Multinational pats on the head for *VideoHound*:

"There are so many mediocre movie books in the field it may require some time for the buffs to recognize the quality of your volume."
—H. Belson, Downers Grove, IL

"I'd like to congratulate you on your excellent reference work. My family and I derive a lot of pleasure from using it."
—R.W. Greenham, Maidstone, England

"I wanted to write and say how much I've enjoyed your book. I have read many video books, but yours has been by far the best book yet." —S. Courville, Bay City, MI

"Where else can you find 20,000 movies broken down into such useful categories as 'Yuppie Nightmares,' 'Nazis and Other Paramilitary Slugs,' 'Negative Utopia,' 'Strained Suburbia,' and 'Sea Critter Attack'?" —*Myrtle Beach Sun-News*

"Videos are encapsulated in lively, pithy, paragraph-length reviews. Tremendous!"—Detroit *Metro Times*

"*VideoHound's Golden Movie Retriever* is the best book of its type that I've seen. In addition to the review of the movies, the list of actors and their films is outstanding! It is exactly what I've been looking for for a long time." —B.D. Richardson, Ottawa, Ontario

Also from **Visible Ink Press:**

Les Brown's Encyclopedia of Television

> "A trivia buff's dream." —*Chicago Sun Times*

Completely revised third edition of the classic comprehensive guide to TV's programs, events, watershed moments, and controversial issues facing the business. 7¼ x 9¼ Paperback. 750 pp. $22.95

Actors, Artists, Authors, & Attempted Assassins: The Almanac of Famous & Infamous People

Brief, lively biographical data on more than 13,000 extraordinary people—both good and up-to-no-good. 7½ x 9¼ Paperback. 590 pp. $13.95

Earl Blackwell's Entertainment Celebrity Register

The world's foremost information bureau on celebrities ropes in 500 of the most provocative figures in the entertainment field in this compulsively readable volume of superstar profiles. 6 x 9 Paperback. 610 pp. $19.95.

VIDEOHOUND'S
GOLDEN
Movie
RETRIEVER

VideoHound's Golden Movie Retriever 1993

Sequel to successful 1991 edition. 22,000 videos. Now an annual. Completely revised, featuring two new indexes (*Series* and *Kibbles*), enlarged cast and director indexing, more categories, thousands of videos new to the *Hound*, and expanded reviews. Distributor guide permits instant tracking of wayward videos. Written on location. Don't leave for the video store without it.

1993. (PG) 1,464p. Beth Fhaner, Christine Tomassini, Carolyn Braun, Marie MacNee, Charles Cassady, Carol A. Schwartz, Kathleen J. Edgar, Nikki Ellert, Thomas Wiloch, Les Stone, Ethan Casey, Mark F. Mikula, David M. Galens *Ed:* Martin Connors, Julia Furtaw. **$17.95**

ADDITIONAL CREDITS

Program Architect: *David Trotter*
Programmer of Certain Indexes: *Don Dillaman*
Cover and Page Design: *Kathleen Hourdakis*
Illustrations: *Terry Colon*
Production: *Mary Beth Trimper, Dorothy Maki, Mary Winterhalter*
Hound Typesetter: *Michael Boyd, ATLIS Publishing Services, Inc.*
Copy Guy: *David Collins*
Visible Ink Trauma Team : *Diane Dupuis, Christa Brelin*

Supporting cast: Julie Bilenchi, Cheryl McDonald, Gwen Turecki, Sara Tal Waldorf, Kim Faulkner, Sandra Doran, Kristin Kahrs, Christine Mathews, Joanna Zakalik, Tony Gerring, Camille A. Killens, Chris Kasic, Evelyn Sullen, Christopher P. Scanlon, Matthew E. Haran, Kathy Lopez, John Krol, Kevin Hillstrom, Janine Eggertson, James Craddock, Kerrie Hurford, Kathleen Maki, Kathy Nemeh, Deborah Burek, Holly Selden, Caren Troshynski-Thomas, Diane Cooper, Melanye Johnson, Margaret Lockwood, Susie Martin, Grant Eldridge, Christine Abraam, Jeffrey Peters, Jackie Barrett, Jared Gulian, Jane Malonis, and Jane Connors.

Art Chartow, Susan Edgar, Kim Forster, Keith Jones, Christine Kesler, Michael Kroll, Marie Rose Napierkowski, and Leslie Norback.

Plus: Jane M. Kelly, Denise E. Kasinec, Barbara Sayles, Mary P. LaBlanc, Betz DesChenes, Kevin Hile, Susan Reicha, Thomas Pendergast, Pam Shelton, Roger Matuz, Steven Carey, Donna Batten, Deborah Stanley, and Promark Ltd.

Key Grips: Kenneth D. Benson, Jr., Yolanda A. Johnson, Tara Y. McKissack, Virgil L. Burton III, Mary Daniels, LySandra C. Davis, and Mauraleen A. Gollob.

Hearty bark for minding the kennel and looking other way when necessary: Dedria Bryfonski, Rod Gauvin, Dennis Poupard, Barb Eschner, Kathi Gruber, Brigitte Darnay, Amy Lucas, Julie Towell, Cindy Baldwin, Theresa Rocklin, James Lesniak, Donna Olendorf, Karen Backus, Gwen Tucker, and Benita Spight.

Thanks to Hector H. Gonzalez for his many corrections.

A special tail-wag to the people who sell the book (including Chuck Lewis at the downstairs Doubleday) and the folks who buy the book and often write in suggesting better ways to do the book.

VIDEOHOUND'S

GOLDEN
Movie
RETRIEVER

VISIBLE
INK
PRESS

DETROIT   WASHINGTON D.C.   LONDON

VideoHound's Golden Movie Retriever 1993

Published by **Visible Ink Press™**
a division of Gale Research Inc.
835 Penobscot Building
Detroit, MI 48226-4094

Visible Ink Press and **VideoHound** are registered trademarks
of Gale Research Inc.

ISBN 0-8103-9425-1

10 9 8 7 6 5 4 3 2 1

C O N T E N T S

Introduction

S ince the last edition, *VideoHound* has been pacing the yard, restless in his quest to build the best possible video guide. He's read the many cards and letters that came his way (thanks, Mom); chewed on a few bones while reviewing every entry; and pulled on the leash after adding thousands of new video descriptions. Tail twitching, many times he began to think almost deeply: what makes a dog-gone good video guide? In his darkest hours, he howled at passing cars and wondered in his existential canine way: why me? And then it came to him: movies serve to entertain. A guide to those videos should do likewise.

Sure, a film may sometimes inform and enlighten as well, but generally a video offers the viewer two hours or so of escapism and the vicarious thrill of being swept to a different place out *there*. Video viewing is a significant commitment of time and resources; the *Hound* believes in making the most of his video opportunities. With those thoughts in mind, *VideoHound 1993* weighs in as a massive movie reference guide with an attitude. Careful research, powerful indexing, and sheer numbers make the *Retriever* useful for determining which video to select; enthusiasm, a fresh view, and a heartening determination to avoid the usual film review cliches (unless the video begs for cliche bludgeoning) keep it engaging. From picks to pans, whether tongue-in-cheek or in earnest, *VideoHound* offers a unique perspective on movies.

During the last 18 months, the *Hound* received and studied hundreds of postcards and letters from devoted video buffs. Perhaps the most popular request was for information on tracking the elusive video. "I've been looking endlessly for a movie and then read the review in your book, so now I know it's out on video but I still can't find it anywhere. Do you know where I might turn?" Of course he does. Always a faithful retriever, the *Hound* has added codes for up to three distributors for each title. The handy distributor guide in the back of the book lists the addresses and telephone numbers (as well as the toll-free and fax numbers when available) for more than 500 distributors. That hard-to-find video may now be only a telephone call away.

The *Golden Movie Retriever* is, of course, a vital data hound, devoted to quickly finding the right video for movie lovers armed with only a fragment of information. Video-store blackout is all-too familiar—Ms. Video Renter is forced to ask 16-year-old clerk who started yesterday if he knows name of movie starring Bergis Winfeld concerning space aliens impersonating accountants in poorly conceptualized attempt to evade persecution from dim-witted authorities. But with the *Hound* and his nose for film you'll face movie-selection gridlock no more. The *Retriever*'s motto: "Don't be stuck again." Armed with seven indexes, the *VideoHound* browser can avoid those video store blues and coast home happily with cassette in hand.

The video-retrieval tools at your disposal include the **Category Index**, listing 350 terms ranging from "Black Comedy" to "Time Travel" and enabling you to scan quickly for a desired flick. Thanks to reader suggestions and staff ingenuity, more than two dozen new categories have been added to this edition, including *Buddy Cops, Detective Spoofs, Loner Cops, Killer Toys, Mad Scientists, Phone Terror, Submarines, Treasure Hunt,* and, of course, *Ghosts, Ghouls, and Goblins.*

The giant **Cast & Director Index** is a compilation of 30,000 performer and director listings citing more than 100,000 credits. Your chances of finding the right video go up dramatically with that kind of movie micro-sizing. At a glance you can discover the range of Al Pacino's acting career or find all of John Huston's movies available on video. Everyone, from major figures in filmdom to obscure participants in forgettable direct-to-video releases, is given his or her due in this comprehensive index. John Wayne, with 123 video listings, zooms to the top of the credit hit list. Stars of tomorrow with one video apiece include Kathleen Bailey and Rula Lenska. In addition to helping you find a movie by searching on performer and director names, the index also serves as a handy look-it-up for all manner of bar bets and trivia questioning. For instance, care to name six features starring Sigourney Weaver? Five with Jodie Foster? Three Joan Crawfords? Tie-buster: four movies starring Ricardo Montalban.

Completely revised, *VideoHound* is bigger and more comprehensive than ever, listing 3,000 movies new to this edition and thousands of videos you won't find in any other guide. Two new indexes (**Series** and **Kibbles**) have been added, as well as a **Music Video Guide**. Thousands of video reviews have been expanded, with additional credits and greater detail, including nominated Academy Awards, alternate titles, remakes, sequels, titles and writers of source books or plays, colorization, special editions, and telling trivia (first feature for now-prominent cast member, etc.). The *Golden Retriever*'s love for unearthing vital detail is gainfully employed, as the *Hound* digs deep to provide information on format (Beta, VHS, or laserdisc), country produced in, price, year released, length, prominent awards, and other video facts. And then there's the exclusive *VideoHound* rating system: one (a likely waste of time) to four (very important movie) bones, with the mega-bad cinematic disasters deserving of a *Woof!*, an emphatic paws down from the *Hound*.

Determined to create the best video guide on the block, *VideoHound* welcomes readers' comments and suggestions. Errors and omissions do occur, and the repentant retriever is always grateful to receive word on them. *VideoHound 1993* could not exist without the many readers who first used their capitalistic moxie to buy the first edition and then wrote in, offering praise, criticism, and ideas. Amid much tail-wagging, the Golden One sincerely thanks them for their interest; like all good dogs, the *Hound* is absolutely loyal to the people he serves. Please contact *VideoHound* at Visible Ink Press, P.O. Box 33477, Detroit, MI 48232–5477. *VideoHound* is a dynamic effort put together by a dedicated pack and sponsored by you.

"How to Get Your Money's Worth with *VideoHound*"

VideoHound enhances your chances of finding the right video. Faced with hundreds of new releases each year (many released directly to video and never reviewed in the mainstream media) and thousands of previously released titles, the average video consumer may be frustrated by the amount of information needed to find a certain video or seek out new and interesting flicks. Well, spin your wheels no more. True to his canine roots, *VideoHound* is a sophisticated information retrieval tool experienced at relieving those occasional video mental blocks. Hundreds of movie classifications break this massive volume of information down to usable bits, enabling the *Hound* to quickly point, like all good retrievers, to the desired video. Thousands of hours have been devoted to creating the system, and work continues as *VideoHound* fine-tunes and devises new retrieval methods. Armed with just the name of a performer or director, producing country, or a vague idea of the subject matter or genre, you can use *VideoHound*'s indexes to quickly find the desired title. Or you can peruse the lists at your leisure, referring back to the review section to find interesting viewing opportunities.

A key to using *VideoHound* effectively is to refer to the exhaustive **Cast & Director Index**. Video title escapes you? You need only remember the director or one cast member to locate a title within the review section. Interested in seeing more movies by a particular performer? All currently available videos are listed under the performer's name. The **Cast & Director Index** includes 30,000 classifications by cast member or director in a straight alphabetical format. More than 100,000 credits are cited. Every cast member and director credited in the main review section is indexed here, creating an intriguing array of movie-making lists and a sure-fire locator for those hard-to-place videos.

And where to turn when title, director, and performers are unknown? To the **Category Index**. Think topically, and scan the lists. Many titles are listed under more than one category, ranging from the wide-reaching genre (comedy) to the thematic (buddy cops) to the specific (chases).

VH's LIST OF CATEGORIES

The **CATEGORY INDEX** contains 350 plot and scene classifications of videos. We provide the following list to give you some insight into VideoHound's thought processes in categorizing videos.

Action-Adventure: Adrenaline testers

Adoption & Orphans: Cute kids lacking permanent authority figures

Advertising: Often full of corporate shenanigans

Africa: Set on the dark continent

AIDS: Trickle of films may turn into minor tide

Airborne: Aeronautics & airplanes

Alien Beings—Benign: Friendly space visitors

Alien Beings—Vicious: Not-so-friendly space visitors

American South: Look away, look away

Amnesia: Who am I?

Angels: Visitors from above

Animated Musicals: Boys, girls, dogs, ducks, birds, mice, and monkeys croon

Animation & Cartoons: Daffy, Donald, Bugs, Micky, and the rest of the gang

Anthology: More than one story to a package

Anti-War War Movies: Often satirical, sometimes extremely grim

Apartheid: Afrikaans term for racial segregation in South Africa

Art & Artists: They paint, pause, propagate

Asia: Set on the big continent

Assassins & Assassination: Lone gunman theory and more

Avant-garde: Ahead of its time or completely bewildering to most in the audience

Babysitting: Not a well-paying job

Ballet: On your toes

Ballooning: Up, up and away

Baseball: Action on the diamond

Basketball: Roundball thrillers

Beach Blanket Bingo: Annette, Frankie, etc.

Behind the Scenes: Peeks at what keeps the illusion alive

Bicycling: Pedaling on location

Big Battles: Big-budget clash of large, opposing military forces

Big Budget Stinkers: Millions spent for no apparent reason

Big Daddies: Usually troublesome, with Freudian overtones

Big Digs: Anthropology and archaeology, including Indiana Jones

Big Ideas: Philosophy, ideology and other pursuits of the mind, decoded for film

Bikers: Usually with a mean streak and traveling in packs

Birds: Beaks—The Movie

Bisexuality: Going both ways

Black Comedy: Funny in a despairing sort of way

Black Culture: Dominant black themes, including blaxploitation

Black Gold: That's oil

Blindness: Can't see or sight impaired

Bloody Mayhem: Expanded scenes of arterial explosions

Boxing: Hitting people with gloves on

Bringing Up Baby: Little ones who sound like Bruce Willis

Buddies: People you can borrow money from

Buddy Cops: Couple of police hanging together

Campus Capers: What really goes on at college

Canada: Above the 49th Parallel or North of the Border

Cannibalism: People who eat people are the luckiest people...

Carnivals & Circuses: Greatest show or sideshow on earth

Cats: Lesser vertebrate nonetheless much beloved by Hollywood

Central America: Strife-ridden area topic of strife-ridden flicks

Chases: Someone's after someone else, usually in high-speed vehicles

Child Abuse: Not very funny at all

Childhood Visions: Stories told from the childhood perspective

China: A very big country

Christmas: Reindeer, Santa, children make appearance amid much sentimentality

City Lights: Films that use various urban sprawls to good advantage

Civil Rights: Fighting for equality

Civil War: The Blue and the Gray

Classic Horror: Boris Karloff, Vincent Price

Classics: Precedent setting, memorable scenes, or just damn good

Cold Spots: Story set in a frostbitten locale with lots of shivering

Comedy: General funny business

Comedy Drama: You'll laugh, you'll weep

Comedy Performance: Stand-Ups

Coming of Age: The hard-fought battle for adulthood

Comic Adventure: Romancing the Stone

Computers: Modern technology plays a part

Cops: Police work

Corporate Shenanigans: Big Business runs amuck

Crime & Criminals: Bad guys and gals

Crop Dusters: Down on the farm

Cult Items: Recognized for that something different

Cults: Something like a gang but more intense and usually governed by a state of mind similar to irrationality

Dance Fever: Whole lotta foot-tapping going on

Deafness: Can't hear or hearing impaired

Death & The Afterlife: Could be ghosts, could be voices from the beyond, could be any number of post-dead things

Deep Blue: The sea around us

Demons & Wizards: Swords, sorcery and wrinkled old men with wands

Dental Mayhem: Tooth-pickin' uproars

Detective Spoofs: Putting together clues in humorous fashion

Detectives: Clue happy but often grizzled and cynical

Dinosaurs: Terrible lizards ruled the Earth

Disaster Flicks: Natural and man-made, including 47 "Airport" flicks

Disease of the Week: Bulimia, polio, cancer, and so on

Divorce: Breaking up is hard to do

Doctors & Nurses: Men and women in white concerned about health of complete strangers

Docudrama: Thin Blue Line between documentaries and drama

Documentary: Real life manipulated on film

Down Under: Set in Australia or New Zealand, mate

Drama: Conflict, tension, climax, resolution

Dream Girls: Product of mid-life male crisis

Ethics & Morals: See Woody Allen or any big European director

Etiquette: Mama always said it was better to be polite

Exploitation: Rigged to take advantage of viewer

Fairy Tales: Age-old children tales

Family Ties: Blood bonds or it's in the genes

Family Viewing: Probably will not embarrass parents if viewing with children

Fantasy: Tales of the imagination

Fast Cars: Racing in the street or on the track

FBI Stories: Men and women of the Bureau

Femme Fatale: She done him and him and him wrong

Film & Film History: Exploring film history or historically important films

Film Noir: Dark and moody

Film Stars: Movies or bios about cinema celebs

Fires: Burning

Flower Children: Peace, luv, and understanding

Folklore & Mythology: Age-old adult tales

Football: Big men crack helmets

Foreign Intrigue: Overseas mystery

France: I love Paris in the springtime...

Front Page: Journalism

Gambling: Lucky seven and eleven

Game Show: C'mon down!

Gangs & Gangsters: Teens and adults running in packs for criminal enterprise

Gender Bending: Boys will be girls and vice versa

Genetics: Fooling with the double helix

Germany: Now one big happy country

Ghosts, Ghouls, & Goblins: Happy Halloween

Giants: Very big guys and gals

Go Fish: Casting a line

Golf: Fore!

Good Earth: Ecology, environment, nature, sod

Grand Hotel: Checkout time is noon

Great Britain: Island off Europe

Great Death Scenes: Signing off with style

Great Depression: The era, not the state of mind

Great Escapes: Seizing the day for freedom

Growing Older: Facing the inevitable

Hard Knock Life: Poverty and bad luck

Heaven Sent: Visits or returns from the place where good souls go

Hell High School: Place where adolescents gather against their will

Historical Drama: Usually based on an historic incident

Hockey: Poetry with a puck

Holidays: Deals with Easter, Thanksgiving, etc. (but not Christmas)

Holocaust: Fate of Jews in Nazi Germany

Homeless: Street people

Homosexuality: Gay and lesbian themes

Horror: Modern cut 'em up scare theater

Horses: Whoa boy

Hospitals & Medicine: Action set within an institutional health environment

Hunting: Big and small game

Identity: Who am I?, part 2

Immigration: Melting pot stories

India: Another very big country

Insects: A straight look at bugs

Interviews: When asked questions, people talk

Inventors & Inventions: Those wacky folks and their newfangled machines

Ireland: A green and rocky place

Islam: Religious themes
Island Fare: Surrounded by water
Israel: The promised land
Italy: Stories set in Chianti country
Japan: Place where many killer beasts got their start
Judaism: Concerning that particular faith
Killer Beasts: Big, mean, and hungry
Killer Bugs: Is that a giant spider crawling up your leg? Aiieeee!
Killer Plants: Too much fertilizer
Killer Toys: Demented play things
King of Beasts (Dogs): Need we say more?
Korean War: M*A*S*H and friends
Labor & Unions: Look for the union label
Late Bloomin' Love: Very mature people diddle with romance
Law & Lawyers: Legal themes
Loner Cops: Men with badges who want to be alone
Macho Men: Pulsating testosterone case studies
Mad Scientists: Eureka! I've sewn his head back on!
Made for Television: Features first shown on small screen
Mafia: The making of irresistible offers
Magic: Hocus pocus
Marriage: Wedding bells and the follow-up
Martial Arts: Head-kicking and rib-crunching
Martyred Pop Icons: For us, they gave their all
Mass Media: Television and other aspects of the electronic age
Medieval Romps: Stories set from roughly 500 A.D. to 1500, often filled with peasants and ox-drawn carts
Meltdown: Nuclear problems
Men: Where "what is man?" is dominant theme
Mental Retardation: Situations dealing with condition
Mexico: South of the Rio Grande
Middle East: Action often set in desert
Military Comedy: Marching to a different drummer
Miners & Mining: Helmets with the little flashlights are nifty
Missing Persons: Kidnapping, other forms of vanishing
Misspelled Titles: Not proper Anglish
Mistaken Identity: A favorite film device
Monsters, General: Not killer beasts, bugs, plants, aliens, toys, giants, vampires, werewolves, or politicians
Motor Vehicle Dept.: Car problems
Mountaineering: Climbing on rocks
Movie & TV Trailers: Brief snippets to entice
Music: Amadeus; movies about music
Musical: Lots of singing and dancing, often for no particular reason
Musical Comedy: Laughter, singing, and dancing
Musical Drama: Singing and dancing, with tension
Musical Fantasy: Singing and dancing, with imagination
Mystery & Suspense: Thrillers, cliffhangers, whodunits
Native Americans: American Indians

Nazis & Other Paramilitary Slugs: WWII types plus David Duke adaptations
Negative Utopia: Things seemed so perfect until...
New Age: New word for day dream
New Black Cinema: Spike is Godfather
Nightclubs: Smokey places that serve drinks and have "entertainment"
Occult: Put a spell on you; witches
Oldest Profession: Prostitution
On the Rocks: Alcoholism, alcohol, barflies, moonshining, Prohibition
Only the Lonely: 50 ways to play solitaire
Opera: Shouting in a melodic way while in costume
Order in the Court: Courtroom tales
Outtakes & Bloopers: Inadvertent laugh riots
Pacific Islands: Hawaii, Tahiti, Philippines, etc.
Parades & Festivals: Includes pageants
Parenthood: Moms, dads
Patriotism & Paranoia: Rambo, Patton
Peace: In opposition to war
Performing Arts: Theater, concerts, etc.
Period Piece: Costume epics
Persian Gulf War: Smart bombs away
Phone Terror: Wrong numbers
Physical Problems: Manifesting into theme
Pill Poppin': Drugs, mostly illegal or overdosed
Poetry: Roses are red...
Poisons: Arsenic and Old Lace
Politics: Local, national, and international
Pool: What Fast Eddie does for a living
Pornography: Selling of naked skin
Post Apocalypse: No more convenience stores
Postwar: After the war, generally WWI or WWII
Presidency: Abe Lincoln leads way
Price of Fame: What goes up...
Propaganda: Deliberately stretching the truth
Psycho-Thriller: It's all in your mind
Puppets: Usually with strings
Pure Ego Vehicles: Big stars doing their big star thing
Race Relations: Black and white in color
Radio: With pictures
Rags to Riches: Grit, determination, and hustling lead to fortune
Rape: Victims and often their revenge
Rebel With a Cause: Bucking the establishment for a reason
Rebel Without a Cause: Bucking the establishment just because it's the establishment
Red Scare: Cold War and Communism (now outdated concepts)
Religion: Generally dealing with faith, churches, etc.
Religious Epics: Charlton Heston parting the sea
Restored Footage: Usually found in a vault or attic
Revenge: A key motivating factor

Revolutionary War: Fought over tea

Road Trip: Escapism courtesy of two- and four-wheel vehicles

Robots/Androids: Mechanical but fascinating

Rock Stars on Film: But can they act?

Rodeos: Rope tricks with steers

Role Reversal: Vice versa; empathy test

Romantic Comedy: Falling in love is often funny

Romantic Drama: Then again, sometimes it's not

Royalty: Kings, queens, crowns, and scepters

Running: Jogging, panting, collapsing

Russia: Back in what used to be the U.S.S.R.

Sail Away: Vessels on the water

Sanity Check: Inmates running the asylum; also deviant mental states

Satanism: The Devil you say!

Satire & Parody: Biting social comment or send-ups

Savants: Half-minded geniuses

Scams, Stings & Cons: The hustle

Schmaltz: Extra cloying

School Daze: Education and school but usually not college and often not high school

Science & Scientists: Madame Curie, Albert Einstein and their work

Sci Fi: Imagination fueled by science

Screwball Comedy: Snappy repartee between sexes dueling in impossible situation

Scuba: Wet suits

Sea Critter Attack: Monsters from the deep surface

Serials: Heyday was in the 1930s

Sex & Sexuality: Focus is on lust

Sexploitation: Softcore epics lacking plot but big on skin

Sexual Abuse: Victims and their stories

Shower & Bath Scenes: Encompassing bathtubs, hottubs, showers, sinks and toilets

Shrinks: Psychiatry

Shutter Bugs: Photographers

Silent Films: No small talk/no big talk/no talk talk

Skateboarding: Teens on wheels

Skating: Roller and ice

Skiing: Downhill or cross-country

Slapstick Comedy: Physically oriented humor

Slavery: Usually the American black experience

Slice of Life: Charming small stories of life the way it is

Soccer: Exciting outside U.S.

South America: Big on soccer

Space Exploration: Going where no spam has gone before

Spaghetti Western: Clint with a squint

Spain: No air-conditioning in Olympic Village

Speculation: In search of...

Spousal Abuse: Usually husbands beating wives

Spies & Espionage: Trench coats, dark glasses

Stepparents: Marrying in to existing clan

Storytelling: Usually directed at children

Strained Suburbia: Neighborhood is not what it seems

Submarines: Deep sea diving

Suicide: Premature ends

Summer Camp: Where children go to misbehave

Supernatural Horror: Forces from beyond terrorize

Surfing: Awesome wave, dude!

Survival: Sometimes an obsession

Swashbucklers: Crossed swords, rope swinging

Swimming: A few laps around the pool

Tearjerkers: Melodrama directed at hankie crowd

Technology—Rampant: Machines that wreak havoc

Teenage Angst: Oh no, it's a pimple!

Tennis: Anyone?

Terrorism: Love affairs with hidden bombs

This Is Your Life: Biography and autobiography

3-D Flicks: Bring your special glasses

Time Travel: Fast forward or reverse

Torrid Love Scenes: Steamy and/or sticky

Toys: R Us

Track & Field: Running in circles

Tragedy: The fate of humankind, magnified

Trains: Rhythm of the clackity clack

Trash: The upper plateau of garbage

Travel: Wanderlust

Treasure Hunt: Looking for hidden riches

Trees & Forests: Can't see one for the other, that wilderness paradox

True Stories: Based on real-life events

Vampires: Blood suckers

Vampire Spoof: Comedic blood suckers

Variety: Little bit of everything

Veterans: When Johnny comes marching home...

Vietnam War: Retrospectives proliferate

Volcanoes: Mountain blowing off steam

War, General: Anything involving war that does not fit under other specific conflict categories

Werewolves: Full moon wonders

Western Comedy: Gags and horses

Westerns: Cowboys, cowgirls, horses and jingle jangling spurs

Wild Kingdom: Animals on their own

Wilderness: More trees than a forest, plus wild critters

Women: Dealing with female issues

Women Cops: Female officers of the law

Women in Prison: The things that go on behind bars

Wonder Women: Females with an edge

World War I: Started in 1914

World War II: The Last Big One

Wrestling: Choreographed sports

Wrong Side of the Tracks: Often involves relationship with someone on the right side

Yuppie Nightmares: Mismatched designer socks

Zombies: Arms straight out, eyes bulging

New Series Index

The **Series Index** provides nearly 50 concise listings of major movie series, ranging from James Bond to National Lampoon to the Three Mesquiteers. It follows the arrangement of the **Category Index**, with series names arranged alphabetically.

How to Commercially Locate the Elusive Video

Each video review lists at least one and as many as three distributor codes. In the reviews, these are located immediately before the critical rating at the bottom of the review. The full address as well as phone, toll-free, and fax numbers of the 550 distributors listed can be found in the **Distributor Guide** in the back of the book. Those reviews with the code *OM* are on moratorium (distributed at one time, though not currently). Since a title enjoying such status was once distributed, it may well linger on your local video-store shelf. When the distributor is not known, the code *NO* appears.

Hound Kibbles

Kibbles is where the miscellaneous heart of the *Hound* lies. The **Kibbles Index** is a potpourri of movie trivia, with more than 100 lists of books-to-film, producers (including Roger Corman and Steven Spielberg) and composers such as John Barry and John Williams. It also provides lists of annual box-office winners (that are now on video) since 1939, classic movies, four-bone delights, and of course, the bottom-of-the-barrel *Woof!* features that are so bad they just might make for an amusing evening. Or maybe not. We'll be carefully scanning the mail for your suggestions to add to the kibbles dish.

Adapted Screenplay
"Alien" & Rip-Offs
BBC-TV Productions
Books to Film:
 Agatha Christie
 Alexandre Dumas
 Alistair MacLean
 Charles Dickens
 Edgar Allan Poe
 Edgar Rice Burroughs
 F. Scott Fitzgerald
 H.G. Wells
 H.P. Lovecraft
 Ian Fleming
 John Steinbeck
 Jules Verne
 Larry McMurtry
 Mark Twain
 Neil Simon
 Ray Bradbury
 Robert Louis Stevenson
 Stephen King
 Tennessee Williams
 William Burroughs
 William Shakespeare
Disney Animated Movies
Disney Family Movies
4-Bone Movies
"Gremlins" & Rip-Offs
Hammer Films: Horror
Hammer Films: Sci Fi & Fantasy
Merchant–Ivory Productions
Musical Score:
 Angelo Badalamenti
 Ennio Morricone
 Jack Nitzsche
 Jerry Goldsmith
 John Barry
 John Williams
Producers:
 Andy Warhol
 Coppola/American Zoetrope
 George Lucas
 Robert Altman
 Roger Corman/New World

Steven Spielberg
Val Lewton
William Castle
"Road Warrior" & Rip-Offs
Special FX Extravaganzas
Special FX Extravaganzas: Make-Up
Special FX Wizards:
 Anton Furst
 Dick Smith
 Douglas Trumball
 Herschell Gordon Lewis
 Ray Harryhausen
 Rick Baker
 Rob Bottin
 Tom Savini
Top Grossing Films of 1939-1991
Troma Films
Woofs!

Other Items of Interest

The **Music Video Guide** alphabetically lists the titles of more than 3,000 music videos and music-performance recordings. Due to space limitations and the remarkable sameness of most of the reviews (Music star sings, dances, songs include...), these videos are not covered elsewhere in the book. These videos are typically prefaced by the name of the artist for quick identification.

And last, but not least, the **Captioned Index** identifies more than 2,500 videos available with captioning for hearing-impaired viewers provided by either the National Captioning Institute (Falls Church, VA) or the Caption Center (Los Angeles, CA).

THE REVIEWS

Each *VideoHound* review contains up to 14 items of information, ranging from the title to the review to awards received to the critical rating. Information of interest to the film buff or casual viewer includes precise length of the video, whether in color or black and white, availability in VHS or Beta formats (as well as laserdisc), year first released, and list price. The review also contains information on colorized versions, alternate titles, sequels, remakes, and sources of adaptations.

1 **Casablanca** **2** Can you see George Raft as Rick? Jack Warner did, but producer Hal Wallis wanted Bogart. Considered by many to be the best film ever made and one of the most quoted movies of all time, it rocketed Bogart from gangster roles to romantic leads as he and Bergman (who never looked lovelier) sizzle on screen. Bogart runs a gin joint in Morocco during the Nazi Occupation, and meets up with Bergman, an old flame, but romance and politics do not mix, especially in Nazi-occupied French Morocco. Greenstreet, Lorre, and Rains all create memorable characters, as does Dooley Wilson, the piano player to whom Bogart says, "Play it, Sam," which is often misquoted. Without a doubt, the best closing scene ever written; it was scripted on the fly during the end of shooting. Written by Julius Epstein, Philip Epstein, and Howard Koch from an unproduced play. See it in the original black and white. The laserdisc edition features restored imaging and sound and commentary on audio track 2 by film historian Ronald Haver about the film's production, the play it was originally based on, and the famed evolution of the screenplay.

3 1943 **4** (PG-13) **5** 102m **6** B **7** *GB* **8** Humphrey Bogart, Ingrid Bergman, Paul Henreid, Claude Rains, Peter Lorre, Sydney Greenstreet, Conrad Veidt, S.Z. Sakall, Dooley Wilson, Marcel Dalio, John Qualen **9** *Dir:* Michael Curtiz. **10** Academy Awards '43: Best Picture, Best Director, Best Screenplay; National Board of Review Awards '45: 10 Best Films of the Year; American Film Institute Survey '77: 3rd Best American Film Ever Made. **11** Beta, VHS, LV **12** $29.95 **13** *CAB* **14** ✓✓✓✓

1 Title

2 Description/review

3 Year released

4 MPAA Rating

5 Length

6 Black and white **(B)** or color **(C)**

7 Country produced in (faked for purposes of display)

8 Cast

9 Director

10 Awards

11 Format **(Beta, VHS,** and/or **LV**-laserdisc)

12 Price (at highest retail level)

13 Distributor code

14 One- to four-bone rating (or *Woof!*)

Alphabetization of Video Titles

VideoHound alphabetically arranges titles word by word, ignoring articles and prepositions if not integral to the title. Leading articles (A, An, The) are also ignored in English-language titles, while foreign-language articles are considered (*The Abyss* appears under "**A**" while *Les Miserables* appears under "**L**").

Other points to keep in mind:
- Acronyms appear alphabetically as regular words.
- Proper names in titles are alphabetized beginning with the individual's first name.
- Numeric-oriented titles (*2001: A Space Odyssey*) sort as if spelled out under the appropriate letter, in this case, "Two Thousand One." However, if a numeric title finds itself in close company with other numeric titles (*2000 Year Old Man, 2001: A Space Odyssey, 2069: A Sex Odyssey*) these titles will be ordered in a more common-sensical numeric sequence.

Foreign Producers

The **Foreign Film Index** classifies movies by the country in which the feature was produced. Since the producers may hail from more than one country, some videos will be listed under more than one country. The codes appear in the review immediately following the color or black and white designation near the top of the review.

Key to Country Origination Tags

AR	Argentinean	FR	French	NZ	New Zealand
AU	Australian	GE	German	NI	Nicaraguan
BE	Belgian	GR	Greek	NO	Norwegian
BR	Brazilian	HK	Hong Kong	PL	Polish
GB	British	HU	Hungarian	PT	Portuguese
CA	Canadian	IN	Indian	RU	Russian
CH	Chinese	IR	Irish	SA	South African
CZ	Czech	IS	Israeli	SP	Spanish
DK	Danish	IT	Italian	SW	Swedish
NL	Dutch	JP	Japanese	SI	Swiss
PH	Filipino	LI	Lithuanian	TU	Turkish
FI	Finnish	MX	Mexican	YU	Yugoslavian

Awards Listed

Care to watch an award-winner tonight? The **Awards Index** lists 1,700 films honored by nine national and international award bodies, representing some 80 categories of competition. This information is also contained in individual video's review, following the credits. Only features available on video are listed in this index; movies not yet released on video are not covered. As award-winning films find their way to video, they will be added to the review section and covered in this index.

American Academy Awards
Australia Film Institute
British Academy of Film and Television Arts
Cannes Film Festival
Canadian Genie
Golden Globe
Directors Guild of America
Edgar Allan Poe Awards
MTV Awards

THE RATING SCALE:
One Dog's Opinion

Woof! **(No bones)**
Junk-food video. May be redeemed by stretches so bad they hold interest, like a freeway accident at rush hour.

(One bone)
Poor use of camera, film, sets, script, actors, and studio vehicles.

(One bone and a half)
Toying with respectability, while still lurching in gutter. Often lacking in standard cinematic devices, like plot and performances.

(Two bones)
May be perfectly delightful for certain tastes. A waste of time for others.

(Two bones and a half)
A very good average film, certain to be enshrined in the hall of mediocrity.

(Three bones)
Above average. Worth investigating. May want to recommend to friends.

(Three bones and a half)
Excellent. Wonderful. Exhalted. The stuff video dreams are made of.

(Four bones)
Masterful cinematic expression. Could confidently recommend to complete strangers.

The *Hound* generally doesn't rate non-features, including cartoons and television series.

Hound Trivia

Number of video titles beginning with the word "blood" (or some bloody variation thereof): *99*

Number of maniacs in slaughter-film director Herschell Gordon Lewis' premier achievement in undiluted gore: *2000.*

Year Alfred Hitchcock's first sound film was released: *1929* ("Blackmail")

Year Alfred Hitchcock made his first on-screen cameo: *1929* (in "Blackmail")

Number of Oscars awarded for Best Actress in 1968: *2*
(to Barbra Streisand in "Funny Girl" and Katherine Hepburn in "The Lion in Winter")

Number of video versions of "Huckleberry Finn": *8*

Dimension Buckaroo Banzai travels through in his jet-propelled Ford Fiesta: *8th*

Number of years Norman Bates spent in a mental institution: *22*

Examples of wild and restless beings featured in "Attack of" flicks:
killer tomatoes, beast creatures, a 50-foot woman, a killer refrigerator, mushroom people, a Mayan mummy, robots, super monsters, and a swamp creature.

Aaron Loves Angela Puerto Rican girl falls in love with a black teen amidst the harsh realities of the Harlem ghetto. "Romeo and Juliet" meets "West Side Story" in a cliched comedy drama. Music by Jose.
1975 (R) 99m/C Kevin Hooks, Irene Cara, Moses Gunn, Robert Hooks, Jose Feliciano; **Dir:** Gordon Parks Jr. **VHS, Beta** $59.95 *COL* 𝅘𝅥𝅘𝅥

Abbott and Costello in Hollywood Bud and Lou appear as a barber and porter of a high class tonsorial parlor in Hollywood. A rather sarcastic look at backstage Hollywood, Abbott & Costello style. Ball makes a guest appearance.
1945 83m/B Bud Abbott, Lou Costello, Frances Rafferty, Warner Anderson, Lucille Ball; **Dir:** Sylvan Simon. **VHS, Beta** $19.98 *MGM, CCB* 𝅘𝅥𝅘𝅥

Abbott and Costello Meet Captain Kidd With Captain Kidd on their trail, Abbott and Costello follow a treasure map. Bland A&C swashbuckler spoof with a disinterested Laughton impersonating the Kidd. One of the duo's few color films.
1952 70m/C Bud Abbott, Lou Costello, Charles Laughton, Hillary Brooke, Fran Warren, Bill Shirley, Leif Erickson; **Dir:** Charles Lamont. **VHS, Beta, LV** $29.95 *NOS, UHV* 𝅘𝅥½

Abbott and Costello Meet Dr. Jekyll and Mr. Hyde Abbott and Costello take on evil Dr. Jekyll, who has transformed himself into Mr. Hyde, and is terrorizing London. A lame attempt at recapturing the success of "Abbott and Costello Meet Frankenstein."
1952 77m/B Bud Abbott, Lou Costello, Boris Karloff, Craig Stevens, Helen Westcott, Reginald Denny; **Dir:** Charles Lamont. **VHS, Beta** $14.95 *MCA* 𝅘𝅥𝅘𝅥

Abbott and Costello Meet Frankenstein Big-budget A&C classic with subsequent big box office. Tells the sad story of Chick and Wilbur, two unsuspecting baggage clerks delivering to a wax museum a crate containing the last but not quite dead remains of Dracula and Dr. Frankenstein's monster. There the fiends are revived, with the help of the vampire's assistant who feigns love for Lou but secretly wants to implant his brain in Frankie Big Shoes. Seems the monster has become too intelligent, so Dracula is seeking to downsize his synapses to make him more controllable. Enter Lou. The Wolf Meister makes a special appearance to warn Chick and Wilbur that trouble looms. Last film to use the Universal creature pioneered by Karloff in 1931, and one of A&C's best efforts.

1948 83m/B Bud Abbott, Lou Costello, Lon Chaney Jr., Bela Lugosi, Glenn Strange, Lenore Aubert, Jane Randolph; **Dir:** Charles T. Barton; **Voices:** Vincent Price. **VHS, Beta, LV** $14.95 *MCA* 𝅘𝅥𝅘𝅥𝅘𝅥

Abbott & Costello Meet the Invisible Man Abbott and Costello play newly graduated detectives who take on the murder case of a boxer (Franz) accused of killing his manager. Using a serum that makes people invisible, the boxer helps out Costello in a prizefight that will frame the real killers, who killed the manager because the boxer refused to throw a fight. Great special effects and hilarious gags make this one of the best from the crazy duo.
1951 82m/B Bud Abbott, Lou Costello, Nancy Guild, Adele Jergens, Sheldon Leonard, William Frawley, Gavin Muir, Arthur Franz; **Dir:** Charles Lamont. **VHS** $14.98 *MCA* 𝅘𝅥𝅘𝅥𝅘𝅥

Abbott and Costello Meet the Killer, Boris Karloff Unremarkable Abbott and Costello murder mystery. Karloff plays a psychic who tries to frame Lou for murder. Pleasant enough but not one of their best.
1949 84m/B Bud Abbott, Lou Costello, Boris Karloff, Lenore Aubert, Gar Moore, Donna Martell, Alan Mowbray, James Flavin, Roland Winters; **Dir:** Charles T. Barton. **VHS, Beta** $9.95 *MCA* 𝅘𝅥𝅘𝅥

Abbott and Costello Scrapbook Rare glimpses of the comedy duo from screen tests, previews, and flubs.
197? 60m/B Bud Abbott, Lou Costello. **VHS, Beta** $29.95 *DVT*

ABC Stage 67: Truman Capote's A Christmas Memory A television drama special narrated by Truman Capote about his remembered childhood experience of baking dozens of fruitcakes for friends at Christmas with his grandmother. Page is quiet, lovely and true in this sensitive portrait of why we give and what we have to be thankful for. Adapted from a Capote short story by Capote and Eleanor Perry.
1966 51m/B Geraldine Page, Donnie Melvin; **Dir:** Frank Perry; **Nar:** Truman Capote. **VHS, Beta, 8mm** *VYY*

ABC's of Love & Sex, Australia Style An Australian-made sex-education parody.
1986 87m/C *AU* Brigittia Almsrom, Rettima Borer, Ian Broadbent. **VHS, Beta** $19.95 *ACA* 𝅘𝅥

Abducted A woman jogger is abducted by a crazed mountain man in the Canadian Rockies. Weird and unbelievably tedious film is nevertheless highlighted by some spectacular wilderness footage of the Vancouver

area, a travelogue bonus for those in the mood.
1986 (PG) 87m/C *CA* Dan Haggerty, Roberta Weiss, Lawrence King Phillips; **Dir:** Boon Collins. **VHS, Beta** $79.95 *PSM* 𝅘𝅥½

Abduction Exploitative account of the Patty Hearst kidnapping loosely adapted from a novel by Harrison James written before the kidnapping. A young woman from a wealthy capitalist family is kidnapped by black radicals and held for an unusual ransom. Oh, and she tangles with lesbians, too.
1975 (R) 100m/C Judith-Marie Bergan, David Pendleton, Gregory Rozakis, Leif Erickson, Dorothy Malone, Lawrence Tierney; **Dir:** Joseph Zito. **VHS, Beta** $59.95 *MED* 𝅘𝅥½

The Abduction of Allison Tate A rich developer takes the land belonging to a group of Native Americans. Three young members of the tribe retaliate by kidnapping the developer's daughter. But when the kidnappers plans go awry and one of the trio is killed, Allison finds herself sympathizing more with them than with her father's ambitions.
1992 (R) 95m/C Leslie Hope, Bernie White; **Dir:** Paul Leder. **VHS** $79.95 *MNC* 𝅘𝅥½

The Abduction of Kari Swenson An account of the true-life (it really happened) kidnapping of Olympic biathalon hopeful Swensen by a pair of mischievous Montana mountain men with matrimony in mind. The movie details the abduction of Swenson by the scruffy father and son duo and the massive manhunt. As the put-upon Kari, Pollan exceeds script expectations in this made-for-television exercise in stress avoidance.
1987 100m/C Joe Don Baker, M. Emmet Walsh, Ronny Cox, Michael Bowen, Geoffrey Blake, Dorothy Fielding, Tracy Pollan; **Dir:** Stephen Gyllenhaal. **VHS** $79.95 *VMK* 𝅘𝅥½

Abduction of St. Anne An almost interesting made-for-television mystical thriller about a detective and a bishop trying to track down a gangster's daughter, who may have nifty supernatural healing powers the Church would be very interested in having documented.
1975 78m/C Robert Wagner, E.G. Marshall, William Windom, Lloyd Nolan; **Dir:** Harry Falk. **VHS, Beta** $49.95 *WOV, GKK* 𝅘𝅥𝅘𝅥

The Abductors Caffaro's super-agent takes on international white slavery, a worthy target for any exploitation effort. While the novelty is a tough and intelligent on-screen heroine, sufficient sleaze and violence bring it all down to the proper level of swampland video. Sequel to the never-to-be-forgotten "Ginger."

1972 **(R)** 90m/C Cheri Caffaro, William Grannell, Richard Smedley, Patrick Wright, Jennifer Brooks; **Dir:** Don Schain. **VHS, Beta $39.95** MON *♂*

Abdulla the Great A dissolute Middle Eastern monarch falls for a model, who spurns him for an army officer. While distracted by these royal shenanigans, the king is blissfully unaware of his subjects' disaffection—until they revolt. Dares to lampoon Egypt's dead King Farouk, going against conventional Hollywood wisdom ("Farouk in film is box office poison"). In some parts of the world known as "Abdullah's Harem."
1956 89m/C *GB* Gregory Ratoff, Kay Kendall, Sydney Chaplin; **Dir:** Gregory Ratoff. **VHS, Beta $29.95** VYY *♂♂*

Abe Lincoln: Freedom Fighter Historical drama documents a turning point in the young life of Abe Lincoln, the 16th President of the United States and honored throughout the country with sales of coats and appliances on his birthday.
1978 54m/C Allen Williams, Andrew Prine, Brock Peters; **Dir:** Jack B. Hively. **VHS, Beta $9.98** VID, LME *♂♂*

Abe Lincoln in Illinois Massey considered this not only his finest film but a part he was "born to play." Correct on both counts, this Hollywood biography follows Lincoln from his log cabin days to his departure for the White House. The Lincoln-Douglass debate scene and Massey's post-presidential election farewell to the citizens of Illinois are nothing short of brilliant. From Robert Sherwood's Pulitzer-Prize winning play. Contrasted with the well-known "Young Mr. Lincoln" (Henry Fonda), its relative anonymity is perplexing. Oscar nominations for Massey and cinematographer James Wong Howe.
1940 110m/B Raymond Massey, Gene Lockhart, Ruth Gordon, Mary Howard, Dorothy Tree, Harvey Stephens, Minor Watson, Alan Baxter, Howard da Silva, Maurice Murphy, Clem Bevans, Herbert Rudley; **Dir:** John Cromwell. **VHS, Beta $19.95** MED, RKO *♂♂♂♂*

The Abe Lincoln of Ninth Avenue All-American tale of a poor young man making good in New York. His role model is Abraham Lincoln. Cooper is exceptional, supported by effective performances by the rest of the cast. Also known as "Streets of New York."
1939 68m/B Jackie Cooper, Martin Spelling, Marjorie Reynolds, Dick Purcell, George Cleveland, George Irving; **Dir:** William Nigh. **VHS, Beta, 8mm $29.95** VYY *♂♂ ½*

Abilene Town In a post-Civil War Kansas town far from the freeway, Scott is the tough marshall trying to calm the conflict between cattlemen and homesteaders. He also finds time to participate in a romantic triangle with dance hall vixen Dvorak and heart-of-gold Fleming. Snappy pace keeps it interesting. Based on a novel by Ernest Haycox.
1946 90m/B Randolph Scott, Ann Dvorak, Edgar Buchanan, Rhonda Fleming, Lloyd Bridges; **Dir:** Edwin L. Marin. **VHS, Beta $9.95** NEG, MRV, NOS *♂♂♂*

The Abominable Dr. Phibes After being disfigured in a freak car accident that killed his wife, an evil genius decides that the members of a surgical team let his wife die and shall each perish by a different biblical plague. High camp with the veteran cast in top form.

1971 **(PG)** 90m/C *GB* Vincent Price, Joseph Cotten, Hugh Griffith, Terry-Thomas, Virginia North, Susan Travers, Alex Scott, Caroline Munro; **Dir:** Robert Fuest. **VHS, Beta, LV $14.98** VES, LIV *♂♂♂*

The Abomination After a 5,000-year-old creature possesses him during a nightmare, a boy goes on an eye-gouging frenzy. Only the audience gets hurt.
1988 **(R)** 100m/C Van Connery, Victoria Chaney, Gaye Bottoms, Suzy Meyer, Jude Johnson, Blue Thompson, Scott Davis; **Dir:** Max Raven. **VHS, Beta** DMP *♂*

About Last Night... Semi-realistic comedy-drama which explores the ups and downs of one couple's relationship. Mostly quality performances, especially Perkins and Belushi as friends of the young lovers. Based on David Mamet's play "Sexual Perversity in Chicago," but considerably softened so that more people would buy tickets at the box office, the film acts as a historical view of contemporary mating rituals before the onset of the AIDS crisis.
1986 **(R)** 113m/C Rob Lowe, Demi Moore, Elizabeth Perkins, James Belushi; **Dir:** Edward Zwick. **VHS, Beta, LV $14.95** COL *♂♂♂*

Above and Beyond Documentary about five veterans who reveal their experiences as POWs in Vietnam.
1987 58m/C **VHS, Beta, 3/4U** CEN *♂♂♂*

Above the Law In his debut, Seagal does his wooden best to portray a tough Chicago police detective planning an enormous drug bust of one of the biggest felons in the state. Unfortunately, the FBI has ordered him to back off and find another bust. The reasons are almost as complex as Seagal's character, and like most details of the flick, stupid. However, people don't watch these movies for the acting or the plot, but for the fight scenes, which are well-choreographed and violent. Watch it with someone you love.
1988 **(R)** 99m/C Steven Seagal, Pam Grier, Henry Silva, Sharon Stone, Ron Dean, Daniel Faraldo; **Dir:** Andrew Davis. **VHS, Beta, LV, 8mm $19.95** WAR, TLF, PIA *♂♂*

Above Us the Waves During WWII, the British navy immobilizes a huge German battleship off the coast of Norway. Effectively dramatizes the British naval preparations for what seemed a suicidal mission: using midget submarines to plant underwater explosives on the hull of the German vessel and detonating them before the Germans could detect the danger.
1956 92m/C *GB* John Mills, John Gregson, Donald Sinden, James Robertson Justice, Michael Medwin, James Kenney, O.E. Hasse, William Russell, Thomas Heathcote, Lee Patterson, Theodore Bikel, Lyndon Brook, Anthony Newley; **Dir:** Ralph Thomas. **VHS** REP *♂♂♂*

Abraham Lincoln D.W. Griffith's first talking movie takes Abraham Lincoln from his birth through his assassination. This restored version includes the original slavery sequences which were believed to be lost, but obviously were not. Musical score included.
1930 97m/B Walter Huston, Una Merkel, Kay Hammond, E. Allen Warren, Hobart Bosworth, Fred Warren, Henry B. Walthall, Russell Simpson; **Dir:** D.W. Griffith. **VHS, Beta, LV $19.95** NOS, MRV, CCB *♂♂ ½*

Abroad with Two Yanks Two Marine buddies on furlough exhibit slapstick tendencies while competing for the same girl. Along the way a big chase ensues with the two

soldiers in drag. Typical wartime shenanigans likely to incite only weak chuckling or inspired snoozing.
1944 81m/B William Bendix, Dennis O'Keefe, Helen Walker, John Loder, George Cleveland, Janet Lambert, James Flavin, Arthur Hunnicutt; **Dir:** Allan Dwan. **VHS, Beta $12.95** IVE *♂♂*

Absence of Malice High-minded story about the harm that the news media can inflict. Field is the earnest reporter who, after being fed some facts by an unscrupulous federal investigator, writes a story implicating Newman in a murder he didn't commit. Field hides behind journalistic confidentiality privilege to put off the outraged Newman, who loses a friend to suicide during the debacle. Interesting performances by Field and Newman.
1981 **(PG)** 116m/C Paul Newman, Sally Field, Bob Balaban, Melinda Dillon, Luther Adler, Barry Primus, Josef Sommer, John Harkins, Don Hood, Wilford Brimley; **Dir:** Sydney Pollack. **VHS, Beta, LV $14.95** COL *♂♂ ½*

The Absent-Minded Professor Classic dumb Disney fantasy of the era. A professor accidentally invents an anti-gravity substance called flubber, causing inanimate objects and people to become airborne. Great sequence of the losing school basketball team taking advantage of flubber during a game. MacMurray is convincing as the absent-minded genius in this newly colored version. Followed by "Son of Flubber."
1961 97m/C Fred MacMurray, Nancy Olson, Keenan Wynn, Tommy Kirk, Leon Ames, Ed Wynn; **Dir:** Robert Stevenson. **VHS, Beta, LV $69.95** DIS *♂♂♂ ½*

Absolute Beginners Fervently stylish camp musical exploring the lives of British teenagers in the 1950s never quite gets untracked, although MTV video moments fill out a spare plot line. Based on a novel by Colin MacInnes.
1986 **(PG)** 107m/C *GB* David Bowie, Ray Davies, James Fox, Eddie O'Connell, Patsy Kensit, Anita Morris, Sade Adu; **Dir:** Julien Temple. **VHS, Beta, LV $79.95** HBO *♂♂*

Absolution Two English boys trapped in a Catholic boarding school conspire to drive a tyrannical priest over the edge of sanity. As a result, bad things (including murder) occur. Burton is interesting in sadistic character study. Not released in the U.S. until 1988 following Burton's death, maybe due to something written in the will.
1981 **(R)** 105m/C *GB* Richard Burton, Dominic Guard, Dai Bradley, Andrew Keir, Billy Connolly, Willoughby Gray; **Dir:** Anthony Page. **VHS, Beta, LV $9.95** SIM, TWE, FCT *♂♂*

The Abyss Underwater sci-fi adventure about a team of oil-drilling divers pressed into service by the navy to locate and disarm an inoperative nuclear submarine. A high-tech thriller with fab footage underwater and pulsating score.
1989 **(PG-13)** 140m/C Ed Harris, Mary Elizabeth Mastrantonio, Todd Graff, Michael Biehn, John Bedford Lloyd, J.C. Quinn, Leo Burmester, Kidd Brewer Jr., Kimberly Scott, Adam Nelson, George Robert Kirk, Chris Elliott; **Dir:** James Cameron. Academy Awards '89: Best Visual Effects. **VHS, Beta, LV $19.98** FOX *♂♂♂*

Academy Award Winners Animated Short Films A collection of six short films that includes Jimmy Picker's "Sundae in New York," "The Hole," "Munro" and "Closed Mondays."
1984 60m/C Academy Award '83: Best Animated Short Film ("Sundae in New York"). **VHS, Beta, LV $59.98** LIV, VES *♂♂♂*

Accatone! Accatone (Citti), a failure as a pimp, tries his luck as a thief. Hailed as a return to Italian neo-realism, this is a gritty, despairing, and dark look at the lives of the street people of Rome. Pasolini's first outing, adapted by the director from his novel, ''A Violent Life.'' Pasolini served as mentor to Bernardo Bertolucci, listed in the credits as an assistant director.
1961 116m/B *IT* Franco Citti, Franca Pasut, Roberto Scaringelli, Adele Cambria, Paolo Guidi, Silvana Corsini; *Dir:* Pier Paolo Pasolini. **VHS** **$79.95** *CON, WBF, APD* 🐾🐾🐾

Acceptable Risks Toxic disaster strikes when a plant manager is ordered to cut costs and sacrifice safety at the Citichem plant. Predictable plot stars Dennehy as the plant manager who risks all and fights politicians to reinstate the safety standards. Meanwhile, Tyson is the city manager who tries to warn the town of a possible chemical accident that could have devastating effects on the community. Made for television drama that shamelessly preys on audience fears.
1986 (R) 97m/C Brian Dennehy, Cicely Tyson, Kenneth McMillan, Christine Ebersole, Beah Richards, Richard Gilliland; *Dir:* Rick Wallace. **VHS, Beta $79.95** *PSM* 🐾🐾½

Access Code Government agents attempt to uncover a private organization that has gained control of nuclear weapons for the purpose of world domination. A ragged patchwork of disconnected scenes meant to test the virtue of patience.
1984 90m/C Martin Landau, Michael Ansara, MacDonald Carey; *Dir:* Mark Sobel. **VHS, Beta $79.95** *PSM* 🐾

Accident A tangled web of guilt, remorse, humor and thwarted sexuality is unravelled against the background of the English countryside in this complex story of an Oxford love triangle. An inside view of English repression at the university level adapted for the screen by Harold Pinter from the novel by Nicholas Mosley. Long-winded but occasionally engrossing character study with interesting performances.
1967 100m/C *GB* Dirk Bogarde, Michael York, Stanley Baker, Jacqueline Sassard, Delphine Seyrig, Alexander Knox, Vivien Merchant, Freddie Jones, Harold Pinter; *Dir:* Joseph Losey. Cannes Film Festival '67: Special Jury Prize; National Board of Review Awards '67: 10 Best Films of the Year. **VHS, Beta $59.95** *HBO* 🐾🐾½

Accident Skate for your life. A hockey game turns into a nightmare when the roof over the arena collapses under the weight of too much ice and snow. One of the few hockey disaster films.
1983 104m/C Terence Kelly, Fiona Reid, Frank Perry; *Dir:* Donald Britnain. **VHS, Beta $59.95** *TWE* 🐾½

The Accidental Tourist A bittersweet and subtle story, adapted faithfully from Anne Tyler's novel, of an introverted, grieving man who learns to love again after meeting an unconventional woman. After his son's death and subsequent separation from wife Turner, Macon Leary (Hurt) avoids emotional confrontation, burying himself in routines with the aid of his obsessive-compulsive siblings. Kooky dog-trainer Muriel Pritchett (an exuberant Davis) wins his attention, but not without struggle. Hurt effectively uses small gestures to describe Macon's emotional journey, while Davis grabs hearts with her open performance. Outstanding supporting cast. Heavily Oscar-nominated.

1988 (PG) 121m/C William Hurt, Geena Davis, Kathleen Turner, Ed Begley Jr., David Ogden Stiers, Bill Pullman, Amy Wright; *Dir:* Lawrence Kasdan. Academy Awards '88: Best Supporting Actress (Davis); New York Film Critics Awards '88: Best Picture. **VHS, Beta, LV, 8mm $89.95** *WAR, BTV* 🐾🐾🐾

Accidents A scientist discovers that his invention has been stolen and is going to be used to cause worldwide havoc. He becomes concerned and spends the remainder of the movie trying to relieve himself of anxiety.
1989 (R) 90m/C Edward Albert, Leigh Taylor-Young, Jon Cypher; *Dir:* Gideon Amir. **VHS, Beta $79.75** *TWE* 🐾½

The Accused Young is cast against character in this story of a college professor who is assaulted by one of her students and kills him in self-defense. A courtroom drama ensues, where Young is defended by the dead man's guardian. Film noir with nice ensemble performance.
1948 101m/B Loretta Young, Robert Cummings, Wendell Corey, Sam Jaffe, Douglas Dick, Sara Allgood, Ann Doran; *Dir:* William Dieterle. **VHS, Beta $89.95** *PAR* 🐾🐾½

The Accused Provocative treatment of a true story involving a young woman gang raped in a bar while onlookers cheer. McGillis is the assistant district attorney who takes on the case and must contend with the victim's questionable past and a powerful lawyer hired by a wealthy defendant's parents. As the victim with a past, Foster gives an Oscar-winning performance that won raves for its strength and complexity.
1988 (R) 110m/C Jodie Foster, Kelly McGillis, Bernie Coulson, Steve Antin, Leo Rossi; *Dir:* Jonathan Kaplan. Academy Awards '88: Best Actress (Foster); National Board of Review Awards '88: Best Actress (Foster). **VHS, Beta, LV, 8mm $14.95** *PAR, FCT, BTV* 🐾🐾🐾

Ace of Aces An American sculptor is reviled, particularly by his girlfriend, when he does not join fellows in enlisting in what becomes WWI. Out to prove he's not lacking testosterone, he becomes a pilot in France, but is embittered by his experiences. Dated but well-acted war melodrama.
1933 77m/B Richard Dix, Elizabeth Allan, Theodore Newton, Ralph Bellamy, Joseph Sawyer, Frank Conroy, William Cagney; *Dir:* J. Walter Ruben. **VHS, Beta $19.98** *TTC* 🐾🐾

Ace Crawford, Private Eye Bumbling private eye struggles with confusing case of murder.
19?? 71m/C Tim Conway, Billy Barty. **VHS** *MCG* 🐾🐾

Ace Drummond The complete 13-chapter serial about a murder organization that tries to stop several countries from forming a worldwide clipper ship air service.
1936 260m/B John ''Dusty'' King, Jean Rogers, Noah Beery; *Dir:* Ford Beebe. **VHS, Beta $29.95** *NOS, HHT, VCN* 🐾🐾

Ace of Hearts A seedy story about a rich guy in the South Pacific who pays heavily to have himself killed.
197? 90m/C Mickey Rooney, Chris Robinson, Pilar Velasquez. **VHS, Beta $39.95** *VCD* 🐾

Ace High Spaghetti western about a ruthless outlaw named Cat Stevens trying to save himself from the noose. Patterned after the famous Sergio Leone-Clint Eastwood westerns, with less of a budget and more camp tendencies.

1968 120m/C *IT* Eli Wallach, Terence Hill, Bud Spencer, Brock Peters, Kevin McCarthy; *Dir:* Giuseppe Colizzi. **VHS, Beta $14.95** *PAR, FCT* 🐾½

Aces and Eights Standard western detailing the final days of Wild Bill Hickok, who was holding aces and eights when shot in the back (he never did finish the game). McCoy does his usual best as the sanitized Billy.
1936 62m/B Tim McCoy, Jimmy Aubrey, Luana Walters, Wheeler Oakman, Earl Hodgins, Frank Glennon; *Dir:* Sam Newfield. **VHS $19.95** *NOS, DVT, RXM* 🐾🐾

Aces: Iron Eagle 3 Colonel ''Chappy'' Sinclair returns once again in this air adventure. He's been keeping busy flying in air shows when he stumbles across the nefarious activities of a Peruvian drug baron working out of a remote village. Sinclair recruits a team of maverick air circus pilots and they ''borrow'' a fleet of WWII vintage aircraft to raid the village, coming up against a fellow Air Force officer who turns out to be another villain. Lots of action and a stalwart cast.
1992 (R) 98m/C Louis Gossett Jr., Rachel McLish, Paul Freeman, Horst Buchholz, Christopher Cazenove, Sonny Chiba, Fred Dalton Thompson, Mitchell Ryan, Robert Estes, J.E. Freeman; *Dir:* John Glen. **VHS** *COL* 🐾🐾

Aces Wild Another poker title. Outlaws menace an honest newspaper until the lawmen show up and send them on their way.
1937 62m/B Harry Carey, Gertrude Messinger, Edward Cassidy, Roger Williams; *Dir:* Harry Fraser. **VHS, Beta, LV $19.95** *WGE* 🐾½

Acorn People An unemployed teacher takes a summer job at a camp for handicapped children and learns sensitivity. Made-for-television weeper adapted by Tewkesbury from the book by Ron Jones.
1982 97m/C Ted Bessell, Cloris Leachman, LeVar Burton, Dolph Sweet, Cheryl Anderson; *Dir:* Joan Tewkesbury. **VHS, Beta** *TLF* 🐾🐾

Acqua e Sapone A young innocent model goes to Rome under the watchful eyes of an appointed priest-chaperon, but finds love, fun and other sinful things. Sudsy/romantic comedy.
1983 (PG) 100m/C *IT* Carlo Verdone, Natasha Hovey, Florinda Bolkan, Elena Bolkan; *Dir:* Carlo Verdone. **VHS, Beta $59.95** *RCA, TAM* 🐾½

Across 110th Street Gritty, violent cop thriller in the blaxploitation genre. Both the Mafia and the cops hunt down three black hoods who, in a display of extremely bad judgement, knocked over a mob-controlled bank while disguised as police. Lots of bullets create buckets of blood. Filmed on location in Harlem.
1972 (R) 144m/C Anthony Quinn, Yaphet Kotto, Anthony (Tony) Franciosa, Paul Benjamin, Ed Bernard, Antonio Fargas, Tim O'Connor; *Dir:* Barry Shear. **VHS, Beta $59.95** *FOX* 🐾🐾

Across the Bridge A man on the run from Scotland Yard for stealing a fortune flees to Mexico, in the process killing a man and assuming his identity. An ironic twist and his love for a dog seal his final destiny. Steiger's psychological study as the fugitive is compelling. Based on a novel by Graham Greene.
1957 103m/B *GB* Rod Steiger, David Knight, Marla Landi, Noel William, Bernard Lee; *Dir:* Ken Annakin. **VHS, Beta** *LCA* 🐾🐾½

Across the Great Divide Two orphans must cross the rugged snow-covered Rocky Mountains in 1876 in order to claim their inheritance—a 400-acre plot of land in Salem, Oregon. Pleasant coming-of-age tale with majestic scenery.
1976 (G) 102m/C Robert F. Logan, George Flower, Heather Rattray, Mark Hall; *W/Dir:* Stewart Raffill. VHS, Beta $9.98 *MED* ♫♫½

Across the Pacific Classic Bogie and Huston made on the heels of ''The Maltese Falcon.'' Bogie is an American Army officer booted out of the service on false charges of treason. When no other military will accept him, he sails to China (via the Panama Canal) to offer his services to Chiang Kai-Shek. On board, he meets a variety of seedy characters who plan to blow up the canal. Huston again capitalizes on the counterpoint between the rotundly acerbic Greenstreet, who plays a spy, and stiff-lipped Bogart, who's wooing Astor. Great Bogie moments and fine direction make this an adventure classic. When he departed for the service just prior to filming the final scenes, Huston turned over direction (with Bogie reportedly in an impossible situation) to Vincent Sherman, who did his best to resolve it. Also available colorized.
1942 97m/B Humphrey Bogart, Mary Astor, Sydney Greenstreet, Charles Halton, Victor Sen Yung, Roland Got, Keye Luke, Richard Loo; *Dir:* John Huston. VHS $19.98 *MGM, FCT* ♫♫♫½

Across the Plains Predictable oater about two brothers who are raised separately after their parents are murdered by outlaws. One is brought up by Indians, the other by the outlaws responsible for wiping out the folks, who tell the boy that Indians did in his ma and pa. Eventually the brothers meet and fight. Odds favor the good one winning, aided by Indian pals.
1939 59m/B Jack Randall, Frank Ysconelli, Joyce Bryant, Hal Price, Dennis Moore, Glenn Strange, Bud Osborne; *Dir:* Spencer Gordon Bennett. VHS, Beta $19.95 *NOS, GPV, DVT* ♫½

Across the Tracks Two brothers, one a juvie-home rebel, the other a straight-A jock, are at odds when the black sheep is pressured into selling drugs. In an attempt to save his brother from a life of crime, the saintly one convinces his brother to join him on the school track team, and the brothers are forced to face off in a big meet. Fairly realistic good guy/bad guy who's really a good guy teen drama. Also available in an ''R'' rated version. Music by Joel Goldsmith; written by director Tung.
1989 (PG-13) 101m/C Rick Schroder, Brad Pitt, Carrie Snodgress; *W/Dir:* Sandy Tung. VHS $19.95 *ACA* ♫♫

Across the Wide Missouri Pioneer epic stars Gable as a rugged fur trapper who marries an Indian woman (Marques) so he can trap beaver pelts on her people's rich land. On the journey to the Indian territory however, the trapper truly falls in love with his bride. Superior historical drama is marred slightly by the use of narration (provided by Howard Keel). Look for lively performances from Menjou as a French tippler and Naish as the quirky Indian Chief. Beautiful scenery filmed in the spectacular Rocky Mountains.
1951 78m/C Clark Gable, Ricardo Montalban, John Hodiak, Adolphe Menjou, Maria Elena Marques, J. Carroll Naish, Jack Holt, Alan Napier; *Dir:* William A. Wellman; *Nar:* Howard Keel. VHS, Beta $19.98 *MGM, CCB* ♫♫

The Act A muddled satire about political double dealing and union corruption further muddled by a twangy musical score manufactured by folksy John Sabastin.
1982 (R) 90m/C Jill St. John, Eddie Albert, Pat Hingle, Robert Ginty, Sarah Lagenfeld, Nicolas Surovy; *Dir:* Sig Shore. VHS, Beta $69.95 *VES* ♫

Act of Aggression When a Parisian man finds his wife and daughter murdered at a summer resort, he takes the law into his own hands, with predictable results.
1973 (R) 100m/C *FR* Jean-Louis Trintignant, Catherine Deneuve, Claude Brasseur. VHS, Beta $59.95 *FCT, KOV* ♫♫

Act of Passion: The Lost Honor of Kathryn Beck A woman meets a man at a party and has the proverbial one-night stand. Her privacy is shattered when she discovers that he's a terrorist under surveillance by the police and press. Strong performances by Thomas and Kristofferson help turn this into an interesting American television remake of the German ''Lost Honor of Katharina Blum,'' based loosely on a novel by Heinrich Boll.
1983 100m/C Marlo Thomas, Kris Kristofferson, George Dzundza, Jon De Vries; *Dir:* Simon Langton. VHS, Beta $69.95 *VMK* ♫♫½

Act of Piracy A bankrupt contractor reunites with his estranged wife to track down the brutal terrorists who have kidnapped their son.
1989 (R) 105m/C Ray Sharkey, Gary Busey. VHS, Beta, LV $89.98 *WAR* ♫½

Act of Vengeance A group of women band together to hunt down and exact revenge on the man who raped them. Exploitative action-filled thriller. Also called ''Rape Squad.''
1974 (R) 90m/C Jo Ann Harris, Peter Brown, Jennifer Lee, Lisa Moore, Connie Strickland, Pat Estrin; *Dir:* Bob Kelljan. VHS, Beta $19.99 *HBO* ♫♫

Act of Vengeance Drama about Jock Yablonski, a United Mine Workers official who challenged the president, Tony Boyle. Made for cable, based on fact, showing the events that led up to the murder of Yablonski and his family. Intriguing story lacking cinematic drive.
1986 97m/C Charles Bronson, Ellen Burstyn, Wilford Brimley, Hoyt Axton, Robert Schenkkan, Ellen Barkin; *Dir:* John MacKenzie. VHS, Beta $19.99 *HBO* ♫♫½

Action This is a behind-the-scenes look at how special effects for ''The Terminator,'' ''Life Force'' and ''Missing in Action'' were done by the Hollywood masters of the art.
1985 60m/C Arnold Schwarzenegger. VHS, Beta $39.98 *LHV* ♫♫½

Action in Arabia A newsman uncovers a Nazi plot to turn the Arabs against the Allies while investigating a colleague's murder in Damascus. The desert teems with spies, double agents, and sheiks as suave Sanders goes about his investigative business. Quintessential wartime B-movie.
1944 75m/C George Sanders, Virginia Bruce, Lenore Aubert, Gene Lockhart, Robert Armstrong, H.B. Warner, Alan Napier, Michael Ansara; *Dir:* Leonide Moguy. VHS, Beta $19.98 *RKO* ♫♫

Action Jackson Power-hungry auto tycoon Nelson tries to frame rebellious black police sergeant Weathers for murder. Being a graduate of Harvard and a tough guy, the cop doesn't go for it. Nelson eats up the screen as the heavy with no redeeming qualities, while Weathers is tongue-in-cheek as the resourceful good guy who keeps running afoul of the law in spite of his best efforts. Lots of action, violence, and a few sexy women help cover the plot's lack of common sense.
1988 (R) 96m/C Carl Weathers, Vanity, Craig T. Nelson, Sharon Stone; *Dir:* Craig R. Baxley. VHS, Beta, LV $19.98 *LHV, WAR* ♫♫

Action in the North Atlantic Massey and Bogart are the captain and first mate of a Merchant Marine vessel running the lone supply route to the Soviet Union. Eventually they wind up locking horns with a Nazi U-boat. Plenty of action and strenuous flag waving in this propagandorama. Gordon fans won't want to miss her as Massey's wife. Also available colorized.
1943 126m/B Humphrey Bogart, Raymond Massey, Alan Hale Jr., Julie Bishop, Ruth Gordon, Sam Levene, Dane Clark; *Dir:* Lloyd Bacon. VHS $19.98 *MGM, FCT* ♫♫

Action for Slander A British army officer sporting the typical stiff upper lip is accused of cheating during a card game, and the slander mars his reputation until the case is taken to court. Dryly earnest and honest in its depiction of class differences in pre-war England. Based on a novel by Mary Borden.
1938 84m/B *GB* Clive Brook, Ann Todd, Margaretta Scott, Ronald Squire, Francis L. Sullivan, Felix Aylmer, Googie Withers; *Dir:* Tim Whelan. VHS, Beta, 8mm $29.95 *VYY* ♫♫½

Action U.S.A. A young woman witnesses the murder of her boyfriend by gangsters, who then pursue her to make sure she will never tell what she saw. Throughout Texas she rambles with the mob sniffing at her heels, grateful for the opportunity to participate in numerous stunts and car crashes.
1989 90m/C Barri Murphy, Greg Cummins, William Knight, William Smith, Cameron Mitchell; *Dir:* John Stewart. VHS, Beta $79.95 *IMP* ♫½

Actor: The Paul Muni Story Musical biography of Paul Muni, from his beginnings as a traveling actor in Hungary to his New York theatre and movie career.
1978 105m/C Herschel Bernardi, Georgia Brown, Harold Gould. VHS, Beta $39.95 *IVE* ♫♫

Actors and Sin Two-part film casting a critical eye toward actors and Hollywood. ''Actor's Blood'' is the melodramatic story of Shakespearean actor Robinson and his unhappy actress daughter. She commits suicide and he sets out to prove it was murder. Heavy going. Lighter and more entertaining is ''Woman of Sin,'' which relates a Hollywood satire involving a theatrical agent and his newest client, a precocious nine-year-old.
1952 82m/B Edward G. Robinson, Eddie Albert, Marsha Hunt, Alan Reed, Dan O'Herlihy, Tracey Roberts; *Dir:* Lee Games, Ben Hecht. VHS, LV $19.95 *SVS* ♫♫½

Adam Television docu-drama based on a tragic, true story. John and Reve Williams (Travanti and Williams) desperately search for their 6-year-old son abducted during an outing. During their long search and struggle, they lobby Congress for use of the FBI's crime computer. Eventually their efforts led to the creation of the Missing Children's Bureau. Sensitive, compelling performances by Travanti and Williams as the agonized, courageous parents.
1983 100m/C Daniel J. Travanti, JoBeth Williams, Martha Scott, Richard Masur, Paul Regina, Mason Adams; *Dir:* Michael Tuchner. VHS, Beta $59.95 *USA* ♫♫♫½

Adam at 6 A.M. Douglas is a young college professor who decides to spend a summer laboring in Missouri, where life, he thinks, is simpler. Of course, he learns that life in the boonies has its own set of problems, but unfortunately it takes him the entire movie before he catches the drift.
1970 (PG) 100m/C Michael Douglas, Lee Purcell, Joe Don Baker, Charles Aidman, Marge Redmond, Louise Latham, Grayson Hall, Dana Elcar, Meg Foster, Richard Derr, Anne Gwynne; **Dir:** Robert Scheerer. **VHS, Beta $59.98** FOX 𝄞𝄞

Adam Had Four Sons Satisfying character study involving the typical turn-of-the-century family nearly consumed by love, jealousy, and hatred. In the early part of the century, a goodly governess (Bergman in her second U.S. film) watches sympathetically over four sons of an American businessman after their mother dies. Economic necessity separates Bergman from the family for several years. Upon her return, she tangles with scheming bride-to-be Hayward, a bad girl intent on dividing and conquering the family before walking down the aisle with one of the sons. Based on a novel by Charles Bonner.
1941 81m/B Ingrid Bergman, Warner Baxter, Susan Hayward, Fay Wray, Richard Denning, June Lockhart, Robert Shaw; **Dir:** Gregory Ratoff. **VHS, Beta $59.95** COL 𝄞𝄞½

Adam's Rib Classic war between the sexes as Tracy and Hepburn portray married attorneys on opposite sides of the courtroom in the trial of blonde bombshell Holliday, charged with attempted murder of the lover of her philandering husband. The battle in the courtroom soon takes its toll at home. Sharp, snappy dialogue by Ruth Gordon and Garson Kanin with superb direction by Cukor. Perhaps the best of the nine movies pairing Tracy and Hepburn. Also available in colorized format.
1949 101m/B Spencer Tracy, Katharine Hepburn, Judy Holliday, Tom Ewell, Jean Hagen, Polly Moran, David Wayne; **Dir:** George Cukor. **VHS, Beta, LV $19.95** MGM, VYG 𝄞𝄞𝄞

The Addams Family Everybody's favorite family of ghouls hits the big screen, with something lost in the translation. An imposter claiming to be long-lost Uncle Fester (Lloyd), who says he was in the Bermuda Triangle for 25 years, shows up at the Addams' home to complete a dastardly deed—raid the family's immense fortune. Although Fester's plan is foiled, a series of plot twists highlight the ghoulish family's eccentricities. Darkly humorous but eventually disappointing. Nominated for an Oscar for Best Costume Design.
1991 (PG-13) 102m/C Anjelica Huston, Raul Julia, Christopher Lloyd, Christina Ricci, Jimmy Workman, Judith Malina, Carel Struycken, Elizabeth Wilson, Dan Hedaya; **Dir:** Barry Sonnenfeld. **VHS, Beta** PAR 𝄞𝄞

Adios Amigo Offbeat western comedy has ad-libbing Pryor hustling as a perennially inept con man. Script and direction (both provided by Williamson) are not up to Pryor's level, although excessive violence and vulgarity are avoided in a boring attempt to provide good clean family fare.
1975 (PG) 87m/C Fred Williamson, Richard Pryor, Thalmus Rasulala, James Brown, Robert Philips, Mike Henry; **W/Dir:** Fred Williamson. **VHS $9.95** SIM, VMK, BUR 𝄞𝄞

Adios, Hombre An innocent man who was imprisoned for murder escapes from prison and seeks revenge. You'll be saying adios as well.
1968 90m/C IT Craig Hill, Giulia Rubini; **Dir:** Mario Caiano. **VHS, Beta $49.95** UNI 𝄞

The Admiral was a Lady Four ex-GIs try to get by in life without going to work. Hendrix walks into their lives as a winning ex-Wave gifted with a knack for repartee who is disgusted by their collective lack of ambition. Nevertheless, she is pursued by the zany quartet with predictable results.
1950 87m/B Edmond O'Brien, Wanda Hendrix, Rudy Vallee, Steve Brodie; **Dir:** Albert Rogell. **VHS, Beta $14.98** NOS, FRH, VHE 𝄞½

Adoption The third of Meszaros's trilogy, involving a middle-aged Hungarian woman who longs for a child and instead forms a deep friendship with a 19-year-old orphan. In Hungarian with English subtitles. Hungarian title: "Orkobefogadas."
1975 89m/B HU GB Marta Meszaros. Berlin Film Festival '75: Best Film. **VHS, Beta $69.95** KIV, TAM 𝄞𝄞𝄞

The Adorable Cheat A low-budget, late-silent melodrama about a young woman who tries to get involved in the family business, despite her father's refusal of her help. Once in the business, she finds love.
1928 76m/B Lila Lee, Cornelius Keefe, Burr McIntosh; **Dir:** Burton King. **VHS, Beta, 8mm $16.95** GPV, VYY 𝄞𝄞

Adorable Julia When an actress takes on a lover many years younger than herself, the laughs begin to fly in this sex comedy based on Somerset Maugham's novel.
1962 97m/C FR Lilli Palmer, Charles Boyer; **Dir:** Alfred Weidenmann. **VHS $29.95** FCT, TIM 𝄞𝄞

The Adultress When a husband and wife cannot satisfy their desire to have a family, they hire a young man to help them in this dumb made-for-television melodrama.
1977 (R) 85m/C Tyne Daly, Eric Braeden, Gregory Morton; **Dir:** Norbert Meisel. **VHS, Beta $24.95** AHV 𝄞

Adventure of the Action Hunters A dying sailor leaves a tourist couple a message leading to a treasure, and they vie for it along with gangsters, mercenaries and other unsavory types.
1987 (PG) 81m/C Ronald Hunter, Sean Murphy, Joe Cimino; **Dir:** Lee Bonner. **VHS, Beta $79.98** LTG 𝄞

Adventure Island Enroute to Australia, a small ship stops at a remote island for supplies. The crew is greeted by a crazed, tyrannical leader who makes their lives difficult. Dull low-budget remake of 1937's "Ebb Tide."
1947 66m/C Rory Calhoun, Rhonda Fleming, Paul Kelly, John Abbott; **Dir:** Peter Stewart. **VHS, Beta $16.95** SNC 𝄞½

The Adventurer An escaped convict saves two wealthy women from death. They mistake him for a gallant sportsman and bring him home. Early Chaplin silent with music track.
1917 20m/B Charlie Chaplin; **Dir:** Charlie Chaplin. **VHS, Beta** CAB, FST 𝄞𝄞½

The Adventurers At the turn of the century, two Boers and an English officer set out to recover stolen jewels hidden in the jungles of South Africa. On the way, greed and anger take their toll a la "The Treasure of Sierra Madre," only on a less convincing scale. Alternate title: "Fortune in Diamonds."
1952 82m/B GB Dennis Price, Jack Hawkins, Siobhan McKenna, Peter Hammond, Bernard Lee; **Dir:** David MacDonald. **VHS, Beta $19.95** NOS, MON 𝄞𝄞

The Adventurers Sleazy Harold Robbins novel retains its trashy aura on film. Unfortunately, this turkey is also long and boring. Set in South America, it tells the tale of a rich playboy who uses and destroys everyone who crosses his path. His vileness results from having seen his mother murdered by outlaws, but his obsession is to avenge his father's murder. Blood, gore, revolutions and exploitive sex follow him everywhere. Watch and be amazed at the big-name stars who signed on for this one.
1970 (R) 171m/C Candice Bergen, Olivia de Havilland, Bekim Fehmiu, Charles Aznavour, Alan Badel, Ernest Borgnine, Leigh Taylor-Young, Fernando Rey, Thommy Berggren, John Ireland, Sydney Tafler, Rossano Brazzi, Anna Moffo, Christian Roberts, Yorgo Voyagis, Angela Scoular, Yolande Donlan, Ferdinand "Ferdy" Mayne, Jaclyn Smith, Peter Graves, Roberta Haynes; **Dir:** Lewis Gilbert. **VHS $29.95** PAR Woof!

Adventures in Babysitting A babysitter and her charges leave peaceful suburbia for downtown Chicago in order to rescue a friend in desperate trouble. After having a flat tire on the freeway, trouble takes on new meaning. The group survives encounters with gangsters and prostitutes and a trip through a blues bar and the office building where their parents are having a party. But will they ever make it home?
1987 (PG-13) 102m/C Elizabeth Shue, Keith Coogan, Maia Brewton, Anthony Rapp, Calvin Levels, Vincent D'Onofrio, Penelope Ann Miller, George Newbern, John Ford Noonan; **Dir:** Chris Columbus. **VHS, Beta, LV, 8mm $89.95** TOU 𝄞𝄞½

The Adventures of Baron Munchausen From the director of "Time Bandits," "Brazil," and "The Fisher King" comes an ambitious, imaginative, chaotic, and under-appreciated marvel based on the tall (and often confused) tales of the Baron. Munchausen encounters the King of the Moon (Williams in a cameo), Venus, and other odd and fascinating characters during what might be described as an circular narrative in which flashbacks dovetail into the present and place is never quite what it seems. Wonderful special effects and visually stunning sets occasionally dwarf the actors and prove what Gilliam can do with a big budget.
1989 (PG) 115m/C John Neville, Eric Idle, Sarah Polley, Robin Williams, Valentina Cortese, Oliver Reed, Uma Thurman, Sting, Jonathan Pryce, Bill Paterson; **Dir:** Terry Gilliam. **VHS, Beta, LV, 8mm $19.95** COL 𝄞𝄞𝄞½

Adventures Beyond Belief An irreverent motorcyclist is chased across Europe for a murder he didn't commit. Firmly within the boundaries of belief. Alternative title, "Neat and Tidy."
1987 95m/C Elke Sommer, Jill Whitlow, Graham Stark, Stella Stevens, Larry Storch, Thick Wilson, Skyler Cole, Edie Adams, John Astin; **Dir:** Marcus Thompson. **VHS, Beta $79.95** SVS 𝄞

The Adventures of Buckaroo Banzai A man of many talents, Buckaroo Banzai (Weller) travels through the eighth dimension in a jet-propelled Ford Fiesta to battle Planet 10 aliens led by the evil Lithgow. Buckaroo incorporates his vast knowledge of medicine, science, music, racing, and foreign relations to his advantage. Offbeat and often humorous cult sci-fi trip also known as "Buckaroo Banzai."
1984 (PG) 100m/C Peter Weller, Ellen Barkin, Jeff Goldblum, Christopher Lloyd, John Lithgow, Lewis Smith, Rosalind Cash, Robert Ito, Pepe Serna, Vincent Schiavelli, Dan Hedaya, Yakov

Smirnoff, Jamie Lee Curtis; **Dir:** W.D. Richter. **VHS, Beta, LV** $9.99 *VES, LIV* 🐾🐾🐾

The Adventures of Bullwhip Griffin

A rowdy, family-oriented comedy-adventure set during the California Gold Rush. Light Disney farce catches Russell at the tail end of his teenage star days. Pleshette and McDowall embark upon an ocean trip from Boston to San Francisco to find her brother, Russell, who's out west digging for gold. Assorted comedic adventures take place.
1966 110m/C Roddy McDowall, Suzanne Pleshette, Karl Malden, Harry Guardino; **Dir:** James Neilson. **VHS, Beta** $69.95 *DIS* 🐾🐾½

Adventures of Captain Fabian

When the captain of the "China Sea" learns that a beautiful woman has been falsely imprisoned, he comes to her rescue. Not one of Flynn's better swashbucklers, with typically low-quality Republic production.
1951 100m/B Errol Flynn, Vincent Price, Agnes Moorehead, Micheline Prelle; **Dir:** William Marshall. **VHS, Beta** $39.95 *NO* 🐾½

Adventures of Captain Marvel

A 12-episode cliff-hanging serial based on the comic book character. Details the adventures of klutzy Billy Batson, who transforms into superhero Captain Marvel by speaking the magic word, "Shazam!"
1941 240m/B Tom Tyler, Frank Coghlan Jr., Louise Currie; **Dir:** William Witney. **VHS, LV** $29.98 *VCN, REP, MLB* 🐾🐾½

Adventures in Dinosaur City

A group of modern-day pre-teens are transported back in time to the stone age. There they hook up with a group of partying, leather-clad dinosaurs and help them solve prehistoric crimes.
1992 (PG) 88m/C Omri Katz, Shawn Hoffman, Tiffanie Poston, Pete Koch, Megan Hughes; **Dir:** Brett Thompson. **VHS, LV** $89.98 *REP* 🐾🐾

Adventures of Don Juan

Flynn's last spectacular epic features elegant costuming and loads of action. Don Juan saves Queen Margaret from her evil first minister. He then swashbuckles his way across Spain and England in an effort to win her heart. Grand, large-scale fun and adventure with Flynn at his self-mocking best. Special laserdisc edition includes the original trailer.
1949 111m/C Errol Flynn, Viveca Lindfors, Robert Douglas, Romney Brent, Alan Hale Jr., Raymond Burr, Aubrey Mather, Ann Rutherford; **Dir:** Vincent Sherman. Academy Awards '49: Best Costume Design (Color). **VHS, Beta, LV** $59.95 *MGM, PIA, MLB* 🐾🐾🐾½

Adventures of Eliza Fraser

A young shipwrecked couple move from bawdy pastoralism to cannibalism after being captured by aborigines.
1976 114m/C Susannah York, Trevor Howard, Leon Lissek, Abigail, Noel Ferrier, Carole Skinner; **Dir:** Tim Burstall. **VHS, Beta, LV** $19.95 *STE, NWV* 🐾½

Adventures of Felix the Cat

In this collection of lighthearted TV episodes, Felix is joined by a whole array of amazing characters-the devious and evil Professor, the absent-minded Poindexter and the Strongman Rock Bottom.
1960 56m/C **VHS, Beta** $9.95 *MED* 🐾½

The Adventures of the Flying Cadets

An early aerial war adventure serial in thirteen complete chapters.
1944 169m/B Johnny Downs, Regis Toomey; **Dir:** Ray Taylor, Lewis D. Collins. **VHS, Beta** $49.95 *NOS, VCN, VYY* 🐾½

The Adventures of Ford Fairlane

The Diceman plays an unusual detective specializing in rock 'n' roll cases. When a heavy metal singer dies on stage, he takes the case in his own inimitable fashion, pursuing buxom gals, sleazy record executives and even his ex-wife. Not surprisingly, many of his stand-up bits are worked into the movie. Clay, the ever-so-controversial comic in his first (and likely last) starring role haplessly sneers his way through this rock 'n roll dud of a comedy thriller. A quick effort to cash in on Clay's fading star. Forget about it.
1990 (R) 101m/C Andrew Dice Clay, Wayne Newton, Priscilla Presley, Morris Day, Lauren Holly, Maddie Corman, Gilbert Gottfried, David Patrick Kelly, Brandon Call, Robert Englund, Ed O'Neill, Sheila E, Kari Wuhrer, Tone Loc; **Dir:** Renny Harlin. **VHS, Beta, LV** $19.98 *FXV* Woof!

The Adventures of Frontier Fremont

A rough and tumble story of a man who leaves the city, grows a beard, and makes the wilderness his home (and the animals his friends). Mountain life, that's the life for me. Almost indistinguishable from Haggerty's "Grizzly Adams," with the usual redeeming panoramic shots of majestic mountains.
1975 95m/C Dan Haggerty, Denver Pyle; **Dir:** Richard Friedenberg. **VHS, Beta** $41.00 *UHV, LME* 🐾½

The Adventures of Gallant Bess

The time-honored story of a rodeo man torn between his girl and his talented horse (the Bess of the title).
1948 73m/C Cameron Mitchell, Audrey Long, Al "Fuzzy" Knight, James Millican; **Dir:** Lew Landers. **VHS** *VEC, TIM* 🐾🐾

The Adventures of Grizzly Adams at Beaver Dam

Fearing that a dam will flood his valley, Grizzly tries desperately to convince a misplaced family of beavers to build their home elsewhere. An episode of the television series.
198? 60m/C Dan Haggerty. **VHS, Beta** $14.98 *UHV* 🐾🐾

Adventures of Hairbreadth Harry

Three short spoofs of the early melodramas cranked out by the movie industry. Includes "Sawdust Baby," "Fearless Harry," and "Rudolph's Revenge."
192? 65m/B Billy West. **VHS, Beta** *GPV* 🐾🐾

The Adventures of Huckleberry Finn

Mark Twain's classic story about a boy who runs away and travels down the Mississippi on a raft, accompanied by a runaway slave, is done over in MGM-style. Rooney is understated as Huck (quite a feat), while the production occasionally floats aimlessly down the Mississippi. An entertaining follow-up to "The Adventures of Tom Sawyer."
1939 89m/B Mickey Rooney, Lynne Carver, Rex Ingram, William Frawley, Walter Connolly; **Dir:** Richard Thorpe. **VHS, Beta** $24.95 *MGM, VID* 🐾🐾🐾

The Adventures of Huckleberry Finn

A lively adaptation of the Twain saga in which Huck and runaway slave Jim raft down the Mississippi in search of freedom and adventure. Miscasting of Hodges as Huck hampers the proceedings, but Randall shines as the treacherous King. Strong supporting cast includes Keaton as a lion-tamer and boxing champ Moore as Jim.
1960 107m/C Tony Randall, Eddie Hodges, Archie Moore, Patty McCormack, Neville Brand, Mickey Shaughnessy, Judy Canova, Andy Devine, Sherry Jackson, Buster Keaton, Finlay Currie, Josephine Hutchinson, Parley Baer, John

Carradine, Royal Dano, Sterling Holloway, Harry Dean Stanton; **Dir:** Michael Curtiz. **VHS, Beta** $19.98 *MGM, FCT* 🐾🐾½

The Adventures of Huckleberry Finn

The classic adventure by Mark Twain of an orphan boy and a runaway slave done again as a TV movie and starring F-Troop. Lacks the production values of earlier versions.
1978 100m/C Forrest Tucker, Larry Storch, Kurt Ida, Mike Mazurki, Brock Peters; **Dir:** Jack B. Hively. **VHS, Beta** $9.98 *VID, CVL* 🐾🐾

The Adventures of Huckleberry Finn

An animated version of the Mark Twain classic novel about the friendship between a young boy and a runaway slave.
1984 48m/C **Dir:** Peter Hunt. **VHS, Beta** $29.98 *SUE* 🐾🐾

The Adventures of Huckleberry Finn

An adaptation of the Mark Twain story about the adventures encountered by Huckleberry Finn and a runaway slave as they travel down the Mississippi River. Top-notch cast makes this an entertaining version. Originally made for television in a much longer version for PBS's "American Playhouse."
1985 121m/C Sada Thompson, Lillian Gish, Richard Kiley, Jim Dale, Barnard Hughes, Patrick Day, Frederic Forrest, Geraldine Page, Butterfly McQueen; **Dir:** Peter Hunt. **VHS, Beta** $19.95 *MCA* 🐾🐾½

The Adventures of Ichabod and Mr. Toad

Disney's wonderfully animated versions of Kenneth Grahame's "The Wind in the Willows" and "The Legend of Sleepy Hollow" by Washington Irving. Rathbone narrates the story of Mr. Toad, who suffers from arrogance and eventually must defend himself in court after being charged with driving a stolen vehicle (Disney did take liberties with the story). Crosby provides all the voices for "Ichabod," which features one of the all-time great animated sequences—Ichabod riding in a frenzy through the forest while being pursued by the headless horseman. A treat for all ages.
1949 68m/C **Dir:** Clyde Geronomi, James Algar; **Voices:** Jack Kinney, Bing Crosby; **Nar:** Basil Rathbone. **VHS** *DIS* 🐾🐾🐾½

The Adventures of the Little Rascals

Six classic comedies from the Little Rascals, complete and uncut.
19?? 100m/B **VHS** *REP* 🐾🐾🐾½

The Adventures of Mark Twain

A clay-animated fantasy based on, and radically departing from, the life and work of Mark Twain. Story begins with Twain flying into outer space in a blimp with stowaways Huck Finn, Tom Sawyer and Becky Thatcher and takes off from there. Above average entertainment for kids and their folks.
1985 86m/C **Dir:** Will Vinton; **Voices:** James Whitmore. **VHS, Beta, LV** *KRT* 🐾🐾½

The Adventures of the Masked Phantom

The exciting adventures of the Masked Phantom.
1938 56m/B Monte Rawlins, Betty Burgess, Larry Mason, Sonny Lamont, Jack Ingram; **Dir:** Charles Abbott. **VHS, Beta** $19.95 *NOS, VYY, RXM* 🐾½

The Adventures of Milo & Otis

A delightful Japanese children's film about a farm-dwelling dog and cat and their odyssey after the cat is accidentally swept away on a river. A record-breaking success in its home-

land. Well received by U. S. children. Narrated by Dudley Moore.
1989 (G) 89m/C *Dir:* Masanori Hata; *Nar:* Dudley Moore. **VHS, Beta, LV $19.95** *COL, RDG* ✓✓✓

The Adventures of Nellie Bly Made for television "Classics Illustrated" story of Nellie Bly, a strong-willed female reporter doing her best to expose wrongdoings in the late 19th century. A decent performance by Purl is overshadowed by the general lack of direction.
1981 100m/C Linda Purl, Gene Barry, John Randolph, Raymond Buktencia, J.D. Cannon, Elayne Heilveil, Cliff Osmond; *Dir:* Henning Schellerup. **VHS, Beta $59.98** *MAG* ✓

Adventures in Paradise Champion surfers Mike Ho, Chris Lassen, and Bobby Owens go on a worldwide search for the perfect wave. Available in Beta hi-fi and VHS Stereo.
1982 79m/C VHS, Beta $39.95 *MON* ✓

The Adventures of Picasso A Swedish satire on the life of Picasso, dubbed in English. Don't look for art or facts here, or, for that matter, many laughs.
1980 88m/C *SW* Gosta Ekman, Lena Nyman, Hans Alfredson, Margareta Krook, Bernard Cribbins, Wilfrid Brambell; *Dir:* Tage Danielsson. **VHS, Beta $59.95** *NSV* ✓✓

Adventures of a Plumber's Helper Seems this stud-muffin plumber's helper is no Maytag man when it comes to house calls: when he stumbles on some stolen moolah whilst romping in the boudoir, he finds his chosen profession to be even more lucrative than the trade school brochures promised.
19?? 72m/C VHS $29.95 *ACA* ✓

Adventures of Prince Achmed Three years in the making using the cutout/silhouette animation method. Based on the Arabian Nights stories.
1927 50m/C *Dir:* Lotte Reiniger. **VHS, Beta** *GVV* ✓

Adventures of a Private Eye A self-mocking British detective farce about an inept private eye who takes his time tracking down a beautiful girl's blackmailer, bedding down with all the women he meets along the way.
1987 96m/C *GB* Christopher Neil, Suzy Kendall, Irene Handl; *Dir:* Stanley Long. **VHS, Beta $29.95** *ACA* ✓

Adventures of Red Ryder The thrills of the rugged West are presented in this twelve-episode serial. Based on the then-famous comic strip character.
1940 240m/B Donald (Don "Red") Barry, Noah Beery; *Dir:* William Witney. **VHS, Beta $80.00** *VCN, MLB* ✓✓

The Adventures of Rex & Rinty An adventure serial in 12 chapters, featuring one of the last appearances by the original Rin-Tin-Tin.
1935 156m/B Kane Richmond, Harry Woods, Rin Tin Tin Jr; *Dir:* Ford Beebe, B. Reeves Eason. **VHS, Beta $49.95** *NOS, VCN, DVT* ✓✓

The Adventures of Rin Tin Tin Chronicles the adventures of that crime-fighting dog, Rin-Tin-Tin.
1947 65m/B Rin Tin Tin, Bobby Blake. **VHS $9.99** *CCB, TAV* ✓✓½

The Adventures of Robin Hood A great, rollicking, Technicolor version of the legendary outlaw, following the justice-minded Saxon knight as he battles the Normans, outwits the evil Prince John, and gallantly romances Maid Marian. With grand castle sets and lush forest photography, "Adventures" displays ample evidence of its huge (at the time) budget of $2 million plus and why it's regarded as the swashbuckler standard-bearer. Just entering his prime, Flynn enthusiastically performs most of his own stunts, including the intricate sword play and various forms of advanced tree and wall climbing. Flynn, fresh from fencing his way through "Captain Blood," brings to life a Robin brimming with charm and bravura, the enthusiastic protector of poor Saxons everywhere and the obvious undeclared king of Sherwood forest. The rest of the cast likewise attacks with zest, ranging from de Havilland as the cold but eventually sympathetic Maid Marian to Rains' fey performance as the dastardly Prince John (the direct predecessor to Alan Rickman's over-the-top spin on the Sheriff in Costner's remake) to Rathbone's conniving Sir Guy to Robin's band of very merry men (led by Hale's Little John and Pallette as Friar Tuck). Oscar nominated for best picture, it's based on various Robin Hood legends as well as Sir Walter Scott's "Ivanhoe" and the opera "Robin Hood" by De Koven-Smith. Also available in letter-boxed format.
1938 102m/C Errol Flynn, Olivia de Havilland, Basil Rathbone, Alan Hale Jr., Una O'Connor, Claude Rains, Patric Knowles, Eugene Pallette; *Dir:* Michael Curtiz. Academy Awards '38: International Prize, Best Film Editing, Best Original Score. **VHS, Beta, LV $19.98** *MGM, FOX, CRC* ✓✓✓✓

The Adventures of Rocky & Bullwinkle: Birth of Bullwinkle From a series of animated stories which chronicle the antics of Rocky the Squirrel and Bullwinkle the Moose. This episode is entitled "Birth of Bullwinkle." A timeless treasure that parents can enjoy as much as kids.
1991 38m/C Voices: June Foray, William Conrad, Bill Scott. **VHS, Beta, LV $12.99** *BVV, FCT* ✓✓✓✓

The Adventures of Rocky & Bullwinkle: Blue Moose Rocky and Bullwinkle are back again in another installment of the animated series. This episode is entitled "Blue Moose."
1991 41m/C Voices: June Foray, Bill Scott; *Nar:* William Conrad. **VHS, Beta, LV $12.99** *BVV, FCT* ✓✓✓✓

The Adventures of Rocky & Bullwinkle: Canadian Gothic This time, Rocky and Bullwinkle travel to Canada in a episode entitled "Canadian Gothic." This cartoon is one in a series released for home video consumption in February 1991, having been aired previously on television.
1991 39m/C Voices: June Foray, Bill Scott, William Conrad. **VHS, Beta, LV $12.99** *BVV, FCT* ✓✓✓✓

The Adventures of Rocky & Bullwinkle: La Grande Moose Another in the series of cartoons featuring the antics of Rocky and Bullwinkle. This installment is entitled "La Grande Moose."
1991 46m/C Voices: June Foray, Bill Scott; *Nar:* William Conrad. **VHS, Beta, LV $12.99** *BVV, FCT* ✓✓✓✓

The Adventures of Rocky & Bullwinkle: Mona Moose Another in a series of stories involving Rocky and Bullwinkle featuring incomparable cartoon images. This epic is entitled "Mona Moose."
1991 46m/C Voices: June Foray, Bill Scott; *Nar:* William Conrad. **VHS, Beta, LV $12.99** *BVV, FCT* ✓✓✓✓

The Adventures of Rocky & Bullwinkle: Norman Moosewell More cartoons from the less-than-bright moose and the spunky squirrel. Features "Wossamatta U."
1991 45m/C VHS, Beta $12.99 *BVV* ✓✓✓✓

The Adventures of Rocky & Bullwinkle: Vincent Van Moose Another in the series of Rocky and Bullwinkle cartoons. Art lovers beware! This episode is entitled "Vincent Van Moose."
1991 44m/C Voices: June Foray, Bill Scott; *Nar:* William Conrad. **VHS, Beta, LV $12.99** *BVV, FCT* ✓✓✓✓

The Adventures of Roger Ramjet Five episodes from the semi-satirical cartoon series about Ramjet and his heroic cohorts battling N.A.S.T.Y., The National Association of Spies, Traitors, and Yahoos.
1965 30m/C VHS, Beta $9.95 *RHI* ✓✓✓✓

The Adventures of Sherlock Holmes The immortal Sherlock Holmes and his assistant Dr. Watson conflict with Scotland Yard as they both race to stop arch-criminal Professor Moriarty. The Yard is put to shame as Holmes, a mere amateur sleuth, uses his brilliant deductive reasoning to save the damsel in distress and to stop Moriarty from stealing the Crown Jewels. Second in the series and the last of Rathbone's appearances as the brilliant detective.
1939 83m/B Basil Rathbone, Nigel Bruce, Ida Lupino, George Zucco, E.E. Clive, Mary Gordon; *Dir:* Alfred Werker. **VHS, Beta $19.95** *FOX* ✓✓½

The Adventures of Sherlock Holmes: Hound of the Baskervilles Holmes tackles the mysterious and long-lived curse of the Baskervilles. Made for British television. Fine acting, detailed production. A pleasure.
1989 120m/C *GB* Jeremy Brett, Edward Hardwick. **VHS $39.95** *MPI* ✓✓✓

The Adventures of Sherlock Holmes' Smarter Brother The unknown brother of the famous Sherlock Holmes takes on some of his brother's more disposable excess cases and makes some hilarious moves. Written, directed, and starring Wilder, with moments of engaging farce borrowing from the Mel Brooks school of parody (and parts of the Brooks ensemble as well).
1978 (PG) 91m/C Gene Wilder, Madeline Kahn, Marty Feldman, Dom DeLuise, Leo McKern, Roy Kinnear, John Le Mesurier, Douglas Wilmer, Thorley Walters; *W/Dir:* Gene Wilder. **VHS, Beta** *FOX* ✓✓✓

The Adventures of Sinbad The adventures of the Arabian mythic hero in an animated film version.
1979 47m/C VHS, Beta $29.98 *MGM, LIV* ✓✓✓

The Adventures of Sinbad the Sailor Sinbad receives a map to a treasure island where fabulous stores of jewels are hidden, and, in his search, falls in love with the King's daughter. Animated adventure.
1973 88m/C VHS, Beta $59.98 *LTG, LIV, FHE* ✓✓✓

Adventures of Smilin' Jack World War II flying ace Smilin' Jack Martin comes to life in this action-packed serial. Character from the Zack Moseley comic strip about air force fighting over China.
1943 90m/B Tom Brown, Sidney Toler; *Dir:* Ray Taylor. **VHS, Beta $26.95** *SNC, NOS, VAN ♂♂♂*

The Adventures of Superman (1942-1943) Seven cartoons from 1942 and 1943 are included in this package: "Underground World," "Terror on the Midway," "Volcano," "Destruction Inc.," "Secret Agent," "Billion Dollar Limited," and "Showdown." Some black and white.
1943 55m/C **VHS, Beta $9.95** *VYY ♂♂♂*

The Adventures of Tartu A British secret agent, sent to blow up a Nazi poison gas factory in Czechoslovakia, poses as a Romanian. Alternate title: "Tartu." One of Donat's lesser films, in the style of "The 39 Steps."
1943 103m/C *GB* Robert Donat, Valerie Hobson, Glynis Johns; *Dir:* Harold Bucquet. **VHS, Beta $19.95** *HHT, VCN, TIM ♂♂♂*

The Adventures of Tarzan The screen's first Tarzan in an exciting jungle thriller. Silent.
1921 153m/B Elmo Lincoln. **VHS, Beta $27.95** *VYY, DNB ♂♂*

Adventures of a Taxi Driver Cabbie finds sex, crime, sex, adventure and sex on the road in this off-duty comedy.
1976 89m/C *GB* Barry Evans, Judy Geeson, Adrienne Posta, Diana Dors, Liz Fraser; *Dir:* Stanley Long. **VHS, Beta $29.95** *ACA Woof!*

The Adventures of Tom Sawyer The vintage Hollywood adaptation of the Mark Twain classic, with music by Max Steiner and art direction by William Cameron Menzies. Not a major effort from the Selznick studio, but quite detailed and the best Tom so far.
1938 91m/C Tommy Kelly, Walter Brennan, Victor Jory, May Robson, Victor Kilian, Jackie Moran, Donald Meek, Anne Gillis, Marcia Mae Jones, Clara Blandick, Margaret Hamilton; *Dir:* Norman Taurog. **VHS, Beta, LV $14.98** *FOX, FCT, MLB ♂♂♂*

The Adventures of Tom Sawyer Tom Sawyer is a mischievous Missouri boy who gets into all kinds of trouble in this white-washed, made for TV adaptation of the Mark Twain classic.
1973 76m/C Jane Wyatt, Buddy Ebsen, Vic Morrow, John McGiver, Josh Albee, Jeff Tyler. **VHS, Beta $39.95** *MCA ♂♂½*

The Adventures of Ultraman The adventures of Ultraman, a futuristic hero from a distant planet who is able to get quite large when necessary.
1981 90m/C **VHS, Beta $24.95** *FHE ♂♂½*

The Adventures of Walt Disney's Alice Three early Disney shorts featuring Lewis Carroll's cast of characters: "Alice's Egg Plant," "Alice's Orphan" and "Alice the Toreador." Silent.
1925 35m/B **VHS, Beta $29.95** *JEF ♂♂½*

The Adventures of the Wilderness Family The story of a modern-day pioneer family who becomes bored with the troubles of city life and heads for life in the wilderness. There, they find trouble in paradise. Family-oriented adventure offering pleasant scenery. Followed by "The Wilderness Family, Part 2."

1976 (G) 100m/C Robert F. Logan, Susan Damante Shaw; *Dir:* Stewart Raffill. **VHS, Beta, LV $9.98** *MED ♂♂*

The Adventurous Knights An athlete learns he is the heir to a Transylvanian throne.
1935 60m/B David Sharpe, Mary Kornman, Mickey Daniels, Gertrude Messinger; *Dir:* Edward Roberts. **VHS, Beta, LV** *WGE Woof!*

Advise and Consent An interesting political melodrama with a fascinating cast, based upon Allen Drury's novel. The President chooses a candidate for the Secretary of State position which divides the Senate and causes the suicide of a senator. Controversial in its time, though somewhat turgid today. Laughton's last film.
1962 139m/B Don Murray, Charles Laughton, Henry Fonda, Walter Pidgeon, Lew Ayres, Burgess Meredith, Gene Tierney, Franchot Tone, Paul Ford, George Grizzard, Betty White, Peter Lawford, Edward Andrews; *Dir:* Otto Preminger. National Board of Review Awards '62: Best Supporting Actor (Meredith). **VHS, Beta $14.95** *QNE, KAR ♂♂♂*

Aelita: Queen of Mars Though the title has blockbuster potential, "Aelita" is a little-known silent Soviet sci-fi flick destined to remain little known. After building a rocket to fly to Mars, a Russian engineer finds its no Martian holiday on the fourth planet from the sun, with the Martians in the midst of a revolution. Silent, with a piano score.
1924 113m/B *RU Dir:* Yakov Protazanov. **VHS $29.95** *KIV, FCT, SNC ♂♂*

The Affair Songwriter/polio victim Wood falls in love for the first time with attorney Wagner. Delicate situation handled well by a fine cast in this television movie.
1973 74m/C Natalie Wood, Robert Wagner, Bruce Davison, Kent Smith, Frances Reid, Pat Harrington; *Dir:* Gilbert Cates. **VHS, Beta $39.98** *CGI ♂♂♂*

An Affair in Mind A British television movie about a professional writer who falls in love with a beautiful woman who tries to convince him to assist her in murderering her husband.
1989 88m/C *GB* Amanda Donohoe, Stephen Dillon, Matthew Marsh, Kean-Laurent Cochot; *Dir:* Michael Baker. **VHS, Beta $39.98** *FOX ♂♂*

The Affair of the Pink Pearl The husband-and-wife private investigation team of Tommy and Tuppence must find the culprit who stole a valuable pink pearl within 24 hours. Based on the Agatha Christie series of short stories. Made for British television.
1984 60m/C *GB* James Warwick, Francesca Annis. **VHS, Beta $14.95** *PAV ♂½*

An Affair to Remember A nightclub singer and wealthy bachelor fall in love on an ocean liner and suffer bad luck later on when they try to get together on shore. Lesser melodramatic remake of McCarey's "Love Affair."
1957 115m/C Cary Grant, Deborah Kerr, Richard Denning, Cathleen Nesbitt, Neva Patterson; *Dir:* Leo McCarey. **VHS, Beta, LV $19.98** *FOX, FCT ♂♂½*

Affair in Trinidad Fun in the tropics as nightclub singer Hayworth enlists the help of brother-in-law Ford to find her husband's murderer. The trail leads to international thieves and espionage in a romantic thriller that reunites the stars of "Gilda." Hayworth sings (with Jo Ann Greer's voice) "I've Been Kissed Before."

1952 98m/B Rita Hayworth, Glenn Ford, Alexander Scourby, Torin Thatcher, Valerie Bettis, Steve Geray; *Dir:* Vincent Sherman. **VHS, LV, SVS $19.95** *COL, PIA, CCB ♂♂♂*

The Affairs of Annabel The first of the popular series of Annabel pictures Lucy made in the late 1930s. Appealing adolescent is zoomed to movie stardom by her press agent's stunts. A behind-the-scenes satire on Hollywood, stars, and agents.
1938 68m/B Lucille Ball, Jack Oakie, Ruth Donnelly, Fritz Feld, Bradley Page; *Dir:* Ben Stoloff. **VHS, Beta $29.95** *CCB, RKO, IME ♂♂½*

Africa Screams Abbott and Costello go on an African safari in possession of a secret map. Unheralded independent A&C film is actually quite good in the stupid vein, with lots of jungle slapstick, generally good production values and a supporting cast of familiar comedy faces.
1949 79m/B Lou Costello, Bud Abbott, Shemp Howard, Hillary Brooke, Joe Besser, Clyde Beatty; *Dir:* Charles T. Barton. **VHS, Beta, LV $9.95** *KAR, MRV, NEG ♂♂½*

Africa Texas Style An East African rancher hires an American cowboy and his Navajo sidekick to help run his wild game ranch. Decent family adventure which served as the pilot for the short-lived TV series "Cowboy in Africa." Features lots of wildlife footage and a cameo appearance by Hayley Mills.
1967 109m/C Hugh O'Brian, John Mills, Nigel Green, Tom Nardini; *Dir:* Andrew Marton. **VHS $19.98** *REP ♂♂*

African Dream A period tale about a black man and a white woman fighting against repression in South Africa.
1990 (R) 94m/C Kitty Aldridge, John Kani, Dominic Jephcott, John Carson, Richard Haines, Joy Stewart Spence; *Dir:* John Smallcombe. **VHS, Beta, LV $29.95** *HMD, HBO, IME ♂½*

African Journey A moving, cross-cultural drama of friendship. A young black American goes to Africa for the summer to be with his divorced father who is working in the diamond mines. There he meets a young black African like himself; they overcome cultural clashes and learn respect for one another. Beautiful scenery, filmed in Africa. Part of the "Wonderworks" series.
1991 165m/C Jason Blicker, Pedzisai Sithole. **VHS $79.95** *PME ♂♂½*

The African Queen After Hepburn's missionary brother is killed by Germans in World War I Africa, a hard-drinking steamer captain (Bogart) offers the bible-thumping spinster safe passage. Not satisfied with sanctuary, she persuades the dissolute captain to try to destroy a German gunboat blocking the British advance. Only thing is, they have to navigate down uncharted rivers to reach the gunboat, anchored in a large central Africa lake. When not battling myriad aquatic obstacles, the two spend more time fighting each other than they do the Germans. Time alone on an African river in the middle of nowhere has a way of changing a man and a woman (as many philosophers have noted), and Bogie and Hepburn are no exceptions. Mistrust and aversion turn to love, a transition effectively counterpointed by the continuing suspense of their daring mission. A classic that earned Bogart his second Academy Award and brought in a host of other nominations, including Best Actress (Hepburn), Director, and Screenplay. Huston and James Agee adapted the screenplay, which makes wonderful use of natural

dialogue and humor, from C.S. Forester's novel. Shot on location in Africa.
1951 105m/C Humphrey Bogart, Katharine Hepburn, Robert Morley, Theodore Bikel, Peter Bull, Walter Gotell; *Dir:* John Huston. Academy Awards '51: Best Actor (Bogart); American Film Institute's Survey '77: 4th Best American Film Ever Made. **VHS, Beta, LV $19.98** *FOX, FCT, TLF* 𝄞𝄞𝄞𝄞

African Rage Little known release about an aging male nurse (yes, Quinn) who discovers he's dying of an incurable disease. With nothing left to lose, he plans the kidnapping of an African leader, hoping that the ransom will support his family. Decent performances help move along the improbable plot.
197? 105m/C Anthony Quinn, John Phillip Law, Simon Sabela; *Dir:* Peter Collinson. **VHS, Beta $59.95** *MPI* 𝄞𝄞

Afrita Hanen A rare Egyptian fantasy starring the famous Mid-Eastern actor, Farrid El Atrache. In Arabic with titles.
1947 97m/B Farrid El Atrache. **VHS, Beta $39.95** *FCT* 𝄞𝄞𝄞

After Darkness Slow-moving psycho-suspenser about a man obsessed with trying to remedy his twin brother's schizophrenia.
1985 104m/C *SI GB* John Hurt, Julian Sands, Victoria Abril, Pamela Salem; *Dir:* Dominique Othenin-Girard. **VHS, Beta $79.95** *CEL* 𝄞𝄞

After the Fall of New York Dimwitted post-apocalyptic tale set in New York after the fall of the ''Big Bomb.'' A man, driven to search for the last normal woman, has reason to believe she is frozen alive and kept in the heart of the city. His mission: locate her, thaw her, engage in extremely limited foreplay with her, and repopulate the planet. A poorly dubbed dating allegory.
1985 (R) 95m/C *IT FR* Michael Sopklw, Valentine Monnier, Anna Kanakis, Roman Geer, Edmund Purdom, George Eastman; *Dir:* Martin Dolman. **VHS, Beta $79.98** *VES, LIV* Woof!

After the Fall of Saigon Soldiers in Vietnam go on a killing rampage. Also known as ''Raiders of the Doomed Kingdom.'' English-dubbed and created in a country of unknown origin.
198? 100m/C Danny Aswatep, Peter Ramwa, Rayak Ramnate, Krao Spartacus. **VHS, Beta** *VCD* 𝄞

After the Fox Sellers is a con artist posing as a film director to carry out a bizarre plan to steal gold from Rome. Features occasional backhand slaps at Hollywood, with Mature turning in a memorable performance as the has-been actor starring in Sellers' movie. Though the screenplay was co-written by Neil Simon, the laughs are marginal. Italian title is ''Caccia alla Volpe.''
1966 103m/C *GB IT* Peter Sellers, Victor Mature, Martin Balsam, Britt Ekland; *Dir:* Vittorio DeSica. **VHS, Beta $59.98** *FOX* 𝄞𝄞

After Hours An absurd, edgy black comedy that's filled with novel twists and turns and often more disturbing than funny. An isolated uptown New York yuppie (Dunne) takes a late night stroll downtown and meets a sexy woman in an all-night coffee shop. From there he wanders through a series of threatening and surreal misadventures, leading to his pursuit by a vigilante mob stirred by ice cream dealer O'Hara. Something like ''Blue Velvet'' with more Catholicism and farce. Or similar to ''Something Wild'' without the high school reunion. Great cameos from the large supporting cast, including Cheech and Chong as burglars. A dark view of a small hell-hole in the Big Apple.

1985 (R) 97m/C Griffin Dunne, Rosanna Arquette, John Heard, Teri Garr, Catherine O'Hara, Verna Bloom, Linda Fiorentino, Cheech Marin, Thomas Chong, Dick Miller, Bronson Pinchot; *Dir:* Martin Scorsese. Cannes Film Festival '86: Best Director (Scorsese). **VHS, Beta, LV $19.98** *WAR, PIA* 𝄞𝄞𝄞

After Julius Twenty years after Julius Grace's death, his memory still hovers over his wife's and daughters' lives. Soap opera moves with glacier-like speed.
19?? 150m/C Faith Brook, John Carson. **VHS $59.95** *SVS* 𝄞𝄞

After Midnight An aspiring playwright tries to get his work produced, only to meet with dead ends. Eventually, he hooks up with a talented young actress with whom he finds romance and success. Also known as ''I Have Lived.''
1933 69m/B Alan Dinehart, Anita Page, Allen Vincent, Gertrude Astor; *Dir:* Richard Thorpe. **VHS, Beta $16.95** *SNC* 𝄞

After Midnight Suspended in a central story about an unorthodox professor who preys upon the deepest fears of his students, a trio of terror tales come to life. From the writers of ''The Fly II'' and ''Nightmare on Elm Street, Part IV.'' Some chills, few thrills.
1989 (R) 90m/C Marg Helgenberger, Marc McClure, Alan Rosenberg, Pamela Segall, Nadine Van Der Velde, Ramy Zada, Jillian McWhirter, Billy Ray Sharkey; *Dir:* Jim Wheat. **VHS, Beta $89.98** *FOX, MGM* 𝄞𝄞

After Pilkington British television thriller about an uptight Oxford professor who runs into his bewitching childhood sweetheart after many years. She persuades him to help search for a missing archaeologist.
1988 100m/C *GB* Bob Peck, Miranda Richardson; *Dir:* Christopher Morahan. **VHS, Beta $39.98** *FOX* 𝄞𝄞

After the Promise During the Depression, a poor carpenter tries to regain custody of his four sons following the death of his wife. Maudlin television melodrama based on a true story.
1987 100m/C Mark Harmon, Diana Scarwid, Rosemary Dunsmore, Donnelly Rhodes, Mark Hildreth, Trey Ames, Richard Billingsley; *Dir:* David Greene. **VHS $19.95** *STE, NWV* 𝄞𝄞

After the Rehearsal Two actresses, one young, the other at the end of her career, challenge their director with love and abuse. Each questions his right to use them on stage and off. A thoughtful discussion of the meaning and reason for art originally made for Swedish television. Swedish with English subtitles.
1984 (R) 72m/C *SW* Erland Josephson, Ingrid Thulin, Lena Olin; *Dir:* Ingmar Bergman. **VHS, Beta $59.95** *COL, TAM* 𝄞𝄞𝄞

After School Sex is prominent as a priest and one of his students find forbidden love while he prepares for a televised debate with an athiest. It could happen.
1988 (R) 90m/C Sam Bottoms, Edward Binns, Page Hannah, Renee Coleman, Dick Cavett; *Dir:* William Olsen. **VHS, Beta $79.95** *ACA* 𝄞

After the Shock Documentary-like dramatization of the San Francisco-Oakland earthquake of October, 1989 and, of course, its aftermath. Incorporates actual footage of the disaster. Written by director Sherman. Made for cable.
1990 (PG) 92m/C Yaphet Kotto, Rue McClanahan, Jack Scalia, Scott Valentine; *W/Dir:* Gary Sherman. **VHS, Beta $79.95** *PAR* 𝄞𝄞

After the Thin Man Second in a series of six ''Thin Man'' films, this one finds Nick, Nora and Asta, the lovable terrier, seeking out a murderer from Nora's own blue-blooded relatives. Fast-paced mystery with a witty script and the popular Powell/Loy charm. Sequel to ''The Thin Man,'' followed by ''Another Thin Man.''
1936 113m/B William Powell, Myrna Loy, James Stewart, Elissa Landi, Joseph Calleia, Jessie Ralph, Alan Marshal; *Dir:* W.S. Van Dyke. **VHS, Beta $19.98** *MGM* 𝄞𝄞𝄞

Afterburn When Ted, her Air Force pilot husband, is killed in a crash of his F-16 fighter, Janet Harduvel learns the official explanation is pilot error. Convinced that something was wrong with his plane, Janet sets out to investigate, and eventualy sue, military contractor General Dynamics. Dern is completely believable as the tough widow determined to clear her husband's name. Based on a true story; made for cable television.
1992 (R) 103m/C Laura Dern, Robert Loggia, Vincent Spano, Michael Rooker; *Dir:* Robert Markowitz. **VHS $89.99** *HBO* 𝄞𝄞𝄞

Aftermath Three astronauts return to Earth and are shocked to discover that the planet has been ravaged by a nuclear war. Quickly they make new plans.
1985 96m/C Steve Barkett, Larry Latham; *Dir:* Ted V. Mikels. **VHS, Beta $9.99** *STE, PSM* 𝄞½

Aftershock A beautiful alien and a mysterious stranger battle the Earth's repressive, evil government.
1988 (R) 90m/C Jay Roberts Jr., Elizabeth Kaitan, Chris Mitchum, Richard Lynch, John Saxon, Russ Tamblyn, Michael Berryman, Chris De Rose, Chuck Jeffreys; *Dir:* Frank Harris. **VHS, Beta $79.95** *PSM* 𝄞½

Afterward Made-for-British-television horror story about a family that moves into a supposedly haunted house with deadly results.
1985 60m/C *GB* Michael J. Shannon, Kay Harper, John Grillo. **VHS, Beta $59.95** *PSM* 𝄞

Against A Crooked Sky A young boy sets out with an elderly trapper to find his sister, who was captured by the Indians. Similiar story to ''The Searchers,'' but no masterpiece.
1975 (G) 89m/C Richard Boone, Stewart Peterson, Clint Richie, Geoffrey Land, Jewel Blanch; *Dir:* Earl Bellamy. **VHS, Beta $29.95** *VES, CTP* 𝄞𝄞

Against All Flags An enjoyable Flynn swashbuckler about a British soldier slashing his way through the Spanish fleet at the turn of the 18th century. Though the story has been told before, tight direction and good performances win out. O'Hara is a tarty eyeful as a hot-tempered pirate moll.
1952 81m/C Errol Flynn, Maureen O'Hara, Anthony Quinn, Mildred Natwick; *Dir:* George Sherman. **VHS, Beta, LV $14.95** *KRT, MCA* 𝄞𝄞𝄞

Against All Odds An interesting love triangle evolves when an ex-football player, played by Bridges, travels to Mexico to find his good friend's girlfriend. Contains complicated plot, numerous double crosses, sensual love scenes, and a chase scene along Sunset Boulevard. As the good friend sans conscience, Woods stars. A remake of 1947's ''Out of the Past.''
1984 (R) 122m/C Jeff Bridges, Rachel Ward, James Woods, Alex Karras, Jane Greer, Richard Widmark, Dorian Harewood, Swoosie Kurtz, Bill McKinney, Saul Rubinek; *Dir:* Taylor Hackford. **VHS, Beta, LV, 8mm $12.95** *COL* 𝄞𝄞½

Against the Drunken Cat Paws A blind young martial arts student seeks to regain her sight and avenge the death of her father.
1975 94m/C Chia Ling, Ou-Yang Ksiek. **VHS, Beta $39.95** UNI ✂

Against the Wind A motley crew is trained for a mission into Nazi Germany to blow up records and rescue a prisoner. The first half of the film focuses on the group's training, but despite its intensity they win only a pyrrhic victory. A well-done production with solid performances from the cast.
1948 96m/B Robert Beatty, Jack Warner, Simone Signoret, Gordon Jackson, Paul Dupuis, Peter Illing; **Dir:** Charles Crichton. **VHS $19.95** NOS, TIM ✂✂½

Agatha A speculative period drama about Agatha Christie's still unexplained disappearance in 1926, and a fictional American reporter's efforts to find her. Beautiful but lackluster mystery. Unfortunately, Hoffman and Redgrave generate few sparks.
1979 (PG) 98m/C Dustin Hoffman, Vanessa Redgrave, Timothy Dalton, Helen Morse, Tony Britton, Timothy West, Celia Gregory; **Dir:** Michael Apted. **VHS, Beta $64.95** WAR ✂✂½

L'Age D'Or Bunuel's first full-length film and his masterpiece of sex, repression and of course, death. A hapless man is prevented from reaching his beloved by middle-class morality, forces of the establishment, the Church, government and every bastion of modern values. Banned for years because of its anti-religious stance. Co-scripted by Salvador Dali. A mercilessly savage, surreal satire, the purest expression of Bunuel's wry misanthropy. In French with English subtitles.
1930 62m/B FR Gaston Modot, Pierre Prevert, Lya Lys, Marie Berthe Ernst, Paul Eluard; **Dir:** Luis Bunuel. New York Film Festival '64: Best Director; New York Film Festival '74: Best Director. **VHS, Beta, LV $59.95** FCT, IME ✂✂✂✂

Age Isn't Everything An appalling clumsy comedy about a recent graduate who abruptly becomes an old man while retaining his youthful exterior. He looks the same, but walks slowly and talks with a thick Yiddish accent, get it? The cast just marks time until an inexplicable ending.
1991 (R) 91m/C Jonathan Silverman, Rita Moreno, Paul Sorvino, Robert Prosky. **VHS $89.99** VES ✂

Age Old Friends Crusty octogenarian Cronyn must choose. His daughter (played by real-life offspring Tandy) wants him to move out of a retirement home and into her house. But he's struggling to keep neighbor and increasingly senile friend Gardenia from slipping into "zombieland." An emotional treat with two fine actors deploying dignity and wit in the battle against old age. Originally adapted from HBO from the Broadway play, "A Month of Sundays" by Bob Larbey.
1989 89m/C Vincent Gardenia, Hume Cronyn, Tandy Cronyn, Esther Rolle, Michelle Scarabelli; **Dir:** Allen Kroeker. **VHS, Beta $79.99** HBO ✂✂✂½

The Agency An advertising agency attempts to manipulate public behavior and opinion through the use of subliminal advertising. A good premise is bogged down by a dull script and plodding performances by all concerned. Also known as "Mind Games" and based on a Paul Gottlieb novel.
1981 (R) 94m/C CA Robert Mitchum, Lee Majors, Valerie Perrine, Saul Rubinek, Alexandra Stewart, George Kaczender; **Dir:** George Kaczender. **VHS, Beta $69.95** VES, MTX, SIM ✂✂

Agent on Ice Hockey team is stalked by lawyers. No, wait. An ex-CIA agent is stalked for cover-up purposes by the agency and the mob. One slippery fellow.
1986 (R) 96m/C Tom Ormeny, Clifford David, Louis Pastore, Matt Craven; **Dir:** Clark Worswick. **VHS, Beta $19.95** STE, NWV ✂✂

Agnes of God Stage to screen translation of John Pielmeier's play loses something in the translation. Coarse chain-smoking psychiatrist Fonda is sent to a convent to investigate whether young nun Tilly is fit to stand trial. Seems that the nun may have given birth to and then strangled her baby, although she denies ever having sexual relations and knows nothing about an infant. Naive Tilly is frightened by probing Fonda, while worldly mother-superior Bancroft is distrusting. Melodramatic stew of Catholicism, religious fervor, and science features generally good performances, although Fonda often seems to be acting (maybe it's the cigarettes).
1985 (PG-13) 99m/C Jane Fonda, Anne Bancroft, Meg Tilly, Anne Pitoniak, Winston Reckert, Gratien Gelinas; **Dir:** Norman Jewison. **VHS, Beta, LV $12.95** COL ✂✂

The Agony and the Ecstacy A big-budget ($12 million, still a lot of money in 1965) adaptation of the Irving Stone book, recounting the conflict between Michelangelo and Pope Julius II after His Holiness directs the artist to paint the Sistine Chapel. We follow the tortured artist through the unpredictable creative process and the hours (it seems literal due to movie length) of painting flat on his back. Heston exudes quiet strength in his sincere interpretation of the genius artist who was of course short and homosexual while Harrison has a fling as the Pope, although the slow script is not up to the generally fine performances. A disappointment at the box office, but certainly worth a look on the small screen for the sets alone.
1965 136m/C Charlton Heston, Rex Harrison, Harry Andrews, Diane Cilento, Alberto Lupo, Adolfo Celi; **Dir:** Carol Reed. **VHS, Beta, LV $19.98** FOX, IGP ✂✂½

Aguirre, the Wrath of God Herzog at his best, combining brilliant poetic images and an intense narrative dealing with power, irony, and death. Spanish conquistadors in 1590 search for the mythical city of gold in Peru. Instead, they descend into the hell of the jungle. Kinski is fabulous as Aguirre, succumbing to insanity while leading a continually diminishing crew in this compelling, extraordinary drama shot in the jungles of South America. Both English- and German-language versions available.
1972 94m/C GE Klaus Kinski, Ruy Guerra, Del Negro, Helena Rojo, Cecilia Rivera; **W/Dir:** Werner Herzog. **VHS $59.95** NYF, APD ✂✂✂✂

Ain't No Way Back Two hunters stumble upon a feudin' bunch of moonshiners and must leave their city ways behind if they plan to survive.
1989 90m/C VHS, Beta $69.95 RAE ✂

Air America It's the Vietnam War and the CIA is operating a secret drug smuggling operation in Southeast Asia to finance the effort. Flyboys Gibson and Downey drop opium and glib lines all over Laos. Big-budget Gibson vehicle with sufficient action but lacking much of a story, which was adapted from a book by Christopher Robbins.
1990 (R) 113m/C Mel Gibson, Robert Downey Jr., Marshall Bell, Nancy Travis, David Marshall Grant, Tim Thomerson, Lane Smith; **Dir:** Roger Spottiswoode. **VHS, Beta, LV $19.95** LIV, IME ✂✂

Air Force One of the finest of the WWII movies, Hawks' exciting classic has worn well through the years, in spite of the Japanese propaganda. It follows the hazardous exploits of a Boeing B-17 bomber crew who fight over Pearl Harbor, Manila, and the Coral Sea. Extremely realistic dogfight sequences and powerful, introspective real guy interfacing by the ensemble cast are masterfully combined by Hawks. Nominated for two Oscars in writing and photography.
1943 124m/B John Garfield, John Ridgely, Gig Young, Arthur Kennedy, Charles Drake, Harry Carey, George Tobias; **Dir:** Howard Hawks. Academy Awards '43: Best Film Editing. **VHS** FOX ✂✂✂½

Air Hawk Australian made for TV release details the adventures of an outback pilot involved with stolen diamonds.
1984 90m/C AU Eric Oldfield, Louise Howitt, Ellie MacLure, David Robson, David Baker; **Dir:** David Baker. **VHS, Beta $59.95** VCD ✂

Airplane! A classic lampoon of disaster flicks that's both dumb and hysterical. Remember, without this, there'd be no "Naked Gun." A former pilot who's lost both his girl (she's the attendant) and his nerve takes over the controls of a passenger plane when the crew is hit with food poisoning. The passengers become increasingly crazed and ground support more surreal as our hero struggles to land the plane. Clever, extremely fast, and often very funny parody mangles every Hollywood cliche within reach. The word and sight gags are so furiously paced that if one bombs, you'll scarcely notice. Cameos worth watching include Beaver's mom and Kareem Abdul-Jabbar. The film that launched Leslie Nielsen's second career as a comic actor (now it can be told). And it ain't over till it's over: don't head for the refrigerator or bathroom until the end of the amusing final credits. Followed in 1982 by "Airplane II: The Sequel," which was just as dumb but not as funny.
1980 (PG) 88m/C Robert Hays, Julie Hagerty, Lloyd Bridges, Peter Graves, Robert Stack, Kareem Abdul-Jabbar, Leslie Nielsen, Stephen Stucker, Ethel Merman, Barbara Billingsley; **Dir:** Jerry Zucker, Jim Abrahams. **VHS, Beta, LV, 8mm $14.95** PAR, PIA ✂✂✂½

Airplane 2: The Sequel Not a Zucker, Abrahams and Zucker effort, and sorely missing their slapstick and script finesse. The first passenger space shuttle has taken off for the moon and there's a mad bomber on board. Given the number of stars mugging, it's more of a loveboat in space than a fitting sequel to "Airplane." Nonetheless, some funny laughs and gags.
1982 (PG) 84m/C Robert Hays, Julie Hagerty, Lloyd Bridges, Raymond Burr, Peter Graves, William Shatner, Sonny Bono, Chuck Connors, Chad Everett, Stephen Stucker, Rip Torn, Ken Finkleman, Sandahl Bergman; **Dir:** Ken Finkleman. **VHS, Beta, LV $14.95** PAR, PIA ✂✂

Airport Old-fashioned disaster thriller built around an all-star cast, fairly moronic script, and an unavoidable accident during the flight of a passenger airliner. A box-office hit that paved the way for many lesser disaster flicks (including its many sequels) detailing the reactions of the passengers and crew as they cope with impending doom. Considered to be the best of the "Airport" series and adapted by director Seaton from the Arthur

Hailey novel. Oscar nominations for Best Picture, Adapted Screenplay, Cinematography, Art Direction, Set Decoration, Sound, Original Score, Film Editing and Costume Design, as well as the award garnered by Hayes.
1970 (G) 137m/C Dean Martin, Burt Lancaster, Jean Seberg, Jacqueline Bisset, George Kennedy, Helen Hayes, Van Heflin, Maureen Stapleton, Barry Nelson, Lloyd Nolan; *Dir:* George Seaton. Academy Awards '70: Best Supporting Actress (Hayes). **VHS, Beta, LV $14.95** MCA, BTV ✗✗✗

Airport '75 After a mid-air collision, a jumbo 747 is left pilotless. Airline attendant Black must fly da plane. She does her cross-eyed best in this absurd sequel to "Airport" built around a lesser "all-star cast." Safe on the ground, Heston tries to talk the airline hostess/pilot into landing, while the impatient Kennedy continues to grouse as leader of the foam-ready ground crew. A slick, insincere attempt to find box office magic again (which unfortunately worked, leading to two more sequels).
1975 (PG) 107m/C Charlton Heston, Karen Black, George Kennedy, Gloria Swanson, Helen Reddy, Sid Caesar, Efrem Zimbalist Jr., Susan Clark, Dana Andrews, Linda Blair, Myrna Loy; *Dir:* Jack Smight. **VHS, Beta $59.95** GKK ✗✗

Airport '77 A converted passenger jet owned by billionaire Stewart is filled with priceless art and sets off to Palm Beach for a museum opening. Stewart and friends are joined by an uninvited gang of hijackers. Lemmon takes over the driving from Heston. The plane crash-lands in the ocean and all aboard must fight for survival. Twist to this in-flight disaster is that the bad time in the air occurs underwater, a novel (and some might say, desperate) twist to the old panic in the plane we're all gonna die formula. With a cast of familiar faces, some of them stars and some of them just familiar faces, this is yet another sequel to "Airport" and another box-office success, leading to the last of the series in 1979.
1977 (PG) 114m/C Jack Lemmon, James Stewart, Lee Grant, Brenda Vaccaro, Joseph Cotten, Olivia de Havilland, Darren McGavin, Christopher Lee, George Kennedy, Kathleen Quinlan; *Dir:* Jack Smight. **VHS, Beta, LV $59.95** MCA ✗✗

Akermania, Vol. 1 This compilation of director Akerman's shorter works includes "I'm Hungry, I'm Cold," the story of two Belgian girls on the loose in Paris, and "Saute Ma Ville," Akerman's first film. Both are in French with English subtitles. Also included is "Hotel Monterey," a silent experimental film.
1992 89m/C FR Maria De Medeiros, Pascale Salkin, Chantal Akerman; *Dir:* Chantal Akerman. **VHS $59.95** WAC ✗✗✗

Akira Secret government experiments on children with ESP go awry, resulting in an atomic bomb drop on Tokyo. The city in turn builds itself up into a Megalopolis and the experiments continue. Combines animation with live-action.
1989 124m/C JP **VHS** STP ✗✗

Akira Production Report Otomo, artist and creator of "Akira," the popular Japanese graphic novel and animated feature, is the focus of this documentary by Streamline Pictures. Noted for his technical adroitness, he explains the production techniques that complement the well-storied plot of "Akira." It takes place some hundred years in the future, after the world has suffered and survived a nuclear war, and the governmen-

tal powers that be are attempting to re-engender a primal power known as Akira. Rather than utilize rotoscoping, where the artist traces over live-action figures, Otomo employs a more precise and technically advanced manner of creating movement and perspective. In English.
1989 52m/C JP Katsuhiro Otomo. **VHS $24.95** STP, WTA ✗✗

Al Capone Film noir character study of one of the most colorful gangsters of the Roaring Twenties. Sort of an underworld "How to Succeed in Business." Steiger chews scenes and bullets as they fly by, providing the performance of his career. Plenty of gangland violence and mayhem and splendid cinematography keep the fast-paced period piece sailing.
1959 104m/C Rod Steiger, Fay Spain, Murvyn Vye, Nehemiah Persoff, Martin Balsam, James Gregory, Joe De Santis; *Dir:* Richard Wilson. **VHS, Beta $19.98** FOX, FCT ✗✗✗

The Al Jolson Collection The famous star of early sound film is honored through seven volumes of his works. Films featured are "The Jazz Singer," "The Singing Fool," "Say it with Songs," "Mammy," "Big Boy," "Wonder Bar," "Go Into Your Dance" and "The Singing Kid."
1992 100m/B Al Jolson. **LV $149.98** MGM ✗✗✗

Alabama's Ghost A musician steals a dead master magician's secrets, incurring the wrath of the paranormal underworld.
1972 (PG) 96m/C Christopher Brooks, E. Kerrigan Prescott; *Dir:* Fredric Hobbs. **VHS, Beta $29.95** IVE ✗ 1/2

Aladdin A comedic Italian modernization of the Aladdin fable.
1986 (PG) 97m/C IT Bud Spencer, Luca Venantini, Janet Agren, Julian Voloshin, Umberto Raho; *Dir:* Bruno Corbucci. **VHS, Beta $9.95** MED ✗

Aladdin and His Wonderful Lamp The story of Aladdin, a young man who finds a magical oil lamp when he is trapped in a tiny cave. From the "Faerie Tale Theatre" series and director of "Beetlejuice" and "Batman" Tim Burton.
1984 60m/C Valerie Bertinelli, Robert Carradine, Leonard Nimoy, James Earl Jones; *Dir:* Tim Burton. **VHS, Beta $14.98** KUI, FOX, FCT ✗✗ 1/2

The Alamo Wayne's directorial effort is an old-fashioned patriotic battle epic recounting the real events of the 1836 fight for independence in Texas. The usual band of diverse and somewhat contentious personalities, including Wayne as a coonskin-capped Davy Crockett, defend a small fort against a very big Mexican raiding party outside of San Antonio. Before taking on the Mexicans and meeting mythic death, the men in the besieged fort fight with each other, learn the meaning of life, and ultimately come to respect each other. Just to make it more entertaining, Avalon sings. The big-budget production features an impeccable musical score by Dmitri Tiomkin and an impressive 7,000 extras for the Mexican army alone. Wayne reportedly received directorial assistance from John Ford, particularly during the big massacre finale.
1960 161m/C John Wayne, Richard Widmark, Laurence Harvey, Frankie Avalon, Richard Boone, Carlos Arruza, Chill Wills; *Dir:* John Wayne. Academy Awards '60: Best Sound. **VHS, Beta, LV $19.98** FOX, MGM, TLF ✗✗✗

Alamo Bay A slow-moving but sincere tale of contemporary racism. An angry Vietnam veteran and his red-neck buddies feel threatened by Vietnamese refugees who want to go into the fishing business. Set in Texas, filled with Texas-sized characters, and based on a true Texas story, as interpreted by the French Malle.
1985 99m/C Ed Harris, Ho Nguyen, Amy Madigan, Donald Moffatt, Cynthia Carle; *Dir:* Louis Malle. **VHS, Beta, LV $79.95** COL ✗✗ 1/2

The Alamo: Thirteen Days to Glory The legendary Davy Crockett (Keith), Colonel William Travis (Baldwin), and Jim Bowie (Arness) overcome personal differences to unite against the Mexican Army, vowing to hold down the fort or die. It takes a true Texan fully to appreciate the merits of this rather pedestrian retelling of a familiar story, but the battle scenes are pretty heady (although they lose some of their froth on the small screen). If nothing else, the ever-versatile Julia is worth seeing as Santa Anna in this made-for-TV rendering of J. Lon Tinkle's "Thirteen Days to Glory." You may not want to remember the Alamo this way.
1987 180m/C James Arness, Lorne Greene, Alec Baldwin, Brian Keith, Raul Julia; *Dir:* Peter Werner. **VHS, Beta $24.98** FRH ✗✗

Alamut Ambush A federal agent is stalked by assassins, and decides to hunt them in return. Sequel to "Cold War Killer."
1986 94m/C GB Terence Stamp, Michael Culver; *Dir:* Ken Grieve. **VHS, Beta $79.95** WES ✗

Alan & Naomi In 1944, Alan Silverman is a 14 year-old Brooklyn boy more concerned with his stickball team than the war raging across Europe. But when the Silverman's upstairs neighbor offers refuge to a French-Jewish mother and her young daughter, Alan is asked by his father to try to befriend the girl. Naomi lives in a world of silence since witnessing the death of her father by the Nazis and it's hoped Alan can break through her defensive wall. Together the two build a sweet, if unlikely, friendship. Based on the novel by Myron Levoy.
1992 (PG) 95m/C Lukas Haas, Vanessa Zaoui, Michael Gross, Amy Aquino, Zohra Lampert; *Dir:* Sterling Van Wagenen. **VHS, LV $89.95** COL ✗✗ 1/2

The Alaska Wilderness Adventure The tale of a family struggling for survival in the Arctic wilderness. Baby, it's cold out there.
1978 90m/C VHS, Beta PGN ✗✗

Albino Albino chief leads African terrorists into murdering white settlers, one of whom is an ex-policeman's fiancee. So the ex-cop pursues the bad guys, who are fairly easy to identify (just look for the big group hanging with the albino). Danning, Lee, and Howard only briefly show faces and collect checks.
1976 (R) 85m/C Christopher Lee, Trevor Howard, Sybil Danning, Horst Frank, James Faulkner; *Dir:* Jurgen Goslar. **VHS, Beta $54.95** MED Woof!

Alcatraz Breakout The story of John Grant, the first man to successfully escape from Alcatraz Prison.
1975 61m/C Marland T. Stewart. **VHS, Beta $59.95** WES Woof!

The Alchemist See humans transformed into murderous zombies! A bewitched man seeks revenge upon the evil magician who placed a curse on him, causing him to live like an animal. Painfully routine, with a few chills along the way. Amonte is an

alias for Charles Band. Filmed in 1981 and released four years later.
1981 (R) 86m/C Robert Ginty, Lucinda Dooling, John Sanderford, Viola Kate Stimpson, Bob Glaudini; *Dir:* James Amonte. **VHS, Beta $79.98** *VES, LIV* Woof!

Alex in Wonderland A semi-autobiographical and satirical look at Hollywood from the standpoint of a young director who's trying to follow up his recent hit (in real life, "Bob & Carol & Ted & Alice") with a picture of some integrity that will keep the mass audience away. The confused plot provides obvious parallels to Fellini (who appears in a cameo) and some sharp, often bitter insights into the Hollywood of the early '60s. Strong performances by Sutherland and Burstyn compensate somewhat for the patience-trying self-indulgent arty whining of the script, written by Muzursky and producer Larry Tucker.
1971 (R) 109m/C Donald Sutherland, Ellen Burstyn, Federico Fellini, Jeanne Moreau, Paul Mazursky; *Dir:* Paul Mazursky. **VHS, Beta** *MGM* ✍✍½

Alexa: A Prostitute's Own Story A prostitute falls for a playwright and decides she wants to change careers. Doesn't sound like much of a switch to us.
1988 (R) 81m/C Christine Moore, Kirk Baily, Ruth Collins; *Dir:* Sean Delgado. **VHS, Beta, LV $29.95** *ACA* ✍

Alexander the Great A lavish epic about the legendary Greek conqueror of the fourth century B.C., which provides Burton a rare chance at an adventure role. Here we find Alexander is the product of a dysfunctional royal family who hopes to create an idealized world modeled after Greek culture to make up for the love he lacks from daddy. This he does by conquering everything before dying at the age of 33. The great cast helps to overcome the sluggish pacing of the spectacle, while numerous battle scenes featuring loads of spears and arrows are staged effectively.
1955 135m/C Richard Burton, Fredric March, Claire Bloom, Harry Andrews, Peter Cushing, Danielle Darrieux; *Dir:* Robert Rossen. **VHS, Beta $59.95** *MGM* ✍✍½

Alexander Nevsky A story of the invasion of Russia in 1241 by the Teutonic Knights of Germany and the defense of the region by good old Prince Nevsky. Eisenstein's first completed project in nearly ten years, it was widely regarded as an artistic disappointment upon release and as pro-war propaganda for the looming conflict with the Nazis. Fabulous Prokofiev score illuminates the classic battle scenes, which used thousands of Russian army regulars. Dubbed in English.
1938 107m/B *RU* Nikolai Cherkassov, N.P. Okholopkov, Al Abrikossov; *Dir:* Sergei Eisenstein. National Board of Review Awards '39: 5 Best Foreign Films of the Year. **VHS, Beta, LV, 8mm, 3/4U $29.95** *MVD, MRV, NOS* ✍✍✍½

Alexander: The Other Side of Dawn Tired made-for-television sequel to the television movie "Dawn: Portrait of a Teenage Runaway." A young man turns to prostitution to support himself on the street of Los Angeles.
1977 100m/C Leigh McCloskey, Eve Plumb, Earl Holliman, Juliet Mills, Jean Hagen, Loni Chapman; *Dir:* John Erman. **VHS** *MOV* ✍✍

Alfie What's it all about, Alfie? Caine, in his first starring role, carries along the fairly routine story and made this a box office hit. As the British playboy out of control in mod London, Caine/Alfie is a despicable, unscrupulous and vile sort of guy who uses woman after woman to fulfill his basic needs and then casts them aside until...tragedy strikes. Though it was seen as a sophisticated take on current sexual mores upon release, it now seems a dated but engaging comedy, notable chiefly for its performances. From the Bill Naughton play and reflective of its stage roots. Oscar nominations for Best Picture, best actor Caine, supporting actress Merchant, screenplay, and Best Song, a top ten hit written by Hal David and Burt Bacharach and sung by Miss Dionne Warwick.
1966 (PG) 113m/C *GB* Michael Caine, Shelley Winters, Millicent Martin, Vivien Merchant, Julia Asher; *Dir:* Lewis Gilbert. National Board of Review Awards '66: 10 Best Films of the Year, Best Supporting Actress (Merchant). **VHS, Beta $59.95** *PAR* ✍✍✍

Alfred Hitchcock Presents Four episodes from the renowned director's classic television series. Episodes are: "Lamb to the Slaughter," "The Case of Mr. Pelham," "Banquo's Chair" and "Back for Christmas."
1966 106m/B *Dir:* Alfred Hitchcock. **LV $29.98** *MCA* ✍✍✍

Alfredo, Alfredo Hoffman plays a mild-mannered bank clerk who regrets marrying a sexy woman. Lightweight domestic comedy. In Italian with English subtitles (Hoffman's voice was dubbed).
1973 (R) 97m/C *IT* Dustin Hoffman, Stefania Sandrelli; *Dir:* Pietro Germi. **VHS** *PAR* ✍✍

Algiers Nearly a scene-for-scene Americanized remake of the 1937 French "Pepe Le Moko" about a beautiful rich girl (Lamarr) who meets and falls in love with a notorious thief (Boyer, then a leading sex symbol). Pursued by French police and hiding in the underworld-controlled Casbah, Boyer meets up with Lamarr in a tragically fated romance done in the best tradition of Hollywood. Boyer provides a measured performance as Le Moko, while Lamarr is appropriately sultry in her American film debut (which made her a star). Received Oscar nominations for Best Actor (Boyer), Best Supporting Actor (Lockhart), Cinematography, and Interior Decoration. Later remade as the semi-musical "Casbah."
1938 96m/B Charles Boyer, Hedy Lamarr, Sigrid Gurie, Gene Lockhart, Joseph Calleia, Alan Hale Jr.; *Dir:* John Cromwell. **VHS, Beta $9.95** *NEG, MRV, NOS* ✍✍✍

Ali: Fear Eats the Soul A widow cleaning woman in her sixties has a love affair with a Moroccan man 30 years her junior. To no one's surprise, both encounter racism and moral hypocrisy in West Germany. Serious melodrama from Fassbinder, who wrote it and appears as the squirmy son-in-law. Also known as "Fear Eats the Soul." In German with English subtitles.
1974 68m/C *GE* Brigitte Mira, El Hedi Ben Salem, Irm Hermann; *Dir:* Rainer Werner Fassbinder. Cannes Film Festival '74: International Critics Award. **VHS $79.95** *NYF, FCT* ✍✍✍

Alias John Law The Good Guy fights for oil rights against the bad guys. Confused program western with an especially convoluted plot.
1935 54m/B Bob Steele; *Dir:* Robert N. Bradbury. **VHS, Beta** *WGE* ✍

Alias John Preston Yet another one of those pseudo-psychological to sleep perchance to dream movies. Lee plays a man haunted by dreams in which he's a murderer, and soon starts to question whether his dreams might not imitate life. It's been done before, it's been done since, and it's been done better.
1956 66m/B *GB* Betta St. John, Alexander Knox, Christopher Lee, Sandra Dorne, Patrick Holt, Betty Ann Davies, John Longden, Bill Fraser, John Stuart; *Dir:* David MacDonald. **VHS $19.98** *SNC* ✍½

Alice A twist on the "Alice in Wonderland" tale. Alice witnesses an attempted murder, faints, and awakens in a weird, yet strangely familiar environment. Adapted from the stage production.
1986 80m/C Sophie Barjae, Susannah York, Jean-Pierre Cassel, Paul Nicholas. **VHS, Beta $59.95** *LHV, WAR* ✍✍

Alice An acclaimed surreal version of Lewis Carroll's already surreal "Alice in Wonderland," with the emphasis on Carroll's obsessiveness. Utilizing animated puppets and a live actor for Alice, Czech director Svankmajer injects grotesque images and black comedy into Wonderland. Not for the kids.
1988 84m/C *GB* Kristina Kohoutova; *Dir:* Jan Svankmajer. **VHS, 3/4U $59.95** *ICA* ✍✍✍½

Alice Farrow is "Alice," a woman plagued with doubts about her lifestyle, her religion, and her happiness. Her perfect children, husband, and apartment don't prevent her backaches, and she turns to an Oriental "herbalist" for aid. She finds his methods unusual and the results of the treatments surprising. Lightweight fairytale of Yuppiedom gone awry. Fine performances, beautiful to look at, and lovely music, but superficial and pointed story that may leave the viewer looking for more. (Perhaps that's Allen's point.) Farrow was nominated for Golden Globe Best Actress in a Musical/Comedy. Oscar nominated for Best Screenplay.
1990 (PG-13) 106m/C Mia Farrow, William Hurt, Joe Mantegna, Keye Luke, Alec Baldwin, Cybill Shepherd, Blythe Danner, Gwen Verdon, Bernadette Peters, Judy Davis; *W/Dir:* Woody Allen. National Board of Review Awards '90: Best Actress (Farrow). **VHS, Beta, LV $94.98** *ORI, FCT* ✍✍✍

Alice Adams Based on the classic Booth Tarkington novel about a poor girl from a small Midwestern town who falls in love with a man from the upper level of society. She tries desperately to fit in and nearly alienates her family and friends. The sets may be dated, but the insight on human behavior is timeless. Received Oscar Nominations for Best Picture and Best Actress (Hepburn).
1935 99m/B Katharine Hepburn, Fred MacMurray, Evelyn Venable, Hattie McDaniel, Fred Stone; *Dir:* George Stevens. National Board of Review Awards '35: 10 Best Films of the Year. **VHS, Beta, LV $14.98** *CCB, MED, NOS* ✍✍½

Alice in the Cities American and German culture are compared and contrasted in this early Wenders road work about a German journalist in the USA on assignment who suddenly finds himself custodian of a worldly nine-year-old girl abandoned by her mother. Together they return to Germany and search for the girl's grandmother. Along the way they learn about each other, share many distinctive and graceful Wenders moments.

1974 110m/B Ruediger Vogler, Yella Rottlaender, Elisabeth Kreuzer, Edda Kochi; *Dir:* Wim Wenders. **VHS, Beta $29.95** *PAV, FCT, APD* ♫♫♫½

Alice Doesn't Live Here Anymore
Scorsese marries road opera with pseudo-feminist semi-realistic melodrama and produces uneven but interesting results. A woman's husband dies suddenly, leaving her to care for their eleven-year-old son. On her way to a new life in California, she runs out of money and is stranded in Phoenix. Down to her last few dollars, she takes a job as a waitress in a diner where she meets kindly rancher Kristofferson. In addition to its female point of view, it's notable for being the basis of the once popular television show "Alice," which took the basic structure ("woman works in diner, has no husband and struggles for self-respect, while being victimized by annoying youngster threatening to sprout into pubescence") into sitcom land of no return. Burstyn and Ladd lend key performances, while Kristofferson is typically kinda wooden.
1974 105m/C Ellen Burstyn, Kris Kristofferson, Diane Ladd, Jodie Foster, Harvey Keitel, Vic Tayback, Billy "Green" Bush; *Dir:* Martin Scorsese. Academy Awards '74: Best Actress (Burstyn); British Academy Awards '75: Best Actress (Burstyn), Best Film, Best Supporting Actress (Ladd). **VHS, Beta, LV $19.98** *WAR, PIA, FCT* ♫♫½

Alice Goodbody
A sex spoof of a lonely girl's misadventures in Hollywood.
1974 (R) 83m/C Sharon Kelly, Daniel Kauffman, Keith McConnell, Arem Fisher, Norman Field, C.D. LaFleure; *Dir:* John Sturges. **VHS, Beta $19.95** *MED* Woof!

Alice to Nowhere
Concerns a running argument in the Outback over a fortune in gems unknowingly carried by a young woman. Made for Australian television and based on a novel by Evan Green.
1986 210m/C *AU* Rosie Jones, Steve Jacobs, John Waters, Ruth Cracknell; *Dir:* John Power. **VHS, Beta $59.95** *PAR* ♫♫

Alice Sweet Alice
A mediocre and gory who-killed-her, best remembered as the debut of Brooke Shields (in a small role). Also known as "Holy Terror" and "Communion."
1976 112m/C Brooke Shields, Linda Miller, Paula Sheppard; *Dir:* Alfred Sole. **VHS, Beta $29.95** *REP, MRV, BTV* ♫½

Alice Through the Looking Glass
Based on Lewis Carroll's classic adventure. Follows the further adventures of young Alice. After a chess piece comes to life, it convinces Alice that excitement and adventure lie through the looking glass.
1966 72m/C Judi Rolin, Ricardo Montalban, Nanette Fabray, Robert Coote, Agnes Moorehead, Jack Palance, Jimmy Durante, Tom Smothers, Roy Castle, Richard Denning; *Dir:* Alan Handley. **VHS, Beta $14.98** *SUE, NEG, KAR* ♫½

Alice in Wonderland
Another version of the Lewis Carroll classic which combines the usage of Lou Bunin's puppets and live action to tell the story. Released independently to cash in on the success of the Disney version. Takes a more adult approach to the story and is worth viewing on its own merits.
1950 83m/C *FR* Carol Marsh, Stephen Murray, Pamela Brown, Felix Aylmer, Ernest Milton; *Dir:* Dallas Bower. **VHS, Beta $39.95** *MON* ♫♫½

Alice in Wonderland
Classic Disney dream version of Lewis Carroll's famous children's story about a girl who falls down a rabbit hole into a magical world populated by strange creatures. Beautifully animated with some startling images, but served with a strange dispassion warmed by a fine batch of songs, including "I'm Late," "A Very Merry Un-Birthday" and the title song. Wynn's Mad Hatter and Holloway's Cheshire Cat are among the treats in store.
1951 (G) 75m/C *Dir:* Clyde Geronimi; *Voices:* Kathryn Beaumont, Ed Wynn, Sterling Holloway, Jerry Colonna, Hamilton Luske, Wilfred Jackson. **VHS, Beta, LV $24.99** *DIS, FCT, KUI* ♫♫♫

Alice in Wonderland
Adult version of the classic tale starring an assortment of Playboy alumnae in various stages of undress.
1977 (R) 76m/C Kristine DeBell; *Dir:* Bud Townsend. **VHS, Beta $54.95** *MED* ♫♫♫

Alice in Wonderland
Musical version of the Lewis Carroll's classic tale performed by the Children's Theatre Company and School.
1982 81m/C Annie Enneking, Solvieg Olsen, Wendy Lehr, Jason McLean, Gary Briggle, Elizabeth Fink. **VHS, Beta $39.95** *MCA* ♫♫½

Alice's Adventures in Wonderland
Musical adaptation of Lewis Carroll's timeless fantasy of a young girl who meets new friends when she falls into a rabbit hole.
1972 96m/C *GB* Fiona Fullerton, Michael Crawford, Ralph Richardson, Flora Robson, Peter Sellers, Robert Helpmann, Dudley Moore, Michael Jayston, Spike Milligan, Michael Hordern; *W/Dir:* William Sterling. **VHS, Beta** *CVL, VES, IPI* ♫♫½

Alice's Restaurant
Based on the popular and funny 20-minute Arlo Guthrie song "Alice's Restaurant Massacre" about a Flower Child during the Last Big War who gets hassled for littering, man. Step back in time and study the issues of the hippie era, including avoiding the draft, dropping out of college, and dealing with the local pigs. Sort of a modern movie in the cinematic ambling genre, in that nothing really happens. Director Penn won an Oscar nomination.
1969 (PG) 111m/C Arlo Guthrie, James Broderick, Pat Quinn, Geoff Outlaw; *Dir:* Arthur Penn. **VHS, Beta $19.95** *MVD, MGM* ♫♫½

Alien
Terse direction, stunning sets and special effects, and a well-seasoned cast save this from being another "Slimy monster from Outerspace" story and instead create a grisly rollercoaster of suspense and fear (and a huge box office hit). An intergalactic freighter's crew is invaded by an unstoppable carnivorous alien life-form intent on picking off the crew one by one. While the cast has nothing much to do other than bitch and banter while awaiting the horror of their imminent departure, Weaver is exceptional as Ripley, a self-reliant survivor who goes toe to toe with the Big Ugly. The futuristic, in the belly of the beast visual design creates a vivid sense of claustrophobic doom enhanced further by Jerry Goldsmith's ominous score. Oscar-winning special effects include the classic baby alien busting out of the crew guy's chest routine, a rib-splitting ten on the gore meter. Successfully followed by "Aliens" and "Alien 3."
1979 (R) 116m/C Tom Skerritt, Sigourney Weaver, Veronica Cartwright, Yaphet Kotto, Harry Dean Stanton, Ian Holm, John Hurt; *Dir:* Ridley Scott. Academy Awards '79: Best Visual Effects; British Academy Awards '79: Best Art Direction/Set Decoration, Best Sound. **VHS, Beta, LV $19.98** *FOX* ♫♫♫½

Alien Contamination
The tale of two astronauts who return to Earth from an expedition on Mars carrying some deadly bacterial eggs. Controlled by a Martian intent on conquering the world, the eggs squirt a gloppy juice that makes people explode on contact (a special effect). A cheap and sloppy attempt to cash in on the success of "Alien" written and directed by Cozzi. Dubbed.
1981 (R) 90m/C *IT* Ian McCulloch, Louise Monroe, Martin Mase, Samuel Rauch, Lisa Hahn; *Dir:* Lewis (Luigi Cozzi) Coates. **VHS, Beta** *PGN* Woof!

Alien Dead
The teenage victims of a bizarre meteor crash reincarnate as flesh-eating ghouls anxious for a new supply of human food in this extremely low-budget sleep inducer.
1985 (R) 87m/C Buster Crabbe, Linda Lewis, Ray Roberts; *Dir:* Fred Olen Ray. **VHS, Beta $19.95** *ACA, GHV, AVD* ♫

The Alien Factor
Another low-budget crazed critter from outerspace dispatch, this one featuring multiple aliens, one of whom is good, who have the misfortune of crash landing near Baltimore. The grotesque extra-terrestrials jolt a small town out of its sleepy state by wreaking havoc (except for the good one, of course). Decent special effects.
1978 (PG) 82m/C Don Leifert, Tom Griffith, Mary Mertens, Dick Dyszel; *Dir:* Donald M. Dohler. **VHS, Beta $59.95** *UHV* ♫½

Alien from L.A.
Awesomely inept comedy about a California girl who unwittingly stumbles onto the famed continent of Atlantis and can't find a yogurt stand. Weakly plotted and acted and filmed. Like, really.
1987 (PG) 88m/C Kathy Ireland, Thom Mathews, Don Michael Paul, Linda Kerridge, William R. Moses; *Dir:* Albert Pyun. **VHS, Beta, LV $79.95** *MED* Woof!

Alien Massacre
One of the worst films of all time—five short horror stories about zombies and vampires. Inexplicably retitled here, it is more commonly known as "Dr. Terror's Gallery of Horrors" or "Return from the Past" or "The Blood Suckers."
1967 77m/C Lon Chaney Jr., John Carradine, Rochelle Hudson, Roger Gentry, Mitch Evans; *Dir:* David L. Hewitt. **VHS, Beta $49.95** *REG, DCH* Woof!

Alien Nation
A few hundred thousand alien workers land accidentally on Earth and slowly become part of its society, although widely discriminated against. One of the "newcomers" teams with a surly and bigoted human cop to solve a racially motivated murder. An inconsistent and occasionally transparent script looks at race conflicts and includes some humorous parallels with contemporary American life. Basis for the television series. Producer Hurd was also the force behind "The Terminator" and "Aliens."
1988 (R) 89m/C James Caan, Mandy Patinkin, Terence Stamp, Kevyn Major Howard, Peter Jason, Jeff Kober; *Dir:* Graham Baker. **VHS, Beta, LV $19.98** *FOX* ♫♫½

Alien Predators
Three friends encounter a malevolent alien in this dull reworking of the plot of "The Andromeda Strain" with laughable special effects tossed in for those outwitted by the script.
1980 (R) 92m/C Dennis Christopher, Martin Hewitt, Lynn-Holly Johnson, Luis Prendes; *Dir:* Deran Sarafian. **VHS, Beta $79.95** *VTR* Woof!

Alien Prey
Two lesbians are making love when they are unexpectedly devoured by a hungry and indiscreet alien. No safe sex there. Graphic sex, violence, and cannibalism

abound. Interesting twist to the old eat 'em and leave 'em genre.
1983 (R) 85m/C Barry Stokes, Sally Faulkner; **Dir:** Norman J. Warren. **VHS, Beta $39.98** CGH ♫♫½

Alien Private Eye An extraterrestrial detective searches Los Angeles for a missing magic disk while investigating an intergalactic crime ring. Also known as "Alien P.I."
1987 90m/C Nikki Fastinetti; **Dir:** Nik Rubenfeld. **VHS $69.95** RAE ♫♫

Alien Seed Aliens kidnap a woman and impregnate her. Estrada is the government scientist hot on her trail. (How far could a woman carrying alien offspring wander?)
1989 88m/C Erik Estrada, Heidi Paine, Steven Blade; **Dir:** Bob James. **VHS $79.95** AIP ♫

Alien Space Avenger A spaceship piloted by four convicts crash lands in New York City. Stalked by an intergalatic bounty hunter whose job is to kill them, the aliens attack and hide inside human bodies to avoid discovery. Its a race against time as the preservation of the human race depends on the avenger's ability to seek and destroy the alien invaders. Bland ripoff of "Aliens."
1991 ?m/C Robert Prichard, Angela Nicholas, Gina Mastrogiacomo, Mike McClerie, Charity Staley; **Dir:** Richard W. Haines. **VHS $89.95** AIP ♫½

Alien Warrior An extraterrestrial fights a street pimp to save a crime-ridden Earth neighborhood. Low-brain rip-off of Superman.
1985 (R) 92m/C Brett Clark, Pamela Saunders; **Dir:** Edward Hunt. **VHS, Beta $79.98** VES, LIV Woof!

Alien Women A soft-core British science fiction yarn about scantily clad alien babes and the special agent who's trying to uncover their secret. It's all very sketchy, silly and campy, though pleasantly cast with plenty of lovely British actresses. Originally titled "Zeta One" and also called "The Love Factor."
1969 (R) 86m/C GB James Robertson Justice, Charles Hawtrey, Robin Hawdon, Anna Gael, Brigitte Skay, Dawn Addams, Valerie Leon, Yutte Stensgaard, Wendy Lingham, Rita Webb, Caroline Hawkins; **Dir:** Michael Cort. **VHS, Beta $59.95** PSM, SNC, TAV ♫

Alienator In the improbable future, an unstoppable android killer is sent after an intergalactic villain. An intentional "Terminator" rip-off.
1989 (R) 93m/C Jan-Michael Vincent, John Phillip Law, Ross Hagen, Dyana Ortelli, Dawn Wildsmith, P.J. Soles, Teagan Clive, Robert Clarke, Leo Gordon, Robert Quarry, Fox Harris, Hoke Howell, Jay Richardson; **Dir:** Fred Olen Ray. **VHS, Beta, LV $79.95** PSM ♫

Aliens The bitch is back, 50 some years later. The popular sequel to "Alien" amounts to non-stop, ravaging combat in space. Contact with a colony on another planet has mysteriously stopped. Fresh from deep space sleep, Ripley and a slew of pulsar-equipped Marines return to the planet where the deadly alien was first found and confront the mother alien at her nest, which is also inhabited by a whole bunch of the nasty critters spewing for a fight. Something's gotta give, and the Oscar-winning special effects are especially inventive (and messy) in the alien demise department. Dimension (acting biz talk) is given to our hero Ripley, as she discovers maternal instincts lurking within her space suit while looking after a young girl, the lone survivor of the colony, aban-

doned in the monsters' wake. A special tension-filled gore blaster.
1986 (R) 138m/C Sigourney Weaver, Michael Biehn, Lance Henriksen, Bill Paxton, Paul Reiser, Carrie Henn, Jenette Goldstein; **Dir:** James Cameron. Academy Awards '86: Best Sound Effects Editing, Best Visual Effects. **VHS, Beta, LV $19.98** FOX ♫♫♫½

Aliens Are Coming A spaceship crash lands on Earth, and its devious denizens begin invading human bodies. TV movie that's a dull echo of "Invasion of the Body Snatchers."
1980 100m/C Tom Mason, Melinda Fee, Max Gail, Eric Braeden, Matthew Laborteaux; **Dir:** Harvey Hart. **VHS, Beta $9.95** WOV, GKK ♫½

Aliens from Spaceship Earth Are strange, celestial forces invading our universe? If they are, is man prepared to defend his planet against threatening aliens of unknown strength? Lame docudrama featuring the Hurdy Gurdy man himself, Donovan.
1977 107m/C Donovan, Lynda Day George; **Dir:** Don Como. **VHS, Beta $59.95** GEM ♫½

Alison's Birthday A teenage girl learns that some of her family and friends are Satan worshipers at a terrifying birthday party. Meanwhile, the ghost of her dad hovers about, asking for more lines.
1979 99m/C AU Joanne Samuel, Lou Brown, Bunny Brooke; **Dir:** Ian Coughlan. **VHS, Beta $19.98** VID ♫♫

All About Eve One of the wittiest (and most cynical) flicks of all time follows an aspiring young actress (Baxter) as she ingratiates herself with a prominent group of theater people so that she may become known on Broadway without the years of work usually required. The not-so-innocent babe becomes secretary to aging star Davis and ruthlessly uses everyone around her in her climb to the top, much to Davis' initial disbelief and eventual displeasure. A satirical, darkly funny view of the world of the theater with an all-star cast featuring exceptional work by Davis, Sanders, and Ritter. Knowing of what he wrote, Mankiewicz penned the screenplay, which was based on "The Wisdom of Eve" by Mary Orr. Later staged as the musical "Applause."
1950 138m/B Bette Davis, Anne Baxter, Gary Merrill, Celeste Holm, George Sanders, Thelma Ritter, Gregory Ratoff, Marilyn Monroe; **Dir:** Joseph L. Mankiewicz. Academy Awards '50: Best Director (Mankiewicz), Best Picture, Best Screenplay, Best Sound, Best Supporting Actor (Sanders); Cannes Film Festival '51: Special Jury Prize, Best Actress (Davis); National Board of Review Awards '50: 10 Best Films of the Year; New York Film Critics Awards '50: Best Actress (Davis), Best Director (Mankiewicz), Best Film. **VHS, Beta, LV $19.98** FOX, BTV ♫♫♫♫

All-American Murder A rebellious young man is enrolled in a typical, all-American college for one last shot at mainstream life. Things start out okay, as he meets an attractive young coed. Hours later, he finds himself accused of her grisly murder. The youth is then given 24 hours to prove his innocence by a canny homicide detective. Can he find the real killer before it's too late?
1991 (R) 94m/C Christopher Walken, Charlie Schlatter, Joanna Cassidy, Josie Bisset, Amy Davis, Richard Kind, Woody Watson; **Dir:** Anson Williams. **VHS, Beta $89.95** PSM ♫♫

All Creatures Great and Small Taken from James Herriot's best selling novels, this is a delightful, quiet drama of a veterinarian's apprentice in rural England. Fine performance by Hopkins. Followed by

"All Things Bright and Beautiful" and a popular British television series.
1974 92m/C GB Simon Ward, Anthony Hopkins, Lisa Harrow, Brian Stirner, Freddie Jones, T.P. McKenna; **Dir:** Claude Whatham. **VHS, Beta $29.97** RDG, FOX ♫♫♫

All Dogs Go to Heaven Somewhat heavy-handed animated musical (Reynolds sings) about a gangster dog who is killed by his partner in business. On the way to Heaven, he discovers how to get back to Earth to seek his revenge. When he returns to Earth, he is taken in by a little girl and learns about something he missed in life the first time around: Love. Expertly animated, but the plot may not keep the grown-ups engrossed, and the kids may notice its lack of charm.
1989 (G) 85m/C **Dir:** Don Bluth, Don Bluth; **Voices:** Burt Reynolds, Judith Barsi, Dom DeLuise, Vic Tayback, Charles Nelson Reilly, Melba Moore, Candy Devine, Loni Anderson. **VHS, Beta, LV, 8mm $24.98** MGM, RDG ♫♫

All Fall Down A young man (de Wilde) idolizes his callous older brother (Beatty) until a tragedy forces him to grow up. Saint plays the older woman taken in by the brothers' family, who is seduced and abandoned. When she finds herself pregnant and alone, she commits suicide causing the younger brother, who loved her from afar, to vow to kill his older sibling. A well-acted melodrama.
1962 111m/B Eva Marie Saint, Brandon de Wilde, Warren Beatty, Karl Malden, Angela Lansbury, Constance Ford, Barbara Baxley; **Dir:** John Frankenheimer. **VHS $19.98** MGM ♫♫½

All God's Children Made for television drama about the controversial forced busing issue and two families who must face it. Top-notch cast is occasionally mislead by meandering script attempting to stay true to a sensitive issue.
1980 100m/C Richard Widmark, Ned Beatty, Ossie Davis, Ruby Dee, Mariclare Costello, George Spell, Trish Van Devere, Ken Swofford; **Dir:** Jerry Thorpe. **VHS, Beta** COM ♫♫½

All the Kind Strangers Traveling photographer Keach picks up a young hitchhiker and takes him to the boy's home. He discovers six other children there who want him as a father, with the alternative being death. He and "Mother" Eggar plot their escape from the dangerous orphans. Made-for-television thriller short on suspense also known as "Evil in the Swamp."
1974 72m/C Stacy Keach, Robby Benson, John Savage, Samantha Eggar; **Dir:** Burt Kennedy. **VHS, Beta $59.98** FOX ♫♫

All the King's Men Set in the Depression, a grim and graphic classic which follows the rise of a Louisiana farm-boy from angry and honest political hopeful to powerful but corrupt governor. Loosely based on the life (and death) of Huey Long and told by a newsman who's followed his career (Ireland). Willy Stark (Crawford), in his break-through role, is the politician who, while appearing to improve the state, rules dictatorially, betraying his friends and constituents, proving once again that power corrupts. In her first major role, McCambridge delivers a powerful award-winning performance as the cunning political aide. A potent morality play based on the Robert Penn Warren book, with a script by Rossen, who also produced and directed. In addition to best picture, actor, and supporting actress awards, it received Oscar nominations for Best Supporting Actor (Ireland), Best Director, Screenplay, and Film Editing.

1949 109m/B Broderick Crawford, Mercedes McCambridge, John Ireland, Joanne Dru, John Derek, Anne Seymour, Sheppard Strudwick; *Dir:* Robert Rossen. Academy Awards '49: Best Actor (Crawford), Best Picture, Best Supporting Actress (McCambridge); New York Film Critics Awards '49: Best Actor (Crawford), Best Film. **VHS, Beta, LV $19.95** *COL, BTV* 🎞🎞🎞🎞

All the Lovin' Kinfolk Two recent Hillbilly High graduates decide to take on the big city, but find themselves in dire situations. For those with time to kill.
1989 (R) 80m/C Mady Maguire, Jay Scott, Anne Ryan, John Denis, Donna Young; *Dir:* John Hayes. **VHS $29.95** *AVD* 🎞

All the Marbles A manager of two beautiful lady wrestlers has dreams of going to the top. Aldrich's last film, and an atypical one, with awkward pacing and a thin veil of sex exploitation. Falk provides needed grace and humor as the seedy manager, with Young contributing his usual competent bit as a hustling promoter dabbling in criminal activity. One of the few tag-team women wrestling pictures, it builds to a rousing finale match in the "Rocky" tradition, although the shift in tone and mounting cliches effectively body slam the intent. Also known in certain regions of the world as "The California Dolls."
1981 (R) 113m/C Peter Falk, Burt Young, Richard Jaeckel, Vicki Frederick; *Dir:* Robert Aldrich. **VHS, Beta, LV** *MGM, OM* 🎞🎞

All of Me A wealthy woman (Tomlin) dies and her guru accidentally transfers her soul to the right side of her lawyer's body, a modern version of existential hell. Lawyer Martin indulges in some funny slapstick as he discovers that his late client is waging an internal war for control of his body. Flat and cliched at times, but redeemed by the inspired clowning of Martin and witty Martin/Tomlin repartee. Based on the novel "Me Too" by Ed Davis.
1984 (PG) 93m/C Steve Martin, Lily Tomlin, Victoria Tennant, Madolyn Smith, Richard Libertini, Dana Elcar, Selma Diamond; *Dir:* Carl Reiner. **VHS, Beta, LV $19.98** *HBO, FCT* 🎞🎞½

All Mine to Give The sad saga of a Scottish family of eight who braved frontier hardships, epidemics, and death in the Wisconsin wilderness more than a century ago. Midway through, mom and dad die, leaving the oldest child struggling to keep the family together. A strange, though often effective, combination of pioneer adventures and tearjerking moments that avoids becoming hopelessly soapy due to fine performances. Unless you're pretty weathered, you'll need some hankies. Based on the reminiscences of Dale and Katherine Eunson as detailed in a "Cosmopolitan" magazine article entitled "The Day They Gave the Babies Away," which was also the original British film title.
1956 102m/C Glynis Johns, Cameron Mitchell; *Dir:* Allen Reisner. **VHS, Beta $15.95** *UHV* 🎞🎞½

All My Good Countrymen A lyrical, funny film about the eccentric denizens of a small Moravian village soon after the socialization of Czechoslovakia in 1948. Completed during the Soviet invasion of 1968 and immediately banned. In Czech with English subtitles. Also known as "All Good Citizens."
1968 115m/C *CZ* Vladimir Mensik, Radoslav Brozobohaty, Pavel Pavlovsky; *Dir:* Voltech Jasny. Cannes Film Festival '69: Best Director (Jasny). **VHS, Beta $59.95** *FCT* 🎞🎞½

All My Sons A wealthy family is distraught when their eldest son is listed as missing-in-action during WWII. They must cope with guilt, as well as grief, because the father's business reaped profits from the war. Adapted for television from the acclaimed Arthur Miller play.
1986 122m/C James Whitmore, Aidan Quinn, Joan Allen, Michael Learned; *Dir:* John Power. **VHS, Beta $39.95** *MCA* 🎞🎞🎞

All Night Long Offbeat middle-age crisis comedy about a burned-out and recently demoted drugstore executive in L.A. who leaves his wife, takes up with his fourth cousin by marriage, and begins a humorous rebellion, becoming an extremely freelance inventor while joining the drifters, weirdos and thieves of the night. An obscure, sometimes uneven little gem with Hackman in top form and an appealing supporting performance by Streisand. Scripted by W.D. Richter, who also wrote "Invasion of the Body Snatchers" and highlighted by delightful malapropisms and sataric inversion of the usual cliches.
1981 (R) 100m/C Gene Hackman, Barbra Streisand, Diane Ladd, Dennis Quaid, Kevin Dobson, William Daniels; *Dir:* Jean-Claude Tramont. **VHS, Beta $14.95** *GKK* 🎞🎞🎞

All Night Long/Smile Please Two classic comedies by the most overlooked of the great silent clowns, both produced by slapstick progenitor Mack Sennett.
1924 66m/B Harry Langdon, Natalie Kingston, Jack Cooper, Alberta Vaughn. **VHS, Beta $29.95** *VYY* 🎞🎞🎞

All in a Night's Work The founder of a one-man publishing empire is found dead with a smile on his face. His nephew inherits the business and finds himself caught in a series of big and small business misunderstandings. He's also falling in love with the woman he suspects was responsible for his uncle's grin. Nicely paced sex and business comedy with warm performances.
1961 94m/C Dean Martin, Shirley MacLaine, Cliff Robertson, Charlie Ruggles; *Dir:* Joseph Anthony. **VHS, Beta $19.95** *STE, FOX* 🎞🎞½

All for Nothing Two brothers are out to get revenge for their murdered family.
1975 98m/C Ferdinando Almada, Irma Lozano, Pedro Armendariz Jr. **VHS, Beta $49.98** *MAD* 🎞

All Over Town Two vaudevillians with a trained seal find themselves involved in a murder when they are kidnapped by a gang of thugs. A lesser Olsen and Johnson comedy with scattered funny moments.
1937 52m/B Ole Olsen, Chic Johnson, Mary Howard, Franklin Pangborn, James Finlayson; *Dir:* James W. Horne. **VHS, Beta $19.95** *NOS, MRV, VYY* 🎞🎞½

All the President's Men Based on the best-selling book by Washington Post reporters Bob Woodward and Carl Bernstein, this is the true story of the Watergate break-in that led to the political scandal of the decade. The politically based plot and investigation provides for an intriguing, terse thriller that will keep you on the edge of your seat even though you know the ending. Expertly paced by Pakula with a script by William Goldman. Features standout performances by Hoffman and Redford as the reporters who slowly uncover and connect the seemingly isolated facts leading to criminal indictments of the Nixon Administration. Deep Throat Holbrook and Robards as executive editor Ben Bradlee lend more authenticity to the endeavor, which realistically portrays the stop and go of journalistic investigations.
1976 (PG) 135m/C Robert Redford, Dustin Hoffman, Jason Robards Jr., Martin Balsam, Jane Alexander, Hal Holbrook, F. Murray Abraham, Stephen Collins, Lindsay Crouse; *Dir:* Alan J. Pakula. Academy Awards '76: Best Adapted Screenplay, Best Art Direction/Set Decoration, Best Sound, Best Supporting Actor (Robards); New York Film Critics Awards '76: Best Director (Pakula), Best Film, Best Supporting Actor (Robards). **VHS, Beta $19.98** *WAR, BTV* 🎞🎞🎞½

All Quiet on the Western Front The extraordinary anti-war epic based upon the Erich Maria Remarque novel realistically follows seven patriotic German youths as they go together from school to the battlefields of World War I. There they experience the horrors of war first-hand, stuck in the trenches and facing gradual extermination. The story centers around the experiences of one of the youths, Paul Baumer, who changes from enthusiastic war endorser to an extremely battle-weary veteran in an emotionally exact performance by Ayres. Made with a gigantic budget for the time of $1.25 million, it features more than 2,000 extras swarming about battlefields set up on ranchland in California. Relentless in its anti-war messaging and emotionally draining, it startles with both graphic shots and haunting visual poetry. Extremely controversial in the U.S. and Germany upon release, the original version was 140 minutes long (some versions are available with restored footage) and featured ZaSu Pitts as Ayres mother (later reshot with Beryl Mercer replacing her). Remarque, who had fought and been wounded on the Western Front, was eventually forced to leave Germany for the U.S. due to the film's ongoing controversy.
1930 103m/B Lew Ayres, John Wray, Louis Wolheim, Russell Gleason, Slim Summerville, Raymond Griffith; *Dir:* Lewis Milestone. Academy Awards '30: Best Director (Milestone), Best Picture; National Board of Review Awards '30: 10 Best Films of the Year. **VHS, Beta, LV $19.95** *MCA, FCT, GLV* 🎞🎞🎞🎞

All Quiet on the Western Front A big-budget television remake of the 1930 masterpiece starring John Boy. Sensitive German youth Thomas plunges excitedly into WWI and discovers its terror and degradation. Nowhere near the original's quality.
1979 150m/C Richard Thomas, Ernest Borgnine, Donald Pleasence, Patricia Neal; *Dir:* Delbert Mann. **VHS, Beta, LV $14.95** *FOX* 🎞🎞½

All the Right Moves Cruise is the high school football hero hoping for a scholarship so he can vacate pronto the dying Pennsylvania mill town where he grew up. At least he thinks that's what he wants to do. Further mixing up his own mixed feelings are his pushy, ambitious coach (Nelson, doing what he does best), his understanding dad (Cioffi) and supportive girlfriend (Thompson, in a notable early role). Strong performances push the relatively cliched melodrama into field goal range. Cinematographer Chapman's directorial debut.
1983 (R) 90m/C Tom Cruise, Lea Thompson, Craig T. Nelson, Christopher Penn; *Dir:* Michael Chapman. **VHS, Beta, LV $14.98** *FOX, FXV* 🎞🎞½

All the Rivers Run First made-for-cable miniseries. A young Australian girl spends her inheritance on a river boat and becomes the first female river captain in Australian history. Directed by George Miller of "Man from Snowy River" fame. Four parts.

1984 400m/C *AU* Sigrid Thornton, John Waters, Diane Craig, Charles Tingwell, Gus Mercurio; *Dir:* George Miller, Pina Amenta. **VHS** $79.95 *HBO* 🎃🎃½

All Screwed Up A group of young immigrants come to Milan and try to adjust to city life; they soon find that everything is in its place, but nothing is in order. Wertmuller in a lighter vein than usual.
1974 104m/C *IT* Luigi Diberti, Lina Polito; *Dir:* Lina Wertmuller. **VHS, Beta** *FOX* 🎃🎃🎃

All-Star Cartoon Parade A collection of popular vintage cartoons, starring such favorites as Little Lulu, Casper the Friendly Ghost, Betty Boop and Raggedy Ann and Andy.
19?? 54m/C **VHS** $29.98 *REP* 🎃🎃🎃

All That Jazz Fosse's autobiographical portrait with Schneider fully occupying his best role as the obsessed, pill-popping, chain-smoking choreographer/director dancing simultaneously with love and death. But even while dying, he creates some great dancing. Vivid and imaginative with exact editing, and an eerie footnote to Fosse's similar death almost ten years later. Egocentric and self-indulgent for sure, but that's entertainment. In addition to Oscars awarded, nominated for Best Picture, Scheider's acting, Fosse's directing, Original Screenplay, and Cinematography.
1979 (R) 120m/C Roy Scheider, Jessica Lange, Ann Reinking, Ben Vereen, John Lithgow, Wallace Shawn, Cliff Gorman, Leland Palmer, Sandahl Bergman; *Dir:* Bob Fosse. Academy Awards '79: Best Art Direction/Set Decoration, Best Costume Design, Best Film Editing, Best Original Score; British Academy Awards '79: Best Cinematography. **VHS, Beta, LV** $19.98 *FOX, FCT* 🎃🎃🎃½

All This and Heaven Too When a governess arrives at a Parisian aristocrat's home in the 1840s, she causes jealous tension between the husband and his wife. The wife is soon killed/murdered. Based on Rachel Field's best-seller. Oscar nominations: Best Picture, Best Supporting Actress (O'Neil), Black & White Cinematography.
1940 141m/B Charles Boyer, Bette Davis, Barbara O'Neil, Virginia Weidler, Jeffrey Lynn, Helen Westley, Henry Daniell, Harry Davenport; *Dir:* Anatole Litvak. **VHS, Beta** $19.95 *MGM* 🎃🎃🎃

All This and Tex Avery, Too! Fourteen cartoons from the great Tex Avery are featured here, including "What Price Freedom," "Doggone Tired," "Wags to Riches," "The Screwy Truant," "King-Size Canary," "House of Tomorrow" and eight others.
1992 120m/C **LV** $34.98 *MGM* 🎃🎃🎃

All Through the Night A very funny spy spoof as well as a thrilling crime story with Bogart playing a gambler who takes on a Nazi spy ring. Features memorable double-talk and a great auction scene that inspired the one with Cary Grant in "North by Northwest." Suspense builds throughout the film as Lorre appears in a sinister role as Pepi and Veidt gives a fine performance as the spymaster.
1942 107m/B Humphrey Bogart, Conrad Veidt, Karen Verne, Jane Darwell, Frank McHugh, Peter Lorre, Judith Anderson, William Demarest, Jackie Gleason, Phil Silvers, Barton MacLane, Martin Kosleck; *Dir:* Vincent Sherman. **VHS** $19.98 *MGM* 🎃🎃🎃

All the Way, Boys Two inept adventurers crash-land a plane in the Andes in the hope of discovering slapstick, but find none. "Trinity" cast up to no good.
1973 (PG) 105m/C *IT* Terence Hill, Bud Spencer, Cyril Cusack, Michel Antoine; *Dir:* Giuseppe Colizzi. **VHS, Beta** $59.98 *SUE* 🎃½

All the Way Down A ski chalet in the Alps is the setting for much fireside cavorting.
1987 80m/C Judith Fritsch, Franz Muxeneder. **VHS, Beta** $29.95 *MED* 🎃

All You Need is Cash "The Rutles" star in this parody of The Beatles' legend, from the early days of the "Pre-Fab Four" in Liverpool to their worldwide success. A marvelous pseudo-documentary, originally shown on NBC-TV and with various SNL alumni, (including Weiss) that captures the development of the Beatles and '60s rock with devastating effect. Served as the inspiration for "This Is Spinal Tap."
1978 70m/C Eric Idle, Neil Innes, Rikki Fataar, Dan Aykroyd, Gilda Radner, John Belushi, George Harrison, Gary Weiss; *Dir:* Eric Idle. **VHS, Beta, LV** $14.95 *MVD, PAV* 🎃🎃🎃

All the Young Men Fairly powerful men-uniting-in-battle story with an interesting cast, highlighted by Poitier in an early role. A tiny marine squadron overrun by the Chinese in the Korean War attempts to resist the numerous attackers. In their spare time, the men confront racial prejudice when a black man (guess who) takes command. Written and directed by Bartlett.
1960 86m/B Alan Ladd, Sidney Poitier, James Darren, Glenn Corbett, Mort Sahl; *Dir:* Hall Bartlett. **VHS, Beta** $14.99 *COL* 🎃🎃🎃

Allan Quartermain and the Lost City of Gold While trying to find his brother, Quartermain discovers a lost African civilization, in this weak adaptation of an H. Rider Haggard adventure. An ostensible sequel to the equally shallow "King Solomon's Mines."
1986 (PG) 100m/C Richard Chamberlain, Sharon Stone, James Earl Jones; *Dir:* Gary Nelson. **VHS, Beta** $9.99 *FHE, MED* 🎃½

Allegheny Uprising Set in 1759, the story of a frontiersman who clashes with a British military commander in order to stop the sale of firearms to Indians. The stars of "Stagecoach" are back on board in this lesser effort.
1939 81m/B John Wayne, Claire Trevor, George Sanders, Brian Donlevy, Chill Wills, Moroni Olsen; *Dir:* William A. Seiter. **VHS, Beta, LV** $19.98 *CCB, RKO, BUR* 🎃🎃½

Allegro Non Troppo An energetic and bold collection of animated skits set to classical music in this Italian version of Disney's "Fantasia." Watch for the evolution of life set to Ravel's Bolero, or better yet, watch the whole darn movie. Features Nichetti (often referred to as the Italian Woody Allen, particularly by people in Italy) in the non-animated segments, who went on to write, direct, and star (he may have sold concessions in the lobby as well) in "The Icicle Thief."
1976 75m/C *IT* Maurizio Nichetti; *Dir:* Bruno Bozzetto. **VHS, LV** $29.95 *BMG, IME* 🎃🎃🎃

Alley Cat A woman uses martial arts to fight back against a street gang that attacked her.
1984 (R) 82m/C Karin Mani, Robert Torti, Brit Helfer, Michael Wayne, Jon Greene; *Dir:* Edward Victor. **VHS, Beta** $69.98 *VES, LIV* 🎃

Alligator Dumped down a toilet 12 long years ago, lonely alligator Ramon resides in the city sewers, quietly eating and sleeping. In addition to feasting on the occasional stray human, Ramon devours the animal remains of a chemical plant's experiment involving growth hormones and eventually begins to swell at an enormous rate. Nothing seems to satisfy Ramon's ever-widening appetite: not all the people or all the buildings in the whole town, but he keeps trying, much to the regret of the guilt-ridden cop and lovely scientist who get to know each other while trying to nab the gator. Mediocre special effects are only a distraction in this witty eco-monster take penned by John Sayles.
1980 (R) 94m/C Robert Forster, Lewis Teague, Jack Carter, Henry Silva, Robin Riker, Dean Jagger; *Dir:* Lewis Teague. **VHS, Beta** $14.98 *LTG, LIV* 🎃🎃🎃

Alligator 2: The Mutation Not a sequel to 1980's surprisingly good "Alligator," but a bland rehash with a decent cast. Once again a toxic alligator grows to enormous size and menaces a community. A Donald-Trump-like villain and pro wrestlers (!) bring this up to date, but it's all on the level of a TV disaster movie; even the PG-13 rating is a bit too harsh.
1990 (PG-13) 92m/C Steve Railsback, Dee Wallace Stone, Joseph Bologna, Woody Brown, Bill Daily, Brock Peters. **VHS** $89.95 *COL* 🎃🎃

Alligator Alley When two young divers witness a major drug deal, they become entangled in a web of danger they never expected.
1972 92m/C Steve Alaimo, John Davis Chandler, Willie Pastrano, Jeremy Slate, Cece Stone; *Dir:* William Grefe. **VHS, Beta** $79.95 *PSM* 🎃

Alligator Eyes A stranger wearing trouble like a cheap perfume enters the midst of a vacationing trio of New Yorkers. The three travelers, who include a boozy, recently jilted young Lothario and an estranged couple attempting to kindle a reconciliation on the road, have their vulnerabilities exposed by a young hitchhiker sporting a slinky polka dot ensemble and way cool sunglasses. Not until the hitchhiker has insinuated herself into their vacation plans (not to mention their private lives) do the three realize the woman is blind and full of manipulative and vindictive tricks, thanks to the usual brutal childhood. A psycho-sexo-logical thriller with some fine performances. Begins promisingly before fizzling into celluloid cotton candy.
1990 (R) 101m/C Annabelle Larsen, Roger Kabler, Mary McLain, Allen McCullough, John MacKay; *W/Dir:* John Feldman. **VHS** $19.95 *ACA* 🎃🎃½

The Allnighter A college coed searches through the hypersexed beach-party milieu of her senior year for Mr. Right. Bangle Hoffs is directed by her mom, to no avail.
1987 (PG-13) 95m/C Susanna Hoffs, John Terlesky, Joan Cusack, Michael Ontkean; *Dir:* Tamar Simon Hoffs. **VHS, Beta** $79.95 *MCA* 🎃

Allonsanfan Early Taviani, in which a disillusioned Jacobin aristocrat in 1816, after Napoleon has fallen, struggles with his revolutionary ideals and his accustomed lifestyle. In Italian with English subtitles. Exciting score by Ennio Morricone.
1973 115m/C *IT* Marcello Mastroianni, Laura Betti, Renato de Carmine, Lea Massari, Mimsy Farmer, Claudio Cassinelli; *Dir:* Paolo Taviani. **VHS, Beta** $59.95 *FCT, WBF, TAM* 🎃🎃🎃½

All's Fair Silly and predictable comedy about executives who take on their spouses at weekend war games. An unfortunate waste of a good cast.
1989 89m/C George Segal, Sally Kellerman, Robert Carradine, Jennifer Edwards, Jane Kaczmarek, John Kapelos, Lou Ferrigno; *Dir:* Rocky Lane. **VHS, Beta, LV** $79.95 *MED 𝆑*

Almonds & Raisins: A History of the Yiddish Cinema The phenomenon of Yiddish cinema is remembered by the actors, directors and producers who made it happen. Excerpts from the great filmworks of Maurice Schwartz, Molly Picon, Moishe Oysher and others provide a glimpse of this lost era.
1983 90m/B *Dir:* Russ Karel; *Nar:* Orson Welles. **VHS, Beta** $79.95 *TAM, ERG 𝆑*

Almos' a Man Richard Wright's story of a black teenage farm worker in the late 1930's who must endure the pains of growing up. From the "American Short Story" series.
1978 (G) 51m/C LeVar Burton. **VHS, Beta** $24.95 *MON, MTI, KAR 𝆑*

...Almost Arquette plays a curiously giddy bookworm with a vivid imagination. When her husband disappears on their anniversary, the man of her dreams shows up to sweep her off her feet. Is he real or is he simply another daydream? Almost a good time.
1990 (PG) 87m/C *AU* Rosanna Arquette, Bruce Spence; *Dir:* Michael Pattinson. **VHS, LV** $89.98 *MAG 𝆑𝆑*

Almost an Angel Another in a recent spat of angels and ghosts assigned back to earth by the head office. A life-long criminal (Hogan) commits a heroic act and finds himself a probationary angel returned to earth to gain permanent angel status. He befriends a wheelchair-bound man, falls in love with the guy's sister, and helps her out at a center for potential juvenile delinquents. Melodramatic and hokey in places, relying too much on Hogan's crocodilian charisma.
1990 (PG) 98m/C Paul Hogan, Linda Kozlowski, Elias Koteas, Doreen Lang, Charlton Heston; *Dir:* John Cornell. **VHS, LV** $19.95 *PAR, PIA 𝆑1/2*

Almost Angels Two boys romp in Austria as members of the Vienna Boys Choir. Lesser sentimental Disney effort that stars the actual members of the Choir; not much of a draw for today's Nintendo-jaded young viewers.
1962 85m/C Vincent Winter, Peter Weck, Hans Holt; *Dir:* Steve Previn. **VHS, Beta** $69.95 *DIS 𝆑𝆑*

Almost Human An Italian Mafia gorefest about a second-class don who kidnaps a businessman's daughter, and then has trouble trying to collect a ransom. For aficionados of Italian Mafia gorefests only (check your weapons at the door).
1979 (R) 90m/C *IT* Henry Silva, Tomas Milian, Laura Belli; *Dir:* Umberto Lenzi. **VHS, Beta** $79.95 *PSM Woof!*

Almost Partners Grandpa's ashes are in trouble. His urn and remains have been stolen. Fortunately, his teenage granddaughter and a detective are on the case.
1987 58m/C Paul Sorvino, Royana Black, Mary Wickes; *Dir:* Alan Kingsberg. **VHS** $29.95 *PME, FCT 𝆑1/2*

An Almost Perfect Affair Taxing romantic comedy about an ambitious independent American filmmaker who, after finishing a movie about an executed murderer, travels to the Cannes festival and proceeds to fall in love or lust with the wife of an Italian producer. Numerous inside jokes and capable performances nearly overcome script lethargy.
1979 (PG) 92m/C Keith Carradine, Monica Vitti, Raf Vallone, Christian de Sica, Dick Anthony Williams; *Dir:* Michael Ritchie. **VHS, Beta** $59.95 *PAR 𝆑𝆑*

Almost Pregnant Linda Anderson (Roberts) desperately wants a baby, but her husband, Charlie, is unable to get her pregnant. Dead set against artificial insemination, Linda decides to take a lover. Although Charlie's not too keen on the idea, he doesn't want to lose her, so he goes along with the proposed solution. Problems arise when Linda falls in love with the new guy and discovers he's had a vasectomy. Still wanting to get pregnant, Linda takes yet another lover while continuing to see the other two guys. Hilarious complications abound in this outrageous quest for a baby. An unrated version containing explicit scenes is also available.
1991 (R) 90m/C Tanya Roberts, Jeff Conaway, Joan Severance, Dom DeLuise; *Dir:* Michael DeLuise. **VHS, LV** $89.95 *COL 𝆑1/2*

Almost Royal Family All kinds of problems arise when a family inherits an island on the St. Lawrence River embargoed by the United States and Canada.
1984 52m/C Sarah Jessica Parker, John Femia. **VHS, Beta** $39.95 *SCL, WAR 𝆑*

Almost You Normal marital conflicts and uncertainties grow exponentially when a wealthy New York City thirty-something couple hires a lovely young nurse to help care for the wife after a car accident. An unsentimental marital comedy with a good cast that still misses.
1985 (R) 91m/C Brooke Adams, Griffin Dunne, Karen Yahng, Marty Watt, Christine Estabrook, Josh Mostel, Laura Dean, Dana Delany, Miguel Pinero, Joe Silver, Suzzy Roche, Spalding Gray; *Dir:* Adam Brooks. **VHS, Beta** $79.98 *FOX 𝆑𝆑*

Aloha, Bobby and Rose A mechanic and his girlfriend in L.A. become accidentally involved in an attempted robbery and murder and go on the run for Mexico, of course. Semi-satisfying drama in the surf, with fine location photography.
1974 (PG) 90m/C Paul LeMat, Dianne Hull, Robert Carradine, Tim McIntire, Edward James Olmos, Leigh French; *Dir:* Floyd Mutrux. **VHS, Beta** $49.95 *MED 𝆑𝆑1/2*

Aloha Summer Six surfing teenagers of various ethnic backgrounds learn of love and life in 1959 Hawaii while riding the big wave of impending adulthood, with a splash of Kung Fu thrown in for good measure. Sensitive and bland.
1988 (PG) 97m/C Chris Makepeace, Lorie Griffin, Don Michael Paul, Sho Kosugi, Yuji Okumoto, Tia Carrere; *Dir:* Tommy Lee Wallace. **VHS, Beta, LV** $79.95 *LHV, WAR 𝆑𝆑*

Alone Against Rome A muscle-bound warrior takes on the forces of Rome to avenge himself against a scornful woman.
1962 100m/B *IT* Lang Jeffries, Rossana Podesta, Phillippe LeRoy; *Dir:* Herbert Wise. **VHS, Beta** $16.95 *SNC 𝆑1/2*

Alone in the Dark Slash and dash horror attempt featuring four escaped patients from a mental hospital who decide that they must kill their doctor because they don't like him. Conveniently, a city-wide blackout provides the opportunity, as the good doctor defends home and family against the aging stars intent on chewing up as much scenery as possible.
1982 (R) 92m/C Jack Palance, Donald Pleasence, Martin Landau, Dwight Schultz; *Dir:* Jack Sholder. **VHS, Beta** $59.95 *COL 𝆑𝆑*

Alone in the Neon Jungle A glamorous big-city police captain is assigned to clean up the most corrupt precinct in town. Pleshette is untypically cast but still manages to make her role believable in a serviceable TV cop drama with more dialogue than action.
1987 90m/C Suzanne Pleshette, Danny Aiello, George Stanford Brown, Frank Converse, Joe Morton; *Dir:* George Stanford Brown. **VHS** $79.95 *COL 𝆑𝆑*

Alone in the T-Shirt Zone A maniacal T-shirt designer lands in a mental institution.
1986 81m/C Michael Barrack, Taylor Gilbert, Bill Barron. **VHS, Beta** $19.95 *STE, NWV 𝆑*

Along Came Jones Cowboy Cooper, who can't handle a gun and is saddled with grumpy sidekick Demarest, is the victim of mistaken identity as both the good guys and the bad guys pursue him thinking he is a vicious killer. Young is the woman who rides to his defense. Cooper produced and Nunnally Johnson scripted this offbeat and charming western parody based on a novel by Alan le May.
1945 93m/B Gary Cooper, Loretta Young, Dan Duryea, William Demarest; *Dir:* Stuart Heisler. **VHS, Beta** $59.98 *FOX 𝆑𝆑𝆑*

Along the Great Divide A U.S. marshal and his deputy battle pursuing vigilantes and the untamed frontier to bring a falsely accused murderer to trial (and of course, find the real bad guy). Douglas' first western has the usual horse opera cliches supported by excellent cinematography.
1951 88m/B Kirk Douglas, John Agar, Walter Brennan, Virginia Mayo; *Dir:* Raoul Walsh. **VHS, Beta** $19.98 *WAR, TLF 𝆑𝆑*

Along the Sundown Trail Standard western escapades as cowboy G-men round up the villains.
1942 59m/B William Boyd, Art Davis, Lee Powell, Julie Duncan, Kermit Maynard, Charles King; *Dir:* Sam Newfield. **VHS, Beta** *VCN 𝆑1/2*

Alpha Beta A stage production set to film studying the break-up of a marriage, where all the husband and wife have in common is the children. Finney and Roberts efficiently carry the load.
1973 70m/C Albert Finney, Rachel Roberts; *Dir:* Anthony Page. **VHS, Beta** $49.95 *CCI 𝆑𝆑1/2*

The Alpha Incident A frightening doomsday drama about an alien organism with the potential to destroy all living things. Ho-hum, not again!
1976 (PG) 86m/C Ralph Meeker, Stafford Morgan, John Goff, Carol Irene Newell, John Alderman; *Dir:* Bill Rebane. **VHS, Beta** $9.95 *MED 𝆑1/2*

Alphabet City A drug kingpin who runs New York's Lower East Side has decided to turn over a new leaf, but first he must survive his last night as a criminal while figuring out a way to pay off his large debts. Very stylish and moody, but light on content and plot.
1984 (R) 85m/C Vincent Spano, Michael Winslow, Kate Vernon, Jami Gertz, Zohra Lampert, Ray Serra, Kenny Marino, Daniel Jordano, Miguel Pinero; *Dir:* Amos Poe. **VHS, Beta** $79.98 *FOX 𝆑𝆑*

The Alphabet Murders Picture this: Randall playing Agatha Christie's famous Belgian sleuth, Hercule Poirot, and Rutherford—in a cameo—as Miss Marple. As if that wouldn't be enough to make Dame Agatha roll over in her grave, there's plenty of cloying wisecracking and slapsticking throughout. An adaptation of ''The ABC Murders'' in which Poirot stalks a literate killer who snuffs out his victims in alphabetical order. Hardly a must-see, unless you're hellbent on viewing the entire Randall opus.
1965 90m/B Tony Randall, Anita Ekberg, Robert Morley, Maurice Denham, Guy Rolfe, Sheila Allen, Margaret Rutherford; *Dir:* Frank Tashlin. **VHS, Beta $19.98** *MGM, FCT* ⅃⅃

Alphaville Engaging and inimitable Godard attempt at science fiction mystery. P.I. Lemmy Caution searches for a scientist in a city (Paris as you've never seen it before) run by robots and overseen by a dictator. The futuristic techno-conformist society must be upended so that Caution may save the scientist as well as nonconformists everywhere. In French with subtitles and also known as ''Alphaville, a Strange Case of Lemmy Caution'' (or ''Alphaville, Une Etrange Aventure de Lemmy Caution'').
1965 100m/B *FR* Eddie Constantine, Anna Karina, Akim Tamiroff; *Dir:* Jean-Luc Godard. Berlin Film Festival '65: Best Film; New York Film Festival '65: Best Film. **VHS, Beta $24.95** *NOS, SNC, MRV* ⅃⅃⅃

Alpine Fire Coming of age story about an adolescent girl and her deaf-mute brother living an isolated life in the Swiss Alps. When life on the mountain overwhelms them, they turn to each other for love. Sharply observed with a naturalistic style, elevating the proceedings above mere voyeurism.
1989 (R) 119m/C *SI* Thomas Knock, Johanna Lier, Dorothea Moritz, Rolf Illig; *Dir:* Fredi M. Murer. **VHS, Beta, LV $79.98** *VES, LIV* ⅃⅃½

Alsino and the Condor An acclaimed Nicaraguan drama about a young boy, caught in war-torn Nicaragua between the Somoza government and the Sandinista rebels, who dreams of flying above the human strife. In Spanish with English subtitles.
1983 (R) 89m/C *NI* Dean Stockwell, Alan Esquivel; *Dir:* Miguel Littin. **VHS, Beta, LV $29.95** *PAV, FCT, TAM* ⅃⅃⅃

Altered States Obsessed with the task of discovering the inner man, Hurt's ambitious researcher ignores his family while consuming hallucinogenic drugs and floating in an immersion tank. He gets too inner, slipping way back through the evolutionary order and becoming a menace in the process. A confusing script by Paddy Chayesfsky (alias Sidney Aaron) based upon a confusing novel by Chayesfsky is supported by great special effects and the usual self-indulgent, provocative, and confusing Russell direction (aided by a $15 million budget). Chayefsky eventually washed his hands of the project after artistic differences with the producers. Others who departed from the film include initial director William Penn and special effects genius John Dkystra (relieved ably by Bran Ferren). Hurt's a solemn hoot in his first starring role.
1980 (R) 103m/C William Hurt, Blair Brown, Bob Balaban, Charles Haid, Dori Brenner, Drew Barrymore; *Dir:* Ken Russell. **VHS, Beta, LV $19.98** *WAR* ⅃⅃⅃

Alternative A female magazine editor finds herself unmarried and quite pregnant. She is caught in a tug-of-war between the baby's father and her current lover. Dated liberated woman and career snoozer.
1976 90m/C Wendy Hughes, Peter Adams, Mary Mackie, Carla Hoogeveen, Tony Bonner. **VHS, Beta $34.95** *KOV* ⅃⅃

Alvarez Kelly Offbeat western with Holden as the Mexican-Irish Kelly who has just sold a herd of cattle to the North during the Civil War. Confederate officer Widmark kidnaps Holden in an effort to have the cattle redirected to the South. Aided by the traditional women in the midst of men intent on double-crossing each other, a fierce hatred develops between the two, erupting into violence. Sleepy performance by Holden is countered by an intensive Widmark. Based on a true Civil War incident, the script occasionally wanders far afield with the cattle, who cleverly heighten the excitement by stampeding.
1966 116m/C William Holden, Richard Widmark, Janice Rule, Patrick O'Neal, Harry Carey Jr.; *Dir:* Edward Dmytryk. **VHS, Beta $12.95** *GKK* ⅃⅃½

Alvin Purple Pedestrian comedy about a Mr. Purple, an ordinary Aussie who sells waterbeds and is for some reason constantly being pursued by throngs of sexually insatiable women. Fortunately, he too enjoys sex, even though complications abound. Sexual situations, double entendres, and a script that aims for cleverness (but rarely attains it) somehow made the lust romp a hit in its native Australia, while worldwide it helped establish a market for Down Under cinema. Written by playwright Alan Hopgood, shot in Melbourne, and followed by ''Alvin Rides Again.''
1973 (R) 97m/C *AU* Graeme Blundell, George Whaley, Ellie MacLure, Jacki Weaver; *Dir:* Tim Burstall. **VHS, Beta, LV $19.95** *STE, NWV* ⅃½

Alvin Rides Again The sexually insatiable Alvin Purple is asked to impersonate an American gangster who was accidentally killed. Another Aussie sex farce co-written by Tim Burstall, Alan Hopgood, and Alan Finney, who conspire to create a decent sequel to ''Alvin Purple,'' with much of the same cast and crew.
1974 89m/C Robin Copping, Graeme Blundell, Alan Finney, Brionny Behets, Frank Thring, Jeff Ashby; *Dir:* David Bilcock. **VHS, Beta $19.95** *STE, NWV* ⅃⅃

Always Jaglom fictionally documents his own divorce and reconciliation with Patrice Townsend: set in the director's home and starring his friends and family, the film provides comic insight into the dynamics of married/about-to-be-married/used-to-be-married relationships. Set at a Fourth of July barbecue, this bittersweet romantic comedy is a veritable feast for Jaglom fans, but not everyone will find the director's free-form narrative to their taste.
1985 (R) 105m/C Henry Jaglom, Patrice Townsend, Bob Rafelson, Melissa Leo, Andre Gregory, Michael Emil, Joanna Frank; *Dir:* Henry Jaglom. **VHS, Beta, LV $79.98** *VES, LIV* ⅃⅃⅃

Always A hotshot slurry pilot (Dreyfuss) meets a fiery end and finds that his spirit is destined to become a guardian angel to the greenhorn fire-fighting flyboy (Johnson) who steals his girl's heart. Warm remake of ''A Guy Named Joe,'' one of Spielberg's favorite movies. Sparks between Dreyfuss and Hunter eventually ignite, and Goodman delivers the most heat. Hepburn makes an appearance as the angel who guides Dreyfuss. An

old-fashioned tree-burner romance that includes actual footage of the 1988 Yellowstone fire.
1989 (PG) 123m/C Holly Hunter, Richard Dreyfuss, John Goodman, Audrey Hepburn, Brad Johnson, Marg Helgenberger, Keith David, Roberts Blossom; *Dir:* Steven Spielberg. **VHS, Beta, LV $19.95** *MCA, FCT* ⅃⅃½

Amadeus An entertaining adaptation by Peter Shaffer of his play about the intense rivalry between 18th century composer Antonio Salieri and Wolfgang Amadeus Mozart. Abraham's Salieri is a man who desires greatness but is tortured by envy and sorrow. His worst attacks of angst occur when he comes into contact with Hulce's Mozart, an immature, boorish genius who, despite his gifts, remains unaffected and delighted by the beauty he creates while irking the hell out of everyone around him. Terrific period piece filmed on location in Prague, with an excellent musical score, beautiful sets, nifty billowy costumes, and realistic American accents for the 18th century Europeans. In addition to the four big Oscars awarded (including best adapted screenplay), the film recieved nominations for best actor (Hulce), photography, editing, and art direction.
1984 (PG) 158m/C F. Murray Abraham, Tom Hulce, Elizabeth Berridge, Simon Callow, Roy Dotrice, Christine Ebersole, Jeffrey Jones, Kenny L. Baker, Cynthia Nixon, Vincent Schiavelli; *Dir:* Milos Forman. Academy Awards '84: Best Actor (Abraham), Best Adapted Screenplay, Best Art Direction/Set Decoration, Best Costume Design, Best Director (Forman), Best Makeup, Best Picture, Best Sound; Directors Guild of America Awards '84: Best Director (Forman); Golden Globe Awards '85: Best Film—Drama. **VHS, Beta, LV $19.99** *HBO, GLV, BTV* ⅃⅃⅃½

Amarcord Semi-autobiograpical Fellini fantasy which takes place in the village of Rimini, his birthplace. Focusing on the young Zanin's impressions of his town's colorful slices of life, Fellini takes aim at facism, family life, and religion in 1930's Italy. Visually ripe, delivering a generous, occasionally uneven mix of satire, burlesque, drama, and tragicomedic lyricism. Considered by people in the know as one of Fellini's best films and the topic of meaningful discussions among art film students everywhere.
1974 (R) 124m/C *IT* Magali Noel, Bruno Zanin, Pupella Maggio, Armando Brancia; *Dir:* Federico Fellini. Academy Awards '74: Best Foreign Language Film; National Board of Review Awards '74: Best Foreign Film; New York Film Critics Awards '74: Best Director (Fellini), Best Film. **VHS, Beta** *APD* ⅃⅃⅃½

The Amateur A computer technologist for the CIA dives into a plot of international intrigue behind the Iron Curtain when he investigates the death of his girlfriend, murdered by terrorists. Confused and ultimately disappointing spy drama cursed with a wooden script written by Robert Littell and Diana Maddox and based on a novel by Littell.
1982 (R) 112m/C John Savage, Christopher Plummer, Marthe Keller, Arthur Hill, Ed Lauter; *Dir:* Charles Jarrott. **VHS, Beta $59.98** *FOX* ⅃½

Amateur Night Sloppy musical comedy about the backstage bickering occuring during the amateur night at a famous nightclub.
1985 91m/C Geoffrey Deuel, Dennis Cole, Allen Kirk. **VHS, Beta $69.98** *VES, LIV* ⅃

Amazing Adventure A millionaire wins a bet when he rises from a chauffeur's position to the executive board room without using his wealth. Though not particularly amazing, lightweight English comedy has

Grant working hard to charm over and above the demands of the dated formula, a performance he undertook during a vacation from Hollywood. Also known as ''Romance and Riches'' and ''The Amazing Quest of Ernest Bliss'' and adapted from a novel by E. Phillips Oppenheim.

1937 63m/C Cary Grant, Mary Brian, Henry Kendall, Leon M. Lion, Ralph Richardson; *Dir:* Alfred Zeisler. **VHS, Beta $19.95** *NOS, KRT, DVT* ♫♫

The Amazing Dobermans Family-oriented pooch performance piece featuring Astaire in one of his lesser roles. Ex-con man Astaire and his five trained Dobermans assist an undercover agent in foiling a small-time criminal's gambling and extortion racket. The last in a series that includes ''The Daring Dobermans'' and ''The Doberman Gang.''

1976 (G) 96m/C Fred Astaire, Barbara Eden, James Franciscus, Jack Carter, Billy Barty; *Dir:* Byron Ross Chudnow. **VHS, Beta $9.95** *MED* ♫♫

Amazing Grace Some righteous mothers led by Moms Mabley go up against corrupt city politics in this extremely dated comedy that was cast with a sense of the absurd.

1974 99m/C Moms Mabley, Slappy White, Moses Gunn, Rosalind Cash, Dolph Sweet, Butterfly McQueen, Stepin Fetchit; *Dir:* Stan Lathan. **VHS** *TAV* ♫♫

Amazing Grace & Chuck Perhaps the only anti-war/sports feature. Upon visiting a Minuteman missile site in Montana, a 12-year-old Little Leaguer (Chuck) learns of the dangers of nuclear arms. The concerned youth begins a protest against nuclear weapons by refusing to play until the nations come to a peace agreement. Upon hearing of the young man's stand for his beliefs, pro basketball star Amazing Grace Smith (Denver Nugget English) quits the team and joins the boy in protest. Turned on by Amazing Grace's sudden surge of social conscience, sports players across the country and the world soon show their support by also refusing to play and by moving into the boy's barn to organize the protest. The movement escalates, forcing the leaders, including President Peck, to take notice. Capra fantasy with good intentions but lacking coherency and plausibility. Big-budget score by Elmer Bernstein.

1987 (PG) 115m/C Jamie Lee Curtis, Gregory Peck, William L. Petersen, Joshua Zuehlke, Alex English; *Dir:* Mike Newell. **VHS, Beta, LV $19.99** *HBO Woof!*

Amazing Howard Hughes Reveals the full story of the legendary millionaire's life and career, from daring test pilot to inventor to Hollywood film producer to isolated wealthy paranoiac with a germ phobia-lingering on the rich guy with big problems theme. Big-budget, made for television drama with a nice performance by Jones.

1977 119m/C Tommy Lee Jones, Ed Flanders, James Hampton, Tovah Feldshuh, Lee Purcell; *Dir:* William A. Graham. **VHS, Beta $19.95** *HBO, FCT* ♫♫½

Amazing Mr. Blunden Solid kidvid about two youngsters aided by a ghost who travel back in time to save the lives of two murdered children. Adapted by Jeffries from Antonia Barber's novel, ''The Ghosts.''

1972 (G) 100m/C Laurence Naismith, Lynne Frederick, Garry Miller, Marc Granger, Rosalyn London, Diana Dors; *Dir:* Lionel Jeffries. **VHS, Beta $19.95** *MED* ♫♫½

The Amazing Mr. X When a woman's husband dies, she tries to contact him via a spiritualist. Things are not as they seem however, and the medium may just be part of an intricate scheme to defraud the woman. Also known as ''The Spiritualist.''

1948 79m/B Turhan Bey, Lynn Bari, Cathy O'Donnell, Richard Carlson, Donald Curtis, Virginia Gregg; *Dir:* Bernard Vorhaus. **VHS, Beta $19.95** *NOS, SNC, RXM* ♫♫½

The Amazing Spider-Man The Marvel Comics superhero's unique powers are put to the test when he comes to the rescue of the government by preventing an evil scientist from blackmailing the government for big bucks. The wall-walking web-slinger has his origins probed (a grad student bit by a radioactive spider develops super-human powers) in his live-action debut, which was made for television and led to a short-lived series. Also known as just plain ol' ''Spider-Man.''

1977 94m/C Nicholas Hammond, David White, Lisa Eilbacher, Michael Pataki; *Dir:* E.W. Swackhamer. **VHS, Beta, LV $24.98** *FOX, IME* ♫♫

Amazing Stories, Book 1 From the Steven Spielberg produced television series. ''The Mission'' sees a World War II tail gunner trapped in an unusual predicament. In ''The Wedding Ring,'' a man gives his wife a strange ring and bizarre situations ensue.

1985 70m/C Kevin Costner, Casey Siemaszko, Kiefer Sutherland, Danny DeVito, Rhea Perlman; *Dir:* Danny DeVito, Steven Spielberg. **VHS, LV $79.95** *MCA, FCT* ♫♫

Amazing Stories, Book 2 Two episodes from the short-lived ''Amazing Stories'' television series. In ''Go To the Head of the Class'' (1986), a cruel English teacher is subjected to black magic by a pair of vengeance-minded students. ''Family Dog'' (1987), is an animated feature designed by Tim Burton about the life of a suburban family from the canine perspective.

1986 71m/C Christopher Lloyd, Scott Coffey, Mary Stuart Masterson; *Dir:* Robert Zemeckis, Brad Bird; *Voices:* Stan Freberg, Annie Potts, Mercedes McCambridge. **VHS, LV $79.95** *MCA* ♫♫

Amazing Stories, Book 3 Three stories from the short-lived series produced by Stephen Spielberg. In ''Life on Death Row,'' Patrick Swayze is a criminal who acquires a great new power. Gregory Hines is ''The Amazing Falsworth,'' a nightclub psychic who uncovers a mass murderer by accident. Then an army outcast, played by Charlie Sheen, performs a miracle to save his fellow soldiers from certain death in ''No Day at the Beach.''

1986 73m/C Patrick Swayze, James Callahan, Gregory Hines, Richard Masur, Charlie Sheen, Ralph Seymour, Philip McKeon, Steven Gethers, Tom Hodges; *Dir:* Mick Garris, Peter Hyams, Lesli Linka Glatter. **VHS, LV $79.95** *MCA, PIA* ♫♫

Amazing Stories, Book 4 Three fantastic tales from the Spielberg-produced television show. ''Mirror, Mirror'' (1985; directed by Martin Scorsese) finds a director of horror films tormented by the very monsters he created. ''Blue Man Down'' (1986; directed by Paul Michael Glaser) tells of a cop who is helped by a mysterious young woman after his partner is shot dead. In ''Mr. Magic'' (1985; directed by Donald Petrie) an aging magician has his powers revitalized thanks to a strange deck of cards.

1985 71m/C Sam Waterston, Helen Shaver, Dick Cavett, Max Gail, Kate McNeil, Chris Nash, Sid Caesar, Leo Rossi; *Dir:* Martin Scorsese, Paul Michael Glaser, Donald Petrie. **VHS, LV $79.95** *MCA, PIA* ♫♫

Amazing Stories, Book 5 More Spielberg produced tales from the television series. ''The Pumpkin Competition'' is a comedy about a parsimonious widow who enters a pumpkin growing contest. A young girl vanishes for two decades in ''Without Diana,'' and a gang of space travellers collide head on with a trio of high school students working on a science project in ''Fine Tuning.''

1985 73m/C June Lockhart, Polly Holliday, J.A. Preston, Billy ''Green'' Bush, Dianne Hull, Gennie James, Matthew Laborteaux, Gary Riley, Jimmy Gatherum, Milton Berle; *Dir:* Norman Reynolds, Lesli Linka Glatter, Bob Balaban. **VHS, LV $79.95** *MCA* ♫♫

The Amazing Transparent Man A mad scientist makes a crook invisible in order to steal the radioactive materials he needs. The crook decides to rob banks instead. Shot at the Texas State Fair for that elusive futuristic look. For Ulmer fans only.

1960 58m/B Douglas Kennedy, Marguerite Chapman, James Griffith, Ivan Triesault; *Dir:* Edgar G. Ulmer. **VHS $16.95** *NOS, SNC Woof!*

The Amazing Transplant A sleazebag psycho has a ''love enhancing'' transplant, much to the pleasure of his sexual partners. Much to their and our dismay, however, he then kills them and bores us.

1970 90m/C Juan Fernandez, Linda Southern; *Dir:* Doris Wishman. **VHS, Beta $49.95** *AVD Woof!*

Amazon An adventure movie filmed in the Brazilian rainforest with an environmental message. A businessman being chased by the police in the Amazon jungle is rescued by a bush pilot who dreams of mining the Amazon's riches. A Brazilian woman enters the picture and persuades the businessman to help save the rainforest. Portions of the proceeds from the sale of this film go to the Rainforest Action Network.

1990 (R) 88m/C Kari Vaananen, Robert Davi, Rae Dawn Chong; *Dir:* Mika Kaurismaki. **VHS, LV $89.98** *LIV* ♫

Amazon Jail Scantily clad women go over the wall and promptly get caught by devil worshiping men in the jungle. They should have known better. Redeemed only by lingerie selection.

1985 94m/C Elisabeth Hartmann, Mauricio Do Valle, Sondra Graffi; *Dir:* Oswald De Oliveira. **VHS, Beta $59.98** *CON* ♫

Amazon Women On the Moon A plotless, irreverent media spoof, depicting the programming of a slipshod television station as it crams weird commercials and shorts around a comical '50s' science fiction film. Inconsistent, occasionally funny anthology hangs together very loosely. Produced by Landis, with the usual amount of in-joke cameos and allusions to his other works of art.

1987 (R) 85m/C Rosanna Arquette, Steve Guttenberg, Steve Allen, B.B. King, Michelle Pfeiffer, Arsenio Hall, Andrew Dice Clay, Howard Hesseman, Lou Jacobi, Carrie Fisher, Griffin Dunne, Sybil Danning, Henny Youngman, Monique Gabrielle, Paul Bartel, Kelly Preston; *Dir:* John Landis, Joe Dante, Carl Gottlieb, Peter Horton, Robert Weiss. **VHS, Beta, LV $79.95** *MCA* ♫♫

The Amazons Goofy made for television drama about a beautiful doctor who discovers an underground organization of Amazon-descended women bent on taking over the world or at least making life hard for men while investigating a Congressman's mysterious death.
1984 100m/C Tamara Dobson, Jack Scalia, Stella Stevens, Madeleine Stowe, Jennifer Warren; ***Dir:*** Paul Michael Glaser. **VHS, Beta $79.95** *WES* 🦴½

Amazons Tall, strong women who occasionally wander around nude search for a magical talisman that will overthrow an evil magician.
1986 (R) 76m/C Windsor Taylor Randolph, Penelope Reed, Joseph Whipp, Willie Nelson, Danitza Kingsley; ***Dir:*** Alex Sessa. **VHS, Beta $79.95** *MGM* 🦴

The Ambassador An American ambassador (Mitchem) is sent to the Middle East to try to solve the area's deep political problems. He quickly becomes the target of terrorist attacks, and is blamed for the nation's unrest. To make matters worse, his wife is having an affair with a PLO leader. The President ignores him, forcing the ambassador to fend for himself. Talk about your bad days. Hudson's last feature. Based on Elmore Leonard's "52 Pickup," and remade a year later under its own title.
1984 97m/C Robert Mitchum, Ellen Burstyn, Rock Hudson, Fabio Testi, Donald Pleasence; ***Dir:*** J. Lee Thompson. **VHS, Beta $79.95** *MGM* 🦴🦴½

Ambassador Bill An Oklahoma rancher is appointed ambassador to a country in revolt. Rogers, of course, saves the day with his rustic witticisms. Based on the story "Ambassador from the United States" by Vincent Sheean.
1931 68m/B Will Rogers, Marguerite Churchill, Greta Nissen; ***Dir:*** Sam Taylor. **VHS $19.98** *FOX, FCT* 🦴🦴

The Ambassador's Daughter The Parisian adventures of an American ambassador's daughter De Havilland and soldier Forsythe, who, unaware of her position, falls in love with her. Faltering comedy written by Krasna is supported by an expert cast (although nearly 40, De Havilland is charming as the young woman) with especially good performances from Menjou and Loy.
1956 102m/C Olivia de Havilland, John Forsythe, Myrna Loy, Adolphe Menjou, Edward Arnold, Francis Lederer, Tommy Noonan, Minor Watson; ***Dir:*** Norman Krasna. **VHS** *MOV* 🦴🦴

Amber Waves Credible made for television drama about a generation-gap conflict between a Midwestern farmer (Weaver) and an irresponsible male model (Russell) who has coasted through life on con and charm.
1982 98m/C Dennis Weaver, Kurt Russell; ***Dir:*** Joseph Sargent. **VHS, Beta** *TLF* 🦴🦴½

Ambition Scriptwriter/star Phillips gets a bone for chutzpah by taking an unsavory lead role; his character torments a paroled psycho so that the killer will kill again and inspire a true-crime bestseller. The plot looks good on paper, but onscreen it's padded and unconvincing.
1991 (R) 99m/C Lou Diamond Phillips, Clancy Brown, Cecilia Peck, Richard Bradford, Willard Pugh, Grace Zabriskie, Haing S. Ngor; ***Dir:*** Scott Goldstein. **VHS, Beta, LV $92.98** *FXV, MED* 🦴½

The Ambulance A New York cartoonist witnesses a mysterious ambulance at work and decides to investigate. His probings uncover a plot to sell the bodies of dying diabetics. A surprisingly good no-money feature from low-budget king Cohen. Includes an appearance by Marvel Comics' Stan Lee as himself.
1990 ?m/C Eric Roberts, James Earl Jones, Megan Gallagher, Red Buttons, Eric Braeden, Laurene Landon, Jill Gatsby; ***W/Dir:*** Larry Cohen. **VHS** *COL* 🦴🦴½

The Ambush Murders True story of a stalwart white attorney defending a black activist accused of killing two cops. Not the compelling television it could be, but still enjoyable. From Ben Bradlee, Jr.'s novel.
1982 100m/C James Brolin, Dorian Harewood, Alfre Woodard, Louis Giambalvo, John McLiam, Teddy Wilson, Antonio Fargas, Amy Madigan; ***Dir:*** Steven Hilliard Stern. **VHS, Beta $49.95** *FRH, IVE* 🦴🦴½

Ambush at Tomahawk Gap Three released prisoners go in search of hidden loot and tempers rise when the goods don't turn up. Then the Apaches show up.
1953 73m/C John Hodiak, John Derek, David Brian, Maria Elena Marques, Ray Teal; ***Dir:*** Fred F. Sears. **VHS, Beta** *GKK* 🦴🦴

The Ambushers Martin's third Matt Helm farce finds him handling a puzzling case involving the first United States spacecraft. When the craft is hijacked with Rule on board, it's Matt to the rescue, regaining control before unfriendly forces can take it back to Earth. Tired formula seems to have worn Martin out while the remainder of the cast goes to camp. Followed by "The Wrecking Crew."
1967 102m/C Dean Martin, Janice Rule, James Gregory, Albert Salmi, Senta Berger; ***Dir:*** Henry Levin. **VHS, Beta, LV $14.95** *COL Woof!*

Amenaza Nuclear A Mexican farce about two idiots who try to stop mysterious Chinamen from stealing nuclear arms.
19?? 106m/C *MX* Emeterio Y. Felipes, Lydia Zamora, Ki Jeong Lee. **VHS, Beta $59.95** *MAD Woof!*

America New York cable station receives worldwide fame when their signal bounces off of the moon. Uninspired piece of fluff from otherwise talented director Downey.
1986 (R) 83m/C Zack Norman, Tammy Grimes, Michael J. Pollard, Monroe Arnold, Richard Belzer, Liz Torres, Howard Thomashefsky, Laura Ashton, Robert Downey Jr.; ***Dir:*** Robert Downey. **VHS $79.95** *SVS* 🦴½

America 3000 Hundreds of years in the holocaust-torn future, men rebel against a brutal, overpowering race of women, with predictable results.
1986 (PG-13) 94m/C Chuck Wagner, Laurene Landon; ***Dir:*** David Engelbach. **VHS, Beta $79.95** *MGM* 🦴

America First A group of seven travelers try to build an "Eden" with the inhabitants of an Appalachian hollow.
1970 90m/C Michael Kennedy, Walter Keller, Pat Estrin, Lois McGuire. **VHS, Beta** *UTM* 🦴½

America Live in Central Park Conceptual sequences combine with stirring performances as the folk-rock group America offers this concert of hits including "Tin Man," "Ventura Highway," "Horse with No Name." In stereo.
1981 53m/C Beta, LV $24.95 *PIA* 🦴½

America at the Movies Scenes from more than 80 of the finest American motion pictures fly by in an effort to tell the story of the cinema and provide a portrait of America as it has been seen on screen for half a century. Scenes from "The Birth of a Nation," "Citizen Kane," "Dr. Strangelove," "East of Eden," "The French Connection," and "From Here to Eternity," are among the many included. Some black and white scenes; produced by the American Film Institute.
1976 116m/C John Wayne, Orson Welles, Peter Sellers, James Dean, Gene Hackman, Burt Lancaster, Julie Harris, Deborah Kerr, Al Pacino, Robert De Niro. **VHS, Beta, LV $59.95** *COL, IME* 🦴🦴½

America/The Fall of Babylon A double feature containing abridged versions; in "America," a Boston patriot and the daughter of an aristocratic Virginia Tory fall in love during the Revolutionary War; "The Fall of Babylon" is one of the stories in D. W. Griffith's "Intolerance." Silent.
1924 56m/B Neil Hamilton, Lionel Barrymore, Constance Talmadge, Elmer Clifton, Alfred Paget. **VHS, Beta $29.98** *CCB* 🦴🦴🦴

The American Angels: Baptism of Blood Three beautiful young women, each with a personal dream, strive to make it in the world of professional wrestling. Hackneyed plot devices, but those simply watching for the wrestling scenes won't be disappointed.
1989 (R) 99m/C Jan MacKenzie, Tray Loren, Mimi Lesseos, Trudy Adams, Patricia Cavoti, Susan Sexton, Jean Kirkland, Jeff Lundy, Lee Marshall; ***Dir:*** Beverly Sebastian, Ferd Sebastian. **VHS, Beta, LV $79.95** *PAR* 🦴

American Anthem A young gymnast must choose between family responsibilities or the parallel bars. Olympic gymnast Gaylord makes his movie debut but doesn't get the gold. Good fare for young tumblers, but that's about it. Followed by two of the films' music videos and tape-ads featuring Max Headroom.
1986 (PG-13) 100m/C Mitch Gaylord, Janet Jones, Michelle Phillips, Michael Pataki; ***Dir:*** Albert Magnoli. **VHS, Beta, LV $19.98** *LHV, WAR* 🦴

American Aristocracy A silent comic romp in which Fairbanks, an old moneyed dandy, wreaks havoc on an island resort whose clientele is composed of the nouveau well-to-do.
1917 52m/B Douglas Fairbanks Sr., Jewel Carmen, Albert Parker; ***Dir:*** Lloyd Ingraham. **VHS, Beta $16.95** *GPV* 🦴🦴½

American Autobahn A low-budget independent actioner about a journalist who discovers an underworld weapons ring that hits the highway in pursuit of him.
1984 90m/C Jan Jalenak, Michael von der Goltz, Jim Jarmusch; ***Dir:*** Andre Degas. **VHS, Beta $79.95** *SVS* 🦴½

American Blue Note Loosely plotted, bittersweet account of a struggling jazz quartet in the early 1960s, as its leader (MacNichol) must decide if their fruitless tours of sleazy bars and weddings are still worth it. The debut of director Toporoff.
1991 (PG-13) 96m/C Peter MacNicol, Charlotte d'Amboise, Trini Alvarado, Carl Capotorto, Tim Guinee; ***Dir:*** Ralph Toporoff. **VHS, LV $89.95** *COL* 🦴🦴

American Born Murder Inc. returns to the dismay of one idealist who embarks on a battle he doesn't intend to lose. The mob had better look out.
1989 90m/C Joey Travolta, Andrew Zeller. **VHS** *PMH* 🦴½

American Boyfriends An ostensible sequel to "My American Cousin," in which two Canadian girls go to California and discover innocent romance and friendship. **1989 (PG-13) 90m/C** *CA* Margaret Langrick, John Wildman, Jason Blicker, Lisa Repo Martell; *Dir:* Sandy Wilson. **VHS, Beta, LV $79.95** *IVE* 🦴🦴

An American Christmas Carol Charles Dickens' classic story is retold with limited charm in a made for television effort. This time a greedy American financier (Winkler) learns about the true meaning of Christmas. **1979 98m/C** Henry Winkler, David Wayne, Dorian Harewood; *Dir:* Eric Till. **VHS, Beta $14.98** *VES, LIV* 🦴🦴

American Commandos An ex-Green Beret slaughters the junkies who killed his son and raped his wife, and then joins his old buddies for a secret, Rambo-esque mission in Vietnam providing a tired rehash of Vietnam movie cliches. **1984 (R) 96m/C** Chris Mitchum, John Phillip Law, Franco Guerrero; *Dir:* Bobby Suarez. **VHS, Beta $79.98** *LTG, LIV* 🦴

American Dream A midwestern family leaves the suburbs and moves into a Chicago inner-city neighborhood. Good TV-movie pilot for the short-lived series that was Emmy nominated for direction and writing. **1981 90m/C** Stephen Macht, Karen Carlson, John Karlen, Andrea Smith, John Malkovich, John McIntire; *Dir:* Mel Damski. **VHS, Beta $29.95** *UNI* 🦴🦴🦴

American Dreamer A housewife wins a trip to Paris as a prize from a mystery writing contest. Silly from a blow on the head, she begins living the fictional life of her favorite literary adventure. Sporadic comedy with a good cast wandering about courtesy of a clumsy screenplay. **1984 (PG) 105m/C** JoBeth Williams, Tom Conti, Giancarlo Giannini, Coral Browne, James Staley; *Dir:* Rick Rosenthal. **VHS, Beta $19.98** *FOX* 🦴🦴

American Drive-In A guffaw-laden teenage comedy about a suburban drive-in theater. The movie being shown at the theater looks suspiciously like the director's previous film, "Hard Rock Zombies." Never released theatrically, for good reason. **1988 92m/C** Emily Longstreth, Joel Bennett, John Rice, Pat Kirton, Rhonda Snow; *Dir:* Krishna Shah. **Beta $79.98** *VES, LIV* 🦴

American Eagle A veteran goes crazy and seeks sadistic, bloody revenge on his war buddies. Now his war buddies are the only ones who can stop him. **1990 (R) 92m/C** Asher Brauner, Robert F. Lyons, Vernon Wells, Kai Baker; *Dir:* Robert J. Smawley. **VHS, LV $89.95** *VMK* 🦴½

American Empire Two Civil War heroes struggle to build a cattle empire in Texas and are hampered by rustlers, one of whom was their partner. A fine, veteran cast and a tight script keep things moving, including the cattle. **1942 82m/B** Preston Foster, Richard Dix, Frances Gifford, Leo Carrillo; *Dir:* William McGann. **VHS, Beta $19.95** *NOS, MRV, DVT* 🦴🦴½

American Film Institutes Life Achievement Awards: Alfred Hitchcock The American Film Institute honors Hitchcock's varied film-making career. Includes excerpts from "The 39 Steps," "Vertigo," and "Psycho." **1989 72m/C VHS $19.95** *WOV*

American Film Institutes Life Achievement Awards: Bette Davis The American Film Institute honors Davis' long and colorful film career. Excerpts include "Now Voyager," "Dark Victory," "Forever Amber," and others. **1989 68m/C VHS $19.95** *WOV*

American Film Institutes Life Achievement Awards: Henry Fonda The American Film Institute honors Henry Fonda for his exceptional film career. Some of Hollywood's biggest celebrities tell of their great admiration for this remarkable man. **1989 97m/C VHS, Beta $19.95** *WOV*

American Film Institutes Life Achievement Awards: James Cagney The American Film Institute honors James Cagney for his wide and exceptional film career. Includes excerpts from "Yankee Doodle Dandy," "The Seven Little Foys," and "Love Me or Leave Me." **1989 74m/C VHS $19.95** *WOV*

American Film Institutes Life Achievement Awards: Jimmy Stewart A tribute to one of America's best loved actors, featuring film clips and interviews with many of Hollywood's biggest names. Includes excerpts from "It's a Wonderful Life," "Harvey," "Vertigo," and others. **1989 71m/C** *Hosted:* Henry Fonda. **VHS, Beta $19.95** *WOV*

American Flyers Two competitive brothers train for a grueling three-day bicycle race in Colorado while tangling with personal drama, including the spector that one of them may have inherited dad's tendency for cerebral aneurisms and is sure to drop dead during a bike race soon. Written by bike movie specialist Steve Teisch ("Breaking Away") with a lot of the usual cliches (the last bike ride, battling bros, eventual understanding), which are gracefully overridden by fine bike-racing photography. Interesting performances, especially Chong as a patient girlfriend and Amos as the trainer. **1985 (PG-13) 113m/C** Kevin Costner, David Grant, Rae Dawn Chong, Alexandra Paul, John Amos, Janice Rule, Robert Townsend, Jennifer Grey; *Dir:* John Badham. **VHS, Beta $19.95** *WAR* 🦴🦴½

American Friend A tribute to the American gangster film that helped introduce Wenders to American moviegoers. A young Hamburg picture framer who thinks he has a terminal disease is set up by American expatriate (Hopper) to become a hired assassin in West Germany. The lure is a promise of quick money that the supposedly dying man can then leave his wife and child. After the first assasination, the two bond. Hopper is the typical Wenders protagonist, a strange man in a strange land looking for a connection, in this case the picture framer and his family. A great, creepy thriller from the acclaimed director adapted from Patricia Highsmith's novel "Ripley's Game." Watch for directors Fuller and Ray as gangsters. **1977 127m/C** *GE* Bruno Ganz, Dennis Hopper, Samuel Fuller, Nicholas Ray, Wim Wenders; *Dir:* Wim Wenders. National Board of Review Awards '77: 5 Best Foreign Films of the Year; New York Film Festival '77: 5 Best Foreign Films of the Year. **VHS, Beta, LV $29.95** *PAV, FCT, GLV* 🦴🦴🦴🦴

The American Gangster Documentary look at the beginnings of organized crime and the rise of the gangster in American society. Includes vintage film and photographs. Profiles include Bugsy Siegel, Al Capone, John Dillinger, Pretty Boy Floyd, and Lucky Luciano. **1992 45m/C VHS $14.95** *COL* 🦴🦴🦴🦴

American Gigolo A Los Angeles loner who sexually services the rich women of Beverly Hills becomes involved with the wife of a California state senator and is then framed for a murder he did not commit. A highly stylized but empty view of seamy low lives marred by a contrived plot, with decent romp in the hay readiness displayed by Gere. **1979 (R) 117m/C** Richard Gere, Lauren Hutton, Hector Elizondo, Paul Schrader; *Dir:* Paul Schrader. **VHS, Beta, LV $59.95** *PAR* 🦴🦴

American Gothic Three couples headed for a vacation are instead stranded on an island and captured by a demented family headed by Steiger and De Carlo, a scary enough proposition in itself. Even worse, Ma and Pa have three middle-aged moronic offspring who still dress as children and are intent on killing the thwarted vacationers (who are none too bright themselves) one by bloody one. A stultifying career low for all involved. **1988 (R) 89m/C** Rod Steiger, Yvonne de Carlo, Michael J. Pollard, Sarah Torgov, Fiona Hutchinson, Willaim Wright; *Dir:* John Hough. **VHS, Beta, LV $89.95** *VMK Woof!*

American Graffiti Lucas' poignant autobiographical take on the early 60s is an atmospheric, episodic look at growing up in the innocence of America before the Kennedy assassination and the Vietnam War. It all takes place on one hectic but typical night in the life of a group of Californian high school friends who've just graduated and are unsure of what the next big step in life is. So they spend their time cruising, girl chasing, listening to Wolfman Jack, and meeting at the drive-in. Successfully creates a slice-of-life for the era with a prudent script, great set design, authentic soundtrack, and consistently fine performances by the young cast, many of whom went on to stardom. Garnered five Oscar nominations but strangely did not win any, although responsible for catapulting Dreyfuss to immediate stardom, branding Lucas a hot directorial commodity with enough leverage to launch "Star Wars," and leading Howard and Williams to find continued age-of-innocence nirvana on "Happy Days." A box-office success that created a minor industry for 60's inspired flicks (including "Diner" and 1979's "More American Graffiti," which had some of the same cast but was not Lucas led), it was produced by Francis Ford Coppola and Gary Kurtz and written by Lucas, Gloria Katz, and William Huyck. **1973 (PG) 112m/C** Richard Dreyfuss, Ron Howard, Cindy Williams, MacKenzie Phillips, Paul LeMat, Charles Martin Smith, Suzanne Somers, Candy Clark, Harrison Ford, Bo Hopkins, Joe Spano, Kathleen Quinlan, Wolfman Jack; *Dir:* George Lucas. Golden Globe Awards '74: Best Film—Musical/Comedy; New York Film Critics Awards '73: Best Screenplay; National Society of Film Critics Awards '73: Best Screenplay. **VHS, Beta, LV $14.95** *MCA, FCT* 🦴🦴🦴½

American Justice Two cops, one of whom looks suspiciously like a Simon of "Simon and Simon" fight political corruption and white slavery near the Mexican border. The chief movie slaver bears a full resemblance to the other Simon. Sufficient action

but less than original. Also known as "Jack-als."
1986 (R) 96m/C Jameson Parker, Gerald McRaney, Wilford Brimley, Jack Lucarelli; *Dir:* Gary Grillo. **VHS, Beta $79.98** *LTG, LIV* 🎞🎞

American Kickboxer 1 A novel story: two fighters kick and make funny noises for honor and revenge. Most excellent fight scenes.
1991 (R) 93m/C Keith Vitali, John Barrett, Terry Norton, Ted Leplat, Brad Morris; *Dir:* Frans Nel. **VHS, Beta $89.99** *CAN* 🎞½

American Me A violent and brutal film depicting more than thirty years of gang wars and drugs in East Los Angeles. Santana has founded and led a street gang since his teenage years. He's spent the last 18 years in Folsom State Prison, where he and his cohorts are the bosses of the so-called Mexican Mafia, which includes the drugs, scams, murders, and violence that are an everyday fact of prison life. Released from Folsom, Santana goes back to his old neighborhood and attempts to distance himself from his old life but finds his gang ties are stronger than any other alliance. Unsparing and desolate. Olmos' directorial debut.
1992 (R) 119m/C Edward James Olmos, William Forsythe, Pepe Serna, Danny De La Paz, Evelina Fernandez, Daniel Villarreal; *Dir:* Edward James Olmos. **VHS, LV** *MCA* 🎞🎞🎞

American Nightmare A young man searches for his missing sister against a background of pornography, drug peddling and prostitution in the slums of a city. The usual titillating squalid urban drama.
1981 85m/C Lawrence S. Day, Lora Staley, Lenore Zann, Michael Ironside, Alexandra Paul; *Dir:* Don McBrearty. **VHS, Beta $54.95** *MED* 🎞½

American Ninja American Dudikoff is G.I. Joe, a martial-arts expert stationed in the Philippines who alienates most everyone around him (he's a rebel). A deadly black-belt war begins with Joe confronting the army which is selling stolen weapons to the South American black market. Aided by one faithful pal, Joe uses his head-kicking martial arts skills to stop hundreds of Ninja combatants working for the corrupt arms dealer. In his spare time he romances the base chief's daughter. Efficient rib-crunching chop-socky action wrapped in no-brainer plot and performed by nonactors. Mercilessly followed by at least three sequels. Also known as "American Warrior."
1985 (R) 90m/C Michael Dudikoff, Guich Koock, Judie Aronson, Steve James; *Dir:* Sam Firstenberg. **VHS, Beta $19.95** *MGM* 🎞

American Ninja 2: The Confrontation Soldiers Dudikoff and James are back again using their martial arts skills (in lieu of any acting) to take on a Caribbean drug-lord. Apparently he has been kidnapping Marines and taking them to his island, where he genetically alters them to become fanatical ninja assassins eager to do his dirty work. The script by actors Conway and James Booth hardly gets in the way of the rib-crunching action, but is an improvement upon Ninja Number One.
1987 (R) 90m/C Michael Dudikoff, Steve James, Larry Poindexter, Gary Conway; *Dir:* Sam Firstenberg. **VHS, Beta, LV $19.95** *MED* 🎞🎞

American Ninja 3: Blood Hunt Third sequel in the American Ninja series. Bradley replaces Dudikoff as the martial arts good guy fighting the martial arts bad guys. More of the same...only worse.

1989 (R) 90m/C David Bradley, Steve James, Marjoe Gortner, Michele Chan, Calvin Jung; *Dir:* Cedric Sundstrom. **VHS, Beta, LV $19.98** *CAN* Woof!

American Ninja 4: The Annihilation More ninja action as Dudikoff ("American Ninja 1 & 2") and Bradley ("American Ninja 3") team up to beat up the bad guys.
1991 (R) 99m/C Michael Dudikoff, David Bradley. **VHS, Beta, LV $89.99** *CAN, LDC* Woof!

An American in Paris Lavish and imaginative musical production scripted by Alan Jay Lerner, with a sweeping score by Ira and George Gershwin and knockout choreography by Kelly. Kelly is an ex-G.I. who stays on in Paris after the war to study painting. Rich American Foch decides to support his efforts, in the hopes of acquiring a little extra attention. But Kelly falls for lovely French lady Caron, who unfortunately is scheduled to marry an older gent. Kelly tries desperately (while dancing up a storm) to win her heart. Highlight is an astonishing 17-minute ballet choreographed by Kelly, holding the record for longest movie dance number (and one of the most expensive at more than a half million for a month of filming). For his efforts, the dance king won a special Oscar citation. Winner of six Academy Awards in all, it also garnered Minnelli a nomination as best director and a nod for film editing. While it sure looks like Paris, most of it was filmed in MGM studios. Songs include "S Wonderful," "I Got Rhythm," "Embraceable You," and "Love is Here to Stay."
1951 113m/C Gene Kelly, Leslie Caron, Oscar Levant, Nina Foch, Georges Guetary; *Dir:* Vincente Minnelli. *Academy Awards '51:* Best Costume Design (Color), Best Art Direction/Set Decoration (Color), Best Color Cinematography, Best Picture, Best Story & Screenplay, Best Musical Score; *Golden Globe Awards '52:* Best Film—Musical/Comedy; *National Board of Review Awards '51:* 10 Best Films of the Year. **VHS, Beta, LV, 8mm $19.98** *MGM, TLF, BTV* 🎞🎞🎞🎞

American Pluck Before he can inherit anything, a playboy millionaire's son must go out into the world and prove he can make his own way. He meets a beautiful princess, who is pursued by a villainous count. Silent with original organ music.
1925 91m/B George Walsh, Wanda Hawley. **VHS, Beta, 8mm $29.99** *VYY* 🎞🎞

American Roulette Garcia is the exiled president of a Latin American country living in London in this thin political thriller. A plot riddled with weaknesses and poor direction make this potentially interesting film dull and lifeless.
1988 (R) 102m/C *GB AU* Andy Garcia, Kitty Aldridge, Robert Stephens, Al Matthews, Susannah York; *Dir:* Maurice Hatton. **VHS, LV $79.95** *VIL* 🎞

The American Scream An innocent family vacationing in the mountains stumbles on to a satanic cult. And that can really ruin a vacation.
198? 85m/C Jennifer Darling, Pons Marr, Blackie Cammett, Kimberly Kramer, Jean Sapienza. **VHS, Beta $59.95** *GHV* 🎞½

The American Soldier Fassbinder's homage to the American gangster film tells the story of Ricky, a charismatic hit man. Ricky always wears a gun in a shoulder holster, sports a fedora and a white double-breasted suit, and drinks Scotch straight from the bottle. He also carries out his assigned murders with complete efficiency

and no emotion. In German with English subtitles.
1970 80m/C *GE* Karl Scheydt, Elga Sorbas, Jan George; *W/Dir:* Rainer Werner Fassbinder. **VHS $79.95** *NYF* 🎞🎞½

An American Summer A Chicago kid spends a summer with his aunt in beach-rich Los Angeles in this coming-of-age tale filled with 90's teen idols.
1990 100m/C Brian Austin Green, Joanna Kerns, Michael Landes. **VHS $89.95** *SVS* 🎞🎞

An American Tail While emigrating to New York in the 1880s, a young Russian mouse (Fievel) is separated from his family. He matures as he learns to live on the Big Apple's dirty boulevards. The bad guys are of course cats. Excellent animation and a high-minded though sentimental and stereotypical plot keep it interesting for adults. Produced by Spielberg and the first big hit for the Bluth factory, a collection of expatriate Disney artists. Aggressively marketed upon release in the hopes of creating Fievel Fever, the flick includes the Oscar-nominated song (and platinum seller) "Somewhere Out There." Knowing better than to let a money-making mouse tale languish, Bluth followed with an "An American Tale: Fievel Goes West."
1986 (G) 81m/C *Dir:* Don Bluth; *Voices:* Dom DeLuise, Madeline Kahn, Phillip Glasser, Christopher Plummer, Nehemiah Persoff, Will Ryan. **VHS, Beta, LV $24.98** *MCA, PIA, FCT* 🎞🎞½

An American Tail: Fievel Goes West Fievel and the Mousekewitz family continue their pursuit of the American dream by heading West, where the intrepid mouse seeks to become a famous lawman while his sister looks to make it big as a dance hall singer. Songs are "Somewhere Out There," "Way Out West," "Rawhide," "Dreams to Dream," and "The Girl I Left Behind," all performed by the London Symphony Orchestra. The laser edition is letterboxed and features chapter stops.
1991 (G) 75m/C *Dir:* Phil Nibblink, Simon Wells; *Voices:* John Cleese, Dom DeLuise, Phillip Glasser, Amy Irving, Jon Lovitz, Cathy Cavadini, Nehemiah Persoff, Erica Yohn, James Stewart. **VHS, Beta, LV $24.95** *MCA* 🎞🎞

American Tickler A thigh-slappin' (or is it head-whacking?) series of satirical pastiches in the grand style of "Kentucky Fried Movie," only more sophomoric. A tasteless collection of yearning to be funny sketches about American institutions that's also known as "Draws."
1976 (R) 77m/C W.P. Dremak, Joan Sumner, Marlow Ferguson, Jeff Alin; *Dir:* Chuck Vincent. **VHS, Beta $59.95** *GEM* 🎞

American Tiger The collegiate hero of "American Tiger" is, like many college students, framed for murder. Using his new study skills, he applies himself to clearing his sullied name. Somewhere during this process, he finds himself in the middle of a battle between good and evil on a football field. About what you'd expect from a supernatural kung-fu teen-action drama.
1989 (R) 93m/C Mitch Gaylord, Donald Pleasence, Daniel Greene, Victoria Prouty; *Dir:* Martin Dolman. **VHS $19.95** *ACA* 🎞🎞

An American Werewolf in London Although not widely popular with American film critics, this strange and darkly humorous retelling of the classic man into wolf horror tale became a cult hit. It follows two American college students (Naughton

and Dunne) backpacking through England. One night they land at the Slaughtered Lamb Pub, whereupon leaving they are attacked viciously by a werewolf. Dunne is killed and Naughton seriously wounded. While recovering with the aid of a friendly nurse, Naughton is visited by his decomposing friend, who presents him with warnings of impending werewolfdom when the moon is full and advises suicide. Seat-jumping horror and gore, highlighted by intensive metamorphosis sequences orchestrated by Rick Baker, are offset by wry humor, though the shifts in tone don't always work. Landis being Landis, he also manages to mangle a great many automobiles for no particular reason. Great moon songs permeate the soundtrack, including CCR's "Bad Moon Rising" and Van Morrison's "Moondance." Naughton may seem familiar as the first Who's a Pepper We're A Pepper Dr. Pepper Guy.
1981 (R) 97m/C David Naughton, Griffin Dunne, Jenny Agutter, Frank Oz, Brian Glover; *Dir:* John Landis. Academy Awards '81: Best Makeup. **VHS, Beta, LV $29.98** *LIV, VES ♫♫♫*

Americana A troubled Vietnam vet tries to restore himself by rebuilding a merry-go-round in a small midwestern town, while dealing with opposition from the local residents. Offbeat, often effective post-Nam editorial that was produced and directed in 1973 by Carradine and then shelved. Hershey and Carradine were a couple back then.
1981 (PG) 90m/C David Carradine, Barbara Hershey, Bruce Carradine, John Blythe Barrymore, Michael Greene; *Dir:* David Carradine. **VHS, Beta $69.98** *VES, LIV ♫♫½*

The Americanization of Emily A happy-go-lucky American naval officer (Garner) with no appetite for war discovers to his horror that he may be slated to become the first casualty of the Normandy invasion as part of a military PR effort in this black comedy-romance. Meanwhile, he spreads the charisma in an effort to woo and uplift Emily (Andrews), a depressed English woman who has suffered the loss of her husband, father, and brother during the war. A cynical, often funny look at military maneuvers and cultural drift that was adapted by Paddy Chayefsky from William Bradford Huie's novel. Also available colorized.
1964 117m/B James Garner, Julie Andrews, Melvyn Douglas, James Coburn, Joyce Grenfell, Keenan Wynn, Judy Carne, Liz Fraser, Edward Binns; *Dir:* Arthur Hiller. **VHS $19.95** *MGM ♫♫♫*

The Americano Ever-suave Fairbanks frees a South American politician locked in a dungeon, returns him to political success, and captures the heart of his beautiful daughter. Based on Eugene P. Lyle, Jr.'s "Blaze Derringer."
1917 58m/B Douglas Fairbanks Sr., Alma Rubens, Spottiswoode Aitken, Lillian Langdon, Carl Stockade, Tom Wilson; *Dir:* John Emerson. **VHS, Beta $16.95** *GPV ♫♫½*

Americano A cowboy travelling to Brazil with a shipment of Brahma bulls discovers the rancher he's delivering them to has been murdered. An odd amalgam of western cliches and a South American setting.
1955 85m/C Glenn Ford, Frank Lovejoy, Abbe Lane, Cesar Romero; *Dir:* William Castle. **VHS $19.98** *REP ♫♫*

The Americano/Variety The two silent classics in abridged form: "The Americano," (1917) about a soldier-of-fortune who aids a revolt-ridden Caribbean country, and "Variety," (1925) a classic German masterpiece about the career of a trapeze artist .

1925 54m/B Douglas Fairbanks Sr., Emil Jannings; *Dir:* John Emerson, E.A. Dupont. **VHS, Beta $39.95** *CCB ♫♫*

America's Favorite Jokes A series of on-the-street interviews with a cross-section of the American public, all telling their favorite jokes.
1987 30m/C Beta $16.95 *RHI ♫♫*

Americathon It is the year 1998 and the United States is almost bankrupt, so President Chet Roosevelt decides to stage a telethon to keep the country from going broke. Interesting satiric premise with a diverse cast, but a poor script and slack pacing spoil all the fun. The soundtrack features music by The Beach Boys and Elvis Costello while the narration is by George Carlin. Tommy LaSorda and Peter Marshall do the cameo thing.
1979 (PG) 85m/C Peter Riegert, John Ritter, Nancy Morgan, Harvey Korman, Fred Willard, Meatloaf, Elvis Costello, Chief Dan George, Howard Hesseman, Jay Leno; *Dir:* Neal Israel. **VHS, Beta $59.95** *LHV, WAR Woof!*

Amin: The Rise and Fall Excessively violent and ultimately pointless dramatization of Idi Amin's eight-year reign of terror in Uganda, which resulted in the deaths of a half million people and the near ruin of a nation.
1982 101m/C Joseph Olita, Geoffrey Keen; *Dir:* Sharad Patel. **VHS, Beta $59.95** *EMI, HBO Woof!*

The Amityville Horror Sometimes a house is not a home. Ineffective chiller that became a box-office biggie, based on a supposedly real-life occurrence in Amityville, Long Island. The Lutz family moves into the house of their dreams only to find it full of nightmares. Once the scene of a grisly mass murder, the house takes on a devilish attitude, plunging the family into supernatural terror. Pipes and walls ooze icky stuff, flies manifest in the strangest places, and doors mysteriously slam while exorcist Steiger staggers from room to room in scene-chewing prayer. Based on the Jay Anson book and followed by a number of sequels.
1979 (R) 117m/C James Brolin, Margot Kidder, Rod Steiger, Don Stroud, Murray Hamilton, Helen Shaver, Amy Wright; *Dir:* Stuart Rosenberg. **VHS, Beta, LV** *WAR, OM ♫♫*

Amityville 2: The Possession More of a prequel than a sequel to "The Amityville Horror" (1979). Relates the story of the house's early years as a haven for demonic forces intent on driving a father to beat the kids, a mother to prayer, and a brother to lust after his sister before he murders them all. Young etc. portray an obnoxious family that you're actually glad to see wasted by the possessed son. A stupid, clumsy attempt to cash in on the success of the first film, which was also stupid and clumsy but could at least claim novelty in the bad housing development genre. Followed by "Amityville III: The Demon" in 1983.
1982 (R) 110m/C James Olson, Burt Young, Andrew Prine, Moses Gunn, Rutanya Alda; *Dir:* Damiano Damiani. **VHS, Beta, LV $9.98** *SUE ♫*

Amityville 3: The Demon America's worst real-estate value dupes another funky buyer. The infamous Amityville house is once again restless with terror and gore, though supported with even less plot than the usual smidgin. Cynical reporter Roberts moves in while trying to get to the bottom of the story by way of the basement. Courtesy of 3-D technology, monsters sprang at theater patrons but the video version is strictly two-

dimensional, forcing the viewer to press his or her face directly onto the television screen in order to derive the same effect. Also known as "Amityville 3D."
1983 (R) 98m/C Tony Roberts, Tess Harper, Robert Joy, Candy Clark, John Beal, Leora Dana, John Harkins, Lori Loughlin, Meg Ryan; *Dir:* Richard Fleischer. **VHS, Beta, LV $79.98** *VES, LIV ♫½*

Amityville 4 It's an unusual case of house-to-house transference as the horror from Amityville continues, now lodged in a Californian residence. The usual good house gone bad story has the place creating a lot of unusual creaks and rattles before deciding to use its inherited powers to attack and possess a little girl. Special effects from Richard Stutsman ("The Lost Boys" and "Jaws"). Made for television.
1989 (R) 95m/C Patty Duke, Jane Wyatt, Norman Lloyd, Frederic Lehne, Brandy Gold; *Dir:* Sandor Stern. **VHS $89.95** *VMK ♫½*

Amityville 1992: It's About Time Time is of the essence in the sixth installment of the Amityville flicks. A vintage clock, which is of course from Amityville, causes the creepy goings-on in a family's house. Actually a halfway decent horror film with high-grade special effects, and definitely much better than previous Amityville sequels.
1992 (R) 95m/C Stephen Macht, Shawn Weatherly, Megan Ward, Damon Martin, Nita Talbot; *Dir:* Tony Randel. **VHS, LV $89.98** *REP ♫♫*

The Amityville Curse The possessed house is yet again purchased by a pathetically uninformed family. The usual ghostly shenanigans occur with low-budget regularity in this fifth film in the series. Never released theatrically, for good reason.
1990 91m/C *CA* Kim Coates, Dawna Wightman, Helen Hughes, David Stein, Anthony Dean Rubes, Cassandra Gava, Jan Rubes; *Dir:* Tom Berry. **VHS, LV $89.95** *VMK Woof!*

Among the Cinders Sometimes interesting coming of age drama about a sixteen-year-old New Zealand boy who runs away to his grandfather's farm to forget about his friend's accidental death. At the farm, he meets an older woman and compromises his virtue.
1983 103m/C *GE NZ* Paul O'Shea, Derek Hardwick, Rebecca Gibney, Yvonne Lawley, Amanda Jones; *Dir:* Rolf Haedrich. **VHS, Beta, LV $9.95** *NWV, STE ♫♫*

Amor Bandido When cab drivers in Rio are turning up dead, a young prostitute is torn between her detective father and her prime suspect boyfriend. Drama based on a true story.
1979 95m/C *BR* Paulo Gracindo, Cristina Ache, Paulo Guarniero; *Dir:* Bruno Barreto. **VHS, LV $79.95** *FXL, IME, TAM ♫♫*

The Amorous Adventures of Moll Flanders An amusing romp set in 18th century England focusing on a poor orphan girl who seeks wealth. Moll plots to get ahead through an advantageous series of romances and marriages. Her plan is ruined when she falls in love and he turns out to be a wanted highwayman, landing her in prison. Not surprisingly, love (and money) conquers all. Based on the novel by Daniel Defoe. Novak tries in this female drivative of "Tom Jones," but this period piece isn't her style.
1965 126m/C Kim Novak, Richard Johnson, Angela Lansbury, Vittorio DeSica, Leo McKern, George Sanders, Lilli Palmer; *Dir:* Terence Young. **VHS $19.95** *PAR ♫♫*

Amos Douglas is Amos, an aging baseball coach confined most reluctantly to a nursing home. He's disturbed by recent suspicious events there, his concern causing him to take on Montgomery, the staunch head nurse. Well-acted drama produced by Douglas's son Peter, offering an echo of "One Flew Over the Cuckoo's Nest," which Dad starred in on Broadway and son Michael helped produced as a movie classic.
1985 100m/C Kirk Douglas, Elizabeth Montgomery, Dorothy McGuire, Pat Morita, James Sloyan, Ray Walston; ***Dir:*** Michael Tuchner. **VHS $29.95** *SVS* 🦴🦴½

The Amphibian Man What does a scientist do once he's created a young man with gills? Plunge him in the real world of aqua pura to experience life and love, albeit underwater. Trouble is, the protagonist, who's come to be known as the Sea Devil, takes a dive for a young pretty he's snatched from the jaws of death. A 60's Soviet sci-fi romance originally seen on American television.
1961 96m/C *RU* K. Korieniev, M. Virzinskaya; ***Dir:*** Y. Kasancki. **VHS $19.98** *SNC* 🦴🦴

The Amsterdam Connection A film company acts as a cover for prostitution and drug smuggling, with the girls acting as international couriers.
197? 90m/C *HK* Chen Shing, Jason Pai Piu, Kid Sherrif. **VHS, Beta $54.95** *MAV* 🦴

The Amsterdam Kill A washed-up ex-agent of the U.S. Drug Enforcement Agency is hired by a desperate U.S. Drug Enforcement Agency to hunt down the kingpin of a narcotics syndicate in Hong Kong. Tedious pace with occasional spells of violence fails to awaken somnambulent Mitchum. Shot on location in Hong Kong.
1978 (R) 93m/C *HK* Robert Mitchum, Richard Egan, Keye Luke, Leslie Nielsen, Bradford Dillman; ***Dir:*** Robert Clouse. **VHS, Beta, LV $9.95** *GKK* 🦴🦴

Amsterdamned A crime thriller taking place on the canals of Amsterdam featuring a serial skindiver killer who surfaces periodically to slash and splash. A steely detective fishes for clues. Dubbed.
1988 (R) 114m/C *NL* Monique Van De Ven, Huub Stapel; ***Dir:*** Dick Maas. **VHS, Beta, LV $89.98** *VES, LIV* 🦴🦴

Amy Set in the early 1900s, the story follows the experiences of a woman after she leaves her well-to-do husband to teach at a school for the deaf and blind. Eventually she organizes a football game between the handicapped kids and the other children in the neighborhood. Good Disney family fare.
1981 (G) 100m/C Jenny Agutter, Barry Newman, Kathleen Nolan, Margaret O'Brien, Nanette Fabray, Chris Robinson, Lou Fant; ***Dir:*** Vincent McEveety. **VHS, Beta $69.95** *DIS* 🦴🦴½

Anastasia Bergman won her second Oscar, and deservedly so, for her classic portrayal of the amnesia victim chosen by Russian expatriate Brynner to impersonate Anastasia, the last surviving member of the Romanoff dynasty. As such, she becomes part of a scam to collect millions of rubles deposited in a foreign bank by her supposed father, the now-dead Czar. But is she just impersonating the princess? Brynner as the scheming White General and Hayes as the Grand Duchess who needs to be convinced turn in fine performances as well. Based on Marcelle Maurette's play and Oscar-nominated for Best Score of a Drama or Comedy.
1956 105m/C Ingrid Bergman, Yul Brynner, Helen Hayes, Akim Tamiroff; ***Dir:*** Anatole Litvak. Academy Awards '56: Best Actress (Bergman). **VHS, LV** *TCF* 🦴🦴🦴½

Anastasia: The Mystery of Anna Irving stars as Anna Anderson, a woman who claimed to be the Grand Duchess Anastasia, the sole surviving daughter of Russian Czar Nicholas II. This powerful epic relives her experience of royalty, flight from execution, and struggle to retain her royal heritage. The story of Anastasia still remains as one of the greatest dramatic mysteries of the 20th Century. Adapted from the book "Anastasia: The Riddle of Anna Anderson" by Peter Kurth and originally aired on cable television.
1986 190m/C Amy Irving, Olivia de Havilland, Jan Niklas, Nicolas Surovy, Susan Lucci, Elke Sommer, Edward Fox, Claire Bloom, Omar Sharif, Rex Harrison; ***Dir:*** Marvin J. Chomsky. **VHS $79.95** *CAF* 🦴🦴½

Anatomy of a Murder Tersely and cleverly directed despite its length. Considered by many to be the best courtroom drama ever made, with a score by Duke Ellington to boot (and the Duke has a cameo). A small-town lawyer in northern Michigan faces an explosive case as he defends an army officer who has killed a man he suspects was his philandering wife's rapist. A realistic, cynical portrayal of the court system not especially concerned with the guilt or innocence of the officer, but instead focusing on the interplay between the various courtroom characters. It's a classic performance by Stewart as the down home but brilliant defense lawyer who matches wits with Scott, the sophisticated prosecutor. Though tame by today's standards, the language used in the courtroom was controversial for its time. Filmed in upper Michigan and based on the bestseller by Robert Traver, who was a judge in the area. Oscar nominated for Best Picture, Best Actor (Stewart), Best Supporting Actor (O'Connell & Scott), Screenplay, Cinematography, and Film Editing.
1959 161m/B James Stewart, George C. Scott, Arthur O'Connell, Ben Gazzara, Lee Remick, Duke Ellington, Eve Arden, Orson Bean; ***Dir:*** Otto Preminger. National Board of Review Awards '59: 10 Best Films of the Year; New York Film Critics Awards '59: Best Actor (Stewart), Best Writing (Wendell Mayes). **VHS, Beta, LV $19.95** *RCA* 🦴🦴🦴🦴

Anatomy of a Psycho A man plans to avenge his gas chambered brother by committing mass murder. Very cheaply and poorly produced.
1961 75m/B Ronnie Burns, Pamela Lincoln, Darrell Howe, Russ Bender; ***Dir:*** Brooke L. Peters. **VHS, Beta $16.95** *SNC, TIM* 🦴

Anatomy of a Seduction Television formula melodrama of a middle-aged divorcee's affair with her son's best friend. Age knows no boundaries when it comes to television lust.
1979 100m/C Susan Flannery, Jameson Parker, Rita Moreno, Ed Nelson, Michael Le Clair; ***Dir:*** Steven Hilliard Stern. **VHS, 8mm $79.99** *HBO* 🦴🦴

Anatomy of Terror A made for television drama about an army vet going bonkers (what? really?) and revealing espionage secrets in his home life.
1974 73m/C Paul Burke, Polly Bergen, Dinsdale Landen, Basil Henson. **VHS, Beta $29.95** *IVE* 🦴

Anchors Aweigh Snappy big-budget (for then) musical about two horney sailors, one a girl-happy dancer and the other a shy singer. While on leave in Hollywood they return a lost urchin to his sister. The four of them try to infiltrate a movie studio to win an audition for the girl from maestro Iturbi. Kelly's famous dance with Jerry the cartoon Mouse (of "Tom and Jerry" fame) is the second instance of combining live action and animation. The young and handsome Sinatra's easy crooning and Grayson's near operatic soprano are blessed with music and lyrics by Jule Styne and Sammy Cahn. Songs include "We Hate to Leave," "I Fall in Love Too Easily," "The Charm of You" and "The Worry Song." Lots of fun, with conductor-pianist Iturbi contributing and Hollywood style Little Mexico also in the brew. Oscar nominated for Best Picture, Best Actor (Kelly), Best Color Cinematography, and Best Song—"I Fall in Love..."
1945 139m/C Frank Sinatra, Gene Kelly, Kathryn Grayson, Jose Iturbi, Sharon McManus, Dean Stockwell, Carlos Ramirez, Pamela Britton; ***Dir:*** George Sidney. Academy Awards '45: Best Musical Score. **VHS, Beta, LV $19.95** *MGM* 🦴🦴🦴

And Another Honky Tonk Girl Says She Will A Tennessee farm girl's dreams are destroyed when she moves to Nashville to become a big star.
1990 30m/C ***Dir:*** Michelle Paymar. **VHS $59.95** *CIG* 🦴½

And Baby Makes Six An unexpected pregnancy creates new challenges for a couple with grown children. Dewhurst is excellent as usual. A television film, scripted by Shelley List. Followed by "Baby Comes Home."
1979 100m/C Colleen Dewhurst, Warren Oates, Maggie Cooper, Mildred Dunnock, Timothy Hutton, Allyn Ann McLerie; ***Dir:*** Waris Hussein. **VHS, 8mm $79.99** *HBO* 🦴🦴½

And God Created Woman Launching pad for Bardot's career as a sex siren, as she flits across the screen in a succession of scanty outfits and hangs out at the St. Tropez beach in what is euphemistically known as a swimsuit while turning up the heat for the males always in attendance. The plot concerns an 18-year-old nymphomaniac who is given a home by a local family with three handsome young sons. A cutting-edge sex film in its time that was boffo at the box office. In French with English subtitles and known originally as "Et Dieu Crea la Femme."
1957 (PG) 93m/C *FR* Brigitte Bardot, Curt Jurgens, Jean-Louis Trintignant, Christian Marquand; ***Dir:*** Roger Vadim. **VHS, Beta $29.95** *VES* 🦴🦴½

And God Created Woman A loose, dull remake by Vadim of his own 1957 softcore favorite about a free-spirited woman dodging men and the law while yearning for rock and roll stardom. DeMornay is a prisoner who hopes to marry one of the local hunks (Spano) so she can be paroled. They marry, she strips, he frets, while political hopeful Langella smacks his lips in anticipation. Available in an unrated 100-minute version.
1988 (R) 98m/C Rebecca DeMornay, Vincent Spano, Frank Langella, Donovan Leitch, Judith Chapman, Thelma Houston; ***Dir:*** Roger Vadim. **VHS, Beta, LV $89.98** *VES, LIV* 🦴½

And God Said to Cain Another Biblically titled western from the prolific Kinski, about a put-upon gunman who must fight for his life.
196? 95m/C Klaus Kinski. **VHS, Beta $29.95** *UNI* 🦴

And Hope to Die Moody, muddled crime drama with a good cast about a fleeing Frenchman who joins a gang of hardened criminals in Canada. They go ahead with their plans to kidnap a retarded girl even though she is already dead. Standard caper film is enhanced by arty camera work and some unusual directorial touches.
1972 95m/C *FR* Robert Ryan, Jean-Louis Trintignant, Aldo Ray, Tisa Farrow; *Dir:* Rene Clement. **VHS, Beta** $59.95 *MON ♂♂½*

And I Alone Survived Made-for-television adaptation of the true story of a woman who survives a leisure-plane crash in the Sierra Nevadas and her struggle to get to the village below. Brown does her best to elevate the proceedings above the usual I hope I don't die cliches.
1978 90m/C Blair Brown, David Ackroyd, Vera Miles, G.D. Spradlin; *Dir:* William A. Graham. **VHS, Beta** $79.95 *PSM ♂♂*

And If I'm Elected The Smothers Brothers satirize famous presidential television commercials.
1984 53m/C Dick Smothers, Tom Smothers. **VHS, Beta** $29.95 *TWE ♂♂*

And Justice for All Extremely earnest attorney Pacino questions the law and battles for justice in and out of the courtroom. He's hired to defend a detested judge from a rape charge, while dealing with a lost-soul caseload of eccentric and tragedy-prone clients (the suicidal transvestite, the young man sent to prison for a minor traffic violation, etc.). Overly melodramatic, with an odd mix of satire, cynicism, and seemingly sincere drama that tends to use a club when a stick will do. Working with a script by Barry Levinson and Valerie Curtin, Jewison aims for a black surrealism, permitting both Pacino and Warden (portraying a judge losing his sanity) to veer into histrionics, to the detriment of what is essentially a gripping behind-the-scenes story. An excellent cast creates sparks, including Lahti in her film debut (Pacino was Oscar nominated). And Baltimore never looked lovelier.
1979 (R) 120m/C Al Pacino, Jack Warden, Christine Lahti, Thomas G. Waites, Craig T. Nelson, John Forsythe, Lee Strasberg; *Dir:* Norman Jewison. **VHS, Beta, LV, 8mm** $14.95 *COL ♂♂½*

And Nothing But the Truth A multinational corporation is out to ruin the investigative television report team trying to do a story on the company. Well-intended drama originally titled "Giro City."
1982 90m/C *GB* Glenda Jackson, Jon Finch, Kenneth Colley, James Donnelly; *Dir:* Karl Francis. **VHS, Beta** $39.95 *MON ♂♂½*

And Now Miguel Plodding tale of a young boy who wants to take over as head shepherd of his family's flock. Filmed in New Mexico, but the over-long outdoor shots make it drag a bit in spite of Cardi's competent performance.
1966 95m/C Pat Cardi, Michael Ansara, Guy Stockwell, Clu Gulager, Joe De Santis, Pilar Del Rey, Buck Taylor; *Dir:* James B. Clark. **VHS** $24.95 *NOS ♂♂*

And Now My Love A French couple endeavor to maintain their romance despite interfering socio-economic factors—she's a millionaire and he's an ex-con filmmaker. Keller plays three roles spanning three generations, as Lelouch invests autobiographical details to invent a highly stylized, openly sentimental view of French folks in love with love. Along the way he comments on social

mores and changing attitudes through the years. Dubbed in English.
1974 (PG) 121m/C *FR* Marthe Keller, Andre Dussollier, Carla Gravina; *Dir:* Claude Lelouch. **VHS, Beta** $29.98 *SUE, TAM ♂♂½*

And Now the Screaming Starts The young bride-to-be of the lord of a British manor house is greeted by bloody faces at the window, a severed hand, and five corpses. Then Cushing shows up to investigate. Good-looking, sleek production with genuine chills.
1973 (R) 87m/C *GB* Peter Cushing, Herbert Lom, Patrick Magee, Ian Ogilvy, Stephanie Beacham, Geoffrey Whitehead, Guy Rolfe; *Dir:* Roy Ward Baker. **VHS, Beta** $79.95 *PSM, MED ♂♂½*

And Now for Something Completely Different A compilation of skits from BBC-television's "Monty Python's Flying Circus" featuring Monty Python's own weird, hilarious brand of humor. Sketches include "The Upper Class Twit of the Year Race," "Hell's Grannies" and "The Townswomen's Guild Reconstruction of Pearl Harbour." A great introduction to Python for the uninitiated, or a chance for the converted to see their favorite sketches again.
1972 (PG) 89m/C *GB* John Cleese, Michael Palin, Eric Idle, Graham Chapman, Terry Gilliam, Terry Jones; *Dir:* Ian McNaughton. **VHS, Beta, LV** $19.95 *COL ♂♂♂*

And the Ship Sails On On the eve of WWI, a group of devoted opera lovers take a luxury cruise to pay their respects to a recently deceased opera diva. Also on board is a group of feisty Serbo-Croation freedom fighters. A charming and absurd autumnal homage-to-life by Fellini shot entirely in the studio.
1984 (PG) 130m/C *IT* Freddie Jones, Barbara Jefford, Janet Suzman, Peter Cellier; *Dir:* Federico Fellini. **VHS, Beta, LV** $59.55 *COL, APD, TAM ♂♂♂*

And Soon the Darkness One of two vacationing young nurses disappears in France where a teenager was once murdered and the search is on. Predictable, ineffective suspenser.
1970 (PG) 94m/C *GB* Pamela Franklin, Michele Dotrice, Sandor Eles, John Nettelton, Claire Kelly, Hanna-Marie Pravda; *Dir:* Robert Fuest. **VHS, Beta** $59.99 *HBO ♂♂*

And Then There Were None An all-star cast makes up the ten colorful guests invited to a secluded estate in England by a mysterious host. What the invitations do not say, however, is the reason they have been specifically chosen to visit—to be murdered, one by one. Cat and mouse classic based on Agatha Christie's play with an entertaining mix of suspense and black comedy. Remade in 1966 and again in 1975 as "Ten Little Indians," but lacking the force and gloss of the original.
1945 97m/C Louis Hayward, Barry Fitzgerald, Walter Huston, Roland Young, C. Aubrey Smith, Judith Anderson, Mischa Auer, June Duprez; *Dir:* Rene Clair. **VHS, Beta, LV** $19.95 *NOS, UHV ♂♂♂½*

And Then You Die An intense crime drama about a Canadian drug lord who amasses a fortune from the cocaine and marijuana trade. His empire is threatened as the Mafia, Hell's Angels, and the police try to bring him down.
1988 (R) 115m/C *CA* Ken Welsh, R.H. Thompson, Wayne Robson, Tom Harvey, George Bloomfield, Graeme Campbell; *Dir:* Francis Mankiewicz. **VHS, Beta** $79.95 *VMK ♂♂♂*

And When She Was Bad... A young vamp gets her stepdad and his girlfriend involved in sexual misadventures. Softcore.
1984 90m/C Heather Vale, Lyllah Torena. **VHS, Beta** $29.95 *MED ♂½*

And You Thought Your Parents Were Weird! A pair of introverted, whiz kid brothers invent a lovable robot to provide fatherly guidance as well as companionship for their widowed mother. Surprisingly charming, sentimental film is only slightly hampered by low-budget special effects.
1991 (PG) 92m/C Marcia Strassman, Joshua Miller, Edan Gross, Sam Behrens; *W/Dir:* Tony Cookson; *Voices:* Alan Thicke. **VHS** $92.95 *VMK ♂♂½*

The Anderson Tapes Newly released from prison, an ex-con assembles his professional pals and plans the million-dollar robbery of an entire luxury apartment house on NYC's upper east side. Of course, he's unaware that a hoard of law men from federal, state, and local agencies are recording their activities for a wide variety of reasons, though none of the surveillance is coordinated and it has nothing to do with the planned robbery. Based on the novel by Lawrence Sanders, the intricate caper is effectively shaped by Lumet, who skillfully integrates broad satire with suspense. Score by Quincy Jones and shot on location in New York City. Walken is The Kid, his first major role.
1971 (PG) 100m/C Sean Connery, Dyan Cannon, Martin Balsam, Christopher Walken, Alan King, Ralph Meeker, Garrett Morris, Margaret Hamilton; *Dir:* Sidney Lumet. **VHS, Beta, LV** $9.95 *GKK, COL ♂♂♂*

The Andersonville Trial Details the atrocities experienced by captured Union soldiers who were held in the Confederacy's notorious Andersonville prison during the American Civil War. Provides an interesting account of the war-crimes trial of the Georgia camp's officials, under whom more than 14,000 prisoners died. Moving, remarkable television drama based on the Pulitzer-honored book by MacKinlay Kantor.
1970 150m/C Martin Sheen, William Shatner, Buddy Ebsen, Jack Cassidy, Richard Basehart, Cameron Mitchell; *Dir:* George C. Scott. **VHS, Beta, LV** $59.95 *IVE ♂♂♂½*

Andrei Roublev A 15th-century Russian icon painter must decide whether to record history or participate in it as the Tartar invaders make life miserable. During the black and white portion, he becomes involved in a peasant uprising, killing a man in the process. After a bout of pessimism and a vow of silence, he goes forth to create artistic beauty as the screen correspondingly blazes with color. A brilliant historical drama censored by Soviet authorities until 1971, it was co-written by acclaimed director Tarkovsky.
1966 185m/C *RU* Anatoli Solonitzin, Ivan Lapikov, Nikolai Grinko, Nikolai Sergueiev; *Dir:* Andrei Tarkovsky. **VHS** $59.99 *TAM, NOS, FCT ♂♂♂♂*

Androcles and the Lion Stagebound Hollywood version of the George Bernard Shaw story about a tailor in Imperial Rome who saves Christians from a hungry lion he had previously befriended. Shavian dialogue and a plot that's relatively (within the bounds of Hollywood) faithful help a great play become a semi-satisfying cinematic morsel. Harpo Marx was originally cast as Androcles, but was fired by producer Howard Hughes five weeks into the shooting.

1952 105m/B Jean Simmons, Alan Young, Victor Mature, Robert Newton, Maurice Evans; *Dir:* Chester Erskine. **VHS, Beta $19.95** *COL, SUE* ♂♂½

Android When an android who has been assisting a quirky scientist in space learns that he is about to be permanently retired, he starts to take matters into his own synthetic hands. Combines science fiction, suspense and cloned romance. A must for Kinski fans. **1982** (PG) 80m/C Klaus Kinski, Don Opper, Brie Howard; *Dir:* Aaron Lipstadt. **VHS, Beta, LV $69.95** *MED* ♂♂½

The Andromeda Strain A satellite falls back to earth carrying a deadly bacteria that must be identified in time to save the population from extermination. The tension inherent in the bestselling Michael Crichton novel is talked down by a boring cast. Also available in letter-box format. **1971** (G) 131m/C Arthur Hill, David Wayne, James Olson, Kate Reid, Paula Kelly; *Dir:* Robert Wise. **VHS, Beta, LV $59.95** *MCA* ♂♂½

Andy and the Airwave Rangers Andy is whisked into the TV!! He finds adventure and excitement—car chases, intergalactic battles, and cartoons. **1989** 75m/C Dianne Kay, Vince Edwards, Bo Svenson, Richard Thomas, Erik Estrada; *Dir:* Deborah Brock. **VHS, LV $79.95** *COL* ♂♂½

Andy Hardy Gets Spring Fever Andy falls for a beautiful acting teacher, and then goes into a funk when he finds she's engaged. Judge Hardy and the gang help heal the big wound in his heart. A lesser entry (and the seventh) from the popular series. **1939** 88m/B Mickey Rooney, Lewis Stone, Ann Rutherford, Fay Holden, Cecilia Parker, Sara Haden, Helen Gilbert; *Dir:* W.S. Van Dyke. **VHS, Beta $19.95** *MGM* ♂♂

Andy Hardy Meets Debutante Seems like there should be an article in that title. Garland's second entry in series, wherein Andy meets and falls foolishly for glamorous debutante Lewis with Betsy's help while family is on visit to New York. Judy/Betsy sings "I'm Nobody's Baby" and "Singing in Rain." Also available with "Love Finds Andy Hardy" on laser disc. **1940** 86m/B Mickey Rooney, Judy Garland, Lewis Stone, Ann Rutherford, Fay Holden, Sara Haden, Cecilia Parker, Diana Lewis, Tom Neal; *Dir:* George B. Seitz. **VHS, Beta, LV $19.95** *MGM* ♂♂½

Andy Hardy's Double Life In this entertaining installment from the Andy Hardy series, Andy proposes marriage to two girls at the same time and gets in quite a pickle when they both accept. Williams makes an early screen splash. **1942** 91m/B Mickey Rooney, Lewis Stone, Ann Rutherford, Fay Holden, Sara Haden, Cecilia Parker, Esther Williams, William Lundigan, Susan Peters, Robert Blake; *Dir:* George B. Seitz. **VHS, Beta $19.95** *MGM, MLB* ♂♂½

Andy Hardy's Private Secretary After Andy fails his high school finals he gets help from a sympathetic faculty member. As the secretary, Grayson makes a good first impression in one of her early screen appearances. The Hardy series was often used as a training ground for new MGM talent. **1941** 101m/B Mickey Rooney, Kathryn Grayson, Lewis Stone, Fay Holden, Ian Hunter, Gene Reynolds, Sara Haden; *Dir:* George B. Seitz. **VHS, Beta $19.95** *MGM* ♂♂

Andy Kaufman: Sound Stage A comedy special devised by Kaufman, using his peculiar brand of humor, abetted by comedienne Elayne Boosler. **1983** 59m/C **VHS, Beta $59.95** *LTG* ♂♂

Andy Kaufman Special Comedy special, with skits and Kaufman's inimitable stand-up routines. **1985** 59m/C Howdy Doody, Andy Kaufman, Cindy Williams. **VHS, Beta $29.95** *MED* ♂♂

Andy Warhol's Bad In the John Waters' school of "crime is beauty," a Queens housewife struggles to make appointments for both her home electrolysis clinic and her all-female murder-for-hire operation, which specializes in children and pets (who are thrown out of windows and knived, respectively). Her life is further complicated by a boarder (King) who's awaiting the go-ahead for his own assignment, an autistic child unwanted by his mother. One of Warhol's more professional-appearing films, and very funny if your tastes run to the tasteless. **1977** (R) 100m/C Perry King, Carroll Baker, Susan Tyrrell, Stefania Cassini; *Dir:* Jed Johnson. **VHS, Beta, LV $59.95** *SUE, MLB* ♂♂♂

Andy Warhol's Dracula Sex and camp humor, as well as a large dose of blood, highlight Warhol's treatment of the tale. As Dracula can only subsist on the blood of pure, untouched maidens ("weregins"), gardener Dallesandro rises to the occasion in order to make as many women as he can ineligible for Drac's purposes. Very reminiscent of Warhol's "Frankenstein," but with a bit more spoofery. Look for Roman Polanski in a cameo peek as a pub patron. Also known as "Blood for Dracula" and "Young Dracula." **1974** (R) 106m/C *IT FR* Udo Kier, Arno Juergling, Maxine McKendry, Joe Dallesandro, Vittorio De Sica, Roman Polanski; *Dir:* Paul Morrissey. **VHS, Beta $79.95** *TRI, GEM* ♂♂♂

Andy Warhol's Frankenstein A most outrageous parody of Frankenstein, featuring plenty of gore, sex, and bad taste in general. Baron von Frankenstein (Kier) derives sexual satisfaction from his corpses (he delivers a particularly thought-provoking philosphy on life as he lustfully fondles a gall bladder); his wife seeks her pleasure from the monster himself (Dallesandro). Originally made in 3-D, this is one of Warhol's campiest outings. **1974** (R) 95m/C *IT GE FR* Udo Kier, Dalia di Lazzaro, Monique Van Vooren, Joe Dallesandro; *Dir:* Paul Morrissey. **VHS, Beta $79.95** *TRI, MLB* ♂♂½

Angel Low-budget leerer about a 15-year-old honor student who attends an expensive Los Angeles private school during the day and by night becomes Angel, a streetwise prostitute making a living amid the slime and sleaze of Hollywood Boulevard. But wait, all is not perfect. A psycho is following her, looking for an opportunity. **1984** (R) 94m/C Donna Wilkes, Cliff Gorman, Susan Tyrrell, Dick Shawn, Rory Calhoun, John Diehl, Elaine Giftos, Ross Hagen; *Dir:* Robert Vincent O'Neil. **VHS, Beta $19.95** *STE, HBO* ♂½

Angel 3: The Final Chapter Former hooker Angel hits the streets to save her newly discovered sister from a life of prostitution. Trashy sequel with a better cast to tepid "Avenging Angel," which was the inept 1985 follow-up to 1984's tasteless "Angel." **1988** (R) 100m/C Maud Adams, Mitzi Kapture, Richard Roundtree, Mark Blankfield, Kin Shriner, Tawny Fere, Toni Basil; *Dir:* Tom DeSimone. **VHS, Beta, LV $19.95** *STE, NWV, IME* Woof!

Angel Baby A mute girl struggles to redefine her faith when she is cured by preacher, but then sees him fail with others. Fine performances all around, notably Reynolds in his screen debut, with cinematography by Haskell Wexler. Adapted from "Jenny Angel" by Elsie Oaks Barber. **1961** 97m/B George Hamilton, Salome Jens, Mercedes McCambridge, Joan Blondell, Henry Jones, Burt Reynolds; *Dir:* Paul Wendkos, Hubert Cornfield. **VHS** *AVC* ♂♂½

Angel and the Badman When notorious gunslinger Wayne is wounded during a shoot-out, a pacifist family takes him in and nurses him back to health. While he's recuperating, the daughter in the family (Russell) falls for him. She begs him not to return to his previous life. But Wayne, though smitten, thinks that a Duke's gotta do what a Duke's gotta do. And that means finding the dirty outlaw (Cabot) who killed his pa. Predictable but nicely done, with a good cast and script. Wayne provides one of his better performances (and also produced). **1947** 100m/B John Wayne, Gail Russell, Irene Rich, Harry Carey, Bruce Cabot; *W/Dir:* James Edward Grant. **VHS, LV $19.98** *REP, NEG, MRV* ♂♂♂

Angel City A Florida labor camp is the setting for this made-for-TV drama. A family of rural West Virginia migrant workers find themselves trapped inside the camp and exploited by the boss-man. Adapted from Patricia Smith's book by scriptwriter James Lee Barrett. **1980** 90m/C Ralph Waite, Paul Winfield, Jennifer Warren, Jennifer Jason Leigh, Mitchell Ryan; *Dir:* Philip Leacock. **VHS $59.95** *XVC, HFE* ♂♂½

Angel City An experimental parody of film mysteries where a detective investigates the nature of visual truth in and about Hollywood. **1984** 75m/C Bob Glaudini; *Dir:* Jon Jost. **VHS, Beta** *FCT* ♂♂

Angel of Death A small mercenary band of Nazi hunters attempt to track down Josef Mengele in South America. Stupid entry in the minor "let's find the darn Nazi before he really causes trouble" genre. Director Franco is also known as A. Frank Drew White. **1986** (R) 92m/C Chris Mitchum, Fernando Rey, Susan Andrews; *Dir:* Jess (Jesus) Franco. **VHS, Beta $19.95** *STE, NWV* Woof!

Angel Heart An exotic and controversial look at murder, voodoo cults, and sex in 1955 New Orleans that enabled Bonet to defiantly shed her image as a young innocent (no more Cosby Show for you, young lady). Rourke is marginal NYC private eye Angel, hired by the devilish DeNiro to track a missing big band singer who violated a "contract." His investigation leads him to the bizarre world of the occult in New Orleans, where the blood drips to a different beat. Visually stimulating, with a provocative sex scene between Bonet and Rourke, captured in both R-rated and unrated versions. Adapted by Parker from "Falling Angel" by William Hjortsberg. **1987** (R) 112m/C Mickey Rourke, Robert De Niro, Lisa Bonet, Charlotte Rampling; *W/Dir:* Alan Parker. **VHS, Beta, LV $14.95** *IVE, FCT* ♂♂½

Angel of H.E.A.T. Porn-star Chambers is Angel, a female super-agent on a mission to save the world from total destruction. Sex and spies abound with trashy nonchalance.

1982 (R) 90m/C Marilyn Chambers, Mary Woronov, Stephen Johnson; *Dir:* Helen Sanford, Myrl A. Schreibman. **VHS, Beta, LV $69.95** *VES Woof!*

Angel on My Shoulder A murdered convict makes a deal with the Devil (Rains) and returns to earth for revenge as a respected judge who's been thinning Hell's waiting list. Occupying the good judge, the murderous Muni has significant problems adjusting. Amusing fantasy with Muni in a rare and successful comic role. Co-written by Harry Segall, who scripted "Here Comes Mr. Jordan," in which Rains played an angel. Remade in 1980.
1946 101m/B Paul Muni, Claude Rains, Anne Baxter, Onslow Stevens; *Dir:* Archie Mayo. **VHS, Beta $9.95** *NEG, MRV, NOS* 𝄞𝄞𝄞

Angel on My Shoulder A small-time hood wrongfully executed for murder comes back as district attorney. He owes the devil, but he's finding it tough to be evil enough to repay his debt. Okay television remake of the better 1946 film starring Paul Muni.
1980 96m/C Peter Strauss, Richard Kiley, Barbara Hershey, Janice Paige; *Dir:* John Berry. **VHS, Beta $69.98** *SUE* 𝄞𝄞

An Angel at My Table A New Zealand television mini-series chronicling the life of Janet Frame, New Zealand's premiere writer/poet. At once whimsical and tragic, the film tells of how a mischievous, free-spirited young girl was wrongly placed in a mental institution for eight years, yet was ultimately able to cultivate her incredible storytelling gifts, achieving success, fame and happiness. Highly acclaimed the world over, winner of over 20 major international awards.
1989 (R) 157m/C *NZ* Kerry Fox, Alexia Keogh, Karen Fergusson, Iris Churn, K.J. Wilson, Martyn Sanderson; *Dir:* Jane Campion. Chicago Film Critics Awards '91: Best Foreign Film. **VHS $89.95** *SVS* 𝄞𝄞𝄞𝄞

Angel Square A Canadian production from the director of the acclaimed "Bye Bye Blues." When the father of a neighborhood boy is brutally attacked, the community bands together to search for the culprit.
1992 106m/C Ned Beatty; *Dir:* Anne Wheeler. **VHS** *CGV* 𝄞𝄞

Angel in a Taxi A six-year old boy in an orphanage decides to choose his own mother, a beautiful ballerina he sees in a magazine, when an ugly couple try to adopt him. Italian film dubbed in English.
1959 89m/B *IT* Wera Cecova, Ettore Manni, Vittorio De Sica, Marietto, Gabriel Ferzetti. **VHS, Beta, 8mm $29.95** *VYY, FCT* 𝄞𝄞

Angel Town A foreign exchange student, who happens to be a champion kick-boxer, is forced into combat with LA street gangs.
1989 90m/C Olivier Gruner, Theresa Saldana, Frank Aragon, Tony Valentino, Peter Kwong, Mike Moroff; *Dir:* Eric Karson. **VHS, Beta $89.95** *IMP* 𝄞

Angela Twisted love story about a young man (Railsback) who jumps the bones of an older woman, unaware that she is the mother he's been separated from for 23 long years. Seems that way back when, the boy was kidnapped by crime boss Huston from mom Loren, an ex-prostitute, who then turned in the boy's dad, who was something of a criminal. Mom thought son was dead, dad vowed revenge from prison, and son went about his unwitting business. It all comes together insipidly, at the expense of the cast and viewer.

1977 91m/C *CA* Sophia Loren, Steve Railsback, John Huston, John Vernon; *Dir:* Boris Sagal. **VHS, Beta $69.98** *SUE, SIM* 𝄞

Angele A lovely, naive country girl is lured to the city by a cunning pimp who knows that she wants to escape her oppressive father. With her illegitimate baby, she's discovered in a whorehouse and taken in disgrace back home to dad, who promptly locks her in the barn. One special guy however, understands her purity and plots to rescue her. What he lacks in material resources he makes up for in character. An overlong but moving story of lost innocence and intolerance. In French with English subtitles. Based on the novel "Un de Baumugnes" by Jean Giono.
1934 130m/B *FR* Orane Demazis, Fernandel, Henri Puopon, Edouard Delmont; *Dir:* Marcel Pagnol. **VHS $59.95** *INT, TAM* 𝄞𝄞𝄞

Angelo My Love Compassionate docudrama about New York's modern gypsy community. Follows the adventures of 12-year-old Angelo Evans, the streetwise son of a fortune teller, who with a fresh view, explores the ups and downs of his family's life. Duvall financed the effort and cast non-professional actors in this charming tale of reality and fairy-tale.
1983 (R) 91m/C Angelo Evans, Micahel Evans, Steve "Patalay" Tsiginoff, Cathy Kitchen, Millie Tsiginoff; *Dir:* Robert Duvall. **VHS, Beta, LV $59.95** *COL* 𝄞𝄞𝄞

Angel's Brigade Seven models get together to stop a big drug operation. Drive-in vigilante movie fare fit for a rainy night.
1979 (PG) 97m/C Jack Palance, Peter Lawford, Jim Backus, Arthur Godfrey; *Dir:* Greydon Clark. **VHS, Beta $69.98** *LTG, LIV, TIM* 𝄞

Angels of the City What's a girl got to do to join a sorority? A house prank turns vicious when two coeds take a walk on the wild side and accidentally observe a murder, making them the next targets. Credible exploitation effort, if that's not an oxymoron.
1989 90m/C Lawrence Hilton-Jacobs, Cynthia Cheston, Kelly Galindo, Sandy Gershman; *Dir:* Lawrence Hilton-Jacobs. **VHS, Beta $29.95** *PMH* 𝄞𝄞

Angels Die Hard Novel biker story with the cyclists as the good guys intent on helping a town during a mining disaster. Grizzly Adams makes an early film appearance.
1984 (R) 86m/C Tom Baker, R.G. Armstrong, Dan Haggerty, William Smith; *Dir:* Richard Compton. **VHS, Beta, LV $19.95** *STE, NWV* 𝄞𝄞

Angels with Dirty Faces The rousing classic with the memorable Cagney twitches and the famous long walk from the cell to the chair. Two young hoodlums grow up on NYC's lower East Side with diverse results—one enters the priesthood and the other opts for crime and prison. Upon release from the pen, Cagney sets up shop in the old neighborhood, where Father O'Brien tries to keep a group of young toughs (the Dead End Kids) from idolizing the famed gangster and following in his footsteps. Bogart's his unscrupulous lawyer and Bancroft a crime boss intent on double-crossing Cagney. Reportedly they were blasting real bullets during the big shootout, not helping Cagney's intensity. Oscar nominated for Best Actor (Cagney), Best Director and Best Original Story. Score by Max Steiner and adapted by John Wexley and Warren Duff from a story by Rowland Brown.

1938 97m/B James Cagney, Pat O'Brien, Humphrey Bogart, Ann Sheridan, George Bancroft, Billy Halop, Leo Gorcey, Huntz Hall, Bobby Jordan, Dan Jesse, Gabriel Dell, J. Frank Burke, William Tracy; *Dir:* Michael Curtiz. **VHS, Beta, LV $29.98** *MGM, FOX, MLB* 𝄞𝄞𝄞𝄞

Angels Hard as They Come Opposing Hell's Angels leaders clash in a hippie-populated ghost town. A semi-satiric spoof of the biker genre's cliches co-written and produced by Demme. Features an early Glenn appearance and Busey's film debut.
1971 (R) 86m/C Gary Busey, Scott Glenn, James Inglehart, Gary Littlejohn, Sharon Peckinpah; *Dir:* Joe Viola. **VHS, Beta $24.98** *SUE* 𝄞𝄞

Angels from Hell Early application in the nutso 'Nam returnee genre. Disillusioned Vietnam veteran forms a massive biker gang for the sole purpose of wreaking havoc upon the Man, the Establishment and anyone else responsible for sending him off to war. The big gang invades a town, with predictably bloody results. Sort of a follow-up to "Hell's Angels on Wheels."
1968 86m/C Tom Stern, Arlene Martel, Ted Markland, Stephen Oliver, Paul Bertoya James Murphy, Jack Starrett, Pepper Martin, Luana Talltree; *Dir:* Bruce Kessler. **VHS $39.95** *TRY* 𝄞

Angels One Five A worm's-eye view of British air power in WWII. What little "excitement" there is, is generated by flashing lights, plotting maps and status boards. Hawkins and Denison are the only bright spots.
1954 97m/B Jack Hawkins, Michael Denison, Dulcie Gray, John Gregson, Cyril Raymond, Veronica Hurst, Harold Goodwin, Geoffrey Keen, Vida Hope. **VHS $19.95** *NOS* 𝄞𝄞

Angels Over Broadway Sliok, fact paced black comedy about con man Fairbanks, who plans to hustle suicidal thief Qualen during a poker game, but has a change of heart. With the help of call-girl Hayworth and drunken playwright Mitchell, he helps Qualen turn his life around. Ahead of its time with an offbeat morality, a delight in the '90s.
1940 80m/B Douglas Fairbanks Jr., Rita Hayworth, Thomas Mitchell, John Qualen, George Watts; *Dir:* Ben Hecht, Lee Garmes. **VHS, Beta, LV $19.95** *RCA, FCT, COL* 𝄞𝄞𝄞

Angels' Wild Women From the man who brought you "Dracula vs. Frankenstein" comes an amalgamation of hippies, motorcycle dudes, evil desert gurus and precious little plot.
1972 85m/C Kent Taylor, Regina Carrol, Ross Hagen, Maggie Bemby, Vicki Volante; *Dir:* Al Adamson. **VHS** *VID Woof!*

AngKor: Cambodia Express An American journalist travels back to Vietnam to search for his long lost love.
1981 96m/C Robert Walker, Christopher George; *Dir:* Lek Kitiparaporn. **VHS, Beta $69.98** *VES, LIV* 𝄞

The Angry Dragon A martial arts guy seeks revenge on the underworld pimps for the murder of his girlfriend's father.
1979 (R) 96m/C VHS, Beta *GEM* 𝄞

Angry Harvest During the World War II German occupation of Poland, a gentile farmer shelters a young Jewish woman on the run, and a serious, ultimately interdependent relationship forms. Acclaimed; Holland's first film since her native Poland's martial law imposition made her an exile to Sweden. In German with English subtitles. Contains nudity and violence.

1985 102m/C *GE* Armin Mueller-Stahl, Elisabeth Trissenaar, Wojciech Pszoniak; *Dir:* Agnieszka Holland. Montreal World Film Festival '85: Best Actor (Mueller-Stahl). **VHS, Beta $69.95** *FCT, GLV, TAM* 🎞🎞🎞

Angry Joe Bass Contemporary Native American Joe Bass faces government officials who continually usurp his fishing rights. 1976 82m/C Henry Bal, Molly Mershon; *Dir:* Thomas G. Reeves. **VHS, Beta** *PGN* 🎞

The Angry Red Planet An unintentionally amusing sci-fi adventure about astronauts on Mars fighting off aliens and giant, ship-swallowing amoebas. Filmed using bizarre ''Cinemagic'' process, which turns almost everything pink. Wild effects have earned the film cult status. 1959 83m/C Gerald Mohr, Les Tremayne, Jack Kruschen; *Dir:* Ib Melchoir. **VHS, Beta $19.95** *HBO* 🎞🎞

Anguish Well-done horror thriller about a lunatic who, inspired to duplicate the actions of an eyeball-obsessed killer in a popular film, murders a movie audience as they watch the movie. Violence and gore abound. 1988 (R) 85m/C *SP* Zelda Rubenstein, Michael Lerner, Talia Paul; *Dir:* Bigas Luna. **VHS, Beta, LV $79.98** *FOX* 🎞🎞½

Animal Behavior An animal researcher and a music professor fall in love on a college campus. Jenny Bowen's (the director's real name) comedy was filmed in 1985 and edited/shelved for over 4 years, with good reason. 1989 (PG) 79m/C Karen Allen, Armand Assante, Holly Hunter, Josh Mostel, Richard Libertini; *Dir:* Riley H. Anne. **VHS, Beta, LV $89.99** *HBO* 🎞

Animal Called Man Two crooked buddies join a big western gang in pillaging a small town and get in too deep. 1987 83m/C Vassilli Karis, Craig Hill. **VHS, Beta $59.95** *TWE* 🎞

Animal Crackers The second and possibly the funniest of the thirteen Marx Brothers films, ''Animal Crackers'' is a screen classic. Groucho is a guest at the house of wealthy matron Margaret Dumont and he, along with Zeppo, Chico, and Harpo, destroy the tranquility of the house. Complete with the Harry Ruby music score-including Groucho's ''Hooray for Captain Spaulding'' with more quotable lines than any other Marx Brothers film: ''One morning I shot an elephant in my pajamas. How he got into my pajamas, I'll never know.'' Based on a play by George S. Kaufman. 1930 (G) 98m/B Groucho Marx, Chico Marx, Harpo Marx, Zeppo Marx, Lillian Roth, Margaret Dumont; *Dir:* Victor Neerman. **VHS, Beta, LV $19.95** *MCA, FCT* 🎞🎞🎞🎞

Animal Farm An animated version of George Orwell's classic political satire about a barnyard full of animals who parallel the growth of totalitarian dictatorships. Not entirely successful, but probably best translation of Orwell to film. 1955 73m/C *Dir:* John Halas, Joy Batchelor. **VHS, Beta, LV, 8mm $29.98** *KUI, AMV, VYY* 🎞🎞🎞

The Animal Kingdom A romantic triangle develops when Howard, married to Loy, has an affair with Harding. The problem is Harding acts more like a wife and Loy a mistress. Intelligently written and directed, with a marvelous performance from veteran character actor Gargan.

1932 95m/B Ann Harding, Leslie Howard, Myrna Loy, Neil Hamilton, William Gargan, Henry Stephenson, Ilka Chase; *Dir:* Edward H. Griffith. **VHS $29.95** *RKO, FCT* 🎞🎞🎞

Animation of the 1930s A collection of nine cartoons from the studios of Warner Brothers and Max Fleischer, including ''Crosby, Columbo and Vallee,'' ''Three's a Crowd,'' ''Hollywood Capers,'' ''Songs You Like to Sing-Margie,'' ''Grampy's Indoor Outing,'' ''Betty in Blunderland,'' ''Let's Sing with Popeye,'' ''Happy You and Me,'' and ''Sinkin' in the Bathtub.'' 1930 (G) 57m/B **VHS, Beta $39.95** *VYY* 🎞🎞🎞

Animation Wonderland A collection of rarely seen cartoons including ''The Early Birds,'' ''The Tinder Box,'' ''Petroushka,'' ''The Chocolate Princess,'' ''King Midas'' and many others. 1990 62m/C *Nar:* Jonathan Winters, Bill Cosby. **VHS $9.95** *SIM* 🎞🎞🎞

Animators from Allegro Non Troppo Live action is combined with Bruno Bozzetto's animation in this feature-length parody of Disney's Fantasia. 1979 70m/C **VHS, Beta** *TEX* 🎞🎞🎞

Anita, Dances of Vice German avant-garde film celebrates Anita Berber, the ''most scandalous woman in 1920s Berlin.'' Berber was openly bisexual, used drugs, and danced nude in public. In von Praunheim's film, she rises from the dead and creates yet more scandal. Women. 1987 85m/C *GE* Lotti Huber, Ina Blum, Mikhael Honesseau; *Dir:* Rosa von Praunheim. **VHS, Beta, 3/4U $49.95** *ICA, GLV, TAM* 🎞🎞

Anita de Montemar A man is torn between his wife and his child-bearing girlfriend. 196? 127m/C *SP* Amparo Rivelles, Fernando Soler. **VHS, Beta $59.95** *UNI* 🎞

The Ann Jillian Story Jillian stars as herself in this television melodrama recounting her battle with breast cancer. Several musical numbers are included. 1988 96m/C Ann Jillian, Tony LoBianco, Viveca Lindfors, Leighton Bewley; *Dir:* Corey Allen. **VHS, Beta $79.95** *JTC* 🎞½

Ann Vickers A dashing young army captain wins over the heart of a dedicated social worker. Okay adaptation of a Sinclair Lewis novel, with Dunne suffering more than usual. 1933 76m/B Irene Dunne, Walter Huston, Bruce Cabot, Conrad Nagel, Edna May Oliver; *Dir:* John Cromwell. **VHS, Beta $19.98** *RKO* 🎞🎞½

Anna Age, envy, and the theatrical world receive their due in an uneven but engrossing drama about aging Czech film star Anna, making a sad living in New York doing commercials and trying for off-Broadway roles. She takes in Krystyna, a young Czech peasant girl who eventually rockets to model stardom. Modern, strongly acted ''All About Eve'' with a story partially based on a real Polish actress. Kirkland drew quite a bit of flak for shamelessly self-promoting for the Oscar. She still lost. 1987 (PG-13) 101m/C Sally Kirkland, Paulina Porizkova, Robert Fields; *Dir:* Yurek Bogayevycz. **VHS, Beta, LV $79.98** *VES, LIV* 🎞🎞🎞

Anna and Bella Two elderly sisters spend an afternoon pouring over the family photo album. 1986 8m/C Academy Awards '86: Best Animated Short Film. **VHS, Beta, 3/4U $150.00** *DCL* 🎞🎞🎞

Anna Christie Garbo is the ex-prostitute who finds love with sailor Bickford. Bickford is unaware of his lover's tarnished past and she does her best to keep it that way. Garbo's first sound effort was advertised with the slogan ''Garbo Talks.'' Adapted from the classic Eugene O'Neill play, the film is a slow but rewarding romantic drama. 1930 90m/B Greta Garbo, Marie Dressler, Charles Bickford, George F. Marion; *Dir:* Clarence Brown. **VHS, Beta $24.95** *MGM* 🎞🎞½

Anna to the Infinite Power Sci fi based on the book of the same name follows a young girl with telepathic powers. When the girl discovers that she has sisters as the result of a strange scientific experiment, she sets out to find them, drawing on her own inner strength. 1984 101m/C Dina Merrill, Martha Byrne, Mark Patton; *Dir:* Robert Wiemer. **VHS, Beta $59.95** *COL* 🎞🎞½

Anna Karenina Cinematic Tolstoy with Garbo as the sad, moody Anna willing to give up everything to be near Varonsky (March), the cavalry officer she's obsessed with. A classic Garbo vehicle with March and Rathbone providing excellent support. Interestingly, a remake of the Garbo and John Gilbert silent, ''Love.'' Photography received an Academy Award nomination. 1935 85m/B Greta Garbo, Fredric March, Freddie Bartholomew, Maureen O'Sullivan, May Robson, Basil Rathbone, Reginald Owen, Reginald Denny; *Dir:* Clarence Brown. **VHS, Beta $19.98** *NOS, MGM* 🎞🎞🎞½

Anna Karenina Stiff version of Tolstoy's passionate story of illicit love between a married woman and a military officer. In spite of exquisite costumes, and Leigh and Richardson as leads, still tedious. Made in Britain. 1948 123m/C Vivien Leigh, Ralph Richardson, Kieron Moore; *Dir:* Julien Duvivier. **VHS, LV $39.95** *GKK, FCT, MRV* 🎞🎞

Anna Karenina Made for television version of Tolstoy's novel of betrayal, intrigue, and forbidden love. Anna, defying all social practices of the time, falls into the arms of a dashing count. Features an excellent performance from Paul ''A Man For All Seasons'' Scofield. 1985 (PG) 96m/C Jacqueline Bisset, Christopher Reeve, Paul Scofield; *Dir:* Simon Langton. **VHS, Beta, LV $89.95** *VES* 🎞🎞½

Anna Russell's Farewell Special Russell parodies Wagnerian operas and Gilbert and Sullivan tunes. Songs include ''Wind Instruments I Have Known,'' ''How to Become a Singer,'' and ''Three Parody Fold Tunes.'' 1988 90m/C **VHS, Beta, 3/4U $39.95** *MTP* 🎞🎞½

Annabel Takes a Tour/Maid's Night Out A delightful comedy double feature. In ''Annabel Takes a Tour,'' a fading movie star falls in love with a writer while on a tour to boost her career, and in ''Maids Night Out,'' an heiress and a millionaire fall in love. 1938 131m/B Lucille Ball, Jack Oakie, Ruth Donnelly, Joan Fontaine, Allan Lane, Hedda Hopper; *Dir:* Lew Landers. **VHS, Beta $34.95** *RKO, MLB* 🎞🎞½

Annapolis Life's in the pink for two guys at the U.S. naval academy until jealousy rears its loathsome head when they fall for the same gal. Just when it's beginning to look like love and honor is an either-or proposition, the lovelorn rivals find that although love transcends all, a guy's gotta do what a guy's gotta do, even if it means

sticking up for his romantic nemesis. Lots of sub-par male bonding.
1928 63m/B Johnny Mack Brown, Hugh Allan, Hobart Bosworth, William Bakewell, Charlotte Walker, Jeanette Loff; *Dir:* Christy Cabanne. **VHS, Beta $19.95** *GPV, FCT 🐾*

The Annapolis Story Cliche-ridden WWII drama about two naval cadets romancing the same lucky girl. Low-rent time for director Siegel which wastes a decent cast.
1955 81m/C John Derek, Kevin McCarthy, Diana Lynn, Pat Dooley, L.Q. Jones; *Dir:* Don Siegel. **VHS, Beta $19.98** *FOX, FCT 🐾 1/2*

Anne of Avonlea The equally excellent mini-series sequel to "Anne of Green Gables" in which the romantic heroine grows up and discovers romance. Lavishly filmed on picturesque Prince Edward Island. On two tapes. Also known as "Anne of Green Gables: The Sequel." Based on the characters from L.M. Montgomery's classic novel series. CBC, PBS, and Disney worked together on this "Wonderworks" production.
1987 224m/C *CA* Megan Follows, Colleen Dewhurst, Wendy Hiller, Frank Converse, Patricia Hamilton, Schuyler Grant, Jonathan Crombie, Rosemary Dunsmore; *Dir:* Kevin Sullivan. **VHS, Beta, LV $29.95** *KUI, BVV, DIS 🐾🐾🐾 1/2*

Anne of Green Gables Warm adaptation of the charming book by Lucy Maud Montgomery about a lonely couple who adopt an orphan who keeps them on their toes with her animated imagination, and wins a permanent place in their hearts.
1934 79m/B Anne Shirley, Tom Brown, O.P. Heggie; *Dir:* George Nicholls. **VHS, Beta $19.98** *RKO, TTC 🐾🐾🐾*

Anne of Green Gables A splendid production of the famous Lucy Maud Montgomery classic about a young girl growing to young adulthood with the help of a crusty pair of adoptive parents. The characters come to life under Sullivan's direction, and the movie is enhanced by the beautiful Prince Edward Island scenery. One of the few instances where an adaptation lives up to (if not exceeds) the quality of the original novel. A WonderWorks presentation that was made with the cooperation of the Disney channel, CBC, and PBS. Followed by "Anne of Avonlea." On two tapes.
1985 197m/C *CA* Colleen Dewhurst, Richard Farnsworth, Megan Follows, Patricia Hamilton, Schuyler Grant, Jonathan Crombie; *Dir:* Kevin Sullivan. **VHS, Beta, LV $29.95** *KUI, BVV, IGP 🐾🐾🐾 1/2*

Anne of the Thousand Days Lavish re-telling of the life and loves of Henry the VIII. Henry tosses aside his current wife for the young and devastatingly beautiful Anne Boleyn. Burton's performance of the amoral king garnered him an Oscar nomination. Watch for Elizabeth Taylor as a masked courtesan at the costume ball. Oscar nominations for Best Picture, actress Bujold, supporting actor Quayle, Adapted Screenplay, Cinematography, Sound, and Score.
1969 (PG) 145m/C Richard Burton, Genevieve Bujold, Irene Papas, Anthony Quayle, John Colicos, Michael Hordern; *Dir:* Hal B. Wallis. Academy Awards '69: Best Costume Design; Golden Globe Awards '70: Best Film—Drama. **VHS $59.95** *MCA 🐾🐾🐾 1/2*

L'Annee des Meduses Bare-chested French nymphettes cavort and scheme against each other and a slimy gigolo on the Mediterranean beaches. Dubbed.
1986 (R) 110m/C Valerie Kaprisky, Barnard Giradeau. **VHS, Beta $69.95** *FCT 🐾 1/2*

Annie Stagy big-budget adaption of the Broadway musical, which was an adaptation of the comic strip. A major financial disaster upon release, still it's an entertaining enterprise curiously directed by Huston and engagingly acted by Finney and Quinn. Oscar nominated for art direction and original song score,.
1981 (PG) 128m/C Aileen Quinn, Carol Burnett, Albert Finney, Bernadette Peters, Ann Reinking, Tim Curry; *Dir:* John Huston. **VHS, Beta, LV, 8mm $19.95** *COL, FCT 🐾🐾*

Annie Hall Planned originally to be based strictly on Allen's life, but expanded upon at the advice of editor Rosenblum, this is the acclaimed coming-of-cinematic-age film for Allen. His love affair with Hall/Keaton is chronicled as an episodic, wistful comedy commenting on family, love, loneliness, communicating, maturity, driving, city life, careers and various other topics. Abounds with classic scenes, including Goldblum and his mantra at a cocktail party; Allen and the lobster pot; and Allen, Keaton, a bathroom, a baseball bat and a spider. The film operates on many levels, as does Keaton's wardrobe, which started a major fashion trend. Bit parts to look for include Weaver as Woody's date. Expertly shot by Gordon Willis and cowritten by Marshall Brickman. Won several Oscar awards, plus an acting nomination for Allen.
1977 (PG) 94m/C Woody Allen, Diane Keaton, Tony Roberts, Carol Kane, Paul Simon, Jeff Goldblum, Janet Margolin, Colleen Dewhurst, Sigourney Weaver, Marshall McLuhan, Beverly D'Angelo; *Dir:* Woody Allen. Academy Awards '77: Best Actress (Keaton), Best Director (Allen), Best Original Screenplay, Best Picture; British Academy Awards '77: Best Actress (Keaton), Best Director (Allen), Best Film, Best Film Editing, Best Screenplay; Directors Guild of America Awards '77: Best Director (Allen); Los Angeles Film Critics Association Awards '77: Best Screenplay; National Board of Review Awards '77: 10 Best Films of the Year, Best Supporting Actress (Keaton); National Society of Film Critics Awards '77: Best Actress (Keaton), Best Picture, Best Screenplay. **VHS, Beta, LV $19.98** *MGM, FOX, VYG 🐾🐾🐾🐾*

Annie Oakley Energetic biographical drama based on the life and legend of sharpshooter Annie Oakley and her on-off relationship with Wild Bill Hickok. Stanwyck makes a great Oakley. Later musicalized as "Annie Get Your Gun."
1935 90m/B Barbara Stanwyck, Preston Foster, Melvyn Douglas, Pert Kelton, Andy Clyde; *Dir:* George Stevens. **VHS, Beta $19.98** *RKO, MED, NOS 🐾🐾🐾*

The Annihilators A group of Vietnam vets band together in an extremely violent manner to protect their small town from a gang of thugs.
1985 (R) 87m/C Gerrit Graham, Lawrence Hilton-Jacobs, Paul Koslo, Christopher Stone, Andy Wood, Sid Conrad, Dennis Redfield; *Dir:* Charles E. Sellier. **VHS, Beta $9.95** *NWV, STE Woof!*

Another Country Julian Mitchell's adaptation of his play about the true story of two English boarding school friends who wind up as spies for the Soviet Union. Based on the lives of Guy Burgess and Donald MacLean, whose youth is rather fancifully depicted in this loving recreation of 1930s' English life. Although the film is inferior to the award-winning play, director Kanievska manages to transform a piece into a solid film, and Rupert Everett's performance as Guy Burgess is outstanding.

1984 90m/C *GB* Rupert Everett, Colin Firth, Michael Jenn, Robert Addie, Anna Massey, Betsy Brantley, Rupert Wainwright, Cary Elwes; *Dir:* Marek Kanievska. **VHS, Beta, LV $24.98** *SUE 🐾🐾 1/2*

Another 48 Hrs. Continuing chemistry between Nolte and Murphy is one of the few worthwhile items in this stodgy rehash. Any innovation by Murphy seems lost, the story is redundant of any other cop thriller, and violence and car chases abound. Pointlessly energetic and occasionally fun for only the true devotee.
1990 (R) 98m/C Eddie Murphy, Nick Nolte, Brion James, Kevin Tighe, Bernie Casey, David Anthony Marshall, Ed O'Ross; *Dir:* Walter Hill. **VHS, Beta, LV, 8mm $14.95** *PAR 🐾 1/2*

Another Man, Another Chance Remake of Lelouch's "A Man and a Woman," set in the turn-of-the-century American West, pales by comparison to the original. Slow-moving tale casts widow Bujold and widower Caan as lovers.
1977 (PG) 132m/C *FR* James Caan, Genevieve Bujold, Francis Huster, Jennifer Warren, Susan Tyrrell; *Dir:* Claude Lelouch. **VHS, Beta $14.95** *WKV 🐾🐾*

Another Pair of Aces: Three of a Kind Nelson and Kristofferson team up to clear the name of Taylor, a Texas Ranger accused of murder. Made for television but not for the kiddies due to added erotic footage.
1991 93m/C Willie Nelson, Kris Kristofferson, Joan Severance, Rip Torn; *Dir:* Bill Bixby. **VHS $89.95** *VMK 🐾🐾🐾*

Another Thin Man Powell and Loy team up for the third in the delightful "Thin Man" series. Slightly weaker series entry takes its time, but has both Powell and Loy providing stylish performances. Nick Jr. is also introduced as the newest member of the sleuthing team. Sequel to "After the Thin Man;" followed by "Shadow of the Thin Man."
1939 105m/B William Powell, Myrna Loy, Virginia Grey, Otto Kruger, C. Aubrey Smith, Ruth Hussey; *Dir:* W.S. Van Dyke. **VHS $24.95** *MGM 🐾🐾 1/2*

Another Time, Another Place Sappy melodrama about an American journalist who suffers an emotional meltdown when her married British lover is killed during WWII. So she heads for Cornwall to console the widow and family.
1958 98m/B *GB* Lana Turner, Barry Sullivan, Glynis Johns, Sean Connery, Terrence Langdon; *Dir:* Lewis Allen. **VHS, Beta $19.95** *KRT, PAR 🐾🐾*

Another Time, Another Place Bored young Scottish housewife married to an older fella falls in love with an Italian prisoner-of-war who works on her farm during WWII. Occasionally quirky, always finely crafted view of wartime Britain and the little-known life of POWs in England.
1983 (R) 101m/C *GB* Phyllis Logan, Giovanni Mauriello, Gian Luca Favilla, Paul Young, Tom Watson; *Dir:* Michael Radford. **VHS, Beta, LV $24.98** *SUE 🐾🐾🐾*

Another Way The director of 1971's much-lauded "Love", sets this politically charged love story in Hungary in 1958. Opening with a view of a female corpse, the story flashes backward to look at the woman's journalistic career and her relationship with a women colleague. Candid love scenes between women in Hungary of 1958, considered a cinematic novelty in many places

outside Hungary. In Hungarian with English subtitles.
1982 100m/C HU Jadwiga Jankowska Cieslak, Grazyna Szapolowska, Josef Kroner, Hernadi Judit, Andorai Peter; **Dir:** Karoly Makk. **VHS, Beta $59.95** EVD, FCT, TAM ♫♫♫

Another Woman The study of an intellectual woman whose life is changed when she begins to eavesdrop. What she hears provokes her to examine every relationship in her life, finding things quite different than she had believed. Heavy going, with Rowlands effective as a woman coping with an entirely new vision of herself. Farrow plays the catalyst. Although Allen's comedies are more popular than his dramas, this one deserves a look.
1988 (PG) 81m/C Gena Rowlands, Gene Hackman, Mia Farrow, Ian Holm, Betty Buckley, Martha Plimpton, Blythe Danner, Harris Yulin, Sandy Dennis, David Ogden Stiers, John Houseman; **Dir:** Woody Allen. **VHS, Beta, LV $89.98** ORI ♫♫♫

Antarctica Due to unfortunate circumstances, a group of scientists must leave their pack of huskies behind on a frozen glacier in the Antarctic. The film focuses on the dogs' subsequent struggle for survival. Dubbed. Music by Vangelis.
1984 112m/C JP Ken Takakura, Masako Natsume, Keiko Oginome; **Dir:** Koreyoshi Kurahara. **VHS, Beta $59.98** FOX ♫♫

Anthony Adverse March is a young man in the 19th century who searches for manhood across America and Mexico. He grows slowly as he battles foes, struggles against adversity and returns home to find his lover in this romantic swashbuckler. With a star-studded cast, lush costuming, and an energetic musical score. Highly acclaimed in its time, but now seems dated. Based on the novel by Hervey Allen. Oscar nominated for Best Picture, Best Cinematography, Best Interior Decoration, Best Assistant Director, Best Film Editing and Best Score.
1936 137m/B Fredric March, Olivia de Havilland, Anita Louise, Gale Sondergaard, Claude Rains, Louis Hayward; **Dir:** Mervyn LeRoy. Academy Awards '36: Best Cinematography, Best Film Editing, Best Supporting Actress (Sondergaard), Best Musical Score. **VHS, Beta $24.95** MGM, BTV, FHE ♫♫½

Antonia and Jane Enjoyable film that tells the story of a longstanding, heavily tested friendship between two women. From the very beginning, they are a study in contrasts—Jane as rather plain, frumpy, and insecure; Antonia as glamorous, elegant, and successful. Both believe that each other's lives are more interesting and exciting than their own and through this smart witty comedy, we are offered an honest look into the complex world of adult friendships.
1991 (R) 75m/C GB Saskia Reeves, Imelda Staunton, Alfred Hoffman, Richard Hope, Brenda Bruce, Bill Nighy; **Dir:** Beeban Kidron. **VHS, Beta** PAR ♫♫♫

Antonio A Texas millionaire on the run from his wife and her divorce lawyer alights in a small Chilean village and turns it upside down.
1973 89m/C Larry Hagman, Trini Lopez, Noemi Guerrero, Pedro Becker; **Dir:** Claudio Guzman. **VHS, Beta $29.95** TAV, TIM ♫♫

Antonio Das Mortes Antonio is a savage mercenary hired to kill rebels against the Brazilian government. Belatedly he realizes he sympathizes with the targets and turns his guns the other way for an incredible shootout finale. A visually lavish political polemic, espousing revolutionary guerrilla action within the format of a South American western. In Portuguese with English subtitles.
1968 100m/C Mauricio Do Valle, Odete Lara; **W/ Dir:** Glauber Rocha. **VHS $59.95** FST ♫♫½

Antony and Cleopatra Heston wrote, directed, and starred in this long, dry adaptation of the Shakespeare play that centers on the torrid romance between Mark Antony and Cleopatra.
1973 (PG) 150m/C Charlton Heston, Hildegarde Neff, Fernando Rey, Eric Porter; **Dir:** Charlton Heston. **VHS, Beta $24.98** SUE ♫♫

Ants A mad bug parable for the planet-obsessed 90s. Insecticide-infected ants turn militant and check into a local hotel to vent their chemically induced foul mood on the unsuspecting clientele. The guest register includes a gaggle of celebrities who probably wish they'd signed on the Love Boat instead. Made for television (an ant farm would probably be just too horrible on the big screen), this bug blitz is also known as "Panic at Lakewood Manor" and "It Happened at Lakewood Manor."
1977 100m/C Suzanne Somers, Robert Foxworth, Myrna Loy, Lynda Day George, Gerald Gordon, Bernie Casey, Barry Van Dyke, Karen Lamm, Anita Gillette, Moosie Drier, Steve Franken, Brian Dennehy, Bruce French, Stacy Keach, Rene Enriquez, James Storm; **Dir:** Robert Scheerer. **VHS, Beta** IVE ♫½

Any Gun Can Play Typical spaghetti western. Three men (banker, thief, and bounty hunter) compete for a treasure of gold while wandering about the Spanish countryside. Alternate title: "For a Few Bullets More."
1967 103m/C IT SP Edd Byrnes, Gilbert Roland, George Hilton, Kareen O'Hara, Pedro Sanchez, Gerard Herter; **Dir:** Enzo G. Castellari. **VHS** HHE ♫

Any Man's Death Savage is a globetrotting reporter on the trail of a worldwide conspiracy who accidentally uncovers a Nazi war criminal in Africa. Well-meaning but confused tale.
1990 (R) 105m/C John Savage, William Hickey, Mia Sara, Ernest Borgnine, Michael Lerner; **Dir:** Tom Clegg. **VHS, Beta, LV** PSM ♫½

Any Wednesday Okay sex farce about powerful industrialist Robards' use of his mistress's apartment as a tax write-off. When a young company executive learns of the "company" apartment, he meets Robards' nonchalant wife for a tryst of their own. Based on Muriel Resnik's Broadway comedy; similar to the 1960 "The Apartment."
1966 110m/C Jane Fonda, Jason Robards Jr., Dean Jones, Rosemary Murphy; **Dir:** Robert Ellis Miller. **VHS, Beta $19.98** WAR ♫♫½

Any Which Way You Can Bad brawler Philo Beddoe and his buddy Clyde, the orangutan, are back again in the sequel to "Every Which Way But Loose." This time Philo is tempted to take part in a big bout for a large cash prize. Clyde steals scenes, brightening up the no-brainer story.
1980 (PG) 116m/C Clint Eastwood, Sondra Locke, Ruth Gordon, Harry Guardino; **Dir:** Buddy Van Horn. **VHS, Beta, LV $19.98** WAR ♫♫

Anything for a Thrill Two kids save a cameraman's career, make friends with a millionairess, and foil some crooks.
1937 59m/B Frankie Darro, Kane Richmond; **Dir:** S.A. Grandi. **VHS $24.95** NOS, DVT ♫

Anzacs: The War Down Under Well-made Australian TV miniseries about the Australian & New Zealand Army Corps during WWI. Follows the men from the time they enlist to the campaigns in Gallipoli and France.
1985 165m/C AU Paul Hogan, Andrew Clarke, Jon Blake, Megan Williams; **Dir:** George Miller. **VHS, Beta $79.95** CEL ♫♫♫

Anzio The historic Allied invasion of Italy during WWII as seen through the eyes of American war correspondent Mitchum. Fine cast waits endlessly to leave the beach, though big battle scenes are effectively rendered. Based on the book by Wynford Vaughan Thomas. Also known as "The Battle for Anzio" and "Lo Sbarco di Anzio."
1968 117m/C Robert Mitchum, Peter Falk, Arthur Kennedy, Robert Ryan, Earl Holliman, Mark Damon, Reni Santoni, Patrick Magee; **Dir:** Edward Dmytryk. **VHS, Beta $9.95** RCA ♫♫

Apache Lancaster is the only Indian in Geronimo's outfit who refuses to surrender in this chronicle of a bitter battle between the Indians and the U.S. cavalry in the struggle for the West. First western for Aldrich is a thoughtful piece for its time that had the original tragic ending reshot (against Aldrich's wishes) to make it more happy. Adapted from "Bronco Apache" by Paul I. Wellman.
1954 91m/C Burt Lancaster, John McIntire, Jean Peters; **Dir:** Robert Aldrich. **VHS, Beta $19.98** MGM, FOX, FCT ♫♫½

Apache Blood An Indian Brave, the lone survivor of an Indian massacre by the U.S. Army, squares off with a calvary scout in the forbidding desert. Also known as "Pursuit."
1975 (R) 92m/C Ray Danton, DeWitt Lee, Troy Nabors, Diane Taylor, Eva Kovacs, Jason Clark; **Dir:** Thomas Quillen. **VHS $19.99** PSM, TAV ♫½

Apache Chief Two Apache tribe leaders, one good, the other evil clash. Ultimately, and perhaps predictably, they face each other in hand-to-hand combat and peace prevails.
1950 60m/B Alan Curtis, Tom Neal, Russell Hayden, Carol Thurston; **Dir:** Frank McDonald. **VHS, Beta, LV** WGE ♫

Apache Kid's Escape The Apache Kid leads the cavalry on a wild chase across the plains in this saga of the old west.
1930 60m/B Jack Perrin, Buzz Barton; **VHS, Beta $19.95** NOS, DVT ♫

Apache Rose A gambling boat owner plots to gain control of oil found on Vegas Ranch. The original, unedited version of the film.
1947 75m/B Roy Rogers, Dale Evans, Olin Howland, George Meeker; **Dir:** William Witney. **VHS, Beta, LV $9.99** NOS, CPB, VCN ♫

Aparajito The second of the Apu trilogy, about a boy growing up in India, after "Pather Panchali," and before "The World of Apu." Apu is brought to Benares and his education seriously begins. The work of a master; in Bengali with English subtitles. Alternative title, "The Unvanquished."
1957 108m/B IN Pinaki Sen Gupta, Karuna Banerjee, Kanu Banerjee; **Dir:** Satyajit Ray. National Board of Review Awards '59: 5 Best Foreign Films of the Year; Venice Film Festival '57: Best Film. **VHS, Beta $29.95** FCT, MRV, TIM ♫♫♫♫

The Apartment A lowly insurance clerk tries to climb the corporate ladder by "loaning" his apartment out to executives having affairs. Problems arise, however, when he unwittingly falls for the most recent girlfriend of the boss. Highly acclaimed social satire and winner of five Academy Awards. Co-written by Wilder and I.A.L. Diamond (who also wrote ''Some Like It Hot''). Additional Oscar nominations for Best Actor (Lemmon), Best Actress (MacLaine), Supporting Actor (Kruschen), Cinematography, and Sound. **1960 125m/B** Jack Lemmon, Shirley MacLaine, Fred MacMurray, Ray Walston, Jack Kruschen, Joan Shawlee, Edie Adams, Hope Holiday, David Lewis; *Dir:* Billy Wilder. Academy Awards '60: Best Art Direction/Set Decoration (B & W), Best Director (Wilder), Best Film Editing, Best Picture, Best Story & Screenplay; British Academy Awards '60: Best Actor (Lemmon), Best Actress (MacLaine), Best Picture; Directors Guild of America Awards '60: Best Director (Wilder); Golden Globe Awards '61: Best Film—Musical/Comedy; National Board of Review Awards '60: 10 Best Films of the Year; New York Film Critics Awards '60: Best Director (Wilder), Best Picture, Best Writing. **VHS, Beta, LV $19.98** *MGM, BTV* 🎞🎞🎞½

Apartment Zero A decidedly weird, deranged black comedy about the parasite/host-type relationship between two roommates in downtown Buenos Aires: one, a obsessive British movie nut, the other, a sexually mesmerizing stud who turns out to be a cold-blooded psycho. **1988 (R) 124m/C** *GB* Hart Bochner, Colin Firth, Fabrizio Bentivoglio, Liz Smith; *Dir:* Martin Donovan. **VHS, Beta, LV $14.95** *ACA, FCT. IME* 🎞🎞🎞½

The Ape When his daughter dies of a crippling disease, Karloff becomes fixated with the mission to cure paralysis. Obviously distraught, he begins donning the hide of an escaped circus ape whose spinal fluid is the key to the serum. Hide-bedecked, he slays unknowing townspeople to tap them of their spinal fluid and cure his latest patient. **1940 62m/B** Boris Karloff, Maris Wrixon, Henry Hall, Gertrude Hoffman; *Dir:* William Nigh. **VHS, Beta $19.95** *NOS, SNC, VYY* 🎞🎞

A*P*E* A*P*E* is thirty-six feet tall and ten tons of animal fury who destroys anything that comes between him and the actress he loves. Cheap rip-off of Kong. **1976 (PG) 87m/C** Rod Arrants, Joanna DeVarona, Alex Nicol; *Dir:* Paul Leder. **VHS, Beta $19.95** *STE. NWV Woof!*

The Ape Man With the aid of a secret potion, a scientist turns himself into a murderous ape. The only way to regain his human side is to ingest human spinal fluid. Undoubtedly inspired by Boris Karloff's 1940 film, ''The Ape.'' **1943 64m/B** Wallace Ford, Bela Lugosi; *Dir:* William Beaudine. **VHS, Beta $19.95** *NOS, MRV, SNC* 🎞🎞

Aphrodite Steamy drama based on Pierre Louy's masterpiece of erotic literature. **1983 89m/C** Valerie Kaprisky; *Dir:* Robert Fuest. **VHS, Beta, LV $69.98** *VES, LIV* 🎞

Apocalipsis Joe A lonely gunfighter looks to kick an evil criminal out of a small town. **1965 90m/C** Eduardo Fajardo, Mary Paz Pondal; Fernando Ceruli, Fernando Bilbao. **VHS, Beta $34.90** *MAD* 🎞🎞½

Apocalypse Now Its merits still the object of debate, Coppola's $40 million epic vision of the Vietnam War was inspired by Joseph Conrad's novella ''Heart of Darkness.'' Disillusioned Army captain Sheen travels upriver into Cambodia to assassinate overweight renegade colonel Brando. His trip upstream is punctuated by sur-

realistic battles and a terrifying descent into a land where human rationality seems to have slipped away. Considered by some to be the definitive picture of war in its overall depiction of chaos and primal bloodletting; by others, over-wrought and unrealistic. A cinematic experience that may not translate as well to the small screen, yet worth seeing if for nothing more than Duvall's ten minutes of scenery chewing as a battle-obsessed general (''I love the smell of napalm in the morning!''), a study in manic machismo. Stunning photography by Vittorio Storaro, awe-inspiring battle scenes, and effective soundtrack montage. Both Sheen and Coppola suffered emotional breakdowns during the prolonged filming, and that's a very young Fishburne in his major film debut. Oscar nominations for Best Picture, Duvall, Coppola, Adapted Screenplay, Art Direction, Set Decoration and Film Editing. Available in widescreen format on laserdisc and in a remastered version in letterbox on VHS with a remixed soundtrack that features Dolby Surround stereo. **1979 (R) 153m/C** Marlon Brando, Martin Sheen, Robert Duvall, Martin Sheen, Sam Bottoms, Scott Glenn, Albert Hall, Larry Fishburne, Harrison Ford, G.D. Spradlin, Dennis Hopper, Colleen Camp; *Dir:* Francis Ford Coppola. Academy Awards '79: Best Cinematography, Best Sound; British Academy Awards '79: Best Director (Coppola); Cannes Film Festival '79: Best Film Co-Winner, International Critics Prize; National Board of Review Awards '79: 10 Best Films of the Year; National Society of Film Critics Awards '79: Best Supporting Actor (Forrest). **VHS, Beta, LV $29.95** *PAR. LDC* 🎞🎞🎞🎞

Apology A psychotic killer stalks Warren, an experimental artist, in Manhattan while a detective stalks the killer. Made for cable, so expect a little more sex and violence than network TV movies. Written by Mark Medhoff, author of ''Children of a Lesser God.'' **1986 98m/C** Lesley Ann Warren, Peter Weller, John Glover, George Loros, Jimmie Ray Weeks, Christopher North, Harvey Fierstein; *Dir:* Robert Bierman. **VHS $19.95** *HBO* 🎞🎞

Appaloosa A lamenting loner who decides to begin anew by breeding Appaloosas is ripped off by a desperate woman who steals his horse in order to get away from her abusive amour. Brando falls in love with the girl and the two amazingly survive a wealth of obstacles in their battle against Mexican bandits. **1966 99m/C** Marlon Brando, Anjanette Comer, John Saxon; *Dir:* Sidney J. Furie. **VHS, Beta, LV $14.95** *MCA* 🎞🎞½

Appassionata Titillating Italian drama about a dentist with a helplessly neurotic wife. He's seduced by his daughter and her best friend. Dubbed. **1974 100m/C** *IT* Ornella Muti, Gabriel Ferzetti, Valentina Cortese, Elenora Giorgi; *Dir:* Gian Luigi Calderone. **VHS, Beta $59.95** *MED* 🎞🎞

Applause Morgan plays a down-and-out burlesque star trying to protect her fresh from the convent daughter. Definitely dated, but a marvelous performance by Morgan. Film buffs will appreciate this early talkie. **1929 78m/B** Helen Morgan, Joan Peers, Fuller Mellish Jr., Henry Wadsworth, Dorothy Cumming; *Dir:* Rouben Mamoulian. **VHS** *PAR* 🎞🎞🎞

The Apple A futuristic musical filmed in Berlin that features a young, innocent, folk-singing couple who nearly become victims of the evil, glitzy record producer who tries to recruit the couple into a life of sex and drugs.

1980 (PG) 90m/C Catherine Mary Stewart, Alan Love, Grace Kennedy, Joss Ackland; *Dir:* Menahem Golan. **VHS, Beta** *PGN* 🎞½

The Apple Dumpling Gang Three frisky kids strike it rich and trigger the wildest bank robbery in the gold-mad West. Unmistakably Disney, a familial subplot and a wacky duo are provided. Mediocre yet superior to its sequel, ''The Apple Dumpling Gang Rides Again.'' **1975 (G) 100m/C** Bill Bixby, Susan Clark, Don Knotts, Tim Conway, David Wayne, Slim Pickens, Harry Morgan; *Dir:* Norman Tokar. **VHS, Beta $19.99** *DIS* 🎞

The Apple Dumpling Gang Rides Again Two lovable hombres terrorize the West in their bungling attempt to go straight. **1979 (G) 88m/C** Tim Conway, Don Knotts, Tim Matheson, Kenneth Mars, Harry Morgan, Jack Elam; *Dir:* Vincent McEveety. **VHS, Beta $69.95** *DIS* 🎞

The Applegates Bugs-on-a-rampage flick cross-meshing with the ''there's something rotten in suburbia'' genre from the director of ''Heathers.'' A bunch of ecologically correct insects in the Amazon are more than a little miffed about the slash-and-burn tactics in the wild, and they decide to implement a plan to establish a kinder, gentler habitat. The head bug orders Begley and his brood to transform into an average American family, set up house in Ohio, and annihilate a neighboring atomic energy generating station. Trouble is, the bug-people find they've grown accustomed to their decadent Homo sapien ways: even insects aren't immune to the lure of sex, drugs, and cable shopping networks. Altogether an imaginative, often quite funny mono joker that should've been shorter. Sometimes you'll wonder whether the screenwriters were given to bouts of narcolepsy. Also known as ''Meet the Applegates.'' Screenplay by director Lehmann and Redbeard Simmons. **1989 (R) 90m/C** Ed Begley Jr., Stockard Channing, Dabney Coleman, Cami Cooper, Bobby Jacoby; *Dir:* Michael Lehmann. **VHS, Beta, LV $89.98** *MED. FOX, NVH* 🎞🎞½

The Appointment A supernatural force enters the bodies and minds of people and suddenly everyone begins going crazy. **1982 90m/C** Edward Woodward, Jane Merrow; *Dir:* Lindsey C. Vickers. **VHS, Beta, 8mm $39.95** *SVS Woof!*

Appointment with Crime After serving a prison sentence, an ex-con sets out to avenge himself against the colleagues who double crossed him. Well done, highlighted by superior characterizations. Based on the story by Michael Leighton. **1945 91m/B** *GB* William Hartnell, Raymond Lovell, Robert Beatty, Herbert Lom, Joyce Howard, Alan Wheatley, Cyril Smith; *W/Dir:* John Harlow. **VHS, Beta $16.95** *SNC* 🎞🎞½

Appointment with Death Disappointing Agatha Christie mystery with Hercule Poirot solving the murder of a shrewish widow in 1937 Palestine. Music by Pino Donaggio. **1988 (PG) 103m/C** Peter Ustinov, Lauren Bacall, Carrie Fisher, John Gielgud, Piper Laurie, Hayley Mills, Jenny Seagrove, David Soul; *Dir:* Michael Winner. **VHS, Beta, LV $19.98** *WAR, FCT* 🎞½

Appointment with Fear A tough detective investigates a murder and all of the clues lead him mysteriously to a comatose asylum inmate.

1985 (R) 95m/C Michael Wyle, Michelle Little, Garrick Dowhen; *Dir:* Alan Smithee. **VHS, Beta** $14.95 *IVE* 🎬

Appointment in Honduras
An adventurer goes on a dangerous trek through the Central American jungles to deliver funds to the Honduran President.
1953 79m/C Glenn Ford, Ann Sheridan, Zachary Scott; *Dir:* Jacques Tourneur. **VHS, Beta** *BVV, OM* 🎬🎬

Apprentice to Murder
A small Pennsylvania Dutch town is shaken by a series of murders, thought to be associated with a bizarre local mystic and healer. Based on a true story, sort of.
1988 (PG-13) 97m/C Donald Sutherland, Mia Sara, Chad Lowe, Eddie Jones; *Dir:* Ralph L. Thomas. **VHS, Beta, LV** $19.95 *STE, NWV, HHE* 🎬🎬

The Apprenticeship of Duddy Kravitz
Young Jewish man in Montreal circa 1948 is driven by an insatiable need to be a "somebody." Everybody has always told him he will be. A series of get-rich-quick schemes backfire in different ways, and he becomes most successful at driving people away. Young Dreyfuss is at his best. Made in Canada with thoughtful detail, and great cameo performances. Oscar-nominated script by Mordecai Richler, from his novel.
1974 (PG) 121m/C *CA* Richard Dreyfuss, Randy Quaid, Denholm Elliott, Jack Warden, Micheline Lanctot, Joe Silver; *Dir:* Ted Kotcheff. Berlin Film Festival '74: Best Film. **VHS, Beta** $69.95 *PAR* 🎬🎬🎬½

April Fools
A bored stockbroker falls in love with a beautiful woman who turns out to be married to his boss. Marvin Hamlisch's score resounds.
1969 (PG) 95m/C Jack Lemmon, Catherine Deneuve, Sally Kellerman, Peter Lawford, Charles Boyer, Myrna Loy, Harvey Korman; *Dir:* Stuart Rosenberg. **VHS, Beta** $19.98 *FOX* 🎬🎬

April Fool's Day
College kids vacationing at a secluded island home are systematically murdered by an unknown assailant.
1986 (R) 90m/C Jay Baker, Pat Barlow, Lloyd Berry, Deborah Foreman; *Dir:* Fred Walton. **VHS, Beta, LV** $19.95 *PAR* 🎬

April In Paris
Dynamite Jackson (Day), a chorus girl accidentally sent by the State Department to perform in Paris, meets S. Winthrop Putnam (Bolger), a timid fellow trapped in an unpleasant marriage. They eventually sing and dance their way to warm feelings as they begin a lifelong romance and live happily ever after.
1952 100m/C Doris Day, Ray Bolger, Claude Dauphin, Eve Miller, George Givot, Paul Harvey; *Dir:* David Butler. **VHS** $29.98 *WAR, CCB* 🎬🎬

Arabesque
A college professor is drawn into international espionage by a beautiful woman and a plot to assassinate an Arab prince. Stylish and fast moving, with music by Henry Mancini. From the novel "The Cipher" by Gordon Votler.
1966 105m/C Gregory Peck, Sophia Loren, George Coulouris; *Dir:* Stanley Donen. **VHS, Beta** $14.95 *MCA, FCT* 🎬🎬½

Arabian Nights
The third of Pasolini's epic, explicit adaptations of classic portmanteau, featuring ten of the old Scheherazade favorites adorned by beautiful photography, explicit sex scenes and homoeroticism. In Italian with English subtitles; available dubbed.
1974 130m/C *IT* Ninetto Davoli, Franco Merli, Ines Pellegrini, Luigina Rocchi; *Dir:* Pier Paolo Pasolini. Cannes Film Festival '74: Special Jury Prize. **VHS, Beta** $11.95 *WBF, TAM, APD* 🎬🎬🎬

Arachnophobia
Erstwhile Spielberg producer Frank Marshall's directorial debut is a mega-budget thrill-com version of a big bug horror story. Lethal bugs—South American spiders this time—wind up in a white picket fence California community somewhere off the beaten track. There they wreak eight-legged antenna-less havoc on their two-legged antagonists, including an utterly arachnophobic town doctor (Daniels) and a gung-ho insect exterminator (Goodman). The script's a bit yawn-inspiring but the cast and effects will keep you from dozing off. Screenplay by Don Jakoby and Wesley Strick, based on a story by Jakoby and Al Williams.
1990 (PG-13) 109m/C Jeff Daniels, John Goodman, Harley Jane Kozak, Julian Sands, Roy Brocksmith; *Dir:* Frank Marshall. **VHS, Beta, LV** $19.99 *BVV* 🎬🎬½

Arch of Triumph
In Paris, an Austrian refugee doctor falls in love just before the Nazis enter the city. Big-budget box-office loser featuring fine cast but sluggish pace. Co-written by Milestone and Harry Brown. Based on the Erich Maria Remarque novel and remade for television.
1948 120m/B Ingrid Bergman, Charles Boyer, Charles Laughton, Louis Calhern; *Dir:* Lewis Milestone. **VHS** $129.98 *REP* 🎬🎬½

Archer: The Fugitive from the Empire
A young warrior battles the forces of evil in this television movie. Lots of strange names to learn even if you already know who's going to win the final battle.
1981 97m/C Lane Caudell, Belinda Bauer, George Kennedy, Victor Campos, Kabir Bedi, George Innes, Marc Alaimo, Allen Rich, John Hancock, Priscilla Pointer, Richard Dix, Sharon Barr; *Dir:* Nick Corea. **VHS, Beta** $39.95 *MCA* 🎬

Archer's Adventure
An Australian family film based on a true story. A horsetrainer's young apprentice delivers a prize racehorse to Melbourne, through 600 miles of tough frontier, devious bush rangers, and disaster.
1985 120m/C *AU* Brett Como; *Dir:* Denny Lawrence. **VHS, Beta** $14.95 *NWV, STE* 🎬½

Are Parents People?
Lighthearted silent comedy about a young girl's successful attempts to reunite her feuding parents. It all begins as she runs away and spends the night in the office of a doctor she has grown to like. After a frantic night of searching, her parents are reunited through their love for her.
1925 60m/B Betty Bronson, Adolphe Menjou, Florence Vidor. **VHS** $27.95 *PAR, DNB* 🎬🎬

Are You in the House Alone?
Made for television adaptation of Richard Peck's award-winning novel. Story of a high school coed who becomes the target of a terror campaign.
1978 100m/C Blythe Danner, Kathleen Beller, Tony Bill, Scott Colomby; *Dir:* Walter Grauman. **VHS, Beta** $49.95 *WOV, GKK* 🎬🎬

Are You Lonesome Tonight
Suspense thriller starring Seymour as a wealthy socialite who discovers her husband is having an affair with a phone-sex girl. After her husband mysteriously disappears, she hires a private detective (Stevenson) to help her embark on a search that opens her eyes to the startling truth.
1992 (PG-13) 91m/C Jane Seymour, Parker Stevenson, Beth Broderick, Joel Brooks; *Dir:* E.W. Swackhamer. **VHS, Beta** $19.95 *PAR* 🎬🎬

Arena
Remember old boxing melodramas about good-natured palookas, slimy opponents, gangsters and dames? This puts those cliches in a garish sci-fi setting, with handsome Steve Armstrong battling ETs and the astro-mob to be the first human pugilistic champ in decades. A really cute idea (from the screenwriters of "The Rocketeer"), but it conks out at the halfway point. Worth a look for buffs.
1988 (PG-13) 97m/C *IT* Paul Satterfield, Claudia Christian, Hamilton Camp, Marc Alaimo, Armin Shimerman, Shari Shattuck, Jack Carter; *Dir:* Peter Manoogian. **VHS, LV** $79.95 *COL, PIA* 🎬🎬

Aria
Ten directors were given carte blanche to interpret ten arias from well-known operas. Henry and D'Angelo star in Julian Temple's rendition of Verdi's "Rigoletto." In Fonda's film debut, she and her lover travel to Las Vegas and eventually kill themselves in the bathtub, just like "Romeo & Juliet." Jarman's piece (a highlight) shows an aged operatic star at her last performance remembering an early love affair. "I Pagliacci" is the one aria in which the director took his interpretation in a straightforward manner.
1988 (R) 90m/C *GB* Theresa Russell, Anita Morris, Bridget Fonda, Beverly D'Angelo, Buck Henry, John Hurt; *Dir:* Ken Russell, Charles Sturridge, Robert Altman, Bill Bryden, Jean-Luc Godard, Bruce Beresford, Nicolas Roeg, Franc Roddam, Derek Jarman, Julien Temple. **VHS, Beta** $29.95 *ACA, FCT* 🎬🎬

Ariel
Refreshing, award-winning offbeat Finnish comedy by much-praised newcomer Kaurismaki (who earlier directed "Leningrad Cowboys Go American"). Hoping to find work in Southern Finland, an out-of-work miner from Northern Finland (Pajala) jets off in his white Cadillac convertible given to him in a cafe by a friend, who promptly shoots himself. There's no linear progression toward a happy ending, although antiheroic subject does find employment and romances a meter maid. Mostly, though, he's one of those it's hell being me guys who wouldn't have any luck if it weren't for bad luck. Strange slice-of-life sporting film noir tendencies, although essentially anti-stylistic. Kaurismaki also scripted.
1989 74m/C *FI* Susanna Haavisto, Turo Pajala; *Dir:* Aki Kaurismaki. National Society of Film Critics Awards '90: Best Foreign Film. **VHS** $79.95 *KIV, FCT, TAM* 🎬🎬🎬

Arizona
A group of illegal aliens struggle for survival after they cross the border into the harsh desert.
1986 94m/C *MX* Roberto "Flaco" Guzman, Juan Valentin. **VHS, Beta** *MED* 🎬½

Arizona Badman
There is a bad man in Arizona, and he must be gotten rid of.
1942 60m/B Reb Russel; *Dir:* S. Roy Luby. **VHS, Beta** $19.95 *NOS, DVT* 🎬

Arizona Bound
Mesa City is infested with a villain and our "Rough Rider" trio must rid the town of him.
1941 57m/B Buck Jones, Tim McCoy, Raymond Hatton, Dennis Moore, Luana Walters; *Dir:* Spencer Gordon Bennet. **VHS, Beta** $19.95 *NOS, VCN, DVT* 🎬

Arizona Cowboy
An ex-G.I., now the rodeo's big attraction, gets involved in a robbery.
1949 57m/B Rex Allen, Gordon Jones, Roy Barcroft; *Dir:* R.G. Springsteen. **VHS, Beta** $29.95 *DVT* 🎬

Arizona Days
Cowboys join a minstrel group and rescue the show when a group of toughs try to break it up.

A

1937 56m/B Tex Ritter, Eleanor Stewart, Syd Saylor, Snub Pollard; *Dir:* John English. **VHS, Beta** $19.95 *NOS, UHV, VDM* 🎬½

Arizona Gangbusters Below-average oater has real-life cowboy McCoy fighting city hall in order to fight other baddies.
1940 57m/B Tim McCoy, Pauline Hadden, Forrest Taylor, Julian Rivero, Curly Dresden; *Dir:* Peter Stewart. **VHS** $19.95 *DVT* 🎬

Arizona Gunfighter A young cowhand seeks revenge against the man who murdered his father in this western.
1937 60m/B Bob Steele, Ted Adams; *Dir:* Sam Newfield. **VHS, Beta** $19.95 *NOS, DVT* 🎬

Arizona Heat A violent cop is teamed up with a tough, but tender female cop in this all-too-familiar tale of two cops chasing a cop killer.
1987 (R) 91m/C Michael Parks, Denise Crosby, Hugh Farrington; *Dir:* John Thomas. **VHS** $79.98 *REP* 🎬🎬

Arizona Kid Another sagebrush saga featuring Roy in singin' and fightin' action.
1939 54m/B Roy Rogers, Dale Evans; *Dir:* Horace Carpenter. **VHS, Beta** $9.99 *NOS, DVT, BUR* 🎬

Arizona Raiders Arizona rangers hunt down killers who have been terrorizing the territory.
1965 88m/C Audie Murphy, Buster Crabbe, Gloria Talbott; *Dir:* William Whitney. **VHS, Beta** $19.95 *NOS, UHV, VCN* 🎬🎬

Arizona Roundup Good vs. bad amid tumbleweed, bleached-white chaps, bloodless shoot-outs and happy endings.
1942 54m/B Tom Keene, Sugar Dawn, Jack Ingram; *Dir:* Robert Tansey. **VHS, Beta** $19.95 *NOS, VDM* 🎬½

Arizona Stagecoach The Range Busters set out to bust a notorious, guiltless, devil-may-care outlaw gang.
1942 58m/B Ray Corrigan, Max Terhune, Kermit Maynard, Charles King, John "Dusty" King; *Dir:* S. Roy Luby. **VHS, Beta** $19.95 *NOS, CPB, BUR* 🎬

Arizona Terror Our hero is on a quest for vengeance, seeking the posse that killed his partner.
1931 64m/B Ken Maynard. **VHS, Beta** *VCN* 🎬

Arizona Whirlwind Our intrepid heroes must battle torrents of gunfire in order to prevent a stage hold-up in this western saga.
1944 59m/B Ken Maynard, Hoot Gibson, Bob Steele. **VHS, Beta** $19.95 *NOS, DVT* 🎬

Ark of the Sun God Another adventurer battles the Nazis and nutsies for a 2000-year-old ark buried in the Sahara.
1982 95m/C David Warbeck, John Steiner, Susie Sudlow, Alan Collins, Riccardo Palacio; *Dir:* Anthony M. Dawson. **VHS, Beta** $59.95 *TWE* 🎬

Armed for Action Estevez stars as Sgt. Phil Towers in this exciting action thriller. Towers gets more than he bargained for when his prisoner, Mafia hitman David Montel escapes while en route from New York to Los Angeles. Montel and his henchman flee to a quiet small town with Towers hot on their trail. When he finally catches up with them, Towers leads a small army of locals on a brutal assault.
19?? ?m/C Joe Estevez, Rocky Patterson, Barri Murphy, David Harrod, J. Scott Guy; *Dir:* Shane Spaulding. **VHS** $89.95 *AIP* 🎬

Armed and Dangerous Candy and Levy are incompetent security guards assigned to a do-nothing job. Things get spiced up when a mobster tries to run a crime ring under their nose. Candy catches on and winds up in a full-fledged chase. Not as funny as it sounds, though occasionally has moments of genuine comedy.
1986 (PG-13) 88m/C John Candy, Eugene Levy, Kenneth McMillan, Brion James, Robert Loggia, Meg Ryan, Don Stroud, Jonathan Banks, Steve Railsback, Bruce Kirby, Tony Burton, Larry Hankin, Judy Landers, David Wohl; *Dir:* Mark L. Lester. **VHS, Beta, LV** $14.95 *COL* 🎬🎬

Armed Response Carradine leads a group of mercenaries in a battle against Chinatown mobsters. They race to locate a priceless jade statue before it can fall into the wrong hands.
1986 (R) 86m/C David Carradine, Lee Van Cleef, Mako, Lois Hamilton, Ross Hagen, Brent Huff; *Dir:* Fred Olen Ray. **VHS, Beta, LV** $12.95 *RCA* 🎬🎬

Armored Car Robbery Talman and his buddies plot to rob an armored car but are foiled by McGraw and his crimefighters. Surprisingly good B-crime drama.
1950 68m/B Charles McGraw, Adele Jergens, William Talman, Steve Brodie; *Dir:* Richard Fleischer. **VHS** *RKO* 🎬🎬

Armored Command A beautiful German spy infiltrates an American outpost during the Battle of the Bulge. Tepid WWII fare made too long after the fact.
1961 105m/B Burt Reynolds, Tina Louise, Howard Keel, Earl Holliman, Warner Anderson, Carleton Young; *Dir:* Byron Haskin. **VHS, Beta** $19.98 *FOX, FCT* 🎬½

Army Brats In this Dutch film a military family goes bloodily and comically to war with itself. Even in a welfare state, parents can't control their wee ones.
1984 105m/C NL Akkemay, Frank Schaafsma, Peter Faber; *Dir:* Ruud Van Hemert. **VHS, Beta** $59.95 *WAR* 🎬½

Arnold Outrageous black comedy involving a woman who marries a cadaver to gain his large inheritance. The dead man's brother is behind the scheme for personal reasons. Lots of bizarre and creative deaths in this horror spoof. Unusual wedding scene is a must-see.
1973 (PG) 96m/C Stella Stevens, Roddy McDowall, Elsa Lanchester, Victor Buono, Bernard Fox, Farley Granger, Shani Wells; *Dir:* Georg Fenady. **VHS, Beta** $69.98 *LTG, LIV* 🎬🎬½

Around the World in 80 Days Niven is the unflappable Victorian Englishman who wagers that he can circumnavigate the earth in four-score days. With his faithful manservant Cantinflas they set off on a spectacular journey. A perpetual favorite providing ample entertainment. Star-gazers will particularly enjoy the more than 40 cameo appearances by many of Hollywood's biggest names. Written by James Poe, John Farrow, and S.J. Perelman from the Jules Verne novel. Oscar nominations for director, color art direction and set design, and color costume design.
1956 (G) 178m/C David Niven, Shirley MacLaine, Frank Sinatra, Marlene Dietrich, Robert Newton, Cantinflas, Charles Boyer, Joe E. Brown, John Carradine, Charles Coburn, Ronald Colman, Noel Coward, Andy Devine, Reginald Denny, Fernandel, John Gielgud, Hermione Gingold, Cedric Hardwicke, Trevor Howard, Glynis Johns, Buster Keaton, Evelyn Keyes, Peter Lorre, Mike Mazurki, Victor McLaglen, John Mills, Robert Morley, Jack Oakie, George Raft, Cesar Romero, Gilbert Roland, Red Skelton, Martine Carol,

Melville Cooper; *Dir:* Michael Anderson Sr. Academy Awards '56: Best Adapted Screenplay, Best Color Cinematography, Best Film Editing, Best Picture, Best Original Score; Golden Globe Awards '57: Best Film—Drama; National Board of Review Awards '56: Best Picture; New York Film Critics Awards '56: Best Picture, Best Writing. **VHS, Beta, LV** $29.98 *WAR, BTV* 🎬🎬🎬

Around the World in 80 Days Monty Python alumnus Palin recreates the globe-trotting journey of Jules Verne's character Phineas Fogg, using the same route and modes of travel from the 1872 novel and accompanied by a four-member film crew. An visual feast for armchair travelers and accidental tourists, laced with Palin's wit. The set of four cassettes runs nearly six hours, but never becomes boring.
1990 330m/C Michael Palin. **VHS** $49.95 *BFS, PME* 🎬🎬🎬

Around the World in 80 Ways Sometimes clever, sometimes crude Australian comedy about an aging man rescued from a nursing home and taken on a phony trip around the world by his sons. Odd, but genuinely funny at times.
1986 (R) 90m/C AU Philip Quast, Alan Penney, Diana Davidson, Kelly Dingwall, Gosia Dobrowolska; *Dir:* Stephen MacLean. **VHS, LV** $19.98 *SUE, CHA* 🎬🎬½

Around the World Under the Sea Bunch of men and one woman scientist plunge under the ocean in an experiment to predict earthquakes. They plant earthquake detectors along the ocean floor and discover the causes of tidal waves. They have men-women battles. They see big sea critters.
1965 111m/C David McCallum, Shirley Eaton, Gary Merrill, Keenan Wynn, Brian Kelly, Lloyd Bridges; *Dir:* Andrew Marton. **VHS, Beta** $59.95 *MGM* 🎬🎬

The Arousers Cult item starring hunk Hunter as a handsome California psycho. Tab travels the coast searching for a woman he is able to make love to; those who fail to fully arouse him come to tragic, climatic ends. Definitely underground and moderately interesting. Originally released as "Sweet Kill." Alternate title: "A Kiss from Eddie."
1970 (R) 85m/C Tab Hunter, Nadyne Turney, Roberta Collins, Isabel Jewell, John Aprea, Angel Fox; *Dir:* Curtis Hanson. **VHS, Beta** $24.98 *SUE* 🎬🎬½

The Arrangement Veteran advertising executive Douglas attempts suicide and then sets out to search for the meaning of life. Along the way he attempts to patch up his "arrangements" with his wife, his mistress and his father. Forced, slow, and self-conscious, though well acted. Adapted by Kazan from the director's own novel.
1969 (R) 126m/C Kirk Douglas, Faye Dunaway, Deborah Kerr, Richard Boone, Hume Cronyn, Elia Kazan; *Dir:* Elia Kazan. **VHS, Beta** $59.95 *WAR* 🎬½

Arrest Bulldog Drummond Captain Drummond is accused of killing the inventor of a futuristic detonator machine and must track down the real killers. Part of the "Bulldog Drummond" series.
1938 57m/B John Howard, Heather Angel, George Zucco, H.B. Warner, E.E. Clive, Reginald Denny, John Sutton; *Dir:* James Hogan. **VHS, Beta** $16.95 *PAR, RXM, SNC* 🎬

The Arrival An never-seen alien parasite turns an old man into a vampiric young stud after female blood. Plot and characterizations never do arrive. Horror director Stuart Gordon cameos as a hairy biker.

1990 (R) 107m/C John Saxon, Joseph Culp, Robert Sampson, Michael J. Pollard; *W/Dir:* David Schmoeller. **VHS, LV** $79.95 PSM *⚏*

Arrowhead A long-running argument between a tough Cavalry scout and an Apache chief pits the cowboys against the Indians in this western fantasy. The personal battles that become all-out wars turn back to fist-fights before the matter is finally settled. **1953 105m/C** Charlton Heston, Jack Palance, Katy Jurado, Brian Keith, Milburn Stone; *Dir:* Charles Marquis Warren. **VHS, Beta, LV** $14.95 PAR *⚏⚏½*

Arrowsmith A small-town medical researcher battles his conscience as he juggles his selfish and unselfish motivations for the work he does. He travels to the West Indies to confront the issues of his life and come to terms with himself once and for all. A talented cast takes their time. Based on the classic Sinclair Lewis novel. Two edited versions available (99 and 89 minutes), both of which delete much of Loy. Oscar nominated for best picture, director, and photography. **1931 95m/B** Ronald Colman, Helen Hayes, Myrna Loy; *Dir:* John Ford. **VHS, Beta** $19.98 SUE *⚏⚏½*

Arsenal Classic Russian propagandist drama about strikes affecting the Russian home front during WWI, marking Dovshenko's first great achievement in the realm of Eisenstein and Pudovkin. Silent. **1929 70m/B** *RU* Semyon Svashenko; *Dir:* Alexander Dovzhenko. **VHS, Beta, 3/4U** $24.95 NOS, MRV, IHF *⚏⚏⚏*

The Arsenal Stadium Mystery Inspector Banks of Scotland Yard tracks down the killer of a football star in this clever but unassuming murder mystery. **1939 85m/B** *GB* Leslie Banks, Greta Gynt, Ian MacLean, Liane Linden, Anthony Bushell, Esmond Knight; *Dir:* Thorold Dickinson. **VHS** $16.95 SNC *⚏⚏½*

Arsenic and Old Lace Set-bound but energetic adaptation of the classic Joseph Kesselring play about easy-going drama critic Grant caught in a sticky situation when he learns of his aunts' favorite pastime. Apparently the kind, sweet, lonely spinsters lure gentlemen to the house and serve them elderberry wine with a touch of arsenic. Even more devastating to Grant is that the women have been burying the bodies in the cellar—a cellar which also serves as the Panama Canal for Grant's crazy cousin who thinks he's Theodore Roosevelt. Massey and Lorre excel in their sinister roles. One of the best madcap black comedies of all time—a must-see. Shot in 1941 and released a wee bit later, and written by the team of Julius J. and Philip G. Epstein. Available colorized. **1944 158m/B** Cary Grant, Josephine Hull, Jean Adair, Raymond Massey, Jack Carson, Priscilla Lane, John Alexander, Edward Everett Horton, John Ridgely, James Gleason, Peter Lorre; *Dir:* Frank Capra. **VHS, Beta, LV** $19.95 MGM, VYG, CCE *⚏⚏⚏½*

The Art of Crime A gypsy/detective is drawn into a homicide case when one of his fellow antique dealers is charged with murder. That certain gypsy feeling provides an atmospheric edge over others of the crime art genre. A pilot for a prospective television series based on the novel "Gypsy in Amber." (Script by Bill Davidson and Martin Smith.) **1975 72m/C** Ron Leibman, Jose Ferrer, David Hedison, Jill Clayburgh; *Dir:* Richard Irving. **VHS** MOV *⚏⚏½*

The Art of Dying A loony videophile decides to start staging productions of his all-time favorite scenes. Trouble is, his idea of a fabulous film moment calls for lots of blood and bile as he lures teenage runaways to his casting couch. Director Hauser stars as the cop who's none too impressed with the cinematic remakes, while cult favorite Pollard is his partner. If you like a little atmosphere and psychological depth in your slashers, you'll find this to be the stuff that populates film noir nightmares. **1990 90m/C** Wings Hauser, Michael J. Pollard, Sarah Douglas, Kathleen Kinmont, Sydney Lassick; *Dir:* Wings Hauser. **VHS** PMH *⚏⚏⚏*

Arthur Spoiled, alcoholic billionaire Moore stands to lose everything he owns when he falls in love with a waitress. He must choose between wealth and a planned marriage, or poverty and love. Surprisingly funny, with an Oscar for Gielgud as Moore's valet, and great performance from Minnelli. Music by Burt Bacharach, whose title song won an Oscar, too. Arguably the best role Moore's ever had, and he makes the most of it, taking the one-joke premise to a nomination for Oscar actor. Screenplay by Gordon was also nominated. **1981 (PG) 97m/C** Dudley Moore, Liza Minnelli, John Gielgud, Geraldine Fitzgerald, Stephen Elliott, Jill Eikenberry, Lou Jacobi; *Dir:* Steve Gordon. Academy Awards '81: Best Song ("Arthur's Theme"), Best Supporting Actor (Gielgud). **VHS, Beta, LV, 8mm** $19.98 WAR, FCT, BTV *⚏⚏⚏*

Arthur 2: On the Rocks When Arthur finally marries his sweetheart, it may not be "happily ever after" because the father of the girl he didn't marry is out for revenge. When Arthur discovers that he is suddenly penniless, a bit of laughter is the cure for the blues and also serves well when the liquor runs out. A disappointing sequel with few laughs. **1988 (PG) 113m/C** Dudley Moore, Liza Minnelli, John Gielgud, Geraldine Fitzgerald, Stephen Elliott, Ted Ross, Barney Martin, Jack Gilford; *Dir:* Bud Yorkin. **VHS, Beta, LV, 8mm** $19.98 WAR, FCT *⚏½*

Arthur's Hallowed Ground A cricket field caretaker battles the board of directors over the fate of his favorite plot of sod. **1984 75m/C** Jimmy Jewel, Jean Bolt, Michael Elphick; *Dir:* Freddie Young. **VHS, Beta** $59.95 MGM *⚏*

Artists and Models Martin is a struggling comic book artist and Lewis his idiot roommate. The pair become mixed up in both romance and intrigue when Lewis begins talking in his sleep about spies and such. One of the duo's more pleasant cinematic outings. **1955 109m/C** Dean Martin, Jerry Lewis, Shirley MacLaine, Dorothy Malone, Eddie Mayehoff, Eva Gabor, Anita Ekberg, George Winslow, Jack Elam, Herbert Rudley, Nick Castle; *Dir:* Frank Tashlin. **VHS** $14.95 PAR, MLB *⚏⚏½*

Artist's Studio Secrets An artist from Greenwich Village only gets aroused by clothed models, so his wife jealously arranges for him to only work with nude women. Also known as "The Story of an Artist's Studio Secrets" and "Artist Nude Secrets." **1964 78m/B VHS, Beta** $19.95 VDM *⚏*

As Is Two gay New Yorkers deal with a troubled romance and AIDS. A cable television adaptation from the William M. Hoffman play.

1985 86m/C Jonathan Hadary, Robert Carradine; *Dir:* Michael Lindsay-Hogg. **VHS, Beta** $59.95 LHV, WAR *⚏½*

As Summers Die Louisiana attorney Glenn fights in the late 1950s to protect the rights of a black family, against the wishes of a powerful local clan. Glenn finds support in surprising places, though. From Winston Groom's acclaimed novel. A cable television movie. **1986 100m/C** Scott Glenn, Jamie Lee Curtis, Penny Fuller, Bette Davis, John Randolph, Beah Richards, Ron O'Neal, John McIntire; *Dir:* Jean-Claude Tramont. **VHS, Beta** $14.95 NEG *⚏⚏*

As You Desire Me Garbo plays an amnesia victim who returns to a husband she doesn't even remember after an abusive relationship with a novelist. An interesting, if not down-right bizarre movie, due to the pairing of the great Garbo and the intriguing von Stroheim. An adaption of Luigi Pirandello's play. **1932 71m/B** Greta Garbo, Melvyn Douglas, Erich von Stroheim, Owen Moore, Hedda Hopper; *Dir:* George Fitzmaurice. **VHS, Beta** $19.98 MGM, FCT *⚏⚏⚏*

As You Like It A Duke's banished daughter poses as a man to win the attentions of one of her father's attendants in this highly stylized Shakespearean comedy adapted by J.M. Barrie and Robert Cullen. Early Shakespearean Olivier. **1936 96m/B** *GB* Elisabeth Bergner, Laurence Olivier, Henry Ainley, Felix Aylmer; *Dir:* Paul Czinner. **VHS, Beta, LV** $19.95 NOS, MRV, VYY *⚏⚏½*

As You Like It Shakespeare's lighthearted comedy about all kinds of love—physical and intellectual, sentimental and cynical, enduring love between friends, and romantic love at first sight. Play features the "Seven Ages of Man" speech. Part of the television series "The Shakespeare Plays." **1979 150m/C** *GB* Helen Mirren, Brian Stirner, Richard Pasco. **VHS, Beta** TLF *⚏⚏⚏*

As You Were A girl with a photographic memory enlists in the Army, becoming both a nuisance and comedic victim to her sergeant. **1951 57m/B** Joe Sawyer, William Tracy, Sondra Rogers, Joan Vohs. **VHS, Beta, LV** WGE *⚏*

As Young As You Feel A 65-year-old man forced to retire from his job at a printing company seeks revenge. He poses as the head of the conglomerate that owns the company and gets them to repeal their retirement policy. He then gains national publicity when he makes a speech about the dignity of man. Although the printing company finds out he's a fake, they still give him his old job back for fear of a negative public opinion. Watch for Monroe as the boss's secretary. Fine comic performances. Based on a story by Paddy Chayefsky. **1951 77m/C** Monty Woolley, Thelma Ritter, David Wayne, Jean Peters, Constance Bennett, Marilyn Monroe, Allyn Joslyn, Albert Dekker, Clinton Sundberg, Minor Watson; *Dir:* Harmon Jones. **VHS** $14.98 FXV *⚏⚏*

Ash Wednesday Taylor endures the pain of cosmetic surgery in an effort to rescue her floundering union with Fonda. Another undistinguished performance by Liz. Fonda is especially slimy as the philandering husband, but only appears in the latter stages of the film. **1973 (R) 99m/C** Elizabeth Taylor, Henry Fonda, Helmut Berger, Keith Baxter, Margaret Blye, Maurice Teynac, Monique Van Vooren; *Dir:* Larry Peerce. **VHS** $14.95 PAR, FCT, CCB *⚏*

Ashanti, Land of No Mercy Caine of the week movie with Michael portraying a doctor acting as a missionary in South Africa who finds himself alone in a battle to rescue his wife from a band of slave traders. The chase spans many Middle Eastern countries and begins to look bleak for our man. Talented cast and promising plot are undone by slow pace. Also known as "Ashanti" and based on the novel "Ebano" by Alberto Vasquez-Figueroa.
1979 117m/C Michael Caine, Omar Sharif, Peter Ustinov, Rex Harrison, William Holden, Beverly Johnson; *Dir:* Richard Fleischer. **VHS, Beta** **$19.98** TWE *♂♂*

Ashes and Diamonds In the closing days of WWII, a Polish resistance fighter assassinates the wrong man, tries to find love with the right women, and questions the meaning of struggle. A seminal Eastern European masterpiece that defined a generation of pre-solidarity Poles. Available in Polish with English subtitles or dubbed into English. The last installment of the trilogy that includes "A Generation" and "Kanal" and based on a novel by Jerzy Andrzeewski. Also known as "Popiol i Diament."
1959 105m/B *PL* Zbigniew Cybulski, Eva Krzyzewska, Adam Pawlikowski; *Dir:* Andrzej Wajda. Venice Film Festival '59: International Film Critics Award; British Film Critics Guild '59: Best Foreign Film. **VHS, Beta** **$19.95** TAM, SUE, MRV *♂♂♂♂*

Ashes and Embers A black Vietnam vet in Los Angeles has trouble fitting into society, eventually running afoul of the police. Ethiopian-born director Gerima endows vital subject matter with a properly alienated mood.
1982 120m/C John Anderson, Evelyn Blackwell; *Dir:* Halle Gerima. **VHS** **$59.95** FCI *♂♂*

Asphalt Jungle An aging criminal emerges from his forced retirement (prison) and assembles the old gang for one final heist. A very realistic storyline and a superb cast make this one of the best crime films ever made. Highly acclaimed; based on the work of W. R. Burnett. Oscar nominated for Best Supporting Actor (Jaffe), Director (Huston), Screenplay, and Black & White Cinematography.
1950 112m/B Sterling Hayden, Sam Jaffe, Louis Calhern, Jean Hagen, Marilyn Monroe, James Whitmore, John McIntire, Marc Lawrence; *Dir:* John Huston. Edgar Allan Poe Awards '50: Best Screenplay; Venice Film Festival '50: Best Actor; National Board of Review Awards '50 '50: Best Director, 10 Best Films of the Year. **VHS, Beta, LV** **$19.98** MGM, VYG *♂♂♂♂*

The Asphyx Nineteenth century doctor Stephens is studying death when he discovers The Asphyx, an aura that surrounds a person just before they die. Stephens delves deeper into his research and finds the keys to immortality. However, his irresponsibility in unleashing the obscure supernatural power on the world brings a swarm of unforeseen and irreversible troubles. High-class sci fi; also on video as, "Spirit of the Dead."
1972 (PG) 98m/C *GB* Robert Stephens, Robert Powell, Jane Lapotaire, Alex Scott, Ralph Arliss, Fiona Walker, John Lawrence; *Dir:* Peter Newbrook. **VHS, Beta** **$19.95** IGV, UHV, MAG *♂♂*

Assassin An athletic karate fighter poses as an underworld figure and infiltrates the largest gang in Japan where he is hired as a hit man.
1979 86m/C Sonny Chiba. **VHS, Beta** **$59.98** FOX *♂*

Assassin Mongolian hordes try to invade China and wind up getting kicked for their efforts.
1979 96m/C VHS, Beta **$29.95** UNI *♂*

Assassin Made for TV drama about a mad scientist who creates a bionic killer for a bizarre plot to take over the world. He programs the cyborg to assassinate the President and other key people to help carry out his plan. A retired CIA operative emerges to stop the scientist by trying to destroy the robot.
1986 (PG-13) 94m/C Robert Conrad, Karen Austin, Richard Young, Jonathan Banks, Robert Webber; *Dir:* Sandor Stern. **VHS, Beta** **$29.95** ACA *♂♂*

Assassin Fairly lame thriller about a CIA agent protecting a Senator who falls under suspicion when his charge is shot by an assassin. In investigating the killing the agent discovers the usual governmental conspiracy.
1989 (R) 92m/C Steve Railsback, Nicholas Guest, Xander Berkeley, Elpidia Carrillo; *Dir:* Jon Hess. **VHS, Beta** **$79.99** HBO *♂♂*

Assassin of Youth A girl is introduced to marijuana and soon becomes involved in "the thrills of wild parties," and the horrors of the "killer weed." Camp diversion.
1935 70m/B Luana Walters, Arthur Gardner; *Dir:* Elmer Clifton. **VHS, Beta** **$16.95** SNC, NOS, TIM *♂ ♦½*

Assassination A serious threat has been made to First Lady Ireland and no one is taking it lightly. Secret Service agent Bronson has been called as Ireland's personal bodyguard and suddenly they are both the target of terrorist attacks. Strangely though, the attacks seem to be directed from inside the White House. Bronson as you've seen him many times before.
1987 (R) 93m/C Charles Bronson, Jill Ireland, Stephen Elliott, Michael Ansara; *Dir:* Peter Hunt. **VHS, Beta, LV** **$9.98** MED *♂*

The Assassination of JFK Chronicles the individuals and events surrounding the assassination of President John F. Kennedy. Includes conspiracy theories involving the FBI, CIA, organized crime, Cuba, the Vietnam war, the Warren Commission, and decisions made by the Kennedys themselves.
1992 78m/C *Dir:* Dennis Mueller. **VHS** **$79.98** MPI *♂*

The Assassination Run A retired British spy is involved against his will in an intricate plot of terrorism, counter-terrorism and espionage.
1980 111m/C *GB* Malcolm Stoddard, Mary Tamm; *Dir:* Ken Hannam. **VHS, Beta** **$59.98** FOX *♂*

Assassination of Trotsky Middling attempt to dramatize the last days of the Russian Revolutionary leader in Mexico before he's done in with an ice pick.
1972 113m/C *FR GB IT* Richard Burton, Alain Delon, Romy Schneider, Valentina Cortese, Jean Desailly; *Dir:* Joseph Losey. **VHS, Beta, LV** **$19.98** REP, PIA, FCT *♂*

Assault Violent sex murders in a girl's school have the police baffled. The school's pretty art teacher offers to act as bait in order to catch the murderer. Also known as "In The Devil's Garden" and "Tower of Terror." Also on video as, "The Creepers."
1970 89m/C *GB* Suzy Kendall, Frank Finlay, Freddie Jones, James Laurenson, Lesley-Anne Down, Tony Beckley; *Dir:* Sidney Hayers. **VHS, Beta** **$59.95** SUE *♂½*

The Assault Powerful and disturbing drama about a Dutch boy who witnesses the arbitrary murder of his family by Nazis. The memory tortures him and leaves him empty as he matures. Years later he meets other victims and also the perpetrators of the incident, each of them changed forever by it. Thought-provoking consideration of WWII and the horrors of living in Nazi Germany from many points of view. Based on a novel by Harry Mulisch. Dutch language dubbed into English. Foreign title "De Aanslag."
1986 (PG) 126m/C *NL* Derek De Lint, Marc Van Uchelen, Monique Van de Van; *Dir:* Fons Rademakers. Academy Awards '86: Best Foreign Language Film. **VHS, Beta** **$79.95** MGM *♂♂♂½*

Assault on Agathon Amid the scenic Greek isles, an "executed" WWII guerilla leader returns to lead a revolution, and bloodshed and bombings ensue.
1975 (PG) 90m/C *GB IT* Nina Van Pallandt, Marianne Faithfull, John Woodvine, Nico Minardos; *Dir:* Laslo Benedek. **VHS, Beta** **$59.95** GEM, MRV, PSM *♂♂*

Assault with a Deadly Weapon When the police budget is cutback, crime runs rampant in an unnamed American city.
1982 86m/C Sandra Foley, Richard Holliday, Lamont Jackson; *Dir:* Arthur Kennedy. **VHS, Beta** **$39.95** IVE *♂*

The Assault of Final Rival A young beggar learns the art of kung-fu and that the inner strength actually lies in the hair. So when a rival cuts his hair, the beggar must strengthen his kung-fu skills so his hair will grow back.
1985 87m/C VHS, Beta **$14.95** NEG *♂½*

Assault of the Killer Bimbos Five foolish girls manage to seduce, then kill, a town's criminal element.
1987 90m/C Karen Nielsen, Debi Thibeault, Lisa Schmidt, Simone; *Dir:* Gorman Bechard. **Beta** **$79.95** UCV *♂½*

Assault and Matrimony Real-life married people Tucker and Eikenberry play a married couple fighting tooth and nail. Lots of slapstick and general nonsense. Television film based on James Anderson's novel.
1987 100m/C John Hillerman, Michelle Phillips, Joe Cortese, Michael Tucker, Jill Eikenberry; *Dir:* James Frawley. **VHS, Beta** **$89.95** VMK *♂½*

Assault of the Party Nerds Nerds throw a wild party to try and attract new members to their fraternity, while a jock frat plots against them. Sound familiar? Little more than a ripoff of "Revenge of the Nerds" made especially for video.
1989 (R) 82m/C Linnea Quigley, Troy Donahue, Richard Gabai, C. Paul Demsey, Marc Silverberg, Robert Mann, Richard Rifkin, Deborah Roush; *Dir:* Richard Gabai. **VHS, Beta** **$79.95** PSM *♂*

Assault on Precinct 13 Urban horror invades LA. A sleepy police station in Los Angeles is suddenly under siege from a violent youth gang. Paranoia abounds as the police are attacked from all sides and can see no way out. Carpenter's musical score adds much to the setting of this unique police exploitation story that somehow stands as Carpenter's adaptation of Howard Hawks' "Rio Bravo." Semi-acclaimed and very gripping.
1976 91m/C John Carpenter, John Carpenter; *Dir:* John Carpenter. **VHS, Beta** **$54.95** MED *♂♂♂*

Assault on a Queen Foolish Sinatra vehicle about a group of con men who get together to rob the Queen Mary on one of her trips to Nassau. Their attack vessel is a renovated World War II German U-boat. The producers tried to capitalize on the popularity of "Ocean's Eleven," but they didn't succeed.
1966 106m/C Frank Sinatra, Virna Lisi, Anthony (Tony) Franciosa, Richard Conte, Reginald Denny; **Dir:** Jack Donohue. **VHS $14.95** PAR ✂️½

Assault of the Rebel Girls A reporter gets involved with smuggling in Castro's Cuba. Flynn's last film, saving the worst for last. Usually titled "Cuban Rebel Girls" or "Attack of the Rebel Girls."
1959 66m/B Errol Flynn, Beverly Aadland, John McKay, Jackie Jackler, Marie Edmund; **Dir:** Barry Mahon. **VHS, Beta $59.95** MED ✂️½

The Assignment The assassination of a high-ranking officer in an uneasy political Latin American nation spurs violence and political instability. A Swedish diplomat is assigned the tremendous task of restoring peace and stability between the political factions.
1978 92m/C SW Christopher Plummer, Thomas Hellberg, Carolyn Seymour, Fernando Rey; **Dir:** Mats Arehn. **VHS, Beta $19.95** STE, NWV ✂️✂️

Assignment Outer Space A giant spaceship with bytes for brains is on a collision course with Earth. A team of astronauts is sent to save the world from certain peril. Seems they take the task lightly, though, and their mission (and hence the plot) revolves more around saving sexy sultress Farinon from certain celibacy. If you're into stultifying Italian space operas with a gratuitous sex sub-plot then look up this assignment, but don't say we didn't warn you. Narrated by Jack Wallace; director Margheriti is also known as Anthony Davies.
1961 79m/B IT Rik von Nutter, Gabriella Farinon, Archie Savage, Dave Montresor, Alan Dijon; **Dir:** Antonio Margheriti; **Nar:** Jack Wallace. **VHS $19.98** NOS, SNC ✂️

Assignment Skybolt When terrorists get their hands on a hydrogen bomb, a special mission is detached to stop them before they wreak havoc.
198? 109m/C **Dir:** Greg Tallas. **VHS, Beta $19.99** BFV ✂️½

The Assisi Underground The true but boringly told story of how the Catholic Church helped to save several hundred Italian Jews from being executed by the Nazis during the 1943 German occupation of Italy. Edited from 178 minutes, a good-will gesture from the producers.
1984 115m/C James Mason, Ben Cross, Maximilian Schell, Irene Papas; **Dir:** Alexander Ramati. **VHS, Beta $79.95** MGM ✂️✂️

Associate A French farce about a penniless financial consultant who invents a fictitious partner in order to get his business rolling. English subtitles.
1982 (R) 93m/C FR Michel Serrault, Claudine Auger, Catherine Alric, Julien Pardot; **Dir:** Rene Gainville. **VHS, Beta $39.98** SUE ✂️✂️

The Association A man and a woman go undercover to bust an international white slavery ring.
198? 92m/C Carter Huang, Angela Mao; **Dir:** Cheung Wo Chang. **VHS, Beta $59.98** CGH, PEV ✂️½

Assunta Spina A very early, very rare Italian epic about ancient Rome, the prototype for the Italian silent epics to come.

1915 70m/B IT VHS, Beta **$39.98** FCT ✂️✂️

The Astounding She-Monster How can you not love a movie with a title like this? A bad script and snail-paced plot are a good start. A geologist wanting only to be left alone with his rocks survives a brush with the kidnappers of a wealthy heiress only to happen upon an alien spacecraft that's crashed nearby. At the helm is a very tall, high-heeled fem-alien in an obligatory skin-tight space outfit. Excellent, our rock jock thinks, but it seems she kills with the slightest touch. For connoisseurs of truly bad movies.
1958 60m/B Robert Clarke, Kenne Duncan, Marilyn Harvey, Jeanne Tatum, Shirley Kilpatrick, Ewing Miles Brown; **Dir:** Ronnie Ashcroft. **VHS $19.98** SNC, MRV, CNM **Woof!**

The Astro-Zombies A contender as one of the worst movies of all time. Carradine plays a mad scientist creating zombies who eat people's guts. Cult favorite Tura Satana stars. Co-written and co-produced by Wayne Rogers of "M*A*S*H" fame.
1967 83m/C Tura Satana, Wendell Corey, John Carradine, Tom Pace, Joan Patrick, Rafael Campos; **W/Dir:** Ted V. Mikels. **VHS, Beta $49.95** UHV **Woof!**

Astroboy Two episodes of Japan's first animated television show are featured on each cassette. Ten untitled volumes.
1992 60m/C VHS **$24.95** CPM **Woof!**

Asylum Four strange and chilling stories weave together in this film. A murderer's victim seeks retribution. A tailor seems to be collecting his bills. A man who makes voodoo dolls...only to become one later on. A doctor visiting the asylum tells each tale. Horrifying and grotesque, not as humorless as American horror films. Reissued as a shorter version in 1980 called "House of Crazies."
1972 (PG) 100m/C GB Peter Cushing, Herbert Lom, Britt Ekland, Barbara Parkins, Patrick Magee, Barry Morse; **Dir:** Roy Ward Baker. **VHS, Beta, LV $9.99** STE, PSM, MED ✂️✂️½

Asylum of Satan A beautiful concert pianist is savagely tortured by a madman in the Asylum of Satan. Filmed on location in Louisville, Kentucky.
1972 87m/C Charles Kissinger, Carla Borelli, Nick Jolly, Sherry Steiner; **W/Dir:** William Girdler. **VHS, Beta $19.95** UHV ✂️

At the Circus Marx Brothers invade the circus to save it from bankruptcy and cause their usual comic insanity, though they've done it better before. Beginning of the end for the Marxes, a step down in quality from their classic work, though frequently darn funny. Groucho sings, "Lydia the Tattooed Lady" and "Step Up and Take a Bow."
1939 87m/B Groucho Marx, Chico Marx, Harpo Marx, Margaret Dumont, Kenny L. Baker, Florence Rice, Eve Arden, Nat Pendleton, Fritz Feld; **Dir:** Edward Buzzell. **VHS, Beta, LV $19.95** MGM ✂️✂️½

At Close Range Based on the true story of Bruce Johnston Sr. and Jr. in Brandywine River Valley, Pennsylvania. Father, Walken, tempts his teen-aged son, Penn, into pursuing criminal activities with talk of excitement and high living. Penn soon learns that his father is extremely dangerous and a bit unstable, but he's still fascinated by his wealth and power. Sometimes overbearing and depressing, but good acting and fancy camera work. A young cast of stars includes Masterson as the girl Penn tries to impress. Features Madonna's "Live to Tell."

1986 (R) 115m/C Sean Penn, Christopher Walken, Christopher Penn, Mary Stuart Masterson, Crispin Glover, Kiefer Sutherland, Candy Clark, Tracy Walter; **Dir:** James Foley. **VHS, Beta, LV $79.98** VES, LIV ✂️✂️✂️

At the Earth's Core A Victorian scientist invents a giant burrowing machine, which he and his crew use to dig deeply into the Earth. To their surprise, they discover a lost world of subhuman creatures and prehistoric monsters. Based on Edgar Rice Burrough's novels.
1976 (PG) 90m/C GB Doug McClure, Peter Cushing, Caroline Munro; **Dir:** Kevin Connor. **VHS, Beta** WAR, OM ✂️✂️

At Gunpoint A store owner becomes the town hero when, by accident, he shoots and kills a bank robber.
1955 81m/C Fred MacMurray, Dorothy Malone, John Qualen, Walter Brennan; **Dir:** Alfred Werker. **VHS, Beta $39.95** REP, HHE ✂️✂️

At Gunpoint A no-account bank robber spends his six-year stint in the slammer plotting his revenge. Having gone thoroughly stir-crazy, the vengeful criminal stalks the lawman who put him away, like so many vengeful criminals before him. Director Harris also scripted this run-of-the-mill addition to big list of bad guy hunting for revenge flicks.
1990 90m/C Frank Kanig, Tain Bodkin, Scott Claflin; **W/Dir:** Steven Harris. **VHS $79.95** CAF ✂️½

At Play in the Fields of the Lord A thoughtful epic that never quite lives up to its own self-importance—or length. Two yankee missionary couples try to evangelize a fearsome tribe of the Brazilian rainforest dwellers. One of the Christian families suffers a crisis of faith that's well-acted but not as powerful as a co-plot regarding a modern American Indian (Berenger) who joins the jungle natives with calamitous results. Based on the novel by Peter Matthiessen.
1991 (R) 186m/C Tom Berenger, Aidan Quinn, Kathy Bates, John Lithgow, Darryl Hannah, Tom Waits; **Dir:** Hector Babenco. **VHS, Beta, LV $44.98** MCA ✂️✂️✂️

At Sword's Point Adventure tale based on Alexandre Dumas' "The Three Musketeers," wherein the sons of the musketeers vie to save their queen from the evil plots of Duke Douglas.
1951 81m/C Cornel Wilde, William Smith, Micheline Lanctot, Maureen O'Hara, Henry Beckman, Brian Patrick Clark, John McFadyen, Ken James; **Dir:** Paul Lynch, Lewis Allen. **VHS, Beta, LV $19.95** MED, TTC, FCT ✂️✂️

At War with the Army Serviceable comedy from Martin and Lewis in their first starring appearance, as the recruits get mixed up in all kinds of wild situations at their army base. Based on the play by James Allardice.
1950 93m/B Dean Martin, Jerry Lewis, Polly Bergen, Mike Kellin; **Dir:** Hal Walker. **VHS, Beta $24.95** NOS, MRV, CAB ✂️✂️

Atalia Relates the love between Atalia, a war widow, and the younger man she loves. The problem stems from the fact that she lives in a Kibbutz, and the lifestyle contradicts sharply from that of her beliefs in love, forcing her to eventually make a momentous decision.
1985 90m/C Michal Bat-Adam, Yftach Katzur, Dan Toren; **Dir:** Akiva Tevet. **VHS $79.95** ERG ✂️✂️

Atlantic City A small-time, aging Mafia hood falls in love with a young clam bar waitress, and they share the spoils of a big score against the backdrop of Atlantic City. Wonderful character study that becomes something more, a piercing declaration about a city's transformation and the effect on the people who live there. Lancaster, in a sterling performance, personifies the city, both of them fading with time. Oscar nominated for best picture, screenplay, actor (Lancaster), actress (Sarandon), and director. Also known as ''Atlantic City U.S.A.''
1981 (R) 104m/C CA FR Burt Lancaster, Susan Sarandon, Kate Reid, Michel Piccoli, Hollis McLaren; **Dir:** Louis Malle. Genie Awards '81: Best Art Direction/Set Decoration, Best Supporting Actress (Reid). **VHS, Beta, LV $14.95** PAR 🎬🎬🎬½

Atlas The mighty Atlas takes on massive armies, one of which includes director Corman, in a bid to win the hand of a princess. About as cheap as they come, although it is one of the few Sword & Sandal epics that isn't dubbed.
1960 84m/C Michael Forrest, Frank Wolff, Barboura Morris, Walter Maslow; **Dir:** Roger Corman. **VHS $19.95** DVT, SNC 🎬

Atlas in the Land of the Cyclops Atlas takes on a hideous one-eyed monster to save a baby from an evil queen. Not a divorce custody drama. Originally titled ''Atlas Against the Cyclops.''
1961 100m/C IT Mitchell Gordon, Chelo Alonso, Vira Silenti; **Dir:** Antonio Leonuiola. **VHS $19.98** SNC, TIM Woof!

Atoll K Laurel and Hardy inherit an paradisiacal island, but their peace is disturbed when uranium is discovered. Final screen appearance of the team is diminished by poor direction and script. Also known as ''Utopia'' and ''Robinson Crusoeland.''
1951 82m/B FR Stan Laurel, Oliver Hardy, Suzy Delair, Max Elloy; **Dir:** Leo Joannon. **VHS, Beta** QNE, MRV, VYY 🎬½

Atom Age Vampire A mad scientist falls in love with a woman disfigured in an auto crash. To remove her scars, he treats her with a formula derived from the glands of freshly killed women. English dubbed. Not among the best of its kind (a low-rent district if ever there was one), but entertaining in a mischievous, boy is this a stupid film sort of way.
1961 71m/B IT Alberto Lupo, Susanne Loret, Sergio Fantoni; **Dir:** Albert Magnoli. **VHS, Beta $16.95** SNC, NOS, VYY Woof!

Atomic Attack A nuclear bomb is dropped on New York City and a family living fifty miles away must escape. Gentlemen, start your engines.
1950 50m/B Walter Matthau. **VHS, Beta, 3/4U $29.95** IHF 🎬

The Atomic Brain An old woman hires a doctor to transplant her brain into the body of a beautiful young girl. Of the three girls who are abducted, two become homicidal zombies and the third starts to act catty when she is given a feline brain. A must-see for bad-brain movie fans. Also known as ''Monstrosity.''
1964 72m/B Frank Gerstle, Erika Peters, Judy Bamber, Marjorie Eaton, Frank Fowler, Margie Fisco; **Dir:** Joseph Mascelli. **VHS $19.98** NOS, SNC Woof!

The Atomic Cafe A chillingly humorous compilation of newsreels and government films of the 1940's and 1950's that show America's preoccupation with the A-Bomb. Some sequences are in black and white.
1982 92m/C **Dir:** Kevin Rafferty. **VHS, Beta $59.95** HBO 🎬🎬🎬

The Atomic Kid A man survives an atomic blast because of a peanut butter sandwich he was eating. As a result, he himself becomes radioactive and discovers that he has acquired some strange new powers which get him into what pass for hilarious predicaments. Screenplay by Blake Edwards.
1954 86m/B Mickey Rooney, Robert Strauss, Elaine Davis, Bill Goodwin; **Dir:** Leslie Martinson. **VHS $19.98** REP 🎬½

The Atomic Man Owing to radioactive experimentation, a scientist exists for a short time in the future. Once there, both good and evil forces want to use him for their own purposes.
1956 78m/B GB Gene Nelson, Faith Domergue, Joseph Tomelty, Peter Arne; **Dir:** Ken Hughes. **VHS $16.95** NOS, SNC 🎬½

Atomic Submarine Futuristic sci fi plots government agents against alien invaders. The battle, however, takes place in the ocean beneath the Arctic and is headed by an atomic-powered submarine clashing with a special alien underwater saucer. We all live on the atomic submarine: fun for devotees.
1959 80m/B Arthur Franz, Dick Foran, Bob Steele, Brett Halsey, Joi Lansing; **Dir:** Spencer Gordon Bennet. **VHS, Beta $39.95** MON, SNC 🎬🎬

The Atonement of Gosta Berling A priest, forced to leave the priesthood because of his drinking, falls in love with a young married woman. Garbo shines in the first role which brought her critical acclaim; Hanson's performance also makes this a memorable drama. Adapted from the novel by Selma Lagerlof. Alternate Titles: ''Gosta Berling's Saga,'' ''The Legend of Gosta Berling,'' and ''The Story of Gosta Berling.'' Silent, with English subtitles.
1924 91m/B SW Lars Hanson, Greta Garbo, Ellen Cederstrom, Mona Martenson, Jenny Hasselquist, Gerda Lundequist; **Dir:** Mauritz Stiller. **VHS** VYY, MRV, DVT 🎬🎬🎬

Ator the Fighting Eagle Styled after ''Conan The Barbarian'' this mythical action fantasy stars O'Keeffe as Ator, son of Thorn. Ator must put an end to the tragic Dynasty of the Spiders, thereby fulfilling the legend of his family at the expense of the viewer. Goofy sword and sandal stuff. Followed by ''The Blade Master.''
1983 (PG) 98m/C IT Miles O'Keefe, Sabrina Siani, Ritza Brown, Edmund Purdom, Laura Gemser; **Dir:** David Hill. **VHS, Beta $69.99** HBO Woof!

Attack of the Beast Creatures The survivors of a wrecked ocean liner are stranded on a desert island overrun by savage creatures. This makes them anxious.
1985 82m/C Robert Nolfi, Robert Langyel, Julia Rust, Lisa Pak; **Dir:** Michael Stanley. **VHS, Beta $19.95** WES Woof!

Attack of the 50 Foot Woman A beautiful, abused housewife has a frightening encounter with a giant alien, causing her to grow to an enormous height. Then she goes looking for hubby. Perhaps the all-time classic '50s sci fi, a truly fun movie highlighted by the sexy, 50-foot Hayes in a giant bikini. Has intriguing psychological depth and social commentary done in a suitably cheesy

manner. Also available with ''House on Haunted Hill'' on laser disc.
1958 72m/B Allison Hayes, William Hudson, Roy Gordon; **Dir:** Nathan Hertz. **VHS, Beta, LV $14.98** FOX, FCT 🎬🎬½

Attack Force Z An elite corps of Australian military is Force Z: volunteers chosen for a dangerous mission: find the plane that crashed somewhere in the South Pacific and rescue the defecting Japanese government official on board, all before the end of WWII and the feature. Talented cast is effectively directed in low-key adventure featuring young Gibson.
1984 84m/C AU Sam Neill, Chris Haywood, Mel Gibson, John Phillip Law, John Waters; **Dir:** Tim Burstall. **VHS, Beta $19.99** MCG, MTX 🎬🎬

Attack of the Giant Leeches Cheapo Corman fare about giant leeches in a murky swamp who suddenly decide to make human flesh their new food supply. Perturbed inn keeper plays along by forcing his wife and lover into the murk. Leeches frolic. Sometimes tedious, sometimes chilling, always low budget and slimy. Although the special effects aren't top notch, this might be a fine choice for a late night scare/laugh. Alternate title: ''The Giant Leeches.''
1959 62m/B Ken Clarke, Yvette Vickers, Gene Roth, Bruno Ve Sota; **Dir:** Bernard L. Kowalski. **VHS $19.95** NOS, SNC 🎬

Attack of the Killer Refrigerator Chiller featuring a group of sleazy college students having a wild party. In the process, they abuse a hapless refrigerator. Fed up, the vengeful appliance goes on a rampage of murder and destruction. Certain to make you view kitchen appliances in a new light. Planned sequels in the newfound kitchen-utility horror genre include ''Refrigerator II: Brutally Defrosted'' and ''Bloody, Bloody Coffee Maker.''
1990 ?m/C **VHS** BFA Woof!

Attack of the Killer Tomatoes Candidate for worst film ever made, deliberate category. Horror spoof that defined ''low budget'' stars several thousand ordinary tomatoes that suddenly turn savage and begin attacking people. No sci-fi cliche remains untouched in this dumb parody. A few musical numbers are performed in lieu of an actual plot. Followed by ''Return of the Killer Tomatoes.''
1977 (PG) 87m/C **Dir:** John DeBello. **VHS, Beta $19.95** MED Woof!

Attack of the Mayan Mummy A greedy doctor gets his patient to channel her former self so that she can show him where to find an ancient tomb that is filled with treasure. Good idea.
1963 77m/B MX Richard Webb, Nina Knight, Norman Burton, Steve Conte; **Dir:** Jerry Warren. **VHS $16.95** NOS, LOO, SNC Woof!

Attack of the Mushroom People A secluded island is the sight where people eating mysterious mushrooms have been turning into oversized, killer 'shrooms' themselves. Trouble is, the only witness to this madness has gone insane. Will anyone believe him before it's too late?
1963 70m/B JP Akira Kubo, Kenji Sahara, Yoshio Tsuchiya, Hiroshi Koizumi; **Dir:** Inoshiro Honda. **VHS $20.00** SMW 🎬🎬

Attack from Outer Space Explores the possibility of extraterrestrial life, particularly on the set.
1980 90m/C **VHS, Beta $9.95** UHV, MRV Woof!

Attack of the Robots A spy spoof about powerful government officials who are being killed off by a mad scientist and replaced with evil robots.
1966 88m/B Eddie Constantine, Fernando Rey; *Dir:* Jess (Jesus) Franco. **VHS, Beta $29.95** VYY *♫ ½*

Attack Squadron Alternately entitled "Kamikaze," this is the story of Japan's suicidal WWII pilots.
197? 120m/C **VHS, Beta $69.95** VCD *♫ ½*

Attack of the Super Monsters A group of prehistoric monsters are developing dastardly plots below the earth to ruin the human race.
1984 85m/C **VHS, Beta $39.95** TWE *♫*

Attack of the Swamp Creature A deranged scientist transforms himself into a swamp critter and terrorizes a small town.
1975 96m/C Frank Crowell, David Robertson; *Dir:* Arnold Stevens. **VHS, Beta $29.95** THR, IVE *Woof!*

The Attic Psycho-drama about an overbearing invalid father and his insecure and unmarried daughter. The girl learns to escape her unhappy life by hiding in the attic. Not horrifying, but a clear analytical look into the game of control.
1980 (R) 92m/C Carrie Snodgress, Ray Milland, Rosemary Murphy, Ruth Cox, Francis Bay, Marjorie Eaton; *Dir:* George Edwards. **VHS, Beta $59.95** FOX, MON *♫♫ ½*

Attica Tense made-for-television depiction of the infamous Attica prison takeover in 1971 and the subsequent bloodbath as state troops were called in. Although edited due to the searing commentary by Rockefeller, it remains powerful and thought-provoking. Adapted from the Tom Wicker bestseller, "A Time to Die."
1980 97m/C George Grizzard, Charles Durning, Anthony Zerbe, Roger E. Mosley; *Dir:* Marvin J. Chomsky. **VHS, Beta $9.98** SUE, CHA *♫♫♫*

Au Revoir Les Enfants During the Nazi occupation of France in the 1940s, the headmaster of a Catholic boarding school hides Jewish boys among the other students by altering their names and identities. One of the boys inadvertently reveals the secret to the Nazis. Based on an incident from director Malle's childhood and considered to be his best film to date, brilliantly told. In French with English subtitles.
1988 (PG) 104m/C FR Gaspard Manesse, Raphael Fejto, Francine Racette, Stanislas Carre de Malberg, Philippe Morier-Genoud, Francois Berleand, Peter Fitz; *Dir:* Louis Malle. Los Angeles Film Critics Association Awards '87: Best Foreign Film; Venice Film Festival '87: Best Film. **VHS, Beta, LV $19.98** TAM, ORI, FCT *♫♫♫♫*

Audrey Rose The parents of a young girl are terrified when their darling daughter is having dreadful dreams. Mysterious friend Hopkins cements their fears when he declares that his dead daughter has been reincarnated in their child. The nightmares continue suggesting that none other than Lucifer could be at work. Slow-moving take-off on "The Exorcist" with a weak staged ending. Based on the book by Frank DeFelitta who also wrote the screenplay.
1977 (PG) 113m/C Marsha Mason, Anthony Hopkins, John Beck, John Hillerman; *Dir:* Robert Wise. **VHS, Beta $59.95** MGM *♫ ½*

Augustine of Hippo One of Rosselini's later historical epics, depicting the last years of St. Augustine and how they exemplify the growing conflicts between Church and State, Christian ethic and societal necessity. In Italian with subtitles.
1972 120m/C IT *Dir:* Roberto Rossellini. **VHS, Beta $59.95** FCT *♫♫♫*

Auntie Entertainer must dress and act like his madam aunt who has died, in order to fool her former employees in a house of prostitution.
1980 (R) 85m/C **VHS, Beta $24.95** NO *Woof!*

Auntie Mame A young boy is brought up by his only surviving relative—flamboyant and eccentric Auntie Mame. Mame is positive that "life is a banquet and most poor suckers are starving to death." Based on the Patrick Dennis novel about his life with "Auntie Mame." Oscar nominated for Best Picture, Best Actress (Russell), Supporting Actress (Cass), Color Cinematography, Art Direction and Set Decoration, and Film Editing. Part of the "A Night at the Movies" series, this tape simulates a 1958 movie evening, with a Road Runner cartoon, "Hook, Line and Stinker," a newsreel and coming attractions for "No Time for Sergeants" and "Chase a Crooked Shadow."
1958 161m/C Rosalind Russell, Patric Knowles, Roger Smith, Peggy Cass, Forrest Tucker, Coral Browne; *Dir:* Morton DaCosta. **VHS, Beta, LV $19.95** HNB, WAR *♫♫♫*

The Aurora Encounter Aliens surreptitiously infiltrate a small town in 1897, and spread benevolence everywhere. Family fare.
1985 (PG) 90m/C Jack Elam, Peter Brown, Carol Bagdasarian, Dottie West, George "Spanky" McFarland; *Dir:* Jim McCullough. **VHS, Beta $14.95** NWV, STE, HHE *♫♫*

Australia Now A look at the people and the music from the lands down under, Australia and New Zealand.
1984 60m/C Little River Band, INXS, Men at Work, Midnight Oil, Split Enz, Moving Pictures. **VHS, Beta $14.95** MVD, MSM *♫♫*

Author! Author! Sweet, likable comedy about playwright Pacino who is about to taste success with his first big hit. Suddenly his wife walks out, leaving him to care for her four children and his own son. His views shift as he begins to worry about who will watch the obnoxious kids on opening night. He soon learns to juggle being father and writer without compromising either one. Written by Israel Horovitz.
1982 (PG) 100m/C Al Pacino, Tuesday Weld, Dyan Cannon, Alan King, Andre Gregory; *Dir:* Arthur Hiller. **VHS, Beta $59.98** FOX *♫♫ ½*

Autobiography of Miss Jane Pittman The history of blacks in the South is seen through the eyes of a 110-year-old former slave. From the Civil War through the Civil Rights movement, Miss Pittman relates every piece of black history, allowing the viewer to experience the injustices. Tyson is spectacular in moving, highly acclaimed television drama. Received nine Emmy awards; adapted by Tracy Keenan Wynn from the novel by Ernest J. Gaines.
1974 110m/C Cicely Tyson, Odetta, Joseph Tremice, Richard Dysart, Michael Murphy, Katherine Helmond; *Dir:* John Korty. **Beta $14.95** KUI, PSM, RRP *♫♫♫ ½*

Autopsy A rash of unexplainable, spontaneous suicides cause a college student to become emotionally upset. Dubbed.
1978 (R) 89m/C IT Mimsy Farmer, Barry Primus, Angela Goodwin, Ray Lovelock; *Dir:* Armando Crispino. **VHS, Beta $19.95** MPI, PSM *♫*

Autopsy A young gold-digger and a millionaire marry, and then cheat on each other, provoking blackmail and murder.
197? 90m/C Fernando Rey, Gloria Grahame, Christian Hey. **VHS, Beta $59.95** MGL *♫*

Autopsy A war correspondent in Vietnam discovers the gruesome secret behind the doctors' autopsy methods at a military hospital.
1986 90m/C **VHS, Beta $19.95** ASE *♫ ½*

An Autumn Afternoon Yasujiro Ozu's final film is a beautiful expression of his talent. In postwar Tokyo, an aging widower loses his only daughter to marriage and begins a life of loneliness. A heart-wrenching tale of relationships and loss. In Japanese with English subtitles.
1962 112m/C JP Chishu Ryu, Shima Iwashita, Shin-Ichiro Mikami, Mariko Okada, Keiji Sada; *W/ Dir:* Yasujiro Ozu. **VHS $69.95** NYF *♫♫♫ ½*

Autumn Born A young heiress is abducted by her guardian and imprisoned while she's taught to obey his will. Ill-fated ex-Playmate Dorothy Stratton's first film.
1979 (R) 76m/C Dorothy Stratton. **VHS, Beta $39.95** MON *♫*

Autumn Leaves Crawford plays a middle-aged typist grasping at her last chance for love. She marries a younger man who's been romancing her, then finds him more and more unstable and violent. Weak story material that could turn melodramatic and tawdry, but doesn't because of Crawford's strength.
1956 108m/B Cliff Robertson, Joan Crawford, Vera Miles, Lorne Greene; *Dir:* Robert Aldrich. Berlin Film Festival '56: Best Director. **VHS, Beta $9.95** GKK *♫♫ ½*

Autumn Marathon To say that this is one of the better Russian movies of the past three decades is sort of faint praise, given the state of Soviet cinema. Written by playwright Alexander Volodin, it's just another paint-by-number version of the philandering man who's really an OK Joe who loves his kids comedy. In Russian with English subtitles.
1979 100m/C *Dir:* Georgi Daniela. **VHS $59.96** IFF *♫♫ ½*

Autumn Sonata Nordic family strife as famed concert pianist Bergman is reunited with a daughter she has not seen in years. Bergman's other daughter suffers from a degenerative nerve disease and had been institutionalized until her sister brought her home. Now the three women settle old scores, and balance the needs of their family. Excellent performance by Bergman in her last film, earned her an Oscar nomination (script was also nominated).
1978 97m/C SW Ingrid Bergman, Liv Ullman, Halvar Bjork, Lena Nyman, Gunnar Bjornstrand; *Dir:* Ingmar Bergman. National Board of Review Awards '78: Best Actress (Bergman), Best Director, Best Foreign Film; New York Film Critics Awards '78: Best Actress (Bergman); National Society of Film Critics Awards '78: Best Actress (Bergman). **VHS, Beta, LV** FOX *♫♫♫*

Avalanche Vacationers at a new winter ski resort are at the mercy of a monster avalanche which leaves a path of terror and destruction in its wake. Talented cast is hemmed in by weak material, producing an snow-bound adventure yarn.
1978 (PG) 91m/C Rock Hudson, Mia Farrow, Robert Forster, Rick Moses; *Dir:* Corey Allen. **VHS, Beta $9.98** SUE *♫ ½*

Avalon Powerful but quite portrait of the break-up of the family unit as seen from the persepective of a Russian family settled in Baltimore at the close of WWII. Initally, the family is unified in their goals, ideologies and social lives. Gradually, all of this disinte-grates; members move to the suburbs and television replaces conversation at holiday gatherings. Levinson based his film on ex-periences within his own family of Russian Jewish immigrants.
1990 (PG) 126m/C Armin Mueller-Stahl, Aidan Quinn, Elizabeth Perkins, Joan Plowright, Lou Jacobi, Leo Fuchs; **W/Dir:** Barry Levinson. **VHS, Beta, LV, 8mm $19.95** COL, PIA, FCT ♪♪♪

Avenged A vigilante saga of no special merit. Crude thugs violate and kill a young girl. Acquitted as a result of a slip in the arrest, they hit the streets again, thinking they are home free. However, the victim's brother shows up demanding they pay dearly for what they have done. His technique of penance is indeed outside the law.
19?? 90m/C VHS $39.95 CON, NSV ♪♪

The Avenger The story of a criminal who cuts off the heads of people and mails them off makes for a shocker. Graphic violence will appeal to those who like a good mail-order gorefest and are not employed by the post office.
1960 102m/B GE Ingrid van Bergen, Heinz Drache, Ina Duscha, Mario Litto, Klaus Kinski; **Dir:** Karl Anton. **VHS $16.95** SNC ♪♪½

The Avenger This time Reeves plays Aeneas and he leads the Trojans in battle against the Etruscans. Also known as "The Last Glory of Troy."
1962 108m/C IT FR Steve Reeves, Giacomo Rossi Stuart, Carla Marlier, Gianno Garko, Liana Orfei; **Dir:** Giorgio Rivalta. **VHS $19.95** SNC, NOS ♪

Avenging Angel Law student Molly "Angel" Stewart is back on the streets to retaliate against the men who killed the policeman who saved her from a life of prostitution. Worthless sequel to 1984's "Angel," exploiting the original's exploitative intent. Followed listlessly by "Angel III: The Final Chapter."
1985 (R) 94m/C Betsy Russell, Rory Calhoun, Susan Tyrrell, Ossie Davis, Barry Pearl, Ross Hagen, Karin Mani, Robert Tessier; **Dir:** Robert Vincent O'Neil. **VHS, Beta, LV $14.95** NWV, STE Woof!

Avenging Conscience An early eerie horror film, based on tales of Edgar Allen Poe. D.W. Griffith's first large-scale feature. Silent. Also known as "Thou Shall Not Kill."
1914 78m/B Henry B. Walthall, Blanche Sweet; **Dir:** D.W. Griffith. **VHS, Beta $29.95** VYY ♪½

Avenging Disco Godfather Moore parodies the "Godfather" and martial arts movies. Also known as "Avenging Godfa-ther."
1976 93m/C Rudy Ray Moore, Lady Reed, Carol Speed, Jimmy Lynch; **Dir:** Rudy Ray Moore. **VHS, Beta $24.95** AHV, IGV ♪½

Avenging Force Okay actioner about retired CIA agent Dudikoff returning to the force to help colleague James run for political office. A group of right-wing terrorists called "Pentangle" threatens James, so Dudikoff adds his name to their wanted list. Conflict leads to a forest manhunt where the aveng-ing force does its avenging.
1986 (R) 104m/C Michael Dudikoff, Steve James, John P. Ryan; **Dir:** Sam Firstenberg. **VHS, Beta, LV $19.95** MED ♪♪

Avenging Ninja A martial arts expert goes on a destructive rampage in Paris when his father and daughter are kidnapped by a European crime syndicate.
198? 83m/C John Liu. **VHS, Beta $59.95** MAG ♪

The Average Woman Gritty journalist consults plain Jane for "Modern Woman" story and finds a major rewrite in order when he falls with a thud for Miss Jane.
1924 52m/B Pauline Garon, David Powell, Burr McIntosh, Harrison Ford, De Sacia Mooers; **Dir:** William Cabanne. **VHS, Beta $14.95** GPV ♪♪

The Aviator Pilot Reeves, haunted by the memory of a fatal crash, tries to find a new line of work. Large sums of money persuade him to transport spoiled Arquette to Wash-ington. When the biplane crashes in the mountain wilderness, the two fall in love between scavenging for food and fighting wild animals. From the director of "Man From Snowy River."
1985 (PG) 98m/C Christopher Reeve, Rosanna Arquette, Jack Warden, Tyne Daly, Marcia Strassman, Sam Wanamaker, Scott Wilson; **Dir:** George Miller. **VHS, Beta $79.95** MGM ♪♪

The Awaken Punch Wherever there's rape, murder, or anything else that demands vengeance, human weapon Yu Young will be there.
19?? ?m/C VHS $16.50 OCE ♪

Awakening An archeologist discovers the tomb of a murderous queen, but upon open-ing the coffin, the mummy's spirit is trans-ferred to his baby daughter, born at that instant. They call that bad luck.
1980 (R) 101m/C Charlton Heston, Susannah York, Stephanie Zimbalist; **Dir:** Mike Newell. **VHS, Beta $19.95** WAR ♪½

Awakening of Candra Based on a real 1975 incident, a young couple honey-mooning in the mountains is assaulted by a lunatic fisherman. The psycho kills the hus-band and rapes the girl, then he brainwashes poor Candra into thinking it was all an accident. Intense subject matter should be horrifying, but falls short of the mark in this television flop.
1981 96m/C Blanche Baker, Cliff DeYoung, Richard Jaeckel; **Dir:** Paul Wendkos. **VHS, Beta $59.95** GEM Woof!

Awakenings Following on the somewhat belated heels of "Big" success, Marshall's first non-comedic film is based on the true story of Dr. Oliver Sacks. Details the doctor's experimentation with the drug L-dopa in the chronic care ward of a Bronx hospital, which inspired the "awakening" of a number of catatonic patients, some of whom had been in a sleeping state for as long as thirty years. Although guilty of lapsing into over-sentimen-tality, the story provides a poignant look at both the "awakened" patients—who find themselves confronted with lost opportuni-ties and faded youth—and at the doctor who awakened them, only to watch their exquisite suffering as they slipped back into their illness. De Niro's performance as Leonard Lowe, the youngest of the group and first to awaken, is heart-rending. Williams offers a subdued and moving performance as the Doctor. At once heart-breaking and uplifting, calculated to provoke thought rather than knee-jerk emotionalism. Oscar nominations for Best Picture, Best Actor (De Niro) and Best Adapted Screenplay (based on Sacks' book of the same title). Production design by the late Anton Furst, best known for his illusionist production work on "Batman" and

laser effects for "Star Wars" and "Alien." Music by Randy Newman.
1990 (PG-13) 120m/C Robin Williams, Robert De Niro, John Heard, Julie Kavner, Penelope Ann Miller, Max von Sydow, Anne Meara; **Dir:** Penny Marshall. **VHS, Beta, LV, 8mm $92.98** COL, FCT ♪♪♪½

Away All Boats The true story of one Captain Hawks, who led a crew of misfits to victory in WWII Pacific aboard transport USS Belinda. Battle scenes are well done; look for early (and brief) appearance by young Clint Eastwood.
1956 114m/B Jeff Chandler, George Nader, Richard Boone, Julie Adams, Keith Andes, Lex Barker; **Dir:** Joseph Pevney. **VHS, Beta $19.95** MCA ♪♪½

The Awful Truth After discarding their marriage made in heaven, a young couple go their separate ways in search of happiness. Meticulously sabotaging each others' new relationships, they discover that, underneath it all, they were made for each other. One of the great screwball comedies; master of comic timing Grant is at his most charming while Dunne is brilliant as his needling ex-wife. The scene in which Dunne poses as Grant's prodigal declasse fan-dancing sister who pays a surprise cocktail-hour visit to the family of his stuffy, haute-mondain bride-to-be is among the most memorable screwball vignettes of all time. And don't miss the custody battle they have over the family dog. Screenplay, based on Arthur Richman's play, by Vina Delmar. Preceded by 1925 and 1929 versions and remade in 1953 as "Let's Do it Again." Oscar-winning direction and Oscar nominations for Best Actress (Dunne), Best Supporting Actor (Bellamy), Best Screenplay and Best Picture.
1937 92m/B Irene Dunne, Cary Grant, Ralph Bellamy, Alexander D'Arcy, Cecil Cunningham, Mary Forbes; **Dir:** Leo McCarey. Academy Awards '37: Best Director (McCarey). **VHS, Beta, LV $19.95** MLB, BTV ♪♪♪♪

Ay, Carmela! During the Spanish Civil War, two traveling entertainers with strong anti-Franco views are captured by Franco forces and sentenced to execution. They are reprieved when a theater-loving Lieutenant offers to spare their lives if they will entertain the troops. But, will Carmela perform in order to save her life - if it means betraying her beliefs?
1990 105m/C SP IT Carmen Maura, Andres Pajares, Gabino Diego, Maurizio De Razza, Miguel Rellan, Edward Zentara, Jose Sancho, Antonio Fuentes; **Dir:** Carlos Savrat. **VHS $89.98** HBO ♪♪♪

Babar: The Movie The lovable Babar, king of the elephants, must devise a plan to outwit an angry hoard of attacking rhinos. Based on the characters of Jean and Laurent de Brunhoff.
1988 (G) 75m/C **W/Dir:** Alan Bunce; **Voices:** Gavin Magrath, Gordon Pinsent, Sarah Polley, Chris Wiggins. **VHS, Beta, LV, 8mm $24.95** FHE, IME ♪♪½

Babe! A fine television move about the life of one of America's most famous woman athletes, Babe Didrickson. Adapted by Joan-na Lee from Didrickson's autobiography, "The Life I've Led." The movie was nominat-ed for Outstanding Special of 1975-76 and Clark won an Emmy for her work.
1975 120m/C Susan Clark, Alex Karras, Slim Pickens, Jeanette Nolan, Ellen Geer, Ford Rainey; **Dir:** Buzz Kulik. **VHS** MET ♪♪♪½

The Babe Follows the life of legendary baseball player Babe Ruth from the time he was labeled incorrigible as a child and sent off to an orphanage to the last day he played baseball. In between, the film portrays Ruth (Goodman) as a sloppy drunkard whose appetites for food, drink, and sex were as large as he was. It also portrays him as an unfaithful husband, although he showered his first wife (Alvarado) with lavish gifts. McGillis plays Ruth's showgirl mistress who eventually became his second wife. Goodman is excellent as Ruth, and he looks like the baseball player, but even his fine job can't make up for a lackluster script with lots of holes.
1991 (PG-13) 115m/C John Goodman, Kelly McGillis, Trini Alvarado, Bruce Boxleitner, Peter Donat, J.C. Quinn, Richard Tyson, James Cromwell, Joe Ragno; *Dir:* Arthur Hiller. **VHS, LV** MCA *♫♫*

Babe Ruth Story An overly sentimental biography about the famed baseball slugger. Bendix is miscast as the Bambino, but the actual film clips of the Babe are of interest. A movie to be watched during those infrequent bouts of sloppy baseball mysticism.
1948 107m/B William Bendix, Claire Trevor, Charles Bickford, William Frawley, Sam Levene, Gertrude Niesen; *Dir:* Roy Del Ruth. **VHS, Beta** **$19.98** FOX *♫*

Babes in Arms The children of several vaudeville performers team up to put on a show to raise money for their financially impoverished parents. The Rodgers and Hart score features "Where or When" and "I Cried For You." Oscar nominations: Best Actor (Rooney), Best Score.
1939 91m/B Judy Garland, Mickey Rooney, Charles Winninger, Guy Kibbee, June Preisser; *Dir:* Busby Berkeley. **VHS, Beta, LV** **$24.95** MGM *♫ ½*

Babes on Broadway Mickey and Judy put on a show to raise money for a settlement house. Nearly the best of the Garland-Rooney efforts, with imaginative numbers staged by Berkeley. Songs include "How About You," "Hoedown," and "Waitin' for the Robert E. Lee."
1941 118m/B Mickey Rooney, Judy Garland, Fay Bainter, Richard Quine, Virginia Weidler, Ray Macdonald, Busby Berkeley; *Dir:* Busby Berkeley. **VHS, Beta, LV** **$29.95** MGM *♫♫ ½*

Babes in Toyland A lavish Disney production of Victor Herbert's timeless operetta, with Toyland being menaced by the evil Barnaby and his Bogeymen. Yes, Annette had a life after Mickey Mouse and before the peanut butter commercials. Somewhat charming, although the roles of the lovers seem a stretch for both Funicello and Kirk. But the flick does sport an amusing turn by Wynn.
1961 105m/C Annette Funicello, Ray Bolger, Tommy Sands, Ed Wynn, Tommy Kirk; *Dir:* Jack Donohue. **VHS, Beta** **$19.99** DIS *♫♫*

Babes in Toyland A young girl must save Toyland from the clutches of the evil Barnaby and his monster minions. Bland TV remake of the classic doesn't approach the original. Music by Leslie Bricusse.
1986 96m/C Drew Barrymore, Pat Morita, Richard Mulligan, Eileen Brennan, Keanu Reeves, Jill Schoelen, Googy Gress; *Dir:* Clive Donner. **VHS, LV** ORI *♫ ½*

Babette's Feast A simple, moving pageant-of-life fable. A group of religiously zealous villagers are taught love and forgiveness through a lavish banquet prepared by one of their housemaids, who turns out to be a world-class chef. Adapted from a tale by Isak Dinesen. In French and Danish with English subtitles.
1987 102m/C DK Stephane Audran, Bibi Andersson, Bodil Kjer, Birgitte Federspiel, Jean-Philippe LaFont, Ebbe Rode, Jarl Kulle; *Dir:* Gabriel Axel. Academy Awards '87: Best Foreign Language Film. **VHS, Beta, LV** **$19.98** ORI, FCT, APD *♫♫♫ ½*

The Baby Bizarre story of a social worker who resorts to an ax to cut the apron strings of a retarded man-child from his over-protective mother and sisters (like, he's still in diapers). Low-budget production looks and feels like a very bad TV movie, but party-scenes-in-'70s-movies fans will need to check this one out, should their other plans fall through.
1972 (PG) 85m/C Anjanette Comer, Ruth Roman, Marianna Hill, Suzanne Zenor, David Manzy, Michael Pataki; *Dir:* Ted Post. **VHS, Beta** **$39.95** KOV, WES *♫*

The Baby and the Battleship The old baby out of (on the?) water plot. While on liberty in Italy, a sailor, after a series of complications, becomes custodian of a baby and attempts to hide the tyke aboard his battleship (hence the title). More complications ensue. Some funny moments. Great cast.
1956 96m/C GB John Mills, Richard Attenborough, Andre Morell, Bryan Forbes, Lisa Gastoni, Michael Hordern, Lionel Jeffries, Gordon Jackson, John Le Mesurier; *Dir:* Jay Lewis. **VHS** **$19.95** NOS, MOV, TIM *♫♫ ½*

Baby Boom When a hard-charging female executive becomes the reluctant mother of a baby girl (a gift from a long-lost relative), she adjusts with great difficulty to motherhood and life outside the rat race and New York City (and Sam Shepard as the love interest). A fairly harmless collection of cliches bolstered by Keaton's usual nervous performance as a power-suited yuppie ad queen saddled with a noncareer-enhancing baby, moving from manic yuppie ad maven to jelly-packing Vermont store-owner/mom. To best appreciate flick, see it with a bevy of five- and six-year-olds (a good age for applauding the havoc that a baby creates).
1987 (PG) 103m/C Diane Keaton, Sam Shepard, Harold Ramis, Sam Wanamaker, James Spader; *Dir:* Charles Shyer. **VHS, Beta, LV** **$19.95** FOX, FCT *♫♫ ½*

Baby Doll Suggestive sex at its best, revolving around the love of cotton in Mississippi. Nubile Baker is married to slow-witted Malden, who runs a cotton gin. His torching of Wallach's cotton gin begins a cycle of sexual innuendo and tension, brought to exhilarating life on screen, without a single filmed kiss. Performers and sets ooze during the steamy exhibition, which was considered highly erotic when released. Excellent performances from entire cast, with expert pacing by director Kazan. Screenplay is based on Tennessee Williams' "27 Wagons Full of Cotton." Oscar nominations for Adapted Screenplay, actress Baker, Black & White Cinematography, and supporting actress Dunnock.
1956 115m/B Eli Wallach, Carroll Baker, Karl Malden, Mildred Dunnock, Rip Torn; *Dir:* Elia Kazan. **VHS, Beta** **$19.98** WAR, FCT *♫♫♫*

Baby Face A small town girl moves to the city when her father dies. There she gets a job at a bank and sleeps her way to the top of the business world, discarding used men left and right. The Hays Office was extremely upset with the then risque material and forced Warner to trim the first cut.
1933 70m/B Barbara Stanwyck, George Brent, Donald Cook, John Wayne, Henry Kolker, Margaret Lindsay, Douglas Dumbrille, James Murray; *Dir:* Alfred E. Green. **VHS, Beta** **$19.98** MGM, FCT *♫♫*

Baby Face Harry Langdon Two silent shorts by the limpid-eyed comedian: "Saturday Afternoon" (1926), and "Lucky Stars" (1925). Both were written by Frank Capra and Arthur Ripley.
1926 60m/B Harry Langdon, Alice Ward, Natalie Kingston, Vernon Dent; *Dir:* Harry Edwards. **VHS, Beta, 8mm** **$24.95** VYY *♫♫ ½*

Baby Face Morgan Poor comedy about gangsters who attempt to take advantage of the FBI's preoccupation with saboteurs and spies by muscling in on an insurance firm.
1942 60m/B Mary Carlisle, Richard Cromwell, Robert Armstrong, Chick Chandler, Charles Judels, Warren Hymer, Vince Barnett, Ralf Harolde; *Dir:* Arthur Dreifuss. **VHS** **$16.95** SNC *♫♫*

Baby Huey the Baby Giant A compilation of the Baby Huey cartoons.
1962 60m/C **VHS, Beta** **$5.99** WOV *♫♫*

Baby It's You The relationship of a smart, attractive Jewish girl who yearns to be an actress and a street-smart Catholic Italian boy puzzles their family and friends. Arquette's Jill Rosen is headed for college while Spano's the Shiek is busy dropping out of high school and looking for something that looks like a future. Set in New Jersey, "Baby" is about listening to your heart as well as your head. It all works due to Arquette's strong acting and Sayles' script, which explores adolescent dreams, the transition to adulthood, class differences, and the late 1960s with insight and humor. Interesting period soundtrack (Woolly Bully and, for some reason, Bruce Springsteen) helps propel the film, a commercial job which helped finance Sayles' more independent ventures.
1982 (R) 105m/C Rosanna Arquette, Vincent Spano, Jack Davidson, Joanna Merlin, Nick Ferrari, Leora Dana, Robert Downey Jr., Tracy Pollan, Matthew Modine; *Dir:* John Sayles. **VHS, Beta, LV** **$79.95** PAR *♫♫♫*

Baby Love A softcore fluff-fest about a trollop seducing a doctor's family. The doctor may or may not be her father. Hayden tantalizes the doctor, the doctor's son, and the doctor's wife, as well as the neighbors, leaving only those in adjoining communities untouched. Based on the novel by Tina Chad Christian.
1969 (R) 98m/C GB Ann Lynn, Keith Barron, Linda Hayden, Derek Lamden, Diana Dors, Patience Collier; *Dir:* Alastair Reid. **VHS, Beta** **$59.95** CHA Woof!

Baby Love Nerds get revenge against the freshmen who run the local frat house.
1983 80m/C **VHS, Beta** **$59.95** MGM Woof!

The Baby Maker A couple who cannot have children because the wife is sterile decides to hire a woman to have a child for them. However, the relationship between the husband and the surrogate progresses beyond what either of them wanted. Hershey stars as the free-love surrogate mama (just before she underwent the supreme 60s transformation into Barbara Seagull) and Bridges makes his directorial debut. Flick is interesting as a combo critique/exploitation of those wild and groovy 1960s.

1970 (R) 109m/C Barbara Hershey, Colin Wilcox-Horne, Sam Groom, Scott Glenn, Jeannie Berlin; *Dir:* James Bridges. **VHS** $59.95 *WAR* 𝕀𝕀

Baby, the Rain Must Fall A rockabilly singer, paroled from prison after serving time for a knifing, returns home to his wife and daughter, but his outbursts of violence make the reunion difficult. Unsentimental with realistic performances, but script is weak (although written by Horton Foote, based on his play, "The Traveling Lady"). Theme song was a Top 40 hit.
1964 100m/B Steve McQueen, Lee Remick, Don Murray; *Dir:* Robert Mulligan. **VHS, Beta, LV** $9.95 *GKK* 𝕀𝕀

Baby Snakes: The Complete Version Concert footage of Zappa is interspersed with clay animation by Bruce Bickford.
1979 166m/C Frank Zappa, Joey Psychotic, Ron Delsener, Donna U. Wanna; *Dir:* Frank Zappa. **VHS, Beta** $79.95 *MVD, MPI* 𝕀𝕀

Baby, Take a Bow Temple's first starring role. As a cheerful Pollyanna-type she helps her father, falsely accused of theft, by finding the true thief.
1934 76m/B Shirley Temple, James Dunn, Claire Trevor, Alan Dinehart; *Dir:* Harry Lachman. **VHS, Beta** $19.98 *FOX* 𝕀 ½

The Babylon Story from "Intolerance" The lavishly produced section of Griffith's "Intolerance" as it was released on its own in 1919, in order to recoup the studio losses incurred by that grandest of cinematic follies. Silent with organ score, tinted color.
1916 25m/C Constance Talmadge, Alfred Paget; *Dir:* D.W. Griffith. **VHS, Beta** $19.95 *VYY* 𝕀𝕀𝕀𝕀

Baby...Secret of the Lost Legend A sportswriter and his paleontologist wife risk their lives to reunite a hatching brontosaurus with its mother in the African jungle. Although this Disney film is not lewd in any sense, beware of several scenes displaying frontal nudity and some violence.
1985 (PG) 95m/C William Katt, Sean Young, Patrick McGoohan, Julian Fellowes; *Dir:* B.W.L. Norton. **VHS, Beta, LV** $79.95 *TOU* 𝕀𝕀

The Babysitter A family hires the mysterious but ingratiating Johanna (Zimbalist) as live-in help without checking on her references (who'd all checked out). The babysitter is the answer to all their problems—Mom's an alcoholic, Dad's a workaholic, and their daughter's just plain maladjusted—but once Johanna's gained their trust (or, in Dad's case, lust), she sets out to manipulate and exploit the family for her own psychotic purposes. Houseman plays the nosy neighbor who's on to her evil plan. Fair made-for-TV treatment of a common suspense plot.
1980 96m/C William Shatner, Patty Duke, Stephanie Zimbalist, Quinn Cummings, John Houseman, David Wallace; *Dir:* Peter Medak. **VHS, Beta, LV** $19.99 *HBO* 𝕀𝕀

Bacall on Bogart Bacall talks about her life with her husband, Humphrey Bogart. Copious film clips are interspersed, most in black and white.
1988 87m/C Humphrey Bogart, Lauren Bacall, Katharine Hepburn, John Huston. **VHS, Beta** $59.98 *RKO, TTC* 𝕀𝕀

The Bacchantes Poorly dubbed account of a ballerina, her life and loves. Based on the play by Euripides.
1963 100m/B *IT* Taina Elg, Pierre Brice, Alessandra Panaro, Alberto Lupo, Akim Tamiroff; *Dir:* Giorgio Ferroni. **VHS** $39.95 *FCT, TIM* 𝕀𝕀

Bachelor Apartment Once at the leading edge of the bachelor on the loose genre, this one's hopelessly dated. The scandalous womanizing of a wealthy thirties Lothario just doesn't have the same impact on the men just don't understand generation. Nevertheless, as vintage if-the-walls-could-talk fluff, it's good for a giggle.
1931 77m/B Lowell Sherman, Irene Dunne, Norman Kerry, Claudia Dell, Noel Francis, Charles Coleman; *Dir:* Lowell Sherman. **VHS** $19.98 *TTC, FCT* 𝕀𝕀 ½

Bachelor Bait A marriage license clerk who's tired of just handing out licenses opens a matrimonial service for men.
1934 75m/B Stuart Erwin, Rochelle Hudson, Pert Kelton, Skeets Gallagher. **VHS, Beta** *CCB* 𝕀 ½

The Bachelor and the Bobby-Soxer Playboy Grant is brought before Judge Loy for disturbing the peace and sentenced to court his teen-age sister Temple. Cruel and unusual punishment? Maybe, but the wise Judge hopes that the dates will help Temple over her crush on handsome Grant. Instead, Loy and Grant fall for each other. Oscar winning screenplay by Sidney Sheldon.
1947 95m/B Cary Grant, Myrna Loy, Shirley Temple, Rudy Vallee, Harry Davenport, Ray Collins; *Dir:* Irving Reis. Academy Awards '47: Best Original Screenplay. **VHS, Beta, LV** $19.95 *CCB, MED, RKO* 𝕀𝕀𝕀

Bachelor of Hearts Sophomoric British comedy set at Cambridge University, with a German exchange student whose difficulty with English brings him to date several women in one evening. Notable for horror fans as the debut of femme fright fave Barbara Steele.
1958 94m/C *GB* Hardy Kruger, Sylvia Syms, Ronald Lewis; *Dir:* Wolf Rilla. **VHS** $19.95 *NOS* 𝕀 ½

Bachelor Mother A single salesgirl causes a scandal when she finds an abandoned baby and is convinced by her boss to adopt the child. Smart, witty comedy with nice performance by Rogers.
1939 82m/B Ginger Rogers, David Niven, Charles Coburn; *Dir:* Garson Kanin. **VHS, Beta, LV** $19.98 *CCB, MED, FCT* 𝕀𝕀𝕀

Bachelor Party He's silly, cute, and poor. She's intelligent, beautiful, and rich. It must be a marriage made in heaven, because no one in their right mind would put these two together. All is basically well, except that her parents hate him and his friends dislike her. Things are calm until right before the big event, when bride-to-be Kitaen objects to Hank's traditional pre-nuptial partying. Light, entertaining film with scattered laughs.
1984 (R) 105m/C Tom Hanks, Tawny Kitaen, Adrian Zmed, George Grizzard, Robert Prescott, William Tepper, Wendie Jo Sperber, Barry Diamond, Michael Dudikoff, Deborah Harmon, John Bloom, Toni Alessandra, Monique Gabrielle, Angela Aames, Rosanne Katon; *Dir:* Neal Israel. **VHS, Beta, LV** $14.98 *FXV, FOX* 𝕀 ½

Bachelorette Party This is an affirmative-action party tape, featuring scantily dressed men for women's viewing pleasure instead of vice versa.
1987 60m/C VHS, Beta $24.95 *AHV* 𝕀 ½

Back to Back A beautiful young vigilante embarks on a rampage to clear her family's name and make her town's redneck crooks pay for their crimes. Never released in theaters.

1990 (R) 95m/C Bill Paxton, Todd Field, Apollonia, Luke Askew, Ben Johnson, David Michael-Standing, Susan Anspach, Sal Landi; *Dir:* John Kincade. **VHS, Beta** $79.95 *MGM* 𝕀

Back to Bataan Colonel forms guerrilla army to raid Japanese in the Philippines and to help Americans landing on Leyte. Also available in a colorized version.
1945 95m/B John Wayne, Anthony Quinn, Beulah Bondi, Fely Franquelli, Richard Loo, Philip Ahn, Lawrence Tierney; *Dir:* Edward Dmytryk. **VHS, Beta, LV** $19.98 *TTC, BUR* 𝕀𝕀 ½

Back to the Beach Frankie and Annette return to the beach as self-parodying, middle-aged parents with rebellious kids, and the usual run of sun-bleached, lovers' tiff comedy ensues. Plenty of songs, and guest appearances from television past.
1987 (PG) 92m/C Frankie Avalon, Annette Funicello, Connie Stevens, Jerry Mathers, Bob Denver, Barbara Billingsley, Tony Dow, Pee-Wee Herman, Edd Byrnes, Dick Dale, Don Adams, Lori Loughlin; *Dir:* Lyndall Hobbs. **VHS, Beta, LV** $19.95 *PAR* 𝕀 ½

Back Door to Heaven Traces the path of a young boy who is born into a poor family and the reasons for his turning to a life of crime. A grim and powerful drama with many convincing performances.
1939 85m/B Wallace Ford, Aline MacMahon, Stuart Erwin, Patricia Ellis, Kent Smith, Van Heflin, Jimmy Lydon; *Dir:* William K. Howard. **VHS** $19.95 *NOS, MOV* 𝕀𝕀𝕀 ½

Back from Eternity Eleven survivors of a plane crash are stranded in a headhunter region of South America's jungle. Remake of "Five Came Back" (1939), which was also directed by Farrow.
1956 97m/B Robert Ryan, Rod Steiger, Anita Ekberg, Phyllis Kirk, Keith Andes, Gene Barry; *Dir:* John Farrow. **VHS, Beta, LV** $19.95 *UHV* 𝕀𝕀

Back to the Future When the neighborhood mad scientist constructs a time machine from a DeLorean, his youthful companion, Marty, accidentally transports himself to 1955. There, Marty must do everything he can to bring his parents together, elude the local bully, and get back...to the future. The soundtrack features Huey Lewis and the News. Lewis also makes a cameo appearance early in the film. Lloyd sparkles and Fox is well-cast as the boy out of his element. Nice close with Fox introducing rock 'n roll to the '50s high schoolers. Followed by two sequels.
1985 (PG) 116m/C Michael J. Fox, Christopher Lloyd, Lea Thompson, Crispin Glover, Wendie Jo Sperber, Marc McClure, Thomas F. Wilson, James Tolken, Casey Siemaszko, Billy Zane, Courtney Gains; *Dir:* Robert Zemeckis. **VHS, Beta, LV** $19.95 *MCA, FCT, TLF* 𝕀𝕀𝕀

Back to the Future, Part 2 The Doc and Marty are time-hopping again. The Doc has discovered that the future has changed because of the time-traveling done in the past. Three generations of Marty's family are visited by the pair while they try to set things straight in the present, past, and future. Just trying to catch your breath between the scenes is difficult as you are led to a cliff-hanger ending, the setting for Part III.
1989 (PG) 107m/C Michael J. Fox, Christopher Lloyd, Lea Thompson, Thomas F. Wilson, Harry Waters Jr., Charles Fleischer, Joe Flaherty, Elizabeth Shue, James Tolkan, Casey Siemaszko, Jeffrey Weissman, Flea; *Dir:* Robert Zemeckis. **VHS, Beta, LV** $19.95 *MCA* 𝕀𝕀

Back to the Future, Part 3 The young time-traveling hero reads an article about his friend Doc Brown's death, and goes back to the Wild West to warn him. Filmed at the same time as Part II, with music by Alan Silvestri and ZZ Top. Complete trilogy is available as a boxed set.
1990 (PG) 118m/C Michael J. Fox, Christopher Lloyd, Mary Steenburgen, Thomas F. Wilson, Lea Thompson, Elizabeth Shue, Matt Clark, Richard Dysart, Pat Buttram, Harry Carey Jr., Dub Taylor, James Tolken; *Dir:* Robert Zemeckis. **VHS, Beta, LV $19.95** MCA �️✐✐

Back Roads A Southern hooker meets a down-on-his-luck boxer and both head out for a better life in California, finding love along the way. Field and Jones should be funnier than this; the story has been told many times before.
1981 (R) 94m/C Sally Field, Tommy Lee Jones, David Keith; *Dir:* Martin Ritt. **VHS, Beta $69.98** FOX ✐ ½

Back to School Dangerfield plays an obnoxious millionaire who enrolls in college to help his wimpy son, Gordon, achieve campus stardom. His motto seems to be "if you can't buy it, it can't be had." At first, his antics embarrass his shy son, but soon everyone is clamoring to be seen with the pair as Gordon develops his own self confidence.
1986 (PG-13) 96m/C Rodney Dangerfield, Keith Gordon, Robert Downey Jr., Sally Kellerman, Burt Young, Paxton Whitehead, Adrienne Barbeau, M. Emmet Walsh, Severn Darden, Ned Beatty, Sam Kinison, Kurt Vonnegut Jr., Robert Picardo, Danny Elfman; *Dir:* Alan Metter. **VHS, Beta, LV $19.98** HBO ✐✐ ½

Back Street The forbidden affair between a married man and a beautiful fashion designer carries on through many anxious years to a tragic end. The lavish third film version of the Fanny Hurst novel.
1961 107m/C Susan Hayward, John Gavin, Vera Miles; *Dir:* David Miller. **VHS, Beta $59.95** MCA, MLB ✐✐

Back Track Foster co-stars in this thriller about an artist who accidentally witnesses a mob hit. The mob puts a hitman (Hopper) on her trail, and after studying her background and listening to audio tapes she recorded, he soon finds himself falling in love.
1991 (R) 102m/C Dennis Hopper, Jodie Foster, Joe Pesci, Dean Stockwell, Fred Ward, Vincent Price, Bob Dylan, John Turturro; *Dir:* Dennis Hopper. **VHS $89.98** LIV ✐✐✐

Back in the USSR Danger follows two lovers caught up in the Moscow underworld. A young American touring Russia unwittingly gets involved with a beautiful art thief. The only thing that can save them is to retrieve the stolen art as they are pursued by the Russian mob, the KGB, and art smugglers. Fast-paced action. The first American film shot entirely on location in Moscow.
1991 (R) ?m/C Frank Whaley, Natalia Negoda, Roman Polanski; *Dir:* Deran Sarafian. **VHS $94.98** TCF ✐✐ ½

Backdraft High action story of Chicago fireman has some of the most stupendous incendiary special effects ever filmed. But then there's that plot, B-movie hokum about a mystery arsonist torching strategic parts of the community with the finesse of an expert and a brother-against-brother conflict. Straight-forward performances from the cast in spite of both. Also available in a letter-boxed version.
1991 (R) 135m/C William Baldwin, Kurt Russell, Scott Glenn, Jennifer Jason Leigh, Donald Sutherland, Rebecca De Mornay, Robert De Niro, J.T. Walsh; *Dir:* Ron Howard. **VHS, Beta, LV $19.98** MCA, PIA ✐✐ ½

Backfire If you're gonna rob a bank, you shouldn't let anyone hear you plan it, and if you're not gonna rob a bank, you shouldn't let anyone hear you plan one. This vintage "Lightning" Carson crime western has Carson and friend suspected of bank robbery because someone heard them planning one. The sheriff follows their every footstep as they search for the real perpetrators.
1922 56m/B Jack Hoxie, George Sowards, Lew Meehan, Florence Gilbert; *Dir:* Alvin J. Neitz. **VHS, Beta $9.95** GPV ✐✐

Backfire A mysterious stranger enters the lives of a disturbed 'Nam vet and his discontented wife, setting a pattern of murder and double-cross in motion.
1988 (R) 90m/C Karen Allen, Keith Carradine, Jeff Fahey, Bernie Casey, Dinah Manoff, Dean Paul Martin; *Dir:* Gilbert Cates. **VHS, Beta, LV $29.95** VMK ✐✐

Backlash An aborigine barmaid is raped. When her assailant turns up dead, she's charged with murder and winds up in the custody of two police officers on a trip across the outback. Holes in the plot undermine this interesting, although graphic, drama with racial overtones.
1986 (R) 88m/C AU David Argue, Gia Carides, Lydia Miller, Brian Syron, Anne Smith; *Dir:* Bill Bennett. **VHS, Beta $79.95** MCG ✐✐ ½

Backstab A spellbinding tale of work, lust, and murder. Architect Brolin finds himself unable to get over the death of his wife, until a seductive and mysterious woman helps him over his grief. They spend the night together, engulfed in passion, but in the morning he wakes to find himself sleeping with the corpse of his boss. Only the first twist in this intriguing thriller.
1990 (R) 91m/C James Brolin, Meg Foster, Isabelle Truchon; *Dir:* Jim Kaufman. **VHS, LV $89.98** MED ✐✐ ½

Backstairs An obscure German silent film about urban degradation and familial strife.
1921 44m/B GE Henny Porten, Fritz Kortner, William Dieterle; *Dir:* Leopold Jessner. **VHS, Beta $39.95** TAM, NOS, DVT ✐✐

Backstreet Dreams The young parents of an autistic child find themselves torn apart due to their feelings of guilt. The father has an affair with a specialist hired to help the boy, causing further strife. Interesting story possibilities never get far.
1990 (R) 104m/C Brooke Shields, Jason O'Malley, Sherilyn Fenn, Tony Fields, Burt Young, Anthony (Tony) Franciosa, Nick Cassavetes, Ray "Boom Boom" Mancini; *Dir:* Rupert Hitzig. **VHS $79.98** VMK ✐✐

Backwoods Two campers wish they had never encountered a mountain man when he begins to stalk them with murder in mind.
1987 (R) 90m/C Jack O'Hara, Dick Kreusser, Brad Armacot; *Dir:* Dean Crow. **VHS, Beta $79.98** NSV ✐✐

The Bad and the Beautiful The story of a Hollywood producer, told from the eyes of an actress, a writer, and a director. Featuring what some consider Turner's best performance ever. Winner of five Oscars, a splendid drama. Douglas received a Best Actor nomination.
1952 118m/B Kirk Douglas, Lana Turner, Dick Powell, Gloria Grahame, Barry Sullivan, Walter Pidgeon, Gilbert Roland; *Dir:* Vincente Minnelli. Academy Awards '52: Best Art Direction/Set Decoration (B & W), Best Black and White Cinematography, Best Costume Design, Best Screenplay, Best Supporting Actress (Grahame). **VHS, LV $19.95** MGM, BTV ✐✐✐ ½

Bad Blood The true story of Stan Graham, who went on a killing spree in the New Zealand bush when his farm was foreclosed and his life ruined.
1987 104m/C NZ Jack Thompson, Carol Burns, Dennis Lill; *Dir:* Mike Newell. **VHS, Beta $29.95** ACA ✐

Bad Boy A country boy goes to the big city, and succumbs to urban evils and temptations, but is eventually saved by motherly love.
1939 60m/B John Downs, Helen MacKellar, Rosalind Keith, Holmes Herbert; *Dir:* Kurt Neumann. **VHS, Beta, LV** WGE ✐ ½

Bad Boys When a gang member's little brother is killed in a rumble, the teen responsible (Penn, who else?) goes to a reformatory, where he quickly (though somewhat reluctantly) takes charge. Meanwhile, on the outside, his rival attacks Penn's girlfriend (Sheedy, in her feature film debut) in retaliation, is incarcerated, and ends up vying with Penn for control of the cell block. Backed into a corner by their mutual hatred and escalating peer pressure, the two are pushed over the brink into a final and shattering confrontation. Not as violent as it could be, to its credit; attempts to communicate a message.
1983 (R) 123m/C Sean Penn, Esai Morales, Reni Santoni, Jim Moody, Eric Gurry, Ally Sheedy, Clancy Brown; *Dir:* Rick Rosenthal. **VHS, Beta $14.99** HBO ✐✐✐

Bad Bunch A white liberal living in Watts tries to befriend a ruthless black street gang, but is unsuccessful.
1976 82m/C Greydon Clark, Tom Johnigam, Pamela Corbett, Jacqulin Cole, Aldo Ray, Jock Mahoney; *Dir:* Greydon Clark. **VHS, Beta $59.95** UHV ✐✐

B.A.D. Cats Two members of a police burglary auto detail team chase after a group of car thieves who are planning a million-dollar gold heist.
1980 74m/C Asher Brauner, Michelle Pfeiffer, Vic Morrow, Jimmie Walker, Steve Hanks, LaWanda Page; *Dir:* Bernard L. Kowalski. **VHS, Beta, LV $59.95** LHV ✐✐

Bad Channels It's radio gone awry as the female listeners of station KDUL are shrunk and put in specimen jars by a way-out disc jockey and a visiting alien, who plans to take the women back to his planet. Features a soundtrack composed and performed by Blue Oyster Cult. Also available with Spanish subtitles.
1992 88m/C Paul Hipp, Martha Quinn, Aaron Lustig, Ian Patrick Williams, Charlie Spradling; *Dir:* Ted Nicolaou. **VHS, Beta** PAR ✐✐

Bad Charleston Charlie Dud of a gangster comedy with terrible acting. A comedy?
1973 (PG) 91m/C Ross Hagen, John Carradine; *Dir:* Ivan Nagy. **VHS $49.95** HCC Woof!

Bad Company Thoughtful study of two very different Civil War draft dodgers roaming the Western frontier and eventually turning to a fruitless life of crime. Both the cast and script are wonderful in an entertaining film that hasn't been given the attention it's due.

1972 (PG) 94m/C Jeff Bridges, Barry Brown, Jim Davis, John Savage; **Dir:** Robert Benton. New York Film Festival '72: Best Supporting Actress. **VHS, Beta, LV** $14.95 *PAR* ✶✶✶½

Bad Day at Black Rock Story of a one-armed man uncovering a secret in a Western town. Wonderful performances from all concerned, especially Borgnine. Fine photography, shot using the new Cinemascope technique. Based on the novel by Howard Breslin. Oscar nominations for Best Actor (Tracy), Director, and Screenplay. 1954 81m/C Spencer Tracy, Robert Ryan, Anne Francis, Dean Jagger, Walter Brennan, John Ericson, Ernest Borgnine, Lee Marvin; **Dir:** John Sturges. **VHS, LV** $19.98 *MGM, VYG, CRC* ✶✶✶½

Bad Dreams The only surviving member of a suicidal religious cult from the '60s awakens in 1988 from a coma. She is pursued by the living-dead cult leader, who seeks to ensure that she lives up (so to speak) to the cult's pact. Blood begins flowing as her fellow therapy group members begin dying, but the only bad dreams you'd get from this flick would be over the money lost on the video rental. 1988 84m/C Bruce Abbott, Jennifer Rubin, Richard Lynch, Harris Yulin, Dean Cameron, Elizabeth Daily, Susan Ruttan, Charles Fleischer; **Dir:** Andrew Fleming. **VHS, Beta, LV** $89.98 *FOX* ✶✶½

Bad Georgia Road A New Yorker inherits a moonshine operation from her uncle, and fights off the syndicate for its profits. 1977 (R) 85m/C Gary Lockwood, Carol Lynley, Royal Dano, John Wheeler, John Kerry; **Dir:** John Broderick. **VHS, Beta** $49.95 *UHV* ✶✶½

Bad Girls A sophisticated erotic French film wherein a young girl, a world-weary playboy and an aging courtesan play games in a house in St. Tropez. Dubbed. 1969 (R) 97m/C FR Jean-Louis Trintignant, Jacqueline Sassard, Stephane Audran; **Dir:** Claude Chabrol. **VHS, Beta** $19.95 *STE, NWV* ✶✶

Bad Girls Dormitory At the New York Female Juvenile Reformatory, suicide seems a painless and welcome escape. 1984 (R) 95m/C Carey Zuris, Teresa Farley; **Dir:** Tim Kincaid. **VHS, Beta** $49.95 *AHV* ✶

Bad Girls Go to Hell From the sultana of sleaze, Wishman, comes this winning entry into Joe Bob Briggs' "Sleaziest Movies in the History of the World" series. A ditsy-but-sexy housewife accidentally commits murder and what follows is a plethora of perversion involving hirsute men and gender-bending women who are hell-bent on showing her how hot it is where bad girls go. 1965 98m/B W/Dir: Doris Wishman. **VHS** $19.98 *SVI Woof!*

Bad Girls from Mars "B" movie sleaze-o-rama in which everyone is murdered, either before, after, or during sex, just like in real life. When the director of the film within this film hires an actress who is, shall we say, popular, to be the heroine of his latest sci-fier, the fun, slim as it is, begins. 1990 86m/C Edy Williams, Brinke Stevens, Jay Richardson, Oliver Darrow; **Dir:** Fred Olen Ray. **VHS** $89.95 *VMK Woof!*

Bad Girls in the Movies A compilation of film clips depicting sleazy, criminal, female characters from women's prison films and other seedy sub-genres. 1986 56m/C David Carradine, Gene Autry, Yvonne de Carlo. **VHS, Beta** $59.98 *LTG, LIV* ✶

Bad Guys A goulash western about the outlaw days of Hungary in the 1860s. A gang of bandits terrorizes the Transdanubian countryside. In Hungarian with English subtitles. For those seeking the Eastern European Wild West experience. 1979 93m/C HU Janos Derzsi, Djoko Rosic, Mari Kiss, Gyorgy Dorner, Laszlo Szabo, Miklos Benedek; **Dir:** Gyorgy Szomjas. **VHS** $69.95 *EVD, FCT, TAM* ✶✶

Bad Guys An inane comedy about two ridiculous policemen who decide to take the wrestling world by storm after being kicked off the police force. Featuring scenes with many of the world's most popular wrestlers. 1986 (PG) 86m/C Adam Baldwin, Mike Jolly, Michelle Nicastro, Ruth Buzzi, James Booth, Gene LeBell, Norman Burton; **Dir:** Joel Silberg. **VHS** $79.95 *IVE* ✶

Bad Influence A lackluster effort in the evil-doppelganger school of psychological mystery, where a befuddled young executive (Spader) is led into the seamier side of life by a mysterious stranger (Lowe). 1990 (R) 99m/C Rob Lowe, James Spader, Lisa Zane, Christian Clemenson, Kathleen Wilhoite; **Dir:** Curtis Hanson. **VHS, Beta, LV** $14.95 *COL* ✶½

Bad Jim A cowpoke buys Billy the Kid's horse and, upon riding it, becomes an incorrigible outlaw himself. First feature film for Hollywood legend Clark Gable's son. 1989 (PG) 110m/C James Brolin, Richard Roundtree, John Clark Gable, Harry Carey Jr., Ty Hardin, Pepe Serna; **Dir:** Clyde Ware. **VHS, Beta, LV** $79.95 *COL* ✶½

Bad Lands A small cowboy posse finds themselves trapped by a band of Apache Indians in the Arizona desert. A remake of "The Lost Patrol." 1939 70m/B Robert Barratt, Guinn Williams, Douglas Walton, Andy Clyde, Addison Richards, Robert Coote, Paul Hurst, Noah Beery Jr.; **Dir:** Lew Landers. **VHS, Beta** $19.98 *TTC* ✶✶½

Bad Man of Deadwood A man-with-a-past joins a circus as a sharp-shooter, and is threatened with disclosure. 1941 54m/B Roy Rogers, George "Gabby" Hayes. **VHS, Beta** $24.95 *DVT, NOS, RXM* ✶

Bad Manners When an orphan is adopted by a wealthy but entirely selfish couple, a group of his orphan friends try to free him from his new home and lifestyle. 1984 (R) 85m/C Martin Mull, Karen Black, Anne De Salvo, Murphy Dunne, Pamela Segall, Edy Williams, Susan Ruttan, Richard Deacon; **Dir:** Bobby Houston. **VHS, Beta** $19.95 *STE, HBO* ✶

Bad Man's River A Mexican revolutionary leader hires a gang of outlaws to blow up an arsenal used by the Mexican Army. 1972 92m/C IT SP Lee Van Cleef, James Mason, Gina Lollobrigida; **Dir:** Gene Martin. **VHS, Beta** $24.95 *WES, BUR* ✶✶

Bad Medicine A youth who doesn't want to be a doctor is accepted by a highly questionable Latin American school of medicine. Remember that it was for medical students like these that the U.S. liberated Grenada. 1985 (PG-13) 97m/C Steve Guttenberg, Alan Arkin, Julie Hagerty, Bill Macy, Curtis Armstrong, Julie Kavner, Joe Grifasi, Robert Romanus, Taylor Negron, Gilbert Gottfried; **Dir:** Harvey Miller. **VHS, Beta** $79.98 *FOX Woof!*

The Bad News Bears Family comedy about a misfit Little League team that gets whipped into shape by a cranky, sloppy, beer-drinking coach who recruits a female pitcher. O'Neal and Matthau are top-notch. The kids' language is rather adult, and very titillating for younger viewers. Spawned two sequels and a television series. 1976 (PG) 102m/C Walter Matthau, Tatum O'Neal, Vic Morrow, Joyce Van Patten, Jackie Earle Haley; **Dir:** Michael Ritchie. **VHS, Beta, LV, 8mm** $14.95 *PAR* ✶✶✶

The Bad News Bears in Breaking Training With a chance to take on the Houston Toros for a shot at the little league baseball Japanese champs, the Bears devise a way to get to Texas to play at the famed Astrodome. Sequel to "The Bad News Bears"; followed by "The Bad News Bears Go to Japan" (1978). 1977 (PG) 105m/C William Devane, Clifton James, Jackie Earle Haley, Jimmy Baio; **Dir:** Michael Pressman. **VHS, Beta, LV** $14.95 *PAR* ✶½

The Bad News Bears Go to Japan The second sequel, in which the famed Little League team goes to the Little League World Series in Tokyo. Comic adventure features Curtis as a talent agent out to exploit the team's fame. 1978 (PG) 102m/C Tony Curtis, Jackie Earle Haley, Tomisaburo Wayakama, Lonny Champman, George Wyner; **Dir:** John Berry. **VHS, Beta, LV** $19.95 *PAR* ✶

Bad Ronald No, not a political biography of Ronald Reagan... Fascinating thriller about a disturbed teenager who kills a friend after being harassed repeatedly. The plot thickens after the boy's mother dies, and he is forced to hide out in a secret room when an unsuspecting family with three daughters moves into his house. The story is accurately recreated from the novel by John Holbrook Vance. 1974 78m/C Scott Jacoby, Pippa Scott, John Larch, Dabney Coleman, Kim Hunter, John Fiedler; **Dir:** Buzz Kulik. **VHS** $49.95 *IVE* ✶✶✶

The Bad Seed A mother makes the torturous discovery that her cherubic eight-year-old daughter harbors an innate desire to kill. Based on Maxwell Anderson's powerful Broadway stage play. Oscar nominations for actress Kelly, supporting actresses Heckart and McCormack, Black & White Cinematography. Remade for television in 1985. 1956 129m/B Patty McCormack, Nancy Kelly, Eileen Heckart, Henry Jones, Evelyn Varden, Paul Fix; **Dir:** Mervyn LeRoy. **VHS, Beta** $59.95 *WAR* ✶✶½

The Bad Seed Television remake of the movie with the same name. Story about a sadistic little child who kills for her own evil purposes. Acting is not up to par with previous version. 1985 100m/C Blair Brown, Lynn Redgrave, David Carradine, Richard Kiley, David Ogden Stiers, Carrie Wells; **Dir:** Paul Wendkos. **VHS** $59.95 *NO* ✶✶

The Bad Sleep Well Japanese variation of the 1940 Warner Brothers crime dramas. A tale about corruption in the corporate world as seen through the eyes of a rising executive. 1960 135m/B JP Toshiro Mifune, Masayuki Kato, Masayuki Mori, Takashi Shimura, Akira Nishimura; **Dir:** Akira Kurosawa. **VHS, Beta** $24.95 *SVS, APD, TAM* ✶✶✶½

Bad Taste A definite pleaser for the person who enjoys watching starving aliens devour the average, everyday human being. Alien fast-food manufacturers come to earth in hopes of harvesting all of mankind. The earth's fate lies in the hands of the govern-

ment who must stop these rampaging creatures before the whole human race is gobbled up. Terrific make-up jobs on the aliens add the final touch to this gory, yet humorous cult horror flick.
1988 90m/C *NZ* Peter Jackson, Pete O'Herne, Mike Minett, Terry Potter, Craig Smith, Doug Wren, Dean Lawrie; *Dir:* Peter Jackson. **VHS, LV $79.98** *MAG* ✗✗✗

The Badge After many failed attempts at a career, a young man finally joins the police force to appease his domineering father.
1990 30m/C *Dir:* Robert Spara. **VHS $59.95** *CIG* ✗✗½

Badge 373 In the vein of "The French Connection," a New York cop is suspended and decides to battle crime his own way.
1973 (R) 116m/C Robert Duvall, Verna Bloom, Eddie Egan; *Dir:* Howard W. Koch. **VHS, Beta $14.95** *PAR* ✗½

Badge of the Assassin A television movie based on the true story of a New York assistant DA who directed a campaign to catch a pair of cop-killers from the '70s.
1985 96m/C James Woods, Yaphet Kotto, Alex Rocco, David Harris, Pam Grier, Steven Keats, Richard Bradford, Rae Dawn Chong; *Dir:* Mel Damski. **VHS, Beta, LV $29.95** *VMK* ✗✗

The Badlanders A western remake of the crime drama "The Asphalt Jungle." Ladd and Borgnine plan a gold robbery against Smith, who cheated them out of their share in a gold mine. Cross and double-cross follow the partners as does romance.
1958 85m/C Alan Ladd, Ernest Borgnine, Kent Smith, Katy Jurado, Claire Kelly, Nehemiah Persoff, Adam Williams; *Dir:* Delmer Daves. **VHS $19.98** *MGM* ✗✗✗

Badlands Based loosely on the Charlie Starkweather murders of the 1950s, this impressive debut by director Malik recounts a slow-thinking, unhinged misfit's killing spree across the midwestern plains, accompanied by a starry-eyed 15-year-old schoolgirl. Sheen and Spacek are a disturbingly numb, apathetic, and icy duo.
1974 (PG) 94m/C Martin Sheen, Sissy Spacek, Warren Oates; *Dir:* Terence Malick. New York Film Festival '73: Best Supporting Actress. **VHS, Beta, LV $39.98** *WAR* ✗✗✗½

Badman's Territory A straight-shooting marshal has to deal with such notorious outlaws as the James and Dalton boys in a territory outside of government control.
1946 79m/B Randolph Scott, Ann Richards, George "Gabby" Hayes, Steve Brodie; *Dir:* Tim Whelan. **VHS, Beta $19.98** *RKO* ✗✗

Badmen of Nevada The early days of Nevada before law and order are featured.
1933 57m/B Kent Taylor, Gail Patrick. **VHS, Beta $29.95** *VCN* ✗✗½

Baffled Nimoy is a race car driver who has visions of people in danger. He must convince an ESP expert (Hampshire) of the credibility of his vision, and then try to save the lives of the people seen with his sixth sense. Made for television.
1972 90m/C Leonard Nimoy, Susan Hampshire, Vera Miles, Rachel Roberts, Jewel Branch, Christopher Benjamin; *Dir:* Philip Leacock. **VHS** *FOX* ✗✗½

Bagdad Cafe A large German woman, played by Sagebrecht, finds herself stranded in the Mojave desert after her husband dumps her on the side of the highway. She encounters a rundown cafe where she becomes involved with the off-beat residents. A hilarious story in which the strange people

and the absurdity of their situations are treated kindly and not made to seem ridiculous. Spawned a short-lived TV series with Whoopi Goldberg.
1988 (PG) 91m/C *GE* Marianne Sagebrecht, CCH Pounder, Jack Palance; *Dir:* Percy Adlon. **VHS, Beta $79.95** *VIR* ✗✗✗

Bahama Passage Trite story of one lady's efforts to win the affection of a macho Bahamas stud.
1942 83m/C Madeleine Carroll, Sterling Hayden, Flora Robson, Leo G. Carroll; *Dir:* Edward H. Griffith. **VHS** *PAR, MLB* ✗✗½

Bail Jumper A story of love and commitment against incredible odds; some of which happen to be a swarm of locusts, a tornado, and falling meteorites. Joe and Elaine are small-time hoods escaping their dreary lives in Murky Springs Missouri by heading for that great bastion of idyllism and idealism—New York City. But even as the world starts to crumble around them, get the message that love prevails.
1989 90m/C Eszter Balint, B.J. Spalding, Tony Askin, Joie Lee; *Dir:* Christian Faber. **VHS, Beta $79.95** *JCI, FXL* ✗✗

Bail Out Three bounty hunters, armed to the teeth, run a car-trashing police gauntlet so they may capture a valuable crook.
1990 (R) 88m/C David Hasselhoff, Linda Blair, John Vernon, Tom Rosales, Charles Brill; *Dir:* Max Kleven. **VHS, Beta, LV $89.98** *VES, LIV* ✗

Bail Out at 43,000 Movie about the lifestyles and love affairs of your average, everyday parachutist.
1957 80m/B John Payne, Karen Steele, Paul Kelly; *Dir:* Francis D. Lyon. **VHS** *MGM* ✗✗

Baja Oklahoma A made-for-cable-television film about a country barmaid with dreams of being a country singer. Songs by Willie Nelson, Emmylou Harris and Billy Vera.
1987 110m/C Lesley Ann Warren, Peter Coyote, Swoosie Kurtz, Willie Nelson, Julia Roberts; *Dir:* Bobby Roth. **VHS, Beta, LV $89.95** *LHV* ✗

Baker's Hawk A young boy befriends a red-tailed hawk and learns the meaning of family and caring.
1976 98m/C Clint Walker, Diane Baker, Burl Ives; *Dir:* Lyman Dayton. **VHS, Beta $39.95** *LGC, BPG, HHE* Woof!

The Baker's Wife (La Femme du Boulanger) There's a new baker in town, and he brings with him to the small French village an array of tantalizing breads, as well as a discontented wife. When she runs off with a shepherd, her loyal and naive husband refuses to acknowledge her infidelity; however, in his loneliness, the baker can't bake, so the townspeople scheme to bring his wife back. Panned as overly cute Marcel Pagnol peasant glorification, and hailed as a visual poem full of wit; you decide. In French with subtitles. Also available for French students without subtitles; a French script booklet is also available.
1933 101m/B *FR* Raimu, Ginette LeClerc, Charles Moulton, Charpin, Robert Vattier; *Dir:* Marcel Pagnol. National Board of Review Awards '40: Best Foreign Film; New York Film Critics Awards '40: Best Foreign Film. **VHS, Beta $39.95** *INT, MRV, DVT* ✗✗✗½

Bakery/Grocery Clerk A package of shorts featuring crazy comedian Larry Semon, getting tangled up in mayhem and molasses. Silent with piano score.
1921 55m/B Larry Semon, Oliver Hardy, Lucille Carlisle; *Dir:* Larry Semon. **VHS, Beta $19.98** *CCB* ✗✗

Balboa Set on sun-baked Balboa Island, this is a melodramatic tale of high-class power, jealousy, and intrigue. Never-aired pilot for a television mini-series, in the nighttime soap tradition (even features Steve Kanaly from TV's "Dallas"). Special appearance by Cassandra Peterson, also known as horror hostess Elvira; and if that interests you, look for Sonny Bono, as well.
1982 92m/C Tony Curtis, Carol Lynley, Chuck Connors, Sonny Bono, Steve Kanaly, Jennifer Chase, Lupita Ferrer, Martine Beswick, Henry Jones, Cassandra Peterson; *Dir:* James Polakof. **VHS, Beta $69.98** *VES, LIV* Woof!

The Balcony A film version of the great Jean Genet play about a surreal brothel, located in an unnamed, revolution-torn city, where its powerful patrons act out their fantasies. Scathing and rude.
1963 87m/B Peter Falk, Shelley Winters, Lee Grant, Kent Smith, Peter Brocco, Ruby Dee, Jeff Corey, Leonard Nimoy, Joyce Jameson; *Dir:* Joseph Strick. **VHS, Beta $29.95** *MFV, TAM* ✗✗✗

Balkan Express A crew of unlikely slobs become heroes in war-ravaged Europe.
1983 102m/C **VHS, Beta $19.95** *STE, NWV* ✗✗½

Ball of Fire A gang moll hides out with a group of mundane professors, trying to avoid her loathsome boyfriend. The professors are busy compiling an encyclopedia and Stanwyck helps them with their section on slang in the English language. Cooper has his hands full when he falls for this damsel in distress and must fight the gangsters to keep her. Stanwyck takes a personal liking to naive Cooper and resolves to teach him more than just slang. Oscar nominations: Best Actress (Stanwyck), Sound Recording and Best Musical Score of a Dramatic Film.
1942 111m/B Gary Cooper, Barbara Stanwyck, Dana Andrews, Gene Krupa, Oscar Homolka, Dan Duryea, S.Z. Sakall, Henry Travers; *Dir:* Howard Hawks. **VHS, Beta, LV $19.98** *SUE* ✗✗✗

Ballad of Billie Blue A top country singer, whose wife betrayed him, has his heart broken and dreams shattered. Fortunately, he finds that God's love eases the pain.
1972 90m/C Jason Ledger, Marty Allen, Ray Danton, Erik Estrada; *Dir:* Kent Osborne. **VHS, Beta** *VGD* Woof!

Ballad in Blue Real life story of Ray Charles and a blind child. Tearjerker also includes some of Ray Charles song hits, such as "I Got a Woman" and "What'd I Say?" Also known as "Blues for Lovers."
1966 89m/B *GB* Ray Charles, Tom Bell, Mary Peach, Dawn Addams, Piers Bishop, Betty McDowall; *Dir:* Paul Henreid. **VHS, Beta $39.95** *IVE* ✗✗

Ballad of Cable Hogue A prospector, who had been left to die in the desert by his double-crossing partners, finds a waterhole. A surprise awaits his former friends when they visit the remote well. Not the usual violent Peckinpah horse drama, but a tongue-in-cheek comedy romance mixed with tragedy. Obviously offbeat and worth a peek.
1970 (R) 122m/C Jason Robards Jr., Stella Stevens, David Warner, L.Q. Jones, Strother Martin, Slim Pickens; *Dir:* Sam Peckinpah. **VHS, Beta, LV $19.98** *WAR* ✗✗✗

Ballad of Gregorio Cortez Tragic story based on one of the most famous manhunts in Texas history. A Mexican cowhand kills a Texas sheriff in self-defense and tries to elude the law, all because of a

misunderstanding of the Spanish language. Olmos turns in a fine performance as Cortez.
1983 99m/C Edward James Olmos, James Gammon, Tom Bower, Alan Vint, Barry Corbin, Rosana De Soto; *Dir:* Robert M. Young. **VHS, Beta, LV $14.98** *SUE, TAM* 🎬🎬🎬

Ballad of a Gunfighter A feud between two outlaws reaches the boiling point when they both fall in love with the same woman.
1964 84m/C Marty Robbins, Bob Barron, Joyce Redd, Nestor Paiva, Laurette Luez; *Dir:* Bill Ward. **VHS, Beta $9.99** *FHE, PSM* 🎬

The Ballad of Narayama Director Imamura's subtle and vastly moving story takes place a vague century ago. In compliance with village law designed to control population among the poverty-stricken peasants, a healthy seventy-year-old woman must submit to solitary starvation atop a nearby mountain. We follow her as she sets into motion the final influence she will have in the lives of her children and grandchildren, a situation described with detachment and without imposing a tragic perspective. In Japanese with English subtitles.
1983 129m/C *JP* Ken Ogata, Sumiko Sakamoto, Takejo Aki, Tonpei Hidari, Shoichi Ozawa; *W/Dir:* Shohei Imamura. Cannes Film Festival '83: Grand Prize. **VHS $59.95** *HMV, FCT* 🎬🎬🎬🎬

The Ballad of Paul Bunyan The good-humored, legendary American giant is pitted against a powerful lumber camp boss who likes to pick on the little guy.
1972 30m/C *Dir:* Arthur Rankin Jr., Jules Bass. **VHS $6.95** *STE, PSM* 🎬🎬🎬🎬

The Ballad of the Sad Cafe An unusual love story starring Redgrave as Miss Amelia, an outcast in a small Southern town during the Depression. When a distant relation of Miss Amelia's shows up, the everyday lives of the townspeople are suddenly transformed. The hunchbacked Cousin Lymon even changes Amelia into a friendly person with enough self-esteem to open a cafe. However, when Miss Amelia's one-time husband of ten days returns to town with revenge on his mind, tragedy is sure to follow. This moving story portrays both sides of love and its power to enhance and destroy simultaneously. Based on the critically acclaimed novella by Carson McCullers.
1991 (PG-13) 100m/C Vanessa Redgrave, Keith Carradine, Rod Steiger, Cork Hubbert, Earl Hindman, Anne Pitoniak; *Dir:* Simon Callow. **VHS, LV $89.95** *SVT* 🎬🎬🎬

Ballad of a Soldier As a reward for demolishing two German tanks, a nineteen-year-old Russian soldier receives a six-day pass so he can see his mother; however, he meets another woman. Well directed and photographed, while avoiding propaganda.
1960 88m/B *RU* Vladimir Ivashov, Shanna Prokhorenko; *Dir:* Grigori Chukrai. **VHS, Beta, 3/4U $35.95** *HTV, NOS, HHT* 🎬🎬🎬½

Ballbuster Cops take on gangs in an all-out high-stakes battle to win back the streets.
19?? 100m/C Ivan Rogers, Bonnie Paine, W. Randolph Galvin, Bill Shirk, Brenda Banet; *W/Dir:* Eddie Beverly Jr. **VHS $39.95** *XVC* 🎬🎬🎬½

Ballet Shoes The classic children's tale of three sisters living a very sheltered life with their guardian Sylvia, who is working to support all four of them. A story of hope and possibilities.
19?? 120m/C *GB* **VHS $29.95** *BFS* 🎬🎬

The Balloonatic/One Week "The Balloonatic" (1923) features Keaton in several misadventures. In "One Week" (1920), Keaton and his new bride, Seely, receive a new home as a wedding gift—the kind you have to assemble yourself. Silent.
192? 48m/B Buster Keaton, Phyllis Haver, Sybil Seely; *Dir:* Buster Keaton. **VHS, Beta $17.95** *CCB* 🎬🎬

Ballyhoo Baby When a couple traveling cross-country pick up a young hitchhiker, the sexual tension between the three builds to a frightening climax.
1990 30m/C *Dir:* Paul Young. **VHS $59.95** *CIG* 🎬½

Baltic Deputy An early forerunner of Soviet historic realism, where an aging intellectual deals with post-revolution Soviet life. In Russian with subtitles.
1937 95m/B *RU* Nikolai Cherkassov; *Dir:* Yosif Heifitz. National Board of Review Awards '37: 10 Best Foreign Films of the Year. **VHS, Beta, 3/4U $35.95** *FCT, IHF* 🎬🎬🎬½

The Baltimore Bullet Two men make their living traveling through the country as pool hustlers, bilking would-be pool sharks. Features ten of the greatest pool players in the world.
1980 (PG) 103m/C James Coburn, Omar Sharif, Bruce Boxleitner, Ronee Blakley, Jack O'Halloran; *Dir:* Robert Ellis Miller. **VHS, Beta $9.98** *SUE* 🎬🎬

Bambi A true Disney classic, detailing the often harsh education of a newborn anthropomorphic deer. One of the greatest children's films of all time, and a genuine perennial from generation to generation. Based very loosely on the book by Felix Salten.
1942 (G) 69m/C *Dir:* David Hand. **VHS, Beta, LV $26.99** *DIS, APD, KUI* 🎬🎬🎬🎬

Bambi Meets Godzilla A spoof on endless film credits, after which the title bout lasts about six seconds. Must be seen to be appreciated.
1969 2m/B *Dir:* Marv Newland. **VHS, Beta $9.95** *PYR* 🎬🎬½

Bambi Meets Godzilla and Other Weird Cartoons Bambi meets Godzilla, plus many classic animated shorts, including "Crazy Town" with Betty Boop, and Max Fleischer creations.
1987 30m/B VHS $9.95 *RHI* 🎬🎬½

The Bamboo Saucer Russian and American scientists race to find a U.F.O. in Red China. Also known as "Collision Course."
1968 103m/C Dan Duryea, John Ericson, Lois Nettleton, Nan Leslie; *Dir:* Frank Telford. **VHS $19.98** *REP* 🎬½

Banana Monster Horror flick spoof about an ape-man who proceeds through town committing murder by banana. The first feature film from the director of "Animal House" and "Spies Like Us"; this move was previously entitled "Schlock" (need we say more?) and features Landis himself in the gorilla suit.
1972 (PG) 80m/C John Landis, Saul Kahan, Eliza Garrett; *Dir:* John Landis. **VHS, Beta $69.95** *LTG* 🎬🎬½

Bananas Intermittently hilarious pre-Annie Hall Allen fare is full of the director's signature angst-ridden philosophical comedy. A frustrated product tester from New York runs off to South America, where he volunteers his support to the revolutionary force of a shaky Latin-American dictatorship and

winds up the leader. Don't miss cameos by Stallone and Garfield. Wittily scored by Marvin Hamlisch.
1971 (PG) 82m/C Woody Allen, Louise Lasser, Carlos Montalban, Howard Cosell, Sylvester Stallone, Allen Garfield, Charlotte Rae, Conrad Bain; *Dir:* Woody Allen. **VHS, Beta, LV $19.98** *FOX, FCT* 🎬🎬🎬

Bananas Boat A captain takes a man and his daughter away from the collapsing government of their banana republic. The cast and director take a few risks but fail in this would-be comedy.
1978 91m/C *GB* Doug McClure, Hayley Mills, Lionel Jeffries, Dilys Hamlett, Warren Mitchell; *Dir:* Sidney Hayers. **VHS $79.95** *UHV* Woof!

Band of the Hand A "Miami Vice" type melodrama about five convicts who are trained to become an unstoppable police unit. The first feature film by Glaser, last seen as Starsky in "Starsky & Hutch."
1986 (R) 109m/C Stephen Lang, Michael Carmine, Lauren Holly, Leon Robinson; *Dir:* Paul Michael Glaser. **VHS, Beta, LV $79.95** *RCA* 🎬

Band of Outsiders A woman hires a pair of petty criminals to rip off her aunt; Godard vehicle for exposing self-reflexive comments on modern film culture. In French with English subtitles.
1964 97m/B *FR* Sami Frey, Anna Karina, Claude Brasseur, Louisa Colpeyn; *Dir:* Jean-Luc Godard. New York Film Festival '64: 10 Best Foreign Films of the Year. **VHS, Beta $49.95** *TAM, NOS, MRV* 🎬🎬½

The Band Reunion The Band is together again performing their greatest hits at a concert taped in Vancouver, Canada.
1984 87m/C VHS, Beta $19.95 *MSM* 🎬🎬½

Band Wagon A Hollywood song-and-dance man finds trouble when he is persuaded to star in a Broadway musical. Songs by Howard Dietz and Arthur Schwartz include "That's Entertainment," and "Dancing in the Dark." Charisse has been called Astaire's most perfect partner, perhaps by those who haven't seen Rogers.
1953 112m/C Fred Astaire, Cyd Charisse, Oscar Levant, Nanette Fabray, Jack Buchanan; *Dir:* Vincente Minnelli. **VHS, Beta, LV $19.95** *MGM, FCT* 🎬🎬🎬

Bandera Bandits An escaped convict, while evading a ruthless sheriff, marries a young girl and together they romp off into the sunset.
19?? 98m/C Charlie Chaplin, Mabel Normand, Mack Swain, Mack Sennett. **VHS, Beta $59.95** *TWE* Woof!

Bandit Queen A spanish girl forms a band to stop seizure of Spanish possessions by lawless Californians.
1951 71m/B Barbara Britton, Willard Parker, Philip Reed, Jack Perrin; *Dir:* William Berke. **VHS, Beta** *WGE, RXM* 🎬½

Bandits Three cowboys team up with a band of Mexican outlaws to fight a Mexican traitor.
1973 83m/C Robert Conrad, Jan-Michael Vincent, Roy Jenson; *Dir:* Robert Conrad, Alfredo Zacharias. **VHS, Beta $19.95** *UNI, TAV* 🎬

Bandits of Orgosolo An acclaimed, patient drama about a Sardinian shepherd who shelters a band of thieves and is thereafter pursued as one of them. English dubbed.
1961 98m/B *IT* Michele Cossu, Peppeddu Cuccu; *Dir:* Vittorio de Seta. Venice Film Festival '61: San Giorgio Award. **VHS, Beta $24.95** *NOS, FCT* 🎬🎬½

Bandolero! In Texas, Stewart and Martin are two fugitive brothers who run into trouble with their Mexican counterparts.
1968 (PG) 106m/C James Stewart, Raquel Welch, Dean Martin, George Kennedy, Will Geer, Harry Carey Jr., Andrew Prine; *Dir:* Andrew V. McLaglen. **VHS, Beta** $19.98 FOX ♪♪½

Bang Bang Kid A western spoof about a klutzy gunfighter defending a town from outlaws.
1967 (G) 78m/C Tom Bosley, Guy Madison, Sandra Milo; *Dir:* Stanley Prager. **VHS, Beta** $59.95 WES ♪♪

Bang the Drum Slowly The original television adaptation of a Mark Harris novel about baseball. A ball player stricken by a terminal illness strikes an unlikely friendship with a teammate. Interesting role for an actor (Newman) who claims to have been driven to acting by running away from the sporting goods business.
1956 60m/B Paul Newman, George Peppard, Albert Salmi. **VHS** $14.95 NOS, WKV ♪♪

Bang the Drum Slowly The touching story of a major league catcher who discovers that he is dying of Hodgkins disease and wants to play just one more season. De Niro is the weakening baseball player and Moriarty is the friend who helps him see it through. Based on a novel by Mark Harris. Gardenia earned an Oscar nomination.
1973 (PG) 97m/C Robert De Niro, Michael Moriarty, Vincent Gardenia, Phil Foster, Ann Wedgeworth, Heather MacRae, Selma Diamond, Danny Aiello; *Dir:* John Hancock. National Board of Review Awards '73: 10 Best Films of the Year; New York Film Critics Awards '73: Best Supporting Actor (De Niro). **VHS, Beta, LV** $14.95 PAR, FCT ♪♪♪

The Bank Dick W.C. Fields wrote the screenplay and stars in this comedy about a man who accidentally trips a bank robber and winds up as a guard. Side two of the laserdisc is in CAV format, which allows single frame access to the frenetic cops and robbers chase sequence.
1940 73m/B W.C. Fields, Cora Witherspoon, Una Richard Purcell, Jack Norton, Franklin Pangborn, Una Merkel, Shemp Howard; *Dir:* Eddie Cline. **VHS, Beta, LV** $29.95 MCA ♪♪♪♪

Bank Shot An hilarious comedy about a criminal who plans to rob a bank by stealing the entire building. Based on the novel by Donald Westlake, and the sequel to "The Hot Rock."
1974 83m/C George C. Scott, Joanna Cassidy, Sorrell Booke, G. Wood, Clifton James, Bob Balaban, Bibi Osterwald; *Dir:* Gower Champion. **VHS, CV** $14.95 WKV ♪♪♪

The Banker A cop, played by Forster, suspects a wealthy, highly influential banker of brutal serial killings.
1989 (R) 90m/C Robert Forster, Jeff Conaway, Leif Garrett, Duncan Regehr, Shanna Reed, Deborah Richter, Richard Roundtree, Teri Weigel, E.J. Peaker; *Dir:* William Webb. **VHS, Beta, LV** $79.95 VIR ♪♪

Banzai Runner A cop whose brother was killed in an exclusive desert-highway race decides to avenge by joining the race himself.
1986 88m/C Dean Stockwell, John Shepherd, Charles Dierkop; *Dir:* John Thomas. **VHS, Beta** $19.95 VMK ♪♪

Bar-20 Stage hold-ups and jewel robberies abound in this Hopalong Cassidy western.

1943 54m/B George Reeves, Douglas Fowley, William Boyd, Andy Clyde, Robert Mitchum, Victor Jory; *Dir:* Lesley Selander. **VHS, Beta** $19.95 NOS, MRV, GVV ♪♪½

Barabbas Barabbas, a thief and murderer, is freed by Pontius Pilate in place of Jesus. He is haunted by this event for the rest of his life. Excellent acting, little melodrama, lavish production make for fine viewing. Based on the novel by Lagerkvist.
1962 144m/C Anthony Quinn, Silvana Mangano, Arthur Kennedy, Jack Palance, Ernest Borgnine, Katy Jurado; *Dir:* Richard Fleischer. National Board of Review Awards '62: 5 Best Foreign Films of the Year. **VHS, Beta, LV** $59.95 COL ♪♪½

Barbarella Based on the popular French sci-fi comic strip drawn by Jean-Claude Forest, this cult classic details the bizarre adventures of a space nymphette (Fonda) encountering fantastic creatures and super beings. You'll see sides of Fonda you never saw before (not even in the workout videos). Notorious in its day; rather silly, dated camp now. Don't miss the elbow-sex scene. Terry Southern contributed to the script.
1968 (PG) 98m/C Jane Fonda, John Phillip Law, David Hemmings, Marcel Marceau, Anita Pallenberg, Milo O'Shea; *Dir:* Roger Vadim. **VHS, Beta, LV** $59.95 PAR ♪♪½

Barbarian and the Geisha The first US diplomat in Japan undergoes culture shock as well as a passionate love affair with a geisha, circa 1856. Music by Hugo Friedhofer.
1958 104m/C John Wayne, Eiko Ando, Sam Jaffe, So Yamamura; *Dir:* John Huston. **VHS, Beta** $19.98 FOX ♪♪

Barbarian Queen Female warriors seek revenge for the capturing of their tribe's men in this sword-and-sorcery epic. Available also in an unrated version.
1985 (R) 71m/C Lana Clarkson, Frank Zagarino, Katt Shea, Dawn Dunlap, Susana Traverso; *Dir:* Hector Olivera. **VHS, Beta** $29.98 VES, LIV Woof!

Barbarian Queen 2: The Empress Strikes Back Beautiful Princess Athalia and her evil brother Ankaris just don't get along. When he strips her of her power and has her imprisoned, she escapes and joins a band of female rebels, leading them into a battle to overthrow her brother. Basically a low-grade remake of the "Conan" movies.
199? (R) 87m/C Lana Clarkson, Greg Kramer, Rebecca Wood, Roger Cundy; *Dir:* Joe Finley. **VHS** $89.98 LIV ♪

The Barbarians Two bodybuilder siblings in animal skins battle wizards and warlords in this dumb-but-fun U.S./Italian co-production.
1987 (R) 88m/C IT David Paul, Peter Paul, Richard Lynch, Eva LaRue, Virginia Bryant, Sheeba Alahani, Michael Berryman; *Dir:* Ruggero Deodato. **VHS, Beta, LV** $19.95 MED ♪

Barbarosa Offbeat western about an aging, legendary outlaw constantly on the lam who reluctantly befriends a naive farmboy and teaches him survival skills. Nelson and Busey are a great team, solidly directed. Lovely Rio Grande scenery.
1982 (PG) 90m/C Willie Nelson, Gilbert Roland, Gary Busey, Isela Vega; *Dir:* Fred Schepisi. **VHS, Beta, LV** $19.95 FOX, JTC ♪♪♪

Barbary Coast A ruthless club owner tries to win the love of a young girl by building her into a star attraction during San Francisco's gold rush days.

1935 90m/B Edward G. Robinson, Walter Brennan, Brian Donlevy, Joel McCrea, Donald Meek, David Niven, Miriam Hopkins; *Dir:* Howard Hawks. **VHS, Beta, LV** $14.98 SUE ♪♪♪

The Barbary Coast A turn-of-the-century detective sleuths the streets of San Francisco in this average made-for-television movie.
1974 100m/C William Shatner, Dennis Cole, Lynda Day George, John Vernon, Charles Aidman, Michael Ansara, Neville Brand, Bill Bixby; *Dir:* Bill Bixby. **VHS** SUE, ORI ♪♪

Barber of Seville Sutherland and some of her puppet friends relate the tale of this famous opera to young people and adults. Sections of the opera are performed featuring the London Symphony Orchestra.
1973 30m/C Joan Sutherland. **VHS, Beta** $29.95 BFA ♪

The Barber Shop Fields portrays the bumbling, carefree barber Cornelius O'Hare, purveyor of village gossip and problem solver. Havoc begins when a gangster enters the shop and demands that Cornelius change his appearance.
1933 21m/B W.C. Fields, Elise Cavanna, Harry Watson, Dagmar Oakland, Frank Yaconelli. **VHS, Beta** $24.98 CCB ♪♪♪

The Barcelona Kill When a journalist and her boyfriend get in too deep with the Barcelona mob, their troubles begin.
197? 86m/C Linda Hayden, John Austin, Simon Andrew, Maximo Valverde; *Dir:* Jose Antonio De La Loma. **VHS, Beta** $59.95 VCL ♪♪

Bare Essentials A made-for-television yuppie Club Med nightmare in which a high-strung couple vacationing from New York find themselves marooned on a desert isle with only two other inhabitants. With no cellular telephoning ability and an absence of large-ticket consumer goods to purchase, the two turn their reluctant sights on each other, focusing on the bare essentials, as it were. Great soul-searching scenes for the inarticulate.
1991 94m/C Gregory Harrison, Mark Linn-Baker, Lisa Hartman, Charlotte Lewis; *Dir:* Martha Coolidge. **VHS** $89.98 REP ♪

Bare Hunt Detective Max T. Unimportant is dragged into the underground world of prospective nude actresses in this supposedly funny "nudie." Also known as "My Gun Is Jammed" and "The Bear Hunt."
1963 69m/B **VHS, Beta** $19.95 VDM ♪

Bare Knuckles The adventures of a low-rent bounty hunter.
1977 (R) 90m/C Robert Vihard, Sherry Jackson, Michael Heit, Gloria Hendry, John Daniels; *Dir:* Don Edmonds. **VHS, Beta** $59.95 TWE ♪

Bare Necessities The narrators travel the world in search of the most expensive women and assorted pleasures.
1985 86m/C Robert Egan, Hilary Scott, Paul Wheeler, Bonnie Paine. **VHS, Beta** $49.95 AHV ♪

Barefoot in Athens Television presentation from "George Schaefer's Showcase Theatre" chronicles the last years of the philosopher Socrates who, barefoot and unkempt, an embarrassment to his wife, and a dangerous critic to the corrupt Athenian leaders, nevertheless believes that democracy and truth are all-important in his city.
1966 76m/C Peter Ustinov, Geraldine Page, Anthony Quayle; *Dir:* George Schaefer. Emmy Award '66: Outstanding Single Performance/Actor (Ustinov). **VHS, Beta** FHS ♪♪

Barefoot Contessa The story, told in flashback, of a Spanish dancer's rise to Hollywood stardom, as witnessed by a cynical director. Shallow Hollywood self-examination also written by Mankiewicz, well-performed. Oscar nominations for Story and Screenplay.
1954 128m/C Ava Gardner, Humphrey Bogart, Edmond O'Brien, Valentina Cortese, Rossano Brazzi; *Dir:* Joseph L. Mankiewicz. Academy Awards '54: Best Supporting Actor (O'Brien). **VHS, Beta $19.98** *MGM, FOX, FCT* 𝄞𝄞𝄞

Barefoot Executive A mailroom boy who works for a national television network finds a chimpanzee that can pick hit shows in this Disney family comedy.
1971 92m/C Kurt Russell, John Ritter, Harry Morgan, Wally Cox, Heather North; *Dir:* Robert Butler. **VHS, Beta** *DIS, OM* 𝄞𝄞

Barefoot in the Park Neil Simon's Broadway hit translates well to screen. A newly wedded bride (Fonda) tries to get her husband (Redford, reprising his Broadway role) to loosen up and be as free-spirited as she is. Natwick received an Oscar nomination.
1967 106m/C Robert Redford, Jane Fonda, Charles Boyer, Mildred Natwick; *Dir:* Gene Saks. **VHS, Beta, LV $14.95** *PAR* 𝄞𝄞𝄞

Barfly A semi-autobiographical screenplay by Charles Bukowski. A talented writer chooses to spend his time as a lonely barfly, hiding his literary abilities behind glasses of liquor. Dunaway's character is right on target as the fellow alcoholic.
1987 (R) 100m/C Mickey Rourke, Faye Dunaway, Alice Krige, Frank Stallone, J.C. Quinn; *Dir:* Barbet Schroeder. New York Film Festival '87: Best Supporting Actor. **VHS, Beta, LV, 8mm $10.08** *WAR, ГОТ* 𝄞𝄞𝄞

The Bargain Hart's first feature, in which he portrays a bandit desperately trying to go straight. Original titles with musical score.
1915 50m/B William S. Hart. **VHS, Beta $19.95** *NOS, GPV, DVT* 𝄞

Baring It All A group therapy collective shares their fantasies and arouses each other.
1985 78m/C Paul Bartel. **VHS, Beta $29.95** *MED Woof!*

Baritone Concerned Polish drama about a prominent opera singer who promises to deliver a grand concert upon returning to his small town, only to lose his voice just before the show is to start. Proof positive that the opera ain't over 'til the fat lady sings.
1985 100m/C *PL* Zbigniew Zapassiewicz; *Dir:* Janusz Zaorski. **VHS $49.95** *FCT* 𝄞𝄞

Barkleys Canine cabdriver Arnie Barkley gets into hot water with his family over his outrageous opinions in this collection of two episodes from the series.
1972 44m/C VHS, Beta $39.95 *TWE* 𝄞𝄞

Barkleys of Broadway The famous dancing team's last film together; they play a quarreling husband/wife showbiz team. Written by Adolph Green and Betty Comden. Songs by Ira Gershwin and Harry Warren include "They Can't Take That Away From Me," "Sabre Dance," "SwingTrot," "Manhattan Downbeat," and "A Weekend in the Country."
1949 109m/C Fred Astaire, Ginger Rogers, Gale Robbins, Oscar Levant, Jacques Francois, Billie Burke; *Dir:* Charles Walters. **VHS, Beta, LV $29.95** *MGM* 𝄞𝄞𝄞

Barn Burning William Faulkner's story of the late nineteenth-century South. The son of a tenant farmer is torn between trying to win his father's acceptance and his aversion to his father's unrelenting and violent nature. Part of the "American Short Story II" series.
1980 41m/C Tommy Lee Jones, Diane Kagan. **VHS, Beta $24.95** *MON, KAR* 𝄞𝄞

Barn of the Naked Dead Prine plays a sicko who tortures women while his radioactive monster/ dad terrorizes the Nevada desert. Rudolph's first film, directed under the pseudonym, Gerald Comier. Also on video as "Terror Circus," and "Nightmare Circus."
1973 86m/C Andrew Prine, Manuella Thiess, Sherry Alberoni, Gylian Roland, Al Cormier, Jennifer Ashley; *Dir:* Alan Rudolph. **VHS $39.95** *REG Woof!*

Barnaby and Me Australian star Barnaby the Koala Bear joins an international con-man in this romantic adventure. The mob is chasing the con-man when he meets and falls for a lovely young woman and her daughter.
1977 90m/C Sid Caesar, Juliet Mills, Sally Boyden; *Dir:* Norman Panama. **VHS, Beta $14.95** *ACA* 𝄞𝄞

Barnum P.T. Barnum's life is focused upon in this biography about the man who helped to form 'The Greatest Show On Earth.'
1986 100m/C Burt Lancaster, Hanna Schygulla, Jenny Lind, John Roney; *Dir:* Lee Philips. **VHS, Beta, LV $29.95** *ACA* 𝄞𝄞½

Baron Vengeance is the name of the game when an underworld boss gets stiffed on a deal. Fast-paced no-brainer street drama.
19?? 88m/C Calvin Lockhart, Richard Lynch. **VHS $14.99** *PGN* 𝄞

Baron of Arizona Land office clerk almost succeeds in convincing the U.S. that he owned the state of Arizona.
1951 99m/B Vincent Price, Ellen Drew, Beulah Bondi, Reed Hadley, Vladimir Sokoloff; *Dir:* Samuel Fuller. **VHS, Beta** *WGE, BUR, RXM* 𝄞𝄞𝄞

Baron and the Kid A pool shark finds out that his opponent at a charity exhibition game is his long-lost son. Based upon Johnny Cash's song.
1984 100m/C Johnny Cash, Darren McGavin, June Carter Cash, Richard Roundtree; *Dir:* Gary Nelson. **VHS, Beta $59.98** *FOX* 𝄞

Baron Munchausen The German film studio UFA celebrated its 25th anniversary with this lavish version of the Baron Munchausen legend, starring a cast of top-name German performers at the height of the Third Reich. Filmed in Agfacolor; available in English subtitled or dubbed versions.
1943 120m/C *GE* Hans Albers, Kaethe Kaack, Hermann Speelmanns, Leo Slezak; *Dir:* Josef von Baky. **VHS, Beta $59.95** *TAM, VCD, GLV* 𝄞𝄞𝄞

Barracuda Lots of innocent swimmers are being eaten by crazed killer barracudas.
1978 (R) 90m/C Wayne Crawford, Jason Evers, Roberta Leighton; *Dir:* Harry Kerwin. **VHS, Beta $19.98** *VID Woof!*

The Barretts of Wimpole Street The moving, almost disturbing, account of poetess Elizabeth Barrett, an invalid confined to her bed, with only her poetry and her dog to keep her company. She is wooed by poet Robert Browning, in whose arms she finds true happiness and a miraculous recovery. Multi-faceted drama expertly played by all. Also known as "Forbidden Alliance."

1934 110m/B Fredric March, Norma Shearer, Charles Laughton, Maureen O'Sullivan, Katherine Alexander, Una O'Connor, Ian Wolfe; *Dir:* Sidney Franklin. **VHS $19.98** *MGM, CCB* 𝄞𝄞𝄞

Barry Lyndon A ravishing adaptation of the classic Thackerey novel about the adventures of an Irish gambler moving from innocence to self-destructive arrogance in the aristocracy of 18th Century England. Visually opulent. Kubrick got excellent performances from all his actors, and a stunning display of history, but the end result still overwhelms. O'Neal has seldom been better. Oscar nominations for Best Picture, director, and Adapted Screenplay.
1975 (PG) 185m/C Ryan O'Neal, Marisa Berenson, Patrick Magee, Hardy Kruger, Guy Hamilton; *Dir:* Stanley Kubrick. Academy Awards '75: Best Art Direction/Set Decoration, Best Cinematography, Best Costume Design, Best Original Score; British Academy Awards '75: Best Cinematography, Best Director (Kubrick); National Society of Film Critics Awards '75: Best Cinematography; Harvard Lampoon Awards '75: Worst Film of the Year. **VHS, Beta, LV $29.95** *WAR, PIA* 𝄞𝄞𝄞½

Barry McKenzie Holds His Own In this sequel to "The Adventures of Barry McKenzie," we find that after a young man's aunt is mistaken for the Queen of England, two emissaries of Count Plasma of Transylvania kidnap her to use as a Plasma tourist attraction. Based on the 'Private Eye' comic strip, this crude Australian film is as disappointing as the first of the Barry McKenzie stories.
1974 93m/C *AU* Barry Humphries, Barry Crocker, Donald Pleasence; *Dir:* Bruce Beresford. **VHS, Beta $14.98** *VID* 𝄞

Bartleby A new version of the classic Herman Melville short story. McEnery is Bartleby the clerk, who refuses to leave his job even after he's fired; Scofield is his frustrated boss.
1970 79m/C Paul Scofield, John McEnery, Colin Jeavons, Thorley Walters; *Dir:* Anthony Friedman. **VHS $29.95** *WST, KUL, FCT* 𝄞𝄞½

Barton Fink This eerie comic nightmare comes laden with awards (including the Palm D'Or from Cannes) but only really works if you care about the time and place. Fink is a trendy New York playwright staying in a seedy Hollywood hotel in the 1940s, straining to write a simple B-movie script. Macabre events, both real and imagined, compound his writer's block. Superb set design from Dennis Gassner complements an unforgettable cast of grotesques.
1991 (R) 116m/C John Turturro, John Goodman, Judy Davis, Michael Lerner, John Mahoney, Jon Polito; *W/Dir:* Joel Coen. Cannes Film Festival '91: Best Actor (Turturro), Best Director (Coen), Best Film; Chicago Film Critics Awards '91: Best Cinematography; Los Angeles Film Critics Association Awards '91: Best Supporting Actor (Lerner); New York Film Critics Awards '91: Best Supporting Actress (Davis); National Society of Film Critics Awards '91: Best Cinematography. **VHS $94.98** *FXV, CCB* 𝄞𝄞𝄞

Basic Training Three sexy ladies wiggle into the Pentagon in their efforts to clean up the government.
1986 (R) 85m/C Ann Dusenberry, Rhonda Shear, Angela Aames, Walter Gotell; *Dir:* Andrew Sugarman. **VHS, Beta $79.98** *LIV, VES Woof!*

Basileus Quartet The replacement for a violinist in a well-established quartet creates emotional havoc. Beautiful music, excellent and evocative photography.

1982 118m/C *FR IT* Pierre Malet, Hector Alterio, Omero Antonutti, Michel Vitold, Alain Cuny, Gabriel Ferzetti, Lisa Kreuzer; *Dir:* Fabio Carpi. **VHS, LV $39.98** *CGI* 🎬🎬🎬

Basin Street Revue A compilation of "soundies," the music videos of their day, featuring a healthy cross section of legendary '40s performers.
1955 41m/B Sarah Vaughan, Lionel Hampton, Cab Calloway, Nat King Cole. **VHS, Beta $29.95** *NO* 🎬🎬🎬

Basket Case A gory horror film about a pair of Siamese twins—one normal, the other gruesomely deformed. The pair is surgically separated at birth, and the evil disfigured twin is tossed in the garbage. Fraternal ties being what they are, the normal brother retrieves his twin—essentially a head atop shoulders—and totes him around in a basket (he ain't heavy). Together they begin twisted and deadly revenge, with the brother-in-a-basket in charge. Very entertaining, if you like this sort of thing. Followed by two sequels, if you just can't get enough.
1982 89m/C Kevin Van Hentenryck, Terri Susan Smith, Beverly Bonner; *Dir:* Frank Henenlotter. **VHS, Beta, LV $19.95** *MED* 🎬🎬½

Basket Case 2 Surgically separated teenage mutant brothers Duane and Belial are back! This time they've found happiness in a "special" family—until they're plagued by the paparazzi. Higher production values make this sequel slicker than its low-budget predecessor, but it somehow lacks the same charm.
1990 (R) 90m/C Kevin Van Hentenryck, Annie Ross, Kathryn Meisle, Heather Rattray, Jason Evers, Ted Sorel; *Dir:* Frank Henenlotter. **VHS, Beta, LV $89.95** *SHG, IME* 🎬🎬½

Basket Case 3: The Progeny In this sequel to the cult horror hits "Basket Case" and "Basket Case 2," Belial is back and this time he's about to discover the perils of parenthood as the mutant Mrs. Belial delivers a litter of bouncing mini-monsters. Everything is fine until the police kidnap the little creatures and chaos breaks out as Belial goes on a shocking rampage in his newly created mechanical body. Weird special effects make this a cult favorite for fans of the truly outrageous.
1991 (R) 90m/C Kevin Van Hentenryck, Gil Roper, Annie Ross; *W/Dir:* Frank Henenlotter. **VHS** *MCA* 🎬🎬½

Bass on Titles Academy Award winner Saul Bass, translates the themes of major motion pictures into graphic animation.
1977 32m/C **VHS, Beta** *PYR* 🎬

The Bat A great plot centering around a murderer called the Bat, who kills hapless victims by ripping out their throats when he isn't busy searching for a million dollars worth of securities stashed in the old house he is living in. Adapted from the novel by Mary Roberts Rinehart.
1959 80m/B Vincent Price, Agnes Moorehead, Gavin Gordon, John Sutton, Lenita Lane, Darla Hood; *Dir:* Crane Wilbur. **VHS, Beta $16.95** *NOS, SNC, DVT* 🎬🎬

The Bat People Less-than-gripping horror flick in which Dr. John Bech is bitten by a bat while on his honeymoon. He then becomes a sadistic bat creature, compelled to kill anyone who stumbles across his path. The gory special effects make for a great movie if you've ever been bitten by that sort of thing. Also known as "It Lives By Night."

1974 (R) 95m/C Stewart Moss, Marianne McAndrew, Michael Pataki, Paul Carr; *Dir:* Jerry Jameson. **VHS, Beta $14.95** *HBO* 🎬

Bat 211 Hackman, an American officer, is stranded in the wilds of Vietnam alone after his plane is shot down. He must rely on himself and Glover, with whom he has radio contact, to get him out. Glover and Hackman give solid performances in this otherwise average film.
1988 (R) 112m/C Gene Hackman, Danny Glover, Jerry Reed, David Marshall Grant, Clayton Rohner, Erich Anderson, Joe Dorsey; *Dir:* Peter Markle. **VHS, Beta, LV $19.98** *MED* 🎬🎬

The Bat Whispers A masked madman is stalking the halls of a creepy mansion; eerie tale that culminates in an appeal to the audience to keep the plot under wraps. Unusually crafted film for its early era. Comic mystery based on the novel and play by Mary Roberts Rinehart and Avery Hopwood.
1930 82m/B Chester Morris, Chance Ward, Richard Tucker, Wilson Benge, DeWitt Jennings, Una Merkel, Spencer Charters; *W/Dir:* Roland West. **VHS, LV $24.95** *NOS, DVT, MLB* 🎬🎬½

Bataan A rugged war-time combat drama following the true story of a small platoon in the Philippines endeavoring to blow up a pivotal Japanese bridge. Also available in a colorized version.
1943 115m/B Robert Taylor, George Murphy, Thomas Mitchell, Desi Arnaz Sr., Lee Bowman, Lloyd Nolan, Robert Walker; *Dir:* Tay Garnett. **VHS, Beta $59.95** *MGM, TLF* 🎬🎬½

Bathing Beauty This musical stars Skelton as a pop music composer with the hots for college swim teacher Williams. Rathbone is a music executive who sees the romance as a threat to Skelton's career and to his own profit margin. Full of aquatic ballet, Skelton's shtick, and wonderful original melodies. The first film in which Williams received star billing.
1944 101m/C Red Skelton, Esther Williams, Basil Rathbone, Bill Goodwin, Jean Porter, Carlos Ramirez, Donald Meek, Ethel Smith, Helen Forrest; *Dir:* George Sidney. **VHS $19.98** *MGM, TTC, FCT* 🎬🎬½

Batman Holy television camp, Batman! Will the caped crusader win the Bat-tle against the combined forces of the Joker, the Riddler, the Penguin, and Catwoman? Will Batman and Robin save the United World Security Council from dehydration? Will the Bat genius ever figure out that Russian journalist Miss Kitka and Catwoman are one in the same? Biff! Thwack! Socko! Don't confuse this with the Michael Keeton version of the Dark Knight; this is the pot-bellied Adam West Batman, teeming with Bat satire.
1966 104m/C Burt Ward, Adam West, Burgess Meredith, Cesar Romero, Frank Gorshin, Lee Meriwether; *Dir:* Leslie Martinson. **VHS, Beta, LV $19.98** *FOX* 🎬🎬½

Batman The blockbuster fantasy epic that renewed Hollywood's faith in media blitzing. The Caped Crusader is back in Gotham City, where even the criminals are afraid to walk the streets alone. There's a new breed of criminal in Gotham, led by the infamous Joker. Their random attacks via acid-based make-up are just the beginning. Keaton is surprisingly good as the dual personality hero though Nicholson steals the show. Marvelously designed and shot. Music by Prince and Danny Elfman. Much better on the big screen. Followed in 1992 by "Batman Returns."

1989 (PG-13) 126m/C Michael Keaton, Jack Nicholson, Kim Basinger, Robert Wuhl, Tracy Walter, Billy Dee Williams, Pat Hingle, Michael Gough, Jack Palance, Jerry Hall; *Dir:* Tim Burton. **VHS, Beta, LV, 8mm $24.98** *FOX, WAR* 🎬🎬🎬½

Batman and Robin: Volume 1 The first movie serial appearance of the Caped Crusader with his partner, Robin. The first seven parts of fifteen part series.
1949 ?m/B **VHS, Beta** *GKK, MLB* 🎬🎬🎬½

Batman and Robin: Volume 2 The second half of the original movie serial, containing the last eight chapters.
1949 ?m/B **VHS, Beta** *GKK, MLB* 🎬🎬🎬½

Battered Wife beating and its effects on three couples are dramatized in this movie, which also tells of the agencies now available to help these women.
1978 98m/C Karen Grassle, Mike Farrell, Joan Blondell, Howard Duff, LeVar Burton, Chip Fields; *Dir:* Peter Werner. **VHS, Beta $49.95** *LCA, PSM* 🎬½

*batteries not included As a real estate developer fights to demolish a New York tenement, the five remaining residents are aided by tiny metal visitors from outer space in their struggle to save their home. Each resident gains a renewed sense of life in this sentimental, wholesome family film produced by Spielberg. Cronyn and Tandy keep the schmaltz from getting out of hand. Neat little space critters.
1987 (PG) 107m/C Hume Cronyn, Jessica Tandy, Frank McRae, Michael Carmine; *Dir:* Matthew Robbins. **VHS, Beta, LV $19.95** *MCA* 🎬🎬½

Battle of Algiers A famous, powerful, award-winning film depicting the uprisings against French Colonial rule in 1954 Algiers. A seminal documentary-style film which makes most political films seem ineffectual by comparison in its use of non-professional actors, gritty photography, realistic violence, and a boldly propagandistic sense of social outrage. Music by Ennio Morricone. Oscar nomination for director Pontecorvo.
1966 123m/B *FR* Brahim Haggiag, Jean Martin, Yacef Saadi, Tommaso Neri; *Dir:* Gillo Pontecorvo. New York Film Festival '67: Best Film; Venice Film Festival '66: Best Film. **VHS, Beta, LV, 3/4U $29.95** *TAM, IHF, AXV* 🎬🎬🎬🎬

The Battle of Austerlitz Lovely and sad French film about Napoleon. Two very different cousins die for the same woman. Fine direction and strong performances. Also known as "Austerlitz."
1960 (PG) 123m/C *FR IT YU* Claudia Cardinale, Martine Carol, Leslie Caron, Vittorio De Sica, Jean Marais, Ettore Manni, Jack Palance, Orson Welles; *Dir:* Abel Gance. **VHS $69.95** *IVE* 🎬🎬

Battle Beneath the Earth The commies try to undermine democracy once again when American scientists discover a Chinese plot to invade the U.S. via a series of underground tunnels. Perhaps a tad jingoistic.
1968 112m/C *GB* Kerwin Mathews, Peter Arne, Viviane Ventura, Robert Ayres; *Dir:* Montgomery Tully. **VHS, Beta $59.95** *MGM* 🎬🎬

Battle Beyond the Stars The planet Akir must be defended against alien rapscallions in this intergalactic Roger Corman creation. John Sayles authored the screenplay and co-authored the story on which it was based.

1980 (PG) 105m/C Richard Thomas, Robert Vaughn, George Peppard, Sybil Danning, Sam Jaffe, John Saxon, Darlanne Fluegel; **Dir:** Jimmy T. Murakami. **VHS, Beta, LV $19.98** VES, LIV 🎬🎬½

Battle Beyond the Sun
Former Russian movie "Nebo Zowet" is Americanized. Everyone is trying to send a mission to Mars. Roger Corman was the producer, Thomas Colchart is a pseudonym for Francis Ford Coppola.
1963 75m/C Edd Perry, Arla Powell, Bruce Hunter, Andy Stewart; **Dir:** Francis Ford Coppola. **VHS $16.95** NOS, SNC 🎬🎬

Battle of Blood Island
Two G.I.s, one Christian and one Jewish, face death at the hands of the Japanese during World War II. They bicker incessantly, and finally pull together to save themselves.
1960 64m/B Richard Devon, Ron Kennedy; **Dir:** Joel Rapp. **VHS, Beta $16.95** NOS, SNC 🎬

Battle of the Bombs
A collection of excerpts from the worst films of all time.
1985 60m/C VHS, Beta $39.95 RHI 🎬

The Battle of Britain
A powerful retelling of the most dramatic aerial combat battle of World War II, showing how the understaffed Royal Air Force held off the might of the German Luftwaffe.
1969 (G) 132m/C Michael Caine, Laurence Olivier, Trevor Howard, Kenneth More, Christopher Plummer, Robert Shaw, Susannah York, Ralph Richardson, Curt Jurgens, Michael Redgrave, Nigel Patrick, Edward Fox; **Dir:** Guy Hamilton. **VHS, Beta, LV $29.95** MGM, FOX 🎬🎬½

Battle of the Bulge
A recreation of the famous offensive by Nazi Panzer troops on the Belgian front during 1944-45, an assault that could have changed the course of World War II.
1965 141m/C Henry Fonda, Robert Shaw, Robert Ryan, Dana Andrews, Pier Angeli; **Dir:** Ken Annakin. **VHS, Beta, LV $59.95** WAR 🎬🎬

Battle of the Bullies
A made-for-television special about an unsalvageable nerd who plots a high-tech revenge upon a slew of high school bullies.
1985 45m/C Manny Jacobs, Christopher Barnes, Sarah Inglis. **VHS, Beta $19.95** NWV Woof!

Battle Circus
Sentimental drama with Bogart portraying a surgeon working at a MASH unit in the Korean War. Allyson plays a combat nurse who comes to Korea to serve at Bogart's hospital and ends up finding love amongst the harsh reality of a war zone. Bogart was badly miscast and his performance proves it. A weak script and uninspired performances don't help this depressing story.
1953 90m/B Humphrey Bogart, June Allyson, Keenan Wynn, Robert Keith, William Campbell; **W/Dir:** Richard Brooks. **VHS $19.98** MGM 🎬½

Battle of the Commandos
A tough Army colonel leads a group of convicts on a dangerous mission to destroy a German-built cannon before it's used against the Allied Forces.
1971 94m/C IT Jack Palance, Curt Jurgens, Thomas Hunter, Robert Hunter; **Dir:** Umberto Lenzi. **VHS $39.98** REP 🎬

Battle Cry
A group of U.S. Marines train, romance, and enter battle in World War II. Based on the novel by Leon Uris. Part of the "A Night at the Movies" series, this tape simulates a 1955 movie evening, with a cartoon, "Speedy Gonzales," a newsreel, and coming attractions for "Mr. Roberts" and "East of Eden."

1955 169m/C Van Heflin, Aldo Ray, Mona Freeman, Tab Hunter, Dorothy Malone, Anne Francis, James Whitmore, Raymond Massey; **Dir:** Raoul Walsh. **VHS, Beta, LV $69.95** WAR, PIA 🎬🎬🎬

Battle of the Eagles
Follows the true adventures of the "Partisan Squadron," the courageous airmen known as the "Knights of the Sky" during World War II in Yugoslavia.
1981 102m/C Bekim Fehmiu, George Taylor, Gloria Samara, Frank Phillips; **Dir:** Tom Raymonth. **VHS, Beta** PGN 🎬

The Battle of El Alamein
An action filled movie about the alliance of Italy and Germany in a war against the British, set in a North African desert in the year 1942.
1968 105m/C IT FR Frederick Stafford, Ettore Manni, Robert Hossein, Michael Rennie, George Hilton, Ira Furstenberg; **Dir:** Calvin Jackson Padget. **VHS, Beta $59.95** PSM, MRV, GEM 🎬🎬

Battle of Elderbush Gulch
An ancient pioneering western short famous for innocently employing the now-established cliches of bad guy, good guy, and helpless frontier heroine. One of Gish's first films.
1913 22m/B Lillian Gish; **Dir:** D.W. Griffith. **VHS, Beta $29.95** VCN, PYR 🎬🎬

Battle Hell
The true story of how a British ship was attacked by the Chinese Peoples Liberation Army on the Yangtze River in 1949.
1956 112m/B GB Richard Todd, Akim Tamiroff, Keye Luke; **Dir:** Michael Anderson Sr. **VHS, Beta $19.98** VID 🎬🎬

Battle of the Japan Sea
A Japanese epic, dubbed in English, centering around the historic battle.
197? 120m/C JP Toshiro Mifune. **VHS, Beta $69.95** VCD 🎬🎬

Battle of Neretva
During WWII, Yugoslav partisans are facing German and Italian troops and local Chetniks as they battle for freedom. Big budget war film lost continuity with U.S. cut.
1971 106m/C YU Yul Brynner, Curt Jurgens, Orson Welles, Hardy Kruger, Franco Nero, Sergei Bondarchuk; **Dir:** Veljko Bulajic. **VHS $39.98** TAM, REP 🎬🎬

Battle for the Planet of the Apes
A tribe of human atomic bomb mutations are out to make life miserable for the peaceful ape tribe in the year 2670 A.D. Final chapter in the five-movie simian saga.
1973 (G) 96m/C Roddy McDowall, Lew Ayres, John Huston, Paul Williams, Claude Akins, Severn Darden, Natalie Trundy; **Dir:** J. Lee Thompson. **VHS, Beta $19.98** FOX 🎬🎬

The Battle of the Sexes
Sophisticated British comedy has mild-mannered Sellers trying to prevent a business takeover by the brash American Cummings. A good supporting cast and the impeccable Sellers make this one unique. Adapted from the James Thurber short story "The Catbird Seat."
1960 88m/B GB Peter Sellers, Robert Morley, Constance Cummings, James B. Clark; **Dir:** Charles Crichton. **VHS, Beta** CGI, MRV 🎬🎬🎬

Battle Shock
A painter is accused of murdering a cantina waitress while on his honeymoon in Mexico. Originally titled "A Woman's Devotion."
1956 88m/C Ralph Meeker, Janice Rule, Paul Henreid; **Dir:** Paul Henreid. **VHS $39.95** REP 🎬½

Battle of Valiant
Thundering hordes of invading barbarians trample the splendor of ancient Rome beneath their grimy sandals.
1963 90m/C Gordon Mitchell, Ursula Davis, Max Serato; **Dir:** John Gentil. **VHS, Beta $69.95** FOR 🎬

Battle of the Worlds
Typical low-budget science fiction. A scientist tries to stop an alien planet from destroying the Earth. Even an aging Rains can't help this one. Poorly dubbed in English.
1961 84m/C IT Claude Rains, Maya Brent, Bill Carter, Marina Orsini, Jacqueline Derval; **Dir:** Anthony M. Dawson. **VHS, CV $16.95** SNC, MRV 🎬

Battleforce
Exciting battle scenes lose their power in the confusion of this mixed-up World War II film about Rommel's last days. Dubbed sequences, news-reel vignettes and surprise performances by big name stars are incomprehensibly glued together. Also on video as "Battle of the Mareth Line."
1978 97m/C GE YU Henry Fonda, Stacy Keach, Helmut Berger, Samantha Eggar, Giuliano Gemma, John Huston; **Dir:** Umberto Lenzi; **Nar:** Orson Welles. **VHS $9.99** CON, NSV, PLV 🎬

Battleground
A tightly-conceived post-World War II character drama, following a platoon of American soldiers through the Battle of the Bulge. Available in a Colorized version. Oscar nominations for Best Picture, Supporting Actor (Whitmore), Wellman, Film Editing.
1949 118m/B Van Johnson, John Hodiak, James Whitmore, George Murphy, Ricardo Montalban, Marshall Thompson, Jerome Courtland, Don Taylor, Bruce Cowling, Leon Ames, Douglas Fowley, Richard Jaeckel; **Dir:** William A. Wellman. Academy Awards '49: Best Black and White Cinematography, Best Story & Screenplay. **VHS, Beta $19.98** MGM 🎬🎬🎬

The Battleship Potemkin
Eisenstein's masterpiece documents mutiny aboard the Russian battleship "Potemkin" in 1905. Exceptionally interesting and beautiful cinematic compositions, especially the Odessa sequence.
1925 71m/B RU Alexander Antonov, Grigori Alexandrov, Vladimir Barsky, Mikhail Gomorov; **Dir:** Sergei Eisenstein. Sight & Sound Survey '52: #3 of the Best Films of All Time; Cinematheque Belgique '52: #1 of the Best Films of All Time; Brussels World's Fair '58: #1 of the Best Films of All Time; Sight & Sound Survey '62: #6 of the Best Films of All Time; Sight & Sound Survey '72: #3 of the Best Films of All Time; Sight & Sound Survey '82: #6 of the Best Films of All Time. **VHS, Beta, LV $19.98** REP, MRV, NOS 🎬🎬🎬

Battlestar Galactica
The pilot episode of the sci-fi television series which was later released in the theaters. The crew of the spaceship Galactica must battle their robot enemies in an attempt to reach Earth. Contains special effects designed by John "Star Wars" Dykstra.
1978 (PG) 125m/C Lorne Greene, Dirk Benedict, Karen Jensen, Jane Seymour, Patrick Macnee, Terry Carter, John Colicos, Richard A. Colla, Laurette Spang, Richard Hatch; **Dir:** Richard A. Colla. **VHS, Beta, LV $19.95** MCA 🎬½

Battling Amazons
Spandex-bedecked women fight, kick-boxing style, though not very well.
1981 60m/C VHS, Beta $19.95 AHV 🎬½

Battling with Buffalo Bill
Twelve episodes of the vintage serial concerning the exploits of the legendary Indian fighter.
1931 180m/B Tom Tyler, Rex Bell, Franklin Farnum. **VHS, Beta $29.95** VCN 🎬½

B

Battling Bunyon A wily youngster becomes a comedy boxer for profit, and eventually gets fed up and battles the champ. Silent.
1924 71m/B Chester Conklin, Wesley Barry, Molly Malone, Jackie Fields. **VHS, Beta $29.95** VYY ⅔ ½

Battling Butler Rich young Keaton tries to impress a young lady by impersonating a boxer. All goes well until he has to fight the real thing. Mostly charming if uneven; one of Keaton's more unusual efforts, thought to be somewhat autobiographical. Silent.
1926 70m/B Buster Keaton, Sally O'Neil, Snitz Edwards, Francis McDonald, Mary O'Brien, Tom Wilson, Walter James; *Dir:* Buster Keaton. **VHS $49.95** EJB ⅔ ½

Battling Marshal Fast-moving action western starring Sunset Carson.
1948 52m/B Sunset Carson, Lee Roberts; *Dir:* Oliver Drake. **VHS, Beta $29.95** VCN ⅔

Battling Outlaws Western featuring the American cowboy Bob Steele.
194? 64m/B Bob Steele. **VHS, Beta $19.95** NOS, VCN ⅔

The Bawdy Adventures of Tom Jones An exploitive extension of the Fielding novel about the philandering English lad, with plenty of soft-core skin and lewdness.
1976 (R) 89m/C GB Joan Collins, Trevor Howard, Terry-Thomas, Arthur Lowe, Murray Melvin; *Dir:* Cliff Owen. **VHS, Beta $59.95** MCA ⅔ ½

Baxter A bull terrier lives his life with three different sets of masters. He examines all of humankind's worst faults and the viewer quickly realizes that Baxter's life depends on his refusal to obey like a good dog should. Based on the novel by Ken Greenhall. Funny, sometimes erotic, quirky comedy. In French with English subtitles.
1989 82m/C FR Lisa Delamare, Jean Mercure, Jacques Spiesser, Catherine Ferran, Jean-Paul Roussillon, Sabrina Leurquin; *Dir:* Jerome Boivin. **VHS $89.95** FXL ⅔⅔ ½

Bay of Blood Four vacationing youngsters are relentlessly pursued by a homicidal maniac bent on decapitation. Known also as "Last House on the Left Part II," "Carnage," or "Twitch of the Death Nerve." Supposedly the inspiration for "Friday the 13th;" very violent and gory.
1971 (R) 85m/C IT Claudine Auger, Isa Miranda, Claudio Volonte, Chris Avran, Luigi Pistilli; *Dir:* Mario Bava. **VHS, Beta** MPI ⅔⅔

The Bay Boy Set in the 1930s in Nova Scotia, this period piece captures the coming-of-age of a rural teenage boy. Young Sutherland's adolescent angst becomes a more difficult struggle when he witnesses a murder, and is tormented by the secret.
1985 (R) 107m/C CA Liv Ullman, Kiefer Sutherland, Peter Donat, Matthieu Carriere, Joe MacPherson, Isabelle Mejias, Alan Scarfe, Chris Wiggins, Leah K. Pinsent; *Dir:* Daniel Petrie. Genie Awards '85: Best Art Direction/Set Decoration, Best Costume Design, Best Picture, Best Screenplay, Best Supporting Actor (Scarfe). **VHS, Beta, LV $79.98** ORI ⅔⅔ ½

Bayou Romance A painter inherits a Louisiana plantation, moves in, and falls in love with a young gypsy.
1986 90m/C Louis Jourdan; *Dir:* Alan Myerson. **Beta $14.95** PSM ⅔

B.C. Rock A cave man learns how to defend himself and has some prehistoric fun with a tribe of female amazons. Animated.
1984 (R) 87m/C **VHS, Beta $69.98** LIV, VES ⅔⅔

B.C.: The First Thanksgiving The caveman B.C. and his friends are trying to find a turkey to flavor their rock soup in this animated featurette.
1984 25m/C **VHS, Beta, LV $14.98** SUE ⅔⅔

Be My Valentine, Charlie Brown The Peanuts gang endures the follies of the heart on St. Valentine's Day. Animated.
1975 30m/C *Dir:* Phil Roman. **VHS, Beta $9.98** MED, VTR ⅔⅔

Beach Blanket Bingo A group of sun-worshipping teens who are fascinated by skydiving and sizzling beach parties become involved in a kidnapping plot.
1965 (G) 96m/C Frankie Avalon, Annette Funicello, Linda Evans, Don Rickles, Buster Keaton, Paul Lynde, Harvey Lembeck; *Dir:* William Asher. **VHS, Beta, LV $19.99** HBO ⅔

Beach Boys: An American Band An in-depth look at the lives and music of the Beach Boys with a soundtrack that features over forty of their songs.
1985 (PG-13) 103m/C Brian Wilson, Carl Wilson, Dennis Wilson, Mike Love, Al Jardine, Bruce Johnson; *Dir:* Malcolm Leo. **VHS, Beta, LV $29.95** MVD, LIV, VES ⅔

Beach Girls Three voluptuous coeds intend to re-educate a bookish young man and the owner of a beach house.
1982 (R) 91m/C Debra Blee, Val Kline, Jeana Tomasina, Adam Roarke; *Dir:* Patrice Townsend. **VHS, Beta, LV $39.95** PAR Woof!

The Beach Girls and the Monster Here's one on the cutting edge of genre bending: while it meticulously maintains the philosophical depth and production values of sixties beach bimbo fare, it manages to graft successfuly with the heinous critter from the sea genre to produce a hybrid horror with acres o' flesh. All that, and music by Frank Sinatra, Jr.
1965 70m/B Jon Hall, Sue Casey, Walker Edmiston; *Dir:* Jon Hall. **VHS $19.98** SNC ⅔

Beach House In this boring film, adolescents frolic on the beach, get inebriated and listen to rock 'n' roll.
1982 76m/C Kathy McNeil, Richard Duggan, Ileana Seidel, John Cosola, Spence Waugh, Paul Anderson, John A. Gallagher. **VHS, Beta $59.99** HBO ⅔ ½

Beach Party A scientist studying the mating habits of teenagers intrudes on a group of surfers, beach bums and motorcyclists.
1963 101m/C Frankie Avalon, Annette Funicello, Harvey Lembeck, Robert Cummings, Dorothy Malone; *Dir:* William Asher. **VHS, Beta $19.98** NO ⅔ ½

Beachballs From band to beach to dreamgirl's bed, this film chronicles the aspirations of a beach guy.
1988 (R) 79m/C Phillip Paley, Heidi Helmer. **VHS, Beta $79.95** MED Woof!

Beachcomber Comedy set in the Dutch East Indies about a shiftless beachcomber (Laughton) who falls in love with a missionary's prim sister (Lanchester), as she attempts to reform him. The real-life couple of Laughton and Lanchester are their usual pleasure to watch. First released as "Vessel of Wrath." Remade in 1954. Story by W. Somerset Maugham.
1938 88m/B GB Charles Laughton, Elsa Lanchester, Robert Newton, Tyrone Guthrie; *Dir:* Erich Pommer. **VHS, Beta $19.95** NOS, MRV, DVT ⅔⅔⅔

Beaches Based on the novel by Iris Rainer Dart about two girls whose friendship survived the test of time. The friendship is renewed once more when one of the now middle-aged women learns that she is dying slowly of a fatal disease.
1988 (PG-13) 123m/C Bette Midler, Barbara Hershey, John Heard, Spalding Gray, Lainie Kazan, James Read; *Dir:* Garry Marshall. **VHS, Beta, LV, 8mm $19.99** TOU, JCF ⅔⅔⅔ ½

Beaks: The Movie Two television reporters try to figure out why birds of prey are suddenly attacking humans. Owes nothing to Hitchcock's "The Birds." Alternate title: "Birds of Prey."
1987 (R) 86m/C Christopher Atkins, Michelle Johnson; *Dir:* Rene Cardona Jr. **VHS, Beta $14.95** IVE Woof!

Beamship Meier Chronicles This UFO documentary focuses on Eduard Meier, a Swiss everyman around whom the most well-documented UFO controversy revolves. Supposedly, alien beings from somewhere deep in the Pleiades contacted earth: Meier has photographs, film footage, tape recordings, and unique metal fragments, as well as collaborative testimony from individuals who claim to have seen the Beamships.
1990 90m/C **VHS, Beta $59.95** ESP, WSH Woof!

Beamship Metal Analysis An in-depth, scientific analysis of the metal fragments said to have been given to Eduard Meier by the Beamship aliens. The metal has such amazing qualities that top national scientists from IBM have concluded that, considering man's present technological limitations, the metal could not have been made on Earth. Well, okay.
1990 45m/C **VHS, Beta $39.95** ESP, WSH Woof!

Beamship Movie Footage Eduard Meier offers photographic proof of alien visitation in the form of 8mm film clips, in slow motion and freeze-frame, of the Pleiadians' ships. Touch football scene is compelling.
1990 60m/C **VHS, Beta $49.95** ESP, WSH Woof!

The Bear A breathtaking, effortlessly entertaining family film about an orphaned bear cub tagging after a grown male and dealing with hunters. The narrative is essentially from the cub's point of view, with very little dialogue. A huge money-maker in Europe. Based on a novel by James Oliver Curwood.
1989 (PG) 89m/C FR Jack Wallace, Tcheky Karyo, Andre Lacombe; *Dir:* Jean-Jacques Annaud. **VHS, Beta, LV $19.95** COL, RDG ⅔⅔⅔

Bear/All the Troubles of the World Classic stories from William Faulkner and Isaac Asimov, for younger viewers.
1987 60m/C **VHS $19.95** KUI ⅔⅔⅔

Bear Island A group of secret agents disguising themselves as U.N. weather researchers converge upon Bear Island in search of a Nazi U-Boat.
1980 (PG) 118m/C Donald Sutherland, Richard Widmark, Barbara Parkins, Vanessa Redgrave, Christopher Lee, Lloyd Bridges; *Dir:* Don Sharp. **VHS, Beta $19.95** STE ⅔ ½

The Bears & I A young Vietnam vet helps Indians regain their land rights while raising three bear cubs. Beautiful photography in this Disney production.
1974 (G) 89m/C Patrick Wayne, Chief Dan George, Andrew Duggan, Michael Ansara; *Dir:* Bernard McEveety. **VHS $69.95** DIS ⅔⅔

The Beast A violent war drama about a wild Russian tank officer who becomes lost in the Afghan wilderness.
1988 (R) 93m/C George Dzundza, Jason Patric, Steven Bauer, Stephen Baldwin; *Dir:* Kevin Reynolds. **VHS, Beta $89.95** *COL* 𝄢½

The Beast from 20,000 Fathoms Atomic testing defrosts a giant dinosaur in the Arctic; the hungry monster proceeds onwards to its former breeding grounds, now New York City. Oft-imitated saurian-on-the-loose formula is still fun, brought to life by Ray Harryhausen special effects. Based loosely on the Ray Bradbury story "The Foghorn."
1953 80m/B Paul Christian, Paula Raymond, Cecil Kellaway, Kenneth Tobey, Donald Woods, Lee Van Cleef, Steve Brodie, Mary Hill; *Dir:* Eugene Lourie. **VHS, Beta, LV $59.95** *WAR, PIA, FCT* 𝄢𝄢½

Beast in the Cellar Every family has something to hide, and in the case of two spinster sisters, it's their murderous inhuman brother, whom they keep chained in the cellar. Like all brothers, however, the "beast" rebels against his sisters' bossiness and escapes to terrorize their peaceful English countryside. The sisters' (Reid and Robson) performances aren't bad, but it's this disappointing movie that should be locked away in the basement.
1970 (R) 85m/C *GB* Beryl Reid, Flora Robson, T.P. McKenna; *Dir:* James Kelly. **VHS, Beta $59.95** *PGN* 𝄢½

The Beast from Haunted Cave Gold thieves hiding in a wilderness cabin encounter a spiderlike monster. Surprisingly good performances from Sinatra (Frank's nephew) and Carol. Produced by Gene Corman, Roger's brother.
1960 64m/B Michael Forest, Sheila Carol, Frank Wolff, Richard Sinatra, Wally Campo; *Dir:* Monte Hellman. **VHS $19.98** *DVT, MRV, SNC* 𝄢𝄢

The Beast Within A young woman is raped by an unseen creature in a Mississippi swamp. Seventeen years later, her son begins to act strangely, forcing a return to the rape scene. The first film to use the air "bladder"-type of prosthetic make-up popularized in later, and generally better, horror films. Contains some choice cuts in photo editing: the juxtaposition of hamburger and human "dead meat" is witty. Based on Edward Levy's 1981 novel.
1982 (R) 98m/C Ronny Cox, Bibi Besch, L.Q. Jones, Paul Clemens, Don Gordon; *Dir:* Philippe Mora. **VHS, Beta $69.95** *MGM* 𝄢½

Beast of the Yellow Night A dying soldier sells his soul to Satan at the close of World War II. Years later, existing without aging, he periodically turns into a cannibal monster. Although the first half is tedious, the monster turns things around when he finally shows up. Decent gore effects.
1970 (R) 87m/C *PH* John Ashley, Mary Wilcox, Eddie Garcia, Vic Diaz; *Dir:* Eddie Romero. **VHS, Beta $19.95** *NOS, UHV* Woof!

The Beast of Yucca Flats A really cheap, quasi-nuclear protest film. A Russian scientist is chased by communist agents into a nuclear testing area and is caught in an atomic blast. As a result, he turns into a club-weilding monster. Voice over narration is used in lieu of dialogue as that process proved too expensive.
1961 53m/B Tor Johnson, Douglas Mellor, Larry Aten, Barbara Francis; *W/Dir:* Coleman Francis. **VHS $12.95** *SNC, CNM, MLB* Woof!

Beastmaster Adventure set in a wild and primitive world. The Beastmaster is involved in a life-and-death struggle with overwhelming forces of evil. Campy neanderthal flesh flick.
1982 (PG) 119m/C Marc Singer, Tanya Roberts, Rip Torn, John Amos, Josh Milrad, Billy Jacoby; *Dir:* Don A. Coscarelli. **VHS, Beta, LV $19.99** *MGM* 𝄢

Beastmaster 2: Through the Portal of Time This time the laughs are intentional as the Beastmaster follows an evil monarch through a dimensional gate to modern-day L.A., where the shopping is better for both trendy clothes and weapons. Fun for genre fans, with a behind-the-scenes featurette on the tape.
1991 (PG-13) 107m/C Marc Singer, Kari Wuhrer, Sarah Douglas, Wings Hauser; *Dir:* Sylvio Tabet. **VHS, LV $92.98** *REP* 𝄢½

Beasts A young couple's plans for a romantic weekend in the Rockies are slightly changed when the pair are savagely attacked by wild beasts.
1983 92m/C Tom Babson, Kathy Christopher, Vern Potter. **VHS, Beta $49.95** *PSM* 𝄢

The Beat Unrealistic film about a bookish new kid who intercedes in the tension between two rival street gangs, changing their lives in his literary way.
1988 101m/C John Savage, Kara Glover, Paul Dillon, David Jacobson, William McNamara; *Dir:* Peter Mones. **VHS, Beta, LV $79.98** *LIV, VES* 𝄢

Beat the Devil Each person on a slow boat to Africa has a scheme to beat the other passengers to the uranium-rich land that they all hope to claim. An unusual black comedy which didn't fare well when released, but over the years has come to be the epitome of spy-spoofs.
1953 89m/C Humphrey Bogart, Gina Lollobrigida, Peter Lorre, Robert Morley, Jennifer Jones; *Dir:* John Huston. **VHS, Beta, LV $9.95** *NEG, MRV, NOS* 𝄢𝄢𝄢

Beat Girl An architect's rebellious teenage daughter sinks to the bottom and gets involved in murder along the way. Originally titled, "Wild for Kicks."
1960 85m/B *GB* David Farrar, Noelle Adam, Christopher Lee, Gillian Hills, Shirley Anne Field, Oliver Reed; *Dir:* Edmond T. Greville. **VHS $29.95** *SNC, FCT, TIM* 𝄢

Beat Street This quasi-documentary about break dancing features the music of Ruben Blades, Afrika Bambaata and the Soul Sonic Force, and Grand Master Melle Mel and the Furious Five.
1984 (PG) 106m/C Rae Dawn Chong, Leon Grant, Saundra Santiago, Robert Taylor, Guy Davis, Jon Chardiet, Duane Jones, Kadeem Hardison; *Dir:* Stan Lathan. **VHS, Beta, LV $79.98** *VES, LIV* 𝄢𝄢½

Beatlemania! The Movie Boring look at the Fab Four. Not the real Beatles, but impersonators who do a very inadequate job. Based on the equally disappointing stage show. Alternate title: "Beatlemania: The Movie."
1981 86m/C Mitch Weissman, Ralph Castelli, David Leon, Tom Teeley; *Dir:* Joseph Manduke. **VHS, Beta, LV** *USA, IME* 𝄢

Beatrice In France during the Middle Ages, a barbaric soldier of the Hundred Years' War returns to his estate that his daughter has maintained, only to brutalize and abuse her. In French with English subtitles.

1988 (R) 132m/C *FR* Julie Delphy, Barnard Pierre Donnadieu, Nils Tavernier; *Dir:* Bertrand Tavernier. **VHS, Beta, LV $79.95** *VIR* 𝄢𝄢½

Beau Brummel The famous silent adaptation of the Clyde Fitch play about an ambitious English dandy's rise and fall.
1924 80m/B John Barrymore, Mary Astor, Willard Louis, Irene Rich, Carmen Myers, Alec B. Francis, William Humphreys; *Dir:* Harry Beaumont. **VHS, Beta $19.95** *NOS, HHT, VYY* 𝄢𝄢½

Beau Geste The classic Hollywood adventure film based on the Percival Christopher Wren story about three brothers who join the Foreign Legion after nobly claiming responsibility for a jewel theft they didn't commit. Once enlisted, they must face desert wars and the most despicable commanding officer ever, before clearing their family name. A rousing, much-copied epic. Oscar nominations: Best Supporting Actor (Donlevy), Interior Decoration.
1939 114m/B Gary Cooper, Ray Milland, Robert Preston, Brian Donlevy, Donald O'Connor, J. Carroll Naish, Susan Hayward, James Stephenson; *Dir:* William A. Wellman. **VHS, Beta $29.95** *MCA* 𝄢𝄢𝄢½

Beau Pere Bittersweet satiric romp from Blier about the war zone of modern romance, wherein a 30-year-old widower falls madly in love with his 14-year-old stepdaughter. Sharp-edged and daring. In French with subtitles.
1981 125m/C *FR* Patrick Dewaere, Nathalie Baye, Ariel Besse, Maurice Ronet; *Dir:* Bertrand Blier. **VHS, Beta $59.95** *MED, TAM* 𝄢𝄢𝄢

Beau Revel Critically acclaimed romantic drama of the silent era. A passionate dancing girl played by Vidor, one of the twenties' more prolific romantic leads (and erstwhile wife of director King Vidor) is the object of romantic interest of a father and son, which leaves the threesome lovelorn, suicidal, and emotionally scarred. You might recognize Stone from his later role as Judge Hardy in the MGM "Hardy Family" series.
1921 70m/B Lewis Stone, Florence Vidor, Lloyd Hughes, Katherine Kirkham, William Conklin; *Dir:* John Wray. **VHS, Beta $24.95** *GPV, FCT* 𝄢𝄢𝄢½

Beauties and the Beast Young nymphomaniacs seek the company of an unlikely suitor—a giant, lonely ape-creature. The real predators, however, are revealed in this comedy.
19?? 83m/C **VHS $39.95** *ACF* 𝄢𝄢𝄢½

The Beautiful Blonde from Bashful Bend Charming comedy-western gets better with age. Grable is the pistol packing mama mistaken for the new school teacher. Fun performances by all, especially Herbert.
1949 77m/C Betty Grable, Cesar Romero, Rudy Vallee, Olga San Juan, Hugh Herbert, Porter Hall, Sterling Holloway, El Brendel; *Dir:* Preston Sturges. **VHS, LV $19.98** *FOX* 𝄢𝄢½

Beauty for the Asking It sounds like a workable Lucille Ball vehicle—a beautician develops a bestselling skin cream—and even the title sounds like something you'd like to love Lucy in. But somehow the numerous plot implausibilities managed to get by the story's five writers, and the idea of a jilted woman making millions while being financed by her ex's wife just doesn't fly (perhaps it was an idea ahead of its time). Die-hard Lucy fans may find this interesting.
1939 68m/B Lucille Ball, Patric Knowles, Donald Woods, Frieda Inescort, Frances Mercer; *Dir:* B.P. Fineman. **VHS $19.98** *TTC, FCT* Woof!

Beauty on the Beach A comedy about a mad psychologist, his bizarre experiments with women, and his eventual descent into insanity.
196? 90m/C Valeria Fabrizi. **VHS, Beta $49.95** *INV ⅆ*

Beauty and the Beast An ethereal, original film based on the classic medieval fairy tale, and the definitive version. The story's themes and famous set-pieces are used by Cocteau at his most cohesive in creating a captivating, quasi-surreal paean to romantic love. In French with subtitles.
1946 90m/B *FR* Jean Marais, Josette Day, Marcel Andre; *Dir:* Jean Cocteau. **VHS, Beta, LV $19.95** *TAM, SUE, MLB ⅆⅆⅆⅆ*

Beauty and the Beast From "Faerie Tale Theatre" comes the story of a Beauty who befriends a Beast and learns a lesson about physical appearance and true love.
1983 60m/C Susan Sarandon, Klaus Kinski; *Dir:* Roger Vadim. **VHS, Beta, LV $14.98** *KUI, FOX, FCT ⅆⅆⅆⅆ*

Beauty and the Beast Includes the pilot episode of the fantasy television series, "Once Upon a Time in New York" and the episode "A Happy Life." A young lawyer who has been attacked and left for dead awakens in a society of outcasts who live under the New York City subway system. She has been rescued and befriended by one of them, a charismatic man/beast named Vincent.
1987 100m/C Linda Hamilton, Ron Perlman. **VHS, Beta, LV $14.98** *REP ⅆⅆⅆⅆ*

Beauty and the Beast Beautiful Belle lives with her eccentric inventor father in a small village, much preferring the pleasures of reading to the unwelcome attentions of muscle-bound bully Gaston, who wants to marry her. When her father becomes lost in the woods he finds himself at a neglected castle and becomes the prisoner of the fearsome Beast. The Beast was once a handsome prince and now is under a terrible curse and if he doesn't find true love soon he will be doomed to be a beast forever. When Belle arrives at the castle she agrees to be the Beast's companion if only he will let her father go. The castle servants, who have all been turned into talking, singing, dancing household objects, delight in Belle's arrival and try to turn their grisly master into a civilized object of Belle's affections. But don't forget Gaston! He decides to "rescue" Belle and kill the Beast! Features the award-ing-winning title song, as well as "Belle," "Something There," and "Be Our Guest." Destined to become a Disney classic, beloved of generations of children. A deluxe video version features a second video that includes a work-in-progress rough film cut, a compact disc of the soundtrack, a lithograph depicting a scene from the film, and an illustrated book.
1991 (G) 84m/C *Dir:* Kirk Wise, Gary Trousdale; *Voices:* Paige O'Hara, Robby Benson, Angela Lansbury, Jerry Orbach, David Ogden Stiers, Richard White, Rex Everhart, Jo Anne Worley. Academy Awards '91: Best Song ("Beauty and the Beast"), Best Musical Score; Golden Globe Awards '92: Best Film—Musical/Comedy; National Board of Review Awards '91: Special Animation Award. **VHS $24.99** *DIS ⅆⅆⅆ½*

Beauty and the Beast: Though Lovers Be Lost The love affair of Vincent and Catherine takes a tragic turn when she is kidnapped by a crime lord. Feature-length drama from the television series.

1989 90m/C Linda Hamilton, Ron Perlman. **VHS, LV $19.98** *REP ⅆⅆⅆ½*

Because of the Cats A police inspector uncovers an evil cult within his seaside village while investigating a bizarre rape and burglary.
1974 90m/C Bryan Marshall, Alexandra Stewart, Alex Van Rooyen, Sylvia Kristel, Sebastian Graham Jones; *Dir:* Fons Rademakers. **VHS, Beta $49.95** *PSM ⅆ½*

Becket The adaptation of Jean Anouilh's play about the tumultuous friendship between Henry II of England and the Archbishop of Canterbury Thomas Becket. Becket views his position in the church of little relation to the sexual and emotional needs of a man, until he becomes archbishop. His growing concern for religion and his shrinking need of Henry as friend and confidant eventually cause the demise of the friendship and the resulting tragedy. Flawless acting from every cast member, and finely detailed artistic direction make up for the occasional slow moment. Oscar nominations for Best Picture, Burton, Geilgud, director, Adapted Screenplay, Cinematography, Art Direction, Set Decoration, Sound, and Costume Design.
1964 148m/C Richard Burton, Peter O'Toole, John Gielgud, Donald Wolfit; *Dir:* Peter Glenville. Academy Awards '64: Best Adapted Screenplay; British Academy Awards '64: Best Art Direction/Set Decoration (Color), Best Costume Design (Color); Golden Globe Awards '65: Best Film—Drama; National Board of Review Awards '64: Best Film. **VHS, Beta, LV $59.95** *KUI, MPI, UHV ⅆⅆⅆ*

Becky Sharp This premiere Technicolor film tells the story of Becky Sharp, a wicked woman who finally performs one good deed. Hopkins received a Best Actress Oscar Nomination.
1935 83m/C Miriam Hopkins, Frances Dee, Cedric Hardwicke, Billie Burke, Nigel Bruce, Pat Nixon; *Dir:* Rouben Mamoulian. **VHS, Beta $19.95** *NOS, MRV, DVT ⅆⅆ*

Bed and Sofa Adultery, abortion, and women's rights are brought about by a housing shortage which forces a man to move in with a married friend. Famous, ground-breaking Russian silent.
1927 73m/B *RU* Nikolai Batalov, Vladimar Fogel; *Dir:* Abram Room. **VHS, Beta, 3/4U $49.95** *TAM, FST, IHF ⅆⅆⅆ½*

Bedazzled In the process of hanging himself, Stanley, a short-order cook is approached by the devil with an offer: seven wishes in exchange for his soul. Each of Stanley's wishes is granted with surprising consequences. Cult comedy is a sometimes uneven, but thoroughly entertaining and funny retelling of the Faustian story.
1968 (PG) 107m/C *GB* Dudley Moore, Peter Cook, Eleanor Bron, Michael Bates, Raquel Welch; *Dir:* Stanley Donen. **VHS, Beta, LV** *FOX ⅆⅆⅆ*

Bedford Incident The U.S.S. Bedford discovers an unidentified submarine in North Atlantic waters. The Bedford's commander drives his crew to the point of exhaustion as they find themselves the center of a fateful controversy.
1965 102m/B Richard Widmark, Sidney Poitier, James MacArthur, Martin Balsam, Wally Cox, Donald Sutherland, Eric Portman; *Dir:* James B. Harris. **VHS, Beta, LV $14.95** *COL, IME ⅆⅆ½*

Bedknobs and Broomsticks A novice witch and three cockney waifs ride a magic bedstead and stop the Nazis from invading England during World War II. Celebrated for its animated art.
1971 (G) 117m/C Angela Lansbury, Roddy McDowall, David Tomlinson, Bruce Forsyth, Sam Jaffe; *Dir:* Robert Stevenson. Academy Awards '71: Best Visual Effects. **VHS, Beta, LV $24.99** *DIS ⅆⅆ½*

Bedlam Val Lewton produced this creeper set in the famed asylum in 18th-century London. A woman, wrongfully committed, tries to stop the evil doings of the chief (Karloff) of Bedlam, and endangers herself. A fine horror film co-written by Lewton.
1945 79m/B Jason Robards Sr., Ian Wolfe, Glenn Vernon, Boris Karloff, Leyland Hodgson, Billy House, Richard Fraser; *Dir:* Mark Robson. **VHS, Beta, LV $19.95** *TTC, MED, FCT ⅆⅆⅆ*

Bedroom Eyes A successful businessman becomes a voyeur by returning nightly to a beautiful woman's window, until she is killed and he is the prime suspect. Part comic, part disappointing thriller.
1986 (R) 90m/C Kenneth Gilman, Dayle Haddon, Christine Cattall; *Dir:* William Fruet. **VHS, Beta, LV $29.95** *FOX ⅆⅆ*

Bedroom Eyes 2 After discovering his wife has had an affair, a stockbroker takes a lover. When she turns up dead, he and his wife become murder suspects. Provides some suspenseful moments.
1989 (R) 85m/C Wings Hauser, Kathy Shower, Linda Blair, Jane Hamilton, Jennifer Delora; *Dir:* Chuck Vincent. **VHS, LV $89.95** *VMK, IME ⅆⅆ*

The Bedroom Window During an illicit liaison with her husband's co-worker (Guttenberg), a beautiful woman (Huppert) witnesses an assault on another woman (McGovern) from his bedroom window. To keep the affair secret, he reports it as his own account. Since it is secondhand, his story is flawed, and he becomes a suspect. Then a short distance away the body of a woman is found. Now he's in really big trouble. Only her truth or his finding of the assailant can clear him. Semi-tight thriller in the Hitchcock mode.
1987 (R) 113m/C Steve Guttenberg, Elizabeth McGovern, Isabelle Huppert, Wallace Shawn, Paul Shenar; *Dir:* Curtis Hanson. **VHS, Beta, LV $9.99** *FHE, VES, LIV ⅆⅆ½*

Bedtime for Bonzo A professor adopts a chimp to prove that environment, not hereditary, determines a child's future. Fun, lighthearted comedy that stars a future president. Followed by "Bonzo Goes to College."
1951 83m/B Ronald Reagan, Diana Lynn, Walter Slezak, Jesse White, Bonzo the Chimp; *Dir:* Fred DeCordova. **VHS, Beta $14.95** *MCA, FCT ⅆⅆ½*

Bedtime Story Two con artists attempt to fleece an apparently wealthy woman and each other on the French Riviera. Re-made in 1988 as "Dirty Rotten Scoundrels." One of Brando's deservedly few forays into comedy.
1963 99m/C Marlon Brando, David Niven, Shirley Jones, Dody Goodman, Marie Windsor; *Dir:* Ralph Levy. **VHS, Beta $59.95** *MCA ⅆⅆ*

Beehive Although it sounds like a sequel to "Hairspray," this experimental film is teeming with dancing and animation. A worker bee is accidently turned into a queen bee by a drone. Beautiful and fascinating.
1985 16m/C *W/DIR* Frank Moore, Jim Self. **VHS $29.95** *FCT, MWF ⅆⅆ*

B

Beer A female advertising executive devises a dangerous sexist campaign for a cheap beer, and both the beer and its nickname become nationwide obsessions. Not especially amusing.
1985 (R) 83m/C Loretta Swit, Rip Torn, Dick Shawn, David Alan Grier, William Russ, Kenneth Mars, Peter Michael Goetz; *Dir:* Patrick Kelly. **VHS, Beta $19.99** *HBO* 🎬

The Bees A strain of bees have ransacked South America and are threatening the rest of the world. The buzz is that no one is safe. Viewers should avoid "The Bees" as well—if it sounds interesting, catch "The Swarm" instead.
1978 (PG) 93m/C John Saxon, John Carradine, Angel Tompkins, Claudio Brook, Alicia Encinas; *Dir:* Alfredo Zarcharias. **VHS, Beta $29.98** *WAR Woof!*

Beethoven Startling biography of the musical genius, filled with opulent, impressionistic visuals.
1936 116m/B *FR* Harry Baur, Jean-Louis Barrault, Marcel Dalio. **VHS $59.95** *CVC Woof!*

Beethoven A St. Bernard puppy escapes from dognappers and wanders into the home of the Newtons, who, over dad's objections, adopt him. Beethoven grows into a huge, slobbering dog who makes all the standard doggie messes (the basis for most of the laughs in this movie). To make things complicated, two sets of villains also wreak havoc on the Newton's lives. Evil veterinarian Dr. Varnick plots to steal Beethoven for lab experiments, and yuppie couple Brad and Brie plot to take over the family business. Of course Beethoven saves the day. This film has an unoriginal plot similar to "101 Dalmations;" good for the younger set.
1992 (PG) 89m/C Charles Grodin, Bonnie Hunt, Dean Jones, Nicholle Tom, Christopher Castile, Oliver Platt, David Duchovny, Patricia Heaton, Sarah Rose Karr; *Dir:* Brian Levant. **VHS $24.98** *MCA* 🎬🎬

Beethoven Cycle Quartets No. 9 in C Major, Op. 59 and No. 11 in F Minor, Op. 95 are performed by the Guarneri String Quartet. Taped at Old Westbury Gardens.
1988 58m/C *Hosted by:* Hal Linden; *Performed by:* Guarneri String Quartet. **VHS, Beta, LV $29.95** *PRS, PIA* 🎬🎬

Beethoven Lives Upstairs In 19th century Vienna 10 year-old Christoph's life is turned upside-down when the family's eccentric new tenant turns out to be composer Ludwig Van Beethoven. In time, Christoph comes to appreciate the beauty of the music and the tragedy of the composer's deafness. Features more than 25 excerpts of Beethoven's works.
1992 52m/C Neil Munro, Illya Woloshyn, Fiona Reid, Paul Soles, Sheila McCarthy; *Dir:* David Devine. **VHS $19.98** *BMG* 🎬🎬½

Beethoven's Nephew A tepid pseudo-historical farce about Beethoven's strange obsession with his only nephew.
1988 103m/C *FR* Wolfgang Reichmann, Ditmar Prinz, Jane Birkin, Nathalie Baye; *Dir:* Paul Morrissey. **VHS, Beta, LV $9.99** *NWV, STE* 🎬½

Beetlejuice The after-life is confusing for a pair of ultra-nice novice ghosts who are faced with chasing an obnoxious family of post-modern art lovers who move into their house. Then they hear of a poltergeist who promises to rid the house of all trespassers for a price. Things go from bad to impossible when the maniacal Keaton (as the demonic "Betelgeuse") works his magic. The calypso scene is priceless. A cheesy, funny, surreal

farce of life after death with inventive set designs popping continual surprises. Ryder is striking as the misunderstood teen with a death complex, while O'Hara is hilarious as the yuppie art poseur.
1988 (PG) 92m/C Michael Keaton, Geena Davis, Alec Baldwin, Sylvia Sidney, Catherine O'Hara, Winona Ryder, Jeffrey Jones, Dick Cavett; *Dir:* Tim Burton. **VHS, Beta, LV, 8mm $19.98** *WAR, FCT, TLF* 🎬🎬🎬½

Before I Hang When a doctor invents a youth serum from the blood of a murderer, he'll stop at nothing to keep his secret. Karloff himself stands the test of time, and is satisfying as the mad scientist, giving this horror flick its appeal.
1940 60m/B Boris Karloff, Evelyn Keyes, Bruce Bennett, Edward Van Sloan; *Dir:* Nick Grinde. **VHS, Beta $59.95** *COL* 🎬🎬🎬

Before Morning Police officer poses as a blackmailer to find out which of the two women in the murder victim's life are guilty.
1933 68m/B Leo Carrillo, Lora Baxter, Taylor Holmes. **VHS, Beta, LV** *WGE* 🎬½

Before the Revolution One of Bertolucci's first films. Love and politics are mixed when a young man who is dabbling in communism is also flirting with his young aunt. Striking and powerful film that has yet to lose its effect despite the times.
1965 115m/C *IT* Francesco Barilli, Adrianna Asti, Alain Midgette, Morando Morandini, Domenico Alpi; *W/Dir:* Bernardo Bertolucci. **VHS** *NYF* 🎬½

Beggars in Ermine A handicapped, impoverished man organizes all the beggars in the world into a successful corporation. Unusual performance from Atwill.
1934 70m/B Lionel Atwill, Henry B. Walthall, Betty Furness, Jameson Thomas, James Bush, Astrid Allwyn, George "Gabby" Hayes; *Dir:* Phil Rosen. **VHS, Beta $16.95** *SNC* 🎬🎬

Beginner's Luck A young man convinces his neighbors to engage in swinging activities, leading to predictable and raunchy situations.
1984 (R) 85m/C Sam Rush, Riley Steiner, Charles Homet, Kate Talbot; *Dir:* Frank Mouris. **VHS, Beta $19.95** *STE, NWV Woof!*

Beginning of the End Produced the same year as "The Deadly Mantis," Gordon's effort adds to 1957's harvest of bugs on a rampage "B"-graders. Giant, radiation-spawned grasshoppers attack Chicago causing Graves to come to the rescue. Easily the best giant grasshopper movie ever made.
1957 73m/B Peggy Castle, Peter Graves, Morris Ankrum, Richard Benedict, James Seay; *Dir:* Bert I. Gordon. **VHS** *VTR* 🎬½

Begone Dull Care Filmmaker Norman McLaren's unusual work was created by ignoring the confines of frame division and drawing fluid lines and brilliant colors on a length of film. Restless shapes and lines cavort as music is played.
1949 9m/C **VHS, Beta** *IFB* 🎬½

The Beguiled During the Civil War a wounded Union soldier is taken in by the women at a girl's school in the South. He manages to seduce both a student and a teacher, and jealousy and revenge ensue. Decidedly weird psychological melodrama from action vets Siegel and Eastwood.
1970 (R) 109m/C Clint Eastwood, Geraldine Page, Elizabeth Hartman; *Dir:* Don Siegel. **VHS, Beta, LV $19.95** *MCA* 🎬🎬🎬

Behave Yourself! A married couple adopts a dog who may be the key for a million-dollar hijacking setup by a gang of hoodlums.
1952 81m/B Shelley Winters, Farley Granger, William Demarest, Lon Chaney Jr., Hans Conried, Elisha Cook Jr., Francis L. Sullivan; *Dir:* George Beck. **VHS, Beta $15.95** *NOS, UHV, TIM* 🎬½

Behind Enemy Lines A Special Forces soldier goes on a special mission to eliminate a possible Nazi spy. Made for television.
1985 (R) 83m/C Hal Holbrook, Ray Sharkey, David McCallum, Tom Isbell, Anne Twomey, Robert Patrick; *Dir:* Sheldon Larry. **VHS, Beta $79.95** *MED* 🎬½

Behind the Front Two friends tumble in and out of trouble in this comic Army film. One of the most profitable films of the late twenties.
1926 60m/B Wallace Beery, Raymond Hatton, Richard Arlen, Mary Brian, Chester Conklin; *Dir:* Edward Sutherland. **VHS** *GPV* 🎬🎬½

Behind Locked Doors A journalist fakes mental illness to have himself committed to an asylum, where he believes a crooked judge is in hiding. A superior "B" mystery/suspense feature, with some tense moments.
1948 61m/B Lucille Bremer, Richard Carlson, Tor Johnson, Douglas Fowley, Herbert Heyes, Ralf Harolde; *Dir:* Budd Boetticher. **VHS $19.95** *VDM* 🎬🎬½

Behind Prison Walls A steel tycoon and his son are sent to prison, and the son tries to convert his dad to socialism. A fun, light hearted film, the last for veteran Tully.
1943 64m/B Alan Baxter, Gertrude Michael, Tully Marshall, Edwin Maxwell, Jacqueline Dalya, Matt Willis; *Dir:* Steve Sekely. **VHS $16.95** *SNC* 🎬🎬

Behind the Rising Sun A Japanese publisher's political views clash with those of his American-educated son in 1930s Japan. Well done despite pre-war propaganda themes.
1943 88m/B Tom Neal, J. Carroll Naish, Robert Ryan, Mike Mazurki, Margo, Gloria Holden, Don Douglas; *Dir:* Edward Dmytryk. **VHS, Beta $19.98** *RKO* 🎬🎬½

Behold the Man! The Gospels are narrated over a performance of the final days of Jesus, with all the high points.
1950 66m/B **VHS, Beta $29.95** *VYY* 🎬🎬½

Behold a Pale Horse A Spanish police captain attempts to dupe Peck into believing that his mother is dying and he must visit her on her deathbed. Loyalist Spaniard Peck becomes privy to the plot against him, but goes to Spain anyway in this post Spanish Civil War drama.
1964 118m/B Gregory Peck, Anthony Quinn, Omar Sharif, Mildred Dunnock; *Dir:* Fred Zinneman. **VHS, Beta $9.95** *GKK* 🎬🎬½

The Being People in Idaho are terrorized by a freak who became abnormal after radiation was disposed in the local dump. Another dull monster-created-by-nuclear-waste flick. The only interest is seeing Buzzi in this plot; however, you'd be more entertained (read: amused) by Troma's "The Toxic Avenger," which takes itself much (much!) less seriously.
1983 (R) 82m/C Ruth Buzzi, Martin Landau, Jose Ferrer; *Dir:* Jackie Kong. **VHS, Beta $19.99** *HBO* 🎬

Being There A feeble-minded gardener whose entire knowledge of life comes from watching television is sent out into the real world when his employer dies. Equipped with his prize possession, his remote control unit, the gardener unwittingly enters the world of politics and is welcomed as a mysterious sage. Oscar nominated Sellers is wonderful in this satiric treat adapted by Jerzy Kosinski from his novel.
1979 (PG) 130m/C Peter Sellers, Shirley MacLaine, Melvyn Douglas, Jack Warden, Richard Dysart, Richard Basehart; *Dir:* Hal Ashby. British Academy Awards '80: Best Screenplay (Jerzy Kosinski); Los Angeles Film Critics Association Awards '79: Best Supporting Actor (Douglas); National Board of Review Awards '79: 10 Best Films of the Year, Best Actor (Sellers). **VHS, Beta, LV $19.98** *FOX, FCT, WAR* 🎏🎏🎏

Bela Lugosi Meets a Brooklyn Gorilla Two men who look like Dean Martin and Jerry Lewis get lost in the jungle, where they meet a mad scientist. Worse than it sounds. Also called "The Boys From Brooklyn."
1952 74m/B Bela Lugosi, Duke Mitchell, Sammy Petrillo; *Dir:* William Beaudine. **VHS, Beta $19.95** *NOS, SNC, AOV* 🎏

Bela Lugosi Scrapbook A compilation of Lugosi appearances, outtakes, flubs, and trailers.
197? 60m/B Bela Lugosi. **VHS, Beta $29.95** *DVT, RXM* 🎏½

Bela Lugosi: The Forgotten King An examination of the horror film star's life and career including vintage interviews, clips from a large selection of his films, stills, and much more.
1988 55m/B **VHS $19.98** *MOV* 🎏½

The Belarus File The bald-headed detective of prime-time television fame searches for a neo-Nazi who is killing Holocaust survivors in New York. Made for television.
1985 95m/C Telly Savalas, Suzanne Pleshette, Max von Sydow, George Savalas; *Dir:* Robert Markowitz. **VHS, Beta $39.95** *MCA* 🎏🎏

Belfast Assassin A British anti-terrorist agent goes undercover in Ireland to find an IRA assassin who shot a British cabinet minister.
1984 130m/C Derek Thompson, Ray Lonnen. **VHS, Beta $49.95** *PSM* 🎏

The Believers Tense horror mystery set in New York city about a series of gruesome, unexplained murders. A widowed police psychologist investigating the deaths unwittingly discovers a powerful Santeria cult that believes in the sacrifice of children. Without warning he is drawn into the circle of the "Believers" and must free himself before his own son is the next sacrifice. Gripping (and grim), unrelenting horror.
1987 (R) 114m/C Martin Sheen, Helen Shaver, Malick Bowens, Harris Yulin, Robert Loggia, Jimmy Smits, Richard Masur, Harley Cross, Elizabeth Wilson, Lee Richardson; *Dir:* John Schlesinger. **VHS, Beta, LV $19.99** *HBO* 🎏🎏½

Belizaire the Cajun 19th-century Louisiana love story. White prejudice against the Cajuns is rampant and violent, but that doesn't stop sexy faith healer Assante from falling in love with the inaccessible Cajun wife (Youngs) of a rich local. Made with care on a tight budget. Worthwhile, though uneven.

1986 (PG) 103m/C Armand Assante, Gail Youngs, Will Patton, Stephen McHattie, Michael Schoeffling, Robert Duvall; *Dir:* Glen Pitre. **VHS, Beta $79.98** *FOX* 🎏🎏½

Bell, Book and Candle A young witch makes up her mind to refrain from using her powers. When an interesting man moves into her building, she forgets her decision and enchants him with a love spell.
1959 106m/C James Stewart, Kim Novak, Jack Lemmon, Elsa Lanchester, Ernie Kovacs, Hermione Gingold; *Dir:* Richard Quine. **VHS, Beta, LV $19.95** *COL, FCT, MLB* 🎏🎏½

Bell Diamond A study by the experimental filmmaker of a Vietnam vet surviving in a bankrupt mining town. Created without a script using non-professional actors.
1981 96m/C *Dir:* Jon Jost. **VHS, Beta** *FCT* 🎏🎏

Bell from Hell A tale of insanity and revenge, wherein a young man, institutionalized since his mother's death, plots to kill his aunt and three cousins.
1974 80m/C Viveca Lindfors, Renaud Verley, Alfredo Mayo; *Dir:* Claudio Guerin Hill. **VHS, Beta $49.95** *UNI* 🎏🎏

The Bell Jar Based on poet Sylvia Plath's acclaimed semi-autobiographical novel, this is the story of a young woman who becomes the victim of mental illness. Not for the easily depressed. Rather disjointed and disappointing adaptation of Plath's work.
1979 (R) 113m/C Marilyn Hassett, Julie Harris, Barbara Barrie, Anne Bancroft, Robert Klein, Anne Jackson; *Dir:* Larry Peerce. **VHS, Beta $29.98** *KUI, VES* 🎏½

Bellamy A murderous madman massacres masseuses, and Bellamy is the cop out to get him.
1980 92m/C John Stanton, Timothy Elston, Sally Conabere; *Dir:* Colin Eggleston, Pina Amenta. **VHS, Beta** *PGN* 🎏

Bellas de Noche A Mexican pimp gets in trouble with the mob. In Spanish.
19?? 110m/C Jorge Rivero, Lalo El Mimo, Rafael Inclan. **VHS, Beta $53.95** *MAD* 🎏

The Bellboy Lewis makes his directorial debut in this plotless but clever outing. He also stars as the eponymous character, a bellboy at Miami's Fountainbleau Hotel. Cameos from Berle and Winchell are highlights.
1960 72m/B Jerry Lewis, Alex Gerry, Bob Clayton, Sonny Sands, Milton Berle; *W/Dir:* Jerry Lewis. **VHS, Beta, LV $59.95** *IVE* 🎏🎏½

The Bellboy and the Playgirls Early Coppola effort adds new film footage to a 1958 German movie. Stars Playboy playmate June "The Body" Wilkinson, with other centerfolds of the time.
1962 93m/C June Wilkinson; *W/Dir:* Francis Ford Coppola. **VHS $59.95** *SVS* Woof!

The Belle of New York A turn-of-the-century bachelor falls in love with a Salvation Army missionary in this standard musical. Songs include "Naughty But Nice," "Baby Doll," "Oops," "I Wanna Be a Dancing Man," and "Seeing's Believing."
1952 82m/C Fred Astaire, Vera-Ellen, Marjorie Main, Keenan Wynn, Alice Pearce, Gale Robbins, Clinton Sundberg; *Dir:* Charles Walters. **VHS, Beta $19.98** *MGM, FHE* 🎏🎏

Belle of San Fernando During the era of Spanish brutality in the California colonies, a Spanish spitfire and Irish immigrant fall in love.
1948 74m/B Donald Woods, Gloria Warren. **VHS, Beta, LV** *WGE* 🎏

Belle Starr Story The career of Wild West outlaw Belle Starr is chronicled.
197? 90m/C Elsa Martinelli, Robert Wood; *Dir:* John Alonzo. **VHS, Beta $39.95** *VCD* 🎏½

Belles of St. Trinian's Sim is priceless in a dual role as the eccentric headmistress of a chaotic, bankrupt girls' school and her bookie twin brother who scheme the school into financial security. The first in a series of movies based on a popular British cartoon by Ronald Searles about a girls' school and its mischievous students. Followed by "Blue Murder at St. Trinian's," "The Pure Hell of St. Trinian's," and "The Great St. Trinian's Train Robbery."
1953 86m/B *GB* Alastair Sim, Joyce Grenfell, Hermione Baddeley, George Cole, Eric Pohlmann, Renee Houston, Beryl Reid; *Dir:* Frank Launder. **VHS, Beta $39.95** *HBO* 🎏🎏

Bellhop/The Noon Whistle Two comedic shorts back to back, separately featuring the famous comics.
1922 46m/B Stan Laurel, Oliver Hardy. **VHS, Beta $49.95** *CCB* 🎏🎏🎏

Bellissima A woman living in an Italian tenement has unrealistic goals for her plain but endearing daughter when a famous director begins casting a role designed for a child. The mother's maternal fury and collision with reality highlight a poignant film. English subtitles.
1951 130m/B *IT* Anna Magnani; *Dir:* Luchino Visconti. **VHS, 3/4U, Special order formats $49.95** *CIG, APD* 🎏🎏🎏

Bellissimo: Images of the Italian Cinema A panoramic film history of Italian cinema, featuring film clips, behind-the-scenes footage and interviews with the industry's leading figures. Clips from "Open City," "Divorce Italian Style," "8 1/2," and "Seven Beauties" are included. In English and Italian with subtitles.
1987 110m/C *IT* Sophia Loren, Marcello Mastroianni, Bernardo Bertolucci, Dino de Laurentiis, Marco Ferreri, Roberto Rossellini, Vittorio De Sica, Pier Paolo Pasolini, Franco Zeffirelli, Federico Fellini, Lina Wertmuller, Monica Vitti, Giancarlo Giannini, Toto, Anna Magnani; *Dir:* Gianfranco Mingozzi. **VHS, Beta $59.95** *WBF, FCT, TAM* 🎏🎏🎏

Bellman and True Rewarding, but sometimes tedious character study of a mild mannered computer whiz who teams with a gang of bank robbers. Fine performances, especially subtle dangerous gang characters.
1988 (R) 112m/C *GB* Bernard Hill, Kieran O'Brien, Richard Hope, Frances Tomelty, Derek Newark, John Kavanagh; *Dir:* Richard Loncraine. **VHS, Beta $79.95** *VES, CAN* 🎏🎏½

Bells of Coronado Rogers and Evans team up to expose the murderer of a profitable uranium mine. A gang of smugglers trying to trade the ore to foreign powers is thwarted. The usual thin storyline, but filled with action and riding stunts.
1950 67m/B Roy Rogers, Dale Evans, Pat Brady, Grant Withers; *Dir:* William Witney. **VHS $12.98** *REP* 🎏🎏

Bells of Death A young martial arts student infiltrates the gang that killed his family and kidnapped his sister, finishing them off one by one.
1990 91m/C Chun Ping, Chiang Yi, Chao Hsin Yen; *Dir:* Yueh Fung. **VHS $69.95** *SOU* 🎏½

Bells are Ringing A girl who works for a telephone answering service can't help but take an interest in the lives of the clients, especially a playwright with an inferiority

complex. Based on Adolph Green and Betty Comden's Broadway musical. Special letter-boxed edition along with the original trailer available on laserdisc format.
1960 126m/C Judy Holliday, Dean Martin, Fred Clark, Eddie Foy Jr., Jean Stapleton; *Dir:* Vincente Minnelli. **VHS, Beta, LV $29.95** MGM, PIA *ₐₐ½*

Bells of Rosarita Roy helps foil an evil plan to swindle Evans out of the circus she inherited.
1945 54m/B Roy Rogers, Dale Evans, George "Gabby" Hayes, Sunset Carson, Adele Mara, Grant Withers, Roy Barcroft, Addison Richards; *Dir:* Frank McDonald. **VHS, Beta $19.95** NOS, MRV, VCN *ₐ½*

The Bells of St. Mary's An easy-going priest finds himself in a subtle battle of wits with the Mother Superior over how the children of St. Mary's school should be raised. It's the sequel to "Going My Way." Songs include the title tune and "Aren't You Glad You're You?" Oscar nominations: Best Picture, Best Actor (Crosby), Best Actress (Bergman), and Best Film Editing. Also available in a colorized version.
1945 126m/B Bing Crosby, Ingrid Bergman, Henry Travers; *Dir:* Leo McCarey. Academy Awards '45: Best Sound; New York Film Critics Awards '45: Best Actress (Bergman). **VHS, Beta, LV $19.98** REP, IGP *ₐₐₐ½*

Bells of San Angelo Roy foils thieves' attempts to steal a girl's inherited ranch.
1947 54m/C Roy Rogers, Dale Evans; *Dir:* William Witney. **VHS, Beta $19.95** NOS, DVT, BUR *ₐ*

Bells of San Fernando An Irish craftsman and evil Spanish ranch overseer fight for the right to marry a young girl in old California.
1947 75m/B Donald Woods, Gloria Warren, Monte Blue, Byron Foulger. **VHS, Beta, 8mm $19.95** NOS, VYY *ₐ*

The Belly of an Architect A thespian feast for the larger-than-life Dennehy as blustering American architect whose personal life and health both crumble as he obsessively readies an exhibition in Rome. A multi-tiered, carefully composed tragicomedy from the ideosyncratic filmmaker Greenaway, probably his most accessible work for general audiences.
1990 (R) 119m/C *GB IT* Brian Dennehy, Chloe Webb, Lambert Wilson; *W/Dir:* Peter Greenaway. **VHS $19.95** HMD *ₐₐₐ½*

Beloved Enemy A romantic tragedy set in Civil War-torn Ireland in the 1920s. A rebel leader and a proper English lady struggle to overcome the war's interference with their burgeoning relationship.
1936 86m/B David Niven, Merle Oberon, Brian Aherne, Karen Morley, Donald Crisp; *Dir:* H.C. Potter. **VHS, Beta $19.95** NOS, CAB *ₐₐₐ*

The Beloved Rogue Crosland—who gained a reputation for innovation, having directed "Don Juan" the previous year (noted for the debut of synchronized music), and "The Jazz Singer" the following year (noted for the first talkie bits)—mounted this well-designed and special effect-laden medieval costumer. Barrymore (whom Crosland had directed through 127 kisses in "Don Juan") is in high profile as a swashbuckling caricature of French poet Francois Villon, who battles mightily (and verbally) with Louis XI (played by Veidt, in his first US film). Marked by the excesses typical of Barrymore's prestige days, it flouts history, taking generous

poetic license when the facts ain't fab enough. In short, it's great entertainment.
1927 110m/B John Barrymore, Conrad Veidt, Marceline Day, Henry Victor, Mack Swain, Slim Summerville; *Dir:* Alan Crosland. **VHS, Beta, LV $27.95** GPV, FCT, VYG *ₐₐₐ*

Below the Belt A street-smart woman from New York City becomes part of the blue-collar "circus" of lady wrestlers.
1980 (R) 98m/C Regina Baff, Mildred Burke, John C. Becher; *Dir:* Robert Fowler. **VHS, Beta $59.99** HBO *ₐ*

Below the Border The Rough Riders go undercover to straighten out some cattle rustlers.
1942 57m/B Buck Jones, Tim McCoy, Raymond Hatton, Linda Brent, Roy Barcroft, Charles King; *Dir:* Howard Bretherton. **VHS, Beta $9.99** NOS, DVT, VDM *ₐ*

The Belstone Fox An orphaned fox goes into hiding, and is hunted by the hound he has befriended and his former owner.
1973 103m/C Eric Porter, Rachel Roberts, Jeremy Kemp; *Dir:* James Hill. **VHS, Beta** SUE *ₐ½*

Ben Sequel to "Willard" finds police Detective Kirtland on the hunt for a killer rat pack led by Ben, king of the rodents. The title song was a number 1 pop hit by Michael Jackson.
1972 (PG) 95m/C Joseph Campanella, Lee Montgomery, Arthur O'Connell, Rosemary Murphy, Meredith Baxter Birney; *Dir:* Phil Karlson. **VHS, Beta, LV $14.95** PSM, PAR *ₐ½*

Ben and Eddie A modern-day spiritual series built around orphan home director Ben, and Eddie, one of the orphans at the home.
1990 45m/C VHS $14.95 SPW *ₐ½*

Ben-Hur Early version of the renowned story of Jewish and Christian divisiveness in the time of Jesus. Battle scenes and chariot races still look good, in spite of age. At a cost of over $4,000,000, it was the most expensive film of its time.
1926 148m/B Ramon Novarro, Francis X. Bushman, May McAvoy, Betty Bronson, Claire McDowell; *Dir:* Fred Niblo. **VHS, Beta, LV $29.95** MGM *ₐₐₐₐ*

Ben-Hur One of the best versions of the Lew Wallace classic stars Heston in the role of a Palestinian Jew battling the Roman empire at the time of Christ. Won a record 11 Oscars, and nominated Best Screenplay. Don't miss the breathtaking chariot race. Perhaps one of the greatest pictures of all time. Also available in letterbox format.
1959 212m/C Charlton Heston, Jack Hawkins, Stephen Boyd, Haya Harareet, Hugh Griffith, Martha Scott, Sam Jaffe, Cathy O'Donnell; *Dir:* William Wyler. Academy Awards '59: Best Actor (Heston), Best Art Direction/Set Decoration (Color), Best Color Cinematography, Best Costume Design, Best Director (Wyler), Best Picture, Best Special Effects, Best Supporting Actor (Griffith), Best Original Score; British Academy Awards '59: Best Picture; Directors Guild of America Awards '59: Best Director (Wyler); Golden Globe Awards '60: Best Film—Drama; New York Film Critics Awards '59: Best Picture. **VHS, Beta, LV $29.98** MGM, APD, BTV *ₐₐₐₐ*

Ben Turpin: An Eye for an Eye Two silent Ben Turpin comedies: "He Looked Crooked, or Why Ben Bolted" (1917) and "Ten Dollars or Ten Days" (1924).
1924 77m/B Ben Turpin, Harry Gribbon. **VHS, Beta, 8mm $29.95** VYY *ₐₐₐₐ*

Bend of the River A haunted, hardened guide leads a wagon train through Oregon territory, pitting himself against Indians, the wilderness and a former comrade-turned-hijacker.
1952 91m/C James Stewart, Arthur Kennedy, Rock Hudson, Harry Morgan, Royal Dano; *Dir:* Anthony Mann. **VHS, Beta $19.95** MCA *ₐₐₐ*

Beneath Arizona Skies/Paradise Canyon A package of John Wayne westerns filled with lots of rumble tumble action.
1935 109m/B John Wayne. **VHS, Beta $19.95** REP, BUR *ₐ½*

Beneath the Planet of the Apes In the first sequel, another Earth astronaut passes through the same warp and follows the same paths as Taylor, through Ape City and to the ruins of bomb-blasted New York's subway system, where warhead-worshiping human mutants are found. The strain of sequelling shows instantly, and gets worse through the next three films; followed by "Escape from the Planet of the Apes."
1970 (G) 108m/C Charlton Heston, James Franciscus, Kim Hunter, Maurice Evans, James Gregory, Natalie Trundy, Jeff Corey, Linda Harrison, Victor Buono; *Dir:* Ted Post. **VHS, Beta, LV $19.98** FOX *ₐₐ½*

Beneath the 12-Mile Reef Two rival groups of divers compete for sponge beds off the Florida coast. Lightweight entertainment notable for underwater photography and early Cinemascope production, as well as Moore in a bathing suit.
1953 102m/C Robert Wagner, Terry Moore, Gilbert Roland, Richard Boone, Peter Graves, J. Carroll Naish; *Dir:* Robert D. Webb. **VHS, Beta $19.95** NOS, WES, GEM *ₐₐ*

Beneath the Valley of the Ultra-Vixens Sex comedy retread directed by the man with an obsession for big. Scripted by Roger Ebert, who also scripted the cult classic "Beyond the Valley of the Dolls." Explicit nudity.
1979 90m/C Francesca "Kitten" Natavidad, Ann Marie, Ken Kerr, Stuart Lancaster; *Dir:* Russ Meyer. **VHS $89.95** RMF *ₐ*

The Beniker Gang Five orphans, supported by the eldest who writes a syndicated advice column, work together as a family. A made-for-television film.
1983 (G) 87m/C Andrew McCarthy, Jennie Dundas, Danny Pintauro, Charlie Fields; *Dir:* Ken Kwapis. **VHS, Beta $59.95** SCL, WAR *ₐₐ½*

Benjamin Argumedo A true story of the struggle to survive a revolution, and what a person has to endure after it is over.
1978 97m/C *SP* Antonio Aguilar, Flor Silvestre, Jose Galvez, Irma Lozano; *Dir:* Mario Hernandez. **VHS, Beta $19.95** WCV *ₐₐ*

Benji In the loveable pooch's first feature-length movie, he falls in love with a female named Tiffany, and saves two young children from kidnappers.
1973 (G) 87m/C Benji, Peter Brek, Christopher Connelly, Patsy Garrett, Deborah Walley, Cynthia Smith; *Dir:* Joe Camp. **VHS, Beta $19.99** VES, VTK, FCT *ₐ½*

Benji the Hunted The heroic canine, shipwrecked off the Oregon coast, discovers a litter of orphaned cougar cubs, and battles terrain and predators to bring them to safety.
1987 (G) 89m/C Benji, Red Steagall, Frank Inn; *Dir:* Joe Camp. **VHS, Beta, LV $89.95** DIS *ₐ½*

Benji Takes a Dive at Marineland/Benji at Work Television's Adam Rich goes to Marineland with Wonder-dog Benji, and chronicles the canine's busy work schedule.
1982 60m/C VHS, Beta $19.98 *CVL* ⅃⅃

Benji's Very Own Christmas Story Benji and his friends go on a magic trip and meet Kris Kringle and learn how Christmas is celebrated around the world. Also included: "The Phenomenon of Benji," a documentary about Benji's odyssey from the animal shelter to international stardom.
1983 60m/C VHS, Beta $14.99 *CVL, BFV* ⅃⅃

The Benny Goodman Story The life and music of Swing Era bandleader Benny Goodman is recounted in this popular bio-pic. Covering Benny's career from his child prodigy days to his monumental 1938 Carnegie Hall Jazz Concert, the movie's soggy plot machinations are redeemed by a non-stop music track featuring the real Benny and an all-star lineup. Tunes include "Don't Be That Way," "Memories of You," "Sing, Sing, Sing," and "Slipped Disc."
1955 116m/C Steve Allen, Donna Reed, Gene Krupa, Lionel Hampton, Kid Ory, Ben Pollack, Harry James, Stan Getz, Teddy Wilson, Martha Tilton; *Dir:* Valentine Davies. **VHS, Beta** $19.95 *MCA, FCT* ⅃⅃

Benny Hill's Crazy World The British comedian delivers some of his best material with the help of "Hill's Angels."
1974 55m/C *GB* Benny Hill. **VHS, Beta** $19.99 *HBO* ⅃⅃

Bergonzi Hand A drama about two art scoundrels who are astonished when an immigrant admits to painting a forged "Bergonzi."
1970 62m/C Keith Mitchell, Gordon Jackson, Martin Miller. **VHS, Beta** $24.95 *VYY* ⅃

The Berlin Affair A sordid tale from "The Night Porter" director about a Japanese woman seducing various parties in pre-World War II Germany.
1985 (R) 97m/C *IT GE* Mio Takaki, Gudrun Landgrebe, Kevin McNally; *Dir:* Liliana Cavani. **VHS, Beta** $79.95 *MGM, TAM* ⅃⅃

Berlin Alexanderplatz (episodes 1-2) Fassbinder's 15 1/2-hour epic, originally produced for German television, follows the life, death, and resurrection of Franz Biberkof, a former transit worker who has just finished a lengthy prison term. With the Berlin of the 1920s as a backdrop, Fassbinder has contrived a melodramatic parable with Biblical overtones.
1980 230m/C *GE* Gunter Lamprecht, Hanna Schygulla, Barbara Sukowa, John Gottfried; *Dir:* Rainer Werner Fassbinder. **VHS, Beta** $400.00 *FCT, MGM* ⅃⅃⅃⅃

Berlin Alexanderplatz (episodes 3-4) The third and fourth episodes of the German television mini series which tells the grand tale of Franz Biberkof, a parolee struggling in the harsh social atmosphere of 1920s Germany. Adapted from a story by writer Alfred Doblin. In German with English subtitles.
1983 230m/C *GE* Hanna Schygulla, Gunter Lamprecht, Barbara Sukowa, John Gottfried; *Dir:* Rainer Werner Fassbinder. **VHS, Beta** *MGM, FCT* ⅃⅃⅃⅃

Berlin Alexanderplatz (episodes 5-6) The fifth and sixth installments of the lengthy adaptation of a story by Alfred Doblin about a prison parolee's painful adjustment

to life in Germany in the 1920s. Originally aired as a mini-series on German television. In German with English subtitles.
1983 230m/C *GE* Gunter Lamprecht, Hanna Schygulla, Barbara Sukowa, John Gottfried; *Dir:* Rainer Werner Fassbinder. **VHS, Beta** *MGM, FCT* ⅃⅃⅃⅃

Berlin Alexanderplatz (episodes 7-8) The final two chapters in the continuing saga of a former inmate's struggle to survive beyond prison walls in 1920s Berlin. Adapted from a tale by Alfred Doblin. Originally produced as a mini-series for German television. In German with English subtitles.
1983 230m/C *GE* Gunter Lamprecht, Hanna Schygulla, Barbara Sukowa, John Gottfried; *Dir:* Rainer Werner Fassbinder. **VHS, Beta** *FCT, MGM* ⅃⅃⅃⅃

Berlin Blues A nightclub singer is torn between two men.
1989 (PG-13) 90m/C Julia Migenes-Johnson, Keith Baxter; *Dir:* Richardo Franco. **VHS, LV** $79.95 *CAN* ⅃⅃⅃⅃

The Berlin Conspiracy This espionage/action potboiler does an imaginative job of setting its action against the fall of the Berlin Wall and the end of the Cold War. A CIA agent forms a shaky alliance with his East German spymaster rival to prevent germ warfare technology from falling into terrorist hands.
1991 (R) 83m/C Marc Singer, Mary Crosby, Stephen Davies, Richard Leparmentier, Terence Henry; *Dir:* Terence H. Winkless. **VHS, LV** $89.95 *COL* ⅃⅃

Berlin Express Battle of wits ensues between the Allies and the Nazis who are seeking to keep Germans divided in post-WWII Germany. Espionage and intrigue factor heavily.
1948 86m/B Robert Ryan, Merle Oberon, Paul Lukas, Charles Korvin; *Dir:* Jacques Tourneur. **VHS, Beta, LV** $19.95 *MED, TTC, FCT* ⅃⅃⅃ ½

Berlin Tunnel 21 Based on the novel by Donald Lindquist, five American soldiers attempt a daring Cold War rescue of a beautiful German girl. The plan is to construct a tunnel under the Berlin Wall. A better-than-average film that was made for television.
1981 150m/C Richard Thomas, Jose Ferrer, Horst Buchholz; *Dir:* Richard Michaels. **VHS, Beta** $59.99 *HBO* ⅃ ½

Bernadette A French-made version of the legend of St. Bernadette, who endured persecution after claiming to have seen the Virgin Mary. Beautiful in its simplicity, but overly long.
1990 120m/C *FR* Sydney Penny, Roland LeSaffre, Michele Simonnet, Bernard Dheran, Arlette Didier; *Dir:* Jean Delannoy. **VHS, Beta** *IGP* ⅃⅃

Bernice Bobs Her Hair An ugly and shy girl's cousin revamps her into a seductress. From the director of "Chilly Scenes of Winter," part of the American Short Story Collection. Adaptation of the F. Scott Fitzgerald story.
1976 49m/C Shelley Duvall, Henry Fonda, Veronica Cartwright, Bud Cort; *Dir:* Joan Micklin Silver. New York Film Festival '76: Best Picture. **VHS, Beta** $24.95 *MON, KAR* ⅃⅃⅃

Berserk! A seedy traveling circus is beset by a series of murders. Not heralded as one of Crawford's best pieces.
1967 95m/C Joan Crawford, Diana Dors, Judy Geeson, Ty Hardin; *Dir:* James O'Connolly. **VHS, Beta** $69.95 *RCA* ⅃ ½

Berserker Six camping college students are attacked by a bloodthirsty psychotic out of a Nordic myth, who takes the shape of a badder-than-the-average bear.
1987 (R) 85m/C Joseph Alan Johnson, Valerie Sheldon, Greg Dawson; *Dir:* Jef Richard. **VHS, Beta** $9.99 *STE, PSM* ⅃

Bert Rigby, You're a Fool A starstruck British coal miner finds his way to Hollywood singing old showtunes, only to be rebuffed by a cynical industry. Reiner also wrote screenplay. Available with Spanish subtitles.
1989 (R) 94m/C Robert Lindsay, Robbie Coltrane, Jackie Gayle, Bruno Kirby, Cathryn Bradshaw, Corbin Bernsen, Anne Bancroft; *Dir:* Carl Reiner. **VHS, Beta, LV** $19.98 *WAR* ⅃⅃ ½

Besame Intrigue, espionage, and romance get mixed together in this movie.
1987 101m/C Sara Montiel, Maurice Ronet, Franco Fabrizi. **VHS, Beta** $53.95 *MAD* ⅃⅃

Bessie Smith and Friends Three jazz shorts are presented on this tape: "St. Louis Blues" (1929), the only film appearance of blues singer Bessie Smith; "Bye Bye Blackbird" (1932), starring Nina Mae McKinney and the dancing Nicholas Brothers with Eubie Blake and His Orchestra; and "Boogie Woogie Dream" (1941), featuring a youthful Lena Horne, boogie woogie pianists Albert Ammons and Pete Johnson, and Teddy Wilson's Band.
1986 39m/B Bessie Smith, Nicholas Brothers, Eubie Blake, Nina Mae McKinney, Lena Horne, Teddy Wilson, Albert Ammons, Pete Johnson. **VHS, Beta** $19.95 *MVD, AFE, AMV* ⅃

The Best of Abbott & Costello Live A compilation of vintage live-television routines by the duo, naturally including "Who's on First," along with their other celebrated bits.
1954 58m/B Bud Abbott, Lou Costello, George Raft, Charles Laughton. **VHS, Beta** $19.88 *WAR* ⅃⅃

Best of the Badmen A whole bunch of outlaws, although seemingly quite nice, are brought together by an ex-Union general who is being framed. Too much talk, not enough action.
1950 84m/B Robert Ryan, Claire Trevor, Jack Buetel, Robert Preston, Walter Brennan, Bruce Cabot, John Archer, Lawrence Tierney; *Dir:* William D. Russell. **VHS, Beta** $19.98 *TTC* ⅃⅃ ½

Best of the Benny Hill Show: Vol. 1 Each volume in this series of tapes contains the raucous, bawdy and unique humor of British comedian Benny Hill, excerpted from his television series.
1985 100m/C VHS, Beta $19.95 *HBO*

Best of the Benny Hill Show: Vol. 2 Another compilation of humorous sketches from "The Benny Hill Show."
1981 115m/C Benny Hill. **VHS, Beta** $19.95 *HBO*

Best of the Benny Hill Show: Vol. 3 British funny man Benny Hill mugs and jokes his way through a new collection of comedy sketches from his popular television series.
1983 110m/C Benny Hill. **VHS, Beta** $19.95 *HBO*

Best of the Benny Hill Show: Vol. 4 Here's another collection of zany sketches from British funnyman Benny Hill.
1984 95m/C *GB* **VHS, Beta** $19.95 *HBO*

Best of the Benny Hill Show: Vol. 5 Here is another volume of madcap comedy from British funnyman Benny Hill. Available in VHS and Beta Hi Fi.
1985 97m/C VHS, Beta $19.95 *HBO* 🦴🦴½

Best of the Best An interracial kick-boxing team strives to win a world championship.
1989 (PG-13) 95m/C Eric Roberts, Sally Kirkland, Christopher Penn, Phillip Rhee, James Earl Jones, John P. Ryan, John Dye, David Agresta, Tom Everett, Louise Fletcher, Simon Rhee; ***Dir:*** Bob Radler. **VHS, Beta, LV $14.95** *COL, SVS* 🦴½

Best of Betty Boop A collection of nine Boop classics, including the immortal "Betty Boop's Ker-choo."
1935 56m/B *Dir:* Dave Fleischer; ***Voices:*** Mae Questel. **VHS, Beta, 8mm $29.95** *VYY* 🦴½

Best of Betty Boop, Vol. 1 Sweet Betty Boop sashays through 11 of her classic cartoon adventures in this collection of original shorts. Mastered from the original negatives.
1939 90m/C *Voices:* Mae Questel. **VHS $14.98** *REP* 🦴½

Best of Betty Boop, Vol. 2 Another collection of original cartoons starring the "Boop-Oop-a-Doop" girl, assisted by Bimbo and Koko the Clown. These black-and-white cartoons have been recolored for this release.
1939 85m/C *Voices:* Mae Questel. **VHS, Beta $14.98** *REP* 🦴½

The Best of Bing Crosby: Part 1 Rosemary Clooney honors the legendary crooner's life with clips of his music and interviews.
1991 50m/C Bing Crosby; ***Hosted:*** Rosemary Clooney. **VHS** *ATV* 🦴½

The Best of Bogart A two-cassette gift pack of the Bogie classics "The Maltese Falcon" and "Casablanca."
1943 204m/B Humphrey Bogart, Ingrid Bergman, Mary Astor, Sydney Greenstreet. **VHS, Beta $39.95** *MGM* 🦴🦴🦴

Best of Bugs Bunny & Friends This collection includes classics from great cartoon stars Bugs Bunny, Daffy Duck, Tweety Pie and Porky Pig. Includes "Duck Soup to Nuts," "A Feud There Was" and "Tweetie Pie."
1940 53m/C *Dir:* Friz Freleng, Friz Freleng, Chuck Jones, Robert McKimson. Academy Awards '47: Best Animated Short Film ("Tweetie Pie"). **VHS, Beta** *MGM* 🦴🦴🦴

Best Buns on the Beach Sexploitation movie about 10 beautiful girls in a bikini contest, that ends up with them going topless. Ridiculous plot, for mature (??) audiences.
1987 60m/C VHS, LV $29.95 *CON, NSV, IME* ***Woof!***

Best of Chaplin Three Chaplin comedies when he worked with Mutual. Includes "The Rink," "One A.M.," and "The Floorwalker."
19?? 60m/B Charlie Chaplin; ***Dir:*** Charlie Chaplin. **VHS $59.95** *MVC* 🦴🦴

Best of Chevy Chase A compilation of Chase's funniest skits and routines from the first season of "Saturday Night Live," including the Land Shark and Chase's Gerald Ford impersonation.
1987 60m/C Chevy Chase, John Belushi, Dan Aykroyd, Bill Murray, Laraine Newman, Gilda Radner, Jane Curtin, Garrett Morris, Richard Pryor, Candice Bergen, Jill Clayburgh. **Beta, LV, Q $19.98** *LHV, WAR* 🦴½

The Best Christmas Pageant Ever The meanest kids in town just decided to participate in the town's Christmas pageant and something special happens when the curtain finally goes up.
1986 60m/C Dennis Weaver, Karen Grassle. **VHS** *IVC* 🦴½

Best of Comic Relief All-star comedic array of stand-up acts was taped "live" in concert to raise money for America's homeless and released on cable television.
1986 120m/C Whoopi Goldberg, Robin Williams, Billy Crystal, Carl Reiner, Harold Ramis, Martin Short, Jerry Lewis, Howie Mandel, Eugene Levy, Steve Allen, Doc Severinsen, Sid Caesar, John Candy, George Carlin. **VHS, Beta, LV $39.95** *LHV, WAR* 🦴½

Best of Comic Relief '90 More than 40 stars make this two hours a real treat, and the hosts aren't so bad themselves: Whoopi Goldberg, Billy Crystal and Robin Williams. Proceeds from this event, and subsequent tape sales go to Comic Relief, Inc. to benefit the homeless in America. Stars include: Bobcat Goldthwait, Elayne Boosler, George Carlin, Louis Anderson, Dennis Miller, The Simpsons.
1990 120m/C *Hosted:* Whoopi Goldberg, Billy Crystal, Robin Williams; ***Performed by:*** Bob Goldthwait, Elayne Boosler, George Carlin, Louie Anderson, Dennis Miller. **VHS, Beta $29.95** *RHI* 🦴½

Best of Dan Aykroyd A collection of Aykroyd's funniest skits and characters from his five-year stint on "Saturday Night Live," including the Bass-O-Matic Salesman, Irwin Mainway, Julia Child, E. Buzz Miller, the Coneheads, Leonard Pinth-Garnell, and the Czechoslovakian Brothers.
1986 57m/C Dan Aykroyd, John Belushi, Laraine Newman, Chevy Chase, Steve Martin, Bill Murray, Jane Curtin, Madeline Kahn, Shelley Duvall, Gilda Radner. **VHS, Beta $19.98** *WAR* 🦴½

The Best of Dark Shadows A collection of highlights from the cult classic daytime serial of the late sixties and early seventies featuring some of the show's most memorable characters.
196? 30m/C Jonathan Frid, Joan Bennett, David Selby, Lara Parker, Kate Jackson, Mitchell Ryan, Alexandre Molthe, Louis Edmonds, Mark Allen. **VHS, Beta $9.98** *MPI* 🦴½

Best Defense A U.S. Army tank operator is sent to Kuwait to test a new state-of-the-art tank in a combat situation. Although the cast is popular, the movie as a whole is not funny and the story frequently is hard to follow.
1984 (R) 94m/C Dudley Moore, Eddie Murphy, Kate Capshaw, George Dzundza, Helen Shaver; ***Dir:*** Willard Huyck. **VHS, Beta, LV $14.95** *PAR* 🦴

Best of Eddie Murphy A compilation of Murphy's funniest bits from his three seasons on "Saturday Night Live," including appearances as Stevie Wonder, Buckwheat, Bill Cosby and Little Richard.
1989 78m/C Eddie Murphy, Joe Piscopo, Robin Duke, Tim Kazurinsky, Mary Gross, Gary Kroeger, Brad Hall, Julia Louis-Dreyfus. **VHS, Beta $14.95** *PAR* 🦴½

Best Enemies An English film about a man suffering the trials of the 1960s, including Vietnam, and how it affects his relationships with his friends and wife.
198? 96m/C *GB* Sigrid Thornton, Paul Williams, Judy Morris, Brandon Burke; ***Dir:*** David Baker. **VHS, Beta $29.95** *ACA* 🦴½

Best of Ernie Kovacs A compilation of some of the best moments from Ernie Kovacs' television programs that features such memorable characters as Percy Dovetonsils and the Nairobi Trio.
1956 60m/B Ernie Kovacs, Edie Adams, Ernie Hatrak, Bill Wendell, Peter Hanley. **VHS, Beta $19.95** *KUL, WST, DVT* 🦴½

Best of Everything Goes Incorrect answers to the queries of adult humorists result in the contestants removing their clothing. Made for television.
1983 56m/C Jaye P. Morgan, Unknown Comic, Pat McCormick, Dick Shawn, Miss Miller, Kip Addotta. **VHS, Beta $39.95** *AHV* 🦴½

Best Foot Forward A vintage musical about a movie star who agrees to accompany a young cadet to a military ball. Based on the popular Broadway show; songs include "Buckle Down, Winsocki," "The Three B's (Barrelhouse, Boogie Woogie, and the Blues)," "Alive and Kicking," and Harry James' "Two O'Clock Jump." The film debuts of Walker, Allyson, and DeHaven.
1943 95m/C Lucille Ball, June Allyson, Tommy Dix, Nancy Walker, Virginia Weidler, Gloria DeHaven, William Gaxton, Harry James; ***Dir:*** Edward Buzzell. **VHS, Beta $19.98** *MGM* 🦴🦴½

Best Friends A pair of screenwriters decide to marry after years of living and working together. Story based on the lives of screenwriters Barry Levinson and Valerie Curtin.
1982 (PG) 109m/C Goldie Hawn, Burt Reynolds, Jessica Tandy, Barnard Hughes, Audra Lindley, Keenan Wynn, Ron Silver; ***Dir:*** Norman Jewison. **VHS, Beta, LV $19.98** *WAR* 🦴🦴

Best of George Pal George Pal's "Puppetoon" series produced between 1938 and 1943 is highlighted. Black and white selections include "Sleeping Beauty," "Captain Kidding," "Cavalcade of Music," and "Sky Pirates." In color are "Ship of the Ether," "Phillips Broadcast of 1938," "Tubby the Tuba," "John Henry," "The Little Broadcast," "Jasper in a Jam," and "Jasper in the Haunted House."
1979 75m/C VHS, Beta $19.95 *VDM* 🦴🦴🦴

Best of Gilda Radner A posthumously released collection of skits featuring the comedienne from "Saturday Night Live," including appearances as RoseAnn Rosannadanna, Lisa Lupner, Baba Wawa and more.
1989 59m/C Gilda Radner, Dan Aykroyd, John Belushi, Chevy Chase, Jane Curtin, Bill Murray, Garrett Morris, Laraine Newman, Madeline Kahn, Buck Henry, Steve Martin, Candice Bergen. **VHS, Beta $19.98** *WAR* 🦴🦴🦴

Best of John Belushi Compilation of sixteen of John Belushi's funniest "Saturday Night Live" routines including "Samurai Deli," "Godfather Therapy" and "Star Trek."
1985 60m/C John Belushi, Dan Aykroyd, Chevy Chase, Jane Curtin, Rob Reiner, Laraine Newman, Bill Murray, Garrett Morris, Gilda Radner, Elliott Gould, Buck Henry, Robert Klein. **VHS, Beta $19.98** *WAR* 🦴🦴🦴

The Best of John Candy on SCTV Some of Candy's most memorable characters, including Johnny LaRue, Yosh Schmenge, and the grown up Theodore "Beaver" Cleaver, are featured on this compilation of his years on SCTV (Second City Television).
19?? 62m/C John Candy, Eugene Levy, Rick Moranis, Catherine O'Hara, Harold Ramis, Martin Short. **VHS, LV** $39.95 COL *Ⅱ Ⅱ Ⅱ*

Best of Judy Garland Judy Garland shines brightly in this television concert that features such standards as "The Man That Got Away," "Over The Rainbow" and "Swanee."
1985 85m/B Judy Garland. **VHS, Beta** COL, OM

The Best of the Kenny Everett Video Show Scenes from the popular British late-night television show combining new wave rock music, outrageous dancing and innovative video special effects.
1981 104m/C Kenny Everett. **VHS, Beta** $29.98 HBO *Ⅱ*

Best Kept Secrets A feisty woman discovers corruption and blackmail in the police department where her husband is an officer.
1988 104m/C Patty Duke, Frederic Forrest, Peter Coyote; *Dir:* Jerrold Freedman. **VHS, Beta** $79.95 VMK *Ⅱ Ⅱ ½*

Best Legs in the 8th Grade A handsome attorney bumps into the gorgeous homeroom angel who had ignored his nebbishly dull countenance in grade school.
1984 60m/C Tim Matheson, James Belushi, Annette O'Toole, Kathryn Harrold; *Dir:* Tom Patchett. **VHS, Beta** $59.98 LTG *Ⅱ ½*

The Best Little Girl in the World Exceptional made for TV tale of an apparently perfect teenager suffering from anorexia. Slow starvation is her only cry for help. Fine performances from Durning and Saint.
1981 96m/C Charles Durning, Eva Marie Saint, Jennifer Jason Leigh, Melanie Mayron, Viveca Lindfors, Jason Miller, David Spielberg, Lisa Pelikan, Ally Sheedy; *Dir:* Sam O'Steen. **VHS, Beta** $39.98 CGI *Ⅱ Ⅱ Ⅱ*

Best of Little Lulu Mischief-prone Little Lulu returns in this special collection of cartoon adventures.
19?? 60m/C **VHS** $19.95 REP *Ⅱ Ⅱ Ⅱ*

The Best of the Little Rascals A collection of six classic "Our Gang" two reelers featuring everyone's favorite rascals Spanky, Alfalfa, Stymie and Buckwheat.
193? 103m/B **VHS, Beta** $19.98 REP *Ⅱ Ⅱ Ⅱ*

The Best Little Whorehouse in Texas Parton is the buxom owner of The Chicken Ranch, a house of ill-repute that may be closed down unless Sheriff-boyfriend Reynolds can think of a way out. Strong performances don't quite make up for the erratically comic script. Based on the long-running Broadway musical, in turn based on a story by Larry McMurtry. Available in a letterboxed edition on laserdisc.
1982 (R) 115m/C Dolly Parton, Burt Reynolds, Dom DeLuise, Charles Durning, Jim Nabors, Lois Nettleton; *Dir:* Colin Higgins. **VHS, Beta, LV** $19.95 MCA, FCT, PIA *Ⅱ Ⅱ*

The Best Man An incisive, darkly satiric political tract, based on Gore Vidal's play, about two presidential contenders who vie for the endorsement of the aging ex-president, and trample political ethics in the process. Tracy, as the ex-president, was Oscar-nominated.
1964 104m/B Henry Fonda, Cliff Robertson, Lee Tracy, Margaret Leighton, Edie Adams, Kevin McCarthy, Ann Sothern, Gene Raymond, Shelley Berman, Mahalia Jackson; *Dir:* Franklin J. Schaffner. **VHS, Beta** $19.98 MGM *Ⅱ Ⅱ Ⅱ ½*

The Best of Mister Magoo Four of Mister Magoo's most hilarious misadventures, entitled " Bearfooted Flat Foot," "Bungled Bungalow," "Bwana Magoo," and "Destination Magoo."
195? 27m/C **VHS, Beta** $9.98 GKK *Ⅱ Ⅱ Ⅱ ½*

Best of the Monsters The Canadian comedy team offers a televised look at movie monsters. There is also plenty of political humor mixed in, which might be difficult or elusive for Americans.
196? 45m/B Wayne & Shuster. **VHS, Beta** $24.95 NOS, DVT *Ⅱ Ⅱ Ⅱ ½*

The Best of Not Necessarily the News From the TV show, combining comedy with actual news film footage, a satirical look at political developments and other happenings.
1988 60m/C Anne Bloom, Danny Breen, Stuart Pankin. **VHS** HBO, TLF *Ⅱ Ⅱ Ⅱ ½*

Best of Popeye Eight classic Popeye cartoons are included in this compilation.
1983 56m/C **VHS, Beta** $29.95 MGM *Ⅱ Ⅱ Ⅱ ½*

Best Revenge Two aging hippies engage in a Moroccan drug deal in order to free a kidnapped friend from a sleazy gangster. They get caught by the police, but escape, searching for the engineers of the frame-up.
1983 (R) 92m/C John Heard, Levon Helm, Alberta Watson, John Rhys-Davies, Moses Znaimer; *Dir:* John Trent. **VHS, Beta** $59.95 LHV, WAR *Ⅱ Ⅱ*

Best Seller An interesting, subtext-laden thriller about a cop/bestselling author with writer's block, and the strange symbiotic relationship he forms with a slick hired killer, who wants his own story written. Dennehy is convincing as the jaded cop, and is paired well with the psychotic Woods.
1987 (R) 112m/C James Woods, Brian Dennehy, Victoria Tennant, Paul Shenar, Seymour Cassel, Allison Balson; *Dir:* John Flynn. **VHS, Beta, LV** $14.98 VES *Ⅱ Ⅱ Ⅱ*

Best of Terrytoons This is a compilation of cartoons featuring Mighty Mouse, Heckle and Jeckle, Deputy Dawg, Gandy Goose, Dinky Duck, Terry Bears and Little Roquefort.
1983 60m/C **VHS, Beta** $29.95 CVL *Ⅱ Ⅱ ½*

The Best of Times Slim story of two grown men who attempt to redress the failures of the past by reenacting a football game they lost in high school due to a single flubbed pass. With this cast, it should have been better.
1986 (PG) 105m/C Robin Williams, Kurt Russell, M. Emmet Walsh, Pamela Reed, Holly Palance, Donald Moffatt, Margaret Whitton, Kirk Cameron; *Dir:* Roger Spottiswoode. **VHS, Beta, LV, 8mm** $14.98 SUE *Ⅱ Ⅱ*

Best of the Two Ronnies A selection of skits from the well-loved BBC comedy series.
197? 45m/C GB Ronnie Barker, Ronnie Corbett. **VHS, Beta** $14.98 FOX *Ⅱ Ⅱ*

Best of Warner Brothers: Volume 1 Classic cartoons from the Warner Brothers archives featuring Porky Pig, Daffy Duck, Bugs Bunny and the gang. Titles include: "Get Rich Quick," "Robinson Crusoe, Jr.," "Porky's Railroad," "Porky's Preview," "I Wanna Be a Sailor," "An Itch in Time," and "Corny Concerto." (Some black and white).
194? 55m/C *Dir:* Friz Freleng, Chuck Jones, Robert McKimson. **VHS, Beta** $19.95 NO *Ⅱ Ⅱ*

Best of Warner Brothers: Volume 2 A collection of eight vintage Warner Brothers cartoons featuring Daffy Duck and Porky Pig, all from 1936-1942.
193? 55m/C *Dir:* Friz Freleng, Chuck Jones, Robert McKimson; *Voices:* Mel Blanc. **VHS, Beta** NO *Ⅱ Ⅱ*

The Best Way Two summer camp counselors discover they might be gay and desirous of each other. Miller's first film; in French with English subtitles.
1982 85m/C FR Patrick Dewaere, Patrick Bouchitey, Christine Pascal, Claude Pieplu; *Dir:* Claude Miller. **VHS, Beta** $79.95 FCT *Ⅱ Ⅱ ½*

Best of W.C. Fields Three of W.C. Field's Mack Sennett shorts are presented in their complete, uncut form: "The Dentist," "The Fatal Glass of Beer," and "The Golf Specialist."
1933 58m/B W.C. Fields, Elise Cavanna. **VHS** $19.95 REP *Ⅱ Ⅱ ½*

The Best Years of Our Lives The all-American classic about three World War II vets returning home to try to pick up the threads of their lives. A film that represented a large chunk of American society and helped it readjust to the modern post-war ambience. Won seven of the eight Academy Awards it was nominated for. Supporting actor Harold Russell, an actual veteran, holds a record for winning two Oscars for a single role.
1946 170m/B Fredric March, Harold Russell, Myrna Loy, Dana Andrews, Teresa Wright, Gladys George, Virginia Mayo, Hoagy Carmichael, Roman Bohnen; *Dir:* William Wyler. Academy Awards '46: Best Actor (March), Best Director (Wyler), Best Film Editing, Best Picture, Best Screenplay, Best Supporting Actor (Russell), Best Original Score; British Academy Awards '47: Best Film; Golden Globe Awards '47: Best Film—Drama; National Board of Review Awards '46: 10 Best Films of the Year, Best Director; New York Film Critics Awards '46: Best Director, Best Picture. **VHS, Beta, LV** $14.98 HBO, SUE, FCT *Ⅱ Ⅱ Ⅱ Ⅱ*

Bestia Nocturna A psychopath stalks the freeways, mutilating his victims. He is hunted by one man, who relies on his wits to hunt the madman.
1986 93m/C SP Raymundo Capetillo, Laura Flores, Eric Del Castillo, Roberto Canedo; *Dir:* Humberto Martinez Mijares. **VHS, Beta** $79.95 MED *Ⅱ*

Bethune The life story of a Canadian doctor who started a practice in Communist China.
1977 88m/C Donald Sutherland, Kate Nelligan, David Gardner, James Hong; *Dir:* Eric Till. **VHS, Beta** $59.95 TWE *Ⅱ Ⅱ ½*

Betrayal A telefilm based on the book by Lucy Freeman and Julie Roy about a historic malpractice case involving a psychiatrist and one of his female patients. The doctor convinced the female patient that sex with him would serve as therapy.
1978 95m/C Lesley Ann Warren, Rip Torn, Ron Silver, Richard Masur, Stephen Elliott, John Hillerman, Peggy Ann Garner; *Dir:* Paul Wendkos. **VHS, Beta** $59.95 VCL *Ⅱ ½*

Betrayal An unusual adult drama, beginning at the end of a seven-year adulterous affair and working its way back in time to finally end at the start of the betrayal of a husband by his wife and his best friend. Kingsley and Irons make Pinter's adaptation of his own play work.
1983 (R) 95m/C *GB* Ben Kingsley, Patricia Hodge, Jeremy Irons; *Dir:* David Jones. **VHS, Beta $59.98** *FOX ♂♂½*

Betrayal A lonely widow hires a companion not realizing her treacherous plans.
19?? 78m/C Amanda Blake, Dick Haymes, Tisha Sterling, Sam Groom; *Dir:* Gordon Hessler. **VHS $79.95** *JEF ♂♂½*

Betrayal from the East A carnival barker saves the Panama Canal from the vicious Japanese war machine in this rather silly wartime drama.
1944 82m/B Lee Tracy, Nancy Kelly, Richard Loo, Abner Biberman, Regis Toomey, Philip Ahn, Addison Richards, Sen Yung, Drew Pearson; *Dir:* William Berke. **VHS, Beta $19.98** *TTC ♂♂*

Betrayal and Revenge The story of the "Monkey King," a Kung Fu warrior determined to avenge his father's murder.
1989 104m/C VHS, Beta $24.95 *JCI ♂*

Betrayed Bombshell Turner and strongman Gable star in this story of World War II intrigue. Suspected of being a Nazi informer, Turner is sent back to Holland for a last chance at redemption. Her cover as a sultry nightclub performer has the Nazis drooling and ogling (can you spell h-o-t?), but her act may be blown by an informant. Can luscious Lana get out of this one intact?
1954 107m/C Clark Gable, Lana Turner, Victor Mature, Louis Calhern, O.E. Hasse, Wilfrid Hyde-White, Ian Carmichael, Niall MacGinnis, Nora Swinburne, Roland Culver; *Dir:* Gottfried Reinhardt. **VHS, Beta $19.98** *MGM, CCB ♂♂♂*

Betrayed A rabid political film, dealing with an implausible FBI agent infiltrating a white supremacist organization via her love affair with a handsome farmer who turns out to be a murderous racist. Winger is memorable in her role as the FBI agent, despite the film's limitations, and admirers of Coasta-Gravas's directorial work and political stances will want to see how the director botched this one.
1988 (R) 112m/C Tom Berenger, Debra Winger, John Mahoney, John Heard, Albert Hall; *Dir:* Constantin Costa-Gavras. **VHS, Beta, LV $19.95** *MGM ♂♂*

The Betsy A story of romance, money, power and mystery centering around the wealthy Hardeman family and their automobile manufacturing business. Loosely patterned after the life of Henry Ford as portrayed in the Harold Robbins' pulp-tome. Olivier is the redeeming feature.
1978 (R) 132m/C Laurence Olivier, Kathleen Beller, Robert Duvall, Lesley-Anne Down, Edward Herrmann, Tommy Lee Jones, Katharine Ross, Jane Alexander; *Dir:* Daniel Petrie. **VHS, Beta $59.98** *FOX ♂♂*

Betsy's Wedding Betsy wants a simple wedding. But her father has other, grander ideas. Then there's the problem of paying for it. Dad tries to take care of that in a not-so-typical manner. Will Betsy make it down the aisle? Watch a fine ensemble cast to find out.
1990 (R) 94m/C Molly Ringwald, Alan Alda, Joey Bishop, Madeline Kahn, Catherine O'Hara, Joe Pesci, Ally Sheedy, Burt Young, Anthony LaPaglia; *W/Dir:* Alan Alda. **VHS, Beta, LV $89.95** *TOU ♂♂½*

Bette Midler Show Midler, accompanied by the Harlettes, jokes, dances and belts out a medley of songs ranging from the Andrew Sisters' "Boogie Woogie Bugle Boy" to "Friends."
1976 84m/C Bette Midler. **VHS, Beta, LV** *SUE, OM ♂♂½*

Better Late Than Never Fun television tale of senior citizens in revolt at an old-age home. Fine characters portrayed by some of the best in the business.
1979 100m/C Harold Gould, Tyne Daly, Strother Martin, Harry Morgan, Victor Buono, George Gobel, Lou Jacobi, Donald Pleasence, Larry Storch; *Dir:* Richard Crenna. **VHS $59.98** *FOX ♂♂*

Better Late Than Never Two penniless old fools vie for the acceptance of a bratty 10-year-old millionairess, who must choose one as her guardian. Niven's last film.
1983 95m/C David Niven, Art Carney, Maggie Smith, Kimberly Partridge, Catherine Hicks, Melissa Prophet; *Dir:* Bryan Forbes. **VHS, Beta $59.98** *FOX ♂½*

Better Off Dead A compulsive teenager's girlfriend leaves him and he decides to end it all. After several abortive attempts, he decides instead to out-ski his ex-girlfriend's obnoxious new boyfriend. Uneven but funny.
1985 (PG) 97m/C John Cusack, Curtis Armstrong, Diane Franklin, Kim Darby, David Ogden Stiers, Dan Schneider, Amanda Wyss, Taylor Negron; *Dir:* Steve Holland. **VHS, Beta, LV $79.98** *FOX ♂♂½*

Betty Blue A vivid, intensely erotic film about two young French lovers and how their inordinately strong passion for each other destroys them, leading to poverty, violence, and insanity. English subtitles. From the director of "Diva."
1986 (R) 121m/C *FR* Beatrice Dalle, Jean-Hughes Anglade, Gerard Darmon; *Dir:* Jean-Jacques Beineix. Montreal Film Festival '86: Grand Prix; Boston Society of Film Critics '86: Best Foreign Film. **VHS, Beta, LV $79.94** *FOX, TAM ♂♂♂*

Betty Boop Cartoons featuring Betty Boop, Daffy Duck and others are featured here. The titles are "The Hot Air Salesman," "Case of the Missing Hare," "There's Good Boos Tonight," and "Yankee Doodle Daffy."
193? 30m/C *Voices:* Mae Questel, Mel Blanc. **VHS, LV** *NEG ♂♂♂*

Betty Boop Cartoon Festival Max Fleischer produced and directed these campy, racy cartoons of the 1930s including such titles as "Baby Be Good," "Betty Boop with Grampy," "Betty Boop with Henry," "Candid Candidate," "Ding Dong Doggie," and "Ker-choo." Mae Questel provides the voice of the popular flapper.
193? 55m/B Max Fleischer; *Voices:* Mae Questel. **VHS, Beta $34.98** *HHT, DVT ♂♂♂*

Betty Boop Classics A compilation of Betty Boop and her cartoon pals most fun-filled escapades.
1939 60m/C *Voices:* Mae Questel. **VHS $19.98** *REP ♂♂♂*

Betty Boop Festival: No. 1 An hour of classic Betty Boop cartoons featuring "Is My Palm Red," "Betty's Rise to Fame," "So Does an Automobile," "SOS," "Making Friends," and several others.
193? 60m/B *Voices:* Mae Questel. **VHS, Beta $19.98** *NO ♂♂♂*

Betty Boop Festival: No. 2 A compilation of classic Betty Boop cartoons, including "Baby Be Good," "A Little Soap and Water," "Kerchoo," "Be Human," "Not Now," "Betty and Jimmy," "Betty in Blunderland," and "Crazy Town."
193? 55m/B *Voices:* Mae Questel. **VHS, Beta $19.98** *NO ♂♂♂*

Betty Boop Festival: No. 3 Nine cartoons of the ageless gal with the great voice: "Happy You and Merry Me," "No, No, A Thousand Times No," "My Friend the Monkey," "A Song a Day," "Training Pigeons," "Scared Crows," "Whoop's, I'm a Cowboy," "Musical Mountaineers," and "We Did It."
193? 60m/B Max Fleischer; *Dir:* Max Fleischer. **VHS, Beta $19.95** *NO ♂♂♂*

Betty Boop Program A compilation of five Betty Boop cartoons of the 1930s: "Betty in Blunderland," "Betty and the Little King," "Betty Boop with Henry," "Betty and Little Jimmy," and "Candid Candidate."
193? 50m/B *Voices:* Mae Questel. **VHS, Beta** *GVV ♂♂♂*

Betty Boop Special Collector's Edition: Volume 1 Betty Boop returns in this collection of vintage cartoons, presented in their original black-and-white form, with appearances by jazz stars Cab Calloway and Don Redman.
1935 90m/B *Dir:* Dave Fleischer, Max Fleischer; *Voices:* Louis Armstrong, Mae Questel. **VHS, LV $19.98** *REP ♂♂♂*

Betty Boop Special Collector's Edition: Volume 2 That boop-a-doop girl is back in this collection of classic cartoons that features Cab Calloway and his orchestra.
1937 90m/B *Voices:* Mae Questel. **VHS, LV $19.98** *REP ♂♂♂*

Between Fighting Men The orphaned daughter of a shepherd is adopted by the cowpoke who caused her father's death. The cowpoke's sons fall in love with, and compete for the love of the girl.
1932 62m/B Ken Maynard, Ruth Hall, Josephine Dunn, Wallace MacDonald; *Dir:* Forrest Sheldon. **VHS $19.95** *NOS, DVT ♂½*

Between Friends Two women with only their respective divorces in common, meet and become fast friends. The rapport between Burnett and Taylor makes for a touching drama. Adapted for cable television from the book "Nobody Makes Me Cry," by Shelley List, one of the producers.
1983 105m/C Elizabeth Taylor, Carol Burnett; *Dir:* Lou Antonio. **VHS, Beta $69.95** *VES ♂♂½*

Between God, the Devil & a Winchester A violent western with lots of shooting, vultures, and dust.
197? 94m/C Gilbert Roland, Richard Harrison; *Dir:* Dario Silvester. **VHS, Beta $49.95** *UNI ♂*

Between Heaven and Hell Prejudiced Southern gentleman Wagner finds how wrong his misconceptions are, as he attempts to survive WWII on a Pacific Island. Ebsen is exceptional, making this rather simplistic story into a meaningful classic.
1956 94m/C Robert Wagner, Terry Moore, Broderick Crawford, Buddy Ebsen, Robert Keith, Brad Dexter, Mark Damon, Kenneth Clark, Harvey Lembeck, Frank Gorshin, Scatman Crothers; *Dir:* Richard Fleischer. **VHS $14.98** *FXV ♂♂♂*

Between the Lines A witty, wonderfully realized ensemble comedy about the staff of a radical post-'60s newspaper always on the brink of folding, and its eventual sell-out.
1977 (R) 101m/C John Heard, Lindsay Crouse, Jeff Goldblum, Jill Eikenberry, Stephen Collins, Lewis J. Stadlen, Michael J. Pollard, Marilu Henner; *Dir:* Joan Micklin Silver. **VHS, Beta** $69.95 VES ✓✓✓½

Between Men A father kills a man he believes killed his son and flees. Later in life, the son meets his father. For father and son, this causes some major concern.
1935 59m/B Johnny Mack Brown, Beth Marion, William Farnum; *Dir:* Robert N. Bradbury. **VHS, Beta** $19.95 NOS, DVT, WGE ✓

Between Two Women A television movie about a wife's rocky relationship with her bossy mother-in-law, until the latter has a stroke and needs care. Dewhurst won an Emmy for her portrayal of the mother-in-law.
1986 95m/C Farrah Fawcett, Michael Nouri, Colleen Dewhurst, Steven Hill, Bridgette Anderson, Danny Corkill; *Dir:* Willis Goldbeck. **VHS, Beta** $79.95 WAR ✓✓

Between Wars A young doctor in Australia's Medical Corps encounters conflict when he tries to introduce Freud's principles into his work.
1974 97m/C Corin Redgrave, Arthur Dingham, Judy Morris, Patricia Leehy, Gunter Meisner; *Dir:* Michael Thornhill. **VHS, Beta** $19.98 VID ✓½

Beulah Land A television mini-series about 45 years in the lives of a Southern family.
1980 280m/C Lesley Ann Warren, Michael Sarrazin, Don Johnson, Meredith Baxter Birney, Dorian Harewood, Eddie Albert, Hope Lange, Paul Rudd. **VHS, Beta, LV** $14.95 COL ✓½

Beverly Hills Bodysnatchers A mad scientist and a greedy mortician plot to get rich, but their plan backfires when they bring a Mafia godfather back to life and he terrorizes Beverly Hills.
1989 (R) 85m/C Vic Tayback, Frank Gorshin, Brooke Bundy, Seth Jaffe, Art Metrano; *Dir:* Jon Mostow. **VHS, LV** $39.95 SOU, IME ✓½

Beverly Hills Brats A spoiled, rich Hollywood brat hires a loser to kidnap him, in order to gain his parents' attention, only to have both of them kidnapped by real crooks.
1989 (PG-13) 90m/C Martin Sheen, Burt Young, Peter Billingsley, Terry Moore; *Dir:* Dimitri Sotirakis. **VHS, Beta, LV** $89.95 IVE, IME ✓

Beverly Hills Call Girls This racy picture deals with prostitutes in Beverly Hills and their "respectable" clients.
1986 60m/C VHS, Beta $24.95 AHV Woof!

Beverly Hills Cop When a close friend of smooth-talking Detroit cop Axle Foley is brutally murdered in L.A., he traces the murderer to the posh streets of Beverly Hills. There he must stay on his toes to keep one step ahead of the killer and two steps ahead of the law. Better than average Murphy vehicle.
1984 (R) 105m/C Eddie Murphy, Judge Reinhold, John Ashton, Lisa Eilbacher, Ronny Cox, Steven Berkoff, James Russo, Jonathan Banks, Stephen Elliott, Bronson Pinchot, Paul Reiser, Damon Wayans, Rick Overton; *Dir:* Martin Brest. **VHS, Beta, LV, 8mm** $14.95 PAR ✓✓½

Beverly Hills Cop 2 The highly successful sequel to the first profitable comedy, with essentially the same plot, this time deals with Foley infiltrating a band of international munitions smugglers.

1987 (R) 103m/C Eddie Murphy, Judge Reinhold, Juergen Prochnow, Ronny Cox, John Ashton, Brigitte Nielsen, Allen Garfield, Paul Reiser, Dean Stockwell; *Dir:* Tony Scott. **VHS, Beta, LV, 8mm** $14.95 PAR ✓✓½

Beverly Hills Knockouts A bevy of beautiful boxers pummel each other with both fists.
1989 60m/C VHS, Beta $39.95 CEL ✓½

Beverly Hills Madam In this routine plot, a stable of elite call girls struggle with their lifestyle and their madam, played by Dunaway. Made for television.
1986 (PG-13) 97m/C Faye Dunaway, Louis Jourdan, Donna Dixon, Robin Givens, Marshall Colt, Melody Anderson, Terry Farrell; *Dir:* Harvey Hart. **VHS, Beta, LV** $79.98 ORI ✓

Beverly Hills Vamp A madame and her girls are really female vampires with a penchant for hot-blooded men.
1988 (R) 88m/C Britt Ekland, Eddie Deezen, Debra Lamb; *Dir:* Fred Olen Ray. **VHS, Beta** $79.95 VMK ✓✓

Beware, My Lovely Taut chiller. Lonely widow Lupino hires a new handy-man. He's great with screen doors and storm windows, but has a problem with sharp tools. Intense and gripping with fine performances.
1952 77m/B Ida Lupino, Robert Ryan, Taylor Holmes, O.Z. Whitehead, Barbara Whiting, Dee Pollock; *Dir:* Harry Horner. **VHS** $19.98 REP ✓✓½

Beware of Pity A crippled baroness thinks she's found true love with a military officer, but it turns out his marriage proposal grew out of pity for her, not passion. A quality but somber British-made historical drama, based on a novel by Stefan Zweig.
1946 1005m/B Lilli Palmer, Albert Lieven, Cedric Hardwicke, Gladys Cooper, Ernest Thesiger, Freda Jackson; *Dir:* Maurice Elvey. **VHS** $19.95 NOS ✓✓½

Bewitched A made-for-British-television horror story about an elderly couple ostensibly plagued by the witch-like conniving of a dead woman.
1985 60m/C GB Eileen Atkins, Alfred Lynch; *Dir:* Edmund Oboler. **VHS, Beta** $59.95 PSM ✓✓½

Beyond Atlantis An ancient underwater tribe is discovered when it kidnaps land-lubbin' women with which to mate.
1973 (PG) 91m/C PH John Ashley, Patrick Wayne, George Nader; *Dir:* Eddie Romero. **VHS, Beta** $49.95 UHV Woof!

Beyond Belief Footage of such phenomena as ESP, automatic writing, telekinesis, etc., featuring Geller fraudulently bending flatware (without touching it).
1976 94m/C Uri Geller. **VHS, Beta** $59.95 UHV Woof!

Beyond the Bermuda Triangle Unfortunate and flat television flick. Businessman MacMurray, now retired, doesn't have enough to do. He begins an investigation of the mysterious geometric island area when his friends and fiancee disappear. Silly.
1975 78m/C Fred MacMurray, Sam Groom, Donna Mills, Suzanne Reed, Dana Plato, Woody Woodbury; *Dir:* William A. Graham. **VHS** MAG ✓½

Beyond the Call of Duty Renegade U.S. Army Commander Len Jordan (Vincent) is after a particularly deadly Vietcong enemy. Aided by the head of a special forces naval unit he tracks his quarry through the notorious Mekong River Delta. But Jordan's mission may be hindered, both personally and

professionally, by a beautiful American journalist after a hot story.
1992 (R) 92m/C Jan-Michael Vincent, Eb Lottimer, Jillian McWhirter. **VHS** $89.98 NHO ✓½

Beyond the Door A San Francisco woman finds herself pregnant with a demonic child. One of the first "Exorcist" ripoffs; skip this one and go right to the sequel, "Beyond the Door 2." In Italian; dubbed.
1975 (R) 97m/C IT Juliet Mills, Richard Johnson, David Colin Jr.; *Dir:* Oliver Hellman. **VHS, Beta, LV** $54.95 MED, COL Woof!

Beyond the Door 2 Better treatment of the possession theme, but this time the door is to the house of a new family: Colin (from the original) plays a boy possessed by his dead father, who seeks revenge on his widow and her new husband. Director Bava improved the sequel.
1979 (R) 90m/C IT John Steiner, Daria Nicolodi, David Colin Jr., Ivan Rassimov; *Dir:* Mario Bava. **VHS, Beta** $59.95 MED ✓½

Beyond the Door 3 Fool American students in Yugoslavia board a hellish locomotive which speeds them toward a satanic ritual. Demonic disaster-movie stuff (with poor miniatures) isn't as effective as the on-location filming; Serbian scenery and crazed peasants impart an eerie pagan aura. What this has to do with earlier "Beyond the Door" movies only the marketing boys can say. Some dialogue in Serbo-Croat with English subtitles.
1991 (R) 94m/C Mary Kohnert, Sarah Conway Ciminera, William Geiger, Renee Rancourt, Alex Vitale, Victoria Zinny, Savina Gersak, Bo Svenson; *Dir:* Jeff Kwitny. **VHS, LV** $89.95 COL, PIA ✓½

Beyond the Doors Schlockmeister Buchannan directed this feeble attempt to convince the American public that the American government, in an effort to wipe out rock music, murdered Jimi Hendrix and Janis Joplin, and scared Jim Morrison into hiding. Also known as "Down on Us."
1983 60m/C Gregory Allen Chatman, Riba Meryl, Brian Wolf, Sandy Kenyon; *Dir:* Larry Buchanan. **VHS, Beta** UNI, HHE ✓

Beyond Dream's Door A young, All-American college student's childhood nightmares come back to haunt him, making dreams a horrifying reality.
1988 86m/C Nick Baldasare, Rick Kesler, Susan Pinsky, Norm Singer; *Dir:* Jay Woelfel. **VHS, Beta** $29.98 VID ✓

Beyond Erotica After being cut out of his father's will, a sadistic young man takes out his rage on his mother and a peasant girl.
1979 90m/C David Hemmings, Alida Valli, Andrea Rau. **VHS** $59.95 PSM ✓

Beyond Evil A newlywed couple moves into an old mansion despite rumors that the house is haunted. The wife becomes possessed by the vengeful spirit of a woman murdered 200 years earlier, and a reign of terror begins.
1980 (R) 98m/C John Saxon, Lynda Day George, Michael Dante, Mario Milano; *Dir:* Herb Freed. **VHS, Beta** $49.95 MED ✓

Beyond Fear Compelling drama about a man forced to aid a gang in robbery while they hold his wife and son captive.
1975 92m/C Michael Boquet. **VHS, Beta** $49.95 PSM, CCI ✓✓½

Beyond the Forest Camp diva Davis, in her last role for Warner Bros., really turns on the histrionics as a big-city gal married to a small-town guy (Cotten) and bored out of her mind. Although the most memorable line, "What a dump," has become larger-than-life, the film itself is rather small and muddled (interestingly, Vidor directed "The Fountainhead" the same year.) A trashy melodrama full of ennui, envy, unwanted pregnancy and murder, it's definitely high on camp and low on art.
1949 96m/B Bette Davis, Joseph Cotten, David Brian, Ruth Roman, Minor Watson, Regis Toomey; **Dir:** King Vidor. **VHS $19.95** *MGM, FCT* ♫♫ ½

Beyond Innocence A 17-year-old lusts after a mature married woman. Ostensibly based on Raymond Radiguet's "Devil in the Flesh."
1987 87m/C Keith Smith, Katia Caballero, John Morris; **Dir:** Scott Murray. **VHS, Beta $79.95** *SVS* ♫

Beyond the Law Spaghetti western with Van Cleef as the too smart bad guy. He becomes sheriff, picks up the stack of silver at the depot, and disappears. Humorous and clever, with fine location photography. Also known as "Bloodsilver."
1968 91m/C *IT* Lee Van Cleef, Antonio Sabato, Lionel Stander, Bud Spencer, Gordon Mitchell, Ann Smyrner; **Dir:** Giorgio Stegani. **VHS $9.99** *QVD, IMP* ♫♫

Beyond the Limit The story of an intense and darkly ominous love triangle which takes place in the South American coastal city of Corrientes. Based on Graham Greene's novel "The Honorary Consul."
1983 (R) 103m/C Michael Caine, Richard Gere, Bob Hoskins, Elpidia Carrillo; **Dir:** John MacKenzie. **VHS, Beta $59.95** *PAR* ♫♫ ½

Beyond the Next Mountain A missionary in China attempts to convert all those he meets.
197? 97m/C Alberto Isaac, Bennett Ohta, James F. Collier, Rolf Forsberg. **VHS, Beta $79.95** *UHV, VGD* ♫

Beyond Obsession A melodramatic love triangle between an Italian ex-diplomat imprisoned in North Africa, a tempestuous young girl, and a lustful American.
1989 (R) 116m/C *IT* Marcello Mastroianni, Tom Berenger, Elenora Giorgi; **Dir:** Liliana Cavani. **VHS, Beta $9.95** *SIM, VID, HHE* ♫♫

Beyond the Poseidon Adventure A sequel to the 1972 film in which salvage teams and ruthless looting vandals compete for access to the sinking ocean liner. Sinking ships should be abandoned.
1979 (PG) 115m/C Michael Caine, Sally Field, Telly Savalas, Peter Boyle, Jack Warden, Slim Pickens, Shirley Knight, Shirley Jones, Karl Malden, Mark Harmon; **Dir:** Irwin Allen. **VHS, Beta $59.95** *WAR* ♫

Beyond Reason A psychologist uses unorthodox methods by treating the criminally insane with dignity and respect.
1977 88m/C Telly Savalas, Laura Johnson, Diana Muldaur, Marvin Laird, Priscilla Barnes; **Dir:** Telly Savalas. **VHS, Beta $59.95** *MED* ♫

Beyond a Reasonable Doubt In order to get a behind-the-scenes glimpse at the judicial system, a man plays the guilty party to a murder. Alas, when he tries to vindicate himself, he is the victim of his own folly. Not as interesting as it sounds on paper.
1956 80m/B Dana Andrews, Joan Fontaine, Sidney Blackmer, Philip Bourneuf, Barbara Nichols, Sheppard Strudwick, Arthur Franz, Edward Binns; **Dir:** Fritz Lang. **VHS $69.95** *VID* ♫♫

Beyond Reasonable Doubt A chilling true-life murder mystery which shatters the peaceful quiet of a small New Zealand town and eventually divides a country.
1980 127m/C *NZ* David Hemmings, John Hargreaves; **Dir:** John Laing. **VHS, Beta $69.95** *VID* ♫♫ ½

Beyond the Silhouette It starts out as another sleazy video sex thriller, as the lawyer heroine discovers her sensuality and poses a lot in her underclothes. Then in the third act it become a hyper-paranoid political-conspiracy assassination-o-rama. Pretty weird, Canadian-made junk, also known as "Ultimate Desires."
1990 90m/C Tracy Scoggins, Marc Singer, Brion James; **Dir:** Lloyd A. Simandl. **VHS, Beta** *NVH* ♫

Beyond the Stars A sci-fi adventure directed by the author of "Cocoon," wherein a whiz-kid investigates the NASA cover-up of a deadly accident that occurred on the moon during the Apollo 11 landing. Unfortunately, the interesting cast can't make up for the script.
1989 94m/C Martin Sheen, Christian Slater, Olivia D'Abo, F. Murray Abraham, Robert Foxworth, Sharon Stone; **Dir:** David Saperstein. **VHS, Beta, LV $89.95** *IVE* ♫

Beyond Therapy A satire on modern psychotherapy, from the play by Christopher Durang, about a confused, crazily neurotic couple and their respective, and not any saner, analysts. Unfortunately, comes off as disjointed and confused.
1986 (R) 93m/C Jeff Goldblum, Tom Conti, Julie Hagerty, Glenda Jackson, Christopher Guest; **Dir:** Robert Altman. **VHS, Beta $19.95** *STE, NWV* ♫ ½

Beyond the Time Barrier Air Force test pilot gets more than he bargained for when his high speed plane carries him into the future. There he sees the ravages of an upcoming plague, to which he must return.
1960 75m/B Robert Clarke, Darlene Tompkins, Arianne Arden, Vladimir Sokoloff; **Dir:** Edgar G. Ulmer. **VHS, Beta $16.95** *SNC, CNM, SMW* ♫♫ ½

Beyond Tomorrow Young romance is guided from the spirit world during the Christmas season, as two "ghosts" come back to help young lovers.
1940 84m/B Richard Carlson, C. Aubrey Smith, Jean Parker, Charles Winninger, Harry Carey, Maria Ouspenskaya; **Dir:** Edward Sutherland. **VHS, Beta $16.95** *SNC, NOS, DVT* ♫♫

Beyond the Valley of the Dolls Russ Meyer ("Faster, Pussycat! Kill! Kill!") directed this Hollywood parody ("BVM," as it came to be known) about an all-girl rock combo and their search for stardom. Labeled the first "exploitation horror camp musical"—how can you pass that up? Screenplay by film critic Roger Ebert, from an original story by Ebert and Meyer.
1970 109m/C Edy Williams, Dolly Reed; **Dir:** Russ Meyer. **VHS, Beta, LV** *FOX* ♫

Beyond the Walls The opposing factions of a hellish Israeli prison unite to beat the system. Brutal with good characterizations. Available in both subtitled and dubbed versions.
1985 (R) 104m/C *IS* Arnon Zadok, Muhamad Bakri; **Dir:** Uri Barbash. Venice Film Festival '84: Best Film. **VHS, Beta $79.95** *WAR, TAM* ♫♫ ½

Bible...In the Beginning The book of Genesis is dramatized, including the stories of Adam and Eve, Cain and Abel, and Noah and the flood.
1966 155m/C *IT* George C. Scott, Richard Harris, Stephen Boyd, John Huston, Peter O'Toole, Franco Nero, Ava Gardner; **Dir:** John Huston. National Board of Review Awards '66: 10 Best Films of the Year. **VHS, Beta, LV $29.98** *FOX* ♫♫♫

The Bicycle Thief A world classic and indisputable masterpiece about an Italian workman who finds a job, only to have the bike he needs for work stolen; he and his son search Rome for it. A simple story that seems to contain the whole of human experience, and the masterpiece of Italian neorealism. In Italian with English subtitles.
1948 90m/B *IT* Lamberto Maggiorani, Enzo Staiola, Lianella Carell; **Dir:** Vittorio DeSica. Academy Awards '49: Best Foreign Language Film; British Academy Awards '49: Best Film; National Board of Review Awards '49: Best Director, Best Film; New York Film Critics Awards '49: Best Foreign Film; Cinematheque Belgique Survey '52: 3rd Best Film of All Time; Sight & Sound Survey '52: 1st Best Film of All Time. **VHS, Beta, LV $69.95** *TAM, NOS, MRV* ♫♫♫♫

Big A 13-year-old boy makes a wish at a carnival fortune-teller to be "big." When he wakes up the next morning he finds that he suddenly won't fit into his clothes and his mother doesn't recognize him. Until he finds a cure, he must learn to live in the real world. Perkins is wonderful as a warming cynic while Hanks is totally believable as the little boy inside a man's body. Marshall directs with authority and the whole thing clicks from the beginning.
1988 (PG) 98m/C Tom Hanks, Elizabeth Perkins, John Heard, Robert Loggia, Jared Rushton, David Moscow, Jon Lovitz, Mercedes Ruehl; **Dir:** Penny Marshall. **VHS, Beta, LV $19.98** *FOX, BUR* ♫♫♫ ½

Big Bad John Some good ol' boys ride around in trucks as they get into a variety of shootouts. The soundtrack includes music by Willie Nelson, The Charlie Daniels Band, and others.
1990 (PG-13) 91m/C Jimmy Dean, Ned Beatty, Jack Elam, Bo Hopkins, Romy Windsor, Doug English. **VHS, Beta, LV $89.98** *MAG, IME* ♫ ½

Big Bad Mama A tough, intelligent, pistol-packing mother moves her two teenage daughters out of Texas during the Depression, and they all turn to robbing banks as a means of support.
1974 (R) 83m/C Angie Dickinson, William Shatner, Tom Skerritt, Susan Sennett, Robbie Lee, Sally Kirkland, Noble Willingham; **Dir:** Steve Carver. **VHS, Beta $39.98** *NO* ♫

Big Bad Mama 2 Belated Depression-era sequel to the 1974 Roger Corman gangster film, where the pistol-packin' matriarch battles a crooked politician with the help of her two daughters.
1987 (R) 85m/C Angie Dickinson, Robert Culp, Danielle Brisebois, Julie McCullough; **Dir:** Jim Wynorski. **VHS, Beta $79.95** *MGM* ♫ ½

The Big Bet High school sex comedy about a guy who is challenged by the school bully to get the gorgeous new girl into bed. Energetic romp is for adults.
1985 90m/C Sylvia Kristel, Kimberly Evenson, Ron Thomas; **Dir:** Bert I. Gordon. **VHS, Beta $79.98** *CGH* ♫

Big Bird Cage Several females living out prison terms in a rural jail decide to defy their homosexual guards and plan an escape. They are aided by revolutionaries led by a Brooklynese expatriate and his lover. Two of the girls survive the escape massacre. Sequel to "The Big Doll House."
1972 (R) 93m/C Pam Grier, Sid Haig, Anitra Ford, Candice Roman, Teda Bracci, Carol Speed, Karen McKevic; *Dir:* Jack Hill. **VHS, Beta** $39.98 *WAR* ✂✂

The Big Blue A semi-true tale about a professional deep sea diver, who dives without the aid of any kind of breathing apparatus. Chronicles his life and the loves weathered by his efforts to be the champion.
1988 (PG) 122m/C Rosanna Arquette, Jean Reno, Jean-Marc Barr, Paul Shenar, Sergio Castellitto, Marc Duret, Griffin Dunne; *Dir:* Luc Besson. **VHS, Beta, LV** $89.95 *COL* ✂✂½

Big Bluff Disappointing result from an interesting premise; fatally ill woman finds love, but when she surprisingly recovers, her new husband decides to help her back along the path to death. Uneven and melodramatic.
1955 70m/B *SI* John Bromfield, Martha Vickers, Robert Hutton, Rosemary Bowie; *Dir:* W. Lee Wilder. **VHS** $19.99 *NOS, MOV* ✂✂

Big Boss A hoodlum climbs to the top of a crime syndicate. An Italian film previously titled "Mr. Scarface."
1977 (R) 90m/C *IT* Jack Palance, Edmund Purdom, Al Cliver, Harry Baer, Gisela Hahn. **Beta** *MFI* ✂✂½

The Big Brawl A Chicago gangster recruits a martial arts expert to fight in a free-for-all match in Texas.
1980 (R) 95m/C Jackie Chan, Jose Ferrer, Mako, Rosalind Chao; *Dir:* Robert Clouse. **VHS, Beta** $19.98 *WAR* ✂½

Big Bus The wild adventures of the world's first nuclear-powered bus as it makes its maiden voyage from New York to Denver. Clumsy disaster-movie parody.
1976 (PG) 88m/C Joseph Bologna, Stockard Channing, Ned Beatty, Ruth Gordon, Larry Hagman, John Beck, Jose Ferrer, Lynn Redgrave, Sally Kellerman, Stuart Margolin, Richard Mulligan, Howard Hesseman, Richard B. Shull; *Dir:* James Frawley. **VHS, Beta** $49.95 *PAR* ✂✂

Big Business Strained high-concept comedy about two sets of identical twins, each played by Tomlin and Midler, mismatched at birth by a near-sighted country nurse. The city set of twins intends to buy out the factory where the country set of twins work. So the country twins march up to the big city to stop the sale and destruction of their beloved home. Both set of twins stay in the Plaza Hotel and zany consequences ensue. Essentially a one-joke outing with some funny moments, but talented comediennes Midler and Tomlin are somewhat wasted. Great technical effects.
1988 (PG) 98m/C Bette Midler, Lily Tomlin, Fred Ward, Edward Herrmann, Michele Placido, Barry Primus, Michael Gross, Mary Gross, Daniel Gerroll, Roy Brocksmith; *Dir:* Jim Abrahams. **VHS, Beta, LV, 8mm** $89.95 *TOU* ✂✂

Big Bust Out Several female convicts escape from prison only to be sold into slavery and face additional torture. Nothing redeeming about this exploitative film.
1973 (R) 75m/C Vonetta McGee, Monica Taylor, Linda Fox, Karen Carter, Gordon Mitchell; *Dir:* Karen Carter. **VHS, Beta** $24.98 *SUE* Woof!

Big Calibre A rancher is inches away from being lynched for his dad's murder before he is found to be innocent.
1935 59m/B Bob Steele, Bill Quinn, Earl Dwire, Peggy Campbell, John Elliott, Georgia O'Dell; *Dir:* Robert N. Bradbury. **VHS, Beta** $12.95 *SNC, DVT, WGE* ✂

The Big Cat A mountain valley in Utah is ravaged by a cougar, and the tense atmosphere is heightened by the hatred between two feuding men.
1949 75m/C Lon McCallister, Peggy Ann Garner, Preston Foster, Forrest Tucker; *Dir:* Phil Karlson. **VHS, Beta** $19.95 *NOS, MRV, DVT* ✂

Big Chase A rookie cop chases down a mob of payroll thieves in this early action film.
1954 60m/B Glenn Langan, Adele Jergens, Lon Chaney Jr., Jim Davis. **VHS, Beta, LV** *WGE* ✂

The Big Chill Seven former '60s radicals, now coming upon middle age and middle class affluence, reunite on the occasion of an eighth friend's suicide and use the occasion to re-examine their past relationships and commitments. A beautifully acted, immensely enjoyable ballad to both the counter-culture and its Yuppie descendants. Great period music. Oscar nominated performance from Close. Kevin Costner is the dead man whose scenes never made it to the big screen.
1983 (R) 108m/C Tom Berenger, Glenn Close, Jeff Goldblum, William Hurt, Kevin Kline, Mary Kay Place, Meg Tilly, JoBeth Williams; *Dir:* Lawrence Kasdan. **VHS, Beta, LV, 8mm** $14.95 *COL, VYG, CRC* ✂✂✂½

Big Combo A gangster's ex-girlfriend helps a cop to smash a crime syndicate. Focuses on the relationship between Wilde's cop and the gangster Conte in an effective film noir, with some scenes of torture that were ahead of their time.
1955 87m/B Cornel Wilde, Richard Conte, Jean Wallace, Brian Donlevy, Earl Holliman, Lee Van Cleef, Helen Walker; *Dir:* Joseph H. Lewis. **VHS, Beta** $19.95 *PSM, WES, KOV* ✂✂

Big Country A sprawling epic in which a sea captain who returns from the Pacific with his wife finds himself without moorings in the midst of a family squall. Classic musical score by Jerome Moross was Oscar nominated.
1958 168m/C Gregory Peck, Charlton Heston, Burl Ives, Jean Simmons, Carroll Baker, Chuck Connors; *Dir:* William Wyler. Academy Awards '58: Best Supporting Actor (Ives). **VHS, Beta** $29.98 *MGM, BTV* ✂✂✂

The Big Crimewave A cast of unknowns in a comedy about a loner who takes a bus to Kansas City (Kansas City?) to become a screenwriter, with comic adventures along the way. A feast of jabs at genre films.
1986 80m/C *CA* John Paizs, Eva Covacs, Darrel Baran; *Dir:* John Paizs. **VHS, Beta** *CGH, PEV, HHE* ✂✂

Big Deadly Game A vacationing American gets caught up in a complicated espionage plot by helping a mysterious wartime buddy.
1954 63m/B Lloyd Bridges, Simone Silva, Finlay Currie. **VHS, Beta, LV** *WGE* ✂½

Big Deal on Madonna Street A band of inept crooks plan to make themselves very rich when they attempt to rob a pawnshop. In Italian with English subtitles.
1960 90m/B *IT* Marcello Mastroianni, Vittorio Gassman, Toto, Claudia Cardinale; *Dir:* Mario Monicelli. **VHS, Beta** $29.95 *TAM, CVC, MRV* ✂✂✂½

The Big Doll House Roger Corman-produced prison drama about a group of tormented female convicts who decide to break out. Features vintage Grier, and a caliber of women's-prison sleaziness that isn't equalled in today's films. Also on video as "Women in Cages."
1971 (R) 93m/C Judy Brown, Roberta Collins, Pam Grier, Brooke Mills, Pat Woodell, Sid Haig, Christianne Schmidtmer, Kathryn Loder; *Dir:* Jack Hill. **VHS, Beta** $24.98 *SUE, CHA* ✂✂

The Big Easy A terrific thriller. A slick New Orleans detective uncovers a heroin-based mob war while romancing the uptight assistant DA who's investigating corruption on the police force. An easy, Cajun-flavored mystery, a fast-moving action-comedy, a very sexy romance, and a serious exploration of the dynamics of corruption.
1987 (R) 101m/C Dennis Quaid, Ellen Barkin, Ned Beatty, John Goodman, Ebboe Roe Smith, Charles Ludlum; *Dir:* Jim McBride. **VHS, Beta, LV** $14.99 *HBO, FCT* ✂✂✂½

The Big Fix Private investigator Moses Wine finds himself in an ironic situation: searching for a fugitive alongside whom he'd protested in the 60s. Based on a Roger Simen novel.
1978 (PG) 108m/C Richard Dreyfuss, Susan Anspach, Bonnie Bedelia, John Lithgow, F. Murray Abraham, Fritz Weaver, Mandy Patinkin; *Dir:* Jeremy Paul Kagan. **VHS, Beta** $59.95 *MCA* ✂✂✂

Big Foot Even genre devotees will be disappointed with this one. Sasquatch, who has procreation on his mind, searches rather half-heartedly for a human mate. A horror flick that forgot to include the horror.
1972 92m/C Chris Mitchum, Joi Lansing, John Carradine, John Mitchum; *Dir:* Bob Slatzer. **VHS, Beta** $39.95 *WES* Woof!

The Big Gag An international group of comedians travel the world and pull gags on people. Lame.
1987 (R) 84m/C **VHS, Beta** $79.95 *TWE, HHE* ✂

The Big Grab A dark big caper thriller about a couple of aging French thugs determined to rip off a casino at Cannes. In French with English subtitles. Also called "Any Number Can Win."
1963 118m/B *FR* Jean Gabin, Alain Delon, Viviane Romance, Maurice Biraud, Carla Marlier; *Dir:* Henri Verneuil. **VHS, Beta** $19.95 *FCT* ✂✂½

A Big Hand for the Little Lady Fonda and Woodward, playing two Westheaded country bumpkins, get involved in a card game in Laredo against high rollers Robards and McCarthy. Fonda risks their savings, finds himself stuck with a losing hand, and has a bit of heart trouble; that's where the little lady comes in. Fine performances and a nifty twist ending don't entirely compensate for the overly-padded script (which evolved from a 48-minute TV play drafted by Sydney Carroll).
1966 95m/C Henry Fonda, Joanne Woodward, Jason Robards Jr., Charles Bickford, Burgess Meredith, Kevin McCarthy; *Dir:* Fielder Cook. **VHS, Beta** $59.95 *WAR, FCT* ✂✂½

Big Heat When detective Ford's wife (played by Jocelyn Brando, sister of Marlon) is killed in an explosion meant for him, he pursues the gangsters behind it and uncovers a police scandal. His appetite is whetted after this discovery and he pursues the criminals even more vigorously with the help of gangster moll Gloria Grahame. Definitive film noir.

1953 90m/B Glenn Ford, Lee Marvin, Gloria Grahame, Jocelyn Brando, Alexander Scourby, Carolyn Jones; *Dir:* Fritz Lang. Edgar Allan Poe Awards '53: Best Screenplay. **VHS, Beta, LV $59.95** COL 🎬🎬½

The Big Hurt A reporter investigating a bizarre double murder discovers a secret government agency involved in torture and mind-control.
1987 (R) 90m/C David Bradshaw, Lian Lunson, Simon Chilvers, Nick Waters; *Dir:* Barry Peak. **VHS, Beta, LV $79.98** MAG 🎬½

Big Jake An aging Texas cattle man who has outlived his time swings into action when outlaws kidnap his grandson and wound his son. He returns to his estranged family to help them in the search for Little Jake. O'Hara is once again paired up with Wayne and the chemistry is still there.
1971 (PG) 90m/C John Wayne, Richard Boone, Maureen O'Hara, Patrick Wayne, Chris Mitchum, Bobby Vinton; *Dir:* George Sherman. **VHS, Beta $19.98** FOX, BUR 🎬🎬

Big Jim McLain Wayne and Arness are federal agents working on behalf of the House Un-American Activities Committee to eliminate communist terrorism in Hawaii. And there's a suspicious psychiatrist, too: Wayne falls for a babe whose boss is a shrink who doesn't quite seem on the level. Definitely not a highpoint in the Duke's career.
1952 90m/C John Wayne, Nancy Olson, James Arness, Veda Ann Borg; *Dir:* Edward Ludwig. **VHS, LV $19.98** WAR, FCT 🎬🎬

Big Lift Two G.I.'s assigned to the Berlin airlift ally themselves in counter-intelligence when they discover that their mutual girlfriend is a spy.
1060 110m/B Montgomery Clift, Paul Douglas, Cornell Borchers, Bruni Lobel, O.E. Hasse; *Dir:* George Seaton. **VHS, Beta $19.95** NOS, MRV, DVT 🎬🎬½

Big Man on Campus A modern-day Quasimodo makes his home in an affluent university's belltower. Of course he falls in love with one of the pretty young co-eds and races out of his tower only to be captured by the psychology department. Really, really bad.
1989 (PG-13) 102m/C Tom Skerritt, Corey Parker, Allan Katz, Cindy Williams, Melora Hardin, Jessica Harper; *Dir:* Jeremy Paul Kagan. **VHS, Beta, LV $33.96** VES 🎬

The Big Man: Crossing the Line A down on his luck British miner joins cahoots with a gangster and takes part in an illegal bare-knuckled boxing match. But will it cost him his wife and children? Based on the novel by William McIlvanney.
1991 (R) 93m/C Liam Neeson, Joanne Whalley-Kilmer, Ian Bannen, Billy Connolly; *Dir:* David Leland. **VHS, LV $89.95** COL 🎬½

Big Meat Eater A musical gore-comedy about extraterrestrials using radioactive butcher's discards for ship fuel. Deliberate camp that is so bad its funny!
1985 81m/C CA George Dawson, Big Miller, Andrew Gillies; *Dir:* Chris Windsor. **VHS, Beta $49.95** MED 🎬🎬

Big Mo The true story of the friendship that developed between Cincinnati Royals basketball stars Maurice Stokes and Jack Twyman after a strange paralysis hit Stokes. Also released under the title "Maurie."
1973 (G) 110m/C Bernie Casey, Bo Svenson, Stephanie Edwards, Janet MacLachlan; *Dir:* Daniel Mann. **VHS, Beta $69.95** VES 🎬

Big Mouth A dopey fisherman gets ahold of a treasure map and is pursued by cops and gangsters. Standard Lewis fare, with the requisite infantile histrionics; must be French to appreciate.
1967 107m/C Jerry Lewis, Jeannine Riley, Harold J. Stone, Charlie Callas, Buddy Lester, Susan Bay; *Dir:* Jerry Lewis. **VHS, Beta $49.98** COL 🎬

Big News Based on the play "For Two Cents" by George S. Brooks, this early talkie uses sound to great advantage. Fired for going after a gangster who's a big advertiser (Hardy), reporter Armstrong nevertheless keeps after the crook. When the intrepid reporter pushes too far, murder enters the picture.
1929 75m/B Robert Armstrong, Carole Lombard, Tom Kennedy, Warner Richmond, Wade Boteler, Sam Hardy, Lew Ayres; *Dir:* Gregory La Cava. **VHS $19.95** NOS, TIM 🎬🎬

The Big Parade Wonderful WWI film. Gilbert and Adoree are exceptional as the lovers torn apart by war. Interesting and thoughtful picture of the trauma and trouble brought to men and their loved ones in wartime. Battle scenes are compelling and intense.
1925 141m/B John Gilbert, Renee Adoree, Hobart Bosworth, Claire McDowell, Claire Adams, Karl Dane; *Dir:* King Vidor. **VHS, LV $29.95** MGM 🎬🎬🎬🎬

The Big Picture A hilarious, overlooked comedy by and starring a variety of Second City/National Lampoon alumni, about a young filmmaker who is contracted by a big studio, only to see his vision trampled by formula-minded producers, crazed agents, hungry starlets, and every other variety of Hollywood predator.
1989 (PG-13) 95m/C Kevin Bacon, Jennifer Jason Leigh, Martin Short, Michael McKean, Emily Longstreth, J.T. Walsh, Eddie Albert, Richard Belzer, John Cleese, June Lockhart, Stephen Collins, Roddy McDowall; *Dir:* Christopher Guest. **VHS, Beta, LV $89.95** COL 🎬🎬½

The Big Push Originally entitled "Timber Tramps." A motley bunch of Alaskan lumberjacks get together to save a poor widow's logging camp from a pair of greedy mill owners.
1975 (PG) 98m/C Joseph Cotten, Claude Akins, Cesar Romero, Tab Hunter, Roosevelt Grier, Leon Ames, Stubby Kaye, Patricia Medina; *Dir:* Tay Garnett. **VHS, Beta $59.95** GEM, STC 🎬½

The Big Rascal Two brothers grow up, learn martial arts, and protect their village amid much kicking and grunting.
19?? ?m/C VHS $29.98 BFV 🎬½

Big Red Set amid the spectacular beauty of Canada's Quebec Province, an orphan boy protects a dog which later saves him from a mountain lion.
1962 89m/C Walter Pidgeon, Gilles Payant; *Dir:* Norman Tokar. **VHS, Beta $69.95** DIS 🎬🎬½

The Big Red One Fuller's harrowing, intense semi-autobiographical account of the U.S. Army's famous First Infantry Division in World War II, the "Big Red One." A rifle squad composed of four very young men, led by the grizzled Marvin, cut a fiery path of conquest through North Africa to the liberation of the concentration camp at Falkenau, Czechoslovakia. In part a tale of lost innocence, the film scores highest by bringing the raw terror of war down to the individual level.

1980 (PG) 113m/C Lee Marvin, Robert Carradine, Mark Hamill, Stephane Audran; *Dir:* Samuel Fuller. **VHS, Beta, LV $19.98** FOX, WAR 🎬🎬🎬½

The Big Scam Criminal mastermind Niven recruits ex-con Jordan to pull off a massive bank heist. Originally titled, "A Nightingale Sang in Berkeley Square" and also on video as "The Mayfair Bank Caper."
1979 102m/C Richard Jordan, David Niven, Oliver Tobias, Elke Sommer, Gloria Grahame, Hugh Griffith, Richard Johnson; *Dir:* Ralph Thomas. **VHS $9.95** SIM, VID 🎬🎬½

Big Score When a policeman is falsely accused and dismissed from the Chicago Police Department, he goes after the men who really stole the money from a drug bust. Script was originally intended to be a Dirty Harry flick; too bad it wasn't.
1983 (R) 88m/C Fred Williamson, John Saxon, Richard Roundtree, Nancy Wilson, Ed Lauter, Ron Dean, D'Urville Martin, Michael Dante, Joe Spinell; *Dir:* Fred Williamson. **VHS, Beta $69.95** VES 🎬

Big Shots Two kids, one naive and white, the other black and streetwise, search for a stolen watch.
1987 (PG-13) 91m/C Ricky Busker, Darius McCrary, Robert Joy, Paul Winfield, Robert Prosky, Jerzy Skolimowski; *Dir:* Robert Mandel. **VHS, Beta, LV $19.98** LHV, WAR 🎬

Big Show A western adventure about the making of a western adventure, with Autry playing a duel role.
1937 54m/B Gene Autry, Smiley Burnette; *Dir:* Mack V. Wright. **VHS, Beta $19.95** NOS, VCN, CPB 🎬½

Big Showdown A young man uses his martial arts skills against the man who murdered his sister and father.
197? 90m/C Shiang Hwa Chyang, Jin Fu War. **VHS, Beta $29.95** UNI 🎬

The Big Sky It's 1830, and a rowdy band of fur trappers embark upon a back breaking expedition up the uncharted Missouri River. Based on the A.B. Guthrie Jr. novel, it's an effortlessly enjoyable and levelheaded Hawksian American myth, with a streak of gentle gallows humor. Oscar nominations: Best Actor (Hunnicutt), Black & White Cinematography.
1952 122m/C Kirk Douglas, Dewey Martin, Arthur Hunnicutt, Elizabeth Threatt, Buddy Baer, Steve Geray, Jim Davis; *Dir:* Howard Hawks. **VHS, Beta $19.98** RKO, FCT 🎬🎬🎬

The Big Sleep Private eye Philip Marlowe, hired to protect a young woman from her own indiscretions, falls in love with her older sister while uncovering murders galore. A dense, chaotic thriller that succeeded in defining and setting a standard for its genre. William Faulkner was one of the script's co-authors. The very best Raymond Chandler on film.
1946 114m/B Humphrey Bogart, Lauren Bacall, Martha Vickers, Elisha Cook Jr., Bob Steele, Dorothy Malone; *Dir:* Howard Hawks. **VHS, Beta, LV $19.95** MGM, FOX, PIA 🎬🎬🎬🎬

The Big Sleep A tired remake of the Raymond Chandler potboiler about exhausted Los Angeles private dick Marlowe and his problems in protecting a wild young heiress from her own decadence and mob connections. Mitchum appears to need a rest.
1978 (R) 99m/C Robert Mitchum, Sarah Miles, Richard Boone, Candy Clark, Edward Fox, Joan Collins, John Mills, James Stewart, Oliver Reed, Harry Andrews, James Donald, Colin Blakely,

Richard Todd; **Dir:** Michael Winner. **VHS, Beta, LV** $19.98 COL, PIA ✂✂

The Big Slice Two would-be crime novelists want to improve their fiction. One masquerades as a cop, the other as a crook, and they infiltrate the underworld from both ends. Clever comedy premise, but vaudeville-level jokes fall flat.
1991 (R) 86m/C Casey Siemaszko, Leslie Hope, Justin Louis, Ken Welsh, Nicholas Campbell, Heather Locklear; **W/Dir:** John Bradshaw. **VHS** $19.95 ACA ✂ ½

The Big Sombrero Autry takes a stand against the marriage between an unsuspecting, wealthy Mexican girl and the fortune-seeking bridegroom who wants her land.
1949 77m/C Gene Autry, Elena Verdugo, Steve Dunne, George Lewis; **Dir:** Frank McDonald. **VHS** $39.98 NO ✂✂½

Big Steal An Army officer recovers a missing payroll and captures the thieves after a tumultuous chase through Mexico.
1949 72m/B Robert Mitchum, William Bendix, Jane Greer, Ramon Novarro, Patric Knowles; **Dir:** Don Siegel. **VHS, Beta, LV** MED ✂✂✂

Big Store Late Marx Brothers in which they are detectives in a large metropolitan department store, foiling a hostile takeover and preventing a murder. Their last MGM effort, with some good moments between the Tony Martin song numbers which include "If It's You" and the immortal "Tenement Symphony." Groucho also leads the "Sing While You Sell" number.
1941 83m/B Groucho Marx, Harpo Marx, Chico Marx, Tony Martin, Margaret Dumont, Virginia Grey, Virginia O'Brien; **Dir:** Charles Riesner. **VHS, Beta** $19.95 MGM, CCB ✂✂½

Big Street A timid busboy, in love with a disinterested nightclub singer, gets to prove his devotion when she is crippled in a fall. Based on a Damon Runyon story, "Little Pinks."
1942 88m/B Henry Fonda, Lucille Ball, Agnes Moorehead, Louise Beavers, Barton MacLane, Eugene Pallette, Ozzie Nelson; **Dir:** Irving Reis. **VHS, Beta, LV** $19.95 MED, RKO ✂✂½

The Big Sweat Maybe you've heard this one before: a man who has been framed by the mob escapes from prison and heads for Mexico. But first, he's got to get past the mobsters and police who are hot on his trail. And maybe you've seen some of it before: some of the same stock footage appears in Lommel's "Cold Heat."
1991 85m/C Robert Z'Dar, Ken Letner, Steve Molone, Kevin McBride, Cheri Caspari, Joanne Watkins; **Dir:** Ulli Lommel. **VHS** $79.95 AIP ✂

Big Switch A gambler is framed for murder and becomes embroiled in the plot to reinstate an old gangster kingpin.
1970 (R) 68m/C Sebastian Breaks, Virginia Wetherell, Erika Raffael; **Dir:** Pete Walker. **VHS, Beta** $59.95 MON ✂½

Big Top Pee Wee Pee Wee's second feature film following the success of "Pee Wee's Big Adventure." This time Pee Wee owns a farm, has a girlfriend(!) and lives the good life until a weird storm blows a traveling circus on to his property. Cute, but not the manic hilarity the first one was. Music by Danny Elfman.
1988 (PG) 86m/C Pee-Wee Herman, Kris Kristofferson, Susan Tyrrell, Penelope Ann Miller, Pee-Wee Herman; **Dir:** Randal Kleiser. **VHS, Beta, LV, 8mm** $14.95 PAR ✂✂

Big Town A farmboy, lucky with dice, hits Chicago to claim his fortune where he meets floozies, criminals, and other streetlife. Look for Lane's strip number.
1987 (R) 109m/C Matt Dillon, Diane Lane, Tommy Lee Jones, Bruce Dern, Tom Skerritt, Lee Grant, Suzy Amis; **Dir:** Ben Bolt. **VHS, Beta, LV** $89.98 VES, LIV ✂✂

Big Town After Dark Daring journalists search for the bottom-line story on a gang of criminals. They find themselves caught behind the firing lines when the sun goes down. Slightly better than average thriller. Also known as "Underworld After Dark."
1947 69m/B Philip Reed, Hillary Brooke, Richard Travis, Anne Gillis; **Dir:** William C. Thomas. **VHS** $16.95 NOS, SNC, RXM ✂✂

Big Trail This pioneering effort in wide-screen cinematography was Wayne's first feature film. A wagon train on the Oregon trail encounters Indians, buffalo, tough terrain, and romantic problems.
1930 110m/B John Wayne, Marguerite Churchill, El Brendel, Tully Marshall, Tyrone Power Sr., Ward Bond, Helen Parrish; **Dir:** Raoul Walsh. **VHS, Beta** $19.98 FOX ✂✂✂

Big Trees A ruthless lumberman attempts a takeover of the California Redwood Timberlands that are owned by a group of peaceful homesteaders.
1952 89m/C Kirk Douglas, Patrice Wymore, Eve Miller, Alan Hale Jr., Edgar Buchanan; **Dir:** Felix Feist. **VHS, Beta** $19.95 NOS, MRV, QNE ✂✂✂

Big Trouble An insurance broker endeavors to send his three sons to Yale by conspiring with a crazy couple in a fraud scheme that goes awry in every possible manner.
1986 (R) 93m/C Alan Arkin, Peter Falk, Beverly D'Angelo, Charles Durning, Robert Stack, Paul Dooley, Valerie Curtin, Richard Libertini; **Dir:** John Cassavetes. **VHS, Beta, LV** $29.95 RCA ✂✂

Big Trouble in Little China A trucker plunges beneath the streets of San Francisco's Chinatown to battle an army of spirits. An uproarious comic-book-film parody with plenty of action and a keen sense of sophomoric sarcasm.
1986 (PG-13) 99m/C Kurt Russell, Suzze Pai, Dennis Dun, Kim Cattrall; **Dir:** John Carpenter. **VHS, Beta, LV** $14.98 FXV, FOX ✂✂½

Big Valley The western television series's two feature-length episodes: "Explosion," and "Legend of a General."
1968 90m/C Richard Long, Linda Evans, Barbara Stanwyck, Lee Majors. **VHS, Beta** $69.98 VID ✂✂½

Big Wednesday Three California surfers from the early sixties get back together after the Vietnam war to reminisce about the good old days and take on the big wave.
1978 (PG) 120m/C Jan-Michael Vincent, Gary Busey, William Katt, Lee Purcell, Patti D'Arbanville; **Dir:** John Milius. **VHS, Beta, LV** $59.95 WAR, PIA ✂✂½

The Big Wheel Old story retold fairly well. Rooney is young son determined to travel in his father's tracks as a race car driver, even when dad buys the farm on the oval. Good acting and direction keep this a cut above average.
1949 92m/B Mickey Rooney, Thomas Mitchell, Spring Byington, Mary Hatcher, Allen Jenkins, Michael O'Shea; **Dir:** Edward Ludwig. **VHS** GPV, UHV, TIM ✂✂½

Big Zapper Violent P.I. Marlowe and masochistic assistant Rock work together in this British comic strip film.
1973 94m/C GB Linda Marlowe, Gary Hope, Sean Hewitt, Richard Monette, Penny Irving; **Dir:** Lindsay Shonteff. **VHS** VCD ✂½

The Bigamist Have you heard the one about the traveling salesman in this movie? He has one wife in Los Angeles, another in San Francisco, and they inevitably find out about each other. A maudlin soap opera with a do-it-yourself ending, only shows why bigamy was done better as farce in the later "Micki and Maude."
1953 79m/B Edmond O'Brien, Joan Fontaine, Ida Lupino, Edmund Gwenn, Jane Darwell, Kenneth Tobey; **Dir:** Ida Lupino. **VHS** $19.98 DVT, NOS, TIM ✂½

Biggles Time-travel fantasy in which a young businessman from present-day New York City is inexplicably transferred into the identity of a 1917 World War I flying ace. He suddenly finds himself aboard a fighter plane over Europe during World War I.
1985 (PG) 100m/C GB Neil Dickson, Alex Hyde-White, Peter Cushing; **Dir:** John Hough. **VHS, Beta, LV** $19.95 STE, NWV ✂✂

Bikini Beach The surfing teenagers at Bikini Beach and a visitor, British recording star The Potato Bug, join forces to keep their beach from being turned into a retirement community. Songs include "Bikini Drag," "Love's a Secret Weapon," and "Because You're You."
1964 100m/C Annette Funicello, Frankie Avalon, Martha Hyer, Harvey Lembeck, Don Rickles, Stevie Wonder; **Dir:** William Asher. **VHS, Beta** $9.98 SUE ✂✂

The Bikini Car Wash Company Babes in bikinis in Los Angeles. A young man is running his uncle's carwash when he meets a business major who persuades him to let her take over the business for a cut of the profits. She decides that a good gimmick would be to dress all the female employees in the tiniest bikinis possible. The story is of course secondary to the amount of flesh on display. Also available in an unrated version.
1992 (R) ?m/C Joe Dusic, Neriah Napaul, Suzanne Browne, Kristie Ducati; **Dir:** Ed Hansen. **VHS** $89.95 IMP Woof!

Bikini Island Beautiful swimsuit models gather on a remote tropical island for a big photo shoot, each vying to be the next cover girl of the hottest swimsuit magazine. Before long, scantily clad lovelies are turning up dead and full out funky madness ensues. Will the mystery be solved before they run out of models, or will the magazine have no choice but to grace its cover with a bikinied cadaver?
1991 (R) 85m/C Holly Floria, Alicia Anne, Jackson Robinson, Shannon Stiles, Cyndi Pass; **Dir:** Tony Markes. **VHS, Beta** $79.95 PSM ✂✂

Bikini Story An examination of the history of scanty swimwear upon its 40th birthday, with lots of semi-nude models.
1985 40m/C **Dir:** Jef Rademakers. **VHS, Beta** $69.98 LTG, VES, LIV ✂✂

Bikini Summer Laughs, music, and skin are the order of the day as two nutty guys and a beautiful girl form an unlikely friendship on the beach. Michelle was Julia Robert's "Pretty Woman" body double.
1991 90m/C David Millburn, Melinda Armstrong, Jason Clow, Shelley Michelle, Alex Smith, Kent Lipham; **Dir:** Carmen Santa Maria. **VHS** $79.95 PMH ✂

Bilitis A young girl from a private girls' school is initiated into the pleasures of sex and the unexpected demands of love. One of director David Hamilton's exploitative meditations on nudity.
1977 (R) 95m/C *FR* Patti D'Arbanville, Bernard Giraudeau, Mona Kristensen; *Dir:* David Hamilton. VHS, Beta $54.95 *MED* ♫♫ 1/2

Bill A made-for-television movie based on a true story about a mentally retarded man who sets out to live independently after 44 years in an institution. Rooney gives an affecting performance as Bill and Quaid is strong as the filmmaker who befriends him. Awarded Emmys for Rooney's performance and the well written script. Followed by "Bill: On His Own."
1981 97m/C Mickey Rooney, Dennis Quaid, Largo Woodruff, Harry Goz; *Dir:* Anthony Page. VHS, Beta $59.95 *IVE* ♫♫♫

Bill and Coo An unusual love story with a villian and hero using an all bird cast.
1947 61m/C *Dir:* Dean Riesner. VHS, Beta, 8mm $24.95 *NOS, MRV, VYY* ♫♫

Bill Cosby: 49 The veteran comedian ponders his life and his experiences. Recorded live at the Chicago Theater.
1989 60m/C Bill Cosby. VHS, Beta, 8mm $19.95 *EKC, WKV, KAR* ♫♫

Bill Cosby, Himself Cosby shares his hilarious observations on marriage, drugs, alcohol, dentists, child-bearing and child-rearing in this performance. Recorded at Toronto's Hamilton Place Performing Arts Center.
1981 (PG) 104m/C Bill Cosby; *Dir:* Bill Cosby. VHS, Beta, LV $19.98 *FOX* ♫♫

A Bill of Divorcement Hepburn's screen debut, as the daughter of a shell-shocked World War I vet, who requires her care after her mother decides to divorce him. Creaky early talker.
1932 76m/B John Barrymore, Katharine Hepburn, Billie Burke, Henry Stephenson, David Manners, Paul Cavanagh, Elizabeth Patterson; *Dir:* George Cukor. National Board of Review Awards '32: 10 Best Films of the Year. VHS, Beta, LV $39.98 *FOX* ♫♫ 1/2

Bill Murray Live from the Second City Prior to his "Saturday Night Live" stardom, Murray performs a few of his comedy acts with his Second City pals.
197? 60m/C Bill Murray, Tim Kazurinsky, Mary Gross, George Wendt, Danny Breen. VHS, Beta $19.98 *TWE* ♫♫ 1/2

Bill: On His Own Rooney is again exceptional in this sequel to the Emmy-winning TV movie "Bill." After 46 years in an institution, a mentally retarded man copes more and more successfully with the outside world. Fine supporting cast and direction control the melodramatic potential.
1983 100m/C Mickey Rooney, Helen Hunt, Teresa Wright, Dennis Quaid, Largo Woodruff, Paul Leiber, Harry Goz; *Dir:* Anthony Page. VHS $59.95 *IVE* ♫♫♫

Bill & Ted's Bogus Journey The big-budget sequel to B & T's first movie has better special effects but about the same quota of laughs, and non-fans still won't think much of it. Slain by lookalike robot duplicates from the future, the airhead heroes pass through heaven and hell before tricking the Grim Reaper into bringing them back for a second duel with their heinous terminators. A most excellent closing-credit montage.
1991 (PG) 98m/C Keanu Reeves, Alex Winter, William Sadler, Joss Ackland, Pam Grier, George Carlin, Amy Stock-Poynton, Annette Azcuy, Chelcie Ross, Taj Mahal; *Dir:* Pete Hewitt. VHS $92.98 *ORI* ♫♫

Bill & Ted's Excellent Adventure Excellent premise: when the entire future of the world rests on whether or not two '80s dudes pass their history final, Rufus comes to the rescue in his time-travelling telephone booth. Bill and Ted share an adventure through time as they meet and get to know some of history's most important figures.
1989 (PG) 105m/C Keanu Reeves, Alex Winter, George Carlin, Bernie Casey, Dan Shor, Robert Barron, Amy Stock-Poynton, Ted Steedman; *Dir:* Stephen Herek. VHS, Beta, LV, 8mm $19.95 *COL, SUE* ♫♫ 1/2

Billie Duke stars as a tomboy athlete who puts the boys' track team to shame. Some amusing but very predictable situations, plus a few songs from Miss Duke. Based on Ronald Alexander's play "Time Out for Ginger."
1965 86m/C Patty Duke, Jim Backus, Jane Greer, Warren Berlinger, Billy DeWolfe, Charles Lane, Dick Sargent, Susan Seaforth Hayes, Ted Bessell, Richard Deacon; *Dir:* Don Weiss. VHS, Beta $19.98 *MGM* ♫ 1/2

A Billion for Boris Boris' TV gives a sneak preview of the future and Boris plans to make some money off of it. Zany comedy in the vein of "Let It Ride." Written by Mary Rogers (author of Freaky Friday).
1990 89m/C Lee Grant, Tim Kazurinsky. VHS, Beta $79.95 *IMP* ♫♫ 1/2

The Billion Dollar Hobo A poor, unsuspecting heir of a multimillion dollar fortune must duplicate his benefactor's experience as a hobo during the Depression in order to collect his inheritance.
1978 (G) 96m/C Tim Conway, Will Geer, Eric Weston, Sydney Lassick; *Dir:* Stuart E. McGowan. VHS, Beta $59.98 *FOX* ♫

Billy Bathgate Uneven but well acted drama set in 1935 New York. A street-wise young man decides that getting ahead during the Depression means gaining the attention of mobster Dutch Schultz and joining his gang. As Billy becomes the confidant of the racketeer he learns that the criminal life is filled with suspicion and violence and in order to stay alive he must rely on every trick he's learned to save his own skin. Willis has a cameo role as a rival mobster who gets fitted for cement overshoes. Kidman garnered a Best Supporting Actress Golden Globe nomination for her role as Dutch's girlfriend while Hill earned a Best Supporting Actor nomination from the New York Film Critics Association for his role as the gang's number man. Based on the novel by E.L. Doctorow.
1991 (R) 115m/C Dustin Hoffman, Loren Dean, Nicole Kidman, Steven Hill, Bruce Willis; *Dir:* Robert Benton. VHS, Beta $92.95 *TOU* ♫♫ 1/2

Billy Budd The classic Melville good-evil allegory adapted to film, dealing with a British warship in the late 1700s, and its struggle between evil master-at-arms and innocent shipmate. Stamp's screen debut as the naive Billy who is tried for the murder of the sadistic first mate. Well directed and acted. Oscar nomination for supporting actor Stamp.
1962 123m/B *GB* Terence Stamp, Peter Ustinov, Robert Ryan, Melvyn Douglas, Paul Rogers, John Neville, Ronald Lewis, David McCallum, John Meillon; *Dir:* Peter Ustinov. VHS, Beta, LV $59.98 *FOX* ♫♫♫

Billy Crystal: A Comic's Line Comedian Billy Crystal performs a one-man show, highlighting spoofs, of rock videos and other stand-up comics.
1983 59m/C Billy Crystal. VHS, Beta, LV $19.95 *PAR* ♫♫♫

Billy Crystal: Don't Get Me Started Crystal hosts a comedy television special featuring his standard routines from "Saturday Night Live" and numerous guest stars.
1986 60m/C Billy Crystal, Sammy Davis Jr., Eugene Levy, Christopher Guest, Rob Reiner. VHS, Beta, LV $14.98 *VES, LIV* ♫♫♫

Billy Crystal: Midnight Train to Moscow Recorded live at Moscow's Pushkin Theatre, Crystal travels to Russia, his ancestral home, and shows that comedy is an international language. A breakthrough in glasnost.
1989 72m/C Billy Crystal; *Dir:* Paul Flaherty. VHS $19.98 *HBO* ♫♫♫

Billy Galvin A bullheaded ironworker tries to straighten out the turbulent relationship he has with his rebellious son.
1986 (PG) 95m/C Karl Malden, Lenny Von Dohlen, Joyce Van Patten, Toni Kalem, Keith Szarabajka, Alan North, Paul Guilfoyle, Barton Heyman; *Dir:* John Gray. VHS, Beta $79.95 *LTG, LIV, VES* ♫♫ 1/2

Billy Jack On an Arizona Indian reservation, a half-breed ex-Green Beret with pugnacious martial arts skills (Laughlin) stands between a rural town and a school for runaways. Laughlin stars with his real-life wife Delores Taylor. Features the then-hit song "One Tin Soldier," sung by Coven Editors. The movie and its marketing by Laughlin inspired a "Billy Jack" cult phenomenon. A Spanish-dubbed version of this film is also available. Followed by a sequel in 1974, "Trail of Billy Jack," which bombed.
1971 (PG) 112m/C Tom Laughlin, Delores Taylor, Clark Howat; *Dir:* T.C. Frank. VHS, Beta, LV $19.98 *WAR, PIA* ♫♫

Billy the Kid Returns While trying to clean up a town of its criminal element, Rogers is mistaken for the legendary outlaw, Billy the Kid.
1938 60m/B Roy Rogers, George "Gabby" Hayes, Smiley Burnette, Lynne Roberts; *Dir:* Joseph Kane. VHS, Beta $19.95 *NOS, HHT, VCN* ♫ 1/2

Billy the Kid in Texas The famed outlaw takes on trouble and makes sure the Texans never forget that he has been there.
1940 52m/B Bob Steele. VHS, Beta $19.95 *NOS, VCN* ♫ 1/2

Billy the Kid Trapped Billy and his partner are rescued from hanging by an outlaw band.
1942 59m/B Buster Crabbe, Al "Fuzzy" St. John, Bud McTaggart, Anne Jeffreys, Glenn Strange; *Dir:* Sherman Scott. VHS $19.95 *NOS, MRV, DVT* ♫ 1/2

Billy the Kid Versus Dracula The title says it all. Dracula travels to the Old West, anxious to put the bite on a pretty lady ranch owner. Her fiance, the legendary outlaw Billy the Kid, steps in to save his girl from becoming a vampire herself. A Carradine camp classic.
1966 95m/C Chuck Courtney, John Carradine, Melinda Plowman, Walter Janovitz, Harry Carey Jr., Roy Barcroft, Virginia Christine, Bing Russell; *Dir:* William Beaudine. VHS, Beta $19.95 *NOS, SUE, VYY* ♫♫

B

Billy Liar A young Englishman dreams of escaping from his working class family and dead-end job as an undertaker's assistant. Parallels James Thurber's story, "The Secret Life of Walter Mitty."
1963 94m/B *GB* Tom Courtenay, Julie Christie, Finlay Currie; *Dir:* John Schlesinger. **VHS, Beta** $39.95 *HBO* ✓✓✓

Billy: Portrait of a Street Kid A ghetto youngster tries pry himself up and out of his bleak surroundings through education, but complications arise when he gets his girlfriend pregnant.
1977 96m/C LeVar Burton, Tina Andrews, Ossie Davis, Michael Constantine; *Dir:* Steven Gethers. **VHS, Beta** $34.95 *WOV* ✓✓✓

Billy Rose's Jumbo Better-than-average update of the circus picture. Durante and Raye are terrific, as are the Rodgers and Hart songs. Fun, with lively production numbers in the inimitable Berkeley style.
1962 125m/C Doris Day, Stephen Boyd, Jimmy Durante, Martha Raye, Dean Jagger; *Dir:* Charles Walters. **VHS, LV** $19.98 *MGM* ✓✓✓

Billyboy Coming of age film with the world of boxing as background. Excellent fight scenes.
1979 94m/C Duane Bobrick, Kim Braden. **VHS, Beta** *BFV* ✓✓✓

Biloxi Blues Eugene Morris Jerome has been drafted and sent to boot camp in Biloxi, Mississippi where he encounters a troubled drill sergeant, hostile recruits, and a skillful prostitute. Walken is the drill sergeant from hell. Some good laughs from the ever-wry Broderick. A sequel to Neil Simon's "Brighton Beach Memoirs" and adapted by Simon from his play.
1988 (PG-13) 105m/C Matthew Broderick, Christopher Walken, Casey Siemaszko, Matt Mulern, Corey Parker, Penelope Ann Miller; *Dir:* Mike Nichols. **VHS, Beta, LV** $14.95 *MCA* ✓✓½

Bim/Dream of the Wild Horses Two famous French shorts, the first of which is dubbed into English.
1952 110m/B *FR* **VHS, Beta** $39.95 *TAM, DVT* ✓✓½

Bimini Code Two female adventurers accept a dangerous mission where they wind up on Bimini Island in a showdown with the mysterious Madame X.
1984 95m/C Vickie Benson, Krista Richardson, Frank Alexander, Rosanna Simanaitis; *Dir:* Barry Clark. **VHS, Beta** $49.95 *PSM* ✓

Bingo The Hound is understandably outraged that someone would make a spoof of hero-dog movies—outraged that they didn't do a better job. The heroic title mutt roams from Denver to Green Bay in search of his absent-minded master, with numerous absurd adventures en route. Some cute moments, but sometimes Bingo is just lame-o.
1991 (PG) 90m/C Cindy Williams, David Rasche; *Dir:* Matthew Robbins. **VHS, LV** *COL, PIA* ✓✓

Bingo Inferno A short narrative film about a precocious kid who connives to break his mother of a compulsive Bingo addiction.
1987 13m/C Beta *OWM* ✓✓

Bingo Long Traveling All-Stars & Motor Kings Set in 1939, this film follows the comedic adventures of a lively group of black ball players who have defected from the old Negro National League. The All-Stars travel the country challenging local white teams.

1976 (PG) 111m/C Billy Dee Williams, James Earl Jones, Richard Pryor, Stan Shaw; *Dir:* John Badham. **VHS, Beta** $49.95 *MCA* ✓✓✓

Bio Hazard A toxic monster needs human flesh to survive and consequently goes on a rampage.
1985 (R) 84m/C Angelique Pettyjohn, Carroll Borland, Richard Hench; *Dir:* Fred Olen Ray. **VHS, Beta** $39.98 *CGI, MTX* ✓

The Bionic Woman A sky-diving accident leaves tennis pro Jaimie Somers crippled and near death. Her bionic buddy, Steve Austin, gets his friends to rebuild her and make her better than she was before. The pilot for the television series.
1975 96m/C Lindsay Wagner, Lee Majors, Richard Anderson, Alan Oppenheimer; *Dir:* Richard Moder. **VHS, Beta** $39.95 *MCA* ✓

The Birch Interval A chronicle of a young Amish girl growing up and experiencing adult passions and fears when she visits her kin in their isolated Pennsylvania community.
1978 (PG) 104m/C Eddie Albert, Rip Torn, Ann Wedgeworth; *Dir:* Delbert Mann. **VHS, Beta** $9.95 *MED* ✓✓

Bird The richly-textured, though sadly one-sided biography of jazz sax great Charlie Parker, from his rise to stardom to his premature death via extended heroin use. A remarkably assured, deeply imagined film from Eastwood that never really shows the Bird's genius of creation. The soundtrack features Parker's own solos re-mastered from original recordings.
1988 (R) 160m/C Forest Whitaker, Diane Venora, Michael Zelniker, Samuel E. Wright, Keith David, Michael McGuire, James Handy, Damon Whitaker, Morgan Nagler; *Dir:* Clint Eastwood. Academy Awards '88: Best Sound; Cannes Film Festival '88: Best Actor (Whitaker); New York Film Critics Awards '88: Best Supporting Actress (Venora). **VHS, Beta, LV** $19.98 *MVD, WAR, FCT* ✓✓✓

Bird with the Crystal Plumage An American writer living in Rome witnesses a murder. He becomes involved in the mystery when the alleged murderer is cleared because the woman believed to be his next victim is shown to be a psychopathic murderer. Vintage Argento mayhem.
1970 (PG) 98m/C *IT* Tony Musante, Suzy Kendall, Eva Renzi; *Dir:* Dario Argento. **VHS, Beta, LV** $19.95 *UHV, MRV* ✓✓½

Bird of Paradise An exotic South Seas romance in which an adventurer is cast onto a remote Polynesian island when his yacht haphazardly sails into a coral reef. There he becomes enamored of an exotic island girl, and nature seems to disapprove.
1932 80m/B Joel McCrea, Dolores Del Rio, Lon Chaney Jr.; *Dir:* King Vidor. **VHS, Beta, 8mm** $19.95 *NOS, MRV, VYY* ✓✓½

Bird on a Wire After suddenly being forced to emerge from a prolonged period under the Witness Protection Program, an ex-hood and his ex-girlfriend are pursued by old enemies.
1990 (PG-13) 110m/C Mel Gibson, Goldie Hawn, David Carradine, Bill Duke, Stephen Tobolowsky; *Dir:* John Badham. **VHS, Beta, LV** $19.98 *MCA* ✓✓

Birdman of Alcatraz Robert Stroud, convicted of two murders and sentenced to life imprisonment on the Island, becomes an internationally accepted authority on birds. Lovingly told, with stunning performance from Lancaster, and exceptionally fine work from the supporting cast. The confinement of

Stroud's prison cell makes the film seem claustrophobic and tedious at times, just as the imprisonment must have been. Oscar nominations for actor Lancaster, supporting actress Ritter as Stroud's mother who never stops trying to get him out of prison, supporting actor Savalas and cinematography.
1962 143m/B Burt Lancaster, Karl Malden, Thelma Ritter, Betty Field, Neville Brand, Edmond O'Brien, Hugh Marlowe, Telly Savalas; *Dir:* John Frankenheimer. **VHS** $19.98 *COL, FCT* ✓✓✓

The Birds Hitchcock attempted to top the success of "Psycho" with this terrifying tale of Man versus Nature, in which Nature alights, one by one, on the trees of Bodega Bay to stage a bloody act of revenge upon the civilized world. Only Hitchcock can twist the harmless into the horrific while avoiding the ridiculous; this is perhaps his most brutal film, and one of the cinema's purest, horrifying portraits of apocalypse. Based on a short story by Daphne DuMaurier; screenplay by novelist Evan Hunter (aka Ed McBain).
1963 120m/C Rod Taylor, Tippi Hedren, Jessica Tandy, Veronica Cartwright, Suzanne Pleshette; *Dir:* Alfred Hitchcock. **VHS, Beta, LV** $19.95 *MCA* ✓✓✓½

Birds & the Bees A millionaire falls in love with an alluring card shark, and then calls it all off when he learns of her profession, only to fall in love with her again when she disguises herself. A remake - and poor shade - of Preston Sturge's 1941 classic "The Lady Eve."
1956 94m/C Mitzi Gaynor, David Niven, George Gobel, Reginald Gardiner, Hans Conried; *Dir:* Norman Taurog. **VHS, Beta** $19.95 *KRT* ✓✓

Birds of Paradise Three attractive women who inherit a yacht travel to the Florida Keys to find romance.
1984 (R) 90m/C **VHS, Beta** $44.95 *KOV* Woof!

Birds of Prey Made-for-television action film pits an ex-World War II army pilot against a group of kidnapping thieves in an airborne chopper chase.
1972 81m/C David Janssen, Ralph Meeker, Elayne Heilveil; *Dir:* William A. Graham. **VHS, Beta** $49.95 *PSM* ✓✓

Birds of Prey A tough urban cop stalks a breed of cold, ruthless assassins in a large city. In HiFi.
1985 90m/C **VHS, Beta** $19.95 *STE, NWV* ✓

Birdy An adaptation of the William Wharton novel about two Philadelphia youths, one with normal interests, the other obsessed with birds, and their eventual involvement in the Vietnam War, wrecking one physically and the other mentally. A hypnotic, evocative film, with a compelling Peter Gabriel soundtrack.
1984 (R) 120m/C Matthew Modine, Nicolas Cage, John Harkins, Sandy Baron, Karen Yahng, Bruno Kirby; *Dir:* Alan Parker. Cannes Film Festival '85: Jury Prize. **VHS, Beta, LV** $79.95 *COL* ✓✓✓½

Birgitt Haas Must Be Killed A ruthless secret agent (Noiret) plots to murder a German female terrorist and make it appear that her boyfriend was the killer. Never quite hits the mark, despite novel premise and strong cast.
1983 105m/C Philippe Noiret, Jean Rochefort, Lisa Kreuzer; *Dir:* Laurent Heynemann. **VHS, Beta** $39.95 *WES, GLV* ✓✓½

Birth of a Legend A documentary showing the on and off antics of Miss Pickford and Mr. Fairbanks when they reigned as King and Queen of the silver screen in 1926.
1984 25m/B Mary Pickford, Douglas Fairbanks Sr. **VHS, Beta $19.95** *CCB* 🎞🎞½

Birth of a Nation A lavish Civil War epic in which Griffith virtually invented the basics of film grammar. Gish and Walthall in some of the most moving scenes ever filmed as well as masterful battle choreography that brought the art of cinematography to new heights. Griffith's attitude toward the KKK provokes argument to this day. Based on "The Clansman" by Thomas Dixon, it is still a rouser, and of great historical interest. Silent with music score. Also available in a 124 minute version.
1915 175m/B Lillian Gish, Henry B. Walthall, Mae Marsh, Wallace Reid, Donald Crisp, Raoul Walsh, Erich von Stroheim, Eugene Pallette; *Dir:* D.W. Griffith. **VHS, LV $49.95** *REP, MRV, VYY* 🎞🎞🎞🎞

Birth Right A cheap cautionary film about VD. "A nighttime of pleasure... a lifetime of regret!!!"
1939 52m/B Ethel Moses, Alec Lovejoy, Carmen Newsome, Laura Bowman. **VHS, Beta $24.95** *JLT* Woof!

The Birth of Soviet Cinema Directors Eisenstein, Pudovkin and Dovzhenko are linked to bring forth sequences of their brilliant masterpieces.
1972 49m/C VHS, Beta *FHS* Woof!

Birthday Boy A cable comedy about a buffoonish salesman's 30th birthday, on which he takes an ill-fated business trip. Written by Belushi.
1985 33m/C James Belushi; *Dir:* Claude Conrad. **VHS, Beta $29.98** *VES, LIV* 🎞🎞

The Bishop's Wife An angel comes down to earth at Christmas to help a young bishop, his wife and his parishioners. Excellent performances by the cast make this an entertaining outing. Oscar nominations for Best Picture, Best Director, Score of a Comedy or Drama, and Film Editing.
1947 109m/B Cary Grant, Loretta Young, David Niven, Monty Woolley, Elsa Lanchester, James Gleason, Gladys Cooper, Regis Toomey; *Dir:* Henry Koster. Academy Awards '47: Best Sound. **VHS, Beta, LV $14.98** *HBO, SUE, FCT* 🎞🎞🎞

The Bitch High-camp follies are the rule in this lustful continuation of "The Stud" as it follows the erotic adventures of a beautiful divorcee playing sex games for high stakes on the international playgrounds of high society. A collaborative effort by the sisters Collins: written by Jackie, with sister Joan well cast in the title role.
1978 (R) 90m/C *GB* Joan Collins, Kenneth Haigh, Michael Coby, Ian Hendry, Carolyn Seymour, Sue Lloyd, John Ratzenberger; *Dir:* Gerry O'Hara. **VHS, Beta $19.99** *HBO* 🎞

Bite the Bullet Moralistic western tells of a grueling 600-mile horse race where the participants reluctantly develop respect for one another. Unheralded upon release and shot in convincing epic style by Harry Stradling. Excellent cast.
1975 (PG) 131m/C Gene Hackman, James Coburn, Candice Bergen, Dabney Coleman, Jan-Michael Vincent, Ben Johnson, Ian Bannen, Paul Stewart, Sally Kirkland; *Dir:* Richard Brooks. **VHS, Beta, LV $12.95** *RCA* 🎞🎞🎞½

Bits and Pieces A crazed mutilating killer strikes fear in the hearts of the city dwellers.
1977 86m/C VHS, Beta $59.95 *TWE* 🎞

Bitter Harvest Emmy-nominated made-for-television movie concerning a dairy farmer frantically trying to discover what is mysteriously killing off his herd. Howard is excellent as the farmer battling the bureaucracy to find the truth. Based on a true story.
1981 98m/C Ron Howard, Art Carney, Tarah Nutter, Richard Dysart, Barry Corbin, Jim Haynie, David Kneel; *Dir:* Roger Young. **VHS, Beta $14.95** *FRH, IVE* 🎞🎞🎞

Bitter Sweet Tragic tale of a woman who finally marries the man she loves, only to find that he is a compulsive gambler. Written by Noel Coward, adapted from his operetta.
1933 76m/B *GB* Anna Neagle, Fernand Graavey, Esme Percy, Clifford Heatherly, Hugh Williams; *Dir:* Herbert Wilcox. **VHS $19.95** *NOS, MRV, DVT* 🎞🎞½

Bitter Sweet The second version of the Noel Coward operetta, about young romance in 1875 Vienna. Creaky and overrated, but the lush Technicolor and Coward standards "Ziguener" and "I'll See You Again" help to compensate.
1941 94m/C Jeanette MacDonald, Nelson Eddy, George Sanders, Felix Bressart, Ian Hunter, Sig Rumann, Herman Bing, Fay Holden, Curt Bois, Edward Ashley; *Dir:* W.S. Van Dyke. **VHS, Beta $29.95** *MGM* 🎞🎞

Bittersweet Love Two young people fall in love and marry, only to have the bride's mother and the groom's father confess a 30-year-old affair, disclosing that the two newlyweds are actually half-siblings.
1976 (PG) 92m/C Lana Turner, Robert Lansing, Celeste Holm, Robert Alda, Meredith Baxter Birney; *Dir:* David Miller. **VHS, Beta $59.98** *SUE* 🎞½

Bizarre When a wife escapes her perverse, psychologically threatening marriage, she finds her husband still haunts her literally and figuratively, and plots psychological revenge. Dubbed.
1987 93m/C *IT* Florence Guerin, Luciano Bartoli, Robert Egon Spechtenhauser, Stefano Sabelli. **VHS, Beta $79.95** *PSM* 🎞

Bizarre Bizarre A mystery writer is accused of murder and disappears, only to return in disguise to try to clear his name. Along the way, a number of French comedians are introduced with a revue of comedy-farce sketches that include slapstick, burlesque, black humor, and comedy of the absurd. In French with English subtitles.
1939 90m/B *FR* Louis Jouvet, Michel Simon, Francoise Rosay; *Dir:* Marcel Carne. **VHS, Beta $24.95** *NOS, HHT, FST* 🎞🎞🎞

Black Adder 3 Three episodes from the popular British sitcom about an aristocrat-turned-butler: "Dish and Dishonesty," "Ink and Incapability," and "Nob and Nobility."
1989 89m/C *GB* Hugh Laurie, Rowan Atkinson. **VHS, Beta $14.98** *FOX* 🎞🎞🎞

Black Adder The Third (Part 2) Three episodes from the popular British television sitcom: "Amy and Amiability," "Sense and Senility," and "Duel and Duality."
1989 (PG) 89m/C *GB* Rowan Atkinson, Hugh Laurie; *Dir:* Mandie Fletcher. **VHS, Beta $19.98** *FOX* 🎞🎞🎞

The Black Arrow Original adventure film of the famous Robert Louis Stevenson novel. Upon return from 16th century's War of the Roses, a young man must avenge his father's murder by following a trail of clues in the form of black arrows. Well made and fun.
1948 76m/B Louis Hayward, Janet Blair, George Macready, Edgar Buchanan, Paul Cavanagh; *Dir:* Gordon Douglas. **VHS, Beta $69.95** *DIS* 🎞🎞🎞

Black Arrow An exiled bowman returns to England to avenge the injustices of a villainous nobleman. Made-for-cable version of the Robert Louis Stevenson medieval romp is not as well done as the 1948 adaptation.
1984 93m/C Oliver Reed, Benedict Taylor, Georgia Slowe, Stephan Chase, Donald Pleasence; *Dir:* John Hough. **VHS, Beta $69.95** *DIS* 🎞½

Black Beauty In this adaptation of Anna Sewell's familiar novel, a young girl develops a kindred relationship with an extraordinary horse.
1946 74m/B Mona Freeman, Richard Denning, Evelyn Ankers; *Dir:* Max Nosseck. **VHS, Beta $9.99** *CCB, MED* 🎞🎞

Black Beauty International remake of the classic horse story by Anna Sewell.
1971 (G) 105m/C *GB GE SP* Mark Lester, Walter Slezak; *Dir:* James Hill. **VHS, Beta, LV $14.95** *PAR, FCT* 🎞🎞

Black Beauty/Courage of Black Beauty Two feature-length films based on Anna Sewell's novel about the love of a child for a black stallion. "Black Beauty" was first released in 1946; "Courage of Black Beauty" in 1957. Excellent for children who love horses, of course.
1957 145m/C Mona Freeman, Johnny Crawford; *Dir:* Max Nosseck, Harold Schuster. **VHS, Beta $12.95** *MED* 🎞🎞

Black Belt Two masters of Kung Fu clash when both attempt to steal a shipment of gold bullion.
1973 (R) 92m/C VHS, Beta $59.95 *HHT* 🎞

Black Belt Fury Chang Li uses his superior kung fu technique and his pureness of soul to fight the evil forces of darkness.
1981 85m/C Chang Li. **VHS, Beta** *NO* 🎞

Black Belt Jones A martial arts expert fights the mob to save a school of self-defense in Los Angeles' Watts district.
1974 (R) 87m/C Jim Kelly, Gloria Hendry, Scatman Crothers; *Dir:* Robert Clouse. **VHS, Beta $19.98** *WAR* 🎞

Black Bikers from Hell Black gang-members infiltrate and wreak havoc on their rivals. Who can stop these brutal young men? Cast with real bikers and biker chicks. Originally titled "Black Angels."
1970 (R) 87m/C King John III, Des Roberts, Linda Jackson, James Whitworth, James Young-El, Clancy Syrko, Beverly Gardner; *W/Dir:* Lawrence Merrick. **VHS, Beta $59.95** *TPV, JEF* 🎞½

Black Bird In this satiric "sequel" to "The Maltese Falcon," detective Sam Spade, Jr. searches for the mysterious black falcon statuette that caused his father such trouble. Features appearances by Elisha Cook Jr. and Lee Patrick, who starred in the 1941 classic "The Maltese Falcon" with Humphrey Bogart.
1975 (PG) 98m/C George Segal, Stephane Audran, Lionel Stander, Lee Patrick, Elisha Cook Jr., Connie Kreski; *Dir:* David Giler. **VHS, Beta $9.95** *GKK* 🎞🎞

Black Box Affair An American secret agent must find a black box lost in a B-52 plane crash before it falls into the wrong hands.
1966 95m/C Craig Hill, Teresa Gimpera, Luis Martin, Jorge Rigaud. **VHS, Beta $19.95** ASE 🎬½

Black Brigade Pryor and Williams star in this low budget movie as leaders of an all-black outfit assigned to a suicide mission behind Nazi lines during World War II. Their force wreaks havoc and earns them the respect of military higher-ups. Lots of action and climactic finish.
19?? 90m/C Richard Pryor, Billy Dee Williams, Stephen Boyd. **VHS** QVD 🎬🎬

Black Caesar A small-time hood climbs the ladder to be the head of a Harlem crime syndicate. Music by James Brown. Followed by the sequel "Hell Up in Harlem."
1973 (R) 92m/C Fred Williamson, Julius W. Harris, Val Avery, Art Lund, Gloria Hendry; *Dir:* Larry Cohen. **VHS, Beta, LV $19.98** ORI 🎬🎬

The Black Cat The first of the Boris and Bela pairings stands up well years after release. Polished and taut, with fine sets and interesting acting. Confrontation between architect and devil worshipper acts as plot, with strange twists. Worth a look. Also available with "The Raven" (1935) on Laser Disc.
1934 65m/B Boris Karloff, Bela Lugosi, David Manners, Jacqueline Wells, Lucille Lund, Henry Armetta; *Dir:* Edgar G. Ulmer. **VHS $9.95** MED, MLB, MCA 🎬🎬🎬½

The Black Cat Spaghetti splatter-meister Fulci, best known for his unabashed ripoffs "Zombie" and "Gates of Hell," tones down the gore this time in a vaguely Poe-ish tale of a medium with some marbles loose (Magee) whose kitty provides the temporary habitat for spirits its master calls up. The dreary English village setting and the downright myopic camerawork add up to an oppressive viewing experience. Written by Fulci.
1981 92m/C IT GB Patrick Magee, Mimsy Farmer, David Warbeck, Al Cliver, Dagmar Lassander; *W/Dir:* Lucio Fulci. **VHS $14.98** MED Woof!

Black Cat Filmmakers find lots of action in a haunted house. Chock full of references to the works of spaghetti horror dons Mario Bava and Dario Argento.
1990 ?m/C IT Caroline Munro, Brett Halsey; *W/Dir:* Lewis (Luigi Cozzi) Coates. **VHS, Beta** COL 🎬½

Black Christmas A college sorority is besieged by an axe-murderer over the holidays. Also known as "Silent Night, Evil Night" and "Stranger in the House."
1975 (R) 98m/C CA Olivia Hussey, Keir Dullea, Margot Kidder, John Saxon; *Dir:* Bob Clark. **VHS, Beta $19.98** WAR 🎬½

Black Cobra A lesbian exacts revenge for her lover's snake-bite murder by trapping the guilty party with her own snakes.
1983 (R) 97m/C Laura Gemser, Jack Palance. **VHS, Beta $79.95** GEM Woof!

The Black Cobra After photographing a psychopath in the process of killing someone, a beautiful photographer seeks the help of a tough police sergeant to protect her. The leader of the Black Cobras gang gives chase.
1987 (R) 90m/C IT Fred Williamson, Karl Landgren, Eva Grimaldi. **VHS, Beta $79.95** TWE Woof!

Black Cobra 2 A mis-matched team of investigators tracks a notorious terrorist. They find him holding a school full of children as hostage.
1989 (R) 95m/C Fred Williamson. **VHS $14.95** HMD, SOU 🎬🎬

Black Cobra 3: The Manila Connection Interpol turns to police lieutenant Robert Malone (Williamson) when a team of high-tech weapons thieves threatens the world. Malone attacks like a cyclone on the terrorists' jungle haven. They won't know what hit 'em!
1990 (R) 92m/C Fred Williamson, Forry Smith, Debra Ward; *Dir:* Don Edwards. **VHS $14.95** HMD, SOU 🎬½

The Black Coin A condensation of the "Black Coin" movie serial.
1936 80m/B Ralph Graves, Ruth Mix; *Dir:* Elmer Clifton. **VHS, Beta $69.95** DVT, MLB, RXM 🎬½

Black Devil Doll from Hell This shot-on-video movie deals with a nasty little voodoo doll that likes to kill its owners.
1984 70m/C Shirley Jones, Rickey Roach, Marie Sainvilvs; *Dir:* Chester Turner. **VHS, Beta $49.95** HHT 🎬

The Black Devils of Kali Adventurers in the Indian jungle discover a lost race of idol-worshipping primitives. Racist garbage produced near the end of Republic Pictures' existence. Also known as "Mystery of the Black Jungle," based on a novel by Emillio Salgari.
1955 72m/B Lex Barker, Jane Maxwell, Luigi Tosi, Paul Muller; *Dir:* Ralph Murphy. **VHS $16.95** SNC, MLB 🎬

Black Dragon A young farm boy moves to the city and teams up with an undercover narcotics agent to infiltrate and destroy the underworld syndicate.
1974 (R) 93m/C Ron Van Cliff, Jason Rai Pow, Jorge Estraga, Nancy Veronica. **VHS, Beta $54.95** SUN 🎬

The Black Dragons A weird war drama involving sabotage by the Japanese, with Lugosi playing a plastic surgeon who enables them to pass as Americans. Also available colorized.
1942 62m/B Bela Lugosi, Joan Barclay, George Pembroke, Clayton Moore; *Dir:* William Nigh. **VHS, Beta $24.95** NOS, MRV, SNC 🎬

Black Dragon's Revenge The Black Dragon arrives in Hong Kong to investigate the death of a great kung fu master.
1975 (R) 90m/C Ron Van Cliff, Charles Bonet, Jason Rai Pow. **VHS, Beta $52.95** SUN 🎬

Black Eagle Pre-Glasnost, anti-Soviet tale of two high-kicking spies. CIA and KGB agents race to recover innovative equipment in the Mediterranean.
1988 (R) 93m/C Bruce Doran, Jean-Claude Van Damme, Sho Kosugi; *Dir:* Eric Karson. **VHS, Beta $89.98** NVH 🎬🎬

Black Eliminator A black cop struggles to stop a maniacal secret agent who plans to destroy the world.
197? 84m/C Jim Kelly. **VHS, Beta $69.95** UNI 🎬

Black Emmanuelle Emmanuelle travels to Africa on an assignment, but work turns to play as she becomes a willing partner.
1976 121m/C Karin Schubert, Angelo Infanti, Don Powell. **VHS, Beta $59.98** FOX 🎬

Black Force Brothers who fight crime with violent actions, are called for assistance in the recovery of an African artifact. Originally known as "Force Four."
1975 82m/C Malachi Lee, Warhawk Tanzania, Owen Watson, Judie Soriano; *Dir:* Michael Fink. **VHS, Beta $9.95** SIM, TAV 🎬🎬

Black Force 2 The brothers are back on the scene in another violent, cartilage-shattering adventure.
1978 (R) 90m/C Terry Carter, James B. Sikking. **VHS $29.95** SIM, TAV 🎬🎬

Black Fury A coal miner's efforts to protest working conditions earn him a beating by the company goons who also kill his friend. He draws national attention to this brutal plight of the workers when he barricades himself inside the mine. Muri's carefully detailed performance adds authenticity to this powerful drama, but it proved too depressing to command a big box office.
1935 95m/B Paul Muni, Barton MacLane, Henry O'Neill, John Qualen, J. Carroll Naish; *Dir:* Michael Curtiz. **VHS, Beta $59.95** FOX 🎬🎬🎬

Black Gestapo A black-exploitation film about a vigilante army taking over a Los Angeles ghetto, first to help residents, but later to abuse them. Extremely violent.
1975 (R) 89m/C Rod Perry, Charles P. Robinson, Phil Hoover, Ed Gross. **VHS, Beta $59.95** UNI Woof!

Black Glove A trumpet star defends himself against charges of murdering a Spanish singer by tracking down the real killer.
1954 84m/B Alex Nicol, Eleanor Summerfield. **VHS, Beta, LV $19.95** NOS, WGE 🎬

Black God & White Devil Another Brazilan socio-political commentary by Rocha, an oft-incendiary filmmaker whose left-leaning ideals are steeped in mysticism, obscure folklore and powerful images. An impoverished man transforms from a religious zealot to a bandit, his tale underscored by conflict between poor masses and wealthy landowners.
1964 102m/C BR *Dir:* Glauber Rocha. **VHS $59.95** FST 🎬🎬

Black Godfather The grueling story of a hood clawing his way to the top of a drug-selling mob. Features an all-black cast.
1974 (R) 90m/C Rod Perry, Damu King, Don Chastain, Jimmy Witherspoon, Diane Summerfield; *Dir:* John Evans. **VHS, Beta $59.98** MAG 🎬

The Black Hand Kelly plays well against character in this atmospheric turn-of-the-century thriller. The evil society of the Black Hand murders his father, and he seeks revenge. Well made drama.
1950 93m/B Gene Kelly, J. Carroll Naish, Teresa Celli, Marc Lawrence, Frank Puglia; *Dir:* Richard Thorpe. **VHS $59.98** MAG 🎬🎬🎬

Black Hand An unemployed Italian immigrant is drawn into a web of murder and betrayal after he is attacked by an Irish gang.
1976 90m/C Lionel Stander, Mike Placido. **VHS, Beta $59.98** MAG 🎬½

Black Hole A high-tech, computerized Disney space adventure dealing with a mad genius who attempts to pilot his craft directly into a black hole. Except for the top quality special effects, a pretty creaky vehicle.
1979 (G) 97m/C Maximilian Schell, Anthony Perkins, Ernest Borgnine, Yvette Mimieux, Joseph Bottoms, Robert Forster; *Dir:* Gary Nelson. **VHS, Beta, LV** DIS, OM 🎬½

The Black Klansman A black man masquerades as a white extremist in order to infiltrate the KKK and avenge his daughter's murder. In the interest of racial harmony, he seduces the Klan leader's daughter. As bad as it sounds.
1966 88m/B Richard Gilden, Rima Kutner, Harry Lovejoy; *Dir:* Ted V. Mikels. VHS, Beta $39.95 UNI 𝒵

The Black Lash Two lawmen go undercover to break a silver hijacking gang.
1952 55m/B Lash LaRue, Al "Fuzzy" St. John, Peggy Stewart, Kermit Maynard; *Dir:* Ron Ormond. VHS, Beta $19.95 NOS, DVT 𝒵

Black Lemons While in prison, a convict is stalked by the Mafia because of what he knows. He eventually spills the beans to the cops, putting himself in unavoidable jeopardy.
1987? 93m/C Peter Carsten, Antonio Sabato. VHS, Beta $59.95 UNI 𝒵

Black Like Me Based on John Howard Griffin's successful book about how Griffin turned his skin black with a drug and traveled the South to experience prejudice firsthand. Neither the production nor the direction enhance the material.
1964 107m/B James Whitmore, Roscoe Lee Browne, Will Geer, Walter Mason, John Marriott, Clifton James, Dan Priest; *Dir:* Carl Lerner. VHS, Beta $19.95 UHV, FCT, RRP 𝒵 1/2

Black Lizard A camp spectacle set in the Japanese underworld. The Lizard is the glamorous queen of Tokyo crime who plots to steal a famous diamond by first kidnapping the owner's daughter, however, complications arise when she falls for a detective. The Lizard is played by female impersonator Maruyama. Yukio Mishima, who wrote the original drama and the screenplay, has a cameo as an embalmed corpse. Style is all. In Japanese with English subtitles.
1968 112m/C JP Akihiro Maruyama, Isao Kimura; *Dir:* Kinji Fukasaku. VHS $79.95 CCN 𝒵𝒵

Black Magic Cagliostro the magician becomes involved in a plot to supply a double for Marie Antoinette.
1949 105m/B Orson Welles, Akim Tamiroff, Nancy Guild, Raymond Burr; *Dir:* Gregory Ratoff. VHS, Beta $12.95 MED 𝒵𝒵 1/2

Black Magic Terror Jilted by her lover, the old queen of black magic has everybody under her spell. Trouble starts, however, when she turns her back on one of her subjects.
1979 85m/C JP Suzanna, W.D. Mochtar, Alan Nuary; *Dir:* L. Sudjio. VHS, Beta $59.95 TWE 𝒵

Black Magic Woman An art gallery owner has an affair with a beautiful and exotic woman but starts to get cold feet when he's plagued by inexplicable phenomena. Seems that black magic woman done put a voodoo spell on him. Listless companion to director Warren's "Blood Spell."
1991 (R) 91m/C Apollonia, Mark Hamill, Amanda Wyss; *Dir:* Deryn Warren. VHS $89.95 VMK Woof!

The Black Maid A scantily-clad, sassy black maid cleans, cooks, cusses and drives her rascist employer to the brink of insanity.
1974 90m/C IT VHS, Beta $39.95 VCD Woof!

Black Marble A beautiful policewoman is paired with a policeman who drinks too much, is divorced, and is ready to retire. Surrounded by urban craziness and corrup-

tion, they eventually fall in love. Based on the Joseph Wambaugh novel.
1979 (PG) 110m/C Paula Prentiss, Harry Dean Stanton, Robert Foxworth, James Woods, Michael Dudikoff; *Dir:* Harold Becker. VHS, Beta $9.98 SUE 𝒵𝒵𝒵

Black Market Rustlers The Range Busters are at it again. This time they break up a cattle rustling syndicate.
1943 60m/B Ray Corrigan, Dennis Moore, Max Terhune; *Dir:* S. Roy Luby. VHS, Beta $19.95 VCN, VDM, RXM 𝒵

Black Moon Rising Based on an idea by John Carpenter ("Halloween"), this film deals with the theft of a new jet-powered car and its involvement in an FBI investigation. Solid performances and steady action enhance this routine effort.
1986 (R) 100m/C Tommy Lee Jones, Linda Hamilton, Richard Jaeckel, Robert Vaughn; *Dir:* Harley Cokliss. VHS, Beta, LV $19.95 STE, NWV 𝒵 1/2

Black Narcissus A group of Anglican nuns attempting to found a hospital and school in the Himalayas confront native distrust and human frailties amid beautiful scenery. Adapted from the novel by Rumer Godden.
1947 101m/C GB Deborah Kerr, Jean Simmons, Flora Robson, Sabu, David Farrar; *Dir:* Michael Powell. Academy Awards '47: Best Art Direction/Set Decoration (Color), Best Color Cinematography; New York Film Critics Awards '47: Best Actress (Kerr). VHS, Beta, LV $29.95 VID, VYG, MLB 𝒵𝒵𝒵𝒵

Black Oak Conspiracy Based on a true story, this film deals with a mining company conspiracy discovered by an inquisitive stuntman.
1977 (R) 92m/C Jesse Vint, Karen Carlson, Albert Salmi, Seymour Cassel, Robert F. Lyons; *Dir:* Bob Kelljan. VHS, Beta $59.98 CHA 𝒵𝒵

Black Orchid A businessman and a crook's widow fall in love and try to persuade their children it can work out.
1959 96m/B Sophia Loren, Anthony Quinn, Ina Balin, Mark Richmond, Jimmie Baird; *Dir:* Martin Ritt. VHS, Beta $14.95 PAR 𝒵𝒵 1/2

Black Orpheus The legend of Orpheus and Eurydice unfolds against the colorful background of the carnival in Rio de Janeiro. In the black section of the city, Orpheus is a street-car conductor and Eurydice, a country girl fleeing from a man sworn to kill her. Dancing, incredible music, and black magic add to the beauty of this film. Dubbed in English.
1958 103m/C BR FR PT Breno Mello, Marpessa Dawn, Lourdes De Oliveira; *Dir:* Marcel Camus. Academy Awards '59: Best Foreign Language Film; Cannes Film Festival '59: Grand Prize Winner. VHS, Beta, LV $29.95 CVC, FOX, VYG 𝒵𝒵𝒵𝒵

Black Panther True story of psycho-killer Donald Neilson, who murdered heiress Lesley Whittle in England in the 1970s.
1977 90m/C Donald Sumpter, Debbie Farrington; *Dir:* Ian Merrick. VHS, Beta $69.95 VES, LIV 𝒵 1/2

The Black Pearl A young pearl diver comes across the ultimate prize pearl while diving off Baja. Trouble is, it's protected by a feisty manta ray. Adapted from a novel by Scott O'Dell.
19?? 37m/C VHS $70.00 AFR 𝒵 1/2

The Black Pirate A shipwrecked mariner vows revenge on the pirates who destroyed his father's ship. Quintessential Fairbanks, this film features astounding athletic

feats and exciting swordplay. Silent film with music score. Also known as "Rage of the Buccaneers." Also available in color.
1926 122m/B IT Douglas Fairbanks Sr., Donald Crisp, Billie Dove; *Dir:* Albert Parker. VHS, Beta, LV $19.95 VYY, MRV, REP 𝒵𝒵𝒵

The Black Planet The distant planet of Terre Verte is rapidly running out of energy. To keep the remaining fuel, warhawks Senator Calhoun and General McNab think their part of the Planet should blow up the other part. Will the entire planet be destroyed?
1982 78m/C AU *Dir:* Paul Williams. VHS, Beta $39.98 SUE 𝒵

Black Rain Erstwhile Ozu assistant Imamura directs this powerful portrait of a post-Hiroshima family five years after the bombing. Tanaka plays a young woman who, having been caught in a shower of black rain (radioactive fallout) on an ill-timed visit to Hiroshima, returns to her village to find herself ostracized by her peers and no longer considered marriage-worthy. Winner of numerous awards (including one at Cannes and five Japanese Academy Awards). In Japanese with English subtitles; also known as "Kuroi Ame."
1988 123m/B JP Kazuo Kitamura, Yoshiko Tanaka; *Dir:* Shohei Imamura. VHS $79.95 FXL, FCT, TAM 𝒵𝒵𝒵

Black Rain Douglas portrays a ruthless American cop chasing a Japanese murder suspect through gang-controlled Tokyo. Loads of action and stunning visuals from the man who brought you "Blade Runner."
1989 (R) 125m/C Michael Douglas, Andy Garcia, Kate Capshaw, Ken Takakura; *Dir:* Ridley Scott. VHS, Beta, LV, 8mm $14.95 PAR 𝒵𝒵 1/2

Black Rainbow A fino, monacing mood haunts most of this terribly familiar premise, made for cable TV by writer/director Hodges. A pretty young medium in a North Carolina carnival sideshow accurately predicts real-life murder, drawing the attention of a reporter—and then the police.
1991 (R) 103m/C Rosanna Arquette, Jason Robards Sr., Tom Hulce; *W/Dir:* Mike Hodges. VHS, Beta, LV $89.98 MED, FXV 𝒵𝒵 1/2

The Black Raven An action film that combines several plots into one. The Black Raven is an inn that sees more excitement than any other—not the least of which is murder!
1943 64m/B George Zucco, Wanda McKay, Glenn Strange, I. Stanford Jolley; *Dir:* Sam Newfield. VHS, Beta $16.95 SNC, MLB, RXM 𝒵 1/2

Black Robe In 1634 a young Jesuit priest journeys across the North American wilderness to bring the word of God to Canada's Huron Indians. The winter journey is brutal and perilous and the young man begins to question his mission after seeing the strength of the Indians native ways. Brian Moore adapted his own novel for the screen. Winner of six Canadian Genie awards in 1991.
1991 (R) 101m/C Lothaire Bluteau, August Schellenberg; *Dir:* Bruce Beresford. VHS $94.91 VMK 𝒵𝒵𝒵

The Black Room As an evil count lures victims into his castle of terror, the count's twin brother returns to fulfill an ancient prophecy. Karloff is wonderful in his dual role as the twin brothers.
1935 70m/B Boris Karloff, Marian Marsh, Robert Allen, Katherine DeMille, John Buckler, Thurston Hall; *Dir:* Roy William Neill. Beta, LV, Q $9.95 GKK, MLB 𝒵𝒵𝒵

The Black Room Couples are lured to a mysterious mansion where a brother and his sister promise to satisfy their sexual desires.
1982 (R) 90m/C Linnea Quigley, Stephen Knight, Cassandra Gaviola, Jim Stathis; *Dir:* Norman Thaddeus Vane. **VHS, Beta, LV** $69.95 *VES, LIV* ⅃

Black Roses A disgusting band of rockers shows up in a small town, and the local kids start turning into monsters. Coincidence?
1988 (R) 90m/C Carmine Appice, Sal Viviano, Carla Ferrigno, Julie Adams, Ken Swofford, John Martin; *Dir:* John Fasano. **VHS, Beta** $79.95 *IMP* ⅃

Black Sabbath An omnibus horror film with three parts, climaxing with Karloff as a Wurdalak, a vampire who must kill those he loves. Also available with "Black Sunday" on Laser Disc.
1964 99m/C *IT* Boris Karloff, Jacqueline Pierreux, Michele Mercier, Lidia Alfonsi, Susy Anderson, Mark Damon, Rika Dialina; *Dir:* Mario Bava. **VHS, Beta** $69.99 *HBO* ⅃⅃⅃

Black Samurai When his girlfriend is held hostage, a martial arts warrior will stop at nothing to destroy the organization that abducted her.
1977 (R) 84m/C Jim Kelly; *Dir:* Al Adamson. **VHS, Beta** $39.98 *CON* ⅃

The Black Scorpion Two geologists in Mexico unearth a nest of giant scorpions living in a dead volcano. Eventually one of the oversized arachnids escapes to wreak havoc on Mexico City.
1957 85m/B Richard Denning, Mara Corday, Carlos Rivas; *Dir:* Edward Ludwig. **VHS** $18.00 *FRG, MLB* ⅃ ½

Black Shampoo A black hairdresser on the Sunset Strip fights the mob with a chainsaw.
1976 90m/C John Daniels, Tanya Boyd, Joe Ortiz; *Dir:* Greydon Clark. **VHS, Beta** $59.95 *UHV* Woof!

Black Sister's Revenge A black Georgia girl gets involved with crime, college fraternities and machine guns in Los Angeles. The film was originally titled "Emma Mae."
197? 100m/C Jerri Hayes, Ernest Williams II; *Dir:* Jamaa Fanaka. **VHS, Beta** $59.95 *UNI* Woof!

The Black Six Six black Vietnam veterans are out to punish the white gang who killed the brother of one of the black men.
1974 (R) 91m/C Gene Washington, Carl Eller, Lem Barney, Mercury Morris, Joe "Mean Joe" Greene, Willie Lanier, Rosalind Miles, John Isenbarger, Ben Davidson, Maury Wills; *Dir:* Matt Cimber. **VHS, Beta** $49.90 *UNI* ⅃

The Black Stallion A young boy and a wild Arabian Stallion are the only survivors of a shipwreck, and they develop a deep affection for each other. When rescued, they begin training for an important race. Exceptionally beautiful first half. Rooney plays a horse trainer, again. Great for adults and kids. Music by Carmine Coppola.
1979 (PG) 120m/C Kelly Reno, Mickey Rooney, Teri Garr, Clarence Muse; *Dir:* Carroll Ballard. Academy Awards '79: Best Sound Effects Editing; Los Angeles Film Critics Association Awards '79: Best Musical Score, Best Cinematography; New York Film Festival '79: Best Musical Score; National Society of Film Critics Awards '79: Best Cinematography (Caleb Deschanel). **VHS, Beta, LV** $19.98 *MGM, FOX, TLF* ⅃⅃⅃

The Black Stallion Returns Sequel to "The Black Stallion" follows the adventures of young Alec as he travels to the Sahara to search for his beautiful horse, which was stolen by an Arab chieftain. Unfortunately lacks much of the charm that was present in the first film. Adapted from the stories by Walt Farley.
1983 (PG) 103m/C Kelly Reno, Teri Garr, Vincent Spano; *Dir:* Robert Dalva. **VHS, Beta, LV** $19.98 *FOX* ⅃⅃ ½

Black Starlet A girl from the projects of Chicago travels to Hollywood in search of fame. She winds her way through a world of sleaze and drugs in order to make it to the top. Also on video as "Black Gauntlet."
1974 (R) 90m/C Juanita Brown, Eric Mason, Rockne Tarkington, Damu King, Diane Holden; *Dir:* Chris Munger. **VHS** $59.95 *GEM* ⅃ ½

Black Sunday Executed one hundred years previously, a witch returns with her vampire servant to do the bidding of Satan. A must see for horror fans; firsts for Steele as star and Bava as director. Also available with "Black Sabbath" on Laser Disc.
1960 83m/B *IT* Barbara Steele, John Richardson, Ivo Garrani, Andrea Cecchi, Arturo Dominici; *Dir:* Mario Bava. **VHS** $19.98 *SNC, MRV, MOV* ⅃⅃⅃

Black Sunday An Arab terrorist group plots to steal the Goodyear Blimp and load it with explosives. Their intent is to explode it over a Miami Super Bowl game to assassinate the U.S. president and to kill all the fans. Based on Thomas Harris' novel.
1977 (R) 143m/C Robert Shaw, Bruce Dern, Marthe Keller, Fritz Weaver, Steven Keats, Michael Gazzo, William Daniels; *Dir:* John Frankenheimer. **VHS, Beta, LV** $24.95 *PAR* ⅃⅃ ½

Black & Tan/St. Louis Blues Two early jazz two-reelers are combined on this tape: "Black and Tan" is the first film appearance of Duke Ellington's Orchestra, featuring Cootie Williams and Johnny Hodges. "St. Louis Blues" is the only surviving film made by legendary blues singer Bessie Smith. She is backed by the Hall Johnson Choir, members of the Fletcher Henderson band directed by James P. Johnson, and dancer Jimmy Mordecai.
1929 36m/B Duke Ellington, Fredi Washington, Bessie Smith, Hall Johnson. **VHS, Beta** $17.95 *CCB* ⅃⅃ ½

Black Terrorist Terrorists take over a ranch and slay the inhabitants. They keep a young boy alive and the mother tries to rescue him.
1985 81m/C Allan Granville, Vera Jones; *Dir:* Neil Hetherington. **VHS, Beta** *RAE* ⅃⅃ ½

Black Tights Chevalier introduces four stories told in dance by Roland Petit's Ballet de Paris company: "The Diamond Crusher," "Cyrano de Bergerac," "A Merry Mourning," and "Carmen." Also known as "Un, Deux, Trois, Quartre!" A keeper for dance fans. Shearer's Roxanne in "Cyrano" was her last performance before retirement.
1960 120m/C *FR* Cyd Charisse, Zizi Jeanmarie, Moira Shearer, Dirk Sanders, Roland Petit; *Dir:* Terence Young; *Nar:* Maurice Chevalier. **VHS, Beta** $29.95 *MVD, NOS, VAI* ⅃⅃ ½

Black Tower An interesting murder mystery telling the story of a desperately impoverished medical student.
1950 54m/B Peter Cookson, Warren Williams, Anne Gwynne, Charles Calvert. **VHS, Beta** $24.95 *VYY* ⅃⅃

Black Vampire Jones plays a professor of African studies who is turned into a vampire by another vampire. He then marries the villain's ex-wife, played by Clark, a veteran of Russ Meyer films. Not much of a story but the acting is actually quite good and there are some creepy moments thanks mainly to an interesting score. Originally titled "Ganja and Hess," also released as "Double Possession." Also on video under the title "Blood Couple."
1973 (R) 83m/C Duane Jones, Marlene Clark, Bill Gunn, Sam Waymon, Leonard Jackson, Candece Tarpley, Mabel King; *W/Dir:* Bill Gunn. **VHS, Beta** $9.95 *SIM, GEM, IPI* ⅃ ½

A Black Veil for Lisa A man attempts to exact revenge upon his unfaithful wife, but things go horribly awry.
1968 87m/C *IT GE* John Mills, Lucianna Paluzzi, Robert Hoffman; *Dir:* Massimo Dallamano. **VHS** *REP* ⅃⅃

Black Venus A soft-core epic, starring the former Miss Bahamas, Josephine Jacqueline Jones, about the 18th-century French aristocracy. Laughably based upon the stories of Balzac. European film dubbed in English.
1983 (R) 80m/C Josephine Jacqueline Jones, Emiliano Redondo; *Dir:* Claude Mulot. **VHS, Beta** $79.95 *MGM* Woof!

Black Water Gold Television movie about a search for sunken Spanish gold.
1969 75m/C Ricardo Montalban, Keir Dullea, Lana Wood, Bradford Dillman, France Nuyen; *Dir:* Alan Landsburg. **VHS** *TAV* ⅃⅃ ½

Black Werewolf A millionaire sportsman invites a group of men and women connected with bizarre deaths or the eating of human flesh to spend the cycle of a full moon at his isolated lodge. Also available under its original title, "The Beast Must Die."
1975 (PG) 93m/C *GB* Peter Cushing, Calvin Lockhart, Charles Gray, Anton Diffring; *Dir:* Paul Annett. **VHS, Beta, LV** $9.95 *SIM, MED, PSM* ⅃⅃

Black and White in Color An award-winning satire about a French soldier at an African outpost, who, upon hearing the news of the beginning of World War I, takes it upon himself to attack a neighboring German fort. Calamity ensues. In French, with English subtitles.
1976 (PG) 100m/C *FR* Jean Carmet, Jacques Dufilho, Catherine Rouvel, Jacques Spiesser, Dora Doll, Jacques Perrin; *Dir:* Jean-Jacques Annaud. Academy Awards '76: Best Foreign Language Film. **VHS, Beta, LV** $39.98 *LHV, WAR, TAM* ⅃⅃⅃

Black & White Swordsmen A running feud between two families provides impetus for much inspired kicking and chopping.
1985 92m/C Jiang Bin, Chen Hong Lei. **VHS, Beta** $29.95 *MAV* ⅃

The Black Widow A fortune-teller plots to steal scientific secrets and take over the world. Serial in thirteen episodes. Also known as "Sombra, the Spider Woman."
1947 164m/B Bruce Edwards, Carol Forman, Anthony Warde; *Dir:* Spencer Gordon Bennet. **VHS** $19.98 *VCN, REP, MLB* ⅃⅃

Black Widow A federal agent pursues a beautiful murderess who marries rich men and then kills them, making the deaths look natural. The agent herself becomes involved in the final seduction. The two women are enticing and the locations picturesque.

1987 (R) 97m/C Debra Winger, Theresa Russell, Sami Frey, Nicol Williamson, Terry O'Quinn, Dennis Hopper; *Dir:* Bob Rafelson. **VHS, Beta, LV** $19.98 *FOX* 🎬🎬🎬

The Black Windmill An English spy is caught between his service and the kidnapping of his family by rival spies.
1974 (PG) 102m/C Michael Caine, Donald Pleasence, Delphine Seyrig, Clive Revill, Janet Suzman, John Vernon; *Dir:* Don Siegel. **VHS, Beta** $59.95 *MCA* 🎬🎬½

Blackbeard the Pirate The 18th-century buccaneer is given the full-blooded, Hollywood treatment.
1952 99m/C Robert Newton, Linda Darnell, Keith Andes, William Bendix, Richard Egan; *Dir:* Raoul Walsh. **VHS, Beta, LV** $19.98 *RKO* 🎬🎬

Blackbeard's Ghost Disney comedy in which the famed 18th-century pirate's spirit (Ustinov) summoned to wreak havoc in order to prevent an old family home from being turned into a casino.
1967 107m/C Peter Ustinov, Dean Jones, Suzanne Pleshette, Elsa Lanchester, Richard Deacon; *Dir:* Robert Stevenson. **VHS, Beta** $19.99 *DIS* 🎬🎬

Blackboard Jungle Well-remembered urban drama about an idealistic teacher in a slum area who fights doggedly to connect with his unruly students. Bill Hailey's "Rock Around the Clock" over the opening credits was the first use of rock music in a mainstream feature film. Based on Evan Hunter novel.
1955 101m/B Glenn Ford, Anne Francis, Louis Calhern, Sidney Poitier, Vic Morrow, Richard Kiley, Margaret Hayes, John Hoyt, Warner Anderson; *Dir:* Richard Brooks. **VHS, Beta** $19.95 *MGM* 🎬🎬🎬½

Blackenstein A doctor restores a man's arms and legs, but a jealous assistant causes the man to turn into a monster who starts attacking people.
1973 (R) 87m/C John Hart, Ivory Stone, Andrea King, Liz Renay, Joe DiSue; *Dir:* William A. Levey. **VHS, Beta** $29.95 *MED* Woof!

Blackjack Las Vegas is the scene for action and excitement as the mob puts the hit on tough guy William Smith!
1978 (R) 104m/C William Smith, Tony Burton, Paris Earl, Damu King, Diane Summerfield, Angela May; *Dir:* John Evans. **VHS** *TPV* 🎬🎬

Blacklash Two cops, male and female, transport an accused murderess to trial and get involved with all kinds of passionate backbiting. Music by Michael Atkinson.
1987 (R) 89m/C David Argue, Gia Carides, Lydia Miller; *Dir:* Bill Bennett. **VHS, Beta** $79.95 *VIR* 🎬½

Blackmail This first sound film for Great Britain and director Hitchcock features an early visualization of some typical Hitchcockian themes. The story follows the police investigation of a murder, and a detective's attempts to keep his girlfriend from being involved. Look for Hitchcock's very first on-screen cameo. Made as a silent, this was reworked to become a talkie.
1929 86m/B *GB* Anny Ondra, John Longdon, Sara Allgood, Charles Paton, Cyril Ritchard; *Dir:* Alfred Hitchcock. **VHS, Beta** $9.95 *NEG, MRV, NOS* 🎬🎬🎬

Blackmail Two con-artists and a detective are caught up in a deadly con against a rich gangster's wife. Based on a short story by Bill Crenshaw.

1991 (R) 87m/C Susan Blakely, Dale Midkiff, Beth Toussaint, John Saxon, Mac Davis; *Dir:* Rubin Preuss. **VHS, Beta** *PAR* 🎬🎬½

Blackout A blind man recovers his sight and finds that the brother of his girlfriend, once thought dead, is actually alive and well and running a smuggling ring. Routine.
1950 73m/B *GB* Maxwell Reed, Dinah Sheridan, Patric Doonan, Eric Pohlmann; *Dir:* Robert S. Baker. **VHS, Beta** $16.95 *SNC* 🎬½

Blackout A drunken private eye is offered a murder case, and is subsequently framed for the crime.
1954 87m/B Dane Clark, Belinda Lee, Betty Ann Davis; *Dir:* Terence Fisher. **VHS, Beta, LV** $89.98 *WGE, MLB* 🎬🎬

Blackout Four killers terrorize an office building during the 1977 New York electrical blackout. Soon the police enter, confront them, and the fun starts. Comic touches provide some relief from the violence here.
1978 (R) 86m/C Jim Mitchum, Robert Carradine, Ray Milland, June Allyson, Jean-Pierre Aumont, Belinda J. Montgomery; *Dir:* Eddy Matalon. **VHS, Beta** $59.95 *CHA* 🎬🎬

Blackout Made-for-cable-TV thriller. An aging police chief suspects a disfigured amnesiac is responsible for past killings, in spite of the fact that he now leads a subdued family life.
1985 99m/C Richard Widmark, Keith Carradine, Kathleen Quinlan, Michael Beck; *Dir:* Douglas Hickox. **VHS, Beta** $79.95 *MED* 🎬🎬½

Blackout Strange memories from childhood come back to her as a woman fights for her life.
1988 (R) 91m/C Carol Lynley, Gail O'Grady, Michael Keys Hall, Joseph Gian, Joanna Miles; *Dir:* Doug Adams. **VHS, Beta, LV** $80.08 *MAG* 🎬🎬

The Blacksmith/The Balloonatic Two Buster Keaton shorts: "The Blacksmith" (1922) features Buster as the local blacksmith's apprentice who suddenly finds himself in charge. "The Balloonatic" (1923) presents Buster trapped in a runaway hot air balloon. Both films include music score.
1923 57m/B Buster Keaton, Virginia Fox, Phyllis Haver; *Dir:* Buster Keaton. **VHS, Beta** $29.95 *VYY* 🎬🎬

The Blacksmith/The Cops "The Blacksmith" is a burlesque of Longfellow's famous poem "The Village Blacksmith." In "Cops" Buster tries a new business venture to win his girl's hand. Chaos ensues. Silent comedy at its most adroit.
1922 38m/B Buster Keaton, Virginia Fox. **VHS, Beta** $19.95 *CCB* 🎬🎬

Blackstar Blackstar fights the forces of evil in three animated adventures.
1981 60m/C VHS, Beta $29.95 *FHE* 🎬🎬

Blackstar Volume 2 John Blackstar and his friends on the planet Sagar continue their never-ending battle against the cruel and ruthless Overlord in three animated adventures.
1981 60m/C VHS, Beta $29.95 *FHE* 🎬

Blackstar Volume 3 John Blackstar and his friends take on the evil overlord in three adventures: "The Mermaid of Serpent Sea," "Lightning City of the Clouds," and "The Airwhales of Anchar."
1981 60m/C VHS, Beta $29.95 *FHE* 🎬

Blackstone on Tour Highlights from magician Harry Blackstone's cross country tour.

1984 60m/C Harry Blackstone Jr. **VHS, Beta** $39.95 *RKO* 🎬

Blacula The African Prince Mamuwalde stalks the streets of Los Angeles trying to satisfy his insatiable desire for blood. Mildly successful melding of blaxploitation and horror that spawned a sequel, "Scream, Blacula, Scream."
1972 (PG) 92m/C William Marshall, Thalmus Rasulala, Denise Nicholas, Vonetta McGee; *Dir:* William Crain. **VHS, Beta, LV** $19.99 *HBO* 🎬🎬

Blade An honest cop challenges a dirty cover-up in killer-stalked New York.
1972 (PG) 79m/C Steve Landesburg, John Schuck, Kathryn Walker; *Dir:* Ernie Pintoff. **VHS, Beta** $59.95 *GEM, MRV* 🎬🎬

A Blade in the Dark A young man composing a score for a horror film moves into a secluded villa and is inspired and haunted by the mysterious murder he witnesses.
1983 90m/C Andrea Qchhipinti, Anny Papa; *Dir:* Lamberto Bava. **VHS, Beta** $69.98 *VES, LIV* 🎬🎬

Blade in Hong Kong Made-for-TV action outing. Investigator travels to Hong Kong, finds trouble and romance in the underbelly of the city. Pilot for un-sold series.
1985 100m/C Terry Lester, Keye Luke, Mike Preston, Jean-Marie Hon, Leslie Nielsen, Nancy Kwan, Anthony Newley, Ellen Regan; *Dir:* Reza Badiyi. **VHS** $59.98 *CGI* 🎬🎬½

Blade Master In this sequel to "Ator the Fighting Eagle," O'Keeffe as Ator is back as the Blade Master. Ator defends his people and his family name in a battle against the "Geometric Nucleus": a primitive bomb. His quest leads him and his small band of men to the castle of knowledge.
1984 (PG) 92m/C Miles O'Keefe, Lisa Foster; *Dir:* David Hills. **VHS, Beta** $19.95 *MED* 🎬½

Blade of the Ripper A madman armed with a razor slashes his way through the lovelies of the international jet set.
1984 90m/C George Hilton, Edwige Fenech. **VHS, Beta** $49.95 *REG* Woof!

Blade Runner In Los Angeles of the 29th century, a world-weary cop tracks down a handful of murdering, renegade "replicants" (synthetically produced human slaves who, with only days left of life, search madly for some way to extend their prescribed lifetimes). A moody, beautifully photographed dark thriller with sets from an architect's dream. Music by Vangelis. Based on "Do Androids Dream of Electric Sheep" by Philip K. Dick. Laser edition features restored footage, information about special effects and production sketches.
1982 (R) 122m/C Harrison Ford, Rutger Hauer, Sean Young, Darryl Hannah, M. Emmet Walsh, Edward James Olmos, Joe Turkel, Brion James, Joanna Cassidy; *Dir:* Ridley Scott. **VHS, Beta, LV, SVS, 8mm** $9.95 *COL, SUE, SUP* 🎬🎬🎬½

Blades Another junk heap from Troma, dealing with the efforts of three golfers who try to stop a maniacal power mower that's been grinding duffers with regularity.
1989 (R) 101m/C Robert North, Jeremy Whelan, Victoria Scott, Jon McBride. **VHS, Beta, LV** $79.98 *MED* 🎬

Blades of Courage A made-for-television biography of Olympic skater Lori Larouche. Choreographed by Debbi Wilkes. Originally known as, "Skate!"
1988 98m/C *CA* Lynn Nightingale, Christianne Hirt, Colm Feore, Rosemary Dunsmore; *Dir:* Randy Bradshaw. **VHS, Beta** $29.95 *ACA* 🎬🎬

Bladestorm Ultra-cheapo fantasy/warrior rehash.
1986 87m/C VHS, Beta $59.95 *MGL* 🦴

Blaise Pascal Another of Rossellini's later historical portraits, detailing the life and times of the 17th-century philosopher, seen as a man whose scientific ideas conflicted with his own religious beliefs. Made for Italian television. In Italian with subtitles.
1971 131m/C *IT* ***Dir:*** Roberto Rossellini. **VHS, Beta $79.95** *FCT* 🦴🦴🦴½

Blake of Scotland Yard Blake, the former Scotland Yard inspector, battles against a villain who has constructed a murderous death ray.
1936 70m/B Ralph Byrd, Herbert Rawlinson, Joan Barclay, Lloyd Hughes; ***Dir:*** Robert Hill. **VHS, Beta $29.95** *NOS, DVT, VCN* 🦴🦴½

Blame It on the Bellboy Wild farce set in Venice about a hotel bellboy who confuses three similarly named visitors—sending the wrong ones to meet corparate bigwigs, date women, or even kill. The brisk tale devolves into chase scenes.
1991 (PG-13) 79m/C Dudley Moore, Bryan Brown, Richard Griffiths, Patsy Kensit, Bronson Pinchot, Penelope Wilton, Andreas Katsulas; ***W/ Dir:*** Mark Herman. **VHS, Beta $94.95** *HPH* 🦴🦴½

Blame It on the Night A rock star gets to take care of the military cadet son he never knew after the boy's mother suddenly dies. Mick Jagger helped write the story. Available in VHS and Beta Hi-Fi.
1984 (PG-13) 85m/C Nick Mancuso, Byron Thames, Leslie Ackerman, Billy Preston, Merry Clayton; ***Dir:*** Gene Taft. **VHS, Beta $79.98** *FOX* 🦴🦴

Blame It on Rio A middle-aged man has a ridiculous fling with his best friend's daughter while on vacation with them in Rio de Janeiro. Caine and Johnson are amusing, but the script is somewhat weak. Remake of the French film "One Wild Moment."
1984 (R) 90m/C Michael Caine, Joseph Bologna, Demi Moore, Michelle Johnson, Valerie Harper; ***Dir:*** Stanley Donen. **VHS, Beta, LV $29.95** *VES, LIV* 🦴🦴

The Blasphemer A well-crafted, engrossing drama about misfortune that befalls someone who disobeys one of the sacred Commandments. Silent music score.
1921 89m/B VHS, Beta $29.95 *VYY* 🦴🦴

Blastfighter After local hoodlums kill his daughter, an ex-con cop goes on a spree of violence and revenge. Director Bava uses the pseudonym "John Old, Jr.," as his father Mario Bava occasionally credited himself as John Old. Italian film shot in Atlanta.
1985 (R) 93m/C *IT* Michael Sopkiw, Valerie Blake, George Eastman; ***Dir:*** Lamberto Bava. **VHS, Beta $69.95** *VES, LIV Woof!*

Blat A gangster of Italian-Russian descent tries to pull a sting on merry old England's financial world in this comedy made for British television.
19?? 90m/C *GB* Robert Hardy, Adrienne Corri, Alfred Molina. **VHS $49.95** *FCT* 🦴½

Blaxploitation Cartoons A collection of seldom seen racist cartoons featuring Amos 'n' Andy, Little Black Sambo and other stereotyped characters. Some black and white sequences.
194? 46m/C VHS, Beta $24.95 *DVT, PTB* 🦴½

Blaze The true story of Louisiana governor Earl Long who became involved with a stripper, Blaze Starr, causing a political scandal of major proportions. Robust, good-humored bio-pic featuring a fine character turn by Newman.
1989 (R) 117m/C Paul Newman, Lolita Davidovich, Jerry Hardin, Robert Wuhl, Gailard Sartain, Jeffrey DeMunn, Richard Jenkins; ***Dir:*** Ron Shelton. **VHS, Beta, LV $89.95** *TOU* 🦴🦴½

Blaze Glory The satire of an "Old West" hero. Sight gags, exaggeration of stereotypes, and pixillation (a form of stop-action animation in which people move like cartoon characters) contribute to the film's humor.
1968 10m/C VHS, Beta *PYR* 🦴½

Blaze Starr: The Original The enormously "gifted" soft-porn star of the sixties joins a nudist camp. Wacky exploits ensue.
1963 80m/C VHS $39.95 *FXL* 🦴🦴

Blazing Guns A marshal, masquerading inexplicably as a tenderfoot, enters a lawless town and is immediately hired as a deputy.
1950 54m/B Raymond Hatton, Russell Hayden, Al "Fuzzy" Knight, Betty Adams. **VHS, Beta, LV $19.95** *NOS, WGE* 🦴

The Blazing Ninja Head-kicking jaw-crunching duels abound in this tale of master versus student.
19?? 86m/C VHS $14.98 *VTR* 🦴½

Blazing Saddles A wild, wacky spoof by Mel Brooks of every cliche in the western film genre. Little is Black Bart, a convict offered a reprieve if he will become a sheriff and clean up a nasty frontier town; the previous recipients of this honor have all swiftly ended up in shallow graves. A crazy, silly film with a cast full of loveable loonies including comedy greats Wilder, Kahn, and Korman. Watch for the Count Basie Orchestra. Was the most-viewed movie in its first year of release on HBO cable.
1974 (R) 90m/C Cleavon Little, Harvey Korman, Madeline Kahn, Gene Wilder, Mel Brooks, John Hillerman, Alex Karras, Dom DeLuise, Liam Dunn; ***Dir:*** Mel Brooks. **VHS, Beta, LV $19.98** *WAR* 🦴🦴🦴½

Blazing Stewardesses The Hound salutes the distributor for truth in advertising, as they stamped this as one of the world's worst videos. Lusty, busty stewardesses relax at a western guest ranch under siege from hooded riders and the aging gags of the Ritz Brothers.
1975 (R) 95m/C Yvonne de Carlo, Bob Livingston, Donald (Don "Red") Barry, The Ritz Brothers, Regina Carrol; ***Dir:*** Al Adamson. **VHS $14.98** *VID Woof!*

Bleak House The British television mini-series adaptation of the Charles Dickens tome about the decadent, criminal ruling class of 19th-century England. On three tapes.
1985 391m/C *GB* Denholm Elliott, Diana Rigg, Philip Franks, Peter Vaughan, T.P. McKenna, Charles Drake; ***Dir:*** Ross Devenish. **VHS, Beta $29.98** *FOX* 🦴🦴½

Bless the Beasts and Children A group of six teenage boys at a summer camp attempt to save a herd of buffalo from slaughter at a national preserve. Treacly Kramer backwater. Based on the novel by Glendon Swarthout.
1971 (PG) 109m/C Billy Mumy, Barry Robins, Miles Chapin, Darel Glaser, Bob Kramer, Ken Swofford, Jesse White; ***Dir:*** Stanley Kramer. **VHS, Beta $59.95** *COL* 🦴🦴½

Blind Ambition Docu-drama television mini-series traces the career of John Dean, special counsel to President Richard M. Nixon. Focuses on his fractured personal life while touching on virtually all of the Watergate headlines.
1982 95m/C Martin Sheen, Rip Torn. **VHS, Beta** *TLF* 🦴½

Blind Date A blind man agrees to have a visual computer implanted in his brain in order to help the police track down a psychopathic killer. Violent scenes may be disturbing to some.
1984 (R) 100m/C Joseph Bottoms, Kirstie Alley, Keir Dullea, James Daughton, Lana Clarkson, Marina Sirtis; ***Dir:*** Nico Mastorakis. **VHS, Beta, LV $79.95** *LTG, VES, LIV* 🦴½

Blind Date A blind date between a workaholic yuppie and a beautiful blonde starts off well, but when she drinks too much at dinner, things get out of hand. In addition to embarrassing her date and destroying the restaurant, she has a jealous ex-boyfriend who must be dealt with.
1987 (PG-13) 95m/C Kim Basinger, Bruce Willis, John Larroquette, William Daniels, George Coe, Mark Blum, Phil Hartman, Stephanie Faracy, Alice Hirson, Graham Stark; ***Dir:*** Blake Edwards. **VHS, Beta, LV $14.95** *COL* 🦴🦴

Blind Faith Action and adventure take a turn for the horrific when several men find themselves in captivity.
1989 (R) ?m/C Eric Gunn, Kevin Yon, Lynne Browne; ***W/Dir:*** Dean Wilson. **VHS $19.95** *AVP, HHE* 🦴

Blind Fear A blind woman is stalked by three killers in an abandoned country inn.
1989 (R) 98m/C Shelley Hack, Jack Langedijk, Kim Coates, Jan Rubes, Heidi von Palleske; ***Dir:*** Tom Berry. **VHS, Beta, LV $29.95** *ACA* 🦴½

Blind Fists of Bruce A rich, idle young man learns the techniques of kung fu from a former warrior.
197? 92m/C Bruce Li. **VHS, Beta $54.95** *MAV* 🦴½

Blind Fools A scathing indictment of children neglected by ambitious parents.
1940 66m/B Herbert Rawlinson, Claire Whitney, Russell Hicks, Miriam Battista. **VHS, Beta, LV** *WGE* 🦴½

Blind Fury A blind Vietnam vet enlists the aid of a Zen master and a sharpshooter to tackle the Mafia.
1990 (R) 86m/C Rutger Hauer, Terry O'Quinn, Lisa Blount, Randall "Tex" Cobb, Noble Willingham, Meg Foster, Sho Kosugi; ***Dir:*** Phillip Noyce. **VHS, Beta, LV $89.95** *COL, FCT* 🦴🦴½

Blind Husbands An Austrian officer is attracted to the pretty wife of a dull surgeon. Controversial in its day, this lurid, sumptuous melodrama instigated many stubborn Hollywood myths, including the stereotype of the brusque, jodhpur-clad Prussian director. This was von Stroheim's first outing as director.
1919 98m/B Erich von Stroheim, Fay Wray; ***Dir:*** Erich von Stroheim. **VHS, Beta $27.95** *VYY, DNB* 🦴🦴🦴½

Blind Justice An innocent man is identified as a rapist, and the accusation ruins his life. Made for television.
1986 94m/C Tim Matheson, Lisa Eichhorn, Mimi Kuzyk, Philip Charles MacKenzie, Tom Atkins; ***Dir:*** Rod Holcomb. **VHS, Beta $59.98** *FOX* 🦴🦴

Blind Man's Bluff A blind professor is accused of murdering his neighbor, but as he tries to solve the mystery, evidence points to his girlfriend who may have framed him for the murder.
1991 (PG-13) 86m/C Robert Urich, Lisa Eilbacher, Patricia Clarkson, Ken Pogue, Ron Perlman; *Dir:* James Quinn. VHS, Beta PAR 🎬🎬

Blind Rage When the United States government transports fifteen million dollars to the Philippines, five blind kung fu masters want a piece of the action.
1978 (R) 81m/C PH D'Urville Martin, Leo Fong, Tony Ferrer, Dick Adair, Darnell Garcia, Leila Hermosa, Fred Williamson; *Dir:* Efren C. Pinon. VHS, Beta $59.95 MGM 🎬

Blind Vengeance When a close friend is killed by white supremacists, a man sets out to exact revenge.
1990 (R) 93m/C Gerald McRaney, Marg Helgenberger, Thalmus Rasulala, Lane Smith, Don Hood, Grand Bush; *Dir:* Lee Philips. VHS $79.95 MCA 🎬🎬🎬

Blind Vengeance Notorious bounty hunter tracks a brutal outlaw for murdering his family. His partner in vengeance is blinded and the hunter's shooting hand gets wounded. The two must rely on teamwork to overcome their disadvantages.
1990 90m/C Robert Lansing, Patrick Wayne, Slim Pickens, Gloria Talbott; *Dir:* Michael Moore. VHS API 🎬

Blind Vision Von Dohlen stars as William Dalton, a mail clerk who is in love with his beautiful neighbor Leanne (Shelton). Suffering from extreme shyness, Dalton can only spy on her through a telephoto lens. Things start to get complicated for him when one of Leanne's boyfriends turns up dead outside her apartment. Soon afterwards, their landlady and a local police detective uncover a shocking sexual secret that forces Dalton into a deadly game of obsession and desire.
1991 92m/C Robert Vaughn, Lenny Von Dohlen, Deborah Shelton, Louise Fletcher, Ned Beatty; *Dir:* Shuki Levy. VHS $89.95 WOV 🎬½

Blinded by the Light A sister tries to save her brother from his attachment to a quasi-religious cult, The Light of Salvation. Real-life brother and sister McNichol team up in this made-for-television drama, one of their earlier examinations of cult behavior.
1982 90m/C Kristy McNichol, Jimmy McNichol, Anne Jackson, Michael McGuire; *Dir:* John Alonzo. VHS, Beta TLF 🎬🎬

Blindside A surveillance hobbyist who owns a motel spies on his tenants until a murder involves him in a big-scale drug war.
1988 (R) 98m/C Harvey Keitel, Lori Hallier, Lolita David, Alan Fawcett, Michael Rudder; *Dir:* Paul Lynch. VHS, Beta, LV $14.98 CHA 🎬🎬

Bliss A savage, surreal Australian comedy about an advertising executive who dies suddenly for a few minutes, and upon his awakening he finds the world maniacally, bizarrely changed. Based on the Peter Carey novel, and one of the most inspired absurdist films of the decade.
1985 (R) 112m/C AU Barry Otto, Lynette Curran, Helen Jones; *Dir:* Ray Lawrence. Australian Film Institute '85: Best Film; Australian Academy Awards '86: Best Picture. VHS, Beta, LV $19.95 NWV, STE 🎬🎬🎬½

The Bliss of Mrs. Blossom Three's a crowd in this light-hearted romp through the machinations of a brassiere manufacturer (Attenborough) and his neglected wife (MacLaine). Mrs. Blossom finds sewing machine repairman Booth so appetizing that she hides him in the attic of the Blossom home. He reads books and redecorates, until, several plot twists later, Attenborough discovers the truth. Witty and wise, with fine supporting cast and excellent pacing.
1968 93m/C GB Shirley MacLaine, Richard Attenborough, James Booth, Freddie Jones, John Cleese; *Dir:* Joseph McGrath. VHS $14.95 PAR, MOV 🎬🎬🎬

Blithe Spirit Charming and funny adaptation of Coward's famed stage play. A man remarries and finds his long-dead wife is unhappy enough about it to come back and haunt him. Clever supporting cast, with Rutherford exceptional as the medium. Received Oscar for its Special Effects.
1945 96m/C GB Rex Harrison, Constance Cummings, Kay Hammond, Margaret Rutherford, Hugh Wakefield, Joyce Carey; *Dir:* David Lean. Academy Awards '46: Best Special Effects. VHS MLB 🎬🎬🎬½

The Blob McQueen's first starring role. Science fiction thriller about a small town's fight against a slimy jello invader from space. McQueen's slightly rebellious character redeems himself when he saves the town with quick action. Low-budget, horror/teen-fantasy that became a camp classic. Other names considered for the movie included The Glob, The Glob That Girdled the Globe, The Meteorite Monster, the Molten Meteorite, and The Night of the Creeping Dead. Followed by a worthless sequel in 1972, "Son of Blob" (directed by Larry Hagman), and a worthwhile remake in 1988. LV edition contains an additional fifteen minutes of McQueen interviews and the trailers for both the original and the 1988 remake.
1958 83m/C Steve McQueen, Aneta Corseaut, Olin Howland, Earl Rowe; *Dir:* Irvin S. Yeawarth Jr. VHS, Beta, LV $19.95 COL, GEM, VYG 🎬🎬½

The Blob A hi-tech remake of the 1958 camp classic about a small town beset by a fast-growing, man-eating mound of glop shot into space by scientists, irradiated into an unnatural being, and then returned to earth. Well-developed characters make this an excellent tribute to the first film.
1988 (R) 92m/C Kevin Dillon, Candy Clark, Joe Seneca, Shawnee Smith, Donovan Leitch, Jeffrey DeMunn; *Dir:* Chuck Russell. VHS, Beta, LV $89.95 COL 🎬🎬🎬

Block-heads Twenty years after the end of World War I, soldier Stan is found, still in his foxhole, and brought back to America, where he moves in with old pal Ollie. Also includes a 1934 Charley Chase short "I'll Take Vanilla."
1938 75m/B Stan Laurel, Oliver Hardy, Billy Gilbert, Patricia Ellis, James Finlayson, Charley Chase; *Dir:* John Blystone. VHS, Beta $19.95 CCB, MED, MLB 🎬🎬🎬

Blockhouse Based on Jean Paul Cleberts' novel "Le Blockhaus." Tells the story of four men entombed in a subterranean stronghold for six years after the D-Day invasion of Normandy.
1973 88m/C GB Peter Sellers, Charles Aznavour, Per Oscarsson, Peter Vaughan, Leon Lissek, Alfred Lynch, Jeremy Kemp; *Dir:* Clive Rees. VHS, Beta $42.95 PGN 🎬½

Blonde in Black Leather A bored Italian housewife takes up with a leather-clad lady biker, and together they cavort about.
1977 88m/C Claudia Cardinale, Monica Vitti; *Dir:* Carlo DiPalma. VHS, Beta $59.95 CHA 🎬½

B

Blonde Cobra/The Doctor's Dream Here are two short films by Ken Jacobs: Underground film star Jack Smith reminisces about his career in "Blonde Cobra" and a country doctor saves the life of a little girl in "Doctor's Dream."
1984 57m/C VHS, Beta $60.00 NEW 🎬½

Blonde Savage Charting African territories, an adventurer encounters a white jungle queen swinging amongst the vines. Essentially a cheap, distaff "Tarzan."
1947 62m/B Leif Erickson, Gale Sherwood, Veda Ann Borg, Douglas Dumbrille, Frank Jenks, Matt Willis, Ernest Whitman; *Dir:* S.K. Seeley. VHS, Beta $16.95 SNC 🎬

Blonde Venus A German cafe singer marries an Englishman, but their marriage hits the skids when he contracts radiation poisoning and she gets a nightclub job to pay the bills. Sternberg's and Dietrich's fourth film together, and characteristically beautiful, though terribly strange. Dietrich's cabaret number "Hot Voodoo," in a gorilla suit and blonde afro, attains new heights in early Hollywood surrealism.
1932 92m/B Marlene Dietrich, Herbert Marshall, Cary Grant, Dickie Moore, Hattie McDaniel, Sidney Toler; *Dir:* Josef von Sternberg. VHS, Beta $29.95 MCA 🎬🎬🎬½

Blondie: Eat to the Beat The multi-million seller platinum album, "Eat to the Beat," taped on location and in a studio. The program contains twelve songs, including the hit singles "Dreaming" and "The Hardest Part."
1980 60m/C Blondie. VHS, Beta $14.95 MVD, WAR 🎬🎬🎬½

Blood Alley A seasoned Merchant Marine captain takes on a cargo of refugee Chinese to smuggle through enemy territory. Middling, mid-career Wayne vehicle.
1955 115m/C John Wayne, Lauren Bacall, Paul Fix, Joy Kim, Berry Kroeger, Mike Mazurki, Anita Ekberg; *Dir:* William A. Wellman. VHS, Beta, LV $19.98 WAR, PIA, FCT 🎬🎬

Blood Avenger Vengeance becomes one man's obsession when he witnesses the brutal rape of his sister and murder of his father.
1984 ?m/C VHS $29.95 OCE 🎬½

Blood Beach A group of teenagers are devoured by menacing sand, which keeps people from getting to the water by swallowing them whole. Weak parody with some humorous moments; more silly than scary.
1981 (R) 92m/C David Huffman, Marianna Hill, John Saxon, Burt Young, Otis Young; *Dir:* Jeffrey Bloom. VHS, Beta $54.95 MED 🎬

Blood Beast Terror An entomologist transforms his own daughter into a Deaths-head Moth and she proceeds to terrorize and drink innocent victims' blood.
1967 81m/C GB Peter Cushing, Robert Flemyng, Wanda Ventham; *Dir:* Vernon Sewell. VHS, Beta $39.98 MON 🎬½

Blood and Black Lace Beautiful models are being brutally murdered and an inspector is assigned to the case, but not before more gruesome killings occur. Bava is, as usual, violent and suspenseful. Horror fans will enjoy this flick.
1964 90m/C IT FR GE Cameron Mitchell, Eva Bartok, Mary Arden; *Dir:* Mario Bava. VHS, Beta $59.95 MED 🎬

Blood Bride A lonely young woman finally finds happiness with her new husband but her world comes crashing about her with soul-mangling ferocity when she discovers he is actually a bloodthirsty maniac.
19?? 90m/C Ellen Barber, Philip English. **VHS, Beta $69.95** *MAG* ⚔

Blood Brothers A young man who dreams of becoming a lawyer is disturbed when he discovers that his family has mafia ties.
1974 (R) 148m/C *IT* Claudia Cardinale, Franco Nero, Lina Polito; *Dir:* Pasquale Squitieri. **VHS, Beta $69.95** *LTG* ⚔

Blood Brothers James Grizzly Adams, who lives in the mountains of the American wilderness, relives the story of how he met his Indian blood brother, Nakuma. Beautiful scenery.
1977 50m/C Dan Haggerty. **VHS, Beta $14.98** *MCG* ⚔⚔

Blood Clan Based on the true story of Katy Bane, daughter of notorious Scottish cult leader, Sawney Bane, in whose lair were found the remains of over 1000 killed and cannibilized followers. When found, Bane's entire family was sentenced to death with the exception of Katy, who left to make a new start in Canada. When a rash of mysterious deaths break out in Katy's new home, she must defend herself from rumors that her father's murderous cult is resurfacing.
1991 91m/C Gordon Pinsent, Michelle Little, Robert Wisden; *Dir:* Charles Wilkinson. **VHS $79.95** *MNC, FXL* ⚔½

Blood & Concrete: A Love Story Bizarre, violent and stylish film-noir spoof, definitely not for all tastes. The innocent hero gets drawn into a maelstrom of intrigue over a killer aphrodisiac drug. Beals, an addicted punk rocker, gets to perform a few songs.
1990 (R) 97m/C Billy Zane, Jennifer Beals, Darren McGavin; *W/Dir:* Jeff Reiner. **VHS $89.95** *COL* ⚔⚔

Blood Cult A bizarre series of murder-mutilations take place on a small midwestern campus. Contains graphic violence that is not for the squeamish. This film was created especially for the home video market.
1985 92m/C Chuck Ellis, Julie Andelman, Jim Vance, Joe Hardt. **VHS, Beta $24.95** *UHV* ⚔

Blood Debts A father is out for revenge after he saves his daughter from some hunters who raped her and also killed her boyfriend.
1983 91m/C Richard Harrison, Mike Manty, Jim Gaines, Anne Jackson, Anne Milhench; *Dir:* Teddy Page. **VHS, Beta $59.98** *CON* ⚔

Blood of Dracula They don't make 1950s rock'n'roll girls' school vampire movies like this anymore, for which we may be grateful. Hypnotism, an amulet, and a greasepaint makeup job turn a shy female student into a bloodsucker.
1957 71m/B Sandra Harrison, Louise Lewis, Gail Ganley, Jerry Blaine, Heather Ames, Malcolm Atterbury; *Dir:* Herbert L. Strock. **VHS $29.95** *COL, MLB* ⚔½

Blood of Dracula's Castle A couple inherit an allegedly deserted castle but upon moving in, discover Mr. and Mrs. Dracula have settled there. The vampires keep young women chained in the dungeon for continual blood supply. Also present are a hunchback and a werewolf. Awesome Adamson production is highlighted by the presence of the gorgeous Volante.

1969 84m/C John Carradine, Alexander D'Arcy, Paula Raymond, Ray Young, Vicki Volante, Robert Dix, John Cardos; *Dir:* Jean Hewitt, Al Adamson. **VHS, Beta $29.95** *UHV* ⚔½

Blood of the Dragon This film has martial-arts expert Wang Yu killing everyone he meets. One scene has him fighting 100 men at the same time.
1973 (R) 90m/C Wang Yu. **VHS, Beta $39.95** *MAG* ⚔½

Blood of the Dragon Peril/ Revenge of the Dragon Two oriental action masterworks, together on one tape for the first time.
1981 172m/C VHS, Beta $7.00 *VTR*

Blood Feast The first of Lewis' gore-fests, in which a demented caterer butchers hapless young women to splice them together in order to bring back an Egyptian goddess. Dated, campy, and gross; reportedly shot in four days (it shows).
1963 70m/C Connie Mason, Thomas Wood, Malcolm Arnold; *Dir:* Herschell Gordon Lewis. **VHS, Beta $29.95** *RHI, SVI* ⚔

Blood Feud In Italy preceding Europe's entry into World War II, a young widow is in mourning over the brutal murder of her husband by the Sicilian Mafia. In the meantime she must contend with the rivalry between Mastroianni as a lawyer and Giannini as a small-time crook both vying for her affections. Dubbed.
1979 (R) 112m/C *IT* Sophia Loren, Marcello Mastroianni, Giancarlo Giannini; *Dir:* Lina Wertmuller. **VHS, Beta $59.95** *FOX* ⚔⚔

Blood Freak An absolutely insane anti-drug, Christian splatter film. A Floridian biker is introduced to drugs by a young woman who eventually turns into a poultry-monster who drinks the blood of junkies. Narrated by a chain smoker who has a coughing fit. Don't miss it.
1972 86m/C Steve Hawkes, Dana Culliver, Randy Grinter Jr., Tina Anderson, Heather Hughes; *Dir:* Steve Hawkes, Brad Grinter. **VHS, Beta $49.95** *REG* Woof!

Blood Frenzy Psychologist's patients take a therapeutic trip to the desert. The sun and heat take their toll, and everyone gets violent. This kind of therapy we don't need, and the movie's a waste, as well. Made for video. Loring was Wednesday on TV's "The Addams Family."
1987 90m/C John Montero, Lisa Loring, Hank Garrett, Wendy MacDonald; *Dir:* Hal Freeman. **VHS, Beta $79.95** *HFE* ⚔

Blood Games Buxom baseball team bats 1000 against the home team and the winning babes find out just how poor losers can be. Diamonds aren't always a girl's best friend.
1990 (R) 90m/C Gregory Cummings, Laura Albert, Shelly Abblett, Luke Shay, Ross Hagen; *Dir:* Tanya Rosenberg. **VHS, LV $79.95** *COL* Woof!

Blood of Ghastly Horror A young man thinks he has a new lease on life when he is the happy recipient of a brain transplant, but his dreams are destroyed when he evolves into a rampaging killer. This movie is so awful it hides behind numerous aliases, including "Psycho-A-Go-Go" (our favorite).
1972 87m/C John Carradine, Kent Taylor, Tommy Kirk, Regina Carrol; *Dir:* Al Adamson. **VHS $14.95** *VID, MOV* Woof!

Blood and Guns Romance, revenge, and action abound in post-revolutionary Mexico.
1979 (R) 96m/C Orson Welles, Tomas Milian, John Steiner; *Dir:* Giulio Petroni. **VHS, Beta** *MRV* ⚔½

The Blood of Heroes A post-apocalyptic action flick detailing the adventures of a battered team of "juggers," warriors who challenge small village teams to a brutal sport (involving dogs' heads on sticks) that's a cross between jousting and football.
1989 (R) 97m/C Rutger Hauer, Joan Chen, Vincent D'Onofrio, Anna Katarina; *Dir:* David Peoples. **VHS, Beta, LV $89.99** *HBO* ⚔⚔

Blood Hook A self-parodying teenage-slasher film about kids running into a backwoods fishing tournament while on vacation, complete with ghouls, cannibalism and grotesquerie.
1986 85m/C Mark Jacobs, Lisa Todd, Patrick Danz; *Dir:* James Mallon. **VHS, Beta $14.95** *PSM, PAR* ⚔½

Blood Island A couple inherit an old house on a remote New England island where the woman grew up. They discover that this old house needs more than a paint job to make it livable; seems there's something evil in them there walls. Based on an H.P. Lovecraft story, it was originally titled "The Shuttered Room." Greene, who's best known for later directing "Godspell," and the solid cast fail to animate the inert script.
1968 100m/C *GB* Gig Young, Carol Lynley, Oliver Reed, Flora Robson; *Dir:* David Greene. **VHS $29.98** *ACE* ⚔⚔

Blood Lake A psychotic killer stalks teenagers, to the detriment of all.
1987 90m/C VHS $29.98 *UHV* Woof!

Blood Legacy Four heirs must survive a night in a lonely country estate to collect their money; what do you think happens? Average treatment of the haunted house theme. Also titled "Legacy of Blood."
1973 (R) 77m/C John Carradine, John Russel, Faith Domergue, Merry Anders; *Dir:* Carl Monson. **VHS, Beta $59.95** *GEM, MRV* ⚔

Blood Lust Dr. Jekyll returns to London to wreak havoc upon the human race.
198? 90m/C VHS, Beta *KOV* ⚔

Blood Mania A retired surgeon's daughter decides to murder her father to collect her inheritance prematurely, but soon learns that crime doesn't pay as well as medicine. Low-budget, low-interest flick.
1970 (R) 90m/C Peter Carpenter, Maria de Aragon, Alex Rocco; *Dir:* Robert Vincent O'Neil. **VHS, Beta $19.95** *UHV* ⚔

Blood Money A dying ex-criminal returns to Australia to redeem his name and die with dignity.
1980 (PG) 64m/C *AU* Bryan Brown, John Flaus, Chrissie James; *Dir:* Christopher Fitchett. **VHS, Beta $69.95** *SVS* ⚔½

Blood Money Man battles woman and both battle world as they struggle for the rights to a stash of loot in this lugubrious greedfest. Yet another illustrious entry in the prolific career of Ms. Black.
1991 (R) ?m/C Wings Hauser, Karen Black, Robert Z'Dar. **VHS, LV $79.95** *PMH* ⚔⚔

Blood Money - The Story of Clinton and Nadine A small-time smuggler and an ex-hooker get seriously involved with each other and Contra weapons trafficking while in Florida. Made for

cable television. Also known as "Clinton & Nadine."
1988 95m/C Ellen Barkin, Andy Garcia, Morgan Freeman; *Dir:* Jerry Schatzberg. **VHS, Beta** $79.95 *JTC, FCT* 🐾🐾

Blood on the Moon Well-acted film about a cowboy's involvement in a friend's underhanded schemes. Based on a Luke Short novel, this dark film revolves around a western land dispute between cattlemen and homesteaders.
1948 88m/B Robert Mitchum, Robert Preston, Walter Brennan, Barbara Bel Geddes; *Dir:* Robert Wise. **VHS, Beta** $19.95 *MED* 🐾🐾½

Blood on the Mountain After escaping from prison, Jim plots revenge against his accomplice, only to kill an innocent man instead.
1988 71m/C Stracker Edwards, Tim Jones, Paula Preston, Cliff Turknett, Rich Jury; *Dir:* Donald W. Thompson. **VHS, Beta** *MIV* 🐾

Blood Orgy of the She-Devils Exploitative gore nonsense about female demons, beautiful witches, and satanic worship. Some movies waste all their creative efforts on their titles.
1974 (PG) 73m/C Lila Zaborin, Tom Pace, Leslie McRae, Ray Myles. **VHS, Beta** $19.95 *WES* 🐾

The Blood of Others A driveling adaptation of the Simone de Beauvoir novel about a young French woman at the outbreak of World War II torn between her absent boyfriend in the Resistance and a kind, wealthy German. Made for French television, it stars Jodie Foster and Michael Ontkean as the Gallic pair.
1984 130m/C *FR* Jodie Foster, Sam Neill, Michael Ontkean, Stephane Audran, Lambert Wilson, John Vernon, Kate Reid, Jean-Pierre Aumont; *Dir:* Claude Chabrol. **VHS, Beta** $9.99 *STE, PSM* 🐾

Blood of a Poet Cocteau's first film, a practically formless piece of poetic cinema, detailing all manner of surreal events occurring in the instant a chimney falls. In French with English subtitles.
1930 55m/B Jean Cocteau; *Dir:* Jean Cocteau. National Board of Review Awards '33: 10 Best Foreign Films of the Year. **VHS, Beta** $24.95 *NOS, MRV, HHT* 🐾

Blood Rage A maniacal twin goes on a murderous rampage through his brother's neighborhood. AKA "Nightmare at Shadow Woods." Only for die-hard "Mary Hartman" fans.
1987 (R) 87m/C Louise Lasser, Mike Soper; *Dir:* John Grissmer. **VHS, Beta** $79.95 *PSM Woof!*

Blood Red In 1895 Northern California, an Italian immigrant and his family give bloody battle to a powerful industrialist who wants their land in wine-growing country. Watch for the scenes involving veteran Eric Roberts and his then-newcomer sister, pretty woman Julia.
1988 (R) 91m/C Eric Roberts, Dennis Hopper, Giancarlo Giannini, Burt Young, Carlin Glynn, Lara Harris, Susan Anspach, Julia Roberts, Elias Koteas, Frank Campanella, Aldo Ray, Horton Foote Jr.; *Dir:* Peter Masterson. **VHS, Beta, LV, 8mm** $14.95 *COL, SUE* 🐾½

Blood Relations A woman is introduced to her fiance's family only to find out that they, as well as her fiance, are murdering, perverted weirdos competing for an inheritance.
1987 (R) 88m/C *CA* Jan Rubes, Ray Walston, Lydie Denier, Kevin Hicks, Lynne Adams; *Dir:* Graeme Campbell. **VHS, Beta, LV** $19.98 *SUE* 🐾½

Blood Relatives Suspenseful thriller with good cast. A family is too close for comfort - incestuously and murderingly close! Release in the U.S. was delayed until 1981 due to the subject matter.
1978 (R) 100m/C *CA FR* Donald Sutherland, Lisa Langlois, Donald Pleasence, Laurent Malet, David Hemmings; *Dir:* Claude Chabrol. **Beta, 8mm, EJ** $79.75 *UAV* 🐾🐾🐾

Blood and Roses A girl who is obsessed with her family's vampire background becomes possessed by a vampire and commits numerous murders. The photography is good, but the plot is hazy and only effective in certain parts. Based on the book "Carmilla" by Sheridan LeFanu. Later remade as "The Vampire Lovers" and "The Blood-Spattered Bride."
1961 74m/C Mel Ferrer, Elsa Martinelli, Annette Vadim, Marc Allegret; *Dir:* Roger Vadim. **VHS** *PAR* 🐾🐾

Blood Salvage A crazy junkman kidnaps beautiful girls, selling their organs to the highest bidder. He meets his match when a potential target refuses to become a victim in spite of her wheelchair. Interesting plot twists keep this Grade B thriller above average.
1990 (R) 90m/C Evander Holyfield, Lori Birdsong, Danny Nelson, John Saxon, Roy Walston; *Dir:* Tucker Johnson. **VHS, Beta, LV** $89.98 *MAG, TTC* 🐾

Blood and Sand Vintage romance based on Vincente Blasco Ibanez's novel about the tragic rise and fall of a matador, and the women in his life. The film that made Valentino a star. Remade in 1941. Silent.
1922 87m/B Rudolph Valentino, Nita Naldi, Lila Lee, Walter Long; *Dir:* Fred Niblo. **VHS** $19.95 *REP, NOS, CAB* 🐾🐾½

Blood and Sand Director Mamoulian "painted" this picture in the new technicolor technique, which makes it a veritable explosion of color and spectacle. Power is the matador who becomes famous and then falls when he is torn between two women, forsaking his first love, bullfighting. This movie catapulted Hayworth to stardom, primarily for her dancing, but also for her sexiness and seductiveness (and of course, her acting!). Nominated for Best Color Interior Decoration Oscar.
1941 123m/C Tyrone Power Sr., Linda Darnell, Rita Haymorth, Anthony Quinn, J. Carroll Naish, John Carradine; *Dir:* Rouben Mamoulian. Academy Awards '41: Best Color Cinematography. **VHS, Beta, LV** $29.99 *FOX* 🐾🐾🐾🐾

Blood and Sand A bullfighter on the verge of super stardom risks it all when he falls under the spell of a sexy, seductive woman. Will he destroy his one opportunity for fame? Interesting for people who actually enjoy watching the "sport" of bullfighting. Originally made in 1922 and remade in 1941.
1989 (R) 96m/C Christopher Rydell, Sharon Stone, Ana Torrent, Jose-Luis De Villalonga, Simon Andrew; *Dir:* Javier Elorrieta. **VHS** $89.95 *VMK* 🐾🐾

Blood and Sand/Son of the Sheik A double feature containing two abridged versions; in "Blood and Sand," an idolized matador meets another woman just before his wedding; in "Son of the Sheik," a man believes he has been betrayed by a dancing girl, and he abducts her to seek revenge. Silent.
1926 56m/B Rudolph Valentino, Lila Lee, Vilma Banky. **VHS, Beta** $29.98 *CCB* 🐾🐾

The Blood on Satan's Claw Graphic tale centering on the Devil himself. Townspeople in an English village circa 1670 find the spirit of Satan taking over their children, who begin to practice witchcraft. Well made with lots of attention to period details. Not for the faint-hearted.
1971 90m/C *GB* Patrick Wymark, Linda Hayden, Barry Andrews, Michele Dotrice; *Dir:* Piers Haggard. **VHS, Beta** $59.95 *PGN* 🐾🐾🐾

Blood Screams Nosy people drop in on an old town in Mexico and attempt to uncover the secrets hidden there. But bizarre entities throw out the unwelcome mat and terrorize them. Lots of blood and screaming.
1988 (R) 75m/C Ran Sands, James Garnett. **VHS, Beta** $19.98 *WAR Woof!*

Blood for a Silver Dollar An action-filled western, littered with murder, revenge and romance.
1966 98m/C Montgomery Wood, Evelyn Stewart. **VHS, Beta** $29.95 *VYY* 🐾

Blood Simple A jealous husband hires a sleazy private eye to murder his adulterous wife and her lover. A dark, intricate, morbid morality tale that deviates imaginatively from the standard murder mystery thriller. Written by Joel and Ethan Coen; their first film.
1985 (R) 96m/C John Getz, M. Emmet Walsh, Dan Hedaya, Frances McDormand; *Dir:* Joel Coen. **VHS, Beta, LV** $79.95 *MCA* 🐾🐾🐾½

Blood Sisters Sorority babes intend to spend a giggle-strewn night in a haunted house but end up decapitated, butchered, and cannibalized.
1986 (R) 85m/C Amy Brentano, Marla MacHart, Brigete Cossu, Randy Mooers; *Dir:* Roberta Findlay. **VHS, Beta** $14.95 *COL, SVS Woof!*

Blood Song A patient (yester-decade teen throb Frankie Avalon) escapes into the night from a mental institution after murdering an attendant. He takes his only possession with him, a carved wooden flute. A young woman sees him burying his latest victim, and now he's on a hunt to play his "blood song" for her. Pretty bad, but fun to see Avalon play a less-than-squeaky-clean role.
1982 90m/C Frankie Avalon, Donna Wilkes, Richard Jaeckel, Dane Clark, Antoinette Bower; *Dir:* Alan J. Levi. **VHS, Beta** *API* 🐾

The Blood Spattered Bride A newlywed couple honeymoon in a remote castle in southern Spain. They are visited by a mysterious young woman who begins to influence the bride in the ways of bloodsucking. Okay 70s Euro-eroti-horror based on LeFanu's "Carmilla." Also known as "Blood Castle."
1972 82m/C *SP* Simon Andrew, Maribel Martin, Alexandra Bastedo, Dean Selmier, Rosa Ma Rodriguez, Montserrat Julio; *Dir:* Vicente Aranda. **VHS, Beta** $59.95 *MPI* 🐾

Blood on the Sun Newspaperman in Japan uncovers plans for world dominance as propaganda, violence, and intrigue combine in this action-adventure. Also available colorized.
1945 98m/B James Cagney, Sylvia Sydney, Robert Armstrong, Wallace Ford; *Dir:* Frank Lloyd. Academy Awards '45: Best Art Direction/Set Decoration (B & W). **VHS, LV** $19.95 *NOS, MRV, HHT* 🐾🐾½

Blood on the Sun Martial arts action highlights this tale of adventure.
1975 (R) 81m/C Ching Ching Chang, Peng Tien. **VHS, Beta, LV** $59.95 *GEM* 🐾

Blood at Sundown A man returns home after the Civil War to find his wife has been kidnapped by a group of Mexican outlaws who have also taken over the village. 19?? 92m/C VHS, Beta $59.95 *IMP* ℤ

Blood Tide A disgusting, flesh-eating monster disrupts a couple's vacation in the Greek isles. Beautiful scenery, good cast, bad movie.
1982 (R) 82m/C James Earl Jones, Jose Ferrer. VHS, Beta $39.98 *PLV* ℤ

Blood Ties An American naval engineer in Sicily gets involved with the Mob in order to save his father's life.
1987 98m/C *IT* Brad Davis, Tony LoBianco, Vincent Spano, Barbara de Rossi, Maria Conchita Alonso; *Dir:* Giacomo Battiato. **VHS, Beta** $69.95 *TWE* ℤ½

Blood Tracks A woman kills her abusive husband, then hides out in the mountains with her kids until they turn into cannibalistic savages. Years later, people show up to shoot a music video. Blood runs in buckets.
1986 82m/C Jeff Harding, Michael Fitzpatrick, Naomi Kaneda; *Dir:* Mike Jackson. **Beta** $79.95 *VHV* ℤ½

Blood of the Vampire A Transylvanian doctor is executed for being a vampire and his hunchbacked assistant brings him back to life.
1958 84m/C *GB* Donald Wolfit, Vincent Ball, Barbara Shelley, Victor Madden; *Dir:* Henry Cass. **VHS, Beta** $59.95 *MPI, MLB* ℤ½

Blood Vows: The Story of a Mafia Wife TV's Laura Ingalls (Melissa Gilbert) leaps from the prairie to modern-day mafia in this warped Cinderella story. She meets the man of her dreams, whom she slowly finds is her worst nightmare as she becomes trapped within the confines of her new-found "family." Made for television.
1987 100m/C Melissa Gilbert, Joe Penny, Eileen Brennan, Talia Shire, Anthony (Tony) Franciosa; *Dir:* Paul Wendkos. **VHS** $39.95 *FRH* ℤ

Blood Voyage A crewman aboard a pleasure yacht must find out who is killing off his passengers one by one.
1977 (R) 80m/C Jonathon Lippe, Laurie Rose, Midori, Mara Modair; *Dir:* Frank Mitchell. **VHS, Beta** $59.95 *MON Woof!*

Blood Wedding (Bodas de Sangre) A wonderfully passionate dance film from Saura and choreographed by Antonio Gades, based on the play by famed author Federico Garcia Lorca. If you like flamenco, there are two more: "Carmen" and "El Amor Brujo." Subtitled; English language title is "Blood Wedding."
1981 71m/C *SP* Antonio Gades, Christina Hoyos, Marisol, Carmen Villena; *Dir:* Carlos Savrat. **VHS, Beta** $24.95 *XVC, CTH, MED* ℤℤℤ

Bloodbath Drugs, sex and terrorism - Dennis Hopper is typecast as an American degenerate who, along with his expatriate friends, is persecuted by local religious cults who need sacrifices.
19?? 89m/C Dennis Hopper, Carroll Baker, Richard Todd, Faith Brook; *Dir:* Silvio Narizzano. **VHS, Beta** $79.98 *GHV, HHE* ℤℤℤ

Bloodbath at the House of Death Price and his compatriots fight a team of mad scientists in this parody of popular horror films. Fun to see Price spoofing his own genre; die-hard horror camp fans will be satisfied. Best line: "Wanna fork?"
1985 92m/C *GB* Kenny Everett, Pamela Stephenson, Vincent Price, Gareth Hunt; *Dir:* Ray Cameron. **VHS, Beta** $59.95 *MED* ℤℤ

Bloodbath in Psycho Town A film crew is marked for death by a hooded man when it enters a remote little village.
1989 (R) 87m/C Ron Arragon, Donna Baltron, Dave Elliott; *W/Dir:* Allessandro DeGaetano. **VHS** $29.95 *AVD* ℤ

Bloodbeat A supernatural being terrorizes a family as they gather at their country home to celebrate Christmas.
1987 84m/C Helen Benton, Terry Brown, Claudia Peyton. **VHS, Beta** $59.98 *TWE* ℤ

Bloodbrothers Portrayal of working-class Italian men's lives—if that's possible without the benefit of Italian writers, producers, or director. Still, lots of cussing and general intensity, as Gere's character struggles between staying in the family construction business and doing what he wants to do—work with children. Re-cut for television and re-titled "A Father's Love."
1978 (R) 120m/C Richard Gere, Paul Sorvino, Tony LoBianco, Kenneth MacMillan, Marilu Henner, Danny Aiello; *Dir:* Robert Mulligan. **VHS** $69.95 *WAR* ℤℤ½

Bloodfight Exciting kick-boxing and martial arts tournament turns sour when champ offends the sense of honor of the master. Now the champ must battle the master, and only one will remain standing!
1990 (PG-13) ?m/C Bolo Yeung. **VHS** *IMP* ℤ

Bloodfist A kickboxer tears through Manila searching for his brother's killer. A Roger Corman production.
1989 (R) 85m/C Don "The Dragon" Wilson, Rob Kaman, Billy Blanks, Kris Aguilar, Riley Bowman, Michael Shaner; *Dir:* Terence H. Winkless. **VHS, Beta** $89.95 *MGM* ℤℤ

Bloodfist 2 Six of the world's toughest martial artists find themselves kidnapped and forced to do the bidding of the evil Su. The mysterious recluse stages a series of incredible fights between the experts and his own army of drugged warriors.
1990 (R) 85m/C Don "The Dragon" Wilson, Maurice Smith, James Warring, Timothy Baker, Richard Hill, Rina Reyes; *Dir:* Andy Blumenthal. **VHS, Beta** $89.98 *MGM* ℤℤ½

Bloodfist 3: Forced to Fight Just as long as nobody's forced to watch, as real-life world champion kickboxer Wilson thrashes his way through another showdown-at-the-arena plot. Better-than-average fight choreography.
1991 90m/C Don "The Dragon" Wilson, Richard Roundtree, Laura Stockman, Richard Paul, Rick Dean, Peter "Sugarfoot" Cunningham. **VHS** $89.98 *NHO* ℤ½

Bloodhounds of Broadway A musical tribute to Damon Runyon, detailing the cliched adventures of an assortment of jazz-age crooks, flappers, chanteuses, and losers.
1989 (PG) 90m/C Madonna, Rutger Hauer, Randy Quaid, Matt Dillon, Jennifer Grey, Julie Hagerty, Esai Morales, Anita Morris, Josef Sommer, William S. Burroughs, Ethan Phillips, Stephen McHattie, Dinah Manoff, Googy Gress, Tony Azito, Tony Longo; *Dir:* Howard Brookner. **VHS, Beta, LV** $19.95 *COL* ℤ½

Bloodlink A well-to-do doctor has a recurring nightmare about killing an elderly woman. This prompts him to explore his past, discovering that he was separated at birth from a twin brother. Naturally, only by searching frantically for his long-lost sibling can the good doctor hope to solve the mystery of the recurring nightmare. A somnolent tale indeed.
1986 (R) 98m/C Michael Moriarty, Penelope Milford, Geraldine Fitzgerald, Cameron Mitchell, Sarah Langenfeld; *Dir:* Alberto De Martino. **VHS, Beta** $24.95 *SUE* ℤ½

Bloodmatch A whodunit, martial-arts-style: the 'sleuth' kidnaps all the suspects in a corruption case and kickboxes each to death until somebody confesses. Not exactly Agatha Christie, and artsy camera moves fail to exploit the fancy footwork.
1991 (R) 85m/C Thom Mathews, Hope Marie Carlton, Benny "The Jet" Urquidez, Marianne Taylor, Dale Jacoby, Thunder Wolf, Vincent Klyn, Peter "Sugarfoot" Cunningham, Hector Pena, Michel Qissi; *Dir:* Albert Pyun. **VHS, LV, Q** $89.99 *HBO, PIA* ℤ

Bloodmoon The setting is Australia but the sleazy story's all too familiar: an insane killer employs knives and other sharp objects to prevent sex-crazed students from getting past third base.
1991 104m/C *AU* Leon Lissek, Christine Amor, Helen Thomson, Ian Patrick Williams; *Dir:* Alec Mills. **VHS, Beta** $89.98 *LIV Woof!*

Bloodshed A psychotic harbors a girl's dead body, and systematically kills anyone who discovers his secret.
1983 (R) 88m/C Laslo Papas, Beverly Ross; *Dir:* Richard Cassidy. **VHS, Beta** $49.95 *REG* ℤ

Bloodspell A student with an evil power unleashes it on those who cross his path.
1987 (R) 87m/C Anthony Jenkins, Aaron Teich, Alexandra Kennedy, John Reno; *Dir:* Deryn Warren. **VHS, Beta** $79.98 *MCG* ℤ

Bloodsport American soldier Van Damme endeavors to win the deadly Kumite, an outlawed martial arts competition in Hong Kong. Lots of kick-boxing action and the sound of bones cracking.
1988 (R) 92m/C Jean-Claude Van Damme, Leah Ayres, Roy Chiao, Donald Gibb, Bolo Yeung, Norman Burton, Forest Whitaker; *Dir:* Newton Arnold. **VHS, Beta, LV** $19.98 *WAR, TLF, PIA* ℤℤ½

Bloodstalkers Two vacationers in Florida meet up with a band of slaughtering, swamp-based lunatics.
1976 91m/C Kenny Miller, Celea Ann Cole, Jerry Albert; *Dir:* Robert W. Morgan. **VHS, Beta** $39.95 *VMK Woof!*

Bloodstone A couple honeymooning in the Middle East unexpectedly become involved in a jewel heist when they discover a valuable ruby amongst their luggage. Non-stop action and humor.
1988 (PG-13) 90m/C Charlie Brill, Christopher Neame, Jack Kehler, Brett Stimely; *Dir:* Dwight Little. **VHS, Beta** $79.98 *MCG* ℤℤ

The Bloodsuckers A British horror tale set on a Greek Island. An Oxford don is seduced into an ancient vampire cult. Director Michael Burrowes replaces Hartford-Davis in the credits due to a dispute over post-production editing.
1970 90m/C *GB* Patrick Macnee, Peter Cushing, Patrick Mower, Edward Woodward; *Dir:* Robert Hartford-Davis. **VHS, Beta** $19.98 *VCL, SNC* ℤ

Bloodsuckers from Outer Space Via an alien invasion, Texas farmers becoming bloodsucking zombies.
1983 80m/C Thom Meyer, Laura Ellis, Billie Keller, Kim Braden; *Dir:* Glenn Coburn. **VHS, Beta** $19.98 *LHV, WAR Woof!*

Bloodsucking Freaks A virtually plotless Troma gagfest full of torture, cannabalistic dwarfs, and similar debaucheries, all played·out on a Grand Guignol stage (horror shows that alledgedly contained real torture and death). Features "The Caged Sexoids," if that tells you anything. AKA "Incredible Torture Show."
1975 (R) 89m/C Seamus O'Brian, Niles McMaster; *Dir:* Joel M. Reed. **VHS, Beta $59.98** *VES Woof!*

Bloodsucking Pharoahs of Pittsburgh Pittsburgh is plagued by crazed cannibals who think eternal life is in Pennsylvania. Two detectives on the case are mystified.
1990 (R) 89m/C Jake Dengel, Joe Sharkey, Susann Fletcher, Beverly Penberthy, Shawn Elliott, Pat Logan, Jane Hamilton; *Dir:* Alan Smithee. **VHS, Beta $79.95** *PAR, FCT Woof!*

Bloody Avenger A trio of detectives search for a murderer in the streets of Philadelphia.
198? 100m/C Jack Palance, George Eastman, Jenny Tamburi; *Dir:* Al Bradley. **VHS, Beta $59.95** *VMK* ℐ

Bloody Birthday Three youngsters, bound together by their eerie birth during an eclipse (you know what that means), kill everyone around them that ever gave them problems. Typical "and the fun continues" ending; standard fare.
1980 92m/C Susan Strasberg, Jose Ferrer, Lori Lethin, Melinda Cordell, Joe Penny, Ellen Geer, Julie Brown, Michael Dudikoff, Billy Jacoby; *Dir:* Edward Hunt. **VHS, Beta $9.99** *STE, PSM* ℐ½

Bloody Che Contra A rebel leader is trying to flee the Bolivian mountains. He talks of the danger in the jungle, the goodwill of his men, and the threat of government forces.
19?? 89m/C John Ireland, Francisco Rabal. **VHS, Beta** *MFI* ℐ

The Bloody Fight Two young people learn the cost of courage and the high price of justice when they avenge a murderer.
197? 90m/C Alan Tang, Yu In Yin, Tan Chin; *Dir:* N.G. Tien Tsu. **VHS, Beta $54.95** *MAV* ℐ

Bloody Fist An exciting story set in China with outstanding martial arts scenes.
197? (R) 90m/C **VHS, Beta $39.95** *MAV* ℐ

Bloody Mama Roger Corman's violent and trashy story of the infamous Barker Gang of the 30s, led by the bloodthirsty and sex-crazed Ma Barker (Shelley Winters, can't you just picture it?) and backed by her four perverted sons (Dern—ditto, DeNiro, Walden, and Stroud).
1970 90m/C Shelley Winters, Robert De Niro, Don Stroud, Pat Hingle, Bruce Dern, Diane Varsi, Robert Walden, Clinton Kimbrough; *Dir:* Roger Corman. **VHS, Beta, LV $69.95** *VES, LIV* ℐ½

Bloody Moon Tourists are being brutally attacked and murdered during a small Spanish village's Festival of the Moon.
1983 84m/C Olivia Pascal, Christopher Brugger; *Dir:* Jess (Jesus) Franco. **VHS, Beta $59.95** *TWE Woof!*

Bloody New Year Corpses stalk the living as a group of teens happen upon an impromptu New Year's Eve party on a deserted island. Auld acquaintance shouldn't be forgot, just this flick.
1987 (R) 90m/C *Dir:* Norman J. Warren. **VHS, Beta $29.95** *ACA* ℐ

The Bloody Pit of Horror While wife Jayne Mansfield was in Italy filming "Primitive Love," bodybuilder Hargitay starred in this sado-horror epic. He owns a castle that is visited by a group of models for a special shoot. While in the dungeon, Hargitay becomes possessed by the castle's former owner, a sadist, and begins torturing the models. Supposedly based on the writings of the Marquis de Sade.
1965 87m/B *IT* Mickey Hargitay, Louise Barrett, Walter Brandi, Moa Thai, Ralph Zucker, Albert Gordon; *Dir:* Max (Massimo Pupillo) Hunter. **VHS $19.98** *SNC* ℐ½

The Bloody Tattoo (The Loot) A suspicious woman has a map tattooed on her body telling where a treasure is buried. The robbers who drew the map are now looking for her so they can find the treasure. A story full of killing and greed.
19?? 90m/C Ko Fay, David Keung, Tsui Siu Keung, Lily Lee. **VHS $49.95** *OCE* ℐ½

Bloody Trail A Union soldier who chooses the recently pummeled South as the venue for postwar R-and-R is, for some reason, pursued by Confederates. And his good ol' boy pursuers don't have a sudden change of heart when he teams up with a former slave. Much violence and nudity; little plot and entertainment.
19?? 91m/C Paul Harper, John Mitchum; *Dir:* Richard Robinson. **VHS $19.95** *ACA* ℐ

Bloody Wednesday It's sanity checkout time when a hotel caretaker is driven mad by tormentors...or is he driving himself mad? What is it about vacant hotels that make men lose their minds? If you can remove "The Shining" from yours, this flick's worthwhile.
1987 89m/C Raymond Elmendort, Pamela Baker, Navarre Perry; *Dir:* Mark Gilhuis. **VHS, Beta $79.95** *PSM* ℐℐ½

Blossoms in the Dust The true story of Edna Gladney is told as she starts the Texas Children's Home and Aid Society of Fort Worth. Major league Garson tear-jerker, Oscar nominated for Best Picture, Best Actress (Garson), and Color Cinematography.
1941 100m/B Greer Garson, Walter Pidgeon, Felix Bressart, Marsha Hunt, Fay Holden, Samuel S. Hinds, Kathleen Howard; *Dir:* Mervyn LeRoy. Academy Awards '41: Best Art Direction/Set Decoration (B & W). **VHS, Beta $19.98** *CRR* ℐℐ

The Blot A story about a poorly paid professor and his family, whose lifestyle contrasts with that of an affluent neighbor, a butcher. Silent.
1921 55m/B Louis Calhern, Claire Windsor; *Dir:* Lois Weber. **VHS, Beta $79.95** *GVV, DVT* ℐℐ

Blow Out When a prominent senator is killed in a car crash, a sound effects engineer becomes involved in political intrigue as he tries to expose a conspiracy with the evidence he has gathered. An intricate mystery and homage to Antonioni's "Blow-Up."
1981 (R) 108m/C John Travolta, Nancy Allen, John Lithgow, Dennis Franz; *Dir:* Brian DePalma. **VHS, Beta, LV** *WAR, OM* ℐℐℐ

Blow-Up A young London photographer takes some pictures of a couple in the park, and finds out he may have recorded evidence of a murder. Though marred by badly dated 1960s modishness, this is Antonioni's most accessible film, a sophisticated treatise on perception and the film-consumer-as-voyeur, brilliantly assembled and wrought. The director was Oscar nominated.

1966 111m/C *GB IT* David Hemmings, Vanessa Redgrave, Sarah Miles, Jane Birkin, Veruschka; *Dir:* Michelangelo Antonioni. Cannes Film Festival '67: Best Film; National Society of Film Critics Awards '66: Best Director, Best Film. **VHS, Beta, LV $19.98** *MGM, VYG, FCT* ℐℐℐ½

Blowing Wild Filmed in Mexico, this Quinn-Stanwyck-Cooper love triangle, set in the early thirties, speaks of lust and vengeance, rashness and greed. Stanwyck, married to oil tycoon Quinn, lusts after one-time lover wildcatter Cooper.
1954 92m/B Gary Cooper, Barbara Stanwyck, Anthony Quinn, Ruth Roman, Ward Bond; *Dir:* Hugo Fregonese. **VHS $19.98** *REP, FCT* ℐ½

Blown Away A mafia wife goes up against her husband in order to retrieve her kidnapped child.
1990 (PG-13) 92m/C Loni Anderson, John Heard, James Naughton; *Dir:* Michael Miller. **VHS, Beta $29.95** *ACA* ℐ

Blue A dull western about an American boy (Stamp) raised by Mexicans who trusts no one until he finds himself face to face with his former gang, led by his adoptive father (Montalban).
1968 113m/C Terence Stamp, Joanna Pettet, Karl Malden, Ricardo Montalban, Joe DeSantis, Sally Kirkland; *Dir:* Silvio Narizzano. **VHS** *PAR* ℐ

The Blue Angel The tale of a man stripped of his dignity. A film classic reeking of sensuality and decay, which made Dietrich a European star and prompted her invitation to Hollywood. When a repressed professor (Jannings) goes to a nightclub hoping to catch some of his students in the wrong, he finds that he too is taken by Lola, the sultry singer portrayed by Dietrich. After spending the night with her, losing his job, and then marrying her, he goes on tour with the troupe, peddling indiscreet photos of his wife. Versions were shot in both German and English, with the German version sporting English subtitles. Songs include "Falling in Love Again" and "They Call Me Wicked Lola."
1930 90m/B *GE* Marlene Dietrich, Emil Jannings, Kurt Gerron, Rosa Valetti; *Dir:* Josef von Sternberg. **VHS, Beta, 8mm $24.95** *NOS, MRV, KRT* ℐℐℐℐ

Blue Angels: A Backstage Pass Behind the scenes view of the Blue Angels with interviews of team members, fantastic footage from the early days of piston engine F8F Bearcats to the in-cockpit footage of today's F18 jets. Also features the music of Tom Petty & the Heartbreakers, Georgia Satellites, Van Halen, Huey Lewis & The News, and Lionel Richie.
1990 30m/C **VHS $9.95** *CAF* ℐℐℐℐ

The Blue Bird A weird, dark fantasy about two children who search for the blue bird of happiness in various fantasy lands, but find it eventually at home. Overlooked and impressively fatalistic.
1940 98m/C Shirley Temple, Gale Sondergaard, Johnny Russell, Eddie Collins, Nigel Bruce, Jessie Ralph, Spring Byington, Sybil Jason; *Dir:* Walter Lang. **VHS, Beta $19.98** *FOX, MLB* ℐℐℐ

Blue Canadian Rockies Autry's employer sends him to Canada to discourage his daughter from marrying a fortune hunter. The daughter has turned the place into a dude ranch and wild game preserve. When Autry arrives, he encounters some mysterious killings.
1952 58m/B Gene Autry, Pat Buttram, Gail Davis, Ross Ford, Tom London; *Dir:* George Archainbaud. **VHS, Beta $39.95** *CCB* ℐ

Blue City A young man returns to his Florida hometown to find his father murdered, and subsequently vows to solve and avenge the matter. Based on a Ross MacDonald thriller.
1986 (R) 83m/C Judd Nelson, Ally Sheedy, Paul Winfield, Anita Morris, David Caruso, Julie Carmen, Scott Wilson; **Dir:** Michelle Manning. **VHS, Beta, LV** $79.95 *PAR ♪♪ ½*

Blue Collar An auto assembly line worker, tired of the poverty of his life, hatches a plan to rob his own union. A study of the working class and the robbing of the human spirit.
1978 (R) 114m/C Richard Pryor, Harvey Keitel, Yaphet Kotto, Ed Begley Jr.; **Dir:** Paul Schrader. **VHS, Beta, LV** $69.95 *MCA ♪♪♪*

Blue Country A joyful romantic comedy about a pair of free souls who leave their stagnant lives behind to seek out a more idyllic existence. Subtitled in English.
1978 (PG) 104m/C *FR* Brigitte Fossey, Jacques Serres. **VHS, Beta** $59.95 *RCA ♪♪ ½*

Blue De Ville Two young women buy a '59 Cadillac and journey from St. Louis to New Mexico, having adventures on the way. The rambling, free-spirited TV movie was a pilot for a prospective series that never set sail—but when "Thelma & Louise" hit big this superficially similar item was hauled out on video.
1986 (PG) 96m/C Jennifer Runyon, Kimberly Pistone, Mark Thomas Miller, Alan Autry, Robert Prescott; **Dir:** Jim Johnston. **VHS** $89.98 *VMK ♪♪*

Blue Desert A woman thinks she's left crime behind her when she moves from New York's concrete jungle to the sandy desert, where she puts the unpleasantries of her past behind her and lives happily ever after. Well no, not really. Seems she has two men pursuing her romantically, and one of them might be a killer. Will true romance triumph?
1991 (R) 98m/C Courtney Cox, D.B. Sweeney, Craig Sheffer; **Dir:** Bradley Battersby. **VHS** $19.95 *ACA ♪♪*

Blue Fin When their tuna boat is shipwrecked, and the crew disabled, a young boy and his father learn lessons of love and courage as the son tries to save the ship.
1978 (PG) 93m/C *AU* Hardy Kruger, Greg Rowe; **Dir:** Carl Schultz. **VHS, Beta** $59.98 *SUE ♪♪*

Blue Fire Lady The heartwarming story of a young girl and her love of horses which endures even her father's disapproval. Good family fare.
1978 96m/C Cathryn Harrison, Mark Holden, Peter Cummins. **VHS, Beta** $19.95 *MED ♪♪*

The Blue and the Grey Epic television mini-series about love and hate inflamed by the Civil War. Keach plays a Pinkerton's secret service agent in this loosely based historical romance.
1985 245m/C Gregory Peck, Lloyd Bridges, Colleen Dewhurst, Stacy Keach, John Hammond, Sterling Hayden, Warren Oates; **Dir:** Andrew V. McLaglen. **VHS, Beta, LV** $14.95 *RCA ♪♪*

Blue Hawaii A soldier, returning to his Hawaiian home, defies his parents by taking a job with a tourist agency. Presley sings "Can't Help Falling in Love."
1962 101m/C Elvis Presley, Angela Lansbury, Joan Blackman, Roland Winters, Iris Adrian, John Archer, Steve Brodie; **Dir:** Norman Taurog. **VHS, Beta, LV** $14.98 *FOX, MVD ♪♪ ½*

Blue Heaven A couple struggles through problems with their marriage and alcohol abuse.
1984 100m/C Leslie Denniston, James Eckhouse; **Dir:** Kathleen Dowdey. **VHS, Beta** $69.95 *VES, LIV ♪*

Blue Iguana An inept bounty hunter travels south of the border to recover millions from a crooked South American bank, and meets up with sexy women, murderous thugs, and corruption.
1988 (R) 88m/C Dylan McDermott, Jessica Harper, James Russo, Dean Stockwell, Pamela Gidley, Tovah Feldshuh; **Dir:** John Lafia. **VHS, Beta, LV** $89.95 *PAR ♪♪*

Blue Jeans A young French boy experiences sexual awakening and humiliation in a British school. In French with English subtitles.
1978 80m/C *FR* **Dir:** Hugues des Roziers. **VHS, Beta** $29.95 *FCT ♪♪ ½*

Blue Jeans and Dynamite A stuntman is hired to lift the "Golden Mask of the Duct Tomb" and is followed on land, water, and air. Great chase scenes and action-filled finale.
19?? 90m/C Robert Vaughn. **VHS** $29.95 *AVD ♪♪*

The Blue Knight Kennedy brings energy and care to this basically standard story. Policeman waiting for retirement searches for his partner's killer. Unexceptional treatment made palatable by actors. Made for television movie.
1975 72m/C George Kennedy, Alex Rocco, Glynn Turman, Verna Bloom, Michael Margotta, Seth Arnold; **Dir:** J. Lee Thompson. **VHS** $49.95 *IVE ♪♪ ½*

The Blue Lagoon Useless remake of 1949 film of the same name. An adolescent boy and girl marooned on a desert isle discover love (read: sex) without the restraints of society. Not too explicit, but nonetheless intellectually offensive. Gorgeous photography of island paradise is wasted on this Shields vehicle.
1980 (R) 105m/C Brooke Shields, Christopher Atkins, Leo McKern, William Daniels; **Dir:** Randal Kleiser. **VHS, Beta, LV** $84.95 *COL ♪ ½*

The Blue Lamp Action-adventure fans familiar with the hoary plot where a cop must avenge the wrongful death of his partner will appreciate this suspenseful British detective effort. It's one of the very first in the genre to explore buddy cop revenge in a very British sort of way. Also sports a concluding chase scene which has stood the test of time. Led to the long-running British TV series "Dixon of Dock Green."
1949 84m/B *GB* Dirk Bogarde, Jimmy Hanley, Jack Warner, Bernard Lee, Robert Flemyng, Patric Doonan, Bruce Seton, Frederick Piper, Betty Ann Davies, Peggy Evans; **Dir:** Basil Dearden. **VHS** $29.95 *FCT ♪♪♪*

The Blue Light Co-written by Riefenstahl and Bela Balazs, a fairy tale love story, based on an Italian fable about a mysterious woman, thought to be a witch, and a painter. Riefenstahl's first film, and it brought her to the attention of Adolf Hitler, who requested she make films glorifying the Nazi Party. In German with English subtitles.
1932 77m/B *GE* Leni Riefenstahl, Matthias Wieman, Max Holsboer; **W/Dir:** Leni Riefenstahl. **VHS, Beta** $29.95 *NOS, MRV, FCT ♪♪♪*

The Blue Lightning Investigator Elliott travels to Australia to retrieve the priceless Blue Lightening gem. He must fight the crime lord in his Aussie encampment. Nice scenery, but unexceptional television story.
1986 95m/C Sam Elliott, Rebecca Gilling, Robert Culp, John Meillon, Robert Coleby, Max Phipps; **Dir:** Lee Philips. **VHS** $39.95 *FRH ♪♪ ½*

The Blue Max During World War I a young German, fresh out of aviation training school, competes for the coveted "Blue Max" flying award with other members of a squadron of seasoned flyers from aristocratic backgrounds. Based on a novel by Jack D. Hunter.
1966 155m/C George Peppard, James Mason, Ursula Andress; **Dir:** John Guillermin. **VHS, Beta, LV** $29.98 *FOX ♪♪ ½*

Blue Money A wild, comedic caper film about a cab-driving nightclub impressionist who absconds with a briefcase packed with cash and is pursued by everyone, even the I.R.A. Originally made for British television.
1984 82m/C *GB* Tim Curry, Dabby Bishop, Billy Connolly, Frances Tomelty; **Dir:** Colin Bucksey. **VHS, Beta** $39.95 *SVS ♪♪*

Blue Monkey A mysterious alien plant impregnates a man, who gives birth to a huge, man-eating insect larva. It subsequently grows up into a giant bug, and roams around a quarantined hospital looking for patients to eat. What made you think it had anything to do with monkeys?
1984 (R) 97m/C Steve Railsback, Susan Anspach, Gwynyth Walsh, John Vernon, Joe Flaherty; **Dir:** William Fruet. **VHS, Beta** $79.95 *COL ♪*

Blue Movies A couple of jerks try their hand at making porn films with predictable results.
1988 (R) 92m/C Larry Linville, Lucinda Crosby, Steve Levitt, Darien Mathias, Larry Poindexter; **Dir:** Paul Koval. **VHS, Beta** $29.95 *ACA Woof!*

Blue Murder at St. Trinian's The second of the madcap British comedy series (based on cartoons by Ronald Searle) about an incredibly ferocious pack of schoolgirls. This time they travel to the European continent and make life miserable for a jewel thief. Highlight: fantasy sequence in ancient Rome showing the girls thrown to the lions—and scaring the lions.
1956 86m/B *GB* Joyce Grenfell, Terry-Thomas, George Cole, Alastair Sim, Lionel Jeffries, Thorley Walters; **Dir:** Frank Launder. **VHS** $29.95 *VDM ♪♪♪*

Blue Paradise A depiction of Adam and Eve as they peruse the primal landscape and each other.
1986 90m/C Mark Gregory, Andrea Goldman. **VHS, Beta** $59.95 *TWE, MLB Woof!*

Blue Ribbon Babes A filmed record of a female athletic competition in Florida, featuring wet T-shirt contests, mud wrestling, and powder-puff boxing.
1986 60m/C **VHS, Beta** $24.95 *AHV Woof!*

Blue Skies Again A spunky young woman determined to play major league baseball locks horns with the chauvinistic owner and the gruff manager of her favorite team.
1983 (PG) 91m/C Robyn Barto, Harry Hamlin, Mimi Rogers, Kenneth McMillan, Dana Elcar, Andy Garcia; **Dir:** Richard Michaels. **VHS, Beta** $69.95 *WAR ♪ ½*

Blue Steel A young Wayne saves a town from financial ruin by leading the citizens to a gold strike. Also available with "Man From Utah" on Laser Disc.
1934 59m/B John Wayne, George "Gabby" Hayes; *Dir:* Robert N. Bradbury. VHS, Beta, LV $8.95 *NEG, NOS, MRV* 🎬🎬

Blue Steel Director Bigelow's much-heralded, proto-feminist cop thriller. A serious female rookie's gun falls into the hands of a Wall Street psycho who begins a killing spree. Action film made silly with over-anxious sub-text and patriarchy-directed rage.
1990 (R) 102m/C Jamie Lee Curtis, Ron Silver, Clancy Brown, Louise Fletcher, Philip Bosco, Elizabeth Pena, Tom Sizemore; *Dir:* Kathryn Bigelow. VHS, Beta, LV, 8mm $19.99 *MGM* 🎬🎬

Blue Sunshine A certain brand of L.S.D. called Blue Sunshine starts to make its victims go insane.
1978 (R) 94m/C Zalman King, Deborah Winters, Mark Goddard, Robert Walden, Charles Siebert, Ann Cooper, Ray Young, Alice Ghostley; *W/Dir:* Jeff Lieberman. VHS, Beta $69.95 *VES* 🎬🎬½

Blue Thunder A police helicopter pilot is chosen to test an experimental high-tech chopper that can see through walls, record a whisper, and level a city block.
1983 (R) 110m/C Roy Scheider, Daniel Stern, Malcolm McDowell, Candy Clark, Warren Oates; *Dir:* John Badham. VHS, Beta, LV $12.95 *COL* 🎬½

Blue Tornado An eerie bright light, emitted from a mountain, makes supersonic jets disappear into thin air. A beautiful researcher and a cocky pilot set out to solve the mystery. Better-than-average aerial sequences don't make up for goofy story and ludicrous dialogue.
1990 (PG-13) 96m/C Dirk Benedict, Patsy Kensit, Ted McGinley, David Warner; *Dir:* Tony B. Dobb. VHS $89.95 *VMK* 🎬

Blue Velvet A disturbing, unique exploration of the dark side of American suburbia by Lynch, involving an innocent college youth who discovers a severed ear in an empty lot, and is thrust into a turmoil of depravity, murder, and sexual deviance. Brutal, grotesque, and unmistakably Lynch; an immaculately made, fiercely imagined film that is unlike any other. Mood is enhanced by the Angelo Badalamenti soundtrack. Graced by splashes of Lynchian humor, most notably the movie's lumber theme. Hopper is riveting as the chief sadistic nutcase and Twin Peaks' MacLachlan is a study in loss of innocence. Written by Lynch; cinematography by Frederick Elmes.
1986 (R) 120m/C Kyle MacLachlan, Isabella Rossellini, Dennis Hopper, Laura Dern, Hope Lange, Jack Nance, Dean Stockwell, George Dickerson, Brad Dourif; *Dir:* David Lynch. Montreal World Film Festival '86: Best Actor (Hopper); National Society of Film Critics '86: Best Picture, Best Director, Best Supporting Actor (Hopper), Best Cinematography; Boston Society of Film Critics '86: Best Film, Best Director, Best Supporting Actor (Hopper), Best Cinematography. VHS, Beta, LV $19.98 *LHV, WAR* 🎬🎬🎬½

The Blue Yonder In this made-for-video feature, a young boy travels back in time to meet the grandfather he never knew, risking historical integrity. Alternate title: "Time Flyers."

1986 89m/C Art Carney, Peter Coyote, Huckleberry Fox; *Dir:* Mark Rosman. VHS, Beta $69.95 *DIS* 🎬

Bluebeard Tormented painter with a psychopathic urge to strangle his models is the basis for this effective, low-budget film. One of Carradine's best vehicles.
1944 73m/B John Carradine, Jean Parker, Nils Asther; *Dir:* Edgar G. Ulmer. VHS, Beta $19.95 *NOS, MRV, SNC* 🎬🎬½

Bluebeard Written by Francoise Sagan, this is a French biography of Henri-Desire Landru, who seduced and murdered eleven women and was subsequently beheaded. Dubbed into English. Also known as "Landru."
1963 108m/C *FR* Charles Denner, Danielle Darrieux, Michele Morgan, Hildegarde Neff; *Dir:* Claude Chabrol. VHS, Beta $59.98 *CHA* 🎬🎬🎬

Bluebeard Lady killer Burton knocks off series of beautiful wives in soporific remake of the infamous pirate story.
1972 (R) 128m/C Richard Burton, Raquel Welch, Joey Heatherton, Nathalie Delon, Virna Lisi, Sybil Danning; *Dir:* Edward Dmytryk. VHS $59.95 *IVE* 🎬½

Blueberry Hill While her mother struggles with the grief over husband's death, a young girl in a small town learns about life, music, and her late father from a jazz singer. Good soundtrack.
1988 (R) 93m/C Jennifer Rubin, Carrie Snodgress, Margaret Avery; *Dir:* Strathford Hamilton. VHS, Beta $79.98 *FOX* 🎬½

Blueblood A demonic butler inflicts nightmares upon a family to gain possession of its mansion.
198? 90m/C *GB* Oliver Reed, Derek Jacobi, Fiona Lewis. VHS, Beta $59.95 *VCL* 🎬

The Blues Brothers As an excuse to run rampant on the city of Chicago, Jake and Elwood Blues attempt to raise $5,000 for their childhood orphanage by putting their old band back together. Good music, and lots of cameos.
1980 (R) 133m/C John Belushi, Dan Aykroyd, James Brown, Cab Calloway, Ray Charles, Frank Oz, Steven Spielberg, Twiggy, Henry Gibson, Aretha Franklin, Carrie Fisher, John Candy, Pee-Wee Herman; *Dir:* John Landis. VHS, Beta, LV $19.95 *MCA* 🎬🎬½

Blues Busters Another entry in the "Bowery Boys" series. When Sach emerges from a tonsillectomy with the velvety voice of a crooner, Slip cashes in by turning Louie's Sweet Shop into a nightclub, the better to showcase his buddy's talents. Sach's singing voice provided by John Lorenz.
1950 68m/B Leo Gorcey, Huntz Hall, Adele Jergens, Gabriel Dell, Craig Stevens, Phyllis Coates, Bernard Gorcey, David Gorcey; *Dir:* William Beaudine. VHS $14.98 *WAR* 🎬🎬½

Bluffing It A made-for-television movie about an older man who has been disguising his illiteracy for years.
1987 120m/C Dennis Weaver, Janet Carroll, Michelle Little, Robert Sean Leonard, Cleavant Derricks, Victoria Wauchope; *Dir:* James Sadwith. VHS $89.95 *OHC* 🎬🎬

The Blum Affair In post-World War I Germany, a Jewish man is framed for the murder of an accountant and uncovers in his capture and prosecution a rat's nest of corruption and anti-Semitism. In German with English subtitles.

1948 109m/B *GE* Hans-Christian Blech, Gisela Trowe, Kurt Ehrhardt, Paul Bildt, Klaus Becker; *Dir:* Erich Engel. VHS, Beta $24.95 *NOS, FCT, APD* 🎬🎬

Blume in Love An ironic comedy/drama about a man who falls hopelessly in love with his ex-wife who divorced him for cheating on her while they were married.
1973 (R) 115m/C George Segal, Susan Anspach, Kris Kristofferson, Shelley Winters, Marsha Mason; *Dir:* Paul Mazursky. VHS, Beta $19.98 *WAR* 🎬🎬🎬

BMX Bandits Three adventurous Aussie teens put their BMX skills to the test when they witness a crime and are pursued by the criminals. Much big air.
1983 92m/C Nicole Kidman, David Argue, John Ley, Angelo D'Angelo; *Dir:* Brian Trenchard-Smith. VHS $79.95 *IMP* 🎬🎬

Bo-Ru the Ape Boy A pioneering documentary shot in Africa that portrays the relationship between natives and the wildlife.
193? 60m/B *Dir:* C. Court Treatt. VHS, Beta $24.95 *NOS, DVT* 🎬

Boarding House Residents of a boardinghouse discover sinister doings in the basement.
1983 90m/C Hank Adly, Kalassu, Alexandra Day; *Dir:* John Wintergate. VHS, Beta *PGN, MRV* 🎬

Boarding School Students at a proper Swiss boarding school for girls devise a plan to seduce the boys at a nearby school by posing as prostitutes. Kinski stars as the American girl who masterminds the caper.
1983 (R) 100m/C *GE* Nastassia Kinski; *Dir:* Andre Farwagi. VHS, Beta $69.95 *VES, LIV* 🎬½

Boat Is Full A group of refugees pose as a family in order to escape from Nazi Germany as they seek asylum in Switzerland. Available in German with English subtitles or dubbed into English.
1981 104m/C *SI* Tina Engel, Curt Bais, Renate Steiger, Mathias Gnaedinger; *Dir:* Markus Imhoof. Berlin Film Festival '81: Best Direction and Screenplay (Imhoof), Cidaic Jury Prize, Fipresci Critic Association Special Mention. VHS, Beta $34.95 *FRI, SUE, GLV* 🎬🎬🎬

The Boatniks An accident-prone Coast Guard ensign finds himself in charge of the "Times Square" of waterways: Newport Harbor. Adding to his already "titanic" problems is a gang of ocean-going jewel thieves who won't give up the ship!
1970 (G) 99m/C Robert Morse, Stefanie Powers, Phil Silvers, Norman Fell, Wally Cox, Don Ameche; *Dir:* Norman Tokar. VHS, Beta $69.95 *DIS* 🎬🎬½

Bob & Carol & Ted & Alice Two California couples have attended a trendy therapy session and, in an attempt to be more in touch with sexuality, resort to applauding one another's extramarital affairs and swinging. Wacky and well-written, this is a farce on free love and psychospeak. Paul Mazursky's directorial debut. Gould and Cannon earned Oscar nominations.
1969 (R) 104m/C Natalie Wood, Robert Culp, Dyan Cannon, Elliott Gould; *Dir:* Paul Mazursky. New York Film Festival '69: Fipresci Critic Association Special Mention. VHS, Beta, LV $9.95 *GKK* 🎬🎬½

Bob le Flambeur Wonderful film noir of a compulsive gambler who decides to take a final fling by robbing the casino at Deauville. In French, subtitled in English. English title is "Bob the Gambler."

1955 106m/B *FR* Roger Duchesne, Isabel Corey, Daniel Cauchy; *Dir:* Jean-Pierre Melville. **VHS, Beta, LV** $59.95 COL ♂♂♂

Bob Goldthwait: Is He Like That All the Time? "Bobcat" Goldthwait performs at San Francisco's American Music Hall. Documentary style footage of the comedian's "Meat Bob '88" concert tour is included.
1989 54m/C Bob Goldthwait. **VHS, Beta** $59.99 HBO ♂♂♂

Bob & Ray, Jane, Laraine & Gilda The whimsical world of Bob & Ray is transferred to video. Featuring members of the cast of "Saturday Night Live."
1983 75m/C Jane Curtin, Laraine Newman, Gilda Radner, Willie Nelson, Leon Russell. **VHS, Beta** $59.95 PAV ♂♂♂

Bobbie Jo and the Outlaw A woman who wants to be a country singer and a man who emulates Billy the Kid are fugitives from the law.
1976 (R) 89m/C Lynda Carter, Marjoe Gortner, Jesse Vint; *Dir:* Mark L. Lester. **VHS, Beta** $69.98 VES, LIV ♂ ½

Bobby Deerfield A cold-blooded Grand Prix driver comes face to face with death each time he races, but finally learns the meaning of life when he falls in love with a critically ill woman.
1977 (PG) 124m/C Al Pacino, Marthe Keller, Anny Duperey, Romolo Valli; *Dir:* Sydney Pollack. **VHS, Beta** $19.98 WAR ♂

Bobo A lousy bullfighter tries to lure a gorgeous woman into romance. Filmed in Spain and Italy.
1967 103m/C *GB* Peter Sellers, Britt Ekland, Rossano Brazzi, Adolfo Celi; *Dir:* Robert Parrish. **VHS, Beta, LV** $19.98 WAR, PIA, FCT ♂ ½

Boccaccio '70 Three short bawdy/comedy/pageant-of-life films inspired by "The Decameron," each pertaining to ironically twisted sexual politics in middle class life. A fourth story, "Renzo and Luciana," by Mario Monichelli, has been cut. Dubbed.
1962 145m/C *IT* Anita Ekberg, Romy Schneider, Tomas Milian, Sophia Loren; *Dir:* Vittorio DeSica, Luchino Visconti, Federico Fellini. **VHS, Beta** FCT, MLB ♂♂

Body Beat This movie follows the integration of jazz and rock dancers into a previously classical ballet academy. While the students become fast friends, the teachers break off into two rival factions.
1988 (PG) 90m/C Tony Dean Fields, Galyn Gorg, Scott Grossman, Eliska Krupka. **VHS, Beta** $29.95 VMK ♂

The Body Beneath A living-dead ghoul survives on the blood of innocents and is still preying on victims today.
1970 85m/C Gavin Reed, Jackie Skarvellis, Susan Heard, Colin Gordon; *Dir:* Andy Milligan. **VHS, Beta** $19.95 WES Woof!

Body Chemistry A married sexual-behavior researcher starts up a passionate affair with his lab partner. When he tries to end the relationship, his female associate becomes psychotic. You've seen it all before in "Fatal Attraction." And you'll see it again in "Body Chemistry 2."
1990 (R) 100m/C Marc Singer, Mary Crosby, Lisa Pescia, Joseph Campanella, David Kagen; *Dir:* Kristine Peterson. **VHS, Beta, LV** $19.95 COL ♂

Body Chemistry 2: Voice of a Stranger An ex-cop (Harrison) obsessed with violent sex gets involved with a talk-radio psychologist (Pescia) whose advice could prove deadly in this erotic sequel.
1991 (R) 84m/C Gregory Harrison, Lisa Pescia, Morton Downey Jr., Robin Riker; *Dir:* Adam Simon. **VHS, LV** $89.95 COL ♂ ½

Body Count A weird and wealthy family will stop at nothing, including murder and cannibalism, to enhance their fortune. Rather than bodies, count the minutes 'til the movie's over. Also titled "The 11th Commandment."
1987 90m/C Marilyn Hassett, Dick Sargent, Steven Ford, Greg Mullavey, Thomas Ryan, Bernie White; *Dir:* Paul Leder. **VHS, Beta** $79.98 MCG ♂ ½

Body Double A voyeuristic unemployed actor peeps on a neighbor's nightly disrobing and sees more than he wants to. A grisly murder leads him into an obsessive quest through the world of pornographic films.
1984 (R) 114m/C Craig Wasson, Melanie Griffith, Greg Henry, Deborah Shelton, Guy Boyd, Dennis Franz, David Haskell, Rebecca Stanley, Barbara Crampton; *Dir:* Brian De Palma. **VHS, Beta, LV** $14.95 COL ♂♂♂

Body Heat During a Florida heat wave, a none-too-bright lawyer becomes involved in a steamy love affair with a mysterious woman and then in a plot to kill her husband. Hurt and Turner (in her film debut) became stars under Kasdan's direction (the three would reunite for "The Accidental Tourist"). Hot love scenes supplement a twisting mystery with a suprise ending. Rourke's arsonist and Danson's soft shoe shouldn't be missed.
1981 (R) 113m/C William Hurt, Kathleen Turner, Richard Crenna, Ted Danson, Mickey Rourke; *Dir:* Lawrence Kasdan. **VHS, Beta, LV** $19.98 WAR ♂♂♂ ½

The Body in the Library A British-television-produced Miss Marple TV mystery, based on Agatha Christie's 1942 novel, involving the septuagenarian detective investigating the murder of a young woman in a wealthy British mansion.
1987 155m/C *GB* Joan Hickson; *Dir:* Silvio Narizzano. **VHS, Beta** $29.98 FOX ♂ ½

Body Moves Romantic entanglements heat up the competition when two couples enter a steamy dance contest. Not apt to move you.
1990 (PG-13) 98m/C *IT* Kirk Rivera, Steve Messina, Dianne Granger, Linsley Allen, Philip Spruce, Nicole Kolman, Susan Gardner; *Dir:* Gerry Lively. **VHS, Beta** $89.95 PSM ♂♂

Body Parts For fans of the demented dismemberer niche, Body Parts, a sleazy skin club, loses some of its star talent when the strippers start turning up in cameo video appearances. Seems there's a psycho killer on the loose who videotapes the dismemberment of his stripper-victims. The police decide they've got to meet this guy when he sends them a sample of his work.
1990 90m/C Teri Lee, Dick Monda, Johnny Mandell. **VHS, LV** $69.95 RAE ♂

Body Parts A crime psychologist loses his arm in an auto accident and receives a transplant from an executed murderer. Does the limb have an evil will of its own? Poorly paced horror goes off the deep end in gore with its third act. Based on the novel "Choice Cuts" by French writers Pierre Boileau and Thomas Narcejac, whose work inspired some of the greatest suspense films.

1991 (R) 88m/C Jeff Fahey, Brad Dourif; *Dir:* Eric Red. **VHS, Beta, LV** $89.98 PAR ♂♂

Body Rock A Brooklyn breakdancer (Lamas) deserts his buddies to work at a chic Manhattan nightclub. Watching Lamas as the emcee/breakdancing fool is a hoot.
1984 (PG-13) 93m/C Lorenzo Lamas, Vicki Frederick, Cameron Dye, Michelle Nicastro, Ray Sharkey, Grace Zabriskie, Carole Ita White; *Dir:* Marcelo Epstein. **VHS, Beta** STE, HBO ♂

The Body Shop An unorthodox love story in which a man decides to piece together his relationship with his fiancee by piecing together her dismembered body. For lovers only.
19?? 75m/C Don Brandon, Jenny Diggers. **VHS** $39.98 PGN ♂

Body Slam A small-time talent monger hits it big with a rock and roll/ professional wrestling tour. Piper's debut; contains some violence and strong language.
1987 (PG) 100m/C Roddy Piper, Captain Lou Albano, Dirk Benedict, Tanya Roberts, Billy Barty, Charles Nelson Reilly, John Astin, Wild Samoan, Tonga Kid, Barry Gordon; *Dir:* Hal Needham. **VHS, Beta** $19.98 SUE ♂ ½

The Body Snatcher Based on Robert Louis Stevenson's story about a grave robber who supplies corpses to research scientists. Set in Edinburgh in the 19th century, this Lewton production is superior. One of Karloff's best vehicles.
1945 77m/B Edith Atwater, Russell Wade, Rita Corday, Boris Karloff, Bela Lugosi, Henry Daniell; *Dir:* Robert Wise. **VHS, Beta, LV** $19.95 MED, RKO ♂♂♂ ½

Body and Soul A young boxer fights his way unscrupulously to the top. Vintage '40s boxing film that defines the genre. Remade in 1981. Oscar nominations for Best Actor (Garfield), Original Screenplay.
1947 104m/B John Garfield, Lilli Palmer, Hazel Brooks, Anne Revere, William Conrad, Canada Lee; *Dir:* Robert Rossen. Academy Awards '47: Best Film Editing. **VHS, LV** $19.98 REP ♂♂♂ ½

Body & Soul A hard hitting film about a boxer who loses his perspective in the world of fame, fast cars and women.
1981 (R) 109m/C Leon Isaac Kennedy, Jayne Kennedy, Peter Lawford, Muhammad Ali; *Dir:* George Bowers. **VHS, Beta** $19.95 MGM ♂ ½

Body Waves Teenage beach comedy starring Calvert as a teenager who bets his father that he can raise money on his own. In typical teen movie fashion, he invents a sex cream that drives boys and girls wild. Brain candy featuring lots of skimpy bikinis.
1991 (R) 80m/C Bill Calvert, Leah Lail, Larry Linville, Dick Miller, Jim Wise; *Dir:* P.J. Pesce. **VHS** $89.98 NHO ♂ ½

The Bodyguard The Yakuza, Japan's mafia, and New York's big crime families face off in this martial arts extravaganza.
1976 (R) 89m/C Sonny Chiba, Aaron Banks, Bill Louie, Judy Lee; *Dir:* Maurice Sarli. **VHS, Beta** $29.95 MED ♂

Bog A boggy beast from the Arctic north awakens to eat people. Scientists mount an anti-monster offensive.
1984 (PG) 90m/C Gloria DeHaven, Marshall Thompson, Leo Gordon, Aldo Ray. **VHS, Beta** $49.95 PSM ♂

Boggy Creek II The continuing saga of the eight-foot-tall, 300 pound monster from Texarkana. Also known as "The Barbaric Beast of Boggy Creek, Part II." Third in a series of low-budget movies including "The

Legend of Boggy Creek'' and ''Return to Boggy Creek.''
1983 (PG) 93m/C Charles B. Pierce, Cindy Butler, Serene Hedin; *Dir:* Charles B. Pierce. VHS, Beta, LV $59.95 MED *‖½*

Bogie Television biography of Humphrey Bogart, populated by almost-lookalikes who can almost act.
1980 100m/C Kevin J. O'Connor, Kathryn Harrold, Ann Wedgeworth, Patricia Barry, Alfred Ryder, Donald May, Richard Dysart, Arthur Franz; *Dir:* Vincent Sherman. VHS, Beta $14.95 FRH *‖½*

Bohemian Girl The last of Laurel and Hardy's comic operettas finds them as guardians of a young orphan (Hood, famous for her roles in the Our Gang comedies), whom no one realizes is actually a kidnapped princess.
1936 74m/B Stan Laurel, Oliver Hardy, Mae Busch, Thelma Hood, Jacqueline Wells, Thelma Todd, James Finlayson; *Dir:* James W. Horne. VHS, Beta $19.95 MED, CCB *‖‖‖*

Boiling Point Lawman proves once again that justice always triumphs.
1932 67m/B Hoot Gibson. VHS, Beta $19.95 NOS, UHV, DVT *‖*

The Bold Caballero Rebel chieftain Zorro overthrows oppressive Spanish rule in the days of early California.
1936 69m/B Bob Livingston, Heather Angel, Sig Rumann, Robert Warwick; *Dir:* Wills Root. VHS, Beta $19.95 NOS, VDM, VCN *‖*

Boldest Job in the West A western about bankrobbers and gamblers.
197? 100m/C Mark Edwards, Frank Sancho, Carmen Sevilla; *Dir:* Joseph Loman. VHS, Beta $19.95 ASE *‖*

Bolero Beginning in 1936, this international epic traces the lives of four families across three continents and five decades, highlighting the music and dance that is central to their lives.
1982 173m/C James Caan, Geraldine Chaplin, Robert Hossein, Nicole Garcia, Jacques Villeret; *Dir:* Claude Lelouch. VHS, Beta $19.95 VES *‖‖½*

Bolero What sounds like a wet dream come true is really just a good snooze: Bo Derek plays a beautiful young woman who goes on a trip around the world in hopes of losing her virginity; in Spain, she meets a bullfighter who's willing to oblige. Too bad Bo cannot act as good as she looks.
1984 (R) 106m/C Bo Derek, George Kennedy, Andrea Occhipinti, Anna Obregon, Olivia D'Abo; *W/Dir:* John Derek. VHS, Beta $19.95 IVE *Woof!*

Bolo The authorities pardon two tough prison inmates, setting the scene for revenge.
197? 90m/C Yang Sze; *Dir:* Bill Raymond. VHS, Beta $47.95 MAV *‖*

Bomb at 10:10 An American pilot escapes from a German POW camp and plots to assassinate a camp commandant.
1967 87m/C George Montgomery, Rada Popovic, Peter Banicevic; *Dir:* Charles Damic. VHS, Beta $14.95 GEM *‖‖*

Bombardier A group of cadet bombardiers discover the realities of war on raids over Japan during World War II.
1943 99m/B Pat O'Brien, Randolph Scott, Robert Ryan, Eddie Albert, Anne Shirley, Barton MacLane; *Dir:* Richard Wallace. VHS, Beta $19.98 RKO *‖‖½*

Bombay Talkie A naive American novelist visits India for ''experience,'' and gets romantically involved with an Indian movie star.
1970 (PG) 110m/C *IN* Jennifer Kendal, Zia Mohyeddi, Shashi Kapoor, Aparna Sen; *Dir:* James Ivory. VHS, Beta $29.95 TAM, SUE *‖‖‖*

Bombs Away! An atomic bomb is mistakenly shipped to a seedy war surplus store in Seattle, and causes much chicanery.
1986 90m/C Michael Huddleston, Pat McCormick; *Dir:* Bruce Wilson. VHS, Beta $59.95 CHA *‖*

Bombshell Wry insightful comedy into the Hollywood of the 1930s. Harlow plays a naive young actress manipulated by her adoring press agent. He thwarts her plans until she finally notices and begins to fall in love with him. Brilliant satire with Harlow turning in perhaps the best performance of her short career. Also known as ''Blonde Bombshell.''
1933 96m/B Jean Harlow, Lee Tracy, Pat O'Brien, Una Merkel, C. Aubrey Smith, Franchot Tone; *Dir:* Victor Fleming. VHS, Beta $19.98 MGM, BTV, FCT *‖‖‖*

Bon Voyage! A family's long-awaited European ''dream'' vacation turns into a series of comic misadventures in this very Disney, very family, very predictable comedy.
1962 131m/C Fred MacMurray, Jane Wyman, Deborah Walley, Michael Callan, Tommy Kirk, Jessie Royce Landis; *Dir:* James Neilson. VHS, Beta, LV $69.95 DIS *‖½*

Bon Voyage, Charlie Brown The comic strip group from ''Peanuts'' become exchange students in Europe, led by Charlie Brown, Linus, Peppermint Patty, Marcie, and the irrepressible beagle, Snoopy.
1980 (G) 76m/C Charlie Brown; *Dir:* Bill Melendez. VHS, Beta, LV $14.95 PAR *‖‖‖*

The Bone Yard A weird mortuary is the setting for strange goings on when a murder is investigated.
1990 (R) 98m/C Ed Nelson, Deborah Rose, Norman Fell, Jim Eustermann, Denise Young, Willie Stratford Jr., Phyllis Diller; *W/Dir:* James Cummins. VHS, Beta $89.98 PSM *‖*

The Bonfire of the Vanities If you read and liked Tom Wolfe's viciously satirical novel, chances are you won't like De Palma's version at all. If you didn't read the book, chances are you still won't like it. Miscast and stripped of the book's gutsy look inside its characters, the film's sole attribute is Vilmos Zsigmond's photography. Hanks is all wrong as wealthy Wall Street trader Sherman McCoy who, taking a wrong turn, finds himself lost in the back streets of the Bronx, where his panic results in the accidental death of a black youth. Newspapers, lawmakers, and ethnic activists are out to use him for a scapegoat and he has neither the brains, nor the skullduggery, to protect himself. Willis' drunken journalist/narrator, Griffith's blonde mistress, and Freeman's righteous judge are all awkward and thinly written. If still awake, look for F. Murray Abraham's cameo as the Bronx D.A.
1990 (R) 126m/C Tom Hanks, Melanie Griffith, Bruce Willis, Morgan Freeman, Alan King, Kim Cattrall, Saul Rubinek, Clifton James, Donald Moffatt, Richard Libertini, Andre Gregory, Robert Stephens; *Dir:* Brian De Palma. VHS, Beta, LV, 8mm $19.98 WAR, FCT *‖‖*

Bongo Originally a part of the Disney cartoon anthology ''Fun and Fancy Free.'' Follows the adventures of a circus bear who flees the big time for the wonders of the

forests. Narrated by Dinah Shore. Also available with ''Ben and Me'' on Laser Disc.
1947 36m/C Bill Roberts, Hamilton Luske; *Dir:* Jack Kinney. VHS, Beta $12.99 DIS, BVV *‖‖*

Bonjour Tristesse An amoral French girl conspires to break up her playboy father's upcoming marriage to her stuffy godmother in order to maintain her decadent freedom. Based on the novel by Francoise Sagan; screenplay by Arthur Laurents. Preminger attempted, unsuccessfully, to use this soaper to catapult Seberg to stardom.
1957 94m/C *FR* Deborah Kerr, David Niven, Jean Seberg, Mylene Demongeot, Geoffrey Horne, Walter Chiari, Jean Kent; *Dir:* Otto Preminger. VHS, Beta $69.95 COL *‖‖‖*

Bonnie & Clyde Based on the biographies of the violent careers of Bonnie Parker and Clyde Barrow, who roamed the Southwest robbing banks. In the Depression era, when any job, even an illegal one, was cherished, money, greed and power created an unending cycle of violence and fury. Highly controversial and influential, with pronounced bloodshed that spurred mainstream cinematic proliferation. Established Dunaway as a star; produced by Beatty in one of his best performances. Oscar nominated for best picture, the scriptwriting of David Newman and Robert Benton, direction, actor (Beatty), actress (Dunaway) and supporting actors Hackman and Pollard.
1967 111m/C Warren Beatty, Faye Dunaway, Michael J. Pollard, Gene Hackman, Estelle Parsons; *Dir:* Arthur Penn. Academy Awards '67: Best Cinematography, Best Supporting Actress (Parsons); New York Film Critics Awards '67: Best Screenplay; National Society of Film Critics Awards '67: Best Screenplay, Best Supporting Actor (Hackman). VHS, Beta, LV $19.98 WAR, BTV *‖‖‖½*

Bonnie Prince Charlie Historical epic opening in 1745 and romanticizing the title pretender to the British throne, who united Scottish clans in a doomed campaign against King George. Talky and rather slow-moving except for stirring battle scenes. A notorious box-office flop in its native Britain, where the original running time was 140 minutes.
1948 135m/C *GB* David Niven, Margaret Leighton, Judy Campbell, Jack Hawkins, Morland Graham, Finlay Currie, Elwyn Brook-Jones, John Laurie; *Dir:* Anthony Kimmins. VHS $19.95 NOS *‖‖½*

Bonnie Scotland Laurel & Hardy accidentally join an India-bound Scottish regiment. Laughs aplenty.
1935 80m/B Stan Laurel, Oliver Hardy, James Finlayson, Daphne Pollard, June Lang; *Dir:* James W. Horne. VHS, Beta $19.98 MGM, MLB *‖‖½*

Bonnie's Kids A story about two sisters who become involved in murder, sex, and stolen money.
1973 (R) 107m/C Tiffany Bolling, Robin Mattson, Scott Brady, Alex Rocco; *Dir:* Arthur Marks. VHS, Beta $39.95 MON *‖½*

Boob Tube Television station KSEX is out to change their audience's viewing habits with their unusual programming.
1975 (R) 74m/C John Alderman, Sharon Kelly, Lois Lane. VHS, Beta $19.95 GEM *‖‖*

The Boogey Man Through the reflection in a mirror, a girl witnesses her brother murder their mother's lover. Twenty years later this memory still haunts her; the mirror is now broken, revealing its special powers. Who will be next? Murder menagerie; see footage from this flick mirrored in ''The Boogey Man 2.''

1980 (R) 93m/C John Carradine, Suzanna Love, Ron James; *Dir:* Ulli Lommel. **VHS, Beta, LV $59.95** UHV, HHE ⊘⊘

Boogey Man 2 The story continues; same footage, new director (but Lommel, the director of th e original ''Boogey Man,'' co-wrote this script and appears in the film).
1983 90m/C John Carradine, Suzanna Love, Shannah Hal, Ulli Lommel, Sholto Von Douglas; *Dir:* Bruce Starr. **VHS, Beta $29.95** VHE ⊘

Book of Love ''Zany'' hijinks as a '50s teenager struggles with friendship, girls and those all-important hormones. Rehash of every 50's movie and TV show cliche in existence. Surprise! There's a classic rock-and-roll soundtrack.
1991 (PG-13) 88m/C Chris Young, Keith Coogan, Michael McKean, Tricia Leigh Fisher, Josie Bisset, Aeryk Egan; *Dir:* Robert Shaye. **VHS, LV, 8mm $19.95** COL, PIA Woof!

Boom in the Moon Keaton fares poorly in this sci-fi comedy. He's trapped on a space ship to the moon. Poor production, with uneven direction and acting.
1946 83m/C MX Buster Keaton, Angel Garasa, Virginia Serret, Fernando Soto, Luis Barreiro; *Dir:* Jaime Salvador. **VHS $39.95** USA Woof!

Boom Town A lively vintage comedy/drama/romance about two oil-drilling buddies competing amid romantic mix-ups and fortunes gained and lost. Oscar-nominated for cinematography.
1940 120m/B Clark Gable, Spencer Tracy, Claudette Colbert, Hedy Lamarr, Frank Morgan, Lionel Atwill, Chill Wills, Curt Bois; *Dir:* Jack Conway. **VHS, Beta $19.98** MGM, CCB ⊘⊘ ½

Boomerang A father rescues his wrongly convicted son from prison. Dubbed.
1976 101m/C IT Alain Delon, Carla Gravina, Dora Doll; *Dir:* Jose Giovanni. **VHS, Beta $59.95** UNI ⊘ ½

The Boost A feverish, messy melodrama about a young couple's spiraling decline from yuppie-ish wealth in a haze of cocaine abuse.
1988 (R) 95m/C James Woods, Sean Young, John Kapelos, Steven Hill, Kelle Kerr, John Rothman, Amanda Blake, Grace Zabriskie; *Dir:* Harold Becker. **VHS, Beta, LV $19.95** HBO ⊘ ½

Boot Hill Two guys mess with western baddies and wild women in spaghetti oater.
1969 (PG) 97m/C IT Terence Hill, Bud Spencer, Woody Strode, Victor Buono, Lionel Stander; *Dir:* Giuseppe Colizzi. **VHS, Beta $9.95** NO ⊘ ½

Boot Hill Bandits ''Crash'' Corrigan fights on the side of right in this classic western as he helps corral Wells Fargo bandits.
1942 58m/B Ray Corrigan, John ''Dusty'' King, Max Terhune, Jean Brooks, John Merton, Glenn Strange; *Dir:* S. Roy Luby. **VHS $19.95** NOS, HEG ⊘⊘ ½

Boothill Brigade Former footballer Brown rides to the rescue when criminals steal land from homesteaders.
1937 58m/B Johnny Mack Brown, Claire Rochelle, Dick Curtis, Horace Murphy, Frank LaRue, Edward Cassidy, Bobby Nelson, Frank Ball, Steve Clark; *Dir:* Sam Newfield. **VHS $19.95** NOS, MRV, ASE ⊘⊘

Bootleg A detective is on a case that leads to crooked politics and espionage.
1989 82m/C Ray Meagher; *Dir:* John Prescott. **VHS $79.95** AIP ⊘

Boots Malone An old down-on-his-luck gambling addict and a young, rich kid fascinated by the sordid atmosphere of the racetrack stumble upon one another and

form a symbiotic relationship. All goes well for while, but there'd be no movie unless a collection of obstacles suddenly threatens their success, friendship, and even their lives. A rather melodramatic buddy tale, but Holden and Stewart hold interest.
1952 103m/C William Holden, John Stewart, Ed Begley Sr., Harry Morgan, Whit Bissell; *Dir:* William Dieterle. **VHS** GKK ⊘⊘ ½

Boots & Saddles A young English lord wants to sell the ranch he has inherited but Autry is determined to make him a real Westerner.
1937 54m/B Gene Autry, Judith Allen, Smiley Burnette; *Dir:* Joseph Kane. **VHS, Beta $19.95** NOS, MRV, VYY ⊘

Bordello A made-for-television western spoof about cowboys and shady ladies.
1979 90m/C Chuck Connors, Michael Conrad, John Ireland, Isela Vega, George Rivero. **VHS, Beta $79.95** PSM ⊘

The Border A border guard faces corruption and violence within his department and tests his own sense of decency when the infant of a poor Mexican girl is kidnapped. Excellent cast, fine cinematography, unusual Nicholson performance.
1982 (R) 107m/C Jack Nicholson, Harvey Keitel, Valerie Perrine, Warren Oates, Elpidia Carrillo; *Dir:* Tony Richardson. **VHS, Beta, LV $39.95** MCA ⊘⊘⊘

Border Badmen It's just another tired old western, even if it does appropriate the classic mystery-thriller plot; following the reading of a siler baron's will, someone starts killing off the relatives.
1945 59m/B Buster Crabbe, Lorraine Miller, Charles King, Ray Bennett, Archie Hall, Budd Buster, Bud Osborne; *Dir:* Sam Newfield. **VHS, Beta** GPV, MRV, TIM ⊘ ½

Border Bandits Plodding oater has Brown going after the bad guys. Not only is it a dirty job, but apparently pretty boring.
1946 57m/B Johnny Mack Brown, Raymond Hatton, John Merton, Frank LaRue, Steve Clark, Charles Stevens, Bud Osborne; *Dir:* Lambert Hillyer. **VHS $19.95** DVT ⊘

Border Devils A boy is pursued by a ruthless gang of outlaws, until he is saved by the good guys.
1932 60m/B Harry Carey, Art Mix, George ''Gabby'' Hayes; *Dir:* William Nigh. **VHS, Beta, LV** WGE ⊘

Border Feud LaRue endeavors to prevent someone who, by instigating a family feud, will claim large gold mine rights.
1947 54m/B Lash LaRue, Al ''Fuzzy'' St. John, Bob Duncan; *Dir:* Ray Taylor. **VHS, Beta $12.95** COL, SVS ⊘

Border Heat A man and his female partner try to save a group of Mexican laborers from the horrible working conditions their cruel employers inflict upon them.
1988 93m/C John Vernon, Darlanne Fluegel, Michael J. Moore; *Dir:* Max Kleven. **VHS, Beta $79.98** MCG ⊘

Border Phantom Steele (whom you might recognize as Canino in ''The Big Sleep''), takes on a suave Chinese businessman who heads a mail order business; seems he's smuggling mail-order brides from Mexico in this vintage ''B''-grade horse opera.
1937 60m/B Bob Steele, Harley Wood, Don Barclay, Carl Hackett, Horace Murphy, Miki Morita; *Dir:* S. Roy Luby. **VHS $19.99** NOS, MRV, ASE ⊘⊘

Border Radio A rock singer decides to steal a car and try to outrun some tough thugs hot on his trail.
1988 88m/C VHS $19.95 PAV, FCT ⊘⊘

Border Rangers A man joins the Rangers in order to avenge the murders of his brother and sister-in-law.
1950 57m/B Robert Lowery, Donald (Don ''Red'') Barry, Lyle Talbot, Pamela Blake; *Dir:* William Berke. **VHS, Beta, LV $29.95** WGE ⊘

Border Romance Three Americans have their horses stolen by bandits while riding through Mexico. Trouble with the locals follows in this early sound western.
1930 58m/B Don Terry, Armida, Marjorie Kane. **VHS, Beta $24.95** VYY ⊘ ½

Border Roundup The ''Lone Rider'' uncovers a gang of thieving crooks.
1941 57m/B George Houston, Al ''Fuzzy'' St. John, Dennis Moore; *Dir:* Sam Newfield. **VHS, Beta $19.95** VDM ⊘

Border Saddlemates A U.S. government agent becomes involved in a crooked deal when he is asked to doctor Canadian silver foxes.
1952 67m/B Rex Allen. **VHS, Beta $29.95** DVT ⊘

Border Shootout A trigger-happy sheriff battles a rich young cattle rustler.
1990 (PG) 110m/C Glenn Ford, Charlene Tilton, Jeff Kaake, Michael Horse, Russell Todd, Cody Glenn, Sergio Calderone, Michael Ansara; *Dir:* C.T. McIntyre. **VHS, Beta $79.98** TTC ⊘⊘

Border Street This film shows how a group of Polish Jews fought off the German occupation forces in the Warsaw ghetto. In Polish with English subtitles.
1948 110m/B PL Venice Film Festival '48: Best Supporting Actor. **VHS, Beta, 3/4U $39.95** IHF ⊘⊘ ½

Border of Tong It was the most heinous killing in the history of Chinatown, and the only suitable response was revenge!
1990 ?m/C David Heavener. **VHS** PMH ⊘

Borderline Bronson is a border patrol guard in pursuit of a murderer in this action flick. Meanwhile, he gets caught up in trying to help an illegal alien and her young child.
1980 (PG) 106m/C Charles Bronson, Wilford Brimley, Bruno Kirby, Benito Morales, Ed Harris, Kenneth McMillan; *Dir:* Jerrold Freedman. **VHS, Beta $59.98** FOX ⊘⊘

Bordertown Gunfighters A cowboy breaks up a vicious lottery racket and falls in love in the process.
1943 56m/B Bill Elliott, Anne Jeffreys. **VHS, Beta $29.95** DVT ⊘ ½

B.O.R.N. Hagen uncovers an underground network of doctors who kill people and sell the body parts and organs.
1988 (R) 90m/C Ross Hagen, P.J. Soles, William Smith, Hoke Howell, Russ Tamblyn, Amanda Blake, Clint Howard, Rance Howard, Debra Lamb; *Dir:* Ross Hagen. **VHS, Beta $79.95** PSM ⊘ ½

Born in America A young orphan is adopted by an undertaker. The boy grows up enveloped by his guardian's business and, when things are slow, works to bring in business himself. Also known as ''Dead Aim.''
1990 90m/C James Westerfield, Glen Lee, Virgil Frye, Venetia Vianello; *Dir:* Jose Antonio Bolanos. **VHS $59.95** AVD, FCT ⊘

Born American Pre-Glasnost flick about teenagers vacationing in Finland who ''accidentally'' cross the Russian border. There they battle the Red Plague. Melodramatic and heavily politicized. Original release banned in Finland.
1986 (R) 95m/C Mike Norris, Steve Durham, David Coburn, Thalmus Rasulala, Albert Salmi; **Dir:** Renny Harlin. **VHS, Beta $79.98** *CGI, MAG* ⚜

Born to Be Bad Fontaine, an opportunistic woman who was b-b-born to be b-b-bad, attempts to steal a wealthy business man away from his wife while carrying on an affair with a novelist. Ray's direction (he's best known for ''Rebel Without a Cause,'' filmed five years later) prevents the film from succumbing to standard Hollywood formula.
1950 93m/B Joan Fontaine, Robert Ryan, Zachary Scott, Joan Leslie, Mel Ferrer; **Dir:** Nicholas Ray. **VHS, Beta, LV $19.98** *TTC, FCT* ⚜⚜½

Born to Be Wild A pair of truck drivers are commissioned to deliver a shipment of dynamite. The dynamite will be used to destroy a dam, preventing the surrounding land from falling into the hands of unscrupulous land barons. Good action picture thanks to the cast.
1938 66m/B Ralph Byrd, Doris Weston, Ward Bond, Robert Emmett Keane, Bentley Hewlett, Charles Williams; **Dir:** Joseph Kane. **VHS, Beta $16.95** *SNC, RXM* ⚜⚜

Born Beautiful Made for TV. Beautiful young women come to New York, hoping to strike it rich in the world of modeling. At the same time, a still beautiful, but older and overexposed model comes to grips with a change in her career. Well-told, but sugarcoated.
1982 100m/C Erin Gray, Ed Marinaro, Polly Bergen, Lori Singer, Ellen Barber, Judith Barcroft, Michael Higgins; **Dir:** Harvey Hart. **VHS** *WAR* ⚜⚜½

Born to Boogie A concert, a chronicle and a tribute, ''Born To Boogie'' is about Marc Bolan and his band T. Rex. The film, which features concert footage from a 1972 concert at the Wembley Empire Pool in London, chronicles the glitter rock era and the next wave of British rock and roll, and pays tribute to Mark Bolan, who died in 1977 just as he was starting a comeback. Ringo Starr and Elton John join T. Rex in the studio for ''Children of the Revolution'' and ''Tutti-Frutti'' as well as a psychadelic soiree where Bolan plays acoustic versions of his hits. This film was never released in the United States.
1972 75m/C Marc Bolan, T. Rex, Elton John, Ringo Starr; **Dir:** Ringo Starr. **VHS $29.98** *MPI* ⚜

Born to Dance A quintessential MGM 1930s dance musical, wherein a beautiful dancer gets a sailor and a big break in a show. Songs by Cole Porter include ''Love Me Love My Pekinese,'' ''I've Got You Under My Skin,'' ''Hey Babe Hey,'' and ''Easy to Love,'' sung by a gangly James Stewart. Powell's first starring vehicle.
1936 108m/B Eleanor Powell, James Stewart, Virginia Bruce, Una Merkel, Frances Langford, Sid Silvers, Raymond Walburn, Reginald Gardiner, Buddy Ebsen; **Dir:** Roy Del Ruth. **VHS, Beta, LV $19.95** *MGM, MLB* ⚜⚜½

Born in East L.A. Marin brings his Bruce Springsteen-parody song to life as he plays a mistakenly deported illegal alien who struggles to return to the U.S.
1987 (R) 85m/C Cheech Marin, Daniel Stern, Paul Rodriguez, Jan-Michael Vincent, Kamala Lopez, Tony Plana; **Dir:** Cheech Marin. **VHS, Beta, LV $14.95** *MCA, FCT* ⚜½

Born Famous An engaging commentary about growing up with famous parents including members of the Kennedy, Conrad, Landon, and Van Patten families.
1990 60m/C VHS $39.95 *FRH* ⚜½

Born of Fire A flutist investigating his father's death seeks the Master Musician; his search leads him to the volcanic (and supernatural) mountains of Turkey.
1987 (R) 84m/C Peter Firth, Susan Crowley, Stephan Kalipha; **Dir:** Jamil Dehlavi. **VHS, Beta, LV $79.95** *VMK* ⚜½

Born in Flames An acclaimed futuristic feminist film about women who band together in an effort to gain control of the state owned media.
1983 90m/C Dir: Lizzie Borden. **VHS, 3/4U** *ICA* ⚜

Born on the Fourth of July A riveting meditation on American life affected by the Vietnam War, based on the real-life, bestselling experiences of Ron Kovic, though some facts are subtly changed. The film follows him as he develops from a naive recruit to an angry, wheelchair-bound paraplegic to an active antiwar protestor. Well-acted and generally lauded; Kovic co-wrote the screenplay and appears as a war veteran in the opening parade sequence. Cruise was Oscar-nominated in this, his most difficult role to date.
1989 (R) 145m/C Tom Cruise, Kyra Sedgwick, Raymond J. Barry, Jerry Levine, Tom Berenger, Willem Dafoe, Frank Whaley, John Getz, Caroline Kava, Dryan Larkin, Abbie Hoffman, Stephen Baldwin, Josh Evans; **Dir:** Oliver Stone. Academy Awards '89: Best Director (Stone), Best Film Editing; Golden Globe Awards '90: Best Film—Drama. **VHS, Beta, LV, 8mm $14.95** *MCA, FCT, BTV* ⚜⚜⚜½

Born Free The touching story of a game warden in Kenya and his wife raising Elsa the orphaned lion cub. When the cub reaches maturity, they work to return her to life in the wild. Great family entertainment based on Joy Adamson's book. Theme song became a hit.
1966 95m/C Virginia McKenna, Bill Travers; **Dir:** James Hill. Academy Awards '66: Best Song (''Born Free''), Best Original Score; National Board of Review Awards '66: 10 Best Films of the Year. **VHS, Beta, LV $14.95** *COL* ⚜⚜⚜

Born Innocent As if ''The Exorcist'' weren't bad enough, Blair is back for more abuse-on-film, this time as a 14-year-old runaway from a dysfunctional family who lands in a reform school for girls. There, she must struggle to be as brutal as her peers in order to survive. Fairly tame by today's standards, but controversial at its made-for-TV premiere, chiefly due to a rape scene involving a broom handle.
1974 92m/C Linda Blair, Joanna Miles, Kim Hunter, Richard Jaeckel; **Dir:** Donald Wrye. **VHS, Beta $49.95** *WES* ⚜½

Born Invincible A Kung Fu master avenges the kick-death of a student by randomly kick-killing suspects.
1976 90m/C Lo Lieh, Carter Wong, Nancy Yan; **Dir:** Joseph Kuo. **VHS, Beta $59.95** *GEM* ⚜

Born to Kill A ruthless killer marries a girl for her money. Minor tough-as-nails noir exemplifying Tierney's stone-faced human-devil film persona.

1947 92m/B Lawrence Tierney, Claire Trevor, Walter Slezak, Elisha Cook Jr.; **Dir:** Robert Wise. **VHS, Beta $19.98** *TTC, KOV* ⚜⚜⚜

Born Killer Teenagers in the woods meet up with murderous escaped convicts.
1989 90m/C Francine Lapensee, Fritz Matthews, Ted Prior, Adam Tucker; **Dir:** Kimberly Casey. **VHS, Beta $79.95** *AIP* ⚜½

Born Losers The original ''Billy Jack'' film in which the Indian martial arts expert takes on a group of incorrigible bikers.
1967 (PG) 112m/C Tom Laughlin, Elizabeth James, Jeremy Slate, William Wellman Jr., Robert Tessier, Jane Russell; **Dir:** T.C. Frank. **VHS, Beta, LV $59.95** *VES, LIV* ⚜

Born to Race In the world of competitive auto racing, a beautiful engineer is kidnapped for her new controversial engine design.
1988 (R) 95m/C Joseph Bottoms, George Kennedy, Marc Singer, Marla Heasly; **Dir:** James Fargo. **VHS, Beta $79.98** *FOX* ⚜

Born to Run A young Australian boy dreams of restoring his grandfather's run-down horse farm to its former glory. Filmed on location.
1977 87m/C VHS, Beta $69.95 *DVT* ⚜⚜

Born to Win A New York hairdresser with an expensive drug habit struggles through life in this well made comedy drama. Excellent acting from fine cast, well-paced, and interestingly photographed. Not well received when first released, but worth a look. Also on video as ''Addict.''
1971 (R) 90m/C George Segal, Karen Black, Paula Prentiss, Hector Elizondo, Robert De Niro, Jay Fletcher; **Dir:** Ivan Passer. **VHS $9.99** *QVD, MGM, IAV* ⚜⚜½

Born Yesterday Ambitious junk dealer Harry Brock is in love with smart but uneducated Billie Dawn. He hires newspaperman Paul Verrall to teach her the finer points of etiquette. During their sessions, they fall in love and Billie realizes how she has been used by Brock. She retaliates against him and gets to deliver that now-famous line: ''Do me a favor, drop dead.'' Based on the Broadway play. Oscar nominations for Best Picture, Best Director, Screenplay, and Black & White Costume Design.
1950 103m/B Judy Holliday, Broderick Crawford, William Holden; **Dir:** George Cukor. Academy Awards '50: Best Actress (Holliday). **VHS, Beta, LV $19.95** *COL, FCT, BTV* ⚜⚜⚜½

Borneo Explorer and naturalist Martin Johnson investigates the many unusual sights and inhabitants of Borneo in this pioneering documentary.
1937 76m/B Martin Johnson, Osa Johnson; **Nar:** Lowell Thomas, Lew Lehr. **VHS, Beta $29.95** *VYY, DVT* ⚜⚜½

The Borrower An exiled unstable mutant insect alien serial killer (you know the type) must switch heads with human beings regularly to survive. This colorful, garish gorefest has humor and attitude, but never develops behind the basic grossout situation.
1989 (R) 97m/C Rae Dawn Chong, Don Gordon, Antonio Fargas, Tom Towles; **Dir:** John McNaughton. **VHS, LV $89.99** *CAN, LDC* ⚜⚜

Boss Blaxploitation western parody with Williamson and Martin as a couple of bounty hunters tearing apart a town to find a fugitive. Originally titled, ''Boss Nigger,'' Williamson wrote the relatively non-violent script. Also known as ''The Black Bounty Killer.''

1974 (PG) 87m/C Fred Williamson, D'Urville Martin, R.G. Armstrong, William Smith, Carmen Hayworth, Barbara Leigh; *Dir:* Jack Arnold. **VHS** *BLC, VTK* 🎬 ½

Boss of Big Town Gangsters try to infiltrate the milk industry in this standard crime drama.
1943 65m/B John Litel, Florence Rice, H.B. Warner, Jean Brooks, John Miljan, Mary Gordon, John Maxwell; *Dir:* Arthur Dreifuss. **VHS, Beta $16.95** *SNC, RXM* 🎬

Boss Cowboy A cowboy shows who wears the chaps in his family when he kidnaps his girlfriend to prevent her from moving East.
1935 51m/B Buddy Roosevelt, Frances Morris, Sam Pierce, Fay McKenzie, Lafe McKee; *Dir:* Victor Adamson. **VHS, Beta** *WGE* 🎬

Boss Foreman A languorous, entwining comedy of errors about a construction foreman and his misadventures with the mob.
1939 58m/B Henry Armetta. **VHS, Beta $24.95** *VYY Woof!*

Boss Is Served A mother and three daughters, racked with debt, take in boarders in the person of a lascivious man and teenage son.
196? 95m/C Senta Berger, Maurizio Arena, Bruno Zanin. **VHS, Beta** *INV* 🎬

Boss of Rawhide The Texas Rangers fight for law and order in the old west despite odds against them.
1944 60m/B Dave O'Brien, James Newill. **VHS, Beta $19.95** *NOS, DVT* 🎬

Boss' Son Worthwhile coming-of-age tale about a young man who learns more than he bargained for about life when he goes to work for his father in the family carpet business. Dad feels he should work his way up through the ranks and makes him a delivery man. The young man meets reality head on and must deal with the injustices of the world around him. Independently filmed by Roth who also wrote the screenplay.
1979 97m/C Rita Moreno, James Darren. **VHS, Beta $69.95** *VES, LIV* 🎬🎬 ½

The Boss' Wife A young stockbroker attempts to fix what's wrong with his life by maneuvering sexually with the boss' wife, and complications set in.
1986 (R) 83m/C Daniel Stern, Arielle Dombasle, Christopher Plummer, Martin Mull, Melanie Mayron, Lou Jacobi; *Dir:* Ziggy Steinberg. **VHS, Beta $79.98** *FOX* 🎬🎬

The Boston Strangler Based on Gerold Frank's bestselling factual book about the deranged killer who terrorized Boston for about a year and a half. Traces events from first killing to prosecution. Curtis, going against type, is compelling in title role.
1968 116m/C Tony Curtis, Henry Fonda, George Kennedy, Murray Hamilton, Mike Kellin, George Voskovec, William Hickey, James Brolin, Hurd Hatfield, William Marshall, Jeff Corey, Sally Kellerman; *Dir:* Richard Fleischer. **VHS, Beta $19.98** *FOX* 🎬🎬 ½

The Bostonians A faith healer's daughter is forced to choose between the affections of a militant suffragette and a young lawyer in 19th Century Boston. Based on Henry James' classic novel.
1984 (PG) 120m/C Christopher Reeve, Vanessa Redgrave, Madeleine Potter, Jessica Tandy, Nancy Marchand, Wesley Addy, Linda Hunt, Wallace Shawn; *Dir:* James Ivory. **VHS, Beta, LV $79.98** *VES, LIV* 🎬🎬 ½

Botany Bay A rousing costumer about a convict ship in the 1700s that, after a trying voyage, lands in Australia, wherein a framed doctor conquers the local plague.
1953 99m/C Alan Ladd, James Mason, Patricia Medina, Cedric Hardwicke; *Dir:* John Farrow. **VHS, Beta $19.95** *KRT* 🎬🎬

Bottoms Up Three nymphomaniacs seduce a virginal mechanic and turn him into an insatiable womanizer. Skin aplenty.
1987 90m/C Adam Janas, Sparky Abbrams, Kathleen Johnson. **VHS, Beta $9.99** *STE, PSM Woof!*

Bottoms Up '81 A screen recreation of this long-running comedy revue, combining slapstick, satire, beautiful showgirls, and lavish musical numbers. Taped on location at Harrah's Lake Tahoe.
1981 75m/C VHS, Beta $49.95 *PAR Woof!*

Boudu Saved from Drowning A suicidal tramp completely disrupts the wealthy household of the man that saves him from drowning. A gentle but sardonic farce from the master filmmaker. Also known as "Boudu Sauve des Eaux." Remade in 1986 as "Down and Out In Beverly Hills."
1932 87m/B **FR** Michel Simon, Charles Grandval, Jean Daste; *Dir:* Jean Renoir. **VHS, Beta $59.95** *INT, MRV, DVT* 🎬🎬🎬 ½

Boulevard Nights A young Latino man tries to escape his neighborhood's street-gang violence, while his older brother is sucked into it. Music by Lalo Schifrin, known for the "Mission: Impossible" theme.
1979 (R) 102m/C Richard Yniguez, Danny De La Paz, Marta Du Bois, Carmen Zapata, Victor Millan; *Dir:* Michael Pressman. **VHS, Beta $59.95** *WAR* 🎬🎬

Bound for Glory The award-winning biography of American folk singer Woody Guthrie set against the backdrop of the Depression. Superb portrayal of the spirit and feelings of the period featuring many of his songs encased in the incidents that inspired them. Haskell Wexler's award-winning camera work is superbly expressive. Oscar nominations for Best Picture, Adapted Screenplay, and Costume Design.
1976 (PG) 149m/C David Carradine, Ronny Cox, Melinda Dillon, Randy Quaid; *Dir:* Hal Ashby. Academy Awards '76: Best Adapted Score, Best Cinematography; National Board of Review Awards '76: Best Actor (Carradine); National Society of Film Critics Awards '76: Best Cinematography. **VHS, Beta, LV $59.95** *MGM* 🎬🎬🎬 ½

The Bounty A new version of "Mutiny on the Bounty," with emphasis on a more realistic relationship between Fletcher Christian and Captain Bligh—and a more sympathetic portrayal of the captain, too. The sensuality of Christian's relationship with a Tahitian beauty also receives greater importance.
1984 (PG) 130m/C Mel Gibson, Anthony Hopkins, Laurence Olivier, Edward Fox, Daniel Day Lewis, Bernard Hill, Philip Davis, Liam Neeson; *Dir:* Roger Donaldson. **VHS, Beta, LV $29.98** *VES, LIV* 🎬🎬🎬

Bounty Hunters A man hunts down his friend's killer.
1989 (R) 91m/C Robert Ginty, Bo Hopkins, Loeta Waterdown; *Dir:* Robert Ginty. **VHS, Beta $79.95** *AIP* 🎬🎬

The Bounty Man A bounty hunter brings an outlaw in, robbing him of his freedom and his girlfriend.

1972 74m/C Clint Walker, Richard Basehart, Margot Kidder, Arthur Hunnicutt, John Ericson, Gene Evans; *Dir:* John Llewellyn Moxey. **VHS, Beta $19.98** *FOX* 🎬🎬

Bouquet of Barbed Wire Three part mini-series which concerns incest and infidelity.
19?? 330m/C Frank Finlay, Sheila Allen. **VHS $79.95** *SVS* 🎬

Bowery Blitzkrieg This lesser "East Side Kids" entry has Gorcey opting to enter the boxing ring rather than turn to crime. Needs more Huntz.
1941 62m/B Leo Gorcey, Huntz Hall, Bobby Jordan, Warren Hull, Charlotte Henry, Keye Luke; *Dir:* Wallace Fox. **VHS $19.95** *NOS, VEC, TIM* 🎬

Bowery Buckaroos The Bowery Boys take their act West in search of gold, meeting up with the usual amounts of goofy characters and hilarious misunderstandings.
1947 66m/B Leo Gorcey, Huntz Hall, Bobby Jordan, Gabriel Dell, Billy Benedict, David Gorcey, Julie Briggs, Bernard Gorcey, Chief Yowlachie, Iron Eyes Cody; *Dir:* William Beaudine. **VHS $14.98** *WAR* 🎬🎬

Bowery at Midnight Lugosi plays a man who recycles criminals into zombies that commit crimes for his benefit. Bela and an intense mood can't quite save this pic.
1942 60m/B Bela Lugosi, Tom Neal, Dave O'Brien, Wanda McKay; *Dir:* Wallace Fox. **VHS, Beta $19.95** *NOS, MRV, SNC* 🎬

Bowery at Midnight/Dick Tracy vs. Cueball On one tape, the vintage Lugosi grade-C thriller about zombies in the Bowery, and a standard Dick Tracy entry.
1942 125m/B Bela Lugosi, Morgan Conway, Dick Wessel, Skelton Knaggs. **VHS, Beta** *AOV*

Boxcar Bertha Scorsese's vivid portrayal of the South during the 1930s' Depression has Hershey portraying a woman who winds up in cahoots with an anti-establishment train robber. Based on the book "Sister of the Road" by Boxcar Bertha Thomson.
1972 (R) 90m/C Barbara Hershey, David Carradine, John Carradine; *Dir:* Martin Scorsese. **VHS, Beta, LV $69.95** *VES, LIV* 🎬🎬 ½

The Boxer and the Death A new-wave Czech film about a boxer in a Nazi concentration camp who interests the boxing-obsessed commandant in a match. In Czech with subtitles.
1963 120m/B Stefan Kvietik; *Dir:* Peter Solan. **VHS, Beta $24.95** *NOS, FCT, DVT* 🎬🎬🎬

Boy in Blue A "Rocky"-esque biography of Canadian speed-rower Ned Hanlan, who set aside lackluster pursuits to turn to rowing.
1986 (R) 97m/C **CA** Nicolas Cage, Christopher Plummer, David Naughton; *Dir:* Charles Jarrott. **VHS, Beta $79.98** *FOX* 🎬 ½

Boy, Did I Get a Wrong Number! A real estate agent gets more than he bargained for when he accidentally dials a wrong number. Zany comedy persists as Hope gets entangled in the life of sexy starlet Sommer.
1966 100m/C Bob Hope, Phyllis Diller, Marjorie Lord, Elke Sommer, Cesare Danova; *Dir:* George Marshall. **VHS, Beta $14.95** *WKV* 🎬

The Boy Friend Russell pays tribute to the Busby Berkeley Hollywood musical. Lots of charming dance numbers and clever parody of plot lines in this adaptation of Sandy Wilson's stage play. Fun!

1971 (G) 135m/C *GB* Twiggy, Christopher Gable, Moyra Fraser, Max Adrian, Vladek Sheybal, Georgina Hale, Tommy Tune; *Dir:* Ken Russell. **VHS $19.95** *MGM* ✂✂✂

The Boy God Rocco here describes his trials and everyday shenanigans; he the young demi-God, battling evil on a fantasy land.
1986 100m/C Nino Muhlach; *Dir:* Erastheo J. Navda. **VHS, Beta $39.95** *VCD* ✂

Boy with the Green Hair When he hears that his parents were killed in an air raid, a boy's hair turns green. The narrow-minded members of his community suddenly want nothing to do with him and he becomes an outcast. Thought-provoking social commentary.
1948 82m/C Pat O'Brien, Robert Ryan, Barbara Hale, Dean Stockwell; *Dir:* Joseph Losey. **VHS, Beta, LV $29.95** *KOV, IME* ✂✂✂

A Boy and His Dog In the post-holocaust world of 2024, a young man (Johnson) and his talking canine cohort search for food and sex. They happen upon a community that drafts Johnson to repopulate their largely impotent race; Johnson is at first ready, willing, and able, until he discovers the mechanical methods they mean to employ. Based on a short story by Harlan Ellison. The dog was played by the late Tiger of "The Brady Bunch."
1975 (R) 87m/C Don Johnson, Suzanne Benton, Jason Robards Jr., Charles McGraw, Alvy Moore; *Dir:* L.Q. Jones. **VHS, Beta, LV $9.99** *CCB, MED, MRV* ✂✂½

A Boy Named Charlie Brown Charlie Brown enters the National Spelling Bee in New York and makes the final rounds with one other contestant. Based on Charles Schultz's popular comic strip characters from "Peanuts." Music by the Vince Guaraldi trio is, as always, a treat.
1969 86m/C *Dir:* Bill Melendez. **VHS, Beta, LV $14.98** *FOX* ✂✂✂

Boy in the Plastic Bubble Well-made, sensitive television drama about a young man born with immunity deficiencies who must grow up in a specially controlled plastic environment. Travolta is endearing as the boy in the bubble.
1976 100m/C John Travolta, Robert Reed, Glynnis O'Connor, Diana Hyland, Ralph Bellamy, Anne Ramsey; *Dir:* Randal Kleiser. **VHS, Beta $59.95** *PSM* ✂✂½

Boy Takes Girl A little girl finds it hard adjusting to life on a farming cooperative during a summer recess.
1983 93m/C Gabi Eldor, Hillel Neeman, Dina Limon; *Dir:* Michal Bat-Adam. **VHS, Beta $59.95** *MGM* ✂½

Boy of Two Worlds Because he is of a lineage foreign to his late father's town, a boy is exiled to the life of a junior Robinson Crusoe.
1970 (G) 103m/C Jimmy Sternman, Edvin Adolphson; *Dir:* Astrid Henning Jensen. **VHS, Beta $29.95** *GEM* ✂

Boy! What a Girl Moore (in drag) is mistaken for a woman, and is fought over by a couple of suitors.
1945 60m/B Tim Moore, Duke Williams, Sheila Guise, Beth Mays, Elwood Smith; *Dir:* Arthur Leonard. **VHS, Beta $57.95** *NOS, GVV, DVT* ✂

Boy Who Could Fly After a plane crash kills his parents, a boy withdraws into a fantasy land where he can fly. The young daughter of a troubled family makes friends with him and the fantasy becomes real. A

sweet film for children, charming though melancholy for adults, too. Fine cast, including Savage, Dewhurst and Bedelia keep this from becoming sappy.
1986 (PG) 120m/C Lucy Deakins, Jay Underwood, Bonnie Bedelia, Colleen Dewhurst, Fred Savage, Fred Gwynne, Louise Fletcher; *Dir:* Nick Castle. **VHS, Beta, LV $19.98** *LHV, WAR* ✂✂✂

The Boy Who Left Home to Find Out About the Shivers From "Faerie Tale Theatre" comes this tale about a boy who learns about fear.
1981 60m/C Peter MacNicol, Christopher Lee, Vincent Price; *Dir:* Graeme Clifford. **VHS, Beta, LV $14.98** *KUI, FOX* ✂✂✂

The Boy Who Loved Trolls Paul's fantastic dreams come true when he meets Ofeoti, a real, live troll. The only problem is Ofeoti only has a day to live, and Paul must find a way to save him. Aired on PBS as part of the WonderWorks Family Movie television series.
1984 58m/C Sam Waterston, Susan Anton, Matt Dill; *Dir:* Harvey Laidman. **VHS $29.95** *PME, FCT* ✂✂✂

Boyfriends & Girlfriends Another one of Rohmer's "Comedies and Proverbs," in which two girls with boyfriends fade in and out of interest with each, casually reshuffling their relationships. Typical, endless-talks-at-cafes Rohmer, the most happily consistent of the aging French New Wave. In French with subtitles.
1988 (PG) 102m/C *FR* Emmanuelle Chaulet, Sophie Renoir, Eric Viellard, Francois-Eric Gendron; *Dir:* Eric Rohmer. **VHS, Beta, LV $79.98** *ORI, TAM* ✂✂½

Boys in the Band A group of gay friends get together one night for a birthday party. A simple premise with a compelling depiction of friendship, expectations, and lifestyle. One of the first serious cinematic presentations to deal with the subject of homosexuality. The film was adapted from Mart Crowley's play using the original cast.
1970 (R) 120m/C Frederick Combs, Cliff Gorman, Laurence Luckinbill, Kenneth Nelson, Leonard Frey; *Dir:* William Friedkin. **VHS, Beta, LV $59.98** *FOX* ✂✂½

The Boys from Brazil Based on Ira Levin's novel, a thriller about Dr. Josef Mengeles endeavoring to reconstitute the Nazi movement from his Brazilian sanctuary by cloning a brood of boys from Hitler's genes. Olivier earned an Oscar nomination.
1978 (R) 123m/C Gregory Peck, James Mason, Laurence Olivier, Uta Hagen, Steve Guttenberg, Denholm Elliott, Lilli Palmer; *Dir:* Franklin J. Schaffner. National Board of Review Awards '78: Best Actor (Laurence Olivier). **VHS, Beta, LV $59.98** *FOX* ✂✂½

Boys of the City The East Side Kids solve a murder in an eerie mansion. Also know as, "The Ghost Creeps."
1940 63m/B Leo Gorcey, Bobby Jordan; *Dir:* Joseph H. Lewis. **VHS, Beta $24.95** *NOS, SNC, DVT* Woof!

The Boys in Company C A frank, hard-hitting drama about five naive young men involved in the Vietnam War.
1977 (R) 127m/C Stan Shaw, Andrew Stevens, James Canning, Michael Lembeck, Craig Wasson; *Dir:* Sidney J. Furie. **VHS, Beta $59.95** *COL* ✂✂

The Boys Next Door Two California lads kill and go nuts during a weekend in Los Angeles. Sheen overacts as a budding psychotic, not an easy thing to do. Apparently is

trying to show what a particular lifestyle can lead to. Violent.
1985 (R) 90m/C Maxwell Caulfield, Charlie Sheen, Christopher McDonald, Hank Garrett, Patti D'Arbanville, Moon Zappa; *Dir:* Penelope Spheeris. **VHS, Beta, LV $19.95** *STE, NWV* ✂✂½

Boys Night Out Kinkiness is the main theme of this video as whipped cream, wrestling, and handcuffs play major roles.
1988 60m/C Teri Peake. **VHS, Beta, LV $39.95** *CEL* Woof!

Boy's Reformatory Brothers are framed. One takes the rap, goes to the title institution, breaks out, and hunts the real bad guys. Creaky and antiquated.
1939 62m/B Frankie Darro, Grant Withers, David Durand, Warren McCollum; *Dir:* Howard Bretherton. **VHS $15.95** *NOS, LOO* ✂½

The Boys from Termite Terrace The animators from Warner Brothers who created Bugs Bunny, Daffy Duck, and Porky Pig are interviewed in this series of two untitled programs.
1975 30m/C VHS, Beta *CMR, NYS* ✂½

Boys Town A righteous portrayal of Father Flanagan and the creation of Boys Town, home for juvenile soon-to-be-ex-delinquents. Oscar nominated for Best Picture, Best Screenplay and Best Director.
1938 93m/B Spencer Tracy, Mickey Rooney, Henry Hull, Gene Reynolds, Sidney Miller, Frankie Thomas; *Dir:* Norman Taurog. Academy Awards '38: Best Actor (Tracy), Best Original Screenplay. **VHS, Beta, LV $19.98** *MGM, BTV* ✂✂✂½

Boyz N the Hood Singleton's debut as a writer and director is an astonishing picture of young black men, four high school students with different backgrounds, aims and abilities trying to survive Los Angeles gangs and bigotry. Excellent acting throughout, with special nods to Fishburne and Gooding, Jr. Violent outbreaks outside theaters where this ran only proves the urgency of its passionately nonviolent, pro-family message. Hopefully those viewers scared off at the time will give this a chance in the safety of their VCRs. Singleton appears on the cassette on behalf of the United Negro College Fund; he received Oscar nominations for Best Original Screenplay and Best Director, making him the youngest director ever so honored. The Criterion Collection laserdisc version includes two extra scenes and an interview with Singleton.
1991 (R) 112m/C Ice Cube, Cuba Gooding Jr., Morris Chestnut, Larry Fishburne; *W/Dir:* John Singleton. Chicago Film Critics Awards '91: Most Promising Actor (Ice Cube); National Board of Review Awards '91: 10 Best Films of the Year; MTV Movie Awards '92: Best New Filmmaker Award (Singleton). **VHS, Beta, LV, 8mm** *COL, CRC* ✂✂✂✂

Braddock: Missing in Action 3 The battle-scarred, high-kicking 'Nam vet battles his way into the jungles once more, this time to rescue his long-lost Vietnamese family. A family effort, Chuck co-wrote the script; his brother directed.
1988 (R) 104m/C Chuck Norris, Aki Aleong, Roland Harrah III; *Dir:* Aaron Norris. **VHS, Beta, LV $9.98** *MED, CCB* ✂

Brady's Escape An American World War II pilot is shot down in the Hungarian countryside and befriended by Hungarian csikos (cowboys) in this made-for-cable television movie.
1984 92m/C *HU* John Savage, Kelly Reno; *Dir:* Pal Gabor. **VHS, Beta, LV $9.95** *CAF* ✂½

B

The Brain A scientist learns that a mind isn't always a terrible thing to waste when he finds himself being manipulated by the brain of a dead man he's trying to keep alive. Remake of "Donovan's Brain" (1953).
1962 83m/B *GB GE* Anne Heywood, Peter Van Eyck, Bernard Lee, Cecil Parker, Jack MacGowran; *Dir:* Freddie Francis. **VHS, Beta, LV $19.95** *NOS, SNC, MON* ♪♪½

The Brain A good international cast holds together this comedy about a multi-million-dollar train heist being planned by Niven.
1969 (G) 100m/C *FR* David Niven, Jean-Paul Belmondo, Bourvil, Eli Wallach, Silvia Monti, Fernand Valois; *Dir:* Gerard Oury. **VHS** *PAR* ♪♪

The Brain Dr. Blake, host of a popular T.V. talk-show, is in league with a power-hungry alien brain. Viewers of his show kill themselves and others in the midst of many special effects.
1988 (R) 94m/C Tom Breznahan, Cyndy Preston, David Gale; *Dir:* Edward Hunt. **VHS, Beta $14.95** *IVE* ♪½

Brain of Blood Also known as "The Creature's Revenge." Deals with a scientist who transplants the brain of a politician into the body of a deformed idiot. The change is minimal.
1971 107m/C Kent Taylor, John Bloom, Regina Carrol, Angelo Rossitto, Grant Williams, Reed Hadley, Vicki Volante, Zandor Vorkov; *Dir:* Al Adamson. **VHS, Beta $49.95** *MAG, REG Woof!*

Brain Damage A tongue-in-bloody-cheek farce about a brain-sucking parasite. The parasite in question, Aylmer, addicts our dubious hero to the euphoria induced by the blue liquid the parasite injects into his brain, paving the way for the bloody mayhem that follows. Poor shadow of Henenlotter's far-superior "Basket Case"—in fact, it even includes an inside-joke cameo by Kevin van Hentenryck, reprising his "Basket Case" character; look for him on the subway.
1988 (R) 89m/C Rick Herbst, Gordon MacDonald, Jennifer Lowry; *Dir:* Frank Henenlotter. **VHS, Beta $14.95** *PAR* ♪♪

Brain Dead Low-budget but brilliantly assembled puzzle-film about a brain surgeon who agrees to perform experimental surgery on a psychotic to retrieve some corporately valuable data—his first mistake, which begins a seemingly endless cycle of nightmares and identity alterations. A mind-blowing sci-fi feast from ex-"Twilight Zone" writer Charles Beaumont.
1989 (R) 85m/C Bill Pullman, Bill Paxton, Bud Cort, Patricia Charbonneau, Nicholas Pryor, George Kennedy; *Dir:* Adam Simon. **VHS, Beta $79.98** *MGM* ♪♪♪½

Brain Donors Goofy, uneven film starring Turturro as a sleazy lawyer trying to take over the Oglethorpe Ballet Company by sweet-talking its aged patronness. He is helped by two ecentric friends, and together the three crack a lot of bad but witty jokes. The film culminates into a hilarious ballet scene featuring someone giving CPR to the ballerina playing the dying swan, an actor in a duck suit, duck hunters, and a pack of hounds. The trio may remind you of second-rate Marx Brothers (or the Three Stooges) but the movie has enough funny moments to be worth a watch.
1992 (PG) 79m/C John Turturro, Bob Nelson, Mel Smith, Nancy Marchand, John Savident, George De La Pena, Juli Donald, Spike Alexander; *Dir:* Dennis Dugan. **VHS, Beta** *PAR* ♪♪½

The Brain Eaters A strange ship from inside the Earth invades a small town, and hairy monsters promptly attach themselves to people's necks in a daring bid to control the planet. The imaginative story compensates somewhat for the cheap special effects. Watch for Nimoy before he grew pointed ears.
1958 60m/B Edwin Nelson, Alan Frost, Jack Hill, Joanna Lee, Jody Fair, Leonard Nimoy; *Dir:* Bruno Ve Sota. **VHS $14.95** *NOS, CNM* ♪♪

The Brain Machine A device that probes thoughts menaces a woman scientist.
1972 90m/C James Best, Barbara Burgess, Gil Peterson, Gerald McRaney, Anne Latham; *Dir:* Joy Houck Jr. **VHS, Beta** *PGN* ♪ -

The Brain from Planet Arous Lassie meets Alien when an evil alien brain appropriates the body of a scientist in order to take over planet Earth. His plans are thwarted, however, by a good alien brain that likewise inhabits the body of the scientists' dog. High camp and misdemeanors.
1957 80m/B John Agar, Joyce Meadows, Robert Fuller, Henry Travis; *Dir:* Nathan Juran. **VHS, Beta $19.95** *RHI, SNC, CNM* ♪

The Brain That Wouldn't Die Love is a many-splattered thing when a brilliant surgeon keeps the decapitated head of his fiancee alive after an auto accident while he searches for a suitable body onto which to transplant the head. A member of the trash film genre; much of the gore was slashed for the video, however.
1963 92m/B Herb Evers, Virginia Leith, Adele Lamont; *Dir:* Joseph Green. **VHS, Beta $19.98** *SNC, MRV, FCT* ♪

Braingames Test your knowledge of history, art and music with this collection of games designed to challenge the mind.
1985 60m/C VHS, Beta $12.95 *HBO* ♪

The Brainiac A sorcerer sentenced for black magic returns to strike dark deeds upon the descendants of those who judged him. He turns himself into a hideous monster, feeding on his victims' brains and blood.
1961 75m/B *MX* Abel Salazar, Ariadne Welter, Mauricio Garces, Rosa Maria Gallardo. **VHS, Beta $54.95** *SNC, MRV, HHT* ♪♪½

Brainstorm A husband-and-wife team of scientists invents headphones that can record dreams, thoughts, and fantasies (VCR-style) and then allow other people to experience them by playing back the tape. Their marriage begins to crumble as the husband becomes more and more obsessed with pushing the limits on the technology; things get worse when the government wants to exploit their discovery. Special effects and interesting camerawork punctuate this sci-fi flick. Wood's last film; in fact, she died before production was completed.
1983 (PG) 106m/C Natalie Wood, Christopher Walken, Cliff Robertson, Louise Fletcher; *Dir:* Douglas Trumbull. **VHS, Beta, LV $79.95** *MGM* ♪♪½

Brainwash A dramatic commentary on the capitalist system and the lengths to which people will go to acquire wealth and power. Also released with the title "Circle of Power."
1982 (R) 98m/C *GB* Yvette Mimieux, Christopher Allport, John Considine, Cindy Pickett; *Dir:* Bobby Roth. **VHS, Beta $59.95** *MED* ♪½

Brainwaves A young woman has disturbing flashbacks after her brain is electrically revived following a car accident. Curtis is the demented doctor who jump starts her.
1982 (R) 83m/C Suzanna Love, Tony Curtis, Keir Dullea, Vera Miles; *Dir:* Ulli Lommel. **VHS, Beta $24.98** *SUE* ♪♪

Brand of Hate A good cowboy rustles up a gang of rustlers.
1934 61m/B Bob Steele, George "Gabby" Hayes, Lucille Brown, James Flavin; *Dir:* Lewis Collins. **VHS, Beta $19.95** *NOS, WGE, RXM* ♪

Brand New Life A childless couple in their 40's are unexpectedly confronted by the wife's first pregnancy. Both have careers and well-ordered lives that promise to be disrupted.
1972 74m/C Cloris Leachman, Martin Balsam, Wilfrid Hyde-White, Mildred Dunnock; *Dir:* Sam O'Steen. **VHS, Beta** *LCA* ♪

Brand of the Outlaws A cowboy becomes an unwitting aid to rustlers in this early western.
1936 56m/B Bob Steele, Margaret Marquis, Virginia True Boardman, Jack Rockwell; *Dir:* Robert N. Bradbury. **VHS, Beta $19.95** *NOS, DVT, WGE* ♪

Branded Ladd (pre-"Shane") impersonates the long gone son of rich rancher Bickford, with unusual results. Nicely balanced action and love scenes makes this old Max Brand story a better-than-average western. Filmed in Technicolor.
1950 95m/C Alan Ladd, Mona Freeman, Charles Bickford, Joseph Calleia, Milburn Stone; *Dir:* Rudolph Mate. **VHS, LV $14.95** *PAR* ♪♪½

Branded a Coward A man whose parents were killed, seeks revenge via his trusty six-shooters.
1935 56m/B Johnny Mack Brown, Billie Seward, Yakima Canutt; *Dir:* Sam Newfield. **VHS, Beta $19.95** *NOS, DVT, WGE* ♪

Brandy Sheriff A former outlaw tries to make up for the time he lost in prison. In Spanish.
1978 90m/C *SP* Alex Nicol, Maria Contrera, Antonio Casas. **VHS, Beta $59.95** *HHT* ♪½

Brannigan The Duke stars in the somewhat unfamiliar role of a rough and tumble Chicago police officer. He travels across the Atlantic to arrest a racketeer who has fled the States rather than face a grand jury indictment. Humor and action abound.
1975 (PG) 111m/C John Wayne, John Vernon, Mel Ferrer, Daniel Pilon, James Booth, Ralph Meeker, Lesley-Anne Down, Richard Attenborough; *Dir:* Douglas Hickox. **VHS, Beta $19.95** *FOX, BUR* ♪♪½

Brass Made for TV. First post-Archie Bunker role for O'Connor, who plods through this average police drama at a snail's pace, dragging everything else down with him. Pilot for a series that never aired.
1985 (PG) 94m/C Carroll O'Connor, Lois Nettleton, Jimmie Baird, Paul Shenar, Vincent Gardenia, Anita Gillette; *Dir:* Corey Allen. **VHS, LV $19.98** *ORI* ♪½

Brass Target A hypothetical thriller about a plot to kill George Patton in the closing days of World War II for the sake of $250 million in Nazi gold.
1978 (PG) 111m/C Sophia Loren, George Kennedy, Max von Sydow, John Cassavetes, Patrick McGoohan, Robert Vaughn, Bruce Davison, Edward Herrmann, Ed Bishop; *Dir:* John Hough. **VHS, Beta, LV $59.95** *MGM* ♪♪½

The Bravados A rough, distressing Western revenge tale, wherein a man is driven to find the four men who murdered his wife. In tracking the perpetrators, Peck realizes that he has been corrupted by his vengeance.
1958 98m/C Gregory Peck, Stephen Boyd, Joan Collins, Albert Salmi, Henry Silva, Lee Van Cleef, George Voskovec, Barry Coe; *Dir:* Henry King. National Board of Review Awards '58: Best Supporting Actor (Salmi). **VHS, Beta $39.98** FOX ⅊⅊⅊

The Brave Bunch A courageous Greek soldier tries to save his men and country during World War II.
1970 110m/C GR Peter Funk, John Miller; *Dir:* Dacosta Carayan. **VHS, Beta $39.95** IVE ⅊½

The Brave One A love story between a Spanish boy and the bull who saves his life. The animal is later carted off to the bull ring. Award-winning screenplay by the then-blacklisted Dalton Trumbo, as "Robert Rich."
1956 100m/C Michael Ray, Rodolpho Moyos, Joi Lansing; *Dir:* Irving Rapper. Academy Awards '56: Best Original Screenplay. **VHS, Beta, LV $49.95** UHV ⅊⅊⅊

Brave Rifles Actual battle footage is used to tell the story of the Battle of the Bulge. Nominated for an Academy Award.
1966 51m/B *Dir:* Laurence E. Mascott; *Nar:* Arthur Kennedy. **VHS, Beta, LV** WGE ⅊½

Brazil The acclaimed nightmare comedy by Gilliam about an Everyman trying to survive in a surreal paper-choked bureaucratic society. There are copious references to "1984" and "The Trial," fantastic mergings of glorious fantasy and stark reality, and astounding visual design. Gilliam co-scripted with Tom Stoppard ("Rosencrantz and Guildenstern are Dead").
1986 (R) 131m/C GB Jonathan Pryce, Robert De Niro, Michael Palin, Katherine Helmond, Kim Greist, Bob Hoskins, Ian Holm, Peter Vaughan, Ian Richardson; *Dir:* Terry Gilliam. Los Angeles Film Critics Association Awards '85: Best Picture. **VHS, Beta, LV $19.95** MCA, FCT ⅊⅊⅊½

Bread (BBC) From the BBC, one of their most popular soap operas. Watch Billy, Joey, Roxy and Adrian go through the hilarious motions of their daily lives.
19?? 85m/C GB Peter Sellars. **VHS $29.95** BFS ⅊⅊⅊½

Break of Hearts Hepburn marries Boyer, who becomes an alcoholic, and then more troubles arise. Very soapy, but well-made.
1935 78m/B Katharine Hepburn, Charles Boyer, Jean Hersholt, John Beal, Sam Hardy; *Dir:* Phillip Moeller. **VHS, Beta, LV $19.98** TTC ⅊⅊½

Breakdown A boxer is set up by his girlfriend's father to take the rap for a murder. Lackluster effort.
1953 76m/B Ann Richards, William Bishop, Anne Gwynne, Sheldon Leonard, Wally Cassell, Richard Benedict; *Dir:* Edmond Angelo. **VHS, Beta $16.95** NOS, SNC ⅊

Breaker! Breaker! A convoy of angry truck drivers launch an assault on the corrupt and sadistic locals of a small Texas town.
1977 (PG) 86m/C Chuck Norris, George Murdock; *Dir:* Don Hulette. **VHS, Beta $9.98** SUE ⅊

Breaker Morant In 1901 South Africa, three Australian soldiers are put on trial for avenging the murder of several prisoners. Based on a true story which was then turned into a play by Kenneth Ross, this riveting, popular antiwar statement and courtroom drama heralded Australia's film renaissance.

Rich performances by Woodward and Waters.
1980 (PG) 107m/C AU Edward Woodward, Jack Thompson, John Waters, Bryan Brown; *Dir:* Bruce Beresford. Australian Film Institute '80: Best Actor (Thompson), Best Film. **VHS, Beta, LV, 8mm $9.99** CCB, COL ⅊⅊⅊½

The Breakfast Club Five students from different cliques at a Chicago suburban high school spend a day together in detention. Rather well done teenage culture study; these characters delve a little deeper than the standard adult view of adolescent stereotypes. One of John Hughes' best movies. Soundtrack features Simple Minds and Wang Chung.
1985 (R) 97m/C Ally Sheedy, Molly Ringwald, Judd Nelson, Emilio Estevez, Anthony Michael Hall; *Dir:* John Hughes. **VHS, Beta, LV $19.95** MCA, FCT ⅊⅊⅊

Breakfast in Hollywood A movie about the popular morning radio show of the 1940s hosted by Tom Breneman, a coast-to-coast coffee klatch.
1946 93m/B Tom Breneman, Bonita Granville, Eddie Ryan, Beulah Bondi, Billie Burke, ZaSu Pitts, Hedda Hopper, Spike Jones; *Dir:* Harold Schuster. **VHS, Beta $29.95** VYY ⅊½

Breakfast in Paris Two American professionals in Paris, both crushed from past failures in the love department, find each other.
1981 85m/C AU Rod Mullinar, Barbara Parkins, Jack Lenoir; *Dir:* John Lamond. **VHS, Beta $19.95** STE, NWV ⅊

Breakfast at Tiffany's Truman Capote's amusing story of an endearingly eccentric New York City playgirl and her shaky romance with a young writer. Hepburn lends Holly Golightly just the right combination of naivete and worldly wisdom with a dash of melancholy. A wonderfully offbeat romance with music by Henry Mancini.
1961 114m/C Audrey Hepburn, George Peppard, Patricia Neal, Buddy Ebsen, Mickey Rooney, Martin Balsam, John McGiver; *Dir:* Blake Edwards. Academy Awards '61: Best Original Score, Best Song ("Moon River"). **VHS, Beta, LV, 8mm $14.95** PAR, FCT ⅊⅊⅊½

Breakheart Pass A governor, his female companion, a band of cavalrymen, and a mysterious man travel on a train through the mountains of Idaho in 1870. The mystery man turns out to be a murderer. Based on a novel by Alistair MacLean.
1976 (PG) 92m/C Charles Bronson, Ben Johnson, Richard Crenna, Jill Ireland, Charles Durning, Ed Lauter, Archie Moore; *Dir:* Tom Gries. **VHS, Beta, LV $19.98** MGM, FOX ⅊⅊

Breakin' This program presents the movie about the dance phenomenon break dancing along with the hit songs that accompany the film.
1984 87m/C Lucinda Dickey, Adolfo "Shabba Doo" Quinones, Michael "Boogaloo Shrimp" Chambers, Ben Lokey, Christopher McDonald, Phineas Newborn III; *Dir:* Joel Silberg. **VHS, Beta, LV $79.95** MGM ⅊⅊

Breakin' 2: Electric Boogaloo Breakdancers hold a benefit concert to preserve an urban community center.
1984 (PG) 94m/C Lucinda Dickey, Adolfo "Shabba Doo" Quinones, Michael "Boogaloo Shrimp" Chambers, Susie Bono; *Dir:* Sam Firstenberg. **VHS, Beta, LV $79.95** MGM ⅊

Breakin' Through The choreographer of a troubled Broadway-bound musical decides to energize his shows with a troupe of street dancers.

1984 73m/C Ben Vereen, Donna McKechnie, Reid Shelton; *Dir:* Peter Medak. **VHS, Beta $69.95** DIS ⅊

Breaking All the Rules Comedy about two teenage lovers who find themselves embroiled in a slapdash jewel robbery at an amusement park during the last day of summer break.
1985 (R) 91m/C CA Carolyn Dunn, Carl Marotte, Thor Bishopric, Rachel Hayward; *Dir:* James Orr. **VHS, Beta $14.95** NWV, STE ⅊

Breaking Away A light-hearted coming-of-age drama about a high school graduate's addiction to bicycle racing, whose dreams are tested against the realities of a crucial race. An honest, open look at present Americana with tremendous insight into the minds of average youth; shot on location at Indiana University. Great bike-racing photography. Dennis Quaid, Barbara Barrie, and Dennis Christopher give exceptional performances. Basis for television series. Oscar nominations for Best Picture, Barrie, director Yates, and Original Score.
1979 (PG) 100m/C Dennis Christopher, Dennis Quaid, Daniel Stern, Jackie Earle Haley, Barbara Barrie, Paul Dooley, Amy Wright; *Dir:* Peter Yates. Academy Awards '79: Best Original Screenplay; British Academy Awards '79: Most Promising Actor (Christopher); Golden Globe Awards '80: Best Film—Musical/Comedy; National Board of Review Awards '79: 10 Best Films of the Year, Best Supporting Actor (Dooley); National Society of Film Critics Awards '79: Best Picture, Best Screenplay. **VHS, Beta, LV $59.98** FOX ⅊⅊⅊½

Breaking Glass A "New Wave" musical that gives an insight into the punk record business and at the same time tells of the rags-to-riches life of a punk rock star.
1980 (PG) 94m/C GB Hazel O'Connor, Phil Daniels, Jon Finch, Jonathan Pryce; *Dir:* Brian Gibson. **VHS, Beta $49.95** PAR ⅊½

Breaking Home Ties Norman Rockwell's illustration is the basis for this coming of age story set in a small Texas town. A young man is eager to be off to college and lessen his family ties but his mother's serious illness may prevent both. Good performances help overcome the sentimentality. Originally made-for-television.
1987 95m/C Jason Robards Jr., Eva Marie Saint, Doug McKeon, Claire Trevor, Erin Gray; *Dir:* John Wilder. **VHS $79.95** CAF ⅊⅊

Breaking the Ice An unlikely mixture of Mennonites and a big city ice skating show. Musical numbers abound with backing by Victor Young and his Orchestra.
1938 79m/B Bobby Breen, Charlie Ruggles, Dolores Costello, Billy Gilbert, Margaret Hamilton. **VHS, Beta $19.95** NOS, VYY, DVT ⅊

Breaking In A semi-acclaimed comedy about a professional thief who takes a young amateur under his wing and shows him the ropes. Under-estimated Reynolds is especially charming. Script by John ("Brother from Another Planet") Sayles is witty and innovative, with intelligent direction from Forsyth.
1989 (R) 95m/C Burt Reynolds, Casey Siemaszko, Sheila Kelley, Lorraine Toussaint, Albert Salmi, Harry Carey, Maury Chaykin, Stephen Tobolowsky, David Frisberg; *Dir:* Bill Forsyth. **VHS, Beta, LV $89.99** HBO ⅊⅊⅊

Breaking Loose A surfer is on the run from a vicious motorcycle gang that kidnapped his girlfriend, but turns to fight eventually, board in hand.
1990 (R) 88m/C Peter Phelps, Vince Martin, Abigail, David Ngcombujarra; *Dir:* Rod Hay. **VHS $14.95** HMD, SOU ⅊½

Breaking with Old Ideas A film made in the waning years of Mao's Cultural Revolution about the construction of a college of workers. In Chinese with subtitles.
1975 126m/C CH *Dir:* Li Wen-Hua. **VHS, Beta** $39.95 FCT 🐾🐾

Breaking Up A woman fights to rediscover her personal identity after her fifteen-year marriage crumbles. Made for television. Emmy nominated.
1978 90m/C Lee Remick, Granville Van Dusen, David Stambaugh; *Dir:* Delbert Mann. **VHS, Beta** TLF 🐾🐾½

Breaking Up Is Hard to Do Six men leave their wives and shack up together on Malibu for a summer of partying and introspection. Made for television; edited down from its original 201 minute length.
1979 96m/C Billy Crystal, Bonnie Franklin, Ted Bessell, Jeff Conaway, Tony Musante, Robert Conrad, Trish Stewart, David Ogden Stiers, George Gaynes; *Dir:* Lou Antonio. **VHS, Beta, LV** $9.99 VMK, STE 🐾

Breakout The wife of a man imprisoned in Mexico hires a Texas bush pilot to help her husband escape. Duvall is the prisoner and Bronson the pilot.
1975 (PG) 96m/C Charles Bronson, Jill Ireland, Robert Duvall, John Huston, Sheree North, Randy Quaid; *Dir:* Tom Gries. **VHS, Beta, LV** $64.95 COL 🐾🐾

Breakthrough German soldier, broken at the end of World War II, sets out to negotiate with the Allies. Average at best.
1978 (PG) 96m/C Richard Burton, Robert Mitchum, Rod Steiger, Michael Parks, Curt Jurgens; *Dir:* Andrew V. McLaglen. **VHS, Beta** $9.95 WOV 🐾🐾

Breath of Scandal An American diplomat in Vienna rescues a princess when she is thrown off a horse; he falls for her like a ton o' bricks. Viennese politics complicate things. Based on a play by Molnar.
1960 98m/C Sophia Loren, John Gavin, Maurice Chevalier, Angela Lansbury; *Dir:* Michael Curtiz. **VHS, Beta** $19.95 KRT 🐾🐾

Breathing Fire A Vietnamese teenager and his American brother find out their ex-GI father is behind an armed bank robbery and murder. They join together to protect the only eyewitness—a young girl—against the ruthless gang of criminals and their own father. Lots of kickboxing action for martial arts fans.
19?? (R) ?m/C Bolo Yeung, Ke Huy Quan, Jerry Trimble; *Dir:* Lou Kennedy. **VHS** $89.98 IMP 🐾

Breathless A car-thief-turned-cop-killer has a torrid love affair with a French student studying in Los Angeles as the police slowly close in. A glossy, smarmy remake of the Godard classic that concentrates on the thin plot rather than any attempt at revitalizing the film syntax.
1983 (R) 105m/C Richard Gere, Valerie Kaprisky, Art Metrano, John P. Ryan; *Dir:* Jim McBride. **VHS, Beta, LV** $29.98 VES 🐾½

Breathless (A Bout de Souffle) Godard's first feature catapulted him to the vanguard of French filmmakers. A carefree Parisian crook, who emulates Humphrey Bogart, falls in love with an American girl with tragic results. Wonderful scenes of Parisian life. Established Godard's Brechtian, experimental style. Belmondo's film debut. Mistitled "Breathless" for American release, the film's French title actually means "Out of Breath"; however, the fast-paced, erratic musical score leaves you breathless. French

with English subtitles. Remade with Richard Gere in 1983 with far less intensity.
1959 90m/B FR Jean-Paul Belmondo, Jean Seberg, Jean-Pierre Melville; *Dir:* Jean-Luc Godard. Berlin Film Festival '60: Best Director. **VHS, Beta** $24.95 NOS, MRV, CVC 🐾🐾🐾

A Breed Apart An on-going battle over a rare eagle's eggs strikes sparks between a reclusive conservationist and a scheming adventurer hired by a billionaire collector.
1984 (R) 95m/C Kathleen Turner, Rutger Hauer, Powers Boothe, Donald Pleasence, Brion James, John Dennis Johnston; *Dir:* Philippe Mora. **VHS, Beta** $19.95 HMD, HBO 🐾🐾

Breed of the Border Tough-fisted cowboys must take the law into their own hands.
1933 60m/B Bob Steele, Marion Byron, George "Gabby" Hayes, Ernie Adams; *Dir:* Robert N. Bradbury. **VHS, Beta** $19.95 NOS, DVT 🐾

Brewster McCloud Altman's first picture after M*A*S*H brings back much of the cast, combining fantasy, black comedy, and satire in telling the story of a young man whose head is in the clouds or at least is trying to inhabit the upper reaches of the Houston Astrodome. Brewster (Cort) lives covertly in the Dome and dreams of flying. Not only does he dream, he practices. While his frustrated girlfriend masturbates, he works on his wings. He also has a guardian angel (Kellerman) who watches over him and may actually be killing people who give him a hard time, after splattering the victims with bird crap. Murphy is a cop driving a Trans-Am who is obsessed with catching the killer. He is assisted by Schuck, the world's dumbest assistant. And there's a circus allegory as well. Hard to figure what it all means and offbeat as they come, but for certain tastes, exquisite.
1970 (R) 101m/C Bud Cort, Sally Kellerman, Shelley Duvall, Michael Murphy, William Windom, Rene Auberjonois, Stacy Keach, John Schuck, Margaret Hamilton; *Dir:* Robert Altman. **VHS, Beta** $69.95 MGM 🐾🐾🐾½

Brewster's Millions If Brewster, an ex-GI, can spend a million dollars in one year, he will inherit a substantially greater fortune. Originally a 1902 novel, this is the fifth of seven film adaptations.
1945 79m/B Dennis O'Keefe, June Havoc, Eddie Anderson, Helen Walker, Gail Patrick, Mischa Auer; *Dir:* Allan Dwan. **VHS, Beta** $19.95 MED 🐾🐾½

Brewster's Millions An aging minor league baseball player must spend 30 million dollars in order to collect an inheritance of 300 million dollars. He may find that money can't buy happiness. Seventh remake of the story.
1985 (PG) 101m/C Richard Pryor, John Candy, Lonette McKee, Stephen Collins, Jerry Orbach, Pat Hingle, Tovah Feldshuh, Hume Cronyn, Rick Moranis; *Dir:* Walter Hill. **VHS, Beta, LV** $19.95 MCA 🐾½

Brian's Song The story of the unique relationship between Gale Sayers, the Chicago Bears' star running back, and his teammate Brian Piccolo. The friendship between the Bears' first interracial roommates ended suddenly when Brian Piccolo lost his life to cancer. Made for television. Incredibly well-received in its time.
1971 73m/C James Caan, Billy Dee Williams, Jack Warden, Shelley Fabares, Judy Pace; *Dir:* Buzz Kulik. **VHS, Beta, LV** $64.95 COL 🐾🐾🐾🐾

The Bride Re-telling of "The Bride of Frankenstein." Sting's Dr. Frankenstein has much more success in creating his second monster (Beals). She's pretty and intelligent and he may even be falling in love with her. When she begins to gain too much independence, though, the awful truth about her origins comes out. Fans of the two leads will want to see this one, but for a true classic see Elsa Lanchester's bride.
1985 (PG-13) 118m/C Sting, Jennifer Beals, Anthony Higgins, David Rappaport, Geraldine Page, Clancy Brown, Phil Daniels, Veruschka; *Dir:* Franc Roddam. **VHS, Beta, LV** $9.95 FOX 🐾½

Bride & the Beast While on an African safari honeymoon, a big game hunter's new bride is carried off by a gorilla. Also goes by the name "Queen of the Gorillas."
1958 78m/B Charlotte Austin, Lance Fuller, William Justine; *Dir:* Adrian Weiss. **VHS, Beta** WGE Woof!

The Bride Came C.O.D. Rough'n'tough pilot Cagney is hired by a rich Texas oil man to prevent his daughter (the one with the Bette Davis eyes) from marrying cheesy bandleader Carson. The payoff: $10 per pound if she's delivered unwed. What a surprise when Cagney faces the timeworn dilemma of having to choose between love and money. A contemporary issue of "Time" magazine trumpeted: "Screen's most talented tough guy roughouses one of screen's best dramatic actresses." That tells you right there it's a romantic comedy. Cagney and Davis—a likely pairing, you'd think—are essentially fish out of water, bothered and bewildered by a weak script. Bette had already been Oscared for her Jezebel, and Cagney was on top of the world as America's favorite "Public Enemy" (although a decade earlier he lamented that he was sick of carrying guns and beating up women).
1941 92m/B Bette Davis, James Cagney, Stuart Erwin, Jack Carson, George Tobias, Eugene Pallette, William Frawley; *Dir:* William Keighley. **VHS** $19.95 MGM, FCT 🐾🐾🐾

The Bride of Frankenstein The classic sequel to the classic original in which Dr. F. seeks to build a mate for his monster. More humor than the first, but also more pathos, including the monster's famous but short-lived friendship with a blind hermit. Lanchester plays both the bride and Mary Shelley in the opening sequence.
1935 75m/B Boris Karloff, Elsa Lanchester, Ernest Thesiger, Colin Clive, Una O'Connor, Valerie Hobson, Dwight Frye, John Carradine, E.E. Clive, O.P. Heggie; *Dir:* James Whale. **VHS, Beta, LV** $14.95 MCA, MLB 🐾🐾🐾🐾

Bride of the Gorilla Burr travels to the jungle where he finds a wife, a plantation, and a curse in this African twist on the werewolf legend. Chaney is the local policeman on his trail. Burr's physical changes are fun to watch. Siodmak wrote the screenplay.
1951 76m/B Raymond Burr, Barbara Payton, Lon Chaney Jr., Tom Conway, Paul Cavanagh; *Dir:* Robert Siodmak. **VHS** $16.95 NOS, MRV, SNC 🐾🐾

The Bride Is Much Too Beautiful Bardot, playing a French farm girl, becomes a renowned model. Dubbed.
1958 90m/C FR Brigitte Bardot, Louis Jourdan, Micheline Presle, Marcel Amont; *Dir:* Fred Surin. **VHS, Beta** $29.95 NOS, FCT 🐾🐾½

Bride of the Monster Lugosi stars as a mad scientist trying to create a race of giants. Classic Woodian badness.

1956 70m/B Bela Lugosi, Tor Johnson, Loretta King, Tony McCoy; *Dir:* Edward D. Wood Jr. **VHS, Beta** $19.95 *NOS, SNC, VYY* 🎬

Bride of Re-Animator Herbert West is back, and this time he not only re-animates life but creates life - sexy female life - in this sequel to the immensely popular "Re-Animator." High camp and blood curdling gore make this a standout in the sequel parade. Available in a R-rated version as well.
1989 99m/C Bruce Abbott, Claude Earl Jones, Fabiana Udenio, Jeffrey Combs, Kathleen Kinmont, David Gale; *Dir:* Brian Yuzna. **VHS, Beta, LV** $89.95 *LIV* 🎬🎬½

Bride Walks Out Newlywed crisis: a woman with rich taste learns how to adjust to living on her husband's poor salary, but not before she samples the life of the wealthy. Interesting in a sociological sort of way.
1936 81m/B Barbara Stanwyck, Gene Raymond, Robert Young, Ned Sparks, Willie Best, Helen Broderick, Hattie McDaniel; *Dir:* Leigh Jason. **VHS, Beta** $19.98 *RKO* 🎬🎬

The Bride Wore Red Crawford stars as a cabaret singer who masquerades as a mysterious socialite in an attempt to break into the upper crust. When a wealthy aristocrat invites her to spend two weeks at a posh resort in Tyrol, Crawford plays her part to the hilt, managing to charm both a rich gentleman and the village postman.
1937 103m/B Joan Crawford, Franchot Tone, Robert Young, Billie Burke, Reginald Owen, Lynne Carver, George Zucco; *Dir:* Dorothy Arzner. **VHS** $19.98 *MGM* 🎬🎬

Brides of the Beast Filipino radiation monsters get their jollies by eating beautiful young women. A newly-arrived research scientist and his bride oppose this custom. Originally titled "Brides of Blood," also known as "Grave Desires" and "Island of the Living Horror."
1968 85m/C *PH* John Ashley, Kent Taylor, Beverly Hills, Eva Darren, Mario Montenegro; *Dir:* Eddie Romero, Gerardo (Gerry) De Leon. **VHS, Beta** $49.95 *REG Woof!*

The Brides of Dracula A young French woman unknowingly frees a vampire. He wreaks havoc, sucking blood and creating more of the undead to carry out his evil deeds. One of the better Hammer vampire films.
1960 86m/C *GB* Peter Cushing, Martita Hunt, Yvonne Monlaur, Freda Jackson, David Peel, Mona Washbourne; *Dir:* Terence Fisher. **VHS** $14.98 *MCA* 🎬🎬½

The Brides Wore Blood Four prospective brides are mysteriously murdered, but one is brought back to life and becomes a vampire's mate.
1984 86m/C **VHS, Beta** $49.95 *REG Woof!*

Brideshead Revisited The acclaimed British mini-series based on the Evelyn Waugh classic about an Edwardian young man who falls under the spell of a wealthy aristocratic family and struggles to retain his integrity and values. On six tapes.
1981 540m/C *GB* Jeremy Irons, Anthony Andrews, Diana Quick, Laurence Olivier, John Gielgud, Claire Bloom, Stephane Audran, Mona Washbourne, John Le Mesurier; *Dir:* Charles Sturridge, Michael Lindsey Hogg. **VHS, Beta** $29.95 *VIR*

Bridge to Hell A group of allied P.O.W.s try to make their way to the American front during World War II in Yugoslavia. A heavily guarded bridge occupied by Nazi troops stands between them and freedom. A special

introduction by Michael Dudikoff doing martial arts.
1987 94m/C Jeff Connors, Francis Ferre, Andy Forrest, Paky Valente; *Dir:* Umberto Lenzi. **VHS** $59.95 *CAN* 🎬½

The Bridge to Nowhere Five city kids go hunting and back-packing in the New Zealand wilderness, and are hunted by a maniacal backwoodsman.
1986 82m/C *NZ* Bruno Lawrence, Alison Routledge, Margaret Umbers, Philip Gordon; *Dir:* Ian Mune. **VHS, Beta** $19.98 *CHA* 🎬🎬

The Bridge at Remagen Based on the true story of allied attempts to capture a vital bridge before retreating German troops destroy it. For war-film buffs.
1969 (PG) 115m/C George Segal, Robert Vaughn, Ben Gazzara, Bradford Dillman, E.G. Marshall; *Dir:* John Guillermin. **VHS, Beta** $14.95 *WKV* 🎬🎬½

The Bridge on the River Kwai The award-winning adaptation of the Pierre Bouelle novel about the battle of wills between a Japanese POW camp commander and a British colonel over the construction of a rail bridge, and the parallel efforts by escaped prisoner Holden to destroy it. Holden's role was originally cast for Cary Grant. Memorable too for whistling "Colonel Bogey March." Received 8 Oscar nominations, winning all but supporting actor for Hayakawa.
1957 161m/C *GB* William Holden, Alec Guinness, Jack Hawkins, Sessue Hayakawa, James Donald; *Dir:* David Lean. Academy Awards '57: Best Actor (Guinness), Best Adapted Screenplay, Best Color Cinematography, Best Director (Lean), Best Film Editing, Best Picture, Best Original Score; British Academy Awards '57: Best Actor (Guinness), Best Film, Best Screenplay; Directors Guild of America Awards '57: Best Director (Lean); Golden Globe Awards '58: Best Film—Drama; National Board of Review Awards '57: Best Actor (Guinness), Best Director (Lean), Best Film, Best Supporting Actor (Hayakawa). **VHS, Beta, LV** $19.95 *COL, BTV* 🎬🎬🎬🎬

Bridge of San Luis Rey A priest investigates the famous bridge collapse in Lima Peru that left five people dead. Based upon the novel by Thorton Wilder.
1944 89m/B Lynn Bari, Francis Lederer, Louis Calhern, Akim Tamiroff, Donald Woods, Alla Nazimova, Blanche Yurka; *Dir:* Rowland V. Lee. **VHS, Beta, LV** $19.95 *STE, NWV, IME* 🎬🎬🎬

Bridge to Silence A young hearing-impaired mother's life begins to crumble following the death of her husband in a car crash. Her mother tries to get custody of her daughter and a friend applies romantic pressure. Made for TV melodrama features Matlin in her first television speaking role.
1989 95m/C Marlee Matlin, Lee Remick, Josef Sommer, Michael O'Keefe, Allison Silva, Candice Brecker; *Dir:* Karen Arthur. **VHS** $59.95 *FRH* 🎬🎬

Bridge to Terabithia Annette O'Toole and a young boy learn the joys of friendship in a magic kingdom called Terabithia. Originally aired by PBS for the WonderWorks television series.
1985 58m/C Annette O'Toole, Juliah Coutts, Julie Beaulieu; *Dir:* Eric Till. **VHS** $29.95 *PME* 🎬🎬

A Bridge Too Far A meticulous recreation of one of the most disastrous battles of World War II, the Allied defeat at Arnhem in 1944. Misinformation, adverse conditions, and overconfidence combined to prevent the Allies from capturing six bridges that connected Holland to the German border.

1977 (PG) 175m/C *GB* Sean Connery, Robert Redford, James Caan, Michael Caine, Elliott Gould, Gene Hackman, Laurence Olivier, Ryan O'Neal, Liv Ullman, Dirk Bogarde, Hardy Kruger, Arthur Hill, Edward Fox, Anthony Hopkins; *Dir:* Richard Attenborough. **VHS, Beta, LV** $29.98 *FOX* 🎬🎬

The Bridges at Toko-Ri Based on the James A. Michener novel, the rousing war-epic about a lawyer being summoned by the Navy to fly bombing missions during the Korean War. A powerful anti-war statement. Screenplay by Valentine Davis.
1955 103m/C William Holden, Grace Kelly, Fredric March, Mickey Rooney, Robert Strauss, Earl Holliman, Keiko Awaji, Charles McGraw; *Dir:* Mark Robson. Academy Awards '55: Best Special Effects. **VHS, Beta, LV** $14.95 *PAR* 🎬🎬🎬½

Brief Encounter Based on Noel Coward's "Still Life" from "Tonight at 8:30," two middle-aged, middle-class people become involved in a short and bittersweet romance in World War II England. Intensely romantic, underscored with Rachmaninoff's Second Piano Concerto. Oscar nominations for Best Actress (Johnson), Best Director (Lean), and Best Screenplay.
1946 86m/B *GB* Celia Johnson, Trevor Howard, Stanley Holloway; *Dir:* David Lean. National Board of Review Awards '46: 10 Best Films of the Year; New York Film Critics Awards '46: Best Actress (Johnson); Cinematheque Belgique '52: #9 of the Best Films of All Time (tie); Sight & Sound Survey '52: #10 of the Best Films of All Time. **VHS, Beta** $19.95 *PAR, TAM* 🎬🎬🎬🎬

The Brig A film by Jonas Mekas documenting the Living Theatre's infamous performance of Kenneth H. Brown's experimental play. Designed by Julian Beck.
1964 65m/B Adolfas Mekas, Jim Anderson, Warren Finnerty, Henry Howard, Tom Lillard, James Tiroff, Gene Lipton; *Dir:* John Mekas. Venice Film Festival '64: Best Documentary Feature. **VHS, Beta** $29.95 *MFV, FCT* 🎬🎬🎬

Brigadoon The story of a magical, 18th century Scottish village which awakens once every 100 years and the two modern-day vacationers who stumble upon it. Highlights of Lerner and Loewe's score. Songs include: "Heather on the Hill," "Almost Like Being in Love," "I'll Go Home with Bonnie Jean," and "Wedding Dance."
1954 108m/C Gene Kelly, Van Johnson, Cyd Charisse; *Dir:* Vincente Minnelli. **VHS, Beta, LV, 8mm** $19.95 *MGM, TLF* 🎬🎬🎬

Bright Angel A young woman attempts to keep her brother out of jail by paying off the man who would testify against him. Her plans go haywire however when the informant turns up dead and the money is stolen. She hits the road with a young boxer in a race against death. Can she clear her name, and free her brother, before it's too late?
1991 (R) 94m/C Dermot Mulroney, Lili Taylor, Valerie Perrine, Bill Pullman, Mary Kay Place, Burt Young, Sam Shepard; *Dir:* Michael Fields. **VHS** $92.99 *HBO* 🎬🎬½

Bright Eyes Shirley Temple stars as an adorable orphan caught between foster parents.
1934 84m/B Shirley Temple, James Dunn, Lois Wilson, Jane Withers, Judith Allen; *Dir:* David Butler. **VHS, Beta** $19.98 *FOX* 🎬🎬

Bright Lights, Big City Based on Jay McInerney's popular novel, Fox plays a contemporary yuppie working in Manhattan as a magazine journalist. As his world begins to fall apart, he embarks on an endless cycle of drugs and nightlife. Fox is not well cast, and his character is hard to care for as he becomes more and more dissolute. Although McInerney wrote his own screenplay, the intellectual abstractness of the novel can't be captured on film. **1988 (R) 108m/C** Michael J. Fox, Kiefer Sutherland, Phoebe Cates, Frances Sternhagen, Swoosie Kurtz, Tracy Pollan, Jason Robards Jr., John Houseman, Dianne Wiest, Charlie Schlatter, William Hickey; *Dir:* James Bridges. **VHS, Beta, LV $19.95** MGM &&

Bright Smiler A made-for-British-television mystery about a writer who retires to her favorite spa and falls into the hands of a malevolent masseuse. **1985 60m/C** GB Janet Suzman. **VHS, Beta $59.95** PSM &&½

Brighton Beach Memoirs The film adaptation of the popular (and semiautobiographical) Neil Simon play. Poignant comedy/drama about a young Jewish boy's coming of age in Depression-era Brooklyn. Followed by ''Biloxi Blues'' and ''Broadway Bound.'' **1986 (PG-13) 108m/C** Blythe Danner, Bob Dishy, Judith Ivey, Jonathan Silverman; *Dir:* Gene Saks. **VHS, Beta, LV $19.95** MCA, FCT &&½

Brighton Rock Sterling performances highlight this seamy look at the British underworld. Attenborough is Pinkie Brown, a small-time hood who ends up committing murder. He manipulates a waitress to get himself off the hook, but things don't go exactly as he plans. Based on the novel by Graham Greene. **1947 92m/B** GB Richard Attenborough, Hermione Baddeley, William Hartnell, Carol Marsh, Nigel Stock, Wylie Watson, Alan Wheatley, George Carney, Reginald Purdell; *Dir:* John Boulting. **VHS** DVT &&&

Brighton Strangler An actor who plays a murderer assumes the role of the strangler after suffering from a concussion. Decent psychodrama. **1945 67m/B** John Loder, June Duprez, Miles Mander; *Dir:* Max Nosseck. **VHS, Beta $34.95** RKO &&

Brighton Strangler Before Dawn A mystery double feature: In ''The Brighton Strangler,'' an actor takes his part too seriously as he murders Londoners at night; in ''Before Dawn'' a brilliant scientist turns to a life of murderous crime. **1945 128m/B** John Loder, June Duprez, Miles Mander, Stuart Oland, Dorothy Wilson. **VHS, Beta $39.95** RKO &&

Brighty of the Grand Canyon The spunky donkey Brighty roams across the Grand Canyon in search of adventure. He finds friendship with a gold-digging old prospector who hits pay dirt. **1967 90m/C** Joseph Cotten, Pat Conway, Dick Foran, Karl Swenson; *Dir:* Norman Foster. **VHS, Beta $24.95** AHV, VTK, KAR &&½

Brimstone & Treacle Weird, obsessive psychodrama written by Dennis Potter, in which a young rogue (who may or may not be an actual agent of the Devil) infiltrates the home of a staid British family caring for their comatose adult daughter. **1982 (R) 85m/C** GB Sting, Denholm Elliott, Joan Plowright, Suzanna Hamilton; *Dir:* Richard Loncraine. Montreal World Film Festival '82: Best Film. **VHS, Beta $59.95** MGM &&&

Bring Me the Head of Alfredo Garcia Peckinpah falters in this poorly paced outing. American piano player on tour in Mexico finds himself entwined with a gang

of bloodthirsty bounty hunters. Bloody and confused. **1974 (R) 112m/C** Warren Oates, Isela Vega, Gig Young, Robert Webber, Helmut Dantine, Emilio Fernandez, Kris Kristofferson; *Dir:* Sam Peckinpah. **VHS $59.95** MGM &&

Bring Me the Vampire A wealthy baron must endure a night of terror in order to receive an inheritance. **1961 100m/B** MX Maria Eugenia San Martin, Hector Godoy, Joaquin Vargas. **VHS $16.95** SNC, MOV &

Bringing Up Baby The quintessential screwball comedy, featuring Hepburn as a giddy socialite with a ''baby'' leopard, and Grant as the unwitting object of her affections. One ridiculous situation after another add up to high speed fun. Hepburn looks lovely, the supporting actors are in fine form, and director Hawke manages the perfect balance of control and mayhem. From a story by Hagar Wilde, who helped Nichols with the screenplay. Also available in a colorized version. **1938 103m/B** Katharine Hepburn, Cary Grant, May Robson, Charlie Ruggles, Walter Catlett, Fritz Feld, Jonathan Hale, Barry Fitzgerald; *Dir:* Howard Hawks. **VHS, Beta, LV $19.98** TTC &&&&

Brink of Life Three pregnant women in a hospital maternity ward await the impending births with mixed feelings. Early Bergman; in Swedish with English subtitles. **1957 82m/B** SW Eva Dahlbeck, Bibi Andersson, Ingrid Thulin, Babro Ornas, Max von Sydow, Erland Josephson, Gunnar Sjoberg; *Dir:* Ingmar Bergman. Cannes Film Festival '58: Best Actress (collective). **VHS, Beta $29.95** VYY, TAM &&½

Brink's Job Recreates the ''crime of the century,'' Tony Pino's heist of 2.7 million dollars from a Brink's truck. The action picks up five days before the statute of limitations is about to run out. **1978 (PG) 103m/C** Peter Falk, Peter Boyle, Warren Oates, Gena Rowlands, Paul Sorvino, Sheldon Leonard, Allen Garfield; *Dir:* William Friedkin. **VHS, Beta $59.95** MCA &&&

Britannia Hospital This is a portrait of a hospital at its most chaotic: the staff threatens to strike, demonstrators surround the hospital, a nosey BBC reporter pursues an anxious professor, and the eagerly-anticipated royal visit degenerates into a total shambles. **1982 (R) 111m/C** GB Malcolm McDowell, Leonard Rossiter, Graham Crowden, Joan Plowright, Mark Hamill, Alan Bates; *Dir:* Lindsay Anderson. **VHS, Beta $69.95** HBO &&½

British Intelligence Silly American-made film about British espionage. Boris Karloff plays a butler (who is also a spy) trapped by an agent who visits the home of a British bureaucrat. Half-baked story that doesn't hold up. **1940 62m/B** Boris Karloff, Margaret Lindsay, Maris Wrixon, Holmes Herbert, Leonard Mudie, Bruce Lester; *Dir:* Terry Morse. **VHS $16.95** NOS, SNC, MLB &½

A Brivele der Mamen (A Letter to Mother) A Jewish mother does her best to hold her fragile family together, in spite of the ravages of war and poverty. Her travails take her and her family from the Polish Ukraine to New York City. In Yiddish with English subtitles. **1938 90m/C** PL Berta Gersten, Lucy Gerhman, Misha Gerhman, Edmund Zayenda; *Dir:* Joseph Green. **VHS, Beta $79.95** ERG, TAM &&&

Broadcast News The acclaimed, witty analysis of network news shows, dealing with the three-way romance between a driven career-woman producer, an ace nebbish reporter and a brainless, popular on-screen anchorman. Incisive and funny, though often simply idealistic. Written by James L. Brooks. **1987 (R) 132m/C** William Hurt, Albert Brooks, Holly Hunter, Jack Nicholson, Joan Cusack, Robert Prosky, Lois Chiles, John Cusack; *Dir:* James L. Brooks. National Board of Review Awards '87: Best Actress (tie) (Hunter), 10 Best Films of the Year. **VHS, Beta, LV $19.98** FOX &&&½

Broadway Bill A man decides to abandon his nagging wife and his job in her family's business for the questionable pleasures of owning a racehorse known as Broadway Bill. This racetrack comedy was also remade by Frank Capra in 1951 as ''Riding High.'' **1934 90m/C** Warner Baxter, Myrna Loy, Walter Connolly, Helen Vinson, Margaret Hamilton; *Dir:* Frank Capra. **VHS $19.95** PAV &&&

Broadway to Cheyenne A cowboy comes to the big city to follow his dreams. But when his dreams do not become realities, he wonders if it was ever to be. **1932 48m/B** Rex Bell, George ''Gabby'' Hayes, Marceline Day; *Dir:* Harry Fraser. **VHS, Beta** VDM, RXM &

Broadway Danny Rose One of Woody Allen's best films, a hilarious, heart-rending anecdotal comedy about a third-rate talent agent involved in one of his client's infidelities. The film magically unfolds as show business veterans swap Danny Rose stories at a delicatessen. Allen's Danny Rose is pathetically lovable. **1984 (PG) 85m/B** Woody Allen, Mia Farrow, Nick Apollo Forte, Sandy Baron, Milton Berle, Howard Cosell; *Dir:* Woody Allen. **VHS, Beta, LV $19.98** VES &&&½

The Broadway Drifter A silent Jazz Age drama about a playboy who repents his decadent ways by opening a girls' health school. Complications ensue. **1927 90m/B** George Walsh, Dorothy Hall, Bigelow Cooper, Arthur Donaldson; *Dir:* Bernard McEveety. **VHS, Beta, 8mm $29.95** VYY &&

The Broadway Melody Early musical in which two sisters hope for fame on Broadway, and encounter a wily song and dance man who traps both their hearts. Dated, but still charming, with lovely score by Arthur Freed and Nacio Herb Brown. Considered the great granddaddy of all MGM musicals. **1929 104m/B** Bessie Love, Anita Page, Charles King, Jed Prouty, Kenneth Thomson, Edward Dillon, Mary Doran; *Dir:* Harry Beaumont. Academy Awards '29: Best Picture. **VHS, Beta $14.95** BTV, MGM &&&

Broadway Melody of 1936 Exceptional musical comedy with delightful performances from Taylor and Powell. Benny is a headline-hungry columnist who tries to entrap Taylor by using Powell. Great music by Freed/Brown team. Nominated for 3 Oscars: Best Picture, Best Writing (Original Story) and Best Dance Direction. **1935 110m/B** Jack Benny, Eleanor Powell, Robert Taylor, Una Merkel, Sid Silvers, Buddy Ebsen; *Dir:* Roy Del Ruth. **VHS, Beta $19.98** MGM, MLB, CCB &&&

Broadway Melody of 1938 A vintage backstage musical with an all-star cast, and songs written by Nacio Herb Brown & Arthur Freed, including ''I'm Feeling Like a

Million," "A New Pair of Shoes" and "Yours and Mine." Judy Garland sings her famous "Dear Mr. Gable."
1937 110m/B Eleanor Powell, Sophie Tucker, George Murphy, Judy Garland, Robert Taylor, Buddy Ebsen; *Dir:* Roy Del Ruth. **VHS, Beta $29.95** *MGM* 🎬🎬½

Broadway Melody of 1940 The best of the "Broadway Melody" series features the only screen teaming of Astaire and Powell. The flimsy plot is just an excuse for a potent series of Cole Porter musical numbers, including "I Concentrate on You," "I've Got My Eye on You," and the famous "Begin the Beguine" duet.
1940 103m/B Fred Astaire, Eleanor Powell, George Murphy, Frank Morgan, Ian Hunter; *Dir:* Norman Taurog. **VHS, Beta, LV $19.98** *MGM, TTC, FCT* 🎬🎬½

Broken Arrow A U.S. scout befriends Cochise and the Apaches, and helps settlers and Indians live in peace together in the 1870s. Acclaimed as the first Hollywood film to side with the Indians, and for Chandler's portrayal of Cochise. Oscar nominations for Best Supporting Actor (Chandler), Best Screenplay, and Color Cinematography.
1950 93m/C James Stewart, Jeff Chandler, Will Geer, Debra Paget, Basil Ruysdael, Arthur Hunnicutt, Jay Silverheels; *Dir:* Delmer Daves. Writers Guild of America Awards '50: Best Written American Western. **VHS, Beta $19.98** *FOX* 🎬🎬½

Broken Badge Crenna stars as a macho, chauvinistic cop whose attitude towards rape victims changes dramatically after he himself is raped by a couple of thugs. Excellent TV movie originally titled "The Rape of Richard Beck."
1985 100m/C Richard Crenna, Meredith Baxter Birney, Pat Hingle, Frances Lee McCain, Cotter Smith, George Dzundza, Joanna Kerns; *Dir:* Karen Arthur. Emmy Award '85: Best Actor (Crenna). **VHS $9.95** *SIM* 🎬🎬🎬

Broken Blossoms One of Griffith's most widely acclaimed films, photographed by Billy Bitzer, about a young Chinaman in London's squalid Limehouse district hoping to spread the peaceful philosophy of his Eastern religion. He befriends a pitiful street waif who is mistreated by her brutal father, resulting in tragedy. Silent. Revised edition contains introduction from Gish and a newly-recorded score.
1919 102m/B Lillian Gish, Richard Barthelmess, Donald Crisp; *Dir:* D.W. Griffith. **VHS, Beta, LV $19.95** *NOS, MRV, HHT* 🎬🎬🎬½

Broken Lance A western remake of "House of Strangers" that details the dissolution of a despotic cattle baron's family. Beautifully photographed. Oscar nomination for supporting actress Jurado.
1954 96m/C Spencer Tracy, Richard Widmark, Robert Wagner, Jean Peters, Katy Jurado, Earl Holliman, Hugh O'Brian, E.G. Marshall; *Dir:* Edward Dmytryk. Academy Awards '54: Best Story. **VHS, Beta, LV $39.98** *FOX* 🎬🎬🎬½

The Broken Mask A doctor performs remarkable plastic surgery on a patient, but when the patient becomes attracted to his sweetheart, he throws the Hippocratic oath to the wind, and his patient's spiffy new look has got to go. A ground-breaker amid plastic surgery-gone-amok pieces.
1928 58m/B Cullen Landis, Barbara Bedford, Wheeler Oakman, James Marcus; *Dir:* James Hogan. **VHS, Beta $19.95** *GPV, FCT* 🎬🎬

Broken Melody An opera singer sent to prison by mistake must escape in order to return to the woman of his dreams.

1934 62m/B *GB* John Garrick, Margot Grahame, Merle Oberon, Austin Trevor, Charles Carson, Harry Terry; *Dir:* Bernard Vorhaus. **VHS $29.95** *FCT* 🎬½

Broken Strings An all-black musical in which a concert violinist must come to terms with himself after an auto accident limits the use of his left hand.
1940 50m/B Clarence Muse, Sybil Lewis, William Washington, Matthew "Stymie" Beard; *Dir:* Bernard B. Ray. **VHS, Beta $24.95** *NOS, VYY, VCN* 🎬

Bromas S.A. A group of teenagers get into a bunch of funny situations.
1987 104m/C *SP* Mauricio Garces, Gloria Marin, Manuel Valdez, Hermanos King. **VHS, Beta $54.95** *MAD* 🎬🎬

Bronco Billy A Wild West Show entrepreneur leads his ragged troupe from one improbable adventure to the next.
1980 (PG) 117m/C Clint Eastwood, Sondra Locke, Bill McKinney, Scatman Crothers, Sam Bottoms; *Dir:* Clint Eastwood. **VHS, Beta, LV $19.98** *WAR, TLF, RXM* 🎬🎬½

Bronson Lee, Champion Kung-fu and karate highlight this action adventure.
1978 (PG) 81m/C Tadashi Yamashita. **VHS, Beta $19.98** *NO* 🎬

Bronson's Revenge Two frontier soldiers face constant danger as they attempt to transport death row prisoners, a large quantity of gold and a stranded woman across the Rocky Mountains territory.
1979 (R) 90m/C Robert Hundar, Roy Hill, Emma Cohen. **VHS, Beta $52.95** *SUN* 🎬½

Bronx Cheers Young Italian-American man returns from WWII, falls prey to the misguidance of a has-been boxer. 1990 Academy Award nominee for Best Live Action Short.
1989 30m/C *Dir:* Raymond De Felitta. **VHS $59.95** *CIG* 🎬½

The Bronx Executioner Android, robot, and human interests clash in futuristic Manhattan and all martial arts hell breaks loose. Special introduction by martial arts star Michael Dudikoff.
1986 88m/C Rob Robinson, Margie Newton, Chuck Valenti, Gabriel Gori; *Dir:* Bob Collins. **VHS, Beta $59.95** *CAN* 🎬

The Bronx War Rival gangs take to the streets in this film from the director of "Hangin' with the Homeboys." A malicious gang leader tricks his gang into going to war over a girl he wants. Unrated version available.
1991 (R) 91m/C Joseph P. Vasquez, Fabio Urena, Charmaine Cruz, Andre Brown, Marlene Forte, Francis Colon, Miguel Sierra, Kim West; *Dir:* Joseph P. Vasquez. **VHS $89.95** *ACA* 🎬🎬

Bronze Buckaroo Director Kahn also scripted this all African-American horse opera in which a cowpoke seeks revenge for his pa's death.
1939 57m/B Herbert Jeffries, Artie Young, Rellie Hardin, Spencer Williams, Clarence Brooks, F.E. Miller; *W/Dir:* Richard C. Kahn. **VHS $24.95** *NOS, MRV, TPV* 🎬🎬

The Brood Cronenberg's inimitable biological nightmares, involving an experimentally malformed woman who gives birth to murderous demon-children that kill every time she gets angry. Extremely graphic. Not for all tastes.
1979 (R) 92m/C *CA* Samantha Eggar, Oliver Reed; *Dir:* David Cronenberg. **VHS, Beta, LV $19.98** *SUE, EMB* 🎬🎬½

Broth of a Boy A hungry British producer decides to film a birthday party of the oldest man in the world. The old Irishman wants a portion of the profits. Well made and thought provoking.
1959 77m/B *IR* Barry Fitzgerald, Harry Brogan, Tony Wright, June Thorburn, Eddie Golden; *Dir:* George Pollock. **VHS $19.95** *NOS* 🎬🎬½

The Brother from Another Planet An black alien escapes from his home planet and winds up in Harlem, where he's pursued by two alien bounty hunters. Independently made morality fable by John Sayles before he hit the big time; features Sayles in a cameo as an alien bounty hunter.
1984 109m/C Joe Morton, Dee Dee Bridgewater, Ren Woods, Steve James, Maggie Renzi, David Strathairn, John Sayles; *Dir:* John Sayles. **VHS, Beta $19.98** *FOX, FCT* 🎬🎬🎬

Brother, Can You Spare a Dime? A compilation of documentary film footage from the 1930's. Hollywood in its heyday, Dillinger vs. the G-men, bread liners, and other memorabilia.
1975 103m/C *Dir:* Philippe Mora. **VHS, Beta $39.95** *UHV, FHS* 🎬🎬🎬

Brother Future T.J., a black, streetsmart city kid who thinks school and helping others is all a waste of time gets knocked out in a car accident. As he's lying unconscious, he is transported back in time to a slave auction block in the Old South. There the displaced urbanite is forced to work on a cotton plantation, and watches the stirrings of a slave revolt. T.J. sees the light and realizes how much opportunity he's been wasting in his own life. He comes to just a few moments later, but worlds away from who he was before. Part of the "Wonderworks" series.
1991 110m/C Phill Lewis, Frank Converse, Carl Lumbly. **VHS $29.95** *PME* 🎬🎬🎬

Brother John An early look at racial tensions and labor problems. An angel goes back to his hometown in Alabama to see how things are going.
1970 (PG) 94m/C Sidney Poitier, Will Geer, Bradford Dillman, Beverly Todd, Paul Winfield; *Dir:* James Goldstone. **VHS, Beta $14.95** *COL* 🎬🎬🎬

Brother Orchid Mobster puts a henchman in charge of his gang while he vacations in Europe. Upon his return, he is deposed and wounded in an assassination attempt. Hiding out in a monastary, he plots to regain control of the gang, leading to fish outta water episodes and a change in his outlook on life. Fine cast fans through farce intelligently.
1940 87m/B Edward G. Robinson, Humphrey Bogart, Joan Blondell, Donald Crisp, Ralph Bellamy, Allen Jenkins, Charles D. Brown, Cecil Kellaway; *Dir:* Lloyd Bacon. **VHS, Beta $19.98** *MGM, CCB* 🎬🎬🎬

Brother Sun, Sister Moon Post-'60s costume epic depicting the trials of St. Francis of Assisi as he evaluates his beliefs in Catholicism. Music by Donovan.
1973 (PG) 120m/C Graham Faulkner, Judi Bowker, Alec Guinness, Leigh Lawson, Kenneth Cranham, Lee Montague, Valentina Cortese; *Dir:* Franco Zeffirelli. **VHS, Beta $44.95** *PAR, APD* 🎬🎬½

The Brotherhood Two hot-headed brothers in a Mafia syndicate clash over old vs. new methods and the changing of the Family's guard. Music by Lalo Schifrin.

1968 96m/C Kirk Douglas, Alex Cord, Irene Papas, Luther Adler, Susan Strasberg, Murray Hamilton; *Dir:* Martin Ritt. **VHS, Beta, LV $14.95** *PAR* 𝄞𝄞½

Brotherhood of Death Three black Vietnam veterans return to their southern hometown to get even with the Klansmen who slaughtered all of the townspeople.
1976 (R) 85m/C Roy Jefferson, Larry Jones, Mike Bass, Le Tari, Haskell V. Anderson; *Dir:* Bill Berry. **VHS, Beta $59.95** *MPI* 𝄞

Brotherhood of Justice Young men form a secret organization to rid their neighborhood of drug dealers and violence. As their power grows, their propriety weakens, until all are afraid of the "Brotherhood of Justice." Exciting made for television feature.
1986 97m/C Keanu Reeves, Kiefer Sutherland, Billy Zane, Joe Spano, Darren Dalton, Evan Mirand, Don Michael Paul; *Dir:* Charles Braverman. **VHS $89.95** *VMK* 𝄞𝄞

The Brotherhood of Satan In an isolated southern town, a satanic coven persuades children to join in their devil-may-care attitude. Worthwhile.
1971 (PG) 92m/C Strother Martin, L.Q. Jones, Charles Bateman, Anna Capri, Charles Robinson, Alvy Moore, Geri Reischl; *Dir:* Bernard McEveety. **VHS, Beta, LV $9.95** *GKK, IME* 𝄞𝄞½

Brotherly Love "Good twin/bad twin" made-for-TV mystery about an escaped psychopath who's out to get his businessman twin brother (Hirsch in both roles).
1985 94m/C Judd Hirsch, Karen Carlson, George Dzundza, Barry Primus, Lori Lethin; *Dir:* Jeff Bleckner. **VHS, Beta $59.98** *FOX* 𝄞½

Brothers in Arms Savage mountainmen have developed a weird religion which requires them to mercilessly hunt down and massacre human prey.
1988 (R) 95m/C Todd Allen, Jack Starrett, Dedee Pfeiffer; *Dir:* George J. Bloom III. **VHS, LV $79.95** *REP, PIA* 𝄞

The Brothers Grimm Fairy Tales Versions of "Little Red Riding Hood" and "The Seven Ravens" are featured on this tape.
1979 35m/C **VHS, Beta $14.95** *NWV* 𝄞𝄞

The Brothers Karamazov Hollywood adaptation of the classic novel by Dostoevsky in which four 19th-Century Russian brothers struggle with their desires for the same beautiful woman and the father who brutalizes them. Incredible performances from every cast member, especially Cobb (he received an Academy Award Best Supporting Actor nomination). Long and extremely intense, with fine direction from Brooks. Marilyn Monroe tried desperately to get Schell's part.
1957 147m/C Yul Brynner, Claire Bloom, Lee J. Cobb, William Shatner, Maria Schell, Richard Basehart; *Dir:* Richard Brooks. **VHS, Beta $24.95** *KUI, MGM, TAM* 𝄞𝄞𝄞

Brothers Lionheart The Lion brothers fight for life, love and liberty during the Middle Ages.
1985 (G) 120m/C Staffan Gotestam, Lars Soderdahl, Allan Edwall; *Dir:* Olle Hellbron. **VHS, Beta** *PAV* 𝄞𝄞½

Brothers O'Toole The misadventures of a pair of slick drifters who, by chance, ride into a broken-down mining town in the 1890s.
1973 94m/C John Astin, Steve Carlson, Pat Carroll, Hans Conried, Lee Meriwether; *Dir:* Richard Erdman. **VHS, Beta $49.95** *UHV* 𝄞

Brothers of the West A cowboy saves his brothers from being lynched by proving the guilt of the real outlaws.
1937 56m/B Tom Tyler, Bob Terry, Lois Wilde, Dorothy Short. **VHS, Beta $19.95** *NOS, VCN, DVT* 𝄞½

Browning Version A lonely, unemotional classics instructor at a British boarding school realizes his failure as a teacher and as a husband. From the play by Terrence Rattigan.
1951 89m/B *GB* Michael Redgrave, Jean Kent, Nigel Patrick, Wilfrid Hyde-White, Bill Travers; *Dir:* Anthony Asquith. National Board of Review Awards '51: 5 Best Foreign Films of the Year. **VHS, Beta** *LCA* 𝄞𝄞𝄞½

Brubaker A sanctimonious drama about a reform warden who risks his life to replace brutality and corruption with humanity and integrity in a state prison farm. Powerful prison drama.
1980 (R) 131m/C Robert Redford, Jane Alexander, Yaphet Kotto, Murray Hamilton, David Keith, Morgan Freeman, Matt Clark, M. Emmet Walsh, Everett McGill; *Dir:* Stuart Rosenberg. **VHS, Beta, LV $69.98** *FOX* 𝄞𝄞𝄞

Bruce Conner Films 1 Five films by the daring experimental filmmaker, all dealing in one way or another with the limits, meaning and effects of the film medium itself: "Ten Second Film," "Permian Strata," "Mongoloid," "America Is Waiting" and "A Movie."
1978 24m/C *Dir:* Bruce Conner. **VHS, Beta** *FCT* 𝄞𝄞𝄞

Bruce Conner Films 2 Four films by Conner, each expressing questions about the nature of film while constructing/deconstructing various cultural objects. Includes "Breakaway," "Vivian," "The White Rose" and "Marilyn Times Five."
1980 30m/C *Dir:* Bruce Conner. **VHS, Beta** *FCT* 𝄞𝄞𝄞

Bruce Is Loose Bruce Le fights the evil Master. More foot-in-the-face follies for fun-loving folks.
1984 76m/C Bruce Le. **VHS, Beta $29.99** *BFV* 𝄞½

The Bruce Lee Collection A collection of Lee's greatest hits including "Fists of Fury," "Chinese Connection," "Return of the Dragon," and "Game of Death."
19?? 397m/C Bruce Lee. **VHS $79.95** *FOX* 𝄞½

Bruce Lee Fights Back From the Grave Bruce Lee returns from the grave to fight the Black Angel of Death and to wreak vengeance on the evil ones who brought about his untimely demise, with his feet.
1976 (R) 97m/C Bruce Lee, Deborah Chaplin, Anthony Bronson; *Dir:* Umberto Lenzi. **VHS, Beta $49.95** *MED* 𝄞

Bruce Lee: The Legend Tribute to the king of karate films features rare footage and interviews with many of the star's closest friends.
1984 88m/C Bruce Lee, Steve McQueen, James Coburn; *Nar:* James B. Nicholson. **VHS, Beta $19.98** *FOX* 𝄞

Bruce Lee: The Man/The Myth A dramatization of the life and times of martial arts master Bruce Lee.
1984 90m/C Bruce Lee. **VHS, Beta $59.95** *LTG, VES, LIV* 𝄞

Bruce Lee's Ways of Kung Fu Copious head-kicking film in honor of the long-dead star.
198? 87m/C Bruce Lei, Philip Cheung, Pearl Lin. **VHS, Beta $39.95** *UNI* 𝄞

Bruce Le's Greatest Revenge Kung-fu action and martial arts fighting highlight this film, in which a martial arts student gets involved in a clash between Chinese and a discriminatory European Club.
1980 (R) 94m/C Bruce Le. **VHS, Beta $24.95** *GEM* 𝄞

Bruce Li the Invincible Cheng, the Kung-Fu killer, is pursued by the vengeance-minded Bruce.
1980 93m/C Bruce Li, Ho Chung Dao, Chen Sing. **VHS $19.95** *OCE, MRV, GEM* 𝄞

Bruce Li in New Guinea The tribe of a remote island worships the legendary Snake Pearl. Two masters of Kung-fu visit the isle and discover they must defend the daughter of the murdered chief against a cruel wizard.
1980 (PG) 98m/C Bruce Li, Ho Chung Dao, Chen Sing; *Dir:* C.Y. Yang. **VHS, Beta $19.95** *OCE, GEM, MRV* 𝄞

Bruce and Shaolin Kung Fu Bruce wins a Kung Fu match against a general's son and now the father's henchmen are after him.
1988 85m/C Bruce Le. **VHS $14.98** *BFV* 𝄞

Bruce the Superhero Chop-socky action as the coattails-riding Bruce Le takes on the Black Dragon Society.
1979 88m/C Bruce Le. **VHS, Beta $19.99** *BFV, MRV* 𝄞

Bruce vs. Bill Bruce is Bruce Le, a minor martial arts hero. Nobody's quite sure who Bill is, but Bruce beats up on him in this movie.
1983 (R) 90m/C Bruce Le, Bill Louie. **VHS, Beta $19.99** *BFV* 𝄞

Bruce's Deadly Fingers Bruce Lee kept a book of deadly finger techniques for Kung fu killing. When a vicious gangster kidnaps the Master's ex-girlfriend, one of his followers must rescue her and maintain the book's integrity.
198? (R) 90m/C Bruce Lee, Lo Lieh, Chan Wai Man, Nora Miao, Rose Marie, Yuen Man Chi, Chang Leih, Young Zee, Wuk Ma No Hans. **VHS, Beta $24.95** *GEM* 𝄞

Bruce's Fists of Vengeance Le and Lee guard master Bruce Lee's kicking secrets from an evil adversary.
1984 87m/C Bruce Le, Jack Lee, Ken Watanabe, Romano Kristoff, Don Gordon. **VHS, Beta $57.95** *GEM* 𝄞

Brujo Luna Evil ritualists kill and eat innocent people. In Spanish.
19?? 90m/C *SP* Laurie Walters, Joe Spano. **VHS, Beta $29.95** *UNI* Woof!

Brussels Transit The story of a Yiddish family's arrival and adjustment to the new world of Brussels, Belgium, after World War II. In Yiddish with English subtitles.
1980 80m/B *Dir:* Samy Szlingerbaum. **VHS $69.95** *WAC* 𝄞𝄞𝄞

Brutal Glory Set in New York City in 1918 and based on the true story of the boxer known as "The Real McCoy." Fascinating and detailed history of a period, a champion, and the man who made it happen.
1989 96m/C Robert Vaughn, Timothy Brantley, Leah K. Pinsent; *Dir:* Koos Roets. **VHS** *QUE* 𝄞

The Brute Man A young man who had been disfigured by his school mates goes out on a trail of revenge. Hatton is convincing in the title role, as in real life he was afflicted with acromegaly, an ailment that produces an enlargment of the bones in the face, hands, and feet.
1946 62m/B Rondo Hatton, Tom Neal, Jane Adams; *Dir:* Jean Yarborough. **VHS, Beta, LV** $14.95 *SVS, AOV, MLB* 🎞🎞

Brutes and Savages A collection of footage of primitive societies surviving in the wild, devised for voyeuristic viewing.
1977 91m/C *Dir:* Arthur Davis. **VHS, Beta** $79.95 *MPI* 🎞

The Buccaneer A swashbuckling version of the adventures of pirate Jean LaFitte and his association with President Andrew Jackson during the War of 1812. Remake of Cecille B. DeMille's 1938 production.
1958 121m/C Yul Brynner, Charlton Heston, Claire Bloom, Inger Stevens, Charles Boyer, Henry Hull, E.G. Marshall, Lorne Greene; *Dir:* Anthony Quinn. **VHS, Beta, LV** $14.95 *KRT, FCT, RXM* 🎞🎞

Buck and the Preacher A trail guide and a con man preacher join forces to help a wagon train of former slaves who are seeking to homestead out West. Poitier's debut as a director.
1972 (PG) 102m/C Sidney Poitier, Harry Belafonte, Ruby Dee, Cameron Mitchell, Denny Miller; *Dir:* Sidney Poitier. **VHS, Beta** $14.95 *COL* 🎞🎞 ½

Buck Privates Abbott and Costello star as two dim-witted tie salesmen, running from the law, who become buck privates during World War II. The duo's first great success, and the film that established the formula for each subsequent film.
1941 84m/B Bud Abbott, Lou Costello, Shemp Howard, Lee Norman, Alan Curtis, The Andrews Sisters; *Dir:* Arthur Lubin. **VHS, Beta, LV** $14.95 *MCA* 🎞🎞 ½

Buck Privates Come Home Abbott and Costello return to their "Buck Privates" roles as two soldiers trying to adjust to civilian life after the war. They also try to help a French girl sneak into the United States. Funny antics culminate into a wild chase scene.
1947 77m/B Bud Abbott, Lou Costello, Tom Brown, Joan Shawlee, Nat Pendleton, Beverly Simmons, Don Beddoe, Don Porter, Donald MacBride; *Dir:* Charles T. Barton. **VHS** $14.98 *MCA* 🎞🎞🎞

Buck Rogers in the 25th Century An American astronaut, preserved in space for 500 years, is brought back to life by a passing Draconian flagship. Outer space adventures begin when he is accused of being a spy from Earth. Based on the classic movie serial. Made for TV movie that began the popular series.
1979 (PG) 90m/C Gil Gerard, Pamela Hensley, Erin Gray, Henry Silva; *Dir:* Daniel Haller. **VHS, Beta, LV** $19.95 *MCA* 🎞🎞

Buck Rogers Cliffhanger Serials, Vol. 1 The original series in which the hero awakens 200 years in the future. Volume one contains episodes 1-6, volume two, 7-12.
1939 122m/B Buster Crabbe, Constance Mour, Jackie Moran; *Dir:* Barry Sarecky. **VHS, Beta** $9.99 *UAV, MLB* 🎞🎞

Buck Rogers Cliffhanger Serials, Vol. 2 The original movie serial in which our hero awakens 200 years in the future to battle evil forces.

1939 122m/B Buster Crabbe, Constance Mour, Jackie Moran; *Dir:* Barry Sarecky. **VHS, Beta** $9.99 *UAV, MLB* 🎞🎞

Buck Rogers Conquers the Universe The story of Buck Rogers, written by Phil Nolan in 1928, was the first science fiction story done in the modern super-hero space genre. Many of the "inventions" seen in this movie have actually come into existence—spaceships, ray guns (lasers), anti-gravity belts—a testament to Nolan's almost psychic farsightedness.
1939 91m/B Buster Crabbe, Constance Moore, Jackie Moran; *Dir:* Ford Beebe, Saul Goodkind. **VHS, Beta** $29.95 *FOX* 🎞🎞

Buck Rogers: Planet Outlaws Drama of the world as it might exist in the 25th century. Compiled from the "Buck Rogers" serial.
1938 70m/B Buster Crabbe, Constance Moore, Jackie Moran. **VHS, Beta** $29.95 *VYY* 🎞🎞

A Bucket of Blood Cult favorite Dick Miller stars as a sculptor with a peculiar "talent" for lifelike artwork. Corman fans will see thematic similarities to his subsequent work, "Little Shop of Horrors" (1960). "Bucket of Blood" was made in just five days, while "Little Shop of Horrors" was made in a record breaking two days. Corman horror/spoof noted for its excellent "beatnik" atmosphere.
1959 66m/B Dick Miller, Barboura Morris, Antony Carbone, Julian Burton, Ed Nelson, Bert Convy; *Dir:* Roger Corman. **VHS, Beta** $19.95 *NOS, RHI, SNC* 🎞🎞🎞

Buckeye and Blue A hero turned outlaw, his 14 year-old female sidekick, and a gang known as the McCoys are all on the lam from the law in this western adventure.
1987 (PG) 94m/C Robin Lively, Jeffery Osterhage, Rick Gibbs, Will Hannah, Kenneth Jensen, Patrick Johnston, Stuart Rogers, Michael Horse; *Dir:* J.C. Compton. **VHS, Beta** $29.95 *ACA* 🎞 ½

Buckskin Frontier Story is built around Western railroad construction and cattle empires in the 1860s.
1943 75m/B Richard Dix, Jane Wyatt, Lee J. Cobb; *Dir:* Lesley Selander. **VHS, Beta** $9.95 *NOS, MRV, IND* 🎞 ½

Bucktown A black man who reopens his murdered brother's bar fights off police corruption and racism in a Southern town.
1975 (R) 95m/C Fred Williamson, Pam Grier, Bernie Hamilton, Thalmus Rasulala, Art Lund; *Dir:* Arthur Marks. **VHS, Beta** $59.98 *ORI* 🎞 ½

Bud and Lou Comedy/drama recounts Abbott & Costello's rise in Hollywood. While the story, based on the book by Bob Thomas, is interesting, the two funnymen in the leads can't pull off the old classic skits.
1978 98m/C Harvey Korman, Buddy Hackett, Michele Lee, Arte Johnson, Robert Reed; *Dir:* Robert C. Thompson. **VHS, Beta, LV** $29.95 *JTC* 🎞 ½

Buddha Assassinator A young Kung-Fu enthusiast challenges a Ming dynasty lord to a duel to the death.
1982 93m/C Hwang Jang Lee, Mang Hai, Lung Fei, Chien Yueh Sheng. **VHS, Beta** $59.95 *WVE* 🎞

Buddy Buddy A professional hitman's well-ordered arrangement to knock off a state's witness keeps being interrupted by the suicide attempts of a man in the next hotel room.

1981 (R) 96m/C Jack Lemmon, Walter Matthau, Paula Prentiss, Klaus Kinski; *Dir:* Billy Wilder. **VHS, Beta** $79.95 *MGM* 🎞🎞 ½

Buddy Holly Story An acclaimed biography of the famed 1950s' pop star, spanning the years from his meteoric career's beginnings in Lubbock to his tragic early death in the now famous plane crash. Busey performs Holly's hits himself and earned an Oscar nomination.
1978 (PG) 113m/C Gary Busey, Don Stroud, Charles Martin Smith, Conrad Janis, William Jordan; *Dir:* Steve Rash. Academy Awards '78: Best Adapted Score; National Society of Film Critics Awards '78: Best Actor (Busey). **VHS, Beta, LV** $19.95 *COL* 🎞🎞🎞 ½

The Buddy System A tale of contemporary love and the modern myths that outline the boundaries between lovers and friends.
1983 (PG) 110m/C Richard Dreyfuss, Susan Sarandon, Jean Stapleton, Nancy Allen, Wil Wheaton, Ed Winter, Keene Curtis; *Dir:* Glenn Jordan. **VHS, Beta** $59.98 *FOX* 🎞🎞

Buffalo Bill A light, fictionalized account of the life and career of Bill Cody, from frontier hunter to showman.
1944 89m/C Joel McCrea, Maureen O'Hara, Linda Darnell, Thomas Mitchell, Edgar Buchanan, Anthony Quinn, Moroni Olsen, Sidney Blackmer; *Dir:* William A. Wellman. **VHS, Beta** $39.98 *FOX* 🎞🎞

Buffalo Bill & the Indians A perennially underrated Robert Altman historical pastiche, portraying the famous Wild West character as a charlatan and shameless exemplar of encroaching imperialism. Great all-star cast amid Altman's signature mise-en-scene chaos.
1976 (PG) 135m/C Paul Newman, Geraldine Chaplin, Joel Grey, Will Sampson, Harvey Keitel, Burt Lancaster, Kevin McCarthy; *Dir:* Robert Altman. **VHS, Beta** $19.98 *FOX* 🎞🎞🎞

Buffalo Bill Rides Again With an Indian uprising on the horizon, Buffalo Bill is called in. Mr. Bill finds land swindlers pitting natives against ranchers, but there's precious little action or interest here.
1947 68m/C Richard Arlen, Jennifer Holy, Edward Cassidy, Edmund Cobb, Charles Stevens; *Dir:* Bernard B. Ray. **VHS, Beta** *GPV* 🎞 ½

Buffalo Jump An independent woman, working as a lounge singer in Toronto, returns to her home in Alberta when her father dies. To her surprise he has left her the family ranch and, to the surprise of everyone else, she decides to stay and run it. She hires a good-looking local man to help her out and they both discover that they want more than a working relationship. However, the fireworks really start when he proposes a marriage of convenience. Engaging performances and beautiful scenery help raise this romantic tale of opposites above the average.
1990 97m/C *CA* Wendy Crewson, Paul Gross, Marion Gilsenan, Kyra Harper, Victoria Snow; *Dir:* Eric Till. **VHS** $89.95 *ACA* 🎞🎞🎞

Buffalo Rider An adventure film depicting the real-life experiences of C.J. "Buffalo" Jones who worked to save the American buffalo from extinction.
1978 90m/C Rick Guinn, John Freeman, Pricilla Lauris, George Sager, Rich Scheeland; *Dir:* George Lauris. **VHS, Beta** $39.95 *ATP* 🎞 ½

Buffalo Stampede Our heroes are involved in a plot to round up buffalo to sell for meat.

1933 60m/B Randolph Scott, Buster Crabbe, Harry Carey, Noah Beery, Raymond Hatton. **VHS, Beta $19.95** *NOS, VCN, MRV ♂*

Buffet Froid Surreal black comedy about a group of bungling murderers. First rate acting and directing makes this film a hilarious treat. From the director of "Menage." In French with English subtitles.
1979 95m/C *FR* Gerard Depardieu, Bernard Blier, Jean Carmet, Genevieve Page, Denise Gence, Carol Bouquet, Jean Benguigui, Michel Serrault; *Dir:* Bertrand Blier. **VHS, LV $59.95** *INT, TAM ♂♂♂½*

Bug The city of Riverside is threatened with destruction after a massive earth tremor unleashes a super-race of ten-inch mega-cockroaches that belch fire, eat raw meat, and are virtually impervious to Raid. Produced by gimic-king William Castle, who wanted to install windshield wiper-like devices under theatre seats that would brush against the patrons' feet as the cockroaches crawled across the screen; unfortunately, the idea was squashed flat.
1975 (PG) 100m/C Bradford Dillman, Joanna Miles, William Castle; *Dir:* Jeannot Szwarc. **VHS, Beta $19.95** *PAR ♂*

Bugles in the Afternoon Life in the army during Custer's last days, with a love triangle, revenge and the Little Big Horn for added spice.
1952 85m/C Ray Milland, Hugh Marlowe, Helena Bonham Carter, Forrest Tucker, Barton MacLane, George Reeves; *Dir:* Roy Rowland. **VHS $14.95** *REP, FCT ♂♂*

Bugs! Seven of the wascally wabbit's finest.
1988 60m/C Bugs Bunny; *Voices:* Mel Blanc. **VHS $14.95** *MGM*

Bugs Bunny Cartoon Festival The animated antics of the nefarious rabbit in four early cartoons.
1944 34m/C *Dir:* Bob Clampett, Chuck Jones, Friz Freleng. **VHS, Beta $14.95** *MGM*

Bugs Bunny Cartoon Festival: Little Red Riding Rabbit Five classic Bugs shorts, including "Old Grey Hare," "Jack Wabbit and the Beanstalk" and the title cartoon are featured.
1943 37m/C *Dir:* Chuck Jones, Friz Freleng. **VHS, Beta $14.95** *MGM*

Bugs Bunny & Elmer Fudd Cartoon Festival Seven classic Bugs and Elmer shorts, including "Wabbitt Twouble," "Stage Door Cartoon," and "The Big Snooze."
1944 54m/C *Dir:* Chuck Jones, Tex Avery, Bob Clampett, Friz Freleng, Robert McKimson. **VHS, Beta $19.95** *MGM*

Bugs Bunny: Hollywood Legend Six episodes featuring the wascally wabbit, including "Hair-Raising Hare," "A Hare Grows in Manhattan," and "Herr Meets Hare."
1990 60m/C *Voices:* Mel Blanc. **VHS $12.98** *MGM, FCT*

Bugs Bunny in King Arthur's Court The Looney Tuners spoof Mark Twain's masterpiece of classic American literature.
1989 25m/C Bugs Bunny; *W/Dir:* Chuck Jones; *Voices:* Mel Blanc. **VHS $14.95** *WAR ♂♂*

Bugs Bunny/Road Runner Movie A compilation of classic Warner Brothers cartoons, starring Bugs Bunny, Daffy Duck, Elmer Fudd, the Road Runner, Wile E. Coy-

ote, Porky Pig, and Pepe Le Pew, including some all-new animated sequences. Also available with Spanish dubbing.
1979 (G) 98m/C *Dir:* Phil Monroe, Chuck Jones. New York Film Festival '79: Best Actor. **VHS, Beta $19.98** *WAR, FCT, APD ♂♂*

Bugs Bunny Superstar A collection of nine animated classics starring everybody's favorite "wascally wabbitt." Includes fascinating interviews and behind the scenes moments featuring the men and women who brought Bugs to life.
1975 (G) 91m/C *Dir:* Larry Jackson; *Voices:* Mel Blanc; *Nar:* Orson Welles. **VHS, LV $19.95** *MGM ♂♂*

Bugs Bunny's 3rd Movie: 1,001 Rabbit Tales A compilation of old and new classic cartoons featuring Bugs, Daffy, Sylvester, Porky, Elmer, Tweety, Speedy Gonzalez and Yosemite Sam.
1982 (G) 74m/C *Dir:* Chuck Jones, Robert McKimson, Friz Freleng. **VHS, Beta $19.98** *WAR, FCT*

Bugs Bunny's Comedy Classics A collection of classic wabbit tales including "Easter Yeggs," "Racketeer Rabbit," "The Hair-Brained Hypnotist," "Acrobatty Rabbit," "Falling Hare," and "Haredevil Hare."
1988 60m/C *Voices:* Mel Blanc. **VHS $12.98** *MGM, FCT*

Bugs Bunny's Hare-Raising Tales Six standouts from the furry varmint's vault of classics; includes "Rabbitson Crusoe" and "Rabbit Hood."
1989 45m/C Bugs Bunny; *Dir:* Abe Levitow; *W/Dir:* Chuck Jones, Robert McKimson; *Voices:* Mel Blanc. **VHS $14.95** *WAR*

Bugs Bunny's Looney Christmas Tales Join Bugs as he and his little nephew re-enact two of the world's most beloved Christmas stories; "A Christmas Carol" and "'Twas the Night Before Christmas."
1979 25m/C *Dir:* Chuck Jones, Friz Freleng; *Voices:* Mel Blanc. **VHS $9.95** *WAR, APD*

Bugs Bunny's Wacky Adventures Here is a collection of that rascally rabbit's classic cartoons featuring "Ali Baba Bunny," "Hare Do," "Long Haired Hare," "Bunny Hugged," "The Grey Hounded Hare," "Roman Legion Hare," and "Duck, Rabbit, Duck!"
1957 59m/C *Dir:* Chuck Jones, Friz Freleng; *Voices:* Mel Blanc. **VHS, Beta $12.95** *WAR*

Bugs & Daffy: The Wartime Cartoons A compilation of Daffy and Bugs classics made during World War II, often dominated by propaganda and Axis-baiting.
1945 120m/C *Dir:* Robert McKimson, Chuck Jones, Friz Freleng; *Nar:* Leonard Maltin. **VHS, Beta, LV $19.98** *MGM*

Bugs & Daffy's Carnival of the Animals A host of favorite animated critters are introduced by the wabbit and the duck.
1989 26m/C Bugs Bunny; *Dir:* Chuck Jones; *Voices:* Mel Blanc. **VHS $14.95** *WAR*

Bugs vs. Elmer Seven of the best Bugs/Elmer battles: "Slick Hare," "Unruly Hare," "Hare Remover," "Stage Door Cartoon," "Wabbit Twouble," "The Big Snooze" and "Fresh Hare."
1953 60m/C *Dir:* Frank Tashlin, Chuck Jones, Friz Freleng, Robert McKimson, Bob Clampett. **VHS, Beta, LV $14.95** *MGM*

Bugsy Beatty stars as Benjamin "Bugsy" Siegel, the Forties gangster who built the Flamingo Hotel in Las Vegas when it was nothing but a virtual desert, but which grew to become a gambling mecca. Bening is perfect as Bugsy's moll, Virginia Hill, who inspired him to carry out his dream of building the Flamingo. Nominated for 10 Academy Awards, including Best Picture, "Bugsy" tells the tale of a very flamboyant man with an extraordinary vision.
1991 (R) 135m/C Warren Beatty, Annette Bening, Harvey Keitel, Ben Kingsley, Joe Mantegna; *Dir:* Barry Levinson. Academy Awards '91: Best Art Direction/Set Decoration, Best Costume Design; Chicago Film Critics Awards '91: Best Supporting Actor (Keitel); Golden Globe Awards '92: Best Film—Drama; Los Angeles Film Critics Association Awards '91: Best Director (Levinson), Best Picture; National Board of Review Awards '91: 10 Best Films of the Year, Best Actor (Beatty). **VHS, Beta, LV, 8mm** *COL ♂♂♂½*

Bugsy Malone An all-children's cast highlights this spoof of 1930s' gangster movies. A musical featuring songs by Paul Williams.
1976 (G) 94m/C *GB* Jodie Foster, Scott Baio, Florrie Augger, John Cassisi, Martin Lev; *Dir:* Alan Parker. **VHS, Beta, LV $14.95** *PAR ♂♂½*

Bull Durham A lovable American romantic comedy, dealing with a very minor minor-league team and three of its current constituents: an aging baseball groupie that beds one player each season; a cocky, foolish new pitcher; and the older, weary catcher brought in to wise the rookie up. The scene in which Sarandon tries poetry out on the banal rookie is a hoot. Highly acclaimed, the film sears with Sarandon and Costner's love scenes.
1988 (R) 107m/C Kevin Costner, Susan Sarandon, Tim Robbins, Trey Wilson, Robert Wuhl, Jenny Robertson; *Dir:* Ron Shelton. **VHS, Beta, LV $19.98** *ORI ♂♂♂½*

Bull of the West Cattlemen battle for land in the wide open spaces of the old West.
1989 90m/C Charles Bronson, Lee J. Cobb, Brian Keith, George Kennedy, DeForest Kelley, Doug McClure. **VHS $14.95** *AVD ♂½*

Bulldance At a gymnastic school in Crete, a girl's obsession with Greek mythological ritual leads to murder. Also known as "Forbidden Son."
1988 105m/C Lauren Hutton, Cliff DeYoung, Renee Estevez; *Dir:* Zelda Barron. **VHS, Beta** *ACA ♂*

Bulldog Courage Young man is out to avenge his father's murder.
1935 66m/B Tim McCoy, Lois January, Joan Woodbury, John Elliott; *Dir:* Sam Newfield. **VHS, Beta $19.95** *NOS, VCN, DVT ♂♂*

Bulldog Drummond A World War I vet, bored with civilian life, is enlisted by a beautiful woman to help her father in various adventures. The first in the long-standing series, based on the detective novels of Herman Cyril McNeile.
1929 85m/B Ronald Colman, Joan Bennett, Montagu Love; *Dir:* F. Richard Jones. **VHS, Beta $19.98** *SUE, RXM ♂♂½*

Bulldog Drummond in Africa/ Arrest Bulldog Drummond Two mid-series Bulldog Drummond films, with two new villains.
1938 115m/B John Howard, H.B. Warner, George Zucco. **VHS, Beta $19.98** *SUE ♂♂*

Bulldog Drummond Comes Back Drummond, aided by Colonel Nielson, rescues his fiancee from the hands of desperate kidnappers in this segment of the serial.
1937 119m/B John Howard, John Barrymore, Louise Campbell, Reginald Denny, Guy Standing; *Dir:* Louis King. **VHS, Beta $16.95** *SNC, NOS, DVT &&*

Bulldog Drummond Double Feature "Bulldog Drummond Comes Back" (1937), in which Drummond is on the trail of a clever criminal who is looking to get at him by kidnapping his fiancee, and "Bulldog Drummond's Bride" (1939), wherein Drummond decides to marry, but not until after bomb explosions, a bank robbery, and a rooftop chase.
193? 119m/B John Howard, John Barrymore, Guy Standing, Reginald Denny, Louis King, James Hogan, H.B. Warner, Eduardo Ciannelli. **VHS, Beta $54.95** *HHT, RXM &&*

Bulldog Drummond Escapes Drummond, aided by his side-kick and valet, rescues a beautiful girl from spies. He then falls in love with her.
1937 67m/B Ray Milland, Heather Angel, Reginald Denny, Guy Standing, Porter Hall, E.E. Clive; *Dir:* James Hogan. **VHS, Beta $16.95** *SNC, NOS, VYY &&*

Bulldog Drummond's Bride Ace detective Bulldog Drummond has to interrupt his honeymoon in order to pursue a gang of bank robbers across France and England. The last of the Bulldog Drummond film series.
1939 69m/B John Howard, Heather Angel, H.B. Warner, E.E. Clive, Reginald Denny, Eduardo Ciannelli; *Dir:* James Hogan. **VHS, Beta $14.95** *QNE, DVT, MRV &&*

Bulldog Drummond's Peril Murder and robbery drag the adventurous Drummond away from his wedding and he pursues the villains until they are behind bars. One in the film series.
1938 77m/B John Barrymore, John Howard, Louise Campbell, Reginald Denny, E.E. Clive, Porter Hall; *Dir:* James Hogan. **VHS, Beta $16.95** *SNC, CAB, RXM &&*

Bulldog Drummond's Revenge The Bulldog Drummond character, created by "Sapper" (Herman Cyril McNeile), underwent a number of different incarnations from twenties' silents through the forties, and even occasionally in the fifties and sixties. A suave ex-British officer (a precursor to the glib shaken-not-stirred gentleman-spy variety), Bulldog has been played by the likes of Ronald Coleman, Ralph Richardson and Tom Conway, among others. Bulldog's "Revenge" was made in the late thirties, during the series' heyday, the second to star John Howard as Drummond; John "the Profile" Barrymore revels in the character of Inspector Neilson of Scotland Yard, and Reginald Denny plays Algy. In this typically fast-paced installment, suave sleuth Drummond stalks the master criminal responsible for stealing the formula to an explosive.
1937 55m/B John Barrymore, Louise Campbell, John Howard, Reginald Denny, Frank Puglia, Nydia Westman, Lucien Littlefield; *Dir:* Louis King. **VHS $16.95** *SNC, MOV & 1/2*

Bulldog Drummond's Revenge/ Bulldog Drummond's Peril Once again, Bulldog Drummond becomes entangled with diamond counterfeiters, secret explosives, and other crime and criminals. Two episodes in a series.

1938 118m/B John Howard, John Barrymore, Louise Campbell, Reginald Denny, E.E. Clive, Porter Hall. **VHS, Beta $19.98** *SUE &&*

Bulldog Drummond's Secret Police The 15th Drummond film, adapted from Herman Cyril McNeile's famed detective novels featuring a million-pound treasure stashed in the Drummond manor and the murderous endeavors to retrieve it.
1939 54m/B John Howard, Heather Angel, H.B. Warner, Reginald Denny, E.E. Clive, Leo G. Carroll; *Dir:* James Hogan. **VHS, Beta $16.95** *SNC, VYY, RXM &&*

Bulldog Edition Two newspapers engage in an all-out feud to be number one in the city, even going so far as to employ gangsters for purposes of sabotage.
1936 57m/B Ray Walker, Evelyn Knapp, Regis Toomey, Cy Kendall, Billy Newell, Oscar Apfel, Betty Compson, Robert Warwick; *Dir:* Charles Lamont. **VHS, Beta $16.95** *SNC &&*

Bulldog Jack Fine entry in the series has Drummond's colleague battling it out with Richardson, the leader of gang of jewel thieves. Also known as, "Alias Bulldog Drummond."
1935 62m/B *GB* Fay Wray, Jack Hulbert, Claude Hulbert, Ralph Richardson, Paul Graetz, Gibb McLaughlin; *Dir:* Walter Forde. **VHS, Beta $16.95** *NOS, SNC, TIM &&*

Bullet for the General An American mercenary joins with rebel forces during the Mexican Revolution.
1968 95m/C *IT* Martine Beswick, Lou Castel, Gian Marie Volonte, Klaus Kinski; *Dir:* Damiano Damiani. **VHS, Beta $59.98** *CHA &*

A Bullet Is Waiting A diligent sheriff finally catches his man only to be trapped in a blinding snowstorm with the hardened criminal.
1954 83m/C Jean Simmons, Rory Calhoun, Stephen McNally, Brian Aherne; *Dir:* John Farrow. **VHS, Beta** *GKK &&*

Bullet for Sandoval An ex-Confederate renegade loots and pillages the north Mexican countryside on his way to murder the grandfather of the woman he loves.
1970 (PG) 96m/C *IT SP* Ernest Borgnine, George Hilton; *Dir:* Julio Buchs. **VHS, Beta $19.95** *UHV &&*

Bulletproof An unstoppable ex-CIA agent battles to retrieve a high-tech nuclear tank from terrorist hands.
1988 (R) 93m/C Gary Busey, Darlanne Fluegel, Henry Silva, Thalmus Rasulala, L.Q. Jones, R.G. Armstrong, Rene Enriquez; *Dir:* Steve Carver. **VHS, Beta, LV $14.95** *COL &*

Bullets or Ballots Tough New York cop goes undercover to join the mob in order to get the goods on them. Old-fashioned danger and intrigue follow, making for some action-packed thrills.
1938 82m/B Edward G. Robinson, Humphrey Bogart, Barton MacLane, Joan Blondell, Frank McHugh, Louise Beavers; *Dir:* William Keighley. **VHS, Beta $19.98** *MGM, CCB &&&*

Bullfighter & the Lady An American goes to Mexico to learn the fine art of bullfighting in order to impress a beautiful woman.
1950 87m/B Robert Stack, Gilbert Roland, Virginia Grey, Katy Jurado; *Dir:* Budd Boetticher. **VHS, LV $19.98** *REP, PIA && 1/2*

Bullfighters Stan and Ollie are in hot pursuit of a dangerous criminal which leads them to Mexico where Stan winds up in a bull ring.

1945 61m/B Stan Laurel, Oliver Hardy, Margo Wood, Richard Lane; *Dir:* Mal St. Clair. **VHS, Beta $29.98** *FOX, MLB &&*

Bullies A woodland-transplanted young man decides to fight back against an ornery mountain clan who've raped his mother, tortured his father, and beat up his girlfriend. Brutal and unpalatable.
1986 (R) 96m/C *CA* Janet Laine Greene, Dehl Berti, Stephen Hunter; *Dir:* Paul Lynch. **VHS, Beta, LV $79.95** *MCA Woof!*

Bullitt A detective assigned to protect a star witness for 48 hours senses danger; his worst fears are confirmed when his charge is murdered. Based on the novel, "Mute Witness" by Robert L. Pike, and featuring one of filmdom's most famous car chases.
1968 (PG) 105m/C Steve McQueen, Robert Vaughn, Jacqueline Bisset, Don Gordon, Robert Duvall, Norman Fell, Simon Oakland; *Dir:* Peter Yates. Academy Awards '68: Best Film Editing; Edgar Allan Poe Awards '68: Best Screenplay. **VHS, Beta, LV $19.98** *WAR &&&*

Bullseye! Knockabout farce is a letdown considering the talents involved. Shady scientists (Moore and Caine) pursue/are pursued by lookalike con-artists (Caine and Moore), who are pursued in turn by international agents. Full of inside jokes and celebrity cameos, but nothing exceptional.
1990 (PG-13) 95m/C Michael Caine, Roger Moore, Sally Kirkland, John Cleese, Patsy Kensit, Jenny Seagrove; *Dir:* Michael Winner. **VHS, LV $89.95** *COL, PIA & 1/2*

Bullshot Zany English satire sends up the legendary Bulldog Drummond. In the face of mad professors, hapless heroines, devilish Huns and deadly enemies, our intrepid hero remains... distinctly British.
1983 (PG) 84m/C *GB* Alan Shearman, Diz White, Ron House, Frances Tomelty, Michael Aldridge, Ron Pember, Christopher Good; *Dir:* Dick Clement. **VHS, Beta $59.99** *HBO && 1/2*

Bullshot Crummond The Low Moan Spectacular comedy troupe performs this satire that trashes oldtime adventure serials.
1984 90m/C VHS, Beta, LV $39.95 *RKO && 1/2*

Bullwhip A man falsely accused of murder saves himself from the hangman's noose by agreeing to a shotgun wedding.
1958 80m/C Guy Madison, Rhonda Fleming, James Griffith, Peter Adams; *Dir:* Harmon Jones. **VHS $19.98** *REP &&*

Bummer A rock band's wild party turns into tragedy when the bass player goes too far with two groupies.
1973 (R) 90m/C Kipp Whitman, Dennis Burkley, Carol Speed, Connie Strickland; *Dir:* William Allen Castleman. **VHS, Beta $59.98** *MAG Woof!*

Bunco Two policemen working for the Los Angeles Police Department's Bunco Squad discover a college for con artists complete with tape-recorded lessons and on-the-job training.
1983 60m/C Tom Selleck, Robert Urich, Donna Mills, Will Geer, Arte Johnson, Alan Feinstein. **VHS, Beta $59.95** *LHV, WAR &*

Bundle of Joy A salesgirl, who saves an infant from falling off the steps of a foundling home, is mistaken for the child's mother. Remake of 1939's "Bachelor Mother."
1956 98m/C Debbie Reynolds, Eddie Fisher, Adolphe Menjou; *Dir:* Norman Taurog. **VHS, Beta, LV $19.95** *UHV & 1/2*

Bunnicula: Vampire Rabbit Strange things happen to a family after they adopt an abandoned bunny. The family dog and cat team up to prove that their furry friend is a vampire in disguise. An animated tale.
1982 23m/C VHS, Beta $19.95 WOV, GKK ♫ 1/2

The 'Burbs A tepid satire about suburbanites suspecting their creepy new neighbors of murderous activities. Well-designed and sharp, but light on story.
1989 (PG) 101m/C Tom Hanks, Carrie Fisher, Rick Ducommun, Corey Feldman, Brother Theodore, Bruce Dern, Gale Gordon, Courtney Gains; Dir: Joe Dante. VHS, Beta, LV $14.95 MCA, CCB ♫ 1/2

Burden of Dreams The landmark documentary chronicling the berserk circumstances behind the scenes of Werner Herzog's epic "Fitzcarraldo." Stuck in the Peruvian jungles, the film crew was subjected to every disaster imaginable while executing Herzog's vision, including disease, horrendous accident, warring local tribes and the megalomaniacal director himself. Considered better than "Fitzcarraldo," although both films discuss a man's obsession.
1982 94m/C Klaus Kinski, Mick Jagger, Jason Robards Jr., Werner Herzog; Dir: Les Blank. VHS, Beta, LV $59.95 NEW, FFM ♫ 1/2

Burgess, Philby, and MacLean: Spy Scandal of the Century An intriguing, true spy story about three double agents who were able to infiltrate British Intelligence and pass on U.S. atomic secrets to the Soviet Union. Originally aired on PBS.
1986 83m/C VHS, Beta MTI ♫ 1/2

Burglar A cat burglar moves to the right side of the law when she tries to solve a murder case. Whoopi's always a treat, but this movie's best forgotten; fans of Whoopi should watch "Jumpin' Jack Flash" again instead. Co-star Bob "Bobcat" Goldthwaite elevates the comedic level a bit.
1987 (R) 103m/C Whoopi Goldberg, Bob Goldthwait, Lesley Ann Warren, John Goodman; Dir: Hugh Wilson. VHS, Beta, LV $19.98 WAR ♫♫

Burglar Not to be confused with the Whoopi Goldberg vehicle, this Russian film (also known as "Vzlomshchik") relates a story no less American than apple pie and teenage angst. Senka and would-be punk star Kostya are two neglected and disaffected brothers whose father is a drunken womanizer whose presence mitigates not at all his absentee paternalism. When Howmuch, a serious heavy metalloid, pressures Kostya to steal a synthesizer, brother Senka steps in to steal the Community Center's property himself, with not a little consternation. In Russian with English subtitles, the solid performances hold their own against the heavy musical content.
1988 101m/C RU Dir: Valery Orgorodnikov. VHS $59.95 FCT, IFF ♫♫♫

Burial Ground A classically grisly splatter film in which the hungry dead rise and proceed to kill the weekend denizens of an isolated aristocratic mansion.
1985 85m/C Karen Well, Peter Bark; Dir: Andrew Bianchi. VHS, Beta $79.95 VES, LIV ♫

Buried Alive A man is sent to prison on trumped up charges. Only the prison nurse believes he is innocent as a crooked politician strives to keep him behind bars.
1939 74m/B Beverly Roberts, Robert Wilcox, George Pembroke, Ted Osborne, Paul McVey, Alden Chase; Dir: Victor Halperin. Beta, CV $16.95 NOS, MRV, LOO ♫

Buried Alive Director d'Amato reaps his finest gore-fest to date, incorporating the well-established taxidermist gone loony motif in his repertoire of appalling bad taste. Not for the squeamish, this bloodier-than-thou spaghetti spooker is chock full of necrophilia, cannibalism, and more.
1981 90m/C IT Kieran Canter, Cinzia Monreale, Franca Stoppi; Dir: Joe D'Amato. VHS THR ♫♫

Buried Alive Once Ravenscroft Hall was an asylum for the incurably insane. Now, the isolated mansion is a school for troubled teenage girls, run by a charismatic psychiatrist. Captivated by his charm, a young woman joins the staff. Soon, she is tormented by nightmare visions of the long-dead victims of a nameless killer. When the students begin to disappear, she realizes he still lives...and she may be his next victim. Carradine's last role.
1989 (R) 97m/C Robert Vaughn, Donald Pleasence, Karen Witter, John Carradine, Ginger Lynn Allen; Dir: Gerard Kikione. VHS, LV $14.95 COL ♫♫

Buried Alive One of the many horror flicks entitled "Buried Alive," this one (written by Patrick Carducci, and made for cable) is not bad, injecting a bit of levity into the time-worn genre. Schemestress Leigh and her paramour poison her husband. Or so they think, only to discover that he's not quite dead.
1990 (PG-13) 93m/C Tim Matheson, Jennifer Jason Leigh, William Atherton, Hoyt Axton; Dir: Frank Darabont. VHS $79.95 MCA ♫♫ 1/2

Burke & Wills A lush, big-budgeted true story of the two men who first crossed Australia on foot, and explored its central region. Popular Australian film.
1985 (PG-13) 120m/C AU Jack Thompson, Nigel Havers, Greta Scacchi; Dir: Graeme Clifford. VHS, Beta, LV $19.98 CHA, TAM ♫♫ 1/2

Burlesque of Carmen A silent pastiche of the Bizet opera, with Charlie as Don Schmose. Interesting early work.
1916 30m/B Charlie Chaplin, Ben Turpin, Edna Purviance; Dir: Charlie Chaplin. VHS, Beta $19.95 NOS, DVT, TIM ♫♫

The Burmese Harp At the end of World War II, a Japanese soldier is spiritually traumatized and becomes obsessed with burying the masses of war casualties. A searing, acclaimed anti-war statement, in Japanese with English subtitles. Remade by Ichikawa in 1985.
1956 115m/B JP Shoji Yasui, Rentaro Mikuni, Tatsuya Mihashi, Tanie Kitabayashi, Yunosuke Ito; Dir: Kon Ichikawa. New York Film Festival '66: San Giorgio Award; Venice Film Festival '56: San Giorgio Award. VHS, Beta $29.95 CVC, TAM, APD ♫♫♫♫

Burn! An Italian-made indictment of imperialist control by guerrilla-filmmaker Pontecorvo, depicting the efforts of a 19th century British ambassador to put down a slave revolt on a Portuguese-run Caribbean island. Great Brando performance; also titled "Queimada!"
1970 112m/C IT Marlon Brando, Evarist Marquez, Renato Salvatori; Dir: Gillo Pontecorvo. VHS, Beta, LV $19.98 VYG, CRC, FCT ♫♫♫ 1/2

Burn 'Em Up Barnes Twelve episodes each depicting the adventures of "Burn'Em Up" Barnes, a racer, and his buddy, Bobbie.
1934 40m/B Frankie Darro, Lola Lane, Jack Mulhall. VHS, Beta $26.95 SNC, NOS, VCN ♫♫♫ 1/2

Burn Witch, Burn! A college professor proudly finds himself rapidly rising in his profession. His pride turns to horror though when he discovers that his success is not due to his own abilities, but to the efforts of his witchcraft practicing wife. Excellent, atmospheric horror with a genuinely suspenseful climax.
1962 87m/B GB Peter Wyngarde, Janet Blair, Margaret Johnston, Anthony Nicholls, Colin Gordon, Kathleen Byron, Reginald Beckwith, Jessica Dunning; Dir: Sidney Hayers. VHS $18.00 FRG ♫♫♫

Burndown Victims of a serial killer in a town with a large nuclear reactor are themselves radioactive, which leads the Police Chief and a beautiful reporter to the truth and the hidden conspiracy.
1989 (R) 97m/C Cathy Moriarty, Peter Firth; Dir: James Allen. VHS, Beta, LV $79.95 VIR ♫

The Burning A story of macabre revenge set in the dark woods of a seemingly innocent summer camp.
1982 (R) 90m/C Brian Matthews, Leah Ayres, Holly Hunter; Dir: Tony Maylam. VHS, Beta $19.99 HBO ♫

Burning Bed A made-for-television dramatic expose (based on a true story) about wife-beating. Fawcett garnered accolades for her performance as the battered wife who couldn't take it anymore. Highly acclaimed and Emmy-nominated.
1985 95m/C Farrah Fawcett, Paul LeMat, Penelope Milford, Richard Masur; Dir: Robert Greenwald. VHS, Beta $79.98 FOX ♫♫♫ 1/2

The Burning Hills Unexceptional cow flick based on a Louis L'Amour novel. Hunter hides from cattle thieves in a barn and, eventually, in the arms of a half-breed Mexican girl (Wood). Tedious and unsurprising.
1956 94m/C Tab Hunter, Natalie Wood, Skip Homeier, Eduard Franz; Dir: Stuart Heisler. VHS MLB ♫ 1/2

Burning Rage A blazing, abandoned coal mine threatens to wreak havoc in the Appalachians. The government sends geologist Mandrell to help prevent a disaster. Made for television.
1984 100m/C Barbara Mandrell, Tom Wopat, Bert Remsen, John Pleshette, Carol Kane, Eddie Albert; Dir: Gilbert Cates. VHS, LV $19.95 JTC, AVD ♫

Burning Secret After WWI, a widow in Austria (Dunaway) meets a baron (Brandauer) who has befriended her son. Mutual seduction ensues but the sparks don't fly. Adapted from Stefan Zweig story.
1989 (PG) 107m/C GB Faye Dunaway, Klaus Maria Brandauer, Ian Richardson, David Eberts; Dir: Andrew Birkin. VHS, Beta, LV $89.98 VES, IME, LIV ♫♫

Burning Vengeance A DEA agent enters the deadly world of drugs and single-handedly annihilates the drug smugglers and destroys their kingpin.
1989 120m/C VHS, Beta AIP ♫♫

Burnt Offerings A family rents a house for the summer and they become affected by evil forces that possess the house.
1976 (PG) 116m/C Oliver Reed, Karen Black, Bette Davis, Burgess Meredith; Dir: Don Curtis. VHS, Beta $59.95 MGM ♫♫

Bury Me an Angel A female biker sets out to seek revenge against the men who killed her brother.

1971 85m/C Dixie Peabody, Terry Mace, Clyde Ventura, Dan Haggerty, Stephen Whittaker, Gary Littlejohn; *Dir:* Barbara Peeters. **VHS, Beta, LV** **$19.95** *STE, NWV, HHT* ⅃

The Bus Is Coming A love story entwined with the problems of blacks in a small town.
1972 95m/C Mike Sims, Stephanie Faulkner, Morgan Jones; *Dir:* Horace Jackson. **VHS, Beta** **$59.95** *VCD* ⅃⅃

Bus Station A group of travellers look for sex and adventure while taking a bus trip.
1988 89m/C Lucha Villa, Sergio Ramos, Maribel Fernandez, Guillermo Rivas, Charly Valentino. **VHS, Beta** **$79.95** *MED* ⅃

Bus Stop Don Murray plays a naive cowboy who falls in love with Monroe, a barroom singer, and decides to marry her without her permission. Considered by many to be the finest performance by Marilyn Monroe; she sings "That Old Black Magic" in this one. Very funny with good performances by all. Oscar nomination for supporting actor Murray. Based on the William Inge play.
1956 96m/C Marilyn Monroe, Arthur O'Connell, Hope Lange, Don Murray, Betty Field, Casey Adams, Hans Conried, Eileen Heckart; *Dir:* Joshua Logan. National Board of Review Awards '56: 10 Best Films of the Year. **VHS, Beta, LV** **$14.98** *FOX* ⅃⅃⅃

The Busher All-American pitcher's romance is interrupted by a gratuitous Mr. Moneybags. Minor-league conflict with little real hardball.
1919 54m/B Charles Ray, Colleen Moore, John Gilbert. **VHS, Beta $19.95** *GPV, FCT* ⅃½

The Bushido Blade An action-packed samurai thriller of adventure and betrayal set in medieval Japan.
1980 92m/C *JP* Richard Boone, James Earl Jones, Frank Converse; *Dir:* Tom Kotani. **VHS, Beta $69.99** *HBO* ⅃⅃½

The Bushwackers Ireland plays a veteran of the Confederate army who wishes only to put his violent past behind him. He is forced to reconsider his vow when old-west bullies threaten his family.
1952 70m/B Dorothy Malone, John Ireland, Wayne Morris, Lawrence Tierney, Jack Elam, Lon Chaney Sr., Myrna Dell; *Dir:* Rod Amateau. **VHS $16.95** *NOS, LOO, SNC* ⅃½

Business as Usual A meek boutique manager in Liverpool defends a sexually harassed employee, is fired, fights back, and creates a national furor over her rights as a woman.
1988 (PG) 89m/C Glenda Jackson, Cathy Tyson, John Thaw; *Dir:* Lezli-Ann Barrett. **VHS, Beta** **$79.98** *WAR* ⅃⅃

Busted Up A young, hard-luck urban street-fighter battles for the sake of his family and neighborhood against local crime-lords.
1986 (R) 93m/C *CA* Irene Cara, Paul Coufos, Tony Rosato, Stan Shaw, Gordon Judges; *Dir:* Conrad Palmisano. **VHS, Beta $79.95** *MCA* ⅃

Buster The story of Buster Edwards, the one suspect in the 1963 Great Train Robbery who evaded being captured by police. Phil Collins makes his screen debut as one of Britain's most infamous criminals. Collins also performs on the film's soundtrack, spawning two hit singles, "Two Hearts" and "Groovy Kind of Love."

1988 (R) 102m/C *GB* Phil Collins, Julie Walters, Larry Lamb, Stephanie Lawrence, Ellen Beaven, Michael Atwell, Ralph Brown, Christopher Ellison, Sheila Hancock; *Dir:* David Green. **VHS, Beta, LV $89.95** *HBO* ⅃⅃½

Buster and Billie Tragedy ensues when a popular high school student falls in love with the class tramp in rural Georgia in 1948. Serious, decently appointed period teen drama.
1974 (R) 100m/C Jan-Michael Vincent, Joan Goodfellow, Clifton James, Pamela Sue Martin, Robert Englund; *Dir:* Daniel Petrie. **VHS, Beta** **$59.95** *COL* ⅃½

Buster Keaton Festival: Vol. 1 A collection of three Keaton silent shorts from 1921-1922: "Paleface," "The Blacksmith," and "Cops." All shorts have a newly composed musical score.
1922 55m/B Buster Keaton. **VHS, Beta $19.95** *VDM* ⅃½

Buster Keaton Festival: Vol. 2 Three rare silent shorts featuring the great comic: "The Boat," "Frozen North," and "Electric House."
1923 55m/B Buster Keaton. **VHS, Beta $19.95** *VDM* ⅃½

Buster Keaton Festival: Vol. 3 Three more rare silent shorts with Keaton: "Daydreams," "The Balloonatic," and "The Garage," also featuring Fatty Arbuckle.
1920 54m/B Buster Keaton. **VHS, Beta $19.95** *VDM* ⅃½

Buster Keaton Rides Again/The Railroader "The Railroader" is a silent comedy short that returns Buster Keaton to the type of slapstick he made famous in his legendary 1920's films. "Buster Keaton Rides Again" (in black-and-white) is a documentary-style look at Keaton filmed during the making of "The Railroader." Besides scenes of Keaton at work, there is also a capsule rundown of his career.
1965 81m/C Buster Keaton. **VHS, Beta $29.95** *GPV, VYY* ⅃½

Buster Keaton: The Golden Years Features three classic Buster Keaton silent shorts; "The Pale Face," "Daydreams," and "The Blacksmith," all from 1921.
1921 60m/B Buster Keaton. **VHS, Beta $29.95** *DVT* ⅃½

Buster Keaton: The Great Stone Face Program from the Rohauer Collection contains footage of Buster Keaton in "Fatty at Coney Island," "Cops," "Balloonatics," "Day Dreams," and "The General."
1968 60m/B Buster Keaton; *Nar:* Harry Morgan. **VHS, Beta $57.95** *MAS, SLF* ⅃½

Bustin' Loose A fast-talking ex-con reluctantly drives a bus load of misplaced kids and their keeper cross-country.
1981 (R) 94m/C Richard Pryor, Cicely Tyson, Robert Christian, George Coe, Bill Quinn; *Dir:* Oz Scott. **VHS, Beta, LV $14.95** *MCA* ⅃⅃⅃

Busting Gould and Blake play a pair of slightly off-the-wall L.A. cops. The pair are forced "to bust" local addicts and prostitutes instead of the real crime bosses because much of the police department is on the take. Plenty of comedy and action as well as highly realistic drama.
1974 92m/C Elliott Gould, Robert Blake, Allen Garfield, Antonio Fargas, Michael Lerner, Sid Haig, Cornelia Sharpe; *Dir:* Peter Hyams. **VHS** *MCA* ⅃⅃½

But Not for Me Gable stars as a middled-aged entertainment executive who thinks he can forestall maturity by carrying on with his youthful secretary.
1959 105m/B Clark Gable, Carroll Baker, Lilli Palmer, Lee J. Cobb, Barry Coe, Thomas Gomez, Charles Lane; *Dir:* Walter Lang. **VHS** **$14.95** *PAR, FCT* ⅃⅃½

But Where Is Daniel Wax? At a class reunion, a doctor and a popular singing star reminisce about their youth and Daniel Wax, their one-time hero. Hebrew with English subtitles.
1974 95m/C Lior Yaeni, Michael Lipkin, Esther Zevko; *Dir:* Avram Heffner. **VHS, Beta $79.95** *ERG* ⅃⅃

Butch Cassidy and the Sundance Kid A couple of legendary outlaws at the turn of the century take it on the lam with a beautiful, willing ex-school teacher. With a clever script, humanly fallible characters, and warm, witty dialogue, this film was destined to become a box-office classic. Featured the hit song, "Raindrops Keep Falling on My Head" and renewed the buddy film industry, as Newman and Redford trade insult for insult. Look for the great scene where Newman takes on giant Ted Cassidy in a fist fight. Oscar nominated for Best Picture, supporting actor Hill, and Sound.
1969 (PG) 110m/C Paul Newman, Robert Redford, Katharine Ross, Jeff Corey, Strother Martin, Cloris Leachman, Kenneth Mars, Ted Cassidy, Sam Elliott; *Dir:* George Roy Hill. Academy Awards '69: Best Cinematography, Best Song ("Raindrops Keep Fallin' on My Head"), Best Story & Screenplay, Best Original Score. **VHS, Beta, LV $19.98** *FOX, TLF* ⅃⅃⅃⅃

Butch and Sundance: The Early Days Traces the origins of the famous outlaw duo. It contains the requisite shoot-outs, hold-ups and escapes. A "prequel" to "Butch Cassidy and the Sundance Kid."
1979 (PG) 111m/C Tom Berenger, William Katt, John Schuck, Jeff Corey, Jill Eikenberry, Brian Dennehy, Peter Weller; *Dir:* Richard Lester. **VHS, Beta $19.98** *FOX* ⅃⅃½

The Butcher's Wife The charming tale of a young psychic who brings romance to a Greenwich Village neighborhood. Moore stars as the clairvoyant whose mystical powers bring magic into the lives of everyone around her, including the local psychiatrist (Daniels), who falls under her spell.
1991 (PG-13) 107m/C Demi Moore, Jeff Daniels, George Dzundza, Frances McDormand, Margaret Colin, Mary Steenburgen; *Dir:* Terry Hughes. **VHS, Beta** *PAR* ⅃⅃

Butler's Dilemma A jewel thief and a playboy both claim the identity of a butler who never existed, with humorous results.
1943 75m/B *GB* Richard Hearne, Francis L. Sullivan, Hermione Gingold, Ian Fleming. **VHS, Beta $29.95** *VYY, TIM* ⅃⅃

Butterfield 8 A seedy film of the John O'Hara novel about a prostitute that wants to go straight and settle down. Taylor won an Oscar, perhaps because she was ill and had lost in the two previous years in more deserving roles. Also Oscar nominated for Color Cinematography.
1960 108m/C Elizabeth Taylor, Laurence Harvey, Eddie Fisher, Dina Merrill, Mildred Dunnock, Betty Field, Susan Oliver, Kay Medford; *Dir:* Daniel Mann. Academy Awards '60: Best Actress (Taylor). **VHS, Beta, LV $19.98** *MGM, PIA, BTV* ⅃⅃⅃

B

Butterflies Ria, a middle-aged housewife with two lazy sons, is beginning to feel her age, and like a butterfly, feeling like she has too much to do and not enough time. Then Leonard appears in her life—divorced, lonely and romantic. Made for British television.
1975 60m/C GB VHS $29.95 BFS ✔✔✔

Butterflies Are Free Fast-paced humor surrounds the Broadway play brought to the big screen. Blind youth Albert is determined to be self-sufficient. A next-door-neighbor actress helps him gain independence from his over-protective mother (Heckart). Oscar nominations for Cinematography and Sound.
1972 (PG) 109m/C Goldie Hawn, Edward Albert, Eileen Heckart, Michael Glaser; **Dir:** Milton Katselas. Academy Awards '72: Best Supporting Actress (Heckart). **VHS, Beta $9.95** GKK ✔✔✔

Butterfly Based on James M. Cain's novel about an amoral young woman (Pia, who'd you think?) who uses her beauty and sensual appetite to manipulate the men in her life, including her father. Set in Nevada of the 1930s, father and daughter are drawn into a daring and forbidden love affair by their lust and desperation.
1982 (R) 105m/C Pia Zadora, Stacy Keach, Orson Welles, Edward Albert, James Franciscus, Lois Nettleton, Stuart Whitman, June Lockhart; **Dir:** Matt Cimber. **VHS, Beta, LV $79.95** VES, LIV ✔✔ ½

Butterfly Affair A beautiful singer, involved in a scheme to smuggle two million dollars worth of gems, plots to double-cross her partners in crime.
1971 (PG) 75m/C Claudia Cardinale, Henri Charriere, Stanley Baker; **Dir:** Jean Herman. **VHS, Beta** PGN ✔ ½

Butterfly Ball This retelling of the 19th century classic combines the rock music of Roger Glover, live action and animation by Halas and Batchelor. Only for the brain-cell depressed.
1976 85m/C GB Twiggy, Ian Gillian, David Coverdale; **Dir:** Tony Klinger; **Nar:** Vincent Price. **VHS, Beta $39.95** VCL ✔

Buxom Boxers An array of semi-nude female boxers whack away at each other in examples of a increasingly popular quasi-sport.
1986 57m/C VHS, Beta $24.95 AHV ✔

Buy & Cell A stockbroker attempts to set up a business while serving a prison sentence. An assortment of amusing inmates doesn't help matters much. Fans of Malcolm McDowell will enjoy his performance as the shady warden.
1989 (R) 95m/C Robert Carradine, Michael Winslow, Malcolm McDowell, Ben Vereen, Lise Cutter, Randall "Tex" Cobb, Roddy Piper; **Dir:** Robert Boris. **VHS, LV $19.95** STE, NWV ✔ ½

Buying Time Two teenagers are arrested when they try to pry their money away from a dishonest bookie, and the police use their connections to solve a drug-related murder.
1989 (R) 97m/C Dean Stockwell, Jeff Schultz, Michael Rudder, Tony De Santis, Leslie Toth, Laura Cruickshank; **Dir:** Mitchell Gabourie. **VHS, Beta $79.95** FOX ✔

By the Blood of Others In a small French town, a psycho holds two women hostage in a farmhouse, and the townspeople try to figure out a solution without getting anyone killed. In French with English subtitles.

1973 90m/C FR Mariangela Melato; **Dir:** Marc Simenon. **VHS, Beta $59.95** FCT ✔✔✔

By Dawn's Early Light Made-for-cable-TV thriller about two Air Force pilots who must decide whether or not to drop the bombs that would begin World War III.
1989 100m/C Powers Boothe, Rebecca DeMornay, James Earl Jones, Martin Landau, Rip Torn, Darren McGavin; **Dir:** Jack Sholder. **VHS, Beta, LV $89.99** HBO ✔✔

By Design Two cohabiting women want to have a baby, so they embark on a search for the perfect stud. Not one of Astin's better performances.
1982 (R) 90m/C CA Patty Duke, Sara Botsford; **Dir:** Claude Jutra. **VHS, Beta $59.99** HBO ✔

By Hook or Crook Zany Mexican comedy/mystery featuring the "Super-Mexican" detective solving a series of high-society murders.
1985 94m/C MX Rafael Inclan, Maribel Guardia; **Dir:** Julio Ruiz Llaneza. **VHS, Beta $72.95** MED ✔

By Love Possessed Can a seemingly typical, quiet, New England town stay quiet when Lana Turner is your neighbor? Vintage romantic melodrama about an attorney who is drawn into an affair when he realizes his home life is not all it could be.
1961 115m/C Lana Turner, Efrem Zimbalist Jr., Jason Robards Jr., George Hamilton, Thomas Mitchell; **Dir:** John Sturges. **VHS $14.95** WKV ✔✔ ½

Bye Bye Baby Two luscious young women cavort amid lusty Italian men on various Mediterranean beaches.
1988 (R) 80m/C IT Carol Alt, Brigitte Nielsen, Jason Connery, Luca Barbareschi; **Dir:** Enrico Oldoini. **VHS, Beta, LV $79.95** PSM, PAR ✔

Bye, Bye, Birdie Energized and sweet film version of the Broadway musical about a teen rock and roll idol (Pearson doing Elvis) coming to a small town to see one of his fans before he leaves for the army. Songs include "Put on a Happy Face" and "We Love You Conrad." The film that made Ann-Margret a star.
1963 112m/C Dick Van Dyke, Janet Leigh, Ann-Margret, Paul Lynde, Bobby Rydell, Maureen Stapleton, Ed Sullivan, Trudi Ames; **Dir:** George Sidney. **VHS, Beta, LV $19.95** COL, LDC ✔✔✔

Bye Bye Blues The lives of the Cooper family are disrupted when the husband is called to service during WWII. In need of money, wife Daisy joins a local swing band and begins charming all who watch her. A sweet-spirited tale of love, loyalty and the search for inner strength.
1991 (PG) 110m/C CA Rebecca Jenkins, Luke Reilly, Stuart Margolin, Michael Ontkean, Wayne Robson, Robyn Stevan, Kate Reid; **W/Dir:** Anne Wheeler. Genie Awards '90: Best Actress (Jenkins), Best Supporting Actress (Stevan). **VHS** MNC ✔✔ ½

Bye Bye Brazil A changing Brazil is seen through the eyes of four wandering gypsy minstrels participating in a tent show traveling throughout the country. Lots of Brazilian charm. Subtitled in English.
1979 100m/C BR Jose Wilker, Betty Faria, Fabio Junior, Zaira Zambello; **Dir:** Carlos Diegues. New York Film Festival '80: Best Supporting Actress. **VHS, Beta, LV $79.95** FXL, TAM ✔✔ ½

The C-Man Customs agent Jagger finds work isn't so dull after all. Murder and theft on an international scale make for a busy week as he follows jewel smugglers from

Paris to New York. Docu-style though routine crime story.
1949 75m/B Dean Jagger, John Carradine, Harry Landers, Rene Paul; **Dir:** Joseph Lerner. **VHS $16.95** SNC ✔

Cabaret Hitler is rising to power, racism and anti-Semitism are growing, and the best place to hide from it all is the cabaret. With dancing girls, androgynous master of ceremonies Grey and American expatriate and singer Minnelli, you can laugh and drink and pretend tomorrow will never come. Minnelli does just that, meanwhile unknowingly sharing her English lover with a homosexual German baron. Face to face with the increasing horrors of Nazism, she persists in the belief that the "show must go on." Based on the John Kander's hit Broadway musical (taken from the Christopher Isherwood stories), the film is even more impressive with excellent direction and cinematography. Songs include "Money Makes the World Go Around," "Wilkommen," and "Mein Herr." It picked up additional Oscar nominations for Best Picture and Adapted Screenplay.
1972 (PG) 119m/C Liza Minnelli, Joel Grey, Michael York, Marisa Berenson; **Dir:** Bob Fosse. Academy Awards '72: Best Actress (Minnelli), Best Art Direction/Set Decoration, Best Cinematography, Best Director (Fosse), Best Film Editing, Best Sound, Best Supporting Actor (Grey), Best Musical Score; British Academy Awards '72: Best Actress (Minnelli), Best Art Direction/Set Decoration, Best Director (Fosse), Best Film, Best Musical Score; Golden Globe Awards '73: Best Film—Musical/Comedy; National Board of Review Awards '72: Best Director (Fosse), Best Film, Best Supporting Actor (Grey), Best Supporting Actress (Berenson). **VHS, Beta, LV $19.98** FOX, BTV, CCB ✔✔✔ ½

Cabin in the Sky A poor woman fights to keep her husband's soul out of the devil's clutches. Based on a Broadway show and featuring an all-Black cast. Lively dance numbers and a musical score by Harold Arlen and E.Y. Harburg, with contributions from Duke Ellington. Songs include "Happiness Is Just a Thing Called Joe," "Taking a Chance on Love," and "Life's Full of Consequences." Minnelli's first feature film.
1943 99m/C Ethel Waters, Eddie Anderson, Lena Horne, Rex Ingram, Louis Armstrong, Duke Ellington; **Dir:** Vincente Minnelli. **VHS, Beta, LV $19.98** MGM, MLB, FHE ✔✔✔

The Cabinet of Dr. Caligari A pioneering film in the most extreme expressionistic style about a hypnotist in a carnival and a girl-snatching somnambulist. Highly influential in its approach to lighting, composition, design and acting. Much imitated. Silent.
1919 52m/B GE Conrad Veidt, Werner Krauss, Lil Dagover; **Dir:** Robert Wiene. Brussels World's Fair '58: #12 of the Best Films of All Time. **VHS, Beta, LV $16.95** SNC, REP, MRV ✔✔✔✔

Cabiria The pioneering Italian epic about a Roman and a slave girl having a love affair in Rome during the Second Punic War. Immense sets and set-pieces; an important influence on Griffith and DeMille. Silent.
1914 123m/B IT Giovanni Pastrone. **VHS, Beta, 8mm $39.95** KIV, FCT, TAM ✔✔ ½

Cabo Blanco A bartender and a variety of other characters, including an ex-Nazi and a French woman searching for her lover, assemble in Peru after WWII. Nazi Robards controls police chief Rey, while American Bronson runs the local watering hole and eyes French woman Sanda. Hey, this sounds familiar. Everyone shares a common interest:

finding a missing treasure of gold, lost in a ship wreck. Remaking "Casablanca" via "The Treasure of Sierra Madre" is never easy.
1981 (R) 87m/C Charles Bronson, Jason Robards Jr., Dominique Sanda, Fernando Rey, Gilbert Roland, Simon MacCorkindale; *Dir:* J. Lee Thompson. **VHS, Beta** $59.95 *MED ♂♂*

Caches de Oro A routine arrest turns into the bloodiest case of a federal judge's career.
19?? 90m/C MX Alvaro Zermeno, Fernando Casanova, Tony Bravo. **VHS, Beta** $79.95 *MED ♂♂*

Cactus Melodrama about a young French woman, separated from her husband, who faces the reality of losing her sight after a car accident. She experiences a growing relationship with a blind man and contemplates the thought of risky surgery which may improve her eyesight or cause complete blindness. Countering all that blindness and tangled romance is a camera that pans lush Australian landscapes and humorously focuses on the small telling details of daily life.
1986 95m/C AU Isabelle Huppert, Robert Menzies, Monica Maughan, Sheila Florence; *Dir:* Paul Cox. **VHS, Beta** $19.98 *LHV ♂♂½*

Cactus Flower Good cast doesn't quite suffice to make this adaptation of a Broadway hit work. A middle-aged bachelor dentist gets involved with a kookie mistress, refusing to admit his real love for his prim and proper receptionist. Hawn's big leap to stardom.
1969 (PG) 103m/C Walter Matthau, Goldie Hawn, Ingrid Bergman, Jack Weston, Rick Lenz; *Dir:* Gene Saks. Academy Awards '69: Best Supporting Actress (Hawn). **VHS, Beta, LV** $14.95 *COL ♂♂*

Cactus in the Snow A virginal Army private serving during the Vietnam War era tries to find sex on a 72-hour leave and instead discovers love. Some guys have all the luck. Melodrama never grabs where it ought to.
1972 90m/C Richard Thomas, Mary Layne, Lucille Benson; *Dir:* Martin Zweiback. **VHS, Beta, LV** *WGE ♂½*

Cada Noche un Amor (Every Night a New Lover) A singer is deceived into working as a harem girl for rich Arabs.
1975 90m/C SP Sara Montiel, Fernando Graney. **VHS, Beta** *MAD ♂*

Caddie Based on an autobiographical story of a woman who leaves her unfaithful husband in 1930s Australia. Struggling to raise her two children, she works as a waitress in a bar where she finds romance with one of the regulars. Average script bolstered by Morse's performance.
1981 107m/C Helen Morse, Jack Thompson; *Dir:* Donald Crombie. **VHS, Beta** $59.95 *HBO ♂♂½*

The Caddy Lewis plays frantic caddy prone to slapstick against Martin's smooth professional golfer with a bent toward singing. Mostly a series of Martin and Lewis sketches that frequently land in the rough. Introduces several songs, including a classic Martin and Lewis rendition of "That's Amore." Look for cameos by a host of professional golfers.
1953 95m/C Dean Martin, Jerry Lewis, Donna Reed, Barbara Bates, Joseph Calleia, Marshall Thompson, Fred Clark; *Dir:* Norman Taurog. **VHS, Beta** $14.95 *PAR ♂♂*

Caddyshack Inspired performances by Murray and Dangerfield drive this sublimely moronic comedy onto the green. Does for golf what "Major League" tried to do for baseball. The action takes place at Bushwood Country Club, where young clean-cut caddy O'Keefe is bucking to win the club's college scholarship. Characters involved in various sophomoric set pieces include obnoxious club president Knight, a playboy who is too laid back to keep his score (Chase), a loud, vulgar, extremely rich golfer (Dangerfield), and Murray as a filthy gopher-hunting groundskeeper. Occasional dry moments are followed by scenes of pure (and tasteless) anarchy, so watch it with someone immature. Gopher holes will never again be the same.
1980 (R) 99m/C Chevy Chase, Rodney Dangerfield, Ted Knight, Michael O'Keefe, Bill Murray, Sarah Holcomb, Brian Doyle-Murray; *W/ Dir:* Harold Ramis. **VHS, Beta, LV** $19.98 *WAR ♂♂♂½*

Caddyshack 2 The obligatory sequel to "Caddyshack," although Bill Murray wisely avoided further encroachment of gopher holes and director Harold Ramis opted for screenwriting chores only. Mason is the star of the show as a crude self-made millionaire who tangles with the snobs at the country club. Although it occasionally earns a side-splitting chuckle, "Shack 2" has significantly fewer guffaws than the original, proving once again that funny guys are always undone by lousy scripts and weak direction.
1988 (PG) 103m/C Jackie Mason, Chevy Chase, Dan Aykroyd, Dyan Cannon, Robert Stack, Dina Merrill, Randy Quaid, Jessica Lundy, Jonathan Silverman, Chynna Phillips; *Dir:* Allan Arkush. **VHS, Beta, LV, 8mm** $19.98 *WAR ♂♂*

Cadence The directorial debut of actor Martin Sheen, this fitful melodrama stars son Charlie as an unruly trooper on a 1960s army base. Placed in an all-black stockade for punishment, he bonds with his brother prisoners by defying the hardcase sergeant (played by Sheen the elder). Characters and situations are intriguing but not rendered effectively. Based on the novel "Count a Lonely Cadence" by Gordon Weaver.
1989 (PG-13) 97m/C Charlie Sheen, Martin Sheen, Larry Fishburne, Michael Beach, Ramon Estevez; *Dir:* Martin Sheen. **VHS, LV** $92.98 *REP, PIA ♂♂*

Cadet Rousselle Two-dimensional puppets sing the old French song of the period between the French Revolution and the Napoleonic Era.
1983 7m/C **VHS, Beta, 3/4U, Special order formats** *IFB ♂♂*

Cadillac Man Williams is the quintessential low-life car salesman in this rather disjointed comedy. A lesser comedic talent might have stalled and been abandoned, but Williams manages to drive away despite the flat script and direction. One storyline follows his attempt to sell 12 cars in 12 days or lose his job, while another follows his confrontation with a gun-toting, mad-as-hell cuckolded husband. Williams and Robbins are often close to being funny in a hyperkinetic way, but the situations are dumb enough to rob most of the scenes of their comedy. Watch for a movie-stealing bit by the spunky waitress at the local Chinese restaurant.
1990 (R) 95m/C Robin Williams, Tim Robbins, Pamela Reed, Fran Drescher, Zack Norman, Annabella Sciorra, Lori Petty, Paul Guilfoyle; *Dir:* Roger Donaldson. **VHS, Beta** $14.98 *ORI ♂♂*

Caesar and Cleopatra Based on the classic George Bernard Shaw play. Caesar meets the beautiful Cleopatra in ancient Egypt and helps her gain control of her life. Remains surprisingly true to Shaw's adaptation, unlike many other historical films of the same era.
1945 135m/C GB Claude Rains, Vivien Leigh, Stewart Granger, Flora Robson, Francis L. Sullivan, Cecil Parker; *Dir:* Gabriel Pascal. **VHS, Beta** $29.95 *VID ♂♂½*

Cafe Express A con artist (he sells coffee illegally aboard an Italian train) stays one step ahead of the law as he raises money to help his ailing son. As in "Bread and Chocolate," Manfredi is again the put-upon working class hero comedically attempting to find a better way of life in this bittersweet tale.
1983 90m/C IT Nino Manfredi, Gigi Reder, Adolfo Celi, Vittorio Mezzogiorno; *Dir:* Nanni Loy. **VHS, Beta** *PGN ♂♂½*

Cafe Romeo Sweet story of friendship that revolves around a New York City neighborhood coffeehouse. Stewart stars as Lia, a waitress who dreams of becoming a fashion designer, but is held back from her aspirations by her small-minded husband.
1991 (R) 93m/C Catherine Mary Stewart, Jonathan Crombie; *Dir:* Rex Bromfield. **VHS, LV** $89.98 *REP ♂♂*

The Cage Gangsters enlist a brain-damaged Vietnam vet played by Ferrigno to participate in illegal "cage fights," enclosed wrestling matches fought to the death. Crude and annoying.
1989 (R) 101m/C Lou Ferrigno, Reb Brown, Michael Dante, Mike Moroff, Marilyn Tokuda, James Shigeta, Al Ruscio; *Dir:* Hugh Kelley. **VHS, Beta, LV** $14.98 *ORI ♂*

The Cage Sent to prison for murdering his girlfriend, Jive has no idea what the pen has in store for him. The inmates stage a mock trial, the defendant is found guilty and the sentence is too horrible to imagine. From a stage play by the San Quentin Drama Workshop and co-starring Hayes, author of "Midnight Express."
1990 90m/C Rick Cluchey, William Hayes; *W/Dir:* Rick Cluchey. **VHS, Beta** $59.95 *LGC ♂♂♂*

Caged Fury American women being held captive in Southeast Asia are brainwashed into becoming walking time bombs. Yes, that dame's gonna blow. Made in the Philippines with the best of intentions.
1980 (R) 90m/C PH Bernadette Williams, Taffee O'Connell, Jennifer Laine; *Dir:* Cirio H. Santiago. **VHS, Beta** $39.95 *WPM ♂*

Caged Fury Two Los Angeles women allegedly commit sexual crimes and are sent to prison. Cheap, exploitive women's prison rehash.
1990 85m/C Erik Estrada, Richie Barathy, Roxanna Michaels, Paul Smith, James Hong, Greg Cummins; *Dir:* Bill Milling. **VHS, Beta, LV** $79.95 *COL ♂*

Caged Heart According to some cinematic code, the male lead always gets in lots of trouble whenever he tries to help a beautiful stranger. This time, Berry, as Bruno Windler, winds up arrested for shoplifting when he attempts to help Abril. Jailed and on the verge of parole, Windler unwittingly becomes involved in a prison break and is accused of helping a crime don mastermind the escape and of shooting a guard. This is where that gallic flare for capturing the sado-masochistic underbelly of humanity comes in,

as Winkler is viciously pursued by the malevolent guard. Also titled "The Patsy."
1985 (R) 85m/C Richard Berry, Victoria Abril, Richard Bohringer, Farid Chopel, Fabrice Eberhard; *Dir:* Denis Amar. **VHS, Beta $19.95** STE, NWV ♂♂½

Caged Heat A low-budget babes-behind-bars film that's been screened by The Museum of Modern Art and the New York Collective for the Living Cinema? And touted as the best sexploitation film of the day, produced by New World, a renowned sex'n'violence shop? Not to mention that it's got a theme dubbed by William ("Naked Lunch") Burroughs as "technological psychiatry;" now that's something wild. Demme scripted and made his directorial debut with this genre-altering installment in Roger Corman's formulaic cellblock Cinderella cycle. What's new isn't the plot—an innocent woman is put behind bars, where she loses some of her innocence—but in the treatment of an historically misogynistic genre (not to mention that this all takes place in the US of A, not some vaguely third world never-neverland). These babes may wear hot pants and gratuitously bare their midriffs, but they're not brainless bimbos whose fate is dictated by he-men. Rather, they're strong individuals who work (brace yourself) together to liberate themselves. Not that Demme didn't make any artistic compromises: some major lapses in artistic integrity occur. But the film owes its cult status to its ability to set the genre akimbo and to its ironic refusal (well, most of the time) to be exploitative. Blues music by John Cale; also known as "Renegade Girls." Cult diva Steele returned to the big screen after a six-year hiatus to play the wheelchair-wielding prison warden, a part written specifically for her.
1974 (R) 83m/C Juanita Brown, Erica Gavin, Roberta Collins, Barbara Steele, Ella Reid, Cheryl "Rainbeaux" Smith; *W/Dir:* Jonathan Demme. **VHS** SUE, NWV ♂♂½

Caged in Paradiso A convicted criminal and his wife are dumped onto a remote, prisoner's isle to live out their lives. When Cara loses contact with her husband, she must fend for herself. Survival tale boringly told.
1989 (R) 90m/C Irene Cara; *Dir:* Michael Snyder. **VHS, Beta, LV $79.95** VMK, IME ♂½

Caged Terror Two urbanites hit the countryside for a weekend and meet a band of crazy rapists who ravage the wife and set the husband raging with bloodthirsty revenge. Squalid stroll through the ruins.
1972 76m/C Percy Harkness, Elizabeth Suzuki, Leon Morenzie. **VHS, Beta $59.95** NWV Woof!

Caged Women An undercover journalist enters a women's prison and gets an eyeful. Also on video as "Women's Penitentiary IV."
1984 (R) 97m/C FR IT Laura Gemser, Gabriele Tinti, Lorraine De Selle, Maria Romano; *Dir:* Vincent Dawn. **VHS, Beta $69.95** VES, LIV, PHX ♂

Cahill: United States Marshal The aging Duke in one of his lesser moments, portraying a marshal who comes to the aid of his sons. The boys are mixed up with a gang of outlaws, proving that no matter how good a parent you are, sometimes the kids just lean toward the wayward. Turns out though, that the boys harbor a grudge against the old man due to years of workaholic neglect. Will Duke see the error of his ways and reconcile with the delinquent boys? Will he catch the outlaw leader? Will he go on to star in other Duke vehicles?
1973 (PG) 103m/C John Wayne, Gary Grimes, George Kennedy, Neville Brand, Marie Windsor, Harry Carey Jr.; *Dir:* Andrew V. McLaglen. **VHS, Beta, LV $19.98** WAR, PIA, TLF ♂♂

The Caine Mutiny A group of naval officers revolt against a captain they consider mentally unfit. Bogart is masterful as Captain Queeg, obsessed with cleanliness while onboard and later a study in mental meltdown during the court-martial of a crew member who participated in the mutiny. Based on the Pulitzer-winning novel by Herman Wouk, the drama takes a close look at the pressure-filled life aboard ship during WWII. Garnered Oscar nominations for Best Picture, Best Actor (Bogart), Best Supporting Actor (Tully), Screenplay, Sound Recording, Score for Comedy & Drama, and Film Editing.
1954 125m/C Humphrey Bogart, Jose Ferrer, Van Johnson, Fred MacMurray, Lee Marvin, Claude Akins, E.G. Marshall; *Dir:* Edward Dmytryk. **VHS, Beta, LV, 8mm $19.95** COL ♂♂♂♂

The Caine Mutiny Court Martial A young lieutenant is up for a court-martial after taking control of the U.S.S. Caine in the midst of a typhoon. In order to save him, his lawyer must discredit the paranoid Commander Queeg. As the events of the mutiny unfold, it becomes clear that Queeg's obsession with discipline had become a threat to everyone aboard. Good performances, particularly by Bogosian as the lawyer. Based on the Pulitzer Prize-winning novel by Herman Wouk. Made for television.
1988 (PG) 100m/C Jeff Daniels, Eric Bogosian, Brad Davis, Peter Gallagher; *Dir:* Robert Altman. **VHS, LV $89.95** VMK ♂♂♂

Cain's Cutthroats A former Confederate army captain and a bounty hunting preacher team up to settle the score with soldiers on a gang-raping, murdering spree.
1971 (R) 87m/C Scott Brady, John Carradine, Robert Dix. **VHS, Beta $19.95** GHV, TIM ♂

Cajon Pass A group of trains challenges the terrain and try and go through Cajon Pass into Los Angeles.
1988 60m/C VHS, Beta $49.95 VRL ♂♂

Cal Set in Northern Ireland, a young Catholic man is recruited into the Irish Republican Army. He falls in love with an older Protestant widow, whose husband, a policeman, he helped kill, acting as the get-away driver for his fellow Republicans. Thoughtful and tragic, with excellent performances. Musical score by Mark Knopfler.
1984 (R) 104m/C IR Helen Mirren, John Lynch, Donal McCann, Kitty Gibson, Ray McAnally, John Kavanagh; *Dir:* Pat O'Connor. Cannes Film Festival '84: Best Actress (Mirren). **VHS, Beta $19.98** WAR, FCT, TAM ♂♂♂

Calacan A non-human settlement filled with culture and tradition fears only the living in this family musical.
19?? ?m/C MX Mauro Mendonca, Sylvia Guevara, Dora Montiel. National Board of Review Awards '79: Best Foreign Film. **VHS, Beta** WCV ♂♂♂

Calamity Jane In one of her best Warner musicals, Day stars as the rip-snortin', gun-totin' Calamity Jane of Western lore. Her on-again, off-again romance with Wild Bill Hickok is set to the songs of Fain and Webster, including "Just Blew in from the Windy City," "The Black Hills of Dakota," "The Deadwood Stage (Whip-Crack-Away!)" and "Secret Love."
1953 101m/C Doris Day, Howard Keel, Allyn Ann McLerie; *Dir:* David Butler. Academy Awards '53: Best Song ("Secret Love"). **VHS, Beta, LV $19.98** WAR, PIA, FCT ♂♂♂

Calamity Jane A made-for-television biography of the famous lady crackshot. Alexander is cast as the tough-as-nails woman who considered herself on a par with any man.
1982 96m/C Jane Alexander, Frederic Forrest, Ken Kercheval, Talia Balsam, David Hemmings; *Dir:* James Goldstone. **VHS, Beta $19.98** FOX ♂♂½

Caledonian Dreams Sequel to "Oriental Dreams" follows the intimate escapades of three beautiful women as they explore the exotic lifestyle of the South Seas. Music by Norio Maeda and Windbreakers.
1982 46m/C JP *Dir:* Shoji Otake. **VHS, Beta, LV $27.98** PNR ♂♂½

California Casanova Mirthless, semi-musical farce about a stagehand learning to be a great lover to win a beautiful singer's heart. Nice cast; nobody's home.
1991 (R) 94m/C Jerry Orbach, Audrey Landers, Tyrone Power Jr., Bryan Genesse, Ken Kercheval; *Dir:* Nathaniel Christian. **VHS $19.95** ACA Woof!

California Dreaming Nerdy young man heads west to California where he tries to fit in with the local beach crowd. Reminiscent of the popular beach movies of the 60s.
1979 (R) 93m/C Dennis Christopher, Tanya Roberts, Glynnis O'Connor, John Calvin, Seymour Cassel; *Dir:* John Hancock. **VHS, Beta $69.95** VES, LIV ♂♂

California Girls A radio station stages a beauty contest and three sexy ladies prove to be tough competition. Features women in little swimsuits, minimal plot, and a soundtrack by the Police, Kool & the Gang, Blondie, Queen, and 10cc.
1984 83m/C Al Music, Mary McKinley, Alicia Allen, Lantz Douglas, Barbara Parks; *Dir:* Rick Wallace. **VHS, Beta $59.95** VCL ♂

California Gold Rush Hays portrays young aspiring writer, Bret Harte, who, in search of adventure in 1849, arrives in Sutter's Fort and takes on a job at the local sawmill. When gold is found, Sutter's Fort is soon overrun with fortune hunters whose greed, violence and corruption threaten to tear apart the peaceful community. Made for television.
1981 100m/C Robert Hays, John Dehner, Dan Haggerty, Ken Curtis; *Dir:* Jack B. Hively. **VHS, Beta $29.98** MAG ♂♂

California Joe California is being eyed by Confederate sympathizers and also by their leader, who wants it for his own plans. Enter California Joe, a Union officer in disguise, who will save the day. Standard western.
1943 55m/B Donald (Don "Red") Barry, Lynn Merrick, Helen Talbot. **VHS, Beta $24.95** DVT ♂

California Manhunt Prime California beefcake parades before the camera in a search for the "perfect man."
1989 60m/C VHS, Beta $24.95 AHV ♂

The California Raisins: Meet the Raisins Claymation video gives the complete history of The California Raisins, including some of their best songs, and includes veggie commercials too.
1989 30m/C The California Raisins. **VHS, Beta $12.95** TMG ♂

California Straight Ahead Denny—whose niche in the twenties was silent action comedies, and who had teamed with director Pollard earlier in the decade to produce "The Leatherpushers" series—stars in this silent actioner which features a cross-country road trip, zoo animals-on-a-rampage, and a car-racing conclusion (sounds like an action formula ahead of its time). Audiences didn't realize until the talkies that the All-American manly man was in fact played by a Brit. For those who revel in trivia, Denny appeared in the 1961 "Batman."
1925 77m/B Reginald Denny, Gertrude Olmsted, Tom Wilson, Charles Gerrard, Lucille Ward, John Steppling; *Dir:* Harry Pollard. VHS, Beta $24.95 *GPV, FCT* ♬♬

California Suite The posh Beverly Hills Hotel is the setting for four unrelated Neil Simon skits, ranging from a battling husband and wife to feuding friends. Simonized dialogue is crisp and funny. Jazz score by Claude Bolling.
1978 (PG) 103m/C Alan Alda, Michael Caine, Bill Cosby, Jane Fonda, Walter Matthau, Richard Pryor, Maggie Smith, Elaine May; *Dir:* Herbert Ross. Academy Awards '78: Best Supporting Actress (Smith). VHS, Beta, LV $14.95 *COL* ♬♬♬

Caligari's Curse Palazzolo's first film is a satiric, non-narrative stab at his Catholic upbringing via German expressionism and modern performance art.
1974 30m/C *Dir:* Tom Palazzolo. VHS, Beta *FCT* ♬♬

Caligula Infamous, extremely graphic and sexually explicit adaptation of the life of the mad Roman emperor, Caligula. Scenes of decapitation, necrophilia, rape, bestiality, and sadomasochism abound. Biggest question is why Gielgud, O'Toole, and McDowell lent their talents to this monumentally abhorred film (not to mention Gore Vidal on the writing end). Adult magazine publisher Bob Guccione coproduced. Also available in a censored, "R" rated version.
1980 143m/C *IT* Malcolm McDowell, John Gielgud, Peter O'Toole, Helen Mirren; *Dir:* Tinto Brass. VHS, Beta, LV $79.95 *VES, HHE Woof!*

Call of the Canyon A crooked agent for a local meat packer won't pay a fair price, so Gene goes off to talk to the head man to set him straight.
1942 71m/B Gene Autry, Smiley Burnette, Ruth Terry, Thurston Hall, Pat Brady; *Dir:* Joseph Santley. VHS, Beta $29.98 *CCB, VCN* ♬

Call of the Forest Western action with good guys and bad guys galore.
1949 74m/B Robert Lowery, Ken Curtis, Martha Sherrill, Chief Thundercloud, Charles Hughes; *Dir:* John F. Link. VHS $19.95 *NOS, DVT* ♬

Call to Glory Pilot for the critically praised television series. An Air Force pilot and his family face turbulent times during the Cuban missile crisis.
1984 96m/C Craig T. Nelson, Cindy Pickett, Gabriel Damon, Keenan Wynn, Elizabeth Shue, G.D. Spradlin. VHS, Beta, LV $14.95 *PAR* ♬♬½

Call Him Mr. Shatter A hired killer stalks a tottering Third World president and becomes embroiled in international political intrigue. Also known as "Shatter."
1974 (R) 90m/C *GB* Stuart Whitman, Peter Cushing, Anton Diffring; *Dir:* Michael Carreras. VHS, Beta $59.98 *CHA* ♬

Call It Murder A jury foreman is the deciding vote that sends a young girl to the electric chair. Criticized by one and all, he says he would do it again. He gets his chance when his own daughter is accused of almost the identical crime. Originally released under the title "Midnight."
1934 74m/B Humphrey Bogart, Sidney Fox, O.P. Heggie, Henry Hull, Margaret Wycherly, Lynne Overman, Richard Whorf; *Dir:* Chester Erskine. VHS, Beta $16.95 *SNC, NOS, NEG* ♬♬½

Call Me Psycho-drama about a lusty young woman who responds positively to an obscene telephone caller until she witnesses him murder another person. Doubt creeps into the relationship.
1988 (R) 98m/C Patricia Charbonneau, Patti D'Arbanville, Sam Freed, Boyd Gaines, Stephen McHattie; *Dir:* Sollace Mitchell. VHS, Beta, LV $79.98 *VES* ♬♬

Call Out the Marines A group of army buddies re-enlist to break up a spy ring. Songs include "Call Out the Marines," "Zana Zaranda," and "The Light of My Life." McLaglen and Lowe's last comedy together.
1942 67m/B Robert Smith, Frank Ryan, Victor McLaglen, Binnie Barnes, Paul Kelly, Edmund Lowe, Franklin Pangborn; *Dir:* Frank Ryan. VHS, Beta $34.95 *RKO* ♬♬

Call of the Wild Jack London's famous story about a man whose survival depends upon his knowledge of the Alaskan wilderness comes almost to life. Filmed in Finland.
1972 (PG) 105m/C Charlton Heston, Michele Mercier, George Eastman; *Dir:* Ken Annakin. VHS, Beta $14.95 *KUI, SIM, MPI* ♬♬

Call of the Wild Adaptation of the Jack London story about a dog's trek across the Alaskan tundra.
1983 68m/C VHS, Beta $29.98 *VES, LIV* ♬♬

The Caller A strange man enters the house of a lone woman and sets off a long night of suspense and an almost longer evening of inept movie-making.
1987 (R) 90m/C Malcolm McDowell, Madolyn Smith; *Dir:* Arthur Seidelman. VHS, Beta $89.95 *TWE* ♬

Callie and Son Details the sordid story of a waitress who works her way up to become a Dallas socialite and her obsessive relationship with her illegitimate son. Written by Thomas Thompson and made for television. Also known as "Rags to Riches."
1981 97m/C Lindsay Wagner, Dabney Coleman, Jameson Parker, Andrew Prine, James Sloyan, Michelle Pfeiffer; *Dir:* Waris Hussein. VHS, Beta $59.95 *SVS, WAR, SIM* ♬♬

Calling Paul Temple The wealthy patients of a nerve doctor have been dying and a detective is called in to investigate.
1948 92m/B *GB* John Bentley, Dinah Sheridan, Margaretta Scott, Abraham Sofaer, Celia Lipton, Alan Wheatley, Wally Patch; *Dir:* Maclean Rogers. VHS, Beta $16.95 *SNC* ♬♬½

Cambio de Cara (A Change of Face) Two reporters investigate corporate corruption. Dubbed into Spanish.
1971 90m/C *IT* Adolfo Celi, Geraldine Hooper. VHS, Beta $59.95 *JCI*

Cambodia: Year Zero and Year One Program, available in two parts, looks first at the strife caused by the overthrown Pol-Pot regime. The second report reviews the advances made during the year after the change in government, thanks largely to $35 million in relief aid.
1980 120m/C John Pilger; *Dir:* David Munro. VHS, Beta $19.95 *AHV, FCT Woof!*

Came a Hot Friday Two cheap con-men arrive in a 1949 southern New Zealand town, run various scams and pursue women.
1985 (PG) 101m/C *NZ* Peter Bland, Philip Gordon, Billy T. James; *Dir:* Ian Mune. VHS, Beta $19.98 *CHA* ♬♬

Camel Boy True story of a young Arabian boy who befriends a camel and their treacherous trek across the desert.
1984 78m/C *AU Dir:* Yoram Gross. VHS, Beta $69.95 *VES, LIV* ♬½

Camelia After much emotional pain, beautiful Camelia learns to love again from the young, warmhearted Rafael. But her new love will not protect her from fate.
197? 108m/B *MX* Maria Felix, Jorge Mistral. VHS, Beta $54.95 *MED* ♬♬

Camelot The long-running Lerner and Loewe Broadway musical about King Arthur, Guinevere, and Lancelot was adapted from T.H. White's book, "The Once and Future King." Redgrave and Nero have chemistry as the illicit lovers, Harris is strong as the king struggling to hold together his dream, but muddled direction undermines the effort. Songs include "If Ever I Would Leave You," "How to Handle a Woman" and "Camelot." Laserdisc edition contains 28 minutes of previously edited footage, trailers and backstage info.
1967 150m/C Richard Harris, Vanessa Redgrave, David Hemmings, Franco Nero, Lionel Jeffries; *Dir:* Joshua Logan. Academy Awards '67: Best Art Direction/Set Decoration, Best Costume Design, Best Musical Score. VHS, Beta, LV $29.98 *WAR, PIA, RDG* ♬♬♬

The Cameraman In 1928, after moving to MGM, Keaton made his first feature with a major studio, giving up the total artistic control he had enjoyed in his previous films. Spared from the vilification of studio politics (as wasn't the case with later Keaton films) "The Cameraman" enjoyed both critical and popular success. Keaton's inept tintype portrait-maker has a heart that pitter-patters for an MGM office girl. He hopes to impress her by joining the ranks of the newsreel photographers. Fortuitously poised to grab a photo scoop on a Chinese tong war, he is forced to return empty-handed when an organ-grinder's monkey absconds with his firsthand footage. Will our hero find both romance and career success? Silent with a musical score.
1928 78m/B Buster Keaton, Marceline Day, Harold Goodwin, Harry Gribbon, Sidney Bracy, Edward Brophy, Vernon Dent, William Irving; *Dir:* Edward Sedgwick. VHS, LV $29.95 *MGM, FCT* ♬♬♬♬

Cameron's Closet Every child's nightmare comes true. A young boy is convinced that a monster lives in his closet due to his perverse father's psychological tortures. Only this time the monster is real!
1989 (R) 86m/C Cotter Smith, Mel Harris, Scott Curtis, Chuck McCann, Leigh McCloskey, Kim Lankford, Tab Hunter; *Dir:* Armand Mastroianni. VHS, LV $14.95 *COL, SVS* ♬½

Camila The true story of the tragic romance between an Argentinean socialite and a Jesuit priest in 1847. Available in Spanish with English subtitles or dubbed into English. Academy Award nomination for Best Foreign Film.
1984 105m/C *AR SP* Susu Pecoraro, Imanol Arias, Hector Alterio, Elena Tasisto; *Dir:* Maria-Luisa Bemberg. VHS, Beta $19.98 *TAM, SUE* ♬♬

Camille The classic Alexandre Dumas story about a dying French courtesan falling in love with a shallow young nobleman. A clean, romantic adaptation that somehow escapes the cliches and stands as one of the most

telling monuments to Garbo's unique magic and presence on film. She was Oscar nominated for Best Actress.
1936 108m/B Greta Garbo, Robert Taylor, Lionel Barrymore, Henry Daniell, Elizabeth Allan, Rex O'Malley, Lenore Ulric, Laura Hope Crews; **Dir:** George Cukor. National Board of Review Awards '37: 10 Best Films of the Year; New York Film Critics Awards '36: Best Actress (Garbo). **VHS, Beta $24.95** *MGM* ⅔½

Camille Claudel A lushly romantic version of the art world at the turn of the century, when art was exploding in new forms and independence for women was unheard of. Sculptor Claudel's unrequited love for art, Rodin, and independence clash, costing her sanity. Very long, it requires an attentive and thoughtful viewing. Adjani won the French Oscar for her performance. In French with English subtitles.
1989 (R) 149m/C *FR* Isabelle Adjani, Gerard Depardieu, Laurent Grevill, Alain Cuny; **Dir:** Bruno Nuytten. **VHS, Beta, LV $79.98** *ORI, IME, TAM* ⅔⅔⅔

Camorra: The Naples Connection A prostitute and an American drug dealer find themselves embroiled in the murders of Neapolitan Mafia heads. Violent, overblown, minor Wertmuller.
1985 (R) 94m/C *IT* Harvey Keitel, Angela Molina, Lorraine Bracco, Francisco Rabal; **Dir:** Lina Wertmuller. **VHS, Beta $79.95** *MGM* ⅔⅔

Camouflage A training film by the Walt Disney studios for the Armed Forces, in which Yehudi the Chameleon teaches camouflage to young Air Corps fliers. Contains some condemning attitudes toward Japan.
194? 20m/C VHS, Beta $19.95 *VYY* ⅔½

Camp Classics #1 Five corny classics spanning four decades make up this collection: "Mystery of the Leaping Fish" with Douglas Fairbanks, "Foreign Press Awards," "Musical Beauty Shop," "Edsel Commercial," and "Nixon's Checkers Speech."
195? 80m/B Douglas Fairbanks Sr. **VHS, Beta $37.00** *WFV*

Camp Cucamonga: How I Spent My Summer Vacation Zany antics at summer camp abound when Camp Cucamonga's owner mistakes the new handyman for the camp inspector. Stars from TV's "Cheers," "The Jefferson s," "Wonder Years," and "The Love Boat" are featured in this made for television flick.
1990 100m/C John Ratzenberger, Sherman Hemsley, Josh Saviano, Danica McKellar, Chad Allen, Dorothy Lyman, Lauren Tewes, G. Gordon Liddy; **Dir:** Roger Duchonwny. **VHS $9.99** *GKK* ⅔

Camp Double Feature Two cautionary anti-pot films, now universally recognized as high camp: "Reefer Madness" and "Marihuana." Both are available individually on video.
1939 120m/B Harley Wood, Hugh McArthur, Paul Ellis, Dave O'Brien, Dorothy Short. **VHS, Beta $34.95** *JEF* ⅔

Campaign in Poland This documentary chronicles the German invasion of Poland during World War II.
1940 80m/B VHS, Beta $29.95 *NOS, DVT* ⅔⅔

The Campus Corpse Young man stumbles into deadly college frat hazing and discovers he and rest of cast are utterly devoid of acting ability.

1977 (PG) 92m/C Charles Martin Smith, Jeff East, Brad Davis; **Dir:** Douglas Curtis. **VHS, Beta $69.95** *VES* ⅔

Campus Knights If you're a serious student of the campus caper film, this is one of the earliest (though not one of the best) of the genre (if such a distinction can be made). The story involves twin brothers—one a tweedy high-browed professor, the other a bon vivant man about town—who wreak fraternal chaos on the quads.
1929 70m/B Raymond McKee, Shirley Palmer, Marie Quillen, Jean Laverty, Sybil Grove; **Dir:** Albert Kelly. **VHS, Beta $24.95** *GPV, FCT* ⅔

Campus Man An entrepreneurial college student markets a beefcake calendar featuring his best friend, until the calendar's sales threaten his friend's amateur athletic status.
1987 (PG) 94m/C John Dye, Steve Lyon, Kim Delaney, Miles O'Keefe, Morgan Fairchild, Kathleen Wilhoite; **Dir:** Ron Casden. **VHS, Beta, LV $14.95** *PAR* ⅔

Can-Can Lackluster screen adaptation of the Cole Porter musical bears little resemblance to the stage version. MacLaine is a cafe owner who goes to court to try and get the "Can-Can," a dance considered risque in gay Paree at the end of the nineteenth century, made legal. Love interest Sinatra happens to be a lawyer. Songs include "C'est Manifique," "Let's Do It" and "I Love Paris."
1960 131m/C Frank Sinatra, Shirley MacLaine, Maurice Chevalier, Louis Jourdan, Juliet Prowse, Marcel Dalio, Leon Belasco; **Dir:** Walter Lang. **VHS, Beta, LV $19.98** *FOX, FCT* ⅔⅔

Can I Do It...Till I Need Glasses? More to point, can you stay awake till the end. Prurient juvenile junk. Brief Williams footage was grafted to this mess after Mork fame.
1977 72m/C Robin Williams, Roger Behr, Debra Klose, Moose Carlson, Walter Olkewicz; **Dir:** I. Robert Levy. **VHS, Beta $19.95** *MED* ⅔

Can She Bake a Cherry Pie? Two offbeat characters meet and fall in love in an odd sort of way. Slow-moving and talky but somewhat rewarding. One of Black's better performances.
1983 90m/C Karen Black, Michael Emil, Michael Margotta, Frances Fisher, Martin Frydberg; **Dir:** Henry Jaglom. **VHS, Beta $39.95** *MON, PAR, FCT* ⅔⅔

Can You Feel Me Dancing? Family-happy lightweight entertainment coproduced by Kent Bateman and starring Bateman progeny Jason and Justine as brother and sister, a novel premise that only a father could love. A blind independence-impaired teenager tries to liberate herself from her overbearing family, and finds the secret to conquering her fears and to standing up for herself when she falls in love. Written by spouse-team Steven and J. Miyoko Hensley.
1985 95m/C Jason Bateman, Justine Bateman. **VHS $79.95** *MNC, TPV* ⅔⅔

Can You Hear the Laughter? The Story of Freddie Prinze Heartstring tugging biography of the late Puerto Rican comedian whose troubled life lead to suicide in spite of his apparent success. He was most noted for his starring role in "Chico and the Man."
1979 100m/C Ira Angustain, Kevin Hooks, Randee Heller, Devon Ericson, Julie Carmen, Stephen Elliott; **Dir:** Burt Brinckerhoff. **VHS, Beta $79.95** *GEM* ⅔⅔

Canadian Capers...Cartoons: Volume I A collection of innovative animated animated shorts produced by the National Film Board of Canada. Titles include: "The Great Toy Robbery," "The Animal Movie," "The Story of Christmas," "The Energy Carol," "The Bear's Christmas," "Carousel" and "TV Sale."
19?? 61m/C VHS, Beta $29.95 *VYY*

Canadian Capers...Cartoons: Volume II A second collection of thought-provoking cartoons from the National Film Board of Canada, including "Spinnolio," "Doodle Film," "Hot Stuff," "The Cruise," "The Specialists" and "No Apple for Johnny."
19?? 58m/C VHS, Beta $29.95 *VYY*

Cancel My Reservation New York talk show host Hope sets out for a vacation on an Arizona ranch, but winds up in trouble due to a mysterious corpse, a rich rancher, and an enigmatic mystic. Even more muddled than it sounds. Based on the novel "Broken Gun" by Louis L'Amour, with pointless cameos by Crosby, Wayne, and Wilson.
1972 (G) 99m/C Bob Hope, Eva Marie Saint, Ralph Bellamy, Anne Archer, Forrest Tucker, Keenan Wynn, John Wayne, Bing Crosby, Flip Wilson, Doodles Weaver, Pat Morita, Chief Dan George; **Dir:** Paul Bogart. **VHS, Beta $19.95** *COL, FCT* ⅔

Cancion en el Alma (Song from the Heart) A talented boy singer is stuck in a greedy custody battle over his potential revenues in the wide world of Puerto Rican entertainment.
19?? 70m/C Luis Fernando Ramirez, Claudia Osuna. **VHS, Beta** *MAV*

Candid Candid Camera, Vol. 1 Each volume in this series is a compilation of funny hidden camera clips which were deemed too risque for general television audiences.
1985 55m/C Hosted: Allen Funt. **VHS, Beta $59.95** *VES*

Candid Candid Camera, Vol. 2 (More Candid Candid Camera) Allen Funt presents this collection of uncensored "Candid Camera" classics that the censors would not allow on television.
1983 60m/C Hosted: Allen Funt. **VHS, Beta $59.98** *VES, LIV*

Candid Hollywood A compilation of Ken Murray home movies featuring dozens of Hollywood stars at rest and play.
1962 115m/B Bob Hope, Gloria Swanson, Jimmy Durante, Clark Gable, Gary Cooper, Desi Arnaz Sr. **VHS, Beta, 3/4U $24.95** *SHO*

The Candidate Realistic, satirical look at politics and political campaigning. A young, idealistic lawyer is talked into trying for the Senate seat and learns the truth about running for office. Ritchie's other efforts include "Downhill Racer" and "Bad News Bears."
1972 (PG) 105m/C Robert Redford, Peter Boyle, Don Porter, Allen Garfield, Karen Carlson, Melvyn Douglas, Michael Lerner; **Dir:** Michael Ritchie. Academy Awards '72: Best Story & Screenplay; National Board of Review Awards '72: 10 Best Films of the Year. **VHS, Beta, LV $19.98** *WAR, PIA* ⅔⅔⅔

Candide Moderately engaging BBC production of Voltaire's eighteenth century satire on human aspirations and weaknesses. The eponymous hero learns a lesson in optimism from Professor Pangloss. Every-

C

thing happens for the best, even mediocre made for television dramas.
1976 95m/C *GB* Frank Finlay. **VHS, 3/4U, Special order formats** *TLF* ✍✍

Candles at Nine An innocent showgirl must spend a month in her late uncle's creepy mansion in order to inherit it, much to the malevolent chagrin of the rest of the family who want the place and the loot for themselves. Uninspired.
1944 84m/B *GB* Jessie Matthews, John Stuart, Reginald Purdell; *Dir:* John Harlow. **VHS, Beta $16.95** *SNC, VYY, MLB* ✍ ½

Candleshoe A Los Angeles street urchin poses as an English matron's long lost granddaughter in order to steal a fortune hidden in Candleshoe, her country estate, where Niven butles. Somewhat slapschticky Disney fare.
1978 (G) 101m/C Vivian Pickles, Helen Hayes, David Niven, Jodie Foster, Leo McKern; *Dir:* Norman Tokar. **VHS, Beta $19.99** *DIS* ✍✍

Candy Mountain Guitar playin' O'Connor roadtrips across America and Canada in search of a legendary guitar maker Yulin. Occasional interest derives from musician cameos from the likes of Buster Poindexter, Dr. John and Redbone.
1987 (R) 90m/C *CA FR SI* Kevin J. O'Connor, Harris Yulin, Tom Waits, Bulle Ogier, David Johansen, Leon Redbone, Joe Strummer, Dr. John, Rita MacNeil, Roberts Blossom, Laurie Metcalf; *Dir:* Robert Frank. **VHS $14.98** *REP, FCT* ✍✍

Candy Stripe Nurses Even hard-core Roger Corman fans might find his final installment in the nursing comedy pentad (preceded by "The Young Nurses") to be a lethargic exercise in gratuitous "sexual situations." Also on video as "Sweet Candy." Bet those uniforms don't meet hospital standards.
1974 (R) 80m/C Candice Rialson, Robin Mattson, Maria Rojo, Kimberly Hyde, Dick Miller, Stanley Ralph Ross, Monte Landis, Tom Baker; *Dir:* Allan Holleb. **VHS, Beta $59.95** *SUE* ✍

Candy Tangerine Man Respectable businessman leads a double life as loving father and LA pimp.
1975 88m/C John Daniels, Tom Hankerson, Eli Haines, Marva Farmer, George Flower; *Dir:* Matt Cimber. **VHS, Beta $49.95** *UNI* ✍

Canine Commando Three wartime Pluto cartoons from Disney: "The Army Mascot," "Dog Watch" and "Canine Patrol."
1945 (G) 23m/C *Dir:* Walt Disney. **VHS, Beta $14.95** *DIS* ✍

Cannery Row Baseball has-been Nolte lives anonymously among the downtrodden in the seemy part of town and carries on with working girl girlfriend Winger. Based on John Steinbeck's "Cannery Row" and "Sweet Thursday."
1982 (PG) 120m/C Nick Nolte, Debra Winger, Audra Lindley, M. Emmet Walsh, James Keane, Sunshine Parker; *Dir:* David S. Ward; *Nar:* John Huston. **VHS, Beta $79.95** *MGM* ✍½

Cannibal Campout Crazed orphans with eating disorders make square meal of coed babes getting back to nature.
1988 89m/C Carrie Lindell, Richard Marcus, Amy Chludzinski, Jon McBride; *Dir:* Jon McBride, Tom Fisher. **VHS, Beta** *DMP* ✍

Cannibal Women in the Avocado Jungle of Death Tongue-in-cheek cult classic features erstwhile playmate Tweed as feminist anthropologist who searches with

ditzy student and mucho macho male guide for lost tribe of cannibal women who dine on their mates.
1989 (PG-13) 90m/C Shannon Tweed, Adrienne Barbeau, Karen Mistal, Barry Primus, Bill Maher; *Dir:* J.D. Athens. **VHS, Beta $79.95** *PAR* ✍✍½

Cannon Ball/Eyes Have It Conklin stars in "The Cannon Ball" (1915), as an explosives expert in the Boom Powder Factory. "The Eyes Have It" (1928), stars Turpin involved in another misunderstanding with wife and mother-in-law. Silent.
192? 41m/B Chester Conklin, Ben Turpin, Georgia O'Dell, Helen Gilmore, Jack Lipson. **VHS, Beta $19.98** *CCB* ✍✍½

Cannon Movie Tales: Red Riding Hood Bright re-telling of the little girl in red and that nasty wolf.
1989 (G) 84m/C Craig T. Nelson, Isabella Rossellini; *Dir:* Adam Brooks. **VHS** *WAR* ✍✍½

Cannon Movie Tales: Sleeping Beauty Fairchild stars in this version of the sleeping princess and the evil witch.
1989 (G) 90m/C Morgan Fairchild, Tracy Welch, Nicholas Clay, Sylvia Miles, Kenny L. Baker, David Holliday; *Dir:* David Irving. **VHS** *WAR* ✍✍½

Cannon Movie Tales: Snow White The incomparable Rigg stars in this witty retelling of the beautiful girl and her seven little friends.
1989 85m/C Diana Rigg, Sarah Peterson, Billy Barty; *Dir:* Michael Berz. **VHS** *WAR* ✍✍✍

Cannon Movie Tales: The Emperor's New Clothes Caesar stars in this version of vain despot and the crafty tailors.
1989 85m/C Sid Caesar, Robert Morse, Clive Revill; *Dir:* David Irving. **VHS** *WAR* ✍✍½

Cannonball Assorted ruthless people leave patches of rubber across the country competing for grand prize in less than legal auto race. Not top drawer New World but nonetheless a cult fave. Inferior to Bartel's previous cult classic, "Death Race 2000." Most interesting for plethora of cult cameos, including Scorsese, Dante and grandmaster Corman.
1976 (PG) 93m/C David Carradine, Bill McKinney, Veronica Hamel, Gerrit Graham, Robert Carradine, Sylvester Stallone, Martin Scorsese, Roger Corman, Joe Dante, Jonathan Kaplan; *Dir:* Paul Bartel. **VHS, Beta $39.98** *NO* ✍½

Cannonball/Dizzy Heights and Daring Hearts Two classic comedies featuring Conklin, the Keystone Kops, and wacky aeroplanes from the irrepressible Sennett.
1916 71m/B Chester Conklin; *Dir:* Mack Sennett. **VHS, Beta $29.95** *VYY* ✍½

Cannonball Run So many stars, so little plot. Reynolds and sidekick DeLuise disguise themselves as paramedics to foil cops while they compete in cross-country Cannonball race. Shows no sign of having been directed by an ex-stuntman. One of 1981's top grossers: go figure. Followed by equally languid sequel "Cannonball Run II."
1981 (PG) 95m/C Burt Reynolds, Farrah Fawcett, Roger Moore, Dom DeLuise, Dean Martin, Sammy Davis Jr., Jack Elam, Adrienne Barbeau, Peter Fonda, Molly Picon, Bert Convy, Jamie Farr; *Dir:* Hal Needham. **VHS, Beta, LV $29.98** *VES* ✍

Cannonball Run II More mindless cross country wheel spinning with gratuitous star cameos. Director Needham apparently subscribes to the two wrongs make a right school of sequels.
1984 (PG) 109m/C Burt Reynolds, Dom DeLuise, Jamie Farr, Marilu Henner, Shirley MacLaine, Jim Nabors, Frank Sinatra, Sammy Davis Jr., Dean Martin, Telly Savalas, Susan Anton, Catherine Bach, Richard Kiel, Tim Conway, Sid Caesar, Don Knotts, Henry Silva, Ricardo Montalban, Jack Elam, Molly Picon; *Dir:* Hal Needham. **VHS, Beta, LV $19.98** *WAR* ✍

The Canon Operation An ostensibly true story about Korean War espionage.
1983 90m/C **VHS, Beta $59.95** *VCD* ✍

Can't Buy Me Love Unpopular high school nerd Dempsey buys a month of dates with teen babe Peterson for $1000 in order to win friends and influence people. Previously known as "Boy Rents Girl."
1987 (PG-13) 94m/C Patrick Dempsey, Amanda Peterson, Dennis Dugan, Courtney Gains; *Dir:* Steve Rash. **VHS, Beta, LV $19.99** *TOU* ✍½

Can't Stop the Music A retired model invites friends from Greenwich Village to a party to help the career of her roommate, an aspiring composer. Walker is best known for her paper towel commercials as Rosy ("the quicker picker upper"). Features two of the top hits by the Village People, "Macho Man" and "YMCA."
1980 (PG) 120m/C Valerie Perrine, Bruce Jenner, Steve Guttenberg, Paul Sand, Leigh Taylor-Young, Village People; *Dir:* Nancy Walker. **VHS, Beta $29.95** *HBO* ✍

A Canterbury Tale Writer-director team Powell and Pressburger have loosely modeled a retelling of Chaucer's famous tale of pilgrimage to the cathedral in Canterbury. Set in Nazi-threatened Britain in 1944, the story follows the pilgrimage of three Brits and an American GI to the eponymous cathedral. Strange, effective, worth looking for. The 95-minute American version, with added footage of Kim Hunter, is inferior to the 124-minute original.
1944 124m/B *GB* Eric Portman, Sheila Sim, Dennis Price, Esmond Knight, Charles Hawtrey, Hay Petrie; *Dir:* Michael Powell, Emeric Pressburger; *Nar:* Esmond Knight. **VHS $59.95** *HMV* ✍✍✍

The Canterbury Tales Four Chaucer tales, most notably "The Merchant's Tale" and "The Wife of Bath," are recounted by travelers, with director Pasolini as the bawdy poet. Deemed obscene by the Italian courts, it's the second entry in Pasolini's medieval "Trilogy of Life," preceded by "The Decameron" and followed by "The Arabian Nights." In Italian with English subtitles.
1971 109m/C *IT* Laura Betti, Ninetto Davoli, Pier Paolo Pasolini, Hugh Griffith, Josephine Chaplin, Michael Balfour, Jenny Runacre; *Dir:* Pier Paolo Pasolini. Berlin Film Festival '72: Best Film. **VHS, Beta $79.95** *WBF, TAM, APD* ✍✍

The Canterville Ghost Laughton, a 300-year old ghost with a yellow streak, is sentenced to spook a castle until he proves he's not afraid of his own shadow. American troops stay at the castle during WWII and, as luck would have it, soldier Young is distantly related to spunky young keeper of the castle O'Brien, ghost Laughton's descendant. Once Young is acquainted with his cowardly ancestor, he begins to fear a congenital yellow streak, and both struggle to be brave despite themselves. Vaguely derived from an Oscar Wilde tale.

1944 95m/B Charles Laughton, Robert Young, Margaret O'Brien, William Gargan, Reginald Owen, Rags Ragland, Una O'Connor, Peter Lawford, Mike Mazurki; **Dir:** Jules Dassin. **VHS, Beta** $19.98 *MGM* ✂✂½

Cantonen Iron Kung Fu The ten tigers of Quon Tung perfect martial arts skills known as Cantonen Iron Kung Fu.
197? 90m/C Liang Jia Ren. **VHS, Beta** $54.95 *MAV, WVE* ✂

The Cantor's Son (Dem Khann's Zindl) Real-life story based on Moishe Oysher's life, vaguely influenced by Jolson's "Jazz Singer." Runaway Oysher joins troupe of performers as a youth, travels to the shores of America, and finds fame once his beautiful voice is discovered. Returning to his mother country to celebrate his parents' golden anniversary, he encounters his childhood sweetheart and falls in love, but another woman is written into the plot to complicate matters. Poorly acted, laughable staging. In Yiddish with English subtitles.
1937 90m/B Judith Abarbanel, Florence Weiss, Moishe Oysher, Isadore Cashier; **Dir:** Ilya Motyleff. **VHS, Beta** $79.95 *ERG* ✂½

Cape Fear Former prosecutor turned small-town lawyer Peck and his family are plagued by the sadistic attentions of criminal Mitchum, who just finished a six year sabbatical at the state pen courtesy of prosecutor Peck. Taut and creepy; Mitchum's a consummate psychopath. Based on (and far superior to) John MacDonald's "The Executioners." Don't pass this one up in favor of the Scorsese remake.
1961 106m/B Gregory Peck, Robert Mitchum, Polly Bergen, Martin Balsam, Telly Savalas, Jack Kruschen; **Dir:** J. Lee Thompson. **VHS, Beta** $19.95 *MCA* ✂✂✂½

Cape Fear Scorsese takes on this terrifying tale of brutality and manipulation (previously filmed in 1961) and cranks it up a notch as a paroled convict haunts the lawyer who put him away. Great cast, a rollercoaster of suspense. Note the cameos by Mitchum and Peck, stars of the first version. Original source material was "The Executioners" by John D. MacDonald.
1991 (R) 128m/C Robert De Niro, Nick Nolte, Jessica Lange, Joe Don Baker, Juliette Lewis, Robert Mitchum, Gregory Peck, Martin Balsam; **Dir:** Martin Scorsese. Chicago Film Critics Awards '91: Most Promising Actress (Lewis). **VHS, LV** $94.95 *MCA* ✂✂✂

Caper of the Golden Bulls Former bank robber Boyd is blackmailed into joining a group of safecrackers who plan to assault the Royal Bank of Spain during the annual Santa Maria bull run. More like siesta of the golden bulls.
1967 104m/C Stephen Boyd, Yvette Mimieux, Giovanna Ralli, Walter Slezak, Vito Scotti; **Dir:** Russel Rouse. **VHS, Beta** $59.98 *CHA* ✂

Capricorn One Astronauts Brolin, Simpson and Waterston follow Mission Controller Holbrook's instructions to fake a Mars landing on a fabricated soundstage when their ship is discovered to be defective. When they find out they're supposed to expire in outer space so that the NASA scam won't become public knowledge, they flee to the desert, while reporter Gould sniffs out the cover up.
1978 (R) 123m/C Elliott Gould, James Brolin, Brenda Vaccaro, O.J. Simpson, Hal Holbrook, Sam Waterston, Karen Black, Telly Savalas; **Dir:** Peter Hyams. **VHS, Beta, LV** $59.98 *FOX* ✂✂✂

Captain America Captain America battles a mad scientist in this fifteen-episode serial based on the comic book character.
1944 240m/B Dick Purcell, Adrian Booth, Lionel Atwill; **Dir:** John English. **VHS, Beta** $109.95 *VCN, MLB* ✂✂

Captain America Marvel Comic character steps into feature film and flounders. The patriotic superhero son of WWII hero fights bad guy with contraband nuclear weapon. Made for television. Followed the same year with "Captain America II: Death Too Soon."
1979 98m/C Reb Brown, Len Birman, Heather Menzies, Steve Forrest, Robin Mattson, Joseph Ruskin, Michael McManus; **Dir:** Rod Holcomb. **VHS, Beta** $39.95 *MCA* ✂½

Captain America Based on the Marvel Comics superhero. It's 1941 and Steve Rogers has just been recruited to join a top secret experimental government program after flunking his army physical. Injected with a serum, Steve becomes super strong, fast, and smart but is matched in all three by an evil Nazi counterpart, Red Skull. The two battle to a World War II standstill and while Red Skull goes on with his evil plots, the next 40 years finds Captain America fast frozen in the Alaskan tundra. Finally, our hero is thawed in time to do a final battle with his evil nemesis. This one is ridiculous even by comic book standards but it may amuse the kids.
1989 (PG-13) 103m/C Matt Salinger, Scott Paulin, Ronny Cox, Ned Beatty, Darren McGavin, Melinda Dillon; **Dir:** Albert Pyun. **VHS, LV** $89.95 *COL* ✂½

Captain America 2: Death Too Soon Terrorists hit America where it hurts, threatening to use age accelerating drug. Sequelized superhero fights chronic crow lines and series dies slow, painful death. Made for television.
1979 98m/C Reb Brown, Connie Sellecca, Len Birman, Christopher Lee, Katherine Justice, Lana Wood, Christopher Carey; **Dir:** Ivan Nagy. **VHS, Beta** $39.95 *MCA* ✂

Captain Apache Union intelligence officer Van Cleef investigates murder of Indian commissioner and discovers fake Indian war landscam. As clever as the title.
1971 95m/C Lee Van Cleef, Carroll Baker, Stuart Whitman; **Dir:** Alexander Singer. **VHS, Beta** $19.95 *KOV, MRV, PSM* ✂

Captain Blackjack All star cast craps out in wannabe thriller about drug smuggling on the French Riviera. Social butterfly Moorehead directs drug traffic, detective Marshall undercovers as doctor and Sanders looks bored in a British sort of way. Also known as "Black Jack."
1951 90m/B *FR* George Sanders, Herbert Marshall, Agnes Moorehead, Patricia Roc, Marcel Dalio; **Dir:** Julien Duvivier. **VHS** *VEC, TIM* ✂

Captain Blood Sabatini adventure story launched then unknown 26-year old Flynn and 19-year old De Havilland to fame in perhaps the best pirate story ever. Exiled into slavery by a tyrannical governor, Irish physician Peter Blood is forced to piracy but ultimately earns a pardon for his swashbuckling ways. Love interest De Havilland would go on to appear in seven more features with Flynn, who took the part Robert Donat declined for health reasons. Cleverly budgeted using ship shots from silents, and miniature sets where possible. First original film score by composer Korngold. Also available colorized. Received Oscar nominations for Best Picture and Best Sound Recording.

1935 120m/B Errol Flynn, Olivia de Havilland, Basil Rathbone, J. Carroll Naish, Guy Kibbee, Lionel Atwill; **Dir:** Michael Curtiz. **VHS, Beta** $19.98 *FOX, FCT, MLB* ✂✂✂½

Captain Caution Young girl throws caution to the wind when dad dies during the War of 1812, assisting young Mature to take over the old man's ship to do battle with the British. Watch for then unknown sailor Ladd.
1940 84m/B Victor Mature, Louise Platt, Bruce Cabot, Alan Ladd, Robert Barratt, Vivienne Osborne; **Dir:** Richard Wallace. **VHS, Beta** *MED* ✂✂

Captain Future in Space The outerspace adventures of Captain Future and his crew aboard the spaceship Comet, fighting evil and making the universe safe for mankind.
197? 54m/C **VHS, Beta** *MED, OM* ✂✂

Captain Gallant, Foreign Legionnaire Two episodes of the vintage television serial: "Flaming Hoop" and "Shifting Sands." Includes original television commercials.
1956 60m/B Buster Crabbe, Cullen Crabbe, Al "Fuzzy" Knight. **VHS, Beta** $19.95 *NOS, DVT*

Captain Horatio Hornblower A colorful drama about the life and loves of a British sea captain during the Napoleonic wars. Peck is rather out of his element as the courageous, swashbuckling hero (Errol Flynn was originally cast) but there's enough fast-paced derring-do to make this a satisfying saga. Based on the novel by C.S. Forester.
1951 117m/C *GB* Gregory Peck, Virginia Mayo, Robert Beatty, Denis O'Dea, Christopher Lee; **Dir:** Raoul Walsh. **VHS, Beta, LV** $59.99 *WAR* ✂✂✂

Captain January Crusty old lighthouse keeper rescues little orphan girl with curly hair from drowning and everyone breaks into cutsey song and dance, interrupted only when the authorities try to separate the two. Temple performs "At the Codfish Ball" with sailor Ebsen.
1936 74m/B Shirley Temple, Guy Kibbee, Buddy Ebsen, Slim Summerville, Jane Darwell, June Lang, George Irving, Si Jenks; **Dir:** David Butler. **VHS, Beta** $19.98 *FOX* ✂✂½

Captain Kidd Laughton huffs and puffs and searches for treasure on the high seas, finds himself held captive with rest of cast in anemic swashbuckler.
1945 83m/B Charles Laughton, John Carradine, Randolph Scott, Reginald Owen, Gilbert Roland, Barbara Britton, John Qualen, Sheldon Leonard; **Dir:** Rowland V. Lee. **VHS, Beta** $9.95 *NEG, MRV, NOS* ✂½

Captain from Koepenick Popular true life comedy about a Berlin cobbler in 1906 who rebels against military bureaucracy by impersonating a Prussian officer and wreaking havoc on his town, arresting authorities and capturing soldiers, only to enjoy folk herodom when his ruse is discovered. Remake of Richard Oswald's 1931 classic. In German with subtitles.
1956 93m/B *GE* Heinz Ruehmann, Hannelore Schroth, Martin Held, Erich Schellow; **Dir:** Helmut Kaeutner. **VHS, Beta** $24.95 *NOS, FCT, TIM* ✂✂½

Captain Kronos: Vampire Hunter Captain Kronos fences thirsty foes in Hammer horror hybrid. Artsy, atmospheric and atypical, it's written and directed with tongue firmly in cheek by Clemens, who penned many an "Avengers" episode.

C

1974 (R) 91m/C *GB* Horst Janson, John Carson, Caroline Munro, Ian Hendry, Shane Briant, Wanda Ventham, Ian Hendry; *Dir:* Brian Clemens. VHS, Beta $44.95 PAR ☾☾☾

Captain Newman, M.D. Three army guys visit stiff shrink Peck during the final months of WWII in VA ward for the mentally disturbed. Much guilt and agonizing, with comic relief courtesy of Curtis. Peck is sub par, the direction flounders and there's something unsettling about quicksilver shifts from pathos to parody. Nonetheless touching, with an Oscar-nominated performance from Darin as guilt ridden hero. Based on the novel by Leo Rosten.
1963 126m/C Gregory Peck, Bobby Darin, Tony Curtis, Angie Dickinson, Eddie Albert, James Gregory, Jane Withers, Larry Storch, Robert Duvall; *Dir:* David Miller. VHS, Beta $59.95 MCA ☾☾½

Captain Scarlett Formulaic swashbuckler has nobleman Greene and highway guy Young fighting nasty French Royalists who've been putting the pressure on impecunious peasants. Runaway Spanish damsel in distress courtesy of Amar. Most novel aspect of the production is that the post-Napoleon French terrain has that vaguely south of the border feel.
1953 75m/C Richard Greene, Leonora Amar, Isobel Del Puerto, Nedrick Young, Manolo Fabregas, Isobel Del Puerto; *Dir:* Thomas Carr. VHS, Beta $19.95 NOS, MRV, WES ☾½

Captain Swagger A man on the brink of thievery experiences some changes that keep him on the right side of the law.
1925 50m/B Rod La Rocque, Sue Carrol. VHS, Beta $14.95 GPV ☾½

Captains Courageous Rich brat Bartholomew takes a dip sans life jacket while leaning over an ocean liner railing to relieve himself of the half dozen ice cream sodas imprudently consumed at sea. Picked up by a Portugese fishing boat, he at first treats his mandatory three month voyage as an unscheduled cab ride, but eventually, through a deepening friendship with crewman Tracy, develops a hitherto unheralded work ethic. The boy's filial bond with Tracy, of course, requires that the seaman meet with watery disaster. Based on the Rudyard Kipling novel. Director Fleming went on to "Gone With the Wind" and "The Wizard of Oz." Oscar nominated for Best Picture and Best Screenplay.
1937 116m/B Spencer Tracy, Lionel Barrymore, Freddie Bartholomew, Mickey Rooney, Melvyn Douglas, Charley Grapewin, John Carradine, Jack LaRue; *Dir:* Victor Fleming. Academy Awards '37: Best Actor (Tracy); National Board of Review Awards '37: 10 Best Films of the Year. VHS, Beta $19.98 MGM, BMV, BTV ☾☾☾

Captain's Paradise Golden Fleece captain Guinness chugs between wives in Gibraltar and North Africa, much to the adulation of chief officer Goldner. While Gibraltar little woman Johnson is homegrown homebody, little woman de Carlo is paint the town red type, allowing Guinness to have cake and eat it too, it seems, except that he's inconveniently positioned in front of a firing squad at movie's start.
1953 89m/B *GB* Alec Guinness, Yvonne de Carlo, Celia Johnson, Miles Malleson, Nicholas Phipps, Ferdinand ''Ferdy'' Mayne, Sebastian Cabot; *Dir:* Anthony Kimmins. VHS, Beta $19.98 HBO, FCT ☾☾☾

The Captain's Table Former cargo vessel captain Gregson is given luxury liner to command, and fails to revise his cargo captain style to fit new crew and clientele. British cast saves unremarkable script from mediocrity.
1960 90m/C *GB* John Gregson, Peggy Cummins, Donald Sinden, Nadia Gray; *Dir:* Jack Lee. VHS $19.95 PAR ☾☾

Captive Spoiled heiress is kidnapped by terrorist trio and brainwashed into anti-establishment Hearst-like creature.
1987 (R) 98m/C *GB* Oliver Reed, Irina Brook, Xavier DeLuc, Hiro Arai; *Dir:* Paul Mayersberg. VHS, Beta $24.95 VIR ☾½

Captive Heart Czech soldier Redgrave assumes the identity of a dead British officer in order to evade Nazis in WWII. Captured and imprisoned in camp reserved for British POWs, his stalagmates think they smell a spy, but he manages to convince them he's an OK Joe. Meanwhile, he's been writing letters home to the little missus, which means he's got a little explaining to do when he's released from prison. Especially fine Redgrave performance.
1947 86m/B *GB* Michael Redgrave, Basil Radford, Jack Warner, Jimmy Hanley, Rachel Kempson, Mervyn Johns; *Dir:* Basil Dearden. VHS, Beta $19.95 NOS, DVT, BUR ☾☾☾

Captive Hearts Well frayed story holds cast captive in sushi romance. Two American flyers are shot down and taken prisoner in isolated Japanese mountain village, and one is shot by Cupid's arrow.
1987 (R) 97m/C Pat Morita, Michael Sarrazin, Chris Makepeace; *Dir:* Paul Almond. VHS, Beta $79.98 FOX ☾

Captive Planet Bargain basement FX and really atrocious acting hold audience captive in routine earth on the verge of obliteration yarn.
1978 95m/C Sharon Baker, Chris Auram, Anthony Newcastle; *Dir:* Al Bradley. VHS, Beta $59.95 MGL ☾

Captive Rage South American general Reed hijacks planeful of girlies to encourage US to release his son, who's in trouble because he swaps money for white powder. Violence lives up to title, all else disappoints.
1988 (R) 99m/C Oliver Reed, Robert Vaughn, Claudia Udy; *Dir:* Cedric Sundstrom. VHS, Beta, CV $79.99 MCG ☾

The Capture of Bigfoot Barefoot monster tracks footprints around town after 25 years of peace, and evil businessman attempts to capture creature for personal gain.
1979 (PG) 92m/C Stafford Morgan, Katherine Hopkins, Richard Kennedy, Otis Young, George Flower, John Goff; *Dir:* Bill Rebane. VHS, Beta $29.95 AHV ☾

The Capture of Grizzly Adams Framed for murder, Adams and his ever-faithful companion Ben the bear must not only clear his name but outwit a band of outlaws who are holding his young daughter captive. Made-for-television.
1982 96m/C Dan Haggerty, Chuck Connors, June Lockhart, Kim Darby, Noah Beery, Keenan Wynn. VHS $89.95 WOV ☾½

Captured in Chinatown Dog chases bad guys in Chinatown. Bow wcw.
1935 53m/B Marion Shilling, Charles Delaney, Philo McCullough, Robert Ellis, Robert Walker, Tarzan the Wonder Dog; *Dir:* Elmer Clifton. VHS, Beta, LV $12.95 SNC, WGE ☾

Car Crash Organized crime hits stock car racing head on to produce crashing bore.
19?? 103m/C Joey Travolta, Anna Obregon. VHS $69.95 SVS ☾

Car Trouble Young English husband buys new Jaguar and wife's not so minor car trouble causes major marital trouble. Funny-bone-tickling pairing of Walters and Charleson.
1986 (R) 93m/C *GB* Julie Walters, Ian Charleson; *Dir:* David Green. VHS, Beta, LV $79.95 VIR, MCG ☾☾½

Car Wash L.A. carwash is soap opera-esque setting for disjointed comic bits about owners of dirty cars and people who hose them down for a living: Econo budget and lite plot, but serious comic talent. A sort of disco carwash version of ''Grand Hotel.'' Written by Joel Schumacher, with music by Rose Royce. Director Schultz known for ''Cooley High.''
1976 (PG) 97m/C Franklin Ajaye, Sully Boyar, Richard Brestoff, George Carlin, Richard Pryor, Melanie Mayron, Ivan Dixon, Antonio Fargas; *Dir:* Michael Schultz. VHS, Beta $59.95 MCA ☾☾½

Caravaggio Controversial biography of the late Renaissance painter famous for his bisexuality, fondness for prostitute models, violence and depravity. Photography by Gabriel Beristain reproduces the artist's visual style. British-made.
1986 97m/C *GB* Nigel Terry, Sean Bean, Tilda Swinson; *Dir:* Derek Jarman. VHS, Beta $79.95 KIV, FCT ☾☾☾

Caravan to Vaccares French Duke hires young American to sneak Eastern European scientist into the States, and little suspense ensues. Based on Alistair MacLean novel.
1974 (PG) 98m/C David Birney, Charlotte Rampling; *Dir:* Geoffrey Reeve. VHS, Beta $69.95 MED ☾

Carbon Copy Successful white executive has life turned inside out when his seventeen-year old illegitimate son, who happens to be black, decides it's time to look up dear old dad. Typical comedy-with-a-moral written by Stanley Shapiro.
1981 (PG) 92m/C George Segal, Susan St. James, Jack Warden, Paul Winfield, Dick Martin, Vicky Dawson, Tom Poston, Denzel Washington; *Dir:* Michael Schultz. VHS, Beta, LV $9.95 COL, SUE ☾☾

Cardiac Arrest Lunatic eviscerates victims in trolley town. They left their hearts in San Francisco.
1974 (PG) 95m/C Garry Goodrow, Mike Chan, Max Gail; *Dir:* Murray Mintz. VHS, Beta $69.95 MED ☾

Cardinal Priestly young Tryon rises through ecclesiastical ranks to become Cardinal, struggling through a plethora of tests of faith, none so taxing as the test of the audience's patience. Had Preminger excised some sixty minutes of footage, he might have had a compelling portrait of faith under fire, but as it stands, the cleric's life is epic confusion. Fine acting, even from Tryon, who later went on to bookish fame, and from Huston who's normally on the other side of the camera. McNamara's final performance. Based on the Henry Morton Robinson novel. Huston garnered an Oscar, while Cinematography, Art Direction, Set Decoration and Film Editing and Costume Design received nominations.

1963 175m/C Tom Tryon, Carol Lynley, Dorothy Gish, Maggie MacNamara, Cecil Kellaway, John Huston, John Saxon, Burgess Meredith; *Dir:* Otto Preminger. Golden Globe Awards '64: Best Film—Drama; National Board of Review Awards '63: 10 Best Films of the Year. **VHS, Beta, LV $59.95** QNE, IME, HRS *♫♫½*

The Care Bears Movie The Care Bears leave their cloud home in Care-a-lot to try and teach earthlings how to share their feelings of love and caring for each other.
1984 (G) 75m/C *Dir:* Arna Selznick; *Voices:* Mickey Rooney, Georgia Engel, Harry Dean Stanton. **VHS, Beta $14.98** LIV, VES, VTR *♫♫*

Career An overwrought, depressing drama about the trials and tribulations of an actor trying to make it on Broadway. He'll try anything to succeed. Good direction, but so-so acting.
1959 105m/B Dean Martin, Anthony (Tony) Franciosa, Shirley MacLaine, Carolyn Jones, Joan Blackman, Robert Middleton, Donna Douglas; *Dir:* Joseph Anthony. **VHS** PAR *♫♫½*

Career Opportunities A "Home Alone" clone for teenagers, from the John Hughes factory. Whaley plays an unsuccessful con-artist who finally gets a job as the night janitor of the local department store. He fools around at company expense until he finds the town's beauty asleep in a dressing room. The pair then play make-believe until they must thwart some small-time thieves. Unexciting and unrealistic in the worst way.
1991 (PG-13) 83m/C Frank Whaley, Jennifer Connelly, Dermot Mulroney; *Dir:* Bryan Gordon. **VHS, LV $19.95** MCA, LDC *♫½*

Carefree Dizzy radio singer Rogers can't make up her mind about beau Bellamy, so he sends her to analyst Astaire. Seems she can't even dream a little dream until shrink Astaire prescribes that she ingest some funny food, which causes her to dream she's in love with the Fredman. Au contraire, says he, it's a Freudian thing, and he hypnotically suggests that she really loves Bellamy. The two line up to march down the aisle together, and Fred stops dancing just long enough to realize he's in love with Ginger. A screwball comedy with music; the Irving Berlin score includes "I Used to be Color Blind," "The Night Is Filled With Music," and "Change Partners."
1938 83m/B Fred Astaire, Ginger Rogers, Ralph Bellamy, Jack Carson, Franklin Pangborn, Hattie McDaniel; *Dir:* Mark Sandrich. **VHS, Beta, LV $14.98** TTC, RKO, MED *♫♫♫*

Careful, He Might Hear You Abandoned by his father, six-year-old P.S. becomes a pawn between his dead mother's two sisters, one working class and the other wealthy, and his worldview is further overturned by the sudden reappearance of his prodigal father. Set in Depression-era Australia, Schultz's vision is touching and keenly observed, and manages a sort of child's eye sense of proportion. Based on a novel by Sumner Locke Elliott.
1984 (PG) 113m/C *AU* Nicholas Gledhill, Wendy Hughes, Robyn Nevin, John Hargreaves; *Dir:* Carl Schultz. Australian Film Institute '83: Best Actress (Hughes), Best Film. **VHS, Beta $59.98** FOX *♫♫♫*

Caribe Caribbean travelogue masquerades as spy thriller. Arms smuggling goes awry, and neither voodoo nor bikinied blondes can prevent audience from dozing. Never released theatrically.
1987 96m/C John Savage, Kara Glover, Stephen McHattie, Sam Malkin; *Dir:* Michael Kennedy. **VHS, Beta, LV $79.98** LIV, VES *♫½*

The Cariboo Trail Two prospecting men seek their fortune in British Columbia, the golden West of Canada in the 1890s. But they find themselves opposed by a ruthless rancher and claim-jumpers. Actually filmed in Colorado with excellent cinematography and solid performances from some big names, it's still a run-of-the-mill entry. Made the same year that the Gabby Hayes' show first aired.
1950 80m/C Angela Douglas, Randolph Scott, George "Gabby" Hayes, Bill Williams, Victor Jory; *Dir:* Gerald Thomas, Edwin L. Marin. **VHS, LV** FRH, IME *♫♫*

Carlton Browne of the F.O. Bumbling Brit diplomat Thomas visits tiny Pacific island of Gallardia, forgotten by the mother country for some fifty years, to insure tenuous international agreement after the island's king dies. Not sterling Sellers but some shining moments. Alternate title: "Man in a Cocked Hat."
1959 88m/C *GB* Peter Sellers, Terry-Thomas, Lucianna Paluzzi, Ian Bannen; *Dir:* Roy Boulting, Jeffrey Dell. **VHS, Beta $19.98** HBO, FCT *♫♫½*

Carmen Choreographer casts Carmen and finds life imitates art when he falls under the spell of the hotblooded Latin siren. Bizet's opera lends itself to erotically charged flamenco context. Well acted, impressive scenes including cigarette girls' dance fight and romance between Carmen and Don Jose. In Spanish with English subtitles.
1983 (R) 99m/C *SP* Antonio Gades; Laura Del Sol, Paco DeLucia, Christina Hoyos; *Dir:* Carlos Savrat. **VHS, Beta $29.98** CTH, MED, TAM *♫♫♫*

Carmen This version of the famous Bizet opera was shot on location in Italy. Sunny villages and dramatic bullrings give the film a look that has, until now, been absent in most versions of this opera. Terrific soundtrack enhances the already beautiful photography.
1984 151m/C *FR IT* Julia Migenes-Johnson, Placido Domingo, Faith Esham, Ruggero Raimondi; *Dir:* Francesco Rosi. **VHS, Beta $19.95** TAM, COL *♫♫♫*

Carmen Jones Bizet's tale of fickle femme fatale Carmen heads South with an all black cast and new lyrics by Hammerstein II. Soldier Belafonte falls big time for factory working belle Dandridge during the war, and runs off with miss thang after he kills his C.O. and quits the army. Tired of prettyboy Belafonte, Dandridge's eye wanders upon prize pugilist Escamillo, inspiring ex soldier beau to wring her throaty little neck. Film debuts of Carroll and Peters. More than a little racist undertone to the direction. Actors' singing is dubbed. Oscar nominations for Dandridge (Best Actress) and Scoring of a Musical Picture.
1954 105m/C Dorothy Dandridge, Harry Belafonte, Pearl Bailey, Roy Glenn, Diahann Carroll, Brock Peters; *Dir:* Otto Preminger. Golden Globe Awards '55: Best Film—Musical/Comedy. **VHS, LV** MLB *♫♫♫*

Carmilla An adaptation of the Sheridan Le Fanu lesbian vampire tale. From Shelley Duvall's cable television series "Nightmare Classics."
1989 60m/C Meg Tilly, Ione Skye, Roddy McDowall, Roy Dotrice; *Dir:* Gabrielle Beaumont. **VHS, Beta $59.95** NO *♫♫♫*

Carnage Hungry house consumes inhabitants.
1984 91m/C Leslie Den Dooven, Michael Chiodo, Deeann Veeder; *Dir:* Andy Milligan. **VHS, Beta $49.95** MED *♫*

Carnal Crimes Well-acted upscale softcore trash about a sensuous woman, ignored by her middle-aged lawyer husband and drawn to a young stud photographer with a shady past and S&M tendencies. Available in a sexy unrated version also.
1991 (R) 103m/C Martin Hewitt, Linda Carol, Rich Crater, Yvette Stefens, Paula Trickey, Alex Kubik; *Dir:* Alexander Gregory Hippolyte. **VHS $79.98** MAG *♫*

Carnal Knowledge Carnal knowledge of the me generation. Three decades in the sex-saturated lives of college buddies Nicholson and Garfunkel, chronicled through girlfriends, affairs and marriages. Controversial upon release, it's not a flattering anatomy of Y chromosome carriers. Originally written as a play. Ann-Margret received an Oscar nomination; Kane's debut.
1971 (R) 96m/C Jack Nicholson, Candice Bergen, Art Garfunkel, Ann-Margret, Rita Moreno, Carol Kane; *Dir:* Mike Nichols. **VHS, Beta, LV, 8mm $14.98** SUE, RCA, EMB *♫♫♫*

Carnegie Hall at 100: A Place of Dreams The one hundredth birthday of the legendary performance hall draws a host of stars, both classical and popular, to celebrate this musical monument. Archival footage and modern performances are juxtaposed in this touching, sentimental documentary. Performances by Leonard Bernstein, Victor Borge, Ray Charles, Van Cliburn, Aaron Copland, Liza Minnelli, Vladimir Horowitz, Frank Sinatra, Isaac Stern, Yo-Yo Ma, and many others make this a special tribute.
1991 ?m/B *Performed by:* Leonard Bernstein, Victor Borge, Ray Charles, Van Cliburn, Aaron Copland, Jascha Heifetz, Marilyn Horne, Vladimir Horowitz, Yo-Yo Ma, Liza Minnelli, Frank Sinatra, Gregor Piatigorsky, Isaac Stern, Arturo Toscanini. **VHS, LV $34.98** BMG, FCT

Carnival of Animals Using the verse of Ogden Nash and the music of Saint-Saens, this film for children explores the animals in the zoo.
1985 30m/C *Dir:* Milos Forman; *Nar:* Gary Burghoff. **VHS, Beta $12.98** LCA, MTT, TTE *♫♫♫*

Carnival of Blood Boring talky scenes punctuated by Coney Island murder mayhem followed by more boring talky scenes. Young's debut, not released for five years. The question's not why the delayed release, but why bother at all. A carnival of cliches.
1987 (PG) 80m/C Earle Edgerton, Judith Resnick, Martin Barlosky, John Harris; *Dir:* Leonard Kirtman. **VHS, Beta** MAV *♫*

Carnival in Flanders (La Kermesse Heroique) A 17th century Gallic village threatened by Spanish invaders decides to roll over and play dead and throws a carnival to end all carnivals to welcome the vanquishers-to-be. Classic French farce won gold medal at the Venice International Exposition of Cinematography and the Grand Prix Du Cinema Francais. In French with English subtitles. Originally titled "La Kermesse Heroique."
1935 90m/B *FR* Francoise Rosay, Louis Jouvet, Jean Murat, Andre Aleme, Micheline Cheirel; *Dir:* Jacques Feyder. National Board of Review Awards '36: Best Foreign Film; New York Film Critics Awards '36: Best Foreign Film; Venice Film Festival '36: Best Director; New York Times Ten Best List '36: Best Film. **VHS, Beta $24.95** NOS, MRV, HHT *♫♫♫½*

Carnival Lady Silver spooner Vincent is scheduled to tie the knot until little lady-to-be discovers he's a bit out of pocket after the stock market takes a wee dip. Jilted and

C

impecunious, he heads for the big top, where he horns in on the high diver's turf, and it's all downhill from there.
1933 66m/B Boots Mallory, A. Vincent. **VHS, Beta $29.95** *VCN* ⚅

Carnival Rock Story of love triangular in seedy nightclub. Club owner Stewart loves chanteuse Cabot who loves card playin' Hutton. Who cares? Who knows. Maybe hard-core Corman devotees. Good tunes from the Platters, the Blockbusters, Bob Luman and David Houston.
1957 80m/C Susan Cabot, Brian Hutton, David J. Stewart, Dick Miller; *Dir:* Roger Corman. **VHS, Beta $9.95** *RHI, AOV, FCT* ⚅ 1/2

Carnival of Souls Cult-followed zero budget zombie opera has young Hilligoss and girlfriends take wrong turn off bridge into river. Mysteriously unscathed, Hilligoss rents room and takes job as church organist, but she keeps running into dancing dead people, led by director Harvey. Spooky, very spooky. Laserdisc version offers restored picture quality and a refined audio track, as well as introductory feature on the movie's background.
1962 80m/B Candace Hilligoss, Sidney Berger, Frances Feist, Stan Levitt, Art Ellison, Herk Harvey; *Dir:* Herk Harvey. **VHS, Beta, LV $19.99** *VID, MRV, SNC* ⚅⚅⚅

Carnival Story Yet another melodramatic cliche about love triangular under the big top. German girl joins American-owned carnival and two guys start acting out unbecoming territorial behavior. Filmed in Germany.
1954 94m/C Anne Baxter, Steve Cochran, Lyle Bettger, George Nader; *Dir:* Kurt Neumann. **VHS, Beta $19.95** *NOS, GEM, DVT* ⚅ 1/2

The Carnivores The flesh-eaters of the animal kingdom—bears, lions, tigers, and others—and their survival instincts are the focus of this nature documentary.
1983 90m/C VHS, Beta $69.95 *DIS* ⚅⚅

Carny Hothead carnival bozo Busey and peacemaker Robertson experience friendship difficulties when runaway Foster rolls in hay with one and then other. Originally conceived as a documentary by "Derby" filmmaker Kaylor, it's a candid, unsavory, behind-the-scenes anatomy. Co-written by "The Band" member Robertson.
1980 (R) 102m/C Gary Busey, Robbie Robertson, Jodie Foster, Meg Foster, Kenneth McMillan, Elisha Cook Jr., Craig Wasson; *Dir:* Robert Kaylor. **VHS, Beta $19.98** *FOX, WAR* ⚅⚅⚅

Caroline? Fifteen years previously a wealthy young woman is presumed dead in a plane crash. Now a stranger appears at the family home claiming to be Caroline—and wanting her inheritance. Is it possible that what she says is true or is it all an elaborate ruse? An especially good performance by Zimbalist. Made for television as a "Hallmark Hall of Fame" presentation. Based on the novel "Father's Arcane Daughter" by E.L. Konigsburg.
1990 (PG) 100m/C Stephanie Zimbalist, Pamela Reed, George Grizzard, Dorothy McGuire, Patricia Neal; *Dir:* Joseph Sargent. Emmy Awards '90: Outstanding Drama/Comedy Special; Emmy Award '90: Outstanding Directing in a Miniseries or a Special (Sargent). **VHS, LV $89.98** *REP* ⚅⚅⚅

Carousel Much-loved Rodgers & Hammerstein musical based on Ferenc Molnar's play "Liliom" (filmed by Fritz Lang in 1935) about a swaggering carnival barker (MacRae) who tries to change his life after he falls in love with a good woman. Killed while at-

tempting to foil a robbery he was supposed to help commit, he begs his heavenly hosts for the chance to return to the mortal realm just long enough to set things straight with his teenage daughter. Songs include "If I Loved You," "Soliloquy," and "You'll Never Walk Alone." Now indisputably a classic, the film lost $2 million when it was released.
1956 128m/C Gordon MacRae, Shirley Jones, Cameron Mitchell, Gene Lockhart, Barbara Ruick, Robert Rounseville, Richard Deacon, Tor Johnson; *Dir:* Henry King. **VHS, Beta, LV $19.98** *FOX, RDG* ⚅⚅⚅

The Carpathian Eagle Police wonder why murdered victims have hearts ripped out while audience wonders what weird title has to do with anything. Routine evisceration fest, part of Elvira's Thriller Video.
1981 60m/C Suzanne Danielle, Sian Phillips, Pierce Brosnan, Anthony Valentine; *Dir:* Francis Megahy. **VHS, Beta $29.95** *THR, IVE* ⚅

The Carpenter Post-nervous breakdown woman receives nightly visits from guy who builds stuff with wood. Very scary stuff. Also available in slightly longer unrated version.
1989 (R) 85m/C Wings Hauser, Lynne Adams, Pierce Lenoir, Barbara Ann Jones; *Dir:* David Wellington. **VHS, LV $89.98** *REP, PIA* ⚅

The Carpetbaggers Uncannily Howard Hughesian Peppard wallows in wealth and women in Hollywood in the 1920s and 1930s. Spayed version of the Harold Robbins novel. Ladd's final appearance. Followed by the prequel "Nevada Smith." Introduced by Joan Collins.
1964 (PG) 150m/C George Peppard, Carroll Baker, Alan Ladd, Elizabeth Ashley, Lew Ayres, Martha Hyer, Martin Balsam, Robert Cummings; *Dir:* Edward Dmytryk. National Board of Review Awards '64: Best Supporting Actor (Balsam). **VHS, Beta $59.95** *PAR* ⚅⚅

Carrie In a part turned down by Cary Grant, Olivier plays a married American who self destructs as the woman he loves scales the heights to fame and fortune. The manager of a posh epicurean mecca, Olivier deserts wife Hopkins and steals big bucks from his boss to head east with paramour Jones, a country bumpkin transplanted to Chicago. Once en route to thespian fame in the Big Apple, Jones abandons her erstwhile beau, who crumbles pathetically. Adapted from Theodore Dreiser's "Sister Carrie," it's mega melodrama, but the performances are above reproach.
1952 118m/B Laurence Olivier, Jennifer Jones, Miriam Hopkins, Eddie Albert, Basil Ruysdael, Ray Teal, Barry Kelley, Sara Berner, William Reynolds, Mary Murphy; *Dir:* William Wyler. **VHS, Beta, LV $19.98** *PAR, FCT* ⚅⚅⚅

Carrie Overprotected by religious fanatic mother Laurie and mocked by the in-crowd, shy, withdrawn high school girl Carrie White is asked to the prom. Realizing she's been made the butt of a joke, she unleashes her considerable telekinetic talents. Travolta, Allen, and Irving are teenagers who get what they deserve. Based on the Stephen King novel. Oscar nominations for Spacek and Laurie. Laserdisc version features the original movie trailer, an interview with screenwriter Lawrence D. Cohen, publicity shots, a study of De Palma's film-making techniques, and commentary by film expert Laurent Bouzereau.
1976 (R) 98m/C Sissy Spacek, Piper Laurie, John Travolta, William Katt, Amy Irving, Nancy Allen, Edie McClurg, Betty Buckley; *Dir:* Brian De Palma. **VHS, Beta, LV $19.95** *FOX, MGM, CRC* ⚅⚅⚅

The Carrier Smalltown Sleepy Rock is ideal family-raising turf until plague mysteriously blights inhabitants, and townspeople are out to exterminate all potential carriers. Silverman is standout as local spiritual leader; well orchestrated crowd scenes. Best park your IQ before watching. Filmed on location in Manchester, Michigan.
1987 (R) 99m/C Gregory Fortescue, Steve Dixon, Paul Silverman; *Dir:* Nathan J. White. **VHS, Beta $29.95** *MAG* ⚅⚅

Carrington, V.C. British army major Niven is brought up for a court-martial on embezzlement charges because he arranges, without official permission, to be reimbursed for money owed him. A former war hero, he decides to defend himself in court, and, once an affair comes to light, his vindictive wife joins the opposition. Good cast, superlative Niven performance. Alternate U.S. title: "Court Martial."
1955 100m/B *GB* David Niven, Margaret Leighton, Noelle Middleton, Laurence Naismith; *Dir:* Anthony Asquith. **VHS, Beta $39.95** *MON* ⚅⚅⚅

Carrott Gets Rowdie Popular English comedian Jasper Carrott comments upon the differences between British and American people in this concert taped in Tampa Bay, Florida.
1984 60m/C VHS, Beta $59.95 *PAV* ⚅⚅⚅

Carry Me Back Two brothers carry dead dad back to Australian ranch in order to inherit mucho dinero.
1982 93m/C *AU* Grant Tilly, Kelly Johnson. **VHS, Beta $39.95** *IVE* ⚅ 1/2

Carry On Behind "Carry On" crew heads for archeological dig and find themselves sharing site with holiday caravan.
1975 95m/C *GB* Sid James, Kenneth Williams, Elke Sommer, Joan Sims; *Dir:* Gerald Thomas. **VHS, Beta $59.98** *SUE* ⚅

Carry On Cleo "Carry On" spoof of Shakespeare's "Antony and Cleopatra."
1965 91m/C *GB* Sidney James, Amanda Barrie, Kenneth Williams, Kenneth Connor, Jim Dale, Charles Hawtrey, Joan Sims; *Dir:* Gerald Thomas. **VHS, Beta $39.99** *HBO* ⚅⚅ 1/2

Carry On Cowboy "Carry On" Western parody of "High Noon," also known as "Rumpo Kid."
1966 91m/C *GB* Sidney James, Kenneth Williams, Jim Dale. **VHS, Beta $69.99** *HBO* ⚅⚅

Carry On Cruising "Carry On" gang attacks sailing world with low humor and raunchiness.
1962 89m/C *GB* Sidney James, Kenneth Williams, Liz Fraser; *Dir:* Gerald Thomas. **VHS, Beta $59.99** *HBO* ⚅

Carry On Dick What made the seemingly endless "Carry On" series of super-low-budget British farces such a hit is a mystery not to be solved. Low production values, scripts that peter out midway, and manifest humor don't normally a classic make; and yet the gang has its following. "Dick," a detective movie spoof, was preceded by some twenty-odd carryings on; suffice it to say that the series, which began in 1958 with "Carry on Sergeant," has not improved with age in subsequent incarnations.
1975 95m/C *GB* Sid James, Joan Sims; *Dir:* Gerald Thomas. **VHS** *TPV* ⚅⚅ •

Carry On Doctor British series continues as characters of questionable competence join the medical profession. Hospital staff gets caught in battle over secret weight loss formula.
1968 95m/C *GB* Frankie Howerd, Kenneth Williams, Jim Dale, Barbara Windsor; *Dir:* Gerald Thomas. **VHS** $19.95 *PAR* 🎬🎬🎬

Carry On Emmanuelle Bedroom diplomat Emmanuelle enters international politics by taking on foreign legion. Part of the British "Carry On" series.
1978 104m/C *GB* Suzanne Danielle, Kenneth O'Connor, Kenneth Williams, Beryl Reid; *Dir:* Gerald Thomas. **VHS, Beta** *PGN* 🎬

Carry On England Mercifully, the gang didn't carry on much beyond this entry, in which a WWII antiaircraft fleet captain bumbles through the usual pranks and imbroglios.
1976 89m/C *GB* Kenneth Connor, Patrick Mower, Judy Geeson; *Dir:* Gerald Thomas. **VHS** *TAF* 🎬

Carry On Nurse Men's ward in a British hospital declares war on nurses and the rest of the hospital. The second of the "Carry On" series.
1959 86m/B *GB* Shirley Eaton, Kenneth Connor, Hattie Jacques, Wilfrid Hyde-White; *Dir:* Gerald Thomas. **VHS, Beta** $19.95 *NOS, VYY, HBO* 🎬

Carry On 'Round the Bend Williams, who played in the original ("Carry On Sergeant"), and Carry On regulars (charpei-mugged James and ever-zaftig Jacques), go 'round the bend in yet another installment of the British spoof.
1972 89m/C *GB* Sid James, Kenneth Williams, Hattie Jacques; *Dir:* Gerald Thomas. **VHS** *TPV* 🎬½

Carry On Screaming "Carry On" does horror. A pair of goofy detectives trail monsters suspected in kidnapping.
1966 97m/C *GB* Harry H. Corbett, Kenneth Williams, Fenella Fielding, Joan Sims, Charles Hawtrey, Jim Dale, Angela Douglas, Jon Pertwee; *Dir:* Gerald Thomas. **VHS** $19.98 *MOV* 🎬🎬

Carry On at Your Convenience Low-brow humor in and around a toilet factory.
1971 86m/C *GB* Sidney James, Kenneth Williams, Charles Hawtrey, Joan Sims, Kenneth Cope; *Dir:* Gerald Thomas. **VHS, Beta** $19.95 *AXV, TAM* 🎬½

Cars That Ate Paris Parasitic town in the Parisian (Australia) outback preys on car and body parts generated by deliberate accidents inflicted by wreck-driving wreckless youths. Weir's first film released internationally, about a small Australian town that survives economically via deliberately contriving car accidents and selling the wrecks' scrap parts. A broad, bitter black comedy with some horror touches.
1974 (PG) 91m/C *AU* Terry Camillieri, Kevin Miles, John Meillon, Melissa Jaffe; *Dir:* Peter Weir. **VHS, Beta** $19.95 *COL* 🎬

Carson City Kid Typical western which has Rogers trying to exact revenge on the man who killed his brother.
1940 54m/B Roy Rogers, Dale Evans. **VHS, Beta** $19.95 *NOS, VCN, MED* 🎬

Cartel O'Keeffe, "B" actor extraordinaire, plays the wrong man to pick on in this rancid dope opera. Hounded by drug lord Stroud and framed for murder, pilot O'Keeffe decides the syndicate has gone too doggone

far when they kill his sister. Exploitive and otherwise very bad.
1990 (R) 106m/C Miles O'Keeffe. **VHS, LV** *SHG, IME* Woof!

Carthage in Flames A graphic portrayal of the destruction of ancient Carthage in a blood and guts battle for domination of the known world. A tender love story is a welcome aside in this colorful Italian-made epic.
1960 96m/C *IT* Anne Heywood, Jose Suarez, Pierre Brasseur; *Dir:* Carmine Gallone. **VHS, Beta** $29.95 *VCD* 🎬🎬

Cartier Affair A beautiful television actress unwittingly falls in love with the man who wants to steal her jewels. Collins designed her own wardrobe. Made for television.
1984 120m/C Joan Collins, David Hasselhoff, Telly Savalas, Ed Lauter, Joe La Due; *Dir:* Rod Holcomb. **VHS, Beta** $39.98 *LHV, WAR* 🎬

Cartoon All-Stars to the Rescue Cartoon characters from Winnie the Pooh to the Ninja Turtles help kids understand the dangers of drug abuse.
1990 60m/C **VHS** *BVV* 🎬

Cartoon Carnival: No. 1 Includes "Song of the Birds" (Max Fleischer), "Jerky Turkey" (MGM), "The Talking Magpies" (Terrytoons), "Jasper in the Haunted House" (George Pal), "It's a Hap Hap Happy Day" (Max Fleischer), "Boy Meets Dog" (Walter Lantz), and "The Friendly Ghost" (Harvey Cartoons).
193? 51m/C **VHS, Beta** $19.95 *HHT* 🎬

Cartoon Carnival: No. 2 A compilation of 8 classic cartoons: "Pincushion Man," "Mary's Little Lamb" and "Jack Frost," all by Ub Iwerks; Max Fleisher's "Cobweb Hotel;" "Farm Frolics," a Merrie Melodies cartoon; Porky Pig in "Timid Toreador;" "Pantry Panic," starring Woody Woodpecker; and a Looney Tunes cartoon, "Hollywood Capers."
193? 55m/C **VHS, Beta** $30.00 *HHT* 🎬

Cartoon Cavalcade Here are three classic cartoons from Burt Gillette and Ub Iwerkes: "Molly Moo Cow and the Indians," "The Three Bears," and "Molly Moo Cow and Rip Van Winkle."
1935 23m/C **VHS, Beta** $19.98 *CCB* 🎬

Cartoon Classics Three vintage cartoons, in black & white and color: "Baby Be Good," a Betty Boop classic, "Secret Agent," anti-Nazi Superman agitprop, and "King for a Day," an adventure set in Swift's Lilliput.
1940 30m/C **VHS, Beta** $59.95 *JEF* 🎬

Cartoon Classics of the 1930s Eight cartoon classics of the 1930s, including "Felix the Cat," "Daffy and the Dinosaur," and "Bold King Cole."
193? 58m/C **VHS, Beta** $29.95 *MED* 🎬

Cartoon Classics in Color, No. 1 Eight cartoon classics filled with color, movement and lots of fun comprise this tape: "Little Black Sambo," "Jack Frost," "Sinbad the Sailor," "Simple Simon," "Ali Baba," "Molly Moo Cow and the Butterflies," "The Toonerville Trolley," and "Somewhere in Dreamland." Reflects the social consciousness of the time, which some may find offensive.
1934 60m/C **VHS, Beta** $24.95 *VYY* 🎬

Cartoon Classics in Color, No. 2 Presents eight animated Warner Brothers favorites. Includes: Daffy Duck in "Yankee Doodle Daffy," Porky Pig in "Ali Baba Bound," and Bugs Bunny in "Falling Hare."
194? 60m/C *Dir:* Chuck Jones, Robert McKimson, Friz Freleng. **VHS, Beta** $24.95 *VYY* 🎬

Cartoon Classics in Color, No. 3 Various Warner Bros. Bugs/Daffy/Elmer/etc. cartoons from the war years: "A Corny Concerto," "Foney Fables," "The Wacky ," "Fifth Column Mouse," "To Duck Or Not To Duck," "The Early Worm Gets the Bird," and "Daffy the Commando."
1943 60m/C *Dir:* Bob Clampett, Chuck Jones, Friz Freleng, Robert McKimson. **VHS, Beta, 8mm** $24.95 *VYY* 🎬

Cartoon Classics in Color, No. 4 Another collection of Warner cartoons, featuring Daffy and Bugs: "The Wabbit Who Came To Supper," "A Tale of," "Wackiki Wabbitt," "Daffy Duck and the Dinosaur," and "Fresh Hare."
1948 54m/C *Dir:* Chuck Jones, Robert McKimson, Friz Freleng, Bob Clampett. **VHS, Beta, 8mm** $24.95 *VYY* 🎬

Cartoon Classics, No. 1: Looney Tunes & Merrie Melodies A collection of early sound cartoons each based on an extended jazz number: "The Queen Was in the Parlor," "Freddy the Freshman," "Red Headed Baby," "Battling Bosko," "You're Too Careless with Your Kisses," "It's Got Me Again," "Moonlight for Two," and "You Don't Know What You're Doin'."
1933 56m/B **VHS, Beta, 8mm** $24.95 *VYY* 🎬

Cartoon Classics, No. 3: The Early Pioneers A compilation of six silent cartoons, including pioneering efforts in stop-motion animation by Willis O'Brien, creator of "King Kong."
1921 52m/B *Dir:* Dave O'Brien, Ub Iwerks. **VHS, Beta** $9.95 *VYY* 🎬

Cartoon Classics, No. 4: Cobweb Hotel A collection of 8 classic early cartoons. Porky Pig makes a few of his first appearances.
1936 55m/B *Dir:* Friz Freleng, Bob Clampett, Paul Terry. **VHS, Beta** $9.98 *VYY* 🎬

Cartoon Classics, No. 5: The Other Studios Another collection of rare, classic cartoons, featuring Bosko and Felix the Cat.
1934 60m/B *Dir:* Walter Lantz. **VHS, Beta** $9.98 *VYY* 🎬

Cartoon Classics, No. 6: Early Animation Seven early cartoons, including two featuring then-famous comic strip stars Mutt and Jeff ("Invisible Revenge" and "Cramps'") and Winsor McCay's famous "Gertie," the first animated short ever. Other titles are: "Dog-Gone," "Goodrich Dirt," "Cowpuncher," "Willi's Nightmare," and "The Animated Grouch Chaser."
1914 56m/B **VHS, Beta, 8mm** $24.95 *VYY* 🎬

Cartoon Classics, Vol. 1: Here's Mickey! Mickey Mouse and Donald Duck in some of the best cartoons ever made by Walt Disney, including "Orphan's Benefit," "Mickey's Garden" and "Mickey's Birthday Party."
1941 27m/C *Dir:* Walt Disney. **VHS, Beta** $12.99 *DIS* 🎬

Cartoon Classics, Vol. 2: Here's Donald! Donald Duck stars in three of his earliest cartoons from the Disney studios: "Wide Open Spaces," "Donald's Ostrich," and "Crazy With the Heat."
1939 22m/C *Dir:* Walt Disney. VHS, Beta $12.99 DIS

Cartoon Classics, Vol. 3: Here's Goofy! That lovable hound Goofy appears in three Disney cartoons: "Whom the Bulls Toll," "Lion Down" and "A Knight for a Day."
1939 22m/C *Dir:* Walt Disney. VHS, Beta $12.99 DIS

Cartoon Classics, Vol. 4: Silly Symphonies! Three early Disney cartoons are included on this tape: "Three Little Wolves," "Toby Tortoise Returns" and "Water Babies."
1939 22m/C VHS, Beta $12.99 DIS

Cartoon Classics, Vol. 5: Here's Pluto! Pluto is the star in these Disney classics. Includes: "Mail Day," "Pantry Pirate" and "Springtime for Pluto."
1939 23m/C VHS, Beta $12.99 DIS

Cartoon Classics, Vol. 6: Starring Mickey & Minnie The beloved mice star in, "The Little Whirlwind," "Hawaiian Holiday" and "Brave Little Tailor."
1988 22m/C VHS, Beta $12.99 DIS

Cartoon Classics, Vol. 7: Starring Donald & Daisy The courtin' fowl star in "Don Donald," "Donald's Double Trouble" and "Donald's Diary."
1939 22m/C VHS, Beta $12.99 DIS

Cartoon Classics, Vol. 8: Starring Silly Symphonies (Animals 2 By 2) More early cartoons from the Disney studios: "Father Noah's Ark," "Peculiar Penguins" and "The Tortoise and the Hare."
1939 27m/C VHS, Beta $12.99 DIS

Cartoon Classics, Vol. 9: Starring Chip 'n' Dale Three more installments of the cartoon antics of the inseparable chipmunk pranksters. These wild episodes are entitled "Donald Applecore," "Dragon Around" and "Working for Peanuts."
1951 22m/C VHS, Beta $12.99 DIS

Cartoon Classics, Vol. 10: Starring Pluto & Fifi Three cartoons from the Disney studios: "Society Dog Show," "Pluto's Blue Note" and "Pluto's Own Puplets."
1939 24m/C VHS, Beta $12.99 DIS

Cartoon Classics, Vol. 11: Mickey & the Gang Mickey stars with Donald, Goofy, Pluto and other favorites in, "Boat Builders," "Canine Caddy" and "Moose Hunters."
1990 25m/C VHS, Beta $12.99 DIS

Cartoon Classics, Vol. 12: Nuts About Chip 'n' Dale Those zany animated chipmunks return to wreak havoc on the lives of veteran Disney performers. In "Trailer Horn," the partners in pranks terrorize Donald Duck. In "Food for Feudin'," they harrass Pluto. And in "Two Chips and a Miss," the boys duel for the affections of a beautiful lass.
1990 22m/C VHS, Beta $12.99 DIS

Cartoon Classics, Vol. 13: Donald's Scary Tales Halloween fun as Donald and friends star in "Donald Duck and the Gorilla," "Duck Pimples" and "Donald's Lucky Day."
1991 22m/C VHS, Beta $12.99 DIS

Cartoon Classics, Vol. 14: Halloween Haunts Disney's best-loved characters star in the Halloween-themed cartoons "Pluto's Judgment Day," "Lonesome Ghost" and "Trick or Treat."
1991 22m/C VHS, Beta $12.99 DIS

Cartoon Collection, No. 1: Porky in Wackyland A collection of sixteen classic cartoons from the thirties, forties, and fifties. Included are Bugs Bunny in "All This and Rabbit Stew," Daffy Duck in "Scrap Happy Daffy," Popeye in "Eugene the Jeep" and "Poop Deck Pappy," and Betty Boop in "Minnie the Moocher." Some in black and white.
194? 115m/C *Dir:* Chuck Jones, Robert McKimson, Max Fleischer, Bob Clampett. VHS, Beta $24.95 SHO

Cartoon Collection, No. 2: Classic Warner Brothers A collection of sixteen favorite Warner Brothers cartoons from the forties and fifties including Bugs Bunny in "Fresh Hare" and "Falling Hare," Daffy Duck in "The Daffy Commando" and "Daffy's Southern Exposure," and Porky Pig in "Notes to You" and "Porky's Midnight Matinee."
194? 115m/C *Dir:* Friz Freleng, Chuck Jones, Robert McKimson. VHS, Beta $24.95 SHO

Cartoon Collection, No. 3: Coal Black & de Sebben Dwarfs Another package of 16 Warner Bros. cartoons from the 1930s and 40s, featuring Bugs Bunny, Daffy Duck and Porky Pig. Titles include "Coal Black and de Sebben Dwarfs," "Calling Dr. Porky," "Tom Turk and Daffy" and "Daffy Doc."
194? 115m/B *Dir:* Chuck Jones, Friz Freleng, Robert McKimson. VHS, Beta $24.95 SHO

Cartoon Collection, No. 4: Warner Brothers & Fleischer Sixteen golden classics from the Depression era, all in their original form in glorious black and white.
193? 110m/B VHS, Beta $24.95 SHO

Cartoon Collection, No. 5: Racial Cartoons A compilation of classic cartoons from the '30s and '40s including "Bugs Bunny Bond Rally," "Little Black Sambo," "Congo Jazz" and "The Japoteurs."
194? 120m/C *Dir:* Chuck Jones, Robert McKimson, Bob Clampett. VHS, Beta $24.95 SHO

Cartoon Collection, No. 6: The Ducktators Sixteen cartoons from the 30's and 40's, featuring Little Lulu, Bosko, Porky Pig, Felix the Cat and Heckle & Jeckle.
1943 115m/C VHS, Beta $24.95 SHO

Cartoon Collection, No. 7: Toyko Jokio Vintage cartoons of the 1930s and the 1940s are shown in color and black and white. Cartoons include: "Yankee Doodle Daffy," "A Tale of Two Kitties," "Unruly Hare," and "Moonlight for Two."
1945 115m/C VHS, Beta SHO

Cartoon Collection, No. 8 Cartoons featuring Little Lulu, Betty Boop, and Porky Pig. Private Snafu, the Warner Brothers World War II creation, is included.
194? 120m/B VHS, Beta $24.95 SHO

Cartoon Fun A compilation of cartoons featuring Casper the Ghost, Betty Boop, Little Lulu and other notables.
1958 40m/C VHS $14.95 REP

Cartoon Magic Each tape in this series contains four or five classic MGM cartoons from the 1930s and 1940s, and feature characters such as Barney Bear, Doctor D and Screwball Squirrel.
1985 45m/C VHS, Beta MGM

Cartoon Moviestars: Bugs! Bugs Bunny performs solo. Included are Jones' "Bugs Bunny and the Three Bears," Clampett's "Bugs Bunny Gets the Boid," and McKimson's "Gorilla My Dreams."
1948 60m/C *Dir:* Bob Clampett, Friz Freleng, Chuck Jones, Robert McKimson. VHS MGM

Cartoon Moviestars: Daffy! Daffy takes center stage. High points: Clampett's "The Great Piggybank Robbery," and "Book Revue," Freleng's "Yankee Doodle Daffy," and Tex Avery's "Daffy Duck and the Egghead," featuring an early appearance of the character who would mature into Elmer Fudd.
1948 60m/C *Dir:* Tex Avery, Bob Clampett, Arthur Davis, Freleng Friz, Robert McKimson. VHS MGM

Cartoon Moviestars: Elmer! Follow the progress of Elmer Fudd. High points: Freleng's "The Hare-Brained Hypnotist," Davis' "What Makes Daff Duck?" and Jones' "A Pest in the House."
1948 60m/C *Dir:* Arthur Davis, Friz Freleng, Chuck Jones. VHS MGM

Cartoon Moviestars: Porky! Porky Pig takes the spotlight, and with him are some of the funniest stories of all time. Freleng's "I Haven't Got a Hat," Jones' "My Favorite Duck," Clampett's "Baby Bottleneck," and McKimson's "Daffy Doodles."
1947 60m/C *Dir:* Tex Avery, Bob Clampett, Friz Freleng, Chuck Jones. Robert McKimson, Frank Tashlin. VHS MGM

Cartoon Parade: No. 1 A collection of cartoons starring Bugs Bunny, Daffy Duck, Porky Pig, Popeye, Superman, and more.
194? 120m/C *Dir:* Friz Freleng. VHS, Beta $19.95 MED

Cartoon Parade: No. 2 Laugh with some of your favorite cartoon characters. Included are Bugs Bunny in "Wabbit Who Came to Supper," Popeye in "Popeye Meets Ali Baba," Superman in "Terror on the Midway," Little Lulu in "Bored of Education," and many more.
194? 117m/C VHS, Beta $19.95 MED

Cartoon Parade: No. 3 A collection of cartoon classics including "Falling Hare" with Bugs Bunny, "Cheese Burglar" starring Herman and Katnip, "Somewhere in Dreamland," "Robin Hood Makes Good," and others.
194? 110m/C VHS, Beta $19.95 MED

Cartoon Parade: No. 4 Another collection of popular cartoons from the 30's and 40's, featuring Max Fleischer's Bouncing Ball, Little Lulu, Superman, Bugs Bunny, and

Porky Pig. Some cartoons are in black and white.
1942 120m/C VHS, Beta *MED*

Cartoonal Knowledge: Confessions of Farmer Gray A collection of 7 silent cartoons by Terry, including "Day at the Park," "The Window Washers," "A Cat's Life," and "Chemistry Lesson."
1923 54m/B *Dir:* Paul Terry. **VHS, Beta $24.95** *VYY*

Cartoonal Knowledge: Farmer Gray Looks at Life Directed by Paul Terry, this compilation features seven Gray silent shorts, with a musical score.
1926 55m/B VHS, Beta $24.95 *VYY*

Cartoonal Knowledge: Farmer Gray and the Mice Seven vintage silent cartoons featuring the irascible Farmer Gray and his adventures with barnyard beasts. Directed by Paul Terry. Silent.
1929 55m/B VHS, Beta $24.95 *VYY*

Cartoonies: Featuring Betty Boop A collection of Max Fleischer's best loved cartoons featuring America's favorite cartoon vamp.
193? 46m/C VHS $14.98 *REP*

The Cartoons Go to War A compilation of cartoons made during World War II, which hilariously propound caricatured propagandistic messages.
1943 58m/C VHS, Beta $29.95 *VIC*

Carve Her Name with Pride The true story of Violette Szabo who at age nineteen became a secret agent with the French Resistance in World War II.
1958 119m/B Virginia McKenna, Paul Scofield. **VHS, Beta** *LCA* ♫♫½

The Cary Grant Collection Three of Cary Grant's most memorable films: "Father Goose," Oscar-winning script); "Indiscreet" (w/ Ingrid Bergman); and "Operation Petticoat" (Tony Curtis and Grant are quite a team in this military comedy! Some great gags.)
1990 336m/C Cary Grant, Ingrid Bergman, Leslie Caron, Trevor Howard, Tony Curtis, Dina Merrill, Gavin McLeod, Madlyn Rhue; *Dir:* Ralph Nelson, Stanley Donen, Blake Edwards. **VHS, Beta $59.95** *REP*

Casablanca Can you see George Raft as Rick? Jack Warner did, but producer Hal Wallis wanted Bogart. Considered by many to be the best film ever made and one of the most quoted movies of all time, it rocketed Bogart from gangster roles to romantic leads as he and Bergman (who never looked lovelier) sizzle on screen. Bogart runs a gin joint in Morocco during the Nazi occupation, and meets up with Bergman, an old flame, but romance and politics do not mix, especially in Nazi-occupied French Morocco. Greenstreet, Lorre, and Rains all create memorable characters, as does Wilson, the piano player to whom Bergman says the oft-misquoted, "Play it, Sam." Without a doubt, the best closing scene ever written; it was scripted on the fly during the end of shooting, and actually shot several ways. Written by Julius Epstein, Philip Epstein, and Koch from an unproduced play. See it in the original black and white. The laserdisc edition features restored imaging and sound and commentary on audio track 2 by film historian Ronald Haver about the film's production, the play it was originally based on, and the famed evolution of the screenplay. Oscar

nominations: Best Actor (Bogart), Best Supporting Actor (Rains), Black & White Cinematography, Best Score for a Dramatic Picture, Film Editing.
1942 102m/B Humphrey Bogart, Ingrid Bergman, Paul Henreid, Claude Rains, Peter Lorre, Sydney Greenstreet, Conrad Veidt, S.Z. Sakall, Dooley Wilson, Marcel Dalio, John Qualen; *Dir:* Michael Curtiz. Academy Awards '43: Best Director (Curtiz), Best Picture, Best Screenplay; National Board of Review Awards '45: 10 Best Films of the Year; American Film Institute's Survey '77: 3rd Best American Film Ever Made. **VHS, Beta, LV, 8mm** *MGM, OM* ♫♫♫♫

Casablanca Express Nazi commandos hijack Churchill's train in this action-adventure drama.
1989 90m/C Glenn Ford, Donald Pleasence, Jason Connery; *Dir:* Sergio Martino. **VHS, Beta $79.95** *TRY* ♫

Casanova '70 Mastroianni plays a handsome soldier who has a knack for enticing liberated women in this comic rendition of the much-cinematized legendary yarn. Trouble is, he's in the mood only when he believes that he's in imminent danger. In Italian with English subtitles.
1965 113m/C *IT* Marcello Mastroianni, Virna Lisi, Michele Mercier, Guido Alberti, Margaret Lee, Bernard Blier, Liana Orfei; *Dir:* Mario Monicelli. **VHS** *FOX* ♫♫½

Casanova's Big Night Classic slapstick comedy stars Hope masquerading as Casanova to determine the nature of Fontaine's intentions in marrying the real Casanova. The all-star cast provides one hilarious scene after another in this dated film. Price has a cameo as the "real" Casanova.
1954 85m/C Bob Hope, Vincent Price, Joan Fontaine, Basil Rathbone, John Carradine, Raymond Burr; *Dir:* Norman Z. McLeod. **VHS, LV $14.95** *PAR, CCB* ♫♫½

Case of the Baby Sitter A private eye investigates jewel thieves posing as nobility.
1947 40m/B Tom Neal, Allen Jenkins, Pamela Blake. **VHS, Beta, LV $16.95** *SNC, WGE* ♫

A Case of Deadly Force Based on the true story of a black man wrongfully killed by Boston's Tactical Police Force who mistook him for a robber. The police investigation calls it self-defense but the family and the attorney they hire fight to change the verdict. Excellent cast, great dramatic story. Originally made-for-television.
1986 95m/C Richard Crenna, John Shea, Lorraine Touissant, Frank McCarthy, Tom Isbell; *Dir:* Michael Miller. **VHS $79.95** *CAF* ♫♫♫

The Case of the Frightened Lady A homicidal family does its collective best to keep its dark past a secret in order to collect some inheritance money. Watch, if you're still awake, for the surprise ending.
1939 80m/B *GB* Marius Goring, Helen Haye, Penelope Dudley Ward, Felix Aylmer, Patrick Barr; *Dir:* George King. **VHS, Beta $16.95** *SNC* ♫

A Case of Libel A made for television dramatization based on attorney Louis Nizer's account of Quentin Reynolds' libel suit against columnist Westbrook Pegler.
1983 90m/C Daniel J. Travanti, Ed Asner; *Dir:* Eric Till. **VHS, Beta $59.95** *IVE* ♫♫½

The Case of the Lucky Legs Yes, there was a Perry Mason before Raymond Burr. Erle Stanley Gardner's sleuthing litigator's first screen appearance came in 1934 with "The Case of the Howling Dog," which initiated Warner Brothers' "A"-grade (but

soon-to-be "B"-grade) series. More akin to Nick Charles than to Gardner's character or to his later TV incarnation, the "Lucky Legs" Perry is a high-living, interminably hungover tippler who winces and wisecracks as he unravels the case of the corpse of a crooked con man. William, who left shortly after the series was downgraded to "B" status, plays the esquire; Tobin plays his smart-mouthed secretary, and Ellis is the tomato suspected of murder. Mayo, by the way, went on to direct "The Petrified Forest" the following year.
1935 76m/B Warren William, Genevieve Tobin, Patricia Ellis, Lyle Talbot, Allen Jenkins, Barton MacLane, Peggy Shannon, Porter Hall; *Dir:* Archie Mayo. **VHS, Beta $16.95** *SNC, RXM* ♫♫½

The Case of the Missing Lady A famous Arctic explorer asks Tommy and Tuppence to find his missing fiancee. Based on the Christie story.
1983 51m/C *GB* Francesca Annis, James Warwick. **VHS, Beta $14.95** *PAV* ♫½

The Case of the Mukkinese Battle Horn The cast of Britain's "The Goon Show" takes over in a zany, slapdash featurette-length mystery spoof.
1956 27m/C *GB* Peter Sellers, Spike Milligan, Dick Emery. **VHS, Beta, 8mm $19.95** *VYY* ♫♫

Casey Jones and Fury (Brave Stallion) An episode from each of these early adventure series.
1957 60m/B Alan Hale Jr., Peter Graves. **VHS, Beta $24.95** *NOS, DVT* ♫♫

Casey's Shadow The eight-year-old son of an impoverished horse trainer raises a quarter horse and enters it in the world's richest horse race.
1978 (PG) 116m/C Walter Matthau, Alexis Smith, Robert Webber, Murray Hamilton; *Dir:* Martin Ritt. **VHS, Beta $59.95** *COL* ♫♫

Cash A bankrupt financier uses counterfeit money to promote a new company. Also called "For Love or Money."
1934 63m/B *GB* Robert Donat, Wendy Barrie, Edmund Gwenn; *Dir:* Zoltan Korda. **VHS, Beta $29.95** *NOS, MRV, DVT* ♫♫

Casimir the Great A dramatized biography of the popular ruler who was instrumental in Poland's development during the period of transition between ancient and modern times. Presented in two segments. In Polish with English subtitles.
19?? ?m/C *PL* Krzysztof Chamiec, Wladyslaw Hanoza. **VHS $69.95** *FCT* ♫½

Casino A made-for-television thriller about back-stabbings and ritzy romance aboard a high-priced gambling liner.
1980 100m/C Mike Connors, Lynda Day George, Bo Hopkins, Gary Burghoff, Joseph Cotten, Robert Reed, Barry Sullivan; *Dir:* Don Chaffey. **VHS, Beta $79.95** *PSM* ♫½

Casino Royale The product of five directors, three writers and a mismatched cast of dozens, this virtually plotless spoof of James Bond films can stand as one of the low-water marks for 1960s' comedy. And yet, there are some marvelous bits within, scenes of bizarre hilarity. Welles and Sellers literally couldn't stand the sight of one another, and their scenes together were filmed separately, with stand-ins.
1967 130m/C David Niven, Woody Allen, Peter Sellers, Ursula Andress, Orson Welles, Jacqueline Bisset, Deborah Kerr, Peter O'Toole, Jean-Paul Belmondo, Charles Boyer, Joanna Pettet, John Huston, William Holden, George Raft; *Dir:* John

Huston, Ken Hughes, Robert Parrish, Val Guest, Joseph McGrath. **VHS, Beta, LV $19.95** *MRV* ✓

Casper the Friendly Ghost Compilation of classic Casper cartoons from the long-running children's show.
1953 60m/C VHS, Beta $9.95 *WOV* ✓

Cass A disenchanted filmmaker returns home to Australia to experiment with alternative lifestyles.
1978 90m/C Michelle Fawden, John Waters, Judy Morris, Peter Carroll. **VHS, Beta $19.98** *VID* ✓

Cassandra A fragile young woman has dreams that foretell the future—specifically, a series of grisly murders.
1987 (R) 94m/C Shane Briant, Brionny Behets, Tessa Humphries, Kit Taylor, Lee James; *Dir:* Colin Eggleston. **VHS, Beta, LV $24.95** *VIR* ✓

The Cassandra Crossing A terrorist with the plague causes havoc on a transcontinental luxury train. Turgid adventure filmed in France and Italy.
1976 (R) 129m/C *GB* Sophia Loren, Richard Harris, Ava Gardner, Burt Lancaster, Martin Sheen, Ingrid Thulin, Lee Strasberg, John Phillip Law, Lionel Stander, O.J. Simpson, Ann Turkel, Alida Valli; *Dir:* George P. Cosmatos. **VHS, Beta** *FOX* ✓ ½

Cassie Follows the rise of Cassie, a successful country and western singer, and documents all her trials along the way.
1987 (R) 75m/C Marilyn Chambers; *Dir:* Godfrey Daniels. **VHS, Beta $69.95** *API* ✓

Cast a Deadly Spell A bubbly, flavorful witch's brew of private-eye and horror cliches, set in a fantasy version of 1948 Los Angeles where sorcery and voodoo abound, but gumshoe Harry P. Lovecraft uses street smarts instead of magic to track down a stolen Necronomicon—and if you know what that is you'll want to watch. Wild creatures and f/x wizardry complement this made-for-cable-TV trick and treat.
1991 (R) 93m/C Fred Ward, David Warner, Julianne Moore, Clancy Brown, Alexandra Powers; *Dir:* Martin Campbell. **VHS $89.99** *HBO* ✓✓✓

Cast the First Stone Made for TV soaper based on the true story of a former nun who becomes a small town schoolteacher. After being raped, Diane Martin discovers she's pregnant and when she decides to keep the baby the school board dismisses her. The plot centers around her fight to regain her job and dignity—while trying to convince everybody that she doesn't deserve to lose her job simply because she is an unwed mother. Eikenberry gives a strong performance, but it can't carry this lackluster film.
1989 94m/C Jill Eikenberry, Richard Masur, Joe Spano, Lew Ayres, Holly Palance; *Dir:* John Korty. **VHS, LV $59.95** *COL* ✓

Cast a Giant Shadow Follows the career of Col. David Marcus, hero of the Arab-Israeli War.
1966 138m/C Kirk Douglas, Senta Berger, Angie Dickinson, John Wayne, James Donald, Chaim Topol, Frank Sinatra, Yul Brynner; *Dir:* Melville Shavelson. **VHS, Beta $59.98** *FOX* ✓✓

Castaway Based on the factual account by Lucy Irvine. The story of Michael Wilmington, who placed an ad for a young woman to spend a year on a Pacific atoll with him, and the battle of the sexes that followed.
1987 (R) 118m/C Oliver Reed, Amanda Donohoe, Tony Rickards, Georgina Hale; *Dir:* Nicolas Roeg. **VHS, Beta $19.98** *WAR* ✓✓½

Castaway Cowboy A shanghaied cowboy becomes partners with a widow when she turns her Hawaiian potato farm into a cattle ranch. Good, clean "family fare."
1974 (G) 91m/C James Garner, Robert Culp, Vera Miles; *Dir:* Vincent McEveety. **VHS, Beta** *DIS, OM* ✓✓

Casting Call Eighteen would-be Hollywood starlets undress and cavort for the camera, in hopes of getting real film work.
1988 60m/C VHS, Beta, CV $19.95 *AHV* ✓✓

Castle A man is summoned by the seemingly invisible occupants of a castle to their village. All his efforts to meet with those inhabiting the castle are futile and the task gradually becomes his obsession. Both wonderfully acted and shot, this is a well-executed adaptation of Franz Kafka's novel.
1968 90m/C *GE SI* Maximilian Schell, Cordula Trantow, Trudik Daniel, Helmut Qualtinger; *Dir:* Rudolf Noelte. **VHS, Beta $79.98** *CGH* ✓✓✓

Castle of Blood Staying overnight in a haunted castle, a poet is forced to deal with a number of creepy encounters. Cult favorite Steele enhances this atmospheric chiller. Dubbed in English. Also on video as "Castle of Terror."
1964 85m/B *IT FR* Barbara Steele, George Riviere, Margrete Robsahm, Henry Kruger, Montgomery Glenn, Sylvia Sorente; *Dir:* Anthony Dawson. **VHS $16.95** *SNC* ✓✓

Castle of the Creeping Flesh A surgeon's daughter is brutally murdered. Vowing to bring her back he begins ripping out the organs of innocent people and transplanting them into her body.
19?? 90m/C Adrian Hoven; *Dir:* Percy G. Parker. **VHS, Beta $59.95** *MAG* ✓

Castle in the Desert Charlie Chan investigates murder and other strange goings-on in a spooky old castle. One of the last in the series.
1942 62m/B Sidney Toler, Arleen Whelan, Richard Derr, Douglas Dumbrille, Henry Daniell, Victor Sen Young; *Dir:* Harry Lachman. **VHS $19.98** *FOX* ✓✓½

Castle of Evil A group of heirs gathers on a deserted isle to hear the reading of a will. One by one, they fall victim to a humanoid being.
1966 81m/C Scott Brady, Virginia Mayo, Lisa Gaye, David Brian, Hugh Marlowe, William Thourlby, Shelly Morrison; *Dir:* Francis D. Lyon. **VHS $14.98** *REP, FCT* Woof!

The Castle of Fu Manchu The final chapter in a series starring Lee as the wicked doctor. This time, Lee has developed a gadget which will put the earth into a deep freeze, and at his mercy. To fine tune this contraption, he enlists the help of a gifted scientist by abducting him. However, the helper/hostage has a bad ticker, so Lee must abduct a heart surgeon to save his life, and thus, the freezer project. Most critics felt this was the weakest installment in the series. Also known as "Assignment Istanbul" or "Die Folterkammer des Dr. Fu Manchu."
1968 (PG) 92m/C *GE SP IT GB* Christopher Lee, Richard Greene, H. Marion Crawford, Tsai Chin, Gunther Stoll, Rosalba (Sara Bay) Neri, Maria Perschy; *Dir:* Jess (Jesus) Franco. **VHS $19.95** *MRV* ✓

Castle of the Living Dead Evil Count Drago's hobbies include mummifying a traveling circus group visiting his castle. Lee is as evil as ever, but be sure to look for Sutherland's screen debut. In a dual role, he

plays not only the bumbling inspector, but also a witch, in drag.
1964 90m/B *IT FR* Christopher Lee, Gala Germani, Phillippe LeRoy, Jacques Stanislawsky, Donald Sutherland; *Dir:* Herbert Wise. **VHS $16.95** *SNC, MRV* ✓✓

Casual Sex? Two young women, looking for love and commitment, take a vacation at a posh resort where they are confronted by men with nothing on their minds but sex, be it casual or the more formal black tie variety. Supposedly an examination of safe sex in a lightly comedic vein, though the comic is too light and the morality too limp. Adapted from the play by Wendy Goldman and Judy Toll.
1988 (R) 87m/C Lea Thompson, Victoria Jackson, Stephen Shellan, Jerry Levine, Mary Gross, Andrew Dice Clay; *Dir:* Genevieve Robert. **VHS, Beta, LV $19.95** *MCA* ✓ ½

Casualties of War A Vietnam war morality play about army private Fox in the bush who refuses to let his fellow soldiers and commanding sergeant (Penn) skirt responsibility for the rape and murder of a native woman. Fox achieves his dramatic breakthrough. Based on the true story by Daniel Lang.
1989 (R) 105m/C Sean Penn, Michael J. Fox, Don Harvey, Thuy Thu Le; *Dir:* Brian DePalma. **VHS, Beta, LV, 8mm $14.95** *COL* ✓✓✓

A Casualty of War Made-for-TV adaptation of Frederic Forsyth's thriller about modern-day arms smuggling (Libya to Ireland) and espionage. Well-acted, with the exception of Hack.
1990 (R) 96m/C Shelley Hack, David Threlfall, Richard Hope, Alan Howard, Clarke Peters; *Dir:* Tom Clegg. **VHS** *NVH* ✓✓½

The Cat A boy and a tamed mountain lion become friends on the run from a murderous poacher.
1966 95m/C Peggy Ann Garner, Roger Perry, Barry Coe; *Dir:* Ellis Kadisan. **VHS, Beta $59.98** *CHA* ✓

Cat Ballou At the turn of the century, a schoolmarm turns outlaw with the help of a drunken gunman. Marvin played Kid Shelleen and his silver-nosed evil twin Tim Strawn in this cheery spoof of westerns. Oscar nominated for score and film editing. Cole and Kaye sing the narration in a one of a kind Greek chorus.
1965 96m/C Jane Fonda, Lee Marvin, Michael Callan, Dwayne Hickman, Reginald Denny, Nat King Cole, Stubby Kaye; *Dir:* Elliot Silverstein. Academy Awards '65: Best Actor (Marvin); Berlin Film Festival '65: Best Actor (Marvin); National Board of Review Awards '65: 10 Best Films of the Year, Best Actor (Marvin). **VHS, Beta $14.95** *COL, BTV* ✓✓✓½

Cat in the Cage A young man finds many things have changed at home while he was in a mental institution: his dad has remarried, the housekeeper is practicing witchcraft, the chauffeur is after his mistress, and the cat is gone.
1968 96m/C Colleen Camp, Sybil Danning, Mel Novak, Frank De Kova; *Dir:* Tony Zarin Dast. **VHS, Beta $24.95** *GHV* ✓ ½

The Cat Came Back Incredibly funny animated short from Canada's Cordell Barker featuring a tiny, but destructive kitty that simply will not leave. If you have a cat, see this. Includes a guide to the film.
1989 8m/C VHS, Special order formats $225.00 *PYR* ✓ ½

The Cat and the Canary A great silent film about the ghost of a madman that wanders nightly through the corridors of an old house. Remade twice, once in 1939 and again in 1979.
1927 70m/B Laura LaPlante, Creighton Hale, Tully Marshall, Gertrude Astor; **Dir:** Paul Leni. **VHS, Beta $19.95** *NOS, MRV, VYY* 🐾🐾🐾

The Cat and the Canary A stormy night, a gloomy mansion, and a mysterious will combine to create an atmosphere for murder. An entertaining remake of the 1927 silent film.
1979 (PG) 96m/C *GB* Carol Lynley, Michael Callan, Wendy Hiller, Olivia Hussey, Daniel Masey, Honor Blackman, Edward Fox, Wilfrid Hyde-White; **Dir:** Radley Metzger. **VHS, Beta** *COL, OM* 🐾🐾

Cat Chaser Weller walks listlessly through the role of an ex-soldier in Miami who has an affair with the wife of an exiled—but still lethal—military dictator. Surpisingly low-key, sometimes effective thriller that saves its energy for sex scenes, also available in a less steamy, 90-minute ''R'' rated version. Based on an Elmore Leonard novel.
1990 97m/C Kelly McGillis, Peter Weir, Charles Durning, Frederic Forrest, Tomas Milian, Juan Fernandez; **Dir:** Abel Ferrara. **VHS $89.98** *VES* 🐾🐾

Cat City A feline James Bond type adventure following the exploits of two hip cats.
1990 90m/C VHS, CV $39.95 *JTC* 🐾

Cat Girl A young bride on her honeymoon finds she has inherited the family curse—she has a psychic link to a ferocious leopard. Numerous murders ensue. Poor production value and a weak script, despite Shelley's fine acting, make this an unworthwhile film.
1957 69m/B Barbara Shelley, Robert Ayres, Kay Callard, Paddy Webster; **Dir:** Alfred Shaughnessy. **VHS $9.95** *PAR* 🐾🐾½

Cat on a Hot Tin Roof Tennessee Williams' powerful play about greed and deception in a patriarchal Southern family. Big Daddy (Ives) is dying. Members of the family greedily attempt to capture his inheritance, tearing the family apart. Taylor is a sensual wonder as Maggie the Cat, though the more controversial elements of the play were toned down for the film version. Intense, believable performances from Ives and Newman. Oscar nominations for Best Picture, Best Actor (Newman), Director, Adapted Screenplay, Color Cinematography.
1958 108m/C Paul Newman, Burl Ives, Elizabeth Taylor, Jack Carson, Judith Anderson; **Dir:** Richard Brooks. National Board of Review Awards '58: 10 Best Films of the Year. **VHS, Beta, LV $19.95** *MGM, PIA, FCT* 🐾🐾½

Cat on a Hot Tin Roof In this adaptation of the Tennessee Williams play, fights over the family inheritance tear a family apart.
1984 122m/C Jessica Lange, Tommy Lee Jones, Rip Torn, David Dukes; **Dir:** Jack Hofsiss. **VHS, Beta, LV $19.98** *MGM, LIV, VES* 🐾🐾

Cat and Mouse A very unorthodox police inspector is assigned to investigate a millionaire's mysterious death. Who done it? French dialogue with English subtitles.
1978 (PG) 107m/C *FR* Michele Morgan, Serge Reggiani, Jean-Pierre Aumont; **Dir:** Claude Lelouch. **VHS, Beta $59.95** *COL* 🐾🐾🐾½

The Cat o' Nine Tails A blind detective and a newsman team up to find a sadistic killer. A gory murder mystery with music by Ennio Morricone.
1971 (PG) 112m/C *GE FR IT* Karl Malden, James Franciscus, Catherine Spaak, Cinzia de Carolis, Carlo Alighiero; **Dir:** Dario Argento. **VHS, 8mm** *SIM* 🐾

Cat from Outer Space An extraterrestrial cat named Jake crashes his spaceship on Earth and leads a group of people on endless escapades. Enjoyable Disney fare.
1978 (G) 103m/C Ken Berry, Sandy Duncan, Harry Morgan, Roddy McDowall, McLean Stevenson; **Dir:** Norman Tokar. **VHS, Beta** *DIS, OM* 🐾🐾

Cat People A young dress designer is the victim of a curse that changes her into a deadly panther who must kill to survive. A classic among the horror genre with unrelenting terror from beginning to end. First horror film from producer Val Lewton.
1942 73m/B Jane Randolph, Elizabeth Russell, Jack Holt, Alan Napier, Simone Simon, Kent Smith, Tom Conway; **Dir:** Jacques Tourneur. **VHS, Beta, LV $19.95** *MED, KOV, RKO* 🐾🐾🐾

Cat People A beautiful young woman learns that she has inherited a strange family trait—she turns into a vicious panther when sexually aroused. The only person with whom she can safely mate is her brother, a victim of the same genetic heritage. Kinski is mesmerizing as the innocent, sensual woman. Remake of the 1942 film.
1982 (R) 118m/C Nastassia Kinski, Malcolm McDowell, John Heard, Annette O'Toole, Ruby Dee, Ed Begley Jr., John Larroquette; **Dir:** Paul Schrader. **VHS, Beta, LV $14.98** *MCA* 🐾🐾½

Cat Women of the Moon Scientists land on the moon and encounter an Amazonlike force of female chauvinists. Also released under the title ''Rocket to the Moon.'' Remade as ''Missile to the Moon.'' Featuring the Hollywood Cover Girls as various cat women. Available in its original 3-D format.
1953 65m/B Sonny Tufts, Victor Jory, Marie Windsor, Bill Phipps, Douglas Fowley, Carol Brewster, Suzanne Alexander, Susan Morrow; **Dir:** Arthur Hilton. **VHS, Beta $12.95** *RHI, SNC, MWP* 🐾

Cataclysm A swell flick about a sadistic demon who spends his time either finding people willing to join him or killing the people who won't. Also on video as ''Satan's Supper.''
1981 (R) 94m/C Cameron Mitchell, Marc Lawrence, Faith Clift, Charles Moll; **Dir:** Tom McGowan, Greg Tallas, Philip Marshak. **VHS, Beta $49.95** *FOX, ACA* 🐾🐾½

Catacombs An investigative monk and a beautiful photographer stumble across a mysterious, centuries-old evil power.
1989 (R) 112m/C Timothy Van Patten, Laura Schaefer, Ian Abercrombie, Jeremy West; **Dir:** David Schmoeller. **VHS, Beta $89.95** *TWE* 🐾

The Catamount Killing The story of a small town bank manager and his lover. They decide to rob the bank and run for greener pastures only to find their escape befuddled at every turn.
1974 (PG) 82m/C *GE* Horst Buchholz, Ann Wedgeworth; **Dir:** Krzysztof Zanussi. **VHS, Beta $29.98** *VID* 🐾½

Catastrophe A documentary which features news footage of natural and man-made disasters, including the destructive fury of Hurricane Camille, car crashes at the Indy 500, the Hindenburg disaster, and the sinking of the Andrea Doria.
1977 90m/C *Nar:* William Conrad. **VHS, Beta $24.98** *SUE, IRN* 🐾½

Catch-22 Buck Henry's adaptation of Joseph Heller's black comedy about a group of fliers in the Mediterranean during WWII. A biting anti-war satire portraying the insanity of the situation in both a humorous and disturbing manner. Perhaps too literal to the book's masterfully chaotic structure, causing occasional problems in the ''are you following along department?'' Arkin heads a fine and colorful cast.
1970 121m/C Alan Arkin, Martin Balsam, Art Garfunkel, Jon Voight, Richard Benjamin, Buck Henry, Bob Newhart, Paula Prentiss, Martin Sheen, Charles Grodin, Anthony Perkins, Orson Welles, Jack Gilford; **Dir:** Mike Nichols. **VHS, Beta, LV $14.95** *PAR* 🐾🐾🐾

Catch as Catch Can An Italian male model is comically besieged by animals of every type, making a mess of his life and career. Dubbed.
1968 95m/C *IT* Vittorio Gassman, Martha Hyer, Gila Golan; **Dir:** Franco Indovina. **VHS, Beta $59.98** *CHA* 🐾

Catch the Heat Alexander is an undercover narcotics agent sent to infiltrate Steiger's South American drug operation.
1987 (R) 90m/C Tiana Alexandra, David Dukes, Rod Steiger; **Dir:** Joel Silberg. **VHS, Beta $79.95** *MED* 🐾

Catch Me a Spy A foreign agent attempts to lure an innocent man into becoming part of a swap for an imprisoned Russian spy. Also known as ''To Catch a Spy.''
1971 94m/C *GB FR* Kirk Douglas, Tom Courtenay, Trevor Howard, Marlene Jobert, Bernard Blier, Patrick Mower, Bernadette LaFont; **Dir:** Dick Clement. **VHS, Beta $59.95** *PSM* 🐾½

Catch Me...If You Can High school class president Melissa doesn't want to see the school torn down. To raise fast cash, she teams up with a drag racer and the fun begins.
1989 (PG) 105m/C Matt Lattanzi, Loryn Locklin, M. Emmet Walsh, Geoffrey Lewis; **Dir:** Stephen Sommers. **VHS, Beta, LV $89.95** *MCG* 🐾🐾

Catch a Rising Star's 10th Anniversary Some of the biggest names in comedy and music got together to celebrate the tenth anniversary of this New York night club.
1983 66m/C Pat Benatar, Billy Crystal, Gabe Kaplan, Joe Piscopo, Robin Williams. **VHS, Beta, LV $39.95** *COL, PIA* 🐾🐾

The Catered Affair Davis, anti-typecast as a Bronx housewife, and Borgnine, as her taxi-driving husband, play the determined parents of soon-to-be-wed Reynolds set on giving her away in a style to which she is not accustomed. Based on Paddy Chayefsky's teleplay, the catered affair turns into a familial trial, sharing the true-to-life poignancy that marked ''Marty,'' the Oscar-winning Chayefsky drama of the previous year. Scripted by Gore Vidal, the dialogue rings true, too.
1956 92m/B Bette Davis, Ernest Borgnine, Debbie Reynolds, Barry Fitzgerald, Rod Taylor, Robert Simon; **Dir:** Richard Brooks. **VHS $19.98** *MGM, FCT, CCB* 🐾🐾🐾

Catherine & Co. A lonely, penniless girl arrives in Paris and ''opens shop'' on the streets of Paris. As business booms, she takes a cue from the big corporations and sells stock in herself.
1976 (R) 91m/C *FR IT* Jane Birkin, Patrick Dewaere, Jean-Pierre Aumont, Jean-Claude Brialy; **Dir:** Michel Boisrond. **VHS, Beta $59.95** *VID* 🐾

Catholic Hour A dramatized version of the life story of Thomas Frederick Price who co-founded the Maryknoll Missionaries in the early 1900's.
1960 30m/B Edward Cullen, Arthur Gary. **VHS, Beta** $19.95 VYY ✰✰

Catholics A sensitive exploration of contemporary mores and changing attitudes within the Roman Catholic church. Sheen is sent by the Pope to Ireland to reform some priests. Based on Brian Moore's short novel and made for television. Also known as "The Conflict."
1973 86m/C Martin Sheen, Trevor Howard; **Dir:** Jack Gold. **VHS, Beta** $39.95 CFV ✰✰✰

Cathy's Curse The spirit of her aunt, who died as a child, possesses a young girl in this Canadian-French collaboration. Tries to capitalize on the popularity of "The Exorcist," but falls seriously short.
1989 90m/C CA FR Alan Scarfe, Beverly Murray; **Dir:** Eddy Matalon. **VHS, Beta** CGI ✰

Cat's Eye An anthology of three Stephen King short stories connected by a stray cat who wanders through each tale.
1985 (PG-13) 94m/C Drew Barrymore, James Woods, Alan King, Robert Hays, Candy Clark, Kenneth McMillan, James Naughton; **Dir:** Lewis Teague. **VHS, Beta, LV** $19.99 FOX, FCT ✰✰

The Cats and Mice of Paul Terry Seven vintage silent Terrytoon cartoons: "The Wild West," "Short Vacation," "China Doll," "Sunny Italy," "Stars of the Circus," "Sharpshooter" and "Ship Ahoy."
1928 55m/B **VHS, Beta, 8mm** $24.95 VYY

Cat's Play A widowed music teacher makes a ceremonial occassion of a weekly dinner with an old flame. Then an old friend from her youth suddenly reappears and begins an affair with the gentleman causing self-destructive passions to explode. In Hungarian with English subtitles.
1974 115m/C HU Margit Dayka, Margit Makay, Elma Bulla; **Dir:** Karoly Makk. **VHS** $59.95 CVC ✰✰

Cattle Queen of Montana Reagan stars as an undercover federal agent investigating livestock rustlings and Indian uprisings.
1954 88m/C Ronald Reagan, Barbara Stanwyck, Jack Elam, Gene Evans, Lance Fuller, Anthony Caruso; **Dir:** Allan Dwan. **VHS, Beta, LV** $34.95 WGE, BVV ✰✰½

Caught A woman marries for wealth and security and is desperately unhappy. She runs away and takes a job with a struggling physician, and falls in love with him. Her husband finds her, forcing her to decide between a life of security or love.
1949 90m/B James Mason, Barbara Bel Geddes, Robert Ryan, Curt Bois, Natalie Schafer, Art Smith; **Dir:** Max Ophuls. **VHS** $14.98 REP ✰✰✰

Cauldron of Blood A blind sculptor uses human skeletons as the framework of his art pieces. Also released as "Blind Man's Bluff."
1967 (PG) 95m/C Boris Karloff, Viveca Lindfors, Jean-Pierre Aumont, Rosenda Monteros, Rueben Rojo, Dianik Zurakowska; **Dir:** Edward Mann. **VHS** $14.98 REP, VDC, FCT ✰✰½

Cause for Alarm A jealous husband recovering from a heart attack begins to lose his mind. He wrongly accuses his wife of having an affair and attempts to frame her for his own murder. A fast-paced thriller with a nifty surprise ending.

1951 74m/B Loretta Young, Barry Sullivan, Bruce Cowling, Margalo Gillmore, Irving Bacon, Carl "Alfalfa" Switzer; **Dir:** Tay Garnett. **VHS** $16.95 NOS, SNC, MGM ✰✰½

Cavalcade of the West A typical western: two brothers, separated by a kidnapping, grow up on opposite sides of the law. Later they meet, and the outlaw is reformed and reunited with his happy family.
1936 70m/B Hoot Gibson, Rex Lease, Marion Shilling, Earl Dwire; **Dir:** Harry Fraser. **VHS, Beta** $19.95 GPV, DVT, VCN ✰

The Cavalier A Chinese cavalier faces love, danger and intrigue in this martial arts film.
198? 90m/C Tang Wei, Lung Fei, Yeh Yuen, Tseng Tsao. **VHS, Beta** $29.95 UNI ✰

Cavalier of the West An Army captain is the only negotiating force between the white man and a primitive Indian tribe.
1931 66m/B Harry Carey, Kane Richmond, George "Gabby" Hayes; **Dir:** John P. McCarthy. **VHS, Beta, LV** $19.95 NOS, WGE ✰

Cavalry A Union Army Lieutenant is reunited with his family after the Civil War, bringing them happiness and joy.
1936 60m/B Bob Steele, Frances Grant, Karl Hackett, Hal Price; **Dir:** Robert N. Bradbury. **VHS, Beta** $19.90 NOS, DVT ✰½

Cavalry Command Good will and integrity characterize the U.S. soldiers called into a small village to quiet a guerrilla rebellion.
1963 77m/C John Agar, Richard Arlen, Myron Healy, Alicia Vergel, William Phipps, Eddie Infante; **W/Dir:** Eddie Romero. **VHS, Beta** $59.95 PGN, RHI ✰½

Cave Girl After falling through a time-warp during a high-school field trip, a social pariah makes a hit with a pre-historic honey. Sexist teen exploitation film, with few original ideas and a not-so-hot cast.
1985 (R) 85m/C Daniel Roebuck, Cindy Ann Thompson, Saba Moor, Jeff Chayette; **W/Dir:** David Oliver. **VHS, Beta, LV** $79.95 COL Woof!

Cave Girls A critically acclaimed experimental short by a group of female filmmakers depicting how a society of stone-age women prior to the evolution of thought processes would interact with each other and relate to their surroundings.
198? 28m/C **Dir:** Cara Brownell. **VHS** $29.95 FCT, MWF ✰✰✰

Cave of the Living Dead Villagers summon Inspector Doren of Interpol to solve a rash of nasty murders that they've blamed on green-eyed, cave dwelling vampires. Filmed in Sepiatone. Also known as "Der Fluch Der Gruenen Augen."
1965 87m/B GE YU Adrian Hoven, Erika Remberg, Carl Mohner, Wolfgang Preiss, Karin Field, Akos V. Ratony, John Kitzmiller, Carl Mohner; **Dir:** Akos Von Rathony. **VHS, Beta** $39.95 MSP Woof!

Caveman Starr stars in this prehistoric spoof about a group of cavemen banished from different tribes who band together to form a tribe called "The Misfits."
1981 (PG) 92m/C Ringo Starr, Barbara Bach, John Matuszak, Dennis Quaid, Jack Gilford, Shelley Long, Cork Hubbert; **W/Dir:** Carl Gottlieb. **VHS, Beta** $59.98 FOX ✰✰

Cease Fire The story of a troubled Vietnam vet who finds solace in a veterans' therapy group. Adapted from the play by George Fernandez.

1985 (R) 97m/C Don Johnson, Robert F. Lyons, Lisa Blount; **Dir:** David Nutter. **VHS, Beta** $79.99 HBO ✰✰

Celebrity In this television film, a tragic childhood secret must be confronted by three friends when one is charged with murder. Based on the Thomas Thompson novel.
1985 313m/C Michael Beck, Ben Masters, Joseph Bottoms; **Dir:** Paul Wendkos. **VHS, Beta** $69.95 RCA ✰½

Celeste Percy Adlon (who later made the offbeat comedy "Sugarbaby") directed this longish but finely detailed and beautifully photographed biographical look at French writer Marcel (Remembrance of Things Past) Proust. Based on the memoirs of Proust's housekeeper, Celeste Albaret, an uneducated farmgirl, portraying the woman's devoted relationship ('til death did them part) with the middle-aged, homosexual author. It does so with wit, poignancy and insight, but is not entirely successful in its attempt to render Proust's verbal literary style into a visual medium. In German with enhanced English subtitles.
1981 107m/C GE Eva Mattes, Jurgen Arndt, Norbert Wartha, Wolf Euba; **Dir:** Percy Adlon. **VHS** $79.95 NYF, FCT, APD ✰✰✰

Celia: Child Of Terror Set in the 1950s, this film deals with the awful results of a young girl's inability to handle disappointment.
1989 110m/C Rebecca Smart, Nicholas Eadie, Victoria Longley, Mary-Anne Fahey; **Dir:** Ann Turner. **VHS, Beta** $79.95 TRY ✰✰

The Cellar A young boy finds an ancient Comanche monster spirit in the basement of his home. His parents, as usual, don't believe him, so he must battle the monster alone.
1990 (PG-13) 90m/C Patrick Kilpatrick, Suzanne Savoy, Chris Miller, Ford Rainey; **Dir:** Kevin S. Tenney. **VHS** $14.95 HMD, SOU ✰½

Cellar Dweller A cartoonist moves into an old house and soon discovers it's haunted by a demonic cartoonist who killed himself thirty years earlier. What a coincidence. Low-budget scare-'em-upper filmed on one set and lit by a floodlamp.
1987 78m/C Pamela Bellwood, Deborah Muldowney, Brian Robbins, Vince Edwards, Jeffrey Combs, Yvonne de Carlo; **Dir:** John Carl Buechler. **VHS, Beta, LV** $19.95 STE, NWV Woof!

Cemetery High Beautiful high school girls decide to lure the local boys into a trap and kill them.
1989 80m/C Debi Thibeault, Karen Nielsen, Lisa Schmidt, Ruth Collins, Simone, Tony Kruk, David Coughlin, Frank Stewart; **Dir:** Gorman Bechard. **VHS, Beta** $79.95 UNI, HHE ✰

Center of the Web John Phillips is a victim of mistaken identity—someone thinks he's a professional hit man. After surviving an apparent mob attempt on his life, Phillips is persuaded by a CIA operative to go along with the deception in order to capture a potential political assassin. At least that's what Phillips is told, but he soon realizes that the deeper he gets into his new role, the deadlier the plot becomes. Davi is one of the best bad guys around and the fast-paced stunts and plot twists make this watchable.
1992 (R) 88m/C Ted Prior, Robert Davi, Tony Curtis, Charlene Tilton, Bo Hopkins, Charles Napier; **Dir:** David A. Prior. **VHS** $89.95 AIP ✰✰

Centerfold The side of the centerfold model we never see—in the dressing room preparing for a shooting session. Model Martha Thomsen talks candidly about her conflicts with men, how she felt posing nude for the first time, and her years growing up "plain."
1980 60m/C VHS, Beta $29.95 *VID* ✂✂

Centerfold Confidential Claims to be an uninhibited look at centerfold modeling, through the eye of the photographer's camera. However, that's just the excuse to show naked women running around for an hour.
1987 73m/C VHS, Beta $39.95 *AHV* ✂✂

The Centerfold Girls Thin storyline involves a deranged man who is determined to kill all the voluptuous young women who have posed nude for a centerfold.
1974 (R) 93m/C Andrew Prine, Tiffany Bolling, Aldo Ray, Jeremy Slate, Ray Danton, Francine York; *Dir:* John Peyser. **VHS, Beta $54.95** *MED* ✂

The Centurions A full-length animated film featuring universe-saving robot characters.
1986 90m/C VHS, Beta $29.95 *CVL* ✂

Certain Fury Two timid women go on the lam when they are mistaken for escaped prostitutes who shot up a courthouse. The sooner they get caught, the better.
1985 (R) 88m/C Tatum O'Neal, Irene Cara, Peter Fonda, Nicholas Campbell, Moses Gunn; *Dir:* Stephen Gyllenhaal. **VHS, Beta, LV $14.95** *NWV, STE Woof!*

Certain Sacrifice Nineteen-year-old Madonna expresses herself in her film debut. Seeking revenge on the man who raped her, she murders him in a strange ritualistic manner underneath the Brooklyn Bridge. Poor is the man whose pleasures depend on the permission of another?
1980 60m/C Madonna, Jeremy Pattnosh, Charles Kurtz; *Dir:* Stephen Lewicki. **VHS, Beta $59.95** *VMV, WOV* ✂

Cesar This is the third and most bittersweet part of Pagnol's famed trilogy based on his play depicting the lives and loves of the people of Provence, France. Marius returns after a 20-year absence to his beloved Fanny and his now-grown son, Cesariot. The first two parts of the trilogy are "Marius" and "Fanny" and were directed by Alexander Korda and Marc Allegret respectively. In French with English subtitles.
1936 117m/B *FR* Raimu, Pierre Fresnay, Orane Demazis, Charpin, Andre Fouche; *Dir:* Marcel Pagnol. **VHS, Beta $24.95** *NOS, MRV, INT* ✂✂✂✂

Cesar & Rosalie Acclaimed French comedy depicts the love triangle between a beautiful divorcee, her aging live-in companion and a younger man. Engaging portrait of how their relationship evolves over time. In French with English subtitles.
1972 (PG) 110m/C *FR* Romy Schneider, Yves Montand, Sami Frey, Umberto Orsini; *Dir:* Claude Sautet. **VHS, Beta, LV $29.95** *AXV, TAM* ✂✂✂

Chain Gang Killings A pair of shackled prisoners, one black, the other white, escape from a truck transporting them to prison.
1985 99m/C Ian Yule, Ken Gampu. **VHS, Beta $59.95** *VCL* ✂

Chain Gang Women Sordid violence. Two escaped convicts plunder, rob, and rape until a victim's husband comes looking for revenge.

1972 (R) 85m/C Robert Lott, Barbara Mills, Michael Stearns, Linda York; *Dir:* Lee Frost.
VHS $39.95 *ACN Woof!*

Chain Lightning Bogart stars as a bomber pilot who falls in love with a Red Cross worker (Parker) while fighting in Europe in 1943. They lose touch after the war until Bogie goes to work as a test pilot for the same shady airplane manufacturer (Massey) where Parker works. He is given the chance to test a new plane, which has already cost the life of one of his friends. Bogart has more success, along with rekindling the flames of romance. Average script but the flying sequences are well-done.
1950 94m/B Humphrey Bogart, Eleanor Parker, Raymond Massey, Richard Whorf; *Dir:* Stuart Heisler. **VHS $19.98** *MGM* ✂✂

Chain Reaction When a nuclear scientist is exposed to radiation after an accident at an atomic power plant, he must escape to warn the public of the danger. Also known as "Nuclear Run."
1980 87m/C Steve Bisley, Ross Thompson; *Dir:* Ian Barry. **VHS, Beta, LV $59.98** *EMB* ✂½

Chained Heat A seamy tale of the vicious reality of life for women behind bars. Little more than another exploitation film, but it has its fans. Sequel to the 1982 film "Concrete Jungle."
1983 (R) 97m/C Linda Blair, Stella Stevens, Sybil Danning, Tamara Dobson, Henry Silva, John Vernon, Nita Talbot, Louisa Moritz; *Dir:* Paul Nicholas. **VHS, Beta, LV $29.98** *VES Woof!*

Chained for Life Daisy and Violet Hilton, the real life Siamese twins, star in this old-fashioned "freak" show. When a gigolo deserts one twin on their wedding night, the other twin shoots him dead. The twins go on trial and the judge asks the viewer to hand down the verdict.
1951 81m/B Daisy Hilton, Violet Hilton; *Dir:* Harry Fraser. **VHS, Beta $16.95** *SNC, NOS, FST Woof!*

Chains A Chicago gangland feud entraps two couples in it web, and the innocents must take care of themselves when the situation goes from bad to worse.
1989 (R) ?m/C Jim Jordan, Michael Dixon, John L. Eves; *Dir:* Roger J. Barski. **VHS $89.95** *IMP* ✂✂

The Chair The penitentiary where an evil superintendent was fried in his own electric chair during an inmate uprising is reopened after two decades. The new warden (Benedict), who used to be a subordinate to the dead man, believes in rigid control of the inmates and locks horns with the big-house shrink (Coco), who on the surface is an intelligent humanitarian, but may have a darker side. This acrimonious dispute takes a back seat, when it appears that the spirit of the late warden has come back to make the inmates pay for his untimely death.
1987 (R) ?m/C James Coco, Paul Benedict, Stephen Geoffreys, Trini Alvarado, John Bentley; *Dir:* Waldemar Korzeniowsky. **VHS $89.95** *IMP* ✂✂

The Chalk Garden A woman with a mysterious past takes on the job of governess for an unruly 14-year-old girl, with unforseen consequences. An excellent adaptation of the Enid Bagnold play although not as suspenseful as the stage production. Dame Evans received a Best Supporting Actress Oscar nomination.

1964 106m/C *GB* Deborah Kerr, Hayley Mills, Edith Evans, John Mills, Elizabeth Sellars, Felix Aylmer; *Dir:* Ronald Neame. National Board of Review Awards '64: 10 Best Films of the Year, Best Supporting Actress (Evans). **VHS, Beta $29.95** *MCA* ✂✂✂

The Challenge Story of the courageous party of explorers who conquered the Matterhorn. Incredible avalanche scenes.
1938 77m/B Luis Trenker, Robert Douglas. **VHS, Beta $19.98** *DVT, VYY, TIM* ✂✂✂

Challenge When his entire family is killed, dad decides to seek revenge with his shotgun. Very bloody and violent and very nearly plotless.
1974 (PG) 90m/C Earl Owensby, William T. Hicks, Katheryn Thompson, Johnny Popwell; *Dir:* Martin Beck. **VHS $59.98** *FOX Woof!*

The Challenge A contemporary action spectacle which combines modern swordplay with the mysticism and fantasy of ancient Samurai legends. John Sayles co-wrote the script with Richard Maxwell.
1982 (R) 108m/C Scott Glenn, Toshiro Mifune; *Dir:* John Frankenheimer. **VHS, Beta $59.95** *FOX* ✂✂½

Challenge to Be Free Action adventure geared toward a young audience depicting the struggles of a man being pursued by 12 men and 100 dogs across a thousand miles of frozen wilderness. The last film directed by Garnett, who has a cameo as Marshal McGee. Released in 1972 as "Mad Trapper of the Yukon."
1974 (G) 90m/C Mike Mazurki, Jimmy Kane; *Dir:* Tay Garnett. **VHS, Beta $9.98** *MED* ✂

Challenge of the Dragon Simple martial-arts tale of a Kung-Fu hero defending a town against bad guys.
1982 (R) 87m/C Tarng Long; *Dir:* Liou-Xiao Ling. **VHS, Beta** *PLV* ✂

Challenge to Lassie When Lassie's Scottish master dies, the faithful pup remains at his grave. An unsympathetic policeman orders Lassie to leave the townsfolk, inspiring a debate among the townsfolk as to the dog's fate.
1949 76m/C Edmund Gwenn, Donald Crisp, Geraldine Brooks, Reginald Owen, Alan Webb, Henry Stephenson, Alan Napier, Sara Allgood; *Dir:* Richard Thorpe. **VHS, Beta $19.98** *MGM, FCT* ✂✂½

Challenge of a Lifetime Marshall stars as a divorcee whose dream is to compete in the Hawaiian Ironman Triathlon. Cult favorite; Woronov has a substantial supporting role.
1985 100m/C Penny Marshall, Richard Gilliland, Jonathan Silverman, Mary Woronov, Paul Gleason, Mark Spitz, Cathy Rigby; *Dir:* Russ Mayberry. **VHS $59.98** *TTC, MOV* ✂✂½

Challenge of the Masters A young Kung-Fu artist wins a new tutor in a contest. Soon he's fighting for a friend to protect his family's honor.
1989 97m/C Liu Chia-Hui, Liu Chia-Yung, Chen Kuan-Tai, Chiang Yang; *Dir:* Lu Chia-Liang. **VHS $69.95** *SOU* ✂

Challenge of McKenna Run-of-the-prairie horse opera has drifter Ireland stumble upon danger and intrigue when he enters a mysterious town.
1983 90m/C John Ireland, Robert Woods. **VHS** *PMH* ✂✂

Challenge the Ninja One talented fighter defends a town from high-kicking hordes.

19?? 92m/C *Hosted:* Sho Kosugi. **VHS, Beta** $59.95 *TWE* 🎬

Challenge to White Fang
A courageous dog prevents a scheming businessman from taking over an old man's gold mine.
1986 (PG) 89m/C Harry Carey Jr., Franco Nero; *Dir:* Lucio Fulci. **VHS, Beta** $69.95 *TWE* 🎬½

Chamber of Fear
A madman inhabits a castle where he practices mental torture on innocent victims. A discredit to Karloff's memory, also known as ''Fear Chamber'' and ''Torture Zone.''
1968 88m/C Boris Karloff, Isela Vega, Julissa, Carlos East; *Dir:* Juan Ibanez, Jack Hill. **VHS, Beta** $57.95 *UNI* 🎬

Chamber of Horrors
A family is brought together at an English castle to claim a fortune left by an aristocrat. However, there's a catch—seven keys may open the vault with the fortune, or leave the key turner dead. Also known as ''The Door With Seven Locks.'' Based on the work by Edgar Wallace.
1940 80m/B *GB* Leslie Banks, Lilli Palmer; *Dir:* Norman Lee. **VHS, Beta** $16.95 *NOS, MRV, SNC* 🎬🎬½

The Chambermaid's Dream
Softcore whimsy about young French maid who fantasizes about her employer. In French, dubbed.
1986 86m/C *FR* Michel Lemoine, Elizabeth Tessier. **VHS, Beta** $29.95 *MED* 🎬

Chameleon
A freeform dramatic examination of a day in the life of a Los Angeles pusher.
1981 90m/C Bob Glaudini; *Dir:* Jon Jost. **VHS, Beta** *FCT* 🎬

The Champ
A washed up boxer dreams of making a comeback and receives support from no one but his devoted son. Minor classic most notorious for jerking the tears and soiling the hankies, this was the first of three Beery/Cooper screen teamings.
1931 87m/B Wallace Beery, Jackie Cooper, Irene Rich, Roscoe Ates, Edward Brophy, Hale Hamilton, Jesse Scott, Marcia Mae Jones; *Dir:* King Vidor. Academy Awards '32: Best Actor (Beery), Best Original Screenplay. **VHS, Beta, LV** $29.98 *MGM, FCT, BTV* 🎬🎬🎬

The Champ
An ex-fighter with a weakness for gambling and drinking is forced to return to the ring in an attempt to keep the custody of his son. Excessive sentiment may cause cringing. Earned an Academy Award nomination for Best Score. Remake of the 1931 classic.
1979 (PG) 121m/C Jon Voight, Faye Dunaway, Rick Schroder, Jack Warden; *Dir:* Franco Zeffirelli. **VHS, Beta, LV** $59.95 *MGM* 🎬

Champ Against Champ
A one-legged ninja masters the 18 kicks and hops his way to ninja superstardom.
198? 90m/C **VHS, Beta** $39.95 *TWE* 🎬

Champagne
A socialite's father fakes bankruptcy to teach his irresponsible daughter a lesson. Early, silent endeavor from Hitchcock is brilliantly photographed.
1928 93m/B *GB* Betty Balfour, Gordon Harker, Ferdinand von Alten, Clifford Heatherly, Jack Trevor; *Dir:* Alfred Hitchcock. **VHS** *GPV, MLB* 🎬🎬

Champagne for Breakfast
Sex comedy follows the fun-filled adventures of ''Champagne,'' a free-spirited beauty living life to the fullest.

1979 (R) 69m/B Mary Carlisle, Hardie Albright, Joan Marsh, Lila Lee, Sidney Toler, Bradley Page, Emerson Tracy; *Dir:* Melville Brown. **VHS** $49.95 *WES* 🎬

Champagne for Caesar
The laughs keep coming in this comedy about self-proclaimed genius on every subject goes on a television quiz show and proceeds to win everything in sight. The program's sponsor, in desperation, hires a femme fatale to distract the contestant before the final program. Wonderful spoof of the game-show industry.
1950 90m/B Ronald Colman, Celeste Holm, Vincent Price, Art Linkletter; *Dir:* Richard Whorf. **VHS, Beta** $39.95 *UHV* 🎬🎬🎬

Champion
An ambitious prizefighter alienates the people around him as he desperately fights his way to the top. When he finally reaches his goal, he is forced to question the cost of his success. From a story by Ring Lardner. Certainly one of the best films ever made about boxing, with less sentiment than ''Rocky'' but concerned with sociological correctness. Oscar nominated for Best Actor (Douglas), Supporting Actor (Kennedy), Screenplay, Black & White Cinematography, Scoring for a Comedy or Drama.
1949 99m/B Kirk Douglas, Arthur Kennedy, Marilyn Maxwell, Ruth Roman, Lola Albright; *Dir:* Mark Robson. Academy Awards '49: Best Film Editing. **VHS** $19.98 *REP* 🎬🎬🎬½

The Champion/His New Job
Two classic silent shorts by Chaplin, with music; the first is alternately entitled ''Champion Charlie.''
1915 53m/B Charlie Chaplin, Edna Purviance, Ben Turpin, Gloria Swanson, Mack Swain, Billy Anderson; *Dir:* Charlie Chaplin. **VHS, Beta, 8mm** $24.95 *VYY* 🎬🎬🎬

Champion Operation
Hong Kong heavies pose as police officers. The action follows them from the slums to the red-light district.
197? 100m/C Tong Chun Chung, Yu Ka Hei, Tang Ho Kwong. **VHS** $49.95 *OCE* 🎬

Champions
Moving but cliched story of Bob Champion, a leading jockey who overcame cancer to win England's Grand National Steeplechase. A true story, no less.
1984 113m/C *GB* John Hurt, Gregory Jones, Mick Dillon, Ann Bell, Jan Francis, Peter Barkworth, Edward Woodward, Ben Johnson, Kirstie Alley, Alison Steadman; *Dir:* John Irvin. **VHS, Beta, LV** $9.98 *SUE* 🎬🎬

Chan is Missing
Two cab drivers try to find the man who stole their life savings. Wry, low-budget comedy filmed in San Francisco's Chinatown was an art-house smash. The first full-length American film produced exclusively by an Asian-American cast and crew.
1982 80m/B Wood Moy, Marc Hayashi, Laureen Chew, Judy Mihei, Peter Wang, Presco Tabios; *Dir:* Wayne Wang. **VHS** $79.95 *CEG, NYF* 🎬🎬🎬

Chance
With over a million dollars in diamonds missing, Haggerty and Jacobs throw out all the stops to recover them in this action thriller.
1989 90m/C Dan Haggerty, Lawrence Hilton-Jacobs, Addison Randall, Roger Rudd, Charles Gries, Pamela Dixon; *Dir:* Addison Randall, Charles Kanganis. **VHS** *PMH* 🎬½

Chances Are
After her loving husband dies in a chance accident, a pregnant woman remains unmarried, keeping her husband's best friend as her only close male companion. Years later, her now teenage daughter brings a friend home for dinner, but due to an

error in heaven, the young man begins to realize that this may not be the first time he and this family have met. A wonderful love-story hampered only minimally by the unbelievable plot.
1989 (PG) 108m/C Cybill Shepherd, Robert Downey Jr., Ryan O'Neal, Mary Stuart Masterson, Josef Sommer, Christopher MacDonald, Joe Grifasi, James Noble; *Dir:* Emile Ardolino. **VHS, Beta, LV, 8mm** $19.95 *COL, FCT* 🎬🎬🎬

Chandu on the Magic Island
Chandu the Magician takes his powers of the occult to the mysterious lost island of Lemuria to battle the evil cult of Ubasti. Sequel to ''Chandu the Magician'' and just as campy.
1934 67m/B Bela Lugosi, Maria Alba, Clara Kimball Young. **VHS, Beta** $19.95 *NOS, SNC, VYY* 🎬½

Chandu the Magician
Bad guy searches desperately for the secret of a powerful death ray so he can (surprise!) destroy civilization. Not well received in its day, but makes for great high-camp fun now. One of Lugosi's most melodramatic performances.
1932 70m/B Edmund Lowe, Bela Lugosi, Irene Ware, Henry B. Walthall; *Dir:* William Cameron Menzies, Marcel Varnel. **VHS, Beta** *RXM* 🎬🎬

Chanel Solitaire
Uninspiring biography follows the fabulous (simply fabulous) career of dress designer Gabrielle ''Coco'' Chanel. Ah, go sew something.
1981 (R) 124m/C Karen Black, Marie-France Pisier, Rutger Hauer, Timothy Dalton, Brigitte Fossey; *Dir:* George Kaczender. **VHS, Beta** $19.95 *MED* 🎬🎬

Chang: A Drama of the Wilderness
A farmer and his family has settled a small patch of ground on the edge of the jungle and must struggle for survival against numerous wild animals. The climactic elephant stampede is still thrilling. Shot on location in Siam.
1927 67m/B **VHS** $39.95 *MIL* 🎬🎬🎬½

Change of Habit
Three novitiates undertake to learn about the world before becoming full-fledged nuns. While working at a ghetto clinic a young doctor forms a strong, affectionate relationship with one of them. Presley's last feature film.
1969 (G) 93m/C Elvis Presley, Mary Tyler Moore, Barbara McNair, Ed Asner, Ruth McDevitt, Regis Toomey; *Dir:* William A. Graham. **VHS, Beta** $14.95 *MCA, GKK* 🎬🎬

A Change of Seasons
One of them so-called sophisticated comedies that look at the contemporary relationships and values of middle-class, middle-aged people who should know better. The wife of a college professor learns of her husband's affair with a seductive student and decides to have a fling with a younger man. The situation reaches absurdity when the couples decide to vacation together.
1980 (R) 102m/C Shirley MacLaine, Bo Derek, Anthony Hopkins, Michael Brandon, Mary Beth Hurt; *Dir:* Richard Lang. **VHS, Beta** $59.98 *FOX* 🎬🎬

The Changeling
A music teacher moves into an old house and discovers that a young boy's ghostly spirit is his housemate. The ghost wants revenge against the being that replaced him upon his death. Scary ghost story with some less than logical leaps of script.

1980 (R) 114m/C *CA* George C. Scott, Trish Van Devere, John Russell, Melvyn Douglas, Jean Marsh; *Dir:* Peter Medak. Genie Awards '80: Best Art Direction/Set Decoration, Best Cinematography, Best Picture; Genie Awards '81: Best Screenplay. **VHS, Beta, LV** $19.98 *VES* 🎬🎬½

Changes A young man leaves home and thumbs across the California coast in order to find himself. On the road again.
1969 (PG) 103m/C Kent Lane, Michele Carey, Jack Albertson, Marcia Strassman; *Dir:* Hall Bartlett. **VHS, Beta** $79.95 *PSM* 🎬

The Channeler The traditional dopey students probe an old Colorado mine and encounter black-robed, demon whazzits. A spunky cast and nice scenery didn't stop the Hound from switching channels on this one.
1989 90m/C Dan Haggerty, Richard Harrison, Jay Richardson, David Homb, Oliver Darrow, Robin Sims, Charles Solomon; *Dir:* Grant Austin Waldman. **VHS, Beta** $79.98 *MAG* 🎬

Chapayev Striking propagandist drama dealing with the exploits of a Red Army commander during the 1919 battles. English subtitles.
1934 101m/B *RU* Boris Babochkin, Leonid Kmit; *Dir:* Sergei, Gregory Vasiliev. National Board of Review Awards '35: Best Foreign Film. **VHS, Beta, 3/4U** $35.95 *IHF* 🎬🎬🎬

Chaplin: A Character Is Born The evolution of Chaplin's "Little Tramp" is shown in scenes from his early classics, "The Pawnshop," "The Rink," and "The Immigrant."
1979 40m/B Charlie Chaplin; *Nar:* Keenan Wynn. **VHS, Beta** *SLF* 🎬🎬🎬

Chaplin: A Character Is Born/ Keaton: The Great Stone Face Wynn narrates "A Character Is Born," with scenes from "Little Champ," "The Pawnshop," "The Rink" and "The Immigrant." Buttons presents "The Great Stone Face" featuring scenes from "Cops," "The Playhouse," "The Boat" and "The General."
19?? 90m/C Charlie Chaplin; *Nar:* Keenan Wynn, Red Buttons. **VHS, Beta** $59.95 *HAR* 🎬½

Chaplin Essanay Book 1 Two of Chaplin's early shorts: "The Tramp" and "The Champion."
1915 51m/B Charlie Chaplin, Edna Purviance, Ben Turpin. **VHS, Beta** $19.98 *CCB* 🎬½

Chaplin Mutuals Contains three classic silent comedy shorts, including "The Pawnshop," "One A.M.," and "Behind the Screen."
1916 83m/B Charlie Chaplin, Edna Purviance; *Dir:* Charlie Chaplin. **VHS, Beta, 8mm** *VYY* 🎬½

The Chaplin Review The "Revue," put together by Chaplin in 1958, consists of three of his best shorts: "A Dog's Life" (1918), the World War II comedy "Shoulder Arms" (1918), and "The Pilgrim," in which a convict hides out in a clerical guise (1922). Chaplin added self-composed score, narration, and some documentary on-the-set material.
1958 121m/B Charlie Chaplin, Edna Purviance, Sydney Chaplin, Mack Swain; *Dir:* Charlie Chaplin. **VHS, Beta** $19.98 *FOX* 🎬½

Chapter in Her Life An early silent film by Lois Weber which examines the relationships between an orphaned child and the adults in her life.
1923 87m/B *Dir:* Lois Weber. **VHS, Beta** $19.95 *GVV* 🎬🎬½

Chapter Two Loosely based on Neil Simon's marriage to Mason and his Broadway hit of the same name. A writer, grief-stricken over the death of his first wife, meets the woman who will become his second. Witty dialogue in first half deteriorates when guilt strikes. Mason was Oscar nominated.
1979 (PG) 124m/C James Caan, Marsha Mason, Valerie Harper, Joseph Bologna; *Dir:* Robert Moore. **VHS, Beta, LV** $69.95 *COL* 🎬🎬

Charade After her husband is murdered, a young woman finds herself on the run from crooks and double agents who want the $250,000 her husband stole during WWII. Hepburn and Grant are charming and sophisticated as usual in this stylish intrigue filmed in Paris. Music by Henry Mancini.
1963 113m/C Cary Grant, Audrey Hepburn, Walter Matthau, James Coburn, George Kennedy; *Dir:* Stanley Donen. Edgar Allan Poe Awards '63: Best Screenplay. **VHS, Beta** $8.95 *NEG, MRV, MCA* 🎬🎬🎬½

Charge of the Light Brigade A British army officer stationed in India deliberately starts the Balaclava charge to even an old score with Surat Khan, who's on the other side. Still an exciting film, though it's hardly historically accurate. De Havilland is along to provide the requisite romance with Flynn. Also available colorized.
1936 115m/B Olivia de Havilland, Errol Flynn, David Niven, Nigel Bruce, Patric Knowles, Donald Crisp, C. Aubrey Smith, J. Carroll Naish, Henry Stephenson, E.E. Clive; *Dir:* Michael Curtiz, Jack Sullivan. **VHS, Beta, LV** $19.98 *FOX, FCT* 🎬🎬🎬

The Charge of the Model T's Set during World War I, this comedy is about a German spy who tries to infiltrate the U.S. army.
1976 (G) 90m/C Louis Nye, John David Carson, Herb Edelman, Carol Bagdasarian, Arte Johnson; *Dir:* Jim McCullough. **VHS, Beta** $9.99 *SUE* 🎬½

Chariots of Fire A lush telling of the parallel stories of Harold Abraham and Eric Liddell, English runners who competed in the 1924 Paris Olympics. One was compelled by a hatred of anti-Semitism, the other by the love of God. Outstanding performances by the entire cast.
1981 (PG) 123m/C *GB* Ben Cross, Ian Charleson, Nigel Havers, Ian Holm, Alice Krige, Brad Davis, Dennis Christopher, Patrick Magee, Cheryl Campbell, John Gielgud, Lindsay Anderson, Nigel Davenport; *Dir:* Hugh Hudson. Academy Awards '81: Best Costume Design, Best Original Screenplay, Best Picture, Best Musical Score; British Academy Awards '81: Best Costume Design, Best Film, Best Supporting Actor (Holm). **VHS, Beta, LV** $19.98 *WAR, FCT, BTV* 🎬🎬🎬½

Charles and Diana: For Better or For Worse A chronicle of the ten-year marriage of Prince Charles and Princess Diana. Includes footage of successful royal tours and public appearances. Also delves into the reported difficulties in their marriage and how they have managed to cope with the strains of public life.
1991 60m/C **VHS** $19.98 *MPI* 🎬🎬🎬½

Charles et Lucie A has-been couple are ripped off, pursued, persecuted and saddled with very bad luck in the South of France. Essentially a character study; semi-acclaimed. In French with subtitles.
1979 96m/C *FR* Daniel Ceccaldi, Ginette Garcin, Jean-Marie Proslier, Samson Fainsilber; *Dir:* Nelly Kaplan. **VHS, Beta** *FCT* 🎬🎬🎬

Charles Pierce at the Ballroom Famous female impersonator Pierce gives an imaginary performance, playing both audience and the on-stage personalities. His impressions of Katharine Hepburn, Bette Davis, Mae West, and others are heckled by Pierce's Joan Collins, Barbara Stanwyck, and Tallulah Bankhead.
1987 70m/C Charles Pierce. **Beta** *CKV* 🎬🎬🎬

Charleston Renoir's third film, a simple erotic dance fantasy that caused much controversy in its day. Silent, with jazz score and historical introduction.
1926 30m/B Catherine Hessling; *Dir:* Jean Renoir. **VHS, Beta** *FCT* 🎬🎬🎬

Charley and the Angel Touching story of a man who changes his cold ways with his family when informed by an angel that he hasn't much time to live. Amusing Disney movie set in the Great Depression.
1973 (G) 93m/C Fred MacMurray, Cloris Leachman, Harry Morgan, Kurt Russell, Vincent Van Patten, Kathleen Cody; *Dir:* Vincent McEveety. **VHS** *DIS, OM* 🎬🎬½

Charley Chase Festival: Vol. 1 Six classic Charley Chase comedies: "At First Sight" (1923), "Fighting Fluid" (1924), "Young Oldfield" (1924), "Ten Minute Egg" (1924), "Stolen Goods" (1924), and "Long Fliv the King" (1926).
192? 70m/B Charley Chase. **VHS, Beta** $19.95 *VDM*

Charley Varrick Matthau, a small-town hood, robs a bank only to find out that one of its depositors is the Mob. Baker's the vicious hit-man assigned the job of getting the loot back. A well-paced, on-the-mark thriller.
1973 (PG) 111m/C Walter Matthau, Joe Don Baker, Felicia Farr, John Vernon, Sheree North, Norman Fell, Andrew Robinson; *Dir:* Don Siegel. **VHS, Beta** $49.95 *MCA* 🎬🎬🎬

Charley's Aunt Amusing Victorian farce in which two young male students convince an older fellow student to dress up as their female chaperon so they can pitch woo to two local lovelies. Fun performances and well-paced direction by Sidney. Based on the Brandon Thomas farce, the movie was remade in 1930, starring Charles Ruggles, and again in 1941, with Jack Benny.
1925 75m/B Sydney Chaplin, Ethel Shannon, Lucien Littlefield, Alec B. Francis; *Dir:* Scott Sidney. **VHS, Beta** $24.95 *GPV, FCT, DNB* 🎬🎬½

Charlie Boy The new owner of an ancient African idol enters the world of the supernatural when he casts a death spell on six people.
1981 60m/C Leigh Lawson, Angela Bruce, Marius Goring; *Dir:* Robert M. Young. **VHS, Beta** $29.95 *IVE* 🎬½

Charlie Bravo A commando group is commanded to rescue a captured nurse in Vietnam. Dubbed.
198? (R) 94m/C *FR* Bruno Pradal, Karen Veriler, Jean-Francois Poron; *Dir:* Rene Demoulin. **VHS, Beta** $59.95 *GEM* 🎬

A Charlie Brown Celebration The Peanuts gang in an extended adventure in which Lucy courts Schroeder, Patty goes to obedience school, and Charlie Brown finally kicks the football.
1984 49m/C *Dir:* Bill Melendez. **VHS, Beta** $14.95 *KRT* 🎬

A Charlie Brown Christmas Disillusioned by the commercialization of the holiday season, Charlie Brown sets out to find the true meaning of Christmas.

1965 30m/C **Dir:** Bill Melendez. **VHS, Beta** $9.95 *MED, SHV, VTR* ♂

A Charlie Brown Thanksgiving
The Peanuts gang learns that Thanksgiving is more than just turkey and stuffing in this family favorite.
1981 25m/C **Dir:** Bill Melendez, Phil Roman. **VHS, Beta** $14.95 *KRT, KAR* ♂

Charlie Brown's All-Stars
Charlie Brown and the whole Peanuts gang is here to delight young and old alike.
1966 26m/C **Dir:** Bill Melendez. **VHS** $9.98 *MED, WPB, VTR* ♂

Charlie Chan and the Curse of the Dragon Queen
The famed Oriental sleuth confronts his old enemy the Dragon Queen, and reveals the true identity of a killer.
1981 (PG) 97m/C Peter Ustinov, Lee Grant, Angie Dickinson, Richard Hatch, Brian Keith, Roddy McDowall, Michelle Pfeiffer, Rachel Roberts; **Dir:** Clive Donner. **VHS, Beta** $9.95 *MED* ♂½

Charlie Chan at the Opera
The great detective investigates an amnesiac opera star (Karloff) who may have committed murder. Considered one of the best of the series. Interesting even to those not familiar with Charlie Chan.
1936 66m/B Warner Oland, Boris Karloff, Keye Luke, Charlotte Henry, Thomas Beck, Gregory Gaye, Nedda Harrigan, William Demarest; **Dir:** H. Bruce Humberstone. **VHS, LV** $19.98 *FOX* ♂♂♂

Charlie Chan in Paris
Chan scours the city of lights to track down a trio of counterfeiters. Top-notch plot and plenty of suspense will please all.
1935 72m/B Warner Oland, Mary Brian, Thomas Beck, Erik Rhodes, John Miljan, Minor Watson, John Qualen, Keye Luke, Henry Kolker; **Dir:** Lewis Seiler. **VHS, LV** $19.98 *FOX* ♂♂½

Charlie Chan in Rio
Local police call on Chan to help solve a double murder. One of the series' weaker entries but with fine setting and music.
1941 60m/B Sidney Toler, Mary Beth Hughes, Cobina Wright Jr., Ted North, Victor Jory, Harold Huber, Sen Yung; **Dir:** Harry Lachman. **VHS, LV** $19.98 *FOX* ♂♂

Charlie Chan at the Wax Museum
Charlie and number one son must find a criminal hiding out in a wax museum. Chills abound when the fugitive heads for the chamber of horrors.
1940 63m/B Sidney Toler, Sen Yung, C. Henry Gordon, Marc Lawrence, Joan Valerie, Marguerite Chapman, Ted Osborn, Michael Visaroff; **Dir:** Lynn Shores. **VHS, LV** $19.98 *FOX* ♂½

Charlie Chan's Secret
Chan must solve the murder of the heir to a huge fortune. A good, logical script with plenty of suspects to keep you guessing.
1935 72m/B Warner Oland, Rosina Lawrence, Charles Quigley, Henrietta Crosman, Edward Trevor, Astrid Allwyn; **Dir:** Gordon Wiles. **VHS** $19.98 *FOX* ♂♂

Charlie Chaplin Carnival, Four Early Shorts
This tape features: "The Vagabond," "The Fireman," "The Count," and "Behind the Screen." Sound effects and music added. Silent.
1916 80m/B Charlie Chaplin. **VHS, Beta** $44.95 *HHT*

Charlie Chaplin Cavalcade, Mutual Studios Shorts
A compilation tape which features Chaplin's: "One A.M.," "Pawnshop," "Floorwalker," and "The Rink."
1916 81m/B Charlie Chaplin. **VHS, Beta** $19.95 *VDM, HHT*

Charlie Chaplin Centennial Collection
Chaplin is up to his stunts again in an anthology of comedy shorts.
191? 87m/B Charlie Chaplin; **Dir:** Charlie Chaplin. **VHS** $19.98 *FOX, BUR*

Charlie Chaplin Festival
This video features four classic Chaplin shorts: "The Immigrant," "The Adventurer," "The Cure," and "Easy Street."
1917 80m/B Charlie Chaplin, Edna Purviance, Henry Bergman, Eric Campbell; **Dir:** Charlie Chaplin. **VHS, Beta** *PSM*

Charlie Chaplin Festival
Four Chaplin shorts: "The Immigrant," "The Fireman," "Behind the Screen" and "The Adventurer."
1917 60m/B Charlie Chaplin. **VHS, Beta** $24.95 *DVT*

Charlie Chaplin, the Mutual Studio Years
Anthology of three shorts from the ingenious comic.
1916 83m/B Charlie Chaplin; **Dir:** Charlie Chaplin. **VHS** $29.95 *VYY*

Charlie Chaplin: Night at the Show
Chaplin, who wrote and directed, plays two different mugs, both out for a night on the town. Mr. Pest, a sharp-dressed upper-cruster, clings to his disgusting habits, and Mr. Rowdy is his working-class doppelganger. Both obnoxious Chaplins-in-disguise collaborate to wreak havoc on a local theater. Contains original organ score.
1915 ?m/B Charlie Chaplin, Edna Purviance; **W/ Dir:** Charlie Chaplin. **VHS, Beta, Special order formats** *VYY* ♂♂♂

Charlie Chaplin: Rare Chaplin
A collection of three Chaplin shorts: "The Bank," "Shanghaied," and "A Night in the Snow."
1915 60m/B Charlie Chaplin, Edna Purviance, Bud Jamison, Billy Armstrong; **Dir:** Charlie Chaplin. **VHS, Beta** $29.95 *KIV* ♂♂½

Charlie Chaplin: The Early Years, Volume 1
This video package includes Chaplin's "The Immigrant," "The Count," and "Easy Street." Silent.
1917 62m/B Charlie Chaplin, Edna Purviance, John Rand, Eric Campbell, James T. Kelley. **VHS, LV** $19.98 *REP, PIA, FCT*

Charlie Chaplin: The Early Years, Volume 2
This video package includes Chaplin's "The Pawnshop" (1916), "The Adventurer" (1917), and "One A.M." (1916).
1917 61m/B Charlie Chaplin, Edna Purviance, Eric Campbell, Albert Austin, Henry Bergman, John Rand. **VHS, LV** $19.98 *REP, PIA, FCT*

Charlie Chaplin: The Early Years, Volume 3
This video package includes Chaplin's "The Cure" (1916), "The Floorwalker" (1916), and "The Vagabond."
1916 64m/B Charlie Chaplin, Edna Purviance, Eric Campbell, Albert Austin. **VHS, LV** $19.98 *REP, PIA, FCT*

Charlie Chaplin: The Early Years, Volume 4
This video package includes Chaplin's "Behind the Screen," "The Fireman" (1916), and "The Rink" (1916).

1916 63m/B Charlie Chaplin, Edna Purviance, Eric Campbell, Lloyd Bacon, Albert Austin, James T. Kelly. **VHS, Beta, LV** $19.98 *REP, PIA, FCT*

Charlie Chaplin...Our Hero!
Two shorts from 1915 written and directed by Chaplin, "Night at the Show" and "In the Park," as well as a 1914 film, "Hot Finish," in which Charlie plays the villain.
1915 58m/B Charlie Chaplin, Edna Purviance, Lloyd Bacon, Mabel Normand, Chester Conklin, Mack Sennett; **Dir:** Charlie Chaplin. **VHS, Beta, 8mm** $24.95 *VYY*

Charlie Chaplin's Keystone Comedies #1
Six one-reelers which Chaplin filmed in 1914, his first movie-making year: "Making a Living," Chaplin's first, in which he plays a villain; "Kid Auto Races," in which Charlie now sports baggy pants, bowler hat and cane; "A Busy Day," featuring Charlie in drag; "Mabel's Married Life;" "Laughing Gas;" and "The New Janitor." Silent with musical score.
1914 59m/B Charlie Chaplin, Mabel Normand, Mack Swain. **VHS, Beta** $24.95 *VYY*

Charlie Chaplin's Keystone Comedies #2
Three hilarious early Chaplin shorts, "His Trysting Place," "Getting Acquainted," and "The Fatal Mallet." Silent with musical score.
1914 60m/B Charlie Chaplin, Mabel Normand, Mack Swain, Mack Sennett. **VHS, Beta** $29.95 *VYY*

Charlie Chaplin's Keystone Comedies #3
Three more vintage early Chaplin shorts: "Caught in a Cabaret," "The Masquerader," and "Between Showers." Silent, includes a musical score.
1914 60m/B Charlie Chaplin, Mack Swain, Mabel Normand, Chester Conklin; **Dir:** Charlie Chaplin. **VHS, Beta** $29.95 *VYY*

Charlie Chaplin's Keystone Comedies #4
Four classic silent shorts featuring Chaplin in 1914, Including "Caught in the Rain (in the Park)," "His Million Dollar Job," "Mabel's Busy Day," and "The Face on the Bar Room Floor." With musical score.
1914 54m/B Charlie Chaplin, Mack Swain, Alice Davenport, Fatty Arbuckle, Mabel Normand, Chester Conklin, Slim Summerville, Cecile Arnold, Alice Howell. **VHS, Beta, 8mm** $29.95 *VYY*

Charlie Chaplin's Keystone Comedies #5
Four classic silent features: "Mabel's Strange Predicament," "The Rounders," "Tango Tangles," and "The Star Border."
1914 57m/B Charlie Chaplin; **Dir:** Charlie Chaplin. **VHS** $24.95 *VYY*

Charlie Chase and Ben Turpin
Three silent comedy classics, "All Wet," "Publicity Pays," and "A Clever Dummy."
1921 67m/B Ben Turpin, Charley Chase, Wallace Beery, Juanita Hansen. **VHS, Beta** $29.95 *VYY*

Charlie Grant's War
The true story of a Canadian businessman who hears of Nazi brutality and determines to save European Jews from persecution during the Holocaust. Not among the more distinguished works of its kind.
1980 130m/C *GB* Jan Rubes, R.H. Thompson. **VHS, Beta** $59.95 *ONE* ♂♂

Charlie and the Great Balloon Chase
When a grandfather takes his grandson on a cross-country balloon trip, they are hotly pursued by the boy's mother, her stuffy fiance (who wants the boy in

military school), the FBI, a reporter, and the Mafia. Will they make it to Virginia? Made for television.
1982 98m/C Jack Albertson, Adrienne Barbeau, Slim Pickens, Moosie Drier; **Dir:** Larry Elikann. **VHS, Beta $59.95** LTG, TLF ✍✍

Charlie, the Lonesome Cougar Life for a rugged logger will never be the same after he adopts an orphaned cougar. From the Disney factory.
1967 (G) 75m/C Linda Wallace, Jim Wilson, Ron Brown, Brian Russell, Clifford Peterson; **Dir:** Winston Hibler; **Nar:** Rex Allen. **VHS, Beta $69.95** DIS ✍✍

Charlotte Forten's Mission: Experiment in Freedom During the Civil War, a wealthy black woman, determined to prove to President Lincoln that blacks are equal to whites, journeys to a remote island off the coast of Georgia. There she teaches freed slaves to read and write.
1985 120m/C Melba Moore, Mary Alice, Ned Beatty, Carla Borelli, Micki Grant, Moses Gunn, Anna Maria Horsford, Bruce McGill, Glynn Turman, Roderick Wimberly; **Dir:** Barry Crane. **VHS $59.95** FCT, PTB ✍✍✍

Charlotte's Web E.B. White's classic story of a friendship between a spider and a pig is handled only adequately by Hanna-Barbera studios. Some okay songs.
1979 94m/C **Dir:** Charles A. Nichols; **Voices:** Debbie Reynolds, Agnes Moorehead, Paul Lynde, Henry Gibson. National Board of Review Awards '73: Special Citation. **VHS, Beta, LV $14.95** KUI, PAR, FCT ✍✍

Charly A retarded man becomes intelligent after brain surgery, then romances a kindly caseworker before slipping back into retardation. Moving account is well served by leads Roberston and Bloom. Adapted by Sterling Silliphant from the Daniel Keyes novel "Flowers for Algernon."
1968 103m/C Cliff Robertson, Claire Bloom, Lilia Skala, Leon Janney, Dick Van Patten, William Dwyer; **Dir:** Ralph Nelson. Academy Awards '68: Best Actor (Robertson); National Board of Review Awards '68: 10 Best Films of the Year, Best Actor (Robertson). **VHS, Beta, LV $19.98** FOX, BTV ✍✍✍

Charm of La Boheme A vintage German musical based on the Puccini opera. With English subtitles.
1936 90m/B GE Jan Keipura, Martha Eggerth, Paul Kemp; **Dir:** Geza von Bolvary. **VHS, Beta $29.95** NOS, DVT, APD ✍

Charro! Elvis Presley in a straight role as a reformed bandit hounded by former gang members. Western fails on nearly all accounts, with Presley hopelessly acting outside his acting limits. He only sings one song. Score by Hugo Montenegro.
1969 (G) 98m/C Elvis Presley, Ina Balin, Victor French, Lynn Kellogg, Barbara Werle, Paul Brinegar, James B. Sikking; **Dir:** Charles Marquis Warren. **VHS $19.98** WAR, AVD ✍

The Chase Not realizing that his boss-to-be is a mobster, Cummings takes a job as a chauffer. Naturally, he falls in love with the gangster's wife (Morgan), and the two plan to elope. Somewhat miffed, the cuckolded mafioso and his bodyguard (Lorre) pursue the elusive couple as they head for Havana. The performances are up to snuff, but the story's as unimaginative as the title, with intermittent bouts of suspense.
1946 86m/B Robert Cummings, Michele Morgan, Peter Lorre, Steve Cochran, Lloyd Corrigan, Jack Holt, Don "The Dragon" Wilson; **Dir:** Arthur Ripley. **VHS, Beta, LV $16.95** NOS, MRV, SNC ✍✍½

The Chase A southern community is undone when rumors circulate of a former member's prison escape and return home. Excellent cast only partially shines. Brando is, as usual, outstanding as the beleaguered, honorable sheriff, and Duvall makes a splash in the more showy role as a cuckold who fears the escapee is returning home to avenge a childhood incident. Reliable Dickinson also makes the most of her role as Brando's loving wife. Fonda, however, was no yet capable of fashioning complex characterizations, and Redford is under-utilized as the escapee. Written by Lillian Hellman from the play by Horton Foote. Notorious conflicts between producer, director, and writer kept it from being the winner it could have been.
1966 135m/C Marlon Brando, Robert Redford, Angie Dickinson, E.G. Marshall, Jane Fonda, James Fox, Janice Rule, Robert Duvall, Miriam Hopkins, Martha Hyer; **Dir:** Arthur Penn. **VHS, Beta, LV $14.95** RCA ✍✍½

Chase A big-city lawyer returns to her small hometown to defend a killer and ends up at odds with the town, including an ex-lover. Routine made for television drama.
1985 90m/C Jennifer O'Neill, Michael Parks, Richard Farnsworth; **Dir:** Rod Holcomb. **VHS, Beta $59.98** FOX ✍½

Chasing Dreams Sickly melodrama about a farmboy who finds fulfillment as a baseball player. Lame film promoted as Costner baseball vehicle, but the star of "Bull Durham" and "Field of Dreams" appears only briefly in a secondary role.
1981 (PG) 96m/C David G. Brown, John Fife, Jim Shane, Lisa Kingston, Matt Clark, Kevin Costner; **Dir:** Sean Roche, Therese Conte. **VHS, Beta $79.95** PSM ✍½

Chasing Those Depression Blues Four '30's comedy shorts: "Money on Your Life," "Dental Follies," "Any Day in Hollywood" and "Art in the Raw."
1935 58m/B Danny Kaye, Pinky Lee, Edgar Kennedy, Ben Turpin. **VHS, Beta $24.95** VYY ✍½

Chaste & Pure A young woman is torn between her blooming sexuality and her vows of chastity. Lesser known Antonelli effort.
1977 90m/C IT Laura Antonelli. **VHS, Beta $59.98** MAG ✍

Chato's Land An Indian is tracked by a posse eager to resolve a lawman's death. Conventional violent Bronson vehicle is bolstered by presence of masterful Palance.
1971 (PG) 100m/C Charles Bronson, Jack Palance, Richard Basehart, James Whitmore, Simon Oakland, Richard Jordan, Ralph Waite, Victor French, Lee Patterson; **Dir:** Michael Winner. **VHS, Beta $34.95** MGM, AVD ✍✍½

Chattahoochee A man suffering from post-combat syndrome lands in a horrifying institution. Strong cast, with Oldman fine in the lead, and Hopper memorable in the extended cameo. Fact-based film is, unfortunately, rather conventionally rendered.
1989 97m/C Ned Beatty, Gary Oldman, Dennis Hopper, Frances McDormand, Pamela Reed; **Dir:** Mick Jackson. **VHS, LV $89.99** HBO, IME, FCT ✍✍½

Chattanooga Choo Choo A football team owner must restore the Chattanooga Choo Choo and make a 24-hour run from Chattanooga to New York in order to collect one-million dollars left to him in a will. The train never leaves the station.
1984 (PG) 102m/C Barbara Eden, George Kennedy, Melissa Sue Anderson, Joe Namath, Bridget Hanley, Christopher McDonald, Clu Gulager, Tony Azito; **Dir:** Bruce Bilson. **VHS, Beta $59.99** HBO ✍½

Chatterbox A starlet's life and career are severely altered when her sex organs begin speaking. So why didn't they warn her about movies like this? Plenty of B-queen Rialson on view. Good double bill with "Me and Him."
1976 73m/C Candice Rialson, Larry Gelman, Jean Kean, Perry Bullington; **Dir:** Tom DeSimone. **VHS, Beta $59.98** LIV, VES ✍

The Cheap Detective Neil Simon's parody of the "Maltese Falcon" gloriously exploits the resourceful Falk in a Bogart-like role. Vast supporting cast—notable Brennan, DeLuise, and Kahn—equally game for fun in this consistently amusing venture.
1978 (PG) 92m/C Peter Falk, Ann-Margret, Eileen Brennan, Sid Caesar, Stockard Channing, James Coco, Dom DeLuise, Louise Fletcher, John Houseman, Madeline Kahn, Fernando Lamas, Marsha Mason, Phil Silvers, Vic Tayback, Abe Vigoda, Paul Williams, Nicol Williamson; **Dir:** Robert Moore. **VHS $19.95** CSM ✍✍✍

Cheap Shots Cheap best describes this. Two guys running a sleazy New York State resort motel plot to videotape patrons having sex, but capture an apparent mob murder instead. More of a character study than sexploitation, but who needs these characters?
1991 (PG-13) 90m/C Louis Zorich, David Patrick Kelly, Mary Louise Wilson, Patience Moore, Michael Twaine; **Dir:** Jeff Ureles. **VHS $89.95** CHV ✍½

Cheaper to Keep Her Supposed comedy about a private detective hired to track likely alimony welches. Loathsome, repellent fare, with Davis hopeless in the lead.
1980 (R) 92m/C Mac Davis, Tovah Feldshuh, Jack Gilford, Rose Marie; **Dir:** Ken Annakin. **VHS, Beta $54.95** MED Woof!

The Cheat Ward plays a frivolous socialite heavily indebted to Hayakawa as the Japanese money lender. Hayakawa makes Ward pay with her honor and her flesh by branding her. A dark and captivating drama. Silent with piano score.
1915 55m/B Fanny Ward, Sessue Hayakawa; **Dir:** Cecil B. DeMille. **VHS, Beta $14.95** GPV ✍✍✍

Cheaters Two middle class couples are having affairs with each other's spouses. Complications arise when their respective children decide to marry each other.
1984 103m/C Peggy Cass, Jack Kruschen. **VHS, Beta $39.95** RKO ✍½

The Cheaters A young gambler runs away with his boss's son's girlfriend, and is pursued therein.
198? (R) 91m/C Dayle Haddon, Luc Merenda; **Dir:** Sergio Martino. **VHS, Beta $59.95** PSM ✍½

Check & Double Check Radio's original Amos 'n' Andy (a couple of blackfaced white guys) help solve a lover's triangle in this film version of the popular radio series. Interesting only as a novelty. Duke Ellington's band plays "Old Man Blues" and "Three Little Words."
1930 85m/B Freeman Gosden, Charles Correll, Duke Ellington; **Dir:** Melville Brown. **VHS, Beta $19.95** NOS, MRV, VYY ✍

Check is in the Mail Lame comedy about a financier determined to make his home entirely independent from the rat-race of capitalist society. Dependables Dennehy and Archer are wasted.
1985 (R) **83m/C** Brian Dennehy, Anne Archer, Dick Shawn; *Dir:* Joan Darling. **VHS, Beta** **$79.95** *MED* 🎞🎞½

Checking Out Black comedy about a manic hypochondriac who is convinced that his demise will soon occur. You'll pray that he's right. Daniels can't make it work, and supporters Mayron and Magnuson also have little chance in poorly conceived roles. Written by Joe Esterhaus.
1989 (R) **95m/C** Jeff Daniels, Melanie Mayron, Michael Tucker, Kathleen York, Ann Magnuson, Allen Havey, Jo Harvey Allen, Felton Perry, Alan Wolfe; *Dir:* David Leland. **VHS, Beta, LV $89.95** *VIR* 🎞½

Cheech and Chong: Still Smokin' Cheech and Chong travel to Amsterdam to raise funds for a bankrupt film festival group by hosting a dope-a-thon. Lots of concert footage used in an attempt to hold the slim plot together. Only for serious fans of the dopin' duo.
1983 (R) **91m/C** Cheech Marin, Thomas Chong; *Dir:* Thomas Chong. **VHS, Beta, LV $14.95** *PAR* 🎞

Cheech and Chong: Things Are Tough All Over Cheech and Chong are hired by two rich Arab brothers (also played by Cheech and Chong) and unwittingly drive a car loaded with money from Chicago to Las Vegas. Fourth in the series.
1982 (R) **87m/C** Cheech Marin, Thomas Chong, Shelby Fiddis, Rikki Marin, Evelyn Guerrero, Rip Taylor; *Dir:* Tom Avildsen. **VHS, Beta $9.95** *COL* 🎞

Cheech and Chong's Next Movie A pair of messed-up bumblers adventure into a welfare office, massage parlor, nightclub, and flying saucer, while always living in fear of the cops. Kinda funny, like. Sequel to "Up in Smoke."
1980 (R) **95m/C** Cheech Marin, Thomas Chong, Evelyn Guerrero, Edie McClurg, Pee-Wee Herman; *Dir:* Thomas Chong. **VHS, Beta, LV $14.95** *MCA* 🎞🎞

Cheech and Chong's Nice Dreams The spaced-out duo are selling their own "specially mixed" ice cream to make cash and realize their dreams. Third in the series.
1981 (R) **97m/C** Cheech Marin, Thomas Chong, Evelyn Guerrero, Stacy Keach, Pee-Wee Herman; *Dir:* Thomas Chong, Timothy Leary. **VHS, Beta, LV $14.95** *COL* 🎞🎞

Cheech and Chong's, The Corsican Brothers Wretched swashbuckler features the minimally talented duo in a variety of worthless roles. Dumas would vomit in his casket if he knew about this. Fifth in the series.
1984 (PG) **91m/C** Cheech Marin, Thomas Chong, Roy Dotrice, Rae Dawn Chong, Shelby Fiddis, Rikki Marin, Edie McClurg; *Dir:* Thomas Chong. **VHS, LV $79.98** *LIV, LTG* Woof!

Cheech and Chong's Up in Smoke A pair of free-spirited burn-outs team up for a tongue-in-cheek spoof of sex, drugs, and rock and roll. First and probably the best of the dopey duo's cinematic adventures. A box-office bonanza when released and still a cult favorite.

1979 (R) **87m/C** Cheech Marin, Thomas Chong, Stacy Keach, Tom Skerritt, Edie Adams, Strother Martin, Cheryl "Rainbeaux" Smith; *Dir:* Lou Adler. **VHS, Beta, LV $14.95** *PAR* 🎞🎞½

The Cheerful Fraud Denny, who appeared in some two hundred-odd films in his career, masquerades as a butler to be near the girl he loves. A fool for love is a cheerful fraud. A zanily romantic silent comedy.
1927 **64m/B** Reginald Denny, Gertrude Olmsted, Emily Fitzroy, Gertrude Astor; *Dir:* William A. Seiter. **VHS, Beta $24.95** *GPV* 🎞🎞½

Cheering Section High school football teams compete for the grand prize—the privilege of ravishing the losing school's cheerleaders. Rah, rah, rah.
1973 (R) **84m/C** Rhonda Foxx, Tom Leindecker, Gregg D'Jah, Patricia Michelle, Jeff Laine; *Dir:* Harry Kerwin. **VHS $29.95** *VCI* Woof!

Cheerleader Camp Nubile gals are stalked by a psychopath while they cavort semi-clad at summer camp. Inexcusable. Released theatrically as "Bloody Pom-poms."
1988 (R) **89m/C** Betsy Russell, Leif Garrett, Lucinda Dickey, Lorie Griffin, George Flower, Teri Weigel, Rebecca Ferratti; *Dir:* John Quinn. **VHS, Beta $14.95** *PSM, PAR* Woof!

The Cheerleaders Dimwitted exploitation effort has the usual suspects—moronic jocks and lamebrained gals—cavorting exuberantly. Followed by "Revenge of the Cheerleaders."
1972 **84m/C** Stephanie Fondue, Denise Dillaway, Jovita Bush, Debbie Lowe, Sandy Evans; *Dir:* Paul Glickler. **VHS, Beta, LV $9.99** *STE, PSM, IME* Woof!

Cheerleaders' Beach Party Good-natured gals undermine the effectiveness of opposing football teams by sexually bewitching enemy players. By standards of the genre, a near masterpiece.
1977 (R) **85m/C** Stephanie Hastings, Linda Jenson, Mary Lou Loredan, Denise Upson; *Dir:* Alex E. Goiten. **VHS, Beta** *PGN* 🎞

Cheerleaders' Wild Weekend A group of cheerleaders are held captive by a disgruntled former football star. Pom poms wave.
1985 (R) **87m/C** Jason Williams, Kristine DeBell; *Dir:* Jeff Werner. **VHS, Beta $69.98** *LIV, VES* Woof!

Cheers for Miss Bishop The story of a woman who graduates from college then teaches at the same institution for the next 50 years. Somewhat moving. Based on Bess Streeter Aldrich's novel. Score by Edward Ward was Oscar nominated.
1941 **95m/B** Martha Scott, William Gargan, Edmund Gwenn, Sterling Holloway, Rosemary DeCamp; *Dir:* Tay Garnett. **VHS, Beta $19.95** *NOS, MRV, VYY* 🎞🎞½

Cheetah Two California kids visiting their parents in Kenya embark on the adventure of their lives when, with the help of a young Masai, they adopt and care for an orphaned cheetah. Usual Disney kids and animals story.
1989 (G) **80m/C** Keith Coogan, Lucy Deakins, Collin Mothupi; *Dir:* Jeff Blyth. **VHS $89.98** *DIS* 🎞🎞

Cherry 2000 Futuristic flick concerns a man who short-circuits his sex-toy robots and embarks on a search for replacement parts across treacherous territory, only to meet a real female—Griffith. Offbeat, occasionally funny.

1988 (PG-13) **94m/C** Melanie Griffith, David Andrews, Ben Johnson, Tim Thomerson, Michael C. Gwynne; *Dir:* Steve DeJarnatt. **VHS, Beta, LV $19.98** *ORI* 🎞🎞

Cherry Hill High Five teen girls compete for former-virgin status. Pass on this one.
1976 (R) **92m/C** Linda McInerney, Carrie Olsen, Nina Carson, Lynn Hastings, Gloria Upson, Stephanie Lawlor; *Dir:* Alex E. Goiten. **VHS, Beta $59.95** *MCA* Woof!

Chesty Anderson USN Ultra-lame sexploitation about female naval recruits. Minimal nudity. Also known as "Anderson's Angels."
1976 **90m/C** Shari Eubank, Dorri Thompson, Rosanne Katon, Marcie Barkin, Scatman Crothers, Frank Campanella, Fred Willard; *Dir:* Ed Forsyth. **VHS, Beta $49.95** *UNI, WES* 🎞

Cheyenne Autumn The newly restored version of the ambitious, ultimately hit-and-miss western epic about three hundred Cheyenne Indians who migrate from Oklahoma to their Wyoming homeland in 1878. The cavalry, for once, are not the good guys. Widmark is strong in the lead. Stewart is memorable in his comic cameo as Wyatt Earp. Last film from genre master Ford, capturing the usual rugged panoramas (cinematographer William H. Clothier picked up an Oscar nomination). Based on a true story as told in the Mari Sandoz novel.
1964 **156m/C** James Stewart, Edward G. Robinson, Sal Mineo, Richard Widmark, Carroll Baker, Dolores Del Rio, Gilbert Roland, Arthur Kennedy, Victor Jory, John Carradine, John Qualen, Mike Mazurki, George O'Brien, Karl Malden, Ricardo Montalban; *Dir:* John Ford. **VHS, Beta $19.98** *WAR, FCT* 🎞🎞½

Cheyenne Kid A bronco buster is up against a gang of bad guys.
1933 **40m/B** Tom Keene, Mary Mason, Roscoe Ates. **VHS, Beta $19.95** *NOS, DVT* 🎞

Cheyenne Rides Again Cheyenne poses as an outlaw to hunt a gang of rustlers.
1938 **60m/B** Tom Tyler, Lucille Browne, James Fox. **VHS, Beta $19.95** *NOS, VYY, VCN* 🎞

The Cheyenne Social Club Stewart inherits a brothel and Fonda helps him operate it. Kelly directs, sort of. Some laughs, but this effort is beneath this trio.
1970 (PG) **103m/C** Henry Fonda, James Stewart, Shirley Jones, Sue Ann Langdon, Elaine Devry; *Dir:* Gene Kelly. **VHS $59.95** *WAR* 🎞🎞

Cheyenne Takes Over Lash sniffs out the bad guy in the case of a murdered ranch heir.
1947 **56m/B** Lash LaRue, Al "Fuzzy" St. John, George Cheseboro; *Dir:* Ray Taylor. **VHS, Beta $12.95** *COL, SVS* 🎞

Chicago Blues The evolution of blues music from its origins in the rural South to the contemporary electric sound is examined through archive footage, interviews, and performances. Featured are Muddy Waters, Junior Wells, Floyd Jones, J.B. Hutto, Dick Gregory, and many others.
1972 **50m/C** Muddy Waters, Junior Wells, Floyd Jones, J.B. Hutto, Dick Gregory. **VHS, Beta $29.95** *MVD, RHP, TLF*

Chicago Joe & the Showgirl A London-based GI befriends a loopy showgirl and helps her in various crimes during WWII. Appealing Lloyd overwhelms Sutherland, who shows only limited acting ability here. Kensit and Pigg are fine in smaller roles.

1990 (R) 105m/C *GB* Emily Lloyd, Kiefer Sutherland, Patsy Kensit, Keith Allen, Liz Fraser, Alexandra Pigg, Ralph Nossek, Colin Bruce; *Dir:* Bernard Rose. VHS, LV $89.95 *LIV, IME* 🐾🐾

The Chicken Chronicles An affluent high school dude longs to get horizontal with his dream girl. Set in the 1960s. First feature film appearance for Guttenberg.
1977 (PG) 94m/C Phil Silvers, Ed Lauter, Steve Guttenberg, Lisa Reeves, Meredith Baer; *Dir:* Francis Simon. VHS, Beta $14.95 *COL, SUE* 🐾

Chicken Ranch Documentary focuses on the women who work at and the men who frequent "The Chicken Ranch," the country's best-known legal brothel.
1983 (R) 84m/C Sandi Sissel, Nick Broomfield. VHS, Beta $59.95 *VES*

Chicken Thing A tongue-in-cheek interpretation of Hollywood's "things that go bump in the night" movies.
1987 12m/C VHS, Beta, 3/4U $100.00 *DCL* 🐾

Chiefs A police chief in a Southern town discovers a series of murders concealed by town big wigs. Made for TV film boasts a surprisingly strong cast, but provides otherwise conventional entertainment.
1983 200m/C Charlton Heston, Paul Sorvino, Keith Carradine, Brad Davis, Billy Dee Williams, Wayne Rogers, Stephen Collins, Tess Harper, Victoria Tennant; *Dir:* Jerry London. VHS, Beta $29.95 *STE, NWV* 🐾🐾

Child Bride An edited version of the mildly infamous exploitation quickie about an Ozark hillbilly and his 12-year-old wife.
1939 52m/B *Dir:* Harry Revier. VHS, Beta $16.95 *SNC, NOS, JLT* Woof!

Child Bride of Short Creek A young Korean war veteran returns to his community where polygamy is allowed, to find that his father intends to marry again, this time to a 15-year-old. Conflict ensues. Made for television.
1981 100m/C Diane Lane, Conrad Bain, Christopher Atkins, Kiel Martin, Helen Hunt; *Dir:* Robert Lewis. VHS, Beta $59.95 *PSM* 🐾🐾½

Child of Darkness, Child of Light A priest is sent to investigate two virgin pregnancies and discovers that one child will be a savior and the other the Anti-Christ. Totally mindless brain candy.
1991 (PG-13) 85m/C Anthony Denison, Sela Ward, Brad Davis, Paxton Whitehead, Claudette Nevins, Sydney Penny, Alan Oppenheimer, Eric Christmas, Viveca Lindfors; *Dir:* Marina Sargenti. VHS, Beta *PAR* Woof!

Child of Glass A young boy's family moves into a huge New Orleans mansion, and soon after encounters a young girl's ghost and lost treasure from the Civil War. A Disney made for TV movie.
1978 93m/C Barbara Barrie, Biff McGuire, Anthony Zerbe, Nina Foch, Steve Shaw, Katy Kurtzman, Olivia Barash; *Dir:* John Erman. VHS, Beta $69.95 *DIS* 🐾½

Child in the Night A shaken eight-year old boy is a witness to murder in his family, and a psychologist with plenty of her own problems must get him to testify. Middling TV-movie suspense.
1990 (R) 93m/C Tom Skerritt, JoBeth Williams, Darren McGavin, Season Hubley; *Dir:* Mike Robe. VHS, Beta $79.95 *TRI* 🐾🐾

Child of the Prairie An early Mix film about a young cowboy and his city-loving wife down on their luck in a western town. Silent with music score.

1918 63m/B Tom Mix. VHS, Beta, 8mm $29.95 *VYY* 🐾

A Child is Waiting Poignant and provocative story of teachers in an institution for the mentally retarded. Fine performances include Lancaster as the institution's administrator and Garland as a teacher who cares too much for everyone's good. Cassavetes incorporates footage using handicapped children as extras—not entirely seamlessly—providing a sensitive and candid edge.
1963 105m/B Burt Lancaster, Judy Garland, Gena Rowlands, Steven Hill, Bruce Ritchey; *Dir:* John Cassavetes. VHS, Beta $19.98 *MGM* 🐾🐾🐾

The Children After a schoolbus passes through radioactive fog, the children inside assume the ability to incinerate whoever they hug. Not as unusual as you might hope it would be.
1980 (R) 89m/C Martin Shaker, Gale Garnett, Gil Rogers; *Dir:* Max Kalmanowicz. VHS, Beta $14.95 *VES, FCT* 🐾½

The Children Kingsley stars as a middle-aged engineer who finds himself caring for seven children in this touching love story from novelist Edith Wharton. Shot on location in Venice, the cinematography is excellent and the cast is superb.
1991 90m/C Ben Kingsley, Kim Novak, Karen Black, Geraldine Chaplin, Britt Ekland, Rupert Graves, Rosemary Leach, Donald Sinden, Robert Stephens, Joe Don Baker; *Dir:* Tony Palmer. VHS $89.95 *HMD*

Children of An Lac Three women attempt to rescue hundreds of Vietnamese orphans before Saigon falls to the communists. The film derives actress Balin's actual wartime experience. Made for television.
1980 100m/C Shirley Jones, Ina Balin, Beulah Quo, Alan Fudge, Ben Piazza, Lee Paul, Kieu Chinh, Vic Diaz. VHS, Beta $14.95 *FRH* 🐾🐾½

The Children Are Watching Us Sobering drama of a family dissolution as seen by a child. Worthy example of Italian neo-realism marks the first collaboration between Zavattini and De Sica. In Italian with subtitles.
1944 92m/B *IT* Luciano de Ambrosis, Isa Pola, Emilio Cigoli; *Dir:* Vittorio DeSica. VHS, Beta $79.95 *FCT* 🐾🐾🐾

Children of the Corn A young couple lands in a small Iowa town where children appease a demon by murderously sacrificing adults. Time to move or dispense with some major spankings. Infrequently scary. Another feeble attempt to translate the horror of a Stephen King book to film.
1984 (R) 93m/C Peter Horton, Linda Hamilton, R.G. Armstrong, John Franklin, Courtney Gains, Robbie Kiger; *Dir:* Fritz Kiersch. VHS, Beta, LV $9.98 *SUE, STE* 🐾½

Children in the Crossfire When youngsters from war-torn Northern Ireland spend a summer in America with children from the Republic of Ireland, both sides find their nationalistic prejudices falling away. But a great challenge awaits them when they return home from their summer of fun. Nicely done drama made for television.
1984 96m/C Charles Haid, Karen Valentine, Julia Duffy, David Hoffman; *Dir:* George Schaefer. VHS, LV $79.98 *LIV, VES* 🐾🐾½

Children of Divorce Made-for-television drama shows divorce from the children's point of view. Standard fare.
1980 96m/C Barbara Feldon, Lance Kerwin, Stacey Nelkin, Billy Dee Williams; *Dir:* Joanna Lee. VHS, Beta $39.95 *IVE* 🐾½

Children of the Full Moon A young couple find themselves lost in a forest that is the home of a family of werewolves. Earns praise only for brevity. Even bad movie aficionados will want to skip this howler.
1984 60m/C *GB* Christopher Cazenove, Celia Gregory, Diana Dors, Jacof Witken, Robert Urquhart; *Dir:* Tom Clegg. VHS, Beta $29.95 *IVE* 🐾

Children of a Lesser God Based upon the play by Mark Medoff, this sensitive, intelligent film deals with an unorthodox speech teacher at a school for the deaf, who falls in love with a beautiful and rebellious ex-student. Inarguably romantic; the original stage production won the Best Play Tony in 1980. Hurt and Matlin reportedly continued their romance off-screen as well. Took one Oscar and was nominated for Best Picture, Actor (Hurt), and Supporting Actress (Laurie).
1986 (R) 119m/C William Hurt, Marlee Matlin, Piper Laurie, Philip Bosco, E. Katherine Kerr; *Dir:* Randa Haines. Academy Awards '86: Best Actress (Matlin); Berlin Film Festival '86: Silver Bear. VHS, Beta, LV, 8mm $19.95 *PAR, BTV, JCF* 🐾🐾🐾½

Children of Paradise (Les Enfants du Paradis) Considered by many to be the greatest film ever made, certainly one of the most beautiful. In the Parisian theatre district of the 1800s an actor falls in love with a seemingly unattainable woman. Although circumstances keep them apart, their love never dies. Produced in France during WWII right under the noses of Nazi occupiers; many of the talent (including the writer, poet Jacques Prevert) were active resistance fighters. Laserdisc version features: interview with director Marcel Carne, production photos, analysis of influences of French painting on the film, and audio essay by film authority Brian Stonehill.
1944 188m/B *FR* Jean-Louis Barrault, Arletty, Pierre Brasseur, Maria Casares; *Dir:* Marcel Carne. VHS, LV $79.95 *HMV, VYG, CRC* 🐾🐾🐾🐾

Children of Rage High-minded and long-winded drama dealing with the Israeli/Palestinian war and terrorism from the viewpoint of an Israeli doctor yearning for peace.
1975 (PG) 106m/C *GB IS* Simon Ward, Cyril Cusack, Helmut Griem, Olga Georges-Picot; *Dir:* Arthur Seidelman. VHS, Beta $69.95 *ACA* 🐾🐾½

Children of Sanchez Mexican man attempts to provide for his family with very little except faith and love. U.S./Mexican production based on Oscar Lewis' novel. Music by Chuck Mangione.
1979 103m/C *MX* Anthony Quinn, Dolores Del Rio, Katy Jurado, Lupita Ferrer; *Dir:* Hall Bartlett. VHS, Beta $39.95 *MON* 🐾🐾

Children Shouldn't Play with Dead Things A band of foolhardy hippie filmmakers on an island cemetary skimp on special effects by using witchcraft to revive the dead. The plan works. Soon the crew has an island full of hungry ghouls to contend with. Film strives for yucks, frequently succeeds. A late night fave, sporting some excellent dead rising from their grave scenes as well as a selection of groovy fashions. Screenwriter/star Ormsby went on to write the remake of "Cat People," while director Clark would eventually helm "Porky's."

1972 85m/C Alan Ormsby, Valerie Mamches, Jeff Gillen, Anya Ormsby, Paul Cronin, Jane Daly, Roy Engelman, Robert Philip, Bruce Solomon, Alecs Baird, Seth Sklarey; *W/Dir:* Benjamin (Bob) Clark. **VHS, Beta $29.95** *MPI* 🐾🐾½

The Children of Theatre Street
Absorbing documentary follows three children attending the renowned Kirov Ballet School. A show for all ages that may be especially enjoyed by kids who are interested in ballet. Kelly's last film appearance; Oscar nominated for Best Documentary.
1978 92m/C Kirov Ballet; *Dir:* Robert Dornhelm, Earle Mack; *Nar:* Grace Kelly. **VHS, Beta $39.95** *MVD, PBC, KUL* 🐾🐾🐾½

The Children of Times Square
A mother pursues her son who has fled home and joined a gang of drug dealers in New York City's seedy Times Square. Ordinary television film despite strong casting of Cassidy and Rollins.
1986 95m/C Joanna Cassidy, Howard E. Rollins Jr., Brandon Douglas, David Ackroyd, Griffin O'Neal, Danny Nucci, Larry B. Scott; *Dir:* Curtis Hanson. **VHS, Beta $14.95** *FRH* 🐾🐾

Children of the Wild
Adventure story about a pack of dogs surviving in the Rocky Mountain wilderness. Splendid scenery and some interesting sequences.
1937 65m/B Patsy Moran. **VHS** *VYY* 🐾🐾½

Children's Carol
The winter solstice brings no special joy to Walton's mountain; WWII has taken many men, with short wave reports indicating the Nazi terror spreading across Europe. But huddled in the glow of Walton's barn, the children rediscover the true meaning of Christmas.
1980 94m/C Judy Norton-Taylor, Jon Walmsley, Mary McDonough, Eric Scott, Kami Cotler, Joe Conely, Ronnie Clare, Leslie Winston, Peggy Rea; *Dir:* Lawrence Dobkin. **VHS, Beta $59.95** *LHV* 🐾🐾

The Children's Hour
The teaching careers of two women are ruined when girls begin circulating vicious rumors. Only an occasionally taut drama despite forceful handling of a lesbian theme. Updated version of Lillian Hellman's play (adapted by Hellman) is more explicit, but less suspenseful in spite of excellent performances from the talented cast. Remake of Wyler's own "These Three." Oscar nominations for Bainter, costume design and sound.
1962 107m/C Shirley MacLaine, Audrey Hepburn, James Garner, Miriam Hopkins, Veronica Cartwright; *Dir:* William Wyler. **VHS $19.98** *MGM, BTV, FCT* 🐾🐾

Children's Island
A two-volume saga, sold in one package, about children who, after being shipwrecked on an island during World War II, begin a democratic mini-society.
1984 100m/C VHS, Beta $59.95 *SVS* 🐾🐾

Child's Christmas in Wales
A film of Dylan Thomas's Christmas story about an old, traditional Christmas in Wales circa 1910.
1988 55m/C Mathonwy Reeves; *Dir:* Don McBrearty; *Nar:* Denholm Elliott. **VHS, Beta $14.98** *LIV, VES* 🐾🐾

Child's Play
A boy discovers that his new doll named Chucky is actually the embodiment of a deranged killer. His initially skeptical mom and a police officer come around after various killings. Exciting, if somewhat moronic, fare, with fine special effects. Followed by "Child's Play 2."
1988 (R) 95m/C Catherine Hicks, Alex Vincent, Chris Sarandon, Dinah Manoff, Brad Dourif; *Dir:* Tom Holland. **VHS, Beta, LV $19.95** *MGM* 🐾🐾½

Child's Play 2
Chuckie lives. The basic doll-on-a-rampage story, a metaphor for the Reagan years, lives on in the sequel (you remember: somehow guy-doll Chuckie made it past the quality control people with a highly inflammable temper). A little dotty from playing with dolls, young Vincent finds himself fostered by two new parents, and plagued by an obnoxious and very animated doll that fosters ill will toward all. What's worse is the doll is transmigratory, craving the boy's body as his next address. Chef d'effects Kevin Yagher's new toy is bad, real bad, and so are the other part two FX. Little Chuckie's saga, however grows a tad tiresome. A bad example for small children.
1990 (R) 84m/C Alex Vincent, Jenny Agutter, Gerrit Graham, Christine Elise, Grace Zabriskie; *Dir:* John Lafia; *Voices:* Brad Dourif. **VHS, Beta, LV $19.98** *MCA* Woof!

Child's Play 3
The possessed doll Chucky returns to life again in search of a new child to control, and finds one at a military school filled with the usual stereotypes. Gory sequel in a mostly awful series. "Don't F#@k With the Chuck" was its catch phrase; that should indicate the level of this crap.
1991 (R) 89m/C Justin Whalin, Perrey Reeves, Jeremy Sylvers, Peter Haskell, Dakin Matthews, Travis Fine, Dean Jacobson, Matthew Walker, Andrew Robinson; *Dir:* Jack Bender; *Voices:* Brad Dourif. **VHS, Beta, LV $91.95** *MCA* 🐾

Chill Factor
A television reporter for an investigative news show uncovers evidence of an international conspiracy. Plot twists abound.
1990 (R) ?m/C Paul Williams, Patrick Macnee, Andrew Prine, Carrie Snodgress, Patrick Wayne, Gary Crosby; *Dir:* David L. Stanton. **VHS** *QUE* 🐾🐾

Chillers
Travellers waiting for a bus are besieged by carnivorous zombies and voracious vampires. Yikes.
1988 90m/C Jesse Emery, Marjorie Fitzsimmons, Laurie Pennington, Jim Wolfe, David Wohl; *W/Dir:* Daniel Boyd. **VHS, Beta $9.95** *SIM, PSM* 🐾

The Chilling
Corpses preserved in a deep freeze come alive and plague Kansas City as flesh-chomping zombies. Utterly worthless film doesn't even provide convincing effects.
1989 (R) 91m/C Linda Blair, Dan Haggerty, Troy Donahue, Jack A. De Rieux, Ron Vincent; *Dir:* Deland Nuse, Jack A. Sunseri. **VHS $89.95** *CHV* Woof!

Chilly Scenes of Winter
Quirky comedy with a cult following about a man obsessed with regaining the love of a former girlfriend who is now married. He must also deal with an insane mom and various other problems. Strong, subtle performances from Heard and Riegert. Hurt is somewhat less satisfying as the supposedly fascinating woman. Released with a different, inferior ending as "Head Over Heels." Adapted by Joan Micklin Silver from Ann Beatty's first novel. Watch for Beatty's cameo as a waitress.
1979 (PG) 96m/C John Heard, Mary Beth Hurt, Peter Riegert, Griffin Dunne; *Dir:* Joan Micklin Silver. **VHS, Beta $59.95** *MGM* 🐾🐾🐾

Chimes at Midnight
Classic tragedy—derived by Welles from five Shakespeare plays—about a corpulent blowhard and his friendship with a prince. Crammed with classic sequences, including a battle that is both realistic and funny. The love scene between massive Welles and a nonetheless willing Moreau also manages to be both sad and amusing. Great performances all around, but Welles understandable dominates. The film's few flaws (due to budget problems) are inconsequential before considerable strengths. This one ranks among Welles', and thus the entire cinema's, very best. Also known as "Falstaff."
1967 115m/B *SP SI* Orson Welles, Keith Baxter, Jeanne Moreau, John Gielgud, Marina Vlady, Margaret Rutherford, Norman Rodway, Fernando Rey; *W/Dir:* Orson Welles. **VHS, Beta** *FCT* 🐾🐾🐾🐾

China Beach
The television movie/pilot that launched the acclaimed series, concerning American military women behind the lines during the chaos of the Vietnam War. Nurse McMurphy has only seven days left on her first tour of duty, and in spite of her hate of the brutality of the war, she has mixed feelings about leaving her friends and her work. A USO singer on a one week tour finds her search for men brings the war too close for comfort. Cherry, a new Red Cross volunteer, shows up at the Beach. Skillfully introduces the viewer to all the characters, their problems and joys, without melodrama or redundancy.
1988 95m/C Dana Delaney, Chloe Webb, Robert Picardo, Nan Woods, Michael Boatman, Marg Helgenberger, Tim Ryan, Concetta Tomei, Jeff Kober, Brian Wimmer; *Dir:* Rod Holcomb. **VHS, Beta $79.95** *WAR* 🐾🐾🐾

China Gate
A band of multi-national troops follows a French officer against a communist stronghold in Indochina. Conventional fare bolstered considerably by director Fuller's flair for action. Weak male leads, but Dickinson shines.
1957 97m/B Gene Barry, Angie Dickinson, Nat King Cole, Paul Dubov, Lee Van Cleef, George Givot, Marcel Dalio; *W/Dir:* Samuel Fuller. **VHS $19.98** *REP* 🐾🐾🐾

China Girl
An Italian-American boy and a Chinese-American girl romance despite a war between gangs in their respective NYC communities. This often violent drama is enriched by director Ferrara's slick, high-energy approach.
1987 (R) 90m/C Richard Panebianco, James Russo, Sari Chang, Russell Wong, David Caruso, Joey Chin; *Dir:* Abel Ferrara. **VHS, Beta, LV $79.98** *VES* 🐾🐾½

The China Lake Murders
A made for cable crazy cop on the loose fable in which a serial killer disguised as a highway patrolman practices population control on the inhabitants of a small town in the Mojave desert. Meanwhile the local sheriff (Skerritt), unwittingly befriends the killer-patrolman. Written by N.D. Schreiner.
1990 (PG-13) 100m/C Tom Skerritt, Michael Parks, Nancy Everhard, Lauren Tewes, Bill McKinney, Loni Chapman; *Dir:* Alan Metzger. **VHS, Beta $79.95** *MCA* 🐾🐾½

China O'Brien
A gorgeous police officer with martial arts expertise returns home for a little R & R, but finds she has to kick some major butt instead. Violent, dimwitted action drama proves only that cleavage can be macho too.
1988 (R) 90m/C Cynthia Rothrock, Richard Norton, Patrick Adamson, David Blackwell, Steven Kerby, Robert Tiller, Lainie Watts, Keith Cooke; *Dir:* Robert Clouse. **VHS $89.95** *IMP, BTV* 🐾½

China O'Brien 2 The unbreakable China, now a sheriff, battles the standard-issue Vietnam-vet druglord who invades her little town. Made back-to-back with the first film (note that both pics' bad guys have the same gang of henchman!), and in both films Rothrock is basically invincible, so there's no suspense.
1989 (R) 85m/C Cynthia Rothrock, Richard Norton, Keith Cook; **Dir:** Robert Clouse. **VHS $89.95** IMP 🐾½

China Seas The captain of a commercial steamship on the China route has to fight off murderous Malay pirates, a spurned woman, and a raging typhoon to reach port safely. Fast-moving romantic action taken from Crosbie Garstin's novel.
1935 89m/B Clark Gable, Jean Harlow, Wallace Beery, Rosalind Russell, Lewis Stone, C. Aubrey Smith, Dudley Digges, Robert Benchley; **Dir:** Tay Garnett. **VHS, Beta $19.98** MGM 🐾🐾🐾

China Sky An American doctor fights alongside Chinese guerillas against the Japanese during WWII. Drama, to its credit, often opts for character conflict instead of warfare, but this makes it merely dull instead of cliched. Adapted from a Pearl S. Buck novel.
1944 78m/B Randolph Scott, Ellen Drew, Ruth Warrick, Anthony Quinn, Carol Thurston, Richard Loo, Philip Ahn; **Dir:** Ray Enright. **VHS, Beta $19.98** TTC 🐾🐾

The China Syndrome A somewhat unstable executive at a nuclear plant uncovers evidence of a concealed accident, takes drastic steps to publicize the incident. Lemmon is excellent as the anxious exec, while Fonda and Douglas are scarcely less distinguished as a sympathetic TV journalist and camera operator, respectively. Tense, prophetic thriller that ironically preceded the Three Mile Island accident by just a few months. Fonda and Lemmon received Oscar nominations, as did the adapted screenplay, art direction and set decoration. Produced by Douglas.
1979 (PG) 123m/C Jane Fonda, Jack Lemmon, Michael Douglas, Scott Brady, James Hampton, Peter Donat, Wilford Brimley, James Karen; **Dir:** James Bridges. British Academy Awards '79: Best Actor (Lemmon), Best Actress (Fonda); National Board of Review Awards '79: 10 Best Films of the Year. **VHS, Beta, LV $12.98** COL, PIA 🐾🐾🐾½

Chinatown A private detective finds himself overwhelmed in a scandalous case involving the rich and powerful of Los Angeles. Gripping, atmospheric mystery excels in virtually every aspect, with strong narrative drive and outstanding performances from Nicholson, Dunaway, and Huston. Director Polanski also appears in a suitable unsettling cameo. Fabulous. A sneaky, snaking delight filled with seedy characters and plots-within-plots. Screenplay by Robert Towne grabbed an Oscar, while the movie was nominated for Best Picture, Director, Actor (Nicholson), Actress (Dunaway), Music (Jerry Goldsmith), and Photography (John A. Alonso). Followed more than 15 years later by "The Two Jakes."
1974 (R) 131m/C Jack Nicholson, Faye Dunaway, John Huston, Diane Ladd, John Hillerman, Burt Young, Perry Lopez, Darrell Zwerling; **Dir:** Roman Polanski. Academy Awards '74: Best Original Screenplay; British Academy Awards '74: Best Actor (Nicholson), Best Director (Polanski); Edgar Allan Poe Awards '74: Best Screenplay; Golden Globe Awards '75: Best Film—Drama; National Board of Review Awards '74: 10 Best Films of the Year; New York Film Critics Awards '74: Best Actor (Nicholson); National Society of Film Critics

Awards '74: Best Actor (Nicholson). **VHS, Beta, LV, 8mm $19.95** PAR 🐾🐾🐾🐾

Chinatown After Dark Ludicrous, inane melodrama about a white girl raised by the Chinese.
1931 50m/B Rex Lease, Barbara Kent, Carmen Myers; **Dir:** Ralph M. Like. **VHS $12.95** SNC, NOS, DVT 🐾

Chinatown Connection For some inexplicable reason, local drug dealers start selling poisoned cocaine. Two renegade cops discover the super-secret organization doing the poisoning, arm themselves heavily, and kill lots of people while putting an end to the poisonings. Violent, foolish film with dull performances.
1990 (R) 94m/C Bruce Ly, Lee Majors, Pat McCormick, Art Camacho, Susan Frailey, Scott **Dir:** Jean-Paul Ouellette. **VHS $79.95** SOU 🐾

The Chinatown Kid Chased by attackers in his native Hong Kong, a young man decides to try the safety of San Francisco.
1990 90m/C VHS $69.98 SOU 🐾

Chinese Boxes An American man in Berlin is framed in a murder and becomes caught in an international web of crime and intrigue. Low budget but interesting thriller.
1984 87m/C GE GB Will Patton, Adelheid Arndt, Robbie Coltrane; **Dir:** Christopher Petit. **VHS, Beta $69.98** LIV, VES 🐾🐾½

Chinese Connection A martial arts expert tracks sadistic brutes who slew his instructor. Wild action sequences provide a breathtaking view of Lee's skill. Dubbed.
1973 (R) 90m/C CH Bruce Lee, James Tien, Robert Baker; **Dir:** Lo Wei. **VHS, Beta $19.98** FOX, MRV, VCD 🐾🐾½

Chinese Connection 2 A martial-arts expert learns that the school where he trained is now run by an unappealing master. Conflict ensues. Bruce Li is not Bruce Lee, but he is fairly good in action portions of this otherwise dull, sloppy venture.
1984 96m/C Bruce Li. **VHS, Beta $39.95** TWE 🐾🐾

Chinese Gods Animated story of Chinese mythology depicts the battles and rivalries occurring circa 1000 B.C. in the period of the Shang Dynasty between Cruel King Cheo's troops and the Marquis Hsi-pa.
1980 (G) 90m/C VHS, Beta $19.95 GEM 🐾🐾

Chinese Roulette A host of unappealing characters convene at a country house for sexual shenanigans and a cruel game masterminded by a sadistic crippled girl. Cold effort from German master Fassbinder. In German with English subtitles.
1986 96m/C GE Anna Karina, Margit Carstensen, Ulli Lommel, Brigitte Mira; **W/Dir:** Rainer Werner Fassbinder. **VHS, Beta, LV $79.95** CVC 🐾🐾½

Chinese Stuntman An American in Hong Kong gets caught up in a web of intrigue, and kicks his way out.
197? 85m/C Bruce Li, Ho Chung Tao, John Ladalski, Dan Inosanto. **VHS, Beta $39.95** TWE 🐾

Chinese Web A Spiderman adventure in which Spidey becomes entwined in international intrigue and corrupt officials.
1978 95m/C Nicholas Hammond, Robert F. Simon, Rosalind Chao, Ted Danson; **Dir:** Donald McDougall. **VHS, Beta $24.98** FOX 🐾

Chino A half-Indian horse rancher struggles to maintain his livelihood in this spaghetti western. Not among Bronson's stronger—that is, more viscerally effective—films. Adapted from Lee Hoffman's novel. Also released as "The Valdez Horses."
1975 (PG) 97m/C IT Charles Bronson, Jill Ireland, Vincent Van Patten; **Dir:** John Sturges. **VHS, Beta $59.95** MPI, MRV 🐾🐾

Chip 'n' Dale & Donald Duck Another entry in the Disney series of cartoon collections. Includes "Working for Peanuts," "Donald Applecare," and "Dragon Around."
1939 48m/C Dir: Walt Disney. **VHS, Beta** DIS 🐾

The Chipmunk Adventure The three animated chipmunks, Simon, Theodore and Alvin, get involved with a diamond smuggling ring. Original songs. In HiFi Stereo.
1987 76m/C Dir: Janice Karman. **VHS, Beta, LV** LHV, WAR 🐾🐾

The Chisholms If you're into oaters, here's a high fiber six-hour serial: Chisolm (Preston) leads the family as they head west and find trouble with a capital "T" en route from Virginia to Californ-i-a. Originally a network miniseries, it's now a two-cassette rentable; adapted by Evan Hunter from his novel.
1979 300m/C Robert Preston, Rosemary Harris, Brian Keith, Ben Murphy, Charles Frank; **Dir:** Mel Stuart. **VHS, Beta $69.95** USA, IVE 🐾🐾½

Chisum Cattle baron faces various conflicts, including a confrontation with Billy the Kid. Lame Wayne vehicle contributes nothing to exhausted western genre.
1970 (G) 111m/C John Wayne, Forrest Tucker, Geoffrey Deuel, Christopher George, Ben Johnson, Bruce Cabot, Patric Knowles, Richard Jaeckel, Glenn Corbett; **Dir:** Andrew V. McLaglen. **VHS, Beta $19.98** WAR, TLF, BUR 🐾½

Chitty Chitty Bang Bang An eccentric inventor spruces up an old car and, in fantasy, takes his children to a land where the evil rulers have forbidden children. Songs include "Chitty Chitty Bang Bang," "Hushabye Mountain," and "Truly Scrumptious." Poor special effects and forgettable score stall effort. Loosely adapted by Roald Dahl and Hughes from an Ian Fleming story. Title song was Oscar nominated.
1968 (G) 142m/C GB Dick Van Dyke, Sally Ann Howes, Lionel Jeffries, Gert Frobe, Anna Quayle, Benny Hill; **Dir:** Ken Hughes. **VHS, Beta, LV $19.98** FOX, MGM, TLF 🐾🐾

Chloe in the Afternoon A married man finds himself inexplicably drawn to an ungainly young woman. Sixth of the "Moral Tales" series is typical of director Rohmer's talky approach. Not for all tastes, but rewarding for those who are drawn to this sort of thing. In French with English subtitles.
1972 (R) 97m/C FR Bernard Verley, Zouzou, Francoise Verley, Daniel Ceccaldi, Malvina Penne, Babette Ferrier, Suze Randall, Marie-Christine Barrault; **Dir:** Eric Rohmer. National Board of Review Awards '72: 5 Best Foreign Films of the Year; New York Film Festival '72: Best Actor. **VHS, Beta $59.95** TAM, FCT 🐾🐾🐾

Chocolat A woman recalls her childhood spent in French West Africa and the unfulfilled sexual tension between her mother and black servant. Vivid film provides a host of intriguing characters and offers splendid panoramas of rugged desert landscapes. Profound, if somewhat personal filmmaking from novice director Denis. In French with English subtitles.

1988 (PG-13) 105m/C Mireille Perrier, Emmet Judson Williamson, Cecile Ducasse, Giulia Boschi, Francois Cluzet, Isaach de Bankole, Kenneth Cranham; *Dir:* Claire Denis. VHS, Beta, LV $79.98 ORI, TAM 🎞🎞🎞½

The Chocolate Soldier Dull musical in which an opera star tests his wife's fidelity. Lovers of the musical genre will find that this one has too much talking, not enough singing. Lovers of fine films will realize that more singing would hardly improve things. Oscar nominated for photography and music. Based loosely on Molnar's play "The Guardsman."
1941 102m/B Nelson Eddy, Rise Stevens, Nigel Bruce, Florence Bates, Dorothy Gilmore, Nydia Westman; *Dir:* Roy Del Ruth. VHS $29.95 MGM 🎞

The Chocolate War An idealistic student and a hardline headmaster butt heads at a Catholic boys' school over an unofficial candy business in this tense, unsettling drama. Glover is notable in his familiar villain role, and Gordon is effective in his first effort as director. Based on the Robert Cormier novel.
1988 (R) 95m/C John Glover, Jenny Wright, Wally Ward, Bud Cort, Ilan Mitchell-Smith; *Dir:* Keith Gordon. VHS, Beta, LV $89.98 MCG 🎞🎞🎞

The Choice A young woman must make the choice between aborting or keeping her baby in this fairly insipid television drama.
1981 96m/C Susan Clark, Jennifer Warren, Mitchell Ryan, Largo Woodruff; *Dir:* David Greene. VHS, Beta $39.95 IVE 🎞🎞

Choice of Arms A gangster's rural retirement is undone by a dimwitted criminal in this sometimes gripping drama. Montand and Deneuve are serviceable in undemanding roles of the gangster and his glamorous wife; Depardieu is more impressive as the troublemaking punk. In French with English subtitles.
1983 114m/C *FR* Gerard Depardieu, Catherine Deneuve, Yves Montand; *Dir:* Alain Corneau. VHS, Beta $59.95 TAM, MED 🎞🎞½

Choice of Weapons Ex-cop trying to solve murders focuses on a curious group of 20th-century men who live within a 12th-century fantasy—jousting for sport and chivalrous honor. Offbeat lance thruster notable for incongruity and all-star cast. Released as "A Dirty Knight's Work," and also known as "Trial by Combat."
1976 88m/C *GB* David Birney, Peter Cushing, Donald Pleasence, Barbara Hershey, John Mills, Margaret Leighton; *Dir:* Kevin Connor. VHS, Beta $42.95 PGN 🎞🎞

Choices A hearing-impaired athlete suffers alienation when banned from the football squad. What? Controversial covergirl Moore is a supporting player in this, her first film.
1981 90m/C Paul Carafotes, Victor French, Lelia Goldoni, Val Avery, Dennis Patrick, Demi Moore; *Dir:* Rami Alon. VHS, Beta $49.95 VHE, HHE 🎞½

The Choirboys Thoroughly mediocre production about overbearing L.A. cops and their off-hours handling of job stress. Few of the stron cast emerge unscathed. Based on the Joseph Wambaugh novel.
1977 (R) 120m/C Charles Durning, Louis Gossett Jr., Perry King, Clyde Kusatsu, Stephen Macht, Tim McIntire, Randy Quaid, Chuck Sacci, Don Stroud, James Woods, Burt Young, Robert Webber, Barbara Rhoades, Vic Tayback, Blair Brown, Charles Haid, Jim Davis; *Dir:* Robert Aldrich. VHS, Beta $79.95 MCA 🎞

Choke Canyon An environmentalist thwarts the henchmen of an industrialist eager to coverup a nuclear dump in a picturesque canyon. Good intentions do not necessarily make for a good movie.
1986 (PG) 95m/C Stephen Collins, Bo Svenson, Lance Henriksen; *Dir:* Chuck Bail. VHS, Beta $79.95 MED 🎞🎞

C.H.O.M.P.S. Comedy in which a youthful inventor and a popular robot guard dog become the target of a business takeover. Harmless but unfunny and unfun.
1979 (PG) 90m/C Jim Backus, Valerie Bertinelli, Wesley Eure, Conrad Bain, Chuck McCann, Red Buttons; *Dir:* Don Chaffey. VHS, Beta $19.98 NO 🎞½

Choose Me Comedy-drama about sad, lonely, and often quirky characters linked to an unlikely L.A. radio sex therapist. Moody, memorable fare features especially strong playing from Bujold as a sexually inexperienced sex therapist and Warren as one of her regular listeners. Typically eccentric fare from director Rudolph.
1984 (R) 106m/C Keith Carradine, Genevieve Bujold, Lesley Ann Warren, Rae Dawn Chong, John Larroquette, John Considine; *W/Dir:* Alan Rudolph. VHS, Beta, LV $9.98 MED 🎞🎞🎞½

Chopper Chicks in Zombietown Tough but sexy Chopper Chicks show up in that American vacation mecca, Zombietown, for a little rest and relaxation. Little do they know that a mad mortician has designs on turning our hot heroines into mindless zombie slaves. Can the buxom biker babes thwart the evil embalmer before it's too late, or will they abandon their Harleys to shuffle about in search of human flesh? From the Troma Team.
1991 (R) 86m/C Jamie Rose, Catherine Carlen, Lycia Naff, Vicki Frederick, Kristina Loggia, Martha Quinn, Don Calfa; *W/Dir:* Dan Hoskins. Fatasporto Film Festival of Portugal & Spain '91: Special Jury Prize. VHS, LV $79.95 COL 🎞🎞

The Choppers Naw, not a fable about false teeth. Teen punk Hall operates a car theft ring made up of fellow punksters. Rock'n'roll tunes by the Hall-meister include the much-overlooked "Monkey in my Hatband."
1961 66m/B Arch Hall Jr., Marianne Gaba, Robert Paget, Tom Brown, Rex Holman, Bruno Ve Sota; *Dir:* Leigh Jason. VHS, Beta $16.95 SNC 🎞

Chopping Mall A freak electric storm unleashes killer security robots on a band of teens trapped inside the mall. Nobody shops. Premise undone by obscure humor, lack of flairful action, or horror. Originally titled "Killbots." Updated imitation of the 1973 TV movie "Trapped."
1986 (R) 77m/C Paul Bartel, Mary Woronov, Barbara Crampton, John Terlesky, Dick Miller; *Dir:* Jim Wynorski. VHS, Beta $19.98 LIV, LTG 🎞½

A Chorus of Disapproval Adaptation of prolific Alan Ayckbourn's play about a withdrawn, somewhat dimwitted British widower who attempts social interaction by joining community theater, then finds himself embroiled in romantic shenanigans and theatrical intrigue. Irons is fine in the lead, but Hopkins sparkles in the more spectacular role of the musical production's demanding but beleaguered director. Seagrove is impressive as an amoral sexpot. Sharper focus from director Winner would have improved this one, but the film is fun even when it isn't particularly funny.
1989 (PG) 105m/C *GB* Jeremy Irons, Anthony Hopkins, Jenny Seagrove, Lionel Jefferies, Patsy Kensit, Gareth Hunt, Prunella Scales, Sylvia Syms, Richard Briers, Barbara Ferris; *W/Dir:* Michael Winner. VHS, Beta $19.98 SOU 🎞½

A Chorus Line A range of performers reveal their insecurities and aspirations while auditioning before a hardnosed director in this adaptation of the popular, overblown Broadway musical. Singing and dancing is rarely rousing. Director Attenborough probably wan't the right choice for this one. Music by Marvin Hamlisch. Songs include "Dance 10, Looks 3," "What I Did For Love," "At the Ballet," "I Can Do That," and the rousing finale, "One." Oscar nominated for editing.
1985 (PG-13) 118m/C Michael Douglas, Audrey Landers, Gregg Burge, Alyson Reed, Janet Jones, Michael Blevins, Terrence Mann, Cameron English, Vicki Fredrick, Nicole Fosse, Michelle Johnston; *Dir:* Richard Attenborough. VHS, Beta, LV, SVS, 8mm $14.98 SUE, SUP 🎞🎞

The Chosen The executive of a nuclear power facility realizes that his son is the Anti-Christ bent on the world's destruction. This truly horrible film provides nothing in terms of entertainment. It's rarely even laughably bad. Woof! Original title: "Holocaust 2000."
1977 (R) 102m/C Kirk Douglas, Simon Ward, Agostina Belli, Anthony Quayle, Virginia McKenna; *Dir:* Alberto DeMartino. VHS, Beta $59.98 FOX Woof!

Chris Elliott: FDR, A One-Man Show/Action Family David Letterman alumni Elliott stars in parodies of one-man theater pieces and television sitcoms.
1989 60m/C Chris Elliott, Marv Albert, Bob Elliott; *Dir:* Gary Weis. VHS, Beta $59.99 HBO

Christ Stopped at Eboli Subdued work about an anti-Fascist writer exiled to rural Italy in the 1930s. Excellent performances from the lead Volante and supporting players Paps and Cuny. Slow, contemplative film is probably director Rosi's masterpiece. Also known as "Eboli." Adapted from Carlo Levi's book. In Italian with English subtitles.
1979 118m/C *IT FR* Gian Marie Volonte, Irene Papas, Paolo Bonacelli, Francois Simon, Alain Cuny, Lea Massari; *Dir:* Francesco Rosi. National Board of Review Awards '80: 5 Best Foreign Films of the Year; Moscow Film Festival '79: Grand Prize. VHS, Beta $59.95 COL, APD, TAM 🎞🎞🎞

Christabel A condensed version of the BBC mini-series based on the true-life World War II exploits of Christabel Bielenberg, a British woman who battled to save her German husband from the horrors of Ravensbruck concentration camp.
1989 148m/C Elizabeth Hurley, Stephen Dillon, Nigel le Vaillant, Geoffrey Palmer. VHS, Beta $39.98 FOX, TAM 🎞🎞

Christian the Lion The true story of Christian, a lion cub raised in a London zoo, who is transported to Africa to learn to live with other lions. With the principals from "Born Free."
1976 (G) 87m/C *GB* Virginia McKenna, Bill Travers, George Adamson, James Hill; *Dir:* Bill Travers. VHS, Beta $29.95 UNI 🎞🎞½

Christiane F. Gripping, visually impressive story of a bored German girl's decline into drug use and prostitution. Based on a West German magazine article. Sobering and dismal look at a milieu in which innocence and youth have run amok. The film's impact is only somewhat undermined by poor dubbing. Bowie appears in a concert sequence.

1982 (R) 120m/C *GE* Natja Brunkhorst, Thomas Haustein, David Bowie; *Dir:* Uli Edel. **VHS, Beta** **$69.95** *MED, GLV* 🎞🎞🎞

Christina Gloomy suspense fare in which a beautiful woman pays a forlorn fellow $25,000 to marry her, then disappears as he begins to actually fall in love with her. His search takes him to various gothic settings. Intriguing, but not really fulfilling. Parkins, however, is appropriately mysterious in the lead.
1974 (PG) 95m/C *CA* Barbara Parkins, Peter Haskell; *Dir:* Paul Krasny. **VHS, Beta $59.95** *MPI* 🎞🎞 ½

Christine An unassuming teen gains possession of a classic auto equipped with a murderous will. The car more than returns the care and consideration its owner provides it. Are you listening GMC? Based on a novel by Stephen King.
1984 (R) 110m/C Keith Gordon, John Stockwell, Alexandra Paul, Robert Prosky, Harry Dean Stanton, Kelly Preston; *Dir:* John Carpenter. **VHS, Beta, LV, 8mm $14.95** *COL* 🎞🎞 ½

A Christmas Carol An early version of Dickens' classic tale about miser Scrooge, who is instilled with the Christmas spirit after a grim evening with some ghosts. Good playing from Owen as Scrooge and Lockhart as the hapless Bob Cratchett. Scary graveyard sequence too.
1938 70m/B Reginald Owen, Gene Lockhart, Terence Kilburn, Leo G. Carroll, Lynne Carver, Ann Rutherford; *Dir:* Edwin L. Marin. **VHS, Beta $19.95** *MGM* 🎞🎞🎞

A Christmas Carol A fine retelling of the classic tale about a penny-pinching holiday hater who learns appreciation of Christmas following a frightful, revealing evening with supernatural visitors. Perhaps the best rendering of the Dickens classic. "And God bless Tiny Tim!" Also known as "Scrooge."
1951 86m/B *GB* Alastair Sim, Kathleen Harrison, Jack Warner, Michael Hordern, Patrick Macnee, Mervyn Johns, Hermione Baddeley, Clifford Mollison, George Cole, Carol Marsh, Miles Malleson, Ernest Thesiger, Hattie Jacques, Peter Bull; *Dir:* Brian Desmond-Hurst. **VHS, Beta $19.95** *VCI, MLB* 🎞🎞🎞

Christmas Carol Musical version of the Dickens' classic about a stingy old man who is visited by three ghosts on Christmas Eve.
1954 54m/B Fredric March, Basil Rathbone, Ray Middleton, Bob Sweeney, Christopher Cook. **VHS, Beta** *CFV* 🎞🎞

Christmas Carol An animated version of the Charles Dickens classic. Hard to beat the Alastair Sim version, however.
1984 72m/C **VHS, Beta $19.95** *CVL* 🎞🎞

Christmas Carol Animated version of Charles Dickens' classic tale. The story of how Ebenezer Scrooge changed from a tyrant to a joyous human being one Christmas Eve.
1984 23m/C *Dir:* Chuck Jones. **VHS, Beta $19.95** *CHI* 🎞🎞 ½

Christmas Coal Mine Miracle Syrupy made-for-television story about miners struggling for survival in a collapsed mine on Christmas Eve. Also titled "Christmas Miracle in Caulfield, U.S.A."
1978 98m/C Kurt Russell, Melissa Gilbert, Andrew Prine, Mitchell Ryan, John Carradine, Barbara Babcock; *Dir:* Judd Taylor. **VHS, Beta $59.98** *FOX* 🎞🎞 ½

Christmas Comes to Willow Creek Mutually antagonistic brothers are enlisted to deliver Christmas gifts to an isolated Alaskan community. Can brotherly love be far off? Two leads played together in the rowdy TV series "Dukes of Hazzard." This is hardly an improvement.
1987 96m/C John Schneider, Tom Wopat, Hoyt Axton, Zachary Ansley, Kim Delaney; *Dir:* Richard Lang. **VHS, Beta $79.95** *JTC* 🎞🎞

Christmas in Connecticut Lightweight comedy about a housekeeping magazine's successful columnist who isn't quite the expert homemaker she presents herself to be. When a war veteran is invited to her home as part of a publicity gimmick, she must master the ways of housekeeping or reveal her incompetence. Stanwyck is winning in the lead role. Also available in a colorized version.
1945 101m/B Barbara Stanwyck, Reginald Gardiner, Sydney Greenstreet, Dennis Morgan, S.Z. Sakall, Una O'Connor, Robert Shayne, Joyce Compton; *Dir:* Peter Godfrey. **VHS, Beta, LV $19.95** *MGM, PIA* 🎞🎞🎞

A Christmas Gift Short animated films from Canada about the December holiday. Includes "The Great Toy Robbery," "The Energy Carol," Oscar nominated "Christmas Cracker" and more!
1990 65m/C *CA* **VHS, Beta $29.95** *WTA, FCT, EXP* 🎞🎞🎞

Christmas in July A young man goes on a spending spree when he thinks he's won a sweepstakes. Things take a turn for the worse when he finds out that it was all a practical joke. Powell provides a winning performance in this second film from comic master Sturges.
1940 67m/B Dick Powell, Ellen Drew, Raymond Walburn, William Demarest, Franklin Pangborn; *Dir:* Preston Sturges. **VHS, Beta, LV $29.95** *MCA, PIA, FCT* 🎞🎞🎞 ½

The Christmas Kid A desperado struggles to determine his true identity while stopping in a frontier town. Gunplay ensues. Don't bother with this one.
1968 87m/C *SP* Jeffrey Hunter, Louis Hayward, Gustavo Rojo, Perla Cristal, Luis Prendes; *Dir:* Sidney Pink. **VHS $59.95** *ACF* 🎞 ½

Christmas Lilies of the Field This made-for-television sequel to "Lilies of the Field" follows an ex-soldier who returns to the church he helped build. This time he sets out to build an orphanage.
1984 98m/C Billy Dee Williams, Maria Schell, Fay Hauser, Judith Piquet; *Dir:* Ralph Nelson. **VHS, Beta $29.95** *MPI* 🎞🎞 ½

The Christmas Messenger Brilliantly-colored animation is used to portray tales of Christmas folklore. A Reader's Digest presentation.
1975 25m/C *Voices:* Richard Chamberlain. **VHS, Beta $195.00** *PYR* 🎞🎞 ½

Christmas Raccoons The Raccoons have their very own special animated Christmas story to tell.
1984 30m/C *Nar:* Rich Little. **VHS, Beta $24.98** *SUE* 🎞🎞 ½

A Christmas to Remember Depression-era Minnesota farmer who has lost his son in WWI brings his city-bred grandson to the farm for a holiday visit. Somber, occasionally poignant film buoyed by Robards presence, and Saint's as well. Made for television.

1978 (G) 96m/C Jason Robards Jr., Eva Marie Saint, Joanne Woodward; *Dir:* George Englund. **VHS, Beta $69.95** *LTG, VES* 🎞🎞 ½

A Christmas Story Unlikely but winning comedy of a boy's single-minded obsession to acquire a Red Ryder BB-gun for Christmas. Particularly great sequence involving an impatient department-store Santa. Fun for everyone. Based on a story by Jean Shepherd.
1983 (PG) 95m/C Peter Billingsley, Darren McGavin, Melinda Dillon, Ian Petrella; *Dir:* Bob Clark. Genie Awards '84: Best Director (Clark), Best Screenplay. **VHS, Beta $19.95** *MGM* 🎞🎞🎞

The Christmas That Almost Wasn't Loathsome humbug decides to destroy Christmas forever by removing Santa Claus from the North Pole. Crude Italian-made children's film nonetheless remembered fondly by a generation of kids.
1966 (G) 95m/C *IT* Rossano Brazzi, Paul Tripp, Lidia Brazzi, Sonny Fox, Mischa Auer; *Dir:* Rossano Brazzi. **VHS, Beta $9.99** *HBO* 🎞 ½

The Christmas Tree Manipulative sudser about the bond that develops between an extremely wealthy man and his mortally ill son. Not among Holden's better filmsw. Also released as "When Wolves Cry."
1969 (G) 110m/C William Holden, Virna Lisi, Andre Bourvil, Brook Fuller; *Dir:* Terence Young. **VHS, Beta $14.95** *UHV, VCI* 🎞 ½

The Christmas Tree The story of a fir tree, played by mime Julian Chagrin, that gets cut down to be sold as a Christmas tree.
1975 12m/C Julian Chagrin. **VHS, Beta $11.95** *PYR* 🎞 ½

The Christmas Wife A lonely man pays a woman to be his holiday companion at a mountain retreat. Sturdy performances by Robards and Harris manage to keep this one from going to the dogs. Based on a story by Helen Norris. Made for cable television.
1988 73m/C Jason Robards Jr., Julie Harris, Don Francks, Patricia Hamilton, Deborah Grover; *Dir:* David Jones. **VHS, Beta $79.99** *HBO* 🎞🎞 ½

Christmas Without Snow A made-for-television film about a lonely divorced woman finding communal happiness within a local church choir which is led by a crusty choir master.
1980 96m/C John Houseman, Michael Learned, Valerie Curtin; *Dir:* John Korty. **VHS, Beta $59.98** *FOX* 🎞🎞 ½

Christopher Columbus Step-by-step biography of the fifteenth-century explorer, his discovery of America, the fame that first greeted him, and his last days.
1949 103m/C *GB* Fredric March, Florence Eldridge, Francis L. Sullivan; *Dir:* David MacDonald. **VHS, Beta** *LCA* 🎞🎞 ½

Christopher Columbus The man who explored the New World is shown in all his flawed complexity in this film that takes a contemporary approach to the discoveries and character of Christopher Columbus. Shot on location in Spain, Malta and the Dominican Republic, the film features an exact replica of Columbus' flagship, the Santa Maria. Originally aired as a six-hour television miniseries.
1991 128m/C Gabriel Byrne, Faye Dunaway, Oliver Reed, Max von Sydow, Eli Wallach, Nicol Williamson, Jose Ferrer, Virna Lisi, Raf Vallone; *Dir:* Alberto Lattuada. **VHS $89.95** *WAR* 🎞🎞 ½

Christopher Strong Interesting Hepburn turn as a daredevil aviatrix who falls in love with a married British statesman.
1933 77m/B Katharine Hepburn, Billie Burke, Colin Clive, Helen Chandler; *Dir:* Dorothy Arzner. VHS, Beta, LV $19.98 *RKO* 𝄞𝄞½

Chrome Hearts A rebel biker rescues a buxom gal from a fate worse than death, than vies for control of the gang. Laughable, with Namath hopeless in his film debut. Ann-Margaret, in continual disarray, is the only redeemable aspect of this one. Also on video under its original title "C.C. and Company."
1970 (R) 91m/C Joe Namath, Ann-Margret, William Smith, Jennifer Billingsley, Teda Bracci, Greg Mullavey, Sid Haig, Bruce Glover; *Dir:* Seymour Robbie. VHS, Beta $9.95 *SIM, SUE* 𝄞

Chrome and Hot Leather A Green Beret is out for revenge after vicious bikers kill his fiance. Conventional, tasteless genre fare notable only as Gaye's first film.
1971 (PG) 91m/C William Smith, Tony Young, Michael Haynes, Peter Brown, Marvin Gaye, Michael Stearns, Katherine Baumann, Wes Bishop, Herbert Jeffries; *Dir:* Lee Frost. VHS $39.95 *TRY* 𝄞½

Chronicles of Narnia Exceptional BBC production of the C.S. Lewis fantasy about four brave children who battle evil in a mythical land where the animals talk and strange creatures roam the countryside. Three volumes, each with a different episode: "The Lion, the Witch and the Wardrobe", "Prince Caspian and the Volyage of the Dawn Treader", and "The Silver Chair". Aired on PBS as part of the WonderWorks Family Movie television series.
1989 180m/C Barbara Kellerman, Jeffery Perry, Richard Dempsey, Sophie Cook, Jonathan Scott, Sophie Wilcox, David Thwaites, Tom Baker; *Dir:* Alex Kirby. VHS $29.95 *HMV, PME* 𝄞½

Chronopolis A brilliant animated feature film which was five years in the making. The citizens of Chronopolis are immmortals who relieve the boredom of perpetual existence by creating time. In French with English subtitles.
1982 70m/C *FR Dir:* Piotr Kamler. VHS $49.95 *FCT* 𝄞𝄞𝄞½

Chu Chu & the Philly Flash A has-been baseball player and a lame dance teacher meet while hustling the same corner; he sells hot watches, she's a one-woman band. A briefcase full of government secrets soon involves them with the feds, the mob, and a motley collection of back-alley bums. Insipid comedy is actually worse than its title, with Burnett and Arkin both trying too hard. Supporters Warden, Aiello, and Glover don't help either.
1981 (PG) 102m/C Alan Arkin, Carol Burnett, Jack Warden, Danny Aiello, Ruth Buzzi, Danny Glover, Lou Jacobi; *Dir:* David Lowell Rich. VHS, Beta $59.98 *FOX* 𝄞

Chuck Amuck: The Movie An in-depth look at legendary animator/director, Chuck Jones.
1991 51m/C Chuck Jones. VHS $19.98 *WAR, FCT*

Chuck Berry: Hail! Hail! Rock 'N' Roll Engaging, energetic portrait of one of rock's founding fathers, via interviews, behind-the-scenes footage and performance clips of Berry at 60. Songs featured: "Johnny B. Goode," "Roll Over Beethoven," "Maybelline," and more. Appearances by Eric Clapton, Etta James, John and Julian Lennon, Roy Orbison, Linda Ronstadt,

Bo Diddley and Bruce Springsteen among others.
1987 (PG) 121m/C Chuck Berry, Eric Clapton, Etta James, Robert Cray, Julian Lennon, Keith Richards, Linda Ronstadt, John Lennon, Roy Orbison, Bo Diddley, Jerry Lee Lewis, Bruce Springsteen, Kareem Abdul-Jabbar; *Dir:* Taylor Hackford. VHS, Beta $19.95 *MCA, MVD* 𝄞𝄞𝄞½

C.H.U.D. Cannibalistic Humanoid Underground Dwellers are what it's about. Exposed to toxic wastes, a race of flesh-craving, sewer-dwelling monstrosities goes food shopping on the streets of New York. Don't be fooled by the presence of real actors, this one is inexcusable. Followed by a sequel.
1984 (R) 90m/C John Heard, Daniel Stern, Christopher Curry, Kim Greist, John Goodman; *Dir:* Douglas Cheek. VHS, Beta, LV $9.98 *MRV* Woof!

C.H.U.D. 2: Bud the Chud Asinine teens steal a corpse that is actually a zombie cannibal capable of passing the trait to anyone it bites (but doesn't eat entirely). Graham excels as the kidnapped corpse, but this horror-comedy is consistently repellent. Contains one of Jagger's few film appearances, a situation of no despair.
1989 (R) 84m/C Brian Robbins, Bill Calvert, Gerrit Graham, Tricia Leigh Fisher, Bianca Jagger, Robert Vaughn, Larry Cedar; *Dir:* David Irving. VHS, Beta, LV $14.98 *LIV, VES, IME* Woof!

Chuka A gunfighter tries to resolve a conflict between Indians and unlikeable troops while simultaneously romancing the fort's beautiful occupant. Pedestrian western features convincing playing from the always reliable Taylor.
1967 105m/C Rod Taylor, Ernest Borgnine, John Mills, Lucianna Paluzzi, James Whitmore, Angela Dorian, Louis Hayward, Michael Cole, Hugh Reilly; *Dir:* Gordon Douglas. VHS $14.95 *PAR, FCT* 𝄞𝄞½

A Chump at Oxford Two street cleaners foil a bank robbery and receive an all-expenses-paid education at Oxford as their reward. This loopy Laurel and Hardy vehicle provides regular amusement in detailing the duo's exploits in Britain.
1940 63m/B Stan Laurel, Oliver Hardy, James Finlayson, Wilfrid Lucas, Peter Cushing, Charlie Hall; *Dir:* Alfred Goulding. VHS, Beta, LV $14.98 *MED, CCB, FCT* 𝄞𝄞½

The Church Italian thrill-meister Dario Argento scripted and produced this ecclesiastical gorefest directed by Michele Soavi. A gargoyle-glutted gothic cathedral which happens to stand on the site of a gruesome mass murder is renovated, and the kirk-cleaning turns into a special-effects loaded demonic epiphany. It'll have you muttering your pater noster. Unrated, it's also available in an R-rated version.
1990 110m/C Tomas Arana, Hugh Quarshine, Feodor Chaliapin; *Dir:* Michele (Michael) Soavi. VHS $89.98 *SOU* 𝄞𝄞𝄞

Churchill and the Generals The true story of how Winston Churchill led England away from the bleak Dunkirk battle and rallied the Allied generals to a D-Day victory. Based upon Churchill's memoirs.
1981 180m/C Timothy West, Joseph Cotten, Arthur Hill, Eric Porter, Richard Dysart; *Nar:* Eric Sevareid. VHS, Beta $14.99 *STE, PSM* 𝄞𝄞

Ciao Federico! Fellini Directs Satyricon A documentary of Frederico Fellini's filming of Petronius' "Satyricon." Portrays the Italian filmmaker's larger-than-life directorial approach and relationship with

his actors. In English and Italian, with English subtitles.
1969 60m/C Federico Fellini, Martin Potter, Hiram Keller, Roman Polanski, Sharon Tate; *Dir:* Gideon Bachmann. VHS, Beta $29.95 *MFV, TAM* 𝄞𝄞

Cigarette Blues Oakland musician Sonny Rhodes addresses three subjects: death, cigarette smoking and the blues.
1989 6m/C Sonny Rhodes; *Dir:* Les Blank, Alan Govenar. VHS, Beta $29.95 *FFM* 𝄞𝄞

The Cigarette Girl of Mosselprom A lowly cigarette girl is thrust into the world of moviemaking in this sharp Russian satire. Silent with orchestral score.
1924 78m/B *RU Dir:* Yuri Zhelyabuzhsky. VHS $29.95 *KIV, FCT, TAM* 𝄞𝄞½

Cimarron Hopelessly overblown saga of an American frontier family from 1890 to 1915. Hokey, cliched, with only sporadic liveliness. How did this one win an Oscar? An adaptation of Edna Ferber's novel, featuring Dunne in an early major role. Remade in 1960.
1930 130m/B Richard Dix, Irene Dunne, Estelle Taylor, Nance O'Neill, William Collier Jr., Roscoe Ates, George E. Stone, Stanley Fields, Edna May Oliver, Dennis O'Keefe; *Dir:* Wesley Ruggles. Academy Awards '31: Best Adapted Screenplay, Best Interior Decoration, Best Picture; National Board of Review Awards '31: 10 Best Films of the Year. VHS, Beta $19.98 *MGM, BTV* 𝄞𝄞

Cincinnati Kid Cardsharks gather for a major stakes poker game, with some romance on the side. Conventional fare helped along by serviceable performances and some stunning cinematography. Sam Peckinpah was dismissed as director from this one. Guess he took the script with him.
1965 104m/C Steve McQueen, Edward G. Robinson, Ann-Margret, Tuesday Weld, Karl Malden, Joan Blondell, Rip Torn, Jack Weston, Cab Calloway; *Dir:* Norman Jewison. VHS, Beta, LV $19.98 *MGM* 𝄞𝄞½

Cinderella Classic Disney animated fairy-tale about the slighted beauty who outshines her evil stepsisters at a royal ball, then returns to her grim existence before the handsome prince finds her again. Engaging film, with a wicked stepmother, kindly fairy godmother, and singing mice. Oscar nominated for Best Song ("Bibbidy Bobbidy Boo") and for musical score.
1950 76m/C *Dir:* Wilfred Jackson; *Voices:* Ilene Woods, William Phipps, Verna Felton, James MacDonald. Berlin Film Festival '51: Best Musical; Venice Film Festival '50: Special Jury Prize. VHS, Beta $26.99 *DIS, KUI* 𝄞𝄞½

Cinderella A made-for-television rendition of the fairy tale as scored by Rodgers and Hammerstein. Songs include "The Loveliness of Evening," "Ten Minutes Ago" and "Do I Love You (Because You're Beautiful?)."
1964 83m/C Lesley Ann Warren, Ginger Rogers, Walter Pidgeon, Stuart Damon, Celeste Holm; *Dir:* Charles S. Dubin. VHS, Beta, LV $19.98 *FOX, FCT* 𝄞𝄞

Cinderella A soft-core musical adaptation of the fairy tale. Also available unrated, in a longer and more graphic version. Followed by "Fairy Tales."
1977 (R) 87m/C Cheryl "Rainbeaux" Smith, Kirk Scott, Brett Smiley, Sy Richardson; *Dir:* Michael Pataki. VHS, Beta $69.98 *LIV, LTG* 𝄞

C

Cinderella From the "Faerie Tale Theatre" comes the classic tale of a poor girl who goes to a ball to meet the man of her dreams, despite her nasty stepmother and stepsisters.
1984 60m/C Jennifer Beals, Jean Stapleton, Matthew Broderick, Eve Arden; *Dir:* Mark Cullingham. VHS, Beta, LV $14.98 KUI, FOX, FCT ✍✍

Cinderella 2000 There must be something about the Cinderella story that cries out for a soft-core musical telling of the "if the shoe fits the girl it must be true love" fairy tale (Pataki directed a 1977 Cinderella bawdorama). Or maybe it's only logical that the director of "Brain of Blood" and "Satan's Sadists" should turn his talents to a futuristic song-and-dance version of the Prince Charming story. It's the year 2047 and sex is outlawed, except by computer. Strains of Sparky Sugarman's score, including "Doin' Without" and "We All Need Love," set the stage for Erhardt's Cinderella to meet her Prince Charming at that conventional single prince romance venue, a sex orgy. Trouble is, it wasn't a shoe Cinderella lost before she fled, and the charming one must interface, as it were, with the local pretenders to the throne in order to find his lost princess. Touching.
1978 (R) 86m/C Catharine Erhardt, Jay B. Larson, Kirk Calloway, Vaughn Armstrong; *Dir:* Al Adamson. VHS $59.95 SVD ✍

Cinderella Liberty Bittersweet romance in which a kindly sailor falls for a brash hooker with a son. Sometimes funny, sometimes moving, with sometimes crude direction overcome by truly compelling performances from Caan and Mason. Story written by by Darryl Ponicsan from his novel. Mason earned an Oscar nomination.
1973 (R) 117m/C James Caan, Marsha Mason, Eli Wallach, Kirk Calloway, Burt Young, Bruce Kirby, Dabney Coleman, Sally Kirkland; *Dir:* Mark Rydell. VHS, Beta $59.98 FOX ✍✍✍

Cinderella/Thumbelina Animated versions of these two classic children's stories. The first is about a poor girl transformed into a princess by her fairy godmother. "Thumbelina" is about a little girl who is no bigger than your thumb.
1983 50m/C VHS, Beta $7.00 VTR ✍✍✍

Cinderfella This twist on the classic children's fairy tale features Lewis as the hapless buffoon guided by his fairy godfather. Somewhat overdone, with extended talking sequences and gratuitous musical interludes. Lewis, though, mugs effectively.
1960 88m/C Jerry Lewis, Ed Wynn, Judith Anderson, Anna Maria Alberghetti, Henry Silva, Count Basie, Robert Hutton; *W/Dir:* Frank Tashlin. VHS, Beta $59.95 IVE ✍✍½

Cinema Paradiso Memoir of a boy's life working at a movie theatre in small-town Italy after World War II. Film aspires to both majestic sweep and stirring poignancy, but only occasionally hits its target. Still manages to move the viewer, and it features a suitably low-key performance by the masterful Noiret. Autobiographically inspired script written by Tornatore. The version shown in America is approximately a half-hour shorter than the original Italian form.
1988 123m/C IT Philippe Noiret, Jacques Perrin, Salvatore Cascio, Mario Leonardi, Agnes Nano, Leopoldo Trieste; *W/Dir:* Giuseppe Tornatore. Academy Awards '89: Best Foreign Language Film; Cannes Film Festival '89: Special Jury Prize. VHS, LV, 8mm $89.99 HBO, APD, TAM ✍✍✍

Cinema Shrapnel A tour through the wonderful world of screen hype and bad movies, with scenes from terrible trailers, maudlin fundraising shorts, and glitzy movie premieres.
1987 60m/C Hosted: Bob Shaw. VHS, Beta $34.95 NO ✍✍✍

Cinemagic A compilation of four short horror and science fiction films including "Nightfright," "Dr. Dobermind," "Illegal Alien" and "The Thing in the Basement."
1985 60m/C VHS, Beta $39.95 MPI ✍✍✍

Circle Canyon In a remote canyon two opposing outlaw gangs battle to the death.
1933 48m/B Buddy Roosevelt. VHS, Beta WGE ✍

Circle of Death Average western about a white boy who is raised as an Indian. Later he falls in love with a white girl and leaves the tribe.
1936 55m/B Monte Montana, Yakima Canutt. VHS, Beta $19.95 NOS, VDM ✍

Circle of Fear An ex-hitman who worked for the Mob upends the black market for sex and drugs in the Philippines while looking for his daughter who was sold into a sex slave ring. El cheapo exploiter.
1989 (R) 90m/C PH Patrick Dollaghan, Welsey Pfenning, Joey Aresco, Vernon Wells; *Dir:* Clark Henderson. VHS, Beta $44.98 MGM ✍½

Circle of Iron Plenty of action and martial arts combat abound in this story of one man's eternal quest for truth. Originally written by Bruce Lee, James Coburn, and Stirling Silliphant as a rib-crunching vehicle for Lee, who died before production began and was replaced by Kung-Fu Carradine. A cut above most chop-socky actioners. Filmed in Israel and released in that country as "The Silent Flute."
1978 (R) 102m/C GB Jeff Cooper, David Carradine, Roddy McDowall, Eli Wallach, Christopher Lee; *Dir:* Richard Moore. VHS, Beta $24.98 SUE ✍✍½

Circle of Love Episodic melodrama drifts from romance to romance in contemporary Paris. A somewhat pretentious remake of Ophuls classic "La Ronde," which was in turn adapted from Arthur Schnitzler's play. Credible performers are undone by Vadim's strained direction. Written by Jean Anouilh.
1964 110m/C FR Jane Fonda, Francine Berge, Marie DuBois, Jean-Claude Brialy, Catherine Spaak, Claude Giraud; *Dir:* Roger Vadim. VHS, Beta $79.95 MFI, JEF ✍✍

Circle Man A made-for-video action drama about the underground sporting phenomenon, bare-knuckle fighting.
1987 90m/C William Sanderson. VHS, Beta $79.95 ACA ✍

Circle of Two A platonic friendship between an aging artist and a young girl is misunderstood by others. Well, of course they do meet in a porno theater. Pedestrian fare adapted from a story by Marie Therese Baird.
1980 (PG) 90m/C CA Richard Burton, Tatum O'Neal, Kate Reid; *Dir:* Jules Dassin. VHS, Beta $69.95 VES ✍✍

Circuitry Man Post-apocalyptic saga of future American life as a woman tries to deliver a briefcase full of computer chips to the underground Big Apple. Along the way she runs into Plughead, a humanoid with electrical outlets that allow him to "plug in" to other people's fantasies. Intelligent retelling of a standard tale with an inspired

soundtrack by Deborah Holland. Witty and original.
1990 (R) 85m/C Jim Metzler, Dennis Christopher, Dana Wheeler-Nicholson, Vernon Wells; *W/Dir:* Steven Lovy. VHS, LV $89.95 COL ✍✍½

The Circus Classic comedy silent details the tramp's exploits as a member of a traveling circus, including a romance with the bareback rider. Hilarious, less sentimental than most of Chaplin's feature films. Chaplin won a special Academy Award for "versatility and genius in writing, acting, directing and producing" for this one. Outrageous final scenes.
1919 105m/B Charlie Chaplin, Merna Kennedy; *Dir:* Charlie Chaplin. VHS $19.98 FOX ✍✍✍½

Circus Angel The renowned director of "The Red Balloon" creates a fantasy about a klutzy burglar transformed into a found nightgown into an angel. He begins to serve the dreams and actions of an odd lot of characters. Subtitled in English.
1965 80m/B FR Philippe Avron, Mirielle Negre; *Dir:* Albert Lamorisse. Cannes Film Festival '65: Best Visual Effects, Jury Prize. VHS, Beta $14.98 SUE ✍✍✍

Circus of Fear A travelling troupe is stalked by a murderer. The unedited version is occasionally scary. Also released as "Psycho-Circus."
1967 92m/C GB GE Christopher Lee, Leo Genn, Anthony Newlands, Heinz Drache, Eddi Arent, Klaus Kinski, Margaret Lee, Suzy Kendall; *Dir:* John Llewellyn Moxey. VHS, Beta $19.95 INC, SNC, NOS ✍½

Circus of Horrors Nip 'n tuck horror about a plastic surgeon who takes over a circus to escape a disfigured patient bent on revenge. The circus is staffed by former patients with new faces who, one by one, fall victim in fine circus style. A bloody one-ring extravaganza.
1960 87m/C GB Donald Pleasence, Anton Diffring, Erika Remberg, Yvonne Monlaur, Jane Hylton, Kenneth Griffith; *Dir:* Sidney Hayers. VHS, Beta $59.99 HBO ✍½

Circus World A circus boss tries to navigate a reckless, romancing crew through a European tour while searching for aerialist he loved 15 years before and whose daughter he has reared. Too long and too familiar, but nonetheless well done with an excellent finale. Written by Ben Hecht, Julian Halevy, and James Edward Grant.
1964 132m/C John Wayne, Rita Hayworth, Claudia Cardinale, Lloyd Nolan, Richard Conte; *Dir:* Henry Hathaway. VHS, Beta $29.98 UHV, LTG ✍✍½

Citadel From the A.J. Cronin novel, the intelligent and honest Hollywood drama about a young British doctor who is morally corrupted by his move from a poor mining village to a well-off practice treating wealthy hypochondriacs. Somewhat hokey but still consistently entertaining, with Donat fine in the lead. Oscar nominated for Best Picture, Best Director, and Best Screenplay.
1938 114m/C GB Robert Donat, Rosalind Russell, Rex Harrison, Ralph Richardson, Emlyn Williams, Penelope Dudley Ward; *Dir:* King Vidor. National Board of Review Awards '38: Best Picture; New York Film Critics Awards '38: Best Picture. VHS, Beta $24.95 MGM ✍✍✍

Citizen Kane Extraordinary film is an American tragedy of a newspaper tycoon (based loosely on William Randolph Hearst) from his humble beginnings to the solitude of his final years. One of the greatest films ever made, and a stunning tour-de-force in virtual-

ly every aspect, from its fragmented narration to its breathtaking, deep-focus cinematography; from its vivid soundtrack to its fabulous ensemble acting. Oscar winning script was written by the 25-year-old Welles and Herman J. Mankiewicz. Curiously, it won a single Oscar, though nominated for Best Picture, Best Actor (Welles), Best Director, Black & White Cinematography, Black & White Interior Decoration, Sound Recording, Film Editing and Best Musical Score of a Dramatic Picture. The laser edition, a three-disc set, was reproduced from a superior negative and features extensive liner notes and running commentary from film historian Robert J. Carringer.
1941 119m/B Orson Welles, Joseph Cotten, Everett Sloane, Dorothy Comingore, Ruth Warrick, George Coulouris, Ray Collins, William Alland, Paul Stewart, Erskine Sanford, Agnes Moorehead, Alan Ladd, Gus Schilling, Philip Van Zandt, Harry Shannon, Sonny Bupp; **Dir:** Orson Welles. Academy Awards '41: Best Original Screenplay; National Board of Review Awards '41: Best Film; New York Film Critics Awards '41: Best Film; Sight & Sound Survey '62: #1 of the Best Films of All Time; Sight & Sound Survey '72: #1 of the Best Films of All Time; American Film Institute's Survey '77: 2nd Best American Film Ever Made; Sight & Sound Survey '82: #1 of the Best Films of All Time. **VHS, Beta, LV, 8mm $19.98** RKO, KOV, VYG ✝✝✝

Citizen Welles The films and life of director/actor Orson Welles are profiled in this series of two untitled programs.
197? 30m/B VHS, Beta CMR, NYS ✝✝✝✝

Citizens Band Episodic, low-key comedy about people united by their CB use in a midwestern community. Notable performance from Clark as a soft-voiced guide for truckers passing through. Demme's first comedy is characteristically idiosyncratic. Written by Paul Brickman and also known as "Handle With Care."
1977 (PG) 98m/C Paul LeMat, Candy Clark, Ann Wedgeworth, Roberts Blossom, Charles Napier, Marcia Rodd, Bruce McGill, Ed Begley Jr., Alix Elias; **Dir:** Jonathan Demme. New York Film Festival '77: Best Supporting Actress; National Society of Film Critics Awards '77: Best Supporting Actress (Wedgeworth). **VHS, Beta $59.95** PAR ✝✝✝

The City Police desperately search for a psychotic determined to kill a country singer. He shouldn't be too hard to find. Made for television.
1976 78m/C Don Johnson, Robert Forster, Ward Costello, Jimmy Dean, Mark Hamill; **Dir:** Harvey Hart. **VHS, Beta $49.95** WOV ✝

City of Blood A South African doctor is embroiled in mystery with racial overtones when prostitutes are being killed with a spiked club. Packaged as a horror film, this movie actually deals with the questions of South African racial tensions in a sophisticated, ultimately tragic manner.
1988 (R) 96m/C SA Joe Stewardson, Ian Yule, Susan Coetzer; **Dir:** Darrell Roodt. **VHS, Beta $79.98** MAG ✝✝ 1/2

City for Conquest Two lovers go their separate ways to pursue individual careers. He attempts to become a boxing champ, but is blinded in the ring and ends up selling newspapers. She takes a shot at a dancing career but hooks up with an unscrupulous partner. Will the ill-fated pair find happiness again?
1940 101m/B James Cagney, Ann Sheridan, Frank Craven, Donald Crisp, Arthur Kennedy, Frank McHugh, George Tobias, Anthony Quinn, Blanche Yurka, Elia Kazan, Bob Steele; **Dir:**

Anatole Litvak. **VHS, Beta $19.98** MGM, CCB ✝✝✝

The City and the Dogs A brutal, cynical adaptation of Mario Vargas Llosa's novel. A young military recruit rebels against the authority establishments around him. In Spanish with English subtitles.
1985 135m/C SP **Dir:** Francisco J. Lombardi Pery. **VHS, Beta $34.95** FCT ✝✝✝

City in Fear Stellar cast lights up a made-for-television drama about a tired newspaper columnist (Janssen) who communicates directly with a murdering psychotic Rourke, as publisher Vaughn applauds and hypes. Rourke's first role, Janssen's last, and Smithee is actually Jud Taylor.
1980 135m/C David Janssen, Robert Vaughn, Mickey Rourke, William Daniels, Perry King, Susan Sullivan, William Prince; **Dir:** Allen Smithee. **VHS, Beta $69.98** LIV, LTG, HHE ✝✝✝

City on Fire An arsonist starts a fire that could engulf an entire city. Flaming bore.
1978 (R) 101m/C CA Barry Newman, Susan Clark, Shelley Winters, Leslie Nielsen, Henry Fonda, Ava Gardner; **Dir:** Alvin Rakoff. **VHS, Beta $9.98** SUE Woof!

City Girl Friedrich Murnau, who directed the silent vampire classic "Nosferatu," was removed from the director's chair before "City Girl" was completed and it shows. But so do the marks of his inimitable camera direction. The story concerns a Minnesota grain grower who visits the Windy City and returns with a waitress as his wife. Frustratingly inconsistent, leading you to wonder what could have been had Murnau remained behind the camera (he died the following year). Silent; also known as "Our Daily Bread."
1930 90m/B Charles Farrell, Mary Duncan; **Dir:** F. W. Murnau. **VHS, Beta $39.95** GPV, FCT, DNB ✝✝ 1/2

City of Gold/Drylanders Combines two riveting Canadian documentaries: "City of Gold" (1957), which is about the Klondike Gold Rush of the 1890's, and "Drylanders" (1962), the story of a city family's attempt to live on a lonely Sakatchewan farm.
1957 92m/B VHS, Beta $29.95 VYY ✝✝ 1/2

City Heat A hardnosed cop and a plucky private eye berate each other while opposing the mob in this overdone comedy. Both Eastwood and Reynolds spoof their screen personas, but the results are only slightly satisfactory. Good back-up, though, from Alexander and Kahn as the dames.
1984 (PG) 98m/C Clint Eastwood, Burt Reynolds, Jane Alexander, Irene Cara, Madeline Kahn, Richard Roundtree, Rip Torn, Tony LoBianco, William Sanderson; **Dir:** Richard Benjamin. **VHS, Beta, LV $19.98** WAR ✝✝ 1/2

City of Hope The picture that "Bonfire of the Vanities" wanted to be, an eventful few days in the fictional metropolis of Hudson: an ugly racial incident threatens to snowball, the corrupt mayor pushes a shady real-estate deal, and a botched robbery has profound implications. Some of the subplots resolve too easily, but this cynical, crazy-quilt of urban life is worthy of comparison with "American Graffiti" and "Nashville" as pure Americana.
1991 (R) 132m/C Vincent Spano, Tony LoBianco, Joe Morton, Todd Graff, David Strathairn, Anthony Denison, Barbara Williams, Angela Bassett, Gloria Foster, Lawrence Tierney, John Sayles; **W/Dir:** John Sayles. **VHS, LV, 8mm** COL ✝✝✝

City Killer A lunatic tries to blackmail the girl he loves into going out on a few dates by committing huge destructive acts of inner-city terrorism (flowers won't do). Made for television.
1987 120m/C Heather Locklear, Gerald McRaney, Terence Knox, Peter Mark Richman, John Harkins; **Dir:** Robert Lewis. **VHS, Beta $9.99** STE, PSM ✝

City Lights Masterpiece that was Chaplin's last silent film. The "Little Tramp" falls in love with a blind flower seller. A series of lucky accidents permits him to get the money she needs for a sight-restoring surgery. One of the most eloquent movies ever filmed, due to Chaplin's keen balance between comedy and tragedy.
1931 86m/B Charlie Chaplin, Virginia Cherrill, Florence Lee, Hank Mann, Harry Myers, Henry Bergman, Jean Harlow; **Dir:** Charlie Chaplin. National Board of Review Awards '31: 10 Best Films of the Year; Cinematheque Belgique '52: #4 of the Best Films of All Time (tie); Sight & Sound Survey '52: #2 of the Best Films of All Time. **VHS, Beta $19.98** FOX ✝✝✝✝

City Limits In a post-apocalyptic city, gangs of young people on choppers clash.
1985 (PG-13) 85m/C John Stockwell, Kim Cattrall, Darrell Larson, Rae Dawn Chong, Robby Benson, James Earl Jones, Jennifer Balgobin; **Dir:** Aaron Lipstadt. **VHS, Beta, LV $79.98** VES, LIV, HHE ✝

City in Panic A detective and radio talk show host try to catch a psychotic mass murderer busy offing homosexuals throughout the city. Violent with no redeeming social qualities.
1986 85m/C Dave Adamson, Ed Chester, Leeann Westegard; **Dir:** Robert Bouvier. **VHS, Beta $19.98** TWE ✝

City of Shadows Twins separated at birth are reunited by a criminal mind with evil intentions. Double trouble? Music by Tangerine Dream.
1986 (R) 97m/C John P. Ryan, Paul Coufos, Tony Rosato; **Dir:** David Mitchell. **VHS, Beta $19.95** STE, NWV ✝✝

City Slickers Three men with mid-life crises leave New York City for a cattle-ranch vacation that turns into an arduous, character-building stint. Many funny moments supplied by leads Crystal, Stern, and Kirby, but Palance steals the film as a salty, wise cowpoke. Slater is also fetching as the lone gal vacationer on the cattle drive. Box office winner notable for a realistic calf birthing scene, one of few in cinema history. From an idea by Crystal, who also produced.
1991 (PG-13) 114m/C Billy Crystal, Daniel Stern, Bruno Kirby, Patricia Wettig, Helen Slater, Jack Palance, Noble Willingham, Tracy Walter, Josh Mostel, Bill Henderson, Jeffrey Tambor, Phill Lewis, Kyle Secor, Yeardley Smith, Jayne Meadows; **Dir:** Ron Underwood. Academy Awards '91: Best Supporting Actor (Palance); Golden Globe Awards '91: Best Supporting Actor (Palance); MTV Movie Awards '92: Best Comedic Performance (Crystal). **VHS, Beta, LV, SVS, 8mm $19.95** COL, PIA ✝✝✝

City That Never Sleeps A Chicago cop must decide whether or not to run off with an entertainer, or continue his life with his family. Decisions, decisions. Moody but dated melodrama that's finely acted and indecisively directed, with good use of dim city lights. Nothing new here.
1953 90m/B Gig Young, Mala Powers, Edward Arnold, Paula Raymond, Chill Wills, Marie Windsor; **Dir:** John H. Auer. **VHS $19.98** REP ✝✝ 1/2

City of the Walking Dead Eco-misery as a community must contend with prowling, radiation-zapped zombies who enjoy chewing through human flesh. Also entitled ''Nightmare City.''
1983 92m/C *IT SP* Mel Ferrer, Hugo Stiglitz, Laura Trotter, Francisco Rabal; *Dir:* Umberto Lenzi. VHS $19.95 *NSV* ✂

City of Women Visually stunning, but otherwise thin fantasy/drama about a man smothered with women. A journalist wanders through a feminist theme park replete with a roller rink and screening room. Some worthwhile adventures are mixed in with rather dull stretches. Not among the best from either Mastroianni, who is under utilized, or Fellini, who here seems incapable of separating good ideas from bad. Plenty of buxom babes, but otherwise undistinguished.
1981 (R) 140m/C *IT* Marcello Mastroianni, Ettore Manni, Anna Prucnall, Bernice Stegers; *W/Dir:* Federico Fellini. VHS, Beta $79.95 *NYF, FCT* ✂✂

The City's Edge A young man becomes involved with the mysterious residents of a boarding house on the edge of the ocean.
1983 86m/C *AU* VHS, Beta $59.95 *MGM* ✂½

The Civil War: Episode 1: The Cause: 1861 The blowout PBS hit mini-series, five years in the making by author and filmmaker Ken Burns, is a staggering historical achievement. Episode 1 focuses on the causes of the Civil War from the schism created by Lincoln's election, John Brown's assault on Harper's Ferry, and the firing on Fort Sumpter. Also available as a set for $179.95.
1990 99m/C *W/Dir:* Ken Burns. VHS, 3/4U $79.95 *PBS* ✂✂✂✂

The Civil War: Episode 2: A Very Bloody Affair: 1862 Second in the enormously popular PBS series looks at the unexpectedly extreme costs in human life of the War Between the States.
1990 69m/C *W/Dir:* Ken Burns. VHS, 3/4U $59.95 *PBS* ✂✂✂✂

The Civil War: Episode 3: Forever Free: 1862 Third volume in the PBS Emmy-nominated mini-series covers Antietam, Robert E. Lee and Stonewall Jackson planning Confederate strategy, and Lincoln's decision to free the slaves.
1990 76m/C *W/Dir:* Ken Burns. VHS, 3/4U $19.95 *PBS* ✂✂✂✂

The Civil War: Episode 4: Simply Murder: 1863 Union forces meet disaster at Fredericksburg, and Gen. Lee wins a victory but loses Stonewall Jackson in the fourth episode of the Emmy-nominated PBS min-series.
1990 62m/C *W/Dir:* Ken Burns. VHS, 3/4U $59.95 *PBS* ✂✂✂✂

The Civil War: Episode 5: The Universe of Battle: 1863 Gettysburg, the battleground which claimed 150,000 American lives, is examined in the 5th episode of the Emmy-nominated PBS mini-series.
1990 95m/C *W/Dir:* Ken Burns. VHS, 3/4U $79.95 *PBS* ✂✂✂✂

The Civil War: Episode 6: Valley of the Shadow of Death: 1864 Profiles of Grant and Lee are offered, plus Sherman's assault on Atlanta, in this volume from the Emmy-nominated PBS mini-series.
1990 70m/C *W/Dir:* Ken Burns. VHS, 3/4U $59.95 *PBS* ✂✂✂✂

The Civil War: Episode 7: Most Hallowed Ground: 1864 The nation re-elects Abraham Lincoln, and the North turns General Lee's mansion into Arlington National Cemetery in this episode of the Emmy-nominated PBS mini-series.
1990 72m/C *W/Dir:* Ken Burns. VHS, 3/4U $59.95 *PBS* ✂✂✂✂

The Civil War: Episode 8: War is All Hell: 1865 Sherman marches to the sea and Lee surrenders to Grant in this episode from the Emmy-nominated PBS mini-series.
1990 69m/C *W/Dir:* Ken Burns. VHS, 3/4U $59.95 *PBS* ✂✂✂✂

The Civil War: Episode 9: The Better Angels of Our Nature: 1865 In this final episode of the Emmy-nominated PBS mini-series, Lincoln's assasination is examined, and the central characters of the war are summarized.
1990 68m/C *W/Dir:* Ken Burns. VHS, 3/4U $59.95 *PBS* ✂✂✂✂

Civilization A silent epic about the horrors and immorality of war as envisioned by the ground-breaking film pioneer Ince, who was later murdered. Famous scene involves Christ walking through body-ridden battlefields.
1916 102m/B Howard Hickman, Enid Markey, Lola May; *Dir:* Thomas Ince. VHS, Beta $27.95 *VYY, DNB* ✂✂✂

Claire's Knee A grown man about to be married goes on a holiday and develops a fixation on a young girl's knee. Another of Rohmer's Moral Tales exploring sexual and erotic obsessions. Lots of talk, little else. You'll either find it fascinating or fail to watch more than 10 minutes. Most, however, consider it a classic. Sophisticated dialogue, lovely visions of summer on Lake Geneva. In French with English subtitles.
1971 105m/C *FR* Jean-Claude Brialy, Aurora Cornu, Beatrice Romand; *Dir:* Eric Rohmer. National Board of Review Awards '71: Best Foreign Film; National Society of Film Critics Awards '71: Best Picture. VHS, Beta, LV $59.95 *TAM, MED, FCT* ✂✂✂½

Clambake Noxious Elvis vehicle about a rich man's son who wants success on his own terms, so he trades places with a water-skiing teacher. Inane, even in comparison to other Elvis ventures. Contains nine songs, including ''Clambake,'' ''Hey, Hey, Hey,'' and ''The Girl I Never Loved.''
1967 98m/C Elvis Presley, Shelley Fabares, Bill Bixby, James Gregory, Gary Merrill, Will Hutchins, Harold Peary; *Dir:* Arthur Nadel. VHS, Beta $19.95 *MGM* ✂

The Clan of the Cave Bear A scrawny cavegirl is taken in by Neanderthals after her own parents are killed. Hannah's lifeless as a primitive gamine, and the film is similarly DOA. Ponderous and only unintentionally funny. Based on the popular novel by Jean M. Auel with a screenplay by John Sayles.
1986 (R) 100m/C Darryl Hannah, James Remar, Pamela Reed, John Doolittle, Thomas G. Waites; *Dir:* Michael Chapman. VHS, Beta, LV $14.98 *FXV, FOX* ✂½

Clara's Heart A Jamaican maid enriches the lives of her insufferable, bourgeois employers and their particularly repellent son. A kinder, gentler waste of film and Goldberg; sentimental clap-trap which occasionally lapses into comedy.
1988 (PG-13) 108m/C Whoopi Goldberg, Michael Ontkean, Kathleen Quinlan, Neil Patrick Harris, Spalding Gray, Beverly Todd, Hattie Winston; *Dir:* Robert Mulligan. VHS, Beta, LV $19.98 *WAR, FCT* ✂

Clarence, the Cross-eyed Lion Follows the many adventures of a cross-eyed lion and his human compatriots. Great family viewing from the creator of ''Flipper.''
1965 98m/C Marshall Thompson, Betsy Drake, Richard Haydn, Cheryl Miller, Rockne Tarkington, Maurice Marsac; *Dir:* Andrew Marton. VHS $19.98 *MGM, FCT* ✂✂✂

Clarence Darrow Riveting one-man stage performance shot for television and based on Darrow's writings, adapted from the play by Irving Stone.
1978 81m/C Henry Fonda; *Dir:* John Rich. VHS, Beta, LV $39.95 *ALL, KAR, IME* ✂✂✂

Clark & McCullough The zany duo, a hit in burlesque and night clubs, transferred some of their crazy antics onto film. Their comedy shorts include ''Fits in a Fiddle'' (1933), ''Love and Kisses'' (1934), ''Alibi Bye Bye'' (1935), and a rehearsal of ''Mademoiselle from New Rochelle'' from ''Strike Up the Band'' (with George Gershwin on piano).
193? 53m/B Bobby Clark, Paul McCullough, George Gershwin. VHS, Beta $57.95 *GVV, DVT* ✂✂½

Clash by Night A wayward woman marries a fisherman, then beds his best friend. Seamy storyline is exploited to the hilt by master filmmaker Lang. An utterly unflinching melodrama. Early Monroe shines in a supporting role too. Based on the Clifford Odets play.
1952 105m/B Barbara Stanwyck, Paul Douglas, Marilyn Monroe, Robert Ryan, J. Carroll Naish; *Dir:* Fritz Lang. VHS, Beta $15.95 *UHV* ✂✂✂½

Clash of the Ninja Two ninja warriors, one good, one evil, battle to the death.
1986 90m/C Paul Torcha, Louis Roth, Eric Neff, Bernie Junker, Joe Redner, Klaus Mutter, Eddie Chan, Max Kwan, Tom Allen, Stanley Tong; *Dir:* Wallace Chan. VHS, Beta $59.95 *TWE* ✂

The Clash: Rude Boy ''Rude Boy'' depicts the rise of The Clash, a top British punk band. Live concert footage features The Clash performing such hits as ''White Riot'' and ''I Fought the Law.'' Included are rare films of early Clash shows.
1980 127m/C The Clash, Ray Gange; *Dir:* Jack Hazan. David Mingay. VHS, Beta $19.95 *MVD, FOX*

Clash of the Titans Mind-numbing fantasy derived from Greek legends about heroic Perseus, who slays the snake-haired Medusa and rescues a semi-clad maiden from the monstrous Kraken. Wooden Hamlin plays Perseus and fares better than more accomplished Olivier and Smith, who seem in need of enemas as they lurch about Mt. Olympus. Only Bloom seems truly godlike in a supporting role. Some good, some wretched special effects from pioneer Harryhausen.
1981 (PG) 118m/C *GB* Laurence Olivier, Maggie Smith, Claire Bloom, Ursula Andress, Burgess Meredith, Harry Hamlin, Sian Phillips, Judi Bowker; *Dir:* Desmond Davis. VHS, Beta, LV $69.95 *MGM, MLB* ✂½

Class A prep school student discovers that his mother is the lover whom his roommate has bragged about. Lowe is serviceable as the stunned son of sexy Bisset, who woos McCarthy, even in an elevator. Too ludicrous to be enjoyed, but you may watch just to see what happens next.

1983 (R) 98m/C Jacqueline Bisset, Rob Lowe, Andrew McCarthy, Cliff Robertson, John Cusack, Stuart Margolin, Casey Siemaszko; *Dir:* Lewis John Carlino. **VHS, Beta, LV $29.98** *VES* 🦴🦴

Class of '44
The sequel to "Summer of '42" finds insufferable boys becoming insufferable men at miltary school. Worse than its predecessor.
1973 (PG) 95m/C Gary Grimes, Jerry Houser, Oliver Conant, Deborah Winters, William Atherton, Sam Bottoms; *Dir:* Paul Bogart. **VHS, Beta $19.98** *WAR* 🦴½

Class of '63
An unfulfilled woman finally discovers the lover she lost nearly ten years ago at her college reunion. Nearly average made for television drama.
1973 74m/C James Brolin, Joan Hackett, Cliff Gorman, Ed Lauter; *Dir:* John Korty. **VHS, Beta $59.95** *LHV* 🦴🦴

Class of 1984
Teacher King must face a motley crew of teenagers in the classroom. Ring leader and student psychopath Van Patten leads his groupies on a reign of terror through the high school halls, stopping to gang rape King's wife. Teacher attempts revenge. Bloody and thoughtless update of "Blackboard Jungle." Early Fox appearance. Followed by "Class of 1999."
1982 (R) 93m/C *CA* Perry King, Roddy McDowall, Timothy Van Patten, Michael J. Fox, Merrie Lynn Ross, Stefan Arngrim, Al Waxman; *Dir:* Mark L. Lester. **VHS, Beta, LV $29.98** *VES* 🦴

Class of 1999
Freewheeling sci-fi set in the near future where teen gangs terrorize seemingly the entire country. A high school principal determines to enforce law and order by installing human-like robots with rocket launchers for arms. Violent, crude entertainment. Class dismissed. Sequel to "Class of 1984" and available in an unrated version.
1990 (R) 98m/C Bradley Gregg, Traci Lin, Malcolm McDowell, Stacy Keach, Patrick Kilpatrick, Pam Grier, John P. Ryan, Darren E. Burrows, Joshua Miller; *Dir:* Mark L. Lester. **VHS, Beta** *VES, LIV* 🦴

Class Action
1960s versus 1990s ethics clash when a father and daughter, both lawyers, wind up on opposing sides of a litigation against an auto manufacturer. Hackman and Mastrantonio give intense, exciting performances, almost surmounting the melodramatic script.
1991 (R) 110m/C Mary Elizabeth Mastrantonio, Gene Hackman, Joanna Merlin, Colin Friels; *Dir:* Michael Apted. **VHS, Beta, LV $19.98** *FXV* 🦴🦴🦴

The Class of Miss MacMichael
A dedicated instructor inherits and ultimately inspires a class of high school misfits and malcontents. Derivative venture is beneath considerable talents of Jackson. Based on a novel by Sandy Hutson.
1978 (R) 95m/C *GB* Glenda Jackson, Oliver Reed, Michael Murphy, Rosalind Cash; *Dir:* Silvio Narizzano. **VHS, Beta $9.95** *FOX, MED* 🦴🦴

Class of Nuke 'Em High
Team Troma once again experiments with the chemicals, with violent results. Jersey high school becomes a hotbed of mutants and maniacs after a nuclear spill occurs. Good teens Chrissy and Warren succumb, the school blows, life goes on. High camp, low budget, heavy gore. Followed by "Class of Nuke 'Em High Part 2."
1986 (R) 84m/C Janelle Brady, Gilbert Brenton; *Dir:* Richard W. Haines. **VHS, Beta $9.98** *MED* 🦴½

Class of Nuke 'Em High Part 2: Subhumanoid Meltdown
The Troma team brings us back into the world of the strange. In this adventure, the evil Nukama-ma Corporation holds secret experiments at their "college" and create sub-humanoids as slave labor, swelling unemployment and wrecking the economy. This does not fare too well with the rest of society, including our heroes Roger the reporter, Professor Holt, and the scantily clad sub-humanoid Victoria.
1991 (R) 96m/C Lisa Gaye, Brick Bronsky, Leesa Rowland. **VHS, Beta, LV $89.98** *FXV, MED* 🦴

Class Reunion
Sexual highjinks are the order of the day at a 10-year high school reunion. Insipid softcore effort. It's cheaper just to cut out heads from yearbooks and paste 'em to skin mags torsos. Not to be confused with "National Lampoon's Class Reunion," equally bad but nonethelss distinct.
1987 (X) 90m/C Marsha Jordan, Renee Bond, Terry Johnson. **VHS, Beta, LV $29.95** *MED* Woof!

Class Reunion Massacre
Typically lame-brained horror about the mysterious deaths of former school jerks who've reconvened for the 10-year reunion. Worthless flick stars someone named Finkbinder. Contains graphic violence that is not for the squeamish. Also titled "The Redeemer."
1977 (R) 87m/C T.K. Finkbinder, Damien Knight, Nick Carter. **VHS, Beta $39.98** *UHV* Woof!

Classic Comedy Video Sampler
A collection of classic comedy shorts featuring Amos and Andy, The Three Stooges, and Abbott and Costello.
1949 78m/B Bud Abbott, Lou Costello, Moe Howard, Larry Fine, Curly Howard. **VHS, Beta $29.95** *UHV*

Claudia
A wealthy woman who's part of London's high society falls for a lowly musician while trying to escape her dominating husband. Melodramatic meanderings not likely to move you.
1985 88m/C Deborah Raffin, Nicholas Ball; *Dir:* Anwar Kawadri. **VHS $19.95** *TTC* 🦴🦴

Claws
A woodsman, a game commissioner and an Indian band together to stop a grizzly bear who is killing residents of a small Alaskan town.
1977 (PG) 100m/C Leon Ames, Jason Evers, Anthony Caruso, Glenn Sipes, Carla Layton, Myron Healey; *Dir:* Richard Bansbach, R.E. Pierson. **VHS, Beta $59.95** *GEM* 🦴

Claws
A lone farmboy is subjected to repeated attacks by feline mutants. Meow, baby.
1985 84m/C Jason Roberts, Brian O'Shaughnessy, Sandra Prinsloo. **VHS, Beta $59.95** *WES* Woof!

The Clay Pigeon
Seaman comes to in the hospital after a long coma and discovers he's about to be court-martialed for treason and murder. Effective suspense as he goes after the man who incriminated him.
1949 63m/B Bill Williams, Barbara Hale, Richard Loo; *Dir:* Richard Fleischer. **VHS, Beta $34.95** *RKO* 🦴🦴½

Clayton County Line
Plucky buddies decide to end a cruel sheriff's reign of terror. Derivative and unredeeming.
19?? 80m/C Kelly Bradish, Vince Csapos, Michael Heinz, Donald Kenney, Kathy Kenney, Gregg Schultz; *Dir:* Dean Wilson. **VHS** *IEC* 🦴

Clean and Sober
A drug addict hides out at a rehabilitation clinic and actually undergoes treatment. A serious, subtle, and realistic look at the physical/emotional detoxification of an obnoxious, substance-abusing real estate broker; unpredictable and powerful without moralizing. Keaton is fine in unsympathetic lead, with both Baker and Freeman excelling in lesser roles. Not for all tastes, but its certainly a worthwhile work. Caron, creator of TV's "Moonlighting," debuts here as director.
1988 (R) 124m/C Michael Keaton, Kathy Baker, Morgan Freeman, M. Emmet Walsh, Claudia Christian, Pat Quinn, Ben Piazza, Brian Benben, Luca Bercovici, Tate Donovan, Henry Judd Baker, Mary Catherine Martin; *Dir:* Glenn Gordon Caron. National Society of Film Critics Awards '88: Best Actor (Keaton). **VHS, Beta, LV, 8mm $19.98** *WAR, FCT* 🦴🦴🦴

Clearcut
A progressive lawyer finds his liberalism and his survival skills tested among modern Canadian Indians, when a militant native resorts to kidnapping the businessman who threatens their land. Well-acted, brutal drama that asks tough questions; it includes what purports to be the first authentic sweat-lodge ceremony ever filmed.
1992 98m/C Graham Greene, Ron Lea, Michael Hogan; *Dir:* Richard Bugajski. **VHS** *CGV* 🦴🦴½

Clearing the Range
Hoot wants the bad guy who knocked off his brother so he poses as a cringing coward in the hopes of uncovering the villain. Eilers is the love interest as well as his real-life wife.
1931 64m/B Hoot Gibson, Sally Eilers. **VHS, Beta $19.95** *NOS, VCN, DVT* 🦴½

Cleo from 5 to 7
A singer strolls through Paris for 90 minutes, and reconsiders her life while awaiting the results of medical tests for cancer. Typical documentary-like effort from innovative filmmaker Varda, who constructed the film in real time. Look for a brief appearance of master director Jean-Luc Godard. In French with English subtitles.
1961 90m/B *FR* Corinne Marchand, Antoine Bourseillor, Dorothee Blanck, Michel Legrand, Jean-Claude Brialy, Jean-Luc Godard, Anna Karina, Eddie Constantine, Sami Frey; *Dir:* Agnes Varda. **VHS, Beta $79.95** *FCT* 🦴🦴🦴½

Cleo/Leo
Crude and sexist jerk Leo gets chased into the East River by a gun-toting feminist. Reincarnated, Leo returns as Cleo and endures the same Neanderthal remarks and attitudes that "he" used to dish out. Tasteless, exploitive comedy only cynically explores the sex reversal theme.
1989 (R) 94m/C Jane Hamilton, Scott Thompson Baker, Kevin Thomas, Alan Naggar, Ginger Lynn Allen; *Dir:* Chuck Vincent. **VHS $19.95** *STE, MED* 🦴

Cleopatra
The early Hollywood DeMille version of the Egyptian temptress's lust for Marc Antony after Julius Caesar's death. Intermittently interesting extravaganza that was Oscar nominated for best picture. Colbert seems to be enjoying herself in the lead role in this hokey, overdone epic. Includes the original theatrical trailer on laser-track 2. Remade in 1963.
1934 100m/B Claudette Colbert, Henry Wilcoxon, Warren William, Gertrude Michael, Joseph Schildkraut, C. Aubrey Smith; *Dir:* Cecil B. DeMille. Academy Awards '34: Best Cinematography. **VHS, Beta, LV $14.95** *MCA* 🦴🦴

Cleopatra
And we thought DeMille's version was extravagant. After the death of Julius Caesar, Cleopatra, Queen of Egypt, becomes infatuated with Mark Anthony.

Costly four-hour epic functions like a blimp-sized, multi-colored sleeping tablet. Historical characters are utterly dwarfed by the film's massive scope, and audiences are ben-umbed by a spectacle of crowd scenes and opulent, grotesque interiors. Taylor looks and often acts like a sex bomb ruler, while Harrison has some notion of Caesar's majesty. Burton, however, is hopelessly wooden. Hard to believe this came from director Mankiewicz. Oscar nominations for Best Picture, actor Harrison, sound, original score and film editing in addition to the awards it won. **1963 246m/C** Elizabeth Taylor, Richard Burton, Rex Harrison, Roddy McDowall, Martin Landau, Pamela Brown, Michael Hordern, Kenneth Haigh, Andrew Keir, Hume Cronyn, Carroll O'Connor; **Dir:** Joseph L. Mankiewicz. Academy Awards '63: Best Art Direction/Set Decoration (Color), Best Cinematography, Best Costume Design, Best Visual Effects; National Board of Review Awards '63: Best Actor (Harrison); Harvard Lamopoon Awards '63: Worst Film of the Century. **VHS, Beta, LV $29.98** FOX *I*

Cleopatra Jones Lean and lethal government agent with considerable martial arts prowess takes on loathsome drug lords. Dobson is fetching as the lead performer in this fast-paced, violent flick. Followed by "Cleopatra Jones and the Casino of Gold." **1973 (PG) 89m/C** Tamara Dobson, Shelley Winters, Bernie Casey, Brenda Sykes; **Dir:** Jack Starrett. **VHS, Beta $19.98** WAR *II*

Cleopatra Jones & the Casino of Gold Dobson returns as the lethal, physically imposing federal agent to Hong Kong to take on a powerful druglord in this sequel to "Cleopatra Jones." Watch for sexy Stevens as the Dragon Lady. **1975 (R) 96m/C** Tamara Dobson, Stella Stevens, Norman Fell; **Dir:** Chuck Bail. **VHS, Beta $19.98** WAR *II*

The Clergyman's Daughter A clergyman's daughter calls upon detectives Tommy and Tuppence to investigate some murders at the family's country house. Made for British television and based on the Agatha Christie story. **1984 60m/C** GB James Warwick, Francesca Annis. **VHS, Beta $14.95** PAV *I 1/2*

Cliffhangers Two volumes of collected movie serial coming attractions from the heyday of Saturday matinees, including "Blackhawk," "Gangbusters," "Captain America," "Mysterious Dr. Satan" and "Tim Tyler's Luck." **194? 57m/C VHS, Beta** CPB

A Climate for Killing Arizona police are baffled when a woman murdered 16 years earlier is again found dead. Case open? Hunks Bauer and Beck are earnest in leads, and Ross and Sara are foxy. Passable entertainment. **1991 (R) 104m/C** Steven Bauer, John Beck, Katharine Ross, Mia Sara; **W/Dir:** J.S. Cardone. **VHS, Beta, LV $89.98** MED *II 1/2*

Climax Blues Band: Live from London The British blues band performs at London's Marquee Club, featuring "Gotta Have More Love," and "Couldn't Get it Right." **1985 60m/C VHS, Beta $29.95** RCA *II 1/2*

The Climb German mountain climbers attempt to conquer the Himalayan Nanga Parbat, the world's fifth highest peak. Sometimes compelling film based on 1953 expedition in which only one member succeeded in reaching the summit.

1987 (PG) 90m/C Bruce Greenwood, James Hurdle, Ken Welsh; **Dir:** Donald Shebib. **VHS, Beta, LV $19.95** VIR, IME *II 1/2*

The Clinic Day in the life farce follows the sometimes humorous events at a VD clinic featuring a homosexual doctor, assisted by a curious student intern. **1983 95m/C** AU Chris Haywood, Simon Burke, Gerda Nicolson, Rona McLeod, Suzanne Raylance, Veronica Lang, Pat Evison; **Dir:** David Stevens. **VHS, Beta $29.98** VID *II*

Clipped Wings The Bowery Boys are at their best as they inadvertantly join the army while visiting a friend and in the process of their usual bumblings, uncover a Nazi plot. **1953 62m/B** Leo Gorcey, Huntz Hall, Bernard Gorcey, David Condon, Bennie Bartlett, June Vincent, Mary Treen, Philip Van Zandt, Elaine Riley, Jeanne Dean, Lyle Talbot; **Dir:** Edward L. Bernds. **VHS $14.98** WAR *II 1/2*

Cloak and Dagger An American physicist joins the secret service during WWII and infiltrates Nazi territory to release a kidnapped scientist who is being forced to build a nuclear bomb. Disappointing spy show with muted anti-nuclear tone. Written by Ring Lardner Jr. and Albert Maltz. **1946 106m/B** Gary Cooper, Lilli Palmer, Robert Alda, James Flavin, Vladimir Sokoloff, J. Edward Bromberg, Marc Lawrence, Ludwig Stossel; **Dir:** Fritz Lang. **VHS, LV $19.95** REP, FCT *II*

Cloak & Dagger A young boy, last seen befriending E.T., depends upon his imaginary super-friend to help him out when some real-life agents are after his video game. Coleman is particularly fun as both dad and fantasy hero in an interesting family adventure. **1984 (PG) 101m/C** Dabney Coleman, Henry Thomas, Michael Murphy, John McIntire, Jeanette Nolan; **Dir:** Richard Franklin. **VHS, Beta, LV $19.95** MCA *II 1/2*

The Clock Appealing romance about an office worker who meets and falls in love with a soldier on two-day leave in New York City. Charismatic Walker and likeable Garland make a fine screen couple, and Wynn is memorable as the drunk. The original theatrical trailer is included on the laserdisc format. **1945 91m/B** Judy Garland, Robert Walker, James Gleason, Marshall Thompson, Keenan Wynn; **Dir:** Vincente Minnelli. **VHS, LV $19.98** BTV, PIA, MGM *III*

The Clockmaker Contemplative drama about a clockmaker whose life is shattered when his son is arrested as a political assassin. Tavernier regular Noiret excels in the lead. In French with English subtitles. **1976 105m/C** FR Philippe Noiret, Jean Rochefort, Jacques Denis, William Sabatier, Christine Pascal; **Dir:** Bertrand Tavernier. National Board of Review Awards '76: 5 Best Foreign Films of the Year. **VHS, Beta, LV $34.95** FCT, LUM, PIA *III 1/2*

Clockwise Monty Python regular Cleese is a teacher preoccupied by punctuality. His neurosis proves his undoing when he falls victim to misadventure while traveling to deliver a speech. Cleese is acceptable, dialogue and story is less so, though sprinkled with a fair amount of humor. **1986 (PG) 96m/C** GB John Cleese, Penelope Wilton, Alison Steadman, Stephen Moore, Sharon Maiden; **Dir:** Christopher Morahan. **VHS, LV $79.99** HBO *II 1/2*

A Clockwork Orange In the Britain of the near future, a sadistic punk leads a gang on nightly rape and murder sprees, then is captured and becomes the subject of a grim experiment to eradicate his violent tenden-

cies in this extraordinary adaptation of Anthony Burgess's controversial novel. The film is an exhilerating experience, with an outstanding performance by McDowell as the funny, fierce psychopath. Many memorable, disturbing sequences, including a rape conducted while assailant McDowell belts "Singing in the Rain." Truly outstanding, provocative work from master filmmaker Kubrick. Oscar nominated for Best Picture, Director, Adapted Screenplay, and Film Editing. **1971 (R) 137m/C** GB Malcolm McDowell, Patrick Magee, Adrienne Corri, David Prowse; **W/Dir:** Stanley Kubrick. New York Film Critics Awards '71: Best Direction, Best Film; Harvard Lampoon Awards '71: Worst Film of the Year. **VHS, Beta, LV $19.98** WAR, PIA, FCT *IIII*

Clodhopper A drama of a young man traveling to New York to find fame and fortune. Music by R. Cameron Menzies. **1917 47m/B** Charles Ray, Margery Wilson. **VHS, Beta $29.98** CCB, DVT *II*

The Clones A doctor discovers a government experiment engineered to murder him with a perfect clone. **1973 (PG) 90m/C** Michael Greene, Gregory Sierra; **Dir:** Paul Hunt, Lamar Card. **VHS, Beta $69.98** LIV, LTG, HHE *II*

Clones of Bruce Lee It's a martial arts free-for-all as bold masters from throughout the world duel for supremacy in the realm of self defense. If you like martial arts movies, you'll probably like this one. If you don't know if you like martial arts movies, find out with one of the actual Bruce Lee's flicks. **1980 (R) 87m/C** Dragon Lee, Bruce Le, Bruce Lai, Bruce Thai; **Dir:** Joseph Kong. **VHS, Beta $49.95** MED *II*

The Clonus Horror A scientist discovers a government plot to clone the population by freezing bodies alive and using their parts in surgery. Also known as "Parts: The Clonus Horror." **1979 90m/C** Tim Donnelly, Keenan Wynn, Peter Graves, Dick Sargent, Paulette Breen; **Dir:** Robert S. Fiveson. **VHS, Beta $69.98** LIV, LTG *I I*

Close Encounters of the Third Kind Middle-American strangers become involved in the attempts of benevolent aliens to contact earthlings. Despite the sometime mundane nature of the characters, this Spielberg epic is a stirring achievement. Studded with classic sequences, and the ending is an exhilerating experience of special effects and peace-on-earth feelings. Dreyfuss and Dillon excel as friends who are at once bewildered and obsessed by the alien presence, and French filmmaker Truffaut is also strong as the stern, ultimately kind scientist. Grabbed two Oscars and was nominated for actress Dillon, director Spielberg, Art Direction, Set Decoration, Sound, Original Score, Film Editing, and Visual Effects. Laserdisc includes: formerly edited scenes, live interviews with Spielberg, special effects man Douglas Trumbull, and composer John Williams, publicity materials, and over 1,000 production photos. **1980 (PG) 152m/C** Richard Dreyfuss, Teri Garr, Melinda Dillon, Francois Truffaut, Bob Balaban; **Dir:** Steven Spielberg. Academy Awards '77: Best Cinematography, Best Sound Effects Editing; British Academy Awards '78: Best Art Direction; National Board of Review Awards '77: 10 Best Films of the Year. **VHS, Beta, LV $14.95** COL, CRC *IIII*

Close to Home A runaway fleeing a home life of abuse and neglect befriends a TV journalist doing a story on similar girls. She turns to him for support after she is arrested and returned home. Merely mediocre.
1986 93m/C *CA* Daniel Allman, Jillian Fargey, Anne Petrie; *Dir:* Rick Beairsto. VHS, Beta $79.95 TWE 🎬🎬

Close My Eyes A hot, steamy summer arouses vibrant passions in a brother and sister, and neither marriage nor blood relation can halt the inevitable.
1991 (R) 105m/C Alan Rickman, Clive Owen, Saskia Reeves; *Dir:* Stephen Poliakoff. VHS $89.95 ACA 🎬🎬

Closely Watched Trains A novice train dispatcher attempts to gain sexual experience in German-occupied Czechoslovakia during WWII. Many funny scenes in this film regarded by some as a classic. Based upon the classic Czech novel by Bohumil Hrabal and subtitled in English.
1966 89m/C *CZ* Vaclav Neckar, Jitka Bendova, Vladimir Valenta, Josef Somr; *Dir:* Jiri Menzel. Academy Awards '67: Best Foreign Language Film. VHS, Beta, LV $59.95 TAM, FXL, COL 🎬🎬🎬½

The Closer Aiello is "The Closer," a high-powered salesman whose world is falling apart. Somewhere between the money and the power, he's lost touch with his family and the truly important things in life. Is it too late to get them back?
1991 (R) 87m/C Danny Aiello, Michael Pare, Joe Cortese, Justine Bateman, Diane Baker, James Karen, Rick Aiello, Michael Lerner; *Dir:* Dimitri Logothetis. VHS $89.95 ACA 🎬🎬½

Closet Cases of the Nerd Kind In this comical takeoff on "Close Encounters of the Third Kind," a zany assortment of mortals meet up with extraterrestrials. All of the memorable scenes of the Spielberg film are recreated and spoofed.
1980 12m/C VHS, Beta PYR 🎬🎬½

Closet Land Severe, stylized political allegory aims high but fails to convincingly distill totalitarian repression into just two characters—a government inquisitor and his captive, a woman subjected to hideous mental and physical torture. Child-molesting emerges as an ill-advised subtheme. A better video on the same subject: "Interrogation," from Poland.
1990 (R) 95m/C Madeleine Stowe, Alan Rickman; *Dir:* Radha Bharadwaj. VHS, Beta, LV $92.98 FXV 🎬½

Cloud Dancer A self-absorbed acrobat uses and abuses those around him, including his girlfriend. Carradine and O'Neill fail to distinguish themselves in this mediocre film.
1980 (PG) 107m/C David Carradine, Jennifer O'Neill, Joseph Bottoms, Colleen Camp; *Dir:* Barry Brown. VHS, Beta $9.99 STE, PSM 🎬🎬

Cloud Waltzing The screen version of a Harlequin romance about a journalist who arrives in France to interview a vinter and finds romance and danger. Melodrama bolstered by the presence of photogenic Beller.
1987 103m/C Kathleen Beller, Francois-Eric Gendron, Paul Maxwell, Therese Liotard, Claude Gensac, David Baxt; *Dir:* Gordon Flemyng. VHS, Beta $19.95 PAR 🎬🎬½

Clouds Over Europe A test pilot and a man from Scotland Yard team up to find out why new bomber planes are disappearing. Also known as "Q Planes."

1939 82m/B Laurence Olivier, Valerie Hobson, Ralph Richardson; *Dir:* Tim Whelan. VHS, Beta, LV $49.95 SUE 🎬🎬🎬

Clown A broken-down funny man with a worshipful son obtains one last career opportunity. This "Champ"-derived tearjerker features credible playing from Skelton and Considine as father and son, respectively.
1953 91m/B Red Skelton, Jane Greer, Tim Considine, Steve Forrest; *Dir:* Robert Z. Leonard. VHS, Beta $49.95 MGM 🎬🎬½

Clown House Young brothers are stalked in their home by three demonic clowns.
1988 (R) 95m/C Nathan Forrest Winters, Brian McHugh, Sam Rockwell, Viletta Skillman, Timothy Enos, Tree; *Dir:* Victor Salva. VHS, Beta, LV $89.95 COL 🎬

Clown Murders A posh Halloween party is undone when a cruel group fakes a kidnapping to ruin a too-prosperous pal's business deal. Awkward thriller boasts little suspense. Candy plays it straight, mostly, in this one.
1983 94m/C John Candy, Al Waxman, Susan Keller, Lawrence Dane; *Dir:* Martyn Burke. VHS, Beta $59.95 TWE 🎬½

Clowning Around Simon, who has lived in foster homes all his life, dreams of becoming a famous circus clown. When he's sent to a new home, his new foster parents think his idea is silly, so he runs away and joins the circus. There he meets Anatole, a European clown has-been who can no longer perform because of injuries. Simon believes Anatole can teach him about clowning, so he follows Anatole to Paris to help make his dream come true. Includes a viewers' guide.
199? 165m/C VHS $29.95 PME 🎬½

The Clowns An idiosyncratic documentary about circus clowns. Director Fellini has fashioned an homage that is sincere, entertaining, and personal. Contains some truly poignant sequences. Subtitled, this was made for Italian television.
1971 90m/C *FR IT Dir:* Federico Fellini. National Board of Review Awards '71: 5 Best Foreign Films of the Year. VHS, Beta $59.95 HTV, NOS, MRV 🎬🎬½

The Club A soccer coach tries to train and motivate a mediocre squad into on worthy of playing for—and perhaps even winning—the league cup. Conventional sports story is significantly improved by Thompson's rendering of the coach. Also known as "Players."
1981 (PG) 99m/C *AU* Jack Thompson, Graham Kennedy, Frank Wilson, John Howard, Alan Cassell; *Dir:* Bruce Beresford. VHS, Beta $19.95 ACA 🎬🎬🎬

Club Extinction A near-future thriller in which Berlin is plagued by a wave of suicides. An investigator suspects a spa representative and a media tycoon of being involved. Features a fine international cast, but the film is a lesser venture from prolific master Chabrol.
1989 (R) 105m/C *GE* Alan Bates, Andrew McCarthy, Jennifer Beals, Jan Niklas; *Dir:* Claude Chabrol. VHS, Beta $89.95 PSM, FCT 🎬🎬½

Club Fed A rigid prison warden's plot to impose greater discipline is undone by inmate highjinks. Crude laughs and one-dimensional play from a cast of recognizable names and faces.
1991 (PG-13) 93m/C Judy Landers, Sherman Hemsley, Karen Black, Burt Young, Rick Schmidt, Allen Garfield, Joseph Campanella, Lyle Alzado, Mary Woronov, Debbie Lee Carrington;

Dir: Nathaniel Christian. VHS, Beta $79.95 PSM 🎬½

Club Life A hardened biker finds sex, drugs, and violence on the Hollywood streets. What did he expect? Peace, love, and understanding?
1986 (R) 93m/C Tony Curtis, Dee Wallace Stone, Tom Parsekian, Michael Parks, Jamie Barrett; *Dir:* Norman Thaddeus Vane. VHS, Beta $9.99 STE, PSM 🎬

Club Med An insecure comedian and his goofy friend try to make the most of a ski vacation. Perhaps your only chance to see Thicke, Killy, and Coolidge together.
1983 (PG) 60m/C Alan Thicke, Jim Carrey, Jean-Claude Killy, Rita Coolidge, Ronnie Hawkins; *Dir:* David Mitchell, Bob Giraldi. VHS, Beta $39.95 AHV 🎬🎬

Club Paradise A Chicago fireman flees the big city for a faltering tropical resort and tries to develop some night life. Somewhat disappointing with Williams largely playing the straight man. Most laughs provided by Martin, particularly when she is assaulted by a shower, and Moranis, who gets lost while windsurfing.
1986 (PG-13) 96m/C Robin Williams, Peter O'Toole, Rick Moranis, Andrea Martin, Jimmy Cliff, Brian Doyle-Murray, Twiggy, Eugene Levy, Adolph Caesar, Joanna Cassidy, Mary Gross, Carey Lowell; *Dir:* Harold Ramis. VHS, Beta, LV $19.98 WAR 🎬🎬½

Clue The popular boardgame's characters must unravel a night of murder at a spooky Victorian mansion. The entire cast seems to be subsisting on suger, with wild eyes and frantic movements the order of the day. Butler Curry best survives the uneven script and direction. Warren is appealing too. The theatrical version played with three alternative endings, and the video version shows all three successively.
1985 (PG) 96m/C Lesley Ann Warren, Tim Curry, Martin Mull, Madeline Kahn, Michael McKean, Christopher Lloyd, Eileen Brennan, Howard Hesseman, Lee Ving, Jane Wiedlin, Colleen Camp; *Dir:* Jonathan Lynn. VHS, Beta, LV $19.95 PAR 🎬🎬½

The Clue According to Sherlock Holmes Sherlock and Dr. Watson host a mystery about a young boy and an older woman who team up to unravel the clues. Made for TV.
19?? 50m/C Dody Goodman, Keith Mitchell; *Dir:* Murray Golden. VHS GEM 🎬½

The Clutching Hand The Clutching Hand seeks a formula that will turn metal into gold and detective Craig Kennedy is out to prevent him from doing so. A serial in 15 chapters on three cassettes.
1936 268m/B Jack Mulhall, Rex Lease. VHS, Beta $26.95 SNC, NOS, VDM 🎬🎬

Coach A sexy woman is unintentionally hired to coach a high school basketball team and mold rookies into lusty young champions. Low-grade roundball fever.
1978 (PG) 100m/C Cathy Lee Crosby, Michael Biehn, Keenan Wynn, Sidney Wicks; *Dir:* Bud Townsend. VHS, Beta $9.95 MED 🎬½

Coal Miner's Daughter A strong bio of country singer Loretta Lynn, who rose from Appalachian poverty to Nashville riches. Spacek is perfect in the lead, and she even provides acceptable rendering of Lynn's tunes. Band drummer Helm shines as Lynn's father, and Jones is strong as Lynn's down-home husband. Uneven melodrama toward the end, but the film is still a good one.

1980 (PG) 125m/C Sissy Spacek, Tommy Lee Jones, Levon Helm, Beverly D'Angelo; *Dir:* Michael Apted. Academy Awards '80: Best Actress (Spacek); Golden Globe Awards '81: Best Film—Musical/Comedy; Los Angeles Film Critics Association Awards '80: Best Actress (Spacek); National Board of Review Awards '80: 10 Best Films of the Year, Best Actress (Spacek); National Society of Film Critics Awards '80: Best Actress (Spacek). **VHS, Beta, LV** $14.95 *MCA, PIA, BTV* 🎙🎙🎙

Coast to Coast A wacky woman escapes from a mental institution and teams with a pugnacious trucker for a cross-country spree. Ostensibly freewheeling comedy is only occasionally worthwhile despite appropriate playing from leads Cannon and Blake, who deserve better.
1980 (PG) 95m/C Dyan Cannon, Robert Blake, Quinn Redeker, Michael Lerner, Maxine Stuart, Bill Lucking; *Dir:* Joseph Sargent. **VHS, Beta, LV** $49.95 *PAR* 🎙🎙

The Coast Patrol B. Reeves Eason, responsible for the chariot scene in the silent "Ben Hur" and the action scenes in "The Charge of the Light Brigade," directed Fay in one of her earliest parts, eight years before the role that prompted her to say "I didn't realize that King Kong and I were going to be together for the rest of our lives, and longer..." Much Eason-style high action when smugglers meet the coast patrol.
1925 63m/B Kenneth McDonald, Claire De Lorez, Fay Wray; *Dir:* B. Reeves Eason. **VHS, Beta** $19.95 *NOS, GPV, FCT* 🎙🎙½

Cobham Meets Bellson A thrilling performance by the two leading big band drummers highlights this concert taped in Switzerland.
1983 36m/C Billy Cobham, Louis Bellson. **VHS, Beta** $29.95 *VVV, MBP* 🎙🎙½

Cobra To pay off the debts incurred by his profligate playbuoyance, Valentino takes a job with an antique dealer, falls for the dealer's secretary and arranges an assignation with the dealer's wife, who is inconveniently killed at the designated rendezvous. Complications abound as Valentino finds himself Shiek out of luck. Not a prequel to the flag-wagging Stallone vehicle, this lesser Valentino effort—directed by Henabery (who played Lincoln in "The Birth of a Nation")—was released a year before the famous heartthrob's untimely death, and enjoyed a rather acrid critical reception. Which isn't to say the public didn't flock to see the Italian stallion flare his nostrils.
1925 70m/B Rudolph Valentino, Nita Naldi, Casson Ferguson, Gertrude Olmsted; *Dir:* Joseph Henabery. **VHS, Beta** $27.95 *GPV, FCT, DNB* 🎙

The Cobra A tightlipped U.S. agent ogles a voluptuous siren when not fighting opium smugglers from the Middle East. A must for all admirers of Ekberg.
1968 93m/C Dana Andrews, Anita Ekberg; *Dir:* Mario Sequi. **VHS** $59.95 *HAR* 🎙

Cobra A hired killer destroys most of a mob family and becomes the target of revenge for the lone survivor. Moody thriller boasts an international cast. Foreign title is "Le Sant de l'Ange."
1971 93m/C *FR IT* Sterling Hayden, Senta Berger, Jean Yanne; *Dir:* Yves Boisset. **VHS, Beta** $59.95 *UNI* 🎙🎙½

Cobra A cold-blooded cop protects a model from a gang of deranged killers. Lowbrow, manipulative action fare void of feeling. Truly exploitive, with little expression from

the leads. Highlight is the extended chase sequence.
1986 (R) 87m/C Sylvester Stallone, Reni Santoni, Brigitte Nielsen, Andrew Robinson; *Dir:* George P. Cosmatos. **VHS, Beta, LV** $19.98 *WAR* 🎙

Cobra Two martial-arts masters duel to the death. Isn't that what always happens in these chop-sock affairs?
19?? 90m/C Bruce Le. **VHS, Beta** $19.99 *BFV* 🎙

Cobra Against Ninja Once again, two martial arts experts try to kill each other. Can't be many of these fellas left.
19?? 88m/C Jimmy Bosco, Paul Branney, Gary Carter, Alan Friss, Alfred Pears; *Dir:* Joseph Lai. **VHS** $59.95 *IMP* 🎙

The Cobra Strikes Nosey thief gets more than he bargained for when he works his way through an inventor's studio. Lowbudget and unexceptional.
1948 62m/B Sheila Ryan, Richard Fraser, Leslie Brooks, Herbert Heyes; *Dir:* Charles F. Reisner. **VHS** $19.95 *NOS, MOV, RXM* 🎙½

The Cobweb Shady goings on are uncovered at a psychiatric ward run by a bunch of neurotic administrators. A strong cast, director and producer do not add much to this dull outing.
1955 125m/C Lauren Bacall, Richard Widmark, Gloria Grahame, Charles Boyer, Lillian Gish, John Kerr, Susan Strasberg, Oscar Levant, Tommy Rettig, Paul Stewart, Adele Jergens, Bert Freed, Sandy Descher, Fay Wray, Virginia Christine; *Dir:* Vincente Minnelli. **LV** $39.98 *MGM* 🎙

The Coca-Cola Kid A smug sales executive treks to Australia to improve regional sales and becomes embroiled in sexual and professional shenanigans. Roberts is strong in the difficult lead role, and Scacchi is compelling in an awkwardly constructed part. Ambitious satire is somewhat scattershot, with more storylines than it needs. Still, filmmaker Makavejev is usually capable of juggling the entire enterprise.
1984 (R) 94m/C *AU* Eric Roberts, Greta Scacchi, Bill Kerr; *Dir:* Dusan Makavejev. **VHS, Beta, LV** $19.95 *LIV, VES, FCT* 🎙🎙½

Cocaine Cowboys Lame rockers support themselves between gigs by peddling drugs then find out they've run afoul of the mob. Palance is the only worthwhile aspect of this rambling venture produced at Andy Warhol's home.
1979 (R) 90m/C Jack Palance, Andy Warhol, Tom Sullivan, Suzanna Love; *Dir:* Ulli Lommel. **VHS, Beta** *MED, OM* 🎙½

Cocaine Fiends Drug use leads siblings into a squalid life of addiction, crime, prostitution, and eventually suicide. Ostensibly straight morality tale functions better as loopy camp. Features memorable slang.
1936 74m/B Lois January, Noel Madison; *Dir:* W.A. Conner. **VHS, Beta** $16.95 *SNC, NOS, MED* 🎙🎙

Cocaine: One Man's Seduction Made-for-television film documents one man's degeneration into drug addiction. Weaver snorts convincingly in lead.
1983 97m/C Dennis Weaver, Karen Grassle, Pamela Bellwood, David Ackroyd; *Dir:* Paul Wendkos. **VHS, Beta** $14.95 *FRH, CHI* 🎙½

Cocaine Wars An undercover U.S. agent kills dozens while trying to rescue his kidnapped girlfriend from an evil South American drug tycoon. Violent, but otherwise unaffecting.

1986 (R) 82m/C John Schneider, Kathryn Witt, Royal Dano; *Dir:* Hector Olivera. **VHS, Beta** $19.95 *MED* 🎙

Cockeyed Cavaliers Wheeler and Woolsey are stockaded for stealing the Duke's horses and carriage. To escape jail they swap clothes with some drunken royalty.
1934 70m/B Bert Wheeler, Robert Woolsey, Dorothy Lee, Thelma Todd; *Dir:* Mark Sandrich. **VHS, Beta** $34.98 *CCB* 🎙🎙

Cockfighter A unique, grim portrait of a cockfighting trainer. Oates, in the lead, provides voice over narrations, but the film otherwise remains silent until the end. Strong support from Stanton. Interesting, violent fare. Alternate titles: "Gamblin' Man," "Born to Kill," and "Wild Drifters."
1974 (R) 84m/C Warren Oates, Harry Dean Stanton, Richard B. Shull, Troy Donahue, Millie Perkins; *Dir:* Monte Hellman. **VHS, Beta** $59.98 *SUE* 🎙🎙🎙

Cocktail A smug young man finds fame and fortune as a proficient, flashy bartender, charming the ladies with his bottle and glass juggling act. Slick, superficial film boasts a busy soundtrack and serviceable exchanges between male leads Cruise and Brown. There's less chemistry between Cruise and love interest Shue. Filmed in a high-tech rock video style.
1988 (R) 103m/C Tom Cruise, Bryan Brown, Elizabeth Shue, Lisa Banes, Laurence Luckinbill, Kelly Lynch, Gina Gershon, Ron Dean, Paul Benedict; *Dir:* Roger Donaldson. **VHS, Beta, LV, 8mm** $19.99 *TOU* 🎙

The Cocoanuts In their film debut, the Marx Brothers create their trademark, indescribable mayhem. Stagey, technically crude comedy nonetheless delights with zany, free-for-all exchanges, antics. Includes famous "viaduct" exchange. Songs by Irving Berlin; written by George S. Kaufman and Morrie Ryskind.
1929 96m/B Groucho Marx, Chico Marx, Harpo Marx, Zeppo Marx, Margaret Dumont, Kay Francis, Oscar Shaw, Mary Eaton; *Dir:* Robert Florey, Joseph Santley. **VHS, Beta, LV** $19.95 *MCA* 🎙🎙½

Cocoon Humanist fantasy in which senior citizens discover their fountain of youth is actually a breeding ground for aliens. Heartwarming, one-of-a kind drama showcases elderly greats Ameche, Brimley, Gilford, Cronyn, and Tandy. A commendable, recommendable venture. Based on David Saperstein's novel and followed by "Cocoon: The Return."
1985 (PG-13) 117m/C Wilford Brimley, Brian Dennehy, Steve Guttenberg, Don Ameche, Tahnee Welch, Jack Gilford, Hume Cronyn, Jessica Tandy, Gwen Verdon, Maureen Stapleton, Tyrone Power Jr., Barret Oliver, Linda Harrison. Academy Awards '85: Best Supporting Actor (Ameche), Best Visual Effects. **VHS, Beta, LV** $19.98 *FOX, FCT, BTV* 🎙🎙🎙

Cocoon: The Return Old timers who left with aliens in "Cocoon" return to earth and face grave problems. Less compelling sequel misses guiding hand of earlier film's director, Ron Howard. Still, most of cast from the original is on board here, and the film has its moments.
1988 (PG) 116m/C Don Ameche, Wilford Brimley, Steve Guttenberg, Maureen Stapleton, Hume Cronyn, Jessica Tandy, Gwen Verdon, Jack Gilford, Tahnee Welch, Courtney Cox, Brian Dennehy, Barret Oliver; *Dir:* Daniel Petrie. **VHS, Beta, LV** $19.98 *FOX*

C

C.O.D. Bosom buddies must develop advertising campaign for brassiere producer or find their careers are bust. Premise milked for all its worth, so you may want to keep abreast of this one. On the other hand, you may not give a hoot. Oh well, tit for tat.
1983 (PG) 96m/C Chris Lemmon, Olivia Pascal, Jennifer Richards, Corinne Alphen, Teresa Ganzel, Carole Davis; *Dir:* Chuck Vincent. **VHS, Beta** $69.95 *VES, LIV* ⅟

Code of the Fearless A cowboy wins the heart of a beautiful maiden while bringing evildoers to justice.
1939 56m/B Fred Scott, Claire Rochelle, John Merton, Harry Harvey, Walter McGrail, Roger Williams; *Dir:* Raymond K. Johnson. **VHS** $19.95 *NOS, DVT* ⅟

Code of Honor Prominent businessman embarks on mission of revenge when police fail to apprehend his wife's rapist. No surprise here. You either enjoy this sort of thing or you lead a reasonably fulfilling life.
19?? 90m/C Cameron Mitchell, Mark Sabin. **VHS** $29.95 *AVD* ⅟

Code Name Alpha You don't have to have a federal case of intelligence to figure out where this FBI story's going and how it's going to get there. Someone's smuggling sophisticated electrical components to the communists, and a team of intelligence agents, including Branger and Schiaffino, hike over to Hong Kong to break up the party. Originally titled "Red Dragon."
1967 89m/C *IT GE* Stewart Granger, Rosanna Schiaffino, Harald Juhnke, Paul Klinger, Helga Sommerfeld, Horst Frank; *Dir:* Ernest Hofbauer. **VHS** $14.99 *VTR* ⅟

Code Name: Dancer A retired spy leaves her peaceful life, husband and family to take care of business in Cuba.
1987 93m/C Kate Capshaw, Jeroen Krabbe, Gregory Sierra, Cliff De Young. **VHS** $89.95 *VMK* ⅟⅟

Code Name: Diamond Head In Hawaii, a secret agent is assigned to retrieve a deadly chemical that has been hidden by a master of disguises. Made for television.
1977 78m/C Roy Thinnes, France Nuyen, Zulu, Ian McShane; *Dir:* Jeannot Szwarc. **VHS, Beta** $49.95 *WOV* ⅟

Code Name: Emerald WWII drama in which an American agent must stop an enemy spy with knowledge of impending D-Day invasion. Conventional TV fare boasts a fine performance by von Sydow, but Stolz is hopelessly miscast as an adult. Perhaps the only film featuring both Bucholz and Berger. First feature film by NBC-TV network.
1985 (PG) 95m/C Ed Harris, Max von Sydow, Eric Stoltz, Horst Buchholz, Helmut Berger, Cyrielle Claire, Patrick Stewart, Graham Crowden; *Dir:* Jonathan Sanger. **VHS, Beta, LV** $79.98 *FOX* ⅟

Code Name: Zebra The Zebra Force helps the cops put bad guys away. When the bad guys get out, they want revenge. The result: lots of violence and action. Wasn't released theatrically.
1984 96m/C Jim Mitchum, Michael Lane, Timmy Brown, Joe Dante, Deana Jurgens, Frank Sinatra Jr.; *Dir:* Joe Tornatore. **VHS, Beta** $79.95 *TWE* ⅟½

Code of Silence A police loner must contend with both a gang war and police department corruption. Hectic, violent action-drama with some wild stunts.

1985 (R) 100m/C Chuck Norris, Henry Silva, Bert Remsen, Molly Hagan, Nathan Davis, Dennis Farina; *Dir:* Andy Davis. **VHS, Beta, LV** $14.99 *HBO, PIA* ⅟⅟

Codename: Foxfire An attractive secret agent is framed for wrongdoing and must clear her name by tracking the enemy spy lurking within the intelligence network. Always interesting Cassidy is an asset, but this by-the-numbers TV film is otherwise a misappropriation of the term ''intelligence'' in relation to espionage. Also known as "Slay It Again, Sam."
1985 99m/C Joanna Cassidy, John McCook, Sheryl Lee Ralph, Henry Jones, Luke Andreas; *Dir:* Corey Allen. **VHS, Beta** $39.95 *MCA* ⅟⅟

Codename: Icarus A young mathematical genius, enrolled at a special school for accelerated study uncovers evil government plots to use the students for espionage. Made for British television.
1985 106m/C *GB* Barry Angel, Jack Galloway; *Dir:* Marilyn Fox. **VHS, Beta** $59.98 *FOX* ⅟⅟

Codename Kyril Edward ("The Equalizer") Woodward plays a British spy pitted against Soviet assassin Kyril (Charleson) who's been sent to England by the KGB to rout out a double agent. Turns out there's more to it than Kyril knows, and nobody loves an assassin. Based on John Trenhaile's "A Man Called Kyril," John Hopkins' script is slow from the gun, but eventually builds something akin to momentum. Made for cable TV.
1991 (R) 115m/C Edward Woodward, Ian Charleson, Denholm Elliott, Joss Ackland, Richard E. Grant; *Dir:* Ian Sharp. **VHS** $79.98 *TTC* ⅟⅟½

Codename: Terminate Title says it all in this tender exploration of the human condition. Jungle-trained mercenaries grab their guns and go after an American pilot shot down behind enemy lines.
1990 84m/C Robert Mason, Jim Gaines. **VHS** $79.95 *ATL* ⅟½

Codename: Vengeance A macho fellow tries to stop a killer holding the shah's wife and son hostage. Dated.
1987 (R) 97m/C Robert Ginty, Cameron Mitchell, Shannon Tweed; *Dir:* David Winters. **VHS** $79.98 *MCG* ⅟

Codename: Wildgeese The big-name cast of this meandering mercenary macho-rama probably wish they'd been credited under code names; solid histrionics cannot a silly script save. A troop of commandos-for-hire are engaged by the Drug Enforcement Administration to obliterate an opium operation in Asia's infamous Golden Triangle, and much mindless agitation ensues.
1984 (R) 101m/C *GE IT* Lewis Collins, Lee Van Cleef, Ernest Borgnine, Klaus Kinski, Mimsy Farmer; *Dir:* Anthony M. Dawson. **VHS** $19.95 *STE, NWV* ⅟½

Cody A former bronco buster wants his son "take it like a man" when the boy's dog dies. Then a mystical goose arrives to mellow the old man. That's really the plot.
1967 82m/C Tony Becker, Terry Evans. **VHS, Beta** *PGN* ⅟

A Coeur Joie Sixties set piece about a gorgeous woman who must choose between her much older lover or her newfound stud. Bardot and U.K. scenery keep this one from sinking entirely. In French with English subtitles. Also known as "Two Weeks in September."

1967 100m/C *FR* Brigitte Bardot, Laurent Terzieff; *Dir:* Serge Bourguignon. **VHS, Beta** $19.95 *KRT* ⅟⅟

Coffy A beautiful woman feigns drug addiction to discover and destroy the evil dealers responsible for her sister's death. Grier is everything in this exploitative flick full of violence and nudity.
1973 (R) 91m/C Pam Grier, Booker Bradshaw, Robert DoQui, Allan Arbus; *Dir:* Jack Hill. **VHS, Beta, LV** $19.98 *ORI, FCT* ⅟⅟

Cohen and Tate Two antagonistic mob hitmen kidnap a 9-year-old who witnessed his parent's recent murder by the mob. In order to survive, the boy begins to play one psycho off the other.
1988 (R) 113m/C Roy Scheider, Adam Baldwin, Harley Cross, Cooper Huckabee; *Dir:* Eric Red. **VHS, Beta, LV** $19.98 *SUE* ⅟⅟

Cold Blood A small dubbed Dutch film about a pair of kidnap victims who turn the tables on their captors.
197? (R) 90m/C *NL* Rutger Hauer. **VHS, Beta** $9.95 *SIM, WES* ⅟½

Cold Blooded Murder A martial arts master assists the cops in a hunt for a Jack-the-Ripper-style killer. Among Lee's more obscure appearances.
197? 96m/C Lau Dan, Bruce Lee, Stephen Leung, Chan Wai Man, Carter Wong; *Dir:* Albert Law. **VHS** *MFI* ⅟

Cold Comfort A father arranges his teen daughter's romance, but finds his plans going awry in this Canadian thriller. Nominated for several Canadian television awards, including Best Actress, Best Actor, Best Picture and Best Musical Score.
1990 (R) 90m/CA Margaret Langrick, Maury Chaykin; *Dir:* Vic Sarin. **VHS** $89.98 *REP* ⅟⅟½

Cold Eyes of Fear A man is besieged by a raving convict whom his father, a judge, had put away years before. Music by Ennio Morricone.
197? 88m/C Fernando Rey. **VHS, Beta** $59.95 *TWE* ⅟

Cold Feet Light romantic comedy about a television director just recovering from a failed marriage and his fall for a lab researcher who just went through a break-up herself.
1984 (PG) 96m/C Griffin Dunne, Marissa Chibas, Blanche Baker, Mark Cronogue; *Dir:* Bruce Van Dusen. **VHS, Beta** $29.95 *FOX* ⅟⅟½

Cold Feet Modern-day western in which a trio of loopy desperados smuggle jewels inside a racehorse's stomach. Quirky comedy offers wild performances from Waits and sex-bomb Kirkland. Could use more laughs, though. Written by Thomas McGuane and Jim Harrison and filmed largely on McGuane's ranch.
1989 (R) 94m/C Keith Carradine, Tom Waits, Sally Kirkland, Rip Torn, Kathleen York, Bill Pullman; *Dir:* Robert Dornhelm. **VHS, Beta** $89.95 *IVE* ⅟⅟½

Cold Front A hired assassin turned serial killer is hunted by two dedicated cops.
1989 (R) 94m/C Martin Sheen, Michael Ontkean, Beverly D'Angelo, Kim Coates; *Dir:* Paul Bnarbic. **VHS, Beta, LV** $89.99 *HBO* ⅟½

Cold Heat When a custody battle rocks a crime boss' world, he pulls out all the stops in the war against his wife.
1990 85m/C John Phillip Law, Britt Ekland, Robert Sacchi, Roy Summerset, Joanne Watkins, Chance Michael Corbitt; *Dir:* Ulli Lommel. **VHS, Beta** $79.95 *AIP* ⅟⅟

Cold Justice An ex-prizefighter (Daltrey) and his friends welcome an English priest into their tough southside Chicago neighborhood. The good father seems too good to be true as he befriends the locals—even raising funds for the children's hospital. When the funds never make it to the hospital and the neighborhood begins to experience a number of violent tragedies the neighborhood decides to take its own revenge.
1989 (R) 106m/C Dennis Waterman, Roger Daltrey, Penelope Milford, Ron Dean, Bert Rosario; *W/Dir:* Terry Green. **VHS, LV $89.95** COL *♫♫*

Cold River An experienced Adirondacks guide takes his two children on an extended trip through the Adirondacks. For the children, it's a fantasy vacation until their father succumbs to a heart attack in the chilly mountains. "Cold River" is a journey of survival, and an exploration of human relationships.
1981 (PG) 94m/C Pat Petersen, Richard Jaeckel, Suzanne Weber; *Dir:* Fred G. Sullivan. **VHS, Beta $59.98** FOX *♫♫½*

Cold Room A college student on vacation with her father in East Berlin discovers the horrors hidden in an antiquated hotel room next to hers. Made for cable television.
1984 95m/C George Segal, Renee Soutendijk, Amanda Pays, Warran Clarke, Anthony Higgins; *Dir:* James Dearden. **VHS, Beta $59.95** FOX *♫½*

Cold Sassy Tree Endearing romance of a scandalous May-December marriage as perceived by the younger woman's teenage son. Dunaway and Widmark shine, and small town pettiness is vividly rendered. Adapted from the books by Olive Ann Burns. A superior made for cable television production.
1989 95m/C Faye Dunaway, Richard Widmark, Neil Patrick Harris, Frances Fisher, Lee Garlington, John M. Jackson; *Dir:* Joan Tewkesbury. **VHS $14.98** TTC *♫♫♫½*

Cold Steel Standard revenge drama about a hardnosed Los Angeles cop tracking his father's disfigured psycho-killer. Cast includes the always intense Davis, ex-pop star Ant, and screen scorchstress Stone. Still it's predictable, low-grade fare.
1987 (R) 91m/C Brad Davis, Jonathan Banks, Adam Ant, Sharon Stone; *Dir:* Dorothy Ann Puzo. **VHS, Beta $79.95** COL *♫½*

Cold Steel for Tortuga An Italian epic about a mercenary rescuing his woman and his gold from an evil governor. Dubbed.
1965 95m/C *IT* Guy Madison, Rick Battaglia, Ingeborg Schoener; *Dir:* Luigi Capuano. **VHS, Beta $69.95** LTG *♫*

Cold Sweat A brutal drug trader takes his ultra-violent revenge after his wife is captured by a drug boss' moronic henchmen. Typical Bronson flick boasts superior supporting cast of Ullman and Mason. Writing and direction, however, is mediocre.
1974 (PG) 94m/C *IT FR* Charles Bronson, Jill Ireland, Liv Ullman, James Mason; *Dir:* Terence Young. **VHS, Beta $59.95** GEM, MRV *♫♫*

Cold Turkey Often witty satire about what happens when an entire town tries to stop smoking for a contest. Van Dyke is fine as the anti-smoking minister; Bob and Ray are riotous as newscasters. Wholesome, somewhat tame fare.
1971 (PG) 99m/C Dick Van Dyke, Pippa Scott, Tom Poston, Bob Newhart, Vincent Gardenia, Barnard Hughes, Jean Stapleton, Graham Jarvis; *Dir:* Norman Lear. **VHS, Beta $19.95** MGM *♫♫♫*

Cold War Killers Soviet and British agents vie for control of prized cargo plane recently retrieved from the sea floor.
1986 (PG) 85m/C *GB* Terence Stamp; *Dir:* William Brayne. **VHS, Beta $79.95** WES *♫½*

Coldfire Two rookie cops work to undo the havoc unleashed on the streets of LA because of a deadly new drug sought for its powerful high. Not many surprises here.
1990 90m/C Wings Hauser, Kamar Reyes, Robert Vihard, Gary Swanson; *Dir:* Wings Hauser. **VHS** PMH *♫½*

The Colditz Story Prisoners of war from the Allied countries join together in an attempt to escape from Colditz, an allegedly escape-proof castle-prison deep within the Third Reich.
1955 93m/B John Mills, Eric Portman, Lionel Jeffries, Bryan Forbes, Ian Carmichael, Anton Diffring; *Dir:* Guy Hamilton. **VHS, Beta $59.99** HBO *♫♫½*

Cole Justice An older man, haunted by the memory of his girlfriend's rape and murder thirty-five years earlier, takes to the streets to protect innocent citizens from lowlife criminals. Somewhat unusual revenge pic filmed in and around Tulsa, Oklahoma.
1989 90m/C Carl Bartholomew, Keith Andrews, Mike Wiles, Nick Zickefoose; *Dir:* Carl Bartholomew. **VHS, Beta $9.95** SIM, RAE *♫♫*

The Collector Compelling adaptation of the John Fowles novel about a withdrawn butterfly collector who decides to add to his collection by kidnapping a beautiful girl he admires. He locks her in his cellar hoping she will fall in love with him. Chilling, unsettling drama with Stamp unnerving, yet sympathetic in lead. Eggar received an Oscar nomination for Best Actress.
1965 119m/C *GB* Terence Stamp, Samantha Eggar, Maurice Dallimore, Mona Washbourne; *Dir:* William Wyler. **VHS, Beta, LV $59.95** COL *♫♫♫*

Collector's Item Two lovers reunite after 16 years and a flurry of memories temporarily rekindles their passion, which comes dangerously close to obsession. Casting of the physically bountiful Antonelli helps this effective erotic drama.
1989 99m/C *IT* Tony Musante, Laura Antonelli, Florinda Bolkan; *Dir:* Giuseppe Patroni Griffi. **VHS, Beta $79.95** NSV *♫½*

College A high school valedictorian tries out for every sport in college, hoping to win the girl. Vintage Keaton antics, including disaster as a soda jerk, an attempt to be a track star, and the pole vault through a window to rescue the damsel in distress. Musical score by John Muri.
1927 60m/B Buster Keaton, A. Cornwall, Harold Goodwin; *Dir:* James W. Horne. **VHS, Beta, LV $19.95** DVT, MRV, CCB *♫♫♫½*

Collision Course A wise-cracking cop from Detroit teams up with Japan's best detective to nail a ruthless gang leader. Release was delayed until 1992 due to a lawsuit, but it was resolved in time to coordinate the release with Leno's debut as the host of "The Tonight Show." Pretty marginal, but diehard fans of Leno will appreciate it. Filmed on location in Detroit.
1989 (PG) 99m/C Pat Morita, Jay Leno, Chris Sarandon, Al Waxman. **VHS, Beta $89.99** HBO *♫½*

The Colombian Connection America's best undercover agent battles a powerful cocaine empire deep in the jungles of the Amazon, where slave labor and political corruption abound.
1991 90m/C Miles O'Keefe, Henry Silva. **VHS, Beta** NVH *♫½*

Colonel Effingham's Raid A retired army colonel uses military tactics to keep an old historical courthouse open and defeat some crooked politicians in the process.
1945 70m/B Joan Bennett, Charles Coburn, William Eythe, Donald Meek; *Dir:* Irving Pichel. **VHS, Beta $19.95** NOS, TIM *♫♫½*

Colonel Redl Absorbing, intricately rendered psychological study of an ambitious officer's rise and fall in pre-World War I Austria. Brandauer is excellent as the vain, insecure homosexual ultimately undone by his own ambition and his superior officer's smug loathing. Muller-Stahl and Landgrebe are particularly distinguished among the supporting players. The second in Szabo/Brandauer's trilogy, after "Mephisto" and before "Hanussen." In German with English subtitles.
1984 114m/C *GE HU* Klaus Maria Brandauer, Armin Mueller-Stahl, Gudrun Landgrebe; *Dir:* Istvan Szabo. **VHS, Beta $29.95** PAV, FCT, GLV *♫♫♫♫*

Colonel Wolodyjowski An adaptation of the monumental novels by Henryk Sinkiewicz chronicling the attack on Poland's eastern border by the Turks in 1668. In Polish with English subtitles.
1969 160m/C *PL* Tadeusz Loniski, Magdalena Zawadzka, Daniel Olbrychski; *Dir:* Jerzy Hoffman. **VHS $49.95** FCT *♫♫*

Color Adventures of Superman Seven cartoon adventures of the Man of Steel animated by the Fleischer Studio between 1941 and 1943. Titles include "Superman," "The Mechanical Monsters," "The Magnetic Telescope," "The Japoteurs," "The Bulleteers," "Jungle Drums," and "The Mummy Strikes."
194? 52m/C **VHS, Beta, 8mm $24.95** VYY *♫♫½*

Color Me Blood Red An artist decides that the red in his paintings is best rendered with human blood. He even manages to continue his art career—when not busy stabbing and mutilating the unsuspecting citizenry. Short, shoddy. Also known as "Model Massacre."
1964 74m/C Don Joseph, Sandi Conder; *Dir:* Herschell Gordon Lewis. **VHS, Beta $29.95** NO Woof!

Color Me Dead A victim of an extremely slow-acting poison frantically spends his final days trying to uncover his killer. Another inferior remake of B-thriller "D.O.A." Rent that one instead.
1969 (R) 97m/C *AU* Tom Tryon, Carolyn Jones, Rick Jason, Patricia Connolly, Tony Ward; *Dir:* Eddie Davis. **VHS** REP, TIM *♫½*

Color of Money Flashy, gripping drama about former pool hustler Fast Eddie Felsen (Newman) who, after years off the circuit, takes a brilliant but immature pool shark (Cruise) under his wing. Strong performances by Newman as the grizzled veteran, Cruise as the showboating youth, and Mastrantonio and Shaver as the men's worldly girlfriends. Worthy sequel to 50's classic "The Hustler."

1986 (R) 119m/C Paul Newman, Tom Cruise, Mary Elizabeth Mastrantonio, Helen Shaver, John Turturro, Forest Whitaker; *Dir:* Martin Scorsese. Academy Awards '86: Best Actor (Newman). **VHS, Beta, LV $19.95** *TOU, BTV* 🎦🎦½

The Color Purple Celie is a poor black girl who fights for her self-esteem when she is separated from her sister and forced into a brutal marriage. Spanning 1909 to 1947 in a small Georgia town, the movie chronicles the joys, pains, and people in her life. Adaptation of Alice Walker's acclaimed book features strong lead from Goldberg (her screen debut), Glover, Avery, and talk-show host Winfrey (also her film debut). It's hard to see director Speilberg as the most suited for this one, but he acquits himself nicely, avoiding the facileness that sometimes flaws his pics. Brilliant photography by Allen Davian and musical score by Quincy Jones (who co-produced) compliment this strong film. **1985 (PG-13) 154m/C** Whoopi Goldberg, Danny Glover, Oprah Winfrey, Margaret Avery, Adolph Caesar, Rae Dawn Chong, Willard Pugh, Akosua Busia; *Dir:* Steven Spielberg. Directors Guild of America '85: Best Director (Spielberg); National Board of Review Awards '85: 10 Best Films of the Year, Best Actress (Goldberg). **VHS, Beta, LV $19.98** *WAR, FCT* 🎦🎦🎦½

Colorado Roy tries to find his brother, a Union deserter, during the Civil War. **1940 54m/B** Roy Rogers, George "Gabby" Hayes, Milburn Stone; *Dir:* Joseph Kane. **VHS, Beta $9.99** *NOS, MED, DVT* 🎦

Colorado Sundown A conniving brother and sister attempt to cheat a man out of his inheritance. **1952 67m/B** Rex Allen, Mary Ellen Kay, Slim Pickens, June Vincent, Koko; *Dir:* William Witney. **VHS $19.95** *NOS, DVT, TIM* 🎦

Colors Vivid, realistic cop drama pairs sympathetic veteran Duvall and trigger-tempered rookie Penn on the gang-infested streets of East Los Angeles. Fine play from leads is one of the many assets in this controversial, unsettling depiction of deadly streetlife. Colorful, freewheeling direction from the underrated Hopper. Rattling rap soundtrack too. Additional footage has been added for video release. **1988 (R) 120m/C** Sean Penn, Robert Duvall, Maria Conchita Alonso, Trinidad Silva, Randi Brooks, Grand Bush, Don Cheadle, Rudy Ramos; *Dir:* Dennis Hopper. **VHS, Beta, LV $19.98** *ORI, FCT* 🎦🎦🎦½

Colossus and the Amazon Queen Idle gladiators are unwillingly recruited for service to Amazons. Doesn't seem like the worst way to make a living. But lead actor Taylor—so notable in comedies, thrillers, and action flicks—is too good for this sort of thing. Dubbed. **1964 94m/C** *IT* Rod Taylor, Dorian Gray, Ed Fury. **VHS, Beta, 8mm $29.95** *VYY* 🎦½

Colossus: The Forbin Project A computer designed to manage U.S. defense systems teams instead with its Soviet equal and they attempt world domination. Wiretight, suspenseful film seems at once dated yet timely. Based on the novel by D.F. Jones. **1970 100m/C** Eric Braeden, Susan Clark, Gordon Pinsent; *Dir:* Joseph Sargent. **VHS, Beta $59.95** *MCA* 🎦🎦🎦

Colt is My Law (Mi Revolver es la Ley) Bandits steal a railroad-funding gold shipment, and are pursued relentlessly by mysterious masked men. **19?? 95m/C** Anthony Clark, Lucy Gilly, Michael Martin. **VHS, Beta $59.95** *JCI Woof!*

Columbia Pictures Cartoon Classics Eight award-winning films from Columbia Pictures, including "Gerald McBoing-Boing" (1950), "Rooty Toot Toot" (1951), "Madeline" (1952), and "Magoo's Puddle Jumper" (1956). **1952 50m/C** *Voices:* Jim Backus. Academy Awards '50: Best Animated Short Film (Gerald McBoing-Boing). **VHS, Beta $19.95** *COL*

Columbo: Murder By the Book The rumpled, cigar-smoking television detective investigates the killing of a mystery writer. Scripted by Steven Bochco of "Hill Street Blues" and "L.A. Law" fame. **1971 79m/C** Peter Falk, Jack Cassidy, Rosemary Forsyth, Martin Milner; *Dir:* Steven Spielberg. **VHS, Beta $39.95** *MCA* 🎦🎦½

Columbo: Prescription Murder Peter Falk's debut as the raincoat-clad lieutenant investigating the death of a psychiatrist's wife. Rich with subplots. Good mystery fare. **1967 99m/C** Peter Falk, Gene Barry, Katherine Justice, William Windom, Nina Foch, Anthony James, Virginia Gregg; *Dir:* Richard Irving. **VHS, Beta** *MCA* 🎦🎦🎦

Coma A female doctor, who discovers murder and corpse-nabbing at her Boston hospital, defies her male bosses and determines to find out what's going on. Exciting, suspenseful fare, with Bujold impressive in lead. Based on the novel by Robin Cook. **1978 (PG) 113m/C** Genevieve Bujold, Michael Douglas, Elizabeth Ashley, Rip Torn, Richard Widmark, Lois Chiles, Harry Rhodes, Tom Selleck; *Dir:* Michael Crichton. **VHS, Beta, LV $59.95** *MGM* 🎦🎦🎦

Comancheros Texas Ranger Wayne and his prisoner fight with the Comancheros, an outlaw gang who is supplying guns and liquor to the dreaded Comanche Indians. Elmer Bernstein's musical score adds flavor. Last film by Curtiz. **1961 108m/C** John Wayne, Ina Balin, Stuart Whitman, Nehemiah Persoff, Lee Marvin, Bruce Cabot; *Dir:* Michael Curtiz. **VHS, Beta $19.98** *FOX* 🎦🎦🎦

Combat Academy Weak comedy about two goofballs sent to a military academy to straighten out, but instead turn the academy on its ear with their antics. Unremarkable. **1986 96m/C** Keith Gordon, Jamie Farr, Sherman Hemsley, John Ratzenberger, Bernie Kopell, Charles Moll; *Dir:* Neal Israel. **VHS, LV $19.95** *STE, NWV* 🎦

Combat Bulletin A collection of World War II action newsreels produced for G.I. viewing. **194? 115m/B** **VHS, Beta $22.00** *WFV* 🎦

Combat Killers A World War II Army captain grabs for individual glory on the battlefield, and jeopardizes the lives of his platoon. **1980 96m/C** Paul Edwards, Marlene Dauden, Claude Wilson; *Dir:* Ken Loring. **VHS, Beta $59.95** *MED* 🎦🎦

Combat Shock A Vietnam veteran returns home and decides to rid his city of undesireables by killing them. Familiar plot handled in typically crude, conventional manner. **1984 (R) 85m/C** Ricky Giovinazzo, Nick Nasta, Veronica Stork; *Dir:* Buddy Giovnazzo. **VHS, Beta $9.99** *STE, PSM* 🎦

Come Along with Me A made-for-television adaptation of Shirley Jackson's unfinished novel about a woman who sells all and leaves her hometown when her husband dies, determined to start a new career as a seer. **1984 60m/C** Estelle Parsons; *Dir:* Joanne Woodward. **VHS, Beta $24.95** *MON, KAR* 🎦🎦½

Come Back to the 5 & Dime Jimmy Dean, Jimmy Dean Five women convene at a run-down Texas drugstore for a 20-year reunion of a local James Dean fan club. The women recall earlier times and make some stunning revelations. Altman's filming of Ed Graczyk's sometimes funny, sometimes wrenching play proves fine vehicle for the actresses. Cher (in her screen debut) is probably the most impressive, but Dennis and Black are also memorable. **1982 (PG) 109m/C** Sandy Dennis, Cher, Karen Black, Kathy Bates, Sudie Bond; *Dir:* Robert Altman. **VHS, Beta, LV $19.98** *SUE* 🎦🎦🎦

Come Back, Africa A classic documentary about a Zulu family surviving under apartheid in South Africa. **1960 83m/B** **VHS, Beta $29.95** *MFV, FCT* 🎦🎦🎦

Come Back Little Reba A semi-educational film about a young black boy and his lovable aging dog. **1986 60m/C** **VHS, Beta, LV** *REB, PAR, PIA* 🎦

Come Back, Little Sheba Unsettling drama about a worn-out housewife, her abusive, alcoholic husband and a comely boarder who causes further marital tension. The title refers to the housewife's despairing search for her lost dog. Booth, Lancaster, and Moore are all excellent. Based on the play by William Inge, this film still packs an emotional wallop. **1952 99m/B** Burt Lancaster, Shirley Booth, Terry Moore, Richard Jaeckel, Philip Ober, Liza Golm, Walter Kelley; *Dir:* Daniel Mann. Academy Awards '52: Best Actress (Booth). **VHS, Beta, LV $19.95** *PAR, FCT, BTV* 🎦🎦🎦½

Come Blow Your Horn Neil Simon's first major Broadway success is a little less successful in its celluloid wrapper, suffering a bit from a familiar script. Sinatra's a playboy who blows his horn all over town, causing his close-knit New York Jewish family to warp a bit. Dad's not keen on his son's pledge of allegiance to the good life, and kid brother Bill would like to be his sibling's understudy in playboyhood. **1963 115m/C** Frank Sinatra, Lee J. Cobb, Tony Bill, Molly Picon, Barbara Rush, Jill St. John; *Dir:* Bud Yorkin. **VHS, Beta $14.95** *PAR* 🎦🎦🎦

Come and Get It A classic adaptation of the Edna Ferber novel about a lumber king battling against his son for the love of a woman. Farmer's most important Hollywood role. **1936 99m/B** Frances Farmer, Edward Arnold, Joel McCrea, Walter Brennan, Andrea Leeds; *Dir:* William Wyler, Howard Hawks. Academy Awards '36: Best Supporting Actor (Brennan). **VHS, Beta $19.98** *HBO, SUE, BTV* 🎦🎦🎦

Come On, Cowboys The Three Mesquiteers rescue an old circus friend from certain death. Part of "The Three Mesquiteers" series. **1937 54m/B** Bob Livingston, Ray Corrigan. **VHS, Beta $25.00** *MED, DVT* 🎦

Come On Rangers Also called "Come On Ranger." Rogers and his friends go to Texas to avenge the death of a comrade. **1938 54m/B** Roy Rogers, George "Gabby" Hayes. **VHS, Beta $19.95** *NOS, DVT, RXM* 🎦

Come On Tarzan Maynard and his trusty horse Tarzan save wild horses from the bad guys who want to turn them into dog food.
1932 60m/B Ken Maynard. **VHS, Beta $19.95** NOS, VDM, DVT ♂

Come and See Harrowing, unnerving epic which depicts the horrors of war as a boy soldier roams the Russian countryside during the Nazi invasion. Some overwhelming sequences, including tracer-bullets flashing across an open field. War has rarely been rendered in such a vivid, utterly grim manner. Outstanding achievement from Soviet director Klimov.
1985 142m/C RU Alexei Kravchenko, Olga Mironova, Lubomiras Lauciavicus; *Dir:* Elem Klimov. **VHS $59.95** TPV, FCT ♂♂♂♂

Come See the Paradise Just before WWII, union organizer Jack McGurn (Quaid) falls in love with Lily (Tomita), who's from Los Angeles' Little Tokyo. They marry (out of state because California law prohibited interracial marriages) before the bombing of Pearl Harbor and the subsequent internment of Japanese-Americans. Not a film with much big screen history, but effectively presented for the little theater of the living room. Told in flashback, the story offers a candid look at the racism implicit in the Asian-American relocations and at the hypocrisy that often lurks beneath the surface of the pursuit of liberty and justice for all. British writer-director Parker's most successful efforts (such as "Birdy," "The Commitments" and "Midnight Express") have been those he has directed but not penned. Typically, the "Paradise" script indulges a bit too much in screenwriter's license and concedes to the obvious. Nonetheless, the cast, apart from Quaid's misguided attempt at seriousness, is excellent, and the subject has not yet been cast into the vast bin of Hollywood cliches.
1990 (R) 135m/C Dennis Quaid, Tamlyn Tomita, Sab Shimono; *W/Dir:* Alan Parker. **VHS, Beta, LV $19.98** FOX, FCT, CCB ♂♂♂

The Comeback A singer attempting a comeback in England finds his wife murdered. Also known as "The Day the Screaming Stopped."
1977 100m/C GB Jack Jones, Pamela Stephenson, David Doyle, Bill Owen, Sheila Keith, Richard Johnson; *Dir:* Pete Walker. **VHS, Beta $19.98** LHV ♂

Comeback The story of a disillusioned rock star, played by Eric Burdon (the lead singer of the Animals), who gives up his life in the fast lane and tries to go back to his roots.
1983 105m/C Eric Burdon; *Dir:* Christel Bushmann. **VHS, Beta $24.95** MGM ♂♂½

Comeback Kid An ex-big league baseball player is conned into coaching an urban team of smarmy street youths, and falls for their playground supervisor. Made for television.
1980 97m/C John Ritter, Susan Dey; *Dir:* Peter Levin. **VHS, Beta $9.98** SUE, CHA ♂

Comedy Classics Three early comedy shorts: "Double Trouble," with Pollard; "So and Sew," with Ball; and "Technocrazy," directed by Al Christie.
1935 60m/B Snub Pollard, Oliver Hardy, Lucille Ball, Billy Bevan, Monty Collins. **VHS, Beta $59.95** JEF ♂

Comedy Classics of Mack Sennett and Hal Roach Three comedy silent films: "Love, Loot and Crash;" "Looking for Trouble;" and "A Desperate Scoundrel."
1915 51m/B VHS, Beta $29.95 VYY ♂

Comedy Festival #1 Collection of three comic films: "Speed in the Gay Nineties," with Clyde; "Disorder in the Court," with the Three Stooges; "Hail Brother," with Gilbert.
193? 53m/B Andy Clyde, Billy Gilbert, Curly Howard, Moe Howard, Larry Fine. **VHS, Beta $34.95** HHT ♂

Comedy Festival #2 Collection of comedy films: "Bashful Romeo," with Lamb; "Super Snooper," with Clyde; "Super Stupid," with Gilbert.
193? 53m/B Gil Lamb, Andy Clyde, Billy Gilbert. **VHS, Beta $34.95** HHT ♂

Comedy Festival #3 Collection of comic films: "Big Flash," with Langdon; "Half A Hero," and "Pardon My Papa," with Temple.
193? 53m/B Harry Langdon, Shirley Temple. **VHS, Beta $34.95** HHT ♂

Comedy Festival #4 A collection of comic films: "His Weak Moment," "Fainting Lover," and "Shopping with the Wife."
193? 53m/B Andy Clyde. **VHS, Beta $34.95** HHT ♂

Comedy Reel, Vol. 1 Contains three W.C. Fields shorts: "Golf Specialist," "Fatal Glass of Beer," and "The Dentist."
19?? 60m/B W.C. Fields. **VHS, Beta $23.95** CAB ♂♂½

Comedy Reel, Vol. 2 These comedy shorts feature the Three Stooges in "Disorder in the Court," Edgar Kennedy and Franklin Pangborn in "Next Door Neighbors," and W. C. Fields in "The Barbershop."
19?? 60m/B Moe Howard, Edgar Kennedy, Franklin Pangborn, W.C. Fields. **VHS, Beta $23.95** CAB ♂♂½

The Comedy of Terrors Comedy in which some deranged undertakers take a hands-on approach to insuring their continued employment. Much fun is supplied by the quartet of Price, Lorre, Karloff, and Rathbone, all veterans of the horror genre.
1964 84m/C Vincent Price, Peter Lorre, Boris Karloff, Basil Rathbone, Joe E. Brown, Joyce Jameson; *Dir:* Jacques Tourneur. **VHS, LV $79.98** MOV ♂♂♂

Comedy Tonight Los Angeles' "Improv" club is the scene for this night of stand-up comedy. Andy Kaufman performs his infamous Tony Clifton routine. Robin Williams reveals his X-rated side, and Gallagher offers his "Sledge-O-Matic" routine.
1977 76m/C Hosted: David Steinberg, Andy Kaufman, Robin Williams, Gallagher, Ed Bluestone, Richard Libertini, MacIntyre Dixon. **VHS, Beta $69.95** VES, LIV ♂♂♂

Comes a Horseman Robards is a cattle baron, attempting to gobble up all the oil-rich land his neighbor owns. Neighbor Fonda has the courage to stand up to him, with the help of a World War II veteran and the local old timer Farnsworth, in this slow-moving but intriguing western drama. Farnsworth received an Oscar nomination for his supporting role.
1978 (PG) 119m/C James Caan, Jane Fonda, Jason Robards Jr., George Grizzard, Richard Farnsworth, Jim Davis, Mark Harmon; *Dir:* Alan J. Pakula. National Board of Review Awards '78: 10 Best Films of the Year, Best Supporting Actor (Farnsworth). **VHS, Beta** FOX ♂♂½

Comfort and Joy After his kleptomaniac girlfriend deserts him, a Scottish disc jockey is forced to reevaluate his life. He becomes involved in an underworld battle between two mob-owned local ice cream companies. Another odd comedy gem from Forsyth, who did "Gregory's Girl" and "Local Hero." Music by Dire Straits guitarist Mark Knopfler.
1984 (PG) 93m/C GB Bill Paterson, Eleanor David, C.P. Grogan, Alex Norton, Patrick Malahide, Rikki Fulton, Roberto Berrardi; *Dir:* Bill Forsyth. **VHS, Beta $69.95** MCA ♂♂♂½

The Comfort of Strangers An atmospheric psychological thriller. Mary and Colin (Richardson and Everett), a handsome young British couple, take a Venetian holiday to rediscover the passion in their relationship. Lost in the aqua-riddled labyrinthine city, they chance, it seems, upon Robert (Walken), who wears an impeccable white suit and tells them of his father. A sort of Vergil in an Italian suit, Robert later reappears, and, abetted by his wife Caroline (Mirren), gradually leads the couple on an eerie tour of urbane decadence that hints at danger. Psychologically tantalizing and horrifyingly erotic, the movie's based on Ian McEwan's novel, with an adroit screenplay by Harold Pinter.
1991 (R) 102m/C Christopher Walken, Natasha Richardson, Rupert Everett, Helen Mirren; *Dir:* Paul Schrader. **VHS, Beta $79.95** PAR, FCT ♂♂♂

The Comic Van Dyke is terrific as a silent screen comedian (a composite of several real-life screen comics) whose ego destroys his career. Recreations of silent films, and the blend of comedy and pathos, are especially effective. Written by Carl Reiner.
1969 (PG) 96m/C Dick Van Dyke, Mickey Rooney, Michele Lee, Cornel Wilde, Nina Wayne, Pert Kelton, Jeannine Riley; *Dir:* Carl Reiner. **VHS, Beta, LV $59.95** COL ♂♂♂

The Comic In a future police state, a young comedian kills the star and takes his place in the show. His future rises, but who cares?
1985 90m/C Steve Munroe, Bernard Plant, Jeff Pirie; *Dir:* Richard Driscoll. **VHS, Beta $59.95** MAG ♂

Comic Book Kids Two youngsters enjoy visiting their friend's comic strip studio, since they have the power to project themselves into the cartoon stories. Viewers will be less impressed.
1982 (G) 90m/C Joseph Campanella, Mike Darnell, Robyn Finn, Jim Engelhardt, Fay De Witt. **VHS, Beta** GEM ♂

Comic Cabby A day in the life of an introvert cabby seems to last forever. From "New York City Cab Driver's Joke Book" by Jim Pietsch.
1987 59m/C Bill McLaughlin, Al Lewis. **VHS, Beta $59.95** VES, LIV Woof!

Comic Relief 2 A non-stop stand-up comedy telethon which raised money for the nation's homeless. Each comic gets only a few minutes, so it's not as satisfying as it might have been. Made for cable television.
1987 120m/C Robin Williams, Billy Crystal, Whoopi Goldberg, Shelley Long, Harry Anderson, Dudley Moore, Peter Cook, Michael J. Fox, Arsenio Hall, Bob Goldthwait, Jon Lovitz, Dabney Coleman, Andrea Martin, Robert Klein, Penn Jillette, Catherine O'Hara. **VHS, Beta $19.95** AXV, SIM

Comic Relief 3 A bevy of comedy favorites appear in this live concert feature, designed to raise money for the homeless. Originally broadcast on cable television.
1989 120m/C Robin Williams, Billy Crystal, Whoopi Goldberg, Arsenio Hall, Bob Goldthwait, Shelley Long, Paul Reiser, Steven Wright. **VHS, Beta, LV $39.95** *KRT*

Coming to America An African prince (Murphy) decides to come to America in search of a suitable bride. He lands in Queens, and quickly finds American women to be more confusing than he imagined. Sometimes overly-cute entertainment relieved by clever costume cameos by Murphy and Hall. Later lawsuit resulted in columnist Art Buchwald being given credit for story.
1988 (R) 116m/C Eddie Murphy, Arsenio Hall, James Earl Jones, John Amos, Madge Sinclair, Shari Headley, Don Ameche, Louie Anderson; *Dir:* John Landis. **VHS, Beta, LV, 8mm $14.95** *PAR* ♫♫♫

The Coming of Amos A wonderful send-up of all those Fairbanks-style melodramas, with La Rocque as an Aussie on vacation on the French Riviera falling for and eventually rescuing Goudal, as the Russian princess held captive by the lustful Beery.
1925 60m/B Rod La Rocque, Jetta Goudal, Noah Beery, Florence Vidor; *Dir:* Paul Sloane. **VHS, Beta $36.95** *DNB* ♫♫♫

Coming Home A look at the effect of the Vietnam War on the American home front. The wife of a gung-ho Marine officer volunteers as an aide in a Veteran's Hospital, befriends and eventually falls in love with a Vietnam veteran, paraplegic from war injuries. His attitudes, pain, and his first-hand knowledge of the war force her to re-examine her previously automatic responses. Honest look at the everyday life of disabled veterans, unusual vision of the possibilities of simple friendship between men and women. Fonda and Voight are great; Dern's character suffers from weak scriptwriting late in the film. Critically acclaimed; Fonda, Voight, and the scriptwriters won Oscars, other nominations for Best Picture, supporting actor Dern, supporting actress Milford and film editing. Compelling score from late '60s music.
1978 (R) 127m/C Jane Fonda, Jon Voight, Bruce Dern, Penelope Milford, Robert Carradine, Robert Ginty; *Dir:* Hal Ashby. Academy Awards '78: Best Actor (Voight), Best Actress (Fonda), Best Screenplay; Cannes Film Festival '78: Best Actor (Voight); Los Angeles Film Critics Association Awards '78: Best Actor (Voight), Best Actress (Fonda), Best Film. **VHS, Beta, LV $19.98** *FOX, MGM, FCT* ♫♫♫½

Coming Next Week: Those Great Movie Trailers Movie previews from the 30's to the 80's covering all film genres. Some trailers are in black and white.
1984 120m/C **VHS, Beta** *AOV* ♫♫♫½

Coming Out Alive Tense thriller about a woman and her hired mercenary who try to rescue her kidnapped son from his estranged father - a radical is involved in an assassination plot. Hylands is especially compelling.
1984 73m/C Helen Shaver, Scott Hylands, Michael Ironside, Monica Parker. **VHS, Beta $59.95** *TWE* ♫♫½

Coming Out of the Ice Engrossing true story of Victor Herman, an outstanding American athlete who worked in Russia in the 1930s. He was imprisoned in Siberia (for 38 years!) for refusing to renounce his American citizenship. Made for television.

1982 (PG) 100m/C John Savage, Willie Nelson, Ben Cross, Francesca Annis, Peter Vaughan; *Dir:* Waris Hussein. **VHS, Beta $59.98** *FOX* ♫♫♫

Coming Soon 50 excerpts from the "previews of coming attractions" of the most famous and infamous of the horror films. Quickie thrills.
1983 55m/C *Dir:* John Landis; *Nar:* Jamie Lee Curtis. **VHS, Beta $14.95** *MCA* ♫♫♫

Coming Up Roses Pleasant comedy about the residents of a small mining village in the South of Wales who fight to keep their local movie house from being shut down. In Welsh with English subtitles.
1987 (PG) 95m/C *GB* Dafydd Hywel, Iola Gregory, Olive Michael, Mari Emlyn, Bill Paterson; *Dir:* Stephen Bayly. **VHS, Beta $29.95** *PAV, FCT, TAM* ♫♫½

Command Decision Upon realizing he must send his men on missions-of-no-return to destroy German jet production, Gable's WWII flight commander becomes tactically at odds with his political superior, Pidgeon (who's not keen to have his precision bombing plans placed in an unflattering light). Based on the stage hit by William Wister Haines, it's a late war pic by Wood, who earlier directed such diverse efforts as "A Night at the Opera," "The Devil and Miss Jones," and "Our Town." Vintage war-is-hell actioner.
1948 112m/B Clark Gable, Walter Pidgeon, Van Johnson, Brian Donlevy, Charles Bickford, John Hodiak; *Dir:* Sam Wood. **VHS $19.98** *MGM, FCT* ♫♫½

Commando An ex-commando leader's daughter is kidnapped in a blackmail scheme to make him depose a South American president. He fights instead, and proceeds to rescue his daughter amid a torrential flow of falling bodies. Violent action spiced with throwaway comic lines.
1985 (R) 90m/C Arnold Schwarzenegger, Rae Dawn Chong, Dan Hedaya, Vernon Wells, James Olson, David Patrick Kelly, Alyssa Milano; *Dir:* Mark L. Lester. **VHS, Beta, LV $19.98** *FOX, TLF* ♫♫♫

Commando Attack A tough sergeant leads a group of misfit soldiers to blow up a German radio transmitter the day before D-Day.
1967 90m/C Michael Rennie, Monica Randall, Bob Sullivan. **VHS, Beta $59.95** *UNI* ♫½

Commando Cody A specially edited feature version of the classic 12-chapter science-fiction television serial launches the original Rocket-Man into action in "Radar Men from the Moon."
1991 85m/B **VHS $14.98** *WEM* ♫½

Commando Invasion When an army captain is accused of murder, he hunts down the man he knows is responsible for the crime. Jungle revenge mayhem is confusing. Set in Viet Nam.
1987 90m/C Jim Gaines, Michael James, Gordon Mitchell, Carol Roberts, Pat Vance, Ken Watanabe; *Dir:* John Gale. **VHS, Beta $59.95** *QHV* ♫

Commando Squad During World War II, soldiers go to battle to prevent the development of a device that re-animates flesh. The script stays dead.
1976 (R) 82m/C Chuck Alford, Peter Owen, April Adams; *Dir:* Charles Nizet. **VHS, Beta $24.95** *GHV* ♫

Commando Squad Story of drug agents undercover in Mexico, and a female agent who rescues her kidnapped lover. Firepower is wasted, but Shower is a former "Playboy" Playmate of the Year.
1987 (R) 90m/C Brian Thompson, William Smith, Kathy Shower, Sid Haig, Robert Quarry, Ross Hagen, Mel Welles, Marie Windsor, Russ Tamblyn; *Dir:* Fred Olen Ray. **VHS, Beta $79.95** *TWE* ♫

Commandos Commando operation against Rommel's forces in North Africa during World War II. Too much said, not enough action. Also know as "Sullivan's Marauders."
1973 (PG) 100m/C *GE IT* Lee Van Cleef, Jack Kelly; *Dir:* Armando Crispino. **VHS, Beta $49.95** *GEM, MRV, PSM* ♫♫½

Commandos Strike at Dawn Norwegian villagers, including Muni, fight the invading Nazis, and eventually help the British Navy battle them over an Arctic supply line. Okay script helped along by veteran actors Muni and Gish. Written by Irwin Shaw, based on a C.S. Forester story.
1942 100m/B Paul Muni, Lillian Gish, Cedric Hardwicke, Anna Lee, Ray Collins, Robert Coote, Alexander Knox, Rosemary DeCamp; *Dir:* John Farrow. **VHS, Beta $69.95** *COL* ♫♫½

The Commies are Coming, the Commies are Coming Cult classic will leave viewers red from laughing in disbelief. Jack Webb narrates this anti-communist movie about the Russians taking over the United States. Filmed in a documentary style, it captures the paranoia of the times. Re-released in 1984, just before the Evil Empire collapsed.
1957 60m/B Jack Webb, Andrew Duggan, Robert Conrad, Jack Kelly, Peter Brown; *Dir:* George Waggner. **VHS, Beta $39.95** *RHI* ♫♫

Commissar Before the Soviet Union ended up in the ashcan of history, this film was labeled as "treason" and shelved in Red Russia. Now, even Americans can view the story of a female Soviet soldier who becomes pregnant during the civil war of 1922. The Soviet military has no policy regarding pregnancy, so the woman is dumped on a family of outcast Jews to complete her pregnancy. This film makes the strong statement that women were just as discriminated against in the U.S.S.R. as were many races or creeds, especially Jews. In Russian with English subtitles. Released in the U.S. in 1988.
1967 105m/C *RU* Nonna Mordyukova, Rolan Bykov; *Dir:* Alexander Askoldov. **VHS $59.95** *IFF* ♫♫♫♫

Commission An experimental video work by Vasulka which explores the relationship between Hector Berlioz and violinist Paganini through electronic breaks with traditional narrative.
197? 30m/C *Dir:* Woody Vasulka, Woody Vasulka. **VHS, Beta $39.95** *FCT* ♫♫♫♫

The Commitments Convinced that they can bring soul music to Dublin, a group of working-class youth form a band. Based on the book by Roddy Doyle, this high-energy production paints an interesting, un-romanticized picture of modern Ireland and refuses to follow standard showbiz cliches, even though its lack of resolution hurts. Honest, whimsical dialog laced with poetic obscenities, delivered by a cast of mostly unknowns. The very successful soundtrack features Wilson Pickett, James Brown, Otis Redding, Aretha Franklin, Percy Sledge and

others, and received a Grammy nomination. much as the band is for their characters.
1991 (R) 116m/C Andrew Strong, Bronagh Gallagher, Glen Hansard, Michael Aberne, Dick Massey, Ken McCluskey, Robert Arkins, Dave Finnegan, Johnny Murphy, Angeline Ball, Felim Gormley, Maria Doyle; *Dir:* Alan Parker. **VHS, Beta $94.98** *FXV, CCB* ☂☂

Committed The story of stage and screen actress Frances Farmer, who went from stardom to a mental institution and a lobotomy. A more subtle, incisive version of the story than the Hollywood biopic "Frances." An independent production directed by McLaughlin and L. Tillman.
1984 77m/C *Dir:* Sheila McLaughlin, Lynne Tillman. **VHS $59.95** *FCT* ☂☂☂

Committed Schlock-meister Levey's latest directorial effort is a step above his earlier "Slumber Party" and "Happy Hooker Goes to Washington," a statement ringing with the faintest of praise. Thinking she's applying for a job at an asylum, a nurse discovers she has committed herself. Try as she may, not even a committed nurse can snake her way out of the pit of madness they call "The Institute."
1991 (R) 93m/C Jennifer O'Neill, William Windom, Robert Forster, Ron Palillo, Sydney Lassick; *Dir:* William A. Levey. **VHS $89.98** *MED* ☂½

The Committee A comedy film of the seminal comedy troupe "The Committee," specialists in short, punchy satire. Dated, but of interest to comedy buffs.
1968 (PG) 88m/C Wolfman Jack, Howard Hesseman, Barbara Bosson, Peter Bonerz, Garry Goodrow, Carl Gottlieb; *Dir:* Jack Del. **VHS, Beta $59.95** *PAV* ☂

Common Bonds After being put together in an experimental outreach program, two mismatched partners struggle to break free from a system that threatens to destroy them both.
1991 (PG) 109m/C Rae Dawn Chong, Michael Ironside, Brad Dourif; *Dir:* Allan Goldstein. **VHS, Beta $89.95** *ACA* ☂☂½

Common Threads: Stories from the Quilt The heavily awarded, made-for-cable-TV documentary about the Quilt: an immense 14-acre blanket created by the survivors of AIDS victims as a monument to their loved ones' premature deaths. All proceeds from the sale of this tape go to the NAMES Project Foundation in support of the Quilt and AIDS service providers. Sincerely touching and extremely well-done film.
1989 79m/C *Dir:* Robert Epstein, Jeffrey Friedman; *Nar:* Dustin Hoffman. Academy Awards '89: Best Feature Documentary. **VHS, Beta, LV $49.95** *IME, HBO*

Communion A serious adaptation of the purportedly nonfictional bestseller by Whitley Strieber about his family's abduction by extraterrestrials. Spacey New Age story is overlong and hard to swallow.
1989 (R) 103m/C Christopher Walken, Lindsay Crouse, Frances Sternhagen, Joel Carlson, Andreas Katsulas, Basil Hoffman; *Dir:* Philippe Mora. **VHS, Beta, LV $89.95** *VIR* ☂☂

Como Dos Gotas de Agua (Like Two Drops of Water) Twin sisters trade places with each other in this limp musical comedy.
197? 104m/C *SP* Pili & Mili, Isabel Garces, Luis Davila, Manolo Moran. **VHS, Beta $49.95** *MAD* ☂

Company Limited The story of a young Indian sales executive working for a British owned firm, set in contemporary Calcutta. English subtitles.
197? 112m/B *IN* **VHS, Beta, 3/4U, Special order formats** *CIG* ☂☂

The Company of Wolves A young girl on the threshold of womanhood listens to her grandmother tell fairy tells and dreams of a medieval fantasy world inhabited by wolves and werewolves. An adult "Little Red Riding-hood" type comedy heavy on Freudian symbolism. Not really for kids.
1985 (R) 95m/C Angela Lansbury, David Warner, Micha Bergese, Tusse Silberg, Sarah Patterson, Brian Glover, Danielle Dax, Graham Crowden, Stephen Rea; *Dir:* Neil Jordan. **VHS, Beta, LV $19.98** *VES, LIV* ☂☂☂

Competition Two virtuoso pianists meet at an international competition and fall in love. Will they stay together if one of them wins? Can they have a performance career and love too? Is he trying to distract her with sex, so he can win? Dreyfuss and Irving are fine (they practiced for four months to look like they were actually playing the pianos), and Remick's character has some interesting insights into the world of art.
1980 (PG) 125m/C Richard Dreyfuss, Amy Irving, Lee Remick, Sam Wanamaker, Joseph Cali, Ty Henderson, Priscilla Pointer, James B. Sikking; *Dir:* Joel Oliansky. **VHS, Beta, LV $19.95** *COL* ☂☂

The Compleat Beatles Music, interviews, film clips, animation and live performances make up this "rockumentary" on the Beatles. New interviews are featured, as well as vintage film clips and studio footage with narration by Malcolm McDowell. The first U.S. press conference, legendary Hamburg footage, and an in-depth interview with George Martin are highlights.
1982 120m/C John Lennon, Paul McCartney, Ringo Starr, George Harrison, George Martin, Brian Epstein, Billy Preston, Milt Oken, Bruce Johnston, Roger McGuinn, Mike McCartney, Mick Jagger; *Dir:* Patrick Montgomery; *Nar:* Malcolm McDowell. **VHS, Beta, LV $19.95** *MVD, MGM* ☂☂☂

Complex Cute tale of a clever dog (Aren't they all?) who annoys his swinish, beer-guzzling master by learning how to play chess, speak French, and read Dostoevsky.
1985 12m/C VHS, Beta, 3/4U $160.00 *BFA* ☂☂☂

Compromising Positions A philandering dentist is killed on Long Island, and a bored housewife begins to investigate, uncovering scandal after scandal. Black comedy mixed unevenly with mystery makes for unsatisfying brew. However, Ivey is a standout as Sarandon's best friend. Screenplay was written by Susan Isaacs, based on her novel of the same title.
1985 (R) 99m/C Susan Sarandon, Raul Julia, Edward Herrmann, Judith Ivey, Mary Beth Hurt, Joe Mantegna, Josh Mostel, Anne De Salvo; *Dir:* Frank Perry. **VHS, Beta, LV $19.95** *PAR* ☂☂

Computer Beach Party A couple of surf-head college computer hackers foil their mayor's plans to develop their favorite beach. The title, at least, is original.
1988 97m/C Hank Amigo, Stacey Nemour, Andre Chimene; *Dir:* Gary A. Troy. **VHS, Beta $69.98** *VES, LIV* ☂

Computer Dreams Various computer animated shorts have been compiled, including "Pencil Test," "Mars Rover," and "Nestor Sextone."
1987 60m/C VHS $19.95 *DVE, MPI* ☂

Computer Wizard A boy genius builds a powerful electronic device. His intentions are good, but the invention disrupts the town and lands him in big trouble. Not the "Thomas Edison Story."
1977 (G) 91m/C Henry Darrow, Kate Woodville, Guy Madison, Marc Gilpin; *Dir:* John Florea. **VHS, Beta $69.95** *VCL* ☂

The Computer Wore Tennis Shoes Disney comedy (strictly for the kids) about a slow-witted college student who turns into a genius after a "shocking" encounter with the campus computer. His new brains give the local gangster headaches. Sequel: "Now You See Him, Now You Don't."
1969 (G) 87m/C Kurt Russell, Cesar Romero, Joe Flynn, William Schallert, Allan Hewitt, Richard Bakalayan; *Dir:* Robert Butler. **VHS, Beta** *DIS, OM* ☂½

The Con Artists A con man (Quinn) recently sprung from prison and his protege set up a sting operation against the beautiful Capucine. Not clever or witty enough by far.
1980 86m/C *IT* Anthony Quinn, Adriano Celentano, Capucine, Corinne Clery; *Dir:* Sergio Corbucci. **VHS, Beta $19.98** *VID, HHE* ☂½

Con Mi Mujer No Puedo A compulsive Don Juan's wife decides to match her husband's carnal activities tryst for tryst. Naughty boredom.
198? 101m/C *SP* Guillermo Bredeston, Leonore Benedetto, Javier Portales. **VHS, Beta $57.95** *UNI* ☂½

Conagher A lyrical, if poorly plotted Western about a veteran cowboy who takes the whole movie to decide to end up in the arms of the pretty widow lady. Cable-TV adaptation of Louis L'Amour's novel doesn't have much intensity, though Elliott captures his character well.
1991 94m/C Sam Elliott, Katharine Ross, Barry Corbin; *Dir:* Reynaldo Villalobos. **VHS, Beta $89.98** *TTC* ☂☂½

Conan the Barbarian A fine sword and sorcery tale featuring brutality, excellent production values, and a rousing score. Conan's parents are killed and he's enslaved. But hardship makes him strong, so when he is set free he can avenge their murder and retrieve the sword bequeathed him by his father. Sandahl Bergman is great as The Queen of Thieves, and Schwarzenegger maintains an admirable sense of humor throughout. Jones is dandy, as always. Based on the character created by Robert E. Howard. Sequel: "Conan the Destroyer."
1982 (R) 115m/C Arnold Schwarzenegger, James Earl Jones, Max von Sydow, Sandahl Bergman, Mako, Ben Davidson, Valerie Quennessen, Cassandra Gaviola, William Smith; *Dir:* John Milius. **VHS, Beta, LV $14.95** *MCA* ☂☂☂

Conan the Destroyer Conan is manipulated by Queen Tamaris into searching for a treasure. In return she'll bring Conan's love Valeria back to life. On his trip he meets Jones and Chamberlain, who later give him a hand. Excellent special effects, good humor, camp fun, somewhat silly finale. Sequel to the better "Conan the Barbarian."
1984 (PG) 101m/C Arnold Schwarzenegger, Grace Jones, Wilt Chamberlain, Sarah Douglas, Mako, Olivia D'Abo, Jeff Corey; *Dir:* Richard Fleischer. **VHS, Beta, LV $14.95** *MCA* ☂☂½

The Concentratin' Kid An unusual horse opera sans shootouts and fisticuffs. Hoot's fiancee is kidnapped and taken to the range so that the renegade cowpokes can,

uh, benefit from a female perspective on home decorating. Despite being outnumbered, Hoot concentrates and sets the rescue in motion, taking on the gang of rustlers without firing a shot.
1930 54m/B Hoot Gibson, Katherine Crawford, Duke Lee, Robert E. Homans, James Mason; *Dir:* Arthur Rosson. **VHS, Beta $9.95** *GPV* 🎞🎞

Concert for Bangladesh 1971 concert was a benefit for the people of Bangladesh.
1972 90m/C George Harrison, Bob Dylan, Ringo Starr, Billy Preston, Eric Clapton, Ravi Shankar, Klaus Voorman. **VHS, Beta $19.95** *MVD, HBO* 🎞🎞

The Concorde: Airport '79 A supersonic film in the "Airport" tradition has the Concorde chased by missiles and fighter aircraft before it crashes in the Alps. Incredibly far-fetched nonsense with an all-star cast. Does not fly. Also known as "Airport '79."
1979 (PG) 103m/C Alain Delon, Susan Blakely, Robert Wagner, Sylvia Kristel, John Davidson, Charo, Sybil Danning, Jimmie Walker, Eddie Albert, Bibi Andersson, Monica Lewis, Andrea Marcovicci, Martha Raye, Cicely Tyson, Mercedes McCambridge, George Kennedy, David Warner; *Dir:* David Lowell Rich. **VHS, Beta $59.95** *MCA* 🎞

Concrete Angels A poignant glimpse at the influence of rock music in its infancy. A group of deprived teens in Toronto form a band and audition to open for the Beatles. Lack of nostalgia is refreshing, but too much of this is amateurish.
1987 (R) 97m/C *CA* Joseph Dimambro, Luke McKeehan, Omie Craden, Dean Bosacki. **VHS, Beta $29.95** *ACA* 🎞

Concrete Beat A newspaper reporter simultaneously searches for a murderer, tries to win his ex-wife back, and writes front page stories for his editor/ex-father-in-law. Tough assignment. Made for television.
1984 74m/C Kenneth McMillan, John Getz, Darlanne Fluegel, Rhoda Gemignani; *Dir:* Robert Butler. **VHS, Beta $59.95** *PSM* 🎞½

The Concrete Cowboys Two bumbling backwoodsmen come to the metropolis of Nashville and promptly turn detective to foil a blackmail scheme and locate a missing singer. Barbara Mandrell and Roy Acuff play themselves. Silly made-for-TV spinoff of short-lived series.
1979 100m/C Jerry Reed, Tom Selleck, Morgan Fairchild, Claude Akins, Gene Evans, Roy Acuff, Barbara Mandrell; *Dir:* Burt Kennedy. **VHS** *FOX* 🎞½

Concrete Jungle After being set up by her no-good boyfriend, a woman is sent to a correctional facility for drug smuggling. Typical woman's prison film, has cute babes in revealing outfits. Followed by "Chained Heat."
1982 (R) 106m/C Tracy Bregman, Jill St. John, Barbara Luna, Peter Brown, Aimee Eccles, Nita Talbot, Sondra Currie; *Dir:* Tom DeSimone. **VHS, Beta $59.95** *COL* 🎞

Condemned to Hell A tough-as-iron street punk gets thrown into a women's prison, where she must fight for her existence. Low-brow violence galore.
198? 97m/C Leonore Benedetto. **VHS, Beta $59.95** *TWE* Woof!

Condemned to Live Mild-mannered doctor Ralph Morgan and his fiancee Maxine Doyle seem like your average early 20th Century Middle-European couple...but the doctor's hunchbacked servant Mischa Auer is a tipoff that this is a horror movie of some

sort. Like many men, the doctor has suffered from a vampire curse all his life. And, like many men, he is unaware of his blood-sucking habit, thanks to the concealment efforts of his loyal (hunchback) servant. Doyle discovers her fiance's sanguine secret, but not before she finds out she really loves Gleason, anyway. Answers the question: just how do you jilt a vampire?
1935 68m/B Ralph Morgan, Maxine Doyle, Russell Gleason, Pedro de Cordoba, Mischa Auer, Lucy Beaumont, Carl Stockade; *Dir:* Frank Strayer. **VHS $19.98** *SNC, TIM* 🎞🎞

Condorman Woody Wilkins, an inventive comic book writer, adopts the identity of his own character, Condorman, in order to help a beautiful Russian spy defect. A Disney film, strictly for the small fry.
1981 (PG) 90m/C Michael Crawford, Oliver Reed, Barbara Carrera, James Hampton, Jean-Pierre Kalfon, Dana Elcar; *Dir:* Charles Jarrott. **VHS, Beta $69.95** *DIS* 🎞🎞½

Conduct Unbecoming Late nineteenth century India is the setting for a trial involving the possible assault of a British officer's wife. Ambitious production based on a British stage play suffers from claustrophobic atmosphere but is greatly redeemed by the first rate cast.
1975 (PG) 107m/C *GB* Michael York, Richard Attenborough, Trevor Howard, Stacy Keach, Christopher Plummer, Susannah York, James Faulkner, Michael Culver, Persis Khambatta; *Dir:* Michael Anderson Sr. **VHS** *CRV* 🎞🎞½

Conexion Mexico Not-bad actioner about the Mexican police fighting against drug traffickers who are aided by a group of vicious women.
1987 85m/C *MX* Sergio Goyri, Victor Junco, Silvia Manriquez, Sergio Bustamante. **VHS, Beta $75.00** *MAD* 🎞🎞½

Conexion Oriental (East Connection) A journalist investigates an Oriental crime syndicate. Not much mystery to the mystery. Dubbed into Spanish.
1970 90m/C Dale Robertson, Lucianna Paluzzi, Calisto Calisti. **VHS, Beta $59.95** *JCI* Woof!

Coney Hatch This Canadian band electrifies head-banging hordes with their Ozzy Osborne-influenced tunes. Includes "Shake It" and "Devils Deck."
1985 17m/C Coney Hatch. **VHS, Beta, LV $9.95** *SVS*

The Confessional A mad priest unleashes a monster from his confessional to wreak havoc upon the world. Pray for your VCR. Also known as "House of Mortal Sin."
1975 (R) 108m/C *GB* Anthony Sharp, Susan Penhaligon, Stephanie Beacham, Norman Eshley, Sheila Keith; *Dir:* Pete Walker. **VHS, Beta $59.95** *PSM* 🎞

Confessions of a Police Captain A dedicated police captain tries to wipe out the bureaucratic corruption that is infecting his city. Balsam gives a fine performance in a heavy-going tale.
1972 (PG) 104m/C *IT* Martin Balsam, Franco Nero, Marilu Tolo; *Dir:* Damiano Damiani. **VHS, Beta $59.98** *SUE* 🎞🎞

Confessions of a Stardreamer A young actress talks about her struggles in the world of show business, while animated images comment on, mirror, and probe hidden meanings in her words.
1979 9m/C **VHS, Beta** *BFA* Woof!

Confessions of Tom Harris Somewhere between all the prize fighting, leg breaking for the mob, and jail terms for rape, Tom Harris finds time for a life-changing encounter with love. Murray wrote the script, based on true story. Also known as "Tale of the Cock" and "Childish Things."
1972 (PG) 90m/C Don Murray, Linda Evans, David Brian, Gary Clarke, Logan Ramsey, Angelique Pettyjohn; *Dir:* John Derek, David Nelson. **VHS, Beta $39.95** *MON* 🎞

Confessions of a Train Spotter This video follows Monty Python's Palin as he travels on British Rail from Euston to the Kyle of Lochalsh. Stays on track.
1987 60m/C Michael Palin. **VHS, Beta $29.95** *PEN, IUF* 🎞

Confessions of a Vice Baron Vice Baron Lombardo makes it big as a drug dealer and flesh peddler, then loses it all when he falls in love. Sleazy exploitation.
1942 70m/B Willy Castello; *Dir:* Harvey Thew. **VHS, Beta $16.95** *SNC, JLT* 🎞

Confessions of a Young American Housewife Routine sexploiter about a recent divorcee who moves in with two younger couples and experiences sexual liberation.
1978 (R) 85m/C Jennifer Wells, Rebecca Brooke, Chris Jordan. **VHS, Beta $29.95** *MED* 🎞

Confidential A G-man goes under cover to infiltrate a crime ring. Well made and full of action, fun array of character actors. Naish particularly good as nasty killer.
1935 67m/B Donald Cook, Evelyn Knapp, Warren Hymer, J. Carroll Naish, Herbert Rawlinson, Morgan Wallace, Kane Richmond, Theodore von Eltz, Reed Howes; *Dir:* Edward L. Cahn. **VHS $79.98** *MAG* 🎞🎞½

Confidential A '40s detective tries to solve a decades-old axe murder, only to disappear himself. Horrible fare from our friends north of the border.
1986 (R) 95m/C *CA* Neil Munroe, August Schellenberg, Chapelle Jaffe, Tom Butler; *Dir:* Bruce Pittman. **VHS, Beta $79.98** *MAG* Woof!

Confidentially Yours Truffaut's homage to Hitchcock, based on Charles Williams' "The Long Saturday Night," deals with a small-town real estate agent who is framed for a rash of murders, while his secretary tries to clear his name. Truffaut's last fils is lightly entertaining. French title: "Vivement Dimanche!"
1983 (PG) 110m/B *FR* Fanny Ardant, Jean-Louis Trintignant, Philippe Morier-Genoud, Philippe Laudenbach, Caroline Sihol; *Dir:* Francois Truffaut. **VHS, Beta $59.98** *FOX, TAM* 🎞🎞🎞

Conflict When Bogart falls for his sister-in-law, he asks his wife for a divorce. When she refuses he plots her murder and thinks he has the perfect alibi (he injures his leg in an accident and supposedly can't walk). When the police fail to notify him of her death Bogart is forced to report his wife missing. But is she dead? Her guilty husband smells her perfume, thinks he sees her walking down the street, and when he goes back to check the scene of the crime his wife's body is missing. Suspenseful melodrama also features Greenstreet as a psychologist/family friend who suspects Bogart knows more than he's telling.
1945 86m/B Humphrey Bogart, Alexis Smith, Sydney Greenstreet, Rose Hobart, Charles Drake, Grant Mitchell; *Dir:* Curtis Bernhardt. **VHS $19.98** *MGM* 🎞🎞🎞

The Conformist Character study of young Italian fascist, plagued by homosexual feelings, who must prove his loyalty by killing his old professor. Decadent and engrossing story is brilliantly acted. Based on the novel by Alberto Moravia, scripted by Bertolucci. **1971 (R) 108m/C** *IT FR GE* Jean-Louis Trintignant, Stefania Sandrelli, Dominique Sanda, Pierre Clementi, Gastone Moschin; *Dir:* Bernardo Bertolucci. National Board of Review Awards '71: 5 Best Foreign Films of the Year; New York Film Festival '70: 5 Best Foreign Films of the Year; National Society of Film Critics '71: Best Director, Best Cinematography. **VHS, Beta, LV $44.95** *TAM, PAR, APD* 🎬🎬🎬🎬

Congress Dances (Der Kongress taenzt) A rare German musical about a romance in old Vienna. Popular in its day, it was subsequently banned by the Nazis in 1937. In German with English subtitles. **1931 92m/B** *GE* Lilian Harvey, Conrad Veidt, Willy Fritsch; *Dir:* Erik Charell. **VHS, Beta $44.95** *TAM, NOS, HHT* 🎬🎬

A Connecticut Yankee A charming, if somewhat dated, version of the popular Mark Twain story, "A Connecticut Yankee in King Arthur's Court." Rogers is a radio shop owner who dreams his way back to the Knights of the Round Table. Story rewritten to fit Rogers' amiable style and to make then-current wisecracks. Great cast overcomes weak points in the script. **1931 96m/B** Will Rogers, Myrna Loy, Maureen O'Sullivan, William Farnum; *Dir:* David Butler. **VHS $19.98** *FOX, FCT* 🎬🎬🎬

A Connecticut Yankee in King Arthur's Court A pleasant version of the famous Mark Twain story about a 19th-century man transported to Camelot and mistaken for a dangerous wizard. Easy going Bing sings "Once and For Always" and "Busy Doin' Nothin'." This was the third film version of the classic, which was later remade as "Unidentified Flying Oddball", a made-for-television movie and an animated feature. **1949 108m/C** Bing Crosby, Rhonda Fleming, William Bendix, Cedric Hardwicke, Henry Wilcoxon, Murvyn Vye, Virginia Field; *Dir:* Tay Garnett. **VHS, Beta, LV $29.95** *MCA* 🎬🎬½

A Connecticut Yankee in King Arthur's Court An animated version of the classic Mark Twain story. **1970 74m/C** *Dir:* Zoran Janjic. **VHS, Beta $19.95** *MGM* 🎬🎬

The Connection The Living Theatre's ground-breaking performance of Jack Gelber's play about heroin addicts waiting for their connection to arrive, while a documentary filmmaker hovers nearby with his camera. **1961 105m/B** Warren Finnerty, Carl Lee, William Redfield, Freddie Redd Quartet, Roscoe Lee Browne, Garry Goodrow, James Anderson, Jackie McLean; *Dir:* Shirley Clarke. **VHS, Beta $29.95** *MFV, FCT* 🎬🎬🎬½

The Connection Durning steps in to work a deal between hotel jewel thieves and the insurance company. Complex plot handled deftly by all concerned. Made for television. **1973 74m/C** Charles Durning, Ronny Cox, Dennis Cole, Zohra Lampert, Heather MacRae, Dana Wynter; *Dir:* Tom Gries. **VHS, Beta $59.95** *VMK* 🎬🎬½

The Conqueror Wayne is horribly miscast as Genghis Khan in this tale of the warlord's early life and involvement with the kidnapped daughter of a powerful enemy. Expensive yet forgettable film with stilted, unbelievable dialogue. Listed in (The Fifty Worst Films of All Time). Unfortunately, the filming took place near a nuclear test site. Later, many members of the cast and crew were diagnosed with cancer. **1956 111m/C** John Wayne, Susan Hayward, William Conrad, Agnes Moorehead, Lee Van Cleef, Pedro Armendariz, Thomas Gomez, John Hoyt; *Dir:* Dick Powell. **VHS, Beta $19.95** *MCA* 🎬

Conqueror & the Empress Low-budget tale of English explorers who roust an island prince from his throne, motivating him to engage in inspired blood-letting. **1964 89m/C** Guy Madison, Ray Danton, Mario Petri, Alberto Farnese. **VHS, Beta $29.95** *FOR* Woof!

Conqueror of the World The adventures of a caveman named Bog. Really. Based on the novel by Albert Cavallone. Available in Spanish. **1985 90m/C** **VHS, Beta $59.95** *MGL* Woof!

The Conqueror Worm Price turns in a fine performance portraying the sinister Matthew Hopkins, a real-life 17th century witch-hunter. No "ham" in this low-budget, underrated thriller. The last of three films for director Reeves, who died from an accidental overdose in 1969. Originally titled "The Witchfinder General." Also available with "Tomb of Ligeia" on Laser Disc. **1968 95m/C** *GB* Vincent Price, Ian Ogilvy, Hilary Dwyer, Rupert Davies, Robert Russell, Patrick Wymark, Wilfrid Brambell; *Dir:* Michael Reeves. **VHS $59.99** *HBO, FCT* 🎬🎬🎬

Conquest Garbo, as the Polish countess Marie Walewska, tries to persuade Napoleon (Boyer) to free her native Poland from the the Russian Tsar. Garbo, Boyer, and Ouspenskaya are outstanding, while the beautiful costumes and lavish production help, but the script is occasionally weak. "Conquest" was a box office flop in the United States and ended up costing MGM more than any movie it had made up until that time. Also known as "Marie Walewska." Oscar nomination for Best Actor (Boyer). **1932 90m/B** Greta Garbo, Charles Boyer, Reginald Owen, Alan Marshal, Henry Stephenson, Leif Erickson, May Whitty, Maria Ouspenskaya, Vladimir Sokoloff, Scotty Beckett; *Dir:* Clarence Brown. **VHS, Beta $19.98** *MGM, FCT* 🎬🎬🎬

Conquest Sword and sorcery tale of two mighty warriors against an evil sorceress who seeks world domination. High point is the score provided by Claudio Simonetti, who scores Dario Argento's horror films. **1983 (R) 92m/C** *IT SP MX* Jorge Rivero, Andrea Occhipinti, Violeta Cela, Sabrina Siani; *Dir:* Lucio Fulci. **VHS, Beta $59.95** *MED* 🎬

Conquest of Cochise Mediocre oater about a cavalry officer who must stop the war between the Apache and Comanche tribes and a group of Mexicans in the Southwest of the 1850s. **1953 70m/C** John Hodiak, Robert Stack, Joy Page; *Dir:* William Castle. **VHS, Beta $14.99** *GKK* 🎬🎬

Conquest of Mycene Hercules battles an evil queen and this time he may actually lose! Muscleman mayhem will entertain the kids. **1963 102m/C** *IT FR* Gordon Scott, Rosalba (Sara Bay) Neri, Jany Clair; *Dir:* Giorgio Ferroni. **VHS $16.95** *SNC* 🎬½

Conquest of the Normans During the Norman invasion of England, Oliver is accused of kidnapping the King. To save his life, Oliver must find the true identity of the abductors. With so much plot, there should have been some suspense. **1962 83m/C** Cameron Mitchell. **Beta $39.95** *AIP* 🎬

Conquest of the Planet of the Apes The apes turn the tables on the human Earth population when they lead a revolt against their cruel masters in the distant year of 1990. Sure, there's plenty of cliches - but the story drags you along. The 4th film in the series. Followed by "Battle for the Planet of the Apes." **1972 (PG) 87m/C** Roddy McDowall, Don Murray, Ricardo Montalban, Natalie Trundy, Severn Darden, Hari Rhodes; *Dir:* J. Lee Thompson. **VHS, Beta, LV $19.98** *FOX* 🎬🎬½

Conrack The true story of Pat Conroy, who tried to teach a group of illiterate Black children in South Carolina by using common-sense teaching techniques to inspire their interest. Voight is convincing as the earnest teacher. Based on Conroy's novel "The Water Is Wide." **1974 (PG) 111m/C** Jon Voight, Paul Winfield, Madge Sinclair, Hume Cronyn, Martin Ritt; *Dir:* Martin Ritt. **VHS, Beta $59.98** *FOX* 🎬🎬🎬

Consenting Adults Made-for-television movie about a distressed family coming to terms with a favorite son's homosexuality. Restrained exploration of the controversial subject. Based on Laura Z. Hobson's novel. **1985 100m/C** Marlo Thomas, Martin Sheen, Barry Tubb, Talia Balsam, Ben Piazza, Corinne Michaels; *Dir:* Gilbert Cates. **VHS $14.95** *KBV* 🎬🎬½

Consolation Marriage A couple meet and marry after being jilted by others, then must make a choice when their old flames return. Dunne and O'Brien make a charming couple in this early talkie. One of Dunne's early starring roles. **1931 82m/B** Irene Dunne, Pat O'Brien, John Halliday, Matt Moore, Myrna Loy, Lester Vail; *Dir:* Paul Sloane. **VHS, Beta $19.98** *TTC* 🎬🎬½

Conspiracy Defense Secretary William Baine is haunted by a past sexual encounter, and a top secret team moves in to avert any hint of what could become a "sex scandal." Based on a true story. **1989 (R) 87m/C** James Wilby, Kate Hardie, Glyn Houston; *Dir:* Christopher Barnard. **VHS, Beta $29.95** *ACA* 🎬🎬

Conspiracy of Hearts A convent of nuns hide Jewish children in 1943 Italy despite threats to their personal safety. Suspenseful tale despite the familiar plot. **1960 113m/B** Lilli Palmer, Yvonne Mitchell, Sylvia Syms, Ronald Lewis; *Dir:* Ralph Thomas. **VHS, Beta** *LCA* 🎬🎬🎬

A Conspiracy of Love A made-for-television movie about three generations of a divorce-torn family trying to pull itself back together. Sugary domestic tale tries to be uplifting, succeeds in being cliche-ridden. **1987 93m/C** Robert Young, Drew Barrymore, Glynnis O'Connor, Elizabeth Wilson, Michael Laurence, John Fujioka, Alan Fawcett; *Dir:* Noel Black. **VHS $19.95** *STE, NWV* 🎬

Conspiracy of Terror Married detectives investigate a series of murders and unearth a grisly cult in a quiet suburb. Standard fare with a few too cute lines concerning the leads' mixed-faith marriage. Made for television pilot.

1975 78m/C Michael Constantine, Barbara Rhoades, Mariclare Costello, Roger Perry, David Opatoshu, Logan Ramsey; *Dir:* John Llewellyn Moxey. VHS *IVE ✂✂*

Conspiracy: The Trial of the Chicago Eight Courtroom drama focuses on the rambunctious trial of the Chicago Eight radicals, charged with inciting a riot at the Democratic National Convention of 1968. Dramatized footage mixed with interviews with the defendants. Imaginative reconstruction of history.
1987 118m/C Peter Boyle, Elliott Gould, Robert Carradine, Martin Sheen, David Clennon, David Kagen, Michael Lembeck, Robert Loggia; *Dir:* Jeremy Paul Kagan. VHS, Beta $79.95 *HBO ✂✂✂*

Constructing Reality: A Film on Film An explanation of "film reality" and a discussion of the five key elements of film construction: time, space, perspective, rhythm, and sound.
1972 18m/C VHS, Beta *JOU ✂✂✂*

Consuming Passions A ribald, food-obsessed English comedy about a young idiot who rises within the hierarchy of a chocolate company via murder. You'll never guess what the secret ingredient in his wonderful chocolate is. Based on a play by Michael Palin and Terry Jones (better known as part of the Monty Python troupe), the film is sometimes funny, more often gross, and takes a single joke far beyond its limit.
1988 (R) 95m/C *GB* Vanessa Redgrave, Jonathan Pryce, Tyler Butterworth, Freddie Jones, Prunella Scales, Sammi Davis, Thora Hird; *Dir:* Giles Foster. VHS, Beta, LV $19.95 *VIR ✂✂½*

Contagion Innocent real estate agent meets a reclusive, eccentric millionaire who offers him unlimited wealth and beautiful women, but only if he kills someone. Interesting premise loses its way.
1987 90m/C *AU* John Doyle, Nicola Bartlett, Roy Barrett, Nathy Gaffney, Pamela Hawksford; *Dir:* Karl Zwicky. VHS, Beta $14.95 *SVS ✂½*

Contempt A film about the filming of a new version of "The Odyssey," and the rival visions of how to tell the story. Amusing look at the film business features Fritz Lang playing himself, Godard as his assistant. Bardot is pleasant scenery. Adapted from Moravia's "A Ghost at Noon," and also known as "Le Mepris."
1964 102m/C *IT FR* Brigitte Bardot, Jack Palance, Fritz Lang, Georgia Moll, Michel Piccoli; *W/Dir:* Jean-Luc Godard. VHS, Beta, LV $19.95 *COL, TAM, SUE ✂✂✂½*

Continental Divide A hard-nosed political columnist takes off for the Colorado Rockies on an "easy assignment"—interviewing a reclusive ornithologist, with whom he instantly falls in love. A city slicker, he first alienates her, but she eventually falls for him, too. But it's not exactly a match made in heaven. Story meanders to a conclusion of sorts. Probably the most normal Belushi ever was on screen. Scripted by Lawrence Kasdan.
1981 (PG) 103m/C John Belushi, Blair Brown, Allen Garfield, Carlin Glynn, Tony Ganios, Val Avery, Tim Kazurinsky; *Dir:* Michael Apted. VHS, Beta, LV $14.95 *MCA, PIA ✂✂*

Contra Company A teenage girl and an embittered ex-terrorist narrowly escape from a terrorist infested region in this confusing film about politics, drugs, and sinister dealings.
1988 (R) 110m/C VHS, Beta *CLV ✂*

Contraband The leader of a smuggling gang escapes from an ambush in which his brother was murdered. He searches for a haven of safety while his cronies seek brutal revenge. Average crime yarn lacks much excitement. Dubbed into English.
1986 95m/C *IT* Fabio Testi, Ivana Monti; *Dir:* Lucio Fulci. VHS, Beta $59.95 *MGL ✂½*

Control Tedious tale of fifteen volunteers for a 20-day fallout shelter habitation experiment who become trapped when a real nuclear emergency occurs. Filmed in Rome. Made for cable TV.
1987 83m/C *IT* Burt Lancaster, Kate Nelligan, Ben Gazzara, Andrea Ferreol, Lavinia Segurini, Andrea Occhipinti, Cyrielle Claire, Jean Benguigui, Kate Reid, Erland Josephson, Ingrid Thulin; *Dir:* Giuliano Montaldo. VHS, Beta $79.99 *HBO ✂½*

The Conversation Nominated for two Academy Awards, this is one of the best movies of the '70s. Hackman plays a surveillance expert increasingly uneasy about his job as he begins to suspect he is an accomplice to murder. Powerful statement about privacy, responsibility and guilt. Written by Coppola.
1974 (PG) 113m/C Gene Hackman, John Cazale, Frederic Forrest, Allen Garfield, Cindy Williams, Robert Duvall, Teri Garr, Michael Higgins, Elizabeth McRae, Harrison Ford; *Dir:* Francis Ford Coppola. British Academy Awards '74: Best Film Editing, Best Musical Score; Cannes Film Festival '74: Best Film; National Board of Review Awards '74: Best Actor (Hackman), Best Director (Coppola), Best Film; National Society of Film Critics Awards '74: Best Director (Coppola). VHS, Beta, LV $19.95 *PAR ✂✂✂½*

Conversation Piece An aging art historian's life is turned upside down when a Countess and her daughters rent out the penthouse in his estate. Sometimes-talky examination of scholarly pretensions.
1975 (R) 112m/C *IT FR* Burt Lancaster, Silvana Mangano, Helmut Berger, Claudia Cardinale; *Dir:* Luchino Visconti. New York Film Festival '75: Best Director. VHS, Beta $69.99 *HBO ✂✂✂*

Convicted Lean thriller about a woman cleared of a murder charge, but only when a second, identical murder occurs while she is jailed.
1932 57m/B Aileen Pringle, Jameson Thomas, Harry Myers, Dorothy Christy, Richard Tucker; *Dir:* Christy Cabanne. VHS, Beta, LV $19.95 *NOS, WGE ✂✂½*

Convicted Solid story of innocent man imprisoned for rape, and his wife's determined efforts to free him. Larroquette does well in unexpected role. Based on a true story. Made for television.
1986 94m/C John Larroquette, Carroll O'Connor, Lindsay Wagner; *Dir:* David Lowell Rich. VHS, Beta $69.98 *TTC ✂✂½*

Convicted: A Mother's Story Jillian embezzles from her employer, all for the sake of a bum of a boyfriend, and goes to jail. Odd casting hinders a ho-hum plot. Made for television.
1987 95m/C Ann Jillian, Kiel Martin, Fred Savage, Gloria Loring, Christa Denton; *Dir:* Richard T. Heffron. VHS, Beta $79.95 *JTC ✂*

Convicts at Large An escaped convict steals the clothes of a inept architect. Left with only prison duds, the architect stumbles into the bad side of town and gets involved with the mob. Inane.
1938 58m/B Ralph Forbes, Paula Stone. VHS, Beta, LV *WGE ✂*

Convoy Life aboard a convoy ship in the North Sea during WWII. The small cruiser is picked on by a German U-Boat. Will a rescuer appear? Noted for technical production values.
1940 95m/B Clive Brook, John Clements, Edward Chapman, Judy Campbell, Edward Rigby, Stewart Granger, Michael Wilding, George Benson; *Dir:* Pen Tennyson. VHS $19.95 *NOS ✂✂½*

Convoy A defiant trucker leads an indestructible truck convoy to Mexico to protect high gasoline prices. Lightweight stuff was inspired by the song "Convoy" by C.W. McCall.
1978 (R) 106m/C Kris Kristofferson, Ali MacGraw, Ernest Borgnine, Burt Young, Madge Sinclair, Franklin Ajaye, Cassie Yates; *Dir:* Sam Peckinpah. VHS, Beta $79.95 *HBO ✂✂*

Coogan's Bluff An Arizona deputy sheriff (Eastwood) travels to New York in order to track down a killer on the loose. First Eastwood, Siegel teaming is tense actioner. The television series "McCloud" was based on this film.
1968 (PG) 100m/C Clint Eastwood, Lee J. Cobb, Tisha Sterling, Don Stroud, Betty Field, Susan Clark, Tom Tully; *Dir:* Don Siegel. VHS, Beta, LV $19.95 *MCA ✂✂✂*

The Cook, the Thief, His Wife & Her Lover An exclusive restaurant houses four disturbing characters. Greenaway's powerful vision of greed, love, and violence may be too strong for some tastes. Available in several different versions: the standard unrated theatrical release, the unrated version in a letterboxed format, an R-rated cut which runs half an hour shorter, and a Spanish-subtitled version.
1990 (R) 123m/C *GB* Richard Boringer, Michael Gambon, Helen Mirren, Alan Howard, Tim Roth; *Dir:* Peter Greenaway. VHS $89.95 *VMK ✂✂✂½*

Cookie A light comedy about a Mafia don's daughter trying to smart mouth her way into the mob's good graces. A character-driven vehicle, this plot takes a backseat to casting. Wiest is superb as Falk's moll. The screenplay is by Nora Ephron and Alice Arlen.
1989 (R) 93m/C Emily Lloyd, Peter Falk, Dianne Wiest, Jerry Lewis, Brenda Vaccaro, Ricki Lake, Lionel Stander, Michael Gazzo, Adrian Pasdar, Bob Gunton; *Dir:* Susan Seidelman. VHS, Beta, LV, 8mm $19.95 *WAR ✂✂½*

Cool As Ice "Rapper" Vanilla Ice makes his feature film debut as a rebel with an eye for the ladies, who motors into a small, conservative town. Several so-so musical segments. For teenage girls only.
1991 (PG) 92m/C Vanilla Ice, Kristin Minter, Michael Gross, Sydney Lassick, Dody Goodman, Naomi Campbell, Candy Clark; *Dir:* David Kellogg. VHS, Beta, LV $79.95 *MCA ✂*

Cool Blue An unsuccessful painter meets the woman of his dreams, has a brief affair, then tries to locate her in the greater Los Angeles metro area. Good luck. Watch for Penn's cameo.
1988 (R) 93m/C Woody Harrelson, Hank Azaria, Ely Pouget, Sean Penn, John Diehl; *Dir:* Mark Mullin, Richard Shepard. VHS, Beta, LV $89.95 *COL ✂½*

Cool it Carol A girl and her boyfriend leaves a small town for the excitement of city life. After she works as a prostitute and a model, they return to their quiet home. There's no place like home.
1968 101m/C Robin Askwith, Janet Lynn, Jess Conrad, Stubby Kaye. VHS, Beta $59.95 *MON* Woof!

Cool Cats: 25 Years of Rock 'n' Roll Style The effect of rock music on contemporary style and mores is the subject of this "rockumentary" which features performance clips and interviews by thirty-four rock trendsetters, including Elvis, The Beatles, Culture Club, David Bowie and many others.
1983 90m/C VHS, Beta, LV **$59.95** *MVD, MGM*

Cool Hand Luke One of the last great men-in-chains films. A man (Newman) sentenced to sweat out a term on a prison farm refuses to compromise with authority. Martin shines in his supporting role as the oily warden, uttering that now-famous phrase, "What we have here is a failure to communicate." Kennedy's performance as leader of the chain gang won him an Oscar. Based on the novel by Donn Pearce. Oscar nominations for Newman, Adapted Screenplay, and Original Score.
1967 126m/C Paul Newman, George Kennedy, J.D. Cannon, Strother Martin, Dennis Hopper, Anthony Zerbe, Lou Antonio, Wayne Rogers, Harry Dean Stanton, Ralph Waite, Joe Don Baker, Richard Davalos, Jo Van Fleet; *Dir:* Stuart Rosenberg. Academy Awards '67: Best Supporting Actor (Kennedy). **VHS, Beta, LV $19.98** *WAR, BTV ♂♂♂½*

Cool Mikado A jazzy modernization of the Gilbert-Sullivan operetta, in which an American soldier is kidnapped by the gangster fiance of the Japanese girl he loves.
1963 81m/C *GB* Stubby Kaye, Frankie Howerd, Dennis Price; *Dir:* Michael Winner. VHS, Beta **$79.95** *JEF ♂♂*

The Cool World Tough-talking docudrama, set on the streets of Harlem, focuses on a 15-year-old black youth whose one ambition in life is to own a gun and lead his gang.
1963 107m/B Gloria Foster, Hampton Clanton, Carl Lee; *Dir:* Shirley Clarke. VHS, Beta **$49.95** *VYY ♂♂½*

Cop Left by his wife and child, a ruthless and work-obsessed detective goes after a twisted serial killer. Woods' exceptional ability to play sympathetic weirdos is diluted by a script—based on James Ellroy's novel "Blood on the Moon"—that warps the feminist theme, is violent, and depends too heavily on coincidence.
1988 (R) 110m/C James Woods, Lesley Ann Warren, Charles Durning, Charles Haid, Raymond J. Barry, Randi Brooks, Annie McEnroe, Victoria Wauchope; *Dir:* James B. Harris. **VHS, Beta, LV $19.95** *PAR ♂♂½*

Cop in Blue Jeans Undercover cop goes after a mob kingpin in this violent, badly dubbed crime yarn.
1978 92m/C Tomas Milian, Jack Palance, Maria Rosaria Omaggio, Guido Mannari; *Dir:* Bruno Corbucci. VHS, Beta **$24.95** *VMK ♂*

Cop & the Girl A cop falls for a teenage runaway, ends up on the lam. Confused story overpowered by camera work. Dubbed.
198? 94m/C *GE* Juergen Prochnow, VHS, Beta **$79.95** *WES ♂*

Cop-Out A cop is framed. His brother is mad. Teamed up with a sexy lawyer, he exposes the filth that is corrupting the city. Okay melodrama.
1991 102m/C David Buff, Kathryn Luster, Dan Ranger, Reggie DeMorton, Lawrence L. Simeone; *W/Dir:* Lawrence L. Simeone. **VHS, Beta $79.95** *AIP ♂♂*

Copacabana A shady theatrical agent (Groucho) books a nightclub singer into two shows at the same time at New York's ritzy nightclub. Lots of energy spent trying to enliven a routine script. Songs include "Tico, Tico" and "Stranger Things Have Happened." Also available colorized.
1947 91m/B Groucho Marx, Carmen Miranda, Steve Cochran, Gloria Jean, Andy Russell, Earl Wilson; *Dir:* Alfred E. Green. **VHS $19.98** *REP, FCT, MLB ♂♂*

Copenhagen Nights A young Danish girl discovers sex in a local orgy-club. Softcore, dubbed, and retitled.
198? 80m/C *DK* Lilli Carati, Ajita Wilson. VHS, Beta **$29.95** *MED Woof!*

Copkillers Two men desperate for an easy way to make money embark on a murderous shooting spree. Mostly blanks.
1977 (R) 93m/C Jason Williams, Bill Osco. VHS, Beta **$59.95** *TWE Woof!*

Copper Canyon Milland plays a Confederate Army officer who heads West after the Civil War, meets up with Lamarr, and sparks a romance. Good chemistry between the leads in this otherwise standard western.
1950 84m/B Ray Milland, Hedy Lamarr, MacDonald Carey, Mona Freeman, Harry Carey Jr., Frank Faylen, Taylor Holmes, Peggy Knudsen; *Dir:* John Farrow. **VHS $14.95** *PAR, FCT ♂♂½*

Copperhead A group of copperhead snakes attack a family who possess a stolen Incan gold necklace. This one bites.
1984 (R) 90m/C Jack Renner, Gretta Ratliff, David Fritts, Cheryl Nickerson. VHS, Beta **$59.95** *UHV ♂*

Cops/One Week Two vintage Keaton shorts, with original organ score.
1922 51m/B Buster Keaton; *Dir:* Buster Keaton, Eddie Cline. VHS, Beta, 8mm *VYY ♂*

Cops and Robbers Two cops use a Wall Street parade for returning astronauts as a cover for a multi-million dollar heist. Exceptional caper film thanks to the likable leads and genuine suspense.
1973 (PG) 89m/C Joseph Bologna, Dick Ward, Sheppard Strudwick, John P. Ryan, Ellen Holly, Dolph Sweet, Joe Spinell, Cliff Gorman; *Dir:* Aram Avakian. **VHS $9.95** *WKV ♂♂½*

Cord & Dischords/Naughty Nurse Two vintage silent shorts featuring some lesser known slapstick clowns cavorting in blackface.
1928 50m/B Fred Parker, Jack Cooper, Ella MacKenzie, Spencer Williams. VHS, Beta **$14.95** *VYY ♂♂*

Cordelia A dramatization of an actual court case in turn-of-the century Quebec, in which an "overly" vivacious woman is accused of murdering her husband.
1987 115m/C VHS, Beta *NFC ♂♂*

Cordell Hull: The Good Neighbor The story of the man known as the Father of the United Nations, Cordell Hull. Hull was appointed Secretary of State by Franklin Roosevelt and in 1943 wrote the United Nations Declaration. His efforts were awarded by his winning of the Nobel Peace Prize. Classic newsreel footage.
1954 15m/B VHS, Beta **$49.95** *TSF ♂♂*

Coriolanus, Man without a Country When the Romans begin to abuse their rights, the Plebians call upon the brave and mighty Coriolanus to lead them into

battle against their oppressors. Predictable muscle man outing.
1964 ?m/C *IT* Gordon Scott, Alberto Lupo, Lilla Brignone. VHS, Beta **$16.95** *SNC ♂♂*

Corleone Limp tale about two boyhood friends who grow up to fight the evil landowners who dominate Italy. To accomplish this, one becomes a mobster, the other a politician.
1979 (R) 115m/C *IT* Claudia Cardinale, Giuliano Gemma; *Dir:* Pasquale Squitieri. Montreal World Film Festival '79: Best Actor (Gemma). VHS, Beta **$79.95** *MGM ♂*

Corn is Green Touching story of a school teacher in a poor Welsh village who eventually sends her pet student to Oxford. Davis makes a fine teacher, though a little young, while the on-site photography provides atmosphere. Based on play by Emlyn Williams. Music by Max Steiner. Remade in 1979. Oscar nominations for Best Supporting Actor (Dall), and Best Supporting Actress (Loring).
1945 115m/B Bette Davis, John Dall, Nigel Bruce, Joan Lorring, Arthur Shields, Mildred Dunnock, Rhys Williams, Rosalind Iven; *Dir:* Irving Rapper. VHS, Beta **$19.95** *MGM, FCT ♂♂♂*

Cornbread, Earl & Me A high school basketball star from the ghetto is mistaken for a murderer by cops and shot, causing a subsequent furor of protest and racial hatred. Superficial melodrama.
1975 (R) 95m/C Moses Gunn, Rosalind Cash, Bernie Casey, Tierre Turner, Madge Sinclair, Keith Wilkes, Antonio Fargas; *Dir:* Joseph Manduke. VHS, Beta **$59.95** *HBO ♂*

Cornered Tough Powell plays an airman released from a German prison camp who pursues a Nazi war criminal to avenge the death of his wife and child.
1945 102m/B Dick Powell, Walter Slezak, Micheline Cheirel, Luther Adler; *Dir:* Edward Dmytryk. VHS, Beta, LV **$19.98** *RKO ♂♂♂*

Coroner Creek Scott's fiancee commits suicide during an Indian attack on her stagecoach, and the couple's dowry is stolen. Scott sets out to trace the loot and exact his revenge. A superior production, notable for being one of the first westerns with an adult theme.
1948 89m/B Randolph Scott, Marguerite Chapman, George Macready, Sally Eilers, Edgar Buchanan, Wallace Ford, Forrest Tucker; *Dir:* Ray Enright. VHS *EBE ♂♂½*

The Corpse Grinders Low-budget bad movie classic in which a cardboard corpse-grinding machine makes nasty cat food that makes cats nasty. Sets are cheap, gore effects silly, and cat attacks ridiculous.
1971 (R) 73m/C Sean Kenney, Monika Kelly, Sandford Mitchell; *W/Dir:* Ted V. Mikels. VHS, Beta **$19.95** *WES Woof!*

The Corpse Vanishes Lugosi at his chilling best as a diabolical scientist who snatches young brides and drains their blood in an effort to keep his 70-year-old wife eternally youthful. Not for newlyweds!
1942 64m/B Bela Lugosi, Luana Walters, Tristram Coffin, Elizabeth Russell, Vince Barnett, Joan Barclay, Angelo Rossitto; *Dir:* Wallace Fox. VHS, Beta **$19.95** *NOS, MRV, SNC ♂♂½*

Corregidor A love triangle develops between doctors treating the wounded during the World War II battle. A poor propaganda piece which contains only shaky stock footage for its "action" sequences.

1943 73m/B Otto Kruger, Elissa Landi, Donald Woods, Rick Vallin, Frank Jenks, Wanda McKay, Ian Keith; *Dir:* William Nigh. VHS $19.95 *NOS, MOV* 🎬🎬

Corrida Interdite Slow-motion sequences reveal the beauty of the bullfighting ritual without neglecting the deadly nature of the sport.
1958 10m/C VHS, Beta *TEX* 🎬🎬

The Corridor: Death Two Americans are involved in a car accident in Israel which triggers a debate on the Jewish perspective of death and the afterlife. Filmed on location at Hadassah Hospital on St. Scopus, Jerusalem. The final part of the trilogy preceded by "The Eighth Day: Circumcision" and "The Journey: Bar Mitzvah."
19?? 25m/C VHS, Beta $39.95 *ERG* 🎬🎬

Corridors of Blood Karloff is a doctor, in search of a viable anesthetic, who accidently becomes addicted to drugs, then turns to grave robbers to support his habit. Karloff plays usual threatening doctor to perfection. Also known as "The Doctor from Seven Dials."
1958 86m/C *GB* Boris Karloff, Betta St. John, Finlay Currie, Christopher Lee, Francis Matthews, Adrienne Corri, Nigel Green; *Dir:* Robert Day. VHS, Beta $39.95 *MPI* 🎬🎬½

Corrupt Bad narcotics cop goes after a murderer, but mysterious Lydon gets in the way. An acquired taste, best if you appreciate Lydon, better known as Johnny Rotten of the Sex Pistols. Alternate titles: "Order of Death" and "Cop Killers."
1984 (PG) 99m/C *IT* Harvey Keitel, John (Johnny Rotten) Lydon, Sylvia Sidney, Nicole Garcia, Leonard Mann; *Dir:* Roberto Faenza. VHS, Beta $69.95 *HBO* 🎬🎬

Corrupt Ones Everyone's after a photographer who has a medallion that will lead to buried treasure in China. Great characters hampered by pedestrian script.
1967 87m/C Robert Stack, Elke Sommer, Nancy Kwan, Christian Marquand, Werner Peters; *Dir:* James Hill. VHS, Beta $49.95 *SUE* 🎬🎬½

Corruption Mostly soft-porn. Alleged plot: The Mob is interested in having Shea work for them, and makes it worth his while.
1989 (R) 80m/C Mary McGee, Patrick Shea. VHS *AVD* Woof!

Corsair Actioner about a gorgeous debutante and a handsome gangster who find themselves caught up with a gang of bootlegging pirates. Thelma Todd is always worth watching.
1931 73m/B Chester Morris, Thelma Todd, Frank McHugh, Ned Sparks, Mayo Methot; *Dir:* Roland West. VHS, Beta $19.95 *NOS, KRT, RXM* 🎬

Corsican Brothers Alexander Dumas' classic about twins who, although separated, remain spiritually tied throughout their various adventures. Fairbanks takes on dual roles in this entertaining swashbuckler. Remade for television in 1984.
1941 111m/B Douglas Fairbanks Jr., Akim Tamiroff, Ruth Warrick, J. Carroll Naish, H.B. Warner, Henry Wilcoxon; *Dir:* Gregory Ratoff. VHS, Beta, LV $12.95 *MED* 🎬🎬

Corvette Summer After spending a semester restoring a Corvette in his high school shop class, an L.A. student must journey to Las Vegas to recover the car when it is stolen. There he meets a prostitute, falls in love, and steps into the "real world" for the first time. Potts intriguing as

the low-life love interest, but she can't save this one.
1979 (PG) 104m/C Mark Hamill, Annie Potts, Eugene Roche, Kim Milford, Richard McKenzie, William Bryant; *Dir:* Matthew Robbins. VHS, Beta $59.95 *MGM* 🎬

Cosh Boy Vicious male youths stalk women on the streets of London in this low-budget Jack the Ripper rip-off.
1961 90m/B Joan Collins, James Kennedy. VHS, Beta *MON* Woof!

Cosi Fan Tutte A performance of the Mozart Lyric Opera at Sweden's 18th-century Drohningham Court Theatre.
1985 139m/C Ann Christine Biel, Maria Hoglind, Lars Tibell. VHS, Beta $39.95 *HMV, HBO, MVD* Woof!

The Cosmic Eye Animated tale of three musicians from outer space who come to earth to spread the message of worldwide peace and harmony. Sticky sweet and moralizing, but the animation will impress the kids.
1971 71m/C *Dir:* Faith Hubley; *Voices:* Dizzy Gillespie, Maureen Stapleton. VHS, Beta $49.95 *DIS* 🎬🎬🎬

The Cosmic Man An alien arrives on Earth with a message of peace and restraint. He is regarded with suspicion by us nasty Earthlings. Essentially "The Day the Earth Stood Still" without the budget, but interesting nonetheless.
1959 72m/B Bruce Bennett, John Carradine, Angela Greene, Paul Langton, Scotty Morrow; *Dir:* Herbert Greene. VHS, Beta $14.95 *RHI, SNC* 🎬½

The Cosmic Monsters A scientist accidentally pops a hole in the ionosphere during a magnetism experiment. That's when huge, mean insects arrive to plague mankind. You figure it out. Original title: "The Strange World of Planet X."
1958 75m/B Forrest Tucker, Gaby Andre, Alec Mango, Hugh Latimer, Martin Benson; *Dir:* Gilbert Gunn. VHS, Beta $14.95 *RHI, MRV, SNC* 🎬½

Cosmos: War of the Planets The ultimate battle for survival is fought in outer space; but not very well. Special effects especially laughable. Alternate titles: "Cosmo 2000: Planet Without a Name" and "War of the Planets."
1980 (PG) 90m/C *IT* Katia Christine, West Buchanan, John Richardson, Yanti Somer; *Dir:* Al Bradley. VHS, Beta *PGN* 🎬

Cottage to Let Propaganda thriller about a Nazi plot to kidnap the inventor of a new bombsight. Cast does the best it can with lackluster script. Also known as "Bombsight Stolen."
1941 90m/B *GB* Leslie Banks, Alastair Sim, John Mills, Jeanne de Casalis, Carla Lehmann, George Cole, Michael Wilding, Frank Cellier, Wally Patch, Catherine Lacey; *Dir:* Anthony Asquith. VHS, Beta $16.95 *SNC* 🎬½

Cotter A Native American rodeo clown feels responsible for a cowboy's death, returns home to reflect on his life. Fair story, best at the small town feeling.
1972 94m/C Don Murray, Carol Lynley, Rip Torn, Sherry Jackson; *Dir:* Paul Stanley. VHS, Beta $39.95 *XVC, AVD, UHV* 🎬

Cotton Candy Follows the faintly interesting trials and tribulations of a high school senior who tries to form a rock band. At first he meets failure, but ultimately he is successful with the band and romance. Made for television. Ron Howard, of "Happy Days"

fame co-wrote the script with his brother, Clint, and also directed.
1982 97m/C Charles Martin Smith, Clint Howard, Leslie King; *Dir:* Ron Howard. VHS, Beta *TLF* 🎬

The Cotton Club With $50 million in his pocket, Francis reaches for an epic and delivers: handsome production, lots of dance, bit of singing, confused plot, uneven performances, tad too long. A musician playing at The Cotton Club falls in love with gangster Dutch Schultz's girlfriend. A black tap dancer falls in love with a member of the chorus line who can pass for white. These two love stories are told against a background of mob violence and music. Songs featured include "Crazy Rhythm," "Am I Blue," "Cotton Club Stomp," "Jitter Bug," and "I ll Wind." Excellent performances by Hoskins and Gwynne.
1984 (R) 121m/C Diane Lane, Richard Gere, Gregory Hines, Lonette McKee, Bob Hoskins, Fred Gwynne, James Remar, Nicolas Cage, Lisa Jane Persky, Allen Garfield, Gwen Verdon, Joe Dallesandro, Jennifer Grey, Tom Waits, Diane Venora; *Dir:* Francis Ford Coppola. VHS, Beta, LV, 8mm $14.98 *SUE* 🎬🎬🎬

The Couch Trip Aykroyd is an escapee from a mental institution who passes himself off as a radio psychologist and becomes a media sensation. There are a few laughs and some funny characters but for the most part this one falls flat.
1987 (R) 98m/C Dan Aykroyd, Walter Matthau, Charles Grodin, Donna Dixon, Richard Romanus, Arye Gross, David Clennon, Mary Gross; *Dir:* Michael Ritchie. VHS, Beta, LV $19.98 *ORI* 🎬½

Could It Happen Here? A lurid Italian spectacle about a metropolis thrown into a state of chaotic martial law by terrorism. Dubbed.
198? 90m/C *IT* Luc Merenda, Marcella Michelangeli; *Dir:* Massimo Pirri. VHS, Beta *MGL* 🎬

The Count Chaplin pretends to be the secretary of his boss, who in turn is posing as a count. Silent with musical soundtrack added.
1916 20m/B Charlie Chaplin; *Dir:* Charlie Chaplin. VHS, Beta *CAB, FST* 🎬🎬½

Count/Adventurer Two Chaplin two-reelers: "The Count (The Phoney Nobleman)" (1916), in which Charlie impersonates a count at the home of Miss Moneybags, and "The Adventurer" (1917), Chaplin's final film for Mutual, in which he plays an escaped convict with the law relentlessly on his trail. Silent with musical score.
1917 52m/B Charlie Chaplin, Eric Campbell, Edna Purviance, Frank Coleman; *Dir:* Charlie Chaplin. VHS, Beta $39.95 *VYY* 🎬🎬½

Count Dracula Passable version of the Dracula legend (based on the novel by Bram Stoker) has Lee as the thirsty count on the prowl for fresh blood.
1971 (R) 90m/C *SP GE IT* Christopher Lee, Herbert Lom, Klaus Kinski, Frederick Williams, Maria Rohm, Soledad Miranda; *Dir:* Jess (Jesus) Franco. VHS $14.98 *REP, FCT* 🎬🎬½

Count of Monte Cristo One of the first full length features made starring popular stage stars of the day. The first truly American feature, it was based on the classic tale of revenge by Alexander Dumas. Silent. Remade many times.
1912 90m/B James O'Neill. VHS, Beta $19.95 *NOS, VYY, DVT* 🎬½

Count of Monte Cristo A true swashbuckling revenge tale about Edmond Dantes, who unjustly spends years prison. After escaping, he gains ever so sweet and served quite cold revenge. Adaptation of the Alexandre Dumas classic.
1934 114m/B Robert Donat, Elissa Landi, Louis Calhern, Sidney Blackmer, Irene Hervey, Raymond Walburn, O.P. Heggie; *Dir:* Rowland V. Lee. **VHS, Beta $9.99** CCB, MED *ЛЛЛ*

Count of Monte Cristo The Alexander Dumas classic about an innocent man (Chamberlain) who is imprisoned, escapes, and finds the treasure of Monte Cristo, which he uses to bring down those who wronged him. Good version of the historical costumer. Originally shown in theaters in Europe, but broadcast on television in the U.S.
1974 104m/C *GB* Richard Chamberlain, Kate Nelligan, Donald Pleasence, Alessio Orano, Tony Curtis, Louis Jourdan, Trevor Howard, Taryn Power; *Dir:* David Greene. **VHS, Beta $59.98** FOX *ЛЛ ½*

The Count of the Old Town Feisty townsfolk settle the score after a gang of booze smugglers trick them out of their money. In Swedish with English subtitles.
1934 90m/B *SW* Ingrid Bergman. **VHS, Beta $29.95** WAC, CRO *ЛЛ*

Count Yorga, Vampire The vampire Count Yorga is practicing his trade in Los Angeles, setting up a coven and conducting seances. Good update of the traditional character, with Quarry suitable solemn and menacing. Followed by "The Return of Count Yorga." Also available with "Cry of the Banshee" on Laser Disc.
1970 (PG) 90m/C Robert Quarry, Roger Perry, Michael Murphy, Michael Macready, Donna Anders, Judith Lang; *Dir:* Bob Kelljan. **VHS, Beta $19.99** HBO *ЛЛ ½*

Countdown Documentary-type fictional look at the first moon mission and its toll on the astronauts and their families. Timely because the U.S. was trying to send a man to the moon in 1968. Interesting as a look at the developing Altman style.
1968 102m/C James Caan, Robert Duvall, Michael Murphy, Ted Knight, Joanna Moore, Barbara Baxley, Charles Aidman, Steve Ihnat, Robert Altman; *Dir:* Robert Altman. **VHS, Beta $59.95** WAR *ЛЛ ½*

Counter Attack Story of murder and insurance fraud takes place on the back lots of the Hong Kong film industry. As good as these get.
1984 105m/C Bruce Li, Dan Inosanto, John Ladalski, Young Kong; *Dir:* Bruce Li. **VHS, Beta** WVE *Л*

Counter Destroyer A woman retreats to a secluded island to write a film script. But then she is attacked by an army of disgruntled Ninja-zombies. Who can work with all that commotion?
19?? 90m/C VHS, Beta TWE *Л*

Counter Measures A helicopter pilot stumbles on to a bizarre chain of crimes and disasters.
19?? 98m/C Monte Markham. **VHS, Beta** PGN *Л*

Counter Punch Blake is a boxer framed for the murder of his manager who sets out to clear his name by finding the killers. Fair rendition of a familiar plot. Also titled "Ripped Off" and "The Boxer."
1971 (R) 72m/C *IT* Robert Blake, Ernest Borgnine, Gabriel Ferzetti, Catherine Spaak, Tomas Milian; *Dir:* Franco Prosperi. **VHS, Beta $9.95** SIM, LIV, LTG *ЛЛ ½*

Counterblast A Nazi spy assumes the role of a British scientist in order to gain classified information for the Fatherland. The trouble begins when he refuses to carry out an order to execute a pretty young assistant. Appealing performances distract from the somewhat confusing plot.
1948 99m/B *GB* Robert Beatty, Mervyn Johns, Nova Pilbeam, Margaretta Scott, Sybilla Binder, Marie Lohr, Karel Stepanek, Alan Wheatley; *Dir:* Paul Stein. **VHS, Beta $16.95** SNC, RXM *ЛЛ*

The Counterfeit Traitor Suspense thriller with Holden playing a double agent in Europe during World War II. Based on the true adventures of Eric Erickson, the top Allied spy of World War II, who was captured by the Gestapo but escaped.
1962 140m/C William Holden, Lilli Palmer, Hugh Griffith, Werner Peters, Eva Dahlbeck; *Dir:* George Seaton. **VHS, Beta $29.95** PAR *ЛЛЛ ½*

Counterforce A rough group of mercenaries is hired to guard a Mideastern leader being chased by his country's ruling despot. Mild action with predictable story.
1987 (R) 98m/C George Rivero, George Kennedy, Andrew Stevens, Isaac Hayes, Louis Jourdan, Kevin Bernhardt, Hugo Stiglitz, Robert Foster, Susana Dosamantes; *Dir:* D.J. Anthony Loma. **VHS, Beta, LV $89.95** IVE *Л*

Countrified A compilation of new and old names in country music singing some really good tunes. Features Hank Williams Jr. /"My Name Is Bocephus," k.d. lang and the reclines / "Turn Me Round," Dwight Yoakam/ "Guitars, Cadillacs," The Nitty Gritty Dirt Band/ "Partners, Brothers, and Friends," Crystaly Gale/ "Nobody's Angel," John Anderson / "Countrified," Tony Perez/ "Oh How I Love You," and Randy Travis/ "I Told You So."
1989 30m/C Performed by: k.d. lang, Dwight Yoakam, Hank Williams Jr. **VHS $16.98** MVD

Country Strong story of a farm family in crisis when the government attempts to foreclose on their land. Good performances all around and an excellent portrayal of the wife by Lange, which earned her an Oscar nomination. "The River" and "Places In the Heart," both released in 1984, also dramatized the plight of many American farm families in the early 1980s.
1984 (PG) 109m/C Jessica Lange, Sam Shepard, Wilford Brimley, Matt Clark, Therese Graham, Levi L. Knebel; *Dir:* Richard Pearce. **VHS, Beta, LV $79.95** TOU *ЛЛЛ*

Country Gentlemen Vaudeville team, Olsen and Johnson, play fast-talking conmen who sell shares in a worthless oil field to a bunch of World War I veterans. What a surprise when oil is actually found there. Not one of the duo's better performances.
1936 54m/B Ole Olsen, Chic Johnson, Joyce Compton, Lila Lee, Ray Corrigan, Donald Kirke, Pierre Watkin; *Dir:* Ralph Staub. **VHS, Beta $24.95** NOS, MRV, VYY *ЛЛ*

Country Girl In the role that completely de-glamorized her (and won her an Oscar), Grace Kelly plays the wife of alcoholic singer Bing Crosby who tries to make a comeback with the help of director William Holden. One of Crosby's four dramatic parts, undoubtedly one of his best. Seaton won an Oscar for his adaptation of the Clifford Odets play. Oscar nominated for Best Picture, Best Actor (Crosby), Director, Black & White Cinematography, Set Decoration and Art Direction. Remade in 1982.

1954 104m/B Bing Crosby, Grace Kelly, William Holden, Gene Reynolds, Anthony Ross; *Dir:* George Seaton. Academy Awards '54: Best Actress (Kelly), Best Screenplay; National Board of Review Awards '54: 10 Best Films of the Year, Best Actress (Kelly). **VHS, Beta $14.95** PAR, BTV *ЛЛЛ ½*

Country Girl An aging, alcoholic actor is desperate for a comeback. His director blames his fiercely loving wife for the downfall. Poor cable TV remake of 1954 version with Bing Crosby.
1982 137m/C Dick Van Dyke, Faye Dunaway, Ken Howard. **VHS, Beta $19.99** HBO *Л*

The Country Kid A freckle-faced teenager and his two brothers are beset by a wicked uncle out to steal their inheritance. One of Barry's best-remembered films.
1923 60m/B Wesley Barry, "Spec" O'Donnell, Bruce Guerin, Kate Toncray, Helen Jerome Eddy, George Nichols; *Dir:* William Beaudine. **VHS, Beta $27.95** DNB *ЛЛЛ*

Country Lovers, City Lovers Two adaptations of short stories by South African novelist Nadine Gordimer about interracial love.
1972 121m/C VHS, Beta $69.95 MGM *ЛЛ ½*

Countryman A dope-smuggling woman crasher her plane in Jamaica and is rescued by Countryman, a rasta super hero. Not very good, but least there's great music by Bob Marley and others.
1983 (R) 103m/C Hiram Keller, Kristine Sinclair; *Dir:* Dickie Jobson. **VHS, Beta $59.95** MVD, MED *ЛЛ*

County Fair This early silent film is an adaptation of Neil Burgess' play about life in New England.
1920 60m/B Helen Jerome Eddy, David Butler. **VHS, Beta $24.95** NOS, DVT *ЛЛ*

Coup de Grace Engaging political satire about a wealthy aristocratic woman in Latvia during the 1919-1920 Civil War, and how she attempts to maintain her lifestyle as German soldiers are housed on her estate. Her unrequited love for a German officer adds to ther troubles. In German with English subtitles. Co-written by von Trotta and based on the novel by Marguerite Yourcenar.
1978 96m/B *GE FR* Margarethe von Trotta, Matthias Habich, Rudiger Kirschstein, Matthieu Carriere; *Dir:* Volker Schlondorff. **Beta $59.95** COL, APD, GLV *ЛЛ ½*

Coup de Torchon Set in 1938 French West Africa, Noiret plays corrupt police chief Lucien Cordier who is consistently harrassed by his community, particularly by the town pimp. He usually overlooks the pimp's crimes, but when Cordier catches him and a friend shooting at plague victims' bodies floating down the river he decides to murder them in cold blood. Based on the novel "POP 1280" by Jim Thompson. Nominated for a Best Foreign Film Oscar in 1981. In French with English subtitles.
1981 128m/C *FR* Philippe Noiret, Isabelle Huppert, Guy Marchand, Stephane Audran, Eddy Mitchell, Jean-Pierre Marielle; *Dir:* Bertrand Tavernier. **VHS $59.95** FRI *ЛЛЛ*

Coupe de Ville Three brothers are forced by their father to drive mom's birthday present from Detroit to Florida in the summer of '63. Period concerns and music keep it interesting.
1990 (PG-13) 98m/C Patrick Dempsey, Daniel Stern, Arye Gross, Joseph Bologna, Alan Arkin, Annabeth Gish, Rita Taggart, James Gammon; *Dir:* Joe Roth. **VHS, Beta, LV $19.95** MCA *ЛЛ ½*

Couples and Robbers A rich couple and a poor couple are unexpectedly brought together.
1984 29m/C VHS, Beta *DCL* &

Courage Based on fact, a television movie about a Hispanic mother in New York City who, motivated by her drug-troubled children, goes undercover and exposes a multimillion-dollar drug ring. Loren is great in a decidedly non-glamorous role.
1986 141m/C Sophia Loren, Billy Dee Williams, Hector Elizondo, Val Avery, Dan Hedaya, Ron Rifkin, Jose Perez; *Dir:* Jeremy Paul Kagan. **VHS, Beta $29.95** *STE, NWV* &&&

Courage of Black Beauty Another version of the perennial heart warmer about a boy and his horse.
1957 80m/C Johnny Crawford, Mimi Gibson, John Bryant, Diane Brewster, J. Pat O'Malley; *Dir:* Harold Schuster. **VHS, Beta $12.95** *IVE* &&½

Courage Mountain A Swiss adventure story involving Johanna Spyri's character Heidi as a teenager in boarding school taken over by the military during World War I. Unsatisfying sequel to the classic tale will appeal to kids despite what the critics say.
1989 (G) 92m/C Charlie Sheen, Leslie Caron, Juliette Caton, Jan Rubes, Yorgo Voyagis; *Dir:* Christopher Leitch. **VHS, Beta, LV $89.95** *COL, FCT* &&

Courage of the North The Mounties break up a fur-stealing ring with the help of Captain Dog and Dynamite Horse. Bound to set your hair on end.
1935 55m/B John Preston, June Love, William Desmond, Tom London, Jimmy Aubrey, Dynamite Horse, Captain Dog **VHS, Beta** *WGE* &½

Courage of Rin Tin Tin Rusty and his faithful dog Rin Tin Tin help the cavalry soldiers of Fort Apache keep law and order in a small Arizona town. The perennial kid-pleasing German Shepherd comes through again.
1983 90m/C James Brown, Lee Aaker. **VHS, Beta $24.95** *MON* &&

Courageous Avenger A sheriff brings a treacherous ore mine foreman to justice.
1935 59m/B Johnny Mack Brown. **VHS, Beta, LV** *WGE, MLB* &

Courageous Dr. Christian Dr. Christian is faced with an epidemic of meningitis among the inhabitants of a shanty town. Typical entry in the "Dr. Christian" series.
1940 66m/B Jean Hersholt, Dorothy Lovett, Tom Neal, Robert Baldwin, Maude Eburne; *Dir:* Bernard Vorhaus. **VHS, Beta $19.95** *NOS, VYY, DVT* &&

The Courageous Mr. Penn Slow-moving British film, outlining the achievements of William Penn, the Quaker founder of Pennsylvania. "Penn of Pennsylvania" was the original British title.
1941 79m/B *GB* Clifford Evans, Deborah Kerr, Dennis Arundell, Aubrey Mallalieu, D.J. Williams, O.B. Clarence, Charles Carson, Henry Oscar, J.H. Roberts; *Dir:* Lance Comfort. **VHS, Beta $19.95** *NOS, VYY, TIM* &&

The Courier A former drug addict seeks revenge on the dealers who killed his friend. Only thing that saves it are the songs by U2, The Pogues and Hothouse Flowers. Musical score by Elvis Costello.
1988 (R) 85m/C *IR* Gabriel Byrne, Ian Bannen, Padraig O'Loingsigh, Cait O'Riordan, Patrick Bergin; *Dir:* Joe Lee, Frank Deasy. **VHS, Beta, LV $79.98** *VES, LIV* &&

Courier of Death A courier is embroiled in a mob war/struggle over a locked briefcase with mysterious contents.
1984 77m/C Joey Johnson, Barbara Garrison; *Dir:* Tom Shaw. **VHS, Beta $69.95** *LTG, VES, LIV* &

Court Jester Swashbuckling comedy stars Danny Kaye as a former circus clown who teams up with a band of outlaws trying to dethrone a tyrant king. Kaye poses as the court jester so he can learn of the evil king's intentions. Filled with more color, more song, and more truly funny lines than any three comedies put together, this is Kaye's best performance.
1956 101m/C Danny Kaye, Glynis Johns, Basil Rathbone, Angela Lansbury, Cecil Parker, John Carradine, Mildred Natwick, Robert Middletown; *Dir:* Norman Panama, Melvin Frank. **VHS, Beta, LV $14.95** *PAR* &&&½

The Court Martial of Billy Mitchell Terrific courtroom drama depicts the secret trial of Billy Mitchell, head of the Army Air Service in the 1920s, who predicted the role of airpower in subsequent warfare and the danger of war with Japan. Mitchell incurred the wrath of the military by publicly faulting the lack of U.S. preparedness for invasion. Steiger is outstanding as the attorney; Cooper is great as Mitchell. Debut for Elizabeth Montgomery.
1955 100m/C Gary Cooper, Charles Bickford, Ralph Bellamy, Rod Steiger, Elizabeth Montgomery, Fred Clark, James Daley, Jack Lord, Peter Graves, Darren McGavin, Robert Simon, Jack Perrin, Charles Dingle; *Dir:* Otto Preminger. **VHS, LV $19.98** *REP* &&&

The Court Martial of Jackie Robinson True story of a little known chapter in the life of the famous athlete. During his stint in the Army, Robinson refused to take a back seat on a bus and subsequntly faced the possibilty of court martial. Originally made for cable television.
1990 (R) 94m/C Andre Braugher, Daniel Stern, Ruby Dee, Stan Shaw, Paul Dooley, Bruce Dern; *Dir:* Larry Peerce. **VHS, Beta $79.98** *TTC, FCT* &&&½

Courtesans of Bombay Gritty docudrama, set in Pavanpul, the poverty-stricken brothel section of Bombay, shows how the women of the area must live.
1985 74m/C *GB* Kareem Samar, Zohra Segal, Saeed Jaffrey; *Dir:* Ismail Merchant. **VHS, Beta $29.98** *SUE, TAM* &&

The Courtney Affair An aristocratic young Britisher causes a stir when he marries an Irish maid. As a result, the couple must face terrific social ostracism. Classy soap opera family saga was a big money maker in England.
1947 112m/B *GB* Anna Neagle, Michael Wilding, Gladys Young, Michael Medwin, Coral Browne, Jack Watling, Bernard Lee; *Dir:* Herbert Wilcox. **VHS** *NO* &&½

Courtship From renowned playwright Horton Foote comes this touching story about a sheltered, upper-crust young girl who shocks her family and friends by eloping with a traveling salesman. Made for television.
1987 84m/C Hallie Foote, William Converse-Roberts, Amanda Plummer, Rochelle Oliver, Michael Higgins; *Dir:* Howard Cummings. **VHS, Beta $19.98** *LHV, WAR* &&&

The Courtship of Eddie's Father A clever nine-year-old boy plays matchmaker for his widowed dad in this rewarding family comedy-drama (the inspiration for the TV series). Some plot elements are outdated,

but young Howard's performance is terrific; he would later excel at direction. Based on the novel by Mark Toby.
1962 117m/C Glenn Ford, Shirley Jones, Stella Stevens, Dina Merrill, Ron Howard, Jerry Van Dyke; *Dir:* Vincente Minnelli. **VHS $19.98** *MGM, FCT* &&&

Cousin, Cousine Pleasant French comedy about distant cousins who meet at a round of family parties, funerals, and weddings and become friends, but their relationship soon becomes more than platonic. Remade in the U.S. in 1989 as "Cousins." Oscar nomination for Barrault and the screenplay.
1976 (R) 95m/C *FR* Marie-Christine Barrault, Marie-France Pisier, Victor Lanoux, Guy Marchand, Ginette Garcin, Sybil Maas; *Dir:* Jean-Charles Tacchella. National Board of Review Awards '76: 5 Best Foreign Films of the Year. **VHS, Beta, LV $29.95** *AXV, IME, FOX* &&&½

The Cousins Set against the backdrop of Parisian student life, two very different cousins (one twisted, one saintly) vie for the hand of Mayniel. This country mouse, city mouse adult fable ultimately depicts the survival of the fittest. Chabrol's lovely but sad second directorial effort.
1959 112m/B *FR* Jean-Claude Brialy, Gerard Blain, Juliette Mayniel, Claude Cerval, Genevieve Cluny; *Dir:* Claude Chabrol. **VHS** *PAR* &&&

Cousins An American remake of "Cousin, Cousine," in which two distant cousins-by-marriage meet at a wedding and, due to their respective spouses' infidelities, fall into each other's arms. A gentle love story with a humorous and biting look at the foibles of extended families.
1989 (PG-13) 110m/C Isabella Rossellini, Sean Young, Ted Danson, William L. Petersen, Norma Aleandro, Lloyd Bridges, Keith Coogan; *Dir:* Joel Schumacher. **VHS, Beta, LV $19.95** *PAR, FCT* &&&½

Cover Girl A vintage wartime musical about a girl who must decide between a nightclub career and a future as a cover model. Hayworth is beautiful; Kelly dances like a dream; Silvers and Arden are hilarious. Songs by Ira Gershwin and Jerome Kern. Oscar-nominated cinematography by Rudolph Mate.
1944 107m/C Rita Hayworth, Gene Kelly, Phil Silvers, Otto Kruger, Lee Bowman, Jinx Falkenberg, Eve Arden, Edward Brophy, Anita Colby; *Dir:* Charles Vidor. Academy Awards '44: Best Musical Score. **VHS, Beta, LV $19.95** *COL* &&&

Cover-Up Terrorist attacks on U.S. military bases in Israel are just a smokescreen for a greater threat. The below-par actioner stands out only for the hokey religious symbolism in its Good Friday climax.
1991 (R) 89m/C Dolph Lundgren, Louis Gossett Jr., John Finn; *Dir:* Manny Coto. **VHS, LV $39.95** *LIV, IME* &½

The Covered Wagon Prototypical silent Western began the genre. Wagon train moves cross-country, battling weather and wild Indians. By today's standards, somewhat bucolic and uneventful, but the location photography holds up quite well. Big budget film in its day, and enormously popular.
1923 98m/B Warren Kerigan, Lois Wilson, Alan Hale Sr., Ernest Torrence; *Dir:* James Cruze. **VHS, LV $14.95** *PAR* &&½

Covered Wagon Days The Three Mesquiteers set out after silver smugglers who have framed Rico's brother for murder.

1940 54m/B Bob Livingston, Raymond Hatton, Duncan Renaldo, John Merton, Reed Howes; *Dir:* George Sherman. **VHS $9.95** REP ✔✔

Covergirl A wealthy man decides to relieve his boredom by molding a young woman into a supermodel. She finds the glamorous life isn't always what it's cracked up to be.
1983 (R) 98m/C *CA* Jeff Conaway, Irena Ferris, Cathie Shirriff, Roberta Leighton, Deborah Wakeham; *Dir:* Jean-Claude Lord. **VHS, Beta $19.99** HBO, STE ✔½

Covert Action A drug kingpin who desires to protect his business tries to frame a captain for a crime he did not commit.
1988 85m/C Rick Washburne, John Christian, Stuart Garrison Day, Amanda Zinsser, Johnny Stumper; *Dir:* J. Christian Ingvordsen. **VHS, Beta $79.98** MCG ✔½

Cow Town A range war results when ranchers begin fencing in their land to prevent cattle rustling. Autry's 72nd film has some good songs and a dependable cast.
1950 70m/B Gene Autry, Gail Davis, Jock Mahoney, Harry Shannon; *Dir:* John English. **VHS, Beta $29.98** CCB ✔✔

Coward of the County A devout pacifist is put to the test when his girlfriend is raped. Good performances by all concerned. Based on the lyrics of Kenny Rogers' hit song of the same name. Made for television.
1981 115m/C Kenny Rogers, Frederic Lehne, Largo Woodruff, Mariclare Costello, Ana Alicia, Noble Willingham; *Dir:* Dick Lowry. **VHS, Beta $59.95** LHV, WAR ✔✔½

Cowboy Film chronicling the determination of cattle-herders in the West.
1954 69m/C *Nar:* Tex Ritter, William Conrad, John Dehner. **VHS, Beta, LV** WGE, RXM ✔½

Cowboy & the Ballerina A television movie about an aging rodeo star who falls in love with a petite ballerina defecting from a Russian Dance Company. Good cast makes it watchable.
1984 96m/C Lee Majors, Leslie Wing, Christopher Lloyd, Anjelica Huston, George De La Pena; *Dir:* Jerry Jameson. **VHS, Beta $79.95** PSM ✔✔

Cowboy & the Bandit A cowboy comes to the aid of a widow being victimized by outlaws.
1935 58m/B Rex Lease, Bobby Nelson, William Desmond; *Dir:* Albert Herman. **VHS, Beta $19.95** NOS, WGE ✔

Cowboy Counselor Cowboy Hoot Gibson helps an enchanting girl clear her innocent brother's name.
1933 60m/B Hoot Gibson. **VHS, Beta** UHV ✔

Cowboy Millionaire An Englishwoman comes to an American dude ranch and falls in love with one of the ranch hands. Entertaining Western comedy.
1935 65m/B George O'Brien, Evalyn Bostock, Edgar Kennedy, Alden Chase; *Dir:* Edward Cline. **VHS $19.95** NOS, DVT, RXM ✔✔½

Cowboy Previews #1 A collection of theatrical trailers from over thirty-five "B" westerns. Titles include "Riders in the Sky," "King of the Bullwhip," "Utah Wagon Train," "Trail of Robin Hood," "Sioux City Sue" and many others.
194? 60m/B **VHS, Beta** MED ✔✔½

Cowboy & the Senorita A cowboy solves the mystery of a missing girl and wins her lovely cousin. The first film that paired Rogers and Evans, who went on to become a winning team, on and offscreen.
1944 56m/B Roy Rogers, John Hubbard, Dale Evans. **VHS, Beta $9.95** NOS, MRV, DVT ✔✔

The Cowboys Wayne stars as a cattle rancher who is forced to hire eleven schoolboys to help him drive his cattle 400 miles to market. Clever script makes this one of Wayne's better Westerns. Carradine's film debut. Inspired the television series.
1972 (PG) 128m/C John Wayne, Roscoe Lee Browne, A. Martinez, Bruce Dern, Colleen Dewhurst, Slim Pickens, Robert Carradine; *Dir:* Mark Rydell. **VHS, Beta $19.98** WAR, TLF, BUR ✔✔✔

Cowboys Don't Cry When a modern cowboy's wife is killed in an automobile accident, he must change his immature ways and build a solid future with his teenage son or perish in loneliness.
1988 96m/C Rebecca Jenkins, Ron White, Janet-Laine Green. **VHS, Beta** MCA ✔✔

Cowboys of the Saturday Matinee The history of B-movie westerns is traced from the beginning of the sound era.
1984 75m/B Gene Autry, Tex Ritter, Roy Rogers; *Hosted:* James Coburn. **VHS, Beta $19.98** LHV, FCT, WAR ✔✔

Cowboys from Texas The Three Mesquiteers bring about a peaceful settlement to a fight between cattlemen and homesteaders. Part of "The Three Mesquiteers" series.
1939 54m/B Bob Livingston, Raymond Hatton, Duncan Renaldo, Carole Landis. **VHS, Beta** MED ✔

Coyote Trails Typical western adventure has cowboy Tyler helping a damsel in distress.
1935 68m/B Tom Tyler. **VHS, Beta $19.95** NOS, VCN, RXM ✔

Crack House A man seeks revenge for the murder of a relative by drug dealers. Average exploitation flick.
1989 (R) 91m/C Jim Brown, Richard Roundtree, Anthony Geary, Angel Tompkins, Greg Gomez Thomsen, Clyde R. Jones, Cheryl Kay; *Dir:* Michael Fischa. **VHS, Beta, LV $89.95** WAR, OM ✔✔

Crack Shadow Boxers Through a series of misadventures, Wu Lung and Chu San battle to protect the inhabitants of a small village from the onslaught of relentless bandits.
197? 91m/C Ku Feng, Chou Li Lung. **VHS, Beta $54.95** OCE, MAV ✔

Crack-Up Largely ignored at the time of its release, this tense film is now regarded as a minor classic of film noir. An art expert suffers a blackout while investigating a possible forgery and must piece together the missing hours to uncover a criminal conspiracy.
1946 70m/B Pat O'Brien, Claire Trevor, Herbert Marshall, Ray Collins, Wallace Ford, Damian O'Flynn, Erskine Sanford; *Dir:* Irving Reis. **VHS, Beta $19.98** TTC ✔✔½

The Cracker Factory Wood is impressive as a woman committed to a mental institution who attempts to charm her way out of treatment. Made for television.

1979 90m/C Natalie Wood, Perry King, Shelley Long, Vivian Blaine, Juliet Mills, Peter Haskell; *Dir:* Burt Brickerhoff. **VHS, Beta $59.95** VCL ✔✔½

Crackers The offbeat story of two bumbling thieves who round up a gang of equally inept neighbors and go on the wildest crime spree you have ever seen. A would-be comic remake of "Big Deal on Madonna Street."
1984 (PG) 92m/C Donald Sutherland, Jack Warden, Sean Penn, Wallace Shawn, Larry Riley, Trinidad Silva, Christine Baransci, Charlaine Woodard, Tasia Valenza; *Dir:* Louis Malle. **VHS, Beta, LV $59.95** MCA ✔✔

Cracking Up An accident-prone misfit's mishaps on the road to recovery create chaos for everyone he meets. Lewis plays a dozen characters in this overboard comedy with few laughs. Also known as "Smorgasbord."
1983 (PG) 91m/C Jerry Lewis, Herb Edelman, Foster Brooks, Milton Berle, Sammy Davis Jr., Zane Buzby, Dick Butkus, Buddy Lester; *Dir:* Jerry Lewis. **VHS, Beta $19.98** WAR ✔½

The Crackler During the '20s, a husband-and-wife private investigation team (Tommy and Tuppence) set out to find a gang of forgers operating in high society circles. Based on a story by Agatha Christie. Made for British television.
1984 60m/C *GB* James Warwick, Francesca Annis. **VHS, Beta $14.95** PAV ✔✔

The Cradle Will Fall A made-for-television adaptation of the pulpy bestselling mystery by Mary Higgins Clark about a woman who cannot convince anyone she witnessed a murder. The cast of "The Guiding Light" appear. Poor script and poor direction do not a thriller make.
1983 103m/C Lauren Hutton, Ben Murphy, James Farentino, Charita Bauer, Peter Simon; *Dir:* John Llewellyn Moxey. **VHS, Beta $59.95** LHV, WAR Woof!

Craig's Wife A classic soap opera about a pitiful woman driven to total ruin by her desire for social acceptance and material wealth. Russell makes her surprisingly sympathetic. Based on a Pulitzer Prize-winning George Kelly play. Remake of a silent film, was also remade as "Harriet Craig."
1936 75m/B Rosalind Russell, John Boles, Alma Kruger, Jane Darwell, Billie Burke; *Dir:* Dorothy Arzner. **VHS, Beta $59.95** RCA ✔✔✔

Crainquebille The classic satire based on the Anatole France story about a street merchant in Paris unfairly imprisoned and eventually surviving peacefully as a tramp. Silent.
1923 50m/B *FR* Maurice Feraudy; *Dir:* Jacques Feyder. **VHS, Beta $34.95** FCT, GPV ✔✔✔

The Cranes are Flying When her lover goes off to fight during World War II, a girl is seduced by his cousin. Touching love story is free of politics. Filmed in Russia; English subtitles.
1957 91m/B *RU* Tatyana Samoilova, Alexei Batalov, Vasily Merkuryev, A. Shvorin; *Dir:* Mikhail Kalatozov. Cannes Film Festival '58: Best Film. **VHS, Beta, 3/4U $35.95** IHF ✔✔✔½

Crash and Burn Rebels in a repressive future police state reactivate a huge, long-dormant robot to battle the establishment's army of powerful androids. Special effects by David Allen make the mayhem interesting.
1990 (R) 85m/C Ralph Waite, Paul Ganus, Eva LaRue, Bill Moseley; *Dir:* Charles Band. **VHS, Beta, LV $19.95** PAR ✔½

The Crash of Flight 401 A plane, full of recognizable TV stars, crashes in the Florida Everglades. "Airport" for a smaller screen. Based on a real event.
1978 97m/C William Shatner, Adrienne Barbeau, Eddie Albert, Brooke Bundy, Christopher Connelly, Lorraine Gary, Ron Glass, Sharon Gless, Brett Halsey, George Maharis, Gerald S. O'Loughlin; *Dir:* Barry Shear. **VHS, Beta $14.95** *FRH* ✍½

Crashout Six men break out of prison in this entertaining melodrama. Bendix plays the gang's leader to nasty perfection.
1955 82m/B William Bendix, Gene Evans, Arthur Kennedy, Luther Adler, William Talman, Marshall Thompson, Beverly Michaels; *Dir:* Lewis R. Foster. **VHS $19.98** *REP* ✍✍½

The Crater Lake Monster The dormant egg of a prehistoric creature hatches after a meteor rudely awakens the dozing dino. He's understandably miffed and begins a revenge campaign. Prehistoric yawner.
1977 (PG) 85m/C Richard Cardella, Glenn Roberts, Mark Siegel, Bob Hyman, Kacey Cobb; *Dir:* William R. Stromberg. **VHS, Beta $19.95** *UHV, MRV* **Woof!**

The Craving Naschy returns as El Hombre Lobo for the umpteenth time and once again battles a female vampire (see "Werewolf vs. the Vampire Woman"). Although continental Europe's biggest horror star, this film was such a box office disaster that Naschy went bankrupt. He was then forced to turn to Japan for financing (see "The Human Beasts"). Naschy also directed under a pseudonym.
1980 (R) 93m/C *MX* Paul Naschy, Julie Saly; *Dir:* Jack Molina. **VHS, Beta $69.95** *VES, LIV* ✍

The Crawling Eye Hidden in a radioactive fog on a mountaintop, the crawling eye decapitates its victims and returns these humans to Earth to threaten mankind. Average acting, but particularly awful special effects. Based on the British TV series "The Trollenberg Terror."
1958 87m/B *GB* Forrest Tucker, Laurence Payne, Janet Munro, Jennifer Jayne, Warren Mitchell; *Dir:* Quentin Lawrence. **VHS, Beta, LV $19.95** *SNC, MRV, MED* ✍✍

The Crawling Hand An astronaut's hand takes off without him, on an unearthly spree of stranglings. Silly stuff is a hands-down loser.
1963 98m/B Alan Hale Jr., Rod Lauren, Richard Arlen, Peter Breck, Kent Taylor, Arlene Judge, Allison Hayes; *Dir:* Herbert L. Strock. **VHS, Beta $9.95** *NOS, RHI, GEM* ✍

Crawlspace Beautiful girls lease rooms from a murdering, perverted doctor who spies on them, then kills. You may find Kinski amusing but not terrifying.
1986 (R) 86m/C Klaus Kinski, Talia Balsam, Joyce Van Patten, Sally Brown, Barbara Whinnery; *Dir:* David Schmoeller. **VHS, Beta $19.98** *LTG, VES, LIV* ✍

Craze Toungue-in-cheek tale of a crazed antique dealer who slays a number of women as sacrifices to an African idol named Chuku. Also known as "The Infernal Idol" and "Demon Master."
1974 (R) 96m/C Jack Palance, Diana Dors, Julie Ege, Suzy Kendall, Michael Jayston, Edith Evans, Hugh Griffin, Trevor Howard; *Dir:* Freddie Francis. **VHS, Beta** *MRV, MFI* ✍✍

Crazed Dull plodder about a psychopath who keeps a dead girl's body in his boarding house and kills all intruders to keep his secret.

1982 88m/C Laslo Papas, Belle Mitchell, Beverly Ross; *Dir:* Richard Cassidy. **VHS, Beta $49.95** *TWE* **Woof!**

Crazed Cop A man's mind snaps when his wife is found raped and murdered, and nothing will stop him on his quest for revenge. Low budget and violent.
1988 85m/C Ivan Rogers, Sandy Brooke, Rich Sutherlin, Doug Irk, Abdulah the Great; *Dir:* Paul Kyriazi. **VHS, Beta $69.95** *UNI* ✍

The Crazies A poisoned water supply makes the residents of a small town go on a chaotic, murderous rampage. When the army is called into to quell the anarchy, a small war breaks out. Message film about the military is muddled and derivative. Also known as "Code Name: Trixie."
1976 (R) 103m/C Lane Carroll, W.G. McMillan, Harold W. Jones, Lloyd Hollar, Lynn Lowry; *Dir:* George A. Romero. **Beta $79.95** *VHV* ✍✍

Crazy Fat Ethel II A fat, hungry, homicidal female psychopath gets released from the asylum and goes on a cannabilistic rampage. Offensive junk. Sequel to "Criminally Insane."
1985 90m/C Priscilla Alden, Michael Flood, Jane Lambert, Robert Copple; *Dir:* Nick Phillips. **VHS, Beta $14.95** *VCD* **Woof!**

Crazy from the Heart Sweet made-for-cable TV tale of a high school principal in a small south Texas town who discovers the romance of her life one weekend with a Mexican janitor. Strong, nuanced performances from Lahti and Blades make this chestnut of a story work beautifully, with able help from supporting actors.
1991 104m/C Christine Lahti, Ruben Blades, Mary Kay Place, Brent Spiner, William Russ, Louise Latham, Tommy Muntz, Robin Lively, Bibi Besch, Kamala Lopez; *Dir:* Thomas Schlamme. **VHS, Beta $89.98** *TTC* ✍✍✍

Crazy Horse and Custer: "The Untold Story" Long before Little Big Horn, two legendary enemies find themselves trapped together in deadly Blackfoot territory. George Armstrong Custer and Crazy Horse are forced to form a volatile alliance in their life or death struggle against the murderous Blackfoot Tribe. History takes a back seat to Hollywood scriptwriting.
1990 (R) 120m/C Slim Pickens, Wayne Maunder, Mary Ann Mobley, Michael Dante. **VHS** *ALE* ✍

Crazy Kung Fu Master Fighting like you've never seen before as lots and lots of martial arts guys kick and rip 1920's Shanghai to pieces.
19?? 90m/C Kent Cheng, Wong Ching, Wang Yu. **VHS, Beta $19.95** *OCE* ✍

Crazy for Love A village idiot stands to inherit the town inn if he can get a diploma within a year. With English subtitles.
1952 80m/B *FR* Brigitte Bardot, Bourvil, Jane Marken, Nadine Basile. **VHS, Beta $24.95** *DVT* ✍✍½

Crazy Mama Three women go on a crime spree from California to Arkansas, picking up men and having a hoot. Crime and comedy in a campy romp. Set in the 1950s and loaded with period kitsch.
1975 (PG) 81m/C Cloris Leachman, Stuart Whitman, Ann Sothern, Jim Backus, Linda Purl; *Dir:* Jonathan Demme. **VHS, Beta $69.98** *SUE* ✍✍½

Crazy Moon A rich high school nerd falls in love with a deaf girl, and must struggle against his domineering father's and older brother's prejudices. The viewer must strug-

gle against the romantic cliches and heavy-handed message to enjoy a basically tender tale of romance. Also known as "Huggers."
1987 (PG-13) 89m/C Kiefer Sutherland, Vanessa Vaughan, Peter Spence, Ken Pogue, Eve Napier; *Dir:* Allan Eastman. **VHS, Beta, LV $14.95** *SUE* ✍✍½

Crazy People An advertising executive devises writes commercials that describe products with complete honesty, and is committed to a mental hospital as a result. A tepid box-office blunder that sounds funnier than it is.
1990 (R) 91m/C Dudley Moore, Darryl Hannah, Paul Reiser, Mercedes Ruehl, J.T. Walsh, Ben Hammer, Floyd Vivino, John Terlesky; *Dir:* Tony Bill. **VHS, Beta, LV $14.95** *PAR* ✍

The Crazy Ray The classic silent fantasy about a mad scientist who endeavors to put the whole population of Paris in a trance. Vintage Clair nonsense. Alternate title: "Paris Qui Dort."
1922 60m/B *FR* Henri Rollan, Albert Prejean, Marcel Vallee, Madeleine Rodrigue; *Dir:* Rene Clair. **VHS, Beta $16.95** *SNC, FCT, MRV* ✍✍✍

The Creation of the Humanoids Set in the familiar post-holocaust future, this is a tale of humans outnumbered by androids and the resulting struggle for survival. Slow and silly low-budget sets. For some reason, Andy Warhol was reported to love this film.
1962 84m/C Don Megowan, Frances McCann, Erica Elliot, Don Doolittle, Dudley Manlove; *Dir:* Wesley E. Barry. **VHS, Beta $39.95** *MON, MRV, MDS* **Woof!**

Creator A Frankenstein-like scientist plans to clone a being based on his wife, who died thirty years ago. As his experiments begin to show positive results, his romantic attention turns towards his beautiful lab assistant. O'Toole as the deranged scientist almost saves this one. Based on a novel by Jeremy Leven.
1985 (R) 108m/C Peter O'Toole, Mariel Hemingway, Vincent Spano, Virginia Madsen, David Ogden Stiers, John Dehner, Karen Kopins, Jeff Corey; *Dir:* Ivan Passer. **VHS, Beta, LV $69.99** *HBO* ✍✍

Creature A two thousand-year-old alien life form is killing off astronauts exploring the planet Titan. "Alien" rip-off has its moments, but not enough of them. Kinski provides some laughs. Also known as "Titan Find."
1985 (R) 97m/C Klaus Kinski, Stan Ivar, Wendy Schaal, Lyman Ward, Annette McCarthy, Diane Salinger; *Dir:* William Malone. **VHS, Beta, LV $9.98** *MED* ✍

Creature from the Black Lagoon An anthropological expedition in the Amazon stumbles upon the Gill-Man, a prehistoric humanoid fish monster who takes a fancy to fetching Adams, a coed majoring in "science," but the humans will have none of it. Originally filmed in 3-D, this was one of the first movies to sport top of the line underwater photography and remains one of the most enjoyable monster movies ever made. The score by Joseph Gershenson became a "Creature Features" standard. Sequels: "Revenge of the Creature" and "The Creature Walks Among Us."
1954 79m/B Richard Carlson, Julie Adams, Richard Denning, Antonio Moreno, Whit Bissell, Nestor Paiva, Ricou Browning; *Dir:* Jack Arnold. **VHS, Beta, LV $14.95** *MCA, GKK* ✍✍✍

Creature from Black Lake Two anthropology students from Chicago travel to the Louisiana swamps searching for the creature from Black Lake. Predictably, they

find him. McClenny, incidentally, is Morgan Fairchild's sister.
1976 (PG) 95m/C Jack Elam, Dub Taylor, Dennis Fimple, John David Carson, Bill Thurman, Catherine McClenny; ***Dir:*** Joy Houck Jr. **VHS, Beta $59.95** *LTG, MRV, VES* 🦴

Creature with the Blue Hand A German horror film based on a passable Edgar Wallace story about a man unjustly convicted of murders actually committed by a lunatic. Dubbed.
1970 92m/C *GE* Klaus Kinski, Harald Leopold, Hermann Leschau, Diana Kerner, Carl Lang, Ilse Page; ***Dir:*** Alfred Vohrer. **VHS, Beta $19.95** *TAV* 🦴🦴

Creature from the Haunted Sea A monster movie satire set in Cuba shortly after the revolution and centering around an elaborate plan to loot the Treasury and put the blame on a strange sea monster. Corman comedy is predictably low budget but entertaining. "Wain" is really Robert Towne, winner of an Oscar for screenwriting. Remake of "Naked Paradise."
1960 76m/B Antony Carbone, Betsy Jones-Moreland, Beach Dickerson, Edward Wain, E.R. Alvarez, Robert Bean; ***Dir:*** Roger Corman, Monte Hellman. **VHS, Beta $19.95** *NOS, MRV, SNC* 🦴½

Creature of the Walking Dead Scientist brings his grandfather back to life with horrifying results for the cast and the audience. Made cheaply and quickly in Mexico in Mexico in 1960, then released with added footage directed by Warren.
1965 74m/B *MX* Rock Madison, Ann Wells, George Todd, Willard Gross, Bruno Ve Sota; ***Dir:*** Frederic Corte, Jerry Warren. **VHS $16.95** *NOS, MRV, SNC* 🦴½

Creatures the World Forgot A British-made bomb about two tribes of cavemen warring over power and a cavewoman. No special effects, dinosaurs, or dialogue.
1970 96m/C Julie Ege, Robert John, Tony Bonner, Rosalie Crutchley, Sue Wilson; ***Dir:*** Don Chaffey. **VHS, Beta $69.95** *COL, MLB Woof!*

The Creeper After experimenting with a variety of serums, a doctor turns into a murderous monster with feline paws.
1948 65m/B Eduardo Ciannelli, Onslow Stevens, June Vincent, Ralph Morgan; ***Dir:*** Jean Yarborough. **VHS, Beta $14.95** *KBV, MLB* 🦴🦴

Creepers A young girl talks to bugs and gets them to follow her instructions, which comes in handy when she battles the lunatic who is killing her school chums. Argento weirdness—and graphic gore—may not be for all tastes.
1985 (R) 82m/C *IT* Jennifer Connelly, Donald Pleasence, Daria Nicolodi, Elenora Giorgi, Dalia di Lazzaro, Patrick Bauchau, Fiore Argento; ***Dir:*** Dario Argento. **VHS, Beta $79.95** *MED* 🦴🦴½

The Creeping Flesh A scientist decides he can cure evil by injecting his patients with a serum derived from the blood of an ancient corpse. Some truly chilling moments will get your flesh creeping.
1972 (PG) 89m/C *GB* Peter Cushing, Christopher Lee, Lorna Heilbron, George Benson; ***Dir:*** Freddie Francis. **VHS, Beta, LV $49.95** *COL, IME, MLB* 🦴🦴½

Creeping Terror A gigantic alien carpet monster (look for the tennis shoes sticking out underneath) devours slow-moving teenagers. Partially narrated because some ot the original soundtrack was lost, the film features bad acting, a worse script, laughable sets, and a ridiculous monster. Beware of the thermometer scene! So bad it's good. Also released as "The Crawling Monster."
1964 81m/B Vic Savage, Shannon O'Neal, William Thourlby, Louise Lawson, Robin James; ***Dir:*** Art J. Nelson. **VHS, Beta $54.95** *UHV Woof!*

Creepozoids In the near future, a monster at an abandoned science complex stalks army deserters hiding out there. Violent nonsense done better by others.
1987 72m/C Linnea Quigley, Ken Abraham, Michael Aranda; ***Dir:*** David DeCoteau. **Beta $69.95** *UCV* 🦴

Creepshow Stephen King's tribute to E.C. Comics, those pulp horror comic books of the 1950s that delighted in grizzly, grotesque, and morbid humor, the film tells five horror tales. Features King himself in one segment, as a none-too-bright farmer who unknowingly cultivates a strange, alien-origin moss. With despicable heroes and gory monsters, this is sure to delight all fans of the horror vein. Those easily repulsed by cockroaches beware!
1982 (R) 120m/C Hal Holbrook, Adrienne Barbeau, Viveca Lindfors, E.G. Marshall, Stephen King, Leslie Nielsen, Carrie Nye, Fritz Weaver, Ted Danson, Ed Harris; ***Dir:*** George A. Romero. **VHS, Beta, LV $19.98** *WAR* 🦴🦴½

Creepshow 2 Romero adapted three Stephen King stories for this horror anthology which presents gruesome looks at a hit-and-run driver, a wooden Indian, and a vacation gone wrong. Gory and childish stuff from two masters of the genre. Look for King as a truck driver.
1987 (R) 92m/C Lois Chiles, George Kennedy, Dorothy Lamour, Tom Savini, Domenick John, Frank S. Salsedo, Holt McCallany, David Holbrook, Page Hannah, Daniel Beer, Stephen King; ***Dir:*** Michael Gornick. **VHS, Beta, LV $19.95** *STE, NWV* 🦴½

The Cremators A meteorite carrying an alien lands at a seaside resort and everyone begins bursting into flames. Low budget effort from the scripter of "It Came from Outer Space."
1972 90m/C Marvin Howard, Maria Di Aragon; ***Dir:*** Harry Essex. **VHS, Beta $59.95** *WES Woof!*

Cria The award-winning story of a 9-year-old girl's struggle for maturity in a hostile adult world.
1976 (PG) 115m/C *SP* Geraldine Chaplin, Ana Torrent, Conchita Perez; ***Dir:*** Carlos Savrat. Cannes Film Festival '76: Special Jury Prize; National Board of Review Awards '77: 5 Best Foreign Films of the Year. **VHS, Beta $59.95** *INT, MAV, APD* 🦴🦴🦴

Cricket on the Hearth An adaptation of Charles Dickens' short story about a mail carrier and his bride, who find the symbol of good luck, a cricket on the hearth, when they enter their new home. Silent with organ score.
1923 68m/B Paul Gerson, Virginia Brown Faire, Paul Moore, Joan Standing. **VHS, Beta $24.98** *CCB* 🦴🦴

Cries and Whispers As a woman dies slowly of tuberculosis, three women care for her: her two sisters, one sexually repressed, the other promiscuous, and her servant. The sisters remember family love and closeness, but are too afraid to look death in the face to aid their sister. Only the servant can touch her in her dying and only the servant believes in God and his will. Beautiful imagery, focused through a nervous camera, which lingers on the meaningless and whisks away from the meaningful. Absolute mastery of cinematic art by Bergman. Oscar nominations for Best Picture, Director, Original Screenplay, and Costume Design.
1972 (R) 91m/C *SW* Harriet Andersson, Ingrid Thulin, Liv Ullman, Kary Sylway, Erland Josephson, Henning Moritzen; ***W/Dir:*** Ingmar Bergman. Academy Awards '73: Best Cinematography; New York Film Critics Awards '72: Best Actress (Ullman), Best Director (Bergman), Best Film; National Society of Film Critics Awards '72: Best Cinematography, Best Screenplay (Bergman). **VHS, Beta $59.95** *WAR, TAM* 🦴🦴🦴

Crime Busters Two guys attempt to pull off a bank heist but accidentally join the Miami police force instead. Hill and Spencer are enjoyable as usual but the film suffers from poor dubbing. Remake of "Two Supercops."
1978 (PG) 114m/C *IT* Terence Hill, Bud Spencer, Laura Gemser, Luciana Catenacci, David Huddleston; ***Dir:*** E.B. Clucher. **VHS, Beta $19.98** *WAR* 🦴½

The Crime of Dr. Crespi A doctor takes his revenge on a man who is after his girlfriend by injecting him with suspended animation serum. Extremely campy performance from von Stroheim; Frye's biggest role in terms of screen time.
1935 64m/B Erich von Stroheim, Dwight Frye, Paul Guilfoyle, Harriett Russell, John Bohn; ***Dir:*** John H. Auer. **VHS, Beta $16.95** *NOS, SNC, MLB* 🦴🦴

Crime of Honor A European executive goes public with his company's corruption, and his family suffers predictable ruin as a result.
198? 95m/C Maria Schneider, David Suchet; ***Dir:*** John Goldschmidt. **VHS, Beta $29.95** *ACA* 🦴

Crime Killer Confusing shenanigans about an explosive ex-CIA agent enlisted by the FBI to battle a brutal crime syndicate shipping guns to nasty Arabs.
1985 90m/C George Pan-Andreas, Leo Morrell, Althan Karras; ***Dir:*** George Pan-Andreas. **VHS, Beta $19.95** *STE, NWV* 🦴

The Crime of Monsieur Lange A charming French socialist fantasy written by Jacques Prevert wherein the workers of a publishing company turn the business into a thriving cooperative while their evil boss is gone. When he returns, they plot to kill him. Rather talky, but humorous. In French with subtitles. French script booklet also available.
1936 90m/B *FR* Rene Lefevre, Jules Berry, Florelle, Sylvia Bataille, Jean Daste, Nadia Sibirskaia; ***Dir:*** Jean Renoir. **VHS, Beta $29.95** *FCT, INT, APD* 🦴🦴🦴

Crime & Passion Two scheming lovers plan to get rich by having the woman marry a multimillionaire and sue for a quick divorce. The multimillionaire is no patsy, however, and seeks a deadly revenge. Three scripts by six writers were combined to create this story. That explains the many problems.
1975 (R) 92m/C Omar Sharif, Karen Black, Joseph Bottoms, Bernhard Wicki; ***Dir:*** Ivan Passer. **VHS, Beta $69.95** *VES, LIV* 🦴½

Crime and Punishment Original French production (preceding the American version by just one week) of Dostoyevsky's novel, the tale of a young murderer, his crime, and ultimately, his confession. Generally considered better than the American version, the French throw out all that is not dramatic, keeping the psychology and suspense. Remade in 1958.
1935 110m/B Harry Baur, Pierre Blanchar, Madeleine Ozeray, Marcelle Geniat, Lucienne Lemarchand; ***Dir:*** Pierre Chenal. **VHS $59.95** *FCT, TIM* 🦴🦴🦴🦴

Crime and Punishment Ponderous, excruciatingly long adaptation of the Dostoyevsky classic which involves a haunted murderer and the relentless policeman who seeks to prove him guilty. In Russian with English subtitles.
1975 200m/C *RU* Georgi Taratorkin, Victoria Fyodorova; *W/Dir:* Lev Kulijanov. **VHS $59.95** *TAM, DVT ✂✂*

Crime Story The full-length television pilot for the Michael Mann cop show, about the police cracking down on the Mob in 1940s Chicago. Tough stuff with moody camera-work and original music by Todd Rundgren.
1986 96m/C Dennis Farina, Anthony Denison, Stephen Lang, Darlanne Fluegel; *Dir:* Abel Ferrara. **VHS, Beta, LV $14.95** *NWV, STE ✂✂✂*

Crime Story: The Complete Saga All 42 episodes of the 1986-1988 NBC cops versus mobsters series set first in Chicago in 1963 and later moving to the action of Las Vegas. The series was created by Michael Mann, the executive producer of "Miami Vice." Del Shannon updated the lyrics to his hit "Runaway" for the series theme. Available as a boxed set of ten tapes or individually.
1986 ?m/C Dennis Farina, Anthony Denison, Bill Smitrovich, Steve Ryan, Bill Campbell, Paul Butler, Stephen Lang, Darlanne Fluegel. **VHS $14.98** *STE ✂✂✂*

Crime Zone In a totalitarian, repressive, future society, two young lovers try to beat the system and make it on their own. A well-made, if occasionally muddled, low-budget film shot in Peru, of all places.
1988 (R) 93m/C David Carradine, Peter Nelson, Sherilyn Fenn, Orlando Sacha, Don Manor; *Dir:* Luis Llosa. **VHS $79.95** *MGM ✂✂½*

Crimebusters A man exacts revenge against a government conspiracy that kidnapped his loved ones.
198? (PG) 90m/C Henry Silva, Anthony Sabato; *Dir:* Michael Tarantini. **VHS, Beta $59.95** *MGL ✂½*

Crimes at the Dark House In this campy, melodramatic adaptation of Wilkie Collins' novel "The Woman in White," a man kills his rich wife and puts a disguised mental patient in her place. Later remade using the book's title.
1939 69m/B *GB* Tod Slaughter, Hilary Eaves, Sylvia Marriott, Hay Petrie, Geoffrey Wardwell; *Dir:* George King. **VHS $19.95** *NOS, MRV, SNC ✂✂✂*

Crimes of the Heart Based on Beth Henley's acclaimed play. A few days in the lives of three very strange Southern sisters, one of whom has just been arrested for calmly shooting her husband after he chased her black lover out of town. Spacek as the suicidal sister is a lark. A tart, black comedy which works better as a play than a film.
1986 (PG-13) 105m/C Sissy Spacek, Diane Keaton, Jessica Lange, Sam Shepard, Tess Harper, Hurd Hatfield; *Dir:* Bruce Beresford. **VHS, Beta, LV $19.98** *LHV, WAR ✂✂½*

Crimes & Misdemeanors One of Allen's most mature films, exploring a whole range of moral ambiguities through the parallel and eventually interlocking stories of a nebbish filmmaker agreeing to make a profile of a smug Hollywood television comic and sabotaging it, and an esteemed ophthalmologist who is being threatened with exposure by his neurotic mistress. Intriguing mix of drama and comedy few directors could pull

off. Look for Daryl Hannah in an unbilled cameo. Heavily Oscar-nominated.
1989 (PG-13) 104m/C Martin Landau, Woody Allen, Alan Alda, Mia Farrow, Joanna Gleason, Anjelica Huston, Jerry Orbach, Sam Waterston, Claire Bloom, Jenny Nichols, Caroline Aaron, Darryl Hannah; *Dir:* Woody Allen. **VHS, Beta, LV $89.98** *ORI ✂✂✂*

Crimes of Passion Vintage whacked-out Russell, not intended for the kiddies. A business-like fashion designer becomes a kinky prostitute at night. A disturbed street preacher makes her the heroine of his erotic fantasies. A dark terrifying vision of the underground sex world and moral hypocrisy. Sexually explicit and violent, with an extremely black comedic center. Music score by Rick Wakeman. Turner's portrayal is honest and believable, Perkins overacts until he nearly gets a nosebleed, but it's all for good effect. Cut for "R" rating to get it in the theatres; this version restores some of excised footage. Also available in rated version.
1984 101m/C Kathleen Turner, Anthony Perkins, Annie Potts, John Laughlin, Bruce Davison, Norman Burton; *Dir:* Ken Russell. **VHS, Beta, LV $14.95** *NWV, STE ✂✂✂*

The Crimes of Stephen Hawke The world knows Stephen Hawke as a big-hearted money lender. What they don't know is his favorite hobby is breaking peoples' spines. An entertaining thriller set in the 1800s.
1936 65m/B *GB* Tod Slaughter, Marjorie Taylor, D.J. Williams, Eric Portman, Ben Soutten; *Dir:* George King. **VHS, Beta $16.95** *SNC, MRV, HEG ✂✂*

Crimewave Zany spoof about serial killers co-written by Ethan and Joel Coen and directed by "Darkman" Raimi. Set in Detroit (Raimi's hometown), it's a rhapsody to comic book style, short on plot and long on style. With such credentials you expect more.
1986 (PG-13) 83m/C Louise Lasser, Paul Smith, Brion James, Bruce Campbell, Reed Birney, Sheree J. Wilson, Edward R. Pressman, Julius W. Harris; *Dir:* Sam Raimi. **VHS, Beta $19.98** *CHA ✂✂*

Criminal Act A newspaper editor hits the street to prove she's still a tough reporter, but uncovers a dangerous scandal in the process.
1988 94m/C Catherine Bach, Charlene Dallas, Nicholas Guest, John Saxon, Vic Tayback; *Dir:* Mark Byers. **VHS, Beta $79.95** *PSM ✂½*

Criminal Code Aging melodrama about a young man who's jailed for killing in self-defense. His life worsens at the hands of a prison warden when the head guy's daughter falls for him. Remade as "Penitentiary" and "Convicted."
1931 98m/B Walter Huston, Phillips Holmes, Boris Karloff, Constance Cummings, Mary Doran, DeWitt Jennings, John Sheehan; *Dir:* Howard Hawks. **VHS, Beta $59.95** *COL, MLB ✂✂½*

Criminal Court A man accused of murdering his blackmailer hires the real killer as his attorney. Hard to follow, with an interesting premise that turns flat.
1946 63m/B Tom Conway, Martha O'Driscoll, Robert Armstrong, Addison Richards, June Clayworth, Pat Gleason, Steve Brodie; *Dir:* Robert Wise. **VHS, Beta** *RKO ✂✂*

Criminal Justice A black man is accused of a crime by a woman with whom he was involved. Is it justice or revenge? Strong cast in important story. Made for cable TV.
1990 (R) 92m/C Forest Whitaker, Jennifer Grey, Rosie Perez, Anthony LaPaglia; *W/Dir:* Andy Wolk. **VHS $89.95** *HBO ✂✂✂*

Criminal Law An ambitious young Boston lawyer gets a man acquitted for murder, only to find out after the trial that the man is guilty and renewing his killing spree. Realizing he's the only one privy to the killer's trust, the lawyer decides to stop him himself. A white-knuckled thriller burdened by a weak script.
1989 (R) 113m/C Kevin Bacon, Gary Oldman, Karen Yahng, Joe Don Baker, Tess Harper; *Dir:* Martin Campbell. **VHS, Beta $89.98** *HBO ✂✂*

The Criminal Life of Archibaldo de la Cruz Seeing the death of his governess has a lasting effect on a boy. He grows up to be a demented cretin whose failure with women leads him to conspire to kill every one he meets, a task at which he also fails. Hilarious, bitter Bunuelian diatribe. In Spanish with English subtitles.
1955 95m/B *MX SP* Ernesto Alonso, Ariadne Welter, Rita Macedo, Rodolfo Landa, Andrea Palma, Miroslava Stern; *Dir:* Luis Bunuel. **VHS, Beta $39.95** *FCT ✂✂✂*

Criminally Insane Authorities release 250 pounds of cleaver-wielding maniacal fury from the loony bin. She seeks food and human blood. Followed by "Crazy Fat Ethel II."
1975 61m/C Priscilla Alden, Michael Flood; *Dir:* Nick Phillips. **VHS, Beta $59.95** *WES ✂*

Criminals Within A scientist working on a top secret formula is murdered and a detective tries to nail the spy ring responsible. Very confusing, poor usage of some fine female cast members.
1941 67m/B Eric Linden, Ann Doran, Constance Worth, Donald Curtis, Weldon Heyburn, Ben Alexander; *Dir:* Joseph H. Lewis. **VHS, Beta $16.95** *SNC ✂*

The Crimson Ghost The Crimson Ghost plots to enslave the world by stealing an atomic weapon. Serial in twelve episodes. Also available in a 93 minute colorized edition.
1946 100m/B Charles Quigley, Clayton Moore, Linda Stirling, I. Stanford Jolley; *Dir:* William Witney, Fred Brannon. **VHS $29.98** *REP, VCN, MLB ✂*

Crimson Pirate An 18th-century buccaneer pits his wits and brawn against the might of a ruthless Spanish nobleman. Considered by many to be one of the best swashbucklers, laced with humor and enthusiastically paced. Showcase for Lancaster and Cravat's acrobatic talents.
1952 104m/C *GB* Burt Lancaster, Eva Bartok, Torin Thatcher, Christopher Lee, Nick Cravat; *Dir:* Robert Siodmak. **VHS, Beta, LV $19.98** *WAR ✂✂✂*

Crimson Romance Two unemployed American pilots in Europe decide to join the German airforce for want of anything better to do. Conflicts revolve around a warmongering Commandant and his pilot.
1934 70m/B Ben Lyon, Sari Maritza, James Bush, Erich von Stroheim, Jason Robards Sr., Herman Bing, Vince Barnett; *Dir:* David Howard. **VHS, Beta, 8mm $19.95** *NOS, VYY, RXM ✂✂*

Crisis at Central High A dramatic television recreation of the events leading up to the 1957 integration of Central High in Little Rock, Arkansas. The screenplay is based on teacher Elizabeth Huckaby's journal. Woodward was nominated for an Emmy for her performance as Huckaby.
1980 120m/C Joanne Woodward, Charles Durning, William Ross, Henderson Forsythe; *Dir:* Lamont Johnson. **VHS, Beta $29.98** *LTG ✂✂✂*

Criss Cross A classic grade-B film noir, in which an armored car driver is suckered into a burglary by his ex-wife and her hoodlum husband. Multiple back-stabbings and double-crossings ensue. Written by Daniel Fuchs.
1948 98m/B Burt Lancaster, Yvonne de Carlo, Dan Duryea, Stephen McNally, Richard Long, Tony Curtis; **Dir:** Robert Siodmak. **VHS, Beta** $59.95 MCA ✓✓✓

Crisscross It's 1969 and Tracy Cross is a divorced mom whose ex-husband is a traumatized Vietnam vet who has left her alone, trying to raise their 12 year-old son Chris during some hard times. She works two jobs, as a waitress and, unbeknownst to her son, as a stripper in a local club. When Chris sneaks into the club and sees her working act he takes drastic measures to help out— by selling drugs. Film is sluggish and relies too much on voice-overs to explain thoughts but Hawn gives a convincing and restrained performance as the mother.
1992 (R) 100m/C Goldie Hawn, David Arnott, Arliss Howard, James Gammon, Keith Carradine, J.C. Quinn; **Dir:** Chris Menges. **VHS, LV** $94.99 MGM ✓✓½

Critical Condition During a blackout, a criminal masquerades as a doctor in a city hospital. Limp comedy embarrassing for Pryor.
1986 (R) 99m/C Richard Pryor, Rachel Ticotin, Ruben Blades, Joe Mantegna; **Dir:** Michael Apted. **VHS, Beta, LV** $19.95 PAR ✓

Critters A gang of furry, razor-toothed aliens escapes from its prison ship to Earth with their bounty hunters right behind them. They make it to a small town in Kansas where they begin to attack anything that moves. Not just a thrill-kill epic, but a sarcastic other-worldly thrill-kill epic.
1986 (PG-13) 86m/C Dee Wallace Stone, M. Emmet Walsh, Billy "Green" Bush, Scott Grimes; **Dir:** Stephen Herek. **VHS, Beta, LV** $14.95 COL ✓✓½

Critters 2: The Main Course Sequel to the hit horror-comedy, wherein the voracious alien furballs return in full force from eggs planted two years before.
1988 (PG-13) 93m/C Scott Grimes, Liane Curtis, Don Opper, Barry Corbin, Terrence Mann; **Dir:** Mick Garris. **VHS, Beta, LV** $14.95 COL ✓½

Critters 3 The not-so-loveable critters are back, terrorizing the occupants of a tenement building.
1991 (PG-13) 86m/C Francis Bay, Aimee Brooks, Leonardo DiCaprio, Don Opper; **Dir:** Kristine Peterson. **VHS** $89.95 COL ✓✓

Critters 4 The ghoulish critters are back, and this time a strain of genetically engineered mutant critters (what's the difference?) wants to take over the universe.
1991 (PG-13) 94m/C Don Opper, Paul Whitthorne, Angela Bassett, Brad Dourif, Terrence Mann; **Dir:** Rupert Harvey. **VHS** $89.95 NLC, COL ✓✓

Crocodile A giant crocodile attacks a beach town, killing and devouring dozens of people. The locals sit around trying to figure a way to stop the critter.
1981 (R) 95m/C Nat Puvanai, Tiny Tim; **Dir:** Herman Cohen. **VHS, Beta** $59.99 HBO Woof!

Crocodile Dundee New York reporter Sue Charlton is assigned to the Outback to interview living legend Mike Dundee. When she finally locates the man, she is so taken with him that she brings him back to New York with her. There, the naive Aussie wan-

ders about, amazed at the wonders of the city and unwittingly charming everyone he comes in contact with, from high-society transvestites to street hookers. One of the surprise hits of 1986.
1986 (PG-13) 98m/C AU Paul Hogan, Linda Kozlowski, John Meillon, David Gulpilil, Mark Blum; **Dir:** Peter Faiman. **VHS, Beta, LV, 8mm** $14.95 PAR ✓✓✓½

Crocodile Dundee 2 In this sequel Mike Dundee, the loveable rube, returns to his native Australia looking for new adventure, having "conquered" New York City. He inadvertently gets involved in stopping a gang of crooks active in Australia and New York. Lacks the charm and freshness of the first film.
1988 (PG) 110m/C Paul Hogan, Linda Kozlowski, Ken Welsh; **Dir:** John Cornell. **VHS, Beta, LV** $14.95 PAR ✓✓½

Cromwell A lavish British-made spectacle about the conflict between Oliver Cromwell and Charles I, and the British Civil War. History is twisted as Cromwell becomes the liberator of the oppressed. Harris gives a commanding performance and the battle scenes are stunners.
1970 (G) 139m/C GB Richard Harris, Alec Guinness, Robert Morley, Frank Finlay, Patrick Magee, Timothy Dalton; **Dir:** Ken Hughes. Academy Awards '70: Best Costume Design. **VHS, Beta, LV** $19.95 COL ✓✓½

Crooked Hearts This dysfunctional-family drama means well but lays it on too thick, as a father-son rivalry threatens to sunder the tight-knit Warrens. Based on a novel by Robert Boswell.
1991 (R) 113m/C Peter Coyote, Jennifer Jason Leigh, Peter Berg, Cindy Pickett, Vincent D'Onofrio; **Dir:** Michael Bortman. **VHS, Beta, LV** $89.98 MGM, LDC ✓✓

Crooked Trail A lawman and bad guy become friends.
1936 58m/B Johnny Mack Brown. **VHS, Beta, LV** WGE ✓

Crooks & Coronets Two-bit crooks plot to rob the country estate of an eccentric British dowager. Alternate title: "Sophie's Place."
1969 (PG) 106m/C GB Cesar Romero, Telly Savalas, Warren Oates, Edith Evans; **Dir:** James O'Connolly. **VHS, Beta** $59.95 WAR ✓½

Cross Country Action revolves around the brutal murder of a call girl, with initial suspicion falling on a television advertising director involved with the woman. The story twists and turns from the suspect to the investigating detective.
1983 (R) 95m/C CA Richard Beymer, Nina Axelrod, Michael Ironside; **Dir:** Paul Lynch. **VHS, Beta** $9.98 SUE, STE, CHA ✓½

Cross Creek Based on the life of Marjorie Kinnan Rawlings (The Yearling) who, after 10 years as a frustrated reporter/writer, moves to the remote and untamed Everglades. There she meets colorful local characters and receives the inspiration to write numerous bestsellers. Well acted though overtly sentimental at times.
1983 (PG) 115m/C Mary Steenburgen, Rip Torn, Peter Coyote, Dana Hill, Alfre Woodard, Malcolm McDowell; **Dir:** Martin Ritt. **VHS, Beta** $19.99 HBO ✓✓½

Cross Examination A boy is charged with his father's murder, and a defense attorney struggles to acquit him.
1932 61m/B H.B. Warner, Sally Blaine, Sarah Padden. **VHS, Beta, LV** WGE ✓½

Cross of Iron During WWII, two antagonistic German officers clash over personal ideals as well as strategy in combatting the relentless Russian attack. Followed by "Breakthrough."
1976 (R) 120m/C James Coburn, Maximilian Schell, James Mason, David Warner, Senta Berger; **Dir:** Sam Peckinpah. **VHS, Beta** $19.95 MED, MED ✓✓½

Cross Mission Predictable actioner about a photographer and a soldier who are captured by the enemy. Packed with a bit of ninja, ammo, and occult.
1989 (R) 90m/C Richard Randall; **Dir:** Al Bradley. **VHS** $59.95 CAN ✓

Cross My Heart Light comedy about two single people with complicated, post-divorce lives who go on a date and suffer accordingly. Will true love prevail?
1988 (R) 91m/C Martin Short, Annette O'Toole, Paul Reiser, Joanna Kerns; **Dir:** Armyan Bernstein. **VHS, Beta** $79.95 MCA ✓✓

The Cross & the Switchblade An idealistic priest tries to bring the message of religion to the members of a vicious street gang. They don't wanna listen.
1972 (PG) 105m/C Pat Boone, Erik Estrada, Jackie Giroux, Jo-Ann Robinson; **Dir:** Don Murray. **VHS, Beta** $79.95 LGC, VGD, VBL ✓½

Crossbar Aaron Kornylo is determined to reach Olympic qualifications in the high jump despite having only one leg. Inspired by a true story, this program dramatically shows how far determination and work can take a person.
1979 77m/C John Ireland, Brent Carver, Kate Reid; **Dir:** John Trent. **VHS, Beta** $49.95 TWE, LCA ✓✓

Crossfire A Jewish hotel guest is murdered and three soldiers just back from Europe are suspected of the crime. The first Hollywood film that explored racial bigotry. Due to the radical nature of its plot, the director and the producer were eventually black-listed for promoting "un-American" themes. Loosely based on Richard Brooks' "The Brick Foxhole." Oscar nominations for Best Picture, Best Supporting Actor (Ryan), Best Supporting Actress (Grahame), Best Director, Screenplay.
1947 86m/B Robert Young, Robert Mitchum, Robert Ryan, Gloria Grahame, Paul Kelly; **Dir:** Edward Dmytryk. Edgar Allan Poe Awards '47: Best Screenplay; National Board of Review Awards '47: 10 Best Films of the Year. **VHS, Beta** $19.95 MED ✓✓✓½

Crossfire Two bandits are almost hung in the midst of the Mexican revolution. Also known as "The Bandit." Released in the U.S. in 1979.
1967 (PG) 82m/C MX Robert Conrad, Jan-Michael Vincent, Manuel Lopez Ochoa, Roy Jenson; **Dir:** Robert Conrad, Alfredo Zacharias. **VHS, Beta** $59.95 ACA ✓

Crossfire A rescue team heads to Vietnam to pick up their MIA comrades. With no help from the government and the enemy at every turn, can they possibly succeed?
1989 (R) 99m/C Richard Norton, Michael Meyer, Daniel Dietrich, Don Pemrick, Eric Hahn, Wren Brown, Steve Young; **Dir:** Anthony Maharaj. **VHS, LV** $19.98 SUE ✓

Crossing Delancey A Jewish woman (Bozyk), in old world style, plays matchmaker to her independent young granddaughter. Charming modern-day New York City fairy tale that deftly manipulates cliches and stereotypes. Lovely performance from Irving as

the woman whose heart surprises her. Riegert is swell playing the gentle but never wimpy suitor. Bozyk was a star on the Yiddish vaudeville stage. Appealing music by the Roches, with Suzzy Roche giving a credible performance as Irving's friend.
1988 (PG) 97m/C Amy Irving, Reizl Bozyk, Peter Riegert, Jeroen Krabbe, Sylvia Miles, Suzzy Roche, George Martin; *Dir:* Joan Micklin Silver. **VHS, Beta, LV, 8mm $19.98** *WAR, FCT, JCF* 🎗️🎗️🎗️

Crossing the Line Two motocross racers battle it out for the championship.
1990 (R) 90m/C Jon Stafford, Rick Hearst, Paul Smith, Cameron Mitchell, Vernon Wells, Colleen Morris, John Saxon; *Dir:* Gary Graver. **VHS, Beta, LV $79.95** *COL* 🎗️

Crossover A devoted male nurse works the graveyard shift in the psychiatric ward and neglects his personal life.
1982 (R) 96m/C James Coburn, Kate Nelligan; *Dir:* John Guillermin. **VHS, Beta $69.95** *LTG, VES, LIV* 🎗️½

Crossover Dreams Actor and musician Blades moves this old story along. Sure that he has at last made the international Big Time, a salsa artist becomes a self-important, back-stabber after cutting an album. Great music. Blades's first film.
1985 86m/C Ruben Blades, Shawn Elliott, Elizabeth Pena, Virgilio Marti, Tom Signorelli, Frank Robles, Joel Diamond, Amanda Barber, John Hammil; *Dir:* Leon Ichaso. **VHS, Beta $14.95** *HBO, NEG* 🎗️🎗️½

Crossroads A blues-loving young white man, classically trained at Juilliard, befriends an aging black blues-master. After helping the old man escape from the nursing home, the two hop trains to the South where it's literally a duel with the devil. Fine performances by Macchio, Gertz, and Morton. Wonderful music, mostly by Ry Cooder.
1986 (R) 100m/C Ralph Macchio, Joe Seneca, Jami Gertz, Joe Morton; *Dir:* Walter Hill. **VHS, Beta, LV $19.95** *RCA* 🎗️🎗️½

Crow Hollow A Victorian mansion inhabited by three whacko sisters, Crow Hollow is the site of a murder committed in an attempt to cash in on an inheritance. When a young woman investigates, eyebrows are raised.
1952 69m/B *GB* Donald Houston, Nastasha Parry, Nora Nicholson, Esma Cannon, Melissa Stribling; *Dir:* Michael McCarthy. **VHS, Beta $16.95** *SNC, RXM* 🎗️🎗️

The Crowd A look at the day-to-day trials of a working-class family set against the backdrop of a wealthy society. True-to-life, it's peppered with some happy moments, too. One of the best silent films. Available on laserdisc as a double feature with another '20's classic, ''The Wind.''
1928 104m/B Eleanor Boardman, James Murray, Bert Roach, Daniel G. Tomlinson, Dell Henderson, Lucy Beaumont; *Dir:* King Vidor. **VHS, Beta, LV $29.95** *MGM* 🎗️🎗️🎗️🎗️

Crucible of Horror Two women plot the murder of an unsuspecting husband, but he turns the tables on them.
1969 91m/C Michael Gough, Yvonne Mitchell, Sharon Gurney; *Dir:* Viktors Ritelis. **VHS, Beta $59.95** *PGN* 🎗️🎗️

Crucible of Terror A mad sculptor covers beautiful models with hot wax, then imprisons them in a mold of bronze.
1972 95m/C Mike Raven, Mary Maude, James Bolam. **VHS, Beta $49.95** *PSM, MRV, GEM* 🎗️

The Crucifer of Blood A disappointing Sherlock Holmes yarn, made for cable TV. Heston is adequate as the Baker Street sleuth, but the mystery—about two retired British soldiers who share an accursed secret and a vengeful comrade—unreels clumsily in the form of flashbacks that give away most of the puzzle from the start. Interesting only in that Dr. Watson has a love affair, more or less. Adapted from a play (and looking like it) by Paul Giovanni, inspired by Arthur Conan Doyle's ''The Sign of Four.''
1991 105m/C Charlton Heston, Richard Johnson, Susannah Harker, John Castle, Clive Wood, Simon Callow, Edward Fox; *Dir:* Fraser Heston. **VHS, Beta $89.98** *TTC* 🎗️🎗️½

The Crucified Lovers A shy scrollmaker falls in love with his master's wife. Excellent Japanese tragedy with fine performances all around. Original title: ''Chikamatsu Monogatari.''
1954 100m/B *JP* Kazuo Hasegawa, Kyoko Kagawa, Yoko Minamida, Eitaro Shindo, Sakae Ozawa; *Dir:* Kenji Mizoguchi. **VHS** *FCT* 🎗️🎗️🎗️½

Cruel Sea Well-made documentary-like account of a Royal Navy corvette on convoy duty in the Atlantic during WWII.
1953 121m/B *GB* Jack Hawkins, Stanley Baker, Denholm Elliott, Virginia McKenna; *Dir:* Charles Frend. **VHS, Beta $29.95** *BVG, HBO, BMV* 🎗️🎗️🎗️

Cruel Story of Youth A teenage girl and her criminal boyfriend use sex to get money out of rich, middle-aged men in this controversial look at the disillusionment of youth and the breaking of old values and traditions in Japan after World War II. In Japanese with English subtitles.
1960 96m/C *JP* Yusuke Kawazu, Miyuki Kuwano, Yoshiko Kuga; *W/Dir:* Nagisa Oshima. **VHS $79.95** *NYF* 🎗️🎗️🎗️

Cruise A cartoon fable about the precision of music, with elements of mystery and fantasy thrown in for good measure.
198? 8m/C **VHS, Beta** *NFC* 🎗️🎗️🎗️

Cruise Into Terror A sarcophagus brought aboard a pleasure cruise ship unleashes an evil force which starts to slowly kill off the ship's passengers, most of whom likely thought they were on the Love Boat. Made for television.
1978 100m/C Ray Milland, Hugh O'Brian, John Forsythe, Christopher George, Dirk Benedict, Frank Converse, Lynda Day George, Stella Stevens; *Dir:* Bruce Kessler. **VHS, Beta $49.95** *PSM* Woof!

Cruise Missile A unique task force is on a mission to keep the world from nuclear holocaust.
1978 100m/C Peter Graves, Curt Jurgens, Michael Dante; *Dir:* Ted V. Mikels. **VHS, Beta $39.95** *MON* 🎗️

Cruisin' High Two city street gangs battle it out.
1975 (R) 109m/C David Kyle, Kelly Yaegermann, Rhodes Reason. **VHS, Beta $69.95** *LTG, VES, LIV* 🎗️

Cruising A rookie cop goes undercover to investigate the bizarre murders of homosexuals in New York's West Village. Sexually explicit but less than suspenseful mystery. Many gay rights groups disapproved of the gay scene represented in this film while others feared copy-cat crimes. Excellent cinematography; music by Jack Nitzsche.
1980 (R) 102m/C Al Pacino, Paul Sorvino, Karen Allen, Powers Boothe; *Dir:* William Friedkin. **VHS, Beta $19.98** *WAR* 🎗️½

Crusader Rabbit vs. the Pirates Early endeavor from Jay Ward (creator of ''Rocky & Bullwinkle'') starring a quick-witted rabbit and his sidekick, Rags the Tiger. This cartoon was unique in the fact that it was the first animated series produced exclusively for television.
1949 80m/B *Nar:* Roy Whaley. **VHS $12.95** *RHI, FCT* 🎗️½

Crusader Rabbit vs. the State of Texas Crusader Rabbit, Jay Ward's pre-''Bullwinkle'' hero, and his sidekick, Rags, take it to the Lone Star state when they begin deporting Crusader's rabbit relatives.
1949 60m/B *Nar:* Roy Whaley. **VHS $12.95** *RHI, FCT* 🎗️½

Crusoe A lushly photographed version of the Daniel Defoe classic. Crusoe is an arrogant slave trader stranded on a desert island he soon finds is populated by unfriendly natives. Themes of prejudice, fear, and choice appear in this never-padded, thoughtful film. Quinn gives an excellent performance as the stranded slave-trader.
1989 (PG-13) 94m/C Aidan Quinn, Ade Sapara, Jimmy Nail, Timothy Spall; *Dir:* Caleb Deschanel. **VHS, Beta, LV $89.95** *VIR* 🎗️🎗️½

Cry-Baby A homage and spoof of '50s teen-rock melodramas by the doyen of cinematic Bad Taste, involving a terminal bad-boy high schooler who goes with a square blond and starts an inter-class rumble. Musical numbers, throwaway gags and plenty of knee-bending to Elvis, with a weak story supported by offbeat celeb cameos.
1990 (PG-13) 85m/C Johnny Depp, Amy Locane, Polly Bergen, Traci Lords, Ricki Lake, Iggy Pop, Susan Tyrrell, Patty Hearst, Kim McGuire, Darren E. Burrows, Troy Donahue, Willem Dafoe, David Nelson; *Dir:* John Waters. **VHS, Beta, LV $19.95** *MCA* 🎗️🎗️🎗️

Cry of the Banshee Witch-hunter Price and family are tormented by Satanic powers seeking revenge. Superior horror period piece. Also available with ''Count Yorga, Vampire'' on Laserdisc.
1970 (PG) 87m/C *GB* Vincent Price, Elisabeth Bergner, Essy Persson, Hugh Griffith, Hilary Dwyer, Sally Geeson, Patrick Mower; *Dir:* Gordon Hessler. **VHS, Beta $79.99** *HBO* 🎗️🎗️½

Cry of Battle Anxious for the challenges of manhood, a well-heeled young man chooses to join a guerrilla militia in the Philippines.
1963 99m/B Van Heflin, Rita Moreno, James MacArthur; *Dir:* Irving Lerner. **VHS, Beta $49.95** *PSM, MRV, KOV* 🎗️🎗️½

Cry, the Beloved Country A black country minister travels to Johannesberg to be with his son after the youth is accused of killing a white man. Through the events of the trial, the horror, oppression and destruction of South Africa's apartheid system are exposed. Startling and moving, the first entertainment feature set against the backdrop of apartheid. Still trenchant; based on the novel by Alan Paton.
1951 111m/B *GB* Canada Lee, Charles Carson, Sidney Poitier, Joyce Carey, Geoffrey Keen; *Dir:* Zoltan Korda. **VHS $69.95** *MON, FCT* 🎗️🎗️🎗️½

Cry of the Black Wolves A trapper wrongly accused of murder saves the life of a bounty hunter hired to capture him.
197? 90m/C Ron Ely, Catherine Conti. **VHS, Beta** *AHV* 🎗️

Cry Danger A falsely imprisoned man is released from jail, and he searches for those who framed him.

1951 80m/B Dick Powell, Rhonda Fleming, William Conrad, Richard Erdman; *Dir:* Robert Parrish. **VHS** $19.98 *REP* 🎞🎞🎞

A Cry in the Dark Tight film story of the infamous Australian murder trial of Lindy Chamberlain, who was accused of killing her own baby, mostly because of the intensely adverse public opinion, aroused by vicious press, that surrounded the case. Chamberlain blamed the death on a wild dingo dog. Near-documentary style, with Streep excellent as the religious, unknowable mother. This case was also detailed in the film "Who Killed Baby Azaria?"
1988 (PG-13) 120m/C *AU* Meryl Streep, Sam Neill, Bruce Myles, Charles Tingwell, Nick Tate, Neil Fitzpatrick, Maurice Fields, Lewis Fitzgerald; *Dir:* Fred Schepisi. Cannes Film Festival '88: Best Actress (Streep); New York Film Critics Awards '88: Best Actress (Streep). **VHS, Beta, LV, 8mm** $19.98 *WAR, FCT* 🎞🎞🎞½

Cry Freedom A romantic look at the short life of South African activist Steven Biko, and his friendship with the white news editor, Donald Woods. The film focuses on Woods' escape from Africa while struggling to bring Biko's message to the world. Based on a true story.
1987 (PG) 157m/C Kevin Kline, Denzel Washington, Penelope Wilton; *Dir:* Richard Attenborough. British Academy Awards '87: Best Sound. **VHS, Beta, LV** $29.95 *KUI, MCA, FCT* 🎞🎞½

Cry of the Innocent An action-packed thriller about a Vietnam veteran who is out to find a group of Irish terrorists that killed his family.
1980 93m/C Cyril Cusack, Alexander Knox, Rod Taylor, Joanna Pettet, Nigel Davenport; *Dir:* Michael O'Herlihy. **VHS, Beta** $59.95 *VCL* 🎞🎞🎞

A Cry for Love An amphetamine addict and an alcoholic meet, fall in love, and help each other recover. Based on the best seller by Jill Schary Robinson, "Bedtime Story."
1980 96m/C Susan Blakely, Powers Boothe, Gene Barry, Charles Siebert, Herb Edelman, Fern Fitzgerald, Lainie Kazan; *Dir:* Paul Wendkos. **VHS, Beta** $14.95 *FRH* 🎞🎞

Cry Panic A man is thrown into a strange series of events after accidentally running down a man on a highway. Made for television.
1974 74m/C John Forsythe, Anne Francis, Earl Holliman, Claudia McNeil, Ralph Meeker; *Dir:* James Goldstone. **VHS, Beta** $49.95 *PSM* 🎞🎞

Cry of a Prostitute: Love Kills A former prostitute joins with a professional assassin in an effort to pacify rival gangsters in Italy.
1972 (R) 86m/C Henry Silva, Barbara Bouchet; *Dir:* Andrew Bianchi. **VHS, Beta** $59.95 *PSM* 🎞

A Cry from the Streets Drama about the plight of orphan children and dedicated social workers in England.
1957 100m/C *GB* Max Bygraves, Barbara Murray, Kathleen Harrison, Colin Petersen; *Dir:* Lewis Gilbert. **VHS** $29.95 *FCT* 🎞🎞

Cry Terror Two escaped convicts take two beautiful women hostage hoping to gain their own freedom. Only a quick thinking undercover police officer can save the girls.
1974 71m/C *GB* Bob Hoskins, Susan Hampshire, Gabrielle Drake. **VHS, Beta** $14.95 *KBV* 🎞🎞

Cry Uncle Comic account of a private eye who investigates a blackmailing case involving film of orgies in which, much to his chagrin, he has participated.

1971 85m/C Allen Garfield, Paul Sorvino; *Dir:* John G. Avildsen. **VHS, Beta** $79.95 *PSM* 🎞🎞½

Cry Vengeance A falsely imprisoned detective gets out of jail and searches for the people who framed him and killed his family.
1954 83m/B Mark Stevens, Joan Vohs, Martha Hyer, Skip Homeier; *Dir:* Mark Stevens. **VHS, Beta** $19.98 *REP* 🎞🎞½

The Crying Red Giant A puppet animation story about a lonely giant who scares away potential friends because of his size.
1989 18m/C VHS, Beta, 3/4U $50.00 *AIM* 🎞🎞

Crypt of Dark Secrets A Vietnam veteran recovering from wounds in the Louisiana swamps encounters a friendly Indian spirit who saves him from death.
1976 (R) 100m/C Maureen Chan, Ronald Tanet, Wayne Mack, Herbert G. Jahncke; *Dir:* Jack Weis. **VHS, Beta** *VHE* 🎞

Crypt of the Living Dead An undead woman from the 13th century makes life miserable for visitors on Vampire Island.
1973 (PG) 75m/C Andrew Prine, Mark Damon, Teresa Gimpera, Patty Sheppard, Francisco (Frank) Brana; *Dir:* Ray Danton. **VHS, Beta** $49.95 *VHE, JLT* Woof!

Crystal Heart A medically isolated songwriter with an incurable disease falls in love with a lovely rock singer.
1987 (R) 103m/C Lee Curreri, Tawny Kitaen, Lloyd Bochner; *Dir:* Gil Bettman. **VHS, Beta** $19.95 *STE, NWV* 🎞

Crystal Triangle An anthropologist seeks a message of God and discovers the Crystal Triangle, which will give him the means to do so. But he isn't the only one who wants the Crystal—because whoever possesses the message of God holds the fate of the world in their hands. Animated; in Japanese with English subtitles.
1987 86m/C *JP* VHS $39.95 *CPM* 🎞🎞½

Crystalstone A wooden cross leads a pair of orphans on a dangerous search for the legendary Crystalstone.
1988 (PG) 103m/C Frank Grimes, Kamlesh Gupta, Laura Jane Goodwin, Sydney Bromley; *Dir:* Antonio Pelaez. Academy of Family Films '88: Award of Merit for Outstanding Achievement; Film Advisory Board '88: Award of Excellence. **VHS, Beta** $79.98 *MCG* 🎞🎞

Cthulhu Mansion The oldest-looking juvenile delinquents you've ever seen release evil spirits when they take over a magician's estate. Not-very-special effects and abominable dialogue; it claims to be based on H.P. Lovecraft stories, but that's a load of shoggoth dung.
1991 (R) 95m/C Frank Finlay, Marcia Layton, Brad Fisher, Melanie Shatner, Luis Fernando Alves, Kaethe Cherney, Paul Birchard, Francisco (Frank) Brana; *W/Dir:* J.P. Simon. **VHS** $89.98 *REP* Woof!

Cuba Cynical, satirical adventure/love story. Mercenary soldier rekindles an old affair during the Castro revolution. Charismatic casting and good directing make for an entertaining, if overlooked, Connery vehicle.
1979 (R) 121m/C Sean Connery, Brooke Adams, Jack Weston, Hector Elizondo, Denholm Elliott, Chris Sarandon, Lonette McKee; *Dir:* Richard Lester. **VHS, Beta** $59.99 *FOX* 🎞🎞🎞

Cujo A rabid dog goes berserk and attacks a mother and her child who are trapped inside a broken-down Pinto. Frighteningly realistic film is based on Stephen King's bestseller.

1983 (R) 94m/C Dee Wallace Stone, Daniel Hugh-Kelly, Danny Pintauro, Ed Lauter, Christopher Stone; *Dir:* Lewis Teague. **VHS, Beta, LV** $19.98 *WAR* 🎞🎞

Culpepper Cattle Co. A young, starry-eyed yokel wants to be a cowboy and gets himself enlisted into a cattle drive, where he learns the harsh reality of the West.
1972 (PG) 92m/C Gary Grimes, Billy "Green" Bush, Bo Hopkins, Charles Martin Smith, Geoffrey Lewis; *Dir:* Dick Richards. **VHS, Beta** $19.98 *FOX* 🎞🎞

Cult People Interviews and film clips featuring a host of filmmaking's favorite cult personalities.
1989 60m/C Patrick Macnee, Cameron Mitchell, James Karen, Russ Meyer, Curtis Harrington, Waris Hussein, Michael Sarne. **VHS** $19.95 *FCT* 🎞

Cure Chaplin arrives at a spa to take a rest cure, accompanied by a trunk full of liquor that somehow gets dumped into the water at the resort. Silent with music track.
1917 20m/B Charlie Chaplin. **VHS, Beta** *CAB, FST* 🎞½

Curfew A young woman rushes home so she doesn't break her curfew, only to find two killers with time on their hands waiting for her. Violent.
1988 (R) 86m/C John Putch, Kyle Richards, William A. Wellman, Bert Remsen; *Dir:* Gary Winnick. **VHS, Beta, LV** $19.95 *STE, NWV* 🎞🎞

Curley High-spirited youngsters play pranks on their schoolteacher. A part of the Hal Roach Comedy Carnival.
1947 53m/C Larry Olsen, Frances Rafferty, Eilene Janssen, Walter Abel; *Dir:* Bernard Carr. **VHS, Beta** $19.95 *DVT, NOS, HHT* 🎞🎞

Curley & His Gang in the Haunted Mansion An eccentric scientist gets involved in a haunted mansion mystery.
1947 54m/C Larry Olsen. **VHS, Beta** $19.95 *UNI, MRV* 🎞

Curly Sue Adorable, homeless waif Porter and her scheming, adoptive father Belushi plot to rip off single, career-minded female attorney Lynch in order to take in some extra cash. Trouble is, all three heartstrings are tugged, and the trio develop a warm, caring relationship. A throwback to the Depression era's Shirley Temple formula films met with mixed reviews. Undiscriminating younger audiences should have a good time, though. Available in widescreen format on laserdisc.
1991 (PG) 102m/C James Belushi, Kelly Lynch, Alison Porter, John Getz, Fred Dalton Thompson; *W/Dir:* John Hughes. **VHS, Beta, LV, 8mm** $94.99 *WAR, CCB* 🎞🎞

Curly Top Orphan Temple helps land a husband for her beautiful sister. Along the way, she sings "Animal Crackers in My Soup." Remake of the silent "Daddy Long Legs," which was remade again in 1955.
1935 74m/B Shirley Temple, John Boles, Rochelle Hudson, Jane Darwell, Esther Dale, Arthur Treacher, Rafaela Ottiano; *Dir:* Irving Cummings. **VHS, Beta** $19.98 *FOX* 🎞🎞½

The Curse After a meteorite lands near a small farming community and contaminates the area, a young boy tries to prevent residents from turning into slime-oozing mutants. Remake of "Die, Monster, Die." Also known as "The Farm."

1987 (R) 92m/C Wil Wheaton, Claude Akins, Malcolm Danare, Cooper Huckabee, John Schneider, David Keith, Amy Wheaton, David Chaskin; *Dir:* David Keith. **VHS, Beta, LV $14.95** MED, VTR, MHV 🎬

Curse 2: The Bite Radiation affected snakes are transformed into deadly vipers whose bites change their unsuspecting victims into horrible creatures.
1988 (R) 97m/C Jill Schoelen, J. Eddie Peck, Jamie Farr, Savina Gersak, Bo Svenson; *Dir:* Fred Goodwin. **VHS, Beta $89.95** TWE 🎬

Curse 3: Blood Sacrifice An African sugar cane plantation becomes host to a horrible nightmare when a voodoo curse is placed on the owners. The monstrous God of the Sea is summoned to avenge the accidental death of a baby, and the hellish journey into insanity begins.
1990 (R) 91m/C Christopher Lee, Jenilee Harrison, Henry Cele; *Dir:* Sean Barton. **VHS, LV $79.95** COL 🎬

Curse of the Alpha Stone A professor experiments with an ancient formula on a student, creating a babbling maniac who slaughters freshman girls.
1985 90m/C **VHS, Beta $49.95** UHV Woof!

The Curse of the Aztec Mummy A mad scientist schemes to rob a Mayan pyramid of its treasure but the resident mummy will have none of it. Sequel to "Robot vs. the Aztec Mummy" and followed by "Wrestling Women vs. the Aztec Mummy."
1959 65m/B MX Ramon Gay, Rosita Arenas, Crox Alvarado; *Dir:* Rafael Portillo. **VHS $16.95** SNC 🎬

Curse of the Black Widow An investigator follows a trail of brutal murders to the lair of a supernatural gigantic spider in the middle of Los Angeles. Made for television. Also known as "Love Trap."
1977 100m/C Patty Duke, Anthony (Tony) Franciosa, Donna Mills, June Lockhart, Sid Caesar, Vic Morrow, June Allyson, Roz Kelly, Jeff Corey; *Dir:* Dan Curtis. **VHS, Beta $39.98** CGI 🎬🎬

Curse of the Blue Lights Mysterious lights begin appearing at the local romantic spot. Little do the young lovers know, it hails the arrival of a ghoul, intent on raising the dead. Also available in an unedited version. Unfortunately, nothing can resurrect this film.
1988 (R) 93m/C Brent Ritter; *Dir:* John H. Johnson. **VHS, Beta $79.98** MAG Woof!

Curse of the Cat People A young sensitive girl is guided by the vision of her dead mother. Sequel to "Cat People." Available in a colorized version.
1944 70m/B Simone Simon, Kent Smith, Jane Randolph, Elizabeth Russell, Ann Carter; *Dir:* Robert Wise, Gunther Von Fristch. **VHS, Beta, LV $19.95** MED, TTC 🎬

The Curse of the Crying Woman An unknowing descendant of a witch is lured to her aunt's home to perform the act that will revive the monstrous crying woman and renew a reign of evil.
1961 74m/B MX Rosita Arenas, Abel Salazar, Rita Macedo, Carlos Lopez Moctezuma. **VHS, Beta $59.95** SNC, HHT, TIM 🎬½

Curse of the Demon A famous psychologist investigates a colleague's mysterious death and enters a world of demonology and the occult, climaxing in a confrontation with a cult's patron demon. Superb thriller based upon the story "Casting the Runes,"

by M.R. James, and alternately titled "Night of the Demon."
1957 81m/B GB Dana Andrews, Peggy Cummins, Niall MacGinnis, Maurice Denham; *Dir:* Jacques Tourneur. **VHS, Beta, LV $9.95** GKK, MLB 🎬🎬🎬½

Curse of the Devil This time Naschy is turned into a werewolf by annoyed gypsies whose ancestors were slain by his.
1973 (R) 73m/C MX SP Paul Naschy, Maria Silva, Patty Shepard, Fay Falcon; *Dir:* Carlos Aured. **VHS $19.95** SNC, UHV 🎬½

The Curse of Frankenstein Young Victor Frankenstein reenacts his father's experiments with creating life from the dead resulting in a terrifying, hideous creature. The first in Hammer's Frankenstein series and followed by "Revenge of Frankenstein." From the Shelley story. Make-up by Jack Pierce, who also created the famous make-up for Universal's Frankenstein monster.
1957 83m/C GB Peter Cushing, Christopher Lee, Hazel Court, Robert Urquhart, Valerie Gaunt; *Dir:* Terence Fisher. **VHS, Beta $19.98** WAR, MLB 🎬🎬½

Curse of King Tut's Tomb Made-for-television movie is set in 1922 where archaeologists have just opened Tutankhamen's tomb. The curse of the boy king is unleashed as tragic events bring the adventurers uncommon gloom.
1980 98m/C Robin Ellis, Harry Andrews, Eva Marie Saint, Raymond Burr, Wendy Hiller; *Dir:* Philip Leacock; *Nar:* Paul Scofield. **VHS, Beta $9.95** GKK 🎬½

Curse of the Living Corpse A millionaire comes back to rotting life to kill his negligent relatives. Scheider's first film.
1964 84m/C Candace Hilligoss, Roy Scheider, Helen Warren, Margot Hartman; *Dir:* Del Tenney. **VHS, Beta $39.95** PSM 🎬

Curse of the Living Dead Two siblings are forced to eat flesh for eternity after their witch-doctor mother puts a curse on them.
1989 90m/C *Dir:* Lawrence Foldes. **VHS, Beta** ELV, HHE 🎬

Curse of the Mummy/Robot vs. the Aztec Mummy A vengeful mummy of an Aztec warrior stalks those who attempt to steal a fabulous treasure. A mad criminal scientist wants the treasure and uses gangsters and a human robot to get what he wants.
196? 130m/B Ramon Gay, Rosita Arenas, Crox Alvarado, Luis Aceves Cantaneda; *Dir:* Rafael Portillo. **VHS, Beta** HHT Woof!

Curse of the Pink Panther Clifton Sleigh, an inept New York City detective played by Wass, is assigned to find the missing Inspector Clouseau. His efforts are complicated by an assortment of gangsters and aristocrats who cross paths with the detective. So-so attempt to keep popular series going after Seller's death. Niven's last film.
1983 (PG) 110m/C Ted Wass, David Niven, Robert Wagner, Herbert Lom, Joanna Lumley, Capucine, Robert Loggia, Harvey Korman, Leslie Ash, Denise Crosby; *Dir:* Blake Edwards. **VHS, Beta $79.95** MGM 🎬🎬

Curse of the Screaming Dead An unpredictable zombie yarn.
1984 90m/C **VHS, Beta $59.95** MGL Woof!

Curse of the Stone Hand A pair of stone hands causes folks to do suicidal things. A mutation of a Mexican and a Chilean horror film, each purchased and monster-mashed into one by Warren.
1964 72m/B MX John Carradine, Ernest Walch, Sheila Bon, Katherine Victor, Lloyd Nelson; *Dir:* Jerry Warren. **VHS $15.95** NOS, LOO Woof!

Curse of the Swamp Creature A mad scientist in the Everglades attempts to create half human/half alligator monsters. In turn, a geologic expeditionary force attempts to stop his experimentation. Low budget thrills courtesy of Larry Buchanan.
1966 80m/C John Agar, Francine York, Shirley McLine, Bill Thurman, Jeff Alexander; *Dir:* Larry Buchanan. **VHS $19.98** MOV, VDM 🎬

The Curse of the Werewolf Horror film about a 19th-century European werewolf that is renowned for its ferocious departure from the stereotypical portrait of the beast.
1961 91m/C GB Oliver Reed, Clifford Evans, Yvonne Romain, Catherine Feller, Anthony Dawson, Michael Ripper; *Dir:* Terence Fisher. **VHS, Beta $14.98** MCA, TLF, MLB 🎬🎬½

Curse of the Yellow Snake Written by Edgar Wallace, this voluminous yarn features a running battle over an ancient Chinese artifact, with crazed Chinese cultists running through foggy London streets.
1963 98m/C GE Joachim Berger, Werner Peters; *Dir:* Franz Gottlieb. **VHS, Beta $29.95** VYY 🎬🎬

Curtain Up "Little theatre" dramatics and the exasperating temperaments of amateur theatricals are exposed in this adaptation of the play "On Monday Next," by Philip King.
1953 82m/C GB Robert Morley, Margaret Rutherford, Olive Sloane; *Dir:* Ralph Smart. **VHS, Beta** LCA 🎬

Curtains A director has a clash of wills with a film star that spells "Curtains" for a group of aspiring actresses.
1983 (R) 90m/C CA John Vernon, Samantha Eggar; *Dir:* Jonathan Stryker. **VHS, Beta $69.95** VES 🎬

Custer's Last Fight A re-enactment of the American West period when the Sioux and Cheyenne Indian tribes bitterly opposed the white man. Silent.
1912 50m/B *Dir:* Francis Ford. **VHS, Beta $19.95** NOS, DVT 🎬½

Custer's Last Stand A feature-length version of the Mascot serial recounting the last days of the famous General.
1936 70m/B Frank McGlynn, Rex Lease, Nancy Caseell, Lona Andre, William Farnum, Reed Howes, Jack Mulhall, Josef Swickard, Ruth Mix; *Dir:* Elmer Clifton. **VHS, Beta $69.95** VCN, GPV, TIM

Cut and Run Two journalists follow a lead to the former South American home of Jim Jones, and are instantly captured by local guerrillas.
1985 (R) 87m/C Michael Berryman, Willie Aames, Lisa Blount, Karen Black. **VHS, Beta $19.95** STE, NWV 🎬

The Cut Throats Americans stumble upon an isolated Nazi outpost stocked with gold and beautiful women. Exploitation at its worst.
1989 (R) 80m/C Jay Scott, Joanne Douglas, Jeff Letham, Pat Michaels, Barbara Lane; *W/Dir:* John Hayes. **VHS $39.95** AVD 🎬

Cutter's Way An embittered and alcoholic disabled Vietnam vet and his small-time crook/gigolo friend wrestle with justice and morality when the drifter uncovers a murder

but declines to get involved. An unusually cynical mystery, it was originally titled "Cutter and Bone," from the novel by Newton Thorburg.
1981 (R) 105m/C Jeff Bridges, John Heard, Lisa Eichhorn, Ann Dusenberry, Stephen Elliott, Nina Van Pallandt, George Dickerson; *Dir:* Ivan Passer. Edgar Allan Poe Awards '81: Best Screenplay. **VHS, Beta, LV $69.95** MGM *♫♫½*

Cutting Class Murders proliferate in a high school, where a student with a history of mental illness is number one on the suspect list. Tongue-in-cheek mayhem.
1989 (R) 91m/C Jill Schoelen, Roddy McDowall, Donovan Leitch, Martin Mull, Brad Pitt; *Dir:* Rospo Pallenberg. **VHS, LV $9.98** REP *♫♫*

The Cutting Edge Take one spoiled figure skater who can't keep a partner, pair her with a cocky ex-hockey player who thinks the only sport on ice is hockey and you've got a winning combination. Rather lame figure skating plot can't be taken too seriously, but the sparks fly between Kelly and Sweeney, and they are the reason this movie works.
1992 (PG) 101m/C D.B. Sweeney, Moira Kelly, Roy Dotrice, Terry O'Quinn. **VHS** MGM *♫♫½*

Cyberpunk Documentary exploring William Gibson's literary and cultural phenomena of Cyberpunk. Includes interviews with Gibson, Jaron Lanier, Timothy Leary and Michael Synergy.
1991 60m/C William Gibson, Timothy Leary, Jaron Lanier, Michael Synergy; *Dir:* Marianne Trench. **VHS $29.95** MFV, FCT *♫♫½*

Cyborg In a deathly, dirty, post-holocaust urban world, an able cyborg battles a horde of evil mutant thugs. Poorly-made action flick.
1989 (R) 95m/C Jean-Claude Van Damme, Deborah Richter, Vincent Klyn, Dayle Haddon; *Dir:* Albert Pyun. **VHS, Beta, LV** PIA *♫*

Cycle Psycho Two girls are abused by a crazed motorcycle gang. Also known as "Numbered Days."
19?? (R) 80m/C Joe Turkel, Tom Drake, Stephen Oliver. **VHS, Beta $59.95** ACA Woof!

Cycle Vixens Three girls jump on their hogs and head from Colorado to California. Much leather, motor revving and other obligatory motorcycle-trash trimmings. Originally titled "The Young Cycle Girls."
1979 (R) 90m/C Loraine Ferris, Daphne Lawrence, Deborah Marcus, Lonnie Pense, Kevin O'Neill, Bee Lechat, Billy Bullet; *Dir:* Peter Perry. **VHS $9.95** SIM, IPI Woof!

Cyclone The girlfriend of a murdered scientist must deliver a secretly devised motorcycle into righteous government hands, much to the dismay of evil agents and corrupt officials. Good fun, sparked by a stellar "B" cast.
1987 (R) 89m/C Heather Thomas, Jeffrey Combs, Ashley Ferrare, Dar Robinson, Martine Beswick, Robert Quarry, Martin Landau, Huntz Hall, Troy Donahue, Michael Reagan, Dawn Wildsmith, Bruce Fairbairn, Russ Tamblyn; *Dir:* Fred Olen Ray. **VHS, Beta, LV $79.95** COL *♫♫*

Cyclone Cavalier A dauntless hero travels through Central America on an adventurous spree.
192? 58m/B Reed Howes; *Dir:* Albert Rogell. **VHS, Beta $16.95** GPV *♫*

Cyclone of the Saddle Typical Western about a range war.
1935 53m/B Rex Lease, Bobby Nelson, William Desmond, Yakima Canutt; *Dir:* Elmer Clifton. **VHS, Beta** WGE *♫*

Cyclops When an expedition party searches throughout Mexico for a woman's long lost brother, they are shocked when they find out that radiation has turned him into a one-eyed monster.
1956 72m/B Tom Drake, Gloria Talbott, Lon Chaney Jr., James Craig; *Dir:* Bert I. Gordon. **VHS, Beta $29.95** IVE, MLB *♫♫*

Cyclotrode "X" The Crimson Ghost attempts to kidnap the inventor of the title machine, which would enable him to rule the world. Edited from what was originally a 12-part serial. Moore (The Lone Ranger) plays a bad guy.
1946 100m/B Charles Quigley, Linda Stirling, I. Stanford Jolley, Clayton Moore, Kenne Duncan; *Dir:* William Witney, Fred Brannon. **VHS, Beta, 3/4U $29.98** VCN *♫♫½*

Cyrano de Bergerac Edmund Rostand's famous story of a large nosed yet poetic cavalier, who finds himself too ugly to be loved. He bears the pain of his devotion to Roxanne from afar, and helps the handsome but tongue-tied Christian to romance her. Ferrer became famous for this role, which won him an Oscar. Also available colorized.
1950 112m/B Jose Ferrer, Mala Powers, William Prince, Elena Verdugo, Morris Carnovsky; *Dir:* Michael Gordon. Academy Awards '50: Best Actor (Ferrer); National Board of Review Awards '50: 10 Best Films of the Year. **VHS, LV $9.95** NEG, MRV, NOS *♫♫♫♫*

Cyrano de Bergerac Depardieu brings to exhilarating life to Rostand's well-loved play about the brilliant but grotesque-looking swordsman/poet, afraid of nothing—except declaring his love to the beautiful Roxanne. One of France's costliest modern productions, a multi-award winner for its cast, costumes, music and sets. English subtitles (by Anthony Burgess) are designed to capture the intricate rhymes of the original French dialogue.
1989 (PG) 135m/C FR Gerard Depardieu, Jacques Weber, Anne Brochet, Vincent Perez, Roland Bertin, Josiane Stoleru, Phillipe Volter, Philippe Morier-Genoud, Pierre Maguelon; *Dir:* Jean-Paul Rappeneau. Academy Awards '90: Best Costume Design. **VHS, LV $89.98** ORI, FCT *♫♫♫♫*

D-Day on Mars The Purple Monster is Earth's only hope against alien invaders bent on taking over the planet. Originally a 15-part Republic serial titled "The Purple Monster Strikes."
1945 100m/B Dennis Moore, Linda Stirling, Roy Barcroft, James Craven, Bud Geary, Mary Moore; *Dir:* Spencer Gordon Bennet, Fred Brannon. **VHS $29.98** REP *♫♫*

D-Day, the Sixth of June An American soldier has an affair with a Englishwoman weeks before D-Day, where he unhappily finds himself fighting side by side with her husband.
1956 106m/C Richard Todd, Dana Wynter, Robert Taylor, Edmond O'Brien, John Williams, Jerry Paris, Richard Stapley; *Dir:* Henry Koster. **VHS, Beta $19.98** FOX *♫♫♫*

Da A middle-aged man returns to Ireland for his father's funeral. As he sorts out his father's belongings, his father returns as a ghostly presence to chat with him about life, death, and their own relationship. Based on the Hugh Leonard play with Hughes recreating his Tony-award winning role.
1988 (PG) 102m/C Barnard Hughes, Martin Sheen; *Dir:* Matt Clark. **VHS, Beta, LV $79.95** VIR *♫♫½*

Dad Hoping to make up for lost time, a busy executive rushes home to take care of his father who has just had a heart attack. What could have easily become sappy is made bittersweet by the convincing performances of Lemmon and Danson. Based on the novel by William Wharton.
1989 (PG) 117m/C Jack Lemmon, Ted Danson, Ethan Hawke, Olympia Dukakis, Kathy Baker, Zakes Mokae, J.T. Walsh, Kevin Spacey, Chris Lemmon; *Dir:* Gary David Goldberg. **VHS, Beta, LV $19.95** MCA *♫♫½*

Daddy Long Legs Far from the great musicals, but enjoyable. An eccentric millionaire glimpses a French orphan and becomes her anonymous benefactor. Her musing over his identity spawns some surreal (often inexplicable) dance numbers, but love conquers all, even lesser Johnny Mercer songs—most prominently "Something's Got to Give." From a story by Jean Webster, also done in 1919 with Mary Pickford, 1931 with Janet Gaynor and 1935 with Shirley Temple as "Curly Top."
1955 126m/C Fred Astaire, Leslie Caron, Terry Moore, Thelma Ritter, Fred Clark, Charlotte Austin, Larry Keating; *Dir:* Jean Negulesco. **VHS $19.98** FXV, FCT *♫♫♫*

Daddy-O Drag racer Daddy-O traps the killers of his best friend and lands the blonde bombshell Jana.
1959 74m/C Dick Contino, Sandra Giles; *Dir:* Lou Place. **VHS $9.95** COL *♫♫*

Daddy's Boys Since farming during the Depression is a rather low-paying endeavor, this family turns to thieving instead. But all-for-one and one-for-all is not the credo for one of the sons who decides to branch out on his own. Produced by Roger Corman.
1987 (R) 90m/C Daryl Haney, Laura Burkett, Raymond J. Barry, Dan Shor; *Dir:* Joe Minion. **VHS, Beta, LV $69.95** COL *♫½*

Daddy's Dyin'...Who's Got the Will? Based on the critically acclaimed play, this bittersweet comedy stars Bridges and D'Angelo as two members of the spiteful Turnover clan. When Daddy is on his deathbed, the entire family uses the opportunity to stab each other in the back. Non-stop humor and deep-hearted honesty carries this film quickly from beginning to end. The cast of well-known screen actors makes this a delightful adaptation.
1990 (PG-13) 100m/C Beau Bridges, Beverly D'Angelo, Tess Harper, Judge Reinhold, Amy Wright, Keith Carradine; *Dir:* Jack Fisk. **VHS, Beta, LV $89.98** MGM *♫♫½*

Daddy's Gone A-Hunting It's an eye for an eye, a baby for a baby in this well-done psychological thriller. A happily married woman is stalked by her deranged ex-boyfriend, whose baby she aborted years before. Now he wants the life of her child as just compensation for the loss of his.
1969 (PG) 108m/C Carol White, Paul Burke, Scott Hylands, Rachel Ames, Mala Powers; *Dir:* Mark Robson. **VHS $19.98** WAR *♫♫♫*

Daffy! The corny quackster meets Duck Twacy and Little Wed Widing Hood.
1988 60m/C **VHS $14.95** MGM

Daffy Duck Cartoon Festival: Ain't That Ducky There's wacky down-filled humor galore in this compilation tape. The cartoons include "Ain't That Ducky," "Daffy Duck Slept Here," "Hollywood Daffy," "Conrad the Sailor" and "The Wise Cracking Duck." Made between 1942 and 1947.

194? 35m/C Dir: Chuck Jones, Friz Freleng, Robert McKimson, Bob Clampett. **VHS, Beta** $14.95 *MGM*

Daffy Duck Cartoon Festival: Daffy Duck & the Dinosaur
Five classic Daffy shorts, including ''Plane Daffy,'' ''Nasty Quacks'' and the title cartoon are featured.
1945 36m/C Dir: Chuck Jones, Chuck Jones, Robert McKimson, Bob Clampett. **VHS, Beta** $14.95 *MGM, MRV*

Daffy Duck & Company
Seven Daffy classics, including a few earlier entries: ''To Duck or Not to Duck,'' ''Conrad the Sailor,'' ''Daffy Duck and the Dinosaur,'' ''The Dover Boys,'' ''House Hunting Mice'' and ''Hep Cat.''
1950 60m/C Dir: Chuck Jones, Robert McKimson, Friz Freleng, Bob Clampett; **Voices:** Jack Benny, Peter Lorre. **VHS, Beta** $14.95 *MGM*

Daffy Duck: The Nuttiness Continues...
Daffy Duck runs amuck in this collection of cartoon classics that include ''Beanstalk Bunny,'' ''Deduce You Say,'' ''Rabbit Fire,'' ''Dripalong Daffy,'' ''Porky's Duck Hunt,'' ''Duck Amuck,'' ''The Daffy Duck,'' and ''The Scarlet Pumpernickel.''
1956 59m/C Dir: Chuck Jones, Bob Clampett, Tex Avery. **VHS, Beta** $12.95 *WAR*

Daffy Duck's Madcap Mania
Six of the billed-comedian's best.
1989 45m/C Dir: Chuck Jones; **W/Dir:** Robert McKimson. **VHS** $14.95 *WAR*

Daffy Duck's Movie: Fantastic Island
A compilation of classic Warner Brothers cartoons, starring Daffy Duck, Speedy Gonzales, Bugs Bunny, Porky Pig, Sylvester, Tweety, Pepe Le Pew, Pirate Sam, Granny, Foghorn Leghorn and the Tasmanian Devil.
1983 (G) 78m/C Dir: Friz Freleng. **VHS, Beta, LV** $19.98 *WAR, FCT*

Daffy Duck's Quackbusters
Daffy, with help from pals Bugs and Porky, sets up his own ''ghostbusting'' service. Good compilation of old classics such as ''Night of the Living Duck,'' plus a new feature, ''The Duxcorcist.'' Video is also available in Spanish.
1989 (G) 79m/C Daffy Duck, Bugs Bunny, Porky Pig; **Dir:** Greg Ford, Terry Lennon; **Voices:** Mel Blanc. **VHS, Beta, LV** $19.98 *WAR, FCT, APD*

Dagger Eyes
A political assassin at work is inadvertently captured on film by a photographer. The assassin's mob employers will go to any lengths to destroy the film. Say cheeeeeese!
198? 84m/C Carol Bouquet, Philip Coccioletti, John Steiner; **Dir:** Carlo Vanzina. **VHS, Beta** $79.95 *VHV* ✂½

Daggers 8
The Kung Fu masters are back, battling their opponents with deadly daggers.
1980 90m/C VHS, Beta $39.95 *TWE* ✂

Dagora, the Space Monster
A giant, slimy, pulsating mass from space lands on Earth and begins eating everything in sight. Scientists join together in a massive effort to destroy the creature. Believe it or not, it's sillier than most of the Japanese sci-fi genre.
1965 80m/C Yosuke Natsuki, Yoko Fujiyama. **VHS, Beta** $29.95 *VYY Woof!*

The Dain Curse
In 1928, private eye Hamilton Nash must recover stolen diamonds, solve a millionaire's suicide, avoid being murdered, and end a family curse. Based on the novel by Dashiell Hammett. Made-for-television as a 312-minute miniseries.
1978 118m/C James Coburn, Jason Miller, Jean Simmons, Beatrice Straight, Hector Elizondo; **Dir:** E.W. Swackhamer. **VHS, Beta** $69.98 *SUE, MTX* ✂✂

Daisy Miller
Shepherd portrays the title character in this adaptation of the Henry James novella about a naive young American woman travelling through Europe and getting a taste of the Continent during the late 19th century. Though it is intelligently written and has a good supporting cast, the film seems strangely flat, due in large part to Shepherd's hollow performance.
1974 (G) 93m/C Cybill Shepherd, Eileen Brennan, Cloris Leachman, Mildred Natwick; **Dir:** Peter Bogdanovich. National Board of Review Awards '74: 10 Best Films of the Year. **VHS, Beta** $59.95 *PAR* ✂✂½

Dakota
A man battles the railroad in a land dispute in this sub-par Wayne western saga. In the meantime, he finds time to fall in love.
1945 82m/B John Wayne, Vera Ralston, Walter Brennan, Ward Bond, Ona Munson; **Dir:** Joseph Kane. **VHS, LV** $14.98 *REP, PIA* ✂½

Dakota
A filming of the 50th Anniversary DC-3 Fly-In, featuring scads of zooming C-47's.
197? 60m/C VHS, Beta $14.95 *VCD* ✂½

Dakota
A troubled half-breed teenager works for a rancher, romances his daughter and befriends his crippled 12-year-old son. A well-meaning, but predictable drama.
1988 (PG) 96m/C Lou Diamond Phillips, Dee Dee Norton, Eli Cummins, Herta Ware; **Dir:** Fred Holmes. **VHS, Beta, LV** $89.99 *HBO* ✂✂

Dakota Incident
Decent western about stagecoach passengers travelling through Dakota Territory who must defend themselves against an Indian attack.
1956 88m/C Dale Robertson, Ward Bond, Linda Darnell, John Lund; **Dir:** Lewis R. Foster. **VHS** $19.98 *REP* ✂✂

Daleks Invasion Earth 2150 A.D.
A sequel to ''Dr. Who and the Daleks,'' wherein the popular British character endeavors to save the Earth from a robotic threat. Also known as ''Invasion Earth 2150 A.D.''
1966 81m/C *GB* Peter Cushing, Andrew Keir, Jill Curzon, Ray Brooks; **Dir:** Gordon Flemyng. **VHS, Beta** $79.99 *HBO* ✂✂½

Dam Busters
This well-done and exciting film details the efforts of British scientists trying to devise a method of destroying a strategic dam in Nazi Germany during World War II. Definitely one of the better ones in the war movie genre.
1955 119m/B *GB* Michael Redgrave, Richard Todd, Ursula Jeans, Basil Sydney; **Dir:** Michael Anderson Sr. **VHS, Beta** $59.99 *HBO* ✂✂✂

Dames
A millionaire with fanatically religious beliefs tries to stop the opening of a Broadway show. Songs from this movie include ''I Only Have Eyes for You'' and ''When You Were a Smile on Mother's Lips and a Twinkle in Your Daddy's Eye.'' In the last of the grand budget-breaking spectacles before the ''production code'' came into being, distinguished choreographer Busby Berkeley took his imagination to the limit creating this 90 minute extravaganza.

1934 95m/B Dick Powell, Joan Blondell, Ruby Keeler, ZaSu Pitts, Guy Kibbee; **Dir:** Ray Enright. **VHS, Beta, LV** $29.95 *FOX* ✂✂✂

Damien: Omen 2
Sequel to ''The Omen'' about Damien, a young boy possessed with mysterious demonic powers, who kills those people who anger him. Followed by ''The Final Conflict.''
1978 (R) 110m/C William Holden, Lee Grant, Lew Ayres, Robert Foxworth, Sylvia Sidney, Lance Henriksen; **Dir:** Don Taylor. **VHS, Beta, LV** $19.98 *FOX* ✂✂

Damien: The Leper Priest
True story of a Roman Catholic priest who doomed himself by voluntarily serving on a leper colony one hundred years ago. Somewhat disappointing considering the dramatic material. Originally a vehicle for David Janssen, who died prior to filming. Also known as ''Father Damien: The Leper Priest.''
1980 100m/C Ken Howard, Mike Farrell, Wilfrid Hyde-White, William Daniels, David Ogden Stiers; **Dir:** Steven Gethers. **VHS, Beta** $59.95 *IVE* ✂✂

Damn the Defiant
Adventure abounds when Guinness, as captain of the HMS Defiant during the Napoleonic wars, finds himself up against not only the French but his cruel second-in-command (Bogarde) and a mutinous crew as well. In the end, both a fleet-wide mutiny and a French invasion of England are avoided. Much attention is paid to period detail in this well-crafted film.
1962 101m/C *GB* Alec Guinness, Dirk Bogarde, Maurice Denham, Anthony Quayle; **Dir:** Lewis Gilbert. **VHS, Beta** $14.95 *COL, FCT* ✂✂✂

Damn Yankees
Musical feature adapted from the Broadway hit. A baseball fan frustrated by his team's lack of success makes a pact with the devil to become the team's new star. Verdon is dynamite as the devil's accomplice. Great Bob Fosse choreography. Songs include ''What Ever Lola Wants'' and ''(You Gotta Have) Heart.''
1958 110m/C Gwen Verdon, Ray Walston, Tab Hunter, Jean Stapleton, Russ Brown; **Dir:** George Abbott, Stanley Donen. **VHS, Beta, LV** $19.98 *WAR, FCT* ✂✂✂

Damnation Alley
A warrior, an artist, and a biker set off in search of civilization after Armageddon. A so-so attempt to revive the post nuclear holocaust genre which was successfully accomplished by the ''Mad Max'' series.
1977 (PG) 87m/C George Peppard, Jan-Michael Vincent, Paul Winfield, Dominique Sanda, Jackie Earle Haley; **Dir:** Jack Smight. **VHS, Beta** $59.98 *FOX* ✂½

The Damned
Depicts the Nazi takeover of a German industrialist family and their descent into greed, lust, and madness. Comes in dubbed or subtitled formats.
1969 (R) 146m/C *IT GE* Dirk Bogarde, Ingrid Thulin, Helmut Griem, Charlotte Rampling, Helmut Berger; **Dir:** Luchino Visconti. National Board of Review Awards '69: 5 Best Foreign Films of the Year. **VHS, Beta** $59.95 *WAR, APD, TAM* ✂✂½

The Damned: Live 79
Some of England's original punkers get destructive as they perform songs such as ''No Fun,'' ''Born to Kill,'' ''Pretty Vacant,'' ''New Rose,'' ''Jet Boy, Jet Girl'' and more live from San Francisco.
1979 40m/B Performed by: The Damned. **VHS, Beta** $39.98 *MVD, TGV* ✂✂½

D

Damned River Four friends take a guided tour down Zimbabwe's Zambezi River, only to find midway down that their guide is a murderous psycho. Survival becomes a priority.
1989 (R) 93m/C Stephen Shellan, Lisa Aliff, John Terlesky, Marc Poppel, Bradford Bancroft; **Dir:** Michael Schroeder. **VHS, Beta $89.98** FOX ♫♫½

Damsel in Distress Astaire falls for an upper-class British girl, whose family wants her to have nothing to do with him. George and Ira Gershwin's memorable songs include "A Foggy Day," "Nice Work If You Can Get It," "Stiff Upper Lip" and "Put Me to the Test."
1937 101m/B Fred Astaire, Joan Fontaine, George Burns, Gracie Allen, Ray Noble, Montagu Love, Reginald Gardiner; **Dir:** George Stevens. **VHS, Beta, LV $14.98** CCB, MED ♫♫♫

Dan Candy's Law A Canadian mountie becomes a driven hunter, and then desperate prey, when he tries to track down the Indian who killed his partner. Also called "Allen Thunder."
1973 90m/C CA Donald Sutherland, Gordon Tootoosis, Chief Dan George, Kevin McCarthy; **Dir:** Claude Fournier. **VHS, Beta $79.95** PSM, SIM ♫½

Dance Two ballet dancers struggle to find love amidst the backstage treacheries and demanding schedules of their profession.
1990 90m/C John Revall, Ellen Troy, Carlton Wilborn, Charlene Campbell; **Dir:** Robin Murray. **VHS, Beta** AIP ♫½

Dance of the Damned A case where low-budget isn't synonymous with bomb. Fascinating noir-ish plot concerning a vampire who wants to learn more about the life of his next victim, a deep-thinking stripper, who has lost the will to live it. Surprisingly well-done and acted; above-par for this genre.
1988 (R) 83m/C Cyril O'Reilly, Starr Andreeff; **Dir:** Katt Shea Rubin. **VHS, Beta, LV $79.95** VIR ♫♫

Dance of Death Just before his death, Karloff agreed to appear in footage for four Mexican cheapies that were practically thrown together. If they had mixed the footage instead of matched it, the flicks couldn't be any worse. This one concerns a lunatic toy-maker whose toys kill and maim his heirs. Also known as "House of Evil" and "Macabre Serenade."
1969 89m/C MX Boris Karloff, Julissa, Andres Garcia, Jack Hill; **Dir:** Juan Ibanez. **VHS, Beta $57.95** UNI Woof!

Dance of Death A girl learns kung fu and kills many warriors by kicking them in the ribs and head.
1984 81m/C Mao Ying, Shih Tien, Shiao Bou-Lo, Chien Pay, Chia Kai, Noh Tikki. **VHS, Beta $39.95** UNI ♫

Dance with Death When strippers turn up brutally murdered, a young journalist goes undercover to solve the case.
1991 (R) 90m/C Maxwell Caulfield, Barbara Alyn Jones, Martin Mull, Drew Snyder, Catya Sasoon; **Dir:** Charles Philip Moore. **VHS $89.98** HBO ♫

Dance or Die Just goes to show that in Las Vegas you can't have your dance and drugs too, because if the Mob doesn't get you, the Feds will. Made for video (though maybe it shouldn't have been made at all). In HiFi Stereo.
1987 90m/C Ray Kieffer, Rebecca Barrington; **Dir:** Richard W. Munchkin. **VHS, Beta** CLV ♫

Dance Fools Dance Fast-paced drama has Crawford and Bakewell as a pair of spoiled rich kids who are face poverty when the stock market crashes. He meets up with Gable, who's producing liquor illegally, while she gets a job at a newspaper. When Gable arranges something akin to the St. Valentine's Day Massacre, Bakewell's investigative reporting of the situation produces fatal results. The Hays Office had a problem with Crawford and friends appearing in their underwear. Cast notes: Gable was just starting out at MGM, which is why he was billed sixth; the William Holden here is not THE William Holden, and Edwards went on to provide the voice of Jiminy Cricket in "Pinocchio."
1931 82m/B Joan Crawford, Lester Vail, Cliff Edwards, William Bakewell, William Holden, Clark Gable; **Dir:** Harry Beaumont. **VHS, Beta $19.98** MGM, FCT ♫♫♫

Dance, Girl, Dance The private lives and romances of a wartime nightclub dance troupe. Standard Hollywood potpourri of dance, drama, comedy.
1940 88m/B Maureen O'Hara, Louis Hayward, Lucille Ball, Virginia Field, Ralph Bellamy, Maria Ouspenskaya, Mary Carlisle, Katherine Alexander, Edward Brophy; **Dir:** Dorothy Arzner. **VHS, Beta, LV $19.98** RKO ♫♫

Dance Hall A Pennsylvania dance hall owner is propelled into romantic confusion with the introduction of a sultry blond dancer into his club.
1941 73m/B Cesar Romero, Carole Landis, William Henry, June Storey; **Dir:** Irving Pichel. **VHS, Beta $19.95** NOS, KRT, TIM ♫½

Dance Hall Racket A sleazy, stilted expose of dance hall vice featuring a brief appearance by Bruce and his wife, Harlow.
1958 60m/B Lenny Bruce, Honey Harlow; **Dir:** Phil Tucker. **VHS, Beta $19.98** DVT, JLT ♫♫

Dance Macabre Englund takes on a dual role as a man with a split personality—one of which is deadly. The "good" guy is the protector of a group of dancers studying in Russia but is the "bad" personality their killer?
1991 (R) ?m/C Robert Englund, Michelle Zeitlin, Marianna Moen, Julene Renee, Alexander Sergeyev; **W/Dir:** Greydon Clark. **VHS, LV $89.95** COL ♫

Dance with Me, Henry When Lou becomes involved with Bud's gambling debts and the local district attorney turns up dead, Lou's not only wanted by the law, but by the mafia as well. This was the great comedy duo's last picture together and it's clear the pair are not happy about working with each other, even on this mediocre effort.
1956 80m/B Bud Abbott, Lou Costello, Gigi Perreau, Rusty Hamer, Mary Wickes, Ted de Corsia; **Dir:** Charles T. Barton. **VHS $19.98** MGM ♫♫½

Dance with a Stranger The engrossing and factual story of Ruth Ellis who gained notoriety as the last woman hanged in Great Britain. This emotional and sometimes violent film mirrors the sensationalism produced in the 1950s. The film follows Ellis's pre-trial life and her perpetual struggle to maintain her independence. Newcomer Richardson portrays the Marilyn Monroe look-alike barmaid.
1985 (R) 101m/C GB Miranda Richardson, Rupert Everett, Ian Holm, Joanne Whalley-Kilmer; **Dir:** Mike Newell. Cannes Film Festival '85: Best Film. **VHS, Beta, LV $79.95** VES, LIV ♫♫♫½

Dancers During the filming of the ballet "Giselle" in Italy, a famous, almost-over-the-hill dancer (Baryshnikov, ten years after his role in "The Turning Point") coaches a young, inexperienced starlet. He hopes to revitalize his life and his dancing. Features dancers from the Baryshnikov-led American Ballet Theatre.
1987 (PG) 99m/C Mikhail Baryshnikov, Leslie Browne, Julie Kent, Mariangela Melato, Alessandra Ferri, Lynn Seymour, Victor Barbee, Tommy Rall; **Dir:** Herbert Ross. **VHS, Beta, LV $19.98** CTH, WAR, FCT ♫½

Dances with Wolves The story of a U.S. Army soldier, circa 1870, whose heroism in battle allows him his pick of posts. His choice, to see the West before it disappears, changes his life. He meets, understands and eventually becomes a member of a Lakota Sioux tribe in the Dakotas. Costner's first directorial attempt proves him a talent of vision and intelligence. This sometimes too objective movie lacks a sense of definitive character, undermining its gorgeous scenery and interesting perspective on the plight of Native Americans. Lovely music and epic proportions.
1990 (PG-13) 181m/C Kevin Costner, Mary McDonnell, Graham Greene, Rodney Grant, Floyd Red Crow Westerman, Tantoo Cardinal, Robert Pastorelli, Charles Rocket, Maury Chaykin, Jimmy Herman, Nathan Lee Chasing His Horse; **Dir:** Kevin Costner. Academy Awards '90: Best Adapted Screenplay, Best Cinematography, Best Director (Costner), Best Film Editing, Best Picture, Best Sound, Best Musical Score; Golden Globe Awards '90: Best Director (Costner), Best Film—Drama, Best Screenplay; National Board of Review Awards '90: Best Director (Costner), Best Picture. **VHS, Beta, LV $49.95** ORI, FCT, IME ♫♫♫½

Dancing in the Dark An interesting but slow-moving Canadian film about the perfect housewife who learns that after 20 years of devotion to him, her spouse has been unfaithful. Realizing how her life has been wasted she murders her husband and, in the end, suffers a mental breakdown.
1986 (PG-13) 93m/C CA Martha Henry, Neil Munroe, Rosemary Dunsmore; **Dir:** Leon Marr. Genie Awards '87: Best Actress (Henry), Best Art Direction/Set Decoration, Best Screenplay. **VHS, Beta $19.95** STE, NWV ♫♫½

Dancing Lady Rarely seen film is MGM's answer to "42nd Street," with Crawford as a small-time hoofer trying to break into Broadway. The screen debuts of Astaire and Eddy; the hodgepodge of musical numbers include "Everything I Have Is Yours," "Heigh-Ho! The Gang's All Here" and "Rhythm of the Day."
1933 93m/B Clark Gable, Joan Crawford, Fred Astaire, Franchot Tone, Nelson Eddy, Ted Healy, Moe Howard, Shemp Howard, Larry Fine; **Dir:** Robert Z. Leonard. **VHS, Beta $19.98** MGM, TTC, FCT ♫♫½

Dancing Man A gigolo romances both mother and daughter and winds up implicated in murder as a result. Dismissable.
1933 65m/B Judith Allen, Reginald Denny, Natalie Moorhead; **Dir:** Albert Ray. **VHS, Beta, 8mm $29.95** VYY ♫½

Dancing Mothers A fast-living woman becomes involved with her mother's roguish boyfriend in this silent film with accompanying musical score.
1926 85m/B Clara Bow, Alice Joyce; **Dir:** Herbert Brenon. **VHS, Beta $19.95** NOS, MRV, DVT ♫♫½

Dancing Pirate A Boston dance instructor is kidnapped by pirates and jumps ship in Mexico, where he romances a mayor's daughter. The silly story boasts some Rodgers and Hart tunes, plus the early use of Technicolor, exploited here for all it was worth and then some.
1936 83m/C Charles Collins, Frank Morgan, Steffi Duna, Luis Alberni, Victor Varconi, Jack LaRue; *Dir:* Lloyd Corrigan. **VHS $19.95** *NOS, LOO, TIM ⅋⅋*

Dancing Princesses When five naughty princesses wear the soles of their slippers out every night, their father the King must foot the bill. He soon tires of this expense and offers a reward to the person who can discover how this happens. A handsome prince becomes invisible to follow the lovely ladies and discover their secret. From the "Faerie Tale Theatre" series.
1984 60m/C Lesley Ann Warren, Peter Weller, Sachi Parker, Roy Dotrice; *Dir:* Peter Medak. **VHS, Beta $14.98** *KUI, FOX, FCT ⅋*

Dandelions This German soft-core flick stars Hauer in an early role as a man jilted by his wife. To make himself feel better he decides to explore the sleazier side of life. Meaningless film evokes no sympathy but some may appreciate the titillation. Dubbed.
1974 92m/C *GE* Rutger Hauer, Dagmar Lassander; *Dir:* Adrian Hoven. **VHS, Beta $79.95** *WES ⅋*

Dandy in Aspic A double-agent is assigned to kill himself in this hard-to-follow British spy drama. The film is based on Derek Marlowe's novel. Mann died midway through shooting and Harvey finished direction.
1968 (R) 107m/C *GB* Laurence Harvey, Tom Courtenay, Lionel Stander, Mia Farrow, Harry Andrews, Peter Cook, Per Oscarsson; *Dir:* Anthony Mann. **VHS, Beta, LV $69.95** *COL ⅋½*

Dangaio Two samples of Japanese animation, with English subtitles, for mature viewers. "Dangaio" follows the adventures of four youngsters gifted with ESP who search to learn their real identities. In "Gunbuster," the daughter of a famous space captain killed in battle fights to join the Earth Space Force and defend Earth against extra-terrestrial invaders.
198? 45m/C *Dir:* Toshihiro Hirano, Hideaki Anno. **VHS $34.95** *CPM, FCT ⅋⅋½*

Danger Ahead! Comedy short featuring an exciting train chase.
1926 15m/B VHS, Beta $49.95 *IUF ⅋⅋½*

Danger Ahead Newill stars again as Renfrew of the Mounties—this time he tries to rid the north woods of the baddies who are trying to take over.
1940 57m/B Jim Newill, Dorothea Kent, Guy Usher, Dave O'Brien, Bob Terry; *Dir:* Ralph Staub. **VHS $19.95** *DVT ⅋*

Danger on the Air A radio program sponsor—a much-despised misogynist—is murdered, and the sound engineer attempts to solve the mystery. Features Cobb in one of his earliest roles; Garrett also directed "Lady in the Morgue" that year.
1938 70m/B Donald Woods, Nan Grey, Berton Churchill, Jed Prouty, William Lundigan, Skeets Gallagher, Edward Van Sloan, George Meeker, Lee J. Cobb, Johnny Arthur, Linda Hayes, Louise Stanley; *Dir:* Otis Garrett. **VHS, Beta $16.95** *SNC ⅋⅋*

Danger: Diabolik A superthief called Diabolik (Law) continuously evades the law while performing his criminal antics.

1968 (PG-13) 99m/C *IT* John Phillip Law, Marisa Mell, Michel Piccoli, Terry Thomas; *Dir:* Mario Bava. **VHS** *PAR ⅋*

Danger Lights Depicts the railroads and the railroad men's dedication to each other and their trains.
1930 73m/B Jean Arthur, Louis Wolheim, Robert Armstrong; *Dir:* George B. Seitz. **VHS, Beta $19.95** *PEN, MRV, NOS ⅋½*

Danger Man Great spy stories that inspired the "Secret Agent" series.
1961 120m/B Patrick McGoohan. **VHS $79.80** *MPI*

Danger in the Skies An alcoholic airline pilot tries to straighten out his life when he nearly loses everything while drinking on the job. Previously titled "The Pilot." Made for cable television.
1979 (PG) 99m/C Cliff Robertson, Diane Baker, Dana Andrews, Gordon MacRae, Milo O'Shea, Frank Converse, Edward Binns; *Dir:* Cliff Robertson. **VHS, Beta $79.95** *UHV ⅋½*

Danger Trails Rehashed western programmer. Features the hero who must save the rancher's honor as well as the ranch, while fighting detestable hombres with the traditional fare as an arsenal.
1935 55m/B Guinn Williams, Marjorie Gordon, Wally Wales, John Elliott, Ace Cain, Edmund Cobb; *Dir:* Bob Hill. **VHS $19.95** *NOS, DVT ⅋½*

Danger Zone A man is hired to bid for a saxophone at an auction and then has the instrument stolen from him. He discovers it contained stolen jewelry and tries to recover it.
1951 56m/B Hugh Beaumont, Tom Neal, Richard Travis, Virginia Dale. **VHS, Beta, LV $19.98** *WGE ⅋⅋*

Danger Zone A low-budget, low-brow flick about an all-female rock band whose bus breaks down in the middle of the desert, leaving the girls to defend themselves against a merciless motorcycle gang. You'll wish this one was a mirage.
1987 (R) 90m/C Robert Canada, Jason Williams, Kriss Braxton, Dana Dowell, Jamie Ferreira; *Dir:* Henry Vernon. **VHS, Beta $19.98** *CHA Woof!*

Danger Zone 2 A pernicious biker is released from prison on a technicality and decides to pay a social call to the undercover cop who set him up.
1989 (R) 95m/C Jason Williams, Robert Random; *Dir:* Geoffrey G. Bowers. **VHS $79.98** *MCG ⅋*

Danger Zone 3: Steel Horse War The "Danger Zone" dude is back and this time he's taken on a bevy of biker-types in a most gruesome battle to make our deserts safe.
1990 91m/C Jason Williams, Robert Random, Barne Suboski, Juanita Ranney, Rusty Cooper, Giles Ashford; *Dir:* Douglas Bronco. **VHS, Beta $79.98** *PMR ⅋*

Dangerous Davis won her first Oscar for her rather overdone portrayal of a has-been alcoholic actress reformed by a smitten architect who recognizes her from her days as a star.
1935 78m/B Bette Davis, Franchot Tone, Margaret Lindsay, Alison Skipworth, John Eldridge, Dick Foran; *Dir:* Alfred E. Green. Academy Awards '35: Best Actress (Davis). **VHS, Beta, LV $19.98** *MGM, BTV ⅋⅋½*

Dangerous Charter A group of fishermen discover a deserted yacht and man its helm. But their good fortune takes a decided turn for the worst as they become caught in mob-propelled danger.
1962 76m/C Chris Warfield, Sally Fraser, Richard Foote, Peter Foster, Wright King; *Dir:* Robert Gottschalk. **VHS, Beta $19.95** *NOS, ACA, TIM ⅋½*

Dangerous Company Murder, sex, infidelity, and expensive living are highlighted in this made-for-video production. What, you want something more in a film?
1988 (R) 98m/C Cliff DeYoung, Tracy Scoggins, Chris Mulkey; *Dir:* Rubin Preuss. **VHS, Beta $29.95** *MCG ⅋*

Dangerous Curves Two friends are assigned to deliver a new Porsche to a billionaire's daughter - one of them talks the other into taking a little detour, and the trouble begins. This one went straight to video.
1988 (PG) 93m/C Robert Stack, Tate Donovan, Danielle von Zernaeck, Robert Klein, Elizabeth Ashley, Leslie Nielsen; *Dir:* David Lewis. **VHS, Beta $79.98** *VES, LIV ⅋½*

Dangerous Game A computer hacker leads his friends through a department store's security system, only to find they can't get out until morning. They soon discover they are not alone when one of the group turns up dead.
1990 (R) 102m/C Miles Buchanan, Sandy Lillingston, Kathryn Walker, John Polson; *Dir:* Stephen Hopkins. **VHS, Beta $29.95** *ACD, QCM ⅋½*

Dangerous Holiday A young violin prodigy would rather be just "one of the boys" than be forced to practice all the time. He decides to run away but has everyone in an uproar thinking he's been kidnapped.
1937 54m/B Hedda Hopper, Guinn Williams, Jack La Rue, Franklin Pangborn, Grady Sutton. **VHS, Beta $24.95** *VYY ⅋*

Dangerous Hours An anti-communist propaganda film with Hughes as the innocent young college boy who gets duped by the evil Marxists until he realizes the truth of his all-American upbringing. Silent with original organ score.
1919 88m/B Lloyd Hughes; *Dir:* Fred Niblo. **VHS, Beta, 8mm $24.95** *VYY, FCT ⅋⅋*

Dangerous Liaisons A couple take each other to the brink of destruction via their insatiable desire for extra-marital affairs. Vadim attempted to repeat his success with Brigitte Bardot by featuring wife Annette as a new sex goddess, but lightning didn't strike twice. Originally titled "Les Liasons Dangereuses" and also known as "Dangerous Love Affairs."
1960 111m/B *FR IT* Gerard Philippe, Jeanne Moreau, Jeanne Valeri, Annette Vadim, Simone Renant, Jean-Louis Trintignant, Nikolas Vogel; *Dir:* Roger Vadim. **VHS $59.95** *INT, APD, TAM ⅋⅋*

Dangerous Liaisons Stylish and absorbing, this adaptation of the Laclos novel and the Christopher Hampton play centers around the relationship of two decadent members of 18th-century French nobility. The two, Close and Malkovich, spend their time testing and manipulating the loves of others. They find love often has a will of its own. Possibly the best of the several versions available and heavily Oscar-nominated. Interesting to comparison-view with director Milos Forman's version of this story, 1989's "Valmont."

1989 (R) 120m/C John Malkovich, Glenn Close, Michelle Pfeiffer, Uma Thurman, Keanu Reeves, Swoosie Kurtz, Mildred Natwick, Peter Capaldi; *Dir:* Stephen Frears. British Academy Awards '88: Best Supporting Actress (Pfeiffer). **VHS, Beta, LV, 8mm** $19.98 *WAR, FCT* 🎬🎬🎬

Dangerous Life Explosive drama brings to the screen the story of the Philippine uprising. It shows the horrible events that lead to the fall of Marcos and permitted the rise of Corazon Aquino.
1989 163m/C Gary Busey. **VHS, Beta, LV** $79.95 *JTC* 🎬🎬

Dangerous Love Advice to the lovelorn: Avoid this one. Female members of a video dating service are being filmed in a death scene—their own.
1987 87m/C Lawrence Monoson, Brenda Bakke, Peter Marc, Elliott Gould, Anthony Geary; *Dir:* Marty Ollstein. **VHS, Beta** $79.95 *MED* Woof!

Dangerous Mission A young woman witnesses a mob murder in New York and flees to the Midwest, pursued by killers and the police. Good cast couldn't do much with this one.
1954 75m/B Victor Mature, Piper Laurie, Vincent Price, William Bendix; *Dir:* Louis King. **VHS, Beta** $19.95 *MED* 🎬🎬½

Dangerous Moonlight Polish concert pianist becomes a bomber pilot for the British in World War II, though his wife wants him to stay at the piano. Great battle and music sequences; the piece ''Warsaw Concerto'' became a soundtrack hit. Also known as ''Suicide Squadron.''
1941 97m/B *GB* Anton Walbrook, Sally Gray, Derrick DeMarney, Cecil Parker, Percy Parsons, Kenneth Kent, Guy Middleton, John Laurie, Frederick Valk; *Dir:* Brian Desmond Hurst. **VHS, Beta** $19.98 *TTC* 🎬🎬🎬

Dangerous Moves A drama built around the World Chess championship competition between a renowned Russian master and a young, rebellious dissident. The chess game serves as both metaphor and background for the social and political tensions it produces. With English subtitles.
1984 (PG) 96m/C *SI* Liv Ullman, Michel Piccoli, Leslie Caron, Alexandre Arbatt; *Dir:* Richard Dembo. Academy Awards '84: Best Foreign Language Film. **VHS, Beta, LV** $59.95 *LHV, IME* 🎬🎬🎬

Dangerous Obsession A woman kidnaps the doctor she blames for her boyfriend's death and instead of exacting revenge, she makes him her personal sex slave. In the case of this flick, it's really a fate worse than death.
1988 81m/C *IT* Corinne Clery, Brett Halsey; *Dir:* Lucio Fulci. **VHS** $79.95 *AIP* 🎬

Dangerous Orphans Not completely awful tale of three brothers, orphaned as boys when their father is murdered before their very eyes, and the vendetta they have against the killer.
198? (R) 90m/C *NZ* Peter Stevens, Peter Bland, Ian Mune, Ross Girven, Jennifer Ward-Lealand; *Dir:* John Laing. **VHS, Beta** $29.95 *ACA* 🎬½

Dangerous Passage A ne'er-do-well inherits 200 grand but gets in trouble before he can collect it. He ends up falling in with assorted misfits aboard a tramp steamer.
1944 60m/B Robert Lowery, Phyllis Brooks, Jack La Rue; *Dir:* William Berke. **VHS, Beta, 8mm** $24.95 *VYY, RXM* 🎬🎬

Dangerous Pursuit A woman discovers that a man she slept with years ago is an assassin. Knowing too much, she finds herself next on his hit list. Decent psycho-thriller that was never released theatrically.
1989 95m/C Gregory Harrison, Alexandra Powers, Scott Valentine, Brian Wimmer, Elena Stiteler; *Dir:* Sandor Stern. **VHS, Beta, LV** $89.99 *PAR* 🎬🎬

Dangerous Relations Not everyone can be a hero, but with a bit of comedy and suspense you can make a half-way decent film.
1973 90m/C Brooke Bundy, Eileen Weston, Kenneth Varnum, Sam Bell. **VHS, Beta** *TVS* 🎬½

Dangerous Summer In Australia, an American businessman building a resort is the victim of elaborate arson/murder insurance schemes.
1982 100m/C *AU* Tom Skerritt, James Mason, Ian Gilmour, Wendy Hughes; *Dir:* Quentin Masters. **VHS, Beta** $24.95 *VIR, VCL* 🎬🎬

Dangerous When Wet A typical Williams water-musical. A farm girl dreams of fame by swimming the English Channel. One famous number pairs Williams with cartoon characters Tom & Jerry in an underwater frolic.
1953 96m/C Esther Williams, Fernando Lamas, Charlotte Greenwood, William Demarest, Jack Carson; *Dir:* Charles Walters. **VHS, Beta** $19.98 *MGM* 🎬🎬

Dangerously Close A group of anti-crime high school students organize a hall monitoring gang that becomes a group of neo-fascist disciplinarian elite.
1986 (R) 96m/C John Stockwell, Carey Lowell, J. Eddie Peck; *Dir:* Albert Pyun. **VHS, Beta** $9.98 *MED* 🎬½

Daniel The children of a couple who were executed for espionage (patterned after the Rosenburgs) struggle with their past in the dissident 1960s. So-so adaptation of E.L. Doctorow's ''The Book of Daniel.''
1983 (R) 130m/C Timothy Hutton, Amanda Plummer, Mandy Patinkin, Lindsay Crouse, Ed Asner, Ellen Barkin; *Dir:* Sidney Lumet. **VHS, Beta** $59.95 *PAR* 🎬🎬½

Daniel Boone Daniel Boone guides a party of settlers from North Carolina to the fertile valleys of Kentucky, facing Indians, food shortages and bad weather along the way. O'Brien turns in a fine performance as the early American frontier hero.
1934 75m/B George O'Brien, Heather Angel, John Carradine; *Dir:* David Howard. **VHS, Beta** $19.95 *NOS, MRV, QNE* 🎬🎬

Daniel Boone: Trail Blazer Low-budget, though surprisingly well-acted rendition of frontiersman's heroics which was filmed in Mexico.
1956 75m/B Bruce Bennett, Lon Chaney Jr., Faron Young, Damian O'Flynn, Fred Kohler Jr., Claudio Brook, Kem Dibbs; *Dir:* Ismael Rodriguez, Albert C. Gannaway. **VHS** $19.95 *NOS, MRV, REP* 🎬🎬½

Danny Charming story of a lonely young girl who receives an injured horse that was no longer fit for the spoiled daughter of the rich owners.
1979 (G) 90m/C Rebecca Page, Janet Zarish, Barbara Jean Earhardt, Gloria Maddox, George Luce; *Dir:* Gene Feldman. **VHS, Beta** $59.95 *MON, WOM, HHE* 🎬🎬½

Danny Boy Veteran dog, returning from the war, has difficulty adjusting to life back home. Things get worse for him and his young master when Danny Boy is assumed to be shell-shocked and dangerous.
1946 67m/B **VHS, Beta** $29.95 *VYY* 🎬½

Danny Boy Takes place in Ireland where a young saxaphonist witnesses a murder and, in an effort to understand it, sets out to find the killer. Thought-provoking film is meant to highlight the continuing struggles in Ireland.
1984 (R) 92m/C *IR* Stephen Rea, Veronica Quilligan, Honor Heffernan, Marie Kean, Donal McCann, Ray McAnally; *Dir:* Neil Jordan. **VHS, Beta** $59.95 *COL* 🎬🎬🎬

Dante's Inferno Abandon all hope you who watch this film. A one-time friend sends his ruthless capitalist nemesis a copy of Dante's ''Inferno,'' and the man reads the book and dreams he's gone to hell (would that be the dress circle?) A silent film that was controversial in its day: the body-stocking actors were believed to be nude (which would have made it the first nudie comedy).
1924 54m/B Lawson Butt, Howard Gaye, Ralph Lewis, Pauline Starke, Josef Swickard; *Dir:* Henry Otto. **VHS** $10.95 *CNM* 🎬🎬🎬

Danton A sweeping account of the reign of terror following the French Revolution. Focuses on the title character (wonderfully portrayed by Depardieu) and is directed with searching parallels to modern-day Poland by that country's premier filmmaker, Andrzej Wajda. Well-done period sets round out a memorable film. In French with English subtitles.
1982 (PG) 136m/C *PL FR* Gerard Depardieu, Wojciech Pszoniak, Patrice Chereau, Angela Winkler; *Dir:* Andrzej Wajda. Montreal World Film Festival '83: Best Actor (Depardieu). **VHS, Beta, LV** $59.95 *COL, TAM* 🎬🎬🎬🎬

Darby O'Gill & the Little People Set in Ireland, a roguish old story teller tumbles into a well and visits the land of leprechauns who give him three wishes in order to rearrange his life. When he tries to tell his friends what happened they think that it is only another one of his stories. A wonderful Disney production (despite its disappointing box-office performance) with wit, charm and an ounce or two of terror.
1959 (G) 93m/C Albert Sharpe, Janet Munro, Sean Connery, Estelle Winwood; *Dir:* Robert Stevenson. **VHS, Beta, LV** $19.99 *DIS* 🎬🎬🎬½

Daredevils of the Red Circle Three young men set out to free a man held prisoner by an escaped convict. A serial in twelve chapters.
1938 195m/B Charles Quigley, Herman Brix, Carole Landis; *Dir:* John English, William Witney. **VHS** $59.95 *REP, VCN, MLB* 🎬🎬🎬½

Daring Dobermans In this sequel to ''The Doberman Gang,'' the barking bank-robbers have a new set of crime-planning masters. A young Indian boy who loves the dogs enters the picture and may thwart their perfect crime.
1973 (PG) 88m/C Charles Robinson, Tim Considine, David Moses, Claudio Martinez, Joan Caulfield; *Dir:* Byron Ross Chudnow. **VHS, Beta** $59.98 *FOX* 🎬🎬

Daring Game A group of scuba divers nicknamed the Flying Fish attempt to rescue a woman's husband and daughter from an island dictatorship. Failed series pilot by producer Ivan Tors, better known for ''Sea Hunt'' and ''Flipper.''

D

1968 100m/C Lloyd Bridges, Brock Peters, Michael Ansara, Joan Blackman; ***Dir:*** Laslo Benedek. **VHS, Beta $39.98** *NO* ₰₰

Dario Argento's World of Horror A look at horror filmmaker Argento and his films "Suspiria," "Demons," "Creepers," "Inferno," and "Tenebrae."
1985 76m/C James Franciscus, Karl Malden, Donald Pleasence, Jessica Harper; ***Dir:*** Dario Argento. **VHS, Beta, LV $39.95** *VMK, IME* ₰₰

The Dark A supernatural beast commits a string of gruesome murders. Also known as "The Mutilator."
1979 (R) 92m/C William Devane, Cathy Lee Crosby, Richard Jaeckel, Keenan Wynn, Vivian Blaine; ***Dir:*** John Cardos. **VHS, Beta $9.95** *MED, SIM* ₰½

Dark Age An Australian conservationist must track down a rampaging, giant, semi-mythical alligator before it is killed by bounty hunters. Based on the novel (and Aboriginal legend) "Numunwari."
1988 (R) 90m/C *AU* John Jarratt, David Gulpilil, Max Phipps, Ray Meagher, Nikki Coghill, Burnam Burnam; ***Dir:*** Arch Nicholson. **VHS, Beta, LV $19.98** *CHA* ₰₰½

Dark August A New Yorker drops out and transplants to rural Vermont, where he accidentally kills a young girl and then suffers numerous horrors as a result of a curse put on him by the girl's grandfather.
1976 (PG) 87m/C J.J. Barry, Carole Shelyne, Kim Hunter, William Robertson; ***Dir:*** Martin Goldman. **VHS, Beta $59.95** *LTG, VES, LIV* ₰₰

Dark Before Dawn American farmers and Vietnam veterans team up against corrupt government officials and ruthless businessmen in this violent film about the plight of the underdog.
1989 (R) 95m/C Doug McClure, Sonny Gibson, Ben Johnson, Billy Drago, Rance Howard, Morgan Woodward, Buck Henry, Jeffery Osterhage, Red Steagall, John Martin; ***Dir:*** Robert Totten. **VHS, Beta, LV $79.98** *VES, LIV* ₰½

Dark Command The story of Quantrell's Raiders who terrorized the Kansas territory during the Civil War until one man came along to put a stop to it. Colorful and talented cast add depth to the script. Also available colorized.
1940 95m/B John Wayne, Walter Pidgeon, Claire Trevor, Roy Rogers, Marjorie Main, George "Gabby" Hayes; ***Dir:*** Raoul Walsh. **VHS, Beta $14.98** *REP, FCT* ₰₰₰

Dark Corner Gripping, intricate film noir about a detective, already framed once, who may go up the river yet again—this time for the murder of his ex-partner, the guy who framed him the first time. Loyal secretary Ball helps to extricate him from this tangled web.
1946 99m/B Lucille Ball, Mark Stevens, Clifton Webb, William Bendix, Cathy Downs, Reed Hadley, Constance Collier, Kurt Kreuger; ***Dir:*** Henry Hathaway. **VHS, Beta $59.98** *FOX* ₰₰₰

The Dark Crystal Jen and Kira, two of the last surviving Gelflings, attempt to return a crystal shard (discovered with the help of a sorceress) to the castle where the Dark Crystal lies, guarded by the cruel and evil Skeksis. Designed by Brian Froud. From the creators of the Muppets.
1982 (PG) 93m/C ***Dir:*** Jim Henson. **VHS, Beta, LV $19.99** *HBO* ₰₰₰

Dark Eyes An acclaimed Italian film based on a several short stories by Anton Chekov. Mastroianni is a weak-willed Italian, trapped in a marriage of convenience, who falls in love with a mysterious, also married, Russian beauty he meets in a health spa. He embarks on a journey to find her and, perhaps, his lost ideals. Hailed as Mastroianni's consummate performance. In Italian with English subtitles.
1987 118m/C *IT* Marcello Mastroianni, Silvana Mangano, Elena Sofonova, Marthe Keller; ***Dir:*** Nikita Mikhailkov. Cannes Film Festival '87: Best Actor (Mastroianni). **VHS, Beta $79.98** *GHV, FOX* ₰₰₰½

Dark Forces A faith-healer promises to help a senator's dying son and finds the politician's wife also desires his assistance. This Australian film, originally titled "Harlequin," is uneven and predictable. Though rated "PG," beware of two rather brief, but explicit, nudity scenes.
1983 (PG) 96m/C *AU* Robert Powell, David Hemmings, Broderick Crawford, Carmen Duncan; ***Dir:*** Simon Wincer. **VHS, Beta $9.95** *MED* ₰₰

Dark Habits An early Almodovar farce about already-demented nuns in a failing convent trying to raise funds with the help of a nightclub singer who is on the run. Although certainly irreverant, it doesn't quite have the zing of his later work. In Spanish with subtitles.
1984 116m/C *SP* Carmen Maura, Christina Pascual, Julieta Serrano, Marisa Paredes; ***Dir:*** Pedro Almodovar. **VHS, Beta, LV $79.95** *TAM, CCN* ₰₰½

The Dark Hour Two detectives team up to solve a murder in which multiple suspects are involved. Based on Sinclair Gluck's "The Last Trap."
1936 72m/B Ray Walker, Irene Ware, Berton Churchill, Hedda Hopper, Hobart Bosworth, E.E. Clive; ***Dir:*** Charles Lamont. **VHS, Beta $16.95** *NOS, LOO, SNC* ₰½

Dark Journey World War I Stockholm is the setting for a love story between a double agent and the head of German Intelligence. Clever, sophisticated production; Leigh is stunning.
1937 82m/B *GB* Vivien Leigh, Conrad Veidt, Joan Gardner, Anthony Bushell, Ursula Jeans; ***Dir:*** Victor Saville. **VHS, Beta, 3/4U $19.95** *NOS, MRV, CAB* ₰₰₰

Dark Mirror A psychologist and a detective struggle to determine which twin sister murdered a prominent physician. Good and evil siblings finely-acted by de Havilland.
1946 85m/B Olivia de Havilland, Lew Ayres, Thomas Mitchell, Garry Owen; ***Dir:*** Robert Siodmak. **VHS $14.98** *REP* ₰₰₰

Dark Mountain A forest ranger rescues his gal from the clutches of a hardened criminal.
1944 56m/B Robert Lowery, Ellen Drew, Regis Toomey, Eddie Quillan, Elisha Cook Jr.; ***Dir:*** William H. Pine, William C. Thomas. **VHS $24.95** *NOS, DVT, TIM* ₰½

Dark of the Night A young woman, new to the city, purchases a used Jaguar and finds she must share it with the car's former owner—a woman murdered inside the Jag. Decent psycho-thriller with some suspenseful and amusing moments.
1985 88m/C *NZ* Heather Bolton, David Letch, Gary Stalker; ***Dir:*** Gaylene Preston. **VHS, Beta $79.98** *LIV, LTG* ₰₰½

Dark Night of the Scarecrow Thriller with a moral. Prejudiced townspeople execute a retarded man who was innocently befriended by a young girl. After his death unusual things begin to happen. Slow to start but effective.
1981 100m/C Charles Durning, Tanya Crowe, Larry Drake; ***Dir:*** Frank DeFelitta. **VHS, Beta $59.98** *FOX, HHE* ₰₰½

Dark Obsession Byrne portrays a husband who lives out his dark erotic fantasies with his wife (Donohoe). Driven by a mysterious guilt, madness slowly begins to take him. Based on a true scandal, also known as "Diamond SKulls." Available in an "NC-17" rated version.
1990 (R) 97m/C Gabriel Byrne, Amanda Donohoe, Michael Hordern, Judy Parfitt; ***Dir:*** Nick Broomfield. **VHS, Beta $89.95** *PSM* ₰₰½

Dark Passage Bogart plays a convict who escapes from San Quentin to prove he was framed for the murder of his wife. He undergoes plastic surgery and is hidden and aided by Bacall as he tries to find the real killer. Stars can't quite compensate for a far-fetched script and so-so direction.
1947 107m/B Humphrey Bogart, Lauren Bacall, Agnes Moorehead, Bruce Bennett, Tom D'Andrea; ***Dir:*** Delmer Daves. **VHS, Beta, LV $19.95** *FOX* ₰₰½

Dark Past A crazy, escaped convict holds a psychologist and his family hostage, and the two men engage in psychological cat-and-mouse combat. An underrated thriller with off-beat casting.
1949 75m/B William Holden, Lee J. Cobb, Nina Foch, Adele Jergens; ***Dir:*** Rudolph Mate. **VHS, Beta $69.95** *COL* ₰₰₰

Dark Places Masquerading as a hospital administrator, a former mental patient inherits the ruined mansion of a man who had killed his wife and children and died insane. As he lives in the house, the spirit of its former owner seems to overcome him and the bizarre crime is repeated.
1973 (PG) 91m/C *GB* Joan Collins, Christopher Lee, Robert Hardy; ***Dir:*** Don Sharp. **VHS, Beta $24.98** *SUE* ₰½

The Dark Power Ex-cowboy LaRue and his trusty whip provide this flick with its only excitement. As sheriff, he must deal with the dead Mexican warriors who rise to wreak havoc when a house is built on their burial ground.
1985 87m/C Lash LaRue, Anna Lane Tatum; ***Dir:*** Phil Smoot. **VHS, Beta $79.95** *MAG* ₰

Dark Ride Lunatic picks up women with the intention of raping and killing them. Ugly thriller based on the evil deeds of serial murderer Ted Bundy.
198? 83m/C James Luisi, Susan Sullivan, Martin Speer; ***Dir:*** Jeremy Hoenack. **VHS, Beta $69.95** *MED* ₰½

Dark Rider Dark Rider is a mysterious, motorcycle riding hero who comes to the rescue of a small town. Besieged by a gangster who buys up all their land, the townspeople must abandon their homes and indeed, their very dreams. That is until Dark Rider rolls into to town to enact his own unique brand of vengeance.
1991 94m/C Joe Estevez, Doug Shanklin, Alicia Kowalski, David Shark, Chuck Williams. **VHS $79.95** *AIP* ₰½

Dark River: A Father's Revenge When his daughter is killed in a toxic waste dump accident, a man begins revenge on the company and its managers. Tense, with important message.
1990 95m/C Helen Hunt, Mike Farrell, Tess Harper. **VHS $79.98** *TTC* ₰₰

The Dark Room A disturbed young man becomes obsessed with his father's voluptuous mistress and longs to possess her.
1984 90m/C VHS, Beta $59.95 *VCL* ♫ ½

Dark Sanity Uninvolving attempt at horror has a recently de-institutionalized woman envisioning death and destruction (something you'll wish the film would do) while everyone else just thinks she's losing her marbles. Problem is we don't really care. Also known as "Straight Jacket."
1982 89m/C Aldo Ray, Kory Clark, Andy Gwyn, Bobby Holt; *Dir:* Martin Greene. **VHS, Beta** $59.95 *PSM Woof!*

The Dark Secret of Harvest Home An urban couple confront a pagan cult in the New England town into which they've moved. Based upon the Thomas Tryon novel. Made for television.
1978 118m/C Bette Davis, Rosanna Arquette, David Ackroyd, Rene Auberjonois, Michael O'Keefe; *Dir:* Leo Penn. **VHS, Beta** $59.95 *MCA* ♫♫ ½

Dark Side of Love A young girl gets involved with the rougher side of New Orleans society, gets pregnant, and realizes how she's been degraded.
1979 94m/C James Stacy, Glynnis O'Connor, Jan Sterling, Mickey Rooney. **VHS, Beta** $59.95 *GEM* ♫ ½

Dark Side of Midnight A super-detective tracks down a psychopathic killer.
1986 (R) 108m/C James Moore, Wes Olsen, Sandy Schemmel, Dave Bowling. **VHS, Beta** $79.95 *PSM* ♫ ½

Dark Side of the Moon Members of a space ship sent on a routine mission to the far side of the moon, discover an unknown force that feeds on human emotion and consumes the soul. Lukewarm science fiction/horror.
1990 (R) 96m/C William Bledsoe, Alan Blumenfeld, John Diehl, Robert Sampson, Wendy MacDonald, Camilla More, Joe Turkel; *Dir:* D.J. Webster. **VHS, LV** $89.95 *VMK* ♫♫

Dark Star John Carpenter's directorial debut is a low-budget, sci-fi satire which focuses on a group of scientists whose mission is to destroy unstable planets. During their journey, they battle their alien mascot (who closely resembles a walking beach ball), as well as a "sensitive" and intelligent bombing device which starts to question the meaning of its existence. Enjoyable early feature from John "Halloween" Carpenter and Dan "Aliens" O'Bannon. Fun, weird, and unpredictable.
1974 (G) 95m/C Dan O'Bannon, Brian Narelle; *Dir:* John Carpenter. **VHS, Beta, LV** $19.95 *HHT, IME, UHV* ♫♫♫

Dark Tower Decent cast is wasted in yet another inept attempt at horror. An architect is dismayed to learn that an evil force is inhabiting her building and it might just be the ghost of her dearly departed husband.
1987 (R) 91m/C Michael Moriarty, Jenny Agutter, Theodore Bikel, Carol Lynley, Anne Lockhart, Kevin McCarthy; *Dir:* Ken Barnett. **VHS, Beta, LV** $79.98 *FOR, IME* ♫

Dark Victory A spoiled young heiress discovers she is dying from a brain tumor. She attempts to pack a lifetime of parties into a few months, but is rescued by her doctor, with whom she falls in love. Classic final scene with Davis at the top of her form. Bogart plays an Irish stable hand, but not especially well. Also available in a colorized

version. Oscar nominated for Best Picture, Best Actress (Davis), and Best Original Score.
1939 106m/B Bette Davis, George Brent, Geraldine Fitzgerald, Humphrey Bogart, Ronald Reagan, Henry Travers; *Dir:* Edmund Goulding. **VHS, Beta, LV** $19.95 *MGM* ♫♫♫ ½

Dark Waters The drowning death of her parents has left a young woman mentally unstable. She returns to her family home in the backwaters of Louisiana with her peculiar aunt and uncle to serve as guardians. It eventually becomes apparent that someone is trying to drive her insane. This one tends to be rather murky and it's not simply due to the plentiful scenes of misty swampland.
1944 93m/B Merle Oberon, Franchot Tone, Thomas Mitchell, Fay Bainter, Elisha Cook Jr., John Qualen, Rex Ingram; *Dir:* Andre de Toth. **VHS, Beta, LV** $19.95 *STE, NWV* ♫♫

Darker Than Amber John D. MacDonald's houseboat-dwelling detective Travis McGee (Taylor) rescues a girl (Kendall) he's fallen for, and soon discovers that the mugs who thugged her were part of a collection racket. A violent action melodrama upgraded from its original "R" rating.
1970 96m/C Rod Taylor, Suzy Kendall, Theodore Bikel, Jane Russell, James Booth, Janet MacLachlan, William Smith, Anna Capri, Chris Robinson; *Dir:* Robert Clouse. **VHS** *CTY* ♫♫

Darkman Raimi's disfigured-man-seeks-revenge suspenser is comicbook kitsch cross-pollinated with a strain of gothic horror. Neeson plays a scientist who's on the verge of discovering the key to cloning body parts; brutally attacked by the henchmen of a crooked politico, his lab is destroyed and he's left for dead. Turns out he's not dead—just horribly disfigured and a wee bit chafed—and he stalks his deserving victims from the shadows, using his lab know-how to disguise his rugged bad looks. Exquisitely violent, it's got tunes by hipster Danny Elfman. Montage by Pablo Ferro.
1990 (R) 95m/C Liam Neeson, Frances McDormand, Larry Drake, Colin Friels, Nelson Mashita, Jenny Agutter, Rafael H. Robledo; *Dir:* Sam Raimi. **VHS, Beta, LV** $19.95 *MCA, CCB* ♫♫♫

Darkroom An unstable young man devises a scheme to photograph his father in bed with his mistress, and then use the pics to blackmail dear ol' dad.
1990 90m/C Jill Pierce, Jeffrey Allen Arbaugh, Sara Lee Wade, Aaron Teich; *Dir:* Terrence O'Hara. **VHS** $79.95 *QUE* ♫ ½

The Darkside In this frightening drama, a young prostitute and an innocent cabbie attempt to escape the clutches of a maniacal film producer with a secret he won't let them reveal.
1987 (R) 95m/C Tony Galati, Cyndy Preston; *Dir:* Constantino Magnatta. **VHS** $79.95 *VMK* ♫ ½

Darktown Strutters Effort to satirize racial stereotypes is humorless and ineffective. Black female motorcycle gang searches for the kidnapped mother of one of the members.
1974 (PG) 85m/C Trina Parks, Roger E. Mosley, Shirley Washington; *Dir:* William Witney. **VHS, Beta** $59.98 *CHA* ♫

Darling A young amoral model tries to hold boredom at bay by having a number of love affairs. She joins the international jet set and manages to reach the top of European society by marrying a prince. She then learns what an empty life she has. Christie won an

Oscar for her portrayal of the disillusioned, cynical young woman. Academy Award winning screenplay by Frederic Raphael. Also Oscar nominated for Best Picture.
1965 122m/B *GB* Julie Christie, Dirk Bogarde, Laurence Harvey; *Dir:* John Schlesinger. Academy Awards '65: Best Actress (Christie), Best Costume Design (B & W), Best Story; British Academy Awards '65: Best Black & White Art Direction, Best Actor (Bogarde), Best Actress (Christie), Best Screenplay; National Board of Review Awards '65: 10 Best Films of the Year, Best Actress (Christie), Best Director (Schlesinger). **VHS, Beta, LV** $69.98 *SUE, VYG, BTV* ♫♫♫ ½

D.A.R.Y.L. The little boy found by the side of the road is too polite, too honest and too smart. His friend explains to him the necessity of imperfection (If you don't want the grown-ups to bother you too much) and he begins to become more like a real little boy. But the American military has a top-secret interest in this child, since he is in fact the combination of a cloned body and a computer brain. More interesting when it's involved with the human beings and less so when it focuses on science.
1985 (PG) 100m/C Mary Beth Hurt, Michael McKean, Barret Oliver, Colleen Camp; *Dir:* Simon Wincer. **VHS, Beta, LV** $14.95 *PAR* ♫♫

Das Boot Superb detailing of life in a German U-boat during World War II. Intense, claustrophobic atmosphere complemented by nail-biting action provides a realistic portrait of the stressful conditions that were endured on these submarines. Excellent performances, especially from Prochnow. With English subtitles. Originally a 6-hour special made for German television. Also available in a dubbed version under the title "The Boat."
1981 (R) 150m/C *GE* Juergen Prochnow, Herbert Gronemeyer, Klaus Wennemann, Hubertus Bengsch, Martin Semmelrogge, Bernd Tauber, Erwin Leder; *Dir:* Wolfgang Petersen. **VHS, Beta, LV** $29.95 *COL, APD* ♫♫♫♫

Das Testament des Dr. Mabuse The third, and only sound, Mabuse film by Lang. The infamous crime lord/megalomaniac plots the world's destruction from the confines of an asylum cell. In German, with English subtitles. Released as "The Last Will of Dr. Mabuse" in the U.S. in 1943.
1933 122m/B *GE* Rudolf Klein-Rogge, Otto Wernicke, Gustav Diesl, Karl Meixner; *Dir:* Fritz Lang. **VHS, Beta** $29.98 *SNC, SUE, GLV* ♫♫♫ ½

Date with an Angel When aspiring musician Knight fishes a beautiful angel out of the swimming pool, he is just trying to rescue her. But the beauty of the angel overwhelms him and he finds himself questioning his upcoming wedding to Cates, a cosmetic mogul's daughter. Too cute and too sentimental, though Beart's beauty is other-wordly.
1987 (PG) 114m/C Emmanuelle Beart, Michael E. Knight, Phoebe Cates, David Dukes, Bibi Besch, Albert Macklin, David Hunt, Michael Goodwin; *Dir:* Tom McLoughlin. **VHS, Beta, LV** $19.99 *HBO* ♫ ½

Date Bait When a teen couple decides—much to their parents' chagrin—to elope, they find themselves on a date with danger when their post-nuptials are plagued by pushers and assorted other bad guys.
1960 71m/B Gary Clarke, Marlo Ryan, Richard Gering, Danny Logan; *Dir:* O'Dale Ireland. **VHS, Beta** $16.95 *SNC* ♫

A Date with Judy Standard post-war musical dealing with teenage mix-ups in and around a big high school dance. Choreography by Stanley Donen. Songs include "It's a Most Unusual Day."
1948 114m/C Jane Powell, Elizabeth Taylor, Carmen Miranda, Wallace Beery, Robert Stack, Xavier Cugat, Selena Royle, Leon Ames; *Dir:* Richard Thorpe. **VHS, Beta, LV $19.95** *MGM, MLB* ♪♪

Daughter of Death A little mentally off-center since she witnessed the gruesome rape and murder of her mother, a teenage girl doesn't bond well with her new mom when dad remarries. Originally titled "Julie Darling."
1982 (R) 100m/C *CA GE* Anthony (Tony) Franciosa, Isabelle Mejias, Sybil Danning, Cindy Gurling, Paul Hubbard, Benjamin Schmoll; *Dir:* Paul Nicholas. **VHS $59.99** *FCT* ♪

The Daughter of Dr. Jekyll The doc's daughter believes she may have inherited her father's evil curse when several of the locals are found dead.
1957 71m/B John Agar, Arthur Shields, John Dierkes, Gloria Talbott; *Dir:* Edgar G. Ulmer. **VHS, Beta $39.98** *FOX* ♪½

Daughter of Horror A young woman finds herself involved with a porcine mobster who resembles her abusive father. Trouble is, Dad's dead, and daughter dearest abetted his departure. Obscure venture into expressionism that was initially banned by the New York State Board of Censors. Shot on a low budget (how low was it? So low that McMahon narrated because shooting with sound was too expensive). The 55 minutes tend to lag, although the film should be intriguing to genre enthusiasts and to fans of things pseudo-Freudian. Originally titled "Dementia."
1955 60m/B Adrienne Barrett, Bruno Ve Sota, Ben Roseman; *W/Dir:* John Parker; *Nar:* Ed McMahon. **VHS $19.98** *NOS, SNC, MRV* ♪♪

Daughter of the Tong FBI agent gets tong twisted when he tries to put a lid on a smuggling ring headed by a woman (who's beautiful, of course). Something that resembles acting is wasted in a mess of a movie.
1939 56m/B Evelyn Brent, Grant Withers, Dorothy Short, Dave O'Brien; *Dir:* Raymond K. Johnson. **VHS, Beta $16.95** *NOS, SNC* ♪

Daughters of Darkness Newlyweds on their way to England stop at a posh French hotel. There they meet a beautiful woman whom the hotel owner swears had been there forty years ago, even though she hasn't aged a bit. When she introduces herself as Countess of Bathory (the woman who bathed in the blood of virgins to stay young) folks begin to wonder. A really superb erotic vampire film charged with sensuality and a sense of dread.
1971 (R) 87m/C *BE GE IT FR* Delphine Seyrig, John Karlen, Daniele Ouimet, Andrea Rau, Paul Esser, Georges Jamin, Joris Collet, Fons Rademakers; *Dir:* Harry Kumel. **VHS, Beta $39.98** *CGI* ♪♪♪

Daughters of the Dust A Gullah family living on the Sea Islands off the Georgia coast in 1902 contemplate moving to the mainland in this emotional tale of change. The Gullah are descendants of West African slaves and their isolation has kept their superstitions and native dialect (a mixture of Western African, Creole, and English) intact. Five women of the Peazant family reflect on their lives and family as they gather together for a final family picnic. They include 88-year-old Nana, the matriarch of the family;

Yellow Mary, a prosperous ex-prostitute who has returned to the family fold; Eula, whose pregnancy may be the result of a rape; Baptist missionary, Viola; and daughter-in-law Haagar, who can't wait to escape what she sees as the backwardness of the island. Family bonds and memories are celebrated with a quiet narrative and beautiful cinematography in Dash's feature-film directorial debut.
1991 113m/C Cora Lee Day, Barbara O, Alva Rogers, Kaycee Moore, Cheryl Lynn Bruce, Adisa Anderson; *W/Dir:* Julie Dash. **VHS $79.95** *KIV* ♪♪♪½

Daughters of Satan Selleck, in an early role as a virile museum buyer, antagonizes a coven of witches when he purchases a painting. His wife, played by Grant, becomes a target of the witches' revenge and salacious shenanigans ensue.
1972 (R) 96m/C Tom Selleck, Barra Grant, Paraluman, Tani Phelps Guthrie; *Dir:* Hollingsworth Morse. **VHS, Beta $14.95** *WKV* ♪♪

David A haunting portrait of the survival of a Jewish teenager in Berlin during the Nazi reign of terror, based on the novel by Joel Koenig. universally acclaimed, this was the first film about the Holocaust made in Germany by a German Jew. In German with English subtitles.
1979 106m/C *GE* Mario Fischel, Walter Taub, Irene Vrkijan, Torsten Hentes, Eva Mattes; *Dir:* Peter Lilienthal. Berlin Film Festival '79: Best Film. **VHS, Beta $79.95** *KIV, GLV, TAM* ♪♪♪½

David and Bathsheba The Bible story comes alive in this lush and colorful Fox production. Peck and Hayward are great together and Peck is properly concerned about the wrath of God over his transgressions. Terrific costumes and special effects, lovely music (Oscar nominated) and a fine supporting cast keep this a notch above other Biblical epics.
1951 116m/C Gregory Peck, Susan Hayward, Raymond Massey, Kieron Moore, James Robertson Justice, Jayne Meadows, John Sutton, Dennis Hoey, Francis X. Bushman, George Zucco; *Dir:* Henry King. **VHS, LV $19.98** *FXV, FOX* ♪♪♪

David Bowie David Bowie at his best performing "Let's Dance," "China Girl" and "Modern Love."
1983 14m/C David Bowie. **VHS, Beta, LV, 8mm $9.95** *MVD, SVS, PIA* ♪♪♪

David Copperfield Superior adaptation of Charles Dickens' great novel. An orphan grows to manhood in Victorian England with a wide variety of help and harm. Terrific acting by Bartholomew, Fields, Rathbone, and all the rest. Lavish production, lovingly filmed. Received Oscar nominations for Best Picture, Best Assistant Director and Best Film Editing.
1935 132m/B Lionel Barrymore, W.C. Fields, Freddie Bartholomew, Maureen O'Sullivan, Basil Rathbone, Lewis Stone; *Dir:* George Cukor. National Board of Review Awards '35: 10 Best Films of the Year. **VHS, Beta, LV $19.98** *MGM* ♪♪♪♪

David Copperfield This British made-for-television production is more faithful to the Dickens classic than any of its predecessors. The added material, however, fails to highlight any one character as had the successful 1935 MGM version. Still, the exceptional (and largely stage-trained) cast do much to redeem the production.

1970 118m/C Richard Attenborough, Cyril Cusack, Edith Evans, Pamela Franklin, Susan Hampshire, Wendy Hiller, Ron Moody, Laurence Olivier; *Dir:* Delbert Mann. **VHS** *FOX* ♪♪½

David Copperfield An animated adaptation of the Dickens classic about a young boy growing up in 19th-century England.
1983 72m/C VHS, Beta **$24.95** *CVL, KUI* ♪♪♪♪

David and Lisa Director Perry was given an Oscar for this independently-produced sensitive film. Adapted from the novelization of Theodore Isaac Rubin of a true case history concerning a young man and woman who fall in love while institutionalized for mental illness. Dullea and Margolin are excellent in the title roles of this sleeper.
1963 94m/B Keir Dullea, Janet Margolin, Howard da Silva, Neva Patterson, Clifton James; *Dir:* Frank Perry. **VHS, Beta, LV $69.95** *COL, MRV* ♪♪♪

David and Moses The animated history of David and Moses from the Old Testament is brought to life in this family program.
197? (G) 46m/C VHS, Beta **$39.95** *VGD* ♪♪♪

David Sanborn: Love and Happiness The saxophonist performs "Run for Cover," "Lisa," "Straight to the Heart," "Smile," "Hideaway," and "Love & Happiness," with his band in a New York rehearsal studio.
1985 37m/B VHS, Beta, LV **$19.95** *MVD, WRV, PIA* ♪♪♪

David Steinberg: In Concert Comedian Steinberg performs some of his best routines in Toronto.
1984 60m/C David Steinberg. **VHS, Beta $39.95** *RKO* ♪♪♪

Davy Crockett, King of the Wild Frontier Three episodes of the popular Disney television series are blended together here to present the life and some of the adventures of Davy Crockett. These include his days as an Indian fighter and his gallant stand in defense of the Alamo. Well-done by a splendid cast, the film helped to spread Davy-mania among the children of the 1950s.
1955 89m/C Fess Parker, Buddy Ebsen, Hans Conried, Ray Whiteside, Pat Hogan; *Dir:* Norman Foster. **VHS, Beta** *DIS, OM*

Davy Crockett and the River Pirates Another Disney splice and dice of two episodes from the television series, chronicling the further adventures of our frontier hero. Davy meets up with Mike Fink, the King of the Ohio River, and the two engage in a furious keelboat race, and then unite to track down a group of thieves masquerading as Indians and threatening the peace.
1956 (G) 81m/C Fess Parker, Buddy Ebsen, Jeff York; *Dir:* Norman Foster. **VHS, Beta, LV** *DIS, OM*

Dawn! The true story of Dawn Fraser, an Olympic champion swimmer and an unfulfilled woman willing to fight for her happiness.
1983 114m/C Bronwyn MacKay-Payne, Tom Richards. **VHS, Beta $19.98** *VID* ♪½

Dawn of the Dead Romero's gruesome sequel to his "Night of the Living Dead." A mysterious plague causes the recently dead to rise from their graves and scour the countryside for living flesh. Very violent, gory, graphic, and shocking, yet not without humor. Gives interesting consideration to the violence created by the living humans in their efforts to save themselves. **1978 126m/C** David Emgee, Ken Foree, Gaylen Ross; *Dir:* George A. Romero. **VHS, Beta, LV $14.99** HBO *ℐℐℐ* ½

Dawn Express Nazi spies infiltrate the U.S. in search of a secret formula designed to enhance the power of gasoline. Another film made to contribute to the war effort that doesn't make much sense. **1942 63m/B** Michael Whalen, Anne Nagel, William Bakewell, Constance Worth, Hans von Twardowski, Jack Mulhall, George Pembroke, Kenneth Harlan, Robert Frazer; *Dir:* Albert Herman. **VHS, Beta $16.95** SNC, TIM *ℐ* ½

Dawn on the Great Divide Jones' last film before his tragic and untimely death trying to save people from a fire at the Coconut Grove in Boston. Wagon train has to battle not only Indians, but bad guys as well. Plot provides more depth than the usual western fare. **1942 57m/B** Buck Jones, Tim McCoy, Raymond Hatton, Mona Barrie, Robert Lowery, Betty Blythe, Jan Wiley, Harry Woods, Roy Barcroft; *Dir:* Howard Bretherton. **VHS, Beta $19.95** NOS, MRV, VCN *ℐℐ* ½

Dawn of the Mummy Lousy plot and bad acting—not to mention gore galore—do this one in. A photographer and a bevy of young fashion models travel to Egypt for a special shoot. They unwittingly stumble upon an ancient tomb, teeming with vengeance-minded mummies. **1982 93m/C** Brenda King, Barry Sattels, George Peck, Joan Levy; *Dir:* Frank Agrama. **VHS, Beta $69.99** HBO Woof!

Dawn Patrol Flynn plays a flight commander whose nerves are shot in this story of the British Royal Flying Corps during World War I. The focus is on the effects that the pressures and deaths have on all those concerned. Fine performances from all in this well-done remake of the 1930 film. **1938 103m/B** Errol Flynn, David Niven, Basil Rathbone, Donald Crisp, Barry Fitzgerald, Melville Cooper; *Dir:* Edmund Goulding. **VHS, Beta $19.98** FOX *ℐℐℐ*

Dawn of the Pirates (La Secta de los Tughs) An adventurer stumbles upon a murderous tribe of Tughs and the beautiful white woman they captured. **19?? 95m/C** SP **VHS, Beta $59.95** JCI *ℐ* ½

Dawn Rider Formula western has all the right ingredients: Wayne as a cowboy out to get revenge on the gang that murdered his father. Features stuntman Canutt in rare acting role. **1935 60m/B** John Wayne, Marion Burns, Yakima Canutt; *Dir:* Robert N. Bradbury. **VHS $8.95** NEG, REP, MRV *ℐ* ½

Dawn Rider/Frontier Horizon A double feature of two of Wayne's earliest western films. **1939 114m/B** John Wayne, Jennifer Jones, Marion Burns; *Dir:* Robert N. Bradbury, George Sherman. **VHS, Beta $34.95** REP *ℐ* ½

Dawson Patrol Royal Canadian Mounty dog-sled race turns into a dramatic battle for survival. **1978 75m/C** George R. Robinson, Tim Henry, James B. Douglas; *Dir:* Peter Kelly. **VHS, Beta $49.95** TWE *ℐ* ½

Day After A powerful drama which graphically depicts the nuclear bombing of a midwestern city and its after-effects on the survivors. Made for television, and very con-troversial when first shown, gathering huge ratings and vast media coverage. **1983 126m/C** Jason Robards Jr., JoBeth Williams, John Lithgow, Steve Guttenberg; *Dir:* Nicholas Meyer. **VHS, Beta, LV $9.98** SUE *ℐℐℐ*

Day of the Animals Nature gone wild. It seems a depleted ozone layer has exposed the animals to the sun's radiation, turning Bambi and Bugs into brutal killers. A group of backpackers trek in the Sierras, unaware of the transformation. Far-fetched and silly (we hope). Also on video as "Something Is Out There." **1977 (PG) 95m/C** Christopher George, Leslie Nielsen, Lynda Day George, Richard Jaeckel, Michael Ansara, Ruth Roman, Jon Cedar, Susan Backlinie, Andrew Stevens, Gil Lamb; *Dir:* William Girdler. **VHS $59.95** MED, ACF *ℐ*

Day of the Assassin A Mideast shah's yacht explodes, sending a huge fortune to the bottom of a murky bay. This sets off a rampage of treasure hunting by a variety of ruthless mercenaries. **1981 94m/C** Glenn Ford, Chuck Connors, Richard Roundtree, Henry Silva, Jorge Rivero; *Dir:* Brian Trenchard-Smith. **VHS, Beta $79.95** PSM *ℐ*

Day the Bookies Wept Decent yarn about cabbies tricked into buying a "race-ehorse." Turns out she's old as the hills and hooked on alcohol to boot, but they enter her in the big race anyway. **1939 50m/B** Betty Grable, Joe Penner, Tom Kennedy, Richard Lane; *Dir:* Leslie Goodwins. **VHS, Beta $34.98** CCB *ℐℐ*

Day of the Cobra A corrupt narcotics bureau official hires an ex-cop to find a heroin kingpin on the back streets of Genoa, Italy. **1984 95m/C** IT Franco Nero, Sybil Danning, Mario Maranzana, Licinia Lentini, William Berger; *Dir:* Enzo G. Castellari. **VHS, Beta $49.95** MED *ℐ* ½

Day with Conrad Green Set in 1927, this program presents a somewhat farcical look at a successful Broadway producer as he attempts to cope with a full schedule of personal and business problems on the day of the funeral of his longtime assistant. **1980 60m/C** Fred Gwynne. **VHS, Beta** MTP *ℐ* ½

A Day in the Country The son of the famed painter gives us the moving tale of a young woman's sudden and intense love for a man she meets while on a picnic with her family. Beautifully adapted from a story by Guy de Maupassant. Renowned photographer Henri Cartier-Bresson contributed to wonderful cinematography. French with English subtitles. **1946 36m/B** FR Sylvia Bataille, Georges Darnoux, Jane Marken, Paul Temps; *Dir:* Jean Renoir. **VHS, Beta $24.95** NOS, MRV, DVT *ℐℐℐ* ½

Day of the Dead The third in Romero's trilogy of films about flesh-eating zombies taking over the world. Romero hasn't thought up anything new for the ghouls to do, and the humans are too nasty to care about this time around. For adult audiences. **1985 91m/C** Lori Cardillo, Terry Alexander, Joseph Pilato, Jarleth Conroy; *Dir:* George A. Romero. **VHS, Beta, LV $19.95** MED, IME *ℐ* ½

A Day in the Death of Joe Egg Based on the Peter Nichols play, an unlikely black comedy about a British couple trying to care for their retarded/autistic child, nearly destroying their marriage in the process. They begin to contemplate euthanasia as a solution. Well-acted, but potentially offensive to some. Written by David Deutsch. **1971 (R) 106m/C** GB Alan Bates, Janet Suzman, Elizabeth Robillard, Peter Bowles, Joan Hickson, Sheila Gish; *Dir:* Peter Medak. **VHS, Beta $69.95** COL *ℐℐℐ*

A Day at Disneyland A colorful souvenir of the attractions at Disneyland. Highlights include a ride down Main Street, a visit to Sleeping Beauty's castle and trips through Adventureland, Frontierland, Fantasyland and Tomorrowland. **1982 39m/C VHS, Beta $49.95** DIS

The Day of the Dolphin Research scientist, after successfully working out a means of teaching dolphins to talk, finds his animals kidnapped; espionage and assassination are involved. Dolphin voices by Buck Henry, who also wrote the screenplay. **1973 (PG) 104m/C** George C. Scott, Trish Van Devere, Paul Sorvino, Fritz Weaver, Jon Korkes, John Dehner, Edward Herrmann, Severn Darden; *Dir:* Mike Nichols. National Board of Review Awards '73: 10 Best Films of the Year. **VHS, Beta, LV $14.98** SUE *ℐℐ*

The Day the Earth Caught Fire The earth is knocked out of orbit and sent hurtling toward the sun when nuclear testing is done simultaneously at both the North and South Poles. Realistic and suspenseful, this is one of the best of the sci-fi genre. **1962 100m/B** Janet Munro, Edward Judd, Leo McKern; *Dir:* Val Guest. **VHS, Beta $14.95** NEG, HBO, KAR *ℐℐℐ* ½

The Day the Earth Stood Still A gentle alien lands on Earth to deliver a message of peace and a warning against experimenting with nuclear power. He finds his views echoed by a majority of the population, but not the ones who are in control. In this account based loosely on the story of Christ, Rennie is the visitor backed by the mighty robot Gort. One of the greatest science fiction films of all time. Score by Bernard Herrmann. **1951 92m/B** Michael Rennie, Patricia Neal, Hugh Marlowe, Sam Jaffe, Frances Bavier, Lock Martin, Bobby Gray; *Dir:* Robert Wise. **VHS, Beta, LV $14.98** FOX, FCT, MLB *ℐℐℐ* ½

Day and the Hour A woman becomes accidentally involved in the resistance movement during the Nazi occupation of France in World War II. **1963 110m/C** FR IT Simone Signoret, Stuart Whitman, Genevieve Page; *Dir:* Rene Clement. **VHS, Beta $24.95** TAM, MON *ℐℐ* ½

The Day It Came to Earth Completely silly and unbelievable sci-fi flick has meteor crashing into the watery grave of a mobster. The decomposed corpse is revived by the radiation and plots to take revenge on those who fitted him with cement shoes. **1977 (PG) 89m/C** Roger Manning, Wink Roberts, Bob Ginnaven, Rita Wilson, Delight de Bruine; *Dir:* Harry Z. Thomason. **VHS, Beta** PGN Woof!

The Day of the Jackal Frederick Forsyth's best-selling novel of political intrigue is splendidly brought to the screen by Zinnemann. A brilliant and ruthless assassin hired to kill Charles de Gaulle skirts the international intelligence pool, while intuitive police work to stop him. Tense, suspenseful, beautiful location photography. Excellent acting by Fox, Cusack and Britton. **1973 (PG) 142m/C** Edward Fox, Alan Badel, Tony Britton, Derek Jacobi, Cyril Cusack, Olga Georges-Picot; *Dir:* Fred Zinnemann. **VHS, Beta, LV $59.95** MCA *ℐℐℐ* ½

Day of Judgment A mysterious stranger arrives in a town to slaughter those people who violate the Ten Commandments.
1981 101m/C William T. Hicks, Harris Bloodworth, Brownlee Davis; ***Dir:*** C.D.H. Reynolds. **VHS, Beta $69.99** *HBO* 🦴½

Day of the Locust Compelling adaptation of Nathaniel West's novel concerning the dark side of 1930s' Hollywood. A no-talent amoral actress's affair with a meek accountant leads to tragedy and destruction. Told from the view of a cynical art director wise to the ways of Hollywood. Oscar nomination for Meredith.
1975 (R) 140m/C Donald Sutherland, Karen Black, Burgess Meredith, William Atherton, Billy Barty, Bo Hopkins, Richard Dysart, Geraldine Page; ***Dir:*** John Schlesinger. British Academy Awards '75: Best Costume Design; National Board of Review Awards '75: 10 Best Films of the Year. **VHS, Beta, LV $49.95** *PAR* 🦴🦴🦴½

Day of the Maniac Psychotic drug addict will stop at nothing to support his growing habit.
1977 (R) 89m/C George Hilton, Susan Scott; ***Dir:*** Sergio Martino. **VHS, Beta $69.95** *REP* 🦴½

Day for Night A wryly affectionate look at the profession of moviemaking—its craft, character, and the personalities that interact against the performances commanded by the camera. French with English subtitles. Earned Oscar nominations for Cortese's acting and Truffaut's direction and screenplay.
1973 (PG) 116m/C *FR* Jean-Pierre Leaud, Jacqueline Bisset, Jean-Pierre Aumont, Valentina Cortese, Francois Truffaut; ***Dir:*** Francois Truffaut. Academy Awards '73: Best Foreign Language Film; British Academy Awards '73: Best Director (Truffaut), Best Picture, Best Supporting Actress (Cortese); New York Film Critics Awards '73: Best Director (Truffaut), Best Film, Best Supporting Actress (Cortese); National Society of Film Critics Awards '73: Best Director (Truffaut), Best Picture, Best Supporting Actress (Cortese). **VHS, Beta $19.95** *TAM, WAR, APD* 🦴🦴🦴🦴

Day of the Panther The Panthers—the world's most formidable martial artists—are mighty torqued when panther-Linda is pithed, and they're not known to turn the other cheek. Much chop-socky kicking and shrilling.
1988 86m/C John Stanton, Eddie Stazak; ***Dir:*** Brian Trenchard-Smith. **VHS, Beta $79.95** *CEL* 🦴½

A Day at the Races Though it seems labored at times, the brilliance of the brothers Marx still comes through in this rather weak tale of a patient in a sanitorium who convinces horse doctor Groucho to take on running the place.
1937 109m/B Groucho Marx, Harpo Marx, Chico Marx, Sig Rumann, Douglas Dumbrille, Margaret Dumont, Allan Jones, Maureen O'Sullivan; ***Dir:*** Sam Wood. **VHS, Beta, LV $19.95** *MGM, CCB* 🦴🦴🦴½

Day that Shook the World Slow-moving but intriguing account of events surrounding the assassination of Archduke Ferdinand of Austria that triggered World War I. Garnered an R-rating due to some disturbingly graphic scenes.
1978 (R) 111m/C Christopher Plummer, Florinda Bolkan, Maximilian Schell; ***Dir:*** Veljko Bulajic. **VHS $29.95** *VID* 🦴🦴

The Day the Sky Exploded Sci-fi disaster drama doesn't live up to the grandiose title, as a runaway rocket ship hits the sun, unleashing an asteroid shower that threatens Earth with tidal waves, earthquakes, heat waves and terrible dialogue. The highlight of this Franco-Italian effort is the cinematography by horror director Mario Bava.
1957 80m/B *FR IT* Paul Hubschmid, Madeleine Fischer, Fiorella Mari, Ivo Garrani, Dario Michaelis; ***Dir:*** Richard Benson. **VHS $29.95** *NOS, FCT, TIM* 🦴½

Day Time Ended A pair of glowing UFO's streaking across the sky and an alien mechanical device with long menacing appendages are only two of the bizarre phenomena witnessed from a lone house in the desert. Good special effects. Also released as "Time Warp" (1978).
1980 80m/C Chris Mitchum, Jim Davis, Dorothy Malone; ***Dir:*** John Cardos. **VHS, Beta** *MED, MRV* 🦴½

Day of the Triffids The majority of Earth's population is blinded by a meteor shower which also causes plant spores to mutate into giant carnivores. Well-done adaptation of John Wyndham's science fiction classic.
1963 94m/C Howard Keel, Janet Scott, Nicole Maurey, Janet Scott, Kieron Moore, Mervyn Johns; ***Dir:*** Steve Sekely. **VHS, Beta, LV $29.95** *MED, MRV, PSM* 🦴🦴🦴

Day of Triumph A dynamic version of the life of Christ as seen through the eyes of the apostles Andrew and Zadok, the leaders of the Zealot underground.
1954 110m/C Lee J. Cobb, Robert Wilson, Ralph Freud; ***Dir:*** Irving Pichel. **VHS $39.95** *REP* 🦴🦴🦴

The Day Will Dawn Suspenseful British-made World War II thriller about a reporter and a young Norwegian girl who plot to sabotage a secret Nazi U-boat base near the girl's town. Alternate title: "The Avengers." Co-written by Terence Rattigan.
1942 100m/B *GB* Deborah Kerr, Hugh Williams, Griffith Jones, Ralph Richardson, Francis L. Sullivan, Roland Culver, Niall MacGinnis, Finlay Currie; ***Dir:*** Harold French. **VHS, Beta, 8mm $29.95** *VYY* 🦴🦴🦴

The Day the Women Got Even Four suburban homemakers take up the life of vigilantes to save unsuspecting actresses from a talent agent's blackmail scheme.
1980 98m/C JoAnn Pflug, Tina Louise, Georgia Engel, Barbara Rhoades, Julie Hagerty, Ed O'Neill; ***Dir:*** Burt Brinckerhoff. **VHS, Beta** *TLF* 🦴½

Day of Wrath An involving psychological thriller based on records of witch trials from the early 1600s. An old woman, burned at the stake as a witch, puts a curse on the local parson and his family. Grim and unrelentingly pessimistic, moving from one horrific scene to another, creating a masterpiece of terror. Also known as "Vredens Dag" and based on a play by Hans Wiers Jenssen.
1943 110m/B *DK* Thorkild Roose, Sigrid Neiiendam, Lisbeth Movin; ***Dir:*** Carl Theodor Dreyer. National Board of Review Awards '48: 10 Best Films of the Year; Venice Film Festival '47: Special Homage. **VHS, Beta $49.95** *TAM, NOS, SNC* 🦴🦴🦴🦴

Daydreamer By combining live action and animation in this delightful feature-length film, young Hans Christian Andersen comes face-to-face with many of the same characters he would later write about. Imaginative and finely-crafted.
1966 98m/C Paul O'Keefe, Ray Bolger, Jack Gilford, Margaret Hamilton; ***Dir:*** Jules Bass; ***Voices:*** Tallulah Bankhead, Boris Karloff, Burl Ives, Terry-Thomas. **VHS, Beta $19.98** *SUE* 🦴🦴🦴

The Daydreamer Richard, who also wrote the screenplay and directed the film, stars as a bumbling fool let loose in the French corporate world with comical results. In French with subtitles.
1977 90m/C *FR* Pierre Richard, Bernard Blier, Maria Pacome, Marie-Christine Barrault; ***W/Dir:*** Pierre Richard. **VHS, Beta $19.95** *FCT* 🦴🦴½

Days of Glory Peck's screen debut finds him as a Russian peasant bravely fighting the Nazi blitzkrieg almost single-handedly.
1943 86m/B Tamara Toumanova, Gregory Peck, Alan Reed, Maria Palmer, Lowell Gilmore, Hugo Haas; ***Dir:*** Jacques Tourneur. **VHS, Beta $19.98** *TTC* 🦴🦴½

Days of Heaven A drifter (Gere), his younger sister, and a woman he claims is his sister become involved with a Texas sharecropper. Told through the eyes of the younger girl, this is a sweeping vision of the U.S. before the first World War. Loss and loneliness, deception, frustration, and anger haunt these people as they struggle to make the land their own. Deservedly awarded an Oscar for breathtaking cinematography.
1978 (PG) 95m/C Richard Gere, Brooke Adams, Sam Shepard, Linda Manz, Stuart Margolin; ***Dir:*** Terence Malick. Academy Awards '78: Best Cinematography; British Academy Awards '79: Best Musical Score; Cannes Film Festival '79: Best Director (Malick); Los Angeles Film Critics Association Awards '78: Best Cinematography; National Board of Review Awards '78: Best Picture; New York Film Critics Awards '78: Best Director (Malick); National Society of Film Critics Awards '78: Best Cinematography, Best Director (Malick). **VHS, Beta, LV $39.95** *PAR* 🦴🦴🦴½

Days of Hell Forgettable tale of four mercenaries who are hired to rescue a doctor and his daughter held captive by guerrillas somewhere in Afghanistan only to find out they're pawns in an even bigger scheme.
197? 90m/C Conrad Nichols, Richard Raymond, Stephen Elliott, Kinako Harada; ***Dir:*** Anthony Richard. **VHS, Beta $59.95** *MGL* 🦴

Days of Thrills and Laughter Delightful compilation of clips from the era of silent films, showcasing the talents of the great comics as well as daring stuntman.
1961 93m/B Buster Keaton, Charlie Chaplin, Harold Lloyd, Stan Laurel, Oliver Hardy, Douglas Fairbanks Sr.; ***Dir:*** Robert Youngson. **VHS, Beta $29.95** *MPI*

Days of Thunder "Top Gun" in race cars! Cruise follows the same formula he has followed for several years now (with the notable exception of "Born on the Fourth of July.") Cruise and Towne co-wrote the screenplay concerning a young kid bursting with talent and raw energy who must learn to deal with his mentor, his girlfriend, and eventually the bad guy. First film that featured cameras that were actually on the race cars. If you like Cruise or race cars then this is the movie for you.
1990 (PG-13) 108m/C Tom Cruise, Robert Duvall, Randy Quaid, Nicole Kidman, Cary Elwes, Michael Rooker; ***Dir:*** Tony Scott. **VHS, Beta, LV, SVS $92.95** *PAR, SUP* 🦴🦴

Days of Wine and Roses The original "Playhouse 90" television version of J.P. Miller's story which was adapted for the big screen in 1962. An executive on the fast track and his young wife, initially only social drinkers, find themselves degenerating into alcoholism. A well-acted, stirring drama.
1958 89m/B Cliff Robertson, Piper Laurie; ***Dir:*** John Frankenheimer. **VHS, Beta $14.95** *WKV, MGM* 🦴🦴½

Days of Wine and Roses A harrowing tale of an alcoholic advertising man who gradually drags his wife down with him into a life of booze. Part of the "A Night at the Movies" series, this tape simulates a 1962 movie evening with a Bugs Bunny cartoon, "Martian Through Georgia," a newsreel, and coming attractions for "Gypsy" and "Rome Adventure."
1962 138m/B Jack Lemmon, Lee Remick, Charles Bickford, Jack Klugman, Jack Albertson; **Dir:** Blake Edwards. Academy Awards '62: Best Song ("Days of Wine and Roses"). **VHS, Beta, LV $19.98** WAR, PIA 🎬🎬🎬½

Dayton's Devils Nielsen is the leader of a motley crew of losers who plan to rob a military base bank in this action-packed heist flick.
1968 107m/C Rory Calhoun, Leslie Nielsen, Lainie Kazan, Hans Gudegast, George Stanford Brown; **Dir:** Jack Shea. **VHS, Beta** TIM 🎬🎬

D.C. Cab A rag-tag group of Washington, D.C. cabbies learn a lesson in self-respect in this endearing comedy. Though not without flaws, it's charming all the same.
1984 (R) 100m/C Mr. T, Leif Erickson, Adam Baldwin, Charlie Barnett, Irene Cara, Anne De Salvo, Max Gail, Gloria Gifford, Gary Busey, Jill Schoelen, Marsha Warfield; **Dir:** Joel Schumacher. **VHS, Beta, LV $19.95** MCA 🎬🎬½

The Dead The poignant final film by Huston, based on James Joyce's short story. At a Christmas dinner party in turn-of-the-century Dublin, a man discovers how little he knows about his wife when a song reminds her of a cherished lost love. Beautifully captures the spirit of the story while providing Huston an opportunity to create a last lament on the fickle nature of time and life.
1987 (PG) 82m/C GB Anjelica Huston, Donal McCann, Marie Kean, Donal Donnelly, Dan O'Herlihy, Helen Carroll, Frank Patterson; **Dir:** John Huston. National Board of Review Awards '87: 10 Best Films of the Year; National Society of Film Critics Awards '87: Best Film. **VHS, Beta, LV $89.98** LIV, VES 🎬🎬🎬½

Dead Again Branagh's first film since his brilliant debut as the director/star of "Henry V" again proves him a visionary force on and off camera. As the smart L.A. detective hired to discover the identity of a beautiful, but mute woman, he's realistic, clever, tender and cynical. He finds that he's apparently trapped in a nightmarish cycle of murder begun years before. Literate, lovely to look at, suspenseful, with a sense of humor to match its high style. Cameo by Robin Williams.
1991 (R) 107m/C Kenneth Branagh, Andy Garcia, Derek Jacobi, Hanna Schygulla, Emma Thompson, Robin Williams, Campbell Scott; **Dir:** Kenneth Branagh. National Board of Review Awards '91: 10 Best Films of the Year. **VHS, LV $94.99** CCB, PAR 🎬🎬🎬½

Dead Aim A dud about Soviet spies who introduce even more drugs into New York City, ostensibly to bring about the downfall of the U.S. Obviously a completely far-fetched plot.
1987 (R) 91m/C Ed Marinaro, Isaac Hayes, Corbin Bernsen; **Dir:** William Vanderkloot. **VHS, LV $89.98** LIV, HHE Woof!

Dead or Alive Tex and the Texas Rangers pretend to be outlaws in order to join a gang terrorizing a town.
1944 56m/B Tex Ritter, Dave O'Brien. **VHS, Beta $19.98** VCN, RXM 🎬½

Dead Bang A frustrated cop uncovers a murderous white supremacist conspiracy in L.A. Frankenheimer's deft directorial hand shapes a somewhat conventional cop plot into an effective vehicle for Johnson.
1989 (R) 102m/C Don Johnson, Bob Balaban, William Forsythe, Penelope Ann Miller, Tim Reid, Frank Military, Michael Higgins, Michael Jeter; **Dir:** John Frankenheimer. **VHS, Beta, LV, 8mm $19.98** WAR 🎬🎬½

Dead and Buried Farentino is the sheriff who can't understand why the victims of some pretty grisly murders seem to be coming back to life. Eerily suspenseful.
1981 (R) 95m/C James Farentino, Jack Albertson, Melody Anderson, Lisa Blount, Bill Quinn, Michael Pataki; **Dir:** Gary Sherman. **VHS, Beta $19.98** VES 🎬🎬½

Dead Calm A taut Australian thriller based on a novel by Charles Williams. A young married couple is yachting on the open seas when they happen upon the lone survivor of a sinking ship. They take him on board only to discover he's a homicidal maniac. Makes for some pretty suspenseful moments that somewhat make up for the weak ending.
1989 (R) 97m/C AU Sam Neill, Billy Zane, Nicole Kidman; **Dir:** Phillip Noyce. **VHS, Beta, LV, 8mm $19.98** WAR, FCT 🎬🎬🎬

Dead for a Dollar A Colonel, a con man, and a mysterious woman team up in the Old West to search for the two hundred thousand dollars they stole from a local bank.
197? 92m/C John Ireland, George Hilton, Piero Vida, Sandra Milo. **VHS, Beta $49.95** UNI 🎬½

The Dead Don't Die Unbelievable plot set in the 1930s has Hamilton as a detective trying to prove his brother was wrongly executed for murder. He ultimately clashes with the madman who wants to rule the world with an army of zombies. Perhaps if they had cast Hamilton as Master of the Zombies.
1975 74m/C George Hamilton, Ray Milland, Linda Cristal, Ralph Meeker, Joan Blondell, James McEachin; **Dir:** Curtis Harrington. **VHS, Beta $49.95** WOV, GKK 🎬

Dead as a Doorman A young writer takes a part-time post as a doorman and becomes the target of a murderous lunatic.
1985 83m/C Bradley Whitford, Sharon Schlarth, Bruce Taylor; **Dir:** Gary Youngman. **VHS, Beta $69.98** LIV, VES 🎬½

Dead Easy Livin' ain't easy for a Sydney sleaze-club owner, his working-girl girlfriend and a slandered cop when they're caught between machete-wielding enemy gangs (it seems somebody made the crime boss REALLY mad). The three flee in that quintessential escape vehicle—a two-ton truck—to qualify the effort as an Australian contender for greatest chase scene ever (heavy chassis category).
1982 92m/C AU Scott Burgess, Rosemary Paul; **Dir:** Bert Deling. **VHS, Beta $59.95** MCG 🎬🎬

Dead End Sidney Kingsley play, adapted for the big screen by Lillian Hellman, traces the lives of various inhabitants of the slums on New York's Lower East Side as they try to overcome their surroundings. Gritty drama saved from melodrama status by some genuinely funny moments. Film launched the Dead End Kids.
1937 92m/B Sylvia Sidney, Joel McCrea, Humphrey Bogart, Wendy Barrie, Claire Trevor, Allen Jenkins, Marjorie Main, Leo Gorcey; **Dir:** William Wyler. **VHS, Beta, LV $19.98** MGM, SUE, MLB 🎬🎬🎬½

Dead End City A street-wise resident of L.A. organizes fellow citizens into a vigilante force to fight the gangs for control of the city.
1988 88m/C Dennis Cole, Greg Cummins, Christine Lunde, Robert Z'Dar, Darrell Nelson, Alena Downs; **Dir:** Peter Yuval. **VHS, Beta $39.95** AIP 🎬🎬

Dead End Drive-In In a surreal, grim future a man is trapped at a drive-in theater-cum-government-run concentration camp, where those considered to be less than desirable members of society are incarcerated.
1986 (R) 92m/C AU Ned Manning, Natalie McCurray, Peter Whitford; **Dir:** Brian Trenchard-Smith. **VHS, Beta $19.95** STE, NWV 🎬🎬

Dead End Street A young prostitute attempts to change her self-destructive lifestyle.
1983 86m/C IS Anat Atzman, Yehoram Gaon. **VHS, Beta $59.95** TWE 🎬½

Dead Eyes of London Blind old German men are dying to lower their premiums in this geriatric thriller: someone (perhaps the director of the home for the blind?) is killing off the clientele for their insurance money, and a Scotland Yard inspector aims to expose the scam. Relatively early-vintage Kinski, it's a remake of the eerie 1939 Lugosi vehicle, "The Human Monster" (originally titled "Dark Eyes of London"), adapted from Edgar Wallace's "The Testament of Gordon Stuart."
1961 95m/B GE Joachim Fuchsberger, Karin Baal, Dieter Borsche, Ady Berber, Klaus Kinski, Eddi Arent, Wolfgang Lukschy; **Dir:** Alfred Vohrer. **VHS $19.98** SNC 🎬🎬½

Dead Girls A rock group whose lyrics focus on suicide goes on vacation. The members find themselves stalked by a serial killer who decides to take matters out of the musicians' hands and into his own. Graphic violence.
1990 (R) 105m/C Diana Karanikas, Angela Eads; **Dir:** Dennis Devine. **VHS** RAE, HHE 🎬

Dead Heat Some acting talent and a few funny moments unfortunately don't add up to a fine film, though they do save it from being a complete woofer. In this one even the cops are zombies when one of them is resurrected from the dead to help his partner solve his murder and rid the city of the rest of the undead.
1988 (R) 86m/C Joe Piscopo, Treat Williams, Lindsay Frost, Darren McGavin, Vincent Price, Keye Luke, Clare Kirkconnell; **Dir:** Mark Goldblatt. **VHS, Beta, LV $19.95** STE, NWV 🎬½

Dead Heat on a Merry-Go-Round Coburn turns in a great performance as the ex-con who masterminds the heist of an airport safe. Intricately woven plot provides suspense and surprises. The film is also notable for the debut of Ford in a bit part (he has one line as a hotel messenger).
1966 (R) 108m/C James Coburn, Camilla Sparv, Aldo Ray, Nina Wayne, Robert Webber, Rose Marie, Todd Armstrong, Marian Moses, Severn Darden, Harrison Ford; **Dir:** Bernard Girard. **VHS, Beta $69.95** COL 🎬🎬🎬

Dead Lucky A novice tries gambling and wins a fortune, but his luck leads him into an entanglement with an assassin, a scam artist, and a beautiful woman. Based on the novel "Lake of Darkness."
1960 91m/B GB Vincent Ball, Betty McDowall, John Le Mesurier, Alfred Burke, Michael Ripper; **Dir:** Montgomery Tully. **VHS, Beta $19.98** FCT 🎬½

Dead Man Out Both Glover and Blades turn in exceptional performances in this thought-provoking drama. A convict on Death Row (Blades) goes insane and therefore cannot be executed. The state calls in a psychiatrist (Glover) to review the case and determine whether he can be cured so that the sentence can be carried out. Powerful and riveting morality check.
1989 87m/C Danny Glover, Ruben Blades, Tom Atkins, Larry Block, Sam Jackson, Maria Ricossa; *Dir:* Richard Pearce. VHS, Beta, LV $89.99 HBO ♂♂♂

Dead Man Walking In a post-holocaust future, half the population has been stricken with a deadly plague. Hauser is a mercenary, dying from the disease, who is hired to rescue a young woman who was kidnapped and is being held in the plague zone.
1988 (R) 90m/C Wings Hauser, Brion James, Pamela Ludwig, Sy Richardson, Leland Crooke, Jeffrey Combs; *Dir:* Gregory Brown. VHS $79.98 REP ♂♂

Dead Mate A woman marries a mortician after a whirlwind romance only to discover that, to her husband, being an undertaker is not just a job, it's a way of life.
1988 93m/C Elizabeth Mannino, David Gregory, Lawrence Bockus, Adam Wahl; *Dir:* Straw Weisman. VHS, Beta $79.95 PSM ♂

Dead Men Don't Die An anchorman is slain by criminals but resurrected as a shambling voodoo zombie. Few viewers notice the difference in this inoffensive but repetitious farce. The living-dead Gould appears to be having a lot of fun.
1991 (PG-13) 94m/C Elliott Gould, Melissa Anderson, Mark Moses, Philip Bruns, Jack Betts, Mabel King; *W/Dir:* Malcolm Marmorstein. VHS $89.95 ACA ♂♂½

Dead Men Don't Wear Plaid Martin is hilarious as a private detective who encounters a bizarre assortment of suspects while trying to find out the truth about a scientist's death. Ingeniously interspliced with clips from old Warner Brothers films. Features Humphrey Bogart, Bette Davis, Alan Ladd, Burt Lancaster, Ava Gardner, Barbara Stanwyck, Ray Milland and others. Its only flaw is that the joke loses momentum by the end of the film.
1982 (PG) 89m/B Steve Martin, Rachel Ward, Reni Santoni, George Gaynes, Frank McCarthy, Carl Reiner; *Dir:* Carl Reiner. VHS, Beta, LV $14.95 MCA, FCT ♂♂½

Dead Men Walk A decent, albeit low budget, chiller about twin brothers (Zucco in a dual role); one a nice, well-adjusted member of society, the other a vampire who wants to suck his bro's blood. Sibling rivalry with a bite.
1943 65m/B George Zucco, Mary Carlisle, Dwight Frye, Nedrick Young, Al "Fuzzy" St. John, Dwight Frye; *Dir:* Sam Newfield. VHS, Beta $14.95 NOS, MRV, SNC ♂♂

Dead on the Money Adequate made-for-TV suspenser about a woman who believes her lover plots a dire fate for her cousin. But nothing is what it seems here. Confusing, droopy in spots, but fun.
1991 92m/C Corbin Bernsen, Amanda Pays, John Glover, Eleanor Parker, Kevin McCarthy; *Dir:* Mark Cullingham. VHS, Beta $89.98 TTC ♂♂½

Dead of Night The template for episodic horror films, this suspense classic, set in a remote country house, follows the individual nightmares of five houseguests. Redgrave

turns in a chillingly convincing performance as a ventriloquist terrorized by a demonic dummy. Not recommended for light-sleepers. Truly spine-tingling.
1945 102m/B GB Michael Redgrave, Sally Ann Howes, Basil Radford, Naunton Wayne, Mervyn Johns, Roland Culver, Googie Withers, Frederick Valk; *Dir:* Basil Dearden, Robert Hamer, Alberto Cavalcanti, Charles Crichton. VHS, Beta $59.99 HBO, RXM ♂♂♂♂

Dead of Night This trilogy features Richard Matheson's tales of the Supernatural: "Second Chance," "Bobby," and "No Such Thing As a Vampire," all served up by your hostess, Elvira.
1977 76m/C Joan Hackett, Ed Begley Jr., Patrick Macnee, Anjanette Comer; *Dir:* Dan Curtis. VHS, Beta, LV $39.95 IVE ♂♂½

Dead Pit Leave this one in the pit it crawled out of. Twenty years ago a mad scientist was killed as a result of horrible experiments he was conducting on mentally ill patients at an asylum. A young woman stumbles across his experiments, awakening him from the dead.
1989 (R) 90m/C Jeremy Slate, Steffen Gregory Foster; *Dir:* Brett Leonard. VHS, Beta $89.95 IMP Woof!

Dead Poets Society An idealistic English teacher inspires a group of boys in a 1950s' prep school to pursue individuality and creative endeavor, resulting in clashes with school and parental authorities. Williams shows he can master the serious roles as well as the comic with his portrayal of the unorthodox educator for which he received an Oscar nomination. Big box office hit occasionally scripted with a heavy hand in order to elevate the message. The ensemble cast is excellent.
1989 (PG) 128m/C Robin Williams, Ethan Hawke, Robert Sean Leonard, Josh Charles, Gale Hansen, Kurtwood Smith, James Waterson, Dylan Kussman; *Dir:* Peter Weir. VHS, Beta, LV, 8mm $19.99 TOU ♂♂♂½

The Dead Pool Dirty Harry Number Five features a new twist on the sports pool: a list of celebrities is distributed and bets are placed on who will be first to cross the finish line, literally. Unfortunately someone seems to be hedging their bet by offing the celebs themselves. When Harry's name appears on the list, he decides to throw the game.
1988 (R) 92m/C Clint Eastwood, Liam Neeson, Patricia Clarkson, Evan C. Kim, David Hunt, Michael Currie, Michael Goodwin; *Dir:* Buddy Van Horn. VHS, Beta, LV, 8mm $19.98 WAR, TLF ♂½

Dead Reckoning Bogart and Prince are two World War II veterans en route to Washington when Prince disappears. Bogart trails Prince to his Southern hometown and discovers he's been murdered. Blackmail and more murders follow as Bogie tries to uncover the truth. Suspenseful with good performances from all, especially Bogart.
1947 100m/B Humphrey Bogart, Lizabeth Scott, Morris Carnovsky, Charles Cane, Wallace Ford, William Prince, Marvin Miller; *Dir:* John Cromwell. VHS, Beta, LV $19.98 CCB, COL ♂♂♂

Dead Reckoning A plastic surgeon and his lovely wife embark on a cruise aboard his new luxury yacht. The intrigue begins with a storm, an isolated island, and the discovery that the captain is the wife's former lover.
1989 (R) 95m/C Cliff Robertson, Susan Blakely, Rick Springfield; *Dir:* Robert Lewis. VHS, LV $79.95 MCA ♂½

Dead Right A black man is unjustly put into prison in the Deep South. He manages to escape in order to try and prove his innocence.
1988 111m/C Raymond St. Jacques, Dana Wynter, Kevin McCarthy, Barbara McNair. VHS, Beta $79.95 PSM ♂½

Dead Ringers A stunning, unsettling chiller, based loosely on a real case and the bestseller by Bari Wood and Jack Geasland. Irons, in an excellent dual role, is effectively disturbing as the twin gynecologists who descend into madness when they can no longer handle the fame, fortune, drugs, and women in their lives. Acutely upsetting film made all the more so due to its graphic images and basis in fact.
1988 (R) 117m/C CA Jeremy Irons, Genevieve Bujold, Heidi von Palleske; *Dir:* David Cronenberg. Genie Awards '89: Best Actor (Irons), Best Art Direction/Set Decoration, Best Cinematography, Best Director (Cronenberg), Best Picture, Best Screenplay, Best Musical Score; Los Angeles Film Critics Association Awards '88: Best Actress (Bujold), Best Director; New York Film Critics Awards '88: Best Actor (Irons). VHS, Beta, LV $9.99 CCB, MED ♂♂♂

Dead Sleep A nurse discovers that comatose patients at a private clinic have been used as guinea pigs by an overzealous doctor. This minor-league chiller from Down Under is said to have been inspired by an actual medical scandal.
1991 (R) 95m/C Linda Blair, Tony Bonner; *Dir:* Alec Mills. VHS $89.98 VES ♂♂

Dead Space Dead space lay between the ears of whoever thought we needed this remake of 1982's "Forbidden World." At a lab on a hostile planet an experimental vaccine mutates into a prickly puppet monster who menaces the medicos. Needed: a vaccine against cheapo "Alien" ripoffs.
1990 72m/C Marc Singer, Laura Tate, Bryan Cranston, Judith Chapman; *Dir:* Fred Gallo. VHS, LV $79.95 COL ♂

Dead in the Water Brown plays Charlie Deegan, a big-time lawyer with a wife that he'd rather see dead. Charlie and his mistress plot the perfect murder to dispose of his spouse and collect her money. When plans go awry, he ends up as a suspect for the wrong murder in this story of infidelity and greed.
1991 (PG-13) 90m/C Bryan Brown, Teri Hatcher, Anne DeSalvo, Veronica Cartwright; *Dir:* Bill Condon. VHS, Beta $79.95 MCA ♂½

Dead of Winter A young actress is suckered into a private screen test for a crippled old director, only to find she is actually being remodeled in the guise of a murdered woman. Edge-of-your-seat suspense as the plot twists.
1987 (R) 100m/C Mary Steenburgen, Roddy McDowall, Jan Rubes, Ken Pouge, William Russ; *Dir:* Arthur Penn. VHS, Beta, LV $79.98 FOX ♂♂♂

Dead Women in Lingerie When beautiful models are turning up dead, a detective is called in to solve the mystery. Perhaps they committed suicide to avoid being in this picture any longer than necessary. Title sounds like the next episode of "Geraldo."
1990 (R) 87m/C June Lockhart, Lyle Waggoner, John Romo, Jerry Orbach, Maura Tierney; *Dir:* Erika Fox. VHS $79.98 MNC Woof!

Dead Wrong An undercover agent falls in love with the drug smuggler she's supposed to bring to justice.

1983 93m/C Britt Ekland, Winston Rekert, Jackson Davies. **VHS, Beta** $49.95 *MED* ℐ½

Dead Wrong: The John Evans Story The true story of Evans' life of crime and eventual execution in the electric chair, using footage of an interview taped before his death.
1984 45m/C **VHS, Beta, 3/4U** *AEF* ℐ½

Dead Zone A man gains extraordinary psychic powers following a near-fatal accident. He is forced to decide between seeking absolute seclusion in order to escape his frightening visions, or using his "gift" to save mankind from impending evil. A good adaptation of the Stephen King thriller.
1983 (R) 104m/C Christopher Walken, Brooke Adams, Tom Skerritt, Martin Sheen, Herbert Lom, Anthony Zerbe, Colleen Dewhurst; *Dir:* David Cronenberg. **VHS, Beta, LV** $14.95 *PAR* ℐℐℐ

The Deadliest Art: The Best of the Martial Arts Films The masters of bare-handed combat team up in this look at the best of the martial arts movies. Action packed.
1990 ?m/C Yuen Biao, Jackie Chan, Samo Hung, Kareem Abdul-Jabbar, Sho Kosugi, Bruce Lee, Angela Mao, Cynthia Rothrock; *W/Dir:* Sandra Weintraub; *Nar:* John Saxon. **VHS** $89.98 *FXV* ℐℐℐ

Deadline A journalist must find out the truth behind a minor earthquake in Australia.
1981 94m/C Barry Newman, Trisha Noble, Bill Kerr. **VHS, Beta** $7.50 *WOV* ℐ½

Deadline Eerie tale of horror writer's life as it begins to reflect his latest story in this play on the "truth is stranger than fiction" adage.
1982 85m/C Stephen Young, Sharon Masters, Cindy Hinds, Phillip Leonard; *Dir:* Mario Azzopardi. **VHS, Beta** $42.95 *PGN* ℐℐ

Deadline Walken is a cynical American journalist assigned to cover the warring factions in Beirut. He finds himself becoming more and more involved in the events he's supposed to report when he falls in love with a German nurse who is aiding the rebel forces. Tense, though sometimes murky drama.
1987 (R) 110m/C *GE* Christopher Walken, Hywel Bennett, Marita Marschall; *Dir:* Nathaniel Gutman. **VHS, Beta, LV** $24.95 *VIR* ℐℐ

Deadline Assault Kate McSweeny is a beautiful young reporter who is brutally attacked and raped by a gang of savage youths. Kate must now face the unsettling facts; they're still on the streets and they may be stalking her again.
1990 (R) 90m/C Elizabeth Montgomery, James Sloyan, Sean Frye, Biff McGuire, Michael Goodwin, Linden Chiles; *Dir:* Paul Wendkos. **VHS** *ELV, HHE* ℐℐ

Deadline Auto Theft A fun lovin' guy drives fast cars and meets beautiful women. Nudge, nudge, wink, wink.
1983 90m/C H.B. Halicki, Hoyt Axton, Marion Busia, George Cole, Judi Gibbs, Lang Jefferies; *Dir:* H.B. Halicki. **VHS, Beta** $79.95 *IMP* ℐ

Deadline at Dawn Hayward is an aspiring actress who tries to help a sailor (Williams) prove that he is innocent of murder. Screenplay by Clifford Odets was based on a novel by Cornell Woolrich.
1946 82m/B Bill Williams, Susan Hayward, Lola Lane, Paul Lukas, Joseph Calleia; *Dir:* Harold Clurman. **VHS, Beta** $19.98 *RKO, MLB* ℐℐ

Deadly Alliance Two impecunious filmmakers (who happen to be hunks) and a babe (for good measure) poke their noses in where they're not wanted (in the shady business of an international oil cartel). "Deadly Dull" would be a more appropriate moniker.
1978 90m/C Kathleen Arc, Tony de Fonte, Michele Marsh, Walter Prince; *Dir:* Paul S. Parco. **VHS, Beta** $59.98 *CGI* ℐ

The Deadly and the Beautiful Dr. Tsu sends her "deadly but beautiful" task force to kidnap the world's prime male athletes for use in her private business enterprise. Originally titled "Wonder Women."
1973 (PG) 82m/C *PH* Nancy Kwan, Ross Hagen, Maria de Aragon, Roberta Collins, Tony Lorea, Sid Haig, Vic Diaz, Shirley Washington, Gale Hansen; *Dir:* Robert Vincent O'Neil. **VHS, Beta** $29.95 *MED* ℐ

Deadly Bet A made-for-video saga of a kickboxer who loses his money and his girl when he loses the big match. Lots of fight sequences as he works to regain his career.
1991 (R) 93m/C Jeff Wincott, Charlene Tilton, Steven Leigh; *Dir:* Richard W. Munchkin. **VHS** $89.95 *PMH* ℐ½

Deadly Blessing A young, recently widowed woman is visited in her rural Pennsylvania home by some friends from the city. Something's not quite right about this country life and they become especially suspicious after meeting their friend's very religious in-laws.
1981 (R) 104m/C Maren Jensen, Susan Buckner, Sharon Stone, Ernest Borgnine, Karen Jensen, Jeff East, Lisa Hartman, Lois Nettleton; *Dir:* Wes Craven. **VHS, Beta** $24.98 *SUE* ℐℐ

Deadly Breed Beware your friendly neighborhood cop. A band of white supremacists has infiltrated the local police force, intent on causing violence, not keeping the peace.
1989 90m/C William Smith, Addison Randall, Blake Bahner, Joe Vance; *Dir:* Charles Kanganis. **VHS** $29.95 *PMH* ℐ½

Deadly Business A hard-nosed collection man for a lethal loan shark finds he is the one with the debt to pay as he is stalked through the city street by hit men.
19?? 95m/C David Peterson, John Lazarus, Leo Gabrowski. **VHS, Beta** $59.95 *AVC* ℐ½

Deadly Chase (La Persecucion Mortal) Police investigate a series of violent acts by an elusive baron in a presumably antiquarian European country.
19?? 95m/C **VHS, Beta** $59.95 *JCI* ℐ

Deadly Commando When terrorists begin kidnapping diplomats, only the Deadly Commandos have a chance of saving them.
19?? 88m/C **VHS, Beta** $59.95 *MAG* ℐ

Deadly Companions Keith is an ex-gunslinger who agrees to escort O'Hara through Apache territory in order to make up for inadvertently killing her son. Director Peckinpah's first feature film.
1961 90m/C Maureen O'Hara, Brian Keith, Chill Wills, Steve Cochran; *Dir:* Sam Peckinpah. **VHS, Beta, LV** $19.95 *STE, NWV* ℐℐ½

Deadly Dancer It's up to a lone cop to find out who's been offing the hoofers in L.A.
1990 88m/C Adolfo "Shabba Doo" Quinones, Smith Wordes, Walter W. Cox, Steve Jonson; *Dir:* Kimberly Casey. **VHS** $79.95 *AIP* ℐ½

Deadly Darling Muddled, completely awful tale centering on two different rapes with no discernible attempt made to link the two or establish a plot.
1985 91m/C Fonda Lynn, Warren Chan, Bernard Tsui, Cherry Kwok; *Dir:* Karen Yahng. **VHS, Beta** $49.95 *UNI* Woof!

Deadly Diamonds A cop is on the trail of diamond smugglers who leave a trail of death in their wake.
1991 90m/C Dan Haggerty, Troy Donahue, Eli Rich, Kathleen Kane, Kenna Grob, Nicholas Mercer; *Dir:* Thomas Atcheson. **VHS, Beta** $48.00 *AMI, GNS, IPI* ℐℐ

Deadly Dreams A young boy's parents are killed by a maniac who then commits suicide. Years later, he dreams that he is being stalked anew by the same killer. Nightmares turn into reality in this decent horror/suspense flick.
1988 (R) 79m/C Mitchell Anderson, Xander Berkeley, Thom Babbes, Juliette Cummins; *Dir:* Kristine Peterson. **VHS, Beta, LV** $79.95 *VIR* ℐℐ

Deadly Embrace Sordid tale of wealthy man (Vincent) who wants to get rid of his wife—permanantly. He enlists the aid of a beautiful young coed. Soft-core sleaze with nary a redeemable quality.
1988 82m/C Jan-Michael Vincent, Jack Carter, Ty Randolph, Linnea Quigley, Michelle Bauer, Ken Abraham; *Dir:* Ellen Cabot. **VHS, Beta** $79.95 *PSM* Woof!

Deadly Encounter Thrilling actioner in which Hagman portrays a helicopter pilot—a former war ace—who allows an old flame to talk him into helping her escape from the mobsters on her trail. Terrific edge-of-your-seat aerial action.
1972 90m/C Larry Hagman, Susan Anspach, Michael C. Goetz, James Gammon; *Dir:* William A. Graham. **VHS, Beta** $19.95 *VCL* ℐℐℐ

Deadly Encounter Lacking any redeemable qualities, this film shows just how low some people go. Merrill hosts a dinner to garner financial support for a budding playwright, but what we get is a look at the lifestyles of the rich and perverse.
1978 90m/C Dina Merrill, Carl Betz, Leon Ames, Vicki Powers, Mark Rasmussen, Susan Logan; *Dir:* R. John Hugh. **VHS, Beta** $19.95 *NEG* Woof!

Deadly Eyes After eating grain laced with steroids a group of rats grows to mammoth proportions. Their appetites grow accordingly—now they crave humans. Not to mention they can bench press 500 lbs.
1982 (R) 93m/C Sam Groom, Sara Botsford, Scatman Crothers, Lisa Langlois, Cec Linder; *Dir:* Robert Clouse. **VHS, Beta** $19.98 *WAR* ℐ½

Deadly Fieldtrip Four female students and their teacher are abducted by a biker gang and held hostage on a deserted farm.
1974 91m/C Zalman King, Brenda Fogarty, Robert Porter; *Dir:* Earl Barton. **VHS, Beta** $79.95 *MNC* ℐ½

Deadly Force An ex-cop turned private detective stalks a killer in Los Angeles who has left an "X" carved in the forehead of each of his 17 victims.
1983 (R) 95m/C Wings Hauser, Joyce Ingalls, Paul Shenar, Al Ruscio, Arlen Dean Snyder, Lincoln Kilpatrick; *Dir:* Paul Aaron. **VHS, Beta** $14.98 *SUE* ℐ½

Deadly Friend A well-meaning horror flick? That's what Wes Craven has tried to provide for us. When the girlfriend of a lonely teenage genius is accidentally killed, he decides to insert his robot's "brain" in her

body, though the results aren't entirely successful. The same can be said for the film which was based on Diana Henstell's more effective novel "Friend."
1986 (R) 91m/C Matthew Laborteaux, Kristy Swanson, Michael Sharrett, Anne Twomey, Richard Marcus, Anne Ramsey; *Dir:* Wes Craven. **VHS, Beta, LV $19.98** WAR 🎞🎞

Deadly Game A reunion at a remote hotel leads to an ordeal of psychological terror and murderous intrigue.
1982 108m/C George Segal, Robert Morley, Trevor Howard, Emlyn Williams, Alan Webb; *Dir:* George Schaefer. **VHS, Beta $59.98** SUE 🎞🎞

Deadly Game Much maligned and preyed upon witness is protected by a weary cop, who speaks with the use of English subtitles.
1983 90m/C GE Mel Ferrer, Helmut Berger, Barbara Sukowa; *Dir:* Karoly Makk, Dieter Geissler. **VHS, Beta $39.95** VCD, GLV 🎞½

Deadly Game Horribly burned, a vengeful millionaire invites seven people he believes have wronged him to spend the weekend on his isolated island. He has promised them each a reward for their good deeds, but instead plans to have a big-game hunt with his "guests" as the prey.
1991 (R) 93m/C Roddy McDowall, Jenny Seagrove, Marc Singer, Michael Beck; *Dir:* Tom Wright. **VHS, Beta** PAR 🎞½

Deadly Games Ineffective plot involving a series of murders in a small town is seemingly played out by the principal characters in a horror board game. Film is saved from being a complete waste by Groom's good performance as the local cop.
1980 94m/C Sam Groom, Jo Ann Harris, Steve Railsback, Dick Butkus, June Lockhart, Colleen Camp; *Dir:* Scott Mansfield. **VHS, Beta $39.95** MON 🎞½

Deadly Harvest Scientists' worst fears have been realized—due to ecological abuse and over-development of the land, food has become extremely scarce. This in turn has caused people to become a bit savage. They are particularly nasty to a farmer and his family. Not a bad plot, but a poorly-acted film.
1972 (PG) 86m/C Clint Walker, Nehemiah Persoff, Kim Cattrall, David G. Brown, Gary Davies; *Dir:* Timothy Bond. **VHS, Beta $19.95** STE, NWV 🎞

Deadly Hero Gripping tale of a young woman who becomes the prey of a New York City cop after questioning the brutal methods he employed while saving her from being assaulted, and ultimately killing her attacker. Williams made his film debut in this chilling suspenser. You just can't trust anyone these days.
1975 (R) 102m/C Don Murray, James Earl Jones, Diahn Williams, Lilia Skala, Conchata Ferrell, George S. Irving, Treat Williams, Josh Mostel; *Dir:* Ivan Nagy, Ivan Nagy. **VHS, Beta $39.98** SUE 🎞🎞🎞

Deadly Illusion Williams is a private detective who manages to get himself tangled in a complex web of intrigue and winds up framed for murder. Decent outing for all. Also known as "Love You to Death."
1987 (R) 95m/C Billy Dee Williams, Morgan Fairchild, Vanity, John Beck, Joe Cortese; *Dir:* William Tannen, Larry Cohen. **VHS, Beta, LV $79.95** COL 🎞🎞

Deadly Impact Las Vegas casinos are targeted for rip-off by means of a computer. Svenson and Williamson give good performances, but that's not enough to save this one.
1984 90m/C IT Bo Svenson, Fred Williamson, Marcia Clingan, John Morghen, Vincent Conti; *Dir:* Larry Ludman. **VHS, Beta $29.98** LIV, VES 🎞½

Deadly Innocence A shy, lonely girl works at an isolated gas station. After her father's death, she takes in a boarder—a mysterious woman being chased by her past.
1988 (R) 90m/C Mary Crosby, Andrew Stevens, Amanda Wyss; *Dir:* John D. Patterson, Hugh Parks. **VHS** QUE 🎞½

Deadly Intent The widow of a murdered archeologist becomes the target of fortune hunters when they try to get their hands on the priceless jewel her husband brought back from a dig.
1988 83m/C Lisa Eilbacher, Steve Railsback, Maud Adams, Lance Henriksen, Fred Williamson, Persis Khambatta; *Dir:* Nigel Dick. **VHS $29.95** FRH 🎞🎞

Deadly Intruder A quiet vacation spot is being terrorized by an escapee from a mental institution. So much for a little R and R.
1984 86m/C Chris Holder, Molly Creek. **VHS, Beta $69.99** HBO 🎞½

Deadly Kick A kick-boxer generously demonstrates his kicking abilities to all.
1986 93m/C **VHS, Beta $19.95** UHV 🎞½

Deadly Mission Five soldiers in World War II France are convicted of crimes against the Army, only to escape from lock-up and become focal points in a decisive battle.
1978 (R) 99m/C Bo Svenson, Peter Hooten, Fred Williamson. **VHS, Beta $69.98** LIV, LTG 🎞½

Deadly Obsession A disfigured lunatic stalks a young co-ed hoping to extort $1 million from the wealthy dean of the school.
1988 (R) 93m/C Jeffrey R. Iorio; *Dir:* Jeno Hodl. **VHS $14.98** REP, FCT 🎞½

Deadly Passion A James Cain-like thriller about a private eye getting carnally involved with a beautiful and treacherous woman who is manipulating her late husband's estate through murder and double-crossing.
1985 (R) 100m/C SA Brent Huff, Ingrid Boulting, Harrison Coburn, Lynn Maree; *Dir:* Larry Larson. **VHS, Beta $79.98** LIV, LTG 🎞½

Deadly Possession A college student plays amateur detective when she tries to clear her ex-husband's name after he is accused of the horrible murder of a fellow student.
1988 99m/C AU Penny Cook, Anna-Maria Winchester, Liddy Clark, Olivia Hamnett; *Dir:* Craig Lahiff. **VHS, Beta, LV $79.98** LIV, VES 🎞🎞

Deadly Prey Prior is up against a fully outfitted army of bloodthirsty mercenaries who are using innocent people as "live targets" at their secret bootcamp. Plot lifted from "The Most Dangerous Game" has more than its share of gruesome violence.
1987 (R) 87m/C Cameron Mitchell, Troy Donahue, Ted Prior, Fritz Matthews; *W/Dir:* David A. Prior. **VHS, Beta $79.95** SVS Woof!

Deadly Reactor In a post-nuclear future a lone preacher turns into an armed vigilante to protect the people of a town being terrorized by a vicious motorcycle gang.
1989 88m/C David Heavener, Stuart Whitman, Darwyn Swalve, Allyson Davis; *Dir:* David Heavener. **VHS, Beta** AIP 🎞🎞

Deadly Recruits The Russians did it! The Russians did it! Stamp is convinced that the downfall of several top students at Oxford, amid scandal and rumor, is due to a KBG conspiracy.
1986 92m/C GB Terence Stamp, Michael Culver, Carmen DuSautoy, Robin Sachs; *Dir:* Roger Tucker. **VHS, Beta $79.95** WES 🎞🎞

Deadly Revenge Action-adventure pic with not much adventure and even less action. Small-town reporter finds himself up against a mob boss when he takes over the nightclub business of a friend killed by the mobsters. Dubbed.
1983 (R) 90m/C IT Rodolfo Ranni, Julio de Grazia, Silvia Montanari, Fred Commoner; *Dir:* Juan Carlos Sesanzo. **VHS, Beta $59.95** UNI 🎞

Deadly Rivals Quirky tale of a youngster who, in order to deal with his jealousy, plots against the man his mother is going to marry. Also known as "Rivals."
1972 103m/C Joan Hackett, Robert Klein, Scott Jacoby; *Dir:* Krishna Shah. **VHS, Beta $39.98** CGI 🎞🎞

Deadly Sanctuary Censors stopped production of this film several times. The Marquis de Sade's writings provided the inspiration (though there's nothing inspired about it) for this tale of two recently orphaned young women who get caught up in prostitution, an S&M club, and murder.
1968 92m/C Jack Palance, Mercedes McCambridge, Sylva Koscina, Klaus Kinski, Akim Tamiroff, Romina Power; *Dir:* Jess (Jesus) Franco. **VHS, Beta $39.95** MON Woof!

Deadly Sting Silva devises an intricate scheme to gain revenge on the men who cheated him out of his share of a gold bullion heist.
1982 90m/C FR Henry Silva, Philip Clay, Andre Pousse; *Dir:* Jean-Claude Roy. **VHS, Beta $59.95** MGL 🎞½

Deadly Stranger Fluegel's talents could have been put to much better use than as the selfish mistress of a plantation owner who is taking advantage of the migrant workers toiling on his land. Moore is a drifter and hired hand who rallies the workers to stand up to the owner.
1988 93m/C Darlanne Fluegel, Michael J. Moore, John Vernon, Ted White; *Dir:* Max Kleven. **VHS** MCG 🎞

Deadly Strangers Two people meet up on the road and decide to share a ride. We discover that one of them is an insane killer who's escaped from a mental institution. Suspense builds nicely, but the ending is flawed.
1982 89m/C GB Hayley Mills, Simon Ward, Sterling Hayden, Ken Hutchison; *Dir:* Sidney Hayers. **VHS, Beta** PGN 🎞🎞

Deadly Strike Lee leads a martial arts team to derring-do. A challenge for Lee, but not for the viewer.
1981 92m/C Bruce Lee. **VHS, Beta $29.99** BFV, MRV 🎞½

The Deadly Summer A beautiful woman swears violent vengeance on the cretins who raped her mother. In French with subtitles.
1983 102m/C FR Isabelle Adjani; *Dir:* Jean Becker. Cesar Awards '83: Best Actress (Adjani). **VHS, Beta $59.95** FCT 🎞🎞½

Deadly Sunday A family's weekly Sunday drive turns terrifying when desperate jewel thieves detour them and hold them hostage.

1982 85m/C Dennis Ely, Henry Sanders, Gylian Roland, Douglas Alexander; **Dir:** Donald M. Jones. **VHS, Beta** $59.98 *LIV, LTG* ℤ 1/2

Deadly Surveillance Two tough policemen investigate a series of drug-related murders and stake out a possible suspect— a beautiful woman who's the lover of one of the lawmen. Tension in this Canadian-made cop opera evaporates in the third act for a standard buddy-buddy action finale.
1991 (R) 92m/C Michael Ironside, Christopher Bondy, Susan Almgren, Vlasta Vrana, David Carradine; **Dir:** Paul Ziller. **VHS** $89.98 *REP* ℤℤ

Deadly Thief A former jewel thief comes out of retirement to challenge those who are now tops in the profession to try and steal the world's most precious ruby from him. Of course all they have to lose is their lives. What could have been an exciting adventure is made curiously boring. Originally titled "Shalimar."
1978 90m/C *IN* Rex Harrison, John Saxon, Sylvia Miles, Dharmendra, Zenat Aman, Shammi Kapoor; **Dir:** Krishna Shah. **VHS, Beta** $49.95 *PSM* ℤℤ

The Deadly Trackers If gorey oaters are your thing, this one's definitely for you; if not, make tracks. Harris plays a spiteful sheriff who heads south of the border to get his pound of flesh from the outlaws who slew his family in a bank robbery. Overlong revenge-o-rama based on Samuel Fuller's short story, "Riata"; might have been bearable if Fuller hadn't been bumped from the director's chair (as it is, he and other contributors refused to be listed in the credits).
1973 (PG) 104m/C Richard Harris, Rod Taylor, Al Lettieri, Neville Brand, William Smith, Paul Benjamin, Pedro Armendariz Jr.; **Dir:** Barry Shear. **VHS, LV** $59.95 *WAR, PIA, FCT* ℤ

The Deadly Trap Confusing thriller that couldn't. Dunaway is the mentally unstable wife of an ex-spy. His former employers have targeted him with plans to use her and her fragile state-of-mind to their advantage. Also called "Death Scream."
1971 (PG) 96m/C *FR IT* Faye Dunaway, Frank Langella, Barbara Parkins, Maurice Ronet; **Dir:** Rene Clement. **VHS** *FCT* ℤ

Deadly Twins After being beaten and gang-raped, twin sister singers are out to get revenge. Nothing new here; don't waste your time.
1985 87m/C Audrey Landers, Judy Landers, Ellie Russell, Wayne Allison; **Dir:** Joe Oaks. **VHS, Beta** $79.95 *PSM* Woof!

Deadly Vengeance Strictly low budget flick in which the mob kills Jones's boyfriend, forcing her to mount a vengeful offensive against the crime syndicate's bigwigs.
1981 84m/C Grace Jones, Alan Marlowe, Arthur Roberts. **VHS, Beta** $29.95 *AHV* ℤ

Deadly Weapon It's a bit of nerdish wish fulfillment when 15-year-old geek Eastman finds a secret anti-matter weapon conveniently lost by the military in an Arizona stream. You won't like the kid any more than you'll like the story of this low rent boy-gets-back-at-bullies sci fi fiasco.
1988 (PG-13) 89m/C Rodney Eastman, Gary Frank, Michael Horse, Ed Nelson, Kim Walker; **Dir:** Michael Miner. **VHS** $79.95 *TWE, HHE* ℤ

Deadly Weapons Chesty Morgan, she of the 73-inch bustline, takes on the mob using only her God-given abilities. One of Joe Bob Brigg's "Sleaziest Movies in the History of the World" series, and he is welcome to it.

1970 (R) 90m/C Chesty Morgan, Harry Reems; **Dir:** Doris Wishman. **VHS** *VCI* ℤ

Deadman's Curve Decent bio pic detailing the lives of Jan Berry and Dean Torrence who, as Jan and Dean, started the surf music wave only to wipe out after an almost-fatal car wreck. Shows how they deal with and recover from the tragedy.
1978 100m/C Richard Hatch, Bruce Davison, Pamela Bellwood, Susan Sullivan, Dick Clark, Wolfman Jack; **Dir:** Richard Compton. **VHS** $59.95 *GEM* ℤℤ

Deadtime Stories Mother Goose it's not. Fairy tales "Little Red Riding Hood" and "Goldilocks and the Three Bears," among others, are presented like you've never before seen them—as low-budget horror tales.
1986 93m/C Michael Mesmer, Brian DePersia, Scott Valentine, Phyllis Craig, Melissa Leo, Nicole Picard; **Dir:** Jeffrey Delman. **VHS, Beta, LV** $39.95 *CON* ℤ 1/2

Deadwood In the untamed West, a young cowboy is mistaken for the notorious Billy the Kid. Cinematography by Vilmos Zsigmond.
1965 100m/C Arch Hall Jr., Jack Lester, Melissa Morgan, Robert Dix; **Dir:** James Landis. **VHS, Beta** $29.95 *RHI, FCT* ℤ 1/2

Deal of the Century A first-rate hustler and his cohorts sell second-rate weapons to third-world nations; unfortunately, their latest deal threatens to blow up in their faces—literally.
1983 (PG) 99m/C Chevy Chase, Sigourney Weaver, Gregory Hines, Richard Libertini, Wallace Shawn; **Dir:** William Friedkin. **VHS, Beta, LV** $19.98 *WAR* ℤℤ

Dealers Two money-hungry stock traders mix business and pleasure, then set their sights on one final all or nothing score.
1989 (R) 92m/C *GB* Rebecca DeMornay, Paul McGann, Derrick O'Connor; **Dir:** Colin Bucksey. **VHS, Beta, LV** $29.95 *ACA, IME* ℤ 1/2

Dear America: Letters Home from Vietnam Acclaimed documentary featuring pictures and film of the war with voice-overs by dozens of Hollywood stars reading the words of American GIs. Initially made for cable television.
1988 (PG) 84m/C **Dir:** Bill Couturie. Emmy '88: Best Information Special; Television Critics Association '88: Program of the Year; Emmy '88: Best Director (Couturie); Booklist '89: Best of the Decade. **VHS, Beta, 3/4U** $89.98 *HBO, AMB*

Dear Brigitte A young American kid ("Lost in Space" tyke Mumy) writes a love letter to Brigitte Bardot and travels to Paris to find and meet her in person.
1965 100m/C James Stewart, Billy Mumy, Glynis Johns, Fabian, Cindy Carol, John Williams, Jack Kruschen, Brigitte Bardot, Ed Wynn, Alice Pearce; **Dir:** Henry Koster. **VHS, Beta** $19.98 *FOX* ℤ 1/2

Dear Dead Delilah Moorehead is Delilah, the matriarch of a southern family, who is on her deathbed. Her heirs are fighting to the bitter—and bloody—end to be the first to find Delilah's money which is buried somewhere on her land.
1972 90m/C Agnes Moorehead, Will Geer, Michael Ansara, Patricia Carmichael, Dennis Patrick; **Dir:** John Farris. **VHS, Beta** $59.98 *SUE* ℤ 1/2

Dear Detective Vaccaro is the head of the police homicide unit who is trying to track down the killer of several politicians. In the midst of her investigation, she falls for an old friend, a professor of Greek language.
1979 90m/C Brenda Vaccaro, Arlen Dean Snyder, Michael MacRae, John Dennis Johnston, Jack Ging, Stephen McNally, M. Emmet Walsh, Constance Forslund, R.G. Armstrong; **Dir:** Dean Hargrove. **VHS** *USA* ℤℤ

Dear Mark A light, dadaist look at the erection of a large abstract sculpture by Mark di Suvero, set to a Gene Autry soundtrack.
1981 15m/B **Dir:** Danny Lyon. **VHS, Beta** $59.00 *FCT, BBV* ℤℤ

Dear Wife Comedy about those ties that bind. Even though her father is running for reelection to the State Senate, Freeman wants her brother-in-law (Holden) to run against him. Sequel to "Dear Ruth."
1949 88m/B William Holden, Joan Caulfield, Edward Arnold, Billy DeWolfe, Mona Freeman; **Dir:** Richard Haydn. **VHS, Beta** $19.95 *KRT* ℤℤ 1/2

Death of Adolf Hitler A British made-for-television film depicting Hitler's last ten days in an underground bunker. Drug-addled, suicidal and Eva Braun-haunted, he receives the news of the fall of the Third Reich. While the film avoids cliche, it lacks emotional depth.
1984 107m/C *GB* Frank Finlay, Caroline Mortimer; **Dir:** Rex Firlin. **VHS, Beta** $59.95 *LHV, TLF, WAR* ℤℤ 1/2

Death of an Angel When her crippled daughter runs away to join a cult in Mexico, a recently ordained priest (Bedelia) follows her, only to become caught up herself with the leader of the group.
1986 95m/C Bonnie Bedelia, Nick Mancuso, Pamela Ludwig, Alex Colon; **Dir:** Petru Popescu. **VHS, Beta, LV** $36.95 *NSV* ℤ 1/2

Death Be Not Proud Based on the book by John Gunther detailing the valiant battle fought by his son against the brain tumor that took his life at the age of 17. Wonderfully acted by the three principals, especially Alexander, this film leaves us feeling hopeful despite its subject.
1975 74m/C Arthur Hill, Robby Benson, Jane Alexander, Linden Chiles, Wendy Phillips; **Dir:** Donald Wrye. **VHS, Beta** *LCA* ℤℤℤ

Death Beach Karate and judo versus Chinese martial arts. A group of Japanese hoodlums invade a Chinese village in search of a magical herb.
198? 83m/C Chen Kuan-Tai, Chen Sing. **VHS, Beta** $59.95 *SVS* ℤ 1/2

Death Before Dishonor Formula actioner stars Dryer as a tough Marine sergeant who battles ruthless Middle Eastern terrorists after they slaughter his men and kidnap his commanding officer (Keith).
1987 (R) 95m/C Fred Dryer, Brian Keith, Joanna Pacula, Paul Winfield; **Dir:** Terry J. Leonard. **VHS, Beta, LV** $19.95 *NWV, STE* ℤ 1/2

Death Blow The victims of a violent rapist team together to stop the criminal after he repeatedly skirts conviction.
1987 (R) 90m/C Martin Landau, Frank Stallone, Jerry Van Dyke, Terry Moore, Henry Darrow, Jack Carter, Peter Lapis, Don Swayze, Donna Denton; **Dir:** Raphael Nussbaum. **VHS, Beta** $79.95 *NSV* ℤ 1/2

Death of a Bureaucrat A Cuban hero dies, and in tribute, the Communist government buries him with his union card. His widow, however, needs the card in order to collect a pension, and she enlists her nephew's help to retrieve it. The nephew soon finds himself buried in red tape, and digs himself in ever-deeper as he seeks to disinter his late uncle. A wry satire of the communist bureaucracy in Cuba, brimming with comedic tributes to Harold Lloyd, Buster Keaton and Laurel & Hardy. Alea's film was only briefly released in 1966 and was promptly banned, eventually finding its way to the U.S. in 1979. In Spanish with English subtitles.
1966 87m/B Salvador Wood, Silvia Planas, Manuel Estanillo, Gaspar de Santelices, Carlos Ruiz de la Tejera, Omar Alfonso, Ricardo Suarez, Luis Romay, Elsa Montero; ***Dir:*** Thomas Gutierrez Alea. **VHS, Beta $69.95** *NYF, FCT* 🦴🦴🦴

Death of a Centerfold A drama based on the life and tragic death of Playboy model and actress Dorothy Stratten. More effectively handled in Bob Fosse's "Star 80."
1981 96m/C Jamie Lee Curtis, Bruce Weitz, Robert Reed, Mitchell Ryan, Bibi Besch; ***Dir:*** Gabrielle Beaumont. **VHS, Beta $59.95** *MGM* 🦴½

Death Challenge Gangs battle without weapons, motive or discretion in this kung fu extravaganza.
1980 94m/C Steve Leving, Susan Wong. **VHS, Beta $39.95** *UNI* 🦴½

Death Chase An average Joe, just jogging down the street, is thrown into a desperate game of cat-and-mouse when a body lands at his feet.
1987 86m/C William Zipp, Paul Smith, Jack Starrett, Bainbridge Scott; ***Dir:*** David A. Prior. **VHS, Beta $39.95** *NSV* 🦴½

Death Code Ninja Ninjas battle each other sword-to-foot for the possession of a map on which the peace of the world depends.
198? 90m/C John Wilford, Mike Abbott, Judy Barnes, Edgar Fox. **VHS, Beta $59.95** *TWE* 🦴½

Death Collector In the future, when insurance companies run the world, if you don't pay your premium you die.
1989 90m/C Daniel Chapman, Ruth Collins; ***Dir:*** Tom Gniazdowski. **VHS, Beta** *REA* 🦴½

Death Cruise The six winners of a free cruise find that there is a fatal catch to the prize in this made-for-TV movie.
1974 74m/C Kate Jackson, Celeste Holm, Tom Bosley, Edward Albert, Polly Bergen, Michael Constantine, Richard Long; ***Dir:*** Ralph Senesky. **VHS, Beta $39.95** *AHV* 🦴½

The Death Curse of Tartu Another aspiring cult classic. (Read: A low-budget flick so bad it's funny). Students accidentally disturb the burial ground of an Indian medicine man who comes to zombie-like life with a deadly prescription.
1966 84m/C Fred Pinero, Doug Hobart, William Grefe. **VHS, Beta $39.95** *AHV Woof!*

Death in Deep Water A sensuous woman lures a man into a plot to kill her incredibly rich husband.
1974 71m/C Bradford Dillman, Suzan Farmer. **VHS, Beta $29.95** *IVE* 🦴🦴

Death by Dialogue Muddled story of some teenagers who discover an old film script from a project never produced because it was haunted by tragic accidents.
1988 (R) 90m/C Laura Albert, Ken Sagoes; ***Dir:*** Tom Dewier. **VHS, Beta $69.95** *CLV* 🦴

Death from a Distance A reporter and a detective put the moves on a group of astronomers who may be responsible for a murder.
1936 73m/B Russell Hopton, Lola Lane, George F. Marion, John St. Polis, Lee Kohlmar, Lew Kelly, Wheeler Oakman, Robert Frazer, Cornelius Keefe; ***Dir:*** Frank Strayer. **VHS, Beta $16.95** *SNC* 🦴½

Death Driver In order to make a comeback, a stuntman attempts to do a stunt that was responsible for ending his career ten years earlier.
1978 93m/C Earl Owensby, Mike Allen, Patty Shaw, Mary Ann Hearn; ***Dir:*** Jimmy Huston. **VHS, Beta $59.99** *HBO* 🦴½

Death Drug Cliched alarmist film about a talented musician ruining his career by becoming addicted to the drug angel dust.
1983 73m/C Philip Michael Thomas, Vernee Watson, Rosalind Cash; ***Dir:*** Oscar Williams. **VHS, Beta $29.95** *ACA* 🦴½

Death Duel of Mantis A martial arts film featuring Chin Yin Fei.
1984 90m/C VHS, Beta $29.98 *UNI* 🦴½

Death Feud A man tries to save the woman he loves from the man who would destroy her—her former pimp.
1989 98m/C Karen Mayo Chandler, Chris Mitchum, Frank Stallone. **VHS, Beta $14.95** *HMD, SOU* 🦴½

Death Force When a Vietnam veteran comes to New York City, he becomes a hitman for the Mafia.
1978 (R) 90m/C Jayne Kennedy, Leon Isaac Kennedy, James Iglehardt, Carmen Argenziano; ***Dir:*** Cirio H. Santiago. **VHS, Beta $33.95** *KOV* 🦴½

Death Game A businessman allows two innocent-looking young women into his home to make a phone call. Once inside, they proceed to wreak havoc on his mind, body, and property. Also known as "The Seducers."
1976 (R) 91m/C Sondra Locke, Colleen Camp, Seymour Cassel; ***Dir:*** Beth Brickell, Peter S. Traynor. **VHS, Beta $59.95** *UHV* 🦴½

Death Games Two young men shooting a documentary about an influential music promoter ask too many wrong questions, causing the powers-that-be to want them out of the picture for good.
1982 78m/C Lou Brown, David Clendenning, Jennifer Cluff. **VHS, Beta $29.98** *VID* 🦴½

Death Goes to School A strangler is loose in a girl's school, and the music teacher has a hunch as to who it is.
1953 65m/B *GB* Barbara Murray, Gordon Jackson, Pamela Allan, Jane Aird, Beatrice Varley; ***Dir:*** Stephen Clarkson. **VHS, Beta $16.95** *SNC* 🦴🦴

Death of a Gunfighter A western town courting eastern investors and bankers seeks a way to kill their ex-gunslinger sheriff.
1969 (PG) 94m/C Richard Widmark, Lena Horne, Carroll O'Connor, John Saxon, Larry Gates; ***Dir:*** Don Siegel. **VHS, Beta $9.95** *MCA* 🦴🦴

Death of a Hooker An ex-boxer, angered by the indifference to a prostitute's murder, decides to find the killer himself. Also known as "Who Killed Mary What's 'Er Name?"
1971 (PG) 90m/C Red Buttons, Sam Waterston, Sylvia Miles, Conrad Baw; ***Dir:*** Ernie Pintoff. **VHS, Beta $59.95** *GEM* 🦴🦴

Death Hunt A man unjustly accused of murder pits his knowledge of the wilderness against the superior numbers of his pursuers.
1981 (R) 98m/C Charles Bronson, Lee Marvin, Ed Lauter, Andrew Stevens, Carl Weathers, Angie Dickinson; ***Dir:*** Peter R. Hunt. **VHS, Beta, LV $14.98** *FXV, FOX* 🦴🦴

Death of the Incredible Hulk Scientist David Banner's new job just may provide the clues for stopping his transformation into the monstrous Incredible Hulk. But first there are terrorists after the Hulk who need to be defeated and Banner's new romance to contend with. Made for television.
1990 96m/C Bill Bixby, Lou Ferrigno, Elizabeth Gracen, Philip Sterling; ***Dir:*** Bill Bixby. **VHS $79.95** *RHI* 🦴½

Death Is Called Engelchen An influential, important work of the Czech new wave. A survivor of World War II remembers his experiences with the SS leader Engelchen in a series of flashbacks. In Czech with subtitles.
1963 111m/B *CZ* ***Dir:*** Jan Kadar, Elmar Klos. **VHS, Beta $59.95** *FCT* 🦴🦴🦴½

Death Journey Williamson portrays Jesse Crowder who is hired by the New York D.A. to escort a key witness cross-country.
1976 (R) 90m/C Fred Williamson, D'Urville Martin, Bernie Kuby, Heidi Dobbs, Stephanie Faulkner; ***Dir:*** Fred Williamson. **VHS, Beta $49.95** *UNI* 🦴

The Death Kiss Creepy thriller about eerie doings at a major Hollywood film studio where a sinister killer does away with his victims while a cast-of-thousands movie spectacular is in production.
1933 72m/B Bela Lugosi, David Manners, Adrienne Ames, Edward Van Sloan, Vince Barnett; ***Dir:*** Edwin L. Marin. **VHS, Beta, 8mm $14.95** *NOS, SNC, CAB* 🦴🦴½

Death Kiss When a man hires a psychopath to murder his wife, his plan doesn't proceed exactly as expected.
1977 (R) 90m/C Larry Daniels, Dorothy Moore. **VHS, Beta $49.95** *PSM* 🦴½

Death at Love House A screen writer becomes obsessed when he is hired to write the life story of a long dead silent movie queen. Filmed at the Harold Lloyd estate. Made for television.
1975 74m/C Robert Wagner, Kate Jackson, Sylvia Sydney, Joan Blondell, John Carradine; ***Dir:*** E.W. Swackhamer. **VHS, Beta $59.95** *PSM* 🦴½

Death Machines A young karate student must face the "Death Machines," a team of deadly assassins who are trained to kill on command.
1976 (R) 93m/C Ron Marchini, Michael Chong, Joshua Johnson; ***Dir:*** Paul Kyriazi. **VHS, Beta $59.98** *VID* 🦴½

Death Mask of the Ninja Two princes, separated at birth, reunite and kick everyone in the head. Dubbed.
198? 95m/C Joey Lee, Tiger Tung, Jon Chen. **VHS, Beta $54.95** *MAV* 🦴½

The Death Merchant Foundering attempt at a nuclear thriller. A modern mad man hopes to sell a computer chip to third world countries but is thwarted by a secret agent. Confusing, illogical plot highlighted by mediocre performances.
1991 90m/C Lawrence Tierney, Martina Castle, Melody Munyan, Monika Schnarre; ***Dir:*** Jim Winburn. **VHS $79.95** *AIP* 🦴

Death on the Nile Agatha Christie's fictional detective, Hercule Poirot, is called upon to interrupt his vacation to uncover who killed an heiress aboard a steamer cruising down the Nile. Anthony Schaffer's screenplay boasts Ustinov's first stint as the Belgian sleuth. Anthony Powell's costume design won an Oscar.
1978 (PG) 135m/C *GB* Peter Ustinov, Jane Birkin, Lois Chiles, Bette Davis, Mia Farrow, David Niven, Olivia Hussey, Angela Lansbury, Jack Warden, Maggie Smith, George Kennedy, Simon McCorkindale, Harry Andrews, Jon Finch; *Dir:* John Guillermin. Academy Awards '78: Best Costume Design; British Academy Awards '78: Best Costume Design; National Board of Review Awards '78: Best Supporting Actress (Lansbury). **VHS, Beta, LV $19.99** *HBO* 𝟚𝟚½

Death Nurse Fresh from "Crazy Fat Ethel II," aspiring Queen of Camp Alden takes takes up residence in a health care establishment, much to the dismay of the patients who give up their lives in exchange for their money.
1987 80m/C Priscilla Alden, Michael Flood; *Dir:* Nick Phillips. **VHS $29.98** *VCD* 𝟚𝟚

Death by Prescription After adjusting his patients' wills to benefit himself, an evil physician sees to their mysterious deaths.
1987 75m/C Timothy West. **VHS, Beta $29.95** *ACA* 𝟚½

Death Promise In this ho-hum excuse for a horror flick, a murderous landlord tries to evict his tenants.
1978 (R) 90m/C Charles Bonet. **VHS, Beta** *PGN* 𝟚

Death Race 2000 In the 21st century five racing car contenders challenge the national champion of a cross country race in which drivers score points by killing pedestrians. Gory fun. Based on the 1956 story by Ib Melchior. Followed by "Deathsport." Made for television. Also known as "T. P. Sloane."
1975 (R) 80m/C David Carradine, Simone Griffeth, Sylvester Stallone, Mary Woronov; *Dir:* Paul Bartel. **VHS, Beta $19.95** *VDM, MRV* 𝟚𝟚½

Death Rage A hitman comes out of retirement to handle the toughest assignment he has ever faced: search for and kill the man who murdered his brother. He finds out he's the victim of a Mafia doublecross.
1977 (R) 92m/C *IT* Yul Brynner, Martin Balsam; *Dir:* Anthony M. Dawson. **VHS, Beta $14.98** *VID* 𝟚½

Death Raiders When an Asian ruler's family is kidnapped it's up to the Death Raiders to kick their way to a peaceful solution.
19?? 86m/C **VHS, Beta $59.95** *MAG* 𝟚½

Death of the Rat A serious/comic meditation on the tragedy of modern technological life.
1989 6m/C *Dir:* Pascal Aubier. **VHS, Beta** *FFM* 𝟚½

Death Ray It's a race to the finish as terrorists threaten to unleash the world's most powerful weapon.
19?? 93m/C Gordon Scott. **VHS, Beta $19.99** *BFV, MRV* 𝟚½

Death Ray 2000 A pair of futuristic spies search out the whereabouts of a stolen military device that could kill all life on Earth.
1981 100m/C Robert F. Logan, Ann Turkel, Maggie Cooper, Dan O'Herlihy; *Dir:* Lee H. Katzin. **VHS, Beta $49.95** *WOV* 𝟚½

Death of Richie Gazzara does a fine job as a father driven to kill his drug-addicted teenage son, portrayed by Benson. Occasionally lapses into melodrama, but all-in-all a decent adaptation of Thomas Thompson's book "Ritchie" which was based on a true story. Also know as "Ritchie."
1976 97m/C Ben Gazzara, Robby Benson, Eileen Brennan, Clint Howard; *Dir:* Paul Wendkos. **VHS, Beta** *LCA* 𝟚𝟚

Death Rides the Plains Western with a twist. A man lures prospective buyers to his ranch, kills them, and steals their money. It's up to our heroes to ride to the rescue.
1944 53m/B Bob Livingston, Al "Fuzzy" St. John, Nica Doret, Ray Bennet; *Dir:* Sam Newfield. **VHS, Beta $24.95** *VYY, VDM* 𝟚𝟚

Death Rides the Range Russian spies start a range war to cover a spy operation.
1940 57m/B Ken Maynard, Fay McKinzie, Ralph Peters, Charles King, John Elliott, Kenneth Rhodes, Bud Osborne; *Dir:* Sam Newfield. **VHS, Beta $19.95** *NOS, DVT, VCN* 𝟚½

Death Row Diner A movie mogul, executed for a crime he didn't commit, is brought back to life by a freak electrical storm. Nothing will stop him from exacting revenge.
1988 90m/C Jay Richardson, Michelle McClellan, John Content, Tom Schell, Dennis Mooney, Frank Sarcinello Sr., Dana Mason. **VHS, Beta** *CAM* 𝟚½

Death of a Salesman A powerful made-for-television adaptation of the famous Arthur Miller play. Hoffman won an Emmy (as did Malkovich) for his stirring portrayal of Willy Loman, the aging salesman who realizes he's past his prime and tries to come to grips with the life he's wasted and the family he's neglected. Reid also turns in a fine performance as his long-suffering wife.
1986 135m/C Dustin Hoffman, John Malkovich, Charles Durning, Stephen Lang, Kate Reid, Louis Zorich; *Dir:* Volker Schlondorff. **VHS, Beta, LV $19.98** *LHV, FCT, WAR* 𝟚𝟚𝟚½

Death of a Scoundrel A womanizing entrepreneur is murdered and the culprit could be any one of his many jealous romantic conquests. Also known as "Loves of a Scoundrel."
1956 119m/B George Sanders, Zsa Zsa Gabor, Yvonne de Carlo, Victor Jory; *Dir:* Charles Martin. **VHS, Beta, LV $15.95** *UHV* 𝟚𝟚½

Death Scream Drama based on the New York murder of Kitty Genovese whose cries for help while being assaulted went ignored by her neighbors. Good casting but the pacing deteriorates. Written by Stirling Silliphant.
1975 100m/C Raul Julia, John P. Ryan, Lucie Arnaz, Ed Asner, Art Carney, Diahann Carroll, Kate Jackson, Cloris Leachman, Tina Louise, Nancy Walker, Eric Braeden, Allyn Ann McLerie, Tony Dow, Sally Kirkland; *Dir:* Richard T. Heffron. **VHS** *ACE* 𝟚𝟚

Death Screams Attractive college coeds have a party and get hacked to pieces for their troubles by a machete-wielding maniac.
1983 (R) 88m/C Susan Kiger, Jennifer Chase, Jody Kay, William T. Hicks, Martin Tucker. **VHS, Beta** *GEM* 𝟚

Death Sentence When a woman juror on a murder case finds out that the wrong man is on trial, she is stalked by the real killer. Made for television. Also known as "Murder One."

1974 74m/C Cloris Leachman, Laurence Luckinbill, Nick Nolte, William Schallert; *Dir:* E.W. Swackhamer. **VHS, Beta $49.95** *PSM* 𝟚½

Death in the Shadows A young woman searches for her mother while trying to avoid a killer. Filmed entirely in Holland.
19?? 96m/C *Dir:* Vivian Pieters. **VHS $39.98** *CON* 𝟚½

Death Ship A luxury liner is destroyed by an ancient, mysterious freighter on open seas, leaving the survivors to confront the ghost ship's inherent evil in this slow-going horror flick.
1980 (R) 91m/C *CA GB* George Kennedy, Richard Crenna, Nick Mancuso, Sally Ann Howes, Saul Rubinek, Kate Reid; *Dir:* Alvin Rakoff. **VHS, Beta $19.98** *SUE* 𝟚

Death Shot Two plainclothes detectives (who seem just as sleazy as the criminals they're trying to bust) try to break up a drug ring and find that, with a little coercion, their best sources of information are the pimps and junkies they want to stop. Senseless attempt to make an action pic.
1973 90m/C Richard C. Watt, Frank Himes; *Dir:* Mitch Brown. **VHS $59.95** *SVS* 𝟚

Death of a Soldier During World War II, the uneasy U.S.-Australian alliance explodes when a psychotic American soldier murders three Melbourne women. The lawyer hired to defend him has to fight political as well as legal influences. Based on a true story.
1985 (R) 93m/C James Coburn, Reb Brown, Maurie Field, Belinda Darey; *Dir:* Philippe Mora. **VHS, Beta $79.95** *FOX* 𝟚𝟚

Death Spa A health club is possessed by the spirit of a vengeful woman. The fitness craze takes on a campy new twist.
1987 (R) 87m/C William Bumiller, Brenda Bakke, Merritt Butrick; *Dir:* Michael Fischa. **VHS $79.98** *MPI* Woof!

Death Sport In this sequel to the cult hit "Death Race 2000," a group of humans play a game of death in the year 3000. The object is to race cars and kill as many competitors as possible. Not as good as the original, but still watchable.
1978 (R) 83m/C David Carradine, Claudia Jennings, Richard Lynch, William Smithers, Will Walker, David McLean, Jesse Vint; *Dir:* Henry Suso, Allan Arkush. **VHS, Beta** *WAR, OM* 𝟚𝟚

The Death Squad A police commissioner hires an ex-cop to find a group of vigilante cops who are behind a series of gangland-style executions. Good cast, though that's about all this flick has got going for it.
1973 74m/C Robert Forster, Melvyn Douglas, Michelle Phillips, Mark Goddard, Bert Remsen, Claude Akins; *Dir:* Harry Falk. **VHS, Beta $59.95** *PSM* 𝟚½

Death Stalk Dream vacation turns nightmarish for two couples when the wives are taken hostage by several escaped convicts.
1974 90m/C Vince Edwards, Vic Morrow, Anjanette Comer, Robert Webber, Carol Lynley, Neville Brand, Norman Fell; *Dir:* Robert Day. **VHS, Beta $59.95** *GEM, WES, HHE* 𝟚½

Death Target Three former mercenaries team up for one last mission that gets sidetracked when one of the soldiers-of-fortune falls for a worker with a habit.
1983 72m/C Jorge Montesi, Elaine Lakeman; *Dir:* Peter Hyams. **VHS, Beta $49.95** *PSM* 𝟚½

Death Train A dead man seems to have been struck by a train, but investigator Ted Morrow cannot figure out where the train came from. **1979 96m/C** Hugh Keays-Byrne, Ingrid Mason, Max Meldrum; *Dir:* Igor Auzins. **VHS, Beta** PGN ⌐½

Death Valley Greed turns gold prospectors in Death Valley against each other. **1946 70m/C** Robert Lowery, Helen Gilbert, Sterling Holloway; *Dir:* Lew Landers. **VHS, Beta** $29.95 WGE, TIM ⌐½

Death Valley Good cast, lousy plot. Youngster (Billingsley) gets caught up with a psycho cowpoke while visiting mom in Arizona. **1981 (R) 90m/C** Paul LeMat, Catherine Hicks, Peter Billingsley, Wilford Brimley, Edward Herrmann, Stephen McHattie; *Dir:* Dick Richards. **VHS, Beta $79.95** MCA ⌐½

Death Valley Manhunt A marshal hired to protect a man's oil wells discovers that the company manager is keeping the profits for himself. **1943 55m/B** Bill Elliott, George "Gabby" Hayes, Anne Jeffreys; *Dir:* John English. **VHS, Beta** $29.95 DVT ⌐⌐

Death Valley Rangers The hero pretends to be an outlaw to expose a gang robbing gold-laden stagecoaches. **1944 59m/B** Ken Maynard, Hoot Gibson, Bob Steele, Linda Brent, Kenneth Harlan; *Dir:* Robert Tansey. **VHS $19.95** NOS, DVT ⌐½

Death in Venice A lush, decadent adaptation of the Thomas Mann novella about an aging artist, here suggested to be Gustav Mahler, tragically obsessed with ideal beauty as personified in a young boy. Music by Mahler. **1971 (PG) 124m/C** *IT* Dirk Bogarde, Mark Burns, Bjorn Andresen, Marisa Berenson, Silvana Mangano; *Dir:* Luchino Visconti. British Academy Awards '71: Best Art Direction/Set Decoration, Best Cinematography, Best Costume Design, Best Musical Score; Cannes Film Festival '71: Special Prize. **VHS, Beta $59.95** WAR, TAM ⌐⌐⌐½

Death Warmed Up A crazed brain surgeon turns ordinary people into bloodthirsty mutants and a small group of young people travel to his secluded island to stop him. **1985 83m/C** Michael Hurst, Margaret Umbers, David Letch; *Dir:* David Blyth. **VHS, Beta $69.98** LIV, VES ⌐½

Death Warrant Van Damme whams and bams a little less than usual in this cop-undercover-in-prison testosterone fest. As a Royal Canadian Mounty undercover in prison—where inmates are perishing under mysterious circumstances—the pectoral-perfect Muscles from Brussels is on the brink of adding two plus two when an inmate transferee threatens his cover. Contains the requisite gratuitous violence, prison bromide, and miscellaneous other Van Dammages. **1990 (R) 99m/C** Jean-Claude Van Damme, Robert Guillaume, Cynthia Gibb, George Dickerson, Patrick Kilpatrick; *Dir:* Deran Sarafian. **VHS, LV $92.98** MGM, PIA ⌐⌐

Death Watch In the future, media abuse is taken to an all time high as a terminally ill woman's last days are secretly filmed by a man with a camera in his head. Intelligent, adult science fiction with an excellent cast. **1980 (R) 128m/C** *FR GE* Romy Schneider, Harvey Keitel, Harry Dean Stanton, Max von Sydow; *Dir:* Bertrand Tavernier. **VHS, LV $29.98** SUE ⌐⌐⌐

Death Weekend A woman is stalked by a trio of murderous, drunken hoodlums who seek to spoil her weekend. Also titled "House by the Lake." **1976 89m/C** *CA* Brenda Vaccaro, Don Stroud, Chuck Shamata, Richard Ayres, Kyle Edwards; *Dir:* William Fruet. **VHS, Beta $69.98** LIV, VES ⌐½

Death Wish A middle-aged businessman turns vigilante after his wife and daughter are violently attacked and raped by a gang of hoodlums (one is Goldblum in his film debut). He stalks the streets of New York seeking revenge on other muggers, pimps, and crooks, making the neighborhood safer for those less macho. Music by Herbie Hancock. **1974 (R) 93m/C** Charles Bronson, Vincent Gardenia, William Redfield, Hope Lange, Jeff Goldblum, Stuart Margolin, Olympia Dukakis; *Dir:* Michael Winner. **VHS, Beta, LV $14.95** PAR ⌐⌐½

Death Wish 2 Bronson recreates the role of Paul Kersey, an architect who takes the law into his own hands when his family is victimized once again. Extremely violent sequel to the successful 1974 movie. Followed by "Death Wish" III (1985) and IV (1987) in which Bronson continues to torture the street scum and the viewers as well. **1982 (R) 89m/C** J.D. Cannon, Charles Bronson, Jill Ireland, Vincent Gardenia, Anthony (Tony) Franciosa; *Dir:* Michael Winner. **VHS, Beta, LV $19.98** WAR, OM Woof!

Death Wish 3 Once again, Charles Bronson blows away the low lifes who have killed those who were dear to him and were spared in the first two films. **1985 (R) 100m/C** Charles Bronson, Martin Balsam, Deborah Raffin, Ed Lauter, Alex Winter, Marina Sirtis; *Dir:* Michael Winner. **VHS, Beta, LV $14.98** MGM ⌐½

Death Wish 4: The Crackdown The four-times-weary urban vigilante hits crack dealers this time, hard. **1987 (R) 100m/C** Charles Bronson, John P. Ryan, Kay Lenz; *Dir:* J. Lee Thompson. **VHS, Beta, LV $19.95** MED ⌐

Death Wish Club Young woman gets involved in a club with an unusual mission—suicide. Alternate title: "Carnival of Fools." **1983 93m/C** Meridith Haze, Rick Barns, J. Martin Sellers, Ann Fairchild; *Dir:* John Carr. **VHS, Beta $49.95** REG ⌐½

Deathcheaters The Australian Secret Service offers two stuntmen a top secret mission in the Phillipines. **1976 (G) 96m/C VHS, Beta $69.98** VES ⌐½

Deathdream In this reworking of the "Monkey's Paw" tale, a mother wishes her dead son would return from Vietnam. He does, but he's not quite the person he used to be. Gripping plot gives new meaning to the saying, "Be careful what you wish for—it might come true." **1972 98m/C** *CA* John Marley, Richard Backus, Lynn Carlin; *Dir:* Bob Clark. **VHS, Beta $59.95** MPI ⌐⌐½

Deathhead Virgin An evil virgin spirit is waiting to possess one of two men eager to find a sunken fortune. The fate of the fortunate one is only death! **1974 (R) 94m/C** Jock Gaynor, Larry Ward, Diane McBain, Vic Diaz. **VHS, Beta $59.95** AHV ⌐½

Deathmask A four-year-old boy's corpse is found buried in a cardboard box. Detective Douglas Andrews begins an obsessive investigation into the murder.

1969 102m/C Farley Granger, Ruth Warrick, Danny Aiello; *Dir:* Richard S. Friedman. **VHS, Beta $69.95** PSM ⌐⌐

Deathmoon A businessman plagued by recurring werewolf nightmares goes on a Hawaiian vacation to try and forget his troubles. **1978 90m/C** Robert Foxworth, Joe Penny, Debralee Scott, Dolph Sweet, Charles Haid; *Dir:* Bruce Kessler. **VHS, Beta** VCL ⌐½

Deathrow Gameshow Condemned criminals can either win their freedom or die in front of millions on a new television show that doesn't win the host many friends. **1988 78m/C** John McCafferty, Robin Bluthe, Beano, Mark Lasky; *Dir:* Mark Pirro. **VHS, Beta, LV** MED ⌐½

Deathstalker Deathstalker sets his sights on seizing the evil wizard Munkar's magic amulet so he can take over Munkar's castle. The only excuse for making such an idiotic film seems to have been to fill it with half-naked women. Filmed in Argentina and followed by two sequels which were an improvement. **1984 (R) 80m/C** Richard Hill, Barbi Benton, Richard Brooker, Victor Bo, Lana Clarkson; *Dir:* John Watson. **VHS, Beta, LV $29.98** LIV, VES Woof!

Deathstalker 2: Duel of the Titans Campy fantasy comically pits the lead character against an evil wizard. Sword-and-sorcery spoof doesn't take itself too seriously. **1987 (R) 85m/C** John Terlesky, Monique Gabrielle, John Lazar, Toni Naples, Maria Socas, Queen Kong; *Dir:* Jim Wynorski. **VHS, Beta, LV $29.98** LIV, VES ⌐⌐

Deathstalker 3 Another humorous entry in the Deathstalker saga. This time there's action, romance, magic and the Warriors From Hell. **1989 (R) 85m/C** John Allen Nelson, Carla Herd, Terri Treas, Thom Christopher; *Dir:* Alfonso Corona. **VHS, Beta, LV $79.98** LIV, VES, HHE ⌐½

Deathstalker 4: Clash of the Titans The sword and sorcery epic continues, with our hero revealing yet more musculature in his efforts to defeat an evil queen and her legion of stone warriors. **1991 (R) 85m/C** Richard Hill, Maria Ford, Michelle Moffett, Brent Baxter Clark; *Dir:* Howard R. Cohen. **VHS $89.98** NHO ⌐

Deathtrap A creatively-blocked playwright, his ailing rich wife, and a former student who has written a surefire hit worth killing for, are the principals in this compelling comedy-mystery. Cross and double-cross are explored in this film based on the Broadway hit by Ira Levin. **1982 (PG) 118m/C** Henry Jones, Michael Caine, Christopher Reeve, Dyan Cannon, Irene Worth; *Dir:* Sidney Lumet. **VHS, Beta $19.98** WAR ⌐⌐

Debbie Does Las Vegas Debbie Reynolds performs her knockout Las Vegas show that features impressions of Dolly Parton and Mae West and such show-stoppers as "Tammy" and "Broadway Melody." **1985 55m/C** Debbie Reynolds. **VHS, Beta $59.95** LTG ⌐⌐

The Decameron Pasolini's first epic pageant in his "Trilogy of Life" series. An acclaimed, sexually explicit adaptation of a handful of the Boccaccio tales. In Italian with English subtitles.

1970 111m/C *GE IT FR* Franco Citti, Ninetto Davoli, Angela Luce, Patrizia Capparelli, Jovan Jovanovich, Silvana Mangano, Pier Paolo Pasolini; *Dir:* Pier Paolo Pasolini. Berlin Film Festival '71: Second Prize; New York Film Festival '71: Second Prize. **VHS, Beta** $79.95 *WBF, TAM, APD* ⅃⅃½

Decameron Nights Story of Boccaccio's pursuit of a recently-widowed young women is interwoven amongst the three of the fourteenth-century Italian writer's bawdy tales.
1953 87m/C *GB* Louis Jourdan, Joan Fontaine, Binnie Barnes, Joan Collins, Marjorie Rhodes; *Dir:* Hugo Fregonese. **VHS, Beta** $19.95 *NOS, MRV, QNE* ⅃⅃½

The Decathalon A history of the grueling Olympic event.
197? 60m/C **VHS, Beta** $14.95 *DVT* ⅃⅃½

Deceived A successful career woman with a passionate husband and a young daughter feels she has it all until her husband is apparently killed in a bizarre tragedy. But just how well did she know the man she married? Who was he really and what secrets are hidden in his past? As she struggles to solve these mysteries, her own life becomes endangered in this psychological thriller.
1991 (PG-13) 115m/C Goldie Hawn, John Heard, Ashley Peldon; *Dir:* Damien Harris. **VHS, Beta** $92.95 *TOU* ⅃⅃½

The Deceivers A British officer infiltrates a deadly sect in colonial India. The sect murders and robs unwary travelers. Interesting premise falters somewhat due to rather slow pacing.
1988 (PG-13) 112m/C *GB* Pierce Brosnan, Saeed Jaffrey, Shashi Kapoor, Keith Michell; *Dir:* Nicholas Meyer. **VHS, Beta, LV** $89.95 *WAR* ⅃⅃

December Four prep-school students in New Hampshire must confront the reality of war when the Japanese bomb Pearl Harbor. A touching story of courage and friendship that takes place in one night during the suprise-attack.
1991 (PG) 92m/C Wil Wheaton, Chris Young, Brian Krause, Balthazar Getty, Jason London; *W/Dir:* Gabe Torres. **VHS, LV** $89.95 *SVT* ⅃⅃

December 7th: The Movie Banned for 50 years by the U.S. Government, this planned Hollywood explanation to wartime audiences of the Pearl Harbor debacle offers such "offensive" images as blacks fighting heroically alongside whites, loyal Japanese-Americans, and Uncle Sam asleep on the morning of the attack. The Chief of Naval Operations confiscated the original film, claiming it demeaned the Navy, but now it's available to all. The battle scenes were so realistic they fooled even documentarians. This isn't the most incisive video on the event—just an unforgettable snapshot.
1991 82m/B Walter Huston, Harry Davenport; *Dir:* John Ford. **VHS** $19.95 *KIT, CPM, TIM* ⅃⅃½

December Flower A young woman arrives at her aunt's estate so she can nurse her back to health, but she finds her aunt a bit down in the mouth: seems someone is trying to kill her, and she could be next in line.
19?? 65m/C Jean Simmons, Mona Washbourne, Bryan Forbes. **VHS** $19.99 *FCT* ⅃⅃½

Deception Davis is a pianist torn between two loves: her intensely jealous sponsor (Rains) and her cellist boyfriend (Henreid). Plot in danger of going over the melodramatic edge is saved by the very effective performances of the stars.

1946 112m/B Bette Davis, Paul Henreid, Claude Rains, John Abbott, Benson Fong; *Dir:* Irving Rapper. **VHS, Beta** $19.95 *MGM* ⅃⅃⅃

Deceptions A risque, semi-erotic thriller about a wealthy society woman (Sheridan) who kills her husband. Sheridan claims that she acted in self-defense, but macho cop Hamlin has a different view. He interrogates her, grows increasingly attracted, and is drawn into her web of seduction.
1990 (R) 105m/C Harry Hamlin, Nicolette Sheridan. **VHS, LV** $89.98 *REP* ⅃⅃

Declassified: The Plot to Kill President Kennedy A revisionist look at the Kennedy assassination, using newly uncovered evidence, computer photograph enhancement and never-before-seen interviews, propounding the conspiracy theory.
1987 60m/C **VHS, Beta** $14.95 *VTK, VID* ⅃⅃

The Decline of the American Empire The critically acclaimed French-Canadian film about eight academic intellectuals who spend a weekend shedding their sophistication and engaging in intertwining sexual affairs. Examines the differing attitudes of men and women in regards to sex and sexuality. Subtitled in English.
1986 (R) 102m/C *CA* Dominique Michel, Dorothee Berryman, Louise Portal, Genevieve Rioux; *Dir:* Denys Arcand. Genie Awards '87: Best Director (Arcand), Best Picture, Best Screenplay, Best Supporting Actor (Arcand), Best Supporting Actress (Portal). **VHS, Beta** $79.95 *MCA, TAM* ⅃⅃⅃½

Decline of Western Civilization 1 The L.A. hard core punk scene. Music by X, Circle Jerks, Black Flag, Fear and more.
1981 100m/C X, Circle Jerks, Black Flag, Fear, Germs, Catholic Discipline; *Dir:* Penelope Spheeris. **VHS, Beta, LV** $49.95 *MVD, MSM* ⅃⅃⅃

The Decline of Western Civilization 2: The Metal Years Following Spheeris's first documentary about hardcore punk, she in turn delves into the world of heavy metal rock. We are given a look at some of the early rockers, as well as some of the smaller bands still playing L.A. clubs. Features appearances by Alice Cooper, Ozzy Osbourne, Poison, Gene Simmons and Megadeth.
1988 (R) 90m/C Alice Cooper, Ozzy Osbourne, Poison, Gene Simmons, Megadeth, Lizzie Borden, Faster Pussycat; *Dir:* Penelope Spheeris. **VHS, Beta** $19.95 *MVD, COL, FCT* ⅃⅃⅃

Dedee D'Anvers A gritty dockside melodrama about a drunken prostitute abused by her pimp. A sailor and a local barkeep don't like it. In French with subtitles.
1949 95m/B Simone Signoret, Marcel Dalio, Bernard Blier, Marcel Pagliero, Jane Marken; *Dir:* Yves Allegret. **VHS, Beta** $24.95 *NOS, FCT, TAM* ⅃⅃½

A Dedicated Man A lonely woman agrees to act as a wife for a businessman to lend him credibility. A British "Romance Theater" presentation.
1986 60m/C *GB* Alec McCowan, Joan Plowright; *Dir:* Robert Knights. **VHS** $11.95 *PSM* ⅃⅃½

The Deep An innocent couple find themselves in an underwater search for a shipwreck. There's more to find than sunken treasure and they quickly find themselves in over their heads. Gorgeous photography manages to keep this slow mover afloat. Famous for Bisset's wet T-shirt scene. Based on the novel by Peter Benchley. The laserdisc edition offers widescreen format.

1977 (PG) 123m/C Nick Nolte, Jacqueline Bisset, Robert Shaw, Louis Gossett Jr., Eli Wallach; *Dir:* Peter Yates. **VHS, Beta, LV** $14.95 *GKK, PIA*

Deep Cover A man dares to go beyond the walls of a mysterious English manor in order to expose the secrets that lie within the manor.
1988 (R) 81m/C Tom Conti, Donald Pleasence, Denholm Elliott, Kika Markham, Phoebe Nicholls; *Dir:* Richard Loncraine. **VHS, Beta** $79.95 *PSM, PAR* ⅃⅃

Deep End A 15-year-old boy working as an attendant in a bath house falls in love with a 20-year-old woman in this tragic tale of a young man obsessed. Good British cast and music by Cat Stevens.
1970 88m/C *GB GE* Jane Asher, John Moulder-Brown, Diana Dors; *Dir:* Jerzy Skolimowski. **VHS** *FCT* ⅃⅃⅃

Deep in the Heart When a young teacher is raped at gunpoint on a second date, she takes the law into her own hands. Just off the target film tries to take a stand against the proliferation of guns in the U.S. Thought-provoking all the same.
1984 (R) 99m/C *GB* Karen Yahng, Clayton Day; *Dir:* Tony Garnett. **VHS, Beta** $69.99 *HBO* ⅃⅃

Deep in My Heart A musical biography of the life and times of composer Sigmund Romberg, with guest appearances by many MGM stars.
1954 132m/C Jose Ferrer, Merle Oberon, Paul Henreid, Walter Pidgeon, Helen Traubel, Rosemary Clooney, Jane Powell, Howard Keel, Cyd Charisse, Gene Kelly, Ann Miller; *Dir:* Stanley Donen. **VHS, Beta** $19.98 *MGM* ⅃⅃½

Deep Red: Hatchet Murders A stylish but gruesome rock music-driven tale of a composer who reads a book on the occult which happens to relate to the sadistic, sangfroid murder of his neighbor. When he visits the book's author, he discovers that she has been horribly murdered as well. Also known as "The Hatchet Murders." The Italian version, "Profondo Rosso," is longer and even bloodier.
1975 100m/C David Hemmings, Daria Nicolodi; *Dir:* Dario Argento. **VHS, Beta** $19.99 *HBO* ⅃⅃½

Deep Six A World War II drama that examines the conflict between pacifism and loyalty. A staunch Quaker is called to active duty as a lieutenant in the U.S. Navy, where his beliefs put him into disfavor with shipmates.
1958 110m/C Alan Ladd, William Bendix, James Whitmore, Keenan Wynn, Efrem Zimbalist Jr., Joey Bishop; *Dir:* Rudolph Mate. **VHS, Beta** $54.95 *UHV* ⅃⅃½

Deep Space A flesh-eating alien lands on earth and, after devouring a cop, is stalked by his partner. An "Alien" rip-off. Some humorous moments keep us from thinking too hard about how much the monster resembles our friend from "Alien."
1987 90m/C Charles Napier, Ann Turkel, Ron Glass, Bo Svenson, Julie Newmar; *Dir:* Fred Olen Ray. **VHS, Beta** $79.95 *TWE* ⅃½

Deepstar Six When futuristic scientists try to set up an undersea research and missile lab, a group of subterranean monsters get in the way.
1989 (R) 97m/C Taurean Blacque, Nancy Everhard, Greg Evigan, Miguel Ferrer, Matt McCoy, Nia Peeples, Cindy Pickett, Marius Weyers; *Dir:* Sean S. Cunningham. **VHS, Beta, LV** $9.99 *CCB, IVE* ⅃½

The Deer Hunter A powerful and vivid portrait of Middle America with three steel-working friends who leave home to face the Vietnam War. Controversial, brutal sequences in Vietnam are among the most wrenching ever filmed; the rhythms and rituals of home are just as purely captured. Neither pro nor anti-war, but rather the perfect evocation of how totally and forever altered these people are by the war. Emotionally shattering; not to be missed. Oscar nominations for De Niro, Streep, Original Screenplay, and Cinematography.
1978 (R) 183m/C Robert De Niro, Christopher Walken, Meryl Streep, John Savage, George Dzundza, John Cazale, Chuck Aspegren, Rutanya Alda, Shirley Stoler, Amy Wright; **Dir:** Michael Cimino. Academy Awards '78: Best Director (Cimino), Best Film Editing, Best Picture, Best Sound, Best Supporting Actor (Walken); Directors Guild of America Awards '78: Best Director (Cimino); Los Angeles Film Critics Association Awards '78: Best Director (Cimino); New York Film Critics Awards '78: Best Picture, Best Supporting Actor (Walken). **VHS, Beta, LV $14.95** *MCA, FCT, BTV* 🐾🐾🐾🐾

Deerslayer Based on the classic novel by James Fenimore Cooper. Frontiersman Hawkeye and his Indian companion Chingachgook set out to rescue a beautiful Indian maiden and must fight bands of hostile Indians and Frenchmen along the way.
1978 98m/C Steve Forrest, Ned Romero, John Anderson, Joan Prather; **Dir:** Richard Friedenberg. **VHS, Beta $19.95** *STE, UHV, MAG* 🐾½

Def-Con 4 Three marooned space travelers return to a holocaust-shaken Earth to try to start again, but some heavy-duty slimeballs are in charge and they don't want to give it up. Good special effects bolster a weak script.
1985 (R) 85m/C Maury Chaykin, Kate Lynch, Tim Choate; **Dir:** Paul Donovan. **VHS, Beta $14.95** *NWV, STE* 🐾½

Def by Temptation A potent horror fantasy about a young black theology student who travels to New York in search of an old friend. There he meets an evil woman who is determined to seduce him and force him to give in to temptation. Great soundtrack.
1990 (R) 95m/C James Bond III, Kadeem Hardison, Bill Nunn, Samuel L. Jackson, Minnie Gentry, Rony Clanton; **Dir:** James Bond III. **VHS, Beta, LV $89.98** *SGE, IME* 🐾🐾

The Defender CBS "Studio One" presentation of a live courtroom drama. First shown in 1957.
1957 112m/B Steve McQueen, William Shatner, Ralph Bellamy, Martin Balsam. **VHS, Beta $29.95** *VYY, FCT* 🐾🐾½

Defending Your Life Brooks' cockeyed way of looking at the world travels to the afterlife in this uneven comedy/romance. In Judgement City, where everyone goes after death, past lives are examined and judged. If you were a good enough person (and didn't let your fears control your life) you get to stay in heaven (where you wear funny robes and eat all you want without getting fat). If not, it's back to earth for another go-round. Brooks plays a L.A. advertising executive who crashes his new B.M.W. and finds himself defending his life. When he meets and falls in love with Streep, his interest in staying in heaven multiplies. Occasionally charming, seldom out-right hilarious. A funny premise taken too seriously. Available in widescreen on laserdisc.

1991 (PG) 112m/C Albert Brooks, Meryl Streep, Rip Torn, Lee Grant, Buck Henry; **W/Dir:** Albert Brooks. **VHS, LV, 8mm $19.98** *WAR, LDC* 🐾🐾🐾

The Defense Never Rests Perry Mason, the lawyer who never loses a case, aids his friend and secretary, Della Street, in this made-for-television movie.
1990 95m/C Raymond Burr, Barbara Hale, Al Freeman Jr., William Katt, Patrick O'Neal. **VHS, Beta $29.98** *MCG* 🐾🐾

Defense Play Two teenagers are unwittingly stuck in the middle of a Soviet plot to steal the plans that created a technologically advanced helicopter.
1988 (PG) 95m/C David Oliver, Susan Ursitti, Monte Markham, William Frankfather, Patch MacKenzie; **Dir:** Monte Markham. **VHS, Beta $19.98** *TWE* 🐾🐾½

Defense of the Realm A British politician is accused of selling secrets to the KGB through his mistress and only a pair of dedicated newspapermen believe he is innocent. In the course of finding the answers they discover a national cover-up conspiracy. An acclaimed, taut thriller.
1985 (PG) 96m/C *GB* Gabriel Byrne, Greta Scacchi, Denholm Elliott, Ian Bannen, Bill Paterson, Fulton Mackay, Robbie Coltrane; **Dir:** David Drury. **VHS, Beta, LV $14.98** *SUE* 🐾🐾🐾½

Defenseless Hershey tries hard in an unplayable part as a giddy attorney who finds herself trapped in a love-affair/legal case that turns murderous. Good cast and interesting twists contend with a sexist subtext, which proves a woman can't "have it all."
1991 (R) 106m/C Barbara Hershey, Sam Shepard, Mary Beth Hurt, J.T. Walsh, Sheree North, **Dir:** Martin Campbell. **VHS, LV $92.98** *LIV* 🐾🐾

Defiance A former merchant seaman moves into a tenement in a run-down area of New York City. When a local street gang begins terrorizing the neighborhood, he decides to take a stand. Familiar plotline handled well in this thoughtful film.
1979 (PG) 101m/C Jan-Michael Vincent, Art Carney, Theresa Saldana, Danny Aiello; **Dir:** John Flynn. **VHS, Beta $69.98** *LIV, VES* 🐾🐾

Defiant Two rival street gangs clash when one admits a young orphan girl.
1970 93m/C Kent Lane, John Rubenstein, Tisha Sterling; **Dir:** Hall Bartlett. **VHS, Beta $69.98** *LIV, LTG* 🐾½

Defiant Ones Thought-provoking story about racism revolves around two escaped prisoners (one black, one white) from a chain gang in the rural south. Their societal conditioning to hate each other dissolves as they face constant peril together. Critically acclaimed. Oscar nominated for Best Picture, Best Actor (Curtis and Poitier), Supporting Actor (Bikel), Supporting Actress (Williams), Director, Original Screenplay, and Film Editing.
1958 97m/B Tony Curtis, Sidney Poitier, Theodore Bikel, Lon Chaney Jr., Charles McGraw, Cara Williams; **Dir:** Stanley Kramer. Academy Awards '58: Best Black and White Cinematography, Best Original Screenplay; Edgar Allan Poe Awards '58: Best Screenplay; Golden Globe Awards '58: Best Film—Drama. **VHS, Beta $19.95** *MGM, FOX* 🐾🐾🐾½

Deja Vu A lame romantic thriller about the tragic deaths of two lovers and their supposed reincarnation 50 years later.
1984 (R) 95m/C *GB* Jaclyn Smith, Nigel Terry, Claire Bloom, Shelley Winters; **Dir:** Anthony Richmond. **VHS, Beta $79.95** *MGM* 🐾

The Delicate Delinquent Lewis's first movie without Dean Martin finds him in the role of delinquent who manages to become a policeman with just the right amount of slapstick.
1956 101m/B Jerry Lewis, Darren McGavin, Martha Hyer, Robert Ivers, Horace McMahon; **Dir:** Don McGuire. **VHS, Beta, LV $24.95** *PAR* 🐾🐾½

A Delicate Thread A collection of four award-winning cartoons from the Hubley Studio, including "Eggs," (1970) "The Hat," (1964) "Children of the Sun," (1960) and "Second Chance: Sea" (1976).
1986 49m/C *Voices:* Dudley Moore, Dizzy Gillespie. **VHS, Beta $49.95** *DVT* 🐾🐾½

Delightfully Dangerous Often Deadly Dull. A farfetched musical rooted in yesteryear about mismatched sisters, one a 15-year-old farm girl, the other a New York burlesque dancer, in competition on Broadway. Songs include "Only Teasin'," "In a Shower of Stars," "Mynah Bird," "Through Your Eyes to Your Heart," "Delightfully Dangerous."
1945 92m/B Jane Powell, Ralph Bellamy, Constance Moore, Arthur Treacher, Louise Beavers; **Dir:** Arthur Lubin. **VHS $19.95** *DVT, NOS, VDM* 🐾

Delinquent Daughters After a high school girl commits suicide, a cop and a reporter try to find out why so many kids are getting into trouble. Slow-paced drama seems lacking in direction.
1944 71m/B June Carlson, Fifi D'Orsay, Teala Loring; **Dir:** Albert Herman. **VHS, Beta $16.95** *SNC, VYY* 🐾

Delinquent Parents A wayward girl gives up her baby for adoption. She turns her life around and becomes a juvenile court judge. Years later, who should come up before her for misdeeds but her long-lost daughter! The Hound cannot improve on a contemporary critic who renamed this "Delinquent Producers."
1938 62m/B Doris Weston, Maurice Murphy, Helen MacKellar, Terry Walker, Richard Tucker, Morgan Wallace; **Dir:** Nick Grinde. **VHS $15.95** *NOS, LOO* 🐾½

Delinquent School Girls Three escapees from an asylum get more than they bargained for when they visit a Female Correctional Institute to fulfill their sexual fantasies. Also known as "Bad Girls," the film buys into just about every conceivable stereotype.
1984 (R) 89m/C Michael Pataki, Bob Minos, Stephen Stucker; **Dir:** Gregory Corarito. **VHS, Beta $69.98** *LIV, VES, SIM* Woof!

Delirium A homicidal maniac goes on a killing spree and has just one thing in mind—women. An angry group of citizens inadvertantly include the demonic villian in their vigilante club. Nothing redeeming here.
1977 94m/C Turk Cekovsky, Debi Shanley, Terry Ten Brock; **Dir:** Peter Maris. **VHS, Beta $19.95** *PGN* Woof!

Deliverance A terrifying exploration of the primal nature of man and his alienation from nature, based on the novel by James Dickey, which he adapted for the film (and in which he makes an appearance as a sheriff). Four urban professionals, hoping to get away from it all for the weekend, canoe down a southern river, encounter crazed backwoodsmen, and end up battling for survival. Excellent performances all around, especially by Voight. Debuts for Beatty and Cox. "Dueling Banjos" scene and tune are memorable as is scene where the backwoods boys promise to make the fellows squeal like pigs. **1972 (R) 109m/C** Jon Voight, Burt Reynolds, Ronny Cox, Ned Beatty, James Dickey, Bill McKinney; **Dir:** John Boorman. National Board of Review Awards '72: 10 Best Films of the Year. **VHS, Beta, LV $19.98** *WAR, FCT* ✍✍✍✍

Delivery Boys The "Delivery Boys," three breakdancers aiming to win the $10,000 New York City Break-Off, find unusual perils that may keep them from competing.
1984 (R) 94m/C Joss Marcano, Tom Sierchio, Jim Soriero, Mario Van Peebles, Samantha Fox; **Dir:** Ken Handley. **VHS, Beta $19.95** *STE, NWV* ✍✍½

Delos Adventure A geological expedition in South America stumbles upon covert Russian activities and must battle to survive in this overly-violent actioner.
1986 (R) 98m/C Roger Kerns, Jenny Neumann, Kevin Brophy; **Dir:** Joseph Purcell. **VHS, Beta $79.95** *TWE* ✍

Delta Force Based on the true hijacking of a TWA plane in June 1985. Arab terrorists take over an airliner; the Delta Force, led by Lee Marvin and featuring Norris as its best fighter, rescue the passengers in ways which cater directly to our nationalistic revenge fantasies. Average thriller, exciting and tense at times, with fine work from Marvin, Norris, and Forster.
1986 (R) 125m/C Lee Marvin, Chuck Norris, Shelley Winters, Martin Balsam, George Kennedy, Hanna Schygulla, Susan Strasberg, Bo Svenson, Joey Bishop, Lainie Kazan, Robert Forster, Robert Vaughn, Kim Delaney; **Dir:** Menahem Golan. **VHS, Beta, LV $9.99** *MED* ✍✍

Delta Force 2: Operation Stranglehold Delta Force is back with martial artist and military technician Norris at the helm. Action-packed and tense.
1990 (R) 110m/C John P. Ryan, Chuck Norris, Billy Drago, Richard Jaeckel, Paul Perri; **Dir:** Aaron Norris. **VHS, Beta $89.98** *MED, BTV* ✍½

Delta Force Commando Two U.S. Fighter pilots fight against terrorism in the deadly Nicaraguan jungle.
1987 (R) 90m/C Fred Williamson, Bo Svenson. **VHS $79.99** *VES* ✍½

Delta Force Commando 2 The celluloid was hardly dry on the first "Delta Force Commando" before the resourceful Italians began grinding out this follow-up, with the lead commando getting the Force entangled in a deadly international conspiracy.
1990 (R) 100m/C *IT* Richard Hatch, Fred Williamson, Giannina Facio, Van Johnson; **Dir:** Frank Valenti. **VHS $89.95** *LIV* ✍

Delta Fox Delta Fox is carrying a million dollars for the mob, but the mob is carrying a grudge and the chase is on.
1977 90m/C Priscilla Barnes, Richard Lynch, Stuart Whitman, John Ireland; **Dir:** Beverly Sebastian, Ferd Sebastian. **VHS, Beta $30.06** *IND* ✍✍

Deluge Tidal waves causd by earthquakes have destroyed most of New York (though some may think this is no great loss) in this early sci-fi pic.
1933 72m/B Edward Van Sloan, Peggy Shannon, Sidney Blackmer, Fred Kohler Jr., Matt Moore, Samuel S. Hinds, Lane Chandler; **Dir:** Felix Feist. **VHS, Beta** *MWP* ✍

Delusion A gothic triller in which invalid Cotten and family are harassed by a possibly supernatural killer as told by Cotten's nurse. Originally titled "The House Where Death

Lives" and filmed in 1980. Ending is a let down.
1984 (R) 93m/C Patricia Pearcy, David Hayward, John Dukakis, Joseph Cotten, Simone Griffeth; **Dir:** Alan Beattie. **VHS, Beta $24.98** *SUE* ✍

Delusion A yuppie with an embezzled fortune is held up in the Nevada desert by a psycho hood with a showgirl lover. But who's really in charge here? The snappy, hip, film-noir thriller takes a few unlikely twists and has an open Lady-or-the-Tiger finale that may infuriate. Available in a widescreen edition on laserdisc.
1991 (R) 100m/C Jim Metzler, Jennifer Rubin, Kyle Secor, Robert Costanzo, Tracy Walter, Jerry Orbach; **Dir:** Carl Colpaert. **VHS, LV $89.98** *COL, LDC* ✍✍½

Delusions of Grandeur A French comedy of court intrigue set in 17th-century Spain. In French with English subtitles. Based loosely on Victor Hugo's "Ruy Blas."
1976 85m/C *FR* Yves Montand, Louis de Funes, Alice Sapritch, Karin Schubert, Gabriele Tinti; **Dir:** Gerard Oury. **VHS, Beta $24.95** *FCT* ✍✍½

Demented A beautiful and talented woman is brutally gang-raped by four men, but her revenge is sweet and deadly as she entices each to bed and gives them a big dose of their own medicine. All-too-familiar plot offers nothing new.
1980 (R) 92m/C Sallee Elyse, Bruce Gilchrist. **VHS, Beta $54.95** *MED* ✍

Dementia 13 This eerie thriller, set in a creepy castle, is an early Coppola film about the members of an Irish family who are being offed by an axe murderer one-by-one.
1963 75m/B William Campbell, Luana Anders, Bart Patton, Patrick Magee; **W/Dir:** Francis Ford Coppola. **VHS, Beta $19.95** *NOS, MRV, SNC* ✍✍½

Demetrius and the Gladiators A sequel to "The Robe," wherein the holy-robe-carrying slave is enlisted as one of Caligula's gladiators and mixes with the trampy empress Messalina.
1954 101m/C Victor Mature, Susan Hayward, Michael Rennie, Debra Paget, Anne Bancroft, Jay Robinson, Barry Jones, Richard Egan, William Marshall, Ernest Borgnine; **Dir:** Delmer Daves. **VHS, Beta, LV $19.98** *FOX* ✍✍

The Demi-Paradise Tongue-in-cheek look at people's perceptions of foreigners has Olivier as a Russian inventor (?!) who comes to England with some trepidation. Though the Brits are plenty quirky, he manages to find romance with Ward in this charming look at British life.
1943 110m/B *GB* Laurence Olivier, Penelope Dudley Ward, Margaret Rutherford, Wilfrid Hyde-White; **Dir:** Anthony Asquith. **VHS, Beta $19.95** *NOS, VDM, DVT* ✍✍✍

Demolition An ex-international courier finds that the "simple" task he has consented to do for his old employers is deceptively hazardous.
1977 90m/C John Waters, Belinda Giblin, Fred Steele, Vincent Ball; **Dir:** Kevin Dobson. **VHS, Beta $59.95** *PGN* ✍½

The Demon A small town may be doomed to extinction, courtesy of a monster's thirst for the blood of its inhabitants.
1981 (R) 94m/C Cameron Mitchell, Jennifer Holmes. **VHS, Beta $14.98** *HBO, UHV, VID* ✍½

Demon Barber of Fleet Street Loosely based on an actual event, this film inspired the 1978 smash play "Sweeney Todd." Slaughter stars as a mad barber who doesn't just cut his client's hair. He happens

to have a deal cooked up with the baker to provide him with some nice 'juicy' filling for his meat pies. Manages to be creepy and funny at the same time.
1936 68m/B *GB* Tod Slaughter, Eve Lister, Bruce Seton; **Dir:** George King. **VHS, Beta $19.95** *NOS, MRV, SNC* ✍✍½

The Demon Fighter A mystical super-being suddenly appears to kick, punch and smash anyone who gets in his way. What he wants, no one knows.
1988 ?m/C VHS $29.95 *OCE* ✍✍½

Demon with a Glass Hand In the vein of "The Terminator," a man goes back two hundred years in time to escape the evil race that rules the planet.
1964 63m/B Robert Culp, Arlene Martel, Abraham Sofaer. **VHS, Beta $14.95** *MGM* ✍½

Demon Hunter Terror reigns while a deranged killer stalks his next victim.
1988 90m/C George Ellis, Erin Fleming, Marrianne Gordon; **Dir:** Massey Cramer. **VHS, Beta** *CAM* ✍½

The Demon Lover Leader of a satanic cult calls forth the devil when he doesn't get his way. Poorly acted and badly produced. Features comic book artists Val "Howard the Duck" Mayerick and Gunnar "Leatherface" Hansen. Also on video as "Devil Master" and "Master of Evil."
1977 (R) 87m/C Christmas Robbins, Val Mayerick, Gunnar Hansen, Tom Hutton, David Howard; **W/Dir:** Donald G. Jackson, Jerry Younkins. **VHS, Beta $9.95** *UNI, REG, PMR* Woof!

Demon Lust A couple's lovely country vacation is turned into a nightmare when two drifters terrorize them.
19?? 76m/C Dir: Bernard Buys. **VHS, Beta $59.95** *GHV* ✍

Demon of Paradise Dynamite fishing off the coast of Hawaii unearths an ancient, man-eating lizard-man. Unveven, unexciting horror attempt.
1987 (R) 84m/C Kathryn Witt, William Steis, Leslie Huntly, Laura Banks; **Dir:** Cirio H. Santiago. **VHS, Beta $19.98** *WAR* Woof!

Demon Queen A vampirish woman seduces, then murders, many men.
198? 70m/C Mary Fanaro, Dennis Stewart, Cliff Dance; **Dir:** Donald Farmer. **VHS, Beta $59.95** *MGL* ✍

Demon Rage A neglected housewife drifts under the spell of a phantom lover. Originally titled "Satan's Mistress."
1982 (R) 98m/C Britt Ekland, Lana Wood, Kabir Bedi, Don Galloway, John Carradine, Sherry Scott; **Dir:** James Polakof. **VHS, Beta $59.95** *HAR, HHE* ✍½

Demon Seed When a scientist and his wife separate so he can work on his computer, the computer takes over the house and impregnates the wife. Bizarre and bad.
1977 (R) 97m/C Julie Christie, Fritz Weaver; **Dir:** Donald Cammell; **Voices:** Robert Vaughn. **VHS, Beta, LV $59.95** *MGM* ✍✍✍

Demon for Trouble Outlaws are murdering land buyers in this routine western.
1934 58m/B Bob Steele. **VHS, Beta $19.95** *NOS, WGE, RXM* ✍

Demon Warrior An Indian shaman curses a land-stealing white man and revenge comes in the form of the Demon Warrior, who can appear but once every ten years.
19?? 82m/C VHS $79.98 *MNC* ✍✍

D

Demon Wind A gateway to Hell opens up on a secluded farm and various heroic types try to close it. Meanwhile, demons attempt to possess those humans in their midst. Zombies abound in this near woof.
1990 (R) 97m/C Eric Larson, Francine Lapensee, Rufus Norris; *Dir:* Charles Philip Moore. **VHS, Beta** $79.95 *PAR*

Demon Witch Child A wrongly accused old woman commits suicide and possesses a young girl who takes revenge on everyone.
1975 90m/C **VHS, Beta** $9.95 *NO Woof!*

Demonic Toys The possessed play things attack a bunch of unfortunates in a warehouse, and pumped-up lady cop Scoggins deserves an award for keeping a straight face. Skimpily scripted gore from the horror assembly-line at Full Moon Productions. Far more entertaining are the multiple behind-the-scenes featurettes on the tape.
1991 (R) 86m/C Tracy Scoggins, Bentley Mitchum, Michael Russo, Jeff Weston, Daniel Cerney, Pete Schrum; *Dir:* Peter Manoogian. **VHS, Beta, LV** *PAR, PIA* 🐾½

Demonoid, Messenger of Death The discovery of an ancient temple of Satan worship drastically changes the lives of a young couple when the husband becomes possessed by the Demonoid, which intially takes on the form of a severed hand. Poor script produces a number of unintentionally laughable moments. Also called "Macabra."
1981 (R) 85m/C Samantha Eggar, Stuart Whitman, Roy Cameron Jenson; *Dir:* Alfredo Zacharias. **VHS, Beta** $9.95 *MED* 🐾

The Demons As she slips into death, a tortured woman curses her torturers.
1974 (R) 90m/C Anne Libert, Britt Nichols, Doris Thomas. **VHS, Beta** $49.95 *UNI* 🐾

Demons A horror film in a Berlin theater is so involving that its viewers become the demons they are seeing. The new monsters turn on the other audience members. Virtually plotless, very explicit. Rock soundtrack by Accept, Go West, Motley Crue, and others. Revered in some circles, blasted in others. Followed by "Demons 2."
1986 (R) 89m/C IT Urbano Barberini, Natasha Hovey, Paolo Cozza, Karl Zinny, Fiore Argento, Fabiola Toledo, Nicoletta Elni; *Dir:* Lamberto Bava. **VHS, Beta, LV** $19.95 *STE, NWV* 🐾½

Demons 2 Lamberto Bava's inferior sequel to "Demons" has been compared to the works of Romero, Raimi, and even Cronenberg, but never, ever, favorably. The son of horror-meister Mario Bava, Lamberto collaborated with Italian auteur Dario Argento (who co-wrote and produced) to create a nonsequitur of a sequel (storywise); although a hit in Italy, "Demons 2" seems to have been edited by some kind of crazed cutting room slasher, and the dubbing is equally hackneyed. We find the residents of a chi-chi high rise watching a documentary about the events of "Demons"—a sort of high-tech play-within-a-play ploy—when a demon emerges from a TV set, spreads his creepy cooties, and causes the tenants to sprout fangs and claws and nasty tempers. Much blood dripping from ceilings, plumbing fixtures and various body parts.
1987 88m/C IT David Knight, Nancy Brill, Coralina Cataldi Tassoni, Bobby Rhodes, Asia Argento, Virginia Bryant; *Dir:* Lamberto Bava. **VHS** $29.99 *IMP* 🐾½

Demons in the Garden Centers on the disintegration of a family after the end of the Spanish Civil War, due to sibling rivalries and a variety of indiscretions. Problems mount as two brothers become involved with their stepsister. Story takes on poignancy as it is told by a young boy. Beautiful cinematography. In Spanish with English subtitles.
1982 100m/C SP Angela Molina, Ana Belen, Encarna Paso, Imanol Arias; *Dir:* Manuel Gutierrez Aragon. **VHS** $79.95 *CVC, NVH, TAM* 🐾🐾🐾

Demons of Ludlow Demons attend a small town's bicentennial celebration intent on raising a little hell of their own.
1975 83m/C Paul von Hauser, Stephanie Cushna, James Robinson, Carol Perry; *Dir:* Steven Kuether. **VHS, Beta** $59.95 *TWE* 🐾

Demons of the Mind A sordid psychological horror film about a baron in 19th-century Austia who imprisons his children, fearing that mental illness is a family trait. Also on video as "Blood Evil."
1972 (R) 85m/C GB Michael Hordern, Patrick Magee, Yvonne Mitchell, Robert Hardy, Gillian Hills, Virginia Wetherell, Shane Briant, Paul Jones; *Dir:* Peter Sykes. **VHS, Beta** $59.99 *HBO, ACA, MLB* 🐾🐾

Demonstone A television reporter becomes possessed by a Filipino demon and carries out an ancient curse against the family of a corrupt government official.
1989 (R) 90m/C Lee Ermey, Jan-Michael Vincent, Nancy Everhard; *Dir:* Andrew Prowse. **VHS, Beta, LV** $29.95 *FRH* 🐾

Demonstrator A father and son are at odds when dad heads an Asian military coalition and junior protests. Notable for being one of the first major Australian films to use an entirely native cast.
1971 82m/C AU Joe James, Irene Inescort, Gerard Maguire, Wendy Lingham, Harold Hopkins; *Dir:* Warwick Freeman. **VHS** $19.95 *ACA* 🐾🐾½

Demonwarp A vengeance-minded hunter journeys into an evil, primeval forest to kill the monsters that abducted his daughter.
1987 91m/C George Kennedy, Emmett Alston. **VHS, Beta** $79.95 *VMK Woof!*

Dempsey In this made-for-television film based on Jack Dempsey's autobiography, Williams plays the role of the famed world heavyweight champ. His rise through the boxing ranks as well as his personal life is chronicled. A bit slow-moving considering the pounding fists and other action.
1983 110m/C Treat Williams, Sam Waterston, Sally Kellerman, Victoria Tennant, Peter Mark Richman, Jesse Vint, Bonnie Bartlett, James Noble; *Dir:* Gus Trikonis. **VHS, Beta** *MED* 🐾🐾

Denial: The Dark Side of Passion A strong-willed woman tries to put a destructive love affair behind her—but if she succeeded there'd be no movie. Torrid stuff with a watchable cast.
1991 (R) 103m/C Robin Wright, Jason Patric, Barry Primus, Christina Harnos, Rae Dawn Chong; *W/Dir:* Eric Dignam. **VHS, LV** $89.98 *REP* 🐾🐾

The Dentist Fields treats several oddball patients in his office. After watching the infamous tooth-pulling scene, viewers will be sure to brush regularly.
1932 22m/B W.C. Fields, Elise Cavanna, Babe Kane, Bud Jamison, Zedna Farley. **VHS, Beta** $19.98 *CCB, FST* 🐾🐾🐾½

The Denver & Rio Grande Action-packed western about two rival railroads competing to be the first to complete a line through the Royal Gorge. Features a climactic head-on train collision.
1951 89m/C Edmond O'Brien, Sterling Hayden, Dean Jagger, ZaSu Pitts, J. Carroll Naish; *Dir:* Byron Haskin. **VHS** *PAR* 🐾🐾½

Departamento Compartido A notorious ladies' man and his shy friend become unlikely roommates after their wives kick them out of their homes.
1985 100m/C SP Alberto Olmedo, Tato Bores, Graciela Alfano, Camila Perisse; *Dir:* Hugo Sofovich. **VHS, Beta** $29.95 *MED* 🐾

The Deputy Drummer A second-rate early British musical, notable as a later vehicle for silent-era star Lane. He plays a penniless composer who impersonates an aristocrat to attend a party, and catches jewel robbers in the act.
1935 71m/B Lupino Lane, Jean Denis, Kathleen Kelly, Wallace Lupino, Margaret Yarde, Syd Crossley. **VHS** $19.95 *NOS* 🐾½

Deputy Marshall Hall trails a pair of bankrobbers, fights off landgrabbers, and finds romance as well.
1950 75m/B Jon Hall, Frances Langford, Dick Foran. **VHS, Beta** *WGE, RXM* 🐾

Der Purimshpiler (The Jester) A drifter embarks on a quest for happiness, which takes him to various small towns. In one, he meets and falls in love with a shoemaker's daughter. In Yiddish with English subtitles.
1937 90m/B PL Miriam Kressyn, Zygmund Turkow, Hymie Jacobsen; *Dir:* Joseph Green. **VHS, Beta** $70.05 *ENG* 🐾🐾½

Deranged Of the numerous movies based on the cannibalistic exploits of Ed Gein ("Psycho," "Texas Chainsaw Massacre," etc.), this is the most accurate. A dead-on performance by Blossom and a twisted sense of humor help move things along nicely. The two directors, Jeff Gillen and Alan Ormsby, previously worked together on the classic "Children Shouldn't Play with Dead Things," as did producer Bob Clark. An added attraction is the early special effect work of gore wizard Tom Savini.
1974 (R) 82m/C CA Roberts Blossom, Cosette Lee, Robert Warner, Marcia Diamond, Brian Sneagle; *Dir:* Jeff Gillen, Alan Ormsby; *Nar:* Leslie Carlson. **VHS** $18.00 *FRG, MRV* 🐾🐾🐾

Deranged A mentally unstable, pregnant woman is attacked in her apartment after her husband leaves town. She spends the rest of the movie engaged in psychotic encounters, real and imagined. Technically not bad, but extremely violent and grim. Not to be confused with 1974 movie of the same name.
1987 (R) 85m/C Jane Hamilton, Paul Siederman, Jennifer Delora, James Gillis, Jill Cumer, Gary Goldman; *Dir:* Chuck Vincent. **VHS** $14.95 *REP* 🐾½

Derby The documentary story of the rise to fame of roller-derby stars on the big rink.
1971 (R) 91m/C Charlie O'Connell, Lydia Gray, Janet Earp, Ann Colvello, Mike Snell; *Dir:* Robert Kaylor. **VHS, Beta** $49.95 *PSM* 🐾🐾

Dersu Uzala An acclaimed, photographically breathtaking film about a Russian surveyor in Siberia who befriends a crusty, resourceful Mongolian. They begin to teach each other about their respective worlds. Produced in Russia; one of Kurosawa's stranger films.

1975 124m/C *JP RU* Yuri Solomin, Maxim Munzuk; *Dir:* Akira Kurosawa. Academy Awards '75: Best Foreign Language Film; New York Film Festival '76: Best Foreign Language Film. **VHS, Beta $29.98** *SUE* 𝄞𝄞𝄞½

Descending Angel The premise is familiar: a swastika-friendly collaborator is forced out of the closet. But the cast and scripting make this made-for-cable there's a Nazi in the woodwork suspenser better than average, despite its flawed finale. Scott plays a well-respected Romanian refugee—active in the community, in the church, and in Romanian-American activities—whose daughter's fiance (Roberts) suspects him of Nazi collusion.
1990 (R) 96m/C George C. Scott, Diane Lane, Eric Roberts, Mark Margolis, Vyto Ruginis, Amy Aquino; *Dir:* Jeremy Paul Kagan; *W/Dir:* George C. Scott. **VHS $89.99** *HBO* 𝄞𝄞𝄞

Desde el Abismo After the birth of her son, a young mother takes to drink while in the throes of post-partum depression.
19?? 115m/C *SP* Thelma Biral, Alberto Argibay, Olga Zubarry. **VHS, Beta $29.95** *MED* 𝄞½

Desert Bloom On the eve of a nearby nuclear bomb test, a beleaguered alcoholic veteran and his family struggle through tensions brought on by a promiscuous visiting aunt and the chaotic, rapidly changing world. Gish shines as the teenage daughter through whose eyes the story unfolds. From a story by Corr and Linda Ramy.
1986 (PG) 103m/C Jon Voight, JoBeth Williams, Ellen Barkin, Annabeth Gish, Allen Garfield, Jay Underwood; *Dir:* Eugene Corr. **VHS, Beta, LV $9.95** *RCA* 𝄞𝄞𝄞

Desert Commandos When the Allies appear to be winning World War II, the Nazis devise a plan to eliminate all of the opposing forces' leaders at once.
196? ?m/C Kenneth Clark, Horst Frank, Jeanne Valeri, Carlo Hinterman, Gianni Rizzo. **VHS** *VTP* 𝄞𝄞

The Desert Fox Big-budgeted portrait of German Army Field Marshal Rommel, played by Mason, focuses on the soldier's defeat in Africa during World War II and his subsequent, disillusioned return to Hitler's Germany. Mason played Rommel again in 1953's "Desert Rats."
1951 87m/B James Mason, Cedric Hardwicke, Jessica Tandy, Luther Adler; *Dir:* Henry Hathaway. **VHS, Beta, LV $19.98** *FOX, TLF* 𝄞𝄞𝄞

Desert Gold A fierce Indian chief battles a horde of greedy white men over his tribe's gold mine. Based on a novel by Zane Grey.
1936 58m/B Buster Crabbe, Robert Cummings, Marsha Hunt, Tom Keene, Raymond Hatton, Monte Blue, Leif Erickson; *Dir:* Charles T. Barton. **VHS, Beta $19.95** *NOS, DVT* 𝄞½

Desert Hearts An upstanding professional woman travels to Reno, Nevada in 1959 to obtain a quick divorce, and slowly becomes involved in a lesbian relationship with a free-spirited casino waitress.
1986 (R) 93m/C Helen Shaver, Audra Lindley, Patricia Charbonneau, Andra Akers, Dean Butler, Jeffrey Tambor, Denise Crosby, Gwen Welles; *Dir:* Donna Deitch. **VHS, Beta, LV $29.98** *LIV, VES* 𝄞𝄞𝄞

Desert Phantom Villains threaten to rob a woman of her ranch until a stranger rescues her.
1936 66m/B Johnny Mack Brown. **VHS, Beta, LV** *WGE* 𝄞½

The Desert Rats A crusty British captain (Burton) takes charge of an Australian division during World War II. Thinking they are inferior to his own British troops, he is stiff and uncaring to the Aussies until a kind-hearted drunk and the courage of the division win him over. Crack direction from Wise and Newton's performance (as the wag) simply steals the movie. Mason reprises his role as Germany Army Field Marshal Rommel from "The Desert Fox."
1953 88m/B Richard Burton, Robert Newton, Robert Douglas, Torin Thatcher, Chips Rafferty, Charles Tingwell, James Mason; *Dir:* Robert Wise. **VHS, Beta $19.98** *FXV, FCT* 𝄞𝄞𝄞

Desert Snow Here's a non-formulaic dope opera oater: cowboys 'n' drug runners battle it out for an out-of-the-way, underpopulated western town. About as clever as the title.
1989 90m/C Frank Capizzi, Flint Carney, Shelley Hinkle, Sam Incorvia, Carolyn Jacobs; *Dir:* Paul de Cruccio. **VHS, Beta** *RHV* 𝄞𝄞

The Desert of the Tartars Story of a young soldier who dreams of war and discovers that the real battle for him is with time.
1982 (PG) 140m/C Jacques Perrin. **VHS, Beta $59.95** *SUE* 𝄞

Desert Tigers (Los Tigres del Desierto) Allied soldiers escape from a German POW camp and battle for survival across the African deserts.
19?? 95m/C **VHS, Beta $39.98** *JCI*

Desert Trail Wayne is a championship rodeo rider accused of bank robbery.
1935 57m/B John Wayne, Paul Fix, Mary Kornman; *Dir:* Robert N. Bradbury. **VHS $19.95** *COL, REP, MRV* 𝄞½

Desert Warrior Earth becomes a waste site after a nuclear war. Ferrigno stars as a post-nuke hero in this low-budget action film. Very bad acting.
1988 (PG-13) 89m/C Lou Ferrigno, Shari Shattuck, Kenneth Peer, Anthony East; *Dir:* Jim Goldman. **VHS, Beta $79.95** *PSM* 𝄞

Deserters Sergeant Hawley, a Vietnam-era hawk, hunts deserters and draft-dodgers in Canada. There, he confronts issues of war and peace head-on.
198? 110m/C Alan Scarfe, Dermot Hennelly, Jon Bryden, Barbara March. **VHS, Beta $59.98** *MAG* 𝄞

Designing Woman Bacall and Peck star in this mismatched tale of romance. She's a chic high-fashion designer, he's a rumpled sports writer. The fun begins when they try to adjust to married life together. Neither likes the other's friends or lifestyle. Things get even crazier when Bacall has to work with her ex-lover Helmore on a fashion show and Peck's former love Gray shows up as well. And as if that weren't enough, Peck is being hunted by the mob because of a boxing story he's written. It's a fun, quick, witty tale that is all entertainment and no message. Bacall's performance is of note because Bogart was dying of cancer at the time.
1957 118m/C Gregory Peck, Lauren Bacall, Dolores Gray, Sam Levene, Tom Helmore, Mickey Shaughnessy, Jesse White, Chuck Connors, Edward Platt, Alvy Moore, Jack Cole; *Dir:* Vincente Minnelli. Academy Awards '57: Best Story & Screenplay (Wells). **VHS, Beta, LV $19.98** *MGM* 𝄞𝄞𝄞

Desire and Hell at Sunset Motel Low-budget thriller that takes place at the Sunset Motel in 1950s Anaheim. Fenn stars as the bombshell wife of a toy salesman who's in town for a sales meeting, while she just wants to visit Disneyland. She's soon fooling around with another guy and all chaos breaks out, as her husband hires a psychotic criminal to spy on her. Very confusing plot which isn't worth figuring out. Film's only redeeming quality is the imaginative and creative work used in the visuals.
1992 (PG-13) 90m/C Sherilyn Fenn, Whip Hubley, David Hewlett, David Johansen, Paul Bartel; *W/Dir:* Alien Castle. **VHS, Beta, LV $89.98** *FXV* 𝄞½

Desire Under the Elms Ives, the patriarch of an 1840s New England farming family, takes a young wife (Loren) who promptly has an affair with her stepson. Loren's American film debut. Based on the play by Eugene O'Neill.
1958 114m/B Sophia Loren, Anthony Perkins, Burl Ives, Frank Overton; *Dir:* Delbert Mann. **VHS, Beta, LV $19.95** *KRT, FCT, PAR* 𝄞𝄞½

Desiree A romanticized historical epic about Napoleon and his 17-year-old mistress, Desiree. Based on the novel by Annemarie Selinko. Slightly better than average historical fiction piece.
1954 110m/C Marlon Brando, Jean Simmons, Merle Oberon, Michael Rennie, Cameron Mitchell, Elizabeth Sellars, Cathleen Nesbitt; *Dir:* Henry Koster. **Beta, LV, CDV $19.98** *FOX, IME* 𝄞𝄞½

Desk Set One of the later and less dynamic Tracy/Hepburn comedies, about an efficiency expert who installs a giant computer in an effort to update a television network's female-run reference department. Still, the duo sparkle as they bicker, battle, and give in to love. Based on William Marchant's play.
1957 103m/C Spencer Tracy, Katharine Hepburn, Joan Blondell, Gig Young, Dina Merrill, Neva Patterson; *Dir:* Walter Lang. **VHS, Beta, LV $19.98** *FOX, FCT* 𝄞𝄞𝄞

Despair A chilling and comic study of a victimized chocolate factory owner's descent into madness, set against the backdrop of the Nazis rise to power in the 1930s. Adapted from the Nabokov novel by Tom Stoppard.
1979 120m/C *GE* Dirk Bogarde, Andrea Ferreal, Volker Spengler, Klaus Lowitsch; *Dir:* Rainer Werner Fassbinder. New York Film Festival '78: Best Story & Screenplay. **VHS, Beta** *GLV* 𝄞𝄞𝄞

Despedida de Casada After their divorce, a man and a woman concentrate on making the other's life miserable.
1987 101m/C *MX* Ana Louisa Peluffo, Mauricio Garces, Hector Suarez. **VHS, Beta $55.95** *MAD* 𝄞𝄞

Desperados After the Civil War, a murderous renegade and two of his sons go on a looting rampage, eventually kidnapping the child of the third son who just wants to live in peace. Made in Spain.
1969 (PG) 90m/C *SP* Jack Palance, Vince Edwards, Christian Roberts, George Maharis, Neville Brand, Sylvia Syms; *Dir:* Henry Levin. **VHS, Beta $14.95** *COL* 𝄞𝄞½

Desperate An honest truck driver witnesses a mob crime and must escape with his wife in this minor film noir. Eventually, the law is on his tail, too.
1947 73m/B Steve Brodie, Audrey Long, Raymond Burr, Jason Robards Sr., Douglas Fowley, William Challee, Ilka Gruning, Nan Leslie; *Dir:* Anthony Mann. **VHS, Beta $19.98** *TTC, RXM* 𝄞𝄞½

Desperate Cargo Two showgirls stranded in a Latin American town manage to get aboard a clipper ship with hoodlums who are trying to steal the vessel's cargo.
1941 69m/B Ralph Byrd, Carol Hughes, Jack Mulhall. **VHS, Beta $24.95** *NOS, DVT* 🦴

Desperate Hours A tough, gritty thriller about three escaped convicts taking over a suburban home and holding the family hostage. Plenty of suspense and fine acting. Based on the novel and play by Joseph Hayes.
1955 112m/B Humphrey Bogart, Fredric March, Martha Scott, Arthur Kennedy, Gig Young, Dewey Martin, Mary Murphy, Robert Middleton, Richard Eyer; ***Dir:*** William Wyler. Edgar Allan Poe Awards '55: Best Screenplay; National Board of Review Awards '55: Best Director (Wyler). **VHS, Beta $14.95** *PAR* 🦴🦴🦴

Desperate Hours In this remake of the 1955 thriller, an escaped prisoner holes up in a suburban couple's home, waiting for his lawyer/accomplice to take him to Mexico. Tensions heighten between the separated couple and the increasingly nerve-wracked criminals. Terrific, if not downright horrifying, performance by Rourke. Hopkins, Rogers, Crouse, and Lynch work well in their roles, too.
1990 (R) 105m/C Mickey Rourke, Anthony Hopkins, Mimi Rogers, Kelly Lynch, Lindsay Crouse, Elias Koteas, David Morse; ***Dir:*** Michael Cimino. **VHS, Beta, LV $19.98** *FHE, MGM* 🦴🦴½

Desperate Lives High school siblings come into contact with drugs and join their guidance counselor in the war against dope. Average made-for-TV fare.
1982 100m/C Diana Scarwid, Doug McKeon, Helen Hunt, William Windom, Art Hindle, Tom Atkins, Sam Bottoms, Diane Ladd, Dr. Joyce Brothers; ***Dir:*** Robert Lewis. **VHS, Beta $49.95** *IVE* 🦴🦴

Desperate Living Typical John Waters trash. A mental patient (Stole) is released and becomes paranoid that her family may be out to kill her. After aiding in the murder of her husband (the hefty maid, Hill, suffocates him by sitting on him), Stole and Hill escape to Mortville, a town populated by outcasts such as transsexuals, murderers, and the woefully disfigured.
1977 90m/C Jean Hill, Mink Stole, Edith Massey, Mary Vivian Pearce, Liz Renay, Susan Lowe; ***Dir:*** John Waters. **VHS, Beta $39.98** *CGI* 🦴🦴½

Desperate Moves An Oregon-transplanted geek in San Francisco sets out to make himself over in order to win his dream girl. Sometimes effective, often sappy treatment of his coming to terms with the big city.
1986 90m/C Isabel Sanford, Steve Tracy, Paul Benedict, Christopher Lee, Eddie Deezen; ***Dir:*** Oliver Hellman. **VHS, Beta $19.98** *TWE* 🦴🦴

Desperate Scoundrel/The Pride of Pikeville Two classic comedy films back to back featuring the best of the early silent comics.
1922 44m/B Ford Sterling, Ben Turpin, The Keystone Cops. **VHS, Beta $19.98** *CCB* 🦴🦴

Desperate Target The story of courageous people fighting against all odds to survive.
1980 90m/C Chris Mitchum. **VHS, Beta** *PGN* 🦴

Desperate Teenage Lovedolls A trashy view of the formation of a tacky all-girl punk group.
1984 60m/C Jennifer Schwartz, Hilary Rubens, Steve McDonald, Tracy Lea; ***Dir:*** David Markey. **VHS, Beta $28.95** *HHT* 🦴

Desperate Women Made-for-television western about three unjustly accused female convicts rescued en route to prison by an ex-hired gun.
1978 98m/C Susan St. James, Dan Haggerty, Ronee Blakley, Ann Dusenberry, Susan Myers; ***Dir:*** Earl Bellamy. **VHS, Beta $59.95** *IVE* 🦴½

Desperately Seeking Susan A bored New Jersey housewife gets her kicks reading the personals. When she becomes obsessed with a relationship between two lovers who arrange their meetings through the columns, she decides to find out for herself who they are. But when she is involved in an accident the housewife loses her memory and thinks she is Susan, the free-spirited woman in the personals. Unfortunately, Susan is in a lot of trouble with all sorts of unsavory folk and our innocent housewife finds herself caught in the middle. Terrific characters, with special appeal generated by Arquette and Madonna. Quinn winningly plays the romantic interest.
1985 (PG-13) 104m/C Rosanna Arquette, Madonna, Aidan Quinn, Mark Blum, Robert Joy, Laurie Metcalf, Steven Wright, John Turturro, Richard Hell, Annie Golden, Ann Magnuson; ***Dir:*** Susan Seidelman. **VHS, Beta, LV $14.99** *HBO, FCT* 🦴🦴

Destination Moon Story of man's first lunar voyage contains Chesley Bonstell's astronomical artwork and a famous Woody Woodpecker cartoon. Includes previews of coming attractions from classic science fiction films.
1950 91m/C Warner Anderson, Tom Powers, Dick Wesson, Erin O'Brien Moore; ***Dir:*** Irving Pichel. Academy Awards '50: Best Special Effects. **VHS, Beta, LV $19.95** *SNC, MED* 🦴🦴½

Destination Moonbase Alpha In the 21st century, an explosion has destroyed half the moon, causing it to break away from the Earth's orbit. The moon is cast far away, but the 311 people manning Alpha, a research station on the moon, must continue their search for other life forms in outer space. The pilot for the television series "Space: 1999."
1975 93m/C Martin Landau, Barbara Bain, Barry Morse. **VHS, Beta** *FOX* 🦴½

Destination Saturn Buck Rogers awakens from suspended animation in the twenty-fifth century.
1939 90m/B Buster Crabbe, Constance Moore; ***Dir:*** Ford Beebe. **VHS, Beta $19.95** *NOS, CAB* 🦴🦴

Destination Tokyo A weathered World War II submarine actioner, dealing with the search-and-destroy mission of a U.S. sub sent into Tokyo harbor. Available in a colorized version.
1943 135m/B Cary Grant, John Garfield, Alan Hale Jr., Dane Clark, John Ridgely, Warner Anderson, William Prince, Robert Hutton, Tom Tully, Peter Whitney, Faye Emerson, John Forsythe; ***Dir:*** Delmer Daves. **VHS, Beta $19.98** *MGM* 🦴🦴

Destroy All Monsters When alien babes take control of Godzilla and his monstrous colleagues, it looks like all is lost for Earth. Adding insult to injury, Ghidra is sent in to take care of the loose ends. Can the planet possibly survive this madness? Classic Toho monster slugfest also features Mothra, Rodan, Son of Godzilla, Angila, Varan, Baragon, Spigas and others.
1968 (G) 88m/C *JP* Akira Kubo, Jun Tazaki, Yoshio Tsuchiya, Kyoko Ai, Yukiko Kobayashi, Kenji Sahara, Andrew Hughes; ***Dir:*** Inoshiro Honda. **VHS $18.00** *FRG* 🦴🦴½

Destroyer Trials and tribulations aboard a WWII destroyer result in tensions, but when the time comes for action, the men get the job done.
1943 99m/B Edward G. Robinson, Glenn Ford, Marguerite Chapman, Edgar Buchanan, Leo Gorcey, Regis Toomey, Edward Brophy; ***Dir:*** William A. Seiter. **VHS, Beta $79.99** *VIR* 🦴🦴🦴

Destroyer A small-budget film crew goes to an empty prison to shoot, and are stalked by the ghost/zombie remains of a huge serial killer given the electric chair 18 months before.
1988 (R) 94m/C Anthony Perkins, Deborah Foreman, Lyle Alzado; ***Dir:*** Robert Kirk. **VHS, Beta, LV $79.95** *VIR* 🦴

The Destructors An American narcotics enforcement officer in Paris seeks the help of a hitman in order to catch a druglord. Also called "The Marseille Contract."
1974 (PG) 89m/C *GB* Anthony Quinn, Michael Caine, James Mason, Maureen Kerwin, Alexandra Stewart; ***Dir:*** Robert Parrish. **VHS, Beta $69.98** *LIV, VES, HHE* 🦴🦴

Destry Rides Again An uncontrollably lawless western town is whipped into shape by a peaceful, unarmed sheriff. A vintage Hollywood potpourri with Dietrich's finest post-Sternberg moment; standing on the bar singing "See What the Boys in the Back Room Will Have." The second of three versions of this Max Brand story. First was released in 1932; the third in 1954.
1939 94m/B James Stewart, Marlene Dietrich, Brian Donlevy, Charles Winninger, Mischa Auer, Irene Hervey, Una Merkel, Billy Gilbert, Jack Carson, Samuel S. Hinds; ***Dir:*** George Marshall. **VHS, Beta, LV $19.95** *MCA, FCT* 🦴🦴🦴🦴

Details of a Duel: A Question of Honor A butcher and a teacher collide in this comedy, and prodded by the church, militia and town officals, must duel before the entire town. In Spanish with English subtitles.
1989 97m/C *SP* ***Dir:*** Sergio Cabrera. **VHS** *FCT* 🦴🦴½

The Detective Based on G.K. Chesterton's "Father Brown" detective stories, a slick, funny English mystery in which the famous priest tracks down a notorious, endlessly crafty antique thief.
1954 91m/B *GB* Alec Guinness, Peter Finch, Joan Greenwood, Cecil Parker, Bernard Lee; ***Dir:*** Robert Hamer. **VHS, Beta $69.95** *COL* 🦴🦴🦴

Detective A New York detective investigating the mutilation murder of a homosexual finds political and police department corruption. Fine, gritty performances prevail in this suspense thriller.
1968 114m/C Frank Sinatra, Lee Remick, Ralph Meeker, Jacqueline Bisset, William Windom, Robert Duvall, Tony Musante, Jack Klugman, Al Freeman Jr.; ***Dir:*** Gordon Douglas. **VHS, Beta $59.98** *FOX* 🦴🦴🦴

Detective Sadie & Son A television movie about an elderly detective and her young son cracking a case.
1984 96m/C Debbie Reynolds, Sam Wanamaker, Brian McNamara. **VHS, Beta $79.95** *JTC* 🦴

Detective School Dropouts Since they couldn't pass detective school, are they smart enough to outwit a kidnapper? Find out and get a few laughs at the same time.
1985 (PG) 92m/C Lorin Dreyfuss, David Landsberg, Christian de Sica, George Eastman; ***Dir:*** Filippo Ottoni. **VHS, Beta $79.95** *MGM* 🦴🦴

Detective Story Intense drama about a New York City police precinct with a wide array of characters led by a disillusioned and bitter detective (Douglas). Excellent casting is the strong point, as the film can be a bit dated. Based on Sydney Kingsley's Broadway play.
1951 103m/B Kirk Douglas, Eleanor Parker, Horace McMahon, William Bendix, Lee Grant, Craig Hill, Cathy O'Donnell, Bert Freed, George Macready, Joseph Wiseman, Gladys George, Frank Faylen, Warner Anderson; *Dir:* William Wyler. Edgar Allan Poe Awards '51: Best Screenplay. **VHS** *PAR ♫♫♫½*

Detour Considered to be the creme de la creme of "B" movies, a largely unacknowledged but cult-followed noir downer that's well-designed, stylish, and compelling, if not a bit contrived and sometimes annoyingly shrill. Shot in six days with six indoor sets by "B" film swami Ulmer. Neal plays a down-on-his-luck pianist hitching cross-country to rejoin his singer fiancee after both have abandoned hopes of stardom. His first wrong turn involves the accidental death of the man who picked him up from hitchhiking, and after that, he's en route to Destiny with a capital "D" when he picks up fatal femme Ann Savage, who's as vicious a vixen as ever ruined a pretty good man. Told in flashback, it's also been called the most despairing of all "B"-pictures. It's as noir as they get.
1946 67m/B Tom Neal, Ann Savage, Claudia Drake, Edmund MacDonald; *Dir:* Edgar G. Ulmer. **VHS, Beta, LV $19.95** *KRT, MRV, SNC ♫♫♫*

Detour to Danger Two young men set out on a fishing expedition and run into crooks and damsels in distress. Interesting primarily because it was filmed in the three-color Kodachrome process.
1945 56m/C Britt Wood, John Day; *Dir:* Richard Talmadge. **VHS, Beta, 8mm $24.95** *VYY ♫*

Detroit 9000 A pair of Detroit policemen investigate a robbery that occurred at a black congressman's fundraising banquet. Also on video as "Detroit Heat."
1973 (R) 106m/C Alex Rocco, Scatman Crothers, Hari Rhodes, Lonette McKee, Herbert Jefferson Jr.; *Dir:* Arthur Marks. **VHS, Beta $69.99** *HBO, SIM ♫½*

Deutschland im Jahre Null The acclaimed, unsettling vision of post-war Germany as seen through the eyes of a disturbed boy who eventually kills himself. Lyrical and grim, in German with subtitles.
1947 75m/B *GE* Franz Gruger; *Dir:* Roberto Rossellini. **VHS, Beta $29.95** *FCT, GLV ♫♫♫½*

Devastator A Vietnam vet exacts violent revenge on the murderer of an old army buddy.
1985 (R) 79m/C Richard Hill, Katt Shea, Crofton Hardester; *Dir:* Cirio H. Santiago. **VHS, Beta $79.95** *MGM ♫*

Devi A minor film in the Ray canon, it is, nonetheless, a strange and compelling tale of an Indian farmer convincing his daughter-in-law that she is a goddess, and therein driving her mad. English subtitles.
1960 96m/B *IN* Chhabi Biswas, Sharmila Tagore, Soumitra Chatterjee; *Dir:* Satyajit Ray. **VHS, Beta $49.95** *TAM, WFV, DVT ♫♫♫½*

The Devil The last remaining humans try to fight off demonic lifeforms which need to feed on their flesh.
1984 90m/C VHS, Beta $39.95 *VCD ♫*

Devil at 4 O'Clock An alcoholic missionary and three convicts work to save a colony of leper children from a South Seas volcano. Quality cast barely compensates for mediocre script.
1961 126m/B Spencer Tracy, Frank Sinatra, Kerwin Mathews, Jean-Pierre Aumont; *Dir:* Mervyn LeRoy. **VHS, Beta, LV $39.95** *GKK, IME ♫♫½*

The Devil Bats Madman Lugosi trains a swarm of monstrous blood-sucking bats to attack whenever they smell perfume. Also titled "Killer Bats" and followed by the sequel "Devil Bat's Daughter."
1941 67m/B Bela Lugosi, Dave O'Brien, Suzanne Kaaren, Yolande Donlan; *Dir:* Jean Yarborough. **VHS, Beta $19.95** *NOS, SNC, PSM ♫♫*

The Devil Bat's Daughter A young woman, hoping to avoid becoming insane like her batty father, consults a psychiatrist when she starts to have violent nightmares. Unsuccessful sequel to "Devil Bats."
1946 66m/B Rosemary La Planche, Michael Hale, John James, Molly Lamont; *Dir:* Frank Wisbar. **VHS, Beta $14.95** *COL, SVS, MRV ♫½*

Devil & Daniel Webster A young farmer, who sells his soul to the devil, is saved from a trip to Hell when Daniel Webster steps in to defend him. Also known as "All That Money Can Buy," this classic fantasy is visually striking and contains wonderful performances. Huston received an Oscar nomination for Best Actor.
1941 109m/B James Craig, Edward Arnold, Walter Huston, Simone Simon, Gene Lockhart, Jane Darwell, Anne Shirley, John Qualen, H.B. Warner; *Dir:* William Dieterle. **VHS, Beta, LV $14.95** *COL, SUE, VYG ♫♫♫½*

Devil of the Desert Against the Son of Hercules The grandson of Zeus ventures into the wasteland to take on a feisty foe.
1962 ?m/C *IT* Kirk Morris, Michele Girardon; *Dir:* Riccardo Freda. **VHS, Beta $16.95** *SNC ♫♫½*

Devil Dog: The Hound of Hell A family has trouble with man's best friend when it adopts a dog that is the son of the "Hound of Hell" in this made-for-television movie.
1978 95m/C Richard Crenna, Yvette Mimieux, Kim Richards, Victor Jory; *Dir:* Curtis Harrington. **VHS, Beta $69.95** *LTG ♫*

Devil Doll An escaped convict uses a mad scientist's human-shrinking formula to evil ends—he sends out miniature assassins in the guise of store-bought dolls. Rarely seen horror oldie. Co-written by Erich Von Stroheim.
1936 80m/B Lionel Barrymore, Maureen O'Sullivan, Frank Lawton; *Dir:* Tod Browning. **VHS, Beta $19.95** *MGM, MLB ♫♫♫*

Devil Doll A ventriloquist's dummy, which contains the soul of a former performer, eyes a beautiful victim in the crowd.
1964 80m/B *GB* Bryant Holiday, William Sylvester; *Dir:* Lindsay Shonteff. **VHS, Beta $59.95** *MPI ♫*

Devil in the Flesh Acclaimed drama about a French soldier's wife having a passionate affair with a high school student while her husband is away fighting in World War I. From the novel by Raymond Radiguet. Dubbed. Updated and remade in 1987.
1946 112m/B *FR* Gerard Philippe, Micheline Presle, Denise Grey; *Dir:* Claude Autant-Lara. **VHS, Beta $24.95** *NOS, FCT, APD ♫♫♫*

Devil in the Flesh A angst-ridden, semi-pretentious Italian drama about an obsessive older woman carrying on an affair with a schoolboy, despite her terrorist boyfriend and the objections of the lad's psychiatrist father, who had treated her. In Italian with subtitles. Famous for a graphic sex scene, available in the unrated version. Updated remake of the 1946 French film.
1987 (R) 110m/C *IT* Maruschka Detmers, Federico Pitzalis; *Dir:* Marco Bellochio. **VHS, Beta, LV $79.98** *ORI, TAM ♫♫*

Devil Girl from Mars A sexy female from Mars and her very large robot arrive at a small Scottish inn to announce that a Martian feminist revolution has occurred. The distaff aliens then undertake a search of healthy Earth males for breeding purposes. Believe it or not, the humans don't want to go and therein lies the rub. A somewhat enjoyable space farce.
1954 76m/B *GB* Hugh McDermott, Hazel Court, Patricia Laffan, Peter Reynolds, Adrienne Corri, Joseph Tomelty, Sophie Stewart, John Laurie, Anthony Richmond; *Dir:* David MacDonald. **VHS, Beta $16.95** *NOS, MRV, SNC ♫½*

Devil Horse A boy's devotion to a wild horse marked for destruction as a killer leads him into trouble. A serial in twelve, thirteen-minute chapters.
1932 156m/B Frankie Darro, Harry Carey, Noah Beery; *Dir:* Otto Brower, Richard Talmadge. **VHS, Beta $49.95** *NOS, VCN, DVT ♫♫*

Devil on Horseback A miner's son journeys to London to become a jockey in this comedy/drama. As he tries to bully his way into the racing circuit, he succeeds, but not in the way he anticipated.
1954 88m/B *GB* Googie Withers, John McCallum, Jeremy Spenser, Liam Redmond; *Dir:* Cyril Frankel. **VHS, Beta, 8mm $29.95** *VYY ♫♫*

Devil In Silk A composer marries a woman, not knowing that she is psychotically jealous. After she commits suicide, he must prove to the police that he did not kill her. Original dialogue in German.
1956 102m/B *GE* Lilli Palmer, Curt Jurgens, Winnie Markus; *Dir:* Rolf Hansen. **VHS, Beta $16.95** *NOS, SNC ♫♫½*

Devil Kiss Gothic horror including zombies, virgins, sorcery, and the inevitable living dead.
1977 93m/C Oliver Mathews, Evelyn Scott. **VHS, Beta $59.95** *HMV ♫*

Devil & Leroy Basset Keema Gregwolf kills a deputy, breaks from jail with the Basset brothers, hijacks a church bus, kidnaps a family, and much more in his posse-eluding cross-country adventure.
1973 (PG) 85m/C Cody Bearpaw, John Goff, George Flower; *Dir:* Robert E. Pearso. **VHS, Beta $59.95** *PSM ♫*

Devil & Max Devlin The recently deceased Max Devlin strikes a bargain with the devil. He will be restored to life if he can convince three mortals to sell their souls. Music by Marvin Hamlisch. A Disney film.
1981 (PG) 95m/C Elliott Gould, Bill Cosby, Susan Anspach, Adam Rich, Julie Budd; *Dir:* Steven Hilliard Stern. **VHS, Beta** *DIS, OM ♫½*

Devil & Miss Jones Engaging romantic comedy finds a big business boss posing as an ordinary salesclerk to weed out union organizers. He doesn't expect to encounter the wicked management or his beautiful co-worker, however. Oscar nominations for Best Supporting Actor (Coburn) and Best Original Screenplay.

1941 90m/B Jean Arthur, Robert Cummings, Charles Coburn, Edmund Gwenn, Spring Byington, William Demarest, S.Z. Sakall; *Dir:* Sam Wood. **VHS $19.98** *REP ✗✗✗½*

Devil Monster A world-weary traveler searches for a girl's fiance in the South Pacific, and is attacked by a large manta.
1946 65m/B Barry Norton, Blanche Mehaffey; *Dir:* S. Edward Graham. **VHS, Beta, LV $16.95** *SNC, WGE ✗*

Devil Rider A monster from Hell, the Devil Rider, once rode the plains in search of human fodder. In the present day, a pleasure ranch is built on his old territory, and people begin dying in the most horrific manners. Could this mean the return of the Devil Rider?
1991 90m/C VHS $59.98 *MAG ✗½*

Devil Thumbs a Ride A naive traveller picks up a hitchhiker, not knowing he's wanted for murder. Will history repeat itself in this interesting noir?
1947 63m/B Ted North, Lawrence Tierney, Nan Leslie; *Dir:* Felix Feist. **VHS, Beta $34.95** *RKO ✗✗✗*

Devil Thumbs a Ride/Having Wonderful Crime A mystery double feature. In "Devil Thumbs a Ride", a ruthless killer-bank robber hitches a ride from an inebriated travelling salesman. In "Having Wonderful Crime", a criminal lawyer investigates the disappearance of a magician.
1947 132m/B Lawrence Tierney, Pat O'Brien, George Murphy, Carole Landis, Nan Leslie, George Zucco; *Dir:* Felix Feist, Edward Sutherland. **VHS, Beta $39.95** *RKO ✗✗✗*

Devil Times Five To take revenge for being incarcerated in a mental hospital, five children methodically murder the adults who befriend them.
1982 (R) 87m/C Gene Evans, Sorrell Booke, Shelly Morrison; *Dir:* Sean McGregor. **VHS, Beta $54.95** *MED ✗✗½*

Devil Wears White A student's vacation is disrupted when he becomes involved in a war with an insane arms dealer who wants to take over a Latin American Republic.
1986 (R) 92m/C Robert Livesy, Jane Higginson, Guy Ecker, Anthony Cordova; *Dir:* Steven A. Hull. **VHS, Beta $79.95** *TWE ✗*

The Devil on Wheels Inspired by his dad's reckless driving, a teenager becomes a hot rodder and causes a family tragedy. A rusty melodrama that can't be described as high-performance.
1947 67m/B James B. Cardwell, Noreen Nash, Darryl Hickman, Jan Ford, Damian O'Flynn, Lenita Love. **VHS $15.95** *NOS, LOO ✗✗*

Devil Woman A Filipino Gorgon-woman seeks reptilian revenge on the farmers who killed her family. Bad news.
1976 (R) 79m/C *PH* Rosemarie Gil; *Dir:* Albert Yu, Felix Vilars. **VHS $19.98** *SNC Woof!*

Devilfish A small seaside community is ravaged by berserk manta rays in this soggy saga.
1984 92m/C *IT Dir:* Lamberto Bava. **VHS, Beta $59.95** *VMK ✗*

The Devils In 1631 France, a priest is accused of commerce with the devil and sexual misconduct with nuns. Since he is also a political threat, the accusation is used to denounce and eventually execute him. Based on Aldous Huxley's "The Devils of Loudun," the movie features masturbating nuns and other excesses—shocking scenes typical of film director Russell. Supposedly this was Russell's attempt to wake the public to their desensitization of modern horrors of war. Controversial and flamboyant.
1971 (R) 109m/C *GB* Vanessa Redgrave, Oliver Reed, Dudley Sutton, Max Adrian, Gemma Jones, Murray Melvin; *Dir:* Ken Russell. National Board of Review Awards '71: Best Director. **VHS, Beta $19.98** *WAR, TAM ✗✗✗*

Devil's Angels A motorcycle gang clashes with a small-town sheriff. Cheap 'n' sleazy fare.
1967 84m/C John Cassavetes, Beverly Adams, Mimsy Farmer, Salli Sachse, Nai Bonet, Leo Gordon; *Dir:* Daniel Haller. **VHS $69.95** *TRY ✗*

The Devil's Brother In one of their lesser efforts, Laurel and Hardy star as bumbling bandits in this comic operetta based on the 1830 opera by Daniel F. Auber. Also available with "Laurel & Hardy's Laughing 20's" on laserdisc.
1933 88m/B Stan Laurel, Oliver Hardy, Dennis King, Thelma Todd, James Finlayson, Lucille Browne; *Dir:* Charles "Buddy" Rogers, Hal Roach. **VHS $19.98** *MGM, MLB, CCB ✗✗*

Devil's Canyon An ex-lawman, serving time in an Arizona prison, is beset by a jailed killer seeking vengeance for his own incarceration. Filmed in 3-D.
1953 92m/B Dale Robertson, Virginia Mayo, Stephen McNally, Arthur Hunnicutt, Robert Keith, Jay C. Flippen, Whit Bissell; *Dir:* Alfred Werker. **VHS, Beta $19.94** *TTC ✗✗*

The Devil's Cargo When a man is accused of killing a racetrack operator, he calls in the Falcon to clear his name. The master detective finds this to be a most difficult task however, especially when the accused is found poisoned in his cell. One of the final three "Falcon" films, in which Calvert replaced Tom Conway.
1948 61m/B John Calvert, Rochelle Hudson, Roscoe Karns, Lyle Talbot, Tom Kennedy; *Dir:* John F. Link. **VHS, Beta $16.95** *SNC ✗½*

The Devil's Commandment A lurid melodrama about a Parisian newspaperman's journey into a network of crime and perversion. Cinematography by Mario Bava. Also known as "I, Vampiri."
1960 90m/B *FR* Giana Maria Canale, Dario Michaelis, Carlo D'Angelo, Wandisa Guida, Paul Muller, Renato Tontini; *Dir:* Riccardo Freda. **VHS, Beta, LV $16.95** *WGE ✗✗*

Devil's Crude An adventurous sailor and a young oil heir uncover a conspiracy within a giant corporation.
1971 85m/C *IT* Franco Nero. **VHS, Beta $59.95** *TWE ✗*

The Devil's Daughter A sister's hatred and voodoo ceremonies play an important part in this all-black drama.
1939 60m/B Nina Mae McKinney, Jack Carter, Ida James, Hamtree Harrington. **VHS, Beta $16.95** *SNC, VYY, DVT ✗½*

The Devil's Eye The devil dispatches Don Juan to tempt and seduce a young virgin bride-to-be, a reverand's daughter, no less. Based on the Danish radio play "Don Juan Returns."
1960 90m/B *SW* Bibi Andersson, Jarl Kulle; *Dir:* Ingmar Bergman. **VHS, Beta $29.98** *SUE, TAM ✗✗*

Devil's Gift A young boy's toy is possessed by a demon and havoc ensues.
1984 112m/C Bob Mendlesohn, Vicki Saputo, Steven Robertson; *Dir:* Kenneth Berton. **VHS, Beta $69.98** *LIV, VES Woof!*

The Devil's Hand When Alda finds a doll that represents his ideal woman in a curio shop, the shop's owner (Hamilton, aka Commissioner Gordon) tells him the dream girl who modeled for the doll lives nearby. Trouble is, she's part of a voodoo cult, and guess who's head voodoo-man. Big trouble for Alda; big snooze for you.
1961 71m/B Linda Christian, Robert Alda, Neil Hamilton, Ariadne Welter; *Dir:* William Hole Jr. **VHS $19.98** *NOS, SNC, LOO ✗*

Devil's Harvest A vintage "Reefer Madness"-style cautionary film about pot, the "smoke of Hell."
193? 52m/B VHS, Beta $29.95 *JLT Woof!*

The Devil's Messenger Ask yourself this: would you rent a movie that wouldn't sell as TV episodes? This is a pastiche of unsold episodes of the Swedish-filmed TV fodder "No. 13 Demon St." Chaney—who plays the devil in a sport shirt—sends a pretty young suicide back to earth to act as hell-bait, luring recruits to their Stygean fate. Curt Siodmak contributed to the poorly written script.
1962 72m/B *SW* Lon Chaney Jr., Karen Kadler, Michael Hinn; *Dir:* Herbert L. Strock. **VHS, Beta $16.95** *SNC ✗*

The Devil's Mistress A gang of criminals on a pillaging spree murder a man and rape his Indian wife, only to find out she's a she-demon who won't let bygones be bygones. Written by director Wanzer, this monster-oater lacks essential nutrients.
1968 66m/C Joan Stapleton, Robert Gregory, Forrest Westmoreland, Douglas Warren, Oren Williams, Arthur Rooley; *W/Dir:* Orvillo Wanzer. **VHS $19.98** *SNC ✗½*

The Devil's Nightmare A woman leads seven tourists (representing the seven deadly sins) on a tour of a medieval European castle. There they experience demonic tortures. Lots of creepy moments. Euro-horror/sex star Blanc is fantastic in this otherwise mediocre production. Also on video as "Succubus" and "The Devil Walks at Midnight."
1971 (R) 88m/C *BE IT* Erika Blanc, Jean Servais, Daniel Emilfork, Lucien Raimbourg, Jacques Monseau, Colette Emmanuelle, Ivana Novak, Shirley Corrigan, Frederique Hender; *Dir:* Jean Brismee. **VHS, Beta $59.95** *MON, APP, REG ✗½*

The Devil's Partner Yet another uninspired devil yarn in which an old-timer trades in his senior citizenship by dying and coming back in the form of his younger self. Young again, he takes a new wife and indulges in multiple ritual sacrifices. Noteworthy only by virtue of the cast's later TV notoriety—Buchanan and Foulger would later appear on "Petticoat Junction," Nelson played Dr. Rossi on "Peyton Place," and Crane beached a role on "Hawaiian Eye."
1958 75m/B Ed Nelson, Jean Allison, Edgar Buchanan, Richard Crane, Spencer Carlisle, Byron Foulger, Claire Carleton; *Dir:* Charles R. Rondeau. **VHS $19.98** *SNC, MRV, CNM ✗½*

Devil's Party A tenement boy's reunion party turns into a night of horror when one of the guests is killed. As a result, the childhood friends band together to uncover the murderer's identity.
1938 65m/B Victor McLaglen, Paul Kelly, William Gargan, Samuel S. Hinds; *Dir:* Ray McCarey. **VHS, Beta $24.95** *NOS, MRV, DVT ✗✗*

D

Devil's Playground A beautiful young woman enlists the aid of Hopalong Cassidy and his friends to make sure that a gold fortune falls into the right hands.
1946 59m/B William Boyd, Andy Clyde, Rand Brooks, Elaine Riley; *Dir:* George Archainbaud. **VHS, Beta $79.95** BVV ⚔️

The Devil's Playground Sexual tension rises in a Catholic seminary, distracting the boys from their theological studies. The attentions of the priests only further their sexual confusion. Winner of many Australian film awards. Schepisi wrote the screenplay as well as directed.
1976 107m/C AU Arthur Dignam, Nick Tate, Simon Burke, Charles Frawley, Jonathon Hardy, Gerry Dugan, Thomas Keneally; *Dir:* Fred Schepisi. Australian Film Institute '76: Best Actor (Tate), Best Actor (Burke), Best Film. **Beta $79.95** ALL ⚔️⚔️⚔️

The Devil's Possessed A Middle Ages despot tortures and maims the peasants in his region until they rise up and enact an unspeakable revenge.
1974 90m/C AR SP Paul Naschy; *Dir:* Leon Klimovsky. **VHS, Beta $19.95** ASE ⚔️½

Devil's Rain The rituals and practices of devil worship, possession, and satanism are gruesomely related. Interesting cast.
1975 (PG) 85m/C Ernest Borgnine, Ida Lupino, William Shatner, Eddie Albert, Keenan Wynn, John Travolta, Tom Skerritt; *Dir:* Robert Fuest. **VHS, Beta, LV $19.95** UHV ⚔️½

The Devil's Sleep When a crusading woman sets out to break up a teen narcotics ring, the threatened thugs attempt to draw her daughter into their sleazy affairs, thus assuring the mother's silence.
1951 81m/B Lita Grey Chaplin, Timothy Farrell, John Mitchum, William Thomason, Tracy Lynn; *Dir:* W. Merle Connell. **VHS, Beta $16.95** SNC ⚔️

Devil's Son-in-Law A black stand-up comic makes a deal with a devil.
1977 (R) 95m/C Rudy Ray Moore; *Dir:* Cliff Roquemore. **VHS, Beta $24.95** AHV Woof!

Devil's Triangle Examines the unexplained incidents which have occurred in the Bermuda Triangle off the Florida Coast.
1978 59m/C *Nar:* Vincent Price. **VHS, Beta $59.95** MGM Woof!

The Devil's Undead When a Scottish orphanage is besieged by a rash of cold-blooded murders, the police are summoned to investigate. Their relentless search to determine the truth leads to a shocking climax.
1975 (PG) 90m/C GB Christopher Lee, Peter Cushing; *Dir:* Peter Sasdy. **VHS, Beta $59.95** MON, MLB ⚔️½

Devil's Wanton A young girl tries to forget an unhappy relationship by beginning a new romance with an equally frustrated beau. Gloomy, but hopeful.
1949 80m/B SW Doris Svedlund, Eva Henning, Hasse Ekman; *Dir:* Ingmar Bergman. **VHS, Beta $29.95** NOS, MRV, DVT ⚔️⚔️½

The Devil's Web A demonic nurse infiltrates, manipulates, and corrupts three beautiful sisters.
1974 (PG) 73m/C Diana Dors, Andrea Marcovicci, Ed Bishop, Cec Linder, Michael Culver. **VHS, Beta $29.95** IVE ⚔️½

Devil's Wedding Night An archaeologist and his twin brother fight over a ring that lures virgins into Count Dracula's Transylvanian castle. Vampire queen Bay seduces the

dimwit twins and strips at every chance she gets.
1973 (R) 85m/C Mark Damon, Sara (Rosalba Neri) Bay, Frances Davis; *Dir:* Paul Solvay. **VHS, Beta $49.95** UHV Woof!

Devlin Connection A retired private investigator helps his son learn the ropes as he embarks on a Los Angeles crime fighting career. Hudson's last starring television series, excluding his "Dynasty" role. All thirteen episodes are available separately.
1982 50m/C Rock Hudson, Jack Scalia, Louis Giambalvo, Leigh Taylor-Young, Jack Kruschen; *Dir:* Christian Nyby. **VHS, Beta $39.95** TWE

Devonsville Terror Strange things begin to happen when a new school teacher arrives in Devonsville, a town which has a history of torture, murder and witchcraft. The hysterical townspeople react by beginning a 20th-century witchhunt.
1983 (R) 97m/C Suzanna Love, Donald Pleasence, Deanna Haas, Mary Walden, Robert Walker Jr., Paul Wilson; *Dir:* Ulli Lommel. **VHS, Beta $9.98** SUE ⚔️⚔️

Devotion A lovesick British miss disguises herself as a governess so she can be near the object of her affection, a London barrister.
1931 80m/B Ann Harding, Leslie Howard, Robert Williams, O.P. Heggie, Louise Closser Hale, Dudley Digges; *Dir:* Robert Milton. **VHS $19.98** TTC ⚔️⚔️½

D.I. A tough drill sergeant is faced with an unbreakable rebellious recruit, threatening his record and his platoon's status. Webb's film features performances by actual soldiers.
1957 106m/B Jack Webb, Don Dubbins, Jackie Loughery, Lin McCarthy, Virginia Gregg; *Dir:* Jack Webb. **VHS, Beta $59.95** WAR ⚔️⚔️

The Diabolical Dr. Z When dad dies of cardiac arrest after the medical council won't let him make the world a kinder gentler place with his personality-altering technique, his dutiful daughter—convinced the council brought on dad's demise—is out to change some personalities in a big way.
1965 86m/B SP FR Mabel Karr, Fernando Montes, Estella Blain, Antonio J. Escribano, Howard Vernon; *Dir:* Jess (Jesus) Franco. **VHS $19.98** SNC ⚔️⚔️

Diabolically Yours A French thriller about an amnesiac who struggles to discover his lost identity. Tensions mount when his pretty spouse and friends begin to wonder if it's all a game. Dubbed.
1967 94m/C FR Alain Delon, Senta Berger; *Dir:* Julien Duvivier. **VHS, Beta $59.95** UNI ⚔️⚔️½

Diabolique The mistress and the wife of a sadistic schoolmaster conspire to murder the man, carry it out, and then begin to wonder if they have covered their tracks effectively. Plot twists and double-crosses abound. In French with English subtitles. Remade for television as "Reflections of Murder."
1955 107m/B FR Simone Signoret, Vera Clouzot, Paul Meurisse, Charles Vanel, Michel Serrault; *Dir:* Henri-Georges Clouzot. National Board of Review Awards '55: 5 Best Foreign Films of the Year; New York Film Critics Awards '55: Best Foreign Film (tie). **VHS, Beta, LV, 8mm $29.95** NOS, MRV, SNC ⚔️⚔️⚔️½

Dial Help A lame Italian-made suspenser about a model plagued by ghostly phone calls.

1988 (R) 94m/C IT Charlotte Lewis, Marcello Modugno, Mattia Sbragia, Victor Cavallo; *Dir:* Ruggero Deodato. **VHS, Beta $79.95** PSM ⚔️½

Dial "M" for Murder An unfaithful husband devises an elaborate plan to murder his wife for her money, but when she accidentally stabs the killer-to-be, with scissors no less, he alters his methods. Part of the "A Night at the Movies" series, this tape simulates a 1954 movie evening with a Daffy Duck cartoon, "My Little Duckaroo," a newsreel, and coming attractions for "Them" and "A Star Is Born."
1954 123m/C Ray Milland, Grace Kelly, Robert Cummings, John Williams, Anthony Dawson; *Dir:* Alfred Hitchcock. **VHS, Beta, LV $19.98** WAR, PIA, TLF ⚔️⚔️⚔️

Diamante, Oro y Amor A pair of amateur gangsters find themselves in some crazy situations.
1987 93m/C MX Julio Aleman, Hilda Aguirre, Hector Suarez, Adriana Roel, Chucho Salinas. **VHS, Beta $59.95** MAD ⚔️⚔️

Diamond Head A Hawaiian landowner brings destruction and misery to his family via his stubbornness. Based on the Peter Gilman novel.
1962 107m/C Charlton Heston, Yvette Mimieux, George Chakiris, France Nuyen, James Darren; *Dir:* Guy Green. **VHS, Beta, LV $69.95** COL ⚔️⚔️½

Diamond Ninja Force Hosted by Sho Kosugi, this film pits unrelentless ninjas against mortal enemies and insatiable revenge.
198? 90m/C **VHS, Beta $59.95** TWE ⚔️

Diamond Run A streetwise American expatriate frantically searches for his girlfriend after unwittingly involving her in an assassination plot.
1990 (R) 89m/C William Bell Sullivan, Ava Lazar, Ayu Azhari, David Thornton, Peter Fox; *Dir:* Robert Chappell. **VHS, Beta, LV $79.95** VIR ⚔️½

Diamond Trail East Coast jewel thieves bring their operation to the West, trailed by a New York reporter.
1933 58m/B Rex Bell, Frances Rich, Bud Osborne, Lloyd Whitlock, Norman Feusier; *Dir:* Harry Fraser. **VHS, Beta $19.95** NOS, DVT, TIM ⚔️½

Diamonds A tense film in which the Israel Diamond Exchange is looted by a motley array of criminal heisters. Shaw plays a dual role as twin brothers.
1972 (PG) 108m/C IS Robert Shaw, Richard Roundtree, Barbara Hershey, Shelley Winters; *Dir:* Menahem Golan. **VHS, Beta $59.98** CHA ⚔️⚔️

Diamond's Edge An adolescent private eye and his juvenescent brother snoop into the affairs of the Fat Man, and find out that the opera ain't over until they find out what's in the Fat Man's mysterious box of bonbons. A genre-parodying kid mystery, written by Anthony Horowitz, based on his novel, "The Falcon's Malteser." Also known as "Just Ask for Diamond."
1988 (PG) 83m/C Susannah York, Peter Eyre, Patricia Hodge, Nickolas Grace; *Dir:* Stephen Bayly. **VHS, Beta $79.99** HBO ⚔️½

Diamonds are Forever 007 once again battles his nemesis Blofeld, this time in Las Vegas. Bond must prevent the implementation of a plot to destroy Washington through the use of a space-orbiting laser. Fabulous stunts include Bond's wild drive through the streets of Vegas in a '71 Mach 1. Connery returned to play Bond in this film

after being offered the then record-setting salary of one million dollars.
1971 (PG) 120m/C *GB* Sean Connery, Jill St. John, Charles Gray, Bruce Cabot, Jimmy Dean, Lana Wood, Bruce Glover, Putter Smith, Norman Burton, Joseph Furst, Bernard Lee, Desmond Llewelyn, Laurence Naismith, Leonard Barr, Lois Maxwell, Margaret Lacey, Joe Robinson, Donna Garrat, Trina Parks; *Dir:* Guy Hamilton. **VHS, Beta, LV $19.95** *MGM, TLF* 𝆐𝆐𝆐½

Diamonds of the Night A break-through masterpiece of the Czech new wave, about two young men escaping from a transport train to Auschwitz and scrambling for survival in the countryside. Surreal, powerfully expressionistic film, one of the most important of its time. In Czech with subtitles. Accompanied by Nemec's short "A Loaf of Bread."
1964 71m/B *CZ Dir:* Jan Nemec. **VHS, Beta $59.95** *FCT* 𝆐𝆐𝆐𝆐

Diamonds on Wheels A Disney film about three British teenagers, amid a big road rally, discovering stolen diamonds and getting pursued by gangsters.
1973 84m/C VHS, Beta $69.95 *DIS* 𝆐

The Diary of Anne Frank In June 1945, a liberated Jewish refugee returns to the hidden third floor of an Amsterdam factory where he finds the diary kept by his youngest daughter, Anne. The document recounts their years in hiding from the Nazis. Based on the actual diary of 13-year-old Anne Frank, killed in a death camp during WWII.
1959 150m/B Millie Perkins, Joseph Schildkraut, Shelley Winters, Richard Beymer, Gusti Huber, Ed Wynn, Lou Jacobi, Diane Baker; *Dir:* George Stevens. Academy Awards '59: Best Art Direction/Set Decoration (B & W), Best Black and White Cinematography, Best Supporting Actress (Winters). **VHS, Beta, LV $79.98** *KUI, FOX, BTV* 𝆐𝆐𝆐½

Diary of a Chambermaid A chambermaid wants to marry a rich man and finds herself the object of desire of a poor servant willing to commit murder for her. Excellent comic drama, but very stylized in direction and set design. Produced during Renoir's years in Hollywood. Adapted from a story by Octave Mirbeau, later turned into a play. Remade in 1964 by Luis Bumel.
1946 86m/B Paulette Goddard, Burgess Meredith, Hurd Hatfield, Francis Lederer, Judith Anderson, Florence Bates, Almira Sessions, Reginald Owen; *Dir:* Jean Renoir. **VHS $19.98** *REP, FCT* 𝆐𝆐𝆐

Diary of a Chambermaid Vintage Bunuelian social satire about a young girl taking a servant's job in a provincial French family, and easing into an atmosphere of sexual hypocrisy and decadence. In French with English subtitles. Remake of the 1946 Jean Renoir film.
1964 97m/C *FR* Jeanne Moreau, Michel Piccoli, Georges Geret, Francoise Lugagne, Daniel Ivernel; *Dir:* Luis Bunuel. **VHS, Beta, LV $19.98** *TAM, APD, IME* 𝆐𝆐𝆐½

Diary of a Country Priest With "Balthazar" and "Mouchette," this is one of Bresson's greatest, subtlest films, treating the story of an alienated, unrewarded young priest with his characteristic austerity and Catholic humanism. In French and English subtitles.
1950 120m/B *FR* Claude Layou, Jean Riveyre, Nicole Ladmiral; *Dir:* Robert Bresson. **VHS, Beta $59.95** *DVT, APD, TAM* 𝆐𝆐𝆐½

Diary of the Dead A newlywed kills his aggravating mother-in-law, only to have her repeatedly return from the grave.
1987? (PG) 93m/C Hector Elizondo, Geraldine Fitzgerald, Salome Jens; *Dir:* Arvin Brown. **Beta $79.95** *VHV* 𝆐

Diary of Forbidden Dreams A beautiful young girl finds herself drawn into bizarre incidents that cause her to go insane. Set on the Italian Riviera, the story is an edited version of Roman Polanski's 1973 film "What?;" it's the most offbeat rendition of "Alice in Wonderland" you're apt to find. Unfortunately, the convoluted plot lessens its appeal.
1976 (R) 94m/C *IT* Marcello Mastroianni, Hugh Griffith, Sydne Rome, Roman Polanski; *Dir:* Roman Polanski. **VHS, Beta $19.95** *VDM, TWE* 𝆐𝆐𝆐

Diary of a Hitman A hitman is hired to knock off the wife and child of a commodities broker who claims his wife is a drug addict and the infant is a crack baby and not his. The hired killer wants out of the business, but needs to pull off one more job for a down payment on his apartment. Beset by doubts, he breaks conduct by conversing with the victim and discovers the broker lied. Based on the play "Insider's Price" by Pressman.
1991 (R) 90m/C Forest Whitaker, James Belushi, Sherilyn Fenn, Sharon Stone, Seymour Cassel, Lewis Smith, Lois Chiles; *Dir:* Roy London. **VHS, LV $89.95** *COL* 𝆐𝆐½

Diary of a Lost Girl The second Louise "Lula" Brooks/G. W. Pabst collaboration (after "Pandora's Box") in which a frail but mesmerizing German girl plummets into a life of hopeless degradation. Dark and gloomy, the film chronicles the difficulties she faces, from rape to an unwanted pregnancy and prostitution. Based on the popular book by Margarete Boehme. Silent. Made after flapper Brooks left Hollywood to pursue greater opportunities and more challenging roles under Pabst's guidance.
1929 99m/B *GE* Louise Brooks, Fritz Rasp, Josef Rovensky, Sybille Schmitz; *Dir:* G. W. Pabst. **VHS, Beta $19.95** *GPV, MRV, VDM* 𝆐𝆐𝆐½

Diary of a Mad Housewife Despairing of her miserable family life, a housewife has an affair with a writer only to find him to be even more selfish and egotistical than her no-good husband. Snodgrass plays her character perfectly, hearing the insensitive absurdity of her husband and her lover over and over again, enjoying her martyrdom even as it drives her crazy. She was Oscar nominated.
1970 (R) 94m/C Carrie Snodgress, Richard Benjamin, Frank Langella; *Dir:* Frank Perry. National Board of Review Awards '70: 10 Best Films of the Year, Best Supporting Actor (Langella). **VHS, Beta $14.95** *GKK* 𝆐𝆐𝆐

Diary of a Mad Old Man A wistful drama about an old man who's lost everything except his desire for his daughter-in-law, which sends him into fits of nostalgia.
1988 (PG-13) 93m/C Derek De Lint, Ralph Michael, Beatie Edney. **VHS, Beta $79.95** *MED* 𝆐𝆐

Diary of a Madman Price is once again possessed by an evil force in this gothic thriller. Fairly average Price vehicle, based on Guy de Maupassant's story.
1963 96m/C Vincent Price, Nancy Kovack, Chris Warfield, Ian Wolfe, Nelson Olmstead; *Dir:* Reginald LeBorg. **VHS, Beta $9.95** *WKV* 𝆐𝆐

Diary of a Rebel A fictional account of the rise of Cuban rebel leader Che Guevara.

1984 89m/C John Ireland, Francisco Rabal. **VHS, Beta $49.95** *REG* 𝆐

Diary of a Teenage Hitchhiker A 17-year-old girl ignores family restrictions and police warnings about a homicidal rapist stalking the area and continues to thumb rides to her job at a beach resort. One night she is picked up for a one-way ride to terror.
1982 96m/C Charlene Tilton, Dick Van Patten; *Dir:* Ted Post. **VHS, Beta $69.95** *LTG* 𝆐

Diary of a Young Comic New York comedian searches for the meaning of lunacy. He finds it after Improvisation in Los Angeles.
1979 74m/C Stacy Keach, Dom DeLuise, Richard Lewis, Bill Macy, George Jessel, Gary Muledeer, Nina Van Pallandt; *Dir:* Gary Weis. **VHS, Beta $39.95** *PAV* 𝆐𝆐

Dias de Ilusion The fantasy world in which Lucia and her sister live was created for only one reason, and only Lucia's diary has the secret. In Spanish.
1980 94m/C *SP* Andrea Del Boca, Luisina Brando. **VHS, Beta $29.95** *MED* 𝆐½

Dick Barton, Special Agent Dick Barton is called in when a mad scientist threatens to attack London with germ-carrying bombs. The first film production for Hammer Studios, later to be known for its horror classics.
1948 70m/B *GB* Don Stannard, Geoffrey Ford, Jack Shaw; *Dir:* Alfred Goulding. **VHS, Beta $16.95** *SNC* 𝆐

Dick Barton Strikes Back Sebastian "Mr. French" Cabot has a nuclear weapon and is willing to use it; that is until Dick Barton arrives to put the Frenchman out of commission. Generally considered the best of the Dick Barton film series.
1948 73m/B *GB* Don Stannard, Sebastian Cabot, Jean Lodge; *Dir:* Godfrey Grayson. **VHS, Beta $16.95** *SNC* 𝆐𝆐½

Dick Powell Theatre: Price of Tomatoes Falk stars as a truck driver who picks up a pregnant hitchhiker (Stevens). Falk won an Emmy for his portrayal.
1962 60m/C Peter Falk, Inger Stevens. Emmy '62: Best Actor (Falk). **VHS, Beta $29.95** *RKO* 𝆐𝆐

Dick Tracy Serial, based on the comic strip character, in fifteen chapters. Tracy tries to find his kidnapped brother as he faces the fiend "Spider." The first chapter is thirty minutes and each additional chapter is twenty minutes.
1937 310m/B Ralph Byrd, Smiley Burnette, Irving Pichel, Jennifer Jones; *Dir:* John English. **VHS, Beta $29.95** *SNC, VYY, UHV* 𝆐½

Dick Tracy Beatty, wearing the cap of producer, director and star, performs admirably on all fronts in one of the biggest movies of 1990. Beatty's Tracy is somewhat flat in comparison to the outstanding performances and makeup of the unique villains, especially Pacino, but altogether, the movie is very entertaining. Stylistically, Beatty hits his mark. Shot in only 7 colors, using timeless sets that capture the essence rather than the reality of the city, the film successfully brings the comic strip to life. Madonna turns in a fine performance as the seductive Breathless Mahoney, belting out Stephen Sondheim like she was born to do it. People expecting the gothic technology of "Batman" may be disappointed, but moviegoers searching for a memory made real will be thrilled. Nominated for 7 Academy Awards: Best Supporting Actor (Pacino), Best Song, Best Cinematog-

raphy, Best Sound, Best Art Direction, Best Costume Design, and Best Make-up.
1990 (PG) 105m/C Warren Beatty, Madonna, Charlie Korsmo, Glenne Headly, Al Pacino, Dustin Hoffman, James Caan, Mandy Patinkin, Paul Sorvino, Charles Durning, Dick Van Dyke, R.G. Armstrong, Catherine O'Hara, Estelle Parsons, Seymour Cassel, Michael J. Pollard, William Forsythe, Kathy Bates; *Dir:* Warren Beatty. Academy Awards '90: Best Art Direction/Set Decoration, Best Makeup, Best Song ("Sooner or Later"). **VHS, Beta, LV $79.95** *TOU, FCT, IME* ✻✻✻½

Dick Tracy Detective The first Dick Tracy feature film, in which Splitface is on the loose, a schoolteacher is murdered, the Mayor is threatened, and a nutty professor uses a crystal ball to give Tracy the clues needed to connect the crimes.
1945 62m/B Morgan Conway, Anne Jeffreys, Mike Mazurki, Jane Greer, Lyle Latelle; *Dir:* William Burke. **VHS, Beta $16.95** *SNC, MED, VYY* ✻✻

Dick Tracy, Detective/The Corpse Vanishes On one tape, the first Dick Tracy feature film and Lugosi's minor epic about a mad scientist using virgin's blood to restore youth to an aging countess.
1942 126m/B Bela Lugosi, Morgan Conway, Luana Walters, Tristram Coffin, Mike Mazurki, Jane Greer, Lyle Latelle; *Dir:* William Berke. **VHS, Beta** *AOV, BUR* ✻½

Dick Tracy Double Feature #1 Mystery-adventure double feature presents Chester Gould's popular comic strip hero in live action. Includes "Dick Tracy, Detective" and "Dick Tracy's Dilemma."
1945 122m/B Ralph Byrd, Lyle Latelle, Morgan Conway, Anne Jeffreys. **VHS, Beta $54.95** *UHV* ✻½

Dick Tracy Double Feature #2 Chester Gould's famous comic strip character is personified in "Dick Tracy Meets Gruesome" and "Dick Tracy vs. Cueball," a double feature videocassette featuring Dick Tracy battling two of his arch enemies.
1947 127m/B Boris Karloff, Ralph Byrd, Morgan Conway, Anne Jeffreys. **VHS, Beta $54.95** *UHV* ✻✻

Dick Tracy Meets Gruesome Gruesome and his partner in crime, Melody, stage a bank robbery using the secret formula of Dr. A. Tomic. Tracy has to solve the case before word gets out and people rush to withdraw their savings, destroying civilization as we know it.
1947 66m/B Boris Karloff, Ralph Byrd, Lyle Latelle; *Dir:* John Rawlins. **VHS, Beta $14.95** *NOS, MRV, SNC* ✻✻

Dick Tracy and the Oyster Caper A collection of cartoons based upon the comic-strip character.
1961 60m/C VHS, Beta $29.95 *MED* ✻✻

Dick Tracy Returns 15-chapter serial. Public Enemy Paw Stark and his gang set out on a wave of crime that brings them face to face with dapper Dick.
1938 100m/B Ralph Byrd, Charles Middleton; *Dir:* William Witney. **VHS, Beta, LV $79.95** *UHV, IME, MLB* ✻✻

Dick Tracy: The Original Serial, Vol. 1 The first seven parts of the original popular serial. The comic strip character comes to life and tries to stop the spider, a dangerous criminal.
1937 150m/B Ralph Byrd, Smiley Burnette, Kay Hughes, Francis X. Bushman, Lee Van Atta; *Dir:* Ray Taylor. **VHS, Beta $24.95** *GKK* ✻✻✻

Dick Tracy: The Original Serial, Vol. 2 The last eight chapters of the highly acclaimed serial. Tracy is still tracking down the Spider in this action-packed adventure.
1937 160m/B Ralph Byrd, Smiley Burnette, Kay Hughes, Francis X. Bushman, Lee Van Atta; *Dir:* Ray Taylor. **VHS, Beta $24.95** *GKK* ✻✻✻

Dick Tracy: The Spider Strikes Dick Tracy, his bumbling sidekick, and his beautiful, brainy assistant go up against the master criminal, The Spider.
1937 60m/B Ralph Byrd, Morgan Conway, Francis X. Bushman. **VHS, Beta $9.95** *SIM, KAR* ✻✻✻

Dick Tracy vs. Crime Inc. Dick Tracy encounters many difficulties when he tries to track down a criminal who can make himself invisible. A serial in 15 chapters.
1941 100m/B Ralph Byrd, Ralph Morgan, Michael Owen; *Dir:* William Witney. **VHS, Beta $79.95** *UHV, IME, MLB* ✻✻

Dick Tracy vs. Cueball Dick Tracy has his work cut out for him when the evil gangster Cueball appears on the scene. Based on Chester Gould's comic strip.
1946 62m/B Morgan Conway, Anne Jeffreys; *Dir:* Gordon Douglas. **VHS, Beta $14.95** *NOS, QNE, UHV* ✻✻

Dick Tracy's Dilemma The renowned police detective Dick Tracy tries to solve a case involving the Claw. Based on the Chester Gould comic strip.
1947 60m/B Ralph Byrd, Lyle Latelle, Kay Christopher, Jack Lambert, Ian Keith, Jimmy Conlin; *Dir:* John Rawlins. **VHS, Beta $14.95** *QNE, MRV, UHV* ✻✻

Dick Turpin In this rare film, Mix plays the famed English highwayman who seeks adventure in historical Britain.
1925 60m/B Tom Mix, Alan Hale Jr.; *Dir:* John Blystone. **VHS, Beta $47.50** *GVV* ✻✻½

Did You Hear the One About ...? Dirty jokes, for all barstool humorists...
1985 60m/C VHS, Beta *ABA* ✻✻½

Didn't You Hear? An alienated college student discovers that dreams have a life of their own when he becomes immersed in his own fantasy world.
1983 (PG) 94m/C Dennis Christopher, Gary Busey, Cheryl Waters, John Kauffman. **VHS, Beta $49.95** *PSM* ✻

Die! Die! My Darling! A young widow visits her mad ex-mother-in-law in a remote English village, and is imprisoned by the mourning woman as revenge for her son's death. Bankhead's last role. Alternate title, "Fanatic."
1965 97m/C *GB* Tallulah Bankhead, Stefanie Powers, Peter Vaughan, Maurice Kaufman, Donald Sutherland, Gwendolyn Watts, Yootha Joyce, Winifred Dennis; *Dir:* Silvio Narizzano. **Beta $9.95** *COL, MLB* ✻½

Die Grosse Freiheit Nr. 7 A man tries to stop his niece's love affair with a sailor. Nazi propaganda minister Goebbels banned this film in Germany, since it showed German soldiers getting drunk and also had Hildebrand cast as a prostitute. After the war, however, it enjoyed great popularity throughout Germany. Albers sings the classic "Auf der Reeperbahn." In German with no English translation.
1945 100m/C *GE* Hans Albers, Hilde Hildebrand. **VHS, Beta, 3/4U $35.95** *IHF* ✻½

Die Hard A high-voltage action thriller about a lone New York cop battling a band of ruthless high-stake terrorists who attack and hold hostage the employees of a large corporation as they celebrate Christmas in a new Los Angeles high rise. Based on the novel by Roderick Thorp. It's just as unbelievable as it sounds, but you'll love it anyway. Rickman, who later charmed audiences as the villain in "Robin Hood: Prince of Thieves," turns in marvelous performance as the chief bad guy.
1988 (R) 114m/C Bruce Willis, Bonnie Bedelia, Alan Rickman, Alexander Godunov, Paul Gleason, William Atherton, Reginald Vel Johnson; *Dir:* John McTiernan. **VHS, Beta, LV $19.98** *FOX* ✻✻✻

Die Hard 2: Die Harder Fast, well-done sequel brings another impossible situation before the wise-cracking, tough-cookie cop. Our hero tangles with a group of terrorists at an airport under siege, while his wife remains in a plane circling above as its fuel dwindles. Obviously a repeat of the plot and action of the first Die Hard, with references to the former in the script. While the bad guys lack the fiendishness of their predecessors, this installment features energetic and finely acted performances. Fairly gory, especially the icicle-in-the-eyeball scene.
1990 (R) 124m/C Bruce Willis, William Atherton, Franco Nero, Bonnie Bedelia, John Amos, Reginald Vel Johnson, Dennis Franz, Art Evans, Fred Dalton Thompson, Bill Sadler; *Dir:* Renny Harlin. **VHS, LV $19.98** *FOX* ✻✻✻

Die Laughing A cab driver unwittingly becomes involved in murder, intrigue, and the kidnapping of a monkey that has memorized a scientific formula that can destroy the world. Benson wrote, produced, scored, and acted in this supposed comedy.
1980 (PG) 108m/C Robby Benson, Charles Durning, Bud Cort, Elsa Lanchester, Peter Coyote; *Dir:* Jeff Werner. **VHS, Beta $19.98** *WAR* Woof!

Die, Monster, Die! A reclusive scientist experiments with a radioactive meteorite and gains bizarre powers. Karloff is great in this adaptation of H. P. Lovecraft's, "The Color Out of Space."
1965 80m/C *GB* Boris Karloff, Nick Adams, Suzan Farmer, Patrick Magee; *Dir:* Daniel Haller. **VHS, LV $19.99** *HBO* ✻✻

Die Screaming, Marianne A girl is on the run from her father, a crooked judge, who wants to kill her before her 21st birthday, when she will inherit evidence that will put him away for life.
1973 81m/C Michael Rennie, Susan George, Karin Dor; *Dir:* Pete Walker. **VHS, Beta $49.95** *UNI* ✻

Die Sister, Die! Thriller about a gothic mansion with an eerie secret in the basement features a battle between a senile, reclusive sister and her disturbed, tormenting brother.
1974 88m/C Jack Ging, Edith Atwater, Antoinette Bower, Kent Smith, Robert Emhardt; *Dir:* Randall Hood. **VHS, Beta $59.95** *MPI* ✻

Different Kind of Love A romantic novel on tape, in modern settings. Love, frustration, tragedy.
1985 60m/C Joyce Redman. **Beta $11.95** *PSM* ✻

Different Story Romance develops when a lesbian real estate agent offers a homosexual chauffeur a job with her firm. Resorts to stereotypes when the characters decide to marry.

1978 (PG) 107m/C Perry King, Meg Foster, Valerie Curtin, Peter Donat, Richard Bull; ***Dir:*** Paul Aaron. **VHS, Beta $19.95** *SUE* 🦴🦴½

Digby, the Biggest Dog in the World A poor comedy-fantasy about Digby, a sheepdog, who wanders around a scientific laboratory, drinks an experimental fluid, and grows and grows and grows.
1973 (G) 88m/C *GB* Jim Dale, Angela Douglas, Spike Milligan, Dinsdale Landen; ***Dir:*** Joseph McGrath. **VHS, Beta $59.95** *PSM* 🦴

Digging Up Business Johnson tries to save the family funeral home business from financial ruin with some unique interment services. This film needs a decent burial.
1991 (PG) 89m/C Lynn-Holly Johnson, Billy Barty, Ruth Buzzi, Murray Langston, Yvonne Craig, Gary Owens; ***Dir:*** Mark Byers. **VHS $79.95** *MNC Woof!*

Dillinger John Dillinger's notorious career, from street punk to public enemy number one, receives a thrilling fast-paced treatment. Tierney turns in a fine performance in this interesting account of the criminal life.
1945 70m/C Lawrence Tierney, Edmund Lowe, Anne Jeffreys, Elisha Cook Jr.; ***Dir:*** Max Nosseck. **VHS, Beta $19.98** *FOX, FCT* 🦴🦴🦴

Dillinger The most colorful period of criminality in America is brought to life in this story of bank-robber John Dillinger, "Baby Face" Nelson, and the notorious "Lady in Red."
1973 (R) 106m/C Warren Oates, Michelle Phillips, Richard Dreyfuss, Cloris Leachman, Ben Johnson, Harry Dean Stanton; ***Dir:*** John Milius. **VHS, Beta, LV $69.98** *LIV, VES* 🦴🦴½

Dim Sum: A Little Bit of Heart The second independent film from the director of "Chan Is Missing." A Chinese-American mother and daughter living in San Francisco's Chinatown confront the conflict between traditional Eastern ways and modern American life. Gentle, fragile picture made with humor and care. In English and Chinese with subtitles.
1985 (PG) 88m/C Laureen Chew, Kim Chew, Victor Wong, Ida F.O. Chong, Cora Miao, John Nishio, Joan Chen; ***Dir:*** Wayne Wang. **VHS, Beta, LV $29.95** *PAV, FCT, IME* 🦴🦴🦴

Dimples When Shirley's pickpocket grandfather is caught red-handed, she steps in, takes the blame, and somehow ends up in show business.
1936 78m/B Shirley Temple, John Carradine, Frank Morgan, Helen Westley, Berton Churchill, Robert Kent, Delma Byron; ***Dir:*** William A. Seiter. **VHS, Beta $19.98** *FOX* 🦴🦴½

Diner A group of old high school friends meet at "their" Baltimore diner to find that more has changed than the menu. A bittersweet look at the experiences of a group of Baltimore post-teenagers growing up, circa 1959. Particularly notable was Levinson's casting of "unknowns" who have since become household names. Features many humorous moments and fine performances.
1982 (R) 110m/C Steve Guttenberg, Daniel Stern, Mickey Rourke, Kevin Bacon, Ellen Barkin, Timothy Daly, Paul Reiser, Michael Tucker; ***Dir:*** Barry Levinson. **VHS, Beta, LV $19.95** *MGM, PIA* 🦴🦴🦴½

Dingaka A controversial drama from the writer and director of "The Gods Must Be Crazy" about a South African tribesman who avenges his daughter's murder by tribal laws, and is then tried by white man's laws. A crusading white attorney struggles to acquit him.
1965 96m/C Stanley Baker, Juliet Prowse, Ken Gampu; ***Dir:*** Jamie Uys. **VHS, Beta $59.98** *CHA* 🦴🦴½

The Dining Room Six performers play more than 50 roles in this adaptation of A. R. Gurney Jr.'s play set in a dining room, where people live out dramatic and comic moments in their lives. From the "American Playhouse" series.
1986 90m/C Remak Ramsay, Pippa Perthree, John Shea, Frances Sternhagen; ***Dir:*** Allan Goldstein. **VHS, Beta $59.95** *PBS, ACA* 🦴🦴½

Dinner at Eight A social climbing woman and her husband throw a dinner party for various members of the New York elite. During the course of the evening, all of the guests reveal something about themselves. Special performances all around, especially Barrymore, Dressler, and Harlow. Superb comedic direction by Cukor.
1933 110m/C John Barrymore, Lionel Barrymore, Wallace Beery, Madge Evans, Jean Harlow, Billie Burke, Marie Dressler, Phillips Holmes, Jean Hersholt; ***Dir:*** George Cukor. **VHS, Beta, LV, 8mm $24.95** *MGM* 🦴🦴🦴🦴

Dinner at Eight A social-climbing romance novelist throws an elegant dinner party. Made for television remake of the 1933 film does not compare well.
1989 100m/C Lauren Bacall, Charles Durning, Ellen Greene, Harry Hamlin, John Mahoney, Marsha Mason, Tim Kazurinsky; ***Dir:*** Ron Lagomarsino. **VHS, Beta $139.98** *TTC* 🦴🦴

Dinner at the Ritz Daughter of a murdered Parisian banker vows to find his killer with help from her fiance.
1937 78m/B *GB* David Niven, Annabella, Paul Lukas; ***Dir:*** Harold Schuster. **VHS, Beta $19.95** *NOS, MRV, VYY* 🦴🦴½

Dino A social worker joins a young woman in helping a seventeen-year-old delinquent re-enter society.
1957 96m/B Sal Mineo, Brian Keith, Susan Kohner; ***Dir:*** Thomas Carr. **VHS, Beta $39.95** *REP* 🦴½

Dino & Juliet When that jerk Loudrock moves next door, he and Fred Flintstone just don't get along! But Dino has goo-goo eyes for the noisy neighbor's poodle. Also includes some Hanna-Barbera trivia!
1989 30m/C ***Voices:*** Alan Reed, Mel Blanc, Jean VanDerPyl. **VHS, Beta $9.95** *HNB* 🦴½

Dinosaurus! Large sadistic dinosaurs appear in the modern world. They eat, burn, and pillage their way through this film. Also includes a romance between a Neanderthal and a modern-age woman.
1960 85m/C Ward Ramsey, Kristina Hanson, Paul Lukather; ***Dir:*** Irvin S. Yeawarth Jr. **VHS, Beta $9.95** *NWV, STE, SNC* 🦴

Diplomaniacs Wheeler and Woolsey, official barbers on an Indian reservation, are sent to the Geneva peace conference to represent the interests of their tribe. Slim plot nevertheless provides quite a few laughs. Fun musical numbers.
1933 62m/B Bert Wheeler, Robert Woolsey, Marjorie White, Hugh Herbert; ***Dir:*** William A. Seiter. **VHS, Beta, LV $19.98** *CCB, IME* 🦴🦴½

Diplomatic Courier Cold-war espionage saga has secret agent Power attempting to re-steal sensitive documents from the hands of Soviet agents. Involved and exciting thriller. Michael Ansara, Charles Bronson and Lee Marvin made brief appearanes.
1952 97m/B Tyrone Power Sr., Patricia Neal, Stephen McNally, Hildegarde Neff, Karl Malden; ***Dir:*** Henry Hathaway. **VHS** *TCF* 🦴🦴🦴

Diplomatic Immunity A vicious killer gets off scot-free due to his diplomatic immunity. That is, until the victim's ex-soldier father follows this low-life to the jungles of Paraguay to exact revenge. A heap o' action.
1991 (R) 95m/C Bruce Boxleitner, Billy Drago, Meg Foster, Robert Forster; ***Dir:*** Peter Maris. **VHS, Beta, LV $89.95** *FRH, IME* 🦴½

Directions '66 A live one-set television drama about a pious couple who take the troubles of local youngsters upon their own shoulders.
1966 29m/B Nancy Marchand, Lawrence Keith. **VHS, Beta $19.95** *VYY* 🦴

Dirkham Detective Agency Three children team up to form the Dirkham Detective Agency and are hired by a veterinarian to recover two dognapped poodles.
1983 60m/C Sally Kellerman, Stan Shaw, John Quade, Gordon Jump, Randy Morton. **VHS, Beta $39.98** *SCL* 🦴½

Dirt Bike Kid A precocious brat is stuck with a used motorbike that has a mind of its own. Shenanigans follow in utterly predictable fashion as he battles bankers and bikers with his bad bike.
1986 (PG) 91m/C Peter Billingsley, Anne Bloom, Stuart Pankin, Patrick Collins, Sage Parker, Chad Sheets; ***Dir:*** Hoite C. Caston. **VHS, Beta, LV, 8mm $19.98** *CHA* 🦴½

Dirt Gang A motorcycle gang terrorizes the members of a film crew on location in the desert.
1971 (R) 89m/C Paul Carr, Michael Pataki, Michael Forest; ***Dir:*** Jerry Jameson. **VHS, Beta $59.95** *MPI* 🦴

Dirty Dancing An innocent 17-year-old (Grey) is vacationing with her parents in the Catskills in 1963. Bored with the program at the hotel, she finds the real fun at the staff dances. Falling for the sexy dance instructor (Swayze), she discovers love, sex and rock and roll dancing. An old story, with little to save it, but Grey and Swayze are appealing, the dance sequences fun, and the music great. Swayze, classically trained in ballet, also performs one of the sound track songs.
1987 (PG-13) 105m/C Patrick Swayze, Jennifer Grey, Cynthia Rhodes, Jerry Orbach; ***Dir:*** Emile Ardolino. Academy Awards '87: Best Song ("The Time of My Life"). **VHS, Beta, LV $19.98** *LIV, VES* 🦴🦴🦴

Dirty Dishes A French film sardonically examining the fruitless life of the average housewife as she confronts a series of bizarre but pointless experiences during the day. Dubbed and subtitled versions available. The director is the American daughter-in-law of the late Luis Bunuel.
1982 (R) 99m/C *FR* Carol Laurie, Pierre Santini; ***Dir:*** Joyce Bunuel. **VHS, Beta $29.98** *SUE* 🦴🦴🦴

The Dirty Dozen A tough Army major is assigned to train and command twelve hardened convicts offered absolution if they participate in a suicidal mission into Nazi Germany in 1944. Well-made movie is a standout in its genre. Cassavetes received an Oscar nomination as did the Sound and Film Editing. Rough and gruff Marvin is good as the group leader.
1967 (PG) 149m/C Lee Marvin, Ernest Borgnine, Charles Bronson, Jim Brown, George Kennedy, John Cassavetes, Clint Walker, Donald Sutherland, Telly Savalas, Robert Ryan, Ralph Meeker, Richard Jaeckel, Trini Lopez; ***Dir:*** Robert Aldrich. Academy Awards '67: Best Sound Effects Editing. **VHS, Beta, LV $19.95** *MGM, TLF* 🦴🦴🦴

The Dirty Dozen: The Next Mission Disappointing made-for-television sequel to the 1967 hit, with Marvin reprising his role as the leader of the motley pack. In this installment, the rag-tag toughs are sent on yet another suicide mission inside Nazi Germany, this time to thwart an assassination attempt of Hitler.
1985 97m/C Lee Marvin, Ernest Borgnine, Richard Jaeckel, Ken Wahl, Larry Wilcox, Sonny Landham, Ricco Ross; *Dir:* Andrew V. McLaglen. VHS, Beta $79.95 *MGM* ♂ ½

Dirty Gertie from Harlem U.S.A. An all-black cast performs a variation of Somerset Maugham's "Rain." Gertie goes to Trinidad to hide out from her jilted boyfriend.
1946 60m/B Gertie LaRue, Spencer Williams, Francine Everett, Don "The Dragon" Wilson. VHS, Beta $24.95 *NOS, DVT, VYY* ♂

Dirty Hands A woman and her lover conspire to murder her husband. A must for fans of the sexy Schneider. Cinematography provides moments that are both interesting and eerie.
1976 102m/C *FR* Rod Steiger, Romy Schneider, Paul Giusti, Jean Rochefort, Hans-Christian Blech; *Dir:* Claude Chabrol. VHS *FCT* ♂♂

Dirty Harry Rock-hard cop Harry Callahan attempts to track down a psychopathic rooftop killer before a kidnapped girl dies. Harry abuses the murderer's civil rights, however, forcing the police to return the criminal to the streets, where he hijacks a school bus and Harry is called on once again. The only answer to stop this vicious killer seems to be death in cold blood, and Harry is just the man to do it. Taut, suspenseful direction by Siegel, who thoroughly understands Eastwood's on-screen character. Features Callahan's famous "Do you feel lucky" line, the precursor to his "Go ahead, make my day."
1971 (R) 103m/C Clint Eastwood, Harry Guardino, John Larch, Andy Robinson, Reni Santoni, John Vernon; *Dir:* Don Siegel. VHS, Beta, LV, 8mm $19.98 *WAR, TLF* ♂♂♂½

Dirty Heroes A band of escaped World War II POWs battle the Nazis for a precious treasure in the war's final days.
1971 (R) 117m/C John Ireland, Curt Jurgens, Adolfo Celi, Daniela Bianchi, Michael Constantine; *Dir:* Alberto DeMartino. VHS, Beta $59.95 *VTR, WES* ♂♂

Dirty Laundry Insipid wreck of a comedy about a klutz who accidentally gets his laundry mixed up with $1 million dollars in drug money. Features olympians Lewis and Louganis. Never released theatrically.
1987 (PG-13) 81m/C Leigh McCloskey, Jeanne O'Brien, Frankie Valli, Sonny Bono, Carl Lewis, Greg Louganis, Nicholas Worth; *Dir:* William Webb. VHS, Beta $79.95 *SVS* Woof!

Dirty Mary Crazy Larry A racecar driver, his mechanic, and a sexy girl hightail it from the law after pulling off a heist. Great action, great fun, and an infamous surprise ending.
1974 (PG) 93m/C Peter Fonda, Susan George, Adam Roarke, Vic Morrow, Roddy McDowall; *Dir:* John Hough. VHS, Beta, LV *FOX* ♂♂♂

Dirty Mind of Young Sally Sally's erotic radio program broadcasts from a mobile studio, which must stay one step ahead of the police.
1972 (R) 84m/C Sharon Kelly. VHS, Beta $39.95 *MON* Woof!

Dirty Partners High-kicking, bone-crunching ninja bank robbers plan to steal a gold bullion shipment.
1987 90m/C VHS, Beta $59.95 *SVS* ♂

Dirty Rotten Scoundrels A remake of the 1964 "Bedtime Story," in which two confidence tricksters on the Riviera endeavor to rip off a suddenly rich American woman, and each other. Caine and Martin are terrific. Martin has some of his best physical comedy ever, and Headly is charming as the prey who's always one step ahead of them. Fine direction from Oz, the man who brought us the voice of Yoda in "The Empire Strikes Back."
1988 (PG) 112m/C Steve Martin, Michael Caine, Glenne Headly; *Dir:* Frank Oz. VHS, Beta, LV $19.98 *ORI* ♂♂♂

Dirty Tricks A history professor fights with bad guys for a letter written by George Washington. Thoroughly forgettable comedy lacking in laughs.
1981 (PG) 91m/C *CA* Elliott Gould, Kate Jackson, Arthur Hill, Rich Little; *Dir:* Alvin Rakoff. VHS, Beta $69.95 *SUE* ♂

Disappearance A hired assassin discovers an ironic link between his new target and his missing wife.
1981 (R) 80m/C *CA* Donald Sutherland, David Hemmings, John Hurt, Christopher Plummer, David Warner, Virginia McKenna; *Dir:* Stuart Cooper. VHS, Beta $19.95 *VES* ♂♂

The Disappearance of Aimee Dramatic recreation of events surrounding the 1926 disappearance of evangelist Aimee Semple McPherson. She claimed she was abducted, but her mother insisted she ran away to have an affair. Excellent made for television movie with an exceptional cast. Originally to star Ann-Margaret.
1976 110m/C Faye Dunaway, Bette Davis, James Sloyan, James Woods, John Lehne, Lelia Goldoni, Barry Brown, Severn Darden. VHS, Beta $59.95 *IVE* ♂♂♂

Disciple Vintage silent western wherein Hart's unsmiling good-bad guy tries to clean up a lawless town and win back his wife, who has shamelessly fallen in with a corrupt saloon-keeper. Co-scripted by Thomas Ince.
1915 80m/B William S. Hart, Dorothy Dalton, Robert McKim, Jean Hersholt; *Dir:* William S. Hart. VHS, Beta, 8mm $29.95 *VYY* ♂½

Disciple of Death Raven plays "The Stranger," a ghoul who sacrifices virgins to the Devil.
1972 (R) 82m/C *GB* Mike Raven, Ronald Lacey, Stephen Bradley, Virginia Wetherell; *Dir:* Tom Parkinson. VHS, Beta $49.95 *UNI* ♂

Disconnected "Sorry, Wrong Number" is updated, with more than ample doses of sex and violence.
1987 81m/C Mark Walker. VHS, Beta $7.00 *VTR* ♂½

Discontent An early silent film by Lois Weber in which a cantankerous old vet goes to live with his wealthy nephew. The harmonious family life is disrupted by his meddling ways.
1916 30m/B *Dir:* Lois Weber. VHS, Beta $42.50 *GVV* ♂½

The Discovery Program Four award-winning short films on one video: "Ray's Male Heterosexual Dance Hall," "Greasy Lake," "The Open Window," and "Hearts of Stone." The subjects range from humorous and offbeat to tragic and deadly serious.
1989 106m/C VHS, Beta $79.95 *JCI, FCT* ♂♂½

The Discreet Charm of the Bourgeoisie Bunuel in top form, satirizing modern society. These six characters are forever sitting down to dinner, yet they never eat. Dreams and reality, actual or contrived, prevent their feast.
1972 (R) 100m/C *FR* Fernando Rey, Delphine Seyrig, Jean-Pierre Cassel, Bulle Ogier, Michel Piccoli, Stephane Audran, Luis Bunuel; *W/Dir:* Luis Bunuel. Academy Awards '72: Best Foreign Language Film; British Academy Awards '73: Best Actress (Seyrig), Best Screenplay; National Board of Review Awards '72: 5 Best Foreign Films of the Year; National Society of Film Critics Awards '72: Best Director (Bunuel), Best Film. VHS, Beta, LV $59.95 *TAM, MED, APD* ♂♂♂♂

Dishonored Lady After her ex-boyfriend is murdered, a female art director finds that she's the number one suspect. When put on trial for the dastardly deed, she takes the fifth in this mediocre melodrama.
1947 85m/B Hedy Lamarr, Dennis O'Keefe, William Lundigan, John Loder; *Dir:* Robert Stevenson. VHS, Beta $16.95 *SNC, NOS, CAB* ♂½

Disney Beginnings A collection of nine silent Disney cartoons, including "Four Musicians of Bremen," "Puss in Boots," "Great Guns" and "Mechanical Cow."
192? 60m/B VHS, Beta $19.95 *VDM*

A Disney Christmas Gift A collection of Christmas scenes from such Disney classics as "Peter Pan," "The Sword in the Stone," and "Cinderella."
1982 47m/C VHS, Beta, LV $12.99 *DIS*

Disney's American Heroes Two Disney tall tales of American folk heroes: "Pecos Bill" and "Paul Bunyan."
1982 39m/C Roy Rogers, The Pioneers. VHS, Beta *DIS*

Disney's Halloween Treat Scenes from Disney classics, such as "Snow White and the Seven Dwarfs," "Fantasia," "Peter Pan" and "The Sword and the Stone" are tied together with a Halloween theme.
1984 47m/C VHS, Beta $39.95 *DIS*

Disney's Storybook Classics A collection of classic children's fables, featuring "Little Toot" (1948, excerpted from "Melody Time"), the story of a little harbor tugboat; "Chicken Little" (1943); "The Grasshopper and the Ant" (1934, a "Silly Symphony"); and "Peter and the Wolf" (1946, excerpted from "Make Mine Music"), a Disneyized version of Prokofiev's famous concert piece.
1982 121m/C Sterling Holloway, The Andrews Sisters. VHS, Beta *DIS*

Disorderlies Members of the popular rap group cavort as incompetent hospital orderlies assigned to care for a cranky millionaire. Fat jokes abound with performances by the Fat Boys.
1987 (PG) 86m/C Fat Boys, Ralph Bellamy; *Dir:* Michael Schultz. VHS, Beta, LV $19.98 *WAR* ♂

Disorderly Orderly When Jerry Lewis gets hired as a hospital orderly, nothing stands upright long with him around. Vintage slapstick Lewis running amuck in a nursing home.
1964 90m/C Jerry Lewis, Glenda Farrell, Everett Sloane, Kathleen Freeman, Susan Oliver; *Dir:* Frank Tashlin. VHS, Beta, LV $14.95 *PAR* ♂½

Disorganized Crime On the lam from the law, Bernsen attempts to organize a group of his ex-con buddies to pull off the perfect heist. Before the boys can organize,

however, the cops are hot on Bernsen's trail, and he must vacate the meeting place. Good cast attempts to lift this movie beyond script.
1989 (R) 101m/C Lou Diamond Phillips, Fred Gwynne, Corbin Bernsen, Ruben Blades, Hoyt Axton, Ed O'Neill, Daniel Roebuck, William Russ; ***Dir:*** Jim Kouf. **VHS, Beta, LV $89.95** *TOU* 🦴🦴

Disparates This program is an experimental film that uses various optical processes, sounds, and images to get its point across.
1974 11m/C VHS, Beta *MDP* 🦴🦴

The Displaced Person In this television adaptation of Flannery O'Connor's story, the inhabitants of a 1940s Georgia farm find their lives disrupted by a Polish refugee family. The strong work ethic practiced by the foreigners makes their less productive, if not lazy, neighbors regard them with caution and contempt. Well-crafted depiction of the prejudice faced by many immigrants aspiring to realize the American dream.
1976 58m/C Irene Worth, John Houseman, Henry Fonda, Shirley Stoler, Lane Smith, Earl Jones; ***Dir:*** Glenn Jordan. **VHS, Beta $24.95** *MON, KAR* 🦴🦴🦴½

Distant Drums A small band of adventurers tries to stop the Seminole War in the Florida Everglades.
1951 101m/C Gary Cooper, Mari Aldon, Robert Barratt, Richard Webb, Ray Teal, Arthur Hunnicutt; ***Dir:*** Raoul Walsh. **VHS $19.98** *REP, FCT* 🦴🦴

Distant Thunder A bitter, unrelenting portrait of a small Bengali neighborhood as the severe famines of 1942, brought about by the "distant thunder" of WWII, take their toll. A mature achievement from Ray; in Bengali with English subtitles.
1973 92m/C *IN* Soumitra Chatterjee, Sandhya Roy, Babita, Gobinda Chakravarty, Romesh Mukerji; ***Dir:*** Satyajit Ray. Berlin Film Festival '73: Best Film; New York Film Festival '73: Best Film. **VHS, Beta $44.95** *FCT* 🦴🦴🦴½

Distant Thunder A scarred Vietnam vet, who has become a recluse, and his estranged son reunite in the Washington State wilderness, causing him to reflect on his isolation. Strong premise and cast watered down by weak script.
1988 (R) 114m/C *CA* John Lithgow, Ralph Macchio, Kerrie Keane, Janet Margolin, Rick Rosenthal; ***Dir:*** Rick Rosenberg. **VHS, Beta, LV $14.95** *PAR* 🦴🦴

Distant Voices, Still Lives A profoundly executed, disturbing film chronicling a British middle-class family through the maturation of the three children, under the dark shadow of their abusive, malevolent father. An evocative, heartbreaking portrait of British life from World War II on, and of the rhythms of dysfunctional families. A film festival favorite.
1988 87m/C *GB* Freda Dowie, Peter Postlethwaite, Angela Walsh, Dean Williams, Lorraine Ashbourne; ***Dir:*** Terence Davies. Cannes Film Festival '89: International Critics Prize. **VHS, Beta, LV, CDV $89.95** *IVE* 🦴🦴🦴

Distortions Olivia Hussey is a widow whose evil aunt (Laurie) is holding her hostage. What the film lacks in suspense throughout most of the movie is made up for in the end.
1987 (PG) 90m/C Piper Laurie, Steve Railsback, Olivia Hussey, Edward Albert, Rita Gam, Terence Knox; ***Dir:*** Armand Mastroianni. **VHS, Beta $29.95** *ACA* 🦴🦴

Disturbance The two personas of a young schizophrenic get him entangled in a mysterious string of murders.
1989 81m/C Timothy Greeson, Lisa Geoffreion; ***Dir:*** Cliff Guest. **VHS, Beta $79.98** *VID* 🦴½

Disturbed Lusty mental hospital director McDowell meets sex starved and suicidal Gidley in this less-than-stellar erotic creeper. Redeemed only by McDowell's performance and more evidence that the droog's career hasn't been just peachy since "Clockwork Orange." Written by director Winkler.
1990 (R) 96m/C Malcolm McDowell, Geoffrey Lewis, Priscilla Pointer, Pamela Gidley, Clint Howard; ***W/Dir:*** Charles Winkler. **VHS, LV $89.95** *LIV* 🦴

Diva While at a concert given by his favorite star, a young French courier secretly tapes a soprano who has refused to record. The film follows the young man through Paris as he flees from two Japanese recording pirates, and a couple of crooked undercover police who are trying to cover-up for the chief who not only has a mistress, but runs a prostitution ring. Brilliant and dazzling photography compliment the eclectic soundtrack.
1982 (R) 123m/C *FR* Frederic Andrei, Roland Bertin, Richard Bohringer, Gerard Darmon, Jacques Fabbri, Wilhelmenia Wiggins Fernandez, Dominique Pinon; ***Dir:*** Jean-Jacques Beineix. **VHS, Beta, LV $29.95** *MGM, APD* 🦴🦴🦴½

The Dive Two North Sea oil-rig workers are trapped far below the waves when the lifeline to their diving bell snaps in rough seas.
1989 (PG-13) 90m/C Bjorn Sundquist, Frank Grimes, Einride Eidsvold, Michael Kitchen; ***Dir:*** Tristan De Vere Cole. **VHS, Beta, LV $89.95** *VIR* 🦴🦴

Divine Two Divine flicks from Waters. First, Divine plays a naughty girl who, not surprisingly, since this one is titled "The Diane Linkletter Story," ends up a successful suicide. "The Neon Woman" is a rare live performance, with Divine a woman who owns a strip joint and has a slew of problems you won't read about in Dear Abbey.
197? 110m/C Divine; ***Dir:*** John Waters. **VHS $29.99** *FCT* 🦴🦴½

The Divine Enforcer A monsignor in a crime-ridden L.A. neighborhood reaches his wits end in trying to stave off perpetrators of injustice. Fortunately, help arrives in the form of a mysterious priest who is equally handy with his fists, nunchakus, and guns. From then on it's no more mister nice guy, and the criminals haven't a prayer.
1991 90m/C Jan-Michael Vincent, Erik Estrada, Jim Brown, Judy Landers, Don Stroud, Robert Z'Dar, Michael Foley, Carrie Chambers, Hiroko; ***Dir:*** Robert Rundle. **VHS, Beta $79.95** *PSM* 🦴½

Divine Madness Bette Midler is captured at her best in a live concert at Pasadena Civic Auditorium.
1980 (R) 87m/C Bette Midler; ***Dir:*** Michael Ritchie. **VHS, Beta, LV $19.98** *WAR* 🦴½

Divine Nymph Charts the erotic adventures of the beautiful and young Manuela, who is dangerously pursued by two cousins.
1971 (R) 100m/C *IT* Laura Antonelli, Marcello Mastroianni, Terence Stamp; ***Dir:*** Giuseppe Patroni Griffi. **VHS, Beta $59.95** *MPI* 🦴½

Diving In Have you heard the one about the acrophobic diver? It seems a paralyzing fear of heights is the only thing between a young diver and Olympic gold. Much splashing about and heartstring-tugging.
1990 (PG-13) 92m/C Burt Young, Matt Adler, Kristy Swanson, Matt Lattanzi, Richard Johnson, Carey Scott, Yolanda Jilot; ***Dir:*** Strathford Hamilton. **VHS, Beta $89.95** *PAR, FCT* 🦴🦴

Diving Sequence This section of Leni Riefenstahl's "Olympia" features the best of the diving action.
1936 4m/C VHS, Beta, 3/4U $85.00 *BFA* 🦴🦴

Divorce Hearing A real-life divorce case is taped and shown, as a couple lambaste each other about their troubled 38 year-marriage. Intended as a diatribe against divorce.
1958 27m/B Dr. Paul Popenoe. **VHS, Beta $19.95** *VYY* 🦴🦴

Divorce His, Divorce Hers The first half of this drama shows the crumbling of a marriage through the husband's eyes. The second half offers the wife's perspective. Made for television.
1972 144m/C Richard Burton, Elizabeth Taylor; ***Dir:*** Waris Hussein. **VHS, Beta $9.95** *VCL* 🦴🦴

Divorce—Italian Style A middle-aged baron bored with his wife begins directing his amorous attentions to a teenage cousin. Since divorce in Italy is impossible, the only way out of his marriage is murder — and the baron finds a little known law which excuses a man from murdering his wife if she is having an affair (since he would merely be defending his honor). A hilarious comedy with a twist ending. Mastroianni's comic performance earned him an Oscar nomination for Best Actor. Available in Italian with English subtitles or dubbed in English.
1962 104m/B *IT* Marcello Mastroianni, Daniela Rocca, Leopoldo Trieste; ***Dir:*** Pietro Germi. Academy Awards '62: Best Screenplay; British Academy Awards '62: Best Actor (Mastroianni); Golden Globe Awards '62: Best Actor (Mastroianni). **VHS $79.95** *HTV* 🦴🦴🦴½

The Divorce of Lady X A spoiled British debutante, in the guise of "Lady X," makes a woman-hating divorce lawyer eat his words through romance and marriage. Based on the play by Gilbert Wakefield.
1938 92m/C *GB* Merle Oberon, Laurence Olivier, Ralph Richardson, Binnie Barnes; ***Dir:*** Tim Whelan. **VHS, Beta $19.95** *NOS, MRV, UNI* 🦴🦴½

The Divorcee Early Leonard film (he'd later direct "The Great Ziegfield" and "Pride and Prejudice") casts Shearer as a woman out to beat her husband at philandering. Married to a journalist, she cavorts with her husband's best friend and a discarded old flame as only a pre-production code gal could. Shearer—who Lillian Hellman described as having "a face unclouded by thought"—grabbed an Oscar, while the Academy also awarded nominations for best film, director, and screenplay. Based on the novel "Ex-Wife" by Ursula Parrott.
1930 83m/B Norma Shearer, Chester Morris, Conrad Nagel, Robert Montgomery, Mary Doran, Tyler Brooke, George Irving, Helen Johnson; ***Dir:*** Robert Z. Leonard. Academy Awards '30: Best Actress (Shearer). **VHS $29.98** *MGM, FCT, BTV* 🦴🦴½

Dixiana A Southern millionaire falls for a circus performer shortly before the start of the Civil War. Part color.
1930 99m/B Bebe Daniels, Bert Wheeler, Robert Woolsey, Dorothy Lamour, Bill Robinson; ***Dir:*** Luther Reed. **VHS, Beta $24.95** *NOS, DVT, TIM* 🦴½

Dixie: Changing Habits A New Orleans madam and a Mother Superior go head to head in this amusing made-for-television movie. In the end all benefit as the nuns discover business sense and pay off a debt, while the former bordello owner cleans up her act. Above average scripting and directing for the medium.

1985 100m/C Suzanne Pleshette, Cloris Leachman, Kenneth McMillan, John Considine, Geraldine Fitzgerald, Judith Ivey; *Dir:* George Englund. **VHS $59.95** IVE *♪♪♪*

Dixie Dynamite The two daughters of a Georgia moonshiner set out to avenge the murder of their father. The music is performed by Duane Eddy and Dorsey Burnette. **1976 (PG)** 88m/C Warren Oates, Christopher George, Jane Anne Johnstone, Kathy McHaley, R.G. Armstrong, Wes Bishop; *Dir:* Lee Frost. **VHS, Beta $49.95** UHV *♪*

Dixie Jamboree A gangster on the lam uses an unusual method to escape from St. Louis—the last Mississippi Showboat. **1944** 69m/B Guy Kibbee, Frances Langford, Louise Beavers, Charles Butterworth; *Dir:* Christy Cabanne. **VHS, Beta $19.95** NOS, VYY, DVT *♪*

Dixie Lanes A relentlessly nostalgic comedy set in a small town at the end of WWII. A woman, troubled by her nephew's restless antics, puts him to work. Unfortunately she adds to his plight, as her business involves the Black Market. **1988** 92m/C Hoyt Axton, Karen Black, Art Hindle, John Vernon, Ruth Buzzi, Tina Louise, Pamela Springsteen, Nina Foch; *Dir:* Don Cato. **VHS, Beta $79.95** CEL *♪*

Django Django is a stranger who arrives in a Mexican-border town to settle a dispute between a small band of Americans and Mexicans. **1968 (PG)** 90m/C Franco Nero, Loredana Nusciak, Angel Alvarez; *Dir:* Sergio Corbucci. **VHS, Beta $29.95** MAG *♪½*

Django Shoots First A colorful western with plenty of action and plot twists. **1974** 96m/C Glenn Saxon, Evelyn Stewart, Alberto Lupo. **VHS, Beta $29.95** VYY *♪½*

Do or Die Former "Playboy" centerfolds Speir and Vasquez team up as a couple of federal agent babes who take on international crime boss Morita. But the crimelord is ready for our scantily clad heroines with a sick game of revenge - and the stakes are their very lives! **1991 (R)** 97m/C Erik Estrada, Dona Speir, Roberta Vasquez, Pat Morita, Bruce Penhall, Carolyn Liu, Stephanie Schick; *W/Dir:* Andy Sidaris. **VHS, LV $79.95** COL *♪½*

Do the Right Thing An uncompromising, brutal comedy about the racial tensions surrounding a white-owned pizzeria in the Bed-Stuy section of Brooklyn on the hottest day of the summer, and the violence that eventually erupts. Ambivalent and, for the most part, hilarious, Lee's coming-of-age. **1989 (R)** 120m/C Spike Lee, Danny Aiello, Richard Edson, Ruby Dee, Ossie Davis, Giancarlo Esposito, Bill Nunn, John Turturro, John Savage, Rosie Perez, Frankie Faison; *W/Dir:* Spike Lee. **VHS, Beta, LV $19.95** MCA *♪♪♪½*

Do You Trust Your Wife/I've Got a Secret A pair of episodes from the early '50s quiz show featuring Edgar Bergen and Charlie McCarthy. **1956** 50m/B Edgar Bergen. **VHS, Beta $24.95** MVC, NOS, DVT *♪♪♪½*

D.O.A. A man is given a lethal, slow-acting poison. As his time runs out, he frantically seeks to learn who is responsible and why he was targeted for death. Dark film noir scored by Dmitri Tiomkin. Remade in 1969 as "Color Me Dead" and in 1988 with Dennis Quaid and Meg Ryan. Also available colorized.

1949 83m/B Edmond O'Brien, Pamela Britton, Luther Adler, Neville Brand, Beverly Garland; *Dir:* Rudolph Mate. **VHS, Beta $16.95** SNC, NOS, VYY *♪♪♪½*

D.O.A. Well-done remake of the 1949 thriller with Quaid portraying a college professor who is poisoned and has only 24 hours to identify his killer. His search for the suspect is further complicated by the fact that he is being sought by the police on phony charges of murder. Directed by the same people who brought "Max Headroom" to television screens. **1988 (R)** 98m/C Dennis Quaid, Meg Ryan, Charlotte Rampling, Daniel Stern, Jane Kaczmarek; *Dir:* Rocky Morton, Annabel Jankel. **VHS, Beta, LV $89.95** TOU *♪♪*

D.O.A.: A Right of Passage Often-depressing documentary about the British punk rock movement of the late '70s. The performance scenes are exciting. Also contains footage of the late Sid Vicious with his girlfriend Nancy Spungen, whom he was later accused of murdering. **1981 (R)** 93m/C The Sex Pistols, X-Ray Specs. **VHS, Beta $54.95** HAR, VES, LIV *♪♪♪*

The Doberman Gang Clever thieves train a gang of Dobermans in the fine art of bank robbery. Sequelled by "The Daring Dobermans." **1972 (PG)** 85m/C Byron Mabe, Hal Reed, Julie Parrish, Simmy Bow, JoJo D'Amore; *Dir:* Byron Ross Chudnow. **VHS, Beta $59.98** FOX *♪♪*

Doc Hollywood A hotshot young physician on his way to a lucrative California practice gets stranded in a small Southern town. Will the wacky woodsy inhabitants persuade the city doctor to stay? There aren't many surprises in this fish-out-of-water comedy, but the cast injects it with considerable charm. **1991 (PG-13)** 104m/C Michael J. Fox, Julie Warner, Woody Harrelson, Barnard Hughes, David Ogden Stiers, Frances Sternhagen, Bridget Fonda, George Hamilton, Roberts Blossom, Helen Martin, Macon McCalman; *Dir:* Michael Caton-Jones. **VHS, Beta, LV, 8mm $94.99** WAR, PIA *♪♪½*

Doc Savage Doc and "The Amazing Five" fight a murderous villain who plans to take over the world. Based on the novels of Kenneth Robeson. **1975 (PG)** 100m/C Ron Ely, Pamela Hensley; *Dir:* Michael Anderson Sr. **VHS, Beta $19.98** WAR *♪♪½*

Docks of New York Von Sternberg made his mark in Hollywood with this silent drama about two dockside losers finding love amid the squalor. A rarely seen, early masterpiece. Score by Gaylord Carter. **1928** 60m/B George Bancroft, Betty Compson, Olga Baclanova; *Dir:* Josef von Sternberg. **VHS, Beta $29.95** PAR *♪♪♪*

The Doctor A hotshot doctor develops throat cancer and gets treated at his own hospital, an ordeal that teaches him a respect and compassion for patients that he formerly lacked. Potential melodrama is saved by fine acting and strong direction. Based on the autobiographical book "A Taste of My Own Medicine" by Dr. Edward Rosenbaum. **1991 (PG-13)** 125m/C William Hurt, Elizabeth Perkins, Christine Lahti, Mandy Patinkin; *Dir:* Randa Haines. **VHS, Beta $92.95** TOU *♪♪½*

Dr. Alien An alien posing as a beautiful scientist turns a college freshman into a sex-addicted satyr. Likewise, the cast turns this flick into a dog.

1988 (R) 90m/C Billy Jacoby, Olivia Barash, Stuart Fratkin, Troy Donahue, Arlene Golonka, Judy Landers; *Dir:* David DeCoteau. **VHS, Beta, LV $19.95** PAR *♪*

Dr. Black, Mr. Hyde Horrifying tale of a black Jekyll who metamorphoses into a white monster with the help of the special potion. Unintended laughs lessen its suspense. Also known by the title "The Watts Monster." **1976 (R)** 88m/C Rosalind Cash, Stu Gilliam, Bernie Casey, Marie O'Henry; *Dir:* William Crain. **VHS, Beta $29.95** UHV, VCI *♪*

Doctor Blood's Coffin The aptly named doctor performs hideous experiments on the unsuspecting denizens of a lonely village. Good cast in this effective chiller. **1962** 92m/C GB Kieron Moore, Hazel Court, Ian Hunter; *Dir:* Sidney J. Furie. **VHS $16.95** SNC, TIM *♪♪½*

Doctor Butcher M.D. A mad doctor's deranged dream of creating "perfect people" by taking parts of one person and interchanging them with another backfires as his monstrocities develop strange side effects. The M.D., by the way, stands for "Medical Deviate." Also known as "Queen of the Cannibals." Dubbed. **1980 (R)** 81m/C IT Ian McCulloch, Alexandra Cole, Peter O'Neal, Donald O'Brien; *Dir:* Frank Martin. **VHS, Beta $39.98** PGN, IVE, IVE Woof!

Dr. Caligari The granddaughter of the original Dr. Caligari promotes "better living through chemistry" in her insane asylum while using her patients as guinea pigs in her bizarre sexual experiments. Nothing outstanding here in terms of story or performance, but "Dick Tracy" has nothing on the visually compelling set designs and make-up. **1989 (R)** 80m/C Madeleine Reynal, Fox Harris, Laura Albert, Jennifer Balgobin, John Durbin, Gene Zerna, David Parry, Barry Phillips; *Dir:* Stephen Sayadian. **VHS, LV $89.98** SHG, FCT *♪♪*

Dr. Christian Meets the Women This entry in the "Dr. Christian" series of films has the small-town physician once again exposing a con man trying to filch the townspeople. **1940** 63m/B Jean Hersholt, Edgar Kennedy, Rod La Rocque, Dorothy Lovett; *Dir:* William McGann. **VHS, Beta, 8mm $29.95** VYY *♪½*

Dr. Cyclops The famous early Technicolor fantasia about a mad scientist miniaturizing a group of explorers who happen upon his jungle lab. Landmark F/X and a slow-moving story. **1940** 76m/C Albert Dekker, Janice Logan, Victor Kilian, Thomas Coley, Charles Halton; *Dir:* Ernest B. Schoedsack. **VHS, Beta, LV $14.98** MCA, FCT, MLB *♪½*

Dr. Death, Seeker of Souls An evil doctor discovered a process for transmigrating his soul into the bodies of people he murdered 1000 years ago. Now he seeks to revive his wife. Final role for 77-year-old "Stooge" Moe Howard. **1973** 93m/C John Considine, Barry Coe, Cheryl Miller, Stewart Moss, Leon Askin, Jo Morrow, Florence Marly, Sivi Aberg, Athena Lorde, Moe Howard; *Dir:* Eddie Saeta. **VHS, Beta $59.95** PSM *♪*

Doctor Detroit Aykroyd is funny in thin film, portraying a meek college professor who gets involved with prostitutes and the mob, under the alias "Dr. Detroit." Aykroyd later married actress Donna Dixon, whom he worked with on this film. Features music by Devo, James Brown, and Pattie Brooks.

1983 (R) 91m/C Dan Aykroyd, Howard Hesseman, Donna Dixon, T.K. Carter, Lynn Whitfield, Lydia Lei, Fran Drescher, Kate Murtagh, George Furth, Andrew Duggan, James Brown, Glenne Headly; *Dir:* Michael Pressman. **VHS, Beta, LV** $14.95 *MCA 𝒥𝒥*

The Doctor and the Devils Based on an old screenplay by Dylan Thomas, this is a semi-Gothic tale about two criminals who supply a physician with corpses to study, either digging them up or killing them fresh.
1985 (R) 93m/C *GB* Timothy Dalton, Julian Sands, Jonathan Pryce, Twiggy, Stephen Rea, Beryl Reid, Sian Phillips, Patrick Stewart; *Dir:* Freddie Francis. **VHS, Beta** $79.98 *FOX 𝒥𝒥½*

Doctor in Distress An aging chief surgeon falls in love for the first time. His assistant tries to further the romance as well as his own love life. Fourth in the "Doctor" series.
1963 102m/C *GB* Dirk Bogarde, James Robertson Justice, Leo McKern, Samantha Eggar; *Dir:* Ralph Thomas. **VHS, Beta** $29.95 *VID 𝒥½*

Doctor Doolittle An adventure about a 19th-century English doctor who dreams of teaching animals to speak to him. Unfortunately suffers from poor script. Based on Hugh Lofting's acclaimed stories. Oscar nomination for Best Picture, Cinematography, Art Direction, Set Decoration, Sound, Original Score, and Film Editing.
1967 144m/C Rex Harrison, Samantha Eggar, Anthony Newley, Richard Attenborough, Geoffrey Holder, Peter Bull; *Dir:* Richard Fleischer. Academy Awards '67: Best Song ("Talk to the Animals"), Best Visual Effects; National Board of Review Awards '67: 10 Best Films of the Year. **VHS, Beta** $19.98 *FOX, FCT 𝒥𝒥*

Doctor of Doom A mad surgeon conducts an insane series of brain transplants, defying law and order. Two lady wrestlers try to stop him.
1962 77m/B *MX* Lorena Velasquez, Armando Silvestre, Elizabeth Campbell, Roberto Canedo; *Dir:* Rene Cardona Jr. **VHS, Beta** $59.95 *SNC, HHT, TIM Woof!*

Doctor Doom Conquers the World Spider Man battles the evil Doctor Doom who uses his scientific skills to conquer the world.
1985 90m/C **VHS, Beta** $9.99 *PSM 𝒥*

Doctor Duck's Super Secret All-Purpose Sauce Conglomeration of video, music and comedy clips directed by the former Monkee done in the style of his popular "Elephant Parts."
1985 90m/C Michael Nesmith, Whoopi Goldberg, Jimmy Buffett, Rosanne Cash, Jay Leno; *Dir:* Michael Nesmith. **VHS, Beta, LV** $9.95 *MVD, PAV 𝒥*

Doctor Faustus A stilted, stagy, but well-meaning adaptation of the Christopher Marlowe classic about Faust, who sold his soul to the devil for youth and the love of Helen of Troy.
1968 93m/C Richard Burton, Andreas Teuber, Elizabeth Taylor, Ian Marter; *Dir:* Richard Burton. **VHS, Beta** $69.95 *COL 𝒥½*

Dr. Frankenstein's Castle of Freaks Dr. Frankenstein and his dwarf assistant reanimate a few Neanderthals that are terrorizing a nearby Rumanian village.
1974 (PG) 87m/C *IT* Rossano Brazzi, Michael Dunn, Edmund Purdom, Christiane Royce; *Dir:* Robert Oliver. **VHS, Beta** $29.98 *MAG, SNC 𝒥*

Dr. Goldfoot and the Bikini Machine A mad scientist employs gorgeous female robots to seduce the wealthy and powerful, thereby allowing him to take over the world. Title song by the Supremes.
1966 90m/C Vincent Price, Frankie Avalon, Dwayne Hickman, Annette Funicello, Susan Hart, Kay Elkhardt, Fred Clark, Deanna Lund, Deborah Walley; *Dir:* Norman Taurog. **VHS** *AIP 𝒥𝒥*

Doctor Gore A demented pediatrician tries to assemble a facsimile of his dead wife with pieces of other women.
1975 90m/C J.G. Patterson, Jenny Driggers, Roy Mehaffey; *Dir:* Pat Patterson. **VHS, Beta** $19.95 *UHV Woof!*

Dr. Hackenstein A doctor tries to revive his wife by "borrowing" parts from unexpected (and unsuspecting) guests.
1988 (R) 88m/C David Muir, Stacy Travis, Catherine Davis Cox, Dyanne DiRosario, Anne Ramsey, Logan Ramsey; *Dir:* Richard Clark. **VHS, Beta** $79.98 *MCG 𝒥½*

Dr. Heckyl and Mr. Hype A naughty, comical version of the Jekyll and Hyde story. Exuberantly wicked performances from Reed and Coogan; stellar cast of "B" veterans.
1980 (R) 100m/C Oliver Reed, Sunny Johnson, Maia Danzinger, Mel Welles, Virgil Frye, Kedrick Wolfe, Jackie Coogan, Corinne Calvert, Dick Miller, Lucretia Love; *W/Dir:* Charles B. Griffith. **VHS, Beta** *PGN 𝒥𝒥*

Doctor in the House Four medical student/roommates seek only to examine lovely women and make lots of cash. In the process, they are also tempted by the evils of drink, but rally to make their grades. A riotous British comedy with marvelous performances all around. Led to six sequels and a television series.
1953 92m/C *GB* Dirk Bogarde, Muriel Pavlow, Kenneth More, Donald Sinden, Kay Kendall, James Robertson Justice, Donald Houston, Geoffrey Keen, George Coulouris, Shirley Eaton, Joan Hickson, Richard Wattis; *Dir:* Ralph Thomas. **VHS** $19.95 *PAR 𝒥𝒥𝒥½*

Dr. Jack/For Heaven's Sake In "Dr. Jack" (1922) and "For Heaven's Sake" (1926) Lloyd acts respectively as an unorthodox doctor and as a debonair millionaire who falls in love with a poor girl. Silent comedy shorts from Lloyd.
1926 88m/B Harold Lloyd, Jobyna Ralston; *Dir:* Sam Taylor. **VHS, Beta** *TLF 𝒥𝒥𝒥*

Dr. Jekyll and Mr. Hyde The first American film version of Robert Louis Stevenson's horror tale about a doctor's experiments which lead to his developing good and evil sides to his personality. Silent. Kino Video's edition also contains the rarely seen 1911 version of the film, as well as scenes from a different 1920 version.
1920 96m/B John Barrymore, Martha Mansfield, Brandon Hurst, Charles Lane, J. Malcolm Dunn, Nita Naldi; *Dir:* John S. Robertson. **VHS, LV** $16.95 *SNC, REP, MRV 𝒥𝒥𝒥*

Dr. Jekyll and Mr. Hyde The hallucinatory, feverish classic version of the Robert Louis Stevenson story, in which the good doctor becomes addicted to the formula that turns him into a sadistic beast. Possibly Mamoulian's and March's best work, and a masterpiece of subversive, pseudo-Freudian creepiness. Eighteen minutes from the original version, lost until recently, have been restored, including the infamous whipping scene.

1931 96m/B Fredric March, Miriam Hopkins, Halliwell Hobbes, Rose Hobart, Holmes Herbert, Edgar Norton; *Dir:* Rouben Mamoulian. Academy Awards '32: Best Actor (March); Venice Film Festival '31: Best Actor (March). **VHS, Beta, LV** $19.98 *MGM, BTV 𝒥𝒥𝒥*

Dr. Jekyll and Mr. Hyde Strangely cast adaptation of the Robert Louis Stevenson story about a doctor's experiment on himself to separate good and evil.
1941 113m/B Spencer Tracy, Ingrid Bergman, Lana Turner, Donald Crisp; *Dir:* Victor Fleming. **VHS, Beta, LV** $24.95 *MGM 𝒥𝒥𝒥*

Dr. Jekyll and Mr. Hyde Yet another production of Robert Louis Stevenson's classic story of a split-personality.
1968 90m/C Jack Palance. **VHS, Beta** $49.95 *IVE 𝒥𝒥𝒥*

Dr. Jekyll and Mr. Hyde Dr. Jekyll discovers a potion that turns him into the sinister Mr. Hyde. Based on the classic story by Robert Louis Stevenson.
1973 90m/C Kirk Douglas, Michael Redgrave, Susan George, Donald Pleasence; *Dir:* David Winters. **VHS, Beta, LV** $59.95 *SVS, IME 𝒥𝒥*

Dr. Jekyll and Sister Hyde A tongue-in-cheek variation on the split-personality theme, which has the good doctor transforming himself into a sultry, knife-wielding woman who kills prostitutes.
1971 (R) 94m/C *GB* Ralph Bates, Martine Beswick, Gerald Sim; *Dir:* Roy Ward Baker. **VHS, Beta** $59.99 *HBO, MLB 𝒥𝒥½*

Dr. Jekyll and the Wolfman Naschy's sixth stint as El Hombre Lobo. This time he visits a mysterious doctor in search of a cure but ends up turning into Mr. Hyde, a man who likes to torture women.
1971 (R) 85m/C *SP* Paul Naschy, Shirley Corrigan, Jack Taylor; *Dir:* Leon Klimovsky. **VHS** $19.98 *SNC 𝒥*

Dr. Jekyll's Dungeon of Death Dr. Jekyll and his lobotomized sister, Hilda, scour the streets of San Francisco looking for human blood to recreate his great-grandfather's secret serum. Also on video as, "Dr. Jekyll's Dungeon of Darkness."
1982 (R) 90m/C James Mathers, Dawn Carver Kelly, John Kearney, Jake Pearson; *Dir:* James Wood. **VHS, Beta** $24.95 *MAG, GHV 𝒥*

Dr. Judym The screen adaptation of writer Stefan Zeromski's story of an impoverished youth who grows up to become an idealistic physician. In Polish with English subtitles.
19?? 94m/C *PL* *Dir:* Wlodzimierz Haupe. **VHS** $49.95 *FCT 𝒥𝒥*

Dr. Kildare's Strange Case Dr. Kildare administers a daring treatment to a man suffering from a mental disorder of a dangerous nature. One in the film series.
1940 76m/B Lew Ayres, Lionel Barrymore, Laraine Day; *Dir:* Harold Bucquet. **VHS, Beta** $19.95 *NOS, HHT, DVT 𝒥½*

Doctor at Large A fledgling doctor seeks to become a surgeon at a hospital for the rich. Humorous antics follow. The third of the "Doctor" series.
1957 98m/C *GB* Dirk Bogarde, Muriel Pavlow, James Robertson Justice, Shirley Eaton, Donald Sinden, Anne Heywood; *Dir:* Ralph Thomas. **VHS, Beta** $29.95 *VID 𝒥𝒥½*

Dr. Mabuse the Gambler The massive, two-part crime melodrama, introducing the raving mastermind/extortionist/villain to the world. The film follows Dr. Mabuse through his life of crime until he finally goes

mad. Highly influential and inventive. Lang meant this to be a criticism of morally bankrupt post-World War I Germany. Followed some time later by ''Testament of Doctor Mabuse'' and ''The Thousand Eyes of Dr. Mabuse.'' Also known as ''Doktor Mabuse der Spieler.'' Silent.
1922 242m/B *GE* Rudolf Klein-Rogge, Aud Egede Nissen, Alfred Abel, Gertrude Welcker, Lil Dagover, Paul Richter; *Dir:* Fritz Lang. **VHS, Beta \$16.95** *SNC, MRV, NOS ♂♂♂ ½*

Dr. Mabuse vs. Scotland Yard A sequel to the Fritz Lang classics, this film features the arch-criminal attempting to take over the world with a mind-controlling camera.
1964 90m/B *GE* Sabine Bethmann, Peter Van Eyck; *Dir:* Paul May. **VHS, Beta \$29.95** *SNC, VYY, GLV ♂♂ ½*

Dr. Minx An account of a professor's sleazy, sordid affairs. Noteworthy only for the exceptionally poor acting of Ms. Williams.
1975 (R) 94m/C Edy Williams, Harvey Jason, Marlene Schmidt, Alvy Moore, William Smith; *Dir:* Hikmet Avedis. **VHS \$59.98** *CGI ♂*

Doctor Mordrid: Master of the Unknown Two immensely powerful sorcerers from the 4th dimension cross over into present time with two very different missions—one wants to destroy the Earth, one wants to save it. Will good or evil triumph?
1992 (R) 102m/C Jeffrey Combs, Yvette Nipar, Jay Acavone, Brian Thompson; *Dir:* Albert Band, Charles Band. **VHS, Beta** *PAR ♂♂*

Dr. No The world is introduced to secret agent James Bond when it is discovered that a mad scientist is sabotaging rocket launchings from his hideout in Jamaica. The first 007 film is far less glitzy than any of its successors but boasts the sexiest ''Bond girl'' of them all in Andress, and promptly made stars of her and Connery. On laserdisc, the film is in wide-screen transfer and includes movie bills, publicity photos, location pictures, and the British and American trailers. The sound effects and musical score can be separated from the dialogue. Audio interviews with the director, writer, and editor are included as part of the disc.
1962 111m/C *GB* Sean Connery, Ursula Andress, Joseph Wiseman, Jack Lord, Zena Marshall, Eunice Gayson, Margaret LeWars, John Kitzmiller, Lois Maxwell, Bernard Lee, Anthony Dawson; *Dir:* Terence Young. **VHS, Beta, LV \$19.95** *MGM, VYG, CRC ♂♂♂*

Dr. Orloff's Monster As if ''The Awful Dr. Orloff'' wasn't awful enough, the doctor is back, this time eliminating his enemies with his trusty killer robot. Awful awful.
1964 88m/B *SP* Jose Rubio, Agnes Spaak; *Dir:* Jess (Jesus) Franco. **VHS \$19.98** *SNC Woof!*

Dr. Otto & the Riddle of the Gloom Beam Ernest is in your face again. Fresh from the TV commercials featuring ''Ernest'' comes Jim Varney playing a villain out to wreck the global economy. He also plays all the other characters, too. Way bizarre, ''Know whut I mean?''
1986 (PG) 92m/C Jim Varney; *Dir:* John R. Cherry III. **VHS, Beta \$59.95** *KWI ♂ ½*

Doctor Phibes Rises Again The despicable doctor contines his quest to revive his beloved wife. Fun, superior sequel to ''The Abominable Dr. Phibes.''
1972 (PG) 89m/C *GB* Vincent Price, Robert Quarry, Peter Cushing, Beryl Reid, Hugh Griffith, Terry-Thomas; *Dir:* Robert Fuest. **VHS, Beta \$19.98** *LIV, VES ♂♂ ½*

Doctor Satan's Robot Indestructible robot is battled by insane Dr. Satan in reedited serial, ''Mysterious Dr. Satan.''
1940 100m/B Eduardo Ciannelli, Robert Wilcox, William Newell, Ella Neal; *Dir:* William Whitney, John English. **VHS \$29.95** *VCN ♂♂ ½*

Doctor at Sea To escape a troublesome romantic entanglement and the stresses of his career, a London physician signs on a cargo boat as a ship's doctor and becomes involved with French bombshell Bardot. The second of the ''Doctor'' series.
1956 93m/C *GB* James Robertson Justice, Dirk Bogarde, Brigitte Bardot; *Dir:* Ralph Thomas. **VHS, Beta \$29.95** *VID ♂♂ ½*

Dr. Seuss' Grinch Grinches the Cat in the Hat/Pontoffel Pock A Dr. Seuss animated double feature: The Cat in the Hat and the Grinch cross paths in ''The Grinch Grinches the Cat in the Hat'' and a young man gains his self confidence in ''Pontoffel Pock.''
1982 49m/C VHS, Beta \$14.98 *FOX ♂♂ ½*

Dr. Seuss' Halloween Is Grinch Night Animated tale about a young boy who must muster up enough courage to save his family and town from the nasty Grinch. Based upon the Dr. Seuss story.
1977 25m/C VHS, Beta \$14.98 *FOX ♂♂ ½*

Doctor Snuggles Animated adventure in which Doctor Snuggles pursues the evil Professor Emerald after a strange series of events occurs.
1984 60m/C VHS, Beta \$49.95 *SUE ♂♂ ½*

Dr. Strange A made-for-television pilot based upon the Marvel Comics character who, with the help of a sorcerer, practices witchcraft in order to fight evil.
1978 94m/C Peter Hooten, Clyde Kusatsu, Jessica Walter, Eddie Benton, John Mills; *Dir:* Philip DeGuere. **VHS, Beta \$39.95** *MCA ♂ ½*

Dr. Strangelove, or: How I Learned to Stop Worrying and Love the Bomb Peter Sellers plays a tour-de-force triple role in Stanley Kubrick's classic black anti-war comedy. While a U.S. President (Sellers) deals with the Russian situation, a crazed general (Hayden) implements a plan to drop the A-bomb on the Soviets. Famous for Pickens' wild ride on the bomb, Hayden's character's ''purity of essence'' philosophy, Scott's gumchewing militarist, a soft drink vending machine dispute, and countless other scenes. Classically written by Terry Southern, based on the novel ''Red Alert'' by Peter George. Oscar nominations for Best Picture, Best Actor (Sellers), Director Kubrick, and Adapted Screenplay.
1964 93m/B *GB* Peter Sellers, George C. Scott, Sterling Hayden, Keenan Wynn, Slim Pickens, James Earl Jones, Peter Bull; *Dir:* Stanley Kubrick. British Academy Awards '64: Best Picture; New York Film Critics Awards '64: Best Director; New York Times Ten Best List '64: Best Film. **VHS, Beta, LV, 8mm \$19.95** *COL ♂♂♂♂*

Dr. Syn The story of a seemingly respectable vicar of Dymchurch who is really a former pirate. The last film of George Arliss.
1937 90m/B *GB* George Arliss, Margaret Lockwood, John Loder; *Dir:* Roy William Neill. **VHS, Beta \$19.95** *NOS, MRV, CAB ♂♂*

Dr. Syn, Alias the Scarecrow A mild-mannered minister is, in reality, a smuggler and pirate avenging King George III's injustices upon the English people. Originally

broadcast in three parts on the Disney television show.
1964 (G) 129m/C Patrick McGoohan, George Cole, Tony Britton, Michael Hordern, Geoffrey Keen, Kay Cole; *Dir:* James Neilson. **VHS, Beta \$69.95** *DIS ♂♂ ½*

Doctor Takes a Wife A fast, fun screwball comedy wherein two ill-matched career people are forced via a publicity mix-up to fake being married. Co-written by George Seaton and Ken Englund.
1940 89m/B Ray Milland, Loretta Young, Reginald Gardiner, Gail Patrick, Edmund Gwenn, George Metaxa; *Dir:* Alexander Hall. **VHS, Beta \$69.95** *COL ♂♂♂*

Dr. Tarr's Torture Dungeon A mysterious man is sent to the forest to investigate the bizarre behavior of Dr. Tarr who runs a torture asylum.
1975 (R) 90m/C Claudio Brook, Ellen Sherman, Robert Dumont. **VHS, Beta \$59.95** *MAG Woof!*

Dr. Terror's House of Horrors On a train, six traveling companions have their fortunes told by a mysterious doctor. Little do they realize that their final destination has changed. Creepy and suspenseful, especially the severed-hand chase sequence.
1965 92m/C *GB* Christopher Lee, Peter Cushing, Donald Sutherland; *Dir:* Freddie Francis. **VHS \$14.95** *REP, MLB ♂♂ ½*

Doctor Who and the Daleks The time-lord Doctor Who and his three grandchildren accidentally transport themselves to a futuristic planet inhabited by the Daleks, who capture the travelers and hold them captive.
1965 78m/C *GB* Peter Cushing, Roy Castle; *Dir:* Gordon Flemyng. **VHS, Beta \$69.99** *HBO ♂♂*

Doctor Who: Death to the Daleks Lord does battle with his arch-foes, the Daleks. From the British television series.
1987 90m/C *GB* Elizabeth Sladen, Jon Pertwee; *Dir:* Michael E. Briant. **VHS, Beta \$19.98** *FOX ♂♂*

Doctor Who: Pyramids of Mars Join Dr. Who and Sarah Jane in another exciting space adventure series. A professor possessed by the God of Darkness must be stopped before he destroys humankind.
1985 91m/C *GB* Tom Baker, Elizabeth Sladen; *Dir:* Paddy Russell. **VHS, Beta \$19.98** *FOX ♂ ½*

Doctor Who: Spearhead from Space Earth is in peril at the hands of murderous mannequins and only Doctor Who can stop them.
19?? 92m/C *GB* Jon Pertwee. **VHS, Beta \$19.98** *FOX, FCT ♂ ½*

Doctor Who: Terror of the Zygons Still more hideous aliens threaten the Earth. Their instrument of destruction? The Loch Ness Monster. Unless of course the Doctor can stop them.
1978 92m/C *GB* Tom Baker. **VHS, Beta \$19.98** *FOX, FCT ♂♂*

Doctor Who: The Ark in Space Giant, intelligent bugs threaten the safety of the universe in this Doctor Who outing.
1978 90m/C *GB* Tom Baker. **VHS, Beta \$19.98** *FOX, FCT ♂♂*

Doctor Who: The Seeds of Death The British time traveler becomes involved in another soul-melting adventure.
1985 137m/B *GB* Patrick Troughton; *Dir:* Michael Ferguson. **VHS, Beta \$19.98** *FOX ♂ ½*

Doctor Who: The Talons of Weng-Chiang The Doctor is faced with a challenge that even he may not survive. From the popular British television series.
1988 140m/C *GB* Tom Baker, Louise Jameson; *Dir:* Michael E. Briant. **VHS, Beta $19.98** FOX ✓✓

Doctor Who: The Time Warrior When scientists begin disappearing, the Doctor pursues them into the Middle Ages, where he finds an unscrupulous Sontaran fleet commander is prepared to exploit his superior technology to no good end.
19?? 90m/C *GB* Jon Pertwee. **VHS, Beta $19.98** FOX, FCT ✓✓

Doctor X An armless mad scientist uses a formula for "synthetic flesh" to grow temporary limbs and commit murder. A classic, rarely-seen horror oldie, famous for its very early use of two-color Technicolor.
1932 77m/C Lionel Atwill, Fay Wray, Lee Tracy, Preston Foster; *Dir:* Michael Curtiz. **VHS, Beta, LV $19.95** MGM, MLB ✓✓½

Doctor Zhivago A sweeping adaptation of the Nobel Prize-winning Boris Pasternak novel about an innocent Russian poet-intellectual caught in the furor and chaos of the Bolshevik Revolution. Essentially a poignant love story filmed as a historical epic. A panoramic film that popularized the song "Lara's Theme." Overlong, with often disappointing performances, but gorgeous scenery. Lean was more successful in "Lawrence of Arabia," where there was less need for ensemble acting. Academy Award nominations for Best Picture, Director, Best Supporting Actor (Courtenay), Sound, and Film Editing.
1965 197m/C Omar Sharif, Julie Christie, Geraldine Chaplin, Rod Steiger, Alec Guinness, Klaus Kinski, Ralph Richardson, Rita Tushingham, Siobhan McKenna, Tom Courtenay; *Dir:* David Lean. Academy Awards '65: Best Adapted Screenplay, Best Art Direction/Set Decoration (Color), Best Color Cinematography, Best Costume Design, Best Original Score; Golden Globe Awards '66: Best Film—Drama; National Board of Review Awards '65: 10 Best Films of the Year, Best Actress (Christie). **VHS, Beta, LV $29.97** MGM, RDG ✓✓✓

Doctors and Nurses A children's satire of soap operas wherein the adults play the children and vice-versa.
1982 90m/C Rebecca Rigg, Drew Forsythe, Graeme Blundell; *Dir:* Maurice Murphy. **VHS, Beta** VID ✓✓

Doctors' Wives An adaptation of the Frank Slaughter potboiler about a large city hospital's doctors, nurses, and their respective spouses, with plenty of affairs, medical traumas, and betrayals.
1970 102m/C Dyan Cannon, Richard Crenna, Gene Hackman, Carroll O'Connor, Rachel Roberts, Janice Rule, Diana Sands, Ralph Bellamy, John Colicos; *Dir:* George Schaefer. **VHS, Beta $69.95** COL ✓

Dodes 'ka-den In this departure from his samurai-genre films, Kurosawa depicts a throng of fringe-dwelling Tokyo slum inhabitants in a semi-surreal manner. Fascinating presentation and content. Nominated for the Best Foreign Film Academy Award in 1971.
1970 140m/C *JP* Yoshitaka Zushi, Junzaburo Ban, Kiyoko Tange; *Dir:* Akira Kurosawa. New York Film Festival '71: Best Actress. **VHS, Beta $19.98** TAM, SUE ✓✓✓½

Dodge City Flynn stars as Wade Hutton, a roving cattleman who becomes the sheriff of Dodge City. His job: to run a ruthless outlaw and his gang out of town. De Havil-

land serves as Flynn's love interest, as she did in many previous films. A broad and colorful shoot-em-up!
1939 104m/C Errol Flynn, Olivia de Havilland, Bruce Cabot, Ann Sheridan, Alan Hale Jr., Frank McHugh, Victor Jory, Henry Travers; *Dir:* Michael Curtiz. **VHS, Beta, LV $19.98** MGM, FOX, FCT ✓✓✓

Dodsworth The lives of a self-made American tycoon and his wife are drastically changed when they take a tour of Europe. The success of their marriage seems questionable as they re-evaluate their lives. Huston excels as does the rest of the cast in this film, based upon the Sinclair Lewis novel. Oscar nominated for Best Picture, Best Actor (Huston), Best Supporting Actress (Ouspenskaya), Best Director, Best Screenplay, Best Sound Recording, and Best Interior Decoration.
1936 101m/B Walter Huston, David Niven, Paul Lukas, John Payne, Mary Astor, Ruth Chatterton, Maria Ouspenskaya; *Dir:* William Wyler. Academy Awards '36: Best Interior Decoration. **VHS, Beta, LV $19.98** HBO, SUE ✓✓✓½

Does This Mean We're Married? A stand-up comedian is trying to find success in the comedy clubs of Paris but without her green card she may soon be deported. She finds a sleazy marriage broker who fixes her up with a womanizing songwriter who needs money. As usual, immigration officials suspect that the marriage is a fake and the mismatched couple must live together to prove them wrong.
1990 93m/C Patsy Kensit, Stephane Freiss; *Dir:* Carol Wiseman. **VHS $89.95** NLC ✓✓

Dog Day An American traitor, who is on the lam from the government, and his cronies, takes refuge on a small farm. A surprise awaits him when the farmers come up with an unusual plan to bargain for his life.
1983 101m/C *FR* Lee Marvin, Miou-Miou, Victor Lanoux; *Dir:* Yves Boisset. **VHS, Beta $79.98** LIV, LTG ✓✓½

Dog Day Afternoon Based on a true story, this taut, yet fantastic thriller centers on a bi-sexual and his slow-witted buddy who rob a bank to obtain money to fund a sex change operation for the ringleader's lover. Pacino is breathtaking in his role as the frustrated robber, caught in a trap of his own devising. Very controversial for its language and subject matter when released, it nevertheless became a huge success. Director Lumet keeps up the pace, fills the screen with pathos without gross sentiment. Oscar nominations for Best Picture, actor Pacino, supporting actor Sarandon, Lumet, and Film Editing.
1975 (R) 124m/C Al Pacino, John Cazale, Charles Durning, James Broderick, Chris Sarandon, Carol Kane, Lance Henriksen, Dick Williams; *Dir:* Sidney Lumet. Academy Awards '75: Best Original Screenplay; British Academy Awards '75: Best Actor (Pacino), Best Film Editing; National Board of Review Awards '75: 10 Best Films of the Year, Best Supporting Actor (Durning). **VHS, Beta, LV $19.98** WAR ✓✓✓

Dog Eat Dog Centering on a heist scheme by Mitchell, this one has nothing to offer except Mansfield's attributes.
1964 86m/B Cameron Mitchell, Jayne Mansfield, Isa Miranda; *Dir:* Ray Nazzaro. **VHS, Beta, 8mm $29.99** VYY ✓

Dog of Flanders A young Dutch boy and his grandfather find a severely beaten dog and restore it to health.

1959 96m/C David Ladd, Donald Crisp, Theodore Bikel; *Dir:* James B. Clark. **VHS, Beta $14.95** PAR ✓✓½

Dog Pound Shuffle Two drifters form a new song-and-dance act in order to raise the funds necessary to win their dog's freedom from the pound. Charming Canadian production. Also known as "Spot."
1975 (PG) 98m/C *CA* Ron Moody, David Soul; *Dir:* Jeffrey Bloom; *W/Dir:* Jeffrey Bloom. **VHS, Beta $59.98** FOX ✓✓✓

The Dog Soldier This "video novel" dramatically depicts a state of future events whereby a lone human in flight struggles to preserve his right to be free.
1972 40m/B VHS, Beta PCV ✓½

Dog Soldier The CIA has discovered a cure for AIDS in this animated Japanese feature. Problem is a bevy of other organizations want the antidote for themselves, including smugglers, ex-Green Berets, and Japan's own intelligence service.
1992 45m/C *JP* **VHS** CPM ✓½

Dog Star Man The silent epic by the dean of experimental American film depicts man's spiritual and physical conflicts through Brakhage's characteristically freeform collage techniques.
1964 78m/C *Dir:* Stan Brakhage. **VHS, Beta $29.95** MFV, FCT ✓✓✓

Dogfight It's 1963 and a baby-faced Marine and his buddies are spending their last night in San Francisco before leaving the U.S. for a tour of duty in Vietnam. The buddies agree to throw a "dogfight," a competition to see who can bring the ugliest date to a party. Birdlace meets a shy, average-looking waitress and the two spend the evening getting to know one another, and though the relationship is at first rocky, they eventually develop a mutual respect which may lead to love. From the director of "True Love."
1991 (R) 94m/C River Phoenix, Lili Taylor, Richard Panebianco, Anthony Clark, Mitchell Whitfield, E.G. Daily, Holly Near; *Dir:* Nancy Savoca. **VHS, Beta, LV, 8mm $92.99** WAR, PIA ✓✓✓

Dogs of Hell The sheriff of an idyllic resort community must stop a pack of killer dogs from terrorizing the residents.
1983 (R) 90m/C Earl Owensby, Bill Gribble, Jerry Rushing; *Dir:* Worth Keeter. **VHS, Beta $59.95** MED ✓½

Dogs in Space A low-budget Australian film about a clique of aimless Melbourne rock kids in 1978, caught somewhere between post-hippiedom and punk, free love and heroin addiction. Acclaimed. Written by Lowenstein; music by Hutchence, Brian Eno, Iggy Pop, and others.
1987 109m/C *AU* Michael Hutchence, Saskia Post, Nique Needles, Tony Helou, Deanna Bond; *Dir:* Richard Lowenstein. **VHS, Beta $79.98** FOX ✓✓

The Dogs of War A graphic depiction of a group of professional mercenaries, driven by nothing but their quest for wealth and power, hired to overthrow the dictator of a new West African nation. Has some weak moments which break up the continuity of the movie. Based on the novel by Frederick Forsythe.
1981 (R) 102m/C *GB* Christopher Walken, Tom Berenger, Colin Blakely, Paul Freeman, Hugh Millais, Victoria Tennant, JoBeth Williams; *Dir:* John Irvin. **VHS, Beta, LV $29.95** MGM ✓✓½

The Dogs Who Stopped the War A Canadian children's film about a dog who puts a halt to a escalating snowball fight between rival gangs.
1984 (G) 90m/C *CA* VHS, Beta $79.95 *HBO* ♂

Doin' Time At the John Dillinger Memorial Penitentiary, the inmates take over the prison under the supervision of warden "Mongo." Silliness prevails.
1985 (R) 80m/C Jeff Altman, Dey Young, Richard Mulligan, John Vernon, Colleen Camp, Melanie Chartoff, Graham Jarvis, Pat McCormick, Eddie Velez, Jimmie Walker, Judy Landers, Nicholas Worth, Mike Mazurki, Muhammad Ali, Melinda Fee, Francesca "Kitten" Natavidad, Ron Palillo; *Dir:* George Mendeluk. VHS, Beta $19.98 *WAR* ♂

Doin' Time on Planet Earth A young boy feels out of place with his family and is convinced by two strange people (aliens themselves?) that he is really an extraterrestrial. Amusing, aimless fun directed by the son of Walter Matthau.
1988 (PG-13) 83m/C Adam West, Candice Azzara, Hugh O'Brian, Matt Adler, Timothy Patrick Murphy, Roddy McDowall, Maureen Stapleton; *Dir:* Charles Matthau. VHS, Beta $19.98 *WAR* ♂♂ 1/2

Doin' What the Crowd Does The tale of the death of Poe's lover Lenore.
1973 (PG) 89m/C Robert Walker Jr., Cesar Romero. VHS, Beta $57.95 *UNI* ♂

Dolemite An ex-con attempts to settle the score with some of his former inmates. He forms a band of kung-fu savvy ladies. Strange combination of action and comedy.
1975 (R) 88m/C Rudy Ray Moore, Jerry Jones; *Dir:* D'Urville Martin. VHS *XVC* ♂ 1/2

Dolemite 2: Human Tornado Nobody ever said Moore was for everybody's taste. But, hey, when blaxploitation movies were the rage, Rudy the standup comic was out there rapping through a series of trashy movies that, when viewed today, have survived the test of time. This one's just as vile, violent and sexist as the day it was released. When Rudy is surprised in bed with a white sheriff's wife, he flees and meets up with a madam and a house of kung-fu-skilled girls who are embroiled in a fight with a local mobster. Also known as "The Human Tornado."
1976 (R) 98m/C Rudy Ray Moore, Lady Reed. VHS *XVC* ♂ 1/2

The Doll A lonely night watchman happens upon two burglars and, in the chase, the thieves knock over a mannequin. The watchman reports the mannequin stolen, brings it home and begins to have conversations with it. Soon, the doll's needs cause him to steal jewelry and clothes, until his brutish neighbor discovers his secrets. In Swedish with English subtitles.
1962 96m/B *SW* Per Oscarsson, Gio Petre, Tor Isedal, Elsa Prawitz; *Dir:* Arne Mattson. VHS $29.95 *FCT, DVT* ♂♂

Doll Face Story of a stripper who wants to go legit and make it on Broadway. Film was adapted from the play "The Naked Genius" by tease queen Gypsy Rose Lee.
1946 80m/B Vivian Blaine, Dennis O'Keefe, Perry Como, Carmen Miranda, Martha Stewart, Reed Hadley; *Dir:* Lewis Seiler. VHS, Beta $19.99 *NOS, HHT, DVT* ♂♂ 1/2

The Doll Squad Three voluptuous special agents fight an ex-CIA agent out to rule the world.

1973 (PG) 93m/C Michael Ansara, Francine York, Anthony Eisley, Tura Satana; *Dir:* Ted V. Mikels. VHS, Beta $39.95 *GHV, MRV* ♂

Dollar The actress, wife of an industrialist, convinced that he is having an affair, follows him to a ski lodge in attempt to catch him in the act. In Swedish with English subtitles.
1938 74m/B *SW* Georg Rydeberg, Ingrid Bergman, Kotti Chave, Tutta Rolf, Hakan Westegren, Elsa Burnett, Edvin Adolphson, Gosta Cederlund, Eric Rosen; *Dir:* Gustaf Molander. VHS, Beta, LV $29.95 *WAC, CRO* ♂♂

Dollars A bank employee and his dizzy call-girl assistant plan to steal the German facility's assets while installing its new security system. Lighthearted fun. Music by Quincy Jones.
1971 (R) 119m/C Warren Beatty, Goldie Hawn, Gert Frobe, Scott Brady, Robert Webber; *Dir:* Richard Brooks. VHS, Beta, LV $9.95 *GKK* ♂♂♂

Dollmaker The excellent made-for-television adaptation of Harriette Arnow's novel. A strong-willed Kentucky mother of five in the 1940s helps move her family to Detroit. Petrie's direction moves the story along and creates a lovely period vision.
1984 140m/C Jane Fonda, Levon Helm, Geraldine Page, Amanda Plummer, Susan Kingsley; *Dir:* Daniel Petrie. VHS, Beta $79.98 *FOX* ♂♂ 1/2

Dollman An ultra-tough cop from an Earth-like planet (even swear words are the same) crashes in the South Bronx—and on this world he's only 13 inches tall. The filmmakers squander a great premise and cast with bloody shootouts and a sequel-ready non-ending.
1991 86m/C Tim Thomerson, Jackie Earle Haley, Kamala Lopez, Humberto Ortiz, Nicholas Guest, Michael Halsey, Eugene Glazer, Judd Omen, Frank Collison, Vincent Klyn; *Dir:* Albert Pyun. VHS, Beta, LV $19.95 *PAR, PIA* ♂ 1/2

The Dolls A tropically located photographer recruits an area beauty into the fashion world with his winning smile and macho charm, only to have native traditions forbid her to follow him.
1983 96m/C Tetchie Agbayani, Max Thayer, Carina Schally, Richard Seward; *Dir:* Hubert Frank. VHS, Beta, LV $69.98 *LIV, VES* Woof!

Dolls A group of people is stranded during a storm in an old, creepy mansion. As the night wears on, they are attacked by hundreds of antique dolls. Tongue-in-cheek.
1987 (R) 77m/C Ian Patrick Williams, Carolyn Purdy-Gordon, Carrie Lorraine, Stephen Lee, Guy Rolfe, Bunty Bailey, Cassie Stuart, Hilary Mason; *Dir:* Stuart Gordon. VHS, Beta $14.98 *LIV, VES* ♂♂ 1/2

Doll's House An all-star cast is featured in this original television production of Henrik Ibsen's classic play about an independent woman's quest for freedom in nineteenth-century Norway.
1959 89m/B Julie Harris, Christopher Plummer, Jason Robards Jr., Hume Cronyn, Eileen Heckart, Richard Thomas. VHS, Beta $39.95 *MGM, FCT* ♂♂ 1/2

Doll's House Jane Fonda plays Nora, a subjugated 19th-century housewife who breaks free to establish herself as an individual. Based on Henrik Ibsen's classic play; some controversy regarding Fonda's interpretation of her role.
1973 (G) 98m/C Jane Fonda, Edward Fox, Trevor Howard, David Warner, Delphine Seyrig; *Dir:* Joseph Losey. New York Film Festival '73: Best Supporting Actor. Beta $9.99 *STE, PSM* ♂♂ 1/2

A Doll's House A Canadian production of the Henrik Ibsen play about a Norwegian woman's search for independence.
1989 (G) 96m/C *GB* Claire Bloom, Anthony Hopkins, Ralph Richardson, Denholm Elliott, Anna Massey, Edith Evans; *Dir:* Patrick Garland. VHS, Beta, LV $9.99 *SOU, STE* ♂♂♂

Dolly Dearest Strange things start happening after an American family takes over a run-down Mexican doll factory. They create a new doll called "Dolly Dearest" with deadly results. In the same tradition as the "Chucky" series.
1992 94m/C Rip Torn, Sam Bottoms, Denise Crosby; *Dir:* Maria Lease. VHS, LV $92.95 *VMK* ♂ 1/2

The Dolphin A dolphin visits a Brazilian fishing village each full moon, turns himself into a man, and casts a spell of seduction over the women. Villagers are both enchanted and angered by the dolphin-man, since his presence creates desire in local women but scares fish from the waters. In Portuguese with English subtitles.
1987 95m/C *BR* Carlos Alberto Riccelli, Cassia Kiss, Ney Latorraca; *W/Dir:* Walter Lima Junior. VHS $79.95 *FXL* ♂♂ 1/2

Dolphin Adventure A slew of dolphin fans set out into tropical waters to meet and observe the sea mammals. Explorers attempt to communicate with dolphins by playing an underwater piano.
1979 58m/C *Dir:* Michael Wiese, Hardy Jones. VHS, Beta $59.98 *LIV, VES, SPV* ♂♂ 1/2

Dominick & Eugene Dominick is a little slow, but he makes a fair living as a garbageman—good enough to put his brother through medical school. Both men struggle with the other's faults and weaknesses, as they learn the meaning of family and friendship. Well-acted, especially by Hulce, never melodramatic or weak.
1988 (PG-13) 96m/C Ray Liotta, Tom Hulce, Jamie Lee Curtis; *Dir:* Robert M. Young. VHS, Beta, LV $19.98 *ORI, FCT* ♂♂♂

Dominion: Tank Police, Act 1 A Japanese cartoon for grown-ups. In this science fiction adventure, it's the tank police versus the sexy evil ladies of the Bauku Gang. First of four parts. Japanese with English subtitles.
1991 40m/C *JP* VHS $34.98 *BMG* ♂♂♂

Dominique is Dead A woman is driven to suicide by her greedy husband; now someone is trying to drive him mad. A.K.A. "Dominique" and "Avenging Spirit."
1979 (PG) 95m/C Cliff Robertson, Jean Simmons, Jenny Agutter, Simon Ward, Ron Moody; *Dir:* Michael Anderson Sr. VHS, Beta $9.95 *SIM, PSM* ♂♂

Domino A beautiful woman and a mysterious guy link up for sex, murder and double-crosses. Dubbed. Tries to be arty and avant-garde, but only succeeds in being a piece of soft-core fluff.
1988 (R) 95m/C *IT* Brigitte Nielsen, Tomas Arana, Daniela Alzone; *Dir:* Ivana Massetti. VHS, Beta, LV $89.95 *IVE* ♂

The Domino Principle Gene Hackman plays a doltish convict plotting to escape from San Quentin prison. He's been offered a new job: working as a political assassin.
1977 (R) 97m/C Gene Hackman, Candice Bergen, Richard Widmark, Mickey Rooney, Edward Albert, Eli Wallach; *Dir:* Stanley Kramer. VHS, Beta $9.99 *CCB, FOX* ♂ 1/2

D

Don Amigo/Stage to Chino A western double feature: The Cisco Kid and Pancho ride together again in "Don Amigo", and in "Stage to Chino", a postal inspector investigates a gang who robs rival stage lines.
194? 122m/B Duncan Renaldo, Leo Carrillo, George O'Brien, Roy Barcroft; *Dir:* Edward Killy. **VHS, Beta** $34.95 *RKO* ✻✻

Don Carlo The spectacular Verdi opera, based on a play by Schiller, is performed at New York City's Metropolitan Opera House in Italian with English subtitles.
1983 214m/C Placido Domingo, Nikolai Ghiaurov, Mirella Freni, Grace Bumbry, Louis Quilico. **VHS, Beta, LV** $39.95 *HMV, PAR, PIA*

Don Daredevil Rides Again Don Daredevil flies into danger in this 12-episode serial.
1951 180m/B Ken Curtis. **VHS, Beta** *MED, MLB* ✻

The Don is Dead A violent Mafia saga wherein a love triangle interferes with Family business, resulting in gang wars.
1973 (R) 96m/C Anthony Quinn, Frederic Forrest, Robert Forster, Al Lettieri, Ina Balin, Angel Tompkins, Charles Cioffi; *Dir:* Richard Fleischer. **VHS, Beta** $59.95 *MCA* ✻✻½

Don Juan Barrymore stars as the swashbuckling Italian duke with Spanish blood who seduces a castleful of women in the 1500s. Many exciting action sequences, including classic sword fights in which Barrymore eschewed a stunt double. Great attention is also paid to the detail of the costumes and settings of the Spanish-Moor period. Noted for employing fledgling movie sound effects and music dubbing technology, which, ironically, were responsible for eclipsing the movie's reputation. Watch for Loy as an Asian vamp and Oland as a pre-Charlie Chan Cesare Borgia.
1926 90m/B John Barrymore, Mary Astor, Willard Louis, Estelle Taylor, Helene Costello, Myrna Loy, June Marlowe, Warner Oland, Montagu Love, Hedda Hopper, Gustav von Seyffertitz; *Dir:* Alan Crosland. **VHS, LV** $29.95 *MGM, FCT* ✻✻✻½

Don Juan My Love Sexy comedy finds the ghost of Don Juan given a chance, after 450 years in Purgatory, to perform a good deed and free his soul. In Spanish with English subtitles.
1990 96m/C *SP* Juan Luis Galiardo, Rossy DePalma, Maria Barranco, Loles Leon; *Dir:* Antonio Mercero. Spanish Press '90: Best Comedy. **VHS** $79.95 *FXL, FCT* ✻✻✻

Don Q., Son of Zorro Zorro's son takes up his father's fight against evil and injustice. Silent sequel to the 1920 classic.
1925 111m/B Douglas Fairbanks Sr., Mary Astor. **VHS, Beta** $19.95 *NOS, WFV, CCB* ✻✻

Don Quixote The lauded, visually ravishing adaptation of the Cervantes classic, with a formal integrity inherited from Eisenstein and Dovshenko. In Russian with English subtitles.
1957 110m/B *SP RU* Nikolai Cherkassov, Yuri Tobubeyev; *Dir:* Grigori Kosintsev. **VHS, Beta** $34.95 *FCT* ✻✻✻½

Don Winslow of the Coast Guard Serial in 13 episodes features comic-strip character Winslow as he strives to keep the waters of America safe for democracy.
1943 234m/B Don Terry, Elyse Knox; *Dir:* Ford Beebe, Ray Taylor. **VHS, Beta** $49.95 *NOS, MED, VDM* ✻✻½

Don Winslow of the Navy Thirteen episodes centered around the evil Scorpion, who plots to attack the Pacific Coast, but is thwarted by comic-strip hero Winslow.
1943 234m/B Don Terry, Walter Sands, Anne Nagel; *Dir:* Ford Beebe, Ray Taylor. **VHS, Beta** $49.95 *NOS, MED, MLB* ✻✻½

Dona Flor & Her Two Husbands A woman becomes a widow when her philandering husband finally expires from drink, gambling, and ladies. She remarries, but her new husband is so boring and proper that she begins fantasizing spouse number one's return. But is he only in her imagination?
1978 106m/C *BR* Sonia Braga, Jose Wilker, Mauro Mendonca; *Dir:* Bruno Barreto. **VHS, Beta, LV** $79.95 *FXL, IME, TAM* ✻✻✻

Dona Herlinda & Her Son A Mexican sex comedy about a mother who manipulates her bisexual son's two lovers (one male, one female), until all four fit together into a seamless unit. In Spanish with English subtitles. Slow, but amusing.
1986 90m/C *MX* Guadalupe Del Toro, Arturo Meza, Marco Antonio Trevino, Leticia Lupersio; *Dir:* Jaime Humberto Hermosillo. **VHS, Beta** $79.95 *KIV* ✻✻

Donald Duck: The First 50 Years Featured highlights from Donald Duck's career, includes "The Wise Little Hen," "Don Donald," and the Oscar-nominated "Rugged Bear."
1953 45m/C *Dir:* Walt Disney. **VHS, Beta, LV** $12.99 *DIS* ✻✻✻

Donald: Limited Gold Edition 1 Features the early adventures of Donald Duck during the World War II period. Includes "Autograph Hound," "The Riveter," "Good Time for a Dime," and "The New Neighbor."
1953 51m/C *Dir:* Walt Disney. **VHS** $12.99 *DIS* ✻✻✻

Donald's Bee Pictures: Limited Gold Edition 2 Includes seven cartoons, featuring Donald Duck and his battles with a honeybee. Stunning artwork in "Window Cleaners," which is the Bee's debut.
1952 50m/C *Dir:* Walt Disney. **VHS, Beta, LV** $12.99 *DIS* ✻✻✻

Donde Duermen Dos...Duermen Tres A Spanish "Mr. Mom"-flavored domestic comedy.
198? 94m/C *SP* Susana Gimenez, Juan Carlos Calabro. **VHS, Beta** $57.95 *UNI* ✻

Donkey Skin (Peau d'Ane) A charming, all-star version of the medieval French fairy tale about a king searching for a suitable wife in a magical realm after his queen dies. In his quest for the most beautiful spouse, he learns that his daughter is that woman. She prefers prince charming, however. English subtitles.
1970 89m/C *FR* Catherine Deneuve, Jean Marais, Delphine Seyrig, Jacques Perrin; *Dir:* Jacques Demy. **VHS, Beta** $19.95 *TAM, SUE* ✻✻✻

Donner Pass: The Road to Survival Tame retelling of the western wagon-train pioneers who were forced to resort to cannibalism during a brutal snowstorm in the Rockies. The tragedy is lightly implied, keeping the film suitable for family viewing.
1984 98m/C Robert Fuller, Diane McBain, Andrew Prine, John Anderson, Michael Callan; *Dir:* James L. Conway. **VHS, Beta** $19.95 *STE, UHV, MAG* ✻✻

Donovan's Brain A scientist maintains the still-living brain of a dead millionaire and begins to be controlled by its powerful force. Based on Curt Siodmak's novel.
1953 85m/B Lew Ayres, Gene Evans, Nancy Davis, Steve Brodie; *Dir:* Felix Feist. **VHS, Beta, LV** $39.95 *IME, MGM, MLB* ✻✻✻

Donovan's Reef Two World War II buddies meet every year on a Pacific atoll to engage in a perpetual bar-brawl, until a stuck-up Bostonian maiden appears to find her dad, a man who has fathered a brood of lovable half-casts. A rollicking, good-natured film from Ford.
1963 109m/C John Wayne, Lee Marvin, Jack Warden, Elizabeth Allan, Dorothy Lamour, Mike Mazurki, Cesar Romero; *Dir:* John Ford. **VHS, Beta, LV** $19.95 *PAR, TLF, BUR* ✻✻✻

Don's Party A rather dark comedy focusing on Australian Yuppie-types who decide to watch the election results on television as a group. A lot more goes on at this party, however, than talk of the returns. Sexual themes surface. Cast members turn fine performances, aided by top-notch script and direction.
1976 90m/C *AU* Pat Bishop, Graham Kennedy, Candy Raymond, Veronica Lang, John Hargreaves; *Dir:* Bruce Beresford. Australian Film Institute '77: Best Actress (Bishop). **VHS, Beta** $14.98 *VID, FCT* ✻✻✻

Don't Answer the Phone A deeply troubled photographer stalks and attacks the patients of a beautiful psychologist talk-show hostess.
1980 (R) 94m/C James Westmoreland, Flo Gerrish, Ben Frank; *Dir:* Robert Hammer. **VHS, Beta** $19.95 *MED* ✻

Don't Be Afraid of the Dark A young couple move into their dream house only to find that demonic little critters are residing in their basement and they want more than shelter. Made for TV with creepy scenes and eerie makeup for the monsters.
1973 74m/C Kim Darby, Jim Hutton, Barbara Anderson, William Demarest, Pedro Armendariz Jr.; *Dir:* John Newland. **VHS, Beta** $49.95 *IVE* ✻✻½

Don't Bother to Knock As a mentally unstable hotel babysitter, Monroe meets a pilot (Widmark) and has a brief rendezvous with him. When the little girl she is babysitting interrupts, Monroe is furious, and later tries to murder the girl. This is one of Monroe's best dramatic roles, and this movie is also Bancroft's film debut as the pilot's girlfriend.
1952 76m/B Richard Widmark, Marilyn Monroe, Anne Bancroft, Elisha Cook Jr., Jim Backus, Lurene Tuttle, Jeanne Cagney, Donna Corcoran; *Dir:* Roy Ward Baker. **VHS** $14.98 *FXV* ✻✻½

Don't Change My World To preserve the natural beauty of the north woods, a wildlife photographer must fight a villainous land developer and a reckless poacher.
1983 (G) 89m/C Roy Tatum, Ben Jones. **VHS, Beta** $59.98 *LIV, CVL* ✻½

Don't Cry, It's Only Thunder A young army medic who works in a mortuary in Saigon becomes involved with a group of Vietnamese orphans and a dedicated army doctor.
1981 (PG) 108m/C Dennis Christopher, Susan St. James; *Dir:* Peter Werner. **VHS, Beta** $59.95 *COL* ✻✻

Don't Drink the Water Based on Woody Allen's hit play, this film places an average Newark, New Jersey family behind the Iron Curtain, where their vacation photo-taking gets them accused of spying.
1969 (G) 100m/C Jackie Gleason, Estelle Parsons, Joan Delaney, Ted Bessell, Michael Constantine; *Dir:* Howard Morris. **VHS, Beta** **$9.95** COL, SUE ♂♂½

Don't Go in the House A long-dormant psychosis is brought to life by the death of a young man's mother.
1980 (R) 90m/C Dan Grimaldi, Robert Osth, Ruth Dardick; *Dir:* Joseph Ellison. **VHS, Beta** **$54.95** MED ♂

Don't Go to Sleep After a fatal car crash, a young girl misses mom and dad so much she returns from the grave to take them where they can be one big happy family again. Better than most made-for-TV junk food fright-fests, written by Keenan Wynn's son, Ned.
1982 93m/C Dennis Weaver, Valerie Harper, Robin Ignico, Kristin Cummings, Ruth Gordon, Robert Webber; *Dir:* Richard Lang. **VHS $29.95** UNI ♂♂

Don't Go in the Woods Four young campers are stalked by a crazed killer.
1981 (R) 88m/C **VHS, Beta $59.98** VES ♂

Don't Look in the Attic A couple finds a haunted house with cows in the attic.
1981 90m/C Beba Longcar, Jean-Pierre Aumont; *Dir:* Carl Ausino. **VHS, Beta $59.95** MGL Woof!

Don't Look Back A famous documentary about Bob Dylan at the beginning of his career: on the road, in performance and during private moments. From the director of "Monterey Pop."
1967 95m/B Bob Dylan, Joan Baez, Donovan; *Dir:* D.A. Pennebaker. **VHS, Beta, LV $19.98** WRV, MVD, PAR ♂♂♂

Don't Look Back: The Story of Leroy "Satchel" Paige Drama of the legendary baseball pitcher who helped break down racial barriers, based on his autobiography. Made-for-TV fare. Gossett hits a home run in the lead, but the overall effort is a ground-rule double.
1981 98m/C Louis Gossett Jr., Beverly Todd, Cleavon Little, Clifton Davis, John Beradino, Jim Davis, Ossie Davis, Hal Williams; *Dir:* George C. Scott, Richard A. Colla. **VHS** ABC ♂♂½

Don't Look in the Basement Things get out of hand at an isolated asylum and a pretty young nurse is caught in the middle. Straight jacketed by a low budget.
1973 (R) 95m/C Rosie Holotik, Anne MacAdams, William Bill McGhee, Rhea MacAdams, Gene Ross, Betty Chandler, Camilla Carr, Robert Dracup, Jessie Kirby, Hugh Feagin, Harryete Warren, Jessie Lee Fulton, Michael Harvey; *Dir:* S.F. Brownrigg. **VHS, Beta $14.98** VID, MPI ♂½

Don't Look Now A psychological thriller about a couple's trip to Venice after their child drowns. Notable for a steamy love scene and a chilling climax. Based on the novel by Daphne du Maurier.
1973 (R) 110m/C Donald Sutherland, Julie Christie, Hilary Mason; *Dir:* Nicolas Roeg. **VHS, Beta, LV $49.95** PAR ♂♂♂

Don't Mess with My Sister! A married, New York junkyard worker falls in love with a belly dancer he meets at a party. The affair leads to murder and subsequently revenge. An interesting, offbeat film from the director of "I Spit on Your Grave."
1985 90m/C Joe Perce, Jeannine Lemay, Jack Gurci, Peter Sapienza, Laura Lanfranchi; *Dir:* Mier Zarchi. **VHS, Beta, LV $29.98** VID ♂♂½

Don't Open the Door! A young woman is terrorized by a killer located inside her house.
1974 (PG) 90m/C Susan Bracken, Gene Ross, Jim Harrell; *Dir:* S.F. Brownrigg. **VHS, Beta** **$24.95** GEM Woof!

Don't Open Till Christmas A weirdo murders various Santa Clauses in assorted gory ways. Best to take this one back to the department store.
1984 86m/C GB Edmund Purdom, Caroline Munro, Alan Lake, Belinda Mayne, Gerry Sundquist, Mark Jones; *Dir:* Edmund Purdom. **VHS, Beta $69.98** LIV, VES Woof!

Don't Raise the Bridge, Lower the River After his wife leaves him, an American with crazy, get-rich-quick schemes turns his wife's ancestral English home into a Chinese discotheque. Domestic farce that comes and goes; if mad for Lewis, rent "The Nutty Professor."
1968 (G) 99m/C Jerry Lewis, Terry-Thomas, Jacqueline Pearce; *Dir:* Jerry Paris. **VHS, Beta** **$59.95** COL ♂♂

Don't Shove/Two Gun Gussie Two early Harold Lloyd shorts feature the comedian's embryonic comic style as a young fellow impressing his date at a skating rink ("Don't Shove") and a city slicker out West ("Two Gun Gussie"). Silent with piano scores.
1919 27m/B Harold Lloyd, Bebe Daniels, Noah Young, Snub Pollard. **VHS, Beta $19.95** CCB ♂♂

Don't Tell Daddy Two less-than-virtuous women get the local men to seduce their virginal older sister.
1990 (R) 90m/C Eva Garden, Elina Moon. **VHS, Beta $39.95** MON ♂

Don't Tell Her It's Me Guttenberg is determined to win the heart of an attractive writer. With the assistance of his sister, he works to become a dream man. Clever premise, promising cast, lame comedy. From the novel by Sarah Bird.
1990 (PG-13) 101m/C Steve Guttenberg, Jami Gertz, Shelley Long, Kyle MacLachlan; *Dir:* Malcolm Mowbray. **VHS, Beta, LV $92.99** HBO, PIA ♂

Don't Tell Mom the Babysitter's Dead Their mother traveling abroad, the title situation leaves a houseful of teenagers with the whole summer to themselves. Eldest daughter Applegate cons her way into the high-powered business world while the metalhead son parties hardy. Many tepid comic situations, not adding up to very much.
1991 (PG-13) 105m/C Christina Applegate, Keith Coogan, Joanna Cassidy, John Getz, Josh Charles, Concetta Tomei, Eda Reiss Merin; *Dir:* Stephen Herek. **VHS, Beta, LV, 8mm $94.99** HBO, PIA ♂½

The Doolins of Oklahoma Scott is the leader of the last of the southwestern outlaw gangs, pursued by lawmen and changing times with the onset of the modern age. Fast-paced and intelligent. Also known as "The Great Manhunt."
1949 90m/B Randolph Scott, George Macready, Louise Allbritton, John Ireland, Virginia Huston, Charles Kemper, Noah Beery Jr.; *Dir:* Gordon Douglas. **VHS, Beta $14.99** GKK ♂♂½

Doom Asylum Several sex kittens wander into a deserted sanatorium and meet up with the grisly beast wielding autopsy instruments. The people involved with this spoof should have been (more) committed.
1988 (R) 77m/C Patty Mullen, Ruth Collins, Kristen Davis, William Hay, Kenny L. Price, Harrison White, Dawn Alvan, Michael Rogan; *Dir:* Richard S. Friedman. **VHS, Beta $29.95** ACA Woof!

Doomed to Die Cargo of stolen bonds leads to a tong war and the murder of a shipping millionaire. Part of Mr. Wong series. Worth a look if only for Karloff's performance.
1940 67m/B Boris Karloff, Marjorie Reynolds, Grant Withers; *Dir:* William Nigh. **VHS, Beta** **$19.95** NOS, MRV, SNC ♂♂

Doomed Love Love is hell, especially when the object of your affection happens to be deceased. A professor of literature in the throes of unrequited love decides to reunite himself with his lost love. Won awards when released in Berlin.
1983 75m/C Bill Rice; *Dir:* Andrew Horn. **VHS** **$59.95** MWF, FCT ♂♂½

The Doomsday Flight Uneven made-for-television thriller scripted by Rod Serling. O'Brien plants an altitude-triggered bomb on a jet in an effort to blackmail the airline. Search for the bomb provides some suspenseful and well-acted moments.
1966 100m/C Jack Lord, Edmond O'Brien, Van Johnson, John Saxon, Katherine Crawford; *Dir:* William A. Graham. **VHS, Beta $39.95** MCA ♂♂

Doomwatch A scientist discovers a chemical company is dumping poison into local waters, deforming the inhabitants of an isolated island when they eat the catch of the day. Unsurprising.
1972 89m/C GB Ian Bannen, Judy Geeson, John Paul, Simon Oates, George Sanders; *Dir:* Peter Sasdy. **VHS, Beta $29.95** EMB, MON, SUE ♂♂½

A Doonesbury Special "Doonesbury" is a humorous view of our changing patterns and lifestyles, showing that change and transition are natural and inevitable parts of the human condition. Features Doonesbury, Zonker, Joanie Caucus, Mike, B. D., Marcus, Jimmy, and other characters from the popular comic strip.
1977 26m/C **VHS, Beta, LV $9.95** PAV, PYR ♂♂½

Door to Door A made-for-cable comedy about two door-to-door salesmen who race to stay one step ahead of the law and their own company. Fairly lightweight, but it has its moments.
1984 93m/C Ron Leibman, Jane Kaczmarek, Arliss Howard, Alan Austin; *Dir:* Patrick Bailey. **VHS, Beta $9.95** MED ♂♂

Door-to-Door Maniac Criminals hold a bank president's wife hostage, unaware that the husband was looking to get rid of her in favor of another woman. Cash's screen debut. Aside from the cast, not particularly worth seeing. Also titled "Five Minutes to Live."
1961 80m/B Johnny Cash, Ron Howard, Vic Tayback, Donald Woods; *Dir:* Bill Karn. **VHS, Beta $19.95** VDM ♂½

The Doors Stone first approached Morrison himself with an early incarnation of the Doors docudrama, but it's hard to believe that even the Lizard King could play himself with any more convincing abandon than does Kilmer. He seems inhabited by the beautiful boy, and the film's strength lies in the truth of Kilmer's performance and of the recreation of an era. Trouble is, Morrison's story—one of drugs, abuse and abject self-indulgence—grows more than a little tiresome, and the

audience, with the exception of the not-to-be-ignored quantities of die-hard Doors fans, are bound to lose sight of whatever sympathy they might have had for the drug-addled singer. Ryan is forgettable as Morrison's hippie-chick wife, MacLachlan sports a funny wig and dabbles on the keyboards as Ray Manszarek, and Quinlan is atypically cast as a sado-masochistic journalist paramour.
1991 (R) 138m/C Val Kilmer, Meg Ryan, Kevin Dillon, Kyle MacLachlan, Frank Whaley, Michael Madsen, Kathleen Quinlan; *Dir:* Oliver Stone. **VHS, LV, 8mm $92.95** *COL, FCT* ✂️✂️½

The Doors: A Tribute to Jim Morrison Interviews and live performance footage capture the power of this famous rock group and its quixotic leader, Jim Morrison. Songs performed include "Light My Fire" and "The End."
1981 60m/C Jim Morrison, Ray Manzarek, Robby Krieger, John Densmore. **VHS, Beta $14.98** *MVD, WAR, FCT* ✂️✂️½

The Doors: Live in Europe, 1968 Footage from The Doors' 1968 European tour through Stockholm, Frankfurt and London, edited into a retrospective documentary. Songs include "Light My Fire," "Break On Through," "Back Door Man," "When the Music's Over," "Hello, I Love You," "Love Me Two Times," "Spanish Caravan," "Unknown Soldier," "Five to One," and "Alabama Song."
1968 58m/C Jim Morrison, Ray Manzarek, Robby Krieger, John Densmore; *Dir:* Ray Manzarek, John Densmore, Paul Justman. **VHS, Beta, LV $19.95** *MVD, AVE, HBO*

The Doors: Live at the Hollywood Bowl A complete performance by the notorious '60s rock band in concert, playing most if not all of their classic songs.
1968 65m/C Jim Morrison, Ray Manzarek, John Densmore, Robby Krieger. **VHS, Beta, LV $19.95** *MCA, MVD*

The Doors: Soft Parade, a Retrospective The Doors' final TV appearance in 1969 forms the centerpiece of this concert-video tie-in to the Oliver Stone film. Fans should a-Door previously unreleased backstage interviews and a notable version of "The Unknown Soldier."
1969 50m/C *Dir:* Ray Manzarek. **VHS, LV $19.95** *MCA, PIA* ✂️✂️½

Dope Mania A potpourri of anti-drug propaganda spanning from the '30's to the '60's, campily detailing the degradation thought to result instantly from pot and cocaine.
1986 60m/C **VHS, Beta $29.95** *RHI* ✂️✂️

Dorf and the First Games of Mount Olympus The video comedy character bungles his way through a parody of the ancient Greek Olympics.
1987 30m/C Tim Conway; *Dir:* Lang Elliott. **VHS, Beta $29.95** *JTC* ✂️

Dorf Goes Auto Racing Wacky "Doozle Dorf" shows us mortals how to race cars through Europe.
1989 30m/C Tim Conway. **VHS, Beta $29.95** *PMR* ✂️

Dorf on Golf Conway's Scandinavian golf expert/moron character gets his own mini-instruction show.
1986 30m/C Tim Conway. **VHS, Beta, LV $19.95** *JTC, VTK, KAR* ✂️

Dorf's Golf Bible Another made-for-tape Dorf golf satire, wherein the pint-sized character mucks up 18 holes as best he can.
1989 45m/C *Hosted:* Tim Conway, Sam Snead. **VHS, Beta $14.95** *JTC, RDG* ✂️

Dorian Gray A modern-day version of the famous tale by Oscar Wilde about an ageless young man whose portrait reflects the ravages of time and a life of debauchery. Not nearly as good as the original version, "The Picture of Dorian Gray."
1970 (R) 92m/C *GE IT* Richard Todd, Helmut Berger, Herbert Lom; *Dir:* Massimo Dallamano. **VHS $14.98** *REP, FCT* ✂️

Dorm That Dripped Blood Five college students volunteer to close the dorm during their Christmas vacation. In a series of grisly and barbaric incidents, the students begin to disappear. As the terror mounts, the remaining students realize that they are up against a terrifyingly real psychopathic killer. Merry Christmas.
1982 (R) 84m/C Laura Lopinski, Stephen Sachs, Pamela Holland; *Dir:* Stephen Carpenter, Jeffrey Obrow. **VHS, Beta $59.95** *MED* ✂️

Dorothy in the Land of Oz The further animated adventures of L. Frank Baum's characters from the "Wizard of Oz."
1981 60m/C *Nar:* Sid Caesar. **VHS, Beta $29.95** *FHE* ✂️

Dorothy Stratten: The Untold Story A look at the life of Dorothy Stratten includes a wealth of footage from Playboy's film and photo vaults, as well as some interviews with former playmates.
1985 85m/C Dorothy Stratten. **VHS, Beta $39.95** *LHV* ✂️

Dos Chicas de Revista A former film actor now recovering from a mental breakdown and drug addiction comes across a book written about him when he was big-time. He meets the girl who wrote the book and they enjoy a happy relationship. In Spanish.
19?? 90m/C *SP* Lina Morgan, Manolo Gomez Bur. **VHS, Beta $19.95** *MED* ✂️½

Dos Esposas en Mi Cama A man finds himself in trouble wherein he must choose between his present wife and the wife he thought was dead. In Spanish.
1987 90m/C *MX* Joaquin Cordero, Teresa Velasquez, Hector Lechuga, Roxana Bellini. **VHS, Beta $57.95** *MAD* ✂️✂️

Dos Gallos Y Dos Gallinas (Two Roosters For ...) Musical focusing on the beauty of the Mexican countryside. In Spanish.
197? 30m/C *MX* Miguel Aceves Mejia, Marco Antonio Muniz, Maria Duval, Rosina Navarro. **VHS, Beta $49.95** *MAD* ✂️½

Dos Postolas Gemelas Two teenage girls sing in the old West. In Spanish.
1965 84m/C *MX* Pili & Milli, Sean Flynn, Jorge Rigaud, Beni Deus. **VHS, Beta $49.95** *MAD* ✂️✂️

Dot & the Bunny The adventures of a spunky young red-haired heroine on her quest for a missing baby kangaroo named Joey. A mix of animation and live action with a subtle moral.
1983 79m/C *AU Dir:* Yoram Gross. **VHS, Beta $14.98** *FOX* ✂️✂️

Dot and the Whale An animated Australian fantasy about a girl who helps a beached whale.
1987 75m/C *AU* **VHS, Beta $14.95** *IVE, FHE* ✂️½

Double Agents Two double agents are sent on a rendezvous to exchange vital government secrets. Dubbed in English.
1959 81m/B *FR* Marina Vlady, Robert Hossein. **VHS, Beta $29.95** *VYY* ✂️

Double Cross Tank Polling (Connery), a British journalist, uses the secrets of an ex-hooker (Donohoe) to get back at a corrupt politician who ruined Polling's life five years earlier. Lots of action and a few graphic murders in this thriller.
199? 96m/C Amanda Donohoe, Peter Wyngarde, Jason Connery; *Dir:* James Marcus. **VHS $79.95** *MNC* ✂️

Double-Crossed In 1984 Barry Seal agrees to inform on Colombia's Medellin Cartel and it s links to the Sandanista government of Nicaragua in exchange for leniency on drug-smuggling charges. Little does he know that the personal consequences will turn deadly when his cover is blown and the government double-crosses him. Hopper's hyped-up performance as drug smuggler turned DEA informant Seal showcases this true-crime tale.
1991 111m/C Dennis Hopper, Robert Carradine, G.W. Dailey, Adrienne Darbeau. **VHS, LV $79.99** *WAR* ✂️✂️

Double Deal Murder, mayhem, and industrial espionage come into play as parties fight over an oil field.
1950 65m/B Marie Windsor, Richard Denning, Taylor Holmes; *Dir:* Abby Berlin. **VHS, Beta $24.95** *DVT* ✂️½

Double Deal An unfaithful woman and her lover plot to steal her husband's priceless opal. Lifeless and silly.
1984 90m/C Louis Jourdan, Angela Punch-McGregor; *Dir:* Brian Kavanagh. **VHS, Beta $69.95** *VCL* ✂️½

Double Dynamite A bank teller is at a loss when his racetrack winnings are confused with the cash lifted from his bank during a robbery. Forgettable, forgotten shambles. Originally filmed in 1948 under the title "It's Only Money."
1951 80m/B Frank Sinatra, Jane Russell, Groucho Marx, Don McGuire, Howard Freeman, Harry Hayden, Nestor Paiva, Lou Nova, Joe Devlin; *Dir:* Irving Cummings. **VHS, Beta $19.98** *RKO, TTC* ✂️½

Double Exposure A young photographer's violent nightmares become the next day's headlines. Unsurprising vision.
1982 (R) 95m/C Michael Callan, James Stacy, Joanna Pettet, Cleavon Little, Pamela Hensley, Seymour Cassel; *Dir:* William B. Hillman. **VHS, Beta $59.98** *VES* ✂️✂️

Double Exposure: The Story of Margaret Bourke-White Beautifully shot but slow biography of a woman who became a well-known professional photographer for "Life" magazine during the 1930s and 1940s, as well as the first official female photojournalist of World War II. Made for

cable television; originally titled "Margaret Bourke-White."
1989 105m/C Farrah Fawcett, Frederic Forrest, Mitchell Ryan, David Huddleston, Jay Patterson, Ken Marshall; **Dir:** Lawrence Schiller. **VHS $9.98** TTC 𝄆𝄆½

Double Face A wealthy industrialist kills his lesbian wife with a car bomb, but she seems to haunt him through pornographic films.
1970 84m/C Klaus Kinski, Annabella Incontrera. **VHS, Beta $49.95** UNI 𝄆½

Double Identity Former college professor turned criminal Mancuso moves to a corn patch called New Hope in hopes of returning to the straight and narrow. He finds a good woman to stand by him, but, alas, crime, like smoking, is easier to start than to quit. Much intrigue and deceit.
1989 (PG-13) 95m/C Nick Mancuso, Leah K. Pinsent, Patrick Bauchau, Anne LeTourneau, Jacques Godin; **Dir:** Yves Boisset. **VHS $79.95** MNC 𝄆𝄆½

Double Impact Van Damme plays twins re-united in Hong Kong to avenge their parents' murder by local bad guys, but this lunkhead kick-em-up doesn't even take advantage of that gimmick; the basic story would have been exactly the same with just one Jean-Claude. Lots of profane dialogue and some gratuitous nudity for the kiddies.
1991 (R) 107m/C Jean-Claude Van Damme, Cory Everson, Geoffrey Lewis; **Dir:** Sheldon Lettich. **VHS, Beta, LV, 8mm $92.00** COL, PIA, CCB 𝄆½

Double Indemnity The classic seedy story of an insurance agent seduced by a deadly woman into killing her husband so they can collect together from his company. Terrific, influential film noir, the best of its kind. Co-written by Wilder and Raymond Chandler; music by Miklos Rozsa. Based on the James M. Cain novel. Oscar nominated for Best Picture, Screenplay, Black & White Cinematography, Best Score of a Drama or Comedy and Best Sound Recording.
1944 107m/B Fred MacMurray, Barbara Stanwyck, Edward G. Robinson, Tom Powers, Porter Hall, Jean Heather, Byron Barr; **Dir:** Billy Wilder. **VHS, Beta, LV $19.95** MCA, FCT 𝄆𝄆𝄆𝄆

A Double Life Ronald Coleman plays a Shakespearean actor in trouble when the characters he plays begin to seep into his personal life and take over. Things begin to look really bad when he is cast in the role of the cursed Othello. Coleman won an Oscar for this difficult role, and the moody musical score garnered another for Miklos Rosza. Oscar nominations for Best Director and Original Screenplay.
1947 103m/B Ronald Colman, Shelley Winters, Signe Hasso, Edmond O'Brien, Ray Collins, Millard Mitchell; **Dir:** George Cukor. Academy Awards '47: Best Actor (Colman), Best Musical Score. **VHS $19.98** REP, FCT, BTV 𝄆𝄆𝄆

The Double Life of Veronique They say everyone has a twin, but this is ridiculous. Two women—Polish Veronika and French Veronique—are born on the same day in different countries, share a singing talent, a cardiac ailment, and, although the two never meet, a strange awareness of each other. Jacob is unforgettable as the two women, and director Krzysztof creates some spellbinding scenes but the viewer has to be willing to forgo plot for atmosphere.

1991 (R) 96m/C Irene Jacob, Phillipe Volter, Sandrine Dumas, Aleksander Bardini, Louis Ducreux, Claude Duneton, Halina Gryglaszewska, Kalina Jedrusik; **W/Dir:** Krzysztof Kieslowski. **VHS, Beta** PAR 𝄆𝄆

The Double McGuffin A plot of international intrigue is uncovered by teenagers - a la the Hardy Boys - when a prime minister and her security guard pay a visit to a small Virginia community. They're not believed, though. Action-packed from the makers of Benji. Dogs do not figure prominently here.
1979 (PG) 100m/C Ernest Borgnine, George Kennedy, Elke Sommer, Ed "Too Tall" Jones, Lisa Whelchel, Vincent Spano; **Dir:** Joe Camp; **Voices:** Orson Welles. **VHS, Beta $69.95** VES 𝄆𝄆

The Double Negative A photojournalist pursues his wife's killer in a confusing story. Based on Ross MacDonald's "The Three Roads." Also known as "Deadly Companion."
1980 96m/C CA Michael Sarrazin, Susan Clark, Anthony Perkins, Howard Duff, Kate Reid, Al Waxman, Elizabeth Shepherd; **Dir:** George Bloomfield. **VHS, Beta** BFV 𝄆½

Double Revenge Two men feel they have a score to settle after a bank heist gone sour leaves two people dead.
1989 90m/C Joe Dallesandro, Nancy Everhard, Leigh McCloskey, Richard Rust, Theresa Saldana; **Dir:** Armand Mastroianni. **VHS $9.98** REP 𝄆½

Double Standard You'll be less concerned with the plot—a community discovers that a prominent judge has maintained two marriages for nearly twenty years—than with the possibility that such dim-witted myopes might actually have existed in the real world. This made-for-TV slice of bigamy is for fans of the oops-I-forgot-I'm-married genre only.
1988 (R) 95m/C Robert Foxworth, Michelle Greene, Pamela Bellwood, James Kee; **Dir:** Louis Rudolph. **VHS $14.95** FRH 𝄆𝄆

Double Suicide From a play by Monzarmon Chikamatsu, a tragic drama about a poor salesman in 18th century Japan who falls in love with a geisha and ruins his family and himself trying to requite the hopeless passion. Stylish with Iwashita turning in a wonderful dual performance. In Japanese with subtitles.
1969 105m/B JP Kichiemon Nakamura, Shima Iwashita, Yusuke Takita, Hosei Komatsu; **Dir:** Masahiro Shinoda. **VHS, Beta $59.95** SVS, TAM 𝄆𝄆𝄆

Double Trouble When rock star Presley falls in love with an English heiress, he winds up involved in an attempted murder. The king belts out "Long Legged Girl" while evading cops, criminals, and crying women. A B-side.
1967 92m/C Elvis Presley, Annette Day, John Williams, Yvonne Romain, The Wiere Brothers, Michael Murphy, Chips Rafferty; **Dir:** Norman Taurog. **VHS, Beta $19.95** MGM 𝄆𝄆

Double Trouble Twin brothers—one a cop, one a jewel thief—team up to crack the case of an international jewel smuggling ring headed by a wealthy and politically well-connected businessman and his righthand man.
1991 (R) 87m/C David Paul, Peter Paul, James Doohan, Roddy McDowall, Steve Kanaly, A.J. Johnson, David Carradine; **Dir:** John Paragon. **VHS, LV $89.95** COL 𝄆

Doubting Thomas Rogers' last film, which was in theatres when he was killed in a plane crash. Silly story of a husband and his doubts about his wife and her amateur acting career. The show goes on in spite of his doubts, forgotten lines, and wardrobe goofs. Remake of 1922's silent film "The Torch Bearers."
1935 78m/B Will Rogers, Billie Burke, Alison Skipworth, Sterling Holloway; **Dir:** David Butler. **VHS $19.98** FOX, FCT 𝄆𝄆½

Doughnuts & Society Two elderly ladies who run a coffee shop suddenly strike it rich and find that life among the bluebloods is not all it's cracked up to be. No sprinkles.
1936 70m/B Louise Fazenda, Maude Eburne, Eddie Nugent, Ann Rutherford, Hedda Hopper, Franklin Pangborn. **VHS, Beta $29.95** VYY 𝄆½

The Dove A hilarious short spoof of the Ingmar Bergman films "Wild Strawberries," "The Seventh Seal," and "The Silence." English sub-titles.
1968 15m/B Madeline Kahn. **VHS, Beta** PYR 𝄆𝄆𝄆

The Dove The true story of a 16-year-old's adventures as he sails around the world in a 23-foot sloop. The trip took him five years and along the way he falls in love with a girl who follows him to Fiji, Australia, South Africa, Panama, and the Galapagos Islands. Stunning location cinematography.
1974 (PG-13) 105m/C Joseph Bottoms, Deborah Raffin, Dabney Coleman; **Dir:** Charles Jarrott. **VHS, Beta $14.95** PAR 𝄆𝄆½

Down Among the Z Men An enlisted man helps a girl save an atomic formula from spies. Funny in spots but weighed down by musical numbers from a female entourage. From the pre-Monty Python comedy troupe "The Goons."
1952 71m/B GB Peter Sellers, Spike Milligan, Harry Secombe, Michael Bentine, Carole Carr; **Dir:** Maclean Rogers. **VHS, Beta, LV $19.95** PAV 𝄆

Down Argentine Way A lovely young woman falls in love with a suave Argentinian horse breeder. First-rate Fox musical made a star of Grable and was Miranda's first American film.
1940 90m/C Don Ameche, Betty Grable, Carmen Miranda, Charlotte Greenwood, J. Carroll Naish, Henry Stephenson, Leonid Kinskey; **Dir:** Irving Cummings. **VHS, Beta, LV $19.98** FOX, MLB 𝄆𝄆𝄆

Down Dakota Way Strange coincidence between a recent murder and a fatal cow epidemic. Rogers helps some locals bring the link to light.
1949 67m/C Roy Rogers, Dale Evans, Pat Brady, Monte Montana, Elisabeth Risdon, Byron Barr, James B. Cardwell, Roy Barcroft, Emmett Vogan; **Dir:** William Witney. **VHS $12.98** REP 𝄆½

Down & Dirty A scathing Italian satire about a modern Roman family steeped in petty crime, incest, murder, adultery, drugs, and arson. In Italian with English subtitles. Italian title: "Brutti, Sporchi, E Cattivi."
1976 115m/C IT Nino Manfredi, Francesco Anniballi, Maria Bosco; **Dir:** Ettore Scola. Cannes Film Festival '76: Best Director (Scola). **VHS, Beta $59.94** MED, TAM 𝄆𝄆𝄆

Down the Drain A broad-as-a-city-block farce about an unscrupulous criminal lawyer and his assortment of crazy clients. A fine first half but someone pulls the plug in the middle of the bath.

1989 (R) 90m/C Andrew Stevens, John Matuszak, Teri Copley, Joseph Campanella, Don Stroud, Stella Stevens, Jerry Mathers, Benny "The Jet" Urquidez; **Dir:** Robert C. Hughes. **VHS, Beta, LV** $89.95 COL ♫♫

Down to Earth Ever-gallant lover Fairbanks overruns a mental hospital to save his sweetheart. Once inside, he's determined to show the patients that their illness is illusion, and reality's the cure. A quixotic gem from Fairbanks' pre-swashbuckling social comic days. Director Emerson's wife was the scenarist, and Victor Fleming, who went on to direct "The Wizard of Oz" and "Gone with the Wind" (to name a few), was the man behind the camera.
1917 68m/B Douglas Fairbanks Sr., Eileen Percy, Gustav von Seyffertitz, Charles P. McHugh, Charles Gerrard, William H. Keith, Ruth Allen, Fred Goodwine; **Dir:** John Emerson. **VHS, Beta** $24.95 GPV, FCT ♫♫♫

Down to Earth A lackluster musical and box office flop with an impressive cast. Hayworth is the Greek goddess of dance sent to Earth on a mission to straighten out Broadway producer Parks and his play that ridicules the Greek gods. Anita Ellis dubs the singing of Hayworth. A parody of "Here Comes Mr. Jordan," remade in 1980 as "Xanadu."
1947 101m/C Rita Hayworth, Larry Parks, Marc Platt, Roland Culver, James Gleason, Edward Everett Horton, Adele Jergens, George Macready, William Frawley, James Burke, Fred F. Sears, Lynn Merrick, Myron Healey; **Dir:** Alexander Hall. **VHS, LV** $19.95 COL, PIA, CCB ♫♫

Down by Law In Jarmusch's follow-up to his successful "Stranger than Paradise," he introduces us to three men: a pimp, an out-of-work disc jockey, and an Italian tourist. When the three break out of prison, they wander through the Louisiana swampland with some regrets about their new-found freedom. Slow moving at times, beautifully shot throughout. Poignant and hilarious, the film is true to Jarmusch form: some will love the film's offbeat flair, and others will find it bothersome. It's a strange and wonderful world. Music by Lurie and Waits.
1986 (R) 107m/C John Lurie, Tom Waits, Roberto Benigni, Ellen Barkin; **Dir:** Jim Jarmusch. **VHS, Beta** $79.98 FOX ♫♫♫

Down Mexico Way Two cowboys come to the aid of a Mexican town whose residents have been hoodwinked by a phony movie company. Very exciting chase on horseback, motorcycle, and in automobiles.
1941 78m/B Gene Autry, Smiley Burnette, Fay McKenzie, Duncan Renaldo, Champion; **Dir:** Joseph Santley. **VHS, Beta** $19.95 CCB ♫♫

Down and Out in Beverly Hills A modern retelling of Jean Renoir's classic "Boudu Sauve Des Eaux" with some nice star turns. A spoiled wealthy couple rescue a suicidal bum from their swimming pool. He, in turn, beds the wife, the maid and the daughter, solves the dog's identity crisis, rescues the man and his son from frustration and finds he doesn't like being a bum. A box-office hit.
1986 (R) 103m/C Nick Nolte, Bette Midler, Richard Dreyfuss, Little Richard, Mike the Dog, Elizabeth Pena; **Dir:** Paul Mazursky. **VHS, Beta, LV** $19.95 TOU ♫♫

Down to the Sea in Ships Clara Bow made her movie debut in this drama about the whalers of 19th-century Massachusetts. Highlighted by exciting action

scenes of an actual whale hunt. Silent with music score and original tinted footage.
1922 83m/B Marguerite Courtot, Raymond McKee, Clara Bow; **Dir:** Elmer Clifton. **VHS, Beta** $19.95 NOS, CCB, DAR ♫♫

Down Texas Way One of the "Rough Riders" series. When one of the boys is accused of murdering his best friend his two companions search for the real killer.
1942 57m/B Buck Jones, Tim McCoy, Raymond Hatton; **Dir:** Howard Bretherton. **VHS, Beta** $9.99 NOS, VCN ♫

Down Twisted A young woman gets involved with a thief on the run in Mexico in this "Romancing the Stone" derivative. The muddled plot deserves such a confusing title.
1989 (R) 89m/C Carey Lowell, Charles Rocket, Trudi Dochtermann, Thom Mathews, Linda Kerridge, Courtney Cox; **Dir:** Albert Pyun. **VHS, Beta** $89.95 MED ♫

Down Under Two gold-hungry beach boys go to Australia looking for riches, and document their adventures on film. Essentially a crudely shot, tongue-in-cheek travelogue narrated by Patrick Macnee.
1986 90m/C Don Atkinson, Donn Dunlop, Patrick Macnee. **VHS, Beta** $79.95 UHV ♫

Downhill Racer An undisciplined American skier has conflicts with his coach and his new-found love while on his way to becoming an Olympic superstar. Character study on film. Beautiful ski and mountain photography keep it from sliding downhill.
1969 (PG) 102m/C Robert Redford, Camilla Sparv, Gene Hackman, Dabney Coleman; **Dir:** Michael Ritchie. **VHS, Beta, LV** $14.95 PAR ♫♫

Downtown Urban comedy about a naive white suburban cop who gets demoted to the roughest precinct in Philadelphia, with a streetwise black partner. Runs a routine beat.
1989 (R) 96m/C Anthony Edwards, Forest Whitaker, Joe Pantoliano, Penelope Ann Miller; **Dir:** Richard Benjamin. **VHS, Beta, LV** $89.98 FOX ♫♫

D.P. A black orphan attaches to the only other black in post WWII Germany. From a story by Kurt Vonnegut Jr.
1985 60m/C Stan Shaw, Rosemary Leach, Julius Gordon. **VHS** $24.95 MON, KAR ♫♫

Drachenfutter (Dragon's Food) Powerful story dealing with a Pakistani immigrant's attempts to enter the Western world. Themes of alienation and helplessness are emphasized. Dialogues occur in 12 languages, predominantly German and Mandarin Chinese; subtitled in English.
1987 75m/B GE SI Bhasker, Ric Young, Buddy Uzzaman; **Dir:** Jan Schutte. **VHS, Beta** $69.95 NYF ♫♫♫

Dracula A vampire terrorizes the countryside in his search for human blood. From Bram Stoker's novel. Lugosi's most famous role. Although short of a masterpiece due to slow second half, deservedly rated a film classic. What would Halloween be like without this movie?
1931 75m/B Bela Lugosi, David Manners, Dwight Frye, Helen Chandler, Edward Van Sloan; **Dir:** Tod Browning. **VHS, Beta, LV** $14.95 MCA, TLF ♫♫♫

Dracula Count on squinty-eyed Palance to shine as the Transylvanian vampire who must quench his thirst for human blood. A made-for-television adaptation of the Bram Stoker novel that really flies.

1973 105m/C Jack Palance, Simon Ward, Fiona Lewis, Nigel Davenport; **Dir:** Dan Curtis. **VHS, Beta** $29.95 IVE ♫♫

Dracula Langella recreates his Broadway role as the count who needs human blood for nourishment. Notable for its portrayal of Dracula as a romantic and tragic figure in history. Overlooked as the vampire spoof "Love at First Bite" came out at the same time.
1979 (R) 109m/C Frank Langella, Laurence Olivier, Kate Nelligan, Donald Pleasence; **Dir:** John Badham. **VHS, Beta, LV** $19.95 MCA ♫♫

Dracula The prince of darkness returns looking for blood in this animated adaptation of Bram Stoker's classic novel.
1984 90m/C **VHS, Beta** $59.98 LIV, VES

Dracula: A Cinematic Scrapbook A look at the blood-sucking monster in both film and legend, featuring rare film trailers and interviews with actors such as Bela Lugosi, Christopher Lee and Lon Chaney, Jr. In color and B&W.
1991 60m/C **W/Dir:** Ted Newsom. **VHS, Beta** $9.95 RHI, FCT ♫♫

Dracula Blows His Cool Three voluptuous models and their photographer restore an ancient castle and open a disco in it. The vampire lurking about the castle welcomes the party with his fangs.
1982 (R) 91m/C John Garco, Betty Verges; **W/Dir:** Carlo Ombra. **VHS, Beta, LV** LUN, OM ♫

Dracula/Garden of Eden The abridged version of the chilling vampire film "Nosferatu" is coupled with "The Garden of Eden," in which Tini Le Brun meets her Prince Charming while vacationing with her Baroness friend. Silent.
1928 52m/B Max Schreck, Alexander Granach, Corrine Griffith, Charles Ray, Louise Dressler. **VHS, Beta** $19.95 CCB ♫♫

Dracula & Son A Dracula spoof, in which the count fathers a son who prefers girls and football to blood. Poor English dubbing and choppy ending drive a stake through the heart. Also known as "Pere et Fils."
1976 (PG) 88m/C FR Christopher Lee, Bernard Menez, Marie Breillat; **Dir:** Edouard Molinaro. **VHS, Beta** $9.95 GKK ♫♫

Dracula (Spanish Version) Filmed at the same time as the Bela Lugosi version of "Dracula," using the same sets and the same scripts, only in Spanish. A real estate agent in Transylvania visits Count Dracula to complete a business transaction, only to discover the Count's evil deeds. He eventually succumbs to the Count, and the two travel to London so Dracula can conquer new blood-sucking territory. However, Dracula doesn't count on Dr. Van Helsing to spoil his plans. Thought to be more visually appealing and more terrifying than the English-language counterpart. Based on the novel by Bram Stoker.
1931 104m/B Carlos Villarias, Lupita Tova, Eduardo Arozamena, Pablo Alvarez Rubio, Barry Norton, Carmen Guerrero; **Dir:** George Melford. **VHS** $14.98 MCA ♫♫

Dracula Sucks A soft-core edit of a hardcore sex parody about Dracula snacking on the usual bevy of screaming quasi-virgins.
1979 (R) 90m/C James Gillis, Reggie Nalder, Annette Haven, Kay Parker, Serena, Seka, John Leslie; **Dir:** Philip Marshak. **VHS, Beta** UNI, OM ♫

D

Dracula: The Great Undead An evocative documentary about vampires—in history and movies. Narated by Vincent Price.
1985 60m/C *Hosted:* Vincent Price. **VHS, Beta** $19.95 *AHV, VTK, FCT* ♫♫

Dracula—Up in Harlem When the Prince of Darkness gets together with the people of the night, the Pentagon rolls out the heavy metal. This may suck your brain out.
1983 90m/C **VHS, Beta** *SFF Woof!*

Dracula vs. Frankenstein An alien reanimates Earth's most infamous monsters in a bid to take over the planet. Rennie's last role. Also on video as, "Assignment: Terror."
1969 91m/C *IT GE SP* Michael Rennie, Paul Naschy, Karin Dor, Patty Shepard; *Dir:* Hugo Fregonese, Tulio Demicheli. **VHS** $29.95 *UAV* ♫

Dracula vs. Frankenstein The Count makes a deal with a shady doctor to keep him in blood. Vampire spoof that's very bad but fun. Last film for both Chaney and Naish. Also known as "Blood of Frankenstein"; features a cameo by genre maven Forrest J. Ackerman.
1971 (R) 90m/C J. Carroll Naish, Lon Chaney Jr., Regina Carrol, Russ Tamblyn, Jim Davis, Anthony Eisley, Zandor Vorkov, John Bloom, Angelo Rossitto; *Dir:* Al Adamson. **VHS** $14.98 *VID, SVS Woof!*

Dracula's Daughter Count Dracula's daughter, Countess Marya Zaleska, heads to London to supposedly find the cure to a mysterious illness. Instead she finds she has a taste for human blood, especially female blood. She also finds a man, who she falls in love with and tries to keep by casting a spell on him. A good script and cast keep this sequel to Bela Lugosi's "Dracula" entertaining.
1936 71m/B Gloria Holden, Otto Kruger, Marguerite Churchill, Irving Pichel, Edward Van Sloan, Nan Grey, Hedda Hopper; *Dir:* Lambert Hillyer. **VHS** $14.98 *MCA* ♫♫♫

Dracula's Great Love Four travellers end up in Dracula's castle for the night, where the horny Count takes a liking to one of the women. Also on video as "Dracula's Virgin Lovers." Left out in the sun too long.
1972 (R) 96m/C *SP* Paul Naschy, Charo Soriano; *Dir:* Javier Aguirre. **VHS, Beta** $59.95 *SNC, MPI Woof!*

Dracula's Last Rites A blood-curdling tale of a sheriff and a mortician in a small town who are up to no good. Technically inept, film equipment can be spotted throughout. Don't stick your neck out for this one. Also called "Last Rites."
1979 (R) 86m/C Patricia Lee Hammond, Gerald Fielding, Victor Jorge; *Dir:* Domonic Paris. **VHS, Beta** $59.95 *PGN, CAN* ♫

Dracula's Widow Countess Dracula, missing her hubby and desperately in need of a substitute, picks innocent Raymond as her victim. His girlfriend and a cynical cop fight to save his soul from the Countess' damnation. Directed by the nephew of Frances Ford Coppola.
1988 (R) 85m/C Sylvia Kristel, Josef Sommer, Lenny Von Dohlen; *Dir:* Christopher Coppola. **VHS, Beta** $14.99 *HBO* ♫♫

Dragnet Sgt. Joe Friday and Officer Frank Smith try to solve a mob slaying but has a rough time. Alexander plays the sidekick in "Dragnet" pre-Morgan days. Just the facts: feature version of the TV show that's suspenseful and well-acted.
1954 88m/C Jack Webb, Ben Alexander, Richard Boone, Ann Robinson; *Dir:* Jack Webb. **VHS, Beta** $14.95 *MCA* ♫♫½

Dragnet This semi-parody of the vintage television cop show has Sgt. Joe Friday's straight-laced nephew and his sloppy partner taking on the seamy crime life of Los Angeles. Neither Aykroyd or Hanks can save the big-budget lackluster spoof that's full of holes.
1987 (PG-13) 106m/C Dan Aykroyd, Tom Hanks, Christopher Plummer, Harry Morgan, Elizabeth Ashley, Dabney Coleman; *Dir:* Tom Mankiewicz. **VHS, Beta, LV** $14.95 *MCA* ♫♫

The Dragon Fist A young man must avenge the death of his father in this martial arts film.
1980 90m/C Jackie Chan. **VHS, Beta** $24.95 *ASE* ♫

Dragon Force Korean martial artists fend off Japanese invaders in the 1930s.
1982 101m/C Kangjo Lee; *Dir:* Key-Nam Nam. **VHS, Beta** *NEG* ♫½

Dragon Lady Ninja/Devil Killer Two Kung-Fu features to shred your sanity, and perhaps your very soul.
19?? 150m/C **VHS, Beta** $49.98 *NSV, CGI* ♫½

Dragon Lee Vs. the Five Brothers Mr. Lee successfully thwarts an attempted overthrow of the Ching Government with his furiously brutal fighting talents.
1981 89m/C Dragon Lee. **VHS, Beta** $54.95 *MAV* ♫

The Dragon Lives Again A martial arts adventure that is dedicated to the memory of Bruce Lee.
198? (R) 90m/C Bruce Leong, Alexander Grand. **VHS, Beta** $57.95 *UNI* ♫♫

Dragon Lord A resident of a small town fights those who would demolish the old ways of life with Kung-Fu.
19?? 90m/C Jackie Chan. **VHS, Beta** $59.98 *CGH, PEV* ♫

Dragon the Odds A man named Mo masters Kung Fu "maniac style" in order to repossess the gymnasium that was willed to him.
198? 90m/C *Hosted:* Sho Kosugi. **VHS, Beta** $39.95 *TWE* ♫

Dragon Princess When the king of karate and his dragon princess confront the blind master of the bloody blades, sparks fly.
197? 90m/C Sonny Chiba, Sue Shiomi. **VHS, Beta** $49.95 *IND* ♫

The Dragon Returns A young man desperately kicks and chops to save the woman he loves from slave traders.
19?? 90m/C Bruce Le, Meng Fei, Lo Lieh. **VHS, Beta** *OCE* ♫

Dragon Rider The Dragon Rider kicks underworld druglords as he goes on a crime-fighting quest.
1985 87m/C Wong Sun. **VHS, Beta** $39.95 *WES* ♫

Dragon Seed The lives of the residents of a small Chinese village are turned upside down when the Japanese invade it. Based on the Pearl S. Buck novel. Lengthy and occasionally tedious, though generally well-made with heart-felt attempts to create Oriental characters, but no Asians in the cast. MacMahon received an Oscar nomination for Best Supporting Actress.
1944 149m/B Katharine Hepburn, Walter Huston, Agnes Moorehead, Akim Tamiroff, Hurd Hatfield, J. Carroll Naish, Henry Travers, Turhan Bey, Aline MacMahon; *Dir:* Jack Conway. **VHS, Beta** $19.98 *MGM, FCT* ♫♫

Dragon from Shaolin The Shaolin monks match wits, chops and kicks with the brutal Ching government.
197? 90m/C Bruce Lee. **VHS, Beta** $54.95 *MAV* ♫½

The Dragon Strikes Back A Chinese immigrant, a master of Kung Fu, stands up to a gang of white toughs who are brutally terrorizing and killing poor Mexican farmers.
1975 (R) 92m/C **VHS, Beta** $59.95 *HHT* ♫

The Dragon That Wasn't (Or Was He?) A bear adopts a baby dragon, then has a difficult time returning the reptile to his home in the Misty Mountains once he grows to dragon-like proportions. Animated.
1983 83m/C **VHS, Beta** $19.95 *MCA*

Dragon vs. Needles of Death It's a battle to the finish against the unstoppable deadly needles in this Kung-Fu adventure.
1982 86m/C Don Chien, Wong Ting. **VHS, Beta** *FOX* ♫

Dragonard Slaves on a West Indies island rebel against their cruel masters.
1988 (R) 93m/C Eartha Kitt, Oliver Reed, Annabel Schofield, Patrick Warburton; *Dir:* Gerard Kikione. **VHS, Beta** $79.95 *WAR* ♫½

Dragonfly Squadron Korean war drama about pilots and their romantic problems. Never gets off the ground.
1954 82m/C John Hodiak, Barbara Britton, Bruce Bennett, Jess Barker; *Dir:* Lesley Selander. **VHS** $39.95 *SVS* ♫♫

Dragon's Claw A young man's parents are murdered, and he trains in shaolin skills in order to exact revenge.
1987 90m/C Jimmy Liu, Hwang Cheng Li; *Dir:* Joseph Kuo. **VHS, Beta** $59.95 *GEM* ♫

The Dragon's Showdown After witnessing the brutal decapitations of his parents, a young man learns Kung-Fu to avenge their deaths.
1983 90m/C Dragon Lee; *Dir:* Godfrey Ho. **VHS, Beta** $59.95 *SVS, GEM* ♫

Dragonslayer A sorcerer's apprentice suddenly finds himself the only person who can save the kingdom from a horrible, fire breathing dragon. Extreme violence but wonderful special effects, smart writing, and a funny performance by Richardson.
1981 (PG) 110m/C Peter MacNicol, Caitlin Clarke, Ralph Richardson, John Hallam, Albert Salmi, Chloe Salaman; *Dir:* Matthew Robbins. **VHS, Beta, LV** $19.95 *PAR, COL* ♫♫

Dragstrip Girl An 18-year-old girl comes of age while burning rubber at the dragstrip—the world of boys, hot rods, and horsepower. A definite "B" movie.
1957 70m/B Fay Spain, Steven Terrell, John Ashley, Frank Gorshin; *Dir:* Edward L. Cahn. **VHS** $9.95 *COL* ♫

The Drake Case A slight courtroom drama made during the waning days of the silent era. You'll probably recognize Brit Lloyd from her later career.
1929 56m/B Robert Frazer, Doris Lloyd, Gladys Brockwell. **VHS, Beta** $19.95 *GPV, FCT* ♫♫

The Draughtsman's Contract A beguiling mystery begins with a wealthy woman hiring an artist to make drawings of her home. Their contract is quite unusual, as is

their relationship. Everything is going along at an even pace until murder is suspected, and things spiral down to darker levels. A simple story turned into a bizarre puzzle. Intense enough for any thriller fan.
1982 (R) 108m/C *GB* Anthony Higgins, Janet Suzman, Anne Louise Lambert, Hugh Fraser; *Dir:* Peter Greenaway. VHS $19.99 *MGM, FCT* 🎬🎬🎬

Draw! Two has-been outlaws warm up their pistols again in this old-fashioned Western. Star power and some smart moments make it worthwhile. Made for TV.
1981 98m/C Kirk Douglas, James Coburn, Alexandra Bastedo, Graham Jarvis; *Dir:* Steven Hilliard Stern. VHS, Beta $69.95 *MED* 🎬🎬

Drawing the Line: A Portrait of Keith Haring An intimate view of the life and art of Keith Haring, the artist whose stick-like figures were seen everywhere from subways to billboards to watches. Included are interviews with gallery owners, curators and the leaders of the international art scene.
1990 30m/C VHS $19.95 *KUL* 🎬🎬

Dreadnaught The White Tiger goes undercover in a circus troupe to escape from his pursuers.
19?? 97m/C Yuen Shun-Yee; *Dir:* Yuen Woo Ping. VHS, Beta $59.98 *CGH, PEV* 🎬

Dreadnaught Rivals A student is taught by his drunken servant how to avenge himself on the culprits who kidnapped his girlfriend. Dubbed.
198? 90m/C Sinon Lin, Kent Ko, Louisa Monk. VHS, Beta $54.95 *MAV* 🎬

Dream to Believe A slice of life of your typical high school coed—mom's dying, stepdad's from hell, sports injury hurts, and, oh yeah, championship gymnastic competitions loom. Designed to make you feel oh-so-good in that MTV kind of way. Keanu is most excellent as the girl's blushing beau.
198? 96m/C Rita Tushingham, Keanu Reeves, Olivia D'Abo; *Dir:* Paul Lynde. VHS, Beta $79.98 *NSV, HHE* 🎬🎬

Dream Boys Revue A slew of female impersonators compete in this colorful contest/ pageant.
1987 74m/C *Hosted:* Lyle Waggoner, Ruth Buzzi. VHS, Beta $19.95 *NWV, STE* 🎬🎬

A Dream Called Walt Disney World This souvenir of Orlando, Florida's Walt Disney World discusses the creation of the magnificent theme park.
1981 25m/C VHS, Beta $49.95 *DIS* 🎬🎬

Dream Chasers A bankrupt old codger and a 11-year-old boy stricken with cancer run away together during the Great Depression.
1984 (PG) 97m/C Harold Gould, Justin Dana. VHS, Beta $79.98 *FOX* 🎬

A Dream for Christmas Earl Hamner, Jr.'s (best known for writing the "The Waltons") moving story of a black minister whose church in Los Angeles is scheduled to be demolished. Made for television.
1973 100m/C Hari Rhodes, Beah Richards, George Spell, Juanita Moore, Joel Fluellen, Robert DoQui, Clarence Muse; *Dir:* Ralph Senesky. VHS $49.95 *IVE* 🎬🎬🎬

The Dream Continues A look at Elvis' fans almost ten years after his death, and how they feel about their idol today.
1985 60m/C VHS, Beta *MNP* 🎬🎬🎬

A Dream of Kings Quinn is exceptional in this Petrakis story of an immigrant working to get his dying son home to Greece. Last film appearance for Stevens before committing suicide at age 36.
1969 (R) 107m/C Anthony Quinn, Irene Papas, Inger Stevens, Sam Levene, Val Avery; *Dir:* Daniel Mann. VHS $59.95 *WAR* 🎬🎬🎬

Dream a Little Dream A strange teen transformation drama about an old man and his wife trying to mystically regain their youth. When they collide bikes with the teenagers down the street, their minds are exchanged and the older couple with the now young minds are transported to a permanent dream-like state. Same old switcheroo made beatable by cast.
1989 (PG-13) 114m/C Corey Feldman, Corey Haim, Meredith Salenger, Jason Robards Jr., Piper Laurie, Harry Dean Stanton, Victoria Jackson, Alex Rocco, Billy McNamara; *Dir:* Marc Rocco. VHS, Beta, LV $89.98 *LIV, VES, HHE* 🎬🎬

Dream Lover Terrifying nightmares after an assault lead McNichol to dream therapy. Treatment causes her tortured unconscious desires to take over her waking behavior, and she becomes a violent schizophrenic. Slow, heavy, and not very thrilling.
1985 (R) 105m/C Kristy McNichol, Ben Masters, Paul Shenar, Justin Deas, Joseph Culp, Gayle Hunnicutt, John McMartin; *Dir:* Alan J. Pakula. VHS, Beta $79.95 *MGM* 🎬🎬

Dream Machine A childish teen comedy, based on that old urban legend of the lucky kid given a free Porshe by the vengeful wife of a wealthy philanderer. The gimmick is that the husband's body is in the trunk; the murderer is in pursuit. The tape includes an anti-drug commerical—but nothing against reckless driving, which the picture glorifies.
1991 (PG) 88m/C Corey Haim, Evan Richards, Jeremy Slate, Randall England, Tracy Fraim, Brittney Lewis, Susan Seaforth Hayes; *Dir:* Lyman Dayton. VHS, LV $89.95 *LIV, PIA* 🎬½

Dream No Evil A mentally disturbed woman is forced to commit bizarre murders to protect her warped fantasy world.
1975 93m/C Edmond O'Brien, Brooke Mills, Marc Lawrence, Arthur Franz; *Dir:* John Hayes. VHS, Beta $19.95 *AHV* 🎬½

A Dream of Passion A woman imprisoned in Greece for murdering her children becomes the object of a publicity stunt for a production of "Medea," and she and the lead actress begin to exchange personalities. Artificial gobbledygook.
1978 (R) 105m/C *GR SI* Ellen Burstyn, Melina Mercouri, Andreas Voutsinas; *Dir:* Jules Dassin. VHS, Beta $9.98 *CHA* 🎬½

Dream Street A weak morality tale of London's lower classes. Two brothers, both in love with the same dancing girl, woo her in their own way, while a Chinese gambler plans to take her by force. Silent with music score. Based on Thomas Burke's "Limehouse Nights."
1921 138m/B Tyrone Power Sr., Carol Dempster, Ralph Graves, Charles Mack; *Dir:* D.W. Griffith. VHS, Beta $49.95 *VYY* 🎬🎬

The Dream Team On their way to a ball game, four patients from a mental hospital find themselves lost in New York City after their doctor is kidnapped by crooks. Some fine moments from a cast of dependable comics.

1989 (PG-13) 113m/C Michael Keaton, Christopher Lloyd, Peter Boyle, Stephen Furst, Lorraine Bracco, Milo O'Shea, Dennis Boutsikaris, Philip Bosco, James Remar; *Dir:* Howard Zieff. VHS, Beta, LV $19.95 *MCA* 🎬🎬½

Dream Trap A young man has an obsession with a girl in his dreams. When he meets a girl who really exists, fantasy louses up every opportunity for the real thing.
1990 90m/C Kristy Swanson, Sasha Jensen, Jeanie Moore; *Dir:* Hugh Parks; *W/Dir:* Tom Logan. VHS $89.95 *QUE* 🎬½

Dreamchild A poignant story of the autumn years of Alice Hargreaves, the model for Lewis Carroll's "Alice in Wonderland." Film follows her on a visit to New York in the 1930s, with fantasy sequences including Wonderland characters created by Jim Henson's Creature Shop invoking the obsessive Reverend Dodgson (a.k.a. Carroll).
1985 (PG) 90m/C *GB* Coral Browne, Ian Holm, Peter Gallagher, Jane Asher, Nicola Cowper; *Dir:* Gavin Millar. VHS, Beta $19.98 *HBO* 🎬🎬🎬

Dreamer The excitement of bowling is exploited for all it's worth in this "strike or die" extravaganza. One of a kind.
1979 (PG) 86m/C Tim Matheson, Susan Blakely, Jack Warden, Richard B. Shull; *Dir:* Noel Nosseck. VHS $59.98 *FOX* 🎬🎬

The Dreaming A young doctor stumbles upon a dimension where reality and fantasy are blurred, and where the supernatural reigns supreme.
1989 87m/C *AU* VHS, Beta *NVH* 🎬

Dreaming Lips An orchestra conductor's wife falls in love with her husband's violinist friend. Tragedy befalls the couple. Bergner outstanding in a familiar script.
1937 70m/B *GB* Raymond Massey, Elisabeth Bergner, Romney Brent, Joyce Bland, Charles Carson, Felix Aylmer; *Dir:* Paul Czinner, Lee Garmes. VHS, Beta $29.95 *CCB, VYY* 🎬🎬½

Dreaming Out Loud Screen debut of popular radio team Lum 'n Abner. The boys get involved in several capers to bring progress to their small Arkansas town. Rural wisecracks make for pleasant outing. First in film series for the duo.
1940 65m/B Frances Langford, Phil Harris, Clara Blandick, Robert Wilcox, Chester Lauck, Norris Goff, Frank Craven, Bob Watson, Irving Bacon; *Dir:* Harold Young. VHS, Beta $24.95 *NOS, DVT* 🎬🎬

Dreams Unfocused film about the lives and loves of two successful women. Subtitled. Also known as "Kvinn odrom" or "Journey into Autumn."
1955 86m/B *SW* Harriet Andersson, Gunnar Bjornstrand, Eva Dahlbeck, Ulf Palme; *Dir:* Ingmar Bergman. VHS, Beta, LV $14.95 *VDM, TAM, SUE* 🎬🎬

Dreams Come True Silly young comedy-romance about two young lovers who discover the trick of out-of-body travel, and have various forms of spiritual contact while their bodies are sleeping.
1984 (R) 95m/C Michael Sanville, Stephanie Shuford; *Dir:* Max Kalmanowicz. VHS, Beta $59.95 *MED* 🎬

Dreams of Desire An engaged couple goes to Hong Kong to enjoy that city's sensual pleasures.
1982 80m/C VHS, Beta *PGN* 🎬

Dreams Lost, Dreams Found A young widow journeys from the U.S. to Scotland in search of her heritage. The third ''Harlequin Romance Movie,'' made for cable television.
1987 102m/C Kathleen Quinlan, Betsy Brantley, Charles Gray; *Dir:* Willi Patterson. VHS $19.95 PAR ✍✍½

Dreamscape When a doctor teaches a young psychic how to enter into other people's dreams in order to end their nightmares, somebody else wants to use this psychic for evil purposes. The special effects are far more convincing than the one man-saves-the-country-with-his-psychic powers plot.
1984 (PG-13) 99m/C Dennis Quaid, Max von Sydow, Christopher Plummer, Eddie Albert, Kate Capshaw, David Patrick Kelly, George Wendt, Jana Taylor; *Dir:* Joseph Ruben. VHS, Beta, LV $19.99 HBO, IME ✍✍½

Dreamwood A myth-packed experimental film by James Broughton with muse-figures, angels, flower-children and the like.
1972 45m/C *Dir:* James Broughton. VHS, Beta $29.95 MFV ✍✍✍

Dressed for Death A tale of true love gone bad as a woman is chased and murdered in her castle by the psychopath she loves. Also known as ''Til Dawn Do Us Part.'' Originally titled, ''Straight on Till Morning.''
1974 (R) 121m/C GB Rita Tushingham, Shane Briant, Tom Bell, Annie Ross, Katya Wyeth, James Bolam, Claire Kelly; *Dir:* Peter Collinson. VHS, Beta $19.95 ACA ✍½

Dressed to Kill Sherlock Holmes finds that a series of music boxes holds the key to plates stolen from the Bank of England. The plot's a bit thin, but Rathbone/Bruce, in their final Holmes adventure, are always a delight.
1946 72m/B Basil Rathbone, Nigel Bruce, Patricia Morison, Edmund Breon, Tom Dillon; *Dir:* Roy William Neill. VHS, LV $9.95 NEG, MRV, REP ✍✍½

Dressed to Kill A contemporary thriller merging bombastic DePalma with tense Hitchcockian. A sexually unsatisfied Dickinson ends up in bed with a man who catches her eye in a museum. After leaving his apartment, she is brutally murdered. Dickinson's son soon teams up with a prostitute to track and lure the killer into their trap. Suspenseful and fast paced. Dickinson's museum scene is wonderfully photographed and edited. DePalma was repeatedly criticized for using a stand-in during the Dickinson shower scene; he titled his next film, ''Body Double'' as a rebuttal.
1980 (R) 105m/C Angie Dickinson, Michael Caine, Nancy Allen, Keith Gordon, Dennis Franz, David Margulies, Brandon Maggart; *Dir:* Brian DePalma. VHS, Beta, LV $19.98 WAR, OM ✍✍½

Dresser The film adaptation of the Ronald Harwood Broadway play about an aging English actor/ manager, his dresser, and their theater company touring England during World War II. Marvelous show biz tale is lovingly told, superbly acted.
1983 (PG) 119m/C Albert Finney, Tom Courtenay, Edward Fox, Michael Gough, Zena Walker, Eileen Atkins, Cathryn Harrison; *Dir:* Peter Yates. VHS, Beta, LV $12.95 COL ✍✍✍½

Dressmaker French comedian Fernandel stars as a man's tailor who designs dresses in secret so as not to compete with his dressmaker wife. Lightweight but entertaining. English subtitled.

1956 95m/B FR Fernandel, Francoise Fabian; *Dir:* Jean Boyer. VHS, Beta $29.95 NOS, FST, DVT ✍✍½

Dressmaker Two sisters in World War II Liverpool must deal with their young niece's romantic involvement with an American soldier. Plowright and Whitelaw turn in fine performances as the two siblings with very different outlooks on life.
1989 90m/C GB Joan Plowright, Billie Whitelaw, Peter Postlethwaite, Jane Horrocks, Tim Ransom; *Dir:* Jim O'Brien. VHS $79.95 CAP, TAM ✍✍½

The Drifter A lumberjack makes sacrifices for the love between his brother and the lumbermill owner's daughter.
1932 56m/B William Farnum, Noah Beery. VHS, Beta, LV $19.95 NOS, WGE ✍✍

The Drifter It's psychos, psychos everywhere as a beautiful young woman learns to regret a one-night stand. A good, low-budget version of ''Fatal Attraction.'' Director Brand plays the cop.
1988 (R) 90m/C Kim Delaney, Timothy Bottoms, Miles O'Keefe, Al Shannon, Larry Brand; *Dir:* Larry Brand. VHS $79.95 MGM ✍✍½

Drifting A controversial Israeli film about a homosexual filmmaker surviving in modern-day Jerusalem. First such movie to be made in Israel. In Hebrew with English titles.
1982 103m/C IS Jonathan Sagalle, Ami Traub, Ben Levine, Dita Arel; *Dir:* Amos Guttman. VHS, Beta $69.95 FCT ✍✍½

Drifting Souls In order to save her father's life, a beautiful young woman marries for money instead of love.
1932 65m/C Lois Wilson, Theodore von Eltz, Shirley Grey, Raymond Hatton, Gene Gowing, Bryant Washburn; *Dir:* Louis King. VHS, Beta GPV ✍✍

Drifting Weeds A remake by Ozu of his 1934 silent film about a troupe of traveling actors whose leader visits his illegitimate son and his lover after years of separation. Classic Ozu in his first color film. In Japanese with English subtitles. Also known as ''Floating Weeds.''
1959 128m/B JP Ganjiro Makamura, Machiko Kyo, Haruko Sugimura, Ayako Wakao; *Dir:* Yasujiro Ozu. VHS, Beta $29.95 FCT, MRV ✍✍✍½

Driller Killer A frustrated artist goes insane and begins to kill off Manhattan residents with a carpenter's drill. The plot is full of holes.
1974 (R) 78m/C Jimmy Laine, Carolyn Marz, Bob DeFrank; *Dir:* Abel Ferrara. VHS, Beta $39.98 MAG Woof!

Drive-In A low-budget bomb showing a night in the life of teenage yahoos at a Texas drive-in.
1976 (PG) 96m/C Lisa Lemole, Glen Morshower, Gary Cavagnaro; *Dir:* Rod Amateau. VHS, Beta $69.95 COL Woof!

Drive-In Massacre Two police detectives investigate a bizarre series of slasher murders at the local drive-in. Honk the horn at this one.
1974 (R) 78m/C Jake Barnes, Adam Lawrence. VHS, Beta $39.98 MAG Woof!

Drive-In Sleaze A mass of grade-Z movie trailers, from the thirties to the seventies, including ''The Girl from S.I.N.,'' ''Mundo Depravados'' and ''The Reluctant Sadist.''
1984 55m/C VHS, Beta $19.95 NO Woof!

Driven to Kill No-name cast assemble in low-rent suspenser portraying the ugly side of love: bent on revenge, a man is out to obliterate the thugs who made his life a horrible experience highlighted by misery and boredom.
1990 ?m/C Jake Jacobs, Chip Campbell, Michele McNeil. VHS $79.95 PMH ✍✍

The Driver A police detective will stop at nothing to catch ''The Driver,'' a man who has the reputation of driving the fastest getaway car around. Chase scenes win out over plot.
1978 (PG) 131m/C Ryan O'Neal, Bruce Dern, Isabelle Adjani, Ronee Blakley, Matt Clark; *Dir:* Walter Hill. VHS, Beta $59.98 FOX ✍½

Driver's Seat Extremely bizarre film with a cult following that was adapted from the novel by Muriel Spark. Liz stars as a deranged woman trying to keep a rendezvous with her strange lover. In the meantime she wears tacky clothes and delivers stupid lines. What's Taylor doing in this? Also on video as, ''Psychotic.''
1973 (R) 101m/C IT Elizabeth Taylor, Ian Bannen, Mona Washbourne, Andy Warhol; *Dir:* Giuseppe Patroni Griffi. VHS, Beta $24.98 SUE, AVD, SIM ✍

Driving Force In an effort to capitalize on the popularity of the ''Road Warrior'' movies, we find a lone trucker battling a gang of roadhogs in another post-holocaust desert. Ultimately, runs off the road.
1988 (R) 90m/C Sam Jones, Catherine Bach, Don Swayze; *Dir:* A.J. Prowse. VHS, Beta, LV $29.95 ACA ✍

Driving Miss Daisy Tender and sincere portrayal of a 25-year friendship between an aging Jewish woman and the black chauffeur forced upon her by her son. Humorous and thought-provoking, skillfully acted and directed, it subtly explores the effects of prejudice in the South. The development of Aykroyd as a top-notch character actor is further evidenced here. Part of the fun is watching the changes in fashion and auto design. Adapted from the play by Alfred Uhry.
1989 (PG) 99m/C Jessica Tandy, Morgan Freeman, Dan Aykroyd, Esther Rolle, Patti LuPone; *Dir:* Bruce Beresford. Academy Awards '89: Best Actress (Tandy), Best Adapted Screenplay, Best Makeup, Best Picture; Golden Globe Awards '90: Best Film—Musical/Comedy; National Board of Review Awards '89: Best Actor (Freeman), Best Picture. VHS, Beta, LV, 8mm $19.98 WAR, FCT, BTV ✍✍✍½

Drop Dead Fred As a little girl, Lizzie Cronin had a manic, imaginary friend named Fred, who protected her from her domineering mother. When her husband dumps her twenty years later, Fred returns to ''help'' as only he can. Although the cast is fine, incompetent writing and direction make this a truly dismal affair. Plus, gutter humor and mean-spirited pranks throw the whole ''heart-warming'' premise out the window. Filmed in Minneapolis.
1991 (PG-13) 103m/C Phoebe Cates, Rik Mayall, Tim Matheson, Marsha Mason, Carrie Fisher, Daniel Gerroll, Ron Eldard; *Dir:* Ate De Jong. VHS, LV $92.95 LIV, PIA Woof!

Drop-Out Mother A woman gives up her executive position to become a housewife and full-time mom; expected domestic turmoil follows. Standard made for television fare and sort of a sequel to ''Drop-Out Father,'' shown in 1982.

1988 100m/C Valerie Harper, Wayne Rogers, Carol Kane, Kim Hunter, Danny Gerard; **Dir:** Charles S. Dubin. VHS **$39.95** *FRH* ✂✂

The Dropkick Erstwhile D.W. Griffith leading man Barthelmess stars in this pigskin whodunnit: a coach's suicide looks like murder, and the prime suspect is the team's most valuable player. Cast includes ten bona fide university football players and Hedda "nobody's interested in sweetness and light" Hopper, while The Duke makes cameo. Webb later went on to direct a musical with Eddie Cantor and Rudy Vallee ("Glorifying the American Girl"). Silent.
1927 62m/B Richard Barthelmess, Barbara Kent, Dorothy Revier, Eugene Strong, Alberta Vaughan, Brooks Benedict, Hedda Hopper; **Dir:** Millard Webb. **VHS, Beta $24.95** *GPV, FCT, DNB* ✂✂✂

Drowning by Numbers Three generations of women, each named Cissie Colpitts, solve their marital problems by drowning their husbands and making deals with a bizarre coroner. Further strange visions from director Greenaway, complemented by stunning cinematography courtesy of Sacha Vierny. A treat for those who appreciate Greenaway's uniquely curious cinematic statements.
1987 (R) 121m/C *GB* Bernard Hill, Joan Plowright, Juliet Stevenson, Joely Richardson; **Dir:** Peter Greenaway. **VHS, LV $89.98** *LIV* ✂✂✂½

The Drowning Pool Newman returns as detective Lew Harper (of "Harper" by Ross MacDonald) to solve a blackmail case. Uneventful script and stodgy direction, but excellent character work from all the cast members keep this watchable. Title is taken from a trap set for Newman, from which he must escape using most of his female companion's clothing.
1975 (PG) 109m/C Paul Newman, Joanne Woodward, Anthony (Tony) Franciosa, Murray Hamilton, Melanie Griffith, Richard Jaeckel; **Dir:** Stuart Rosenberg. **VHS, Beta $19.98** *WAR* ✂✂

Drugstore Cowboy A gritty, uncompromising depiction of a pack of early 1970s drugstore-robbing junkies as they travel around looking to score. Brushes with the law and tragedy encourage them to examine other life-styles, but the trap seems impossible to leave. A perfectly crafted piece that reflects the "me generation" era, though it tends to glamorize addiction. Dillon's best work to date. Based on an unpublished novel by prison inmate James Fogle.
1989 (R) 100m/C Matt Dillon, Kelly Lynch, James Remar, James Le Gros, Heather Graham, William S. Burroughs, Beah Richards, Grace Zabriskie, Max Perlich; **Dir:** Gus Van Sant. National Society of Film Critics Awards '89: Best Director (Van Sant), Best Film, Best Screenplay. **VHS, Beta, LV $19.95** *IVE, FCT* ✂✂✂½

Drum This steamy sequel to "Mandingo" deals with the sordid interracial sexual shenanigans at a Southern plantation. Bad taste at its best.
1976 (R) 101m/C Ken Norton, Warren Oates, Pam Grier, Yaphet Kotto, Fiona Lewis, Isela Vega, Cheryl "Rainbeaux" Smith; **Dir:** Steve Carver. **VHS, Beta $69.98** *LIV, VES* Woof!

Drum Beat An unarmed Indian fighter sets out to negotiate a peace treaty with a renegade Indian leader. Bronson is especially believable as chief. Based on historical incident.
1954 111m/C Alan Ladd, Charles Bronson, Marisa Pavan, Robert Keith, Rodolfo Acosta, Warner Anderson, Elisha Cook Jr., Anthony Caruso; **Dir:** Delmer Daves. **VHS, Beta $54.95** *UHV* ✂✂

Drum Taps Maynard saves the day for a young girl who is being pushed off her land by a group of speculators.
1933 55m/B Ken Maynard, Dorothy Dix, Junior Coughlin, Kermit Maynard; **Dir:** J.P. McGowan. **VHS, Beta $19.95** *NOS, VDM, VYY* ✂✂

Drums A native prince helps to save the British army in India from being annihilated by a tyrant. Rich melodrama with interesting characterizations and locale.
1938 96m/B *GB* Sabu, Raymond Massey, Valerie Hobson, Roger Livesey, David Tree; **Dir:** Zoltan Korda. **VHS, Beta, LV $14.98** *SUE* ✂✂½

Drums Along the Mohawk Grand, action-filled saga about pre-Revolutionary America, detailing the trials of a colonial newly wed couple as their village in Mohawk Valley is besieged by Indians. Based on the Walter Edmonds novel, and vintage Ford. Oscar nominations: Best Supporting Actress (Oliver), Color Cinematography.
1939 104m/C Henry Fonda, Claudette Colbert, Edna May Oliver, Eddie Collins, John Carradine, Dorris Bowdon, Arthur Shields, Ward Bond, Jessie Ralph, Robert Lowery; **Dir:** John Ford. **VHS, Beta, LV $19.98** *FOX* ✂✂✂½

Drums in the Deep South A rivalry turns ugly as two former West Point roommates wind up on opposite sides when the Civil War breaks out. Historical drama hampered by familiar premise.
1951 87m/C James Craig, Guy Madison, Craig Stevens, Barbara Payton, Barton MacLane; **Dir:** William Cameron Menzies. **VHS, Beta $19.95** *NOS, MRV, VGR* ✂✂

Drums of Fu Manchu Fu Manchu searches for the scepter of Genghis Khan, an artifact that would give him domination over the East. Brandon smoothly evil as the Devil Doctor. Originally a serial in 15 chapters.
1940 150m/B Henry Brandon, Robert Kellard, George Cleveland, Dwight Frye, Gloria Franklin, Tom Chatterton; **Dir:** William Witney, John English. **VHS, Beta $69.95** *VCN, MLB* ✂✂✂

Drums of Jeopardy A father wanders through czarist Russia and the U.S. to seek revenge on his daughter's killer. Cheap copy of "Dr. Fu Manchu."
1931 65m/B Warner Oland, June Collyer, Lloyd Hughes, George Fawcett, Mischa Auer; **Dir:** George B. Seitz. **VHS $24.95** *NOS, SNC, DVT* ✂

Drums O'Voodoo A voodoo princess fights to eliminate the town bad guy in this all black feature which is greatly hampered by shoddy production values.
1934 70m/B Laura Bowman, J. Augustus Smith, Edna Barr; **Dir:** Arthur Hoerl. **VHS, Beta $16.95** *SNC* Woof!

Drunken Angel Alcoholic doctor gets mixed up with local gangster. Kurosawa's first major film aided by strong performances. With English subtitles.
1948 108m/B *JP* Toshiro Mifune, Takashi Shimura, Choko Iida; **Dir:** Akira Kurosawa. **VHS, Beta $39.95** *DVT, MRV* ✂✂✂

A Dry White Season A white Afrikaner living resignedly with apartheid confronts the system when his black gardener, an old friend, is persecuted and murdered. A well-meaning expose that, like many others, focuses on white people. Brando was Oscar-nominated.
1989 (R) 105m/C Donald Sutherland, Marlon Brando, Susan Sarandon, Zakes Mokae, Janet Suzman, Juergen Prochnow, Winston Ntshona; **Dir:** Euzhan Palcy. **VHS, Beta, LV $19.98** *FOX* ✂✂✂

Drying Up the Streets Tepid message tale of a police drug squad out to terminate the pattern that has young women turning to a life of prostitution and drugs.
1976 90m/C *CA* Len Cariou, Don Francks, Sarah Torov, Calvin Butler; **Dir:** Robin Spry. **VHS, Beta $69.98** *LIV, VES* ✂

Dual Alibi Identical twin trapeze artists murder a colleague for a winning lottery ticket. Lom plays the brothers in a dual role.
1947 87m/B *GB* Herbert Lom, Terence de Marney, Phyllis Dixey, Ronald Frankau, Abraham Sofaer, Harold Berens, Sebastian Cabot; **Dir:** Alfred Travers. **VHS, Beta $16.95** *NOS, SNC* ✂✂

Dual Flying Kicks A Chinese provincial mayor allies himself to a group of bandits in this martial arts film.
1982 95m/C Chen Hsing, Tan Tao Liang. **VHS, Beta** *WVE* ✂

Dubarry Early sound version of the romantic experiences of Madame Dubarry, the alluring French heroine of David Belasco's "DuBarry." Talmadge's last film.
1930 81m/B Norma Talmadge, Conrad Nagel, William Farnum, Hobart Bosworth, Alison Skipworth; **Dir:** Sam Taylor. **VHS, Beta $24.95** *NOS, DVT* ✂✂½

DuBarry was a Lady Skelton is a washroom attendant who daydreams that he is King Louis XV of France. Dorsey's band gets to dress in period wigs and costumes. Cole Porter score includes "Friendship" and "Do I Love You?"
1943 112m/C Lucille Ball, Gene Kelly, Red Skelton, Virginia O'Brien, Zero Mostel, Dick Haymes, Jo Stafford & the Pied Pipers, Tommy Dorsey & His Orchestra, Rags Ragland, Donald Meek, George Givot, Louise Beavers; **Dir:** Roy Del Ruth. **VHS, Beta $19.98** *MGM, CCB* ✂✂½

Dubeat-E-O A filmmaker races against time to finish making a documentary about the Los Angeles underground hardcore punk scene, and becomes immersed in the subculture. Somewhat of a haphazard exercise in filmmaking that is best appreciated by rock fans.
1984 84m/C Ray Sharkey, Joan Jett, Derf Scratch, Len Lesser, Nora Gaye; **Dir:** Alan Sacks. **VHS, Beta $59.95** *MVD, MED* ✂½

The Duchess and the Dirtwater Fox A period western strung together with many failed attempts at humor, all about a music-hall hooker who meets a bumbling card shark on the make.
1976 (PG) 105m/C George Segal, Goldie Hawn, Conrad Janis, Thayer David; **Dir:** Melvin Frank. **VHS, Beta, LV $19.98** *FOX* ✂½

Duchess of Idaho Williams tries to help her roommate patch up a romance gone bad, but ends up falling in love herself. MGM guest stars liven up an otherwise routine production.
1950 98m/C Esther Williams, Van Johnson, John Lund, Paula Raymond, Amanda Blake, Eleanor Powell, Lena Horne; **Dir:** Robert Z. Leonard. **VHS** *MGM* ✂✂

Duck Soup The Marx Brothers satiric masterpiece (which failed at the box office). Groucho becomes the dictator of Freedonia, and hires Chico and Harpo as spies. Jampacked with the classic anarchic and irreverent Marx shtick; watch for the mirror scene. Zeppo plays a love-sick tenor, in this, his last film with the brothers. Screenplay by Bert Kalmar and Holly Ruby.
1933 70m/B Groucho Marx, Chico Marx, Harpo Marx, Zeppo Marx, Louis Calhern, Margaret Dumont, Edgar Kennedy; **Dir:** Leo McCarey. **VHS, Beta, LV $19.95** *MCA, FCT* ✂✂✂✂

D

DuckTales, the Movie Uncle Scrooge et al embark on a lost-ark quest—sans Harrison Ford—for misplaced treasure (i.e., a lamp that can make the sky rain ice cream). Based on the daily Disney cartoon of the same name, it's more like an extended-version Saturday morning sugar smacks'n'milk 'toon. See it with someone young.
1990 (G) 74m/C VHS, Beta, LV $22.99 *DIS* ♫♫½

Dude Bandit An unscrupulous money-lender tries to gain control of a ranch.
1932 68m/B Hoot Gibson, Gloria Shea. **VHS, Beta** $29.95 *VCN, UHV* ♫

Dude Ranger Well-acted Zane Grey story about an easterner who gets caught up in a range war and cattle rustling when he takes possession of some property out west.
1934 58m/B Smiley Burnette, George O'Brien, Irene Hervey; *Dir:* Eddie Cline. **VHS, Beta** $19.95 *VDM* ♫♫½

Dudes Three city kids head for the desert and run afoul of some rednecks. Tired grade-6 revenge tale.
1987 (R) 90m/C Jon Cryer, Catherine Mary Stewart, Daniel Roebuck, Flea, Lee Ving, Calvin Bartlett, Pete Willcox, Glenn Withrow; *Dir:* Penelope Spheeris. **VHS, Beta, LV** $9.99 *CCB, IVE* ♫

Duel Spielberg's first notable film, a made-for-television exercise in paranoia penned by Robert Bloch. A docile traveling salesman is repeatedly attacked and threatened by a huge, malevolent tractor-trailer on an open desert highway. Released theatrically in Europe.
1971 (PG) 90m/C Dennis Weaver, Lucille Benson, Eddie Firestone, Cary Loftin; *Dir:* Steven Spielberg. **VHS, Beta** $39.95 *MCA* ♫♫♫

Duel of the Brave Ones A man undertakes a dangerous trek through Japan to search for jewel thieves.
1980 87m/C Chang Wu Lang, Tony Wai Shing. **VHS, Beta** $39.95 *TWE* ♫

Duel of Champions It's ancient Rome, and the prodigal gladiator has come home. Now the family must have a duel to decide who will rule.
1961 90m/C *IT SP* Alan Ladd, Francesca Bett; *Dir:* Ferdinando Baldi. **VHS, Beta** $69.95 *FOR* ♫

Duel at Diablo An exceptionally violent film that deals with racism in the old west. Good casting, western fans will enjoy the action.
1966 103m/C James Garner, Sidney Poitier, Bibi Andersson, Dennis Weaver, Bill Travers; *Dir:* Ralph Nelson. **VHS** $14.95 *WKV* ♫♫

Duel-Duo A clarinet and a trumpet have a clash of wills, but harmony is instrumental in bringing them together as friends. No dialogue.
198? 2m/C *CA* **VHS, Beta** *NFC, SAL*

Duel in the Eclipse A strategically timed eclipse permits a gringo to avenge his brother's murder.
19?? 100m/C Fernando Sancho. **VHS, Beta** $39.95 *TWE* ♫

Duel of Fists When the owner of a boxing institute dies, he instructs his son in his will to go searching for his long lost brother in Thailand. All he can go on is the knowledge that his relative is a Thai boxer and an old photograph.
19?? 111m/C David Chiang, Chen Hsing, Ti Lung, Tang Ti. **VHS, Beta** $69.95 *SOU* ♫

Duel of Hearts The beautiful Caroline falls in love with the wealthy Lord Vane Brecon, who unfortunately is accused of murder. When Caroline decides to help him clear his name she discovers he's hiding a number of secrets. Based on the romance novel "Duel of Love" by Barbara Cartland. Originally made for cable television.
1992 (PG-13) 95m/C Alison Doody, Benedict Taylor, Michael York, Geraldine Chaplin, Billie Whitelaw, Richard Johnson, Jeremy Kemp; *Dir:* John Hough. **VHS, Beta** $89.98 *TTC* ♫♫

Duel of the Iron Fist A master of the martial arts must do battle with a brutal gang to save his own group from extinction.
1972 (R) 98m/C David Chiang, Ti Lung, Wang Ping, Yu Hui. **VHS, Beta** $59.95 *HHT, UHV* ♫

Duel of the Masters Two ninja masters fight to the finish and then start again.
1976 95m/C **VHS, Beta** $29.99 *BFV* ♫½

Duel of the Seven Tigers Nine of the world's greatest Kung fu and Karate exponents team up in this martial arts spectacular.
1980 80m/C Cliff Lok, Ka Sa Fa, Chio Chi Ling, Lam Men Wei, Yang Pan Pan, Charlie Chan. **VHS, Beta** $39.95 *TWE* ♫♫

Duel in the Sun A lavish, lusty David O. Selznick production of a minor western novel about a vivacious half-breed Indian girl, living on a powerful dynastic ranch, who incites two brothers to conflict. Selznick's last effort at outdoing his epic success with "Gone With the Wind." Oscar nominations for Best Actress (Jones), Supporting Actress (Gish).
1946 130m/C Gregory Peck, Jennifer Jones, Joseph Cotten, Lionel Barrymore, Lillian Gish, Butterfly McQueen, Harry Carey, Walter Huston, Charles Bickford, Herbert Marshall; *Dir:* King Vidor. **VHS, Beta** $19.98 *FOX* ♫♫♫

The Duellists A beautifully photographed picture about the long running feud between two French officers during the Napoleonic wars. Based on "The Duel" by Joseph Conrad.
1978 (PG) 101m/C *GB* Keith Carradine, Harvey Keitel, Albert Finney, Edward Fox, Tom Conti, Christina Raines, Diana Quick; *Dir:* Ridley Scott. **VHS, Beta, LV** $49.95 *PAR* ♫♫♫

Duelo en el Dorado A white man and Indian fight over the custody of an orphan.
19?? 90m/C Luis Aguilar, Lola Beltran, Emilio Fernandez, German Valdes. **VHS, Beta** $78.95 *MAD*

Duet for One A famous concert violinist learns she is suffering from multiple sclerosis and is slowly losing her ability to play. Convinced that her life is meaningless without music, she self-destructively drifts into bitter isolation. From the play by Tom Kempinski.
1986 (R) 108m/C Julie Andrews, Max von Sydow, Alan Bates, Liam Neeson; *Dir:* Andrei Konchalovsky. **VHS, Beta** $79.95 *MGM* ♫♫

Duke of the Derby A scheming race-horse handicapper bets over his head and ruins his higher-than-means lifestyle. In French with English subtitles.
1962 83m/C *FR* Jean Gabin; *Dir:* Gilles Grangier. **VHS, Beta** $19.95 *FCT, DVT* ♫♫

Duke Ellington & His Orchestra: 1929-1952 Duke Ellington and His Famous Orchestra are featured in two vintage film shorts, "Black and Tan" (1929) and "Symphony in Black"(1934). Also included are three 1952 Telescriptions: "Sophisticat-ed Lady," "Caravan" and "The Hawk Talks."
1986 40m/B Duke Ellington, Billie Holiday, Fredi Washington, Louis Bellson, Cootie Williams, Johnny Hodges, Tricky Sam Nanton. **VHS, Beta** $19.95 *MVD, AFE, AMV* ♫♫

Duke Is Tops In Horne's earliest existing film appearance, she's off to attempt the "big-time," while her boyfriend joins a traveling medicine show.
1938 80m/C Ralph Cooper, Lena Horne, The Basin Street Boys, Rubber Neck Boys, Marie Bryant. **VHS, Beta** $39.95 *NOS, MRV, VCN* ♫♫½

Duke: The Films of John Wayne A gift pack collection of three of Wayne's finest: "Flying Leathernecks," "Fort Apache," and "She Wore a Yellow Ribbon."
1989 332m/C John Wayne. **VHS** $59.95 *TTC* ♫♫½

Dumb Waiter An hour-long mini-play by Harold Pinter about two hitmen awaiting instructions for a job in an empty boarding-house, and getting comically mixed messages.
1987 60m/C John Travolta, Tom Conti; *Dir:* Robert Altman. **VHS, Beta** $79.95 *PSM* ♫♫½

Dumbo Animated Disney classic about a baby elephant growing up in the circus who is ridiculed for his large ears, until he discovers he can fly. The little elephant who could fly then becomes a circus star. Expressively and imaginatively animated, highlighted by the hallucinatory dancing pink elephants sequence. Endearing songs by Frank Churchill, Oliver Wallace, and Ned Washington, including "Baby Mine," "Pink Elephants on Parade," and "I See an Elephant Fly."
1941 63m/C *Dir:* Ben Sharpsteen; *Voices:* Sterling Holloway, Edward Brophy, Verna Felton, Herman Bing, Cliff Edwards. Academy Awards '41: Best Musical Score; National Board of Review Awards '41: 10 Best Films of the Year. **VHS, Beta, LV** $24.99 *DIS, KUI, APD* ♫♫♫♫

The Dummy Talks When a ventriloquist is murdered, a midget goes undercover as a dummy to find the killer. Watch closely and you might see his lips move.
1943 85m/B *GB* Jack Warner, Claude Hulbert, Beryl Orde, Derna-Hazell, Ivy Benson, Manning Whiley; *Dir:* Oswald Mitchell. **VHS, Beta** $16.95 *SNC* ♫

Dune Lynch sci-fi opus based on the Frank Herbert novel boasting great set design, muddled scripting, and a good cast. The story: controlling the spice drug of Arrakis permits control of the universe in the year 10,991. Paul, the heir of the Atreides family, leads the Freemen in a revolt against the evil Harkhonens who have violently seized control of Arrakis, also known as Dune, the desert planet. That's as clear as it ever gets. Music by Brian Eno and Toto.
1984 (PG-13) 137m/C Kyle MacLachlan, Francesca Annis, Jose Ferrer, Sting, Max von Sydow, Juergen Prochnow, Linda Hunt, Freddie Jones, Dean Stockwell, Virginia Madsen, Brad Dourif, Kenneth McMillan, Silvana Mangano, Jack Nance, Sian Phillips, Paul Smith, Richard Jordan, Everett McGill, Sean Young; *Dir:* David Lynch. **VHS, Beta, LV** $24.95 *MCA, FCT* ♫♫

Dune Warriors Earth is a parched planet in the year 2040. Renegade bands cruise the desert, making short rift of civilized people who band together in remote villages. When peaceful farmer Carradine's family falls prey to the pillagers, surely he and the most evil warlord of all must clash. Good action and stunts, if the dummy plot doesn't get you first.

1991 (R) 77m/C David Carradine, Richard Hill, Luke Askew, Jillian McWhirter, Blake Boyd; *Dir:* Cirio H. Santiago. VHS, LV $89.95 COL, PIA ⫙½

Dunera Boys Hoskins shines in this story about Viennese Jews who, at the onset of World War II, escape to England where they are suspected as German spies. Subsequently, the British shipped them all to an Australian prison camp, where they recreated in peace their Viennese lifestyle.
1985 (R) 150m/C AU Bob Hoskins, Joe Spano, Warren Mitchell, Joseph Furst, Moshe Kedem, Dita Cobb, John Meillon, Mary-Anne Fahey, Simon Chilvers, Steven Vidler; *Dir:* Sam Lewin. VHS, Beta, LV $9.99 STE, PSM ⫙⫙⫙

Dungeon of Harrow Unbelievably cheap tale of shipwrecked comrades who encounter a maniacal family on an otherwise deserted island. Not your stranded-on-a-desert-island fantasy come true. Filmed in San Antonio; a harrowing bore.
1964 84m/B Russ Harvey, Helen Hogan, Bill McNulty, Pat Boyette; *Dir:* Pat Boyette. VHS $19.98 SNC ⫙

Dungeonmaster A warlord forces a computer wiz to participate in a bizarre "Dungeons and Dragons" styled game in order to save a girl held captive. Save yourself the trouble.
1985 (PG-13) 80m/C Jeffrey Byron, Richard Moll, Leslie Wing, Danny Dick, John Carl Buechler, Charles Band, David Allen, Steven Ford, Peter Manoogian, Ted Nicolaou; *Dir:* Charles Band, Rosemarie Turko. VHS, Beta, LV $29.98 LIV, LTG ⫙½

The Dunwich Horror Young warlock acquires a banned book of evil spells, starts trouble on the actral plane. Stookwoll ham my. Loosely based on H. P. Lovecraft story.
1970 90m/C Sandra Dee, Dean Stockwell, Lloyd Bochner, Ed Begley Sr., Sam Jaffe, Joanna Moore; *Dir:* Daniel Haller. VHS, Beta $9.98 SUE ⫙⫙

Durango Valley Raiders Sheriff is the leader of the outlaws, and a young cowboy finds out.
1938 55m/B Bob Steele; *Dir:* Sam Newfield. VHS, Beta $19.95 VCN, DVT ⫙½

Dust A white woman in South Africa murders her father when he shows affection for a young black servant. Controversial and intense.
1985 87m/C FR BE Jane Birkin, Trevor Howard, John Matshikiza, Nadine Uwampa; *W/Dir:* Marion Hansel. Venice Film Festival '85: Silver Lion. VHS, Beta $79.95 TAM ⫙⫙⫙

Dust to Dust A short mystery made for British television about an aging gentleman who answers a request for correspondence and discovers ulterior, and evil, motives.
1985 60m/C GB Patricia Hodge, Michael Jayston, Judy Campbell. VHS, Beta $59.95 PSM ⫙⫙⫙

Dusty Touching story of a wild dingo dog raised by an Australian rancher.
1985 89m/C AU Bill Kerr, Noel Trevarthen, Carol Burns, Nicholas Holland, John Stanton; *Dir:* John Richardson. VHS, Beta $19.95 MED ⫙⫙

Dutch Girls Muddled mayhem as a horny high school field hockey team travels through Holland, cavorts about, and discovers the meaning of life.
1987 83m/C GB Bill Paterson, Colin Firth, Timothy Spall; *Dir:* Giles Foster. VHS, Beta $29.95 ACA Woof!

Dutch Treat Two nerds con a sultry, all-girl rock band into thinking they're powerful record company execs. Nobody's buying.
1986 (R) 95m/C Lorin Dreyfuss, David Landsberg; *Dir:* Boaz Davidson. VHS, Beta $79.95 MGM ⫙

D.W. Griffith Triple Feature A collection of three short D.W. Griffith films including "The Battle of Elderbush Gulch," "Iola's Promise," and "The Goddess of Sagebrush Gulch."
1913 50m/B Mae Marsh, Lillian Gish, Mary Pickford, Blanche Sweet, Charles West; *Dir:* D.W. Griffith. VHS, Beta $19.95 KRT, FCT ⫙⫙½

The Dybbuk A man's bride is possessed by a restless spirit. Set in the Polish-Jewish community before WWI and based on the play by Sholom Anski. Considered a classic for its portrayal of Jewish religious and cultural mores. In Yiddish with English subtitles.
1937 123m/B PL Abraham Morewski, Isaac Samberg, Moshe Lipman, Lili Liliana, Dina Halpern; *Dir:* Michal Waszynski. VHS, Beta $39.95 IHF, FCT, NCJ ⫙⫙⫙½

Dying Room Only Travelling through the desert, a woman's husband suddenly disappears after they stop at a secluded roadside diner. Richard Matheson script is a spooker. Made for television.
1973 74m/C Cloris Leachman, Ross Martin, Ned Beatty, Louise Latham, Dana Elcar, Dabney Coleman; *Dir:* Philip Leacock. VHS $49.95 IVE ⫙⫙½

The Dying Truth A wrongfully accused prisoner lies dying in his cell. Trouble is, he can't die in peace until he solves the murder they say he committed. Interesting mix of the supernatural and mystery, although perhaps not one of Carradine's more memorable roles.
1991 80m/C David Carradine, Stephanie Beacham, Stephen Grief, Stephan Chase, Larry Carby, Lesley Dunlop; *Dir:* John Hough. VHS $79.95 IMD ⫙

Dying Young Muted romance with a wealthy leukemia victim hiring a spirited, unschooled beauty as his nurse. They fall in love, but the film is either too timid or too unimaginative to mine emotions denoted by the title. Nobody dies, in fact; a grim ending (faithful to the Marti Leimbach novel on which this was based) got scrapped after testing poorly with audiences. The actors try their best, photography is lovely, and Kenny G's mellow music fills the soundtrack.
1991 (R) 111m/C Julia Roberts, Campbell Scott, Vincent D'Onofrio, Colleen Dewhurst, Ellen Burstyn, David Selby; *Dir:* Joel Schumacher. VHS $19.98 FXV ⫙⫙

Dynamite Romance explodes as two young demolitions experts vie for the same girl. Stand back.
1949 68m/B William Gargan, Virginia Welles, Richard Crane, Irving Bacon; *Dir:* William H. Pine. VHS, Beta $24.95 NOS, DVT ⫙

Dynamite Chicken Melange of skits, songs, and hippie satire is dated. But includes performances by Joan Baez, Lenny Bruce, B. B. King, and others.
1970 75m/C Joan Baez, Richard Pryor, Lenny Bruce, Jimi Hendrix, Sha-Na-Na; *Dir:* Ernie Pintoff. VHS, Beta MON, OM ⫙⫙

Dynamite and Gold Nelson and crew embark on a search for lost gold, battling hostile Indians, the Mexican army and assorted bandits along the way. A mediocre televi-

sion movie originally titled "Where the Hell's the Gold?"
1988 (PG) 91m/C Willie Nelson, Delta Burke, Jack Elam, Alfonso Arau, Gregory Sierra, Michael Wren, Gerald McRaney; *W/Dir:* Burt Kennedy. VHS $89.95 ACA ⫙⫙

Dynamite Pass Offbeat oater about disgruntled ranchers attempting to stop the construction of a new road.
1950 61m/B Tim Holt, Richard Martin, Regis Toomey, Mary Hart, Denver Pyle; *Dir:* Lew Landers. VHS, Beta MED ⫙⫙

Dynamite Ranch Ranchers fight tooth and nail to keep their rights and property in this Maynard oater.
1932 60m/B Ken Maynard. VHS, Beta $19.95 NOS, UHV, DVT ⫙

Dynamo A glamorous advertising executive pursues a kung fu expert to sign a contract with her agency for a kick boxing contest.
1980 (R) 81m/C VHS, Beta $9.98 SUE ⫙

Dynasty A leader of the Ming Dynasty is killed by evil Imperial Court eunuch and his son the monk sets out to avenge him. In 3-D. Run-of-the-mill kung-fu stunts.
1977 94m/C Bobby Ming, Lin Tashing, Pai Ying, Tang Wei, Jin Gang; *Dir:* Zhang Meijun. VHS, Beta $29.95 BFV ⫙⫙

Dynasty Pulitzer Prize-winning author James Michener creates the usual sweeping saga of a family torn by jealousy, deception and rivalry in love and business as husband, wife and brother-in-law seek their fortune in the Ohio frontier of the 1820s.
1982 90m/C Sarah Miles, Harris Yulin, Stacy Keach. VHS, Beta TLF, LME ⫙⫙

Dynasty of Fear Matters get rather sticky at a British boys' school when the headmaster's wife seduces her husband's assistant. Together they conspire to murder her husband and share his fortune. Originally titled, and also on video as, "Fear in the Night" and "Honeymoon of Fear."
1972 (PG) 93m/C GB Peter Cushing, Joan Collins, Ralph Bates, Judy Geeson, James Cossins; *Dir:* Jimmy Sangster. VHS, Beta $19.95 ACA, HBO ⫙½

E. Nick: A Legend in His Own Mind Satire of videos and magazines designed for adults.
1984 75m/C Don Calfa, Cleavon Little, Andra Akers, Pat McCormick; *Dir:* Robert Hegyes. VHS, Beta $39.95 IVE ⫙½

Each Dawn I Die Cagney stars as a reporter who is a ferverent critic of the political system. Framed for murder and imprisoned, is subsequently befriended by fellow inmate, Raft, a gangster. Once hardened by prison life, Cagney shuns his friend and becomes wary of the system. Despite its farfetched second half and mediocre script, the film makes interesting viewing thanks to a stellar performance from Cagney.
1939 92m/B James Cagney, George Raft, George Bancroft, Jane Bryan, Maxie Rosenbloom, Alan Baxter, Thurston Hall, Stanley Ridges, Victor Jory; *Dir:* William Keighley. VHS, Beta $19.95 MGM ⫙⫙½

The Eagle In this tale of a young Cossack "Robin Hood," Valentino assumes the personna of the Eagle to avenge his father's murder. The romantic rogue encounters trouble when he falls for the beautiful Banky much to the chagrin of the scorned Czarina Dresser. Fine performances from Valentino and Dresser. Silent, based on a Alexander

Pushkin story. Released on video with a new score by Carl Davis.
1925 77m/B Rudolph Valentino, Vilma Banky, Louise Dressler, George Nichols, James Marcus; *Dir:* Clarence Brown. **VHS, Beta, LV** $19.95 NOS, HBO, VYY 𝄞𝄞𝄞

Eagle Claws Champion A ninja master accidentally kills a friend, goes to prison, gets released and returns to fight again.
198? 90m/C Sho Kosugi. **VHS, Beta** $29.98 TWE 𝄞

The Eagle Has Landed Duvall, portraying a Nazi colonel in this World War II spy film, commissions Sutherland's Irish, English-hating character to aid him in his mission to kill Prime Minister Winston Churchill. Adapted from the bestselling novel by Jade Higgins.
1977 (PG) 123m/C Michael Caine, Donald Sutherland, Robert Duvall, Larry Hagman, Jenny Agutter, Donald Pleasence, Treat Williams, Anthony Quayle; *Dir:* John Sturges. **VHS, Beta** $19.98 FOX 𝄞𝄞𝄞

Eagles Attack at Dawn After escaping from an Arab prison, an Israeli soldier vows to return with a small commando force to kill the sadistic warden.
1974 (PG) 96m/C IS Rick Jason, Peter Brown, Joseph Shiloal. **VHS, Beta** $39.98 VID 𝄞½

Eagle's Shadow After a poor orphan boy rescues an aged beggar, the grateful old man tutors the lad in Snake-Fist techniques.
1984 101m/C Jackie Chan, Juan Jan Lee, Simon Yuen, Roy Horan; *Dir:* Yuen Woo Ping. **VHS, Beta** WVE 𝄞

Eagle's Wing John Briley, screenwriter of the award-winning epic ''Gandhi,'' attempts to weave the threads of allegory smoothly in this white man vs. red man Western from England. The mediocre story finds Native American Waterston dueling white trapper Sheen in a quest to capture an elusive, exotic white stallion.
1979 111m/C Martin Sheen, Sam Waterston, Harvey Keitel, Stephane Audran, Caroline Langrishe; *Dir:* Anthony Harvey. **VHS, Beta** $19.95 MED 𝄞½

Early Cinema Vol. 1 A compendium of early silent shorts, mostly predating 1920, which includes: Melies' ''Conquest of the North Pole''; Linder's ''Man and the Statue'' and ''Max's Mother In Law''; Drew's ''Auntie's Portrait'' Normand's ''Mabel's Dramatic Career''; Mix's ''An Angelic Attitude'' Sennett and Swain's ''Cowboy Ambrose''; Lloyd's ''Just Neighbors'' and Parrott's ''Take the Air.''
1919 118m/B Mabel Normand, Harold Lloyd, Sidney Drew, Max Linder, Paul Parrott, Tom Mix, Mack Swain; *Dir:* Mabel Normand, Harold Lloyd, Max Linder, Georges Melies, Mack Sennett. **VHS, Beta** FCT

Early Days A cantankerous, salty, once-powerful politician awaits death.
1981 67m/C Ralph Richardson. **VHS, Beta** $49.98 FOX 𝄞𝄞

Early Elvis Clips of a young Elvis as he appears on ''Stage Show'' with the Tommy and Jimmy Dorsey Orchestra, on the ''Steve Allen Show'' in a comedy sketch, and on the ''Ed Sullivan Show.'' The popular rocker performs ten songs in all.
1956 56m/B Elvis Presley, Ed Sullivan, Charles Laughton. **VHS, Beta** $24.95 VYY

Early Films #1 A collection of silent movies from the very early days of the industry, including: ''The Great Train Robbery,'' ''1895 Lumiere Films,'' ''Dreams of a

Rarebit Fiend,'' ''Trip to the Moon,'' and ''Life of an American Fireman.''
1903 45m/B *Dir:* The Lumiere Brothers, Georges Melies, Edwin S. Porter. **VHS, Beta** WFV 𝄞𝄞

Early Frost A suspenseful whodunit centering around a simple divorce investigation that leads to the discovery of a corpse.
1984 95m/C Diana McLean, Jon Blake, Janet Kingsbury, David Franklin; *Dir:* Terry O'Connor. **VHS, Beta, LV** $29.95 VCL 𝄞½

An Early Frost Highly praised, surprisingly intelligent made-for-television drama following the anguish of a successful lawyer who tells his closed-minded family that he is gay and is dying of AIDS. Sensitive performance by Quinn in one of the first TV films to focus on the devastating effects of HIV. Rowlands also adeptly displays her acting talents as the despairing mother.
1985 97m/C Aidan Quinn, Gena Rowlands, Ben Gazzara, John Glover, D.W. Noffett, Sylvia Sydney; *Dir:* John Erman. **VHS, Beta, LV** $69.95 COL 𝄞𝄞𝄞½

Early Sound Films This series of experimental shorts from throughout the silent era depicts the various methods of devising sound film, prior to ''The Jazz Singer'' in 1927.
1928 90m/B **VHS, Beta** $59.50 GVV

Early Summer (Bakushu) A classic from renowned Japanese director Ozu, this film chronicles family tensions in post-World War II Japan caused by newly-independent women rebelling against the social conventions they are expected to fulfill. Perhaps the best example of this director's work. Winner of Japan's Film of the Year Award. In Japanese with titles.
1951 150m/B JP Ichiro Sugai, Setsuko Hara, Chikage Awashima; *Dir:* Yasujiro Ozu. **VHS, Beta** $59.95 SVS, FCT 𝄞𝄞𝄞

Early Warner Brothers Cartoons Eight cartoons from the first days of the Merrie Melodies. Includes ''Sinking in the Bathtub,'' ''Lady Play Your Mandolin,'' ''One More Time,'' ''Freddie the Freshman,'' ''You Don't Know What You're Doin','' ''Red Headed Baby,'' ''I Love a Parade,'' and ''The Shanty Where Santy Lives.''
1931 55m/B **VHS, Beta** $19.95 VDM

The Earrings of Madame De... A simple story about a society woman who sells a pair of diamond earrings that her husband gave her, then lies about it. Transformed by Ophuls into his most opulent, overwrought masterpiece, the film displays a triumph of form over content. Also titled ''Madame de'' or ''The Diamond Earrings.'' In French with English subtitles.
1954 105m/C FR Charles Boyer, Danielle Darrieux, Vittorio De Sica, Lea di Lea, Jean Debucourt; *Dir:* Max Ophuls. **VHS, Beta** $69.95 FCT, MRV 𝄞𝄞𝄞𝄞

Earth Classic Russian silent film with English subtitles. Problems begin in a Ukrainian village when a collective farm landowner resists handing over his land. Outstanding camera work. Kino release runs 70 minutes.
1930 57m/B RU Semyon Svashenko, Mikola Nademsy, Stephan Shkurat; *Dir:* Alexander Dovzhenko. National Board of Review Awards '30: 5 Best Foreign Film of the Year; Brussels World's Fair '58: #10 of the Best Films of All Time. **VHS, Beta, 3/4U** $24.95 VYY, CCB, IHF 𝄞𝄞𝄞𝄞

Earth Entranced (Terra em Transe) A complex political lamentation from Rocha, the premiere director of Brazil's own new wave cinema. Here a writer switches his allegiance from one politician to another, only to find that he (and the masses) lose either way. In Portuguese with English subtitles.
1966 105m/C BR Jose Lewgoy, Paulo Gracindo, Jardel Filho, Glauce Rocha; *W/Dir:* Glauber Rocha. Cannes Film Festival '66: International Critics Award (Terra Em Transe). **VHS** $59.95 FST 𝄞𝄞

Earth Girls are Easy Valley girl Valerie is having a bad week: first she catches her fiancee with another woman, then she breaks a nail, then furry aliens land in her swimming pool. What more could go wrong? When the aliens are temporarily stranded, she decides to make amends by giving them a head-to-toe makeover. Devoid of their excessive hairiness, the handsome trio of fun-loving extraterrestrials set out to experience the Southern California lifestyle. A sometimes hilarious sci-fi/musical, featuring bouncy shtick and a gleeful dismantling of modern culture.
1989 (PG) 100m/C Geena Davis, Jeff Goldblum, Charles Rocket, Julie Brown, Jim Carrey, Damon Wayans, Michael McKean, Angelyne, Larry Linville; *Dir:* Julien Temple. **VHS, Beta, LV** $14.98 LIV, VES 𝄞𝄞½

Earth vs. the Flying Saucers Extraterrestrials land on Earth and issue an ultimatum to humans concerning their constant use of bombs and missiles. Peace is threatened when the military disregards the extraterrestrials' simple warning. Superb special effects by Ray Harryhausen; written by Curt Siodmak.
1956 83m/B Hugh Marlowe, Joan Taylor, Donald Curtis, Morris Ankrum; *Dir:* Fred F. Sears. **VHS, Beta, LV** $9.95 COL, MLB 𝄞𝄞½

Earth vs. the Spider Man-eating giant mutant (teenage ninja?) tarantula makes life miserable for a small town in general and high school partyers in particular. Silly old drive-in fare is agony for many, camp treasure for a precious few.
1958 72m/B Edward Kemmer, June Kennedy, Gene Persson, Gene Roth, Hal Tory, Mickey Finn; *Dir:* Bert I. Gordon. **VHS** $29.95 COL, MLB 𝄞

Earthling A terminally ill Holden helps a young Schroder survive in the Australian wilderness after the boy's parents are killed in a tragic accident. Lessons of life and the power of the human heart are passed on in this panoramic, yet mildly sentimental film.
1980 (PG) 102m/C AU William Holden, Rick Schroder, Jack Thompson, Olivia Hamnett, Alwyn Kurtis; *Dir:* Peter Collinson. **VHS, Beta** $69.98 LIV, VES 𝄞𝄞½

Earthquake This mediocre drama centers on a major earthquake in Los Angeles and its effect on the lives of an engineer, his spoiled wife, his mistress, his father-in-law and a suspended policeman. Filmed in Sensurround—a technique intended to shake up the theater a bit, but one which will likely have little effect in home television viewing rooms. Good special effects, but not enough to compensate for lackluster script.
1974 (PG) 129m/C Charlton Heston, Ava Gardner, George Kennedy, Lorne Greene, Genevieve Bujold, Richard Roundtree, Marjoe Gortner, Barry Sullivan, Victoria Principal, Lloyd Nolan, Walter Matthau; *Dir:* Mark Robson. Academy Awards '74: Best Sound, Best Visual Effects. **VHS, Beta, LV** $19.95 MCA 𝄞

Earthworm Tractors Ambitious tractor salesman Alexander Botts will do anything to make a sale. Brown excels as the fast-talking lead. Comedy is strengthened by its supporting cast. Based on characters first appearing in the Saturday Evening Post.
1936 63m/B Joe E. Brown, June Travis, Guy Kibbee, Dick Foran, Carol Hughes, Gene Lockhart, Olin Howland; *Dir:* Ray Enright. **VHS** $19.95 *NOS, RXM* ⚜⚜½

East of Borneo A tropical adventure involving a "lost" physician whose worried wife sets out to find him in the jungle. She locates her love only to discover he has a prestigious new job tending to royalty and didn't want to be found.
1931 76m/B Charles Bickford, Rose Hobart, Georges Renavent, Noble Johnson; *Dir:* George Melford. **VHS, Beta** $16.95 *SNC, DVT, CAB* ⚜½

East of Eden Steinbeck's contemporary retelling of the biblical Cain and Abel myth receives superior treatment from Kazan and his excellent cast. Dean, in his first starring role, gives a reading of a young man's search for love and acceptance that defines adolescent pain. Though filmed in the 1950s, this story still rivets today's viewers with its emotional message.
1955 115m/C James Dean, Julie Harris, Richard Davalos, Raymond Massey, Jo Van Fleet, Burl Ives, Albert Dekker; *Dir:* Elia Kazan. Academy Awards '55: Best Supporting Actress (Van Fleet); Golden Globe Awards '56: Best Film—Drama; National Board of Review Awards '55: 10 Best Films of the Year. **VHS, Beta** $19.98 *WAR, BTV* ⚜⚜⚜⚜

East of Eden Remade into a television mini-series, this Steinbeck classic tells the tale of two brothers who vie for their father's affection and the woman who comes between them. Rife with biblical symbolism and allusions.
1980 375m/C Jane Seymour, Bruce Boxleitner, Warren Oates, Lloyd Bridges, Anne Baxter, Timothy Bottoms, Soon Tech-Oh, Karen Allen, Hart Bochner, Sam Bottoms, Howard Duff, Richard Masur, Wendell Burton, Nicholas Pryor, Grace Zabriskie, M. Emmet Walsh, Matthew "Stymie" Beard; *Dir:* Harvey Hart. **VHS, Beta, LV** $29.95 *STE, IVE* ⚜⚜½

East of Elephant Rock In 1948, a young first secretary of the British Embassy returns from leave in England to a tense atmosphere in a colony in southeast Asia. Beautiful scenery can't make up for weak plot.
1976 93m/C *GB* John Hurt, Jeremy Kemp, Judi Bowker, Christopher Cazenove; *Dir:* Don Boyd. **VHS, Beta** $59.98 *SUE* ⚜

East End Hustle A high-priced call girl rebels against her pimp and sets out to free other prostitutes as well. Also available in an 88-minute unrated version.
1976 (R) 86m/C **VHS, Beta** $69.98 *LIV, VES* ⚜

East of Kilimanjaro A freelance photographer and a doctor search frantically to find the carrier of a fatal disease near the slopes of the majestic Mt. Kilimanjaro. Routine killer virus flick filmed in Africa. Also know as "The Big Search."
1959 75m/C Marshall Thompson, Gaby Andre, Fausto Tozzi; *Dir:* Arnold Belgard. **VHS** *SUE* ⚜⚜

East L.A. Warriors A mobster, Hilton-Jacobs, manipulates Los Angeles gangs in this non-stop action adventure.
1989 90m/C Tony Bravo, Lawrence Hilton-Jacobs, Kamar Reyes, William Smith; *Dir:* Addison Randall. **VHS, Beta** $69.95 *PAR* ⚜½

East of Ninevah A teleplay by Dr. Jim Peyton about four Kentucky men who try to cleanse themselves of past sins.
1983 90m/C VHS, Beta *KET* ⚜

East Side Kids Early East Side kids. A hoodlum wants to prevent his brother from beginning a similar life of crime.
1940 60m/B Leon Ames, Harris Berger, Dennis Moore, Joyce Bryant; *Dir:* Robert F. Hill. **VHS, Beta** $19.95 *NOS, MRV, DVT* ⚜⚜

East Side Kids/The Lost City This first East Side Kids film, entitled "East Side Kids," introduces some soon familiar faces. Coupled with that is "The Lost City," a composite of all twelve episodes of the serial; edited for brevity.
1935 135m/B Dennis Moore, Vince Barnett, William Boyd, Kane Richmond. **VHS, Beta** *AOV* ⚜⚜

East and West Morris Brown, a worldly New York Jew, returns home to Galicia with his daughter Mollie for a traditional family wedding. She teaches the young villagers to dance and box but meets her romantic match in a young yeshiva scholar who forsakes tradition to win her heart. Picon shines as the exuberant flapper. With English and Yiddish intertitles.
1923 85m/B Molly Picon, Jacob Kalish, Sidney Goldin; *Dir:* Sidney Goldin. **VHS** $72.00 *NCJ* ⚜⚜⚜

Easter Parade A big musical star (Astaire) splits with his partner (Miller) claiming that he could mold any girl to replace her in the act. He tries and finally succeds after much difficulty. Astaire and Garland in peak form, aided by a classic Irving Berlin score.
1948 103m/C Fred Astaire, Judy Garland, Peter Lawford, Ann Miller, Jules Munshin; *Dir:* Charles Walters. Academy Awards '48: Best Musical Score. **VHS, Beta, LV** $19.95 *MGM* ⚜⚜⚜½

Easy Come, Easy Go Elvis, as a Navy frogman, gets excited when he accidentally discovers what he believes is a vast sunken treasure. Music is his only solace when he finds his treasure to be worthless copper coins.
1967 96m/C Elvis Presley, Dodie Marshall, Pat Priest, Elsa Lanchester, Frank McHugh, Pat Harrington; *Dir:* John Rich. **VHS, Beta** $14.95 *PAR* ⚜½

Easy Kill A man takes the law into his own hands to exact revenge on drug lords.
1989 100m/C Jane Badler, Cameron Mitchell, Frank Stallone; *Dir:* Josh Spencer. **VHS, Beta** $79.95 *TRY* ⚜½

The Easy Life Haunting film about a hedonistic man from a small Italian village who takes a mild-mannered student pleasure-seeking. The ride turns sour and the playboy is ultimately destroyed by his own brutality.
1963 105m/B *IT* Vittorio Gassman, Catherine Spaak, Jean-Louis Trintignant, Luciana Angiolillo; *Dir:* Dino Risi. **VHS** $59.95 *EMB, FST* ⚜⚜⚜

Easy Living Compromised melodrama about an over-the-hill football player who must cope with his failing marriage and looming retirement. Based on an Irwin Shaw story.
1949 77m/B Victor Mature, Lucille Ball, Jack Paar, Lizabeth Scott, Sonny Tufts, Lloyd Nolan, Paul Stewart; *Dir:* Jacques Tourneur. **VHS, Beta, LV** $19.95 *MED* ⚜⚜½

Easy to Love Williams is in love with her boss, but he pays her no attention until a handsome singer vies for her affections. Set at Florida's Cypress Gardens, this aquatic

musical features spectacular water ballet productions choreographed by Busby Berkeley, and several songs, including "Easy to Love," "Coquette," and "Beautiful Spring."
1953 96m/C Esther Williams, Van Johnson, Tony Martin, John Bromfield, King Donovan, Carroll Baker; *Dir:* Charles Walters. **VHS** *MGM* ⚜⚜½

Easy Money A basic slob has the chance to inherit $10 million if he can give up his loves: smoking, drinking, and gambling among others. Dangerfield is surprisingly restrained in this harmless, though not altogether unpleasing comedy.
1983 (R) 95m/C Rodney Dangerfield, Joe Pesci, Geraldine Fitzgerald, Jennifer Jason Leigh, Tom Ewell, Candice Azzara, Taylor Negron; *Dir:* James Signorelli. **VHS, Beta, LV** $19.98 *VES* ⚜⚜

Easy Rider Slim-budget, generation-defining movie. Two young men in late 1960s undertake a motorcycle trek throughout the Southwest in search of the real essence of America. Along the way they encounter hippies, rednecks, prostitutes, drugs, Jack Nicholson, and tragedy. One of the highest grossing pictures of the decade, this movie undoubtedly influenced two generations of "youth-oriented dramas," which all tried unsuccessfully, to duplicate the original accomplishment. Some psychedelic scenes and a great role for Nicholson are a added bonus. Look for the graveyard dancing scene in New Orleans. Features one of the best 60s rock scores around, in a league that includes "Mean Streets" and "The Wanderers." Oscar nomination for Nicholson.
1969 (R) 94m/C Peter Fonda, Dennis Hopper, Jack Nicholson, Karen Black, Toni Basil, Robert Walker Jr.; *Dir:* Dennis Hopper. **VHS, Beta, LV, 8mm** $19.95 *COL, CCB* ⚜⚜⚜½

Easy Street Chaplin portrays a derelict who, using some hilarious methods, reforms the residents of Easy Street. Chaplin's row with the town bully is particularly amusing. Silent with musical soundtrack added.
1916 20m/B Charlie Chaplin; *Dir:* Charlie Chaplin. **VHS, Beta** *CAB, FST* ⚜⚜

Easy Virtue Master of suspense Hitchcock directs this adaptation of a Noel Coward play in which the wife of an alcoholic falls for another man—a lover with suicidal tendencies.
1927 79m/B *GB* Isabel Jean, Ian Hunter, Frank Dyall, Eric Bransby Williams, Robin Irvine, Violet Farebrother; *Dir:* Alfred Hitchcock. **VHS, Beta** $19.95 *NOS, MRV, GPV* ⚜⚜½

Easy Wheels Like most decent biker movies, "Wheels" is propelled by bad taste and a healthy dose of existentialist nihilism. But there's an unusual plot twist in this parody. A biker named She-Wolf and her gang kidnap female babies and let wolves rear the children. Their elaborate plan is to create a race of super women who will subdue the troublesome male population. But can this "noble" plan succeed?
1989 94m/C Paul LeMat, Eileen Davidson, Marjorie Bransfield, Jon Menick, Mark Holton, Karen Russell, Jami Richards, Roberta Vasquez, Barry Livingston, George Plimpton; *Dir:* David O'Malley. **VHS, Beta, LV** $29.95 *FRH* ⚜⚜½

Eat a Bowl of Tea Endearing light drama-comedy concerning a multi-generational Chinese family who must learn to deal with the problems of life in America and in particular, marriage, when Chinese women are finally allowed to immigrate with their husbands to the United States following World War II. Adaptation of Louis Chu's story directed by the man who brought us

"Dim Sum." This film enjoyed a particularly favorable response in art theaters.
1989 (PG-13) 102m/C Cora Miao, Russell Wong, Lau Siu Ming, Eric Tsang Chi Wai, Victor Wong, Jessica Harper; *Dir:* Wayne Wang. **VHS, Beta, LV $19.98** *COL* ⅋⅋⅋

Eat My Dust The teenage son of a California sheriff steals the best of stock cars from a race track to take the town's heart-throb for a joy ride. Subsequently he leads the town on the wildest car chase ever filmed. Brainless but fast-paced.
1976 (PG) 89m/C Ron Howard, Christopher Norris, Warren Kemmerling, Rance Howard, Clint Howard, Corbin Bernsen; *Dir:* Charles B. Griffith. **VHS, Beta $9.98** *SUE* ⅋½

Eat the Peach An idiosyncratic Irish comedy about two young unemployed rebels who attempt to break free from their tiny coast town by becoming motorcycle champs after seeing the Elvis Presley film "Roustabout." Together they create the "wall of death,"—a large wooden barrel in which they can perform various biker feats.
1986 90m/C IR Stephen Brennan, Eamon Morrissey, Catherine Byrne, Niall Toibin; *Dir:* Peter Ormrod. **VHS, Beta $79.98** *FOX* ⅋⅋½

Eat the Rich A British farce about a group of terrorists who take over the popular London restaurant, Bastard's. Led by a former, disgruntled transvestite employee, they turn diners into menu offerings. Paul McCartney and Koo Stark appear in cameos; music by Motorhead.
1987 (R) 89m/C GB Nosher Powell, Lanah Pellay, Paul McCartney, Bill Wyman, Koo Stark, Sandra Dorne; *Dir:* Peter Richardson. **VHS, Beta $89.95** *COL* ⅋

Eat and Run A bloody comedy about a 400-pound alien who ravages New York consuming Italians.
1986 (R) 85m/C Ron Silver, R.L. Ryan, Sharon Schlarth; *Dir:* Christopher Hart. **VHS, Beta $19.95** *STE, NWV* ⅋

Eaten Alive A Southerner takes an unsuspecting group of tourists into a crocodile death trap. Director Tobe Hooper's follow-up to "The Texas Chainsaw Massacre." Englund is more recognizable with razor fingernails as Freddy Krueger of the "Nightmare on Elm Street" films.
1976 (R) 96m/C Neville Brand, Mel Ferrer, Carolyn Jones, Marilyn Burns, Stuart Whitman, Robert Englund; *Dir:* Tobe Hooper. **VHS, Beta $59.95** *PSM* ⅋⅋

Eating Raoul The Blands are a happily married couple who share many interests: good food and wine, entrepreneurial dreams, and an aversion to sex. The problem is, they're flat broke. So, when the tasty swinger from upstairs makes a pass at Mary and Paul accidentally kills him, they discover he's got loads of money; Raoul takes a cut in the deal by disposing—or rather recycling—the bodies. This may just be the way to finance that restaurant they've been wanting to open. Wonderful, offbeat, hilarious comedy.
1982 (R) 83m/C Mary Woronov, Paul Bartel, Buck Henry, Ed Begley Jr., Edie McClurg, Robert Beltran, John Paragon; *Dir:* Mary Woronov, Paul Bartel. **VHS, Beta $59.98** *FOX* ⅋⅋⅋½

Ebony Dreams World War II documentary focuses on the experience of female factory workers during and after the war. Five contemporary women relate their personal endeavors and struggles to retain their jobs once the war ended. Excellent use of newsreel footage and interesting interviews with the women.
1980 85m/C Philomena Nowlin; *Dir:* Bill Brame. **VHS $19.95** *LPC* ⅋⅋⅋½

Ebony Tower Based on the John Fowles novel about a crusty old artist, who lives in a French chateau with two young female companions. They are visited by a handsome young man, thereby initiating sexual tension and recognizably Fowlesian plot puzzles. Made for British television. Features some partial nudity.
1986 80m/C GB Laurence Olivier, Roger Rees, Greta Scacchi, Toyah Willcox; *Dir:* Robert Knights. **Beta $79.95** *VHV* ⅋⅋½

Echo Murders A Sexton Blake mystery wherein he investigates a mine owner's mysterious death, opening up a veritable can of murdering, power-hungry worms.
1945 75m/B David Farrar, Dennis Price. **VHS, Beta $19.95** *NOS, VYY* ⅋½

Echo Park An unsung sleeper comedy about three roommates living in Los Angeles' Echo Park: a body builder, a single-mother waitress and an itinerant songwriter. Charts their struggles as they aim for careers in showbiz. Offbeat ensemble effort.
1986 (R) 93m/C AU Tom Hulce, Susan Dey, Michael Bowen, Cheech Marin, Christopher Walker, Shirley Jo Finney, Cassandra Peterson, Yana Nirvana, Timothy Carey; *Dir:* Robert Dornhelm. **VHS, Beta, LV $19.98** *KRT* ⅋⅋½

Echoes A young painter's life slowly comes apart as he is tormented by nightmares that his stillborn twin brother is attempting to murder him.
1983 (R) 90m/C Gale Sondergaard, Mercedes McCambridge, Richard Alferi, Ruth Roman, John Spencer, Nathalie Nell; *Dir:* Arthur Seidelman. **VHS, Beta $19.98** *VID* ⅋

Echoes in the Darkness Television mini-series based on the true-life Joseph Wambaugh bestseller about the "Main Line" murder investigation in Pennsylvania in 1979. Police are baffled when a teacher is murdered and her children vanish. Suspicion falls on a charismatic coworker and the school's principal. On two tapes.
1987 234m/C Peter Coyote, Robert Loggia, Stockard Channing, Peter Boyle, Cindy Pickett, Gary Cole, Zeljko Ivanek, Alex Hyde-White, Treat Williams; *Dir:* Glenn Jordan. **VHS $29.95** *STE, NWV* ⅋⅋½

Echoes of Paradise After a series of earth-shattering events, a depressed woman journeys to an island where she falls for a Balinese dancer. Soon she must choose between returning home or beginning life anew in paradise.
1986 (R) 90m/C AU Wendy Hughes, John Lone, Rod Mullinar, Peta Toppano, Steven Jacobs, Gillian Jones; *Dir:* Phillip Noyce. **VHS, Beta $14.95** *ACA* ⅋½

The Eclipse The last of Antonioni's trilogy (after "L'Avventura" and "La Notte"), wherein another fashionable and alienated Italian woman passes from one lover to another searching unsuccessfully for truth and love. Highly acclaimed. In Italian with subtitles.
1966 123m/B IT Monica Vitti, Alain Delon, Francisco Rabal, Louis Seigner; *Dir:* Michelangelo Antonioni. Cannes Film Festival '62: Special Jury Prize. **VHS, Beta $29.95** *TAM, MRV, FCT* ⅋⅋⅋½

Ecstasy A romantic, erotic story about a young woman married to an older man, who takes a lover. Features Lamarr, then Hedy Kiesler, before her discovery in Hollywood. Film subsequently gained notoriety for Lamarr's nude scenes. Original title: "Extase." Subtitled in English.
1933 90m/B CZ Hedy Lamarr, Jaromir Rogoz; *Dir:* Gustav Machaty. **VHS, Beta, 8mm $29.95** *HTV, NOS, MRV* ⅋⅋½

Ecstasy Soft core fluff about a film director's wife's erotic adventures.
1984 (R) 82m/C Tiffany Bolling, Franc Luz, Jack Carter, Britt Ekland, Julie Newmar. **VHS, Beta $19.95** *MGM Woof!*

Ecstasy Inc. A husband-wife sex therapist team kill their patients.
1988 86m/C Gabriel River, Christina Andress. **VHS, Beta $29.95** *MED* ⅋

Eddie & the Cruisers In the early 1960s, rockers Eddie and the Cruisers score with one hit album. Amid their success, lead singer Pare dies mysteriously in a car accident. Years later, a reporter decides to write a feature on the defunct band, prompting a former band member to begin a search for missing tapes of the Cruisers' unreleased second album. Questions posed at the end of the movie are answered in the sequel. Enjoyable soundtrack by John Cafferty and the Beaver Brown Band.
1983 (PG) 90m/C Tom Berenger, Michael Pare, Ellen Barkin, Joe Pantoliano, Matthew Laurance; *Dir:* Martin Davidson. **VHS, Beta, LV, 8mm $14.95** *MVD, SUE* ⅋⅋

Eddie & the Cruisers 2: Eddie Lives! A sequel to the minor cult favorite, in which a rock star believed to be dead emerges under a new name in Montreal to lead a new band. The Beaver Brown Band again provides the music.
1989 (PG-13) 106m/C CA Michael Pare, Marina Orsini, Matthew Laurance, Bernie Coulson; *Dir:* Jean-Claude Lord. **VHS, Beta, LV $9.99** *CCB, IVE, IME* ⅋½

Eddie Macon's Run Based on a true story, Eddie Macon has been unjustly jailed in Texas and plans an escape to run to Mexico. He is followed by a tough cop who is determined to catch the fugitive. Film debut of Goodman.
1983 (PG) 95m/C Kirk Douglas, John Schneider, Lee Purcell, John Goodman, Lisa Ayres; *Dir:* Jeff Kanew. **VHS, Beta, LV $69.95** *MCA* ⅋⅋

Eddie Murphy "Delirious" Murphy raps about life, sex, childhood etc. in this scathing, scatalogical, stand-up performance. Strong language.
1983 (R) 69m/C Eddie Murphy, Richard Tienken, Robert Wachs; *Dir:* Bruce Gowers. **VHS, Beta, LV $14.95** *PAR* ⅋⅋½

Eddie Murphy: Raw Murphy's second stand-up performance film, featuring many of his woman-bashing observations on life.
1987 (R) 90m/C Eddie Murphy; *Dir:* Robert Townsend. **VHS, Beta, LV, 8mm $14.95** *PAR* ⅋⅋

Eddie Rickenbacker: Ace of Aces The life of Eddie Rickenbacker, America's ace of flying aces, who spent twenty-four days adrift on a raft until rescued, is shown through classic newsreel footage.
1954 15m/B *VHS, Beta $49.95* *TSF* ⅋⅋

The Eddy Duchin Story Glossy tearjerker that profiles the tragic life of Eddy Duchin, the famous pianist/bandleader of the 30's and 40's. Features almost 30 songs, including classics by Cole Porter, George and Ira Gershwin, Hammerstein, Chopin and several others.
1956 123m/C Tyrone Power Sr., Kim Novak, Victoria Shaw, James Whitmore, Rex Thompson; *Dir:* George Sidney. **VHS $19.95** *COL* ⅋⅋

Edgar Kennedy Slow Burn Festival Popular 30s comedian Edgar Kennedy stars in three shorts from his long-running "Average Man" series: "Poisoned Ivory" (1934), "Edgar Hamlet" (1935), and "A Clean Sweep" (1938).
1938 59m/B Edgar Kennedy, Florence Lake, Dot Farley, Jack Rice, Vivien Oakland, Tiny Sandford. VHS, Beta $24.95 VYY ♂♂

Edge of the Axe A small-town is held in the grip of terror by a demented slasher.
1989 91m/C Barton Faulks, Christina Marie Lane, Page Moseley, Fred Hollyday; *Dir:* Joseph Braunstein. VHS, Beta $79.98 MCG ♂♂

Edge of Darkness Originally produced as a mini-series for British television, this mystery involves a police detective who investigates his daughter's murder and uncovers a web of espionage and intrigue.
1986 307m/C GB Joe Don Baker, Bob Peck, Jack Woodson, John Woodvine, Joanne Whalley-Kilmer; *Dir:* Martin Campbell. VHS, Beta $79.98 FOX ♂♂

Edge of Fury When a businessman is unjustly arrested on a drug charge, he enlists the help of his chauffeur to clear his name.
198? 90m/C Bruce Li, Andrew Sage, Michael Danna, Tommy Lee. VHS, Beta $39.95 KOV ♂

Edge of Honor Young Eagle Scouts camping in the Pacific Northwest discover a woodland weapons cache and wage guerilla war against killer lumberjacks out to silence them. A boneheaded, politically correct action bloodbath; you don't have to like the logging industry to disapprove of the broad slurs against its men shown here.
1991 (R) 92m/C Corey Feldman, Meredith Salenger, Scott Reeves, Ken Jenkins, Christopher Neame, Don Swayze; *Dir:* Michael Spence. VHS $89.95 ACA ♂ 1/2

Edge of Sanity An overdone Jekyll-Hyde reprise, with cocaine serving as the villainous substance. The coke-created Hyde is posited as a prototype for Jack the Ripper. Not for the easily queasy. Available in a 90 minute, unrated version.
1989 (R) 85m/C GB Anthony Perkins, Glynis Barber, David Lodge, Sarah Maur-Thorp; *Dir:* Gerard Kikione. VHS, Beta, LV $89.95 VIR ♂ 1/2

Edge of the World Moody, stark British drama of a mini-society in its death throes, expertly photographed on a six-square-mile island in the Shetlands. A dwindling fishing community of fewer than 100 souls agonize over whether to migrate to the mainland; meanwhile the romance of a local girl with an off-islander takes a tragic course. Choral effects were provided by the Glasgow Orpheus Choir.
1937 80m/B GB Finlay Currie, Niall MacGinnis, Grant Sutherland, John Laurie, Michael Powell; *W/Dir:* Michael Powell. VHS $19.95 NOS ♂♂♂

Edie in Ciao! Manhattan The real-life story of Edie Sedgwick, Warhol superstar and international fashion model, whose life in the fast lane led to ruin.
1972 (R) 84m/C Edie Sedgwick, Baby Jane Holzer, Roger Vadim, Paul America, Viva. VHS, Beta $14.95 PAV, FCT ♂

Edipo Re (Oedipus Rex) A freeform, revisionist version of the Sophocles play, newly translated by Pasolini, in which he uses both ancient and modern settings. The play is used to explicitly address the Freudian patterns it has come to symbolize. Director Pasoline also wrote some of the music and stars as the high priest. In Italian with English subtitles.

1967 110m/C IT Franco Citti, Silvana Mangano, Alida Valli, Julian Beck, Pier Paolo Pasolini; *Dir:* Pier Paolo Pasolini. VHS, Beta $79.95 FCT ♂♂♂

Edith & Marcel A fictionalization of the love affair between chanteuse Edith Piaf and boxer Marcel Cerdan (played by his son, Marcel Cerdan, Jr.). In French, with English subtitles.
1983 104m/C FR Evelyne Bouix, Marcel Cerdan Jr., Charles Aznavour, Jacques Villeret; *Dir:* Claude Lelouch. VHS, Beta $59.95 MED ♂ 1/2

Educating Rita Walters and Caine team beautifully in this adaptation of the successful Willy Russell play which finds an uneducated hairdresser determined to improve her knowledge of literature. In so doing, she enlists the aid of a tutor: a disillusioned alcoholic, adeptly played by Caine. Together, the two find inspiration in one another's differences and experiences. Ultimately, the teacher receives a lesson in how to again appreciate his work and the classics as he observes his pupil's unique approach to her studies. Some deem this a "Pygmalion" for the '80s.
1983 (PG) 110m/C GB Michael Caine, Julie Walters, Michael Williams, Maureen Lipman; *Dir:* Lewis Gilbert. British Academy Awards '83: Best Picture. VHS, Beta, LV, 8mm $14.95 COL ♂♂♂ 1/2

Education of Sonny Carson Chilling look at the tribulations of a black youth living in a Brooklyn ghetto amid drugs, prostitution, crime and other forms of vice. Based on Sonny Carson's autobiography. Still pertinent some twenty years after its theatrical release.
1974 (R) 104m/C Rony Clanton, Don Gordon, Paul Benjamin; *Dir:* Michael Campus. VHS, Beta $59.95 MED ♂♂ 1/2

Edward and Mrs. Simpson Made-for-television dramatic reconstruction of the years leading to the abdication of King Edward VII, who forfeited the British throne in 1936 so that he could marry American divorcee Wallis Simpson. Originally aired on PBS.
1980 270m/C Edward Fox, Cynthia Harris; *Dir:* Waris Hussein. VHS, Beta $79.95 HBO ♂♂

Edward Scissorhands Depp's a young man created by loony scientist Price, who unfortunately dies before he can attach hands to his boy-creature. The boy's "rescued" from his lonely existence outside of suburbia (in a gothic castle, not exurbia) by an indefatigably ingratiating lady from Avon. With scissors in place of hands, he has more trouble fitting into suburbia than would most new kids on the block, and he struggles with being different and lonely in a cardboard-cutout world. Burton co-wrote and directed this visually captivating fairy tale full of splash and color. But toward the end—as if Burton suddenly realized his social satire was failing or that box-office needs must be met—the tale resorts to a Hollywood-prefab denouement, sapping whatever moral fiber might have been. Oscar-nominated for Best Make-up.
1990 (PG-13) 100m/C Johnny Depp, Dianne Wiest, Winona Ryder, Alan Arkin, Vincent Price, Anthony Michael Hall; *Dir:* Tim Burton. VHS, Beta, LV $19.98 FOX, FCT ♂♂♂

Edwin S. Porter/Edison Program A compilation of five films made by Edwin S. Porter from 1903-1907: "Life of an American Fireman," "The Great Train Robbery," "Dream of a Rarebit Fiend," "Rescued from the Eagle's Nest," and "The Kleptomaniac."

190? 60m/B *Dir:* Edwin S. Porter, Edwin S. Porter. VHS, Beta $52.95 GVV ♂♂♂

Eegah! Another Arch Hall (Merriwether) epic in which an anachronistic Neanderthal falls in love in '60s California. Reputed to be one of the worst films of all time.
1962 93m/C Marilyn Manning, Richard Kiel, Arch Hall Jr.; *Dir:* Nicholas Merriwether. VHS, Beta $9.95 RHI, FCT Woof!

The Eerie Midnight Horror Show A life-size statue of a crucified thief comes to life after being stolen from a church by a female art student. After centuries of chastity, the possessed statue seeks relief. Also known as "The Sexorcists" and "The Tormented."
1982 (R) 92m/C Stella Carnachia, Chris Auram, Lucretia Love; *Dir:* Mario Gariazzo. VHS, Beta $39.98 CGI Woof!

Egg Striking film from Holland chronicles the life of a quiet, middle-aged baker when he answers a personal ad from a school-teacher. After much correspondence, the teacher visits the baker in his village, and the townspeople take great interest in what happens next. In Dutch with English subtitles.
1988 58m/C NL Johan Leysen, Marijke Vengelers; *Dir:* Danniel Danniel. VHS, LV $29.95 KIV, LUM ♂♂ 1/2

Egg and I Based on the true-life adventures of best-selling humorist Betty MacDonald. A young urban bride agrees to help her new husband realize his life-long dream of owning a chicken farm. A dilapidated house, tempermental stove, and suicidal chickens test the bride's perseverance, as do the zany antics of her country-bumpkin neighbors, Ma and Pa Kettle, who make their screen debut. Oscar nomination for supporting actress Main as the blunt Ma Kettle. Plenty of old-fashioned laughs.
1947 104m/B Claudette Colbert, Fred MacMurray, Marjorie Main, Percy Kilbride, Louise Allbritton, Richard Long, Billy House, Richard Long, Donald MacBride; *Dir:* Chester Erskine. VHS, Beta, LV $14.95 MCA, KRT ♂♂♂

The Egyptian Based on the sword-and-sandal novel by Mika Waltari, this is a ponderous big-budgeted epic about a young Egyptian in Akhnaton's epoch who becomes physician to the Pharaoh.
1954 140m/C Edmund Purdom, Victor Mature, Peter Ustinov, Bella Darvi, Gene Tierney, Henry Daniell, Jean Simmons, Michael Wilding, Judith Evelyn, John Carradine, Carl Benton Reid; *Dir:* Michael Curtiz. VHS, Beta, LV $19.98 FOX, IME ♂♂ 1/2

The Eiger Sanction An art teacher returns to the CII (a fictionalized version of the CIA) as an exterminator hired to assassinate the killers of an American agent. In the process, he finds himself climbing the Eiger. Beautiful Swiss Alps scenery fails to totally compensate for several dreary lapses. Based on the novel by Trevanian.
1975 (R) 125m/C Clint Eastwood, George Kennedy, Vonetta McGee, Jack Cassidy, Thayer David; *Dir:* Clint Eastwood. VHS, Beta, LV $19.95 MCA, TLF ♂♂

Eight Men Out Taken from Eliot Asinof's book, a moving, full-blooded account of the infamous 1919 "Black Sox" scandal, in which members of the Chicago White Sox teamed to throw the World Series for $80,000. A dirge of lost innocence, this is among Sayles' best films. Provides an interesting look at the "conspiracy" that ended "shoeless" Joe Jackson's major-league ca-

reer. The actual baseball scenes are first-rate, and Straithairn, Sweeny and Cusack give exceptional performances. Sayles makes an appearance as Ring Lardner. Enjoyable viewing for even the non-sports fan.
1988 (PG) 121m/C John Cusack, D.B. Sweeney, Perry Lang, Jane Alexander, Bill Irwin, Clifton James, Michael Rooker, Michael Lerner, Christopher Lloyd, Studs Terkel, David Straithairn, Charlie Sheen, Kevin Tighe, John Mahoney, John Sayles, Gordon Clapp, Richard Edson, James Reed, Don Harvey; *Dir:* John Sayles. **VHS, Beta, LV** $19.98 *ORI, FCT* 🎬🎬🎬

8 Million Ways to Die An ex-cop hires himself out to rescue a pimp-bound hooker, and gets knee-deep in a mess of million-dollar drug deals, murder, and prostitution. Slow-moving but satisfying. Based on the book by Lawrence Block.
1985 (R) 115m/C Jeff Bridges, Rosanna Arquette, Andy Garcia, Alexandra Paul; *Dir:* Hal Ashby. **VHS, Beta, LV** $19.98 *FOX* 🎬🎬

8 1/2 The acclaimed Fellini self-portrait of a revered Italian film director struggling with a fated film project wanders through his inter-mixed life, childhood memories and hallucinatory fantasies. Subtitled in English. Fellini received a Best Director Academy Award nomination. Other Oscar nominations for Original Screenplay, Art Direction, Set Decoration, and Costume Design.
1963 135m/B *IT* Marcello Mastroianni, Claudia Cardinale, Anouk Aimee, Sandra Milo, Barbara Steele, Rossella Falk; *Dir:* Federico Fellini. Academy Awards '63: Best Costume Design (B & W), Best Foreign Language Film; National Board of Review Awards '63: Best Foreign Language Film; New York Film Critics Awards '63: Best Foreign Film; Sight & Sound Survey '72: #4 of the Best Films of All Time; Sight & Sound Survey '82: #5 of the Best Films of All Time. **VHS, Beta, LV** $69.95 *VES, MRV, MPI* 🎬🎬🎬🎬

18 Again! After a bump on the head, an 81-year old man and his 18 year-old grandson mentally switch places, giving each a new look at his life. Lightweight romp with Burns in especially good form, but not good enough to justify redoing this tired theme.
1988 (PG) 100m/C George Burns, Charlie Schlatter, Anita Morris, Jennifer Runyon, Tony Roberts, Red Buttons, Miriam Flynn; *Dir:* Paul Flaherty. **VHS, Beta, LV** $19.95 *STE, NWV* 🎬🎬1/2

18 Bronzemen Eighteen martial artists band together to defeat evil.
1979 90m/C Polly Kuan, James Tien, Carter Wong; *Dir:* Joseph Kuo. **VHS, Beta** *IGV* 🎬

Eighteen Jade Arhats A kung fu epic with a blind heroine who nevertheless kicks other people in the eyes.
1984 95m/C Chia Ling, Ou-Yang Ksiek, Sun Chia-Lin, Otae Mofo. **VHS, Beta** $57.95 *UNI* 🎬

18 Weapons of Kung-Fu Ninjas take on their foes in a deadly battle.
19?? ?m/C VHS $29.98 *BFV* 🎬

The Eighties A comic, pseudo-documentary romp through the making of a musical. Not just another dime-a-dozen song-and-dance flick, but plays with the genre with humor and intelligence, treating us to an insider's view of a performance arranged at a shopping plaza, from the rigorous auditions to the tedium of the production of songs and routines. Treated in a light-hearted, sensitive, just on the verge of laughable manner, it concludes with 30 minutes of song and dance. In French with English subtitles.

1983 86m/C *FR* Aurore Clement, Lio, Magali Noel, Pascale Salkin; *Dir:* Chantal Akerman. **VHS** $79.95 *WAC, FCT* 🎬🎬1/2

84 Charing Cross Road A lonely woman in New York and a book-seller in London begin corresponding for business reasons. Over a twenty-year period, their relationship grows into a friendship, and then a romance, though they communicate only by mail. Based on a true story.
1986 (PG) 100m/C Anne Bancroft, Anthony Hopkins, Judy Dench; *Dir:* David Jones. British Academy Awards '87: Best Actress (Bancroft). **VHS, Beta, LV** $19.95 *COL* 🎬🎬🎬

84 Charlie Mopic A widely acclaimed drama about the horrors of Vietnam seen through the eyes of a cameraman assigned to a special front-line unit. Filled with a cast of unknowns, this is an unsettling film which sheds new light on the subject of the Vietnam war. Powerful and energetic; Music by Donovan.
1989 (R) 89m/C Richard Brooks, Christopher Burgard, Nicholas Cascone, Jonathon Emerson, Glen Morshower, Joason Tomlins, Byron Thames; *Dir:* Patrick Duncan. **VHS, Beta, LV** $89.95 *COL* 🎬🎬

Eijanaika (Why Not?) A gripping story of a poor Japanese man who returns to his country after a visit to America in the 1860s. Memorable performances make this an above-average film. In Japanese with English subtitles.
1981 151m/C *JP* Ken Ogata, Shigeru Izumiya; *Dir:* Shohei Imamura. **VHS** $79.95 *KIV, FCT* 🎬🎬🎬1/2

Eisenhower: A Place in History Documents Ike's career as a general in World War II.
195? 52m/B VHS, Beta $9.99 *VTR* 🎬🎬🎬1/2

Eisenstein Well-done biography of Sergei Eisenstein, the famous Russian director. Footage of his early life, first works and masterpieces such as "Potemkin" and "Ivan the Terrible."
1958 48m/B Sergei Eisenstein. **VHS, Beta** $24.95 *VYY, DVT* 🎬🎬🎬1/2

El Ametralladora A retired general tussles with a man known as "The Machine Gun."
1965 90m/C *MX* Pedro Infante, Margarita Mora, Alfredo Varela. **VHS, Beta** $44.95 *MAD* 🎬🎬

El Amor Brujo An adaptation of the work of Miguel de Falla, in which flamenco dancers enact the story of a tragic romance.
1986 100m/C Antonio Gades, Christina Hoyos, Laura Del Sol; *Dir:* Carlos Savrat. **VHS, Beta, LV** $29.95 *PAV, CTH, FCT* 🎬🎬🎬

El Aviador Fenomeno A private eye goes undercover to investigate several murders.
1947 90m/B *MX* Adalberto Martinez, Maria Eugenia San Martin, David Silva, Pedro de Aguillon. **VHS, Beta** $39.95 *MAD* 🎬🎬

El Aviso Inoportuno Two men journey to Mexico City to become actors, but are unable to find jobs in their chosen field.
1987 121m/C *MX* Enrique Cuenca, Eduardo Manzano, Carlos Lopez Moctezuma. **VHS, Beta** $57.95 *MAD* 🎬🎬

El Barbaro In the beginning of civilization, when evil powers rule, a barbarian has two ambitions in life: to seek revenge for his friend's death and to rescue the world from a sorcerer's power.
1984 88m/C *SP* Andrea Occhipinti, Maria Scola, Violeta Cela, Sabrina Sellers. **VHS, Beta** *MED* 🎬

El Bruto (The Brute) A mid-Mexican-period Bunuel drama, about a brainless thug who is used as a bullying pawn in a struggle between a brutal landlord and discontented tenants. In Spanish with English titles.
1952 83m/B *MX* Pedro Armendariz, Katy Jurado, Andres Soler, Rosita Arenas; *Dir:* Luis Bunuel. **VHS, Beta** $59.95 *TAM, FCT, APD* 🎬🎬1/2

El Callao (The Silent One) A Puerto Rican gang in New York City searches for their former leader. When they mistakenly kill a priest who looks like him, trouble begins.
197? 90m/C Jose Albar, Olga Agostini. **VHS, Beta** $29.98 *MAV* 🎬1/2

El Ceniciento An innocent man gets into trouble because he can't stop thinking about a beautiful princess.
1947 90m/B *MX* Tin Tan, Alicia Caro, Andres Soler. **VHS, Beta** $48.95 *MAD* 🎬🎬

El Ciclon An invincible lawman comes to town, looking to clean things up.
1947 120m/B *MX* Miguel Aceves Mejia, Flor Silvestre, Sonia Furio, Raul Ramirez. **VHS, Beta** $45.95 *MAD* 🎬🎬

El Cid Charts the life of Rodrigo Diaz de Bivar, known as El Cid, who was the legendary eleventh-century Christian hero who freed Spain from Moorish invaders. Music by Miklos Roza. Noted for its insanely lavish budget, this epic tale is true to its setting and features elaborate battle scenes.
1961 184m/C Charlton Heston, Sophia Loren, Raf Vallone, Hurd Hatfield, Genevieve Page; *Dir:* Anthony Mann. **VHS, Beta, LV** $29.95 *BFV, LTG, UHV* 🎬🎬🎬

El Cochecito The great Spanish actor, Jose Isbert, stars as the head of a large family whose closest friends are all joined together in a kind of fraternity defined by the fact that they each use a wheelchair. He feels excluded because he does not have or need one, so he goes to great lengths to gain acceptance in this "brotherhood." In Spanish with English subtitles. Also known as "The Wheelchair."
1960 90m/B *SP* Jose Isbert; *Dir:* Marco Ferreri. **VHS, Beta** $69.95 *KIV* 🎬🎬🎬

El Condor Two drifters search for gold buried in a Mexican fort.
1970 (R) 102m/C *SP* Jim Brown, Lee Van Cleef, Patrick O'Neal, Marianna Hill, Iron Eyes Cody, Elisha Cook Jr.; *Dir:* John Guillermin. **VHS** $59.95 *WAR, SIM* 🎬1/2

El Coyote Emplumado Comedian "La India Maria" unwittingly winds up as a detective in Acapulco, pursued by artifact smugglers.
1983 99m/C *SP* Maria Elena Velasco, Miguel Angel Rodriguez. **VHS, Beta** *MED* 🎬1/2

El Cristo de los Milagros After a ruthless thief has a vision of Christ, he changes his criminal ways. In Spanish.
198? 90m/C *SP* Juan Gallardo, Norma Lazarendo, Claudio Lanuza, Julio Cesar. **VHS, Beta** $74.95 *MAD* 🎬

El Cuarto Chino (The Chinese Room) A madman terrorizes the ports of Acapulco, Mexico.
1971 97m/C *MX* Cathy Crosby, Guillermo Murray, Carlos Rivas, Regina Torne. **VHS, Beta** $69.95 *MAD* 🎬

El Derecho de Comer (The Right to Eat) A good natured vagrant finds a purse filled with money and a love letter.

1987 86m/C *MX* Arturo Correa, Findingo, Lissette, Luis Cosme. **VHS, Beta $54.95** *MAD* ♂♂½

El Dia Que Me Quieras The master of the Tango brings you a heartwarming musical drama.
1935 95m/B *MX* Rosita Moreno; *Dir:* John Reinhardt. **VHS, Beta $72.95** *MED* ♂♂½

El Diablo A young man finds the West more wild than he expected. He finds "help" in the shape of Gossett as he tries to free a young girl who's being held by the notorious El Diablo. Made for television. Better than average.
1990 (PG-13) 107m/C Louis Gossett Jr., Anthony Edwards, John Glover, Robert Beltran; *Dir:* Peter Markle. **VHS $89.99** *HBO* ♂♂½

El Diablo Rides A fierce feud between cattlemen and sheepmen develops, with touches of comedy in between.
1939 57m/B Bob Steele, Carleton Young, Kit Guard. **VHS, Beta $19.95** *VDM, VCN* ♂½

El Diputado (The Deputy) A famous politician jeopardizes his career when he has an affair with a young man. In Spanish. Also available with English subtitles.
1978 111m/C *SP* Jose Sacristan, Maria Luisa San Jose, Jose Alonso; *Dir:* Eloy De La Iglesia. **VHS, Beta $79.95** *FCT* ♂♂♂

El Dorado A gunfighter rides into the frontier town of El Dorado to aid a reckless cattle baron in his war with farmers over land rights. Once in town, the hired gun meets up with an old friend—the sheriff—who also happens to be the town drunkard. Switching allegiances, the gunslinger helps the lawman sober up and defend the farmers. This Hawks western displays a number of similarities to the director's earlier "Rio Bravo" (1959), staring Wayne, Dean Martin, and Ricky Nelson—who charms viewers as the young sidekick "Colorado" much like Caan does as "Mississippi" in El Dorado.
1967 126m/C John Wayne, Robert Mitchum, James Caan, Charlene Holt, Ed Asner, Arthur Hunnicutt, Christopher George, R.G. Armstrong, Jim Davis, Paul Fix, Johnny Crawford, Michele Carey; *Dir:* Howard Hawks. **VHS, Beta, LV $14.95** *PAR, TLF, BUR* ♂♂♂

El Fego Baca: Six-Gun Law An Arizona DA tackles a seemingly clear-cut murder case. Compiled from Walt Disney television episodes.
1962 77m/C Robert Loggia, James Dunn, Lynn Bari, Annette Funicello. **VHS, Beta $69.95** *DIS* ♂½

El Fin del Tahur An unwed mother falls in love with a gambler who must face a crooked rival.
1987 90m/C *MX* Luciana, Jaime Moreno, Mario Almada, Gloriella Ruben Benavides. **VHS, Beta $59.95** *MAD* ♂♂

El Forastero Vengador A man arrives in town and challenges an evil woman dictator.
1965 95m/C *MX* Jaime Fernandez, Eleazar Garcia, Ofelia Guillman. **VHS, Beta $39.95** *MAD* ♂♂

El Fugitivo A man convicted of a crime he didn't commit waits for the right moment to exact his revenge.
1965 95m/C *MX* Luis Aguilar, Lucha Villa, Amparo Rivelles, Rita Macedo. **VHS, Beta $58.95** *MAD* ♂♂

El Gallo de Oro (The Golden Rooster) Two rival gamblers bet everything on a big cock fight in Mexico. This film examines the story behind the cockfights and what they mean in Mexican tradition.
1965 94m/C *MX* Ignacio Lopez Tarso, Lucha Villa, Narciso Busquets. **VHS, Beta $68.00** *MAD, APD*

El Gavilan Pollero A singing man in the West enthralls many a beautiful lady in this Western. Packed full with gambling, horse racing, music, and women.
1947 90m/B *MX* Pedro Infante, Antonio Badu, Lilia Prado. **VHS, Beta $44.95** *MAD* ♂♂

El Hombre Contra el Mundo A Mexican comic book hero comes to life so he can battle a variety of evils.
1987 100m/C *MX* Jorge Rivero, Nadia Nilton, Narciso Busquets, Tito Junco. **VHS, Beta $75.95** *MAD* ♂♂½

El Hombre de la Furia (Man of Fury) The son of a rancher seeks to avenge his father's death.
1965 90m/B *MX* Javier Solis, Dacia Gonzalez, Fernando Soto. **VHS, Beta $49.95** *MAD* ♂♂½

El Hombre y el Monstruo When a pianist hears a certain melody, he becomes a killing monster.
1947 90m/B *MX* Abel Salazar, Enrique Rambal, Martha Roth. **VHS, Beta $39.95** *MAD* ♂♂

El Libro de Piedra An isolated ranch harbors dark, horrible secrets. There's no telling what could be causing such strange occurrences.
1968 100m/C *MX* Margo Lopez, Joaquin Cordero; *Dir:* Carlos Enrique Taboada. **VHS, Beta $59.95** *UNI* ♂

El Monte de las Brujas A mad widow burns her daughter alive in a satanistic ritual. In Spanish.
1977? 90m/C *MX* Patty Shepherd. **VHS, Beta $57.95** *UNI* ♂

El Muerto In nineteenth-century Buenos Aires, a young man flees his home after killing an enemy. Arriving in Montevideo, he becomes a member of a smuggling ring. Dialogue in Spanish.
19?? 105m/C *SP* Thelma Biral, Juan Jose Camero, Francisco Rabal; *Dir:* Hector Olivera. **VHS, Beta** *MED* ♂

El Norte Gripping account of a Guatemalan brother and sister, persecuted in their homeland, who make an arduous journey north ("El Norte") to America. Their difficult saga continues as they struggle against overwhelming odds in an attempt to realize their dreams. Passionate, sobering, and powerful. In English and Spanish with English subtitles. Produced in association with the "American Playhouse" series for PBS. Produced by Anna Thomas who also co-wrote the story with Nava.
1983 (R) 139m/C *SP* David Villalpando, Zaide Silvia Gutierrez, Ernesto Cruz, Eracio Zepeda, Stella Quan; *Dir:* Gregory Nava. **VHS, Beta $68.00** *FOX, APD* ♂♂♂♂

El Paso Stampede A cowboy investigates raids on cattle herds used to feed Americans fighting in the Spanish-American War.
1953 50m/B Allan "Rocky" Lane, Eddy Waller, Phyllis Coates; *Dir:* Harry Keller. **VHS, Beta $29.95** *VCN* ♂

El Penon de las Animas Spanish film features a classic premise—two long-feuding families find complications in their bitter relationship as their children fall in love with one another.
1947 90m/B *MX* Maria Felix, George Negrete. **VHS, Beta $44.95** *MAD* ♂♂

El Pirata Negro (The Black Pirate) Italian pirate yarn dubbed into Spanish.
1965 82m/C *SP IT* George Hilton, Claude Dantes, Tony Kendall. **VHS, Beta** *JCI* ♂

El Preprimido (The Timid Bachelor) A newlywed travels to France to learn about sex before approaching his bride. Upon arrival, however, he finds the prostitutes are on strike.
197? 101m/C *SP* Alfredo Landa, Isabel Garces. **VHS, Beta** *MAV* ♂

El Profesor Hippie A freewheeling college history professor befriends a group of students, taking them to a far-off part of the country where adventures befall them. In Spanish.
1969 95m/C *MX* Luis Sandrini, Soledad Silveyra; *Dir:* Fernando Ayala. **VHS, Beta $29.95** *MED* ♂♂

El Rey The story of "The King," a rebel during the Mexican Revolution.
1987 93m/C *MX* Antonio Aguilar, Carmen Montejo, Pancho Cordova, Gerardo Reyes. **VHS, Beta $58.95** *MAD* ♂♂

El Robo al Tren Correo A young man must confront a ruthless band of outlaws to prove that he didn't steal a gold shipment.
1947 90m/B *MX* Carlos Cortez, Noe Murayama, Julissa. **VHS, Beta $39.95** *MAD* ♂♂

El Sabor de la Venganza A man's wife and son vow to avenge his murder.
1965 92m/C *MX* Isela Vega, Jorge Luke, Cameron Mitchell, Mario Almada. **VHS, Beta $59.95** *MAD* ♂♂

El Sheriff Terrible Comedy-western about two drifters trying to make a living by cheating others out of their money. In Spanish.
1963 90m/C *SP* **VHS, Beta** *TSV* ♂½

El Sinaloense A staid family becomes annoyed with their jet-setting son and writes him out of their will.
1984 98m/C *MX* Cruz Infante, Evita Munoz, Roberto Canedo, Fernando Fernandez. **VHS, Beta $59.95** *WCV* ♂½

El Super A Cuban refugee, still homesick after many years, struggles to make a life for himself in Manhattan as an apartment superintendent. In Spanish with English subtitles; an American production shot on location in New York City.
1979 90m/C *SP* Raymundo Hidalgo-Gato, Orlando Jiminez-Leal, Zully Montero, Raynaldo Medina, Juan Granda, Hilda Lee, Elizabeth Pena; *Dir:* Leon Ichaso. **VHS $79.95** *CVW, APD* ♂♂♂

El Tesoro de Chucho el Roto A widow with half of a treasure map must fend off thieves while she searches for the other half.
1947 90m/B *MX* Joaquin Cordero, Ana Berta Lepe, Manuel Lopez Ochoa. **VHS, Beta $44.95** *MAD* ♂♂

El: This Strange Passion Disturbing drama in which a man, convinced he is cuckolded, slowly descends into madness. Bunuel's surreal, almost Freudian look at the

Spanish male's obsession with female fidelity. In Spanish with English subtitles.
1952 82m/B *SP* Arturo de Cordova, Delia Garces, Luis Beristain, Aurora Walker; *Dir:* Luis Bunuel. **VHS $59.95** *HTV, FCT* 🎬🎬🎬

El Zorro Blanco A man in a white mask rescues a kidnapped heiress.
1987 90m/C *MX* Juan Miranda, Hilda Aguirre, Carlos Agosti. **VHS, Beta $59.95** *MAD* 🎬🎬

El Zorro Vengador Zorro's best friend is killed and his bride abducted. And, you better believe he's going to make those baddies pay.
1947 90m/B *MX* Luis Aguilar, Maria Eugenia San Martin. **VHS, Beta $39.95** *MAD* 🎬🎬

Elayne Boosler: Broadway Baby Boosler attacks everything from sex, to sports, to contact lenses in this hilarious solo stand-up performance.
1987 60m/C Elayne Boosler; *Dir:* Steve Gerbson. **VHS $59.98** *LIV, VES* 🎬🎬

Elayne Boosler: Party of One A straightforward comedy evening with Elayne Boosler, featuring many cameos by her friends.
1989 60m/C Elayne Boosler, David Letterman, Dr. Ruth Westheimer, Larry "Bud" Melman. **VHS, Beta $14.98** *LIV, VES*

Eleanor & Franklin An exceptional television dramatization of the personal lives of President Franklin D. Roosevelt and his wife Eleanor. Based on Joseph Lash's book, this Emmy award-winning film features stunning performances by Alexander and Herrmann in the title roles.
1976 208m/C Jane Alexander, Edward Herrmann, Ed Flanders, Rosemary Murphy, MacKenzie Phillips, Pamela Franklin, Anna Lee, Linda Purl, Linda Kelsey, Lindsay Crouse; *Dir:* Daniel Petrie. **VHS, Beta** *TLF* 🎬🎬🎬½

Electra Glide in Blue An Arizona motorcycle cop uses his head in a world that's coming apart at the seams. Good action scenes; lots of violence.
1973 (R) 106m/C Robert Blake, Billy "Green" Bush, Mitchell Ryan, Jeannine Riley, Elisha Cook Jr., Royal Dano; *Dir:* James W. Guercio. **VHS, LV $29.95** *MGM* 🎬🎬🎬

Electric Boogie Documentary capturing the spirit of four boys from the Bronx who perform breakdancing.
1983 30m/C VHS, Beta *FLL* 🎬🎬🎬

Electric Dreams A young man buys a computer that yearns to do more than sit on a desk. First it takes over his apartment, then it sets its sights on the man's cello-playing neighbor—the same woman his owner is courting. To win her affections, the over-eager computer tries to dazzle her with a variety of musical compositions from his unique keyboard. Cort supplies the voice of Edgar the computer in this film which integrates a rock music video format.
1984 (PG) 95m/C Lenny Von Dohlen, Virginia Madsen, Maxwell Caulfield, Bud Cort, Koo Stark; *Dir:* Steven Barron. **VHS, Beta, LV $79.95** *MGM* 🎬🎬½

Electric Grandmother Ray Bradbury's story "I Sing the Body Electric" is the source of this gentle and affecting tale about a grandmother that can be made to exact specifications. An NBC Project Peacock presentation.
1981 49m/C Maureen Stapleton, Edward Herrmann; *Dir:* Noel Black. **VHS, Beta** *NWV, LCA* 🎬🎬🎬

The Electric Horseman Journalist Fonda sets out to discover the reason behind the kidnapping of a prized horse by an ex-rodeo star. The alcoholic cowboy has taken the horse to return it to its native environment, away from the clutches of corporate greed. As Fonda investigates the story she falls in love with rebel Redford. Excellent Las Vegas and remote western settings.
1979 (PG) 120m/C Robert Redford, Jane Fonda, John Saxon, Willie Nelson, Valerie Perrine, Wilford Brimley, Nicholas Coster; *Dir:* Sydney Pollack. **VHS, Beta, LV $19.95** *MCA* 🎬🎬½

Electric Light Voyage Electronic fantasy showing how computer animation can control and change moods of elation and tranquility. The animated visuals are in sync with a mesmerizing soundtrack.
1980 60m/C VHS, Beta $19.95 *MED* 🎬🎬½

The Electronic Monster Insurance claim investigator Cameron looks into the death of a Hollywood starlet and discovers an exclusive therapy center dedicated to hypnotism. At the facility, people vacation for weeks in morgue-like body drawers, while evil Dr. Illing uses an electronic device to control the sleeper's dreams and actions. Eerie. Intriguingly, it is one of the first films to explore the possibilities of brainwashing and mind control. Also known as "Escapement."
1957 72m/B *GB* Rod Cameron, Mary Murphy, Meredith Edwards, Peter Illing; *Dir:* Montgomery Tully. **VHS $16.95** *SNC* 🎬🎬

The Element of Crime In a monochromatic, post-holocaust future, a detective tracks down a serial killer of young girls. Made in Denmark, this minor festival favorite features an impressive directional debut and awaits cult status. Filmed in Sepiatone.
1985 104m/C *DK* Michael Elphick, Esmond Knight, Jerold Wells, Meme Lei, Astrid Henning-Jensen, Preben Leerdorff-Rye, Gotha Andersen; *Dir:* Lars von Trier. **VHS, Beta $79.95** *UNI* 🎬🎬½

Elena and Her Men The romantic entanglements and intrigues of a poor Polish princess are explored in this enjoyable French film. Beautiful cinematography by Claude Renoir. In French with subtitles. Alternate title: "Paris Does Strange Things."
1956 98m/C *FR* Ingrid Bergman, Jean Marais, Mel Ferrer, Jean Richard, Magali Noel, Pierre Bertin, Juliette Greco; *Dir:* Jean Renoir. **VHS, Beta, LV $59.95** *INT, FCT* 🎬🎬🎬

Eleni The true story of New York Times reporter Nicholas Gage and his journey to Athens to discover the truth about his mother's execution by Communists during the Greek rebellions after World War II. Adapted by Steve Tesich from Gage's bestselling book.
1985 (R) 117m/C John Malkovich, Kate Nelligan, Linda Hunt, Oliver Cotton, Ronald Pickup, Dimitra Arliss, Rosalie Crutchley; *Dir:* Peter Yates. **VHS, Beta, LV, 8mm $14.98** *SUE, IGP* 🎬🎬

Elephant Boy An Indian boy helps government conservationists locate a herd of elephants in the jungle. Sabu's first film. Available in digitally remastered stereo.
1937 80m/B *GB* Sabu, Walter Hudd, W.E. Holloway; *Dir:* Robert Flaherty, Zoltan Korda. Venice Film Festival '37: Best Director. **VHS, Beta, LV $19.98** *HBO, SUE* 🎬🎬🎬

Elephant Man A biography of John Merrick, a severely deformed man who, with the help of a sympathetic doctor, moved from freak shows into posh London society. Lynch's first mainstream film, shot in black and white, it presents a startlingly vivid picture of the hypocrisies evident in the social

moves of the Victorian era. Moving performance from Hurt in title role.
1980 (PG) 125m/B Anthony Hopkins, John Hurt, Anne Bancroft, John Gielgud, Wendy Hiller, Freddie Jones, Kenny Baker; *Dir:* David Lynch. British Academy Awards '80: Best Actor (Hurt), Best Art Direction/Set Decoration, Best Picture; National Board of Review Awards '80: 10 Best Films of the Year. **VHS, Beta, LV $14.95** *PAR* 🎬🎬🎬🎬

Elephant Parts A video album by ex-Monkee Michael Nesmith, which contains several amusing comedy sketches and original music.
1981 60m/C Michael Nesmith. Grammy '91: Video of the Year Award. **VHS, Beta, LV $14.95** *MVD, PAV, PIA* 🎬🎬🎬🎬

Elephant Walk The balmy jungles of Sri Lanka provide the backdrop for a torrid triangular story of love on an agricultural estate in this post-prime Dieterle effort. Taylor, ignored by her wealthy drunkard husband, finds solace in the arms of her spouse's sexy right-hand man. As if keeping the affair secret weren't elephantine task enough, Taylor braves an epidemic of cholera and a pack of vengeful elephants who decide to take an unscheduled tour of her humble home. The lethargic performances of the lead characters make this more of a sleep walk, but Sofaer and Biberman's supporting appearances are worth the price of rental. Taylor replaced Vivian Leigh in the early stages of filming after Leigh fell ill, but the final production includes footage of Leigh in faraway shots.
1954 103m/C Elizabeth Taylor, Dana Andrews, Peter Finch, Abraham Sofaer, Abner Biberman; *Dir:* William Dieterle. **VHS $14.95** *PAR, FCT* 🎬🎬½

11 Harrowhouse The Consolidated Selling System at 11 Harrowhouse, London, controls much of the world's diamond trade. Four adventurous thieves plot a daring heist relying on a very clever cockroach. A rather successful stab at spoofing detailed "heist" films. Alternate title: "Anything for Love."
1974 (PG) 95m/C *GB* Charles Grodin, Candice Bergen, James Mason, Trevor Howard, John Gielgud; *Dir:* Aram Avakian. **VHS, Beta $59.98** *FOX* 🎬🎬🎬

Eli Eli An American Yiddish film about a family forced to live apart because of their farm's failure. Lighthearted and fun, with songs by Sholom Secunda. In Yiddish with English titles.
1940 85m/B Esther Feild, Lazar Fried, Muni Serebroff; *Dir:* Joseph Seiden. **VHS, Beta $39.95** *FCT* 🎬🎬½

The Eliminators A cyborgian creature endeavors to avenge himself on his evil scientist creator with help from a kung fu expert and a Mexican. The group uses time travel and other devices to achieve their mission.
1986 (PG) 95m/C Roy Dotrice, Patrick Reynolds, Denise Crosby, Andrew Prine, Conan Lee; *Dir:* Peter Manoogian. **VHS, Beta $79.98** *FOX* 🎬

L'Elisir D'Amore Performance of Donizetti's enjoyable opera.
1985 45m/C VHS $29.95 *PAR*

Elizabeth of Ladymead Four generations of a British family live through their experiences in the Crimean War, Boer War, World War I and World War II. Neagle, playing the wives in all four generations, aptly explores the woman's side of war. Enjoyable performances from husband and wife team Neagle and Wilcox.

1948 97m/C Anna Neagle, Hugh Williams, Bernard Lee, Michael Laurence, Nicholas Phipps, Isabel Jeans; **Dir:** Herbert Wilcox. **VHS, Beta** $24.95 *DVT, TIM* 𝕚𝕚𝕚

Elizabeth, the Queen Historical drama recreates the struggle for power by Robert Devereaux, Earl of Essex, whom the aging Queen Elizabeth I both loved and feared, and whose downfall she finally invoked. Part of "George Schaefer's Showcase Theatre."
1968 76m/C Judith Anderson, Charlton Heston; **Dir:** George Schaefer. **VHS, Beta** *FHS* 𝕚𝕚

Eliza's Horoscope A frail Canadian woman uses astrology in her search for love in Montreal, and experiences bizarre and surreal events. An acclaimed Canadian film.
1970 120m/C *CA* Tommy Lee Jones, Elizabeth Moorman; **Dir:** Gordon Sheppard. **VHS, Beta** $24.95 *AHV* 𝕚𝕚½

Ella Cinders The American mania for breaking into the movies is satirized in this look at a girl who wins a trip to Hollywood in a small-town beauty contest. Silent.
1926 60m/B Harry Langdon, Colleen Moore, Lloyd Hughes; **Dir:** Alfred E. Green. **VHS, Beta** $19.95 *NOS, FST, TIM* 𝕚𝕚𝕚

Elle Voit des Nains Partout A film of France's "Cafe Theatre Show," wherein various fairy tale characters' stories are revised per modern situations. In French with subtitles.
1983 83m/C *FR* **Dir:** Jean-Claude Sussfeld. **VHS, Beta** $59.95 *FCT* 𝕚𝕚½

Ellie A murderous widow's stepdaughter tries to save her father from being added to the woman's extensive list of dearly departed husbands.
1984 (R) 90m/C Shelley Winters, Sheila Kennedy, Pat Paulsen, George Gobel, Edward Albert; **Dir:** Peter Wittman. **VHS, Beta** $79.98 *LIV, VES* 𝕚½

Ellis Island Choreographer-video artist Meredith Monk has fashioned a dance piece to express the ordeal of immigration.
1983 30m/C **VHS, Beta** $79.95 *SVA, TIM*

Elmer Follows the adventures of a temporarily blinded youth and a lovable hound dog who meet in the wilderness and together set off in search of civilization.
1976 (G) 82m/C Elmer Swanson, Phillip Swanson; **Dir:** Christopher Cain. **VHS, Beta** $59.98 *LIV, LTG* 𝕚½

Elmer Fudd Cartoon Festival Fussy Elmer Fudd gets little respect and more than his share of torment from "wascals." Includes "An Itch in Time," "Hardship of Miles Standish," "Elmer's Pet Rabbit," and "Back-Alley Operator." Made between 1940 and 1947.
194? 33m/C **Dir:** Chuck Jones, Robert McKimson, Friz Freleng. **VHS, Beta** $14.95 *MGM*

Elmer Fudd's Comedy Capers Eight classic Fudd cartoons, including "Hare Brush," "The Rabbit of Seville," and "What's Opera, Doc?" The latter short finds Fudd crooning "Kill the Wabbit" to the tune of Wagner's popular composition "Die Valkyrie," while "Seville" features the memorable Bugs the Barber using his toes to lather up Fudd's head.
1955 57m/C **Dir:** Chuck Jones, Robert McKimson, Friz Freleng. **VHS, Beta** $12.95 *WAR*

Elmer Gantry The classic multi-Oscar-winning adaptation of the Sinclair Lewis novel written to expose and denounce the flamboyant, small-town evangelists spreading through America at the time. In the film, Lancaster is the amoral Southern preacher who pursues wealth and power from his congregation, and takes a nun as a mistress. Jones stars as his ex-girlfriend who resorts to a life of prostitution. Oscar nominations for Best Picture and Score.
1960 146m/C Burt Lancaster, Shirley Jones, Jean Simmons, Dean Jagger, Arthur Kennedy, Patti Page, Edward Andrews, John McIntire, Hugh Marlowe, Rex Ingram; **Dir:** Richard Brooks. Academy Awards '60: Best Actor (Lancaster), Best Adapted Screenplay, Best Supporting Actress (Jones); New York Film Critics Awards '60: Best Actor (Lancaster). **VHS, Beta, LV** $19.98 *FOX, MGM, PIA* 𝕚𝕚𝕚½

Elsa & Her Cubs Documentary, photographed by Joy and George Adamson in Kenya, records their five-year friendship with the lioness Elsa, from the time she was a cub until her turn as mother. True animal lovers will remember Elsa from her earlier appearance in 1966's acclaimed film "Born Free" and her later adventures in 1972's "Living Free."
196? 25m/C *Nar:* Joy Adamson, George Adamson. **VHS, Beta** *BMF* 𝕚𝕚𝕚

Elusive Corporal Set in a P.O.W. camp on the day France surrendered to Germany, this is the story of the French and Germans, complete with memories of a France that is no more.
1962 108m/B Jean-Pierre Cassel, Claude Brasseur, O.E. Hasse, Claude Rich; **Dir:** Jean Renoir. **VHS, Beta, LV** $59.95 *INT, VYY, APD* 𝕚𝕚𝕚½

The Elusive Pimpernel Niven sparkles in this otherwise undistinguished adaptation of Baroness Orczy's, "The Scarlet Pimpernel." Originally titled, "The Fighting Pimpernel."
1950 109m/C *GB* David Niven, Margaret Leighton, Jack Hawkins, Cyril Cusack, Robert Coote, Edmund Audran, Danielle Godet, Patrick Macnee; **W/Dir:** Michael Powell, Emeric Pressburger. **VHS, Beta** $59.95 *HMV, FCT* 𝕚𝕚𝕚

Elves A group of possessed, neo-Nazi elves performs serious human harm at Christmas time in this tongue-in-check thriller.
1989 (PG-13) 95m/C Dan Haggerty, Deanna Lund, Julie Austin, Borah Silver; **Dir:** Jeffrey Mandel. **VHS, Beta** $79.95 *AIP* 𝕚

Elvira Madigan Chronicles the 19th-century Swedish romance between a young officer and a beautiful circus tight-rope walker. Based on a true incident. Exceptional direction and photography.
1967 (PG) 89m/C *SW* Pia Degermark, Thommy Berggren; **Dir:** Bo Widerberg. Cannes Film Festival '67: Best Actress (Degermark); National Board of Review Awards '67: Best Foreign Film. **VHS, Beta, LV** $34.95 *HBO* 𝕚𝕚𝕚

Elvira, Mistress of the Dark A manic comedy based on Peterson's infamous B-movie horror-hostess character. The mega-busted terror-queen inherits a house in a conservative Massachusetts town and causes double-entendre chaos when she attempts to sell it.
1988 (PG-13) 96m/C Cassandra Peterson, Jeff Conaway, Susan Kellerman, Edie McClurg, Daniel Greene, W. Morgan Shepherd; **Dir:** James Signorelli. **VHS, Beta, LV** $19.95 *STE, NWV* 𝕚𝕚

Elvis in the '50s A collection of Elvis television appearances from the 1950s. Songs include: "Love Me Tender," "Hound Dog," "I'm Ready," "Don't Be Cruel," and "I Want You, I Need You, I Love You."
195? 60m/B Elvis Presley. **VHS, Beta** *BTV, DVT* 𝕚𝕚

Elvis: Aloha from Hawaii Elvis' 1973 Hawaiian concert appearance features the King performing "Burning Love," "Suspicious Minds," "Blue Suede Shoes," and a smoldering cover of "My Way."
1973 75m/C Elvis Presley. **VHS, Beta, LV** $19.95 *BMG, MED, MVD* 𝕚𝕚

Elvis in Concert 1968 Recorded before the famous singer's television special, this informal concert features Elvis at his best.
1968 60m/B Elvis Presley. **VHS, Beta** $19.95 *NOS, DVT* 𝕚

The Elvis Files Documentary examining the strange details surrounding Elvis's death. Contains some rare footage and interviews in support of Brewer's theory that Elvis's death was faked.
1990 55m/C **VHS** $19.98 *MED, TTC* 𝕚𝕚

Elvis and Me A TV mini-series based on Priscilla Beaulieu Presley's autobiography about her life with the Pelvis. Features many of Presley's biggest hits, sung by country star Ronnie McDowell. Film may aggravate Elvis fans, however, as the King is often depicted negatively from his ex-wife's point of view.
1988 187m/C Dale Midkiff, Susan Walters, Billy "Green" Bush, Linda Miller, Jon Cypher; **Dir:** Larry Peerce. **VHS, Beta** $29.95 *STE, NWV* 𝕚½

Elvis Memories A video scrapbook of Elvis' career, including concert appearances, home movies, and contemporary interviews.
1985 48m/C Elvis Presley, Priscilla Presley, Cybill Shepherd, Dick Clark. **VHS, Beta, LV** $29.95 *MVD, LIV, VES* 𝕚½

Elvis in the Movies All of the King's movies are covered, complete with scenes and highlights from each. A must for Elvis fans. A must for Elvis impersonators. A must for average Elvis look-alikes. Essentially, a must for fans of all types and ages.
1990 ?m/C Elvis Presley. **VHS** *COL* 𝕚½

Elvis: One Night with You The celebrated crooner performs such memorables as "Heartbreak Hotel," "Blue Suede Shoes," and "Are You Lonesome Tonight" in this unedited performance from his 1968 Christmas Special.
1968 53m/C Elvis Presley. **VHS, Beta, LV** $19.95 *MVD, MED, VTK* 𝕚½

Elvis Presley's Graceland Elvis' ex-wife (Priscilla Beaulieu Presley) takes the viewer on a tour of Graceland, the famous Memphis, Tennessee, mansion, which was home to the rock star from 1957 until his death in 1977.
1984 60m/C **Hosted:** Priscilla Presley. **VHS, Beta** $19.95 *NEG, KAR* 𝕚½

Elvis: Rare Moments with the King Footage of Elvis from news conferences, his wedding and media newsreels.
1990 60m/C Elvis Presley. **VHS, Beta** $12.95 *MVD, GKK* 𝕚½

Elvis Stories Collection of humorous shorts dealing with the King, from Mojo Nixon's song "Elvis Is Everywhere" to Cusack as a burger maker who claims to

receive messages from Elvis through his patties.
1989 30m/C John Cusack, Ben Stiller, Joel Murray, Mojo Nixon, Rick Saucedo. **VHS, Beta** $14.95 *MVD, RHI 🎞 ½*

Elvis: That's the Way It Is Elvis performs more than 30 songs during his 1970 tour, interspersed with rehearsal and backstage footage.
1970 (G) 109m/C Elvis Presley; *Dir:* Denis Sanders. **VHS, Beta** $19.95 *MVD, MGM 🎞 ½*

Elvis: The Echo Will Never Die Celebrities chat about the King and how much they miss him, recounting anecdotes about his rise to fame.
1986 50m/C Tom Jones, B.B. King, Sammy Davis Jr., Casey Kasem, Ursula Andress. **VHS, Beta, 3/4U** $29.95 *IHF, MVD, MPI 🎞 ½*

Elvis: The Great Performances, Vol. 1: Center Stage Rare early clips of the King have been digitally remastered for this two-part series. Elvis' first recording, "My Happiness," and his first screen test are included.
1990 52m/B Elvis Presley; *Dir:* Andrew Solt; *Nar:* George Klein. **VHS, Beta, LV, 8mm** $19.99 *MVD, BVV, RDG 🎞🎞*

Elvis: The Great Performances, Vol. 2: Man & His Music Never-before-seen live footage of Presley's groundbreaking early performances. Also includes excerpts from his last concert. Well assembled and nicely narrated by D.J./musician George Klein of Memphis, Tennessee.
1990 54m/C Elvis Presley. **VHS, LV, 8mm** $19.99 *MVD, BVV, RDG 🎞🎞*

Elvis: The Lost Performances A video collection of live performances and rehearsal footage that was discovered in MGM's underground vaults six years ago. Includes outtakes from "Elvis: That's the Way It Is" and "Elvis on Tour." A must for fans and collectors of Elvis memorabilia.
1992 60m/C Elvis Presley. **VHS, LV** $19.98 *WAR 🎞🎞*

Elvis: The Movie A made-for-television biography of the legendary singer, from his high school days to his Las Vegas comeback. Russell gives a convincing performance and lip syncs effectively to the voice of the King (provided by country singer Ronnie McDowell). Also available in 150-minute version.
1979 117m/C Kurt Russell, Season Hubley, Shelley Winters, Ed Begley Jr., Dennis Christopher, Pat Hingle, Bing Russell, Joe Mantegna; *Dir:* John Carpenter. **VHS, Beta** $69.98 *LIV, VES 🎞🎞 ½*

Elvis on Tour A revealing glimpse of Elvis Presley, onstage and off, during a whirlwind concert tour.
1972 93m/C Elvis Presley; *Dir:* Pierre Adidge, Robert Abel. **VHS, Beta, LV** $19.95 *MGM 🎞🎞 ½*

Emanon A coming-of-age tale about a young kid and the Christ-like vagabond, named Emanon (or "no name" if spelled in reverse), he befriends.
1986 (PG-13) 101m/C Stuart Paul, Cheryl Lynn, Jeremy Miller; *Dir:* Stuart Paul. **VHS, Beta** $19.98 *CHA 🎞 ½*

Embassy In this mediocre espionage thriller, von Sydow is a Soviet defector under asylum at the U.S. embassy in Beirut. Colonel Connors, a Russian spy, penetrates embassy security and wounds von Sydow. He is caught, escapes, is captured, escapes,

and is caught again. Great cast, but script is often too wordy and contrived.
1972 90m/C *GB* Richard Roundtree, Chuck Connors, Max von Sydow, Ray Milland, Broderick Crawford, Marie-Jose Nat; *Dir:* Gordon Hessler. **VHS** $79.95 *WVE 🎞🎞*

Embassy A made-for-television movie about an American family in Rome that unknowingly possesses a secret computer chip, and thus is pursued by ruthless agents.
1985 104m/C Nick Mancuso, Blanche Baker, Eli Wallach, Sam Wanamaker, Richard Gilliland, Mimi Rogers, George Grizzard, Richard Masur, Kim Darby; *Dir:* Robert Lewis. **VHS, Beta** $79.95 *WES 🎞🎞*

Embryo An average sci-fi drama about a scientist who uses raw genetic material to artificially produce a beautiful woman, with ghastly results. Also released on video as "Created to Kill."
1976 (PG) 108m/C Rock Hudson, Barbara Carrera, Diane Ladd, Roddy McDowall, Ann Schedeen, John Elerick, Dr. Joyce Brothers; *Dir:* Ralph Nelson. **VHS, Beta** $19.95 *STE, IVE 🎞🎞*

Emerald of Aramata Anyone viewing the most precious and sought after stone in the history of man does not live to tell about it.
19?? 90m/C Rory Calhoun. **VHS** $59.95 *ACF 🎞*

The Emerald Forest Captivating adventure about a young boy who is kidnapped by a primitive tribe of Amazons while his family is traveling through the Brazilian jungle. Boothe is the father who searches 10 years for him. An engrossing look at tribal life in the vanishing jungle. Beautifully photographed and based upon a true story. Features the director's son as the kidnapped youth.
1985 (R) 113m/C Powers Boothe, Meg Foster, Charley Boorman, Dira Pass, Rui Polonah; *Dir:* John Boorman. **VHS, Beta, LV, SVS, 8mm** $14.98 *SUE, SUP 🎞🎞 ½*

Emerald Jungle While searching for her missing sister in the jungle, Agren encounters cannibal tribes and a colony of brainwashed cult followers. Originally titled "Eaten Alive by Cannibals," this Italian feature was cut for its U.S. release.
1980 92m/C *IT* Robert Kerman, Janet Agren, Mel Ferrer, Luciano Martino, Mino Loy; *Dir:* Umberto Lenzi. **VHS, Beta** $39.98 *CGI 🎞 ½*

Emergency Scenes from the Living Theatre's works after their return to America from four years of self-imposed exile, including "Frankenstein," "Mysteries" and "Paradise Now."
1968 29m/C Julian Beck, Judith Malina; *Dir:* Gwen Brown. **VHS, Beta** $29.95 *MFV 🎞🎞🎞*

Emil & the Detectives A German ten-year-old is robbed of his grandmother's money by gangsters, and subsequently enlists the help of pre-adolescent detectives to retrieve it. Good Disney dramatization of the Erich Kastner children's novel. Remake of the 1931 German film starring Rolf Wenkhaus.
1964 99m/C Walter Slezak, Roger Mobley; *Dir:* Peter Tewkesbury. **VHS, Beta** $69.95 *DIS 🎞🎞 ½*

Emiliano Zapata Zapata was a hero of the Mexican Revolution who sacrificed everything because of the high hopes he had for his country.
1987 139m/C *MX* Antonio Aguilar, Jaime Fernandez, Mario Almada, Patricia Aspillaga. **VHS, Beta** $59.95 *MAD 🎞🎞*

Emilienne An artistic young couple toys with the sexual possibilities that exist outside of, and within, a marriage.
1978 (X) 94m/C VHS, Beta $59.95 *VID 🎞🎞*

Emilio Varela vs. Camelia La Texana A pair of criminals encounter trouble when one tries to reform while the other insists on a life of crime.
1987 107m/C *MX* Maria Almada, Jorge Russek, Silvia Manriquez. **VHS, Beta** $59.95 *MAD 🎞🎞*

Emily Returning from her exclusive Swiss finishing school, licentious young Emily discovers mama is a prostitute and deals with the scalding news by delving into a series of erotic encounters at the hands of her willing "instructors." Koo Stark's first venture into the realm of soft porn.
1977 87m/C Koo Stark; *Dir:* Henry Herbert. **VHS, Beta** $59.95 *MGM Woof!*

Eminent Domain A communist party member wakes up to find himself stripped of power in a Kafkaesque purge, not knowing what he is accused of—or why. A potent premise, indifferently handled despite filming on location in Poland. Based on actual events that befell the family of scriptwriter Androej Krakowski.
1991 (PG-13) 102m/C Donald Sutherland, Anne Archer, Paul Freeman, Bernard Hepton, Francoise Michaud, Jodhi May; *Dir:* John Irvin. **VHS** $39.95 *SVS, IME 🎞🎞 ½*

Emissary An American politician and his wife are blackmailed in Africa by the Russians into revealing state secrets that could spark World War III.
1989 (R) 98m/C *SA* Robert Vaughn, Ted Leplat, Terry Norton, Andre Jacobs; *Dir:* Jan Scholtz. **VHS, Beta, LV** $79.95 *VIR 🎞🎞*

Emmanuelle Filmed in Bangkok, a young, beautiful woman is introduced by her husband to an uninhibited world of sensuality. Above-average soft-core skin film, made with sophistication and style. Kristel maintains a vulnerability and awareness, never becoming a mannequin.
1974 92m/C *FR* Sylvia Kristel, Alain Cuny, Marika Green; *Dir:* Just Jaeckin. **VHS, Beta** $84.95 *COL 🎞🎞 ½*

Emmanuelle 4 Emmanuelle flees a bad relationship and undergoes plastic surgery to mask her identity. Ultimately, she becomes a beautiful young model in the form of a different actress, of course.
1984 (R) 95m/C *FR* Sylvia Kristel, Mia Nygren, Patrick Bauchau; *Dir:* Francis Giacobetti. **VHS, Beta** $79.95 *MGM 🎞*

Emmanuelle 5 This time around sexy Emmanuelle is abducted from her yacht at the Cannes Film Festival and forced to join an Arab sheik's harem of slaves. A good example of a series that should have quit when it was ahead.
1987 (R) 78m/C *FR* Monique Gabrielle, Charles Foster; *Dir:* Steve Barnett, Walerian Borowczyk. **VHS** $89.98 *NHO 🎞 ½*

Emmanuelle in Bangkok Emanuelle's exotic, erotic experiences in the Far East include a royal masseuse, suspense, and the Asian art of love.
1978 (R) 94m/C *FR* **VHS, Beta** $29.98 *VID 🎞*

Emmanuelle in the Country Emmanuelle becomes a nurse in an attempt to bring comfort and other pleasures to those in need.
1978 90m/C *FR* Laura Gemser. **VHS, Beta, LV** $59.98 *MAG, IME 🎞*

Emmanuelle & Joanna A mistreated newlywed bride goes to her sister's whorehouse and stumbles across lots of soft-core antics.
1986 90m/C *FR* Sherry Buchanna, Danielle Dublino. **VHS, Beta $29.95** *MED* 🎬

Emmanuelle, the Joys of a Woman The amorous exploits of a sensuous, liberated couple take them and their erotic companions to exotic Hong Kong, Bangkok, and Bali.
1976 92m/C *FR* Sylvia Kristel, Umberto Orsini, Catherine Rivet, Frederic Lagache; *Dir:* Francis Giacobetti. **VHS, Beta, LV, 8mm $69.95** *PAR* 🎬

Emmanuelle, the Queen Seeking revenge, Emanuelle plots the murder of her sadistic husband. The lecherous assassin she hires tries to blackmail her, and she challenges him at his own game of deadly seduction.
1975 90m/C Laura Gemser. **VHS, Beta $29.98** *VID* 🎬

Emmanuelle in Soho Yet another episode of the sleaze-queen's sexual exploits, this time in London's porn-saturated Soho district.
198? (R) 90m/C *FR* **VHS, Beta $59.95** *PSM* 🎬

Emmanuelle on Taboo Island A marooned young man discovers a beautiful woman on his island.
1976 (R) 95m/C *FR* Laura Gemser, Paul Giusti, Arthur Kennedy. **VHS, Beta** *PGN* 🎬

Emmanuelle's Daughter A Greek-made thriller in which a vapidly sensuous heroine cavorts amid episodes of murder, blackmail and rape.
1979 91m/C *GR* Laura Gemser, Gabriele Tinti, Livia Russa, Vagelis Varton, Nadia Neri. **Beta $49.95** *MFI* 🎬

Emma's Shadow In the 1930s, a young girl fakes her own kidnapping to get away from her inattentive family. Winner of the Prix de la Jeunesse at Cannes, it was voted 1988's Best Danish Film.
1988 93m/C *DK* Bjorje Ahistedt, Kruse Line; *Dir:* Soeren Kragh-Jacobsen. Cannes Film Festival '88: Prix de la Jeunesse. **VHS $79.95** *FXL, FCT* 🎬🎬🎬

The Emmett Kelly Circus The wonders and fun of the circus are highlighted.
1981 90m/C **VHS, Beta $9.95** *IND*

Emperor of the Bronx A look at crime, sleaze, violence, and the actions of bad guys in the inner-city.
1989 90m/C William Smith. **VHS, Beta $29.95** *CLV, HHE* 🎬

Emperor Caligula: The Untold Story A raunchy look at the famed Roman's bacchanalian sexual excesses.
198? 100m/C David Cain Haughton. **VHS, Beta $19.98** *TWE* Woof!

Emperor Jones Loosely based on Eugene O'Neill's play, film portrays the rise and fall of a railroad porter whose exploits take him from a life sentence on a chain gang to emperor of Haiti.
1933 72m/B Paul Robeson, Dudley Digges, Frank Wilson, Fredi Washington, Ruby Elzy; *Dir:* Dudley Murphy. **VHS, Beta $12.95** *SNC, NOS, HHT* 🎬🎬½

Emperor Jones/Paul Robeson: Tribute to an Artist A classic, rarely viewed version of the Eugene O'Neill play about a criminal who schemes his way into becoming the king of a Caribbean island.

Robeson's only true dramatic film appearance. The film is accompanied by a documentary profile of Robeson, made in 1979.
1933 102m/B Paul Robeson, Dudley Digges, Frank Wilson; *Dir:* Dudley Murphy; *Nar:* Sidney Poitier. **VHS, Beta $39.98** *SUE*

The Emperor's New Clothes The story of an emperor and the unusual outfit he gets from his tailor. A "Faerie Tale Theatre" presentation.
1984 60m/C Art Carney, Alan Arkin, Dick Shawn; *Dir:* Peter Medak; *Nar:* Timothy Dalton. **VHS, Beta $14.98** *KUI, FOX, FCT* 🎬🎬½

Empire of the Air: The Men Who Made Radio Documentary traces the history of radio from 1906-1955 and profiles its three rival creators: Lee de Forest, Edwin Howard Armstrong and David Sarnoff.
1991 120m/C **VHS $19.95** *PBS*

Empire of the Ants A group of enormous, nuclear, unfriendly ants stalk a real estate dealer and prospective buyers of undeveloped oceanfront property. Story originated by master science-fiction storyteller H. G. Wells.
1977 (PG) 90m/C Joan Collins, Robert Lansing, John David Carson, Albert Salmi, Jacqueline Scott, Robert Pine; *Dir:* Bert I. Gordon. **VHS, Beta $9.98** *SUE* 🎬½

Empire of the Dragon A rip-roaring martial arts adventure with plenty of kicks.
198? 90m/C Chen Tien Tse, Chia Kai, Chang Shan. **VHS, Beta $57.95** *UNI* 🎬

The Empire of Passion A peasant woman and her low-life lover kill the woman's husband, but find the future they planned with each other is not to be. The husband's ghost returns to haunt them, and destroy the passionate bond which led to the murder. Oshima's follow-up to "In the Realm of the Senses." In Japanese with yellow English subtitles.
1976 110m/C *JP* Nagisa Oshima, Kazuko Yoshiyuki, Tatsuya Fuji, Takahiro Tamura, Takuzo Kawatani; *Dir:* Nagisa Oshima. Cannes Film Festival '78: Best Director (Oshima). **VHS $89.95** *FCT* 🎬🎬🎬

Empire of Spiritual Ninja A ninja student gone bad encounters his former teacher.
19?? 90m/C Tom Berlin, Hubert Hays; *Dir:* Bruce Lambert. **VHS $59.95** *TWE* 🎬

Empire State When his friend disappears from a posh London nightclub, a young man searches for him and finds more action and intrigue than he can handle. Even so, this movie's still pretty bland.
1987 (R) 102m/C *GB* Cathryn Harrison, Martin Landau, Ray McAnally; *Dir:* Ron Peck. **VHS, Beta $79.95** *VMK* 🎬

The Empire Strikes Back The second film in the epic "Star Wars" trilogy finds young Luke Skywalker and the Rebel Alliance plotting new strategies as they prepare to battle the evil Darth Vader and the forces of the Dark Side. Luke travels to a distant planet to learn the ways of a Jedi knight from master Yoda, while Han and Leia find time for romance and a few adventures of their own. This installment introduces the charismatic Lando Calrissian, the vulgar and drooling Jabba the Hut, and a mind-numbing secret from Vadar. With the same superb special effects and a hearty plot, this is a true continuation of the excellent tradition set by 1977's "Star Wars." Followed by "Return of the Jedi" in 1983. Also available

on Laserdisc with "The Making of 'Star Wars'."
1980 (PG) 124m/C Mark Hamill, Carrie Fisher, Harrison Ford, Billy Dee Williams, Alec Guinness, David Prowse, Kenny Baker, Frank Oz, Anthony Daniels, Peter Mayhew; *Dir:* Irvin Kershner; *Voices:* James Earl Jones. Academy Awards '80: Best Visual Effects. **VHS, Beta, LV $19.98** *FOX, FCT, BUR* 🎬🎬🎬🎬

Empire of the Sun Spielberg's mature, extraordinarily vivid return to real storytelling, adapted from the best-selling J. G. Ballard novel. Although yearning to be a great film, it experiences occasional flat spots which keep it slightly out of contention. A young, wealthy British boy living in Shanghai is caught in the turbulence when China is invaded by the Japanese at the onset of World War II. Separated from his family, he is thrust into an unfamiliar life of poverty and discomfort when he is interred in a prison camp. A mysterious, breathtaking work, in which Spielberg's heightened juvenile romanticism has a real, heartbreaking context. Written by Tom Stoppard; cinematography by Allen Daviau; cameo by Ballard. Interesting to compare to other films made in 1987 concerning a young boy's recollection of World War II, including Louis Malle's "Au Revoir Les Enfants" and John Boorman's "Hope and Glory."
1987 (PG) 153m/C Christian Bale, John Malkovich, Miranda Richardson, Nigel Havers, Joe Pantoliano; *Dir:* Steven Spielberg. National Board of Review Awards '87: Outstanding Juvenile Performance (Bale), Best Director, Best Picture. **VHS, Beta, LV, 8mm $19.98** *WAR* 🎬🎬🎬

The Empty Beach A tough private detective investigates the disappearance of a wealthy business tycoon.
1985 87m/C *AU* Bryan Brown, Anna Maria Monticelli; *Dir:* Chris Thomson. **VHS, Beta $69.98** *LIV, LTG* 🎬½

An Empty Bed Award-winning independent production depicting Bill Frayne, an older homosexual, and the challenges of his everyday life. Told mostly in flashbacks as Bill encounters people and places during the course of one day, this is a delicate drama with important statements about homosexuality, aging and honesty.
1990 60m/C John Wylie, Mark Clifford Smith, Conan McCarty, Dorothy Stinnette, Kevin Kelly, Thomas Hill, Harriet Bass; *W/Dir:* Mark Gasper. Houston International Film Festival '88: Bronze Plaque; Sinking Creek Film and Video Festival '88: Director's Choice. **VHS $59.99** *YOC, FCT* 🎬🎬🎬

Empty Canvas A spiritually-bankrupt artist becomes obsessively jealous of his mistress who refuses to marry him in the hope that someone better will come along. Based on the Alberto Moravia novel.
1964 118m/B *FR IT* Horst Buchholz, Catherine Spaak, Bette Davis; *Dir:* Damiano Damiani. **VHS, Beta $59.98** *SUE, MLB* 🎬½

Empty Suitcases A woman becomes alienated from society when she commutes between a lover in New York City and a job in Chicago.
1984 55m/C **VHS, Beta** *NEW* 🎬🎬

En Passant A highly detailed and creative example of the pin screen animation technique which was created and perfected by Alexandre Alexeïeff and Claire Parker.
1943 3m/C *CA* **VHS, Beta** *NFC* 🎬🎬½

En Retirada (Bloody Retreat) Inside look at the cruel methods of oppression implemented by the Argentine government.

1984 88m/C *AR* Rodolfo Ranni, Julio de Grazia; *Dir:* Juan Carlos Desanzo. **VHS, Beta $49.95** *MED* 🎬🎬½

Enchanted Cottage Represents Hollywood's ''love conquers all'' fantasy hokum, as a disfigured war vet and a homely girl retreat from the horrors of the world into a secluded cottage, where they both regain youth and beauty. A four-tissue heart-tugger. Adopted from an Arthur Pinero play.
1945 91m/B Dorothy McGuire, Robert Young, Herbert Marshall, Mildred Natwick, Spring Byington, Hillary Brooke, Richard Gaines, Robert Clarke; *Dir:* John Cromwell. **VHS, Beta, LV $19.98** *NOS, RKO* 🎬🎬½

The Enchanted Forest An elderly man teaches a boy about life and the beauty of nature when the boy gets lost in a forest.
1945 78m/C *AR* Harry Davenport, Edmund Lowe, Brenda Joyce; *Dir:* Lew Landers. **VHS, Beta, LV $19.95** *STE, NOS, NWV* 🎬🎬½

Enchanted Island Thinly based upon Herman Melville's ''Typee,'' this movie features a sailor who stops on an island to find provisions and ends up falling in love with a cannibal princess.
1958 94m/C Jane Powell, Dana Andrews, Arthur Shields; *Dir:* Allan Dwan. **VHS, Beta $15.95** *UHV* 🎬½

The Enchanted Journey A full-length animated film about a chimpanzee who escapes from a zoo and journeys to freedom.
197? 90m/C *Voices:* Orson Welles, Jim Backus. **VHS, Beta $29.95** *MED* 🎬

The Enchantress A daring boy experiences his first true loves and passions when he journeys through a mystical land in order to find a beautiful fairy. In Greek with subtitles.
1988 93m/C *GR* Alkis Kourkoulos, Sofia Aliberti, Lily Kokodi, Antogone Amanitou, Vicky Koulianou, Nicols Papaconstantinou, Stratos Pachis; *Dir:* Manoussos Manoussakis. **VHS, Beta $59.95** *FCT* 🎬½

Encore The third W. Somerset Maugham omnibus (following ''Quartet'' and ''Trio'') which includes: ''Winter Cruise,'' ''The Ant and the Grasshopper'' and ''Gigolo and Gigolette.''
1952 85m/B *GB* Nigel Patrick, Kay Walsh, Roland Culver, John Laurie, Glynis Johns, Ronald Squire, Noel Purcell, Peter Graves; *Dir:* Pat Jackson, Anthony Pelissier, Harold French. **VHS, Beta $39.95** *AXV* 🎬🎬🎬

Encounter with Disaster Using authentic footage from some of the most frightening events of the 20th-century, this film explores how tragic events unfolded and how humankind has prevailed.
1979 (PG) 93m/C **VHS, Beta $15.95** *UHV, LME*

Encounter at Raven's Gate Punk rockers and extraterrestrials meet amid hard rock and gallons of gore.
1988 (R) 85m/C Eddie Cleary, Steven Violer; *Dir:* Rolf de Heer. **VHS, Beta, LV $29.95** *HMD, HBO, IME* 🎬½

Encounter with the Unknown Relates three fully documented supernatural events including a death prophesy and a ghost.
1975 90m/C Rosie Holotik, Gene Ross; *Dir:* Harry Z. Thomason; *Nar:* Rod Serling. **VHS, Beta $29.95** *UHV* 🎬🎬½

The End Reynolds plays a young man who discovers he is terminally ill. He decides not to prolonging his suffering and attempts various tried-and-true methods for committing suicide, receiving riotous help from the crazed DeLuise.
1978 (R) 100m/C Burt Reynolds, Sally Field, Dom DeLuise, Carl Reiner, Joanne Woodward, Robby Benson, Kristy McNichol, Norman Fell, Pat O'Brien, Myrna Loy, David Steinberg; *Dir:* Burt Reynolds. **VHS, Beta, LV $34.98** *FOX* 🎬🎬🎬

End of August A spinster of New Orleans Creole aristocracy, circa 1900, breaks out of her sheltered life and experiences new sexual and romantic awareness. Adapted from ''The Awakening'' by Kate Chopin.
1982 (PG) 104m/C Sally Sharp, David Marshall Grant, Paul Roebling; *Dir:* Bob Graham. **VHS, Beta $59.98** *CHA* 🎬🎬

End of Desire Schell discovers that her husband, Marquand, has married her for her money and is having an affair with the maid in this melodrama based on De Maupassant's story, ''Une Vie.'' A period piece that is a little on the slow side. In French with English subtitles.
1962 86m/C *FR* Maria Schell, Christian Marquand, Pascale Petit, Ivan Desny; *Dir:* Alexandre Astruc. **VHS** *CGI* 🎬🎬

The End of Innocence In her attempts to please everyone in her life, a woman experiences a nervous breakdown. Written and directed by Cannon.
1990 (R) 102m/C Dyan Cannon, John Heard, George Coe, Viveka Davis, Rebecca Shaeffer, Billie Bird; *W/Dir:* Dyan Cannon. **VHS, Beta $89.95** *PAR, FCT* 🎬🎬½

End of the Line Two old-time railroad workers steal a locomotive for a cross-country jaunt to protest the closing of the local railroad company. Produced by Steenburgen.
1988 (PG) 103m/C Wilford Brimley, Levon Helm, Mary Steenburgen, Kevin Bacon, Holly Hunter, Barbara Barrie, Bob Balaban, Howard Morris, Bruce McGill, Clint Howard, Trey Wilson; *Dir:* Jay Russell. **VHS, Beta, LV $79.95** *KRT, LHV, WAR* 🎬🎬½

End of the Road Keach is a troubled college professor whose bizarre treatment by his psychologist (Jones) produces tragic results. He eventually enters into an affair with the wife of a co-worker. Adapted from John Barth's story, with a confused script cowritten by Terry Southern. Rated X upon release for adult story and nudity. Fascinating, if uneven.
1970 (X) 110m/C Stacy Keach, James Earl Jones, James Coco, Harris Yulin, Dorothy Tristan; *Dir:* Aram Avakian. **VHS, Beta $59.98** *FOX* 🎬🎬½

End of St. Petersburg A Russian peasant becomes a scab during a workers' strike in 1914. He is then forced to enlist in the army prior to the 1917 October Revolution. Fascinating, although propagandistic film commissioned by the then-new Soviet government. Silent.
1927 75m/B Ivan Chuvelov; *Dir:* Vsevolod Pudovkin. **VHS, Beta $24.95** *NOS, MRV, WFV* 🎬🎬🎬½

End of the World A coffee machine explodes, sending a man through a window and into a neon sign, where he is electrocuted. A priest witnesses this and retreats to a convent where he meets his double and heads for more trouble with alien invaders.

1977 (PG) 88m/C Christopher Lee, Sue Lyon, Lew Ayres, MacDonald Carey, Dean Jagger, Kirk Scott; *Dir:* John Hayes. **VHS, Beta $29.95** *MED* 🎬

Endangered Species A retired New York cop on vacation in America's West helps a female sheriff investigate a mysterious series of cattle killings. Based on a true story.
1982 (R) 97m/C Robert Urich, JoBeth Williams, Paul Dooley, Hoyt Axton, Peter Coyote, Harry Carey Jr., Dan Hedaya, John Considine; *Dir:* Alan Rudolph. **VHS, Beta $69.95** *MGM* 🎬

Endgame Grotesquely deformed survivors of World War III fight their way out of radioactive New York City to seek a better life.
1985 98m/C Al Oliver, Moira Chen, Jack Davis, Bobby Rhodes, Jill Elliot; *Dir:* Steve Benson. **VHS, Beta $59.95** *MED* 🎬

Endless Descent ...at least it seems endless. A group of scientist-types set out in search of a sunken sub, but somehow take a wrong turn into the rift, the deepest chasm at the bottom of the sea. There, they discover unimaginable horrors—that is, if you've never seen ''Alien,'' or just about any other icky-monster flick. A Spanish film, shot in English, by director Simon, whose other works of art include ''Pieces'' and ''Slugs.''
1990 (R) 79m/C Jack Scalia, Lee Ermey, Ray Wise, Deborah Adair, Ely Pouget; *Dir:* J.P. Simon. **VHS, Beta, LV $89.95** *LIV* Woof!

The Endless Game Cold War suspenser has British secret-agent-man Finney attempting to solve the murder of his erstwhile fellow agent-lover, much to the dismay of the government. Written by director Forbes, this made-for-cable would-be thriller is a decidedly mediocre waste of a talented cast. Also of note, this was actor-director Sir Anthony Quayle's final appearance.
1989 (PG-13) 123m/C *GB* Albert Finney, George Segal, Derek De Lint, Monica Guerritore, Ian Holm, John Standing, Anthony Quayle, Kristin Scott Thomas. **VHS $89.95** *PSM, FCT* 🎬🎬

Endless Love Although only 15, David and Jade are in love. Her parents think they are too serious and demand that the two spend time apart. David attempts to win her parents' affection and approval, goes mad in the process, and commits a foolish act (he burns the house down) that threatens their love forever. Based on the novel by Scott Spencer. Of interest only to those with time on their hands or smitten by a love so bad that this movie will seem grand in comparison. Features Cruise's first film appearance.
1981 (R) 115m/C Brooke Shields, Martin Hewitt, Don Murray, Shirley Knight, Beatrice Straight, Richard Kiley, Tom Cruise, James Spader, Robert Moore, Jami Gertz; *Dir:* Franco Zeffirelli. **VHS, Beta, LV $29.98** *LIV, VES* 🎬

Endless Night An adaptation of an Agatha Christie tale. Focuses on a young chauffeur who wants to build a dream house, and his chance meeting with an American heiress.
1971 95m/C *GB* Hayley Mills, Hywel Bennett, Britt Ekland, George Sanders, Per Oscarsson, Peter Bowles; *Dir:* Sidney Gilliat. **VHS, Beta, LV $19.99** *HBO* 🎬🎬

The Endless Summer Director Bruce Brown follows two young surfers around the world in their search for the perfect wave in this wryly narrated documentary.
1966 90m/C Mike Hynson, Robert August; *Dir:* Bruce Brown. **VHS, Beta, LV $24.95** *PAV, FCT*

Endplay An Australian-made crime/horror drama in which two brothers cover for each other in a series of murders involving blonde hitchhikers. Based on a novel by Russell Braddon.
1975 110m/C *AU* George Mallaby, John Waters, Ken Goodlet, Delvene Delaney, Charles Tingwell, Robert Hewett, Kevin Miles; *Dir:* Tim Burstall. **VHS \$14.95** *ACA* ♫♫ ½

Endurance This tape shows highlights from the popular Japanese television show, in which endurance of pain and psychological discomfort, is the name of the game. The men competing for the grand prize are dragged over rocks, bitten by fish, submerged, attacked by bulls and rats, burned by the sun, and subjected to a variety of other tortures.
1985 90m/C **VHS, Beta \$69.95** *NSV*

Enemies, a Love Story A wonderfully resonant, subtle tragedy based on the novel by Isaac Bashevis Singer. A post-Holocaust Jew, living in Coney Island, can't choose between three women—his current wife, his tempestuous lover (who hid him during the war), and his reappearing pre-war wife he presumed dead. A hilarious, confident tale told with grace and patience; heavily Oscar-nominated.
1989 (R) 119m/C Ron Silver, Lena Olin, Anjelica Huston, Margaret Sophie Stein, Paul Mazursky, Alan King, Judith Malina, Rita Karin, Phil Leeds, Elya Baskin, Marie-Adele Lemieux; *Dir:* Paul Mazursky. New York Film Critics Awards '89: Best Director, Best Supporting Actress (Olin). **VHS, Beta, LV \$89.98** *MED* ♫♫♫ ½

Enemy Below Suspenseful World War II sea epic, in which an American destroyer and a German U-Boat chase one another and square off in the South Atlantic.
1957 98m/C Robert Mitchum, Curt Jurgens, David Hedison, Theodore Bikel, Doug McClure, Russell Collins; *Dir:* Dick Powell. Academy Awards '57: Best Special Effects; National Board of Review Awards '57: 10 Best Films of the Year. **VHS, Beta \$19.98** *FOX, FCT* ♫♫♫

Enemy of the Law Texas rangers battle evil in an old frontier town.
1945 59m/B Tex Ritter, Dave O'Brien; *Dir:* Harry Fraser. **VHS, Beta \$19.95** *NOS, VCN, DVT* ♫♫ ½

Enemy Mine A space fantasy in which two pilots from warring planets, one an Earthling, the other an asexual reptilian Drac, crash land on a barren planet and are forced to work together to survive. Warm-hearted family entertainment.
1985 (PG-13) 108m/C Dennis Quaid, Louis Gossett Jr., Brion James, Richard Marcus, Lance Kerwin; *Dir:* Wolfgang Petersen. **VHS, Beta, LV \$19.98** *FOX* ♫♫

Enemy Territory A handful of citizens trapped in a New York City housing project after dark, are stalked by a violent, murderous street gang.
1987 (R) 89m/C Ray Parker Jr., Jan-Michael Vincent, Gary Frank, Frances Foster; *Dir:* Peter Manoogian. **VHS, Beta \$79.98** *FOX* ♫♫

Enemy Unseen A bickering squad of mercenaries slog through the African jungle to find a girl abducted by crocodile-worshipping natives. Dull Jungle-Jim-style adventure, notable for the hilariously phony crocs that occasionally clamp onto the characters.
1991 (R) 90m/C Vernon Wells, Stack Pierce, Ken Gampu, Michael McCabe, Angela O'Neill; *Dir:* Elmo De Witt. **VHS \$79.95** *AIP* ♫

Enemy of Women Chronicles the life and loves of Nazi propagandist Dr. Joseph Goebbels. Also called "The Private Life of Paul Joseph Goebbels."
1944 90m/B H.B. Warner, Claudia Drake, Donald Woods; *Dir:* Alfred Zeisler. **VHS, Beta \$24.95** *NOS, DVT, GLV* ♫♫ ½

The Enforcer A district attorney goes after an organized gang of killers in this film noir treatment of the real-life "Murder, Inc." case.
1951 87m/B Humphrey Bogart, Zero Mostel, Ted de Corsia, Everett Sloane, Roy Roberts, Lawrence Tolan, King Donovan, Bob Steele, Adelaide Klein, Don Beddoe, Tito Vuolo, John Kellogg; *Dir:* Bretaigne Windust, Raoul Walsh. **VHS, LV \$19.95** *REP* ♫♫ ½

The Enforcer Dirty Harry takes on a female partner and a vicious terrorist group that is threatening the city of San Francisco. See how many "punks" feel lucky enough to test the hand of the tough cop.
1976 (R) 96m/C Clint Eastwood, Tyne Daly, Harry Guardino, Bradford Dillman, John Mitchum; *Dir:* James Fargo. **VHS, Beta, LV \$19.98** *WAR* ♫♫ ½

Enforcer from Death Row An ex-con is recruited by a secret international peacekeeping organization to track down and eliminate a band of murderous spies.
1978 87m/C Cameron Mitchell, Leo Fong, Darnell Garcia, Booker T. Anderson, John Hammond, Mariwin Roberts. **VHS, Beta \$69.98** *LIV, LTG, GHV* ♫

An Englishman Abroad During the Cold War a British actress visits Moscow and chances to meet a notorious English defector. Behind their gossip and small talk is a tragicomic portrait of the exiled traitor/spy. He was the infamous Guy Burgess; the actress was Coral Browne, here playing herself in a pointed recreation scripted by Alan Bennett. Made for British TV; nuances may be lost on yank viewers. Winner of several British awards.
1983 63m/C *GB* Alan Bates, Coral Browne, Charles Gray; *Dir:* John Schlesinger. **VHS \$49.95** *PRV, FCT* ♫♫♫

Enid is Sleeping Well-done comedy noir in the now-popular there's a corpse in the closet subgenre. A woman in a mythical New Mexican town tries to hide the body of her sister Enid, who she's accidentally killed. Enid, it turns out, wasn't thrilled to discover her sister sleeping with her police-officer husband. Phillips and Perkins restored the film to its original state after it was ruthlessly gutted by the studio. Written by director Phillips.
1990 ?m/C Elizabeth Perkins, Judge Reinhold; *W/Dir:* Maurice Phillips. **VHS** *VES* ♫♫♫

Enigma Trapped behind the Iron Curtain, a double agent tries to find the key to five pending murders by locating a Russian coded microprocessor holding information that would unravel the assassination scheme.
1982 (PG) 101m/C *FR GB* Martin Sheen, Brigitte Fossey, Sam Neill, Derek Jacobi, Frank Finlay, Michael Lonsdale; *Dir:* Jeannot Szwarc. **VHS, Beta, LV \$69.98** *SUE* ♫♫

Enigma Secret Three Polish mathematicians use their noggins to break the Nazi secret code machine during WWII. Based on a true story; in Polish with English subtitles.
1979 158m/C *PL* Tadeusz Borowski, Piotr Fronczewski, Piotr Garlicki. **VHS \$49.95** *FCT* ♫♫

Enola Gay: The Men, the Mission, the Atomic Bomb Based on the bestselling book by Gordon Thomas and Max Gordon Witts, this made for television drama tells the story of the airmen aboard the B-29 that dropped the first atomic bomb. Details the events during World War II leading up to the decision to bomb Hiroshima and the concerns of the crew assigned the task.
1980 150m/C Patrick Duffy, Billy Crystal, Kim Darby, Gary Frank, Gregory Harrison, Ed Nelson, Robert Walden, Stephen Macht, Robert Pine, James Shigeta, Henry Wilcoxon; *Dir:* David Lowell Rich. **VHS, Beta \$69.95** *PSM* ♫♫

Enormous Changes Three stories about New York City women and their personal relationships. Screenplay co-authored by John Sayles was based on the stories of Grace Paley. Also titled "Enormous Changes at the Last Minute."
1983 115m/C Ellen Barkin, David Strathairn, Ron McLarty, Maria Tucci, Lynn Milgrim, Kevin Bacon; *Dir:* Mirra Bank. **VHS, Beta, LV \$79.95** *VMK* ♫♫♫

Enrapture A chauffeur is the prime suspect when a promiscuous passenger is murdered in his limousine. The driver did it?
1990 (R) 87m/C Ona Simms, Harvey Siegel, Richard Parnes; *Dir:* Chuck Vincent. **VHS, LV \$79.95** *ATL* ♫ ½

Ensign Pulver A continuation of the further adventures of the crew of the U.S.S. Reluctant from "Mister Roberts," which was adapted from the Broadway play.
1964 104m/C Walter Matthau, Robert Walker Jr., Larry Hagman, Jack Nicholson, Millie Perkins, James Coco, James Farentino, Burl Ives, Gerald S. O'Loughlin, Al Freeman Jr.; *Dir:* Joshua Logan. **VHS, Beta \$19.98** *WAR* ♫♫

Enter the Dragon The American film that broke Bruce Lee worldwide combines Oriental conventions with 007 thrills. Spectacular fighting sequences including Karate, Judo, Tae Kwon Do, and Tai Chi Chuan are featured as Lee is recruited by the British to search for opium smugglers in Hong Kong.
1973 (R) 98m/C Bruce Lee, John Saxon, Jim Kelly; *Dir:* Robert Clouse. **VHS, Beta, LV \$19.98** *WAR* ♫♫♫

Enter the Game of Death Bruce Lee stars in this kung fu epic with mucho head kicking and rib punching.
1980 88m/C Bruce Lee. **VHS, Beta \$29.99** *BFV* ♫ ½

Enter Laughing Based on Reiner's semi-autobiographical novel and play, depicts the botched efforts of a Bronx-born schlump to become an actor. Worthwhile, but doesn't live up to the original.
1967 112m/C Reni Santoni, Jose Ferrer, Elaine May, Shelley Winters, Jack Gilford, Don Rickles, Michael J. Pollard, Janet Margolin, Rob Reiner; *Dir:* Carl Reiner. **VHS, Beta \$59.95** *RCA* ♫♫

Enter the Ninja The story of the Ninja warrior's lethal, little-known art of invisibility.
1981 94m/C Franco Nero, Susan George, Sho Kosugi, Christopher George, Alex Courtney; *Dir:* Menahem Golan. **VHS, Beta \$69.95** *MGM* ♫♫

Enter the Panther Amidst, and regardless of, a double-crossing plot featuring a treacherous family and a sought-after gold mine, a young man kicks heads and punches his way to fame.
1979 91m/C Bruce Li, Tsao Chen, Tse Lan. **VHS, Beta \$24.95** *GEM* ♫

Enter Three Dragons Kung Fu artistry abounds in this head-kicking action-adventure film.
1981 (R) 90m/C Dragon Lee, Bruce Lea, Yang Tsze, Samuel Walls, Chang Li, Bruce Li, Jackie Chin. **VHS, Beta** *SUN* 🎬

The Entertainer Splendid drama of a down and out vaudevillian who tries vainly to gain the fame his dying father once possessed. His blatant disregard for his alcoholic wife and his superficial sons brings his world crashing down around him, and he discovers how self-destructive his life has been. Remade in 1975 for television. Adapted from the play by John Osborne. Oscar nomination for Olivier.
1960 97m/B *GB* Laurence Olivier, Brenda de Banzie, Roger Livesey, Joan Plowright, Daniel Massey, Alan Bates, Shirley Anne Field, Albert Finney, Thora Hird; *Dir:* Tony Richardson. **VHS** $39.95 *FCT* 🎬🎬🎬½

Entertaining Mr. Sloane Playwright Joe Orton's masterpiece of black comedy concerning a handsome criminal who becomes the guest and love interest of a widow and her brother.
1970 90m/C Beryl Reid, Harry Andrews, Peter McEnery, Alan Webb; *Dir:* Douglas Hickox. **VHS, Beta** $59.99 *HBO* 🎬🎬½

Entertaining the Troops A look at Hollywood's contribution to WWII fund-raising and entertainment.
1989 53m/C Bud Abbott, Lou Costello, Jack Benny, Jimmy Durante, Eddie Cantor, Lucille Ball, Bob Hope. **VHS, LV** $29.95 *SVS*

Enthusiasm A historic documentary/propaganda piece portraying the coal miners in the Don Basin enthusiastically working out their communist roles; chiefly fascinating for its montage-use of sound. In Russian with English titles.
1931 67m/B *RU* Dir: Dziga Vertov. **VHS, Beta** $29.95 *FCT*

The Entity Based on a true story about an unseen entity that repeatedly raped a woman. Hershey's the victim who eventually ends up at a university for talks with parapsychologist Silver. Pseudo-science to the rescue as the over-sexed creature is frozen dead in its tracks. Violence, gore, and nudity aplenty.
1983 (R) 115m/C Barbara Hershey, Ron Silver, Alex Rocco; *Dir:* Sidney J. Furie. **VHS, Beta** $19.98 *FOX* 🎬🎬½

Entr'acte Clair's famous dada-surrealist classic short, reflecting his sense of humor, as various eccentric characters become involved in a crazy chase. Silent.
1924 21m/B *FR* Dir: Rene Clair. **VHS, Beta** *FST, MRV* 🎬🎬½

Entre-Nous (Between Us) Two attractive, young French mothers find in each other the fulfillment their husbands cannot provide. One of the women was confined in a concentration camp during WWII; the other is a disaffected artist. French dialogue with English subtitles. Alternate title: "Coup de Foudre."
1983 (PG) 110m/C *FR* Isabelle Huppert, Miou-Miou, Guy Marchand; *Dir:* Diane Kurys. **VHS, Beta** $29.95 *MGM* 🎬🎬🎬½

EPIC: Days of the Dinosaurs An animated fable about two babies born into a pack of dingoes in a prehistoric world. They must fight for superiority over the mystical powers of nature.
1987 75m/C *Dir:* Yoram Gross. **VHS, Beta** $39.95 *IVE, FHE* 🎬½

Epic That Never Was A fascinating documentary of the circumstances under which a big-budgeted 1937 adaptation of Graves' "I, Claudius," filmed by Von Sternberg, was shelved after Oberon's auto accident and director/star disagreements. Features both interviews made 30 years later and much of the existing footage, a testament to its unfulfilled potential as a masterpiece.
1965 74m/C Charles Laughton, Flora Robson, Merle Oberon, Emlyn Williams, Robert Newton, Josef von Sternberg, Robert Graves; *Hosted:* Dirk Bogarde. **VHS, Beta, 8mm** $24.95 *NOS, VYY*

Epitaph An average American family's life is disrupted by the mother's habitual axe-murdering.
1987 90m/C *Dir:* Joseph Merhi. **VHS, Beta** $39.95 *CLV* 🎬

Equalizer 2000 A warrior in a post-holocaust future plots to overthrow a dictatorship by using a high-powered gun.
1986 (R) 85m/C Richard Norton, Corinne Wahl, William Steis; *Dir:* Cirio H. Santiago. **VHS, Beta** $79.95 *MGM* 🎬

Equinox Young archaeologists uncover horror in a state forest. The ranger, questing for a book of spells that the scientists have found, threatens them with wonderful special effects, including winged beasts, huge apes, and Satan. Though originally an amateur film, it is deemed a minor classic in its genre.
1971 (PG) 80m/C Edward Connell, Barbara Hewitt, Frank Bonner, Robin Christopher, Jack Woods, Fritz Leiber; *Dir:* Jack Woods. **VHS** *UHV* 🎬🎬🎬

Equinox Flower Ozu's first color film tells the sensitive story of two teenage girls who make a pact to protect each other from the traditional prearranged marriages their parents have set up. Lovely film that focuses on the generation gap between young and old in the Japanese family. In Japanese with English subtitles.
1958 118m/C *JP* Shin Saburi, Kinuyo Tanaka, Ineko Arima, Miyuki Kuwano, Chishu Ryu; *Dir:* Yasujiro Ozu. **VHS** $69.95 *NYF, ART, CNT* 🎬🎬🎬

Equus A psychiatrist undertakes the most challenging case of his career when he tries to figure out why a stable-boy blinded horses. Based upon the successful play by Peter Shaffer, but not well transfered to film. Oscar nominations for Burton and Firth.
1977 (R) 138m/C Richard Burton, Peter Firth, Jenny Agutter, Joan Plowright, Colin Blakely, Harry Andrews; *Dir:* Sidney Lumet. British Academy Awards '77: Best Supporting Actress (Agutter); National Board of Review Awards '77: 10 Best Films of the Year. **VHS, Beta, LV** $69.95 *MGM* 🎬🎬½

Era Notte a Roma An American, Russian and British soldier each escape from a concentration camp in the waning days of WWII and find refuge in the home of a young woman. Rossellini also co-scripted.
1960 145m/C *IT* Dir: Roberto Rossellini. **VHS** $79.95 *TPV* 🎬🎬🎬

Eraserhead The infamous cult classic about a numb-brained everyman wandering through what amounts to a sick, ironic parody of the modern urban landscape, innocently impregnating his girlfriend and fathering a pestilent embryonic mutant. Surreal and bizarre, the film has an inner, completely unpredictable logic all its own. Lynch's first feature-length film (he directed, produced, and wrote the screenplay). Stars Jack Nance, who later achieved fame in Lynch's "Twin Peaks" as Pete the logger.
1978 90m/B Jack Nance, Charlotte Stewart, Jack Fisk, Jeanne Bates; *Dir:* David Lynch. **VHS, Beta, LV** *COL, OM* 🎬🎬🎬

Erendira Based on Gabriel Garcia-Marquez's story about a teenage girl prostituting herself to support her witch-like grandmother after accidentally torching the elder's house. The film follows the two as they travel across the desert trying out new and inventive ways to survive. Unusual ending tops off a creative, if not eclectic, movie. In Spanish with English subtitles.
1983 103m/C *FR MX GE* Irene Papas, Claudia Ohana, Michael Lonsdale, Rufus, Jorge Fegan; *Dir:* Ruy Guerra. **VHS, Beta** $24.95 *XVC, MED, APD* 🎬🎬½

Eric Made-for-television tear-jerker about a young athlete who fights for his life after he's diagnosed with leukemia. Based on the true-life account written by Eric's mother Doris Lund.
1975 100m/C Patricia Neal, John Savage, Claude Akins, Sian Barbara Allen, Mark Hamill, Nehemiah Persoff, Tom Clancy; *Dir:* James Goldstone. **VHS, Beta** $39.95 *IVE* 🎬🎬🎬

Erik An ex-government agent working in Central America finds his loyalties divided when he is approached by a female activist and his old CIA friends. They want him to help expose his current boss as a drug smuggler.
1990 90m/C Stephen McHattie, Deborah Van Valkenburgh. **VHS, Beta** *SVS* 🎬🎬

Erik, the Viking The Norse Warrior discovers not only the New World but traitorous subordinates among his crew, calling for drastic measures.
1972 95m/C Giuliano Gemma, Gordon Mitchell. **VHS, Beta** $59.95 *UNI* 🎬

Erik the Viking A mediocre Monty Pythonesque farce about a Viking who grows dissatisfied with his barbaric way of life and decides to set out to find the mythical Asgaard, where Norse gods dwell. Great cast of character actors is wasted.
1989 (PG-13) 104m/C *GB* Tim Robbins, Terry Jones, Mickey Rooney, John Cleese, Imogen Stubbs, Anthony Sher, Gordon John Sinclair, Freddie Jones, Eartha Kitt; *Dir:* Terry Jones. **VHS, Beta, LV** $89.98 *ORI* 🎬½

The Ernest Film Festival Nothing but ads fill this hour-long video, featuring more than 100 Ernest and Vern commercials.
1986 55m/C Jim Varney; *Dir:* John R. Cherry III. **VHS, Beta** *KWI*

Ernest Goes to Camp A screwball, slapstick summer camp farce starring the character Ernest P. Worrell as an inept camp counselor. When progress threatens the camp, Ernest leads the boys on a turtle-bombing, slop-shooting attack on the construction company. Followed by "Ernest Saves Christmas" and "Ernest Goes to Jail."
1987 (PG) 92m/C Jim Varney, Victoria Racimo, John Vernon, Iron Eyes Cody, Lyle Alzado, Gailard Sartain, Daniel Butler, Hakeem Abdul-Samad; *Dir:* John R. Cherry III. **VHS, Beta** $19.99 *TOU* 🎬½

Ernest Goes to Jail The infamous loon Ernest P. Worrell winds up in the jury box and the courtroom will never be the same again. Jury duty suddenly becomes hard-time in the slammer for poor Ernest when he is mistaken for a big-wig organized

crime boss. Sequel to "Ernest Goes to Camp" and "Ernest Saves Christmas."
1990 81m/C Jim Varney, Gailard Sartain, Randall "Tex" Cobb, Bill Byrge, Barry Scott, Charles Napier; *Dir:* John R. Cherry III. **VHS, LV** $89.95 *BVV, TOU ✗*

Ernest Saves Christmas Ernest P. Worrell is back. When Santa decides that it's time to retire, Ernest must help him recruit a has-been children's show host who is a bit reluctant. For Ernest fans only, and only the most dedicated of those. Second in the series featuring the nimble-faced Varney, the first of which was "Ernest Goes to Camp," followed by "Ernest Goes to Jail."
1988 (PG) 91m/C Jim Varney, Douglas Seale, Oliver Clark, Noelle Parker, Billie Bird; *Dir:* John R. Cherry III. **VHS, Beta, LV** $89.95 *TOU ✗✗½*

Ernest Scared Stupid Pea-brained Ernest P. Worrell returns yet again in this silly comedy. When Ernest accidentally releases a demon from a sacred tomb a 200-year-old curse threatens to destroy his hometown, unless Ernest can come to the rescue. Would you want your town depending on Ernest's heroics?
1992 (PG) 93m/C Jim Varney, Eartha Kitt; *Dir:* John Cherry. **VHS, Beta** $94.95 *TOU ✗✗½*

Ernesto A lushly erotic Italian film depicting the troubled relationship between a young, devil-may-care gay youth and his older, cool-ly seducing lover. In Italian with subtitles.
1979 98m/C *IT* Martin Halm, Michele Placido; *Dir:* Salvatore Samperi. **VHS, Beta** $79.50 *FCT ✗✗½*

Ernie Banks: History of the Black Baseball Player Documents the history of the black athlete in America's favorite pastime.
1990 45m/C VHS $19.95 *FRH*

Ernie Kovacs: Between the Laughter A made for television version of the television comic's life and career, ending with his tragic death in a car accident in 1962. Shows the brand of humor that made Kovacs famous as well as the devastation that he suffered after the tragic kidnapping of his children by his first wife during divorce proceedings. Kovac's second wife, Edie Adams, makes a brief appearance.
1984 95m/C Jeff Goldblum, Cloris Leachman, Melody Anderson, Madolyn Smith, John Glover, Edie Adams; *Dir:* Lamont Johnson. **VHS, Beta** $79.95 *WES ✗✗*

Ernie Kovacs/Peter Sellers A sample of Sellers and Kovacs performing classic routines as memorable characters circa 1956.
1956 60m/B Ernie Kovacs, Peter Sellers. **VHS, Beta** $29.95 *DVT*

The Ernie Kovacs Show The November 24, 1961 episode of Kovacs' comedy show, wherein there was no spoken dialogue, not even in the three commercials.
1961 29m/B Ernie Kovacs. **VHS, Beta, 8mm** $19.95 *VYY*

Ernie Kovacs: TV's Comedy Wizard A compilation of outstanding skits and zany characters dreamed up in the golden years of television by one of the medium's greatest innovators.
1991 70m/B Ernie Kovacs. **VHS** $9.99 *VTR*

Eros, Love & Lies A unique talk on human relationships and the great importance of being honest with ourselves and with others, by Laing.

1990 55m/C R.D. Laing; *Dir:* Mark Elliot. **VHS** $29.95 *MFV, IVA*

Erotic Adventures of Pinocchio Softcore comedy version of the fairy tale. As the ads say, "it's not his nose that grows!" You'll have to look carefully to spot Danning and Brandt; Steckler served as cinematographer.
1971 (R) 77m/C Alex Roman, Dyanne Thorne, Karen Smith, Eduardo Ranez, Carolyn Brandt, Sybil Danning; *Dir:* Corey Allen. **VHS** $49.95 *AVD ✗✗*

Erotic Dreams The Devil tries to persuade a young guy to come to Hell by showing him a tableau of lusty soft-core vignettes.
1987 70m/C VHS, Beta $59.95 *CEL* Woof!

Erotic Escape Soft-core fun about party girls taking over a small French town. Originally in French, but dubbed in English. Check your brain at the door.
1986 86m/C *FR* Marie-Claire Davy, Pauline Larrieu. **VHS, Beta** $29.95 *MED ✗*

Erotic Illusion A strange mother, father and son view a porn film and each becomes obsessed with its heroine. Italian softcore.
1986 90m/C *IT* Silvana Venturelli, Frank Wolff, Paolo Turco. **VHS, Beta** $29.95 *MED ✗*

Erotic Images A beautiful psychology teacher publishes her doctoral thesis on sex and becomes a best-selling author, which threatens her marriage.
1985 (R) 93m/C Britt Ekland, Edd Byrnes, John McCann; *Dir:* Declan Langan. **VHS, Beta** $69.98 *LIV, VES* Woof!

Erotic Taboo A softcore look at taboo-breaking sexual practices around the world.
1988 85m/C VHS, Beta $29.95 *MED*

The Erotic Three An unusual love triangle is the focus of this film.
1983 90m/C VHS, Beta *PGN ✗*

Erotic Touch of Hot Skin Sex, murder, false identities, car crashes, and striptease in the Riviera.
1965 78m/B Fabienne Dali, Sophie Hardy, Jean Valmont, Francois Dryek; *Dir:* Max Pecas. **VHS, Beta** $19.95 *VDM ✗*

Erotica: Fabulous Female Fantasies A made-for-video soft-core feature involving the sexual adventures of five women at a resort hotel.
1986 60m/C *Dir:* Adam Friedman. **VHS, Beta** $29.98 *LIV, VES ✗*

Errand Boy Jerry Lewis' patented babbling schnook hits Hollywood in search of a job. When he lands a position as an errand boy, Hollywood may never be the same again in this prototypical comedy; a must for Lewis fans only.
1961 92m/B Jerry Lewis, Brian Donlevy, Dick Wesson, Howard McNear, Felicia Atkins, Fritz Feld, Sig Rumann, Renee Taylor, Doodles Weaver, Mike Mazurki, Lorne Greene, Michael Landon, Dan Blocker, Pernell Roberts, Snub Pollard, Kathleen Freeman; *Dir:* Jerry Lewis. **VHS, Beta** $59.95 *IVE ✗✗½*

Eruption: St. Helens Explodes A look at the eruption of Mount St. Helens on May 18, 1980. Station KOIN-TV in Portland, Oregon, shot the blast, showing the volcanic peak being torn away. They almost lost a $70,000 remote broadcast truck while shooting.
1980 25m/C VHS, Beta $29.95 *CCB*

Escapade Sons thinking their parents are en route to divorce court create a scheme to achieve peace worldwide. The pacifist father's reaction juxtaposes idealistic youth with the cynicism of adulthood.
1955 87m/B *GB* John Mills, Yvonne Mitchell, Alastair Sim, Jeremy Spenser, Andrew Ray, Marie Lohr, Peter Asher; *Dir:* Philip Leacock. **VHS** $29.95 *MOV ✗✗✗*

Escapade in Florence As two students in Florence paint their way to immortality, an elaborate art-forging ring preys upon their talents.
1962 81m/C Tommy Kirk, Ivan Desny; *Dir:* Steve Previn. **VHS, Beta** $69.95 *DIS ✗*

Escapade in Japan A Japanese youth helps an American boy frantically search the city of Tokyo for his parents. Shot in Japan.
1957 93m/C Cameron Mitchell, Teresa Wright, Jon Provost, Roger Nakagawa, Philip Ober, Clint Eastwood; *Dir:* Arthur Lubin. **VHS, Beta** $15.95 *UHV ✗✗*

Escape A woman attempts to track down her brother's killer and finds an entire town mysteriously controlled by a sadistic and powerful man. Features General Hospital's bad-boy Shriner.
1990 (R) 100m/C Elizabeth Jeager, Kim Richards, Kin Shriner; *Dir:* Richard Styles. **VHS** *QUE, HHE ✗½*

Escape A fun-loving, care-free Irish officer is sent to oversee the toughest POW prison in Scotland. There, he becomes consumed with keeping the facility secure despite the intricate escape plans laid out by a group of rioters.
1990 (PG) 90m/C Brian Keith, Helmut Griem; *Dir:* Lamont Johnson. **VHS** *API ✗✗*

Escape 2000 In a future society where individuality is considered a crime, those who refuse to conform are punished by being hunted down in a jungle. Gross, twisted takeoff of Richard Connell's "The Most Dangerous Game."
1981 (R) 80m/C *AU* Steve Railsback, Olivia Hussey, Michael Craig; *Dir:* Brian Trenchard-Smith. **VHS, Beta** $9.98 *SUE, STE* Woof!

Escape from Alcatraz A fascinating account of the one and only successful escape from the maximum security prison at Alcatraz in 1962. The three men were never heard from again.
1979 (PG) 112m/C Clint Eastwood, Patrick McGoohan, Roberts Blossom, Fred Ward, Danny Glover; *Dir:* Don Siegel. **VHS, Beta, LV** $14.95 *PAR ✗✗✗*

Escape Artist Award-winning cinematographer Deschanel's first directorial effort is this quirky film about a teenage escape artist who sets out to uncover the identity of his father's killers.
1982 (PG) 96m/C Griffin O'Neal, Raul Julia, Teri Garr, Joan Hackett, Desi Arnaz Sr., Gabriel Dell, Huntz Hall, Jackie Coogan; *Dir:* Caleb Deschanel. **VHS, Beta, LV** $79.98 *LIV, VES ✗✗✗*

Escape to Athena A motley group is stuck in a German P.O.W. camp on a Greek island during World War II.
1979 (PG) 102m/C *GB* Roger Moore, Telly Savalas, David Niven, Claudia Cardinale, Richard Roundtree, Stefanie Powers, Sonny Bono, Elliott Gould, William Holden; *Dir:* George P. Cosmatos. **VHS, Beta** $19.98 *FOX ✗✗½*

Escape from the Bronx Invading death squads seek to level the Bronx. Local street gangs cry foul and ally to defeat the uncultured barbarians. Sequel to "1990: The Bronx Warriors."

1985 (R) 82m/C Mark Gregory, Henry Silva, Valeria D'Obici, Timothy Brent, Thomas Moore, Andrea Coppola; *Dir:* Enzo G. Castellari. **VHS, Beta $69.95** *MED* 🎞

Escape to Burma A man on the run for a murder he did not commit finds refuge and romance in an isolated jungle home.
1955 86m/C Barbara Stanwyck, Robert Ryan, Reginald Denny; *Dir:* Allan Dwan. **VHS, Beta $39.95** *DIS, BVV* 🎞½

Escape from Cell Block Three Five escaped female convicts take it on the lam for Mexico and freedom.
1978 82m/C Carolyn Judd, Teri Guzman, Bonita Kalem. **VHS, Beta $9.99** *STE, PSM* 🎞

Escape from Death Row A convicted criminal mastermind, sentenced to die, devises a brilliant and daring plan of escape on the eve of his execution.
1976 (R) 85m/C Lee Van Cleef, James Lane, Barbara Moore, Alice Belios. **VHS, Beta** *PGN* 🎞

Escape from DS-3 Bostwick and "Police Academy" alumnus Smith team up in a familiar story: framed for a serious crime, a man attempts to escape from a maximum security satellite prison. About as good as you'd expect.
1982 88m/C Barry Bostwick, Bubba Smith. **VHS $79.99** *MNC* 🎞🎞

Escape from El Diablo Two juvenile delinquents harass guards at a Mexican prison. Being from California, they use frisbees and skateboards to help their escape.
1983 (PG) 92m/C *SP GB* Jimmy McNichol, Timothy Van Patten, John Ethan Wayne; *Dir:* Gordon Hessler. **VHS, Beta $9.95** *MED* 🎞

Escape from Galaxy Three A pair of space travelers fight off a bevy of evil aliens.
1976 (G) 90m/C James Milton, Cheryl Buchanan. **VHS, Beta $79.95** *PSM* 🎞

Escape from Hell Two scantily clad women escape from a jungle prison and are pursued by their sadistic warden.
198? ?m/C Anthony Steffen, Ajita Wilson; *Dir:* Edward Muller. **VHS $59.95** *HHE* 🎞

Escape from the KGB A CIA agent escapes from a Soviet prison, taking plans for a new space installation with him.
1987 99m/C Thomas Hunter, Ivan Desny, Marie Versini, Walter Barns; *Dir:* Harald Phillipe. **VHS, Beta $59.95** *MAG Woof!*

Escape to Love A beautiful student helps a famous dissident escape from Poland, only to lose him to another heroic venture.
1982 105m/C Clara Perryman, Louis Jourdan; *Dir:* Herbert Stein. **Beta $14.95** *PSM* 🎞½

Escape from New York The ultimate urban nightmare: a ruined, future Manhattan is an anarchic prison for America's worst felons. When convicts hold the President hostage, a disgraced war hero unwillingly attempts an impossible rescue mission. Cynical but largely unexceptional sci-fi action, putting a good cast through tight-lipped peril.
1981 (R) 99m/C Kurt Russell, Lee Van Cleef, Ernest Borgnine, Donald Pleasence, Isaac Hayes, Adrienne Barbeau, Harry Dean Stanton, Season Hubley; *Dir:* John Carpenter. **VHS, Beta, LV $14.98** *SUE* 🎞🎞½

Escape to Paradise The last of Breen's films for RKO has him as a South American motorcycle-taxi driver acting as a guide for tourist Taylor. When Breen sets Taylor up with Shelton, trouble ensues.

Breen sings "Rhythm of the Rio" and several others.
1939 60m/B Bobby Breen, Kent Taylor, Maria Shelton, Joyce Compton, Rosina Galli, Frank Yaconelli; *Dir:* Erle C. Kenton. **VHS $19.95** *NOS, RXM* 🎞½

Escape from the Planet of the Apes Reprising their roles as intelligent, English-speaking apes, McDowall and Hunter flee their world before it's destroyed, and travel back in time to present-day America. In L.A. they become the subjects of a relentless search by the fearful population, much like humans Charlton Heston and James Franciscus were targeted for experimentation and destruction in simian society in the earlier "Planet of the Apes" and "Beneath the Planet of the Apes." Sequelled by "Conquest of the Planet of the Apes" and a television series.
1971 (G) 98m/C Roddy McDowall, Kim Hunter, Sal Mineo, Ricardo Montalban, William Windom, Bradford Dillman, Natalie Trundy, Eric Braeden; *Dir:* Don Taylor. **VHS, Beta, LV $19.98** *FOX* 🎞🎞🎞

Escape from Planet Earth A spaceship is damaged deep in space and only a few of the crew can make it back to Earth. But who will decide who lives or dies? Also known as "The Doomsday Machine."
1967 91m/B Grant Williams, Bobby Van, Ruta Lee, Henry Wilcoxon, Mala Powers, Casey Kasem, Mike Farrell, Harry Hope; *Dir:* Lee Sholem. **VHS $59.95** *ACA* 🎞

Escape from Safehaven Brutal slimeballs rule a mad, sadistic world in the post-apocalyptic future, and a family tries to escape them. In very poor taste.
1988 (R) 87m/C Rick Gianasi, Mollie O'Mara, John Wittenbauer, Roy MacArthur, William Beckwith; *Dir:* Brian Thomas Jones, James McCalmont. **VHS, Beta $79.95** *SVS Woof!*

Escape from Sobibor Nail-biting, true account of the largest successful escape from a Nazi prison camp, adapted from Richard Rashke's book. Originally a made for TV movie.
1987 (PG-13) 120m/C Alan Arkin, Joanna Pacula, Rutger Hauer, Hartmut Becker, Jack Shepherd; *Dir:* Jack Gold. **VHS $89.98** *LIV* 🎞🎞🎞

Escape to the Sun A pair of Russian university students plan to flee their homeland so they can be allowed to live and love free from oppression.
1972 (PG) 94m/C *IS* Laurence Harvey, Josephine Chaplin, John Ireland, Jack Hawkins, Lila Kedrova, Clive Revill; *Dir:* Menahem Golan. **VHS, Beta $39.95** *IVE, MON* 🎞½

Escape to Witch Mountain Two young orphans with supernatural powers find themselves on the run from a greedy millionaire who wants to exploit their amazing gift.
1975 (G) 97m/C Kim Richards, Ike Eisenmann, Eddie Albert, Ray Milland, Donald Pleasence; *Dir:* John Hough. **VHS, Beta, LV** *DIS, OM* 🎞🎞

Escapes In the tradition of "The Twilight Zone," Vincent Price introduces five short thrillers featuring time travel, aliens, and telepathy. Produced with computer assistance for sharper, more contrasted images.
1986 72m/C Vincent Price, Jerry Grisham, Lee Canfield, John Mitchum, Gil Reade; *Dir:* David Steensland. **VHS, Beta $9.99** *STE, PSM* 🎞

Escapist A professional escape artist becomes involved in a perverted corporate plot and must escape to save his life and livelihood.
1983 87m/C Bill Shirk, Peter Lupus; *Dir:* Eddie Beverly Jr. **VHS, Beta $69.98** *LIV, VES Woof!*

Escort Girls The escort girls are a group of actresses, models and secretaries who really know how to show their clients a good time.
1974 77m/C David Dixon, Maria O'Brien, Marika Mann, Gil Barber, Helen Christie. **VHS, Beta $59.95** *GEM Woof!*

E.S.P. A young man is given the amazing power of extra-sensory perception.
1983 96m/C Jim Stafford, George Deaton. **VHS, Beta** *BFV Woof!*

Esposa y Amante While her daughter contemplates suicide, a mother remembers the happy early years of her marriage, followed by the wrong-doings of her husband which caused her to seek comfort in the arms of an old lawyer friend. Her daughter is now suffering for the problems of her own marriage. In Spanish.
197? 102m/C *SP* Ramiro Oliveros, Ricardo Merino, Victoria Abril, Frika Wallner. **VHS, Beta $29.95** *MED* 🎞

Esta Noche Cena Pancho A Spanish comedy about the antics of men at a bachelor party.
1987 90m/C *SP* Alfonso Zayas, Carmen Salinas, Armando Silvestre, Alberto Rojas; *Dir:* Victor Manuel Castro. **VHS, Beta $79.95** *CCN, MED* 🎞

Estate of Insanity An English lord and his second wife become involved in a web of death when a maniac stalks their ancient estate.
1970 90m/C John Turner, Heather Sears, Ann Lunn; *Dir:* Robert Hartford-Davis. **VHS, Beta $9.95** *VCL* 🎞

Et La Tenoresse?...Bordel! A French comedy about a color-blind painter and his kinky fashion photographer roommate living in Paris. In French with subtitles.
1984 100m/C *FR Dir:* Patrick Schulmann. **VHS, Beta $59.95** *FCT* 🎞🎞

E.T.: The Extra-Terrestrial The famous Spielberg fantasy, one of the most popular and highest grossing films in history, portrays a limpid-eyed alien stranded on earth and his special bonding relationship with a young boy. A modern fairy tale providing warmth, humor and sheer wonder.
1982 (PG) 115m/C Henry Thomas, Dee Wallace Stone, Drew Barrymore, Robert MacNaughton, Peter Coyote, C. Thomas Howell, Sean Frye; *Dir:* Steven Spielberg. Academy Awards '82: Best Sound, Best Visual Effects, Best Original Score; British Academy Awards '82: Best Original Score; Golden Globe Awards '83: Best Film—Drama. **VHS, Beta, LV $24.95** *MCA, APD, RDG* 🎞🎞🎞🎞

L'Etat Sauvage This French political thriller finds its heart in the middle of racism. The story unfolds around a newly independent African republic and a love affair between a black cabinet minister and a white Frenchwoman. Corruption abounds in this film based on George Conchon's award-winning novel.
1990 111m/C *FR* Marie-Christine Barrault, Claude Brasseur, Jacques Dutronc, Doura Mane, Michel Piccoli; *Dir:* Francis Girod. **VHS, Beta, LV $59.95** *INT* 🎞🎞🎞½

Eternal Evil A bored television director is taught how to have out-of-body experiences by his devil-worshiping girlfriend. He eventually realizes that when he leaves his body, it runs around killing people.
1987 (R) 85m/C Karen Black, Winston Rekert, Lois Maxwell; *Dir:* George Mihalka. **VHS, Beta, LV $79.98** *LIV, LTG* 🎞

Eternal Return A lush modern retelling of the Tristan/Isolde legend. A young man falls passionately in love with his aunt, only to die of unrequited love. The script was written by Jean Cocteau. In French with English subtitles. French title: ''L'Eternel Retour.''
1943 100m/B FR Jean Marais, Madeleine Sologne; *Dir:* Jean Delannoy. **VHS, Beta $29.98** SUE 🎬🎬½

The Eternal Waltz Overly sentimental chronicle of the life of composer Johann Strauss. Only director Verhoeven's touch holds the viewer's interest, though the production values are excellent.
1959 97m/C Bernhard Wicki, Hilde Krahl, Annemarie Duerringer, Friedl Loor, Eduard Strauss Jr., Gert Froebe, Arnulf Schroeder; *Dir:* Paul Verhoeven. **VHS $24.95** NOS 🎬

Eternally Yours A witty magician's career threatens to break up his marriage.
1939 95m/C David Niven, Loretta Young, Hugh Herbert, Broderick Crawford, C. Aubrey Smith, Billie Burke, Eve Arden, ZaSu Pitts; *Dir:* Tay Garnett. **VHS, Beta $19.95** NOS, MRV, PFI 🎬🎬½

Eternity While trying to uncover corrupt corporate America a television reporter falls in love with a model who works for a media king and puts his credibility on the line. He believes he and the woman shared romance in a past life.
1990 (R) 122m/C Jon Voight, Armand Assante, Wilford Brimley, Eileen Davidson, Kaye Ballard, Lainie Kazan, Joey Villa, Steven Keats, Eugene Roche, Frankie Valli, John P. Ryan; *Dir:* Steven Paul. **VHS $14.95** ACA 🎬

Ethan A missionary in the Philippines falls in love with a woman and exiles himself for his fall from grace.
1071 01m/C Robert Sampoon, Rosa Rosal, Eddie Infante; *Dir:* Michael DuPont. **VHS, Beta $24.95** GHV 🎬½

Eubie! The popular Broadway musical revue based on the life and songs of Eubie Blake is presented in a video transfer. Some of Eubie's best known songs, performed here by members of the original cast, include ''I'm Just Wild About Harry,'' ''Memories of You,'' ''In Honeysuckle Time'' and ''The Charleston Rag.''
1982 100m/C Gregory Hines, Maurice Hines, Leslie Dockery, Alaina Reed, Lynnie Godfrey, Mel Johnson Jr., Jeffrey V. Thompson; *Dir:* Julianne Boyd. **VHS, Beta $19.98** WAR, OM 🎬½

Eureka! A bizarre, wildly symbolic slab of Roegian artifice that deals with the dream-spliced life of a rich, bored gold tycoon who becomes tortured over his daughter's marriage and his own useless wealth. Eventually he is bothered by the Mafia and led to the courtroom by business competitors. From the book by Paul Mayersberg.
1981 (R) 130m/C GB Gene Hackman, Theresa Russell, Joe Pesci, Rutger Hauer, Mickey Rourke, Ed Lauter, Jane Lapotaire; *Dir:* Nicolas Roeg. **VHS, Beta $79.95** MGM 🎬🎬½

Eureka Stockade Four early Australian gold prospectors join forces to fight their governor and the police for the rights to dig on the continent.
1949 103m/B Chips Rafferty, Peter Finch, Jane Barrett, Peter Illing; *Dir:* Harry Watt. **VHS, Beta $24.95** NOS, DVT 🎬

Europa '51 The despairing portrait of post-war malaise, as an American woman, whose son committed suicide, lives in Rome searching for some semblance of meaning, and ends up in an asylum. One of Bergman & Rossellini's least-loved films. In Italian with subtitles. Also titled ''The Greatest Love.''
1952 110m/B Ingrid Bergman, Alexander Knox, Ettore Giannini, Giulietta Masina; *Dir:* Roberto Rossellini. **VHS, Beta $59.95** FCT 🎬🎬½

Europa, Europa The incredible, harrowing and borderline-absurdist true story of Solomon Perel, a young Jew who escaped the Holocaust by passing for German at an elite, Nazi-run academy. Such a sharp evocation of the era that the modern German establishment wouldn't submit it for the Academy Awards; the Hound suggests you follow their 'recommendation' and see it. In German and Russian with English subtitles. because the film has a Polish director and was cofinanced by the French that it violated national content requirements. In German and Russian with English subtitles.
1991 (R) 115m/C GE Marco Hofschneider, Klaus Abramowsky, Michele Gleizer, Rene Hofschneider, Nathalie Schmidt, Delphine Forest; *Dir:* Agnieszka Holland. Golden Globe Awards '92: Best Foreign Language Film; National Board of Review Awards '91: Best Foreign Language Film; New York Film Critics Awards '91: Best Foreign Language Film. **VHS** ORI 🎬🎬🎬½

The Europeans Henry James's satirical novel about two fortune-seeking expatriates and their sober American relations.
1979 (G) 90m/C Lee Remick, Lisa Eichhorn, Robin Ellis, Wesley Addy, Tim Woodward; *Dir:* James Ivory. National Board of Review Awards '79: 10 Best Films of the Year; New York Film Festival '79: 10 Best Films of the Year. **VHS, Beta $59.95** CVC 🎬🎬🎬

Eve of Destruction Hell knows no fury like a cutting-edge android-girl on the war-path. Modeled after her creator, Dr. Eve Simmons, Eve VII has android-babe good looks and a raging nuclear capability. Wouldn't you know, something goes haywire during her trial run, and debutante Eve turns into a PMS nightmare machine, blasting all the good Doctor's previous beaux. That's where military agent Hines comes in, though you wonder why. Dutch actress Soutendijk plays dual Eves in her first American film.
1990 (R) 101m/C Gregory Hines, Renee Soutendijk, Kevin McCarthy; *Dir:* Duncan Gibbins. **VHS, Beta, LV $19.95** COL, ORI, SUE Woof!

Evel Knievel The life of motorcycle stuntman Evel Knievel is depicted in this movie, as portrayed by George Hamilton. Stunts will be appreciated by Evel Knievel fans.
1972 (PG) 90m/C George Hamilton, Bert Freed, Rod Cameron, Sue Lyon; *Dir:* Marvin J. Chomsky. **VHS, Beta $59.95** MPI 🎬🎬

Even Angels Fall One of heartthrob singer Humperdinck's scattered dramatic projects, a mystery thriller about a romance novelist moving into a New York brownstone whose previous inhabitant committed suicide—or was it murder?
1990 ?m/C Morgan Fairchild, Engelbert Humperdinck; *Dir:* Thomas Calabrese. **VHS** FMP 🎬🎬

An Evening with Bobcat Goldthwait: Share the Warmth Goldthwait's complete fevered stand-up routine, as recorded live at New York's Bottom Line nightclub.
1987 54m/C Bob Goldthwait. **VHS, Beta $19.98** VES 🎬🎬

Evening at the Improv All-star performance at the Improv features the decade's funniest men.
1986 60m/C Harry Anderson, Billy Crystal, Michael Keaton, Howie Mandel, Steven Wright. **VHS, Beta $29.98** NO 🎬🎬

An Evening with Robin Williams Robin Williams explodes all over the screen in this live nightclub performance taped at the Great American Music Hall in San Francisco.
1983 92m/C Robin Williams. **VHS, Beta, LV $14.95** PAR 🎬🎬

Evergreen The daughter of a retired British music hall star is mistaken for her mother and it is thought that she has discovered the secret of eternal youth. Music score by Rogers and Hart.
1934 91m/B GB Jessie Matthews, Sonnie Hale, Betty Balfour, Barry Mackey; *Dir:* Victor Saville. **VHS, Beta $39.95** NOS, HMV, VMK 🎬🎬½

Eversmile New Jersey A dentist travels through Patagonia, offering dental care and advice to anyone in need, and a young woman, taken with the dental knight, dumps her boyfriend and stows away with him. When he discovers his admirer, he is less than ecstatic, but a chance tooth extraction on the road reveals the young woman's natural dental talents, and the two bond, as it were. Screenplay by Jorge Goldenberg, Roberto Scheuer and Carlos Sorin; from the director of ''The Official Story.''
1989 (PG) 88m/C AR Daniel Day Lewis, Mirjana Jokovic; *Dir:* Carlos Sorin. **VHS, Beta, LV $89.98** FXV, MED 🎬½

Every Girl Should Be Married A shopgirl sets her sights on an eligible bachelor doctor.
1948 84m/B Cary Grant, Betsy Drake, Diana Lynn, Franchot Tone; *Dir:* Don Hartman. **VHS, Beta, LV $19.98** RKO 🎬🎬½

Every Girl Should Have One A rambunctious comedy about a chase following a million-dollar diamond theft.
1978 90m/C Zsa Zsa Gabor, Robert Alda, Alice Faye, Sandra Vacey, John Lazar. **VHS, Beta $59.95** IGV, GEM 🎬½

Every Man for Himself & God Against All This film tells the story of Kaspar Hauser, a young man who mysteriously appears in a small German town, having been confined since birth. His alternate vision of the world and attempts to reconcile with reality make for a lovely, though demanding film. Also released as ''The Mystery of Kaspar Hauser.'' In German with English subtitles.
1975 110m/C GE Bruno S; *Dir:* Werner Herzog. Cannes Film Festival '75: Ecumenical Prize, International Critics Award, Special Jury Prize; New York Film Festival '75: Ecumenical Prize. **VHS, Beta $59.95** COL, APD, GLV 🎬🎬🎬🎬

Every Man's Law A cowboy who poses as a hired gunman is almost lynched by ranchers who think he is a murderer.
1935 60m/B Johnny Mack Brown. **VHS, Beta $19.95** VCN, WGE 🎬

Every Time We Say Goodbye In 1942, Jerusalem, an American flyboy falls in love with a young Sephardic Jewish girl, whose family resists the match.
1986 (PG-13) 97m/C Tom Hanks, Christina Marsillach, Benedict Taylor; *Dir:* Moshe Mizrahi. **VHS, Beta, LV $79.98** LIV, LTG 🎬🎬½

Every Which Way But Loose A beer guzzling, country music-loving truck driver earns a living as a barroom brawler. He and his orangutan travel to Colorado in pursuit of the woman he loves. Behind him are a motorcycle gang and an L.A. cop. All have been victims of his fists. Sequel is ''Any Which Way You Can.''

1978 (R) 119m/C Clint Eastwood, Sondra Locke, Geoffrey Lewis, Beverly D'Angelo, Ruth Gordon; *Dir:* James Fargo. **VHS, Beta, LV $19.98** WAR ♂½

Everybody Wins A mystery-romance about a befuddled private eye trying to solve a murder and getting caught up with the bizarre prostitute who hired him. Arthur Miller based this screenplay on his stage drama "Some Kind of Love Story." Confused and, given its pedigree, disappointing mystery. **1990 (R) 110m/C** Nick Nolte, Debra Winger, Will Patton, Jack Warden, Kathleen Wilhoite, Frank Converse, Frank Military, Judith Ivey; *Dir:* Karel Reisz. **VHS, Beta, LV $89.98** ORI ♂

Everybody's All American Shallow, sentimental melodrama about a college football star and his cheerleader wife whose lives subsequent to their youthful glories is a string of disappointments and tragedies. Decently acted and based on Frank Deford's novel. **1988 (R) 127m/C** Jessica Lange, Dennis Quaid, Timothy Hutton, John Goodman, Carl Lumbly, Ray Baker, Savannah Smith; *Dir:* Taylor Hackford. **VHS, Beta, LV, 8mm $89.95** WAR ♂♂

Everybody's Dancin' A ballroom proprietor's business is marred by random killings in her establishment as the bands play on. **1950 66m/B** Spade Cooley, Dick Lane, Hal Derwin, The Pioneers; *Dir:* Will Jason. **VHS, Beta, LV** WGE ♂½

Everybody's Fine Mastroianni stars in this bittersweet story of a father on a mission to reunite his five grown children. His journey takes him all over Italy as he tries to bring them together in this touching film about enduring family love. In Italian with English subtitles. **1991 (PG-13) 115m/C** *IT* Marcello Mastroianni, Salvatore Cascio, Valeria Cavalli, Norma Martelli, Marino Cenna, Roberto Nobile, Salvatore Cascio; *Dir:* Giuseppe Tornatore. **VHS, LV $89.95** COL ♂♂½

Everything You Always Wanted to Know About Sex (But Were Afraid to Ask) Satiric comical sketches about sex includes a timid sperm cell, an oversexed court jester, a sheep folly, and a giant disembodied breast. Quite entertaining in its own jolly way. **1972 (R) 88m/C** Woody Allen, John Carradine, Lou Jacobi, Louise Lasser, Anthony Quayle, Geoffrey Holder, Lynn Redgrave, Tony Randall, Burt Reynolds, Gene Wilder, Robert Walden, Jay Robinson; *Dir:* Woody Allen. **VHS, Beta, LV $19.98** MGM, FOX, PIA ♂♂♂

The Evictors A young couple moves into an abandoned, haunted farmhouse in a small Louisiana town. Unfortunately, they don't know anything about its horrible bloody history. **1979 (PG) 92m/C** Vic Morrow, Michael Parks, Jessica Harper, Sue Ann Langdon, Dennis Fimple; *Dir:* Charles B. Pierce. **VHS, Beta $59.98** LIV, VES ♂

The Evil A psychologist must destroy an evil force that is killing off the members of his research team residing at an old mansion. **1978 (R) 80m/C** Richard Crenna, Joanna Pettet, Andrew Prine, Victor Buono, Cassie Yates; *Dir:* Gus Trikonis. **VHS, Beta $19.98** SUE ♂♂

Evil Altar A man controls a small town, but only so long as he offers sacrifices to the devil! **1989 (R) 90m/C** William Smith, Robert Z'Dar, Pepper Martin, Theresa Cooney, Ryan Rao; *Dir:* Jim Winburn. **VHS $89.95** SOU ♂

The Evil Below A couple hits the high seas in search of the lost treasure ship "El Diablo," resting on the ocean floor. In the process they trigger an evil curse and then must attempt to thwart it. **1987 90m/C** *SA* June Chadwick, Wayne Crawford; *Dir:* Jean-Claude Dubois. **VHS, Beta $69.95** RAE ♂

Evil Clutch A young couple vacationing in the Alps encounter several creepy locals when they find themselves in the midst of a haunted forest. The cinematography is extremely amateurish in this Italian gorefest and the English dubbing is atrocious. However, the special makeup effects are outstanding and the musical score adds a touch of class to this otherwise inept horror film. **1989 (R) 88m/C** Coralina Tessoni, Diego Riba, Elena Cantarone, Luciano Crovato, Stefano Molinari; *W/Dir:* Andreas Marfori. **VHS $79.95** UND, RHI ♂½

The Evil Dead Five vacationing college students unwittingly resurrect demons which transform the students into evil monsters. Scripted and filmed by real college students, this might well be one of the goriest films ever made. **1983 126m/C** Bruce Campbell, Ellen Sandweiss, Betsy Baker, Hal Delrich; *Dir:* Sam Raimi. **VHS, Beta, LV $14.99** HBO, IME ♂♂½

Evil Dead 2: Dead by Dawn A gory, tongue-in-cheek sequel/remake of the original festival of gag and gore, in which an ancient book of magic invokes a crowd of flesh-snacking, joke-tossing ghouls. Campbell co-produces. **1987 (R) 84m/C** Bruce Campbell, Sarah Berry, Dan Hicks; *Dir:* Sam Raimi. **VHS, Beta, LV $14.98** LIV, VES ♂♂½

The Evil of Frankenstein The third of the Hammer Frankenstein films, with the mad doctor once again finding his creature preserved in ice and thawing him out. Preceded by "The Revenge of Frankenstein" and followed by "Frankenstein Created Woman." **1948 84m/C** *GB* Peter Cushing, Duncan Lamont, Peter Woodthorpe, Sandor Eles, Kiwi Kingston, Katy Wild; *Dir:* Freddie Francis. **VHS, Beta $14.98** MCA, MLB ♂♂

Evil Judgement A young girl investigates a series of murders and finds the culprit is a psychopathic judge. **1985 93m/C** Pamela Collyer, Jack Langedijk, Nanette Workman; *Dir:* Claude Castravelli. **VHS, Beta $69.95** MED ♂½

Evil Laugh Medical students and their girlfriends party at an abandoned orphanage, until a serial killer decides to join them. **1986 (R) 90m/C** Tony Griffin, Kim McKamy, Jody Gibson, Dominick Brascia; *Dir:* Dominick Brascia. **VHS $79.95** CEL Woof!

The Evil Mind A fraudulent mind reader predicts many disasters that start coming true. Also called "The Clairvoyant." **1934 80m/B** *GB* Claude Rains, Fay Wray, Jane Baxter, Felix Aylmer; *Dir:* Maurice Elvey. **VHS, Beta $16.95** SNC, NOS, KRT ♂♂½

Evil Spawn A fading movie queen takes an experimental drug to restore her youthful beauty, but it only turns her into a giant silverfish. **1987 90m/C** Bobbie Bresee, John Carradine, Drew Godderis, John Terrance, Dawn Wildsmith; *Dir:* Kenneth J. Hall. **VHS, Beta $59.95** CAM ♂

Evil Spirits Boardinghouse tenants are murdered while the crazy landlady cashes their social security checks. This seedy horror cheapie doesn't take itself seriously, and like-minded genre buffs may enjoy the cult-film cast. **1991 (R) 95m/C** Karen Black, Arte Johnson, Virginia Mayo, Michael Berryman, Martine Beswick, Bert Remsen, Yvette Vickers, Robert Quarry, Mikel Angel, Debra Lamb; *Dir:* Gary Graver. **VHS, Beta $79.95** PSM ♂½

The Evil That Men Do A hitman comes out of retirement to break up a Central American government's political torture ring and, in the process, bring a friend's killer to justice. Based on the novel by R. Lance Hill. **1984 (R) 90m/C** Charles Bronson, Theresa Saldana, Joseph Maher, Jose Ferrer, Rene Enriquez, John Glover, Raymond St. Jacques, Antoinette Bower, Enrique Lucero, Jorge Luke; *Dir:* J. Lee Thompson. **VHS, Beta, LV, 8mm $14.95** COL ♂

Evil Toons A quartet of lovely coeds on a cleaning job venture into a deserted mansion. There they accidentally release a vulgar, lustful, animated demon who proceeds to cause their clothes to fall off. Can the girls escape the haunted mansion with their sanity, virtue and wardrobes intact? **1990 (R) 86m/C** David Carradine, Dick Miller, Monique Gabrielle, Suzanne Ager, Stacy Nix, Madison Stone, Don Dowe, Arte Johnson, Michelle Bauer; *Dir:* Fred Olen Ray. **VHS, Beta $79.95** PSM ♂½

Evil Town In this poorly made film, a wandering guy discovers a town overrun with zombies created by a mad doctor. **197? 88m/C** Dean Jagger, James Keach, Robert Walker Jr., Doria Cook, Michele Marsh; *Dir:* Edward Collins. **VHS, Beta $79.95** TWE, HHE ♂

Evil Under the Sun An opulent beach resort is the setting as Hercule Poirot attempts to unravel a murder mystery. Based on the Agatha Christie novel. **1982 (PG) 102m/C** *GB* Peter Ustinov, Jane Birkin, Maggie Smith, Colin Blakely, Roddy McDowall, Diana Rigg, Sylvia Miles, James Mason, Nicholas Clay; *Dir:* Guy Hamilton. **VHS, Beta $19.99** HBO ♂♂

Evils of the Night Teenage campers are abducted by sex-crazed alien vampires. Bloody naked mayhem follows. **1985 85m/C** John Carradine, Julie Newmar, Tina Louise, Neville Brand, Aldo Ray, Karrie Emerson, Bridget Hollman; *Dir:* Mardi Rustam. **VHS, Beta $69.98** LIV, LTG ♂

Evilspeak A bumbling misfit enrolled at a military school is mistreated by the other cadets. He retaliates with satanic power with the help of his computer. **1982 (R) 89m/C** Clint Howard, Don Stark, Lou Gravance, Lauren Lester, R.G. Armstrong, Joe Cortese, Claude Earl Jones, Haywood Nelson; *Dir:* Eric Weston. **VHS, Beta $59.98** FOX Woof!

Evolutionary Spiral A combination of visual imagery and a musical soundtrack by the group Weather Report. **1983 45m/C** Weather Report. **VHS, Beta, LV $9.95** SVS Woof!

The Ewok Adventure Those adorable, friendly and funny characters from "Return of the Jedi" make the jump from film to television in a new adventure from George Lucas. In this installment, the Ewoks save a miraculous child from harm with the help of a young human. This fun-filled adventure has Lucas's thumbprint all over it and great

special effects. Followed by "Ewoks: The Battle for Endor."
1984 (G) 96m/C Warwick Davis, Eric Walker, Aubree Miller, Fionnula Flanagan; *Dir:* John Korty; *Nar:* Burl Ives. **VHS, Beta, LV $19.98** *MGM, FCT* 🎞️🎞️½

The Ewoks: Battle for Endor A children's television movie based on the furry creatures from "Return of the Jedi," detailing their battle against an evil queen to retain their forest home. Preceded by "The Ewok Adventure."
1985 98m/C Wilford Brimley, Warwick Davis, Aubree Miller, Sian Phillips, Paul Gleason, Eric Walker; *Dir:* Jim Wheat, Ken Wheat. **VHS, Beta, LV $19.98** *MGM, FCT* 🎞️🎞️½

Ex-Mrs. Bradford Amateur sleuth Dr. Bradford teams up with his ex-wife to solve a series of murders at the race track. Sophisticated comedy-mystery; witty dialogue.
1936 80m/B William Powell, Jean Arthur, James Gleason, Eric Blore, Robert Armstrong; *Dir:* Stephen Roberts. **VHS, Beta $19.95** *NOS, MED* 🎞️🎞️🎞️

Excalibur A sweeping, visionary retelling of the life of King Arthur, from his conception, to the sword in the stone, to the search for the Holy Grail and the final battle with Mordred. An imperfect, sensationalized version, but still the best yet filmed.
1981 (R) 140m/C Nigel Terry, Nicol Williamson, Nicholas Clay, Helen Mirren, Cherie Lunghi, Paul Geoffrey, Gabriel Byrne, Liam Neeson, Patrick Stewart, Charley Boorman, Corin Redgrave; *Dir:* John Boorman. **VHS, Beta, LV $19.98** *WAR* 🎞️🎞️🎞️½

The Execution A made-for-television film about five female friends who discover that the Nazi doctor who brutalized them in a concentration camp during World War II is now living a normal life in California. Together they plot his undoing.
1985 92m/C Loretta Swit, Valerie Harper, Sandy Dennis, Jessica Walter, Rip Torn, Barbara Barrie, Robert Hooks, Michael Lerner; *Dir:* Paul Wendkos. **VHS, Beta $9.99** *STE, PSM* 🎞️½

The Execution of Private Slovik This quiet powerhouse of a TV movie recounts in straightforward terms the case of Eddie Slovik, a WWII misfit who became the only American soldier executed for desertion since the Civil War. The Richard Levinson/William Link screenplay (based on the book by William Bradford Huie) ends up deifying Slovik, which some might find hard to take. But there's no arguing the impact of the drama, or of Sheen's unaffected lead performance.
1974 122m/C Martin Sheen, Mariclare Costello, Ned Beatty, Gary Busey, Matt Clark, Ben Hammer, Warren Kemmerling; *Dir:* Lamont Johnson. **VHS $79.95** *MCA* 🎞️🎞️🎞️½

Execution of Raymond Graham The lawyers and family of Raymond Graham struggle to keep him from being executed for murder. Based on a true story; made for television.
1985 104m/C Morgan Freeman, Jeff Fahey, Kate Reid, Laurie Metcalf, Josef Sommer; *Dir:* Daniel Petrie. **VHS, Beta** *NSV* 🎞️🎞️½

The Executioner A thriller wherein a British spy must prove that his former colleague is a double agent. Elements of back-stabbing, betrayal, and espionage abound.
1970 (PG) 107m/C GB Judy Geeson, Oscar Homolka, Charles Gray, Nigel Patrick, George Peppard, Joan Collins, Keith Mitchell; *Dir:* Sam Wanamaker. **VHS, Beta $9.95** *COL, GKK* 🎞️🎞️

The Executioner A very cheap, very "Godfather"-like story of a mafia family gone awry. Also known as "Like Father, Like Son" and "Massacre Mafia Style."
1978 (R) 84m/C Duke Mitchell, Vic Caesar, Dominic Micelli, John Strong, Jim Williams, Lorenzo Dodo; *W/Dir:* Duke Mitchell. **VHS, Beta** *GEM, MRV* 🎞️

The Executioner, Part 2: Frozen Scream A brutal feud rocks the Mafia, and a crime kingpin's passionate son seeks revenge on his father's slayers. Not a sequel to any other films bearing similar titles. Strange thing is, no "Executioner, Part I" was ever made. Pretty laughable.
1984 (R) 150m/C Chris Mitchum, Aldo Ray, Antoine John Mottet, Renee Harmon; *Dir:* James Bryant. **VHS, Beta $49.98** *NSV, GEM* 🎞️

Executioner of Venice Marauding pirates swarm in from the Adriatic Sea and attempt to rob the Venetians. The Doge and his godson come to the rescue.
1963 90m/C Guy Madison, Lex Barker, Sandra Panaro; *Dir:* Louis Capauno. **VHS, Beta $69.95** *FOR* 🎞️

The Executioner's Song European version of the television movie based on Norman Mailer's Pulitzer Prize-winner, recounting the life and death of convicted murderer Gary Gilmore. Features adult-minded footage not seen in the U.S. version.
1982 157m/C Tommy Lee Jones, Rosanna Arquette, Eli Wallach, Christine Lahti, Jenny Wright, Jordan Clark, Steven Keats; *Dir:* Lawrence Schiller. **VHS, Beta $29.95** *STE, IVE* 🎞️🎞️½

Executive Action A different look at the events leading to the assassination of JFK. In this speculation, a millionaire pays a professional spy to organize a secret conspiracy to kill President Kennedy. Ryan's final film. Adapted from Mark Lane's "Rush to Judgement."
1973 (PG) 91m/C Burt Lancaster, Robert Ryan, Will Geer, Gilbert Green, John Anderson; *Dir:* David Miller. **VHS, Beta, LV $19.98** *WAR* 🎞️

Executive Suite One of the first dog-eat-dog dramas about high finance and big business. The plot centers on the question of a replacement for the freshly buried owner of a gigantic furniture company.
1954 104m/B William Holden, June Allyson, Barbara Stanwyck, Fredric March, Walter Pidgeon, Louis Calhern, Shelley Winters, Paul Douglas, Nina Foch, Dean Jagger; *Dir:* Robert Wise. **VHS, Beta $29.95** *MGM* 🎞️🎞️🎞️

Exiled to Shanghai A couple of newsreel men invent a television device that revolutionizes the business.
1937 65m/B Wallace Ford, June Travis, Dean Jagger, William Bakewell, Arthur Lake, Jonathan Hale, William Harrigan, Sarah Padden; *Dir:* Armand Schaefer. **VHS $24.95** *NOS, DVT* 🎞️🎞️½

Exit the Dragon, Enter the Tiger Story about the death of karate master Bruce Lee.
1976 84m/C **VHS, Beta $24.95** *UHV, TWE* 🎞️

Exodus Based on the novel by Leon Uris and filmed in Cyprus and Israel chronicles the post WWII partition of Palestine into a homeland for Jews, the anguish of refugees from Nazi concentration camps held on ships in the Mediterranean, the struggle of the tiny nation with the forces dividing it within and destroying it on the outside, and the simple heroic men and women who saw a job needing to be done and doing it. Preminger battled the Israeli government, the studio and

the novel's author to complete this epic. Cost more than $4 million, a phenomenal amount at the time.
1960 213m/C Paul Newman, Eva Marie Saint, Lee J. Cobb, Sal Mineo, Ralph Richardson, Hugh Griffith, Gregory Ratoff, Felix Aylmer, Peter Lawford, Jill Haworth, John Derek; *Dir:* Otto Preminger. Academy Awards '60: Best Original Score. **VHS, Beta, LV $29.98** *FOX* 🎞️🎞️🎞️

Exorcism A satanic cult in a small English village commits a series of gruesome crimes that have the authorities baffled.
1974 90m/C Paul Naschy, Maria Perschy; *Dir:* Juan Bosch. **VHS, Beta $19.95** *ASE, HHE* 🎞️½

The Exorcist Truly terrifying story of a young girl who is possessed by a malevolent demon. Brilliantly directed by Friedkin, with underlying themes of the workings and nature of fate. Impeccable casting and unforgettable, thought-provoking performances. A rare film that remains startling and engrossing with every viewing, it spawned countless imitations and changed the way horror films were made. Based on the bestseller by William Peter Blatty, who won an Oscar for his screenplay. Not for the squeamish. When first released, the film created mass hysteria in theaters, with people fainting and paramedics on the scene.
1973 (R) 120m/C Ellen Burstyn, Linda Blair, Jason Miller, Max von Sydow, Jack MacGowran, Lee J. Cobb, Kitty Winn; *Dir:* William Friedkin. Academy Awards '73: Best Adapted Screenplay, Best Sound; Golden Globe Awards '74: Best Film—Drama. **VHS, Beta, LV $19.98** *WAR* 🎞️🎞️🎞️½

Exorcist 2: The Heretic A sequel to the 1973 hit "The Exorcist." After four years, Blair is still under psychiatric care, suffering from the effects of being possessed by the devil. Meanwhile, a priest investigates the first exorcist's work as he tries to help the head-spinning lass.
1977 (R) 118m/C Richard Burton, Linda Blair, Louise Fletcher, Kitty Winn, James Earl Jones, Ned Beatty, Max von Sydow; *Dir:* John Boorman. **VHS, Beta $19.98** *WAR* 🎞️

The Exorcist 3 Apparently subscribing to the two wrongs make a right school of sequels, novelist Blatty this time sits in the director's chair. The result is slightly better than the first sequel, but still a far cry from the original. It's fifteen years later, and Detective Kinderman (played by Scott) is faced with a series of really gross murders that bear the indubitable mark of a serial killer who was flambeed in the electric chair, it seems, the same night as the exorcism of the pea-soup expectorating devil of the original. With the aid of priests Flanders and Douriff the detective stalks the transmigratory terror without the help of Linda Blair, who was at the time spoofing "The Exorcist" in "Repossessed."
1990 (R) 105m/C George C. Scott, Ed Flanders, Jason Miller, Nicol Williamson, Scott Wilson, Brad Dourif, Nancy Fish, George DiCenzo, Viveca Lindfors. **VHS, Beta, LV $19.98** *FOX, CCB* 🎞️🎞️

The Expendables A rugged captain turns a platoon of criminals and misfits into a tough fighting unit for a particularly dangerous mission from which they might not return.
1989 (R) 89m/C Anthony Finetti, Peter Nelson, Loren Haynes, Kevin Duffis; *Dir:* Cirio H. Santiago. **VHS, Beta, LV $79.95** *MED* 🎞️½

Experience Preferred... But Not Essential An English schoolgirl gets her first job at a resort where she learns about life. Made for British television.
1983 (PG) 77m/C *GB* Elizabeth Edmonds, Sue Wallace, Geraldine Griffith, Karen Meagher, Ron Bain, Alun Lewis, Robert Blythe; *Dir:* Peter Duffell. **VHS, Beta** $59.95 *MGM* ♫♫½

Experiment Perilous A psychologist and a recently widowed woman band together to find her husband's murderer. An atmospheric vintage mystery.
1944 91m/B Hedy Lamarr, Paul Lukas, George Brent, Albert Dekker; *Dir:* Jacques Tourneur. **VHS, Beta** $19.95 *MED* ♫♫♫

Experiment in Terror A psychopath kidnaps a girl in order to blackmail her sister, a bank teller, into embezzling $100,000. Score by Henry Mancini.
1962 123m/B Lee Remick, Glenn Ford, Stefanie Powers, Ross Martin; *Dir:* Blake Edwards. **VHS, Beta, LV** $59.95 *COL, IME* ♫♫♫

Expertos en Pinchazos In this comedy from Spain, Albert and George, experts at giving injections to women, inject a patient with venom by mistake. Now they must find her within 48 hours. In Spanish.
1979 100m/C *SP* Alberto Olmedo, Jorge Porcel. **VHS, Beta** $29.95 *MED* ♫½

The Experts When the KGB needs real Americans for their spies to study, they kidnap two out-of-work New Yorkers who mistakenly believe that they have been hired to open a nightclub in Nebraska. Shot on location in Canada. Directed by former Second City TV MacKenzie Brother, Dave Thomas.
1989 (PG-13) 94m/C John Travolta, Arye Gross, Charles Martin Smith, Kelly Preston, James Keach, Deborah Foreman, Brian Doyle Murray; *Dir:* Dave Thomas. **VHS, Beta, LV, 8mm** $14.95 *PAR* ♫

Explorers Intelligent family fare involving three young boys who use a contraption from their makeshift laboratory to travel to outer space. From the director of "Gremlins," displaying Dante's characteristic surreal wit and sense of irony.
1985 (PG) 107m/C Ethan Hawke, River Phoenix, Jason Presson, Amanda Peterson, Mary Kay Place, Dick Miller, Robert Picardo; *Dir:* Joe Dante. **VHS, Beta, LV** $14.95 *PAR* ♫♫½

Explosion A distraught and disturbed young man evades the draft after losing a brother in Vietnam. Arriving in Canada, he meets another draft-dodger with whom he embarks on a murderous rampage.
1969 (R) 96m/C Don Stroud, Gordon Thomson, Michele Chicione, Richard Conte; *Dir:* Jules Bricken. **VHS, Beta** $59.95 *TWE* ♫½

Exposed High fashion model Kinski falls in with a terrorist gang through a connection with violinist Nureyev. Weak plotting undermines the end of this political thriller. However, Kinski is brilliant, stripping the barrier between performance and audience.
1983 (R) 100m/C Nastassia Kinski, Rudolf Nureyev, Harvey Keitel, Ian McShane, Bibi Andersson; *Dir:* James Toback. **VHS, Beta** $79.95 *MGM* ♫♫

Exposure A rugged American photographer (Coyote) on assignment in Rio turns vigilante to locate the vicious killer of a young prostitute. Coyote and his girlfriend (Pays) get caught up in the deadly underworld of international arms trading and drug running as they search for the murderer.

1991 (R) 99m/C Peter Coyote, Amanda Pays, Tcheky Karyo. **VHS** $92.99 *HBO* ♫

Express to Terror Passengers aboard an atomic-powered train en route to Los Angeles attempt to kill a sleazy theatrical agent. Pilot for the "Supertrain" series.
1979 120m/C Steve Lawrence, George Hamilton, Vic Morrow, Broderick Crawford, Robert Alda, Don Stroud, Fred Williamson, Stella Stevens, Don Meredith; *Dir:* Dan Curtis. **VHS, Beta** $49.95 *PSM* ♫½

Exquisite Corpses An Oklahoma hayseed charges to New York with a new, slick image. He meets the wife of a wealthy man, and together they organize a murderous operation.
1988 95m/C Zoe Tamerlaine Lund, Gary Knox, Daniel Chapman, Ruth Collins; *Dir:* Temistocles Lopez. **VHS** $79.95 *MNC* ♫½

The Exterminating Angel A fierce, funny surreal nightmare, wherein dinner guests find they cannot, for any definable reason, leave the dining room; full of dream imagery and characteristically scatological satire. One of Bunuel's best, in Spanish with English subtitles.
1962 95m/B *MX SP* Silvia Pinal, Enrique Rambal, Jacqueline Andere, Jose Baviera, Augusto Benedico, Luis Beristain; *Dir:* Luis Bunuel. Cannes Film Festival '62: International Critics Award; New York Film Festival '63: International Critics Award; New York Film Festival '74: International Critics Award. **VHS, Beta** $59.95 *TAM, HTV, FCT* ♫♫♫

Exterminator A Vietnam veteran hunts down the gang that assaulted his friend and becomes the target of the police, the CIA and the underworld in this bloody banal tale of murder and intrigue. Followed by "Exterminator II."
1980 (R) 101m/C Christopher George, Samantha Eggar, Robert Ginty; *Dir:* James Glickenhaus. **VHS, Beta, LV** $14.98 *SUE* ♫

Exterminator 2 The Exterminator battles the denizens of New York's underworld after his girlfriend is crippled, then murdered by the ruthless Mr. X. Violence galore.
1984 (R) 88m/C Robert Ginty, Mario Van Peebles, Deborah Geffner, Frankie Faison; *Dir:* Mark Buntzman. **VHS, Beta** $79.95 *MGM* ♫

Exterminators of the Year 3000 The Exterminator and his mercenary girlfriend battle with nuclear mutants over the last remaining tanks of purified water on Earth. Low-budget Road Warrior rip-off.
1983 (R) 101m/C *IT SP* Robert Jannucci, Alicia Moro, Alan Collins, Fred Harris; *Dir:* Jules Harrison. **VHS, Beta** $69.99 *HBO Woof!*

Extra Girl A silent melodrama/farce about a farm girl, brilliantly played by Normand, who travels to Hollywood to be a star. Once in the glamour capital, she gets used and abused for her trouble. Written by Mack Sennett.
1923 87m/B Mabel Normand, Ralph Graves, Vernon Dent; *Dir:* F. Richard Jones. **VHS, Beta, 8mm** $29.95 *VYY* ♫♫

The Extraordinary Adventures of Mr. West in the Land of the Bolsheviks The first achievement from the Kuleshov workshop, a wacky satire on American insularity depicting a naive and prejudiced American visiting Russia and being taken advantage of. Silent.
1924 55m/B Vsevolod Pudovkin, Boris Barnett; *Dir:* Lev Kuleshov. **VHS, Beta** $29.95 *FCT* ♫♫

Extreme Close-Up Drama about spies and their equipment. Screenplay by Michael Crichton; also known as "Sex Through a Window."
1973 (R) 80m/C James McMullan, James A. Watson Jr., Kate Woodville, Bara Byrnes, Al Checco, Antony Carbone; *Dir:* Jeannot Szwarc. **VHS** $69.95 *VES* ♫½

Extreme Prejudice A redneck Texas Ranger fights a powerful drug kingpin along the U.S.-Mexican border. Once best friends, they now fight for justice and the heart of the woman they both love.
1987 (R) 104m/C Nick Nolte, Powers Boothe, Maria Conchita Alonso, Michael Ironside, Rip Torn; *Dir:* Walter Hill. **VHS, Beta, LV** $14.95 *IVE* ♫♫

Extremities An adaptation of the topical William Mastrosimone play about an intended rape victim who turns on her attacker, captures him and plots to kill him. Violent and exploitive.
1986 (R) 83m/C Farrah Fawcett, Diana Scarwid, James Russo, Alfre Woodard; *Dir:* Robert M. Young. **VHS, Beta, LV, 8mm** $19.95 *PAR* ♫♫

The Eye Creatures Alien creatures in the form of eyeballs are fought off by a teenager and his girlfriend. A low-budget, gory, science fiction feature.
1965 80m/B John Ashley, Cynthia Hull, Warren Hammack, Chet Davis, Bill Peck; *Dir:* Larry Buchanan. **VHS** $16.98 *NOS, SNC* ♫

Eye of the Demon A couple moves to a small town in Massachusetts and discovers that the area had once been a haven for witchcraft. To their horror, they soon find that old habits die hard, and the spellcasting continues in a nearby graveyard. Made for cable television.
1987 92m/C Tim Matheson, Pamela Sue Martin, Woody Harrelson, Barbara Billingsley, Susan Ruttan; *Dir:* Carl Schenkel. **VHS** $89.95 *VMK* ♫

Eye of the Eagle A special task force is given a dangerous assignment during the Vietnam war.
1987 (R) 84m/C Brett Clark, Ed Crick, Robert Patrick; William Steis, Cec Verrell; *Dir:* Cirio H. Santiago. **VHS, Beta** $79.95 *MGM* ♫½

Eye of the Eagle 2 When his platoon is betrayed and killed in Vietnam, a surviving soldier joins with a beautiful girl to get revenge.
1989 (R) 93m/C Todd Field, Andy Wood, Ken Jacobson, Ronald Lawrence; *Dir:* Carl Franklin. **VHS, Beta** *SOU* ♫

Eye of the Eagle 3 Filipino-made Vietnam-War shoot-em-up, rack-em-up, shoot-em-up again, with U.S. forces pinned down against seemingly overwhelming odds. Violent.
1991 (R) 90m/C Steve Kanaly, Ken Wright, Peter Nelson, Carl Franklin. **VHS** $89.98 *NHO, CGV* ♫½

An Eye for an Eye A story of pursuit and revenge with Norris as an undercover cop pitted against San Francisco's underworld and high society.
1981 (R) 106m/C Chuck Norris, Christopher Lee, Richard Roundtree, Matt Clark, Mako, Maggie Cooper; *Dir:* Steve Carver. **VHS, Beta** $19.98 *SUE* ♫♫

Eye of the Needle Based on Ken Follett's novel about a German spy posing as a shipwrecked sailor on a deserted English island during World War II. Lonely, sad, yet capable of terrible violence, he is stranded on an isolated island while en route to report to his Nazi commander. He becomes involved with an English woman living on the island, and begins to contemplate his role in the war. **1981 (R) 112m/C** Donald Sutherland, Kate Nelligan, Ian Bannen, Christopher Cazenove, Philip Brown; *Dir:* Richard Marquand. **VHS, Beta** $29.95 *MGM, FOX* ♫♫½

Eye of the Octopus Based on a true story of a boy who traveled to a Pacific Island to learn the ways of the people there. **1982 23m/C VHS, Beta** *JOU* ♫

Eye on the Sparrow A couple (Winningham and Carradine) desperately want to raise a child of their own, but the system classifies them as unfit parents since they are both blind. Together they successfully fight the system in this inspiring made for television movie that was based on a true story. **1991 (PG) 94m/C** Mare Winningham, Keith Carradine, Conchata Ferrell; *Dir:* John Korty. **VHS** $89.98 *REP* ♫♫½

Eye of the Storm At the highway gas station/motel/diner where they live, two young brothers witness their parents murder and the younger brother is also blinded in the incident. Ten years later both brothers are still there and the tragedy may have turned one of them psychotic. When the abusive Gladstone and his young and sexy wife are stranded at the gas station it brings out the worst in everyone, with a violent climax during an equally violent thunderstorm. **1991 (R) 98m/C** Craig Sheffer, Bradley Gregg, Lara Flynn Boyle, Dennis Hopper, Leon Rippy; *Dir:* Yuri Zeltser. **VHS** $89.95 *NLC* ♫♫½

Eye of the Tiger A righteous ex-con battles a crazed, crack-dealing motorcycle gang that terrorized and murdered his wife, and is moving on to infest his town. **1986 (R) 90m/C** Gary Busey, Yaphet Kotto, Seymour Cassel, Bert Remsen, William Smith; *Dir:* Richard Sarafian. **VHS, Beta** $14.95 *IVE* ♫♫

Eye Witness An American attorney goes abroad to free a friend from the British legal system. A book of poems becomes the necessary device in deducing the whereabouts of the witness testifying to his friend's alibi. Originally released as "Your Witness." **1949 104m/B** *GB* Robert Montgomery, Felix Aylmer, Leslie Banks, Michael Ripper, Patricia Wayne; *Dir:* Robert Montgomery. **VHS** $16.95 *NOS, SNC* ♫♫½

Eyeball An intrepid policeman is stumped by a madman who is removing eyeballs from his victims. At least they won't have to watch this movie. **1978 (R) 87m/C** *IT* John Richardson, Martine Brochard; *Dir:* Umberto Lenzi. **VHS, Beta** $59.95 *PSM* Woof!

Eyes of the Amaryllis A young girl becomes involved in a mysterious game when she arrives in Nantucket to care for her insane, invalid grandmother. Based on the story by Natalie Babbitt and filmed on location on Nantucket Island. **1982 (R) 94m/C** Martha Byrne, Ruth Ford, Guy Boyd, Jonathan Bolt, Katherine Houghton; *Dir:* Frederick King Keller. **VHS, Beta** $69.98 *LIV, VES* ♫½

Eyes Behind the Stars A news photographer accidentally gets a few pictures of invading aliens, but nobody takes him seriously, particularly not the government. **1972 95m/C** *IT* Martin Balsam, Robert Hoffman, Nathalie Delon, Sherry Buchanan; *Dir:* Roy Garrett. **VHS, Beta** $49.95 *NEG* ♫

Eyes of Fire In early rural America, a group of pioneers set up camp in the wilderness and are besieged during the night by Indian witchcraft. **1984 (R) 86m/C** Dennis Lipscomb, Rebecca Stanley, Fran Ryan, Rob Paulsen, Guy Boyd, Karlene Crockett; *Dir:* Avery Crounse. **VHS, Beta, LV** $79.95 *VES* ♫♫

Eyes of Laura Mars A photographer exhibits strange powers—she can foresee a murder before it happens through her snapshots. In time she realizes that the person responsible for a series of killings is tracking her. Title song performed by Barbra Streisand. **1978 (R) 104m/C** Faye Dunaway, Tommy Lee Jones, Brad Dourif, Rene Auberjonois, Raul Julia, Darlanne Fluegel, Michael Tucker; *Dir:* Irvin Kershner. **VHS, Beta, LV** $14.95 *GKK* ♫♫½

Eyes, the Mouth A young man has an affair with his dead twin brother's fiancee. Happiness eludes them as they are haunted by the deceased's memory. **1983 (R) 100m/C** *FR IT* Lou Castel, Angela Molina; *Dir:* Marco Bellochio. **VHS, Beta** $59.95 *COL* ♫♫

Eyes of the Panther One of Shelley Duvall's "Nightmare Classics" series, this adaptation of an Ambrose Bierce story concerns a pioneer girl haunted for years by the animal urges of a wild cat. **1990 60m/C** Daphne Zuniga, C. Thomas Howell, John Stockwell; *Dir:* Noel Black. **VHS, Beta, LV** $59.95 *NO* ♫♫

Eyes Right! Based on a story by Ernest Grayman, this portrayal of life in a military prep school has its main character experience struggle, recognition, love, jealousy, and triumph. Silent. **1926 46m/B** Francis X. Bushman. **VHS, Beta** $29.95 *VYY, GPV* ♫½

Eyes of a Stranger A terrifying maniac stalks his female prey by watching their every move. Tewes is cast as a journalist, the stronger of the two sisters in this exploitative slasher. **1981 (R) 85m/C** Lauren Tewes, John Disanti, Jennifer Jason Leigh; *Dir:* Ken Wiederhorn. **VHS, Beta** $19.98 *WAR* Woof!

Eyes of Texas Bryant, a lawyer who uses a pack of killer dogs to get the land she wants, is pursued by a U.S Marshal played decently by Rogers. **1948 54m/B** Roy Rogers, Lynne Roberts, Andy Devine, Nana Bryant, Roy Barcroft; *Dir:* William Witney. **VHS, Beta** $9.99 *NOS, VCN, CPB* ♫

Eyes of Turpin Are Upon You! Two early Turpin silent comedies: "Idle Eyes" and "A Small Town Idol." **1921 56m/B** Ben Turpin, Ramon Novarro, Marie Prevost, James Finlayson. **VHS, Beta** $24.95 *VYY* ♫♫

The Eyes of Youth A young woman searches her soul for answers: to marry or not to marry is the question. A little foresight, in the form of a glimpse into the hypothetical future, helps her make the right choice. Very early Valentino fare in which the sheik plays a cad. **1919 78m/B** Clara Kimball Young, Edmund Lowe, Rudolph Valentino. **VHS, Beta** $29.95 *GPV, FCT* ♫♫

Eyewitness When a murder occurs in the office building of a star-struck janitor, he fabricates a tale in order to initiate a relationship with the television reporter covering the story. Unfortunately the killers think he's

telling the truth, which plunges the janitor and reporter into a dangerous and complicated position, pursued by both police and foreign agents. Somewhat contrived, but Hurt and Weaver are always interesting. Woods turns in a wonderful performance as Hurt's somewhat-psychotic best friend. **1981 (R) 102m/C** William Hurt, Sigourney Weaver, Christopher Plummer, James Woods, Kenneth MacMillan, Pamela Reed, Irene Worth, Steven Hill, Morgan Freeman; *Dir:* Peter Yates. **VHS, Beta, LV** $14.98 *FXV, FOX* ♫♫½

F. Scott Fitzgerald in Hollywood Dramatization begins in the late 1930s when Fitzgerald was lured to Hollywood to work in the burgeoning film industry. It covers the period when, with his wife in a mental institution, he lived with Sheila Graham and struggled against alcoholism. Interesting story made dull and depressing by Miller's uninspired performance, but it's almost saved by Weld's portrayal of Zelda. Made for television. **1976 98m/C** Jason Miller, Tuesday Weld, Julia Foster, Dolores Sutton, Michael Lerner, James Woods; *Dir:* Anthony Page. **VHS, Beta** *LME* ♫♫

F/X A Hollywood special effects expert is contracted by the government to fake an assassination to protect a mob informer. After completing the assignment, he learns that he's become involved in a real crime and is forced to reach into his bag of F/X tricks to survive. Twists and turns abound in this fast-paced story that was the sleeper hit of the year. Followed by a sequel. **1986 (R) 109m/C** Bryan Brown, Cliff DeYoung, Diane Venora, Brian Dennehy, Jerry Orbach, Mason Adams; *Dir:* Robert Mandel. **VHS, Beta, LV** $14.98 *HBO, FCT* ♫♫♫

F/X 2: The Deadly Art of Illusion Weak follow-up finds the special-effects specialist set to pull off just one more illusion for the police. Once again, corrupt cops use him as a chump for their scheme, an over-complicated business involving a stolen Vatican treasure. **1991 (PG-13) 107m/C** Bryan Brown, Brian Dennehy, Rachel Ticotin, Philip Bosco, Joanna Gleason; *Dir:* Richard Franklin. **VHS, LV** $34.95 *ORI, IME* ♫

The Fable of the Beautiful Pigeon Fancier A powerful man who has always gotten what he wanted spies a beautiful young married woman and sets out to make her his own. Based on the Gabriel Gracia Marquee novel "Life in the Time of Cholera." In Spanish with English subtitles. **1988 73m/C** *SP* Ney Lotorraco, Claudia Ohana, Tonia Carrero, Dina Stat, Chico Diaz; *Dir:* Ruy Guerra. **VHS** $79.95 *FXL, FCT* ♫♫½

Fables of the Green Forest Johnny Chuck, Peter Cottontail, Chatter the Squirrel and other memorable Thornton W. Burgess characters come to life in "Whose Footprint Is That?" and "Johnny's Hibernation." Available in English and Spanish versions. **1979 55m/C VHS, Beta** $14.95 *MED* ♫♫½

The Fables of Harry Allard The New England storyteller narrates animated versions of "It's So Nice to Have a Wolf Around the House," and "Miss Nelson Is Missing." **1979 30m/C VHS, Beta** $14.95 *NWV* ♫♫½

The Fabulous Baker Boys Two brothers have been performing a tired act as nightclub pianists for fifteen years. When they hire a sultry vocalist to revitalize the routine, she inadvertently triggers long-sup-

pressed hostility between the "boys." The story may be a bit uneven, but fine performances by the three leading actors, the steamy 1940s atmosphere, and Pfeiffer's classic rendition of "Makin' Whoopee," are worth the price of the rental. Music by Dave Grusin.
1989 (R) 116m/C Michelle Pfeiffer, Jeff Bridges, Beau Bridges, Elie Raab, Jennifer Tilly; *Dir:* Steve Kloves. Los Angeles Film Critics Association Awards '89: Best Actress (Pfeiffer); National Board of Review Awards '89: Best Actress (Pfeiffer); New York Film Critics Awards '89: Best Actress (Pfeiffer). **VHS, Beta, LV, 8mm $19.95** IVE, LIV, FCT 🎬🎬🎬

The Fabulous Dorseys The musical lives of big band leaders Tommy and Jimmy Dorsey are portrayed in this less than fabulous biographical film that's strong on song but weak on plot. Guest stars include Art Tatum, Charlie Barnet, Ziggy Elman, Bob Eberly and Helen O'Connell. Highlights are the many musical numbers, such as "Green Eyes," "Runnin' Wild," and "Marie."
1947 91m/B Tommy Dorsey, Jimmy Dorsey, Janet Blair, Paul Whiteman, Sara Allgood, Arthur Shields; *Dir:* Alfred E. Green. **VHS, LV $19.95** NOS, MRV, QNE 🎬1/2

Fabulous Fleischer Folio, Vol. 3 This volume features six more classic Max and Dave Fleischer cartoons including "Cobweb Hotel" (1939), "The Stork Market" (1949), and "The Little Stranger" (1936).
193? 50m/C *Dir:* Dave Fleischer. **VHS, Beta** DIS, OM 🎬

Fabulous Fred Astaire Contains Fred Astaire's 1958 Emmy Award winning special, "An Evening With Fred Astaire," plus a "Person-to-Person" interview by Edward R. Murrow with Astaire.
1958 70m/B Fred Astaire, Barrie Chase. **VHS, Beta $24.95** NOS, DVT 🎬1/2

Fabulous Joe A talking dog named Joe gets involved in the life of a hen-pecked husband.
1947 54m/C Walter Abel, Donald Meek, Margot Grahame, Marie Wilson; *Dir:* Bernard Carr. **VHS, Beta $19.95** NOS, UNI 🎬🎬

Face of Another A severely burned man gets a second chance when a plastic surgeon makes a mask to hide the disfigurement. Unfortunately this face leads him to alienation, rape, infidelity, and murder.
1966 124m/B JP Tatsuya Nakadai, Machiko Kyo; *Dir:* Hiroshi Teshigahara. **VHS, Beta $59.95** SVS 🎬🎬

A Face in the Crowd Neal is a journalist who discovers a down-home philosopher (Griffith) and puts him on her television show. His aw-shucks personality soon wins him a large following and increasing influence-even political clout. However, off the air he reveals his true nature to be insulting, vengeful, and power-hungry-all of which Neal decides to expose. This marks Griffith's spectacular film debut as a thoroughly despicable character. Also debut of Remick as the pretty cheerleader he takes an interest in. Budd Schulberg wrote the screenplay from his short story "The Arkansas Traveler." He and directory Kazan had collaborated equally well in "On the Waterfront."
1957 126m/B Andy Griffith, Patricia Neal, Lee Remick, Walter Matthau, Anthony (Tony) Franciosa; *Dir:* Elia Kazan. **VHS, Beta $19.98** WAR, FCT 🎬🎬🎬1/2

Face to Face Dramatic adaptation of two short stories, "The Secret Sharer," by Joseph Conrad, and Stephen Crane's, "The Bride Comes to Yellow Sky." The Conrad story features Mason as a sea captain who discovers a fugitive stowaway aboard his ships. The Crane story features Preston as a sheriff threatened by an old-time gunfighter.
1952 92m/B Bretaigne Windust, James Mason, Michael Pate, Robert Preston, Marjorie Steele, Gene Lockhart, Minor Watson; *Dir:* John Brahm. **VHS** RKO 🎬🎬1/2

A Face in the Fog Two newspaper reporters set out to solve a number of murders that have plagued the cast of a play. Also interested is the playwright. Good low-budget thriller.
1936 66m/B June Collyer, Lloyd Hughes, Lawrence Gray, Al "Fuzzy" St. John, Jack Mulhall, Jack Cowell, John Elliott, Sam Flint, Forrest Taylor, Edward Cassidy; *Dir:* Bob Hill. **VHS, Beta $16.95** NOS, SNC 🎬🎬

Face in the Mirror Story of the pain caused by teenage suicide and the answers found through love and faith.
1988 65m/C Michael Mitchell, Brian Park, Scott Wier, Tom Vanderwell, Marlene O'Malley; *Dir:* Russell S. Doughten Jr. **VHS, Beta** MIV 🎬

Face of the Screaming Werewolf Originally a Mexican horror/comedy that was bought for U.S. distribution by Jerry Warren, who earned his spot in the horror hall of fame with "The Incredible Petrified World" and "Teenage Zombies." A scientist brings a mummy back to life, but when he removes the bandages (gasp), the subject turns out to be of the canine persuasion. Suffers from comic evisceration; not much to scream about.
1959 60m/B MX Lon Chaney Jr., Landa Varle, Raymond Gaylord; *Dir:* Gilberto Martinez Solares, Jerry Warren. **VHS $16.95** SNC, MRV 🎬1/2

The Face at the Window Melodramatic crime story of a pair of ne'er-do-well brothers who terrorize Paris to conceal their bank robberies.
1939 65m/B Tod Slaughter, Marjorie Taylor, John Warwick, Robert Adair, Harry Terry; *Dir:* George King. **VHS $14.99** NOS, MRV, HEG 🎬🎬

Faces of Death, Part 1 Morbid, gruesome, "shockumentary" looks at death experiences around the world; uncensored film footage offers graphic coverage of autopsies, suicides, executions, and animal slaughter.
1974 88m/C *Nar:* Frances B. Gross. **VHS, Beta $19.98** MPI 🎬🎬

Faces of Death, Part 2 This sequel to "Faces of Death" further explores violent termination of man by man and by nature in a graphic-grisly gala. Not for the squeamish.
1985 84m/C **VHS, Beta $79.95** MPI Woof!

Fade to Black A young man obsessed with movies loses his grip on reality and adopts the personalities of cinematic characters to seek revenge on people who have wronged him. Thoroughly unpleasant.
1980 (R) 100m/C Dennis Christopher, Tim Thomerson, Linda Kerridge, Mickey Rourke, Melinda Fee; *Dir:* Vernon Zimmerman. **VHS, Beta $9.95** MED, MLB 🎬1/2

Fahrenheit 451 Chilling adaptation of the Ray Bradbury novel about a futuristic society that has banned all reading material and the firemen whose job it is to keep the fires at 451 degrees: the temperature at which paper burns. Werner is a fireman who begins to question the rightness of his ac-

tions when he meets the book-loving Christie-who also plays the dual role of Werner's TV-absorbed wife. Truffaut's first color and English-language film.
1966 112m/C FR Oskar Werner, Julie Christie, Cyril Cusack, Anton Diffring; *Dir:* Francois Truffaut. **VHS, Beta, LV $29.98** MCA, KUI, MLB 🎬🎬🎬

Fail-Safe A nail-biting nuclear age nightmare, in which American planes have been erroneously sent to bomb the USSR, with no way to recall them. An all-star cast impels this bitterly serious thriller, the straight-faced flipside of "Dr. Strangelove."
1964 111m/B Henry Fonda, Dan O'Herlihy, Walter Matthau, Larry Hagman, Fritz Weaver, Dom DeLuise; *Dir:* Sidney Lumet. New York Film Festival '64: Best Actress. **VHS, Beta, LV $14.95** COL 🎬🎬🎬1/2

Fair Game Hedonistic maniacs roam rural areas killing, looting and raping, until a beautiful woman they capture fights back.
1985 (R) 90m/C Kim Trengove, Kerry Mack, Marie O'Loughlina, Karen West; *Dir:* Christopher Fitchett. **VHS, Beta $79.95** IMP Woof!

Fair Game A psychotic but imaginative ex-boyfriend locks his former girlfriend in her apartment with a lethal giant Mamba snake. Understandably uninterested in its serpentine attention, the young woman must trespass against the Hollywood code and keep her wits in the face of danger. Guaranteed not to charm you.
1989 (R) 81m/C Greg Henry, Trudie Styler, Bill Moseley; *Dir:* Mario Orfini. **VHS $89.95** VMK Woof!

Fairytales A ribald musical fantasy for adults. In order to save the kingdom, the prince must produce an heir. The problem is that only the girl in the painting of "Princess Beauty" can "interest" the prince and she must be found. Good-natured smut.
1979 (R) 83m/C Don Sparks, Prof. Irwin Corey, Brenda Fogarty, Sy Richardson, Nai Bonet, Martha Reeves; *Dir:* Harry Tampa. **VHS, Beta $39.95** MED 🎬🎬

Fake Out A nightclub singer is caught between the mob and the police who want her to testify against her gangland lover. A typical vanity outing for Zadora.
1982 89m/C Pia Zadora, Telly Savalas, Desi Arnaz Jr.; *Dir:* Matt Cimber. **VHS, Beta $69.99** HBO 🎬

Falco: Rock Me Falco Four songs are performed by the German dance-music artist.
1985 20m/C GE **VHS, Beta** A&M, OM 🎬

Falcon A brave knight battles for the honor of a beautiful maiden; with or without the Falcon's help.
1974 105m/C Franco Nero. **VHS, Beta $59.95** TWE 🎬

The Falcon in Hollywood The falcon gets caught up with the murder of an actor who was part of a Tinseltown love triangle. The film takes place on RKO's back lot for a behind-the-scenes view of Hollywood. Part of the popular "Falcon" series from the 1940s.
1944 67m/B Tom Conway, Barbara Hale, Veda Ann Borg, Sheldon Leonard, Frank Jenks, Rita Corday, John Abbott; *Dir:* Gordon Douglas. **VHS $59.98** MED 🎬🎬1/2

The Falcon in Mexico When paintings from a supposedly dead artist turn up for sale in New York City the Falcon and the artist's daughter wind up journeying to Mexi-

co to solve the mystery. Part of "The Falcon" series.
1944 70m/B Tom Conway, Mona Maris, Nestor Paiva; *Dir:* William Berke. **VHS, Beta** *MED* 🎬½

The Falcon and the Snowman True story of Daulton Lee and Christopher Boyce, two childhood friends who, almost accidentally, sell American intelligence secrets to the KGB in 1977. Hutton and Penn are excellent, creating a relationship we care about and strong characterizations. Pat Metheny and Lyle Mays perform the musical score.
1985 (R) 110m/C Sean Penn, Timothy Hutton, Lori Singer, Pat Hingle, Dorian Harewood, Richard Dysart, Michael Ironside, Jennifer Runyon; *Dir:* John Schlesinger. **VHS, Beta, LV** $29.98 *LIV, VES* 🎬🎬🎬

Falcon Takes Over/Strange Bargain A mystery double feature. In "The Falcon Takes Over," The Falcon becomes involved in a series of murder and an ex-con's last love, and in "Strange Bargain," an underpaid bookkeeper gets in on an insurance swindle.
1949 131m/B George Sanders, Ward Bond, Allen Jenkins, Martha Scott, Harry Morgan, James Gleason, Anne Revere, Hans Conried; *Dir:* Irving Reis. **VHS, Beta** $34.95 *TTC, RKO* 🎬🎬

Falcon's Adventure/Armored Car Robbery A mystery double feature: The Falcon uncovers a vicious plot to steal a formula for synthetic diamonds in "The Falcon's Adventure," and four participants in an armored car robbery flee after slaying a policeman in "Armored Car Robbery."
1950 122m/B Tom Conway, Madge Meredith, William Talman, Charles McGraw, Adele Jergens; *Dir:* William Berke. **VHS, Beta** $34.95 *TTC, RKO* 🎬½

The Falcon's Brother Enemy agents intent on killing a South American diplomat are foiled by the Falcon's brothers at a steep personal cost. The plot enabled Sanders' real-life brother Conway to take over the title role in "The Falcon" mystery series, a role which he played nine more times.
1942 64m/B Tom Conway, George Sanders, Keye Luke, Jane Randolph; *Dir:* Stanley Logan. **VHS, Beta** *MED* 🎬

The Fall of the Berlin Wall Chronicles the Berlin Wall from its construction in 1963 to its opening in November 1989, when the people of eastern Germany rose up in a peaceful revolution.
1990 49m/C *Hosted:* Hans-Joachim Friedrichs. **VHS, Beta** $59.95 *WAR, FCT* 🎬

The Fall of the House of Usher Lord Roderick Usher is haunted by his sister's ghost in this poor adaptation of the Poe classic.
1952 70m/B *GB* Kay Tendeter, Gwen Watford, Irving Steen, Lucy Pavey; *Dir:* Ivan Barnett. **VHS** $16.95 *SNC, FCT* 🎬

The Fall of the House of Usher The moody Roger Corman/Vincent Price filmization, the first of their eight Poe adaptations, depicting the collapse of the famous estate due to madness and revenge. Usually titled "House of Usher." Terrific sets and solid direction as well as Price's inimitable presence.
1960 85m/C Vincent Price, Myrna Fahey, Mark Damon; *Dir:* Roger Corman. **VHS, Beta** $19.95 *KUI* 🎬🎬🎬

The Fall of the House of Usher Another version of Edgar Allan Poe's classic tale of a family doomed to destruction. Still to the Roger Corman/Vincent Price version to see how it should have been done.
1980 (PG) 101m/C Martin Landau, Robert Hays, Charlene Tilton, Ray Walston; *Dir:* James L. Conway. **VHS, Beta** $19.98 *UHV, LME* 🎬

Fall from Innocence A girl is driven to a life of prostitution and drugs by the sexual abuse inflicted upon her by her father.
1988 81m/C Isabelle Mejias, Thom Haverstock, Amanda Smith, Rob McEwan; *Dir:* Carey Connor. **VHS** $79.95 *MNC* 🎬½

Fall and Rise of Reginald Perrin 1 Starring Leonard Rossiter, this BBC show has Reginald getting out of the rat race.
1978 107m/C *GB* Leonard Rossiter. **VHS** $29.95 *BFS* 🎬🎬🎬

Fall and Rise of Reginald Perrin 2 After part one's escape from the rat race, this time Reginald is back in suburbia.
1978 120m/C *GB* Leonard Rossiter. **VHS** $29.95 *BFS* 🎬🎬🎬

The Fall of the Roman Empire An all-star, big budget extravaganza set in ancient Rome praised for its action sequences. The licentious son of Marcus Aurelius arranges for his father's murder and takes over as emperor while Barbarians gather at the gate. Great sets, fine acting, and thundering battle scenes.
1964 187m/C Sophia Loren, Alec Guinness, James Mason, Stephen Boyd, Christopher Plummer, John Ireland, Anthony Quayle, Eric Porter, Mel Ferrer, Omar Sharif; *Dir:* Anthony Mann. **VHS, Beta** $29.95 *BFV, UHV, LTG* 🎬🎬🎬

Fallen Angel A television movie about a child pornographer and a particular young girl he finds easily exploitable because of her unbalanced home situation.
1981 100m/C Dana Hill, Richard Masur, Melinda Dillon, Ronny Cox; *Dir:* Robert Lewis. Emmy Awards '81: Outstanding Drama Special. **Beta** $59.95 *COL* 🎬½

The Fallen Idol A young boy wrongly believes that a servant he admires is guilty of murdering his wife. Unwittingly, the child influences the police investigation of the crime so that the servant becomes the prime suspect. Richardson as the accused and Henrey as the boy cure notable. Oscar nominated screenplay by Graham Greene from his short story, "The Basement Room." Reed received a Best Director Oscar nomination.
1949 92m/B *GB* Ralph Richardson, Bobby Henrey, Michele Morgan, Sonia Dresdel, Jack Hawkins, Bernard Lee; *Dir:* Carol Reed. British Academy Awards '48: Best Film; National Board of Review Awards '49: 10 Best Films of the Year, Best Actor (Richardson). **VHS, Beta** $19.95 *NOS, MRV, PSM* 🎬🎬🎬

The Fallen Sparrow Garfield gives a superb performance as a half-mad veteran of the Spanish Civil War. Captured and brutalized by the Nazis he never revealed the whereabouts of the valuable possession they seek. Even back in the U.S. he is hounded by Nazi agent Slezak who used the woman Garfield loves to set a trap and finish the job.
1943 94m/B John Garfield, Maureen O'Hara, Walter Slezak, Patricia Morison, Martha O'Driscoll, Bruce Edwards, John Miljan, John Banner, Hugh Beaumont; *Dir:* Richard Wallace. **VHS, Beta** $29.95 *MED, IME* 🎬🎬🎬

Falling from Grace Bud Parks (Mellencamp) is a successful country singer who, accompanied by his wife and daughter, returns to his small Indiana hometown to celebrate his grandfather's 80th birthday. He's tired of both his career and his marriage and finds himself taking up once again with an old girlfriend (Lenz), who is not only married to Bud's brother but is also having an affair with his father. Bud believes he's better off staying in his old hometown but the problems caused by his return may change his mind. Surprisingly sedate, although literate, family drama with good ensemble performances. Actor-director debut for Mellencamp.
1992 (PG-13) 100m/C John Cougar Mellencamp, Mariel Hemingway, Kay Lenz, Claude Akins, Dub Taylor, Brent Huff, Deirdre O'Connell; *Dir:* John Cougar Mellencamp. **VHS, LV** *COL* 🎬🎬🎬

Falling in Love Two married New Yorkers unexpectedly fall in love after a coincidental meeting at the Rizzoli Book Store. Weak but gracefully performed reworking of "Brief Encounter," where no one seems to ever complete a sentence. An unfortunate reteaming for Streep and De Niro after their wonderful work in "The Deer Hunter."
1984 (PG-13) 106m/C Robert De Niro, Meryl Streep, Harvey Keitel, Dianne Wiest, George Martin, Jane Kaczmarek, David Clennon; *Dir:* Ulu Grosbard. **VHS, Beta, LV** $19.95 *PAR* 🎬🎬½

Falling in Love Again A middle-aged dreamer and his realistic wife travel from Los Angeles to their hometown of New York for his high school reunion. The man is filled with nostalgia for his youth, seen in flashback. Tedious, but notable for the debut of Pfeiffer. Also known as "In Love."
1980 (PG) 103m/C Elliott Gould, Susannah York, Michelle Pfeiffer; *Dir:* Steven Paul. **VHS, Beta** $24.98 *SUE, TAV* 🎬½

Falling for the Stars Both stars and their stunt-doubles talk about the perils involved in movie stunts. Using spectacular clips and insightful commentary, the program steps back from the camera to show familiar camaraderie of the stunt trade.
1985 58m/C Richard Farnsworth, Harvey Perry, Polly Burson, Buddy Ebsen, Robert Duvall, Robert Conrad, Betty Thomas. **VHS, Beta** $49.95 *DIS* 🎬½

False Colors Hopalong Cassidy unmasks a crook posing as a ranch heir. This was the 49th Hopalong Cassidy feature.
1943 54m/B William Boyd, Robert Mitchum, Andy Clyde, Jimmy Rogers; *Dir:* George Archainbaud. **VHS, Beta** $9.99 *NOS, MRV, VYY* 🎬

False Faces A ruthless, money-hungry quack is hounded by the law and the victims of his unscrupulous plastic surgery.
1932 80m/B Lowell Sherman, Peggy Shannon, Lila Lee, Joyce Compton, Berton Churchill, David Landau, Eddie Anderson, Ken Maynard; *Dir:* Lowell Sherman. **VHS, Beta** $29.95 *VYY* 🎬½

False Identity A radio psychologist buys a Purple Heart medal at a garage sale and then tries to find out about the original recipient. But her questions are making a number of folks uneasy, including a potentially dangerous stranger.
1990 (PG-13) 97m/C Genevieve Bujold, Stacy Keach, Veronica Cartwright, Tobin Bell, Mimi Maynard; *Dir:* James Keach. **VHS, Beta, LV** $89.99 *PAR* 🎬½

False Note Against a background of vibrant colors and collages, a little, crippled beggar-musician roams through a modern city, playing his small barrel organ which inevitably emits a sour note. He discovers a kindred spirit in an abandoned carousel horse. **1972 10m/C VHS, Beta** *IFB* ✂

Fame Follows eight talented teenagers from their freshmen year through graduation from New York's High School of Performing Arts. Insightful and absorbing, director Parker allows the kids to mature on screen, revealing the pressures of constantly trying to prove themselves. A faultless parallel is drawn between these "special" kids and the pressures felt by high schoolers everywhere. Great dance and music sequences. The basis for the television series. **1980 (R) 133m/C** Irene Cara, Barry Miller, Paul McCrane, Anne Meara, Joanna Merlin, Richard Belzer; *Dir:* Alan Parker. Academy Awards '80: Best Song ("Fame"), Best Original Score. **VHS, Beta, LV $19.99** *MGM, FCT* ✂✂✂

Fame is the Spur A lengthy but interesting look at the way power corrupts, plus an insight into the Conservative versus Labor dynamics of British government. Redgrave is a poor, idealistic worker who decides to help his fellow workers by running for Parliament. There, he falls prey to the trappings of office with surprising consequences. Look hard for Tomlinson, who went on to star in "Mary Poppins" and "Bedknobs and Broomsticks" for Disney. **1947 116m/B** Michael Redgrave, Rosamund John, Bernard Miles, Hugh Burden, Guy Verney, Carla Lehmann, Seymour Hicks, David Tomlinson; *Dir:* Roy Boulting. **VHS $19.95** *NOS, TIM* ✂✂½

The Family As a New Orleans hit-man who resists joining the mob, Bronson initiates an all-out war on the syndicate and its boss, played by Savalas. A poorly dubbed Italian action film. **1973 94m/C** *IT* Charles Bronson, Jill Ireland, Telly Savalas; *Dir:* Sergio Sollima. **VHS, Beta $59.95** *MPI* ✂✂½

The Family An 80-year-old patriarch prepares for his birthday celebration reminiscing about his family's past triumphs, tragedies and enduring love. The charming flashbacks, convincingly played, all take place in the family's grand old Roman apartment. Available either in Italian with subtitles or dubbed. **1988 128m/C** *IT* Vittorio Gassman, Fanny Ardant, Philippe Noiret, Stefania Sandrelli; *Dir:* Ettore Scola. **VHS, Beta, LV $79.98** *LIV, VES* ✂✂✂

Family Business A bright Ivy Leaguer, impressed by the exploits and vitality of his criminal grandfather, recruits him and his ex-con dad to pull off a high-tech robbery, which goes awry. Caper film, with its interest in family relationships and being true to one's nature. Casting Connery, Hoffman, and Broderick as the three leaves a big believability problem in the family department. **1989 (R) 114m/C** Sean Connery, Dustin Hoffman, Matthew Broderick, Rosana De Soto, Janet Carroll, Victoria Jackson, Bill McCutcheon, Deborah Rush, Marilyn Cooper, Salem Ludwig, Rex Everhart, James Tolken; *Dir:* Sidney Lumet. **VHS, Beta, LV, 8mm $19.95** *COL, FCT* ✂✂½

A Family Circus Christmas All the characters of the famous comic strip, "Family Circus," celebrate Christmas with a funny approach. The hysterics begin when one of the children asks Santa for a very unusual present, which he delivers! **1984 30m/C** *Dir:* Al Kouzel. **VHS, Beta $9.95** *FHE* ✂✂½

Family Enforcer A small-time hoodlum is bent on becoming the best enforcer in an underworld society. Also known as "Death Collector." **1977 (R) 82m/C** Joe Cortese, Lou Criscuola, Joe Pesci, Anne Johns, Keith Davis; *Dir:* Ralph De Vito. **VHS, Beta $19.95** *UHV* ✂

Family Game In "The Newlywed Game" fashion, mom and dad return from the isolation booth to guess what their children said about them in their absence. **1967 29m/C** *Hosted:* Bob Barker. **VHS, Beta $59.95** *VYY* ✂

The Family Game An obsessive satire about a poor, belligerent college student hired by a wealthy contemporary Japanese family to tutor their spoiled teenage son. Provides an all-out cultural assault on the Japanese bourgeoisie. In Japanese with English subtitles. **1983 107m/C** *JP* Yusaku Matsuda, Juzo Itami, Saori Yuki, Ichirota Miyagawa; *Dir:* Yoshimitsu Morita. **VHS, Beta $59.95** *SVS, FCT* ✂✂✂

Family Jewels A spoiled child-heiress has to choose among her six uncles to decide which should be her new father. If you like Jerry Lewis, you can't miss this! In addition to playing all six uncles, Lewis plays the chauffeur, as well as serving as producer, director, and coauthor of the script. **1965 100m/C** Jerry Lewis, Donna Butterworth, Sebastian Cabot, Robert Strauss; *Dir:* Jerry Lewis. **VHS, Beta $19.95** *PAR* ✂✂

Family Life A portrait of a 19-year-old girl battling her parents to establish her own identity. She is subjected to increasingly vigorous psychological interventions in this documentary-style effort originally made for British television. **1972 108m/C** *GB* Sandy Ratcliff, Bill Dean, Grace Cave; *Dir:* Ken Loach. **VHS, Beta $59.95** *COL* ✂✂½

The Family Man A brief encounter-between a married Manhattanite (Asner) and a much younger single woman (Baxter Birney) is sensitively handled in this film which explores their passionate affair. Made-for-television. **1979 98m/C** Ed Asner, Meredith Baxter Birney, Paul Clemens, Mary Joan Negro, Anne Jackson, Martin Short, Michael Ironside; *Dir:* Glenn Jordan. **VHS, Beta** *TLF* ✂✂

A Family Matter (Vendetta) A Mafia chieftain adopts a child whose father was killed by hit men. She grows into a beautiful woman bent on revenge. Melodramatic Italian-American co-production, made for TV. **1991 (R) 100m/C** Eric Roberts, Carol Alt, Eli Wallach, Burt Young. **VHS $89.95** *VMK* ✂✂

Family Plot Alfred Hitchcock's last film focuses on the search for a missing heir which is undertaken by a phony psychic and her private-eye boyfriend, with all becoming involved in a diamond theft. Campy, lightweight mystery that stales with time and doesn't fit well into Hitchcock's genre. Screenplay by Ernest Lehman. **1976 (PG) 120m/C** Karen Black, Bruce Dern, Barbara Harris, William Devane, Ed Lauter, Katherine Helmond; *Dir:* Alfred Hitchcock. Edgar Allan Poe Awards '76: Best Screenplay; National Board of Review Awards '76: 10 Best Films of the Year. **VHS, Beta, LV $19.95** *MCA* ✂✂½

Family Reunion A young man brings a pretty hitchhiker to his family reunion where they mistake her for his fiancee (who recently dumped him).

1988 97m/C David Eisner, Rebecca Jenkins, Henry Beckman, Linda Sorensen; *Dir:* Dick Sarin. **VHS, Beta $19.95** *MNC* ✂✂½

Family Secrets A made-for-television drama about a daughter, mother, and grandmother overprotecting and manipulating each other during a weekend together. **1984 96m/C** Maureen Stapleton, Stefanie Powers, Melissa Gilbert, James Spader; *Dir:* Jack Hofsiss. **Beta $79.95** *VHV* ✂✂½

Family Sins A television movie about a domineering father who dotes on his sports-oriented son while willfully neglecting the other. Eventually this leads to calamity. **1987 93m/C** James Farentino, Jill Eikenberry, Andrew Bendarski, Mimi Kuzyk, Brent Spiner, Michael Durrell, Tom Bower; *Dir:* Jerrold Freedman. **VHS $79.98** *REP* ✂✂

Family Upside Down An aging couple fight their separation after the husband has a heart attack and is put into a nursing home. The fine cast received several Emmy nominations, with Astaire the winner. Made-for-television. **1978 100m/C** Helen Hayes, Fred Astaire, Efrem Zimbalist Jr., Patty Duke; *Dir:* David Lowell Rich. **VHS, Beta $59.95** *COL* ✂✂✂

Family Viewing Surrealistic depiction of a family—obsessed with television and video—whose existence is a textbook model of home sweet dysfunctional home. An early, experimental film from Canada's Egoyan, it won considerable praise for its social commentary. **1988 92m/C** *CA* Adian Tierney, Gabrielle Rose; *Dir:* Atom Egoyan. **VHS $79.95** *FXL, FCT* ✂✂✂

Famous Five Get Into Trouble Four precocious youngsters and a dog get involved with a criminal plot and cutely wile their way out of it. Scandinavian; dubbed in English. **1977? 90m/C** Astrid Villaume, Ove Sprogoe, Lily Broberg; *Dir:* Trine Hedman. **VHS, Beta $69.95** *TWE* ✂

The Fan A Broadway star is threatened and her immediate circle cut down when a lovestruck fan feels he has been rejected by his idol. The stellar cast makes this bloody and familiar tale seem better than it is. **1981 (R) 95m/C** Lauren Bacall, Maureen Stapleton, James Garner, Hector Elizondo, Michael Biehn, Griffin Dunne; *Dir:* Edward Bianchi. **VHS, Beta, LV $14.95** *PAR* ✂✂

Fancy Pants Remake of "Ruggles of Red Gap" features Hope, a British actor posing as a butler. Also featuring Ball, an amusing contrast. Fine performances all around. **1950 92m/C** Bob Hope, Lucille Ball, Bruce Cabot, Jack Kirkwood, Lea Penman, Eric Blore, John Alexander, Norma Varden; *Dir:* George Marshall. **VHS, Beta, LV $14.95** *PAR* ✂✂½

Fandango Five college friends take a wild weekend drive across the Texas Badlands for one last fling before graduation and the prospect of military service. Expanded by Reynolds with assistance from Steven Spielberg, from his student film. Provides a look at college and life during the '60s Vietnam crisis. **1985 (PG) 91m/C** Judd Nelson, Kevin Costner, Sam Robards, Chuck Bush, Brian Cesak, Elizabeth Daily, Suzy Amis, Glenne Headly, Pepe Serna; *W/Dir:* Kevin Reynolds. **VHS, Beta $19.98** *WAR* ✂✂½

Fangoria's Weekend of Horrors A documentary about the making of horror films, featuring interviews with the industry's most scream-invoking artists. Copious film clips.
1986 60m/C Rick Baker, Wes Craven, Tobe Hooper, Robert Englund. **VHS, Beta $9.98** *MED* ✂✂½

Fangs Unfriendly reptile-stomping villagers take the life of Mr. Snakey's favorite serpent. He sends his slithering pets on a vengeful and poisonous spree.
1975 (R) 90m/C *SP IT* Les Tremayne, Janet Wood, Bebe Kelly, Marvin Kaplan, Alice Nunn; *Dir:* Vittorio Schiraldi. **VHS, Beta $24.95** *GEM, MRV* ✂½

Fangs of Fate A gang of bad guys battles the town's new marshal.
1925 67m/B Bill Patton, Dorothy Donald. **VHS, Beta, 8mm $29.95** *GPV, VYY* ✂½

Fangs of the Living Dead When a young woman inherits a castle, her uncle, who happens to be a vampire, tries to persuade her to remain among the undead. Also known as "Malenka, the Vampire."
1968 80m/C *SNC IT* Anita Ekberg, Rossana Yanni, Diana Lorys, Fernando Bilbao, Paul Muller, John Hamilton, Julian Ugarte, Andriana Ambesi; *W/Dir:* Armando de Ossorio. **VHS $19.98** *SNC* ✂½

Fanny Second part of Marcel Pagnol's trilogy depicting the lives of the people of Provence, France. The poignant tale of Fanny, a young woman who marries an older man when Marius, her young lover, leaves her pregnant when he goes to sea. Remade several times but the original holds its own very well. "Marius" was first in the trilogy; "Cesar" was third.
1932 128m/B *FR* Raimu, Charpin, Orane Demazis, Pierre Fresnay, Alida Rauffe; *Dir:* Marc Allegret. **VHS, Beta $64.95** *INT, MRV, HHT* ✂✂✂✂

Fanny A young girl falls in love with an adventurous sailor, and finds herself pregnant after he returns to the sea. With the help of the sailor's parents, she finds, marries and eventually grows to love a much older man, who in turn cares for her and adores her son as if he were his own. When the sailor returns, all involved must confront their pasts and define their futures. Beautifully made, with fine performances and a plot which defies age or nationality. Oscar nominations for Best Picture, actor Boyer, color cinematography, score and film editing. Part of the "A Night at the Movies" series, this tape simulates a 1961 movie evening, with a Tweety Pie cartoon, "The Last Hungry Cat," a newsreel and coming attractions for "Splendor in the Grass" and "The Roman Spring of Mrs. Stone."
1961 148m/C *FR* Leslie Caron, Maurice Chevalier, Charles Boyer, Horst Buchholz, Lionel Jeffries; *Dir:* Joshua Logan. National Board of Review Awards '61: 10 Best Films of the Year. **VHS, Beta $19.98** *NOS, WAR* ✂✂✂

Fanny and Alexander The culmination of Bergman's career, this autobiographical film is set in a rural Swedish town in 1907. It tells the story of one year in the lives of the Ekdahl family, as seen by the young children, Fanny and Alexander. Magic and religion, love and death, reconciliation and estrangement are skillfully captured in this carefully detailed, lovingly photographed film. In Swedish with English subtitles.
1983 (R) 197m/C *SW* Pernilla Allwin, Bertil Guve, Gunn Walgren, Allan Edwall, Ewa Froling, Erland Josephson, Harriet Andersson, Jarl Kulle; *Dir:* Ingmar Bergman. Academy Awards '83:

Best Art Direction/Set Decoration, Best Cinematography, Best Costume Design, Best Foreign Language Film. **VHS, Beta, LV $39.98** *SUE* ✂✂✂✂

Fanny Hill A softcore adaptation of the racy Victorian classic.
1983 (R) 80m/C *GB* Lisa Raines, Shelley Winters, Wilfrid Hyde-White, Oliver Reed. **VHS, Beta $79.95** *MGM* ✂✂

Fanny Hill: Memoirs of a Woman of Pleasure The sexual exploits of an innocent in bawdy eighteenth-century London, as directed by the notorious Russ Meyer of "Super Vixen" repute. Inept.
1964 105m/B Miriam Hopkins, Walter Giller, Alexander D'Arcy, Leticia Roman; *Dir:* Russ Meyer. **VHS, Beta** *PGN* Woof!

Fantasia Disney's most personal animation feature first bombed at the box office and irked purists who couldn't take the plotless, experimental mix of classical music and cartoons. It became a cult movie, embraced by more liberal generations of moviegoers. Reissue of the original version, painstakingly restored, has ceased because of a planned remake, so get copies while they last. Now-legendary sequences include the short "The Sorcerer's Apprentice," "Toccata & Fugue in D Minor," "The Cossack Dance," "Pastoral Symphony," "Night on Bald Mountain" and "Ave Maria."
1940 116m/C Mickey Mouse; *Nar:* Deems Taylor; *Performed by:* Philadelphia Symphony Orchestra. **VHS, LV $24.99** *DIS, FCT, RDG* ✂✂✂✂

Fantasies Teenage lovers Derek and Hooten return to their Greek island home and decide to improve their village by turning it into a tourist haven. First collaboration of the Dereks. Bo plays no talent but the obvious one.
1973 81m/C Bo Derek, Peter Hooten, Anna Alexiades; *Dir:* John Derek. **VHS, Beta $59.98** *FOX* Woof!

The Fantasist This sex crime thriller features a deranged killer who makes obscene, yet seductive phone calls to young ladies and murders them.
1989 (R) 98m/C Timothy Bottoms, Christopher Cazenove; *Dir:* Robin Hardy. **VHS $79.98** *REP* ✂

Fantastic Animation Festival Fourteen award-winning animated shorts are combined into one feature-length program. Included are "Closed Mondays," "The Last Cartoon Man," "French Windows," "Moonshadow," and "Cosmic Cartoon."
1977 (PG) 91m/C **VHS, Beta** *MED, OM* ✂

Fantastic Balloon Voyage Three men embark on a journey across the equator in a balloon, experiencing countless adventures along the way.
198? (G) 100m/C Hugo Stiglitz, Jeff Cooper. **VHS, Beta $29.95** *GEM* ✂

Fantastic Planet A critically acclaimed French, animated, sci-fi epic based on the drawings of Roland Topor. A race of small humanoids are enslaved and exploited by a race of giants on a savage planet, until one of the small creatures manages to unite his people and fight for equality.
1973 (PG) 68m/C *FR* *Dir:* Rene Laloux; *Voices:* Barry Bostwick. Cannes Film Festival '73: Grand Prix. **VHS, Beta $39.98** *SNC, MRV, VYY* ✂✂✂

Fantastic Seven The Fantastic Seven is a daredevil team sent to rescue a Hollywood sex symbol who has been kidnapped and held for ransom while shooting on location off the coast of Miami.

1982 96m/C Elke Sommer, Christopher Connelly. **VHS, Beta** *TLF* ✂½

Fantastic Voyage An important scientist, rescued from behind the Iron Curtain, is so severely wounded by enemy agents that traditional surgery is impossible. After being shrunk to microscopic size, a medical team journeys inside his body where they find themselves threatened by the patient's natural defenses. Great action, award-winning special effects.
1966 100m/C Stephen Boyd, Edmond O'Brien, Raquel Welch, Arthur Kennedy, Donald Pleasence, Arthur O'Connell, William Redfield, James Brolin; *Dir:* Richard Fleischer. Academy Awards '66: Best Art Direction/Set Decoration (Color), Best Visual Effects. **VHS, Beta, LV $19.98** *FOX, CCB* ✂✂✂

The Fantastic World of D.C. Collins Gary Coleman plays a daydreaming teenager who thinks he is being pursued by mysterious strangers seeking a videotape that was unknowingly slipped to him. Made for television.
1984 100m/C Gary Coleman, Bernie Casey, Shelley Smith, Fred Dryer, Marilyn McCoo, Phillip Abbott, George Gobel, Michael Ansara; *Dir:* Leslie Martinson. **VHS $19.95** *STE, NWV* ✂

Fantasy in Blue The search for the solution to a sexual stalemate results in a couple's strange experimentation.
197? 81m/C **VHS, Beta** *MED, MRV* ✂

The Fantasy Film Worlds of George Pal A documentary examining the career of the famed producer/director, from the animated Puppetoons to films like "The War of the Worlds," "The Time Machine," "Tom Thumb," and "The Seven Faces of Dr. Lao."
1986 93m/C **VHS, Beta, LV $19.95** *STE, NWV, IME* ✂

Fantasy Island Three people fly ("De plane, boss!") to an island paradise and get to live out their fantasies for a price. Made for television; the basis for the TV series.
1976 100m/C Ricardo Montalban, Bill Bixby, Sandra Dee, Peter Lawford, Carol Lynley, Hugh O'Brian, Eleanor Parker, Dick Sargent, Victoria Principal; *Dir:* Richard Lang. **VHS, Beta $49.95** *PSM* ✂

Fantasy Man A restless middle-aged man tries to relieve his mid-life crisis with the help of three women.
1984 100m/C *AU* Harold Hopkins, Jeanie Drynan, Kerry Mack, Kate Fitzpatrick; *W/Dir:* John Meagher. **VHS, Beta $59.95** *VCD* ✂½

Fantasy Masquerade Young lovelies play dress-up in slinky fantasy costumes that would probably be illegal if worn in public.
1987 60m/C **VHS, Beta $19.95** *AHV* ✂½

Fantasy Mission Force Dastardly double crossing catches a crack commando unit with its pants down.
198? 90m/C Jackie Chan. **VHS, Beta $24.95** *ASE* ✂

Far Away and Long Ago Based on the autobiographical novel by Guillermo Hudson which details the memories which haunted his childhood—the Argentinian pampas, its gauchos, witchcraft, and women. In Spanish with English subtitles.
1974 91m/C *AR* Juan Jose Camero, Leonor Manso; *Dir:* Manuel Antin. **VHS $24.98** *TAM* ✂✂½

Far Country Cattlemen must battle the elements and frontier lawlessness in this classic. Stewart leads his herd to the Yukon in hopes of large profits, but ends up having to kidnap it back from the crooked sheriff and avenging the deaths of his friends. Entertaining and the Yukon setting takes it out of the usual Western arena.
1955 97m/C James Stewart, Ruth Roman, Walter Brennan, Harry Morgan, Corinne Calvert, Jay C. Flippen, John McIntire; *Dir:* Anthony Mann. **VHS, Beta $14.95** KRT ✅✅✅

Far Cry from Home A battered wife attempts to escape from her domineering husband before it's too late.
197? 87m/C Mary Ann McDonald, Richard Monette. **VHS, Beta $49.95** TWE ✅

Far East Two ex-lovers meet in Southeast Asia and join forces to find the woman's missing husband, a reporter.
1985 105m/C Bryan Brown, Helen Morse; *Dir:* John Duigan. **VHS, Beta $59.95** LHV, VIR ✅✅½

Far Frontier Roy Rogers saves the day by thwarting a band of outlaws who are being smuggled across the border in soybean oil cans. Includes musical numbers and horse tricks.
1948 60m/B Roy Rogers, Andy Devine, Clayton Moore. **VHS $9.95** NOS, CPB, REP ✅✅

Far from Home Drew Barrymore is the seductive teen being scoped by a psychotic killer while on vacation with her father. Lots of over-the-top performances by the familiar cast.
1989 (R) 86m/C Matt Frewer, Drew Barrymore, Richard Masur, Karen Austin, Susan Tyrrell, Anthony Rapp, Jennifer Tilly, Andras Jones, Dick Miller; *Dir:* Meiert Avis. **VHS, LV $89.98** LIV, VES ✅✅

Far from the Madding Crowd A lavish, long adaptation of Thomas Hardy's nineteenth century classic about the beautiful Bathsheba and the three very different men who love her. Her first love is a handsome and wayward soldier (Stamp), her second the local noble lord (Finch) and her third the ever-loving and long-patient farmer (Bates). Christie is well cast as the much-desired beauty. Gorgeous cinematography by Nicolas Roeg.
1967 (PG) 165m/C *GB* Julie Christie, Terence Stamp, Peter Finch, Alan Bates, Prunella Ransome; *Dir:* John Schlesinger. National Board of Review Awards '67: Best Actor (Finch), Best Picture. **VHS, Beta, LV $29.98** MGM ✅✅½

Far North Written by Shepard, this is a quirky comedy about a woman returning to her rural family homestead after her father is seriously injured by a horse and trying to deal with her eccentric family's travails.
1988 (PG-13) 96m/C Jessica Lange, Charles Durning, Tess Harper, Donald Moffatt, Ann Wedgeworth, Patricia Arquette; *Dir:* Sam Shepard. **VHS, Beta, LV, 8mm $19.98** ORI ✅✅½

Far Out Man An unalterable middle-aged hippie is sent by his worried family and his psychiatrist on a cross-country journey to rediscover himself. The script makes no sense and has very few laughs.
1989 (R) 105m/C Thomas Chong, Rae Dawn Chong, C. Thomas Howell, Shelby Chong, Martin Mull, Paris Chong, Paul Bartel, Judd Nelson, Michael Winslow, Cheech Marin; *Dir:* Thomas Chong. **VHS, Beta, LV $89.95** COL Woof!

Far Out Space Nuts, Vol. 1 Two NASA ground crewmen accidentally launch a spacecraft propelling themselves into the vastness of outer space.

1975 48m/C Bob Denver, Chuck McCann. **VHS, Beta $29.98** SUE ✅

The Far Pavilions A British officer falls in love with an Indian princess during the second Afghan War. Cross is appropriately noble and stiff-upper-lipped but Irving is miscast as his ethnic love. Based on the romantic bestseller, this lavish production was done for cable television. Also known as "Blade of Steel."
1984 108m/C Ben Cross, Amy Irving, Omar Sharif, Benedict Taylor, Rossano Brazzi, Christopher Lee; *Dir:* Peter Duffell. **VHS, Beta $14.95** NEG, HBO, KAR ✅✅½

Faraway Fantasy A Texan beds down in love with the perky, nubile maids.
1987 86m/C Judith Frisch, Franz Muxeneder. **VHS, Beta $29.95** MED Woof!

A Farewell to Arms The original film version of Ernest Hemingway's novel about the tragic love affair between an American ambulance driver and an English nurse during the Italian campaign of World War I. The novelist disavowed the ambiguous ending, but the public loved the film. Fine performances and cinematography.
1932 85m/B Helen Hayes, Gary Cooper, Adolphe Menjou, Mary Philips, Jack LaRue, Blanche Frederici; *Dir:* Frank Borzage. Academy Awards '32: Best Cinematography, Best Sound; National Board of Review Awards '32: 10 Best Films of the Year. **VHS, Beta, LV $9.95** NEG, MRV, NOS ✅✅✅

Farewell to the King During World War II, a ship-wrecked American deserter becomes the chief of a tribe of Borneo headhunters until his jungle kingdom is caught between the forces of the U.S. and Japanese. With the help of a British officer he helps them defend themselves when the Japanese invade. As old-fashioned war epic with beautiful performances and solid if uninspired-performances by the leads.
1989 (PG-13) 114m/C Nick Nolte, Nigel Havers, Marius Weyers, Frank McRae, Marilyn Tokuda, Elan Oberon, William Wise, James Fox; *Dir:* John Milius. **VHS, Beta, LV $19.98** ORI ✅✅½

Farewell, My Lovely A remake of the 1944 Raymond Chandler mystery, "Murder, My Sweet," featuring private eye Phillip Marlowe hunting for an ex-convict's lost sweetheart in 1941 Los Angeles. Perhaps the most accurate of Chandler adaptations, but far from the best, this film offers a nicely detailed production. Mitchum is a bit too world-weary as the seen-it-all detective. Miles earned an Oscar nomination.
1975 (R) 95m/C *GB* Robert Mitchum, Charlotte Rampling, Sylvia Miles, John Ireland, Anthony Zerbe, Jack O'Halloran, Harry Dean Stanton, Sylvester Stallone, Cheryl "Rainbeaux" Smith; *Dir:* Dick Richards. National Board of Review Awards '75: 10 Best Films of the Year. **VHS, Beta $24.98** SUE ✅✅✅

Fargo Express When the kid brother of Maynard's true love is falsely accused of robbing a stagecoach, Maynard works to clear his name. Routine stuff.
1932 60m/B Ken Maynard, Helen Mack, Paul Fix; *Dir:* Alan James. **VHS, Beta $19.95** NOS, VCN, DVT ✅

Farmer Gray Goes to the Dogs (Cats, Monkeys & Lions) Seven animated silent films by pioneer animator Paul Terry, all featuring Farmer Gray: "Small Town Sheriff," "Cracked Ice," "The Huntsman," "The Medicine Man," "Mouse's

Bride," "Monkey Shines" and "Coast to Coast."
1926 55m/B VHS, Beta, 8mm $24.95 VYY ✅

The Farmer Takes a Wife Henry Fonda's first film, recreating his stage role, as a mid-nineteenth century farmer at odds with the Erie canal builders and struggling to court the woman he loves. Remade as a musical in 1953.
1935 91m/B Janet Gaynor, Henry Fonda, Charles Bickford, Slim Summerville, Jane Withers; *Dir:* Victor Fleming. **VHS** TCF ✅✅½

The Farmer Takes a Wife Poor musical remake of the 1935 drama, about the trials of a struggling 1850s farmer and the Erie canal boat cook he loves. Contains no memorable music and Robertson can't compete with Henry Fonda's original role as the farmer.
1953 81m/C Betty Grable, Dale Robertson, Thelma Ritter, John Carroll, Eddie Foy Jr.; *Dir:* Henry Levin. **VHS, Beta, LV $19.98** FOX ✅✅

The Farmer's Daughter Young portrays Katrin Holmstrom, a Swedish farm girl who becomes a maid to Congressman Cotten and winds up running for office herself (not neglecting to find romance as well). The outspoken and multi-talented character charmed audiences and was the basis of a television series in the 1960s. Oscar nomination for supporting actor Bickford.
1947 97m/B Loretta Young, Joseph Cotten, Ethel Barrymore, Charles Bickford, Harry Davenport, Lex Barker, James Arness, Rose Hobart; *Dir:* H.C. Potter. Academy Awards '47: Best Actress (Young). **VHS, Beta $59.98** FOX, BTV ✅✅✅

The Farmer's Other Daughter A rural comedy about Farmer Brown whose lovely daughter is eyed by all the farmhands. He also has to contend with dastardly Mr. Barksnapper who wants to take his farm away from him until one of his daughter's beaus comes to the rescue.
1965 84m/C Ernest Ashworth, Judy Pennebaker, Bill Michael; *Dir:* John Patrick Hayes. **VHS $19.95** SVS ✅✅

The Farmer's Wife Silent British comedy about a recently widowed farmer searching for a new wife. Meanwhile, his lovely housekeeper would be the perfect candidate. Screenplay written by Hitchcock, based on Eden Philpott's play.
1928 97m/B *GB* Jameson Thomas, Lillian Hall-Davies, Gordon Harker; *Dir:* Alfred Hitchcock. **VHS $16.95** NOS, SNC, MLB ✅✅½

Fashions of 1934 A typical, lightweight thirties musical with impressive choreography by Busby Berkeley. Powell plays a disreputable clothing designer who goes to Paris to steal the latest fashion designs and winds up costuming a musical and fall in love.
1934 78m/B Bette Davis, William Powell, Frank McHugh, Hugh Herbert, Reginald Owen, Busby Berkeley; *Dir:* William Dieterle. **VHS, Beta $19.95** VDM ✅✅

Fass Black Yet another man against the mob flick straight out of Bartlett's familiar plot lines. Amidst gratuitous music and violence, a disco owner locks horns with the syndicate after they try to take over his place. There's a good reason disco is dead.
1977 (R) 105m/C John Poole, Jeanne Bell, Cal Wilson, Harold Nicholas, Nicholas Lewis; *Dir:* D'Urville Martin. **VHS $39.95** XVC ✅✅

Fast Break A New York deli clerk who is a compulsive basketball fan talks his way into a college coaching job. He takes a team of street players with him, with predictable results on and off the court. Kaplan's screen debut.
1979 (PG) 107m/C Gabe Kaplan, Harold Sylvester, Randee Heller; *Dir:* Jack Smight. **VHS, Beta $19.95** *GKK* 🎬½

Fast Bullets Texas Rangers fight a gang of smugglers and rescue a kidnap victim in this typical western.
1944 52m/B Tom Tyler, Rex Lease. **VHS, Beta $19.95** *NOS, VCN, DVT* 🎬

Fast Company The life story of champion race car driver Lonnie Johnson including women, money, and the drag racing sponsors.
1978 90m/C William Smith, John Saxon, Claudia Jennings, Nicholas Campbell, Don Francks; *Dir:* David Cronenberg. **VHS, Beta** *AOV* 🎬

Fast Fists During the reign of the Min Kou Government, the Red Light Net gang of bandits terrorized the countryside. One fateful evening, the gang's leader attempted to kidnap and kill a famous actress. This is the incredible story of what happened.
19?? 90m/C Jimmy Wang Tu. **VHS, Beta $57.95** *OCE, UNI* 🎬

Fast Food A super-cheap, strangulated attempt at low comedy, wherein an entrepreneurial hamburger peddler invents a secret aphrodisiac sauce. Look for former porn-star Traci Lords.
1989 90m/C Clark Brandon, Tracy Griffith, Randal Patrick, Traci Lords, Kevin McCarthy, Michael J. Pollard, Jim Varney; *Dir:* Michael A. Simpson. **VHS, Beta, LV $29.95** *FRH* 🎬

Fast Forward A group of eight teenagers from Ohio learn how to deal with success and failure when they enter a national dance contest in New York City. A break-dancing variation on the old show business chestnut.
1984 (PG) 110m/C John Scott Clough, Don Franklin, Tracy Silver, Cindy McGee; *Dir:* Sidney Poitier. **VHS, Beta, LV $79.95** *COL* 🎬½

The Fast and the Furious On the lam after being falsely charged with murder, Ireland picks up a fast car and a loose woman (or is it a loose car and a fast woman?) and makes a run for the border by entering the Pebble Beach race.
1954 73m/B John Ireland, Dorothy Malone, Bruce Carlisle, Iris Adrian; *Dir:* John Ireland, Edwards Sampson. **VHS $19.95** *NOS, TIM* 🎬🎬

Fast Getaway Chases scenes proliferate when a teen criminal mastermind plots bank heists for his outlaw father. Relatively painless adolescent action-comedy, cleverly acted, quickly forgotten.
1991 (PG) 91m/C Corey Haim, Cynthia Rothrock, Leo Rossi, Ken Lerner, Marcia Strassman; *Dir:* Spiro Razatos. **VHS, LV, 8mm $89.95** *COL, PIA* 🎬🎬

Fast Kill A terrorist plot begins to fall apart when the conspirators fight among themselves.
1973 94m/C Tom Adams, Susie Hampton, Michael Culver, Peter Halliday; *Dir:* Lindsay Shonteff. **VHS, Beta $19.95** *GEM* 🎬

Fast Lane Fever A drag racer challenges a factory worker to a no-holds-barred race.
1982 (R) 94m/C **VHS, Beta $69.95** *MGM* 🎬

Fast Money Three pals can't resist the temptation of big bucks from flying drugs in from Mexico, but they find danger along the way.
1983 92m/C Sammy Allred, Sonny Carl Davis; *Dir:* Doug Holloway. **VHS, Beta $39.95** *IVE* 🎬½

Fast Talking The story of a quick-talking, incorrigible, fifteen-year-old Australian boy's humorous though tragic criminal schemes. The young protagonist slides his way through a life marked by a degenerate home life, a brother who forces him to sell drugs, and a future that holds no promise.
1986 93m/C *AU* Steve Bisley, Tracey Mann, Peter Hehir, Dennis Moore, Rod Zuanic, Toni Allaylis, Chris Truswell; *W/Dir:* Ken Cameron. **VHS, Beta $19.98** *SUE* 🎬🎬

Fast Times at Ridgemont High Teens at a Southern California high school revel in sex, drugs, and rock 'n' roll. A full complement of student types meet at the Mall—that great suburban microcosm percolating with angst-ridden teen trials—to contemplate losing their virginity, plot skipping homeroom, and move inexorably closer to the end of their adolescence. The talented young cast became household names: Sean Penn is most excellent as the California surfer dude who antagonizes teacher, Walston, aka "Aloha Mr. Hand." Based on the best-selling book by Cameron Crowe, it's one of the best of this genre.
1982 (R) 92m/C Sean Penn, Jennifer Jason Leigh, Judge Reinhold, Robert Romanus, Brian Backer, Phoebe Cates, Ray Walston, Scott Thomson, Vincent Schiavelli, Amanda Wyss, Forest Whitaker, Kelli Maroney, Eric Stoltz, Pamela Springsteen; *Dir:* Amy Heckerling. **VHS, Beta, LV $14.95** *MCA* 🎬🎬🎬

Fast Walking A prison guard is offered fifty-thousand dollars to help a militant black leader escape from jail, the same man his cousin has contracted to kill.
1982 (R) 116m/C James Woods, Kay Lenz, M. Emmet Walsh, Robert Hooks, Tim McIntire, Timothy Carey, Susan Tyrrell; *Dir:* James B. Harris. **VHS, Beta $19.98** *FOX, HHE* 🎬🎬

Faster Pussycat! Kill! Kill! It doesn't get any better than this! Three sexy go-go dancers get their after-work kicks by hot-rodding in the California desert. They soon find themselves enveloped in murder, kidnapping, lust and robbery after a particular race gets out of hand. Easily the most watchable, fun and funny production to spring from the mind of Russ Meyer. Those who haven't seen this cannot truly be called "cool."
1966 83m/B Tura Satana, Haji, Lori Williams, Susan Bernard, Stuart Lancaster, Paul Trinka, Dennis Busch, Ray Barlow, Mickey Foxx; *Dir:* Russ Meyer. **VHS $79.95** *RMF, FCT* 🎬🎬

Fastest Guitar Alive Debut of the Man with a Tear in His Voice is strictly for inveterate fans of the legendary Orbison. Rebel operative/crooner during the Civil War, whose rhythm is good but timing is bad, steals a Union gold supply, but the war ends before he makes it back to the land of Dixie. Seems that makes him a common thief. Orbison penned seven of the eight songs.
1968 88m/C Roy Orbison, Sammy Jackson, Margaret Pierce, Joan Freeman; *Dir:* Michael Moore. **VHS, Beta $19.95** *MVD, MGM* 🎬

The Fat Albert Christmas Special Bill Cosby's famous creations star in this animated Christmas story. And if you're not careful, you just may learn something about yourself.

1977 27m/C *Dir:* Hal Sutherland; *Voices:* Bill Cosby, Jan Crawford, Gerald Edwards, Eric Suter. **VHS, Beta $7.00** *VTR* 🎬½

The Fat Albert Easter Special Fat Albert and the Cosby Kids learn lessons and have fun in this animated special.
1982 27m/C *Dir:* Hal Sutherland; *Voices:* Bill Cosby, Jan Crawford, Eric Suter, Gerald Edwards. **VHS, Beta $7.00** *VTR* 🎬½

The Fat Albert Halloween Special More fun and learning from Fat Albert and the gang.
1977 27m/C *Dir:* Hal Sutherland; *Voices:* Bill Cosby, Jan Crawford, Gerald Edwards, Eric Suter, Erika Carroll. **VHS, Beta $7.00** *VTR* 🎬½

Fat City One of Huston's later triumphs, a seedy, street-level drama based on the Leonard Gardner novel about an aging alcoholic boxer trying to make a comeback and his young worshipful protege. Highly acclaimed. Tyrrell earned an Oscar nomination as the boxer's world-weary lover.
1972 (PG) 93m/C Stacy Keach, Jeff Bridges, Susan Tyrrell, Candy Clark, Nicholas Colasanto; *Dir:* John Huston. **VHS, Beta $19.95** *COL* 🎬🎬🎬½

Fat Guy Goes Nutzoid A crude farce about an obese mental patient, who escapes from the mental hospital and joins two teenagers on a wild trip to New York City.
1986 85m/C Tibor Feldman, Peter Linari, John MacKay, Joan Allen; *Dir:* John Golden. **VHS, Beta $79.95** *PSM* Woof!

Fat and the Lean An early short film from Roman Polanski that satirizes government tyranny and absurdity.
1961 16m/C *Dir:* Roman Polanski, Roman Polanski. **VHS, Beta** *IHF, TEX* 🎬🎬🎬

Fat Man and Little Boy A lavish, semi-fictional account of the creation of the first atomic bomb, and the tensions between J. Robert Oppenheimer and his military employer, Gen. Leslie Groves. Overlong but interesting. Cusack, whose character never existed, is especially worthwhile as an idealistic scientist.
1989 (PG-13) 127m/C Paul Newman, Dwight Schultz, Bonnie Bedelia, John Cusack, Laura Dern, John C. McGinley, Natasha Richardson; *Dir:* Roland Joffe. **VHS, Beta, LV, 8mm $14.95** *PAR, SOU, FCT* 🎬🎬½

Fatal Assassin The president of a black African nation is threatened by a hit-man in a South African hospital. Also known as "Target of an Assassin."
1978 90m/C *SA* John Phillip Law, Anthony Quinn, Simon Sabela; *Dir:* Peter Collinson. **VHS, Beta $9.95** *SIM, HHE* 🎬½

Fatal Attraction Couple meet after an auto accident. They begin an affair which turns dark and bizarre. Fine cast doesn't get far with this peculiar premise, although every one tries hard. Also known as "Head On".
1980 (R) 90m/C *CA* Sally Kellerman, Stephen Lack, John Huston, Lawrence Dane; *Dir:* Michael Grant. **VHS, Beta $29.98** *VES* 🎬

Fatal Attraction When a very married New York lawyer is seduced by a beautiful blonde associate, the one-night stand leads to terror as she continues to pursue the relationship. She begins to threaten his family and home with possessive, violent acts. An expertly made, manipulative thriller; one of the most hotly discussed films of the 1980s. A successful change of role for Close as the sexy, scorned, and deadly other woman. Music by Maurice Jarre. Also available in a special "director's series" edition, featuring Lyne's original, controversial ending. **1987 (R) 120m/C** Michael Douglas, Glenn Close, Anne Archer, Stuart Pankin, Ellen Hamilton-Latzen, Ellen Foley, Fred Gwynne, Meg Mundy, J.J. Johnston; *Dir:* Adrian Lyne. **VHS, Beta, LV, 8mm $14.95** *PAR, FCT, PIA* 🎞🎞🎞

Fatal Beauty A female undercover cop in Los Angeles tracks down a drug dealer selling cocaine (from which the title is taken). Elliott is the mob bodyguard who helps her out. Violent and sensational, the film tries to capitalize on the success of "Beverly Hills Cop" and fails miserably. Goldberg is wasted in this effort and the picture tiptoes around the interracial romance aspects that are implied. **1987 (R) 104m/C** Whoopi Goldberg, Sam Elliott, Ruben Blades, Harris Yulin, Cheech Marin, Brad Dourif; *Dir:* Tom Holland. **VHS, Beta, LV $19.95** *MGM* 🎞½

Fatal Chase Van Cleef is a U.S. Marshal assigned to protect ex-hitman Musante from his vengeful former employers. And you thought you didn't get along with your boss. Originally titled "Nowhere to Hide"; typical made-for-TV fodder written by Edward Anhalt, who usually manages to be a trifle more entertaining. **1977 78m/C** Lee Van Cleef, Tony Musante, Charles Robinson, Lelia Goldoni, Noel Fournier, Russell Johnson, Edward Anhalt; *Dir:* Jack Starrett. **VHS $9.95** *SIM, IPI* 🎞🎞

Fatal Confinement Wrenching drama finds Crawford, having lived 15 years in seclusion with her young daughter, forced to sell her land to a giant corporation bent on expansion. Interesting also for the similarities to Crawford's real-life "Mommie Dearest" lifestyle. Silent. **192? 70m/B** *GB* Joan Crawford, Paul Berg, Charles Bickford; *Dir:* Robert Guest. **VHS $39.95** *FCT* 🎞🎞½

Fatal Error A Dutch police thriller about the investigation of a cop killing. Dubbed. Originally titled, "Outsider in Amsterdam" and also released on video as, "The Outsider." **1983 85m/C** *NL* Rutger Hauer, Rijk de Gooyer, Willeke Van Ammelrooy, Donald M. Jones; *Dir:* Wim Verstappen. **VHS, Beta $79.95** *WES* 🎞½

Fatal Exposure The insatiably sanguinary Jack the Ripper lives on in the form of his great grandson, who avails himself of modern technology to capture those magic moments on videotape. Just goes to show home movies don't have to be bland. **1990 ?m/C** Blake Bahner, Ena Henderson, Julie Austin, Dan Schmale, Renee Cline, Gary Wise, Joy Ovington. **VHS** *TPV, HHE* 🎞🎞

Fatal Exposure A divorcee on vacation with her young sons accidentally picks up the wrong snapshots at the developers'. It turns out she has her hands on some incriminating photos, and the subject will resort to anything, even murder, to keep her quiet. **1991 (PG-13) 89m/C** Mare Winningham, Nick Mancuso, Christopher McDonald, Geoffrey Blake; *Dir:* Alan Metzger. **VHS, Beta** *PAR* 🎞½

Fatal Fix A French film expose of urban heroin addiction. Music by the Pretenders. **198? 90m/C** *FR* Helmut Berger, Corinne Clery; *Dir:* Massimo Pirri. **VHS, Beta $59.95** *MGL* 🎞½

Fatal Games Young female athletes are mysteriously disappearing at the Falcon Academy of Athletics and a crazed killer is responsible. **1984 88m/C** Sally Kirkland, Lynn Banashek, Sean Masterson. **VHS, Beta $59.95** *MED* 🎞½

The Fatal Glass of Beer Fields' son returns to his home in the north woods after serving a jail term in this slapstick comedy short. **1933 18m/B** W.C. Fields, Rosemary Theby, George Chandler, Richard Cramer. **VHS, Beta $19.98** *CCB, FST, RXM* 🎞½

The Fatal Hour Karloff is enlisted to aid police in solving the murder of a detective. As Karloff is rounding up suspects, three more murders take place. Feeble. **1940 68m/B** Boris Karloff, Marjorie Reynolds, Grant Withers, Charles Trowbridge, John Hamilton, Frank Puglia, Jason Robards Sr.; *Dir:* William Nigh. **VHS, Beta, 8mm $16.98** *VYY, SNC, TIM* 🎞½

The Fatal Image A mother and daughter on vacation in Paris inadvertently videotape an international mob hit, making them the target of ruthless assassins. Filmed on location in Paris. **1990 96m/C** Michele Lee, Justine Bateman, Francois Dunoyer, Jean-Pierre Cassel, Sonia Petrovna; *Dir:* Thomas J. Wright. **VHS $89.95** *WOV* 🎞🎞½

Fatal Images A camera mysteriously causes all the models who pose for it to die, prompting the photographer to investigate. **1982 90m/C** Lane Coyle. **VHS, Beta $49.95** *AHV* 🎞

Fatal Instinct A tough-guy cop trying to solve a murder instead finds himself a victim of sexual obsession in this erotic thriller. Also available in an unrated version. Not to be confused with (but doesn't everything sound familiar) "Basic Instinct." **1992 (R) 93m/C** Michael Madsen, Laura Johnson, Tony Hamilton; *Dir:* John Dirlam. **VHS $89.95** *NLC, COL* 🎞🎞

Fatal Mission A Vietnam soldier captures a female Chinese guerilla and uses her as his hostage and guide through the jungle. **1989 (R) 84m/C** Peter Fonda, Mako, Tia Carrere. **VHS, Beta, LV $79.98** *MED* 🎞

Fatal Pulse Sorority girls are being killed off in grisly ways. Basic slasher/gore. **1988 90m/C** Michelle McCormick, Ken Roberts, Joe Phelan, Alex Courtney. **VHS, Beta $79.95** *CEL* Woof!

Fatal Skies An evil schemer plots to dump toxic waste in a small town's water supply until a group of teens comes to the rescue. **1990 88m/C** Timothy Leary. **VHS, Beta $139.95** *AIP* 🎞

Fatal Vision The television film version of Joe McGinniss' controversial book about the murder trial of Dr. Jeffrey MacDonald, the Marine corps physician accused of murdering his whole family and blaming the killings on crazed hippies. McGinniss first believed MacDonald may have been innocent but his subsequent investigation, and his book, proved otherwise. The starring debut for Cole. MacDonald still protests his innocence. **1984 192m/C** Gary Cole, Karl Malden, Eva Marie Saint, Andy Griffith, Barry Newman; *Dir:* David Greene. **VHS, Beta, LV $69.95** *COL* 🎞

Fate Sweet comedy. A young man plans his love-affair carefully, but his biggest problem is meeting the girl of his dreams. Cheerful acting from Lynn and Paul. Stylish and charming. **1990 (PG-13) 115m/C** Cheryl Lynn, Stuart Paul, Kaye Ballard, Susannah York; *Dir:* Stuart Paul. **VHS $19.95** *ACA* 🎞🎞½

The Fate of Lee Khan Much blood is shed as a rebellion is led against the Mongols during the Yuan dynasty. **1973 107m/C** Roy Chaid, Angela Mao; *Dir:* Kinghu. **VHS, Beta $39.95** *CGH, PEV, FCT* 🎞🎞½

Father After World War II, a Hungarian youth makes up stories about his dead father's heroism that enhance his own position. Eventually he becomes obsessed with the facts surrounding his father's death at the hands of the enemy and, learning the truth, lays the past to rest. **1967 89m/B** *HU* Andras Balint, Miklos Gabor; *W/Dir:* Istvan Szabo. New York Film Festival '67: Best Actress. **VHS, Beta $175.00** *VYY, DVT* 🎞🎞🎞

Father of the Bride A classic, quietly hilarious comedy about the tribulations of a father preparing for his only daughter's wedding. Tracy is suitably overwhelmed as the loving father and Taylor radiant as the bride. A warm vision of American family life, accompanied by the 1940 MGM short "Wedding Bills." Also available in a colorized version. Followed by "Father's Little Dividend" and later a television series. Oscar nominated for Best Picture, Best Actor (Tracy), Screenplay. Remade in 1991. **1950 94m/C** Spencer Tracy, Elizabeth Taylor, Joan Bennett, Billie Burke, Leo G. Carroll, Russ Tamblyn, Don Taylor, Moroni Olsen, Russ Tamblyn; *Dir:* Vincente Minnelli. **VHS, Beta, LV $19.95** *MGM, PIA* 🎞🎞🎞½

Father of the Bride Welcome to one of the most overextravagant weddings in recent film history in this remake of the 1950 comedy classic. Martin is believable as the reluctant dad who complains about everything from the expenses, to the groom, to the future in-laws, all the while secretly hoping the wedding will be called off. Of course, this is all a front to hide his struggle to accept the fact that his little girl is growing up and getting married. There aren't any surprises here since the plot and characters are predictable, but nothing detracts from the purpose of the film: to be a nice, charming movie. Keaton is little more than window dressing as the mother of the bride and Short steals several scenes as a pretentious and rather irritating wedding coordinator. Williams pulls off a nice film debut—observant television watchers will recognize him as the bride to be in the Sprint ad, calling her dad to tell him she's engaged. **1991 (PG) 105m/C** Steve Martin, Diane Keaton, Martin Short, Kimberly Williams, George Newbern; *Dir:* Charles Shyer. **VHS, Beta $94.95** *BVV* 🎞🎞½

Father Figure When a divorced man attends his ex-wife's funeral, he discovers that he must take care of his estranged sons. Based on young adult writer Richard Peck's novel; a well-done made-for-television movie. **1980 94m/C** Hal Linden, Timothy Hutton, Cassie Yates, Martha Scott, Jeremy Licht; *Dir:* Jerry London. **VHS, Beta $69.95** *LTG* 🎞½

Father Goose During World War II, a liquor-loving plane-spotter stationed on a remote Pacific isle finds himself stuck with a group of French refugee schoolgirls and their teacher. Some predictable gags, romance and heroism fill out the running time pleasantly. Scriptwriters Peter Stone and Frank Tarloff, who were competitors, not collaborators on the project, shared an Oscar. **1964 116m/C** Cary Grant, Leslie Caron, Trevor Howard; *Dir:* Ralph Nelson. Academy Awards '64: Best Story & Screenplay. **VHS $19.98** *REP* 🎞🎞½

Father Guido Sarducci Goes to College The satirical character of the title is shown performing at a campus concert. The tape also includes footage from the padre's campus tours, and a glimpse into Sarducci's secret Vatican film archives.
1985 60m/C Don Novello. **VHS, Beta $19.98** *LIV, VES* ♫♫½

Father Hubbard: The Glacier Priest This program explores the finds of explorer-scientist Father Bernard Hubbard. As a geologist he discovered the secrets behind Alaska's great active volcanoes and as a sociologist, the origins and customs of Eskimos. Presented through classic newsreel footage.
1954 15m/B VHS, Beta $49.95 *TSF* ♫♫½

Father's Little Dividend Tracy expects a little peace and quiet now that he's successfully married off Taylor in this charming sequel to "Father of the Bride." However, he's quickly disillusioned by the news he'll soon be a grandfather - a prospect that causes nothing but dismay. Reunited the stars, director, writers, and producer from the successfull first film
1951 82m/B Spencer Tracy, Joan Bennett, Elizabeth Taylor, Don Taylor, Billie Burke, Russ Tamblyn, Moroni Olsen; *Dir:* Vincente Minnelli. **VHS, Beta, LV $9.95** *NEG, MRV, NOS* ♫♫♫

Fats Domino and Friends Join Fats Domino and his friends Ray Charles, Jerry Lee Lewis, Ron Wood, and Paul Schaffer for some good-time rock & roll. Filmed/ recorded at the Storyville Club in New Orleans. Songs include "Blueberry Hill," "Walking to New Orleans," "The Fat Man" and more.
1988 60m/C Fats Domino; *Dir:* Len Dell'Amico. *Performed by:* Paul Schaffer. **VHS, LV $19.98** *MVD, AVE, HBO* ♫♫♫

Fats Domino Live! The Fats performs in Los Angeles, featuring "Blueberry Hill," "Blue Monday," "Ain't That a Shame" and "I'm Ready."
1985 19m/C VHS, Beta $14.95 *MCA, MVD* ♫♫♫

Fats Waller and Friends The jovial Fats Waller appears in the only four Soundies he made, all from 1941: "Your Feet's Too Big," "Ain't Misbehavin'," "Honeysuckle Rose" and "The Joint Is Jumpin'." Also on this tape are seven other Soundies by an assortment of performers.
1986 29m/B Fats Waller, Tiny Grimes, Dorothy Dandridge, Bob Howard, Mabel Lee, Cook & Brown, Dusty Brooks, Myra Johnson. **VHS, Beta $19.95** *MVD, AFE, AMV* ♫♫♫

Fatso After the shocking death of his obese cousin, an obsessive overeater struggles to overcome his neurosis with the aid of his sister and a self-help group called "Chubby Checkers." Bancroft's first work as both writer and director.
1980 (PG) 93m/C Dom DeLuise, Anne Bancroft, Ron Carey, Candice Azzara; *W/Dir:* Anne Bancroft. **VHS, Beta $79.98** *FOX* ♫½

Fatty Finn A children's gagfest about young kids and bullies during the Depression, based on Syd Nicholls' comic strip.
1984 91m/C *AU* Ben Oxenbould, Bart Newton; *Dir:* Maurice Murphy. **VHS, Beta $14.98** *VID* ♫½

Fatty and Mabel Adrift/Mabel, Fatty and the Law Fatty and Mabel have problems enjoying their wedded bliss in these two silent shorts, which have a newly recorded orchestral score on the soundtrack.
1916 40m/B Fatty Arbuckle, Mabel Normand, Al "Fuzzy" St. John, Minta Durfee, Teddy the Dog. **VHS, Beta $19.98** *CCB* ♫½

Fatty's Tin-Type Tangle/Our Congressman In "Fatty's Tin-Type Tangle" (1915), Fatty and Louise are snapped by a traveling tintyper. In "Our Congressman" (1924), Will Rogers offers an "expose" of political life.
192? 44m/B Fatty Arbuckle, Louise Fazenda, Edgar Kennedy, Frank Hayes, Will Rogers, James Finlayson. **VHS, Beta $19.98** *CCB* ♫½

Faust The classic German silent based upon the legend of Faust, who sells his soul to the devil in exchange for youth. Based on Goethe's poem, and directed by Murnau as a classic example of Germanic expressionism. Remade as "All That Money Can Buy" in 1941.
1926 117m/B *GE* Emil Jannings, Warner Fuetterer, Gosta Ekman, Camilla Horn; *Dir:* F. W. Murnau. **VHS, Beta $24.95** *NOS, MRV, GPV* ♫♫♫½

Faust Joan Sutherland introduces the opera based on Goethe's dramatic poem telling of a man who makes a pact with the devil. Some of the most beautiful moments in French opera are presented. Puppet characters help explain the story and make it accessible to youngsters.
1973 30m/C Joan Sutherland. **VHS, Beta** *BFA* ♫♫½

The Favor, the Watch, & the Very Big Fish Set in a fairytale version of Paris. Hoskins is a photographer of religious subjects searching for a man to pose as Jesus. He discovers his subject in the hirsute Goldblum, a mad bar pianist who does think he's the savior. Richardson plays the object of Hoskin's shy affections, an actress who does dubbing work by providing the moaning and groaning for porno flicks. A messy attempt at screwball-comedy with some brief humorous moments.
1992 (R) 89m/C Bob Hoskins, Jeff Goldblum, Natasha Richardson, Michel Blanc, Angela Pleasence, Jean-Pierre Cassel; *W/Dir:* Ben Lewin. **VHS, LV $92.95** *VMK* ♫♫

Favorite Black Exploitation Cartoons Black stereotype cartoons exhibiting the bigotry of the times in which they were produced.
194? 60m/B VHS, Beta $11.95 *VCD* ♫♫

Favorite Celebrity Cartoons Famous people, events and literature are seen in comic cartoon portrayals.
194? 60m/B VHS, Beta *VCD* ♫♫

Favorite Racist Cartoons A compilation of banned racist cartoons.
194? 60m/B VHS, Beta *VCD* ♫♫

FBI Girl An FBI clerk is used as bait to trap a murderer and break up a gang. Burr, pre-Perry Mason, plays a bad guy.
1952 74m/B Cesar Romero, George Brent, Audrey Totter, Raymond Burr, Tom Drake; *Dir:* William Berke. **VHS, Beta, LV** *WGE, RXM* ♫♫

The FBI Story Mr. Stewart goes to Washington in this anatomy of the Federal Bureau of Investigation. If you're a fan of the gangster genre (LeRoy earlier directed "Little Caesar"), and not especially persnickety about fidelity to the facts, this actioner offers a pseudo-factual (read fictional) glimpse—based on actual cases from the 1920s through the 1950s—into the life of a fictitious agent-family man.
1959 149m/C James Stewart, Vera Miles, Nick Adams, Murray Hamilton, Larry Pennell; *Dir:* Mervyn LeRoy. **VHS, LV $19.98** *WAR, PIA, FCT* ♫♫♫

Fear An impoverished medical student murders the professor he believes is tormenting him and his life suddenly gets better. He falls in love and gets a scholarship but the police are closing in on his crime. Except...did any of it really happen? Or is he suffering from hallucinations. Low-budget psycho-drama with some good paranoid moments.
1946 68m/B Peter Cookson, Warren William, Anne Gwynne, James B. Cardwell, Nestor Paiva; *Dir:* Alfred Zeisler. **VHS** *VES* ♫♫

Fear A vacationing family is plagued by a murderous Vietnam vet and other psychotic cons.
1988 (R) 96m/C Frank Stallone, Cliff DeYoung, Kay Lenz, Robert Factor; *Dir:* Robert A. Ferretti. **VHS, Beta, LV $79.95** *VIR* ♫

Fear A young psychic (Sheedy) aids police by getting into the minds of serial killers and then writes about the cases. But what happens when the next killer is also a psychic and decides to play mind-games with her? Above average suspense sustains this made-for-cable thriller.
1990 (R) 98m/C Ally Sheedy, Lauren Hutton, Michael O'Keefe, Stan Shaw, Dina Merrill, John Agar, Marta DuBois; *Dir:* Rockne S. O'Bannon. **VHS $89.98** *VES* ♫♫½

Fear, Anxiety and Depression A neurotic aspiring playwright in New York has various problems with his love life. Sub-Woody Allen comedy attempt, with a few bright spots. The dregs of New York, with all the violence and degradation, are brought to the screen here.
1989 (R) 84m/C Todd Solondz, Stanley Tucci, *W/Dir:* Todd Solondz. **VHS, Beta $89.95** *VIR* ♫½

The Fear Chamber Hardly a Karloff vehicle. Boris shot the footage for this and three other Mexican "horror" films in LA, an unfortunate swan song to his career, though he was fortunate to be quickly written out of this story. The near plot concerns a mutant rock that thrives on human fear. Doctor Karloff and his assistants make sure the rock is rolling in sacrificial victims (women, of course). A prodigious devaluation of the "B"-grade horror flick, it's so bad it's just bad. Also known as "Torture Zone" and "Chamber of Fear."
1968 88m/C *MX* Boris Karloff, Yerye Beirut, Julissa, Carlos East; *Dir:* Juan Ibanez, Jack Hill. **VHS $19.99** *SNC, MPI* Woof!

Fear in the City The Sicilian Mafia and a black crime gang battle for control of the streets.
1981 90m/C *IT* Michael Constantine, Fred Williamson, Gianni Manera; *Dir:* Gianni Manera. **VHS, Beta $59.95** *MGL* ♫

Fear City Two partners who own a talent agency are after the psychopath who is killing off their prized strippers with the aid of a local cop. Sleazy look at Manhattan low life.
1985 (R) 93m/C Billy Dee Williams, Tom Berenger, Jack Scalia, Melanie Griffith, Rae Dawn Chong, Joe Santos, Maria Conchita Alonso, Rossano Brazzi; *Dir:* Abel Ferrara. **VHS, Beta, LV $79.99** *HBO* ♫½

Fear in the Night Suspenseful tale of a murder committed by a man under hypnosis. Fearing his nightmares are real Kelley talks his detective friend into investigating his "crime", which leads them to a mansion, a

mirrored room, and a plot that takes some clever and unexpected twists. Remade in 1956 as ''Nightmare.''
1947 72m/B Paul Kelly, DeForest Kelley, Ann Doran, Kay Scott, Robert Emmett Keane; *Dir:* Maxwell Shane. **VHS $59.99** SNC, HBO ♂♂♂

Fear No Evil A teenager who is the human embodiment of the demon Lucifer commits acts of demonic murder and destruction. His powers are challenged by an 18-year-old girl, who is the embodiment of the archangel Gabriel. First feature from La Loggia is better than it sounds.
1980 (R) 90m/C Stefan Arngrim, Kathleen Rowe McAllen, Elizabeth Hoffman; *Dir:* Frank Laloggia. **VHS, Beta $9.95** COL, SUE ♂♂

Fear Strikes Out Perkins plays Jimmy Piersall, star outfielder for the Boston Red Sox, and Malden, his demanding father, in the true story of the baseball star's battle for sanity. One of Perkins' best screen performances.
1957 100m/B Anthony Perkins, Karl Malden, Norma Moore, Adam Williams, Perry Wilson; *Dir:* Robert Mulligan. **VHS, Beta, LV $14.95** PAR ♂♂♂

Fearless An Italian detective has found a Viennese banker's daughter, but continues to pursue the unanswered questions of the case, embroiling himself in a web of intrigue and plotting.
1978 89m/C Joan Collins, Maurizio Merli, Franz Antel. **VHS, Beta $19.95** STE, NWV ♂

Fearless Fighters Kung fu action is fast and furious in this film wherein everybody is Kung fu fightin'.
1977 83m/C Lei Peng, Mu Lan. **VHS, Beta $7.00** VTR ♂

Fearless Hyena A young man uses his martial arts skills to get revenge against the men who murdered his grandfather.
1979 97m/C Jackie Chan. **VHS, Beta $24.95** ASE ♂

Fearless Hyena: Part 2 A happy-go-lucky Yin/Yang clan disciple must abandon his gambling and acrobatic endeavors to kick members of a rival clan in the head.
1986 90m/C Jackie Chan. **VHS, Beta $24.95** ASE ♂

The Fearless Vampire Killers Underrated, off-off-beat, and deliberately campy spoof on vampire films in which Tate is kidnapped by some fangy villains. Vampire trackers MacGowran and Polanski pursue the villains to the haunted castle and attempt the rescue. Only vampire movie with a Jewish bloodsucker (''Boy, have you got the wrong vampire,'' he proclaims to a maiden thrusting a crucifix at him). Inside the castle, Polanski is chased by the count's gay vampire son. Highlight is the vampire ball with a wonderful mirror scene. Many other amusing moments. Originally titled ''Fearless Vampire Killers, or Pardon Me, Your Teeth are in My Neck;'' also known as ''Dance of the Vampires.''
1967 98m/C GB Jack MacGowran, Roman Polanski, Alfie Bass, Jessie Robbins, Sharon Tate, Ferdinand ''Ferdy'' Mayne, Iain Quarrier; *Dir:* Roman Polanski. **VHS, Beta $19.98** MGM ♂♂♂

The Fearless Young Boxer A young man trains in martial arts in order to exact revenge for his father's murder.
1973 94m/C VHS, Beta $39.95 UNI ♂

Fearmaker After the mysterious death of her father, the heiress to his fortune is tangled in a web of treachery created by other potential inheritors. Also known as ''House of Fear.''
1989 (R) 90m/C MX Katy Jurado, Paul Picerni, Sonia Amelio, Carlos East; *Dir:* Anthony Carras. **VHS, Beta $59.95** PSM, MHV ♂

Feast for the Devil A woman searches for her missing sister in a mysterious coastal village, only to fall under the occult spell of a mad doctor.
197? 90m/C Krista Nell, Thomas Moore, Teresa Gimpera. **VHS, Beta $59.95** MGL ♂½

Feds Two women enter the FBI Academy and take on the system's inherent sexism with feebly comic results.
1988 (PG-13) 82m/C Rebecca DeMornay, Mary Gross, Ken Marshall, Fred Dalton Thompson, Larry Cedar, James Luisi, Raymond Singer; *Dir:* Dan Goldberg. **VHS, Beta, LV** WAR ♂½

Feel the Motion A female auto-mechanic wants to make it big in the music biz. She sees her chance when she slips her demo tape onto a hit TV-music show.
1986 98m/C Sissy Kelling, Frank Meyer-Brockman, Ingold Locke, Falco, Meatloaf. **VHS, Beta $69.95** VMK ♂

Feel My Pulse Silent comedy about a rich fanatic who leaves everything in his will to his young niece on the stipulation that she lead a germ-free life; when she reaches 21, she moves into the sanitarium she's inherited, not knowing it has become a base for prohibition-era rum runners.
1928 86m/B Bebe Daniels, Richard Arlen, William Powell; *Dir:* Gregory La Cava. **VHS, Beta $27.95** VYY, DNB ♂♂

Feelin' Screwy A couple of misfit dweebs attempt to rid their town of the local drug dealer in order to impress a couple of babes in this limp coming-of-ager.
1990 90m/C Quincy Reynolds, Larry Gamal, Darin McBride, Hassan Jamal, Marsha Carter, Brooks Morales, Anna Fuentes; *W/Dir:* Riffat A. Khan. **VHS $69.95** RAE ♂

Feelin' Up A young man sells all his possessions to come to New York in search of erotic adventures.
1976 (R) 84m/C Malcolm Groome, Kathleen Seward, Rhonda Hansome, Tony Collado, Charles Douglass; *Dir:* David Secter. **VHS, Beta $69.98** LIV, VES Woof!

Feet First A shoe salesman puts on ''upper crust'' airs as he begins a shipboard romance with a girl who thinks he's wealthy. Lloyd's second sound film shows him grappling with technique and has scenes that recall highlights of his silent hits.
1930 85m/B Harold Lloyd, Barbara Kent, Robert McWade; *Dir:* Clyde Bruckman. **VHS, Beta** TLF ♂♂½

Feet Foremost A made-for-British-television horror story about an industrialist haunted by the spirit of a teenage girl.
1985 60m/C GB Jeremy Kemp. **VHS, Beta $59.95** PSM ♂½

Felicity Upon graduating an exclusive finishing school, a young girl travels to the Orient and has a number of sexual experiences.
1983 (R) 91m/C Jody Hanson, Marilyn Rodgers; *Dir:* John Lamond. **VHS, Beta $39.95** LUN ♂

Felix the Cat: The Movie Classic cartoon creation Felix returns in a trite feature. The feline and his bag of tricks enter a dimension filled with He-Man/Mutant Ninja

Turtles leftovers; new-age princess, comic reptiles, robots and a Darth Vader clone who's defeated with ridiculous ease. Strictly for undemanding kids.
1991 83m/C VHS, Beta $19.99 BVV, FCT ♂½

Felix in Outer Space Felix and Poindexter travel through the galaxy to do battle with the evil duo, the Professor and Rock Bottom.
1985 55m/C VHS, Beta $29.95 MED ♂½

Felix's Magic Bag of Tricks The professor is after Felix's Magic Bag of Tricks once again in this collection of cartoon favorites.
1984 60m/C *Dir:* Joseph Oriolo. **VHS, Beta $14.95** MED ♂½

Fellini Satyricon Fellini's famous, garish, indulgent pastiche vision of ancient Rome, based on the novel ''Satyricon'' by Petronius, follows the adventures of two young men through the decadences of Nero's reign. Actually an exposition on the excesses of the 1960s, with the actors having little to do other than look good and react to any number of sexual situations. Crammed with excesses of every variety. In Italian with English subtitles. Also available on laserdisc with additional footage and letterboxing. Oscar nomination for the director.
1969 (R) 138m/C IT Martin Potter, Capucine, Hiram Keller, Salvo Randone, Max Born; *Dir:* Federico Fellini. **VHS, Beta, LV $19.98** MGM, VYG, FCT ♂♂♂

Fellow Traveler Two old friends, an actor and a screenwriter, both successful, get blacklisted for reputed Communist leanings in 1950s Hollywood. Good performances and a literate script. Made for cable television.
1989 97m/C Ron Silver, Hart Bochner, Daniel J. Travanti, Imogen Stubbs, Katherine Borowitz, Jonathan Hyde; *Dir:* Philip Saville. **VHS, Beta $89.95** PAR ♂♂♂

The Female Bunch The man-free world of an all-woman settlement is shattered by the arrival of a handsome stranger. Some of this garbage was filmed on location at the notorious Charles Manson ranch, and that's the least of its flaws.
1969 (R) 86m/C Jennifer Bishop, Russ Tamblyn, Lon Chaney Jr., Nesa Renet, Geoffrey Land, Regina Carrol; *Dir:* Al Adamson, John Cardos. **VHS $16.95** SNC, VTP Woof!

The Female Jungle Below-average whocares whodunnit directed by Roger Corman stock-company actor Ve Sota. Police sergeant Tierney is caught between a rock and a hard place. The prime suspect in a murder case, Tierney discovers a series of clues that implicate his friend Carradine. Interesting only for the screen debut of the nympho-typecast Miss Jayne—of whom Bette Davis said ''Dramatic art in her opinion is knowing how to fill a sweater.'' Also known as ''The Hangover.''
1956 56m/B Lawrence Tierney, John Carradine, Jayne Mansfield, Burt Kaiser, Kathleen Crowley, James Kodl, Rex Thorson, Jack Hill; *Dir:* Bruno Ve Sota. **VHS $29.95** COL, FCT ♂

Female Trouble Divine leads a troublesome existence in this $25,000 picture. She turns to a life of crime and decadence, seeking to live out her philosophy: ''Crime is beauty.'' Look closely at the Divine rape scene where she plays both rapist and victim. Climax of her deviant ways comes with her unusual night club act, for which the

law shows no mercy. Trashy, campy; for die-hard Waters fans.
1975 (R) 95m/C Divine, David Lochary, Mary Vivian Pearce, Mink Stole, Edith Massey, Danny Mills, Cookie Mueller, Susan Walsh; **Dir:** John Waters. **VHS, Beta $39.98** CGI, HHE *ll½*

Femme Fatale When a man's new bride disappears he teams up with an artist pal to track her down. Their search takes them to L.A.'s avant-garde art scene - and deeper into deception and mystery.
1990 (R) 96m/C Colin Firth, Lisa Zane, Billy Zane, Scott Wilson, Lisa Blount; **Dir:** Andre Guttfreund. **VHS, LV $89.98** REP, PIA *ll½*

Femmes de Paris A funny and risque French musical comedy about the comings and goings of the cast of a naughty nightclub shows. In French with English subtitles.
1953 79m/C FR Robert Dhery, Collette Brosset, Louis de Funes, Bluebell Girls. **VHS, Beta $29.95** NO *ll*

Fer-De-Lance A stricken submarine is trapped at the bottom of the sea, with a nest of deadly snakes crawling through the ship. Made for television.
1974 120m/C David Janssen, Hope Lange, Ivan Dixon, Jason Evers; **Dir:** Russ Mayberry. **VHS, Beta $49.95** WOV *l½*

Fernandel the Dressmaker Fernandel dreams of designing exquisite dresses, but when the opportunity arises, he finds his cheating partners get in the way. In French with English subtitles.
1957 95m/B FR Fernandel, Suzy Delair, Francoise Fabian, Georges Chamarat; **Dir:** Jean Boyer. **VHS, Beta, 8mm $29.95** FCT, VYY *ll½*

Ferngully: The Last Rain Forest A rain forest beset by pollution is the setting for this animated musical tale. Crysta is an independent-minded flying sprite, living in the rain forest, who discovers the outside world and becomes smitten with the human Zak, who is helping to cut down the forest. Crysta decides to reduce him to her size and show him the error of his ways aided by fellow sprite, Pips, and a crazy bat who's escaped from a research laboratory. But the humans aren't the only bad guys, there's also an evil woodland spirit (in the shape of a gloppy blob named Hexxus) who likes toxic waste. A so-so script is enhanced by the animation's brilliant colors and its non-violent, environmental message.
1992 (G) 72m/C Dir: Bill Kroyer; **Voices:** Samantha Mathis, Christian Slater, Robin Williams, Tim Curry, Jonathan Ward, Grace Zabriskie. **VHS $24.98** FXV *ll½*

Ferocious Female Freedom Fighters A typical foreign-made action flick with female wrestlers is spoofed by the L.A. Connection comedy troupe which did the totally ridiculous dubbed dialog.
1988 90m/C Eva Arnaz, Barry Prima. **VHS, Beta, LV** MED, IME *l*

Ferris Bueller's Day Off It's almost graduation and if Ferris can get away with just one more sick day—it had better be a good one. He sweet talks his best friend into borrowing his dad's antique Ferrari and sneaks his girlfriend out of school to spend a day in Chicago. Their escapades lead to fun, adventure, and almost getting caught. Broderick is charismatic as the notorious Bueller with Gray amusing as his tattle-tale sister doing everything she can to see him get caught. Led to TV series. One of Hughes' more solid efforts.

1986 (PG-13) 103m/C Matthew Broderick, Mia Sara, Alan Ruck, Jeffrey Jones, Jennifer Grey, Cindy Pickett, Edie McClurg, Charlie Sheen; **Dir:** John Hughes. **VHS, Beta, LV, 8mm $14.95** PAR, TLF *lll*

Ferry to Hong Kong A world-weary and heavy drinking traveler comes aboard the "Fat Annie," a ship skippered by the pompous Captain Hart. The two men clash, until an act of heroism brings them together. An embarrassingly hammy performance by Welles as the ferry skipper.
1959 103m/C Curt Jurgens, Orson Welles, Sylvia Syms; **Dir:** Lewis Gilbert. **VHS, Beta $9.95** SUE *l½*

The Feud A wacky comedy about the ultimate "family feud." The Bullards of Millville and the Bealers of Hornbeck engage in a battle of the witless, and no one in either town is safe. A silly, irreverent comedy of the slapstick variety.
1990 (R) 87m/C Rene Auberjonois, Ron McLarty, Joe Grifasi, David Strathairn, Gale Mayron; **Dir:** Bill D'Elia. **VHS $89.95** VMK *ll*

Feud of the Trail A cowboy saves a range family's gold and falls in love with their daughter in the process. Tyler plays both the hero and the villain here.
1937 56m/B Tom Tyler, Harlin Wood, Guinn Williams; **Dir:** Robert Hill. **VHS, Beta $19.95** NOS, DVT *l*

Feud of the West Old West disagreements settled with guns in this outing for cowboy star Gibson, here portraying a rodeo performer.
1935 60m/B Hoot Gibson. **VHS, Beta $19.95** NOS, UHV, VCN *l*

Fever Controversial Polish film based on Andrzej Strug's novel "The Story of One Bomb." Focusing on a period in Polish history marked by anarchy, violence, resistance and revolution, it was banned before it won eventual acclaim at the Gdansk Film Festival. In Polish with English subtitles.
1981 115m/C PL **Dir:** Agnieszka Holland. **VHS $49.95** FCT *ll*

Fever A once-honest cop trades his good life in for a shot at dealing drugs. Meanwhile, his wife's lover wants to murder him.
1988 (R) 83m/C Bill Hunter, Gary Sweet, Mary Regan, Jim Holt; **Dir:** Craig Lahiff. **VHS, Beta $14.95** ACA *l*

Fever An ex-con and a high-powered lawyer join forces to rescue the woman both of them love from a vicious killer. Made for cable television. Available in Spanish.
1991 (R) 99m/C Armand Assante, Sam Neill, Marcia Gay Harden, Joe Spano; **Dir:** Larry Elikann. **VHS $89.98** HBO *ll*

Fever Mounts at El Pao A minor effort from director Bunuel about the regime of a dictator in an imaginary South American country. In Spanish with English subtitles. Also known as "Los Ambiciosos."
1959 97m/B MX Gerard Philippe, Jean Servais; **Dir:** Luis Bunuel. **VHS, Beta $24.95** NOS, FCT, FST *ll½*

Fever Pitch Sordid story of a sports writer (O'Neal) who becomes addicted to gambling. Very poor script.
1985 (R) 95m/C Ryan O'Neal, Catherine Hicks, Giancarlo Giannini, Bridgette Andersen, Chad Everett, John Saxon, William Smith, Patrick Cassidy, Chad McQueen; **Dir:** Richard Brooks. **VHS, Beta $79.98** FOX *Woof!*

A Few Moments with Buster Keaton and Laurel and Hardy Three short appearances: "The Tree in a Test Tube," a government film with Stan & Ollie about wood's role in the war effort; an episode of "The Ed Sullivan Show," with the duo, and a commercial for Simon Pure Beer featuring Keaton.
1963 17m/B Stan Laurel, Oliver Hardy, Buster Keaton, Ed Sullivan, Pete Smith. **VHS, Beta, 8mm $9.95** VYY *lll*

ffolkes Rufus Excalibur Ffolkes is an eccentric underwater expert who is called upon to stop a madman (Perkins, indulging himself) from blowing up an oil rig in the North Sea. Entertaining farce with Moore playing the opposite of his usual suave James Bond character. Also known as "Assault Force" on television.
1980 (PG) 99m/C Roger Moore, James Mason, Anthony Perkins, David Hedison, Michael Parks; **Dir:** Andrew V. McLaglen. **VHS, Beta $59.95** MCA *ll*

Fiction Makers Roger Moore stars as Simon Templar, also known as "The Saint," a sophisticated detective who is hired to help Amos Klein. Klein turns out to be an alias for a beautiful novelist who is being threatened by the underworld crime ring. Based on the Leslie Chateris' character.
1967 102m/C Roger Moore, Sylvia Syms. **VHS, Beta** FOX *l*

Fictitious Marriage Eldad Ilan, a high school teacher experiencing a mid-life crisis, travels to Israel to get away from his family in order to consider his life's direction. In Hebrew with English subtitles.
1988 90m/C Shlomo Bar-Aba, Irit Sheleg, Ofra Voingarton; **Dir:** Haim Bouzaglo. **VHS $79.95** ERG *l*

Fiddler on the Roof Based on the long-running Broadway musical. The poignant story of Tevye, a poor Jewish milkman at the turn of the century in a small Ukrainian village, and his five dowry-less daughters, his lame horse, his wife, and his companionable relationship with God. Topol, an Israeli who played the role in London, is charming, if not quite as wonderful as Zero Mostel, the Broadway star. Finely detailed set decoration and choreography, strong performances from the entire cast create a sense of intimacy in spite of near epic proportions of the production. The play was based on the Yiddish stories of Tevye the Dairyman, written by Sholem Aleichem, which have been previously filmed as straight drama.
1971 (G) 184m/C Chaim Topol, Norma Crane, Leonard Frey, Molly Picon; **Dir:** Norman Jewison. Academy Awards '71: Best Cinematography, Best Sound, Best Musical Score; Golden Globe Awards '72: Best Film—Musical/Comedy. **VHS, Beta, LV $29.97** FOX, MGM, TLF *lll½*

Fidelio A production of the Beethoven opera taped at the Glyndebourne Festival in Great Britain. With English subtitles.
1979 130m/C Elizabeth Soderstrom, Anton de Ridder, Robert Allman. **VHS, Beta, LV $49.95** MVD, HMV, VAI *lll½*

The Field After an absence from the big screen, intense and nearly over the top Harris won acclaim (and an Oscar nomination) as an iron-willed peasant fighting to retain a patch of Irish land he's tended all his life, now offered for sale to a wealthy American. His uncompromising stand divides the community in this glowing adaptation of John B. Keane's classic play, an allegory of Ireland's internal conflicts.

1990 (PG-13) 113m/C Richard Harris, Tom Berenger, John Hurt, Sean Bean, Brenda Fricker, Frances Tomelty; **W/Dir:** Jim Sheridan. **VHS, LV** $89.95 LIV ☆☆☆½

Field of Dreams Based on W. P. Kinsella's novel "Shoeless Joe," this uplifting mythic fantasy depicts an Iowa corn farmer who, following the directions of a mysterious voice that first speaks to him while he's out tending the farm, cuts a baseball diamond in his corn field. "If you build it and he will come," instructs the voice and soon the ball field is inhabited by the spirit of Joe Jackson and the other ballplayers who were caught up and disgraced in the notorious 1919 "Black Sox" World Series baseball scandal. Jones, portraying a character based on reclusive author J.D. Salinger, is reluctantly pulled into the mystery. It's all about chasing a dream, paying debts, maintaining innocence in spite of adulthood, finding redemption, reconciling the child with the man, and of course, celebrating the mythic lure of baseball. Costner and Madigan are strong, believable characters. Heavily Oscar-nominated.
1989 (PG) 106m/C Kevin Costner, Amy Madigan, James Earl Jones, Burt Lancaster, Ray Liotta, Timothy Busfield, Frank Whaley, Gaby Hoffman; **Dir:** Phil Alden Robinson. **VHS, Beta, LV** $19.95 MCA, FCT ☆☆☆½

Field of Honor A look at the harrowing experience of jungle combat through the eyes of a Dutch infantryman in Korea.
1986 (R) 93m/C Everett McGill, Ron Bradsteder, Hey Young Lee; **Dir:** Hans Scheepmaker. **VHS, Beta, LV** $79.95 MGM ☆☆½

Field of Honor A quiet French anti-war drama set during the Franco-Prussian war in 1869 in which a peasant boy volunteers to fight in place of a rich man's son. He is befriended behind enemy lines by a young boy and both try to avoid capture. In French with English subtitles.
1987 (PG) 87m/C **FR** Cris Campion, Eric Wapler, Pascale Rocard, Frederic Mayer; **Dir:** Jean-Pierre Denis. **VHS, Beta** $79.98 ORI ☆☆½

The Fiend A religious cultist, already unbalanced, grabs a knife and starts hacking away Jack-the-Ripper style.
1971 (R) 87m/C **GB** Ann Todd, Patrick Magee, Tony Beckley, Madeline Hinde, Suzanna Leigh, Percy Herbert; **Dir:** Robert Hartford-Davis. **VHS, Beta** $39.95 MON ☆

Fiend A small-town music teacher feeds parasitically on his students to satisfy his supernatural hunger. His neighbor suspects some discord.
1983 93m/C Don Liefert, Richard Nelson, Elaine White, George Stover; **Dir:** Donald M. Dohler. **VHS, Beta** $49.95 PSM ☆

Fiend Without a Face An isolated air base in Canada is the site for a scientist using atomic power in an experiment to make a person's thoughts materialize. Only his thoughts are evil and reveal themselves as flying brains with spinal cords that suck human brains right out of the skull. Tons of fun for fifties SF fans and anyone who appreciates the sheer silliness of it all.
1958 77m/B Marshall Thompson, Kim Parker, Terence Kilburn, Michael Balfour, Gil Winfield, Shane Cordell; **Dir:** Arthur Crabtree. **VHS, LV** $19.98 CCB, REP, PIA ☆½

Fiendish Plot of Dr. Fu Manchu A sad farewell from Sellers, who in his last film portrays Dr. Fu in his desperate quest for the necessary ingredients for his secret life-preserving formula. Sellers portrays both Dr. Fu and the Scotland Yard detective on his

trail, but it's not enough to save this picture, flawed by poor script and lack of direction.
1980 (PG) 100m/C **GB** Peter Sellers, David Tomlinson, Sid Caesar, Helen Mirren; **Dir:** Piers Haggard. **VHS, Beta, LV** $19.98 WAR, FCT Woof!

Fiesta A girl comes from Mexico City to her father's hacienda where her old boyfriend awaits her return with a proposal of marriage. Full of authentic Mexican dances and music.
1947 44m/C Anne Ayars, George Negrete, Armida; **Dir:** Richard Thorpe. **VHS, Beta** $24.95 VYY, DVT, QNE ☆☆

Fifth Avenue Girl An unhappy millionaire brings home a poor young woman to pose as his mistress to make his family realize how they've neglected him. As this below-par social comedy drones toward its romantic conclusion, the rich folks see the error of their ways and love conquers all.
1939 83m/B Ginger Rogers, Walter Connolly, Tim Holt, James Ellison, Franklin Pangborn; **Dir:** Gregory La Cava. **VHS, Beta** $29.95 RKO, IME ☆½

Fifth Day of Peace In the aftermath of the World War I armistice, two German soldiers find themselves wandering around the Italian countryside without direction until they are caught and tried for desertion by their German commander. An interesting plot that is marred by too obvious plot twists. Based on a true story.
1972 (PG) 95m/C **IT** Richard Johnson, Franco Nero, Larry Aubrey, Helmut Schneider; **Dir:** Giuliano Montaldo. **VHS, Beta** $19.95 PSM ☆☆

The Fifth Floor A college disco dancer overdoses on drugs and winds up in an insane asylum, complete with a menacing attendant and an apathetic doctor. Pathetic, exploitative trash.
1980 (R) 90m/C Bo Hopkins, Dianne Hull, Patti D'Arbanville, Mel Ferrer; **Dir:** Howard Avedis. **VHS, Beta** $59.95 MED Woof!

The Fifth Monkey A man embarks on a journey to sell four monkeys in order to fill a dowry for his bride, but along the way he encounters all sorts of obstacles and adventures. Filmed on location in Brazil.
1990 (PG-13) 93m/C Ben Kingsley; **W/Dir:** Eric Rochat. **VHS, LV** $89.98 COL, FCT ☆½

The Fifth Musketeer A campy adaptation of Dumas's "The Man in the Iron Mask," wherein a monarch's evil twin impersonates him while imprisoning the true king. Also known as "Behind the Iron Mask." A good cast and rich production shot in Austria make for a fairly entertaining swashbuckler.
1979 (PG) 90m/C **GB** Beau Bridges, Sylvia Kristel, Ursula Andress, Cornel Wilde, Ian McShane, Alan Hale Jr., Helmut Dantine, Olivia de Havilland, Jose Ferrer, Rex Harrison; **Dir:** Ken Annakin. **Beta** $59.95 COL ☆☆

50 of the Greatest Cartoons Join many of your cartoon favorites for this once in a lifetime festival including Bugs Bunny, Casper, Mighty Mouse, Woody Woodpecker and Felix the Cat.
1990 360m/C **VHS** $14.99 STE ☆☆

$50,000 Dollar Reward Maynard's first western finds him being victimized by an unscrupulous banker who wants Maynard's land deeds for property on which a new dam is being built.
1925 49m/B Ken Maynard, Esther Ralston, Tarzan the Horse. **VHS, Beta** $19.98 DVT, CCB ☆

52 Pick-Up After a fling, a wealthy industrialist is blackmailed by a trio of repulsive criminals and determines to save himself. First he becomes deeply caught in their web of murder. Based on an Elmore Leonard novel with lots of gruesome violence and good performances.
1986 (R) 111m/C Roy Scheider, Ann-Margret, Vanity, John Glover, Doug McClure, Clarence Williams III, Kelly Preston; **Dir:** John Frankenheimer. **VHS, Beta, LV** $19.95 MED ☆☆½

55 Days at Peking A costume epic depicting the Chinese Boxer Rebellion and the fate of military Britishers caught in the midst of the chaos. Standard fare superbly handled by director Ray and an all-star cast.
1963 150m/C Charlton Heston, Ava Gardner, David Niven, John Ireland, Flora Robson, Paul Lukas, Jacques Sernas; **Dir:** Nicholas Ray. **VHS, Beta** $29.95 BFV, LTG, UHV ☆☆☆

The Fig Tree After her mother passes away, Miranda becomes tormented by the fear of death. Now her aunt must help her cope with this difficult part of life. Aired on PBS as part of the WonderWorks Family Movie television series.
1987 58m/C Olivia Cole, William Converse-Roberts, Doris Roberts, Teresa Wright, Karron Graves; **Dir:** Calvin Skaggs. **VHS** $29.95 PME, FCT ☆☆☆

Fight to the Death An evil businessman hires a group of deadly assassins to assist with the Japanese invasion of China at the start of World War II. Includes self-defense tips segment.
1983 100m/C Chang Li, Ron Van Cliff. **VHS** $39.95 KBV ☆

Fight for Gold In the frozen wastes of the North, a man becomes embroiled in a violent quest for gold.
19?? 107m/C **GE** Doug McClure, Harald Leipnitz, Heinz Reinl, Roberto Bianco, Angelica Ott, Kristina Nel; **Dir:** Harold Reinl. **VHS** $14.95 ACA ☆

A Fight for Jenny Miami Viceroy Thomas and Warren play an interracial couple who fight for custody of the wife's child from a previous marriage. Standard made-for-TV meaningful drama (bite-sized issues served with a modicum of melodrama).
1990 (PG) 95m/C Philip Michael Thomas, Lesley Ann Warren, Jean Smart, Lynne Moody, William Atherton; **Dir:** Gilbert Moses. **VHS** $79.95 PSM ☆☆

Fight for Survival A young female aspirant of kung fu must recover sacred books that were stolen by disguised kung fu masters.
1977 (R) 101m/C Shangkuan Ling-Feng. **VHS, Beta** FOX ☆

Fight for Us A social activist, recently released from prison, takes on the death squads terrorizing the countryside in post-Marcos Philippines. In Spanish with English subtitles.
1989 (R) 92m/C Phillip Salvador, Dina Bonnevie, Gina Alajar, Benbol Roco. **VHS, Beta, LV** $59.95 NO ☆½

Fight for Your Life Three escaped convicts—a white bigot, an Asian, and a Chicano—take a black minister's family hostage. After suffering all manner of vicious torture, the family exacts an equally brutal revenge. Shades of "The Desperate Hours" and "Extremities," but with much more graphic violence and a racial twist.

1979 89m/C CA William Sanderson, Robert Judd, Lela Small. **VHS, Beta** $59.95 MON ℤ

The Fighter A Mexican patriot, involved in a struggle to overthrow a dictator, falls for a co-revolutionist. Flashbacks show the destruction of his family and village, and his pugilistic expertise, which he uses to fight for a huge purse to help the cause. Adapted from Jack London's "The Mexican."
1952 78m/B Richard Conte, Vanessa Brown, Lee J. Cobb, Frank Silvera, Roberta Haynes, Hugh Sanders, Claire Carleton, Martin Garralaga; **Dir:** Herbert Kline. **VHS** DVT ℤℤℤ

Fighter Attack Story of a WWII pilot, based in Corsica, who is shot down during a mission to destroy a Nazi supply station. He meets with a woman of the Italian underground, and succeeds in destroying his target with her help.
1953 80m/C Sterling Hayden, J. Carroll Naish, Joy Page; **Dir:** Lesley Selander. **VHS** $19.98 REP ℤℤ

Fightin' Foxes Women wrestle in mud, oil, and few clothes.
1983 60m/C **VHS, Beta** $19.95 AHV ℤℤ

Fightin' Ranch A famous lawman saves his reputation, previously tainted. A Maynard classic.
1930 60m/B Ken Maynard. **VHS, Beta** UHV ℤℤ

Fighting Ace A young Chinese student sets out to take martial arts revenge on the villains who killed his family. Dubbed.
198? 86m/C Liu Chang Liang, Lung Chun Erl. **VHS, Beta** $54.95 MAV ℤ

The Fighting American A frat boy proposes to a girl on a brotherly dare, and, unflattered by the fraternal proposal, the girl flees to the Far East to return to the bosom of her family. But it turns out the boy really does love the girl, by golly, and he's got just enough of that fighting American spirit to prove it. Predates Oland's Charlie Chan career by seven years.
1924 65m/B Pat O'Malley, Mary Astor, Raymond Hatton, Warner Oland; **Dir:** Tom Forman. **VHS, Beta** $16.95 GPV, FCT ℤℤ

Fighting Back Reactionary tale of an angry resident in a crime-ridden Philadelphia neighborhood who organizes a patrol of armed civilian vigilantes. The police attempt to head off a racial confrontation in this effective drama graced with some fine performances. Also known as "Death Vengeance."
1982 (R) 96m/C Tom Skerritt, Patti LuPone, Michael Sarrazin, Yaphet Kotto; **Dir:** Lewis Teague. **VHS, Beta** $14.95 PAR ℤℤ

Fighting Black Kings Martial arts and karate masters appear in this tale of action.
1976 (PG) 90m/C William Oliver, Charles Martin, Willie Williams, Mas Oyama. **VHS, Beta** $19.98 WAR, OM ℤ

Fighting Caballero The Rough Riders effectively handle a gang of mine-harassing outlaws.
1935 60m/B Rex Lease; **Dir:** Elmer Clifton. **VHS, Beta** WGE ℤ

Fighting Caravans In this early big budget western, based on a story by Zane Grey, a wagon train sets out west from Missouri. Cooper emerges as the hero after warding off an Indian attack that takes the lives of the original leaders.
1932 80m/B Gary Cooper, Ernest Torrance, Tully Marshall, Fred Kohler Jr., Lily Damita; **Dir:** Otto Brower, David Burton. **VHS, Beta** $19.95 NOS, MRV, VYY ℤℤ

Fighting Cowboy Bill settles a tungsten mine dispute with a greedy miner and some outlaws.
1933 50m/B Buffalo Bill Jr.; **Dir:** Denver Dixon. **VHS, Beta, LV** WGE ℤ

Fighting Devil Dogs Two Marine lieutenants are assigned the task of obtaining a deadly secret weapon controlled by crooks. A serial in twelve chapters.
1943 195m/B Eleanor Stewart, Montagu Love, Hugh Sothern; **Dir:** William Witney. **VHS** $29.98 REP, VCN, MLB ℤ

Fighting Duel of Death Subculture evildoers threaten a convict and his lady friend. The two fight back with astounding expertise and effective results.
19?? 88m/C Mah Sah, Richard Chui. **VHS** $14.99 NEG ℤ½

Fighting Father Dunne Mega-hokey Hollywood steamroller about a tough priest caring for St. Louis newsboys in the early 1900s. Based on a true story.
1948 92m/B Pat O'Brien, Darryl Hickman, Charles Kemper, Una O'Connor, Arthur Shields, Harry Shannon, Joe Sawyer, Anna Q. Nilsson, Donn Gift, Myrna Dell; **Dir:** Ted Tetzlaff. **VHS, Beta** RKO ℤℤ

Fighting Fists of Shanghai Joe A fearsome fighting man from the Far East engages a vile American land baron in a battle of honor.
1965 94m/C Klaus Kinski, Gordon Mitchell, Carla Ronanelli, Robert Hundar, Chen Lee; **Dir:** Mario Ciaino. **VHS, Beta** $59.95 TWE ℤ

The Fighting Kentuckian Homeward bound from the Battle of New Orleans in 1814, a Kentucky rifleman lingers in a French settlement in Alabama. His romance with the daughter of a French general is blocked by the father until the American saves the community from an assault by land grabbers. Oliver Hardy as a frontiersman is well worth the view. An action-packed, well-photographed hit.
1949 100m/B John Wayne, Oliver Hardy, Vera Ralston, Marie Windsor, Philip Dorn, John Howard, Hugo Haas, Grant Withers; **Dir:** George Waggner. **VHS** $19.98 MED, REP ℤℤ½

Fighting with Kit Carson Famous guide and Indian fighter lead bands of settlers westward. Action-packed. Twelve chapters, 13 minutes each.
1933 156m/B Johnny Mack Brown, Noah Beery; **Dir:** Armand Schaefer, Colbert Clark. **VHS, Beta** $49.95 NOS, VCN, VDM ℤ½

Fighting Life The tale of two brothers who overcome immense physical and emotional handicaps and become vital members of society. The two stars of the film are both physically handicapped.
1980 90m/C **VHS, Beta** MAV ℤ½

Fighting Mad A singing Mountie tangles with border-crossing robbers and saves the reputation of a woman they exploit. One of the "Renfrew of the Mounties" series.
1939 57m/B James Newill, Milburn Stone, Sally Blane; **Dir:** Sam Newfield. **VHS, Beta** $39.98 VDM ℤ½

Fighting Mad Soldiers leave their buddy for dead in wartime Vietnam. He is alive, but captured by Japanese soldiers who believe they are still fighting in WWII.
1977 (R) 96m/C James Iglehardt, Jayne Kennedy, Leon Isaac Kennedy; **Dir:** Cirio H. Santiago. **VHS** $39.98 CGI ℤ½

Fighting Marines The U.S. Marines are trying to establish an air base on Halfway Island in the Pacific, but are thwarted by the "Tiger Shark," a modern-day pirate. First appeared as a serial.
1936 69m/B Jason Robards Sr., Grant Withers, Ann Rutherford, Pat O'Malley; **Dir:** Joseph Kane, B. Reeves Eason. **VHS, Beta** $29.95 VYY, VCN, VDM ℤ½

Fighting Parson In order to infiltrate a lawless town Gibson, a gunsliging cowboy, dresses as a revivalist preacher.
1935 65m/B Hoot Gibson, Marceline Day, Robert Frazer, Stanley Blystone, Skeeter Bill Robbins, Charles King; **Dir:** Harry Fraser. **VHS, Beta** UHV, GPV ℤ

Fighting Pilot Talmadge is a real fighter in this classic talkie. His assignment is to rescue the plans for a secret aircraft from ill fate.
1935 62m/B Richard Talmadge, Victor Mace, Gertrude Messinger, Eddie Davis, Robert Frazer; **Dir:** Noel Mason. **VHS, Beta** $24.95 NOS, DVT, GPV ℤ

Fighting Prince of Donegal An Irish prince battles the invading British in 16th Century Ireland. Escaping their clutches, he leads his clan in rescuing his mother and his beloved in this Disney swashbuckler. Based on the novel "Red Hugh, Prince of Donegal" by Robert T. Reilly.
1966 110m/C Peter McEnery, Susan Hampshire, Tom Adams, Gordon Jackson, Andrew Keir; **Dir:** Michael O'Herlihy. **VHS, Beta** $69.95 DIS ℤℤ½

Fighting Renegade The search for an Indian burial ground in Mexico prompts two murders and McCoy, disguised as a notorious bandit of the time, is wrongly accused. Justice triumphs in the end.
1939 60m/B Tim McCoy; **Dir:** Sam Newfield. **VHS, Beta** $19.95 NOS, VCN, DVT ℤ

The Fighting Rookie LaRue, better known for playing hoods, is a rookie cop set up and disgraced by the mob. But he fights back to regain his honor and bring the gang to justice.
1934 65m/B Jack LaRue; **Dir:** Spencer Gordon Bennet. **VHS, Beta, 8mm** $29.95 VYY ℤ½

Fighting Seabees As a hot-tempered construction foreman who battles Navy regulations as well as the Japanese, Wayne emerges as a larger-than-life hero in this action-packed saga of the Pacific theater of World War II. Extremely popular patriotic drama depicts the founding of the Seabees, the naval construction corps, amidst the action and the would-be romance with a woman reporter. Oscar-nominated score. Also available colorized.
1944 100m/B John Wayne, Susan Hayward, Dennis O'Keefe, William Frawley; **Dir:** Edward Ludwig. **VHS** $19.98 REP, BUR ℤℤ½

The Fighting Stallion Canutt stars as a drifter hired by a rancher who's determined to capture a beautiful wild stallion (played by Boy the Wonder Horse). Silent with original organ score.
1926 76m/B Yakima Canutt, Neva Gerber, Bud Osborne. **VHS, Beta, 8mm** $16.95 GPV, VYY ℤℤℤ

The Fighting Sullivans The true story of five brothers killed on the Battleship Juneau at Guadalcanal during World War II. The tale tells of the fury felt by the siblings after Pearl Harbor, their enlistment to fight for their country, and their tragic fate in the heat of battle. Truly a stirring tribute to all

lives lost in combat. Originally released as "The Sullivans."
1942 110m/B Anne Baxter, Thomas Mitchell, Selena Royle, Eddie Ryan, Trudy Marshall, James B. Cardwell, Roy Roberts, Ward Bond, Mary McCarty, Bobby Driscoll, Addison Richards, Selmer Jackson, Mae Marsh, Harry Strang, Barbara Brown; *Dir:* Lloyd Bacon. **VHS** IVY, MLB ✍️✍️✍️½

Fighting Texans An entire town invests in an oil well, on the advice of young salesman Bell, that the bankers are sure is a dud. Imagine their surprise when it turns out to be a gusher.
1933 60m/B Rex Bell, Luana Walters, Betty Mack, Gordon DeMain; *Dir:* Armand Schaefer. **VHS $19.95** NOS, DVT ✍️

Fighting Thru Maynard fights off the bad guys and wins the girl after some nifty riding in his first all-talkie western.
1930 60m/B Ken Maynard, Jeanette Loff; *Dir:* William Nigh. **VHS, Beta $19.95** NOS, DVT ✍️

The Fighting Trooper Based on the James Oliver Curwood story, "Footprints," this frontier tale presents a Mountie who goes undercover as a trapper in the Northwest. He is out to catch a murderer.
1934 57m/B Kermit Maynard. **VHS, Beta $9.95** NOS, VDM, DVT ✍️

The Fighting Westerner Scott stars as a mining engineer who becomes entangled in mysterious murders. Mrs. Leslie Carter stars in a rare screen role. Also titled "Rocky Mountain Mystery."
1935 54m/B Randolph Scott, Chic Sale, Ann Sheridan, Charles T. Barton. **VHS, Beta, 8mm $24.95** VYY, MRV ✍️½

Film Firsts Documentary-style look at early film segments from the "History of the Motion Picture" series. Includes the first attempt at science fiction with Melies' "Trip to the Moon" (1902), and the first cartoon and western.
1960 51m/B *Dir:* Georges Melies, Edwin S. Porter. **VHS, Beta $29.98** CCB ✍️½

Film House Fever A compilation of clips from sleazy, cult-type movies, including "Blood Feast" and "Hot Night at the Go Go Lounge."
1986 58m/C Jamie Lee Curtis, James Keach, Lon Chaney Jr., Harvey Korman. **VHS, Beta $59.98** LIV, VES ✍️

Filmmaker: A Diary by George Lucas A portrait by Lucas of his fellow director Francis Ford Coppola during the filming of "The Rain People." Made in 1969, the film was revised by Lucas in 1982.
1982 33m/C Francis Ford Coppola; *Dir:* George Lucas. **VHS, Beta** DCL ✍️

Filmmakers: King Vidor A profile of the veteran film director whose work includes "The Big Parade," "War and Peace," and "Our Daily Bread."
1965 28m/B King Vidor. **VHS, 3/4U, Special order formats** INU ✍️

Filthy Harry Eastwood and Bronson look-alikes send up the vigilante film.
19?? 90m/C VHS $59.95 TPV, HHE ✍️½

Fin de Fiesta When a corpse is found in the swimming pool, the party takes a dive. Spanish pool party fans will lap this one up. In Spanish.
1971 91m/C SP Isela Vega, Hector Suarez, Ana Martin; *Dir:* Mauricio Walerstein. **VHS, Beta $49.95** MED ✍️

The Final Alliance A tough loner takes on a vicious motorcycle gang that's terrorizing a small town, and realizes that they're also responsible for his family's death.
1989 (R) 90m/C David Hasselhoff, John Saxon, Bo Hopkins, Jeanie Moore; *Dir:* Mario Di Leo. **VHS, Beta, LV $79.95** COL ✍️

Final Analysis Glossy thriller starring Gere as a San Francisco psychiatrist who falls for the glamorous sister of one of his patients. Basinger plays the femme fatale and Thurman is Gere's sexually neurotic patient. Although heavily influenced by "Vertigo," this film never comes close to attaining the depth of Hitchock's cinematic masterpiece. Roberts gives the most gripping performance in this slick suspense movie as Basinger's sleazy gangster husband.
1992 (R) 125m/C Richard Gere, Kim Basinger, Uma Thurman, Eric Roberts, Paul Guilfoyle; *Dir:* Phil Joanou. **VHS, LV $94.99** WAR ✍️✍️½

Final Approach Test pilot Jason Halsey (Sikking) crashes in the desert and awakens in the office of psychiatrist Dio Gottlieb (Elizondo). Halsey can't remember anything about his past, or his own name for that matter, but through word association games and psychological tests, he begins to remember as Gottlieb tries to pry information from his brain. Showy computer effects are excellent, but they only serve to fill holes in a story that has an unsatisfying ending. Also available in a letterboxed version.
1992 (R) 100m/C James B. Sikking, Hector Elizondo, Madolyn Smith, Kevin McCarthy, Cameo Kneuer, Wayne Duvall; *Dir:* Eric Steven Stahl. Golden Scroll Award '92: Academy of Science Fiction, Fantasy & Horror Films. **VHS, LV $92.95** VMK ✍️

Final Assignment A Canadian television reporter agrees to smuggle a dissident Soviet scientist's ill granddaughter out of Russia for treatment along with a videotape documenting tragic experiments on children with steroids. She manages to evade the KGB while carrying on with a Russian press officer, and enlists the support of a Jewish fur trader. Location shooting in Canada instead of Russia is just one pitfall of this production. Also on video as, "The Moscow Chronicle."
1980 101m/C CA Genevieve Bujold, Michael York, Burgess Meredith, Colleen Dewhurst; *Dir:* Paul Almond. **VHS, Beta $69.98** LIV, VES ✍️

Final Comedown A black revolutionary attempts to get white radicals behind his war against racism. He fails and starts a racial bloodbath.
1972 (R) 84m/C Billy Dee Williams; *Dir:* Oscar Williams. **VHS, Beta $59.98** CHA, SIM ✍️

Final Conflict The third installment in the "Omen" series, concerning Satan's son Damien. Now 32, and the head of an international conglomerate, he is poised for world domination but fears another savior is born. Several monks and many babies meet gruesome deaths before he gets his comeuppance. The last theatrical release was then followed by a made-for-television movie called "Omen IV: The Awakening" (1991).
1981 (R) 108m/C Sam Neill, Lisa Harrow, Barnaby Holm, Rossano Brazzi; *Dir:* Graham Baker. **VHS, Beta, LV $19.98** FOX ✍️½

The Final Countdown A U.S. nuclear-powered aircraft carrier, caught in a time warp, is transported back to 1941, just hours before the bombing of Pearl Harbor. The commanders face the ultimate decision—leave history intact or stop the incident and

thus avoid World War II. Excellent photography and a surprise ending.
1980 (PG) 92m/C Kirk Douglas, Martin Sheen, Katharine Ross, James Farentino, Charles Durning; *Dir:* Don Taylor. **VHS, Beta, LV $19.98** VES ✍️✍️✍️

Final Cut While filming in a secluded swampland, a crew stumbles on a local sheriff's crooked scheme and one by one, the crew members disappear.
1988 92m/C Carla De Lane, T.J. Kennedy, Joe Rainer, Brett Rice, Jordan Williams; *Dir:* Larry G. Brown. **VHS, Beta $79.95** VMK ✍️½

Final Exam A psychotic killer stalks college students during exam week. This one's too boring to be scary. You root for the psycho to kill everyone off just to have the movie over.
1981 90m/C Cecile Bagdadi, Joel Rice; *Dir:* Jimmy Huston. **VHS, Beta $14.95** COL, SUE Woof!

The Final Executioner A valiant man finds a way to stop the slaughter of innocent people in a post-nuclear world.
1983 95m/C William Mang, Marina Costo, Harrison Muller, Woody Strode; *Dir:* Romolo Guerrieri. **VHS, Beta $59.95** MGM ✍️

Final Extra An ambitious reporter's career gets a boost when he's assigned to investigate the murder of a rival. Silent with music score.
1927 60m/B Grant Withers, Marguerite de la Motte. **VHS, Beta, 8mm $19.95** NOS, VYY, DVT ✍️½

Final Impact Nick Taylor seeks vengeance on reigning kickboxing champ Jake Gerard through his prodigy, Danny Davis. Will sweet revenge for the wicked beating, suffered at the hands of Gerard years earlier, be his?
1991 (R) 99m/C Lorenzo Lamas, Kathleen Kinmont, Mimi Lesseos, Kathrin Lautner, Jeff Langton, Mike Worth; *Dir:* Joseph Merhi, Stephen Smoke. **VHS $89.95** PMH ✍️½

Final Justice A small-town Texan sheriff wages a war against crime and corruption that carries him to Italy and the haunts of Mafia hitmen.
1984 (R) 90m/C Joe Don Baker, Rossano Brazzi, Patrizia Pellegrino; *Dir:* Greydon Clark. **VHS, Beta $79.95** VES ✍️

Final Mission A vengeful one-man army follows the professional hit man who slaughtered his family from L.A. to Laos. His pursuit leads to a jungle showdown.
1984 101m/C Richard Young, John Dresden, Kaz Garaz, Christine Tudor. **VHS, Beta $19.99** HBO ✍️

Final Notice A detective tracking a serial killer has as his only evidence a trail of shredded photos of nude women. Not released theatrically.
1989 91m/C Gil Gerard, Steve Landesburg, Jackie Burroughs, Melody Anderson, Louise Fletcher, David Ogden Stiers, Kevin Hicks; *Dir:* Steven Hilliard Stern. **VHS, Beta, LV $59.99** PAR, PIA ✍️

The Final Option An agent of England's Special Air Services team goes undercover as one of a band of anti-nuclear terrorists that take over the U.S. Embassy in London. Violent, but unconvincing. Also known as "Who Dares Wins."
1982 (R) 125m/C GB Lewis Collins, Judy Davis, Richard Widmark, Robert Webber, Edward Woodward, Ingrid Pitt, Kenneth Griffith, Tony Doyle, John Duttine; *Dir:* Ian Sharp. **VHS, Beta $79.95** MGM ✍️

F

The Final Programme In this futuristic story, a man must rescue his sister—and the world—from their brother, who holds a microfilmed plan for global domination. Meanwhile, he must shield himself from the advances of a bisexual computer programmer who wants to make him father to a new, all-purpose human being. Based on the Michael Moorcock "Jerry Cornelius" stories, the film has gained a cult following. Also known as "The Last Days of Man on Earth."
1973 85m/C *GB* Hugh Griffith, Harry Andrews, Jon Finch, Jenny Runacre, Sterling Hayden, Patrick Magee, Sarah Douglas; *Dir:* Robert Fuest. **VHS, Beta** $59.95 *HBO, SUE* 🎬½

Final Sanction After a nuclear holocaust exhausts their military resources, the U.S. and the Soviets each send one-man armies to battle each other for final control of the world.
1989 90m/C Robert Z'Dar, Ted Prior. **VHS, Beta** *AIP, HHE* 🎬🎬½

The Final Terror A group of young campers is stalked by a mad killer in a desolate, backwoods area. Better-than-average stalked-teens entry is notable for the presence of some soon-to-be stars.
1983 (R) 90m/C John Friedrich, Rachel Ward, Adrian Zmed, Darryl Hannah, Joe Pantoliano, Ernest Harden, Mark Metcalf, Lewis Smith, Cindy Harrel, Akosua Busia; *Dir:* Andrew Davis. **VHS, Beta, LV** $29.98 *LIV, VES* 🎬½

Final Verdict A trial lawyer defends a man he knows is guilty, throwing his life and his family into turmoil. Made for cable television.
1991 93m/C Treat Williams, Glenn Ford, Amy Wright, Olivia Burnette; *Dir:* Jack Fisk. **VHS, Beta** $89.98 *TTC* 🎬🎬½

Final Warning A dramatization of actual events surrounding the 1986 melt-down of the nuclear power plant at Chernobyl. Also known as "Chernobyl: The Final Warning." Made for television.
1990 94m/C Jon Voight, Jason Robards Jr., Sammi Davis; *Dir:* Anthony Page. **VHS** $79.98 *TTC* 🎬🎬½

Find the Lady Candy assumes a supporting role as a bumbling cop who is part of an incompetent police team trying to rescue a kidnapped socialite. Also known as "Call the Cops."
1975 79m/C *CA* John Candy, Peter Cook, Mickey Rooney, Lawrence Dane, Alexandra Bastedo; *Dir:* John Trent. **VHS, Beta** $24.95 *VIR* 🎬½

Finders Keepers A wild assortment of characters on board a train en route from California to Nebraska search for five million dollars hidden in the baggage car.
1984 (R) 96m/C Michael O'Keefe, Beverly D'Angelo, Ed Lauter, Louis Gossett Jr., Pamela Stephenson, Jim Carrey, David Wayne, Brian Dennehy; *Dir:* Richard Lester. **VHS, Beta** $59.98 *FOX* 🎬🎬

Fine Gold A false charge of embezzlement leaves a man without home and family, so he decides to turn the tables on those who betrayed him.
19?? 91m/C Andrew Stevens, Ray Walston, Ted Wass, Stewart Granger, Lloyd Bochner, Jane Badler. **VHS, Beta** $79.99 *MCG* 🎬

A Fine Madness A near-classic comedy about a lusty, rebellious poet thrashing against the pressures of the modern world, and fending off a bevy of lobotomy-happy psychiatrists. Shot on location in New York City, based on Elliot Baker's novel.
1966 104m/C Sean Connery, Joanne Woodward, Jean Seberg, Patrick O'Neal, Colleen Dewhurst, Clive Revill, John Fiedler; *Dir:* Irvin Kershner. **VHS, Beta** $19.98 *WAR* 🎬🎬🎬

A Fine Mess Two buffoons cash in when one overhears a plan to dope a racehorse, but they are soon fleeing the plotters' slapstick pursuit. The plot is further complicated by the romantic interest of a gangster's wife. The television popularity of the two stars did not translate to the big screen; perhaps it's Edwards' fault.
1986 (PG) 100m/C Ted Danson, Howie Mandel, Richard Mulligan, Stuart Margolin, Maria Conchita Alonso, Paul Sorvino; *Dir:* Blake Edwards. **VHS, Beta, LV** $19.95 *COL* 🎬½

Finessing the King Private eyes Tommy and Tuppence follow a mysterious newspaper notice to a masked ball where a murder is about to occur. Based on the Agatha Christie story. Made for British television.
1984 60m/C *GB* James Warwick, Francesca Annis. **VHS, Beta** $14.95 *PAV* 🎬🎬½

Finest Hours Sir Winston Churchill, removed as Lord of the Admiralty after WWI's Dardanelles campaign, later went on to lead Great Britain as Prime Minister through WWII. Documentary recounts the life of Churchill from his early childhood to the remarkable career which thrust him into the center of the great historical events of the twentieth century. Narrated by Orson Welles.
1964 116m/C *Dir:* Peter Baylis, Peter Baylis; *Nar:* Orson Welles. National Board of Review Awards '64: 10 Best Films of the Year. **VHS, Beta** $49.95 *PSM* 🎬🎬½

The Finger Man An ex-con cooperates with the feds rather than return to jail. The deal is that he gets the dirt on an underworld crime boss. Solid performances.
1955 82m/B Frank Lovejoy, Forrest Tucker, Peggy Castle, Timothy Carey, Glenn Gordon; *Dir:* Harold Schuster. **VHS** $19.98 *REP* 🎬🎬½

Finger on the Trigger Veterans from both sides of the Civil War are after a hidden supply of gold, but find that they must band together to fend off an Indian attack. Routine.
1965 89m/C Rory Calhoun, James Philbrook, Todd Martin, Silvia Solar, Brad Talbot; *Dir:* Sidney Pink. **VHS, Beta** $7.00 *VTR* 🎬🎬

Fingerprints Don't Lie A fingerprint expert pins the murder of the town mayor on an innocent guy, then suspects a frame-up.
1951 56m/B Richard Travis, Sheila Ryan, Tom Neal; *Dir:* Sam Newfield. **VHS, Beta, LV** *WGE* 🎬

Fingers Keitel is a mobster's son, reluctantly working as a mob debt-collector, all the while dreaming of his ambitions to be a concert pianist. The divisions between his dreams and reality cause him to crack. Toback's first film generates psychological tension and excellent performances.
1978 (R) 89m/C Harvey Keitel, Tisa Farrow, Jim Brown, Michael Gazzo, Danny Aiello, Marian Seldes, James Toback, Tanya Roberts; *Dir:* James Toback. **VHS, Beta** $49.95 *MED* 🎬🎬🎬

Finian's Rainbow A leprechaun comes to America to steal back a pot of gold taken by an Irishman and his daughter in this fanciful musical comedy based on a 1947 Broadway hit. Both the sprite and the girl find romance; the share-cropping locals are saved by the cash; a bigot learns the error of his ways; and Finian (Astaire) dances off to new adventures. The fine production and talented cast are not used to their best advantage by the director who proved much better suited for "The Godfather." Entertaining, nonetheless.
1968 (G) 141m/C Fred Astaire, Petula Clark, Tommy Steele, Keenan Wynn, Al Freeman Jr., Don Francks, Susan Hancock, Dolph Sweet; *Dir:* Francis Ford Coppola. **VHS, Beta, LV** $19.98 *WAR, PIA* 🎬🎬

Finish Line A high school track star turns to steroids to enhance his performance after being pushed by an overzealous father. The father-and-son Brolins play the pair. Made for cable.
1989 100m/C James Brolin, Josh Brolin, Mariska Hargitay, Kristoff St. John, John Finnegan, Billy Vera, Stephen Lang; *Dir:* John Nicolella. **VHS** $139.96 *TTC* 🎬🎬½

Finishing School Girls' school roomates Rogers and Dee experience heartaches and loves lost while enduring disinterested parents and snobbish peers. Box office bomb when released, despite a strong cast.
1933 73m/B Ginger Rogers, Frances Dee, George Nicholls, Beulah Bondi, Bruce Cabot, Billie Burke, John Halliday, Sara Haden, Jack Norton, Joan Barclay, Jane Darwell; *Dir:* George Nicholls. **VHS, Beta** $19.98 *TTC* 🎬½

Finnegan Begin Again A middle-aged schoolteacher and a grouchy, 65-year-old newspaper editor find romance despite their other obligations. Winning performances add charm to this cable TV movie.
1984 112m/C Mary Tyler Moore, Robert Preston, Sam Waterston, Sylvia Sydney, David Huddleston; *Dir:* Joan Micklin Silver. **VHS, Beta** $14.95 *NEG, HBO, KAR* 🎬🎬

Fiorello La Guardia: The Crusader In this program, classic newsreel footage shows Fiorello La Guardia when he was Mayor of New York.
1954 15m/B **VHS, Beta** $49.95 *TSF* 🎬🎬

Fire A fire started by an escaped convict rages through Oregon timberland in this suspenseful made-for-television Irwin Allen disaster drama.
1977 98m/C Ernest Borgnine, Vera Miles, Patty Duke, Alex Cord, Donna Mills; *Dir:* Earl Bellamy. **VHS, Beta** $59.95 *WAR* 🎬

Fire Alarm A change of scene for Western star Brown as he portrays a fireman who, in-between his work, romances a career girl.
1938 67m/B Johnny Mack Brown, Noel Francis; *Dir:* Karl Brown. **VHS, Beta, LV** $19.95 *NOS, WGE, VYY* 🎬

Fire Birds Army attack helicopters and the people who fly them are used in the war on drugs in South America. Failed to sustain either the exciting flight sequences, the romantic interest or the box office of "Top Gun."
1990 (PG-13) 85m/C Nicolas Cage, Tommy Lee Jones, Sean Young, Bryan Kestner; *Dir:* David Green. **VHS, Beta, LV** $89.95 *TOU* 🎬

Fire Down Below Hayworth is the been-around-the-block beauty who persuades Mitchum and Lemmon, two smalltime smugglers, to take her to a safe haven, no questions asked. Both men fall for her obvious charms, causing them to have a falling out until a life or death situation puts their friendship to the test. An unoriginal melodrama indifferently acted by everyone but Lemmon. Good location work in Trinidad and Tobago.
1957 116m/C *GB* Robert Mitchum, Jack Lemmon, Rita Hayworth, Herbert Lom, Anthony Newley; *Dir:* Robert Parrish. **VHS, Beta** $19.95 *COL, COL, CCB* 🎬½

Fire with Fire A boy at a juvenile detention center and a Catholic school girl fall in love. However, they find themselves on the run from the law when he escapes to be with her. Sheffer and Madsen are appealing in this otherwise unspectacular film.
1986 (PG-13) 103m/C Craig Sheffer, Virginia Madsen, Jon Polito, Kate Reid, Jean Smart; *Dir:* Duncan Gibbins. **VHS $79.95** PAR ♂♂

Fire and Ice An animated adventure film that culminates in a tense battle between good and evil, surrounded by the mystical elements of the ancient past. Designed by Frank Frazetta.
1983 (PG) 81m/C Randy Norton, Cynthia Leake; *Dir:* Ralph Bakshi; *Voices:* Susan Tyrrell, William Ostrander. **VHS, Beta, LV $12.98** COL ♂♂½

Fire and Ice A tale of love on the slopes. Two skiers realize that their feelings for each other are perhaps even stronger than their feelings about skiing. You won't care though, except for some fine ski footage.
1987 (PG) 83m/C Suzy Chaffee, John Eaves; *Dir:* Willy Bogner; *Nar:* John Denver. **VHS, Beta, LV $14.95** SUE ♂

Fire, Ice and Dynamite Moore is onscreen only briefly as an eccentric tycoon who fakes suicide to watch several teams of challengers scramble in a madcap wintersports contest for his millions. A German-made avalanche of crazy stunts, bad jokes, and product plugs for countless European companies, commencing with a literal parade of guests, from astronaut Buzz Aldrin to soul man Isaac Hayes. Numbing; must be seen to be believed.
1991 (PG) 105m/C Roger Moore, Shari Belafonte-Harper, Simon Shepherd, Uwe Ochsenknecht, Marjoe Gortner; *Dir:* Willy Bogner. **VHS $89.98** VES, LIV ♂½

Fire Maidens from Outer Space Fire maidens prove to be true to the space opera code that dictates that all alien women be in desperate need of male company. Astronauts on an expedition to Jupiter's thirteenth moon discover the lost civilization of Atlantis, which, as luck would have it, is inhabited by women only. Possibly an idea before its time, it might've been better had it been made in the sixties, when space-exploitation came into its own.
1956 80m/C GB Anthony Dexter, Susan Shaw, Paul Carpenter, Harry Fowler, Jacqueline Curtiss, Sydney Tafler, Maya Koumani, Jan Holden, Kim Parker; *W/Dir:* Cy Roth. **VHS $16.95** CNM, MLB ♂

Fire in the Night In a small Southern town, a beautiful and sure-footed woman battles the limitless resources of the town's predominant dynastic family.
1985 89m/C Graciela Casillas, John Martin. **VHS, Beta $19.95** STE, NWV ♂½

Fire Over England A young naval officer volunteers to spy at the Spanish court to learn the plans for the invasion of his native England and to identify the traitors among the English nobility. He arouses the romantic interest of his queen, Elizabeth I, one of her ladies, and a Spanish noblewoman who helps with his missions, and later leads the fleet to victory over the huge Spanish Armada. The first on-screen pairing of Olivier and Leigh is just one of the many virtues of this entertaining drama.
1937 81m/B GB Flora Robson, Raymond Massey, Laurence Olivier, Vivien Leigh, Leslie Banks, James Mason; *Dir:* William K. Howard. **VHS, Beta $19.95** NOS, MRV, VYY ♂♂♂

Fire and Rain Based on the real-life crash of a Delta Airlines plane in Dallas and the rescue efforts made following the disaster. Familiar cast-members turn in decent performance. Made for television.
1989 89m/C Angie Dickinson, Charles Haid, Tom Bosley, David Hasselhoff, Robert Guillaume, Susan Ruttan, John Beck, Patti LaBelle, Dean Jones, Lawrence Pressman, Penny Fuller; *Dir:* Jerry Jameson. **VHS, Beta, LV $59.99** PAR, PIA ♂♂

The Fire in the Stone A young boy discovers an opal mine and dreams of using the treasure to reunite his family. However, when the jewels are stolen from him, he enlists his friends to help get them back. Based on the novel by Colin Thiele.
1985 97m/C AU Paul Smith, Linda Hartley, Theo Pertsindis. **VHS, Beta $39.95** SCL, WAR ♂♂

Fire and Sword A Cornish knight must choose between loyalty for king and country and the love of an Irish woman in this medieval drama. A not very inspired retelling of the Tristan and Isolde legend.
1982 84m/C GB Peter Firth, Leigh Lawson, Antonia Preser, Christopher Waitz; *Dir:* Veith von Furstenberg. **VHS, Beta $69.98** LIV, LTG ♂

Fireback A Vietnam vet's wife is kidnapped by the mob, causing him to take up arms and spill blood yet again.
1978 90m/C Bruce Baron, Richard Harrison. **VHS, Beta $39.95** IVE ♂½

Fireballs Sexploitaton pic set at a firehouse, where a batch of female recruits turn up the heat for three stud hose-bearers. The title should give you some idea of the intellectual level.
1990 89m/C Mike Shapiro, Goren Kalezik; *Dir:* Mike Shapiro. **VHS $79.95** ATL ♂½

Firebird 2015 A.D. Dreary action-adventure movie set in a 21st century society where automobile use is banned because of an extreme oil shortage. Private cars are hunted down for destruction by the Department of Vehicular Control. One over-zealous enforcer decides to make it a package deal and throws in the owners as well. The film runs out of gas as well.
1981 (PG) 97m/C Darren McGavin, George Touliatos, Doug McClure; *Dir:* David Robertson. **VHS, Beta $69.98** SUE ♂

Firecracker A female martial arts expert retaliates against the crooks who murdered her sister.
1971 83m/C Jillian Kesner, Darby Hinton. **VHS, Beta $59.95** MON ♂

Firefight After a nuclear holocaust, convicted criminals endeavor to rule the wasteland.
1987 100m/C James Pfeiffer, Janice Carraher, Jack Tucker; *Dir:* Scott Pfeiffer. **VHS, Beta $69.95** TWE ♂

Firefox A retired pilot sneaks into the Soviet Union for the Pentagon to steal a top-secret, ultra-sophisticated warplane and fly it out of the country. Best for the low-altitude flight and aerial battle sequences, but too slow on character and much too slow getting started.
1982 (PG) 136m/C Clint Eastwood, Freddie Jones, David Huffman, Warren Clarke, Ronald Lacey, Kenneth Colley, Nigel Hawthorne, Kai Wulff; *Dir:* Clint Eastwood. **VHS, Beta, LV $19.98** WAR ♂♂½

Firehead A pyrokinetic Soviet defector begins using his powers to destroy American munitions in the name of peace. When a clandestine pro-war organization hears of this, they attempt to capture the man and use him for their own evil purposes.
1990 (R) 88m/C Christopher Plummer, Chris Lemmon, Martin Landau, Gretchen Becker, Brett Porter; *Dir:* Peter Yuval. **VHS, Beta $89.98** AIP ♂♂½

Firehouse Tempers ignite in a lily-white firehouse when a black rookie replaces an expired veteran. March—of "Paper Lion" renown—directed this made-for-TV emergency clone, which ran ever-so-briefly as an adventure series on TV in 1974 (with a largely different cast).
1972 73m/C Richard Roundtree, Vince Edwards, Andrew Duggan, Richard Jaeckel, Sheila Frazier, Val Avery, Paul LeMat, Michael Lerner; *Dir:* Alex March. **VHS $59.95** XVC ♂♂

Firehouse In the style of "Police Academy," three beautiful and sex-starved firefighting recruits klutz up an urban firehouse. A softcore frolic.
1987 (R) 91m/C Barrett Hopkins, Shannon Murphy, Violet Brown, John Anderson; *Dir:* J. Christian Ingvordsen. **VHS, Beta $14.95** ACA ♂

The Fireman Chaplin portrays a fireman who becomes a hero. Silent with musical soundtrack added.
1916 20m/B Charlie Chaplin; *Dir:* Charlie Chaplin. **VHS, Beta $39.95** CAB, FST ♂♂½

Firemen's Ball A critically acclaimed comedy about a small-town ball held for a retiring fire chief. Plans go amusingly awry as beauty contestants refuse to show themselves, raffle prizes and other items—including the gift for the guest of honor—are stolen, and the firemen are unable to prevent an old man's house from burning down. Forman's second film is sometimes interpreted as political allegory; in Czechoslovalcian, with English subtitles.
1968 73m/C CZ Vaclav Stockel, Josef Svet; *Dir:* Milos Forman. New York Film Festival '68: 10 Best Films of the Year. **VHS, Beta $59.95** COL ♂♂♂

Firepower Loren blames her chemist husband's death on a rich industrialist and hires hitman Coburn to take care of the matter. Uninvolving.
1979 (R) 104m/C Sophia Loren, James Coburn, O.J. Simpson, Christopher F. Bean; *Dir:* Michael Winner. **VHS, Beta $59.98** FOX ♂½

Fires on the Plain A grueling Japanese anti-war film about an unhinged private in the Philippines during World War II who roams the war-torn countryside encountering all manner of horror and devastation. Originally called "Nobi." In Japanese with English subtitles.
1959 105m/B JP Eiji Funakoshi, Osamu Takizawa, Mickey Custis, Asao Suno; *Dir:* Kon Ichikawa. **VHS, Beta $29.98** SUE, MRV, HHT ♂♂♂

Firesign Theatre Presents Nick Danger in the Case of the Missing Yolks Firesign characters Nick Danger and Rocky Rococo are featured in this story of a truly interactive family who live through their television set.
1983 60m/C Phil Proctor, Phil Austin, Peter Bergman. **VHS, Beta $29.95** PAV ♂½

Firesign Theatre's Hot Shorts A compilation of the comedy troupe's most hilarious short films, mostly consisting of old serial episodes like "Commander Cody" and

"Black Widow," reedited and laden with farcical sound effects and dialogue.
1984 90m/B VHS, Beta, LV COL ℤ½

Firestarter A C.I.A.-like organization is after a little girl who has the ability to set anything on fire in this filmed adaptation of Stephen King's bestseller. Good special effects help a silly plot.
1984 (R) 115m/C David Keith, Drew Barrymore, Freddie Jones, Martin Sheen, George C. Scott, Heather Locklear, Louise Fletcher, Moses Gunn, Art Carney, Antonio Fargas, Drew Snyder; **Dir:** Mark L. Lester. VHS, Beta, LV $19.95 MCA ℤℤ

Firewalker An "Indiana Jones" clone about three mercenaries endeavoring to capture a fortune in hidden gold. Paper-mache sets and a villain with an eye patch that consistently changes eyes are just some of the gaffes that make this one of the worst edited movies in history.
1986 (PG) 106m/C Chuck Norris, Louis Gossett Jr., Melody Anderson; **Dir:** J. Lee Thompson. VHS, Beta, LV $19.95 MED ℤ

The Firing Line The Central American government hires a mercenary rebel-buster to squash insurgents. Everything's great until he finds out he agrees with the rebel cause, and he trains them to fight the government. Below average renegade-with-a-hidden-heart warpic.
1991 93m/C Reb Brown, Shannon Tweed; **Dir:** John Gale. VHS $79.95 AIP ℤℤ

First Affair A made-for-television film about a young innocent girl undergoing the pressures of freshman life at college, including her first love affair, which is with the husband of a female professor.
1983 95m/C Loretta Swit, Melissa Sue Anderson, Joel Higgins; **Dir:** Gus Trikonis. VHS, Beta $59.98 FOX ℤ

First Blood Stallone portrays a former Green Beret survivor of Vietnam whose nightmares of wartime horrors are triggered by a wrongful arrest in a small town. He escapes to lead his pursuers to all manner of bloody ends and the Army is finally summoned to crush him. Invincible, he succumbs at last to the order of his former commander. Extremely violent and confused story launched the "Rambo" series.
1982 (R) 96m/C Sylvester Stallone, Richard Crenna, Brian Dennehy, Jack Starrett; **Dir:** Ted Kotcheff. VHS, Beta, LV $14.95 HBO, LIV, TLF ℤ½

First Born In search of romance, divorced mother, Garr, finds an intriguing man, Weller. Trouble erupts when her son learns Weller is a cocaine dealer. Talented cast makes most of story.
1984 (PG-13) 100m/C Teri Garr, Peter Weller, Christopher Collet, Corey Haim, Sarah Jessica Parker, Robert Downey Jr.; **Dir:** Michael Apted. VHS, Beta, LV $14.95 PAR ℤℤ½

The First Christmas A shepherd boy is miraculously rewarded for his generosity to nuns. Animated, with voices of Angela Lansbury and Cyril Ritchard.
1975 23m/C **Voices:** Angela Lansbury, Cyril Ritchard. VHS, Beta $12.95 WAR ℤℤ½

The First Deadly Sin A police lieutenant (Sinatra) tracks down a homicidal killer in spite of his wife's (Dunaway) illness and his impending retirement. Read on Lawrence Sanders' bestselling novel; it's a lot more exciting than this movie on which it was based.
1980 (R) 112m/C Frank Sinatra, Faye Dunaway, David Dukes, Brenda Vaccaro, James Whitmore; **Dir:** Brian G. Hutton. VHS, Beta, LV $59.95 WAR ℤ½

First Family A flat satire of life in the White House. The humor is weak and silly at best, despite the excellent comedy cast. Henry's first directional effort.
1980 (R) 100m/C Bob Newhart, Madeline Kahn, Gilda Radner, Richard Benjamin; **Dir:** Buck Henry. VHS, Beta $19.98 WAR ℤ½

First Love A sixteen year old lad falls in love with a slightly older woman only to have his feelings rejected and to discover that she's his father's mistress. Actor Schell's directorial debut.
1970 (R) 90m/C GE SI John Moulder-Brown, Dominique Sanda, Maximilian Schell, Valentina Cortese; **Dir:** Maximilian Schell. VHS, Beta $49.95 NEG ℤℤ½

First Love A story of an idealistic college student who takes love (and especially making love) more seriously than the rest of his peers, including his girlfriend. Based on Harold Brodkey's story "Sentimental Education." Darling's directorial debut.
1977 (R) 92m/C William Katt, Susan Dey, John Heard, Beverly D'Angelo, Robert Loggia; **Dir:** Joan Darling. VHS, Beta $49.95 PAR ℤℤ½

First Man into Space An astronaut returns to Earth covered with strange space dust and with an organism feeding inside him (shades of "Alien"). The alien needs human blood to survive and starts killing in order to get it.
1959 78m/B Marshall Thompson, Marla Landi, Bill Edwards; **Dir:** Robert Day. VHS, Beta $19.98 DVT, RHI, MON ℤℤ

First Men in the Moon A fun, special effects-laden adaptation of the H. G. Wells novel about an Edwardian civilian spacecraft visiting the moon and the creature found there. Visual effects by Ray Harryhausen. Finch makes a brief appearance.
1964 103m/C GB Martha Hyer, Edward Judd, Lionel Jeffries, Erik Chitty, Peter Finch; **Dir:** Nathan Juran. VHS, Beta, LV $14.95 COL, PIA, MLB ℤℤ

First Monday in October A comedy concerning the first woman appointed to the Supreme Court, a conservative, and her colleague, a crusty but benign liberal judge. Though based on a Broadway hit, it ended up seeming to foreshadow the real-life appointment of Sandra Day O'Connor, which occurred at about the time the film was released. The title refers to the date the court begins its sessions.
1981 (R) 99m/C Walter Matthau, Jill Clayburgh, Barnard Hughes, James Stephens, Jan Sterling; **Dir:** Ronald Neame. VHS, Beta, LV $84.95 PAR ℤℤ½

The First Movie Cowboy Three shorts featuring the very first movie cowboy, Bronco Billy.
1913 47m/B G.M. Anderson; **Dir:** G.M. Anderson. VHS, Beta, 8mm $24.95 VYY ℤℤ½

First Name: Carmen Carmen, although posing as an aspiring filmmaker, really is a bank robber and terrorist. She is also such a femme fatale that during a bank robbery one of the guards decides to run away with her. Gooddard cast himself as Carmen's uncle. Amusing late Godard.
1983 95m/C FR Maruschka Detmers, Jacques Bonaffe, Jean-Luc Godard, Myriem Roussel, Christophe Odent; **Dir:** Jean-Luc Godard. VHS, Beta, LV $79.95 CVC ℤℤℤ

The First Nudie Musical A producer tries to save his studio by staging a 1930s style musical, but with a naked cast and risque lyrics. Has attained semi-cult/trash status.
1975 (R) 93m/C Cindy Williams, Stephen Nathan, Diana Canova, Bruce Kimmel; **Dir:** Mark Haggard. VHS, Beta $59.95 MED, HHT ℤℤ

First Olympics: Athens 1896 Recounts the organization and drama surrounding the first modern-day Olympic games and has the inexperienced American team shocked the games with their success. Made for television.
1984 95m/C Angela Lansbury, Louis Jourdan, David Ogden Stiers, Virginia McKenna, Jason Connery, Alex Hyde-White; **Dir:** Alvin Rakoff. VHS, Beta $29.95 COL ℤℤ½

The First Power A detective and psychic join forces to track down a serial killer who, after being executed, uses his satanic powers to kill again.
1989 90m/C Lou Diamond Phillips, Tracy Griffith, Jeff Kober, Mykel T. Williamson, Elizabeth Arlen; **W/Dir:** Robert Resnikoff. Homer '90: Best Horror Film. VHS, Beta, LV, 8mm $14.95 COL, SUE, ORI ℤ½

First Spaceship on Venus Eight scientists from various countries set out for Venus and find the remains of a civilization far in advance of Earth's that perished because of nuclear weapons. A sometimes compelling anti-nuclear sci-fi effort made with German and Polish backing. Originally released at 130 minutes.
1960 78m/C GE PL Yoko Tani, Oldrich Lukes, Ignacy Machowski, Julius Ongewe, Michal Postnikow, Kurt Rackelmann, Gunther Simon, Tang-Hua-Ta, Lucyna Winnicka; **Dir:** Kurt Maetzig. VHS, Beta $19.95 NOS, MRV, SNC ℤℤℤ

First Strike The United States and the USSR engage in nuclear submarine warfare when the Soviets hijack a U.S. sub and aims its weapons at Arab oil fields.
1985 90m/C Stuart Whitman, Persis Khambatta. VHS, Beta $69.95 VCD ℤ

First & Ten A failing football team emphasizes sexual, rather than athletic conquests, under the guidance of their female owner. Made for cable TV and followed by "First & Ten: The Team Scores Again."
1985 88m/C Delta Burke, Geoffrey Scott, Reid Shelton, Ruta Lee, Fran Tarkenton; **Dir:** Donald Kushner. VHS, Beta $69.98 LIV, VES ℤ

First & Ten: The Team Scores Again A sequel to the made-for-cable-television comedy "First and Ten" about a football team's relationship with their female owner.
1985 101m/C Delta Burke, Geoffrey Scott, Ruta Lee, Reid Shelton, Clayton Landey, Fran Tarkenton. VHS, Beta $69.98 LIV, VES ℤ

The First Time It's pre-Porky's zaniness as a vacationing youth understandably wants to spend his "first time" with Bisset. Other than her presence, this film has little to offer.
1969 (R) 90m/C Jacqueline Bisset, Wes Stern, Rick Kelman, Wink Roberts; **Dir:** James Neilson. VHS $9.95 WKV ℤ½

First Time A comedy about a college student who can't quite succeed with women, despite coaching from his roommate and counseling from a psychology professor whose assistance is part of a research project. Also known as "Doin' It."

1982 (R) 96m/C Tim Choate, Krista Errickson, Marshall Efron, Wallace Shawn, Wendie Jo Sperber, Cathryn Damon; *Dir:* Charles Loventhal. **VHS, Beta** $19.99 *HBO, SIM* 🎞🎞½

First Turn On Not to be confused with "The Thomas Edison Story," this sex-comedy follows the adventures of several young campers who decide to die happy when an avalanche leaves them trapped in a cave. A longer unrated version is available.
1983 (R) 84m/C Georgia Harrell, Michael Sanville, Googy Gress, Jenny Johnson, Heide Basset, Vincent D'Onofrio, Sheila Kennedy; *Dir:* Michael Herz, Samuel Weil. **VHS, Beta** $69.98 *LIV, LTG* 🎞

First Twenty Years: Part 1 (Porter pre-1903) Early examples of narrative films produced by the Edison Company and photographed by Edwin S. Porter from 1898 to 1903.
1904 22m/B **VHS, Beta** *PYR* 🎞

First Twenty Years: Part 2 (Porter 1903-1904) These dramas, filmed between 1903 and 1904 by the Edison Company, illustrate the variety of sources which influenced filmmaker Edwin S. Porter.
1904 29m/B **VHS, Beta** *PYR* 🎞

First Twenty Years: Part 3 (Porter 1904-1905) Three prime examples of early films from the Edison Company. Filmed by Edwin S. Porter between the years 1904 and 1905.
1905 29m/B *Dir:* Edwin S. Porter, Edwin S. Porter, Edwin S. Porter, Edwin S. Porter. **VHS, Beta** *PYR* 🎞

First Twenty Years: Part 4 Three Edison Company comedies which illustrate Edwin S. Porter's knowledge of photography and camera techniques.
1905 17m/B *Dir:* Edwin S. Porter, Edwin S. Porter, Edwin S. Porter. **VHS, Beta** *PYR* 🎞

First Twenty Years: Part 5 (Comedy 1903-1904) First of several reels in the history of comedy presents five films produced by the American Mutoscope and Biograph companies between the years 1903 and 1904.
1904 25m/B **VHS, Beta** *PYR* 🎞

First Twenty Years: Part 6 (Comedy 1904-1907) Presents three films produced by the American Mutoscope and Biograph companies between the years 1904 and 1907 which are early examples of the situation comedy and the chase.
1907 25m/B **VHS, Beta** *PYR* 🎞

First Twenty Years: Part 7 (Camera Effects) Four films produced by the American Mutoscope and Biograph companies between the years 1902 and 1908 in which the common element is the use of special camera effects, the matte, stop action, time lapse, etc.
1908 23m/B **VHS, Beta** *PYR* 🎞

First Twenty Years: Part 8 (Camera Effects 1903-1904) These films, produced by the American Mutoscope and Biograph companies between the years 1903 and 1904, are representative of the films of the era, which use delayed suspense, camera movement, and special effects.
1904 29m/B **VHS, Beta** *PYR* 🎞

First Twenty Years: Part 9 (Cameramen) Shows examples of the work of American Mutoscope and Biograph cameramen F. S. Armitage, F. A. Dobson, and C. W. Bitzer as a comparison of their individual styles.
1906 26m/B **VHS, Beta** *PYR* 🎞

First Twenty Years: Part 10 (Special Effects) Presents film documentaries of dramatic incidents (American Mutoscope and Biograph) to show interesting camera techniques, day for night photography, tight close-ups, etc.
1906 23m/B **VHS, Beta** *PYR* 🎞

First Twenty Years: Part 11 (Drama on Location) Examples from American Mutoscope and Biograph of a dramatic story woven into the photography of dangerous occupations shot on location.
1906 18m/B **VHS, Beta** *PYR* 🎞

First Twenty Years: Part 12 (Direction & Scripts) Four pictures produced by the American Mutoscope and Biograph Companies that show how far camera movement, script, and direction have progressed to becoming an integral part of the film.
1908 26m/B **VHS, Beta** *PYR* 🎞

First Twenty Years: Part 13 (D.W.Griffith) Examples of comedies directed by D.W. Griffith and produced by the American Mutoscope and Biograph companies between the years 1908-1912.
1912 27m/B *Dir:* D.W. Griffith. **VHS, Beta** *PYR* 🎞

First Twenty Years: Part 14 (Griffith's Dramas) Dramas produced during the first six months of D.W. Griffith's career as a director (1908-1909), showing the beginnings of techniques recognized in later films.
1909 30m/B *Dir:* D.W. Griffith. **VHS, Beta** *PYR* 🎞

First Twenty Years: Part 15 (Griffith's Editing) Shows how D.W. Griffith edited raw footage by first showing "The Girls and Daddy" in the sequential order in which the scenes were photographed, and then with the scenes rearranged according to the scene numbers.
1909 23m/B *Dir:* D.W. Griffith. **VHS, Beta** *PYR* 🎞

First Twenty Years: Part 16 (Later Griffith) Three dramas directed by D.W. Griffith during his second six months with American Mutoscope and Biograph, showing refinements in his style and techniques.
1909 26m/B *Dir:* D.W. Griffith. **VHS, Beta** *PYR* 🎞

First Twenty Years: Part 17 (Make-up Effects) An early attempt by D.W. Griffith at making longer films. Among other contributions, they are noted for the "aging" of the cast through make-up.
1911 22m/B *Dir:* D.W. Griffith. **VHS, Beta** *PYR* 🎞

First Twenty Years: Part 18 (2-reelers) "Enoch Arden," directed by D.W. Griffith, was the first American Mutoscope and Biograph film released and publicized as a two-reel drama.
1911 25m/B *Dir:* D.W. Griffith. **VHS, Beta** *PYR* 🎞

First Twenty Years: Part 19 (A Temporary Truce) "A Temporary Truce" from American Mutoscope and Biograph, with its large cast, illustrates D.W. Griffith's progress in his craft as he uses the camera to build suspense.

1912 27m/B *Dir:* D.W. Griffith. **VHS, Beta** *PYR* 🎞

First Twenty Years: Part 20 (Lubin films) Four examples from the prodigious number of films produced by the Siegmund Lubin Manufacturing Company, Philadelphia, between the years 1904 and 1909.
19?? 28m/B **VHS, Beta** *PYR* 🎞

First Twenty Years: Part 21 (3 Independent Producers) Five examples of films from independent producers J.H. White, William Seling, and J.B. Kent between the years 1899-1908.
19?? 30m/B **VHS, Beta** *PYR* 🎞

First Twenty Years: Part 22 (Melies) Examples from the northern 1500 films of French filmmaker Georges Melies, magician, artist, and early master of special effects. These selections were produced between 1903 and 1904.
19?? 29m/B *Dir:* Georges Melies. **VHS, Beta** *PYR* 🎞

First Twenty Years: Part 23 (British Comedies) Comedies produced between 1904 and 1905 in Great Britain by the Clarendon, Gaumont, and Hepworth Companies. They were copyrighted in the U.S. and distributed under reciprocal trade agreements.
19?? 22m/B **VHS, Beta** *PYR* 🎞

First Twenty Years: Part 24 Dramas of social-documentary realism produced in Great Britain between 1903 and 1904 and distributed in the U. S.
1904 30m/B **VHS, Beta** *PYR* 🎞

First Twenty Years: Part 25 "The Aviator's Generosity" was produced by the Novdisk Film Company of Copenhagen which had, at this time, one of the largest and best equipped studios in the world.
1912 23m/B **VHS, Beta** *PYR* 🎞🎞½

First Twenty Years: Part 26 (Danish Superiority) This Danish production, "Love and Friendship," was professionally made even by today's standards.
19?? 26m/B **VHS, Beta** *PYR* 🎞🎞½

First Yank into Tokyo An American army pilot who grew up in Japan undergoes plastic surgery in order to infiltrate Japanese lines and get information from an American scientist. Some last minute editing to capitalize on current events made this the first American film to deal with the atomic bomb.
1945 83m/B Tom Neal, Richard Loo, Barbara Hale, Marc Cramer; *Dir:* Gordon Douglas. **VHS, Beta** $19.98 *RKO* 🎞🎞

A Fish Called Wanda An absurd, high-speed farce about four criminals trying to retrieve $20 million they've stolen from a safety deposit box and each other. Meanwhile, barrister Cleese falls in love with the female thief (Curtis). There's some sick, but funny, humor involving Palin's problem with stuttering and very dead doggies. Kline received an Oscar for his dumb and vain character. Written by Cleese and Crichton, who understand that silence is sometimes funnier than speech, and that timing is everything. Wickedly funny.
1988 (R) 98m/C John Cleese, Kevin Kline, Jamie Lee Curtis, Michael Palin, Tom Georgeson, Maria Aitken; *Dir:* Charles Crichton. Academy Awards '88: Best Supporting Actor (Kline). **VHS, Beta, LV** $19.98 *FOX, BTV* 🎞🎞🎞

Fish Hawk When an alcoholic Indian, Fish Hawk, meets a young boy in the forest, they strike up a friendship and he attempts to clean up his act. Attempts to be heartwarming.
1979 (G) 95m/C *CA* Will Sampson, Charlie Fields; *Dir:* Donald Shebib. **VHS, Beta $19.95** *MED* 🎬½

The Fish that Saved Pittsburgh A floundering basketball team hires an astrologer to try and change their luck. She makes sure all the team members' signs are compatible with their star's Pisces sign (the fish). Produced a disco soundtrack with several Motown groups who also performed in the movie.
1979 (PG) 104m/C Jonathan Winters, Stockard Channing, Flip Wilson, Julius Erving, Margaret Avery, Meadowlark Lemon, Nicholas Pryor, James Bond III, Kareem Abdul-Jabbar, Jack Kehoe, Debbie Allen; *Dir:* Gilbert Moses. **VHS, Beta $59.95** *LHV, WAR* 🎬🎬

The Fisher King In derelict-infested Manhattan a down-and-out radio deejay meets a crazed vagabond (Williams) obsessed with medieval history and in search of the Holy Grail. At first the whimsical mix of Arthurian myth and modern urban hell seems amazingly wrongheaded. In retrospect it still does. But while this picture runs it weaves a spell that pulls you in, especially in its quiet moments. Your reaction to the silly ending depends entirely on how well you're bamboozled by a script that equates madness with enlightment and the homeless with holy fools. Filmed on the streets of New York, with many street people playing themselves.
1991 (R) 138m/C Robin Williams, Jeff Bridges, Amanda Plummer, Mercedes Ruehl; *Dir:* Terry Gilliam. Academy Awards '91: Best Supporting Actress (Ruehl); Chicago Film Critics Awards '91: Best Supporting Actress (Ruehl); Golden Globe Awards '91: Best Supporting Actress (Ruehl); Golden Globe Awards '92: Best Actor (Williams); Los Angeles Film Critics Association Awards '91: Best Actress (Ruehl). **VHS, Beta, LV, 8mm $94.99** *CCB, COL* 🎬🎬🎬

Fisherman's Wharf Breen stars as an orphan adopted by a San Francisco fisherman who runs away when his aunt and bratty cousin come to live with them.
1939 72m/B Bobby Breen, Leo Carrillo, Henry Armetta, Lee Patrick, Rosina Galli, Leon Belasco; *Dir:* Bernard Vorhaus. **VHS $19.95** *NOS* 🎬½

F.I.S.T. A young truck driver turns union organizer for idealistic reasons, but finds himself teaming with gangsters to boost his cause. His rise to the top of the union comes at the cost of his integrity, as Stallone does a character resembling Jimmy Hoffa.
1978 (R) 145m/C Sylvester Stallone, Rod Steiger, Peter Boyle, David Huffman, Melinda Dillon, Tony LoBianco, Kevin Conway, Peter Donat, Cassie Yates, Brian Dennehy; *Dir:* Norman Jewison. **VHS, Beta, LV $79.98** *MGM, FOX* 🎬🎬½

Fist A street fighter battles his way through the urban jungle seeking personal freedom and revenge.
1979 (R) 84m/C Richard Lawson, Annazette Chase, Dabney Coleman. **VHS, Beta $59.95** *HAR* 🎬

Fist of Death Two high-kicking kids battle gangs of killers using only their bare fighting skills.
1987 90m/C Jackie Chang, Tong Lung. **VHS, Beta $59.95** *MGL* 🎬

Fist of Fear, Touch of Death Madison Square Garden is the scene for a high stakes martial arts face-off. Standard kung-fu film highlighted by short clips of the late Bruce Lee.
1977 (R) 90m/C Fred Williamson, Ron Van Cliff, Adolph Caesar, Aaron Banks, Bill Louie; *Dir:* Matthew Mallinson. **VHS, Beta $39.95** *IVE* 🎬

Fist Fighter The ups and downs of a professional bare-knuckle fighter as he avenges a friend's murder.
1988 (R) 99m/C George Rivero, Edward Albert, Brenda Bakke, Mike Connors, Simon Andrew, Matthias Hues; *Dir:* Frank Zuniga. **VHS, Beta, LV $9.99** *CCB, IVE* 🎬½

A Fist Full of Talons A sexy maiden, a young warrior, and a goofy thief team up to fight evil in the early Chinese republic.
19?? 100m/C VHS, Beta $34.95 *OCE* 🎬

The Fist, the Kicks, and the Evils A young man who's father is murdered learns kung fu and sets out for revenge.
1980 90m/C Kao Fei, Bruce Liang. **VHS, Beta $29.95** *OCE* 🎬

Fist of Vengeance The East Asia Society hires a samurai to kill a young Chinese officer.
197? 90m/C Shoji Karada, Lu Pi Chen. **VHS, Beta $39.95** *MAV* 🎬

Fistful of Death Kinski portrays a vengefully violent and desperate cowpoke in the days of the old West.
1967 84m/C Klaus Kinski. **VHS, Beta $59.95** *TWE* 🎬

A Fistful of Dollars The epitome of the ''spaghetti western'' pits Eastwood as ''the man with no name'' against two families who are feuding over land. A remake of Kurosawa's ''Yojimbo,'' and followed by Leone's ''For a Few Dollars More'', and ''The Good, The Bad, and The Ugly.''
1964 96m/C *IT* Clint Eastwood, Gian Marie Volonte, Marianne Koch; *Dir:* Sergio Leone. **VHS, Beta, LV $19.98** *MGM, FCT, TLF* 🎬🎬🎬

Fistful of Dynamite A spaghetti western, with Leone's trademark humor and a striking score by Ennio Morricone. An Irish demolitions expert and a Mexican peasant team up to rob a bank during a revolution in Mexico. Also known as ''Duck, You Sucker.''
1972 (PG) 138m/C *IT* James Coburn, Rod Steiger, Romolo Valli; *Dir:* Sergio Leone. **VHS, Beta, LV $59.95** *MGM* 🎬🎬🎬

Fists of Blood A martial arts expert seeks revenge on the drug baron who murdered his friend and kidnapped his girl.
1987 90m/C *AU* Eddie Stazak; *Dir:* Brian Trenchard-Smith. **VHS, Beta $79.95** *CEL* 🎬½

Fists of Bruce Lee An undercover cop cracks a drug-smuggling ring by kicking out in all directions.
198? 90m/C Bruce Lee, Lo Lieh; *Dir:* Bruce Lee. **VHS, Beta $24.95** *VCL* 🎬

Fists of Dragons Vengeful, sacrificial bloodshed abounds in this Chinese boxing extravaganza.
1969 92m/C Huang I Lung, Ou Ti; *Dir:* Yeh Yung Chu; *Hosted:* Sho Kosugi. **VHS, Beta $29.95** *TWE* 🎬

Fists of Fury Bruce Lee stars in this violent Kung Fu action adventure in which he must break a solemn vow to avoid fighting in order to avenge the murder of his teacher by drug smugglers.
1973 (R) 102m/C Bruce Lee, Maria Yi; *Dir:* Lo Wei. **VHS, Beta $19.98** *FOX* 🎬

Fists of Fury 2 Chen Shan must survive the Organizations' onslaughts to kill him. He escapes their perilous plots only to return to battle against them after they kill his mother for her inability to disclose Chen's hiding place. Finally Chen defeats the evil Organization himself.
1980 (R) 90m/C Bruce Li, Ho Chung Do, Shum Shim Po. **VHS, Beta $19.98** *GEM* 🎬½

Fists of the White Lotus A Shaolin warrior seeks revenge for the killing of his pals by the vicious White Lotus.
198? 95m/C *Dir:* Lo Lieh. **Beta $69.95** *VHV* 🎬

Fit for a King A reporter becomes a princess' knight in shining armor when he foils an assassination plot in this screwball romance.
1937 73m/B Joe E. Brown, Leo Carrillo, Helen Mack, Paul Kelly, Harry Davenport; *Dir:* Edward Sedgwick. **VHS, Beta $24.95** *NOS, DVT* 🎬🎬

Fitzcarraldo Although he failed to build a railroad across South America, Fitzcarraldo is determined to build an opera house in the middle of the Amazon jungles and have Enrico Caruso sing there. Based on a true story of a charismatic Irishman's impossible quest at the turn of the century. Of note: No special effects were used in this movie - - everything you see actually occurred during filming, including hauling a large boat over a mountain.
1982 (PG) 157m/C *GE* Klaus Kinski, Claudia Cardinale, Jose Lewgoy, Miguel Angel Fuentes; *Dir:* Werner Herzog. Cannes Film Festival '82: Best Director (Herzog). **VHS, Beta $69.95** *WAR, APD* 🎬🎬🎬🎬

Five Came Back When a plane with twelve passengers crashes in the South American jungle, only five can ride in the patched-up wreck. Since the remainder will be left to face head hunters, intense arguments ensue. Same director remade this as ''Back from Eternity'' in 1956.
1939 93m/B Lucille Ball, Chester Morris, John Carradine, Wendy Barrie, Kent Taylor, Joseph Calleia, C. Aubrey Smith, Patric Knowles; *Dir:* John Farrow. **VHS, Beta $29.95** *RKO* 🎬🎬

Five Card Stud Five members of a lynching party are being killed one by one, and a professional gambler, who tried to prevent the lynching, attempts to ensnare the killer with the aid of a preacher with a gun. The poor script was based on a novel by Ray Gaulden.
1968 (PG) 103m/C Robert Mitchum, Dean Martin, Inger Stevens, Roddy McDowall, Yaphet Kotto, John Anderson, Katherine Justice; *Dir:* Henry Hathaway. **VHS, Beta $14.95** *PAR* 🎬½

Five Corners A quixotic, dramatic comedy about the inhabitants of the 5 Corners section of the Bronx in 1964, centering around a girl being wooed by a psychotic rapist, her crippled boyfriend, and the hero-turned-racial-pacifist who cannot rescue her out of principle.
1988 (R) 92m/C Jodie Foster, John Turturro, Todd Graff, Tim Robbins, Elizabeth Berridge, Rose Gregorio, Gregory Rozakis, Rodney Harvey; *Dir:* Tony Bill. **VHS, Beta $89.95** *NO* 🎬🎬🎬

Five Days One Summer Set in 1932, the story of a haunting and obsessive love affair between a married Scottish doctor and a young woman who happens to be his niece. While on vacation in the Swiss Alps, the doctor must vie for her love with their

handsome young mountain climbing guide. Based on a story by Kay Boyle.
1982 (PG) 108m/C Sean Connery, Betsy Brantley, Lambert Wilson; *Dir:* Fred Zinneman. **VHS, Beta** $19.98 *WAR* ♂♂

Five Easy Pieces Nicholson's superb acting brings to life this character study of a talented musician who has given up a promising career and now works on the oil rigs. After twenty years he returns home to attempt one last communication with his dying father and perhaps, reconcile himself with his fear of failure and desire for greatness. Black, Anspach, and Bush create especially memorable characters. Nicholson ordering toast via a chicken salad sandwich is a classic. Oscar nominated for Best Picture, actor Nicholson, supporting actress Black, and Original Screenplay.
1970 (R) 98m/C Jack Nicholson, Karen Black, Susan Anspach, Lois Smith, Billy "Green" Bush, Fannie Flagg, Ralph Waite; *Dir:* Andrien Joyce, Bob Rafelson. **VHS, Beta, LV, 8mm** $14.95 *COL, VYG* ♂♂♂♂

Five Fighters from Shaolin Five high-kicking fighters endeavor to recapture their temple from an evil duke.
197? 88m/C Chang Shen, Lung Kwan, Dhiu Min Shien; *Dir:* Ko Shu-How. **VHS, Beta** $39.95 *MPI* ♂

Five Forty-Eight A man has an affair with a former employee, then breaks it off. He meets her again on the 5:48 train and the commuter run becomes a dangerous ride through obsession and revenge. From the "Cheever Short Stories" series.
1979 60m/C Laurence Luckinbill, Mary Beth Hurt. **VHS, Beta** *FLI* ♂ 1/2

Five Giants from Texas El cheapo spaghetti western features Madison, television's former Wild Bill Hickock, as a rancher fighting off displaced Mexican peasants and outlaws.
1966 103m/C Guy Madison, Monique Randall; *Dir:* Aldo Florio. **VHS, Beta** $59.95 *TWE* ♂

Five Golden Dragons A typical action film about an American running afoul of ruthless gold trafficking in Hong Kong. Even this stellar cast can't help the script.
1967 92m/C *GB* Robert Cummings, Christopher Lee, Brian Donlevy, Klaus Kinski, George Raft, Dan Duryea, Margaret Lee; *Dir:* Jeremy Summers. **VHS, Beta** $59.98 *NO* ♂

The Five Heartbeats Well told story of five black singers in the 1960s, their successes and failures as a group and as individuals. Although every horror story of the music business is included, the story remains fresh and the acting excellent. Music is fine, but secondary to the people. Well written characters with few cliches. Skillfully directed by Townsend (of "Hollywood Shuffle") who did research by talking to the Dells. Less than memorable showing at the box office but fun and entertaining.
1991 (R) 122m/C Robert Townsend, Tressa Thomas, Michael Wright, Harry J. Lennix, Diahann Carroll, Leon, Hawthorne James, Chuck Patterson, Roy Fegan; *W/Dir:* Robert Townsend. **VHS, Beta, LV** $19.95 *FXV* ♂♂♂

Five for Hell The army picks five of its meanest men for a suicide mission during World War II. They must go behind German lines and find the plans for the enemy offensive.
1967 88m/C Klaus Kinski, John Garko, Nick Jordan, Margaret Lee; *Dir:* Frank Kramer. **VHS, Beta** $59.95 *TWE* ♂

The Five of Me A man with five personalities finds that livin' ain't easy and seeks professional help. Made for television, based on the autobiography of Henry Hawksworth.
1981 100m/C David Birney, Dee Wallace Stone, Mitchell Ryan, John McLiam, James Whitmore Jr., Ben Piazza; *Dir:* Paul Wendkos. **VHS** $39.95 *IVE* ♂♂1/2

Five Miles to Midnight Brett and Danny have to depend on a photographer (Collins) for help to smuggle a gangster out of Italy. From the British, "Persuaders" adventure series.
1980 52m/C *GB* Roger Moore, Tony Curtis, Joan Collins. **VHS** $39.95 *SVS* ♂♂1/2

The Five Pennies Sentimental biography starring Kaye as famed jazzman Red Nichols. Featuring performances by legendary musicians Bob Crosby, Bobby Troup, Ray Anthony and Louis Armstrong. Songs include "Good Night Sleep Tight," "The Five Pennies," "Battle Hymn of the Republic," "When The Saints Go Marching In," "Jingle Bells," and "Carnival of Venice." This movie marked Weld's film debut.
1959 117m/C Danny Kaye, Louis Armstrong, Barbara Bel Geddes, Tuesday Weld, Harry Guardino; *W/Dir:* Melville Shavelson. **VHS, Beta** $14.95 *PAR* ♂♂1/2

Five Weeks in a Balloon This adaptation of the Jules Verne novel follows the often-comic exploits of a 19th-century British expedition that encounters many adventures on a balloon trek across Africa. Pleasant fluff with a good cast.
1962 101m/C Fabian, Peter Lorre, Red Buttons, Cedric Hardwicke, Barbara Eden; *Dir:* Irwin Allen. **VHS, Beta** $19.98 *FOX, FCT* ♂♂

The 5000 Fingers of Dr. T In Dr. Seuss' only non-animated movie, a boy tries to evade piano lessons and runs right into the castle of the evil Dr. Terwilliger, where hundreds of boys are held captive for piano lessons. Worse yet, they're forced to wear silly beanies with "happy fingers" waving on top. Luckily, the trusted family plumber is on hand to save the day through means of an atomic bomb. Wonderful satire, horrible music, mesmerizing Seussian sets. The skating brothers (who are joined at their beards) are a treat.
1953 88m/C Peter Lind Hayes, Mary Healy, Tommy Rettig, Hans Conried; *Dir:* Roy Rowland. **VHS, Beta, LV** $59.95 *XVC, FCT* ♂♂♂

The $5.20 an Hour Dream Lavin plays a divorced mother and factory worker burdened with debt and determined to get and keep a job on the higher-paying, traditionally all-male assembly line. Feminist drama on the heels of "Norma Rae."
1980 96m/C Linda Lavin, Richard Jaeckel, Nicholas Pryor, Pamela McMyler, Mayf Nutter, Taurean Blacque, Robert Davi, Dennis Fimple, Dana Hill, Ernie Hudson; *Dir:* Russ Mayberry. **VHS, Beta** *TLF* ♂

The Fix A group of cocaine smugglers get their just desserts as the federal government catches them in a sting operation.
1984 95m/C Vince Edwards, Richard Jaeckel, Julie Hill, Charles Dierkop, Byron Cherry, Robert Tessier; *Dir:* Will Zens. **VHS, Beta** $39.95 *WES* ♂

The Fixx: Live in the USA New wave rockers The Fixx, perform such hits as "Saved by Zero," "Deeper and Deeper" and "One Thing Leads to Another" in this concert video. Available in Hi-Fi stereo for both formats.

1984 58m/C VHS, Beta $19.95 *MCA, MVD* ♂

The Flame & the Arrow Dardo the Arrow, a Robin Hood-like outlaw in medieval Italy, leads his band of mountain fighters against a mercenary warlord who has seduced his wife and kidnapped his son. Spectacular acrobatics, with Lancaster performing his own stunts, add interest to the usual swashbuckling. Music score by Max Steiner.
1950 88m/C Burt Lancaster, Virginia Mayo, Aline MacMahon, Nick Cravat, Robert Douglas, Frank Allenby; *Dir:* Jacques Tourneur. **VHS, Beta, LV** $19.98 *WAR* ♂♂♂

Flame of the Barbary Coast A rancher from Montana vies with a gambling czar for a beautiful dance hall queen and control of the Barbary Coast district of San Francisco. The great earthquake of 1906 provides the plot with a climax. Also available colorized.
1945 91m/B John Wayne, Ann Dvorak, Joseph Schildkraut, William Frawley; *Dir:* Joseph Kane. **VHS** $14.98 *MED, REP* ♂♂1/2

Flame of the Islands De Carlo plays a smoldering, passionate chanteuse who struggles with love and gangsters for possession of a Bahamian casino in this tropical heat wave.
1955 92m/C Yvonne de Carlo, Howard Duff, Zachary Scott, James Arness, Kurt Kasznar, Barbara O'Neil; *Dir:* Edward Ludwig. **VHS, Beta** $39.98 *NO* ♂♂

Flame is Love A made-for-television adaptation of one of Barbara Cartland's romantic novels. An American heiress falls tragically in love with a Parisian journalist despite her engagement to an Englishman. Takes place at the turn of the century.
1979 98m/C Linda Purl, Timothy Dalton, Shane Briant; *Dir:* Michael O'Herlihy. **VHS, Beta** $69.98 *LTG, LIV* ♂1/2

Flame to the Phoenix A World War II drama about the Polish cavalry forces' decimation to German Panzer tanks.
1985 80m/C *GB* Paul Geoffrey, Ann Firbank, Frederick Treves; *Dir:* William Brayne. **VHS, Beta** $79.95 *WES* ♂1/2

Flaming Frontiers A frontier scout matches wits against gold mine thieves. In fifteen episodes.
1938 300m/B Johnny Mack Brown, Eleanor Hanson, Ralph Bowman; *Dir:* Ray Taylor. **VHS, Beta** $49.95 *NOS, VCN, VYY* ♂

Flaming Lead A hard-drinkin' ranch owner hires a nightclub cowboy to help him get an Army horse contract.
1939 57m/B Ken Maynard, Eleanor Stewart, Walter Long, Tom London; *Dir:* Max Alexander. **VHS** $19.95 *NOS, DVT* ♂1/2

Flaming Signal Featuring Flash the Wonder Labrador, this action-packed film deals with a pilot who crash-lands near a Pacific island, just as the natives, provoked by an exploitative German trader, are rising up against the whites. Both Flash and his master manage to rescue the missionary's pretty daughter.
1933 64m/B Noah Beery, Madeline Day, Carmelita Geraghty, Mischa Auer, Henry B. Walthall. **VHS, Beta, LV** $16.95 *SNC, WGE, RXM* ♂1/2

Flaming Star In 1870s Texas, a family with a white father and an Indian mother is caught in the midst of an Indian uprising. The mixed-blood youth, excellently played by Presley, must choose a side with tragic results for all. A stirring, well-written drama

of frontier prejudice and one of Presley's best films.
1960 101m/C Elvis Presley, Dolores Del Rio, Barbara Eden, Steve Forrest, John McIntire, Richard Jaeckel, L.Q. Jones; *Dir:* Don Siegel. **VHS, Beta $14.98** *FOX, MVD* 🎞🎞🎞

The Flaming Urge A small town is plagued by mysterious fires. Could a pyromaniac be on the loose?
1953 67m/B Harold Lloyd Jr., Cathy Downs. **VHS, Beta $16.95** *SNC* 🎞🎞🎞

The Flamingo Kid A Brooklyn teenager (Dillon) gets a summer job at a fancy beach club on Long Island. His father is a plumber, who remembers how to dream, but is also aware of how rough the world is on dreamers. Suddenly making lots of money at a mostly easy job, the kid is attracted to the high style of the local sports car dealer and gin rummy king, and finds his father's solid life a bore. By the end of the summer, he's learned the true value of both men, and the kind of man he wants to be. Excellent performances all around, nice ensemble acting among the young men who play Dillon's buddies. Great sound track. Film debut of Janet Jones, who seems a little old for her part as a California college sophomore.
1984 (PG-13) 100m/C Matt Dillon, Hector Elizondo, Molly McCarthy, Martha Gehman, Richard Crenna, Jessica Walter, Carole Davis, Janet Jones, Fisher Stevens, Bronson Pinchot; *Dir:* Garry Marshall. **VHS, Beta, LV $29.98** *LIV, VES* 🎞🎞🎞

Flamingo Lead Maynard aids a female cavalry horse-raiser amid various western adversities.
1939 57m/B Ken Maynard. **VHS, Beta** *UHV* 🎞

Flamingo Road A scandalously entertaining melodrama in which Crawford portrays a carnival dancer who intrigues Scott and Brian in a small Southern town where the carnival stops. Crawford shines in a role that demands her to be both tough and sensitive in a corrupt world full of political backstabbing and sleazy characters. Remade as a TV movie and television soap-opera series in 1980.
1949 94m/B Joan Crawford, Zachary Scott, David Brian, Sydney Greenstreet, Gertrude Michael, Gladys George, Virginia Huston, Fred Clark; *Dir:* Michael Curtiz. **VHS $19.98** *MGM* 🎞🎞🎞

Flanagan Boy A prizefighter and his manager's loose wife conspire to drown her husband for more than one reason. Also known as ''Bad Blonde.''
1953 79m/B *GB* Barbara Payton, Tony Wright. **VHS, Beta, LV** *WGE* 🎞

The Flash When police scientist Barry Allen is accidentally doused by chemicals and then struck by lightening the combination makes him into a new superhero. His super-strength and quickness help his quest in fighting crime in Central City where he's aided by fellow scientist Tina McGee (the only other person to know his secret). In this adventure, the Flash seeks out the violent and mesmerizing leader of a biker gang who caused the death of Barry's brother. Based on the DC comic book character, this is the pilot episode for the short-lived television series. The look is dark and stylized and not played for camp.
1990 94m/C John Wesley Shipp, Amanda Pays, Michael Nader. **VHS, Beta, LV $89.99** *WAR* 🎞🎞

Flash Challenger Frank Chan fights to reclaim some stolen gold meant to finance a revolution.

1969 90m/C *Hosted:* Sho Kosugi. **Beta $29.98** *TWE* 🎞

Flash & Firecat A beautiful blonde and a crazy thief steal and race their way across the country in a dune buggy with the police hot on their trail.
1975 (PG) 90m/C Roger Davis, Tricia Sembera, Dub Taylor, Richard Kiel; *Dir:* Ferd Sebastian. **VHS, Beta $39.95** *IND* 🎞

Flash Gordon Camp version of the adventures of Flash Gordon in outer space. This time, Flash and Dale Arden are forced by Dr. Zarkov to accompany him on a mission to far-off Mongo, where Ming the Merciless is threatening the destruction of Earth. Music by Queen.
1980 (PG) 111m/C Sam Jones, Melody Anderson, Chaim Topol, Max von Sydow, Ornella Muti, Timothy Dalton, Brian Blessed; *Dir:* Mike Hodges. **VHS, Beta, LV $39.95** *MCA* 🎞🎞

Flash Gordon Battles the Galactic Forces of Evil Flash Gordon returns to battle with Ming the Merciless in this collection of cartoon adventures.
1979 59m/C VHS, Beta $19.95 *MED* 🎞½

Flash Gordon Conquers the Universe Ravaging plague strikes the earth and Flash Gordon undertakes to stop it. A serial in twelve chapters.
1940 240m/B Buster Crabbe, Carol Hughes, Charles Middleton, Frank Shannon. **VHS, Beta $49.95** *NOS, MRV, SNC* 🎞½

Flash Gordon: Mars Attacks the World The earth is plagued by the evil Ming, but Flash Gordon steps in. From the serial.
1938 87m/B Buster Crabbe, Jean Rogers, Charles Middleton. **VHS, Beta $39.95** *CAB* 🎞½

Flash Gordon: Rocketship Re-edited from the original Flash Gordon serial in which Flash and company must prevent the planet Mongo from colliding with Earth. Good character acting and good clean fun. Also known as ''Spaceship to the Unknown'' and ''Perils from Planet Mongo.''
1936 97m/B Buster Crabbe, Jean Rogers, Frank Shannon, Charles Middleton, Priscilla Lawson, Jack Lipson; *Dir:* Frederick Stephani. **VHS $29.95** *VYY, PSM, CAB* 🎞🎞½

Flash Gordon: Space Adventurer Join Flash Gordon, Dale Arden and Dr. Zarkov as they encounter all kinds of adventures while traveling through outer space.
1979 58m/C VHS, Beta $19.95 *MED* 🎞

Flash Gordon: The Beast Men's Prey Flash and crew take on an alien menace in this brand new adventure from the King Features ''Animated Comics'' series.
1991 30m/C VHS $9.99 *BFV* 🎞

Flash Gordon: Vol. 1 Join Flash Gordon and Dale Arden as they struggle to maintain peace in the galaxy in this pair of episodes from the series.
1953 60m/B Steve Holland, Irene Champlin. **VHS, Beta $29.95** *DVT* 🎞

Flash Gordon: Vol. 2 Flash and Dale continue their quest for intergalactic tranquility in this pair of episodes from the series.
1953 60m/B Steve Holland, Irene Champlin. **VHS, Beta $29.95** *DVT* 🎞

Flash Gordon: Vol. 3 The struggle for intergalactic peace continues as Flash Gordon battles Ming the Merciless in this pair of episodes from the series.
1953 60m/B VHS, Beta $29.95 *DVT* 🎞

A Flash of Green A crooked politician is helping a construction firm exploit valuable waterfront property. He enlists the influence of a hesitant local journalist, who then falls for the woman leading the homeowner's conservation drive against the development plan. Made for American Playhouse and produced by costar Jordon. Based on the work of John D. MacDonald.
1985 122m/C Ed Harris, Blair Brown, Richard Jordan, George Coe; *W/Dir:* Victor Nunez. **VHS, Beta $59.95** *MED* 🎞🎞

Flashback FBI agent Sutherland's assignment sounds easy: escort aging 1960s radical Hopper to prison. But Hopper is cunning and decides not to go without a fight. He uses his brain to outwit the young Sutherland and to turn him against himself. Good moments between the two leads and with Kane, as a woman who never left the sixties behind.
1989 (R) 108m/C Dennis Hopper, Kiefer Sutherland, Carol Kane; *Dir:* Franco Amurri. **VHS, Beta, LV $89.99** *PAR* 🎞🎞🎞

Flashdance 18 year-old Alex wants to dance. She works all day as a welder (and has a hot affair going with her boss), dances at a local bar at night, and hopes someday to get enough courage to audition for a spot at the School of Ballet. Glossy music video redeemed somewhat by exciting choreography with Marine Jahan doing the dancing for Beals. Oscar-winning title song sung by Irene Cara. Inspired the torn-sweatshirt trend in fashion of the period.
1983 (R) 95m/C Jennifer Beals, Michael Nouri, Belinda Bauer, Lilia Skala, Cynthia Rhodes; *Dir:* Adrian Lyne. Academy Awards '83: Best Song (''Flashdance...What a Feeling''). **VHS, Beta, LV, 8mm $14.95** *PAR* 🎞🎞

Flashpoint A pair of Texas border patrolmen discover an abandoned jeep that contains a fortune in cash, apparently from the 1960s. As they try to figure out how it got there, they become prey to those who want to keep the secret. With this cast, flick ought to be better.
1984 (R) 95m/C Treat Williams, Kris Kristofferson, Tess Harper, Rip Torn, Miguel Ferrer, Roberts Blossom; *Dir:* William Tannen. **VHS, Beta, LV $79.99** *HBO* 🎞🎞

Flashpoint Africa When a news team follows a terrorist group's activities it winds up in a power struggle with terrifying consequences.
1984 99m/C Trevor Howard, Gayle Hunnicutt, James Faulkner. **VHS, Beta $69.95** *VCL* 🎞½

Flat Top The training of Navy fighter pilots aboard ''flat top'' aircraft carriers during World War II provides the drama here. A strict commander is appreciated only after the war when the pilots realize his role in their survival. The film makes good use of actual combat footage; fast-paced and effective.
1952 85m/C Sterling Hayden, Richard Carlson, Keith Larsen, John Bromfield; *Dir:* Lesley Selander. **VHS, Beta $19.98** *REP, FCT* 🎞🎞½

Flatbed Annie and Sweetiepie: Lady Truckers A couple of good ol' gals hit the road and encounter a variety of bad guys out to steal their truck. Made for television.
1979 100m/C Annie Potts, Kim Darby, Harry Dean Stanton, Arthur Godfrey, Rory Calhoun, Bill Carter; *Dir:* Robert Greenwald. **VHS, Beta $79.99** *HBO* 🎞½

Flatfoot Tough police officer will let nothing stop him from finding and arresting drug smugglers. Good for laughs as well as lots of action. Also known as "The Knock-Out Cop."
1978 (PG) 113m/C *IT GE* Bodo, Werner Pochat, Bud Spencer, Enzo Cannavale, Dagmar Lassander, Joe Stewardson; *Dir:* Steno. **VHS, Beta $14.95** *LEG* ✂✂

Flatliners A group of medical students begin after-hours experimentation with death and out-of-body experiences. Some standard horror film images but Roberts and Sutherland create an energy that makes it worth watching. Oscar nominated for Best Sound Effects Editing.
1990 (R) 111m/C Kiefer Sutherland, Julia Roberts, William Baldwin, Oliver Platt, Kevin Bacon, Kimberly Scott, Joshua Rudoy; *Dir:* Joel Schumacher. **VHS, Beta, LV, 8mm $19.95** *COL* ✂✂✂

Fleischer Color Classics A compendium of 8 1930's Max Fleischer cartoons, including "An Elephant Never Forgets," "Dancing on the Moon," and "Fresh Vegetable Mystery" and "Greedy Humpty Dumpty."
1939 55m/C **VHS, Beta $19.95** *VDM* ✂✂

Flesh An Andy Warhol-produced seedy urban farce about a bisexual street hustler who meets a variety of drug-addicted, deformed, and sexually deviant people. Dallesandro fans will enjoy his extensive exposure (literally).
1968 90m/C Joe Dallesandro, Geraldine Smith, Patti D'Arbanville; *Dir:* Paul Morrissey. **VHS, Beta $29.95** *MFV, PAR, FCT* ✂✂✂

Flesh & Blood An unjustly convicted lawyer is released from prison to find out his wife has died. He vows revenge on those who falsely imprisoned him and assumes the disguise of a crippled beggar to begin his plot. Silent with musical score.
1922 75m/B Lon Chaney Sr.; *Dir:* Irving Cummings. **VHS, Beta, 8mm $16.95** *SNC, VYY, GVV* ✂✂

Flesh and Blood A rowdy group of 16th Century hellions makes off with a princess who is already spoken for and pillage and plunder their way to revenge. Hauer leads the motley group through sword fights, raids, and the like. Dutch director Verhoeven's first English language film. Not for children; with rape scenes, nudity, and graphic sex. Also known as "The Rose and the Sword."
1985 (R) 126m/C Rutger Hauer, Jennifer Jason Leigh, Tom Burlinson, Susan Tyrrell; *Dir:* Paul Verhoeven. **VHS, Beta, LV $19.99** *LIV, VES, SIM* ✂✂

Flesh and Blood Show Rehearsal turns into an execution ritual for a group of actors at a mysterious Seaside theatre. Blood, gore, and some sex. Shot in 3-D.
1973 (R) 93m/C *GB* Robin Askwith, Candace Glendenning, Tristan Rogers, Ray Brooks, Jenny Hanley, Luan Peters, Patrick Barr; *Dir:* Pete Walker. **VHS, Beta $59.95** *MON* ✂

The Flesh and the Devil Classic Garbo at her seductive best as a woman who causes a feud between two friends. Gilbert is an Austrian officer, falls for the married Garbo and winds up killing her husband in a duel. Banished to the African Cops he asks his best friend (Hanson) to look after his lady love. But Hanson takes his job too seriously, falling for the lady himself. Great silent movie with surprise ending to match. The first Gilbert and Garbo pairing.

1927 112m/B John Gilbert, Greta Garbo, Lars Hanson, Barbara Kent, George Fawcett, Eugenie Besserer; *Dir:* Clarence Brown. **VHS $29.95** *MGM* ✂✂✂½

The Flesh Eaters A claustrophobic low-budget thriller about a film queen and her secretary who crash-land on an island inhabited by your basic mad scientist. His latest experiment is with tiny flesh-eating sea creatures. Shock ending.
1964 87m/C Martin Kosleck, Rita Morley, Byron Sanders, Barbara Wilkin, Ray Tudor; *Dir:* Jack Curtis. **VHS, Beta $39.95** *MON, SNC* ✂½

Flesh Eating Mothers Housewives are transformed into cannibals after a mystery virus hits their town. Their kids must stop the moms from eating any more people.
1989 90m/C Robert Lee Oliver, Donatella Hecht, Valorie Hubbard, Neal Rosen, Terry Hayes; *Dir:* James Aviles Martin. **VHS, Beta $14.95** *ACA* ✂

Flesh Feast Classically horrendous anti-Nazi bosh, in which a mad female plastic surgeon (Lake) rejuvenates Hitler and then tortures him to death with maggots to avenge her mother's suffering. Lake's last film, and the sorriest sign-off any actress ever had.
1969 (R) 72m/C Veronica Lake, Phil Philbin, Heather Hughes, Martha Mischon, Yanka Mann, Dian Wilhite, Chris Martell; *Dir:* Brad F. Ginter. **VHS, Beta $39.95** *WES* Woof!

Flesh Gordon Soft-core spoof of the "Flash Gordon" series. Flesh takes it upon himself to save Earth from the evil Wang's sex ray; Wang, of course, being the leader of the planet Porno. Lackluster special effects and below par story dull an already ridiculous movie. Look for real-life porn starlet Candy Samples in a cameo appearance.
1972 70m/C Jason Willaims, Suzanne Fields, Joseph Hudgins, John Hoyt, Howard Zieff, Candy Samples, Michael Benveniste; *Dir:* Mike Light. **VHS, Beta, LV $19.98** *MED, IME, TAV* ✂

Flesh and the Spur A cowboy tracks the killer of his brother. Future Mannix Mike Connors (nicknamed 'Touch' at the time) joins the search. Unedifying western with oddball touches, notably a theme song by Chipmunks creator Ross Bagdasarian.
1957 78m/C John Agar, Maria English, Mike Connors, Raymond Hatton, Maria Monay, Joyce Meadows, Kenne Duncan; *Dir:* Edward L. Cahn. **VHS $15.95** *NOS, LOO* ✂½

Fleshburn An Indian Vietnam War veteran escapes from a mental institution to get revenge on the four psychiatrists who committed him.
1984 (R) 91m/C Steve Kanaly, Karen Carlson, Sonny Landham, Macon McCalman; *Dir:* George Gage. **VHS, Beta $59.95** *MED* ✂✂

Fletch Somewhat charming comedy. When newspaper journalist Fletch goes undercover to get the scoop on the local drug scene, a wealthy young businessman enlists his help in dying. Something's rotten in Denmark when the man's doctor knows nothing of the illness and Fletch comes closer to the drug scene than he realizes. Throughout the entire film, Chevy Chase assumes a multitude of flippant comic characters to discover the truth. Based on Gregory McDonald's novel.
1985 (PG) 98m/C Chevy Chase, Tim Matheson, Joe Don Baker, Dana Wheeler-Nicholson, M. Emmet Walsh, Kenneth Mars, Geena Davis, Richard Libertini, George Wendt, Kareem Abdul-Jabbar, Alison La Placa; *Dir:* Michael Ritchie. **VHS, Beta, LV $14.95** *MCA* ✂✂

Fletch Lives In this sequel to "Fletch," Chase is back again as the super-reporter. When Fletch learns of his inheritance of a Southern estate he is eager to claim it. During his down-home trip he becomes involved in a murder and must use his disguise skills to solve it before he becomes the next victim. Based on the novels of Gregory MacDonald.
1989 (PG) 95m/C Chevy Chase, Hal Holbrook, Julianne Phillips, Richard Libertini, Lee Ermey, Cleavon Little; *Dir:* Michael Ritchie. **VHS, Beta, LV $14.95** *MCA* ✂✂

Flicks A compilation of skits parodying the tradition of Saturday afternoon matinees, including coming attractions and a cartoon. Never released theatrically.
1985 (R) 79m/C Martin Mull, Joan Hackett, Pamela Sue Martin, Betty Kennedy, Richard Belzer; *Dir:* Peter Winograd. **VHS, Beta $69.95** *MED* ✂✂

The Flight A non-narrative view of glider planes as they soar over the countryside and rivers of the Laurentians and Canadian Rockies. Useful in language arts classes.
1989 96m/C Eli Danker, Sandy McPeak, Lindsay Wagner; *Dir:* Paul Wendkos. **VHS, Beta, LV $79.95** *VES, IME* ✂✂

Flight of Black Angel Wacked-out F-16 pilot fancies himself an angel of death, and, after annihilating a number of trainees, sets out to make Las Vegas a nuked-out ghost town. Squadron commander Strauss, however, is not pleased with his pilot's initiative. Written by Henry Dominic, the script runs out of gas and heads into a nosedive. Made for cable.
1991 (R) 102m/C Peter Strauss, William O'Leary, James O'Sullivan, Michael Keys Hall; *Dir:* Jon Mostow. **VHS $89.95** *VMK* ✂✂½

Flight of Dragons This animated tale takes place between the Age of Magic and the Age of Science, in a century when dragons ruled the skies.
1982 98m/C *Dir:* Arthur Rankin Jr., Jules Bass; *Voices:* John Ritter, Victor Buono, James Earl Jones. **VHS, Beta** *CVL* ✂✂✂✂

Flight of the Eagle Based on the actual ill-fated expedition of Salomon Andree who, with two friends, attempted to fly from Sweden to the North Pole in a hydrogen balloon in 1897. The last half of the film drags somewhat as the three struggle to survive in the frozen north after their balloon crashes. Beautifully photographed adventure was nominated for an Oscar as Best Foreign Film. In Swedish with English subtitles.
1982 115m/C *SW* Max von Sydow, Goran Stangertz, Clement Harari, Sverre Anker; *Dir:* Jan Troell. **VHS, Beta, LV $24.95** *AHV* ✂½

Flight to Fury A cheap independent adventure film, featuring the novice Hellman-Nicholson team, about a few assorted mercenaries searching for a horde of diamonds when their plane crashes in the wilderness of the Philippines. Nicholson also wrote the script from a story by the director and producer.
1966 73m/B Jack Nicholson, Dewey Martin, Fay Spain, Vic Diaz, Jacqueline Hellman; *Dir:* Monte Hellman. **VHS, Beta $59.95** *WAR* ✂½

Flight from Glory A group of pilots fly supplies over the Andes from their isolated base camp to even more isolated mines. Morris is their leader who watches as, one by one, the men are killed on their dangerous flights. To make a bad situation worse, Heflin arrives as a new recruit - along with his pretty wife.

1937 66m/B Chester Morris, Onslow Stevens, Whitney Bourne, Van Heflin; *Dir:* Lew Landers. VHS, Beta *RKO* 𝄞𝄞½

Flight of the Grey Wolf A tame, innocent wolf is mistaken for a killer and must run for his life with the help of his boy-owner.
1976 82m/C Bill Williams, Barbara Hale, Jeff East. VHS, Beta $69.95 *DIS* 𝄞½

Flight of the Intruder Vietnam naval pilots aboard an aircraft carrier don't like the way the war is being handled, so they go rogue and decide to go on a mission to bomb an enemy air base in Hanoi. Loads of male bonding.
1990 (PG-13) 115m/C Danny Glover, Willem Dafoe, Brad Johnson, Rosanna Arquette, Tom Sizemore; *Dir:* John Milius. VHS, Beta, LV $14.95 *PAR* 𝄞½

Flight to Mars An expedition crash lands on the red planet and discovers an advanced underground society that wants to invade earth using the U.S. spacecraft. Includes previews of coming attractions from classic science fiction films. First movie of this genre to be shot in color.
1952 72m/C Cameron Mitchell, Marguerite Chapman, Arthur Franz; *Dir:* Lesley Selander. VHS, Beta, LV $19.95 *MED, IME* 𝄞𝄞

Flight of the Navigator A boy boards an alien spacecraft and embarks on a series of time-travel adventures with a crew of wisecracking extraterrestrial creatures. When he returns home eight years later, a NASA investigation ensues. Paul Reubens, better known as Pee Wee Herman, provides the voice of the robot.
1986 (PG) 89m/C Joey Cramer, Veronica Cartwright, Cliff De Young, Sarah Jessica Parker, Matt Adler, Howard Hesseman; *Dir:* Randal Kleiser. VHS, Beta, LV $29.95 *DIS* 𝄞𝄞

Flight to Nowhere An FBI agent tracks down a stolen map of atomic bomb source material with the help of a charter pilot and a dizzy blonde. Muddled plot and no discernable acting.
1946 74m/B Alan Curtis, Evelyn Ankers, Jack Holt; *Dir:* William Rowland. VHS, Beta, LV $16.95 *SNC, WGE* Woof!

The Flight of the Phoenix A group of men stranded in the Arabian desert after a plane crash attempt to rebuild their plane in order to escape before succumbing to the elements. Big budget, all-star survival drama based on the novel by Elleston Trevor. Bannen received an Oscar nomination, as did the film editing.
1966 147m/C James Stewart, Richard Attenborough, Peter Finch, Hardy Kruger, Dan Duryea, George Kennedy, Ernest Borgnine, Ian Bannen; *Dir:* Robert Aldrich. VHS, Beta, LV $19.98 *FOX* 𝄞𝄞𝄞

A Flight of Rainbirds A repressed cell biologist comes to believe unless he loses his virginity within a week he will die. Aiding him in his quest for a willing woman is his rakish alter ego. In Dutch with English subtitles.
1981 94m/C *NL* Jeroen Krabbe; *Dir:* Ate De Jong. VHS $79.95 *WOR* 𝄞𝄞½

A Flight of Rainbirds In a dual role, Krabbe plays a biologist who must lose his virginity within a week or die and his alter ego who joins him in the search for the perfect woman. In Dutch with English subtitles.
1981 94m/C *NL* Jeroen Krabbe; *Dir:* Ate De Jong. VHS $79.95 *WAC* 𝄞𝄞

Flight from Singapore Transporting desperately needed blood to Malaysia, a flight crew is forced to crash land in the jungle. Nicely done if trite.
1962 74m/B *GB* Patrick Allan, Patrick Holt, William Abney, Harry Fowler; *W/Dir:* Dudley Birch. VHS, Beta $16.95 *SNC* 𝄞½

Flight from Vienna A high-ranking Hungarian security officer, disenchanted with communism, stages a daring escape from his country. In Vienna, he asks the British for political asylum, but is sent back to Hungary to help a scientist escape in order to prove his loyalty. Cold war drama hangs together on strength of Bikel's performance.
1958 54m/B Theodore Bikel, John Bentley, Donald Gray. VHS, Beta $24.95 *VYY* 𝄞𝄞

Flights of Fancy The renowned animators, Faith and John Hubley, produced several whimsical short subjects through the '50s and '60s. This tape includes several, including: "The Adventure of an *," "Moonbird," "Windy Day" and "Zuckerkandl."
1986 45m/C Faith Hubley, John Hubley. Academy Awards '59: Best Animated Short Film ("Moonbird"). VHS, Beta $49.95 *DIS* 𝄞𝄞𝄞

Flights & Flyers Three Fox-Movietone newsreels covering stories about famous flyers such as Will Rogers, Amelia Earhart, Howard Hughes, Eddie Rickenbacker, and Wrong Way Corrigan, to name a few.
193? 30m/B VHS, Beta $29.98 *CCB* 𝄞𝄞𝄞

Flights & Flyers: Amelia Earhart The flying exploits and heroics of Amelia Earhart are chronicled in this program. The determination of a woman who made both cross-country and trans-Atlantic flights, and vowed to follow her failures with successes, is emphasized.
193? 11m/B VHS, Beta $24.98 *CCB* 𝄞𝄞𝄞

Flim-Flam Man A con man teams up with an army deserter to teach him the fine art of flim-flamming as they travel through small southern towns. Love may lead the young man back to the straight and narrow, but not his reprobate mentor. Scott is wonderful; and the slapstick episodes move at a good pace.
1967 104m/C George C. Scott, Michael Sarrazin, Slim Pickens, Sue Lyon, Jack Albertson, Harry Morgan; *Dir:* Irvin Kershner. VHS, Beta $59.98 *FOX* 𝄞𝄞

Flintstones The foibles of the two Stone Age families, the Flintstones and the Rubbles, are chronicled in these two animated episodes of the classic series.
1960 50m/C *Voices:* Alan Reed, Mel Blanc, Jean Vander Pyl, Bea Benadaret. VHS, Beta $29.95 *HNB* 𝄞𝄞

The Flintstones Meet Rockula & Frankenstone Fred, Wilma, Barney and Betty win a trip to Rocksylvania and meet up with Count Rockula and his monster, Frankenstone. "Flinstones' New Neighbors!" is also included.
1980 75m/C VHS, Beta, LV $29.95 *HNB, WOV* 𝄞½

Flipper A fisherman's son befriends an injured dolphin, is persuaded to return him to the wild, and earns the animal's gratitude. Prime kids' fare, as its sequels and television series attest.
1963 87m/C Chuck Connors, Luke Halpin, Kathleen Maguire, Connie Scott; *Dir:* James B. Clark. VHS, Beta $19.98 *MVD, MGM* 𝄞𝄞𝄞

Flipper's New Adventure Believing they are to be separated, Flipper and Sandy travel to a remote island. Little do they know, a British family is being held for ransom on the island they have chosen. It's up to the duo to save the day. An enjoyable, nicely done family adventure.
1964 103m/C Luke Halpin, Pamela Franklin, Tom Helmore, Francesca Annis, Brian Kelly, Joe Higgins, Ricou Browning; *Dir:* Leon Benson. VHS, Beta $19.98 *MGM, FCT* 𝄞𝄞𝄞

Flipper's Odyssey Flipper has disappeared and his adopted family goes looking for him, but when one of the boys gets trapped in a cave, Flipper is his only hope.
1966 77m/C Luke Halpin, Brian Kelly, Tommy Norden; *Dir:* Paul Landres. VHS, Beta $59.95 *PSM* 𝄞𝄞

Flirting with Fate Early Fairbanks-cum-acrobat vehicle. Having hired a hitman to rub himself out, a young man decides he doesn't want to die, after all. Seems there's a girl involved...
1916 51m/B Douglas Fairbanks Sr., Jewel Carmen, Howard Gaye, William E. Lawrence, George Beranger, Dorothy Hadel, Lillian Langdon; *Dir:* Christy Cabanne. VHS, Beta $24.95 *NOS, GPV, FCT* 𝄞𝄞𝄞

Floating Weeds A quiet drama by the master director about a traveling troupe of actors in ancient Japan who visit a remote island where the lead actor had fathered an illegitimate son years before. With English subtitles. Ozu also directed the original silent version ("A Story of Floating Weeds") in 1934. AKA "Drifting Weeds," it is Ozu's first color film.
1959 118m/C *JP* Ganjiro Nakamura, Haruko Sugimura, Machiko Kyo, Ayako Wakao; *Dir:* Yasujiro Ozu. VHS, Beta, LV $29.95 *VDA, MRV, VYG* 𝄞𝄞𝄞𝄞

A Flock of Seagulls This program presents the British band performing "Wishing (If I Had a Photograph of You)," "Nightmares" and "I Ran."
1983 13m/C A Flock of Seagulls. VHS, Beta $9.95 *SVS* 𝄞𝄞𝄞𝄞

Flood! Irwin Allen's first made-for-television disaster film. A dam bursts and devastates a small town, so a helicopter pilot must save the day. Good cast is swept along in a current of disaster-genre cliches.
1976 98m/C Robert Culp, Martin Milner, Barbara Hershey, Richard Basehart, Carol Lynley, Roddy McDowall, Cameron Mitchell, Teresa Wright, Francine York; *Dir:* Earl Bellamy. VHS, Beta $59.95 *WAR* 𝄞𝄞

The Floorwalker Chaplin becomes involved with a dishonest floorwalker in a department store. Silent.
1917 20m/B Charlie Chaplin; *Dir:* Charlie Chaplin. VHS, Beta *CAB, FST* 𝄞𝄞𝄞

The Floorwalker/By the Sea Two early Chaplin short comedies, featuring a hilarious scene on an escalator in the first film. Silent.
1915 51m/B Charlie Chaplin, Edna Purviance, Eric Campbell; *Dir:* Charlie Chaplin. VHS, Beta, 8mm $24.95 *VYY* 𝄞𝄞𝄞

Flor Silvestre (Wild Flower) The son of a rich rancher romances a young woman against the background of the Mexican Revolution.
1958 90m/B *MX* Dolores Del Rio, Pedro Armendariz; *Dir:* Emilio Fernandez. VHS, Beta $64.95 *MAD* 𝄞

Florence Chadwick: The Challenge The story of swimming champion Florence Chadwick, who swam the English Channel in both directions. Classic newsreel footage.
1954 15m/B GB VHS, Beta $49.95 TSF ✗

The Florida Connection An action thriller set in the Florida Swamps with a collection of villains and a vague plot about smuggling.
1974 90m/C Dan Pastorini, June Wilkinson. VHS, Beta $9.95 UNI ✗

Florida Straits A recently released Cuban prisoner hires a few losers and their boat to return to the island, supposedly to find the girl he loves. The real quest is for gold buried during the Bay-of-Pigs invasion in this none-too-original made-for-cable movie that at least offers a good cast.
1987 (PG-13) 98m/C Raul Julia, Fred Ward, Daniel H. Jenkins, Antonio Fargas; *Dir:* Mike Hodges. VHS, Beta, LV $19.98 ORI ✗

Flower Angel The Flower Angel and her friends, a white kitty and a lovable brown dog, are searching for the Flower of Seven Colors. In their travels they help a lonely old man and his beautiful daughter find the love they have for each other.
1980 46m/C VHS, Beta $29.95 FHE ✗✗

Flower Drum Song The Rodgero and Hammerstein musical played better on Broadway than in this overblown adaptation of life in San Francisco's Chinatown. Umeki plays the young girl who arrives from Hong Kong for an arranged marriage. Her intended (Soo) is a fast-living nightclub owner already enjoying the love of singer Kwan. Meanwhile Umeki falls for the handsome Shigeta. Naturally, everything comes right in a happy ending.
1961 133m/C Nancy Kwan, Jack Soo, James Shigeta, Miyoshi Umeki, Juanita Hall; *Dir:* Henry Koster. VHS, Beta, LV $39.98 MCA, FCT, PIA ✗✗

Flowers in the Attic Based on the V.C. Andrews bestseller, a would-be thriller about four young siblings locked for years in their family's old mansion by their grandmother with their mother's selfish acquiescence. A chicken-hearted, clumsy flop that skimps on the novel's trashier themes.
1987 (PG-13) 93m/C Victoria Tennant, Kristy Swanson, Louise Fletcher, Jeb Stuart Anderson; *Dir:* Jeffrey Bloom. VHS, Beta, LV $19.95 NWV, STE ✗ 1/2

The Flowers of St. Francis Rossellini's presentation of St. Francis and his friars' attainment of spiritual harmony. In Italian with English subtitles. This is the British release version, 10-15 minutes longer than the U.S. version.
1950 75m/B IT Federico Fellini; *Dir:* Roberto Rossellini. VHS, Beta $34.95 FCT ✗✗✗

Flowing Falls A continuous picture of a lush cove fed by a majestic waterfall to create a relaxed background.
1978 30m/C VHS, Beta PCC ✗✗✗

Flowing Sea A continuous picture of a rocky cove of cross-breaking waves rolling over a green coral reef to create a relaxed background.
1978 30m/C VHS, Beta PCC ✗✗✗

Flush Unorthodox comedy involving funny noises.
1981 90m/C William Calloway, William Bronder, Jeannie Linero; *Dir:* Andrew J. Kuehn. VHS TPV ✗✗

Flush It While working in the honey factory, Harvey finds a clue that leads him on a wild escapade in search of a buried treasure. Surprises abound along the way.
1976 (R) 90m/C VHS, Beta $59.95 JLT ✗✗

The Flustered Comedy of Leon Errol The popular Depression-era comedian Errol is showcased in three shorts: "Crime Rave," "Man-I-Cured," and "A Panic in the Parlor."
1939 56m/B Leon Errol, Vivian Tobin, Dorothy Granger, Frank Faylen, Virginia Vale. VHS, Beta, 8mm $24.95 VYY ✗✗

The Fly The historic, chillingly original '50s sci-fi tale about a hapless scientist experimenting with teleportation who accidentally gets anatomically confused with a housefly. Campy required viewing; two sequels followed, and a 1986 remake which itself has spawned one sequel.
1958 94m/C Vincent Price, David Hedison, Herbert Marshall, Patricia Owens; *Dir:* Kurt Neumann. VHS, Beta, LV $14.98 FOX ✗✗✗

The Fly A sensitive, humanistic remake of the 1958 horror film about a scientist whose flesh is genetically intermixed with a housefly via his experimental transportation device. A thoughtful, shocking horror film, with fine performances from Goldblum and Davis and a brutally emotional conclusion. Followed by "The Fly II" in 1989.
1986 (R) 96m/C Jeff Goldblum, Geena Davis, John Getz; *Dir:* David Cronenberg. Academy Awards '86: Best Makeup. VHS, Beta, LV $19.98 FOX ✗✗✗

The Fly 2 Inferior sequel to Cronenberg's opus, in which the offspring of Seth Brundle achieves full genius maturity in three years, falls in love, and discovers the evil truth behind his father's teleportation device and the corporate auspices that backed it.
1989 (R) 105m/C Eric Stoltz, Daphne Zuniga, Lee Richardson, John Getz, Harley Cross; *Dir:* Chris Walas. VHS, Beta, LV $19.95 FOX, CCB ✗ 1/2

Fly with the Hawk A troubled teenager gets lost in the wilderness for a year, and learns some important things from nature to take back to civilization.
19?? 90m/C Peter Ferri, Peter Snook, Shelley Lynne Speigel; *Dir:* Robert Tanos. VHS $49.99 NAM ✗✗

Flying Blind Foreign agents are thwarted in their attempt to steal a vital air defense secret. Unconvincing espionage plot cobbled into a story about a Los Angeles-Las Vegas puddle jumper.
1941 69m/B Richard Arlen, Jean Parker, Marie Wilson. VHS, Beta $24.95 NOS, DVT, TIM ✗ 1/2

The Flying Deuces Ollie's broken heart lands Laurel and Hardy in the Foreign Legion. The comic pair escape a firing squad only to suffer a plane crash that results in Hardy's reincarnation as a horse. A musical interlude with a Laurel soft shoe while Hardy sings "Shine On, Harvest Moon" is one of the film's highlights.
1939 65m/B Stan Laurel, Oliver Hardy, Jean Parker, Reginald Gardiner, James Finlayson; *Dir:* Edward Sutherland. VHS, LV $9.95 NEG, MRV, NOS ✗✗✗

Flying Down to Rio The first Astaire-Rogers musical, although they are relegated to supporting status behind Del Rio and Raymond. Still, it was enough to make them stars and a team that epitomizes the height of American musical films. The slim story revolves around singer Del Rio's two suitors and receives a splendid, art deco production, featuring Vincent Voumans' score, which includes "The Carioca." Showgirls dancing on plane wings in flight provide another memorable moment.
1933 89m/B Fred Astaire, Ginger Rogers, Dolores Del Rio, Eric Blore, Gene Raymond, Franklin Pangborn; *Dir:* Thornton Freeland. VHS, Beta $19.95 MED ✗✗ 1/2

Flying Fool Two brothers battle for the charms of a sexy singer with the worldly Boyd winning out.
1929 75m/B William Boyd, Marie Prevost, Russell Gleason; *Dir:* Tay Garnett. VHS, Beta RXM, VYY ✗ 1/2

Flying from the Hawk A 12-year-old boy sees something he shouldn't and soon finds that his friends and relatives are mysteriously vanishing.
1986 110m/C John Ireland, Diane McBain. VHS, Beta $19.95 NEG ✗ 1/2

Flying Karamazov Brothers Comedy Hour The wild and crazy Karamazov brothers show what they can do in this fun-filled video of comedy, acrobatics, and juggling.
1983 59m/C The Flying Karamazov Brothers. VHS, Beta $19.98 MCG ✗ 1/2

Flying Leathernecks Tough squadron leader Wayne fights with his fellow officer Ryan in Guadalcanal when their leadership styles clash. But when the real fighting begins all is forgotten as Wayne leads his men into victorious battle, winning the admiration and devotion of his fliers. Memorable World War II film deals with war in human terms.
1951 102m/C John Wayne, Robert Ryan, Janis Carter; *Dir:* Nicholas Ray. VHS, Beta, LV $19.98 RKO, VID, TLF ✗✗✗

Flying Machine/Coup de Grace/Interlopers Three film versions of short stories by Ray Bradbury, Ambrose Bierce, and Saki (H.H. Munro).
19?? 59m/C VHS, LV KRT ✗✗✗

The Flying Saucer U.S. and Russian scientists clash over their search for a huge flying saucer that is hidden under a glacier. The first movie to deal with flying saucers. The cassette includes animated opening and closing sequences plus previews of coming attractions.
1950 120m/B Mikel Conrad, Pat Garrison, Hanz von Teuffen; *Dir:* Mikel Conrad. VHS, Beta $14.95 RHI, MWP, UHV ✗ 1/2

The Flying Scotsman A fired railroad worker tries to wreck an express train on the engineer's last journey. The daughter of the intended victim saves the day. Silent, with sound added.
1929 60m/B Ray Milland, Pauline Johnson, Moore Marriot; *Dir:* Castleton Knight. VHS, Beta $24.95 NOS, DVT, TIM ✗✗

Flying Tigers Salutes the All-American Volunteer Group which flew for China under General Claire Chennault against the Japanese before the U.S. entered World War II. A squadron leader and a brash new recruit both vie for the affections of a pretty nurse in-between their flying missions. Romance and a few comic touches take a back seat to graphic scenes of aerial battles and dramatization of heroic sacrifice in this rousing war film.
1942 101m/B John Wayne, Paul Kelly, John Carroll, Anna Lee, Mae Clark, Gordon Jones; *Dir:* David Miller. VHS, LV $19.98 REP, BUR ✗✗ 1/2

Flying Wild A gang of saboteurs is out to steal top-secret airplane blueprints. Who else but the Bowery Boys could conceivably stop them? No Satch in this one.
1941 62m/B Leo Gorcey, Bobby Jordan, Donald Haines, Joan Barclay, David Gorcey, Bobby Stone, Sammy Morrison; **Dir:** William West. **VHS** $19.95 NOS, LOO, TIM 🎬½

FM The disc jockeys at an L.A. radio station rebel in the name of rock'n'roll. Despite the promising cast and setting (Mull makes his movie debut as a memorable space case), this is just disjointed and surprisingly unhip; one producer took his name off it due to creative difficulties. The decent soundtrack includes concert footage of Jimmy Buffet and Linda Ronstadt.
1978 (PG-13) 104m/C Eileen Brennan, Alex Karras, Cleavon Little, Martin Mull, Cassie Yates, Linda Ronstadt, Jimmy Buffett; **Dir:** Lamont Johnson. **VHS, LV** $79.95 MCA, PIA 🎬🎬

The Fog John Carpenter's blustery follow-up to his success with "Halloween." An evil fog containing murderous, vengeful ghosts envelops a sleepy seaside town and subjects the residents to terror and mayhem.
1978 (R) 91m/C Hal Holbrook, Adrienne Barbeau, Jamie Lee Curtis, Janet Leigh, John Houseman; **Dir:** John Carpenter. **VHS, Beta, LV, 8mm** $14.95 COL, SUE 🎬🎬🎬

Fog Island Murder and terror lurk after a greedy inventor, who was framed for fraud by his business partner, is released from prison and plots revenge by inviting his foes to his island home.
1945 72m/B George Zucco, Lionel Atwill, Terry Morse, Jerome Cowen; **Dir:** Terry Morse. **VHS, Beta** $19.95 NOS, SNC, VCN 🎬½

Foghorn Leghorn's Fractured Funnies Eight Leghorn classic shorts, including "Plop Goes the Weasel," "Feather Dusted," and "The Foghorn Leghorn." "I say...I say...boy, let go of my leg."
1952 58m/C **Dir:** Robert McKimson, Chuck Jones, Friz Freleng, Bob Clampett. **VHS, Beta** $12.95 WAR 🎬½

The Folk Music Reunion Several folk music stars, including Judy Collins, John Sebastian, Mary Travers, the Kingston Trio, Tom Paxton, and others perform their hits.
1988 80m/C Judy Collins, The Kingston Trio, Tom Paxton. **VHS, Beta** $19.95 JCI, VTK 🎬½

Folks Selleck is a successful Chicago stockbroker who finds his world crashing down on him. His wife and kids have left him, the FBI is after him, and worst of all, his parents have moved in with him. His parents don't want to be a burden, so Selleck decides that the best way to solve his financial woes is to help his parents commit suicide so he can collect on their insurance policies. The parents help too by driving their car into oncoming traffic. In the meantime, Selleck receives numerous injuries, including a black eye, a dog bite, broken fingers, a broken leg, and an amputated toe. This is a tasteless comedy that makes fun of aging and Alzheimer's Disease. Selleck is to sweet and cuddly for his role, but Ameche is good as the senile father, and Ebersole is great as Selleck's unpleasant sister.
1992 (PG-13) 109m/C Tom Selleck, Don Ameche, Anne Jackson, Christine Ebersole, Wendy Crewson, Robert Pastorelli, Michael Murphy, Kevin Timothy Chevalia, Margaret Murphy; **Dir:** Ted Kotcheff. **VHS** FXV 🎬

Follies in Concert A filmed record of the famed Stephen Sondheim musical play, performed at Lincoln Center in New York with an all-star cast. Songs include "I'm Still Here," "Losing My Mind," "Broadway Baby" and "The Ladies Who Lunch."
1985 90m/C Carol Burnett, Lee Remick, Betty Comden, Andre Gregory, Adolph Green, Mandy Patinkin, Phyllis Newman, Elaine Stritch, Jim Walton, Licia Albanese. **VHS, Beta, LV** $19.95 MVD, FRH 🎬

Follies Girl There's folly in expecting that this wartime tuner would hold up today. An army private romances a dress designer and a musical show somehow results.
1943 74m/B Wendy Barrie, Doris Nolan, Gordon Oliver, Anne Barrett, Arthur Pierson; **Dir:** William Rowland. **VHS** $15.95 NOS, LOO 🎬🎬

Follow the Fleet A song-and-dance man joins the Navy and meets two sisters in need of help in this Rogers/Astaire bon-bon. The excellent score by Irving Berlin includes the classic, "Let's Face the Music and Dance" as well as "Get Thee Behind Me, Satan" and "I'm Putting All My Eggs in One Basket." Look for Betty Grable, Lucille Ball, and Tony Martin in minor roles. Hilliard went on to be best known as the wife of Ozzie Nelson in TV's "The Adventures of Ozzie and Harriet."
1936 110m/B Fred Astaire, Ginger Rogers, Randolph Scott, Harriet Hilliard, Betty Grable, Lucille Ball; **Dir:** Mark Sandrich. **VHS, Beta, LV** $29.95 RKO, MED, TTC 🎬🎬🎬

Follow Me Round-the-world odyssey of three moon doggies searching for adventure—and the perfect wave. Spectacular surfing scenes are complemented by songs performed by Dino, Desi, and Billy.
1909 (G) 90m/C Claude Codgen, Mary Lou McGinnis, Bob Purvey; **Dir:** Gene McCabe. **VHS, 3/4U, Q, Special order formats** BEI 🎬½

Follow Me, Boys! A Disney film about a simple man who decides to put down roots and enjoy the quiet life, after one year too many on the road with a ramshackle jazz band. That life is soon interrupted when he volunteers to lead a high-spirited boy scout troop.
1966 120m/C Fred MacMurray, Vera Miles, Lillian Gish, Charlie Ruggles, Elliott Reid, Kurt Russell, Luanna Patten, Ken Murray; **Dir:** Norman Tokar. **VHS, Beta** $69.95 DIS 🎬🎬🎬

Follow Me Quietly Serial strangler who only kills in the rain is stalked by Lundigan. Very good little thriller which packs a punch in less than an hour.
1949 59m/B William Lundigan, Dorothy Patrick, Jeff Corey, Nestor Paiva, Charles D. Brown, Paul Guifoyle; **Dir:** Richard Fleischer. **VHS, Beta** $19.95 TTC, MLB, RXM 🎬🎬🎬

Follow That Camel A Foreign Legion sergeant who invents acts of heroism finally gets a chance to really help out a friend in need. Part of the "Carry On" series.
1967 95m/C GB Phil Silvers, Kenneth Williams, Anita Harris, Jim Dale; **Dir:** Gerald Thomas. **VHS, Beta** $29.95 VID 🎬½

Follow That Car Three southern kids become FBI agents and begin a thigh-slappin', rip-snortin' down-home escapade.
1980 (PG) 96m/C Dirk Benedict, Tanya Tucker, Teri Nunn; **Dir:** Daniel Haller. **VHS, Beta** $59.98 CHA 🎬

Follow That Dream Elvis plays a musical hillbilly whose family is trying to homestead on government land along the sunny Florida coast. Based on Richard C. Powell's novel "Pioneer Go Home." The songs-the only reason to see this movie—include "Angel", and "What a Wonderful Life", and the title track.
1961 111m/C Elvis Presley, Arthur O'Connell, Anne Helm, Simon Oakland, Jack Kruschen, Joanna Moore, Howard McNear; **Dir:** Gordon Douglas. **VHS, Beta** $19.95 MGM 🎬🎬

Follow That Rainbow Believing that her long-lost father is a popular touring singer, a young lass pursues him from Switzerland to South Africa.
1966 90m/C Joe Stewardson, Memory Jane, Joan Bickhill. **VHS, Beta** PGN 🎬

Food of the Gods On a secluded island giant rats, chickens, and other creatures crave human flesh, blood, bones, etc. This cheap, updated version of the H. G. Wells novel suffers from lousy performances and a lack of imagination.
1976 (PG) 88m/C Marjoe Gortner, Pamela Franklin, Ralph Meeker, Ida Lupino, Jon Cypher; **Dir:** Bert I. Gordon. **VHS, Beta** $19.98 LIV, VES Woof!

Food of the Gods: Part 2 The killer beasts and animals of the first film (and/ of H.G. Wells' classic novel) strike again; gigantic rats maim young girls. Easily as bad as the original.
1988 (R) 93m/C Paul Coufos, Lisa Schrage; **Dir:** Damian Lee. **VHS, Beta, LV** $9.99 CCB, IVE Woof!

The Fool Killer In the post-Civil War south, a 12 year-old boy learns that a terrifying legend about an axe-murderer may be all too real. Offbeat film with striking visuals and photography. Debut film for Albert and another choice psycho role for Perkins.
1965 100m/B Anthony Perkins, Edward Albert, Dana Elcar, Henry Hull, Salome Jens, Arnold Moss; **Dir:** Servando Gonzalez. **VHS** $19.98 REP 🎬🎬½

Fool for Love Explores the mysterious relationship between a modern-day drifter and his long-time lover, who may or may not be his half-sister, as they confront each other in a seedy New Mexico motel. Adapted by Shepard from his play; music by George Burt.
1986 (R) 108m/C Sam Shepard, Kim Basinger, Randy Quaid, Harry Dean Stanton; **Dir:** Robert Altman. **VHS, Beta** $79.95 MGM 🎬🎬

A Fool There Was The rocket that blasted Theda Bara to stardom and launched the vamp film genre. One of the few extant Bara films, it tells the now familiar story of a good man whom crumbles from the heights of moral rectitude thanks to the inescapable influence of an unredeemable vamp. Bette Davis described Bara as "divinely, hysterically, insanely malevolent." Much heavy emoting; subtitled (notoriously) "Kiss Me, My Fool!" Based on Rudyard Kipling's "The Vampire."
1914 70m/B Theda Bara, Edward Jose, Runa Hodges, Clifford Bruce; **Dir:** Frank Powell. **VHS, Beta** $27.95 GPV, FCT, DNB 🎬🎬½

Foolin' Around An innocent Oklahoma farm boy arrives at college and falls in love with a beautiful heiress. He will stop at nothing to win her over, including crashing her lavish wedding ceremony.
1980 (PG) 101m/C Gary Busey, Annette O'Toole, Eddie Albert, Tony Randall, Cloris Leachman; **Dir:** Richard T. Heffron. **VHS, Beta** $24.98 SUE 🎬🎬½

Foolish Wives A reconstruction of Von Stroheim's classic depicting the confused milieu of post-war Europe as reflected through the actions of a bogus count and his seductive, corrupt ways. Comes as close as possible to the original film.
1922 107m/B Erich von Stroheim, Mae Busch, Maude George, Cesare Gravina; *Dir:* Erich von Stroheim. VHS, Beta, LV $29.95 *KIV, CCB, HHT* ✍✍✍½

Fools Two lonely people—he an aging horror film actor and she a young woman estranged from her husband—start a warm romance when they meet in San Francisco. The husband reacts violently in a film whose biggest surprise is how awful it is.
1970 (PG) 93m/C Jason Robards Jr., Katharine Ross, Scott Hylands; *Dir:* Tom Gries. VHS, Beta $59.95 *PSM* ✍

Fools of Fortune Adaptation of William Trevor's novel depicting Willie Clinton's childhood and adult experiences of family dramatics played against the backdrop of post-WWI Ireland. During the Irish war for independence, a family is attacked by British soldiers, creating emotional havoc for the survivors. Though well-acted and poignant, it's a rather disjointed and straying shamrock opera.
1991 (PG-13) 104m/C Mary Elizabeth Mastrantonio, Julie Christie, Iain Glen, Michael Kitchen; *Dir:* Pat O'Connor. VHS, LV $89.95 *COL, PIA, FCT* ✍½

Footlight Frenzy The Law Moan Spectacular comedy troupe are at it again as they perform a benefit play where everything goes wrong.
1984 110m/C *GB* Alan Shearman, Diz White, Ron House, Frances Tomelty, Michael Aldridge, Ron Pember. VHS, Beta $39.95 *RKO* ✍✍

Footlight Parade Broadway producer Cagney is out of work. Sound films have scared off his backers until his idea for staging live musical numbers before the cinema features lures them back. Lots of authentic backstage action precedes three spectacular Busby Berkeley-choreographed numbers that climax the film. "By a Waterfall" uses more than 100 girls in a giant water ballet; Cagney himself sings and dances delightfully in the lowdown "Shanghai Lil."
1933 104m/B James Cagney, Joan Blondell, Dick Powell, Ruby Keeler, Guy Kibbee, Ruth Donnelly; *Dir:* Lloyd Bacon. Beta, LV, Q $29.95 *MGM, FOX* ✍✍✍

Footlight Serenade A boxer falls for a beautiful dancer with whom he's costarring in a Broadway play. Unfortunately, she's secretly married to another of the actors and her husband is becoming jealous of the boxer's intentions. Light, fun musical. One of a series of Grable movies made to boost morale during World War II.
1942 80m/B Betty Grable, Victor Mature, John Payne, Jane Wyman, Phil Silvers, James Gleason, Mantan Moreland; *Dir:* Gregory Ratoff. VHS, Beta $19.98 *FOX* ✍✍½

Footloose When a city boy moves to a small Midwestern town, he discovers some disappointing news: rock music and dancing have been forbidden by the local government. Determined to bring some '80s-style life into the town, he sets about changing the rules and eventually enlists the help of the daughter of the man responsible for the law. Rousing music, talented young cast, and plenty of trouble make this an entertaining musical-drama.

1984 (PG) 107m/C Kevin Bacon, Lori Singer, Christopher Penn, John Lithgow, Dianne Wiest, John Laughlin, Sarah Jessica Parker; *Dir:* Herbert Ross. VHS, Beta, LV, 8mm $14.95 *PAR* ✍✍½

For All Mankind A collage of original footage and interviews with the 24 astronauts who were the first humans to land on the moon, spliced together as one mission. The laser disc edition (CLV) includes 500 still pictures from NASA's archives, images of international media coverage of the moon mission from around the world, and an hour-long talk by Apollo 12 astronaut Alan Bean. Interactive laser edition (CAV) available.
1989 90m/C VHS, LV *VYG, CRC* ✍✍½

For Auld Lang Syne: The Marx Brothers Four classic comedy shorts: "Monkey Business Special," "Gai Dimanche," "Hooks and Jabs," and "For Auld Lang Syne."
1932 46m/B Chico Marx, Groucho Marx, Alan Marx, Harpo Marx. VHS, Beta *VYY* ✍✍½

For the Boys Midler stars as Dixie Leonard, a gutsy singer-comedian who hooks up with Eddie Sparks (Caan) to become one of America's favorite USO singing, dancing and comedy teams. The movie spans 50 years and three wars—including Korea and Vietnam—and raises such issues as the blacklist and the role of politics in showbiz. Glitzy Hollywood entertainment on a grand scale. Midler's performance earned her an Oscar nomination for Best Actress.
1991 (R) 120m/C Bette Midler, James Caan, George Segal; *Dir:* Mark Rydell. Golden Globe Awards '92: Best Actress (Midler). VHS $94.98 *FOX* ✍✍

For a Few Dollars More The Man With No Name returns as a bounty hunter who teams up with a gunslinger/rival to track down the sadistic leader of a gang of bandits. Violent. Sequel to "A Fistful of Dollars" (1964) and followed by "The Good, The Bad, and The Ugly". Letterboxed laserdisc edition features the original movie trailer.
1965 (PG) 127m/C Clint Eastwood, Lee Van Cleef, Klaus Kinski, Gian Marie Volonte; *Dir:* Sergio Leone. VHS, Beta, LV $19.95 *FOX, MGM, PIA* ✍✍½

For Heaven's Sake Lloyd's first film for Paramount has him making an accidental donation to a skid row mission, then marrying the preacher's daughter and converting all the neighborhood tough guys.
1926 60m/B Harold Lloyd, Jobyna Ralston, Noah Young, James Mason, Paul Weigel; *Dir:* Sam Taylor. VHS $49.95 *EJB* ✍✍✍

For Heaven's Sake A bumbling basketball team is lent some heavenly assistance in the form of a meddling angel. But can he hit the jumper?
1979 90m/C Ray Bolger, Kent McCord. VHS $49.99 *IVE* ✍½

For Keeps Two high school sweethearts on the verge of graduating get married after the girl becomes pregnant, and suffer all the trials of teenage parenthood. Written by Tim Kazurinsky and Denise DeClure.
1988 (PG-13) 98m/C Molly Ringwald, Randall Batinkoff, Kenneth Mars; *Dir:* John G. Avildsen. VHS, Beta, LV $14.95 *COL* ✍½

For Ladies Only A struggling actor takes a job as a male stripper to pay the rent. Among the leering ladies is Patricia Davis, daughter of former President Reagan. Made for television.

1981 100m/C Gregory Harrison, Patricia Davis, Dinah Manoff, Louise Lasser, Lee Grant, Marc Singer, Viveca Lindfors, Steven Keats; *Dir:* Mel Damski. VHS, Beta $49.95 *USA* ✍✍

For Love Alone A young Australian college co-ed in the 1930s falls in love with a handsome, controversial teacher and follows him to London, only to discover he's not her Mr. Right after all.
1986 102m/C *AU* Helen Buday, Sam Neill, Hugo Weaving; *Dir:* Stephen Wallace. VHS, Beta $69.95 *IVE* ✍✍

For the Love of Angela A pretty young shop clerk involves herself with both the shopkeeper and his son.
1986 90m/C Louis Jourdan. Beta $14.95 *PSM* ✍½

For the Love of Benji In this second "Benji" film, the adorable little dog accompanies his human family on a Greek vacation. He is kidnapped to be used as a messenger for a secret code, but escapes, to have a series of comic adventures in this entertaining family fare.
1977 (G) 85m/C Benji, Patsy Garrett, Cynthia Smith, Allen Finzat, Ed Nelson; *Dir:* Joe Camp. VHS, Beta $19.99 *CVL, FCT, APD* ✍½

For the Love of It Television stars galore try to hold together this farce about car chases in California, stolen secret documents, and, (of course) true love. Outstandingly mediocre.
1980 100m/C Deborah Raffin, Jeff Conaway, Tom Bosley, Norman Fell, Don Rickles, Henry Gibson, Pat Morita, William Christopher, Lawrence Hilton-Jacobs, Adrian Zmed, Barbi Benton, Adam West; *Dir:* Hal Kanter. VHS, Beta $14.95 *FRH* ✍½

For Love of Ivy Poitier is a trucking executive who has a gambling operation on the side. Ivy is the Black maid of a rich white family who is about to leave her job to look for romance. The two are brought together but the road to true love doesn't run smooth. Based on a story by Poitier.
1968 (PG) 102m/C Sidney Poitier, Abbey Lincoln, Beau Bridges, Carroll O'Connor, Nan Martin; *Dir:* Daniel Mann. VHS, Beta $19.98 *FOX* ✍✍

For Love or Money A super-rich hotel owner wants his lawyer to find her three beautiful daughters. However, she doesn't trust him enough not to meddle in the search. Lavish production and a good cast can't overcome the mediocrity of the script and direction.
1963 108m/C Kirk Douglas, Mitzi Gaynor, Thelma Ritter, William Bendix, Julie Newmar, Gig Young, Leslie Parrish, William Windom, Richard Sargent; *Dir:* Michael Gordon. VHS $47.97 *NSV* ✍✍

For Love or Money Made-for-television fare about game-show contestants finding money and romance.
1984 91m/C Jamie Farr, Suzanne Pleshette, Gil Gerard, Ray Walston, Lawrence Pressman, Mary Kay Place; *Dir:* Terry Hughes. VHS, Beta, 3/4U $39.95 *NSV, IME* ✍½

For Love or Money A real estate man with Career on the mind has to choose between the woman in his life and the condo development that stands firmly between them and happily-ever-afterdom.
1988 ?m/C Timothy Daly, Haviland Morris, Kevin McCarthy. VHS $89.99 *HBO* ✍✍

For Me & My Gal In his film debut, Gene Kelly plays an opportunistic song-and-dance man who lures a young vaudevillian (Garland) away from her current partners. World War I interrupts both their career and romance, but you can count on them being reunited. It's fun to watch the talented stars perform such vintage tunes as "Oh You Beautiful Doll," "Pack Up Your Troubles," "How Ya Gonna Keep 'Em Down on the Farm," "After You've Gone," and, of course, the title tune.
1942 104m/B Gene Kelly, Judy Garland, George Murphy, Martha Eggerth, Ben Blue, Richard Quine, Keenan Wynn, Stephen McNally; *Dir:* Busby Berkeley. **VHS, Beta, LV** $19.95 *MGM* ♫♫½

For Pete's Sake Topsy-turvy comedy about a woman whose efforts to get together money to put her husband through school involve her with loan sharks, a madame, and even cattle rustling in New York City. Not one of Streisand's best.
1974 (PG) 90m/C Barbra Streisand, Michael Sarrazin, Estelle Parsons, William Redfield, Molly Picon; *Dir:* Peter Yates. **VHS, Beta** $9.95 *COL* ♫½

For Richer, For Poorer A very rich, successful businessman notices his son is happy just to sit back and let dad earn all the dough while he waits for his share. In order to teach his son a lesson, Dad decides to give all his money away. But the plan backfires when both realize that earning a second fortune may not be as easy as they assumed.
1992 (PG) 90m/C Jack Lemmon, Jonathan Silverman, Talia Shire, Joanna Gleason, Madeline Kahn; *Dir:* Jay Sandrich. **VHS** $89.99 *HBO* ♫♫

For Us, the Living The life and assassination of civil rights activist Medgar Evers are dramatically presented in the production of "American Playhouse" for PBS. Insight into the man is given, not just a recording of the events surrounding his life. Adapted from the biography written by Evers' widow.
1988 84m/C Howard E. Rollins Jr., Rocky Aoki, Paul Winfield, Irene Cara; *Dir:* Michael Schultz. **VHS, Beta** $79.95 *MED, KUI* ♫♫♫

For Your Eyes Only In this James Bond adventure, 007 must keep the Soviets from getting hold of a valuable instrument aboard a sunken British spy ship. Sheds the gadgetry of its more recent predecessors in the series in favor of some spectacular stunt work and the usual beautiful girl and exotic locale. Glen's first outing as director, though he handled second units on previous Bond films. Sheena Easton sang the hit title tune.
1981 (PG) 127m/C *GB* Roger Moore, Carol Bouquet, Chaim Topol, Lynn-Holly Johnson, Julian Glover, Cassandra Harris, Jill Bennett, Michael Gothard, John Wyman, Jack Hedley, Lois Maxwell, Desmond Llewelyn, Geoffrey Keen, Walter Gotell, Charles Dance; *Dir:* John Glen. **VHS, Beta, LV** $19.95 *MGM, FOX, TLF* ♫♫♫

For Your Love Only A beautiful young student falls in love with her teacher and is blackmailed, leading her to murder her tormentor. Soap opera was made for German television. Dubbed in English.
1979 90m/C *GE* Nastassia Kinski, Christian Quadflieg, Judy Winter, Klaus Schwarzkopf; *Dir:* Wolfgang Petersen. **VHS, Beta** $69.00 *PGN, GLV* ♫♫

Forbidden In Nazi Germany, a German countess has an affair with a Jewish intellectual, and she winds up hiding him from the S.S. Slow made-for-cable movie.

1985 116m/C Jacqueline Bisset, Juergen Prochnow, Irene Worth, Peter Vaughan, Amanda Cannings, Avis Bunnage; *Dir:* Anthony Page. **VHS, Beta** $79.95 *IVE* ♫♫½

The Forbidden Dance The first of several quickies released in 1990 applying hackneyed plots to the short-lived Lambada dance craze. The nonsensible plot has a Brazilian princess coming to the U.S. in order to stop further destruction of the rain forest. Instead, she winds up falling for a guy, teaching him to Lambada, and going on TV for a dance contest. Features an appearance by Kid Creole and the Coconuts.
1990 (PG-13) 90m/C Laura Herring, Jeff James, Sid Haig, Richard Lynch, Kid Creole & the Coconuts; *Dir:* Greydon Clark. **VHS, Beta, LV** $19.95 *COL* Woof!

Forbidden Games The famous anti-war drama about two French children play-acting the dramas of war amid the carnage of World War II. Fossey is a young refugee who sees her parents and dog killed. She meets a slightly older boy whose family takes the girl in. The children decide to bury the animals they have seen killed in the same way that people are buried - even stealing the crosses from the cemetery to use over the animal graves. Eventually they are discovered and Fossey is again separated from her new-found home. Acclaimed; available in both dubbed and English-subtitled versions. French title: "Jeux Interdits."
1952 90m/B *FR* Brigitte Fossey, Georges Poujouly, Amedee; *Dir:* Rene Clement. Academy Awards '52: Best Foreign Language Film; British Academy Awards '53: Best Film; National Board of Review Awards '52: 5 Best Foreign Films of the Year; New York Film Critics Awards '52: Best Foreign Film; Venice Film Festival '52: Best Film. **VHS, Beta, LV** $29.98 *NOS, SUE, APD* ♫♫♫♫

Forbidden Impulse A group of teenage girls set out to seduce older men. Softcore.
1985 80m/C Denise Sexton, Crystal Tilley, Martin Hendler. **VHS, Beta** $29.95 *MED* ♫

Forbidden Love A man in his early 20's falls in love with a woman twice his age, much to the chagrin of her daughters. Made for television.
1982 96m/C Andrew Stevens, Yvette Mimieux, Dana Elcar, Lisa Lucas, Jerry Hauser, Randi Brooks, Lynn Carlin, Hildy Brooks, John Considine; *Dir:* Steven Hilliard Stern. **VHS, Beta** *NEG* ♫

Forbidden Passion: The Oscar Wilde Movie A British-made look at Wilde's later years, emphasizing his homosexuality and the Victorian repression that accompanied it.
1987 120m/C *GB* Michael Gambon, Robin Lermitte. **VHS, Beta** $79.95 *VMK* ♫♫½

Forbidden Planet In 2200 A.D., a space cruiser visits the planet Altair-4 to uncover the fate of a previous mission of space colonists. They are greeted by Robby the Robot and discover the only survivors of the Earth colony which has been preyed upon by a terrible space monster. A classic science fiction version of the Shakespearean classic "The Tempest." Laserdisc features formerly "cut" scenes, production and publicity photos, the original screen treatment, and special effects outtakes.
1956 98m/C Walter Pidgeon, Anne Francis, Leslie Nielsen, Warren Stevens, Jack Kelly, Richard Anderson, Earl Holliman, George Wallace; *Dir:* Fred M. Wilcox. **VHS, Beta, LV** $19.95 *MGM, PIA, VYG* ♫♫♫½

Forbidden Sun An Olympic gymnastics coach and her dozen beautiful students go to Crete to train. When one of them is brutally raped, vengeance is meted out by the girls.
1989 (R) 88m/C Lauren Hutton, Cliff DeYoung, Renee Estevez; *Dir:* Zelda Barron. **VHS, Beta, LV** $14.95 *ACA* ♫

Forbidden Trails A Rough Riders adventure set in the old West shows Silver the horse saving his owner from the vengeance of two outlaws. Two more heroes go under cover to bring them to justice.
1941 60m/B Buck Jones, Tim McCoy, Raymond Hatton; *Dir:* Robert N. Bradbury. **VHS, Beta** $9.95 *NOS, VCN, DVT* ♫

Forbidden World The lives of a genetic research team become threatened by the very life form they helped to create: a man-eating organism capable of changing its genetic structure as it grows and matures. A graphically violent rip-off of "Alien."
1982 (R) 82m/C Jesse Vint, Dawn Dunlap, June Chadwick, Linden Chiles; *Dir:* Allan Holzman. **VHS, Beta, LV** $69.98 *SUE* ♫½

Forbidden Zone A sixth dimension kingdom is ruled by the midget, King Fausto, and inhabited by dancing frogs, bikini-clad tootsies, robot boxers and degraded beings of all kinds. Original music by Oingo Boingo, and directed by founding member Elfman.
1980 (R) 75m/B Herve Villechaize, Susan Tyrrell, Kipper Kids, Viva; *Dir:* Richard Elfman. **VHS, Beta** $39.95 *MED* ♫½

Force of Evil A cynical attorney who works for a mob boss and for Wall Street tries to save his brother from the gangster's takeover of the numbers operation. The honorable, though criminal, brother refuses the help of the amoral lawyer, and he is finally forced to confront his conscience. Garfield's sizzling performance and the atmospheric photography have made this a film noir classic.
1949 80m/B John Garfield, Thomas Gomez, Marie Windsor, Sheldon Leonard, Ray Roberts; *Dir:* Abraham Polonsky. **VHS, LV** $19.98 *REP* ♫♫♫

Force of Evil A paroled murderer returns to his hometown to brutalize the family that refused to fabricate an alibi for him. Made for television.
1977 100m/C Lloyd Bridges, Eve Plumb, William Watson, Pat Crowley; *Dir:* Richard Lang. **VHS, Beta, LV** $49.95 *WOV, GKK* ♫

Force Five A group of ex-cons form an anti-crime undercover force turning their skills toward justice. Nothing new in this made for television pilot.
1975 78m/C Gerald Gordon, Nicholas Pryor, James Hampton, David Spielberg, Leif Erickson, Bradford Dillman; *Dir:* Walter Grauman. **VHS, Beta** $54.95 *MED* ♫♫

Force: Five A mercenary gathers a group of like-minded action groupies together to rescue the daughter of a powerful man from a religious cult. All action and no brains.
1981 (R) 95m/C Joe Lewis, Pam Huntington, Master Bong Soo Han; *Dir:* Robert Clouse. **VHS, Beta** $54.95 *MED* ♫

Force of One A team of undercover narcotics agents is being eliminated mysteriously, and karate expert Norris saves the day in this sequel to "Good Guys Wear Black."
1979 (PG) 91m/C Chuck Norris, Bill Wallace, Jennifer O'Neill, Clu Gulager; *Dir:* Paul Aaron. **VHS, Beta** $54.95 *MED* ♫♫

Force 10 from Navarone So-so sequel to Alistair MacLean's ''The Guns of Navarone,'' follows a group of saboteurs whose aim is to blow up a bridge vital to the Nazi's in Yugoslavia. Keep an eye out for Ford, Nero, and Bach. Lots of double-crosses and action sequences, but doesn't quite hang together.
1978 (PG) 118m/C Robert Shaw, Harrison Ford, Barbara Bach, Edward Fox, Carl Weathers, Richard Kiel, Franco Nero; *Dir:* Guy Hamilton. **VHS, Beta, LV $19.98** *WAR, OM ♂♂*

Force on Thunder Mountain A father and son go camping and encounter ancient Indian lore and flying saucers.
1977 93m/C Christopher Cain, Todd Dutson. **VHS, Beta $19.95** *UHV ♂*

Forced Entry A psychopathic killer-rapist hesitates over one of his victims, and she murders him instead. Originally titled ''The Last Victim.''
1975 (R) 92m/C Tanya Roberts, Ron Max, Nancy Allen; *Dir:* Jim Sotos. **VHS, Beta $59.95** *HAR ♂*

Forced March On location in Hungary, an actor portrays a poet who fought in World War II and became a victim of the Holocaust. However, the deeper he gets into his role, the thinner the line between illusion and reality becomes.
1990 104m/C Chris Sarandon, Renee Soutendijk, Josef Sommer, John Seitz; *Dir:* Rick King. **VHS, Beta, LV $89.99** *SGE, MOV, IME ♂♂½*

Forced Vengeance A Vietnam vet pits himself against the underworld in Hong Kong. With Norris in the lead, you can take it for granted there will be plenty of martial arts action, however, there is little else.
1982 (R) 103m/C Chuck Norris, Mary Louise Weller; *Dir:* James Fargo. **VHS, Beta $19.95** *MGM ♂♂½*

Ford Startime An original television drama, ''The Man'' is a psychological study of an unbalanced veteran who visits the mother of an old army buddy and proves impossible to get rid of.
1960 50m/B Audie Murphy, Thelma Ritter. **VHS, Beta $24.95** *VYY Woof!*

Ford: The Man & the Machine An episodic biography of ruthless auto magnate Henry Ford I from the building of his empire to his personal tragedies. The cast doesn't have any spark and Robertson (Ford) is positively gloomy. The only one appearing to have any fun is Thomas (Ford's mistress). Based on the biography by Robert Lacey.
1987 200m/C Cliff Robertson, Hope Lange, Heather Thomas, Michael Ironside, Chris Wiggins, R.H. Thomson; *Dir:* Allan Eastman. **VHS $79.95** *CAF ♂♂½*

Foreign Body An Indian (played by ''Passage to India'' star, Banerjee) visiting London pretends to be a doctor and finds women flocking to him. Excellent overlooked British comedy.
1986 (PG-13) 108m/C *GB* Victor Banerjee, Warren Mitchell, Trevor Howard, Geraldine McEwan, Amanda Donohoe, Denis Quilley, Eve Ferret, Anna Massey; *Dir:* Ronald Neame. **VHS $19.99** *HBO ♂♂½*

Foreign Correspondent A classic Hitchcock tale of espionage and derring-do. A reporter is sent to Europe during World War II to cover a pacifist conference in London, where he becomes romantically involved with the daughter of the group's founder and befriends an elderly diplomat. When the diplomat is kidnapped, the reporter uncovers a Nazi spy-ring headed by his future father-in-law. Oscar nominations: Best Picture, Best Supporting Actor (Basserman), Original Screenplay, Black & White Cinematography, Black & White Interior Decoration, Special Effects.
1940 120m/B Joel McCrea, Laraine Day, Herbert Marshall, George Sanders, Robert Benchley, Albert Basserman; *Dir:* Alfred Hitchcock. National Board of Review Awards '40: 10 Best Films of the Year. **VHS, Beta, LV $59.95** *LTG, WAR, PIA ♂♂♂♂*

Foreign Legionnaire: Court Martial An episode from the television serial about the adventures of Captain Michael Gallant and his ward, Cuffy Sanders, in the French Foreign Legion of North Africa.
1957 30m/B Buster Crabbe, Cullen Crabbe, Al ''Fuzzy'' Knight. **VHS, Beta $29.95** *VCN ♂♂♂♂*

The Foreigner A secret agent who comes to New York City to meet his contact becomes entrapped in a series of mysterious events revolving around the underground club scene.
1978 90m/B Eric Mitchell, Patti Astor, Deborah Harry; *Dir:* Amos Poe. **VHS, Beta $59.95** *NEW, FCT ♂*

The Foreman Went to France An industrial engineer travels to France during World War II to help prevent secret machinery from falling into the hands of the Nazis and their allies. Also known as ''Somewhere in France''.
1942 88m/B *GB* Clifford Evans, Constance Cummings, Robert Morley; *Dir:* Charles Frend. **VHS, Beta $24.95** *NOS, MRV, DVT ♂½*

Foreplay A trilogy of comedy segments involving characters with White House connections. Also known as ''The President's Women,'' since the stories are introduced by a former President (Mostel) discussing his downfall in a television interview. Lame rather than risque and wastes the talents of all involved. Each segment was scripted and directed by a different team.
1975 100m/C Pat Paulsen, Jerry Orbach, Estelle Parsons, Zero Mostel; *Dir:* Bruce Malmuth, John G. Avildsen. **VHS, Beta $59.98** *LIV, VES Woof!*

The Forest Spooks and a cannibalistic killer terrrorize a group of campers. Nothing that hasn't been done countless times before.
1983 90m/C *Dir:* Don Jones. **VHS, Beta $59.95** *STE, PSM ♂*

Forest Duel Revenge is sweet in this rural Kung Fu epic.
1985 83m/C **VHS, Beta $14.95** *SVS ♂*

Forest of Little Bear A man returns to his impoverished family in 1928 and decides to seek the reward put on the head of a one-eared man-eating bear. He kills the bear but faces a dilemma when he sees the bear left a helpless cub behind. In Japanese with English subtitles.
1979 124m/C *JP Dir:* Toshio Goto. **VHS $59.95** *FCT ♂♂♂*

Forever A teenage girl experiences true love for the first time and struggles with its meaning. Made for television adaptation of Judy Blume's' novel.
1978 100m/C Stephanie Zimbalist, Dean Butler, John Friedrich, Beth Raines, Diana Scarwid; *Dir:* John Korty. **VHS $59.95** *GEM ♂♂½*

Forever Darling Mixed effort from the reliable comedy duo sees Desi playing a dedicated chemist who neglects his wife while pursuing the next great pesticide. Lucy calls on her guardian angel (Mason) to help her rekindle her marriage. He advises her to go with Desi when he tests his new bug killer, and a series of hilarious, woodsy calamities occur. Lucy and Desi are always fun to watch, but '60s TV sitcom fans will enjoy the fact that Mrs. Howell (Schaefer) and Jane Hathaway (Kulp) appear in the same picture.
1956 91m/C Lucille Ball, Desi Arnaz Sr., James Mason, Louis Calhern, John Emery, John Hoyt, Natalie Schafer, Nancy Kulp; *Dir:* Alexander Hall. **VHS, Beta $19.98** *MGM, CCB ♂♂*

Forever and a Day Tremendous salute to British history centers around a London manor originally built by an English admiral during the Napoleonic era and the exploits of succeeding generations. The house even manages to survive the blitz of World War II showing English courage during wartime. Once-in-a-lifetime casting and directing.
1943 104m/B Brian Aherne, Robert Cummings, Ida Lupino, Charles Laughton, Herbert Marshall, Ray Milland, Anna Neagle, Merle Oberon, Claude Rains, Victor McLaglen, Buster Keaton, Jessie Matthews, Roland Young, C. Aubrey Smith, Edward Everett Horton, Elsa Lanchester, Edmund Gwenn; *Dir:* Rene Clair, Edmund Goulding, Cedric Hardwicke, Frank Lloyd, Victor Saville, Robert Stevenson, Herbert Wilcox, Kent Smith. **VHS $19.95** *HTV, MLB ♂♂♂♂*

Forever Emmanuelle A sensual young woman finds love and the ultimate erotic experience in the wilds of the South Pacific in this sequel to the porn-with-production-values ''Emmanuelle.''
1982 (R) 89m/C Annie-Belle, Emmanuelle Arsan, Al Cliver; **VHS, Beta, LV $29.98** *LIV, VES ♂*

Forever Evil The vacationing denizens of a secluded cabin are almost killed off by the cult followers of a mythic god.
1987 107m/C Red Mitchell, Tracey Huffman, Charles Trotter, Howard Jacobsen, Kent Johnson. **VHS, Beta $79.95** *UHV ♂*

Forever James Dean A profile of the legendary actor/cult hero, examining his life, short career and remaining legacy. Also includes the original previews of his three feature films. A must for serious Dean fans.
1988 69m/C James Dean, Julie Harris, William Bast, Kenneth Kendall; *Dir:* Ara Chekmayan. **VHS, Beta $29.98** *WAR, FCT ♂*

Forever, Lulu A down-on-her-luck novelist winds up involved with the mob and a gangster's girlfriend. Completely laughless comedy with amateurish direction.
1987 (R) 86m/C Hanna Schygulla, Deborah Harry, Alec Baldwin, Annie Golden, Paul Gleason, Dr. Ruth Westheimer; *Dir:* Amos Kollek. **VHS, Beta, LV $79.95** *COL ♂*

Forever Mary A teacher tries to better the lives of the boys sentenced to a reformatory in Palermo, Sicily, when a teenage transvestite prostitute (the title character) is admitted to the school. In Italian with English subtitles.
1991 100m/C *IT* Michele Placido, Alesandro DiSanzo, Francesco Benigno; *Dir:* Marco Risi. **VHS $79.95** *CCN ♂♂♂*

Forever Young A young boy immerses himself in Catholicism. Admiring a priest, he is unaware that the priest's friend is involved with his mother. Tangled, but delicate handling of the coming-of-age story. Made originally for British television.
1985 85m/C *GB* James Aubrey, Nicholas Gecks, Alec McCowen; *Dir:* David Drury. **VHS, Beta $59.95** *MGM ♂♂*

Forget Mozart On the day of Mozart's death, the head of the secret police gathers a group together for questioning. They include the composer's wife, his lyricist, the head of a Masonic lodge, and Salieri, Mozart's rival. Each has a different story to tell to the investigator about the composer's life and death. The film makes use of a number of sets and costumes from "Amadeus" and was also filmed on location in Prague. In German with English subtitles.
1986 93m/C *GE* Armin Mueller-Stahl, Catarina Raacke; ***Dir:*** Salvo Luther. **VHS $79.95** *WBF* 🦴🦴

The Forgotten After 17 years as POWs, six Green Berets are freed, only to face a treacherous government conspiracy aimed at eliminating them. Produced and written by Carradine, Railsback and James Keach. Made for television.
1989 96m/C Keith Carradine, Steve Railsback, Stacy Keach, Pepe Serna, Don Opper, Richard Lawson; ***Dir:*** James Keach. **VHS, Beta, LV $59.95** *PAR* 🦴🦴

The Forgotten One An uninspired author and a female reporter become mixed up in a century-old murder and the restless ghost it spawned.
1989 (R) 89m/C Kristy McNichol, Terry O'Quinn, Blair Parker, Elisabeth Brooks; ***Dir:*** Phillip Badger. **VHS, Beta $14.95** *ACA* 🦴½

Forgotten Prisoners Celluloid proof that conviction and true grit do not always entertainment make. Greenwald—whose credits range from "Xanadu" to "The Burning Bed"—directed this righteous but stationary drama about an Amnesty International official (Silver) who's assigned to look into the incarceration and egregious mistreatment of Turkish citizens. Made for cable and also known as "Forgotten Prisoners: The Amnesty Files."
1990 (R) 92m/C Ron Silver, Roger Daltrey, Hector Elizondo; ***Dir:*** Robert Greenwald. **VHS, Beta $79.98** *TTC, FCT* 🦴🦴

A Forgotten Tune for the Flute A delightful, Glasnost-era romantic comedy. A bureaucrat caught up in a dull, stuffy existence has a heart attack and winds up falling in love with his nurse. He then must decide between his comfortable, cared-for life or the love which beckons him. In Russian with English subtitles.
1988 131m/C *RU* Leonid Filatov, Tatiana Dogileva, Irina Kupchenko; ***Dir:*** Edgar Ryazanov. **VHS, Beta, LV $29.95** *FRH* 🦴🦴🦴

Forgotten Warrior Vengeance is the name of the game as a Vietnam vet tracks down the fellow officer who tried to kill him and shot his commander.
1986 (R) 76m/C Quincy Frazer, Sam T. Lapuzz, Ron Marchini, Joe Meyer; ***Dir:*** Nick Cacas, Charlie Ordonez. **VHS, Beta $79.95** *MNC* 🦴½

Forlorn River Nevada and Weary, two law-abiding men, seek to return their hometown to peace by eliminating a gang of outlaws. Based on a novel by Zane Grey.
1937 62m/B Buster Crabbe, June Martel, John D. Patterson, Harvey Stephens, Chester Conklin, Lew Kelly, Syd Saylor; ***Dir:*** Charles T. Barton. **VHS $19.95** *NOS, DVT, RXM* 🦴½

The Formula A convoluted story about a secret formula for synthetic fuel that meanders it's way from the end of World War II to the present. A U.S. soldier waylays, and then joins forces with, a German general entrusted with the formula. Years later, after the American is murdered, his friend, a hard-nosed L.A. cop, starts investigating, meeting up with spies and a reclusive oil billionaire. Scott and Brando have one good confrontation scene but the rest of the movie is just hot air.
1980 (R) 117m/C Marlon Brando, George C. Scott, Marthe Keller, G.D. Spradlin, Beatrice Straight, John Gielgud, Richard Lynch; ***Dir:*** John G. Avildsen. **VHS, Beta $19.98** *MGM* 🦴½

Formula for a Murder A very rich but paralyzed woman marries a scheming con artist who plots to kill her and inherit everything.
1985 89m/C Christina Nagy, David Warbeck, Rossano Brazzi. **VHS, Beta $69.98** *LIV, LTG* 🦴

Forsaking All Others Screwball comedy featuring several MGM superstars. Friends since childhood, Gable has been in love with Crawford for 20 years, but she never realizes it and plans to marry Montgomery. Crawford is very funny in this delightful story of a wacky romantic triangle.
1935 84m/B Joan Crawford, Clark Gable, Robert Montgomery, Charles Butterworth, Billie Burke, Rosalind Russell, Frances Drake; ***Dir:*** W.S. Van Dyke. **VHS $19.98** *MGM* 🦴🦴½

Fort Apache The first of director Ford's celebrated cavalry trilogy, in which a fanatical general leads his reluctant men to an eventual slaughter, recalling George Custer at Little Big Horn. In residence: Ford hallmarks of spectacular landscapes and stirring action, as well as many vignettes of life at a remote outpost. Don't forget to catch "She Wore a Yellow Ribbon" and "Rio Grande," the next films in the series. Also available in a colorized version.
1948 125m/B John Wayne, Henry Fonda, Shirley Temple, John Agar, Pedro Armendariz, Victor McLaglen, Ward Bond, Anna Lee; ***Dir:*** John Ford. **VHS, Beta, LV $19.95** *TTC, RKO, VID* 🦴🦴🦴½

Fort Apache, the Bronx A police drama set in the beleaguered South Bronx of New York City, based on the real-life experiences of two former New York cops who served there. Newman is a decent cop who goes against every kind of criminal and crazy and against his superiors in trying to bring law and justice to a downtrodden community.
1981 (R) 123m/C Paul Newman, Ed Asner, Ken Wahl, Danny Aiello, Rachel Ticotin, Pam Grier, Kathleen Beller; ***Dir:*** Daniel Petrie. **VHS, Beta, LV $9.99** *CCB, VES* 🦴🦴🦴

Fortress A teacher and her class are kidnapped from their one-room schoolhouse in the Australian outback. They must use their ingenuity and wits to save their lives in this violent made-for-cable-television suspense drama.
1985 90m/C *AU* Rachel Ward, Sean Garlick, Rebecca Rigg; ***Dir:*** Arch Nicholson. **VHS, Beta $19.99** *HBO* 🦴

Fortress of Amerikka In the not-too-distant future, mercenaries get their hands on a secret weapon that could allow them to take over the USA.
1989 ?m/C Gene LeBrok, Kellee Bradley. **VHS $79.99** *IMP* 🦴½

The Fortune Cookie After receiving a minor injury during a football game, a TV cameraman is convinced by his seedy lawyer brother-in-law to exaggerate his injury and start an expensive lawsuit. A classic, biting comedy by Wilder. First of several Lemmon-Matthau comedies. Oscar nomination for original screenplay. The Letterbox laserdisc version contains the original theatrical trailer.
1966 125m/B Jack Lemmon, Walter Matthau, Ron Rich, Cliff Osmond, Judi West, Lurene Tuttle; ***Dir:*** Billy Wilder. Academy Awards '66: Best Supporting Actor (Matthau). **VHS, Beta, LV $19.95** *FOX, MGM, PIA* 🦴🦴🦴

Fortune Dane Police detective Fortune Dane goes undercover to clear his name after he is framed for murder.
1986 (PG) 83m/C Carl Weathers, Adolph Caesar; ***Dir:*** Nicholas Sgarro. **VHS, Beta $29.95** *VMK* 🦴½

Fortune's Fool A beef king and profiteer marries a younger woman and soon discovers the problems that ambition can cause. Silent.
1921 60m/B *GE* Emil Jannings, Daguey Servaes, Reinhold Schunzel; ***Dir:*** Reinhold Schunzel. **VHS, Beta $29.95** *NOS, MRV, VYY* 🦴🦴

Fortunes of War British professor Guy Pringle arrives with his bride, Harriet, to take up a teaching post in the Balkans in 1939. The idealistic Guy soons becomes enmeshed in anti-fascist politics and involved with the local members of the British embassy. When threatened by war they travel to Athens and then Cairo, where Guy's increasing involvement in the political situation, and his nelgect of Harriet, causes a crisis in their marriage. Slow-moving drama with some self-centered characters is redeemed by the acting skills of those involved. Based on the autobiographial novels of Olivia Manning. Originally shown on "Masterpiece Theater" on PBS.
1990 360m/C *GB* Kenneth Branagh, Emma Thompson, Rupert Graves. **VHS $89.98** *FOX* 🦴🦴🦴

Forty Carats Ullmann plays the just-turned forty divorcee who has a brief fling with the half-her-age Albert while on vacation in Greece. She figures she'll never see him again but, back in New York, he turns up on the arm of her beautiful daughter (Rafflin). Except he's still interested in mom. The rest of the movie is spent trying to convince Ullmann that love can conquer all. Ullmann's very attractive but not well-suited for the part; Raffins's screen debut. Based on the Broadway hit.
1973 (PG) 110m/C Liv Ullman, Edward Albert, Gene Kelly, Binnie Barnes, Deborah Raffin, Nancy Walker; ***Dir:*** Milton Katselas. **VHS, Beta $59.95** *COL* 🦴🦴½

Forty Days of Musa Dagh Based on the true story of Armenia's fight for freedom against Turkey in the 1914 uprising and the dreadful cost. Based on the novel by Franz Werfel.
1985 120m/C Michael Constantine, David Opatoshu. **VHS, Beta $69.95** *VCD* 🦴🦴

The Forty-Ninth Parallel Six Nazi servicemen, seeking to reach neutral American land, are trapped and their U-boat sunk by Royal Canadian Air Force bombers, forcing them into Canada on foot, where they run into an array of stalwart patriots. Dated wartime propaganda made prior to the U.S. entering the war; riddled with entertaining star turns. Also titled "The Invaders."
1941 90m/B *GB* Laurence Olivier, Leslie Howard, Eric Portman, Raymond Massey, Glynis Johns, Finlay Currie, Anton Walbrook; ***Dir:*** Michael Powell. **VHS, Beta, LV $29.95** *VID, VYG* 🦴🦴🦴

Forty-Seven Ronin, Part 1 Turn of the 18th-century epic chronicling the samurai legend. The warriors of Lord Asano set out to avenge their leader, tricked into committing a forced seppuku, or hara-kiri. The photography is generously laden with views of 18th century gardens as well as panoram-

ic vistas. This is the largest and most popular film of the Kabuki version of the story by Seika Mayama. Also known as "The Loyal 47 Ronin." In Japanese with English subtitles.
1942 111m/B *JP Dir:* Kenji Mizoguchi. **VHS** **$59.95** *SVS* ♫♫♫

Forty-Seven Ronin, Part 2 Second half of the film in which the famous Japanese folklore tale of Lord Asano and his warriors is told. The film follows Asano's samurai as they commit themselves to avenging their leader in 1703. In Japanese with English subtitles.
1942 108m/B *JP Dir:* Kenji Mizoguchi. **VHS** **$59.95** *SVS* ♫♫♫

Forty Thousand Horsemen The story of the ANZACS of Australia, created to fight Germany in the Middle East during World War II. This Australian war drama is full of cavalry charges, brave young lads, and "Waltzing Matilda."
1941 86m/B *AU* Chips Rafferty, Grant Taylor, Betty Bryant, Pat Twohill. **VHS, Beta $29.95** *FCT, VYY* ♫♫½

42nd Street A Broadway musical producer faces numerous problems in his efforts to reach a successful opening night. The choreography is by Busby Berkeley; the songs are by Harry Warren and Al Dubin, including "You're Getting to Be a Habit with Me," "Young and Healthy," and "Shuffle Off to Buffalo." A colorized version of the film is also available.
1933 89m/B Warner Baxter, Ruby Keeler, Bebe Daniels, George Brent, Dick Powell, Guy Kibbee, Ginger Rogers, Una Merkel, Busby Berkeley; *Dir:* Lloyd Bacon. **VHS, Beta, LV $29.98** *MGM, FOX* ♫♫♫♫

48 Hours to Live A reporter travels to a nuclear scientist's secluded island only to find the scientist held hostage by nuclear weapon-seeking terrorists.
1960 86m/B Anthony Steel, Ingemar Johannson, Marlies Behrens. **VHS, Beta $39.95** *SVS* ♫

48 Hrs. An experienced San Francisco cop (Nolte) springs a convict (Murphy) from jail for 48 hours to find a vicious murdering drug lord. Murphy's film debut is great and Nolte is perfect as his gruff foil.
1982 (R) 97m/C Nick Nolte, Eddie Murphy, James Remar, Annette O'Toole, David Patrick Kelly, Sonny Landham, Brion James, Denise Crosby; *Dir:* Walter Hill. **VHS, Beta, LV $14.95** *PAR* ♫♫♫

Fotografo de Senoras A novice photographer finds himself being mistaken for a rapist when naked women keep popping up in front of his camera.
1985 110m/C *SP* Jorge Porcel, Graciela Alfano, Tristan. **VHS, Beta** *MED* ♫

Foul Play Bearing no similarity to the Goldie Hawn vehicle of the same name, this Polish crimer is based on a true story, and stars two of Poland's Olympic boxers. A young man, accused of theft, is coerced by the Warsaw police to go undercover in return for having the charges against him dropped. With English subtitles.
1976 98m/C *PL* Marek Powowski, Jan Szcepanski, Jerzy Kulej. **VHS $14.95** *FCT* ♫♫

Foul Play Hawn is a librarian who picks up a hitchhiker which leads to nothing but trouble. She becomes involved with San Francisco detective Chase in an effort to expose a plot to kill the Pope during his visit to the city. Also involved is Moore as an English orchestra conductor with some kinky

sexual leanings. Chase is charming (no mugging here) and Hawn both bubbly and brave. A big winner at the box office; features Barry Manilow's hit tune "Ready to Take a Chance Again."
1978 (PG) 116m/C Goldie Hawn, Chevy Chase, Dudley Moore, Burgess Meredith, Billy Barty, Rachel Roberts, Eugene Roche, Brian Dennehy, Chuck McCann, Bruce Solomon; *Dir:* Colin Higgins. **VHS, Beta, LV, 8mm $19.95** *PAR* ♫♫♫

Found Alive After a painful divorce, a woman kidnaps her son and steals away into the Mexican jungle, with only the help of her faithful butler.
1934 65m/B Barbara Bedford, Maurice Murphy, Robert Frazer, Edwin Cross; *Dir:* Charles Hutchinson. **VHS, Beta $16.95** *SNC* ♫

The Fountainhead Cooper is an idealistic, uncompromising architect who refuses to change his designs. When he finds out his plans for a public housing project have been radically altered he blows up the building and winds up in court defending his actions. Neal is the sub-plot love interest. Based on the novel by Ayn Rand.
1949 113m/C Gary Cooper, Patricia Neal, Raymond Massey, Ray Collins, Henry Hull; *Dir:* King Vidor. **VHS, Beta $19.98** *MGM, FOX* ♫♫♫

Four Adventures of Reinette and Mirabelle Small but touching Rohmer tale of the friendship between two women—one a naive country girl (Miquel, as Reinette), the other a sophisticated Parisian (Forde, as Mirabelle)—who share an apartment and a number of experiences, but who have a very different manner of inhabiting those experiences. A thoroughly French human comedy, one of Rohmer's better studies.
1989 95m/C *FR* Joelle Miquel, Jessica Forde; *W/Dir:* Eric Rohmer. **VHS $69.95** *NYF, FCT* ♫♫♫

Four Bags Full A bitter comedy about two smugglers during World War II who try to get a slaughtered pig to the black market under the Nazis' noses. In French with English subtitles.
1956 82m/C *FR* Jean Gabin, Louis de Funes, Bourvil, Jeanette Batti; *Dir:* Claude Autant-Lara. **VHS, Beta $14.95** *VDM, FCT* ♫♫½

Four Daughters Classic, three-hankie outing in which four talented daughters of music professor Rains fall in love and marry. Garfield shines, in a role tailor-made for him, as the world-weary suitor driven to extremes in the name of love. Great performances from all. Based on the novel "Sister Act" by Fannie Hurst.
1938 90m/B Claude Rains, John Garfield, May Robson, Priscilla Lane, Lola Lane, Rosemary Lane, Gale Page, Dick Foran, Jeffrey Lynn; *Dir:* Michael Curtiz. **VHS $19.98** *MGM, CCB* ♫♫♫

Four Deuces A gang war is underway between the Chico Hamilton mob and Vic Morano and the Four Deuces during the Depression. Comedy and action combine with elements of strong language, strong sex, and strong violence.
1975 87m/C Jack Palance, Carol Lynley, Warren Berlinger, Adam Roarke; *Dir:* William H. Bushnell Jr. **VHS, Beta $59.99** *HBO, TIM* ♫½

Four Faces West McCrea is an honest rancher who nevertheless robs the local bank in order to save his father's ranch from foreclosure. His humanity in helping a diphtheria-ridden family leads to his capture, but the sheriff promises a light sentence, since he is not a typical bad guy. Fine performances strengthen this low-key western.

1948 90m/B Joel McCrea, Frances Dee, Charles Bickford; *Dir:* Alfred E. Green. **VHS $14.95** *REP* ♫♫½

The Four Feathers A grand adventure from a story by A.E.W. Mason. After resigning from the British Army a young man is branded a coward and given four white feathers as symbols by three of his friends and his lady love. Determined to prove them wrong he joins the Sudan campaign of 1898 and rescues each of the men from certain death. They then take back their feathers as does his girl upon learning of his true courage. Excellent performances by Smith and Richardson.
1939 99m/C *GB* John Clements, Ralph Richardson, C. Aubrey Smith, June Duprez, Donald Gray; *Dir:* Zoltan Korda. Venice Film Festival '39: Cup of the Biennial. **VHS, Beta, LV $14.98** *SUE, FCT* ♫♫♫♫

Four Feathers Determined to return the symbols of cowardice, four feathers, to his friends and fiancee, a man courageously saves his friends' lives during the British Sudan campaign and regains the love of his lady. Made-for-television movie is the fifth remake.
1978 95m/C Beau Bridges, Jane Seymour, Simon Ward, Harry Andrews, Richard Johnson, Robert Powell; *Dir:* Don Sharp. **VHS, Beta $59.99** *HBO* ♫♫

Four Friends The magical good and bad dream of the 1960s is remembered in this story of four friends. A young woman and the three men in love with her first come together in high school and then separate, learning from college, war and drug abuse, and each other in this ambitious movie from Steve Tesich, the writer of "Breaking Away." Good performances by all.
1981 (R) 114m/C James Leo Herlihy, Craig Wasson, Jodi Thelen, Michael Huddleston, Jim Metzler, Reed Birney; *Dir:* Arthur Penn. **VHS, Beta $19.98** *WAR, OM* ♫♫♫

The Four Hands of Death: Wily Match Two martial artists are the best of friends, in and out of the ring—until they both fall for the same woman!
19?? 90m/C Cheung Fu Hung, Jimmy Lee, Mo Man Sau. **VHS, Beta $49.95** *OCE* ♫½

The Four Horsemen of the Apocalypse Silent classic and star maker for Valentino concerning an Argentine family torn apart by the outbreak of WWI. Valentino is a painter who moves from his native Argentina to France and is persuaded to enlist by a recruiter who invokes the image of the Biblical riders. His excellence as a soldier, however, proves to be his undoing. The 1962 remake can't hold a candle to original, adapted from a novel by Vicente Blasco-Ibanez.
1921 110m/B Rudolph Valentino, Alice Terry, Pomeroy Cannon, Josef Swickard, Alan Hale Sr., Mabel van Buren, Nigel de Brulier, Bowditch Turner, Wallace Beery; *Dir:* Rex Ingram. **VHS $99.95** *EJB* ♫♫♫½

Four Horsemen of the Apocalypse The members of a German family find themselves fighting on opposite sides during World War II. This remake of the vintage Valentino silent failed at the box office, with complaints about its length, disjointed script, and uninspired performances. The title refers to the horrors of conquest, pestilence, war, and death.
1962 153m/C Glenn Ford, Charles Boyer, Lee J. Cobb, Paul Henreid, Yvette Mimieux; *Dir:* Vincente Minnelli. **VHS, Beta $24.95** *MGM* ♫♫♫

Four Infernos to Cross Martial arts mixes with drama in this story of the struggle the Korean people encountered while under Japanese rule before World War II.
1977 90m/C Musung Kwak, Kyehee Kim. **VHS, Beta $39.95** UNI ⅅ

Four Jacks and a Jill Four musicians are left in the lurch when their band singer leaves them for her gangster boyfriend and they must frantically search for a replacement. Not completely uninteresting, but rather bland and old now. Earlier versions of the story were released as ''Street Girl'' and ''That Girl from Paris.''
1941 68m/B Anne Shirley, Ray Bolger, Desi Arnaz Sr., Jack Durant, June Havoc, Eddie Foy Jr., Fritz Field; **Dir:** Jack Hively. **VHS, Beta $19.98** TTC ⅅⅅ

Four in a Jeep In Vienna in 1945, soldiers from different countries are serving as an international military police force. They clash as a result of political demands and their love for the same woman. Shot on location in Austria.
1951 83m/B Ralph Meeker, Viveca Lindfors, Joseph Yadin, Michael Medwin; **Dir:** Leopold Lindtberg. **VHS, Beta $19.95** NOS, SVS ⅅⅅ½

The Four Musketeers A fun-loving continuation of ''The Three Musketeers,'' reportedly filmed simultaneously. Lavish swashbuckler jaunts between France, England, and Italy, in following the adventures of D'Artagnan and his cohorts. Pictures give an amusing depiction of Lester-interpreted seventeenth-century Europe, with fine performances especially by Dunaway as an evil countess seeking revenge on our heroes and Welch as the scatterbrained object of York's affections.
1975 (PG) 108m/C Michael York, Oliver Reed, Richard Chamberlain, Frank Finlay, Raquel Welch, Christopher Lee, Faye Dunaway, Jean-Pierre Cassel, Geraldine Chaplin, Simon Ward, Charlton Heston, Roy Kinnear, Nicole Calfan; **Dir:** Richard Lester. **VHS, Beta** IVE, OM ⅅⅅⅅ

Four Robbers The ruthless Ma initiates his series of crimes throughout the Far East, somehow staying ahead of the cops. When the final showdown occurs, it's an all out kung fu massacre.
1977 90m/C Lau Chun, Shek Hon, Kong Seng, Lee Wing Shan. **VHS, Beta $49.95** OCE ⅅ

Four Rode Out A Mexican outlaw is pursued by his girlfriend, one-time partner, and the law. Brutal, inferior western.
1969 99m/C Pernell Roberts, Sue Lyon, Leslie Nielsen, Julian Mateos; **Dir:** John Peyser. **VHS** AVD ⅅ

The Four Seasons Three upper-middle-class New York couples share their vacations together, as well as their friendship, their frustrations and their jealousies. Alda's first outing as a film director is pleasant and easy on the eyes.
1981 (PG) 108m/C Alan Alda, Carol Burnett, Sandy Dennis, Len Cariou, Jack Weston, Rita Moreno, Bess Armstrong; **Dir:** Alan Alda. **VHS, Beta, LV $19.95** MCA ⅅⅅ½

Four Sided Triangle Two mad scientists find their friendship threatened when they discover that they are both in love with the same woman. So they do what anyone would do in this situation—they invent a machine and duplicate her.
1953 81m/B GB James Hayter, Barbara Payton, Stephen Murray, John Van Eyssen, Percy Marmont; **Dir:** Terence Fisher. **VHS $19.98** NOS, SNC, MLB ⅅ

The Four Sons A guide to Passover and how to prepare the traditional seder. Theodore Bikel plays Manny who helps three brothers learn the methods and significance of Passover.
197?? 37m/C Theodore Bikel. **VHS, Beta $34.95** ERG ⅅ

Four for Texas Perhaps Aldrich's later success with ''The Dirty Dozen'' can be attributed, in part, to this exercise in how not to make a comic western; poorly made, over long, and far too dependant on the feminine charisma of Ekberg and Andress. Slow-moving Sinatra-Martin vehicle tells the tale of con men in the Old West who battle bandits and bad bankers for a stash of loot.
1963 124m/C Frank Sinatra, Dean Martin, Anita Ekberg, Ursula Andress, Charles Bronson, Victor Buono, Jack Elam; **Dir:** Robert Aldrich. **VHS, Beta, LV $59.95** WAR, FCT ⅅ

Four Ways Out Four average guys are fed up by their lot in life and stage a box-office robbery of their local soccer stadium. They don't get to enjoy their ill-gotten loot. In Italian with subtitles or dubbed.
1957 77m/B IT Gina Lollobrigida, Renalto Baldini, Paul Muller; **Dir:** Pietro Germi. **VHS $16.95** SNC ⅅⅅ½

The 400 Blows The classic, ground-breaking semi-autobiography that initiated Truffaut's career and catapulted him to international acclaim, about the trials and rebellions of a 12-year-old French schoolboy. One of the greatest and most influential of films, and the first of Truffaut's career-long Antoine Doinel series. With English subtitles.
1959 97m/B FR Jean-Pierre Leaud, Claire Maurier, Albert Remy, Guy Decomble, Georges Flamant, Patrick Auffay; **Dir:** Francois Truffaut. Cannes Film Festival '59: Catholic Film Office Award, Best Director (Truffaut). **VHS, Beta, LV $59.98** FOX, MRV, APD ⅅⅅⅅⅅ

The 4D Man A physicist makes two fateful discoveries while working on a special project that gets out of control, leaving him able to pass through matter and see around corners. He also finds that his touch brings instant death. Cheap but effective sci-fier. Young Duke has a small part.
1959 85m/C Robert Lansing, Lee Meriwether, James Congdon, Guy Raymond, Robert Strauss, Patty Duke; **Dir:** Irvin S. Yeaworth Jr. **VHS, Beta, LV $19.95** STE, NWV, MLB ⅅⅅ½

The 4th Man Steeped in saturated colors and jet black comedy, Verhoeven's nouveau noir mystery enjoyed considerable arthouse success but was not released in the US until 1984. The story is decidedly non-linear, the look stylish and symbolic. Krabbe is an alcoholic bisexual Catholic writer who inadvertently becomes the hypotenuse of a love triangle involving Herman, a young man he encounters at a railway station, and his lover Christine, who owns the Sphinx beauty parlor and whose three husbands died, shall we say, mysteriously. Named one of the year's best films by the National Board; atmospheric score by Loek Dikker. In Dutch with English subtitles.
1979 104m/C NL Jeroen Krabbe, Renee Soutendijk, Thom Hoffman, Jon De Vries, Geert De Jong; **Dir:** Paul Verhoeven. **VHS, Beta, LV $24.95** XVC, MED ⅅⅅⅅ½

The Fourth Protocol Well-made thriller based on the Frederick Forsyth bestseller about a British secret agent trying to stop a young KGB agent from destroying NATO and putting the world in nuclear-jeopardy. Brosnan, as the totally dedicated Russkie, gives his best performance to date while

Caine, as usual, is totally believable as he goes about the business of tracking down the bad guys.
1987 (R) 119m/C GB Michael Caine, Pierce Brosnan, Ned Beatty, Joanna Cassidy, Julian Glover, Ray McAnally, Michael Gough, Ian Richardson; **Dir:** John MacKenzie. **VHS, Beta, LV $19.98** LHV, WAR ⅅⅅⅅ

Fourth Story A misfit detective is hired by a beautiful woman to find her missing husband. The investigation reveals some unseemly facts about the gentleman. Characterizations and plot twists keep this made-for-cable mystery interesting.
1990 (PG-13) 91m/C Mark Harmon, Mimi Rogers, Cliff DeYoung, Paul Gleason, M. Emmet Walsh; **Dir:** Ivan Passer. **VHS, Beta, LV $89.98** MED, IME ⅅⅅⅅ

The Fourth War Scheider and Prochnow are American and Russian colonels, respectively, assigned to guard the West German-Czechoslovakian border against each other. With the end of the cold war looming, these two frustrated warriors begin to taunt each other with sallies into the other's territory, threatening to touch off a major superpower conflict.
1990 (R) 109m/C Roy Scheider, Juergen Prochnow, Tim Reid, Lara Harris, Harry Dean Stanton; **Dir:** John Frankenheimer. **VHS, Beta, LV $89.99** HBO ⅅⅅ

Fourth Wise Man A Biblical Easter story about a rich physician searching for Christ in Persia.
1985 72m/C Martin Sheen, Lance Kerwin, Alan Arkin, Harold Gould, Eileen Brennan, Ralph Bellamy, Adam Arkin, Richard Libertini. **VHS, Beta $69.95** LIV, LTG ⅅⅅ

Fourth Wish When a single father learns that his son is dying, he vows to make his son's last months as fulfilling as possible.
1975 107m/C John Meillon, Robert Bettles; **Dir:** Don Chaffey. Australian Film Institute '75: Best Actor (Meillon). **VHS, Beta $49.95** SUE ⅅ

Fowl Play An unlikely threesome encounter various hardships and mishaps on their way to the first cock-fighting Olympics. Let's hope it's the last cock-fighting Olympics.
1983 90m/C Nancy Kwan. **VHS, Beta** PGN Woof!

Fox and His Friends Fassbinder's breakthrough tragi-drama, about a lowly gay carnival barker who wins the lottery, thus attracting a devious, exploiting lover, who takes him for everything he has. In German with English subtitles. Also known as: ''Fist Right of Freedom;'' German title: ''Faustrecht der Freiheit.''
1975 123m/C GE Rainer Werner Fassbinder, Peter Chatel, Karl-Heinz Bohm, Adrian Hoven, Harry Baer, Ulla Jacobsen, Kurt Raab; **Dir:** Rainer Werner Fassbinder. **VHS, Beta $79.95** FCT, NYF ⅅⅅⅅ½

Foxes Four young California girls grow up with little supervision from parents still trying to grow up themselves. They rely on each other in a world where they have to make adult choices, yet are not considered grown-up. They look for no more than a good time and no tragic mistakes.
1980 (R) 106m/C Jodie Foster, Cherie Currie, Marilyn Kagan, Scott Baio, Sally Kellerman, Randy Quaid, Laura Dern; **Dir:** Adrian Lyne. **VHS, Beta $59.98** FOX ⅅⅅ

Foxfire Light A young woman vacationing in the Ozarks is drawn into a romance with a cowboy. But her mother's social ambitions and her own indecision may tear them apart.

1984 (PG) 102m/C Tippi Hedren, Lara Parker, Leslie Nielsen, Barry Van Dyke; *Dir:* Allen Baron. VHS, Beta **$79.95** *PSM* ✰✰

Foxstyle A wealthy nightclub owner struggles with his country roots and his city sophistication.
1973 84m/C Juanita Moore, Richard Lawson, John Taylor, Jovita Bush; *Dir:* Clyde Houston. VHS, Beta **$59.98** *MAG, MOV* ✰ 1/2

Foxtrap An L.A. courier is hired to find a runaway girl and, upon finding her in Europe, discovers he's been led into a trap. A low budget Italian co-production.
1985 89m/C *IT* Fred Williamson, Christopher Connelly, Arlene Golonka; *Dir:* Fred Williamson. VHS, Beta **$69.95** *VES* ✰

Foxtrot A wealthy count isolates himself, his wife, and their two servants, on an island but cannot escape his past or the horrors of World War II. Even the good cast can't save the pretentious script and inadequate production.
1976 (R) 91m/C *MX Sl* Peter O'Toole, Charlotte Rampling, Max von Sydow, Jorge Luke; *Dir:* Arturo Ripstein. VHS, Beta **$59.98** *CHA* ✰✰

Foxy Brown A bitter woman poses as a prostitute to avenge the mob-backed deaths of her drug dealer brother and undercover cop boyfriend. Extremely violent black exploitation flick.
1974 (R) 92m/C Pam Grier, Terry Carter, Antonio Fargas, Kathryn Loder; *Dir:* Jack Hill. VHS, Beta **$19.98** *ORI* ✰

F.P. 1 An artifical island (Floating Platform 1) in the Atlantic is threatened by treason. This slow-moving 1930s technothriller is the English-language version of the German "F.P. 1 Antwortet Nicht" ("F.P. 1 Doesn't Answer"). Both were directed at the same time by Hartl, using different casts.
1933 74m/B *GE* Leslie Fenton, Conrad Veidt, Jill Esmond. VHS **$19.95** *NOS* ✰✰

F.P. 1 Doesn't Answer A mid-Atlantic refueling station (Floating Platform 1) is threatened by treason and a pilot sets out to put things right. Features pre-Hollywood vintage Lorre; Albers was Germany's number one box office draw at the time. In German.
1933 74m/B *GE* Hans Albers, Sybille Schmitz, Paul Hartmann, Peter Lorre, Hermann Speelmanns; *Dir:* Karl Hartl. VHS **$16.95** *SNC* ✰✰

Fraidy Cat This program includes four animated fantasies for children.
1974 45m/C VHS, Beta **$29.95** *PSM* ✰✰

Frame Up A small town sheriff is trying to investigate a murder tied to a fraternity's initiation. However, he runs into opposition from the most powerful man in town.
1991 (R) 90m/C Wings Hauser, Bobby DiCicco, Frances Fisher, Dick Sargent, Robert Picardo; *Dir:* Paul Leder. VHS, LV **$89.98** *REP* ✰✰

Framed A nightclub owner is framed for murder, which understandably irks him. He's determined to get paroled and then seek revenge on the crooked cops responsible for his incarceration. This action melodrama features the writer, director and star of "Walking Tall."
1975 106m/C Joe Don Baker, Gabriel Dell, Brock Peters, Conny Van Dyke, John Marley; *Dir:* Phil Karlson. VHS, Beta **$24.95** *PAR* ✰✰

Framed An art forger gets tripped up by a beautiful con artist. However, when they meet again he is willingly drawn into her latest swindle.

1990 87m/C Jeff Goldblum, Kristin Scott Thomas, Todd Graff. VHS **$89.99** *HBO* ✰✰

Fran A young woman is torn between her desperate need for a male companion and the care of her children. Well made, albeit depressing, Australian production.
1985 92m/C *AU* Noni Hazelhurst, Annie Byron, Alan Fletcher; *Dir:* Glenda Hambly. Australian Film Institute '85: Best Actress (Hazelhurst). VHS, Beta *API* ✰✰

France/Tour/Detour/Deux/ Enfants A twelve-part television series made for French television in which Godard and Mieville chronicle, in their characteristically reflexive fashion, the lives of two French schoolchildren and how their everyday growth is informed by television and the media. In French with English subtitles.
1978 26m/C *FR Dir:* Jean-Luc Godard, Anne-Marie Mieville. VHS, Beta *EAI* ✰✰✰

Frances The tragic story of Frances Farmer, the beautiful and talented screen actress of the 30's and early 40's, who was driven to a mental breakdown by bad luck, drug and alcohol abuse, a neurotic, domineering mother, despicable mental health care, and her own stubbornness. After being in and out of mental hospitals, she is finally reduced to a shadow by a lobotomy. Not nearly as bleak as it sounds, this film works because Lange understands this character from the inside out, and never lets her become melodramatic or weak.
1982 134m/C Jessica Lange, Kim Stanley, Sam Shepard; *Dir:* Graeme Clifford. VHS, Beta, LV **$19.99** *HBO* ✰✰✰

The Franchise Affair A teenage girl in need of an alibi accuses two reclusive women of kidnapping and abusing her. It's up to a lawyer to seek out the truth. Based on the novel by Josephine Tey.
1952 149m/C *GB* Michael Denison, Dulcie Gray, Anthony Nicholls, Marjorie Fielding, Athene Seyler, Ann Stephens, Patrick Troughton; *Dir:* Lawrence Huntington. VHS **$49.95** *FCT* ✰✰

Francis Gary Powers: The True Story of the U-2 Spy A television dramatization of the true experiences of Gary Powers, a CIA spy pilot whose plane was shot down over the Soviet Union in 1960. His capture, trial and conviction are all portrayed in detail, taken from Power's own reminiscences.
1976 120m/C Lee Majors, Noah Beery Jr., Nehemiah Persoff, Lew Ayres, Brooke Bundy; *Dir:* Delbert Mann. VHS, Beta *WOV* ✰✰ 1/2

Francis in the Navy The precocious talking mule, Francis, is drafted, and his buddy, played by O'Connor, comes to the rescue. The loquacious beast proves he has the superior grey matter however, and ends up doing all the thinking. Look for Clint Eastwood in his second minor role.
1955 80m/B Donald O'Connor, Martha Hyer, Jim Backus, Paul Burke, David Janssen, Clint Eastwood, Martin Milner; *Dir:* Arthur Lubin. VHS **$9.95** *MCA* ✰✰

Francisco Oller In Spanish and English versions, a survey of the life and times of the Puerto Rican 19th century master.
1984 25m/C Renee Channey, Ivan Silva. VHS, Beta **$30.00** *MUS* ✰✰

Franck Goldberg Videotape A collection of Goldberg films. Includes "Red Souvenir," about kids in New York City, "No Sellout," a rap music video about Malcom X, "Lynch: The Murder of Michael Stewart," about the police killing of a subway grafitti

artist, and "Promised Land," a filmette starring Zedd.
19?? 58m/C Nick Zedd; *W/Dir:* Franck Goldberg. VHS **$39.95** *MWF* ✰✰

Frank and I A soft-core epic about a Victorian woman who disguises herself as a boy in order to enter the house of a distinguished Englishman.
1983 (R) 75m/C Jennifer Inch, Christopher Pearson; *Dir:* Gerard Kikione. VHS, Beta **$79.95** *MGM* Woof!

Frank, Liza and Sammy: The Ultimate Event Taped live from Detroit, this show includes the stars performing individual hits, as well as a set together. Backstage footage is included.
1989 120m/C Frank Sinatra, Liza Minnelli, Sammy Davis Jr. VHS, Beta, 8mm **$29.95** *EKC* Woof!

Frank Zappa's Does Humor Belong in Music? Lively and irreverent concert tape by Frank Zappa. Songs include "Dancin' Fool," "Dinah - Moe Humm" and "Zoot Allures."
1986 57m/C Frank Zappa. VHS, Beta **$24.95** *MVD, MPI* Woof!

The Franken and Davis Special Writer-performers Al Franken and Tom Davis, once featured on "Saturday Night Live," star in this live comedy concert taped in New Jersey.
1984 60m/C VHS, Beta **$59.95** *PAV*

Franken and Davis at Stockton State Former "Saturday Night Live" writers-stars Al Franken and Tom Davis perform stand-up comedy in this concert taped at Stockton College in New Jersey.
1984 55m/C Al Franken, Tom Davis. VHS, Beta **$59.95** *PAV* Woof!

Frankenhooker Jeffrey Franken is a nice guy; he didn't mean to mow his fiancee down in the front lawn. But sometimes bad things just happen to good people. Luckily Jeff thought to save her head and decides to pair it up with the body of some sexy streetwalkers. Voila! You have Frankenhooker: the girlfriend with more than a heart. The posters say it best, "A Terrifying Tale of Sluts and Bolts." Also available in an "R" rated version.
1990 90m/C James Lorinz, Patty Mullen, Charlotte J. Helmkamp, Louise Lasser; *Dir:* Frank Henenlotter. VHS **$89.98** *SHG, FCT* ✰ 1/2

Frankenstein The definitive expressionistic Gothic horror classic that set the mold. Adapted from the Mary Shelley novel about Dr. Henry Frankenstein, the scientist who creates a terrifying, yet strangely sympathetic monster. Great performance by Karloff as the creation, which made him a monster star. Several powerful scenes, excised from the original version, have been restored. Side two of the laser disc version contains the original theatrical trailer, plus a collection of photos and scenes replayed for study purposes.
1931 71m/B Boris Karloff, Colin Clive, Mae Clarke, John Boles, Dwight Frye, Edward Van Sloan; *Dir:* James Whale. VHS, Beta, LV **$14.95** *MCA, TLF* ✰✰✰✰

Frankenstein A brilliant scientist plays God, unleashing a living monster from the remains of the dead. A TV movie version of the legendary horror story. Good atmosphere provided by producer Dan "Dark Shadows" Curtis.

1973 130m/C Robert Foxworth, Bo Svenson, Willie Aames, Susan Strasberg; *Dir:* Glenn Jordan. VHS, Beta $49.95 *IVE* 🎬🎬½

Frankenstein A remake of the horror classic, closely following the original story, wherein the creature speaks (and waxes philosophical), the doctor sees him as his dark subconscious, and the two die in an arctic confrontation.
1982 81m/C Robert Powell, Carrie Fisher, David Warner, John Gielgud; *Dir:* James Ormerod. VHS, Beta $59.98 *LIV, LTG* 🎬🎬½

Frankenstein '80 A guy named Frankenstein pieces together a monster who goes on a killing spree. Bottom of the barrel Italian production with funky music and lots of gore.
1979 88m/C *IT GE* John Richardson, Gordon Mitchell, Leila Parker, Dada Galloti, Marisa Travers; *Dir:* Mario Mancini. VHS, Beta $59.95 *GRG, MPI* 🎬

Frankenstein: A Cinematic Scrapbook A behind the scenes look at the world's most infamous monster, featuring interviews with Boris Karloff, Bela Lugosi and Lon Chaney, Jr. Also included are original movie trailers. In color and B&W.
1991 60m/C *W/Dir:* Ted Newsom. VHS, Beta $9.95 *RHI, FCT* 🎬

Frankenstein General Hospital A completely laughless horror spoof wherein the twelfth grandson of the infamous scientist duplicates his experiments in the basement of a modern hospital. A must see for Frankenstein fans; considered by some the worst Frankenstein movie ever made.
1988 90m/C Mark Blankfield, Kathy Shower, Leslie Jordan, Irwin Keyes; *Dir:* Deborah Roberts. VHS, Beta, LV $79.95 *NSV* Woof!

Frankenstein Island Four balloonists get pulled down in a storm and end up on a mysterious island. They are greeted by one Sheila Frankenstein and encounter monsters, amazons and other obstacles. Completely inept; Carradine "appears" in a visionary sequence wearing his pajamas.
1981 (PG) 97m/C John Carradine, Andrew Duggan, Cameron Mitchell; *Dir:* Jerry Warren. VHS, Beta $39.95 *MON* 🎬

Frankenstein Meets the Space Monster A classic grade-Z epic about a space robot gone berserk among Puerto Rican disco dancers.
1965 80m/B James Karen, Nancy Marshall, Marilyn Hanold; *Dir:* Robert Gaffney. VHS, Beta $39.95 *PSM* Woof!

Frankenstein Meets the Wolfman The two famous Universal monsters meet and battle it out in this typical grade-B entry, the fifth from the series. The Werewolf wants Dr. Frankenstein to cure him, but only his monster (played by Lugosi) remains.
1942 73m/C Lon Chaney Jr., Bela Lugosi, Patric Knowles, Lionel Atwill, Maria Ouspenskaya; *Dir:* Roy William Neill. VHS, Beta, LV $14.98 *MCA, MLB* 🎬🎬🎬

Frankenstein and the Monster from Hell A young doctor is discovered conducting experiments with human bodies and thrown into a mental asylum run by none other than Dr. Frankenstein himself. They continue their gruesome work together, creating a monster who develops a taste for human flesh. This really lame film was the last of the Hammer Frankenstein series.
1974 (R) 93m/C Peter Cushing, Shane Briant, Madeleine Smith, David Prowse, John Stratton; *Dir:* Terence Fisher. VHS *PAR* 🎬🎬

Frankenstein, Unbound Corman returns after nearly 20 years with a better than ever B movie. Hurt plays a nuclear physicist time traveler who goes back to the 1800s and runs into Lord Byron, Percy and Mary Shelley and their neighbor Dr. Frankenstein and his monster. Great acting, fun special effects, intelligent and subtle message, with a little sex to keep things going.
1990 (R) 86m/C John Hurt, Raul Julia, Bridget Fonda, Jason Patric, Michael Hutchence, Catherine Rabett; *Dir:* Roger Corman. VHS, Beta, LV $89.98 *FOX* 🎬🎬🎬

Frankenstein's Daughter A demented descendant of Dr. Frankenstein sets a den of gruesome monsters loose, including the corpse of a teenaged girl he revitalizes, as he continues the mad experiments of his forefathers.
1958 85m/B John Ashley, Sandra Knight; *Dir:* Richard Cunha. VHS, Beta $9.95 *NOS, MRV, RHI* 🎬

Frankenstein's Great Aunt Tillie An excrutiatingly bad send up of the Frankenstein saga with the good doctor about to be evicted from his estate because of back taxes. Gabor appears for a few seconds in a flashback.
1983 99m/C Donald Pleasence, Yvonne Furneaux, Aldo Ray, June Wilkinson, Zsa Zsa Gabor; *W/Dir:* Myron G. Gold. VHS, Beta $59.95 *VCD* Woof!

Frankenweenie "Frankenweenie" is the tale of a lovable spunky dog named Sparky and his owner, Victor Frankenstein. When Sparky is hit by a car, Victor brings him back to life through the use of electric shock. This affectionate parody of "Frankenstein" launched renowned director Tim Burton's career.
1987? (PG) 27m/B Shelley Duvall, Daniel Stern, Barret Oliver, Paul Bartel; *W/Dir:* Tim Burton. VHS, Beta $14.99 *BVV* Woof!

Frankie and Johnny Based on the song of the same name, famed torch singer Helen Morgan portrays a singer in a bordello who shoots her unfaithful lover. Release was delayed two years by the Hays Office's intervention in how a whorehouse could be portrayed on screen. Too bad they reached an agreement. Not even Morgan's rendition of the title song redeems this.
1936 68m/B Helen Morgan, Chester Morris; *Dir:* Chester Erskine. VHS, Beta $24.98 *CCB* 🎬

Frankie and Johnny Elvis is a riverboat gambler/singer with lousy luck until Kovack changes the odds. This upsets his girlfriend Douglas who shoots him (but not fatally). Songs include: "When the Saints Go Marchin' In" and "Frankie & Johnny." Elvis wears period costumes, but this is otherwise similar to his contemporary films.
1965 88m/C Elvis Presley, Donna Douglas, Harry Morgan, Audrey Christie, Anthony Eisley, Sue Ann Langdon, Robert Strauss, Nancy Kovack, Sue Ann Langdon; *Dir:* Fred DeCordova. VHS, Beta $19.95 *MGM* 🎬🎬½

Frankie & Johnny An ex-con gets a job as a short-order cook and falls for a world-weary waitress. She doesn't believe in romance, but finally gives into his pleas for a chance and finds out he may not be such a bad guy after all. Nothing can make Pfeiffer dowdy enough for this role, but she and Pacino are charming together. In a change of pace role, Nelligan has fun as a fellow waitress who loves men. Based on the play "Frankie and Johnny in the Clair de Lune" by Terrance McNally who also wrote the screenplay.

1991 (R) 117m/C Al Pacino, Michelle Pfeiffer, Kate Nelligan, Hector Elizondo; *Dir:* Garry Marshall. National Board of Review Awards '91: 10 Best Films of the Year, Best Supporting Actress (Nelligan). VHS $94.98 *PAR, CCB* 🎬🎬🎬

Franklin D. Roosevelt: F.D.R. The story of Franklin D. Roosevelt from the time he was governor of New York until 1933, when he became President of the United States. Classic newsreel footage.
1954 15m/B VHS, Beta $19.98 *TSF* 🎬🎬🎬

Frantic From Louis Malle comes one of the first French New Wave film noir dramas. A man kills his boss with the connivance of the employer's wife, his lover, and makes it look like suicide. Meanwhile, teenagers have used his car and gun in the murder of a tourist couple and he is indicted for that crime. Their perfectly planned murder begins to unravel into a panic-stricken nightmare. A suspenseful and captivating drama. Director Malle's first feature film. Musical score by jazz legend Miles Davis. Also known as "Elevator to the Gallows."
1958 92m/B *FR* Maurice Ronet, Jeanne Moreau, Georges Poujouly; *Dir:* Louis Malle. VHS, Beta, 8mm $16.95 *SNC, NOS, VYY* 🎬🎬½

Frantic While in Paris, an American surgeon's wife is kidnapped when she inadvertantly picks up the wrong suitcase. Her kidnappers want their hidden treasure returned, which forces the husband into the criminal underground and into unexpected heroism when he seeks to rescue her. Written by Polanski and Gerard Brach; music by Ennio Morricone. Contrived ending weakens on the whole, but Polanski is still master of the dark film thrillor.
1988 (R) 120m/C Harrison Ford, Betty Buckley, John Mahoney, Emmanuelle Seigner, Jimmie Ray Weeks, Yorgo Voyagis, David Huddleston, Gerard Klein; *Dir:* Roman Polanski. VHS, Beta, LV $19.98 *WAR, FCT* 🎬🎬🎬

Frantic Antics A collection of animated shorts, including "Cookoo's Place," "The Girl That Ate New Jersey," and "The Wild Bus."
1987? 60m/C VHS, Beta $14.95 *VDC* 🎬🎬🎬

Franz A French mercenary and an insecure woman struggle with a doomed love affair. In French with English subtitles.
1972 90m/C *FR* Jacques Brel; *Dir:* Jacques Brel. VHS, Beta $19.95 *KRT* 🎬

Frasier the Sensuous Lion A zoology professor is able to converse with Frasier, a lion known for his sexual stamina and whose potency is coveted by a billionaire (shadow of monkey glands rejuvenation therapy). A fairly typical, if slightly risque, talking critter film. Also known as "Frasier the Lovable Lion."
1973 (PG) 97m/C Michael Callan, Katherine Justice, Frank De Kova, Malachi Throne, Victor Jory, Peter Lorre Jr., Marc Lawrence; *Dir:* Pat Shields. VHS, Beta $9.99 *STE, PSM* 🎬🎬

Fraternity Vacation Two college fraternity men show a nerd the greatest time of his life while he's on vacation in Palm Springs.
1985 (R) 95m/C Stephen Geoffreys, Sheree J. Wilson, Cameron Dye, Leigh McCloskey, Tim Robbins, Matt McCoy, Amanda Bearse, John Vernon, Nita Talbot, Barbara Crampton, Kathleen Kinmont, Max Wright, Julie Payne, Franklin Ajaye, Charles Rocket, Britt Ekland; *Dir:* James Frawley. VHS, Beta, LV $9.95 *NWV, STE* 🎬

F

The Freakmaker A mad professor attempts to breed plants with humans in his lab and has the usual bizarre results. Includes real freaks; released theatrically as "Mutations."
1973 (R) 90m/C *GB* Donald Pleasence, Tom Baker, Brad Harris, Julie Ege, Michael Dunn, Jill Haworth; *Dir:* Jack Cardiff. VHS, Beta $19.98 *VDC* ✍

Freaks The infamous, controversial, cult-horror classic about a band of circus freaks that exact revenge upon a beautiful aerialist and her strongman lover after enduring humiliation and exploitation. Based on Ted Robbins story "Spurs." It was meant to out-horror "Frankenstein" but was so successful that it was repeatedly banned. Browning's film may be a shocker but it is never intended to be exploitative since the "Freaks" are the only compassionate, loyal, and loving people around.
1932 66m/B Wallace Ford, Olga Baclanova, Leila Hyams, Roscoe Ates, Harry Earles; *Dir:* Tod Browning. VHS, Beta, LV $59.95 *MGM* ✍✍✍½

Freaky Friday A housewife and her teenage daughter inadvertently switch bodies and each then tries to carry on the other's normal routine. Mary Rodgers' popular book is brought to the screen with great charm in this above average Disney film.
1976 (G) 95m/C Barbara Harris, Jodie Foster, Patsy Kelly, Dick Van Patten, Ruth Buzzi; *Dir:* Gary Nelson. VHS, Beta $19.99 *DIS* ✍✍½

Freddie Hubbard A top-rate jazz performance by one of the greatest trumpet players of our time, Freddie Hubbard.
1981 59m/C *Performed by:* Freddie Hubbard. VHS, Beta, LV $29.95 *SVS, PIA, MVD* ✍✍½

Freddy's Dead: The Final Nightmare Freddy's daughter journeys to Springwood, Ohio to put a stop to her father's evil ways. Will she be able to destroy this maniac, or is a new reign of terror just beginning? The sixth, and supposedly last, film in the "Nightmare" series.
1991 (R) 96m/C Robert Englund, Lisa Zane, Shon Greenblatt, Leslie Deane, Ricky Dean Logan, Breckin Majer, Yaphet Kotto, Elinor Donahue, Roseanne (Barr) Arnold, Johnny Depp, Alice Cooper; *Dir:* Rachel Talalay. VHS, LV *COL* ✍✍

Free Amerika Broadcasting A drama about an alternate reality in which Richard Nixon was assassinated, leaving Spiro Agnew president. As a result, the country turns to violent revolution.
1989 80m/C Brian Kincaid; *Dir:* Dean Wilson. VHS $59.95 *IEC* ✍✍

Free to Love Screen idol Bow plays a young woman, fresh out of prison, who is taken in by a kindly, affluent patron. But, as the silent drama axiom would have it, there's no escaping the past, and she's forced to defend her former life when her guardian is murdered.
1925 61m/B Clara Bow, Donald Keith, Raymond McKee; *Dir:* Frank O'Connor. VHS, Beta $49.95 *GPV, FCT* ✍✍

Free Ride A kooky kid steals a sportscar to impress a girl, and gets involved with local hoods. They chase each other around.
1986 (R) 92m/C Gary Hershberger, Reed Rudy, Dawn Schneider, Peter DeLuise, Brian MacGregor. VHS, Beta $19.95 *IVE* ✍

A Free Soul Tippling litigator Barrymore helps low-life mobster Gable beat a murder rap, only to discover the hood has stolen his daughter's heart. Bargaining with his ditzy daughter, he vows to eschew the bottle if she'll stay away from good-for-nothing Gable. Directed by Brown, best known for his work with Garbo, and based on the memoirs of Adela Rogers St. John. The final courtroom scene—a cloak-and-gavel classic—cinched Barrymore's Oscar. Also nominated for best director and actress (Shearer). Remade with Liz Taylor as "The Girl Who Had Everything" (although, ironically, the earlier version is the racier of the two).
1931 91m/B Norma Shearer, Leslie Howard, Lionel Barrymore, Clark Gable, James Gleason, Lucy Beaumont; *Dir:* Clarence Brown. Academy Awards '31: Best Actor (Barrymore). VHS $29.95 *MGM, FCT, BTV* ✍✍½

Free, White, and 21 A black motel owner is accused of raping a white civil rights worker. Produced in a "pseudo documentary" form which sometimes drags.
1962 104m/B Frederick O'Neal, Annalena Lund, George Edgely, John Hicks, Hugh Crenshaw, George Russell; *Dir:* Larry Buchanan. VHS $19.95 *ASE* ✍✍

Freebie & the Bean Two San Francisco cops nearly ruin the city in their pursuit of a numbers-running mobster. Top-flight car chases and low-level, bigoted humor combine. Watch for Valerie Harper's appearance as Arkin's wife. Followed by a flash-in-the-pan television series.
1974 (R) 113m/C Alan Arkin, James Caan, Loretta Swit, Valerie Harper, Jack Kruschen, Mike Kellin; *Dir:* Richard Rush. VHS, Beta, LV $19.98 *WAR* ✍½

Freedom A young man finds the price of freedom to be very high when he tries to escape from Australian society in a silver Porsche.
1981 102m/C Jon Blake, Candy Raymond, Jad Capelja, Reg Lye, John Clayton; *Dir:* Joseph Sargent. VHS, Beta $69.95 *VID* ✍½

Freedom Road A made-for-television drama about a Reconstruction Era ex-slave, portrayed by heavyweight champion Ali, who is elected to the U.S. Senate and subsequently killed while trying to obtain total freedom for his race. Based on a novel by Howard Fast.
1979 186m/C Ron O'Neal, Edward Herrmann, John McLiam, Ernest Dixon, Alfre Woodard, Kris Kristofferson, Muhammad Ali; *Dir:* Jan Kadar. VHS, Beta $69.95 *WOV* ✍✍

Freefall A female attorney begins to feel the toll of the '80s when, at the height of her dissatisfaction with her husband and job, an old beau shows up.
1990 28m/C *Dir:* Liz Leshin. VHS $59.95 *CIG* ✍½

Freejack Futuristic thriller set in the year 2009, where pollution, the hole in the ozone layer and the financial gap between the social classes have grown to such horrific proportions that the rich must pillage the past to find young bodies to replace their own. Estevez stars as a young race car driver whose sudden death makes him an ideal candidate for this bizarre type of surgery. Once transported to the future, he becomes a "Freejack" who must run for his life to escape not only the bounty hunter (Jagger), but also the ruthless tycoon (Hopkins), who puts a price on Freejack's head.
1992 (R) 110m/C David Johansen, Amanda Plummer, Emilio Estevez, Mick Jagger, Rene Russo, Anthony Hopkins; *Dir:* Geoff Murphy. VHS, Beta, LV $94.99 *WAR* ✍✍½

Freeway A nurse attempts to find the obsessive killer who shot her husband. The murderer phones a radio psychiatrist from his car, using Biblical quotes, while cruising for new victims. Okay thriller, based on the L. A. freeway shootings.
1988 (R) 91m/C Darlanne Fluegel, James Russo, Billy Drago, Richard Belzer, Michael Callan, Steve Franken, Kenneth Tobey, Clint Howard; *Dir:* Francis Delia. VHS, Beta, LV $14.98 *NWV, STE* ✍✍½

The Freeway Maniac A low budget thriller about an escaped convict who ends up on a movie set where he continues his former career: murder. Music by former "Doors" guitarist Robby Krieger.
1988 (R) 94m/C Loren Winters, James Courtney, Shepard Sanders, Donald Hotton; *Dir:* Paul Winters. VHS, Beta $79.95 *MED* ✍

Freeze-Die-Come to Life First feature film from Kanevski, who spent 8 years in a labor camp before glasnost. The story of two children who overcome the crushing poverty and bleakness of life in a remote mining community with friendship and humor. Beautifully filmed images, fine acting, touching but never overly sentimental story. The title comes from a children's game of tag. In Russian with English subtitles.
1990 105m/C *RU* Pavel Nazarov, Dinara Drukarova; *Dir:* Vitaly Kanevski. Cannes Film Festival '91: Best First Film (Kanevski). VHS $89.95 *FXL, ART, BST* ✍✍½

French Can-Can The dramatically sparse but visually stunning depiction of the can-can's revival in Parisian nightclubs. Gabin plays the theater impressario who discovers laundress Arnoul and decides to turn her into the dancing str of his new revue at the Moulin Rouge. In French with English subtitles. Alternate title, "Only the French Can!"
1955 93m/C *FR* Jean Gabin, Francoise Arnoul, Maria Felix, Jean-Roger Caussimon, Edith Piaf, Patachou; *Dir:* Jean Renoir. VHS, Beta $39.95 *INT, APD* ✍✍✍

The French Connection Two NY hard-nosed narcotics detectives stumble onto what turns out to be one of the biggest narcotics rings of all time. Cat-and-mouse thriller will keep you on the edge of your seat; contains one of the most exciting chase scenes ever filmed. Hackman's portrayal of Popeye Doyle is exact and the teamwork with Scheider special. Won multiple awards. Based on a true story from the book by Robin Moore. Followed in 1975 by "French Connection II." Oscar nominations for Scheider, Cinematography, and Sound.
1971 (R) 102m/C Gene Hackman, Roy Scheider, Fernando Rey, Tony LoBianco, Eddie Egan, Sonny Grosso, Marcel Bozzuffi; *Dir:* William Friedkin. Academy Awards '71: Best Actor (Hackman), Best Adapted Screenplay, Best Director (Friedkin), Best Film Editing, Best Picture; British Academy Awards '71: Best Actor (Hackman), Best Film Editing; Directors Guild of America Awards '71: Best Director (Friedkin); Edgar Allan Poe Awards '71: Best Screenplay; Golden Globe Awards '72: Best Film—Drama; National Board of Review Awards '71: 10 Best Films of the Year, Best Actor (Hackman). VHS, Beta, LV $19.98 *FOX* ✍✍✍½

French Connection 2 New York policeman "Popeye" Doyle goes to Marseilles to crack a heroin ring headed by his arch nemesis, Frog One, who he failed to stop in the United States. Dour, super-gritty sequel to the 1971 blockbuster, and featuring one of Hackman's most uncompromising performances.

1975 (R) 118m/C Gene Hackman, Fernando Rey, Bernard Fresson; *Dir:* John Frankenheimer. **VHS, Beta, LV** $59.98 *FOX* 🎞🎞🎞

The French Detective Two detectives, one an old-time tough guy and the other a young cynic, are after a ruthless politician and his pet hood. Grim but entertaining.
1975 90m/C *FR* Lino Ventura, Patrick Dewaere, Victor Lanoux; *Dir:* Pierre Granier-Deferre. **VHS, Beta** $59.95 *COL* 🎞🎞🎞

French Intrigue International agents track drug lords from the U.S. to France in a so-so spy thriller.
19?? (R) 90m/C Jane Birkin, Curt Jurgens. **VHS, Beta** $19.99 *BFV* 🎞🎞

The French Lesson A romantic British farce from a screenplay written by Posy S. Simmonds, dealing with an English girl who goes to school in Paris and finds love.
1986 (PG) 90m/C *GB* Alexandre Sterling, Jane Snowdon; *Dir:* Brian Gilbert. **VHS, Beta** $69.95 *WAR* 🎞🎞½

The French Lieutenant's Woman Romantic love and tragedy in the form of two parallel stories, that of an 18th-century woman who keeps her mysterious past from the soldier who loves her, and the lead actor and actress in the film of the same story managing an illicit affair on the set. Extraordinary performances and beautifully shot. From a Harold Pinter screenplay, based upon the John Fowles novel.
1981 (R) 124m/C Meryl Streep, Jeremy Irons, Leo McKern, Lynsey Baxter; *Dir:* Karel Reisz. British Academy Awards '81: Best Actress (Streep), Best Sound, Best Musical Score. **VHS, Beta, LV** $19.95 *MGM, FOX* 🎞🎞🎞½

The French Line A millionairess beauty travels incognito while trying to sort out which men are after her money, and which ones aren't. The 3-D presentation of Russell's physique in skimpy costumes earned this the condemnation of the Catholic Legion of Decency, which pumped the box office even higher. Lacks the charm of "Gentlemen Prefer Blonds," which it imitates.
1954 102m/C Jane Russell, Gilbert Roland, Craig Stevens, Kim Novak, Arthur Hunnicutt; *Dir:* Lloyd Bacon. **VHS, Beta** $19.98 *CCB* 🎞🎞

French Postcards Three American students study all aspects of French culture when they spend their junior year of college at the Institute of French Studies in Paris. The same writers penned "American Graffiti" a few years earlier.
1979 (PG) 95m/C Miles Chapin, Blanche Baker, Valerie Quennessen, Debra Winger, Mandy Patinkin, Marie-France Pisier; *Dir:* Willard Huyck. **VHS, Beta** $59.95 *PAR* 🎞🎞

French Quarter A dual story about a young girl in the modern-day French Quarter of New Orleans who is also the reincarnation of a turn-of-the-century prostitute. Everyone gets to play two roles but this film is more curiosity than anything else.
1978 (R) 101m/C Bruce Davison, Virginia Mayo, Lindsay Bloom, Alisha Fontaine, Lance LeGault, Anne Michelle; *Dir:* Dennis Kane. **VHS, Beta** $19.95 *UHV* 🎞🎞½

French Quarter Undercover Two undercover cops in New Orleans thwart a terrorist plot aimed at the World's Fair.
1985 (R) 84m/C Michael Parks, Bill Holiday; *Dir:* Joe Catalanotto. **VHS, Beta** $69.95 *LIV, LTG* 🎞

French Singers The French vocalists perform many of their most famous songs.

195? 60m/B *FR* Edith Piaf, George Brassens, Maurice Chevalier. **VHS, Beta** $39.95 *TAM, DVT* 🎞

The French Touch Fernandel stars in this French farce about a shepherd who decides to open his own clip joint as a hairdresser and finds himself a hit among the Parisiennes. Fair to middling comedy.
1954 84m/B *FR* Fernandel, Renee Devillers, Georges Chamarat; *Dir:* Jean Boyer. **VHS** $24.95 *NOS* 🎞🎞½

The French Way One of the legendary Baker's few films, in which she portrays a Parisian nightclub owner playing matchmaker for a young couple. In French with English subtitles.
1952 72m/B *FR* Josephine Baker, Micheline Presle, Georges Marchal; *Dir:* Jacques de Baroncelli. **VHS, Beta** $29.95 *VDM* 🎞🎞½

The French Woman A soft-core story of blackmail, murder, and sex involving French cabinet ministers mixing passion and politics. From the director of "Emmanuelle".
1979 (R) 97m/C Francoise Fabian, Klaus Kinski; *Dir:* Just Jaeckin. **VHS, Beta** $29.98 *VID* 🎞

Frenchman's Farm A young woman witnesses a killing at a deserted farmhouse. When she tells the police, they tell her that the murder happened forty years before. Average thriller that probably won't keep you on the edge of your seat.
1987 (R) 90m/C *AU* Tracey Tanish, David Reyne, John Meillon, Norman Kaye, Tui Bow; *Dir:* Ron Way. **VHS, Beta** $79.95 *MAG* 🎞½

Frenzy A sculptor, pushed over the brink of sanity when he discovers that his wife is having an affair, turns to his art for solace...and seals his wife's corpse in a statue. Creepy story written by director Sewell; also known as "Latin Quarter."
1946 75m/B *GB* Derrick DeMarney, Frederick Valk, Joan Greenwood, Joan Seton, Valentine Dyall, Martin Miller; *W/Dir:* Vernon Sewell. **VHS** $19.98 *SNC* 🎞🎞½

Frenzy The only film in which Hitchcock was allowed to totally vent the violence and perverse sexuality of his distinctive vision, in a story about a strangler stalking London women in the late '60s. Finch plays the convicted killer-only he's innocent and must escape prison to find the real killer. McGowen is wonderful as the put-upon police inspector. A bit dated, but still ferociously hostile and cunningly executed.
1972 (R) 116m/C *GB* Jon Finch, Barry Foster, Barbara Leigh-Hunt, Anna Massey, Alec McCowen, Vivien Merchant, Billie Whitelaw, Jean Marsh; *Dir:* Alfred Hitchcock. National Board of Review Awards '72: 10 Best Films of the Year. **VHS, Beta, LV** $19.95 *MCA* 🎞🎞🎞

Fresh Horses A wrong-side-of-the-tracks Depression-era romance. McCarthy is the engaged college boy who falls for backwoods girl Ringwald, who turns out to have a destructive secret.
1988 (PG-13) 92m/C Molly Ringwald, Andrew McCarthy, Patti D'Arbanville, Ben Stiller; *Dir:* David Anspaugh. **VHS, Beta, LV** $89.95 *COL* 🎞🎞

Fresh Kill A young guy from Chicago goes to Hollywood, and gets mixed up in drugs and murder. Made for video.
1987 90m/C Flint Keller, Patricia Parks; *Dir:* Joseph Merhi. **VHS, Beta** *CLV* 🎞

The Freshman Country boy Lloyd goes to college and, after many comic tribulations, saves the day with the winning touchdown at the big game and wins the girl of his dreams.

This was one of the comedian's most popular films.
1925 75m/B Harold Lloyd, Jobyna Ralston, Brooks Benedict, James Anderson, Hazwel Keener; *Dir:* Sam Taylor. **VHS, Beta** *TLF* 🎞🎞🎞🎞

The Freshman Brando in an incredible parody of his Don Corleone character makes this work. Broderick is a college student in need of fast cash, and innocent enough to believe that any work is honest. A good supporting cast and a twisty plot keep things interesting. Sometimes heavy handed with its sight gags, but Broderick and Brando push the movie to hilarious conclusion. Don't miss Burt Parks' musical extravaganza.
1990 (PG) 103m/C Marlon Brando, Matthew Broderick, Bert Parks, Frank Whaley, B.D. Wong, Jon Polito, Paul Benedict, Richard Gant, Penelope Ann Miller, Maximilian Schell, Bruno Kirby; *Dir:* Andrew Bergman. **VHS, Beta, LV, 8mm** $19.95 *COL, FCT* 🎞🎞½

Frida The life of controversial Mexican painter Frido Kahlo is told via deathbed flashbacks using the artist's paintings to set the scenes.
1984 108m/C *SP* Ofelia Medina, Juan Jose Gurrola, Max Kerlow; *Dir:* Paul Leduc. **VHS, Beta, LV** $79.95 *CVC, IME, APD* 🎞🎞½

Friday the 13th The first in a very long series of slasher flicks. A New Jersey camp that's been closed for 20 years after a history of "accidental" deaths reopens and the horror begins again. Six would-be counselors arrive to get the place ready. Each are progressively murdered: knifed, speared, and axed. Followed by numerous, equally gory sequels.
1980 (R) 95m/C Betsy Palmer, Adrienne King, Harry Crosby, Laurie Bartrarr, Mark Nelsor, Kevin Bacon; *Dir:* Sean S. Cunningham. **VHS, Beta, LV** $14.95 *PAR* 🎞

Friday the 13th, Part 2 A new group of teen camp counselors are gruesomely executed by the still-undead Jason. Equally as graphic as the first installment; followed by several more gorefests.
1981 (R) 87m/C Amy Steel, John Furey, Adrienne King, Betsy Palmer; *Dir:* Steve Miner. **VHS, Beta, LV** $14.95 *PAR* Woof!

Friday the 13th, Part 3 Yet another group of naive counselors at Camp Crystal Lake fall victim to the maniacal Jason. The 3-D effects actually helped lessen the gory effects but this is still awful. Followed by more disgusting sequels.
1982 96m/C Dana Kimmell, Paul Krata, Richard Brooker; *Dir:* Steve Miner. **VHS, Beta, LV** $14.95 *PAR* Woof!

Friday the 13th, Part 4: The Final Chapter Jason escapes from the morgue to once again slaughter and annihilate teenagers at a lakeside cottage. Preceded by four earlier "Friday the 13th" films, equally as graphic. The title's also a lie since there's several more sequels to look forward to.
1984 (R) 90m/C Erich Anderson, Judie Aronson, Kimberly Beck, Peter Barton, Tom Everett, Corey Feldman, Crispin Glover; *Dir:* Joseph Zito. **VHS, Beta, LV** $14.95 *PAR* Woof!

Friday the 13th, Part 5: A New Beginning Jason rises from the dead to slice up the residents of a secluded halfway house. The sequels, and the sameness, never stop. See Part 6...
1985 (R) 92m/C John Shepherd, Melanie Kinnaman, Shavar Ross, Richard Young, Juliette Cummins, Corey Feldman; *Dir:* Danny Steinmann. **VHS, Beta, LV** $14.95 *PAR* Woof!

Friday the 13th, Part 6: Jason Lives One of the few youths not butchered by Jason digs him up and discovers he's not (and never will be) dead. Carnage ensues...and the sequels continue.
1986 (R) 87m/C Thom Mathews, Jennifer Cooke, David Kagan; *Dir:* Tom McLoughlin. VHS, Beta, LV $14.95 *PAR* Woof!

Friday the 13th, Part 7: The New Blood A young camper with telekinetic powers accidentally unchains Jason from his underwater lair with the now-familiar results. There's still another bloody sequel to go.
1988 (R) 90m/C Lar Park Lincoln, Kevin Blair, Susan Blu; *Dir:* John Carl Buechler. VHS, Beta, LV $14.95 *PAR* Woof!

Friday the 13th, Part 8: Jason Takes Manhattan Yet another sequel (so far the last), with the ski-masked walking slaughterhouse transported to New York. Most of the previous action in the movie takes place on a cruise ship. This one is less gruesome than others in the series.
1989 (R) 96m/C Jensen Daggett, Scott Reeves, Peter Mark Richman, Barbara Bingham; *W/Dir:* Rob Hedden. VHS, Beta, LV $14.95 *PAR* ✠

Friday Foster A beautiful, young photographer investigates an assassination attempt and uncovers a conspiracy against black politicians. Based on the Chicago Tribune comic strip.
1975 90m/C Pam Grier, Yaphet Kotto, Thalmus Rasulala, Carl Weathers, Godfrey Cambridge; *Dir:* Arthur Marks. VHS, Beta $59.98 *ORI* ✠½

Fridays of Eternity A romantic comedy about loyalty, fidelity, love, and lovers all with a tinge of the supernatural. In Spanish with English subtitles.
1981 89m/C *AR* Thelma Biral, Hector Alterio; *Dir:* Hector Olivera. VHS $24.98 *TAM* ✠✠

Fried Green Tomatoes Two stories about four women, love, friendship, Southern charm, and eccentricity are untidily held together by wonderful performances. Evelyn Couch (Bates) is an unhappy, middle-aged woman who meets the talkative 83 year-old Ninny Threadgoode (Tandy) while visiting a nursing home. Ninny reminisces about her life in the town of Whistle Stop, Alabama, during the Depression and the two women, Idgie (Masterson) and Ruth (Parker), who ran the local cafe. The more Evelyn hears about how the unconventional duo remade their lives to suit each other so long ago, the more Evelyn finds the confidence to work on her own life. Back-and-forth narrative as it tracks multiple storylines is occasionally confusing, though strong character development holds interest. Adapted by Fannie Flagg from her novel, "Fried Green Tomatoes at the Whistle Stop Cafe."
1991 (PG-13) 130m/C Kathy Bates, Mary Stuart Masterson, Mary Louise Parker, Jessica Tandy, Cicely Tyson, Chris O'Donnell, Stan Shaw; *Dir:* Jon Avnet. VHS, Beta, LV *MCA* ✠✠✠

Frieda An RAF officer brings his German wife home after the war and she is naturally distrusted by his family and neighbors. An interesting look at post-war bigotry.
1947 97m/B *GB* David Farrar, Glynis Johns, Mai Zetterling, Flora Robson, Albert Lieven; *Dir:* Basil Dearden. VHS $19.95 *NOS, VYY, TIM* ✠✠½

Friendly Fire Based on a true story of an American family in 1970 whose soldier son is killed by "friendly fire" in Vietnam, and their efforts to uncover the circumstances of the tragedy. Touching and powerful, with an excellent dramatic performance by Burnett. Made for television.
1979 146m/C Carol Burnett, Ned Beatty, Sam Waterston, Timothy Hutton; *Dir:* David Greene. VHS, Beta $59.98 *FOX* ✠✠✠

Friendly Persuasion Earnest, solidly acted tale about a peaceful Quaker family struggling to remain true to its ideals in spite of the Civil War which touches their farm life in southern Indiana. Cooper and McGuire are excellent as the parents with Perkins fine as the son worried he's using his religion to hide his cowardice. Based on a novel by Jessamyn West; music by Dimitri Tiomkin. Oscar nominations for Best Picture, supporting actor Perkins, director, Adapted Screenplay, Best Song and Sound Recording.
1956 140m/C Gary Cooper, Dorothy McGuire, Anthony Perkins, Marjorie Main; *Dir:* William Wyler. VHS, Beta, LV $59.98 *FOX* ✠✠✠

Friends, Lovers & Lunatics A weekend turns into a romantic nightmare/laugh-fest when a man visits his ex-girlfriend and her new "friend", and winds up in a new romnce himself. Also known as "Crazy Horse."
1989 87m/C *CA* Daniel Stern, Deborah Foreman, Sheila McCarthy, Page Fletcher, Elias Koteas; *Dir:* Stephen Withrow. VHS, Beta, LV $29.95 *FRH, IME* ✠½

Fright A woman—convinced that she died in 1889 as part of a suicide pact with Prince Rudolph of Austria—seeks help from a psychiatrist who promptly falls in love with her (didn't Freud say it was supposed to be the other way around?). Originally titled "Spell of the Hypnotist."
1956 ?m/B Nancy Malone, Eric Fleming, Frank Marth, Humphrey Davis, Ned Glass, Norman Burton; *Dir:* W. Lee Wilder. VHS $19.98 *SNC* ✠

Fright A baby-sitter is menaced by a mental hospital escapee. He turns out to be the father of the boy she is watching. Tense but violent thriller.
1971 87m/C *GB* Susan George, Honor Blackman, Ian Bannen, John Gregson, George Cole, Dennis Waterman; *Dir:* Peter Collinson. VHS *MOV* ✠✠

Fright House Two stories of terror. "Fright House" features witches preparing an old mansion for a visit from the Devil. "Abandon" explains the prolonged youth of a teacher to her young student.
1989 (R) 110m/C Al Lewis, Duane Jones; *Dir:* Len Anthony. VHS, Beta $79.95 *SED* ✠

Fright Night It's Dracula-versus-the-teens time when Charley suspects that his new neighbor descends from Count Vlad's line. He calls in the host of "Fright Night," the local, late-night, horror-flick series, to help de-ghoul the neighborhood. But they have a problem when the vampire discovers their plans (and nobody belives them anyway). Sarandon is properly seductive as the bloodsucker. Followed by sequel.
1985 (R) 106m/C William Ragsdale, Chris Sarandon, Amanda Bearse, Roddy McDowall, Stephen Geoffreys; *Dir:* Tom Holland. VHS, Beta, LV $19.95 *COL* ✠✠½

Fright Night 2 The sequel to the 1985 release "Fright Night," in which the harassed guy from the original film learns slowly that the vampire's sister and her entourage have come to roost around his college. Not quite as good as the original but the special effects are worth a look.
1988 (R) 108m/C Roddy McDowall, William Ragsdale, Traci Lin, Julie Carmen; *Dir:* Tommy Lee Wallace. VHS, Beta, LV $14.95 *IVE* ✠✠

Frightmare A seemingly quiet British couple do indulge in one strange habit—they're cannibals.
1974 (R) 86m/C *GB* Deborah Fairfax, Kim Butcher, Rupert Davies, Sheila Keith; *Dir:* Pete Walker. VHS, Beta $59.98 *PSM* ✠✠½

Frightmare A great horror star dies, but he refuses to give up his need for adoration and revenge.
1983 (R) 84m/C Ferdinand "Ferdy" Mayne, Luca Bercovici, Nita Talbot, Peter Kastner; *Dir:* Norman Thaddeus Vane. VHS, Beta $59.95 *VES* ✠

Frightshow Four independent short horror films; "Nightfright," "Thing in the Basement," "Dr. Dobermind," and "Illegal Alien."
1984 60m/C VHS $29.98 *MPI* ✠

The Fringe Dwellers An Aborigine family leave their shantytown and move to the white, middle-class suburbs of Australia, encountering prejudice and other difficulties. Well acted and interesting but the ending's a letdown.
1986 (PG) 98m/C *AU* Kristina Nehm, Justine Saunders, Bob Maza, Kylie Belling, Denis Walker, Ernie Dingo; *Dir:* Bruce Beresford. VHS, Beta, LV $79.95 *SVS* ✠✠½

The Frisco Kid An innocent orthodox rabbi (Wilder) from Poland is sent to the wilds of San Francisco during the 1850s gold rush to lead a new congregation. He lands in Philadelphia, joins a wagon train and is promptly robbed and abandoned. He eventually meets up with a not-too-bright robber (Ford) who finds himself unexpectly befriending the man and undergoing numerous tribulations in order to get them both safely to their destination. This isn't a laugh riot and some scenes fall distinctly flat but Wilder is sweetness personified and lends the movie its charm.
1979 (R) 119m/C Gene Wilder, Harrison Ford, Ramon Bieri, George DiCenzo, Penny Peyser, William Smith; *Dir:* Robert Aldrich. VHS, Beta $19.98 *WAR* ✠✠

Fritz the Cat Ralph Bakshi's animated tale for adults about a cat's adventures as he gets into group sex, college radicalism, and other hazards of life in the '60s. Loosely based on the underground comics character by Robert Crumb. Originally X-rated.
1972 77m/C *Dir:* Ralph Bakshi. VHS, Beta *WAR, OM* ✠✠✠

Frog A boy adds a frog to his reptile collection but gets the surprise of his life when he finds it's really an enchanted Prince!
1989 (G) 55m/C Paul Williams, Scott Grimes, Shelley Duvall, Elliott Gould, David Grossman. VHS, Beta, LV $19.98 *ORI, IME* ✠✠½

Frogs An environmental photographer working on a small island in Florida interrupts the birthday celebration of a patriarch. He and the folks at the party soon realize that various amphibians animals in the surrounding area are going berserk and attacking humans. One of the first environmentally-motivated animal-vengeance films, and one of the best to come out of the '70s.
1972 (PG) 91m/C Ray Milland, Sam Elliott, Joan VanArk, Adam Roarke, Judy Pace, Lynn Borden, Mae Mercer, David Gilliam, George Skaff; *Dir:* George McCowan. VHS, Beta *WAR, OM* ✠✠½

Frolics on Ice Pleasant comedy-musical about a family man saving to buy the barber shop he works at. Irene Dare is featured in several ice skating production numbers.
1940 65m/B Roscoe Karns, Lynne Roberts, Irene Dare, Edgar Kennedy. **VHS, Beta $29.95** VYY *ℐℐ*

From Beyond A gruesome, tongue-in-cheek adaptation of the ghoulish H.P. Lovecraft story. Scientists discover another dimension through experiments with the pineal gland. From the makers of "Re-Animator," and just as funny.
1986 (R) 90m/C Jeffrey Combs, Barbara Crampton, Ted Sorel; *Dir:* Stuart Gordon. **VHS, Beta, LV $19.98** LIV, VES *ℐℐℐ*

From Beyond the Grave This horror compendium revolves around a mysterious antique shop whose customers experience various supernatural phenomena, especially when they try to cheat the shop's equally mysterious owner.
1973 (PG) 98m/C GB Peter Cushing, David Warner, Ian Bannen, Donald Pleasence, Margaret Leighton, Lesley-Anne Down, Diana Dors, Ian Ogilvy; *Dir:* Kevin Connor. **VHS, Beta $59.95** WAR *ℐℐ*

From Broadway to Cheyenne Here is an early western with lots of rip roarin' action, stunts and guns.
1932 60m/B Rex Bell, Robert Ellis, Alan Bridge, Matthew Betz, Marceline Day, George "Gabby" Hayes; *Dir:* Harry Fraser. **VHS, Beta $19.95** NOS, DVT *ℐ*

From China with Death After a vicious War Lord wipes out his family, a young man trains under a martial arts master to prepare himself for revenge.
1973 (R) 90m/C **VHS, Beta $59.95** HHT *ℐ*

From D-Day to Victory in Europe Based on Mr. Hastings best-selling novel, this program combines original battlefield footage with computer graphics for a thorough description of the War's turning point.
1985 110m/B Max Hastings. **VHS, Beta $39.98** MPI *ℐ*

From the Earth to the Moon A mad scientist invents a new energy source and builds a rocket which takes him, his daughter, and two other men for some adventures on the moon. Based on the novel by Jules Verne.
1958 100m/C George Sanders, Joseph Cotten, Debra Paget, Don Dubbins; *Dir:* Byron Haskin. **VHS, Beta $15.95** UHV *ℐℐ*

From the Earth to the Moon An animated version of the Jules Verne classic about man's first landing on the moon.
1979 50m/C **VHS, Beta $19.95** MGM *ℐℐ*

From Hell to Borneo A mercenary must defend his secluded island from pirates and gangsters.
1964 (PG) 90m/C George Montgomery, Torin Thatcher, Julie Gregg, Lisa Moreno; *Dir:* George Montgomery. **VHS, Beta $59.95** MON *ℐ½*

From Hell to Victory A group of friends of different nationalities vow to meet each year in Paris on the same date but World War II interrupts their lives and friendships. Lenzi used the alias Hank Milestone when he directed this film. Okay battle-scenes but nothing special.
1979 (PG) 100m/C SP FR George Peppard, George Hamilton, Horst Buchholz, Jean-Pierre Cassel, Capucine, Sam Wanamaker, Anny Duperey, Ray Lovelock; *Dir:* Umberto Lenzi. **VHS, Beta $9.95** SIM, MED, HHE *ℐ½*

From Here to Eternity A complex, hard-hitting look at the on and off-duty life of soldiers (both enlisted and officers) at the Army base in Honolulu in the days before the Pearl Harbor attack. Pruitt wants to be the company bugler but instead is given every dirty detail in camp because he refuses to continue as the company boxer (he won, but also blinded a friend in a fight). His best friend is Maggio, who's always in trouble, and his good-guy top sergeant just happens to be having a torrid affair with the commander's wife. Pruitt, meanwhile, is introduced to a club "hostess" who is a lot more vulnerable than she's willing to admit. This is a movie filled with great performances: Kerr and Reed shine in atypical roles; Clift is sensitive but stubborn; Lancaster, strong and passionate; and Borgnine is the epitome of villainy as a brutal seargant. Sinatra won an Oscar for his portrayal of the doomed Maggio. Still has the most classic waves-on-the-beach love scene in filmdom. Based on the novel by James Jones, which was toned down by the censors. Oscar nominations for Best Actor (Clift and Lancaster), Best Actress (Kerr), Score of a Comedy or Drama, and Black & White Costume Design.
1953 118m/B Burt Lancaster, Montgomery Clift, Frank Sinatra, Deborah Kerr, Donna Reed, Ernest Borgnine, Philip Ober, Jack Warden, Mickey Shaughnessy, George Reeves, Claude Akins; *Dir:* Fred Zinneman. Academy Awards '53: Best Black and White Cinematography, Best Director (Zinneman), Best Film Editing, Best Picture, Best Screenplay, Best Sound, Best Supporting Actor (Sinatra), Best Supporting Actress (Reed); Directors Guild of America Awards '53: Best Director (Zinneman); National Board of Review Awards '53: 10 Best Films of the Year; New York Film Critics Awards '53: Best Actor (Lancaster), Best Director (Zinneman), Best Film. **VHS, Beta, LV, 8mm $19.95** COL, PIA, BTV *ℐℐℐℐ*

From Here to Eternity From here to eternity...and back on the silver screen. Remake of the 1953 classic is based on James Jones' novel about life on a Hawaiian military base just before WWII. Decent actioner, but no Oscar-winner this time around.
1979 (PG-13) 110m/C Natalie Wood, Kim Basinger, William Devane, Steve Railsback, Peter Boyle, Will Sampson, Andy Griffith, Roy Thinnes; *Dir:* Buzz Kulik. **VHS, Beta, LV $89.95** VMK, FCT *ℐℐ½*

From Here to Maternity A made-for-television soap opera-based spoof about modern maternity. Three women want to be pregnant but their significant others aren't interested in paternity.
1985 40m/C Carrie Fisher, Arleen Sorkin, Lauren Hutton, Griffin Dunne, Paul Reiser. **VHS, Beta $59.95** LIV, VES *ℐℐ½*

From the Hip A young lawyer (Nelson) gets the chance of a lifetime when his office assigns a murder case to him. The only problem is that he suspects his client is very guilty indeed and must discover the truth without breaking his code of ethics. Nelson's courtroom theatrics are a bit out of hand, as he flares his nostrils at every opportunity. The movie is encumbered by a weak script and plot as well as a mid-story switch from comedy to drama.
1986 (PG) 112m/C Judd Nelson, Elizabeth Perkins, John Hurt, Ray Walston; *Dir:* Bob Clark. **VHS, Beta, LV $19.98** LHV, WAR *ℐℐ*

From Hollywood to Deadwood A beautiful starlet is kidnapped and the private eye searching for her discovers blackmail and danger.
1989 (R) 90m/C Scott Paulin, Jim Haynie, Barbara Schock; *Dir:* Rex Pickett. **VHS, Beta, LV $19.98** MED, FCT, IME *ℐℐ*

From the Life of the Marionettes A repressed man in a crumbling marriage rapes and kills a young prostitute. Another look at the powers behind individual motivations from Bergman, who uses black and white and color to relate details of the incident.
1980 (R) 104m/C SW Robert Atzorn, Christine Buchegger, Martin Benrath, Rita Russek, Lola Muethel, Walter Schmidinger, Heinz Bennent; *Dir:* Ingmar Bergman. **VHS, Beta, LV $39.95** IVE *ℐℐℐ*

From Russia with Love Bond is back and on the loose in exotic Istanbul looking for a super-secret coding machine. He's involved with a beautiful Russian spy and has the SPECTRE organization after him, including villainess Rosa Klebb (she of the killer shoe). Lots of exciting escapes but not an over-reliance on the gadgetry of the later films. The second Bond feature, thought by many to be the best. The laserdisc edition includes audio interviews with director Terence Young and others on the creative staff. The musical score and special effects can be separated from the actors' dialogue. The laserdisc also features publicity shots, American and British trailers, on-location photos, and movie posters.
1963 118m/C GB Sean Connery, Daniela Bianchi, Pedro Armendariz, Lotte Lenya, Robert Shaw, Eunice Gayson, Walter Gotell, Lois Maxwell, Bernard Lee, Desmond Llewelyn, Nadja Regin, Alizia Gur, Martine Beswick, Leila; *Dir:* Terence Young. **VHS, Beta, LV $19.95** MGM, VYG, CRC *ℐℐℐ½*

From "Star Wars" to "Jedi": The Making of a Saga Using copious amounts of actual footage from the films, this movie follows the production of all three "Star Wars" sagas, concentrating on the special effects involved.
1983 65m/C David Prowse, Harrison Ford, Carrie Fisher, Billy Dee Williams, Alec Guinness; *Dir:* George Lucas; *Nar:* Mark Hamill. **VHS, Beta $19.98** FOX, FLI *ℐℐℐ½*

From the Terrace Newman is a wealthy Pennsylvania boy who goes to New York and marries into even more money and social position when he weds Woodward. He gets a job with her family's investment company and neglects his wife for business. She turns to another man and when he goes home for a visit he also finds a new romance which leads to some emotional soul searching. The explicitness of O'Hara's very long novel was diluted by the censors and Newman's performance is a stilted disappointment. Loy, as Newman's alcoholic mother, earned the best reviews.
1960 144m/C Paul Newman, Joanne Woodward, Myrna Loy, Ina Balin, Felix Aylmer, Leon Ames, George Grizzard, Patrick O'Neal, Barbara Eden, Mae Marsh; *Dir:* Mark Robson. **VHS, Beta, LV $19.98** FOX *ℐℐ½*

The Front Woody is the bookmaker who becomes a "front" for blacklisted writers during the communist witch hunts of the 1950s in this satire comedy. The scriptwriter and several of the performers suffered blacklisting themselves during the Cold War. Based more or less on a true story.
1976 (PG) 95m/C Woody Allen, Zero Mostel, Herschel Bernardi, Michael Murphy, Danny Aiello, Andrea Marcovicci; *Dir:* Martin Ritt. National Board of Review Awards '76: 10 Best Films of the Year. **VHS, Beta, LV $12.95** COL *ℐℐℐ*

F

The Front Page The original screen version of the Hecht-MacArthur play about the shenanigans of a battling newspaper reporter and his editor in Chicago. O'Brien's film debut here is one of several hilarious performances in this breathless pursuit of an exclusive with an escaped death row inmate. **1931 101m/B** Adolphe Menjou, Pat O'Brien, Edward Everett Horton, Mae Clarke, Walter Catlett; *Dir:* Lewis Milestone. National Board of Review Awards '31: 10 Best Films of the Year. **VHS, Beta $19.95** *NOS, MRV, HHT* ♫♫♫½

The Front Page A remake of the Hecht-MacArthur play about the managing editor of a 1920s Chicago newspaper who finds out his ace reporter wants to quit the business of get married. But first an escaped convicted killer offers the reporter an exclusive interview. **1974 (PG) 105m/C** Jack Lemmon, Walter Matthau, Carol Burnett, Austin Pendleton, Vincent Gardenia, Charles Durning, Susan Sarandon; *Dir:* Billy Wilder. **VHS, LV $89.95** *MCA* ♫♫½

Frontera Sin Ley The son of Santo fights against local gangsters' crime and abuse in Pueblo Viejo. **19?? ?m/C** *MX* Mil Mascaras, Eleazar Garcia, Carmen Del Valle, Hijo Del Santo. **VHS, Beta $19.95** *WCV*

Frontera (The Border) If you're looking for action, go to the Border! Anything goes on the dangerous, exciting "Frontera." **19?? ?m/C** *SP* Fernando Allende, Daniela Romo. **VHS, Beta** *VVC*

Frontier Days Cody goes undercover to capture the leader of a gang of stagecoach robbers. He also manages to capture a killer along the way. **1934 60m/B** Bill Cody, Ada Ince, Wheeler Oakman; *Dir:* Sam Newfield. **VHS, Beta $19.95** *NOS, DVT* ♫½

Frontier Horizon Wayne and the Three Mesquiteers help out ranchers whose land is being bought up by crooked speculators. **1939 55m/B** John Wayne, Jennifer Jones, Ray Corrigan, Raymond Hatton; *Dir:* George Sherman. **VHS $19.95** *REP, CAB* ♫

Frontier Justice Gibson is the foolish son of a cattle owner who has been forced into an insane asylum because of a battle over water rights. Naturally, he must take on responsibility and rescue his father. **1935 56m/B** Hoot Gibson, Richard Cramer; *Dir:* Robert McGowan. **VHS, Beta $19.95** *NOS, VCN, DVT* ♫

Frontier Pony Express Roy and Trigger do their best to help the Pony Express riders who are being attacked by marauding gangs in California during the time of the Civil War. Plenty of action here. **1939 54m/B** Roy Rogers; *Dir:* Joseph Kane. **VHS, Beta $19.95** *NOS, MED, DVT* ♫♫

Frontier Scout This action-packed western has Wild Bill Hickock, portrayed by former opera singer Houston, forcing Lee's surrender in the Civil War before pursuing cattle rustlers out West. **1938 60m/B** George Houston, Mantan Moreland; *Dir:* Sam Newfield. **VHS, Beta $19.95** *NOS, DVT, RXM* ♫

Frontier Vengeance Barry is falsely accused of murder and works to clear his name while at the same time romancing a stagecoach driver who's trying to win a race with a rival stagecoach line. **1940 54m/B** Donald (Don "Red") Barry, Nate Watt. **VHS, Beta** *MED, RXM* ♫

Frontline Music Festival An amazing UNICEF benefit concert taped live in Zimbabwe. The music of Harry Belafonte, Miriam Makeba, and Johnny Clegg, combined with the setting of the dark continent, produces a rich and fascinating concert. Includes Maxi Priest, Youssou N'Dour, and more. **1989 80m/C** Harry Belafonte, Miriam Makeba, Johnny Clegg. **VHS, Beta $29.95** *WKV, MVD, KAR* ♫

Frosty the Snowman Classic cartoon about the magic snowman and the children he befriends. **1969 30m/C** *Dir:* Arthur Rankin Jr., Jules Bass; *Voices:* Billy DeWolfe, Jackie Vernon; *Nar:* Jimmy Durante. **VHS $14.95** *FHE, BTV* ♫

Frozen Alive A scientist experiments with suspended animation but, wouldn't you know, someone murders his wife while he's on ice. Apparently being frozen stiff does not an alibi make, and he becomes the prime suspect. Much unanimated suspense. **1964 80m/B** *GB GE* Mark Stevens, Marianne Koch, Delphi Lawrence, Joachim Hansen, Walter Rilla, Wolfgang Lukschy; *Dir:* Bernard Knowles. **VHS $19.98** *MOV* ♫

Frozen Terror A madman resembling both Jack Frost and Jack the Ripper claims victims at random. **1980 (R) 93m/C** Bernice Stegers; *Dir:* Lamberto Bava. **VHS, Beta $69.95** *LIV, VES* ♫

Fu Manchu Villainous arch-criminal Fu Manchu plots to enslave the world in "The Golden God" and "The Master Plan," two episodes from the television series. **1956 60m/B** Glenn Gordon, Laurette Luez, Clark Howat, Lee Matthews. **VHS, Beta $24.95** *DVT* ♫

Fuerte Perdido Settlers battle for their lives against hostile Indians led by Geronimo. In Spanish. Also known as "Fort Lost." **1978 90m/C** *SP* Esther Rojo, German Cobos, Mario Vidal. **VHS, Beta $59.95** *HHT* ♫

The Fugitive Fonda plays a priest devoted to his God and the peasants under his care. He finds himself on the run when all religion is outlawed in this nameless South-of-the-border dictatorship. Despite the danger of capture, he continues to minister to his flock. His eventual martyrdom unites the villagers in prayer, much to the chagrin of the militia. Considered Fonda's best performance and Ford's favorite film. Shot on location in Taxco, Cholula and Cuernavaca, Mexico. Excellent supporting performances. Based on the Graham Greene novel "The Power and the Glory" although considerably cleaned up for the big screen - Greene's priest had lost virtually all of his faith and moral code. Here the priest character is a genuine Ford hero with no moral ambiguities. A gem. **1948 99m/B** Henry Fonda, Dolores Del Rio, Pedro Armendariz, J. Carroll Naish, Leo Carrillo, Ward Bond, Robert Armstrong, John Qualen; *Dir:* John Ford. **VHS, Beta, LV $19.98** *TTC* ♫♫♫

Fugitive Alien Alien and Earth forces battle it out. Dubbed for Japanese. **1986 103m/C** *JP* **VHS, Beta $39.95** *CEL* Woof!

The Fugitive Kind A young drifter walks into a sleepy Mississippi town and attracts the attention of three of its women with tragic results. Based upon the Tennessee Williams' play "Orpheus Descending." good performances but the writing never hangs well-enough together for coherence.

1960 122m/B Marlon Brando, Anna Magnani, Joanne Woodward, Victor Jory, R.G. Armstrong, Maureen Stapleton; *Dir:* Sidney Lumet. **VHS, Beta $19.98** *MGM, FOX* ♫♫½

Fugitive Lovers An innocent guy rescues a gangster's beautiful wife from suicide, and invites his wrath by running away with her. **197? (R) 95m/C** Steve Oliver, Virginia Mayo, Sondra Currie, John Russell. **Beta** *MFI* ♫

Fugitive Samurai A samurai, with his young son, flees a vengeful shogun and seeks the friends who betrayed him. **1984 92m/C** Kinnosuke Yorozuya, Katzutaka Nishikawa. **VHS, Beta, LV $59.95** *SVS* ♫♫

The Fugitive: Taking of Luke McVane An early silent western in which Hart upholds the cowboy code of honor. **1915 28m/B** William S. Hart, Enid Markey. **VHS, Beta $24.95** *VYY* ♫♫

Full Fathom Five Central American militants, angered by America's invasion of Panama, hijack a submarine and threaten a nuclear assault on Houston. Boring and dumb. **1990 (PG) 90m/C** Michael Moriarty, Maria Rangel, Michael Cavanaugh, John Lafayette, Todd Field, Daniel Faraldo; *Dir:* Carl Franklin. **VHS, Beta $79.98** *MGM* ♫½

Full Hearts & Empty Pockets Follows the happy-go-lucky adventures of a young, handsome, impoverished gentleman on the loose in Rome. Dubbed in English. **1963 88m/B** *IT GE* Linda Christian, Gino Cervi, Senta Berger; *Dir:* Camillo Mastrocinque. **VHS, Beta $29.95** *VYY* ♫½

Full Metal Jacket A three-act Vietnam War epic co-written by Kubrick, Michael Herr and the novel's original author, Gustav Hasford, about a single Everyman impetuously passing through basic training then working in the field as an Army photojournalist and fighting at the onset of the Tet offensive. First half of the film is the most realistic bootcamp sequence ever done. Unfocused but powerful. **1987 (R) 116m/C** Matthew Modine, Lee Ermey, Vincent D'Onofrio, Adam Baldwin, Dorian Harewood, Arliss Howard, Kevyn Major Howard; *Dir:* Stanley Kubrick. Boston Society of Film Critics Awards '87: Best Director; London Film Critics Circle '87: Best Director. **VHS, Beta, LV, 8mm $19.98** *WAR, FCT* ♫♫♫½

Full Metal Ninja When his family is kidnapped, a trained martial artist trails their abductors and swears revenge. **1989 90m/C** Pat Allen, Pierre Kirby, Sean Odell, Jean Paul, Renato Sala; *Dir:* Charles Lee. **VHS, Beta $59.95** *IMP* ♫

Full Moon in Blue Water His wife's been dead a year, he owes back taxes on his bar, his father has Alzheimer's, and his only form of entertainment is watching old home movies of his wife: Floyd has problems. Enter Louise, a lonely spinster who feels that it is her personal duty to change his life. If she doesn't do it, the bizarre things that happen after her arrival will. **1988 (R) 96m/C** Gene Hackman, Teri Garr, Burgess Meredith, Elias Koteas; *Dir:* Peter Masterson. **VHS, Beta, LV $19.98** *MED* ♫♫½

Full Moon in Paris A young woman in Paris moves out on her architect lover in order to experience freedom. Through a couple of random relationships, she soon finds that what she hoped for is not what she really wanted.

1984 (R) 101m/C *FR* Pascale Ogier, Tcheky Karyo, Fabrice Luchini; *Dir:* Eric Rohmer. Venice Film Festival '84: Best Actress (Ogier). **VHS, Beta $24.95** *XVC, MED* 𝔅𝔅½

Fuller Brush Man A newly hired Fuller Brush man becomes involved in murder and romance as he tries to win the heart of his girlfriend. A slapstick delight for Skelton fans.
1948 93m/B Red Skelton, Janet Blair, Don McGuire, Adele Jergens; *Dir:* Frank Tashlin. **VHS, Beta, LV $14.95** *COL* 𝔅𝔅½

Fun in Acapulco Former trapeze artist Elvis romances two beauties and acts as a part-time lifeguard and night club entertainer. He must conquer his fear of heights for a climactic dive from the Acapulco cliffs. Features ten musical numbers.
1963 97m/C Elvis Presley, Ursula Andress, Elsa Cardenas, Paul Lukas, Alejandro Rey, Larry Domasin, Howard McNear; *Dir:* Richard Thorpe. **VHS, Beta $14.98** *FOX, MVD* 𝔅𝔅

Fun with Dick and Jane An upper-middle class couple turn to armed robbery to support themselves when the husband is fired from his job. Though it has good performances, the film never develops its intended bite.
1977 (PG) 104m/C George Segal, Jane Fonda, Ed McMahon; *Dir:* Ted Kotcheff. **VHS, Beta $9.95** *COL* 𝔅𝔅

Fun Down There This smalltown boy-comes-to-the-big city is so bad, it's...bad. A naive—and unsympathetic—gay man from upstate New York moves to Greenwich Village and has fun down there (and they don't mean Australia). Written by director Roger Stigliano and Michael Waite.
1988 89m/C Michael Waite, Nickolas Nagurney, Gretschen Somerville, Martin Goldin, Kevin Och; *Dir:* Roger Stigliano. **VHS $59.95** *WBF, FCT* *Woof!*

Fun with the Fab Four Rare Beatles footage of the sixties' youth explosion is captured, including Peter Best and the Shakespearean sketch where John and Paul play lovers.
1990 60m/C John Lennon, Paul McCartney, George Harrison, Ringo Starr. **VHS $10.99** *MVD, GKK* *Woof!*

Fun Factory/Clown Princes of Hollywood Collection of various slapstick situations including Keystone Cops segments. From the "History of Motion Pictures" series.
192? 56m/B Mack Sennett, Charlie Chaplin, Buster Keaton, Charley Chase, Ben Turpin, Stan Laurel. **VHS, Beta $29.98** *CCB* *Woof!*

Fun & Fancy Free Part-animated, part-live-action feature is split into two segments: "Bongo" with Dinah Shore narrating the story of a happy-go-lucky circus bear; and "Mickey and the Beanstalk"—a "new" version of an old fairy tale.
1947 96m/C Edgar Bergen, Charlie McCarthy, Jiminy Cricket, Mickey Mouse, Donald Duck, Goofy, Dinah Shore; *Dir:* Walt Disney. **VHS, Beta** *DIS* *Woof!*

Fun Shows One episode apiece of "Life of Riley" and "The Great Gildersleeve."
1950 60m/B Jackie Gleason, Willard Waterman. **VHS, Beta $24.95** *NOS, DVT*

The Funeral A sharp satire of the clash of modern Japanese culture with the old. The hypocrisies, rivalries and corruption in an average family are displayed at the funeral of its patriarch who also happened to own a house of ill repute. Itami's breakthrough film, in Japanese with English subtitles.
1984 112m/C *JP* Tsutomu Yamazaki, Nobuko Miyamoto, Kin Sugai, Ichiro Zaitsu, Nekohachi Edoya, Shoji Otake; *Dir:* Juzo Itami. **VHS $19.98** *REP, FCT* 𝔅𝔅𝔅½

Funeral for an Assassin A professional assassin seeks revenge for his imprisonment by the government of South Africa, a former client. Planning to kill all of the country's leading politicians, he masquerades as a black man, a cover which is designed to fool the apartheid establishment.
1977 (PG) 92m/C Vic Morrow, Peter Van Dissel; *Dir:* Ivan Hall. **VHS, Beta $59.95** *GEM, MRV* 𝔅½

Funeral in Berlin Second of the Caine-Harry Palmer espionage films (following up "Ipcress File"), in which the deadpan British secret serviceman arranges the questionable defection of a Russian colonel. Good look at he spy biz and postwar Berlin. "Billion Dollar Brain" continues the series.
1966 102m/C Michael Caine, Eva Renzi, Oscar Homolka; *Dir:* Guy Hamilton. **VHS, Beta, LV $14.95** *PAR* 𝔅𝔅𝔅

Funeral Home A terrified teen spends her summer vacation at her grandmother's tourist home, a former funeral parlor. A rip-off of "Psycho." Also known as "Cries in the Night."
1982 90m/C *CA* Lesleh Donaldson, Kay Hawtry; *Dir:* William Fruet. **VHS, Beta** *PGN* 𝔅½

The Funhouse Four teenagers spend the night at a carnival funhouse, witness a nurder, and become next on the list of victims. From the director of cult favorite "Texas Chainsaw Massacre," but nothing that hasn't been seen before.
1981 (R) 96m/C Elizabeth Berridge, Shawn Carson, Cooper Huckabee, Largo Woodruff, Sylvia Miles; *Dir:* Tobe Hooper. **VHS, Beta, LV $14.98** *MCA* 𝔅½

Funhouse Made-for-PBS performance with Bogosian portraying mild-mannered employees describing gruesome feats to unsuspecting clients.
1987 60m/C Eric Bogosian. **VHS, Beta, LV $19.95** *PAV* 𝔅½

Funland In the world's weirdest amusement park a clown goes nuts over the new corporate owners' plans for the place and decides to seek revenge.
1989 (PG-13) 86m/C David Lander, William Windom, Bruce Mahler, Michael McManus; *Dir:* Michael A. Simpson. **VHS, LV $79.98** *LIV, IME, VES* 𝔅

Funnier Side of Eastern Canada with Steve Martin In his first television special, Steve Martin "tours" eastern Canada, taking his viewers to lunch at a fancy French restaurant, to Toronto's city hall, and to other attractions. The zany comedian performs magic tricks, plays the banjo, and juggles.
1974 60m/C VHS, Beta *IND* 𝔅

Funny About Love Fairly absurd tale with dashes of inappropriate black humor about a fellow with a ticking biological clock. Nimoy sheds his Spock ears to direct a star-studded cast in this lame tale of middle-aged cartoonist Wilder who strays from wife Lahti after having problems on the conception end into the welcoming arms of fertile college co-ed Masterson in his quest to contribute to the population count.
1990 (PG-13) 107m/C Gene Wilder, Christine Lahti, Mary Stuart Masterson, Robert Prosky, Stephen Tobolowsky, Anne Jackson, Susan Ruttan, David Margulies; *Dir:* Leonard Nimoy. **VHS, Beta, LV, 8mm $19.95** *PAR* 𝔅½

Funny, Dirty Little War An Argentinian farce about the petty rivalries in a small village erupting into a violent, mini-civil war. In Spanish with English subtitles.
1983 80m/C *AR* Federico Luppi, Julio de Grazia, Miguel Angel Sola; *Dir:* Hector Olivera. **VHS, Beta $24.95** *XVC, MED* 𝔅𝔅½

Funny Face A musical satire on beatniks and the fashion scene also features the May-December romance between Astaire and the ever-lovely Hepburn. He is a high-fashion photographer (based on Richard Avedon); she is a Greenwich Village bookseller fond of shapeless, drab clothing. He decides to take her to Paris and show her what modeling's all about. The elegant musical score features such Gershwin songs as "Bonjour Paris," "How Long Has This Been Going On," "S'Wonderful," and "Funny Face." The laserdisc includes the original theatrical trailer and is available in widescree.
1957 103m/C Fred Astaire, Audrey Hepburn, Kay Thompson, Suzy Parker; *Dir:* Stanley Donen. **VHS, Beta, LV, 8mm $14.95** *PAR, LDC, FCT* 𝔅𝔅𝔅

The Funny Farm A group of ambitious young comics strive to make it in the crazy world of comedy at Los Angeles' famous club, The Funny Farm.
1982 90m/C *CA* Miles Chapin, Eileen Brennan, Peter Aykroyd; *Dir:* Ron Clark. **VHS, Beta $19.95** *STE, HBO* 𝔅½

Funny Farm Chevy Chase as a New York sportswriter finds that life in the country is not quite what he envisioned. The comedy is uneven throughout; with the best scenes coming at the end.
1988 (PG) 101m/C Chevy Chase, Madolyn Smith, Joseph Maher, Jack Gilpin, Brad Sullivan, MacIntyre Dixon; *Dir:* George Roy Hill. **VHS, LV $19.98** *WAR* 𝔅𝔅

Funny Girl Follows the early career of comedian Fanny Brice, her rise to stardom with the Ziegfeld Follies and her stormy romance with gambler Nick Arnstein. The classic songs "People" and "Don't Rain on My Parade" are featured. Streisand's film debut followed her auspicious performance of the role on Broadway. Score was augmented by several tunes sung by Brice during her performances. Excellent performances from everyone, captured beautifully by Wyler in his musical film debut. Fun and funny look at back stage music hall life in the early 1900s. Oscar nominations for Best Picture, supporting actress Medford, Cinematography, Sound, Best Song, Score of a Musical Picture and Film Editing. Followed by "Funny Lady."
1968 (G) 151m/C Barbra Streisand, Omar Sharif, Walter Pidgeon, Kay Medford, Anne Francis; *Dir:* William Wyler. Academy Awards '68: Best Actress (Streisand). **VHS, Beta, LV $19.95** *COL, PIA, BTV* 𝔅𝔅𝔅

Funny Guys & Gals of the Talkies Four short pictures featuring the comedy stars of the early talking pictures: "The Golf Specialist," "Pardon My Pups," "Girls Will Be Boys," and "Band Rally Radio Show."
193? 60m/B W.C. Fields, Shirley Temple, Charlotte Greenwood, Groucho Marx, Marlene Dietrich. **VHS, Beta $24.95** *VYY* 𝔅½

Funny Lady A continuation of "Funny Girl," recounting Fanny Brice's tumultuous marriage to showman Billy Rose in the 1930s and her lingering affection for first husband Nick Arnstein. Songs featured include: "How Lucky Can You Get?," "Great Day," "More

Than You Know'' and ''I Found a Million Dollar Baby in the Five and Ten Cent Store.''
1975 (PG) 137m/C Barbra Streisand, Omar Sharif, James Caan, Roddy McDowall, Ben Vereen, Carole Wells, Larry Gates, Heidi O'Rourke; *Dir:* Herbert Ross. **VHS, Beta, LV** **$19.95** *COL, PIA, FCT* 🎬🎬½

Funny Money Pointed, satirical look at the British habit of financing life with the almighty credit card. Sometimes uneven comedy.
1982 92m/C *GB* Greg Henry, Elizabeth Daily, Gareth Hunt, Derren Nesbitt, Annie Ross; *Dir:* James K. Clarke. **VHS, Beta $69.95** *LIV, LTG* 🎬

A Funny Thing Happened on the Way to the Forum A bawdy Broadway farce set in ancient Rome where a conniving, slave plots his way to freedom by aiding in the romantic escapades of his master's inept son. Terrific performances by Mostel and Gilford. Keaton's second-to-last film role. The Oscar-winning score includes such highlights as ''Comedy Tonight,'' ''Lovely,'' and ''Everybody Ought to Have a Maid.''
1966 100m/C Zero Mostel, Phil Silvers, Jack Gilford, Buster Keaton, Michael Hordern, Michael Crawford, Annette Andre; *Dir:* Richard Lester. Academy Awards '66: Best Musical Score. **VHS, Beta, LV $19.99** *FOX, FCT* 🎬🎬🎬

Funstuff Shirley Temple and friends struggle for stardom. Harold Lloyd plays ''Non-Stop Kid'' and Snub Pollard strives to be an artist.
193? 59m/B Shirley Temple, Harold Lloyd, Snub Pollard. **VHS, Beta $19.98** *CCB* 🎬🎬🎬

Furia Pasional A hot, passionate film about a lusty couple in romantic trouble.
19?? ?m/C *SP* Alfonso Muguia, Hilda Cibar, Manuel Cepeda. **VHS, Beta** *MAD*

The Furious Bruce Lee fights the forces of a villainous drug dealer known as the Cobra.
1981 (PG) 88m/C Bruce Le. **VHS, Beta $29.99** *BFV* 🎬

Furious The sister of a young martial arts apprentice is killed by the master, with a great deal of action and special effects.
1983 (PG) 85m/C Simon Rhee, Howard Jackson, Arlene Montano. **VHS, Beta $49.95** *VHE* 🎬

The Furious Avenger A released convict, on his way to avenging his family's murder, stops a rape and takes on one opponent after another in this rousing kick-fest. Dubbed.
1976 84m/C Hsiung Fei, Fan Ling, Chen Pei Ling, Tien Yeh, Wang Mo Chou. **VHS, Beta $39.95** *UNI* 🎬

Furious Slaughter In the 1930's, a martial artist risks it all in a bold attempt to obliterate a white slave ring.
19?? 90m/C Jimmy Wang Yu. **VHS, Beta $19.95** *OCE* 🎬

The Further Adventures of Tennessee Buck A mercenary adventurer travels through tropical jungles acting as a guide to a dizzy couple. Along the way, they encounter cannibals, headhunters and other bizarre things in this lame take on ''Indiana Jones.'' Shot on location in Sri Lanka.
1988 (R) 90m/C David Keith, Kathy Shower, Brant Van Hoffman; *Dir:* David Keith. **VHS, Beta, LV $19.98** *MED* 🎬

The Fury The head of a government institute for psychic research finds that his own son is snatched by supposed terrorists who wish to use his lethal powers. The father tries to use another a young woman with psychic power to locate him, with bloody results. A real chiller.
1978 (R) 117m/C Kirk Douglas, John Cassavetes, Carrie Snodgress, Andrew Stevens, Amy Irving, Charles Durning, Darryl Hannah; *Dir:* Brian DePalma. **VHS, Beta $19.98** *FOX* 🎬🎬

Fury A man and woman meet in the Amazonian jungle and the sparks begin to fly.
198? 75m/C Stuart Whitman, Laura Gemser. **VHS, Beta $59.95** *MGL* 🎬

Fury to Freedom: The Life Story of Raul Ries A young man raised in the volatile world of abusive and alcoholic parents grows into an abusive and violent adult. One night, however, he meets something he can't beat up, and it changes his life.
1985 78m/C Gil Gerard, John Quade, Tom Silardi. **VHS $39.95** *UHV* 🎬

The Fury of Hercules It's up to the mighty son of Zeus to free an enslaved group of people from an oppressive, evil ruler. Surely all who stand in his way will be destroyed. Another Italian muscleman film.
1961 95m/C *IT* Brad Harris, Brigitte Corey, Mara Berni, Carlo Tamberlani, Serge Gainsbourg, Elke Arendt, Alan Steel; *Dir:* Gianfranco Parolini. **VHS $9.95** *GKK, SNC* 🎬½

Fury of King Boxer The rebel Shaolin warriors fight the evil overlords, as usual.
1983 92m/C Kuo Shu Chung, Wang Yu; *Dir:* Ting Shan Si. **VHS, Beta $29.99** *BFV* 🎬

Fury in the Shaolin Temple A kung fu master kills the thief who stole his temple's secret by kicking him in the head.
198? 89m/C Liu Chia Hui. **VHS, Beta $54.95** *MAV* 🎬

Fury on Wheels A tale of men who race cars and wreak havoc.
1971 (PG) 89m/C Judd Hirsch, Tom Ligon, Paul Sorvino, Logan Ramsey, Colin Wilcox. **VHS, Beta** *PGN* 🎬

The Fury of the Wolfman A murdering werewolf is captured by a female scientist who tries to cure his lycanthropy with drugs and brain implants. Pretty slow going, but hang in there for the wolfman versus werewoman climax.
1970 80m/C *SP* Paul Naschy, Perla Cristal; *Dir:* Jose Maria Zabalza. **VHS, Beta $49.95** *UNI, CHA, LOO* 🎬

Fury is a Woman A Polish version of Shakespeare's Macbeth which ranks with the greatest film translations of his work. In Czarist Russia the passionate wife of a plantation owner begins an affair with a farm hand, and poisons her father-in-law when he finds them out. As her madness grows, she plots the murder of the husband. Letterbox format. In Serbian with English subtitles.
1961 93m/B Olivera Markovic, Ljuba Tadic, Kapitalina Eric; *Dir:* Andrzej Wajda. **VHS $59.95** *HTV* 🎬🎬🎬½

Future Cop A hard-nosed, old-fashioned cop is forced to team with the ultimate partner—a robot. Silly made-for-TV movie, but likable leads.
1976 78m/C Ernest Borgnine, Michael J. Shannon, John Amos, John Larch, Herbert Nelson, Ronnie Clair Edwards; *Dir:* Judd Taylor. **VHS** *PAR* 🎬🎬

Future Force In the crime-filled future cops can't maintain order. They rely on a group of civilian mercenaries to clean up the streets.
1989 (R) 90m/C David Carradine, Robert Tessier, Anna Rapagna, William Zipp. **VHS, Beta $79.95** *AIP* 🎬

Future Hunters In the holocaust-blitzed future, a young couple searches for a religious artifact that may decide the future of the planet.
1988 96m/C Robert Patrick, Linda Carol, Ed Crick, Bob Schott; *Dir:* Cirio H. Santiago. **VHS, Beta, LV $19.98** *LIV, VES* 🎬

Future Kill Anti-nuclear activists battle fraternity brothers in this grim futuristic world. Shaky political alliances form between revenge-seeking factions on both sides.
1985 (R) 83m/C Edwin Neal, Marilyn Burns, Doug Davis; *W/Dir:* Ronald W. Moore. **VHS, Beta, LV $19.98** *LIV, VES* 🎬

Future Zone In an attempt to save his father from being murdered, a young man travels backwards in time.
1990 (R) 88m/C David Carradine, Charles Napier, Ted Prior; *Dir:* David A. Prior. **VHS $79.95** *AIP* 🎬🎬

Futureworld In the sequel to ''Westworld,'' two reporters junket to the new ''Futureworld'' theme park, where they first support a scheme to clone and control world leaders. Includes footage shot at NASA locations.
1976 (PG) 107m/C Peter Fonda, Blythe Danner, Arthur Hill, Yul Brynner, Stuart Margolin; *Dir:* Richard T. Heffron. **VHS, Beta $19.98** *WAR, OM* 🎬🎬

Futz Story from off-Broadway is of a man who loves his pig and the world that can't understand the attraction. Better in original stage production.
1969 90m/C John Pakos, Victor Lipari, Sally Kirkland; *Dir:* Tom O'Horgan. **VHS, Beta $49.95** *IND* 🎬½

Fuzz Reynolds in the Boston cop on the track of a bomber killing policemen, punks setting winos on fire, and some amorous fellow officers. Combines fast action and some sharp-edged humor.
1972 (PG) 92m/C Burt Reynolds, Tom Skerritt, Yul Brynner, Raquel Welch, Jack Weston, Charles Martin Smith; *Dir:* Richard A. Colla. **VHS, Beta $59.98** *FOX* 🎬🎬½

Fyre A young beautiful girl who moves from the midwest to Los Angeles and winds up as a prostitute.
1978 (R) 90m/C Allen Garfield, Lynn Theel, Tom Baker, Cal Haynes, Donna Wilkes, Bruce Kirby; *Dir:* Richard Grand. **VHS, Beta** *MED* 🎬½

''G'' Men Powerful story based loosely on actual events that occurred during FBI operations in the early 1930s, with Cagney on the right side of the law, though still given to unreasonable fits of anger. Raised and educated by a well-known crime kingpin, a young man becomes an attorney. When his friend the FBI agent is killed in the line of duty, he joins the FBI to seek vengeance on the mob. But his mob history haunts him, forcing him to constantly prove to his superiors that he is not under its influence. Tense and thrilling classic.
1935 86m/B James Cagney, Barton MacLane, Ann Dvorak, Margaret Lindsay, Robert Armstrong, Lloyd Nolan, William Harrigan, Regis Toomey; *Dir:* William Keighley. **VHS, Beta $19.98** *MGM, CCB* 🎬🎬🎬½

G-Men Never Forget Moore, former stunt-man and future Lone Ranger, stars in this 12-part serial about FBI agents battling the bad guys. Lots of action. Alternate title, "Code 645." On two cassettes.
1948 167m/B Clayton Moore, Roy Barcroft, Ramsay Ames, Drew Allen, Tommy Steele, Eddie Acuff; **Dir:** Fred Brannon, Yakima Canutt. **VHS** $29.98 REP, MLB ✂✂½

G-Men vs. the Black Dragon Reedited serial of "Black Dragon of Manzanar" stars Cameron as Fed who battles Asian Axis agents during WWII. Action-packed.
1943 244m/B Rod Cameron, Roland Got, Constance Worth, Nino Pipitone, Noel Cravat; **Dir:** William Witney. **VHS, LV** $29.98 REP, MED, MLB ✂✂½

Gabe Kaplan as Groucho Kaplan impersonates the famous Marx Brother and runs through his most renowned routines.
1988 90m/C Gabe Kaplan. **VHS, Beta, LV** $29.95 JTC ✂✂½

Gabriela A sultry romance develops between a Brazilian tavern keeper and the new cook he's just hired. Music by Antonio Carlos. Derived from Jorge Amado's novel. In Portuguese with English subtitles.
1984 (R) 105m/C BR Sonia Braga, Marcello Mastroianni, Nelson Xavier, Antonio Cantafora; **Dir:** Bruno Barreto. **VHS, Beta** $79.95 MGM ✂✂

Gaby: A True Story The story of a woman with congenital cerebral palsy who triumphs over her condition with the help of family and loved ones, and ends up a college graduate and acclaimed author. Based on a true story.
1987 (R) 115m/C Rachel Levin, Liv Ullman, Norma Aleandro, Robert Loggia; **Dir:** Luis Mandoki. **VHS, Beta, LV** $79.95 COL ✂✂✂

Gai Dimanche/Swing to the Left Two early Tati shorts, wherein he takes unknowing tourists on an excursion in a used bus, and slugs it out with a professional boxer.
1936 40m/B **Dir:** Jacques Tati. **VHS, Beta** $49.95 FCT ✂✂✂

Gaiety Explores the lighter side of lion hunting.
1943 40m/B Antonio Moreno, Armida, Anne Ayars. **VHS, Beta** $14.95 QNE ✂½

Gal Young 'Un Set during the Prohibition era, a rich, middle-aged woman living on her property in the Florida backwoods finds herself courted by a much younger man. She discovers she is being used to help him set up a moonshining business. Unsentimental story of a strong woman. Based on a Marjorie Kinnan Rawlings story.
1979 105m/C Dana Preu, David Peck, J. Smith; **Dir:** Victor Nunez. **VHS, Beta** $29.95 ACA ✂✂✂

Galactic Gigolo An alien descends to earth on vacation, and discovers he's irresistible to human women. Intended as science-fiction satire, but there's more sleaze than humor.
1987 80m/C Carmine Capobianco, Debi Thibeault, Ruth Collins, Angela Nicholas, Frank Stewart; **Dir:** Gorman Bechard. **Beta** $69.95 UCV ✂

Galaxina In the 31st century, a beautiful robot woman capable of human feelings is created. A parody of superspace fantasies featuring one of murdered Playmate Stratten's few film appearances.

1980 (R) 95m/C Dorothy Stratten, Avery Schreiber, Stephen Macht; **Dir:** William Sachs. **VHS, Beta, LV** $59.95 MCA ✂✂½

Galaxy Express In this animated adventure, a young boy sets out to find immortality by traveling on The Galaxy Express, an ultra-modern 35th century space train that carries its passengers in search of their dreams. Also known as "Galaxy Express 999."
1982 (PG) 91m/C JP **Dir:** Taro Rin; **Voices:** Booker Bradshaw, Corey Burton. **VHS, Beta** $39.98 SUE ✂✂

Galaxy Invader Chaos erupts when an alien explorer crash lands his spacecraft in a backwoods area of the United States.
1985 (PG) 90m/C Richard Ruxton, Faye Tilles, Don Liefert. **VHS, Beta** $24.95 NEG ✂

Galaxy of Terror Astronauts sent to rescue a stranded spaceship get killed by vicious aliens. An inferior "Alien" imitation without any surprises but lots of gore. Also known as "Mindwarp: An Infinity of Terror" and "Planet of Horrors." Followed by the sequel "Forbidden World."
1981 (R) 85m/C Erin Moran, Edward Albert, Ray Walston; **Dir:** B.D. Clark. **VHS, Beta** $9.98 SUE ✂½

Gallagher in Concert Gallagher is back and funnier than ever in this video release to benefit wildlife preservation.
1988 43m/C Gallagher. **VHS, Beta** $29.95 GEO

Gallagher: Melon Crazy Human sledge-o-matic live on stage! Some of Gallagher's most renowned material.
1985 58m/C Gallagher. **VHS, Beta, LV** $19.95 PAR

Gallagher: Over Your Head The popular comedian expounds on politicians, ancient history, childbearing, and other topical subjects.
1986 58m/C Gallagher. **VHS, Beta** $19.95 PAR

Gallagher: Stuck in the 60's Comic Gallagher describes what his life was like in the 1960's in this concert performance.
1984 58m/C VHS, Beta, LV $19.95 PAR

Gallagher: The Bookkeeper The notorious comic takes a stand-up aim at bureaucracy and the trials of modern life.
1985 58m/C VHS, Beta $19.95 PAR

Gallagher: The Maddest The madcap comedy of Gallagher is captured in this concert.
1984 60m/C VHS, Beta, LV $19.95 PAR

Gallagher's Overboard The nutty comedian bashes even more watermelons...and it just gets funnier every time.
1987 54m/C Gallagher. **VHS** $19.95 PAR

Gallagher's Travels A screwball male/female reporting team chase an animal smuggling ring through the wilds of Australia.
198? 94m/C AU Ivar Kants, Joanne Samuels; **Dir:** Michael Caulfield. **VHS, Beta** $79.95 TWE

The Gallant Hours Biography of Admiral "Bull" Halsey (Cagney) covers five weeks, October 18 through December 1, 1942, in the World War II battle of Guadalcanal in the South Pacific. Director Montgomery forgoes battle scenes to focus on the human elements in a war and what makes a great leader. Fine performance by Cagney.

1960 115m/B James Cagney, Dennis Weaver, Ward Costello, Richard Jaeckel, Les Tremayne, Robert Burton, Raymond Bailey, Karl Swenson, Harry Landers, James T. Goto; **Dir:** Robert Montgomery. **VHS** $19.98 MGM ✂✂✂

Gallery of Horror Series of five horror stories featuring memorable, nightmarish moments. Chaney and Carradine each appear in one segment. Also known as "Dr. Terror's Gallery of Horrors," and "Return from the Past."
1967 90m/C Lon Chaney Jr., John Carradine. **VHS, Beta** $19.95 ACA, MRV ✂

Gallipoli History blends with the destiny of two friends as they become part of the legendary World War I confrontation between Australia and the German-allied Turks. A superbly filmed, gripping commentary on the wastes of war. Haunting score; excellent performances by Lee and a then-unknown Gibson.
1981 (PG) 111m/C AU Mel Gibson, Mark Lee, Bill Kerr, David Argue, Tim McKenzie; **Dir:** Peter Weir. Australian Film Institute '81: Best Actor (Gibson), Best Film. **VHS, Beta, LV** $19.95 PAR ✂✂✂✂

Galloping Dynamite Early western film based on a James Oliver Curwood story. A ranger avenges the death of his brother killed for gold.
1937 58m/B Kermit Maynard; **Dir:** Harry Fraser. **VHS, Beta** $19.95 VDM ✂

The Galloping Ghost A 12-chapter serial starring football great "Red" Grange about big games, gambling, and underworld gangs.
1931 226m/B VHS, Beta $49.95 NOS, VDM, VYY ✂

Galyon A soldier of fortune is recruited by an oil tycoon to find his daughter and son-in-law who have been kidnapped by terrorists in South America.
1977 (PG) 92m/C Stan Brock, Lloyd Nolan, Ina Balin. **VHS, Beta** $59.95 MON ✂

Gambit Caine's first Hollywood film casts him as a burglar who develops a "Topkapi"-style scheme to steal a valuable statue with MacLaine as the lure and Lom as the equally devious owner. Once Caine's careful plan is put into operation, however, everything begins to unravel.
1966 109m/C Shirley MacLaine, Michael Caine, Herbert Lom, Roger C. Carmel; **Dir:** Ronald Neame. **VHS, Beta** $59.95 MCA ✂✂½

The Gamble A young man (Modine) seeks to rescue his father from gambling debts by wagering himself against a lustful countess (Dunaway). When he loses the bet, however, he flees and joins up with a runaway (Beals) with the countess and her henchmen in close pursuit.
1988 (R) 108m/C Matthew Modine, Jennifer Beals, Faye Dunaway; **Dir:** Carlo Vanzina. **VHS, Beta, LV** $89.98 PSM ✂✂½

Gamble on Love A woman returns to her father's Las Vegas casino, and falls for the man who manages the gaming room.
1986 90m/C Louis Jourdan. **Beta** $14.95 PSM ✂

The Gambler College professor Axel Freed has a gambling problem so vast that it nearly gets him killed by his bookies. He goes to Las Vegas to recoup his losses and wins big, only to blow the money on stupid sports bets that get him deeper into trouble. Excellent character study of a compulsive loser on a downward spiral.

1974 (R) 111m/C James Caan, Lauren Hutton, Paul Sorvino, Burt Young, James Woods; *Dir:* Karel Reisz. VHS, Beta, LV $69.95 *PAR* ✂️✂️✂️

The Gambler & the Lady A successful London-based gambler and casino owner endeavors to climb the social ladder by having an affair with a member of the British aristocracy. He not only has to contend with his jilted nightclub singer girlfriend but also with the gangsters who would like to take over his clubs. Dull direction and a mediocre script.
1952 72m/B *GB* Dane Clark, Naomi Chance, Kathleen Byron; *Dir:* Patrick Jenkins. VHS, Beta, LV *WGE* ✂️½

The Gambling Samurai A samurai returns to his village to find it besieged by a ruthless government. Instead of attacking, he bides his time and works to undermine the government—finally achieving his revenge. Also known as "Kunisada Chuji." In Japanese with English subtitles.
1960 101m/C *JP* Toshiro Mifune, Michiyo Aratama; *Dir:* Senkichi Taniguhi. VHS, Beta $59.95 *VDA* ✂️✂️

The Gambling Terror A cowboy goes after a mobster selling protection to hapless ranchers.
1937 53m/B Johnny Mack Brown, Charlie King; *Dir:* Sam Newfield. VHS, Beta $19.95 *NOS, VCN* ✂️✂️

The Game Three friends arrange a puzzle-murder-mystery for a selected group of guilty people, with $1 million as the prize—and death for the losers.
1989 91m/C Tom Blair, Debbie Martin. VHS, Beta $69.95 *TWE* ✂️

Game of Death Bruce Lee's final kung fu thriller about a young martial arts movie star who gets involved with the syndicate. Lee died halfway through the filming of this movie and it was finished with out-takes and a double. Also known as "Goodbye Bruce Lee: His Last Game of Death."
1979 (R) 100m/C Bruce Lee, Dean Jagger, Kareem Abdul-Jabbar, Colleen Camp, Chuck Norris; *Dir:* Robert Clouse. VHS, Beta, LV $19.98 *FOX* ✂️✂️

The Game of Love Olin presides over Henry's bar, the lost and found of lonely hearts and lust, where people go to quench that deep thirst for love.
1990 (PG) 94m/C Ed Marinaro, Belinda Bauer, Ken Olin, Robert Rusler, Tracy Nelson, Max Gail, Janet Margolin, Brynn Thayer; *Dir:* Bobby Roth. VHS $19.95 *ACA* ✂️✂️

The Game is Over A wealthy neglected wife falls in love with her grown stepson, causing her to divorce her husband and shatter her life. A good performance by Fonda, then Vadim's wife. Modern version of Zola's "La Curee."
1966 (R) 96m/C *FR* Jane Fonda, Peter McEnery, Michel Piccoli; *Dir:* Roger Vadim. VHS, Beta $69.95 *MED* ✂️✂️½

Game of Seduction A murdering, womanizing playboy pursues a single, married woman, and gets caught up in a treacherous cat-and-mouse game. Dubbed.
1986 81m/C Sylvia Kristel, Jon Finch, Nathalie Delon; *Dir:* Roger Vadim. VHS, Beta $19.95 *VIR, VCL* ✂️

Game of Survival A young rebel warrior from another planet is sent to earth to battle six of the galaxy's most brutal warriors.

1989 85m/C Nikki Hill, Cindy Coatman, Roosevelt Miller Jr.; *Dir:* Armand Gazarian. VHS, Beta *RAE* ✂️

Game for Vultures Racial unrest in Rhodesia is the backdrop for this tale of a black revolutionary who fights racist sanctions. The noble topic suffers from stereotypes and oversimplification.
1986 (R) 113m/C *GB* Richard Harris, Richard Roundtree, Joan Collins, Ray Milland; *Dir:* James Fargo. VHS, Beta $19.98 *TWE* ✂️½

Gamera This Japanese monster flick features the ultimate nuclear super-turtle who flies around destroying cities and causing panic. First in the series of notoriously bad films, this one is cool in black and white and has some impressive special effects. Dubbed in English. Originally titled, "Gamera, the Invincible."
1966 86m/B *JP* Brian Donlevy, Albert Dekker, Diane Findlay, John Baragrey, Dick O'Neill, Eiji Funakoshi; *Dir:* Noriaki Yuasa. VHS, LV $9.99 *JFK, SNC* ✂️½

Gamera vs. Barugon The monstrous turtle returns to Earth from his outer space prison, now equipped with his famous leg-jets. He soon wishes he had stayed airborne, however, when he is forced to do battle with 130-foot lizard Barugon and his rainbow-melting ray. Tokyo and Osaka get melted in the process. Also known as "The War of the Monsters."
1966 101m/C *JP* Kojiro Hongo, Kyoko Enami; *Dir:* Shigeo Tanaka. VHS, Beta $9.99 *JFK* ✂️½

Gamera vs. Gaos Now fully the good guy, Gamera slugs it out with a bat-like critter named Gaos. Suspense rules the day when Gaos tries to put out the super turtle's jets with his built-in fire extinguisher but luckily for Earth, Gamera has a few tricks of his own up his shell. Also known as "The Return of the Giant Monsters."
1967 87m/C *JP* Kojiro Hongo, Kichijiro Ueda, Naoyuki Abe; *Dir:* Noriaki Yuasa. VHS, Beta, LV $9.99 *JFK* ✂️½

Gamera vs. Guiron Gamera risks it all to take on an evil, spear-headed monster on a distant planet. Highlight is the sexy, leotard clad aliens who want to eat the little Earth kids' brains. Also known as "Attack of the Monsters."
1969 88m/C *JP* Nobuhiro Kashima, Christopher Murphy, Miyuki Akiyama, Yuko Hamada, Eiji Funakoshi; *Dir:* Noriaki Yuasa. VHS, Beta $9.99 *JFK* ✂️

Gamera vs. Zigra Gamera the flying turtle chose an ecological theme for this, his final movie. It seems that the alien Zigrans have come to Earth to wrest the planet from the hands of the pollutive humans who have nearly destroyed it. The aliens kill the staunch turtle, but the love and prayers of children revive him that he may defend Earth once more.
1971 91m/C *JP* Reiko Kasahara, Mikiko Tsubouchi, Koji Fujiyama, Arlene Zoellner, Gloria Zoellner; *Dir:* Noriaki Yuasa. VHS, Beta, LV $9.99 *JFK* ✂️

Games Girls Play The daughter of an American diplomat organizes a contest at a British boarding school to see which of her classmates can seduce important dignitaries. Also known as "The Bunny Caper" and "Sex Play."
1975 90m/C Christina Hart, Jane Anthony, Jill Damas, Drina Pavlovic; *Dir:* Jack Arnold. VHS, Beta $59.95 *MON* ✂️½

Gametime Vol. 1 Here are two vintage game shows from the 1960s: "Video Village," where contestants moved along a giant board, and "Treasure Isle," where married couples competed to find buried treasure in Florida.
196? 60m/C Jack Narz. VHS, Beta $19.95 *NO* ✂️½

The Gamma People A journalist discovers a Balkan doctor shooting children with gamma rays, creating an army of brainless zombies. Cult-renowned grade-Z tripe.
1956 83m/B *GB* Walter Rilla, Paul Douglas, Eva Bartok; *Dir:* John Gilling. VHS, Beta $9.95 *RCA* Woof!

Gandhi The biography of Mahatma Gandhi from the prejudice he encounters as a young attorney in South Africa, to his role as spiritual leader to the people of India and his use of passive resistance against the country's British rulers, and his eventual assassination. Kingsley is a marvel in his Academy Award-winning role and the picture is a generally riveting epic worthy of its eight Oscars.
1982 (PG) 188m/C *GB* Ben Kingsley, Candice Bergen, Edward Fox, John Gielgud, John Mills, Saeed Jaffrey, Trevor Howard, Ian Charleson, Roshan Seth, Athol Fugard, Martin Sheen, Daniel Day Lewis; *Dir:* Richard Attenborough. Academy Awards '82: Best Actor (Kingsley), Best Art Direction/Set Decoration, Best Cinematography, Best Costume Design, Best Director (Attenborough), Best Film Editing, Best Original Screenplay, Best Picture; British Academy Awards '82: Best Film; New York Film Critics Awards '82: Best Film; Directors Guild of America Awards '82: Best Director (Attenborough). VHS, Beta, LV $14.95 *COL, BTV* ✂️✂️✂️½

Gang Bullets A mobster tries to continue his dirty systems in a new town, but finds the legal force incorruptible.
1938 62m/B Anne Nagel, Robert Kent, Charles Trowbridge, Morgan Wallace, J. Farrell MacDonald, Arthur Loft, John Merton; *Dir:* Lambert Hillyer. VHS, Beta $16.95 *SNC* ✂️

Gang Busters Prisoners plan a breakout. Slow and uninteresting adaptation of the successful radio series. Filmed in Oregon.
1955 78m/B Myron Healey, Don Harvey, Sam Edwards, Frank Gerstle; *Dir:* Bill Karn. VHS $19.95 *NOS, SNC* ✂️½

Gang Wars There's a riot going on as Puerto Rican, Black, and Chinese gangs fight it out for control of the city.
1984 (R) 90m/C VHS, Beta $9.95 *SUN* ✂️

Gangbusters Men battle crime in the city in this serial in thirteen episodes based on the popular radio series of the same name.
1938 253m/B Kent Taylor, Irene Hervey, Robert Armstrong. VHS, Beta $49.95 *NOS, CAB, VCN* ✂️✂️

The Gang's All Here Darro joins a trucking firm whose rigs are being run off the road and hijacked by a competitor. Features "Charlie Chan" regulars Luke and Moreland.
1941 63m/B Frankie Darro, Marcia Mae Jones, Jackie Moran, Mantan Moreland, Keye Luke; *Dir:* Jean Yarborough. VHS $19.95 *NOS, RXM* ✂️½

Gangs, Inc. A woman is sent to prison for a hit-and-run accident, actually the fault of a wealthy playboy. When she learns his identity she vows revenge and turns to crime to achieve her evil ends. Okay thriller suffers from plot problems. Ladd has a minor role filmed before he was a star.

1941 72m/B Joan Woodbury, Lash LaRue, Linda Ware, John Archer, Vince Barnett, Alan Ladd; *Dir:* Phil Rosen. **VHS, Beta** $19.95 *NOS, MRV, DVT* 🎬🎬

The Gangster Sullivan plays a small-time slum-bred hood who climbs to the top of the underworld, only to lose his gang because he allows his fears to get the better of him. (There's nothing to fear except when someone knows you're afeared). Noir crimer with an interesting psychological perspective, though a bit sluggish at times.
1947 84m/B **VHS** $19.98 *FOX, FCT* 🎬🎬½

Gangsters A city is ruled by the mob and nothing can stop their bloody grip except "The Special Squad."
1979 90m/C *IT* Michael Gazzo, Tony Page, Vicki Sue Robinson, Nai Bonet; *Dir:* Mac Ahlberg. **VHS, Beta** $39.95 *VCD, HHE, MTX* 🎬

Gangster's Den A gang of renegades scare a woman out of buying a gold-laden piece of land and Billy the Kid comes to her rescue.
1945 56m/B Buster Crabbe, Kermit Maynard, Al "Fuzzy" St. John, Charles King; *Dir:* Sam Newfield. **VHS, Beta** $19.95 *VDM, TIM* 🎬

Gangsters of the Frontier The people of Red Rock are forced to work in the town mines by two prison escapees until Ritter and his friends ride to the rescue. Part of the "Texas Ranger" series.
1944 56m/B Tex Ritter, Dave O'Brien; *W/Dir:* Elmer Clifton. **VHS, Beta** $19.95 *NOS, DVT, RXM* 🎬

Gangster's Law A look at a seedy gangster's downfall. Dubbed.
19?? 89m/C Klaus Kinski. **VHS, Beta** $69.95 *TWF* Woof!

Ganjasaurus Rex A monster spoof about a giant green dinosaur emerging from the pot-laden hills of California.
1987 (R) 100m/C Paul Bassis, Dave Fresh, Rosie Jones; *Dir:* Ursi Reynolds. **VHS, Beta** $59.95 *RHI* 🎬

Gap-Toothed Women This acclaimed Blank documentary features interviews with numerous gap-toothed women including Hutton, O'Connor, Dearman and Wallach. Beginning with a lighthearted tone, the film then probes more deeply into issues such as self-acceptance and feminism.
1989 31m/C Lauren Hutton, Sandra Day O'Connor, Jill Dearman, Katherine Wallach; *Dir:* Les Blank. **VHS, Beta** $49.95 *FFM* 🎬

Gar Wood: The Silver Fox Through classic newsreel footage, this program shows the career of speedboat racer Gar Wood.
1954 15m/B **VHS, Beta** $49.95 *TSF*

The Garbage Pail Kids Movie The disgusting youngsters, stars of bubblegum cards, make their first and last film appearance in this live action dud. The garbage is where it should be thrown.
1987 (PG) 100m/C Anthony Newley, MacKenzie Astin, Katie Barberi; *Dir:* Rod Amateau. **VHS, Beta, LV** $14.95 *PAR, KRT, IME* Woof!

Garbancito de la Mancha A child with magic powers takes his goat on a quest to rescue his friends.
1987 70m/C *MX* **VHS, Beta** $45.95 *MAD* 🎬🎬

Garbo Talks A dying eccentric's last request is to meet the reclusive screen legend Greta Garbo. Bancroft amusingly plays the dying woman whose son goes to

extreme lengths in order to fulfill his mother's last wish.
1984 (PG-13) 104m/C Anne Bancroft, Ron Silver, Carrie Fisher, Catherine Hicks, Steven Hill, Howard da Silva, Dorothy Loudon, Harvey Fierstein, Hermione Gingold; *Dir:* Sidney Lumet. **VHS, Beta** $19.98 *MGM, FOX* 🎬🎬

The Garden of Allah Dietrich finds hyper-romantic encounters in the Algerian desert with Boyer, but his terrible secret may doom them both. This early Technicolor production, though flawed, is an absolute must for Dietrich fans. Yuma, Ariz. substituted for the exotic locale.
1936 85m/C Marlene Dietrich, Charles Boyer, Basil Rathbone, C. Aubrey Smith, Tilly Losch, Joseph Schildkraut, Henry Kleinbach (Brandon), John Carradine; *Dir:* Richard Boleslawski. **VHS, LV** $59.98 *FOX, RXM* 🎬🎬

The Garden of Delights A wicked black comedy about a millionaire, paralyzed and suffering from amnesia after a car accident. His horrendously greedy family tries to get him to remember and reveal his Swiss bank account number. Acclaimed; in Spanish with English subtitles.
1970 95m/C *SP* Jose Luis Lopez Vasquez, Lina Canelajas, Luchy Soto, Francisco Pierra, Charo Soriano; *Dir:* Carlos Savrat. New York Film Festival '70: Best Director. **VHS, Beta** $68.00 *TAM, CVC, FCT* 🎬🎬🎬

The Garden of Eden Milestone—whose "Two Arabian Nights" had won an Academy Award the previous year, and who went on to direct "All Quiet on the Western Front" and "Of Mice and Men"—directed this sophisticated ersatz Lubitsch sex comedy. Griffith, the so-called "Orchid Lady," is a young girl who dreams of diva-dom but falls deep into the underbelly of the seedy side of Budapest. With the aid of a fallen baroness, however, she finds her way to Monte Carlo and has her turn in the limelight. A handsome production designed by William Cameron Menzies, art director for "Gone With the Wind."
1928 115m/B Corrine Griffith, Louise Dressler, Charles Ray, Lowell Sherman; *Dir:* Lewis Milestone. **VHS, Beta, 8mm** $19.95 *NOS, VYY, DNB* 🎬🎬🎬

Garden of the Finzi-Continis The acclaimed film by DeSica about an aristocratic Jewish family living in Italy under the increasing Fascist oppression on the eve of World War II. The garden wall symbolizes the distance between the Finzi-Continis and the Nazi reality about to engulf them. Flawless acting and well-defined direction. Dubbed.
1971 (R) 90m/C *IT* Dominique Sanda, Helmut Berger, Lino Capolicchio, Fabio Testi; *Dir:* Vittorio DeSica. Academy Awards '71: Best Foreign Language Film; Berlin Film Festival '71: Best Film; National Board of Review Awards '71: 5 Best Foreign Films of the Year. **VHS, Beta, LV** $39.98 *LHV, IME, APD* 🎬🎬🎬🎬

Gardens of Stone A zealous young cadet during the Vietnam War is assigned to the Old Guard patrol at Arlington Cemetery, and clashes with the patrol's older officers and various pacifist civilians. Falls short of the mark, although Jones turns in an excellent performance.
1987 (R) 112m/C James Caan, James Earl Jones, D.B. Sweeney, Anjelica Huston, Dean Stockwell, Lonette McKee, Mary Stuart Masterson, Bill Graham, Sam Bottoms, Casey Siemaszko, Larry Fishburne; *Dir:* Francis Ford Coppola. **VHS, Beta, LV** $89.98 *FOX* 🎬🎬

Garringo Lawmen attempt to bring to justice a man who is intent on killing the soldiers responsible for his father's death.

1965 95m/C *MX* Anthony Steffen, Peter Lee Lawrence, Jose Bodalo. **VHS, Beta** $44.95 *MAD* 🎬🎬

Garry Shandling Show, 25th Anniversary Special A comedy special with stand-up comic Shandling saluting his fictional long-running talk show.
1985 57m/C Garry Shandling. **VHS, Beta, LV** $39.95 *PAR* 🎬🎬

Gas This dog has tycoon Hayden buying up all the gas stations in town to create a crisis that will make him richer. Everyone in the cast wastes their time, especially Sutherland as a hip DJ reporting on the gas shortage.
1981 (R) 94m/C *CA* Donald Sutherland, Susan Anspach, Sterling Hayden, Peter Aykroyd, Howie Mandel, Helen Shaver; *Dir:* Les Rose. **VHS, Beta** $79.95 *PAR* Woof!

Gas Pump Girls Five lovely ladies manage a gas station and use their feminine wiles to win the battle against a shady oil sheik.
1979 (R) 102m/C Kirsten Baker, Dennis Bowen, Huntz Hall, Steve Bond, Leslie King, Linda Lawrence; *Dir:* Joel Bender. **VHS, Beta, LV** *VES* Woof!

Gas-s-s-s! A gas main leak in an Alaskan defense plant kills everyone beyond thirty-something and the post-apocalyptic pre-boomer survivors are left to muddle their way through the brume. Trouble is, AIP edited the heck out of the movie, much to Corman's chagrin, and the result is a truncated comedy; Corman was so displeased, in fact, he left to create New World studios. Music by Country Joe and the Fish. Also known as "Gas-s-s-s...or, it May Become Necessary to Destroy the World in Order to Save It."
1970 (PG) 79m/C Robert Corff, Elaine Giftos, Pat Patterson, George Armitage, Alex Wilson, Ben Vereen, Cindy Williams, Bud Cort, Talia Shire, Country Joe & the Fish; *Dir:* Roger Corman. **VHS, Beta** $69.95 *LTG* 🎬🎬½

Gaslight Lavish remake of the 1939 film, based on the Patrick Hamilton play "Angel Street." A man tries to drive his beautiful wife insane while searching for priceless jewels. Her only clue to his evil acts is the frequent dimming of their gaslights. A suspenseful Victorian era mystery. Lansbury's film debut, as the tarty maid. Oscar nominated for Best Picture, Best Supporting Actress (Lansbury), Screenplay, Black & White Cinematography, and Best Score of a Drama or Comedy.
1944 114m/B Charles Boyer, Ingrid Bergman, Joseph Cotten, Angela Lansbury, Terry Moore, May Whitty, Barbara Everest, Emil Rameau, Edmund Breon, Halliwell Hobbes, Tom Stevenson; *Dir:* George Cukor. Academy Awards '44: Best Actress (Bergman), Best Art Direction/Set Decoration (B & W), Best Interior Decoration. **VHS, Beta, LV** $19.98 *MGM, TIM, BTV* 🎬🎬🎬½

The Gate Kids as a whole like to explore and Dorff and Tripp are no exception. When a large hole is exposed in their backyard, it would be a childhood sin not to see what's in it, right? What they don't know is that this hole is actually a gateway to and from Hell. The terror they unleash includes more than just run-of-the-mill demons. Good special effects. Followed by a sequel.
1987 (PG-13) 85m/C Christa Denton, Stephen Dorff, Louis Tripp; *Dir:* Tibor Takacs. **VHS, Beta, LV** $14.98 *LIV, VES* 🎬🎬½

Gate 2 Tripp returns in his role as Terry, a young student of demonology in this lame sequel. Tripp and a few of his teen buddies call up a group of demons that have been confined behind a gate for billions of years.

The demons are used to grant modest wishes, but the wishes end up having very evil effects. Incredibly weak plot saved only by the special monster effects, which include live action and puppetry.
1992 (R) 90m/C Louis Tripp, Simon Reynolds, Pamela Segall, James Villemaire, Andrea Ladanyi; **Dir:** Tibor Takacs. VHS, LV COL *♫♫½*

Gate of Hell Set in 12th-century Japan. A warlord desires a beautiful married woman and seeks to kill her husband to have her. However, he accidentally kills her instead. Filled with shame and remorse, the warlord abandons his life to seek solace as a monk. Heavily awarded and critically acclaimed. In Japanese with English subtitles.
1954 89m/C *JP* Kazuo Hasegawa, Machiko Kyo; **Dir:** Teinosuke Kinugasa. Academy Awards '54: Best Costume Design (Color), Best Foreign Language Film; Cannes Film Festival '54: Best Film; National Board of Review Awards '54: Special Citation (Kyo), 10 Best Foreign Films of the Year; The New York Film Critics Awards '54: Best Foreign Film. VHS, Beta, 8mm $29.98 NOS, SUE, DVT *♫♫♫½*

Gates of Heaven A venerated documentary about pet cemeteries, the proprietors, and the pet owners themselves. A quizzical, cold-eyed look at the third-rail side of American middle-class life. Morris' first film is shockingly gentle, yet telling, allowing the subjects to present their own stories.
1978 85m/C Dir: Errol Morris. VHS, Beta $19.98 COL, FCT *♫♫♫½*

Gates of Hell Salem, Massachusetts, must battle the risen dead in this gore-ridden film that tried to disguise itself under many other titles, including "City of the Living Dead," "The Fear," "Twilight of the Dead," and "Fear in the City of the Living Dead."
1983 90m/C Christopher George, Janet Agren, Katherine MacColl, Robert Sampson; **Dir:** Lucio Fulci. VHS, Beta PGN Woof!

The Gathering A dying man seeks out the wife and family he has alienated for a final Christmas gathering. James Poe authored this well-acted holiday tearjerker that won the 1977-78 Emmy for outstanding TV drama special.
1977 94m/C Ed Asner, Maureen Stapleton, Lawrence Pressman, Stephanie Zimbalist, Bruce Davison, Gregory Harrison, Veronica Hamel, Gail Strickland; **Dir:** Randal Kleiser. VHS, Beta $49.95 WOV *♫♫*

A Gathering of Eagles Hudson portrays a hard-nosed Air Force colonel during peacetime whose British wife must adjust to being a military spouse.
1963 115m/C Rock Hudson, Rod Taylor, Mary Peach, Barry Sullivan, Kevin McCarthy, Henry Silva, Leif Erickson; **Dir:** Delbert Mann. VHS MGM *♫♫½*

The Gathering: Part 2 Sequel in which a widow has a Christmas reunion with her family, but conflict arises when she introduces a new man in her life. Not as effective as the original, also made for TV.
1979 98m/C Maureen Stapleton, Efrem Zimbalist Jr., Jameson Parker, Bruce Davison, Lawrence Pressman, Gail Strickland, Veronica Hamel; **Dir:** Charles S. Dubin. VHS HNB *♫♫*

Gathering Storm Based on the first book of memoirs from Sir Winston Churchill, this drama examines the pre-World War II years.
1974 72m/C Richard Burton, Virginia McKenna, Ian Bannen; **Dir:** Herbert Wise. Beta $14.95 PSM *♫♫*

The Gatling Gun Poor tale of the Cavalry, Apache Indians, and renegades all after the title weapon.
1972 (PG) 93m/C Guy Stockwell, Woody Strode, Patrick Wayne, Robert Fuller, Barbara Luna, John Carradine, Pat Buttram, Phil Harris; **Dir:** Robert Gordon. VHS HHE *♫♫½*

Gator Sequel to "White Lightning" (1973) follows the adventures of Gator (Reynolds), who is recruited to gather evidence to convict a corrupt political boss who also happens to be his friend. Reynolds in his good ole' boy role with lots of chase scenes. Talk show host Michael Douglas made his film debut in the role of the governor. First film Reynolds directed.
1976 (PG) 116m/C Burt Reynolds, Jerry Reed, Lauren Hutton, Jack Weston, Alice Ghostley, Dub Taylor; **Dir:** Burt Reynolds. VHS, Beta $29.95 FOX *♫♫*

Gator Bait The Louisiana swamp is home to the beautiful Desiree, and woe be unto any man who threatens her family. Aims to please fans of the raunchy and violent. Followed by an equally vile sequel.
1973 (R) 91m/C Claudia Jennings, Clyde Ventura, Bill Thurman, Janet Baldwin; **W/Dir:** Ferd Sebastian, Beverly Sebastian. VHS, Beta, LV $19.95 IND, PAR *♫♫½*

Gator Bait 2: Cajun Justice A city girl comes to the Louisiana swamp and turns violent in an effort to exact revenge on the men who have been tormenting her.
1988 (R) 94m/C Jan MacKenzie, Tray Loren, Paul Muzzcat, Brad Kepnick, Jerry Armstrong, Ben Sebastian; **W/Dir:** Ferd Sebastian, Beverly Sebastian. VHS, Beta, LV $19.95 PAR *♫*

The Gauntlet A cop is ordered to Las Vegas to bring back a key witness for an important trial—but the witness turns out to be a beautiful prostitute being hunted by killers. Violence-packed action.
1977 (R) 111m/C Clint Eastwood, Sondra Locke, Pat Hingle, Bill McKinney; **Dir:** Clint Eastwood. VHS, Beta, LV $19.98 WAR, TLF *♫♫*

The Gay Divorcee Astaire pursues Rogers to an English seaside resort, where she mistakes him for the hired correspondent in her divorce case. Songs include "Night and Day," "Don't Let It Bother You," "Needle in a Haystack," and the first Best Song Oscar for "The Continental." Based on the musical play "The Gay Divorce" by Dwight Taylor and Cole Porter. The title was slightly changed for the movie because of protests from the Hays Office.
1934 107m/B Fred Astaire, Ginger Rogers, Edward Everett Horton, Eric Blore, Alice Brady, Erik Rhodes, Betty Grable; **Dir:** Mark Sandrich. Academy Awards '34: Best Song ("The Continental"). VHS, Beta, LV $29.95 MED *♫♫♫*

The Gay Lady Light romantic comedy set in the 1890s about an actress who climbs to stardom in London theater and winds up marrying into the aristocracy. Look for Christopher Lee and Roger Moore in early bit parts.
1949 95m/C *GB* Jean Kent, James Donald, Hugh Sinclair, Lana Morris, Bill Owen, Michael Medwin; **Dir:** Brian Desmond Hurst. VHS MOV *♫♫½*

Gay Purr-ee Delightful tale of feline romance in the City of Lights. For all ages.
1962 85m/C Voices: Judy Garland, Robert Goulet, Red Buttons, Hermione Gingold, Mel Blanc. VHS, LV $19.98 WAR, FCT *♫♫*

The Gay Ranchero Rogers, without Dale Evans, in an average tale of old west meets new crooks. Rogers is the sheriff who foils gangsters trying to take over an airport. The action doesn't stop him from singing a duet with leading lady Frazee on "Wait'll I Get My Sunshine in the Moonlight."
1942 55m/B Roy Rogers, Andy Devine, Tito Guizar, Jane Frazee, Estelita Rodriguez; **Dir:** William Witney. VHS, Beta $9.99 NOS, VCN, CAB *♫♫*

Geek Maggot Bingo It's too bad director Zed waited until mid-"Bingo" to post a sign warning "Leave Now, It Isn't Going Get Any Better!" Conceived by the New York underground's don of the "Cinema of Transgression" to be an off-the-rack cult classic, this horror spoof is too long on in-jokes and short on substance to earn its number in the cult hall of fame. It does, however, boast Death as Scumbalina the vampire queen and Hell as a punk cowboy crooner (before alternative country hit the airwaves). TV horror-meister Zacherle narrates.
1983 70m/C Richard Hell, Donna Death, John Zacherle; **Dir:** Nick Zedd. VHS $39.99 MWF, MOV *♫½*

Geheimakte WB1 (Secret Paper WB1) Historical drama centering on the invention of the U-boat by Sergeant Wilhelm Bauer during Denmark's WWII blockade of the ports of Schleswig-Holstein. In German with no subtitles.
1942 91m/B *GE* Alexander Golling; **Dir:** Herbert Selpin. VHS, Beta, 3/4U $35.95 IHF *♫♫*

The Geisha Boy Jerry is a floundering magician who joins the USO and tours the Far East. His slapstick confrontations with the troupe leader and an officer who dreams of attending his own funeral provide hearty laughs, plus Lewis finds romance with a Japanese widow after her son claims him as a father. Pleshette's film debut, with an appearance by the Los Angeles Dodgers.
1958 98m/C Jerry Lewis, Marie McDonald, Sessue Hayakawa, Barton MacLane, Suzanne Pleshette, Nobu McCarthy; **Dir:** Frank Tashlin. VHS $14.95 PAR *♫♫*

Gello Rama The wonderful world of female gelatin wrestling.
1987 58m/C VHS, Beta $39.95 MED *♫♫*

Gemidos de Placer A young, sex-obsessed couple get involved with murder. In Spanish.
197? 90m/C *SP* Lina Romay. VHS, Beta $57.95 UNI *♫*

Gemini Affair Two women go to Hollywood to become rich and famous but end up being very disappointed.
198? 88m/C Marta Kristen, Kathy Kersh, Anne Seymour. VHS, Beta $49.95 UNI *♫*

Gendarme Desconocido Cantiflas bungles his way into becoming a police officer after accidentally apprehending a gang of criminals in this bumble-filled comedy.
19?? 110m/B *SP* VHS, Beta $49.95 VLA *♫♫*

Gene Autry Matinee Double Feature #1 Includes "Blue Canadian Rockies" (1952), where Autry must investigate a series of murders, and "Last of the Pony Riders" (1953), in which the final days of the pony express yield a sinister conspiracy.
1953 117m/C Gene Autry; **Dir:** George Archainbaud. VHS $39.98 NO *♫♫*

Gene Autry Matinee Double Feature #2 An exciting western double feature: A cowboy goes after the man who stole his rodeo winnings in "Melody Trail" and a territorial ranger investigates a series of Indian raids in "Winning of the West."
1953 117m/B Gene Autry, Champion, Smiley Burnette, Alan Bridge; *Dir:* George Archainbaud, Joseph Kane. VHS, Beta $39.98 *NO* ✂️

Gene Autry Matinee Double Feature #3 Gene stars in "Prairie Moon," in which he teaches some tough kids a lesson about life, and "On Top of Old Smokey," which sees a young woman trying to save her small town.
1953 113m/B Gene Autry, Smiley Burnette; *Dir:* George Archainbaud, Ralph Staub. VHS, Beta $39.98 *NO* ✂️

The Gene Krupa Story The story of the famous jazz drummer and his career plunge after a drug conviction. Krupa recorded the soundtrack, mimed by Mineo, who was too young for a convincing portrayal. Gavin McLeod (of "Love Boat" fame) has a small part as Mineo's dad.
1959 101m/B Sal Mineo, James Darren, Susan Kohner, Susan Oliver, Anita O'Day, Red Nichols; *Dir:* Don Weis. VHS, Beta $19.95 *COL, FCT* ✂️✂️½

The General Keaton's masterpiece and arguably the most formally perfect and funniest of silent comedies. Concerns a plucky Confederate soldier who single-handedly retrieves a pivotal train from Northern territory. Full of eloquent man-vs.-machinery images and outrageous sight gags. Remade as "The Great Locomotive Chase" in 1956 with Fess Parker.
1927 78m/B Buster Keaton, Marion Mack; *Dir:* Buster Keaton, Clyde Bruckman. Sight & Sound Survey '72: #8 of the Best Films of All Time; Sight & Sound Survey '82: #10 of the Best Films of All Time (tie). VHS, Beta, LV, 8mm $29.95 *NOS, HHT, VYY* ✂️✂️✂️✂️

The General Died at Dawn A clever, atmospheric suspense film about an American mercenary in Shanghai falling in love with a beautiful spy as he battles a fierce Chinese warlord who wants to take over the country. Playwright-to-be Odets' first screenplay; he, O'Hara and '30s gossip hound Skolsky have cameos as reporters. Oscar nominated for Best Supporting Actor (Tamiroff), and Best Cinematography.
1936 93m/B Gary Cooper, Madeleine Carroll, Akim Tamiroff, Dudley Digges, Porter Hall, William Frawley, Philip Ahn, John O'Hara, Clifford Odets, Sidney Skolsky; *Dir:* Lewis Milestone. VHS, Beta $29.95 *MCA* ✂️✂️✂️

The General Line Eisenstein's classic pro-Soviet semi-documentary about a poor farm woman who persuades her village to form a cooperative. Transgresses its party-line instructional purpose by vivid filmmaking. The director's last silent film. Also known as "The Old and the New."
1929 90m/B *RU* Marfa Lapkina; *Dir:* Sergei Eisenstein. National Board of Review Awards '30: 5 Best Foreign Films of the Year. VHS, Beta, 3/4U $24.95 *NOS, MRV, DVT* ✂️✂️✂️✂️

General/Slapstick A double feature composed of abridged versions of the Buster Keaton classic comedy of Civil War espionage, and a Mack Sennett anthology of slapstick comedy.
1926 56m/B Buster Keaton, Marion Mack, Charles Murray, Mabel Normand, Fatty Arbuckle, Edgar Kennedy. VHS, Beta $29.98 *CCB, MRV* ✂️✂️✂️✂️

General Spanky The only feature "Our Gang" comedy goofed by transplanting them to a Civil War milieu popular at the time. Confederate kids Spanky and Alfala play soldier, ultimately outsmarting Union troops. The rascals are funny sometimes, but Buckwheat's role as an eager slave is disturbing today.
1936 73m/B George "Spanky" McFarland, Phillips Holmes, Ralph Morgan, Irving Pichel, Rosina Lawrence, Billie "Buckwheat" Thomas, Carl "Alfalfa" Switzer, Louise Beavers; *Dir:* Fred Newmeyer. VHS $19.98 *MGM, FCT* ✂️✂️

Generale Della Rovere A World War II black marketeer is forced by the Nazis to go undercover in a local prison. To find out who the resistance leaders are, he poses as a general. But when prisoners begin to look to him for guidance, he finds the line between his assumed role and real identity diminished, leading to a tragic conclusion. Acclaimed film featuring a bravura lead performance by veteran director De Sica. In Italian with English subtitles.
1960 139m/B *IT* Vittorio De Sica, Otto Messmer, Sandra Milo; *Dir:* Roberto Rossellini. VHS, Beta $29.95 *NOS, CVC, HHT* ✂️✂️✂️½

Generation A very pregnant bride informs everyone she'll give birth at home without doctors, and creates a panic. Based on the Broadway play by William Goodhart.
1969 109m/C David Janssen, Kim Darby, Carl Reiner, Peter Duel, Andrew Prine, James Coco, Sam Waterston; *Dir:* George Schaefer. VHS, Beta $69.98 *SUE* ✂️✂️

Generations: Mind Control— From the Future Max Lang has lost all self-control and is being driven by evil forces beyond time, driven to kill and to try and flee total extinction.
1980 54m/C VHS, Beta *BEI* ✂️

Genevieve A 1904 Darracq roadster is the title star of this picture which spoofs "classic car" owners and their annual rally from London to Brighton. Two married couple challenge each other to a friendly race which becomes increasingly intense as they near the finish line.
1953 86m/C *GB* John Gregson, Dinah Sheridan, Kenneth More, Kay Kendall; *Dir:* Henry Cornelius. Best British Film '53: 5 Best Foreign Films of the Year. VHS, Beta *LCA* ✂️✂️✂️

Genie of Darkness The ashes of Nostradamus himself are retrieved in a bid to destroy his vampiric descendent. Edited from a Mexican serial; if you were able to sit through this one, look for "Curse of Nostradamus" and "The Monster Demolisher."
1960 ?m/B *MX* German Robles; *Dir:* Frederick Curiel. VHS $16.95 *SNC, NOS, LOO* ✂️

Gente Violenta A crook who steals a million dollars worth of diamonds has to flee from both the police and his old gang.
1987 90m/C *MX* Armando Silvestre, Antonio De Hudd, Rebeca Silva, Victor Junco. VHS, Beta $58.95 *MAD* ✂️

Gentle Ben Episodes from the late '60s television show about a friendly family who adopts an orphaned bear. Derived from the film "Gentle Giant."
197? 60m/C Dennis Weaver, Beth Brickell, Clint Howard, Rance Howard, Angelo Rutherford, Burt Reynolds. VHS, Beta $14.95 *NEG* ✂️✂️

Gentle Giant An orphaned bear is taken in by a boy and his family and grows to be a 750 pound giant who must be returned to the wild. "Gentle Ben" TV series was derived from this feature.
1967 93m/C Dennis Weaver, Vera Miles, Ralph Meeker, Clint Howard, Huntz Hall; *Dir:* James Neilson. VHS $14.98 *REP* ✂️✂️

Gentle Savage When an Indian is wrongly accused of raping and beating a white woman, the white community seeks revenge. The vengeful mob, led by the victim's step-father, is unaware that it's really their ringleader who is responsible for the crime. When the accused's brother is slain, the Indian community retaliates. Violent but well-crafted; originally titled "Camper John."
1973 (R) 85m/C William Smith, Gene Evans, Barbara Luna, Joe Flynn; *Dir:* Sean McGregor. VHS $29.95 *IVE, USA* ✂️✂️½

Gentleman Bandit So-so drama about Father Bernard Pagano, a Boston priest mistakenly arrested as a stickup artist. Based on a true story. Made for television.
1981 96m/C Ralph Waite, Julie Bovasso, Jerry Zaks, Joe Grifasi, Estelle Parsons, Tom Aldredge; *Dir:* Jonathan Kaplan. VHS, Beta $59.95 *MTI* ✂️✂️

Gentleman from California The dashing son of a Mexican landowner returns home and finds his people's land stolen from them by greedy tax collectors. Also known as "The Californian."
1937 56m/B Ricardo Cortez, Marjorie Weaver, Katherine DeMille; *Dir:* Gus Meins. VHS, Beta, 8mm *VYY* ✂️

Gentleman Jim A colorful version of the life of oldtime heavyweight boxing great Jim Corbett, transformed from a typical Warner Bros. bio-pic by director Walsh into a fun-loving, anything-for-laughs donnybrook. Climaxes with Corbett's fight for the champion ship against the great John L. Sullivan. One of Flynn's most riotous performances.
1942 104m/B Errol Flynn, Alan Hale Jr., Alexis Smith, Jack Carson, Ward Bond, Arthur Shields, William Frawley; *Dir:* Raoul Walsh. VHS, Beta, LV $34.98 *FOX* ✂️✂️✂️½

The Gentleman Killer A man-with-no-name clears a ravaged border town of murdering bandits.
1978 95m/C Anthony Steffen, Silvia Solar. VHS, Beta $49.95 *UNI* ✂️

Gentleman from Texas Ropin', ridin', and romance figure in this less than dramatic oater.
1946 55m/B Johnny Mack Brown, Raymond Hatton, Claudia Drake, Reno Blair, Christine McIntyre; *Dir:* Lambert Hillyer. VHS $19.95 *NOS, DVT* ✂️

Gentlemen Prefer Blondes Amusing satire involving two show-business girls from Little Rock trying to make it big in Paris. Seeking rich husbands or diamonds, their capers land them in police court. Monroe plays Lorelei Lee, Russell is her sidekick. Despite an occasionally slow plot, the music, comedy, and performances are great fun. Film version of a Broadway adaption of a story by Anita Loos. Songs by Jule Styne include "Diamonds Are a Girl's Best Friend." Followed by (but without Monroe) "Gentlemen Marry Brunettes."
1953 91m/C Marilyn Monroe, Jane Russell, Charles Coburn, Elliott Reid, Tommy Noonan, George Winslow; *Dir:* Howard Hawks. VHS, Beta, LV $14.98 *FOX* ✂️✂️✂️

Gentlemen of Titipu Animated version of Gilbert and Sullivan's "The Mikado."
1982 27m/C VHS, Beta *PGN* ✂️✂️✂️

Genuine Risk Gorgeous young woman turns off her crime boss boyfriend bigtime when she seduces his bodyguard. Genuine tripe.
1989 (R) 89m/C Terence Stamp, Michelle Johnson, Peter Berg, M.K. Harris; *Dir:* Kurt Voss. **VHS, LV** $89.95 COL ✄½

George! A carefree bachelor takes his girlfriend and his 250-pound St. Bernard on a trip to the Swiss Alps where he proves that a dog is not always man's best friend.
1970 (G) 87m/C Marshall Thompson, Jack Mullaney, Inge Schoner; *W/Dir:* Wallace C. Bennett. **VHS, Beta** $39.95 UHV ✄½

George Burns in Concert America's oldest working comedian will make you laugh again and again.
1982 57m/C George Burns. **VHS** $39.95 USA

George Burns: His Wit & Wisdom This day in the life of Burns begins with his comic views on life, and ends with him performing for a live audience.
1988 45m/C George Burns, Red Buttons, Carol Channing, Emma Samms, Morey Amsterdam, Army Archerd, Yakov Smirnoff. **VHS, Beta** $19.98 VID

George Carlin on Campus Carlin performs his classic routines and new material in this concert filmed at UCLA's Wadsworth Theater.
1984 59m/C George Carlin. **VHS, Beta, LV** $19.98 LIV, VES

George Carlin at Carnegie Hall It's George Carlin at his funniest, with his classic routine of the "Seven Words You Can Never Say on Television."
1983 60m/C George Carlin. **VHS, Beta, LV** $14.98 LIV, VES

George Carlin - Live!: What am I Doing in New Jersey? The veteran comedian continues to ponder life's questions in this live concert appearance.
1989 60m/C George Carlin. **VHS, Beta** $14.99 HBO, TLF

George Carlin: Playin' with Your Head At the Beverly Theater in Los Angeles, the renowned stand-up comic performs all-new material. The tape also features a short, "The Envelope," starring Tayback.
1988 57m/C George Carlin, Vic Tayback. **VHS, Beta, LV** $19.98 LIV, VES

George and the Christmas Star George is dissatisfied with the paper star he cut out to put atop his Christmas tree, so he builds a spaceship and goes to catch a real star. Features songs by Paul Anka.
1985 25m/C Paul Anka; *Dir:* Gerald Potterton. **VHS, Beta** $12.95 PAR

George Clinton: Parliament/Funkadelic Clinton performs his biggest hits: "Atomic Dog," "Get Off Your Arse and Jam" and "Night of the Thumpasorus People."
1984 30m/C **VHS, Beta** SVS, OM

George Melies, Cinema Magician A look at film pioneer George Melies who originated the use of special effects in cinema.
1978 17m/C Georges Melies. **VHS, Beta** $19.98 CCB

George Pal Color Cartoon Carnival A collection of eight classic George Pal cartoons: "Jasper in a Jam," "Wolf Wolf," "Mighty Mouse," "Pin Cushion Man," "Mary's Little Lamb," "Molly Moo Coo and Robinson Crusoe," "Little Black Sambo," and "Jasper's Derby."
1944 80m/C *Dir:* George Pal. **VHS, Beta** $24.95 DVT ✄½

George Pal Puppetoons #1 Collection of the animation of George Pal includes "Tubby the Tuba," "Rhapsody in Wood," "A Date with Duke," "Jasper in a Jam," and "John Henry and the Inky Poo." Jazz greats Woody Herman, Duke Ellington, and Peggy Lee appear and contribute their music.
194? 45m/C **VHS, Beta** $52.95 GVV ✄✄✄

George Pal Puppetoons #2 Collection of the animation of George Pal includes "Hot Lips Jasper" with trumpet virtuoso Rafael Mendez, "The Phillips Broadcast of 1938" with Ambrose and his Orchestra, "The Little Broadcast," "The Sky Princess," and "Ship of the Ether."
194? 45m/C **VHS, Beta** $52.95 GVV ✄✄✄

George Stevens: A Filmmaker's Journey Documentary-style tribute to the famed director whose works include "Alice Adams," "Gunga Din," "Shane," and "Giant." Includes home movies and personal insights from his son.
1984 110m/C Katharine Hepburn, Cary Grant, Joel McCrea, Fred Astaire, Ginger Rogers, Warren Beatty; *Dir:* George Stevens. **VHS** $69.95 CON ✄✄✄

George White's Scandals A lightweight musical comedy look at the show biz world and why the show must go on. Based on White's Broadway extravaganzas. Jazz numbers by Gene Krupa and his band. Preceded by "George White's Scandals" of 1934 and 1935.
1945 95m/B Joan Davis, Jack Haley, Jane Greer, Phillip Terry, Margaret Hamilton, Martha Holliday; *Dir:* Felix Feist. **VHS, Beta** $39.98 CCB ✄✄

George's Island When young George is placed with the worst foster parents in the world, his eccentric grandfather helps him escape. They wind up on Oak's Island, where, legend has it, Captain Kidd's ghost and buried treasure reside. They soon find out the legends are true!
1991 (PG) 89m/C Ian Bannen, Sheila McCarthy, Maury Chaykin, Nathaniel Moreau; *Dir:* Paul Donovan. **VHS** $89.95 COL, NLC ✄✄✄

Georgia, Georgia Sands is a black entertainer on tour in Sweden who falls for a white photographer. Her traveling companion, who hates whites, takes drastic action to separate the lovers. Poor acting, script, and direction. Based on a book by Maya Angelou.
1972 (R) 91m/C Diana Sands, Dirk Benedict, Minnie Gentry; *Dir:* Stig Bjorkman. **VHS, Beta** $59.95 PSM ✄

Georgy Girl Redgrave finely plays the overweight ugly-duckling Georgy who shares a flat with the beautiful and promiscuous Meredith (Rampling). Georgy is, however, desired by the wealthy and aging Mason and soon by Meredith's lover (Bates) who recognizes her good heart. When Meredith becomes pregnant, Georgy persuades her to let her raise the baby, leaving Georgy with the dilemma of marrying the irresponsible Bates (the baby's father) or the settled Mason. The film and title song (sung by the Seekers) were both huge hits. Based on the novel by Margaret Foster who also co-wrote the screenplay. Mason, Redgrave, and cinematographer Ken Higgins were Oscar nominees.
1966 100m/B GB Lynn Redgrave, James Mason, Charlotte Rampling, Alan Bates, Bill Owen, Claire Kelly, Rachel Kempson; *Dir:* Silvio Narizzano. National Board of Review Awards '66: 10 Best Films of the Year; New York Film Critics Awards '66: Best Actress (tie) (Redgrave). **VHS, Beta, LV** $19.95 RCA, FCT ✄✄✄

Germicide A scientist tries to warn the world of the threat posed by a horrifying bacterial weapon. Terrorists and his mistress hatch plots against him.
1974 90m/C Rod Taylor, Bibi Andersson. **Beta** $59.95 PSM ✄

Geronimo's Revenge When Geronimo starts a battle with the settlers, a rancher he has befriended finds himself caught in the middle. The fourth soggy Disney TV oater from the "Tales of Texas John Slaughter" series.
1960 77m/C Tom Tryon, Darryl Hickman; *Dir:* Harry Keller. **VHS, Beta** IPI ✄½

Gerry Mulligan This program presents the jazz music of Gerry Mulligan featuring his compositions "K4 Pacific" and "North Atlantic Run."
1981 18m/C Gerry Mulligan. **VHS, Beta** $9.95 SVS

Gertrud The simple story of an independent Danish woman rejecting her husband and lovers in favor of isolation. Cold, dry, minimalistic techniques make up Dreyer's final film. In Danish with subtitles.
1964 116m/B DK Nina Pens Rode, Bendt Rothe, Ebbe Rode; *W/Dir:* Carl Theodor Dreyer. National Board of Review Awards '65: 5 Best Foreign Films of the Year; New York Film Festival '65: International Critics Award; Venice Film Festival '65: International Critics Award. **VHS, Beta** $34.95 FCT, DVT ✄✄✄½

Gerty, Gerty, Gerty Stein Is Back, Back, Back This is a one-woman drama concerning the anecdotes and reminiscenses of Gertrude Stein.
1981 60m/C **VHS, Beta** NO ✄✄✄

Gervaise The best of several films of Zola's "L'Assommoir." A 19th-century middle-class family is destroyed by the father's plunge into alcoholism despite a mother's attempts to save them. Well acted but overwhelmingly depressing. Music by Georges Auric. In French with English subtitles.
1957 89m/B FR Maria Schell, Francois Perier, Suzy Delair, Armand Mestral; *Dir:* Rene Clement. British Academy Awards '56: Best Actor (Perier), Best Film; National Board of Review Awards '57: 5 Best Foreign Films of the Year; New York Film Critics Awards '57: Best Foreign Film; Venice Film Festival '56: International Critics Award, Best Actress (Schell). **VHS, Beta** $29.95 DVT, FST, WFV ✄✄✄

Get Along Gang The Get Along Gang has an opportunity to find the meaning of cooperation and friendship at their annual Scavenger Hunt.
1984 22m/C **VHS, Beta** $19.98 SCL

Get Christie Love! Vintage blaxploitation from the genre's halcyon days, when going undercover meant a carte blanche for skimpy outfits. Based on Dorothy Uhnak's detective novel, the eponymous policewoman goes undercover to bring a thriving drug empire to it knees. The outcome: a TV series. Graves previously appeared on television as one of the "Laugh-In" party girls.

1974 95m/C Teresa Graves, Harry Guardino, Louise Sorel, Paul Stevens, Andy Romano, Debbie Dozier; *Dir:* William A. Graham. **VHS** $59.95 *XVC* ✍✍

Get Crazy The owner of the Saturn Theatre is attempting to stage the biggest rock-and-roll concert of all time on New Year's Eve 1983, and everything is going wrong in this hilarious, off-beat film.
1983 90m/C Malcolm McDowell, Allen Garfield, Daniel Stern, Gail Edwards, Ed Begley Jr., Lou Reed, Bill Henderson; *Dir:* Allan Arkush. **VHS, Beta, LV** $9.98 *SUE* ✍✍½

Get 'Em Off A documentary history of striptease.
197? 60m/C VHS, Beta $39.95 *JEF*

Get Happy Includes Shirley Temple in "Glad Rags to Riches," Weber and Fields in "Beer is Here," and Bessie Smith in "St. Louis Blues." Also a Flip the Frog Cartoon.
193? 59m/B Shirley Temple, Weber & Fields, Bessie Smith. VHS, Beta $19.98 *CCB*

Get Out Your Handkerchiefs Unconventional comedy about a husband desperately attempting to make his sexually frustrated wife happy. Determined to go to any lengths, he asks a Mozart-loving teacher to become her lover. She is now, however, bored by both men. Only when she meets a 13-year-old genius at the summer camp where the three adults are counselors does she come out of her funk and find her sexual happiness. Laure is a beautiful character, but Depardieu and Dewaere are wonderful as the bewildered would-be lovers. Academy Award winner for best foreign film. Also known as "Preparez Vous Monchoirs."
1978 (R) 109m/C *FR* Gerard Depardieu, Patrick Dewaere, Carole Laure; *Dir:* Bertrand Blier. Academy Awards '78: Best Foreign Language Film. VHS, Beta $19.98 *WAR, INT* ✍✍✍½

Get Rita Italian prostitute sets up her gangster boyfriend for a murder she committed and she winds up as mob boss.
19?? 90m/C Sophia Loren, Marcello Mastroianni; *Dir:* Tom Rowe. VHS, Beta $19.95 *CTC, API* ✍✍

Get That Man When a cabby is mistaken for the murdered heir to a vast fortune, his life suddenly becomes complex.
1935 57m/B Wallace Ford, Leon Ames, Lillian Miles, E. Allen Warren; *Dir:* Spencer Gordon Bennett. VHS, Beta, LV $19.95 *NOS, WGE* ✍✍

The Getaway McQueen plays a thief released on a parole arranged by his wife (McGraw) only to find out a corrupt politician wants him to rob a bank. After the successful holdup, McQueen finds out his cohorts are actually in the politician's pocket and trying unsuccessfully to double-cross him. McQueen and McGraw are forced into a feverish chase across Texas to the Mexican border, pursued by the politician's henchmen and the state police. This film is a completely amoral depiction of crime and violence with McQueen taciturn as always and McGraw again showing a complete lack of acting skills. Based on a novel by Jim Thompson.
1972 (PG) 123m/C Steve McQueen, Ali MacGraw, Ben Johnson, Sally Struthers, Al Lettieri, Slim Pickens, Jack Dodson, Dub Taylor, Bo Hopkins; *Dir:* Sam Peckinpah. VHS, Beta *FCT* ✍✍

Getting Even A maniac threatens to poison the entire population of Texas with a deadly gas unless he gets $50 million. An adventurous businessman sets out to stop him. Also known as "Hostage: Dallas."

1986 (R) 90m/C Edward Albert, Joe Don Baker, Audrey Landers; *Dir:* Dwight Little. **VHS, Beta** $79.95 *LIV, VES* ✍1½

Getting It On A high school student uses his new-found video equipment for voyeuristic activity.
1983 (R) 100m/C Martin Yost, Heather Kennedy, Jeff Edmond, Kathy Rockmeier, Mark Alan Ferri; *Dir:* William Olsen. VHS, Beta $69.95 *LIV, VES* ✍

Getting It Right A sweet-natured British comedy about an inexperienced, shy adult male who is quite suddenly pursued by a middle-aged socialite, a pregnant, unstable rich girl, and a modest single mother. Novel and screenplay by Elizabeth Jane Howard.
1989 (R) 101m/C *GB* Jesse Birdsall, Helena Bonham Carter, Lynn Redgrave, Peter Cook, John Gielgud, Jane Horrocks, Richard Huw, Shirley Anne Field, Pat Haywood, Judy Parfitt, Bryan Pringle; *Dir:* Randal Kleiser. VHS, Beta, LV $89.98 *VIR* ✍✍½

Getting Lucky A shrinking violet schoolboy is granted his adolescent fantasy by a pandering little leprechaun who makes him the object of desire of a voluptuous pom-pom girl...Seems it's more complicated than that, though.
1989 (PG-13) ?m/C Lezlie Z. McGraw, Steven Cooke, Rick McDowell; *Dir:* Michael Paul Girard. VHS $59.95 *RAE, HHE* ✍½

Getting Mama Married Special double length episode starring Amos and Andy.
19?? 60m/B Spencer Williams Jr. VHS, Beta $45.95 *VCN*

Getting Over A black promoter is hired as a figurehead by a bigoted record company president and proceeds to try to run the company his own way. He develops an all-girl group called "The Love Machine" and gets involved in a singer's kidnapping and a record company run by gangsters. Filled with cliches and stereotypes.
1981 108m/C John Daniels, Gwen Brisco, Mary Hopkins, John Goff, Andrew "Buzz" Cooper; *W/ Dir:* Bernie Rollins. VHS, Beta $59.95 *UNI* ✍

Getting Physical After being mugged, a secretary decides to get in shape and enters the world of female body building. Crammed with workout scenes and disco music. Made for television.
1984 95m/C Alexandra Paul, Sandahl Bergman, David Naughton; *Dir:* Steven Hilliard Stern. VHS, Beta $59.98 *FOX* ✍

Getting Straight A returning Vietnam soldier (Gould) goes back to his alma mater to secure a teaching degree and gets involved in the lives of his fellow students and the turbulence of the end of the '60s, including campus riots. A now-dated "youth" picture somewhat watchable for Gould's performance.
1970 (R) 124m/C Elliott Gould, Candice Bergen, Jeff Corey, Cecil Kellaway, Jeannie Berlin, Harrison Ford, John Rubinstein, Robert F. Lyons, Max Julien; *Dir:* Richard Rush. VHS, Beta $59.95 *COL* ✍✍

Getting Wasted Set in 1969 at a military academy for troublesome young men, chaos ensues when the cadets meet hippies.
1980 (PG) 98m/C Brian Kerwin, Stephen Furst, Cooper Huckabee. VHS, Beta $19.95 *UHV* ✍½

The Getting of Wisdom A 13-year-old girl from the Australian outback tries to establish her identity, and her individuality, within the restricting confines of a Victorian girl's boarding school. Based on the novel by Henry Handel Richardson.

1977 100m/C *AU* Susannah Fowle, Hilary Ryan, Alix Longman, Sheila Helpmann, Patricia Kennedy, Barry Humphries, John Waters; *Dir:* Bruce Beresford. VHS, Beta *FOX* ✍✍½

The Ghastly Ones Three couples must stay in a haunted mansion to inherit an estate, but they're soon being violently killed off. No budget and no talent. Remade as "Legacy of Blood."
1968 81m/C Don Williams, Maggie Rogers, Hal Belsoe, Veronica Redburn; *Dir:* Andy Milligan. VHS, Beta *VHM* Woof!

Ghetto Blaster A Vietnam vet takes on out-of-control street gangs to make his neighborhood safe.
1989 (R) 86m/C R.G. Armstrong, Richard Hatch, Richard Jaeckel, Harry Caesar, Rose Marie; *Dir:* Alan L. Stewart. VHS, Beta $79.95 *PSM* ✍

Ghetto Revenge An embittered Vietnam vet gets caught between white supremacists and black radical activists in this coming home story. He returns and tries to build a life but succumbs to the desire for vengeance and things are never the same.
19?? ?m/C Mike Sims. VHS $19.99 *IPI* ✍

Ghidrah the Three Headed Monster When a three-headed monster from outer-space threatens the world, humans appeal to the friendly Mothra, Rodan, and Godzilla. Tokyo once again gets trampled. Japanese kiddie fare; badly dubbed.
1965 85m/C *JP* Yosuke Natsuki, Yuriko Hoshi; *Dir:* Inoshiro Honda. VHS, Beta $19.95 *HHT, VCN, VHE* ✍½

The Ghost A woman is driven mad by her supposedly dead husband and their evil housekeeper. Sequel to "The Horrible Dr. Hichcock."
1963 96m/C *IT* Barbara Steele, Peter Baldwin, Leonard Eliott, Harriet White; *Dir:* Robert Hampton. VHS, Beta $14.95 *SNC, LPC* ✍½

Ghost Zucker, known for overboard comedies like "Airplane!" and "Ruthless People," changed tack and directed this undemanding romantic thriller, which was the surprising top grosser of 1990. Swayze is a murdered investment consultant attempting (from somewhere near the hereafter) to protect his lover, Moore, from imminent danger when he learns he was the victim of a hit gone afoul. Goldberg is the medium who suddenly discovers that the powers she's been faking are real. A winning blend of action, special effects (from Industrial Light and Magic) and romance. Snared two Oscars while garnering nominations for best film, score (Maurice Jarre), and film editing.
1990 (PG-13) 127m/C Patrick Swayze, Demi Moore, Whoopi Goldberg, Tony Goldwyn, Vincent Schiavelli; *Dir:* Jerry Zucker. Academy Awards '90: Best Original Screenplay, Best Supporting Actress (Goldberg). VHS, Beta, LV, SVS, 8mm $89.95 *PAR, SUP, FCT* ✍✍

Ghost Chase A young filmmaker desperate for funds inherits his dead relative's clock, from which issues the ghost of the deceased's butler. The ghostly retainer aids in a search for the departed's secret fortune. Neither scary nor funny.
1988 (PG-13) 89m/C Jason Lively, Jill Whitlow, Tim McDaniel, Paul Gleason, Chuck "Porky" Mitchell; *Dir:* Roland Emmerich. VHS, Beta, LV $79.95 *VIR* ✍½

Ghost Chasers The Bowery Boys become mixed up in supernatural hijinks when a seance leads to the appearance of a ghost that only Sach can see.

1951 70m/B Leo Gorcey, Huntz Hall, Billy Benedict, David Gorcey, Buddy Gorman, Bernard Gorcey, Jan Kayne, Philip Van Zandt, Lloyd Corrigan; ***Dir:*** William Beaudine. **VHS $14.98** *WAR* 🦴🦴½

Ghost City Our hero must deal with a masked gang that is "haunting" a town.
1932 60m/B Bill Cody, Helen Foster, Ann Rutherford, John Shelton, Reginald Owen; ***Dir:*** Harry Fraser. **VHS $19.95** *NOS, DVT* 🦴

Ghost Dad A widowed workaholic dad is prematurely killed and returns from the dead to help his children prepare for the future. Cosby walks through doors, walls, and other solid objects for the sake of comedy—only none of it is funny.
1990 (PG) 84m/C Bill Cosby, Denise Nicholas, Ian Bannen, Christine Ebersole, Dana Ashbrook, Arnold Stang; ***Dir:*** Sidney Poitier. **VHS, Beta, LV $19.95** *MCA* 🦴

Ghost Dance A sacred Indian burial ground is violated, with grim results.
1983 93m/C Sherman Hemsley, Henry Ball, Julie Amato; ***Dir:*** Peter Bufa. **VHS, Beta $49.95** *TWE* 🦴

Ghost Fever Two bumbling cops try to evict the inhabitants of a haunted house and get nowhere. A bad rip-off of "Ghostbusters." Alan Smithee is a pseudonym used when a director does not want his or her name on a film—no wonder.
1987 (PG) 86m/C Sherman Hemsley, Luis Avalos; ***Dir:*** Alan Smithee. **VHS, Beta, LV $19.98** *CHA Woof!*

The Ghost Goes West A brash American family buys a Scottish castle and transports the pieces to the United States along with the Scottish family ghost (Donat). Also along is the ghost's modern-day descendant (also played by Donat) who is the new castle caretaker and in love with the new owner's daughter. It turns out to be up to the ghost to get the two lovers together and find his own eternal rest. A rather lovable fantasy written by Robert Sherwood. Rene Clair's first English film. Available in digitally remastered stereo with original movie trailer.
1936 85m/B *GB* Robert Donat, Jean Parker, Eugene Pallette, Elsa Lanchester; ***Dir:*** Rene Clair. **VHS, Beta $19.98** *HBO, SUE* 🦴🦴🦴

The Ghost and the Guest Newlyweds spend their honeymoon in a house in the country only to find that gangsters are using the place as a hideout. Chases involving secret passages ensue.
1943 59m/B James Dunn, Florence Rice, Mabel Todd; ***Dir:*** William Nigh. **VHS, Beta $16.95** *SNC* 🦴🦴½

The Ghost of H.L. Mencken Reporter dreams he is the ghost of writer H.L. Mencken.
1980 60m/C VHS, Beta *NO* 🦴🦴½

Ghost Keeper Three girls are trapped in a mansion with an old hag and various supernatural apparitions.
1980 87m/C Riva Spier, Murray Ord, Georgie Collins; ***Dir:*** Vernon Sewell. **VHS, Beta $59.95** *NWV* 🦴

The Ghost and Mrs. Muir A charming, beautifully orchestrated fantasy about a feisty widow who, with her young daughter, buys a seaside house and refuses to be intimidated by the crabby ghost of its former sea captain owner. When the widow falls into debt, the captain dictates his sea adventures, which she adapts into a successful novel. The ghost also falls in love with the beautiful lady. Tierney is exquisite and Harrison is sharp-tongued and manly. Oscar-nominated for Charles Lang's cinematography. Based on R. A. Dick's novel.
1947 104m/B Gene Tierney, Rex Harrison, George Sanders, Edna Best, Anna Lee, Vanessa Brown, Robert Coote, Natalie Wood, Isobel Elsom; ***Dir:*** Joseph L. Mankiewicz. **VHS, Beta, LV $19.98** *FOX* 🦴🦴🦴

Ghost of the Ninja An evil, masked assassin terrorizes the city but meets his match in the heroic "ghost" ninja.
19?? 114m/C Roy Chiao Hung, Steve Tung-Wai. **VHS, Beta $19.99** *BFV* 🦴🦴🦴

Ghost in the Noonday Sun The crew of a treasure-seeking pirate ship sets sail for high sea silliness in this slapstick adventure film. Not a theatrical release for one obvious reason—it's terrible.
1974 90m/C *GB* Peter Sellers, Spike Milligan, Anthony (Tony) Franciosa, Clive Revill, Rosemary Leach, Peter Boyle; ***Dir:*** Peter Medak. **VHS, Beta $19.95** *VIR, VCL Woof!*

Ghost Patrol A cowboy duo kidnaps the inventor of a ray gun in order to hijack mail planes, but they're stopped by G-man McCoy and friends. An odd hybrid of science fiction and western.
1936 57m/B Tim McCoy, Walter Miller, Wheeler Oakman, Sam Newfield. **VHS, Beta $45.95** *VCN, UHV* 🦴

The Ghost of Rashmon Hall A doctor tries to rid his home of the ghosts of a sailor, his wife and her lover. Also known as "Night Comes Too Soon."
1947 52m/B Valentine Dyall, Anne Howard, Alex Flavorsham, Howard Douglas, Beatrice Marsden, Arthur Brander. **VHS, Beta $16.95** *SNC, MWP* 🦴

Ghost Rider A lawman is aided in his outlaw-nabbing efforts by the ghost of a gunfighter.
1935 56m/B Rex Lease; ***Dir:*** Jack Levine. **VHS, Beta $19.95** *NOS, WGE* 🦴

Ghost Rider Former footballer Brown rides the range once again, this time with an element of the supernatural nipping at his heels.
1943 58m/B Johnny Mack Brown, Raymond Hatton, Tim Seidel, Beverly Boyd, Milburn Morante; ***Dir:*** Wallace Fox. **VHS** *ASE* 🦴🦴

Ghost Ship A young couple is tortured by ghostly apparitions when they move into an old yacht with a dubious past. The murdered wife of the ship's former owner is the poltergeist in this Grade B movie.
1953 69m/B *GB* Dermot Walsh, Hazel Court, Hugh Burden, John Robinson; ***Dir:*** Vernon Sewell. **VHS, Beta $14.95** *NOS, SNC* 🦴

Ghost Stories: Graveyard Thriller A compilation of stories with ghosts, vampires and other creatures of the night.
1986 56m/C ***Dir:*** Lynn Silver. **VHS, Beta $59.98** *LIV, VES*

Ghost Story Four elderly men, members of an informal social club called the Chowder Society, share a terrible secret buried deep in their pasts—a secret that comes back to haunt them to death. Has moments where it's chilling, but unfortunately they're few. Based on the best-selling novel by Peter Straub.
1981 (R) 110m/C Fred Astaire, Melvyn Douglas, Douglas Fairbanks Jr., John Houseman, Craig Wasson, Alice Krige, Patricia Neal; ***Dir:*** John Irvin. **VHS, Beta, LV $19.95** *MCA* 🦴🦴

Ghost Town Good guy Carey saves a mine from claim jumpers, but not before he is mistakenly incarcerated as his friend's murderer, and then proved innocent.
1937 65m/B Harry Carey; ***Dir:*** Harry Fraser. **VHS, Beta, LV $19.95** *WGE, TIM, VDM* 🦴

Ghost Town Fine sleeper about a modern sheriff who must rid a ghost town of its age-old curse.
1988 (R) 85m/C Franc Luz, Catherine Hickland, Jimmie F. Skaggs, Penelope Windust, Bruce Glover, Blake Conway, Laura Schaefer; ***Dir:*** Richard Governor. **VHS, LV $19.95** *STE, NWV* 🦴🦴½

Ghost Town Gold The Three Mesquiteers series continues as the trio outwits bank robbers who have hidden their loot in a ghost town.
1936 53m/B Bob Livingston, Ray Corrigan, Max Terhune; ***Dir:*** Joseph Kane. **VHS, Beta $19.95** *NOS, DVT* 🦴

Ghost Town Law The Rough Riders step in to save the day when a group of outlaws defend their hideout by murdering anyone who approaches.
1942 62m/B Buck Jones, Tim McCoy, Raymond Hatton; ***Dir:*** Howard Bretherton. **VHS, Beta $9.99** *NOS, VCN, CAB* 🦴½

Ghost Town Renegades A pair of government agents prevent a crook from gaining control of a mining town.
1947 57m/B Lash LaRue, Al "Fuzzy" St. John, Jennifer Holt; ***Dir:*** Ray Taylor. **VHS, Beta $12.95** *COL, SVS* 🦴

The Ghost Train Passengers stranded at a train station are told tales of a haunted train. Forced to stay at the station overnight, they encounter a number of comically spooky situations.
1941 85m/B *GB* Arthur Askey, Richard Murdock, Kathleen Harrison; ***Dir:*** Walter Forde. **VHS, Beta $16.95** *NOS, SNC* 🦴🦴

The Ghost Walks A playwright has his new masterpiece acted out in front of an unsuspecting producer, who thinks that a real murder has taken place. But when the cast really does start disappearing, who's to blame?
1934 69m/B John Miljan, June Collyer, Richard Carle, Spencer Charters, Johnny Arthur, Henry Kolker; ***Dir:*** Frank Strayer. **VHS $12.95** *SNC, NOS, DVT* 🦴

Ghost Writer A writer discovers the ghost of a movie star haunting her beach house, and together they solve the dead vamp's murder.
1989 (PG) 94m/C Audrey Landers, Judy Landers, Jeff Conaway, David Doyle, Anthony (Tony) Franciosa, Joey Travolta, John Matuszak, David Paul, Peter Paul; ***Dir:*** Kenneth J. Hall. **VHS, Beta $79.95** *PSM* 🦴

The Ghost of Yotsuya Based on the Japanese legend of a man who must betray his wife to get the power he seeks. When he does, however, what follows is a horrifying revenge. In Japanese with English subtitles.
1958 100m/C *JP* Shigeru Amachi. **VHS, Beta $59.95** *VDA* 🦴½

Ghostbusters After losing their scholastic funding, a group of "para-normal" investigators decide to go into business for themselves, aiding New York citizens in the removal of ghosts, goblins and other annoying spirits. This comedy-thriller about Manhattan being overrun by ghosts, contains great special effects, zany characters, and some of the best laughs of the decade. Also available in a laserdisc version with letterbox-

ing, analysis of all special effects, complete screenplay, and original trailer. Followed by a sequel.
1984 (PG) 103m/C Bill Murray, Dan Aykroyd, Harold Ramis, Rick Moranis, Sigourney Weaver, Annie Potts, Ernie Hudson, William Atherton; *Dir:* Ivan Reitman. **VHS, Beta, LV, 8mm $19.95** COL, PIA, VYG *ᛚᛚᛚ*

Ghostbusters 2 After being sued by the city for the damages they did in the original "Ghostbusters," the boys in khaki are doing kiddie shows at birthday parties. When a river of slime that is actually the physical version of evil is discovered running beneath the city, the Ghostbusters are back in action. They must do battle with a wicked spirit in a painting or the entire world will fall prey to its ravaging whims.
1989 (PG) 102m/C Bill Murray, Dan Aykroyd, Sigourney Weaver, Harold Ramis, Rick Moranis, Ernie Hudson, Peter MacNicol, David Margulies, Wilhelm von Homburg, Harris Yulin, Annie Potts; *Dir:* Ivan Reitman. **VHS, Beta, LV, 8mm $19.95** COL *ᛚᛚ½*

Ghosthouse It looks like an average suburban house, but it's haunted by the ghost of a little girl—and when she comes, evil is sure to follow.
1988 91m/C Lara Wendell, Gregg Scott; *Dir:* Humphrey Humbert. **VHS, Beta $79.95** IMP *ᛚ*

Ghostriders One hundred years after their hanging, a band of ghostly outlaws take revenge on the townfolk's descendents.
1987 85m/C Bill Shaw, Jim Peters, Ricky Long, Cari Powell, Mike Ammons, Arland Bishop; *Dir:* Alan L. Stewart. **VHS, Beta $9.99** STE, PSM *ᛚ*

Ghosts of Berkeley Square The ghosts of two retired soldiers of the early 18th century are doomed to haunt their former home, and only a visit from a reigning monarch can free them.
1947 85m/B Wilfrid Hyde-White, Robert Morley, Felix Aylmer; *Dir:* Vernon Sewell. **VHS, Beta $24.95** VYY *ᛚᛚ½*

Ghosts Can Do It The Hound doesn't like jumping to conclusions, but is it possible this obscurity was titled to be reminiscent of the more publicized Bo Derek vehicle "Ghosts Can't Do it?" Comedy about a man who returns to his sexy wife despite her efforts at killing him.
1990 88m/C Garry McDonald, Pamela Stephenson. **VHS $59.95** ATL, HHE *ᛚ½*

Ghosts Can't Do It Another wet-shirted, Bo-dacious softcore epic, involving a young woman who would like to have the spirit of her virile, but unfortunately dead, husband return to inhabit a living body.
1990 (R) 91m/C Bo Derek, Anthony Quinn, Don Murray, Leo Damian; *Dir:* John Derek. **VHS, Beta, LV $59.95** COL, TTC *ᛚ*

The Ghosts of Hanley House No-budget spooker with no-brain plot. Several unlucky guests spend an evening in a creepy mansion...sound familiar? Shot in Texas.
1968 80m/C Barbara Chase, Wilkie De Martel, Elsie Baker, Cliff Scott. **VHS, Beta $16.95** SNC *ᛚ½*

Ghosts on the Loose The Bowery Boys stumble into nasty Nazi spy Lugosi in a supposedly haunted mansion. Gardner plays Bowery Boy Hall's sister.
1943 67m/B Leo Gorcey, Huntz Hall, Bobby Jordan, Sammy Morrison, Billy Benedict, Stanley Clements, Bobby Stone, Bill Bates, Bela Lugosi, Ava Gardner; *Dir:* William Beaudine. **VHS, Beta $9.95** NEG, MRV, NOS *ᛚ½*

Ghosts of the Sky A documentary about the restoration of World War II B-series planes.
197? 60m/C *Nar:* Glenn Ford. **VHS, Beta $39.95** VCD

Ghosts That Still Walk Spooky phenomena occur. The demons possessing a young lad's soul may be responsible.
1977 92m/C Ann Nelson, Matt Boston, Jerry Jenson, Caroline Howe, Rita Crafts; *Dir:* James T. Flocker. **VHS, Beta $29.95** UHV, MRV *ᛚ*

Ghostwarrior A 16th century samurai warrior's ice-packed body is revived and runs amuck in the streets of modern-day Los Angeles.
1986 (R) 86m/C Hiroshi Fujioka, Janet Julian, Andy Wood, John Calvin; *Dir:* Larry Carroll. **VHS, Beta $79.94** VES *ᛚ*

Ghostwriter Based on Philip Roth's best-selling novel, the story is of a writer coming to terms with his past. Roth co-wrote the screenplay. First seen on PBS.
1984 90m/C Sam Wanamaker, Claire Bloom, Rose Arrick, MacIntyre Dixon, Cecile Mann, Joseph Wiseman, Mark Linn-Baker, Paulette Smit; *Dir:* Tristam Powell. **VHS, Beta $69.95** PBS *ᛚᛚ*

The Ghoul An eccentric English Egyptologist desires a sacred jewel to be buried with him and vows to come back from the grave if the gem is stolen. When that happens, he makes good on his ghostly promise. A minor horror piece with Karloff only appearing at the beginning and end of the film. This leaves the middle very dull.
1934 73m/B *GB* Boris Karloff, Cedric Hardwicke, Ernest Thesiger, Dorothy Hyson, Ralph Richardson, Anthony Bushell; *Dir:* T. Hayes Hunter. **VHS $19.95** NOS, SNC, DVT *ᛚᛚ½*

The Ghoul A defrocked clergyman has a cannibal son to contend with—especially after the drivers in a local auto race begin to disappear. Hurt is the lunatic family gardener.
1975 (R) 88m/C *GB* Peter Cushing, John Hurt, Alexandra Bastedo, Don Henderson, Stewart Bevan; *Dir:* Freddie Francis. **VHS, Beta $19.95** MED, VCL *ᛚᛚ*

Ghoul School A spoof of horror films and college flicks, with lots of scantily clad women and bloody creatures in dark hallways.
1990 (R) 90m/C Joe Franklin, Nancy Siriani, William Friedman. **VHS $79.95** HHE *ᛚ*

Ghoulies A young man gets more than he bargained for when he conjures up a batch of evil creatures when dabbling in the occult. Ridiculous but successful. Followed by two sequels.
1985 (PG-13) 81m/C Lisa Pelikan, Jack Nance, Scott Thompson, Tamara DeTreaux, Mariska Hargitay, Bobbie Breese; *Dir:* Luca Bercovici. **VHS, Beta $29.98** LIV, VES, HHE *ᛚ*

Ghoulies 2 Sequel to "Ghoulies" (1985), wherein the little demons haunt a failing carnival horror house, whose revenues begin to soar. Followed by a second sequel.
1988 (R) 89m/C Damon Martin, Royal Dano, Phil Fondacaro, J. Downing, Kerry Remsen; *Dir:* Albert Band. **VHS, Beta, LV $14.98** LIV, VES *ᛚ*

Ghoulies 3: Ghoulies Go to College Third in the series has the best special effects and the worst storyline. Not satisfied with ripping off "Gremlins," this one imposes Three Stooges personae upon a trio of demons at large on a beer- and babe-soaked campus.

1991 (R) 94m/C Kevin McCarthy, Griffin O'Neal, Evan Mackenzie; *Dir:* John Carl Buechler. **VHS $89.98** VES, LIV *Woof!*

G.I. Blues Three G.I.'s form a musical combo while stationed in Germany. Prowse is the nightclub singer Presley falls for. Songs include "Blue Suede Shoes." Presley's first film after his military service.
1960 104m/C Elvis Presley, Juliet Prowse, Robert Ivers, Leticia Roman, Ludwig Stossel, James Douglas, Jeremy Slate; *Dir:* Norman Taurog. **VHS, Beta $14.98** FOX, MVD *ᛚᛚ*

G.I. Executioner An adventure set in Singapore featuring a Vietnam veteran turned executioner.
1971 (R) 86m/C Tom Kenna, Victoria Racimo, Angelique Pettyjohn, Janet Wood, Walter Hill; *Dir:* Joel M. Reed. **VHS, Beta $69.95** VES *ᛚ*

G.I. Jane A television producer faints when he gets his draft notice and dreams his company is stranded at a desert post with a company of WACs.
1951 62m/B Jean Porter, Tom Neal, Iris Adrian, Jimmy Lloyd, Mara Lynn, Michael Whalen; *Dir:* Reginald LeBorg. **VHS, Beta, LV** WGE *ᛚ*

G.I. Joe: The Movie A full-length animated cartoon based on the famous toy character.
1986 90m/C *Dir:* Don Jurwich; *Voices:* Don Johnson, Burgess Meredith, Sgt. Slaughter. **VHS, Beta $19.95** CEL *ᛚ½*

The G.I. Road to Hell Contains three films of major operations and battles during World War II. The Burma campaign where General Joseph Stilwell, Merrills Marauders, Wingates Chindits, Flying Tigers, RAF and Gurkas combined to recapture Burma and build Burma and Ledo roads is the subject of "The Stilwell Road." In "Fortress of the Sea" American troops battle their way back to the Phillipines in an island hopping campaign led by MacArthur. Bloody battles in a ballpark and an assault on a 16th century fortress are documented in "The Battle of Manila."
1985 106m/B VHS, Beta $29.95 BVG, FGF, AVL *ᛚ½*

Giant Based on the Edna Ferber novel, this epic saga covers two generations of a wealthy Texas cattle baron (Hudson) who marries a strong-willed Virginia woman (Taylor) and takes her to live on his vast ranch. It explores the problems they have adjusting to life together, as well as the politics and prejudice of the time. Dean plays the resentful ranch hand (who secretly loves Taylor) who winds up striking oil and beginning a fortune to rival that of his former boss. Nominated for ten Academy Awards, including Best Picture, Best Director (which it won), Best Actor (Dean and Hudson), Supporting Actress (McCambridge), Adapted Screenplay, Score of a Comedy or Drama, Color Art Direction, Set Decoration, Costume Design and Film Editing. Dean's last movie—he died in a car crash shortly before filming was completed.
1956 201m/C Elizabeth Taylor, Rock Hudson, James Dean, Carroll Baker, Chill Wills, Dennis Hopper, Rod Taylor, Earl Holliman, Jane Withers, Sal Mineo, Mercedes McCambridge; *Dir:* George Stevens. Academy Awards '56: Best Director (Stevens); Directors Guild of America Awards '56: Best Director (Stevens). **VHS, Beta, LV $29.98** WAR, BTV *ᛚᛚᛚᛚ*

The Giant Claw A giant, winged (and stringed) bird attacks from outer space and scientists Morrow, Corday and Ankrum attempt to pluck it. Good bad movie; Corday was Playboy's Miss October, 1958.

1957 76m/B Jeff Morrow, Mara Corday, Morris Ankrum, Louis D. Merrill, Edgar Barrier, Robert Shayne, Morgan Jones, Clark Howat; *Dir:* Fred F. Sears. **VHS** $14.98 MOV ♂♂½

The Giant Gila Monster A giant lizard has the nerve to disrupt a local record hop, foolishly bringing upon it the wrath of the local teens. Rear-projection monster isn't particularly effective, but the film provides many unintentional laughs.
1959 74m/B Don Sullivan, Lisa Simone, Shug Fisher, Jerry Cortwright, Beverly Thurman, Don Flourney, Pat Simmons; *Dir:* Ray Kellogg. **VHS** $9.95 NOS, RHI, SNC ♂♂½

The Giant of Marathon Reeves shrugs off the role of Hercules to play Philippides, a marathoner (who uses a horse) trying to save Greece from invading Persian hordes. Lots of muscle on display, but not much talent.
1960 90m/C IT Steve Reeves, Mylene Demongeot, Daniela Rocca, Ivo Garrani, Alberto Lupo, Sergio Fantoni; *Dir:* Jacques Tourneur. **VHS** $16.95 SNC, MLB ♂♂

The Giant of Metropolis Muscle-bound hero goes shirtless to take on the evil, sadistic ruler of Atlantis (still above water) in 10,000 B.C. Ordinary Italian adventure, but includes interesting sets and bizarre torture scenes.
1961 92m/C IT Gordon Mitchell, Roldano Lupi, Bella Cortez, Liana Orfei; *Dir:* Umberto Scarpelli. **VHS** $19.98 SNC ♂♂

The Giant Spider Invasion A meteorite carrying spider eggs crashes to Earth and soon the alien arachnids are growing to humongous proportions. Notoriously bad special effects, but the veteran "B" cast has a good enough time.
1975 (PG) 76m/C Steve Brodie, Barbara Hale, Leslie Parrish, Robert Easton, Alan Hale Jr., Dianne Lee Hart, Bill Williams, Christianne Schmidtmer; *Dir:* Bill Rebane. **VHS** MOV, VCL ♂

Giant from the Unknown A giant conquistador is revived after being struck by lightning and goes on a murderous rampage. Unbelievably bad.
1958 80m/B Edward Kemmer, Buddy Baer, Bob Steele, Sally Fraser; *Dir:* Richard Cunha. **VHS, Beta** $16.95 NOS, SNC, MWP Woof!

Giants of Rome Rome is in peril at the hands of a mysterious secret weapon that turns out to be a giant catapult. Ludicrous all around.
1963 87m/C IT Richard Harrison, Ettore Manni. **VHS, Beta** $16.95 SNC Woof!

The Giants of Thessaly Jason and Orpheus search for the Golden Fleece, encountering and defeating monsters, wizards, and a scheming witch.
1960 86m/C IT Roland Carey, Ziva Rodann, Massimo Girotti, Alberto Farnese; *Dir:* Riccardo Freda. **VHS** $16.95 SNC ♂♂

Gideon's Trumpet True story of Clarence Earl Gideon, a Florida convict whose case was decided by the Supreme Court and set the precedent that everyone is entitled to defense by a lawyer, whether or not they can pay for it. Based on the book by Anthony Lewis. Made for television.
1980 104m/C Henry Fonda, Jose Ferrer, John Houseman, Dean Jagger, Sam Jaffe, Fay Wray; *Dir:* Robert Collins. **VHS, Beta** $14.95 WOV, VTK ♂♂½

Gidget A plucky, boy-crazy teenage girl (whose nickname means girl midget) discovers romance and wisdom on the beaches of Malibu when she becomes involved with a group of college-aged surfers. First in a series of Gidget/surfer films. Based on a novel by Frederick Kohner about his daughter.
1959 95m/C Sandra Dee, James Darren, Cliff Robertson, Mary Laroche, Arthur O'Connell, Joby Baker; *Dir:* Paul Wendkos. **VHS, Beta, LV** $9.95 COL

Gidget Goes Hawaiian Gidget is off to Hawaii with her parents and is enjoying the beach (and the boys) when she is surprised by a visit from boyfriend "Moondoggie." Sequel to "Gidget."
1961 102m/C Deborah Walley, James Darren, Carl Reiner, Peggy Cass, Michael Callan, Eddie Foy Jr.; *Dir:* Paul Wendkos. **VHS, Beta** $59.95 COL ♂♂

Gidget Goes to Rome Darren returns in his third outing as the boyfriend to yet another actress playing "Gidget" as the two vacation in Rome and find themselves tempted by other romances. Second sequel to "Gidget."
1963 104m/C Cindy Carol, James Darren, Jeff Donnell, Cesare Danova, Peter Brooks, Jessie Royce Landis; *Dir:* Paul Wendkos. **VHS, Beta** $69.95 RCA ♂½

The Gift Story of a 55-year-old man who chooses early retirement in the hopes of changing his dull and boring life. Unbeknownst to him, his co-workers arrange the ultimate retirement gift - a woman. Silly French sexual farce. Dubbed.
1982 (R) 105m/C FR Pierre Mondy, Claudia Cardinale, Clio Goldsmith, Jacques Francois, Cecile Magnet, Remy Laurent; *Dir:* Michael Lang. **VHS, Beta** $19.95 FCT ♂♂½

The Gift Horse A domineering British officer commands an old battleship but finds both the ship and the crew are still seaworthy after a battle with the Germans. Also known as "Glory at Sea."
1952 99m/B GB Trevor Howard, Richard Attenborough, Sonny Tufts, Bernard Lee; *Dir:* Compton Bennett. **VHS, Beta** $39.95 MON, DVT ♂♂

The Gift of Love Marie Osmond's acting debut in this saccharine adaption of O. Henry's Christmas romance "The Gift of the Magi." Made for television.
1990 96m/C James Woods, Marie Osmond, Timothy Bottoms, June Lockhart, David Wayne; *Dir:* Don Chaffey. **VHS** $24.95 MON, KAR ♂½

The Gig In this small, independently-made film, a band of middle-aged amateur jazz musicians give up their stable occupations for a once-in-a-lifetime shot at a two-week gig in the Catskills.
1985 95m/C Wayne Rogers, Cleavon Little, Warren Vache, Joe Silver, Daniel Nalbach; *Dir:* Frank D. Gilroy. **VHS, Beta, LV** $79.95 LHV, WAR ♂♂½

Gigi Based on Colette's story of a young Parisian girl (Caron) trained to become a courtesan to the wealthy Gaston (Jourdan). But he finds out he prefers her to be his wife rather than his mistress. Chevalier is Gaston's rougish father who casts an always admiring eye on the ladies, Gingold is Gigi's grandmother and former Chevalier flame, and Gabor amuses as Gaston's current, and vapid, mistress. One of the first MGM movies to be shot on location, this extravaganza features some of the best tributes to the French lifestyle ever filmed. Musical score by Lerner and Loewe includes "Thank Heaven for Little Girls," "I Remember It Well" and "The Night They Invented Champagne." Winner of nine Academy Awards.
1958 119m/C Leslie Caron, Louis Jourdan, Maurice Chevalier, Hermione Gingold, Eva Gabor, Isabel Jeans; *Dir:* Vincente Minnelli. Academy Awards '58: Best Adapted Screenplay, Best Art Direction/Set Decoration, Best Color Cinematography, Best Costume Design, Best Director (Minnelli), Best Film Editing, Best Picture, Best Song ("Gigi"), Best Musical Score; Directors Guild of America Awards '58: Best Director (Minnelli); Golden Globe Awards '59: Best Film—Musical/Comedy; National Board of Review Awards '58: 10 Best Films of the Year. **VHS, Beta, LV, 8mm** $19.98 MGM, BTV ♂♂♂♂

Gilbert & Sullivan Present Their Greatest Hits The chorus performs songs from the duo's most famous operettas, including "Pirates of Penzance," "The Mikado," and "Iolanthe."
1985 54m/C D'Oyly Carte Singers. **VHS, Beta** $29.95 MVD, LIV, VES

Gilda An evil South American gambling casino owner hires young American Ford as his trusted aide, unaware that Ford and his sultry wife Hayworth have engaged in a steamy affair. Hayworth does a striptease to "Put the Blame on Mame" in this prominently sexual film. This is the film that made Hayworth into a Hollywood sex goddess.
1946 110m/B Rita Hayworth, Glenn Ford, George Macready, Joseph Calleia; *Dir:* Charles Vidor. **VHS, Beta, LV** $19.95 COL ♂♂♂½

Gilda Live A live taping of Radner's stage show at New York's Winter Garden Theater. Gilda presents many of her "Saturday Night Live" characters, including dimwitted Lisa Lubner, loudmouthed Roseanne Roseannadanna and punk rocker Candy Slice.
1980 (R) 90m/C Gilda Radner, Don Novello; *Dir:* Mike Nichols. **VHS, Beta** $19.98 WAR ♂♂♂½

The Gilded Cage Two brothers are falsely accused of art theft and must find the real crooks in order to clear their names.
1954 77m/B GB Alex Nicol, Veronica Hurst, Clifford Evans, Ursula Howells, Elwyn Brook-Jones, John Stuart; *Dir:* John Gilling. **VHS, Beta** $16.95 SNC ♂♂

Gimme an F The handsome cheerleading instructor at Camp Beaver View ruffles a few pom-poms when he discovers that the camp's owner is about to enter into a shady deal with some foreigners. The film certainly deserves an "F" for foolishness if nothing else. Also known as "T & A Academy 2."
1985 (R) 100m/C Stephen Shellan, Mark Keyloun, John Karlen, Jennifer Cooke; *Dir:* Paul Justman. **VHS, Beta** $79.98 FOX, SIM Woof!

Gimme Shelter The sixties ended as violence occurred at a December 1969 free Rolling Stones concert attended by 300,000 people at Altamont, California. This "Woodstock West" became a bitter remembrance in rock history, as Hell's Angels (hired for security) do some ultimate damage to the spectators. A provocative look at an out-of-control situation.
1970 91m/C Mick Jagger, Keith Richards, Charlie Watts, Bill Wyman, Mick Taylor; *Dir:* David Maysles. **VHS, Beta, LV** $29.95 COL

Gin Game Taped London performance of the Broadway play about an aging couple who find romance in an old age home. Two-character performance won awards; touching and insightful.
1984 82m/C Jessica Tandy, Hume Cronyn; *Dir:* Mike Nichols. **VHS, Beta** $39.95 RKO, VTK ♂♂♂

Ginger Fabulous super-sleuth Ginger faces the sordid world of prostitution, blackmail and drugs. Prequel to "The Abductors."

1972 90m/C Cheri Caffaro, William Grannell; *Dir:* Don Schain. VHS, Beta $39.95 MON ⬜

Ginger Ale Afternoon A smutty comedy revolving around a triangle formed by a married couple and their sexy next-door neighbor. The film tries for fizz but only comes up flat. Music by Willie Dixon.
1989 88m/C Dana Anderson, John M. Jackson, Yeardley Smith; *Dir:* Rafal Zielinski. VHS, Beta, LV $29.95 ACA, IME ⬜

Ginger & Fred An acclaimed story of Fellini-esque poignancy about an aging dance team. Years before they had gained success by reverently impersonating the famous dancing duo of Astaire and Rogers. Now, after thirty years, they reunite under the gaudy lights of a high-tech television special. Wonderful performances by the aging Mastroianni and the still sweet Masina. In Italian with English subtitles.
1986 (PG-13) 128m/C IT Marcello Mastroianni, Giulietta Masina, Franco Fabrizi; *Dir:* Federico Fellini. VHS, Beta $19.98 MGM, FCT, APD ⬜

Ginger in the Morning A salesman is enamored of a young hitchhiker whom he picks up on the road. This is the same year Spacek played the innocent-gone-twisted in "Badlands."
1973 90m/C Sissy Spacek, Slim Pickens, Monte Markham, Susan Oliver, Mark Miller; *Dir:* Gordon Wiles. VHS, Beta $39.98 WGE ⬜

Gino Vannelli Gino Vannelli performs his hit songs in concert, including "I Just Wanna Stop," "Living Inside Myself," "Brother to Brother," and others.
1981 60m/C Gino Vannelli. VHS, Beta $29.98 WAR, OM

The Girl The first of Meszaros' trilogy, dealing with a young girl who leaves an orphanage to be reunited with her mother, a traditional country peasant. In Hungarian with English subtitles. Meszaros' first film. Followed by "Riddance" (1973) and "Adoption" (1975).
1968 86m/B HU *Dir:* Marta Meszaros. VHS, Beta $69.95 KIV ⬜⬜⬜

The Girl A wealthy attorney begins an affair with a young schoolgirl eventually leading to murder.
1986 (R) 104m/C Franco Nero, Christopher Lee, Bernice Stegers, Clare Powney; *Dir:* Arne Mattson. VHS, Beta $19.95 STE, NWV ⬜½

Girl in Black The shy daughter of a once wealthy household gets the chance to escape her genteel poverty when she falls in love with the young man boarding with her family. He is attracted to her but a shocking tragedy changes everything. In Greek with English subtitles.
1956 100m/C GR Ellie Lambetti; *Dir:* Michael Cacoyannis. VHS, Beta $59.95 FCT ⬜⬜

Girl in Black Stockings Beautiful women are mysteriously murdered at a remote, Utah hotel. Interesting, little known thriller similar to Hitchcock's "Psycho," but pre-dating it by three years.
1957 75m/B Lex Barker, Anne Bancroft, Mamie Van Doren, Ron Randell, Marie Windsor, John Dehner, John Holland, Diana Van Der Vlis; *Dir:* Howard W. Koch. VHS $18.00 FRG ⬜⬜

The Girl in Blue A lawyer, ill at ease with his current romance, suddenly decides to search for a beautiful woman he saw once, and instantly fell in love with, years before.

1974 (R) 103m/C CA Maud Adams, David Selby, Gay Rowan, William Osler, Diane Dewey, Michael Kirby; *Dir:* George Kaczender. VHS, Beta $79.95 PSM ⬜⬜

Girl from Calgary A female rodeo champ ropes her boyfriends much like she ropes her cows, but not without a knockdown, drag-out fight between the two which forces our cowgirl to make the choice she seems to know is fated. Stock rodeo footage is intercut with the heroine riding an obviously mechanical bucking bronc.
1932 66m/B Fifi D'Orsay, Paul Kelly; *Dir:* Philip H. Whitman. VHS, Beta, LV WGE ⬜

The Girl Can't Help It Satiric rock and roll musical comedy about a retired mobster who hires a hungry talent agent to promote his girlfriend, a wannabe no-talent singer. Mansfield's first starring role. Cameos by Eddie Cochran, Gene Vincent, The Platters, Little Richard and Fats Domino.
1956 99m/B Jayne Mansfield, Tom Ewell, Edmond O'Brien, Julie London, Ray Anthony; *Dir:* Frank Tashlin. VHS, Beta, LV $59.98 FOX ⬜⬜

Girl on a Chain Gang A girl on the run gets caught by police and does her time on a chain gang - an otherwise all-male chain gang. Predictable.
1965 96m/B William Watson, Julie Ange, R.K. Charles; *W/Dir:* Jerry Gross. VHS, Beta $16.95 SNC Woof!

Girl Crazy A wealthy young playboy is sent to an all-boy school in Arizona to get his mind off girls. Once there, he still manages to fall for a local girl who can't stand the sight of him. Songs by George and Ira Gershwin include "I Got Rhythm," "Embraceable You" and "Could You Use Me?" The eighth film pairing for Rooney and Garland.
1943 99m/B Mickey Rooney, Judy Garland, Tommy Dorsey & His Orchestra, Nancy Walker, June Allyson; *Dir:* Norman Taurog. VHS, Beta, LV $19.98 MGM, TTC, FCT ⬜⬜⬜

A Girl in Every Port Navy buddies acquire an ailing racehorse and try to conceal it aboard ship along with a healthy horse they plan to switch it with in an upcoming race. Groucho flies without his brothers in this zany film.
1952 86m/B Groucho Marx, William Bendix, Marie Wilson, Don DeFore, Gene Lockhart; *Dir:* Chester Erskine. VHS, Beta, LV $19.98 CCB ⬜⬜

Girl in Gold Boots An aspiring starlet meets up with a draft evader and a biker on her way to Hollywood where she gets a job as a go-go dancer in a sleazy club. But things don't go smoothly when a murder and a drug theft are revealed.
1969 94m/C Jody Daniels, Leslie McRae, Tom Pace, Mark Herron; *Dir:* Ted V. Mikels. VHS, Beta $39.95 WES ⬜

Girl of the Golden West A musical version of the David Belasco chestnut about a Canadian frontier girl loving a rogue. One of Eddy and MacDonald's lesser efforts but features Ebsen's song and dance talents in "The West Ain't Wild Anymore."
1938 120m/B Jeanette MacDonald, Nelson Eddy, Walter Pidgeon, Leo Carrillo, Buddy Ebsen, Leonard Penn, Priscilla Lawson, Bob Murphy, Olin Howland, Cliff Edwards, Billy Bevin; *Dir:* Robert Z. Leonard. VHS, Beta $29.95 MGM ⬜⬜

Girl Groups: The Story of a Sound Documentary on the "girl group" sound of the early 60s features rare footage and interviews with many of the original

singers, record producers and songwriters of that period. Among the 25 songs performed are "Please Mr. Postman," "Be My Baby," "Chapel of Love," "Baby Love" and "Stop! In the Name of Love."
1983 90m/C The Supremes, Ronettes, Shangri-Las, The Marvelettes, Shirelles. VHS, Beta, LV $19.95 MVD, MGM

A Girl, a Guy and a Gob A shy rich boy falls in love with a girl who has her sights on another. Silly but enjoyable cast.
1941 91m/B Lucille Ball, Edmond O'Brien, George Murphy, Franklin Pangborn, Henry Travers, Lloyd Corrigan; *Dir:* Richard Wallace. VHS, Beta $19.98 RKO ⬜⬜

Girl Happy The King's fans will find him in Fort Lauderdale, Florida this time as the leader of a rock 'n' roll group. His mission: to chaperon the daughter of a Chicago mobster who naturally falls for him. Why not?
1965 96m/C Elvis Presley, Harold J. Stone, Shelley Fabares, Gary Crosby, Nita Talbot, Mary Ann Mobley, Jackie Coogan; *Dir:* Boris Sagal. VHS, Beta $19.95 MGM ⬜⬜

The Girl with the Hat Box Early Russian production about a poor working girl who's paid with a lottery ticket rather than rubles (an ironic pre-communist theme?). When she strikes it rich in the lottery, her boss gives chase and silent antics ensue. Silent with orchestral score.
1927 67m/B RU Anna Sten, Vladimir Fogel, Serafina Birman, Ivan Koval-Samborsky; *Dir:* Boris Barnett. VHS $29.95 KIV, FCT ⬜⬜

A Girl on Her Own A young actress embarking on her first big role in Paris must also deal with family and political considerations in this drama set during the political upheaval of 1935.
1976 112m/C FR Sophie Chemineau, Bruno La Brasca; *Dir:* Philippe Nahoun. VHS $49.95 FCT ⬜⬜

Girl from Hunan Turn-of-the-century China, a 12-year-old girl is married to a 2-year-old boy in a typical arranged marriage. She grows into womanhood treating her toddler husband with sibling affection, but conducts a secret love affair with a farmer. Intolerant village laws against adultry place her life in jeopardy. In Mandarin with English subtitles.
1986 99m/C Na Renhua, Deng Xiaotuang, Yu Zhang; *Dir:* Xie Fei. VHS $79.95 NYF ⬜⬜⬜

Girl Hunters That intrepid private eye, Mike Hammer (played by his creator, Spillane), is caught up with communist spies, wicked women, and a missing secretary. Hammer is his usual judge-and-jury character but the action is fast-paced.
1963 103m/B Mickey Spillane, Lloyd Nolan, Shirley Eaton, Hy Gardner, Scott Peters; *Dir:* Roy Rowland. VHS, Beta $16.95 SNC, NOS, DVT ⬜⬜½

A Girl to Kill for A temptress lures a randy college guy into a plot of murder and betrayal.
1990 (R) 85m/C Sasha Jensen, Karen Austin, Alex Cord, Rod McCary, Karen Medak; *Dir:* Richard Oliver. VHS, Beta, LV $79.95 COL ⬜½

A Girl of the Limberlost A young Indiana farmgirl fights to bring her estranged parents back together, in spite of a meddling aunt and the unexplained phenomena that haunt her. Fine acting and beautiful scenery. Originally broadcast on public television as part of the WonderWorks Family Movie television series, it's best appreciated by

young teens. Adapted from Gene Stratton Porter's much-filmed novel.
1990 120m/C Annette O'Toole, Joanna Cassidy, Heather Fairfield. **VHS** $29.95 PME, HMV ♂♂

The Girl in Lover's Lane A drifter falls in love with a small town girl but becomes a murder suspect when she turns up dead. Interesting mainly for Elam's uncharacteristic portrayal of a village idiot.
1960 78m/B Brett Halsey, Joyce Meadows, Lowell Brown, Jack Elam; **Dir:** Charles R. Rondeau. **VHS, Beta** $16.95 SNC

A Girl in a Million Williams dumps his verbose wife and seeks peace and quiet in a men-only War Department office. All is well until the arrival of a colonel and his daughter, mute from trauma. Finding her silence appealing, he marries her, but another trauma restores her speech. Sexist comedy may have played better in its day.
1946 90m/B GB Hugh Williams, Joan Greenwood, Naunton Wayne, Wylie Watson, Garry Marsh; **Dir:** Francis Searle. **VHS** $19.95 NOS ♂♂

The Girl Most Likely Light musical comedy about a romance-minded girl who dreams of marrying a wealthy, handsome man. She runs into a problem when she finds three prospects. Remake of 1941's "Tom, Dick and Harry" with Ginger Rogers. Choreography by great Gower Champion.
1957 98m/C Jane Powell, Cliff Robertson, Tommy Noonan, Kaye Ballard, Una Merkel; **Dir:** Mitchell Leisen. **VHS, Beta, LV** $15.95 UHV ♂♂½

The Girl on a Motorcycle Singer Faithfull is a bored housewife who dons black leather and hops on her motorcycle to meet up with her lover (Delon), all the while remembering their other erotic encounters. Also known as "Naked Under Leather."
1968 (R) 92m/C GB Alain Delon, Marianne Faithfull, Roger Mutton; **Dir:** Jack Cardiff. **VHS, Beta** $59.95 MON, MLB ♂♂

The Girl from Petrovka A high-spirited Russian ballerina falls in love with an American newspaper correspondent. Their romance is complicated by the suspicious KGB. Bittersweet with a bleak ending.
1974 (PG) 103m/C Goldie Hawn, Hal Holbrook, Anthony Hopkins; **Dir:** Robert Ellis Miller. **VHS, Beta** $59.95 MCA ♂½

The Girl in the Picture A mild British comedy about a young photographer who is happy to lose, but then strives to regain, his live-in girlfriend.
1986 (PG-13) 89m/C GB John Gordon-Sinclair, Irina Brook, David McKay, Catherine Guthrie, Paul Young; **Dir:** Cary Parker. **VHS, Beta** $79.95 VES ♂♂½

The Girl in Room 2A A young woman is trapped in a mansion with the elderly owner and his demented son. Routine horrorfest.
1976 (R) 90m/C Raf Vallone, Daniela Giordano. **VHS, Beta** $59.95 PSM Woof!

Girl Rush Two bad vaudeville comics are stuck in California during the gold rush and are promised sacks of gold if they can persuade some women to come to a rowdy mining town.
1944 65m/B Wally Brown, Alan Carney, Frances Langford, Vera Vague, Robert Mitchum, Paul Hurst, Patti Brill, Sarah Padden, Cy Kendall; **Dir:** Gordon Douglas. **VHS, Beta** RKO ♂

The Girl Said No A sleazy bookie poses as a theatrical producer in order to get even with a dance hall floozy who once stiffed him. Also released as "With Words and Music."
1937 72m/B Robert Armstrong, Irene Hervey; **W/ Dir:** Andrew L. Stone. **VHS, Beta** $19.95 NOS, DVT ♂½

Girl School Screamers School's brightest students are trapped in a deceased benefactor's mansion. The ghost of the benefactor and a host of demons terrorize the girls. Contains nudity and violence, but nothing of merit.
1986 (R) 85m/C Mollie O'Mara, Sharon Christopher, Vera Gallagher; **Dir:** John P. Finegan. **VHS, Beta** $69.98 LTG Woof!

Girl Shy Harold is a shy tailor's apprentice who is trying to get a collection of his romantic fantasies published. Finale features Lloyd chasing wildly after girl of his dreams.
1924 65m/B Harold Lloyd, Jobyna Ralston, Richard Daniels; **Dir:** Fred Newmeyer. **VHS, Beta** TLF ♂♂♂

The Girl in a Swing An Englishman impulsively marries a beautiful German girl and both are subsequently haunted by phantoms from her past - are they caused by her psychic powers? Cliche-ridden silliness. Based on the Richard Adams novel.
1989 (R) 119m/C Meg Tilly, Rupert Frazer, Elspet Gray, Lynsey Baxter, Nicholas Le Prevost, Jean Boht; **Dir:** Gordon Hessler. **VHS, Beta, LV** $89.99 HBO ♂½

The Girl from Tobacco Row A Southern slut involved with a cache of stolen cash fools with both a convict and the sheriff.
1966 90m/C Tex Ritter, Rachel Romen, Earl Richards, Tim Ormond, Rita Faye, Ralph Emery; **Dir:** Ron Ormond. **VHS, Beta** $59.95 NO ♂

Girl-Toy A nerdy guy is afraid of women until a beautiful motorcyclist sweeps him off his feet. Before they can consummate their relationship, he is kidnapped by a lustful erotic cult who find him irresistible.
1984 86m/C Gerry Sont, Lenita Psillakis, Jon Finlayson. **VHS, Beta** $69.95 LTG ♂

Girl Under the Sheet A con-man/archaeologist moves into an ancient castle and begins romancing a beautiful resident ghost.
196? 89m/C Chelo Alonso, Walter Chiari. **VHS, Beta** INV ♂

Girl Who Spelled Freedom A young Cambodian girl, speaking little English, strives to adjust to life in Chattanooga, Tennessee. She faces her challenges by becoming a champion at the national spelling bee. Based on a true story. Made-for-television.
1986 90m/C Wayne Rogers, Mary Kay Place, Jade Chinn; **Dir:** Simon Wincer. **VHS, Beta** $59.95 DIS ♂♂♂

Girlfriend from Hell The Devil inhabits the body of a teenage wallflower and turns her into an uncontrollable vamp.
1989 (R) 95m/C Liane Curtis, Dana Ashbrook, Leslie Deane, James Daughton, Anthony Barrie, James Karen; **W/Dir:** Daniel M. Peterson. **VHS, Beta, LV** $14.95 IVE ♂

Girlfriends The bittersweet story of a young Jewish photographer learning to make it on her own. Directorial debut of Weill reflects her background in documentaries as the true-to-life episodes unfold.
1978 (PG) 87m/C Melanie Mayron, Anita Skinner, Eli Wallach, Christopher Guest, Amy Wright, Viveca Lindfors, Bob Balaban; **Dir:** Claudia Weill. **VHS, Beta** $59.95 WAR ♂♂♂

Girls Are for Loving Undercover agent Ginger faces real adventure when she battles it out with her counterpart, a seductive enemy agent. Her third adventure following "Ginger" (1970) and "The Abductors" (1971).
1973 90m/C Cheri Caffaro, Timothy Brown; **Dir:** Don Schain. **VHS, Beta** $39.95 MON ♂

Girls in Chains Girls in a reformatory, a teacher, a corrupt school official, and the detective trying to nail him. There's also a murder with the killer revealed at the beginning of the film.
1943 72m/B Arlene Judge, Roger Clark, Robin Raymond, Barbara Pepper, Dorothy Burgess, Clancy Cooper; **Dir:** Edgar G. Ulmer. **VHS, Beta** $16.95 NOS, SNC ♂½

Girls of the Comedy Store Taped at the Los Angeles Comedy Store, female stand-up comediennes do their best to break up their audience.
1986 60m/C Pam Matteson, Shirley Hemphill, Karen Haber, Tamayo Otsuki, Carrie Snow. **VHS, Beta** $29.98 LIV, LTG

The Girls of Crazy Horse Saloon Las Vegas' topless cabaret is on display. Titillating trash (no pun intended).
1989 60m/C VHS, Beta, LV $39.95 CEL, IME

Girls of Don Juan A nerd finds that every woman he meets somehow falls in love or lust with him.
197? 88m/C Anne Libert, Monica Vitti. **VHS, Beta** $29.95 MED ♂

A Girl's Folly A country girl falls in love with the leading man of a film shooting on location. She wants to become a rich actress, but fails miserably. The actor offers her all the luxuries she wants if she'll become his mistress. When she finally agrees, her mother shows up. What will our country maid do now?
1917 66m/B Robert Warwick, Doris Kenyon, June Elvidge, Johnny Hines; **Dir:** Maurice Tourneur. **VHS, Beta, 8mm** $19.95 VYY ♂♂½

Girls! Girls! Girls! Poor tuna boat fisherman Elvis moonlights as a nightclub singer to get his father's boat out of hock. He falls for a rich girl pretending to be poor; and after some romantic trials, there's the usual happy ending. Among the songs is "Return to Sender."
1962 106m/C Elvis Presley, Stella Stevens, Laurel Goodwin, Jeremy Slate, Benson Fong, Robert Strauss, Ginny Tiu, Guy Lee, Beulah Quo; **Dir:** Norman Taurog. **VHS, Beta** $14.98 FOX, MVD ♂♂

Girls of Huntington House A teacher in a school for unwed mothers finds herself becoming increasingly absorbed in her students' lives. Made for television.
1973 73m/C Shirley Jones, Mercedes McCambridge, Sissy Spacek, William Windom, Pamela Sue Martin, Darrell Larson; **Dir:** Alf Kjellin. **VHS** $59.95 IVE ♂

Girls Just Want to Have Fun An army brat and her friends pull out all the stops and defy their parents for a chance to dance on a national television program. Based loosely upon Cyndi Lauper's song of the same name.
1985 (PG) 90m/C Sarah Jessica Parker, Helen Hunt, Ed Lauter, Holly Gagnier, Morgan Woodward, Lee Montgomery, Shannen Doherty, Biff Yeager; **Dir:** Alan Metter. **VHS, Beta, LV** $9.95 NWV, STE ♂½

The Girls of Malibu A bevy of girls frolic on the Malibu beaches without clothes.

1986 60m/C VHS, Beta $24.95 *AHV Woof!*

Girls of the Moulin Rouge
A video glimpse into the stripteasing antics of the Moulin Rouge stage performers.
1986 60m/C VHS, Beta $59.98 *LIV, LTG*

Girls Next Door
A bevy of boisterous beauties cavort and cause havoc in buffoon-cluttered suburbia.
1979 85m/C Kirsten Baker, Perry Lang, Leslie Cederquist, Richard Singer; *Dir:* James Hong. VHS, Beta, LV *WGE ⅃*

Girls Night Out
An ex-cop must stop a killer who is murdering participants in a sorority house scavenger hunt and leaving cryptic clues on the local radio station.
1983 (R) 96m/C Hal Holbrook, Rutanya Alda, Julia Montgomery, James Carroll; *Dir:* Robert Deubel. VHS, Beta $69.95 *HBO ⅃*

Girls in Prison
Anne Carson is sent to prison for a bank robbery she didn't commit, and although the prison chaplain believes her story, the other prisoners think she knows where the money is hidden. They plan a prison break, forcing Carson to go with them and get the money, but they soon run across the real robber, also searching for the loot. A below-average "B" prison movie, lacking the camp aspects of so many of these films.
1956 87m/B Richard Denning, Joan Taylor, Adele Jergens, Helen Gilbert, Lance Fuller, Jane Darwell, Raymond Hatton; *Dir:* Edward L. Cahn. VHS $9.95 *COL ⅃½*

Girls Riot
Young, female juvenile delinquents have had enough! They decide to give their warden-like headmistress a taste of her own violence.
1988 100m/C Jocelyne Boisseau, Cornelia Calwer, Angelica Domrose, Ute Freight, *Dir:* Manfred Purzer. VHS, Beta $39.95 *ACN ⅃*

Girls on the Road
Two girls are just out for fun cruising the California coast. But that handsome hitchhiker turns out to be a deadly mistake.
1973 (PG) 91m/C *CA* Kathleen Cody, Michael Ontkean, Dianne Hull, Ralph Waite; *Dir:* Thomas J. Schmidt. VHS, Beta $59.95 *UNI ⅃½*

Girls School Screamers
Six young women and a nun are assigned to spend the weekend checking the contents for sale in a scary mansion bequeathed to their school. Unfortunately, the psychotic killer inhabiting the place doesn't think that's a good idea.
1986 (R) 85m/C Mollie O'Mara, Sharon Christopher, Vera Gallagher. VHS, Beta $69.98 *LIV, LTG ⅃*

Girls of the White Orchid
A naive American girl thinks she's getting a job singing in a Japanese nightclub, but it turns out to be a front for a prostitution ring run by the Japanese Yakuza. A made-for-television movie based on real stories, though the producers concentrate on the seamy side. Also known as "Death Ride to Osaka."
1985 96m/C Ann Jillian, Jennifer Jason Leigh; *Dir:* Jonathan Kaplan. VHS, Beta $59.95 *LHV ⅃*

Girlschool
The all-female metal band rocks raw and bawdy at London's Camden Palace. Includes "Play Dirty," "Rock Me Shock Me" and "Out to Get You."
1984 59m/C Girlschool. VHS, Beta $14.95 *SVS ⅃*

Girly
An English gothic about an excessively weird family that lives in a crumbling mansion and indulges in murder, mental aberration, and sexual compulsion. Originally titled, "Mumsy, Nanny, Sonny, and Girly."

1970 (R) 101m/C *GB* Michael Bryant, Ursula Howells, Pat Heywood, Howard Trevor, Vanessa Howard, Michael Ripper; *Dir:* Freddie Francis. VHS, Beta $79.95 *PSM ⅃*

Git!
A young runaway and his faithful dog are taken in by a wealthy dog-breeder and trained by him into a crack hunting team.
1965 90m/C Jack Chaplain, Richard Webb, Heather North; *Dir:* Ellis Kadisan. VHS, Beta $14.95 *COL, CHA ⅃½*

Git Along Little Dogies
Cattle ranchers versus oil men with Autry caught in the middle of the fight, but he still finds the time to romance the town banker's pretty daughter.
1937 60m/B William Farnum, Gene Autry, Judith Allen, Champion, Smiley Burnette; *Dir:* Joseph Kane. VHS, Beta $19.95 *NOS, VYY, DVT ⅃½*

Give 'Em Hell, Harry!
James Whitmore's one-man show as Harry S. Truman filmed in performance on stage; Whitmore was nominated for a Tony and an Oscar.
1975 103m/C James Whitmore; *Dir:* Steve Binder. VHS, Beta $19.95 *WOV*

Give a Girl a Break
Three talented but unknown babes vie for the lead in a stage production headed for Broadway after the incumbent prima donna quits. After befriending one of the various men associated with the production, each starlet thinks she's a shoe-in for the part. Entertaining but undistinguished musical with appealing dance routines of Fosse and the Champions. Ira Gershwin, collaborating for the first and only time with Burton Lane, wrote the lyrics to "It Happens Every Time" and "Applause, Applause."
1953 84m/C Marge Champion, Gower Champion, Debbie Reynolds, Helen Wood, Bob Fosse, Kurt Kasznar, Richard Anderson, William Ching, Larry Keating, Donna Martell; *Dir:* Stanley Donen. VHS $19.98 *MGM, TTC, FCT ⅃⅃½*

Give My Regards to Broad Street
McCartney film made for McCartney fans. Film features many fine versions of the ex-Beatle's songs that accompany his otherwise lackluster portrayal of a rock star in search of his stolen master recordings.
1984 (PG) 109m/C *GB* Paul McCartney, Bryan Brown, Ringo Starr, Barbara Bach, Tracy Ullman, Ralph Richardson, Linda McCartney; *Dir:* Peter Webb. VHS, Beta, LV $19.95 *MVD, FOX ⅃⅃*

Gizmo!
A hilarious and affectionate tribute to crackpot inventors everywhere, with footage of dozens of great and not-so-great machines and other creations. Some segments in black-and-white.
1977 77m/C *Dir:* Howard Smith. VHS, Beta $19.98 *WAR, OM ⅃⅃½*

The Gladiator
An angry Los Angeles citizen turns vigilante against drunk drivers after his brother is one of their victims. Made for television.
1986 94m/C Ken Wahl, Nancy Allen, Robert Culp, Stan Shaw, Rosemary Forsyth; *Dir:* Abel Ferrara. VHS, Beta, LV $19.95 *STE, NWV, HHE ⅃½*

Gladiator
When suburban Golden Gloves boxing champion Tommy Riley (Marshall) is forced to move to the inner city because of his father's gambling debts, he becomes involved with an evil boxing promotor who thrives on pitting different ethnic races against each other in illegal boxing matches. Eventually Riley is forced to fight his black friend Lincoln (Gooding), even though Lincoln has been warned that another blow to the head could mean his life. Although this film

tries to serve some moral purpose, it falls flat on the mat.
1992 (R) 98m/C Cuba Gooding Jr., James Marshall, Robert Loggia, Ossie Davis, Brian Dennehy, John Heard, Cara Buono; *Dir:* Rowdy Herrington. VHS, LV *COL ⅃⅃*

Gladiator of Rome
A gladiator flexes his pecs to save a young girl from death at the hands of evil rulers.
1963 105m/C *IT* Gordon Scott, Wandisa Guida, Roberto Risso, Ombretta Colli, Alberto Farnese; *Dir:* Mario Costa. VHS $16.95 *SNC ⅃*

The Gladiators
Televised gladiatorial bouts are designed to subdue man's violent tendencies in a futuristic society until a computer makes a fatal error.
1970 90m/C Arthur Pentelow, Frederick Danner; *Dir:* Peter Watkins. VHS, Beta *UHV ⅃*

Glass
A normally complacent man is driven to the brink of savagery when a killer begins stalking his employees.
1990 92m/C Alan Lovell, Lisa Peers, Adam Stone, Natalie McCurray. VHS $59.95 *ATL ⅃½*

The Glass Bottom Boat
A bubbly but transparent Doris Day comedy, in which she falls in love with her boss at an aerospace lab. Their scheme to spend time together get her mistaken for a spy. Made about the time they stopped making them like this anymore, it's innocuous slapstick romance with a remarkable 1960's cast, including a Robert Vaughn cameo as the Man From U.N.C.L.E. Doris sings "Que Sera, Sera" and a few other numbers.
1966 110m/C Doris Day, Rod Taylor, Arthur Godfrey, John McGiver, Paul Lynde, Dom DeLuise, Eric Fleming, Dick Martin, Robert Vaughn, Edward Andrews; *Dir:* Frank Tashlin. VHS, Beta $19.98 *MGM ⅃⅃½*

Glass House
Adapted from a Truman Capote story, this is a tense movie about a power struggle among the inmates of a state prison. Filmed at Utah State Prison, real-life prisoners as supporting cast add to the drama. Still has the power to chill, with Morrow particularly effective as one of the inmate leaders.
1972 92m/C Vic Morrow, Clu Gulager, Billy Dee Williams, Alan Alda, Dean Jagger, Kristopher Tabori; *Dir:* Tom Gries. VHS, Beta $49.95 *WES ⅃⅃⅃*

The Glass Jungle
Will a cab driver save Los Angeles? When armed terrorists and the FBI battle it out in the sunny California city an innocent cabby seems to hold the solution to survival.
1988 94m/C Lee Canalito, Joe Filbeck, Diana Frank, Mark High, Frank Scala; *Dir:* Joseph Merhi. VHS, Beta $69.95 *CLV, HHE ⅃⅃⅃*

The Glass Key
The second version of Dashiell Hammet's novel, a vintage mystery movie concerning a nominally corrupt politician (Donlevy) being framed for murder, and his assistant sleuthing out the real culprit. One of Ladd's first starring vehicles; Lake is the mystery woman who loves him, and Bendix a particularly vicious thug.
1942 85m/B Alan Ladd, Veronica Lake, Brian Donlevy, William Bendix, Bonita Granville, Richard Denning, Joseph Calleia, Moroni Olsen, Dane Clark; *Dir:* Stuart Heisler. VHS, Beta $29.95 *MCA ⅃⅃⅃*

The Glass Menagerie
An aging Southern belle deals with her crippled daughter Laura, whose one great love is her collection of glass animals. The third film adaptation of the Tennessee Williams classic, which preserves the performances of the Broadway revival cast, is a solid-but-not-stel-

G

lar adaptation of the play. Music by Henry Mancini with an all-star acting ensemble.
1987 (PG) 134m/C Joanne Woodward, Karen Allen, John Malkovich, James Naughton; *Dir:* Paul Newman. **VHS, Beta, LV $79.95** KUI, MCA 🎬🎬🎬

The Glass Slipper In this version of the "Cinderella" saga, Caron plays an unglamourous girl gradually transformed into the expected beauty. Winwood is the fairy godmother who inspires the girl to find happiness, rather then simply providing it for her magically. The film is highlighted by the stunning dance numbers, choreographed by Roland Petit and featuring the Paris ballet.
1955 93m/C Leslie Caron, Michael Wilding, Keenan Wynn, Estelle Winwood, Elsa Lanchester, Barry Jones, Amanda Blake, Lurene Tuttle; *Dir:* Charles Walters; *Nar:* Walter Pidgeon. **VHS, Beta $19.98** MGM, FCT 🎬🎬

Glass Tomb At a circus, a man performs the world's longest fast inside a glass cage, and becomes the raison d'etre for murder.
1955 59m/B John Ireland, Honor Blackman, Eric Pohlmann, Tonia Bern, Sidney James; *Dir:* Montgomery Tully. **VHS, Beta, LV** WGE 🎬🎬

Gleaming the Cube A skateboarding teen investigates his brother's murder. Film impresses with its stunt footage only. For those with adolescent interests.
1989 (PG-13) 102m/C Christian Slater, Steven Bauer, Min Luong, Art Chudabala, Le Tuan; *Dir:* Graeme Clifford. **VHS, Beta, LV $14.98** LIV, VES 🎬🎬½

Glen or Glenda An appalling, quasi-docudrama about transvestism, interspersed with meaningless stock footage, inept dream sequences and Lugosi sitting in a chair spouting incoherent prattle at the camera. Directorial debut of Wood, who, using a pseudonym, played the lead; one of the most phenomenally bad films of all time. An integral part of the famous anti-auteur's canon.
1953 70m/B Bela Lugosi, Edward D. Wood Jr., Lyle Talbot, Donald Woods; *Dir:* Edward D. Wood Jr. **VHS, Beta $19.95** NOS, MRV, SNC Woof!

Glen and Randa Two young people experience the world after it has been destroyed by nuclear war. Early Jim McBride, before the hired-gun success of "The Big Easy."
1971 (R) 94m/C Steven Curry, Shelley Plimpton; *Dir:* Jim McBride. **VHS, Beta, LV $59.95** UHV, IME 🎬🎬½

The Glenn Miller Story The music of the Big Band Era lives again in this warm biography of the legendary Glenn Miller, following his life from the late '20s to his untimely death in a World War II plane crash. Among the Miller hits featured are "Moonlight Serenade," "PE 6500," "In the Mood," "Chattanooga Choo Choo," and "St. Louis Blues March." Stewart's likably convincing and even fakes the trombone playing well.
1954 (G) 113m/C James Stewart, June Allyson, Harry Morgan, Gene Krupa, Louis Armstrong, Ben Pollack; *Dir:* Anthony Mann. Academy Awards '54: Best Sound; Film Daily Poll '54: Best Picture of the Year. **VHS, Beta, LV $19.95** MCA, RDG 🎬🎬🎬

Glitch! Two inept burglars fumble an attempted robbery and somehow wind up with a slew of beautiful models, but in big trouble with the mob.
1988 (R) 88m/C Julia Nickson, Will Egan, Steve Donmyer, Dan Speaker, Dallas Cole, Ji Tu Cambuka, Dick Gautier, Ted Lange, Teri Weigel, Fernando Carzon, John Kreng, Lindsay Carr; *Dir:* Nico Mastorakis. **VHS, Beta $29.95** ACA 🎬½

Glitter Dome Two policemen discover the sleazier side of Hollywood when they investigate the murder of a pornographer. Based upon the novel by Joseph Wambaugh. Made for television.
1984 90m/C Stuart Margolin, John Marley, James Garner, John Lithgow, Margot Kidder, Colleen Dewhurst; *Dir:* Stuart Margolin. **VHS, Beta $59.99** HBO 🎬½

Glitz A Miami cop and an Atlantic City lounge singer team up to investigate a call-girl's death and wind up too close to a drug ring. Based on the novel by Elmore Leonard. Made for television.
1988 96m/C Jimmy Smits, John Diehl, Markie Post, Ken Foree, Madison Mason, Robin Strasser; *Dir:* Sandor Stern. **VHS, Beta $79.95** WAR 🎬½

Gloria She used to be a Mafia moll, now she's outrunning the Mob after taking in the son of a slain neighbor. He's got a book that they want, and they're willing to kill to get it. Gritty attempt by director Cassavetes at a mainstream hit.
1980 (PG) 123m/C Gena Rowlands, John Adams, Buck Henry, Julie Carmen; *Dir:* John Cassavetes. Venice Film Festival '80: Best Film. **VHS, Beta, LV $69.95** COL 🎬🎬½

Glorifying the American Girl Eaton's a chorus girl performing in the Ziegfeld Follies, which lends itself to numerous production numbers. The only film Ziegfeld ever produced.
1930 96m/B Mary Eaton, Dan Healey, Eddie Cantor, Rudy Vallee; *Dir:* Millard Webb. **VHS, Beta $19.95** NOS, HHT, DVT 🎬🎬

Glory A racehorse owner finds the money to enter her filly in the Kentucky Derby. No surprises here.
1956 100m/C Margaret O'Brien, Walter Brennan, Charlotte Greenwood; *Dir:* Dick Butler. **VHS, Beta $15.95** UHV 🎬🎬

Glory A rich, historical spectacle chronicling the 54th Massachusetts, the first black volunteer infantry unit in the Civil War. The film manages to artfully focus on both the 54th and their white commander, Robert Gould Shaw. Based on Shaw's letters, the film uses thousands of accurately costumed "living historians" (re-enactors) as extras in this panoramic production that was Oscar-nominated for best picture of 1989. A haunting, bittersweet musical score pervades what finally becomes an anti-war statement. Stunning performances throughout, with exceptional work from Freeman and Washington.
1989 (R) 122m/C Matthew Broderick, Morgan Freeman, Denzel Washington, Cary Elwes, Jihmi Kennedy, Andre Braugher, John Finn, Donovan Leitch, John Cullum, Bob Gunton, Jane Alexander, Raymond St. Jacques; *Dir:* Edward Zwick. Academy Awards '89: Best Cinematography, Best Sound, Best Supporting Actor (Washington). **VHS, Beta, LV, 8mm $19.95** COL, FCT, BTV 🎬🎬🎬½

Glory Boys A secret agent is hired to protect an Israeli scientist who is marked for assassination by the PLO and IRA.
1984 110m/C Rod Steiger, Anthony Perkins, Gary Brown, Aaron Harris; *Dir:* Michael Ferguson. **VHS, Beta $59.95** PSM 🎬½

Glory! Glory! Slashing satire about TV evangelism. Richard Thomas as the meek son of founder of Church of the Champions of Christ brings in lewd female rock singer to save the money-machine ministry from financial ruin. Rock music, sex, and MTV-like camera work make it worth seeing. James Whitmore is a treat.

1990 (R) 152m/C Ellen Greene, Richard Thomas, Barry Morse, James Whitmore; *Dir:* Lindsay Anderson. **VHS, Beta, LV $59.98** ORI, IME 🎬🎬🎬

The Glory Stompers Hopper prepares for his Easy Rider role as the leader of a motorcycle gang who battles with a rival leader over a woman. Very bad, atrocious dialogue, and a "love-in" scene that will be best appreciated by insomniacs.
1967 85m/C Dennis Hopper, Jody McCrea, Chris Noel, Jock Mahoney, Lindsay Crosby, Robert Tessier, Casey Kasem; *Dir:* Anthony M. Lanza. **VHS $59.95** TRY Woof!

The Glory of Their Times Professional baseball during the years 1896 to 1916, with vintage photos and film footage, plus voices of the great players.
1970 60m/C VHS $14.98 CSB, VTK Woof!

Glory Years Three old friends find themselves in charge of the scholarship fund at their 20th high school reunion. Unfortunately, they decide to increase the fund by gambling in Las Vegas and lose it all on a fixed fight!
1987 150m/C George Dzundza, Archie Hahn, Tim Thomerson, Tawny Kitaen, Donna Pescow, Donna Denton; *Dir:* Arthur Seidelman. **VHS, Beta $79.95** HBO 🎬

The Glove An ex-cop turned bounty hunter has his toughest assignment ever. It's his job to bring in a six-and-a-half foot, 250 pound ex-con who's been breaking havoc with an unusual glove—it's made of leather and steel.
1978 (R) 93m/C John Saxon, Roosevelt Grier, Joanna Cassidy, Joan Blondell, Jack Carter, Aldo Ray; *Dir:* Ross Hagen. **VHS, Beta $19.95** MED 🎬½

Gnome-Mobile A lumber baron and his two grandchildren attempt to reunite a pair of forest gnomes with a lost gnome colony. Brennan has a dual role as both a human and a gnome grandfather. Wynn's last film. Based on a children's novel by Upton Sinclair.
1967 84m/C Walter Brennan, Richard Deacon, Ed Wynn, Karen Dotrice, Matthew Garber; *Dir:* Robert Stevenson. **VHS, Beta** DIS, OM 🎬🎬

Gnomes The gnomes are wily little creatures who are in constant battle with their oversized, underwitted adversaries, the trolls. The books of Wil Huggen and Rien Poortvliet inspired the film.
1980 48m/C VHS, Beta $225.00 PYR 🎬🎬

The Go-Between A wonderful tale of hidden love. A young boy (Guard) acts as messenger between the aristocratic Christie and her former lover Bates. But tragedy befalls them all when the lovers are discovered. The story is told as the elderly messenger (now played by Redgrave) recalls his younger days as this go-between and builds to a climax when he is once again asked to be a messenger for the lady he loved long ago. Based on a story by L. P. Hartley and adapted by Harold Pinter. Leighton, as Christie's mother, received an Oscar nomination.
1971 (PG) 116m/C GB Julie Christie, Alan Bates, Dominic Guard, Margaret Leighton, Michael Redgrave, Michael Gough, Edward Fox; *Dir:* Joseph Losey. **VHS** CSM, MGM 🎬🎬🎬½

Go for Broke! Inexperienced officer Johnson heads a special World War II attack force which is made up of Japanese Americans. Sent to fight in Europe, they prove their bravery and loyalty to all. Good, offbeat story.

1951 92m/B Van Johnson, Giana Maria Canale, Warner Anderson, Lane Nakano, George Miki; *Dir:* Robert Pirosh. **VHS, Beta, 3/4U $19.95** *NOS, MRV, VYY* 🐾🐾½

Go for Broke Motorcycle stunts, fights, and car chases highlight this action film. **1990** ?m/C **VHS** *VYY* 🐾🐾½

Go Down Death In this early all-black film, a minister is caught in a moral dilemma literally between Heaven and Hell. Scenes of the afterlife are taken from early silent films. **1941** 63m/B Myra D. Hemmings, Samuel H. James, Eddy L. Houston, Spencer Williams, Amos Droughan; *Dir:* Spencer Williams. **VHS, Beta $29.95** *NOS, DVT, TIM* 🐾½

Go Go Big Beat A collection of simulated concert shorts featuring a plethora of British '60's rock bands, including The Hollies, The Merseybeats, The Animals, and many more. Released in England as three shorts: "Swinging UK," "UK Swings Again" and "Mods and Rockers." **1965** 70m/C **VHS, Beta $29.95** *RHI* 🐾½

Go for the Gold A young marathon runner risks his girlfriend's love when he decides to pursue fame and fortune. **1984** 98m/C James Ryan, Cameron Mitchell, Sandra Horne. **VHS, Beta $69.98** *LIV, LTG* 🐾

Go Go's: Wild at the Greek The rock and roll girl group performs "Head Over Heels," "We Got The Beat," and other favorites in this concert taped at Los Angeles Greek Theatre. **1984** 52m/C Belinda Carlisle, Charlotte Caffey, Kathy Valentine, Jane Wiedlin, Gina Schock. **VHS, Beta $14.95** *MVD, COL* 🐾

Go for It Oddball supercops go after a mad scientist—who has created a deadly "K"-bomb—and they encounter killer whales, karate busboys and malevolent Shirley Temple clones along the way. **1983** (PG) 109m/C *IT* Bud Spencer, Terence Hill, David Huddleston; *Dir:* E.B. Clucher. **VHS, Beta $79.95** *WAR* 🐾🐾

Go, Johnny Go! Rock promoter Alan Freed molds a young orphan into rock sensation "Johnny Melody." Musical performances include Ritchie Valens (his only film appearance), Eddie Cochran, Jackie Wilson. **1959** 75m/B Alan Freed, Sandy Stewart, Chuck Berry, The Cadillacs, Jimmy Clanton, Eddie Cochran, Jackie Wilson, Ritchie Valens; *Dir:* Paul Landres. **VHS, Beta, LV $14.95** *QNE, MSM, AOV* 🐾🐾

Go Kill and Come Back A bounty hunter tracks down a notoriously dangerous train robber. **1968** (PG) 95m/C Gilbert Roland, George Helton, Edd Byrnes; *Dir:* Enzo G. Castellari. **VHS, Beta $24.95** *MON* 🐾

The Go-Masters A historic co-production between Japan and China, centering on the ancient strategy board game "Go." A young competitor becomes obsessed with winning at any price. Episodic tale of the relationships between a Japanese and Chinese family spans 30 years and was critically acclaimed in both China and Japan. In Japanese and Chinese with English subtitles. **1982** 134m/C *JP CH* Rentaro Mikuni, Sun Dao-Lin, Shen Guan-Chu, Misako Honno; *Dir:* Junya Sato, Duan Jishun. Montreal World Film Festival '83: Best Film. **VHS, Beta $59.95** *SVS, TAM, FCT* 🐾🐾🐾

Go Tell the Spartans In Vietnam, 1964, a hard-boiled major is ordered to establish a garrison at Muc Wa with a platoon of burned out Americans and Vietnamese mercenaries. Blundering but politically interesting war epic pre-dating the flood of 1980s American-Vietnam apologetics. Based on Daniel Ford's novel. **1978** (R) 114m/C Burt Lancaster, Craig Wasson, David Clennon, Marc Singer; *Dir:* Ted Post. **VHS, Beta, LV $19.98** *VES* 🐾🐾🐾

Go West The brothers Marx help in the making and un-making of the Old West. Weak, late Marx Bros., but always good for a few yucks. **1940** 82m/B Groucho Marx, Chico Marx, Harpo Marx, John Carroll, Diana Lewis; *Dir:* Edward Buzzell. **VHS, Beta, LV $19.95** *MGM, CCB* 🐾🐾½

The Goalie's Anxiety at the Penalty Kick A chilling landmark film that established Wenders as one of the chief film voices to emerge from post-war Germany. The story deals with a soccer player who wanders around Vienna after being suspended, commits a random murder, and slowly passes over the brink of sanity and morality. A suspenseful, existential adaptation of the Peter Handke novel. In German with English subtitles. **1971** 101m/C *GE* Arthur Brauss, Erika Pluhar, Kai Fischer; *Dir:* Wim Wenders. **VHS, Beta $29.95** *PAV, FCT, GLV* 🐾🐾🐾½

God, Man and Devil A wager between God and Satan begins this allegory of money versus the spirit. A poor Torah scribe, Hershele Dubrovner, has a life that glorifies God until Satan, disguised as a business partner, turns him greedy and dishonest. Dubrovner's success destroys both his religion and his community leaving only betrayal and abandonment. In Yiddish with English subtitles. **1949** 100m/B Mikhal Mikhalesko, Gustav Berger; *Dir:* Joseph Seiden. **VHS $54.00** *NCJ* 🐾🐾½

God Told Me To A religious New York cop is embroiled in occult mysteries while investigating a series of grisly murders. He investigates a religious cult that turns out to be composed of half-human, half-alien beings. A cult-fave Larry Cohen epic, alternately titled "Demon." **1976** (R) 89m/C Tony LoBianco, Deborah Raffin, Sylvia Sidney, Sandy Dennis, Richard Lynch; *W/Dir:* Larry Cohen. **VHS, Beta, LV $19.98** *CHA* 🐾🐾½

Godard/Truffaut Shorts A pair of early shorts directed by two French "new wave" masters: Jean-Luc Godard's "All the Boys Are Called Patrick" from 1957 and Francois Truffaut's "Les Mistons" from 1958. **1958** 39m/B *FR Dir:* Francois Truffaut, Jean-Luc Godard. **VHS, Beta $29.95** *DVT, MRV*

The Goddess Written by Paddy Chayefsky, this is the sordid story of a girl who rises to fame as a celluloid star by making her body available to anyone who can help her career. When the spotlight dims, she keeps going with drugs and alcohol to a bitter end. **1958** 105m/B Kim Stanley, Lloyd Bridges, Patty Duke; *Dir:* John Cromwell. **VHS, Beta, LV $69.95** *RCA* 🐾🐾🐾

The Godfather Coppola's award-winning adaptation of Mario Puzo's novel about a fictional Mafia family in the late 1940s. Revenge, envy, and parent-child conflict mix with the rituals of Italian mob life in America. Minutely detailed, with excellent perfor-

mances by Pacino, Brando, and Caan as the violence-prone Sonny. The horrific horse scene is an instant chiller. Indisputably an instant piece of American culture. Followed by two sequels. **1972** (R) 171m/C Marlon Brando, Al Pacino, Robert Duvall, James Caan, Diane Keaton, John Cazale, Talia Shire, Richard Conte, Richard Castellano, Abe Vigoda, Alex Rocco, Sterling Hayden, John Marley, Al Lettieri; *W/Dir:* Francis Ford Coppola. Academy Awards '72: Best Actor (Brando), Best Adapted Screenplay, Best Picture; British Academy Awards '72: Best Original Score; Directors Guild of America Awards '72: Best Director (Coppola); Golden Globe Awards '73: Best Film—Drama; National Board of Review Awards '72: 10 Best Films of the Year, Best Supporting Actor (Pacino); National Society of Film Critics Awards '72: Best Actor (Pacino). **VHS, Beta, LV, 8mm $29.95** *PAR, BTV* 🐾🐾🐾🐾

Godfather 1902-1959—The Complete Epic Coppola's epic work concerning the lives of a New York crime family. Comprises the first two "Godfather" films, reedited into a chronological framework of the Corleone family history, with much previously discarded footage restored. **1981** 386m/C Marlon Brando, Al Pacino, Robert Duvall, James Caan, Richard Castellano, Diane Keaton, Robert De Niro, John Cazale, Lee Strasberg, Talia Shire, Michael Gazzo, Troy Donahue, Joe Spinell, Abe Vigoda, Alex Rocco, Sterling Hayden, John Marley, Richard Conte, G.D. Spradlin, Bruno Kirby, Harry Dean Stanton, Roger Corman, Al Lettieri; *Dir:* Francis Ford Coppola. **VHS, Beta $99.95** *PAR, FCT* 🐾🐾🐾🐾

The Godfather: Part 2 A continuation and retracing of the first film, interpolating the maintenance of the Corleone family by the aging Michael, and it's founding by the young Vito (DeNiro, in a terrific performance) 60 years before in New York City's Little Italy. Often considered the second half of one film, the two films stand as one of American film's greatest efforts, and as a 1970s high-water mark. Combined into one work for television presentation. **1974** (R) 200m/C Al Pacino, Robert De Niro, Diane Keaton, Robert Duvall, James Caan, Danny Aiello, John Cazale, Lee Strasberg, Talia Shire, Michael Gazzo, Troy Donahue, Joe Spinell, Abe Vigoda, Marianna Hill, Fay Spain, G.D. Spradlin, Bruno Kirby, Harry Dean Stanton, Roger Corman, Kathleen Beller; *Dir:* Francis Ford Coppola. Academy Awards '74: Best Adapted Screenplay, Best Art Direction/Set Decoration, Best Director (Coppola), Best Picture, Best Supporting Actor (De Niro), Best Original Score (Rota); National Society of Film Critics Awards '74: Best Cinematography, Best Director (Coppola). **VHS, Beta, LV, 8mm $29.95** *PAR, BTV* 🐾🐾🐾🐾

The Godfather: Part 3 Don Corleone (Pacino), now aging and guilt-ridden, determines to buy his salvation by investing in the Catholic Church, which he finds to be a more corrupt brotherhood than his own. Meanwhile, back on the homefront, his young daughter discovers her sexuality as she falls in love with her first cousin. While not the best film of the trilogy, it's still a stunning and inevitable conclusion to the story, and Pacino adds some exquisite finishing touches to his trilogy-worn character. Beautifully photographed in Italy by Gordon Willis. Music by Francis' father, Carmine Coppola. Oscar nominations for Best Picture, Best Director, Best Song, Cinematography, Film Editing and Art Direction. Video release contains the final director's cut featuring nine minutes of footage not included in the theatrical release. Also available as part of "The Godfather's Collector's Edition," which contains all three

films, along with a 73-minute chronicle of the filming of each movie.
1990 (R) 170m/C Al Pacino, Diane Keaton, Andy Garcia, Joe Mantegna, George Hamilton, Talia Shire, Sofia Coppola, Eli Wallach, Don Novello, Bridget Fonda, John Savage, Al Martino; **W/Dir:** Francis Ford Coppola. **VHS, Beta, LV, SVS, 8mm $79.98** *PAR, PIA, SUP ♪♪♪*

God's Bloody Acre Three backwoods mountain-dwelling brothers kill construction workers in order to defend their land, and then begin preying on vacationers.
1975 (R) 90m/C Scott Lawrence, Jennifer Gregory, Sam Moree, Shiang Hwa Chyang; **Dir:** Harry Kerwin. **VHS, Beta $69.95** *TWE Woof!*

God's Country A loner kills a man in self-defense and escapes the pursuing law by disappearing into California Redwood country with his dog.
1946 62m/C Robert Lowery, Helen Gilbert, William Farnum, Buster Keaton; **Dir:** Robert Tansey. **VHS, Beta, LV** *WGE ♪♪*

God's Gun A preacher, who was once a gunfighter, seeks revenge on the men who tried to kill him. Also known as "A Bullet from God."
1975 (R) 93m/C Richard Boone, Lee Van Cleef, Jack Palance, Sybil Danning; **Dir:** Frank Kramer. **VHS, Beta $9.99** *CCB, PGN ♪*

God's Little Acre Delves into the unexpectedly passionate lives of Georgia farmers. One man, convinced there's buried teasure on his land, nearly brings himself and his family to ruin trying to find it. Based on the novel by Erskine Caldwell.
1958 110m/B Robert Ryan, Tina Louise, Michael Landon, Buddy Hackett, Vic Morrow, Jack Lord, Aldo Ray; **Dir:** Anthony Mann. **VHS, Beta $19.95** *KOV, PSM, VHE ♪♪♪*

The Gods Must Be Crazy An innocent and charming film. A peaceful Bushman travels into the civilized world to return a Coke bottle "to the gods." Along the way he meets a transplanted schoolteacher, an oafishly clumsy microbiologist and a gang of fanatical terrorists. A very popular film, disarmingly crammed with slapstick and broad humor of every sort. Followed by a weak sequel.
1984 (PG) 109m/C N!xau, Marius Weyers, Sandra Prinsloo, Louw Verwey, Jamie Uys, Michael Thys, Nic de Jager; **Dir:** Jamie Uys. **VHS, Beta, LV $19.39** *FOX, FCT ♪♪♪½*

The Gods Must Be Crazy 2 A slapdash sequel to the original 1981 African chortler, featuring more ridiculous shenanigans in the bush. This time the bushman's children find themselves in civilization and N!xau must use his unique ingenuity to secure their safe return.
1989 (PG) 90m/C N!xau, Lena Farugia, Hans Strydom; **Dir:** Jamie Uys. **VHS, Beta, LV $89.95** *COL, TTC, FCT ♪♪♪*

Gods of the Plague Fassbinder goes noir—or grey with lots of sharp lighting—in this gangster-auteur tale of robbery gone awry. The requisite trappings of the crime genre (guys and dolls and cops and robbers) provide a vague backdrop (and a vague plot) for a moody story full of teutonic angst and alienation, and that certain Fassbinder feeling (which, need we say, isn't to everyone's taste). An early effort by the director, who remade the story later the same year as "The American Soldier" (Fassbinder acts in both). In German with (difficult-to-read) English subtitles; originally titled "Gotter der Pest."

1969 92m/C GE Hanna Schygulla, Harry Bear; **Dir:** Rainer Werner Fassbinder. **VHS $39.99** *INF ♪♪♪*

The Godsend A little girl is thrust upon a couple who, after adopting her, lose their natural family to her evil ways.
1979 (R) 93m/C CA Cyd Hayman, Malcolm Stoddard, Angela Pleasence, Patrick Barr; **Dir:** Gabrielle Beaumont. **VHS, Beta $59.98** *VES ♪*

Godson Emotionless assassin lives a barren life under the surveillance of Paris police who suspect him in a recent killing. He suddenly discovers that his boss has a contract out on his life. Delon's performance raises this to an average crime drama.
1972 103m/C FR Alain Delon, Nathalie Delon, Francois Perier; **Dir:** Jean-Pierre Melville. **VHS** *PSV, AVD ♪♪♪*

Godzilla 1985 Moviedom's reptilian superstar sets out to reverse Japan's trade imbalance by destroying Tokyo.
1985 (PG) 91m/C JP Raymond Burr, Keiju Kobayashi, Ken Takaka, Yasuka Sawaguchi; **Dir:** Kohji Hashimoto, R.J. Kizer. **VHS, Beta, LV $9.95** *NWV, STE ♪♪½*

Godzilla, King of the Monsters The radioactive monster Godzilla attacks Tokyo and terrifies the world. Burr's scenes are intercut in the American version, where he serves as a narrator telling the monster's tale in flashbacks.
1956 80m/B JP Raymond Burr, Takashi Shimura; **Dir:** Terry Morse. **VHS, Beta, LV $19.98** *VES ♪♪½*

Godzilla on Monster Island Even Godzilla himself cannot hope to take on both Ghidra and Gigan alone and hope to succeed. Therefore he summons his pal Angillus for help. Together, they offer Earth its only hope of survival. Though the terror may repel you, the movie is a must for Godzilla fans; it's his first speaking part.
1972 89m/C JP Hiroshi Ishikawa, Tomoko Umeda, Yuriko Hishimi, Minoru Takashima, Zan Fujita; **Dir:** Jun Fukuda. **VHS $9.95** *SNC, MRV, NWV ♪♪*

Godzilla Raids Again Also known as "Gigantis, the Fire Monster"—Warner Bros. had a problem securing rights to Godzilla's name. Yearning for a change of pace, the King of Monsters opts to destroy Osaka instead of Tokyo, but the spiny Angorous is out to dethrone our hero. Citizens flee in terror when the battle royale begins. Also known as "Godzilla's Counter Attack." The first Godzilla sequel.
1959 78m/B JP Hugo Grimaldi, Makayama, Minoru Chiaki; **Dir:** Motoyoshi. **VHS $14.99** *VTR ♪♪*

Godzilla vs. the Cosmic Monster Godzilla's worst nightmares become a reality as he is forced to take on the one foe he cannot defeat - a metal clone of himself! To make matters worse, Earth is in dire peril at the hands of cosmic apes. We all need friends, and Godzilla is never more happy to see his buddy King Seeser, who gladly lends a claw. Also known as "Godzilla Versus the Bionic Monster."
1974 (G) 80m/C JP Masaaki Daimon, Kazuya Aoyama, Reiko Tajima, Barbara Lynn, Akihiko Hirata; **Dir:** Jun Fukuda. **VHS, CV $9.95** *NWV, MRV ♪♪*

Godzilla vs. Gigan The Super-Lizard must save a children's amusement park from Gigan and Ghidra, the monsters from outer space. Also known as "Godzilla on Monster Island."

1972 89m/C JP Hiroshi Ishikawa, Minoru Takashima,, Tomoko Umeda; **Dir:** Jun Fukuda. **VHS, Beta, LV $9.95** *NWV, STE, IME ♪♪*

Godzilla vs. Mechagodzilla Following his daughter's fatal accident, a deranged scientist turns her remains into a robot. She bands together with a group of monsters, and they try to dominate earth.
1975 84m/C JP Kazuya Aoyama, Masaaki Daimon, Barbara Lynn, Reiko Tajima; **Dir:** Jun Fukuda. **VHS, LV $9.95** *NWV, STE, SNC ♪♪*

Godzilla vs. Megalon Godzilla faces the ultimate challenge of his monstrous career as he is forced to take on Megalon, a giant cockroach, and Gigan, a flying metal creature, simultaneously. Fortunately, his giant robot pal Jet Jaguar is on hand to slug it out side by side with Tokyo's ultimate defender.
1976 (G) 80m/C JP Katsuhiko Sasakai, Hiroyuki Kawase, Yutaka Hayashi, Robert Dunham; **Dir:** Jun Fukuda. **VHS $19.95** *NOS, MRV, NWV ♪½*

Godzilla vs. Monster Zero The suspicious denizens of Planet X require the help of Godzilla and Rodan to rid themselves of the menacing Ghidra, whom they refer to as Monster Zero. Will they, in return, help Earth as promised, or is this just one big, fat double cross?
1968 (G) 93m/C JP Nick Adams; **Dir:** Inoshiro Honda. **VHS, Beta $19.95** *PAR, SIM ♪*

Godzilla vs. Mothra The Mighty Mothra is called in to save the populace from Godzilla, who is on a rampage; and he's aided by two junior Mothras who hatch in the nick of time. Also known as "Godzilla vs. the Thing."
1964 88m/C JP Akira Takarada, Yuriko Hoshi, Hiroshi Koizumi; **Dir:** Inoshiro Honda. **VHS, Beta $19.95** *PAR ♪*

Godzilla vs. the Sea Monster The famous giant lizard makes a friend of former enemy Mothra so the two can battle Ebirah, the Sea Monster.
1966 80m/C JP **Dir:** Jun Fukuda. **VHS, Beta $44.95** *DVT, MRV, HHT ♪*

Godzilla vs. the Smog Monster The Big G battles a creature born of pollution, a 400-pound sludge blob named Hedora. Great opening song: "Save the Earth." Dubbed in English. Also known as "Godzilla vs. Hedora."
1972 (G) 87m/C JP Akira Yamauchi, Hiroyuki Kawase, Toshio Shibaki; **Dir:** Yoshimitu Banno. **VHS, Beta $19.98** *ORI, SIM ♪*

Godzilla's Revenge A young boy who is having problems dreams of going to Monster Island to learn from Minya, Godzilla's son. Using the lessons in real life, the boy captures some bandits and outwits a bully. Uses footage from "Godzilla vs. the Sea Monster" and "Son of Godzilla" for battle scenes. One of the silliest Godzilla movies around.
1969 70m/C JP Kenji Sahara, Tomonori Yazaki, Machiko Naka, Sachio Sakai, Chotaro Togin, Yoshibumi Tujima; **Dir:** Inoshiro Honda. **VHS $9.95** *SIM ♪*

Goetz von Berlichingen (Iron Hand) During the reign of Emperor Maximilian, a brave knight fights injustice in the service of the Emperor and becomes a hunted man. In German with English subtitles.
1980 100m/C GE **VHS, Beta $39.95** *VCD, GLV ♪♪*

Goin' All the Way Seventeen-year-old Monica decides that she has to prove her love to her boyfriend, Artie, by going all the way.
1982 85m/C Deborah Van Rhyn, Dan Waldman, Joshua Cadman, Sherie Miller, Joe Colligan; *Dir:* Robert Freedman. **VHS, Beta** $39.95 *MON ∅*

Goin' Coconuts Donny and Marie are miscast as Donny and Marie in this smarmy story of crooks and jewels. Seems someone covets the chanteuse's necklace, and only gratuitous crooning can spare the twosome from certain swindling. Big bore on the big island.
1978 (PG) 93m/C Donny Osmond, Marie Osmond, Herb Edelman, Kenneth Mars, Ted Cassidy, Marc Lawrence, Harold Sakata; *Dir:* Howard Morris. **VHS** *STC ∅*

Goin' South An outlaw is saved from being hanged by a young woman who agrees to marry him in exchange for his help working a secret gold mine. They try to get the loot before his old gang gets wind of it. A tongue-in-cheek western that served as Nicholson's second directorial effort.
1978 (PG) 109m/C Jack Nicholson, Mary Steenburgen, John Belushi, Christopher Lloyd, Veronica Cartwright, Richard Bradford; *Dir:* Jack Nicholson. **VHS, Beta, LV** $19.95 *PAR ∅∅½*

Going Ape! A young man inherits a bunch of orangutans. If the apes are treated well, a legacy of $5 million will follow.
1981 (PG) 87m/C Tony Danza, Jessica Walter, Danny DeVito, Art Metrano, Rick Hurst; *Dir:* Jeremy Joe Kronsberg. **VHS, Beta** $49.95 *PAR ∅½*

Going Back Four friends reunite after college to relive a memorable summer they spent together after graduating from high school.
1983 85m/C Bruce Campbell, Christopher Howe, Perry Mallette, Susan W. Yamasaki; *Dir:* Ron Teachworth. **VHS, Beta** $69.98 *LIV, VES ∅½*

Going Bananas Wacky adventure about a chimp who is being chased by a sinister circus owner. The monkey runs through Africa dragging along a boy, his caretaker, and a guide. The cast says it all.
1988 (PG) 95m/C Dom DeLuise, Jimmie Walker, David Mendenhall, Herbert Lom; *Dir:* Boaz Davidson. **VHS, Beta** $19.95 *MED ∅*

Going Berserk The members of SCTV's television comedy troupe are featured in this comedy which lampoons everything from religious cults to kung fu movies to "Father Knows Best." Lacks structure and humor.
1983 (R) 85m/C *CA* John Candy, Joe Flaherty, Eugene Levy, Paul Dooley, Pat Hingle, Richard Libertini, Ernie Hudson; *Dir:* David Steinberg. **VHS, Beta, LV** $69.95 *MCA ∅½*

Going to Congress/Don't Park There Two typical silent Rogers shorts. With original organ score.
1924 57m/B Will Rogers. **VHS, Beta, 8mm** $24.95 *VYY*

Going for the Gold: The Bill Johnson Story Biographical tale of the American downhill skier who went from hard-luck punk to Olympic champion in the 1984 games at Sarajevo, Yugoslavia.
1985 100m/C Anthony Edwards, Dennis Weaver, Sarah Jessica Parker, Deborah Van Valkenburgh, Wayne Northrop; *Dir:* Don Taylor. **VHS** $14.95 *COL, SVS ∅*

Going Hollywood A documentary that examines the stars and films of the Great Depression of the 1930s.

1983 (G) 75m/C *Nar:* Robert Preston. **VHS, Beta** $24.95 *MON ∅∅*

Going Home During World War I, Canadian soldiers held in British camps awaiting transport back home stage a riot over the endless delays and intolerable conditions. A fact-based story of the events leading up to this tragedy.
1986 100m/C Nicholas Campbell, Paul Maxwell, Eugene Lipinski. **VHS** $79.95 *SVS ∅∅½*

Going My Way A musical-comedy about a progressive young priest assigned to a downtrodden parish who works to get the parish out of debt, but clashes with his elderly curate, who's set in his ways. Songs include "Going My Way," "Ave Maria," "Swinging on a Star," "Too-ra-loo-ra-loo-ra," and "The Day After Forever." Followed by "The Bells of St. Mary's." Fitzgerald's Oscar-winning Supporting Actor performance was also nominated in the Best Actor category. Other Oscar nominations include Film Editing and Black & White Cinematography.
1944 126m/B Bing Crosby, Barry Fitzgerald, Rise Stevens, Frank McHugh, Gene Lockhart, Porter Hall; *W/Dir:* Leo McCarey. Academy Awards '44: Best Actor (Crosby), Best Director (McCarey), Best Picture, Best Screenplay, Best Song ("Swinging on a Star"), Best Story, Best Supporting Actor (Fitzgerald); National Board of Review Awards '44: 10 Best Films of the Year; New York Film Critics Awards '44: Best Actor (Fitzgerald), Best Director (McCarey), Best Film. **VHS, Beta, LV** $14.95 *MCA, IGP, BTV ∅∅∅½*

Going Places A cynical, brutal satire about two young thugs traversing the French countryside raping, looting and cavorting as they please. Moreau has a brief role as an ex-con trying to go straight. An amoral, controversial comedy that established both its director and two stars. Dubbed into English. Also known as "Les Valseuses" or "Making It."
1974 (R) 122m/C *FR* Gerard Depardieu, Patrick Dewaere, Miou-Miou, Isabelle Huppert, Jeanne Moreau, Brigitte Fossey; *Dir:* Bertrand Blier. **VHS, Beta, LV** $59.95 *COL, INT, TAM ∅∅∅*

Going Steady A story of teenage love, set to the beat of fifties rock 'n' roll.
1980 (R) 90m/C **VHS, Beta** *PGN ∅*

Going Straight A documentary is presented about the San Francisco county jail early release parole program. A community-based alternative to incarceration is examined. San Francisco has a recidivism rate of ten percent compared to sixty percent nationwide. This program explains why.
1978 29m/B **VHS, Beta** *VPH*

Going in Style Three elderly gentlemen, tired of doing nothing, decide to liven up their lives by pulling a daylight bank stick-up. They don't care about the consequences because anything is better than sitting on a park bench all day long. The real fun begins when they get away with the robbery. Great cast makes this a winner.
1979 (PG) 91m/C George Burns, Art Carney, Lee Strasberg; *Dir:* Martin Brest. **VHS, Beta** $19.98 *WAR ∅∅∅½*

Going to Town Radio team Lum and Abner in one of their last films, as two conmen who scheme to make money from a phony oil well.
1944 77m/B Barbara Hale, Grady Sutton, Florence Lake. **VHS, Beta** *GVV ∅*

Going Undercover A bumbling private investigator is hired to protect a rich, beautiful, and spoiled young woman on her European vacation. Poor excuse for a comedy.
1988 (PG-13) 90m/C Lea Thompson, Jean Simmons, Chris Lemmon; *Dir:* James Kenelm Clark. **VHS, Beta, LV** $29.95 *VMK ∅*

Gold of the Amazon Women When two explorers set out to find gold, they stumble onto a society of man-hungry women who follow them into the urban jungle of Manhattan. Made for television. Also on video as "Amazon Women" in its "R" rated European version.
1979 94m/C Bo Svenson, Anita Ekberg, Bond Gideon, Donald Pleasence; *Dir:* Mark L. Lester. **VHS, Beta** $59.98 *SUE ∅*

The Gold Coast A continuous picture of a coastline that blazes brilliantly golden as the sun shines through to create a relaxed background.
1978 30m/C **VHS, Beta** *PCC*

Gold Diggers of 1933 In this famous period musical, showgirls help a songwriter save his Busby Berkeley-choreographed show. Songs include "We're in the Money," and "My Forgotten Man." Followed by two sequels.
1933 96m/B Joan Blondell, Ruby Keeler, Aline MacMahon, Dick Powell, Guy Kibbee, Warren Williams, Ned Sparks, Ginger Rogers; *Dir:* Mervyn LeRoy. **VHS, Beta, LV** $59.98 *MGM, FOX, PIA ∅∅∅½*

Gold Diggers of 1935 The second Gold Diggers film, having something to do with a New England resort, romance, and a charity show put on at the hotel. Plenty of Berkeleian large-scale drama, especially the bizarre, mock-tragic "Lullaby on Broadway" number, which details the last days of a Broadway baby.
1935 95m/B Dick Powell, Adolphe Menjou, Gloria Stuart, Alice Brady, Frank McHugh, Glenda Farrell, Grant Mitchell, Hugh Herbert, Wini Shaw; *Dir:* Busby Berkeley. **VHS, Beta, LV** $29.95 *MGM, PIA, MLB ∅∅∅*

The Gold & Glory A young underdog runs and wins an iron man triathalon.
1988 (PG-13) 102m/C Colin Friels, Josephine Smulders, Grant Kenny. **VHS, Beta, LV** $39.95 *IVE ∅*

The Gold of Naples A four-part omnibus film about life in Naples, filled with romance and family tragedies, by turns poignant, funny and pensive. Originally six tales; two were trimmed for U.S. release. In Italian with subtitles.
1954 107m/B *IT* Vittorio De Sica, Eduardo de Filippo, Toto, Sophia Loren, Paolo Stoppa, Silvana Mangano; *Dir:* Vittorio De Sica. **VHS, Beta** $19.95 *NOS, FCT, APD ∅∅∅*

Gold Raiders Larry, Moe, and Shemp team up with an insurance agent to take on the bad guys in the wild west. The Stooges are long past their prime.
1951 56m/B George O'Brien, Moe Howard, Moe Howard, Shemp Howard, Larry Fine, Sheila Ryan, Clem Bevans, Lyle Talbot; *Dir:* Edward L. Bernds. **VHS** *MED ∅*

Gold Raiders A team of secret agents are sent to the Jungles of Laos to find a plane carrying two hundred million dollars worth of gold, and nothing will get in its way.
1984 106m/C Robert Ginty, Sarah Langenfeld, William Steven. **VHS, Beta** $69.95 *MED ∅*

The Gold Rush Chaplin's most critically acclaimed film. The best definition of his simple approach to film form; adept maneuvering of visual pathos. The "Little Tramp" searches for gold and romance in the Klondike in the mid-1800s. Includes the dance of the rolls, pantomime sequence of eating the shoe, and Chaplin's lovely music.
1925 85m/B Charlie Chaplin, Mack Swain, Tom Murray, Georgina Hale; ***Dir:*** Charlie Chaplin. Cinematheque Belgique '52: #2 of the Best Films of All Time; Sight & Sound Survey '52: #2 of the Best Films of All Time; Brussels World's Fair '58: #2 of the Best Films of All Time. **VHS, Beta, LV, 8mm $9.95** *NEG, NOS, CAB* 🦴🦴🦴🦴

Gold Rush/Payday Chaplin drifts along the Arctic tundra, searching for gold and love in "The Gold Rush." In "Payday," he goes out drinking with his buddies.
1925 92m/B Charlie Chaplin, Georgina Hale, Mack Swain. **VHS, Beta $19.98** *FOX* 🦴🦴🦴🦴

Golden Age of Comedy The great comedians of silent cinema are seen in clips from some of their funniest films.
1958 78m/B Ben Turpin, Harry Langdon, Will Rogers, Jean Harlow, Carole Lombard, Stan Laurel, Oliver Hardy, Keystone Kops. **VHS, Beta, LV $14.98** *VID, PIA, MLB*

Golden Age of Hollywood This episode of "Life Goes to the Movies" examines Hollywood as it quickly became the ultimate dream factory, providing escape from the grimness of the depression.
1977 34m/C Henry Fonda, Shirley MacLaine; ***Hosted:*** Liza Minnelli. **VHS, Beta** *TLF*

Golden Boy Holden plays a young and gifted violinist who earns money for his musical education by working as a part-time prizefighter. Fight promoter Menjou has Stanwyck cozy up to the impressionable young man to convince him to make the fight game his prime concern. She's successful but it leads to tragedy. Holden's screen debut, with Cobb as his immigrant father and Stanwyck successfully slinky as the corrupting love interest. Classic pugilistic drama with well-staged fight scenes. Based on Clifford Odet's play with toned down finale.
1939 99m/B William Holden, Adolphe Menjou, Barbara Stanwyck, Lee J. Cobb, Joseph Calleia, Sam Levene, Don Beddoe; ***Dir:*** Rouben Mamoulian. **VHS, Beta, LV $9.95** *GKK* 🦴🦴🦴

The Golden Child When a Tibetan child with magic powers is kidnapped and transported to Los Angeles, Chandler, a professional "finder of lost children" must come to the rescue. The search takes him through Chinatown in a hunt that cliches every Oriental swashbuckler ever made. Good fun.
1986 (PG-13) 94m/C Eddie Murphy, Charlotte Lewis, Charles Dance, Victor Wong, Randall "Tex" Cobb, James Hong; ***Dir:*** Michael Ritchie. **VHS, Beta, LV, 8mm $19.95** *PAR* 🦴🦴

The Golden Coach Based on a play by Prosper Merimee, the tale of an 18th century actress in Spanish South America who takes on all comers, including the local viceroy who creates a scandal by presenting her with his official coach. Rare cinematography by Claude Renoir.
1952 101m/C *FR* Anna Magnani, Odoardo Spadaro, Nada Fiorelli, Dante Rino, Duncan Lamont; ***Dir:*** Jean Renoir. **VHS, Beta $29.95** *FCT* 🦴🦴🦴

Golden Demon An acclaimed Japanese epic, from the novel by Koyo Ozaki, about the destructive powers of wealth on the lives of two star-crossed lovers in medieval Japan. In Japanese with English subtitles.
1953 95m/C *JP* Jun Negami, Fujiko Yamamoto; ***Dir:*** Koji Shima. **VHS, Beta $29.95** *SUE, EMB* 🦴🦴½

Golden Destroyers Hosted by the inimitable Sho Kosugi, this film chronicles the hapless adventures of rampaging golden buddhas.
198? 90m/C VHS, Beta $39.95 *TWE* 🦴

Golden Exterminators Two brothers battle the forces of evil who have killed their martial arts master with the heels of their formidable feet.
197? 86m/C Raymond Chan. **VHS, Beta $39.95** *UNI* 🦴

The Golden Honeymoon Ring Lardner's story about an elderly couple who journey to St. Petersburg to celebrate their 50th anniversary. There they encounter the wife's suitor of 50 years past, who is also vacationing with his spouse. Curiosity draws the couples together, and jealousy, doubt and a simmering competitive spirit momentarily jar the marriage. Part of the American Short Story Collection. Made for television.
1980 52m/C James Whitmore, Teresa Wright; ***Dir:*** Noel Black. **VHS, Beta** *MON, MTI* 🦴

Golden Lady A beautiful woman leads her female gang in a deadly game of international intrigue.
1979 90m/C Christina World, Suzanne Danielle, June Chadwick; ***Dir:*** Joseph Larraz. **VHS, Beta $39.95** *MON* 🦴½

Golden Ninja Invasion Ninja lords battle in a scheme to conquer the world. Their method? Fighting with sticks, naturally.
198? 85m/C Leonard West, Stephanie Burd. **VHS, Beta $59.95** *TWE* 🦴

Golden Ninja Warrior Lady ninjas endeavor to capture a golden statue and gain superiority over chauvinists everywhere.
198? 92m/C ***Hosted:*** Sho Kosugi. **VHS, Beta $59.95** *TWE* 🦴

Golden Rendezvous A tale of treachery aboard a "gambler's paradise" luxury liner which is hijacked and held for ransom. Based on Alistair MacLean's novel.
1977 103m/C Richard Harris, David Janssen, John Carradine, Burgess Meredith, Ann Turkel, John Vernon; ***Dir:*** Ashley Lazarus. **VHS, Beta $69.95** *VES* 🦴🦴

The Golden Salamander Howard shines in otherwise lackluster adventure about archaeologist searching for ancient ruins in Tunisia who must deal with gun runners and their evil leader, while torridly romancing a beautiful Tunisian girl. Filmed on location in Tunis. Based on the novel by Victor Canning.
1951 96m/B *GB* Trevor Howard, Anouk, Herbert Lom, Miles Maleson, Walter Rilla, Jacques Sernas, Wilfrid Hyde-White, Peter Copely, Eugene Deckers, Henry Edwards, Marcel Poncin, Percy Walsh, Sybilla Binder, Kathleen Boutall, Valentine Dyall; ***Dir:*** Ronald Neame. **VHS $29.95** *FCT* 🦴🦴

The Golden Seal Tale of a small boy's innocence put in direct conflict, because of his love of a rare wild golden seal and her pup, with the failed dreams, pride and ordinary greed of adults.
1983 (PG) 94m/C Steve Railsback, Michael Beck, Penelope Milford, Torguil Campbell; ***Dir:*** Frank Zuniga. **VHS, Beta, LV, 8mm $19.98** *SUE* 🦴🦴½

The Golden Stallion Trigger falls in love with a stunning mare. He sees her villainous owners abusing her and kills them. Roy takes the blame for his equine pal, but true love triumphs to save the day.
1949 67m/B Roy Rogers, Dale Evans, Estelita Rodriguez, Pat Brady, Douglas Evans, Frank Fenton, Trigger; ***Dir:*** William Witney. **VHS $12.98** *REP* 🦴½

Golden Sun A young Kung Fu student decides to investigate the death of his hero, Bruce Lee, and comes up against a sinister conspiracy
198? 90m/C Lei Hsiao Lung, Chen Pei Ling, Ou-Yang Chung. **VHS, Beta $29.95** *UNI* 🦴

Golden Tales and Legends These two volumes of the best-loved fairy tale from the Brothers Grimm and Hans Christian Andersen are sure to captivate every child. Volume one includes "The Frog Prince," "Rapunzel" and "Hansel and Gretel;" the second volume features "The Juggler," "The Golden Apple" and "The Iron Mountain."
1985 60m/C VHS, Beta $24.95 *MPI* 🦴

The Golden Triangle Big battle of the drug pushers set in the mountainous region surrounded by Laos, Thailand, and Burma. Rival gangs fight for control of the world's opium crop.
1980 (PG) 90m/C Lo Lieh, Sambat Metanen, Tien Ner. **VHS, Beta $59.98** *OCE, MAG* 🦴½

Golden Voyage of Sinbad In the mysterious ancient land of Lemuria, Sinbad and his crew encounter magical and mystical creatures. A statue of Nirvana comes to life and engages in a sword fight with Sinbad. He later meets up with a one-eyed centaur and a griffin. Ray Harryhausen can once again claim credit for the unusual and wonderful mythical creatures springing to life.
1973 (G) 105m/C John Phillip Law, Caroline Munro, Tom Baker, Douglas Wilmer, Martin Shaw, John Garfield, Gregoire Aslan; ***Dir:*** Gordon Hessler. **VHS, Beta, LV $14.95** *COL, LDC, MLB* 🦴🦴½

Goldengirl A neo-Nazi doctor tries to make a superwoman of his daughter who has been specially fed, exercised, and emotionally conditioned since childhood to run in the Olympics. Anton's starring debut.
1979 (PG) 107m/C Susan Anton, James Coburn, Curt Jurgens, Robert Culp, Leslie Caron, Harry Guardino, Jessica Walter; ***Dir:*** Joseph Sargent. **VHS, Beta $14.95** *COL, SUE* 🦴🦴

Goldenrod A successful rodeo champion is forced to reevaluate his life when he sustains a crippling accident in the ring. Made for television.
1977 100m/C *CA* Tony LoBianco, Gloria Carlin, Donald Pleasence, Donnelly Rhodes; ***Dir:*** Harvey Hart. **VHS, Beta $59.95** *PSM* 🦴½

Goldfinger Ian Fleming's James Bond, Agent 007, attempts to prevent international gold smuggler Goldfinger and his pilot Pussy Galore from robbing Fort Knox. Features villainous assistant Oddjob and his deadly bowler hat. The third in the series is perhaps the most popular. Shirley Bassey sings the theme song. The laserdisc edition includes audio interviews with the director, the writer, the editor, and the production designer; music and sound effects/ dialogue separation; publicity stills, movie posters, trailers,

and on-location photos. A treat for Bond fans.
1964 108m/C GB Sean Connery, Honor Blackman, Gert Frobe, Shirley Eaton, Tania Mallet, Harold Sakata, Cec Linder, Bernard Lee, Lois Maxwell, Desmond Llewelyn, Nadja Regin; *Dir:* Guy Hamilton. Academy Awards '64: Best Sound Effects Editing. **VHS, Beta, LV $19.95** MGM, VYG, PIA 🎬🎬🎬

Goldie Gold & Action Jack Futuristic reporter Goldie Gold and intern Action Jack team up to pursue stories with the help of computerized gadgetry. Animated adventure tale.
1981 40m/C VHS, Beta $19.95 WOV 🎬🎬🎬

Goldie & Kids Goldie Hawn talks to twelve children about such subjects as love, marriage, and divorce in this tape that features Barry Manilow.
1982 52m/C Goldie Hawn, Barry Manilow. **Beta $9.99** STE, PSM

Goldilocks & the Three Bears Entry from "Faerie Tale Theatre" tells the story of Goldilocks, who wanders through the woods and finds the home of three bears—only in this version Goldilocks is a nasty child, and it's the bears who'll get your sympathy.
1983 60m/C Tatum O'Neal, Alex Karras, Brandis Kemp, Donovan Scott, Hoyt Axton, John Lithgow, Carole King; *Dir:* Gilbert Cates. **VHS, Beta, LV $14.98** KUI, FOX, FCT 🎬🎬🎬

Goldwyn Follies Lavish disjointed musical comedy about Hollywood. A movie producer chooses a naive girl to give him advice on his movies. Songs include "Love Walked In" and "Love is Here to Stay" by Ira and George Gershwin, who died during the filming.
1938 115m/C Adolphe Menjou, Vera Zorina, The Ritz Brothers, Helen Jepson, Phil Baker, Bobby Clark, Ella Logan, Andrea Leeds, Edgar Bergen; *Dir:* George Marshall. **VHS, Beta $19.98** SUE 🎬🎬

Goldy 2: The Saga of the Golden Bear The adventures in the High Sierras of an orphan girl and her friend, the Golden Bear.
1985 96m/C VHS, Beta $69.98 LIV, VES 🎬

Goldy: The Last of the Golden Bears An orphaned child and a lonely prospector risk their lives to save a Golden Bear from a circus owner.
1984 91m/C VHS, Beta $69.98 LIV, VES 🎬½

The Golem A huge clay figure is given life by a rabbi in hopes of saving the Jews in the ghetto of medieval Prague. Rarely seen Wegener expressionist myth-opus that heavily influenced the "Frankenstein" films of the sound era.
1920 80m/B RU Paul Wegener, Albert Steinruck, Ernst Deutsch, Syda Salmonava; *Dir:* Paul Wegener. **VHS, Beta, 3/4U $16.95** SNC, NOS, MRV 🎬🎬🎬½

Goliath Against the Giants Goliath takes on Bokan, who has stolen his throne. Goliath must fight Amazons, storms, and sea monsters to save the lovely Elea. Goliath conquers all.
1963 95m/C IT SP Brad Harris, Gloria Milland, Fernando Rey, Barbara Carrol; *Dir:* Guido Malatesta. **VHS, Beta $19.98** SNC 🎬½

Goliath Awaits Harmon plays an oceanographer who discovers a ship that was sunk by German U-boats during WWI. Nothing much new there, except that the survivors still reside on board the vessel, some 40 years after the oceanliner was torpedoed.

And get this: they're living a quasi-Utopian existence. Kind of hard to swallow. Made for TV, it originally aired in two parts.
1981 110m/C Mark Harmon, Robert Forster, Christopher Lee, Eddie Albert, John Carradine, Alex Cord, Emma Samms, Jean Marsh; *Dir:* Kevin Connor. **VHS** VMK 🎬🎬

Goliath and the Barbarians Goliath and his men go after the barbarians who are terrorizing and ravaging the Northern Italian countryside during the fall of the Roman Empire. A basic Reeves muscleman epic.
1960 86m/C IT Steve Reeves, Bruce Cabot; *Dir:* Carlo Campogalliani. **VHS, Beta $59.95** MGM 🎬🎬

Goliath and the Dragon Even Goliath must have doubts as he is challenged by the evil and powerful King Eurystheus. A must for fans of ridiculous movies with ridiculous monsters.
1961 90m/C IT FR Bruce Cabot, Mark Forest, Broderick Crawford, Gaby Andre, Leonora Ruffo; *Dir:* Vittorio Cottafavi. **VHS, Beta $19.98** SNC Woof!

Goliath and the Sins of Babylon Well-sculpted Forest plays yet another mesomorph to the rescue in this poorly dubbed spaghetti legend. Goliath must spare twenty-four virgins whom the evil Crisa would submit as human sacrifice. Forest—who played the mythic Maciste in a number of films—was randomly assigned the identity of Goliath, Hercules or Samson for US viewing; go figure.
1964 80m/C IT Mark Forest, Eleanora Bianchi, Jose Greco, Giuliano Gemma, Paul Muller; *Dir:* Michele Lupo. **VHS $19.98** SNC 🎬🎬

Gondoliers A new version of Gilbert and Sullivan's opera. Entertaining lampoon against class bigotry.
19?? 112m/C Keith Michell. **VHS, Beta $49.98** FOX 🎬🎬

Gone in 60 Seconds A car thief working for an insurance adjustment firm gets double-crossed by his boss and chased by the police. Forty minutes are consumed by a chase scene which destroyed more than 90 vehicles.
1974 105m/C H.B. Halicki, Marion Busia, George Cole, James McIntyre, Jerry Daugirda; *Dir:* H.B. Halicki. **VHS, Beta $9.95** MED 🎬½

Gone are the Days Shaky adaptation of the play, "Purlie Victorious." A black preacher wants to cause the ruin of a white plantation owner. Alda's screen debut.
1963 97m/B Ossie Davis, Ruby Dee, Sorrell Booke, Godfrey Cambridge, Alan Alda, Beah Richards; *Dir:* Nicholas Webster. **VHS, Beta $69.95** DIS 🎬🎬½

Gone are the Days Government agent (Korman) is assigned to protect a family who witnessed an underworld shooting, but the family would like to get away from both the mob and the police. Made-for-cable Disney comedy is well-acted but done in by cliches.
1984 90m/C Harvey Korman, Susan Anspach, Robert Hogan; *Dir:* Gabrielle Beaumont. **VHS, Beta $69.95** DIS 🎬🎬

Gone to Ground A mad killer terrorizes a group of vacationers who are trapped in an isolated beach house.
1976 74m/C Charles Tingwell, Elaine Lee, Eric Oldfield, Marion Johns, Robyn Gibbes, Judy Lynne, Dennis Grosvenor, Alan Penney, Marcus Hale. **VHS, Beta** PGN 🎬

Gone with the West Little Moon and Jud McGraw seek revenge upon the man who stole their cattle.

1972 92m/C James Caan, Stefanie Powers, Sammy Davis Jr., Aldo Ray, Michael Conrad, Michael Walker Jr.; *Dir:* Bernard Gerard. **VHS, Beta $49.95** UNI 🎬½

Gone with the Wind Epic Civil War drama focusing on the life of petulant southern belle Scarlett O'Hara. Starting with her idyllic lifestyle on a sprawling plantation, the film traces her survival through the tragic history of the South during the Civil War and Reconstruction, and her tangled love affairs with Ashley Wilkes and Rhett Butler. Classic Hollywood doesn't get any better than this; one great scene after another, equally effective in its intimate drama and its sweeping spectacle. The train depot scene, one of the more technically adroit shots in movie history, involved hundreds of extras and dummies, and much of the MGM lot was razed to simulate the burning of Atlanta. Beautifully filmed, scored, and acted. Based on Margaret Mitchell's novel. For its 50th anniversary, a 231-minute restored version was released that included the trailer for "The Making of a Legend: GWTW." Won seven Oscars, with additional nominations for Best Actor (Gable), Best Supporting Actress (de Havilland), Best Sound Recording, Best Special Effects and Best Original Score. The laserdisc is available in a limited, numbered edition, fully restored to Technicolor from the original negative, with an enhanced soundtrack and seven minutes of rare footage, including the original trailer. coverage of the 1939 premiere, and the 1961 Civil War centennial screening attended by Leigh.
1939 231m/C Clark Gable, Vivien Leigh, Olivia de Havilland, Leslie Howard, Thomas Mitchell, Hattie McDaniel, Butterfly McQueen, Evelyn Keyes, Harry Davenport, Jane Darwell, Ona Munson, Barbara O'Neil, William Bakewell, Rand Brooks, Ward Bond, Laura Hope Crews, Yakima Canutt, George Reeves, Marjorie Reynolds; *Dir:* Victor Fleming. Academy Awards '39: Best Actress (Leigh), Best Art Direction/Set Decoration, Best Color Cinematography, Best Director (Fleming), Best Film Editing, Best Picture, Best Screenplay, Best Supporting Actress (McDaniel); New York Film Critics Awards '39: Best Actress (Leigh); American Film Institute's Survey '77: Best American Film Ever Made; American Film Institute Survey '77: Best American Film Ever Made; American Film Institute's Survey '77: Best American Film of All Time. **VHS, Beta, LV $89.98** MGM, LDC, BTV 🎬🎬🎬🎬

Gonza the Spearman A gifted samurai lancer is accused of having an affair with the wife of the lord of his province. The lancer must leave to save the woman's honor, because to stay would mean death to them both. In Japanese with English subtitles.
1986 126m/C JP Hiromi Goh, Shima Iwashita; *Dir:* Masahiro Shinoda. **VHS $79.95** KIV 🎬🎬½

Goober and the Ghost Chasers The semi-visible cartoon dog goes ghost-huntin' with his buddies in this cartoon compilation.
19?? 50m/C VHS, Beta $19.95 WOV, GKK

The Good, the Bad, and Huckleberry Hound Journeying out West to begin a ranch, Huckleberry Hound winds up becoming a sheriff in the small town of Two-Bit. Huckleberry stands up against a group of bully brothers who have been terrorizing the town.
1988 94m/C *Dir:* Ray Patterson. *Voices:* Daws Butler. **VHS, Beta $29.95** HNB, IME 🎬🎬½

The Good, the Bad and the Ugly Leone's grandiloquent, shambling tribute to the American Western. Set during the Civil War, it follows the seemingly endless adventures of three dirtbags in search of a cache of Confederate gold buried in a nameless grave. Violent, exaggerated, beautifully crafted, it is the final and finest installment of the "Dollars" trilogy: a spaghetti Western chef d'oeuvre.
1967 161m/C *IT* Clint Eastwood, Eli Wallach, Lee Van Cleef; *Dir:* Sergio Leone. **VHS, Beta, LV** $19.98 *FOX, MGM, FCT* 🎬🎬½

The Good Earth Pearl S. Buck's classic recreation of the story of a simple Chinese farm couple beset by greed and poverty. Outstanding special effects. MGM's last film produced by master Irving Thalberg and dedicated to his memory. Oscar-nominated for Best Picture and Best Director. Rainer won the second of her back-to-back Best Actress Oscars for her portrayal of the self-sacrificing O-Lan.
1937 138m/B Paul Muni, Luise Rainer, Charley Grapewin, Keye Luke, Walter Connolly; *Dir:* Sidney Franklin. Academy Awards '37: Best Actress (Rainer), Best Cinematography. **VHS, Beta** $19.98 *KUI, MGM, BTV* 🎬🎬½

The Good Father An acclaimed British TV movie about a bitter divorced man trying to come to terms with his son, his ex-wife and his own fury by supporting the courtroom divorce battle of a friend.
1987 (R) 90m/C *GB* Anthony Hopkins, Jim Broadbent, Harriet Walter, Frances Viner, Joanne Whalley-Kilmer; *Dir:* Mike Newell. **VHS, Beta** $79.95 *MED, FOX* 🎬🎬🎬

Good Grief, Charlie Brown That lovably misunderstood Charlie Brown is featured again in this Peanuts production.
1983 30m/C *Dir:* Bill Melendez, Sam Jaimes, Phil Roman. **VHS** $14.95 *KRT* 🎬🎬🎬

Good Guys Wear Black A mild-mannered professor keeps his former life as leader of a Vietnam commando unit under wraps until he discovers that he's number one on the CIA hit list. Sequel is "A Force of One."
1978 (PG) 96m/C Chuck Norris, Anne Archer, James Franciscus; *Dir:* Ted Post. **VHS, Beta, LV** $14.98 *VES* 🎬🎬

Good Morning, Babylon The Taviani brothers' first American film. Two young Italian brothers skilled in cathedral restoration come to America and find success building the sets to D. W. Griffith's "Intolerance." Eventually, their fortune is shattered by the onslaught of World War I.
1987 (PG-13) 113m/C Vincent Spano, Joaquim de Almeida, Greta Scacchi, Charles Dance, Desiree Becker, Omero Antonutti; *Dir:* Paolo Taviani, Vittorio Taviani. **VHS, Beta, LV** $79.98 *LIV, VES* 🎬🎬🎬

Good Morning, Vietnam Based on the story of Saigon DJ Adrian Cronauer, although Williams portrayal is reportedly a bit more extroverted than the personality of Cronauer. Williams spins great comic moments that may have been scripted but likely were not as a man with no history and for whom everything is manic radio material. The character adlibs, swoops and swerves, finally accepting adult responsibility. Engaging all the way with an outstanding period soundtrack.
1987 (R) 121m/C Robin Williams, Forest Whitaker, Bruno Kirby, Richard Edson, Robert Wuhl; *Dir:* Barry Levinson. **VHS, Beta, LV, 8mm** $29.95 *TOU, BVV* 🎬🎬🎬

The Good Mother A divorced mother works at creating a fulfilling life for herself and her daughter, with an honest education for her daughter about every subject, including sex. But her ex-husband, unsure of how far this education is being taken, fights her for custody of their 8-year old daughter after allegations of sexual misconduct against the mother's new lover. Based on Sue Miller's bestselling novel. Well acted, weakly edited and scripted.
1988 (R) 104m/C Diane Keaton, Liam Neeson, Jason Robards Jr., Ralph Bellamy, James Naughton, Teresa Wright; *Dir:* Leonard Nimoy. **VHS, Beta, LV** $89.95 *TOU* 🎬🎬½

Good Neighbor Sam A married advertising executive (Lemmon) agrees to pose as a friend's husband in order for her to collect a multi-million dollar inheritance. Complications ensue when his biggest client mistakes the friend for his actual wife and decides they're the perfect couple to promote his wholesome product—milk.
1964 130m/C Jack Lemmon, Romy Schneider, Dorothy Provine, Mike Connors, Edward G. Robinson, Joyce Jameson; *Dir:* David Swift. **VHS, Beta** $9.95 *GKK* 🎬🎬🎬

Good News A vintage Comden-Green musical about the love problems of a college football star, who will flunk out if he doesn't pass his French exams. Revamping of the 1927 Broadway smash features "The Varsity Drag," "Just Imagine," "Pass That Peace Pipe" and the unlikely sight of Lawford in a song-and-dance role.
1947 92m/C June Allyson, Peter Lawford, Joan McCracken, Mel Torme; *Dir:* Charles Walters. **VHS, Beta, LV** $19.98 *MGM, FCT* 🎬🎬½

Good Old Days of Radio A nostalgic look at old time radio, featuring appearances by many former radio stars.
197? 60m/B *Hosted:* Steve Allen. **VHS, Beta** $24.95 *DVT* 🎬🎬½

Good Sam An incurable "Good Samaritan" finds himself in one jam after another as he tries too hard to help people. Lots of missed opportunities for laughs with McCrarey's mediocre direction.
1948 116m/B Gary Cooper, Ann Sheridan, Ray Collins, Edmund Lowe, Joan Lorring, Ruth Roman; *Dir:* Leo McCarey. **VHS** $19.98 *REP* 🎬🎬

Good Time with a Bad Girl A married millionaire visiting Las Vegas experiments with exotic sexual pleasures when he meets up with Sue, a teenage nymphomaniac. Also known as "A Good Time, A Bad Girl."
1967 61m/B VHS, Beta $19.95 *VDM* 🎬

The Good Wife A bored and sexually frustrated wife scandalizes her small Australian town by taking up with the new hotel barman. There's no denying Ward's sexuality, but overall the movie is too predictable.
1986 (R) 97m/C *AU* Rachel Ward, Bryan Brown, Sam Neill; *Dir:* Ken Cameron. **VHS, Beta, LV** *KRT, IME* 🎬🎬½

Goodbye Again A romantic drama based on the novel by Francoise Sagan. Bergman is a middle-aged interior decorator whose lover, Montand, has a roaming eye. She finds herself drawn to the son (Perkins) of a client and begins an affair—flattered by the young man's attentions. Montand is outraged and vows to change his ways and marry Bergman if she will give up the young man. A bit soapy, but a must for Bergman fans.
1961 120m/B Jessie Royce Landis, Ingrid Bergman, Anthony Perkins, Yves Montand, Diahann Carroll; *Dir:* Anatole Litvak. **VHS** $14.95 *WKV* 🎬🎬½

Goodbye Columbus Philip Roth's novel about a young Jewish librarian who has an affair with the spoiled daughter of a nouveau riche family is brought to late-'60s life, and vindicated by superb performances all around. Benjamin and McGraw's first starring roles.
1969 (PG) 105m/C Richard Benjamin, Ali MacGraw, Jack Klugman, Nan Martin, Jaclyn Smith; *Dir:* Larry Peerce. **VHS, Beta, LV** $49.95 *PAR* 🎬🎬🎬

Goodbye Cruel World Black comedy about a suicidal television anchorman who decides to spend his last day filming the relatives who drove him to the brink.
1982 (R) 90m/C Dick Shawn, Cynthia Sikes, Chuck "Porky" Mitchell; *Dir:* David Irving. **VHS, Beta** $69.98 *LIV, LTG* 🎬½

Goodbye Emmanuelle Follows the further adventures of Emmanuelle in her quest for sexual freedom and the excitement of forbidden pleasures. The second sequel to "Emmanuelle."
1979 (R) 92m/C Sylvia Kristel. **VHS, Beta** $79.99 *HBO* Woof!

Goodbye Girl Neil Simon's story of a former actress, her precocious nine-year-old daughter and the aspiring actor who moves in with them. The daughter serves as catalyst for the other two to fall in love. While Mason's character is fairly unsympathetic, Dreyfuss is great and Simon's dialogue witty. Oscar nominations for Best Picture, Actress Mason, Supporting Actress Cummings, and Original Screenplay.
1977 (PG) 110m/C Richard Dreyfuss, Marsha Mason, Quinn Cummings, Barbara Rhoades, Marilyn Sokol; *Dir:* Herbert Ross. Academy Awards '77: Best Actor (Dreyfuss); Golden Globe Awards '78: Best Film—Musical/Comedy; Los Angeles Film Critics Association Awards '77: Best Actor (Dreyfuss). **VHS, Beta, LV** $19.98 *MGM, BTV* 🎬🎬🎬

Goodbye Love Dull comedy about a group of ex-husbands who refuse to pay alimony and end up in jail.
1934 65m/B Charlie Ruggles, Verree Teasdale, Sidney Blackmer, Mayo Methot, Phyllis Barry, Ray Walker, John Kelly, Hale Grace, Luis Alberni; *Dir:* Herbert Ross. **VHS** $29.95 *FCT* 🎬½

Goodbye, Mr. Chips An MGM classic, the sentimental rendering of the James Hilton novel about a shy Latin professor in an English school who marries the vivacious Garson only to tragically lose her. He spends the rest of his life devoting himself to his students and becoming a school legend. Multi award-winning soaper featuring Garson's first screen appearance, which was Oscar nominated. Other Oscar nominations for Best Picture, Best Director, Best Screenplay, and Best Sound Recording, and Best Film Editing. Special two-tape letterboxed laserdisc edition includes the original trailer and an extra-special production featurette. Remade in 1969 as a fairly awful musical starring Peter O'Toole.
1939 115m/B Robert Donat, Greer Garson, Paul Henreid, John Mills; *Dir:* Sam Wood. Academy Awards '39: Best Actor (Donat); National Board of Review Awards '39: 10 Best Films of the Year. **VHS, Beta, LV** $24.98 *MGM, LDC, BTV* 🎬🎬½

Goodbye, Mr. Chips Ross, whose later achievements include "Steel Magnolias" and "My Blue Heaven" debuted as director with this big-budget but inferior musical re-make of the classic James Hilton novel about a gentle teacher at an English all-boys private school and the woman who helps him demonstrate his compassion and overcome his shyness. O'Toole, although excellent, is categorically a non-singer, while Clark, a popular singer at the time ("Downtown"), musters very little talent in front of the camera. The plot is altered unnecessarily and updated to the WWII era for little reason, and the music is thoroughly forgettable. Nonetheless, the unflappable O'Toole received an Oscar nomination. Also nominated was John Williams' musical direction and the music by Leslie Bricusse.
1969 151m/C Peter O'Toole, Petula Clark, Michael Redgrave, George Baker, Sian Phillips, Michael Bryant, Jack Hedley, Herbert Ross; *Dir:* Herbert Ross. **VHS, Beta, LV** *MGM, PIA* ✓✓

Goodbye, My Lady Based on the novel by James Street, this is a tear-jerking film about a young Mississippi farmboy who finds and cares for a special dog that he comes to love but eventually must give up.
1956 95m/B Brandon de Wilde, Walter Brennan, Sidney Poitier, Phil Harris, Louise Beavers; *Dir:* William A. Wellman. **VHS, Beta, LV $19.98** *WAR, FCT* ✓✓

Goodbye, New York A New York yuppie on a vacation finds herself penniless and stranded in Israel. She makes the best of the situation by joining a kibbutz and falling for a part-time soldier. Well, of course.
1985 (R) 90m/C *IS* Julie Hagerty, Amos Kollek, Shmuel Shilo; *Dir:* Amos Kollek. **VHS, Beta, LV $79.95** *VES* ✓✓

Goodbye, Norma Jean A detailed and sleazy recreation of Marilyn Monroe's early years in Hollywood. Followed by "Goodnight, Sweet Marilyn."
1975 (R) 95m/C Misty Rowe, Terrence Locke, Patch MacKenzie; *Dir:* Larry Buchanan. **VHS, Beta $19.99** *HBO* ✓

The Goodbye People Balsam is an elderly man who decides to reopen his Coney Island hot dog stand that folded 22 years earlier. Hirsch and Reed help him realize his impossible dream in this sweetly sentimental film.
1983 104m/C Judd Hirsch, Martin Balsam, Pamela Reed, Ron Silver; *W/Dir:* Herb Gardner. **VHS, Beta, LV $14.95** *COL, SUE* ✓✓½

Goodbye Pork Pie With the police on their trail, two young men speed on a 1000-mile journey in a small, brand-new, yellow stolen car. Remember: Journey of thousand miles always begin in stolen cars!
1981 (R) 105m/C *NZ* Tony Barry, Kelly Johnson; *Dir:* Geoff Murphy. **VHS, Beta $69.98** *SUE* ✓½

Goodfellas Considered a quintessential picture about "wise guys" and their lives, at turns both violent and funny. A young man grows up in the mob, working hard to advance himself through the ranks, enjoying the life of the rich and violent, and oblivious to the horror of which he is a part. Soon his addiction to cocaine and many wise-guy missteps help unravel his climb to the top. Excellent performances all around (particularly Liotta and Pesci, with DeNiro pitching in around the corners), with visionary cinematography and careful pacing. Received Golden Globe nominations for Best Screenplay, Picture, Supporting Actor (Pesci), Supporting Actress (Bracco), and Director. Five Oscar nominations for Best Picture, Director, Supporting Actor (Pesci), Supporting Actress (Bracco), and Film Editing. Based on the life of Henry Hill, the ex-mobster now in the Witness Protection Program. Adapted from the book by Nicholas Pileggi.
1990 (R) 146m/C Robert De Niro, Ray Liotta, Joe Pesci, Paul Sorvino, Lorraine Bracco; *Dir:* Martin Scorsese. Academy Awards '90: Best Supporting Actor (Pesci); Los Angeles Film Critics Association Awards '90: Best Director (Scorsese), Best Picture, Best Supporting Actor (Pesci), Best Supporting Actress (Bracco); New York Film Critics Awards '90: Best Director (Scorsese), Best Picture; National Society of Film Critics Awards '90: Best Director (Scorsese), Best Picture. **VHS, Beta, LV, 8mm $19.98** *WAR, PIA, FCT* ✓✓✓✓

Goodnight God Bless A depressing movie about a mentally disturbed priest who wanders into an elementary school with murder on his mind. He kills the teacher and all the children—except one. Will he find that one too? It's not worth it to find out.
1987 (R) 90m/C **VHS, Beta $79.98** *MAG* ✓

Goodnight, Michelangelo Funny, engaging and somewhat confusing story of the life of an Italian immigrant family as seen through the eyes of its youngest member, eight-year-old Michelangelo. Winner of the 1990 Best Comedy Award at the Greater Fort Lauderdale Film Festival.
1989 (R) 91m/C *IT* Lina Sastri, Kim Cattrall, Giancarlo Giannini, Daniel Desanto; *W/Dir:* Carlo Liconti. **VHS $89.98** *SHG* ✓✓½

Goodnight, Sweet Marilyn Supposed "never-before-told" story of Marilyn Monroe's tragic death. According to this, her death was the result of pre-arranged "mercy-killing." Follow-up to director Buchanan's "Goodbye, Norma Jean" (1977) with Lane as the 1960's Marilyn intercut with scenes of Rowe from the first film as young Marilyn. You'd be better off leaving "Sweet Marilyn" to rest in peace.
1989 (R) 100m/C Paula Lane, Jeremy Slate, Misty Rowe; *Dir:* Larry Buchanan. **VHS, Beta $89.95** *SED* ✓

Goof Balls The tourist denizens of a island golf resort putt and cavort about, and golf isn't all that's on their minds.
1987 (R) 87m/C Ben Gordon, Laura Robinson. **VHS, Beta $29.95** *ACA* Woof!

Goofy Over Sports Goofy stars in this compilation of sports cartoons from the Disney archives: "How to Play Football" (1944), "Double Dribble" (1946), "Art of Skiing" (1941), "How to Swim" (1942), "Art of Self Defense" (1941), and, "How to Ride a Horse" (a segment from the 1941 feature, "The Reluctant Dragon").
194? 46m/C **VHS, Beta $49.95** *DIS*

Goon Movie (Stand Easy) The cast of "The Goon Show," Britain's popular radio comedy series, perform some of their best routines in this, their only film appearance.
1953 75m/B Peter Sellers, Spike Milligan, Harry Secombe. **VHS, Beta $19.95** *VDM* ✓✓

The Goonies Two brothers who are about to lose their house conveniently find a treasure map. They pick up a couple of friends and head for the "X." If they can recover the treasure without getting caught by the bad guys, then all will be saved. Steven Spielberg produced this high-energy action fantasy for kids of all ages. The laserdisc edition comes in a widescreen format.
1985 (PG) 114m/C Sean Astin, Josh Brolin, Jeff B. Cohen, Corey Feldman, Martha Plimpton, John Matuszak, Robert Davi, Anne Ramsey, Mary Ellen Trainer; *Dir:* Richard Donner. **VHS, Beta, LV $19.98** *WAR, PIA* ✓✓½

Gor Sword and sorcery: a magic ring sends a meek college professor to "Gor," a faraway world in which survival goes to the fittest and the most brutal. Followed by "Outlaw of Gor."
1988 (PG) 95m/C Urbano Barberini, Rebecca Ferratti, Jack Palance, Paul Smith, Oliver Reed; *Dir:* Fritz Kiersch. **VHS, Beta $19.98** *WAR* ✓½

Gorath The world's top scientists are racing to stop a giant meteor from destroying the Earth.
1967 77m/C *JP* *Dir:* Inoshiro Honda. **VHS, Beta $19.95** *GEM, PSM* ✓✓

Gordon's War When a Vietnam vet returns to his Harlem home, he finds his wife overdosing on the drugs that have infiltrated his neighborhood. He leads a vigilante group in an attempt to clean up the area, which makes for a lot of action; however, there's also an excessive amount of violence.
1973 (R) 89m/C Paul Winfield, Carl Lee, David Downing, Tony King, Grace Jones; *Dir:* Ossie Davis. **VHS, Beta $59.98** *FOX* ✓✓

Gore-Met Zombie Chef From Hell A demon/vampire opens up a seafood restaurant and slaughters his customers. Deliberately campy and extremely graphic.
1987 90m/C Theo Depuay, Kelley Kunicki, C.W. Casey, Alan Marx, Michael O'Neill; *Dir:* Don Swan. **VHS, Beta** *CAM* Woof!

Gore Vidal's Billy the Kid An unusual treatment of the William Donney legend in which the Kid is merely a misunderstood teenager caught up in the midst of the brutal range wars. Made for cable television.
1989 100m/C Val Kilmer, Duncan Regehr, Wilford Brimley, Julie Carmen, Michael Parks, Rene Auberjonois, Albert Salmi; *Dir:* William A. Graham. **VHS $9.98** *TTC* ✓✓½

The Gorgeous Hussy Crawford stars in this fictionalized biography of Peggy Eaton, Andrew Jackson's notorious belle, who disgraces herself and those around her. Although this film featured a star-studded cast complete with beautiful costumes, it wasn't enough to save this overly long and dull picture.
1936 102m/B Joan Crawford, Robert Taylor, Lionel Barrymore, Melvyn Douglas, James Stewart, Franchot Tone, Louis Calhern; *Dir:* Clarence Brown. **VHS $19.98** *MGM* ✓

Gorgo An undersea explosion off the coast of Ireland brings to the surface a prehistoric sea monster, which is captured and brought to a London circus. Its irate mother appears looking for her baby, creating havoc in her wake.
1961 76m/C *GB* Bill Travers, William Sylvester, Vincent Winter, Bruce Seton, Christopher Rhodes; *Dir:* Eugene Lourie. **VHS, Beta, LV $19.95** *NOS, MRV, UHV* ✓✓½

The Gorgon In pre-World War I Germany, the lovely assistant to a mad brain surgeon moonlights as a snake-haired gorgon, turning men to stone. A professor arrives in the village to investigate, only to become another victim.
1964 83m/C *GB* Peter Cushing, Christopher Lee, Richard Pasco, Barbara Shelley, Michael Goodliffe, Patrick Troughton; *Dir:* Terence Fisher. **VHS, Beta, LV $9.95** *GKK, MLB* ✓✓

The Gorilla The bumbling Ritz Brothers are hired to protect a country gentleman receiving threats from a killer. Lugosi plays the menacing butler. Derived from the play by Ralph Spence.
1939 67m/B The Ritz Brothers, Anita Louise, Patsy Kelly, Lionel Atwill, Bela Lugosi; *Dir:* Allan Dwan. **VHS, Beta, LV $19.95** NOS, SNC, MRV ♫½

Gorilla A rampaging gorilla is sought by the local game warden and a journalist researching the natives. Filmed on location in the Belgian Congo.
1956 79m/B SW Gio Petre, Georges Galley; *Dir:* Sven Nykvist, Lar Henrik Ottoson. **VHS, Beta $19.95** FCT ♫♫

Gorilla Farming This humorous documentary/spoof explores both sides of the marijuana growth industry and its downfall.
1990 85m/C *Dir:* Terry Lewis, Gary Barron. **VHS $39.95** PRM ♫♫

Gorillas in the Mist The life of Dian Fossey, animal rights activist and world-renowned expert on the African Gorilla, from her pioneering contact with mountain gorillas to her murder at the hands of poachers. Weaver is totally appropriate as the increasingly obsessed Fossey, but the character moves away from us, just as we need to see and understand more about her. Excellent special effects. Music by Maurice Jarre.
1988 (PG-13) 117m/C Sigourney Weaver, Bryan Brown, Julie Harris, Iain Cuthbertson, John Omirah Miluwi; *Dir:* Michael Apted. **VHS, Beta, LV $19.95** MCA ♫♫♫

Gorky Park Adaptation of Martin Cruz Smith's bestseller. Three strange, faceless corpses are found in Moscow's Gorky Park. There are no clues for the Russian police captain investigating the incident. He makes the solution to this crime his personal crusade, and finds himself caught in a web of political intrigue. Excellent police procedure yarn.
1983 (R) 127m/C William Hurt, Lee Marvin, Brian Dennehy, Joanna Pacula; *Dir:* Michael Apted. Edgar Allan Poe Awards '83: Best Screenplay. **VHS, Beta, LV $29.98** VES ♫♫♫

Gorp Sex, fun, and lewd, sophomoric humor reign at a summer camp. For the bored and witless.
1980 (R) 90m/C Dennis Quaid, Rosanna Arquette, Michael Lembeck, Philip Casnoff, Frank Drescher; *Dir:* Joseph Ruben. **VHS, Beta $79.99** HBO Woof!

Gospel A rousing musical theatrical tribute to the leading exponents of Gospel singing with the Mighty Clouds of Joy, Shirley Caeser, the Clark Sisters, Walter Hawkins, and Rev. James Cleveland.
1982 92m/C Rev. James Cleveland, Shirley Caeser, Joy, Twinkie Clark & the Clark Sisters, Walter Hawkins & the Hawkins Family; *Dir:* David Leivick. **VHS, Beta $39.95** MVD, MON, VTK

Gospel According to St. Matthew Perhaps Pasolini's greatest film, retelling the story of Christ in gritty, neo-realistic tones and portraying the man less as a divine presence than as a political revolutionary. The yardstick by which all Jesus-films are to be measured; in Italian with subtitles or dubbed in English.
1964 136m/B *IT* Enrique Irazoqui, Susanna Pasolini; *Dir:* Pier Paolo Pasolini. National Board of Review Awards '66: 5 Best Foreign Films of the Year. **VHS, Beta, 8mm $24.95** NOS, MRV, WBF ♫♫♫♫

The Gospel According to Vic A Scottish comedy about a skeptical teacher at a remedial school, who, having survived miraculously from a fall, is taken for proof of the sainthood of the school's patron namesake.
1987 (PG-13) 92m/C GB Tom Conti, Helen Mirren; *Dir:* Charles Gormley. **VHS, Beta $79.98** FOX ♫♫

Gotcha! The mock assassination game "Gotcha!" abounds on the college campus and sophomore Edwards is one of the best. What he doesn't know is that his "assassination" skills are about to take on new meaning when he meets up with a female Czech graduate student who is really an international spy.
1985 (PG-13) 97m/C Anthony Edwards, Linda Fiorentino, Alex Rocco, Nick Corri; *Dir:* Jeff Kanew. **VHS, Beta, LV $19.95** MCA ♫♫½

Gotham A wealthy financier hires private-eye Jones to track down beautiful wife Madsen. It seems that she keeps dunning her husband for money. Sounds easy enough to Jones until he learns that she has been dead for some time. Mystery ventures into the afterlife and back as it twists its way to its surprise ending. Intriguing throwback to '40s detective flicks. Made for television.
1988 100m/C Tommy Lee Jones, Virginia Madsen, Colin Bruce, Kevin Jarre, Denise Stephenson, Frederic Forrest; *Dir:* Lloyd Fonvielle. **VHS, LV $19.98** CAN ♫♫♫

Gothic Mary Shelley, Lord Byron, Percy Bysshe Shelley, and John Polidori spend the night in a Swiss villa telling each other ghost stories and experimenting with laudanum and sexual partner combinations. The dreams and realizations of the night will color their lives ever after. Interesting premise, well-carried out, although burdened by director Russell's typical excesses. Soundtrack by Thomas Dolby.
1987 (R) 87m/C Julian Sands, Gabriel Byrne, Timothy Spall, Natasha Richardson; *Dir:* Ken Russell. **VHS, Beta, LV $14.98** LIV, VES ♫♫♫

Gotta Dance, Gotta Sing Compilation of memorable dance routines from great Hollywood films.
1984 53m/C Fred Astaire, Ginger Rogers, Shirley Temple, Carmen Miranda, Betty Grable. **VHS, Beta $39.95** RKO

Grace Quigley An elderly woman, tired of life, hires a hitman to kill her. Instead they go into business together, bumping off other elderly folk who decide they'd rather be dead. Notoriously inept black comedy noted for being the career embarrassment for both of its stars. Also known as "The Ultimate Solution of Grace Quigley."
1984 (PG) 102m/C Katharine Hepburn, Nick Nolte, Walter Abel, Chip Zien, Elizabeth Wilson, Kit Le Fever, William Duell, Jill Eikenberry; *Dir:* Anthony Harvey. **VHS, Beta $79.95** MGM ♫½

Grad Night The anticipation for graduation in a small town high school brings surprises no one could have predicted, unless they've seen all the other typical adolescent films.
1981 (R) 85m/C Joey Johnson, Suzanna Fagan, Barry Stolze, Sam Whipple, Caroline Bates; *Dir:* John Tenorio. **VHS, Beta $79.95** MNC ♫

The Graduate The famous, influential slice of comic Americana starring Hoffman as Benjamin Braddock, a shy, aimless college graduate who, without any idea of responsibility or ambition, wanders from a sexual liaison with a married woman (the infamous Mrs. Robinson) to pursuit of her engaged daughter. Benjamin's pursuit of Elaine right to her wedding has become a film classic. Extremely popular and almost solely responsible for establishing both Hoffman and director Nichols. The career advice given to Benjamin, "plastics," became a catchword for the era. Music by Simon & Garfunkel. Based on the novel by Charles Webb. Oscar nominations for Best Picture, actor Hoffman, actress Bancroft, supporting actress Ross, adapted screenplay, and cinematography. Laserdisc edition includes screen tests, promotional stills, and a look at the creative process involved in making a novel into a movie. A 25th Anniversary Limited Edition is also available, presented in a wide-screen format and including the original movie trailer and interviews with the cast and crew.
1967 106m/C Dustin Hoffman, Anne Bancroft, Katharine Ross, Murray Hamilton, Brian Avery, Marion Lorne, Alice Ghostley, William Daniels, Elizabeth Wilson, Norman Fell, Buck Henry, Richard Dreyfuss, Mike Farrell; *Dir:* Mike Nichols. Academy Awards '67: Best Director (Nichols); British Academy Awards '68: Most Promising Actor (Hoffman), Best Director (Nichols), Best Film, Best Film Editing, Best Screenplay; Directors Guild of America Awards '67: Best Director (Nichols); Golden Globe Awards '68: Best Film—Musical/Comedy; National Board of Review Awards '67: 10 Best Films of the Year; New York Film Critics Awards '67: Best Director (Nichols). **VHS, Beta, LV, 8mm $14.98** SUE, VYG, CRC ♫♫♫♫

Graduation Day Another teen slasher about the systematic murder of members of a high school track team. Notable mainly for brief appearance by Vanna White.
1981 (R) 85m/C Christopher George, Patch McKenzie, E. Danny Murphy, Vanna White; *Dir:* Herb Freed. **VHS, Beta $69.95** COL ♫

Graffiti Bridge Prince preaches love instead of sex and the result is boring. He plays a Minneapolis nightclub owner who battles with Day over the love of a beautiful woman. Chauvinistic attitudes toward women, but the females in the film don't seem to mind. Visually interesting, although not as experimental as his first try. Music made Top Ten.
1990 (PG-13) 90m/C Prince, Morris Day, Jerome Benton, The Time, Jill Jones, Mavis Staples, George Clinton, Ingrid Chavez; *W/Dir:* Prince. **VHS, Beta, LV $19.98** MVD, WAR ♫½

Graham Parker This program presents Graham Parker performing some of his greatest hits.
1982 60m/C Graham Parker. **VHS, Beta $29.95** SVS

Grain of Sand Based on a true story, an understated French drama about a troubled woman searching her past, and her home town in Corsica, for the man of her dreams. Directorial debut of Meffre. In French with subtitles.
1984 90m/C FR Delphine Seyrig, Genevieve Fontanel, Coralie Seyrig, Michel Aumont; *Dir:* Pomme Meffre. **VHS, Beta $59.95** FCT ♫♫♫

Grambling's White Tiger The true story of Jim Gregory, the first white man to play on Grambling College's all-black football team. Jenner's a little too old, but Belafonte, in his TV acting debut, is fine as legendary coach Eddie Robinson. Based on book by Bruce Behrenberg. Made for television.
1981 98m/C Bruce Jenner, Harry Belafonte, LeVar Burton, Ray Vitte, Byron Stewart; *Dir:* George Stanford Brown. **VHS, Beta $39.95** MCA ♫♫½

Grampa's Monster Movies Grampa (star of "The Munsters" TV show) hosts a tribute to Frankenstein, Dracula, Wolfman and the Mummy, with legends Bela Lugosi, Boris Karloff, Lon Chaney Jr. and Vincent Price.
1990 65m/C *Hosted:* Al Lewis. **VHS, Beta** *AMV* 🎬🎬½

Grampa's Sci-Fi Hits Sequel to Grampa's "Monster Movies," this one carries some of the more fantastic of sci-fi creatures from beyond.
1990 65m/C *Hosted:* Al Lewis. **VHS, Beta** *AMV* 🎬🎬½

Gran Valor en la Facultad de Medicina A Spanish simpleton is mistakenly admitted to medical school.
198? 90m/C *SP* Juan Carlos Calabro, Adriana Aguirre. **VHS, Beta $57.95** *UNI* 🎬

Grand Canyon A group of very diverse characters are thrown together through chance encounters while coping with urban chaos in L.A. The film mainly centers on the growing friendship between an immigration lawyer (Kline) and a tow-truck driver (Glover) who meet when Kline's car breaks down in a crime-ridden neighborhood. The entire cast gives first-rate performances in this heartwarming drama filled with subtle moral messages that hold even more meaning now as the country tries to heal the wounds caused by the recent violence in L.A.
1991 (R) 134m/C Danny Glover, Kevin Kline, Steve Martin, Mary McDonnell, Mary Louise Parker, Alfre Woodard; *W/Dir:* Lawrence Kasdan. National Board of Review Awards '91: 10 Best Films of the Year. **VHS $94.98** *FXV* 🎬🎬🎬

Grand Canyon Trail Our hero is the owner of a played-out silver mine who is the target of an unscrupulous engineer who thinks there's silver to be found if you know where to look. The first appearance of the Riders of the Purple Sage, who replaced the Sons of the Pioneers.
1948 68m/B Roy Rogers, Andy Devine, Charles Coleman, Jane Frazee, Robert Livingston; *Dir:* William Witney. **VHS, Beta $9.99** *NOS, MED, DVT* 🎬

The Grand Duchess and the Waiter Menjou is a French millionaire who disguises himself as a waiter in order to enter the service of a grand duchess with whom he has fallen in love.
1926 70m/B Adolphe Menjou, Florence Vidor, Lawrence Grant; *Dir:* Malcolm St. Clair. **VHS, Beta $27.95** *DNB* 🎬🎬🎬

The Grand Highway A sweet, slice-of-life French film about a young boy's idyllic summer on a family friend's farm while his mother is having a baby. Hubert's son plays the boy Louis. In French with English subtitles.
1988 113m/C *FR* Anemone, Richard Bohringer, Antoine Hubert, Vanessa Guedj, Christine Pascal, Raoul Billerey, Pascale Roberts; *Dir:* Jean-Loup Hubert. **VHS, Beta $29.95** *PAV* 🎬🎬🎬

Grand Hotel A star-filled cast is brought together by unusual circumstances at Berlin's Grand Hotel and their lives become hopelessly intertwined over a twenty-four hour period. Adapted (and given the red-carpet treatment) from a Vicki Baum novel; notable mostly for Garbo's world-weary ballerina. Time has taken its toll on the concept and treatment, but still an interesting star vehicle.

1932 112m/B Greta Garbo, John Barrymore, Joan Crawford, Lewis Stone, Wallace Beery, Jean Hersholt, Lionel Barrymore; *Dir:* Edmund Goulding. Academy Awards '32: Best Picture. **VHS, Beta, LV $19.98** *MGM, FCT, BTV* 🎬🎬🎬½

Grand Illusion The unshakably classic anti-war film by Renoir, in which French prisoners of war attempt to escape from their German captors during World War I. An indictment of the way Old World, aristocratic nobility was brought to modern bloodshed in the Great War. Renoir's optimism remains relentless, but was easier to believe in before World War II. In French, with English subtitles. Received Oscar Nomination for Best Picture. The laserdisc version includes an audio essay by Peter Crowe.
1937 111m/B *FR* Jean Gabin, Erich von Stroheim, Pierre Fresnay, Marcel Dalio, Julien Carette, Gaston Modot, Jean Daste, Dita Parlo; *Dir:* Jean Renoir. National Board of Review Awards '38: Best Foreign Film; New York Film Critics Awards '38: Best Foreign Film; Venice Film Festival '37: Best Artistic Ensemble; Cinematheque Belgique '52: #4 of the Best Films of All Time (tie); Brussels World's Fair '58: #5 of the Best Films of All Time. **VHS, Beta, LV, 8mm $29.95** *NOS, MRV, FOX* 🎬🎬🎬🎬

Grand Master of Shaolin Kung Fu Follow the exploits of Dhama, the very man responsible for the creation of Kung Fu.
19?? 88m/C **VHS, Beta $14.98** *NEG*

Grand Prix A big-budget look at the world of Grand Prix auto racing, as four top competitors circle the world's most famous racing circuits. Strictly for those who like cars racing round and round; nothing much happens off the track. Letterboxed screen is featured in the laserdisc format.
1966 161m/C James Garner, Eva Marie Saint, Yves Montand, Toshiro Mifune, Brian Bedford; *Dir:* John Frankenheimer. Academy Awards '66: Best Film Editing, Best Sound, Best Sound Effects Editing. **VHS, Beta, LV $29.99** *MGM, FCT* 🎬🎬

Grand Theft Auto In Howard's initial directorial effort, a young couple elopes to Las Vegas in a Rolls Royce owned by the bride's father. The father, totally against the marriage and angered by the stolen Rolls, offers a reward for their safe return and a cross-country race ensues.
1977 (PG) 84m/C Ron Howard, Nancy Morgan, Marion Ross, Barry Cahill, Clint Howard; *Dir:* Ron Howard. **VHS, Beta $39.98** *WAR* 🎬🎬

Grandes Amigos (Great Friends) A young boy finds out the meaning of having a true friend in this comedy-drama.
197? 105m/C *SP* Nino Del Arco. **VHS, Beta $49.95** *MAD* 🎬

Grandma Didn't Wave Back A touching story of how a young girl handles her grandmother's senility.
1984 25m/C Molly Picon. **VHS $89.95** *ERG* 🎬

Grandma's House A brother and sister discover that the house in which they live with their grandmother is chock-full of secrets and strange happenings, including madness, incest, and murder.
1988 (R) 90m/C Eric Foster, Kim Valentine, Brinke Stevens, Ida Lee; *Dir:* Peter Rader. **VHS, Beta $79.95** *ACA* 🎬

Grandview U.S.A. A low-key look at low-rent middle America, centering on a foxy local speedway owner and the derby-obsessed boys she attracts.

1984 (R) 97m/C Jamie Lee Curtis, Patrick Swayze, C. Thomas Howell, M. Emmet Walsh, Troy Donahue, William Windom, Jennifer Jason Leigh, Ramon Bieri, John Cusack, Joan Cusack, Jason Court; *Dir:* Randal Kleiser. **VHS, Beta $19.98** *FOX* 🎬🎬½

Grant at His Best Collection of Cary Grant's most charming work. Includes footage from: "Father Goose," "Operation Petticoat," and "Indiscreet." Available on three cassettes.
19?? 336m/C Cary Grant. **VHS, Beta $59.95** *REP*

The Grapes of Wrath John Steinbeck's classic American novel about the Great Depression. We follow the impoverished Joad family as they migrate from the dust bowl of Oklahoma to find work in the orchards of California and as they struggle to maintain at least a little of their dignity and pride. A sentimental but dignified, uncharacteristic Hollywood epic. Written by Nunnally Johnson. Oscar nominations: Best Picture, Best Actor (Fonda), Best Screenplay, Sound Recording, Film Editing.
1940 129m/B Henry Fonda, Jane Darwell, John Carradine, Charley Grapewin, Zeffie Tilbury, Dorris Bowdon, Russell Simpson, John Qualen, Eddie Quillan, O.Z. Whitehead, Grant Mitchell; *Dir:* John Ford. Academy Awards '40: Best Director (Ford), Best Supporting Actress (Darwell); New York Film Critics Awards '40: Best Director (Ford), Best Film; New York Times 10 Best List '40: Best Film. **VHS, Beta, LV $19.98** *KUI, FOX, BTV* 🎬🎬🎬½

The Grass is Always Greener Over the Septic Tank Based on Erma Bombeck's best-seller. A city family's flight to the supposed peace of the suburbs turns out to be a comic compilation of complications. Made for television.
1978 98m/C Carol Burnett, Charles Grodin, Linda Gray, Alex Rocco, Robert Sampson, Vicki Belmonte, Craig Richard Nelson, Anrae Walterhouse, Eric Stoltz; *Dir:* Robert Day. **VHS, Beta $69.95** *LTG* 🎬🎬

The Grass is Greener An American millionaire invades part of an impoverished Earl's mansion and falls in love with the lady of the house. The Earl, who wants to keep his wife, enlists the aid of an old girlfriend in a feeble attempt to make her jealous. The two couples pair off properly in the end.
1961 105m/C Cary Grant, Deborah Kerr, Jean Simmons, Robert Mitchum; *Dir:* Stanley Donen. **VHS $19.98** *REP* 🎬🎬½

The Grateful Dead Movie This rockumentary looks at the lives and the music of the rock band, The Grateful Dead. Songs include "Uncle John's Band," "Sugar Magnolia," "Casey Jones," "Ripple" and "Morning Dew."
1976 131m/C *Dir:* Jerry Garcia. **VHS, Beta, LV $39.95** *MON, VTK, MVD*

Grave Secrets A professor of psychic phenomena teams up with a medium to find out who—or what—is controlling the life of a young woman.
1989 (R) 90m/C Paul LeMat, Renee Soutendijk, David Warner, Olivia Barash; *Dir:* Donald P. Borchers. **VHS, LV $89.95** *SGE* 🎬½

Grave of the Vampire A vampire rapes and impregnates a modern-day girl. Twenty years later, their son, who grows up to be a bitter bloodsucker, sets out to find his father and kill him. Also known as "Seed of Terror."

1972 (R) 95m/C William Smith, Michael Pataki, Lyn Peters, Diane Holden, Jay Adler, Kitty Vallacher, Jay Scott, Lieux Dressler; *Dir:* John Patrick Hayes. **VHS, Beta** $49.95 *UNI* ℐ

The Graveyard A mommy spends many years torturing her little boy, David. In one blood-filled day, the fellow seeks his revenge. Also released as "Terror of Sheba" and "Persecution."
1974 90m/C Lana Turner, Trevor Howard, Ralph Bates, Olga Georges-Picot, Suzan Farmer; *Dir:* Don Chaffey. **VHS, Beta** *VCL, ELV* ℐℐ

Graveyard of Horror His wife and baby dead, a man seeking revenge discovers a horrible secret: his brother, a doctor, is stealing the heads from corpses. Dubbed.
1971 105m/C *SP* Bill Curran, Francisco (Frank) Brana, Beatriz Lacy; *Dir:* Miguel Madrid. **VHS, Beta** $49.95 *IMG* ℐ

Graveyard Shift A New York cabby on the night shift is actually a powerful vampire who uses his fares to build an army of vampires. Not to be confused with the 1990 Stephen King scripted film. Followed by "Understudy: Graveyard Shift II."
1987 (R) 89m/C *IT* Silvio Oliviero, Helen Papas, Cliff Stoker; *Dir:* Gerard Ciccoritti. **VHS, Beta** $79.95 *VIR* ℐ ½

Graveyard Shift Just because Stephen King wrote the original story doesn't mean its celluloid incarnation is guaranteed to stand your hair on end; unless that's your response to abject boredom. When a man takes over the night shift at a recently reopened textile mill, he starts to find remnants...and they aren't fabric. Seems there's a graveyard in the neighborhood.
1990 (R) 89m/C David Andrews, Kelly Wolf, Stephen Macht, Brad Dourif; *Dir:* Ralph S. Singleton. **VHS, Beta, LV** $14.95 *PAR, FCT* ℐ

Gray Lady Down A nuclear submarine sinks off the coast of Cape Cod and with their oxygen running out, a risky escape in an experimental diving craft seems to be their only hope. Dull, all-semi-star suspenser.
1977 (PG) 111m/C Charlton Heston, David Carradine, Stacy Keach, Ned Beatty, Ronny Cox, Christopher Reeve, Michael O'Keefe, Rosemary Forsyth; *Dir:* David Greene. **VHS, Beta** $59.95 *MCA* ℐℐ

Grayeagle When a frontier trapper's daughter is kidnapped by Cheyenne Indians, he launches a search for her recovery. More style than pace in this Western, but worth a look for fans of Johnson.
1977 (PG) 104m/C Ben Johnson, Iron Eyes Cody, Lana Wood, Alex Cord, Jack Elam, Paul Fix, Cindy Butler, Charles B. Pierce; *W/Dir:* Charles B. Pierce. **VHS, Beta** $19.98 *WAR, OM* ℐℐ

Grazuole A nine-year-old girl is mockingly made "Beauty Queen" by her malicious playmates.
1969 71m/B *Dir:* Arunas Zhebrunas. **VHS, Beta** *IHF* ℐℐ

Grease Film version of the hit Broadway musical about summer love. Set in the 1950s, this spirited musical follows a group of high-schoolers throughout their senior year. The story offers a responsible moral: act like a tart and you'll get your guy; but, hey, it's all in fun anyway. Songs, many of which are performed by Newton-John, include "Hopelessly Devoted to You," "We Go Together," and "Summer Nights." Followed by a weak sequel.

1978 (PG) 110m/C John Travolta, Olivia Newton-John, Jeff Conaway, Stockard Channing, Eve Arden, Sha-Na-Na, Frankie Avalon, Sid Caesar; *Dir:* Randal Kleiser. **VHS, Beta, LV** $14.95 *PAR, FCT* ℐ

Grease 2 The saga of the T-Birds, the Pink Ladies and young love at Rydell High continues with a new crop of students in the class of '61. Not only doesn't have the cast, but also doesn't have the good-humor flair of the original.
1982 (PG) 115m/C Maxwell Caulfield, Michelle Pfeiffer, Adrian Zmed, Lorna Luft, Didi Conn, Eve Arden, Sid Caesar, Tab Hunter; *Dir:* Patricia Birch. **VHS, Beta, LV** $14.95 *PAR* ℐ

Greased Lightning The story of the first black auto racing champion, Wendell Scott, who had to overcome racial prejudice to achieve his success. Slightly-better-than-average Pryor comedy vehicle.
1977 (PG) 95m/C Richard Pryor, Pam Grier, Beau Bridges, Cleavon Little, Vincent Gardenia; *Dir:* Michael Schultz. **VHS, Beta** $59.95 *WAR* ℐℐ ½

Greaser's Palace Seaweedhead Greaser, owner of the town's saloon, faces his arch-nemesis in this wide-ranging satiric Christ allegory from the director of "Putney Swope."
1972 91m/C Albert Henderson, Allan Arbus; *Dir:* Robert Downey. **VHS, Beta** *COL, OM* ℐℐ ½

The Great Adventure Two farm boys capture and attempt to train a wild otter. Unfortunately, the otter longs to once again join his kind despite the boys' efforts. Arne Sucksdorff won an Oscar for his arresting view of animal life.
1953 73m/B *SW* Anders Norberg, Kjell Sucksdorff, Arne Sucksdorff; *W/Dir:* Arne Sucksdorff. **VHS, Beta** *MED* ℐℐℐ

Great Adventure In the severe environment of the gold rush days on the rugged Yukon territory, a touching tale unfolds of a young orphan boy and his eternal bond of friendship with a great northern dog. Based on a Jack London story. Paul Elliots is a pseudonym for Gianfranco Baldanello and Fred Romer is actually Fernando Romero.
1975 (PG) 90m/C *IT SP* Jack Palance, Joan Collins, Fred Romer, Elisabetta Virgili, Remo de Angelis, Manuel de Blas; *Dir:* Paul Elliots. **VHS, Beta** $19.95 *MED* ℐℐ

The Great Alligator As if you couldn't guess, this one's about a huge alligator who does what other huge alligators do...terrorizes the people at a resort.
1981 89m/C Barbara Bach, Mel Ferrer, Richard Johnson; *Dir:* Sergio Martino. **VHS, Beta** *MPI* ℐ

The Great American Broadcast Two WWI vets want more than anything to strike it rich. After many failed endeavors, the two try that new-fangled thing called radio. The station takes off, as does the plot. The girlfriend of one vet falls for his partner and numerous other misunderstandings and complications ensue. Charming and crazy musical with an energetic cast.
1941 92m/B Alice Faye, John Payne, Jack Oakie, Cesar Romero, The Four Ink Spots, James Newill, Mary Beth Hughes; *Dir:* Archie Mayo. **VHS** *MLB* ℐℐℐ

Great American Cowboy From the stable to the arena, this documentary tells the story of modern rodeo cowboy Larry Mahan.
1973 89m/C *Dir:* Keith Merrill; *Nar:* Joel McCrea. Academy Awards '73: Best Feature Documentary. **VHS, Beta** $69.95 *DIS* ℐℐℐ

Great American Traffic Jam A made-for-television farce about the humorous variety of characters interacting on the interstate during a massive California traffic jam.
1980 (PG) 97m/C Ed McMahon, Vic Tayback, Howard Hesseman, Abe Vigoda, Noah Beery Jr., Desi Arnaz Jr., John Beck, Shelley Fabares, James Gregory; *Dir:* James Frawley. **VHS, Beta** $59.98 *FOX* ℐℐ ½

The Great Armored Car Swindle An anonymous English businessman becomes involved in international intrigue when he becomes a pawn in a plan to transfer money to a Middle Eastern government under communist rule. But thanks to his wife things don't go off as planned.
1964 58m/B Peter Reynolds, Dermot Walsh, Joanna Dunham, Lisa Gastoni, Brian Cobby; *Dir:* Lance Comfort. **VHS, Beta** $16.95 *SNC* ℐℐ

Great Balls of Fire A florid, comic-book film of the glory days of Jerry Lee Lewis, from his first band to stardom. Most of the drama is derived from his marriage to his 13-year-old cousin. Somewhat overacted but full of energy. The soundtrack features many of the "Killer's" greatest hits re-recorded by Jerry Lee Lewis for the film.
1989 (PG-13) 108m/C Dennis Quaid, Winona Ryder, Alec Baldwin, Trey Wilson, John Doe, Lisa Blount; *Dir:* Jim McBride. **VHS, Beta, LV** $19.98 *ORI* ℐℐ ½

Great Bank Hoax Three bank managers decide to rob their own bank to cover up the fact that all the assets have been embezzled. The film was originally titled "Shenanigans" and is also known as "The Great Georgia Bank Hoax."
1978 (PG) 89m/C Richard Basehart, Ned Beatty, Burgess Meredith, Michael Murphy, Paul Sand, Arthur Godfrey; *Dir:* Joseph Jacoby. **VHS, Beta** $19.98 *WAR* ℐℐ ½

The Great Battle The events that led up to the last great World War II North African battle between the Allies and the German Panzers for the Mareth Line. International mishmash. Also known as "Battleforce."
1979 97m/C *GE YU* Henry Fonda, Stacy Keach, Samantha Eggar, Helmut Berger, John Huston; *Dir:* Humphrey Longan; *Nar:* Orson Welles. **VHS, Beta** *TLF* ℐ

Great Battle of the Volga A fierce World War II Russian front documentary.
1961 75m/B **VHS, Beta** *WFV* ℐ

Great British Striptease Sixteen of England's most fetching young women are featured performing the "Great British Striptease."
1981 60m/C **VHS, Beta** $39.95 *MON*

The Great Caruso The story of opera legend Enrico Caruso's rise to fame, from his childhood in Naples, Italy, to his collapse on the stage of the Metropolitan Opera House. Lanza is superb as the singer, and there are 27 musical numbers to satisfy the opera lover.
1951 113m/C Mario Lanza, Ann Blyth, Dorothy Kirsten; *Dir:* Richard Thorpe. Academy Awards '51: Best Sound; Film Daily Poll '51: Ten Best Pictures. **VHS, Beta** $19.98 *MGM* ℐℐ ½

Great Chase A historical anthology of the funniest and most suspenseful classic movie chases, especially "The General." Also features score by Larry Adler.
1963 79m/B Buster Keaton, Lillian Gish, Pearl White, Noah Beery; *Nar:* Frank Gallop. **VHS, Beta** $24.98 *SUE* ℐℐ ½

The Great Commandment Young man throws himself into a revolution against the overbearing Roman Empire in 30 A.D. His brother joins him but is killed. The man, who has become influenced by the teachings of Christ, decides to give up his revolution and go off with his brother's widow. (He always loved her anyway.)
1941 78m/B John Beal, Maurice Moscovich, Albert Dekker, Marjorie Cooley, Warren McCullum; ***Dir:*** Irving Pichel. **VHS $29.98** *REP* 🦴

Great Dan Patch The story of the great Dan Patch, the horse that went on to become the highest-earning harness racer in history.
1949 92m/B Dennis O'Keefe, Gail Russell, Ruth Warrick; ***Dir:*** Joseph M. Newman. **VHS, Beta $19.95** *WES, TIM* 🦴½

Great Day A soap opera set around the wartime visit of first lady Eleanor Roosevelt to a small British town. It focuses mainly on the wife and daughter of a bitter, alcoholic WWI vet.
1946 62m/B Eric Portman, Flora Robson, Sheila Sim, Isabel Jeans, Walter Fitzgerald, Philip Friend, Marjorie Rhodes, Maire O'Neill, Beatrice Varley; ***Dir:*** Lance Comfort. **VHS, Beta $19.98** *TTC* 🦴½

Great Day in the Morning A rebel sympathizer and a Northern spy are both after Colorado gold to help finance the Civil War. But our Southern gentlemen soon find themselves equally interested in the local ladies.
1956 92m/C Robert Stack, Ruth Roman, Raymond Burr, Virginia Mayo, Alex Nicol, Regis Toomey; ***Dir:*** Jacques Tourneur. **VHS, Beta $15.95** *UHV* 🦴🦴½

The Great Dictator Chaplin's first all-dialogue film, a searing satire on Nazism in which he has dual roles as a Jewish barber with amnesia who is mistaken for a Hitlerian dictator, Adenoid Hynkel. A classic scene involves Hynkel playing with a gigantic balloon of the world. Hitler banned the film's release to the German public due to its highly offensive portrait of him. The film also marked Chaplin's last wearing of the Little Tramp's (and Hitler's equally little) mustache. Oscar nominations: Best Picture, Best Actor (Chaplin), Best Supporting Actor (Oakie), Best Original Screenplay, Original Score.
1940 126m/B Charlie Chaplin, Paulette Goddard, Jack Oakie, Billy Gilbert, Reginald Gardiner, Henry Daniell; ***Dir:*** Charlie Chaplin. National Board of Review Awards '40: 10 Best Films of the Year; New York Film Critics Awards '40: Best Actor (Chaplin- award refused). **VHS, Beta $19.98** *FOX, GLV* 🦴🦴🦴🦴

The Great Escape During WWII, the Nazis gather all of the troubelsome allied P.O.W.s in a escape proof camp. These prisoners join forces in a single mass break for freedom. A true story based on the novel by Paul Brickhill and scripted by James Clavell and W.R. Burnett. Many of the stunts were performed by McQueen himself, particularly the climatic motorcycle ride, against the wishes of director Sturges. Laserdisc version features: an analysis of the true story that inspired the movie, actual photos of the camp, allied prisoners, and German guards, indepth commentaries by director John Sturges, composer Elmer Bernstein, and stuntman Bud Ekins, excerpts from the script by novelist James Clavell, over 250 production photos, and the original movie trailer.
1963 170m/C Steve McQueen, James Garner, Richard Attenborough, Charles Bronson, James Coburn, Donald Pleasence, David McCallum, James Donald, Gordon Jackson; ***Dir:*** John Sturges. National Board of Review Awards '63: 10 Best Films of the Year. **VHS, Beta, LV $29.97** *FOX, MGM, VYG* 🦴🦴🦴½

The Great Escape 2 This flaccid actioner (you'd think that would be a contradiction of terms) can't decide whether it's a remake of the original or a sequel; it pretends to be a sequel yet spends two-thirds of its reel time retelling—poorly—the first great escape. And once it gets around to the "and then what happened" part, it dispenses with sequel protocol: Pleasance, who played an escapee in the original, is a Nazi bad guy in this go-round. What's more, the casting is screwy: Reeves is unconvincing as a German-speaking British officer who rallies the surviving escapees in revenge against their WWII captors. Made for television and originally shown in two parts, it's been mercifully shortened for video.
1988 93m/C Christopher Reeve, Judd Hirsch, Ian McShane, Donald Pleasence, Anthony Denison, Judd Taylor, Charles Haid; ***Dir:*** Judd Taylor, Paul Wendkos. **VHS $89.95** *VMK, TLF* 🦴🦴

Great Expectations Lean's magisterial adaptation of the Dickens tome, in which a young English orphan is graced by a mysterious benefactor and becomes a well-heeled gentleman. Hailed and revered over the years; possibly the best Dickens on film. Well-acted by all, but especially notable is Hunt's slightly mad and pathetic Miss Havisham. Remake of the 1934 film. Oscar nominations for Best Picture, Best Director, Screenplay.
1946 118m/B *GB* John Mills, Valerie Hobson, Anthony Wager, Alec Guinness, Finlay Currie, Jean Simmons, Bernard Miles, Francis L. Sullivan, Martita Hunt; ***Dir:*** David Lean. Academy Awards '47: Best Art Direction/Set Decoration (B & W), Best Black and White Cinematography; National Board of Review Awards '47: 10 Best Films of the Year. **VHS, Beta, LV $79.98** *PAR, LCA* 🦴🦴🦴🦴

Great Expectations Animated version of Charles Dickens' classic is about an orphan's rise to gentleman status due to a mysterious benefactor.
1978 72m/C VHS, Beta $19.98 *CVL*

Great Expectations A British television mini-series adaptation of the Dickens epic about Pip and his mysterious benefactor in Victorian London. On two tapes.
1981 300m/C *GB* Gerry Sundquist, Stratford Johns, Joan Hickson. **VHS, Beta $29.98** *KUI, FOX* 🦴🦴

Great Expectations This tape is based upon the Dickens classic about a young boy's rise from a humble childhood to find fortune and happiness.
1983 72m/C ***Dir:*** Jean Tych. **VHS, Beta $19.95** *VES* 🦴

Great Expectations Dickens classic retold on 3 cassettes. A mysterious benefactor turns a poor orphan boy into a gentleman of means.
1989 325m/C *GB* Jean Simmons, Anthony Hopkins, John Rhys-Davies; ***Dir:*** Kevin Connor. **VHS, Beta $49.95** *DIS* 🦴🦴🦴

Great Flamarion A woman-hating trick-shot artist is nevertheless duped into murdering the husband of his femme fatale assistant after he comes to believe she loves him. He's wrong.
1945 78m/B Dan Duryea, Erich von Stroheim, Mary Beth Hughes; ***Dir:*** Anthony Mann. **VHS, Beta $16.95** *SNC, NOS, DVT* 🦴🦴½

The Great Frost Artist Seymour Chwast captures the dazzling imagery of Virginia Wolfe's prose from "Orlando," the love story of a young nobleman and a mysterious Russian princess. Part of the "Simple Gifts" series.
1979 15m/C VHS, Beta *TLF* 🦴🦴

The Great Gabbo A ventriloquist can express himself only through his dummy, losing his own identity and going mad at the end. Von Stroheim's first talkie. May quite possibly be the first mentally-twisted-ventriloquist story ever put on film. Based on a story by Ben Hecht.
1929 82m/B Erich von Stroheim, Betty Compson, Don Douglas, Marjorie Kane; ***Dir:*** James Cruze. **VHS, Beta $16.95** *SNC, NOS, DVT* 🦴🦴

The Great Gatsby Adaptation of F. Scott Fitzgerald's novel of the idle rich in the 1920s. A mysterious millionaire crashes Long Island society, and finds his heart captured by an impetuous and emotionally impoverished girl. Skillful acting and directing captures the look and feel of this era, but the movie doesn't have the power of the book. Screenplay by Francis Ford Coppola.
1974 (PG) 144m/C Robert Redford, Mia Farrow, Bruce Dern, Karen Black, Patsy Kensit, Howard da Silva, Sam Waterston, Howard da Silva, Edward Herrmann; ***Dir:*** Jack Clayton. Academy Awards '74: Best Adapted Score, Best Costume Design; Harvard Lampoon Awards '73: Worst Film of the Year. **VHS, Beta, LV $19.95** *PAR, FCT* 🦴🦴½

The Great Gildersleeve Throckmorton P. Gildersleeve adds another problem to his list; the sister of a local judge wants to marry him. His niece and nephew won't allow it and his attempts to juggle everyone's desires make for fine fun. Peary was also the voice for the popular radio series of the same name.
1943 62m/B Harold Peary, Jane Darwell, Nancy Gates, Charles Arnt, Thurston Hall; ***Dir:*** Gordon Douglas. **VHS** *MED, RXM* 🦴🦴

Great Gold Swindle A dramatization of the 1982 swindling of the Perth Mint in Australia of over $650,000 worth of gold.
1984 101m/C John Hargreaves, Robert C. Hughes, Tony Rickards, Barbara Llewellyn. **VHS, Beta $69.98** *LIV, LTG* 🦴🦴

The Great Gundown A violent tale set in the Old West. The peace of frontier New Mexico erupts when a half-breed Indian leads a brutal assault on an outlaw stronghold.
1975 98m/C Robert Padilla, Richard Rust, Milila St. Duval; ***Dir:*** Paul Hunt. **VHS, Beta $29.95** *HHT, VID* 🦴½

Great Guns Stan and Ollie enlist in the army to protect a spoiled millionaire's son but wind up being targets at target practice instead. Not one of their better efforts.
1941 74m/B Stan Laurel, Oliver Hardy, Sheila Ryan, Dick Nelson; ***Dir:*** Montague Banks. **VHS, Beta $29.98** *FOX, MLB* 🦴🦴

Great Guy One of Cagney's lesser roles as an ex-boxer turned food inspector who wipes out graft in his town.
1936 50m/B James Cagney, Mae Clarke, Edward Brophy; ***Dir:*** John Blystone. **VHS, Beta $19.95** *NOS, MRV, PSM* 🦴½

The Great Hunter A village militia captain and his girlfriend seek to destroy the men who killed the girl's father.
1975 91m/C Chia Ling, Wang Yu, Hsu Feng. VHS, Beta $29.95 *UNI* 🎬

The Great Impostor The true story of Ferdinand Waldo Demara, Jr., who during the 1950s, hoodwinked people by impersonating a surgeon, a college professor, a monk, a prison warden and a schoolteacher, with the FBI always one step behind him.
1961 112m/C Tony Curtis, Edmond O'Brien, Arthur O'Connell, Gary Merrill, Raymond Massey, Karl Malden, Mike Kellin, Frank Gorshin; *Dir:* Robert Mulligan. VHS, Beta $59.95 *MCA* 🎬🎬½

Great Jesse James Raid Jesse James comes out of retirement to carry out a mine theft. Routine.
1949 73m/C Willard Parker, Barbara Payton, Tom Neal, Wallace Ford; *Dir:* Reginald LeBorg. VHS, Beta $12.95 *AVD, MRV, WGE* 🎬½

Great K & A Train Robbery A railroad detective is hired to find and stop the bandits who have been preying on the K & A Railroad. Silent.
1926 55m/B Tom Mix, Dorothy Dwan, William Walling, Harry Grippe, Carl Miller; *Dir:* Lewis Seiler. VHS, Beta $52.95 *GPV, GVV* 🎬🎬½

Great Kate: The Films of Katharine Hepburn A three tape set of some of Hepburn's finest work. Includes the sparkling comedy "Bringing Up Baby," the Oscar winning "Morning Glory," and a fascinating women's tale "Stage Door."
19?? 278m/B Katharine Hepburn, Douglas Fairbanks Jr., Ginger Rogers, Cary Grant. VHS $59.95 *TCF* 🎬🎬½

The Great Land of Small Two children enter a fantasy world and try to save it from Evil. For kids.
1986 (G) 94m/C Karen Elkin, Michael Blouin, Michael Anderson Jr., Ken Roberts; *Dir:* Vojta Jasny. VHS, Beta $19.95 *STE, NWV, HHE* 🎬🎬

Great Leaders The inspiring stories of two Old Testament heroes, Gideon and Samson, are dramatized in this beautifully constructed film.
197? 105m/C Ivo Garrani, Fernando Rey, Giorgio Ceridni. VHS, Beta $49.95 *UHV* 🎬🎬

The Great Lie A great soaper with Davis and Astor as rivals for the affections of Brent, an irresponsible flyer. Astor is the concert pianist who marries Brent and then finds out she's pregnant after he's presumed dead in a crash. The wealthy Davis offers to raise the baby so Astor can continue her career. But when Brent does return, who will it be to? Sparkling, catty fun. Astor's very short, mannish haircut became a popular trend.
1941 107m/B Bette Davis, Mary Astor, George Brent, Lucile Watson, Hattie McDaniel, Grant Mitchell, Jerome Cowan; *Dir:* Edmund Goulding. Academy Awards '41: Best Supporting Actress (Astor). VHS, Beta, LV $19.95 *MGM, BTV* 🎬🎬🎬

The Great Locomotive Chase During the Civil War, Parker and his fellow soldiers head into Confederate territory to disrupt railroad supply lines. They take over a locomotive, but are pursued by Confederate Hunter, the conductor of another train; and soon a pursuit is underway. Based on the true story of Andrew's Raiders.
1956 85m/C Fess Parker, Jeffrey Hunter, Kenneth Tobey; *Dir:* Francis D. Lyon. VHS, Beta *DIS* 🎬🎬½

The Great Los Angeles Earthquake Made-for-TV disaster movie is the same old story; multiple soap-opera plotlines are spun around terrifying special effects, as L.A. is devastated by three major earth tremors and attendant disasters. Condensed from a two-part miniseries, also known as "The Big One: The Great Los Angeles Earthquake."
1991 106m/C Ed Begley Jr., Joanna Kerns; *Dir:* Larry Elikann. VHS $89.98 *VMK* 🎬½

Great Love Experiment Everyone in a high school learns a valuable lesson about beauty and friendship when four popular students try to change the personality of an awkward classmate.
1984 52m/C Tracy Pollan, Esai Morales, Kelly Wolf, Scott Benderer. VHS, Beta $39.98 *SCL* 🎬

The Great Lover Aboard an ocean liner, a bumbler chaperoning school kids gets involved with a gambler, a duchess and a murder. Vintage Bob Hope.
1949 80m/B Bob Hope, Rhonda Fleming, Roland Young, Roland Culver, George Reeves, Jim Backus; *Dir:* Alexander Hall. VHS, Beta $19.95 *COL, FCT* 🎬🎬½

The Great Man Votes Alcoholic college professor Barrymore falls into a deep depression after his wife's death. His children then plan to create a situation wherein his vote will decide the fate of the new town mayor, thus instilling in him a newly found sense of importance for himself and his children. Satiric and funny.
1938 72m/B John Barrymore, Peter Holden, Virginia Weidler, Katherine Alexander, Donald MacBride, Elisabeth Risdon, Granville Bates, Luis Alberni, J.M. Kerrigan, William Demarest, Roy Gordon; *Dir:* Garson Kanin. VHS, Beta $19.98 *TTC* 🎬🎬🎬

Great Massacre Sets the Robin Hood legend against the backdrop of Mainland China.
1969 95m/C *Hosted:* Sho Kosugi. VHS, Beta $29.98 *TWE* 🎬

The Great McGinty Sturges' directorial debut, about the rise and fall of a small-time, bribe-happy politician. As written by Sturges, an acerbic, ultra-cynical indictment of modern politics that stands, like much of his other work, as bracingly courageous as anything that's ever snuck past the Hollywood censors. Political party boss Tamiroff chews the scenery wih gusto while Donlevy, in his first starring role, is more than his equal as the not-so-dumb political hack.
1940 82m/B Brian Donlevy, Muriel Angelus, Akim Tamiroff, Louis Jean Heydt, Arthur Hoyt, William Demarest; *W/Dir:* Preston Sturges. Academy Awards '40: Best Original Screenplay. VHS, Beta, LV $29.95 *MCA* 🎬🎬🎬½

Great McGonagall Tale of an unemployed Scot trying to become Britain's poet laureate. Sellers adds a semi-bright spot with his portrayal of Queen Victoria!
1975 95m/C *GB* Peter Sellers, Spike Milligan, Julia Foster; *Dir:* Joseph McGrath. VHS $49.95 *MOV* 🎬🎬

Great Missouri Raid Follows the famous adventures of the James/Younger gang and their eventual demise.
1951 81m/C Wendell Corey, MacDonald Carey, Ellen Drew, Ward Bond; *Dir:* Gordon Douglas. VHS, Beta $19.95 *KRT* 🎬🎬

The Great Moment Story of the Boston dentist who discovered ether's use as an anesthetic in 1845. Confusing at times, changing from comedy to drama and con-

taining flashbacks that add little to the story. Surprisingly bland result from ordinarily solid cast and director.
1944 87m/B Joel McCrea, Betty Field, Harry Carey, William Demarest, Franklin Pangborn, Porter Hall; *Dir:* Preston Sturges. VHS, Beta $29.95 *MCA* 🎬🎬

The Great Mouse Detective Animated version of the book "Basil of Baker Street," by Eve Titus. Fun adventure about Sherlock-of-the-mouse-world. The detective is up against the sinister Professor Ratigan—will he solve this mystery??
1986 (G) 74m/C *Dir:* John Musker; *Voices:* Ron Clements, Dave Michener, Burny Mattinson, Vincent Price, Barrie Ingham, Candy Candido, Eve Brenner, Alan Young. VHS, Beta $24.99 *DIS* 🎬🎬🎬

Great Movie Stunts & the Making of Raiders of the Lost Ark Two television specials: "Movie Stunts" demonstrates how major action sequences were designed and executed, and "Making of Raiders" captures the cast and crew as they tackle the many problems created in filming the spectacular scenes.
1981 107m/C Harrison Ford. VHS, Beta, LV $12.95 *PAR*

Great Movie Trailers Coming attractions for many classic films: "My Little Chickadee," "High Noon," "African Queen," "2001," and many more.
197? 60m/B W.C. Fields, Mae West, Gary Cooper, Grace Kelly, Lloyd Bridges, Humphrey Bogart, Katharine Hepburn. VHS, Beta *BTV*

The Great Muppet Caper A group of hapless reporters (Kermit, Fozzie Bear, and Gonzo) travel to London to follow up on a major jewel robbery.
1981 (G) 95m/C Charles Grodin, Diana Rigg, John Cleese, Robert Morley, Peter Ustinov, Frank Oz, Peter Falk, Jack Warden, Jim Henson's Muppets; *Dir:* Jim Henson. VHS, Beta, LV $19.98 *FOX* 🎬🎬🎬

Great Music from Chicago The famous conductor and orchestra perform pieces by Barber, Toch, Borodin, Ravel, and Chabrier.
1957 55m/B Andre Kostelanetz, The Chicago Symphony Orchestra. VHS, Beta $24.99 *VYY* 🎬🎬🎬

The Great Northfield Minnesota Raid The Younger/James gang decides to rob the biggest bank west of the Mississippi, but everything goes wrong. Uneven, offbeat western, but Duvall's portrayal of the psychotic Jesse James and Robertson's cunning Cole Younger are notable.
1972 (PG) 91m/C Cliff Robertson, Robert Duvall, Elisha Cook Jr., Luke Askew, R.G. Armstrong, Donald Moffatt, Matt Clark; *Dir:* Philip Kaufman; *W/Dir:* Philip Kaufman. VHS, Beta $29.95 *MCA* 🎬🎬½

The Great Outdoors A family's peaceful summer by the lake is disturbed by their uninvited, trouble-making relatives. Aykroyd and Candy are two funny guys done in by a lame script and the tale of a giant bear that somehow seems to cast a giant shadow over this movie. May be fun for the kids, however.
1988 (PG) 91m/C Dan Aykroyd, John Candy, Stephanie Faracy, Annette Bening, Chris Young, Lucy Deakins; *Dir:* Howard Deutch. VHS, Beta, LV $14.95 *MCA* 🎬½

Great Race A dastardly villain, a noble hero and a spirited suffragette are among the competitors in an uproarious New York-to-Paris auto race circa 1908, complete with pie fights, saloon brawls, and a confrontation with a feisty polar bear. Overly long and only sporadically funny. **1965 160m/C** Jack Lemmon, Tony Curtis, Natalie Wood, Peter Falk, Keenan Wynn, George Macready; *Dir:* Blake Edwards. Academy Awards '65: Best Sound Effects Editing. **VHS, Beta, LV $19.98** *WAR, PIA, FCT* 🎬🎬½

Great Ride The state police are after two dirt bikers who are riding through areas where bike riding is illegal. **1978 90m/C** Perry Lang, Michael MacRae, Michael Sullivan; *Dir:* Don Hulette. **VHS, Beta $59.95** *MON* 🎬

The Great Riviera Bank Robbery A genius executes a bank robbery on the French Riviera netting fifteen million dollars. Based on a true story, the film chronicles the unfolding of the heist. **1979 98m/C** Ian McShane, Warren Clarke, Stephen Greif, Christopher Malcolm; *Dir:* Francis Megahy. **VHS, Beta $59.95** *MPI* 🎬🎬½

The Great Rupert Durante and family are befriended by a helpful squirrel (a puppet) in obtaining a huge fortune. Good fun; Durante shines. **1950 86m/B** Jimmy Durante, Terry Moore, Tom Drake, Frank Orth, Sara Haden, Queenie Smith; *Dir:* Irving Pichel. **VHS, Beta $19.98** *DVT* 🎬🎬🎬

Great Sadness of Zohara A young Jewish woman embarks on a spiritual journey that leads her to remote Arab lands, and her Orthodox community exiles her. Filmmaker Menkes tells the story via sound, image, and poetry. **1983 40m/C** *Dir:* Nina Menkes. **VHS $39.95** *FCT* 🎬🎬½

Great St. Trinian's Train Robbery Train robbers hide their considerable loot in an empty country mansion only to discover, upon returning years later, it has been converted into a girls' boarding school. When they try to recover the money, the thieves run up against a band of pestiferous adolescent girls, with hilarious results. Based on the cartoon by Ronald Searle. Sequel to "The Pure Hell of St. Trinian's." **1966 90m/C** *GB* Dora Bryan, Frankie Howerd, Reg Varney, Desmond Walter Ellis; *Dir:* Sidney Gilliat, Frank Launder. **VHS, Beta $59.99** *HBO* 🎬🎬

The Great Santini Lt. Col. Bull Meechum, the "Great Santini," a Marine pilot now stationed stateside, fights a war involving his frustrated career goals, his repressed emotions, and his family. His family becomes his company of marines, as he abuses them in the name of discipline, because he doesn't allow himself any other way to show his affection. His teen-aged son is forced to face the realities of his father's shortcomings as well as his own, and the racism he witnesses in a small Southern town in the 1960s. With a standout performance by Duvall, the film successfully blends warm humor and tenderness with the harsh cruelties inherent with dysfunctional families and racism. Based on Pat Conroy's autobiographical novel, the movie was virtually undistributed when first released, and then re-released due to critical acclaim. **1980 (PG) 118m/C** Robert Duvall, Blythe Danner, Michael O'Keefe, Julie Ann Haddock, Lisa Jane Persky, David Keith; *Dir:* Lewis John Carlino. Montreal World Film Festival '80: Best Actor (Duvall); National Board of Review Awards '80: 10 Best Films of the Year. **VHS, Beta, 8mm $19.98** *WAR* 🎬🎬🎬

Great Scout & Cathouse Thursday Marvin and Reed vow to take revenge on their third partner, Culp, who made off with all their profits from a gold mine. The title refers to Marvin's May-December romance with prostitute Lenz. Already-forgotten, unfunny, all-star comedy. **1976 (PG) 96m/C** Lee Marvin, Oliver Reed, Robert Culp, Elizabeth Ashley, Kay Lenz; *Dir:* Don Taylor. **VHS, Beta $69.98** *LIV, VES* 🎬½

Great Serial Prevues A series of movie trailer compilations, with everything from "Radar Men from the Moon" to "Captain America." **198? 60m/B** **VHS, Beta $19.95** *NOS, DVT* 🎬½

The Great Skycopter Rescue Ruthless businessmen hire a motorcycle gang to terrorize and scare away the inhabitants of an oil-rich town. A local teenage flying enthusiast organizes his friends into an attack force to fight back. **1982 96m/C** William Marshall, Aldo Ray, Russell Johnson, Terry Michos, Terry Taylor; *Dir:* Lawrence Foldes. **VHS, Beta $59.95** *MGM* 🎬

Great Smokey Roadblock While in the hospital, a sixty-year-old truck driver's rig is repossessed by the finance company. Deciding that it's time to make one last perfect cross country run, he escapes from the hospital, steals his truck, picks up six prostitutes, heads off into the night with the police on his tail, and becomes a folk hero. Also known as "Last of the Cowboys." **1976 (PG) 84m/C** Henry Fonda, Eileen Brennan, Susan Sarandon, John Byner; *Dir:* John Leone. **VHS, Beta $59.95** *MED* 🎬

The Great Texas Dynamite Chase Two sexy young women drive across Texas with a carload of dynamite. They leave a trail of empty banks with the cops constantly on their trail. **1976 (R) 90m/C** Claudia Jennings, Jocelyn Jones, Johnny Crawford, Chris Pennock, Tara Strohmeier, Miles Watkins, Bart Braverman; *Dir:* Michael Pressman. **VHS, Beta $39.98** *WAR, OM* 🎬½

Great Theatrical Previews Two hours of full length previews of motion pictures from the 30's, 40's, 50's, and 60's. Includes both color and B/W clips. **196? 120m/C** **VHS, Beta $29.95** *DVT*

The Great Trailers Compilation of coming attractions from the top films of the 30's, 40's and 50's. **194? 120m/B** **VHS, Beta $29.95** *DVT* 🎬½

Great Train Robbery A dapper thief arranges to heist the Folkstone bullion express in 1855, the first moving train robbery. A well-designed, fast-moving costume piece based on Crichton's best-selling novel. Music by Jerry Goldsmith. **1979 (PG) 111m/C** *GB* Sean Connery, Donald Sutherland, Lesley-Anne Down, Alan Webb; *Dir:* Michael Crichton. **VHS, Beta, LV $19.98** *MGM, FCT* 🎬🎬🎬

The Great Train Robbery: Cinema Begins Three pioneering film shorts: "Billy Whiskers," a collection of Lumiere footage from their most famous films, and the title short. **1903 42m/B** Billy Anderson; *Dir:* Edwin S. Porter, Auguste Lumiere, Hans A. Spanuth. **VHS, Beta, 8mm $24.95** *VYY*

Great Treasure Hunt Four Western rogues help a blind man steal a wealthy crook's gold.

197? 90m/C Mark Damon, Stan Cooper, Luis Marin; *Dir:* Tonino Ricci. **VHS, Beta $69.95** *TWE* 🎬

The Great Waldo Pepper Low key and (for Hill) less commercial film about a WWI pilot-turned-barnstormer who gets hired as a stuntman for the movies. Features spectacular World War I vintage aircraft flying sequences. **1975 (PG) 107m/C** Robert Redford, Susan Sarandon, Margot Kidder, Bo Svenson, Scott Newman, Geoffrey Lewis, Edward Herrmann; *Dir:* George Roy Hill. **VHS, Beta, LV $19.95** *MCA* 🎬🎬🎬

A Great Wall A Chinese-American family travels to mainland China to discover the country of their ancestry and to visit relatives. They experience radical culture shock. Wang's first independent feature. In English and Chinese with subtitles. **1986 (PG) 103m/C** Peter Wang, Sharon Iwai, Kelvin Han Yee, Lin Qinqin; *Dir:* Peter Wang. **VHS, Beta, LV $29.95** *PAV, FCT, IME* 🎬🎬🎬

Great Wallendas The true story of the tragedies and triumphs of the Wallendas, a seven-person acrobatic family, who were noted for creating a pyramid on the high wire without nets below them. Made for television. **1978 96m/C** Lloyd Bridges, Britt Ekland, Cathy Rigby; *Dir:* Larry Elikann. **VHS, Beta $59.95** *LTG* 🎬🎬

The Great Waltz The first of two musical biographies on the life of Johann Strauss sees the musician quit his banking job to pursue his dream of becoming a successful composer. A fine, overlooked production rich with wonderful music and romantic comedy. **1938 102m/B** Luise Rainer, Fernand Gravet, Milza Korjus, Hugh Herbert, Lionel Atwill, Curt Bois, Leonid Kinskey, Al Shean, Minna Gombell, George Houston, Bert Roach, Herman Bing, Alma Kruger, Sig Rumann; *Dir:* Julien Duvivier. **VHS $29.98** *FOX, FCT* 🎬🎬🎬

The Great War Two Italian soldiers find themselves in the midst of World War I, much against their will. They try a number of schemes to get out of fighting and working, but their circumstance dictates otherwise. Slightly muddled, but generally good acting. **1959 118m/B** *FR IT* Vittorio Gassman, Alberto Sordi, Silvana Mangano, Folco Lulli; *Dir:* Mario Monicelli. **VHS $69.95** *FCT* 🎬🎬

Great White Death Join Glenn Ford as he searches for the Great White Shark in a real life underwater adventure. **1981 (PG) 88m/C** Glenn Ford. **VHS, Beta $59.95** *GEM, MRV* 🎬

The Great White Hope A semi-fictionalized biography of boxer Jack Johnson, played by Jones, who became the first black heavyweight world champion in 1910. Alexander makes her film debut as the boxer's white lover, as both battle the racism of the times. Two Oscar-nominated performances in what is essentially an "opened-out" version of the Broadway play. **1970 (PG) 103m/C** James Earl Jones, Jane Alexander, Lou Gilbert, Joel Fluellen, Chester Morris, Robert Webber, Hal Holbrook, R.G. Armstrong, Moses Gunn, Scatman Crothers; *Dir:* Martin Ritt. **VHS, Beta $39.98** *FOX* 🎬🎬½

The Great Ziegfeld The big, bio-pic of the famous showman; acclaimed in its time as the best musical biography ever done, still considered the textbook for how to make a musical. Look for cameo roles by many famous stars, including Fanny Brice, as well

as a walk on role by future First Lady Pat Nixon. The movie would have stood up as a first-rate biography of Ziegfeld, even without the drop-dead wonderful songs, including "You Gotta Pull Strings," "My Man," "You," and "A Pretty Girl Is Like a Melody." Oscar nominated for Best Director, Best Original Story, Best Interior Decoration, Best Film Editing.
1936 179m/B William Powell, Luise Rainer, Myrna Loy, Frank Morgan, Reginald Owen, Nat Pendleton, Ray Bolger, Fannie Brice, Virginia Bruce, Harriet Hocter, Ernest Cossart, Robert Greig, Gilda Gray, Leon Errol, Dennis Morgan, Mickey Daniels, William Demarest; **Dir:** Robert Z. Leonard. Academy Awards '36: Best Actress (Rainer), Best Picture; New York Film Critics Awards '36: Best Actress (Rainer). **VHS, Beta, LV $29.98** MGM, BTV ✠✠✠½

The Greatest The filmed autobiography of Cassius Clay, the fighter who could float like a butterfly and sting like a bee. Ali plays himself, and George Benson's hit "The Greatest Love of All" is introduced.
1977 (PG) 100m/C Muhammad Ali, Robert Duvall, Ernest Borgnine, James Earl Jones, John Marley, Roger E. Mosley, Dina Merrill, Paul Winfield; **Dir:** Tom Gries. **VHS, Beta $14.95** COL ✠½

The Greatest Clown Acts of All Time A compilation of various renowned circus clown acts.
1987 30m/C VHS, Beta $14.95 IVE

The Greatest Fights of Martial Arts A compilation of foreign filmdom's best head-kicking matches.
1987 50m/C VHS, Beta $9.95 SAT

The Greatest Flying Heroes Its up, up, and away with Superman, Flash Gordon, Rocky Jones and others as these flying heroes take to the skies. Also included is a sequence on man's early attempts at flight.
1991 50m/C VHS $14.98 WEM

Greatest Heroes of the Bible Two Bible stories: Moses receiving the Ten Commandments; and the story of how Samson lost and regained his extraordinary strength.
1979 95m/C VHS, Beta $34.95 VID

The Greatest Man in the World James Thurber's tale of an illiterate, incorrigible lout who, upon becoming the first man to fly non-stop around the world, receives immediate national attention. Part of the American Short Story Collection.
1980 51m/C Brad Davis, William Prince, John McMartin, Howard da Silva, Carol Kane. **VHS, Beta $29.95** MON, MTI, KVI ✠½

Greatest Show on Earth DeMille, who received an Oscar nomination, is in all his epic glory here in a tale of a traveling circus wrought with glamour, romance, mysterious clowns, a tough ringmaster, and a train wreck. Also received a Film Editing nomination.
1952 149m/C Betty Hutton, Cornel Wilde, James Stewart, Charlton Heston, Dorothy Lamour, Lawrence Tierney; **Dir:** Cecil B. DeMille. Academy Awards '52: Best Picture, Best Story; Golden Globe Awards '53: Best Film—Drama. **VHS, Beta, LV $29.95** PAR, LDC, BTV ✠✠✠

The Greatest Story Ever Told Christ's journey from Galilee to Golgotha is portrayed here in true international-all-star-epic treatment by director Stevens. A lackluster version of Christ's life, remarkable only for Heston's out-of-control John the Baptist.

1965 196m/C Max von Sydow, Charlton Heston, Sidney Poitier, Claude Rains, Jose Ferrer, Telly Savalas, Angela Lansbury, Dorothy McGuire, John Wayne, Donald Pleasence, Carroll Baker, Van Heflin, Robert Loggia, Shelley Winters, Ed Wynn, Roddy McDowall; **Dir:** George Stevens. National Board of Review Awards '65: 10 Best Films of the Year; Harvard Lampoon Awards '65: Worst Film of the Year. **VHS, Beta, LV $29.98** FOX, PIA, RDG ✠✠

Greed A wife's obsession with money drives her husband to murder. Highly acclaimed. Although the original version's length (eight hours) was trimmed to 140 minutes, it remains one of the greatest silent films ever made. Effective use of Death Valley locations. Adapted from the Frank Norris novel, "McTeague."
1924 140m/B Dale Fuller, Gibson Gowland, ZaSu Pitts, Jean Hersholt, Chester Conklin; **Dir:** Erich von Stroheim. **VHS, LV $29.95** MGM ✠✠✠✠

Greed & Wildlife: Poaching in America Illegal hunting kills hundreds of innocent and endangered species each year. Film witnesses the tragedy and discusses methods of stopping the carnage.
1990 60m/C Nar: Richard Chamberlain. **VHS $29.98** VES

The Greed of William Hart Reworking of the Burke and Hare legend has grave robbers providing Edinburgh medical students with the requisite cadavers. Also known as "Horror Maniacs."
1948 78m/B GB Tod Slaughter, Henry Oscar, Jenny Lynn, Winifred Melville; **Dir:** Oswald Mitchell. **VHS $19.98** SNC ✠✠

Greedy Terror A cross-country motorcycle trip becomes a real nightmare when two young men realize they're being stalked by a maniac with a thirst for blood.
1978 90m/C Michael MacRae, Perry Lang. **VHS, Beta** NAV ✠

Greek Street The owner of small cafe in London discovers a poor girl singing in the street for food. He takes her in and spotlights her songs in his cafe. Also known as "Latin Love."
1930 51m/B GB Sari Maritza, Arthur Ahmbling, Martin Lewis; **Dir:** Sinclair Hill. **VHS, Beta $19.98** DVT, VYY ✠✠

The Greek Tycoon The widow of an American president marries a billionaire shipping magnate and finds that money cannot buy happiness. A transparent depiction of the Onassis/Kennedy marriage, done to a turn.
1978 (R) 106m/C Anthony Quinn, Jacqueline Bisset, James Franciscus, Raf Vallone, Edward Albert; **Dir:** J. Lee Thompson. **VHS, Beta $59.95** MCA ✠

Green Archer Fifteen episodes of the famed serial, featuring a spooked castle complete with secret passages and tunnels, trapdoors, and the mysterious masked figure, the Green Archer.
1940 283m/B Victor Jory, Iris Meredith, James Craven, Robert Fiske; **Dir:** James W. Horne. **VHS, Beta $49.95** NOS, VYY, VCN ✠✠

The Green Berets Based on Robin Moore's novel. Wayne stars as a Special Forces colonel, leading his troops against the Viet Cong. Painfully insipid pro-war propaganda, notable as the only American film to come out in support of U.S. involvement in Vietnam. Truly embarrassing but did spawn the hit single "Ballad of the Green Beret" by Barry Sadler.

1968 (G) 135m/C John Wayne, David Janssen, Jim Hutton, Aldo Ray, George Takei, Raymond St. Jacques; **Dir:** John Wayne. **VHS, Beta, LV $19.98** WAR, TLF, BUR ✠

Green Card Some marry for love, some for money, others for an apartment in the Big Apple. Refined but socially aware McDowell covets a commodious rent-controlled apartment in Manhattan. The lease stipulates that the apartment be let to a married couple. Enter brusk and burly Depardieu, who covets that elusive green card. To secure said lease and card, they consent to jump the broom. The counterpoint between McDowell, as a stuffy horticulturist, and Depardieu, as a French composer who has none of Cyrano's refinement, works well, and the two adroitly play a couple whose relationship covers the romantic continuum from loveless to love/hate to love less hate. Weir—who earlier directed "Picnic at Hanging Rock" and "Dead Poet's Society," among others—is adept at creating an updated screwball-type comedy (he wrote the screenplay as well, with Depardieu in mind). If "cute story" and "happy ending" are descriptors you look for in a movie, you should find Depardieu's English-language debut entertaining. Nominated for three Golden Globe Awards in Musical/Comedy category: Best Picture, Best Actor (Depardieu) and Best Actress (MacDowell). Also nominated for an Oscar for Best Screenplay.
1990 (PG-13) 108m/C Andie MacDowell, Gerard Depardieu, Bebe Neuwirth, Gregg Edelman; **W/Dir:** Peter Weir. Golden Globe Awards '90: Best Actor (Depardieu), Best Film—Musical/Comedy. **VHS, Beta $19.99** TOU ✠✠½

Green for Danger A detective stages a chilling mock operation to find a mad killer stalking the corridors of a British hospital during World War II. Sim's first great claim to fame (pre-"A Christmas Carol").
1947 91m/B GB Trevor Howard, Alastair Sim, Leo Genn; **Dir:** Sidney Gilliat. **VHS, Beta, LV $59.95** LCA, RXM ✠✠

Green Dolphin Street Flimsy romantic epic set in 19th century New Zealand. A young girl marries the beau she and her sister are battling over in this Oscar-winning special effects show which features one big earthquake. Based on the Elizabeth Goudge novel, this was one of MGM's biggest hits in 1947.
1947 161m/B Lana Turner, Van Heflin, Donna Reed, Edmund Gwenn; **Dir:** Victor Saville. Academy Awards '47: Best Special Effects. **VHS, Beta $19.98** MGM ✠✠½

Green Dragon Inn When trying to escort an accused murderer to trial, a Ming court emissary finds himself battling against Kung Fu masters.
1982 94m/C Yue Hwa, Lo Lei, Shang Kuan Ling Fung. **VHS, Beta** WVE ✠

Green Eyes Winfield, an ex-GI, returns to post-war Vietnam to search for his illegitimate son who he believes will have green eyes. A highly acclaimed, made-for-television drama featuring moving performances from Winfield and Jonathan Lippe.
1976 100m/C Paul Winfield, Rita Tushingham, Jonathon Lippe, Victoria Racimo, Royce Wallace, Claudia Bryar; **Dir:** John Erman. **VHS $29.95** IVE ✠✠✠½

Green Fields A quiet romance based on Peretz Hirschbein's legendary tale of a young scholar of the Talmud who leaves the shelter of the synagogue in order to learn about people in the real world. The search takes him to the countryside where he finds himself

in the middle of a battle between two families who both want him as a tutor and a suitor for their daughters. In Yiddish with English subtitles. Also known as "Grine Felder."
1937 95m/B Michael Goldstein, Herschel Bernardi, Helen Beverly; *Dir:* Jacob Ben-Ami. **VHS, Beta $79.95** ERG, NCJ *ZZ*½

The Green Glove A jewel thief steals a beautiful relic from a tiny church. WWII makes it impossible for him to fence it and the church finds a relentless ally to track the thief. Excellent action sequences lift this standard plot slightly above the ordinary.
1952 88m/B Glenn Ford, Geraldine Brooks, Cedric Hardwicke, George Macready, Gaby Andre, Roger Treville, Juliette Greco, Jean Bretonniere; *Dir:* Rudolph Mate. **VHS $19.95** NOS, FCT, TIM *ZZ*½

Green Grow the Rushes The English government tries to put a stop to brandy smuggling on the coast, but the townspeople drink up the evidence. A young Richard Burton highlights this watered-down comedy, also known as "Brandy Ashore." Based on a novel by co-scripter Harold Clewes.
1951 77m/B GB Roger Livesey, Honor Blackman, Richard Burton, Frederick Leister; *Dir:* Derek Twist. **VHS $19.95** NOS, TIM *ZZ*

Green Horizon A bucolic old codger faces a dilemma when he must choose between his granddaughter's future and his disaster-prone natural environment.
1983 80m/C James Stewart, Philip Sayer, Elenora Vallone. **VHS, Beta $29.95** FHE *Z*½

The Green Hornet The feature version of the original serial, in which the Green Hornet and Kato have a series of crime-fighting adventures.
1940 100m/B Gordon Jones, Keye Luke, Anne Nagel, Wade Boteler, Walter McGrail, Douglas Evans, Cy Kendall. **VHS, Beta $49.95** NOS, GKK, GPV *ZZZ*

Green Ice An American electronics expert gets involved with a brutal South American government dealing with emeralds and plans a heist with his girlfriend and other cohorts. A little bit of romance, a little bit of action, a lot of nothing much.
1981 109m/C GB Ryan O'Neal, Anne Archer, Omar Sharif, John Larroquette; *Dir:* Ernest Day. **VHS, Beta, LV $19.95** JTC *Z*½

Green Inferno A wealthy eccentric who lives deep in the South American jungle incites the natives to rebel. Mercenary agents are then hired to kill him.
198? 90m/C Richard Yesteran, Didi Sherman, Caesar Burner. **VHS, Beta $59.95** MGL *Z*

The Green Man Maurice Allington (Finney), the alcoholic and randy owner of the Green Man Inn in rural England, jokes about the spirits that supposedly haunt the inn. It's not too funny, however, when he discovers the spirits are real and he has to figure out who the spirit of a 17th-century murderer is inhabiting before he kills again. A classy ghost story based on the novel by Kingsley Amis.
1991 150m/C GB Albert Finney, Sarah Berger, Linda Marlowe, Michael Hordern, Nickolas Grace; *Dir:* Elijah Moshinsky. **VHS $89.95** AAE, TAM, ART *ZZZ*

Green Pastures An adaptation of Marc Connelly's 1930 Pulitzer Prize-winning play which attempts to retell Biblical stories in Black English Vernacular of the '30s. Southern theater owners boycotted the controversial film which had an all-Black cast.

1936 93m/C Rex Ingram, Oscar Polk, Eddie Anderson, George Reed, Abraham Graves, Myrtle Anderson, Frank Wilson; *Dir:* William Keighley, Marc Connelly. National Board of Review Awards '36: 10 Best Films of the Year. **VHS, Beta $59.98** FOX *ZZZ*

The Green Promise A well-meaning but domineering farmer and his four motherless children face disaster as a result of the father's obstinacy. But when he is laid up, the eldest daughter takes over, tries modern methods and equipment, and makes the farm a success.
1949 90m/B Walter Brennan, Marguerite Chapman, Robert Paige, Natalie Wood; *Dir:* William D. Russell. **VHS, Beta $24.95** NOS, DVT, TIM *ZZ*

Green Room Truffaut's haunting tale of a man who, valuing death over life, erects a shrine to all the dead he has known. Offered a chance at love, he cannot overcome his obsessions to take the risk. Based on the Henry James story "Altar of the Dead." In French with subtitles.
1978 94m/C FR Francois Truffaut, Nathalie Baye, Jean Daste; *Dir:* Francois Truffaut. New York Film Festival '78: 10 Best Films of the Year. **VHS, Beta** WAR, OM *ZZ*

The Green Wall A city-dwelling young man attempts to homestead with his family in the Amazon. A lovely and tragic film with stunning photography of the Peruvian forests. First Peruvian feature to be shown in the U.S. In Spanish, with English subtitles.
1970 110m/C SP Julio Aleman, Sandra Riva; *W/ Dir:* Armando Robles. **VHS $79.95** FCT, DVT *ZZZ*

Greenstone The green stone is a glowing rock in the forbidden forest that conveys a young boy into an exciting new world.
1985 (G) 48m/C Joseph Corey, John Riley, Kathleen Irvine, Jack Mauck; *Dir:* Kevin Irvine. **VHS, Beta $19.95** AHV *ZZ*

Greetings De Niro stars in this wild and crazy comedy about a man who tries to help his friend flunk his draft physical. One of De Palma's first films, and a pleasantly anarchic view of the late '60s, wrought with a light, intelligent tone. Followed by the sequel "Hi, Mom."
1968 (R) 88m/C Robert De Niro, Gerrit Graham, Allen Garfield; *Dir:* Brian De Palma. **VHS, Beta $79.95** VMK *ZZZ*

Gregory Issacs Live Gregory Issacs is backed by the Riddim Kings as he performs "Love is Overdue," "Turn Me On," "Slave Master," "Out What a Feeling," and more.
1987 60m/C Gregory Issacs. **VHS, Beta $19.95** MVD, KRV *ZZZ*

Gregory's Girl The sweet, disarming comedy that established Forsyth. An awkward young Scottish schoolboy falls in love with the female goalie of his soccer team. He turns to his ten-year-old sister for advice, but she's more interested in ice cream than love. His best friend is no help, either, since he has yet to fall in love. A perfect mirror of teenagers and their instantaneous, raw, and all-consuming loves. Very sweet scene with Gregory and his girl lying on their backs in an open space illustrates the simultaneous simplicity and complexity of young love.
1980 91m/C GB Gordon John Sinclair, Dee Hepburn, Jake D'Arcy, Chic Murray, Alex Norton, John Bett, Claire Grogan; *Dir:* Bill Forsyth. **VHS, Beta $69.98** SUE *ZZZ*½

Gremlins Comedy horror with deft satiric edge. Produced by Spielberg. Fumbling gadget salesman Rand Peltzer is looking for something really special to get his son Billy.

He finds it in a small store in Chinatown. The wise shopkeeper is reluctant to sell him the adorable "mogwai" but relents after fully warning him "Don't expose him to bright light, don't ever get him wet, and don't ever, ever feed him after midnight." Naturally, this all happens and the result is a gang of nasty gremlins who decide to tear up the town on Christmas Eve. Followed by "Gremlins 2, the New Batch" which is less black comedy, more parody, and perhaps more inventive.
1984 (PG) 106m/C Zach Galligan, Phoebe Cates, Hoyt Axton, Polly Holliday, Frances Lee McCain, Keye Luke, Dick Miller, Corey Feldman, Judge Reinhold; *Dir:* Joe Dante. **VHS, Beta, LV, 8mm $19.98** WAR, TLF *ZZZ*

Gremlins 2: The New Batch The sequel to "Gremlins" is superior to the original, which was quite good. Set in a futuristic skyscraper in the Big Apple, director Dante presents a less violent but far more campy vision, paying myriad surreal tributes to scores of movies, including "The Wizard of Oz" and musical extravaganzas of the past. Also incorporates a Donald Trump parody, takes on television news, and body slams modern urban living. Great fun.
1990 (PG-13) 107m/C Phoebe Cates, Christopher Lee, John Glover, Zach Galligan; *Dir:* Joe Dante. **VHS, Beta, LV, 8mm $19.95** WAR, HHE *ZZZ*½

Grendel, Grendel, Grendel An utterly urbane dragon wants to be friends, but for some reason people are just terrified of Grendel. It's true, he bites a head off once in a while, but nobody's perfect! An ingenious, animated retelling of Beowulf.
1982 90m/C AU *Dir:* Alexander Stitt; *Voices:* Peter Ustinov, Arthur Dignam, Julie McKenna, Keith Michell. **VHS, Beta $20.05** FHE *ZZ*

Greta During the second World War, a young Polish boy escapes from a detention camp. Now on the run, several bizarre twists of fate place him in the care of Greta, a young girl whose father is an officer in the S.S. In Polish with English subtitles.
1986 60m/C PL Janusz Grabowski, Agnieszka Kruszewska, Eva Borowik, Tomasz Grochoczki, Andrzej Precig; *Dir:* Krzysztof Gruber. **VHS, Beta $39.50** EVD *ZZZ*

The Grey Fox A gentlemanly old stage-coach robber tries to pick up his life after thirty years in prison. Unable to resist another heist, he tries train robbery, and winds up hiding out in British Columbia where he meets an attractive photographer, come to document the changing West. Farnsworth is perfect as the man who suddenly finds himself in the 20th century trying to work at the only craft he knows. Based on the true story of Canada's most colorful and celebrated outlaw, Bill Miner.
1983 (PG) 92m/C CA Richard Farnsworth, Jackie Burroughs, Wayne Robson, Timothy Webber, Ken Pogue; *Dir:* Phillip Borsos. Genie Awards '83: Best Art Direction/Set Decoration, Best Director (Borsos), Best Picture, Best Screenplay, Best Supporting Actress (Burroughs), Best Musical Score. **VHS, Beta, LV $19.95** MED *ZZZ*½

Grey Gardens For Jackie Kennedy fans, a documentary sure to shock. Filmed in her 79-year-old aunt's rotting Long Island mansion, we're allowed to witness the reclusive Big Edith and her 56-year-old daughter, Little Edie, singing, dancing, and fighting in this most bizarre of documentaries.
1976 94m/C *Dir:* Al Maysles, David Maysles. **VHS $79.95** FCT

Greyfriars Bobby A true story of a Skye terrier named Bobby who, after his master dies, refuses to leave the grave. Even after being coaxed by the local children into town, he still returns to the cemetery each evening. Word of his loyalty spreads, and Bobby becomes the pet of 19th century Edinburgh. Nicely told, with fine location photography and good acting. Great for children and animal lovers.
1961 91m/C Donald Crisp, Laurence Naismith, Kay Walsh; *Dir:* Don Chaffey. **VHS $69.95** DIS 🐾🐾½

Greystoke: The Legend of Tarzan, Lord of the Apes The seventh Earl of Greystoke becomes a shipwrecked orphan and is raised by apes. Ruling the ape-clan in the vine-swinging persona of Tarzan, he is discovered by an anthropologist and returned to his ancestral home in Scotland, where he is immediately recognized by his grandfather. The contrast between the behavior of man and ape is interesting, and Tarzan's introduction to society is fun, but there's no melodrama or cliff-hanging action, as we've come to expect of the Tarzan genre. Due to her heavy southern accent, Andie McDowell (as Jane) had her voice dubbed by Glenn Close.
1984 (PG) 130m/C Christopher Lambert, Ralph Richardson, Ian Holm, James Fox, Andie MacDowell, Ian Charleson, Cheryl Campbell, Nigel Davenport; *Dir:* Hugh Hudson. **VHS, Beta, LV $19.98** WAR 🐾🐾

Grievous Bodily Harm Morris Martin's wife Claudine is missing. When he finds a videotape of Claudine in very compromising positions, he becomes murderously determined to find out where she has gone.
1989 (R) 136m/C John Waters, Colin Friels, Bruno Lawrence; *Dir:* Mark Joffe. **VHS, Beta $29.95** FRH 🐾🐾

Griffin and Phoenix: A Love Story A sentimental film about the mortality of two vital people. Falk deserts his family when he learns he has terminal cancer. Clayburgh, also terminally ill, meets him for a short but meaningful affair. Good performances help overcome the tear-jerker aspects of the story. Made for television.
1976 110m/C Peter Falk, Jill Clayburgh, Dorothy Tristan, John Lehne; *Dir:* Daryl Duke. **VHS $59.98** FOX 🐾🐾½

Griffith Biograph Program Some of the best Biograph shorts are featured: "The Lonely Villa," "The Lonedale Operator," "The New York Hat," "Friends," and "The Unseen Enemy," all produced between 1909 and 1912.
191? 50m/B Mary Pickford, Blanche Sweet, Lillian Gish, Dorothy Gish; *Dir:* D.W. Griffith. **VHS, Beta, LV** GVV

The Grifters Evocative rendering of a terrifying sub-culture. Huston, Cusak, and Bening are con artists, struggling to stay on top in a world where violence, money, and lust are the prizes. Bening and Huston fight for Cusak's soul, while he's determined to decide for himself. Seamless performances and dazzling atmosphere, along with superb pacing, make for a provocative film. Based on a novel by Jim Thompson.
1990 (R) 114m/C Anjelica Huston, John Cusack, Annette Bening, Pat Hingle, J.T. Walsh, Charles Napier, Henry Jones, Gailard Sartain; *Dir:* Stephen Frears. Los Angeles Film Critics Association Awards '90: Best Actress (Huston); National Society of Film Critics Awards '90: Best Actress (Huston), Best Supporting Actress (Bening). **VHS, Beta, LV $92.99** MCA, HBO, PIA 🐾🐾🐾½

Grim Prairie Tales A city slicker and a crazed mountain man meet by a campfire in the high plains one night and pass the time sharing four tales of Western horror. Dourif and Jones are fun, the stories scary, making for an interesting twist on both the horror and the Western genres.
1989 (R) 90m/C Brad Dourif, James Earl Jones, Marc McClure, William Atherton, Scott Paulin; *W/ Dir:* Wayne Coe. **VHS $29.95** ACA, FCT

Grim Reaper Mia Farrow's sister is one of a group of American students vacationing on a Greek island. Farrow learns of the twisted cannibalistic murderer who methodically tries to avenge the tragic deaths of his family.
1981 (R) 81m/C IT Tisa Farrow, George Eastman; *Dir:* Joe D'Amato. **VHS, Beta $39.95** MON 🐾½

The Grim Reaper (La Commare Secca) 22-year-old Bertolucci directed his first feature with this grim and brutal treatment of the investigation of the murder of a prostitute, told via flashbacks and from the disparate perspectives of three people who knew the woman. A commercial bust, the critics were no kinder than the public at the time; not until two years later, with "Before the Revolution," did Bertolucci earn his directorial spurs. The script was written by Pasolini, with whom the erstwhile poet had collaborated on "Accattone" the previous year (as assistant). In Italian with English subtitles; also known as "La Commare Secca."
1962 100m/B IT Francesco Rulu, Giancarlo de Rosa; *Dir:* Bernardo Bertolucci. **VHS, LV $79.95** CVC, FCT 🐾🐾🐾

The Grissom Gang Remake of the 1948 British film "No Orchids for Miss Blandish." Darby is a wealthy heiress kidnapped by a family of grotesques, led by a sadistic mother. The ransom gets lost in a series of bizarre events, and the heiress appears to be falling for one of her moronic captors. Director Robert Aldrich skillfully blends extreme violence with dark humor. Superb camera work, editing, and 1920's period sets.
1971 (R) 127m/C Kim Darby, Scott Wilson, Tony Musante, Ralph Waite, Connie Stevens, Robert Lansing, Wesley Addy; *Dir:* Robert Aldrich. **VHS $29.99** FOX 🐾🐾🐾

Grit of the Girl Telegrapher/In the Switch Tower A pair of silent railroad dramas, which feature thrilling chase sequences, nefarious schemers and virginal heroines. Silent with piano scores by Jon Mirsalis.
1915 47m/B Anna Q. Nilsson, Hal Clements, Walter Edwards, Frank Borzage. **VHS, Beta $19.98** CCB, PEN

Grizzly A giant, killer grizzly terrorizes a state park in this blatant "Jaws" rip-off. From the same folks who produced "Abby," a blatant "Exorcist" rip-off. Filmed in Georgia. Also known as "Killer Grizzly."
1976 (PG) 92m/C Christopher George, Richard Jaeckel, Andrew Prine, Victoria (Vicki) Johnson, Charles Kissinger; *Dir:* William Girdler. **VHS, Beta $9.95** MED 🐾

Grizzly Adams: The Legend Continues A small town is saved from three desperados by Grizzly and his huge grizzly bear pet, Martha.
1990 90m/C Gene Edwards, Link Wyler, Red West, Tony Caruso, Acquanetta, L.Q. Jones; *Dir:* Ken Kennedy. **VHS $79.95** NVH, QUE 🐾½

Grizzly Golfer When Magoo and his nephew Waldo tee off at the Ozark Hollow Golf Club, a bear becomes involved in the game. Part of the "Mister Magoo" series.
1951 7m/C Voices: Jim Backus. **VHS, Beta** CHF

Grizzly Mountain A man moves to the Sierra Mountains and befriends an Indian.
1980 96m/C **VHS, Beta** VDY 🐾

The Groove Tube A television series called "The Groove Tube" is the context for these skits that spoof everything on television from commercials to newscasts. Chevy Chase's movie debut.
1972 (R) 75m/C Lane Sarasohn, Chevy Chase, Richard Belzer, Marcy Mendham, Bill Kemmill; *Dir:* Ken Shapiro. **VHS, Beta $19.95** MED 🐾🐾½

Gross Anatomy Light-weight comedy/drama centering on the trials and tribulations of medical students. Modine is the very bright, but somewhat lazy, future doctor determined not to buy into the bitter competition among his fellow students. His lack of desire inflames Lahti, a professor dying of a fatal disease who nevertheless believes in modern medicine. She pushes and inspires him to focus on his potential, and his desire to help people. Worth watching, in spite of cheap laughs. Interesting cast of up-and-comers.
1989 (PG-13) 107m/C Matthew Modine, Daphne Zuniga, Christine Lahti, John Scott Clough, Alice Carter, Robert Desiderio, Zakes Mokae, Todd Field; *Dir:* Thom Eberhardt. **VHS, Beta, LV $19.99** BVV 🐾🐾½

Gross Jokes The comic producers used Julius Alvin's best-selling book as the inspiration for this joke-fest. Filmed at L.A.'s Improv.
1985 53m/C Tommy Sledge, Barry Diamond, Budd Friedman. **VHS, Beta $29.95** MCA

Grotesque After slaughtering a young woman's family as they vacationed in a remote mountain cabin, a gang of bloodthirsty punks are in turn killed by the family's secret deformed son. The title says it all.
1987 (R) 79m/C Linda Blair, Tab Hunter, Guy Stockwell, Donna Wilkes, Nels Van Patten; *Dir:* Joe Tornatore. **VHS, Beta, LV $79.95** MED Woof!

Groucho/Howdy Doody Show Episodes from television shows featuring Groucho Marx and Howdy Doody are featured.
19?? 60m/B Groucho Marx, Howdy Doody. **VHS, Beta $39.95** VCN

Ground Zero Ground Zero is the term used to describe the site of a nuclear explosion. San Francisco is threatened with just such a new name in this story of nuclear blackmail. Someone plants an atomic bomb on the Golden Gate Bridge. It must be defused and the culprit caught.
1988 85m/C Ron Casteel, Melvin Belli, Yvonne D'Angiers; *Dir:* James T. Flocker. **VHS, Beta $39.95** GHV 🐾½

Ground Zero Political thriller in which a man searches for the reasons for his father's murder in the films of the British nuclear tests in the Australian outback in the fifties.
1988 (R) 109m/C AU Colin Friels, Jack Thompson, Donald Pleasence, Natalie Bate, Simon Chilvers, Neil Fitzpatrick, Bob Maza; *Dir:* Michael Pattinson, Bruce Myles. **VHS, Beta, LV $14.95** IVE 🐾🐾🐾

The Groundstar Conspiracy Spy thriller brings to life L.P. Davies' novel, "The Alien." After an explosion kills all but one space project scientist, Peppard is sent to investigate suspicions of a cover-up. Meanwhile, the surviving scientist (Sarrazin) suffers from disfigurement and amnesia. He pursues his identity while Peppard accuses him of being a spy. Splendid direction by Lamont Johnson. Sarrazin's best role.
1972 (PG) 96m/C *CA* George Peppard, Michael Sarrazin, Christine Belford, Cliff Potts, James Olson, Tim O'Connor, James McEachin, Alan Oppenheimer; *Dir:* Lamont Johnson. **VHS $59.95** *MCA ♂♂♂*

The Group Based upon the novel by Mary McCarthy, the well-acted story deals with a group of graduates from a Vassar-like college as they try to adapt to life during the Great Depression. Has some provocative subject matter, including lesbianism, adultery, and mental breakdowns. Soapy, but a good cast with film debuts of Bergen, Hackett, Pettet, Widdoes, and Holbrook.
1966 150m/C Candice Bergen, Joanna Pettet, Shirley Knight, Joan Hackett, Elizabeth Hartman, Jessica Walter, Larry Hagman, Kathleen Widdoes, Hal Holbrook, Mary-Robin Redd, Richard Mulligan, James Broderick, Carrie Nye; *Dir:* Sidney Lumet. **VHS, Beta $59.98** *FOX ♂♂½*

Group Marriage Six young professionals fall into a marriage of communal convenience and rapidly discover the many advantages and drawbacks of their thoroughly modern arrangement.
1972 (R) 90m/C Victoria Vetri, Aimee Eccles, Zack Taylor, Jeff Pomerantz, Claudia Jennings, Jayne Kennedy, Milt Kamen; *Dir:* Stephanie Rothman. **VHS, Beta $39.95** *UHV, VCI ♂♂½*

Growing Pains A young couple discover that their newly adopted son possesses extraordinary powers.
1982 60m/C Gary Bond, Barbara Keilermann, Norman Beaton. **VHS, Beta $29.95** *IVE ♂½*

Grown Ups A journalist watches his family slowly disintegrating around him in this adaptation of Jules Feiffer's play.
1986 106m/C Jean Stapleton, Martin Balsam, Charles Grodin, Marilu Henner, Kerry Segal; *Dir:* John Madden. **VHS, Beta $59.95** *LHV, WAR ♂♂*

GRP All-Stars from the Record Plant An unforgettable collection of jazz greats assembled in Los Angeles to play live at the Record Plant.
19?? 50m/C *Performed by:* Dave Grusin, Lee Ritenour, Carlos Vega, Dave Valentin, Larry Williams, Abraham Laboriel, Ivan Lins, Diane Schuur. **VHS, Beta, LV $29.95** *GRP, MVD, HMV*

Gruesome Twosome Another guts/cannibalism/mutilation fun-fest by Lewis, about someone who's marketing the hair of some very recently deceased college coeds.
1967 75m/C Elizabeth Davis, Chris Martell; *Dir:* Herschel Gordon Lewis. **VHS, Beta $39.98** *RHI ♂*

Grunt! The Wrestling Movie A spoof of documentaries about the behind-the-scenes world of wrestling. For fans only.
1985 (R) 91m/C Wally Greene, Steven Cepello, Dick Murdoch, John Tolos; *Dir:* Allan Holzman. **VHS, Beta $19.95** *STE, NWV Woof!*

Gryphon Ricky's new teacher can do all sorts of magical things. Hidden among her tricks are lessons about creativity, beauty, and imagination. Aired on PBS as part of the WonderWorks Family Movie television series.

1988 58m/C Amanda Plummer, Sully Diaz, Alexis Cruz; *Dir:* Mark Cullingham. **VHS $29.95** *PME, FCT*

Guadalcanal Diary A vintage wartime flag-waver, with a typical crew of Marines battling the "Yellow Menace" over an important base on the famous Pacific atoll. Based on Richard Tregaskis' first-hand account.
1943 93m/B Preston Foster, Lloyd Nolan, William Bendix, Richard Conte, Anthony Quinn, Richard Jaeckel, Roy Roberts, Minor Watson, Miles Mander, Ralph Byrd, Lionel Stander, Reed Hadley; *Dir:* Lewis Seiler. **VHS, Beta, LV $39.98** *FOX, IME ♂♂½*

The Guardian The residents of a chic New York apartment building hire a security expert/ex-military man to aid them in combatting their crime problem. But one of the liberal tenants thinks their guard's methods may come at too high a cost. Made for cable television.
1984 (R) 102m/C Martin Sheen, Louis Gossett Jr., Arthur Hill; *Dir:* David Greene. **VHS, Beta, LV $69.98** *LIV, VES ♂½*

The Guardian A young couple unwittingly hires a human sacrificing druid witch as a babysitter for their child. Based on the book "The Nanny" by Dan Greenburg.
1990 (R) 92m/C Jenny Seagrove, Dwier Brown, Carey Lowell; *Dir:* William Friedkin. **VHS, Beta, LV $19.98** *MCA ♂*

Guardian of the Abyss A young couple who buy an antique mirror get more then they bargained for when they discover that it is the threshold to Hell and the devil wants to come through.
1982 60m/C Ray Lonnen, Rosalyn Landor, Paul Darrow, Barbara Ewing; *Dir:* Don Sharp. **VHS, Beta $29.95** *IVE ♂*

Guerilla Brigade A rarely seen Ukrainian epic about Russian/Ukrainian solidarity during the Civil War. In Russian with English subtitles. Also called "Riders."
1939 90m/B *RU Dir:* Igor Savchenko. **VHS, Beta $59.95** *FCT*

Guess What We Learned in School Today? Parents in a conservative suburban community protest sex education in the schools. Intended as a satire, but it only reinforces stereotypes.
1970 (R) 85m/C Richard Carballo, Devin Goldenbrry; *Dir:* John G. Avildsen. **VHS, Beta** *PGN Woof!*

The Guess Who Reunion Before a live audience in Toronto, the original Guess Who perform the hits that made them worldwide superstars. Included are performances of "Shakin' All Over," "These Eyes," and "American Woman."
1983 118m/C VHS, Beta $19.95 *MVD, MSM Woof!*

Guess Who: Together Again Canadian band reunites to perform some of their hits, including "American Woman," "These Eyes," "No Time," and "Share the Land."
1990 53m/C VHS, LV $29.95 *PIA*

Guess Who's Coming to Dinner Controversial in its time. A young white woman brings her black fiance home to meet her parents. The situation truly tests their open-mindedness and understanding. Hepburn and Tracy (in his last film appearance) are wonderful and serve as the anchors in what would otherwise have been a rather sugary film (although the screenplay won an Academy Award). Houghton, who portrays the independent daughter, is the real-life

niece of Hepburn who garnered an Oscar for her performance. Other Oscar nominations for Best Picture, Best Actor (Tracy), Supporting Actor (Kellaway), Supporting Actress (Richards), Director, Art Direction, Set Decoration, and Film Editing.
1967 108m/C Katharine Hepburn, Spencer Tracy, Sidney Poitier, Katherine Houghton, Cecil Kellaway; *Dir:* Stanley Kramer. Academy Awards '67: Best Actress (Hepburn), Best Original Screenplay; Harvard Lampoon Awards '67: Worst Film of the Year. **VHS, Beta, LV $14.95** *COL, FCT, COL ♂♂♂*

Guest in the House A seemingly friendly young female patient is invited to stay in the family home of her doctor. She skillfully attempts to dissect the family's harmony in the process.
1944 121m/B Anne Baxter, Ralph Bellamy, Ruth Warrick, Marie McDonald, Margaret Hamilton; *Dir:* John Brahm. **VHS, Beta $19.95** *NOS, CAB ♂♂½*

Guest Wife Determined to impress his sentimental boss, a quick-thinking bachelor talks his best friend's wife into posing as his own wife, but the hoax gets a bit out of hand and nearly ruins the real couple's marriage.
1945 90m/B Claudette Colbert, Don Ameche, Dick Foran, Charles Dingle, Wilma Francis; *Dir:* Sam Wood. **VHS** *REP ♂♂½*

A Guide for the Married Man One suburban husband instructs another in adultery, with a cast of dozens enacting various slapstick cameos. Followed by "A Guide for the Married Woman."
1967 91m/C Walter Matthau, Robert Morse, Inger Stevens, Lucille Ball, Jack Benny, Jeffrey Hunter, Polly Bergen, Sid Caesar, Art Carney, Wally Cox, Jayne Mansfield, Louis Nye, Carl Reiner, Phil Silvers, Terry-Thomas, Sam Jaffe; *Dir:* Gene Kelly. **VHS, Beta $59.98** *FOX ♂♂♂*

A Guide for the Married Woman A bored housewife tries to take the romantic advice of a girlfriend to add a little spice to her life. A poor made-for-television follow-up to "A Guide for the Married Man."
1978 96m/C Cybill Shepherd, Barbara Feldon, Eve Arden, Chuck Woolery, Peter Marshall, Charles Frank; *Dir:* Hy Averback. **VHS, Beta $59.95** *FOX ♂½*

Guido Manuli: Animator (Nightmare at the "Opera") Manuli, collaborator on "Allegro Non Troppo," presents his own intriguing and humorous animation in this superbly done video.
1990 ?m/C *Dir:* Guido Manuli. **VHS $39.95** *EXP*

Guilty as Charged A butcher rigs his own private electric chair and captures and executes paroled killers in his personal quest for justice. But when a politician frames an innocent man for murder how far will this vigilante go?
1991 (R) 95m/C Rod Steiger, Lauren Hutton, Heather Graham, Isaac Hayes; *Dir:* Sam Irvin. **VHS, LV $89.98** *COL ♂½*

Guilty of Innocence A young African-American man finds he's guilty until proven innocent when he runs into the law in Texas. A bold lawyer joins his fight for justice. Based on a true story and made for television.
1987 (PG) 95m/C Dorian Harewood, Dabney Coleman, Hoyt Axton, Paul Winfield, Dennis Lipscomb, Debbi Morgan, Marshall Colt, Victor Love; *Dir:* Richard T. Heffron. **VHS $89.98** *VMK ♂♂*

Guilty by Suspicion Often compelling examination of the 1950s' McCarthy investigations. DeNiro plays a director who has no Communist party connections, but refuses to incriminate anyone else just to get off easily. He re-evaluates his life, his work and his relationships, as the likelihood of their loss becomes more apparent. Strong performance from DeNiro. Supporting characters are not written as thoroughly. Available in widescreen on laserdisc.
1991 (PG-13) 105m/C Robert De Niro, Annette Bening, George Wendt, Patricia Wettig, Sam Wanamaker, Chris Cooper, Barry Primus, Gailard Sartain, Robin Gammell, Brad Sullivan, Ben Piazza, Martin Scorsese, Roxann Biggs, Stuart Margolin, Barry Tubb; **W/Dir:** Irwin Winkler. **VHS, LV, 8mm $92.99** WAR, LDC ♫♫♫

Guilty of Treason A documentary on the life and times of Joszef Cardinal Mindszenty of Hungary, who was imprisoned by the communists as an enemy of the state for speaking out against the totalitarian regime. At the trial, it was revealed that the Cardinal's confession was obtained only by the use of drugs, hypnosis and torture. A realistic look at the dark side of Communism.
1950 86m/B Charles Bickford, Paul Kelly, Bonita Granville, Richard Derr, Berry Kroeger, Elisabeth Risdon, Roland Winters, John Banner; **Dir:** Felix Feist. **VHS $24.95** NOS, MRV, TIM ♫♫½

The Guinness Book of World Records This is a video version of the popular book that chronicles unusual facts about everyone from Jim Brown to the Wright Brothers.
1977 30m/C VHS, Beta $9.98 VID

Guitarras Lloren Guitarras A popular artist gets frustrated when the girl he loves falls for his brother.
1947 111m/B MX Cuco Sanchez, Lucha Villa, Fernando Soto Mantequilla. **VHS, Beta $54.95** MAD ♫♫½

Gulag An American sportscaster is wrongly sentenced to ten years of hard labor in a Soviet prison, and plans his escape from the cruel guards. Good suspense. Made for television.
1985 130m/C GB David Keith, Malcolm McDowell, David Suchet, Warren Clarke; **Dir:** Roger Young. **VHS, Beta $69.95** PSM ♫♫½

Gulliver A farce about an escaped convict who hides out with a community of dwarves, manipulates them, and eventually impels them to murder.
197? 100m/C Fernando Gomez, Yolanda Farr. **VHS, Beta** MAV ♫½

Gulliver's Travels The Fleischer Studio's animated version of Jonathan Swift's classic about the adventures of Gulliver, an English sailor who is washed ashore in the land of Lilliput, where everyone is about two inches tall.
1939 74m/C Dir: Dave Fleischer; **Voices:** Lanny Ross, Jessica Dragonette. **VHS, LV $8.95** NEG, MRV, NOS ♫♫½

Gulliver's Travels In this partially animated adventure the entire land of Lilliput has been constructed in miniature. Cartoon and real life mix in a 3-dimensional story of Dr. Lemuel Gulliver and his discovery of the small people in the East Indies. From the classic by Jonathan Swift.
1977 (G) 80m/C GB Richard Harris, Catherine Schell; **Dir:** Peter Hunt. **VHS, Beta $19.95** UHV, LME, HHE ♫½

Gulliver's Travels An animated version of the Jonathan Swift satire about a sailor whose voyage takes him to an unusual island populated by tiny people.
1979 52m/C VHS, Beta $9.95 WOV, GKK, KUI ♫½

Gumball Rally An unusual assortment of people converge upon New York for a cross country car race to Long Beach, California where breaking the rules is part of the game.
1976 (PG) 107m/C Michael Sarrazin, Gary Busey, Raul Julia, Nicholas Pryor, Tim McIntire, Susan Flannery; **Dir:** Chuck Bail. **VHS, Beta $59.95** WAR ♫♫

Gumby and the Moon Boggles Five episodes featuring the notorious claymation character: "Indian Trouble," "Weight and See," "Mystic Magic," "Gabby Auntie" and the title episode.
1956 30m/C VHS, Beta $14.95 FHE ♫♫

Gumby's Supporting Cast A series featuring the clay-animated adventures of Gumby's buddies, with six 6-minute episodes per program.
1986 42m/C VHS, Beta $14.95 FHE

Gumshoe An homage to American hard-boiled detectives. Finney plays a small-time worker in a Liverpool nightclub who decides to become a detective like his movie heroes. He bumbles through a number of interconnecting cases, but finds his way in the end. Good satire of film noir detectives. Music by Andrew Lloyd Webber.
1972 (G) 85m/C GB Albert Finney, Billie Whitelaw, Frank Finlay, Janice Rule; **Dir:** Stephen Frears. **VHS, Beta $69.95** RCA ♫♫½

The Gumshoe Kid A college-bound kid enters the family detective business to prevent his mother from being evicted. He gets in over his head though, trying to track the fiancee of a wealthy mobster.
1989 (R) 98m/C Jay Underwood, Tracy Scoggins, Vince Edwards, Arlene Golonka, Pamela Springsteen; **Dir:** Joseph Manduke. **VHS, Beta, LV $29.95** ACA ♫

Gun Cargo Muddled story about a conniving shipowner who fires his crew for demanding shore leave. He replaces them with assorted low-lifes, resulting in the usual hijinks. A terrible performance by an otherwise great cast.
1949 49m/C Rex Lease, Smith Ballew, William Farnum, Gibson Gowland. Golden Turkey Award '91: So Bad, It's Good. **VHS, Beta, 8mm $24.95** VYY ♫

Gun Code The white-hatted cowboy-hero puts a stop to small-town racketeering.
1940 52m/B Tim McCoy, Ina Guest; **Dir:** Peter Stewart. **VHS, Beta $19.95** VDM ♫½

Gun Crazy Annie Laurie Starr—an ersatz Annie Oakley in a Wild West show—meets gun-coveting Bart Tare, who says a gun makes him feel good inside, "like I'm somebody." Sparks fly, and the two get married and live happily ever after for a couple of nanoseconds until the money starts to run out and fatal femme Laurie's craving for excitement and violence starts to flare up. She convinces Bart to turn to robbery, and the two become lovebirds-of-the-lam (and inspiring "Bonnie and Clyde"). Needless to say, they find out in no uncertain terms that crime doesn't pay. But not before Laurie realizes (perhaps) that she really does love Bart. The sleeper's original title, "Deadly is the Female," belies its noir streak. Now a cult fave, it's based on a MacKinlay Kantor story that first appeared in the Saturday Evening Post. Ex-stuntman Russell Harlan's photography is daring, and the realism of the impressive robbery scenes owes in part to the technical consultation of octogenarian former train robber Al Jennings. And watch for a Rusty (Russ) Tamblyn as the 14-year-old Bart.
1950 87m/B Peggy Cummins, John Dall, Berry Kroeger, Morris Carnovsky, Anabel Shaw, Nedrick Young, Trevor Bardette; **Dir:** Joseph H. Lewis. **VHS $19.98** FOX, FCT ♫♫♫

Gun Crazy A gambler becomes enmeshed in an arranged marriage with a cursed Mexican family. Lightweight comedy, good cast. Originally titled, "A Talent for Loving."
1969 110m/C Richard Widmark, Chaim Topol, Cesar Romero, Genevieve Page, Judd Hamilton, Caroline Munro; **Dir:** Richard Quine. **VHS $9.95** SIM ♫♫½

Gun Fury A panoramic western about a Civil War veteran pursuing the bandits who kidnapped his beautiful bride-to-be. First screened as a 3-D movie.
1953 83m/C Rock Hudson, Donna Reed, Phil Carey, Lee Marvin; **Dir:** Raoul Walsh. **VHS, Beta, LV $14.95** COL ♫♫½

A Gun in the House While being attacked in her own home, a woman shoots one of her assailants. The police find no conclusive evidence of attack and arrest the woman for murder. Exploitive drama on gun control. Made-for-television.
1981 100m/C Sally Struthers, David Ackroyd, Joel Bailey, Jeffrey Tambor; **Dir:** Ivan Nagy. **VHS, Beta $49.95** VHE, HHE ♫

Gun Law The hero tackles the Sonora Kid, an outlaw terrorizing Arizona folk. One of the few sound westerns with Hoxie, a silent-era star who hung up his spurs when his voice proved non-photogenic.
1933 59m/B Jack Hoxie, Paul Fix, Mary Carr, Edmund Cobb, Robert Burns; **Dir:** Lewis Collins. **VHS, Beta** GPV ♫½

Gun Lords of Stirrup Basin A cattleman and the daughter of a homesteader are in love in the Old West, but a seedy lawyer wants to douse their flame. Bullets and fists fly as a result.
1937 55m/B Bob Steele, Louise Stanley, Karl Hackett, Ernie Adams, Frank LaRue, Frank Ball, Steve Clark; **Dir:** Sam Newfield. **VHS $19.95** NOS, DVT ♫♫

Gun Play Brother and sister venture west to run their father's ranch. Mexican bandits believe the ranch's property houses a revolutionary fortune, and set about recovering it. The brother defends his father's ranch along with a sidekick, who sweeps the sister off her feet. Also known as "Lucky Boots."
1936 59m/B Guinn Williams, Frank Yaconelli, Marion Shilling, Wally Wales, Charles French; **Dir:** Albert Herman. **VHS $19.95** NOS, DVT ♫

The Gun Ranger Our hero plays a Texas Ranger who becomes fed up with soft judges and shady prosecutors and decides to take the law into his own hands.
1934 56m/B Bob Steele, Eleanor Stewart. **VHS, Beta $19.95** NOS, DVT ♫

Gun Riders A gunman must seek out and stop a murderer of innocent people. Variant titles: "Five Bloody Graves," "Lonely Man," "Five Bloody Days to Tombstone."
1969 98m/C Scott Brady, Jim Davis, John Carradine; **Dir:** Al Adamson. **VHS, Beta** CCB ♫

Gun Smugglers/Hot Lead A western double feature: Tim Holt and his sidekick Chito must recover a shipment of stolen guns in "Gun Smugglers," and the duo gallop into action against a gang of train robbers in "Hot Lead."
1951 120m/B Tim Holt, Richard Martin, Martha Hyer. **VHS, Beta $34.95** *RKO* ₰

Gunblast A convict is released from San Quentin and wastes no time in getting involved with mob-powered drug smuggling.
197? 70m/C Lloyd Allan, Christine Cardan, James Cunningham. **VHS, Beta $59.95** *MGL* ₰

Gunbuster 1 Animated feature from Japan aimed at a mature audience. In this science fiction adventure, two girls train to pilot Earth's defense systems against alien invaders. First of three parts. Japanese with English subtitles.
1991 55m/C *JP* **VHS $34.95** *CPM, FCT* ₰

A Gunfight When Cash is stranded in a small Western town, he meets up with fellow old-time gunfighter Douglas. The two strike up a friendship, but discover the town folk expect a gun battle. Needing money, the two arrange a gunfight for paid admission — winner take all. Cash makes his screen debut.
1971 (PG) 90m/C Kirk Douglas, Johnny Cash, Jane Alexander, Karen Black, Dana Elcar, Keith Carradine, Raf Vallone; *Dir:* Lamont Johnson. **Beta $59.95** *ALL, MTX* ₰₰

Gunfight at the O.K. Corral The story of Wyatt Earp and Doc Holliday joining forces in Dodge City to rid the town of the criminal Clanton gang. Filmed in typical Hollywood style, but redeemed by its great stars. Written by Leon Uris.
1957 122m/C Burt Lancaster, Kirk Douglas, Rhonda Fleming, Jo Van Fleet, John Ireland, Kenneth Tobey, Lee Van Cleef, Frank Faylen, DeForest Kelley, Earl Holliman; *Dir:* John Sturges. **VHS, Beta, LV $14.95** *PAR, FCT* ₰₰₰

Gunfight at Sandoval A Texas Ranger investigates the headquarters of a criminal gang while hunting for the bandits who killed his partner. Adapted from the "Texas John Slaughter" episodes of the Disney television series.
1959 72m/C Tom Tryon, Beverly Garland, Dan Duryea. **VHS, Beta $69.95** *DIS* ₰½

Gunfighter A mature, serious, Hollywood-western character study about an aging gunfighter searching for peace and quiet but unable to avoid his reputation and the duel-challenges it invites. One of King's best films.
1950 85m/B Gregory Peck, Helen Westcott, Millard Mitchell, Jean Parker, Karl Malden, Skip Homeier, Mae Marsh; *Dir:* Henry King. **VHS, Beta, LV $19.98** *FOX* ₰₰₰½

The Gunfighters Lackluster pilot of a proposed Canadian series pits three individualistic relatives against a powerful empire-builder.
1987 100m/C *CA* Art Hindle, Reiner Schoene, Tony Addabbo, George Kennedy, Michael Kane, Lori Hallier; *Dir:* Clay Borris. **VHS, LV $19.99** *VMK* ₰½

Gunfire A Frank James look-alike cavorts, robs and loots the countryside until the real James decides to set things straight.
1950 59m/B Donald (Don "Red") Barry, Robert Lowery, Pamela Blake, Wally Vernon; *Dir:* William Berke. **VHS, Beta, LV** *WGE, HHE* ₰½

Gunfire A gunfighter is rescued from the hangman's noose by railroad tycoons who want him to kill a farmer. He doesn't kill the man, but runs off with his wife instead, and the incensed railway honchos send assassins after the double-crossing gunman. Peckinpah's part is pretty puny in this leisurely horse opera. Originally titled, "China 9, Liberty 37," (a clever reference to a signpost).
1978 94m/C *IT* Fabio Testi, Warren Oates, Jenny Agutter, Sam Peckinpah; *Dir:* Monte Hellman. **VHS $9.95** *SIM* ₰₰

Gung Ho! Carlson's Raiders are a specially trained group of Marine jungle fighters determined to retake the Pacific island of Makin during World War II.
1943 88m/B Robert Mitchum, Randolph Scott, Noah Beery Jr., Alan Curtis, Grace McDonald; *Dir:* Ray Enright. **VHS, Beta $9.95** *NEG, MRV, REP* ₰₰

Gung Ho A Japanese firm takes over a small-town U.S. auto factory and causes major cultural collisions. Keaton plays the go-between for employees and management while trying to keep both groups from killing each other. From the director of "Splash" and "Night Shift." Made into a short-lived television series.
1985 (PG-13) 111m/C Michael Keaton, Gedde Watanabe, George Wendt, Mimi Rogers, John Turturro, Clint Howard, Michelle Johnson, So Yamamura, Sab Shimono; *Dir:* Ron Howard. **VHS, Beta, LV, 8mm $14.95** *PAR* ₰₰½

Gunga Din Based loosely on Rudyard Kipling's famous poem. The prototypical "buddy" film. Three veteran British sergeants in India try to suppress a native uprising, but it's their water boy, the intrepid Gunga Din, who saves the day. Friendship, loyalty, and some of the best action scenes ever filmed.
1939 117m/B Cary Grant, Victor McLaglen, Douglas Fairbanks Jr., Sam Jaffe, Eduardo Ciannelli, Montagu Love, Joan Fontaine; *Dir:* George Stevens. **VHS, Beta, LV $19.98** *TTC, RKO, VID* ₰₰₰₰

Gunman from Bodie The second film in the Rough Rider series. Jones, McCoy, and Hatton rescue a baby orphaned by rustlers and then go after the bad guys.
1941 62m/B Tim McCoy, Buck Jones, Raymond Hatton; *Dir:* Spencer Gordon Bennet. **VHS, Beta $9.95** *NOS, MRV, VCN* ₰

Gunners & Guns A vintage western starring Black King.
1934 51m/B Black King, Edmund Cobb, Edna Asetin, Eddie Davis; *Dir:* Ray Nazarro. **VHS, Beta $29.95** *VCN* ₰

Gunplay Two cowboys befriend a boy whose father has been killed and search for the murderer.
1951 61m/B Tim Holt, Joan Dixon, Richard Martin. **VHS, Beta** *MED* ₰

Gunpowder Two Interpol agents endeavor to stop a crime lord from causing the collapse of the world economy.
1984 85m/C David Gilliam, Martin Potter, Gordon Jackson, Anthony Schaeffer. **VHS, Beta $69.98** *LIV, VES* ₰

The Gunrunner Costner plays a Canadian mobster, involved in liquor smuggling, who aids a Chinese rebellion in this 1920s drama. Interesting concept. Received more attention after Costner gained notoriety in later films.
1984 92m/C *CA* Kevin Costner, Sara Botsford, Paul Soles; *Dir:* Nardo Castillo. **VHS, LV $19.95** *STE, HHE* ₰½

Guns An international gunrunner puts the moves on buxom women between hair-raising adventures.
1990 (R) 95m/C Erik Estrada, Dona Speir; *W/ Dir:* Andy Sidaris. **VHS, LV $79.95** *COL* ₰₰½

Guns at Batasi Attenborough plays a tough British Sergeant Major stationed in Africa during the anti-colonial 1960s. Ultimately the regiment is threatened by rebel Africans. Fine performances make this rather predictable film watchable.
1964 103m/B *GB* Richard Attenborough, Jack Hawkins, Mia Farrow, Flora Robson, John Leyton; *Dir:* John Guillermin. **VHS** *TCF* ₰₰½

Guns in the Dark A cowboy vows never to use a gun again when he believes that he accidentally shot his best friend.
1937 60m/C Johnny Mack Brown, Claire Rochelle; *Dir:* Sam Newfield. **VHS, Beta $19.95** *NOS, DVT, RXM* ₰

Guns of Diablo Wagon train master Bronson has to fight a gang that controls the supply depot, all the while showing his young helper (beardless boy Russell) the fascinating tricks of the trade.
1964 91m/C Charles Bronson, Kurt Russell, Susan Oliver; *Dir:* Boris Sagal. **VHS $39.99** *ACE, MTX* ₰₰

Guns for Dollars Mex-Western has four men, including a secret agent and a cossack, tangled up in the Mexican Revolution as they search for a fortune in smuggled jewels. Originally titled, "They Call Me Hallelujah."
1973 94m/C *MX* George Hilton, Charles Southwood, Agata Flory, Roberto Camardiel, Paolo Gozlino, Rick Boyd; *Dir:* Anthony Ascot. **VHS $19.95** *ACA* ₰

Guns of Fury Cisco and Pancho solve the troubles of a small boy in this wild western.
1945 60m/B Duncan Renaldo. **VHS, Beta $19.95** *NOS, UHV, VCN* ₰

The Guns and the Fury Two American oil riggers in the Middle East, circa 1908, are forced to fight off an attack by a vicious Arab sheik and his tribe.
1983 90m/C Peter Graves, Cameron Mitchell, Michael Ansara, Albert Salmi; *Dir:* Tony Zarindast. **VHS, Beta $49.95** *WES* ₰½

Guns of Justice Undercover Colorado rangers stop a swindler from forcing homesteaders off their land. Also known as "Colorado Ranger."
1950 60m/B James Ellison, Russell Hayden, Tom Tyler, Raymond Hatton; *Dir:* Thomas Carr. **VHS, Beta, LV** *WGE* ₰

Guns of the Magnificent Seven The third remake of "The Seven Samurai." Action-packed western in which the seven free political prisoners and train them to kill. The war party then heads out to rescue a Mexican revolutionary being held in an impregnable fortress.
1969 (G) 106m/C George Kennedy, Monte Markham, James Whitmore, Reni Santoni, Bernie Casey, Joe Don Baker, Scott Thomas, Michael Ansara, Fernando Rey; *Dir:* Paul Wendkos. **VHS $19.98** *MGM* ₰₰½

The Guns of Navarone During World War II, British Intelligence in the Middle East sends six men to the Aegean island of Navarone to destroy guns manned by the Germans. Consistently interesting war epic based on the Alistair MacLean novel, with a vivid cast. Academy Award nominations for

Best Picture, Director, Sound, Film Editing, as well as the Special Effects Award it won.
1961 159m/C Gregory Peck, David Niven, Anthony Quinn, Richard Harris, Stanley Baker, Anthony Quayle, James Darren, Irene Papas; *Dir:* J. Lee Thompson. Academy Awards '61: Best Special Effects; Golden Globe Awards '62: Best Film—Drama; Filmdom's Famous Five '61: Best Actor (Peck), Best Supporting Actor (Quinn). **VHS, Beta, LV $19.95** COL 🐾🐾🐾½

Guns of War This is the true story of a group of Yugoslavians formed a partisan army to stop the reign of Nazi terror in their native land.
1975 114m/C YU **VHS, Beta** VCL 🐾🐾🐾½

The Gunslinger A woman marshall struggles to keep law and order in a town overrun by outlaws. Unique western with a surprise ending.
1956 83m/C John Ireland, Beverly Garland, Allison Hayes, Jonathan Haze, Dick Miller, Bruno Ve Sota; *Dir:* Roger Corman. **VHS $19.95** AVD, MLB, TIM 🐾½

Gunslinger Armed to the teeth with weaponry, Sartana shows a mining town's heavies how to sling a gun or two.
197? 90m/C John Garko, Tony Vilar; *Dir:* Anthony Ascot. **VHS, Beta $19.98** VDC 🐾½

Gunsmoke Trail Cowpoke Randall saves an heiress from losing her estate to a phony uncle in low-budget Western. Also features the prolific "Fuzzy" St. John in his cowboy's sidekick stage (interestingly, St. John, a nephew of Fatty Arbuckle, later co-starred in the "Lone Rider" series with Randall's brother, Robert Livingston).
1938 59m/C Jack Randall, Al "Fuzzy" St. John. **VHS $19.99** NOS, MRV, MOV 🐾½

Gus A Disney film about the California Atoms, a football team that has the worst record in the league until they begin winning games with the help of their field goal kicking mule of a mascot, Gus. The competition then plots a donkey-napping. Enjoyable comedy for the whole family.
1976 (G) 96m/C Ed Asner, Tim Conway, Dick Van Patten, Ronnie Schell, Bob Crane, Tom Bosley; *Dir:* Vincent McEveety. **VHS, Beta, LV** DIS, OM 🐾🐾½

The Guy from Harlem He's mad, he's from Harlem, and he's not going to take it any more.
1977 86m/C Loye Hawkins, Cathy Davis, Patricia Fulton, Wanda Starr; *Dir:* Rene Martinez Jr. **VHS** XVC 🐾🐾

A Guy Named Joe A sentimental, patriotic Hollywood fantasy about the angel of a dead World War II pilot guiding another young pilot through battle and also helping him to romance his girl, who's still too devoted to his memory. Screenplay by Dalton Trumbo. Remade in 1989 as "Always."
1944 121m/B Spencer Tracy, Irene Dunne, Van Johnson, Ward Bond, James Gleason, Lionel Barrymore, Esther Williams; *Dir:* Victor Fleming. **VHS, Beta $19.95** MGM 🐾🐾½

The Guyana Tragedy: The Story of Jim Jones Dramatization traces the story of the Reverend Jim Jones and the People's Temple from its beginnings in 1953 to the November 1978 mass suicide of more than 900 people. Boothe is appropriately hypnotic as the Reverend. Made for television.
1980 240m/C Powers Boothe, Ned Beatty, Randy Quaid, Brad Dourif, Brenda Vaccaro, LeVar Burton, Colleen Dewhurst, James Earl Jones; *Dir:* William A. Graham. **VHS, Beta $59.95** VHE, HHE 🐾🐾

Guys & Dolls New York gambler Sky Masterson takes a bet that he can romance a Salvation Army lady. Frank Loesser's great score includes "Luck Be a Lady," "If I were a Bell," and "Sit Down You're Rocking the Boat." Based on the stories of Damon Runyon with Blaine, Kaye, Pully, and Silver recreating their roles from the Broadway hit. Brando's not-always-convincing musical debut.
1955 150m/C Marlon Brando, Jean Simmons, Frank Sinatra, Vivian Blaine, Stubby Kaye, Sheldon Leonard; *Dir:* Joseph L. Mankiewicz. Golden Globe Awards '56: Best Film—Musical/Comedy. **VHS, Beta, LV $19.98** FOX, FCT 🐾🐾½

Gymkata A gymnast (Olympian Thomas) must use his martial arts skills to conquer and secure a military state in a hostile European country.
1985 (R) 89m/C Kurt Thomas, Tetchie Agbayani, Richard Norton, Conan Lee; *Dir:* Robert Clouse. **VHS, Beta $79.95** MGM 🐾

Gypsy The life story of America's most famous striptease queen, Gypsy Rose Lee. Russell gives a memorable performance as the infamous Mama Rose, Gypsy's stage mother. Based on both Gypsy Rose Lee's memoirs and the hit Broadway play by Arthur Laurents. Songs by Jule Styne and Stephen Sondheim include "Everything's Coming Up Roses" and "Let Me Entertain You."
1962 144m/C Rosalind Russell, Natalie Wood, Karl Malden, Ann Jillian, Parley Baer, Paul Wallace, Betty Bruce; *Dir:* Mervyn LeRoy. **VHS, Beta, LV $19.98** WAR 🐾🐾🐾

Gypsy A modern day gypsy-blooded Robin Hood cavorts between heists and romantic trysts. Dubbed.
1975 90m/C Alain Delon, Annie Girardot, Paul Meurisse; *Dir:* Jose Giovanni. **VHS, Beta $59.95** FCT 🐾🐾

Gypsy Blood One of the first screen versions of "Carmen," closer to Prosper Merrimee's story than to the opera, depicting the fated love between a dancing girl and a matador. Lubitsch's first notable film. Silent.
1918 104m/B Pola Negri; *Dir:* Ernst Lubitsch. **VHS, Beta $44.95** FCT, GPV 🐾🐾

The Gypsy Warriors Two Army captains during World War II infiltrate Nazi-occupied France in order to prevent the distribution of a deadly toxin and are aided by a band of Gypsies. Made for television.
1978 77m/C Tom Selleck, James Whitmore Jr., Joseph Ruskin, Lina Raymond; *Dir:* Lou Antonio. **VHS, Beta $39.95** MCA 🐾½

H-Bomb Mitchum stars as a CIA agent who is sent to Bangkok to retrieve two stolen nuclear warheads from a terrorist. It just so happens that his ex-girlfriend's father is the terrorist he must contend with in this stupid movie.
1971 98m/C Olivia Hussey, Chris Mitchum, Krung Savilai; *Dir:* P. Chalong. **VHS, Beta $59.98** CGI Woof!

H-Man A Japanese sci-fi woofer about a radioactive mass of slime festering under Tokyo. Extremely lame special effects get quite a few unintentional laughs. Dubbed.
1959 79m/C JP Koreya Senda, Kenji Sahara, Yumi Shirakawa, Akihiko Hirata; *Dir:* Inoshiro Honda. **VHS, Beta $69.95** COL

Hack O'Lantern Granddad worships Satan, and gets his grandson involved: the result is much bloodshed, violence and evil.
1987 90m/C *Dir:* Jag Mundhra. **VHS, Beta $79.95** LGC 🐾

Hadley's Rebellion A Georgia farm boy adjusts to relocation at an elitist California boarding school by flaunting his wrestling abilities.
1984 (PG) 96m/C Griffin O'Neal, Charles Durning, William Devane, Adam Baldwin, Dennis Hage, Lisa Lucas; *Dir:* Fred Walton. **VHS, Beta $59.98** FOX 🐾½

Hail A biting satire of what-might-have-been if certain key cabinet members had their way. Also called "Hail to the Chief" and "Washington, B.C."
1973 (PG) 85m/C Richard B. Shull, Dick O'Neill, Phil Foster, Joseph Sirola, Dan Resin, Willard Waterman, Gary Sandy; *Dir:* Fred Levinson. **VHS, Beta $39.95** MON 🐾🐾

Hail to the Chieftains Irish traditional musicians, the Chieftains, perform a brief concert in this program.
1976 30m/C **VHS, Beta** CMR, NYS 🐾🐾

Hail the Conquering Hero A slight young man is rejected by the Army. Upon returning home, he is surprised to find out they think he's a hero. Biting satire with Demarest's performance stealing the show.
1944 101m/B Eddie Bracken, Ella Raines, William Demarest, Franklin Pangborn, Raymond Walburn; *Dir:* Preston Sturges, Preston Sturges. **VHS $29.95** MCA 🐾🐾🐾½

Hail, Hero! A dated Vietnam War-era drama about a confused hippie rebelling against his parents and the draft. First film for both Douglas and Strauss. Adapted from the novel by John Weston. Includes music by Gordon Lightfoot.
1969 (PG) 97m/C Michael Douglas, Arthur Kennedy, Teresa Wright, John Larch, Charles Drake, Deborah Winters, Peter Strauss; *Dir:* David Miller. **VHS, Beta $59.98** FOX 🐾½

Hail Mary A modern-day virgin named Mary inexplicably becomes pregnant in this controversial film that discards notions of divinity in favor of the celebration of a lively, intellectual humanism. Godard rejects orthodox narrative structure and bourgeois prejudices. Very controversial, but not up to his breathless beginning work. Also known as "Je Vous Salue Marie."
1985 (R) 107m/C FR SI GB Myriem Roussel, Thierry Rode, Philippe Lacoste, Manon Anderson; *Dir:* Jean-Luc Godard. **VHS, Beta, LV $29.98** LIV, VES 🐾🐾½

Hair Film version of the explosive 1960s Broadway musical about the carefree life of the flower children and the shadow of the Vietnam War that hangs over them. Great music, as well as wonderful choreography by Twyla Tharp help portray the surprisingly sensitive evocation of the period so long after the fact. Forman has an uncanny knack for understanding the textures of American life. Watch for a thinner Nell Carter.
1979 (PG) 122m/C Treat Williams, John Savage, Beverly D'Angelo, Annie Golden, Nicholas Ray; *Dir:* Milos Forman. **VHS, Beta $19.95** MGM 🐾🐾🐾½

Hairspray Waters' first truly mainstream film, if that's even possible, and his funniest, detailing the struggle in 1962 Baltimore among teenagers for the top spot in a local television dance show. Deals with racism and stereotypes, as well as typical "teen" problems (hair-do's and "hair-don'ts"). Filled with refreshingly tasteful, subtle social satire (although not without its yester-Waters touches; the scene with Debbie Harry, her daughter, and a pimple will please die-hard Waters fans). Ricki Lake is lovable and appealing as Divine's daughter; Divine, in his

last film, is likeable as an iron-toting mom. Written and directed by Waters; watch for him in his cameo as the demented shrink. Look also for Divine's appearance as a man. Features some great 60s music, which Waters refers to as "the only known remedy to today's Hit Parade of Hell." Tunes include Gene Pitney's "Town Without Pity" and the Ray Bryant Combo's "Madison Time."
1988 (PG) 94m/C Ricki Lake, Divine, Jerry Stiller, Colleen Fitzpatrick, Sonny Bono, Deborah Harry, Ruth Brown, Pia Zadora; *W/Dir:* John Waters. **VHS, Beta, LV $19.95** *COL, FCT* 𝄢𝄢𝄢

The Hairy Ape Screen adaptation of the Eugene O'Neill play. A beast-like coal stoker becomes obsessed with a cool and distant passenger aboard an ocean liner.
1944 90m/B William Bendix, Susan Hayward, John Loder, Dorothy Comingore, Roman Bohnen, Alan Napier; *Dir:* Alfred Santell. **VHS $19.95** *NOS, MRV, FCT* 𝄢𝄢

The Half-Breed Routine oater with settlers and Apaches fighting, and the obligatory half-breed stuck in the middle.
1951 81m/B Robert Young, Jack Buetel, Janis Carter, Barton MacLane, Reed Hadley, Porter Hall, Connie Gilchrist; *Dir:* Stuart Gilmore. **VHS, Beta $19.98** *TTC* 𝄢𝄢

Half of Heaven A critically popular film about a young Spanish woman who begins to share certain telepathic powers with her wizened grandmother. In Spanish with English subtitles.
1986 95m/C *SP* Angela Molina, Margarita Lozano; *Dir:* Manuel Gutierrez Aragon. **VHS, Beta $29.95** *PBS, PAV, FCT* 𝄢𝄢𝄢

Half Human Dull Japanese monster movie about an ape-human who terrorizes Northern Japan. When released in the U.S., Carradine and Morris were added. Without subtitles, just Carradine's narration. Director Honda made better movies before and after, with Godzilla and Rodan.
1958 78m/B *JP* John Carradine, Akira Takarada, Morris Ankrum; *Dir:* Inoshiro Honda. **VHS $14.95** *RHI, SNC, MLB* 𝄢½

Half a Lifetime The stakes are too high for four friends playing a game of poker.
1986 85m/C Keith Carradine, Gary Busey, Nick Mancuso, Saul Rubinek. **VHS, Beta $79.95** *HBO* 𝄢½

Half a Loaf of Kung Fu The worthy bodyguards of Sern Chuan are called upon to deliver a valuable jade statue. Thwarted again and again by ruthless robbers, they falter. Only Chan carries on, meeting the enemy alone. Fast action, good story.
1985 98m/C Jackie Chan. **VHS, Beta $24.95** *ASE* 𝄢

Half Moon Street A brilliant woman scientist supplements her paltry fellowship salary by becoming a hired escort and prostitute, which leads her into various incidents of international intrigue. Adapted from a story by Paul Theroux, "Dr. Slaughter."
1986 (R) 90m/C Sigourney Weaver, Michael Caine, Keith Buckley, Ian MacInnes; *Dir:* Bob Swaim. **VHS, Beta, LV, 8mm $14.98** *SUE* 𝄢𝄢

Half-Shot at Sunrise Madcap vaudeville comedians play AWOL soldiers loose in 1918 Paris. Continuous one-liners, sight gags, and slapstick nonsense. First film appearance of comedy team Wheeler and Woolsey.
1930 78m/B Bert Wheeler, Robert Woolsey, Dorothy Lee, Robert Rutherford, Edna May Oliver; *Dir:* Paul Sloane. **VHS, Beta $24.95** *VYY, CAB, DVT* 𝄢

Half a Sixpence Former musician Sidney, who earlier directed "Showboat" and "Kiss Me Kate," shows a bit less verve in this production. Based on H.G. Wells' 1905 novel, "Kipps," it sports much of the original cast from its Broadway incarnation. Edwardian orphan Steele, a cloth-dealer-in-training, comes into a large sum of money and proceeds to lose it with great dispatch. Somewhere along the way he loses his girlfriend, too, but she takes him back because musicals end happily.
1967 148m/B Tommy Steele, Julia Foster, Penelope Horner, Cyril Ritchard, Grover Dale; *Dir:* George Sidney. **VHS $19.95** *PAR, FCT* 𝄢𝄢

Half Slave, Half Free This adaptation of "Solomon Northrup's Odyssey" tells the true story of a free black man who is kidnapped and forced into slavery for 12 years. Made for television as part of the "American Playhouse" series on PBS. Also titled "Solomon Northrup's Odyssey."
1985 113m/C Avery Brooks, Mason Adams; *Dir:* Gordon Parks. **VHS, Beta $59.95** *SVS, KAR, KUI* 𝄢𝄢½

Half a Soldier A spoiled girl gets a big surprise when she steals a car and discovers the body of a gangster in the trunk.
1940 59m/B Heather Angel, John "Dusty" King. **VHS, Beta $24.95** *NOS, DVT* 𝄢𝄢

Hallelujah, I'm a Bum A happy-go-lucky hobo reforms and begins a new life for the sake of a woman. Bizarre Depression-era musical with a Rodgers and Hart score and continuously rhyming dialogue. The British version, due to the slang meaning of 'bum,' was titled "Hallelujah, I'm a Tramp." Also known as "The Heart of New York," "Happy Go Lucky," and "Lazy Bones." Script by Ben Hecht and S.N. Behrman.
1933 83m/B Al Jolson, Madge Evans, Frank Morgan, Chester Conklin; *Dir:* Lewis Milestone. **VHS, Beta** *LCA, MLB* 𝄢𝄢½

The Hallelujah Trail A Denver mining town in the late 1800s is about to batten down the hatches for a long winter, and there's not a drop of whiskey to be had. The U.S. Cavalry sends a shipment to the miners, but temperance leader Cora Templeton Massingale (Remick) and her bevy of ladies against liquor—along with a bunch of whiskey-ravening Indians—stand between them and the would-be whistle whetters. A limp Western satire directed by Preston Sturges' brother John, who fared much better when he kept a straight face (he directed "The Great Escape" two years earlier). Based on Bill Gulick's novel, "The Hallelujah Train." Special two-laserdisc set includes additional footage not seen in over 25 years (though it's already overlong at two and three-quarters hours), the original movie trailer, and a letter-boxed screen.
1965 166m/C Burt Lancaster, Lee Remick, Jim Hutton, Pamela Tiffin, Donald Pleasence, Brian Keith, Martin Landau; *Dir:* John Sturges; *Nar:* John Dehner. **VHS, LV $29.99** *MGM, PIA, FCT* 𝄢𝄢

Hallmark Theater (Sometimes She's Sunday) Adult fare about a Portuguese-American fisherman whose daughter becomes engaged to a typical American boy.
1952 26m/B VHS, Beta $19.95 *VYY*

Halloween John Carpenter's horror classic has been acclaimed "the most successful independent motion picture of all time." A deranged youth returns to his hometown with murderous intent after fifteen years in an asylum. Very, very scary—you feel this movie more than see it.

1978 (R) 90m/C Jamie Lee Curtis, Donald Pleasence, Nancy Loomis, P.J. Soles; *Dir:* John Carpenter. **VHS, Beta, LV $19.98** *MED* 𝄢𝄢𝄢½

Halloween 2: The Nightmare Isn't Over! Picking up where "Halloween" left off, the sequel begins with the escape of vicious killer Michael, who continues to murder and terrorize the community of Haddonfield, Illinois. Unlike its predecessor, this is not innovative and creates discomfort, not suspense. Not directed by Carpenter.
1981 (R) 92m/C Jamie Lee Curtis, Donald Pleasence, Jeffrey Kramer, Charles Cyphers, Lance Guest; *Dir:* Rick Rosenthal. **VHS, Beta, LV $19.95** *MCA* 𝄢½

Halloween 3: Season of the Witch A modern druid plans to kill 50 million children with his specially made Halloween masks. Produced by John Carpenter, this second sequel to the 1978 honor classic is not based on the events or characters of its predecessors or successors. Sequelled by "Halloween 4," and "Halloween 5: The Revenge of Michael Myers" (1989).
1982 (R) 98m/C Tom Atkins, Stacey Nelkin, Dan O'Herlihy, Ralph Strait; *Dir:* Tommy Lee Wallace. **VHS, Beta, LV $19.95** *MCA* 𝄢½

Halloween 4: The Return of Michael Myers The third sequel, wherein the lunatic that won't die returns home to kill his niece.
1988 (R) 88m/C Donald Pleasence, Ellie Cornell, Danielle Harris, Michael Pataki; *Dir:* Dwight Little. **VHS, Beta, LV $89.98** *FOX* 𝄢

Halloween 5: The Revenge of Michael Myers Fifth in the series, this Halloween is an improvement over 2, 3, and 4, thanks to a few well-directed scare scenes. Unfortunately, the plot remains the same: a psycho behemoth chases down and kills more teens. An open ending promises yet another installment.
1989 (R) 96m/C Donald Pleasence, Ellie Cornell, Danielle Harris, Don Shanks; *Dir:* Dominique Othenin-Girard. **VHS, Beta, LV $89.98** *FOX* 𝄢½

Halloween with the Addams Family A really pathetic production spawned during the "reunion" craze of the late '70s, sparked only by the lively performance of Coogan's Uncle Fester and the always lovable Lurch.
1979 87m/C John Astin, Carolyn Jones, Jackie Coogan, Ted Cassidy. **VHS, Beta** *GKK Woof!*

Halloween Monster Madness A fun-filled giggle-fest ripe with chills featuring all your favorite Hanna-Barbera stars.
1990 ?m/C VHS $24.95 *HNB*

Halloween Night Ancient evil rises up to destroy the perfect small town.
1990 (R) 90m/C Hy Pyke, Katrina Garner; *Dir:* Emilio P. Miraglio. **VHS $59.95** *ATL* 𝄢𝄢

Halls of Montezuma Large, bombastic post-World War II combat epic. Depicts the Marines fighting the Japanese in the Pacific.
1950 113m/C Richard Widmark, Jack Palance, Reginald Gardiner, Robert Wagner, Karl Malden, Richard Boone, Richard Hylton, Skip Homeier, Jack Webb, Neville Brand, Martin Milner, Bert Freed; *Dir:* Lewis Milestone. **VHS, Beta $14.98** *FOX* 𝄢𝄢½

Hallucination Expatriate acid heads murder an antique dealer under the influence of LSD. Depicts LSD as the catalyst in bringing out the group's criminal behavior. In sepia and color. Also known as "Hallucination Generation."

1967 90m/C George Montgomery, Danny Stone; *Dir:* Edward Mann. **VHS, Beta $19.98** *VDC* Woof!

Hambone & Hillie An elderly woman makes a 3000-mile trek across the United States to search for the dog she lost in an airport.

1984 (PG) 97m/C Lillian Gish, Timothy Bottoms, Candy Clark, O.J. Simpson, Robert Walker Jr., Jack Carter, Alan Hale Jr., Anne Lockhart; *Dir:* Roy Watts. **VHS, Beta $19.95** *STE, HBO* 𝔡𝔡

Hamburger Hill The popular war epic depicting the famous battle between Americans and Viet Cong over a useless hill in Vietnam. Made in the heyday of 1980s Vietnam backlash, and possibly the most realistic and bloodiest of the lot. Written by Jim Carabatsos.

1987 (R) 104m/C Michael Dolan, Daniel O'Shea, Dylan McDermott, Tommy Swerdlow, Courtney B. Vance, Anthony Barille, Michael Boatman, Don Cheadle; *Dir:* John Irvin. **VHS, Beta, LV $19.98** *LIV, VES, FCT* 𝔡𝔡½

Hamburger... The Motion Picture The life and times of students at Busterburger U., the only college devoted to hamburger franchise management.

1986 (R) 90m/C Leigh McCloskey, Dick Butkus, Randi Brooks, Sandy Hackett; *Dir:* Mike Marvin. **VHS, Beta, LV $9.95** *MED* 𝔡

Hamlet Splendid adaptation of Shakespeare's dramatic play. Hamlet vows vengeance on the murderer of his father in this tight version of the four hour stage play. Some scenes were cut out, including all of Rosencrantz and Guildenstern. Beautifully photographed in Denmark. An Olivier triumph; won several Oscars, plus a Best Director nomination for Olivier and Best Supporting Actress nomination for Simmons. Remade in 1991.

1948 153m/B *GB* Laurence Olivier, Basil Sydney, Felix Aylmer, Jean Simmons, Stanley Holloway, Peter Cushing, Christopher Lee; *Dir:* Laurence Olivier; *Voices:* John Gielgud. Academy Awards '48: Best Actor (Olivier), Best Art Direction/Set Decoration (B & W), Best Costume Design, Best Picture. **VHS $19.95** *PAR, BTV* 𝔡𝔡𝔡

Hamlet Williamson brings to the screen his stage performance as the classic Shakespeare character. Something is lost in the process, although there are some redeeming moments including those with Faithfull as Ophelia.

1969 (G) 114m/C *GB* Nicol Williamson, Anthony Hopkins, Marianne Faithfull, Gordon Jackson, Judy Parfitt, Mark Dingham, Anjelica Huston; *Dir:* Tony Richardson. **VHS, Beta $19.95** *COL* 𝔡𝔡½

Hamlet In this tragedy, Shakespeare uses a tense drama of murder, conspiracy, and revenge as a medium for inquiring into the most fundamental problems: justice, guilt, madness, death, and the difficulty of understanding oneself and others. Part of the television series "The Shakespeare Plays."

1979 150m/C *GB* Claire Bloom, Patrick Stewart, Eric Porter, Derek Jacobi. **VHS, Beta $69.95** *TLF* 𝔡𝔡𝔡

Hamlet Zeffirelli—who has much Shakespeare under his belt, having directed "The Taming of the Shrew," "Romeo and Juliet" and a filmed version of the opera "Otello"—creates a surprisingly energetic and accessible interpretation of the Bard's moody play. Gibson brings charm, humor and a carefully calculated sense of violence, not to mention a good deal of solid flesh, to the eponymous role, and he handles the language skillfully

(although if it's a poetic Dane you seek, you'd best stick with Olivier). The cast is uniformly excellent, with exceptional work from Scofield as the ghost of Hamlet's father and Bates as Claudius; Close may seem a tad hysterical at times (not to mention too close to Gibson's age to play his mother), but she brings some insight and nuance to the maternal role. Purists beware: Zeffirelli and co-scriptwriter Christopher DeVore have cut the play to underline the themes of revenge and insanity, and even add a scene at the beginning for the sake of narrative expediency (not that Olivier et al didn't take their liberties). Beautifully costumed and shot on location in Northern Scotland. Oscar nominations for costumes and art direction.

1990 (PG) 135m/C Mel Gibson, Glenn Close, Alan Bates, Paul Scofield, Ian Holm, Helena Bonham Carter; *Dir:* Franco Zeffirelli. **VHS, Beta, LV, 8mm $19.98** *WAR, BTV, PIA* 𝔡𝔡𝔡½

Hammered: The Best of Sledge Contains four episodes from the comedy series about Inspector Sledge Hammer.

1988 104m/C David Rasche, Anne-Marie Martin, Harrison Page, John Vernon; *Dir:* Jackie Cooper, Gary Walkow. **VHS, LV $19.95** *STE, NWV*

Hammersmith is Out A violent lunatic cons an orderly into letting him escape. Chases, romance and craziness follow.

1972 (R) 108m/C Richard Burton, Elizabeth Taylor, Peter Ustinov, Beau Bridges; *Dir:* Peter Ustinov. **VHS, Beta $59.95** *PSM* 𝔡½

Hammett After many directors and script rewrites, Wenders was assigned to this arch neo-noir what-if scenario. Depicts Dashiell Hammett solving a complex crime himself, an experience he uses in his novels. Interesting, but ultimately botched studio exercise, like many from executive producer Francis Coppola, who is said to have reshot much of the film.

1982 (PG) 98m/C Frederic Forrest, Peter Boyle, Sylvia Sidney, Elisha Cook Jr., Marilu Henner; *Dir:* Wim Wenders. **VHS, Beta $69.95** *WAR* 𝔡𝔡

Hamsin A Jewish landowner and an Arab worker encounter difficulties in their relationship when the government announces plans to confiscate Arab land. In Hebrew with English subtitles.

1983 90m/C Shlomo Tarshish, Yasin Shawap, Hemda Levy, Ruth Geler; *Dir:* Daniel Wachsmann. **VHS, Beta $79.95** *ERG* 𝔡𝔡𝔡

The Hand During World War II, three British POWs are tortured by the Japanese. Refusing to give information, two of them have their right hands cut off. The third opts to talk, thereby keeping his hand. After the war, a series of murders occur in London where the victims have their hands amputated. Could there be a connection?

1960 61m/B *GB* Derek Bond, Ronald Lee Hunt, Reed de Rouen, Ray Cooney, Brian Coleman; *Dir:* Henry Cass. **VHS, Beta $16.95** *SNC* 𝔡½

The Hand A gifted cartoonist's hand is severed in a car accident. Soon, the hand is on the loose with a mind of its own, seeking out victims with an obsessive vengeance. Stone's first film is a unique, surreal psychohorror pastiche consistently underrated by most critics.

1981 (R) 105m/C *GB* Michael Caine, Andrea Marcovicci, Annie McEnroe, Bruce McGill, Viveca Lindfors; *W/Dir:* Oliver Stone. **VHS, Beta $19.98** *WAR* 𝔡𝔡𝔡

The Hand that Rocks the Cradle De Mornay is frightening as Peyton Flanders, the nanny from hell who charms her way into a suburban Seattle family's life. Sciorra is the

pregnant and unbelievably naive Claire Bartel, who unwittingly starts a terrifying chain of events when she brings charges against her new obstetrician after he molests her during a check-up. The physician's wife, also pregnant, loses her baby after he commits suicide and begins to plot her revenge. Peyton literally shows up and gets the nanny job and Claire ignores warnings from her best friend until she notices that her husband and daughter are becoming closer to Peyton. The plot is very predictable and preys on the worst fears of viewers: that a nanny or babysitter will betray your trust and harm your child, not to mention that a physician will take advantage of you. Still, the movie is worth it if only to see De Mornay in her fantastic Jeckle and Hyde performance.

1991 (R) 110m/C Rebecca De Mornay, Annabella Sciorra, Matt McCoy, Ernie Hudson, Madeline Zima, Julianne Moore, John de Lancie; *Dir:* Curtis Hanson. MTV Movie Awards '92: Best Villain (DeMornay). **VHS, Beta $94.95** *HPH* 𝔡𝔡𝔡

A Handful of Dust A dry, stately adaptation of the bitter Evelyn Waugh novel about a stuffy young aristocrat's wife's careless infidelity and how it sends her innocent husband to a tragic downfall. A well-meaning version that captures Waugh's cynical satire almost in spite of itself. Set in post-World War I England.

1988 (PG) 114m/C *GB* James Wilby, Kristin Scott Thomas, Rupert Graves, Alec Guinness, Anjelica Huston, Judy Dench; *Dir:* Charles Sturridge. **VHS, Beta $19.95** *COL* 𝔡𝔡𝔡

The Handmaid's Tale A cool, shallow but nonetheless chilling nightmare based on Margaret Atwood's bestselling novel, about a woman caught in the machinations of a near-future society so sterile it enslaves the few fertile women and forces them into being child-bearing "handmaids."

1990 (R) 109m/C Natasha Richardson, Robert Duvall, Faye Dunaway, Aidan Quinn, Elizabeth McGovern, Victoria Tennant, Blanche Baker; *Dir:* Volker Schlondorff. **VHS, Beta, LV $19.98** *HBO, FCT, IME* 𝔡𝔡½

Hands Across the Border A musical western in which Roy helps a woman find the men who killed her father.

1943 72m/B Roy Rogers, Ruth Terry, Guinn Williams, Onslow Stevens, Mary Treen, Joseph Crehan; *Dir:* Joseph Kane. **VHS $19.95** *NOS, DVT* 𝔡½

Hands of Death Thieves use their Ninja skills to stop any one in their path to evil. Action-packed, somewhat contrived.

1988 90m/C Richard Harrison, Mike Abbott; *Dir:* Geoffrey Ho. **VHS, Beta** *IMP* 𝔡

The Hands of Orlac Third remake of Maurice Renard's classic tale. When a concert pianist's hands are mutilated in an accident, he receives a graft of a murderer's hands. Obsession sweeps the musician, as he believes his new hands are incapable of music, only violence. Bland adaptation of the original story. Also known as "Hands of the Strangler."

1960 95m/B *GB FR* Mel Ferrer, Christopher Lee, Felix Aylmer, Basil Sydney, Donald Wolfit, Donald Pleasence; *Dir:* Edmond T. Greville. **VHS $16.95** *SNC* 𝔡𝔡

Hands of the Ripper Jack the Ripper's daughter returns to London where she works as a medium by day and stalks the streets at night. Classy Hammer horror variation on the perennial theme.

1971 (R) 85m/C *GB* Eric Porter, Angharad Rees, Jane Merrow, Keith Bell; *Dir:* Peter Sasdy. VHS, Beta $19.98 *VID, MLB* 🐾🐾½

Hands of Steel A ruthless cyborg carries out a mission to find and kill an important scientist. Terrible acting, and lousy writing: an all-around woofer!
1986 (R) 94m/C Daniel Greene, John Saxon, Janet Agren, Claudio Cassinelli, George Eastman; *Dir:* Martin Dolman. VHS, Beta $79.98 *LIV, LTG* Woof!

Hands of a Stranger Another undistinguished entry in the long line of remakes of "The Hands of Orlac." A pianist who loses his hands in an accident is given the hands of a murderer, and his new hands want to do more than tickle the ivories. Kellerman has a very small role as does former "Sheena" McCalla.
1962 95m/B Paul Lukather, Joan Harvey, James Stapleton, Sally Kellerman, Irish McCalla; *Dir:* Newton Arnold. VHS $19.98 *SNC* 🐾🐾

Hands of a Stranger Policeman's wife is raped during an adulterous tryst. As the painful truth dawns on the cop, he himself dallies with a compassionate lady D.A. while hunting the attacker. About as sordid as a network TV-movie gets without toppling into sleaze; strong acting and a sober script (by dramatist Arthur Kopit) pull it through. But the plot, from Robert Daly's novel, doesn't justify miniseries treatment; the two-cassette package is excessively long.
1987 179m/C Armand Assante, Beverly D'Angelo, Blair Brown, Michael Lerner, Philip Casnoff, Arliss Howard; *Dir:* Larry Elikann. VHS $89.95 *WOV* 🐾🐾½

Hands Up Jaunty Confederate spy (played by much-overlooked American comedian Griffith) attempts, amidst much comic nonsense, to thwart Yankee gold mining during the Civil War. Silent.
1926 50m/B Raymond Griffith, Mack Swain, Marion Nixon, Montagu Love; *Dir:* Clarence Badger. VHS, Beta $19.95 *GPV, DNB* 🐾🐾½

Hands Up Famed Polish film finished in 1967 but not released until 1981 for political reasons (prologue shot in '81 included). Doctors at a medical school reunion reflect upon the effect of Stalinist rule on their education and lives. In Polish with English subtitles and known also as "Rece do Gory."
1981 78m/C *PL Dir:* Jerzy Skolimowski. VHS, Beta $49.95 *FCT* 🐾🐾🐾

Hang 'Em High A cowboy is saved from a lynching and vows to hunt down the gang that nearly killed him in this American-made spaghetti western. Eastwood's first major vehicle made outside of Europe.
1967 114m/C Clint Eastwood, Inger Stevens, Ed Begley Jr., Pat Hingle, James MacArthur; *Dir:* Ted Post. VHS, Beta, LV $19.95 *MGM, TLF* 🐾🐾½

Hang Tough A goofy loser of a teenage slob has no luck in life until the older brother of one of his friends shows him the ropes. That's when the laughs begin.
1989 (R) 105m/C Carl Marotte, Lisa Langlois, Charlaine Woodward, Grand Bush, Vincent Bufano; *Dir:* Daryl Duke. VHS $49.95 *MNC* 🐾½

Hangar 18 A silly sci-fi drama about two scientists who see a spaceship crash-land and keep it in a hangar so that news of its existence doesn't spread while they study it. Shown on TV as "Invasion Force" with an entirely new ending.

1980 (PG) 97m/C Darren McGavin, Robert Vaughn, Gary Collins, Joseph Campanella, James Hampton, Tom Hallick, Pamela Bellwood; *Dir:* James L. Conway. VHS, Beta $19.99 *WOV* 🐾½

The Hanged Man Tale of a gunslinger in the Old West who mystically escapes death by hanging. Reborn, he re-evaluates the meaning of justice.
1974 78m/C Steve Forrest, Cameron Mitchell, Sharon Acker, Dean Jagger, Will Geer, Barbara Luna, Rafael Campos; *Dir:* Michael Caffey. VHS $59.95 *KOV* 🐾🐾

Hangfire When a New Mexican prison is evacuated thanks to a nasty chemical explosion, several prisoners decide it's time for a furlough, and they elect the local sheriff's wife as a traveling companion. Another mediocre actioner from director Maris (scripted by Brian D. Jeffries) it gets an extra bone for stunts kids shouldn't try at home.
1991 (R) 91m/C Brad Davis, Yaphet Kotto, Lee De Broux, Jan-Michael Vincent, George Kennedy, Kim Delaney; *Dir:* Peter Maris. VHS, LV $89.98 *COL* 🐾

Hangin' with the Homeboys One night in the lives of four young men. Although the Bronx doesn't offer much for any of them, they have little interest in escaping its confines, and they are more than willing to complain. Characters are insightfully written and well portrayed, with strongest work from Serrano as a Puerto Rican who is trying to pass himself off as Italian. Lack of plot may frustrate some viewers.
1991 (R) 89m/C Mario Joyner, Doug E. Doug, John Leguizamo, Nestor Serrano; *W/Dir:* Joseph P. Vasquez. Sundance Film Festival '91: Best Screenplay. VHS, LV $89.95 *COL* 🐾🐾🐾

Hanging on a Star A small rock band encounters many comic adventures as it climbs its way up the charts. Raffin stars as the band's persistent and competent road agent.
1978 (PG) 92m/C Deborah Raffin, Lane Caudell, Wolfman Jack, Jason Parker, Danil Torppe; *Dir:* Mike MacFarland. VHS, Beta $79.95 *MGM* 🐾

The Hanging Woman A man is summoned to the reading of a relative's will, and discovers the corpse of a young woman hanging in a cemetery. As he investigates the mystery, he uncovers a local doctor's plans to zombify the entire world. Not bad at all and quite creepy once the zombies are out in force. Euro-horror star Naschy plays the necrophiliac grave digger, Igor. Also on video as "Return of the Zombies" and "Beyond the Living Dead."
1972 (R) 91m/C *SP IT* Stan Cooper, Vickie Nesbitt, Marcella Wright, Catherine Gilbert, Gerard Tichy, Paul Naschy; *Dir:* John Davidson. VHS, Beta $59.95 *WES, UHV, UNI* 🐾🐾½

Hangman's Knot Members of the Confederate Cavalry rob a Union train, not knowing that the war is over. Now facing criminal charges, they are forced to take refuge in a stagecoach stop. Well-done horse opera with a wry sense of humor.
1952 80m/C Randolph Scott, Donna Reed, Claude Jarman Jr., Frank Faylen, Glenn Langan, Richard Denning, Lee Marvin, Jeanette Nolan; *Dir:* Roy Huggins. VHS, Beta *GKK* 🐾🐾½

Hangmen Ex-CIA agents battle it out on the East side of New York. If you're into very violent, very badly-acted films, then this one's for you.
1987 88m/C Jake La Motta, Rick Washburne, Doug Thomas; *Dir:* J. Christian Ingvordsen. VHS, Beta $29.95 *ACA* Woof!

Hanky Panky An insipid comic thriller in which Wilder and Radner become involved in a search for top-secret plans.
1982 (PG) 103m/C Gene Wilder, Gilda Radner, Richard Widmark, Kathleen Quinlan; *Dir:* Sidney Poitier. VHS, Beta, LV $79.95 *COL* 🐾½

Hanna K. The gripping story of divided passions set in the tumultuous state of Israel. An American lawyer tries to settle her personal and political affairs with various Middle-Eastern men. Unpopular for its pro-Palestine stance, it is only a mediocre achievement from both star and director.
1983 (R) 111m/C *FR* Jill Clayburgh, Gabriel Byrne, Jean Yanne, Muhamad Bakri, David Clennon, Oded Kotler; *Dir:* Constantin Costa-Gavras. VHS, Beta $59.95 *MCA* 🐾½

Hannah and Her Sisters Woody Allen's grand epic about a New York show-biz family, its three adult sisters and their various complex romantic entanglements. Excellent performances by the entire cast, especially Caine and Hershey. Classic Allen themes of life, love, death, and desire are explored through Allen's most assured and sensitive film-making ever. Witty, ironic, and heartwarming.
1986 (PG) 103m/C Mia Farrow, Barbara Hershey, Dianne Wiest, Michael Caine, Woody Allen, Maureen O'Sullivan, Lloyd Nolan, Sam Waterston, Carrie Fisher, Max von Sydow, Julie Kavner, Daniel Stern, Tony Roberts, John Turturro; *Dir:* Woody Allen. Academy Awards '86: Best Original Screenplay, Best Supporting Actor (Caine), Best Supporting Actress (Wiest); New York Film Critics Awards '86: Best Film. VHS, Beta, LV $19.99 *HBO, BTV* 🐾🐾🐾½

Hanna's War A moving film about Hanna Senesh, a young Hungarian Jew living in Palestine, who volunteered for a suicide mission behind Nazi lines during World War II. After her capture by the Germans, she was killed, and is now considered by many to have been a martyr. Detmers is very believable as Hanna and Burstyn puts in a good performance as her mother.
1988 (PG-13) 148m/C Ellen Burstyn, Maruschka Detmers, Anthony Andrews, Donald Pleasence, David Warner, Denholm Elliott, Vincenzo Ricotta, Ingrid Pitt; *Dir:* Menahem Golan. VHS, Beta, LV $9.99 *CCB, MED* 🐾🐾½

Hannie Caulder A woman hires a bounty hunter to avenge her husband's murder and her own rape at the hands of three bandits. Excellent casting but uneven direction. Boyd is uncredited as the preacher.
1972 (R) 87m/C Raquel Welch, Robert Culp, Ernest Borgnine, Strother Martin, Jack Elam, Christopher Lee, Diana Dors, Stephen Boyd; *Dir:* Burt Kennedy. VHS $14.95 *PAR, FCT* 🐾🐾

Hanoi Hilton A brutal drama about the sufferings of American POWs in Vietnamese prison camps. Non-stop torture, filth, and degradation.
1987 (R) 126m/C Michael Moriarty, Jeffrey Jones, Paul LeMat, Lawrence Pressman, David Soul; *Dir:* Lionel Chetwynd. VHS, Beta, LV $19.98 *WAR* 🐾½

Hanoi Rocks: The Nottingham Tapes The heavy metal band performs live, 14 of their most popular songs, including "Motorvatin'," "Up Around the Bend," and "I Feel Alright."
1984 60m/C Hanoi Rocks. VHS, Beta $29.95 *MVD* 🐾½

Hanoi Rocks: Video LP These Alice Cooper-influenced musicians bridge the gap between punk and metal. Includes tunes recorded live at London's Marquee, including "Pipeline," "Oriental Beat," "Motorvatin',"

"Tragedy," "I Feel Alright," "Blitzkreig Bop" and 10 more.
1985 55m/C Hanoi Rocks. **VHS, Beta $29.95** SVS 🎬½

Hanover Street An American bomber pilot and a married British nurse fall in love in war-torn Europe. Eventually the pilot must work with the husband of the woman he loves on a secret mission. A sappy, romantic tearjerker.
1979 (PG) 109m/C Harrison Ford, Lesley-Anne Down, Christopher Plummer, Alec McCowan; **W/Dir:** Peter Hyams. **VHS, Beta $59.95** COL 🎬

Hans Brinker Young Hans Brinker and his sister participate in an ice skating race, hoping to win a pair of silver skates. A musical version of the classic tale.
1969 103m/C Robin Askwith, Eleanor Parker, Richard Basehart, Cyril Ritchard, John Gregson; **Dir:** Robert Scheerer. **VHS, Beta $19.95** WAR, OM 🎬🎬½

Hans Christian Andersen Sentimental musical story of Hans Christian Andersen, a young cobbler who has a great gift for storytelling. Frank Loesser's score features "Inchworm" and "Wonderful Copenhagen." Written by Moss Hart. Digitally remastered editions available with stereo sound and original trailer.
1952 112m/C Danny Kaye, Farley Granger, Zizi Jeanmarie, Joey Walsh; **Dir:** Charles Vidor. **VHS, Beta, LV $14.98** HBO, SUE, FCT 🎬🎬

Hansel & Gretel Grimm's fairy tale tells the story of the woodcutter's children who venture into the forest and are caught in the clutches of a wicked old witch. Puppet animation.
1954 82m/C **Voices:** Anna Russell, Mildred Dunnock. **VHS, Beta $19.95** MED 🎬½

Hansel & Gretel From Shelly Duvall's "Faerie Tale Theatre" comes the story of two young children who get more than they bargained for when they eat a gingerbread house.
1982 51m/C Rick Schroder, Joan Collins, Paul Dooley, Bridgette Anderson; **Dir:** James Frawley. **VHS, Beta $14.98** KUI, FOX, FCT 🎬🎬

Hanussen The third of Szabo and Brandauer's evil-of-power trilogy (after "Mephisto" and "Colonel Redl"), in which a talented German magician and clairvoyant collaborates with the Nazis during their rise to power in the 1930s, although he can foresee the outcome of their reign. In German with English subtitles. Nominated for a foreign film Oscar.
1989 (R) 110m/C GE HU Klaus Maria Brandauer, Erland Josephson, Walter Schmidinger; **Dir:** Istvan Szabo. **VHS, Beta, LV $79.95** COL 🎬🎬🎬½

The Happiest Millionaire A Disney film about a newly immigrated lad who finds a job as butler in the home of an eccentric millionaire. Based on the book "My Philadelphia Father," by Kyle Chrichton.
1967 118m/C Fred MacMurray, Tommy Steele, Greer Garson, Geraldine Page, Lesley Ann Warren, John Davidson; **Dir:** Norman Tokar. **VHS, Beta** DIS, OM 🎬🎬

Happily Ever After A Las Vegas starmaker recruits and manipulates two green performers. Show business life in Vegas is given a truthful treatment in this show. Made for television.
1982 95m/C Suzanne Somers, John Rubinstein, Eric Braeden; **Dir:** Robert Scheerer. **VHS, Beta $69.95** TLF 🎬

Happily Ever After A Brazilian housewife meets a bisexual transvestite and has a passionate affair, bringing her into the depths of the criminal underworld. In Portuguese with English subtitles. Contains nudity.
1986 106m/C Regina Duarte, Paul Castelli; **Dir:** Bruno Barreto. **VHS, Beta $69.95** FCT 🎬🎬

Happiness Another zany comedy from the land of the hammer and sickle. Banned in Russia for 40 years, it was deemed to be risque and a bit too biting with the social satire. Silent with orchestral score.
1934 69m/B RU **Dir:** Alexander Medvedkin. **VHS $29.95** KIV, FCT 🎬🎬

Happiness Is... Mike Adams is the perfect boy next door until a series of pranks and lies lead to disaster. He returns to the right path through Christian love.
1988 68m/C Diane Bertouille, Brent Campbell, Eddie Moran, Carleton Smith; **Dir:** Russell S. Doughten Jr. **VHS, Beta** MIV, BPG 🎬

Happy Anniversary 007: 25 Years of James Bond Film clip retrospective of the dashing British spy created by Ian Fleming. Hosted by Roger Moore.
1987 59m/C Sean Connery, Roger Moore, George Lazenby, Timothy Dalton; **Dir:** Mel Stuart. **VHS, Beta $14.98** FOX

Happy Birthday to Me Several elite seniors at an exclusive private school are mysteriously killed by a deranged killer who has a fetish for cutlery.
1981 (R) 108m/C CA Melissa Sue Anderson, Glenn Ford, Tracy Bregman, Jack Blum, Matt Craven; **Dir:** J. Lee Thompson. **VHS, Beta, LV $69.95** COL 🎬

The Happy Ending Woman struggles with a modern definition of herself and her marriage, causing pain and confusion for her family. Simmons is solid as the wife and mother seeking herself. Michel LeGrand's theme song "What Are You Doing With the Rest of Your Life?" was a big hit. Simmons won an Oscar nomination.
1969 (PG) 112m/C Jean Simmons, John Forsythe, Lloyd Bridges, Shirley Jones, Teresa Wright, Dick Shawn, Nanette Fabray, Bobby Darin, Tina Louise; **Dir:** Richard Brooks. **VHS $14.95** WKV 🎬🎬½

Happy Face Cartoon Classics A series of cartoon compilations featuring public domain cartoons from Warner Bros., Max Fleischer and others.
194? 25m/C **VHS, Beta** GHV

The Happy Gigolo Soft-core fun about a gigolo in a European hotel making his nightly rounds. Foreign.
197? 73m/C Peter Ham, Margaret Rose Keil, Eva Gross, Rose Gardner, Al Price; **Dir:** Ilja Nutrof. **VHS, Beta $29.95** ACA 🎬

Happy Go Lovely An unscrupulous theatrical producer casts a chorus girl in a leading role in order to get closer to her boyfriend's fortune. Tired plot, but good performances in this lightweight musical.
1951 95m/C Vera-Ellen, David Niven, Cesar Romero, Bobby Howes; **Dir:** H. Bruce Humberstone. **VHS, Beta $24.95** NOS, MRV, DVT 🎬🎬

Happy Hooker Xaviera Hollander's cheery memoir of her transition from office girl to "working girl" has been brought to the screen with a sprightly air of naughtiness. Followed by "The Happy Hooker Goes to Washington" and "The Happy Hooker Goes Hollywood."
1975 (R) 96m/C Lynn Redgrave, Jean-Pierre Aumont, Nicholas Pryor; **Dir:** Nicholas Sgarro. **VHS, Beta, LV $29.95** CAN, VES 🎬½

Happy Hooker Goes Hollywood The third film inspired by the title of Xaviera Hollander's memoirs, in which the fun-loving Xaviera comes to Hollywood with the intention of making a movie based on her book, but soon meets up with a series of scheming, would-be producers.
1980 (R) 86m/C Martine Beswick, Chris Lemmon, Adam West, Phil Silvers; **Dir:** Alan Roberts. **VHS, Beta, LV $39.98** CAN, MCA 🎬

Happy Hooker Goes to Washington The further adventures of the world's most famous madam find Xaviera Hollander testifying before the U.S. Senate. Second in the trilogy of Happy Hooker pictures, which also include "The Happy Hooker" and "The Happy Hooker Goes Hollywood."
1977 (R) 89m/C Joey Heatherton, George Hamilton, Ray Walston, Jack Carter; **Dir:** William A. Levey. **VHS, Beta $69.95** CAN, VES 🎬

Happy Hour A young brewing scientist discovers a secret formula for beer, and everyone tries to take it from him. Little turns in a good performance as a superspy.
1987 (R) 88m/C Richard Gilliland, Jamie Farr, Tawny Kitaen, Ty Henderson, Rich Little; **Dir:** John DeBello. **VHS, Beta $19.95** IVE 🎬½

Happy New Year Charming romantic comedy in which two thieves plan a robbery but get sidetracked by the distracting woman who works next door to the jewelry store. Available in both subtitled and dubbed versions. Also called "The Happy New Year Caper." Remade in 1987.
1974 114m/C FR IT Francoise Fabian, Lino Ventura, Charles Gerard, Andre Falcon; **Dir:** Claude Lelouch. **VHS, Beta $29.98** SUE 🎬🎬🎬

Happy New Year Two sophisticated thieves plan and execute an elaborate jewel heist that goes completely awry. Remake of the 1974 French film of the same name.
1987 (PG) 86m/C Peter Falk, Wendy Hughes, Tom Courtenay, Charles Durning; **Dir:** John G. Avildsen. **VHS, Beta $79.95** COL 🎬🎬½

Happy New Year, Charlie Brown Strife upsets the Peanuts gang as the New Year's Eve gala becomes a source of tension.
1982 25m/C **Dir:** Bill Melendez, Sam Jaimes. **VHS, Beta $9.95** KRT, KAR 🎬🎬½

Happy Sex An Italian-made farce about nymphomaniac women in a small Italian village.
1986 60m/C IT Marina Frajese, Guia Lauri. **VHS, Beta $29.95** MED 🎬

Happy Together An eager freshman accidentally gets a beautiful, impulsive girl as his roommate. Together they meet the challenges of secondary education.
1989 (PG-13) 102m/C Helen Slater, Patrick Dempsey, Dan Schneider, Marius Weyers, Barbara Babcock; **Dir:** Mel Damski. **VHS, Beta, LV $89.95** IVE 🎬

Hard Body Workout Beautiful girls in an erotic aerobic workout.
1989 60m/C **VHS, Beta, LV $39.95** CEL, IME

Hard-Boiled Mahoney Slip, Sach, and the rest of the Bowery Boys try to solve a mystery involving mysterious women and missing men. The last film in the series for Bobby Jordan, whose career was ended

when he was injured in an accident involving a falling elevator.
1947 64m/B Leo Gorcey, Huntz Hall, Bobby Jordan, Billy Benedict, David Gorcey, Gabriel Dell, Teala Loring, Dan Seymour, Bernard Gorcey, Patti Brill, Betty Compson; *Dir:* William Beaudine. **VHS** $14.98 *WAR ♂♂½*

Hard Choices A 15-year-old Tennessee boy is unjustly charged as an accessory to murder, until a female social worker decides to help him. From then on, nothing is predictable. Excellent work by Klenck, McCleery, and Seitz. Don't miss Sayles as an unusual drug dealer. Intelligent, surprising, and powerful. Based on a true story, this is a low profile film that deserves to be discovered.
1984 90m/C Margaret Klenek, Gary McCleery, John Seitz, John Sayles, Liane Curtis, J.T. Walsh, Spalding Gray; *W/Dir:* Rick King. **VHS, Beta** $79.95 *KRT, LHV, WAR ♂♂♂*

Hard Country Caught between her best friend's success as a country singer and the old values of a woman's place, a small town girl questions her love and her life style. A warm and intelligent rural drama. Kim Basinger's debut. Music by Michael Martin Murphy.
1981 (PG) 104m/C Jan-Michael Vincent, Kim Basinger, Michael Parks, Gailard Sartain, Tanya Tucker, Ted Neely, Darryl Hannah, Richard Moll; *Dir:* David Greene. **VHS, LV** $19.95 *JTC, JWD ♂♂½*

Hard Day's Night The Beatles' first film is a joyous romp through an average "day in the life" of the Fab Four, shot in a pseudo-documentary style with great flair by Lester. Songs include "If I Fell" and "Can't Buy Me Love." Noted as the first music video. Laser version includes original trailer and an interview with Lester.
1964 90m/B *GB* John Lennon, Paul McCartney, George Harrison, Ringo Starr; *Dir:* Richard Lester. **VHS, Beta, LV** $29.95 *MVD, MPI, VYG ♂♂♂½*

Hard Drivin' Rowdy action featuring Southern stock car drivers with well shot race scenes from the Southern 500.
1960 92m/C Rory Calhoun, John Gentry, Alan Hale Jr.; *Dir:* Paul Helmick. **VHS, Beta** *MAG ♂*

Hard Frame An ex-con returns home hoping all will turn to sweetness and light when he attempts to make nice-nice with his stepdad, who wouldn't stick by him during his trial. Reynolds' debut into TV moviedom; could be the hirsute actor was thinking of this one when he said his movies were the kind they show in prisons and airplanes because nobody can leave. Originally titled "Hunters are for Killing."
1970 100m/C Burt Reynolds, Melvyn Douglas, Suzanne Pleshette, Larry Storch, Martin Balsam, Peter Brown, Jill Banner, Donald (Don "Red") Barry, Angus Duncan, Ivor Francis; *Dir:* Bernard Girard. **VHS** $9.95 *SIM, IPI ♂*

Hard to Hold Rockin' Rick's lukewarm film debut where he falls in love with a children's counselor after an automobile accident. Springfield sings "Love Somebody" with music by Peter Gabriel.
1984 (PG) 93m/C Rick Springfield, Janet Eilber, Patti Hansen, Albert Salmi, Monique Gabrielle; *Dir:* Larry Peerce. **VHS, Beta, LV** $69.95 *MCA ♂½*

Hard Hombre Gibson rides and shoots across the screen in one of his best western adventures.
1931 60m/B Hoot Gibson, Lena Basquette, Skeeter Bill Robbins, Jack Byron, Glenn Strange; *Dir:* Otto Brower. **VHS, Beta** $29.95 *GPV, VCN, UHV ♂♂*

Hard to Kill Policeman Seagal is shot and left for dead in his bedroom, but survives against the odds, though his wife does not. After seven years, he is well enough to consider evening the score with his assailants. He hides while training in martial arts for the final battle. Strong outing from Seagal, with good supporting cast.
1989 (R) 96m/C Steven Seagal, Kelly Le Brock, Bill Sadler, Frederick Coffin, Bonnie Burroughs, Zachary Rosencrantz, Dean Norris; *Dir:* Bruce Malmuth. **VHS, Beta, LV, 8mm** $14.00 *WAR, PIA ♂♂*

Hard Knocks An ex-con goes from one tragic situation to another, and is eventually stalked by shotgun-wielding rednecks in this grim, powerful drama. Excellent performance by Mann holds this picture together.
1980 85m/C *AU* Tracey Mann, John Arnold, Bill Hunter, Max Cullen, Tony Barry; *Dir:* Don McLennan. Australian Film Institute '80: Best Actress (Mann), Best Actress (Mann). **Beta** $59.95 *MFI ♂♂½*

Hard Knocks A young man on the run is tormented by an aging actress and her husband. His attempt to leave a past of violence and pain creates even more havoc.
19?? 90m/C Michael Christian, John Crawford, Donna Wilkes, Keenan Wynn. **VHS, Beta** *MFI ♂♂½*

Hard Knox A hard-nosed Marine pilot is dumped from the service and takes up command at a military school filled with undisciplined punks. A made-for-television movie.
1983 96m/C Robert Conrad, Frank Howard, Alan Ruck, Red West, Bill Erwin, Dean Hill, Joan Sweeney; *Dir:* Peter Werner. **VHS, Beta** $59.95 *KRT, LHV ♂*

Hard Part Begins A naive country singer gets sucked into the pratfalls of show business.
1987 87m/C Donnelly Rhodes, Nancy Belle Fuller, Paul Bradley. **VHS, Beta** $49.95 *TWE ♂*

Hard Promises An innocuous domestic comedy-drama about Joey, a man with wandering feet who hasn't been home in so long he doesn't even know his wife has divorced him—until he is accidentally invited to her wedding! Joey hightails it back to discover the intended groom was a hated high-school rival, so he decides to battle to win his ex-wife back. But this time his macho charm may not be enough to smooth out all the marital rough spots.
1992 (PG) 95m/C Sissy Spacek, William L. Petersen, Mare Winningham, Brian Kerwin; *Dir:* Martin Davidson. **VHS, LV, 8mm** *COL ♂♂½*

Hard Rain-The Tet 1968 An American platoon stationed in Saigon must mobilize in the wake of the North Vietnamese Army's Tet offensive. An episode of television's "Tour of Duty." Other episodes are also available on tape.
1989 89m/C Terence Knox, Stephen Caffrey; *Dir:* Bill L. Norton. **VHS** $59.95 *NWV*

Hard Rock Zombies Four heavy metal band members die horribly on the road and are brought back from the dead as zombies. This horror/heavy metal spoof is somewhat amusing in a goofy way.
1985 (R) 90m/C E.J. Curcio, Sam Mann; *Dir:* Krishna Shah. **VHS, Beta** $69.98 *LIV, VES ♂½*

Hard Ticket to Hawaii A shot-on-video spy thriller about an agent trying to rescue a comrade in Hawaii from a smuggling syndicate. Supporting cast includes many Playboy Playmates.
1987 (R) 96m/C Ron Moss, Dona Speir, Hope Marie Carlton, Cynthia Brimhall; *Dir:* Andy Sidaris. **VHS, Beta** $79.95 *KRT, LHV, WAR ♂*

Hard Times A Depression-era drifter becomes a bare knuckle street fighter, and a gambler decides to promote him for big stakes. One of grade-B meister Walter Hill's first genre films. A quiet, evocative drama.
1975 (PG) 92m/C Charles Bronson, James Coburn, Jill Ireland, Strother Martin; *Dir:* Walter Hill. **VHS, Beta, LV** $12.95 *GKK ♂♂♂*

Hard Traveling An unemployed Depression-era farmworker is accused of murder. He must fight a hostile court system to maintain his dignity and salvage the love and respect of his wife. Based on a novel by Alvah Bessie.
1985 (PG) 99m/C J.E. Freeman, Ellen Geer, Barry Corbin, James Gammon, Jim Haynie; *Dir:* Dan Bessie. **VHS, Beta** $19.95 *STE, NWV ♂♂*

Hard Way A world-weary assassin is pressured by his boss to perform one last assignment. A surprise awaits him if he accepts.
1980 88m/C Patrick McGoohan, Lee Van Cleef, Donal McCann, Edna O'Brien; *Dir:* Mihael Dryhurst. **VHS, Beta** $59.95 *TWE ♂♂*

The Hard Way A Hollywood super-star decides to learn the cop business first hand on the mean streets of New York. The cop is in pursuit of a serial killer, but informed by his commanders that he WILL do the necessary training. Not especially innovative script or direction, but Woods and Fox bring such intensity and good humor to their roles that the package works. Fox has fun poking fun at the L.A. life-style, with a terrific cameo appearance by Marshall as his agent. Woods, known for his over the edge acting, goes as far as he can. Silly finale almost destroys the film, but other small vignettes - the bar scene when Fox play-acts as a woman - are terrific.
1991 (R) 111m/C James Woods, Michael J. Fox, Annabella Sciorra, Stephen Lang, Penny Marshall, L.L. Cool J.; *Dir:* John Badham. **VHS, Beta, LV** $19.98 *MCA, PIA ♂♂♂*

Hardbodies Three middle-aged men hit the beaches of Southern California in search of luscious young girls. Followed by a sequel.
1984 (R) 88m/C Grant Cramer, Teal Roberts, Gary Wood, Michael Rapport, Sorrels Pickard, Roberta Collins, Cindy Silver, Courtney Gains, Kristi Somers, Crystal Shaw, Kathleen Kinmont, Joyce Jameson; *Dir:* Mark Griffiths. **VHS, Beta, LV** $79.95 *COL* Woof!

Hardbodies 2 A sequel to the original comedy hit, dealing with film crew in Greece that is distracted by hordes of nude natives. Sophomoric humor dependent on nudity and profanity for laughs.
1986 (R) 89m/C Brad Zutaut, Brenda Bakke, Fabiana Udenio, James Karen; *Dir:* Mark Griffiths. **VHS, Beta, LV** $9.95 *COL* Woof!

Hardcase and Fist A framed cop leaves prison with his mind on revenge. The godfather who used him is just his first target.
1989 (R) 92m/C Maureen Lavette, Ted Prior; *Dir:* Tony Zarindast. **VHS, Beta** $79.98 *MCG ♂♂*

Hardcore A Midwestern businessman who is raising a strict Christian family learns of his daughter's disappearance while she is on a church trip. After hiring a streetwise investigator, he learns that his daughter has become an actress in pornographic films out in California. Strong performance by Scott and glimpse into the hardcore pornography

industry prove to be convincing points of this film, though it exploits what it condemns.
1979 (R) 106m/C George C. Scott, Season Hubley, Peter Boyle, Dick Sargent; *Dir:* Paul Schrader. **VHS, Beta, LV** $9.95 COL &&½

Hardcore, Volume 1 A compilation of live clips of America's and Britain's most notorious hardcore punk groups including Black Flag, Honor, the Sex Pistols, and the Toxic Reasons.
1985 60m/C Black Flag, Honor, The Sex Pistols, The Toxic Reasons. **VHS, Beta** $39.95 MVD, TGV

Hardcore, Volume 2 More live clips of infamous hardcore punk bands. Featuring the songs "No God," "Wasted," "FVK," and "War Dance" from bands such as the Sleepers, the Germs, and the Bad Brains.
1985 30m/C Sleepers, Germs, Bad Brains, TSOL. **VHS, Beta** $29.95 MVD, TGV

The Harder They Come A poor Jamaican youth becomes a success with a hit reggae record after he has turned to a life of crime out of desperation. Songs, which are blended nicely into the framework of the film, include "You Can Get It If You Really Want it" and "Sitting In Limbo."
1972 (R) 93m/C Jimmy Cliff, Janet Barkley, Carl Bradshaw; *Dir:* Perry Henzell. **VHS, Beta, LV** $19.95 IVA, HBO, CRC &&½

Harder They Fall A cold-eyed appraisal of the scum-infested boxing world. An unemployed reporter (Bogart in his last role) promotes a fighter for the syndicate, while doing an expose on the fight racket. Bogart became increasingly debilitated during filming and died soon afterward. Based on Budd Schulberg's novel.
1956 109m/B Humphrey Bogart, Rod Steiger, Jan Sterling, Mike Lane, Max Baer; *Dir:* Mark Robson. **VHS, Beta** $19.98 CCB, COL &&&

Hardhat & Legs Comic complications arise when a New York construction worker falls in love with the woman who's teaching the modern sexuality course in which he's enrolled. Made for television; Ruth Gordon co-scripted this one.
1980 96m/C Sharon Gless, Kevin Dobson, Ray Serra, Elva Josephson, Bobby Short, Jacqueline Brookes, W.T. Martin; *Dir:* Lee Philips. **VHS, Beta** $69.95 LTG &&

Hardly Working A circus clown finds it difficult to adjust to real life as he fumbles about from one job to another. Lewis's last attempt at resuscitating his directorial career. His fans will love it, but others need not bother.
1981 (PG) 90m/C Jerry Lewis, Susan Oliver, Roger C. Carmel, Gary Lewis, Deanna Lund; *Dir:* Jerry Lewis. **VHS, Beta** $59.98 FOX &

Hardware Rag-tag ripoff in which McDermott is a post-apocalyptic garbage picker who gives girlfriend Travis some robot remains he collected (ours is not to ask why), oblivious to the fact that the tidy android was a government-spawned population controller programmed to destroy warm bodies. Seems old habits die hard, and that spells danger, danger, danger, Will Robinson. Much violence excised to avoid x-rating.
1990 (R) 94m/C Dylan McDermott, Stacy Travis, John Lynch, Iggy Pop; *Dir:* Richard Stanley. **VHS, Beta, LV** $92.95 HBO &½

Hardware Wars This parody of the spectacular space epic creates its special effects from household appliances available in any hardware store.
1978 13m/C **VHS, Beta, 3/4U** PYR &½

Hardware Wars & Other Film Farces A collection of four award-winning spoofs of big-budget film epics, featuring "Hardware Wars," "Porklips Now," "Bambi Meets Godzilla," and "Closet Cases of the Nerd Kind." Some black-and-white segments.
1981 (G) 48m/C **VHS, Beta** WAR, OM

Harem A beautiful stockbroker gets kidnapped by a wealthy OPEC oil minister and becomes part of his harem. Kingsley stars as the lonely sheik who longs for the love of a modern woman.
1985 107m/C FR Ben Kingsley, Nastassia Kinski; *Dir:* Alfred Joffe. **VHS, Beta, LV** $79.98 LIV, VES &

Harlan County, U.S.A. The emotions of 180 coal mining families are seen up close in this classic documentary about their struggle to win a United Mine Workers contract in Kentucky. Award-winning documentary.
1976 (PG) 103m/C *Dir:* Barbara Kopple. Academy Awards '76: Best Feature Documentary; National Board of Review Awards '77: 10 Best Films of the Year. **VHS, Beta** $19.98 COL &

Harlem Nights Two Harlem nightclub owners in the 1930s battle comically against efforts by the Mob and crooked cops to take over their territory. High-grossing, although somewhat disappointing effort from Murphy directed, wrote, produced, and starred in this film.
1989 (R) 118m/C Eddie Murphy, Richard Pryor, Redd Foxx, Danny Aiello, Jasmine Guy, Michael Lerner, Arsenio Hall, Della Reese; *Dir:* Eddie Murphy. **VHS, Beta, LV** $14.95 PAR &&

Harlem on the Prairie This first-ever western with an all-black cast has Jeffries helping a young lady find hidden gold and foiling a villain. Even though the technical standards and acting are below par, this is worth watching for the singing and a comedy routine done by a well-known contemporary black team. Jeffries also sings "Romance in the Rain" and the title tune.
1938 54m/B Herbert Jeffries, Mantan Moreland, Flourney E. Miller, Connie Harris, Maceo B. Sheffield, William Spencer Jr.; *Dir:* Sam Newfield. **VHS** $24.95 NOS &

Harlem Rides the Range Jeffries must outsmart villainous Brooks, who means to swindle a radium mine away from its rightful owner. Early oater of interest only because of its atypical all-black cast and crew.
1939 58m/B Herbert Jeffries, Lucius Brooks, Artie Young, F.E. Miller, Spencer Williams, Clarence Brooks, Tom Southern, Wade Dumas, The Four Tunes, Leonard Christmas; *Dir:* Richard C. Kahn. **VHS** $24.95 NOS, MRV, TPV &½

Harley Diamond is Harley, an L.A. thug-on-a-hog, who's on his way to an extended stay at one of the state's luxury institutions when he's sent to a Texas rehabilitation community. There, he bonds with an ex-biker, and may just find his way back down the straight and narrow. And the local longhorn rednecks applaud his efforts...not.
1990 (PG) 80m/C Lou Diamond Phillips, Eli Cummings, DeWitt Jan, Valentine Kim; *Dir:* Fred Holmes. **VHS** $89.95 VMK &

Harley Davidson and the Marlboro Man An awful rehash of "Butch Cassidy and the Sundance Kid" with blatant vulgarity and pointless sci-fi touches. The title duo are near-future outlaws who rob a bank to save their favorite bar, then find

they've stolen mob money. Some action, but it's mostly talk: lewd, meant-to-be-whimsical soul-probing chats between H.D. and M.M. that would bore even the biker crowd—and did.
1991 (R) 98m/C Mickey Rourke, Don Johnson, Chelsea Field, Tom Sizemore, Vanessa Williams, Robert Ginty; *Dir:* Simon Wincer. **VHS, Beta, LV, 8mm** $94.99 MGM, PIA &

Harlow The more lavish of the two Harlow biographies made in 1965, both with the same title. A sensationalized "scandal sheet" version of Jean Harlow's rise to fame that bears little resemblance to the facts of her life.
1965 125m/C Carroll Baker, Martin Balsam, Red Buttons, Michael Connors, Angela Lansbury, Peter Lawford, Raf Vallone, Leslie Nielsen; *Dir:* Gordon Douglas. **VHS, Beta** $59.95 PAR &&½

Harmony Lane A musical semi-biography of Stephen Foster. Including the familiar Foster songs: "Oh Susanna," "My Old Kentucky Home," and "Swanee River."
1935 84m/B Douglass Montgomery, Evelyn Venable, William Frawley, Adrienne Ames, Joseph Cawthorn, Clarence Muse; *Dir:* Joseph Santley. **VHS, Beta** $24.95 DVT &½

Harmony Trail Maynard stars in another rip-roaring western adventure.
1944 60m/B Ken Maynard. **VHS, Beta** $19.95 NOS, VCN, DVT &

Harold and Maude A cult classic pairing Bud Cort as a dead-pan disillusioned 20-year-old obsessed with suicide (his staged attempts are a highlight) and a loveable Ruth Gordon as a fun-loving 80-year-old eccentric. They meet at a funeral (a mutual hobby), and develop a taboo romantic relationship, in which they explore the tired theme of the meaning of life with a fresh perspective. Features music by the pre-Islamic Cat Stevens.
1971 (PG) 92m/C Ruth Gordon, Bud Cort, Cyril Cusack, Vivian Pickles, Charles Tyner, Ellen Geer; *Dir:* Hal Ashby. **VHS, Beta, LV, 8mm** $14.95 PAR &&&

Harper A tight, fast-moving genre piece about cynical LA private eye Newman investigating the disappearence of Bacall's wealthy husband. From a Ross McDonald novel, adapted by William Golding. Also called "The Moving Target." Later sequelled in "The Drowning Pool."
1966 121m/C Paul Newman, Shelley Winters, Lauren Bacall, Julie Harris, Robert Wagner, Janet Leigh, Arthur Hill; *Dir:* Jack Smight. Edgar Allan Poe Awards '66: Best Screenplay. **VHS, Beta, LV** $19.98 WAR, PIA &&&

Harper Valley P.T.A. Barbara Eden raises hell in Harper Valley after the PTA questions her parental capabilities. TV series followed.
1978 (PG) 93m/C Barbara Eden, Nanette Fabray, Louis Nye, Pat Paulsen, Ronny Cox; *Dir:* Richard Bennett. **VHS, Beta** $19.98 VES &&

Harrad Experiment Adaptation of Robert Rimmer's love-power bestseller in which an experiment-minded college establishes a campus policy of sexual freedom. Famous for the Hedren/Johnson relationship, shortly before Johnson married Hedren's real-life daughter Melanie Griffith.
1973 98m/C James Whitmore, Tippi Hedren, Don Johnson, Bruno Kirby, Laurie Walters, Victoria Thompson, Elliot Street, Sharon Taggart, Robert Middleton, Billy Sands, Melanie Griffith; *Dir:* Ted Post. **VHS, Beta** UHV, VES &½

Harrad Summer Sequel to "The Harrad Experiment." College coeds take their sex-education to the bedroom where they can apply their knowledge by more intensive means.
1974 (R) 103m/C Richard Doran, Victoria Thompson, Laurie Walters, Robert Reiser, Bill Dana, Marty Allen; ***Dir:*** Steven Hilliard Stern. **VHS** *UHV* 🦴🦴

Harry Black and the Tiger Jungle adventure involving a one-legged hunter and a man-eating tiger. The hunter initally teams up with the old war buddy who cost him his leg. The motivation grows with each failure to catch the beast and to make matters worse, the friend's wife and son are lost somewhere in the perilous jungle. Filmed in India.
1958 107m/C *GB* Stewart Granger, Barbara Rush, Anthony Steel; ***Dir:*** Hugo Fregonese. **VHS** *TCF* 🦴🦴

Harry and the Hendersons An ordinary American family is on vacation in the Northwest when the family car has a collision with Bigfoot. Thinking that the big guy is dead, they throw him on top of the car and head home. Lo and behold, he revives and starts wrecking the furniture and in the process, endears himself to the at-first frightened family. But the arrival of Harry starts the neighbors talking, and the family decides he must return to the wild for his own good. Trouble is, Harry doesn't want to go so he escapes. And the hunt is on. Nice little tale efficiently told, with Lithgow fine as the frustrated dad trying to hold his Bigfoot-invaded home together.
1987 (PG) 111m/C John Lithgow, Melinda Dillon, Don Ameche, David Suchet; ***Dir:*** William Dear. Academy Awards '87: Best Makeup. **VHS, Beta, LV $19.95** *MCA, APD* 🦴🦴½

Harry & Son A widowed construction worker faces the problems of raising his son. Newman is miscast as the old man and we've all seen Benson play the young role too many times.
1984 (PG) 117m/C Paul Newman, Robby Benson, Ellen Barkin, Wilford Brimley, Judith Ivey, Ossie Davis, Morgan Freeman, Joanne Woodward; ***Dir:*** Paul Newman. **VHS, Beta, LV $79.98** *VES* 🦴🦴

Harry and Tonto A gentle comedy about an energetic septuagenarian who takes a cross-country trip with his cat, Tonto. Never in a hurry, still capable of feeling surprise and joy, he makes new friends and visits old lovers. Carney deserved his Oscar. Mazursky has seldom been better. Oscar nomination for Original Screenplay.
1974 115m/C Art Carney, Ellen Burstyn, Larry Hagman, Geraldine Fitzgerald, Chief Dan George, Arthur Hunnicutt, Josh Mostel; ***Dir:*** Paul Mazursky. Academy Awards '74: Best Actor (Carney); National Board of Review Awards '74: 10 Best Films of the Year. **VHS, Beta $19.98** *FOX, FCT, BTV* 🦴🦴🦴

Harry Tracy Whimsical tale of the legendary outlaw whose escapades made him both a wanted criminal and an exalted folk hero. Alss known as "Harry Tracy, Desperado."
1983 (PG) 111m/C *CA* Bruce Dern, Gordon Lightfoot, Helen Shaver, Michael C. Gwynne; ***Dir:*** William A. Graham. **VHS, Beta $69.95** *VES* 🦴🦴½

Harry & Walter Go to New York At the turn of the century, two vaudeville performers are hired by a crooked British entrepreneur for a wild crime scheme. The cast and crew try their hardest, but it's not enough to save this boring comedy. The vaudeville team of Caan and Gould perhaps served as a model for Beatty and Hoffman in "Ishtar."
1976 (PG) 111m/C James Caan, Elliott Gould, Michael Caine, Diane Keaton, Burt Young, Jack Gilford, Charles Durning, Lesley Ann Warren, Carol Kane; ***Dir:*** Mark Rydell. **VHS, Beta, LV $9.95** *COL* 🦴🦴

Harry's War A middle-class, middle-aged American declares military war on the IRS in this overdone comedy.
1984 (PG) 98m/C Edward Herrmann, Geraldine Page, Karen Grassle, David Ogden Stiers; ***Dir:*** Keith Merrill. **VHS, Beta $59.95** *IMG, TPI* 🦴🦴

Harum Scarum Elvis tune-fest time! When a movie star (Presley) travels through the Middle East, he becomes involved in an attempted assassination of the king and falls in love with his daughter. He also sings at the drop of a veil: "Shake That Tambourine," "Harem Holiday," and seven others.
1965 95m/C Elvis Presley, Mary Ann Mobley, Fran Jeffries, Michael Ansara, Billy Barty, Theo Marcuse, Jay Novello; ***Dir:*** Gene Nelson. **VHS, Beta $19.95** *MGM* 🦴🦴

Harvest A classically Pagnolian rural pageant-of-life melodrama, wherein a pair of loners link up together in an abandoned French town and completely revitalize it and the land around it. From the novel by Jean Giono. In French with English subtitles. Music by Arthur Honegger.
1937 128m/B *FR* Fernandel, Gabriel Gabrio, Orane Demazis, Edouard Delmont; ***Dir:*** Marcel Pagnol. National Board of Review Awards '39: 5 Best Foreign Films of the Year; New York Film Critics Awards '39: Best Foreign Film. **VHS, Beta $69.95** *INT, MRV, FCT* 🦴🦴🦴🦴

Harvest Melody Singing star Gilda Parker travels to farm country for a mere publicity stunt but decides to stay. Corny in more ways than one.
1943 70m/B Rosemary Lane, Johnny Downs, Sheldon Leonard, Charlotte Wynters, Luis Alberni, Claire Rochelle, Syd Saylor; ***Dir:*** Sam Newfield. **VHS $19.95** *NOS, LOO* 🦴🦴

Harvey Straightforward version of the Mary Chase play about a friendly drunk with an imaginary 6-foot rabbit friend named Harvey, and a sister who tries to have him committed. A fondly remembered, charming comedy. Hull is a standout, well deserving her Oscar. Stewart also received an Oscar nomination for Best Actor.
1950 104m/B James Stewart, Josephine Hull, Victoria Horne, Peggy Dow, Cecil Kellaway, Charles Drake, Jesse White, Wallace Ford, Nana Bryant; ***Dir:*** Henry Koster. Academy Awards '50: Best Supporting Actress (Hull). **VHS, Beta, LV $14.95** *MCA, BTV, RDG* 🦴🦴🦴½

The Harvey Girls Lightweight Musical about a restaurant chain that sends its waitresses to work in the Old West. Songs, by Johnny Mercer and Harry Warren, include "On the Atchison, Topeka and the Santa Fe," "In the Valley," and "It's a Great Big World."
1946 102m/C Judy Garland, Ray Bolger, John Hodiak, Preston Foster, Angela Lansbury, Virginia O'Brien, Marjorie Main, Chill Wills, Kenny L. Baker, Selena Royale, Cyd Charisse; ***Dir:*** George Sidney. Academy Awards '46: Best Song ("On the Atchison, Topeka and Santa Fe"). **VHS, Beta, LV $19.95** *MGM, FCT* 🦴🦴🦴

Has Anyone Seen My Pants? A naive farmboy from Texas inherits a hotel in Heidelberg, Germany. Hilarity and sensuality result when he discovers that the hotel is really a brothel.
19?? 86m/C Judith Fritsch, Franz Muxeneder. **VHS $39.95** *MED* 🦴🦴

A Hash House Fraud/The Sultan's Wife "A Hash House Fraud" (1915) features a frenetic Keystone chase. "The Sultan's Wife" (1917) concerns a woman who attracts the unwanted attention of a sultan.
1915 33m/B Louise Fazenda, Hugh Fay, Chester Conklin, Gloria Swanson, Bobby Vernon. **VHS, Beta $19.98** *CCB* 🦴🦴

The Hasty Heart A rowdy, terminally ill Scotsman brings problems and friendships to a makeshift war-time hospital built in the jungles of Burma. Originally filmed in 1949 with Richard Todd and Ronald Reagan in leading roles.
1986 100m/C Gregory Harrison, Cheryl Ladd, Perry King; ***Dir:*** Martin Speer. **VHS, Beta $59.95** *KRT, LHV, WAR* 🦴🦴

Hat Box Mystery A detective's secretary is tricked into shooting a woman and private eye Neal is determined to save her from prison. Although not outstanding, it's worth seeing as perhaps the shortest detective film (a mere 44 min.) to ever make it as a feature.
1947 44m/B Tom Neal, Pamela Blake, Allen Jenkins; ***Dir:*** Lambert Hillyer. **VHS, Beta, LV $16.95** *SNC, WGE, VYY* 🦴🦴

Hatari An adventure-loving team of professional big game hunters ventures to East Africa to round up animals for zoos around the world. Led by Wayne, they get into a couple of scuffs along the way, including one with lady photographer Martinelli who is doing a story on the expedition. Extraordinary footage of Africa and the animals. Fantastic musical score by Henry Mancini.
1962 158m/C John Wayne, Elsa Martinelli, Red Buttons, Hardy Kruger, Gerard Blain, Bruce Cabot; ***Dir:*** Howard Hawks. **VHS, Beta, LV $49.95** *PAR, BUR* 🦴🦴🦴

Hatchet for the Honeymoon A rather disturbed young man goes around hacking young brides to death as he tries to find out who murdered his wife. Typical sick Bava horror flick; confusing plot, but strong on vivid imagery.
1970 90m/C *SP IT* Stephen Forsythe, Dagmar Lassander, Laura Betti, Gerard Tichy; ***Dir:*** Mario Bava. **VHS, Beta $12.95** *MED, TIM* 🦴

Hatfields & the McCoys A made-for-television re-telling of the most famous feud in American history-the legendary mountain war between the Hatfields and the McCoys.
1975 90m/C Jack Palance, Steve Forrest, Richard Hatch, Karen Lamm; ***Dir:*** Clyde Ware. **VHS, Beta $49.95** *WOV* 🦴🦴

Hats Off Clarke and Payne star as opposing press agents in this tedious musical. The only bright spot is the finale, which features white-robed girls on a cloud-covered carousel.
1937 65m/B Mae Clarke, John Payne, Helen Lynd, Luis Alberni, Skeets Gallagher, Franklin Pangborn; ***Dir:*** Boris L. Petroff. **VHS $19.95** *NOS, MLB* 🦴

The Haunted The ancient curse of an Indian woman haunts a present day family by possessing the body of a beautiful girl. Through the girl, a horrible vengeance is carried out.
1979 81m/C Aldo Ray, Virginia Mayo, Anne Michelle, Jim Negele; ***Dir:*** Michael DeGaetano. **VHS, Beta $9.95** *SIM, VHE, HHE* 🦴

The Haunted Castle (Schloss Vogelod) One of Murnau's first films, and a vintage, if crusty, example of German expressionism. Silent, with English titles.
1921 56m/B Arnold Korff, Lulu Keyser-Korff; *Dir:* F. W. Murnau. **VHS, Beta $16.95** *SNC, NOS, VYY* 🐾🐾🐾

Haunted Honeymoon Arthritic comedy about a haunted house and a couple trapped there. Sad times for the Mel Brooks alumni participating in this lame horror spoof.
1986 (PG) 82m/C Gene Wilder, Gilda Radner, Dom DeLuise, Jonathan Pryce, Paul Smith, Peter Vaughan, Bryan Pringle, Roger Ashton-Griffiths, Jim Carter; *Dir:* Gene Wilder. **VHS, Beta, LV $34.95** *HBO* 🐾

The Haunted Palace Price plays both a 17th-century warlock burned at the stake and a descendant who returns to the family dungeon and gets possessed by the mutant-breeding forebearer. The movie has its own identity crisis, with title and ambience from Poe but story from H.P. Lovecraft's "The Case of Charles Dexter Ward." Respectable but rootless chills, also available with the quasi-Lovecraft "Curse of the Crimson Altar" on laser disc.
1963 87m/C Vincent Price, Debra Paget, Lon Chaney Jr., Frank Maxwell, Leo Gordon, Elisha Cook, John Dierkes; *Dir:* Roger Corman. **VHS, Beta $59.98** *HBO, MLB* 🐾🐾½

Haunted Ranch Reno Red has been murdered and a shipment of gold bullion is missing. A gang on the lookout for the gold tries to convince people that Red's ranch is haunted by his ghost.
1943 56m/B John "Dusty" King, David Sharpe, Max Terhune, Rex Lease, Julie Duncan; *Dir:* Robert Tansey. **VHS, Beta $19.95** *NOS, DVT, VYY* 🐾

The Haunted Strangler Boris Karloff is a writer investigating a 20-year old murder who begins copying some of the killer's acts. Also known as "The Grip of the Strangler."
1958 78m/B *GB* Boris Karloff, Anthony Dawson, Elizabeth Allan; *Dir:* Robert Day. **VHS $49.95** *MED, MPI* 🐾🐾½

Haunted Summer A glamorized version of the bacchanalian summer of 1816 spent by Lord Byron, Percy Shelley, Mary Shelley, and John Polidori, and others that resulted in the writing of "Frankenstein." Based on the novel of the same name by Anne Edwards.
1988 (R) 106m/C Alice Krige, Eric Stoltz, Philip Anglim, Laura Dern, Alex Winter; *Dir:* Ivan Passer. **VHS, Beta, LV $79.95** *MED* 🐾½

Haunted: The Ferryman Adapted by Julian Bond from a Kingsley Amis story, the film deals with a novelist who is mysteriously confronted with various events and macabre set-pieces from his own novel, including a dead ferryman rising from the grave.
1974 50m/C *GB* Jeremy Brett, Nastasha Parry, Lesley Dunlop; *Dir:* John Irvin. **VHS, Beta $59.95** *PSM* 🐾🐾

HauntedWeen All trick, no treat, as a guy nailed by a frat house prank twenty years ago comes-back-for-revenge.
1991 ?m/C Brien Blakely, Blake Pickett, Brad Hanks, Bart White, Leslee Lacey, Ethan Adler; *Dir:* Doug Robertson. **VHS $29.50** *CVD* 🐾½

The Haunting A subtle, bloodless horror film about a weekend spent in a monstrously haunted mansion by a parapsychologist, the mansion's skeptic heir, and two mediums. A chilling adaptation of Shirley Jackson's "The Haunting of Hill House," in which the psychology of the heroine is forever in question.

1963 113m/B Julie Harris, Claire Bloom, Russ Tamblyn, Richard Johnson; *Dir:* Robert Wise. **VHS, Beta, LV $19.98** *KUI, MGM* 🐾🐾🐾½

Haunting Fear Poe's "The Premature Burial" inspired this pale cheapie about a wife with a fear of early interment. After lengthy nightmares and graphic sex, her greedy husband uses her phobia in a murder plot.
1991 88m/C Jan-Michael Vincent, Karen Black, Brinke Stevens, Michael Berryman; *Dir:* Fred Olen Ray. **VHS $79.95** *RHI* 🐾

Haunting of Harrington House A tame teenage haunted house film originally produced for television.
1982 50m/C Dominique Dunne, Roscoe Lee Browne, Edie Adams, Phil Leeds; *Dir:* Murray Golden. **VHS, Beta $29.95** *GEM* 🐾

The Haunting of Julia Peter Straub wrote this eerie tale about a grief-stricken young woman whose child has recently died. To overcome her loss, she moves into a new house—which is haunted by the ghost of a long-dead child. Also called "Full Circle."
1977 (R) 96m/C *CA GB* Mia Farrow, Keir Dullea, Tom Conti, Jill Bennett, Robin Gammell; *Dir:* Richard Loncraine. **VHS, Beta $19.95** *MAG, MED* 🐾🐾

The Haunting of Morella Poe-inspired cheapjack exploitation about a an executed witch living again in the nubile body of her teen daughter. Ritual murders result, in between lesbian baths and nude swims. Producer Roger Corman did the same story with more class and less skin in his earlier anthology "Tales of Terror."
1991 (R) 82m/C David McCallum, Nicole Eggert, Maria Ford, Lana Clarkson. **VHS $89.98** *NHO* 🐾

The Haunting Passion Newlywed Seymour moves into a haunted house only to be seduced by the ghost of the former occupant's dead lover. Effective and erotic presentation.
1983 100m/C Jane Seymour, Gerald McRaney, Millie Perkins, Ruth Nelson, Paul Rossilli, Ivan Bonar; *Dir:* John Korty. **VHS, Beta $39.95** *IVE* 🐾🐾

Haunting of Sarah Hardy Ward, as a recently wed heiress, returns to the scene of her unhappy childhood home. In a standard plot, Ward is torn between an apparent haunting and the question of her own sanity. Made for TV film is redeemed by the creditable acting.
1989 92m/C Sela Ward, Michael Woods, Roscoe Born, Polly Bergen, Morgan Fairchild; *Dir:* Jerry London. **VHS, Beta, LV $59.95** *PAR, PIA* 🐾🐾

Haunts A tormented woman has difficulty distinguishing between fantasy and reality after a series of brutal slayings lead police to the stunning conclusion that dead people have been committing the crimes.
1977 (PG) 97m/C Cameron Mitchell, Aldo Ray, May Britt, William Gray Espy, Susan Nohr; *Dir:* Herb Freed. **VHS, Beta $59.95** *MED* 🐾

Haunts of the Very Rich A loose remake of "Outward Bound" in which a group of spoiled people on a vacation find themselves spiritually dead. Made for TV.
1972 72m/C Lloyd Bridges, Anne Francis, Donna Mills, Tony Bill, Robert Reed, Moses Gunn, Cloris Leachman, Ed Asner; *Dir:* Paul Wendkos. **VHS, Beta $59.95** *VMK* 🐾🐾½

Havana During the waning days of the Batista regime, a gambler travels to Havana in search of big winnings. Instead, he meets the beautiful wife of a communist revolutionary. Unable to resist their mutual physical

attraction, the lovers become drawn into a destiny which is far greater than themselves. Reminiscent of "Casablanca."
1990 (R) 145m/C Robert Redford, Lena Olin, Alan Arkin, Raul Julia; *Dir:* Sydney Pollack. **VHS, Beta, LV $92.95** *MCA, FCT* 🐾🐾½

Having It All A wry television comedy about a love triangle. Dyan Cannon stars as a successful business woman with a husband on each coast, who finds dividing her time is more than she bargained for.
1982 100m/C Dyan Cannon, Barry Newman, Hart Bochner, Melanie Chartoff, Sylvia Sidney; *Dir:* Edward Zwick. **VHS, Beta $39.95** *IVE* 🐾🐾

Having Wonderful Crime Madcap comedy-thriller about a newly married couple who combine sleuthing with a honeymoon. With their criminal lawyer friend along, the trio winds up searching for a magician who has vanished. Fast-paced action and quick dialogue.
1945 70m/B Pat O'Brien, George Murphy, Carole Landis, Lenore Aubert, George Zucco; *Dir:* Edward Sutherland. **VHS, LV $19.98** *TTC, FCT, IME* 🐾🐾½

Having Wonderful Time A young girl tries to find culture and romance on her summer vacation at a resort in the Catskills. Film debut of comic Red Skelton, who wasn't always as old as the hills. Based on a Broadway hit by Arthur Kober.
1938 71m/B Ginger Rogers, Lucille Ball, Eve Arden, Red Skelton, Douglas Fairbanks Jr.; *Dir:* Alfred Santell. **VHS, Beta $34.95** *RKO* 🐾

Having Wonderful Time/Carnival Boat A Ginger Rogers double feature: "Having Wonderful Time" (1938)-fun and frolic at an adult summer camp in the Catskills; and "Carnival Boat" (1932)-the lives and loves of rugged loggers in the Pacific Northwest.
1938 134m/B Ginger Rogers, Douglas Fairbanks Jr., Lucille Ball, Ann Miller, Red Skelton, William Boyd. **VHS, Beta $39.95** *RKO* 🐾🐾½

Hawaii James Michener's novel about a New England farm boy who decides in 1820 that the Lord has commanded him to the island of Hawaii for the purpose of "Christianizing" the natives. Filmed on location. Also available in a 181 minute version with restored footage. Oscar nominations for supporting actress Lagarde, color cinematography, sound, original score, special effects and costume design. Available in director's cut version with an additional 20 minutes.
1966 161m/C Max von Sydow, Julie Andrews, Richard Harris, Carroll O'Connor, Bette Midler, Gene Hackman; *Dir:* George Roy Hill. **VHS, Beta, LV $19.95** *FOX, BUR* 🐾🐾🐾

Hawaii Calls Two young stowaways on a cruise ship are allowed to stay on after one of them (Breen) struts his stuff as a singer. The two then turn sleuth to catch a gang of spies. More ham than a can of Spam, but still enjoyable.
1938 72m/B Bobby Breen, Ned Sparks, Irvin S. Cobb, Warren Hull, Gloria Holden, Pua Lani, Raymond Paige, Philip Ahn, Ward Bond; *Dir:* Edward Cline. **VHS $19.95** *DVT, NOS* 🐾🐾

Hawaiian Buckaroo A couple of cowboys fight to save their pineapple plantation from a sneaky foreman.
1938 62m/B Smith Ballew, Evelyn Knapp, Benny Burt, Harry Woods, Pat O'Brien; *Dir:* Ray Taylor. **VHS $19.95** *NOS, DVT* 🐾

Hawk of the Caribbean Minor Spanish pirate film is presented in an English-dubbed version.

H

197? 100m/C SP Johnny Desmond, Yvonne Monlaur, Armando Francioli. **VHS, Beta $39.95** VCD, JCI ♂ .

Hawk and Castile
Bandits murder, pillage, and terrorize the nobility in Medieval Europe until the Hawk comes onto the scene.
197? 95m/C Tom Griffith, Jerry Cobb, Nurla Forway, Mari Real. **VHS, Beta** MFI ♂

Hawk the Slayer
A comic book fantasy. A good warrior struggles against his villainous brother to possess a magical sword that bestows upon its holder great powers of destruction. Violent battle scenes; Palance makes a great villain.
1981 93m/C Jack Palance, John Terry, Harry Andrews, Roy Kinnear, Ferdinand ''Ferdy'' Mayne; *Dir:* Terry Marcel. **VHS, Beta $39.95** IVE ♂ ½

Hawk of the Wilderness
A man, shipwrecked as an infant and reared on a remote island by native Indians, battles modern day pirates. Serial in 12 episodes. Also called ''Lost Island of Kioga.''
1938 195m/B Herman Brix, Mala, William Boyle; *Dir:* William Witney. **VHS, Beta $29.95** VCN, MLB ♂♂

Hawken's Breed
Drifter meets American Indian beauty and hears her tale of woe; the two fall for each other and set out together to avenge the murder of the woman's husband and son. Nothing much to crow about.
1987 (R) 93m/C Peter Fonda, Serene Hedin, Jack Elam; *Dir:* Charles B. Pierce. **VHS, Beta, LV $19.95** STE, VMK, IME ♂♂

Hawkeye
After his best friend is murdered by gangsters in a seedy drug deal, Hawkeye, a rough and tumble Texas cop, and his slick new partner hit the streets of Las Vegas with a vengeance.
1988 90m/C Troy Donahue, Chuck Jeffreys, George Chung, Stan Wertlieb; *Dir:* Leo Fong. **VHS, Beta** VSI ♂ ½

Hawks
Somewhere on the road to black-comedy this film gets waylaid by triviality. Two terminally ill men break out of the hospital, determined to make their way to Amsterdam for some last-minute fun. Mediocre at best.
1989 (R) 105m/C GB Anthony Edwards, Timothy Dalton, Janet McTeer, Jill Bennett, Sheila Hancock, Connie Booth, Camille Coduri; *Dir:* Robert Ellis Miller. **VHS, Beta, LV, 8mm $34.95** PAR ♂ ½

Hawks & the Sparrows
(Uccellacci e Uccellini) Shortly after ''Mr. Ed'' came Paolini's garrulous crow, which follows a father and son's perambulations through Italy spouting pithy bits of politics. A comic Pasolinian allegoryfest full of then-topical political allusions, the story (written by penman Pier) comes off a bit wooden. Worth watching to see Toto, the Italian comic, in his element. Music by the ever-prolific Ennio Morricone, whose credits include ''A Fistful of Dollars,'' Pasolini's ''Decameron,'' and ''Exorcist II.''
1967 91m/B IT Toto, Ninetto Davoli, Femi Benussi; *W/Dir:* Pier Paolo Pasolini. **VHS $79.95** WBF ♂♂

Hawmps!
A Civil War lieutenant trains his men to use camels. When the soldiers and animals begin to grow fond of each other, Congress orders the camels to be set free. Hard to believe this was based on a real life incident.

1976 (G) 98m/C James Hampton, Christopher Connelly, Slim Pickens, Denver Pyle; *Dir:* Joe Camp. **VHS, Beta $79.95** VES ♂ ½

Hawthorne of the USA
Reid stars as Anthony Hamilton Hawthorne, American extraordinaire. After gambling a fortune away in Monte Carlo, Hawthorne heads to mainland Europe for rest and revelry. Instead, he finds the love of a princess and a Communist coup in the making. Silent with orchestral score.
1919 55m/B Wallace Reid, Harrison Ford, Lila Lee, Tully Marshall; *Dir:* James Cruze. **VHS, Beta $19.95** GPV ♂♂ ½

Hay Muertos Que No Hacen Ruido
This comedy centers around a plot to pull off a fake murder.
1947 95m/B MX Su Carnal Marcelo, Amanda Del Llano, Tony Diaz. **VHS, Beta $44.95** MAD ♂♂

Hazel's People
A bitter and hostile college student attends his friend's funeral in Mennonite country. He discovers not only a way of life he never knew existed, but a personal faith in a living Christ. From the Merle Good novel, ''Happy as the Grass Was Green.''
1975 (G) 105m/C Geraldine Page, Pat Hingle, Graham Beckel. **VHS, Beta $29.95** UHV, VGD ♂ ½

He Kills Night After Night After Night
The British police are baffled as they seek the man who has been killing women in the style of Jack the Ripper.
1970 88m/C Jack May, Linda Marlowe, Justine Lord. **VHS, Beta $59.95** MON ♂

He Knows You're Alone
Very lame horror flick that focuses on a psychotic killer terrorizing young girls in his search for a suitable ''bride.''
1980 (R) 94m/C Don Scardino, Elizabeth Kemp, Tom Rolfing, Paul Gleason, Caitlin O'Heaney, Tom Hanks, Patsy Pease; *Dir:* Armand Mastroianni. **VHS, Beta $59.95** MGM ♂

He Lives: The Search for the Evil One
A man who was imprisoned by Nazis as a child attempts to keep Hitler, now living in Buenos Aires, from starting a 4th Reich.
1988 90m/C Lee Patterson. **VHS, Beta** CAM ♂

He is My Brother
Two boys survive a shipwreck, landing on an island that houses a leper colony.
1975 (G) 90m/C Keenan Wynn, Bobby Sherman, Robbie Rist; *Dir:* Edward Dmytryk. **VHS, Beta $59.98** MAG ♂ ½

He Said, She Said
Shifting narrative point of view is old hat in the moving picture biz, but it usually brings into focus subtle and interesting variations in perspective. But put it into the hands of the pop psychology, drunk genderspeak generation, and it becomes a cloying and facile formula for box office bucks. The story—a contrapuntal flip-flop between the guy's point of view and the girl's—tells how the romance between two Baltimore journalists winds up on the rocks, and why (or so they say). Directors Silver and Kwapis each wrote half, and the end result is way too long.
1990 (R) 115m/C Elizabeth Perkins, Kevin Bacon, Sharon Stone, Anthony LaPaglia, Charlaine Woodard, Phil Leeds; *W/Dir:* Marisa Silver, Ken Kwapis. **VHS, LV $14.95** PAR, PIA, CCB ♂♂ ½

He Walked by Night
Los Angeles homicide investigators track down a cop killer in this excellent drama. Based on a true story from the files of the Los Angeles police, this first rate production reportedly inspired Webb to create ''Dragnet.''
1948 80m/B Richard Basehart, Scott Brady, Roy Roberts, Jack Webb, Whit Bissell; *Dir:* Alfred Werker, Anthony Mann. **VHS, Beta $16.95** SNC, NOS, KRT ♂♂♂ ½

He Who Walks Alone
Made-for-television film documents the life of Thomas E. Gilmore, who became the South's first elected black sheriff in the 1960s.
1978 74m/C Louis Gossett Jr., Clu Gulager, Mary Alice, James McEachin, Barton Heyman, Barry Brown, Loni Chapman; *Dir:* Jerrold Freedman. **VHS, Beta $49.95** VCL ♂♂♂ ½

The Head
A scientist comes up with a serum that can keep the severed head of a dog alive. Before too long, he tries the stuff out on a woman, transferring the head from her own hunchbacked body to that of a beautiful stripper. Weird German epic sports poor special effects and unintentional laughs, but is interesting nonetheless.
1959 92m/B GE Horst Frank, Michel Simon, Paul Dahlke, Karin Kernke, Helmut Schmidt; *W/Dir:* Victor Trivas. Emmy Awards '53: Best Public Affairs Program. **VHS $19.98** SNC ♂♂

Head
Written by Jack Nicholson and Bob Rafelson, this is the infamously plotless musical comedy starring the television fab four of the '60s, the Monkees, in their only film appearance. A number of guest stars appear and a collection of old movie clips are also included. Songs include ''Can You Dig It,'' ''Circle Sky,'' and ''Daddy's Song.''
1968 (R) 86m/C Peter Tork, Mickey Dolenz, Davy Jones, Michael Nesmith, Frank Zappa, Annette Funicello, Teri Garr; *Dir:* Bob Rafelson. **VHS, Beta, LV $19.95** MVD, COL ♂♂♂

Head of the Family
After a woman sacrifices her political ideals and career goals for her role as a family matriarch, she eventually falls apart.
1971 (PG) 105m/C IT FR Nino Manfredi, Leslie Caron, Ugo Tognazzi, Claudine Auger; *W/Dir:* Nanni Loy. Venice Film Festival '71: Critic's Prize. **VHS, Beta $59.95** GEM ♂♂ ½

Head Office
A light comedy revolving around the competition between corporate management and the lower echelon for the available position of chairman.
1986 (PG-13) 90m/C Danny DeVito, Eddie Albert, Judge Reinhold, Rick Moranis, Jane Seymour; *Dir:* Ken Finkleman. **VHS, Beta $79.99** HBO ♂♂

Head Over Heels
A model married to an older man has an affair with a younger one and cannot decide between them. In French with subtitles. Also titled ''Two Weeks in September.''
1967 89m/C FR Brigitte Bardot, Laurent Terzieff, Michael Sarne; *Dir:* Serge Bourguignon. **VHS, Beta $19.95** FCT ♂♂♂

Headhunter
Voodoo-esque killers are leaving headless bodies all around Miami. Plenty of prosthetic make-up.
1989 92m/C Kay Lenz, Wayne Crawford, John Fatooh, Steve Kanaly, June Chadwick, Sam Williams; *Dir:* Francis Schaeffer. **VHS, Beta, LV $29.95** ACA, IME ♂ ½

Headin' for the Rio Grande
A cowboy and his sheriff brother drive off cattle-rustlers in this Western saga. Ritter's film debut. Songs include ''Campfire Love Song,'' ''Jailhouse Lament,'' and ''Night Herding Song.''

1936 60m/B Tex Ritter, Eleanor Stewart, Syd Saylor, Warner Richmond, Charles King; *Dir:* Robert Bradbury. **VHS, Beta $19.95** *NOS, DVT* 🎬

Heading for Heaven A mistaken medical report leaves a frazzled realtor believing he has only three months to live, and when he disappears the family suspects the worst. A minor comedy based on a play by Charles Webb.
1947 65m/B Stu Erwin, Glenda Farrell, Russ Vincent, Irene Ryan, Milburn Stone, George O'Hanlon; *Dir:* Lewis D. Collins. **VHS $19.95** *NOS, TIM* 🎬🎬

Headless Eyes Gory entry into the psycho-artist category. An artist is miffed when mistaken by a woman as a burglar he loses an eye. Not one to forgive and forget, he becomes quite the eyeball fetishist, culling donations from unsuspecting women. An eye for an eye ad finitum.
1983 78m/C Bo Brundin, Gordon Raman, Mary Jane Early; *Dir:* Kent Bateman. **VHS** *UHV* 🎬½

Headline Woman An ongoing feud between a police commissioner and a newspaper's city editor causes a reporter to make a deal with a policeman in exchange for news. A good cast makes this otherwise tired story tolerable.
1935 75m/B Heather Angel, Roger Pryor, Jack LaRue, Ford Sterling, Conway Tearle; *Dir:* William Nigh. **VHS, Beta $24.95** *NOS, DVT* 🎬🎬

Headwinds Silent romance in which Mr. Right spares helpless girl against her will from marrying Mr. Wrong.
1928 52m/B House Peters, Patsy Ruth Miller. **VHS, Beta $16.95** *GPV* 🎬🎬

Healer A doctor forgets his pledge to help the lame and becomes a fashionable physician. A young crippled lad helps him eventually remember his original purpose was to serve others. Also known as "Little Pal."
1936 80m/B Mickey Rooney, Ralph Bellamy, Karen Morley, Judith Allen, Robert McWade; *Dir:* Reginald Barker. **VHS, Beta $19.95** *DVT* 🎬½

The Healing John Lucas, a successful doctor, faces personal tragedy after the loss of his wife and unborn child. After falling to the seduction of alcohol, he finds personal salvation in Christianity.
1988 71m/C Brian Collins, Jon Lormer, Erin Blunt; *Dir:* Russell S. Soughten. **VHS, Beta** *MIV, BPG* 🎬

Hear My Song Mickey O'Neill (Dunbar) is an unscrupulous club promoter, who books singers with names like Franc Cinatra, trying to revive his failing nightclub in this charming, hilarious comedy. He books the mysterious Mr. X, who claims he is really the legendary singer Josef Locke who fled England for Ireland because of tax problems, only to find out he is an imposter. To redeem himself with his fiancee and her mother, Locke's former lover, O'Neill goes to Ireland to find the real Locke (Beatty) and bring him to Liverpool to sing. Dunbar is appealing in his role, and Beatty is magnificent as the legendary Locke—so much so that he was nominated for a Golden Globe for Best Supporting Actor.
1991 (R) 104m/C *GB* Ned Beatty, Adrian Dunbar, Shirley Anne Field, Tara Fitzgerald, William Hootkins, David McCallum; *W/Dir:* Peter Chelsom. **VHS, Beta** *PAR* 🎬🎬🎬½

Hear O Israel Three classic short films about the land of Israel and the Jewish faith: "Hear O Israel," "The Changing Land," and "My Holiday in Israel." The first two have

dialogue in English, the last is in Hebrew with no subtitles.
19?? 71m/C **VHS, Beta $29.95** *VYY* 🎬🎬🎬½

The Hearse Incredibly boring horror film in which a young school teacher moves into a mansion left to her by her late aunt and finds her life threatened by a sinister black hearse.
1980 (PG) 97m/C Trish Van Devere, Joseph Cotten; *Dir:* George Bowers. **VHS, Beta $54.95** *MED* 🎬

Heart A down-and-out small-time boxer is given a chance to make a comeback, and makes the Rocky-like most of it.
1987 (R) 93m/C Brad Davis, Jesse Doran, Sam Gray, Steve Buscemi, Frances Fisher; *Dir:* James Lemmo. **VHS, Beta $19.95** *STE, NWV* 🎬½

Heart Beat The fictionalized story of Jack Kerouac (author of "On the Road"), his friend and inspiration Neal Cassady, and the woman they shared, Carolyn Cassady. Based on Carolyn Cassady's memoirs. Strong performances, with Nolte as Cassady and Spacek as his wife, overcome the sometimes shaky narrative.
1980 (R) 105m/C Nick Nolte, John Heard, Sissy Spacek, Ann Dusenbery, Ray Sharkey, Tony Bill, Steve Allen, John Larroquette; *Dir:* John Byrum. **VHS, Beta $59.95** *WAR* 🎬🎬½

Heart of a Champion: The Ray Mancini Story Made-for-TV movie based on the life of Ray "Boom Boom" Mancini, former lightweight boxing champion. Inspired by the career his father never had due to WWII, Mancini fought his way to the top. Sylvester Stallone staged the fight sequences.
1985 94m/C Robert Blake, Doug McKeon, Mariclare Costello; *Dir:* Richard Michaels. **VHS, Beta $59.98** *FOX* 🎬🎬

Heart Condition A deceased black lawyer's heart is donated to a bigoted Los Angeles cop who was stalking the lawyer when he was alive. Soon afterwards the lawyer's ghost returns to haunt the police officer, hoping the officer will help to avenge his murder. Both Hoskins and Washington display fine performances given the unlikely script.
1990 (R) 95m/C Bob Hoskins, Denzel Washington, Chloe Webb, Ray Baker, Ja'net DuBois, Alan Rachins; *Dir:* James D. Parriott. **VHS, Beta, LV $19.95** *COL* 🎬🎬

Heart of Dixie Three college co-eds at a southern university in the 1950s see their lives and values change with the influence of the civil rights movement. College newspaper reporter Sheedy takes up the cause of a black man victimized by racial violence. Lightweight social-conscience fare.
1989 (PG) 96m/C Virginia Madsen, Ally Sheedy, Phoebe Cates, Treat Williams, Kyle Secor, Francesca Roberts, Don Michael Paul, Kurtwood Smith, Richard Bradford; *Dir:* Martin Davidson. **VHS, Beta $19.98** *ORI* 🎬🎬

Heart of the Golden West Roy protects ranchers of Cherokee City from unjust shipping charges.
1942 65m/B Roy Rogers, Smiley Burnette, George "Gabby" Hayes, Bob Nolan, Ruth Terry; *Dir:* Joseph Kane. **VHS, Beta $19.95** *NOS, VYY, VCN* 🎬🎬

Heart Like a Wheel The story of Shirley Muldowney, who rose from the daughter of a country-western singer to the leading lady in drag racing. The film follows her battles of sexism and choosing whether to have a career or a family. Bedelia's

perfomance is outstanding. Fine showings from Bridges and Axton in supporting roles.
1983 (PG) 113m/C Bonnie Bedelia, Beau Bridges, Bill McKinney, Leo Rossi, Hoyt Axton, Dick Miller, Anthony Edwards; *Dir:* Jonathan Kaplan. **VHS, Beta, LV $59.98** *FOX* 🎬🎬½

The Heart is a Lonely Hunter Set in the South, Carson McCuller's tale of angst and ignorance, loneliness and beauty comes to the screen with the film debuts of Keach and Locke. Arkin was nominated for an Academy Award for his instinctive, gentle acting as the deaf mute; Locke's screen debut earned her an Oscar nomination as well.
1968 (G) 124m/C Alan Arkin, Cicely Tyson, Sondra Locke, Stacy Keach, Chuck McCann, Laurinda Barrett; *Dir:* Robert Ellis Miller. New York Film Critics Award '68: Best Actor (Arkin). **VHS, Beta $59.95** *KUI, WAR* 🎬🎬🎬

Heart of Midnight An emotionally unstable young woman inherits a seedy massage parlor. She works hard to turn it, and herself, into something better. Nice performance by Leigh.
1989 (R) 93m/C Jennifer Jason Leigh, Brenda Vaccaro, Frank Stallone, Peter Coyote; *Dir:* Matthew Chapman. **VHS, Beta, LV $89.95** *VIR* 🎬🎬½

Heart of a Nation Saga of the Montmarte family and their life during three wars, beginning with the Franco-Prussian War and ending with the Nazi occupation in France. Although ordered to be destroyed by the Nazis, this film was saved, along with many others, by the French "Cinema Resistance." In French with English subtitles.
1943 111m/B *FR* Louis Jouvet, Raimu, Suzy Prim, Michael Morgan, Renee Devillers; *Dir:* Paul Graetz; *Nar:* Charles Boyer. **VHS $24.95** *NOS, DVT* 🎬🎬🎬

Heart of the Rio Grande A spoiled young rich girl tries to trick her father into coming to her "rescue" at a western dude ranch.
1942 70m/B Gene Autry, Smiley Burnette, Fay McKenzie, Edith Fellows, Joseph Stauch Jr.; *Dir:* William M. Morgan. **VHS, Beta $29.98** *CCB, VCN* 🎬

Heart of the Rockies The Three Mesquiteers stop a mountain family's rustling and illegal game trappers.
1937 54m/B Bob Livingston, Ray Corrigan, Max Terhune, Lynn Roberts, Yakima Canutt, J.P. McGowan; *Dir:* Joseph Kane. **VHS, Beta $9.95** *NOS, VCN, MED* 🎬½

Heart of the Rockies Roy, along with his trusty horse and dog, takes care of a highway construction project.
1951 67m/B Roy Rogers, Penny Edwards, Gordon Jones, Ralph Morgan, Fred Graham, Mira McKinney; *Dir:* William Witney. **VHS $19.95** *NOS, DVT* 🎬½

Heart of the Stag On an isolated sheep ranch in the New Zealand outback, a father and daughter suffer the repercussions of an incestuous relationship when she becomes enamored with a hired hand.
1984 (R) 94m/C *NZ* Bruno Lawrence, Mary Regan, Terence Cooper; *Dir:* Michael Firth. **VHS, Beta, LV $19.95** *STE, NWV* 🎬🎬½

Heart of Texas Ryan The flamboyant cowboy star fights with kidnappers and wins in this silent film.
1916 50m/B Tom Mix. **VHS, Beta $19.95** *NOS, GPV, DVT* 🎬½

Heartaches A reporter tries to track down a murderer who has killed twice.

1947 71m/B Chill Wills, Sheila Ryan, Edward Norris, James Seay. **VHS, Beta $19.95** NOS, DVT, TIM ♫♫

Heartaches A young pregnant woman (Potts) meets up with a crazy girlfriend (Kidder) in this touching film about love and friendship. Although a study in contrasts, the two end up sharing an apartment together in Toronto. Fine performances from Potts and Kidder give life to this romantic comedy. **1982** 90m/C *CA* Margot Kidder, Annie Potts, Robert Carradine, Winston Rekert; *Dir:* Donald Shebib. Genie Awards '82: Best Actress (Kidder), Best Screenplay. **VHS, Beta, LV $69.98** LIV, VES ♫♫♫

Heartbeat If you thought pointless Hollywood remakes of French films were a new phenomenon (see ''Pure Luck,'' for example), then note this lighthearted remake of a 1940 Gallic farce. Rogers becomes the best student in a Parisian school for pickpockets, but when she tries out her skills on a dashing diplomat they fall in love instead. **1946** 100m/B Ginger Rogers, Jean-Pierre Aumont, Adolphe Menjou, Basil Rathbone, Melville Cooper, Mona Maris, Henry Stephenson, Eduardo Ciannelli; *Dir:* Sam Wood. **VHS $19.95** NOS, TIM ♫♫½

Heartbeeps In 1995, two domestic robot servants fall in love and run off together. **1981** (PG) 79m/C Andy Kaufman, Bernadette Peters, Randy Quaid, Kenneth McMillan, Christopher Guest, Melanie Mayron, Jack Carter; *Dir:* Allan Arkush. **VHS, Beta $39.95** MCA ♫½

Heartbreak Hotel Johnny Wolfe kidnaps Elvis Presley from his show in Cleveland and drives him home to his mother, a die-hard Elvis fan. Completely unbelievable, utterly ridiculous, and still a lot of fun. **1988** (PG-13) 101m/C David Keith, Tuesday Weld, Charlie Schlatter, Angela Goethals, Jacque Lynn Colton, Chris Mulkey, Karen Landry, Tudor Sherrard, Paul Harkins; *Dir:* Chris Columbus. **VHS, Beta, LV $89.95** TOU ♫♫½

Heartbreak House A production of George Bernard Shaw's classic play about a captain and his daughter who invite an odd assortment of people into their home for a few days. In the course of their visit, each person shares his ambitions, hopes, and fears. From the ''American Playhouse'' television series. **1986** 118m/C Rex Harrison, Rosemary Harris, Amy Irving; *Dir:* Anthony Page. **VHS, Beta $59.95** ACA, VTK, PBS ♫♫½

The Heartbreak Kid Director May's comic examination of love and hypocrisy. Grodin embroils himself in a triangle with his new bride and a woman he can't have, an absolutely gorgeous and totally unloving woman he shouldn't want. Walks the fence between tragedy and comedy, with an exceptional performance from Berlin. Neil Simon script, based on Bruce Jay Friedman's story. **1972** (PG) 106m/C Charles Grodin, Cybill Shepherd, Eddie Albert, Jeannie Berlin, Audra Lindley, Art Metrano; *Dir:* Elaine May. **VHS, Beta, LV $19.95** MED ♫♫♫

Heartbreak Ridge An aging Marine recon sergeant is put in command of a young platoon to whip them into shape to prepare for combat in the invasion of Grenada. Eastwood whips this old story into shape too, with a fine performance of a man who's given everything to the Marines. The invasion of Grenada, though, is not epic material. (Also available in English with Spanish subtitles.)

1986 (R) 130m/C Clint Eastwood, Marsha Mason, Everett McGill, Arlen Dean Snyder, Bo Svenson; *Dir:* Clint Eastwood. **VHS, Beta, LV, 8mm $19.98** WAR, TLF ♫♫♫

Heartbreaker Eastern Los Angeles explodes with vicious turf wars when Beto and Hector battle for the affection of Kim, the neighborhood's newest heartbreaker. **1983** (R) 90m/C Fernando Allende, Dawn Dunlap, Michael D. Roberts, Robert Dryer, Apollonia; *Dir:* Frank Zuniga. **VHS, Beta $49.95** MED ♫½

Heartbreakers Two male best friends find themselves in the throes of drastic changes in their careers and romantic encounters. On target acting makes this old story new, with a fine performance by Carol Wayne, in her last film before her death. Tangerine Dream performs the musical score. **1984** (R) 98m/C Peter Coyote, Nick Mancuso, Carole Laure, Max Gail, James Laurenson, Carol Wayne, Jamie Rose, Kathryn Harrold; *W/Dir:* Bobby Roth. **VHS, Beta $79.95** VES ♫♫♫

Heartburn Based on Nora Ephron's own semi-autobiographical novel about her marital travails with writer Carl Bernstein, this is a tepid, bitter modern romance between writers already shell-shocked from previous marriages. **1986** (R) 109m/C Meryl Streep, Jack Nicholson, Steven Hill, Richard Masur, Stockard Channing, Jeff Daniels, Milos Forman, Catherine O'Hara, Maureen Stapleton, Karen Ackers, Joanna Gleason, Mercedes Ruehl, Caroline Aaron, Yakov Smirnoff, Wilfrid Hyde-White; *Dir:* Mike Nichols. **VHS, Beta, LV, 8mm $19.95** PAR ♫♫♫

Heartland Set in 1910, this film chronicles the story of one woman's life on the Wyoming Frontier, the hazards she faces, and her courage and spirit. Stunningly realistic and without cliche. Based on the diaries of Elinore Randall Stewart. **1981** (PG) 95m/C Conchata Ferrell, Rip Torn, Barry Primus, Lilia Skala, Megan Folson; *Dir:* Richard Pearce. **VHS, Beta $79.99** HBO ♫♫♫½

Hearts & Armour A holy war between Christians and Moors erupts when a Moorish princess is kidnapped. Based on Ludovico Ariosto's ''Orlando Furioso.'' **1983** 101m/C *IT* Tanya Roberts, Leigh McCloskey, Ron Moss, Rick Edwards, Giovanni Visentin; *Dir:* Giacomo Battiato. **VHS, Beta $19.98** WAR ♫♫

Hearts of Darkness: A Filmmaker's Apocalypse This riveting, critically acclaimed documentary about the making of Francis Ford Coppola's masterpiece ''Apocalypse Now'' is based largely on original footage shot and directed by his wife Eleanor. Also included are recent interviews with cast and crew members including Coppola, Martin Sheen, Robert Duvall, Frederic Forrest and Dennis Hopper. The film received several award nominations, including one by the Director's Guild in the documentary category. **1991** (R) 96m/C Sam Bottoms, Eleanor Coppola, Francis Ford Coppola, Robert Duvall, Larry Fishburne, Frederic Forrest, Albert Hall, Dennis Hopper, George Lucas, John Milius, Martin Sheen; *W/Dir:* Fax Bahr, George Hickenlooper. National Board of Review Awards '91: Best Documentary; Eddie Award '91: Best Editing of a Documentary. **VHS, Beta** PAR ♫♫♫♫

Heart's Desire Opera great Richard Tauber stars in this musical of an unknown Viennese singer who falls in love with an English girl.

1937 79m/B *GB* Richard Tauber, Lenora Corbett, Kathleen Kelly, Paul Graetz; *Dir:* Paul Stein. **VHS, Beta $29.95** VYY ♫½

Hearts of Fire A trio of successful rock 'n roll stars work out their confused romantic entanglements. Barely released and then, not even in the U.S. Sad outing for the cast, especially legendary Dylan. **1987** 90m/C Bob Dylan, Fionna, Rupert Everett, Julian Glover; *Dir:* Richard Marquand. **VHS, Beta, LV $19.95** LHV, WAR ♫

Hearts of the West A fantasy-filled farm boy travels to Hollywood in the 1930s and seeks a writing career. Instead, he finds himself an ill-suited western movie star in this small offbeat comedy-drama that's sure to charm. Also known as ''Hollywood Cowboy.'' **1975** (PG) 103m/C Jeff Bridges, Andy Griffith, Donald Pleasence, Alan Arkin, Blythe Danner; *Dir:* Howard Zieff. National Board of Review Awards '75: 10 Best Films of the Year; New York Film Critics Awards '75: Best Supporting Actor (Arkin); New York Film Festival '75: 10 Best Films of the Year. **VHS, Beta $59.95** MGM ♫♫♫½

Hearts of the World A vintage Griffith epic about a young boy who goes off to World War I and the tribulations endured by both him and his family on the homefront. Overly sentimental but powerfully made melodrama, from Griffith's waning years. Silent with music score. **1918** 152m/B Lillian Gish, Robert Harron, Dorothy Gish, Erich von Stroheim, Ben Alexander, Josephine Crowell, Noel Coward, Mary Gish; *Dir:* D.W. Griffith. **VHS, Beta, LV, 8mm $29.98** VYY, REP, PIA ♫♫♫

Heat Another Andy Warhol-produced journey into drug-addled urban seediness. Features a former child actor/junkie and a has-been movie star barely surviving in a run-down motel. This is one of Warhol's better film productions; even non-fans may enjoy it. **1972** 102m/C Joe Dallesandro, Sylvia Miles, Pat Ast; *Dir:* Paul Morrissey. **VHS, Beta $29.95** MFV, PAR, FCT ♫♫♫

Heat A Las Vegas bodyguard avenges the beating of an old flame by a mobster's son, and incites mob retaliation. Based on the William Goldman novel. **1987** (R) 103m/C Burt Reynolds, Karen Yahng, Peter MacNicol, Howard Hesseman; *Dir:* R.M. Richards. **VHS, Beta, LV $14.95** PAR ♫½

Heat of Desire (Plein Sud) A philosophy professor abandons everything (including his wife) for a woman he barely knows and is conned by her ''relatives.'' With English subtitles. **1984** (R) 90m/C *FR* Clio Goldsmith, Patrick Dewaere, Jeanne Moreau, Guy Marchand; *Dir:* Luc Beraud. **VHS, Beta, LV $59.95** COL ♫♫

Heat and Dust A young bride joins her husband at his post in India and is inexorably drawn to the country and its prince of state. Years later her great niece journeys to modern day India in search of the truth about her scandalous and mysterious relative. More than the story of two romances, this is the tale of women rebelling against an unseen caste system which keeps them second-class citizens. Ruth Jhabvala wrote the novel and screenplay. **1982** (R) 130m/C *GB* Julie Christie, Greta Scacchi, Shashi Kapoor, Christopher Cazenove, Nickolas Grace; *Dir:* James Ivory. **VHS, Beta $59.95** MCA ♫♫♫

Heat of the Flame A vicious woodsman captures and rapes a beautiful girl. Psychological games, eroticism and exploitive nudity ensue.
1982 88m/C Tony Ferrandis, Ellie MacLure, Raymond Young, Anthony Mayans. **VHS, Beta** $39.95 VCD &

Heat Street Two thugs exact revenge on street gangs for the murders of loved ones. Made for video.
1987 90m/C Quincy Adams, Deborah Gibson, Wendy MacDonald, Del Zamora; *Dir:* Joseph Merhi. **VHS, Beta** $39.95 CLV, HHE &

Heat and Sunlight A photographer becomes obsessively jealous of his lover as their relationship comes to an end. Written, directed and starring Nilsson, who used, as with his previous film "Signal 7," a unique improvisational, video-to-film technique.
1987 98m/B Rob Nilsson, Consuelo Faust, Bill Bailey, Don Bajema, Ernie Fosselius; *Dir:* Rob Nilsson. **VHS, Beta** $79.95 CVC & 1/2

Heat Wave A writer gets involved with a "femme fatale" who kills her wealthy husband for the love of a young pianist.
1954 60m/B GB Alex Nicol, Hillary Brooke, Sidney James, Susan Stephen, Paul Carpenter; *W/Dir:* Ken Hughes. **VHS, Beta, LV** $59.99 WGE & 1/2

Heat Wave The Watts ghetto uprising of 1965 is a proving ground for a young black journalist. Excellent cast and fine script portrays the anger and frustration of blacks in Los Angeles and the U.S. in the 1960's and the fear of change felt by blacks and whites when civil rights reform began. Strong drama with minimal emotionalism. Made for cable TV.
1990 (R) 92m/C Blair Underwood, Cicely Tyson, James Earl Jones, Sally Kirkland, Margaret Avery; *Dir:* Kevin Hooks. **VHS, Beta** $79.88 TTC &&&

Heated Vengeance A Vietnam vet returns to the jungle to find the woman he left behind, and runs into some old enemies.
1987 (R) 91m/C Richard Hatch, Michael J. Pollard, Mills Watson, Cameron Dye; *Dir:* Edward Murphy. **VHS, Beta** $79.95 MED &

Heathcliff and the Catillac Cats Heathcliff and his cronies face new adventures in these humorous cartoons.
1985 25m/C **VHS, Beta** $29.95 RCA &

Heathcliff & Marmaduke A program of animated adventures with the hapless dog Marmaduke matching wits with Heathcliff, the cat, his constant nemesis.
1983 60m/C **VHS, Beta** $5.99 WOV &

Heathers A clique of stuck-up schoolgirls named Heather rules the social scene until the newest member decides that enough is enough. She and her outlaw boyfriend embark (accidentally on her part, intentionally on his) on a murder spree disguised as a rash of teen suicides. A dense, take-no-prisoners black comedy with buckets of potent slang, satire and unforgiving hostility. Humor this dark is rare; sharply observed and acted, though the end is out of place.
1989 (R) 102m/C Winona Ryder, Christian Slater, Kim Walker, Shannen Doherty, Lisanne Falk, Penelope Milford, Glenn Shadix, Lance Fenton, Patrick Laborteaux; *Dir:* Michael Lehmann. Edgar Allan Poe Awards '89: Best Screenplay. **VHS, Beta, LV** $19.95 STE, NWV, FCT &&& 1/2

Heatwave Local residents oppose a multi-million dollar residential complex in Australia. Davis portrays a liberal activist, who wages war with the developers, and, in return, becomes involved with a possible murder case.
1983 (R) 92m/C AU Judy Davis, Richard Moir, Chris Haywood, Bill Hunter, John Gregg, Anna Jemison; *Dir:* Phillip Noyce. **VHS, Beta** $59.99 HBO &&&

Heaven An exploration of "heaven" including the idea, the place, and people's views about it. Questions such as "How do you get there?" and "What goes on up there?" are discussed. Offbeat interviews mixed with a collage of celestial images.
1987 (PG-13) 80m/C *Dir:* Diane Keaton. **VHS, Beta, LV** $14.95 PAV, FCT &&

Heaven for Betsy/Jackson & Jill An episode from each of these early comedy series.
19?? 60m/B Jack Lemmon, Cynthia Stone, Todd Karns. **VHS, Beta** $24.95 NOS, DVT &&

Heaven Can Wait Social satire in which a rogue tries to convince the Devil to admit him into Hell by relating the story of his philandering life and discovers that he was a more valuable human being than he thought. A witty Lubitsch treat based on a play called "Birthdays." Oscar-nominated for Best Picture and Best Director.
1943 112m/C Don Ameche, Gene Tierney, Laird Cregar, Charles Coburn, Marjorie Main, Eugene Pallette, Allyn Joslyn, Spring Byington, Signe Hasso, Louis Calhern, Dickie Moore, Florence Bates; *Dir:* Ernst Lubitsch. **VHS, Beta, LV** $19.98 FOX, FCT &&&

Heaven Can Wait A remake of 1941's "Here Comes Mr. Jordan," about a football player who is mistakenly summoned to heaven before his time. When the mistake is realized, he is returned to Earth in the body of a corrupt executive about to be murdered by his wife. Nominated for ten Academy Awards including Best Picture, actor and director Beatty, supporting actor Warden, supporting actress Cannon, cinematography, and Original Score and screenplay. Beatty co-wrote the screenplay with Elaine May. Not to be confused with the 1943 film of the same name.
1978 (PG) 101m/C Warren Beatty, Julie Christie, Charles Grodin, Dyan Cannon, James Mason, Jack Warden; *Dir:* Warren Beatty, Buck Henry. Academy Awards '78: Best Art Direction/Set Decoration; Golden Globe Awards '79: Best Film—Musical/Comedy. **VHS, Beta, LV** $14.95 PAR, FOX &&&

Heaven on Earth The story of two orphaned British children who, along with thousands of orphans shipped from England to Canada between 1867 and 1914, try to make new lives in the Canadian wilderness. Canadian-made.
1989 101m/C CA R.H. Thompson, Sian Leisa Davies, Torquil Campbell, Fiona Reid; *Dir:* Allen Kroeker. **VHS, Beta** $79.95 SVS && 1/2

Heaven & Earth A samurai epic covering the battle for the future of Japan between two feuding warlords. The overwhelming battle scenes were actually filmed on location in Canada. In Japanese with English subtitles.
1990 (PG-13) 104m/C JP Masahiko Tsugawa, Takaai Enoki; *Dir:* Haruki Kadowawa. **VHS** $79.98 LIV && 1/2

Heaven Help Us Three mischievous boys find themselves continually in trouble with the priests running their Brooklyn Catholic high school during the mid-1960s. Realistic and humorous look at adolescent life.
1985 (R) 102m/C Andrew McCarthy, Mary Stuart Masterson, Kevin Dillon, Malcolm Danare, Jennie Dundas, Kate Reid, Wallace Shawn, Jay Patterson, John Heard, Donald Sutherland, Yeardley Smith, Sherry Steiner, Calvert Deforest; *Dir:* Michael Dinner. **VHS, Beta, LV** $19.99 HBO && 1/2

Heaven is a Playground On Chicago's South Side an inner city basketball coach and an idealistic young lawyer are determined to change the fate of a group of high school boys. The men use the incentive of athletic scholarships to keep their team in school and away from drugs and gangs.
1991 (R) 104m/C D.B. Sweeney, Michael Warren, Richard Jordan, Victor Love; *W/Dir:* Randall Fried. **VHS, LV** $89.95 COL && 1/2

Heavenly Bodies A young woman who dreams of owning a health club will stop at nothing to accomplish her goals. When a rival tries to put her out of business, a "dance-down" takes place in this lame low-budget aerobics musical.
1984 (R) 99m/C CA Cynthia Dale, Richard Rebrere, Laura Henry, Stuart Stone, Walter George Alton, Cec Linder; *Dir:* Lawrence Dane. **VHS, Beta** $19.98 FOX Woof!

The Heavenly Kid A leather-jacketed "cool" guy who died in a sixties hot rod crash finally receives an offer to exit limbo and enter heaven. The deal requires that he educate his dull earthly son on more hip and worldly ways. A big problem in this movie is that the soundtrack and wardrobe are 1955, whereas the cocky cool greaser supposedly died seventeen years ago in 1968.
1985 (PG-13) 92m/C Lewis Smith, Jane Kaczmarek, Jason Gedrick, Richard Mulligan; *Dir:* Cary Medoway. **VHS, Beta** $19.99 HBO & 1/2

Heavens Above A sharp, biting satire on cleric life in England. Sellers stars as the quiet, down-to-earth reverend who is appointed to a new post in space.
1963 113m/B GB Peter Sellers, Cecil Parker, Isabel Jeans, Eric Sykes, Ian Carmichael; *Dir:* John Boulting, Roy Boulting. New York Times Ten Best List '63: Best Film. **VHS, Beta** $19.98 HBO, FCT &&&

Heaven's Gate The uncut version of Cimino's notorious folly. A fascinating, plotless, and exaggerated account of the Johnson County cattle war of the 1880s. Ravishingly photographed, the film's production almost single-handedly put United Artists out of business.
1981 (R) 149m/C Kris Kristofferson, Christopher Walken, Isabelle Huppert, John Hurt, Richard Masur, Mickey Rourke, Brad Dourif, Joseph Cotten, Jeff Bridges, Sam Waterston, Terry O'Quinn; *Dir:* Michael Cimino. **VHS, Beta, LV** $29.99 MGM, FCT &&

Heaven's Heroes Dennis Hill, a dedicated policeman, was killed in the line of duty on August 27, 1977. This film recalls his life and deep faith in Jesus.
1988 72m/C David Ralphe, Heidi Vaughn, James O'Hagen; *Dir:* Donald W. Thompson. **VHS, Beta** $29.95 MIV, BPG &&

Heavy Petting A hilarious compilation of "love scene" footage from feature films of the silent era to the sixties, newsreels, news reports, educational films, old TV shows, and home movies.
1989 75m/C *Dir:* Obie Benz. **VHS, Beta, LV** $29.95 SUE, ACA && 1/2

Heavy Traffic Ralph Bakshi's animated fantasy portrait of the hard-edged underside of city life. A young cartoonist draws the people, places, and paranoia of his environment.
1973 76m/C *Dir:* Ralph Bakshi. **VHS, Beta** *WAR, OM* 🎞🎞🎞

Heckle y Jeckle The talking magpies get in and out of mischief. Available in Spanish only.
19?? 90m/C VHS, Beta $9.95 *FOX*

Hector's Bunyip What's a bunyip? People think it's just Hector's scaly imaginary friend. It therefore comes as quite a surprise when Hector gets kidnapped - by his supposed imaginary friend! Aired on PBS as part of the WonderWorks Family Movie television series.
1986 58m/C Scott Bartle, Robert Coleby, Barbara Stephens, Tushka Hose; *Dir:* Mark Callan. **VHS $29.95** *PME, FCT* 🎞🎞½

Hedda A dramatization of the Henrik Ibsen play, "Hedda Gabler," about a middle-class pregnant woman. The story finds her frustrated with her life and manipulating those around her with tragic results. Jackson earned an Oscar nomination.
1975 102m/C *GB* Glenda Jackson, Peter Eyre, Timothy West, Jennie Linden, Patrick Stewart; *Dir:* Trevor Nunn. **VHS, Beta $59.95** *MED* 🎞🎞½

Hedda Hopper's Hollywood Newsreel version of the gossip-hound's column, featuring scoops on Ball, Cooper, Minelli and many more.
1942 58m/B Hedda Hopper, Gary Cooper, Lucille Ball, Gloria Swanson, Liza Minnelli. **VHS $19.95** *REP*

Heidi Johanna Spyri's classic tale puts Shirley Temple in the hands of a mean governess and the loving arms of her Swiss grandfather. Also available colorized. Remade in 1967. Also available with "Poor Little Rich Girl" on Laser Disc.
1937 88m/B Shirley Temple, Jean Hersholt, Helen Westley, Arthur Treacher; *Dir:* Allan Dwan. **VHS, Beta $19.95** *FOX* 🎞🎞½

Heidi Swiss do own version of children's classic by Johann Spyri. A precocious little girl enjoys life in the Swiss Alps with her grandfather, until her stern aunt takes her away to live in the village in the valley.
1952 98m/B *SI* Elsbeth Sigmund, Heinrich Gretler, Thomas Klameth, Elsie Attenoff; *Dir:* Luigi Comencini. **VHS $8.95** *NEG, MRV, KAR* 🎞🎞½

Heidi The classical story from Johanna Spyri's novel is filmed beautifully in the Swiss Alps in Eastmancolor. Heidi is kidnapped by her mean aunt and forced to work as a slave for a rich family. Her kindly old grandfather comes to her rescue. Family entertainment for all, dubbed in English.
1965 95m/C Eva Maria Singhammer, Gustav Knuth, Lotte Ledl; *Dir:* Werner Jacobs. **VHS** *NO* 🎞🎞🎞

Heidi The second adaptation of the classic Johanna Spyri novel tells the story of an orphaned girl who goes to the Swiss Alps to live with her grandfather. Made for television.
1967 100m/C Maximilian Schell, Jennifer Edwards, Michael Redgrave, Jean Simmons; *Dir:* Delbert Mann. **VHS, Beta $19.98** *LIV, VES, GLV* 🎞🎞

Heidi A delightful full-length cartoon based on Johanna Spyri's famous children's story.
1975 93m/C *SP* **VHS, Beta $19.95** *CVS* 🎞🎞

Heidi An orphan girl's optimism brings new life to the residents of a village in the Swiss Alps. From the classic children's story by Johanna Spyri.
1979 93m/C VHS, Beta $14.95 *PAV* 🎞🎞

The Heiress Based on the Henry James novel "Washington Square," about a wealthy plain Jane, her tyrannical father, and the handsome fortune-seeking scoundrel who seeks to marry her. Musical score by Aaron Copland. Oscar nominations for Best Picture, Supporting Actor (Richardson), Best Director, Black & White Cinematography.
1949 115m/B Olivia de Havilland, Montgomery Clift, Ralph Richardson, Miriam Hopkins; *Dir:* William Wyler. Academy Awards '49: Best Actress (de Havilland), Best Art Direction/Set Decoration (B & W), Best Costume Design (B & W), Best Original Score (Copland); New York Film Critics Awards '49: Best Actress (de Havilland). **VHS, Beta $14.95** *MCA, MLB, BTV* 🎞🎞🎞½

The Heist An explosive crime thriller...
1988 (R) 93m/C Lee Van Cleef, Karen Black, Edward Albert, Lionel Stander, Robert Alda. **VHS $12.95** *AVD* 🎞🎞

The Heist A made-for-cable-television film about an ex-con who, upon regaining his freedom, sets out to rip off the crook who framed him. Entertaining enough story with a first-rate cast.
1989 97m/C Pierce Brosnan, Tom Skerritt, Wendy Hughes; *Dir:* Stuart Orme. **VHS, Beta $89.99** *HBO, HHE* 🎞🎞

Held Hostage The true story of Jerry Levin, a reporter kidnapped by terrorists while on assignment in Beirut, and his wife, Sis, who struggled with the State Department for his release.
1991 95m/C Marlo Thomas, David Dukes, G.W. Bailey, Ed Winter, Robert Harper, William Schallert; *Dir:* Roger Young. **VHS** *CGV* 🎞🎞

Held for Murder When a vacationing daughter is accused of murder, her mom lovingly takes the blame. Will the daughter come back to clear her mom, or will she let her fry in the electric chair? Also known as, "Her Mad Night."
1932 67m/B Irene Rich, Conway Tearle, Mary Carlisle, Kenneth Thomson, William B. Davidson; *Dir:* E. Mason Hopper. **VHS, Beta $16.95** *SNC* 🎞

Helen Hayes Relates the life and career of Helen Hayes to the chronological development of American theater since 1900. Includes personal recollections and clips from Hayes' greatest performances.
1974 90m/C Helen Hayes. **VHS, Beta** *BFA*

Helen Wills: Miss Poker Face "Miss Poker Face" is Helen Wills, queen of the tennis courts. At seventeen she was the women's champion of the United States. Classic newsreel footage.
1954 15m/B VHS, Beta $49.95 *TSF*

Helix These outstandingly crazed rockers perform such tunes as "Don't Get Mad Get Even" and "Heavy Metal Love."
1984 14m/C Helix. **VHS, Beta $9.95** *KRT*

Hell on the Battleground A group of inexperienced recruits are led into combat by two battle-hardened veterans. Their only hope for survival is a counterattack by U.S. tanks.
1988 (R) 91m/C William Smith, Ted Prior, Fritz Matthews; *Dir:* David A. Prior. **VHS, Beta $9.95** *AIP* 🎞

Hell Comes to Frogtown In a post-nuclear holocaust land run by giant frogs, a renegade who is one of the few non-sterile men left on earth must rescue some fertile women and impregnate them. Sci-fi spoof is extremely low-budget, but fun. Stars "Rowdy" Roddy Piper of wrestling fame.
1988 88m/C Roddy Piper, Sandahl Bergman, Rory Calhoun, Donald S. Jackson; *Dir:* R.J. Kizer. **VHS, Beta, LV $19.95** *STE, NWV* 🎞

Hell Commandos Soldiers in World War II struggle to prevent the Nazis from releasing a deadly bacteria that will kill millions. Dubbed.
1969 92m/C *IT* Guy Madison, Stan Cooper; *Dir:* J.L. Merino. **VHS, Beta $69.98** *LTG* 🎞½

Hell to Eternity The true story of how World War II hero Guy Gabaldon persuaded 2,000 Japanese soldiers to surrender. Features several spectacular scenes.
1960 132m/B Jeffrey Hunter, Sessue Hayakawa, David Janssen, Vic Damone, Patricia Owens; *Dir:* Phil Karlson. **VHS, Beta $14.98** *FOX* 🎞🎞½

Hell Fire Austin Ken Maynard finds himself mixed up with outlaws in the Old West.
1932 60m/B Ken Maynard, Nat Pendleton, Jack Perrin; *Dir:* Forrest Sheldon. **VHS, Beta $19.95** *NOS, VCN, DVT* 🎞½

Hell on Frisco Bay An ex-waterfront cop, falsely imprisoned for manslaughter, sets out to clear his name. His quest finds him taking on the Mob, in this thirties-style gangster film. Good performances all around, especially Robinson, who steals every scene he's in.
1955 93m/C Alan Ladd, Edward G. Robinson, Joanne Dru, Fay Wray, William Demarest, Jayne Mansfield; *Dir:* Frank Tuttle. **VHS, Beta $54.95** *UHV* 🎞🎞½

Hell Harbor Caribbean love, murder and greed combine to make this early talkie. Exterior shots were filmed on the west coast of Florida and the beauty of the Tampa area in the 1930s is definitely something to see.
1930 65m/B Lupe Velez, Gibson Gowland, Jean Hersholt, John Holland; *Dir:* Henry King. **VHS, Beta $47.50** *GVV* 🎞🎞

Hell is for Heroes McQueen stars as the bitter leader of a small infantry squad outmanned by the Germans in this tight World War II drama. A strong cast and riveting climax make this a must for action fans.
1962 90m/B Steve McQueen, Bobby Darin, Fess Parker, Harry Guardino, James Coburn, Mike Kellin, Nick Adams, Bob Newhart; *Dir:* Don Siegel. **VHS, Beta, LV $14.95** *PAR* 🎞🎞🎞½

Hell High Four high schoolers plan a night of torture and humiliation for an annoying teacher, only to find she has some deadly secrets of her own. Dumber than it sounds.
1986 (R) 84m/C Christopher Stryker, Christopher Cousins, Millie Prezioso, Jason Brill; *Dir:* Douglas Grossman. **VHS, Beta $79.95** *PSM* 🎞

Hell Hounds of Alaska A man goes searching for a kidnapped boy in the gold rush days of Alaska.
197? (G) 90m/C Doug McClure. **VHS, Beta $9.95** *NO* 🎞

Hell Hunters Nazi-hunters foil the plot of an old German doctor to poison the population of L.A. Meanwhile, a daughter avenges the death of her mother by blowing up the doctor's jungle compound.
1987 98m/C Stewart Granger, Maud Adams, George Lazenby, Candice Daly, Romulo Arantes, William Berger. **VHS, Beta $39.95** *NSV* 🎞

Hell Night Several young people must spend the night in a mysterious mansion as part of their initiation into Alpha Sigma Rho fraternity in this extremely dull low-budget horror flick.
1981 (R) 100m/C Linda Blair, Vincent Van Patten, Kevin Brophy, Peter Barton, Jenny Neumann; *Dir:* Tom DeSimone. **VHS, Beta** $19.95 MED ♂½

Hell in the Pacific A marvelously photographed (by Conrad Hall) psycho/macho allegory, about an American and a Japanese soldier stranded together on a tiny island and the mini-war they fight all by themselves. Overly obvious anti-war statement done with style.
1969 (PG) 101m/C Lee Marvin, Toshiro Mifune; *Dir:* John Boorman. **VHS, Beta, LV** $14.98 FOX ♂♂♂

Hell Raiders A force invades a supposedly defenseless Pacific island, and finds itself fiercely attacked.
1968 90m/C John Agar, Richard Webb, Joan Huntington; *Dir:* Larry Buchanan. **VHS, Beta** $79.95 JEF ♂½

Hell River In 1941 Yugoslavia, Yugoslav partisans and Nazis battle it out at a place called Hell River.
1975 (PG) 100m/C Rod Taylor, Adam West; *Dir:* Stole Jankovic. **VHS, Beta** $69.95 VCD ♂♂

Hell Ship Mutiny A man comes to the aid of a lovely island princess, whose people have been forced to hand over their pearls to a pair of ruthless smugglers. An excellent cast in an unfortunately tepid production.
1957 66m/B John Hall, John Carradine, Peter Lorre, Roberta Haynes, Mike Mazurki, Stanley Adams; *Dir:* Lee Sholem, Elmo Williams. **VHS, Beta** $16.95 SNC ♂

Hell Squad Lost in North Africa during WWII, five American GIs wander through the desert. They find numerous pitfalls and unexpected help. Overall, dismal and disappointing.
1958 64m/B Wally Campo, Brandon Carroll; *Dir:* Burt Topper. **VHS** $19.95 NOS ♂½

Hell Squad Unable to release his son from the Middle Eastern terrorists who kidnapped him, a U.S. ambassador turns to the services of nine Las Vegas showgirls. These gals moonlight as vicious commandos in this low-budget action film.
1985 (R) 88m/C Bainbridge Scott, Glen Hartford, Tina Lederman; *W/Dir:* Kenneth Hartford. **VHS, Beta** $59.95 MGM, GHV ♂

Hell Up in Harlem A black crime lord recuperates from an assassination attempt and tries to regain his power. Poor sequel to the decent film ''Black Caesar.''
1973 (R) 98m/C Fred Williamson, Julius W. Harris, Margaret Avery, Gerald Gordon, Gloria Hendry; *Dir:* Larry Cohen. **VHS, Beta** $19.98 ORI Woof!

Hell on Wheels Two successful brothers in the racing industry are torn apart by the same girl. Brotherly love diminishes into a hatred so deep that murder becomes the sole purpose of both.
1967 96m/C Marty Robbins, Jennifer Ashley, John Ashley, Gigi Perreau, Robert Dornan, Connie Smith, Frank Gerstle; *Dir:* Will Zens. **VHS, Beta** $39.95 ACA, ACN ♂♂

Hellbenders A confederate veteran robs a Union train and must fight through acres of Civil War adversity. Dubbed. Poorly directed and acted.

1967 (PG) 92m/C *IT SP* Joseph Cotten, Norman Bengell, Julian Mateos; *Dir:* Sergio Corbucci. **VHS, Beta** $14.95 COL, EMB, SUE ♂½

Hellbent A crazed musician and a revenge obssessed housewife embark on an unusual adventure which challenges their very sanity.
1990 90m/C Phil Ward, Lynn Levand, Cheryl Slean, David Marciano; *W/Dir:* Richard Casey. **VHS, Beta** RAE ♂½

Hellbound: Hellraiser 2 In this, the first sequel to Clive Barker's inter-dimensional nightmare, the traumatized daughter from the first film is pulled into the Cenobites' universe. Gore and weird imagery abound. An uncut, unrated version is also available.
1988 (R) 96m/C *GB* Ashley Laurence, Clare Higgins, Kenneth Cranham, Imogen Boorman, William Hope, Oliver Smith, Sean Chapman, Doug Bradley; *Dir:* Tony Randel. **VHS, Beta, LV** $19.95 STE, NWV ♂♂

Hellcats A mob of sleazy, leather-clad female bikers terrorize small Midwestern towns in this violent girl-gang thriller. Even biker movie fans might find this one unworthy.
1968 (R) 90m/C Ross Hagen, Dee Duffy, Sharyn Kinzie, Sonny West, Bob Slatzer; *Dir:* Bob Slatzer. **VHS, Beta** $19.95 ACA Woof!

Hellcats of the Navy Soggy true saga of the World War II mission to sever the vital link between mainland Asia and Japan. The only film that Ronald and Nancy Reagan starred in together and the beginning of their grand romance.
1957 82m/B Ronald Reagan, Nancy Davis, Arthur Franz, Robert Arthur; *Dir:* Nathan Juran. **VHS, Beta** $9.95 GKK ♂½

Helldorado A lively western starring the singing cowboy, Roy Rogers.
1946 54m/B Roy Rogers, George ''Gabby'' Hayes, Dale Evans, Paul Harvey; *Dir:* William Witney. **VHS, Beta** $19.95 NOS, VCN, CAB ♂♂

Heller in Pink Tights The Louis L'Amour story follows the adventures of a colorful theatrical troupe that visits the 1880s West. Performing amid Indians, bill collectors, and thieves, they are often just one step ahead of the law themselves in this offbeat western.
1960 100m/C Sophia Loren, Anthony Quinn, Margaret O'Brien, Steve Forrest, Edmund Lowe; *Dir:* George Cukor. **VHS, Beta** $14.95 KRT, PAR, FCT ♂♂

Hellfighters Texas oil well fire fighters experience trouble between themselves and the women they love.
1968 (G) 121m/C John Wayne, Katharine Ross, Jim Hutton, Vera Miles, Bruce Cabot, Jay C. Flippen; *Dir:* Andrew V. McLaglen. **VHS, Beta, LV** $14.95 MCA, PIA ♂♂

Hellfire A gambler promises to build a church and follow the precepts of the Bible after a minister sacrifices his life for him.
1948 90m/B William Elliott, Marie Windsor, Forrest Tucker, Jim Davis; *Dir:* R.G. Springsteen. **VHS** $39.98 REP ♂♂

Hellfire A short film about a maniacal TV evangelist who is contacted by God-who turns out to be a woman.
198? 30m/C VHS, Beta $19.95 MPI ♂♂

Hellgate A woman hitchhiking turns out to be one of the living dead. Her benefactor lives to regret picking her up.
1989 (R) 96m/C Abigail Wolcott, Ron Palillo; *Dir:* William A. Levey. **VHS, LV** $89.95 VMK Woof!

Hellhole A young woman who witnesses her mother's murder is sent to a sanitarium where the doctors are perfecting chemical lobotomies. Extremely weak script and poor acting make this a suspenseless thriller.
1985 (R) 93m/C Judy Landers, Ray Sharkey, Mary Woronov, Marjoe Gortner, Edy Williams, Terry Moore, Dyanne Thorne; *Dir:* Pierre De Moro. **VHS, Beta, LV** $14.95 COL ♂

Hello A small animated fable wherein three extraterrestrial musicians send peace messages to earthlings.
1982 9m/C VHS, Beta, Special order formats $195.00 PYR

Hello Again The wife of a successful plastic surgeon chokes to death on a piece of chicken. A year later she returns to life, with comical consequences, but she soon discovers that life won't be the same.
1987 (PG-13) 96m/C Shelley Long, Corbin Bernsen, Judith Ivey, Gabriel Byrne, Sela Ward; *Dir:* Frank Perry. **VHS, Beta, LV** $19.99 TOU ♂♂

Hello, Dolly! Widow Dolly Levi, while matchmaking for her friends, finds a match for herself. Based on the stage musical adapted from Thornton Wilder's play ''Matchmaker.'' Jerry Herman won an Oscar for his score. Lightweight story needs better actors with stronger characterizations. Oscar nominations for Best Picture, Costume Design, and Film Editing.
1969 (G) 146m/C Barbra Streisand, Walter Matthau, Michael Crawford, Louis Armstrong, E.J. Peaker, Marianne McAndrew, Tommy Tune; *Dir:* Gene Kelly. Academy Awards '69: Best Art Direction/Set Decoration, Best Sound, Best Musical Score. **VHS, Beta, LV** $19.98 FOX, FCT ♂♂

Hello Mary Lou: Prom Night 2 A sequel to the successful slasher flick, wherein a dead-for-30-years prom queen relegates the current queen to purgatory and comes back to life in order to avenge herself. Wild special effects.
1987 (R) 97m/C Michael Ironside, Wendy Lyon, Justin Louis, Lisa Schrage, Richard Monette; *Dir:* Bruce Pittman. **VHS, Beta, LV** $19.95 VIR ♂½

Hellraiser A graphic, horror fantasy about a woman who is manipulated by the monstrous spirit of her husband's dead brother. In order for the man who was also her lover to be brought back to life, she must lure and kill human prey for his sustenance. Grisly and inventive scenes keep the action fast-paced; not for the faint-hearted. Screenplay written by Barker.
1987 (R) 94m/C *GB* Andrew Robinson, Clare Higgins, Ashley Laurence, Sean Chapman, Oliver Smith; *W/Dir:* Clive Barker. **Beta, LV, Q** $19.95 STE, NWV ♂♂½

Hellriders A motorcycle gang rides into a small town and subjects its residents to a reign of terror.
1984 (R) 90m/C Adam West, Tina Louise. **VHS, Beta** $59.95 TWE Woof!

Hell's Angels Classic WWI aviation movie created by the fledgling but well-bankrolled studio of Howard Hughes, which had produced ''Two Arabian Nights'' the year before. Sappy and a bit lumbering, it's still an extravagant spectacle with awesome big air scenes. Hughes fired directors Howard Hawks and Luther Reed, spent a hitherto unprecedented $3.8 million and was ultimately credited with the film's direction (although James Whale, who wrote the script, spent some time in the director's chair). Three years in the making, the venture cost three pilots their lives and Hughes lost a bundle.

Harlow—who replaced Swedish Greta Nissen when Hughes gave the word to let there be sound—was catapulted into blond bombshelldom playing a two-timing dame who dallies with two brothers before they head off to war. And the tinted and two-color scenes—restored in 1989—came well before Ted Turner ever wielded a crayola. Oscar nominated for photography.
1930 135m/B Jean Harlow, Ben Lyon, James Hall, John Darrow, Lucien Prival; *Dir:* Howard Hughes. **VHS $19.98** *MCA, MON, MRV* 🎬🎬🎬

Hell's Angels '69 Two wealthy brothers plot a deadly game by infiltrating the ranks of the Hell's Angels in this below average biker film.
1969 (PG) 97m/C Tom Stern, Jeremy Slate, Conny Van Dyke; *Dir:* Lee Madden. **VHS, Beta $49.95** *MED* 🎬

Hell's Angels Forever A revealing ride into the world of honor, violence, and undying passion for motorcycles on the road. Documentary was filmed with cooperation of the Angels. Features appearances by Willie Nelson, Jerry Garcia, Bo Diddley, Kevin Keating and Johnny Paycheck.
1983 (R) 93m/C Willie Nelson, Jerry Garcia, Johnny Paycheck, Bo Diddley; *Dir:* Richard Chase. **VHS, Beta $59.95** *MED* 🎬

Hell's Angels on Wheels A low-budget, two-wheeled Nicholson vehicle that casts him as a gas station attendant who joins up with the Angels for a cross country trip. Laszlo Kovacs is responsible for the photography. This is one of the better 1960's biker films.
1967 95m/C Jack Nicholson, Adam Roarke, Sabrina Scharf, Jana Taylor, John Garwood; *Dir:* Richard Rush. **VHS, Beta $39.95** *VMK, MON* 🎬

Hell's Brigade: The Final Assault Low budget, poorly acted film in which a small band of American commandos is ordered on the most dangerous and important mission of World War II.
1980 (PG) 99m/C Jack Palance, John Douglas, Carlos Estrada; *Dir:* Henry Mankiewirk. **VHS, Beta $59.95** *MPI Woof!*

Hell's Harbor In a tiny Caribbean harbor, the owner of a small trading store witnesses a murder committed by a spitfire's father.
1934 64m/B Lupe Velez, Jean Hersholt. **VHS, Beta, LV** *WGE* 🎬🎬

Hell's Headquarters A startling tale of murder and greed involving a hunt for a fortune in African ivory.
1932 59m/B Jack Mulhall, Barbara Weeks, Frank Mayo; *Dir:* Andrew L. Stone. **VHS $16.95** *SNC, NOS, DVT* 🎬

Hell's Hinges The next to last Hart western, typifying his good/bad cowboy character. Perhaps Hart's best film.
1916 65m/B William S. Hart, Clara Williams, Jack Standing, Robert McKim; *Dir:* William S. Hart. **VHS, Beta $29.95** *GPV, VYY* 🎬🎬½

Hell's House After his mother dies, young lad Durkin goes to the city to live with relatives, gets mixed up with moll Davis and her bootlegger boyfriend O'Brien, and is sent to a brutal reform school. Interesting primarily for early appearances by Davis and O'Brien (before he was typecast as the indefatigable good-guy).
1932 80m/B Junior Durkin, Pat O'Brien, Bette Davis, Junior Coghlan, Charley Grapewin, Emma Dunn; *Dir:* Howard Higgin. **VHS, Beta $19.95** *NOS, MRV, NEG* 🎬🎬

Hell's Wind Staff Two boys uncover a plot to sell their townsmen as slaves to other South East Asian countries.
1982 95m/C Hwang Jang Lee, Mang Yuen Man, Mang Hai. **VHS, Beta** *WVE* 🎬

Hellstrom Chronicle A powerful quasi-documentary about insects, their formidable capacity for survival, and the conjectured battle man will have with them in the future.
1971 (G) 90m/C Lawrence Pressman; *Dir:* Walon Green. Academy Awards '71: Best Feature Documentary. **VHS, Beta** *COL, OM* 🎬🎬🎬

Helltown A western programmer based on Zane Grey's novel.
1938 60m/B John Wayne; *Dir:* Charles T. Barton. **VHS, Beta $19.95** *NOS, DVT, VCN* 🎬½

Helltrain Passengers take the ride of their lives on the ''Helltrain.''
1980 (R) 90m/C *Dir:* Alain Payet. **VHS $59.95** *UHV Woof!*

Help! Ringo's ruby ring is the object of a search by an Arab cult who chase the Fab Four all over the globe in order to acquire the bauble. Songs include the title tune, ''Ticket to Ride,'' ''You've Got to Hide Your Love Away,'' and ''Another Girl.'' The laserdisc version includes a wealth of Beatles memorabilia, rare footage behind the scenes and at the film's premiere, and extensive publicity material.
1965 (G) 90m/C John Lennon, Paul McCartney, Ringo Starr, George Harrison, Leo McKern, Eleanor Bron; *Dir:* Richard Lester. **VHS, Beta, LV $29.95** *MVD, MPI, CRC* 🎬🎬🎬

Help Wanted: Male When a magazine publisher discovers that her fiance cannot have children, she looks for someone else who can do the job right. Made for television.
1982 97m/C Suzanne Pleshette, Gil Gerard, Bert Convy, Dana Elcar, Harold Gould, Caren Kaye; *Dir:* William Wiard. **VHS, Beta $39.95** *WOV* 🎬½

Helter Skelter The harrowing story of the murder of actress Sharon Tate and four others at the hands of Charles Manson and his psychotic ''family.'' Based on the book by prosecutor Vincent Bugliosi, adapted by J. P. Miller. Features an outstanding performance by Railsback as Manson. A TV mini-series.
1976 194m/C Steve Railsback, Nancy Wolfe, George DiCenzo, Marilyn Burns; *Dir:* Tom Gries. **VHS, Beta $59.98** *FOX* 🎬🎬½

The Henderson Monster Experiments of a genetic scientist are questioned by a community and its mayor. A potentially controversial television drama turns into typical romantic fluff.
1980 105m/C Jason Miller, Christine Lahti, Stephen Collins, David Spielberg, Nehemiah Persoff, Larry Gates; *Dir:* Waris Hussein. **VHS, Beta $39.95** *IVE* 🎬🎬½

Hennessy An IRA man plots revenge on the Royal family and Parliament after his family is violently killed. Tense political drama, but slightly far-fetched.
1975 (PG) 103m/C *GB* Rod Steiger, Lee Remick, Richard Johnson, Trevor Howard, Peter Egan, Eric Porter; *Dir:* Don Sharp. **VHS $19.99** *HBO* 🎬🎬½

Henry Fonda: The Man and His Movies Documentary features highlights from Fonda's stage, screen and television career that spanned more than fifty years.
1984 60m/C Henry Fonda; *Nar:* Arthur Hill. **VHS, Beta $39.95** *RKO*

Henry IV The adaptation of the Luigi Pirandello farce about a modern-day recluse who shields himself from the horrors of the real world by pretending to be mad and acting out the fantasy of being the medieval German emperor Henry IV. In Italian, with English subtitles.
1985 94m/C *IT* Marcello Mastroianni, Claudia Cardinale, Leopoldo Trieste, Paolo Bonacelli; *Dir:* Marco Bellochio. **VHS, Beta $59.95** *MED, APD* 🎬🎬½

Henry IV, Part I Against a background of civil war, Prince Hal, the heir to the throne, appears to waste his youth hanging around taverns in London. The political tensions between the king and the rebels lead to the battle of Shrewsbury, at which time the Prince redeems himself. Part of the television series ''The Shakespeare Plays.''
1980 147m/C *GB* Jon Finch, Anthony Quayle, David Gwillim. **VHS, Beta** *TLF* 🎬🎬🎬

Henry IV, Part II A colorful panorama of Medieval English life. The young prince is now preparing himself for leadership. The play ends with the death of Henry IV and the coronation of Prince Hal as Henry V. Part of the television series ''The Shakespeare Plays.''
1980 151m/C *GB* Jon Finch, Anthony Quayle, David Gwillim. **VHS, Beta** *TLF* 🎬🎬🎬

Henry & June Based on the diaries of writer Anais Nin, Kaufman's movie focuses on Nin's relationships with author Henry Miller and his wife June. Set in Paris in the early '30s, the setting moves between the impecunious expatriate's cheap room on the Left bank—filled with artists, circus performers, prostitutes and gypsies—to the conservative and well-appointed home of Nin and her banker husband Hugo. June rejoins husband Miller in Paris, finds him engaged in an affair with Nin, and joins them in a triangular relationship, all of which is fastidiously chronicled in Nin's diaries and provides the erotic backdrop to Miller's ''Tropic of Capricorn.'' Scripted by Kaufman and his wife Rose, the movie captures the heady atmosphere of Miller's gay Paris, and no one plays an American better than Ward (who replaced Alec Baldwin in the part and earlier appeared in Kaufman's ''The Right Stuff''). The film is noted for having prompted the creation of an NC-17 rating for its adult theme, most likely because of scenes of lesbian sex, rather than the violence the censors are used to.
Phillipe Rousselot's camerawork received an Oscar nomination for Best Cinematography.
1990 (X) 136m/C Fred Ward, Uma Thurman, Maria De Medeiros, Richard E. Grant, Kevin Spacey; *Dir:* Philip Kaufman. **VHS, LV $92.95** *MCA, FCT* 🎬🎬🎬½

Henry: Portrait of a Serial Killer Based on the horrific life and times of serial killer Henry Lee Lucas, this film has received wide praise for its straight-forward and uncompromising look into the minds of madmen. The film follows Henry and his roommate Otis as they set out on mindless murder sprees (one of which they videotape). Extremely disturbing and graphic film. Unwary viewers should be aware of the grisly scenes and use their own discretion when viewing this otherwise genuinely moving film.
1990 (X) 90m/C Michael Rooker, Tom Towles, Tracy Arnold; *Dir:* John McNaughton. **VHS, Beta, LV $79.98** *MPI* 🎬🎬🎬🎬

Henry V The classic, epic adaptation of the Shakespeare play, and Olivier's first and most successful directorial effort, dealing with the medieval British monarch that defeated the French at Agincourt. Distinguished by Olivier's brilliant formal experiment of beginning the drama as a 16th Century performance of the play in the Globe Theatre, and having the stage eventually transform into realistic historical settings of storybook color. Music by Sir William Walton. Oscar nominations for Best Picture, Best Actor (Olivier), Color Interior Decoration, Best Score of a Drama or Comedy.
1944 136m/C GB Laurence Olivier, Robert Newton, Leslie Banks, Esmond Knight, Renee Asherson, Leo Genn, George Robey, Ernest Thesiger, Felix Aylmer; *Dir:* Laurence Olivier. Academy Awards '46: Special Prize (Olivier); National Board of Review Awards '46: Best Actor (Olivier), Best Film; New York Film Critics Awards '46: Best Actor (Olivier); Venice Film Festival '46: Special Mention. **VHS, Beta $19.95** *PAR, VTK* 🎬🎬🎬🎬

Henry V England's most admired national hero is shown uniting his people as he embarks on the invasion of France, deals justly with traitors, tirelessly leads his soldiers to victory, and ensures future peace by his marriage to the princess of France. From the "Shakespeare Plays" series.
1980 163m/C GB VHS, Beta $19.95 *TLF* 🎬🎬🎬

Henry V Stirring, expansive retelling of Shakespeare's drama about the warrior-king of England. Branagh stars as Henry, leading his troops and uniting his kingdom against France. A very impressive production that rivals Olivier's 1945 rendering but differs by stressing the high cost of war—showing the ego-mania, doubts and subterfuge that underlie the conflict. A marvelous film-directorial debut for Branagh (who also adapted the screenplay). Wonderful supporting cast includes some of Britain's finest actors. Branagh was Oscar nominated for Best Actor.
1989 138m/C GB Kenneth Branagh, Derek Jacobi, Brian Blessed, Alec McCowen, Ian Holm, Richard Briers, Robert Stephens, Robbie Coltrane, Christian Bale, Judy Dench, Paul Scofield, Michael Maloney, Emma Thompson; *W/ Dir:* Kenneth Branagh. **VHS, LV $19.98** *FOX* 🎬🎬🎬🎬

Henry VI, Part 1 The Shakespeare classic offers a glimpse of the beginning of the long civil wars between the Houses of York and Lancaster.
1982 120m/C Peter Benson, Trevor Peacock. **VHS, Beta** *TLF* 🎬

Henry VIII In "Henry VIII" Shakespeare spins a web of intrigue and betrayal. Production was shot at one of Henry's favorite visiting spots, Leeds Castle. Part of the television series "The Shakespeare Plays."
1979 165m/C GB Claire Bloom, John Stride, Julian Glover. **VHS, Beta** *TLF* 🎬🎬½

Henry VI, Part 3 The conclusion of the Henry VI chronicles, climaxed by the Duke of Gloucester's plan to murder Henry.
1982 120m/C Peter Benson, Julia Foster, Bernard Hill. **VHS, Beta** *TLF* 🎬

Hepburn & Tracy The finest moments from the legendary film careers of Hepburn and Tracy, as a team and on their own. Some sequences in black and white. A must for fans!
1984 45m/C Katharine Hepburn, Spencer Tracy. **VHS, Beta $29.95** *RKO*

Her Alibi When successful murder-mystery novelist Phil Blackwood runs out of ideas for good books, he seeks inspiration in the criminal courtroom. There he discovers a beautiful Romanian immigrant named Nina who is accused of murder. He goes to see her in jail and offers to provide her with an alibi. Narrated by Blackwood in the tone of one of his thriller novels. Uneven comedy, with appealing cast and arbitrary plot.
1988 (PG) 95m/C Tom Selleck, Paulina Porizkova, William Daniels, James Farentino, Hurd Hatfield, Patrick Wayne, Tess Harper, Joan Copeland; *Dir:* Bruce Beresford. **VHS, Beta, LV, 8mm $19.98** *WAR* 🎬🎬

Her Husband's Affairs Tone is an advertising wonder and Ball is his loving wife who always gets credit for his work. Tone does the advertising for an inventor who is searching for the perfect embalming fluid in this rather lifeless comedy.
1947 83m/B Lucille Ball, Franchot Tone, Edward Everett Horton, Gene Lockhart; *Dir:* Sylvan Simon. **VHS $9.95** *GKK* 🎬

Her Life as a Man A female reporter is refused a job as a sportswriter because of her gender. She makes herself up as a man, gets the job, and creates havoc when she has to deal with lustful women on the job. A made-for-television movie.
1983 93m/C Robin Douglas, Joan Collins, Robert Culp, Marc Singer, Laraine Newman; *Dir:* Robert Ellis Miller. **VHS, Beta $59.95** *LHV* 🎬½

Her Silent Sacrifice A silent melodrama about a French girl torn between material wealth and true love.
1918 42m/B Alice Brady, Henry Clive. **VHS, Beta $24.95** *VYY* 🎬½

Her Summer Vacation A beautiful teenager spends her summer vacation on a country ranch which belongs to her father. A sensual story about becoming a woman.
19?? 90m/C *Dir:* Victor Di Mello. **VHS, Beta $39.98** *CON, HHE* 🎬½

Herbie Goes Bananas While Herbie the VW is racing in Rio de Janeiro, he is bothered by the syndicate, a pickpocket, and a raging bull. The fourth and final entry in the Disney "Love Bug" movies, but Herbie later made his way to a television series.
1980 (G) 93m/C Cloris Leachman, Charles Martin Smith, Harvey Korman, John Vernon, Alex Rocco, Richard Jaeckel, Fritz Feld; *Dir:* Vincent McEveety. **VHS, Beta $69.95** *DIS* 🎬½

Herbie Goes to Monte Carlo While participating in a Paris-to-Monte-Carlo race, Herbie the VW takes a detour and falls in love with a Lancia. Third in the Disney "Love Bug" series.
1977 (G) 104m/C Dean Jones, Don Knotts, Julie Sommars, Roy Kinnear; *Dir:* Vincent McEveety. **VHS, Beta $69.95** *DIS* 🎬🎬

Herbie Rides Again In this "Love Bug" sequel, Herbie comes to the aid of an elderly woman who is trying to stop a ruthless tycoon from raising a skyscraper on her property. Humorous Disney fare. Two other sequels followed.
1974 (G) 88m/C Helen Hayes, Ken Berry, Stefanie Powers, John McIntire, Keenan Wynn; *Dir:* Robert Stevenson. **VHS, Beta, LV $69.95** *DIS* 🎬🎬½

Hercules The one that started it all. Reeves is perfect as the mythical hero Hercules who encounters many dangerous situations while trying to win over his true love. Dubbed in English.

1958 107m/C IT Steve Reeves, Sylva Koscina, Fabrizio Mioni, Giana Maria Canale, Arturo Dominici; *Dir:* Pietro Francisci. **VHS, Beta, LV $14.98** *VID, MRV, IME* 🎬🎬½

Hercules Legendary muscleman Hercules fights against the evil King Minos for his own survival and the love of Cassiopeia, a rival king's daughter.
1983 (PG) 100m/C IT Lou Ferrigno, Sybil Danning, William Berger, Brad Harris, Ingrid Anderson; *Dir:* Lewis (Luigi Cozzi) Coates. **VHS, Beta, LV $79.95** *MGM, IME* 🎬

Hercules 2 The muscle-bound demi-god returns to do battle with more evil foes amidst the same stunningly cheap special effects.
1985 (PG) 90m/C IT Lou Ferrigno, Claudio Cassinelli, Milly Carlucci, Sonia Viviani, William Berger, Carlotta Green; *Dir:* Lewis (Luigi Cozzi) Coates. **VHS $79.95** *MGM* 🎬

Hercules Against the Moon Men It's no holds barred for the mighty son of Zeus when evil moon men start killing off humans in a desperate bid to revive their dead queen.
1964 88m/C IT FR Alan Steel, Jany Clair, Anna Maria Polani; *Dir:* Giacomo Gentilomo. **VHS, Beta $16.95** *SNC, MRV* 🎬

Hercules and the Captive Women Hercules' son is kidnapped by the Queen of Atlantis, and the bare-chested warrior goes on an all-out rampage to save the boy. Directed by sometimes-lauded Cottafavi.
1963 93m/C IT Reg Park, Fay Spain, Ettore Manni; *Dir:* Vittorio Cottafavi. **VHS, Beta $9.95** *RHI, MRV, TIM* 🎬🎬½

Hercules in the Haunted World Long before the days of 24-hour pharmacies, Hercules—played yet again by Reeves-clone Park—must journey to the depths of Hell in order to find a plant that will cure a poisoned princess. Better than most muscle operas, thanks to Bava.
1964 91m/C IT Reg Park, Leonora Ruffo, Christopher Lee, George Ardisson; *Dir:* Mario Bava. **VHS, Beta $9.99** *CCB, RHI, MRV* 🎬🎬½

Hercules in New York In his motion picture debut, Arnold Schwarzenegger is a Herculean mass of muscle who becomes a professional wrestling superstar. Two hundred and fifty pounds of lighthearted fun. Previously released on video as "Hercules Goes Bananas."
1970 (G) 93m/C Arnold Schwarzenegger, Arnold Stang, Deborah Loomis, James Karen, Ernest Graves; *Dir:* Arthur Seidelman. **VHS, Beta $59.98** *MPI, UNI* 🎬½

Hercules and the Princess of Troy Hercules is up to his pecs in trouble as he battles a hungry sea monster in order to save a beautiful maiden. Originally a pilot for a prospective television series, and shot in English, not Italian, with Hercules played by erstwhile Tarzan Scott. Highlights are special effects and color cinematography.
1965 ?m/C IT Gordon Scott, Diana Hyland, Paul Stevens, Everett Sloane. **VHS, Beta $16.95** *SNC, MLB* 🎬🎬½

Hercules, Prisoner of Evil Spaghetti myth-opera in which Hercules battles a witch who is turning men into werewolves. Made for Italian television by director Dawson (the nom-de-cinema of Antonio Margheriti).
1964 ?m/C IT Reg Park; *Dir:* Anthony Dawson. **VHS $19.98** *SNC* 🎬

Hercules Unchained In this sequel to "Hercules," the superhero must use all his strength to save the city of Thebes and the woman he loves from the giant Antaeus.
1959 101m/C *IT* Steve Reeves, Sylva Koscina, Silvia Lopel, Primo Carnera; *Dir:* Pietro Francisci. **VHS, Beta, LV** $14.95 *VDM, SUE, MRV* ♂️½

Herculoids Zandor, Tara, Dorno and the towering man of stone, Igoo, encounter adventures on a wild, semi-primitive planet.
196? 60m/C **VHS, Beta** $19.95 *HNB* ♂️½

Herculoids, Vol. 2 A team of futuristic animals attempt to save a king and his planetary community from alien invaders.
1967 60m/C **VHS, Beta** $19.95 *HNB* ♂️½

Here Come the Littles: The Movie A twelve-year-old boy finds many new adventures when he meets The Littles—tiny folks who reside within the walls of people's houses.
1985 76m/C *Dir:* Bernard Deyries. **VHS, Beta** $14.98 *FOX* ♂️½

Here Come the Marines The Bowery Boys accidentally join the marines and wind up breaking up a gambling ring.
1952 66m/B Leo Gorcey, Huntz Hall, Tim Ryan; *Dir:* William Beaudine. **VHS** $19.95 *NOS* ♂️♂️

Here Comes Droopy The sad-eyed sheriff with the creaky voice saves the day in spite of himself.
1990 (G) 60m/C **VHS** $12.98 *MGM*

Here Comes Garfield Garfield's first step from newspaper to television has Odie getting captured by the dog catcher. After much thought and a lot of lasagna, Garfield decides to rescue him. The only problem is that Odie only has until dawn to live.
1990 (G) 24m/C *Voices:* Lorenzo Music. **VHS** $12.98 *FOX*

Here Comes the Groom Late, stale Capra-corn, involving a rogue journalist who tries to keep his girlfriend from marrying a millionaire by becoming a charity worker. Includes a number of cameos.
1951 114m/B Bing Crosby, Jane Wyman, Franchot Tone, Alexis Smith, James Barton, Connie Gilchrist, Robert Keith, Anna Maria Alberghetti; *Dir:* Frank Capra. Academy Awards '51: Best Song ("In the Cool, Cool, Cool of the Evening"). **VHS, Beta, LV** $14.95 *PAR* ♂️♂️

Here Comes Kelly Controlling his temper and hanging on to his job is more than our hero can handle.
1943 63m/C Eddie Quillan, Joan Woodbury, Maxie Rosenbloom; *Dir:* William Beaudine. **VHS, Beta** *GPV* ♂️½

Here Comes Mr. Jordan Montgomery is the young prizefighter killed in a plane crash because of a mix-up in heaven. He returns to life in the body of a soon-to-be murdered millionaire. Rains is the indulgent and advising guardian angel. A lovely fantasy/romance remade in 1978 as "Heaven Can Wait." Laserdisc edition includes actress Elizabeth Montgomery talking about her father Robert Montgomery as well as her television show "Bewitched."
1941 94m/B Robert Montgomery, Claude Rains, James Gleason, Evelyn Keyes, Edward Everett Horton; *Dir:* Alexander Hall. Academy Awards '41: Best Screenplay, Best Story; National Board of Review Awards '41: 10 Best Films of the Year. **VHS, Beta, LV** $19.95 *COL, VYG, CRC* ♂️♂️♂️♂️

Here Comes Santa Claus A young boy and girl travel to the North Pole to deliver a very special wish to Santa Claus.
1984 78m/C Karen Cheryl, Armand Meffre; *Dir:* Christian Gion. **VHS, Beta** $14.95 *NWV, STE* ♂️½

Here It is, Burlesque Male and female striptease, baggy-pants comedians, exotic dancers and classic comedy sketches in this tribute to the living art, burlesque.
1979 88m/C Ann Corio, Morey Amsterdam. **VHS, Beta, LV** $49.95 *VES* ♂️½

Here We Go Again! Fibber McGee and Molly are planning a cross-country trip for their 20th anniversary celebration, but complications abound. Based on the popular NBC radio series.
1942 76m/B Marion Jordan, Jim Jordan, Harold Peary, Gale Gordon, Edgar Bergen, Charlie McCarthy, Mortimer Snerd, Ray Nobel; *Dir:* Allan Dwan. **VHS, Beta** $9.95 *CCB* ♂️♂️

The Hereafter An old church is suddenly overrun by zombies.
1987 92m/C **VHS, Beta** $59.95 *MGL* ♂️

Heritage of the Desert One of the earlier westerns made by the rugged Randolph Scott.
1933 62m/B Randolph Scott, Sally Blane, Guinn Williams; *Dir:* Henry Hathaway. **VHS, Beta** $24.95 *DVT, BUR, TIM* ♂️♂️

Hero A popular soccer player agrees to throw a game for big cash, and then worries about losing the respect of a young boy who idolizes him. Nothing new in this overly sentimental sports drama.
1972 (PG) 97m/C Richard Harris, Romy Schneider, Kim Burfield, Maurice Kaufman; *Dir:* Richard Harris. **VHS, Beta** $59.98 *SUE* ♂️½

Hero Ain't Nothin' But a Sandwich A young urban Black teenager gets involved in drugs and is eventually saved from ruin. Slow-moving, over-directed and talky. However, Scott turns in a fine performance. Based on Alice Childress' novel.
1978 (PG) 107m/C Cicely Tyson, Paul Winfield, Larry B. Scott, Helen Martin, Glynn Turman, David Groh; *Dir:* Ralph Nelson. **VHS, Beta** $14.95 *PAR, FCT* ♂️♂️½

Hero Bunker Hero bunk about a platoon of heroic Greek soldiers who attempt, at the expense of their personal longevity, to defend an important bunker from enemy invasion.
1971 93m/C John Miller, Maria Xenia; *Dir:* George Andrews. **VHS, Beta** $39.95 *IVE* ♂️♂️

Hero at Large An unemployed actor foils a robbery while dressed in a promotional "Captain Avenger" suit, and instant celebrity follows. Lightweight, yet enjoyable.
1980 (PG) 98m/C John Ritter, Anne Archer, Bert Convy, Kevin McCarthy, Kevin Bacon; *Dir:* Martin Davidson. **VHS, Beta** $59.95 *MGM* ♂️♂️½

Hero of Rome Rome has many who could be called "hero," but one stands biceps and pecs above the rest.
1963 ?m/C *IT* Gordon Scott, Gabriella Pallotti. **VHS, Beta** $16.95 *SNC* ♂️♂️½

Hero and the Terror Perennial karate guy Norris plays a sensitive policeman who conquers his fear of a not-so-sensitive maniac who's trying to kill him. Plenty of action and a cheesy subplot to boot.
1988 (R) 96m/C Chuck Norris, Brynn Thayer, Steve James, Jack O'Halloran, Ron O'Neal, Billy Drago; *Dir:* William Tannen. **VHS, Beta, LV** $89.95 *MED* ♂️½

Hero of the Year The characters from Falk's "Top Dog" return in this story of a Polish television personality who loses his job and is forced to face unemployment and jobseeking.
1985 115m/C *PL* *Dir:* Feliks Falk. **VHS** $49.99 *FCT* ♂️♂️

Herod the Great Biblical epic of the downfall of Herod, the ruler of ancient Judea. Scantily clad women abound. Dubbed.
1960 93m/C *IT* Edmund Purdom, Sandra Milo, Alberto Lupo; *Dir:* Arnaldo Genoino. **VHS** $16.95 *SNC* ♂️

Heroes An institutionalized Vietnam vet (Winkler) escapes, hoping to establish a worm farm which will support all his crazy buddies. On the way to the home of a friend he hopes will help him, he encounters Field. Funny situations don't always mix with serious underlying themes of mental illness and post-war adjustment.
1977 (PG) 97m/C Henry Winkler, Sally Field, Harrison Ford; *Dir:* Jeremy Paul Kagan. **VHS, Beta** $59.95 *MCA* ♂️♂️

Heroes of the Alamo Remember the Alamo...but forget this movie, dull, threadbare dramatization of the battle that was scorned even back in '37.
1937 75m/B Rex Lease, Lane Chandler, Roger Williams, Earle Hodgins, Julian Rivero; *Dir:* Harry Fraser. **VHS, Beta** *GPV* ♂️½

Heroes in Blue Two brothers join the police force, but one goes astray and hooks up with gangsters. Plenty of action distracts from the clichéd story.
1939 60m/B Dick Purcell, Charles Quigley, Bernadene Hayes, Edward Keane; *Dir:* William Watson. **VHS, Beta** $16.95 *SNC* ♂️

The Heroes of Desert Storm The human spirit of the people who fought in the high-tech Persian Gulf War is captured in this film that also features real war footage from ABC News.
1991 93m/C Daniel Baldwin, Angela Bassett, Marshall Bell, Michael Alan Brooks, William Bumiller, Michael Champion, Maria Diaz; *Dir:* Don Ohlmeyer. **VHS, Beta** $79.95 *PSM* ♂️♂️

Heroes Die Young A cheap independent feature depicting an American command mission with orders to sabotage German oil fields during World War II.
1960 76m/B Krika Peters, Scott Borland, Robert Getz, James Strother. **VHS, Beta** $19.98 *FOX, FCT* ♂️½

Heroes in Hell Two escaped WWII POWs join the Allied underground in an espionage conspiracy against the Third Reich.
1967 90m/C Klaus Kinski, Ettore Manni. **VHS, Beta** $69.95 *FOR* ♂️½

Heroes of the Hills The Three Mesquiteers back up a plan that would allow trusted prisoners to work for neighboring ranchers. Part of "The Three Mesquiteers" series.
1938 54m/B Bob Livingston, Ray Corrigan, Max Terhune. **VHS, Beta** *MED* ♂️

Heroes in the Ming Dynasty A martial arts period film set in the Ming Dynasty.
1984 90m/C **VHS, Beta** $39.95 *UNI* ♂️

Heroes Stand Alone When a U.S. spy plane is downed in Central America, a special force is sent in to rescue the crew. Another "Rambo" rip-off.

1989 (R) 90m/C Bradford Dillman, Chad Everett, Wayne Grace, Rick Dean; *Dir:* Mark Griffiths. VHS, Beta **$79.99** MGM ♫

Heroes Three When his crewmate is murdered while on leave in in the Far East, a Navy officer enlists the aid of a Chinese detective to track down the killer. **1984 90m/C** Rowena Cortes, Mike Kelly, Laurens C. Postma, Lawrence Tan; *Dir:* S.H. Lau. VHS, Beta **$99.95** IMP ♫♫

Heroes of the West An early sound action western serial. **1932 70m/B** Noah Beery Jr. VHS, Beta **$29.95** VCN, MLB ♫

Hero's Blood A young married couple tries to survive in ancient China despite endless kung-fu marauders attempting to kick the living daylights out of them. **197? 82m/C** Cheang Leung, Kan Mei Chai. VHS, Beta **$54.95** MAV ♫

He's My Girl When a rock singer wins a trip for two to Hollywood, he convinces his agent to dress up as a woman so they can use the free tickets. **1987 (PG-13) 104m/C** David Hallyday, T.K. Carter, Misha McK, Jennifer Tilly; *Dir:* Gabrielle Beaumont. VHS, Beta **$14.95** IVE ♫ ½

He's Your Dog, Charlie Brown! An animated Peanuts adventure wherein Charlie Brown decides to send Snoopy back to the Daisy Hill Puppy Farm for obedience lessons. **1968 30m/C** Charlie Brown, Snoopy. VHS, Beta **$9.98** MED, VTR

Hester Street Set at the turn of the century, the film tells the story of a young Jewish immigrant who ventures to New York City to be with her husband. As she reacquaints herself with her husband, she finds that he has abandoned his Old World ideals. The film speaks not only to preserving the heritage of the Jews, but to cherishing all heritages and cultures. Highly regarded upon release, unfortunately forgotten today. Kane received an Oscar nomination for her performance. **1975 89m/C** Carol Kane, Doris Roberts, Steven Keats, Mel Howard; *Dir:* Joan Micklin Silver. VHS, Beta **$39.95** VES, FRV ♫♫♫

Hey Abbott! A compilation of Abbott and Costello routines from their classic television series. Sketches include "Who's On First," "Oyster Stew," "Floogle Street" and "The Birthday Party." **1978 76m/B** Bud Abbott, Lou Costello, Phil Silvers, Steve Allen, Joe Besser; *Nar:* Milton Berle. VHS, Beta **$14.98** VID ♫♫♫

Hey, Babu Riba A popular Yugoslavian fit of nostalgia about four men convening at the funeral of a young girl they all loved years before, and their happy, Americana-bathed memories therein. In Serbo-Croatian with subtitles. **1988 (R) 109m/C** *YU* Gala Videnovic, Nebojsa Bakocevic, Dragan Bjelogrlic, Marko Todorovic, Goran Radakovic; *Dir:* Jovan Acin. VHS, Beta **$79.98** ORI ♫♫ ½

Hey Good Lookin' Ralph Bakshi's irreverent look at growing up in the 1950s bears the trademark qualities that distinguish his other adult animated features, "Fritz the Cat" and "Heavy Traffic." **1982 87m/C** *Dir:* Ralph Bakshi. VHS, Beta **$39.98** WAR ♫♫ ½

Hey! Hey! A nostalgic, musical and comedic look at the history and heritage of the Cubs.

1985 30m/C The Chicago Cubs. VHS, Beta MPI

Hey There, It's Yogi Bear When Yogi Bear comes out of winter hibernation to search for food, he travels to the Chizzling Brothers Circus. The first feature-length cartoon to come from the H-B Studios, it features the voices of Mel Blanc, J. Pat O'Malley, Julie Bennett and Daws Butler. **1964 98m/C** *Dir:* William Hanna; *Voices:* Daws Butler, James Darren, Mel Blanc, H. Pat O'Malley, Julie Bennett. VHS, Beta, LV **$19.95** WOV, IME ♫♫

Hey Vern! It's My Family Album Ernest P. Worrell finds his family album and portrays various comical ancestors. **1985 57m/C** Jim Varney. VHS, Beta **$19.95** KWI

Hi-De-Ho The great Cab Calloway and his red hot jazz are featured in this film, which has the band caught between rival gangsters. The plot is incidental to the music, anyway. **1947 60m/B** Cab Calloway, Ida James, The Peters Sisters; *Dir:* Josh Binney. VHS, Beta **$19.95** MVD, NOS, VCN ♫♫ ½

Hi-Di-Hi A British television comedy about the shenanigans that inevitably follow when a stuffy Cambridge professor becomes the entertainment manager for a holiday camp. **1988 91m/C** *GB* Simon Cadell, Ruth Madoc, Paul Shane, Jeffrey Holland, Leslie Dwyer. VHS, Beta **$39.98** FOX ♫♫ ½

Hi Diddle Diddle Topsy-turvy comedy features young lovers who long for conventional happiness. Instead they are cursed with con-artist parents who delight in crossing that law-abiding line. **1943 72m/B** Adolphe Menjou, Martha Scott, Dennis O'Keefe, Pola Negri; *Dir:* Andrew L. Stone. VHS, Beta **$24.95** NOS, DVT ♫♫ ½

Hi-Jacked When a parolee trucker's cargo is stolen, he inevitably becomes a suspect and must set out to find the true culprits. **1950 66m/B** Jim Davis, Paul Cavanagh, Marsha Jones. VHS, Beta, LV WGE, RXM ♫

Hi, Mom! Amateur pornographer/movie maker De Niro is advised by a professional in the field (Garfield) of sleazy filmmaking. De Niro films the residents of his apartment building and eventually marries one of his starlets. One of De Palma's earlier and better efforts with De Niro playing a crazy as only he can. Also known as "Confessions of a Peeping John." **1970 (R) 87m/C** Robert De Niro, Charles Durnham, Allen Garfield, Lara Parker, Jennifer Salt, Gerrit Graham; *W/Dir:* Brian De Palma. VHS **$59.95** AVP, FCT, HHE ♫♫♫

Hi-Riders A revenge-based tale about large, mag-wheeled trucks and their drivers. **1977 (R) 90m/C** Mel Ferrer, Stephen McNally, Neville Brand, Ralph Meeker; *Dir:* Greydon Clark. VHS, Beta **$49.95** UHV ♫

The Hidden A seasoned cop (Nouri) and a benign alien posing as an FBI agent (MacLachlan) team up to track down and destroy a hyper-violent alien who survives by invading the bodies of humans, causing them to go on murderous rampages. Much acclaimed, high velocity action film with state-of-the-art special effects. **1987 (R) 98m/C** Kyle MacLachlan, Michael Nouri, Clu Gulager, Ed O'Ross; *Dir:* Jack Sholder. VHS, Beta, LV **$19.95** MED, CVS ♫♫♫

Hidden Agenda A human rights activist and an American lawyer uncover brutality and corruption among the British forces in Northern Ireland. Slow-paced but strong performances. The Northern Irish dialect is sometimes difficult to understand as are the machinations on the British police system. Generally worthwhile. **1990 (R) 108m/C** Frances McDormand, Brian Cox, Brad Dourif, Mai Zetterling, John Benfield, Des McAleer, Jim Norton, Maurice Roeves; *Dir:* Ken Loach. VHS, LV **$92.99** HBO, PIA, FCT ♫♫ ½

Hidden City A film archivist and a statistician become drawn into a conspiracy when a piece of revealing film is spliced onto the end of an innocuous government tape. **1987 112m/C** Charles Dance, Cassie Stuart, Alex Norton, Tusse Silberg, Bill Paterson; *Dir:* Stephen Poliakoff. VHS, Beta **$79.95** SVS ♫♫

Hidden Enemy In World War II, a scientist develops a new metal alloy that is stronger than steel yet lighter than aluminum. His nephew must protect his scientist/uncle from spies who would steal the formula to use against the Allies. **1940 63m/B** Warren Hull, Kay Linaker, William von Brinken, George Cleveland; *Dir:* Howard Bretherton. VHS, Beta **$16.95** SNC, RXM ♫ ½

Hidden Fortress Kurosawa's tale of a warrior who protects a princess from warring feudal lords. An inspiration for George Lucas' "Star Wars" series and deserving of its excellent reputation. In Japanese with English subtitles. The laserdisc edition is letter boxed to retain the film's original integrity. **1958 139m/B** *JP* Toshiro Mifune, Misa Vehara, Minoru Chiaki; *Dir:* Akira Kurosawa. Berlin Film Festival '59: International Critics Prize, Best Director. VHS, Beta, LV **$29.98** MED, VYG, FCT ♫♫♫ ½

Hidden Gold A routine western for Mix, which has him going undercover to win the confidence of three outlaws and discover where they have hidden their loot. **1933 57m/B** Tom Mix. VHS, Beta **$19.95** DVT, VCN, BUR ♫ ½

The Hidden Room A doctor finds out about his wife's affair and decides to get revenge on her lover. He kidnaps him, imprisons him in a cellar, and decides to kill him - slowly. Tense melodrama. **1949 98m/B** *GB* Robert Newton, Sally Gray, Naunton Wayne, Phil Brown, Olga Lindo; *Dir:* Edward Dmytryk. VHS, Beta **$16.95** NOS, SNC, RXM ♫♫♫

Hide and Go Shriek Several high school seniors are murdered one by one during a graduation party. Also available in an unrated, gorier version. **1987 (R) 90m/C** Annette Sinclair, Brittain Frye, Rebunkah Jones; *Dir:* Skip Schoolnik. VHS, Beta, LV **$19.95** NSV ♫

Hide-Out A man running from the law winds up at a small farm, falls in love, and becomes reformed by the good family that takes him in. Good supporting cast includes a very young Rooney. **1934 83m/B** Robert Montgomery, Maureen O'Sullivan, Edward Arnold, Elizabeth Patterson, Mickey Rooney, Edward Brophy; *Dir:* W.S. Van Dyke. VHS MGM ♫♫ ½

Hide in Plain Sight A distraught blue-collar worker searches for his children who disappeared when his ex-wife and her mobster husband are given new identities by federal agents. Caan's fine directorial debut is based on a true story.

1980 (PG) 96m/C James Caan, Jill Eikenberry, Robert Viharo, Kenneth McMillan, Josef Sommer, Danny Aiello; *Dir:* James Caan. **VHS, Beta** $59.95 *MGM* 𝄞𝄞𝄞

Hide and Seek When a high school hacker converts his computer into a nuclear weapon, he finds that it has developed a mind of its own.
1977 60m/C Bob Martin, Ingrid Veninger, David Patrick. **VHS, Beta** $59.95 *LHV* 𝄞

Hide and Seek A study of children living in the land that would soon become Israel. Although they play as normal children, life is particularly difficult. This adversity will lay the foundation for their Israeli citizenry.
1980 90m/C Gila Almagor, Doron Tavori, Chaim Hadaya; *Dir:* Dan Wolman. **VHS, Beta** $79.95 *ERG* 𝄞

The Hideaways A 12-year-old girl and her younger brother run away and hide in the Metropolitan Museum of Art. The girl becomes enamored of a piece of sculpture and sets out to discover its creator. Also known as: "From the Mixed-Up Files of Mrs. Basil E. Frankweiler." Based on the children's novel by M. E. Kerr.
1973 (G) 105m/C Richard Mulligan, George Rose, Ingrid Bergman, Sally Prager, Johnny Doran, Madeline Kahn; *Dir:* Fielder Cook. **VHS, Beta** $59.95 *WAR, VHE, FCT* 𝄞𝄞 ½

Hideous Sun Demon A physicist exposed to radiation must stay out of sunlight or he will turn into a scaly, lizard-like creature. Includes previews of coming attractions from classic science fiction films.
1959 75m/B Robert Clarke, Patricia Manning, Nan Peterson; *Dir:* Robert Clarke. **VHS, Beta** $9.95 *NOS, MHV, HHI* 𝄞 ½

Hider in the House In a peaceful neighborhood a family is about to spend an evening of terror when they discover a psychopath is hiding in the attic of their home.
1990 (R) 109m/C Michael McKean, Gary Busey, Mimi Rogers; *Dir:* Michael Patrick. **VHS, Beta, LV** $89.98 *VES, LIV* 𝄞𝄞 ½

Hiding Out A young stockbroker testifies against the Mafia and must find a place to hide in order to avoid being killed. He winds up at his cousin's high school in Delaware, but can he really go through all these teen-age troubles once again?
1987 (PG-13) 99m/C Jon Cryer, Keith Coogan, Gretchen Cryer, Annabeth Gish, Tim Quill; *Dir:* Bob Giraldi. **VHS, Beta** $19.99 *HBO* 𝄞𝄞

The Hiding Place True story of two Dutch Christian sisters sent to a concentration camp for hiding Jews during World War II. Film is uneven but good cast pulls it through. Based on the Corrie Ten Boom book and produced by Billy Graham's Evangelistic Association.
1975 145m/C Julie Harris, Eileen Heckart, Arthur O'Connell, Jeanette Clift; *Dir:* James F. Collier. **VHS** $49.95 *VBL, REP, COV* 𝄞𝄞

High Adventure A series of over 150 programs in which celebrities and other accomplished guests are interviewed about their profession and credits within that profession. Later, these guests provide an insight into their Christian outlook and discuss the role which Christianity plays in their lives.
19?? 30m/C Debbie Boone, Dale Evans, Susan Stafford, Don Sutton, Dean Jones, Roger Staubach, Tom Landry. **VHS, Beta** *TVS*

High Anxiety Brooks' tries hard to please in this low-brow parody of Hitchcock films employing dozens of references to films like "Psycho," "Spellbound," "The Birds," and "Vertigo." Tells the tale of a height-fearing psychiatrist caught up in a murder mystery. The title song performed a la Sinatra by Brooks is one of the film's high moments. Brooks also has a lot of fun with Hitchcockian camera movements. Uneven (What? Brooks?) but amusing tribute.
1977 (PG) 92m/C Mel Brooks, Madeline Kahn, Cloris Leachman, Harvey Korman, Ron Carey, Howard Morris, Dick Van Patten; *Dir:* Mel Brooks. **VHS, Beta, LV** $19.98 *FOX* 𝄞𝄞

High Ballin' Fonda and Reed team up with lady trucker Shaver to take on the gang trying to crush the independent drivers.
1978 (PG) 100m/C *CA* Peter Fonda, Jerry Reed, Helen Shaver; *Dir:* Peter Carter. **VHS, Beta** $29.98 *LIV, VES* 𝄞 ½

High Command To save his daughter from an ugly scandal, the British general of a Colonial African outpost traps a blackmailer's killer.
1937 84m/B *GB* Lionel Atwill, Lucie Mannheim, James Mason; *Dir:* Thorold Dickinson. **VHS, Beta** $19.95 *NOS, MRV, VYY* 𝄞 ½

High Country Two misfits, one handicapped, the other an ex-con, on the run from society learn mutual trust as they travel through the Canadian Rockies.
1981 (PG) 101m/C *CA* Timothy Bottoms, Linda Purl, George Sims, Jim Lawrence, Bill Berry; *Dir:* Harvey Hart. **VHS, Beta** $79.95 *VES* 𝄞𝄞 ½

High Country Calling Wolf pups escape their man-made home and set off a cross-country chase.
1975 (G) 84m/C *Nar:* Lorne Greene. **VHS, Beta** $29.95 *GEM* 𝄞 ½

High Crime A "French Connection"-style suspense story about the heroin trade. Features high-speed chases and a police commissioner obsessed with capturing the criminals.
1973 91m/C James Whitmore, Franco Nero, Fernando Rey; *Dir:* Enzo G. Castellari. **VHS, Beta** $49.95 *MED* 𝄞𝄞

High Desert Kill A science-fiction tale will keep you guessing. Aliens landing in the New Mexican desert want something totally unexpected, and a trio of friends will be the first to find out what it is. Written for cable by T. S. Cook ("The China Syndrome").
1990 (PG) 93m/C Chuck Connors, Marc Singer, Anthony Geary, Micah Grant; *Dir:* Harry Falk. **VHS, Beta** $79.95 *MCA* 𝄞𝄞

High Frequency Two young men monitoring a satellite relay station witness the murder of a woman. Suspecting whom the killer's next victim will be, they set out to warn her. Lack of development mars what could have been a tidy little thriller.
1988 105m/C *IT* Oliver Benny, Vincent Spano, Isabelle Pasco, Anne Canovos; *Dir:* Faliero Rosati. **VHS, LV** $79.98 *MCG* 𝄞𝄞

High Heels A black comedy about a doctor who believes beauty is only skin deep and marries a plain woman. However, when he meets her beautiful sister, he begins to wonder. In French with English subtitles.
1972 100m/C *FR* Laura Antonelli, Jean-Paul Belmondo, Mia Farrow, Daniel Ivernel; *Dir:* Claude Chabrol. **VHS, Beta** $39.95 *MON* 𝄞𝄞

High Heels An outrageous combination murder-melodrama-comedy from Almodovar. Rebecca is a TV anchorwoman in Madrid whose flamboyant singer/actress mother has returned to the city for a concert. Rebecca happens to be married to one of her mother's not-so-ex-flames. When her husband winds up dead, Rebecca confesses to his murder during her newscast—but is she telling the truth or just covering up for mom? Mix in a drag queen/judge, a dancing chorus of women prison inmates, and a peculiar police detective, and see if the plot convolutions make sense. In Spanish with English subtitles.
1991 (R) 113m/C *SP* Victoria Abril, Marisa Paredes, Miguel Bose; *W/Dir:* Pedro Almodovar. **VHS, Beta** *PAR* 𝄞𝄞𝄞

High Hopes A moving yet nasty satiric comedy about a pair of ex-hippies maintaining their counter-culture lifestyle in Margaret Thatcher's England, as they watch the signs of conservative "progress" overtake them and their geriatric, embittered Mum. Hilarious and mature.
1989 110m/C *GB* Philip Davis, Ruth Sheen, Edna Dore; *Dir:* Mike Leigh. **VHS, Beta** $29.95 *ACA* 𝄞𝄞𝄞

High Ice A forest ranger and a lieutenant colonel are involved in a clash of wills over a rescue mission high in the snow-capped Washington state peaks.
1980 97m/C David Janssen, Tony Musante; *Dir:* Eugene S. Jones. **VHS, Beta** $69.95 *VES* 𝄞 ½

High & Low Fine Japanese film noir about a wealthy business who is being blackmailed by kidnappers who claim to have his son. When he discovers that they have mistakenly taken his chauffeur's son he must decide whether to face financial ruin or risk the life of a young boy. Based on an Ed McBain novel. In Japanese with English subtitles.
1962 (R) 143m/B *JP* Toshiro Mifune, Tatsuya Mihashi, Tatsuya Nakadai; *Dir:* Akira Kurosawa. **VHS, Beta, LV** $29.95 *PAV, IME* 𝄞𝄞𝄞 ½

High Noon The award-winning Western about Hadeyville town marshal Will Kane (Cooper) who must face down four professional killers alone, after being abandoned to his fate by the gutless townspeople who professed to admire him. Cooper has just married his young, beautiful Quaker bride (Kelly) when the town learns that a revenge-minded killer will arrive on the noon train to be met by his three cohorts. The sherriff and his bride are urged to leave town, which they do, but Cooper's sense of honor drives him back for the noon showdown. A landmark Western, with Cooper as the ultimate hero figure, his sheer presence overwhelming. Note the continuing use of the ballad written by Dimitri Tamkin, "Do Not Forsake Me, Oh My Darlin'" (sung by Tex Ritter) to heighten the tension and action throughout the film. Oscar nominations: Best Picture, Director, Screenplay. The laserdisc includes the original theatrical trailer, an appreciative essay by Howard Suber on audio 2, a photo essay of production stills, Carl Foreman's original notes of the film and the complete text of "The Tin Star," the story on which the film is based.
1952 85m/B Gary Cooper, Grace Kelly, Lloyd Bridges, Lon Chaney Jr., Thomas Mitchell, Otto Kruger, Katy Jurado, Lee Van Cleef; *Dir:* Fred Zinneman. Academy Awards '52: Best Actor (Cooper), Best Film Editing, Best Song ("High Noon (Do Not Forsake Me, Oh My Darlin')"), Best Musical Score; National Board of Review Awards '52: 10 Best Films of the Year; New

York Film Critics Awards '52: Best Director (Zinneman), Best Film. VHS, LV $19.98 *REP, VYG, TLF ♫♫♫*

High Noon: Part 2 Majors attempts to fill Gary Cooper's shoes, as Marshal Kane returns a year after the events of the original "High Noon" to remove the corrupt marshal who replaced him. Subtitled "The Return of Will Kane." Made for television.
1980 96m/C Lee Majors, David Carradine, Pernell Roberts, Katherine Cannon; *Dir:* Jerry Jameson. **VHS, Beta $14.95** *FRH ♫*

High Plains Drifter A surreal, violent western focusing on a drifter who defends a town from gunmen intent on meting out death. One of Eastwood's most stylistic directorial efforts, indebted to his "Man with No Name" days with Sergio Leone. Laser format is letterboxed and includes chapter stops and the original theatrical trailer.
1973 (R) 105m/C Clint Eastwood, Verna Bloom, Mitchell Ryan, Marianna Hill; *Dir:* Clint Eastwood. **VHS, Beta, LV $19.95** *MCA, TLF ♫♫1/2*

High Risk Improbable action-adventure of four unemployed Americans who battle foreign armies, unscrupulous gunrunners, and jungle bandits in a harrowing attempt to steal five million dollars from an expatriate American drug dealer living in the peaceful splendor of his Columbian villa. Interesting in its own run-on way.
1976 (R) 94m/C James Brolin, Anthony Quinn, Lindsay Wagner, James Coburn, Ernest Borgnine, Bruce Davison, Cleavon Little; *Dir:* Stewart Raffill. **VHS, Beta $9.98** *SUE ♫1/2*

High Road to China A hard-drinking former WWI air ace is recruited by a young heiress who must find her father before his ex-partner takes over his business. Post-"Raiders of the Lost Ark" thrills and romance, but without that tale's panache.
1983 (PG) 105m/C Bess Armstrong, Tom Selleck, Jack Weston, Robert Morley, Wilford Brimley, Brian Blessed; *Dir:* Brian G. Hutton. **VHS, Beta, LV $19.98** *WAR ♫♫*

High Rolling in a Hot Corvette Two carnival workers leave their jobs and hit the road in search of adventure and excitement, eventually turning to crime.
1977 (PG) 82m/C *AU* Joseph Bottoms, Greg Taylor, Judy Davis, Wendy Hughes; *Dir:* Igor Auzins. **VHS, Beta $69.95** *VES ♫*

High School Caesar A seedy teenage exploitation flick about a rich teenager, with no parental supervision, who starts his own gang to run his high school.
1956 75m/B John Ashley; *Dir:* O'Dale Ireland. **VHS, Beta $9.95** *RHI, SNC ♫*

High School Confidential Teen punk Tamblyn transfers to Santo Bello high school from Chicago and causes havoc among the locals. He soon involves himself in the school drug scene and looks to be the top dog in dealing, but what no one knows is that he's really a narc! A must for midnight movie fans thanks to high camp values, "hep cat" dialogue and the gorgeous Van Doren as Tamblyn's sex-crazed "aunt."
1958 85m/B Russ Tamblyn, Jan Sterling, John Barrymore Jr., Mamie Van Doren, Diane Jergens, Jerry Lee Lewis, Ray Anthony, Jackie Coogan, Charles Chaplin Jr., Burt Douglas, Michael Landon, Jody Fair, Phillipa Fallon, Robin Raymond, James Todd, Lyle Talbot, William Wellman Jr.; *Dir:* Jack Arnold. **VHS, Beta $39.98** *REP ♫♫*

High School USA A high school class clown confronts the king of the prep-jock-bullies. Antics ensue in this made-for-television flick which features many stars from fifties and sixties sitcoms.
1984 96m/C Michael J. Fox, Nancy McKeon, Bob Denver, Angela Cartwright, Elinor Donahue, Dwayne Hickman, Lauri Hendler, Dana Plato, Tony Dow, David Nelson; *Dir:* Rod Amateau. **VHS, Beta $59.95** *LHV, WAR ♫1/2*

High Season A satire about a beautiful English photographer (Bisset) stranded on an idyllic Greek island with neither money nor inspiration, and the motley assembly of characters who surround her during the tourist season. Written by Clare and Mark Peploe.
1988 (R) 95m/C *GB* Jacqueline Bisset, James Fox, Irene Papas, Sebastian Shaw, Kenneth Branagh, Robert Stephens; *Dir:* Clare Peploe. **VHS, Beta, LV, 8mm $19.98** *SUE, FCT ♫♫1/2*

High Sierra Bogart is Roy "Mad Dog" Earle, an aging gangster whose last job backfires, so he's hiding out from the police in the High Sierras. Screenplay by John Huston and W.R. Burnett, on whose novel it's based. Bogart's first starring role. Remade in 1955 as "I Died A Thousand Times." Also available colorized.
1941 96m/B Humphrey Bogart, Ida Lupino, Arthur Kennedy, Joan Leslie, Cornel Wilde, Henry Travers, Henry Hull; *Dir:* Raoul Walsh. National Board of Review Awards '41: 10 Best Films of the Year. **VHS, Beta, LV $19.95** *FOX ♫♫♫*

High Society A wealthy man attempts to win back his ex-wife who's about to be remarried in this enjoyable remake of "The Philadelphia Story." The Cole Porter score includes: "True Love," "Well Did You Evah," "Now You Has Jazz," "Mind If I Make Love to You," and "High Society Calypso." Letterboxed laserdisc format also includes the original movie trailer.
1956 107m/C Frank Sinatra, Bing Crosby, Grace Kelly, Louis Armstrong, Celeste Holm, Sidney Blackmer, Louis Calhern; *Dir:* Charles Walters. **VHS, Beta, LV, 8mm $19.95** *MGM, PIA ♫♫♫*

High Spirits An American inherits an Irish castle and a sexy 200-year-old ghost. A painful, clumsy comedy by the fine British director that was apparently butchered by the studio before its release.
1988 (PG-13) 99m/C Darryl Hannah, Peter O'Toole, Steve Guttenberg, Beverly D'Angelo, Liam Neeson, Martin Ferrero, Peter Gallagher, Jennifer Tilly; *Dir:* Neil Jordan. **VHS, Beta, LV $19.98** *MED ♫♫*

High Stakes Kirkland stars as a prostitute who's battling the Mob and falling in love. Limited release in theaters but another fine performance from Kirkland. Also released as "Melanie Rose."
1989 (R) 102m/C Sally Kirkland, Robert LuPone, Richard Lynch, Sarah Gellar, Kathy Bates, W.T. Martin; *Dir:* Amos Kollek. **VHS, Beta $89.95** *VMK ♫♫*

High Tide A strong and strange drama once again coupling the star and director of "My Brilliant Career." A small-time rock 'n' roll singer who is stranded in a small, beach town fortuitously meets up with her previously-abandoned teenage daughter. Acclaimed.
1987 (PG-13) 120m/C *AU* Judy Davis, Jan Adele, Claudia Karvan, Colin Friels; *Dir:* Gillian Armstrong. Australian Film Institute '87: Best Actress (Davis). **VHS, Beta, LV, 8mm $19.98** *SUE, FCT ♫♫♫*

High Velocity An action-packed adventure of two mercenaries involved in the challenge of a lifetime to rescue a kidnapped executive.
1976 (PG) 105m/C Ben Gazzara, Paul Winfield, Britt Ekland, Keenan Wynn; *Dir:* Remi Kramer. **VHS, Beta $29.95** *MED ♫1/2*

High Voltage A fast-moving comedy adventure set in the High Sierras.
1929 62m/B Carole Lombard, William Boyd, Gwen Moore, Billy Bevan; *Dir:* Howard Higgin. **VHS, Beta $19.95** *NOS, MRV, KRT ♫♫*

Higher Education Andy leaves his small town for a big city college and falls head over heels for a student and a teacher. He didn't know college could be so much fun.
1988 (R) 83m/C Kevin Hicks, Isabelle Mejias, Lori Hallier, Maury Chaykin, Richard Monette; *Dir:* John Sheppard. **VHS $9.95** *SIM, NSV ♫♫*

Higher and Higher Bankrupt aristocrat conspires with servants to regain his fortune. He also tries to marry his daughter into money. Features Sinatra singing "I Couldn't Sleep a Wink Last Night."
1943 90m/B Frank Sinatra, Leon Errol, Michele Morgan, Jack Haley, Mary McGuire; *Dir:* Tim Whelan. **VHS, Beta, LV $19.98** *CCB, IME ♫♫*

The Highest Honor The true story set during World War II of a friendship between an Australian army officer, captured while attempting to infiltrate Singapore, and a Japanese security officer. Too long, but gripping.
1984 (R) 99m/C *AU* John Howard, Atsuo Nakamura, Stuart Wilson, Michael Aitkens, Steve Bisley; *Dir:* Peter Maxwell. **VHS, Beta, LV $19.98** *SUE, STE ♫♫1/2*

Highlander A strange tale about an immortal 16th-century Scottish warrior who has had to battle his evil immortal enemy through the centuries. The feud comes to blows in modern-day Manhattan. Connery makes a memorable appearance as the good warrior's mentor. Spectacular battle and death scenes.
1986 (R) 110m/C Christopher Lambert, Sean Connery, Clancy Brown; *Dir:* Russell Mulcahy. **VHS, Beta, LV $19.99** *HBO ♫♫♫*

Highlander 2: The Quickening The saga of Connor MacLeod and Juan Villa-Lobos continues in this sequel set in the year 2024. An energy shield designed to block out the sun's harmful ultraviolet rays has left planet Earth in perpetual darkness, but there is evidence that the ozone layer has repaired itself. An environmental terrorist and her group begin a sabotage effort and are joined by MacLeod and Villa-Lobos in their quest to save Earth. Stunning visual effects don't make up for the lack of substance.
1991 (R) 90m/C Christopher Lambert, Sean Connery, Virginia Madsen, Michael Ironside; *Dir:* Russell Mulcahy. **VHS, LV, 8mm** *COL ♫1/2*

Highpoint When an unemployed man becomes the chauffeur of a wealthy family, he finds himself in the middle of a mysterious murder and part of an international CIA plot. Hard to figure what's happening in this one.
1980 (PG) 91m/C *CA* Richard Harris, Christopher Plummer, Beverly D'Angelo, Kate Reid, Saul Rubinek; *Dir:* Peter Carter. **VHS, Beta, LV $19.95** *STE, SUE ♫*

Highway 13 When a trucker witnesses an "accidental" death at the loading warehouse, he falls under suspicion. He then sets out to clear his name.

1948 60m/B Robert Lowery, Pamela Blake, Lyle Talbot, Michael Whalen, Maris Wrixon, Clem Bevans; *Dir:* William Berke. **VHS, Beta, LV**
WGE ✠½

Highway 61 Pokey Jones is a shy barber in the small Canadian town of Pickerel Falls who becomes a celebrity when he discovers the frozen corpse of an unknown young man in his backyard. Jackie Bangs is a rock roadie who has stolen her band's stash of drugs and finds herself stranded in the same small town. She quickly proclaims herself the dead man's sister and persuades Pokey to drive her and the corpse to New Orleans for burial, neglecting to mention she has stashed the drugs in the coffin. Amoral Jackie and innocent Pokey begin a wild road trip which also features the sinister Mr. Skin, who says he's the Devil and tries to claim the corpse for himself, and a wasted rock star based on Elvis. Lackluster direction undercuts much of the first-rate acting and amusing story quirks.
1991 (R) 110m/C Valerie Buhagiar, Don McKellar, Earl Pastko, Peter Breck, Art Bergmann; *Dir:* Bruce McDonald. **VHS $89.95**
PAR ✠✠

Highway to Hell Bergen plays the devil in a horror-comedy about going to hell—even if you're not dead yet. Lowe and Swanson are newlyweds on their way to Las Vegas when they're stopped by a "hellcop," who kidnaps the bride and takes her to the netherworld to be the devil's new plaything. Her human hubby doesn't take kindly to this and follows them to hell to rescue his bride. This one's played for laughs and has some good special effects.
1992 (R) 93m/C Patrick Bergin, Chad Lowe, Kristy Swanson, Adam Storke, Pamela Gidley, Richard Farnsworth; *Dir:* Ate De Jong. **VHS $89.99** *HBO* ✠✠

Hijazo De Mi Vidaza (Son of My Life) The history of a family as told by a mother to her son. In Spanish.
197? 90m/C Enrique Cuenca, Eduardo Manzano, Maria Fernanda, Fernando Casanova. **VHS, Beta $49.95** *MAD* ✠

The Hill Contains episodes from the "Tour of Duty" Vietnam War series on television—"Angel of Mercy" and the title episode.
1988 93m/C Terence Knox, Stephen Caffrey, Joshua Maurer, Tony Becker, Ramon Franco; *Dir:* Bill L. Norton, Robert Iscove. **VHS, LV $59.95** *NWV* ✠½

Hill 24 Doesn't Answer Four Zionist soldiers defend a strategic entrance to Jerusalem. While at their posts, they reflect on their lives and devotion to the cause. Made in Israel, in English. The first Israeli feature film.
1955 101m/B *IS* Edward Mulhare, Haya Harareet, Michael Wager; *Dir:* Thorold Dickinson. **VHS, Beta $79.95** *ERG* ✠✠✠

Hill Number One A dramatization of the story of Easter, featuring Dean in one of his first major roles as John the Baptist.
1951 57m/B James Dean, Michael Ansara, Leif Erickson, Ruth Hussey, Roddy McDowall. **VHS, Beta $19.95** *RHI, FCT* ✠✠✠

Hillbillys in a Haunted House Two country and western singers en route to the Nashville jamboree encounter a group of foreign spies "haunting" a house. Features Rathbone's last film performance. Sequel to "Las Vegas Hillbillies" (1966).

1967 88m/C Ferlin Husky, Joi Lansing, Don Bowman, John Carradine, Lon Chaney Jr., Basil Rathbone; *Dir:* Jean Yarborough. **VHS, Beta $29.95** *UHV* ✠½

The Hills Have Eyes A desperate family battles for survival and vengeance against a brutal band of inbred hillbilly cannibals.
1977 (R) 83m/C Susan Lanier, Robert Houston, Martin Speer, Dee Wallace Stone, Russ Grieve, John Steadman, James Whitworth, Michael Berryman; *Dir:* Wes Craven. **VHS, Beta, LV $19.95** *STE, HAR* ✠

Hills Have Eyes, Part 2 Ignorant teens disregard warnings of a "Hills Have Eyes" refugee, and go stomping into the grim reaper's proving grounds.
1984 86m/C Michael Berryman, Kevin Blair, John Bloom, Janus Blythe, John Laughlin, Tamara Stafford; *Dir:* Wes Craven. **VHS, Beta $14.99** *HBO* ✠

Hills of Utah Autry finds himself in the middle of a feud between the local mine operator and a group of cattlemen, while searching for his father's murderer.
1951 70m/B Gene Autry, Pat Buttram, Denver Pyle; *Dir:* John English. **VHS, Beta $29.98** *CCB* ✠½

The Hillside Strangler Made-for-television, crime-of-the-week drama about the investigation to apprehend the serial killer responsible for 10 deaths in 1977-78.
1989 91m/C Richard Crenna, Dennis Farina, Billy Zane, Tony Plana, James Tolken, Tasia Valenza; *Dir:* Steven Gethers. **VHS, Beta $79.95** *FRH* ✠½

Himatsuri (Fire Festival) An acclaimed film detailing the conflicts between man and nature, as a careless lumberjack in a Japanese forest is visited by the ancient gods. English subtitles. Also known as "Fire Festival."
1985 95m/C *JP* Kinya Kitaoji; *Dir:* Mitsuo Yanagimachi. **VHS, Beta $29.95** *LHV, WAR* ✠✠✠

Hips, Hips, Hooray Two supposed "hot shot" salesmen are hired by a cosmetic company to sell flavored lipstick. This lavish musical comedy is one of Wheeler and Woolsey's best vehicles.
1934 68m/B Robert Woolsey, Bert Wheeler, Ruth Etting, Thelma Todd, Dorothy Lee; *Dir:* Mark Sandrich. **VHS, Beta, LV $19.95** *CCB* ✠✠½

Hired Hand Two drifters settle on a farm belonging to one of their wives, only to leave again seeking to avenge the murder of a friend. TV prints include Larry Hagman in a cameo role.
1971 93m/C Peter Fonda, Warren Oates, Verna Bloom, Severn Darden, Robert Pratt; *Dir:* Peter Fonda. **VHS, Beta $14.95** *KRT* ✠✠½

Hired to Kill A pair of Mafia hoods plot against each other over a $6 million drug shipment. Previously titled "The Italian Connection."
1973 (R) 90m/C Henry Silva, Woody Strode, Adolfo Celi, Mario Adorf. **Beta** *MFI* ✠½

Hired to Kill The granite-hewn Thompson plays a male-chauvinist commando grudgingly leading a squad of beautiful mercenary-ettes on an international rescue mission. Not as bad as it sounds (mainly because the ladies aren't complete bimbos), but between senseless plotting, gratuitous catfights, and a love scene indistinguishable from rape, the Hound found plenty to dislike.

1991 (R) 91m/C Brian Thompson, George Kennedy, Jose Ferrer, Oliver Reed, Penelope Reed, Michelle Moffett, Barbara Lee Alexander, Jordana Capra; *Dir:* Peter Rader; *W/Dir:* Nico Mastorakis. **VHS, Beta** *PAR* ✠½

Hiroshima Maiden An American family gains new perspective when they meet a young survivor of the atomic bomb. Originally aired on PBS as part of the WonderWorks Family Movie series.
1988 58m/C Susan Blakely, Richard Masur, Tamlyn Tomita; *Dir:* Joan Darling. **VHS $29.95** *PME, WNE* ✠½

Hiroshima, Mon Amour A profoundly moving drama exploring the shadow of history over the personal lives of a lonely French actress living in Tokyo and a Japanese architect with whom she's having an affair. Presented in a complex network of flashbacks. Highly influential; adapted by Marguerite Duras from her book.
1959 88m/B *JP FR* Emmanuelle Riva, Eiji Okada, Bernard Fresson; *Dir:* Alain Resnais. Cannes Film Festival '59: Film Writers Award, International Critics Award; National Board of Review Awards '60: 5 Best Foreign Films of the Year; New York Film Critics Awards '60: Best Foreign Film. **VHS, Beta, 8mm $29.98** *NOS, MRV, SUE* ✠✠✠✠

Hiroshima: Out of the Ashes The terrible aftermath of the bombing of Hiroshima, August, 1945, as seen through the eyes of American and Japanese soldiers, the Japanese people, and a priest. Carefully detailed and realistic, creates an amazing emotional response. Don't miss the scene in which Nelson and his buddy find themselves surrounded by Japanese—the enemy—only to realize that their "captors" have been blinded by the blast. Created as an homage for the 45th anniversary of the bombing.
1990 (PG) 98m/C Max von Sydow, Judd Nelson, Pat Morita; *Dir:* Peter Werner. **VHS** *VMK* ✠✠✠

His Brother's Ghost When Fuzzy is rubbed out by bandits, Billy Carson (no longer called Billy the Kid thanks to mothers' protestations) convinces his twin brother to impersonate him in order to make the bandits believe they are the victims of hocus pocus from the otherworld. Crabbe—who dressed noir before it was chic—made this during his post sci-fi western period, when he slugged it through some fifty low-rent oaters (usually under the direction of Newfield at PRC studios).
1945 54m/C Buster Crabbe, Al "Fuzzy" St. John, Charles King, Bud Osborne, Karl Hackett, Archie Hall; *Dir:* Sam Newfield. **VHS $19.99** *NOS, MRV, HEG* ✠✠

His Double Life When a shy gentlemen's valet dies, his master assumes the dead man's identity and has a grand time. From the play "Buried Alive" by Arnold Bennett. Remade in 1943 as "Holy Matrimony."
1933 67m/B Roland Young, Lillian Gish, Montagu Love; *Dir:* Arthur Hopkins. **VHS, Beta $19.95** *NOS, MRV, KRT* ✠✠½

His Fighting Blood Northwest Mounties in action. Based on a James Curwood story.
1935 63m/B Kermit Maynard. **VHS, Beta $19.95** *VDM* ✠

His First Command An early talkie wherein a playboy falls in love with the daughter of a cavalry officer, and enlists in order to win her affections.

1929 60m/B William Boyd, Dorothy Sebastian, Gavin Gordon; *Dir:* Gregory La Cava. **VHS, Beta, 8mm $24.95** VYY ⅇⅇ½

His First Flame A classic Langdon silent romantic comedy. Co-written by Frank Capra.
1926 62m/B Harry Langdon, Vernon Dent, Natalie Kingston. **VHS, Beta $19.95** NOS, VYY ⅇⅇⅇ

His Girl Friday The classic, unrelentingly hilarious war-between-the-sexes comedy in which a reporter and her ex-husband editor help a condemned man escape the law while at the same time furthering their own ends as they try to get the big scoop on political corruption in the town. One of Hawks' most furious and inventive screen combats in which women are given uniquely equal (for Hollywood) footing, with staccato dialogue scripted by Charles Lederer and wonderful performances. Based on the Hecht-MacArthur play "The Front Page," which was filmed as the 1931 classic "The Front Page." Remade again in 1974 as "The Front Page" and in 1988 as "Switching Channels." Hawks' masterwork is also available colorized.
1940 92m/B Cary Grant, Rosalind Russell, Ralph Bellamy, Gene Lockhart, John Qualen, Porter Hall, Roscoe Karns, Abner Biberman, Cliff Edwards, Billy Gilbert; *Dir:* Howard Hawks. **VHS, Beta $9.95** NEG, MRV, NOS ⅇⅇⅇⅇ

His Kind of Woman A fall guy, being used to bring a racketeer back to the U.S. from Mexico, discovers the plan and tries to halt it.
1951 120m/B Robert Mitchum, Jane Russell, Vincent Price, Tim Holt, Charles McGraw, Raymond Burr, Jim Backus; *Dir:* John Farrow. **VHS, Beta** KOV ⅇⅇⅇ

His Majesty O'Keefe Lancaster stars as a South Seas swashbuckler dealing in the lucrative coconut-oil trade of the mid-1800s. The natives see him as a god and allow him to marry a beautiful maiden. When his reign is threatened by unscrupulous traders, Lancaster springs into action to safeguard his kingdom. Based on a real-life American adventurer, this was the first movie ever filmed in the Fiji Islands.
1953 92m/C Burt Lancaster, Joan Rice, Benson Fong, Philip Ahn, Grant Taylor; *Dir:* Byron Haskin. **VHS, Beta, LV $59.99** WAR ⅇⅇ½

His Majesty, the Scarecrow of Oz/The Magic Cloak of Oz The creator of the Oz stories, L. Frank Baum, wrote, produced and directed these two stories himself, with all the characters from his best-selling children's books.
1914 80m/B *Dir:* L. Frank Baum. **VHS, Beta $59.95** GVV ⅇⅇ½

His Memory Lives On A documentary about Elvis and his fanatical fans.
1984 60m/C **VHS, Beta $19.95** MNP ⅇⅇ½

His Name was King A man goes after the gang who murdered his brother and raped his young wife in the old West.
1985 90m/C IT Richard Harrison, Klaus Kinski, Anne Puskin; *Dir:* Don Reynolds. **VHS, Beta $14.95** NEG Woof!

His Picture in the Papers An early silent comedy wherein the robust, red-meat-eating son of a health food tycoon get his father's permission to marry by getting his picture favorably in the paper for advertising's sake. Written by Anita Loos and John Emerson.

1916 68m/B Douglas Fairbanks Sr.; *Dir:* John Emerson. **VHS, Beta, 8mm $19.95** NOS, VYY, GPV ⅇⅇ

His Prehistoric Past/The Bank Two of Chaplin's classic early shorts.
1915 54m/B Charlie Chaplin, Mack Swain, Edna Purviance; *Dir:* Charlie Chaplin. **VHS, Beta $24.95** VYY ⅇⅇ

His Private Secretary Wayne plays the jet-setting son of a wealthy businessman who wants his boy to settle down. And he does - after he meets the minister's beautiful daughter.
1933 68m/B John Wayne, Evelyn Knapp, Alec B. Francis, Reginald Barlow, Natalie Kingston, Arthur Hoyt, Al "Fuzzy" St. John; *Dir:* Philip H. Whitman. **VHS, Beta $19.98** DVT ⅇ½

His Royal Slyness/Haunted Spooks Two Harold Lloyd shorts. "His Royal Slyness" (1919) offers Harold impersonating the king of a small kingdom, while in "Haunted Spooks" (1920), Harold gets suckered into living in a haunted mansion. Both films include music score.
1920 52m/B Harold Lloyd; *Dir:* Harold Lloyd. **VHS, Beta $29.95** VYY, CCB ⅇ½

His Wife's Lover (Zayn Vaybs Lubovnik) An actor disguised as an old man wins the heart of a lovely young woman. He decides to test her fidelity by reverting to his handsome young self and attempting to seduce her. An enchanting comedy in Yiddish with English subtitles.
1931 77m/B Ludwig Satz, Michael Rosenberg, Isadore Cashier, Lucy Levine; *Dir:* Sidney Goldin. **VHS, Beta $59.95** ERG, FCT ⅇ½

Historia de un Gran Amor (History of a Great Love) Two children of rival families fall in love against their parents' wishes.
1957 155m/B George Negrete, Sara Garcia, Gloria Marin. **VHS, Beta $64.95** MAD ⅇ½

History is Made at Night A wife seeks a divorce from a jealous husband while on an Atlantic cruise; she ends up finding both true love and heartbreak in this story of a love triangle at sea.
1937 98m/B Charles Boyer, Jean Arthur, Leo Carrillo, Colin Clive; *Dir:* Frank Borzage. **VHS, Beta, LV $19.98** WAR, LTG, FCT ⅇⅇⅇ

The History of White People in America A pseudo-documentary look at the complicated life of a white midwestern family. Succeeds in making hilarious fun of many traditional prejudices. Originally made for cable television.
1985 48m/C Martin Mull, Fred Willard, Mary Kay Place, Teri Garr, Bob Eubanks; *Dir:* Harry Shearer. **VHS, Beta $24.95** MCA ⅇⅇⅇ

The History of White People in America: Vol. 2 Mull's television-special comedy satirizing WASP society in America.
1985 40m/C Martin Mull; *Dir:* Harry Shearer. **VHS, Beta $24.95** MCA ⅇⅇⅇ

History of the World: Part 1 More misses than hits in this Brooks' parody of historic epics. A bluntly satiric vision of human evolution, from the Dawn of Man to the French Revolution told in an episodic fashion. Good for a few laughs.
1981 (R) 90m/C Mel Brooks, Dom DeLuise, Madeline Kahn, Harvey Korman, Cloris Leachman, Gregory Hines, Pamela Stephenson, Paul Mazursky, Bea Arthur, Fritz Feld, John Hurt, Jack Carter, John Hillerman, John Gavin, Barry Levinson, Ron Carey, Howard Morris, Sid

Caesar, Jackie Mason; *Dir:* Mel Brooks; *Nar:* Orson Welles. **VHS, Beta, LV $19.98** FOX ⅇ½

The Hit A feisty young hooker gets mixed-up with two strong-armed hired killers as they escort an unusual stool-pigeon from his exile in Spain to their angry mob bosses. Minor cult-noir from the director of "My Beautiful Launderette" and "Dangerous Liaisons."
1985 (R) 105m/C GB Terence Stamp, John Hurt, Laura Del Sol, Tim Roth, Fernando Rey; *Dir:* Stephen Frears. **VHS, Beta, LV $19.98** SUE ⅇⅇ½

Hit the Deck Second-rate studio musical about sailors on leave and looking for romance. Based on the 1927 Broadway musical.
1955 112m/C Jane Powell, Tony Martin, Debbie Reynolds, Walter Pidgeon, Vic Damone, Gene Raymond, Ann Miller, Russ Tamblyn, J. Carroll Naish, Kay Armen, Richard Anderson; *Dir:* Roy Rowland. **VHS, Beta, LV $19.95** MGM ⅇ½

Hit the Ice Newspaper photographers Bud and Lou are mistaken as gangsters in Chicago and the usual complications ensue. Includes an appearance by Johnny Long & His Orchestra.
1943 89m/C Bud Abbott, Lou Costello, Patric Knowles, Elyse Knox, Ginny Simms, Johnny Long & His Orchestra; *Dir:* Charles Lamont. **VHS, Beta $14.95** MCA ⅇⅇ

Hit Lady An elegant cultured woman becomes a hit lady for the syndicate in this predictable, yet slick gangster movie. Made for television.
1974 74m/C Yvette Mimieux, Dack Rambo, Clu Gulager, Keenan Wynn; *Dir:* Tracy Keenan Wynn. **VHS, Beta $59.95** PSM ⅇⅇ

The Hit List A regular guy fights back brutally when his wife and child are mistakenly kidnapped by the Mafia. Fast-paced action makes up for thin plot.
1988 (R) 87m/C Jan-Michael Vincent, Leo Rossi, Lance Henriksen, Charles Napier, Rip Torn; *Dir:* William Lustig. **VHS, Beta, LV $14.95** COL ⅇⅇ

Hit Man (Il Sicario) A crime drama that profiles the crisis of a man unfit to commit murder but who does for a price.
1961 100m/B **VHS, Beta, 3/4U, Special order formats** CIG ⅇⅇ

Hit Men Double crosses and gun battles highlight this tale of inner-city crime.
1973 90m/C Henry Silva, Mario Adorf, Woody Strode, Adolfo Celi. **VHS, Beta $19.95** VDC ⅇ½

Hit & Run David Marks, a Manhattan cab driver, is haunted by recurring flashbacks of a freak hit-and-run accident in which his wife was struck down on a city street. Also called "Revenge Squad."
1982 96m/C Paul Perri, Claudia Cron; *Dir:* Charles Braverman. **VHS, Beta $59.99** HBO ⅇ

Hit the Saddle The Three Mesquiteers track down a gang involved in capturing wild horses in protected areas although one of three is more interested in the charms of a dance hall girl (played by starlet Rita Cansino who would soon become Rita Hayworth).
1937 54m/B Bob Livingston, Ray Corrigan, Max Terhune, J.P. McGowan, Yakima Canutt, Rita Hayworth; *Dir:* Mack V. Wright. **VHS, Beta $9.99** NOS, VCN, REP ⅇ½

The Hitch-Hiker Two young men off on the vacation of their dreams pick up a psychopathic hitchhiker with a right eye that never closes, even when he sleeps. Taut suspense makes this flick worth seeing.

1953 71m/B Edmond O'Brien, Frank Lovejoy, William Talman, Jose Torvay; *Dir:* Ida Lupino. **VHS $16.95** *SNC* 🎞

Hitchcock Classics Collection: Trailers on Tape Collection of 20 previews for such Hitchcock favorites as "Shadow of a Doubt," "Rear Window," "Vertigo," "North by Northwest" and "Psycho." Some trailers are in black and white.
1985 55m/C *Dir:* Alfred Hitchcock. **VHS, Beta $79.95** *SFR* 🎞🎞🎞

Hitchcock Collection A collector's set containing three films from the master of suspense. Includes: "The Lady Vanishes," "The 39 Steps," "Vintage Hitchcock" - see him from a set designer to a director. In B/W and color.
1987 200m/C Alfred Hitchcock, Michael Redgrave, Paul Lukas, Robert Donat, Madeleine Carrol; *Dir:* Alfred Hitchcock. **VHS, Beta $79.95** *MPI* 🎞🎞🎞

Hitcher A young man picks up a hitchhiker on a deserted stretch of California highway only to be tormented by the man's repeated appearances: is he real or a figment of his imagination? Ferociously funny, sadomasochistic comedy with graphic violence.
1985 (R) 98m/C Rutger Hauer, C. Thomas Howell, Jennifer Jason Leigh; *Dir:* Robert Harmon. **VHS, Beta, LV $14.99** *HBO* 🎞🎞½

Hitchhiker 1 First of a series from HBO's horror anthology series, titles are: "The Curse," "W.G.O.D.," and "Hired Help."
1985 90m/C Harry Hamlin, Gary Busey, Geraldine Page, Karen Black. **VHS, Beta, LV $36.95** *HBO, WAR* 🎞🎞½

Hitchhiker 2 Three more episodes from HBO's horror series: "Dead Man's Curve," "Nightshift," and "Last Scene."
1985 90m/C Peter Coyote, Margot Kidder, Darren McGavin, Susan Anspach. **VHS, Beta, LV $36.95** *HBO, WAR* 🎞🎞½

Hitchhiker 3 Three episodes from the anthology mystery series: "Ghost Writer," "And If We Dream," and "True Believer." Made-for-cable television.
1987 90m/C Ornella Muti, Willem Dafoe, Barry Bostwick, Dayle Haddon, Stephen Collins, Robert Weiss, Tom Skerritt. **VHS, Beta, LV $39.98** *LHV, WAR* 🎞🎞½

Hitchhiker 4 Three episodes of the cable-television suspense series: "Man's Best Friend," about a lonely man whose pet dog tears his enemies to bits. "Face to Face," about the revenge exacted upon a drug-abusing cosmetic surgeon after he mucks up an operation; and "Videodate," about a guy being lured into a sinister video performance piece via his use of video dating services.
1987 90m/C Michael O'Keefe, Sybil Danning, Sonja Smitts, Robert Vaughn, Shannon Tweed, Greg Henry. **VHS, Beta, LV $39.98** *LHV, WAR* 🎞🎞½

Hitchhikers Trash film about a bunch of scantily clad female hitchhikers who rob the motorists who stop to pick them up. Keep right on going when you spot this roadkill.
1972 (R) 90m/C Misty Rowe, Norman Klar, Linda Avery; *Dir:* Ferd Sebastian. **VHS, Beta $39.95** *IND* Woof!

Hitler The true story of the infamous Nazi dictator's rise to power and his historic downfall.
1962 103m/B Richard Basehart, Maria Emo, Cordula Trantow; *Dir:* Stuart Heisler. **VHS, Beta $59.98** *FOX* 🎞🎞½

Hitler: Dead or Alive Three ex-cons devise a scheme to gain a million-dollar reward for capturing Adolf Hitler in this low-budget production.
1943 64m/B Ward Bond, Dorothy Tree, Bruce Edwards, Warren Hymer, Paul Fix, Russell Hicks, Bob Watson; *Dir:* Nick Grinde. **VHS, Beta $19.95** *NOS, MRV, DVT* 🎞🎞

Hitler: The Last Ten Days Based on an eyewitness account, the story of Hitler's last days in an underground bunker gives insight to his madness.
1973 (PG) 106m/C Alec Guinness, Simon Ward, Adolfo Celi; *Dir:* Ennio de Concini. **VHS, Beta $49.95** *PAR, GLV* 🎞🎞

Hitler's Children Two young people, a German boy and an American girl, are caught in the horror of Nazi Germany. He's attracted to Hitler's rant, she's repelled. Exploitative yet engrossing.
1943 83m/B Tim Holt, Bonita Granville, Otto Kruger, H.B. Warner, Irving Reis, Hans Conried, Lloyd Corrigan; *Dir:* Edward Dmytryk. **VHS, Beta $19.95** *MED, IHF* 🎞🎞🎞

Hitler's Daughter Made-for-TV movie following the trials and tribulations of the illegitimate daughter of Adolf Hitler. She's all grown up now...and wants to rule the United States. She secures a position of power during an election and now Nazi-hunters must figure out her identity before she takes over the world!
1990 (R) 88m/C Kay Lenz, Veronica Cartwright, Melody Anderson, Patrick Cassidy, Carolyn Dunn, Lindsay Merrithew, George R. Robertson; *Dir:* James A. Contner. **VHS, Beta** *PAR* 🎞

Hitler's Henchmen Powerful look at the inhumanity of war and the graphic reality of Nazi death camps during World War II.
1984 60m/C **VHS, Beta $39.95** *MPI* 🎞🎞½

The Hitman Norris plays a cop undercover as a syndicate hit man in Seattle, where he cleans up crime by triggering a three-way mob bloodbath between beastly Italian mafiosi, snooty French-Canadian hoods, and fanatical Iranian scum. But he teaches a black kid martial arts, so you know he's politically correct. Action addicts will give this a passing grade; all others need not apply.
1991 (R) 95m/C Chuck Norris, Michael Parks, Al Waxman, Alberta Watson, Salim Grant, Ken Pogue, Marcel Sabourin, Bruno Gerussi, Frank Ferrucci; *Dir:* Aaron Norris. **VHS, LV, 8mm $92.99** *PIA* 🎞🎞

The Hitter A former prizefighter, his girlfriend, and a washed-up promoter are all on the run in this adventure tale.
1979 (R) 94m/C Ron O'Neal, Adolph Caesar, Sheila Frazier; *Dir:* Christopher Leitch. **VHS $59.95** *SVS* 🎞🎞

Hittin' the Trail The Good Cowboy is mistaken for a murdering varmint, and must prove his innocence.
1937 57m/B Tex Ritter, Charles King, Ray Whitley, The Range Ramblers; *Dir:* Robert Bradbury. **VHS, Beta $19.95** *NOS, VDM* 🎞

Hitz In a Los Angeles barrio, Musico, a Loco gang member, kills two rival gang members. Arrested and convicted when a young boy named Pepe is coerced into testifying against him, Musico is killed in the courtroom. But the violence doesn't stop. The judge, who wants to reform the juvenile court system, tries to protect Pepe, and both are targeted for death by the Loco gang.
1992 (R) 90m/C Elliott Gould, Karen Black, Cuba Gooding Jr. **VHS $89.95** *VMK* 🎞½

H.M.S. Pinafore The opera is captured in a live performance.
1982 90m/C Beta, LV $24.95 *PNR* 🎞½

The Hobbit An animated interpretation of J.R.R. Tolkien's novel of the same name. The story follows Bilbo Baggins and his journeys in Middle Earth and encounters with the creatures who inhabit it. He joins comrades to battle against evil beings and creatures. Animator Ralph Bakshi utilizes "live motion" animation to give authentic movement to the characters. Made for television.
1978 76m/C *Dir:* Arthur Rankin Jr., Jules Bass; *Voices:* Orson Bean, John Huston, Otto Preminger, Richard Boone. **VHS, Beta, 8mm $19.95** *WAR, SVS, VTK* 🎞🎞🎞

Hobgoblins Little creatures escape from a studio vault and wreak havoc.
1987? 92m/C Jeffrey Culver, Tom Bartlett; *Dir:* Rick Sloane. **VHS, Beta $79.95** *TWE* 🎞

Hobo's Christmas A man who left his family to become a hobo comes home for Christmas 25 years later. Hughes turns in a delightful performance as the hobo in this made for television special.
1987 94m/C Barnard Hughes, William Hickey, Gerald McRaney, Wendy Crewson; *Dir:* Will MacKenzie. **VHS, Beta $79.95** *NSV* 🎞🎞

Hobson's Choice A prosperous businessman in the 1890s tries to keep his daughter from marrying, but the strong-willed daughter has other ideas. Remade for television in 1983.
1953 102m/B *GB* Charles Laughton, John Mills, Brenda de Banzie; *Dir:* David Lean. Berlin Film Festival '54: Top Audience Award; British Academy Awards '54: Best Film. **VHS, Beta, LV $19.98** *SUE, HBO* 🎞🎞🎞

Hobson's Choice Made-for-television remake of the old family comedy about a crusty, penny-pinching businessman whose headstrong daughter proves to him she'll not become an old maid. Instead she marries one of his employees. Not bad, but not as good as the original.
1983 95m/C Jack Warden, Sharon Gless, Richard Thomas, Lillian Gish; *Dir:* Gilbert Cates. **VHS, Beta $59.98** *FOX* 🎞

Hockey Night A movie for pre- and early teens in which a girl goalie makes the boys' hockey team.
1984 77m/C *CA* Megan Follows, Rick Moranis, Gail Youngs, Martin Harburg, Henry Ramer; *Dir:* Paul Shapiro. **VHS, Beta** *FHE* 🎞🎞½

Hog Wild A clique of high school nerds exact revenge on bullies from the local mean motorcycle gang, and a girl is torn between the two factions.
1980 (PG) 97m/C *CA* Patti D'Arbanville, Tony Rosato, Michael Biehn; *Dir:* Les Rose. **VHS, Beta $59.98** *CHA* 🎞½

The Holcroft Covenant Based upon the complex Robert Ludlum novel. Details the efforts of a man trying to release a secret fund that his Nazi father humanistically set up to relieve the future sufferings of Holocaust survivors. Confusing, and slow, but interesting, nonetheless.
1985 (R) 112m/C *GB* Michael Caine, Victoria Tennant, Anthony Andrews, Lilli Palmer, Mario Adorf, Michael Lonsdale; *Dir:* John Frankenheimer. **VHS, Beta, LV $79.99** *HBO* 🎞🎞

Hold the Dream A made-for-television mini-series sequel to Barbara Taylor Bradford's "A Woman of Substance."
1985 195m/C Jenny Seagrove, Stephen Collins, Deborah Kerr, James Brolin; *Dir:* Don Sharp. **VHS, Beta $14.95** *QNE* 🎞

Hold 'Em Jail Some good laughs with Wheeler and Woolsey starting a competitive football team at Kennedy's Bidemore Prison.
1932 65m/B Bert Wheeler, Robert Woolsey, Edgar Kennedy, Betty Grable, Edna May Oliver; **Dir:** Norman Taurog. **VHS, Beta $34.98** *CCB* ✂️✂️½

Hold That Ghost Abbott and Costello inherit an abandoned roadhouse where the illicit loot of its former owner, a "rubbed out" mobster, is supposedly hidden.
1941 86m/B Bud Abbott, Lou Costello, Joan Davis, Richard Carlson, Mischa Auer, The Andrews Sisters, Ted Lewis & his Band; **Dir:** Arthur Lubin. **VHS, Beta $14.95** *MCA* ✂️✂️✂️

Hole in the Head A comedy-drama about a shiftless but charming lout who tries to raise money to save his hotel from foreclosure and learn to be responsible for his young son. Notable for the introduction of the song "High Hopes."
1959 120m/C Frank Sinatra, Edward G. Robinson, Thelma Ritter, Carolyn Jones, Eleanor Parker, Eddie Hodges, Keenan Wynn, Joi Lansing; **Dir:** Frank Capra. Academy Awards '59: Best Song ("High Hopes"). **VHS, Beta $19.95** *MGM* ✂️✂️½

The Holes (Les Gaspards) A Parisian book shop owner's daughter vanishes along with other local citizens and American tourists, and he decides to investigate when the police won't. In French with subtitles.
1972 (PG) 92m/C *FR* Philippe Noiret, Charles Denner, Michael Serrault, Gerard Depardieu. **VHS, Beta $59.95** *FHE, FCT* ✂️✂️½

Holiday The classically genteel screwball comedy about a rich girl who steals her sister's fiance. A yardstick in years to come for sophisticated, urbane Hollywood romanticism. Based on the play by Philip Barry who later wrote "The Philadelphia Story."
1938 93m/B Cary Grant, Katharine Hepburn, Doris Nolan, Edward Everett Horton, Ruth Donnelly, Lew Ayres, Binnie Barnes; **Dir:** George Cukor. **VHS, Beta, LV $19.95** *RCA* ✂️✂️✂️½

Holiday Affair Hollywood yuletide charmer about a pretty widow being courted by two very different men.
1949 86m/B Robert Mitchum, Janet Leigh, Griff Barnett, Wendell Corey, Esther Dale, Henry O'Neill, Henry Morgan, Larry J. Blake; **Dir:** Don Hartman. **VHS, Beta, LV** *RKO* ✂️✂️✂️

Holiday in Havana Desi Arnaz is back in Cuba winning hearts and dance contests. Watch for Desi's presence.
1949 73m/B Desi Arnaz Sr., Lolita Valdez, Mary Hatcher, Ann Doran; **Dir:** Jean Yarborough. **VHS** *GKK* ✂️✂️

Holiday Hotel A group of wild French vacationers let loose at their favorite little resort hotel along the Brittany coast. Available in French with English subtitles or dubbed into English.
1977 (R) 109m/C *FR* Sophie Barjae, Daniel Ceccaldi; **Dir:** Michael Lang. **VHS, Beta $29.95** *SUE* ✂️✂️

Holiday Inn Fred Astaire and Bing Crosby are rival song-and-dance men who decide to work together to turn a Connecticut farm into an inn, open only on holidays. Songs include: "Happy Holiday," "Be Careful, It's My Heart," and "Easter Parade." Remade in 1954 as "White Christmas."
1942 101m/B Bing Crosby, Fred Astaire, Marjorie Reynolds, Walter Abel, Virginia Dale; **Dir:** Mark Sandrich. Academy Awards '42: Best Song ("White Christmas"). **VHS, Beta, LV $19.95** *MCA* ✂️✂️✂️

Holiday Rhythm A television promoter is knocked out and dreams of a world-wide trip full of entertainment, comedy, and insanity.
1950 61m/B Mary Beth Hughes, Tex Ritter, David Street, Donald MacBride. **VHS, Beta, LV** *WGE* ✂️½

Hollow Gate Trick or Treat?? Four kids find out when they run into a guy with hungry dogs.
1988 90m/C Addison Randall. **VHS, Beta $69.95** *CLV* ✂️½

Hollow Triumph A down-and-out con man assumes the identity of a respected psychiatrist in this film noir. Based on the novel by Murray Forbes. Retitled "The Scar."
1948 83m/B Paul Henreid, Joan Bennett, Eduard Franz, Leslie Brooks; **Dir:** Steve Sekely. **VHS $19.98** *NOS, MOV* ✂️✂️✂️

Hollywood Bloopers Hilarious outtakes from some of Hollywood's finest movies, featuring stars such as Bette Davis, James Cagney, Humphrey Bogart, Rosalind Russell, Errol Flynn, Kirk Douglas, and many more.
195? 85m/B Lauren Bacall, Ronald Reagan, James Stewart, Joan Blondell, Rosalind Russell, Errol Flynn, Kirk Douglas, Barbara Stanwyck. **VHS, Beta $19.95** *VDM, AMV, PPI* ✂️

Hollywood Boulevard Behind-the-scenes glimpse of shoestring-budget movie making offers comical sex, violence, sight gags, one-liners, comedy bits, and mock-documentary footage. Commander Cody and His Lost Planet Airmen are featured.
1976 (R) 93m/C Candice Rialson, Mary Woronov, Rita George, Jonathan Kaplan, Jeffrey Kramer, Dick Miller, Paul Bartel; **Dir:** Joe Dante, Allan Arkush. **VHS, Beta $39.98** *WAR, OM* ✂️✂️½

Hollywood Boulevard 2 A "B" movie studio finds its top actresses being killed off and terror reigns supreme as the mystery grows more intense. Could that psycho-bimbo from outer space have something to do with it?
1989 (R) 82m/C Ginger Lynn Allen, Kelly Monteith, Eddie Deezen, Ken Wright, Steve Vinovich; **Dir:** Steve Barnett. **VHS, Beta $79.95** *MGM* ✂️

Hollywood or Bust The zany duo take their act on the road as they head for Tinsel Town in order to meet Lewis' dream girl Ekberg. This was the last film for the Martin & Lewis team and it's not the swan song their fans would have hoped for.
1956 95m/C Dean Martin, Jerry Lewis, Anita Ekberg, Pat Crowley, Maxie Rosenbloom, Willard Waterman; **Dir:** Frank Tashlin. **VHS $14.95** *PAR, CCB* ✂️✂️½

Hollywood Canteen A star-studded patriotic extravaganza. Just about every Warner Brothers lot actor turns out for this tribute to love and nationalism. Hutton stars as a lovesick G.I. who has fallen for Joan Leslie. He goes to the Hollywood Canteen hoping to meet her. The workers at the canteen, including Bette Davis, set-up a phony raffle so that he can win a date with Leslie. The sparks fly right away, but then he thinks it's all been just a ruse as he boards his train and Leslie's not there to see him off. But, of course she makes it in the nick of time, declares her love for him, and says she'll wait for him. Kind of a lame story but the talent and musical numbers make this picture fly with charm and style. Musical numbers include "Don't Fence Me In," "You Can Always Tell A Yank," "We're Having A Baby," "Tumblin' Tumbleweeds," and many many more. Before the picture was made there were arguments over actors being labled unpatriotic for not participating in this, and other similar movies being produced at the time.
1944 124m/B Robert Hutton, Dane Clark, Janice Paige, Jonathan Hale, Barbara Brown, James Flavin, Eddie Marr, Ray Teal, Bette Davis, Joan Leslie, Jack Benny, Jimmy Dorsey, Joan Crawford; **W/Dir:** Delmer Daves. **VHS $29.98** *FOX, FCT* ✂️✂️✂️

Hollywood Cartoons Go to War Hollywood's top cartoon stars support the war effort by appealing to citizens to purchase War Bonds and to save waste fats, cooking oils and scrap metals. Featuring Bugs Bunny, Donald Duck, Pluto, Minnie, Superman and Private Snafu.
1945 45m/C VHS, Beta *GVV* ✂️✂️✂️

Hollywood Centerfolds, Volume 1 Several bosomy young women bare all for the camera in hopes of landing real film work.
1984 30m/C VHS, Beta $19.95 *AHV* ✂️✂️✂️

Hollywood Centerfolds, Volume 2 More well-endowed starlets shimmy suggestively for your entertainment and their careers.
1985 30m/C VHS, Beta $19.95 *AHV* ✂️✂️✂️

Hollywood Centerfolds, Volume 3 Sleepwear, swimwear, and no wear at all drapes the bodies of a bevy of busty babes, as they show off their assets in hopes of getting jobs in which they can take their clothes off in real movies, and not just videos.
1986 30m/C VHS, Beta $19.95 *AHV* ✂️✂️✂️

Hollywood Centerfolds, Volume 4 Another collection of ambitious, well-built young ladies romping in the buff.
1988 30m/C VHS, Beta $19.95 *AHV* ✂️✂️✂️

Hollywood Chainsaw Hookers A campy, sexy, very bloody parody about attractive prostitutes who dismember their unsuspecting customers.
1988 (R) 90m/C Linnea Quigley, Gunnar Hansen, Jay Richardson, Michelle Bauer; **Dir:** Fred Olen Ray. **VHS, Beta $59.95** *CAM* ✂️✂️

Hollywood Chaos Director Ballantine's stars quit his production before it even starts, so he gets the next best thing - star look-alikes. However, his problems are just beginning. One of his new "stars" is kidnapped, and his reclusive stage manager is hunted as the object of desire for a female motorcycle gang.
1989 90m/C Carl Ballantine, Tricia Leigh Fisher, Kathleen Freeman; **Dir:** Sean McNamara. **VHS $79.95** *WLA* ✂️½

Hollywood Clowns Glenn Ford narrates this tribute to the screen's greatest comedians, with clips from many classic scenes.
1985 60m/C Buster Keaton, W.C. Fields, Charlie Chaplin, Harold Lloyd, Bud Abbott, Lou Costello, Stan Laurel, Oliver Hardy, Groucho Marx, Harpo Marx, Chico Marx, Jerry Lewis, Dean Martin, Red Skelton, Danny Kaye, Betty Hutton, Bob Hope. **VHS, Beta $29.95** *MGM* ✂️½

Hollywood Come Home Ray Milland hosts this look at the history of the movie industry from World War II to the early 60's. Also on this tape is "Hollywood Historama," a collection of video biographies of such luminaries as Mae West and John Barrymore.

H

1962 54m/B Ray Milland. **VHS, Beta, 8mm** $24.99 *VYY* ⅋½

Hollywood Confidential A beautiful Eastern girl goes to Hollywood and suffers irrevocable corruption.
193? 55m/B **VHS, Beta** $29.95 *JLT* ⅋

Hollywood Cop A tough cop in Hollywood goes up against the mob for the sake of a kidnapped boy.
1987 (R) 101m/C Jim Mitchum, David Goss, Cameron Mitchell, Troy Donahue, Aldo Ray; *Dir:* Amir Shervan. **VHS, Beta** $79.95 *CEL* ⅋½

Hollywood Crime Wave An examination of crime and criminals as seen in vintage Hollywood gangster films. Features scenes with Humphrey Bogart, James Cagney and Edward G. Robinson.
196? 60m/B *Nar:* Broderick Crawford. **VHS, Beta** $14.95 *MPI* ⅋½

The Hollywood Detective An average TV movie that spoofs "Kojak." A TV detective tries it out for real, and gets into troubles for which he can't pause "for a message from our sponsor."
1989 (PG) 88m/C Telly Savalas, Helene Udy, George Coe, Joe Dallesandro, Tom Reese; *Dir:* Kevin Connor. **VHS, Beta** $79.95 *MCA* ⅋⅋

Hollywood Erotic Film Festival A package of soft-core shorts, animated as well as live action, from around the world.
1987 75m/C **VHS, Beta** $59.95 *PAR* ⅋

The Hollywood Game Head of Hollywood talent agency charms more than the critics with his casting couch-side manner . That is, until his bevy of starlets discover that they are to be the prize behind door number one in his game show/sleaze fest. Not based on a true story.
1989 92m/C Dick Miller, Thelma Houston. **VHS, Beta** $79.95 *JEF* ⅋½

Hollywood Ghost Stories A view of supernatural activity in Hollywood, on the screen and behind the scenes. Includes interviews with William Peter Blatty, Sir Arthur Conan Doyle, Elke Sommer and others.
1986 75m/C *Nar:* John Carradine. **VHS, Beta** $29.98 *WAR* ⅋½

Hollywood Goes to War Five shorts produced for the entertainment of G.I.'s overseas or for home front bond drives: "The All-Star Bond Rally" features Bob Hope, Fibber McGee and Molly, Bing Crosby, Harry James and His Orchestra, Sinatra, Grable, Harpo Marx, and the Talking Pin-Ups; "Hollywood Canteen Overseas Special" features Dinah Shore, Eddie Cantor, Jimmy Durante, and Red Skelton; "Mail Call (Strictly G.I.)" is a filmed "Mail Call" radio program with Don Wilson announcing guests Dorothy Lamour, Cass Daley, and Abbott and Costello doing "Who's on First"; "G.I. Movie Weekly-Sing with the Stars" features Carmen Miranda and her fruit-basket hat, a Portuguese "follow-the-bouncing-ball' sing-along, Richard Lane, and Mel Blanc's voice; and "Invaders in Greasepaint," the story of "Four Jills in a Jeep'-the North African USO tour by Martha Raye, Carol Landis, Mitzi Mayfair, and Kay Francis. Also includes newsreel footage of the girls touring, plus their memorable rendition of the wartime classic tune "Snafu," which was banned from the airwaves for its racy lyrics.
1954 41m/B Bob Hope, Bing Crosby, Frank Sinatra, Betty Grable, Harpo Marx, Jimmy Durante, Eddie Cantor, Red Skelton, Bud Abbott, Lou Costello, Dinah Shore, Dorothy

Lamour, Carmen Miranda. **VHS, Beta** $24.95 *VYY* ⅋½

Hollywood Harry A private-eye mystery/parody, about an inept detective who searches for a missing porn actress.
1986 (PG-13) 99m/C Robert Forster, Kathrine Forster, Joe Spinell, Shannon Wilcox; *Dir:* Robert Forster. **VHS, Beta** $79.95 *MED* ⅋

Hollywood Heartbreak It's heartbreak in tinseltown for a young writer who's trying to pitch his movie script but winds up defending himself from slimy agents and persistent starlets. Another entry for Bartlett's familiar plotlines.
1989 90m/C Mark Moses, Richard Romanus, James LeGros. **VHS** $69.99 *RAE* ⅋⅋

Hollywood High Four attractive teen couples converge on the mansion of an eccentric silent-film star for an unusual vacation experience in this teenage skin flick. Followed by a sequel.
1977 (R) 81m/C Marcy Albrecht, Sherry Hardin, Rae Sperling, Susanne Kevin Mead; *Dir:* Patrick Wright. **VHS, Beta** $29.98 *LIV, VES* ⅋

Hollywood High Part 2 Ignored by their boyfriends for the lure of sun and surf, three comely high school students try their best to regain their interest.
1981 (R) 86m/C April May, Donna Lynn, Camille Warner, Lee Thornburg; *Dir:* Caruth C. Byrd. **VHS, Beta** $29.98 *LIV, VES* ⅋

Hollywood Home Movies The home movies of more than 60 celebrities are compiled in this program. Offers a nostalgic look at how the demi-gods of yesteryear lived, worked, played, and relaxed.
1987 (PG) 58m/B Marilyn Monroe, John Wayne, Gregory Peck, Robert Mitchum, Ann-Margret. **VHS, Beta, 3/4U** $19.98 *IHF, MPI* ⅋

Hollywood Hot Tubs A teenager picks up some extra cash and a lot of fun as he repairs the hot tubs of the rich and famous.
1984 (R) 103m/C Donna McDaniel, Michael Andrew, Katt Shea, Paul Gunning, Edy Williams, Jewel Shepard; *Dir:* Chuck Vincent. **VHS, Beta** $79.98 *LIV, VES* ⅋

Hollywood Hot Tubs 2: Educating Crystal A softcore comedy about a well-tanned ingenue learning about business in Hollywood during numerous hot tub trysts.
1989 (R) 103m/C Jewel Shepard, Patrick Day, Remy O'Neill, David Tiefen, Bart Braverman; *Dir:* Ken Raich. **VHS, Beta, LV** $9.99 *CCB, IVE* ⅋

Hollywood without Makeup Home movie footage taken over the years by Ken Murray includes candid shots of over 100 stars of the film world.
1965 51m/C **VHS, Beta, 8mm** $24.95 *HHT, MRV, VYY* ⅋

Hollywood Man A Hollywood actor wants to make his own film but when his financial support comes from the mob there's trouble ahead.
1976 (R) 90m/C William Smith, Don Stroud, Jennifer Billingsley, Mary Woronov; *Dir:* Jack Starrett. **VHS, Beta** $39.95 *MON* ⅋⅋

Hollywood My Hometown Hollywood host Ken Murray looks at the stars of the glittering movie world.
19?? 60m/B Ken Murray. **VHS, Beta** $19.95 *VCN, DVT, HHT* ⅋⅋

Hollywood Mystery A publicity agent stops at nothing to promote his would-be starlet girlfriend.

1934 53m/B June Clyde, Frank Albertson, Joe Crespo; *Dir:* Breezy Eason. **VHS** $19.95 *NOS, LOO* ⅋½

Hollywood Nights The dark underbelly of Hollywood glitz is explored, with an emphasis on semi-naked women.
1985 60m/C **VHS, Beta** $24.95 *AHV* ⅋½

Hollywood Outtakes & Rare Footage A collection of rare and unusual film clips and excerpts that feature dozens of Hollywood stars in screen tests, promotional shorts, home movies, outtakes, television appearances and World War II propaganda shorts. Some segments in black and white.
1983 85m/C Marilyn Monroe, James Dean, Joan Crawford, Judy Garland, Humphrey Bogart, Ronald Reagan, Bette Davis, Vivien Leigh, Carole Lombard. **VHS, Beta, LV** $59.95 *COL, RXM* ⅋½

Hollywood Palace A program from the popular comedy-variety show, complete with commercials for Playtex, Bayer, Johnson's Wax, Kool, and Tegrin.
1968 55m/B Dino Desi & Billy, Mitchell Ayres Orchestra, Victor Borge, Steve Allen, Jayne Meadows, The King Family; *Performed by:* The Scots Guard. **VHS, Beta** $29.95 *VYY, DVT* ⅋½

Hollywood Revels A burlesque show including Can-Can dancers, singers and comedians.
1947 58m/B Bill Rose, Hillary Dawn, Peggy Bond, Pat Dorsey, Aleen Dupree, Mickey Lotus Wing. **VHS, Beta** $24.95 *VYY* ⅋½

Hollywood Scandals and Tragedies A look at the skeletons-in-the-closet surrounding the deaths and private turmoils of James Dean, Rock Hudson, Vivien Leigh, Frances Farmer, Sal Mineo, Sharon Tate, Jayne Mansfield, and others.
1988 90m/C **VHS, Beta** $79.95 *MPI* ⅋½

Hollywood Shuffle Townsend's autobiographical comedy about a struggling black actor in Hollywood trying to find work and getting nothing but stereotypical roles. Written, directed, financed by Townsend, who created this often clever and appealing film on a $100,000 budget.
1987 (R) 81m/C Robert Townsend, Anne-Marie Johnson, Starletta DuPois, Helen Martin, Keenan Ivory Wayans, Damon Wayans; *Dir:* Robert Townsend. **VHS, Beta, LV** $24.95 *VIR* ⅋⅋½

Hollywood Stadium Mystery A boxer turns up dead before a big match and a D.A. investigates. Performances hold this one up, especially by Hamilton who later played Commissioner Gordon on TV's "Batman."
1938 66m/B Neil Hamilton, Evelyn Venable, Jimmy Wallington, Barbara Pepper, Lucien Littlefield, Lynn Roberts, Charles Williams, James Spottswood, Reed Hadley, Smiley Burnette; *Dir:* David Howard. **VHS, Beta** $16.95 *SNC* ⅋⅋

Hollywood Strangler Plotless ramble in which a fashion photographer strangles models and a store employee slashes the throats of L.A. bums. Dialogue is almost nonexistent; what little there is consists of voiceovers. Unintentionally amusing because it's so bad. Also known as "The Model Killer." Wolfgang Schmid is an alias for Ray Dennies Stekler.
1982 72m/C Pierre Agostino, Carolyn Brandt, Forrest Duke, Chuck Alford. **VHS, Beta** $24.95 *AHV Woof!*

The Hollywood Strangler Meets the Skid Row Slasher Voiced-over narration, canned music and an unfathomable plot are but a few of this would-be fright fest's finer points. Uninhibited by any narrative connection, two terrors strike fear in the heart of Tinseltown. While a psycho photographer cruises L.A. taking pictures of models he subsequently strangles, a woman working in a bare-bums magazine store takes a stab (with a knife) at lowering the city's derelect population. There's a word for this sort of dribble, and it isn't verisimilitude. Also available on video under the title "The Model Killer."
1979 (R) 72m/C Pierre Agostino, Carolyn Brandt; *Dir:* Ray Dennis Steckler. **VHS $14.99** *AHV, REG* Woof!

Hollywood in Trouble A bumbling middle-aged fool decides to break into the wacky movie-making business.
1987 90m/C Vic Vallard, Jean Levine, Jerry Tiffe, Pamela Dixon, Jerry Cleary. **VHS, Beta** *CLV* ✓

Hollywood Uncensored A history of Hollywood censorship, including long-suppressed scenes from "King Kong," "The Outlaw," "Carnal Knowledge," "Baby Doll," "Easy Rider," "Woodstock," and more.
1987 75m/C *Hosted:* Douglas Fairbanks Jr., Peter Fonda. **VHS, Beta $9.99** *IVE, FCT* ✓

Hollywood Varieties A virtually plotless no-star conglomeration of comedy, music, and burlesque.
1950 60m/B The 3 Rio Brothers, Robert Alda, Glenn Vernon, Eddie Ryan. **VHS, Beta, LV** *WGE* ✓

Hollywood Vice Sqaud Explores the lives of Hollywood police officers and the crimes they investigate. Contains three different story lines involving prostitution, child pornography, and organized crime. The film can't decide to be a comedy send-up of crime stories or a drama and is also crippled by a poor script. Written by the real-life chief of the Hollywood Vice Squad.
1986 (R) 100m/C Trish Van Devere, Ronny Cox, Frank Gorshin, Leon Isaac Kennedy, Carrie Fisher, Ben Frank, Robin Wright; *Dir:* Penelope Spheeris. **VHS, LV $79.98** *CGI, HHE, MTX* ✓

Hollywood at War Collection of wartime shorts featuring "All Star Bond Rally," "Spirit of '43," "Stamp Day for Superman."
194? 60m/C Bob Hope, Donald Duck, Superman. **VHS, Beta $34.95** *HHT* ✓

Hollywood Zap Two losers, one a Southern orphan, the other a video game nut, hit Hollywood and engage in antics that are meant to be funny, but are actually tasteless.
1986 (R) 85m/C Ben Frank, Ivan E. Roth, De Waldron, Annie Gaybis, Chuck "Porky" Mitchell; *Dir:* David Cohen. **VHS, Beta $14.95** *PSM* Woof!

Hollywood's Greatest Trailers Theatrical previews from an assortment of classic films, including "Citizen Kane," "Top Hat," "It's a Wonderful Life," "Fort Apache" and many others. Some are in color.
194? 60m/B VHS, Beta *MED* Woof!

Hollywood's New Blood Young actors making a movie are haunted by a film crew from hell.
1988 90m/C Bobby Johnson, Francine Lapensee; *Dir:* James Shyman. **VHS, Beta $69.95** *RAE* ✓

Holocaust The war years of 1935 to 1945 are relived in this account of the Nazi atrocities, focusing on the Jewish Weiss family, destroyed by the monstrous crimes, and the Dorf family, Germans who thrived under the Nazi regime. Highly acclaimed riveting made-for-television mini-series with and exceptional cast. Won eight Emmys.
1978 475m/C Michael Moriarty, Fritz Weaver, Meryl Streep, James Woods, Joseph Bottoms, Tovah Feldshuh, David Warner, Ian Holm, Michael Beck, Marius Goring; *Dir:* Marvin J. Chomsky. **VHS, Beta $69.95** *WOV* ✓✓✓½

Holocaust 2000 A nuclear facility in the Sahara Desert becomes plagued with a series of accidents, putting the whole world in jeopardy. Also known as "The Chosen."
1983 (R) 101m/C Kirk Douglas, Simon Ward, Virginia McKenna, Anthony Quayle, Alexander Knox; *Dir:* Alberto De Martino. **VHS, Beta $69.98** *LIV, VES* ✓½

Holocaust Survivors... Remembrance of Love A concentration camp survivor and his daughter attend the 1981 World Gathering of Holocaust Survivors in Tel-Aviv and both find romance. Originally titled "Remembrance of Love." Made for television.
1983 100m/C Kirk Douglas, Pam Dawber, Chana Eden, Yoram Gal, Robert Clary; *Dir:* Jack Smight. **VHS, Beta $69.95** *VMK* ✓✓

Holt of the Secret Service A secret service agent runs afoul of saboteurs and fifth-columnists in this 15-episode serial.
1942 225m/B Jack Holt. **VHS, Beta $49.95** *NOS, MED, DVT* ✓

The Holy Father Loves You Pope John Paul II holds a unique meeting with ten thousand children at the Vatican.
1985 26m/C VHS, Beta *DSP* ✓

The Holy Innocents A pleasant family is torn asunder by their feudal obligations to their patrons. In Spanish with English subtitles.
1984 (PG) 108m/C *SP* Alfredo Landa, Francisco Rabal; *Dir:* Mario Camus. Cannes Film Festival '84: Best Actor (Landa), Best Actor (Rabal). **VHS, Beta $68.00** *LHV, APD* ✓✓✓

Holy Terror The story of Florence Nightingale, who shocked Victorian society by organizing a nursing staff to aid British soldiers in the Crimean War, is told in this presentation from "George Schaefer's Showcase Theatre."
1965 76m/C Julie Harris, Denholm Elliott, Torin Thatcher, Kate Reid; *Dir:* George Schaefer. **VHS, Beta** *FHS* ✓

Hombre A white man in the 1880s, raised by a band of Arizona Apaches, is forced into a showdown. In helping a stagecoach full of settlers across treacherous country, he not only faces traditional bad guys, but prejudice as well. Based on a story by Elmore Leonard.
1967 111m/C Paul Newman, Fredric March, Richard Boone, Diane Cilento, Cameron Mitchell, Barbara Rush, Martin Balsam; *Dir:* Martin Ritt. **VHS, Beta, LV $19.98** *FOX* ✓✓✓

Home Alone Eight-year-old Kevin is sent to his attic room without supper for misbehaving, and is forgotten the next day when his entire family (including a cousin played by brother Kieran) rushes to catch a plane to France. That's the plausible part. Left alone and besieged by burglars who are plundering the deserted neighborhood, Culkin turns into a pint-sized Rambo defending his suburban castle with the wile and resources of a boy genius with perfect timing and unlimited wherewithal. That's the implausible part. Cute and sentimental, it's chock-full of feel-good sentiment for youngsters and oldsters alike. As the targets of Macauley's preadolescent wrath, house-breakers Pesci and Stern enact painful slapstick with considerable vigor, while Candy has a small but funny part as the leader of a polka band traveling cross-country with mom O'Hara. With an Oscar-nominated score and theme song, it was the highest-grossing picture of 1990, surpassing "Ghost" and "Jaws" in dollars earned. Scripted and produced by the ever hit-seeking Hughes.
1990 (PG) 105m/C Macauley Culkin, Catherine O'Hara, Joe Pesci, Daniel Stern, John Heard, Roberts Blossom, John Candy; *Dir:* John Hughes. **VHS, Beta, LV $24.98** *FXV, IME, RDG* ✓✓✓

Home Before Midnight A young songwriter falls in love with a 14-year-old, is discovered, and is subsequently charged with statutory rape.
1984 115m/C James Aubrey, Alison Elliot; *Dir:* Pete Walker. **VHS, Beta $59.95** *LHV, WAR* ✓½

Home of the Brave A black soldier is sent on a top secret mission in the South Pacific, but finds that he must battle with his white comrades as he is subjected to subordinate treatment and constant racial slurs. Hollywood's first outstanding statement against racial prejudice.
1949 86m/B Lloyd Bridges, James Edwards, Frank Lovejoy, Jeff Corey; *Dir:* Mark Robson. National Board of Review Awards '49: 10 Best Films of the Year. **VHS, Beta $19.98** *REP, FCT* ✓✓✓

Home Feeling In a black ghetto the residents try to make a viable community.
19?? 60m/C VHS, Beta $49.95 *VSV* ✓✓✓

Home Free All A Vietnam vet turned revolutionary, turned writer, tries to reacquaint himself with childhood friends only to find one has joined the mob and another is settled in suburbia and none of them are dealing well with adulthood.
1984 92m/C Allan Nicholls, Roland Caccavo, Maura Ellyn, Shelley Wyant, Lucille Rivim; *Dir:* Stuart Bird. **VHS, Beta $69.98** *LIV, VES* ✓✓½

Home From the Hill A solemn, brooding drama about a southern landowner and his troubled sons, one of whom is illegitimate. This one triumphs because of casting. Cinematography by Milton Krasner.
1960 150m/C Robert Mitchum, George Peppard, George Hamilton, Eleanor Parker, Everett Sloane, Luana Parker, Constance Ford; *Dir:* Vincente Minnelli. National Board of Review Awards '60: Best Actor (Mitchum). **VHS, Beta, LV $19.98** *MGM* ✓✓✓

Home for the Holidays Four daughters are called home on Christmas by their father who is convinced that his second wife is trying to poison him. Made for television.
1972 74m/C Eleanor Parker, Walter Brennan, Sally Field, Jessica Walter, Julie Harris; *Dir:* John Llewellyn Moxey. **VHS, Beta $24.95** *VMK* ✓✓

Home Movies Brian De Palma and his film students at Sarah Lawrence College devised this loose, sloppy comedy which hearkens back to De Palma's early films. Tells the story of a nebbish who seeks therapy to gain control of his absurdly chaotic life.
1979 (PG) 89m/C Kirk Douglas, Nancy Allen, Keith Gordon, Gerrit Graham, Vincent Gardenia, Harry Davenport; *Dir:* Brian DePalma. **VHS, Beta, LV $69.98** *LIV, VES, IME* ✓½

Home in Oklahoma A boy will be swindled out of his inheritance and a killer will escape justice until Rogers comes to the rescue, as usual.
1947 72m/B Roy Rogers, Dale Evans, Trigger, George "Gabby" Hayes, Carol Hughes; *Dir:* William Witney. **VHS, Beta $9.95** *NOS, CPB, CAB* ⟊

A Home of Our Own An American priest sets up a home in Mexico for orphan boys and changes their lives for the better. A true story of Father William Wasson.
1975 100m/C Jason Miller. **VHS, Beta $49.95** *WOV* ⟊

Home Remedy An introverted New Jersey bachelor's isolated lifestyle is invaded by a flirtatious housewife and her jealous husband.
1988 92m/C Seth Barrish, Maxine Albert; *Dir:* Maggie Greenwald. **VHS, Beta** *MNC, TAM* ⟊⟊

Home Safe A young boy, who has had a long history of disciplinary problems, finds love and salvation in Christian faith.
1988 74m/C Newell Alexander, Anita Jesse, Howard Culver, Michael Hornaday; *Dir:* Donald W. Thompson. **VHS, Beta** *MIV, BPG* ⟊

Home to Stay A farmer who has suffered a stroke is fighting off repeated lapses into senility. His son and daughter have conflicting interests as to whether he should be committed to a nursing home. Made for television.
1979 74m/C Henry Fonda, Frances Hyland, Michael McGuire; *Dir:* Delbert Mann. **VHS, Beta $59.95** *LTG, TLF* ⟊⟊

Home Sweet Home Suggested by the life of John Howard Payne (actor, poet, dramatist, critic, and world-wanderer) who wrote the title song amid the bitterness of his sad life. Silent with musical score.
1914 62m/B Lillian Gish, Dorothy Gish, Henry B. Walthall, Mae Marsh, Blanche Sweet, Donald Crisp, Robert Haron; *Dir:* D.W. Griffith. **VHS, Beta $29.95** *VYY* ⟊⟊½

Home Sweet Home A murdering psychopath escapes from the local asylum, and rampages through a family's Thanksgiving dinner. Also known as "Slasher in the House."
1980 84m/C Jake Steinfeld, Sallee Elyse, Peter de Paula; *Dir:* Nettie Pena. **VHS, Beta $59.95** *MED, TAV* ⟊½

Home Town Story A politician is convinced big business is behind his election loss. Fair drama. Watch for Marilyn Monroe in a bit part.
1951 61m/B Jeffrey Lynn, Donald Crisp, Marjorie Reynolds, Alan Hale Jr., Marilyn Monroe, Barbara Brown, Melinda Plowman; *Dir:* Arthur Pierson. **VHS $19.95** *NOS* ⟊⟊

Home is Where the Hart is A rich 103-year-old is kidnapped by a nurse wanting to marry him and become heir to his estate. It's up to his sons, all in their 70s, to rescue him.
1988 (PG-13) 94m/C Martin Mull, Leslie Nielsen, Valri Bromfield, Stephen E. Miller, Eric Christmas, Ted Stidder; *Dir:* Rex Bromfield. **VHS, Beta, LV $19.95** *PAR, IME* ⟊⟊

The Home and the World Another masterpiece from India's Satyajit Ray, this film deals with a sheltered Indian woman who falls in love with her husband's friend and becomes politically committed in the turmoil of 1907-1908. In Bengali with English subtitles. Based on Rabindranath Tagore's novel.
1984 130m/C *IN* Victor Banerjee, Soumitra Chaterjee; *Dir:* Satyajit Ray. **VHS, Beta $29.98** *SUE* ⟊⟊⟊½

Homebodies Six senior citizens, disgusted when they learn that they will be callously tossed out of their home, become violent murderers in their shocking attempt to solve their problem. An odd little film made in Cincinnati.
1974 (PG) 96m/C Ruth McDevitt, Linda Marsh, William Hansen, Peter Brocco, Frances Fuller; *Dir:* Larry Yust. **VHS, Beta $59.98** *SUE* ⟊⟊

Homeboy A small-time club boxer who dreams of becoming middleweight champ is offered his big break, but the opportunity is jeopardized by his dishonest manager's dealings. Never released theatrically.
1988 (R) 118m/C Mickey Rourke, Christopher Walken, Debra Feuer, Kevin Conway, Antony Alda, Ruben Blades; *Dir:* Michael Seresin. **VHS, Beta, LV, CDV $9.99** *CCB, IVE* ⟊⟊

Homeboy A Black boxer owes his career to the mob. When he tries to break free is wife is murdered - and he seeks revenge.
198? 93m/C Dabney Coleman, Philip Michael Thomas. **VHS** *SIM, TAV* ⟊½

Homeboys Two East L.A. brothers find themselves on opposite sides of the law in this quickie designed to cash in on gang violence and ghetto dramas.
1992 91m/C Todd Bridges, Ron Odriozola, Ken Michaels; *Dir:* Lindsay Norgard. **VHS $89.95** *AIP Woof!*

Homecoming Two German prisoners of war escape from a Siberian lead mine. One reaches home first and has an affair with his comrade's wife. Silent.
1928 74m/B *GE* Dita Parlo, Lars Hanson, Gustav Froehlich; *Dir:* Joe May. **VHS, Beta $29.95** *FST, GLV* ⟊⟊⟊

Homecoming: A Christmas Story The made-for-television heart-tugger that inspired the television series "The Waltons." A depression-era Virginia mountain family struggles to celebrate Christmas although the whereabouts and safety of their father are unknown. Adapted from the autobiographical novel by Earl Hamner, Jr.
1971 98m/C Richard Thomas, Patricia Neal, Edgar Bergen, Cleavon Little, Ellen Corby; *Dir:* Fielder Cook. **VHS, Beta $59.98** *FOX* ⟊⟊½

Homer Comes Home A homely and diligent inventor finds nothing but failure, but sticks it out in hopes of turning his luck around. Silent comedy starring down-home Ray.
1920 55m/B Charles Ray. **VHS, Beta $16.95** *GPV* ⟊⟊

Homer and Eddie A witless road comedy with Goldberg as a dying sociopath and a mentally retarded Belushi. They take off cross-country to make a little trouble, learn a little about life, and create a less than entertaining movie.
1989 (R) 100m/C Whoopi Goldberg, James Belushi, Karen Black; *Dir:* Andrei Konchalovsky. **VHS, Beta, LV $89.95** *HBO* ⟊

Homesteaders of Paradise Valley The Hume brothers oppose Red Ryder and a group of settlers building a dam in Paradise Valley.
1947 54m/B Allan Lane, Bobby Blake. **VHS, Beta $19.95** *NOS, VCN, CAB* ⟊

Hometown U.S.A. Set in Los Angeles in the late 50s' this teenage cruising movie is the brainchild of director Baer who is better known as Jethro from that madcap television series "The Beverly Hillbillies."
1979 (R) 97m/C Brian Kerwin, Gary Springer, David Wilson, Cindy Fisher, Sally Kirkland; *Dir:* Max Baer. **VHS, Beta $69.98** *LIV, VES Woof!*

Homeward Bound A dying teenager tries to reunite his long-estranged father and grandfather. Made for television.
1980 96m/C David Soul, Moosie Drier, Barnard Hughes; *Dir:* Richard Michaels. **VHS, Beta $59.95** *LTG* ⟊

Homework A young man's after-school lessons with a teacher are definitely not part of the curriculum. Yawning sexploitation.
1982 (R) 90m/C Joan Collins, Michael Morgan, Betty Thomas, Shell Kepler, Wings Hauser, Lee Purcell; *Dir:* James Beshears. **VHS, Beta, LV $19.95** *MCA* ⟊

Homicide Terrific police thriller with as much thought as action; a driven detective faces his submerged Jewish identity while probing an anti-Semetic murder and a secret society. Playwright/filmmaker David Mamet creates nail-biting suspense and shattering epiphanies without resorting to Hollywood glitz. Rich (often profane) dialogue includes a classic soliloquy mystically comparing a lawman's badge with a Star of David.
1991 (R) 100m/C Joe Mantegna, W.H. Macy, Natalija Nogulich, Ving Rhames, Rebecca Pidgeon; *W/Dir:* David Mamet. National Board of Review Awards '91: 10 Best Films of the Year. **VHS, LV, 8mm $92.95** *COL* ⟊⟊⟊½

L'Homme Blesse A serious erotic film about an 18-year-old boy, cloistered by an overbearing family, and his sudden awakening to his own homosexuality. In French with Engish Subtitles.
1983 90m/C *FR* Jean-Hughes Anglade, Roland Bertin, Vittorio Mezzogiorno; *Dir:* Patrice Chereau. **VHS, Beta $79.95** *KIV* ⟊⟊⟊

Homo Eroticus/Man of the Year A handsome servant finds it difficult to satisfy the desires of Bergamo, Italy's socialite.
1971 93m/C *IT* Lando Buzzanca, Rossana Podesta. **VHS, Beta $59.95** *HRS, QNE Woof!*

Honey An attractive writer pays an unusual visit to the home of a distinguished publisher. Brandishing a pistol, she demands that he read aloud from her manuscript. As he reads, a unique fantasy unfolds, a dreamlike erotic tale that may be the writer's own sexual awakening.
1981 89m/C *IT SP* Clio Goldsmith, Fernando Rey, Catherine Spaak; *Dir:* Gianfranco Angelucci. **VHS, Beta $69.95** *VES* ⟊½

Honey, I Shrunk the Kids The popular Disney fantasy about a suburban inventor. His shrinking device accidentally reduces his kids to 1/4 inch tall, and he subsequently throws them out with the garbage. Now they must journey back to the house through the jungle that was once the back lawn. Accompanied by "Tummy Trouble," the first of a projected series of Roger Rabbit Maroon Cartoons. Followed by a sequel.
1989 (G) 101m/C Rick Moranis, Matt Frewer, Marcia Strassman, Kristine Sutherland, Thomas Wilson Brown, Jared Rushton; *Dir:* Joe Johnston, Rob Minkoff; *Voices:* Charles Fleischer, Kathleen Turner, Lou Hirsch, April Winchell. **VHS, Beta, LV $19.99** *TOU, IVE, RDG* ⟊⟊½

The Honey Pot A millionaire feigns death to see the reaction of three of his former lovers. Amusing black comedy with fine performances from all. Based on Moliere's "Volpone." Also known as "It Comes Up Murder."
1967 131m/C Rex Harrison, Susan Hayward, Cliff Robertson, Capucine, Edie Adams, Maggie Smith, Adolfo Celi; ***W/Dir:*** Joseph L. Mankiewicz. **VHS $14.95** *WKV* 🦴🦴½

Honeybaby A smooth international soldier of fortune and a bright, sexy American interpreter are entangled in Middle-Eastern turbulence when they must rescue a politician kidnapped by terrorists.
1974 (PG) 94m/C Calvin Lockhart, Diana Sands; ***Dir:*** Michael Schultz. **VHS $19.95** *GEM, MRV* 🦴🦴

Honeyboy Estrada tries to win fame, fortune, and a ticket out of the barrio as an up-and-coming boxer. Outstanding fight scenes, but little else in this made for television drama of the boxing scene.
1982 100m/C Erik Estrada, Morgan Fairchild, James McEachin, Robert Costanzo, Yvonne Wilder; ***Dir:*** John Berry. **VHS, Beta $59.95** *USA, IVE* 🦴½

Honeymoon A young French woman visiting New York learns she will be deported when her boyfriend is arrested for drug-smuggling. To stay, she is set up in a marriage of convenience and told her new husband will never see or bother her. But hubby has other plans!
1987 96m/C John Shea, Nathalie Baye, Richard Berry, Peter Donat; ***Dir:*** Patrick Jamain. **VHS, Beta $79.95** *LHV, WAR* 🦴🦴

Honeymoon Academy Newly married man has lots to learn in this comic outing. His new wife never told him about her unusual line of business—as an undercover agent!
1990 (PG-13) 94m/C Robert Hays, Kim Cattrall, Leigh Taylor-Young; ***Dir:*** Gene Quintano. **VHS, LV $89.99** *HBO, IME* 🦴🦴

Honeymoon Horror Stranded on an island, three honeymooning couples soon realize their romantic retreat is a nightmare. A crazy man with an axe is tracking them down. Minimally effective.
197? 90m/C Cheryl Black, William F. Pecchi, Bob Wagner. **VHS, Beta $59.95** *SVS* 🦴

Honeymoon Killers A grim, creepy filmization of a true multiple-murder case wherein an overweight woman and slimy gigolo living on Long Island seduce and murder one lonely woman after another. Independently made and frankly despairing. Also known as "The Lonely Hearts Killers."
1970 (R) 103m/C Tony LoBianco, Shirley Stoler, Mary Jane Higby, Dortha Duckworth, Doris Roberts, Marilyn Chris, Kip McArdle, Mary Breen, Barbara Cason, Ann Harris, Guy Sorel; ***Dir:*** Leonard Kastle. **VHS, Beta, LV $69.98** *LIV, VES* 🦴🦴🦴

Honeysuckle Rose A road-touring country-Western singer whose life is a series of one night stands, falls in love with an adoring young guitar player who has just joined his band. This nearly costs him his marriage when his wife, who while waiting patiently for him at home, decides she's had enough. Easygoing performance by Nelson, essentially playing himself. Also known as "On the Road Again."
1980 (PG) 120m/C Willie Nelson, Dyan Cannon, Amy Irving, Slim Pickens, Joey Floyd, Charles Levin, Priscilla Pointer; ***Dir:*** Jerry Schatzberg. **VHS, Beta, LV $39.98** *WAR* 🦴🦴½

Hong Kong Nights Two agents attempt to break up a Chinese smuggling ring run by an unscrupulous madman. Good, action-packed low-budget film.
1935 59m/B Tom Keene, Wera Engels, Warren Hymer, Tetsu Komai, Cornelius Keefe; ***Dir:*** E. Mason Hopper. **VHS, Beta $16.95** *SNC* 🦴🦴½

Honky An innocent friendship between a black woman and a white man develops into a passionate romance that has the whole town talking.
1971 (R) 92m/C Brenda Sykes, John Nielson, Maria Donzinger; ***Dir:*** William A. Graham. **VHS, Beta $49.95** *UNI* 🦴½

Honky Tonk A western soap opera in which ne'er do well Clark marries Lana and tries to live a respectable life. In this, the first of several MGM teamings between Gable and Turner, the chemistry between the two is evident. So much so in fact that Gable's then wife Carole Lombard let studio head Louis B. Mayer know that she was not at all thrilled. The public was pleased however, and made the film a hit.
1941 106m/B Clark Gable, Lana Turner, Frank Morgan, Claire Trevor, Marjorie Main, Albert Dekker, Chill Wills, Henry O'Neill, John Maxwell, Morgan Wallace, Betty Blythe, Francis X. Bushman; ***Dir:*** Jack Conway. **VHS, Beta $19.98** *MGM, CCB* 🦴🦴🦴

Honky Tonk Freeway An odd assortment of people become involved in a small town Mayor's scheme to turn a dying hamlet into a tourist wonderland. You can find better things to do with your evening than watch this accident.
1981 (PG) 107m/C Teri Garr, Howard Hesseman, Beau Bridges, Hume Cronyn, William Devane, Beverly D'Angelo, Geraldine Page; ***Dir:*** John Schlesinger. **VHS, Beta $69.99** *HBO* 🦴

Honkytonk Man An unsuccessful change-of-pace Eastwood vehicle, set during the Depression. An aging alcoholic country singer tries one last time to make it to Nashville, hoping to perform at the Grand Ole Opry. This time he takes his newphew (played by Eastwoood's real-life son) with him.
1982 (PG) 123m/C Clint Eastwood, Kyle Eastwood, John McIntire, Alexa Kenin, Verna Bloom; ***Dir:*** Clint Eastwood. **VHS, Beta, LV $19.98** *WAR* 🦴½

Honkytonk Nights A honky-tonkin' romp through an evening in a country/western bar, featuring legendary topless dancer Carol Doda, Georgina ("Devil in Mrs. Jones") Spelvin, and the music of The Hot Licks (minus front man Dan Hicks).
197? (R) 90m/C Carol Doda, Georgina Spelvin, Ramblin' Jack Elliot. **VHS, Beta $59.95** *TPV* 🦴

Honor Among Thieves Two former mercenaries reteam for a robbery that doesn't come off as planned. Also known as "Farewell, Friend." Not as fast paced as some of Bronson's other action-adventure films.
1968 (R) 115m/C *FR IT* Charles Bronson, Alain Delon, Brigitte Fossey, Olga Georges-Picot, Bernard Fresson; ***Dir:*** Jean Herman. **VHS, Beta $9.99** *CCB, MPI, FRH* 🦴½

Honor of the Range The good sheriff's lady is whisked away by his evil twin brother and the sheriff must go undercover to get her back. Odd plot twists for a seemingly normal western.
1934 60m/B Ken Maynard, Cecilia Parker, Fred Kohler Jr.; ***Dir:*** Ken Maynard. **VHS $19.95** *NOS, DVT* 🦴🦴

Honor Thy Father The everyday life of a real-life Mafia family as seen through the eyes of Bill Bonanno, the son of mob chieftain Joe Bonanno. Adapted from the book by Gay Talese. Made for television.
1973 97m/C Raf Vallone, Richard Castellano, Brenda Vaccaro, Joseph Bologna; ***Dir:*** Paul Wendkos. **Beta $14.95** *PSM* 🦴🦴

Hooch Comedy about three members of the New York Mafia who get some unexpected southern hospitality when they try to muscle in on a southern family's moonshine operation.
1976 (PG) 96m/C Gil Gerard, Erika Fox, Melody Rogers, Danny Aiello, Mike Allen, Ray Serra; ***Dir:*** Edward Mann. **VHS, Beta $49.95** *PSM* 🦴½

The Hooded Terror A merciless group of murderers take on a G-man.
1938 70m/B Tod Slaughter, Greta Gynt. **VHS $14.99** *NOS, HEG* 🦴½

The Hoodlum A criminal does his best to rehabilitate and, with a little help from his brother, is able to hold a steady job. However the crime bug dies hard and when he spots an armored car, he just can't resist the temptation. Tierney's real brother, Edward, plays his screen brother.
1951 61m/B Lawrence Tierney, Allene Roberts, Marjorie Riordan, Liza Golm, Edward Tierney; ***Dir:*** Max Nosseck. **VHS, Beta $16.95** *SNC* 🦴🦴

Hoodlum Empire A senator enlists the aid of a former gangster, now a war hero, in his battle against the syndicate. Loosely based on the Kefauver investigations of 1950-51.
1952 98m/B Brian Donlevy, Claire Trevor, Forrest Tucker, Vera Ralston, Luther Adler, John Russell, Gene Lockhart, Grant Withers, Taylor Holmes, Roy Barcroft, Richard Jaeckel; ***Dir:*** Joseph Kane. **VHS $19.98** *REP* 🦴🦴

Hoodoo Ann The story of Hoodoo Ann, from her days in the orphanage to her happy marriage. Silent.
1916 27m/B Mae Marsh, Robert Harron. **VHS, Beta $29.98** *CCB* 🦴🦴

Hook Although he said he was never going to do it, Peter Pan has grown up to be Peter Banning, an uptight, portable phone-toting adult who places his work before his family and has forgotten everything about Neverland and the evil Captain Hook. When Peter and his family visit Wendy in London, Hook kidnaps Peter's children and takes them to Neverland. With the help of Tinkerbell, her magic pixie dust, and happy thoughts, Peter flies to Neverland to rescue his children. He rediscovers his youth while visiting with the Lost Boys, who help him rescue his children. While this is a good fantasy with lots of special effects, it lacks the charm it needs to really fly.
1991 (PG) 142m/C Dustin Hoffman, Robin Williams, Julia Roberts, Bob Hoskins, Maggie Smith, Charlie Korsmo, Caroline Goodall, Amber Scott; ***Dir:*** Steven Spielberg. **VHS, Beta, LV, 8mm $24.95** *COL* 🦴🦴

Hook, Line and Sinker A couple of insurance investigators try to help a young woman restore a hotel, and they find romance and run-ins with crooks. Directed by Cline, who later directed W.C. Fields in "The Bank Dick," "My Little Chickadee," and "Never Give a Sucker an Even Break."
1930 75m/B Bert Wheeler, Robert Woolsey, Dorothy Lee, Jobyna Howland, Ralf Harolde, Natalie Moorhead, George F. Marion, Hugh Herbert; ***Dir:*** Eddie Cline. **VHS, LV $19.98** *NOS, TTC, FCT* 🦴🦴

The Hooked Generation A group of drug pushers kidnap defenseless victims, rape innocent girls, and even murder their own Cuban drug contacts (and all without the aid of a cohesive plot). Many gory events precede their untimely deaths; few will be entertained.
1969 (R) 92m/C Jeremy Slate, Steve Alaimo. **VHS** *AVC* ✶

Hooker This documentary takes a revealing look at the private moments and public times of prostitutes.
1983 79m/C *Dir:* Robert Niemack. **VHS, Beta $69.95** *VES* ✶

Hoomania Award-winning Christian children's film. Hoomania is a board game that fantastically whisks young Kris to a strange world. He meets the most unlikely characters, some of whom aren't too helpful in his journey to Mount Wisdom.
1990 30m/C VHS, Beta, 3/4U $15.95 *SPW* ✶

Hooper A behind-the-scenes look at the world of movie stuntmen. Burt Reynolds is a top stuntman who becomes involved in a rivalry with an up-and-coming young man out to surpass him. Dated good-ole-boy shenanigans.
1978 (PG) 94m/C Burt Reynolds, Jan-Michael Vincent, Robert Klein, Sally Field, Brian Keith, John Marley, Adam West; *Dir:* Hal Needham. **VHS, Beta, LV $59.95** *WAR* ✶ 1/2

Hoosiers In Indiana, where basketball is the sport of the gods, a small town high school basketball team gets a new, but surprisingly experienced coach. He makes the team, and each person in it, better than they thought possible. Classic plot rings true because of Hackman's complex and sensitive performance coupled with Hopper's touching portrait of an alcoholic basketball fanatic.
1986 (PG) 115m/C Gene Hackman, Barbara Hershey, Dennis Hopper, David Neidorf; *Dir:* David Anspaugh. **VHS, Beta, LV $29.98** *LIV, HBO* ✶✶✶

Hootch Country Boys A film packed with moonshine, sheriffs, busty country girls, and chases. Also known as "Redneck County," originally titled, "The Great Lester Boggs."
1975 (PG) 90m/C Alex Karras, Scott MacKenzie, Dean Jagger, Willie Jones, Bob Ridgely, Susan Denbo, Bob Ginnaven, David Haney; *Dir:* Harry Z. Thomason. **VHS, Beta $19.95** *ACA* ✶

Hopalong Cassidy: Borrowed Trouble Hoppy must save a righteous teacher from the clutches of kidnappers who plan to place a saloon right next to the school.
1948 61m/B William Boyd, Andy Clyde, Rand Brooks, Elaine Riley, John Kellogg, Helen Chapman; *Dir:* George Archainbaud. **VHS $9.99** *BVV* ✶ 1/2

Hopalong Cassidy: Dangerous Venture "Dangerous Venture" was one of Hoppy's last ventures, produced the final year of the series' 13-year tenure. Hoppy and the gang hunt for Aztec ruins in the great Southwest, and have to deal with renegade Indians and a band of treacherous looters along the way.
1948 55m/B William Boyd, Andy Clyde, Rand Brooks. **VHS, Beta $9.95** *BVV* ✶✶

Hopalong Cassidy: False Paradise Another late Hopalong has the noir-clad cowboy rushing to the aid of a girl and her dad when control of their ranch is threatened by swindlers with false mining claims.
1948 59m/B William Boyd, Andy Clyde, Rand Brooks; *Dir:* George Archainbaud. **VHS, Beta $9.99** *BVV* ✶✶

Hopalong Cassidy: Hoppy's Holiday Boyd—the only celluloid incarnation of Clarence Mulford's pulp hero—plays the Hopster yet again. This time, the non-drinking, non-smoking (what does he do for kicks?) hero and a passel of Bar-20 cowboys visit Mesa City for some well-earned R and R, but find themselves caught in a web of corruption.
1947 60m/B William Boyd, Andy Clyde, Rand Brooks, Jeff Corey; *Dir:* George Archainbaud. **VHS, Beta $9.99** *BVV* ✶✶

Hopalong Cassidy: Riders of the Deadline Hoppy turns into a baddie in this episode in order to infiltrate a treacherous gang and apprehend their leader.
1943 70m/B William Boyd, Andy Clyde, Jimmy Rogers, Robert Mitchum; *Dir:* Lesley Selander. **VHS, Beta $9.99** *BVV* ✶✶

Hopalong Cassidy: Silent Conflict Lucky is tricked and hypnotized by a traveling medicine man who commands him to rob and kill Hoppy.
1948 61m/B William Boyd, Andy Clyde, Rand Brooks; *Dir:* George Archainbaud. **VHS, Beta $9.99** *BVV* ✶✶

Hopalong Cassidy: Sinister Journey Mysterious accidents are plaguing a railroad where an old friend of Hoppy's works, so Hoppy, California, and Lucky take railroad jobs to find out what gives.
1948 58m/B William Boyd, Andy Clyde, Rand Brooks; *Dir:* George Archainbaud. **VHS, Beta $9.99** *BVV* ✶✶

Hopalong Cassidy: The Dead Don't Dream Hoppy's sidekick Lucky decides to get married, but happily-ever-afterdom is threatened when the father of Mrs. Lucky-to-be is murdered. When Hoppy sets out to find the killer, he finds himself suspected of the dastardly deed.
1948 55m/B William Boyd, Andy Clyde, Rand Brooks, John Parrish; *Dir:* George Archainbaud. **VHS, Beta $9.99** *BVV* ✶✶

Hopalong Cassidy: The Devil's Playground Strange things are afoot in a peaceful valley adjacent to a desolate, forbidding wasteland. When Hoppy and a young woman set out across the wasteland to make a gold delivery, they encounter a band of desperados and begin to unravel the secrets of a dishonest political ring.
1946 65m/B William Boyd, Andy Clyde, Rand Brooks, Elaine Riley; *Dir:* George Archainbaud. **VHS, Beta $9.99** *BVV* ✶✶

Hopalong Cassidy: The Marauders After seeking shelter in an abandoned church one rainy night, Hoppy and the boys must defend the kirk when it is threatened by a marauder and his gang.
1947 64m/B William Boyd, Andy Clyde, Rand Brooks; *Dir:* George Archainbaud. **VHS, Beta $9.99** *BVV* ✶✶

Hopalong Cassidy: Unexpected Guest When a small fortune is left to a California family, the head count starts to dwindle, and Hoppy aims to find out why.
1947 59m/B William Boyd, Andy Clyde, Rand Brooks, Una O'Connor; *Dir:* George Archainbaud. **VHS, Beta $30.88** *BVV* ✶✶ 1/2

Hope and Glory Boorman turns his memories of World War II London into a complex and sensitive film. Father volunteered, mother must deal with awakening sexuality of her teen-aged daughter, keep her son in line, balance the ration books and try to make it to the bomb shelter in the middle of the night. Seen through the boy's eyes, war creates a playground of shrapnel to collect and wild imaginings come true.
1987 (PG-13) 97m/C *GB* Sebastian Rice Edwards, Geraldine Muir, Sarah Miles, Sammi Davis, David Hayman, Derrick O'Connor, Susan Woolridge, Jean-Marc Barr, Ian Bannen, Jill Baker, Charley Boorman, Annie Leon, Katrine Boorman, Gerald James; *Dir:* John Boorman. British Academy Awards '87: Best Director (Boorman), Best Film; Golden Globe Awards '88: Best Film—Musical/Comedy; Los Angeles Film Critics Association Awards '87: Best Director (Boorman), Best Picture, Best Screenplay; National Board of Review Awards '87: 10 Best Films of the Year. **VHS, Beta, LV, 8mm $14.98** *SUE, FCT* ✶✶✶ 1/2

Hoppity Goes to Town Full-length animated feature from the Max Fleischer studios tells the story of the inhabitants of Bugville, who live in a weed patch in New York City. Songs by Frank Loesser and Hoagy Carmichael. Original title: "Mr. Bug Goes to Town."
1941 77m/C *Dir:* Dave Fleischer. **VHS, LV $14.98** *REP, MRV, FCT* ✶✶✶ 1/2

Hopscotch A C.I.A. agent drops out when his overly zealous chief demotes him to a desk job. When he writes a book designed to expose the dirty deeds of the CIA, he leads his boss and KGB pal on a merry chase. Amiable comedy well-suited for Matthau's rumpled talents.
1980 (R) 107m/C Walter Matthau, Glenda Jackson, Ned Beatty, Sam Waterston, Herbert Lom; *Dir:* Ronald Neame. **VHS, Beta, LV $9.95** *COL, SUE* ✶✶✶

Horizons West Two brothers go their separate ways after the Civil War. One leads a peaceful life as a rancher but the other, corrupted by the war, engages in a violent campaign to build his own empire. Ryan's outstanding performance eclipses that of the young Hudson.
1952 81m/C Robert Ryan, Julie Adams, Rock Hudson, Raymond Burr, James Arness, John McIntire, Dennis Weaver, Francis Bavier; *Dir:* Budd Boetticher. **VHS** *MCA* ✶✶ 1/2

The Horn Blows at Midnight A band trumpeter falls asleep and dreams he's a bumbling archangel, on Earth to blow the note bringing the end of the world. But a pretty girl distracts him, and...A wild, high-gloss, well-cast fantasy farce, uniquely subversive in its lighthearted approach to biblical Doomsday. Benny made the film notorious by acting ashamed of it in his later broadcast routines.
1945 78m/B Jack Benny, Alexis Smith, Dolores Moran, Allyn Joslyn, Reginald Gardiner, Guy Kibbee, John Alexander, Margaret Dumont; *Dir:* Raoul Walsh. **VHS $19.98** *MGM, FCT* ✶✶✶

The Horrible Dr. Hichcock A sicko doctor, who accidentally killed his first wife while engaged in sexual antics, remarries to bring the first missus back from the dead using his new wife's blood. Genuinely creepy. Sequelled by "The Ghost."
1962 76m/C *IT* Robert Flemyng, Barbara Steele; *Dir:* Robert Hampton, Riccardo Freda. **VHS $69.95** *SNC* ✶✶ 1/2

Horrible Double Feature This package includes condensed silent versions of both "Dr. Jekyll and Mr. Hyde" and "Hunchback of Notre Dame." From the "History of the Motion Picture" series.
192? 56m/B John Barrymore, Lon Chaney Jr. **VHS, Beta $29.98** *CCB* ♨♨½

Horror of the Blood Monsters John Carradine made a career out of being in bad movies, and this one competes as one of the worst. Also released under four or five other titles, such as "Vampire Men of the Lost Planet," this is an editor's nightmare, made up of black & white film spliced together and colorized. Vampires from outer space threaten to suck all the blood from the people of Earth.
1970 (PG) 85m/C *PH* John Carradine, Robert Dix, Vicki Volante, Jennifer Bishop; *Dir:* Al Adamson, George Joseph. **VHS, Beta** *VID, REP* Woof!

Horror Chamber of Dr. Faustus A wickedly intelligent, inventive piece of Grand Guignol about a mad doctor who kills young girls so he may graft their skin onto the face of his accidentally mutilated daughter. In French with English subtitles. Also known as "Eyes Without a Face."
1959 84m/B *FR* Alida Valli, Pierre Brasseur, Edith Scob, Francois Guerin; *Dir:* Georges Franju. **VHS, Beta $29.95** *SNC* ♨♨♨½

The Horror of Dracula The first Hammer Dracula film, in which the infamous vampire is given a new, elegant and ruthless persona, as he battles Prof. Van Helsing after coming to England. Possibly the finest, most inspired version of Bram Stoker's macabre chestnut, and one that single-handedly revived the horror genre.
1958 82m/C *GB* Peter Cushing, Christopher Lee, Michael Gough, Melissa Stribling, Carol Marsh, John Van Eyssen, Valerie Gaunt; *Dir:* Terence Fisher. **VHS, Beta, LV $19.98** *WAR, LDC, FCT* ♨♨♨½

Horror Express A creature from prehistoric times, that was removed from its tomb, is transported on the Trans-Siberian railroad. Passengers suddenly discover strange things happening—such as having their souls sucked out of them.
1972 (R) 88m/C *SP* Christopher Lee, Peter Cushing, Telly Savalas; *Dir:* Gene Martin. **VHS, Beta $59.95** *SNC, MRV, PSM* ♨♨½

The Horror of Frankenstein Spoof of the standard Frankenstein story features philandering ex-med student Baron Frankenstein, whose interest in a weird and esoteric branch of science provides a shocking and up-to-date rendition of the age-old plot. Preceded by "Frankenstein Must Be Destroyed" and followed by "Frankenstein and the Monster from Hell."
1970 93m/C *GB* Ralph Bates, Kate O'Mara, Dennis Price, David Prowse; *Dir:* Jimmy Sangster. **VHS, Beta $59.99** *HBO, MLB* ♨♨

Horror Hospital Patients are turned into zombies by a mad doctor in this hospital where no anesthesia is used. Those who try to escape are taken care of by the doctor's guards.
1973 (R) 91m/C *GB* Michael Gough, Robin Askwith, Vanessa Shaw, Ellen Pollock, Skip Martin, Dennis Price; *Dir:* Anthony Balch. **VHS, Beta $29.95** *GRG, UHV, MPI* ♨

Horror Hotel A young witchcraft student visits a small Massachusetts town which has historic ties to witch burnings and discovers a new, and deadly, coven is now active. Well done and atmospheric.

1960 76m/B *GB* Christopher Lee, Patricia Jessel, Betta St. John, Dennis Lotis, Venetia Stevenson, Valentine Dyall; *Dir:* John Llewellyn Moxey. **VHS $19.95** *NOS, MRV, SNC* ♨♨½

Horror House on Highway 5 Someone in a Nixon mask kills people in this extremely cheap dud.
1986 90m/C *W/Dir:* Richard Casey. **VHS, Beta $9.95** *NO* Woof!

The Horror of It All A multi-faceted look at popular horror stars, including Karloff, Chaney, Lugosi, Price, and others, plus an examination of the dark side of the imaginative minds that produce horror. Also includes a behind-the-scenes peek at the creation of great horror scenes.
1991 58m/C Boris Karloff, Lon Chaney Sr., Bela Lugosi, John Barrymore, Lionel Atwill, John Carradine, Vincent Price; *Nar:* Jose Ferrer. **VHS $24.98** *MPI, FCT* Woof!

Horror of Party Beach Considered to be one of the all-time worst films. Features a mob of radioactive seaweed creatures who eat a slew of nubile, surf-minded teenagers. Featured song: "Zombie Stomp."
1964 71m/C John Scott, Alice Lyon, Allen Laurel; *Dir:* Del Tenney. **VHS, Beta $39.95** *PSM* Woof!

Horror Planet Capitalizing on the popularity of "Aliens," this is a graphic and sensationalistic film. An alien creature needs a chance to breed before moving on to spread its horror. When a group of explorers disturbs it, the years of waiting are over, and the unlucky mother-to-be will never be the same. Originally titled "Inseminoid."
1980 (R) 93m/C Robin Clarke, Jennifer Ashley, Stephanie Beacham, Judy Geeson, Stephen Grives, Victoria Tennant; *Dir:* Norman J. Warren. **VHS, Beta, LV $9.98** *SUE* Woof!

Horror Rises from the Tomb A 15th century knight and his assistant are beheaded for practicing witchcraft. Five hundred years later, they return to possess a group of vacationers and generally cause havoc.
1972 89m/C *SP* Paul Naschy, Vic Winner, Emma Cohen, Helga Line; *Dir:* Carlos Aured. **VHS, Beta $59.98** *WES, CHA, ASE* ♨♨

The Horror Show Serial killer fries in electric chair, an event that really steams him. So he goes after the cop who brought him in. Standard dead guy who won't die and wants revenge flick.
1989 (R) 95m/C Brion James, Lance Henriksen, Rita Taggart, Dedee Pfeiffer, Aron Eisenberg, Matt Clark; *Dir:* James Isaac. **VHS, Beta $19.98** *MGM* ♨

Horror of the Zombies Bikinied models are pursued by hooded zombies on board a pleasure yacht...the horror. Effete final installment in the "blind dead" trilogy; see also "Tombs of the Blind Dead" and "Return of the Evil Dead."
1974 (R) 90m/C *SP* Maria Perschy, Jack Taylor; *W/Dir:* Armando de Ossorio. **VHS $59.95** *VID, VTO* ♨

Horrors of Burke & Hare A gory tale about the exploits of a bunch of 19th-century grave robbers.
1971 94m/C Derren Nesbitt, Harry Andrews, Yootha Joyce; *Dir:* Vernon Sewell. **VHS, Beta, LV $19.95** *STE, NWV* ♨

Horrors of the Red Planet Cheapo epic sees astronauts crash on Mars, meet its Wizard, and stumble into a few Oz-like creatures. Technically advised by Forrest J.

Ackerman. Also known as "The Wizard of Mars."
1964 81m/C John Carradine, Roger Gentry, Vic McGee; *Dir:* David L. Hewitt. **VHS, Beta $19.99** *GHV, MRV, REP* ♨

The Horse A moving, mature Turkish film about a father and son trying to overcome socioeconomic obstacles and their own frailties in order to make enough money to send the boy to school. In Turkish with English subtitles.
1982 116m/C *TU Dir:* Ali Ozgenturk. **VHS, Beta $69.95** *KIV* ♨♨♨½

Horse Feathers Huxley College has to beef up its football team to win the championship game, and the corrupt new college president knows just how to do it. Features some of the brothers' classic routines, and the songs "Whatever It Is, I'm Against It" and "Everyone Says I Love You."
1932 67m/B Groucho Marx, Chico Marx, Harpo Marx, Zeppo Marx, Thelma Todd, David Landau, Nat Pendleton; *Dir:* Norman Z. MacLeod. **VHS, Beta, LV $19.95** *MCA* ♨♨♨

The Horse in the Gray Flannel Suit Disney comedy portrays an advertising executive who links his daughter's devotion to horses with a client's new ad campaign.
1968 (G) 114m/C Dean Jones, Ellen Janov, Fred Clark, Diane Baker, Lloyd Bochner, Kurt Russell; *Dir:* Norman Tokar. **VHS, Beta** *DIS, OM* ♨♨½

The Horse of Pride Chabrol takes a telling look at the everyday life of the peasants of Breton—both their pleasures and their sorrows. In French with English subtitles. French script booklet available.
1980 118m/C *FR* Jacques Dufilho, Francois Cluzet; *Dir:* Claude Chabrol. **VHS, Beta $59.95** *INT, APD* ♨♨½

The Horse Soldiers An 1863 Union cavalry officer is sent 300 miles into Confederate territory to destroy a railroad junction and is accompanied by a fellow officer who is also a pacifist doctor. Based on a true Civil War incident.
1959 119m/C John Wayne, William Holden, Hoot Gibson, Anna Lee, Russell Simpson, Strother Martin; *Dir:* John Ford. **VHS, Beta $19.95** *FOX, BUR* ♨♨½

The Horse Thief Noted as the first movie from the People's Republic of China to be released on video, this epic tells the tale of Norbu, a man who, exiled from his people for horse thievery, is forced to wander the Tibetan countryside with his family in search of work. His son dies while he is in exile, and he, a devout Buddhist, is ultimately forced to accept tribal work in a ritual exorcism, after which he pleads to be accepted back into his clan. Beautiful and image-driven, offering a rare glimpse into the Tibet you won't see in travel brochures. Filmed on location with locals as actors. In Mandarin with English subtitles; originally titled "Daoma Zei."
1987 88m/C *CH Dir:* Tian Zhuangzhuang. **VHS $59.95** *KIV, FCT* ♨♨♨

The Horse Without a Head Stolen loot has been hidden in a discarded toy horse which is now the property of a group of poor children. The thieves, however, have different plans. Good family fare originally shown on the Disney television show.
1963 89m/C Jean-Pierre Aumont, Herbert Lom, Leo McKern, Pamela Franklin, Vincent Winter; *Dir:* Don Chaffey. **VHS, Beta $69.95** *DIS* ♨♨½

Horsemasters A group of young riders enter a special training program in England to achieve the ultimate equestrian title of horsemaster, with Annette having to overcome her fear of jumping. Originally a two-part Disney television show, and released as a feature film in Europe.
1961 85m/C Tommy Kirk, Annette Funicello, Janet Munro, Tony Britton, Donald Pleasence, Jean Marsh, John Fraser, Millicent Martin; ***Dir:*** William Fairchild. **VHS, Beta $69.95** *DIS* 🦴🦴

Horsemen An Afghani youth enters the brutal buzkashi horse tournament to please his macho-minded father. Beautifully shot in Afghanistan and Spain by Claude Renoir.
1970 (PG) 109m/C Omar Sharif, Leigh Taylor-Young, Jack Palance, David De, Peter Jeffrey; ***Dir:*** John Frankenheimer. **VHS, Beta $69.95** *COL* 🦴½

Horseplayer A loner is drawn into a mysterious triangle of decadence and desire in this riveting psychothriller.
1991 (R) ?m/C Brad Dourif, Sammi Davis, M.K. Harris, Vic Tayback; ***Dir:*** Kurt Voss. **VHS, LV $89.98** *REP* 🦴½

The Horse's Mouth An obsessive painter discovers that he must rely upon his wits to survive in London. A hilarious adaptation of the Joyce Cary novel. The laserdisc version is accompanied by an episode of the British documentary series "The Art of Film" profiling Guinness and his screen career.
1958 93m/C *GB* Alec Guinness, Kay Walsh, Robert Coote, Renee Houston, Michael Gough; ***Dir:*** Ronald Neame. National Board of Review Awards '58: 5 Best Foreign Films of the Year, Best Supporting Actress (Walsh). **VHS, Beta, LV $19.98** *SUE, VYG* 🦴🦴🦴

The Hospital Cult favorite providing savage, unrelentingly sarcastic look at the workings of a chaotic metropolitan hospital beset by murders, witchdoctors, madness, and plain ineptitude. Scott was Oscar nominated. Scripted by Paddy Chayefsky.
1971 (PG) 101m/C George C. Scott, Diana Rigg, Barnard Hughes, Stockard Channing, Nancy Marchand, Richard Dysart; ***Dir:*** Arthur Hiller. Academy Awards '71: Best Story & Screenplay; Berlin Film Festival '72: Second Prize; British Academy Awards '72: Best Screenplay (tie). **VHS, Beta $19.98** *MGM, FOX* 🦴🦴🦴½

Hospital Massacre A psychopathic killer is loose in a hospital. Seems he holds a grudge against the woman who laughed at his valentine when they were children twenty years earlier, and too bad for her that she just checked in as a patient.
1981 (R) 89m/C Barbi Benton, Jon Van Ness, Chip Lucio; ***Dir:*** Boaz Davidson. **VHS, Beta $59.95** *MGM* 🦴

Hospital of Terror A transmigratory religious fanatic who died on the operating table manages to possess a nurse (Jacobson) before he kicks the bucket, and once in his candystripe incarnation, vents his spleen on the doctors who botched his operation. Originally titled "Nurse Sherri," it's got a little more plasma than Adamson's "AstroZombies," considered by many to be a world-class boner.
1978 (R) 88m/C Jill Jacobson, Geoffrey Land, Marilyn Joi, Mary Kay Pass, Prentiss Moulden, Clayton Foster; ***Dir:*** Al Adamson. **VHS $19.99** *SSI* 🦴🦴

The Hostage A young teenager is witness to a gruesome and clandestine burial. The two murderers responsible take it upon themselves to see that their secret is never told.
1967 (PG) 84m/C Don O'Kelly, Harry Dean Stanton, John Carradine, Danny Martins; ***Dir:*** Russell S. Doughten Jr. **VHS $39.95** *ACN* 🦴🦴½

Hostage Arab terrorists hijack a plane and the passengers start to fight back. Filmed in South Africa.
1987 (R) 94m/C Karen Black, Kevin McCarthy, Wings Hauser; ***Dir:*** Hanro Mohr. **VHS, Beta $79.95** *COL* 🦴½

The Hostage Tower In France, a group of international crime figures capture a the visiting U.S. president's mother, hold her hostage in the Eiffel Tower and demand $30 million in ransom. Made for television.
1980 (PG) 97m/C Peter Fonda, Maud Adams, Britt Ekland, Billy Dee Williams, Keir Dullea, Douglas Fairbanks Jr., Rachel Roberts, Celia Johnson; ***Dir:*** Claudio Guzman. **VHS, Beta, LV $69.98** *SUE* 🦴🦴

Hostages A gang of criminals kidnap a family on vacation at a Caribbean island.
1979 93m/C Stuart Whitman, Marisa Mell. **VHS, Beta $49.98** *KOV* 🦴

Hostile Guns A law man discovers a woman he once loved is now an inmate he is transporting across the Texas badlands. Routine with the exception of cameos by veteran actors.
1967 91m/C George Montgomery, Yvonne de Carlo, Tab Hunter, Brian Donlevy, John Russell; ***Dir:*** R.G. Springsteen. **VHS, LV $14.95** *PAR* 🦴🦴½

Hostile Takeover A mild-mannered accountant finally cracks and takes his coworkers hostage. He faces the police in a tense showdown.
1988 (R) 93m/C David Warner, Michael Ironside, Kate Vernon, Jayne Eastwood; ***Dir:*** George Mihalka. **VHS, Beta, LV $89.95** *IVE* 🦴½

Hot Blood Two bank robbers take an innocent woman hostage and thrust her in the midst of a huge family battle. Another lukewarm entry from the director of "The Arrogant."
1989 89m/C Sylvia Kristel, Alicia Moro, Aldo Sambrel, Gaspar Cano; ***Dir:*** Philippe Blot. **VHS, Beta $79.95** *CAN* 🦴½

Hot Box A Filipino-shot, low budget woofer. Prison flick about women who break out and foment a revolution.
1972 (R) 85m/C Andrea Cagan, Margaret Markov, Rickey Richardson, Laurie Rose; ***Dir:*** Joe Viola. **VHS, Beta $24.98** *SUE Woof!*

Hot Bubblegum Three teenagers discover sex, the beach, rock 'n' roll, sex, drag racing and sex.
1981 (R) 94m/C Jonathan Segal, Zachi Noy, Yftach Katzur. **VHS, Beta** *PGN* 🦴

Hot Child in the City A country girl goes to Los Angeles and searches through the city's fleshpots for her sister's killer. Music by Nick Glider, Lou Reed, Billy Idol, and Fun Boy Three.
1987 85m/C Leah Ayres Hendrix, Shari Shattuck, Geof Pryssir; ***Dir:*** John Flores. **VHS, Beta $14.95** *PSM, PAR* 🦴

Hot Chili Four sex-obsessed guys go to Mexico for fun and games, and find just that.
1985 (R) 91m/C Charles Schillaci, Allan J. Kayser, Louisa Moritz; ***Dir:*** William Sachs. **VHS, Beta $79.95** *MGM Woof!*

The Hot, the Cool, and the Vicious A quiet town turns into a bloody, martial arts playground when three hombres trade blows in a tavern.
198? ?m/C Wang Tao, Tommy Lee, Sun Chia-Lin. **VHS $34.95** *OCE Woof!*

Hot Dog...The Movie! There's an intense rivalry going on between an Austrian ski champ and his California challenger during the World Cup Freestyle competition in Squaw Valley. Snow movie strictly for teens or their equivalent.
1983 (R) 96m/C David Naughton, Patrick Houser, Shannon Tweed, Tracy N. Smith; ***Dir:*** Peter Markle. **VHS, Beta** *FOX* 🦴½

Hot Foot/A Happy Married Couple Two silent comedies: "Hot Foot" (1920), is about Dunn's hobo character in the Chaplin mold; "A Happy Married Couple" (1916), is about a young married couple whose puppy, as described in a letter, is mistaken for a baby by their parents.
1920 51m/B Bobbie Dunn, Neal Burns, Betty Compson. **VHS, Beta, 8mm $24.95** *VYY* 🦴½

Hot Lead & Cold Feet Twin brothers (one a gunfighter, the other meek and mild) compete in a train race where the winner will take ownership of a small western town. Dale not only plays both brothers, but also their tough father. Standard Disney fare.
1978 (G) 89m/C Jim Dale, Don Knotts, Karen Valentine; ***Dir:*** Robert Butler. **VHS, Beta $69.95** *DIS* 🦴🦴

Hot Line The world's two superpowers become befuddled when the hot line connecting Washington and Moscow breaks down. Taylor's final film. Also released as "The Day the Hot Line Got Hot."
1969 87m/C Robert Taylor, Charles Boyer, George Chakiris, Dominique Fabre, Gerard Tichy; ***Dir:*** Etienne Perier. **VHS** *REP* 🦴½

Hot Money A small-town drama about an ex-con who poses as an assistant sheriff and plots a million-dollar robbery during the town's 4th of July parade. Made shortly before Welles' death, and kept on the shelf for more than five years.
198? 78m/C Orson Welles, Michael Murphy, Bobby Pickett, Michelle Finney. **VHS, Beta $79.95** *VMK* 🦴

Hot Moves Four high school boys make a pact to lose their virginity before the end of the summer.
1984 (R) 89m/C Michael Zorek, Adam Silbar, Jill Schoelen, Debi Richter, Monique Gabrielle, Tami Holbrook, Virgil Frye; ***Dir:*** Jim Sotos. **VHS, Beta $79.98** *VES* 🦴

Hot Potato Black Belt Jones rescues a senator's daughter from a megalomaniacal general by using skull-thwacking footwork.
1976 (PG) 87m/C Jim Kelly, George Memmoli, Geoffrey Binney; ***Dir:*** Oscar Williams. **VHS, Beta $59.95** *WAR* 🦴

Hot Pursuit A woman who has been framed for murder runs from the law and a relentless hitman.
1984 (PG) 94m/C Mike Preston, Dina Merrill, Kerrie Keane. **VHS, Beta $79.95** *VMK* 🦴½

Hot Pursuit A prep-school bookworm resorts to Rambo-like tactics in tracking down his girlfriend and her family after being left behind for a trip to the tropics.
1987 (PG-13) 93m/C John Cusack, Robert Loggia, Jerry Stiller, Wendy Gazelle; ***Dir:*** Steven Lisberger. **VHS, Beta, LV $79.95** *PAR* 🦴½

Hot Resort Young American lads mix work and play when they sign on as summer help at an island resort.

1985 (R) 92m/C Bronson Pinchot, Tom Parsekian, Michael Berz, Linda Kenton, Frank Gorshin; *Dir:* John Robins. **VHS, Beta $79.95** *MGM* 🐶

The Hot Rock A motley crew of bumbling thieves conspire to steal a huge, priceless diamond; a witty, gritty comedy that makes no moral excuses for its characters and plays like an early-'70s crime thriller gone awry. Written by William Goldman, from a Donald E. Westlake novel. The sequel, 1974's "Bank Shot," stars George C. Scott in the role created by Redford.
1970 (PG) 97m/C Robert Redford, George Segal, Ron Leibman, Zero Mostel, Moses Gunn, William Redfield, Charlotte Rae, Topo Swope, Paul Sand; *Dir:* Peter Yates. **VHS, Beta** *FOX* 🐶🐶🐶

Hot Rod A young California hotrodder battles the evil sponsor of a drag race championship.
1979 97m/C Greg Henry, Robert Culp, Pernell Roberts, Robin Mattson, Grant Goodeve; *Dir:* George Armitage. **VHS, Beta $59.98** *CHA* 🐶

Hot Rod Girl A concerned police officer (Conners) organizes supervised drag racing after illegal drag racing gets out of hand in his community.
1956 75m/B Lori Nelson, Chuck Connors, John W. Smith; *Dir:* Leslie Martinson. **VHS $16.95** *SNC, NOS, MRV* 🐶

Hot Shorts The Firesign Theatre turn their satiric wit on Saturday matinee cliff hanger serials. While familiar characters cross the screen the re-recorded soundtrack features hilarious new dialogue, sound effects, and music.
1984 73m/B Phil Austin, Peter Bergman, Phil Proctor. **VHS, Beta, LV $39.95** *COL, PIA* 🐶

Hot Shot An inspirational tale involving a young soccer player who goes to great lengths to take training from a soccer star, played - of course - by Pele.
1986 (PG) 90m/C Pele, Jim Youngs, Billy Warlock, Weyman Thompson, Mario Van Peebles, David Groh; *Dir:* Rick King. **VHS, Beta $39.98** *WAR* 🐶½

Hot Shots! A lesser feat from "The Naked Gun" team of master movie parodists, this has clever sight gags but the verbal humor often plummets to ground. A spoof of "Top Gun" and similar gung-ho air corps adventures, with Sheen out to avenge the family honor as an ace fighter pilot. Great when you're in the mood for laughs that don't require thought.
1991 (PG-13) 85m/C Charlie Sheen, Lloyd Bridges, Cary Elwes, Valeria Golino, Jon Cryer, Efrem Zimbalist Jr., Kevin Dunn, Bill Irwin; *W/ Dir:* Jim Abrahams. **VHS $19.98** *FXV* 🐶🐶½

Hot Spell In an attempt to repair her marriage and unify her family, a mother plans a reunion-birthday party for her husband. Despair reigns in this sultry, Southern melodrama.
1958 86m/B Shirley Booth, Anthony Quinn, Shirley MacLaine, Earl Holliman, Eileen Heckart; *Dir:* Daniel Mann. **VHS, Beta $19.95** *KRT* 🐶½

The Hot Spot An amoral drifter arrives in a small Texas town and engages in affairs with two women, including his boss's over-sexed wife and a young woman with her own secrets. Things begin to heat up when he decides to plot a bank robbery.
1990 (R) 120m/C Don Johnson, Virginia Madsen, Jennifer Connelly, Charles Martin Smith; *Dir:* Dennis Hopper. **VHS, LV $89.98** *ORI, FCT* 🐶🐶½

Hot Stuff Officers on a burglary task force decide the best way to obtain convictions is to go into the fencing business themselves.
1980 (PG) 91m/C Dom DeLuise, Jerry Reed, Suzanne Pleshette, Ossie Davis; *Dir:* Dom DeLuise. **VHS, Beta, LV $14.95** *COL* 🐶

Hot Summer in Barefoot County A state law enforcement officer on the search for illegal moonshiners finds more than he bargained for.
1974 (R) 90m/C Sherry Robinson, Tonia Bryan, Dick Smith; *Dir:* Will Zens. **VHS, Beta** *PGN* 🐶🐶

Hot Summer Night...with Donna Donna Summer gives an electrifying performance of her hits at a concert held at California's Pacific Amphitheater. Songs include "She Works Hard For the Money," "Bad Girls" and "Hot Stuff."
1984 60m/C Donna Summer. **VHS, Beta, LV $24.95** *COL* 🐶🐶

Hot T-Shirts A small town bar owner needs a boost for business, finding the answer in wet T-shirt contests. Softcore trash.
1979 (R) 86m/C Ray Holland, Stephanie Lawlor, Pauline Rose, Corinne Alphen; *Dir:* Chuck Vincent. **VHS, Beta $39.95** *MCA* *Woof!*

Hot Target An ideal mother becomes obsessed with the stranger who seduced her, even after she finds out he's a thief planning to rob her home.
1985 (R) 93m/C *NZ* Simone Griffith, Steve Marachuk; *Dir:* Denis Lewiston. **VHS, Beta, LV $79.98** *LIV, VES* 🐶

Hot Times Low-budget comedy about a high school boy who, after striking out with the high school girls, decides to go to New York and have the time of his life.
1974 (R) 80m/C Gail Lorber, Amy Farber, Henry Cory; *Dir:* Jim McBride. **VHS, Beta $39.95** *MON* 🐶

Hot Touch An art forger with a large heist is besieged by a slashing maniac, and must discover the culprit.
198? 92m/C Wayne Rogers, Samantha Eggar, Marie-France Pisier, Melvyn Douglas. **VHS, Beta $69.95** *TWE, HHE* *Woof!*

Hot to Trot! A babbling idiot links up with a talking horse in an updated version of the "Francis, the Talking Mule" comedies, by way of Mr. Ed. The equine voice is provided by Candy. Some real funny guys are wasted here.
1988 (PG) 90m/C Bob Goldthwait, Dabney Coleman, Virginia Madsen, Jim Metzler, Cindy Pickett, Tim Kazurinsky, Santos Morales, Barbara Whinnery, Garry Kluger; *Dir:* Michael Dinner; *Voices:* John Candy. **VHS, Beta, LV $19.95** *WAR* 🐶

Hot Under the Collar Jerry is a weasel who tries to seduce the luscious Monica by using hypnosis. But his plans backfire and instead Monica enters a convent and takes a vow of chastity. This isn't going to stop Jerry, he tries posing as a priest and even dressing as a nun in order to get her out. Things get even more complicated when a mobster, looking for a fortune in stolen diamonds, sneaks into the convent.
1991 (R) 87m/C Richard Gabai, Angela Visser, Daniel Friedman, Mindy Clarke, Tane McClure; *Dir:* Richard Gabai. **VHS $89.99** *HBO* 🐶½

Hot Water/Safety Last Two classic Harold Lloyd silent comedies, "Hot Water" (1924) and "Safety Last" (1923). "Safety Last" features one of Lloyd's most famous stunt scenes, where he is hanging from the hands of a giant clock on the side of a building, high above the city.
1924 81m/B Harold Lloyd. **VHS, Beta** *TLF* 🐶½

Hot Wire A tired comedy about the unscrupulous side of the car repossession racket.
197? 92m/C George Kennedy, Strother Martin. **VHS, Beta** *PGN* 🐶½

Hotel A pale rehash of the "Grand Hotel" formula about an array of rich characters interacting in a New Orleans hotel. From the Arthur Hailey potboiler; the basis for the television series.
1967 (PG) 125m/C Rod Taylor, Catherine Spaak, Karl Malden, Melvyn Douglas, Merle Oberon, Michael Rennie, Richard Conte, Kevin McCarthy; *Dir:* Richard Quine. **VHS, Beta $59.95** *WAR* 🐶½

Hotel Colonial A young Italian idealist ventures to Columbia to retrieve the body of his brother, who reportedly killed himself. Once there, he finds the dead man is not his brother, and decides to investigate.
1988 (R) 103m/C *IT* Robert Duvall, John Savage, Rachel Ward, Massimo Troisi; *Dir:* Cinzia Torrini. **VHS, Beta, LV, 8mm $19.98** *SUE* 🐶½

Hotel Du Lac A made-for-British-television adaptation of the Anita Brookner novel about a woman writer languishing at a Swiss lakefront hotel.
1986 75m/C *GB* Anna Massey; *Dir:* Giles Foster. **VHS, Beta $59.98** *FOX* 🐶🐶

The Hotel New Hampshire The witless, amoral adaptation of John Irving's novel about a very strange family's adventures in New Hampshire, Vienna and New York City, which include gang rape, incest, and Kinski in a bear suit.
1984 (R) 110m/C Jodie Foster, Rob Lowe, Beau Bridges, Nastassia Kinski, Wallace Shawn, Wilford Brimley, Amanda Plummer, Anita Morris, Matthew Modine; *Dir:* Tony Richardson. **VHS, Beta, LV $29.98** *VES* 🐶🐶

Hotel Reserve A Nazi spy is reportedly afoot in a French resort during World War II, and tension mounts as the guests attempt to flush him out. Critical response to this film is wildly split; some find it obvious, while others call it well-done and suspenseful. Based on the novel "Epitaph for a Spy" by Eric Ambler.
1944 79m/B *GB* James Mason, Lucie Mannheim, Raymond Lovell, Julien Mitchell, Martin Miller, Herbert Lom, Frederick Valk, Valentine Dyall; *Dir:* Victor Hanbury, Lance Comfort, Max Greene. **VHS, Beta $19.98** *TTC, RXM* 🐶🐶🐶

Hothead A heedless, impulsive soccer player's behavior gets him cut from the team, fired from his mill job, and booted out of his favorite bar. Dubbed. Also known as "Coup de Tete."
1978 98m/C *FR* Patrick Dewaere; *Dir:* Jean-Jacques Annaud. **VHS, Beta $59.95** *COL* 🐶🐶

Hothead (Coup de Tete) A talented soccer player's quick temper causes him to be cut from his team, lose his job and even be banned from the local bar.
1978 90m/C *FR* Patrick Dewaere; *Dir:* Jean-Jacques Annaud. **VHS, Beta $59.95** *COL* 🐶🐶½

Hotline After taking a job answering phones at a crisis center, a woman becomes the next target of a psychotic killer. Made for television.
1982 96m/C Lynda Carter, Steve Forrest, Granville Van Dusen, Monte Markham, James Booth; *Dir:* Jerry Jameson. **VHS $39.95** *USA* 🐶🐶

H.O.T.S. A sex-filled fraternity rivalry film, starring a slew of ex-Playboy Playmates in wet shirts. Also known as "T & A Academy." Screenplay co-written by exploitation star Cheri Caffaro.
1979 (R) 95m/C Susan Kiger, Lisa London, Kimberly Cameron, Danny Bonaduce, Steve Bond; *Dir:* Gerald Seth Sindell. **VHS, Beta** **$69.98** *LIV, VES, SIM* ⟂

Houdini Historically inaccurate but entertaining biopic of the infamous magician and escapologist. Chronicles his rise to stardom, his efforts to contact his dearly departed mother through mediums (a prequel to "Ghost"?), and his heart-stopping logic-defying escapes. First screen teaming of Leigh and Curtis, who had already been married for two years.
1953 107m/C Tony Curtis, Janet Leigh, Torin Thatcher, Angela Clark, Stefan Schnabel, Ian Wolfe, Sig Rumann, Michael Pate, Connie Gilchrist, Mary Murphy, Tor Johnson; *Dir:* George Marshall. **VHS $14.95** *PAR, FCT* ⟂⟂½

The Hound of the Baskervilles The curse of a demonic hound threatens descendants of an English noble family until Holmes and Watson solve the mystery. Also available with "Sherlock Holmes Faces Death" on Laser Disc.
1939 80m/B *GB* Basil Rathbone, Nigel Bruce, Richard Greene, John Carradine, Wendy Barrie, Lionel Atwill, E.E. Clive; *Dir:* Sidney Lanfield. **VHS, Beta $19.95** *KUI, FOX* ⟂⟂⟂

The Hound of the Baskervilles Cushing's not half bad as Sherlock Holmes as he investigates the mystery of a supernatural hound threatening the life of a Dartmoor baronet. Dark and moody.
1959 86m/C *GB* Peter Cushing, Christopher Lee, Andre Morell; *Dir:* Terence Fisher. **VHS, Beta, LV $29.98** *FOX, MLB* ⟂⟂½

The Hound of the Baskervilles Awful spoof of the Sherlock Holmes classic with Cook as Holmes and Moore as Watson (and Holmes' mother). Even the cast can't save it.
1977 84m/C Dudley Moore, Peter Cook, Denholm Elliott, Joan Greenwood, Spike Milligan, Jessie Matthews, Roy Kinnear; *Dir:* Paul Morrissey. **VHS** *ATV* Woof!

Hour of the Assassin A mercenary is hired to assassinate a South American ruler, and a CIA operative is sent to stop him. Shot in Peru.
1986 (R) 96m/C Erik Estrada, Robert Vaughn; *Dir:* Luis Llosa. **VHS, Beta $79.95** *MGM* ⟂½

Hour of Decision When a newspaper columnist turns up dead, a reporter investigates the murder. Things get hairy when he finds the prime suspect is his own wife.
1957 81m/B *GB* Jeff Morrow, Hazel Court, Lionel Jeffries, Anthony Dawson, Mary Laura Wood; *Dir:* C. Pennington Richards. **VHS, Beta $16.95** *SNC* ⟂½

Hour of the Gun Western saga chronicles what happens after the gunfight at the OK Corral. Garner plays the grim Wyatt Earp on the trail of vengeance after his brothers are killed. Robards is excellent as the crusty Doc Holliday.
1967 100m/C James Garner, Jason Robards Jr., Robert Ryan, Albert Salmi, Charles Aidman, Steve Ihnat, Jon Voight; *Dir:* John Sturges. **VHS** *MGM* ⟂⟂

Hour of the Star The poignant, highly acclaimed feature debut by Amaral, about an innocent young woman moving to the city of Sao Paulo from the impoverished countryside of Brazil, and finding happiness despite her socioeconomic failures. Based on Clarice Lispector's novel. In Portuguese with English subtitles.
1985 96m/C *BR* Marcelia Cartaxo; *Dir:* Suzana Amaral. Berlin Film Festival '85: Best Actress. **VHS, Beta $79.95** *KIV* ⟂⟂⟂½

Hour of Stars An early television dramatization of "The Genius."
1958 60m/B ZaSu Pitts, Eddie Bracken, Reginald Denny. **VHS, Beta $29.95** *DVT* ⟂

Hour of the Wolf An acclaimed, surreal view into the tormented inner life of a painter as he and his wife are isolated on a small northern island. Cinematography by Sven Nykvist. In Swedish with English subtitles.
1968 89m/B *SW* Max von Sydow, Liv Ullman, Ingrid Thulin, Erland Josephson, Gertrud Fridh, Gudrun Brost; *Dir:* Ingmar Bergman. National Board of Review Awards '68: Best Actress (Ullman); National Society of Film Critics Awards '68: Best Director. **VHS, Beta $39.95** *FCT, MRV* ⟂⟂⟂½

House A horror novelist moves into his dead aunt's supposedly haunted house only to find that the monsters don't necessarily stay in the closets. His worst nightmares come to life as he writes about his Vietnam experiences and is forced to relive the tragic events, but these aren't the only visions that start springing to life. It sounds depressing, but is actually a funny, intelligent "horror" flick. Followed by several lesser sequels.
1986 (R) 93m/C William Katt, George Wendt, Richard Moll, Kay Lenz; *Dir:* Steve Miner. **VHS, Beta, LV $19.95** *NWV, STE* ⟂⟂⟂

House 2: The Second Story The flaccid sequel to the haunted-house horror flick concerns two innocent guys who move into the family mansion and discover Aztec ghosts. Has none of the humor which helped the first movie along.
1987 (PG-13) 88m/C John Ratzenberger, Arye Gross, Royal Dano, Bill Maher, Jonathan Stark; *Dir:* Ethan Wiley. **VHS, Beta, LV $19.95** *NWV, STE* ⟂

House 4 Another bad real estate epic on a par with the other unusually inept "House" efforts. When her husband (Katt) is killed in an auto accident, a young woman and her daughter move into a mysterious old house. Terror begins to confront them at every turn, threatening to destroy them both. We can only hope.
1991 (R) 93m/C Terri Treas, Scott Burkholder, Melissa Clayton, William Katt, Ned Romero; *Dir:* Lewis Abernathy. **VHS, LV $89.95** *COL* ⟂½

The House of 1,000 Dolls Doleful, not dollful, exploitation thriller as magician Price drugs young girls and sells them to slavery rings. An international production set in Tangiers; dialogue is dubbed.
1967 79m/C *GE SP GB* Vincent Price, Martha Hyer, George Nader, Ann Smyrner, Maria Rohm; *Dir:* Jeremy Summers. **VHS, Beta $59.98** *HBO* ⟂

House Across the Bay Raft, an imprisoned nightclub owner, finds that he is being duped by his attorney who's eager to get his hooks into the lingerie-clad prisoner's wife. But prison isn't going to stop the bitter husband from getting his revenge.
1940 88m/B George Raft, Walter Pidgeon, Joan Bennett, Lloyd Nolan, Gladys George; *Dir:* Archie Mayo. **VHS, Beta $19.95** *MON* ⟂⟂½

The House of the Arrow Investigating the murder of a French widow, who was killed by a poisoned arrow, a detective must sort through a grab bag of potential suspects. Superior adaptation of the A.E.W. Mason thriller.

1953 73m/B *GB* Oscar Homolka, Yvonne Furneaux, Robert Urquhart; *Dir:* Michael Anderson Sr. **VHS, Beta $16.95** *NOS, SNC* ⟂⟂⟂

House of the Black Death Horror titans Chaney and Carradine manage to co-star in this as warring warlocks, yet they share no scenes together! Chaney is the evil warlock (the horns are a giveaway) holding people hostage in the title edifice.
1965 89m/B Lon Chaney Jr., John Carradine, Katherine Victor, Tom Drake, Andrea King; *Dir:* Harold Daniels, Reginald LeBorg. **VHS $15.95** *NOS, LOO* ⟂½

The House that Bled to Death Strange things happen to a family when they move into a run-down house where a murder occurred years before.
1981 60m/C Nicholas Ball, Rachel Davies, Brian Croucher, Pat Maynard, Emma Ridley. **VHS, Beta $29.95** *IVE* ⟂

House Calls A widowed surgeon turns into a swinging bachelor until he meets a witty but staid English divorcee. Wonderful dialogue. Jackson is exceptional. Made into a short-lived TV series.
1978 (PG) 98m/C Walter Matthau, Glenda Jackson, Art Carney, Richard Benjamin, Candice Azzara; *Dir:* Howard Zieff. **VHS, Beta, LV $69.95** *MCA* ⟂⟂

The House on Carroll Street It's New York in 1951, the middle of the McCarthy era. A young woman overhears a plot to smuggle Nazi war criminals into the U.S. But she's lost her job because of accusations of subversion, and it's not easy persuading FBI agent Daniels that she knows what she's talking about. Contrived plot and melodramatic finale help sink this period piece.
1988 (PG) 111m/C Kelly McGillis, Jeff Daniels, Mandy Patinkin, Jessica Tandy; *Dir:* Peter Yates. **VHS, Beta, LV $89.99** *HBO* ⟂½

The House by the Cemetery When a family moves into a house close to a cemetery, strange things start to happen to them.
1983 (R) 84m/C *IT* Katherine MacColl, Paolo Malco; *Dir:* Lucio Fulci. **VHS, Beta $69.98** *LIV, VES* ⟂

The House on Chelouche Street In Tel-Aviv, during the summer of 1946, 15 year-old Sami, an Egyptian immigrant, tries to help support his mother and family during the upheavals of Palestine under the British mandate. A well thought-out story detailing the struggles of a North African Jewish family before the creation of the country of Israel. In Hebrew with English subtitles.
1973 111m/C *IS* Gila Almagor, Michal Bat-Adam, Shai K. Ophir; *Dir:* Moshe Mizrahi. **VHS, Beta $29.95** *VYY, NCJ* ⟂⟂½

House of the Damned A young woman, a mental hospital, a murdered father, a priceless gold statue—mix together and you get a mystery horror movie. A woman tries to solve her father's murder.
19?? 89m/C Donald Pleasence, Michael Dunn. **VHS, Beta $39.98** *NSV* ⟂½

House of Dark Shadows A 150-year-old vampire, accidentally released by a handyman, wreaks havoc on the inhabitants of the Collinswood mansion. Based on the TV daytime soap.
1970 (PG) 97m/C Jonathan Frid, Joan Bennett, Grayson Hall, Kathryn Leigh Scott, Roger Davis, Nancy Barrett, John Carlen, Thayer David; *Dir:* Dan Curtis. **VHS, Beta $19.98** *MGM* ⟂⟂

House of the Dead Man is trapped inside a haunted house. Scary.

1980 100m/C John Erikson, Charles Aidman, Bernard Fox, Ivor Francis. **VHS** *JLT* ✒

House of Death The ex-Playmate of the Year runs in hysterical fear from a knife-wielding lunatic.
1982 (R) 88m/C Susan Kiger, William T. Hicks, Jody Kay, Martin Tucker, Jennifer Chase. **VHS, Beta $59.95** *GEM* ✒

The House of Dies Drear A black family moves into an old, mysterious house. Based on the story by Virginia Hamilton. Also available with an author interview. Aired on PBS as part of the WonderWorks Family Movie television series.
1988 116m/C Howard E. Rollins Jr., Moses Gunn, Shavar Ross, Gloria Foster, Clarence Williams III; *Dir:* Allan Goldstein. **VHS $29.95** *PPM, PME, FCT* ✒

The House that Dripped Blood A Scotland Yard inspector discovers the history of an English country house while investigating an actor's disappearance. Four horror tales written by Robert Bloch comprise the body of this omnibus creeper, following the successful "Tales from the Crypt" mold. Duffell's debut as director.
1971 (PG) 101m/C *GB* Christopher Lee, Peter Cushing, Jon Pertwee, Denholm Elliott, Ingrid Pitt, John Bennett, Tom Adams, Joss Ackland, Chloe Franks; *Dir:* Peter Duffell. **VHS, Beta $59.95** *PSM* ✒✒½

House on the Edge of the Park A frustrated would-be womanizer takes revenge on a parade of women during an all-night party.
1984 91m/C *IT* David Hess, Annie Belle, Lorraine De Selle; *Dir:* Ruggero Deodato. **VHS, Beta $69.98** *LIV, VES* Woof!

House of Evil Karloff operates a torture dungeon and has harmless musical toys turning into mechanical killers. One of his final four films. Like the others, this is a Mexican production but Karloff filmed his scenes in Los Angeles. Also known as "Dance of Death" and "Macabre Serenade."
1968 (PG) 75m/C *MX* Boris Karloff, Andres Garcia; *Dir:* Juan Ibanez. **VHS, Beta $59.95** *SNC, MPI, SVS* Woof!

House of Fear Holmes and Watson investigate the murders of several members of the Good Comrades Club, men who were neither good nor comradely.
1945 69m/B Basil Rathbone, Nigel Bruce, Dennis Hoey, Aubrey Mather, Paul Cavanagh, Gavin Muir; *Dir:* Roy William Neill. **VHS, Beta $19.95** *FOX* ✒✒½

House of Frankenstein An evil scientist (Karloff) escapes from prison and, along with his hunchback manservant, revives Dracula, Frankenstein's monster, and the Wolfman to carry out his dastardly deeds. An all-star cast helps this horror fest along.
1944 71m/B Boris Karloff, J. Carroll Naish, Lon Chaney Jr., John Carradine, Elena Verdugo, Anne Gwynne, Lionel Atwill, Peter Coe, George Zucco, Glenn Strange, Sig Rumann; *Dir:* Erle C. Kenton. **VHS $14.98** *MCA* ✒✒½

House of Games The directorial debut of playwright Mamet is a tale about an uptight female psychiatrist who investigates the secret underworld of con-artistry and becomes increasingly involved in an elaborate con game. She's led along her crooked path by smooth-talking con-master Mantegna. A taut, well plotted psychological suspense film with many twists and turns. Stylishly shot, with dialogue which has that marked Mamet cadence. Most of the leads are from Mamet's theater throng (including his wife, Crouse).
1987 (R) 102m/C Joe Mantegna, Lindsay Crouse, Lilia Skala, J.T. Walsh, Meshach Taylor; *Dir:* David Mamet. **VHS, Beta, LV $19.99** *HBO, FCT* ✒✒✒

The House on Garibaldi Street Spy drama about the capture of Nazi war criminal Adolph Eichmann in Argentina and his extradition to Israel for trial. Based on the book by Isser Harel. Made for television.
1979 96m/C Chaim Topol, Nick Mancuso, Martin Balsam, Janet Suzman, Leo McKern, Charles Gray, Alfred Burke; *Dir:* Peter Collinson. **VHS, Beta $14.95** *FRH* ✒✒½

House on Haunted Hill A wealthy man throws a haunted house party and offers ten thousand dollars to anyone who can survive the night there. Vintage cheap horror from the master of the macabre Castle. Remembered for Castle's in-theater gimmick of dangling a skeleton over the audiences' heads in the film's initial release. Also available with "Attack of the 50-Foot Woman" on laserdisc.
1958 75m/B Vincent Price, Carol Ohmart, Richard Long, Alan Marshal; *Dir:* William Castle. **VHS, Beta, LV $14.98** *FOX, FCT, MLB* ✒✒½

House of Insane Women While running an insane asylum, a man discovers that a woman has lost her mind because as a child, she witnessed her mother's death during an exorcism. Pretty grim but interesting for genre fans. A natural double feature with "House of Psychotic Women."
1974 (R) 93m/C *SP* Amelia Gade, Francisco Rabal, Espartaco Santoni; *Dir:* Rafael Morena Alba. **VHS $16.95** *SNC* ✒

House of the Living Dead Brattling Manor harbors a murderous and flesh-eating secret ready to lure the unsuspecting.
1973 (PG) 87m/C *SA* Mark Burns, Shirley Anne Field, David Oxley; *Dir:* Ray Austin. **VHS, Beta $29.95** *UHV, HHE* ✒

House of the Long Shadows Comedy mixes with gore, as four of the horror screen's best leading men team up for the first time, although they have only minor roles. Might have been a contender, but Arnaz is miscast as author who stays in spooky house on a bet with Todd. Some wit but not enough. Based on the play "Seven Keys to Baldpate" by George M. Cohan (and novel by Earl Derr Biggers).
1982 (PG) 96m/C Vincent Price, Christopher Lee, Peter Cushing, John Carradine, Desi Arnaz Jr., Sheila Keith, Richard Todd; *Dir:* Pete Walker. **VHS, Beta $69.95** *MGM* ✒½

The House of Lurking Death Detectives Tommy and Tuppence find a box of chocolates laced with arsenic at the home of Lois Hargreaves, and the mad old maid is the prime suspect. Based on the Agatha Christie story. Made for British television.
1984 60m/C *GB* James Warwick, Francesca Annis. **VHS, Beta $14.95** *PAV* ✒½

The House in Marsh Road A faithless husband decides to rid himself of She Who Must Be Obeyed, but his wife's ghost-friend intervenes.
1960 70m/B *GB* Tony Wright, Patricia Dainton, Sandra Dorne, Derek Aylward, Sam Kydd; *Dir:* Montgomery Tully. **VHS $19.98** *SNC* ✒✒

House of Mystery For a shot at some priceless jewels stolen by the host from a Hindu princess, several guests gather for a weekend in a spine-tingling mansion. A creepy curse keeps them looking over their shoulders.
1934 62m/C Ed Lowry, Verna Hillie, Brandon Hurst, George "Gabby" Hayes; *Dir:* William Nigh. **VHS $12.95** *SNC, DVT, MLB* ✒✒

House of Mystery Splendid British remake of 1934 film in which the detective does some true sleuthing to catch the culprits. An old widow is killed for her jewels, but the murderers can't locate the goods. After they try a series of fiendish plans to retrieve them, the detective must work to stop the crooks.
1941 61m/B *GB* Kenneth Kent, Judy Kelly, Peter Murray Hill, Walter Rilla, Ruth Maitland; *Dir:* Walter Summers. **VHS $24.95** *NOS, DVT* ✒✒½

House Party A light-hearted, black hip-hop version of a 1950s teen comedy with rap duo Kid 'n' Play. After his father grounds him for fighting, a high-schooler attempts all sorts of wacky schemes to get to his friend's party. Sleeper hit features real-life music rappers and some dynamite dance numbers.
1990 (R) 100m/C Christopher Reid, Christopher Martin, Lawrence Martin, Tisha Campbell, Paul Anthony, Kid 'N' Play, A.J. Johnson, Full Force, Robin Harris; *Dir:* Reginald Hudlin. Sundance Film Festival '90: Best Screenplay. **VHS, Beta, LV $19.95** *COL* ✒✒½

House Party 2: The Pajama Jam Rap stars Kid 'N' Play are back in this hip-hop sequel to the original hit. At Harris Univerity Kid 'N' Play hustle up overdue tuition by holding a campus "jammie jam jam." A stellar cast shines in this rap-powered pajama bash.
1991 (R) 94m/C Christopher Reid, Christopher Martin, Full Force, Tisha Campbell, Iman, Queen Latifah, George Stanford Brown, Martin Lawrence; *Dir:* Doug McHenry, George Jackson. **VHS** *COL* ✒✒½

House of Psychotic Women Naschy, Spain's premier horror movie star, plays a studly drifter who winds up in the home of three sisters, one with a horribly scarred arm, one confined to a wheelchair and the other a nymphomaniac. No problemo thinks the hirsute Spaniard, until a series of bizarre murders—in which the victims eyes are ripped out—spoils his fun. Naschy, who usually sprouts fur and fangs, actually doesn't turn into a werewolf. Catchy but inappropriate score. A must for Naschy/bad movie fans.
1973 (R) 87m/C *SP* Paul Naschy, Maria Perschy, Diana Lorys, Eva Leon, Ines Morales, Tony Pica; *Dir:* Carlos Aured. **VHS, Beta $14.98** *VID* Woof!

House of the Rising Sun A Los Angeles reporter goes under cover for a story on an exclusive brothel only to discover the owner is a psychopath who has very permanent ways of dealing with troublesome employees. Music by Tina Turner and Bryan Ferry.
1987 86m/C John York, Bud Davis, Deborah Wakeham, Frank Annese; *Dir:* Greg Gold. **VHS, Beta $79.95** *PSM, PAR* ✒½

The House of Secrets A Yank travels to Britain to collect an inheritance and stays in a dusty old mansion filled with an odd assortment of characters.
1937 70m/B Leslie Fenton, Muriel Evans, Noel Madison, Sidney Blackmer, Morgan Wallace, Holmes Herbert; *Dir:* Roland D. Reed. **VHS, Beta $16.95** *SNC* ✒✒

The House of Seven Corpses A crew attempts to film a horror movie in a Victorian manor where seven people died in a variety of gruesome manners. Things take a turn for the ghoulish when a crew member becomes possesed by the house's evil spirits. Good, low budget fun with a competent "B" cast. Filmed in what was the Utah governor's mansion.
1973 (PG) 90m/C John Ireland, Faith Domergue, John Carradine, Carole Wells; *Dir:* Paul Harrison. VHS $29.95 GEM, MRV, NEG ⅟₂

House of Shadows A 20-year-old murder comes back to haunt the victim's friends in this tale of mystery and terror.
1983 90m/C John Gavin, Yvonne de Carlo, Leonor Manso; *Dir:* Richard Wulicher. VHS, Beta $49.95 MED ⅟ ½

The House on Skull Mountain The four surviving relatives of a deceased voodoo priestess are in for a bumpy night as they gather at the House on Skull Mountain for the reading of her will.
1974 (PG) 85m/C Victor French, Janee Michelle, Mike Evans, Jean Durand; *Dir:* Ron Honthaner. VHS, Beta $59.98 FOX ⅟ ½

The House on Sorority Row The harrowing story of what happens when seven sorority sisters have a last fling and get back at their housemother at the same time.
1983 90m/C Eileen Davidson, Kathy McNeil, Robin Meloy; *Dir:* Mark Rosman. VHS, LV $79.95 VES ⅟ ½

House of Strangers Conte, in a superb performance, swears vengeance on his brothers, whom he blames for his father's death. Robinson, in a smaller part than you'd expect from his billing, is nevertheless excellent as the ruthless banker father who sadly reaps a reward he didn't count on. Based on Philip Yordan's "I'll Never Go There Again."
1949 101m/B Edward G. Robinson, Susan Hayward, Richard Conte, Luther Adler, Efrem Zimbalist Jr., Debra Paget; *Dir:* Joseph L. Mankiewicz. VHS $39.98 FOX, FCT ⅟⅟⅟½

The House on Straw Hill A successful novelist is intrigued by an attractive woman who lives in an isolated farmhouse. Her presence inspires gory hallucinations, lust and violence. Also known as "Expose."
1976 (R) 84m/C Udo Kier, Linda Hayden, Fiona Richmond; *Dir:* James K. Clarke. VHS, Beta $19.95 STE, NWV, LUN ⅟ ½

House of Terror A woman and her boyfriend plot to kill her rich boss, while they too are being stalked by some undefined horror.
1987 91m/C Jennifer Bishop, Arell Blanton, Jacquelyn Hyde. VHS, Beta $79.95 TWE ⅟

The House of Usher Some serious overacting by Reed (as Roderick) and Pleasance (as his balmy brother) almost elevate this Poe retread to "so bad it's good" status, and there are a few admirably grisly moments. But, without other camp virtues, it's merely dreary. Adapted by Michael J. Murray and filmed in South Africa.
1988 (R) 92m/C Oliver Reed, Donald Pleasence, Romy Windsor; *Dir:* Alan Birkinshaw. VHS, LV $79.95 COL ⅟

The House that Vanished A beautiful young model witnesses a murder in a house but later can't find the place. Her boyfriend is skeptical. Also known as "Scream and Die."
1973 (R) 84m/C GB Andrea Allan, Karl Lanchbury, Judy Matheson; *Dir:* Joseph Larraz. VHS, Beta $14.98 MED ⅟

House of Wax A deranged sculptor (Price, who else?) builds a sinister wax museum which showcases creations that were once alive. A remake of the early horror flick "Mystery of the Wax Museum," and one of the 50s' most popular 3-D films. This one still has the power to give the viewer the creeps, thanks to another chilling performance by Price. Look for a very young Charles Bronson, as well as Carolyn "Morticia Addams" Jones as a victim.
1953 (PG) 88m/C Vincent Price, Frank Lovejoy, Carolyn Jones, Phyllis Kirk, Paul Cavanagh, Charles Bronson; *Dir:* Andre de Toth. VHS, Beta $59.95 WAR, MLB ⅟⅟⅟

House Where Evil Dwells An American family is subjected to a reign of terror when they move into an old Japanese house possessed by three deadly samurai ghosts. The setting doesn't upgrade the creepiness of the plot.
1982 (R) 88m/C Edward Albert, Susan George, Doug McClure, Amy Barett; *Dir:* Kevin Connor. VHS, Beta $59.95 MGM ⅟ ½

House of Whipcord Beautiful young women are kidnapped and tortured in this British gore-o-rama. Awful and degrading.
1975 102m/C GB Barbara Markham, Patrick Barr, Ray Brooks, Penny Irving, Anne Michelle; *Dir:* Pete Walker. VHS, Beta $39.95 MON, IVE, AVD Woof!

House of the Yellow Carpet A couple try to sell an ancient Persian carpet heirloom, and in doing so transgress some unwritten mystical law. Havoc ensues.
1984 90m/C Roland Josephson, Beatrice Romand; *Dir:* Carlo Lizzani. VHS, Beta $69.98 LIV, LTG ⅟ ½

Houseboat Widower Grant with three precocious kids takes up residence on a houseboat with Italian maid Loren who is actually a socialite, incognito. Naturally, they fall in love. Light, fun Hollywood froth. Oscar nominated for the script and the song, "Almost in Your Arms."
1958 110m/C Cary Grant, Sophia Loren, Martha Hyer, Eduardo Ciannelli, Murray Hamilton, Harry Guardino; *Dir:* Melville Shavelson. Beta, LV, CDV $19.95 PAR ⅟⅟½

The Housekeeper Tushingham's portrayal of a deranged housekeeper brings this movie to life even as she kills off her employers. Also known as "A Judgement in Stone."
1986 (R) 97m/C CA Rita Tushingham, Ross Petty, Jackie Burroughs, Tom Kneebone, Shelly Peterson, Jessica Stern, Jonathan Crombie; *Dir:* Ousama Rawi. VHS, Beta, LV $39.98 WAR ⅟⅟

Housekeeping A quiet but bizarre comedy by Forsyth (his first American film). A pair of orphaned sisters are cared for by their newly arrived eccentric, free-spirited aunt in a small and small-minded community in Oregon in the 1950's. Conventional townspeople attempt to intervene, but the sisters' relationship with their offbeat aunt has become strong enough to withstand the coercion of the townspeople. May move too slowly for some viewers. Based on novel by Marilynne Robinson.
1987 (PG) 117m/C Christine Lahti, Sarah Walker, Andrea Burchill; *Dir:* Bill Forsyth. VHS, Beta $89.95 COL ⅟⅟⅟

Housemaster In a British boys' school the housemaster is kind and understanding of his sometimes-unruly charges. The stern new headmaster is not, and the students conspire against him. An okay but dated comedy, featuring characters with names like Bimbo and Button. Based on "Bachelor Born" by Ian Hay, a London stage hit at the time.
1938 95m/B Otto Kruger, Diana Churchill, Phillips Holmes, Rene Ray, Walter Hudd, John Wood, Cecil Parker, Michael Shepley, Jimmy Hanley, Rosamund Barnes; *Dir:* Herbert Brenon. VHS $19.95 NOS, MLB ⅟⅟½

Housewife A vengeful black man holds an unhappily married Beverly Hills couple hostage in their home.
1972 (R) 96m/C Yaphet Kotto, Andrew Duggan, Joyce Van Patten, Jeannie Berlin; *Dir:* Larry Cohen. VHS, Beta $19.95 STE, NWV ⅟⅟

How to Beat the High Cost of Living Three suburban housewives decide to beat inflation by taking up robbery, and they start by planning a heist at the local shopping mall.
1980 (PG) 105m/C Jessica Lange, Susan St. James, Jane Curtin, Richard Benjamin, Eddie Albert; *Dir:* Dabney Coleman. VHS, Beta $69.98 LIV, VES ⅟ ½

How the Best was Won Some of Walt Disney's most popular cartoons ranging from the 1930's to the 1960's are featured in this video that is sure to win the hearts of both the young and the old. Includes, "Ferdinand the Bull," "Three Orphan Kittens," "Funny Little Bunnies," and "Building a Building."
1960 48m/C *Dir:* Walt Disney. VHS, Beta, LV DIS ⅟ ½

How to Break Up a Happy Divorce A divorced woman starts a vigorous campaign to win her ex-husband back by dating another man to make him jealous. Lightweight made-for-television comedy.
1976 74m/C Barbara Eden, Hal Linden, Peter Bonerz, Marcia Rodd, Harold Gould; *Dir:* Jerry Paris. VHS, Beta $34.95 WOV ⅟⅟

How Come Nobody's On Our Side? Two out-of-work stuntmen try astrology to find the best time to make some easy money smuggling marijuana out of Mexico.
1973 (PG) 84m/C Adam Roarke, Larry Bishop, Alexandra Hay, Rob Reiner, Penny Marshall. VHS, Beta $39.95 MON ⅟

How Funny Can Sex Be? Eight risque sketches about love and marriage Italian style.
1976 (R) 97m/C IT Giancarlo Giannini, Laura Antonelli, Dulio Del Prete; *Dir:* Dino Risi. VHS, Beta FOX ⅟⅟

How to Get Ahead in Advertising A cynical, energetic satire about a manic advertising idea man who becomes so disgusted with trying to sell a pimple cream that he quits the business. Ultimately he grows a pimple of his own that talks and begins to take over his life. Acerbic and hilarious.
1989 (R) 95m/C GB Richard E. Grant, Rachel Ward, Susan Wooldridge, Mick Ford, Richard Wilson; *W/Dir:* Bruce Robinson. VHS, Beta, LV $89.95 VIR ⅟⅟⅟

How Green was My Valley Compelling story of the trials and tribulations of a Welsh mining family, from the youthful perspective of the youngest child (played by a 13-year-old McDowall). Spans 50 years, from the turn of the century, when coal mining was a difficult but fair-paying way of life, and ends, after unionization, strikes, deaths, and child abuse, with the demise of a town and its culture. Considered by many to be director Ford's finest work (and he produced a lot of works over his 40-year tenure in sound

films), it was nominated for 10 Oscars. When WWII prevented shooting on location, producer Zanuck built a facsimile Welsh valley in California (although Ford, born Sean Aloysius O'Fearna, was said to have been thinking of his story as taking place in Ireland rather than Wales). Based on the novel by Richard Llewellyn.
1941 118m/C Walter Pidgeon, Maureen O'Hara, Donald Crisp, Anna Lee, Roddy McDowall, John Loder, Sara Allgood, Barry Fitzgerald, Patric Knowles, Rhys Williams; *Dir:* John Ford. Academy Awards '41: Best Black and White Cinematography, Best Director (Ford), Best Interior Decoration, Best Picture, Best Supporting Actor (Crisp). **VHS, Beta $39.98** FOX 𝄞𝄞𝄞𝄞

How I Got Into College A slack, uninspired satire about a high school senior who barely manages to get into a local college so he may pursue the girl of his dreams.
1989 (PG-13) 87m/C Corey Parker, Lara Flynn Boyle, Christopher Rydell, Anthony Edwards, Phil Hartman, Brian Doyle-Murray, Nora Dunn, Finn Carter, Charles Rocket; *Dir:* Steve Holland. **VHS, Beta, LV $89.98** FOX 𝄞𝄞

How I Learned to Love Women A British comedy of errors involving a car-loving youth who learns about women and love.
196? 95m/C *GB* Anita Ekberg. **VHS, Beta** INV 𝄞

How I Won the War An inept officer must lead his battalion out of England into the Egyptian desert to conquer a cricket field. Indulgent, scarcely amusing Richard Lester comedy; Lennon, in a bit part, provides the brightest moments.
1967 (PG) 111m/C *GB* John Lennon, Michael Crawford, Michael Hordern; *Dir:* Richard Lester. **VHS, Beta, LV $59.95** MGM 𝄞𝄞

How to Kill 400 Duponts A bumbling Scotland Yard detective is off to France to stop a master criminal from trying to kill off the rich Dupont family. Why? So he'll be the only one left with any claim to the remaining family fortune.
1965 98m/C Terry-Thomas, Johnny Dorelli, Margaret Lee. **VHS, Beta** VCD 𝄞

How to Make a Monster In-joke from the creators of youth-oriented '50s AIP monster flicks; a mad makeup man's home-brew greasepaint brainwashes the actors. Disguised as famous monsters, they kill horror-hating movie execs. Mild fun if—and only if—you treasure the genre. Conway repeats his role as Teenage Frankenstein, but the studio couldn't yet Michael Landon for the Teenage Werewolf.
1958 73m/B Robert Harris, Paul Brinegar, Gary Conway, Gary Clarke, Malcolm Atterbury, Dennis Cross, John Ashley, Morris Ankrum, Walter Reed, Heather Ames; *Dir:* Herbert L. Strock. **VHS $29.95** COL, MLB 𝄞𝄞

How Many Miles to Babylon? During WWI, two young men become friends despite their different backgrounds. But one is court-martialed for desertion, and the other is supposed to oversee his execution. Made for British television.
1982 106m/C *GB* Daniel Day Lewis, Christopher Fairbank; *Dir:* Moira Armstrong. **VHS, Beta $59.98** FOX 𝄞𝄞

How to Marry a Millionaire Three models pool their money and rent a lavish apartment in a campaign to trap millionaire husbands. Clever performances by three lead women salvage a vehicle intended primarily to bolster Monroe's career. Filmed just after "Gentlemen Prefer Blonds," it has little to add—except that it is the first movie to be

filmed in CinemaScope. Nunnally Johnson wrote the script, based on "The Greeks Had a Word for Them." The opening street scene, with the accompanying theme music by Alfred Newman, were state-of-the-art achievements of screen and sound.
1953 96m/C Lauren Bacall, Marilyn Monroe, Betty Grable, William Powell, David Wayne, Cameron Mitchell; *Dir:* Jean Negulesco. **VHS, Beta, LV $14.98** FOX 𝄞𝄞½

How to Murder Your Wife While drunk, a cartoonist marries an unknown woman and then frantically tries to think of ways to get rid of her — even contemplating murder. Frantic comedy is tailored for Lemmon.
1964 118m/C Jack Lemmon, Terry-Thomas, Virna Lisi, Eddie Mayehoff, Sidney Blackmer, Claire Trevor, Mary Wickes, Jack Albertson; *Dir:* Richard Quine. **VHS, Beta $19.95** MGM 𝄞𝄞𝄞

How to Pick Up Girls Based on Eric Weber's best-seller, this film tells of the small-town boy who moves to New York City and finds the secret to picking up women. Made for television.
1978 93m/C Desi Arnaz Jr., Bess Armstrong, Fred McCarren, Polly Bergen, Richard Dawson, Alan King, Abe Vigoda, Deborah Raffin; *Dir:* Bill Persky. **VHS, Beta $19.98** TWE 𝄞

How to Seduce a Woman A playboy attempts to bed five supposedly unattainable women.
1974 108m/C Angus Duncan, Marty Ingels, Lillian Randolph. **VHS, Beta $39.95** PSM *Woof!*

How to Stuff a Wild Bikini Sixth in the too-long tradition of Frankie and Annette do the beach thing movies. This one features a pregnant Funicello (though this is hidden and not part of the plot) and Brian Wilson of the Beach Boys. Avalon actually has only a small role as the jealous boyfriend trying to see if Annette will remain faithful.
1965 90m/C Frankie Avalon, Annette Funicello, Dwayne Hickman, Beverly Adams, Buster Keaton, Harvey Lembeck, Mickey Rooney, Brian Donlevy; *Dir:* William Asher. **VHS, Beta $19.98** WAR, OM 𝄞½

How to Succeed in Business without Really Trying Classic musical comedy about a window-washer who charms his way to the top of a major company. Robert Morse repeats his Tony winning Broadway role. Loosely based on a non-fiction book of the same title by Shepherd Mead, whose book Morse purchases on his first day of work. Excellent transfer of stage to film, with choreography by Moreda expanding Bob Fosse's original plan. Fine cameo performances. Songs include "Coffee Break," "A Secretary is Not a Toy," "Grand Old Ivy," and the title. Dynamite from start to finish.
1967 121m/C Robert Morse, Michele Lee, Rudy Vallee, Anthony Teague, George Fenneman, Maureen Arthur; *W/Dir:* David Swift. **VHS, LV $19.95** MGM 𝄞𝄞𝄞½

How Sweet It Is! Typical American family takes a zany European vacation. Looks like a TV sitcom and is about as entertaining. National Lampoon's version is a lot more fun.
1968 99m/C James Garner, Debbie Reynolds, Maurice Ronet, Paul Lynde, Erin Moran, Marcel Dalio, Terry-Thomas; *Dir:* Jerry Paris. **VHS $59.95** WAR 𝄞

How the West was Won A panoramic view of the American West, focusing on the trials, tribulations and travels of three generations of one family, set against the

background of wars and historical events. Particularly notable for its impressive cast list and expansive western settings. Oscar nominations for Best Picture, Cinematography, Art Direction and Set Decoration, Original Score and Costume Design, in addition to the awards it won.
1962 165m/C John Wayne, Carroll Baker, Lee J. Cobb, Spencer Tracy, Gregory Peck, Karl Malden, Robert Preston, Eli Wallach, Henry Fonda, George Peppard, Debbie Reynolds, Carolyn Jones, Richard Widmark, James Stewart, Walter Brennan, Andy Devine, Raymond Massey, Agnes Moorehead, Henry Morgan, Thelma Ritter, Russ Tamblyn; *Dir:* John Ford, Henry Hathaway, George Marshall. Academy Awards '63: Best Film Editing, Best Sound, Best Story & Screenplay; National Board of Review Awards '63: 10 Best Films of the Year. **VHS, Beta, LV $19.98** MGM 𝄞𝄞𝄞

Howard the Duck A big-budget Lucasfilm adaptation of the short-lived Marvel comic book about an alien, resembling a cigar-chomping duck, who gets sucked into a vortex, lands on Earth and saves it from the Dark Overlords. While he's at it, he befriends a nice young lady in a punk rock band and starts to fall in love. A notorious box-office bomb and one of the '80s' worst major films, although it seems to work well with children, given the simple story line and good special effects.
1986 (PG) 111m/C Lea Thompson, Jeffrey Jones, Tim Robbins; *Dir:* Willard Huyck. **VHS, Beta, LV $79.95** MCA 𝄞

The Howards of Virginia A lavish Hollywood historical epic, detailing the adventures of a backwoodsman and the rich Virginia girl he marries as their families become involved in the Revolutionary War.
1940 117m/B Cary Grant, Martha Scott, Cedric Hardwicke, Alan Marshal, Richard Carlson, Paul Kelly, Irving Bacon, Tom Drake, Anne Revere, Ralph Byrd, Alan Ladd; *Dir:* Frank Lloyd. **VHS, Beta $69.95** COL 𝄞𝄞½

Howdy Doody Clarabell and Buffalo Bob show movies of Clarabell's recent trip. Princess Summerfall Winterspring and Zippy the Chimp, Flubadub, Dilly Dally, Inspector John, Mr. Bluster, and others also appear. Two complete shows.
19?? 49m/B Buffalo Bob Smith. **VHS, Beta $24.95** NOS, VYY, DVT 𝄞𝄞½

Howie From Maui - Live! Comedian Howie Mandel performs some of his best routines in a live Hawaiian setting. First seen on cable.
1989 60m/C Howie Mandel. **VHS, Beta $14.99** HBO, TLF 𝄞𝄞½

Howie Mandel's North American Watusi Tour Frenetic comedian Howie Mandel keeps the audience in stitches with his zaniness and energetic hilarity.
1986 52m/C Howie Mandel. **VHS, LV $29.95** PAR 𝄞𝄞½

The Howling A pretty television reporter takes a rest at a clinic and discovers slowly that its denizens are actually werewolves. An in-joke-crammed, highly Dantean horror comedy that pioneered the use of the body-altering prosthetic make-up (by Rob Bottin) now essential for on-screen man-to-wolf transformations. Co-scripted by John Sayles. At last count, followed by five sequels.
1981 (R) 91m/C Dee Wallace Stone, Patrick Macnee, Dennis Dugan, Christopher Stone, Belinda Balaski, Kevin McCarthy, John Carradine, Slim Pickens, Elisabeth Brooks, Robert Picardo, Dick Miller; *Dir:* Joe Dante. **VHS, Beta, LV $14.95** COL, SUE 𝄞𝄞

Howling 2: Your Sister Is a Werewolf
A policeman (the brother of one of the first film's victims) investigates a Transylvanian werewolf-ridden castle and gets mangled for his trouble. Followed by four more sequels.
1985 (R) 91m/C *FR IT* Sybil Danning, Christopher Lee, Annie McEnroe, Marsha Hunt, Reb Brown, Ferdinand "Ferdy" Mayne; *Dir:* Philippe Mora. **VHS, Beta, LV $14.99** *HBO* ⅋

Howling 3
Subtitled "The Marsupials;" Australians discover a pouch-laden form of lycanthrope. Third "Howling" (surprise), second from Mora and better than his first.
1987 (PG-13) 94m/C *AU* Barry Otto, Imogen Annesley, Dasha Blahova, Max Fairchild, Ralph Cotterill, Leigh Biolos, Frank Thring, Michael Pate; *Dir:* Philippe Mora. **VHS, Beta, LV $19.95** *IVE* ⅋⅋

Howling 4: The Original Nightmare
A woman novelist hears the call of the wild while taking a rest cure in the country. This werewolf tale has nothing to do with the other sequels.
1988 (R) 94m/C Romy Windsor, Michael T. Weiss, Anthony Hamilton, Susanne Severeid, Lamya Derval; *Dir:* John Hough. **VHS, Beta, LV $14.95** *IVE* ⅋

Howling 5: The Rebirth
The fifth in the disconnected horror series, in which a varied group of people stranded in a European castle are individually hunted down by evil.
1989 (R) 99m/C Philip Davis, Victoria Catlin, Elizabeth She, Ben Cole, William Shockley; *Dir:* Neal Sundstrom. **VHS, Beta, LV $14.95** *IVE* ⅋

Howling 6: The Freaks
Hideous werewolf suffers from multiple sequels, battles with vampire at freak show, and discovers that series won't die. Next up: "Howling 7: The Nightmare that Won't Go Away."
1990 (R) 102m/C Brendan Hughes, Michelle Matheson, Sean Gregory Sullivan, Antonio Fargas, Carol Lynley, Jered Barclay, Bruce Martin Payne; *Dir:* Hope Perello. **VHS, LV $89.95** *LIV, IME* ⅋¹/₂

The Huberman Festival
This series of five concerts features the Israeli Philharmonic and seven prominent violinists performing works by Tchaikovsky, Bach, and Vivaldi. Available in VHS and Beta Hi-Fi.
1984 45m/C VHS, Beta, LV $19.95 *PAV, PIA* ⅋⅋

Huckleberry Finn
The musical version of the Mark Twain story about the adventures a young boy and a runaway slave encounter along the Mississippi River.
1974 118m/C Jeff East, Paul Winfield, Harvey Korman, David Wayne, Arthur O'Connell, Gary Merrill, Natalie Trundy, Lucille Benson, Farrah Fawcett; *Dir:* J. Lee Thompson. **VHS, Beta $19.98** *FOX* ⅋⅋

Huckleberry Finn
Whitewashed version of the Twain classic features post Opie Howard as the incredible Huck and Dano as the sharp witted short storyist. Made for TV. Stars four members of the Howard clan.
1975 (G) 74m/C Ron Howard, Donny Most, Royal Dano, Antonio Fargas, Jack Elam, Merle Haggard, Rance Howard, Jean Howard, Clint Howard, Shug Fisher, Sarah Selby, Bill Erwin; *Dir:* Robert Totten. **VHS, Beta $14.98** *KUI, FXV, FCT* ⅋⅋¹/₂

Huckleberry Finn
A version of the classic Mark Twain novel about the adventures a young boy and a runaway slave encounter as they travel down the Mississippi River.
1981 72m/C VHS, Beta $59.98 *LIV, LTG* ⅋¹/₂

The Hucksters
Account of a man looking for honesty and integrity in the radio advertising world and finding little to work with. All performances are excellent; but Greenstreet, as the tyrannical head of a soap company, is a stand out. Deborah Kerr's American debut.
1947 115m/B Clark Gable, Deborah Kerr, Sydney Greenstreet, Adolphe Menjou, Ava Gardner, Keenan Wynn, Edward Arnold; *Dir:* Jack Conway. **VHS, Beta $19.98** *MGM, FCT* ⅋⅋⅋

Hud
Newman is a hard-driving, hard-drinking, woman-chasing young man whose life is a revolt against the principles of stern father Douglas. Neal is outstanding as the family housekeeper. Excellent photography. Oscar nominations for Newman, director Ritt, Adapted Screenplay, Art Direction and Set Decoration.
1963 112m/B Paul Newman, Melvyn Douglas, Patricia Neal, Brandon de Wilde; *Dir:* Martin Ritt. Academy Awards '63: Best Actress (Neal), Best Black and White Cinematography, Best Supporting Actor (Douglas); New York Film Critics Awards '63: Best Actress (Neal). **VHS, Beta, LV $69.95** *PAR, BTV* ⅋⅋⅋⅋

Hudson Hawk
A big-budget star vehicle with little else going for it. Willis plays a master burglar released from prison, only to find himself trapped by the CIA into one last theft. Everyone in the cast tries to be extra funny, resulting in a disjointed situation where no one is. Weakly plotted, poorly paced.
1991 (R) 95m/C Bruce Willis, Danny Aiello, Andie MacDowell, James Coburn, Sandra Bernhard, Richard E. Grant, Frank Stallone; *Dir:* Michael Lehmann. **VHS, Beta, LV, 8mm $19.95** *COL, PIA* ⅋

Hue and Cry
Nobody believes a young boy when he discovers crooks are sending coded messages in a weekly children's magazine. A detective writer finally believes his story and they set off to capture the crooks.
1947 82m/B *GB* Alastair Sim, Jack Warner, Frederick Piper, Jack Lambert, Joan Dowling; *Dir:* Charles Crichton. **VHS $16.95** *SNC, NOS, FCT* ⅋⅋⅋

Hughes & Harlow: Angels in Hell
The story of the romance that occurred between Howard Hughes and Jean Harlow during the filming of "Hell's Angels" in 1930.
1977 (R) 94m/C Lindsay Bloom, Victor Holchak, David McLean, Royal Dano, Adam Roarke, Linda Cristal; *Dir:* Larry Buchanan. **VHS, Beta $39.95** *MON* ⅋

Hugo the Hippo
Animated musical about a forlorn baby hippo who struggles to survive in the human jungle of old Zanzibar.
1976 (G) 90m/C *Voices:* Paul Lynde, Burl Ives, Robert Morley, Marie Osmond. **VHS, Beta $49.98** *FOX* ⅋⅋

Hula
Unflappable flapper Bow and her gang frolic on a Pacific island without the benefit of sound FX.
1928 64m/B Clara Bow, Clive Brook, Patricia Dupont, Arnold Kent, Agastino Borgato; *Dir:* Victor Fleming. **VHS, Beta $24.95** *GPV, FCT, DNB* ⅋¹/₂

Hullabaloo Over Georgie & Bonnie's Pictures
A light-hearted comedy about valuable paintings in a Rajah's possession and the Englishman who will stop at nothing to get them. Written by Ruth Prawer Jhabvala.
1978 85m/C *GB* Victor Banerjee, Aparna Sen, Larry Pine, Saeed Jaffrey, Peggy Ashcroft; *Dir:* James Ivory. **VHS, Beta $29.98** *SUE* ⅋⅋¹/₂

Human Beasts
Jewel robbers face a tribe of hungry cannibals.
1980 90m/C *SP JP* Paul Naschy, Eiko Nagashima; *W/Dir:* Paul Naschy. **VHS, Beta $19.95** *ASE* ⅋

The Human Comedy
A small-town boy experiences love and loss and learns the meaning of true faith during WWII. Straight, unapologetically sentimental version of the William Saroyan novel. Oscar nominations: Best Picture, Best Actor (Rooney), Best Director, Black & White Cinematography.
1943 117m/B Mickey Rooney, Frank Morgan, James Craig, Fay Bainter, Ray Collins, Donna Reed, Van Johnson, Barry Nelson, Robert Mitchum; *Dir:* Clarence Brown. Academy Awards '43: Best Story. **VHS, Beta $24.95** *KUI, MGM* ⅋⅋⅋¹/₂

The Human Condition: A Soldier's Prayer
A Japanese pacifist escapes from his commanders and allows himself to be captured by Russian troops, hoping for better treatment as a P.O.W. The final part of the trilogy preceded by "Human Condition: No Greater Love" and "Human Condition: Road to Eternity." In Japanese with English subtitles.
1961 190m/B *JP* Tatsuya Nakadai, Michiyo Aratama, Yusuke Kawazu, Tamao Nakamura, Chishu Ryu; *Dir:* Masaki Kobayashi. **VHS, Beta $59.95** *SVS* ⅋⅋⅋

The Human Condition: No Greater Love
First of a three-part series of films. A pacifist is called into military service and subsequently sent to a Siberian military camp. A gripping look at one man's attempt to retain his humanity in the face of war. Followed by "Human Condition: Road to Eternity" and "Human Condition: A Soldier's Prayer." In Japanese with English subtitles.
1958 200m/B *JP* Tatsuya Nakadai, Michiyo Aratama, So Yamamura, Eitaro Ozawa, Akira Ishihama; *Dir:* Masaki Kobayashi. **VHS, Beta $59.95** *SVS* ⅋⅋⅋

The Human Condition: Road to Eternity
A Japanese pacifist is on punishment duty in Manchuria where he is beaten by sadistic officers who try to destroy his humanity. The second part of the trilogy, preceded by "Human Condition: No Greater Love" and followed by "Human Condition: A Soldier's Prayer." In Japanese with English subtitles.
1959 180m/B *JP* Tatsuya Nakadai, Michiyo Aratama, Kokinji Katsura, Jun Tatara, Michio Minami, Keiji Sada; *Dir:* Masaki Kobayashi. **VHS, Beta $59.95** *SVS* ⅋⅋⅋

Human Desire
Femme fatale Grahame and jealous husband Crawford turn out the lights on Crawford's boss (seems there's some confusion about how Miss Gloria managed to convince the guy to let her hubby have his job back). Ford's wise to them but plays see no evil because he's gone crackers for Grahame, who by now has decided she'd like a little help to rid herself of an unwanted husband. Bleak, melodramatic, full of big heat, its Lang through and through. Based on the Emile Zola novel "La bete humaine," which inspired Renoir's 1938 telling as well.
1954 90m/B Glenn Ford, Gloria Grahame, Broderick Crawford, Edgar Buchanan, Kathleen Case; *Dir:* Fritz Lang. **VHS** *GKK* ⅋⅋¹/₂

The Human Duplicators
Kiel is an alien who has come to Earth to make androids out of important folk, thus allowing the "galaxy beings" to take over. Cheap stuff, but earnest performances make this more fun than it should be.

1964 82m/C George Nader, Barbara Nichols, George Macready, Dolores Faith, Hugh Beaumont, Richard Kiel, Richard Arlen; *Dir:* Hugo Grimaldi. **VHS, Beta** $39.95 IVE ♂½

Human Experiments A psychiatrist in a women's prison conducts a group of experiments in which he destroys the "criminal instinct" in the inmates through brute fear.
1979 (R) 82m/C Linda Haynes, Jackie Coogan, Aldo Ray; *Dir:* Gregory Goodell. **VHS, Beta** $39.95 VID ♂

The Human Factor Kennedy stars as a computer expert who tracks down his family's murderers using technology. Bloody and violent.
1975 (R) 96m/C GB IT George Kennedy, John Mills, Raf Vallone, Rita Tushingham, Barry Sullivan, Arthur Franz; *Dir:* Edward Dmytryk. **VHS, Beta** $39.95 IVE ♂½

Human Gorilla A thriller involving a mad scientist, an insane asylum, a reporter, and various other mysterious trappings. Also known as, "Behind Locked Doors."
1948 58m/B Richard Carlson, Lucille Bremer, Tor Johnson. **VHS, Beta** $19.95 NO ♂

Human Hearts A creaky old silent starring "Phantom of the Opera" leading lady Philbin. A family is torn apart by a con man and his femme fatale.
1922 99m/B House Peters, Russell Simpson, Mary Philbin. **VHS, Beta, 8mm** $29.95 VYY ♂♂

The Human Monster Scotland Yard inspector investigates five drownings of the blind patients Lugosi's character is exploiting for insurance money. Originally titled "Dark Eyes of London."
1939 73m/B GB Bela Lugosi, Hugh Williams, Greta Gynt; *Dir:* Walter Summers. **VHS, Beta** $19.95 NOS, MRV, SNC ♂½

The Human Shield A man risks his life on a mission to save his brother during the Persian Gulf War.
1992 (R) 88m/C Michael Dudikoff, Tommy Hinkley, Steve Inwood; *Dir:* Ted Post. **VHS, LV** $89.99 WAR ♂♂

The Human Vapor A normal human being has the ability to vaporize himself at will and use it to terrorize Tokyo.
1968 81m/C JP **VHS, Beta** $19.95 GEM, PSM ♂

The Humanoid On a distant Earth outpost, a scientist has created a beautiful, childlike android named Antoinette. Their peaceful world is disrupted by an evil corporate magnate in search of the ultimate weapon. Will the sexy, synthetic woman find the humanity she so desperately craves before her world comes crashing down around her? In Japanese with English subtitles.
1991 45m/C JP *Dir:* Shin-Ichi Masaki. **VHS** $34.95 CPM ♂♂

Humanoid Defender A made-for-television film about a scientist and an android rebelling against the government that wants to use the android as a warfare prototype.
1985 94m/C Terence Knox, Gary Kasper, Aimee Eccles, William Lucking; *Dir:* William Lucking. **VHS, Beta** $39.95 MCA ♂♂

Humanoid Woman An animated adventure about a female android battling evil on the planet of Dessa.
1981 100m/C **VHS, Beta** $39.98 CEL ♂♂

Humanoids from the Deep Mutated salmon-like monsters rise from the depths of the ocean and decide to chop on some bikinied babes. Violent and bloody.

1980 (R) 81m/C Doug McClure, Ann Turkel, Vic Morrow, Cindy Weintraub, Anthony Penya, Denise Balik; *Dir:* Barbara Peeters. **VHS, Beta** WAR, OM Woof!

Humongous Seemingly brain dead bevy of teens shipwreck on an island where they encounter a deranged mutant giant who must kill to survive. A humongous bore.
1982 (R) 93m/C CA Janet Julian, David Wallace, Janet Baldwin; *Dir:* Paul Lynch. **VHS, Beta** $14.98 SUE ♂

Humoresque A talented but struggling young musician finds a patron in the married, wealthy, and older Crawford. His appreciation is not as romantic as she hoped. Stunning performance from Crawford, with excellent supporting cast, including a young Robert Blake. Fine music sequences (Isaac Stern dubbed the violin), and lush production values. Laserdisc version features the original movie trailer.
1947 123m/B Joan Crawford, John Garfield, Oscar Levant, J. Carroll Naish, Joan Chandler, Tom D'Andrea, Peggy Knudsen, Ruth Nelson, Craig Stevens, Paul Cavanagh, Richard Gaines, John Abbott, Bobby Blake; *Dir:* Jean Negulesco. **VHS, LV** $19.98 MGM, BTV, PIA ♂♂♂

The Hunchback of Notre Dame The first film version of Victor Hugo's novel about the tortured hunchback bellringer of Notre Dame Cathedral, famous for the contortions of Lon Chaney's self-transformations via improvised makeup. Silent.
1923 100m/B Lon Chaney Sr., Patsy Ruth Miller, Norman Kerry, Ernest Torrance; *Dir:* Wallace Worsley. **VHS, LV** $19.98 KIV, MRV, NOS ♂♂♂

The Hunchback of Notre Dame The best Hollywood version of the Victor Hugo classic, infused with sweep, sadness and an attempt at capturing a degree of spirited, Hugoesque detail. Laughton plays Quasimodo, a deformed Parisian bellringer, who provides sanctuary to a young gypsy woman accused by church officials of being a witch. The final scene of the townspeople storming the cathedral remains a Hollywood classic. Great performances all around; the huge facade of the Notre Dame cathedral was constructed on a Hollywood set for this film. Remake of several earlier films, including 1923's Lon Chaney silent. Remade in 1957 and later for television.
1939 117m/B Charles Laughton, Maureen O'Hara, Edmond O'Brien, Cedric Hardwicke, Thomas Mitchell, George Zucco; *Dir:* William Dieterle. **VHS, Beta, LV** $19.95 RKO, MED, VID ♂♂♂♂

Hunchback of Notre Dame It's not often that a classic novel is remade into a classic movie that's remade into a made-for-TV reprise, and survives its multiple renderings. But it's not often a cast so rich in stage trained actors is assembled on the small screen. Hopkins gives a textured, pre-Hannibal Lecter interpretation of Quasimodo, the Hunchback in Hugo's eponymous novel. Impressive model of the cathedral by production designer John Stoll.
1982 (PG) 102m/C Anthony Hopkins, Derek Jacobi, Lesley-Anne Down, John Gielgud, Tim Pigott-Smith, Rosalie Crutchley, Robert Powell; *Dir:* Michael Tuchner. **VHS** $89.95 VMK ♂♂♂

Hundra A warrior queen vows revenge when her all-female tribe is slain by men. Nothing can stop her fierce vendetta—except love.
1985 96m/C IT Laurene London, John Gaffari, Ramiro Oliveros, Marissa Casel; *Dir:* Matt Cimber. **VHS, Beta** $59.95 MED ♂

A Hungarian Fairy Tale Political satire meets fantasy in a magical tale of a fatherless young boy searching for a surrogate dad. Unusual characters as well as an homage to Mozart's "The Magic Flute." In Hungarian with English subtitles.
1987 97m/B HU Arpad Vermes, Maria Varga, Frantisek Husak, Eszter Csakanyi, Szilvia Toth, Judith Pogany, Geza Balkay; *Dir:* Gyula Gazdag. **VHS** $59.95 EVD ♂♂♂

Hungarian Rhapsody In 1911, a Hungarian nobleman joins the ranks of the rebelling peasants in order to oppose his brother, who represents aristocratic repression. One of Jancso's clearest political films, and one that continues the filmmaker's experiments with long, uncut sequences and dynamic mise-en-scene. In Hungarian with English subtitles.
1978 101m/C HU Gyorgy Cserhalmi, Lajos Balaszovits, Gabor Koncz, Bertalan Solti; *Dir:* Miklos Jancso. **VHS, Beta** $59.95 FCT ♂♂♂

Hunger In the late 1800's a starving Norwegian writer, unable to sell his work, rejects charity out of pride, and retains his faith in his talent. Based on a novel "Sult" by Knut Hamsun. In Swedish with English subtitles.
1966 100m/C SW Per Oscarsson. Cannes Film Festival '66: Best Actor (Oscarsson). **VHS** $49.95 FCT ♂♂

The Hunger A beautiful 2000-year-old vampire needs new blood when she realizes that her current lover, Bowie, is aging fast. Visually sumptuous but sleepwalking modern vampire tale, complete with soft-focus lesbian love scenes between Deneuve and Sarandon. Features a letterboxed screen and original movie trailer on the laserdisc format.
1983 (R) 100m/C Catherine Deneuve, David Bowie, Susan Sarandon, Cliff De Young, Ann Magnuson, Dan Hedaya, Willem Dafoe; *Dir:* Tony Scott. **VHS, Beta, LV** $79.95 MGM, PIA ♂♂

Hungry i Reunion A reunion of stars who made the "Hungry i" San Francisco's favorite nightclub of the 50s and 60s. Includes rare footage of Lenny Bruce in performance.
1981 90m/C Bill Cosby, Phyllis Diller, Ronnie Schell, Bill Dana, Mort Sahl, Prof. Irwin Corey, Jackie Vernon, Jonathan Winters, The Kingston Trio, Limelighters, Lenny Bruce. **VHS, Beta** $59.95 PAV ♂♂

Hunk A computer nerd sells his soul to the devil for a muscular, beach-blonde physique. Answers the question, "Was it worth it?"
1987 (PG) 102m/C John Allen Nelson, Steve Levitt, Deborah Shelton, Rebecca Bush, James Coco, Avery Schreiber; *Dir:* Lawrence Bassoff. **VHS, Beta, LV** $79.95 COL ♂♂

Hunt the Man Down An innocent man is charged with murder, and a public defender must find the real killer before time runs out.
1950 68m/B Gig Young, Lynn Roberts, Gerald Mohr; *Dir:* George Archainbaud. **VHS, Beta** $34.95 RKO ♂½

Hunt the Man Down/Smashing the Rackets A double feature: In "Hunt the Man Down" (1950) a public defender tries to clear a captured fugitive's name, and in "Smashing the Rackets" (1938) a special prosecutor takes on the gangsters who have corrupted a big city.
1950 137m/B Gig Young, Lynne Roberts, Chester Morris, Frances Mercer, Willard Parker, Gerald Mohr, Paul Frees; *Dir:* George Archainbaud. **VHS, Beta** $34.95 RKO ♂♂½

The Hunt for the Night Stalker
Another one of those True-Detective quickies so beloved by network TV, following two hardworking L.A. cops who tracked down satanic serial killer Richard Ramirez in the mid-1980s. It originally aired (under the title "Manhunt: The Search for the Night Stalker") on the date of the killer's death-sentence verdict; the videotape lacks that particular timely sparkle.
1991 (PG-13) 95m/C Richard Jordan, A. Martinez, Lisa Eilbacher, Julie Carmen, Alan Feinstein; *Dir:* Bruce Seth Green. **VHS $19.95** ACA *ℤℤ*

The Hunt for Red October Based on Tom Clancy's blockbuster novel, a high-tech Cold War yarn about a Soviet nuclear sub turning rogue and heading straight for U.S. waters, as both the U.S. and the U.S.S.R. try to stop it. Complicated, ill-plotted potboiler that succeeds breathlessly due to the cast and McTiernan's tommy-gun direction. Introduces the character of CIA analyst Jack Ryan who returns in "Patriot Games," though in the guise of Harrison Ford. Oscar nominated for Best Sound, Best Sound Effects Editing, and Best Film Editing.
1990 (PG) 137m/C Sean Connery, Alec Baldwin, Richard Jordan, Scott Glenn, Joss Ackland, Sam Neill, James Earl Jones, Peter Firth, Tim Curry, Courtney B. Vance, Jeffrey Jones, Fred Dalton Thompson; *Dir:* John McTiernan. Academy Awards '90: Best Sound Effects Editing. **VHS, Beta, LV, SVS, 8mm $14.95** PAR, SUP *ℤℤℤ*

Hunted! A young Jewish boy teams up with a resistance fighter to find his mother in Nazi Germany during World War II.
1979 90m/C VHS, Beta $29.95 GEM *ℤℤ*

Hunted WWII prisoner attempts to escape from demented officer who seems to enjoy bugging him. Hunted by the MPs, he relies on unorthodox guide to sustain feature length footage.
1988 (PG) 75m/C Andrew Buckland, Richard Carlson, Ron Smerczak, Mercia Van Wyk; *Dir:* David Lister. **VHS $34.95** ACA *ℤℤ*

The Hunted Lady In this television pilot, Mills is an undercover policewoman who is framed by the mob. Tired and predictable.
1977 100m/C Donna Mills, Robert Reed, Lawrence Casey, Andrew Duggan, Will Sampson, Alan Feinstein; *Dir:* Richard Lang. **VHS $39.95** WOV *ℤ½*

Hunter Pilot for a television series that never emerged. A government agent uncovers an enemy brainwashing scam.
1973 90m/C John Vernon, Steve Ihnat, Fritz Weaver, Edward Binns, Sabrina Scharf, Barbara Rhoades; *Dir:* Leonard Horn. **VHS** LHV *ℤ½*

Hunter An attorney is falsely accused of a crime and imprisoned. Released after eight years, he sets out to even the score with the mysterious millionaire who set him up.
1977 120m/C James Franciscus, Linda Evans, Broderick Crawford. **VHS, Beta $59.95** LHV, WAR *ℤ*

The Hunter An action-drama based on the real life adventures of Ralph (Papa) Thorson, a modern day bounty hunter who makes his living by finding fugitives who jumped bail. McQueen's last film.
1980 (PG) 97m/C Steve McQueen, Eli Wallach, Kathryn Harrold, LeVar Burton; *Dir:* Buzz Kulik. **VHS, Beta, LV $19.95** PAR *ℤ½*

Hunter in the Dark Set in 18th-century Japan, where power, betrayal, and corruption are the dark side of the samurai world. In Japanese with yellow English subtitles.
1980 138m/C JP Tatsuya Nakadai, Tetsuro Tamba, Sonny Chiba; *Dir:* Hideo Gosha. **VHS $79.95** WAC *ℤℤ*

The Hunters A motley crew of pilots learn about each other and themselves in this melodrama set during Korean war. Incredible aerial photography sets this apart from other films of the genre.
1958 108m/C Robert Mitchum, Robert Wagner, Richard Egan, May Britt, Lee Phillips; *Dir:* Dick Powell. **VHS** TCF *ℤℤ½*

Hunter's Blood Five urbanites plunge into the Southern wilderness to hunt deer, and are stalked by maniacal hillbillies.
1987 (R) 102m/C Sam Bottoms, Kim Delaney, Clu Gulager, Mayf Nutter; *Dir:* Robert C. Hughes. **VHS, Beta, LV $19.98** SUE *ℤ½*

Hunters of the Golden Cobra Two American soldiers plot to recover the priceless golden cobra from the Japanese general who stole the prized relic during the last days of WWII.
1982 (R) 95m/C David Warbeck, Almanta Suska, Alan Collins, John Steiner; *Dir:* Anthony M. Dawson. **VHS, Beta $69.98** LIV, VES *ℤ*

The Hunting A married woman falls for a ruthless businessman and is drawn into a web of blackmail and murder in this low-budget erotic thriller.
1992 (R) 97m/C John Savage, Kerry Armstrong, Rebecca Rigg; *W/Dir:* Frank Howson. **VHS, Beta** PAR *ℤℤ*

Hunting the White Rhino Peter Capstick gives us a history of the near extinction of the White Rhino and how the tranplantation program brought them back to a point where they could be hunted once again. He helps a client stalk one and finally bring it down.
1988 70m/C VHS, Beta $29.95 SOF *ℤℤ*

Hurray for Betty Boop Betty Boop gets herself into the true spirit election year by running for President.
1980 81m/C Voices: Tom Smothers, Victoria D'Orzai. **VHS, Beta $19.98** WAR, OM *ℤ½*

Hurricane A couple on the run from the law are aided by a hurricane and are able to build a new life for themselves on an idyllic island. Filmed two years before the Academy's "special effects" award came into being, but displaying some of the best effects of the decade. Oscar nominated for Best Actor (Mitchell). Boringly remade in 1979.
1937 102m/B Jon Hall, Dorothy Lamour, Mary Astor, C. Aubrey Smith, Raymond Massey; *Dir:* John Ford. Academy Awards '37: Best Sound. **VHS, Beta, LV $14.98** SUE *ℤℤℤ*

Hurricane Television's answer to the disaster movie craze of the early 1970s. Realistic hurricane footage and an adequate cast cannot save this catastrophe.
1974 78m/C Larry Hagman, Martin Milner, Jessica Walter, Barry Sullivan, Will Geer, Frank Sutton; *Dir:* Jerry Jameson. **VHS $39.98** CGI *ℤ*

Hurricane Robards is the governor of a tropical island with Farrow as his daughter who falls for a native. Very expensive, very mediocre remake of the 1937 film. Even the hurricane sequences are dull in comparison.
1979 (PG) 120m/C Mia Farrow, Jason Robards Jr., Trevor Howard, Max von Sydow, Timothy Bottoms, James Keach; *Dir:* Jan Troell. **VHS, Beta $14.95** PAR *ℤ*

Hurricane Express Twelve episodes of the vintage serial, in which the Duke pits his courage against an unknown, powerful individual out to sabotage a railroad.
1932 223m/B John Wayne, Joseph Girard, Conway Tearle, Shirley Grey; *Dir:* J.P. McGowan, Armand Schaefer. **VHS, Beta $16.95** SNC, NOS, RHI *ℤℤ½*

Hurricane Smith Weathers stars as a roughneck Texan who travels to Australia's Gold Coast in search of his missing sister. While in the land of Oz, he gets entangled in a Mafia-style drug and prostitution ring around the "Surfer's Paradise" section of the Gold Coast. Packed with mind-blowing stunts and a good performance from Weathers, this fast-paced action thriller won't disappoint fans of this genre.
1992 (R) 86m/C Carl Weathers, Cassandra Delaney, Tony Bonner, Juergen Prochnow; *Dir:* Colin Budds. **VHS $94.99** WAR *ℤ*

Hurricane Sword Thriller sets savage sword play against a background of murder, betrayal, prostitution and reunion.
197? 86m/C Chen Sau Kei, Li Tai Shing. **VHS, Beta $54.95** MAV *ℤ½*

The Hurried Man An international art smuggler like to live life to the extreme. But his obsessions, and a beautiful woman, just may lead to death.
1977 91m/C FR Alain Delon, Christian Barbier, Andre Falcon, Stefano Patrizi. **VHS, Beta $59.95** UNI *ℤ*

Hurry, Charlie, Hurry A husband gets in trouble when he bows out of a trip with his social-climber wife when he tells her he must travel to Washington to see the Vice President.
1941 65m/B Leon Errol. **VHS, Beta $34.95** RKO, MLB *ℤ*

Hurry Up or I'll Be Thirty A Brooklyn bachelor celebrates his thirtieth birthday by becoming morose, depressed, and enraged. His friends try to help; they fail, but he finds love anyway. For those seeking a better life through celluloid.
1973 87m/C Danny DeVito, John Lefkowitz, Steve Inwood, Linda DeCoff, Ronald Anton, Maureen Byrnes, Francis Gallagher; *Dir:* Joseph Jacoby. **VHS, Beta $69.98** LIV, VES *ℤℤ*

Husband Hunting Working girls hit the beach in search of husbands, and find that love is easy to find or hold on to.
196? 90m/C IT Walter Chiari. **VHS, Beta** INV *ℤ*

Husbands and Lovers Controversial film that focuses on a couple's untraditional marriage. Stephen (Sands) and Alina (Pacula) have agreed to be totally honest with each other and when Alina decides to take a lover, Stephen agrees on the one condition that she report back to him every intimate detail of her affair with Paolo. In the confusion and excitement of this kinky love triangle, everyone starts losing control as sexual boundaries are pushed to the limit. Also available in an unrated version.
1992 (R) 91m/C Julian Sands, Joanna Pacula, Tcheky Karyo; *Dir:* Mauro Bolognini. **VHS $89.95** COL *ℤℤ*

Husbands, Wives, Money, and Murder A couple is pushed over the edge by a nosey census taker and then must dispose of the consequences.

1986 92m/C Garrett Morris, Greg Mullavey, Meredith McRae, Timothy Bottoms; *Dir:* Bruce Cook Jr. **VHS, Beta** $79.95 *TWE, HHE* 🎟

Hush, Hush, Sweet Charlotte A fading southern belle finds out the truth about her fiance's murder when the case is reopened thirty-seven years later by her cousin in an elaborate plot to drive her crazy. Grisly, superbly entertaining Southern Gothic horror tale, with vivid performances from the aging leads. Moorehead received an Oscar nomination for her supporting actress performance.
1965 134m/B Bette Davis, Olivia de Havilland, Joseph Cotten, Agnes Moorehead, Mary Astor, Bruce Dern; *Dir:* Robert Aldrich. Edgar Allan Poe Awards '64: Best Screenplay. **VHS, Beta, LV** $19.98 *FOX* 🎟🎟🎟

Hush Little Baby, Don't You Cry A made-for-television drama about a seemingly normal middle-aged family man who is in actuality a child abuser/murderer.
1986 110m/C Emery L. Kedocia, Gary Giem, Tony Grant, Burt Douglas; *Dir:* Don Hawks. **VHS, Beta** $79.95 *PSM* 🎟

Hussy A hooker and her boyfriend find themselves trapped in a web of gangsters and drugs.
1980 (R) 95m/C *GB* Helen Mirren, John Shea, Jenny Runacre; *Dir:* Matthew Chapman. **VHS, Beta, LV** $69.95 *VES* 🎟

Hustle Burt Reynolds plays a detective investigating a young girl's supposed suicide who becomes romantically entangled with a high-priced call girl.
1975 (R) 120m/C Burt Reynolds, Catherine Deneuve; *Dir:* Robert Aldrich. **VHS, Beta** $24.95 *PAR* 🎟½

The Hustler The original story of Fast Eddie Felsen and his adventures in the seedy world of professional pool. Newman plays the naive, talented and self-destructive Felsen perfectly, Laurie is oustanding as his lover, and Gleason epitomizes the pool great Minnesota Fats. Rivetingly atmospheric, and exquisitely photographed. Parent to the reprise "The Color of Money," made 25 years later. Academy Award nominations for Best Picture, actor Newman, supporting actors Gleason and Scott, actress Laurie, director, and screenplay.
1961 134m/B Paul Newman, Jackie Gleason, Piper Laurie, George C. Scott, Myron McCormick; *Dir:* Robert Rossen. Academy Awards '61: Best Art Direction/Set Decoration (B & W), Best Black and White Cinematography; British Academy Awards '61: Best Actor (Newman); National Board of Review Awards '61: 10 Best Films of the Year, Best Supporting Actor (Gleason). **VHS, Beta, LV** $19.98 *FOX, FCT* 🎟🎟🎟

Hustler Squad A U.S. Army major and a Philippine guerrilla leader stage a major operation to help rid the Philippines of Japanese Occupation forces: they have four combat-trained prostitutes infiltrate a brothel patronized by top Japanese officers.
1976 (R) 98m/C John Ericson, Karen Ericson, Lynda Sinclaire, Nory Wright; *Dir:* Ted V. Mikels. **VHS, Beta** $19.95 *UHV, MRV* 🎟

Hustling A reporter writing a series of articles on prostitution in New York City takes an incisive look at their unusual and sometimes brutal world. Notable performance by Remick as the reporter and Clayburgh as a victimized hooker. Made for TV and based on a novel by Gail Sheehy.
1975 96m/C Jill Clayburgh, Lee Remick, Alex Rocco, Monte Markham; *Dir:* Joseph Sargent. **VHS, Beta** $49.95 *WOV* 🎟🎟🎟

Hyper-Sapien: People from Another Star Two cuddly aliens run away from home, are befriended by a lonely farmboy, and are chased all over the wilds of Wyoming.
1986 (PG) 93m/C Sydney Penny, Keenan Wynn, Gail Strickland, Ricky Paull Goldin, Peter Jason, Talia Shire; *Dir:* Peter Hunt. **VHS, Beta, LV** $19.98 *WAR* 🎟½

Hypothesis of the Stolen Painting An art collector guides an interviewer around six paintings by Frederic Tonnerre, an academic 19th century painter, in an attempt to solve the mysterious disappearance of a seventh painting. The film presents the paintings as tableaux vivants in which the actors hold poses as they are examined. In French with English subtitles. Based on the novel "Baphomet" by Pierre Klossowski.
1978 78m/B *W/Dir:* Raul Ruiz. **VHS** $59.95 *WAC* 🎟🎟

Hysteria When an American becomes involved in an accident and has amnesia, a mysterious benefactor pays all his bills and gives the man a house to live in. But a series of murders could mean he's the murderer or the next victim.
1964 (PG) 85m/B *GB* Robert Webber, Susan Lloyd, Maurice Denham; *Dir:* Freddie Francis. **VHS, Beta** $59.95 *MGM, MLB* 🎟½

Hysterical Odd little attempt at a horror flick parody features the Hudson Brothers and involves a haunted lighthouse occupied by the vengeful spirit of a spurned woman.
1983 (PG) 86m/C Brett Hudson, William Hudson, Mark Hudson, Cindy Pickett, Richard Kiel, Julie Newmar, Bud Cort; *Dir:* Chris Bearde. **VHS, Beta, LV** $9.98 *SUE* 🎟

I Accuse My Parents A juvenile delinquent tries to blame a murder and his involvement in a gang of thieves on his mom and dad's failure to raise him properly. Wonder if the producer blamed lukewarm reception on his parents too.
1945 70m/B Mary Beth Hughes, Robert Lowell, John Miljan. **VHS, Beta** $16.95 *SNC, NOS, DVT* 🎟½

I Am a Camera A young English writer develops a relationship with a reckless young English girl in Berlin during the 1930s. Based on the Berlin stories by Christopher Isherwood, later musicalized as "Cabaret."
1955 99m/B *GB* Julie Harris, Shelley Winters, Laurence Harvey, Patrick McGoohan; *Dir:* Henry Cornelius. **VHS, Beta** $24.95 *MON* 🎟🎟🎟

I Am the Cheese An institutionalized boy undergoes psychiatric treatment; with the aid of his therapist (Wagner) he relives his traumatic childhood and finds out the truth about the death of his parents. A bit muddled, but with its moments. Adapted from a Robert Cormier teen novel.
1983 95m/C Robert MacNaughton, Hope Lange, Don Murray, Robert Wagner, Sudie Bond; *Dir:* Robert Jiras. **VHS, Beta** $69.98 *KUI, LIV, VES* 🎟🎟

I Am Curious (Yellow) A woman sociologist is conducting a sexual survey on Swedish society, which leads her to have numerous sexual encounters in all sorts of places. Very controversial upon its U.S. release because of the nudity and sexual content but tame by today's standards. Followed by "I Am Curious (Blue)" which was filmed at the same time. In Swedish with English subtitles.
1967 95m/B *SW* Lena Nyman, Peter Lindgren; *Dir:* Vilgot Sjoman. **VHS** $59.95 *HTV* 🎟🎟½

I Am a Fugitive from a Chain Gang WWI veteran Muni returns home with dreams of traveling across America. After a brief stint as a clerk, he strikes out on his own. Near penniless, Muni meets up with a tramp who takes him to get a hamburger. He becomes an unwilling accomplice when the bum suddenly robs the place. Convicted and sentenced to a Georgia chain gang, he's brutalized and degraded, though he eventually escapes and lives the degrading life of a criminal on the run. Based on the autobiography by Robert E. Burns. Brutal docu-details combine with powerhouse performances to create a classic. Timeless and thought-provoking; Oscar nominated for best picture and actor (Muni).
1932 90m/B Paul Muni, Glenda Farrell, Helen Vinson, Preston Foster; *Dir:* Mervyn LeRoy. National Board of Review Awards '32: Best Film. **VHS, Beta** $19.98 *MGM, FOX, CCB* 🎟🎟🎟🎟

I Am the Law Robinson is given the task of stopping gangster activity in the city. Although regularly a law professor, he is made special prosecutor and starts canvassing for people to testify. His witnesses become murder victims as an enemy on the inside attempts to thwart the game plan. Robinson is dismissed from the case but continues to pursue justice in streetclothes. Fine performances but somewhat predictable.
1938 83m/B Edward G. Robinson, Otto Kruger, John Beal, Barbara O'Neil, Wendy Barrie, Arthur Loft, Marc Lawrence; *Dir:* Alexander Hall. **VHS, Beta, LV** $69.95 *COL, PIA* 🎟🎟🎟

I Am the Law & the Hunter An episode from each of these early television crime shows.
19?? 60m/C George Raft, Barry Nelson. **VHS, Beta** $24.95 *NOS, DVT* 🎟🎟🎟

I Beheld His Glory Story based on the experience of Cornelius, the Roman centurion who guarded Christ's tomb.
1978 53m/C George Macready, Robert Wilson, Virginia Wave. **VHS** $29.95 *REP* 🎟🎟🎟

I Bury the Living A cemetery manager sticks pins in his map of a graveyard and people mysteriously start to die. Well-done suspense film.
1958 76m/B Richard Boone, Theodore Bikel, Peggy Maurer; *Dir:* Albert Band. **VHS** $16.95 *SNC, MRV* 🎟🎟🎟

I Can't Escape A man attempts to piece his life back together after serving time for a crime he didn't commit. To prove himself to his girlfriend, he tries to break up an illegal stock scam.
1934 60m/B Onslow Stevens, Lila Lee, Russell Gleason, Otis Harlan, Hooper Atchley, Clara Kimball Young; *Dir:* Otto Brower. **VHS, Beta** $16.95 *SNC* 🎟🎟

I, Claudius, Vol. 1: A Touch of Murder/Family Affairs Award-winning Masterpiece Theatre series tells the true story of the Roman Empire and the people determined to rule it. In these first two episodes, Augustus tries to control his conniving second wife, the beautiful Livia, who is determined that her son Tiberius will be the next Roman ruler. Her manipulations lead to bloody mayhem. Brilliant acting. Based on the books by Robert Graves. Also available as a seven-volume set.
1991 120m/C *GB* Sian Phillips, Brian Blessed, Derek Jacobi; *Dir:* Herbert Wise. **VHS** $24.95 *PBS, PAV, FCT* 🎟🎟

I, Claudius, Vol. 2: Waiting in the Wings/What Shall We Do With Claudius? Continuation of the Masterpiece Theatre series portraying the true story of the Roman Empire and its rulers. In the third and fourth episodes, Tiberius and his mother Livia attempt to capture the throne with a series of brutal manipulations. Young cousin Claudius, considered an ugly dimwit, watches from afar. Based on the books of Robert Graves. Also available as a seven-volume set.
1980 120m/C Derek Jacobi, Sian Phillips, Brian Blessed; *Dir:* Herbert Wise. **VHS** $24.95 *PBS, PAV, FCT* 🐾🐾

I, Claudius, Vol. 3: Poison is Queen/Some Justice Episodes 5 and 6 see Emperor Augustus murdered and Tiberius on the throne he's coveted for so long. But how long can he stay, hiding the treachery that brought him there? Good question. Agrippina, wife of his latest victim, is determined to see him punished, and the Roman masses, not so stupid after all, have begun to see a pattern in the deaths of their best and brightest. Based on the books of Robert Graves. Also available as a seven-volume set.
1980 120m/C Derek Jacobi, Sian Phillips, Brian Blessed; *Dir:* Herbert Wise. **VHS** $24.95 *PBS, PAV, FCT* 🐾🐾

I, Claudius, Vol. 4: Queen of Heaven/Reign of Terror In episodes 7 and 8, Caligula begins his evil manipulation of Tiberius, involving him in a wide variety of amoral activities and creating a trail of blood. Based on the books of Robert Graves. Also available as a seven-volume set.
1980 120m/C Derek Jacobi, Sian Phillips. **VHS** $24.95 *PBS, PAV, FCT* 🐾🐾

I, Claudius, Vol. 5: Zeus, By Jove!/Hail Who? Caligula's schemes result in his rise to the throne after Tiberius. Debauchery rules and murder is a standard procedure that returns to remove the emperor. At last, patient Claudius, the only remaining heir, thought too foolish to be dangerous, ascends the throne. Award-winning Masterpiece Theatre production created by the BBC. Based on the books of Robert Graves. The first episode on this cassette contains the graphic scene of Caligula's evisceration of his sister, as he enacts the life of Zeus in demented belief that he is the god—be forewarned. Also available as a seven-volume set.
1980 120m/C Derek Jacobi, John Hurt; *Dir:* Herbert Wise. **VHS** $24.95 *PBS, PAV, FCT* 🐾🐾

I, Claudius, Vol. 6: Fool's Luck/A God in Colchester Episodes 11 and 12 see Emperor Claudius wise in the ways of statesmanship, but foolish in the ways of love. His young wife Messalina creates a furor when her sexual practices become known. In spite of his love, the emperor must punish her. Award-winning BBC production, shown in U.S. on Masterpiece Theatre. Based on the books of Robert Graves. Also available as a seven-volume set.
1980 120m/C Derek Jacobi; *Dir:* Herbert Wise. **VHS** $24.95 *PBS, PAV, FCT* 🐾🐾

I, Claudius, Vol. 7: Old King Log Claudius, aging and sad, resigns himself to his destiny, as his wife makes plans for his death. Claudius wants only the time to complete the journal of his life. Brilliant acting from Jacobi in this BBC production shown in the U.S. on Masterpiece Theatre. Based on

the books of Robert Graves. Also available as a seven-volume set.
1980 60m/C Derek Jacobi; *Dir:* Herbert Wise. **VHS** $24.95 *PBS, PAV, FCT* 🐾🐾

I Come in Peace A tough, maverick Texas cop embarks on a one-way ride to Nosebleed City when he attempts to track down a malevolent alien drug czar who kills his victims by sucking their brains. Mindless thrills.
1990 (R) 92m/C Dolph Lundgren, Brian Benben, Betsy Brantley, Jesse Vint, Michael J. Pollard; *Dir:* Craig R. Baxley. **VHS, Beta, LV** $89.95 *MED* 🐾🐾

I Confess When a priest (Clift) hears a murderer's confession, the circumstances seem to point to him as the prime suspect. Tepid, overly serious and occasionally interesting mid-career Hitchcock. Adapted from Paul Anthelme's 1902 play.
1953 95m/B Montgomery Clift, Anne Baxter, Karl Malden, Brian Aherne; *Dir:* Alfred Hitchcock. **VHS, Beta, LV** $19.98 *WAR, FCT, MLB* 🐾🐾½

I Conquer the Sea Two brothers, who works as whalers, cast their affection on the same woman. Drowns in its own melodrama.
1936 68m/B Steffi Duna, Stanley Morner, Dennis Morgan, Douglas Walton, George Cleveland, Johnny Pirrone; *Dir:* Victor Halperin. **VHS, Beta, 8mm** *VYY* 🐾½

I Could Go on Singing An aging American songstress, on a tour in Britain, becomes reacquainted with her illegitimate son and his British father, but eventually goes back to the footlights. Garland's last film. Songs include "By Myself" and "It Never Was You." A must for Garland fans. Letterboxed.
1963 99m/C Judy Garland, Dirk Bogarde, Jack Klugman, Aline MacMahon; *Dir:* Ronald Neame. **VHS, Beta, LV** $19.98 *MGM* 🐾🐾½

I Cover the Waterfront A reporter is assigned to write about a boatman involved in a fishy scheme to smuggle Chinese immigrants into the country wrapped in shark skins. While trying to get the story, the journalist falls in love with the fisherman's daughter. Torrance passed away before its release.
1933 70m/B Claudette Colbert, Ben Lyon, Ernest Torrance, Hobart Cavanaugh; *Dir:* James Cruze. **VHS, Beta** $19.95 *NOS, MRV, VYY* 🐾🐾½

I Died a Thousand Times Aging gangster Mad Dog Earle (pushup prince Palance) plans one last death-defying heist while hiding from police in the mountains. Meanwhile, the hard boiled gangster softens a bit thanks to surgery-needing girlfriend, but moll Winters doesn't seem to think that three's company. A low rent "High Sierra."
1955 109m/C Jack Palance, Shelley Winters, Lori Nelson, Lee Marvin, Earl Holliman, Lon Chaney Jr., Howard St. John; *Dir:* Stuart Heisler. **VHS, LV** $59.99 *WAR, PIA, FCT* 🐾🐾

I Dismember Mama Classless story of an asylum inmate who escapes to kill his mother. Although he hates women, he likes little girls as evidenced by his nine-year-old love interest. Notably lacking in bloody scenes. Also known as "Poor Albert and Little Annie."
1974 (R) 81m/C Zooey Hall, Joanne Moore Jordan, Greg Mullavey, Marlene Tracy; *Dir:* Paul Leder. **VHS, Beta** $24.95 *GEM* Woof!

I Do! I Do! This Los Angeles production of the Broadway musical covers 50 years of a marriage, beginning just before the turn of the century.

1984 116m/C Lee Remick, Hal Linden. **VHS, Beta** $39.95 *RKO* Woof!

I Don't Buy Kisses Anymore Heartwarming story starring Alexander and Peeples as two mismatched lovers who end up realizing that they're made for each other. Alexander plays Bernie Fishbine, an overweight Jewish shoe store owner who falls for a psychology graduate student (Peeples). Little does he know, but Peeples is studying him for her term paper, appropriately titled, "The Psychological Study of an Obese Male." Alexander gives a great performance, as do Kazan and Jacobi who play Bernie's parents.
1992 (PG) 112m/C Jason Alexander, Nia Peeples, Lainie Kazan, Lou Jacobi, Eileen Brennan; *Dir:* Robert Marcarelli. **VHS, Beta** *PAR* 🐾🐾½

I Don't Give a Damn An embittered wounded soldier returns home to his loved ones, rejects them, and becomes more and more uninterested in life itself. Subtitled.
1988 (R) 94m/C *IS* Ika Sohar, Anat Waxman; *Dir:* Shmuel Imberman. **VHS, Beta** $59.95 *TWE* 🐾🐾

I Don't Want to Be Born It's got all the right ingredients for overnight camp: a spurned dwarf, a large, howling baby-thing, slice and dice murder and mayhem, and Collins. It could've been so bad. Instead, this "Rosemary's Baby" rehash is just stupid bad. Also known as "The Devil Within Her."
1975 (R) 90m/C *GB* Joan Collins, Eileen Atkins, Donald Pleasence, Ralph Bates, Caroline Munro; *Dir:* Peter Sasdy. **VHS** *AXV* 🐾

I Dood It Young tailor's assistant Skelton falls hard for young actress Powell working near his shop. She agrees to date, and eventually marry him, but only to spite her boyfriend who has just run off with another woman. All's well however, when Skelton stumbles across a spy ring, is hailed a hero and heps Powell realize she really loves him. Songs include "Star Eyes," "Hola E Pae," "Swing the Jinx Away," and "Taking a Chance on Love."
1943 102m/B Red Skelton, Eleanor Powell, Richard Ainley, Patricia Dane, Sam Levene, Thurston Hall, Lena Horne, Butterfly McQueen, Jimmy Dorsey & His Orchestra; *Dir:* Vincente Minnelli. **VHS** $19.98 *MGM, CCB* 🐾🐾

I Dream of Jeannie The third and worst bio of Stephen Foster has Foster as a bookkeeper-cum-songwriter alternating between writing tunes and chasing Lawrence. When she dumps him, he goes into a funk. Will he be able to complete the title song? The suspense will kill you. Lots of singing, but not much else.
1952 90m/C Ray Middleton, Bill Shirley, Muriel Lawrence, Lynn Bari, Rex Allen; *Dir:* Allan Dwan. **VHS** $19.95 *NOS* 🐾

I Dream Too Much A musical vehicle for opera star Pons, as a French singer who marries an American composer played by Fonda. Songwriter falls into wife's shadow after pushing her into a singing career; she then raises his spirits a couple of octaves by helping him sell a musical comedy. Went into production after a rival company, Columbia, launched opera singer Grace Moore's acting career. Ball later bought this studio, where she was given her second starring role.
1935 90m/B Lily Pons, Henry Fonda, Eric Blore, Lucille Ball, Mischa Auer; *Dir:* John Cromwell. **VHS, Beta** $19.98 *HHT* 🐾🐾

I Drink Your Blood Hippie satanists looking for kicks spike an old man's drink with LSD. To get revenge, the old codger's grandson sells the nasty flower children meat pies injected with the blood of a rabid dog. The hippies then turn into cannibalistic maniacs, infecting anyone they bite. From the man responsible for "I Spit on Your Grave;" originally played on a double bill with "I Eat Your Skin."
1971 (R) 83m/C Bhasker, Jadine Wong, Ronda Fultz, Elizabeth Marner-Brooks, George Patterson, Riley Mills, Iris Brooks, John Damon; *Dir:* David E. Durston. **VHS $18.00** *FRG* 🎞½

I Eat Your Skin Cannibalistic zombies terrorize a novelist and his girlfriend on a Caribbean island. Blood and guts, usually shown on a gourmet double bill with "I Drink Your Blood." Also known as "Zombies" and "Voodoo Blood Bath."
1964 82m/B William Joyce, Heather Hewitt, Betty H. Lindon; *Dir:* Del Tenney. **VHS $9.95** *RHI, SNC, FCT Woof!*

I Give My All Animated Japanese feature for adults. Based on Japanese comic books' "love comedy" genre. A bawdy tale of a rich woman and her daughters' sex adventures. Japanese with English subtitles.
1991 45m/C *JP* **VHS $34.95** *CPM Woof!*

I Hate Blondes A group of criminals use a ghostwriter's burglaring stories as instructional manuals. The writer becomes involved with the gang. Relies heavily on visual humor. Notably among the comedy scenes is the author's search for jewels at a get-together.
1983 90m/C *IT* *Dir:* Giorgio Capitani, Giorgio Capitani. **VHS, Beta** *PGN* 🎞🎞½

I Heard the Owl Call My Name Poignant story of an Anglican priest who is relocated to an Indian fishing village on the ouskirts of Vancouver, British Columbia. Made for television and based on a Margaret Craven book.
1973 74m/C Tom Courtenay, Dean Jagger, Paul Stanley; *Dir:* Daryl Duke. **Beta $14.95** *PSM* 🎞🎞½

I, the Jury A remake of the 1953 Mike Hammer mystery in which the famed PI investigates the murder of his best friend. Assante mopes around as Hammer and looks out of place in this slowed-down version.
1982 (R) 111m/C Armand Assante, Barbara Carrera, Laurene Landon, Alan King, Geoffrey Lewis, Paul Sorvino, Jessica James, Leigh Anne Harris, Lynette Harris; *Dir:* Richard T. Heffron. **VHS, Beta $19.98** *FOX* 🎞🎞

I Killed Rasputin The "Mad Monk" who rose to power before the Russian Revolution lost his life in a bizarre assassination by Felix Youssoupoff. Film deals with the friendship between the men that ended in betrayal. Well-intentioned, but overwrought.
1967 95m/C *FR* *IT* Geraldine Chaplin, Gert Frobe, Peter McEnery; *Dir:* Robert Hossein. **VHS, Beta $59.95** *MPI* 🎞½

I Killed That Man A prisoner scheduled to die in the electric chair is given an early ticket out when he is found poisoned. Evidence points to an unusual group of suspects. Effective low-budget thriller offers a few unexpected surprises.
1942 72m/B Ricardo Cortez, Joan Woodbury, Iris Adrian, George Pembroke, Herbert Rawlinson, Pat Gleason, Ralf Harolde, Jack Mulhall, Vince Barnett, Gavin Gordon, John Hamilton; *Dir:* Phil Rosen. **VHS, Beta $16.95** *SNC* 🎞🎞½

I Know Where I'm Going A young woman (Hiller), who believes that money brings happiness, is on the verge of marrying a rich old man, until she meets a handsome naval officer (Livesey) and finds a happy, simple life. Early on the female lead appears in a dream sequence filmed in the mode of surrealist painter Salvador Deli and avant garde director Luis Bunvelo. Scottish setting and folk songs give a unique flavor. Brown provides a fine performance as a native Scot.
1947 91m/B *GB* Roger Livesey, Wendy Hiller, Finlay Currie, Pamela Brown; *Dir:* Michael Powell. **VHS, Beta $59.95** *LCA* 🎞🎞🎞

I Know Why the Caged Bird Sings A black writer's memories of growing up in the rural South during the 1930s. Strong performances from Rolle and Good. Made-for-television film is based on the book by Maya Angelou.
1979 100m/C Diahann Carroll, Ruby Dee, Esther Rolle, Roger E. Mosley, Paul Benjamin, Constance Good; *Dir:* Fielder Cook. **VHS, Beta $59.95** *KUI, USA* 🎞🎞🎞

I Like Bats Can a psychiatrist help a vampire become a human? Polish dialogue with English subtitles.
1985 90m/C *Dir:* Grzegorz Warchol. **VHS $49.95** *FCT*

I Live in Grosvenor Square Also known as "A Yank In London." An American soldier in Great Britain falls in love with a major's fiancee. Entertaining if a bit drawn out.
1946 106m/B *GB* Anna Neagle, Rex Harrison, Dean Jagger, Robert Morley, Jane Darwell; *Dir:* Herbert Wilcox. **VHS $19.95** *NOS, DVT* 🎞

I Live with Me Dad A vagrant drunk and his son fight the authorities for the right to be together.
1986 86m/C *AU* Peter Hehir, Haydon Samuels; *Dir:* Paul Maloney. **VHS, Beta $79.98** *FOX* 🎞½

I Live My Life Stylish glossy flick with Crawford playing a bored New York debutante who travels to Greece and meets a dedicated archaeologist (Aherne). A love/hate relationship ensues in this typical Crawford vehicle where she is witty and parades around in sophisticated fashions, but there is little substance here.
1935 92m/B Joan Crawford, Brian Aherne, Frank Morgan, Aline MacMahon, Fred Keating, Jessie Ralph, Arthur Treacher, Frank Conroy, Sterling Holloway, Vince Barnett, Hedda Hopper, Lionel Stander; *Dir:* W.S. Van Dyke. **VHS $19.98** *MGM* 🎞½

I Love All of You Deneuve relives experiences with three former lovers - a composer, a well-intentioned nobody, and a rock star - as she enters into a relationship with a widower.
1983 103m/C *FR* Catherine Deneuve, Jean-Louis Trintignant, Gerard Depardieu, Serge Gainsbourg; *Dir:* Claude Berri. **VHS, Beta $59.95** *MON* 🎞🎞½

I Love Melvin Reynolds wants to be a Tinseltown goddess, and O'Connor just wants Reynolds; so he passes himself off as chief lenseman for a famous magazine and promises her a shot at the cover of "Look." Seems he has a little trouble on the follow through. Choreographed by Robert Alton, it's got a best ever football ballet (with Reynolds as pigskin).
1953 77m/C Donald O'Connor, Debbie Reynolds, Una Merkel, Richard Anderson, Jim Backus, Allyn Joslyn, Les Tremayne, Noreen Corcoran, Robert Taylor, Howard Keel; *Dir:* Don Weis. **VHS $19.98** *MGM, TTC, FCT* 🎞🎞½

I Love My...Wife A medical student plays doctor with hospital nurses when he and his wife stop having sex during her pregnancy. The marriage becomes more strained when their son is born and his mother-in-law moves in. A comedy without the proper dosage of laughs.
1970 (R) 98m/C Elliott Gould, Brenda Vaccaro, Angel Tompkins; *Dir:* Mel Stuart. **VHS, Beta $59.95** *MCA* 🎞🎞

I Love N.Y. A young metropolitan couple struggles to find true love amidst disapproving parents and doubting friends. Choppy direction and poor writing contribute to its failure.
1987 100m/C Scott Baio; Kelley Van Der Velden, Christopher Plummer, Jennifer O'Neill, Jerry Orbach, Virna Lisi; *Dir:* Alan Smithee. **VHS, Beta, LV $89.98** *MAG, HHE* 🎞

I Love You A man, down on his luck, mistakenly assumes that a woman he meets is a hooker. She plays along, only to find that they are becoming emotionally involved. Pretentious cat-and-mouse game.
1981 (R) 104m/C Sonia Braga, Paulo Cesar Pereio; *Dir:* Arnaldo Jabor. **VHS, Beta $59.95** *MGM* 🎞½

I Love You, Alice B. Toklas A straight uptight lawyer decides to join the peace and love generation in this somewhat maniacal satire of the hippie culture. Authored by Paul Mazursky and Larry Tucker, and known in Europe as "Kiss My Butterfly." Incidentally, the title's Alice B. Toklas was actually the lifemate of "Lost Generation" author Gertrude Stein.
1968 94m/C Peter Sellers, Jo Van Fleet, Leigh Taylor-Young, Joyce Van Patten; *Dir:* Hy Averback. **VHS, Beta, LV $19.98** *WAR, FCT* 🎞🎞½

I Love You to Death Dry comedy based on a true story, concerns a woman who tries to kill off her cheating husband. Lots of stars, but they never shine. Hurt and Reeves are somewhat amusing as drugged up hit men who struggle with the lyrics to the National Anthem.
1990 (R) 110m/C Kevin Kline, Tracy Ullman, Joan Plowright, River Phoenix, William Hurt, Keanu Reeves, Phoebe Cates; *Dir:* Lawrence Kasdan. **VHS, Beta, LV $19.95** *COL* 🎞🎞½

I Love You Rosa Rosa, a young Jewish widow, wrestles big time with old world values. Required by custom to marry her dearly departed's eldest brother, she's not enamored with her newly betrothed. Not that he's not a nice guy; he seems to plan to take his marital duties very seriously. It's just that he's a tad youthful (eleven years old, to be exact). Much rabbi consulting and soul searching.
1973 90m/C Michal Bat-Adam, Gabi Oterman, Yoseph Shiloah; *W/Dir:* Moshe Mizrahi. **VHS $39.95** *AVD, FCT, NCJ* 🎞🎞½

I Love You...Goodbye A frustrated housewife, fed up with the constant role of wife and mother, leaves her family in an effort to find a more challenging and fulfilling life. A good performance by Lange compensates for some of its muddledness.
1973 74m/C Hope Lange, Earl Holliman. **VHS, Beta** *LCA* 🎞🎞

I, Madman A novel-loving horror actress is stalked by the same mutilating madman that appears in the book she's presently reading. We call that bad luck.
1989 (R) 90m/C Jenny Wright, Clayton Rohner, William Cook; *Dir:* Tibor Takacs. **VHS, Beta, LV $89.95** *MED* 🎞🎞

I Married an Angel The last MacDonald/Eddy film. A playboy is lured away from his usual interests by a beautiful angel. Adapted from Rogers and Hart Broadway play. Strange and less than compelling.
1942 84m/C Jeanette MacDonald, Nelson Eddy, Binnie Barnes, Edward Everett Horton, Reginald Owen, Mona Maris, Janis Carter, Inez Cooper, Douglas Dumbrille, Leonid Kinskey, Marion Rosamond, Anne Jeffreys, Marek Windheim; *Dir:* W.S. Van Dyke. VHS, Beta $29.95 MGM 🎬½

I Married a Centerfold Television movie fluff about a young man's amorous pursuit of a model.
1984 100m/C Teri Copley, Timothy Daly, Diane Ladd, Bert Remsen, Anson Williams; *Dir:* Peter Werner. VHS, Beta, LV $79.95 SVS 🎬🎬

I Married a Monster from Outer Space The vintage thriller about a race of monster-like aliens from another planet who try to conquer earth. Despite its head-shaking title, an effective '50s sci-fi creeper.
1958 78m/B Tom Tryon, Gloria Talbott, Maxie Rosenbloom, Mary Treen, Ty Hardin; *Dir:* Gene Fowler. VHS, Beta $49.95 PAR, MLB 🎬🎬½

I Married a Vampire A country girl in the city is romanced and wed by a dashing vampire. Troma-produced hyper-camp.
1987 85m/C Rachel Gordon, Brendan Hickey; *Dir:* Jay Raskin. VHS, Beta $9.99 STE, PSM 🎬🎬

I Married a Witch A witch burned at the stake during the Salem witch trials comes back to haunt the descendants of her accusers and gets romantic with one of them. Wonderfully played fantasy/comedy.
1942 77m/B Veronica Lake, Fredric March, Susan Hayward, Broderick Crawford; *Dir:* Rene Clair. VHS, Beta, LV $59.95 LTG, MLB, WAR 🎬🎬🎬

I Married a Woman An advertising executive marries a beautiful blonde woman, but finds it very difficult to balance his career and marriage. Low on laughs. Shot in black and white, but includes a fantasy scene in color that features John Wayne. Pay attention to see young Angie Dickinson.
1956 84m/B George Gobel, Diana Dors, Adolphe Menjou, Nita Talbot; *Dir:* Hal Kanter. VHS, Beta, LV $15.95 UHV 🎬🎬

I Met a Murderer A man kills his nagging wife after a bitter argument and flees from her vengeful brother. Fine fugitive drama with a number of interesting twists.
1939 79m/B James Mason, Pamela Kellino, Sylvia Coleridge; *Dir:* Roy Kellino. VHS $16.95 SNC, NOS, DVT 🎬🎬½

I, Mobster Cochran tells a Senate Sub-Committee of his rise in the ranks of the Underworld—from his humble beginning as a bet collector for a bookie to his position as kingpin of the crime syndicate.
1958 80m/B Steve Cochran, Lita Milan, Robert Strauss, Celia Lovsky; *Dir:* Roger Corman. VHS $29.95 KBV 🎬🎬½

I Never Promised You a Rose Garden A disturbed 16-year-old girl spirals down into madness and despair while a hospital psychiatrist struggles to bring her back to life. Based on the Joanne Greenberg bestseller. Compelling and unyielding exploration of the clinical treatment of schizophrenia.
1977 (R) 90m/C Kathleen Quinlan, Bibi Andersson, Sylvia Sidney, Diane Varsi; *Dir:* Anthony Page. VHS, Beta $59.95 WAR, OM 🎬🎬🎬

I Never Sang For My Father A devoted son must choose between caring for his cantankerous but well-meaning father, and moving out West to marry the divorced doctor whom he loves. While his mother wants him to stay near home, his sister, who fell out of her father's favor by marrying out of the family faith, argues that he should do what he wants. An introspective, stirring story based on the Robert Anderson play. Oscar nominations for Douglas, Hackman, and Anderson's Adapted Screenplay.
1970 (PG) 90m/C Gene Hackman, Melvyn Douglas, Estelle Parsons, Dorothy Stickney; *Dir:* Gilbert Cates. National Board of Review Awards '70: 10 Best Films of the Year; Writers Guild of America Awards '70: Best Adapted Drama. VHS, Beta, LV $69.95 COL 🎬🎬🎬½

I Ought to Be in Pictures After hitch-hiking from New York to Hollywood to break into the movies, a teen-aged actress finds her father, a screenwriter turned alcoholic-gambler. Late, desperately unfunny Neil Simon outing, adapted from a Simon play.
1982 (PG) 107m/C Walter Matthau, Ann-Margret, Dinah Manoff, Lance Guest, Michael Dudikoff; *Dir:* Herbert Ross. VHS, Beta $59.98 FOX 🎬🎬

I Posed for Playboy Three women—one a college co-ed, one a stockbroker, and one a 37-year-old mother—quench their private passions by posing for Playboy Magazine. Made for television with additional footage added to give it an "R" rating.
1991 (R) 98m/C Lynda Carter, Michelle Greene, Amanda Peterson, Brittany York; *Dir:* Stephen Stafford. VHS, LV $29.98 REP 🎬🎬

I Remember Mama A true Hollywood heart tugger chronicling the life of a Norwegian immigrant family living in San Francisco during the early 1900s. Dunne plays the mother with a perfect Norwegian accent and provides her family with wisdom and inspiration. A kindly father and four children round out the nuclear family. A host of oddball characters regularly pop in on the household—three high-strung aunts and an eccentric doctor who treats a live-in uncle. Adapted from John Van Druten's stage play, based on Kathryn Forbes memoirs, "Mama's Bank Account;" a TV series came later. Dunne triumphs in her only character role; Oscar nominations went to her for Best Actress, Corby for Best Supporting Actress, Homolka for Best Supporting Actor, and Black & White Cinematography.
1948 95m/B Irene Dunne, Barbara Bel Geddes, Oscar Homolka, Ellen Corby, Cedric Hardwicke, Edgar Bergen, Rudy Vallee, Barbara O'Neil, Florence Bates; *Dir:* George Stevens. VHS, Beta, LV $19.98 CCB, RKO 🎬🎬🎬½

I See a Dark Stranger Cynical yet whimsical postwar British spy thriller about an angry Irish lass who agrees to steal war plans for the Nazis in order to battle her native enemies, the British—then she falls in love with a British officer. Also titled "The Adventuress," this is a sharply performed, decidedly jaded view of nationalism and wartime "heroism."
1947 112m/B GB Deborah Kerr, Trevor Howard, Raymond Huntley; *Dir:* Frank Laudner. VHS, Beta $29.95 VID 🎬🎬🎬

I Sent a Letter to My Love An aging spinster, faced with the lonely prospect of the death of her paralyzed brother, places a personal ad for a companion in a local newspaper, using a different name. Unknown to her, the brother is the one who answers it, and they begin a romantic correspondence. Well-acted, touching account of

relationships that change lives, but it takes its time in telling the story.
1981 102m/C FR Simone Signoret, Jean Rochefort, Delphine Seyrig; *Dir:* Moshe Mizrahi. VHS, Beta $59.95 HBO 🎬🎬🎬

I Shot Billy the Kid Billy the Kid decides to turn over a new leaf and live a decent life. The cards are stacked against him though: He's killed a man for nearly every year of his life, and old habits are hard to break.
1950 58m/B Donald (Don "Red") Barry, Robert Lowery, Tom Neal, Jack Perrin; *Dir:* William Berke. VHS, Beta, LV WGE 🎬🎬

I Shot Jesse James In his first film, director Fuller breathes characteristically feverish, maddened fire into the story of Bob Ford (Ireland) after he killed the notorious outlaw. An essential moment in Fuller's unique, America-as-tabloid-nightmare canon, and one of the best anti-westerns ever made.
1949 83m/B John Ireland, Barbara Britton, Preston Foster, Reed Hadley; *Dir:* Samuel Fuller. VHS, Beta WGE, RXM 🎬🎬🎬

I Spit on Your Corpse A vicious female hired killer engages in a series of terrorist activities. Stars Spelvin in a non-pornographic role. Originally titled "Girls for Rent."
1974 (R) 90m/C Georgina Spelvin, Susan McIver, Kent Taylor; *Dir:* Al Adamson. VHS, Beta $59.95 REP Woof!

I Spit on Your Grave A woman vacationing at an upstate New York lake house is brutally attacked and raped by four men. Left for dead, she recovers and avenges the attack. Violence deserves an "X" rating. Not to be confused with the 1962 film of the same title. Also known as "Day of the Women." Also available in a 102-minute version.
1977 (R) 98m/C Camille Keaton, Eron Tabor, Richard Pace, Anthony Nichols, Gunter Kleeman; *W/Dir:* Mier Zarchi. VHS, Beta, LV $39.95 VES, VID Woof!

I Stand Condemned A Russian officer is tricked into borrowing money from a spy and is condemned for treason. He's saved when a girl who loves him gives herself to a profiteer. Worth seeing only for young Olivier's performance. Also known as "Moscow Nights."
1936 90m/B GB Laurence Olivier, Penelope Dudley Ward, Robert Cochran; *Dir:* Anthony Asquith. VHS, Beta, 8mm $19.95 NOS, MRV, VYY 🎬🎬½

I Vitelloni Fellini's semi-autobiographical drama, argued by some to be his finest work. Five young men grow up in a small Italian town. As they mature, four of them remain in Romini and limit their opportunities by roping themselves off from the rest of the world. The characters are multi-dimensional, including a loafer who is supported by his sister and a young stud who impregnates a local woman. The script has some brilliant insights into youth, adulthood and what's in between. Also known as "The Young and the Passionate," "Vitelloni," and "Spivs."
1953 104m/B IT Alberto Sordi, Franco Interlenghi, Franco Fabrizi, Leopoldo Trieste; *Dir:* Federico Fellini. VHS, LV $68.00 MOV, APD 🎬🎬🎬½

I Wake Up Screaming An actress' promoter is accused of her murder. Entertaining mystery with a surprise ending. Originally titled "Hot Spot" and remade as "Vicki."

1941 82m/B Betty Grable, Victor Mature, Carole Landis, Laird Cregar, William Gargan, Alan Mowbray, Allyn Joslyn; *Dir:* H. Bruce Humberstone. **VHS, Beta** $19.98 *FOX* √√√

I Walked with a Zombie The definitive and eeriest of the famous Val Lewton/Jacques Tourneur horror films. Dee, a young American nurse, comes to Haiti to care for the catatonic matriarch of a troubled family. Local legends bring themselves to bear when the nurse takes the ill woman to a local voodoo ceremony for "healing." Superb, startling images and atmosphere create a unique context for this serious "Jane Eyre"-like story; its reputation has grown through the years.
1943 69m/B Frances Dee, Tom Conway, James Ellison, Christine Gordon, Edith Barrett, Darby Jones, Sir Lancelot; *Dir:* Jacques Tourneur. **VHS, Beta, LV** $19.98 *MED, FCT, MLB* √√√½

I Wanna Be a Beauty Queen Divine hosts the "Alternative Miss World" pageant featuring all sorts of bizarre contestants.
1985 81m/C Divine, Little Nell, Andrew Logan. **VHS, Beta** $19.95 *AHV, FCT, HHE* √√√½

I Wanna Hold Your Hand Teenagers try to crash the Beatles' appearance on the Ed Sullivan show.
1978 (PG) 104m/C Nancy Allen, Bobby DiCicco, Wendie Jo Sperber, Marc McClure, Susan Kendall Newman, Theresa Saldana, Eddie Deezen, Will Jordan; *Dir:* Robert Zemeckis. **VHS, Beta, LV** $19.98 *WAR, FCT* √√½

I Want to Live! Based on a scandalous true story, Hayward gives a riveting, Oscar-winning performance as a prostitute framed for the murder of an elderly woman and sentenced to death in the gas chamber. Producer Walter Wanger's seething indictment of capital punishment. Music by Johnny Mandel. Oscar nominations for director Wise, Adapted Screenplay, Black & White Cinematography, Sound, and Film Editing.
1958 120m/B Susan Hayward, Simon Oakland, Theodore Bikel; *Dir:* Robert Wise. Academy Awards '58: Best Actress (Hayward). **VHS, Beta** $19.98 *MGM, BTV, CCB* √√√

I Want What I Want A young Englishman wants a sex-change, lives as a woman, falls in love, complications ensue. Not as shocking to watch as it might seem, but more a melodramatic gender-bender.
1972 (R) 91m/C *GB* Anne Heywood, Harry Andrews, Jill Bennett; *Dir:* John Dexter. **VHS, Beta** $39.95 *PSM* √√

I Was a Teenage TV Terrorist Two teenagers pull on-the-air pranks at the local cable television, blaming it all on an imaginary terrorist group. Cheaply made and pointless, with bad acting and a script by Kevin McDonough. Also released as "Amateur Hour."
1987 85m/C Adam Nathan, Julie Hanlon; *Dir:* Stanford Singer. **VHS, Beta** $79.98 *LIV, LTG* Woof!

I Was a Teenage Werewolf "Rebel Without a Cause" meets "The Curse of the Werewolf" in this drive-in rock'n'roll horror. A troubled young man (pre-Bonanza Landon, in his first feature film appearance) suffers from teen angst and low production values, and falls victim to shrink Bissel. The good doctor turns out to be a bad hypnotist, and Landon's regression therapy takes him beyond childhood into his, gasp, primal past, where he sprouts a premature beard and knuckle hair. Much misconstrued and full of terrible longings, the hirsute highschooler is under-

stood only by girlfriend Lime and misunderstood hair-sprouting teen viewers. Directorial debut of Fowler, who was once Fritz Lang's editor.
1957 70m/B Michael Landon, Yvonne Lime, Whit Bissell, Tony Marshall, Dawn Richard, Barney Phillips, Ken Miller, Cindy Robbins, Michael Rougas, Robert Griffin, Joseph Mell, Malcolm Atterbury, Eddie Marr, Vladimir Sokoloff, Louise Lewis, S. John Launer, Guy Williams, Dorothy Crehan; *Dir:* Gene Fowler. **VHS** $9.95 *COL, FCT, MLB* √√½

I Was a Teenage Zombie Spoof of high school horror films features a good zombie against a drug-pushing zombie. Forget the story and listen to the music by Los Lobos, the Fleshtones, the Waitresses, Dream Syndicate and Voilent Femmes.
1987 92m/C Michael Rubin, Steve McCoy, Cassie Madden, Allen Rickman; *Dir:* John E. Michalakis. **VHS, Beta, LV** $19.98 *CHA* √

I Was a Zombie for the FBI Intelligence agency toughens its hiring criteria. Much McCarthyian mirth.
1982 105m/B *Dir:* Maurice Penczner. **VHS, Beta** *CGI* √½

I Will Fight No More Forever A vivid recounting of the epic true story of the legendary Chief Joseph who led the Nez Perce tribe on a 1,600-mile trek to Canada in 1877. Disturbing and powerful.
1975 100m/C James Whitmore, Ned Romero, Sam Elliott; *Dir:* Richard T. Heffron. **VHS, Beta** $14.95 *NEG, MRV, GEM* √√√

I Will, I Will for Now Tacky sex clinic comedy with Keaton and Gould trying to save their marriage through kinky therapy.
1976 (R) 109m/C Diane Keaton, Elliott Gould, Paul Sorvino, Victoria Principal, Robert Alda, Warren Bellinger; *Dir:* Norman Panama. **VHS, Beta** $69.95 *MED* √√

I Wonder Who's Killing Her Now? Husband pays to have his wife murdered, then changes his mind. On-again, off-again comedy.
1976 87m/C Bob Dishy, Joanna Barnes, Bill Dana, Vito Scotti; *Dir:* Steven Hilliard Stern. **VHS** $59.95 *IVE* √

Ice Castles A young figure skater's Olympic dreams are dimmed when she is blinded in an accident, but her boyfriend gives her the strength, encouragement, and love necessary to perform a small miracle. Way too schmaltzy.
1979 (PG) 110m/C Robby Benson, Lynn-Holly Johnson, Tom Skerritt, Colleen Dewhurst, Jennifer Warren, David Huffman; *Dir:* Donald Wrye. **VHS, Beta** $64.95 *COL* √√

The Ice Flood Silent family type comedy of a son who proves to the old man that he wrote the book on life, wagering the successful supervision of dad's lumberjacks.
1926 63m/B Kenneth Harlen, Viola Dana; *Dir:* George B. Seitz. **VHS, Beta** $9.95 *GPV* √√

Ice House A sophisticated young lady flees Texas for Hollywood, searching for a better life. When her ex-lover shows up and wants her to go back, the lackless turmoil begins.
1989 81m/C Melissa Gilbert, Bo Brinkman, Andreas Manolikakis, Buddy Quaid; *Dir:* Eagle Pennell. **VHS, Beta** $79.95 *MNC* √√

Ice Palace Two rugged adventurers maintain a lifelong rivalry in the primitive Alaskan wilderness, their relationship dramatizing the development of the 49th state. Silly but entertaining. Based on Edna Ferber's novel; music by Max Steiner.

1960 143m/C Richard Burton, Robert Ryan, Carolyn Jones, Shirley Knight, Martha Hyer, Jim Backus, George Takei; *Dir:* Vincent Sherman. **VHS, Beta** $59.95 *WAR* √√

Ice Pirates Space pirates in the far future steal blocks of ice to fill the needs of a thirsty galaxy. Cool plot has its moments.
1984 (PG) 93m/C Robert Urich, Mary Crosby, Michael D. Roberts, John Matuszak, Anjelica Huston, Ron Perlman, John Carradine, Robert Symonds; *Dir:* Stewart Raffill. **VHS, Beta, LV** $79.95 *MGM* √½

Ice Station Zebra A nuclear submarine races Soviet seamen to find a downed Russian satellite under a polar ice cap. Suspenseful Cold War adventure with music by Michel Legrand. Based on the novel by Alistair MacLean.
1968 150m/C Rock Hudson, Ernest Borgnine, Patrick McGoohan, Jim Brown, Lloyd Nolan, Tony Bill; *Dir:* John Sturges. **VHS, Beta, LV** $29.95 *MGM* √√½

Icebox Murders A crazed, motiveless, faceless killer hangs his victims in a walk-in freezer. Store this one.
198? 90m/C Jack Taylor, Mirta Miller. **VHS, Beta** $59.95 *MGL* Woof!

Iced Wild fun turns to horror on a ski trip as a group of college friends are hunted by a maniac on the loose. It's all downhill.
1988 86m/C Debra Deliso, Doug Stevenson, Ron Kologie, Elizabeth Gorcey, Alan Johnson; *Dir:* Jeff Kwitny. **VHS, Beta** $79.95 *PSM* √

Iceman A frozen prehistoric man is brought back to life, after which severe culture shock takes hold. Underwritten but nicely acted, especially by Lone as the primal man.
1984 (PG) 101m/C *CA* Timothy Hutton, Lindsay Crouse, John Lone, David Straithairn, Josef Sommer, Danny Glover; *Dir:* Fred Schepisi. **VHS, Beta, LV** $69.95 *MCA* √√√

The Icicle Thief "The Naked Gun" for the art-house crowd. As a stark, black-and-white neo-realist tragedy airs on TV, bright color commercials disrupt the narrative. Soon ads invade the the movie itself, and the film's director jumps into his picture to fix the damage. Study up on your postwar Italian cinema and this'll seem hilarious; otherwise the real fun doesn't kick in until the halfway point. In Italian with English subtitles.
1989 93m/C *IT* Maurizio Nichetti, Cateria Sylos Labini, Claudio G. Fava; *Dir:* Maurizio Nichetti. Moscow Film Festival '90: Best Film. **VHS** $89.95 *FXL, FCT, APD* √√√

Icy Breasts A French psychiatrist tries to prevent his beautiful but psychotic patient from continuing her murdering spree. Dubbed.
1975 105m/C *FR* Alain Delon, Mireille Darc; *Dir:* Georges Lautner. **VHS, Beta** $39.95 *FCT* √√√

I'd Give My Life A gangster's ex-wife is presently married to the governor. He tries to use their honest but framed son to blackmail her.
1936 73m/B Guy Standing, Frances Drake, Tom Brown, Janet Beecher; *Dir:* Edwin L. Marin. **VHS, Beta** $29.95 *VYY* √½

Idaho Rogers is on a mission to close down houses of ill repute, teamed up with Autry's old sidekick Burnette.
1943 70m/B Roy Rogers, Harry Shannon, Virginia Grey, Smiley Burnette; *Dir:* Joseph Kane. **VHS, Beta** $9.95 *NOS, CPB, VCN* √

Idaho Transfer Fonda's second directorial effort; Carradine's first screen appearance. An obnoxious group of teens travel through time to Idaho in the year 2044. Environmental wasteland story is dull and confused.
1973 (PG) 90m/C Keith Carradine, Kelley Bohanan; *Dir:* Peter Fonda. **VHS, Beta $19.95** *MPI, FCT* Woof!

Identity Crisis Campy fashion maven and flamboyant rapper switch identities causing much tedious overacting. Lifeless murder comedy from the brothers Van Peebles.
1990 (R) 98m/C Mario Van Peebles, Ilan Mitchell-Smith, Nicholas Kepros, Shelly Burch, Richard Clarke; *Dir:* Melvin Van Peebles. **VHS $19.95** *ACA* ♫

Identity Unknown Shell-shocked veteran goes AWOL to discover who he is, meeting grieving relatives along the way. Interesting premise is sometimes moving.
1945 70m/B Richard Arlen, Cheryl Walker, Roger Pryor, Bobby Driscoll; *Dir:* Walter Colmes. **VHS, Beta, 8mm $29.95** *VYY* ♫♫½

The Idiot Dostoevski's Russian novel is transported by Kurosawa across two centuries to post-war Japan, where the madness and jealousy continue to rage. In Japanese with English subtitles.
1951 166m/B Toshiro Mifune, Masayuki Mori, Setsuko Hara, Yoshiko Kuga; *Dir:* Akira Kurosawa. **VHS $79.95** *NYF* ♫♫♫♫

Idiot's Delight At an Alpine hotel, a song and dance man meets a gorgeous Russian countess who reminds him of a former lover. Incredibly, Gable sings and dances through "Puttin' on the Ritz," the film's big highlight. Based on the Pulitzer Prize-winning play by Robert Sherwood.
1939 107m/B Clark Gable, Norma Shearer, Burgess Meredith, Edward Arnold; *Dir:* Clarence Brown. **VHS, Beta, LV $19.98** *CCB, MGM* ♫♫♫

Idolmaker A conniving agent can make a rock star out of anyone. Well-acted fluff with Sharkey taking a strong lead. The first film by hack-meister Hackford and somewhat based on true-life teen fab Fabian.
1980 (PG) 107m/C Ray Sharkey, Tovah Feldshuh, Peter Gallagher, Paul Land, Joe Pantoliano, Maureen McCormick, John Aprea, Richard Bright, Olympia Dukakis, Steven Apostlee Peck; *Dir:* Taylor Hackford. **VHS, Beta, LV $59.95** *MGM* ♫♫½

If... Three unruly seniors at a British boarding school refuse to conform. A popular, anarchic indictment of staid British society, using the same milieu as Vigo's "Zero de Conduite," with considerably more violence. The first of Anderson and McDowell's trilogy, culminating with "O Lucky Man!" and "Britannia Hospital." In color and black and white.
1969 (R) 111m/C *GB* Malcolm McDowell, David Wood, Christine Noonan, Richard Warwick; *Dir:* Lindsay Anderson. Cannes Film Festival '69: Best Film. **VHS, Beta, LV $49.95** *PAR* ♫♫♫♫

If Ever I See You Again The creator of "You Light Up My Life" stars in the sentimental melodrama. Two adolescent lovers meet again years later and try to rekindle old passions. Done with as much skill as "Light" minus a hit song.
1978 (PG) 105m/C Joseph Brooks, Shelley Hack, Jerry Keller; *Dir:* Joseph Brooks. **VHS, Beta $69.95** *COL* ♫

If I Perish A drama about a young Korean Christian during World War II who preaches gospel in prison. Dubbed.

197? 75m/C VHS, Beta $39.95 *VBL, UHV, VGD* ♫½

If I Were Rich A playboy on the skids must earn a living for the first time in his life. Also known as "Cash" and "For Love or Money."
1933 63m/B *GB* Robert Donat, Wendy Barrie, Edmund Gwenn, Clifford Heatherly. **VHS, Beta, 8mm $29.95** *VYY, FCT* ♫♫

If It's a Man, Hang Up A model is terrified by harrowing telephone calls from a stranger.
1975 71m/C Carol Lynley, Gerald Harper. **VHS, Beta $29.95** *IVE* ♫½

If Looks Could Kill Would-be film noir about a man spying on a female embezzler and falling in love.
1986 (R) 90m/C Alan Fisler, Tim Gail, Kim Lambert, Jeanne Marie; *Dir:* Chuck Vincent. **VHS $79.95** *REP, NRT* ♫

If Looks Could Kill TV stud Grieco makes film debut as a high school class cutup who travels to France with his class to parlez vous for extra credit. Mistaken for a CIA agent, he stumbles into a plot to take over all the money in the whole wide world, and much implausible action and eyelash batting follows. Extra kibbles for supporting characterizations. Oh, and don't be put off by the fact that the Parisian scenes were shot in Montreal.
1991 (PG-13) 89m/C Richard Grieco, Linda Hunt, Roger Rees, Robin Bartlett, Gabrielle Anwar; *Dir:* William Dear. **VHS, Beta, LV, 8mm $19.98** *WAR, PIA* ♫♫½

If Things Were Different A feisty woman struggles to hold her family together after her husband is hospitalized with a nervous breakdown.
1979 96m/C Suzanne Pleshette, Tony Roberts, Arte Johnson, Chuck McCann, Don Murray; *Dir:* Robert Lewis. **VHS, Beta $34.98** *WOV* ♫½

If You Could See What I Hear Irritatingly upbeat true-life story of blind singer-musician, Tom Sullivan. Covers his life from college days to marriage. Sophomoric comedy centers around the hero's irresponsible, and often life-endangering, behavior.
1982 (PG) 103m/C *CA* Marc Singer, R.H. Thompson, Sarah Torgov, Shari Belafonte-Harper; *Dir:* Eric Till. Genie Awards '83: Best Supporting Actor (Thomson). **VHS, Beta, LV $79.95** *VES* ♫½

If You Don't Stop It...You'll Go Blind A series of gauche and tasteless vignettes from various little-known comedians.
1977 (R) 80m/C Pat McCormick, George Spencer, Patrick Wright; *Dir:* Bob Levy, Keefe Brasselle. **VHS, Beta $54.95** *MED* ♫½

If You Knew Susie Two retired vaudeville actors find a letter from George Washington which establishes them as descendants of a colonial patriot and the heirs to a seven billion dollar fortune.
1948 90m/B Eddie Cantor, Joan Davis, Allyn Joslyn, Charles Dingle; *Dir:* Gordon Douglas. **VHS, Beta $34.98** *CCB* ♫♫

Igor & the Lunatics Tasteless tale of a cannibal cult leader released from prison who picks up where he left off.
1985 (R) 79m/C Joseph Eero, Joe Niola, T.J. Michaels; *Dir:* Billy Parolini. **VHS, Beta $69.98** *LIV, LTG* ♫

Iguana The videocassette box art makes this look like a horror movie, but it's really a stiff, solemn period drama about a deformed sailor with lizardlike features. After a life of mistreatment he reigns mercilessly over a handful of island castaways. An international coproduction with mostly English dialogue, some Spanish and Portuguese with subtitles.
1989 ?m/C Everett McGill; *Dir:* Monte Hellman. **VHS $79.99** *IMP* ♫

Ike Duvall is top brass in this made-for-television epic biography tracing Eisenhower's career during World War II. Previously a six-hour mini-series; also titled "Ike: The War Years."
1979 291m/C Robert Duvall, Lee Remick, Darren McGavin, Dana Andrews, Laurence Luckinbill; *Dir:* Melville Shavelson. **VHS, Beta, LV $29.95** *STE, VMK* ♫♫½

Ikiru When a clerk finds out he is dying of cancer, he decides to build a children's playground and give something of himself back to the world. Highly acclaimed, heartbreaking drama from the unusually restrained Kurosawa; possibly his most "eastern" film. In Japanese with English subtitles.
1952 134m/B *JP* Takashi Shimura, Kyoko Seki; *Dir:* Akira Kurosawa. **VHS, Beta, LV** *MED, VYG, CRC* ♫♫♫♫

I'll Cry Tomorrow Hayward brilliantly portrays actress Lillian Roth as she descends into alcoholism and then tries to overcome her addiction. Based on Roth's memoirs. Oscar nominations for Hayward, Black & White Cinematography, Art Direction and Set Decoration.
1955 119m/B Susan Hayward, Richard Conte, Eddie Albert, Jo Van Fleet, Margo, Don Taylor, Ray Danton; *Dir:* Daniel Mann. Academy Awards '55: Best Costume Design; Cannes Film Festival '56: Best Actress (Hayward). **VHS, Beta, LV $19.98** *MGM* ♫♫♫

I'll Get You An FBI agent goes to England on the trail of a kidnapping ring. Also released as "Escape Route."
1953 79m/B George Raft, Sally Gray, Clifford Evans; *Dir:* Seymour Friedman. **VHS, Beta, LV** *WGE, RXM* ♫♫

Illegal Crime melodrama stars Robinson as an attorney who risks all to acquit his assistant of murder. Early Mansfield appearance. Remake of "The Mouthpiece."
1955 88m/B Edward G. Robinson, Nina Foch, Hugh Marlowe, Jayne Mansfield, Albert Dekker, Ellen Corby, DeForest Kelley, Howard St. John; *Dir:* Lewis Allen. **VHS** *NO* ♫♫½

Illegally Yours Miscast comedy about a college student serving on the jury in trial of old girlfriend. Bring down the gavel on this one.
1987 (PG) 94m/C Rob Lowe, Colleen Camp, Kenneth Mars; *Dir:* Peter Bogdanovich. **VHS, Beta $79.98** *FOX* ♫

Illicit Behavior A dangerous manipulative woman (Severance) is out to claim a $2 million inheritance in this seductive police thriller. Included among her victims in this intricate plot are her husband, a Hollywood vice cop suspended for use of excessive force; his partner; her husband's internal affairs adversary; and her own psychiatrist.
1991 (R) 101m/C Joan Severance, Robert Davi, James Russo, Jack Scalia, Kent McCord; *Dir:* Worth Keeter. **VHS $79.95** *PSM* ♫½

The Illusion Travels by Streetcar Odd but enchanting story of two men who restore a streetcar, only to find that it will not be used by the city. In a gesture of defiance,

they steal the streetcar and take it for one last ride, picking up an interesting assortment of characters along the way. Not released in the U.S. until 1977.
1953 90m/B *MX* Lilia Prado, Carlos Navarro, Domingo Soler, Fernando Soto, Agustin Isunza, Miguel Manzano; *Dir:* Luis Bunuel. **VHS $59.95** *CVC, FCT* 𝒵𝒵𝒵

The Illustrated Hitchcock The films of quintessential film director Alfred Hitchcock are examined in this two-volume offering which also includes interviews with the master of the mystery genre.
1972 30m/C Beta *CMR, NYS* 𝒵𝒵𝒵

The Illustrated Man A young drifter meets a tattooed man. Each tattoo causes a fantastic story to unfold. A strange, interesting, but finally limited attempt at literary sci-fi. Based on the story collection by Ray Bradbury.
1969 (PG) 103m/C Rod Steiger, Claire Bloom, Robert Drivas, Don Dubbins; *Dir:* Jack Smight. **VHS, Beta $19.98** *KUI, WAR, FCT* 𝒵𝒵𝒵

Ilsa, Harem Keeper of the Oil Sheiks The naughty Ilsa works for an Arab sheik in the slave trade. More graphic violence and nudity. Plot makes an appearance.
1976 (R) 45m/C Dyanne Thorne, Michael Thayer. **VHS, LV $59.98** *CIC* Woof!

Ilsa, She-Wolf of the SS The torture-loving Ilsa runs a medical camp for the Nazis. Plot is incidental to nudity and violence.
1974 (R) 45m/C Dyanne Thorne, Greg Knoph; *Dir:* Don Edmunds. **VHS, LV $59.98** *CIC* Woof!

Ilsa, the Tigress of Siberia Everyone's favorite torturer has been working her wiles on political prisoners in Russia. After escaping to Canada, her past threatens to catch up with her. As usual, no plot, violence.
1979 85m/C Dyanne Thorne, Michel Morin, Tony Angelo, Terry Coady, Howard Mauer; *Dir:* Jean LaFleur. **VHS $59.98** *AVD* Woof!

Ilsa, the Wicked Warden Ilsa is now a warden of a woman's prison in South America, behaving just as badly as she always has, until the prisoners stage an uprising. Lots of skin, no violence, no acting. Also: no plot. AKA "Ilsa, the Absolute Power" and "Greta the Mad Butcher."
1978 (R) 90m/C Dyanne Thorne, Lina Romay, Jess (Jesus) Franco; *Dir:* Jess (Jesus) Franco. **VHS, LV $59.98** *CIC* Woof!

I'm All Right Jack Sellers plays a pompous communist union leader in this hilarious satire of worker-management relations. Based on Alan Hackney's novel "Private Life."
1969 101m/B Peter Sellers, Ian Carmichael, Terry-Thomas; *Dir:* John Boulting. National Board of Review Awards '60: 5 Best Foreign Films of the Year. **VHS, Beta $19.98** *HBO, FCT* 𝒵𝒵𝒵

I'm from Arkansas The little town of Pitchfork, Arkansas, goes nuts when a pig gives birth to ten piglets.
1944 68m/B El Brendel, Slim Summerville, Iris Adrian, Harry Harvey, Bruce Bennett; *Dir:* Lew Landers. **VHS, Beta $24.95** *NOS, MRV, DVT* 𝒵

I'm Dancing as Fast as I Can A successful television producer hopelessly dependent on tranquilizers tries to stop cold turkey. Good story could be better. Written by David Rabe, based on Barbara Gordon's memoir.

1982 (R) 107m/C Jill Clayburgh, Nicol Williamson, Dianne West, Joe Pesci, Geraldine Page, John Lithgow, Daniel Stern; *Dir:* Jack Hofsiss. **VHS, Beta, LV $69.95** *PAR* 𝒵𝒵

I'm Dangerous Tonight Wallflower Amick (Shelley the Waitress in Twin Peaks) turns into party monster when she dons red dress made from evil ancient Aztec cloak. Many dead soldiers when this girl parties. Based on a Cornell Woolrich short story, it's another in a series of ever more disappointing entries from Hooper, whose cult fave "Texas Chainsaw Massacre" had low as they come production values, but far better scares. Originally cable fare.
1990 (R) 92m/C Madchen Amick, Corey Parker, R. Lee Ermey, Mary Frann, Dee Wallace Stone, Anthony Perkins; *Dir:* Tobe Hooper. **VHS $79.95** *MCA* 𝒵𝒵

I'm a Fool A two-character romantic study involving an act of insincerity between two race-track aficionados. Based on the Sherwood Anderson story. From the American Short Story Collection.
1977 38m/C Ron Howard, Amy Irving; *Dir:* Noel Black. **VHS, Beta $24.95** *MON, KAR* 𝒵𝒵½

I'm the Girl He Wants to Kill A woman sees a murder committed, and the murderer sees her. Now he's after her.
1974 71m/C Julie Sommars, Anthony Steel. **VHS, Beta $29.95** *THR* 𝒵½

I'm Gonna Git You Sucka Parody of "blaxploitation" films popular during the '60s and '70s. Funny and laced with out-right bellylaughs. A number of stars who made "blaxploitation" films, including Jim Brown, take part in the gags.
1988 (R) 89m/C Keenan Ivory Wayans, Bernie Casey, Steve James, Isaac Hayes, Jim Brown, Ja'net DuBois, Anne-Marie Johnson, Antonio Fargas, Eve Plumb; *W/Dir:* Keenan Ivory Wayans. **VHS, Beta $19.95** *MGM* 𝒵𝒵𝒵

I'm on My Way/The Non-Stop Kid Two classic shorts: "I'm on My Way" (1919), in which Lloyd's dreams of an idyllic marriage are shattered; and "The Non-Stop Kid" (1918), in which Lloyd must contend with a rival for the affection of his beloved.
191? 30m/B Harold Lloyd, Snub Pollard, Bebe Daniels; *Dir:* Harold Lloyd. **VHS, Beta $19.98** *CCB* 𝒵𝒵𝒵

I'm the One You're Looking For A famous model obsessively combs Barcelona for the man who raped her. Suspenseful thriller.
1988 85m/C *SP* Patricia Adriani, Chus Lampreave, Ricard Borras, Toni Canto, Angel Alcazar, Marta Fedz, Frank Muro, Miriam DeMaeztu; *Dir:* Jaime Chavarri. **VHS, LV $79.95** *FXL, FCT, IME* 𝒵𝒵½

The Image Anchorman Finney, the unscrupulous czar of infotainment, does a little soul delving when a man he wrongly implicated in a savings and loan debacle decides to check into the hereafter ahead of schedule. Seems the newsmonger's cut a few corners en route to the bigtime, and his public is mad as hell and...oops, wrong movie (need we say there are not a few echoes of "Network"?). Finney's great. Made for cable.
1989 91m/C Albert Finney, John Mahoney, Kathy Baker, Swoosie Kurtz, Marsha Mason; *Dir:* Peter Werner. **VHS $89.95** *HBO* 𝒵𝒵½

Image of Bruce Lee Martial arts fight scenes prevail in this story about a jeweler who is swindled out of $1 million worth of diamonds.

197? (G) 88m/C Bruce Li, Chang Wu Lang, Chang Lei, Dana. **VHS, Beta $29.95** *MED* 𝒵

Image of Death Woman plots to take the life of an old school chum, assume her identity and then blow up the Capitol.
1977 83m/C Cathy Paine, Cheryl Waters, Tony Banner, Barry Pierce, Sheila Helpmann, Pennie Hackforth-Jones; *Dir:* Kevin Dobson. **VHS, Beta** *PGN* 𝒵

Image of Passion Weak account of male stripper/lovely advertising executive romance.
1986 90m/C James Horan, Susan Damante Shaw, Edward Bell, Louis Jourdan. **Beta $14.95** *PSM* 𝒵

Imagemaker Uneven story of a presidential media consultant who bucks the system and exposes corruption.
1986 (R) 93m/C Michael Nouri, Jerry Orbach, Jessica Harper, Farley Granger; *Dir:* Hal Wiener. **VHS, Beta, LV $79.98** *LIV, VES* 𝒵½

Imitation of Life Remake of the successful 1939 Claudette Colbert outing of the same title, and based on Fannie Hurst's novel, with a few plot changes. Turner is a single mother, more determined to achieve acting fame and fortune than function as a parent. Her black maid, Moore, is devoted to her own daughter (Kohner), but loses her when the girl discovers she can pass for Caucasian. When Turner discovers that she and her daughter are in love with the same man, she realizes how little she knows her daughter, and how much the two of them have missed by not having a stronger relationship. Highly successful at the box office, with Oscar nominations for Kohner and Moore.
1959 124m/C Lana Turner, John Gavin, Troy Donahue, Sandra Dee, Juanita Moore, Susan Kohner; *Dir:* Douglas Sirk. **VHS $14.95** *MCA* 𝒵𝒵𝒵

Immediate Family Childless couple contact a pregnant, unmarried girl and her boyfriend in hopes of adoption. As the pregnancy advances, the mother has doubts about giving up her baby. A bit sugar-coated in an effort to woo family trade.
1989 (PG-13) 112m/C Glenn Close, James Woods, Kevin Dillon, Mary Stuart Masterson; *Dir:* Jonathan Kaplan. National Board of Review Awards '89: Best Supporting Actress (Masterson). **VHS, Beta, LV $19.95** *COL* 𝒵𝒵½

The Immigrant Chaplin portrays a newly arrived immigrant who falls in love with the first girl he meets. Silent with musical soundtrack added.
1917 20m/B Charlie Chaplin; *Dir:* Charlie Chaplin. **VHS, Beta** *CAB, FST* 𝒵𝒵𝒵½

Immortal Bachelor Good cast can't overcome the silliness of this one. A juror daydreams about the dead husband of a cleaning woman on trial for killing her adulterous spouse. She decides the dead man seems a lot more exciting than her own dull and tiresome husband.
1980 (PG) 94m/C *IT* Giancarlo Giannini, Monica Vitti, Claudia Cardinale, Vittorio Gassman; *Dir:* Marcello Fondato. **VHS, Beta $69.95** *VID* 𝒵½

Immortal Battalion Entertaining wartime psuedo-documentary follows newly recruited soldiers as they are molded from an ordinary group of carping civilians into a hardened battalion of fighting men by Niven. Uses some actual training and combat footage. Based on idea from Niven and scripted by Ustinov and Eric Ambler. Originally known

as "The Way Ahead," with some prints running to 116 minutes.
1944 89m/B David Niven, Stanley Holloway, Reginald Tate, Peter Ustinov; *Dir:* Carol Reed. **VHS, Beta $14.95** NOS, MRV, QNE *♫♫♫*

Immortal Sergeant After the battle death of the squad leader, an inexperienced corporal takes command of North African troops during World War II. Fonda gives a strong performance. Based on a novel by John Brody and scripted by Lamar Trotti.
1943 91m/B Henry Fonda, Thomas Mitchell, Maureen O'Hara, Allyn Joslyn, Reginald Gardiner, Melville Cooper, Morton Lowry; *Dir:* John M. Stahl. **VHS, Beta $19.98** FOX *♫♫½*

Immortal Sins Michael and Susan inherited an old Spanish Castle, complete with its own curse, which comes in the form of a seductive women that has haunted families for centuries.
1991 80m/C Cliff De Young, Maryam D'Abo. **VHS** CGV *♫*

The Immortalizer A mad doctor transfers people's brains into young bodies for a price, although it's not as simple as it sounds.
1989 (R) 85m/C Ron Kay, Chris Crone, Melody Patterson, Clarke Lindsley, Bekki Armstrong; *Dir:* Joel Bender. **VHS, Beta, LV $79.95** COL *♫*

Impact A woman and her lover plan the murder of her rich industrialist husband, but the plan backfires. More twists and turns than a carnival ride.
1949 83m/B Brian Donlevy, Ella Raines; *Dir:* Arthur Lubin. **VHS, Beta $14.95** NOS, MRV, QNE *♫♫♫*

The Imperial Japanese Empire Epic of the World War II Pacific seen through the Japanese' eyes.
1985 130m/C JP Tetsuro Tamba, Tomokazu Miura, Teruhiko Saigo, Teruhiko Aoi, Saburo Shinoda, Keiko Sekine, Masako Natsume, Akiko Kana. **VHS, Beta $69.95** VCD *♫♫*

Imperial Navy A World War II epic as seen from the Japanese perspective, concentrating on their great navy and its suicide squadrons. Dubbed.
198? 145m/C JP *Dir:* Paul Aaron. **VHS, Beta $59.95** SVS *♫♫*

Imperial Venus Biography of Napoleon's sister, Paolina Bonaparte—with particular focus on her many loves, lusts and tribulations.
1971 (PG) 121m/C Gina Lollobrigida, Stephen Boyd, Raymond Pellegrin; *Dir:* Jean Delannoy. **VHS, Beta $49.95** UHV *♫♫*

Importance of Being Donald Three vintage Disney Donald Duck shorts from the early days of World War II: "Donald's Better Self," "Polar Trappers," and "Timber."
1939 25m/C VHS, Beta $14.95 DIS *♫♫*

Importance of Being Earnest Fine production of classic Oscar Wilde comedy-of-manners. Cast couldn't be better or the story funnier.
1952 95m/C Michael Redgrave, Edith Evans, Margaret Rutherford, Michael Denison, Joan Greenwood; *Dir:* Anthony Asquith. Venice Film Festival '52: Best Decor (Carmen Dillon). **VHS, Beta $19.95** PAR *♫♫♫*

The Impossible Spy True story of Elie Cohen, an unassuming Israeli who doubled as a top level spy in Syria during the 1960s. Cohen became so close to Syria's president that he was nominated to become the Deputy Minister of Defense before his double life was exposed. Well-acted spy thriller made

for cable and is worth a look for espionage fans.
1987 96m/C GB John Shea, Eli Wallach; *Dir:* Jim Goddard. **VHS $79.99** HBO, NCJ *♫♫♫*

The Imposter Improbable tale of a con artist/ex-con who impersonates a high school principal in order to regain his estranged girlfriend, who works as a teacher. Complications ensue when he tackles the school's drug problems. Made for television.
1984 95m/C Anthony Geary, Billy Dee Williams, Lorna Patterson; *Dir:* Michael Pressman. **VHS, Beta $79.95** PSM *♫½*

Imposters A comedy centering on the romantic problems between a self-indulgent young man and the woman who cheats on him.
1984 110m/C Ellen McElduff, Charles Ludlum, Lisa Todd, Peter Evans. **VHS, Beta** NEW *♫♫*

Impromptu A smart, sassy, romantic comedy set in 1880s Europe among the era's artistic greats—here depicted as scalawags, parasites and early beatniks. The film's ace is Davis in a lusty, dynamic role as mannish authoress George Sand, obsessively in love with the dismayed composer Chopin. Though he's too pallid a character for her attraction to be credible, a great cast and abundant wit make this a treat.
1990 (PG-13) 108m/C Judy Davis, Hugh Grant, Mandy Patinkin, Bernadette Peters, Julian Sands, Ralph Brown, Georges Corraface, Anton Rodgers, Emma Thompson; *Dir:* James Lapine. **VHS $19.95** HMD *♫♫♫*

Improper Channels Mediocre comedy about a couple trying to recover their daughter after being accused of child abuse.
1902 (PG) 91m/C Monica Parker, Alan Arkin, Mariette Hartley; *Dir:* Eric Till. **VHS, Beta, LV $69.95** VES *♫½*

Impulse When his wife leaves for a visit with her mother, a young man is thrust into cahoots with gangsters, and wrapped up in an affair with another woman.
1955 81m/B GB Arthur Kennedy, Constance Smith, Joy Shelton; *Dir:* Charles De Latour. **VHS $29.95** FCT *♫½*

Impulse Shatner, a crazed killer, is released from prison and starts to kill again in this low-budget, predictable, critically debased film about child-molesting. Alternative titles, "Want a Ride, Little Girl?," and "I Love to Kill."
1974 (PG) 85m/C William Shatner, Ruth Roman, Harold Sakata, Kim Nicholas, Jennifer Bishop, James Dobbs; *Dir:* William Grefe. **VHS, Beta $79.95** IVE, VES **Woof!**

Impulse Small town residents can't control their impulses, all because of government toxic waste in their milk. A woman worried about her mother's mental health returns to town with her doctor boyfriend to discover most of the inhabitants are quite mad. Starts strong but...
1984 (R) 95m/C Tim Matheson, Meg Tilly, Hume Cronyn, John Karlen, Bill Paxton, Amy Stryker, Claude Earl Jones, Sherri Stoner; *Dir:* Graham Baker. **VHS, Beta, LV $29.98** VES *♫♫*

Impulse A beautiful undercover vice cop poses as a prostitute, and gives in to base desires. Russell gives a fine performance in this underrated sleeper.
1990 (R) 109m/C Theresa Russell, Nicholas Mele, Eli Danker, Charles McCaughan, Jeff Fahey, George Dzundza; *Dir:* Sondra Locke. **VHS, Beta, LV $92.95** WAR *♫♫½*

Impure Thoughts Four friends meet in the afterlife and discuss their Catholic school days. Odd premise never catches fire.
1986 (PG) 87m/C Brad Dourif, Lane Davies, John Putch, Terry Beaver; *Dir:* Michael A. Simpson. **VHS, Beta $19.98** CHA *♫♫*

In the Aftermath Mix of animation and live-action in this post-nuclear wasteland tale never quite works.
1987 85m/C Tony Markes, Rainbow Dolan. **VHS, Beta $19.95** STE, NWV *♫½*

In Between When Margo, Jack, and Amy wake up they expect it to be another average day but instead these three strangers find themselves together in a peculiar house. An angel informs them that the house is a way station between life and the hereafter. The three must look at their pasts, imagine their futures, and decide whether their old lives are worth returning to or should they accept death and go on to an afterlife.
199? 92m/C Wings Hauser, Robin Mattson, Alexandra Paul, Robert Forster. **VHS $79.95** MNC *♫♫½*

In a Cartoon Studio Nine rarely seen cartoon shorts from the Van Beuren Studio, rival to Paul Terry's Aesop Fables Studio in the 1930s: "In a Cartoon Studio (Making 'Em Move)," "Cinderella Blues," "Redskin Blues," "The Ball Game," "Galloping Hoofs (Gallopin' Fanny)," "Gay Gaucho," "Happy Hoboes," "Indian Whoopee," and "Brownie's Victory Garden (How's Crops)."
1934 60m/B VHS, Beta, 8mm $24.95 VYY *♫♫½*

In Celebration Three successful brothers return to the old homestead for their parents' 40th wedding celebration. Intense drama but somewhat claustrophobic. Based on a play by David Storey.
1975 (PG) 110m/C GB CA Alan Bates, James Bolam, Brian Cox; *Dir:* Lindsay Anderson. **VHS** AFF *♫♫♫*

In Cold Blood Truman Capote's supposedly factual novel provided the basis for this hard-hitting docu-drama about two ex-cons who ruthlessly murder a Kansas family in 1959 in order to steal their non-existent stash of money. Blake is riveting as one of the killers. Oscar nominations for director Brooks, Adapted Screenplay, Original Score.
1967 133m/B Robert Blake, Scott Wilson, John Forsythe; *Dir:* Richard Brooks. National Board of Review Awards '67: 10 Best Films of the Year, Best Director. **VHS, Beta, LV $19.95** COL *♫♫♫½*

In the Cold of the Night Another drowsy entry into the "to sleep perchance to have a nightmare genre." A photographer with a vivid imagination dreams he murders a woman he doesn't know, and when said dream girl rides onto his life on the back of a Harley, Mr. Foto's faced with an etiquette quandary: haven't they met before? Cast includes Hedren (Hitchcock's Marnie and Melanie Griffith's mother).
1989 (R) 112m/C Jeff Lester, Adrienne Sachs, Shannon Tweed, David Soul, John Beck, Tippi Hedren, Marc Singer; *Dir:* Nico Mastorakis. **VHS, LV $9.98** REP *♫*

In Country Based on Bobbie Ann Mason's celebrated novel, the story of a young Kentuckian high schooler (perfectly played by Lloyd) and her search for her father, killed in Vietnam. Willis plays her uncle, a veteran still struggling to accept his own survival, and crippled with memories. Moving scene at the Vietnam Veterans Memorial.

1989 (R) 116m/C Bruce Willis, Emily Lloyd, Joan Allen, Kevin Anderson, Richard Hamilton, Judith Ivey, Peggy Rea, John Terry; *Dir:* Norman Jewison. **VHS, Beta, LV, 8mm $19.95** *WAR* 𝄫𝄫½

The In Crowd A bright high school student gets involved with a local television dance show circa 1965, and must choose between uncertain fame and an Ivy League college. A fair nostalgic look at the period. Leitch is the son of psychedelic folk singer Donovan.
1988 (PG) 96m/C Donovan Leitch, Jennifer Runyon, Scott Plank, Joe Pantoliano; *Dir:* Mark Rosenthal. **VHS, Beta, LV $19.98** *ORI* 𝄫𝄫

In the Custody of Strangers A teenager's parents refuse to help when he is arrested for being drunk and he ends up spending the night in jail. Realistic handling of a serious subject. Sheen and Estevez play the father and son, mimicking their real-life relationship. Made for television.
1982 100m/C Martin Sheen, Jane Alexander, Emilio Estevez, Kenneth McMillan, Ed Lauter, Matt Clark, John Hancock; *Dir:* Robert Greenwald. **VHS** *MOV* 𝄫𝄫𝄫

In Dangerous Company Beautiful babe balances beaux and battles abound. A brow lowering experience.
1988 (R) 92m/C Tracy Scoggins, Cliff De Young, Chris Mulkey, Henry Darrow, Steven Keats, Richard Portnow; *Dir:* Rubin Preuss. **VHS $79.98** *ORI* 𝄫

In the Days of the Thundering Herd & the Law & the Outlaw In the first feature, a pony express rider sacrifices his job to accompany his sweetheart on a westward trek to meet her father. In the second show, a fugitive falls in love with a rancher's daughter and risks recognition.
1914 76m/B Tom Mix, Myrtle Stedman; *Dir:* Tom Mix. **VHS, Beta $24.98** *CCB* 𝄫

In Deadly Heat Crime, revenge, murder and urban warfare are done once again.
1987 98m/C William Dame, Catherine Dee, M.R. Murphy; *Dir:* Don Nardo. **VHS, Beta** *JEF* 𝄫

In Desert and Wilderness Two kidnapped children escape and are thrust into the wilds of Africa, where they must survive with wits and courage. Based on a novel by Nobel prize-winning author Henyk Sienkiewicz. In Polish with English subtitles.
1973 144m/C *PL Dir:* Wladyslaw Slesicki. **VHS $49.95** *FCT* 𝄫𝄫½

In Eagle Shadow Fist Japanese invasion forces meet their match when an unconventional resistance fighter sabotages their advance.
198? 90m/C Jackie Chan. **VHS, Beta $24.95** *ASE* 𝄫

In a Glass Cage A young boy who was tortured and sexually molested by a Nazi official seeks revenge. Years after the war, he shows up at the Nazi's residence in Spain and is immediately befriended by his tormentor's family. Confined to an iron lung, the official is at the mercy of the young man who tortuously acts out the incidents detailed in the Nazi's journal. Suspenseful and extremely graphic, it's well crafted and meaningful but may be just too horrible to watch. Sexual and violent acts place it in the same league as Pasolini's infamous "Salo, or The 120 Days in Sodom." Definitely not for the faint of heart. Also known as "Tras el Cristal."
1986 110m/C *SP* Gunter Meisner, David Sust, Marisa Paredes, Gisela Echevarria; *Dir:* Agustin Villaronga. **VHS $79.95** *CCN, FCT* 𝄫𝄫½

In Gold We Trust Vincent leads an elite battalion of fighters against a group of rogue American MIAs (!) laying waste to the Vietnemese countryside.
1991 89m/C Jan-Michael Vincent, Sam Jones; *Dir:* P. Chalong. **VHS $89.98** *AIP* 𝄫½

In the Good Old Summertime This pleasant musical version of "The Shop Around the Corner" tells the story of two bickering co-workers who are also anonymous lovelorn pen pals. Minnelli made her screen debut at 18 months in the final scene. Featured songs include, "I Don't Care" and "Play that Barbershop Chord." The original theatrical trailer is included on the laserdisc format.
1949 104m/C Judy Garland, Van Johnson, S.Z. Sakall, Buster Keaton, Spring Byington, Liza Minnelli, Clinton Sundberg; *Dir:* Robert Z. Leonard. **VHS, Beta, LV $19.95** *MGM, PIA* 𝄫𝄫𝄫

In Harm's Way Overdine story about two naval officers and their response to the Japanese attack at Pearl Harbor. Even the superb cast members (and there are plenty of them) can't overcome the incredible length and overly intricate plot.
1965 165m/B John Wayne, Kirk Douglas, Tom Tryon, Patricia Neal, Paula Prentiss, Brandon de Wilde, Burgess Meredith, Stanley Holloway, Henry Fonda, Dana Andrews, Franchot Tone, Jill Haworth, George Kennedy, Carroll O'Connor, Patrick O'Neal, Slim Pickens, Bruce Cabot, Larry Hagman, Hugh O'Brian, Jim Mitchum, Barbara Bouchet; *Dir:* Otto Preminger. **VHS, Beta, LV $14.95** *PAR, BUR* 𝄫𝄫

In the Heat of the Night A wealthy industrialist in a small Mississippi town is murdered. A black homicide expert is asked to help solve the murder, despite resentment on the part of the town's chief of police. Powerful script with underlying theme of racial prejudice is served well by taut direction and powerhouse performances. Poitier's memorable character Virgil Tibbs appeared in two more pictures, "They Call Me Mister Tibbs" and "The Organization."
1967 109m/C Sidney Poitier, Rod Steiger, Warren Oates, Lee Grant; *Dir:* Norman Jewison. Academy Awards '67: Best Actor (Steiger), Best Picture; Edgar Allan Poe Awards '67: Best Screenplay; Golden Globe Awards '68: Best Film—Drama. **VHS, Beta, LV $19.98** *MGM, FOX, BTV* 𝄫𝄫𝄫½

In the Heat of Passion A low-budget ripoff of "Body Heat" and "Fatal Attraction" starring Kirkland as a married woman who involves a much younger man in a very steamy affair that leads to murder. This movie wants to be a thriller, but is little more than a soft-core sex flick. Also available in an unrated version that contains explicit footage.
1991 (R) 84m/C Sally Kirkland, Nick Corri, Jack Carter, Michael Greene; *W/Dir:* Rodman Flender. **VHS, LV $89.95** *COL* 𝄫½

In Heaven There is No Beer? Documentary presents the history of polka dancing and music, with Blank at his American low-culture best.
1983 51m/C *Dir:* Les Blank. **VHS, Beta $49.95** *NEW, FFM* 𝄫½

In Hot Pursuit Convicted for drug smuggling, two young men stage a daring but not too exciting escape from a southern prison.
1982 90m/C Bob Watson, Don Watson, Debbie Washington. **VHS, Beta** *PGN* 𝄫½

In the King of Prussia Recreation of the trial of the "Plowshares Eight" radicals who were jailed for sabotaging two nuclear missiles at an electric plant in Pennsylvania.
1982 92m/C Martin Sheen, Rev. Daniel Berrigan. **VHS, Beta $59.95** *MPI, BUL, NEW* 𝄫½

In the Land of the Owl Turds The story of a decidely eccentric young man who drives a dada art gallery on wheels while searching for love. Blank's senior thesis film.
1989 30m/C *Dir:* Harrod Blank. **VHS, Beta $39.95** *FFM* 𝄫½

In-Laws A wild comedy with Falk, who claims to be a CIA agent, and Arkin, a dentist whose daughter is marrying Falk's son. The fathers foil a South American dictator's counterfeiting scheme in a delightfully convoluted plot.
1979 (PG) 103m/C Peter Falk, Alan Arkin; *Dir:* Arthur Hiller. **VHS, Beta, LV $19.98** *WAR* 𝄫𝄫𝄫

In Like Flint Sequel to "Our Man Flint" sees our dapper spy confronting an organization of women endeavoring to take over the world. Spy spoofery at its low-level best.
1967 107m/C James Coburn, Lee J. Cobb, Anna Lee, Andrew Duggan, Jean Hall; *Dir:* Gordon Douglas. **VHS, Beta, LV $19.98** *FOX* 𝄫𝄫½

In a Lonely Place Hollywood screenwriter has an affair with a starlet while under suspicion of murder. Bogart is outstanding as the hard-drinking brawler. Written by Andrew Solt.
1950 93m/B Humphrey Bogart, Gloria Grahame, Frank Lovejoy, Carl Benton Reid, Art Smith, Jeff Donnell; *Dir:* Nicholas Ray. **VHS, Beta $69.95** *COL* 𝄫𝄫𝄫½

In Love with an Older Woman Ritter takes fewer pratfalls than usual in this film; after all, it's a romantic comedy. Adapted from "Six Months with an Older Woman," the novel by David Kaufelt.
1982 100m/C John Ritter, Karen Carlson, Jamie Rose, Robert Mandan, Jeff Altman, George Murdock; *Dir:* Jack Bender. **VHS, Beta $14.95** *FRH* 𝄫𝄫

In Love and War Woods plays Navy pilot Jim Stockdale, whose plane was grounded over hostile territory and who endured nearly eight years of torture as a POW. Meanwhile, wife Sybil is an organizer of POW wives back in the States. Low octane rendering of the US Navy Commander's true story. Aaron earlier directed "The Miracle Worker," and you may recognize Ngor from "The Killing Fields."
1991 (R) 96m/C James Woods, Jane Alexander, Haing S. Ngor, Concetta Tomei, Richard McKenzie, James Pax; *Dir:* Paul Aaron. **VHS $89.95** *VMK* 𝄫𝄫½

In Memoriam Tragic and gripping drama about one man's ability, albeit too late, to cope with his unforgettable past. In Spanish with English subtitles.
1976 96m/C *SP* Geraldine Chaplin, Jose Luis Gomez. **VHS $59.95** *FCT* 𝄫𝄫½

In Memoriam A man must come to terms with a past he would rather leave behind, only to learn that his inability to express his feelings has caused a great tragedy. In Spanish with English subtitles.
1976 96m/C *SP* Geraldine Chaplin, Jose Luis Gomez; *Dir:* Enrique Braso. **VHS $24.98** *TAM* 𝄫𝄫½

In the Mood Based on fact, this is the story of teenager Sonny Wisecarver, nicknamed "The Woo Woo Kid," who in 1944, seduced two older women and eventually landed in jail after marrying one of them. Look for Wisecarver in a cameo role as a mailman in the film.
1987 (PG-13) 98m/C Patrick Dempsey, Beverly D'Angelo, Talia Balsam; *Dir:* Phil Alden Robinson. VHS, Beta, LV $19.98 *KRT, LHV, WAR* ♫♫½

In 'n Out An American loser ventures to Mexico to recover a long-lost inheritance, meeting with mildly comic obstacles along the way.
1986 85m/C Sam Bottoms, Pat Hingle. VHS, Beta $19.95 *STE, NWV* ♫½

In Name Only Somber drama about a heartless woman who marries for wealth and prestige and holds her husband to a loveless marriage. Based on "Memory of Love" by Bessie Brewer.
1939 102m/B Carole Lombard, Cary Grant, Kay Francis; *Dir:* John Cromwell. VHS, Beta, LV $19.95 *CCB, RKO, MLB* ♫♫

In the Name of the Pope-King An acclaimed historical epic about the politics and warfare raging around the 1867 Italian wars that made the Pope the sole ruler in Rome. In Italian with English subtitles.
1985 115m/C Nino Manfredi; *Dir:* Luigi Magni. VHS, Beta *WBF, VPI, FCT* ♫♫♫

In the Navy Abbott and Costello join the Navy in one of four military-service comedies they cranked out in 1941 alone. The token narrative involves about a singing star (Powell) in uniform to escape his female fans, but that's just an excuse for classic A & C routines. With the Andrews Sisters.
1941 85m/B Lou Costello, Bud Abbott, Dick Powell, The Andrews Sisters; *Dir:* Arthur Lubin. VHS, LV $14.95 *MCA* ♫♫½

In Old Caliente Rogers battles the bad guys in the frontier town of Caliente.
1939 60m/B Roy Rogers, Mary Hart, George "Gabby" Hayes. VHS, Beta $9.95 *NOS, VCN, DVT* ♫½

In Old California Plucky story of a young Boston pharmacist who searches for success in the California gold rush and runs into the local crime boss. Also available colorized.
1942 89m/B John Wayne, Patsy Kelly, Binnie Barnes, Albert Dekker; *Dir:* William McGann. VHS $14.98 *REP* ♫♫½

In Old Cheyenne Bank holdups, cattle rustling, fist fights, and even car crashes mix with some good ol' guitar plunking in this western.
1941 60m/B Roy Rogers, George "Gabby" Hayes. VHS, Beta $19.95 *NOS, CAB* ♫

In an Old Manor House Residents of an old house are tormented by the vindictive ghost of a murdered adultress whose husband didn't approve of the, uh, affection she showed his son from a previous marriage. Based on the play by Stanislaw Witkiewicz. In Polish with English subtitles.
1984 90m/C PL *Dir:* Jerzy Kotkowski. VHS $49.99 *FCT* ♫♫½

In Old Montana A Cavalry lieutenant tries to settle a cattleman vs. sheep-herding war.
1939 60m/B Fred Scott, Jeanne Carmen, Harry Harvey, John Merton; *Dir:* Raymond K. Johnson. VHS, Beta $19.95 *VDM* ♫

In Old New Mexico The Cisco Kid and Pancho reveal the murderer of an old woman—a mysterious doctor who was after an inheritance.
1945 60m/B Duncan Renaldo, Martin Garralaga, Gwen Kenyon, Pedro de Cordoba. VHS, Beta $19.95 *NOS, UHV, VYY* ♫

In Old Santa Fe Cowboy star Maynard's best western, and the first film ever with Autry and Burnette. Maynard is framed for murder and his pals save him.
1934 60m/B Ken Maynard, Gene Autry, Smiley Burnette, Evelyn Knapp, George "Gabby" Hayes, Kenneth Thomson, Wheeler Oakman, George Cheseboro; *Dir:* David Howard, Joseph Kane. VHS, Beta $19.95 *NOS, VCN, DVT* ♫♫½

In Person Spry comedy about a movie star who disguises herself to take a vacation incognito and falls in love with a country doctor who is unaware of her identity. Songs by Oscar Levant include "Don't Mention Love to Me," "Out of Sight, Out of Mind," and "Got a New Lease on Love."
1935 87m/B Ginger Rogers, George Brent, Alan Mowbray, Grant Mitchell; *Dir:* William A. Seiter. VHS, Beta $19.98 *RKO* ♫♫½

In Praise of Older Women Berenger is just right as a young Hungarian who is corrupted by World War II and a large number of older women. Based on a novel by Stephen Vizinczey.
1978 110m/C CA Karen Black, Tom Berenger, Susan Strasberg, Helen Shaver, Alexandra Stewart; *Dir:* George Kaczender. VHS, Beta, LV *FOX, SUE* ♫♫

In the Realm of the Senses Taboo-breaking story of a woman and man who turn their backs on the militaristic rule of Japan in the mid-1930s by plunging into an erotic and sensual world all their own. Striking, graphic work that was seized by U.S. customs when it first entered the country. Violent with explicit sex, and considered by some critics as pretentious, while others call it Oshima's masterpiece.
1976 105m/C JP FR Tatsuya Fuji, Eiko Matsuda, Aio Nakajima, Meika Seri; *Dir:* Nagisa Oshima. VHS $89.95 *FXL* ♫♫♫

In the Region of Ice Award-winning short reveals the complexities of the relationship between a nun and a disturbed student.
1976 38m/B Fionnula Flanagan, Peter Lempert; *Dir:* Peter Werner. New York Film Festival '76: Best Film—Drama. VHS, Beta *BFA* ♫♫♫

In Search of Anna When a convict is released from jail, he hits the road with a flighty model in search of the girlfriend who corresponded with him in prison. There auto be a law.
1979 90m/C Judy Morris, Richard Moir, Chris Hayward, Bill Hunter; *Dir:* Esben Storm. VHS, Beta $24.95 *AHV* Woof!

In Search of the Castaways Stirring adventure tale of a teenage girl and her younger brother searching for their father, a ship's captain lost at sea years earlier. Powerful special effects and strong cast make this a winning Disney effort. Based on a story by Jules Verne.
1962 98m/C GB Hayley Mills, Maurice Chevalier, George Sanders, Wilfrid Hyde-White, Michael Anderson Jr.; *Dir:* Robert Stevenson. VHS, Beta $19.99 *DIS* ♫♫♫

In Search of a Golden Sky Following their mother's death, a motley group of children move in with their secluded cabin-dwelling uncle, much to the righteous chagrin of the welfare department.

1984 (PG) 94m/C Charles Napier, George Flower, Cliff Osmond; *Dir:* Jefferson Richard. VHS, Beta $79.98 *FOX* ♫♫

In Search of Historic Jesus Docudrama of sorts that attempts to pull together a careful tabulation of data about Jesus Christ, who was hardly known to the historians of his time. Low-budget presentation at odds with serious topic. From the makers of the equally suspect "In Search of Noah's Ark."
1979 (PG) 91m/C John Rubinstein, John Anderson; *Dir:* Henning Schellerup; *Nar:* Brad Crandall. VHS, Beta, LV $79.98 *LIV, UHV, LME* ♫

In the Secret State Inspector Strange finds himself faced with the apparent suicide of a colleague on the day of his own retirement from the force. His private investigation into the case finds him mired in murder and espionage.
19?? 107m/C Frank Finlay. VHS $79.95 *SVS* ♫♫

In the Shadow of Kilimanjaro A pack of hungry baboons attacks innocent people in the wilderness of Kenya. Based on a true story but the characters are contrived and just plain stupid at times. Quite gory as well. Filmed on location in Kenya.
1986 (R) 105m/C GB John Rhys-Davies, Timothy Bottoms, Michele Carey, Irene Miracle, Calvin Jung, Donald Blakely; *Dir:* Raju Patel. VHS, Beta $19.95 *IVE* Woof!

In a Shallow Grave Shallow treatment of James Purdy's novel about a WWII veteran, badly disfigured in combat at Guadalcanal, who returns to Virginia in an attempt to return to the life he once had with the woman he still loves. Handsome Biehn plays the disfigured vet, Mueller's his ex-fiancee, and Dempsey's the drifter who adds a bisexual hypotenuse to the love triangle. Pokey American Playhouse coproduction.
1988 (R) 92m/C Michael Biehn, Maureen Mueller, Patrick Dempsey, Michael Beach, Thomas Boyd Mason; *Dir:* Kenneth Bowser. VHS *WAR* ♫♫

In the Spirit New Age couple move from Beverly Hills to New York, where they meet an air-headed mystic and are forced to hide from a murderer. Falk gives solid performance in sometimes rocky production.
1990 (R) 94m/C Elaine May, Marlo Thomas, Jeannie Berlin, Peter Falk, Melanie Griffith, Olympia Dukakis, Chad Burton, Thurn Hoffman; *Dir:* Sandra Seacat. VHS, Beta $29.95 *ACA* ♫♫½

In the Steps of a Dead Man A family grieving for a fallen soldier is earmarked for evil by a young man who bears an uncanny resemblance to the dead man. A friend of the family has her suspicions and must find out the truth before something terrible happens.
1974 71m/C Skye Aubrey, John Nolan. VHS, Beta $29.95 *THR* ♫½

In This House of Brede A sophisticated London widow turns her back on her worldly life to become a cloistered Benedictine nun. Rigg is outstanding as the woman struggling to deal with the discipline of faith. Made for television and based on the novel by Rumer Godden.
1975 105m/C Diana Rigg; *Dir:* George Schaefer. VHS, Beta *LCA, OM* ♫♫

In This Our Life Histrionic melodrama handled effectively by Huston about a nutsy woman who steals her sister's husband, rejects him, manipulates her whole family and eventually leads herself to ruin. Vintage

star vehicle for Davis, who chews scenery and fellow actors alike. Adapted from Ellen Glasgon's novel by Howard Koch. Score by Max Steiner.
1942 101m/B Bette Davis, Olivia de Havilland, Charles Coburn, Frank Craven, George Brent, Dennis Morgan, Billie Burke, Hattie McDaniel, Lee Patrick, Walter Huston, Ernest Anderson; **Dir:** John Huston. **VHS, Beta $19.95** *MGM ⅃⅃½*

In Too Deep A rock star and a beautiful woman are physically drawn to one another in spite of the fact that any contact between them could have disasterous results.
1990 (R) 106m/C Hugo Race, Santha Press, Rebekah Elmaloglou, John Flaus, Dominic Sweeney; **Dir:** Colin South. **VHS, Beta $89.95** *PAR ⅃½*

In for Treatment A terminally ill cancer patient, lost in the bureaucracy of the hospital, fights to rediscover the joy of life. Strong and disturbing story. In Dutch with English subtitles.
1982 92m/C *NL* Helmut Woudenberg, Frank Groothof, Hans Man Int Veld; **Dir:** Eric Van Zuylen, Marja Kok. **VHS, Beta** *IFE ⅃⅃⅃*

In Trouble Hard-to-follow and frequently aimless story of a pregnant girl who invents a fake rape story that her dull-witted brothers take seriously, embarking on a cross-country search for the culprit.
1967 82m/C *CA* Julie LaChapelle, Katherine Mousseau, Daniel Pilon; **Dir:** Gilles Carle. **VHS, Beta $59.95** *NWV Woof!*

In Which We Serve Much stiff upperlipping in this classic that captures the spirit of the British Navy during WWII. The sinking of the destroyer HMS Torrin during the Battle of Crete is told via flashbacks, with an emphasis on realism that was unusual in wartime flag-wavers. Features the film debuts of Johnson and Attenborough, and the first Lean directorial effort. Coward received a special Oscar for his "outstanding production achievement," having scripted, scored, codirected and costarred. Oscar nominated for Best Picture, Original Screenplay.
1942 114m/B *GB* Noel Coward, John Mills, Bernard Miles, Celia Johnson, Kay Walsh, James Donald, Richard Attenborough; **Dir:** Noel Coward, David Lean. National Board of Review Awards '42: Best Film; New York Times Ten Best List '42: Best Film. **VHS, Beta $19.95** *NOS, EMB, SUE ⅃⅃⅃½*

In the White City A sailor jumps ship in Lisbon and uses a movie camera to record himself and his search through the twisted streets and alleys of the city for something or someone to connect to. In French, Portuguese, and German with English subtitles.
1983 108m/C *PT SI* Bruno Ganz, Teresa Madruga; **Dir:** Alain Tanner. **VHS $79.95** *NYF ⅃⅃⅃*

In Your Face Little known blaxploitation film about a black family harrassed in a white suburb. A black motorcycle gang comes to their rescue. Originally titled "Abar, the First Black Superman."
1977 (R) 90m/C J.Walter Smith, Tobar Mayo, Roxie Young, Tina James; **Dir:** Frank Packard. **VHS $59.95** *XVC, HHE ⅃½*

Inch High Private Eye Animated yarn about Inch High, the world's smallest private eye, who solves mysteries for the Finkerton Organization.
1973 60m/C VHS, Beta $19.95 *HNB ⅃½*

The Incident A gritty, controversial (at-the-time) drama about two abusive thugs who take over a New York subway car late at night, humiliating and terrorizing each

passenger in turn. Film debut for Sheen and Musante as the thugs.
1967 99m/B Martin Sheen, Tony Musante, Beau Bridges, Ruby Dee, Jack Gilford, Thelma Ritter, Brock Peters, Ed McMahon, Gary Merrill, Donna Mills, Jan Sterling, Mike Kellin; **Dir:** Larry Peerce. **VHS, Beta, LV $59.98** *FOX ⅃⅃½*

The Incident Political thriller set during World War II in Lincoln Bluff, Colorado. Matthau stars as the small-town lawyer who must defend a German prisoner of war accused of murder at nearby Camp Bremen. An all-star cast lends powerful performances to this riveting drama.
1989 95m/C Walter Matthau, Susan Blakely, Harry Morgan, Robert Carradine, Barnard Hughes, Peter Firth. **VHS $79.95** *CAF ⅃⅃⅃*

Incident at Channel Q A heavy-metal rock musical about a suburban neighborhood declaring war on a chain-and-leather deejay.
1986 85m/C Al Corley, Bon Jovi, Deep Purple, Motley Crue, The Scorpions. **VHS, Beta, LV $19.95** *MVD, BMG, RCA ⅃½*

Incident at Map-Grid 36-80 A Soviet-made combination of espionage and sea-battle. Gripping and fascinating look at the U.S. from the pre-Glasnost era.
1983 85m/C *RU* **VHS** *TPV ⅃*

Incoming Freshmen Teen sex-venture about a young innocent girl who discovers sex when she enrolls in a liberal co-ed institution.
1979 (R) 84m/C Ashley Vaughn, Debralee Scott, Leslie Blalock, Richard Harriman, Jim Overbey; **Dir:** Glenn Morgan, Eric Lewald. **VHS, Beta $39.95** *MCA ⅃½*

An Inconvenient Woman When an older married man has an affair with a young seductress, there's suicide, murder, a nasty gossip columnist, a reporter, a knowing society wife, and a lot of cover-ups among the rich and famous. From the novel by Dominick Dunne and based on the Alfred Bloomingdale scandal. Originally a made for television mini-series.
1991 126m/C Rebecca DeMornay, Jason Robards Jr., Jill Eikenberry, Peter Gallagher, Roddy McDowall, Chad Lowe. **VHS, Beta $79.95** *ABC, PSM ⅃⅃*

Incredible Detectives A dog, a cat and a crow team up to search for a boy who has been kidnapped by a trio of hoods.
1979 23m/C VHS, Beta $19.95 *WOV ⅃⅃*

The Incredible Hulk Bixby is a scientist who achieves superhuman strength after he is exposed to a massive dose of gamma rays. But his personal life suffers, as does his wardrobe. Ferrigno is the Hulkster. The pilot for the television series is based on the Marvel Comics character.
1977 94m/C Bill Bixby, Susan Sullivan, Lou Ferrigno, Jack Colvin; **Dir:** Kenneth Johnson. **VHS, Beta, LV $19.95** *MCA ⅃½*

The Incredible Hulk Returns The beefy green mutant is back and this time he wages war against a Viking named Thor. Very little substance in this made for television flick, so be prepared to park your brain at the door. Followed by "The Trial of the Incredible Hulk."
1988 100m/C Bill Bixby, Lou Ferrigno, Jack Colvin, Lee Purcell, Charles Napier, Steve Levitt; **Dir:** Nick Corea. **VHS, LV $19.95** *STE, NWV ⅃½*

Incredible Hulk, Volume 1 The secret origin of the Hulk is told in this collection of three episodes from the animated series.

1985 70m/C VHS, Beta, LV $39.95 *PSM ⅃½*

Incredible Hulk, Volume 2 In this trio of cartoons from the television series, the great green man battles gigantic mice, spiders and lizards, the huge and hungry Gammatron monster, and his arch-nemesis Dr. Octopus.
1985 70m/C VHS, Beta $39.95 *PSM ⅃½*

The Incredible Journey A labrador retriever, bull terrier and Siamese cat mistake their caretaker's intentions when he leaves for a hunting trip, believing he will never return. The three set out on a 250 mile adventure-filled trek across Canada's rugged terrain. Entertaining family adventure from Disney taken from Sheila Burnford's book.
1963 80m/C Dir: Fletcher Markle. **VHS, Beta $69.95** *DIS ⅃⅃½*

The Incredible Journey of Dr. Meg Laurel A young doctor returns to her roots to bring modern medicine to the Appalachian mountain people during the 1930's. Made for television featuring Wagner in a strong performance.
1979 143m/C Lindsay Wagner, Jane Wyman, Dorothy McGuire, James Woods, Gary Lockwood; **Dir:** Guy Green. **VHS, Beta $59.95** *COL ⅃⅃⅃*

Incredible Manitoba Animation Prize-winning animated shorts, brought together in honor of the Film Board of Canada's 50th Anniversary. Includes Richard Condie's Oscar-nominated "The Big Snit," and his "Getting Started;" Cordell Barker's "The Cat Came Back" and more fun for the entire family.
1990 51m/C *CA* **VHS, Beta $29.95** *WTA, FCT ⅃⅃⅃*

Incredible Master Beggars Against all odds, the Beggars challenge the Great Iron Master, using "Tam Leg" tactics versus the "Iron Cloth" fighting style.
1982 (R) 88m/C Tan Tao Liang, Ku Feng, Han Kuo Tsai, Li Tang Ming, Li Hai Sheng, Lui I. Fan, Pan Yao Kun. **VHS, Beta $59.95** *GEM ⅃*

Incredible Melting Man Two transformations change an astronaut's life after his return to earth. First, his skin starts to melt, then he displays cannibalistic tendencies. Hard to swallow. Special effects by Rick Baker.
1977 (R) 85m/C Alex Rebar, Burr de Benning, Cheryl "Rainbeaux" Smith; **Dir:** William Sachs. **VHS, Beta $69.98** *LIV, VES Woof!*

The Incredible Mr. Limpet Limp comedy about a nebbish bookkeeper who's transformed into a fish, fulfilling his aquatic dreams. Eventually he falls in love with another fish, and helps the U.S. Navy find Nazi subs during World War II. Partially animated, beloved by some, particularly those under the age of seven. Based on Theodore Pratt's novel.
1964 99m/C Don Knotts, Jack Weston, Carole Cook, Andrew Duggan, Larry Keating, Elizabeth McRae; **Dir:** Arthur Lubin. **VHS, Beta, LV $19.98** *WAR ⅃⅃*

The Incredible Petrified World Divers are trapped in an underwater cave when volcanic eruptions begin. Suffocating nonsense.
1958 78m/B John Carradine, Allen Windsor, Phyllis Coates, Lloyd Nelson, George Skaff; **Dir:** Jerry Warren. **VHS $19.95** *NOS, SNC, MON Woof!*

The Incredible Rocky Mountain Race The townspeople of St. Joseph, fed up with Mark Twain's destructive feud with a neighbor, devise a shrewd scheme to rid the town of the troublemakers by sending them on a road race through the West. Made for television comedy starring the F-Troop.
1977 97m/C Christopher Connelly, Forrest Tucker, Larry Storch, Mike Mazurki; *Dir:* James L. Conway. **VHS, Beta $19.95** STE, MAG 🎬🎬

The Incredible Sarah A limp adapted stage biography from ''Reader's Digest'' of the great actress Sarah Bernhardt. Jackson chews scenery to no avail.
1976 (PG) 105m/C Glenda Jackson, Daniel Massey, Yvonne Mitchell, Douglas Wilmer; *Dir:* Richard Fleischer. **VHS, Beta $19.95** STE, LTG, IME 🎬🎬

The Incredible Shrinking Man Adapted by Richard Matheson from his own novel, the sci-fi classic is a philosophical thriller about a man who is doused with radioactive mist and begins to slowly shrink. His new size means that everyday objects take on sinister meaning and he must fight for his life in an increasingly hostile, absurd environment. A surreal, suspenseful allegory with impressive special effects. Endowed with the tension usually reserved for Hitchcock films.
1957 81m/B Grant Williams, Randy Stuart, April Kent, Paul Langton, Raymond Bailey; *Dir:* Jack Arnold. **VHS, Beta, LV $14.98** MCA, PIA, MLB 🎬🎬🎬½

The Incredible Shrinking Woman ''...Shrinking Man'' spoof and inoffensive social satire finds household cleaners producing some strange side effects on model homemaker Tomlin, slowly shrinking her to doll-house size. She then encounters everyday happenings as big tasks and not so menial. Her advertising exec husband has a hand in the down-sizing. Sight gags abound but the cuteness wears thin by the end.
1981 (PG) 89m/C Lily Tomlin, Charles Grodin, Ned Beatty, Henry Gibson; *Dir:* Joel Schumacher. **VHS, Beta, LV $39.95** MCA 🎬🎬½

The Incredible Two-Headed Transplant Mad scientist Dern has a criminal head transplanted on to the shoulder of big John Bloom and the critter runs amuck. Low-budget special effects guaranteed to give you a headache or two. Watch for onetime Pat ''Marilyn Munster'' Priest in a bikini.
1971 (PG) 88m/C Bruce Dern, Pat Priest, Casey Kasem, Albert Cole, John Bloom, Berry Kroeger; *Dir:* Anthony M. Lanza. **VHS** TRV Woof!

Incredibly Strange Creatures Who Stopped Living and Became Mixed-Up Zombies The infamous, super-cheap horror spoof that established Sheckler (for whom Cash Flagg is a pseudonym), about a carny side show riddled with ghouls and bad rock bands. Assistant cinematographers include the young Laszlo Kovacs and Vilmos Zsigmond. Also known as ''The Teenage Psycho Meets Bloody Mary,'' it's a must-see for connoisseurs of cult and camp.
1963 90m/C Cash Flagg, Carolyn Brandt, Brett O'Hara, Atlas King, Sharon Walsh; *Dir:* Ray Dennis Steckler. **VHS, Beta $39.95** CAM Woof!

Incubus A doctor and his teenaged daughter settle in a quiet New England community, only to encounter the incubus, a terrifying supernatural demon who enjoys sex murders. Offensive trash.

1982 (R) 90m/C CA John Cassavetes, Kerrie Keane, Helen Hughes, Erin Flannery, John Ireland; *Dir:* John Hough. **VHS, Beta, LV $19.98** VES Woof!

An Indecent Obsession Colleen McCullough's bestseller about a wartime asylum on the day the Japanese surrendered. Hughes plays the nurse with a heart of gold.
1985 100m/C AU Wendy Hughes, Bill Hunter, Bruno Lawrence; *Dir:* Lex Marinos. **VHS, Beta** QNE 🎬🎬½

Independence Dramatically recreates the debates, concerns, and events that led up to the U.S. Declaration of Independence in 1776.
1976 30m/C Eli Wallach, Anne Jackson, Pat Hingle, Patrick O'Neal; *Dir:* John Huston; *Nar:* E.G. Marshall. **VHS, Beta $19.98** MPI, FHF, KAR 🎬🎬½

Independence Day Uneven romantic drama about a small-town photographer, yearning for the big city, who falls in love with a racing car enthusiast, whose sister is a battered wife. Strong cast, particularly Wiest as the abused wife, is left occasionally stranded by a script reaching too hard for social significance. Also known as ''Follow Your Dreams.''
1983 (R) 110m/C Kathleen Quinlan, David Keith, Frances Sternhagen, Dianne Wiest, Cliff DeYoung, Richard Farnsworth, Josef Sommer, Cheryl ''Rainbeaux'' Smith; *Dir:* Robert Mandel. **VHS, Beta $19.98** WAR, CCB 🎬🎬

The Indestructible Man Chaney, electrocuted for murder and bank robbery, is brought back to life by a scientist. Naturally, he seeks revenge on those who sentenced him to death. Chaney does the best he can with the material.
1956 70m/B Lon Chaney Jr., Marian Carr, Casey Adams; *Dir:* Jack Pollexfen. **VHS $16.95** NOS, MRV, SNC 🎬

Indian Paint A pleasant children's film about an Indian boy's love for his horse and his rite of passage. Silverheels was Tonto to television's ''The Lone Ranger,'' while Crawford was the ''Rifleman's'' son.
1964 90m/C Jay Silverheels, Johnny Crawford, Pat Hogan, Robert Crawford Jr., George Lewis; *Dir:* Norman Foster. **VHS, Beta $19.95** UHV 🎬🎬

The Indian Runner In Penn's debut as a writer-director, he brings us the story of two brothers in Nebraska during the late '60s, who are forced to change their lives with the loss of their family farm. Joe is the good cop and family man who can't deal with the rage of his brother Frank, who has just returned from Vietnam and is turning to a life of crime. Penn does a decent job of representing the struggle between the responsible versus the rebellious side of human nature. Based on the song ''Highway Patrolman'' by Bruce Springsteen.
1991 (R) 127m/C David Morse, Viggo Mortensen, Sandy Dennis, Charles Bronson, Valeria Golino, Patricia Arquette, Dennis Hopper; *W/Dir:* Sean Penn. **VHS $94.99** MGM 🎬🎬

Indian Uprising An army captain tries to keep local settlers from causing trouble with Geronimo and his people when gold is discovered on their territory.
1951 75m/C George Montgomery, Audrey Long, Carl Benton Reid, John Baer, Joe Sawyer; *Dir:* Ray Nazzaro. **VHS, Beta $14.99** GKK 🎬🎬

Indiana Jones and the Last Crusade In this, the third and last Indiana Jones adventure, we find Jones, the fearless archaeologist, once again up against the Nazis in a race to find the Holy Grail. Introduces Connery as Indy's father and explains the origin of the infamous fedora. More adventures, exotic places, dastardly villains, and daring escapes than ever before; a must for Indy fans and a bit tedious for others.
1989 (PG) 126m/C Harrison Ford, Sean Connery, Denholm Elliott, Alison Doody, Julian Glover, John Rhys-Davies, River Phoenix; *Dir:* Steven Spielberg. **VHS, Beta, LV, SVS, 8mm $24.95** PAR, SUP, TLF 🎬🎬🎬

Indiana Jones and the Temple of Doom Daredevil archaeologist Indiana Jones is back and on the trail of the legendary Ankara Stone and a ruthless cult that has enslaved hundreds of children. Enough action for ten movies, special effects galore, and the usual booming John Williams score make it a cinematic roller coaster ride, with less regard for plot and pacing than the original. The scene where a heart is plucked out of a chest is gory, and may be too much for children. Though second of three Indiana Jones movies, it's actually a prequel to ''Raiders of the Lost Ark.''
1984 (PG) 118m/C Harrison Ford, Kate Capshaw, Ke Huy Quan, Amrish Puri; *Dir:* Steven Spielberg. Academy Awards '84: Best Visual Effects. **VHS, Beta, LV, 8mm $19.95** PAR, APD 🎬🎬🎬

India's Master Musicians This program presents India's master musicians including Imrat Khan, Lakshmi Shankar, Ashish Khan, Alla Rakha, Lakshmi, Viswanathan, Zakir Hussain and Ramanad Raghaven.
1982 60m/C VHS, Beta AUP

Indio A Marine-trained half-breed takes it on himself to save the Amazon rain forest. Not a peaceful demonstrator, his tactics are violent ones.
1990 (R) 94m/C Marvin Hagler, Francesco Quinn, Brian Dennehy; *Dir:* Anthony M. Dawson. **VHS, Beta $89.98** MED 🎬

Indio 2: The Revolt Greedy developers are cutting a highway into the Amazon jungle, destroying everything in their path. All is not lost though as U.S. Marine Sgt. Iron is on top of things, continuing his struggle to unite the Amazon tribes and save the rainforest, no matter what it takes.
1992 (R) 104m/C Marvin Hagler, Charles Napier. **VHS $89.98** LIV 🎬

Indiscreet Empty-headed romantic comedy about a fashion designer whose past catches up with her when her ex-lover starts romancing her sister. Nearly Swanson's swan song as she sings ''If You Haven't Got Love'' and other songs. No relation to the classic 1958 film with Cary Grant and Ingrid Bergman.
1931 81m/B Gloria Swanson, Ben Lyon, Barbara Kent; *Dir:* Leo McCarey. **VHS, Beta $19.95** NOS, MRV, REP 🎬½

Indiscreet A charming American diplomat in London falls in love with a stunning actress, but protects himself by saying he is married. Needless to say, she finds out. Stylish romp with Grant and Bergman at their sophisticated best. Adapted by Norman Krasna from his stage play ''Kind Sir.''
1958 100m/C Cary Grant, Ingrid Bergman, Phyllis Calvert; *Dir:* Stanley Donen. **VHS $19.98** REP, MLB 🎬🎬🎬

Indiscreet Indistinctive made-for-TV rehash of the charming 1958 Grant/Bergman romantic comedy in which a suave American falls for an English actress. Both versions derive from the play "Kind Sir" by Norman Krasna, but this one suffers by comparison.
1988 (PG) 94m/C Robert Wagner, Lesley-Anne Down, Maggie Henderson, Robert McBain, Jeni Barnett; *Dir:* Richard Michaels. **VHS $79.98** REP *&½*

Indiscretion of an American Wife Set almost entirely in Rome's famous Terminal Station, where an ill-fated couple say goodbye endlessly while the woman tries to decide whether to join her husband in the States. Co-written by Truman Capote and trimmed from 87 minutes upon U.S. release. Also known as "Indiscretion" and "Terminus Station."
1954 63m/C *IT* Jennifer Jones, Montgomery Clift, Richard Beymer; *Dir:* Vittorio De Sica. **VHS, Beta $19.95** NOS, FOX *&*

Indomitable Teddy Roosevelt Entertaining mix of newsreel footage and staged scenes traces the career of the feisty president.
1983 93m/C *Dir:* Harrison Engle; *Nar:* George C. Scott. **VHS, Beta** PSM, CHF *&&½*

Infamous Crimes Better known as "Philo Vance's Return." Philo investigates the murders of a playboy and his girlfriends. One of the last of the series whose character was based on the S.S. Van Dine stories.
1947 64m/B Willaim Wright, Terry Austin, Leon Belasco, Clara Blandick, Iris Adrian, Frank Wilcox; *Dir:* William Beaudine. **VHS, Beta, 8mm $29.95** VYY *&&*

Infamous Daughter of Fanny Hill Softcore drama about a Victorian floozy who seduces dozens of men.
1987 73m/C Stacy Walker. **VHS, Beta $29.95** MED *&*

Infernal Trio An attorney comes up with a plan involving two sisters (who are his lovers) which will make them both wealthy; but it will also make them both murderers. Sophisticated black comedy based on a true story and available in dubbed or subtitled versions.
1974 100m/C *FR IT* Romy Schneider, Michel Piccoli, Mascha Gonska; *Dir:* Francis Girod. **VHS $59.95** CVC *&&½*

Inferno Uneven occult horror tale about a young man who arrives in New York to investigate the mysterious circumstances surrounding his sister's death. Score by Keith Emerson. Dubbed.
1978 (R) 83m/C *IT* Leigh McCloskey, Elenora Giorgi, Irene Miracle, Sacha Pitoeff; *Dir:* Dario Argento. **VHS, Beta $59.98** FOX *&&*

Inferno in Paradise Mucho macho firefighter and bimbous gal pal photographer get hot while trailing a murderous pyromaniac. More like purgatory in the living room (that is, unless you thought Forsyth's earlier "Chesty Anderson, US Navy" was divine comedy).
1988 115m/C Richard Young, Betty Ann Carr, Jim Davis, Andy Jarrell, Dennis Chun; *Dir:* Ed Forsyth. **VHS $19.95** ACA *&*

The Informer Based on Liam O'Flaherty's novel about the Irish Sinn Fein Rebellion of 1922, it tells the story of a hard-drinking Dublin man (Mclaglan) who informs on a friend (a member of the Irish Republican Army) in order to collect a 20-pound reward. When his "friend" is killed during capture, he goes on a drinking spree instead of using the money, as planned, for passage to America. Director Ford allowed McLaglen to improvise his lines during the trial scene in order to enhance the realism, leading to excruciating suspense. Wonderful score by Max Steiner. Received Oscar nominations for Best Picture, Best Director, Best Score and Best Actor (McLagen).
1935 91m/B Victor McLaglen, Heather Angel, Wallace Ford, Margot Grahame, Joseph Sawyer, Preston Foster, Una O'Connor; *Dir:* John Ford. Academy Awards '35: Best Actor (McLaglen), Best Director (Ford), Best Screenplay, Best Original Score; National Board of Review Awards '35: Best Film; New York Film Critics Awards '35: Best Director (Ford), Best Film; Venice Film Festival '35: Best Screenplay; New York Times Ten Best List '35: Best Film. **VHS, Beta, LV $19.98** VID, MED, KOV *&&&&*

Infra-Man Infra-man is a bionic warrior who must destroy the Princess Dragon Mom and her army of prehistoric monsters in order to save the galaxy. Outlandish hokum is entertaining stuff for kids.
1976 (PG) 89m/C *JP* Wang Hsieh; *Dir:* Hua-Shan. **VHS, Beta $49.95** PSM *&&*

Inga Story of a teenage girl becoming a woman. Sensitive and well-acted.
1967 86m/B *SW* **VHS, Beta $24.95** VPL *&&*

Ingrid Documentary on the life and accomplishments of actress Ingrid Bergman. Clips from twenty-five of her greatest films, rare home movies from her childhood, and interviews with her peers form a portrait of the star.
1989 70m/B *GB* Liv Ullman, Angela Lansbury, Anthony Quinn, Jose Ferrer, Colleen Dewhurst, Ann Todd, John Gielgud, Dirk Bogarde, Honor Blackman, Felix Aylmer, Hermione Baddeley, Mai Zetterling, Cecil Parker, Wilfrid Hyde-White, Jack Watling, Mervyn Johns, Ernest Thesiger, Basil Radford, Naunton Wayne, Ian Fleming; *Dir:* Ken Annakin, Arthur Crabtree, Harold French. National Board of Review Awards '49: 10 Best Films of the Year. **VHS, Beta, 3/4U $195.00** WOM *&&*

Inherit the Wind Powerful courtroom drama, based on the Broadway play, is actually a fictionalized version of the infamous Scopes "Monkey Trial" of 1925. Tracy is the defense attorney for the schoolteacher on trial for teaching Darwin's Theory of Evolution to a group of students in a small town ruled by religion. March is the prosecutor seeking to put the teacher behind bars and restore religion to the schools. Oscar nominations for Tracy, Adapted Screenplay, Cinematography and Film Editing.
1960 128m/B Spencer Tracy, Fredric March, Florence Eldridge, Gene Kelly, Dick York; *Dir:* Stanley Kramer. Berlin Film Festival '60: Best Actor (March); National Board of Review Awards '60: 10 Best Films of the Year. **VHS, Beta, LV $19.98** FOX, FCT *&&&½*

Inheritance Victorian-era melodrama of a young heiress endangered by her guradian, who plots to murder his charge for her inheritance. Chilling and moody story based on a novel by Sheridan Le Fanu. Original title: "Uncle Silas."
1947 103m/B *GB* Jean Simmons, Katina Paxinou, Derrick DeMarney, Derek Bond; *Dir:* Charles Frank. **VHS, Beta $29.98** DVT *&&&*

The Inheritance Wealthy patriarch becomes sexually involved with scheming daughter-in-law. Tawdry tale is nonetheless engaging.
1976 (R) 121m/C *IT* Anthony Quinn, Fabio Testi, Dominique Sanda; *Dir:* Mauro Bolognini. **VHS, Beta $14.98** VID *&&½*

The Inheritor Set in a small New England town, this mystery centers on the death of a young woman and her sister's attempts to uncover the murderer. The local celebrity, a writer, seems to be at the heart of the mystery and the sister finds herself growing interested. Keeps you guessing.
1990 (R) 83m/C *Dir:* Dan Haggerty; *Dir:* Brian Savegar. **VHS $79.98** VID *&&*

The Inheritors Not-too-subtle tale of a young German boy who becomes involved with a Nazi youth group as his home life deteriorates. Heavy-going anti-fascism creates more message than entertainment. Available in German with English subtitles, or dubbed into English.
1985 89m/C *GE* Nikolas Vogel, Roger Schauer, Klaus Novak, Johanna Tomek; *Dir:* Walter Bannert. **VHS, Beta $29.98** SUE, APD, GLV *&&½*

Inhibition A beautiful heiress has lots of sex with different people in exotic locales. Softcore soap.
1986 90m/C Claudine Beccaire, Ilona Staller. **VHS, Beta $29.95** MED *&*

The Initiation Trying to rid herself of a troublesome nightmare, a coed finds herself face-to-face with a psycho. Gory campus slaughterfest.
1984 (R) 97m/C Vera Miles, Clu Gulager, James Read, Daphne Zuniga; *Dir:* Larry Stewart. **VHS, Beta $19.95** STE, HBO *&*

Initiation of Sarah A college freshman joins a sorority, undergoes alusive initiation, and gains a supernatural revenge. Made-for-television "Carrie" ripoff.
1978 100m/C Kay Lenz, Shelley Winters, Kathryn Crosby, Morgan Brittany, Tony Bill, Tisa Farrow, Robert Hays, Morgan Fairchild; *Dir:* Robert Day. **VHS, Beta $49.95** WOV *&*

Inn of the Damned Detective looks into an inn where no one ever gets charged for an extra day; they simply die. Anderson is watchable.
1974 92m/C *AU* Alex Cord, Judith Anderson, Tony Bonner, Michael Craig, John Meillon; *Dir:* Terry Burke. **VHS, Beta** PGN *&½*

The Inn of the Sixth Happiness Inspiring story of Gladys Aylward, an English missionary in 1930s' China, who leads a group of children through war-torn countryside. Donat's last film. Director Robson received an Academy Award nomination.
1958 158m/C Ingrid Bergman, Robert Donat, Curt Jurgens; *Dir:* Mark Robson. National Board of Review Awards '58: Special Citation (Donat), Best Actress (Bergman). **VHS, Beta** FOX *&&&*

Inn of Temptation A bunch of bowling buddies take off for Bangkok on an impulsive weekend, and meet lots of Oriental women.
1986 72m/C Michael Jacot, Claude Martin. **VHS, Beta $29.95** MED *&*

The Inner Circle A private detective, framed for murder by his secretary, has a limited amount of time to find the real murderer. Confusing.
1946 57m/B Adele Mara, Warren Douglas, William Frawley, Ricardo Cortez, Virginia Christine; *Dir:* Philip Ford. **VHS, Beta $16.95** SNC *&½*

The Inner Circle Ivan Sanshin is a meek, married man working as a movie projectionist for the KGB in 1935 Russia. Sanshin is taken by the KGB to the Kremlin to show movies, primarily Hollywood features, to leader Joseph Stalin, a job he cannot discuss, even with his wife. Under the spell of Stalin's personality Sanshin sees only

what he's told and overlooks the oppression and persecution of the times. Based on the life of the projectionist who served from 1935 until Stalin's death in 1953. Filmed on location at the Kremlin.
1991 (PG-13) 122m/C Tom Hulce, Lolita Davidovich, Bob Hoskins, Alexandre Zbruev; *W/ Dir:* Andrei Konchalovsky. **VHS, LV, 8mm** *COL* 🎗️🎗️🎗️

Inner Sanctum Adequate tale of a young girl accosted by a gypsy who claims that tragedy awaits her. Based on a radio show of the same name.
1948 62m/B Chuck Russell, Mary Beth Hughes, Lee Patrick, Nana Bryant; *Dir:* Lew Landers. **VHS $16.95** *SNC* 🎗️🎗️

Inner Sanctum Cheating husband hires sensuous nurse to tend invalid wife. You can guess the rest; in fact, you have to because the plot ultimately makes no sense. Available in R-rated and sex-drenched unrated editions.
1991 87m/C Tanya Roberts, Margaux Hemingway, Joseph Bottoms, Valerie Wildman, William Butler, Brett Clark; *Dir:* Fred Olen Ray. **VHS, LV $89.98** *COL, PIA* Woof!

Innerspace A space pilot, miniaturized for a journey through a lab rat a la "Fantastic Voyage," is accidentally injected into a nebbish supermarket clerk, and together they nab some bad guys and get the girl. Award-winning special effects support some funny moments between micro Quaid and nerdy Short, with Ryan producing the confused romantic interest.
1987 (PG) 120m/C Dennis Quaid, Martin Short, Meg Ryan, Kevin McCarthy, Fiona Lewis, Henry Gibson, Robert Picardo, John Hora, Wendy Schaal, Orson Bean, Chuck Jones, William Schallert, Dick Miller; *Dir:* Joe Dante. Academy Awards '87: Best Visual Effects. **VHS, Beta, LV, 8mm $19.98** *WAR* 🎗️🎗️½

Innocence Unprotected A film collage that contains footage from the 1942 film "Innocence Unprotected," the story of an acrobat trying to save an orphan from her wicked stepmother, newsreels from Nazi-occupied Yugoslavia, and interviews from 1968 with people who were in the film. Confiscated by the Nazis during final production, "Innocence Unprotected" was discovered by director Makavejev, who worked it into the collage. Filmed in color and black and white; in Serbian with English subtitles.
1968 78m/C *YU* Dragoljub Aleksic, Ana Milosavljevic, Vera Jovanovic; *Dir:* Dusan Makavejev. **VHS $59.95** *FCT* 🎗️🎗️½

The Innocent The setting: turn of the century Rome, ablaze with atheism and free love. The story: Giannini's a Sicilian aristocrat who ignores wife Antonelli and dallies with scheming mistress O'Neill. Discovering his wife's infidelity, he's profoundly disillusioned, and his manly notion of self disintegrates. Visconti's last film, it was an art house fave and is considered by many to be his best. Lavishly photographed, well set and costumed, and true to the Gabrielle D'Annunzio's novel. English language version; originally titled "L'Innocente."
1976 (R) 125m/C *IT* Laura Antonelli, Jennifer O'Neill, Giancarlo Giannini; *Dir:* Luchino Visconti. **VHS, Beta $79.95** *VES, FCT* 🎗️🎗️🎗️½

An Innocent Man Uneven story of Selleck, an ordinary family man and airline mechanic, framed as a drug dealer and sent to prison.
1989 (R) 113m/C Tom Selleck, F. Murray Abraham, Laila Robins, David Rasche; *Dir:* Peter Yates. **VHS, Beta, LV $19.99** *TOU* 🎗️🎗️½

Innocent Prey When a woman finally summons the courage to bid hasta la vista to her abusive husband, it turns out he's not so quick to say ciao. Definitely not a highlight in Balsam's career.
1988 (R) 88m/C *AU* P.J. Soles, Martin Balsam, Kit Taylor, Grigor Taylor, John Warnock, Susan Stenmark, Richard Morgan; *Dir:* Colin Eggleston. **VHS $89.95** *SVS* 🎗️🎗️

Innocent Victim Convoluted tale of revenge and madness. Bacall plays a mad grandmother who kidnaps a boy to take the place of her dead grandson. The kidnapped boy's stepfather then is accused of murdering him, and starts murdering others to get the boy back.
1990 (R) 100m/C Lauren Bacall, Helen Shaver, Paul McGann, Peter Firth; *Dir:* Giles Foster. **VHS $29.95** *ACA* 🎗️

The Innocents Abroad Entertaining adaptation of Mark Twain's novel about a group of naive Americans cruising the Mediterranean. Ensemble cast offers some surprising performances. Originally produced for PBS.
1984 116m/C Craig Wasson, Brooke Adams, David Ogden Stiers. **VHS $19.95** *MCA* 🎗️🎗️½

Innocents in Paris A group of seven Britons visit Paris for the first time. A bit long but enough ooo-la-la to satisfy.
1953 89m/B *GB* Alastair Sim, Laurence Harvey, Jimmy Edwards, Claire Bloom, Margaret Rutherford; *Dir:* Gordon Parry. **VHS, Beta** *MON* 🎗️🎗️½

The Inquiry A new twist on the resurrection of Christ. Carradine senses a cover-up when he's sent from Rome to investigate the problem of Christ's missing corpse. He retraces Christ's final days to solve the case in this interesting, offbeat film. Dubbed.
1987 107m/C *IT* Keith Carradine, Harvey Keitel, Phyllis Logan, Angelo Infanti, Lina Sastri; *Dir:* Damiano Damiani. **VHS $79.99** *HBO* 🎗️🎗️½

Inquisition Naschy is a 16th century witch hunting judge who finds himself accused of witchcraft.
1976 85m/C *SP* Paul Naschy, Daniela Giordano, Juan Gallardo; *W/Dir:* Paul Naschy. **VHS, Beta** *VCD* 🎗️½

Insanity A film director becomes violently obsessed with a beautiful actress.
1982 101m/C Terence Stamp, Fernando Rey, Corinne Cleary. **VHS, Beta $59.95** *MGL* 🎗️

The Insect Woman Chronicles 45 years in the life of a woman who must work with the diligence of an ant in order to survive. Thoughtfully reflects the exploitation of women and the cruelty of human nature in Japanese society. Beautiful performance from Hidari, who ages from girlhood to middle age. In Japanese with English subtitles.
1963 123m/B *JP* Sachiko Hidari, Yitsuko Yoshimura, Hiroyuki Nagaes, Sumie Sasaki; *Dir:* Shohei Imamura. **VHS $29.95** *CVC* 🎗️🎗️🎗️

Inserts A formerly successful director has been reduced to making porno films in this pretentious, long-winded effort set in a crumbling Hollywood mansion in the 1930s. Affected, windbaggish performances kill it.
1976 (R) 99m/C *GB* Richard Dreyfuss, Jessica Harper, Veronica Cartwright, Bob Hoskins; *Dir:* John Byrum. **VHS $59.98** *FOX* 🎗️

Inside Daisy Clover Wood is the self-sufficient, junior delinquent waif who becomes a teenage musical star and pays the price for fame in this Hollywood saga set in the 1930s. Discovered by tyrannical studio head Plummer, Daisy is given her big break and taken under his wing for grooming as Swan Studio's newest sensation. She falls for fellow performer, matinee idol Redford, who has some secrets of his own, and eventually has a breakdown from the career pressure. Glossy melodrama is filled with over-the-top performances but its sheer silliness makes it amusing.
1965 128m/C Natalie Wood, Christopher Plummer, Robert Redford, Ruth Gordon, Roddy McDowall, Katharine Bard; *Dir:* Robert Mulligan. **VHS** *WEA* 🎗️🎗️½

Inside Hitchcock Documentary reveals much about one of America's best-loved film makers. Scenes from his movies and his last interview enliven this informative program.
1974 55m/C Alfred Hitchcock; *Nar:* Cliff Robertson. **VHS, Beta, 3/4U $29.95** *IHF, MPI* 🎗️🎗️½

Inside Information With Tarzan's help, a cop searches out the leader of a ring of jewel thieves.
1934 51m/B Rex Lease, Marion Shilling, Philo McCullough, Tarzan the Wonder Dog. **VHS, Beta, LV** *WGE* 🎗️

Inside the Labyrinth A documentary look at the making of the puppet and special-effects-laden fantasy film "Labyrinth."
1986 57m/C David Bowie, Jennifer Connelly, Jim Henson. **VHS, Beta $19.98** *SUE* 🎗️

Inside the Lines A World War I tale of espionage and counter-espionage.
1930 73m/B Betty Compson, Montagu Love, Mischa Auer, Ralph Forbes. **VHS, Beta $29.95** *VYY* 🎗️🎗️

The Inside Man Double agents struggle to find a submarine-detecting laser device. Based on true incidents in which a soviet sub ran aground in Sweden. Made in Sweden; dubbed.
1984 90m/C *SW* Dennis Hopper, Hardy Kruger, Gosta Ekman, David Wilson; *Dir:* Tom Clegg. **VHS, Beta $9.95** *SIM, CEL* 🎗️🎗️½

Inside Moves A look at handicapped citizens trying to make it in everyday life, focusing on the relationship between an insecure, failed suicide and a volatile man who is only a knee operation away from a dreamed-about basketball career.
1980 (PG) 113m/C John Savage, Diana Scarwid, David Morse, Amy Wright; *Dir:* Richard Donner. **VHS, Beta, LV $19.95** *FOX, JTC* 🎗️🎗️½

Inside Out An ex-GI, a jewel thief and a German POW camp commandant band together to find a stolen shipment of Nazi gold behind the Iron Curtain. To find the gold, they help a Nazi prisoner who knows the secret to break out of prison. Good action caper.
1975 (PG) 97m/C *GB* Telly Savalas, Robert Culp, James Mason, Aldo Ray, Doris Kunstmann; *Dir:* Peter Duffell. **VHS, Beta $59.95** *WAR* 🎗️🎗️½

Inside Out Hiding in his apartment from a world which terrifies him, Jimmy doesn't feel he's missing much - since he has the necessary women and bookies come to him. He soon finds he's not quite as safe as he thought, and the only way out is "out there". Fine performance from Gould, suspenseful pacing.
1991 (R) 87m/C Elliot Gould, Jennifer Tilly, Howard Hesseman, Beah Richards, Timothy Scott, Sandy McPeak, Dana Elcar, Meshach Taylor; *W/Dir:* Robert Taicher. **VHS $79.95** *HMD* 🎗️🎗️½

Inside the Third Reich A made-for-television mini-series detailing the rise to power within the Third Reich of Albert Speer, top advisor to Hitler. Based on Speer's rather self-serving memoires.
1982 250m/C Derek Jacobi, Rutger Hauer, John Gielgud, Blythe Danner, Maria Schell, Ian Holm, Trevor Howard, Randy Quaid, Viveca Lindfors, Robert Vaughn, Stephen Collins, Elke Sommer, Renee Soutendijk; *Dir:* Marvin J. Chomsky. **VHS, Beta, LV $34.95** *SUE, GLV* 🦴🦴½

Insignificance A film about an imaginary night spent in a New York hotel by characters who resemble Marilyn Monroe, Albert Einstein, Joe McCarthy, and Joe DiMaggio. Entertaining and often amusing as it follows the characters as they discuss theory and relativity, the Russians, and baseball among other things.
1985 110m/C *GB* Gary Busey, Tony Curtis, Theresa Russell, Michael Emil, Will Sampson; *Dir:* Nicolas Roeg. **VHS, Beta, LV $59.95** *LHV, WAR* 🦴🦴🦴

Inspector Clouseau Cartoon Festival A series of tapes which features five cartoons each from the Pink Panther canon and the diminutive French detective.
1966 32m/C VHS, Beta $14.95 *MGM* 🦴🦴🦴

The Inspector General Classic Kaye chaziness of mistaken identities with the master comic portraying a carnival medicine man who is mistaken by the villagers for their feared Inspector General. If you like Kaye's manic performance, you'll enjoy this one.
1949 97m/C Danny Kaye, Walter Slezak, Barbara Bates, Elsa Lanchester, Gene Lockhart, Walter Catlett, Alan Hale Jr.; *Dir:* Henry Koster. **VHS, Beta $9.95** *NEG, MRV, NOS* 🦴🦴🦴

The Inspector General The filmed performance of Gogol's great play by the Moscow Art Theatre, wherein a provincial town panics when they mistake a wandering moron for the Inspector come to check up on them. In Russian with subtitles.
1954 128m/B *RU Dir:* Vladimir Petrov. **VHS, Beta $44.95** *FCT* 🦴🦴½

Inspector Hornleigh A pair of bumbling detectives arrive on the scene when a Chancellor's fortune is stolen. Fine comedy, although the Scottish and British accents can be rather thick now and again.
1939 76m/B *GB* Gordon Harker, Alastair Sim, Miki Hood, Wally Patch, Steve Geray, Edward Underdown, Hugh Williams, Gibb McLaughlin; *Dir:* Eugene Forde. **VHS, Beta $16.95** *NOS, SNC* 🦴🦴½

Inspiration of Mr. Budd A clever barber helps police trap a murderer. Based on a short story by Dorothy L. Sayers. Part of EBE's "Orson Welles' Great Mysteries" series.
1975 25m/C Orson Welles. **VHS, Beta** *BRT* 🦴🦴½

Instant Justice A professional Marine jeopardizes his military career by avenging his sister's murder, vigilante style. Exclusive action. Original title: "Marine Issue."
1986 (R) 101m/C *GB* Michael Pare, Tawny Kitaen, Charles Napier; *Dir:* Craig T. Rumar. **VHS, Beta $79.95** *WAR* 🦴½

Instant Karma Would be comedy depicts young man's far fetched attempts to score babewise. Ask no more what happened to erstwhile teen heart throb Cassidy: he suffers from a major bout of bad karma.
1990 94m/C Craig Sheffer, Chelsea Noble, David Cassidy, Alan Blumenfeld, Glen Hirsch, Marty Ingels, Orson Bean; *Dir:* Roderick Taylor. **VHS, Beta $89.98** *MGM* 🦴½

Instant Kung-Fu Man Twin brothers get involved in a case of mistaken identity in this martial arts film.
1984 104m/C John Liu, Hwang Jang Lee, Yeh Fei Yang. **VHS, Beta** *WVE* 🦴

The Instructor The head of a karate school proves the value of his skill when threatened by the owner of a rival school.
1983 91m/C Don Bendell, Bob Chaney, Bob Saal, Lynday Scharnott; *Dir:* Don Bendell. **VHS, Beta $69.98** *LIV, VES* 🦴

Interface A computer game gets out of hand at a university, turning the tunnels beneath the campus into a battleground of good and evil.
1984 88m/C John Davies, Laura Lane, Matthew Sacks. **VHS, Beta $69.98** *LIV, VES* 🦴½

Interiors Ultra serious, Bergmanesque drama about three neurotic adult sisters, coping with the dissolution of their family. When Father decides to leave mentally unbalanced mother for a divorcee, the daughters are shocked and bewildered. Depressing and humorless, but fine performances all the way around, supported by the elegant camera work of Gordon Willis. Oscar nominations for Page, Stapleton, Allen as director, the screenplay (by Allen), and art direction. Original theatrical trailer available on the laserdisc format.
1978 (PG) 95m/C Diane Keaton, Mary Beth Hurt, E.G. Marshall, Geraldine Page, Richard Jordan, Sam Waterston, Kristin Griffith, Maureen Stapleton; *W/Dir:* Woody Allen. **VHS, Beta, LV $79.95** *MGM* 🦴🦴🦴½

Intermezzo Married violinist Ekman meets pianist Bergman and they fall in love. He deserts his family to tour with Bergman, but the feelings for his wife and children become too much. When Bergman sees their love won't last, she leaves him. Shakily, he returns home and as his daughter runs to greet him, she is hit by a truck and he realizes that his family is his true love. One of the great grab-your-hanky melodramas. In Swedish with English subtitles. 1939 English re-make features Bergman's American film debut.
1936 88m/B *SW* Gosta Ekman, Inga Tidblad, Ingrid Bergman, Bullen Berglund; *Dir:* Gustaf Molander. **VHS, LV $29.95** *WAC, CRO, LUM* 🦴🦴🦴

Intermezzo Fine, though weepy, love story of a renowned, married violinist who has an affair with his stunningly beautiful protege (Bergman), but while on concert tour realizes that his wife and children hold his heart, and he returns to them. A re-make of the 1936 Swedish film, it's best known as Bergman's American debut. Howard's violin playing was dubbed by Jascha Heifetz.
1939 70m/B *IT* Ingrid Bergman, Leslie Howard, Edna Best, Ann Todd; *Dir:* Gregory Ratoff. **VHS, Beta, LV $59.98** *FOX, FCT* 🦴🦴🦴

Internal Affairs Wild, sexually charged action piece about an Internal Affairs officer for the L.A.P.D. (Garcia) who becomes obsessed with exposing a sleazy, corrupt street cop (Gere). Drama becomes illogical, but boasts excellent, strung out performances, especially Gere as the creepy degenerate you'd love to bust.
1990 (R) 114m/C Richard Gere, Andy Garcia, Laurie Metcalf, Ron Vawter, Marco Rodriguez, Nancy Travis, Adam Baldwin; *Dir:* Mike Figgis. **VHS, Beta, LV, 8mm $14.95** *PAR* 🦴🦴½

International Crime Radio hero The Shadow solves a tough crime in this adventure film.
1937 72m/B Rod La Rocque. **VHS, Beta $16.95** *SNC, NOS, AOV* 🦴🦴

International House A wacky Dada-esque Hollywood farce about an incredible array of travelers quarantined in a Shanghai hotel where a mad doctor has perfected television. Essentially a burlesque compilation of skits, gags, and routines. Guest stars Rudy Vallee, Baby Rose Marie and Cab Calloway appear on the quaint "television" device, in musical sequences.
1933 72m/B W.C. Fields, Peggy Hopkins Joyce, Rudy Vallee, George Burns, Cab Calloway, Sari Maritza, Gracie Allen, Bela Lugosi, Sterling Holloway, Baby Rose Marie; *Dir:* Edward Sutherland. **VHS, Beta, LV $29.95** *MCA, MLB* 🦴🦴½

International Velvet In this long overdue sequel to "National Velvet," adult Velvet, with live-in companion (Plummer), grooms her orpahned neice (O'Neal) to become an Olympic champanion horsewoman. Stunning photography and good performances by Plummer and Hopkins manage to keep the sentiment at a trot.
1978 (PG) 126m/C *GB* Tatum O'Neal, Anthony Hopkins, Christopher Plummer; *Dir:* Bryan Forbes. **VHS, Beta $59.95** *MGM* 🦴🦴½

Internecine Project Stylistic espionage tale with Coburn's English professor acting as the mastermind behind an unusual series of murders where industrial spies kill each other, so Coburn can garner a top governmet job in D.C. Unexpected ending is worth the viewing. Screenplay by Barry Levinson; based on Mort Elkind's novel.
1973 (PG) 89m/C *GB* James Coburn, Lee Grant, Harry Andrews, Keenan Wynn; *Dir:* Ken Hughes. **VHS, Beta $59.98** *FOX* 🦴🦴½

The Interns Finessed hospital soaper about interns on the staff of a large city hospital, whose personal lives are fraught with trauma, drugs, birth, death, and abortion. Good performances and direction keep the medical melodrama moving in a gurney-like manner. Followed by the "The New Interns" and the basis for a television series. Adapted from a novel by Richard Frede.
1962 120m/C Cliff Robertson, James MacArthur, Michael Callan, Nick Adams, Stefanie Powers, Suzy Parker, Buddy Ebsen, Telly Savalas, Haya Harareet; *Dir:* David Swift. **VHS, Beta, LV $59.95** *COL* 🦴🦴½

Interrogation Grueling depiction of Stalin-era Poland proves that nobody makes anticommunist movies better than those who knew tyranny firsthand. A 1950s cabaret starlet is arrested on false charges and endures years of torment in custody. Banned under Polish martial law in 1982, it circulated illegally in the country until 1989. Janda won an award at Cannes for her transformation from showgirl floozy to defiant heroine. In Polish with English subtitles.
1982 118m/C *PL* Krystyna Janda. Cannes Film Festival '82: Best Actress (Janda). **VHS $79.95** *KIV, FCT* 🦴🦴🦴

Interrupted Journey A married man runs away to start a new life with another woman. When she is killed in a train accident, he becomes the prime suspect in her murder. Lots of action and speed, but a disappointing ending.
1949 80m/B *GB* Richard Todd, Valerie Hobson; *Dir:* Daniel Birt. **VHS $16.95** *NOS, SNC* 🦴🦴½

Interval A middle-aged, globe-trotting woman becomes involved with a young American painter while running from her past. Oberon, 62 at the time, produced and played in this, her last film. Filmed in Mexico.
1973 (PG) 84m/C Merle Oberon, Robert Wolders; *Dir:* Daniel Mann. **VHS, Beta $59.95** CHA 🎬🎬

Interview Two filmmakers relate their perceptions of each other through their respective animation techniques.
1987 14m/C VHS, Beta NFC 🎬🎬

Interzone Humans battle mutants in a post-holocaust world.
1988 97m/C Bruce Abbott. **VHS, Beta $79.95** MED 🎬½

Intimate Contact Top executive discovers he has AIDS after a business trip fling. Powerful drama as he and his wife try to cope. Excellent, understated performances from Massey and Bloom. Written by Alma Cullen and originally made for cable television.
1987 (PG) 140m/C GB Claire Bloom, Daniel Massey, Abigail Cruttenden, Mark Kingston, Sylvia Syms; *Dir:* Waris Hussein. **VHS, Beta $79.99** HBO 🎬🎬🎬

Intimate Games Soft-core nonsense about a professor assigning his class to research sexual fantasies.
198? (R) 90m/C VHS, Beta $59.95 PSM 🎬

Intimate Lessons An erotic tale like no other.
19?? ?m/C VHS TAV, HHE 🎬

Intimate Moments Madame Claude runs an exclusive call-girl operation catering to the upper echelons of power in France, when she discovers that a newspaper is investigating her business.
1982 (R) 82m/C Alexandra Stewart, Dirke Altevogt. **VHS, Beta $69.98** SUE 🎬

Intimate Power True story of a French schoolgirl who is sold into slavery to an Ottoman sultan and becomes the harem queen. She bears an heir and begins to teach her son about his people's rights, in the hope that he will bring about reform in Turkey where he grows up to be a sultan. Good cast supports the exotic storytelling in this made for cable drama.
1989 (R) 104m/C F. Murray Abraham, Maud Adams, Amber O'Shea; *Dir:* Jack Smight. **VHS, Beta, LV $89.99** HBO 🎬🎬½

Intimate Story Intimate story of a husband and wife trying unsuccessfully to have a baby, amid the lack of privacy in a Kibbutz. Each blames the other for the problem and tries to escape the problems by focusing on personal fantasies. Hebrew with English subtitles.
1981 95m/C IS Chava Alberstein, Alex Peleg; *Dir:* Israeli Nadav Levitan. **VHS, Beta $79.95** ERG 🎬🎬½

Intimate Stranger Sexy, would-be rock star by day, Harry sets out in search of a steadier source of income. Rather than wait on tables all day, she chooses instead to become a phone-sex girl. This turns out to be an unwise career move, as she promptly finds herself the target of a phone-psycho who just can't take "no" for an answer. Will Debbie be able to shake this loser, or will he put a permanent end to her life-long dreams?
1991 (R) 96m/C Deborah Harry, James Russo, Tim Thomerson, Paige French, Grace Zabriskie; *Dir:* Allan Holzman. **VHS, Beta $89.95** PAR 🎬½

Intimate Strangers Taut made-for-television expose on battering and its effects on the family. Weaver and Struthers outshine the material, but the finest performance is a cameo by Douglas, Weaver's irascible aged father.
1977 96m/C Dennis Weaver, Sally Struthers, Quinn Cummins, Tyne Daly, Larry Hagman, Rhea Perlman, Melvyn Douglas; *Dir:* John Llewellyn Moxey. **VHS, Beta $39.95** WOV 🎬🎬½

Into the Badlands A mild made-for-cable-TV thriller presents three suspense tales in "Twilight Zone" fashion, linked by the appearance of a mysterious Man in Black. No, it's not Johnny Cash, but Dern as a sinister bounty-hunter in the old west, who inspires strange events on the trail.
1992 89m/C Bruce Dern, Mariel Hemingway, Helen Hunt, Dylan McDermott, Lisa Pelikan, Andrew Robinson; *Dir:* Sam Pillsbury. **VHS $89.95** MCA 🎬🎬

Into the Darkness In the world of fashion models, a maniacal, demented killer stalks his prey.
1986 90m/C Donald Pleasence, Ronald Lacey. **VHS, Beta $49.95** WWV 🎬½

Into the Fire Thriller about a young drifter who, against the backdrop of a Canadian winter, happens upon a roadside lodge and diner, where he gets involved in sex and murder. Title gives a good idea of what to do with this one.
1988 (R) 88m/C CA Susan Anspach, Art Hindle, Olivia D'Abo, Lee Montgomery; *Dir:* Graeme Campbell. **VHS, Beta, LV $79.98** LIV, VES 🎬

Into the Homeland Seasoned ex-cop (Boothe) infiltrates a white supremacist group which had kidnapped his teenage daughter. Failed "message" about neo-nazi evils, but good confrontation between Boothe and his son (Howell) make this so-so viewing. Made foe cable television.
1987 95m/C Powers Boothe, C. Thomas Howell, Paul LeMat; *Dir:* Lesli Linka Glatter. **VHS, Beta, LV $79.95** KRT, LHV, WAR 🎬🎬

Into the Night Campy, off-beat thriller about a middle-aged, jilted deadbeat (Goldblum), who meets a beautiful woman (Pfeiffer) when she suddenly drops onto the hood of his car, with a relentless gang of Arabs pursuing. During their search through L.A. and other parts of California for the one person who can help her out of this mess, a whole crew of Hollywood directors make cameo appearances, making this a delight for film buffs, but occasionally tedious for general audiences. B.B. King sings the title song.
1985 (R) 115m/C Jeff Goldblum, Michelle Pfeiffer, David Bowie, Carl Perkins, Richard Farnsworth, Dan Aykroyd, Paul Mazursky, Roger Vadim, Jim Henson, Paul Bartel, David Cronenberg, Irene Papas, Bruce McGill, Vera Miles, Clu Gulager, Don Steel, Jack Arnold, Jonathan Demme, Lawrence Kasdan, Amy Heckerling, Kathryn Harrold; *Dir:* John Landis. **VHS, Beta, LV $19.95** MCA 🎬🎬

Into the Sun Hall plays a Hollywood star who encounters real-life danger while hanging out with an Air Force pilot (Pare) to prepare for an upcoming role. Spectacular aerial scenes steal the show, which isn't hard to do in this lackluster movie.
1991 (R) 101m/C Anthony Michael Hall, Michael Pare, Terry Kiser, Deborah Maria Moore; *Dir:* Fritz Kiersch. **VHS, LV** VMK 🎬½

Into Thin Air A mother searches tirelessly for her college bound son who disappeared on a cross-country drive. Well-done television drama with a good performance from Burstyn. Written by George Rubio.
1985 100m/C Ellen Burstyn, Robert Prosky, Sam Robards, Tate Donovan; *Dir:* Roger Young. **VHS, Beta, LV $79.95** NBI 🎬🎬🎬

Intolerance Griffith's largest film in every aspect. An interwoven, four-story epic about human intolerance, with segments set in Babylon, ancient Judea, and Paris. One of the cinema's grandest follies, and greatest achievements. Silent with music score. Laserdisc version contains supplemental material including cuts from other Griffith works. Black and white with some restored color footage.
1916 175m/B Lillian Gish, Mae Marsh, Constance Talmadge, Bessie Love, Elmer Clifton, Erich von Stroheim, Eugene Pallette; *Dir:* D.W. Griffith. Cinematheque Belgique '52: #9 of the Best Films of All Time (tie); Sight and Sound Survey '52: #6 of the Best Films of All Time; British Film Institute Poll '52: #5 of the Best Films of All Time (tie); Brussels World's Fair '58: #7 of the Best Films of All Time. **VHS, LV $29.95** NOS, MRV, KIV 🎬🎬🎬🎬

Intrigue Former CIA agent (Loggia) defects to the KGB, but wants to return to the US when he realizes he has a progressive illness. Underground agent and old buddy (Glenn) is recruited with orders to smuggle his former colleague out. Tension continues throughout as murder comes along with the orders. Loads of suspense and twist ending in this made for television nail-biter.
1990 (PG) 96m/C Scott Glenn, Robert Loggia, William Atherton, Martin Shaw; *Dir:* David Drury. **VHS, LV $89.95** VMK 🎬🎬½

Intruder A man and his stepson are terrorized by a stranger in a panel truck as they travel from Rome to Paris.
1977 98m/C Jean-Louis Trintignant, Mireille Darc, Adolfo Celi. **VHS, Beta $59.95** VID 🎬🎬

Intruder A psychotic killer stalks an all-night convenience store shopping for fresh meat (customers beware). Slasher movie fans should get a kick out of the plot twist at the end.
1988 (R) 90m/C Elizabeth Cox, Renee Estevez, Alvy Moore; *Dir:* Scott Spiegel. **VHS, Beta, LV $79.95** PAR 🎬🎬

Intruder in the Dust A small southern community develops a lynch mob mentality when a black man is accused of killing a white man. Powerful, but largely ignored portrait of race relations in the South. Solid performances from the whole cast; filmed in Mississippi. Adapted from a novel by William Faulkner.
1949 87m/B David Brian, Claude Jarman Jr., Juano Hernandez, Porter Hall, Elizabeth Patterson; *Dir:* Clarence Brown. **VHS** MGM 🎬🎬🎬

Intruder Within Half-baked thriller about crew on an isolated oil rig in Antarctica who, while drilling, unearth a nasty creature that goes around terrorizing the frost-bitten men and occasional stray woman.
1981 91m/C Chad Everett, Joseph Bottoms, Jennifer Warren; *Dir:* Peter Carter. **VHS, Beta $69.95** TWE 🎬

The Invaders Two Viking brothers battle each other and the Britons. Originally titled "Erik the Conqueror."
1963 88m/C FR IT Cameron Mitchell, George Ardisson, Andrea Checchi, Francoise Christophe, Joe Robinson; *Dir:* Mario Bava. **VHS, Beta $59.95** TTE 🎬🎬

Invaders from Mars Young boy cries "martian" in this sci-fi cheapy classic. He can't convince the townspeople of this invasion because they've already been possessed by the alien beings. His parents are zapped by the little green things first, making this perhaps an allegory for the missing Eisenhower/Stepford years. Includes previews of coming attractions from classic science fiction films. Remade in 1986 by Tobe Hooper.
1953 78m/C Arthur Franz, Jimmy Hunt, Leif Erickson; *Dir:* William Cameron Menzies. **VHS, Beta $19.95** *MED, MLB* 🎞🎞½

Invaders from Mars A high-tech remake of Menzies' 1953 semi-classic about a Martian invasion perceived only by one young boy and a sympathetic (though hysterical) school nurse, played by mother and son Black and Carson. Instant camp.
1986 (PG) 102m/C Hunter Carson, Karen Black, Louise Fletcher, Laraine Newman, Timothy Bottoms, Bud Cort; *Dir:* Tobe Hooper. **VHS, Beta, LV $39.95** *MED, IME* 🎞½

Invasion A hospital opens its doors to an accident victim, and his attractive female visitors don't seem sympathetic to the idea of a long hospital stay. Turns out he's an escaped alien prisoner, and the alien babes, intent on intergalactic extradition, place a force field around the hospital and demand his return. Early effort of director Bridges, who later did "The Shooting Party." Interesting, creepy, atmospheric, with very cool camera moves.
1965 82m/B *GB* Edward Judd, Yoko Tani, Valerie Gearon, Lyndon Brook, Tsai Chin, Barrie Ingham; *Dir:* Alan Bridges. **VHS $19.98** *MOV* 🎞🎞½

Invasion of the Animal People Hairy monster from outer space attacks Lapland. Narrated by Carradine, the only American in the cast. Pretty silly. Also known as "Terror in the Midnight Sun," "Space Invasion from Lapland," "Space Invasion of Lapland," and "Horror in the Midnight Sun."
1962 73m/C *SW* Robert Burton, Barbara Wilson, John Carradine; *Dir:* Virgil W. Vogel, Jerry Warren; *Nar:* John Carradine. **VHS $19.95** *VMK, SNC, NOS Woof!*

Invasion of the Bee Girls California girls are mysteriously transformed into deadly nymphomaniacs in this delightful campy science fiction romp. Written by Nicholas Meyer who later directed "Star Trek II: The Wrath of Khan" and wrote "The Seven Per Cent Solution."
1973 85m/C Victoria Vetri, William Smith, Anitra Ford, Cliff Osmond, Wright King, Ben Hammer; *Dir:* Denis Sanders. **VHS $69.98** *SUE* 🎞🎞🎞

Invasion of the Blood Farmers In a small New York town, members of an ancient Druidic cult murder young women, taking their blood in the hope of finding the precise, rare blood type to keep their queen alive. A real woofer.
1972 (PG) 86m/C Norman Kelley, Tanna Hunter, Bruce Detrick, Jack Neubeck, Cythia Fleming, Paul Craig Jennings; *Dir:* Ed Adlum. **VHS, Beta $49.95** *REG Woof!*

Invasion of the Body Snatchers The one and only post-McCarthy paranoid sci-fi epic, where a small California town is infiltrated by pods from outer space that replicate and replace humans. A chilling, genuinely frightening exercise in nightmare dislocation. Based upon a novel by Jack Finney; scripted by Sam Peckinpah. Remade in 1978. The laserdisc version contains commentary by Maurice Yacowar, the text of an

interview with Siegel, and the original theatrical trailer. The film itself is presented in its original wide-screen format.
1956 80m/B Kevin McCarthy, Dana Wynter, Carolyn Jones, King Donovan, Don Siegel; *Dir:* Don Siegel. **VHS, LV $14.98** *REP, VYG, CRC* 🎞🎞🎞

Invasion of the Body Snatchers One of the few instances where a remake is an improvement on the original, which was itself a classic. This time, the "pod people" are infesting San Francisco, with only a small group of people aware of the invasion. A ceaselessly inventive, creepy version of the alien-takeover paradigm, with an intense and winning performance by Sutherland.
1978 (PG) 115m/C Donald Sutherland, Brooke Adams, Veronica Cartwright, Leonard Nimoy, Jeff Goldblum, Kevin McCarthy, Don Siegel, Art Hindle; *Dir:* Philip Kaufman. **VHS, Beta, LV $19.95** *MGM* 🎞🎞🎞½

Invasion of the Body Stealers Aliens are thought to be the captors when sky-divers begin vanishing in mid-air. Also known as "Thin Air," and originally titled, "The Body Stealers." Uneven sci-fi with nothing going for it.
1969 (PG) 115m/C *GB* George Sanders, Maurice Evans, Patrick Allen; *Dir:* Gerry Levy. **VHS, Beta $29.95** *IVE Woof!*

The Invasion of Carol Enders A woman, almost killed by a prowler, awakens with an expanded consciousness, and tries to convince her near husband that her death wasn't an accident.
1974 72m/C Meredith Baxter Birney, Christopher Connelly, Charles Aidman. **VHS, Beta $29.95** *IVE* 🎞½

Invasion Earth: The Aliens Are Here! A cheap spoof of monster movies, as an insectoid projectionist takes over the minds of a movie audience. Clips of Godzilla, Mothra and other beasts are interpolated.
1987 84m/C Janice Fabian, Christian Lee; *Dir:* George Maitland. **VHS, Beta, LV $19.95** *STE, NWV* 🎞

Invasion of the Flesh Hunters A group of tortured Vietnam veterans returns home carrying a cannibalistic curse with them. A smorgasbord of sensationalism; not for discriminating tastes.
1982 90m/C *IT* John Saxon, Elizabeth Turner; *Dir:* Anthony Dawson. **VHS, Beta $69.98** *LIV, VES Woof!*

Invasion Force A terrorist army parachutes into a remote region where a film crew is preparing to set up for an action movie. The moviemakers must scare away the terrorists with their smokepots and blanks. Sounds promising, but proves to be as empty as the movie props used to scare the terrorists.
1990 93m/C Richard Lynch, David Shark, Renee Cline, Douglas Hartier, Graham Times, Angie Synodis; *Dir:* David A. Prior. **VHS $79.95** *AIP* 🎞

Invasion of the Girl Snatchers Aliens from another planet, in cahoots with a cult kingpin, subdue young earth girls and force them to undergo bizarre acts that may just rob them of their dignity. Save a bit of your own dignity by staying away. Cheap look, cheap feel, cheap thrills—don't sell yourself short.
1973 90m/C Elizabeth Rush, Ele Grigsby, David Roster; *Dir:* Lee Jones. **VHS, Beta $39.95** *UHV Woof!*

Invasion of the Space Preachers Hideous creatures from outer space arrive on Earth with plans for conquest. Led by the seemingly human Reverend Lash, they prey on the innocent and trusting, fleecing God-fearing folk out of their hard earned cash!
1990 100m/C Jim Wolfe, Guy Nelson, Eliska Hahn, Gary Brown, Jesse Johnson, John Riggs; *W/Dir:* Daniel Boyd. **VHS, Beta $79.95** *RHI Woof!*

Invasion U.S.A. A mercenary defends nothing less than the entire country against Russian terrorists landing in Florida and looking for condos. A record-setting amount of people killed. Count them—it's not like you'll miss anything important. Norris is a bit less animated than a wooden Indian, and the acting is more painful to endure than the violence. Glasnost saved the movie industry from additional paranoia movies of this kind.
1985 (R) 90m/C Chuck Norris, Richard Lynch, Melissa Prophet, Alex Colon, Billy Drago; *Dir:* Joseph Zito. **VHS, Beta, LV $19.95** *MGM* 🎞

Invasion of the Vampires Burrito bat fest has Count Frankenhausen doing his vampire thing in a small, 16th century village. Atmospheric sets are highlight. Not a good substitute for No Doze.
1961 78m/B *MX* Carlos Agosti, Rafael Etienne, Bertha Moss, Tito Junco, Erna Martha Bauman, Fernando Soto, Enrique Garcia Alvarez, David Reynoso; *W/Dir:* Miguel Morayta. **VHS $19.98** *SNC* 🎞

Invasion of the Zombies Bare-fisted wrestler Santo faces of brace of zombies in order to prove that men with short necks can hug other men and still be mucho macho.
1961 85m/B *MX* Santo; *Dir:* Benito Alzraki. **VHS $19.98** *SNC, HHT* 🎞

Investigation A Frenchman plans to murder his wife so he can marry his pregnant mistress, guaranteeing an heir. Subtle study of small town people so concerned with their own welfare that the dirty deeds of the lead character might be overlooked. Fine performances and story development make this worth investigating. In French with English subtitles.
1979 116m/C *FR* Victor Lanoux, Valerie Mairesse; *Dir:* Etienne Perier. **VHS, Beta $29.98** *SUE* 🎞🎞½

Invincible A martial arts student must find and correct another student who has turned evil. For Bruce Lee fans, as if they haven't already memorized all the memorable lines—"hiyaah," "ugh," "oof," etc.
1980 (R) 93m/C Bruce Lee, Chen Sing, Ho Chung Dao. **VHS, Beta** *GEM* 🎞

Invincible Armor The wrong man is accused of murder when an ambitious minister of justice unleashes a sinister plot.
197? 92m/C **VHS, Beta $39.95** *TWE* 🎞

Invincible Barbarian A young man leads a tribe of amazing Amazon warriors in a sneak attack against the tribe that annihilated his native village. One of those movies that presumes "primitivism" can be recreated through bad acting.
1983 92m/C Diana Roy, David Jenkins. **VHS, Beta $59.95** *PSM Woof!*

The Invincible Gladiator A gladiator who saves the evil man ruling a land for a boy-king is given command of an army. He eventually leads the army in revolt when the evil man tries to take over the throne. Plodding with stock characters.

1963 96m/C *IT SP* Richard Harrison, Isabel Corey, Livio Lorenzon; *Dir:* Frank Gregory. **VHS** $16.95 *SNC* 𝄞½

Invincible Gladiators The sons of Hercules labor to aid a prince whose bride had been kidnapped by an evil queen. Laborious to watch.
1964 87m/C Richard Lloyd, Claudia Lange, Tony Freeman; *Dir:* Robert Mauri. **VHS, Beta** $69.95 *LTG* *Woof!*

Invincible Mr. Disraeli Presentation from "George Schaefer's Showcase Theatre" deals with the life and career of Benjamin Disraeli, novelist, philosopher, first Earl of Beaconsfield, statesman, and Prime Minister. Basically a tribute.
1963 76m/C Trevor Howard, Greer Garson, Hurd Hatfield, Kate Reid; *Dir:* George Schaefer. **VHS, Beta** *FHS* 𝄞𝄞½

Invincible Obsessed Fighter Two heavy-handed warriors (Wong and Chong) battle over a stolen treasure.
198? 92m/C Elton Chong, Michael Wong. **VHS, Beta** $54.95 *MAV* 𝄞

The Invincible Six An unlikely band of thieves running from the law winds up protecting a small village from marauding bandits. Plenty of action, as you would expect when thieves take on bandits, but with no scene-stealers or sympathetic characters the story lacks interest. Unremarkable choreography by ballet great Rudolf Nureyev. Filmed in Iran.
1968 96m/C Stuart Whitman, Elke Sommer, Curt Jurgens, Jim Mitchum, Ian Ogilvy; *Dir:* Jean Negulesco. **VHS, Beta** $39.98 *CGI* 𝄞𝄞

Invincible Sword A motley band of zany characters use their collective brain power to save their imprisoned chum, amid much chop-suey action.
1978 93m/C Wang Yu; *Dir:* Hsu Tseng Hung. **VHS** $59.98 *CGH* 𝄞

Invisible Adversaries A photographer uncovers an extra-terrestrial plot to cause excessive aggression in humans. She and her lover attempt to hold on to their crumbling humanity. Movies like this make you mad enough to tear something up. Enjoyed a meek cult following. In German with English subtitles.
1977 112m/C *GE* Susanne Widl, Peter Weibel; *Dir:* Valie Export. **VHS, Beta** $59.95 *FCT* 𝄞𝄞

The Invisible Avenger Based on the vintage radio character "The Shadow." A detective who can make himself invisible investigates the murder of a New Orleans jazz musician. The mysterious element successful on radio is lost on film, and only the Shadow knows shy. Additional footage was added and the film was re-released in 1962 as "Bourbon St. Shadows."
1958 60m/B Richard Derr, Marc Daniels, Helen Westcott, Jeanne Neher; *Dir:* John Sledge, Ben Parker. **VHS, Beta** $16.95 *SNC, SMW* 𝄞

The Invisible Dead So many invisible man movies, so little time. A scientist creates an invisible man, imprisons and tortures him. A bit miffed with his host, he who can't be seen escapes, and vents his invisible spleen.
19?? 90m/C *IT* Howard Vernon, Britt Carva; *Dir:* Jess (Jesus) Franco. **VHS** $59.95 *UHV* 𝄞

The Invisible Ghost A man carries out a series of grisly stranglings while under hypnosis by his insane wife. Typically bad low-budget exploiter about a fun-lovin' crazy couple bringing down property values in the neighborhood.

1941 70m/B Bela Lugosi, Polly Ann Young; *Dir:* Joseph H. Lewis. **VHS, Beta** $24.95 *NOS, MRV, SNC* 𝄞½

The Invisible Kid A shy teen manages to make himself invisible, a feat which ironically permits him to get people to notice him. Now he can visit places he's only dreamed of, including the girls' locker room. For those who have dreamed about girls' locker rooms.
1988 95m/C Karen Black, Jay Underwood, Chynna Phillips, Wally Ward, Brother Theodore; *Dir:* Avery Crounse. **VHS, Beta, LV** *MED* 𝄞

The Invisible Killer Someone is killing folks by sending poison through telephone lines. A detective and a reporter team up to find the scoundrel by setting up an answering machine that automatically plays back messages to the caller. Bad connection of events leads to a static resolution.
1940 61m/B Grace Bradley, Roland Drew, William Newell; *Dir:* Sam Newfield. **VHS, Beta** $16.95 *SNC, MLB* 𝄞½

The Invisible Man The vintage horror-fest based on H. G. Wells' novella about a scientist whose formula for invisibility slowly drives him insane. His mind definitely wandering, he plans plans to use his recipe to rule the world. Rains' first role; though his body doesn't appear until the final scene, his voice characterization is magnificent. The visual detail is excellent, setting standards that are imitated because they are difficult to surpass; with special effects by John P. Fulton.
1933 71m/B Claude Rains, Gloria Stuart, Dudley Digges, William Harrigan, Una O'Connor; *Dir:* James Whale. New York Times 10 Best Films of the Year '33: #7 of the Best Films of All Time. **VHS, Beta, LV** $14.95 *MCA, MLB, RXM* 𝄞𝄞𝄞

The Invisible Man Returns Price stars as the original invisible man's brother. Using the same invisibility formula, Price tries to clear himself after being charged with murder. He reappears at the worst times, and you gotta love that floating gun. Fun sequel to 1933's classic "The Invisible Man."
1940 81m/B Cedric Hardwicke, Vincent Price, John Sutton, Nan Grey; *Dir:* Joe May. **VHS** $14.98 *MCA, MLB* 𝄞𝄞𝄞

The Invisible Maniac A crazy voyeur perfects a serum for invisibility and promptly gets a job as a physics teacher in a high school. There he leers at girls taking showers and slaughters students when he is caught. For voyeurs. Get outta here, you maniac.
1990 (R) 87m/C Noel Peters, Shannon Wilsey, Melissa Moore, Robert Ross, Rod Sweitzer, Eric Champnella, Kalei Shellabarger, Gail Lyon, Debra Lamb; *W/Dir:* Rif Coogan. **VHS** $89.95 *REP* *Woof!*

The Invisible Ray For a change, this horror film features Lugosi as the hero, fighting Karloff, a scientist who locates a meteor that contains a powerful substance. Karloff is poisoned and becomes a murdering megalomaniac. Watching Karloff and Lugosi interact, and the great special effects—including a hot scene where a scientist bursts into flames—helps you ignore a generally hokey script.
1936 82m/B Boris Karloff, Bela Lugosi, Frances Drake, Frank Lawton, Beulah Bondi, Walter Kingsford; *Dir:* Lambert Hillyer. **VHS, Beta, LV** $14.98 *MCA, MLB* 𝄞𝄞½

The Invisible Strangler A death-row murderer can make himself invisible and rubs out witnesses who helped put him away; woman risks her life to expose him when the police fail to see the problem. Very violent

collection of brutal scenes. Not released theatrically until 1984.
1976 (PG) 85m/C Robert Foxworth, Stefanie Powers, Elke Sommer, Sue Lyon, Leslie Parrish, Marianna Hill; *Dir:* John Florea. **VHS, Beta** $59.95 *TWE* 𝄞½

Invitation to the Dance Three classic dance sequences, "Circus," "Ring Around the Rosy," and "Sinbad the Sailor." For dance lovers, featuring excellent performances by Kelly.
1956 93m/C Gene Kelly, Igor Youskevitch, Tamara Toumanova; *Dir:* Gene Kelly. Berlin Film Festival '56: Best Picture. **VHS, Beta** $19.98 *MGM, CCB* 𝄞𝄞½

Invitation to a Gunfighter Small town politics change when a paid assassin ambles into town and creates a lot of talk among the neighbors. Like a long joke in which the teller keeps forgetting the important details, the plot grows confusing though it's not complicated. Brynner is interesting as an educated, half-black/half-creole, hired gun.
1964 92m/C Yul Brynner, George Segal, Strother Martin, William Hickey, Janice Rule, Mike Kellin, Pat Hingle; *Dir:* Richard Wilson. **VHS, Beta** $19.98 *FOX, HHE* 𝄞𝄞

Invitation to Hell A tormented maiden has a devil of a time discovering a secret power that can enable her to gracefully decline a satanic summoning. The cliched evil is tormenting.
1984 100m/C Becky Simpson, Joseph Sheahan, Colin Efford, Stephen Longhurst. **VHS, Beta** $79.95 *WES* *Woof!*

Invitation to Hell "Faust" meets "All My Children" in celluloid suburbia. Never Emmied Lucci is the devil's dirty worker, persuading upwardly mobile suburbanites to join a really exclusive country club in exchange for a little downward mobility. Urich, a space scientist, and family are new in town, and soul searching Lucci's got her devil vixen sights set on space jock and brood. Bound to disappoint both fans of Craven's early independent work ("Last House on the Left") and his later high gloss ("Nightmare on Elm Street") formulas. Made for the small screen.
1984 100m/C Susan Lucci, Robert Urich, Joanna Cassidy, Kevin McCarthy, Patty McCormack, Joe Regalbuto, Soleil Moon Frye; *Dir:* Wes Craven. **VHS, Beta** $79.98 *SVS* 𝄞½

Invitation to Paris A French musical revue set in the streets of Paris which features the girls of the French Can-Can revue. Has a lot of kick.
1960 51m/B Maurice Chevalier, Les Djinns, Patachou, Fernandel, Jean Sablon, George Ulmer, Line Renaud. **VHS, Beta** $24.95 *NOS, VYY, DVT* 𝄞½

Invitation au Voyage Follows the journey of a twin who refuses to accept the death of his sister, a rock singer. The obsession is played fairly well, but it's a no-twin situation, even though the guy can't admit it. Perversity is played to the point of overkill. Subtitled in English.
1983 (R) 90m/C *FR* Laurent Malet, Nina Scott, Aurore Clement, Mario Adorf; *Dir:* Peter Del Monte. **VHS, Beta** $59.95 *COL* 𝄞½

Invitation to the Wedding When the best friend of a bridegroom falls in love with the bride, he stops at nothing to halt the wedding. Not exactly the Gielgud movie of the year, but there are some excellent scenes between him and Richardson.

1973 (PG) 89m/C John Gielgud, Ralph Richardson, Paul Nickolaus, Elizabeth Shepherd; *Dir:* Joseph Brooks. **VHS, Beta** $69.95 *VES* 🎅🎅

The Ipcress File The first of the Harry Palmer spy mysteries that made Caine a star. Based upon the bestseller by Len Deighton, it features the flabby, nearsighted spy investigating the kidnapping of notable British scientists. Solid scenes, including a scary brainwashing session, and tongue-firmly-in-British-cheek humor. Lots of camera play to emphasize Caine's myopia. Two sequels: "Funeral in Berlin" and "Billion Dollar Brain."
1965 108m/C *GB* Michael Caine, Nigel Green, Guy Doleman, Sue Lloyd; *Dir:* Sidney J. Furie. British Academy Awards '65: Best Art Direction/Set Decoration (Color), Best Color Cinematography, Best Film. **VHS, Beta** $29.95 *MCA* 🎅🎅🎅

Iphigenia Based on the classic Greek tragedy by Euripides, this story concerns the Greek leader Agamemnon, who plans to sacrifice his lovely daughter, Iphigenia, to please the gods. Mere mortals protest and start a save-the-babe movement, but Euripides' moral—you can't please 'em all—is devastatingly realized. Fine adaptation that becomes visually extravagant at times, and an equally fine musical score from Mikos Theodorakis. Subtitled.
1977 130m/C *GR* Irene Papas, Costa Kazakos, Tatiana Papamoskou; *Dir:* Michael Cacoyannis. **VHS, Beta** $59.95 *COL, FHS* 🎅🎅🎅

Iran: Days of Crisis An insightful made-for-cable-TV miniseries recounts America's humiliation at the hands of Iranian radicals in 1979, told from the vantage of a U.S. embassy official married to a Tehran woman. Former Carter administration aides Hamilton Jordan and Gerald Rafshoon devised the docudrama, which spares neither them nor their president in showing how bungled and shortsighted policies led to 52 Americans held hostage for over a year.
1991 185m/C Arliss Howard, Alice Krige, George Grizzard, Jeff Fahey, Tony Goldwyn; *Dir:* Kevin Connor. **VHS** $89.98 *TTC* 🎅🎅1/2

Irezumi (Spirit of Tattoo) A Japanese woman has her body elaborately tattooed to please her lover. Things start off well enough. He's an inept lover, so she has directions tattooed, but then he discovers a "Dear John" letter on her foot just as she gives him the boot. Actually proves an interesting study in obsession and eroticism, with the tattoos drawing the lovers deeper and deeper until they get under each other's skin. Subtitled in English.
1983 88m/C *JP* **VHS, Beta, LV** $29.95 *PAV* 🎅🎅

Iris Drama follows the passionate and tragic life of New Zealand writer Iris Wilkinson, an international novelist, poet, and journalist. Wilkinson achieved a great deal of acclaim as a writer, but suffered many personal tragedies, and eventually took her own life at age 33. Well intentioned but clunky treatment of powerful subject matter.
1989 90m/C *NZ* Helen Morse. **VHS, Beta** $160.00 *WOM* 🎅🎅

Irish Cinderella A silent version of the Cinderella legend set in Ireland, with thematic stress on Irish politics and patriotism.
1922 72m/B Pattie MacNamara. **VHS, Beta, 8mm** $29.95 *VYY* 🎅🎅

Irish Luck A bellhop turns detective to figure out the mysterious happenings in his hotel. Becomes terminally cute by playing on a luck o' the Irish theme.

1939 58m/B Frankie Darro, Dick Purcell, Sheila Darcy; *Dir:* Grant Withers, Scott R. Dunlap. **VHS** $24.95 *NOS, DVT* 🎅🎅

The Irishman Set in 1920s Australia, the tale of a proud North Queensland family and their struggles to stay together. The Irish immigrant father is a teamster whose horse-drawn wagons are threatened by progress. His fight to preserve old ways is impressive, but gives way to sentimentality.
1978 108m/C *AU* Lou Brown, Michael Craig, Simon Burke, Robyn Nevin, Bryan Brown; *Dir:* Donald Crombie. **VHS, Beta** $69.95 *VES* 🎅🎅1/2

Irma La Douce A gendarme pulls a one-man raid on a back-street Parisian joint and falls in love with one of the hookers he arrests. Lemmon is great as a well-meaning, incompetent boob, and MacLaine gives her all as the hapless hooker. A plodding pace, however, robs it of the original's zip, though it was a box office smash. Broadway musical is lost without the music. MacLaine received an Oscar nomination, as did the cinematography.
1963 146m/C Jack Lemmon, Shirley MacLaine, Herschel Bernardi; *Dir:* Billy Wilder. Academy Awards '63: Best Original Score. **VHS, Beta, LV** $59.98 *FOX, IME* 🎅🎅1/2

Iron Angel During the Korean War a squadron sets out to silence North Korean guns. They do. Judging by their eternal bickering they must have put the enemy to sleep.
1964 84m/B Jim Davis, Margo Woode, Donald (Don "Red") Barry, L.Q. Jones; *W/Dir:* Ken Kennedy. **VHS** $15.95 *NOS, LOO* 🎅1/2

Iron Bodies Several well-developed men and women discuss the mental and physical sacrifices they've made in order to swell and harden their flesh.
1985 60m/C **VHS, Beta** $39.95 *AHV* 🎅1/2

Iron Cowboy Actor finds romance with film editor while audience stifles yawns. Filmed on the set of "Blue," this pre-"Deliverance" Reynolds comedy never made it to the theaters. (Smithee is actually director Jud Taylor.)
1968 86m/C Burt Reynolds, Barbara Loden, Terence Stamp, Noam Pitlik, Ricardo Montalban, Patricia Casey, Jane Hampton, Joseph Perry; *Dir:* Allen Smithee. **VHS** $39.98 *ACE* 🎅🎅

The Iron Crown A 13th-century legend inspired this sometime violent spectacle involving the title totem, a symbol of justice ignored by a powerful king at his peril. There's wonderful pagentry and a notable Tarzan imitator, but the pace is terribly slow. Italian dialogue with English subtitles.
1941 100m/C *Dir:* Alexandro Blasetti. **VHS** $59.95 *FST* 🎅🎅

Iron Dragon Strikes Back A man seeks out the men who murdered his martial arts instructor.
1984 90m/C Bruce Li. **VHS, Beta** $39.95 *TWE* 🎅

Iron Duke An historical account of the life of the Duke of Wellington.
1934 88m/B George Arliss, Gladys Cooper, Victor Saville. **VHS, Beta** $19.95 *NOS, WFV, KRT* 🎅1/2

Iron Eagle A teenager teams with a renegade fighter pilot to rescue the youth's father from captivity in the Middle East. Predictable but often exciting.
1985 (PG-13) 117m/C Louis Gossett Jr., Jason Gedrick, Tim Thomerson; *Dir:* Sidney J. Furie. **VHS, Beta, LV** $14.98 *FOX* 🎅🎅1/2

Iron Eagle 2 Lower-budget extended adventures of a maverick fighter pilot after he is reinstated in the Air Force. This time he links with an equally rebellious commie fighter and blasts away at nuke-happy Ivan. Yahoo fun for all overt Yankees! Available in Spanish.
1988 (PG) 102m/C *CA IS* Louis Gossett Jr., Mark Humphrey, Stuart Margolin, Alan Scarfe, Maury Chaykin, Sharon H. Brandon; *Dir:* Sidney J. Furie. **VHS, Beta, LV** $14.95 *IVE* 🎅🎅

Iron Horsemen A motorcycle gang leader has a prophetic vision while tripping on LSD, and returns to his former town to challenge its religious establishment. Also available on video under its original title, "J.C."
1971 (R) 97m/C William F. McGaha, Hannibal Penny, Joanna Moore, Burr de Benning, Slim Pickens, Pati Delaney; *Dir:* William F. McGaha. **VHS, Beta** $9.95 *SIM, CHA* 🎅

The Iron Major Frank Cavanaugh, a famous football coach, becomes a hero in World War I. Standard flag-waving biography to increase morale back home, but fairly well-made.
1943 85m/B Pat O'Brien, Ruth Warrick, Robert Ryan, Leon Ames, Russell Wade, Bruce Edwards; *Dir:* Ray Enright. **VHS, Beta** $19.98 *TTC* 🎅🎅1/2

The Iron Mask Early swashbuckling extravaganza with a master swordsman defending the French king from a scheme involving substitution by a lookalike. Still fairly exciting, thanks largely to director Dwan's flair. Based on Alexandre Dumas' "Three Musketeers" and "The Man in the Iron Mask" with talking sequences.
1929 87m/B Douglas Fairbanks Sr., Nigel de Brulier, Marguerite de la Motte; *Dir:* Allan Dwan. **VHS, Beta, LV** $9.98 *NOS, CCB, DVT* 🎅🎅1/2

Iron Maze A fascinating notion forms the center of this dramatic scrapheap; the classic Japanese "Rashomon" plot shifted to a rusting Pennsylvania steel town. When a Tokyo businessman is found bludgeoned, witnesses and suspects (including his American-born wife) tell contradictory stories. It's too convoluted and contrived to work, with a hollow happy ending tacked on. Oliver Stone helped produce.
1991 (R) 102m/C Jeff Fahey, Bridget Fonda, Hiroaki Murakami, J.T. Walsh, Gabriel Damon, John Randolph, Peter Allas; *Dir:* Hiroaki Yoshida. **VHS** $89.95 *ACA* 🎅1/2

Iron & Silk A young American searches for himself while he teaches English and learns martial arts in mainland China. Based on the true story of Salzman's travels. His studies of martial arts and Chinese culture provide a model for his students in their studies of American language and culture. Beautiful photography. Fine performances.
1991 (PG) 94m/C Mark Salzman, Pan Qingfu, Jeanette Lin Tsui, Vivian Wu; *Dir:* Shirley Sun. **VHS, LV** $89.95 *LIV, PIA* 🎅🎅

Iron Thunder Liberty is in peril and only much kicking and pec flexing will make the world a kinder gentler place. A thundering bore.
1989 85m/C Iron "Amp" Elmore, George M. Young, Julius Dorsey. **VHS** $79.95 *XVC, HHE* 🎅1/2

The Iron Triangle A U.S. officer taken prisoner by the Viet Cong forms a bond with one of his captors. Film lends the viewer an opportunity to see things from the other sides perspective.

1989 (R) 94m/C Beau Bridges, Haing S. Ngor, Liem Whatley; *Dir:* Eric Weston. **VHS, Beta, LV** $19.95 *IVE* ⚫⚫

Iron Warrior A barbarian hacks his way through a fantastical world to get to a beautiful princess.
1987 (R) 82m/C Miles O'Keefe, Savina Gersak, Tim Lane; *Dir:* Al Bradley. **VHS, Beta** $19.95 *MED* ⚫

Ironclads Memorable dramatization of the five-hour naval battle between the Confederate's Merrimac and the Union's Monitor during the Civil War. The special effects are the only thing worth watching in this overlong tale that includes standard subplots of loyalty, lost loves, and death from war.
1990 94m/C Virginia Madsen, Alex Hyde-White, Reed Edward Diamond, E.G. Marshall, Fritz Weaver; *Dir:* Delbert Mann. **VHS** $79.98 *TTC* ⚫⚫½

Ironmaster When a primitive tribesman is exiled from his tribe, he discovers a mysteriously power-filled iron staff on a mountainside.
1983 98m/C George Eastman, Pamela Field. **VHS, Beta** $59.95 *PSM* ⚫

Ironweed Grim and gritty drama about bums living the hard life in Depression-era Albany. Nicholson excels as a former ballplayer turned drunk bothered by visions of the past, and Streep is his equal in the lesser role of a tubercular boozer. Tom Waits also shines. Another grim view from "Pixote" director Babenco. Based on William Kennedy's Pulitzer Prize-winning tragedy, and scripted by Kennedy. Both Streep and Nicholson garnered Oscar nominations.
1987 (R) 135m/C Jack Nicholson, Meryl Streep, Tom Waits, Carroll Baker, Michael O'Keefe, Fred Gwynne, Diane Venora, Margaret Whitton; *Dir:* Hector Babenco. **VHS, Beta, LV** $89.98 *LIV, VES* ⚫⚫⚫

Irreconcilable Differences When her Beverly Hills parents spend more time working and fretting than giving hugs and love, a ten-year-old girl sues them for divorce on the grounds of "irreconcilable differences." The media has a field day when they hear that she would rather go live with the maid. Well cast, with well-developed characterizations. The script, by the creators of "Private Benjamin," is humanely comic rather than uproariously funny.
1984 (PG) 112m/C Ryan O'Neal, Shelley Long, Drew Barrymore, Sam Wanamaker, Allen Garfield, Sharon Stone, Luana Anders; *Dir:* Charles Shyer. **VHS, Beta, LV** $29.98 *LIV, VES* ⚫⚫½

Irving Berlin's America A documentary depicting Berlin's legacy of patriotic pop tunes, shown in a variety of film and performance clips, plus interviews with the stars who sang them.
1986 83m/C Ginger Rogers, Mary Martin, Howard Keel, Kate Smith, Fred Astaire; *Nar:* Sandy Duncan. **Beta** *CMR* ⚫⚫½

Is Paris Burning? A spectacularly star-studded but far too sprawling account of the liberation of Paris from Nazi occupation. The script, in which seven writers had a hand, is based on Larrry Collins and Dominique Lapierre's best seller.
1968 173m/C *FR* Jean-Paul Belmondo, Charles Boyer, Leslie Caron, Jean-Pierre Cassel, George Chakiris, Claude Dauphin, Alain Delon, Kirk Douglas, Glenn Ford, Gert Frobe, Daniel Gelin, E.G. Marshall, Yves Montand, Anthony Perkins, Claude Rich, Simone Signoret, Robert Stack, Jean-Louis Trintignant, Pierre Vaneck, Orson Welles, Bruno Cremer, Suzy Delair, Michael

Lonsdale; *Dir:* Rene Clement. **VHS** $29.95 *PAR, FCT* ⚫⚫⚫

Is There Sex After Death? Often funny satire on the sexual revolution is constructed as a behind-the-scenes view of the porn film world. Odd cast includes Henry and Warhol superstar Woodlawn. Originally rated "X."
1971 (R) 97m/C Buck Henry, Alan Abel, Marshall Efron, Holly Woodlawn, Earl Doud; *Dir:* Alan Abel, Jeanne Abel. **VHS, Beta** $59.98 *MAG* ⚫⚫⚫

Is This Goodbye, Charlie Brown? Charlie Brown and the gang struggle to understand their loss when a friend moves away.
1983 25m/C Snoopy, Charlie Brown. **VHS** $14.95 *MED, VTR* ⚫⚫⚫

Isadora A loose, imaginative biography of Isadora Duncan, the cause celebre dancer in the 1920s who became famous for her scandalous performances, outrageous behavior, public love affairs, and bizarre, early death. Redgrave is exceptional in the lead, and Fox provides fine support. Music by Maurice Jarre. Alternate title: "The Loves of Isadora." Restored from its original 131-minute print length by the director. Redgrave was nominated for an Oscar.
1968 153m/C Vanessa Redgrave, Jason Robards Jr., James Fox, Ivan Tchenko, John Fraser, Bessie Love; *Dir:* Karel Reisz. Cannes Film Festival '69: Best Actress (Redgrave); National Board of Review Awards '69: 10 Best Films of the Year; National Society of Film Critics Awards '69: Best Actress (Redgrave). **VHS, Beta** $29.95 *MCA* ⚫⚫⚫

Ishtar Two astoundingly untalented performers have a gig in a fictional Middle Eastern country and become involved in foreign intrigue. A big-budget box-office bomb produced by Beatty offering few laughs, though it's not as bad as its reputation. Considering the talent involved, though, it's a dissappointment, with Beatty and Hoffman laboring to create Hope and Crosby chemistry. Anyone for a slow night?
1987 (PG-13) 107m/C Dustin Hoffman, Warren Beatty, Isabelle Adjani, Charles Grodin; *W/Dir:* Elaine May. **VHS, Beta, LV** $14.95 *COL* ⚫⚫

The Island Poetic examination of a peasant family's daily struggle for existence on a small island. No dialogue in this slow but absorbing work by one of Japan's master directors. Also known as "Hadaka no Shima."
1961 96m/B *JP* Nobuko Otowa, Taiji Tonoyama, Shinji Tanaka, Masanori Horimoto; *Dir:* Kaneto Shindo. National Board of Review Awards '62: 5 Best Foreign Films of the Year. **VHS, Beta, LV, 8mm** $24.95 *IHF, VYY, HHT* ⚫⚫⚫½

Island Tautly paced, the story creates an increasingly complex situation, until a World War I army officer is revealed to be the victim of a shocking and ingenious trap.
1977 30m/C John Hurt; *Dir:* Robert Fuest. **VHS, Beta** *LCA* ⚫

The Island A New York reporter embarks on a Bermuda triangle investigation, only to meet with the murderous descendants of seventeenth-century pirates on a deserted island. Provides little entertainment value or believability. Based on the Peter Benchley novel.
1980 (R) 113m/C Michael Caine, David Warner, Angela Punch McGregor, Frank Middlemass, Don Henderson; *Dir:* Michael Ritchie. **VHS, Beta, LV** $19.95 *MCA* ⚫

Island of Adventure Four children explore an island and find a gang of criminals inhabiting it.
1981 85m/C VHS, Beta $39.98 *SUE* ⚫

Island of Blood Buckets of blood abound on a remote island when a film crew's beset by berserk butcher.
19?? 84m/C Jim Williams, Dean Richards. **VHS** $59.98 *AIP* ⚫½

Island of the Blue Dolphins Based on a popular children's book by Scott O'Dell, this is a true story of a young Native American girl who, with her brother, are stranded on a deserted island when their people accidentally leave them. When the boy is killed, the girl learns to survive by her skills and by befriending the leader of a pack of wild dogs. Good family fare.
1964 99m/C Celia Kaye, Larry Domasin, Ann Daniel, George Kennedy; *Dir:* James B. Clark. **VHS, Beta** $59.95 *MCA* ⚫⚫½

Island of the Burning Doomed A brutal heat wave accompanies invading aliens in this British-made outing. Lee and Cushing carry the picture. Also known as "Island of the Burning Damned" and "Night of the Big Heat."
1967 94m/C *GB* Christopher Lee, Peter Cushing, Patrick Allen, Sarah Lawson, Jane Merrow; *Dir:* Terence Fisher. **VHS** *NSV* ⚫⚫½

Island Claw A group of marine biologists experimenting on a tropical island discover the "Island Claw," who evolved as the result of toxic waste seeping into the ocean. A festering, oozing woofer.
1980 91m/C Barry Nelson, Robert Lansing; *Dir:* Hernan Cardenas. **VHS, Beta** $69.98 *LIV, VES* Woof!

Island of Desire An Army nurse, a doctor, and a young Navy Adonis are trapped on a deserted island. Not surprisingly, a love triangle develops.
1952 93m/C *GB* Tab Hunter, Linda Darnell, Donald Gray; *Dir:* Stuart Heisler. **VHS, Beta** $59.95 *MWP, MLB* ⚫

Island of Dr. Moreau This remake of the H.G. Wells story "Island of the Lost Souls" (1933) is a bit disappointing but worth watching for Lancaster's solid performance as the scientist who has isolated himself on a Pacific island in order to continue his chromosome research—he can transform animals into near-humans and humans into animals. Neat-looking critters.
1977 (PG) 99m/C Burt Lancaster, Michael York, Nigel Davenport, Barbara Carrera, Richard Basehart; *Dir:* Don Taylor. **VHS, Beta** $19.98 *WAR, OM* ⚫⚫½

Island of the Lost An anthropologist's family must fight for survival when they become shipwrecked on a mysterious island.
1968 92m/C Richard Greene, Luke Halpin; *Dir:* John Florea. **VHS** $19.98 *REP, GHV* ⚫½

Island of Lost Girls Young women are taken to an unknown location in the Far East to be sold as part of a white slave ring.
1973 (R) 85m/C Brad Harris, Tony Kendall, Monica Pardo. **VHS, Beta** $16.95 *SNC* Woof!

Island Monster Also called "Monster of the Island," this film deals with ruthless, kidnapping drug-smugglers led by "monster" Karloff, and the efforts to bring them to justice.
1953 87m/B *IT* Boris Karloff, Renata Vicario, Franco Marzi; *Dir:* Robert Montero. **VHS, Beta** $29.95 *SNC, HHT* ⚫½

The Island of Nevawuz A beautiful island is in trouble when J.B. Trumphorn decides to make lots of money by building factories and refineries on it. Will the Island of Nevawuz end up a polluted mess?
1980 50m/C VHS, Beta $39.98 *SUE* ♫ ½

Island Reggae Greats A compilation of concert and studio performances from such reggae legends as Toots and the Maytals and Bob Marley.
1985 28m/C Toots & the Maytals, Black Uhuru, Aswad, Third World, Bob Marley. **VHS, Beta $19.95** *COL* ♫ ½

Island in the Sun Racial tension pulls apart the lives of the residents of a Caribbean island. Good cast but a very poor adaption of Alec Waugh's novel. Marvelous location shots.
1957 119m/C James Mason, Joan Fontaine, Dorothy Dandridge, John Williams, Harry Belafonte; *Dir:* Robert Rossen. **VHS** *TCF* ♫♫

Island of Terror First-rate science fiction chiller about an island overrun by shell-like creatures that suck the bones out of their living prey. Good performances and interesting twists make for prickles up the spine.
1966 90m/C *GB* Peter Cushing, Edward Judd, Carole Gray, Sam Kydd, Niall MacGinnis; *Dir:* Terence Fisher. **VHS $16.95** *SNC, MLB* ♫♫

Island at the Top of the World A rich Englishman, in search of his missing son, travels to the Arctic Circle in 1908. The rescue party includes an American archeologist, a French astronaut, and an Eskimo guide. Astonishingly, they discover an unknown, "lost" Viking kingdom. This Jules Verne-style adventure doesn't quite measure up, but kids will like it.
1974 (G) 93m/C David Hartman, Donald Sinden, Jacques Marin, Mako, David Gwillim; *Dir:* Robert Stevenson. **VHS, Beta $69.95** *DIS* ♫♫½

Island Trader A young lad battles pirates and a crooked tug captain who seek hidden gold that he knows about.
1971 95m/C John Ewart, Ruth Cracknell, Eric Oldfield; *Dir:* Harold Rubie. **VHS, Beta** *PGN* ♫

Island Warriors On a remote island, a beautiful Amazon queen rules an army of kung fu masters.
1986 88m/C VHS, Beta $29.99 *BFV* ♫ ½

The Islander Sixteen-year-old Inga comes of age among the commercial fishermen of Lake Michigan.
1988 99m/C Kit Wholihan, Jeff Weborg, Celia Klehr, Jacob Mills, Julie Johnson, Mary Ann McHugh, Michael Rock, Sheri Parish; *Dir:* Phyllis Berg-Pigorsch. **VHS, Beta $79.95** *NSV* ♫

Islands A made-for-television drama about a middle-aged ex-hippie and a young rebellious punk who clash during a summer together on a secluded island.
1987 55m/C Louise Fletcher. **VHS, Beta $19.95** *NWV* ♫ ½

Islands in the Stream Based on Ernest Hemingway's novel, this film is actually two movies in one. An American painter/ sculptor, Thomas Hudson, lives a reclusive life on the island of Bimini shortly before the outbreak of World War II. The first part is a sensitive story of a broken family and the coming of the artist's three sons to live with their father after a four-year separation. The second part is a second rate action-adventure.

1977 (PG) 105m/C George C. Scott, David Hemmings, Claire Bloom, Susan Tyrrell, Gilbert Roland; *Dir:* Franklin J. Schaffner. **VHS, Beta, LV $29.95** *AIM, PAR* ♫♫½

Isle of the Dead A Greek general is quarantined with seemingly all manner of social vermin on a plague-infested island in the early 1900s. The fear is that vampires walk among them. Characteristically spooky Val Lewton production with some original twists.
1945 72m/B Boris Karloff, Ellen Drew, Marc Cramer; *Dir:* Mark Robson. **VHS, Beta, LV $19.95** *TTC, MED, FCT* ♫♫½

Isle of Forgotten Sins Deep sea divers and an evil ship's captain vie for treasure. Good cast and direction but unremarkable material. Also released under the title "Monsoon."
1943 82m/B John Carradine, Gale Sondergaard, Sidney Toler, Frank Fenton; *Dir:* Edgar G. Ulmer. **VHS, Beta, 8mm $16.95** *SNC, VYY* ♫♫

Isle of Secret Passion A romance novel comes to video life as two lovers thrash out their problems on a Greek island.
1985 90m/C Patch MacKenzie, Michael MacRae, Zohra Lampert. **VHS, Beta $14.95** *PSM* ♫

Isn't Life Wonderful Outstanding film from Griffith depicts the unbearable conditions in post-World War I Germany. Filmed on location.
1924 90m/B Carol Dempster, Neil Hamilton, Lupino Lane, Hans von Schlettow; *W/Dir:* D.W. Griffith. **VHS, Beta $27.95** *DNB* ♫♫♫♫

Istanbul An American charms a Belgian student into taking part in a kidnap scheme.
1985 (R) 90m/C *FR* Brad Dourif, Dominique Deruddere, Ingrid De Vos, Francois Beukelaers; *Dir:* Marc Didden. **VHS, Beta $59.98** *CGI* ♫

Istanbul Believing he has stumbled upon the best story of his career, a journalist soon discovers that he has been pulled into a world of guns and violence.
1990 (PG-13) 88m/C *TU SW* Robert Morley, Timothy Bottoms, Twiggy; *Dir:* Mats Arehn. **VHS, Beta, LV $89.98** *MAG, IME* ♫

It A department store clerk with plenty of sex appeal desires her rich boss. Established Bow as a prominent screen siren. Otherwise, it's not particularly interesting. Silent. Adapted by Elinor Glyn from her own story; Glyn also makes an appearance.
1927 71m/B Clara Bow, Gary Cooper, Antonio Moreno; *Dir:* Clarence Badger. **VHS, Beta, LV $34.98** *CCB, VYG, CRC* ♫♫½

It Came from Beneath the Sea A giant octopus arises from the depths of the sea scouring San Francisco searching for human food. Ray Harryhausen effects are special.
1955 80m/B Kenneth Tobey, Faith Domergue, Ian Keith, Donald Curtis; *Dir:* Robert Gordon. **VHS, Beta, LV $9.95** *GKK, MLB* ♫♫

It Came From Outer Space Aliens take on the form of local humans to repair their spacecraft in a small Arizona town. Good performances and outstanding direction add up to a better science fiction film. Based on the story "The Meteor" by Ray Bradbury, who also wrote the script. Originally filmed in 3-D.
1953 81m/B Richard Carlson, Barbara Rush, Charles Drake, Russell Johnson; *Dir:* Jack Arnold. **VHS** *GKK, MLB* ♫♫♫

It Came from Hollywood A compilation of scenes from "B" horror and science fiction films of the 1950s, highlighting the funny side of these classic schlock movies. Some sequences are in black and white.
1982 (PG) 87m/C *Dir:* Malcolm Leo; *Nar:* Dan Aykroyd, Cheech Marin, Thomas Chong, John Candy, Gilda Radner. **VHS, Beta, LV $59.95** *PAR* ♫½

It Came Upon a Midnight Clear A heavenly miracle enables a retired (and dead) New York policeman to keep a Christmas promise to his grandson. Made for television.
1984 96m/C Mickey Rooney, Scott Grimes, George Gaynes, Annie Potts, Lloyd Nolan, Barrie Youngfellow; *Dir:* Peter Hunt. **VHS, Beta $9.95** *GKK* ♫½

It Conquered the World Vegetable critters from Venus follow a probe satellite back to Earth and drop in on scientist Van Cleef, who'd been their intergalactic pen pal (seems he thought a Venusian invasion would improve the neighborhood). Their travelling companions, little bat creatures, turn earth dwellers into murderous zombies, and it starts to look like the Venusians are trying to make earth their planetary exurbia. Early vintage zero budget Corman, for schlock connoisseurs. Remade for television as "Zontar, the Thing from Venus."
1956 68m/B Peter Graves, Beverly Garland, Lee Van Cleef, Sally Fraser, Charles B. Griffith, Russ Bender, Jonathan Haze, Dick Miller, Karen Kadler, Paul Blaisdell; *Dir:* Roger Corman. **VHS $9.95** *COL, FCT, MLB* ♫♫½

It Could Happen to You Stepbrothers raised by a gentle immigrant take different paths in life. When the one ends up killing someone during a botched robbery, the other - a lawyer - defends him. Standard '30s morality play.
1937 64m/B Alan Baxter, Andrea Leeds, Owen Davis Jr., Astrid Allwyn, Walter Kingsford, Al Shear; *Dir:* Phil Rosen. **VHS $19.95** *NOS, KRT* ♫½

It Could Happen to You A drunken advertising executive is charged with murder after a dead nightclub singer is found in his car. His wife sets out to clear him.
1939 64m/C Stuart Erwin, Gloria Stuart, Raymond Walburn, Douglas Fowley, June Gale, Paul Hurst, Richard Lane; *Dir:* Alfred Werker. **VHS, Beta $19.95** *KRT* ♫♫

It Couldn't Happen Here Fans of the techno-pop duo The Pet Shop Boys can test their loyalty by enduring this feature music-video that actually played a few theaters. The dour pair journey across a surreal England, encountering a sinister blind priest, a philisophical ventriloquist's dummy, and more. Songs, sometimes presented only as muted incidental music, include the title track, "West End Girls," "It's a Sin" and "Always on My Mind."
1988 (PG-13) 87m/C *GB* Neil Tennant, Chris Lowe, Joss Ackland, Neil Dickson, Carmen Du Sautoy, Gareth Hunt; *Dir:* Jack Bond. **VHS, LV $59.95** *COL, MVD* ♫

It Happened in Brooklyn Former sailor with blue eyes bunks with janitor with big nose in Brooklyn. Sailor and cronies encounter many musical opportunities. Lots of songs and falling in love, little entertainment.
1947 104m/B Frank Sinatra, Kathryn Grayson, Jimmy Durante, Peter Lawford, Gloria Grahame; *Dir:* Richard Whorf. **VHS $19.98** *MGM, FCT* ♫♫

It Happened in New Orleans A young boy and a former slave struggle to survive in post-Civil War New Orleans.
1936 86m/B Bobby Breen, May Robson, Alan Mowbray, Benita Hume. **VHS, Beta $29.95** VYY 🎬🎬½

It Happened One Night Classic Capra comedy about an antagonistic couple determined to teach each other about life. Colbert is an unhappy heiress who runs away from her affluent home in search of contentment. On a bus she meets newspaper reporter Gable, who teaches her how "real" people live. She returns the favor in this first of the 1930's screwball comedies. The plot is a framework for an amusing examination of war between the sexes. Colbert and Gable are superb as affectionate foes. Remade as the musicals "Eve Knew Her Apples" and "You Can't Run Away From It."
1934 105m/B Clark Gable, Claudette Colbert, Roscoe Karns, Walter Connolly, Alan Hale Jr., Ward Bond; *Dir:* Frank Capra. Academy Awards '34: Best Actor (Gable), Best Actress (Colbert), Best Adapted Screenplay, Best Director (Capra), Best Picture; National Board of Review Awards '34: Best Picture. **VHS, Beta, LV $19.95** COL, BTV 🎬🎬🎬🎬

It Happened in the Park A rare film exercise for De Sica, wherein the patrons of the Villa Borghese parks are surveyed for a 24-hour period. In French with English dialogue.
1956 72m/B FR *Dir:* Vittorio De Sica. **VHS, Beta $24.95** NOS, FCT 🎬🎬½

It Happened at the World's Fair Fun and light romance comedy has Elvis and a companion (O'Brien) being escorted through the Seattle World's Fair by a fetching Chinese girl. The King seems to be fresh and fun in this film - songs include "One Broken Heart for Sale," "A World of our Own," and "Happy Ending."
1963 105m/C Elvis Presley, Joan O'Brien, Gary Lockwood, Kurt Russell, Edith Atwater, Yvonne Craig; *Dir:* Norman Taurog. **VHS, Beta $19.95** MGM 🎬🎬½

It Rained All Night the Day I Left Two drifters find their fortunes change when they are hired by a wealthy widow. Interesting romance adventure combination.
1978 (R) 100m/C Louis Gossett Jr., Sally Kellerman, Tony Curtis; *Dir:* Nicolas Gessner. **VHS, Beta $29.95** IVE 🎬🎬

It Should Happen to You An aspiring model attempts to boost her career by promoting herself on a New York City billboard. The results, however, are continually surprising. Fine comedy teamwork from master thespians Holliday and Lemmon in this, the latter's first film. Written by Ruth Gordon and Garson Kanin.
1953 87m/B Judy Holliday, Jack Lemmon, Peter Lawford; *Dir:* George Cukor. **VHS, Beta, LV $9.95** GKK 🎬🎬🎬

It Started in Naples Good performances by both Gable and Loren in this comedy-drama about an American lawyer in Italy who, upon preparing his late brother's estate, finds that his brother's nephew is living with a stripper. A custody battle ensues, but love wins out in the end. Loren is incredible in nightclub scenes.
1960 100m/C Clark Gable, Sophia Loren, Marietto, Vittorio De Sica, Paolo Carlini, Claudio Ermelli, Giovanni Filidoro; *Dir:* Melville Shavelson. **VHS $14.95** PAR 🎬🎬🎬

It Takes a Thief Tough cookie Jayne's a buxom gangstress who does the hokey pokey with her criminal minions. A former lover who's been released from the big house seems to think she's been minding the mint for him. Au contraire.
1959 93m/B GB Jayne Mansfield, Anthony Quayle, Carl Mohner, Peter Reynolds, John Bennett; *Dir:* John Gilling. **VHS** BBF 🎬🎬

It Takes Two A young man spends his last ten days of bachelorhood cruising down Texas highways with a beautiful car saleswoman. Meanwhile, back at the altar, his bride waits patiently. Upbeat with some genuinely funny moments. From the director of "Pass the Ammo."
1988 (PG-13) 79m/C George Newbern, Leslie Hope, Kimberly Foster, Barry Corbin; *Dir:* David Beaird. **VHS, Beta $79.98** FOX 🎬🎬

It Was a Short Summer, Charlie Brown The Peanuts gang enjoys the summer until it's time to go back to school.
1969 26m/C *Dir:* Bill Melendez. **VHS $14.95** HIL 🎬🎬

Italian The story of an Italian immigrant family living in the slums of New York during the turn of the century.
1915 78m/B George Beban, Clara Williams, Leo Wills, J. Frank Burke, Fanny Midgley. **VHS, Beta** GVV, DNB 🎬🎬

The Italian Connection A man marked for death by the mob starts his own private war against the organization. Lots of action. Also known as "Manhunt."
1973 92m/C IT Henry Silva, Woody Strode, Mario Adorf, Lucianna Paluzzi; *Dir:* Fernando Di Leo. **VHS** AIP 🎬🎬

The Italian Job Caine and Coward pair up to steal $4 million in gold by causing a major traffic jam in Turin, Italy. During the jam, the pair steals the gold from an armored car. Silliness and chases through the Swiss mountains ensue, culminating in a hilarious ending.
1969 (G) 99m/C GB Michael Caine, Noel Coward, Benny Hill, Raf Vallone, Tony Beckley, Rossano Brazzi, Margaret Blye; *Dir:* Peter Collinson. **VHS** PAR 🎬🎬½

Italian Stallion Stallone plays a man with only one thing on his mind...sex!
1973 90m/C Sylvester Stallone, Henrietta Holm. **VHS, Beta $79.95** JEF 🎬🎬½

Italian Straw Hat Entertaining classic silent about a chain of errors that ensues when a man's horse eats a woman's hat. His vast, unending struggle to replace the hat is a source of continual comedy. From Eugene Labiche's play and originally titled "Un Chapeau de Paille d'Italie."
1927 72m/B FR Albert Prejean; *Dir:* Rene Clair. **VHS, Beta $29.95** VYY, MRV, INT 🎬🎬🎬

It's A Gift A grocery clerk moves his family west to manage orange groves in this classic Fields comedy. Several inspired sequences. The supporting cast shines too. A real find for the discriminating comedy buff. Remake of the silent "It's the Old Army Game."
1934 71m/B W.C. Fields, Baby LeRoy, Kathleen Howard; *Dir:* Norman Z. MacLeod. **VHS, Beta, LV $14.95** MCA, KRT 🎬🎬🎬½

It's an Adventure, Charlie Brown! A collection of six vignettes featuring Charlie Brown and the whole Peanuts gang.
1983 50m/C *Dir:* Bill Melendez. **VHS, Beta $9.98** SHV, VTR 🎬🎬🎬½

It's Alive! Slow-moving dud about a farmer feeding passersby to the area's cave-dwelling lizard man. The ping pong ball-eyed monster puts in a belated appearance that's not worth the wait.
1968 80m/C Tommy Kirk, Shirley Bonne, Bill Thurman; *W/Dir:* Larry Buchanan. **VHS $15.95** NOS, LOO 🎬🎬½

It's Alive Cult film about a mutated baby that is born to a normal Los Angeles couple, who escapes and goes on a bloodthirsty, murderous rampage, is a sight to behold. Fantastic score by Bernard Herrmann makes this chilling film a memorable one.
1974 (PG) 91m/C John P. Ryan, Sharon Farrell, Andrew Duggan, Guy Stockwell, James Dixon, Michael Ansara; *Dir:* Larry Cohen. **VHS, Beta $19.98** WAR, FCT 🎬🎬🎬

It's Alive 2: It Lives Again In this sequel to "It's Alive," the original hellspun baby meets up with two more of the same and all three terrorize the city, murdering everyone they can find. Truly horrendous.
1978 (R) 91m/C Frederic Forrest, Kathleen Lloyd, John P. Ryan, Andrew Duggan, John Marley, Eddie Constantine; *Dir:* Larry Cohen. **VHS, Beta $19.98** WAR, FCT 🎬½

It's Alive 3: Island of the Alive The second sequel to the tongue-in-cheek horror film, in which the infant mutant of the previous films has been left with other mutations to spawn on a desert island.
1987 (R) 94m/C Michael Moriarty, Karen Black, Laurene Landon, Gerrit Graham, James Dixon, Neal Israel, MacDonald Carey; *Dir:* Larry Cohen. **VHS, Beta $19.98** WAR 🎬

It's Always Fair Weather Three World War II buddies meet again at a 10-year reunion and find that they don't like each other very much. A surprisingly cynical film, written by Betty Comden & Adolph Green, with songs by Comden, Green, and Andre Previn.
1955 102m/C Gene Kelly, Cyd Charisse, Dan Dailey, Michael Kidd; *Dir:* Stanley Donen. **VHS, Beta, LV $19.98** MGM, FHE 🎬🎬

It's Arbor Day, Charlie Brown Tribulations aplenty for Charlie Brown as his baseball team turns his field into an Arbor Day garden.
1983 26m/C **VHS, Beta $14.95** KRT 🎬🎬

It's in the Bag A shiftless flea circus owner sells chairs he has inherited, not knowing that a fortune is hidden in one of them.
1945 87m/B Fred Allen, Don Ameche, Jack Benny, William Bendix, Binnie Barnes, Robert Benchley; *Dir:* Richard Wallace. **VHS $19.95** REP, MED 🎬🎬½

It's Called Murder, Baby The famed porn star gives the viewer a lesson in English vocabulary in this tale of a blackmailed movie queen.
1982 (R) 94m/C John Leslie, Cameron Mitchell, Lisa Trego; *Dir:* Sam Weston. **VHS, Beta $29.98** LIV, LTG 🎬

It's a Date A mother and daughter, both actresses, continually get their professional and romantic lives intertwined. It begins when Durbin (the daughter) is offered Kay's (the mother) Broadway role. It continues in Hawaii where the man that Kay is in love with tries to court Durbin. As it is a comedy, of course all is put right in the end. Musical numbers include "Ave Maria," "Loch Lomond," and more.

1940 103m/B Deanna Durbin, Kay Francis, Walter Pidgeon, Samuel S. Hinds, S.Z. Sakall, Henry Stephenson, Charles Lane, Leon Belasco; *Dir:* William A. Seiter. **VHS $29.98** FOX, FCT, MLB 🐾🐾

It's Flashbeagle, Charlie Brown/ She's a Good Skate, Charlie Brown
A Peanuts double header: In "It's Flashbeagle Charlie Brown" Snoopy infects the Peanuts gang with dance fever and in "She's a Good Skate Charlie Brown" Snoopy trains Peppermint Patty to become a figure skating champion.
1984 55m/C *Dir:* Bill Melendez. **VHS, Beta $29.95** SHV 🐾🐾

It's Good to Be Alive
The true story of how Brooklyn Dodgers' catcher Roy Campanella learned to face life after an automobile accident left him a quadriplegic. Nice cast, good performances in this television drama.
1974 100m/C Paul Winfield, Ruby Dee, Louis Gossett Jr.; *Dir:* Michael Landon. **VHS, Beta $59.95** PSM 🐾🐾½

It's a Great Feeling
Another Carson and Day team-up where Carson plays a camera-hogging show-off whom no one wants to direct. He's such an industry piranha that he ends up directing himself! He cons Day, a waitress and would-be actress, by promising her a part in his new film. Some executives hear her sing before she heads back home to Wisconsin, fed up with Hollywood. They make Carson track her down to star in the new picture, but when he arrives she is at the alter with her high school sweetheart. A very funny film just brimming with pure comedy. Many say it's worth it just to see directors Curtiz, Vidor, and Walsh playing around in front of the camera. Appearances by stars Errol Flynn, Joan Crawford, Sydney Greenstreet, Gary Cooper, Danny Kaye, Patricia Neal, Eleanor, Parker, Edwin G. Robinson, Jane Wyman, and Ronald Reagan.
1949 85m/C Dennis Morgan, Doris Day, Jack Carson, Bill Goodwin, Errol Flynn, Gary Cooper, Joan Crawford, Ronald Reagan, Jane Wyman; *Dir:* David Butler. **VHS, LV $29.98** FOX, FCT, MLB 🐾🐾½

It's the Great Pumpkin, Charlie Brown
It's Halloween, and Charlie Brown and his friends are going trick-or-treating, but it is Linus who faithfully awaits the arrival of the Great Pumpkin.
1988 25m/C **VHS, Beta $9.98** MED, WPB, VTR 🐾🐾½

It's a Joke, Son!
Follows the fictional politician, Senator Claghorn from the Fred Allen radio show, during his first run for the U.S. Senate.
1947 67m/B Kenny Delmar, Una Merkel; *Dir:* Ben Stoloff. **VHS $19.95** NOS, MRV, QNE 🐾🐾

It's Love Again
A streetwise chorus girl poses as a fictional socialite in a light music-hall comedy.
1936 83m/B GB Jessie Matthews, Robert Young, Sonnie Hale; *Dir:* Victor Saville. **VHS, Beta $24.95** NOS, DVT 🐾🐾½

It's a Mad, Mad, Mad, Mad World
Overblown epic comedy with a cast of notables desperately seeking the whereabouts of stolen money. Ulotimately exhausting film undone by the length and overbearing performers. Ernest Laszlo's photography was Oscar nominated, as was Ernest Gold's music. Laserdisc contains 20 minutes of supplemental footage including interviews with some of the surviving cast members.
1963 155m/C Spencer Tracy, Sid Caesar, Milton Berle, Ethel Merman, Jonathan Winters, Jimmy Durante, Buddy Hackett, Mickey Rooney, Phil Silvers, Dick Shawn, Edie Adams, Dorothy Provine, Buster Keaton, Terry-Thomas, Moe Howard, Larry Fine, Joe DeRita, Jim Backus, William Demarest, Peter Falk, Leo Gorcey, Edward Everett Horton, Joe E. Brown, Carl Reiner, ZaSu Pitts, Eddie Anderson, Jack Benny, Jerry Lewis; *Dir:* Stanley Kramer. Academy Awards '63: Best Sound Effects Editing. **VHS, Beta, LV $29.99** FOX, MGM, PIA 🐾🐾

It's Magic, Charlie Brown/ Charlie Brown's All Star
In the first of these two stories, Snoopy, "The Great Houndini," makes Charlie Brown disappear. Problems begin when Snoopy finds it difficult conjuring his master back to reality. In the second story, Charlie Brown's baseball team has lost 999 games in a row, and prospects are bleak. New hope arrives when he gets an offer to have the team sponsored-in a real league!
1981 55m/C *Dir:* Phil Roman. **VHS, Beta $14.95** SHV, VTR 🐾🐾

It's My Turn
A mathematics professor (Clayburgh) struggles in her relationship with live-in lover (Grodin) who sells real-estate in Chicago. She meets retired baseball player (Douglas) and falls in love. "Ho-hum" just about describes the rest.
1980 (R) 91m/C Jill Clayburgh, Michael Douglas, Charles Grodin, Beverly Garland, Steven Hill, Dianne Wiest, Daniel Stern; *Dir:* Claudia Weill. **VHS, Beta, LV $64.95** COL 🐾🐾

It's a Mystery, Charlie Brown
An animated Peanuts adventure in which Snoopy interrogates everyone on the disappearence of Woodstock's birdhouse.
1987 30m/C *Dir:* Phil Roman. **VHS, Beta $14.95** KRT 🐾🐾

It's Not Easy Bein' Me
Essentially a series of stand-up comedy acts, framed by Dangerfield and Barr doing sketch comedy.
1987 (R) 59m/C Rodney Dangerfield, Sam Kinison, Jeff Altman, Bob Nelson, Roseanne (Barr) Arnold, Jerry Seinfeld, Robert Townsend. **VHS, Beta, LV $19.98** ORI, IME 🐾🐾

It's Not the Size That Counts
After the Earth's drinking water is contaminated, Lawson is the only man on Earth not to be struck by impotency. Limp sequel to "Percy"—and originally titled "Percy's Progress"- - the British comedy about the first man to receive a penis transplant. Promising comedy that doesn't deliver.
1974 90m/C Leigh Lawson, Elke Sommer, Denholm Elliott, Vincent Price; *Dir:* Ralph Thomas. **VHS $59.98** SUE Woof!

It's the Old Army Game
Some classic Fields gags in three of his best Ziegfeld Follies sketches: "The Drug Store," "A Peaceful Morning," and "The Family Flivver." The beautiful Brooks serves as his comic foil.
1926 75m/B W.C. Fields, Louise Brooks, Blanche Ring, William Gaxton; *Dir:* Edward Sutherland. **VHS, Beta $24.95** GPV, DNB 🐾🐾🐾

It's Three Strikes, Charlie Brown
Misfortunes around the baseball diamond for the Peanuts gang. Animated.
1986 41m/C *Dir:* Bill Melendez, Phil Roman, Robert E. Balzer, Sam Nicholson. **VHS, Beta $14.95** KRT 🐾🐾🐾

It's a Wonderful Life
Endearing American classic concerns a man saved from suicide by a considerate angel, who then shows the hero how important he's been to the lives of loved one. Corny but inspirational and heartwarming, with endearing performance by Travers as the angel. Stewart and Reed are typically wholesome. A perfect film for people who want to feel good, joyfully teetering on the border between Hollywood schmaltz and genuine heartbreak. Also available colorized. The laserdisc version includes production and publicity stills, the theatrical trailer, and commentary by film professor Jeanine Basinger. Oscar nominated for Best Picture, Best Actor (Stewart), Best Director, Sound Recording, Film Editing. Also available in a 160-minute Collector's Edition with original preview trailer, "The Making of 'It's a Wonderful Life,'" and a new digital transfer from the original negative.
1946 125m/B James Stewart, Donna Reed, Henry Travers, Thomas Mitchell, Lionel Barrymore, Samuel S. Hinds, Frank Faylen, Gloria Grahame, H.B. Warner, Ellen Corby, Sheldon Leonard, Beulah Bondi, Ward Bond; *Dir:* Frank Capra. **VHS, LV $9.95** IGP, MRV, NEG 🐾🐾🐾🐾

It's Your First Kiss, Charlie Brown/Someday You'll Find Her Charlie Brown
In "It's Your First Kiss, Charlie Brown," our hero has to escort the school's homecoming queen to the ball. What's worse, he has to actually kiss her in front of everyone! In "Someday You'll Find Her, Charlie Brown," Charlie Brown sees "the most beautiful girl in the world" on television and, after recruiting Linus, embarks upon a door-to-door search to find her.
1977 55m/C *Dir:* Phil Roman. **VHS, Beta $14.95** SHV, VTR 🐾🐾🐾🐾

Ivan the Terrible
Yuri Grigorovich choreographed this ballet to the score Sergei Prokofiev wrote for Eisenstein's historic film of the same name. The Bolshoi dances this ballet of murder and intrigue in the czarist past.
1989 91m/C RU Yuri Vladimirov, Natalia Bessmertnova, Boris Akimov. **VHS, Beta, LV $29.95** KUL 🐾🐾🐾🐾

Ivan the Terrible, Part 1
Contemplative epic of Russia's first czar is a classic, innovative film from early cinema genius Eisenstein. Visually stunning, with a fine performance by Cherkassov. Ivan's struggles to preserve his country are the main concerns of this first half of Eisenstein's masterwork, which he originally planned as a trilogy. In Russian with English subtitles.
1944 96m/B RU Nikolai Cherkassov, Ludmila Tselikovskaya, Serafina Birman, Piotr Kadochnikov; *Dir:* Sergei Eisenstein. Sight & Sound Survey '62: #8 of the Best Films of All Time. **VHS, Beta, LV, 8mm, 3/4U $24.95** NOS, MRV, WST 🐾🐾🐾🐾

Ivan the Terrible, Part 2
Landed gentry conspire to dethrone the czar in this continuation of the innovative epic. More stunning imagery from master Eisenstein, who makes no false moves in this one. Slow going, but immensely rewarding. Russian dialogue with English subtitles; contains color sequences.
1946 84m/B RU Nikolai Cherkassov, Ludmila Tselikovskaya, Serafina Birman, Piotr Kadochnikov; *Dir:* Sergei Eisenstein. Sight & Sound Survey '62: #8 of the Best Films of All Time. **VHS, Beta, LV, 8mm, 3/4U $24.95** NOS, MRV, WST 🐾🐾🐾½

Ivanhoe Knights fight each other and woo maidens in this chivalrous romance derived from the Sir Walter Scott classic. Tayolor is suitably noble, while Sanders is familiarly serpentine. Remade in 1982. Oscar nominations for Best Picture, Color Cinematography, Score of a Comedy or Drama.
1952 107m/C Robert Taylor, Elizabeth Taylor, Joan Fontaine, George Sanders, Finlay Currie, Felix Aylmer; *Dir:* Richard Thorpe. Film Daily Poll '53: Ten Best of Year. VHS, Beta $19.98 *MGM* ♫♫♫

Ivanhoe A version of Sir Walter Scott's classic novel of chivalry and knighthood in 12th century England. Made for television. Remake of the 1953 film classic.
1982 142m/C Anthony Andrews, James Mason, Lysette Anthony, Sam Neill, Olivia Hussey, Michael Hordern, Julian Glover, George Innes, Ronald Pickup, John Rhys-Davies, Chloe Franks; *Dir:* Douglas Camfield. VHS, Beta $19.95 *RCA* ♫♫½

I've Heard the Mermaids Singing Independent Canadian semi-satiric romantic comedy details the misadventures of a klutzy woman who suddenly obtains a desirable job in an art gallery run by a lesbian on whom she develops a crush. Good-natured tone helped considerably by McCarthy's winning performance.
1987 81m/C *CA* Sheila McCarthy, Paule Baillargeon, Ann-Marie MacDonald; *Dir:* Patricia Rozema. Cannes Film Festival '87: Prix de la Jeunesse (Youth Prize); Genie Awards '87: Best Actress (McCarthy), Best Supporting Actress (Baillargeon). VHS, Beta, LV $19.98 *CHA, FCT* ♫♫♫

Ivory Hunters Conservationists team up to thwart a massacre of elephants by poachers. Predictable made-for-cable entertainment co-produced by the National Audubon Society.
1990 94m/C John Lithgow, Isabella Rossellini, James Earl Jones, Tony Todd, Olek Krupa; *Dir:* Joseph Sargent. VHS, Beta $79.95 *TTC* ♫♫½

Izzy & Moe Carney and The Great One are reunited for the last time on screen as former vaudevillians who become federal agents and create havoc amid speak-easies and bathtub gin during Prohibition. A made-for-television movie based on true stories.
1985 92m/C Jackie Gleason, Art Carney, Cynthia Harris; *Dir:* Jackie Cooper. VHS, Beta, LV $79.95 *VMK* ♫♫½

J. Edgar Hoover Made for cable adaptation from "My 30 years in Hoover's FBI" by William G. Sullivan and William S. Brown. Biographical drama deals with Hoover's fifty-five year career with the government. Included are his FBI days.
1987 110m/C Treat Williams, Rip Torn, David Ogden Stiers, Andrew Duggan, Art Hindle, Louise Fletcher; *Dir:* Robert Collins. VHS, Beta, LV $79.98 *MCG* ♫♫♫

J-Men Forever! A spoof on early sci-fi/spy serials in which an alien takeover is attempted via rock-n-roll (your parents warned you). Employs an amusing technique in dubbing footage from dozens of Republic dramas intercut with new film featuring the "Firesign Theater" crew.
1979 (PG) 73m/B Peter Bergman, Phil Proctor; *Dir:* Richard Patterson. VHS, Beta $59.98 *VES* ♫♫♫

Jabberwalk A "Mondo Cane" American-style. A reckless and funny view of the bizarre in U.S. life.
1979 (R) 110m/C VHS, Beta $59.95 *VCD* ♫♫

Jabberwocky Pythonesque chaos prevails in the medieval cartoon kingdom of King Bruno the Questionable, who rules with cruelty, stupidity, lust and dust. Jabberwocky is the big dragon mowing everything down in its path until hero Palin decides to take it on. Uneven, but real funny at times.
1977 (PG) 104m/C *GB* Eric Idle, Michael Palin, Max Wall, Deborah Fallender, Terry Jones; *Dir:* Terry Gilliam. VHS, Beta, LV $19.95 *COL* ♫♫½

J'Accuse A Frenchman creates a device he believes will stop the war but it is confiscated by the government and used as part of the national defense. Subsequently, he goes mad. Gance remade his 1919 silent film and retells his powerful anti-war message with great sensitivity. Banned in Nazi Germany.
1937 125m/B *FR* Victor Francen, Jean Max, Deltaire, Renee Devillers; *Dir:* Abel Gance. VHS $29.95 *CVC, SNC, FCT* ♫♫♫

Jack & the Beanstalk While baby-sitting, Lou falls asleep and dreams he's Jack in this spoof of the classic fairy tale.
1952 78m/C Bud Abbott, Lou Costello, Buddy Baer, Jean Yarborough. VHS, Beta $19.95 *NOS, UHV, AMV* ♫♫

Jack & the Beanstalk An animated musical version of the familiar story of Jack, the young boy who climbs a magic beanstalk up into the clouds, where he meets a fearsome giant.
1976 80m/C *W/Dir:* Peter J. Solmo. VHS, Beta *COL* ♫♫

Jack & the Beanstalk From the "Faerie Tale Theatre" comes this classic tale of Jack, who sells his family's cow for 5 magic beans, then climbs the huge beanstalk that sprouts from them and encounters an unfriendly giant.
1983 60m/C Dennis Christopher, Katherine Helmond, Elliott Gould, Jean Stapleton, Mark Blankfield; *Dir:* Lamont Johnson. VHS, Beta, LV $14.98 *KUI, FOX, FCT* ♫♫♫

Jack Benny Show The cast does a parody of the "The Honeymooners" with Dennis playing Ed Norton and Jack playing Ralph Kramden (with the help of a pillow under his shirt).
1958 25m/B Jack Benny, Dennis Day, Audrey Meadows. VHS, Beta $29.95 *VYY*

Jack Frost Moving biographical drama that follows a man from the sweat factories of London to his eventual rise to success.
1943 89m/B Michael O'Shea, Susan Hayward. VHS, Beta, 8mm *VYY* Woof!

Jack Frost A made-for-television Christmas special filmed with puppet animation.
1979 48m/C *Voices:* Buddy Hackett, Robert Morse, Dave Garroway. VHS, Beta, 8mm $14.98 *LIV, LTG* Woof!

Jack the Giant Killer A young farmer joins a medieval princess on a journey to a distant convent. Along the way they combat an evil wizard, dragons, sea monsters and other mystical creatures. Conversely, they are assissted by leprechauns, a dog and a chimp. Generally considered a blatant rip-off of "The Seventh Voyage of Sinbad," the film nonetheless delivers plenty of fun and excitement. Jim Danforth of "Gumby" fame provided the stop-motion animation.
1962 94m/C Kerwin Mathews, Judi Meredith, Torin Thatcher, Walter Burke, Roger Mobley, Barry Kelley, Don Beddoe, Anna Lee, Robert Gist; *Dir:* Nathan Juran. VHS, Beta $19.98 *MGM, FCT* ♫♫♫

The Jack Knife Man Sentimental silent treatment of the relationship between a boy and a man. A woman, about to check into the great hereafter, hands over her bundle of joy to a kindly old river man, and the two bond in a familial way despite incessant adversity. Made the year following Vidor's debut as a feature film director (Vidor coproduced and coscripted as well).
1920 70m/B Fred Turner, Todd Harry, Bobby Kelso, Willis Marks, Lillian Leighton, James Corrigan; *Dir:* King Vidor. VHS, Beta $24.95 *VYY, GPV, FCT* ♫♫♫

Jack London Jack London's most creative years during his careers as oyster pirate, prospector, war correspondent and author are dramatized. Based on "The Book of Jack London" by Charmian London.
1943 94m/B Michael O'Shea, Susan Hayward, Harry Davenport, Virginia Mayo, Frank Craven; *Dir:* Alfred Santell. VHS, Beta $19.95 *NOS, MRV, DVT* ♫½

Jack the Ripper An American detective joins Scotland Yard in tracking down the legendary and elusive Jack the Ripper. More gory than most. The last scene is in color.
1960 88m/B *GB* Lee Patterson, Betty McDowall, Barbara Burke, John Le Mesurier, George Rose; *Dir:* Robert S. Baker, Monty Berman. VHS $16.95 *SNC, PAR* ♫♫

Jack the Ripper The inimitable Kinski assumes the role of the most heinous criminal of modern history - Jack the Ripper.
1980 (R) 82m/C Klaus Kinski, Josephine Chaplin; *Dir:* Jess (Jesus) Franco. VHS, Beta $59.95 *VES* ♫♫

Jack the Ripper Another retelling of the life of the legendary serial killer. Caine is the Scotland Yard inspector who tracks down the murderer. Ending is based on recent evidence found by director/co-writer Wickes. Extremely well done made-for-television film.
1988 200m/C Michael Caine, Armand Assante, Ray McNally, Susan George, Jane Seymour, Lewis Collins; *Dir:* David Wickes. VHS *VES, LIV* ♫♫♫

Jackie Chan's Police Force Chopsocker Chan's assigned to protect a witness in a drug case. Faced by unsavory thugs, he flies through the air with the greatest of ease. Very cool stunts. Also known as "The Police Story," "Police Force," and "Jackie Chan's Police Story."
1985 (PG-13) 89m/C *CH* Jackie Chan, Bridget Lin, Maggie Cheung; *Dir:* Jackie Chan. VHS $79.98 *NSV, CGH* ♫♫½

Jackie Mason on Broadway The veteran comedian's one-man Broadway show.
1989 60m/C Jackie Mason. VHS, Beta $14.99 *HBO, TLF* ♫♫½

The Jackie Robinson Story Chronicles Robinson's rise from UCLA to his breakthrough as the first black man to play baseball in the major league. Robinson plays himself; the film deals honestly with the racial issues of the time.
1950 76m/C Jackie Robinson, Ruby Dee, Minor Watson, Louise Beavers, Richard Lane, Harry Shannon, Joel Fluellen; *Dir:* Alfred E. Green. VHS $19.95 *NOS, MRV, MOV* ♫♫♫

Jacknife The well-crafted story of a Vietnam veteran who visits his old war buddy and tries to piece together what's happened to their lives since their homecoming. During his visit he discovers anger and hostility from the other veteran, and tenderness from his

friend's sister. A masterfully acted, small-scale drama, adapted by Stephen Metcalfe from his play.
1989 (R) 102m/C Robert De Niro, Kathy Baker, Ed Harris, Loudon Wainwright III; *Dir:* David Jones. **VHS, Beta, LV** $89.98 *HBO, IME* &&&

Jacko & Lise A French lad falls in love with a young girl, and abandons his previously decadent lifestyle to win her. With English subtitles or dubbed.
1982 (R) 92m/C *FR* Laurent Malet, Annie Girardot; *Dir:* Walter Bal. **VHS, Beta** $29.98 *SUE* &&

The Jackpot An average Joe wins a bushel of money from a radio quiz show but can't pay the taxes. Maybe that was funny before the age of read my lips economics, but the all star cast doesn't deliver on its promise. Lang, noted for mostly mediocre pictures, went on to direct "The King and I."
1950 87m/C James Stewart, Natalie Wood, Barbara Hale, James Gleason, Fred Clark; *Dir:* Walter Lang. **VHS** *STC* &&

Jack's Back When a lunatic is killing Los Angeles prostitutes Jack-the-Ripper style, twin brothers get mistakenly involved (Spader in both roles). One of the brothers is accused of the murders and it is up to the other to either clear his name or provide the final evidence of guilt. A well thought out story with enough twists and turns for any suspense buff.
1987 (R) 97m/C James Spader, Cynthia Gibb, Rod Loomis, Rex Ryon, Robert Picardo, Jim Haynie; *Dir:* Rowdy Herrington. **VHS, Beta, LV** $19.95 *PAR* &&&

Jackson County Jail While driving cross-country, a young woman is robbed, imprisoned, and raped by a deputy, whom she kills. Faced with a murder charge, she flees, with the law in hot pursuit. Also known by its later remade-for-television name, "Outside Chance," this is a minor cult film.
1976 (R) 84m/C Yvette Mimieux, Tommy Lee Jones, Robert Carradine, Severn Darden, Howard Hesseman, Mary Woronov; *Dir:* Michael Miller. **VHS, Beta** $39.98 *WAR* &&½

Jackson & Jill A vintage situation comedy episode about a silly suburban couple.
1953 27m/B Todd Karns, Helen Chapman. **VHS, Beta** $19.95 *VYY* Woof!

Jacob: The Man Who Fought with God Jacob's struggle to receive his father's blessing and inheritance rights is depicted, as well as his marriage to Rachel and the return of his brother Esau. From the "Bible" series.
197? 118m/C **VHS, Beta** $19.95 *UHV* Woof!

Jacob's Ladder A man struggles with events he experienced while serving in Vietnam. Gradually, he becomes unable to separate reality from the strange, psychotic world into which he increasingly lapses. His friends and family try to help him before he's lost forever. Great story potential is flawed by too many flashbacks, leaving the viewer more confused than the characters.
1990 (R) 116m/C Tim Robbins, Elizabeth Pena, Danny Aiello, Matt Craven, Macauley Culkin; *Dir:* Adrian Lyne. **VHS, Beta, LV** $92.95 *LIV, FCT* &&½

Jacqueline Bouvier Kennedy A made-for-television biography of the former First Lady, from her childhood to "Camelot," the glorious years with JFK in the White House. Smith bears a physical resemblance to Jackie, but this drama is far too glossy.

1981 150m/C Jaclyn Smith, James Franciscus, Rod Taylor, Donald Moffatt, Dolph Sweet, Stephen Elliott; *Dir:* Steven Gethers. **VHS, Beta** $24.95 *VMK* &&½

The Jagged Edge The beautiful wife of successful newspaper editor, Jack Forester, is killed and the police want to point the guilty finger at Jack. Attorney Teddy Barnes is brought in to defend him and accidentally falls in love. Taut murder mystery will keep you guessing "whodunit" until the very end.
1985 (R) 108m/C Jeff Bridges, Glenn Close, Robert Loggia, Peter Coyote, John Dehner, Leigh Taylor-Young, Lance Henriksen, James Karen; *Dir:* Richard Marquand. **VHS, Beta, LV, 8mm** $14.95 *RCA* &&½

Jaguar Three men leave their homes in Niger in order to find jobs in the cities. When they return to their village, they bring back with them their experiences.
1956 93m/C *Dir:* Jean Rouch. **VHS** *INT, DER* &&½

Jaguar Lives A high-kicking secret agent tracks down bad boy drug kings around the world.
197? (PG) 91m/C Joe Lewis, Barbara Bach, Christopher Lee, Woody Strode, Donald Pleasence, Joseph Wiseman, John Huston, Capucine; *Dir:* Ernie Pintoff. **VHS, Beta** $79.95 *TWE* &

Jail Bait Early Ed Wood film about a group of small-time crooks who are always in trouble with the law; they blackmail a plastic surgeon into using his talents to help them ditch the cops. Not as "bad" as Wood's "Plan 9 From Outer Space," but still bad enough for camp fans to love (check the cheesy score leftover from an equally cheesy Mexican mad-scientist flick).
1954 80m/B Timothy Farrell, Clancy Malone, Lyle Talbot, Steve Reeves, Herbert Rawlinson; *Dir:* Edward D. Wood Jr. **VHS, Beta** $16.95 *SNC, AOV* &½

Jailbait: Betrayed By Innocence A man is on trial for statutory rape. Made-for-television.
1986 (R) 90m/C Barry Bostwick, Lee Purcell, Paul Sorvino, Cristen Kauffman, Isaac Hayes; *Dir:* Elliot Silverstein. **VHS, Beta** $79.95 *MNC, HHE* &½

Jailbird Rock We'd like to tell you that this isn't really a sweet-young-thing-in-prison musical, but it is. When Antin shoots her stepfather (he told her to turn that damn noise down?) she tries to bring down the Big House with song and dance. No kidding.
1988 90m/C Ronald Lacey, Rhonda Aldrich, Robin Antin; *Dir:* Phillip Schuman. **VHS, Beta** $79.95 *TWE* &½

Jailbird's Vacation Two ex-cons pose as lumberjacks in order to exact revenge on the bum that put them away. In French with English subtitles.
1965 125m/C *FR* Lino Ventura; *Dir:* Robert Enrico. **VHS, Beta** $19.95 *FCT* &&

Jailbreakin' A faded country singer and a rebellious youth team up to break out of jail.
1972 90m/C Erik Estrada. **VHS, Beta** $59.95 *GEM* &

Jailhouse Rock While in jail for manslaughter, a teenager learns to play the guitar. After his release, he slowly develops into a top recording star. Songs include "Jailhouse Rock," "Treat Me Nice," "Baby, I Don't Care," and "Young and Beautiful." Probably the only film that captured the magnetic power of the young Elvis Presley;

an absolute must for fans. Also available in a colorized version.
1957 96m/B Elvis Presley, Judy Tyler, Vaughn Taylor, Dean Jones, Mickey Shaughnessy, William Forrest, Glenn Strange; *Dir:* Richard Thorpe. **VHS, Beta, LV** $19.95 *MGM* &&&

Jakarta A love-hardened CIA operative is whisked away to Jakarta for reasons unknown, only to find his thought-to-be-dead lover still alive and caught in a deadly game of espionage.
1988 94m/C Christopher Noth, Sue Francis Pai, Ronald Hunter; *Dir:* Charles Kaufman. **VHS, Beta, LV** $79.95 *VIR* &½

Jake Speed Comic-book mercenary Jake and his loyal associate Remo rescue a beautiful girl from white slave traders. Supposedly a parody of the action adventure genre.
1986 (PG) 93m/C Wayne Crawford, John Hurt, Karen Kopins, Dennis Christopher; *Dir:* Andrew Lane. **VHS, Beta, LV** $14.95 *NWV, STE* &½

Jamaica Inn In old Cornwall, an orphan girl becomes involved with smugglers. Remade in 1982; based on the story by Daphne Du Maurier.
1939 98m/B *GB* Charles Laughton, Maureen O'Hara, Leslie Banks, Robert Newton; *Dir:* Alfred Hitchcock. **VHS, Beta** $16.95 *SNC, NOS, CAB* &&

Jamaica Inn A made-for-television miniseries based on the old Daphne Du Maurier adventure about highwaymen and moor-lurking thieves in Cornwall. Remake of the 1939 Hitchcock film.
1982 192m/C Patrick McGoohan, Jane Seymour; *Dir:* Lawrence Gordon Clark. **VHS, Beta** $29.95 *STE, NWV* &½

James Cagney: That Yankee Doodle Dandy Definitive homage to Cagney's life and work features footage from his films, plus interviews and comments by his friends and co-workers. Portions are in black and white.
1986 73m/C James Cagney; *Nar:* Treat Williams, Pat O'Brien, Donald O'Connor, Milos Forman. **VHS, Beta** $29.95 *MGM* &½

James Dean Dean's friend Bast wrote this behind-the-scenes look at the short life of the enigmatic movie star. Made for television. Also known as "The Legend."
1976 99m/C Stephen McHattie, Michael Brandon, Candy Clark, Amy Irving, Brooke Adams, Dane Clark, Jayne Meadows, Meg Foster; *Dir:* Robert Butler; *W/Dir:* William Bast. **VHS, Beta** $19.95 *STE, NWV, SIM* &&

James Dean Production reflects an exhaustive search through archival film, videotape and photographs, and includes many exclusive interviews with friends of the late actor.
1985 120m/C James Dean; *Dir:* Claudio Masenza. **VHS, Beta** $69.95 *KRT* &&

The James Dean 35th Anniversary Collection A must for James Dean fans. This special five-tape collection contains Dean's movies "East of Eden," "Rebel Without a Cause," and "Giant," plus the documentary "Forever James Dean."
19?? ?m/C **VHS, LV** *WAR* &&

James Dean Story An intimate portrait of James Dean presented by Robert Altman. The program includes never-before-seen outtakes from "East of Eden" and rare footage from the Hollywood premiere of "Giant" and the infamous Highway Public Safety message Dean made for television.

1957 80m/B James Dean; *Dir:* George W. George; *Nar:* Martin Gabel. **VHS, Beta, LV** $14.95 *PAV, FCT*

James Dean: The First American Teenager
A look at the life and legend of the charismatic filmstar, with comments by those who knew him best and scenes from his films.
1976 (PG) 83m/C James Dean, Elizabeth Taylor, Sammy Davis Jr., Rock Hudson, Sal Mineo, Natalie Wood, Julie Harris, Jack Larson, Nicholas Ray. **VHS, Beta** $14.98 *VID* 🐾🐾

James Joyce: A Portrait of the Artist as a Young Man
A moving, lyrical adaptation of the author's autobiography, told through the character of Stephen Dedalus. Joyce's characterizations, words, and scenes are beautifully translated to the medium of film. Excellent casting.
1977 93m/C John Gielgud, T.P. McKenna, Bosco Hogan. **VHS** $29.95 *MFV* 🐾🐾

James Joyce's Women
Adapted from her one-woman stage show as well as produced by Flanagan, this acclaimed film features enacted portraits of three real-life Joyce associates, including his wife (Molly Bloom) and three of his famous characters.
1985 (R) 91m/C Fionnula Flanagan, Timothy E. O'Grady, Chris O'Neill; *Dir:* Michael Pearce. **VHS, Beta** $69.95 *MCA* 🐾🐾🐾

Jane Austen in Manhattan
Two rival producers endeavor to stage a long-lost play by Jane Austen. Katrina Hodiak is the real life daughter of Anne Baxter.
1980 108m/C Anne Baxter, Robert Powell, Michael Wager, Sean Young, Kurt Johnson, Katrina Hodiak; *Dir:* James Ivory. **VHS, Beta** $59.98 *CHA* 🐾½

Jane Doe
Valentine plays a young victim of amnesia who is linked to a series of brutal slayings. Quite suspenseful for a television movie.
1983 100m/C Karen Valentine, William Devane, Eva Marie Saint, Stephen E. Miller, Jackson Davies; *Dir:* Ivan Nagy. **VHS, Beta** $39.95 *IVE* 🐾🐾½

Jane Eyre
Stiff, early version of Charlotte Bronte's classic gothic romance. English orphan grows up to become the governess of a mysterious manor. Notable as the first talkie version. Remade several times.
1934 67m/B Virginia Bruce, Colin Clive, Beryl Mercer, Jameson Thomas, Aileen Pringle, David Torrence; *Dir:* Christy Cabanne. **VHS** $16.95 *SNC* 🐾🐾

Jane Eyre
A made-for-British-television mini-series based upon the famed Charlotte Bronte novel about the maturation of a homeless English waif, her love of the tormented Rochester, and her quest for permanent peace.
1983 239m/C *GB* Timothy Dalton, Zelah Clarke; *Dir:* Julian Aymes. **VHS, Beta, LV** $29.98 *KUI, FOX* 🐾🐾🐾

Jane & the Lost City
A British farce based on the age-old, barely dressed comic-strip character, Jane, as she stumbles on ancient cities, treasures, villains, and blond/blue-eyed heroes. Low-budget camp fun.
1987 (PG) 94m/C *GB* Kristen Hughes, Maud Adams, Sam Jones; *Dir:* Terry Marcel. **VHS, Beta** $14.95 *NWV, STE* 🐾🐾½

Janis: A Film
The famous biographical film about Janis Joplin, from her Texas origins to Woodstock and superstardom, featuring performances with Big Brother and the Holding Company and rare interviews.

1974 (R) 79m/C *Dir:* Howard Alk. **VHS, Beta** $19.95 *MVD, MCA, FCT* 🐾🐾½

The January Man
An unorthodox cop, previously exiled by a corrupt local government to the fire department, is brought back to the force to track down a serial killer. New York City is the locale. Written by the "Moonstruck" guy, John Patrick Shanley, who apparently peaked with the earlier movie.
1989 (R) 97m/C Kevin Kline, Susan Sarandon, Mary Elizabeth Mastrantonio, Harvey Keitel, Rod Steiger, Alan Rickman, Danny Aiello; *Dir:* Pat O'Connor. **VHS, Beta, LV** $19.98 *FXV* 🐾🐾

Japanese Connection
Warring crime chiefs fight furiously with deadly kung-fu action.
1982 96m/C Li Chao, Yang Wei, Wu Ming-Tsai. **VHS, Beta** $59.95 *GEM* 🐾

The Jar
A recluse discovers a disgusting creature in a jar. He keeps it, only to discover that it is out to kill him.
1984 90m/C Gary Wallace, Karen Sjoberg; *Dir:* Bruce Toscano. **VHS, Beta** $79.95 *MAG* 🐾

Jason and the Argonauts
Jason, son of King of Thessaly, sails on the Argo to the land of Colchis, where the Golden Fleece is guarded by a seven-headed hydra. Superb special effects and multitudes of mythological creatures; fun for the whole family.
1963 104m/C *GB* Todd Armstrong, Nancy Kovack, Gary Raymond, Laurence Naismith, Nigel Green, Michael Gwynn; *Dir:* Don Chaffey. **VHS, Beta, LV, 8mm** $14.95 *COL, MLB* 🐾🐾🐾

Java Head
Young Chinese wife commits suicide so her husband can marry an English girl. Set in mid-1800s England. Based on the novel by Joseph Hergesheimer.
1935 70m/B *GB* Anna May Wong, Elizabeth Allan, John Loder, Edmund Gwenn, Ralph Richardson, George Curzan; *Dir:* J. Walter Ruben. **VHS, Beta, 8mm** *VYY* 🐾🐾🐾

Jaws
Early directorial effort by Spielberg from the Peter Benchley potboiler. A tight, very scary, and sometimes hilarious film about the struggle to kill a giant great white shark that is terrorizing an eastern beach community's waters. The characterizations by Dreyfuss, Scheider, and Shaw are much more endurable than the shock effects. Memorable score by John Williams. Sequelled by "Jaws 2" in 1978, "Jaws 3" in 1983, and "Jaws the Revenge" in 1987. Oscar nomination for Best Picture. Look for Benchley in a cameo as a TV reporter.
1975 (PG) 124m/C Roy Scheider, Robert Shaw, Richard Dreyfuss, Lorraine Gary, Murray Hamilton, Carl Gottlieb; *Dir:* Steven Spielberg. Academy Awards '75: Best Film Editing, Best Sound, Best Original Score; British Academy Awards '75: Best Original Score. **VHS, Beta, LV** $19.95 *MCA* 🐾🐾🐾½

Jaws 2
Unsatisfactory sequel to "Jaws." It's been four years since the man-eating shark feasted on the resort town of Amity; suddenly a second shark stalks the waters and the terror returns. Scheider, who must wonder at his luck in attracting large sharks with mayhem and lunch as their sole biological imperatives, battles without his compatriots from the original. And we haven't seen the last of the shark yet; two more sequels follow.
1978 (PG) 116m/C Roy Scheider, Lorraine Gary, Murray Hamilton; *Dir:* Jeannot Szwarc. **VHS, Beta, LV** $19.95 *MCA* 🐾🐾

Jaws 3
Same monster, new setting: in a Sea World-type amusement park, a great white shark escapes from its tank and proceeds to cause terror and chaos. Little

connection to the previous "Jaws" sagas; and this one is in 3-D. The theatrical title was "Jaws 3-D." Followed by one more sequel, "Jaws: The Revenge."
1983 (PG) 98m/C Dennis Quaid, Bess Armstrong, Louis Gossett Jr., Simon MacCorkindale, Lea Thompson; *Dir:* Joe Alves. **VHS, Beta, LV** $19.95 *MCA* 🐾🐾

Jaws of Death
A "Jaws"-like saga of shark terror. Jaeckel strikes out illogically to protect his "friends," the sharks.
1976 91m/C Richard Jaeckel, Harold Sakata, Jennifer Bishop, John Chandler, Buffy Dee; *Dir:* William Grefe. **VHS, Beta** *PGN* 🐾🐾

Jaws of the Dragon
The story of two rival gangs in the Far East.
1976 (R) 96m/C James Nam, Johnny Taylor, Kenny Nam. **VHS, Beta** $24.95 *GEM* 🐾

Jaws of Justice
A Canadian Mountie and his trusty German shepherd track down outlaws in the Canadian Northwest.
1933 55m/B Kazan the Dog, Richard Terry, Lafe McKee. **VHS, Beta, 8mm** $24.95 *VYY* 🐾½

Jaws of Satan
Weaver is terrorized by a slimy snake who is, in actuality, the Devil. Not released for 2 years after it was filmed, and generally considered a bomb of the first caliber. Unfortunate work for usually worthwhile Weaver.
1981 (R) 92m/C Fritz Weaver, Gretchen Corbett, Jon Korkes, Norman Lloyd; *Dir:* Bob Claver. **VHS** $9.95 *WKV* Woof!

Jaws: The Revenge
The third sequel, in which Mrs. Brody is pursued the world over by a seemingly personally motivated Great White Shark. Includes footage not seen in the theatrical release. Each sequel in this series is progressively inferior to the original.
1987 (PG-13) 87m/C Lorraine Gary, Lance Guest, Karen Yahng, Mario Van Peebles, Michael Caine, Judith Barsi; *Dir:* Joseph Sargent. **VHS, Beta, LV** $19.95 *MCA* 🐾

The Jayhawkers
Chandler and Parker battle for power and women in pre-Civil War Kansas.
1959 100m/C Jeff Chandler, Fess Parker, Nicole Maurey, Henry Silva, Herbert Rudley, Frank DeKova, Don Megowan, Leo Gordon; *Dir:* Melvin Frank. **VHS** $14.95 *PAR, FCT* 🐾🐾

The Jayne Mansfield Story
A television movie recounting the blond bombshell's career from her first career exposure, through her marriage to a bodybuilder, to the famous car crash that beheaded her.
1980 97m/C Loni Anderson, Arnold Schwarzenegger, Raymond Buktencia, Kathleen Lloyd, G.D. Spradlin, Dave Shelley; *Dir:* Dick Lowry. **VHS, Beta** $39.95 *IVE* 🐾

Jazz Ball
A compilation of songs and performances by the great jazz stars of the '30s and '40s taken from various movie shorts and features.
1956 60m/B Duke Ellington, Louis Armstrong, Artie Shaw, Cab Calloway, Gene Krupa, Peggy Lee, Buddy Rich, Betty Hutton. **VHS** $19.95 *VYY, REP* 🐾

Jazz in Exile
A profile of several well-known American jazz musicians, including Dexter Gordon and Phil Woods, who have spent much of their careers and made their reputations in Europe.
1982 58m/C Dexter Gordon, Phil Woods, Randy Weston, Richard Davis. **VHS, Beta** $29.95 *MVD, RHP, FST* 🐾

Jazz & Jive Duke Ellington provides early jazz background in "Black and Tan," his first movie. Dance numbers accompany Dewey Brown in "Toot the Trumpet," followed by Major Bowes in "Radio Revels."
193? 60m/B Duke Ellington, Major Bowes, Dewey Brown. VHS, Beta CCB 🎬

Jazz is Our Religion An in-depth look at the lives and music of some of the great jazz musicians of our time. Val Wilmer's quick cut stills and narration stand out, as does her theme that "jazz musicians play the way they live and live the way they play."
1972 50m/B Jo Jones, Dizzy Gillespie, Sunny Murray, Ted Jones, Art Blakey, Johnny Griffin; **Dir:** Val Kilmer. VHS, Beta $19.95 MVD, KIV 🎬

Jazz Singer A Jewish cantor's son breaks with his family to become a singer of popular music. Of historical importance as the first successful part-talkie. Jolson's songs include "Toot Toot Tootsie," "Blue Skies," and "My Mammy." With the classic line "You ain't heard nothing yet!" Remade several times.
1927 89m/B Al Jolson, May McAvoy, Warner Oland, William Demarest, Eugenie Besserer, Myrna Loy; **Dir:** Alan Crosland. VHS, Beta, LV $29.98 FOX, FCT, MLB 🎬🎬½

Jazz Singer Another remake of the 1927 classic about a Jewish boy who rebels against his father and family tradition to become a popular entertainer. Stick with the original, this is little more than a vehicle for Diamond to sing. Unintentionally funny.
1980 (PG) 115m/C Neil Diamond, Laurence Olivier, Lucie Arnaz, Catlin Adams, Franklin Ajaye; **Dir:** Richard Fleischer. VHS, Beta, LV, 8mm $14.95 PAR 🎬

Jazz at the Smithsonian: Art Blakey The legendary jazz drummer and his Great Messengers perform and converse about their history at a Smithsonian jazz festival.
1982 58m/C VHS, Beta, 8mm $29.95 SVS, MVD, KUL 🎬

Jazz at the Smithsonian: Art Farmer A look at the concert work and life of the famed jazz flugelhorn player.
1980 58m/C Art Farmer; **Performed by:** Billy Hart, Dennis Irwin. VHS, Beta, 8mm $29.95 SVS, MVD, KUL 🎬

Jazz at the Smithsonian: Benny Carter The archetypal jazz saxophonist, mentor to Miles Davis and Charlie Parker, performs "A Train," "Honeysuckle Rose" and "Autumn Leaves."
1982 57m/C VHS, Beta, 8mm $29.95 SVS, MVD, KUL 🎬

Jazz at the Smithsonian: Bob Wilber Wilber, in dedication to his mentor Sidney Bechet, plays sax at the Smithsonian: "Down in Honky Tonk," "Coal Cart Blues," "Kansas City Man Blues," and others.
1984 59m/C VHS, Beta $29.95 SVS, MVD, KUL 🎬

Jazz at the Smithsonian: Joe Williams The 1986 winner of the Best Male Jazz Vocalist Grammy performs an array of tunes: "Everyday I Have the Blues," "Once in a While," "Who She Do," "Save that Time for Me," and others.
1984 58m/C VHS, Beta $29.95 SVS, MVD, KUL 🎬

Jazz at the Smithsonian: Mel Lewis The astute jazz drummer performs at the Smithsonian: "One Finger Snap," "Dolphin Dance," "Make Me Smile" and "Eye of the Hurricane."
1984 55m/C VHS, Beta, 8mm $29.95 SVS, MVD, KUL 🎬

Jazz at the Smithsonian: Red Norvo The innovative and ground-breaking jazz vibist performs at the Smithsonian: "All of Me," "Jitterbug Waltz," "School Days," and others.
1984 58m/C VHS, Beta, 8mm $29.95 SVS, MVD, KUL 🎬

Jazz on a Summer's Day Filmed version of the 1958 Newport Jazz Festival. Well-edited. Jazz fans won't want to miss this fabulous production.
1959 85m/C Louis Armstrong, Big Maybelle, Chuck Berry, Thelonious Monk, Anita O'Day; **Dir:** Ken Hughes. VHS $59.95 MVD, FCT, CEG 🎬🎬🎬

J.D.'s Revenge Attorney in training Glynn Turman is possessed by the spirit of a gangster who was murdered on Bourbon Street in the early 1940s and grisly blaxploitation results. Ever so slightly better than others of the genre. Shot in New Orleans, with a soundtrack by a then unknown Prince.
1976 (R) 95m/C Louis Gossett Jr., Glynn Turman; **Dir:** Arthur Marks. VHS $59.95 ORI, FCT 🎬🎬½

je tu il elle A hyperactive young woman desperately seeking the answers to life gradually gains experience and maturity as she travels around France. The directorial debut of Ackerman; in French with English subtitles.
1974 90m/C FR Niels Arestrup, Claire Wauthion; **Dir:** Chantal Akerman. VHS $79.95 WAC, FCT 🎬🎬

Jealousy Three dramatic episodes dealing with jealousy. Made for TV.
1984 95m/C Angie Dickinson, Paul Michael Glaser, David Carradine, Bo Svenson; **Dir:** Jeffrey Bloom. VHS, Beta $79.95 IMP 🎬🎬½

Jean de Florette The first of two films (with "Manon of the Spring") based on Marcel Pagnol's novel. A single spring in drought-ridden Provence, France is cemented up by two scheming countrymen (Montand and Auteuil), while they await the imminent failure of the farm nearby, inherited by a city-born hunchback, whose chances for survival fade without water for his crops. A devastating story with a heartrending performance by Depardieu as the hunchback. Lauded and awarded; in French with English subtitles.
1987 (PG) 122m/C FR Gerard Depardieu, Yves Montand, Daniel Auteuil, Elisabeth Depardieu, Ernestine Mazurowna; **Dir:** Claude Berri. British Academy Awards '87: Best Picture, Best Supporting Actor (Auteuil). VHS, Beta, LV $19.98 ORI, IME, FCT 🎬🎬🎬

Jean Sheperd on Route #1... and Other Thoroughfares Sheperd travels the American highways in search of truth and finds only humor.
1985 60m/C Beta NJN 🎬🎬

Jekyll & Hyde...Together Again A New Wave comic version of the classic story of Dr. Jekyll and Mr. Hyde, with a serious young surgeon who turns into a drug-crazed punk rocker after sniffing a mysterious powder.
1982 (R) 87m/C Mark Blankfield, Bess Armstrong, Krista Errickson, Tim Thomerson; **Dir:** Jerry Belson. VHS, Beta $59.95 PAR 🎬

Jennifer Elaboration of that old "Carrie" theme: a teen-age girl outcast who likes snakes wreaks havoc against her catty classmates in this supernatural thriller.

1978 (PG) 90m/C Lisa Pelikan, Bert Convy, Nina Foch, John Gavin, Wesley Eure; **Dir:** Brice Mack. VHS, Beta $69.98 LIV, VES 🎬½

Jenny A loveless marriage of convenience between a draft-dodger and a film buff gets turned around as they fall in love despite themselves.
1970 (PG) 90m/C Marlo Thomas, Alan Alda, Marian Hailey, Elizabeth Wilson, Vincent Gardenia, Charlotte Rae; **Dir:** Edgar J. Scherick. VHS, Beta $59.98 FOX 🎬

Jenny Lamour A dark mystery thriller about a singer accused of murdering a man he thought was stealing his woman. Acclaimed genre piece, dubbed into English. Alternate title: "Quai des Orfevres."
1947 95m/B FR Louis Jouvet, Suzy Delair, Bernard Blier, Simone Rennant; **Dir:** Henri-Georges Clouzot. VHS, Beta $24.95 NOS, FCT 🎬🎬

Jeremiah Johnson The story of a man who turns his back on civilization, circa 1850, and learns a new code of survival in a brutal land of isolated mountains and hostile Indians. In the process, he becomes part of the mountains and their wildlife. A notable and picturesque film.
1972 107m/C Robert Redford, Will Geer; **Dir:** Sydney Pollack. VHS, Beta, LV $19.98 WAR, FCT 🎬🎬🎬

Jericho Adventures in Africa occur as a court-martialed captain pursues a murderous deserter. Also known as "Dark Sands."
1938 77m/B GB Paul Robeson, Henry Wilcoxon, Wallace Ford; **Dir:** Thornton Freeland. VHS, Beta $44.95 NOS, HHT, DVT 🎬🎬

The Jericho Mile A made-for-television drama about a track-obsessed convicted murderer who is given a chance at the Olympics.
1979 97m/C Peter Strauss, Roger E. Mosley, Brian Dennehy; **Dir:** Michael Mann. VHS, Beta $9.98 CHA, SUE 🎬🎬

The Jerk A jerk tells his rags to riches to rags story in comedic flashbacks, from "I was born a poor Black child," through his entrepreneurial success in his invention of the "Optigrab," to his inevitable decline. Only film in history with a dog named "Shithead." Martin's first starring role, back in his wild and crazy days; his ridiculous misadventures pay tribute to Jerry Lewis movies of the late sixties. TV version released in 1984 as "The Jerk Too," with Mark Blankfield as the Jerk.
1979 (R) 94m/C Steve Martin, Bernadette Peters, Catlin Adams, Bill Macy, Jason Jackson; **Dir:** Carl Reiner. VHS, Beta, LV $14.95 MCA 🎬🎬½

Jesse A heroic nurse ministers to remote Death Valley residents—so state bureaucats hound her for practicing medicine without a license. Although based on a true case the well-acted made-for-TV drama just lacks import and impact.
1988 (PG) 94m/C Lee Remick, Scott Wilson, Leon Rippy, Priscilla Lopez, Albert Salmi; **Dir:** Glenn Jordan. VHS $79.98 REP 🎬🎬

Jesse James A Hollywood biography of the famed outlaw. One of director King's best efforts.
1939 105m/C Henry Fonda, Tyrone Power Sr., Randolph Scott, Henry Hull, Jane Darwell, Brian Donlevy; **Dir:** Henry King. VHS, Beta, LV $19.98 FOX 🎬🎬🎬

Jesse James at Bay Exciting saga of the notorious Jesse James and his fight against the railroads.

1941 54m/B Roy Rogers, George "Gabby" Hayes, Sally Payne; *Dir:* Joseph Kane. **VHS, Beta $19.95** *NOS, MRV, VCN* ♫♫ ½

Jesse James Meets Frankenstein's Daughter
The gunslinger and Frankenstein's granddaughter, Maria, meet up in the Old West in this wacky combination of western and horror genres.
1965 95m/C John Lupton, Cal Bolder, Narda Onyx, Steve Geray, Estelita, Jim Davis, William Fawcett, Nestor Paiva; *Dir:* William Beaudine. **VHS, Beta $19.98** *NOS, SUE* ♫♫ ½

Jesse James Rides Again
Originally a serial, this feature depicts the further adventures of the West's most notorious outlaw.
1947 181m/B Clayton Moore, Linda Stirling, Roy Barcroft, Tristram Coffin; *Dir:* Fred Brannon, Thomas Carr. **VHS $29.98** *REP, MLB* ♫♫

Jesse James Under the Black Flag
An interesting western/pseudo biography/flashback in which Coates and James Jr. (the outlaw's son) play themselves. As Coates attempts to gather information from Jesse's decendants, a stranger appears and falls for Jesse's granddaughter, who shuns him. He proceeds to read the Coates biography and the flashback begins, in the Jesse was really a good guy forced off the straight and narrow vein.
1921 59m/B Jesse James Jr., Franklin Coates, Diana Reed, Marguerite Hungerford; *Dir:* Franklin Coates. **VHS, Beta $24.95** *GPV, FCT* ♫♫ ½

The Jesse Owens Story
The moving story of the four-time Olympic Gold medal winner's triumphs and misfortunes. A made-for-television movie.
1984 174m/C Dorian Harewood, Debbi Morgan, George Stanford Brown, LeVar Burton, George Kennedy, Tom Bosley, Ben Vereen; *Dir:* Richard Irving. **VHS, Beta $69.95** *PAR, KUI* ♫♫

Jessie's Girls
In retaliation for the murder of her husband, an angry young woman frees three female prisoners, and they embark on a bloody course of revenge. Together, they track down the killers, and one by one, they fight to even the score.
1975 (R) 86m/C Sondra Currie, Regina Carrol, Jennifer Bishop; *Dir:* Al Adamson. **VHS, Beta $39.95** *MON* ♫ ½

Jesus
A supposedly authenticated life of Christ, using real artifacts.
1979 (G) 118m/C Brian Deacon; *Dir:* Peter Sykes, John Kirsh. **VHS, Beta** *VBL* ♫♫

Jesus Christ Superstar
A rock opera that portrays, in music, the last seven days in the earthly life of Christ as reenacted by young tourists in Israel. Songs include "I Don't Know How to Love Him," "What's the Buzz," "Blood Money," and "Heaven On Their Minds." Based on the stage play by Tim Rice and Andrew Lloyd Weber, film is sometimes stirring while exhibiting the usual heavy-handed Jewison approach.
1973 (G) 108m/C Ted Neeley, Carl Anderson, Yvonne Elliman, Josh Mostel; *Dir:* Norman Jewison. **VHS, Beta, LV $29.95** *MCA* ♫♫

Jesus of Montreal
A vagrant young actor (stage trained Canadian star Bluteau) is hired by a Montreal priest to produce a fresh interpretation of an Easter passion play. Taking the good book at its word, he produces a contemporized literal telling that captivates audiences, inflames the men of the cloth, and eventually wins the players' faith. Quebecois director Arcand (keep an eye out for him as the judge) tells a compelling, acerbically satirical and haunting story

that never forces its Biblical parallels. Earned an Academy Award nomination for Best Foreign Language Film. French with English subtitles.
1989 (R) 119m/C *FR CA* Gilles Pelletier, Lothaire Bluteau, Catherine Wilkening, Robert Lepage, Johanne-Marie Tremblay, Remy Girard, Marie-Christine Barrault; *W/Dir:* Denys Arcand. Genie Awards '90: Best Actor (Girard), Best Art Direction/Set Decoration, Best Cinematography, Best Costume Design, Best Director (Arcand), Best Picture, Best Screenplay, Best Musical Score. **VHS, LV $79.95** *ORI, FCT* ♫♫♫ ½

Jesus of Nazareth
The earliest known filmed account of the life of the Christian messiah.
1928 85m/B **VHS, Beta, 8mm $29.95** *VYY, FCT* ♫♫ ½

Jesus of Nazareth
An all-star cast portrays the life of Jesus Christ. A made-for-television mini-series.
1977 371m/C Robert Powell, Anne Bancroft, Ernest Borgnine, Claudia Cardinale, James Mason, Laurence Olivier, Anthony Quinn; *Dir:* Franco Zeffirelli. **VHS, Beta $169.95** *CCB, FOX, JER* ♫♫ ½

Jesus: The Nativity
An animated retelling of the birth of Christ.
1987 52m/C **Beta $9.95** *GKK* ♫♫ ½

Jesus: The Resurrection
An animated children's version of Christ's execution and subsequent resurrection.
1987 49m/C **Beta $9.95** *GKK* ♫♫ ½

The Jesus Trip
Hunted motorcyclists take a young nun as hostage in the desert.
1971 86m/C Robert Porter, Tippy Walker; *Dir:* Russ Mayberry. **VHS, Beta $57.95** *UNI* ♫ ½

Jet Attack/Paratroop Command
A double feature of "B" war movies. "Jet Attack" is a bad movie about a scientist whose plane gets shot down over Korea. He amazingly survives, only to be taken to a hospital and brainwashed. Enter a squad of soldiers leading a rescue mission who meet up with a beautiful Russian aide and some life-saving guerillas. This movie is so bad it's funny. In "Paratroop Command," a paratrooper accidentally kills a member of his squadron. In order to regain his dignity, the paratrooper sets off land mines in North Africa to prove himself.
1958 139m/B John Agar, Audrey Totter, Gregory Walcott, James Dobson, Richard Bakalayan, Jack Hogan, Jimmy Murphy, Ken Lynch; *Dir:* Edward L. Cahn, William Witney. **VHS $19.95** *COL* ♫

Jet Benny Show
A Buck Rogers/Star Wars style spoof of Jack Benny's classic TV show.
1986 76m/C Steve Norman, Kevin Dees, Polly MacIntyre, Ted Luedemann; *Dir:* Roger Evans. **VHS, Beta $69.95** *CHA* ♫

Jet Over the Atlantic
A dangerous situation arises as a bomb is discovered aboard a plane en route from Spain to New York.
1959 92m/B Guy Madison, George Raft, Virginia Mayo, Brett Halsey; *Dir:* Byron Haskin. **VHS, Beta $39.95** *BVV* ♫ ½

Jet Pilot
An American Air Force colonel (Wayne) in charge of an Alaskan Air Force base falls in love with a defecting Russian jet pilot (Leigh). They marry, but Wayne suspects Leigh is a spy planted to find out top U.S. secrets. He pretends to defect with her back to Russia to see what he can find out, but they again flee. Ludicrous plot is saved only by spectacular flying scenes, some

performed by Chuck Yeager. Although this was filmed in 1950, it took seven more years to be released because producer Hughes couldn't keep his hands off it.
1957 112m/C John Wayne, Janet Leigh, Jay C. Flippen, Paul Fix, Richard Rober, Roland Winters, Hans Conried, Ivan Triesault; *Dir:* Josef von Sternberg. **VHS $19.98** *MCA* ♫♫

Jet Sets
Three models pose, in and out of scanty outfits, then record music for no apparent reason.
1989 60m/C **VHS, Beta $19.95** *AHV* ♫♫

A Jetson Christmas Carol
Even Mr. Spacely gets in the spirit!
1989 30m/C *Voices:* George O'Hanlon, Penny Singleton, Daws Butler, Mel Blanc. **VHS, Beta, LV $9.95** *HNB, IME* ♫♫

The Jetsons Meet the Flintstones
Trouble arises when young Elroy Jetson's time machine takes he and his family back to the 25th century B.C. There they meet the Flintstones and the Rubbles, who accidentally get transported into the future. Will either family ever get back home?
1988 100m/C *Voices:* Mel Blanc, Daws Butler, Jean VanDerPyl, George O'Hanlon, Penny Singleton. **VHS, Beta, LV $19.95** *HNB, IME* ♫♫

The Jetsons: Movie
The famous outer space family of 60's TV is given a new silver-screened life. George gets a promotion that puts him in charge of an asteroid populated by furry creatures. Ecological concerns are expressed; while this is rare for a cartoon, the story is overall typical.
1990 (G) 82m/C *Dir:* William Hanna, Joseph Barbera; *Voices:* George O'Hanlon, Mel Blanc, Penny Singleton, Tiffany, Patric Zimmerman, Don Messick, Jean VanDerPyl, Ronnie Schell, Patti Deutsch, Dana Hill, Russi Taylor, Paul Kreppel, Rick Dees. **VHS, Beta, LV $22.95** *MCA, APD* ♫

The Jewel in the Crown
Epic saga of the last years of British rule in India from 1942-1947 concentrating on an Indian man and the Englishwoman that he loves. Originally shown on British television, it aired in the United States on the PBS series "Masterpiece Theatre." An excellent mini-series that was shown in fourteen segments. Available on five video cassettes.
1984 750m/C *GB* Charles Dance, Susan Woolridge, Art Malik, Tim Pigott-Smith, Geraldine James, Peggy Ashcroft, Judy Parfitt; *Dir:* Christopher Morahan, Jim O'Brien. Emmy Awards '84: Outstanding Limited Series. **VHS, Beta $199.95** *MTI* ♫♫ ½

Jewel of the Nile
The sequel to "Romancing the Stone," with same cast and new director. Jack endeavors to rescue Joan from the criminal hands of a charming North African president. Not quite up to the "Stone's" charm.
1985 (PG) 106m/C Michael Douglas, Kathleen Turner, Danny DeVito, Avner Eisenberg, The Flying Karamazov Brothers, Spiros Focas; *Dir:* Lewis Teague. **VHS, Beta, LV $19.98** *FOX* ♫♫ ½

The Jeweller's Shop
A highly spiritual tale of two married couples in Poland whose children meet and fall in love much later in Canada. Based on a play by Karol Wojtyla—who later became Pope Jean-Paul II. The film was accordingly adapted under strict Vatican supervision.
1990 90m/C Burt Lancaster, Ben Cross, Olivia Hussey. **VHS $29.95** *IGP* ♫♫

Jezebel
Davis is a willful Southern belle who loses fiance Fonda through her selfish and spiteful ways in this pre-Civil War drama. When he becomes ill, she realizes her cruelty

J

and rushes to nurse him back to health. Davis' role won her an Oscar for Best Actress, and certainly provided Scarlett O'Hara with a rival for most memorable female character of all time. Oscar nominations: Best Picture, Best Cinematography, Best Score.
1938 105m/B Bette Davis, George Brent, Henry Fonda, Margaret Lindsay, Fay Bainter, Donald Crisp, Spring Byington, Eddie Anderson; *Dir:* William Wyler. Academy Awards '38: Best Actress (Davis), Best Supporting Actress (Bainter). **VHS, Beta, LV $19.95** *MGM, FCT, BTV* ♫♫♫ ½

Jezebel's Kiss A sizzling young beauty returns to the town where she grew up and proceeds to destroy any man she pleases.
1990 (R) 96m/C Meg Foster, Malcolm McDowell, Meredith Baxter Birney, Everett McGill, Katherine Barrese; *W/Dir:* Harvey Keith. **VHS $79.95** *COL* ♫♫

JFK Costner stars as former New Orleans district attorney Jim Garrison in Stone's highly controversial examination of the 1963 assassination of President John F. Kennedy. Hotly debated because of Stone's theory of a vast conspiracy behind the assassination, this film has sparked new calls for Congress to open long-sealed government records from the 1977 House Select Committee on Assassinations investigation. Hailed as one of the year's best films, "JFK" features outstanding performances from all-star principal and supporting casts. Even Jim Garrison himself shows up as Chief Justice Earl Warren, chairman of the Warren Commission. Stunning cinematography and excellent editing techniques combine to make "JFK" a true cinematic masterpiece that will leave the viewer wanting to know more.
1991 (R) 189m/C Kevin Costner, Tommy Lee Jones, Kevin Bacon, Laurie Metcalf, Gary Oldman, Michael Rooker, Jay O. Sanders, Sissy Spacek, Sally Kirkland, Ed Asner, Jack Lemmon, Brian Doyle Murray, Joe Pesci, Walter Matthau, John Candy, Donald Sutherland; *Dir:* Oliver Stone. Academy Awards '91: Best Cinematography, Best Film Editing; Golden Globe Awards '92: Best Director (Stone). **VHS, Beta, LV $94.99** *WAR* ♫♫♫♫

The JFK Conspiracy Did Lee Harvey Oswald act alone? Opinions of government officials, eyewitnesses, and director Oliver Stone as well as testimony, photographs, documents, and other evidence work to answer that question. Includes footage from the House Select Committee on Assassination.
1991 98m/C *Hosted:* James Earl Jones. **VHS, LV $24.98** *BMG* ♫♫♫♫

JFK Remembered This look at John F. Kennedy's 1000-day Presidential term, released on the 25th anniversary of his assassination, features interviews, rare film clips, and famous news footage.
1988 54m/C *Hosted:* Peter Jennings. **VHS, Beta $29.98** *LIV, VES* ♫♫♫♫

Jigsaw Crime drama with a smattering of cameos by Hollywood favorites. A newspaper reporter is murdered. The dead man's friend, an assistant district attorney, seeks the punks responsible. They send in a seductress to keep the D.A. busy, but she too is murdered. Action and tension in a well made flick. Also known as "Gun Moll."
1949 70m/B Franchot Tone, Jean Wallace, Myron McCormick, Marc Lawrence, Robert Gist; *Dir:* Fletcher Markle. **VHS $16.95** *SNC, DVT* ♫♫♫

Jigsaw Missing persons lawman is lured into a cover-up more complex than he had bargained for. A tense drama originated as the pilot for a series. Retitled "Man on the Move."
1971 100m/C James Wainwright, Vera Miles, Richard Kiley, Andrew Duggan, Edmond O'Brien; *Dir:* William A. Graham. **VHS** *IVE* ♫♫

Jigsaw A developer is found dead the day after his wedding. His widow and a cop team up to find out who, what where and why. Lukewarm Australian mystery export.
1990 85m/C Rebecca Gibney, Dominic Sweeny, Michael Coard. **VHS $59.95** *ATL* ♫♫

The Jigsaw Man A British-Russian double agent is sent back to England to retrieve a list of Soviet agents which he hid there many years before.
1984 (PG) 90m/C *GB* Michael Caine, Laurence Olivier, Susan George, Robert Powell, Charles Gray, David Kelly, Michael Medwin; *Dir:* Terence Young. **VHS, Beta, LV $79.99** *HBO* ♫♫

The Jilting of Granny Weatherall An adaptation of the Katherine Anne Porter short story about a dying matriarch who is still haunted by the man who jilted her decades ago.
1980 57m/C Geraldine Fitzgerald, Lois Smith, William Swetland; *Hosted:* Henry Fonda. **VHS, Beta $24.95** *MON, KAR* ♫♫

Jim McCann & the Morrisseys A popular Irish solo singer and a group perform live in concert in Ireland.
1988 55m/C Jim McCann, Morrisseys. **VHS, Beta** *RGO* ♫♫

Jim Thorpe: All American The life story of Thorpe, a Native American athlete who gained international recognition for his excellence in many different sports. A must for sports fans.
1951 105m/B Burt Lancaster, Phyllis Thaxter, Charles Bickford, Steve Cochran, Dick Wesson; *Dir:* Michael Curtiz. **VHS, Beta** *ISS, WAR* ♫♫ ½

Jimi Hendrix: Berkeley May 1970 The master guitarist Hendrix electrifies a Berkeley audience with songs "Johnny B. Goode," "Hear My Train A Comin'" "Star Spangled Banner," "Purple Haze," "I Don't Live Today," "Hey Baby," "Lover Man," "Machine Gun," and "Voodoo Chile."
1973 55m/C Jimi Hendrix. **VHS, Beta, LV $19.98** *MVD, WRV, VES* ♫♫ ½

Jimi Hendrix: Johnny B. Goode Rarely-seen footage of Hendrix performing "Are You Experienced?," "All Along the Watchtower" and "Voodoo Child" is featured in this compilation of live concert material.
1967 26m/C Jimi Hendrix. **VHS, Beta, LV $14.95** *SVS* ♫♫ ½

Jimi Hendrix: Live in Monterey, 1967 Hendrix caught live in the unedited footage of his performance in "Monterey Pop," by Pennebaker, the premier rock documentarian. Songs include "Rock Me Baby," "Like a Rolling Stone," "Purple Haze" and "Wild Thing."
1967 48m/C Jimi Hendrix; *Dir:* D.A. Pennebaker, Chris Hegedus. **Beta, LV $19.95** *MVD, HBO* ♫♫ ½

Jimi Hendrix: Rainbow Bridge Jimi Hendrix at his best doing the classics "Purple Haze," "Dolly Dagger," "Foxy Lady," "Star Spangled Banner" and more.
1989 74m/C Jimi Hendrix; *Dir:* Chuck Wein. **VHS, LV $19.95** *MVD, RHI* ♫♫ ½

Jimi Hendrix: Story This documentary about the guitar-playing legend features interviews with Eric Clapton and Pete Townsend, along with concert footage from Jimi's appearances at the Woodstock and Isle of Wight festivals.
1973 (R) 103m/C *Dir:* Joe Boyd, John Hend, Gary Weis; *Voices:* Eric Clapton, Pete Townsend. **VHS, Beta $19.95** *MVD, WAR* ♫♫ ½

Jiminy Cricket's Christmas A Yuletide collection of Disney cartoons, featuring Donald Duck, Chip 'n' Dale, Goofy, Jiminy and the rare "Mickey's Good Deed," (1932).
1932 (G) 47m/C VHS, Beta, LV $12.99 *DIS* ♫♫ ½

Jimmy the Kid A young boy becomes the unlikely target for an improbable gang of would-be crooks on a crazy, "fool-proof" crime caper.
1982 (PG) 95m/C Gary Coleman, Cleavon Little, Fay Hauser, Ruth Gordon, Dee Wallace Stone, Paul LeMat, Don Adams; *Dir:* Gary Nelson. **VHS, Beta $59.99** *HBO Woof!*

Jimmy Valentine An adaptation of the O. Henry story wherein a reformed safecracker must rescue a child from a bank vault and risk revealing his identity. Also available in an edited 30-minute version.
1985 46m/C VHS, Beta *LCA* ♫♫

Jinxed A Las Vegas nightclub singer tries to convince a gullible blackjack dealer to murder her crooked boyfriend, but the plan backfires when the gangster electrocutes himself while taking a shower.
1982 (R) 104m/C Bette Midler, Ken Wahl, Rip Torn; *Dir:* Don Siegel. **VHS, Beta $19.98** *MGM* ♫ ½

The Jitters The murdered dead turn into zombies and seek revenge on their killers.
1988 80m/C Sal Viviano, Marilyn Tokuda, James Hong, Frank Dietz; *Dir:* John Fasano. **VHS, Beta $79.95** *PSM* ♫

Jive Junction A group of patriotic teenagers convert a barn into a canteen for servicemen and name it "Jive Junction."
1943 62m/B Dickie Moore, Tina Thayer, Gerra Young, Johnny Michaels, Jack Wagner, Jane Wiley, Beverly Boyd, William Halligan; *Dir:* Edgar G. Ulmer. **VHS, Beta $24.95** *NOS, VYY* ♫♫

Jive Turkey Cool dude busts heads.
1976 ?m/C Paul Harris, Frank DeKova, Serena; *Dir:* Bill Brame. **VHS** *MOV* ♫ ½

Jo Jo Dancer, Your Life is Calling Pryor directed and starred in this semi-autobiographical price-of-fame story of a comic, hospitalized for a drug-related accident, who must re-evaluate his life. A serious departure from Pryor's slapstick comedies that doesn't quite make it as a drama, but Pryor deserves credit for the honesty he displayed in dealing with his real-life problems. Music by Herbie Hancock.
1986 (R) 97m/C Richard Pryor, Debbie Allen, Art Evans, Fay Hauser, Barbara Williams, Paula Kelly, Wings Hauser, Carmen McRae, Diahnne Abbott, Scoey Mitchell, Billy Eckstine; *Dir:* Richard Pryor. **VHS, Beta, LV $14.95** *COL* ♫♫ ½

Joan of Arc A touching and devout look at the life of Joan of Arc. Perhaps unfortunately, the film is very accurately based on the play by Maxwell Anderson and adds up to too much talk and too little action. Oscar nominations: Best Actress (Bergman), Best Supporting Actor (Ferrer), Color Art Direction, Score for a Drama or Comedy and Film Editing.

1948 100m/C Ingrid Bergman, Jose Ferrer, John Ireland, Leif Erickson; *Dir:* Victor Fleming. Academy Awards '48: Best Color Cinematography, Best Costume Design; National Board of Review Awards '48: 10 Best Films of the Year. **VHS, Beta, LV $14.98** *VID, IGP* ♂♂½

Joan of Paris A French resistance leader dies so that Allied pilots can escape from the Nazis. A well-done, but obviously dated propaganda feature.
1941 91m/B Michele Morgan, Paul Henreid, Thomas Mitchell, Laird Cregar, May Robson, Alexander Granach, Alan Ladd; *Dir:* Robert Stevenson. **VHS, Beta $19.98** *TTC* ♂♂♂½

Jock Jokes Comics and athletes share their funniest jokes and stories revolving around sports.
1989 49m/C Franklin Ajaye, Howie Gold, Mike Guido, Jonathan Soloman, John Witherspoon. **VHS, Beta $29.95** *MED* ♂♂♂½

Jock Peterson A light-hearted story about a blonde hunk who enrolls in college, cavorting and seducing his way to questionable fame and fortune.
1974 (R) 97m/C Jack Thompson, Wendy Hughes; *Dir:* Tim Burstall. **VHS, Beta $59.98** *CHA* Woof!

Jocks A whiz-kid tennis team competes in a pro meet in Las Vegas, and paints the town red.
1987 (R) 90m/C Christopher Lee, Perry Lang, Richard Roundtree, Scott Strader; *Dir:* Steve Carver. **VHS, Beta $79.95** *COL* ♂½

Joe An odd friendship grows between a businessman and a blue-collar worker as they search together for the executive's runaway daughter. Thrust into the midst of the counter-culture, they react with an orgy of violence.
1970 (R) 107m/C Peter Boyle, Susan Sarandon, Dennis Patrick; *Dir:* John G. Avildsen. **VHS, Beta, LV $39.98** *VES* ♂♂½

Joe Bob Briggs Dead in Concert The popular Mid-American comic does his schtick in concert, singing "We are the Weird" and "Drive-In Oath."
1985 60m/C VHS, Beta $29.95 *UHV* ♂♂½

Joe Cocker: Mad Dogs Includes classic tour footage, guests Rita Coolidge and Leon Russell, and legendary hits such as "The Letter," "With a Little Help from My Friends," and "Please Give Peace a Chance."
117m/C VHS, LV $29.95 *PIA* ♂♂½

Joe Kidd A land war breaks out in New Mexico between Mexican natives and American land barons. Eastwood, once again portraying the "mysterious stranger," must decide with whom he should side, all the while falling in love with Garcia. Lackluster direction results in a surprisingly tedious western, in spite of the cast. One of Eastwood's lowest money grossers. The laserdisc edition is letterboxed and features chapter stops, the original trailer and digital sound.
1972 (PG) 88m/C Clint Eastwood, Robert Duvall, John Saxon, Don Stroud, Stella Garcia, James Wainwright, Paul Koslo, Gregory Walcott, Dick Van Patten, Lynn Marta; *Dir:* John Sturges. **VHS, Beta, LV $14.95** *MCA, PIA* ♂♂½

Joe Louis Story The story of Joe Louis' rise to fame as boxing's Heavyweight Champion of the world.
1953 88m/B Coley Wallace, Paul Stewart, Hilda Simms. **VHS, Beta $19.95** *NOS, DVT, FCT* ♂♂

Joe Panther Family drama about a young Seminole Indian who stakes his claim in the Anglo world by wrestling alligators. Montalban plays a wise old chieftain who helps the youth handle the conflict between Indian and white societies.
1976 (G) 110m/C Brian Keith, Ricardo Montalban, Alan Feinstein, Cliff Osmond, A. Martinez, Robert Hoffman; *Dir:* Paul Krasny. **VHS** *IND, WAR* ♂♂♂

Joe Piscopo Joe does some of his best impressions, from Frank Sinatra to Jerry Lewis, and some memorable improvisations from Saturday Night Live in this one hour special.
1984 60m/C Joe Piscopo, Eddie Murphy. **VHS, Beta $19.98** *LIV, VES, TLF* ♂♂♂

Joe Piscopo Live! The "Saturday Night Live" alumnus gives a solo performance.
1989 60m/C Joe Piscopo. **VHS, Beta $14.99** *HBO* ♂♂♂

Joe Piscopo: New Jersey Special A comedic salute to Piscopo's home state, with guest appearances by Eddie Murphy and Danny DeVito.
1987 60m/C Joe Piscopo, Eddie Murphy, Danny DeVito. **VHS, Beta $19.98** *LIV, VES* ♂

Joe Versus the Volcano Expressionistic goofball comedy about a dopey guy who, after finding out he has only months to live, is contracted by a millionaire into leaping into a volcano alive. An imaginative farce with great "Metropolis"-pastiche visuals that eventually fizzle out. Available with Spanish subtitles. Written and directed by the "Moonstruck" Oscar winner.
1990 (PG) 106m/C Tom Hanks, Meg Ryan, Lloyd Bridges, Robert Stack, Amanda Plummer, Abe Vigoda, Dan Hedaya, Barry McGovern, Ossie Davis; *Dir:* John Patrick Shanley. **VHS, Beta, LV, 8mm $19.98** *WAR* ♂½

Joe's Bed-Stuy Barbershop: We Cut Heads Lee's first film takes place inside a neighborhood landmark, the corner barbershop. Lee displays the social awareness and urban humor that he honed in his later, bigger films.
1983 60m/C Leon Errol, Morris Carnovsky, Elyse Knox, Billy House, Trudy Marshall; *Dir:* Spike Lee. **VHS, Beta, 3/4U** *ICA* ♂♂½

Joey Daddy, a former doo-wopper, looks back on his years of musical success as a waste of time. His son takes to the world of rock guitar with blind fervor. Their argument plays against the backdrop of the "Royal Doo-Wopp Show" at New York City's Radio Music Hall. Both old and new rock songs are featured.
1985 (PG) 90m/C Neill Barry, James Quinn. **VHS, Beta $69.95** *VID* ♂½

John F. Kennedy A retrospective of JFK's term in office.
1981 104m/C VHS, Beta, LV $29.95 *FOX* ♂½

John Huston: The Man, the Movies, the Maverick Tribute to the legendary director features home movies, film clips, and interviews with the likes of Lauren Bacall and Paul Newman. A captivating profile of the renaissance man. Robert Mitchum narrates.
1989 128m/C Robert Mitchum, Paul Newman, Lauren Bacall; *Dir:* Frank Martin. **VHS, Beta $29.95** *TTC* ♂½

John Jacob Niles A portrait of John Jacob Niles, balladeer and composer, in concert, at home, and at work arranging his tunes and discussing the historical place of balladry in American music.
1978 32m/C VHS, Beta $31.95 *APL* ♂½

John & Julie Two young people get it together despite many obstacles.
1977 82m/C Colin Gibson, Lesley Dudley. **VHS, Beta $29.95** *FHE* ♂½

John Lee Hooker: Blue Monday Party Hooker plays blues that reflect his Mississippi roots and his Detroit days. Great back-up band and harmonica player help Hooker kick it in.
1981 30m/C Performed by: John Lee Hooker, Charlie Musselwhite, Mark Naftalin. **VHS $24.95** *KIV* ♂½

John Lennon: Imagine A film of Lennon's solo album, featuring psychedelic video images accompanying 13 songs.
1972 55m/C *Dir:* Andrew Solt. **VHS, Beta, LV, 8mm $9.95** *WAR, SVS, MVD* ♂½

John Lennon: Interview with a Legend A television interview with John Lennon, made with Tom Snyder on the "Tomorrow Show," originally aired April 28, 1975. Lennon discusses what it was like to be a Beatle, how he dealt with worldwide popularity, the breakup of the group, and his life in New York during the post-Beatle era.
1981 60m/C John Lennon, Tom Snyder. **VHS, Beta** *LHV, OM* ♂½

John Lennon Live in New York City Lennon performs at Madison Square Garden in this rare live concert performance featuring such songs as "Imagine," "Instant Karma," "Come Together" and "Give Peace a Chance."
1985 55m/C John Lennon, Yoko Ono, Plastic Ono Elephant's Memory Band. **VHS, Beta, LV, 8mm $9.95** *SVS, PIA, MVD* ♂½

John Lennon and the Plastic Ono Band: Live Peace in Toronto, 1969 The original film of the band's first public performance in Toronto. Songs include "Give Peace a Chance," "Money," "Blue Suede Shoes," and "Cold Turkey." Features an appearance by Clapton.
1969 56m/C John Lennon, Yoko Ono, Eric Clapton, Klaus Voorman, Alan White; *Dir:* D.A. Pennebaker. **Beta, LV $19.95** *MVD, HBO* ♂½

John Lennon/Yoko Ono: Then and Now Ono looks back on her life before and after Lennon along with songs performed by both.
1984 56m/C VHS, Beta, LV $9.95 *MVD, MSM, PIA* ♂½

John and the Missus After the discovery of a copper mine threatens to destroy a nearby town, the Canadian government begins a relocation process. However, one man refuses to give up his generations-old home. Canadian Pinsent's directorial debut, adapting his own 1973 novel.
1987 (PG) 98m/C CA Gordon Pinsent, Jackie Burroughs, Randy Follet, Jessica Steen; *Dir:* Gordon Pinsent. Genie Awards '87: Best Actor (Pinsent), Best Musical Score. **VHS, Beta $79.98** *CGH* ♂½

John Wayne: American Hero of the Movies A tribute to the Duke featuring the greatest action scenes from his films.
1990 ?m/C John Wayne. **VHS, Beta** *GKK* ♂½

John Wayne Anthology: Star Packer/Lawless Range In addition to two of Wayne's films, this program captures 40 years of the actor's career.
1934 156m/B John Wayne; *Dir:* Robert Bradbury. VHS **$9.95** *SIM* ✍✍½

John Wayne: Bigger Than Life A documentary of the Duke which revolves around clips of his most famous westerns.
1990 55m/B John Wayne. VHS **$9.99** *VTR, BUR* ✍✍½

John Wayne Previews, Vol. 1 A collection of previews from the Duke's biggest films, including "The Alamo," "Red River," "Rio Bravo" and "The Spoilers."
196? 59m/C John Wayne. VHS, Beta *CPB* ✍✍½

John Wayne: The Duke Lives On A look at the life and films of John Wayne including highlights from "She Wore a Yellow Ribbon" and "The Quiet Man."
1984 48m/C John Wayne. VHS, Beta **$19.98** *RKO* ✍✍½

John & Yoko: A Love Story Appalling television movie version of the rock legend's life and marriage. With original Beatles tunes.
1985 146m/C Mark McGann, Kim Mayori, Kenneth Price, Peter Capaldi, Richard Morant, Philip Walsh; *Dir:* Sandor Stern. VHS, Beta, LV **$79.95** *SVS, IME* ✍

Johnnie Gibson F.B.I. Supposedly based on a true story, the adventures of a beautiful, black FBI agent who falls in love with a man she's investigating.
1987 96m/C Howard E. Rollins Jr., Lynn Whitfield, William Allen Young, Richard Lawson; *Dir:* Bill Duke. VHS, Beta **$29.95** *ACA, PTB* ✍

Johnny Angel Merchant Marine captain unravels mystery of his father's murder aboard a ship. Tense film noire.
1945 79m/B George Raft, Claire Trevor, Signe Hasso, Lowell Gilmore, Hoagy Carmichael; *Dir:* Edwin L. Marin. VHS, Beta, LV **$19.98** *RKO, CCB* ✍✍✍

Johnny Apollo Upright college student Power keeps his nose clean until Dad's sent to jail for playing with numbers. Scheming to free his old man (Arnold) from the big house, he decides crime will pay for the stiff price tag of freedom. Predictable and melodramatic, but well directed.
1940 93m/B Tyrone Power Sr., Dorothy Lamour, Lloyd Nolan, Edward Arnold, Charley Grapewin, Lionel Atwill; *Dir:* Yves Allegret. VHS, Beta **$19.98** *FOX, FCT* ✍✍½

Johnny Barrows, El Malo An ex-Vietnam vet battles crime on urban streets. In Spanish.
197? 90m/C *SP* Fred Williamson. VHS, Beta **$57.95** *UNI* ✍✍½

Johnny Be Good A too-talented high school quarterback is torn between loyalty to his best friend and to his girlfriend amid bribery and schemings by colleges eager to sign him up.
1988 (R) 91m/C Anthony Michael Hall, Robert Downey Jr., Paul Gleason, Uma Thurman, John Pankow, Steve James, Seymour Cassel, Michael Greene, Marshall Bell, Deborah May; *Dir:* Bud Smith. VHS, Beta, LV **$19.98** *ORI* ✍

Johnny Belinda A compassionate physician cares for a young deaf mute woman and her illegitimate child. Tension builds as the baby's father returns to claim the boy. Oscar nominations for Best Picture, Best Actor (Ayres), Best Supporting Actor (Bickford), Best Supporting Actress (Moorehead),

Screenplay, Black & White Cinematography, Black & White Art Direction, Sound recording, Score of a Drama or Comedy, and Film Editing.
1948 103m/B Jane Wyman, Lew Ayres, Charles Bickford, Agnes Moorehead; *Dir:* Jean Negulesco. Academy Awards '48: Best Actress (Wyman); Golden Globe Awards '49: Best Film—Drama. VHS, Beta **$19.98** *MGM, BTV, CCB* ✍✍✍

Johnny Belinda A television movie adaptation of the Elmer Harris weepie about a small town doctor who befriends an abused deaf girl. Remake of the 1948 film.
1982 95m/C Richard Thomas, Rosanna Arquette, Dennis Quaid, Candy Clark, Roberts Blossom, Fran Ryan; *Dir:* Anthony Page. VHS, Beta **$59.95** *IVE* ✍

Johnny Carson Highlights of the fabulous "Tonight Show," starring Johnny Carson. The program shows material not seen in "The Tonight Show" anniversary programs. Also included are bloopers and some early "Who Do You Trust" clips.
197? 60m/C Johnny Carson, Don Rickles, Pearl Bailey, Joey Heatherton, Jerry Van Dyke, Ed McMahon. VHS, Beta **$24.95** *SHO* ✍

Johnny Come Lately An elderly editor helps out an ex-newspaperman with a police charge. The two then team up to expose political corruption despite threats from a rival newspaperman.
1943 97m/B James Cagney, Grace George, Marjorie Main, Marjorie Lord, Hattie McDaniel, Ed McNamara; *Dir:* William K. Howard. VHS, LV **$19.98** *REP, PIA* ✍✍½

Johnny Dangerously A gangster spoof about Johnny Dangerously, who turned to crime in order to pay his mother's medical bills. Now, Dangerously wants to go straight, but competitive crooks would rather see him killed than law-abiding and his mother requires more and more expensive operations. Crime pays in comic ways.
1984 (PG-13) 90m/C Michael Keaton, Joe Piscopo, Danny DeVito, Maureen Stapleton, Marilu Henner, Peter Boyle, Griffin Dunne, Glynnis O'Connor, Dom DeLuise, Danny De Vito, Richard Dimitri, Ray Walston, Dick Butkus, Alan Hale Jr., Bob Eubanks; *Dir:* Amy Heckerling. VHS, Beta **$79.98** *FOX* ✍✍

Johnny Firecloud A modern Indian goes on the warpath when the persecution of his people reawakens his sense of identity.
1975 94m/C Victor Mohica, Ralph Meeker, Frank De Kova, Sacheen Little Feather, David Canary, Christina Hart; *Dir:* William Allen Castleman. VHS, Beta **$49.95** *PSM* ✍

Johnny Frenchman Rival fishing groups battle it out, complicated by a romance across the water. Featured in the cast are real fishermen and villagers from Cornwall, as well as authentic Free French Resistance fighters from WWII.
1946 104m/B *GB* Francoise Rosay, Tom Walls, Patricia Roc, Ralph Michael, Frederick Piper, Pierre Richard, Carroll O'Connor; *Dir:* Charles Frend. VHS **$19.95** *NOS* ✍

Johnny Got His Gun Dalton Trumbo's story of a young war veteran who realizes that his arms and legs have been amputated.
1971 (R) 111m/C Timothy Bottoms, Jason Robards Jr., Donald Sutherland, Diane Varsi, Kathy Field; *Dir:* Dalton Trumbo. VHS, Beta, LV **$59.95** *MED* ✍✍

Johnny Guitar Women strap on six-guns in Nicholas Ray's unintentionally hilarious, gender-bending western. A guitar-playing loner wanders into a small town feud between lovelorn saloon owner Crawford and

McCambridge, the town's resident lynchmob-leading harpy. This fascinating cult-favorite has had film theorists arguing for decades: is it a parody, a political McCarthy-era allegory, or Freudian exercise? The off-screen battles of the two female stars are equally legendary. Stick around for the end credits to hear Peggy Lee sing the title song.
1953 110m/C Joan Crawford, Ernest Borgnine, Sterling Hayden, Mercedes McCambridge, Scott Brady, Ward Bond, Royal Dano, John Carradine; *Dir:* Nicholas Ray. VHS, Beta, LV **$14.98** *REP* ✍✍✍½

Johnny Handsome An ugly, deformed hood, after he's been double-crossed and sent to prison, volunteers for a reconstructive surgery experiment and is released with a new face, determined to hunt down the scum that set him up. A terrific modern B-picture based on John Godey's "The Three Worlds of Johnny Handsome."
1989 (R) 96m/C Mickey Rourke, Ellen Barkin, Lance Henriksen, Elizabeth McGovern, Morgan Freeman, Forest Whitaker, Scott Wilson; *Dir:* Walter Hill. VHS, Beta, LV **$19.95** *IVE* ✍✍✍

Johnny Lightning This musical about the rise and fall and comeback of a rock guitar player features both the glories and pitfalls of wine, women, and ego.
1973 29m/B VHS, Beta *PCV* Woof!

Johnny Nobody A Mysterious stranger murders a writer who has been taunting the residents of a quaint Irish town. A sleeper of a thriller, especially in the treatment of the town's reaction to this murder.
1961 88m/B *GB* William Bendix, Aldo Ray, Nigel Patrick, Yvonne Mitchell, Cyril Cusack; *Dir:* Nigel Patrick. VHS, Beta **$19.95** *MON* ✍✍½

Johnny Shiloh An underage youth becomes an heroic drummer during the Civil War. Originally a two-part Disney television show.
1963 90m/C Kevin Corcoran, Brian Keith, Darryl Hickman, Skip Homeier; *Dir:* James Neilson. VHS, Beta **$69.95** *DIS* ✍✍

Johnny Tiger A teacher has his hands full when he arrives at the Seminole Reservation in Florida to instruct the Indian children.
1966 100m/C Robert Taylor, Geraldine Brooks, Chad Everett, Brenda Scott; *Dir:* Paul Wendkos. VHS, Beta *REP* ✍✍½

Johnny Tremain & the Sons of Liberty The story of the gallant American patriots who participated in the Boston Tea Party.
1958 85m/C Sebastian Cabot, Hal Stalmaster, Luanna Patten, Richard Beymer; *Dir:* Robert Stevenson. VHS, Beta **$69.95** *DIS* ✍✍

Johnny We Hardly Knew Ye Well-made biographical drama recounts John F. Kennedy's first political campaign for local office in Boston in 1946. Made for television.
1977 90m/C Paul Rudd, Kevin Conway, William Prince, Burgess Meredith, Tom Berenger, Brian Dennehy, Kenneth McMillan; *Dir:* Gilbert Gates. VHS, Beta *TLF* ✍✍½

Johnny Winter Live This program presents the best young, white blues artists of the '60's in a live concert appearance.
1984 45m/C Johnny Winter. VHS, Beta, LV **$19.95** *MSM, PIA* ✍✍½

A Joke of Destiny, Lying in Wait Around the Corner Like a Bandit A irreverent satire about a Minister of the Interior (Tognazzi, of "La Cage aux Folles" fame) who becomes trapped in his high-tech limousine before a vital press conference. An

exaggerated vision of Italian bureaucracy. With English subtitles.
1984 (PG) 105m/C *IT* Ugo Tognazzi, Piera Degli Esposti, Gastone Moschin, Renzo Montagnani, Valeria Golino; *Dir:* Lina Wertmuller. **VHS, Beta** $59.95 LHV, APD *♂♂*

Jokes My Folks Never Told Me
A series of blackouts and sketches featuring a bevy of beautiful women.
1977 82m/C Sandy Johnson, Mariwin Roberts; *Dir:* Gerry Woolery. **VHS, Beta** $59.98 SUE *♂*

The Jolly Paupers
A pre-war comedy made in Warsaw about two small-town Jews who try to achieve fame and fortune in the face of setbacks, community quarrels, and insanity. In Yiddish with English subtitles.
1938 62m/B Shimon Dzigan, Yisroel Shumacher, Max Bozyk, Menasha Oppenheim; *Dir:* Zygmund Turkow. **VHS** $89.95 ALD *♂♂½*

Jolson Sings Again
This sequel to "The Jolson Story" brings back Larry Parks as the ebullient entertainer, with Jolson himself dubbing Parks's voice for the songs. Picking up where the other film ended, the movie chronicles Jolson's comeback in the 1940s and his tireless work with the USO overseas during World War II and the Korean War. Songs include "Ma Blushin' Rosie," "Pretty Baby," "I Only Have Eyes for You," "Baby Face" and "Carolina in the Morning."
1949 96m/C Larry Parks, William Demarest, Barbara Hale, Bill Goodwin; *Dir:* Henry Levin. **VHS, Beta, LV** $19.95 COL *♂♂*

The Jolson Story
A smash Hollywood bio of Jolson, from his childhood to superstardom. Features dozens of vintage songs from Jolson's parade of hits. Jolson himself dubbed the vocals for Parks, rejuvenating his own career in the process. Oscar nominations for Best Actor (Parks), Supporting Actor (Demarest), Color Cinematography, and Film Editing.
1946 128m/C Larry Parks, Evelyn Keyes, William Demarest, Bill Goodwin, Tamara Shayne, John Alexander, Jimmy Lloyd, Ludwig Donath; *Dir:* Alfred E. Green. Academy Awards '46: Best Sound, Best Musical Score. **VHS, Beta, LV** $19.95 COL *♂♂♂*

Jonah Who Will Be 25 in the Year 2000
A group of eight friends, former sixties radicals, try to adjust to growing older and coping with life in the seventies. The eccentric octet include a disillusioned journalist turned gambler, an unorthodox teacher, and a grocery store cashier who gives away food to the elderly. Wonderful performances highlight this social comedy. In French with English subtitles.
1976 110m/C *SI* Jean-Luc Bideau, Myriam Meziere, Miou-Miou, Jacques Denis, Rufus, Dominique Labourier, Roger Jendly, Miriam Boyer, Raymond Bussieres, Jonah; *W/Dir:* Alain Tanner. **VHS** $79.95 NYF *♂♂♂½*

Jonathan Livingston Seagull
Based on the best-selling novella by Richard Bach, this film quietly envisions a world of love, understanding, achievement, hope and individuality.
1973 (G) 99m/C James Franciscus, Juliet Mills; *Dir:* Hall Bartlett. **VHS, Beta** $29.95 PAR *♂♂½*

Jonathan Miller's Production of Verdi's Rigoletto
Famed lecturer Dr. Miller examines the tragic opera, its creation and its meaning in today's world.
1982 140m/C **VHS, Beta** $39.95 HBO *♂♂½*

Jonathan Winters on the Edge
Jonathan Winters and his friends show viewers how crazy they can be.

1986 60m/C Jonathan Winters, Robin Williams, Milton Berle, Martin Mull, Phyllis Diller. **VHS, Beta** $19.98 MCG *♂♂½*

Joni
An inspirational story based on the real life of Joni, playing herself, who was seriously injured in a diving accident, and her conquering the odds. A born-again feature film based on the book by Tada.
1979 (G) 75m/C Joni Eareckson Tada, Bert Remsen, Katherine De Hetre, Cooper Huckabee; *Dir:* James F. Collier. **VHS, Beta** $59.95 IVE *♂♂*

Jornada de Muerte
An ex-cop is hired to bring an important anti-Mob witness to trial via transcontinental train, and must fend off Mob attack squads in the process. In Spanish.
198? 90m/C *SP* Fred Williamson, Bernie Kuby, Heidi Dobbs. **VHS, Beta** $49.95 UNI *♂*

Jory
A young man's father is killed in a saloon fight, and he must grow up quickly to survive.
1972 (PG) 96m/C Robby Benson, B.J. Thomas, John Marley; *Dir:* Jorge Fons. **VHS, Beta** $9.95 COL, SUE *♂♂*

Jose Canseco's Baseball Camp
The games' biggest hitter shows how it's done.
1989 60m/C VHS, Beta $19.95 IVE *♂♂*

Jose' Jose' en Acapulco
Spanish singing star Jose Jose performs fifteen classic songs in this concert taped in Acapulco, Mexico. Available in VHS and Beta Hi-Fi Stereo.
1985 60m/C VHS, Beta $19.95 MCA *♂♂*

Joseph Andrews
This adaptation of a 1742 Henry Fielding novel chronicles the rise of Joseph Andrews from servant to personal footman (and fancy) of Lady Booby.
1977 (R) 99m/C *GB* Ann-Margret, Peter Firth, Jim Dale, Michael Hordern, Beryl Reid; *Dir:* Tony Richardson. **VHS, Beta** $49.95 PAR *♂♂*

Joseph & His Brethren
A full-length dramatization of the Old Testament story.
1960 103m/C Geoffrey Horne, Robert Morley; *Dir:* Irving Rapper. **VHS, Beta** $49.95 UHV. VGD *♂*

Joseph & His Brothers
The story of Joseph, sold into slavery by his jealous brothers. Part of the "Greatest Heroes of the Bible" series.
1979 55m/C Sam Bottoms, Walter Brooke, Harvey Jason, Bernie Kopell, Barry Nelson, Carol Rossen, Albert Salmi. **VHS, Beta** $14.95 VGD

Josepha
A husband and wife, both actors, are forced to re-examine their relationship when the wife finds a new love while on a film location. In French with English subtitles.
1982 (R) 114m/C *FR* Miou-Miou, Claude Brasseur, Bruno Cremer; *Dir:* Christopher Frank. **VHS, Beta** $59.95 COL *♂♂♂*

The Josephine Baker Story
Made-for-cable-TV bio of exotic entertainer/activist Josephine Baker, a black woman who found superstardom in pre-WWII Europe but repeated rejection in her native USA. At times trite treatment turns her eventful life into a standard rise-and-fall showbiz tale, but a great cast and lavish scope pull it through. Whitfield recreates Baker's (sometimes topless) dance routines; Carol Dennis dubs her singing.
1990 (R) 129m/C Lynn Whitfield, Ruben Blades, David Dukes, Craig T. Nelson, Louis Gossett Jr.; *Dir:* Brian Gibson. Emmy '91: Best Actress-Miniseries (Lynn Whitfield), Best Director (Gibson). **VHS** $59.99 HBO, FCT *♂♂♂*

Joshua
Western drama about a vigilante who tracks down the group of outlaws that killed his mother.
1976 (R) 75m/C Fred Williamson, Isela Vega; *Dir:* Larry Spangler. **VHS** $79.98 MAG *♂*

Joshua Then and Now
When a Jewish-Canadian novelist (Woods) is threatened by the breakup of his marriage to his WASPy wife, compounded by a gay scandal, he re-examines his picaresque history, including his life with his gangster father (Arkin). A stirring story that is enhanced by a strong performance by Woods, as well as a picturesque Canadian backdrop. Adapted by Mordecai Richler from his own novel.
1985 (R) 102m/C *CA* James Woods, Gabrielle Lazure, Alan Arkin, Michael Sarrazin, Chuck Shamata; *Dir:* Ted Kotcheff. Genie Awards '86: Best Art Direction/Set Decoration, Best Cinematography, Best Costume Design, Best Supporting Actor (Arkin), Best Supporting Actress (Sorensen). **VHS, Beta** $79.98 FOX *♂♂♂*

Jour de Fete
Tati's first film, dealing with a French postman's accelerated efforts at efficiency after viewing a motivational film of the American post service. Wonderful slapstick moments. In French with English subtitles.
1948 79m/B *FR* Jacques Tati, Guy Decomble, Paul Fankeur; *Dir:* Jacques Tati. Venice Film Festival '49: Best Scenario. **VHS, Beta, LV** $19.95 VDM, SUE, LV *♂♂♂½*

Journey
A violent story of a girl who is rescued from the Sagueney River and falls in love with her rescuer. Choosing to remain in the remote pioneer community of this "hero," she brings everyone bad luck and misery.
1977 (PG) 87m/C *CA* Genevieve Bujold, John Vernon; *Dir:* Paul Almond. **VHS, Beta** $54.95 UHV *♂*

Journey Back to Oz
This animated special features Dorothy and Toto returning to visit their friends in the magical land of Oz.
1971 90m/C *Voices:* Liza Minnelli, Ethel Merman, Paul Lynde, Milton Berle, Mickey Rooney, Danny Thomas. **VHS, Beta** $14.95 FHE, KAR, HHE *♂*

Journey Beneath the Desert
Three engineers discover the lost but always found in the movies kingdom of Atlantis when their helicopter's forced down in the sunny Sahara. A poor hostess with a rotten disposition, the mean sub-saharan queen doesn't roll out the welcome mat for her grounded guests, so a beautiful slave babe helps them make a hasty exit. Dawdling Euro production with some sufficient visuals.
1961 105m/C *FR* *IT* Haya Harareet, Jean-Louis Trintignant, Brad Fulton, Amadeo Nazzari, George Riviere, Giulia Rubini, Gabriele Tinti, Gian Marie Volonte; *Dir:* Edgar G. Ulmer, Giuseppe Masini, Frank Borzage. **VHS, Beta** $16.95 SNC *♂♂*

Journey into the Beyond
A brain-incinerating look at the inexplicable world of the supernatural. Includes examinations of voodoo, bare-handed surgery and psychic sex that are guaranteed to fry your soul.
1976 (R) 83m/C *Nar:* John Carradine. **VHS, Beta** FHE *♂♂*

Journey to the Center of the Earth
A scientist and student undergo a hazardous journey to find the center of the earth and along the way they find the lost city of Atlantis. Based upon the Jules Verne novel.

1959 132m/C James Mason, Pat Boone, Arlene Dahl, Diane Baker, Thayer David; *Dir:* Henry Levin. VHS, Beta, LV $19.98 FOX, FCT ✍✍✍

Journey to the Center of the Earth An animated version of the Jules Verne classic adventure about a scientist's quest to reach the earth's core.
1976 50m/C VHS, Beta $19.95 MGM ✍✍✍

Journey to the Center of the Earth A young nanny and two teenage boys discover Atlantis while exploring a volcano.
1988 (PG) 83m/C Nicola Cowper, Paul Carafotes, Ilan Mitchell-Smith; *Dir:* Rusty Lemorande. VHS, Beta, LV $79.95 CAN ✍✍✍

Journey to the Center of Time A scientist and his crew are hurled into a time trap when a giant reactor explodes.
1967 83m/C Lyle Waggoner, Scott Brady, Gigi Perreau, Anthony Eisley; *Dir:* David L. Hewitt. VHS, Beta $19.95 ACA, GHV ✍½

Journey to the Far Side of the Sun Chaos erupts in the Earth's scientific community when it is discovered that a second, identical Earth is on the other side of the Sun. Both planets end up sending out identical exploratory missions. The denouement is worth the journey. Originally known as "Doppelganger."
1969 (G) 92m/C GB Roy Thinnes, Ian Hendry, Lynn Loring, Patrick Wymark, Loni von Friedl, Herbert Lom, Ed Bishop; *Dir:* Robert Parrish. VHS, Beta $29.95 MCA ✍✍✍

Journey into Fear During World War II, an American armaments expert is smuggled out of Istanbul, but Axis agents are close behind. From the novel by Eric Ambler; remade for television in 1974.
1942 71m/B Joseph Cotten, Dolores Del Rio, Orson Welles, Agnes Moorehead, Norman Foster; *Dir:* Norman Foster. VHS, Beta, LV $19.95 RKO, MED, RXM ✍✍✍

Journey into Fear A remake of the 1942 Orson Welles classic about a geologist ensnared in Turkish intrigue and murder.
1974 96m/C CA Sam Waterston, Vincent Price, Shelley Winters, Donald Pleasence, Zero Mostel, Yvette Mimieux, Ian McShane; *Dir:* Daniel Mann. VHS, Beta $59.98 LIV, LTG ✍✍

Journey: Frontiers and Beyond Platinum rock group Journey is profiled on and behind the stage. Includes "Wheel in the Sky," "Stone in Love," "After the Fall" and "Escape."
1983 98m/C Journey. VHS, Beta $19.95 MSM, VTK ✍½

Journey of Honor During a civil war in 17th century Japan, a nearly defeated Shogun sends his son, Lord Mayeda, to Spain to buy guns from King Phillip III. Along the way he must fight shipboard spies and fierce storms, and once Mayeda arrives in Spain, he is confronted by the evil Duke Don Pedro. In defeating Don Pedro, Mayeda saves King Phillip, gets his guns, and wins the love of Lady Cecilia. To get revenge, Don Pedro plots to capture the Japanese ship and enslave everyone on board. Lots of action and fine acting. Based on a story by Sho Kosugi.
1991 (PG-13) 107m/C Sho Kosugi, David Essex, Kane Kosugi, Christopher Lee, Norman Lloyd, Ronald Pickup, John Rhys-Davies, Dylan Kussman, Toshiro Mifune; *Dir:* Gordon Hessler. VHS, Beta, LV MCA ✍✍½

Journey of Hope Powerful drama about Kurdish family that sells its material possessions in hopes of emigrating legally to Switzerland, where life will surely be better. During the perilous journey, smugglers take their money and the family must attempt crossing the formidable slopes of the Swiss Mountains on foot. Based on a true story. In Swiss with English subtbitles.
1990 (PG) 111m/C SI Necmettin Cobanoglu, Nur Surer, Emin Sivas, Yaman Okay, Mathias Gnaedinger, Dietmar Schoenherr; *W/Dir:* Xavier Koller. Academy Awards '90: Best Foreign Language Film. VHS, LV $92.99 HBO ✍✍✍½

Journey to the Lost City An architect living in India happens upon a lost city, the rulership of which is being contested by two brothers. In the midst of fighting snakes and tigers, he falls in love with Paget, a beautiful dancer. This feature is actually a hybrid of two films, "Der Tiger von Eschnapur" and "Das Indische Grabmal," poorly edited together to make one feature length film.
1958 95m/C GE FR IT Debra Paget, Paul Christian, Walter Reyer, Claus Holm, Sabine Bethmann, Valeri Inkizhinov, Rene Deltgen, Lucianna Paluzzi; *Dir:* Fritz Lang. VHS, Beta $16.95 SNC ✍

Journey of Natty Gann With the help of a wolf (wonderfully portrayed by a dog) and a drifter, a 14-year-old girl travels across the country in search of her father in this warm and touching film. Excellent Disney representation of life during the Great Depression.
1985 (PG) 101m/C Meredith Salenger, John Cusack, Ray Wise, Scatman Crothers, Lainie Kazan, Verna Bloom; *Dir:* Jeremy Paul Kagan. VHS, Beta, LV $29.95 DIS ✍✍✍½

Journey Through Rosebud A re-enactment of the 1970s protest at Wounded Knee by Native Americans.
1972 93m/C Robert Forster, Eddie Little Sky; *Dir:* Tom Giles. VHS, Beta $59.98 CHA ✍✍

Journey Together An old black woman teaches a young black girl about life.
1982 25m/C Esther Rolle, Tina Andrews; *Dir:* Paul Asselin. VHS, Beta $19.95 NOS, GEM ✍✍

Journeys from Berlin/1971 This is the true story about a woman who tries to find a solution to her own problems in life.
1984 125m/C VHS, Beta NEW ✍✍½

Joy House An American woman befriends a handsome French playboy when her husband sends gangsters to kill him.
1964 98m/B Jane Fonda, Alain Delon, Lola Albright; *Dir:* Rene Clement. VHS, Beta $39.95 MON ✍✍

Joy of Living Vintage screwball farce about a randy playboy trying to win over a pragmatic songstress. With music by Jerome Kern.
1938 90m/B Irene Dunne, Douglas Fairbanks Jr., Alice Brady, Guy Kibbee, Jean Dixon, Eric Blore, Lucille Ball, Warren Hymer, Billy Gilbert, Frank Milan; *Dir:* Tay Garnett. VHS, Beta, LV $19.98 RKO, TTC ✍✍✍

Joy Ride to Nowhere Two young women steal a Cadillac with two million dollars in the trunk and ride away with the owner hot on their tail.
1978 (PG) 86m/C Leslie Ackerman, Sandy Serrano, Len Lesser, Mel Welles, Ron Ross, Speed Stearns. VHS, Beta $19.95 NEG ✍

Joy of Sex An undercover narcotics agent is sent to Southern California's Richard M. Nixon High School to investigate the school's extracurricular activities. Typical teen sex flick; no relation to the book.
1984 (R) 93m/C Cameron Dye, Michelle Meyrink, Colleen Camp, Christopher Lloyd, Ernie Hudson, Lisa Langlois; *Dir:* Martha Coolidge. VHS, Beta $79.95 PAR ✍

Joy Sticks Something outrageously hilarious, very sexy and thoroughly entertaining is going on at the local video arcade!
1983 (R) 88m/C Joe Don Baker, Leif Green, Jim Greenleaf, Scott McGinnis, Jonathan Gries; *Dir:* Greydon Clark. VHS, Beta, LV $79.95 VES ✍½

Joyful Laughter The dreary life of an Italian movie bit player draws strength from Magnani's powerful emotions as she repeatedly comes near to success only to find disappointment. Set in Rome. With English subtitles.
1954 106m/C IT *Dir:* Mario Monicelli. VHS $59.95 FST ✍✍

Joyless Street Silent film focuses on the dismal life of the middle class in Austria during an economic crisis. Lovely piano score accompanies the film.
1925 96m/B GE Greta Garbo, Werner Krauss, Asta Nielson; *Dir:* G. W. Pabst. VHS, Beta, LV $29.95 VYY, MRV, FST ✍✍✍½

Joyride A union official mistreats them, so three friends steal a car for a joyride and plummet into a life of crime.
1977 (R) 91m/C Desi Arnaz Jr., Robert Carradine, Melanie Griffith, Anne Lockhart; *Dir:* Joseph Ruben. VHS, Beta $29.98 LIV, VES ✍½

Juarez A revolutionary leader overthrows the Mexican government and then becomes President of the country. Based on the true story of Benito Pablo Juarez. Oscar nominations: Best Supporting Actor (Aherne), Cinematography.
1939 132m/B Paul Muni, John Garfield, Bette Davis, Claude Rains, Gale Sondergaard; *Dir:* William Dieterle. VHS, Beta $19.95 MGM, FCT ✍✍

Jubal A rancher (Borgnine) seeks advice from a cowhand (Ford) about pleasing his wife, but another cowhand (Steiger) implies that Ford is "advising" Borgnine's wife, as well. A western take on "Othello."
1956 101m/C Glenn Ford, Rod Steiger, Ernest Borgnine, Felicia Farr, Charles Bronson, Valerie French, Noah Beery Jr.; *Dir:* Delmer Daves. VHS, Beta, LV $9.95 RCA, IME ✍✍

Jubilee Adam Ant stars as a punk who takes over Buckingham Palace and turns it into a recording studio. Music by Eno, Adam and the Ants, and Siouxsie and the Banshees.
1978 103m/C GB Jordan, Toyah Wilcox, Jenny Runacre, Little Nell. VHS, Beta $29.95 MVD, MFV, VCL ✍✍

Jubilee Trail Wagon trail western follows the lives of a group of pioneers from New Orleans to California. A woman goes through a marriage, a baby, and her husband's death. Ralston has some good tunes in this lush picture.
1954 103m/C Vera Ralston, Joan Leslie, Forrest Tucker, John Russell, Ray Middleton, Pat O'Brien; *Dir:* Joseph Kane. VHS $14.95 REP ✍✍½

Jud Society's refusal to understand a young soldier returning from the Vietnam conflict leads him to violence and tragedy. Jennings first film; music by Stu Phillips of Creedence Clearwater Revival.

1971 (PG) 80m/C Joseph Kaufmann, Robert Deman, Alix Wyeth, Norman Burton, Claudia Jennings, Maurice Sherbanee, Victor Dunlap, Bonnie Bittner; *Dir:* Gunther Collins. **VHS, Beta** $49.95 *PSM ♫½*

Jud Suess The classic, scandalous Nazi anti-Semitic tract about a Jew who rises to power under the duchy of Wuerttemberg by stepping on, abusing, and raping Aryans. A film that caused riots at its screenings and tragedy for its cast and crew, and the Third Reich's most notorious fictional expression of policy. In German, subtitled.
1940 100m/B *GE* Ferdinand Marian, Werner Krauss, Heinrich George, Kristina Soderbaum, Eugene Klopfer; *Dir:* Veit Harlan. **VHS, Beta, 3/4U** $24.95 *NOS, VCD, IHF ♫♫♫*

Judex Judex, a sensitive cloaked hero-avenger, fights master criminal gangs. In French with English subtitles. Remake of the silent French serial.
1964 103m/B *FR IT* Channing Pollock, Francine Berge, Jacques Jouanneau; *Dir:* Georges Franju. **VHS, Beta** $24.95 *NOS, SNC, HHT ♫♫½*

The Judge A courtroom crime-drama about the consequences of infidelity. A lawyer sends an acquitted hit man to kill his wife and her lover. Things go amiss and the lawyer winds up dead.
1949 69m/B Milburn Stone, Katherine DeMille, Paul Guilfoyle, Jonathan Hale; *Dir:* Elmer Clifton. **VHS** $16.95 *SNC, EME ♫½*

The Judge and the Assassin Intriguing courtroom drama. A prejudiced judge has his values challenged when he must decide if a child-killer is insane or not. The relationship that develops between the judge and the killer is the film's focus. Excellent direction by Tavernier.
1975 130m/C *FR* Philippe Noiret, Michel Galabru, Isabelle Huppert, Jean-Claude Brialy; *Dir:* Bertrand Tavernier. **VHS, Beta, LV** $79.95 *CVC, IME ♫♫♫½*

Judge Horton and the Scottsboro Boys A courtroom drama focusing on a famous 1931 trial. A courageous judge must battle the South in the case of nine black men charged with gang-raping two white women. Made for television.
1976 100m/C Arthur Hill, Vera Miles, Ken Kercheval, Suzanne Lederer, Tom Ligon; *Dir:* Fielder Cook. **VHS** $59.95 *IVE ♫♫½*

Judge Priest Small-town judge in the old South stirs up the place with stinging humor and common-sense observances as he tangles with prejudices and civil injustices. Funny, warm slice of life is occasionally defeated by racist characterizations. Ford remade it later as "The Sun Shines Bright." Taken from the Irvin S. Cobb stories.
1934 80m/B Will Rogers, Stepin Fetchit, Anita Louise, Henry B. Walthall; *Dir:* John Ford. **VHS, Beta** $19.95 *NOS, MRV, PSM ♫♫♫*

The Judge Steps Out Pleasant, light comedy about a hen-pecked, Bostonian judge who decides he's had enough and heads for balmy California. There he finds a job in a restaurant and more importantly, a sympathetic friend in its female owner. He soon finds himself falling for the woman, and realizes he faces an important decision.
1949 91m/B Alexander Knox, Ann Sothern, George Tobias, Sharyn Moffett, Florence Bates, Frieda Inescort, Myrna Dell, Ian Wolfe, H.B. Warner; *Dir:* Boris Ingster. **VHS** $19.95 *TTC, NVH ♫♫½*

Judgment A devout Christian couple are shocked to learn that their young son has been molested by the popular priest of their small Louisiana parish. When they find out their son is not the first victim, and other families have been coerced into silence, they vow to fight back. Tasteful treatment of a real court case. Made for cable.
1990 (PG-13) 89m/C Keith Carradine, Blythe Danner, Jack Warden, David Strathairn; *W/Dir:* Tom Topor. **VHS** $89.98 *HBO ♫♫♫*

Judgment in Berlin An East German family hijacks a U.S. airliner into West Germany, and then must stand trial in Berlin before a troubled American judge. Based on a true 1978 incident and the book by Herbert J. Stern.
1988 (PG) 92m/C Martin Sheen, Sam Wanamaker, Max Gail, Sean Penn, Heinz Hoenig, Carl Lumbly, Max Volkert Martens, Harris Yulin, Joshua Sinclair, Jutta Speidel; *Dir:* Leo Penn. **VHS, Beta, LV** $29.95 *COL ♫♫½*

Judgment Day Four college buddies pick the wrong New England town for their vacation. It seems that the town made a deal with the Devil in 1689 to save themselves from the plague, and now the DarkMaster is here to collect.
1988 (PG-13) 93m/C Kenneth McLeod, David Anthony Smith, Monte Markham, Gloria Hayes, Peter Mark Richman, Cesar Romero; *Dir:* Ferde Grofe Jr. **VHS, Beta** $79.98 *MAG ♫*

Judgment at Nuremberg It's 1948 and a group of high-level Nazis are on trial for war crimes. Chief Justice Tracy must resist political pressures as he presides over the trials. Excellent performances throughout, especially by Dietrich and Garland. Considers to what extent an individual may be held accountable for actions committed under orders of a superior officer. Consuming account of the Holocaust and World War II; deeply moving and powerful. Based on a "Playhouse 90" television program. Academy Award nominations for Best Picture, actor Tracy, supporting actor Clift, supporting actress Garland, director, cinematography, art direction, set design and film editing.
1961 178m/B Spencer Tracy, Burt Lancaster, Richard Widmark, Montgomery Clift, Maximilian Schell, Judy Garland, Marlene Dietrich, William Shatner; *Dir:* Stanley Kramer. Academy Awards '61: Best Actor (Schell), Best Adapted Screenplay; National Board of Review Awards '61: 10 Best Films of the Year; New York Film Critics Awards '61: Best Actor (Schell). **VHS, Beta, LV** $29.98 *MGM, FOX, BTV ♫♫♫♫*

Judy Garland Scrapbook A compilation of vintage Garland appearances from television specials, film trailers and rare screen tests.
1987 60m/B Judy Garland. **VHS, Beta** $19.95 *DVT ♫♫♫♫*

Juggernaut Not the classic disaster film of the 70s. A young woman hires a sinister doctor (Karloff) to murder her wealthy husband. The doctor, who happens to be insane, does away with the husband and then goes on a poisoning spree. Seems to drag on forever.
1937 64m/B Boris Karloff, Mona Goya, Arthur Margetson. **VHS, Beta** $19.95 *NOS, SNC, VYY ♫½*

Juggernaut Well-done drama about a doomed luxury liner. A madman plants bombs aboard a cruise ship and mocks the crew about his plan over the wireless. The countdown ensues, while the bomb experts struggle to find the explosives. Suspenseful with good direction.

1974 (PG-13) 109m/C *GB* Richard Harris, Omar Sharif, David Hemmings, Anthony Hopkins, Shirley Knight, Ian Holm, Roy Kinnear, Freddie Jones; *Dir:* Richard Lester. **VHS** $59.98 *FOX ♫♫♫*

Juggler of Notre Dame A hobo and a street juggler embark on an unusual journey that will change the lives of the people they visit.
1984 110m/C Carl Carlson, Patrick Collins, Melinda Dillon, Merlin Olsen, Gene Roche. **VHS, Beta** $69.95 *BVV ♫½*

Juice The day-to-day street life of four Harlem youths is depicted as they try to earn respect ("juice") in their neighborhood. Q, an aspiring deejay, is talked into a robbery by his friends but everything takes a turn for the worse when one of the others, Bishop, gets hold of a gun. The gritty look and feel of the drama comes naturally to Dickerson in his directorial debut. Prior to his first film, Dickerson served as cinematographer for Spike Lee's "Do the Right Thing" and "Jungle Fever."
1992 (R) 95m/C Omar Epps, Jermaine Hopkins, Tupac Shakur, Ernest R. Dickerson; *W/Dir:* Ernest R. Dickerson. **VHS** *PAR ♫♫½*

Juke Joint "Amos 'n Andy" star Spencer Williams acted in and directed this tale of two men who arrive in Hollywood with only 25 cents between them. Features an all-black cast and musical numbers.
1947 60m/B Spencer Williams, Judy Jones, Mantan Moreland; *Dir:* Spencer Williams. **VHS, Beta** $39.95 *NOS, VCN, DVT ♫½*

Jules and Jim The story of a friendship between two men, one German and the other French and their twenty-year love for the same woman. Adapted from the novel by Henri-Pierre Roche. English subtitles.
1962 104m/B *FR* Jeanne Moreau, Oskar Werner, Henri Serre, Marie DuBois, Vanna Urbino; *Dir:* Francois Truffaut. Mar Del Plata Festival '62: Best Director; Sight & Sound Survey '82: #14 of the Best Films of All Time (tie). **VHS, Beta, LV** $59.98 *FOX, APD ♫♫♫♫*

Julia A woman tries to lose her virginity while vacationing in the Swiss Alps.
1974 (R) 83m/C Sylvia Kristel, Jean-Claude Bouillon, Terry Torday. **VHS, Beta** $59.95 *GEM Woof!*

Julia The story recounted in Lillian Hellman's fictional memoir "Pentimento." Fonda plays Hellman as she risks her life smuggling money into Germany during World War II for the sake of Julia, her beloved childhood friend (Redgrave) who is working in the Resistance. All cast members shine in their performances; watch for Meryl Streep in her screen debut. Oscar nominations for Best Picture, Fonda, Schell, Director, Cinematography, Score, and Film Editing and Costume Design.
1977 (PG) 118m/C Jane Fonda, Jason Robards Jr., Vanessa Redgrave, Maximilian Schell, Hal Holbrook, Rosemary Murphy, Meryl Streep, Lisa Pelikan; *Dir:* Fred Zinneman. Academy Awards '77: Best Adapted Screenplay, Best Supporting Actor (Robards), Best Supporting Actress (Redgrave); National Board of Review Awards '77: 10 Best Films of the Year; New York Film Critics Awards '77: Best Supporting Actor (Schell). **VHS, Beta, LV** $69.98 *FOX, BTV ♫♫♫½*

Julia and Julia A beautiful American woman in Trieste is tossed between two seemingly parallel dimensions, one in which her husband died six years earlier in a car crash, and the other in which he didn't. A purposefully obscure Italian-made psychological thriller, filmed in high-definition video

and then transferred to film stock, it has a very different look and feel from other films of the genre. Intriguing and engaging most of the time, slow moving and confusing some of the time, it ultimately challenges but fails adequately to reward the viewer.
1987 (R) 98m/C Kathleen Turner, Sting, Gabriel Byrne, Gabriel Ferzetti, Angela Goodwin; *Dir:* Peter Del Monte. **VHS, Beta, LV $19.98** *FOX* 🐾🐾½

Julia has Two Lovers Minimalist comedy-drama about a woman torn between two men—her dull fiancee and a mysterious phone caller. Mostly a long phone conversation; don't expect blistering action.
1991 (R) 87m/C Daphna Kastner, David Duchovny, David Charles; *Dir:* Bashar Shbib. **VHS $89.98** *SOU* 🐾🐾

Juliet of the Spirits Fellini uses the sparse story of a woman (Fellini's real-life wife) deliberating over her husband's possible infidelity to create a wild, often senseless surrealistic film. With a highly symbolic internal logic and complex imagery, Fellini's fantasy ostensibly elucidates the inner life of a modern woman.
1965 142m/C *IT* Giulietta Masina, Valentina Cortese, Sylva Koscina, Mario Pisu; *Dir:* Federico Fellini. National Board of Review Awards '65: Best Foreign Film; New York Film Critics Awards '65: Best Foreign Film. **VHS, Beta $69.95** *CVC, MRV, VDM* 🐾🐾🐾

Julius Caesar The all-star version of the Shakespearean tragedy, heavily acclaimed and deservedly so. Working directly from the original Shakespeare, director Mankiewicz produced a life-like, yet poetic production. Brando was Best Actor Oscar-nominated, as was the film, the cinematography (black & white) and the score.
1953 121m/B James Mason, Marlon Brando, John Gielgud, Greer Garson, Deborah Kerr, Louis Calhern, Edmond O'Brien, George Macready, John Hoyt, Michael Pate; *Dir:* Joseph L. Mankiewicz; *W/Dir:* Joseph L. Mankiewicz. Academy Awards '53: Best Art Direction/Set Decoration (B & W); British Academy Awards '53: Best Actor (Brando), Best Actor (Gielgud). **VHS, Beta $19.95** *MGM* 🐾🐾🐾½

Julius Caesar Subpar adaptation of the Shakespeare play about political greed and corruption within the Roman Empire.
1970 116m/C *GB* Charlton Heston, John Gielgud, Jason Robards Jr., Richard Chamberlain, Robert Vaughn, Diana Rigg; *Dir:* Stuart Burge. **VHS, Beta $19.98** *NOS, REP, APD* 🐾🐾

Julius Caesar A drama of power, assassination, and revenge set in ancient Rome as complicated human beings are shown in conflict with one another and themselves. Part of the television series "The Shakespeare Plays."
1979 161m/C Richard Pasco, Keith Mitchell, Charles Gray. **VHS, Beta, LV $49.95** *TLF* 🐾🐾🐾

Julius Caesar A performance of Handel's opera, which is based on Shakespeare. All the roles are sung by women.
1984 220m/C Janet Baker, Valerie Masterson, Sarah Walker, Della Jones. **VHS, Beta $49.95** *HMV, HBO, MVD* 🐾🐾🐾

July Group A Quaker family's peaceful existence is threatened when kidnappers invade their home and hold them for ransom.
197? 75m/C Nicholas Campbell, Calvin Butler, Maury Chaykin. **VHS, Beta $59.95** *TWE* 🐾½

Jumper An abstracted, theatrical narrative about a cast of tragi-comic characters floundering through their meaningless lives. In German with English subtitles.
1985 90m/C *GE Dir:* Benno Trachtmann. **VHS, Beta $39.95** *FCT* 🐾🐾½

Jumpin' Jack Flash A bank worker is humorously embroiled in international espionage when her computer terminal picks up distress signals from a British agent in Russia. Marshall's directing debut. Good performances from a fun cast (which includes cameo appearances by McKean and Ullman) and a particularly energetic effort by Goldberg are held back by a rather average and predictable script.
1986 (R) 98m/C Whoopi Goldberg, Stephen Collins, Carol Kane, Annie Potts, Jonathan Pryce, James Belushi, Jon Lovitz, Michael McKean, Tracy Ullman; *Dir:* Penny Marshall. **VHS, Beta, LV $19.98** *FOX* 🐾🐾½

Junction 88 This is an early all black musical chock full of great dancing and hot music.
1940 60m/B Noble Sissle, Bob Howard. **VHS, Beta $24.95** *NOS, DVT* 🐾

June Night A woman who was victimized by a shooting incident cannot escape the public eye due to her former promiscuous behavior.
1940 90m/C *SW* Ingrid Bergman, Marianne Lofgren, Gunnar Sjoberg, Olaf Widgren; *Dir:* Per Lindberg. **VHS, Beta, LV $29.95** *WAC, CRO, LUM* 🐾🐾

Jungle A princess and an American adventurer lead an expedition into the Indian wilds to discover the source of recent elephant attacks.
1952 74m/B Rod Cameron, Cesar Romero, Marie Windsor; *Dir:* William Berke. **VHS, Beta, LV** *WGE* 🐾

Jungle Assault Two Vietnam vets with nothing left to lose are called back into the bloodiest action of their lives. Their assignment is to locate and destroy a terrorist base and bring back an American general's brainwashed daughter alive.
1989 86m/C William Smith, Ted Prior, William Zipp; *Dir:* David A. Prior. **VHS, Beta** *AIP* 🐾

Jungle Book A lavish version of Rudyard Kipling's stories about Mowgli, the boy raised by wolves in the jungles of India. Musical score by Miklos Rosza.
1942 109m/C Sabu, Joseph Calleia, Rosemary de Camp, Ralph Byrd, John Qualen; *Dir:* Zoltan Korda. **VHS, LV $9.95** *NEG, MRV, REP* 🐾🐾

The Jungle Book Based on Kipling's classic, a young boy raised by wolves must choose between his jungle friends and human "civilization." Along the way he meets a variety of jungle characters including zany King Louie, kind-hearted Baloo, wise Bagherra and the evil Shere Khan. Great, classic songs including "Trust in Me," "I Wanna Be Like You," and Oscar nominated "Bare Necessities." Last Disney feature overseen by Uncle Walt himself and a must for kids of all ages.
1967 78m/C *Dir:* Wolfgang Reitherman; *Voices:* Phil Harris, Sebastian Cabot, Louis Prima, George Sanders, Sterling Holloway, J. Pat O'Malley, Verna Felton, Darlene Carr. **VHS, Beta $24.99** *DIS, BVV, FCT* 🐾🐾🐾

Jungle Bride Three people are shipwrecked and one of them is a suspected murderer.
1933 66m/B Anita Page, Charles Starrett. **VHS, Beta, LV $16.95** *SNC, NOS, WGE* 🐾🐾½

Jungle Cat This major Disney wildlife film depicts the South American Jaguar in its original habitat.
1959 69m/C *Dir:* James Algar. **VHS, Beta $69.95** *DIS* 🐾🐾½

Jungle Cavalcade Naturalist Frank Buck traveled deep into the jungle to photograph wild animals in their natural surroundings.
1941 80m/B *Dir:* Clyde Elliot. **VHS, Beta $19.95** *NOS, CAB* 🐾🐾½

Jungle Drums of Africa Jungle adventures abound as Moore and Coates encounter lions, wind tunnels, voodoo and enemy agents in deepest Africa. A 12-episode serial re-edited onto two cassettes. Also known as "U-238 and the Witch Doctor."
1953 167m/B Clayton Moore, Phyllis Coates, Roy Glenn, John Cason; *Dir:* Fred Brannon. **VHS REP, MLB** 🐾½

Jungle Fever A married black architect's affair with his white secretary provides the backdrop for a cold look at an interracial love affair. Lee focuses more on the discomfort of the friends and families involved than with the intense world created by the lovers for themselves. The movie provides the quota of humor and fresh insight we expect from Lee, but none of the joyous sexuality experienced by the black lovers in "She's Gotta Have It." In fact, Lee tells us that interracial love is unnatural, never more than skin deep. His lovers have "Jungle Fever," a blind obsession with the allure of the opposite race. A very fine cast but if you don't agree with Lee's viewpoint, a real disappointment as well.
1991 (R) 131m/C Wesley Snipes, Annabella Sciorra, John Turturro, Samuel L. Jackson, Ossie Davis, Ruby Dee, Lonette McKee, Anthony Quinn; *W/Dir:* Spike Lee. Cannes Film Festival '91: Best Supporting Actor (Jackson); National Board of Review Awards '91: 10 Best Films of the Year; New York Film Critics Awards '91: Best Supporting Actor (Jackson). **VHS, Beta, LV $94.95** *MCA, CCB* 🐾🐾🐾

Jungle Goddess Two mercenaries set out for the reward offered for finding of a wealthy heiress last seen in the jungles of Africa.
1949 61m/B George Reeves, Wanda McKay, Ralph Byrd, Armida; *Dir:* Lewis D. Collins. **VHS, Beta, LV** *WGE* 🐾

Jungle Heat While searching for a pygmy tribe, an anthropologist and her pilot run into a pack of anthropoid mutants. Originally released as "Dance of the Dwarfs."
1984 (PG) 93m/C Peter Fonda, Deborah Raffin; *Dir:* Gus Trikonis. **VHS, Beta $59.95** *TWE* 🐾½

Jungle Hell Sabu comes to the rescue of an Indian tribe being harassed by flying saucers, death rays and radioactive debris.
1955 ?m/C Sabu, David Bruce, George E. Stone, K.T. Stevens; *Dir:* Norman A. Cerf. **VHS, Beta $16.95** *SNC* 🐾🐾

Jungle Holocaust One member of a scientific team survives an attack by a cannibal tribe and in an effort of self-preservation becomes one of them.
1985 90m/C **VHS, Beta $59.95** *VCD* 🐾

Jungle Inferno Nature boy, a fugitive from the government, fights lions, tigers and pursuing feds in this story of escape and survival.
197? 90m/C Brad Harris. **VHS, Beta** *WES* 🐾

J

Jungle Master Noble British types hunt for Karzan, the jungle's number one guy, with the aid of a tribal priestess and a lovely photo-journalist. A must see for trash movie fans.
1956 90m/C John Kitzmiller, Simone Blondell, Edward Mann; ***Dir:*** Miles Deem. **VHS, Beta $69.95** *FOR* 🦴

Jungle Patrol In 1943 New Guinea, a squadron of entrapped fliers are confronted with a beautiful USO entertainer. Romance and show tunes follow.
1948 72m/B Kristine Miller, Arthur Franz, Richard Jaeckel, Ross Ford, Tommy Noonan, Gene Reynolds; ***Dir:*** Joseph M. Newman. **VHS, Beta $39.95** *SVS* 🦴½

Jungle Raiders An Indiana Jones-esque mercenary searches the steamy jungles of Malaysia for a valuable jewel, the Ruby of Gloom.
1985 (PG-13) 102m/C Lee Van Cleef, Christopher Connelly, Marina Costa; ***Dir:*** Anthony M. Dawson. **VHS, Beta $79.95** *MGM* 🦴½

Jungle Siren Nazis in Africa try to foment rebellion amongst the black natives against the white residents. Crabbe and a white woman raised in the jungle set things right.
1942 68m/B Ann Corio, Buster Crabbe, Evelyn Wahl, Milton Kibbee; ***Dir:*** Sam Newfield. **VHS, Beta $16.95** *SNC, MLB* 🦴🦴

Jungle Warriors Seven fashion models are abducted by a Peruvian cocaine dealer. To escape him, they must become Jungle Warriors.
1984 (R) 96m/C *GE MX* Sybil Danning, Marjoe Gortner, Nina Van Pallandt, Paul Smith, John Vernon, Alex Cord, Woody Strode, Kai Wulff; ***Dir:*** Ernst R. von Theumer. **VHS, Beta $59.95** *MED* 🦴½

Junior A raving, drooling lunatic cuts up girls with a chainsaw.
1986 80m/C Suzanne DeLaurentis, Linda Singer, Jeremy Ruthford, Michael McKeever. **VHS, Beta $9.99** *STE, PSM Woof!*

Junior Bonner A rowdy modern-day western about a young drifting rodeo star who decides to raise money for his father's new ranch by challenging a formidable bull.
1972 (PG) 100m/C Steve McQueen, Robert Preston, Ida Lupino, Ben Johnson, Joe Don Baker, Barbara Leigh; ***Dir:*** Sam Peckinpah. **VHS, Beta $19.98** *FOX* 🦴🦴🦴

Junior G-Men The Dead End Kids fight Fifth Columnists who are trying to sabotage America's war effort. Twelve episodes.
1940 237m/B Billy Halop, Huntz Hall. **VHS, Beta $49.95** *NOS, VCN, VYY* 🦴🦴

Junior G-Men of the Air The Dead End Kids become teen-age flyboys in this twelve-episode serial adventure.
1942 215m/B Billy Halop, Huntz Hall. **VHS, Beta $26.95** *SNC, NOS, MED* 🦴

Junior High School A comically, on target film detailing the romantic yearnings, petty cruelties, and claustrophobia of 8th grade existence.
1981 60m/C Paula Abdul. **VHS, Beta** *PIC* 🦴🦴½

Junkman A movie maker whose new film is about to be premiered is being chased by a mysterious killer. The ultimate car chase film, this production used and destroyed over 150 automobiles.
1982 (R) 97m/C H.B. Halicki, Christopher Stone, Susan Shaw, Hoyt Axton, Lynda Day George, Freddy Cannon & the Belmonts; ***Dir:*** H.B. Halicki. **VHS, Beta $18.95** *TWE* 🦴½

Juno and the Paycock Perhaps if Hitchcock hadn't been so faithful to O'Casey this early effort would have been less stagey and more entertaining. In Dublin during the civil uprising, a poor family is torn apart when they receive news of an imminent fortune. Hitchcock's reported to have said that "drama is life with the dull bits left out," though here too many dull bits remain.
1930 96m/B Sara Allgood, Edward Chapman, John Longden, John Laurie, Maire O'Neill; ***Dir:*** Alfred Hitchcock; ***Nar:*** Barry Fitzgerald. **VHS $19.98** *NOS, VEC, HHT* 🦴🦴½

Jupiter Menace An examination of speculative theories dealing with the inevitable end of the world. Kennedy predicts a continuing cycle of unnatural occurrences and disasters which will culminate with the tilting of the earth's axis in the year 2000.
1982 84m/C George Kennedy, Greg Michaels; ***Dir:*** Lee Auerbach, Peter Matulavich. **VHS, Beta $69.99** *HBO* 🦴🦴½

Jupiter's Darling This spoof of Hannibal (Keel) and Amytis (Williams) gravely misses the mark in making funny the world of the Roman Empire. Amytis has the job of distracting Hannibal from attacking the Eternal City, and does so through musical interludes and unfunny jokes.
1955 96m/C Esther Williams, Howard Keel, George Sanders, Gower Champion, Marge Champion, Norma Varden, Richard Haydn, William Demarest, Douglas Dumbrille, Michael Ansara, Martha Wentworth, Chris Alcaide, William Tannen; ***Dir:*** George Sidney. **LV $34.98** *MGM* 🦴🦴

Jupiter's Thigh A madcap married pair of treasure-hunters honeymoon in Greece in search of an ancient statue's lost thigh. Available in French with English subtitles, or dubbed into English.
1981 96m/C *FR* Annie Girardot, Philippe Noiret; ***Dir:*** Phillipe DeBroca. **VHS, Beta $29.98** *SUE, FCT* 🦴🦴½

Just Another Pretty Face An aging detective catches a gang of jewel smugglers, and gets involved with their young moll. In French with English subtitles. Also titled "Be Beautiful but Shut Up."
1958 110m/B *FR* Henri Vidal, Mylene Demongeot, Isabelle Miranda; ***Dir:*** Henri Verneuil. **VHS, Beta $19.95** *FCT* 🦴🦴½

Just Around the Corner Temple helps her Depression-poor father get a job after she befriends a cantankerous millionaire. Among others, she sings "This Is a Happy Little Ditty."
1938 70m/B Shirley Temple, Charles Farrell, Bert Lahr, Joan Davis, Bill Robinson, Cora Witherspoon, Franklin Pangborn; ***Dir:*** Irving Cummings. **VHS, Beta $19.98** *FOX* 🦴½

Just Before Dawn Another murderers stalk campers story; humans resort to their animal instincts in their struggle for survival. Not to be confused with the William Castle film (1946) from the "Crime Doctor" series.
1980 (R) 90m/C Chris Lemmon, Deborah Benson, Greg Henry, George Kennedy; ***Dir:*** Jeff Lieberman. **VHS, Beta** *PGN* 🦴🦴

Just Between Friends Two women become friends, not knowing that one is having an affair with the husband of the other. Allan Burns directorial debut.
1986 (PG-13) 110m/C Mary Tyler Moore, Christine Lahti, Sam Waterston, Ted Danson, Jim Mackrell, Jane Greer; ***Dir:*** Allan Burns. **VHS, Beta $79.99** *HBO* 🦴🦴

Just Call Me Kitty A program of fun facts and trivia about cats, big and small, wild and domesticated.
1986 60m/C VHS, Beta $14.95 *UNI* 🦴🦴

Just a Gigolo Bowie stars in this unusual melodrama about a Prussian vet turned male prostitute. He spends most of his time working for the sexiest of women. Splendid cast, but a bit incoherent.
1979 (R) 105m/C *GE* David Bowie, Sydne Rome, Kim Novak, David Hemmings, Maria Schell, Curt Jurgens, Marlene Dietrich; ***Dir:*** David Hemmings. **VHS, LV** *WBF, MOV, LUM* 🦴🦴½

Just for the Hell of It A quartet of teenage punks ruthlessly terrorize their suburban Miami 'hood while an innocent kid gets blamed. More exploitation from schlock king Lewis, who also wrote the theme song, "Destruction Inc" (pic's alternate title). Essentially the same cast as the director's "She Devils on Wheels," filmed simultaneously.
1968 85m/C Rodney Bedell, Ray Sager, Nancy Lee Noble, Agi Gyenes, Steve White; ***Dir:*** Herschell Gordon Lewis. **VHS $20.00** *SMW Woof!*

Just Hold Still Short film collection from Cohen includes "Never Change," set to a poem by Blake Nelson, and "Love Teller," about a carnival machine that reveals a person's level of passion.
19?? ?m/C ***W/Dir:*** Jem Cohen. **VHS** *FCT*

Just Like Us A made-for-television program designed for adolescents, depicting the growing friendship, and problems, of a very rich girl and the daughter of her cook.
1983 55m/C Jennifer Jason Leigh, Karli Michaelson, Marion Ross, Carol Lawrence. **VHS, Beta $39.95** *GEM* 🦴

Just Like Weather Contemporary Hong Kong and the city's looming return to control by mainland China provide the backdrop for the troubled marriage of a young couple. Couple endure abortion, arrest, and veterinarian episodes before departing for on U.S. trip with high hopes.
1986 98m/C *HK* Christine Lee, Lee Chi-Keung, Allen Fong; ***Dir:*** Allen Fong. **VHS $39.95** *FCT* 🦴🦴🦴

Just Me & You An "It Happened One Night" made-for-television tale of an unlikely couple who fall in love with each other when chance brings them together on a cross-country drive. Written by Lasser.
1978 100m/C Louise Lasser, Charles Grodin; ***Dir:*** Charles Erman. **VHS, Beta $59.95** *VCL* 🦴½

Just One of the Guys When the school newspaper refuses to accept the work of an attractive young girl, she goes undercover as a boy to prove that her work is good. She goes on to befriend the school's nerd, and even helps him grow out of his awkward stage, falling for him in the process. Very cute, but predictable.
1985 (PG-13) 100m/C Joyce Hyser, Clayton Rohner, Billy Jacoby, Toni Hudson, Leigh McCloskey, Sherilyn Fenn; ***Dir:*** Lisa Gottlieb. **VHS, Beta, LV $79.95** *COL* 🦴🦴

Just Plain Daffy A compilation of pre-1948 Daffy Duck cartoon favorites.
1948 60m/C ***Dir:*** Bob Clampett, Robert McKimson, Friz Freleng, Frank Tashlin. **VHS, Beta $14.95** *MGM* 🦴🦴🦴½

Just Suppose Prince in waiting decides he'd prefer the simple life, travels to the land of the free in search of honest work and is felled by cupid's arrow. All's well 'til duty calls, and his princely presence is expected back in the royal fold. Features beautiful boy Barthelmess, whom Lillian Gish described as

having "the most beautiful face of any man who ever went before a camera." Silent.
1926 90m/B Richard Barthelmess, Lois Moran, Geoffrey Kerr, Henry Vibart, George Spelvin; *Dir:* Kenneth Webb. **VHS, Beta $19.95** GPV 🎬🎬½

Just Tell Me What You Want A wealthy, self-made married man finally drives his longtime mistress away when he refuses to let her take over the operation of a failing movie studio he has acquired. After she falls for another man, the tycoon does everything he can to win her back. The department store battle between MacGraw and King is priceless.
1980 (R) 112m/C Alan King, Ali MacGraw, Myrna Loy, Keenan Wynn, Tony Roberts; *Dir:* Sidney Lumet. **VHS, Beta $19.98** WAR 🎬🎬½

Just Tell Me You Love Me Three budding con artists plot to make easy money in this Hawaiian romp.
1980 (PG) 90m/C Robert Hegyes, Debralee Scott, Lisa Hartman, Ricci Martin, June Lockhart; *Dir:* Tony Mordente. **VHS, Beta $69.98** LIV, VES 🎬

Just the Way You Are An attractive musician struggles to overcome a physical handicap and winds up falling in love while on vacation in the French Alps.
1984 (PG) 96m/C Kristy McNichol, Robert Carradine, Kaki Hunter, Michael Ontkean, Alexandra Paul, Lance Guest, Timothy Daly, Patrick Cassidy; *Dir:* Edouard Molinaro. **VHS, Beta $79.95** MGM 🎬

Just William's Luck A precocious English brat sneaks into an old mansion, which just happens to be the headquarters for a gang of thieves. Based on British series of children's books.
1947 87m/B GB William A. Graham, Garry Marsh; *Dir:* Val Guest. **VHS, Beta $29.95** VYY 🎬🎬

Justice A crusading attorney tries to keep a waterfront kangaroo court from applying its harsh justice to an admitted killer. A television crime-drama also titled "Flight from Fear."
1955 26m/B William Prince, Jack Klugman, Biff McGuire, Jack Warden. **VHS, Beta $19.95** VYY 🎬🎬

Justice Rides Again Tom Mix tries to uphold law and order, but he's up against some very tough outlaws.
193? 55m/B Tom Mix. **VHS, Beta $29.95** VCN 🎬

Justice of the West The Lone Ranger and his sidekick, Tonto, perform good deeds in the Old West, including helping to retrieve stolen gold, helping to build an Indian school, and giving a blind man a fresh perspective on life.
1961 71m/C Clayton Moore, Jay Silverheels; *Dir:* Earl Bellamy. **VHS, Beta $29.95** MGM 🎬

Justicia para los Mexicanos A gang of outlaws make a big mistake when they try to blame Mexican citizens for their crimes. The righteous folks show they're not pushovers.
19?? ?m/C MX Tony Kendall. **VHS, Beta** MAD 🎬

Justin Morgan Had a Horse The true story of a colonial school teacher in post-Revolutionary War Vermont who first bred the Morgan horse, the first and most versatile American breed.
1981 91m/C Don Murray, Lana Wood, Gary Crosby; *Dir:* Hollingsworth Morse. **VHS, Beta $69.95** DIS 🎬🎬

Justine In the 1930s, a prostitute who marries an Egyptian banker becomes involved with a variety of men and a plot to arm Palestinian Jews in their revolt against English rule. A condensed film version of Lawrence Durrell's, "The Alexandria Quartet."
1969 115m/C IS Anouk Aimee, Michael York, Dirk Bogarde, Philippe Noiret, Michael Constantine, John Vernon, Jack Albertson; *Dir:* George Cukor. **VHS, Beta $59.98** FOX, HHE 🎬🎬

Juve Contre Fantomas A pioneering, highly influential action serial about a super-villain/ master of disguises and the police inspector out to stop him. Filmed and edited with amazing pre-"Birth of a Nation" brilliance, this series has been hailed by the Surrealists and recognized by critics as a foundation in the development of film syntax. Silent.
1914 64m/B Rene Navarre, Edmund Breon, Georges Melchior; *Dir:* Louis Feuillade. **VHS, Beta $24.95** NOS, MRV, FCT 🎬🎬🎬

K-9 After having his car destroyed by a drug dealer, "I work alone" Belushi is forced to take on a partner—a German Shepherd. Together they work to round up the bad guys and maybe chew on their shoes a little. Sometimes amusing one-joke comedy done in by a paper-thin script. Both the dog and Belushi are good, however.
1989 (PG-13) 111m/C James Belushi, Mel Harris; *Dir:* Rod Daniel. **VHS, Beta, LV $19.95** MCA 🎬🎬½

K-9000 The Hound salutes the idea behind this standard sci-fi crime-fighter; a cyberdog fights the forces of evil with the aid of a cop, a lady reporter, and the usual cliches.
1989 96m/C Chris Mulkey, Catherine Oxenberg; *Dir:* Kim Manners. **VHS $64.95** FRH 🎬🎬

K2 Two men, one a skirt-chasing lawyer, the other a happily married physicist, tackle the world's second largest mountain—the K-2 in Kashmir, northern Pakistan. They encounter a number of dangers, including an ascent of sheer rock face, an avalanche, and a fall down perpendicular mountain ice, but even the exciting mountain-climbing scenes can't hold up this watered down movie. The film was actually shot on Canada's Mount Waddington. Based on the play by Patrick Meyers.
1992 (R) 104m/C Michael Biehn, Matt Craven, Raymond J. Barry, Luca Bercovici, Patricia Charbonneau, Julia Nickson-Soul, Hiroshi Fujioka, Jamal Shah; *Dir:* Franc Roddam. **VHS, Beta** PAR 🎬🎬

Kagemusha A thief is rescued from the gallows because of his striking resemblance to a warlord in 16th Century Japan. When the ambitious warlord is fatally wounded, the thief is required to pose as the warlord. In Japanese with English subtitles. Also known as "The Shadow Warrior."
1980 (PG) 160m/C JP Tatsuya Nakadai, Tsutomu Yamazaki, Kenichi Hagiwara; *Dir:* Akira Kurosawa. British Academy Awards '80: Best Costume Design, Best Director (Kurosawa); National Board of Review Awards '80: 5 Best Foreign Films of the Year. **VHS, Beta $29.98** FOX, APD 🎬🎬🎬½

Kajagoogoo Features the band Kajagoogoo performing their songs "Too Shy," "Ooh to Be An," and "Hang on Now."
1983 11m/C Kajagoogoo. **VHS, Beta, LV $9.95** SVS 🎬🎬🎬½

Kameradschaft A great, early German sound film about Germans struggling to free themselves and French miners trapped underground on the countries' border. In German and French with English subtitles.
1931 80m/B GE Ernst Busch, Alexander Granach, Fritz Kampers; *Dir:* G. W. Pabst. National Board of Review Awards '32: 10 Best Foreign Films of the Year. **VHS, Beta, 3/4U $29.98** SUE, MRV, IHF 🎬🎬🎬½

Kamikaze A Japanese World War II action drama about suicide pilots.
1945 100m/B JP **VHS $14.99** VCD, RXM 🎬

Kamikaze '89 German director Fassbinder has the lead acting role (his last) in this offbeat story of a police lieutenant in Berlin, circa 1989, who investigates a puzzling series of bombings. Music by Tangerine Dream. In German with English subtitles.
1983 90m/C GE Rainer Werner Fassbinder, Gunther Kaufman, Boy Gobert; *Dir:* Wolf Gremm. **VHS, Beta $59.95** MGM, FCT, GLV 🎬🎬🎬

Kanako Kanako Higuchi, one of Japan's most famous and adored actresses, is presented by renowned photographer Kishin Shinoyama in this series of nude photographs. Stereo.
19?? ?m/C Beta, LV PNR 🎬🎬🎬

Kanal Wajda's first major success, a grueling account of Warsaw patriots, upon the onset of the Nazis toward the end of the war, fleeing through the ruined city's sewers. Highly acclaimed. Polish; available with English subtitles or dubbed. Also known as "They Loved Life."
1956 96m/B PL Teresa Izewska, Tadeusz Janczar; *Dir:* Andrzej Wajda. Cannes Film Festival '57: Special Prize. **VHS, Beta $29.98** SUE 🎬🎬🎬½

Kandyland An over-the-hill stripper takes a young innocent stripper under her wing.
1987 (R) 94m/C Sandahl Bergman, Kimberly Evenson, Charles Laulette, Bruce Baum; *Dir:* Robert Allen Schnitzer. **VHS, Beta, LV $19.95** STE, NWV 🎬

Kangaroo An Australian adaptation of the semi-autobiographical D. H. Lawrence novel. A controversial English novelist and his wife move to the Outback in 1922, and are confronted with all manner of prejudice and temptation.
1986 (R) 115m/C AU Judy Davis, Colin Friels; *Dir:* Tim Burstall. Australian Film Institute '86: Best Actress (Davis). **VHS, Beta $79.95** MCA 🎬🎬🎬

The Kansan The marshall in a Kansas town won't rest until he's stamped out all traces of corruption.
1943 79m/B Richard Dix, Victor Jory, Albert Dekker; *Dir:* George Archainbaud. **VHS, Beta $19.95** NOS, MRV, IND 🎬

Kansas Live concert by the rock group Kansas.
1982 87m/C Kansas. **VHS, Beta $29.95** SVS

Kansas Two young men, one a lawless rebel, the other a rational loner, stage a bank heist, and then go on the lam from the cops. Lame plot and very weak acting combine to make this film a dud.
1988 (R) 111m/C Matt Dillon, Andrew McCarthy, Leslie Hope, Kyra Sedgwick; *Dir:* David Stevens. **VHS, Beta, LV $19.95** MED 🎬

Kansas City Confidential An ex-cop on the wrong side of the law launches a sophisticated armored car heist. A disgruntled ex-con gets arrested for the crime on

circumstantial evidence. When released, he scours the underworld for the real thieves.
1952 98m/B John Payne, Coleen Gray, Preston Foster, Neville Brand, Lee Van Cleef, Jack Elam; *Dir:* Phil Karlson. **VHS $16.95** NOS, MGM, RXM ♂♂♂

Kansas City Massacre A made-for-television film chronicling the efforts of Melvin Purvis, as he tracked down Depression-era hoods. First called "Melvin Purvis, G-Man."
1975 99m/C Dale Robertson, Sally Kirkland, Bo Hopkins, Scott Brady, Lynn Loring, Matt Clark, Mills Watson, Robert Walden; *Dir:* Dan Curtis. **VHS, Beta $59.95** VMK ♂ 1/2

Kansas Pacific A group of Confederate sympathizers try to stop the Kansas Pacific Railroad from reaching the West Coast in the 1860s.
1953 73m/C Sterling Hayden, Eve Miller, Barton MacLane, Reed Hadley, Douglas Hadley; *Dir:* Ray Nazarro. **VHS, Beta $14.95** NOS, MRV, QNE ♂♂

Kapo A 14-year-old Jewish girl and her family are imprisoned by the Nazis in a concentration camp. There, the girl changes identities with the help of the camp doctor, and rises to the position of camp guard. She proceeds to become taken with her power until a friend commits suicide and jolts the girl back into harsh reality. An Academy Award nominee for Best Foreign Film (lost to "The Virgin Spring").
1959 116m/B IT FR YU Susan Strasberg, Laurent Terzieff, Emmanuelle Riva, Gianno Garko; *W/Dir:* Gillo Pontecorvo. **VHS** DVT ♂♂♂

The Karate Kid A teen-age boy finds out that Karate involves using more than your fists when a handyman agrees to teach him martial arts. The friendship that they develop is deep and sincere; the Karate is only an afterthought. From the director of the original "Rocky," this movie is easy to like.
1984 (PG) 126m/C Ralph Macchio, Pat Morita, Elizabeth Shue, Randee Heller, Martin Kove, Chad McQueen; *Dir:* John G. Avildsen. **VHS, Beta, LV, 8mm $19.95** COL ♂♂♂ 1/2

The Karate Kid: Part 2 Sequel to the first film wherein our high-kicking hero tests his mettle in real-life karate exchanges in Okinawa, and settles a long-standing score. Followed by a second sequel.
1986 (PG) 95m/C Ralph Macchio, Pat Morita, Danny Kamekona, Martin Kove, Tamlyn Tomita; *Dir:* John G. Avildsen. **VHS, Beta, LV, 8mm $19.95** RCA ♂♂ 1/2

The Karate Kid: Part 3 The second sequel, wherein the high-kicking young guy again battles an evil nemesis and learns about himself.
1989 (PG) 105m/C Ralph Macchio, Pat Morita; *Dir:* John G. Avildsen. **VHS, Beta, LV, 8mm $19.95** COL ♂♂♂

Karate Killer A Kung Fu master seeks revenge on a brutal and vicious gang.
1973 (R) 95m/C VHS, Beta $54.95 HHT ♂

Karate Polish Style Polish painter duo paint, plaster and draw blood. A sort of "Dances with Poles" in Warsaw's Mazurian Lake District. In Polish with English subtitles. Not a Polish joke.
1982 91m/C PL *Dir:* Jerzy Wojcik. **VHS $49.95** FCT ♂

Karate Rock Non-stop head-kicking, rib-cracking karate action.
1988 30m/C VHS, Beta KAR, VAL ♂

Karate Warrior Young martial artist is beaten and left for dead by Filipino crime syndicate. Bones mended, he prescribes a dose of their medicine. Plenty of rest and retaliation.
19?? 90m/C Jared Martin, Ken Watanabe. **VHS, Beta $79.98** IMP ♂ 1/2

Karate Warriors Chiba takes his battle against crime to the streets.
197? 90m/C Sonny Chiba. **VHS, Beta $79.95** IND ♂

Kashmiri Run Action adventure about trio on the lam from Chinese communists chasing them through Tibet.
1969 (R) 101m/C SP Pernell Roberts, Alexandra Bastedo, Julian Mateos, Gloria Gamata; *Dir:* John Peyser. **VHS $29.95** IVE ♂♂

Katherine A young heiress rejects her pampered lifestyle and becomes a violent revolutionary, rebelling against social injustices and the system that spawned them. Made for television. Also released on video as "The Radical."
1975 98m/C Sissy Spacek, Art Carney, Jane Wyatt, Henry Winkler, Julie Kavner; *Dir:* Jeremy Paul Kagan. **VHS, Beta, LV $19.95** STE, NWV, IME ♂

Katie's Passion In 19th century Holland, a young girl acquires fame, money, power and love through her wiles. Dubbed.
197? 90m/C NL Rutger Hauer, Monique Van De Ven; *Dir:* Paul Verhoeven. **VHS, Beta $79.95** WES ♂

Kavik the Wolf Dog A heartwarming story of a courageous dog's love and suffering for the boy he loves. Made for television.
1984 99m/C Ronny Cox, Linda Sorensen, Andrew Ian McMillian, Chris Wiggins, John Ireland; *Dir:* Peter Carter. **VHS, Beta $19.95** MED ♂ 1/2

Keaton's Cop Another cheap cop comedy involving the mistaken identity of an important mob witness.
1990 (R) 95m/C Lee Majors, Abe Vigoda, Don Rickles; *Dir:* Bob Burge. **VHS, Beta, LV $89.95** WAR, OM ♂

The Keep At the height of the Nazi onslaught, several German soldiers unleash an unknown power from a medieval stone fortress. Although technically impressive, all other aspects of the film are lacking.
1983 (R) 96m/C Scott Glenn, Alberta Watson, Juergen Prochnow, Robert Prosky, Gabriel Byrne; *Dir:* Michael Mann. **VHS, Beta, LV $79.95** PAR ♂ 1/2

Keep 'Em Flying Bud and Lou star in the this wartime morale-booster that hasn't aged well. The duo follow their barnstorming friend into flight academy; a not-too-taxing plot includes five musical numbers, two Martha Rayes (she plays twins).
1941 86m/B Bud Abbott, Lou Costello, Martha Raye; *Dir:* Arthur Lubin. **VHS, LV $14.95** MCA ♂♂

Keep It Up, Jack! A good for nothing, out-of-work actor winds up with a bordello on his hands. Lucky for him it comes with a bunch of blonde beauties to keep him company.
1975 90m/C Mark Jones, Sue Longhurst, Maggie Burton, Steve Viedor; *Dir:* Tom Parker. **VHS $39.95** ACF ♂♂

Keep My Grave Open A woman lives in an isolated house where a series of strange murders take place. She attributes them to her brother, but does he really exist?

Made cheaply, but not without style; filmed in Harrison County, Texas.
1980 (R) 85m/C Camilla Carr, Gene Ross, Stephen Tobolowsky, Ann Stafford, Sharon Bunn, Chelcie Ross; *Dir:* S.F. Brownrigg. **VHS, Beta $9.95** UNI ♂

Keep Your Mouth Shut An animated human skull is the central character in this wartime anti-gossip film.
1944 3m/C VHS, Beta NFC ♂

Keeper The patients at Underwood Asylum suffer unspeakable horrors while under the care of the Keeper.
1984 96m/C Christopher Lee, Tell Schreiber. **VHS, Beta $49.95** TWE, HHE ♂

Keeper of the City Gossett stars as a tough detective out to get a gangster killer (LaPaglia) who is roaming the streets of Chicago. LaPaglia turns in a good performance as the son of a Mafia operative who wants to rid himself of the Mafia ties that have controlled him. As he pursues his deadly course of action, a newspaper journalist tags him "The Gangster Killer." Coyote plays the crusading journalist who is always interfering in Gossett's business. First-rate performances and an intriguing story line combine to make this action-packed thriller worthwhile.
1992 (R) 95m/C Louis Gossett Jr., Peter Coyote, Anthony LaPaglia, Renee Soutendijk; *Dir:* Bobby Roth. **VHS, Beta, LV $89.98** FXV ♂♂ 1/2

Keeping On Television movie with a pro-union stance. A preacher, who is also a millworker, teams up with an organizer to try to unionize the mill. Originally produced for the PBS "American Playhouse" series.
1981 75m/C Dick Anthony Williams, Carol Kane, James Broderick, Marcia Rodd, Rosalind Cash, Carl Lee, Danny Glover, Guy Boyd; *Dir:* Barbara Kopple. **VHS $59.95** TWE ♂♂

Keeping Track Two tourists witness a murder and robbery and find the stolen $5 million on a New York-bound train. The two are relentlessly pursued by everyone, including the CIA and Russian spies.
1986 (R) 102m/C Michael Sarrazin, Margot Kidder, Alan Scarfe, Ken Pogue; *Dir:* Robin Spry. **VHS, Beta, LV $19.98** CHA ♂

Keetje Tippei A Dutch prostitute works her way out of poverty and enters a world of education and wealth. Subtitled in English.
1975 104m/C NL Monique Van De Ven, Rutger Hauer; *Dir:* Paul Verhoeven. **VHS, Beta, LV $79.95** CVC, IME ♂

Kelly's Heroes A misfit band of crooks are led by Eastwood on a daring mission: to steal a fortune in gold from behind enemy lines. In the process, they almost win World War II. Sutherland is superb, as is McLeod, in his pre-Love Boat days.
1970 (PG) 143m/C Clint Eastwood, Donald Sutherland, Telly Savalas, Gavin McLeod, Don Rickles, Carroll O'Connor, Stuart Margolin, Harry Dean Stanton; *Dir:* Brian G. Hutton. **VHS, Beta, LV $19.95** MGM, TLF ♂♂ 1/2

Kemek An eccentric chemical company owner has his mistress push a new mind-control drug on an American writer, and has them killed after discovering their subsequent affair. Unfortunately, he didn't count on the woman's ex-husband seeking revenge.
1988 (R) 82m/C David Hedison, Helmut Snider, Mary Woronov, Alexandra Stewart, Cal Haynes. **VHS, Beta** GHV ♂

Kennedy The long television mini-series biography of JFK from his inauguration to assassination.
1983 278m/C Martin Sheen, Blair Brown, Vincent Gardenia, Geraldine Fitzgerald, E.G. Marshall, John Shea; *Dir:* Richard Hartley. **VHS, Beta $14.99** STE, PSM 🎞🎞

Kennel Murder Case Debonair detective Philo Vance suspects that a clear-cut case of suicide is actually murder. Fourth Vance mystery starring Powell. Remade as "Calling Philo Vance" in 1940.
1933 73m/B William Powell, Mary Astor, Jack LaRue, Ralph Morgan, Eugene Pallette; *Dir:* Michael Curtiz. **VHS, Beta $16.95** SNC, NOS, HHT 🎞🎞

Kenneth Anger, Vol. 1: Fireworks Compilation of Anger's surrealistic works which merge images of cult figures with erotic fantasies. In "Fireworks," (1947) Anger stars as a man who discovers his growing need for perverse forms of sexual fulfillment. "Rabbit's Moon" (1950) combines an ancient Japanese legend with the interchangeable characters from the 16th Century Harlequin comedies dell'Arte. "Eaux D'Artifice" (1953) involves a strange adventure in the Tivoli Gardens of Denmark.
1947 34m/C Kenneth Anger, Gordon Gray; *Dir:* Kenneth Anger. **VHS $29.99** MFV

Kenneth Anger, Vol. 2: Inauguration of the Pleasure Dome Mystic philosopher Aleister Crowley's writings are the basis for avant gardist Anger's Dionysian fantasy. A group of wizards transform into godlike entities and perform an erotic ceremony. Anais Nin plays a love goddess.
1954 38m/C Marjorie Cameron, Anais Nin; *Dir:* Kenneth Anger. **VHS $29.99** MFV

Kenneth Anger, Vol. 3: Scorpio Rising Anger's shorts, "Kustom Kar," "Kommandos," "Puce Moment," and "Scorpio Rising" are full of machismo, sleaze and tinseltown tales.
1965 37m/C Bruce Byron, Johnny Sapienza; *Dir:* Kenneth Anger. **VHS $29.99** MFV

Kenneth Anger, Vol. 4: Invocation of My Demon Brother Collection of surrealistic visions by explicit filmmaker Kenneth Anger. In "Lucifer Rising," Satan is reincarnated as a deity of light; "Invocation of My Demon Brother" features tunes of Mick Jagger. Sex and violence aplenty.
1980 39m/C Leslie Huggins, Marianne Faithfull; *Dir:* Kenneth Anger. **VHS $29.99** MFV

Kenny Rogers as the Gambler Rogers stars as Brady Hawkes, debonair gambler searching for a son he never knew he had. Based on the Rogers song of the same name. One of the highest rated TV movies ever. Followed by three sequels.
1980 94m/C Kenny Rogers, Christine Belford, Bruce Boxleitner, Harold Gould, Clu Gulager, Lance LeGault, Lee Purcell, Noble Willingham; *Dir:* Dick Lowry. **VHS** WKV 🎞🎞½

Kenny Rogers as the Gambler, Part 2: The Adventure Continues The surprise success of the made-for-TV western based on the popular Kenny Rogers' song spawned this equally popular sequel. Rogers returns as Brady Hawkes, this time searching for his kidnapped son. Followed by two more sequels.

1983 195m/C Kenny Rogers, Bruce Boxleitner, Linda Evans, Harold Gould, David Hedison, Clu Gulager, Johnny Crawford; *Dir:* Dick Lowry. **VHS, Beta, 8mm $24.95** WKV, EKC 🎞🎞½

Kent State A made-for-television drama recounting the tragic events that took place at Kent State University in 1970, when student demonstrators faced National Guardsmen. Goldstone won an Emmy for Outstanding Direction.
1981 120m/C Talia Balsam, Ellen Barkin, Jane Fleiss, John Getz, Keith Gordon; *Dir:* James Goldstone. **VHS, Beta $39.95** MCA 🎞🎞

Kentuckian Burt Lancaster stars as a rugged frontiersman who leaves with his son to go to Texas. On their journey the two are harassed by fighting mountaineers.
1955 104m/C Burt Lancaster, Walter Matthau, Diana Lynn, John McIntire, Dianne Foster, Una Merkel, John Carradine; *Dir:* Burt Lancaster. **VHS, Beta $19.98** FOX 🎞🎞½

Kentucky Blue Streak A young jockey is framed for murder while riding at an "illegal" racetrack. Later, almost eligible for parole, he escapes from jail to ride "Blue Streak" in the Kentucky Derby.
1935 61m/B Eddie Nugent, Junior Coughlin, Patricia Scott, Ben Carter. **VHS, Beta $24.95** VYY 🎞½

Kentucky Fried Movie A zany potpourri of satire about movies, television, commercials, and contemporary society. Written by Jim Abrahams, Jerry Zucker and David Zucker, who were later known for "Airplane!"
1977 (R) 85m/C Bill Bixby, Jerry Zucker, Jim Abrahams, David Zucker, Donald Sutherland, Henry Gibson, George Lazenby; *Dir:* John Landis. **VHS, Beta, LV $59.95** MED 🎞🎞🎞

Kentucky Jubilee At the jubilee, a movie director is kidnapped and the master of ceremonies, among others, decides to find him.
1951 67m/B Jerry Colonna, Jean Porter, James Ellison, Raymond Hatton, Fritz Feld. **VHS, Beta, LV** WGE, RXM 🎞½

Kentucky Kernels A pair of down and out magicians (thirties comic duo Wheeler and Woolsey) happen upon a young boy (Little Rascal Spanky) who happens to be heir to a fortune. The three head for the rascal's Kentucky home, where they're welcomed with southern inhospitality. Much feuding and slapsticking.
1934 75m/B Bert Wheeler, Robert Woolsey, Mary Carlisle, George "Spanky" McFarland, Noah Beery, Lucille LaVerne, Willie Best; *Dir:* George Stevens. **VHS, LV $19.98** TTC, FCT 🎞🎞

Kentucky Rifle A Comanche Indian tribe will let a group of stranded pioneers through their territory only if they agree to sell the Kentucky rifles aboard their wagon.
1955 80m/C Chill Wills, Lance Fuller, Cathy Downs, Jess Barker, Sterling Holloway, Jeanne Cagney; *Dir:* Carl K. Hittleman. **VHS, Beta $29.95** MON, HHE 🎞½

Kerouac Documentary frames the life of the beat poet Jack Kerouac with excerpts from his appearance on the Steve Allen Show in 1959. The film includes interviews with many of Kerouac's contemporaries.
1984 90m/C Lawrence Ferlinghetti, Allen Ginsberg, William S. Burroughs, Steve Allen, Jack Kerouac. **VHS, Beta $39.95** AHV

The Key A long, slow World War II drama about the key to an Italian girl's apartment that gets passed from tugboat skipper to tugboat skipper before dangerous missions.

Ultimately she finds true love, or does she? Based on the novel "Stella" by Jan de Hartog.
1958 134m/B *GB* William Holden, Sophia Loren, Trevor Howard, Oscar Homolka, Kieron Moore; *Dir:* Carol Reed. British Academy Awards '58: Best Actor (Howard). **VHS, Beta, LV $69.95** COL 🎞🎞½

Key Exchange Kevin Scott and Paul Kurta based this contemporary look at love and commitments on Kevin Wade's popular play. Two New York City "yuppies" have reached a point in their relationship where an exchange of apartment keys commonly occurs—but they are hesitant.
1985 (R) 96m/C Brooke Adams, Ben Masters, Daniel Stern, Tony Roberts, Danny Aiello, Annie Golden; *Dir:* Barnet Kellman. **VHS, Beta $79.98** FOX 🎞🎞½

Key Largo A gangster melodrama set in Key West, Florida, where hoods take over a hotel in the midst of a hurricane. Based on a play by Maxwell Anderson.
1948 101m/B Humphrey Bogart, Lauren Bacall, Claire Trevor, Edward G. Robinson, Lionel Barrymore; *Dir:* John Huston. Academy Awards '48: Best Supporting Actress (Trevor). **VHS, Beta, LV $19.98** MGM, FOX, PIA 🎞🎞🎞

The Key Man A radio show host manages to get mixed up with gangsters after recreating a crime on the air. He uses his knowledge of crime to thwart the crooks.
1957 63m/C *GB* Lee Patterson, Hy Hazell, Colin Gordon, Philip Leaver, Paula Byrne; *Dir:* Montgomery Tully. **VHS, Beta $16.95** NOS, SNC 🎞½

The Key to Rebecca The Nazis and the British go head-to-head in war torn North Africa. As the Germans push their way across Egypt, they find an unexpected ally in a half-German, half-Arab killer behind the British lines. Tense, well made thriller, originally a mini-series. On two cassettes.
1985 190m/C David Soul, Cliff Robertson, Robert Culp, Season Hubley, Lina Raymond, Anthony Quayle, David Hemmings; *Dir:* David Hemmings. **VHS $89.95** WOV 🎞🎞½

Key to Vengeance Lone wolf battles bad guy with wealth of weapons. Vengeance is mine saith the guy with the most toys.
19?? 90m/C Patrick L'Argent, Laura Yang. **VHS** AVD 🎞½

The Keys of the Kingdom An earnest adaptation of A. J. Cronin's novel about a young Scottish missionary spreading God's word in 19th Century China. Oscar-nominated for Best Actor (Peck), and Black & White Cinematography.
1944 137m/B Gregory Peck, Thomas Mitchell, Edmund Gwenn, Vincent Price, Roddy McDowall, Cedric Hardwicke, Peggy Ann Garner, James Gleason, Anne Revere, Rose Stradner, Sara Allgood, Abner Biberman, Arthur Shields; *Dir:* John M. Stahl. **VHS, Beta, LV $19.98** FOX, IGP 🎞🎞🎞

The KGB: The Secret War Reheated Cold War fare.
19?? 90m/C Sally Kellerman, Michael Ansara, Michael Billington. **VHS, LV $29.99** MAG, HHE 🎞½

Khartoum A sweeping but talky adventure epic detailing the last days of General "Chinese" Gordon as the title city is besieged by Arab tribes in 1833.
1966 134m/C Charlton Heston, Laurence Olivier, Ralph Richardson, Richard Jordan, Alexander Knox, Hugh Williams, Nigel Green, Michael Hordern, Johnny Sekka; *Dir:* Basil Dearden. **VHS, Beta, LV $19.98** MGM, PIA, CCB 🎞🎞½

Kick of Death: The Prodigal Boxer A young boxer is accused of a murder he didn't commit, and fights a battle to the death to prove it.
198? (R) 90m/C Mang Sei, Suma Wah Lung, Pa Hung; *Dir:* Chai Yang Min. **VHS, Beta** $19.95 GEM 🎬

The Kick Fighter Would you believe the hero has to kickbox to finance his kid sister's operation? Grungy, Bangkok-set chopsocky cheapie leaves no cliche untouched, but fights are well-staged. A strange end credit lauds real-life champ Urquidez, here a bad guy.
1991 92m/C Richard Norton, Benny "The Jet" Urquidez, Glen Ruehland, Franco Guerrero, Erica Van Wagener, Steve Rackman; *Dir:* Anthony Maharaj. **VHS** $79.95 AIP 🎬

Kickboxer The brother of a permanently crippled kickboxing champ trains for a revenge match.
1989 (R) 97m/C Jean-Claude Van Damme, Rochelle Ashana, Dennis Chan, Dennis Alexio; *Dir:* David Worth, Mark DiSalle. **VHS, Beta, LV** $89.99 HBO 🎬½

Kickboxer the Champion The opium trade is in for a kick to the stomach when Archer challenges the big man in charge to a deathly duel.
1991 ?m/C Don Murray, Wayne Archer; *Dir:* Alton Cheung. **VHS** $79.95 WLA 🎬

Kicks Two well-off San Francisco professionals play high-risk games with each other, culminating in a life-or-death hunt. Made for television.
1985 97m/C Anthony Geary, Shelley Hack, Tom Mason, Ian Abercrombie, James Avery; *Dir:* William Wiard. **VHS, Beta** $79.95 WES 🎬½

The Kid Sensitive and sassy film about a tramp who takes home an orphan. Chaplin's first feature. Also launched Coogan as the first child superstar.
1921 60m/B Charlie Chaplin, Jackie Coogan, Edna Purviance; *Dir:* Charlie Chaplin. **VHS** $29.98 FCT, MLB 🎬🎬🎬

Kid Young guy carries grudge for the murder of his parents and seeks pound of flesh. Enter mysterious beautiful woman and copious complications. Good sound FX.
1990 (R) 94m/C C. Thomas Howell, R. Lee Ermey, Brian Austin Green, Sarah Trigger; *Dir:* John Mark Robinson. **VHS, Beta, LV** $89.95 LIV, IME 🎬🎬½

Kid with the 200 I.Q. When an earnest boy genius who enters college at age 13, predictable comic situations arise that involve his attempts at impressing his idolized astronomy professor (Guillaume), as well as an equally unrequited bout of first love. Harmless comedy, one of a series of squeaky-clean made-for-TV family fare starring Coleman.
1983 96m/C Gary Coleman, Robert Guillaume, Harriet Nelson, Dean Butler, Karli Michaelson, Christina Murrull, Mel Stewart; *Dir:* Leslie Martinson. **VHS, Beta** $49.95 IVE 🎬🎬

The Kid with the Broken Halo Coleman and Guillaume are paired again in this unsuccessful pilot for a TV series. A young angel, out to earn his wings, must try to help three desperate families, with the help of an experienced angel.
1982 100m/C Gary Coleman, Robert Guillaume, June Allyson, Mason Adams, Ray Walston, John Pleshette, Kim Fields, George Stanford Brown, Telma Hopkins; *Dir:* Leslie Martinson. **VHS** $49.95 USA 🎬🎬

Kid from Brooklyn A shy, musically inclined milkman becomes a middleweight boxer by knocking out the champ in a street brawl. Remake of "The Milky Way" by Harold Lloyd. Available with digitally remastered stereo and original movie trailer.
1946 113m/C Danny Kaye, Virginia Mayo, Eve Arden, Fay Bainter, Walter Abel; *Dir:* Norman Z. MacLeod. **VHS, Beta, LV** $19.98 HBO, SUE 🎬🎬

The Kid Brother The shy, weak son of a tough sheriff, Harold fantasizes about being a hero like his father and big brothers, falls in love with a carnival lady, and somehow saves the day. Classic silent comedy.
1927 84m/B Harold Lloyd, Walter James, Jobyna Ralston; *Dir:* Ted Wilde. **VHS, Beta** TLF 🎬🎬🎬½

Kid Colter An innocent country boy is attacked, left for dead in the mountains, but lives to pursue his attackers relentlessly.
1985 101m/C Jim Stafford, Jeremy Shamos, Hal Terrance, Greg Ward, Jim Turner. **VHS, Beta** $79.98 FOX 🎬

Kid Courageous An athlete goes west, tracks ore mine thieves and prevents a hot-blooded spitfire from marrying the wrong man.
1935 53m/B Bob Steele. **VHS, Beta** $19.95 NOS, WGE 🎬

Kid Creole and the Coconuts Live The Coconuts perform their famous songs at Carnegie Hall: "Don't Take My Coconuts," "Laughing," and "Lifeboat Party."
1986 60m/C **VHS, Beta** $19.98 SUE 🎬

Kid Dynamite A Bowery Boys series episode. Gorcey is a boxer who is kidnapped to prevent his participation in a major fight. The real fighting occurs when his brother is substituted, and Gorcey is smitten.
1943 73m/B Leo Gorcey, Huntz Hall, Bobby Jordan, Gabriel Dell, Pamela Blake; *Dir:* Wallace Fox. **VHS** $19.95 NOS, MNE 🎬

Kid Galahad The King plays a young boxer who weathers the fight game, singing seven songs along the way.
1962 95m/C Elvis Presley, Lola Albright, Charles Bronson, Ned Glass, Joan Blackman, Ed Asner, Gig Young; *Dir:* Phil Karlson. **VHS, Beta** $19.95 MGM 🎬🎬

Kid from Gower Gulch A singing, non-horse-riding cowboy enters a rodeo through a clever ruse.
1947 56m/B Spade Cooley. **VHS, Beta, LV** $19.95 NOS, WGE 🎬

The Kid/Idle Class The Little Tramp adopts a homeless orphan in "The Kid," Chaplin's first feature-length film. This tape also includes "The Idle Class," a rare Chaplin short.
1921 85m/B Charlie Chaplin, Jackie Coogan, Edna Purviance, Chuck Reisner, Lita Grey; *Dir:* Charlie Chaplin. **VHS, Beta** $19.98 FOX 🎬🎬🎬

Kid and the Killers A young orphan and a hardened criminal band together to pursue a villain.
198? (PG) 90m/C Jon Cypher. **VHS, Beta** $49.95 IND 🎬

The Kid from Left Field A bat boy for the San Diego Padres transforms the team from losers to champions when he passes on the advice of his father, a "has-been" ballplayer, to the team members. A made-for-television remake of the 1953 classic.

1979 80m/C Gary Coleman, Robert Guillaume, Ed McMahon, Tab Hunter; *Dir:* Adell Aldrich. **VHS, Beta** $69.95 VES 🎬

Kid Millions Vintage musical comedy in which a dull-witted Brooklyn boy must travel to exotic Egypt to collect an inherited fortune. The finale is filmed in early three-strip Technicolor. Songs include "Mandy," "Okay, Toots," "An Earful of Music," and "When My Ship Comes In." Lucille Ball appears as a Goldwyn Girl.
1934 90m/B Eddie Cantor, Ethel Merman, Ann Sothern, George Murphy; *Dir:* Roy Del Ruth. **VHS, Beta** $19.98 SUE, HBO 🎬🎬½

Kid 'n' Hollywood and Polly Tix in Washington These two "Baby Burlesks" shorts star a cast of toddlers, featuring the most famous moppet of all time, Shirley Temple, in her earliest screen appearances.
1933 20m/B Shirley Temple. **VHS, Beta** $19.98 CCB 🎬🎬½

Kid from Not-So-Big A family film that tells the story of Jenny, a young girl left to carry on her grandfather's frontier-town newspaper. When two con-men come to town, Jenny sets out to expose them.
1978 87m/C Jennifer McAllister, Veronica Cartwright, Robert Viharo, Paul Tulley; *Dir:* Bill Crain. **VHS, Beta** $39.98 WAR, OM 🎬½

Kid Ranger A ranger shoots an innocent man, but later sets things straight by bringing the real culprit to justice.
1936 57m/B Bob Steele. **VHS, Beta** $24.98 WGE 🎬

Kid Sister A young girl is determined to grab her sister's boyfriend for herself, and enlists the aid of a burglar to do it.
1945 56m/B Roger Pryor, Judy Clark, Frank Jenks, Constance Worth. **VHS, Beta** $24.95 VYY 🎬½

A Kid for Two Farthings An episodic, sentimental portrait of the Jewish quarter in London's East End, centered on a boy with a malformed goat he thinks is a magic unicorn capable of granting wishes and bringing happiness to his impoverished 'hood. Acclaimed adaptation of a novel by screenwriter Wolf Mankowitz, also on tape in a shorter rehash as "The Unicorn."
1955 96m/C Jonathan Ashmore. **VHS** $39.95 HMV, FCT 🎬🎬½

Kid Vengeance After witnessing the brutal slaying of his family, a boy carries out a personal vendetta against the outlaws.
1975 (R) 90m/C Leif Garrett, Jim Brown, Lee Van Cleef, John Marley, Glynnis O'Connor; *Dir:* Joseph Manduke. **VHS, Beta** $59.95 PGN 🎬

Kidco The true story of a money-making corporation headed and run by a group of children ranging in age from nine to sixteen.
1983 (PG) 104m/C Scott Schwartz, Elizabeth Gorcey, Cinnamon Idles, Tristine Skyler; *Dir:* Ronald F. Maxwell. **VHS, Beta** $59.98 FOX 🎬🎬

Kidnap Syndicate Kidnappers swipe two boys, releasing one—the son of a wealthy industrialist who meets their ransom demands. When they kill the other boy, a mechanic's son, the father goes on a revengeful killing spree.
1976 (R) 105m/C *IT* James Mason, Valentina Cortese; *Dir:* Fernando Di Leo. **VHS, Beta** $29.95 VID 🎬½

Kidnapped A young boy is sold by his wicked uncle as a slave, and is helped by an outlaw. A Disney film based on the Robert Louis Stevenson classic.
1960 94m/C Peter Finch, James MacArthur, Peter O'Toole; *Dir:* Robert Stevenson. **VHS, Beta $19.99** *DIS* ✍✍ ½

Kidnapped When a teenage girl is kidnapped by pornographers, her sister seeks help from a tough cop and together they go undercover to find her. Plot is flimsy cover for what is actually an exploitative piece.
1987 (R) 100m/C David Naughton, Barbara Crampton, Lance LeGault, Chick Vennera, Charles Napier; *Dir:* Howard Avedis. **VHS, Beta, LV $19.95** *VIR Woof!*

Kidnapping of Baby John Doe Tragedy strikes a family's newborn, and they must make the terrifying decision between life and death. A doctor and nurse, however, take matters into their own hands to save the baby.
1988 90m/C Helen Hughes, Jayne Eastwood, Janet-Laine Green, Geoffery Boues, Peter Gerretsen. **VHS, Beta $79.98** *PEV Woof!*

The Kidnapping of the President The U.S. president is taken hostage by Third World terrorists. The Secret Service is on the ball trying to recover the nation's leader. Well-integrated subplot involves the vice president in a scandal. Engrossing political thriller. Based on novel by Charles Templeton.
1980 (R) 113m/C *CA* William Shatner, Hal Holbrook, Van Johnson, Ava Gardner, Miguel Fernandez; *Dir:* George Mendeluk. **VHS, Beta $39.98** *CGI* ✍✍ ½

Kid's Auto Race/Mabel's Married Life "Kid's Auto Race" (1914), also known as "Kid Auto Races at Venice," concerns a kiddie-car contest; "Mabel's Married Life" (1915) is about flirtations in the park between married individuals. Piano and organ scores.
1915 21m/B Charlie Chaplin, Mabel Normand, Mack Swain; *Dir:* Mack Sennett. **VHS, Beta $19.98** *CCB*

Kids from Candid Camera A collection of classic children's segments from episodes of "Candid Camera."
1985 60m/C *Hosted:* Allen Funt. **VHS, Beta $59.98** *LIV, VES*

Kids from Fame The cast of the television show "Fame" sings and dances in a live sold-out performance at London's Royal Albert Hall.
1983 75m/C Debbie Allen, Gene Anthony Ray, Lee Curreri, Erica Gimpel, Lori Singer, Carlo Imperato. **VHS, Beta, LV $39.95** *MGM* ✍✍

Kids is Kids A collection of cartoons from 1938-1953. Includes "Good Scouts," "Soups On," "Donald's Fountain of Youth," and "Lucky Number." Donald and his four saucy nephews at their best!
1953 50m/C *Dir:* Walt Disney. **VHS, Beta, LV $49.95** *RCA*

Kid's Last Ride Three tough guys come into town to settle a feud.
1941 55m/B Ray Corrigan, John "Dusty" King, Max Terhune, Luana Walters, Edwin Brian; *Dir:* S. Roy Luby. **VHS $19.95** *NOS, DVT* ✍

The Kill A rough, cynical, hard-boiled, womanizing detective tracks down stolen cash in downtown Macao.
1973 81m/C Richard Jaeckel, Henry Duval, Judy Washington. **VHS, Beta $19.95** *STE, NWV* ✍✍ ½

Kill Alex Kill A Vietnam POW returns to find his family murdered, and uses the crime underworld to take revenge.
1983 88m/C Tony Zarindast, Tina Bowmann, Chris Ponti. **VHS, Beta $59.95** *GEM Woof!*

Kill, Baby, Kill A small Transylvania town is haunted by the ghost of a seven-year-old witchcraft victim, and the town's suicide victims all seem to have hearts of gold (coins, that is). Lots of style and atmosphere in this Transylvanian tale from horror tongue in cheekster Bava. Considered by many genre connoisseurs to be the B man's finest, except that it bears early symptoms of the director's late onset infatuation with the zoom shot. Also known as "Curse of the Living Dead" and "Operacione Paura."
1966 83m/C *IT* Erika Blanc, Giacomo Rossi Stuart, Fabienne Dali, Giana Vivaldi; *Dir:* Mario Bava. **VHS $19.98** *SNC, FRG* ✍✍✍

Kill or Be Killed A martial arts champion is lured to a phony martial arts contest by a madman bent on revenge.
1980 (PG) 90m/C James Ryan, Charlotte Michelle, Norman Combes; *Dir:* Ivan Hall. **VHS, Beta $54.95** *MED* ✍

Kill Castro A Key West boat skipper is forced to carry a CIA agent to Cuba on a mission to assassinate Castro. Low budget and deadly dull.
1980 (R) 90m/C Sybil Danning, Albert Salmi, Michael Gazzo, Raymond St. Jacques, Woody Strode, Stuart Whitman, Robert Vaughn, Caren Kaye; *Dir:* Chuck Workman. **VHS, Beta $59.95** *MON* ✍

Kill Factor A spy thriller with a dash of martial arts. Also known as "Death Dimension."
19?? (R) 91m/C Jim Kelly, George Lazenby, Aldo Ray, Harold Sakata, Terry Moore. **VHS, Beta $19.95** *ACA* ✍

Kill and Go Hide A young girl visits her mother's grave nightly to communicate with and command the ghoul-like creatures that haunt the surrounding woods.
1976 95m/C **VHS, Beta $39.95** *MON* ✍

Kill the Golden Goose Two martial arts masters work on opposite sides of a government corruption and corporate influence peddling case.
1979 (R) 91m/C Brad von Beltz, Ed Parker, Master Bong Soo Han; *Dir:* Elliot Hong. **VHS, Beta $59.95** *GEM* ✍

Kill and Kill Again A martial arts champion attempts to rescue a kidnapped Nobel Prize-winning chemist who has developed a high-yield synthetic fuel. Colorful, tongue-in-cheek, and fun even for those unfamiliar with the genre.
1981 (PG) 100m/C James Ryan, Anneline Kriel, Stan Schmidt, Bill Flynn, Norman Robinson, Ken Gampu, John Ramsbottom; *Dir:* Ivan Hall. **VHS, Beta $19.95** *MED* ✍✍ ½

Kill Line Kim stars as Joe, a street fighter who seeks to clear his name after serving a 10-year prison sentence for a crime he didn't commit. When he finds his brother's family has been murdered by criminals looking for the millions he supposedly stole, Joe wages a one-man martial arts war against a corrupt police force and the gang of thugs who are looking for the money.
1991 (R) 93m/C Bobby Kim, Michael Parker, Marlene Zimmerman, H. Wayne Lowery, C.R. Valdez, Mark Williams, Ben Pfeifer; *W/Dir:* Richard H. Kim. **VHS $89.95** *HMD* ✍✍

Kill Me Again A woman hires a private detective to fake her death, which only gets him targeted by her boyfriend, the Mob, and the cops.
1989 (R) 93m/C Val Kilmer, Joanne Whalley-Kilmer, Michael Madsen, Jonathan Gries, Bibi Besch; *Dir:* John Dahl. **VHS, Beta $89.95** *MGM* ✍ ½

Kill My Wife...Please! Comic adventure of a despicable husband trying to bump off his wife in order to cash in on her insurance policy.
19?? 90m/C Bob Dishy, Joanna Barnes, Bill Dana; *Dir:* Steven Hilliard Stern. **VHS $19.95** *AVD, API, HHE* ✍✍

The Kill Reflex A tough black cop and his beautiful rookie partner battle the mob and corruption.
1989 (R) 90m/C Maud Adams, Fred Williamson, Bo Svenson, Phyllis Hyman; *Dir:* Fred Williamson. **VHS, Beta, LV $79.95** *COL* ✍

Kill Slade When a United Nations food-aid diversion conspiracy is uncovered by a beautiful journalist, a plan to kidnap her is put into action. Romance follows.
1989 (PG-13) 90m/C *SA* Patrick Doolaghan, Lisa Brady, Anthony Fridjhon, Danny Keogh, Alfred Nowke; *Dir:* Bruce McFarlane. **VHS, Beta, LV $19.98** *SUE* ✍

Kill Squad A squad of martial arts masters follow a trail of violence and bloodshed to a vengeful, deadly battle of skills.
1981 (R) 85m/C Jean Claude, Jeff Risk, Jerry Johnson, Bill Cambra, Cameron Mitchell. **VHS, Beta $59.95** *HAR* ✍

Killcrazy On the way to a weekend camping trip, five Vietnam vets released from a mental hospital are slowly massacred by a group of dangerous killers until one of the vets decides to fight back.
1989 (R) 94m/C David Heavener, Danielle Brisebois, Burt Ward, Lawrence Hilton-Jacobs, Bruce Glover, Gary Owens, Rachelle Carson; *W/Dir:* David Heavener. **VHS, LV $89.98** *MED, IME* ✍

Killer An ex-convict sets up an ingenious bank robbery, and gets double-crossed himself.
197? 87m/C Henry Silva. **VHS, Beta $59.95** *MGL* ✍

Killer on Board Folks spending their holiday on a cruise ship are afflicted by a fatal virus and subsequently quarantined. Typical made-for-TV disaster flick.
1977 100m/C Claude Akins, Beatrice Straight, George Hamilton, Patty Duke, Frank Converse, Jane Seymour, William Daniels; *Dir:* Philip Leacock. **VHS, Beta $49.95** *IVE* ✍✍

Killer Commandos Mercenaries go to the rescue of an oppressed African nation to overthrow the sinister white supremacy dictator.
19?? 87m/C Cameron Mitchell, Anthony Eisley. **VHS $39.95** *ACF* ✍

Killer Dill Also killer dull. A meek salesman is misidentified as a death-dealing gangster, and the comic mixups begin.
1947 71m/B Stuart Erwin, Anne Gwynne, Frank Albertson, Mike Mazurki, Milburn Stone, Dorothy Granger; *Dir:* Lewis D. Collins. **VHS $19.95** *NOS* ✍ ½

Killer Diller An all-black musical revue, featuring Nat King Cole and his Trio.
1948 70m/B Dusty Fletcher, Nat King Cole, Butterfly McQueen, Moms Mabley, George Wiltshire; *Dir:* Joe Binney. **VHS, Beta $19.95** *MVD, NOS, DVT* ✍ ½

Killer Elephants A Thai man struggles to save his plantation, his wife, and his baby from the terrorists hired by a land baron to drive him away from his property.
1976 83m/C Sung Pa, Alan Yen, Nai Yen Ne, Yu Chien. **VHS, Beta $29.95** *UNI*

The Killer Elite Straight-ahead Peckinpah fare examining friendship and betrayal. Two professional assassins begin as friends but end up stalking each other when they are double-crossed. This minor Peckinpah effort is murky and doesn't have a clear resolution, but is plenty bloody. Lots of Dobermans roam through this picture too.
1975 (R) 120m/C James Caan, Robert Duvall, Arthur Hill, Gig Young, Burt Young, Mako, Bo Hopkins; ***Dir:*** Sam Peckinpah. **VHS, Beta, LV $59.98** *MGM* 🦴🦴½

A Killer in Every Corner Three psychology students visit a loony professor and succumb to his hair-raising shenanigans.
1974 80m/C Joanna Pettet, Patrick Magee, Max Wall, Eric Flynn. **VHS, Beta $29.95** *IVE* 🦴

Killer Fish A scheme to steal and then hide a fortune in emeralds at the bottom of a tank full of piranhas backfires as the criminals find it impossible to retrieve them...as if that took a lot of foresight. Director Dawson is AKA Antonio Margheriti. Also called "Deadly Treasure of the Piranha," but "Killer Fish" sounds more interesting, doesn't it? Filmed in Brazil.
1979 (PG) 101m/C *BR GB* Lee Majors, James Franciscus, Margaux Hemingway, Karen Black, Roy Brocksmith, Marisa Berenson; ***Dir:*** Anthony M. Dawson. **VHS, Beta $59.98** *FOX* 🦴

Killer Force An adventure of international diamond smuggling.
1975 (R) 100m/C Telly Savalas, Peter Fonda, Maud Adams; ***Dir:*** Val Guest. **VHS, Beta $69.98** *LIV, VES* 🦴½

Killer Image A photographer unwittingly sees more than he should and becomes involved in a lethal political coverup. At the same time a wealthy senator targets him and he must fight for his life. Viewers might feel that they have to fight their boredom.
1992 (R) 97m/C John Pyper-Ferguson, Michael Ironside, M. Emmet Walsh, Krista Errickson; ***Dir:*** David Winning. **VHS, Beta** *PAR* 🦴½

Killer Inside Me The inhabitants of a small Western town are unaware that their mild-mannered deputy sheriff is actually becoming a crazed psychotic murderer. From the novel by Jim Thompson.
1976 (R) 99m/C Stacy Keach, Susan Tyrrell, Tisha Sterling, Keenan Wynn, John Dehner, John Carradine, Don Stroud; ***Dir:*** Burt Kennedy. **VHS, Beta $59.95** *WAR, OM* 🦴🦴

Killer Klowns from Outer Space Bozo-like aliens resembling demented clowns land on earth and set up circus tents to lure Earthlings in. Visually striking, campy but slick horror flick that'll make you think twice about your next visit to the big top. Mood is heightened by a cool title tune by the Dickies. Definitely has cult potential!
1988 (PG-13) 90m/C Grant Cramer, Suzanne Snyder, John Allen Nelson, Royal Dano, John Vernon; ***Dir:*** Stephen Chiodo. **VHS, Beta, LV $19.95** *MED, CVS* 🦴🦴½

Killer Likes Candy An assassin stalks the King of Kafiristan, and a CIA operative tries to stop him.
1978 86m/C Kerwin Mathews, Marilu Tolu; ***Dir:*** Richard Owens. **VHS, Beta $59.95** *GEM, MRV* 🦴

Killer Party Three coeds pledge a sorority and are subjected to a hazing that involves a haunted fraternity house. Standard horror plot; Paul Bartel ("Eating Raoul"; "Lust in the Dust") is the only significant element.
1986 (R) 91m/C Elaine Wilkes, Sherry Willis-Burch, Joanna Johnson, Paul Bartel, Martin Hewitt, Ralph Seymour; ***Dir:*** William Fruet. **VHS, Beta $79.98** *FOX Woof!*

The Killer Shrews Lumet (Sidney's father) creates a serum that causes the humble shrew to take on killer proportions. The creatures are actually dogs in makeup. Goude was 1957's Miss Universe.
1959 70m/B James Best, Ingrid Goude, Baruch Lumet, Ken Curtis; ***Dir:*** Ray Kellogg. **VHS $19.98** *SNC, MRV, HHT* 🦴🦴

Killer Tomatoes Strike Back The third 'Killer Tomatoes' movie isn't in the league of "The Naked Gun" satires, but it's still bright parody for the Mad Magazine crowd, as tomato-mad scientist Astin harnesses the powers of trash-TV in a planned vegetable invasion. Perhaps due to a 'Killer Tomatoes' cartoon series at the time, this isn't as saucy as its predecessors and is acceptable for family audiences.
1990 87m/C John Astin, Rick Rockwell, Crystal Carson, Steve Lundquist, John Witherspoon; ***Dir:*** John DeBello. **VHS $39.98** *FXV, IME* 🦴🦴

Killer with Two Faces A woman is terrorized by the evil twin of her boyfriend. Made for television.
1974 70m/C Donna Mills, Ian Hendry, David Lodge, Roddy McMillian. **VHS, Beta $12.95** *IVE* 🦴

Killer Workout A murderer stalks the clients of a sweat/sex/muscle-filled gym.
1986 (R) 89m/C VHS, Beta $29.95 *ACA* 🦴

The Killers After two hired assassins kill a teacher, they look into his past and try to find leads to a $1,000,000 robbery. Ronald Reagan's last film. Remake of 1946 film of the same name, which was loosely based on a short story by Ernest Hemingway. Originally intended for television, but released to theaters instead due to its violence.
1964 95m/C Lee Marvin, Angie Dickinson, John Cassavetes, Ronald Reagan, Clu Gulager, Claude Akins, Norman Fell; ***Dir:*** Don Siegel. **VHS, Beta $39.95** *MCA* 🦴🦴

Killers The remote jungles of southern Africa are the scene of a military coup.
1988 83m/C Cameron Mitchell, Alicia Hammond, Robert Dix; ***Dir:*** Ewing Miles Brown. **VHS, Beta $39.95** *GHV* 🦴

The Killer's Edge Cop is caught between rock and hard place in L.A. when he's forced to confront criminal who once saved his life in 'Nam. Plenty of gut busting and soul wrenching.
1990 (R) 90m/C Wings Hauser, Robert Z'Dar, Karen Black. **VHS $79.95** *PMH* 🦴🦴½

Killer's Kiss A boxer and a dancer set out to start a new life together when he saves the woman from an attempted rape. Gritty, second feature from Kubrick was financed by friends and family and shows signs of his budding talent.
1955 67m/B Frank Silvera, Jamie Smith, Irene Kane, Jerry Jarret; ***W/Dir:*** Stanley Kubrick. **VHS** *FCT* 🦴🦴½

Killer's Moon Four sadistic psychopaths escape from a prison hospital and unleash their murderous rage on anyone who crosses their path.
1984 90m/C VHS, Beta $49.95 *VCL* 🦴

Killers from Space Cheap sci-fi flick in which big-eyed men from beyond Earth bring a scientist (Graves) back to life to assist them with their evil plan for world domination.
1954 80m/B Peter Graves, Barbara Bestar, James Seay; ***Dir:*** W. Lee Wilder. **VHS, Beta $9.95** *NOS, MRV, RHI Woof!*

The Killing The dirty, harsh, street-level big heist epic that established Kubrick and presented its genre with a new and vivid existentialist aura, as an ex-con engineers the rip-off of a racetrack with disastrous results. Displays characteristic nonsentimental sharp-edged Kubrick vision. Cinematography by Lucien Ballard. Based on the novel "Clean Break" by Lionel White.
1956 83m/B Sterling Hayden, Marie Windsor, Elisha Cook Jr., Jay C. Flippen, Vince Edwards, Timothy Carey, Coleen Gray, Joe Sawyer, Ted de Corsia; ***Dir:*** Stanley Kubrick. **VHS, Beta, LV $19.95** *MGM, VYG* 🦴🦴🦴

A Killing Affair Set in West Virginia, 1943, this is the story of a widow who takes in a drifter who she believes is the man who killed her husband. She begins to fall for him, but cannot be sure if she should trust him. Vague and melodramatic.
1985 (R) 100m/C Peter Weller, Kathy Baker, John Glover, Bill Smitrovich; ***Dir:*** David Saperstein. **VHS, Beta, LV $9.99** *STE, PSM* 🦴

Killing of Angel Street A courageous young woman unwittingly becomes the central character in an escalating nightmare about saving a community from corrupt politicians and organized crime.
1981 (PG) 101m/C *AU* Liz Alexander, John Hargreaves; ***Dir:*** Donald Crombie. **VHS, Beta $14.98** *VID, FCT* 🦴🦴

Killing Cars A German car designer's pet project, a car that runs without gas, is halted by the influence of an Arab conglomerate. He nevertheless tries to complete it, and is hunted down.
198? (R) 104m/C Juergen Prochnow, Senta Berger, William Conrad, Agnes Soral; ***Dir:*** Michael Verhoeven. **VHS, Beta $79.95** *VMK, HHE* 🦴½

The Killing Edge In post-nuclear holocaust Earth, a lone warrior seeks justice in a lawless land.
198? 85m/C VHS $39.98 *VCD, HHE* 🦴½

Killing 'em Softly Segal portrays a down-on-his-luck musician who accidentally murders a music manager during an argument. Cara's boyfriend is accused and to clear his name, the singer moonlights as a detective. She and Segal end up falling in love. Disappointing.
1985 90m/C *CA* George Segal, Irene Cara; ***Dir:*** Max Fischer. **VHS, Beta $79.95** *PSM, HHE* 🦴

The Killing Fields Based on the New York Times' Sydney Schanberg's own account of his friendship with Cambodian interpreter Dith Pran. They separated during the fall of Saigon, when the Western journalists fled, leaving behind countless assistants, secretaries and interpretors who were later accused of collusion with the enemy by the vicious Khmer Rouge and killed or sent to reeducation camps during the bloodbath known as "Year Zero." Schanberg continued to search for Pran through the Red Cross and U.S. government, while Pran struggled to survive, secretly drinking ox blood to supplement his meager food supply until he was able to escape, walking miles to the Thailand border. Ngor's own experiences echoed much of the film's. Debut of Malkovich is intense and watchable. Joffe's direc-

torial debut shows a generally sure hand, with only a bit of melodrama at the end.
1984 (R) 142m/C *GB* Sam Waterston, Haing S. Ngor, John Malkovich, Athol Fugard, Craig T. Nelson, Julian Sands, Spalding Gray, Bill Paterson; ***Dir:*** Roland Joffe. Academy Awards '84: Best Cinematography, Best Film Editing, Best Supporting Actor (Ngor). **VHS, Beta, LV $19.98** *WAR, PIA, BTV* 🦴🦴🦴½

Killing Floor A black sharecropper during World War I travels to Chicago to get work in the stockyards and becomes a voice in the growing labor movement. Made for television.
1985 (PG) 118m/C Damien Leake, Alfre Woodard, Moses Gunn, Clarence Felder, Bill Duke; ***Dir:*** William Duke. **VHS, Beta, LV $19.98** *SUE* 🦴🦴

The Killing Game A husband-wife cartoonist team link up with an unhinged playboy and act out a murder-mystery comic they produce together. In French with subtitles.
1967 90m/C *FR* Jean-Pierre Cassel, Claudine Auger, Michael Duchaussoy, Anna Gaylor; ***Dir:*** Alain Jessua. **VHS, Beta $39.95** *FCT* 🦴🦴½

The Killing Game A slew of Californians kill and betray each other for lustful reasons.
1987 90m/C Chad Hayward, Cynthia Killion, Geoffrey Sadwith; ***Dir:*** Joseph Merhi. **VHS, Beta** *CLV* 🦴

Killing Heat An independent career woman living in South Africa decides to abandon her career to marry a struggling jungle farmer.
1984 (R) 104m/C Karen Black, John Thaw, John Kani, John Moulder-Brown; ***Dir:*** Michael Raeburn. **VHS, Beta $59.98** *FOX* 🦴½

Killing at Hell's Gate A made-for-television movie about a group of white-water rafters being picked off by snipers as they travel down river.
1981 96m/C Robert Urich, Deborah Raffin, Lee Purcell; ***Dir:*** Jerry Jameson. **VHS, Beta $59.98** *FOX* 🦴½

Killing Hour A psychic painter finds that the visions she paints come true in a string of grisly murders. Her ability interests a television reporter and a homicide detective. Also known as "The Clairvoyant."
1984 (R) 97m/C Elizabeth Kemp, Perry King, Norman Parker, Kenneth McMillan; ***Dir:*** Armand Mastroianni. **VHS, Beta $79.98** *FOX* 🦴🦴

The Killing Kind Man released from prison is obsessed with wreaking vengeance on his daft lawyer and his accuser. His vendetta eventually draws his mother into the fray as well. Fine performance by Savage.
1973 95m/C Ann Sothern, John Savage, Ruth Roman, Luana Anders, Cindy Williams; ***Dir:*** Curtis Harrington. **VHS $14.99** *AVD, PGN* 🦴🦴½

The Killing Machine A Japanese World War II army veteran uses his martial arts skills to subdue local black market gangs.
1976 89m/C Sonny Chiba. **VHS, Beta $9.99** *STE, PSM* 🦴

The Killing Mind A young girl witnesses a grisly murder that is never solved. Twenty years later she (Zimbalist) becomes a cop who specializes in trapping psychos. Haunted by her memories, she and a reporter (Bill) team up to find the killer, never suspecting that she is his next victim. Originally made-for-cable-television.
1990 96m/C Stephanie Zimbalist, Tony Bill, Daniel Roebuck; ***Dir:*** Michael Ray Rhodes. **VHS $89.98** *WOV* 🦴

Killing of Randy Webster A father attempts to prove that his son did not die as a criminal when Texas policemen shot him after transporting a stolen van across state lines.
1981 90m/C Hal Holbrook, Dixie Carter, Sean Penn, Jennifer Jason Leigh; ***Dir:*** Sam Wanamaker. **VHS, Beta** *VCL* 🦴½

The Killing of Sister George Racy, sensationalized film based on the Frank Marcus black comedy/melodrama about a lesbian love triangle between a television executive and two soap opera stars. Things get a little uncomfortable when they learn one of their characters is to be written out of the show. Shot in England.
1969 (R) 138m/C Beryl Reid, Susannah York, Coral Browne, Ronald Fraser, Patricia Medina, Hugh Paddick, Cyril Elaine Church, Brandan Dillon; ***Dir:*** Robert Aldrich. **VHS, Beta $79.98** *FOX* 🦴🦴

Killing in a Small Town Candy Morrison seems like the perfect member of her small Texas community—she's married, has children, and even teaches Bible school. But Candy needs a little excitement in her life, which she gets by initiating a brief affair with fellow church-goer Stan. However, when Stan's wife Peggy finds out she confronts Candy—with an axe. Candy defends herself and winds up killing Peggy, by striking her with the axe 41 times (shades of Lizzie Borden)! But was Candy only defending herself? Or did she seek to get her own sort of revenge? Good performances by Hershey as the frustrated Candy and Dennehy as her skeptical lawyer. Based on a true story; originally a television miniseries.
1990 95m/C Barbara Hershey, Brian Dennehy, Hal Holbrook, Richard Gilliland, John Terry; ***Dir:*** Stephen Gyllenhaal. **VHS, LV $89.95** *VMK* 🦴🦴🦴

Killing Stone A freelance writer uncovers a small town sheriff's plot to cover up a scandalous homicide. Made for television and written by Landon.
1978 120m/C Gil Gerard, J.D. Cannon, Jim Davis, Nehemiah Persoff; ***Dir:*** Michael Landon. **VHS, Beta $49.95** *PSM* 🦴🦴

Killing Streets A commando learns that his twin brother is being held hostage in Lebanon and plans a rescue mission. A standard farfetched actioner; noteworthy for a real sense of despair over the endless carnage in the Mideast. Some dialogue in Arabic with subtitles.
1991 109m/C Michael Pare, Lorenzo Lamas; ***Dir:*** Stephen Cornwell. **VHS $89.98** *VES, LIV* 🦴🦴

Killing in the Sun Gripping struggle for prosperous smuggling in the Mediterranean consume mobsters from three countries.
19?? (R) 90m/C Henry Silva, Michael Constantine. **VHS** *API* 🦴

The Killing Time A minor, effective murder thriller about a quiet small resort town suddenly beset by a web of murder, double-crossings, blackmail, and infidelity. They seem to coincide with the appearance of a mysterious stranger posing as the town's deputy sheriff just as the new sheriff is about to take up the badge.
1987 (R) 94m/C Kiefer Sutherland, Beau Bridges, Joe Don Baker, Wayne Rogers; ***Dir:*** Rick King. **VHS, Beta, LV $19.95** *NWV, STE* 🦴🦴

The Killing Zone Convict nephew of onetime Drug Enforcement agent rewrites zoning ordinances to hunt for Mexican drug lord south of the border.
1990 90m/C Daron McBee, James Dalesandro, Melissa Moore, Armando Silvestre, Augustine Beral, Sydne Squire, Deborah Dare; ***Dir:*** Addison Randall. **VHS $79.95** *PMH* 🦴½

Killings at Outpost Zeta A crew of Earthmen investigate a barren planet where previous expeditions have disappeared, and find hordes of aliens.
1980 92m/C Gordon Devol, Jackie Ray, James A. Watson Jr. **VHS, Beta $69.95** *UHV* 🦴

Killjoy A sleazy surgeon's daughter is the prey in this television thriller. The plot twists are led by the array of people who become involved. A clever suspense mystery. Made for television. Also known as "Who Murdered Joy Morgan."
1981 100m/C Kim Basinger, Robert Culp, Stephen Macht, Nancy Marchand, John Rubinstein, Ann Dusenberry, Ann Wedgeworth; ***Dir:*** John Llewellyn Moxey. **VHS, Beta $49.95** *IVE* 🦴🦴½

Killpoint A special task force is assembled to catch the criminals who robbed a National Guard armory for its weapons.
1984 (R) 89m/C Leo Fong, Richard Roundtree, Cameron Mitchell; ***Dir:*** Frank Harris. **VHS, Beta, LV $79.98** *LIV, VES* 🦴½

Killzone A brainwashed Vietnam vet breaks down during a training exercise and embarks on a psychotic killing spree.
1985 86m/C Ted Prior, David James Campbell, Richard Massery; ***Dir:*** David A. Prior. **VHS, Beta $69.98** *VES* 🦴

Kilma, Queen of the Amazons A shipwrecked sailor finds himself on an island populated by man-hating Amazons. When a shipload of lusty sailors arrive to rescue him, carnage ensues.
1986 90m/C VHS, Beta $19.95 *ASE Woof!*

Kim A colorful Hollywood adaptation of the Rudyard Kipling classic about an English boy disguised as a native in 19th Century India, and his various adventures.
1950 113m/C Errol Flynn, Dean Stockwell, Paul Lukas, Cecil Kellaway; ***Dir:*** Victor Saville. **VHS, Beta $19.98** *KUI, MGM* 🦴🦴🦴

Kind Hearts and Coronets Black comedy in which an ambitious young man sets out to bump off eight relatives in an effort to claim a family title. Guinness is wonderful in his role as all eight of the fated relations. Based on Roy Horiman's novel.
1949 104m/B *GB* Alec Guinness, Dennis Price, Valerie Hobson, Joan Greenwood; ***Dir:*** Robert Hamer. National Board of Review Awards '50: 5 Best Foreign Films of the Year. **VHS, Beta, LV $19.98** *HBO, FCT* 🦴🦴🦴½

A Kind of Loving Two North English young people marry rashly as a result of pregnancy, find they really didn't like each other all that much, but manage to adjust.
1962 107m/B *GB* Alan Bates, Thora Hird, June Ritchie; ***Dir:*** John Schlesinger. **VHS, Beta $59.99** *HBO* 🦴🦴🦴

Kindergarten Cop Pectoral perfect cop Kimble (Schwarzenegger) stalks mama's boy/criminal Crisp (Tyson) by locating the drug lord's ex and his six-year-old son. When the pec man's female partner succumbs to a nasty bout of food poisoning, he's forced to take her place as an undercover kindergarten teacher in the drowsy Pacific northwest community where mother and son reside incognito. A cover all the bases Christmas

release, it's got romance, action, comedy and cute. And box office earnings to match Arnie's chest measurements. A bit violent for the milk and cookie set.
1990 (PG-13) 111m/C Arnold Schwarzenegger, Penelope Ann Miller, Pamela Reed, Linda Hunt, Richard Tyson, Carroll Baker; *Dir:* Ivan Reitman. **VHS, Beta, LV $19.95** MCA, CCB *♫♫½*

The Kindred A young student discovers that his mother the biologist has created a hybrid creature using his body tissue. Naturally, he is horrified and begins to search for his test-tube brother; the problem is, this brother likes eating people. Not too bad; boasts some good acting.
1987 (R) 92m/C Rod Steiger, Kim Hunter, David Allan Brooks, Timothy Gibbs, Amanda Pays, Talia Balsam; *Dir:* Stephen Carpenter. **VHS, Beta, LV $14.98** LIV, VES *♫♫½*

King Telefilm follows the life and career of one of greatest non-violent civil rights leaders of all time, Martin Luther King.
1978 272m/C Paul Winfield, Cicely Tyson, Roscoe Lee Browne, Ossie Davis, Art Evans, Ernie Banks, Howard E. Rollins Jr., William Jordan, Cliff DeYoung. **VHS, Beta $69.99** HBO *♫½*

King of America A Greek sailor and a local labor agent battle over who will become King of America.
1980 90m/C *Dir:* Dezso Magyar. **VHS, Beta $59.95** VCL Woof!

King Arthur, the Young Warlord The struggle that was the other side of Camelot—the campaign against the Saxon hordes. Realistic portrayal of Medieval warfare.
1975 (PG) 90m/C Oliver Tobias, Michael Gothard, Jack Watson, Brian Blessed, Peter Firth, *Dir:* Sidney Hayers, Pat Jackson, Peter Sasdy. **VHS, Beta $29.95** GEM *♫*

King Boxer A famous martial arts school is taken over by a menacing gang. When a student-to-be finds this out, he enlists himself to take on the best individual of the gang in a deadly match.
1990 106m/C Lo Lieh, Wang Ping. **VHS $69.95** SOU *♫*

King Boxers Japan's top actor, Yasuka Kurate, stars in this tale of elephant hunts, warding off Triad Society gangs, and personal combat. Also stars Johnny Nainam, Thailand's fists and kick boxing champion.
1980 90m/C Yasuka Kurata, Johnny Nainam. **VHS, Beta $59.95** GEM *♫*

King of the Bullwhip Two undercover U.S. Marshals are sent to Tioga City to stop the killing and looting of a masked bandit, whose whip is as dangerous as his gun.
1951 59m/B Lash LaRue, Al "Fuzzy" St. John, Anne Gwynne, Tom Neal, Jack Holt, Dennis Moore, Michael Whalen, George Lewis; *Dir:* Ron Ormond. **VHS, Beta $19.95** NOS, VCN, DVT *♫*

King of Comedy An unhinged would-be comedian haunts and eventually kidnaps a massively popular Johnny Carson-type television personality. A cold, cynical farce devised by Scorsese seemingly in reaction to John Hinckley's obsession with his film "Taxi Driver." Controlled, hard-hitting performances, especially by De Niro and Lewis.
1982 (PG) 101m/C Robert De Niro, Jerry Lewis, Sandra Bernhard, Tony Randall, Diahnne Abbott, Shelley Hack, Liza Minnelli; *Dir:* Martin Scorsese. **VHS, Beta, LV $19.95** COL, FCT *♫♫♫*

King of the Congo Crabbe, in his last serial, as a pilot who is frantically searching for an important piece of film. In 15 parts.

1952 ?m/B Buster Crabbe, Gloria Dee; *Dir:* Spencer Gordon Bennet, Wallace Grissell. **VHS** MLB *♫½*

King of the Cowboys Roy fights a gang of saboteurs and saves a defense installation.
1943 54m/B Roy Rogers, Smiley Burnette, James Bush. **VHS $9.99** NOS, VCN, DVT *♫*

King Creole The King goes film noir as a teenager with a criminal record who becomes a successful pop singer in New Orleans but is threatened by his ties to crime, represented by Walter Matthau. One of the better Elvis films, based on Harold Robbins' "A Stone for Danny Fisher." Features Elvis' last film appearance before his service in the Army. Songs included "Crawfish," "Hard-Headed Woman," "Trouble," and "Don't Ask Me Why."
1958 115m/B Elvis Presley, Carolyn Jones, Walter Matthau, Dean Jagger, Dolores Hart, Vic Morrow, Paul Stewart; *Dir:* Michael Curtiz. **VHS, Beta $19.95** MVD, FOX *♫♫½*

King David The story of David, the legendary Biblical hero whose acts of bravery paved the way for him to become king of Israel.
1985 (PG-13) 114m/C Richard Gere, Alice Krige, Cherie Lunghi, Hurd Hatfield, Edward Woodward; *Dir:* Bruce Beresford. **VHS, Beta, LV $79.95** PAR *♫½*

King Dinosaur A new planet arrives in the solar system, and a scientific team checks out its giant iguana-ridden terrain.
1955 63m/B Bill Bryant, Wanda Curtis, Patti Gallagher, Douglas Henderson. **VHS, Beta, LV** WGE, MLB *♫½*

The King and Four Queens Gable, on the run from the law, happens upon a deserted town, deserted, that is, except for a woman and her three daughters. Clark soon discovers that the women are looking for $100,000 in gold that one of their missing husbands had stolen. True to form, conniving Clark wastes no time putting the moves on each of them to find the whereabouts of the loot.
1956 86m/C Clark Gable, Eleanor Parker, Jo Van Fleet, Jean Wiles, Barbara Nichols, Sara Shane, Roy Roberts, Arthur Shields, Jay C. Flippen; *Dir:* Raoul Walsh. **VHS, Beta $19.98** MGM, CCB *♫♫½*

King of the Grizzlies The mystical relationship between a Cree Indian and a grizzly cub is put to the acid test when the full grown bear attacks a ranch at which the Indian is foreman.
1969 (G) 93m/C Chris Wiggins, John Yesno; *Dir:* Ron Kelly. **VHS, Beta $69.95** DIS *♫½*

King of the Gypsies Interesting drama. A young man, scornful of his gypsy heritage, runs away from the tribe and tries to make a life of his own. He is summoned home to his grandfather's deathbed where he is proclaimed the new king of the gypsies, thus incurring the wrath of his scorned father. From Peter Maas's best-selling novel.
1978 (R) 112m/C Sterling Hayden, Eric Roberts, Susan Sarandon, Brooke Shields, Shelley Winters, Annie Potts, Annette O'Toole, Judd Hirsch, Michael Gazzo, Roy Brocksmith; *Dir:* Frank Pierson. **VHS, Beta, LV $14.95** PAR, CCB *♫♫½*

King of Hearts In World War I, a Scottish soldier finds a battle-torn French town evacuated of all occupants except a colorful collection of escaped lunatics from a nearby asylum. The lunatics want to make him their king, which is not a bad alternative

to the insanity of war. Bujold is cute as ballerina wannabe; look for Serrault ("La Cage aux Folles") as, not surprisingly, a would-be effeminate hairdresser. Light-hearted comedy with a serious message; definitely worthwhile. Also known as "Le Roi de Coeur."
1966 101m/C FR GB Alan Bates, Genevieve Bujold, Adolfo Celi, Francoise Christophe, Micheline Presle; *Dir:* Phillipe DeBroca. **VHS, Beta, LV $19.95** FOX *♫♫*

The King and I Adaptation of Rogers and Hammerstein's Broadway play based on the novel "Anna and the King of Siam" by Margaret Landon. An English governess is hired to teach the King of Siam's many children and bring them into the 20th century. She has more of a job than she realizes, for this is a king, a country, and a people who value tradition above all else. Her intelligence and good humor eventually begin to convince the King that she'll not destroy the world he loves, and they discover characteristics to appreciate in each other. One of Rodgers and Hammerstein's best-loved scores, with favorites such as "Getting to Know You," "We Kiss in a Shadow," "Shall We Dance," "Something Wonderful," and "Hello, Young Lovers." Brynner made this role his and his alone, playing it over 4,000 times on stage and screen before his death. Kerr's voice was dubbed when she sang; the voice you hear is Marni Nixon, who also dubbed the star's singing voices in "West Side Story" and "My Fair Lady." A wonderful film with important messages. Received a Best Picture Oscar nomination.
1956 133m/C Deborah Kerr, Yul Brynner, Rita Moreno, Martin Benson, Terry Saunders, Rex Thompson, Alan Mowbray, Carlos Rivas; *Dir:* Walter Lang. Academy Awards '56: Best Actor (Brynner), Best Art Direction/Set Decoration (Color), Best Costume Design (Color), Best Sound, Best Musical Score; Golden Globe Awards '57: Best Film—Musical/Comedy. **VHS, Beta, LV $19.98** FOX, BTV, RDG *♫♫♫♫*

King of Jazz A lavish revue built around the Paul Whiteman Orchestra with comedy sketches and songs by the stars on Universal Pictures' talent roster. (Though billed as the King of Jazz, Whiteman was never as good as jazz's real royalty.) Musical numbers include George Gershwin's "Rhapsody in Blue," "Happy Feet," and "It Happened in Monterey." Filmed in two-color Technicolor with a cartoon segment by Walter Lantz.
1930 93m/C Paul Whiteman, John Boles, Jeanette Loff, Bing Crosby; *Dir:* John Murray Anderson. Academy Awards '30: Best Interior Decoration. **VHS, Beta $29.95** MCA, MLB, RXM

The King of the Kickboxers The villains make kung-fu snuff movies with unwitting actors killed on camera. Otherwise, same old junk about a karate cop back in Bangkok to get the dude who squashed his brother. Pretty racist at times.
1991 (R) 90m/C Loren Avedon, Richard Jaeckel, Billy Blanks, Don Stroud, Keith Cooke; *Dir:* Lucas Lowe. **VHS $89.95** IMP Woof!

King of Kings DeMille depicts the life of Jesus Christ in this highly regarded silent epic. The resurrection scene appears in color. Remade by Nicholas Ray in 1961.
1927 115m/B H.B. Warner, Dorothy Cumming, Ernest Torrence, Joseph Schildkraut, Jacqueline Logan, Victor Varconi, William Boyd; *Dir:* Cecil B. DeMille. **VHS, Beta $29.98** MSP *♫♫½*

King of Kings The life of Christ is intelligently told, with an attractive visual sense and a memorable score by Miklos Rosza. Narrated by Orson Welles. Remake of Cecil B. DeMille's silent film, "The King of Kings," released in 1927. Laserdisc format features a special letterboxed edition of this film.
1961 170m/C Jeffrey Hunter, Siobhan McKenna, Hurd Hatfield, Robert Ryan, Rita Gam, Viveca Lindfors, Rip Torn; *Dir:* Nicholas Ray; *Nar:* Orson Welles. **VHS, Beta, LV $29.98** *MGM, PIA, BUR* �́✍✍✍

King Kong The original beauty and the beast film classic tells the story of Kong, a giant ape captured in Africa and brought to New York as a sideshow attraction. Kong falls for Wray, escapes from his captors and rampages through the city, ending up on top of the newly built Empire State Building. Moody Max Steiner score adds color. Willis O'Brien's stop-motion animation still holds up well. Remade numerous times with various theme derivations. Available in a colorized version. The laserdisc, produced from a superior negative, features extensive liner notes and running commentary by film historian Ronald Haver.
1933 105m/B Fay Wray, Bruce Cabot, Robert Armstrong, Frank Reicher, Noble Johnson; *Dir:* Ernest B. Schoedsack. **VHS, Beta, LV, 8mm $19.98** *TTC, RKO, MED* ✍✍✍✍

King Kong An oil company official travels to a remote island to discover it inhabited by a huge gorilla. The transplanted beast suffers unrequited love in classic fashion: monkey meets girl, monkey gets girl and brandishes her while atop the World Trade Center. An updated remake of the 1933 movie classic that also marks the screen debut of Lange. Impressive sets and a believable King Kong romp around New York City in this film.
1976 (PG) 135m/C Jeff Bridges, Charles Grodin, Jessica Lange, Rene Auberjonois, John Randolph, Ed Lauter, Jack O'Halloran; *Dir:* John Guillermin. Academy Awards '76: Best Visual Effects. **VHS, Beta, LV, 8mm $24.95** *PAR* ✍✍

King of Kong Island Intent on world domination, a group of mad scientists implant receptors in the brains of gorillas on Kong Island, and the monster apes run amok.
1978 92m/C *SP* Brad Harris, Marc Lawrence; *Dir:* Robert Morris. **VHS, Beta $54.95** *UHV* Woof!

King Kong Lives Unnecessary sequel to the 1976 remake of "King Kong," in which two scientists get the big ape, now restored after his asphalt-upsetting fall, together with a lady ape his size and type.
1986 (PG-13) 105m/C Brian Kerwin, Linda Hamilton, John Ashton, Peter Michael Goetz; *Dir:* John Guillermin. **VHS, Beta, LV $19.98** *MED, WAR* ✍

King Kong Versus Godzilla The planet issues a collective shudder as the two mightiest monsters slug it out for reasons known only to themselves. Humankind can only stand by and watch in impotent horror as the tide of the battle sways to and fro, until one monster stands alone and victorious.
1963 105m/C *JP* Inoshiro Honda, Michael Keith, Tadao Takashima; *Dir:* Thomas Montgomery. **VHS $9.95** *GKK* ✍✍½

King of the Kongo A handsome young man is sent by the government to Nuhalla, deep in the jungle, to break up a gang of ivory thieves. Serial was made in silent and sound versions.
1929 213m/B Jacqueline Logan, Boris Karloff, Richard Tucker; *Dir:* Richard Thorpe. **VHS, Beta $49.95** *NOS, VYY, MLB* ✍✍

King Kung Fu A karate master raises a gorilla, and sends it from Asia to the U.S. There, two out-of-work reporters decide to release it from captivity and then recapture it so they can get the story and some needed recognition. The background they don't have on the gorilla is that its master taught it kung fu.
1987 (G) 90m/C John Balee, Tom Leahy, Maxine Gray, Bill Schwartz; *Dir:* Bill Hayes. **VHS $29.95** *AVD* ✍

King of Kung-Fu A martial arts adventure featuring plenty of kung-fu kicks.
198? 90m/C Bobby Baker, Nam Chun Pan, Lam Chun Chi. **VHS, Beta $57.95** *UNI* ✍

King Lear Brook's version of Shakespeare tragedy. The king drives away the only decent daughter he has, and when he realizes this, it is too late. Powerful performances and interesting effort at updating the bard.
1971 137m/B *GB* Paul Scofield, Irene Worth, Jack MacGowran, Alan Webb, Cyril Cusack, Patrick Magee; *Dir:* Peter Brook. **VHS $29.95** *NOS, DVT* ✍✍✍½

King Lear Loosely adapted from Shakespeare's tragedy of a king who loses everything. A fragmental French existential reading with gangster undercurrent. This updated version is dark, ominous, wandering, and strange. Scripted by Norman Mailer and not like the other 1984 remake. Or any other remake, for that matter.
1987 91m/C *SI* Peter Sellars, Burgess Meredith, Molly Ringwald, Jean-Luc Godard, Woody Allen, Norman Mailer; *Dir:* Jean-Luc Godard. **VHS** *CAN* ✍½

The King on Main Street A king on vacation in America finds romance.
1925 ?m/B Adolphe Menjou, Bessie Love. **VHS, Beta $27.95** *DNB* ✍½

King of the Mountain The "Old King" and the "New King" must square off in this tale of daredevil road racers.
1981 (PG) 92m/C Richard Cox, Harry Hamlin, Dennis Hopper, Joseph Bottoms, Deborah Van Valkenburgh, Dan Haggerty; *Dir:* Noel Nosseck. **VHS, Beta $24.98** *SUE* ✍½

A King in New York Chaplin plays the deposed king of a European mini-monarchy who comes to the United States in hope of making a new life. Looks critically at 1950s-era America, including Cold War paranoia and over reliance on technology. Containing Chaplin's last starring performance, this film wasn't released in the U.S. until 1973. Uneven but interesting.
1957 105m/B *GB* Charlie Chaplin, Dawn Addams, Michael Chaplin, Oliver Johnston, Maxine Audley, Harry Green; *Dir:* Charlie Chaplin. **VHS, Beta $19.98** *FOX* ✍✍

King of New York Drug czar Frank White (Walken), recently returned from a prison sabbatical, regains control of his New York drug empire with the aid of a loyal network of black dealers. How? Call it dangerous charisma, an inexplicable sympatico. Headquartered in Manhattan's chic Plaza hotel, he ruthlessly orchestrates the drug machine, while funneling the profits into a Bronx hospital for the poor. The motive? As inscrutable as White himself, who's smooth, calculating, driven by some unfathomable vision. Walken, the master of the inscrutable, makes White's power tangible, believable,

yet never fathomable. Scripted by Nicholas St. John, directed by crime stylist/ cult director Ferrara ("China Girl," "Crime Story").
1990 106m/C Christopher Walken, Larry Fishburne, David Caruso, Victor Argo, Wesley Snipes, Janet Julian, Joey Chin, Giancarlo Esposito; *Dir:* Abel Ferrara. **VHS, LV $89.95** *LIV* ✍✍✍

King of the Pecos A young lawyer (who's also adept at the shootin' iron) exacts revenge on his parents' killers in the courtroom. Well made early Wayne outing, with good character development, taut pacing and beautiful photography.
1936 54m/B John Wayne, Muriel Evans, Cy Kendall, Jack Clifford, John Beck, Yakima Canutt; *Dir:* Joseph Kane. **VHS $12.98** *REP, MLB* ✍✍½

King, Queen, Knave Sexy black comedy about a shy, awkward 19-year old boy, keenly aware that his interest in girls is not reciprocated, who has to go live with his prosperous uncle and his much younger wife when his parents are killed. Based on the Vladimir Nabokov novel.
1974 94m/C *GE* Gina Lollobrigida, David Niven, John Moulder-Brown; *Dir:* Jerzy Skolimowski. **VHS, Beta $69.98** *SUE* ✍✍✍

King Ralph When the rest of the royal family passes away in a freak accident, lounge lizard Ralph finds himself the only heir to the throne. O'Toole is the long-suffering valet who tries to train him for the job. Funny in spots and Goodman is the quintessential good sport, making the whole outing pleasant. Sometimes too forced.
1991 (PG) 96m/C John Goodman, Peter O'Toole, Camille Coduri, Natasha Richardson, John Hurt; *Dir:* David S. Ward. **VHS, Beta, LV $19.95** *MCA, LDC, CCB* ✍✍

King Rat Drama set in a World War II Japanese prisoner-of-war camp. Focuses on the effect of captivity on the English, Australian and American prisoners. An American officer bribes his Japanese captors to live more comfortably than the rest. Based on James Clavell's novel.
1965 134m/B George Segal, Tom Courtenay, James Fox, James Donald, Denholm Elliott, Patrick O'Neal, John Mills; *Dir:* Bryan Forbes. **VHS, Beta $59.95** *COL* ✍✍✍

King of the Rocketmen Jeff King thwarts an attempt by traitors to steal government scientific secrets. Serial in twelve episodes. Later released as a feature titled "Lost Planet Airmen."
1949 156m/B Tristram Coffin, Mae Clark, I. Stanford Jolley; *Dir:* Fred Brannon. **VHS, LV $29.98** *MED, VCN, REP* ✍½

King of the Rodeo Cowpolk Hoot wins a rodeo and catches a thief in the Windy City. Hoot's final silent ride into the sunset.
1928 54m/B Hoot Gibson, Katherine Crawford, Slim Summerville, Charles French, Monte Montague, Joseph Girard; *Dir:* Henry MacRae. **VHS, Beta $14.95** *GPV* ✍✍

King Solomon's Mines The search for King Solomon's Mines leads a safari through the treacherous terrain of the desert, fending off sandstorms, Zulus, and a volcanic eruption. Adapted from the novel by H. Rider Haggard and remade twice.
1937 80m/B Cedric Hardwicke, Paul Robeson, Roland Young, John Loder, Anna Lee; *Dir:* Robert Stevenson. **VHS, Beta $16.95** *SNC, NOS, HMV* ✍✍✍

King Solomon's Mines A lavish version of the classic H. Rider Haggard adventures about African explorers searching out a fabled diamond mine. Remake of the 1937 classic; remade again in 1985. Oscar nomination for Best Picture.
1950 102m/C Stewart Granger, Deborah Kerr, Richard Carlson, Hugo Haas, Lowell Gilmore; **Dir:** Compton Bennett. Academy Awards '50: Best Color Cinematography, Best Film Editing. **VHS, Beta $14.98** MGM ♫♫♫

King Solomon's Mines The third remake of the classic H. Rider Haggard novel about a safari deep into Africa in search of an explorer who disappeared while searching for the legendary diamond mines of King Solomon. Updated but lacking the style of the previous two films. Somewhat imperialistic, racist point of view.
1985 (PG-13) 101m/C Richard Chamberlain, John Rhys-Davies, Sharon Stone, Herbert Lom; **Dir:** J. Lee Thompson. **VHS, Beta $14.98** MGM ♫♫ ½

King Solomon's Mines This cartoon version of the H. Rider Haggard novel chronicles the adventures of big game hunter Alan Quartermain as he searches for a missing man and a fantastic lost treasure.
1986 60m/C Voices: Tom Burlinson, Arthur Dignam. **VHS, Beta $19.95** LTG ♫♫ ½

King Solomon's Treasure The great white adventurer takes on the African jungle, hunting for hidden treasure in the Forbidden City. Britt Ekland stars as a Phoenician Queen—need we say more??
1976 90m/C CA GB David McCallum, Britt Ekland, Patrick Macnee, Wilfrid Hyde-White; **Dir:** Alvin Rakoff. **VHS, Beta $59.95** UHV ♫

King of the Stallions A dangerous stallion is a menacing threat to the herd and the cowboys and Indians and other scattered wildlife. Chief Thundercloud is the only one capable of tackling the matter. Largely Indian cast, including Iron Eyes Cody, famed for the anti-litter campaign.
1942 63m/B Chief Thundercloud, Princess Bluebird, Chief Yowlachie, Rick Vallin, Dave O'Brien, Barbara Felker, Iron Eyes Cody; **Dir:** Edward Finney. **VHS $19.95** NOS, DVT ♫ ½

King of the Texas Rangers Tom King, Texas Ranger, finds that his father's killers are a group of saboteurs who have destroyed American oil fields. This twelve episode serial comes on two tapes.
1941 195m/B Sammy Baugh, Neil Hamilton; **Dir:** William Witney, John English. **VHS $59.95** REP, MRV, MLB

King of the Wild Horses Engaging children's story about a wild stallion that comes to the rescue. Terrific action photography of Rex going through his paces. Look for Chase in a rare dramatic role. Silent.
1924 50m/B Rex, Edna Murphy, Leon Bary, Pat Hartigan, Frank Butler, Charley Chase; **Dir:** Fred Jackman. **VHS, Beta $27.95** DNB ♫♫♫

King of the Zombies A scientist creates his own zombies without souls, to be used as the evil tools of a foreign government.
1941 67m/B John Archer, Dick Purcell, Mantan Moreland, Henry Victor, Joan Woodbury; **Dir:** Jean Yarborough. **VHS, Beta $24.95** NOS, MRV, SNC ♫ ½

Kingdom of the Spiders A desert town is invaded by swarms of killer tarantulas, which begin to consume townspeople.
1977 (PG) 90m/C William Shatner, Tiffany Bolling, Woody Strode; **Dir:** John Cardos. **VHS, Beta $19.95** UHV, FCT ♫♫ ½

Kingfisher Caper A power struggle between a businessman, his brother, and a divorced sister is threatening to rip a family-owned diamond empire apart.
1976 (PG) 90m/C Hayley Mills, David McCallum; **Dir:** Dirk DeVilliers. **VHS, Beta $9.95** MED ♫♫

Kingfisher the Killer Charles, the Raging Titan of Ninja skill, stalks the bad gang that killed his mother and hits them with nunchakos.
198? 90m/C Sho Kosugi. **VHS, Beta $39.95** TWE ♫

King's Christmas The renowned Kings' Singers perform some favorite Christmas tunes.
1983 27m/C VHS, Beta $19.95 PRS ♫

Kings and Desperate Men Apprentice terrorists plea their case at a public forum when they take a rakish talk show host hostage.
1983 118m/C GB Patrick McGoohan, Alexis Kanner, Andrea Marcovicci, Robin Spry, Frank Moore; **Dir:** Alexis Kanner. **VHS $79.98** MAG ♫ ½

Kings Go Forth Hormones and war rage in this love triangle set against the backdrop of WWII France. Sinatra loves Wood who loves Curtis. When Sinatra asks for her hand in marriage she refuses because she is mixed—half black, half white. He says it doesn't matter, but she still declines because she's in love with Curtis. When Sinatra tells Curtis that Wood is mixed, Curtis says that it doesn't matter to him either because he'd never planned on marrying her. Meanwhile, the war continues. Not particularly satisfying on either the war or race front. Based on a novel by Joe David Brown.
1958 109m/D Frank Sinatra, Tony Curtis, Natalie Wood, Leora Dana; **Dir:** Delmer Daves. **VHS, Beta $19.98** MGM ♫♫ ½

Kings, Queens, Jokers Package includes a foolish lover and ghosts in "Haunted Spooks," an escaped convict in "Dangerous Females," and a search for lost jewels in "Stolen Jools."
193? 60m/B Harold Lloyd, Marie Dressler, Polly Moran, Edward G. Robinson, Gary Cooper, Joan Crawford. **VHS, Beta** CCB

King's Ransom When the Japanese Emperor's Pearl is stolen, the underworld is abuzz with excitement. But only the wealthiest individuals are invited to bid for the jewel at Cameron King's secret casino. There, untold pleasures, and dangers, await them.
1991 ?m/C Miles O'Keefe, Dedee Pfeiffer, Christopher Atkins; **Dir:** Hugh Parks. **VHS** FMP Woof!

The King's Rhapsody Flynn is an heir who abdicates his throne in order to be with the woman he loves. But when the king dies, Flynn sacrifices love for honor and goes back to marry a princess. Very dated and Flynn is long past his swashbuckling-romantic days.
1955 93m/C GB Errol Flynn, Anna Neagle, Patrice Wymore, Martita Hunt, Finlay Currie, Frank Wolff, Joan Benham, Reginald Tate, Miles Malleson; **Dir:** Herbert Wilcox. **VHS $29.95** FCT ♫♫

Kings of the Road (In the Course of Time) A writer and a film projectionist travel by truck across Germany. The film stands out for its simple and expressive direction, and truly captures the sense of physical and ideological freedom of being on the road. A landmark Wenders epic. In German with English subtitles.

1976 176m/B GE Ruediger Vogler, Hanns Zischler, Lisa Kreuzer; **Dir:** Wim Wenders. **VHS, Beta $29.95** PAV, FCT, GLV ♫♫♫ ½

Kings Row The Harry Bellamann best-selling Middle American potboiler comes to life. Childhood friends grow up with varying degrees of success, in a decidedly macabre town. All are continually dependent on and inspired by Parris (Cummings) a psychiatric doctor and genuine gentleman. Many cast members worked against type with unusual success. Warner held the film for a year after its completion, in concern for its dark subject matter, but it received wide acclaim. Shot completely within studio settings - excellent scenic design by William Cameron Menzies, and wonderful score by Erich W. Korngold. Oscar nominations: Black & White Cinematography, Best Director and Best Picture.
1941 127m/B Ann Sheridan, Robert Cummings, Ronald Reagan, Betty Field, Charles Coburn, Claude Rains, Judith Anderson, Nancy Coleman, Karen Verne, Maria Ouspenskaya, Harry Davenport, Ernest Cossart, Pat Moriarity; **Dir:** Sam Wood. **VHS, Beta $19.98** MGM, FOX, CCB ♫♫♫

Kinjite: Forbidden Subjects A cop takes on a sleazy pimp whose specialty is recruiting teenage girls, including the daughter of a Japanese business man. Slimy, standard Bronson fare.
1989 (R) 97m/C Charles Bronson, Juan Fernandez, Peggy Lipton; **Dir:** Lee Thompson. **VHS, Beta, LV $19.95** CAN ♫♫

A Kink in the Picasso MTV personality Daddo joins a cast of criminals who chase after a Picasso artwork and a counterfeit. Artless Australian production.
1990 84m/C Peter O'Brien, Jon Finlayson, Jane Clifton, Andrew Daddo; **Dir:** Marc Gracie. **VHS $79.95** ATL ♫

Kino Pravda/Enthusiasm Two famous documentaries by Vertov, the first a prime example of his pioneering avant-garde newsreel that helped establish Soviet montage aesthetic, and the second the famous portrait of Don Basin coal miners, augmented by the famous use of sound editing/montage. In Russian with English titles.
1931 120m/B RU Dir: Dziga Vertov. **VHS, Beta $69.95** FCT, FST ♫♫

Kipling's Women Supposedly based on Rudyard Kipling's "The Ladies," only with much more sexual content.
1963 50m/C Margie Sutton, Lisa Gordon, Felix de Cola. **VHS, Beta $19.95** VDM ♫

Kipperbang During the summer of 1948, a 13-year-old boy wishes he could kiss the girl of his dreams. He finally gets the chance in a school play. Not unsatisfying, but falls short of being the bittersweet coming-of-age drama it could have been.
1982 (PG) 85m/C GB John Albasiny, Abigail Cruttenden, Alison Steadman; **Dir:** Michael Apted. **VHS, Beta $59.95** MGM ♫

Kipps Based on H. G. Wells' satirical novel. A young British shopkeeper inherits a fortune and tries to join high society while neglecting his working-class girlfriend. Also known as "The Remarkable Mr. Kipps." Set the stage for the musical "Half a Sixpence."
1941 95m/B GB Michael Redgrave, Phyllis Calvert, Michael Wilding; **Dir:** Carol Reed. **VHS, Beta $19.95** NOS, HHT, CAB ♫♫♫

The Kirlian Witness A woman uses the power of telepathic communication with house plants to solve her sister's murder. Also known as, "The Plants Are Watching."

1978 (PG) 88m/C Nancy Snyder, Joel Colodner, Ted Leplat; *Dir:* Jonathan Sarno. **VHS, Beta** $59.98 *MAG* ℐ

Kismet Original screen version of the much filmed lavish Arabian Nights saga (remade in '30, '44 and '55). A beggar is drawn into deception and intrigue among Bagdad upper-crusters. Glorious sets and costumes; silent with original organ score.
1920 98m/C *Dir:* Louis Gasnier. **VHS** $29.95 *VYY* ℐℐℐ

Kismet A big budget Arabian Nights musical drama of a Baghdad street poet who manages to infiltrate himself into the Wazir's harem. The music was adapted from Borodin by Robert Wright and George Forrest, producing such standards as "Baubles, Bangles, and Beads," "Stranger in Paradise," and "And This Is My Beloved." Original trailer and letterboxed screen available in special laserdisc edition.
1955 113m/C Howard Keel, Ann Blyth, Dolores Gray, Vic Damone; *Dir:* Vincente Minnelli. **VHS, Beta, LV** $19.95 *MGM, PIA, FCT* ℐℐ

The Kiss Garbo, the married object of earnest young Ayres' lovelorn affection, innocently kisses him nighty night since a kiss is just a kiss. Or so she thought. Utterly misconstrued, the platonic peck sets the stage for disaster, and murder and courtroom anguish follow. French Feyder's direction is stylized and artsy. Garbo's last silent and Ayres' first film.
1929 89m/B Greta Garbo, Conrad Nagel, Holmes Herbert, Lew Ayres, Anders Randolph; *Dir:* Jacques Feyder. **VHS** $29.98 *MGM, FCT* ℐℐℐ

The Kiss A kind of "Auntie Mame from Hell" story in which a mysterious aunt visits her teen-age niece in New York, and tries to apprentice her to the family business of sorcery, demon possession, and murder. Aunt Felicity's kiss will make you appreciate the harmless cheek-pinching of your own aunt; your evening would be better spent with her, rather than this movie.
1988 (R) 98m/C Pamela Collyer, Peter Dvorsky, Joanna Pacula, Meredith Salenger, Mimi Kuzyk, Nicholas Kilbertus, Jan Rubes; *Dir:* Pen Densham. **VHS, Beta, LV** $19.95 *COL, IME* ℐ

Kiss and Be Killed Crazy guy with knife cuts short couple's wedding night. Shortchanged widow is mighty miffed. Will she have to return the wedding gifts?
1991 89m/C Caroline Ludvik, Crystal Carson, Tom Reilly, Chip Hall, Ken Norton, Jimmy Baio; *Dir:* Tom Milo. **VHS, Beta, 8mm** $79.95 *MNC* ℐ½

A Kiss Before Dying A botched adaptation of Ira Levin's cunning thriller novel (filmed in 1956), this serves up a burst head in the first few minutes. So much for subtlety. The highlight is Dillon's chilly role as a murderous opportunist bent on marrying into a wealthy family. Young plays two roles (not very well) as lookalike sisters on his agenda. The ending was hastily reshot and it shows.
1991 (R) 93m/C Matt Dillon, Sean Young, Max von Sydow, Diane Ladd, James Russo; *Dir:* James Dearden. **VHS, Beta, LV** $34.98 *MCA, PIA* ℐℐ

Kiss Daddy Goodbye A widower keeps his two children isolated in order to protect their secret telekinetic powers. When he is killed by bikers, the kids attempt to raise him from the dead. Also on video as "Revenge of the Zombie" and "The Vengeful Dead."

1981 (R) 81m/C Fabian Forte, Marilyn Burns, Jon Cedar, Marvin Miller; *Dir:* Patrick Regan. **VHS, Beta** $59.95 *MON, MRV, IVE* ℐ½

Kiss Daddy Goodnight A Danish-made thriller about a beautiful young girl who seduces men, drugs them and takes their money. One man turns the tables on her, however, and decides that she will only belong to him.
1987 (R) 89m/C *DK* Uma Thurman, Paul Dillon, Paul Richards, David Brisbin; *Dir:* P.I. Huemer. **VHS, Beta** $29.95 *ACA* ℐ

Kiss of Death Paroled when he turns state's evidence, Mature must now watch his back constantly. Widmark, in his film debut, seeks to destroy him. Police chief Donlevy tries to help. Filmed on location in New York, this gripping and gritty film is a vision of the most terrifying sort of existence, one where nothing is certain, and everything is dangerous. Excellent. Oscar nominations for supporting actor Widmark and the original story.
1947 98m/B Victor Mature, Richard Widmark, Anne Grey, Brian Donlevy, Karl Malden, Coleen Gray; *Dir:* Henry Hathaway. **VHS** $39.98 *FOX* ℐℐℐ

Kiss and Kill Lee returns in his fourth outing as Fu Manchu. This time the evil one has injected beautiful girls with a deadly poison that reacts upon kissing. They are then sent out to seduce world leaders. Not on par with the previous movies, but still enjoyable. Also known as "Blood of Fu Manchu" and "Against All Odds." Sequel to "Castle of Fu Manchu."
1968 (R) 91m/C Christopher Lee, Richard Greene, Shirley Eaton, Tsai Chin, Maria Rohm, H. Marion Crawford; *Dir:* Jess (Jesus) Franco. **VHS** $14.99 *AVD, MRV* ℐℐ

A Kiss for Mary Pickford A rare, hilarious cinematic oddity, a film formulated from Kuleshov montage techniques from footage of the famous American couple's visit to Russia in 1926, wherein a regular guy tries to win a girl through friendship with the stars. Fairbanks and Pickford didn't know of their role in this film, being spliced in later according to montage theory.
1927 70m/B Mary Pickford, Douglas Fairbanks Sr., Igor Ilinsky; *Dir:* Sergei Komarov. **VHS, Beta** $49.95 *FCT* ℐℐ½

Kiss Me Deadly Aldrich's adaptation of Mickey Spillane's private eye tale has been called the meeting of "art" and pulp literature. At the least, its the meeting of high concept and low brow. Meeker, as Mike Hammer, is a self interested rough and tumble all American dick (detective, that is). When a woman to whom he happened to give a ride is found murdered, he follows the mystery (because he sniffs a golden goose) straight into a nuclear conspiracy. A pretty standard spy story that Aldrich, with tongue deftly in cheek, styles a message through the medium. His pugilistic editorial style (one scene cuts to the next with Hammer falling unconscious and waking), topsy turvy camerawork and rat-a-tat-tat pacing tell volumes about Hammer, the world he orbits, and that special fifties kind of paranoia. Now a cult fave, not always critic raved, it's considered to be the American grandaddy to the French New Wave. Cinematography by Ernest Laszlo.
1955 105m/B Ralph Meeker, Albert Dekker, Paul Stewart, Wesley Addy, Cloris Leachman, Strother Martin, Marjorie Bennett; *Dir:* Robert Aldrich. **VHS, Beta** $19.98 *MGM, FCT* ℐℐℐ½

Kiss Me Goodbye Young widow Fields can't shake the memory of her first husband, a charismatic but philandering Broadway choreographer, who's the antithesis of her boring but devoted professor fiance. She struggles with the charming ghost of her first husband, as well as her domineering mother, attempting to understand her own true feelings. Harmless but two-dimensional remake of "Dona Flor and Her Two Husbands."
1982 (PG) 101m/C Sally Field, James Caan, Jeff Bridges, Paul Dooley, Mildred Natwick, Claire Trevor; *Dir:* Robert Mulligan. **VHS, Beta** $59.98 *FOX* ℐℐ

Kiss Me Kate A married couple can't separate their real lives from their stage roles in this musical-comedy screen adaptation of Shakespeare's "Taming of the Shrew," based on Cole Porter's Broadway show. Bob Fosse bursts from the screen—particularly if you see the 3-D version—when he does his dance number.
1953 110m/C Kathryn Grayson, Howard Keel, Ann Miller, Tommy Rall, Bob Fosse, Bobby Van, Keenan Wynn, James Whitmore; *Dir:* George Sidney. **VHS, Beta, LV** $19.95 *MGM, PIA, FCT* ℐℐℐ

Kiss Me, Kill Me A confused woman is on the run after she may have murdered someone.
1969 (R) 91m/C Carroll Baker, George Eastman, Isabelle DeFunes, Ely Gallo. **VHS, Beta** *PGN* ℐℐ½

Kiss Me a Killer The ancient thriller plot about a young wife and her lover who scheme to kill her middle-aged husband. It's set in L.A.'s Latino community, but extensive shots of salsa music and ethnic cooking hardly raise this above the mediocre.
1991 (R) 91m/C Julie Carmen, Robert Beltran, Guy Boyd, Ramon Franco, Charles Boswell; *Dir:* Marcus De Leon. **VHS, LV** $89.95 *COL* ℐℐ

Kiss Me, Stupid! Once condemned as smut, this lesser Billy Wilder effort now seems no worse than an average TV sitcom. Martin basically plays himself as a horny Vegas crooner stranded in the boondocks. A local songwriter wants Dino to hear his tunes but knows the cad will seduce his pretty wife, so he hires a floozy to pose as the tempting spouse. It gets better as it goes along, but the whole thing suffers from staginess, being an adaptation of an Italian play "L'Ora Della Fantasia" by Anna Bonacci.
1964 (PG) 126m/B Dean Martin, Kim Novak, Ray Walston, Felicia Farr, Cliff Osmond, Barbara Pepper, Doro Merande, Howard McNear, Henry Gibson, John Fiedler, Mel Blanc; *Dir:* Billy Wilder. **VHS, Beta** $19.98 *MGM* ℐℐ

KISS Meets the Phantom of the Park The popular '70s rock band is featured in this Dr. Jekyll-esque Halloween horror tale, interspersed with musical numbers.
1978 96m/C Gene Simmons, Paul Stanley, Peter Criss, Ace Frehley, Anthony Zerbe, Deborah Ryan, John Dennis Johnston; *Dir:* Gordon Hessler. **VHS, Beta** $10.98 *MVD, WOV* ℐℐ½

Kiss My Grits A good ole boy hightails it to Mexico with his girlfriend and son, chased by mobsters and the law.
1982 (PG) 101m/C Bruce Davison, Anthony (Tony) Franciosa, Susan George, Bruno Kirby; *Dir:* Jack Starrett. **VHS, Beta** $49.95 *MED* ℐ

Kiss the Night An unfortunate hooker falls head over heels for a man she had the pleasure of doing business with. To her chagrin, it is much harder to leave the street than she thought. A dramatic and timely presentation.

1987 (R) 99m/C Patsy Stephens, Warwick Moss, Gary Aron Cook; *Dir:* James Ricketson. **VHS** $79.95 *MCG* 🎬

Kiss Shot Goldberg is a struggling single mother who loses her job but still must make the mortgage payments. She takes a job as a waitress but realizes it isn't going to pay the bills so she tries her hand as a pool hustler. Frantz is the promoter who finances her bets and Harewood, the pool-shooting playboy whose romantic advances are destroying her concentration.
199? 88m/C Whoopi Goldberg, Dennis Franz, Dorian Harewood, David Marciano, Teddy Wilson; *Dir:* Jerry London. **VHS, Beta** $89.95 *ACA* 🎬🎬

Kiss of the Spider Woman From the novel by Manuel Puig, an acclaimed drama concerning two cell mates in a South American prison, one a revolutionary, the other a homosexual. Literate, haunting, powerful.
1985 (R) 119m/C *BR* William Hurt, Raul Julia, Sonia Braga, Jose Lewgoy; *Dir:* Hector Babenco. Academy Awards '85: Best Actor (Hurt); Cannes Film Festival '85: Best Actor (Hurt). **VHS, Beta, LV, 8mm** $14.98 *CHA, BTV* 🎬🎬🎬

Kiss of the Tarantula The story of a young girl and her pet spiders as they spin a deadly web of terror.
1975 (PG) 85m/C Eric Mason; *Dir:* Chris Munger. **VHS, Beta** $19.95 *MON, MPI, FCT* 🎬

Kiss Tomorrow Goodbye A brutal, murderous escaped convict rises to crime-lord status before his inevitable downfall. Based on a novel by Horace McCoy.
1950 102m/B James Cagney, Barbara Payton, Ward Bond, Luther Adler, Helena Bonham Carter, Steve Brodie, Rhys Williams; *Dir:* Gordon Douglas. **VHS, LV** $19.98 *REP* 🎬🎬

Kissin' Cousins An Air Force man on a secluded base in the South discovers a local hillbilly is his double. Typical Presley film that includes country tunes such as "Smokey Mountain Boy" and "Barefoot Ballad" as well as the title song.
1963 96m/C Elvis Presley, Arthur O'Connell, Jack Albertson, Glenda Farrell, Pam Austin, Yvonne Craig, Cynthia Pepper, Donald Woods, Tommy Farrell, Beverly Powers; *Dir:* Gene Nelson. **VHS, Beta** $19.95 *MGM* 🎬½

The Kissing Place A woman who abducted her "son" years ago tracks the boy to New York City to prevent him from finding his biological family.
1990 90m/C Meredith Baxter Birney, David Ogden Stiers; *Dir:* Tony Wharmby. **VHS, Beta** $29.95 *PAR* 🎬½

Kit Carson Frontiersman Kit Carson leads a wagon train to California, fighting off marauding Indians all the way.
1940 97m/B Jon Hall, Dana Andrews, Ward Bond, Lynn Bari; *Dir:* George B. Seitz. **VHS, Beta** $14.95 *MED* 🎬🎬

The Kitchen Toto Set in 1950 Kenya as British rule was being threatened by Mau Mau terrorists. A young black boy is torn between the British for whom he works and the terrorists who want him to join them. It is a complex and powerful story of the Kenyan freedom crusade.
1987 (PG-13) 96m/C *GB* Bob Peck, Phyllis Logan, Robert Urquhart, Edward Judd; *Dir:* Harry Hook. **VHS, Beta, LV** $79.95 *WAR* 🎬🎬🎬

Kitty and the Bagman A comedy about two rival madams who ruled Australia in the 1920's.
1982 (R) 95m/C *AU* Liddy Clark; *Dir:* Donald Crombie. **VHS, Beta** $59.98 *SUE* 🎬🎬

Kitty Foyle From the novel by Christopher Morley, Rogers portrays the white-collar working girl whose involvement with a married man presents her with both romantic and social conflicts. Oscar nominations: Best Picture, Best Director, Best Screenplay.
1940 108m/B Ginger Rogers, Dennis Morgan, James Craig, Gladys Cooper, Ernest Cossart, Eduardo Ciannelli; *Dir:* Sam Wood. Academy Awards '40: Best Actress (Rogers). **VHS, Beta** $19.98 *VID, MED, CCB* 🎬🎬🎬

The Klansman Race relations come out on the short end in this film about a sheriff trying to keep the lid on racial tensions in a southern town. Even the big-name cast can't save what comes off as a nighttime soaper rather than a serious drama. Also known as "Burning Cross."
1974 (R) 112m/C Lee Marvin, Richard Burton, Cameron Mitchell, Lola Falana, Lucianna Paluzzi, Linda Evans, O.J. Simpson; *Dir:* Terence Young. **VHS** $14.95 *PAR, FCT* 🎬

Klondike Fever Join the young Jack London as he travels from San Francisco to the Klondike fields during the Great Gold Rush of 1898.
1979 (PG) 118m/C Rod Steiger, Angie Dickinson, Lorne Greene; *Dir:* Peter Carter. **VHS, Beta** $59.95 *LTG* 🎬½

Klute A small-town policeman (Sutherland) comes to New York in search of a missing friend and gets involved with a prostitute/would-be actress (Fonda) being stalked by a killer. Intelligent, gripping drama. Oscar nomination for Original Screenplay.
1971 (R) 114m/C Jane Fonda, Donald Sutherland, Charles Cioffi, Roy Scheider, Rita Gam, Jean Stapleton; *Dir:* Alan J. Pakula. Academy Awards '71: Best Actress (Fonda); New York Film Critics Awards '71: Best Actress (Fonda); National Society of Film Critics '71: Best Actress (Fonda). **VHS, Beta, LV** $19.98 *WAR, BTV* 🎬🎬🎬½

The Klutz A bumbling fool, on the way to visit his girlfriend, becomes accidentally embroiled in a bank robbery and other ridiculous mishaps. Light French comedy; dubbed.
1973 87m/C *FR* Claude Michaud, Louise Portal, Guy Provost; *Dir:* Pierre Rose. **VHS, Beta** $19.95 *STE, NWV* 🎬🎬

Knick Knack The brilliantly funny computer animated short that for its time represented the apex of image technology; a frustrated snowman trapped inside a paperweight tries to get out to pursue a sexy desk object. A capella score by Bobby McFerrin.
1989 4m/C **VHS** $24.95 *DCL* 🎬🎬

Knife in the Water A journalist, his wife and a hitchhiker spend a day aboard a sailboat. Sex and violence can't be far off. Tense psychological drama. Served as director Polanski's debut. In Polish with English subtitles. Originally titled "Noz w Wodzie."
1962 94m/B *PL* Leon Niemczyk, Jolanta Umecka, Zygmunt Malandowicz; *Dir:* Roman Polanski. New York Film Festival '63: International Critics Award; Venice Film Festival '62: International Critics Award. **VHS, Beta, LV** $39.95 *HMV, FOX, VYY* 🎬🎬🎬🎬

Knight Without Armour A journalist opposed to the Russian monarchy falls in love with the daughter of a czarist minister in this classic romantic drama.
1937 107m/B *GB* Marlene Dietrich, Robert Donat; *Dir:* Jacques Feyder. **VHS, Beta** $14.98 *SUE* 🎬🎬🎬

Knightriders The story of a troupe of motorcyclists who are members of a traveling Renaissance Fair.

1981 (R) 145m/C Ed Harris, Gary Lahti, Tom Savini, Amy Ingersoll; *Dir:* George A. Romero. **VHS, Beta** $54.95 *MED* 🎬🎬½

Knights of the City Miami street gangs fight each other over their "turf." One leader decides to opt out, using music.
1986 (R) 87m/C Nicholas Campbell, Stoney Jackson, Leon Isaac Kennedy, John Mengatti, Janine Turner, Michael Ansara; *Dir:* Dominic Orlando. **VHS, Beta** $14.95 *NWV, STE* 🎬

Knights & Emeralds Cross-cultural rivalries and romances develop between members of two high school marching bands in a British factory town as the national band championships draw near.
1987 (PG) 90m/C *GB* Christopher Wild, Beverly Hills, Warren Mitchell; *Dir:* Ian Emes. **VHS, Beta** $19.98 *WAR* 🎬½

Knights of the Range Vintage western.
1940 66m/B Russell Hayden, Victor Jory; *Dir:* Lesley Selander. **VHS, Beta** $29.95 *VCN, MLB* 🎬½

Knights of the Round Table The story of the romantic triangle between King Arthur, Sir Lancelot and Guinevere during the civil wars of sixth century England.
1954 106m/C Robert Taylor, Ava Gardner, Mel Ferrer, Anne Crawford, Felix Aylmer, Stanley Baker; *Dir:* Richard Thorpe. **VHS, Beta, LV** $19.95 *MGM* 🎬🎬½

Knock on Any Door A young hoodlum from the slums is tried for murdering a cop. He is defended by a prominent attorney who has known him from childhood.
1949 100m/B Humphrey Bogart, John Derek, George Macready; *Dir:* Nicholas Ray. **VHS, Beta, LV** $9.95 *COL* 🎬🎬

Knock Outs Samantha not only loses her shirt at a sorority strip poker marathon but she loses her tuition money to a gang of biker chicks. Then Samantha and a bevy of bikini-clad friends decide to pose for a swimsuit calendar to earn some cash but are secretly videotaped in the nude by some local sleaze promoters. Tired of being taken advantage of these lovelies take up martial arts and challenge their nefarious girl biker rivals to a winner-take-all wrestling match, not forgetting about the purveyors of the not-so-secret videotape.
1992 90m/C Chona Jackson, Sindi Rome, Brad Zutaut; *Dir:* John Bowen. **VHS** $79.95 *HMD* 🎬

The Knockout/Dough and Dynamite Two Chaplin shorts: "The Knockout" (1914), in which Charlie referees a big fight; "Dough and Dynamite" (1914), in which a labor dispute at a bake shop leaves Charlie in charge when the regular baker walks out.
1914 54m/B Charlie Chaplin, Fatty Arbuckle, Mabel Normand, The Keystone Cops; *Dir:* Charlie Chaplin, Sydney Chaplin. **VHS, Beta** $29.95 *VYY* 🎬🎬🎬

Knute Rockne: All American Life story of Notre Dame football coach Knute Rockne, who inspired many victories with his powerful speeches. Ronald Reagan, as the dying George Gipp, utters that now-famous line, "Tell the boys to win one for the Gipper."
1940 98m/B Ronald Reagan, Pat O'Brien, Gale Page, Donald Crisp, John Qualen; *Dir:* Lloyd Bacon. **VHS, Beta** $19.98 *FHE, MGM* 🎬🎬🎬

Knute Rockne: The Rock of Notre Dame This program tells of the football career of Knute Rockne, who helped to make Notre Dame's football team the most famous. Classic newsreel footage.
1954 15m/B VHS, Beta $49.95 TSF ☾☾☾

Kojiro A sprawling epic by prolific Japanese director Inagaki about a dashing rogue's bid for power in feudal times. With English subtitles.
1967 152m/C JP Kikunosuke Onoe, Tatsuya Nakadai; *Dir:* Hiroshi Inagaki. VHS, Beta VDA ☾☾☾

Koko Cartoons: No. 1 A collection of Koko's best cartoons, including "Flies," "Koko's Cartoon Factory," "Koko's Thanksgiving," "Sweet Adeline," and "The Fortune Teller."
1927 50m/C *Dir:* Max Fleischer. VHS, Beta GVV

Koko the Clown, Vol. 1 Examples from this silent cartoon series include "The Tantalizing Fly," "Perpetual Motion," "The Clown's Little Brother," "Ouija Board," "Bubbles," "Modeling," "In the Good Old Summertime," and "Koko the Kop."
193? 55m/B VHS, Beta, LV $19.95 VDM

Kolberg The true story of a Prussian town heroically withstanding Napoleon. Produced by Joseph Goebbels in the last days of the Third Reich, it is best remembered as the film whose expensive production and momentous use of real German soldiers, supplies and ammunition eventually helped to fell the Axis war machine. In German with English subtitles.
1945 118m/C GE Kristina Soderbaum, Heinrich George, Horst Caspar, Paul Wegener; *Dir:* Veit Harlan. VHS, Beta, 3/4U $39.95 FCT, IHF, GLV ☾

Konrad Konrad, an "instant" child made in a factory, is accidentally delivered to an eccentric woman. The factory wants him back when the mistake is discovered, but Konrad stands up against it. Inlcudes a viewers' guide.
199? 110m/C Ned Beatty, Polly Holliday. VHS $29.95 PME ☾

Koroshi Secret Agent John Drake is dispatched to Hong Kong to disband a secret society who is killing off international political figures.
1967 100m/C Patrick McGoohan, Kenneth Griffith, Yoko Tani; *Dir:* Michael Truman, Peter Yates. VHS, Beta $59.95 MPI ☾½

Kostas A migrant Greek taxi driver romantically pursues a wealthy Australian divorcee in England.
1979 (R) 88m/C Takis Emmanuel, Wendy Hughes; *Dir:* Paul Cox. VHS, Beta $69.98 LIV, VES ☾☾

Kotch An elderly man resists his children's attempts to retire him. Warm detailing of old age with a splendid performance by Matthau. Lemmon's directorial debut. Matthau was Oscar nominated, as was the sound.
1971 (PG) 113m/C Walter Matthau, Deborah Winters, Felicia Farr; *Dir:* Jack Lemmon. VHS $59.98 FOX ☾☾☾

Kovacs! The television humorist Ernie Kovacs is examined primarily through showing his old comedy routines.
1971 85m/C Ernie Kovacs. VHS, Beta $19.95 RHI ☾☾☾

Kowloon Assignment Comic book superhero Golgo 13 comes to life in this action film dealing with a drug war in the Hong Kong underground.
1977 93m/C Sonny Chiba. VHS, Beta $9.98 SUE

Koyaanisqatsi A mesmerizing film that takes an intense look at modern life (the movie's title is the Hopi word for 'life out of balance'). Without dialogue or narration, it brings traditional background elements, landscapes and cityscapes, up front to produce a unique view of the structure and mechanics of our daily lives. Riveting and immensely powerful. A critically acclaimed score by Philip Glass, and Reggio's cinematography prove to be the perfect match to this brilliant film. Followed by "Powaqqatsi."
1983 87m/C *Dir:* Godfrey Reggio. VHS, Beta, LV $14.95 MVD, PAV, FCT ☾☾☾

Kramer vs. Kramer The highly acclaimed family drama about an advertising executive husband and child left behind when their wife and mother leaves on a quest to find herself, and the subsequent courtroom battle for custody when she returns. Hoffman and Streep give exacting performances as does young Justin Henry. Successfully moves you from tears to laughter and back again. Based on the novel by Avery Corman. Oscar nominated Cinematography and Film Editing, in addition to the several awards it won.
1979 (PG) 105m/C Dustin Hoffman, Meryl Streep, Jane Alexander, Justin Henry, Howard Duff, JoBeth Williams; *Dir:* Robert Benton. Academy Awards '79: Best Actor (Hoffman), Best Adapted Screenplay, Best Director (Benton), Best Picture, Best Supporting Actress (Streep); Los Angeles Film Critics Association Awards '79: Best Actor (Hoffman), Best Director (Benton), Best Picture, Best Screenplay, Best Supporting Actress (Streep); New York Film Critics Awards '79: Best Actor (Hoffman), Best Film, Best Supporting Actress (Streep); Directors Guild of America Awards '79: Best Director (Benton). VHS, Beta, LV $19.95 COL, PIA, BTV ☾☾☾½

The Krays An account of British gangsters Reggie and Ronnie Kray, the brothers who ruled London's East End with brutality and violence, making them bizarre celebrities of the sixties. The leads are portrayed by Gary and Martin Kemp, founders of the British pop group Spandau Ballet.
1990 (R) 119m/C GB Gary Kemp, Martin Kemp, Billie Whitelaw, Steven Berkoff, Susan Fleetwood, Charlotte Cornwell; *Dir:* Peter Medak. VHS, LV, 8mm $19.95 COL, FCT ☾☾☾½

Kriemhilde's Revenge The second film, following "Siegfried," of Lang's "Die Nibelungen," a lavish silent version of the Teutonic legends Wagner's "Ring of the Nibelungen" was based upon.
1924 95m/B GE Paul Richter, Margareta Schoen; *Dir:* Fritz Lang. VHS, LV $19.95 CCB, IHF, REP ☾☾☾☾

Kronos A giant robot from space drains the Earth of all its energy resources. A good example of this genre from the 50s. Includes previews of coming attractions from classic science fiction films.
1957 78m/B Jeff Morrow, Barbara Lawrence, John Emery; *Dir:* Kurt Neumann. VHS, Beta, LV $19.95 MED, MLB ☾☾½

Krull Likeable fantasy adventure set in a world peopled by creatures of myth and magic. A prince embarks on a quest to find the Glaive (a magical weapon) and then rescues his young bride, taken by the Beast of the Black Fortress.
1983 (PG) 121m/C GB Ken Marshall, Lysette Anthony, Freddie Jones, Francesca Annis, Liam Neeson; *Dir:* Peter Yates. VHS, Beta, LV $14.95 GKK ☾☾

Krush Groove The world of rap music is highlighted in this movie, the first all-rap musical.
1985 (R) 95m/C Blair Underwood, Eron Tabor, Kurtis Blow, Fat Boys, Sheila E; *Dir:* Michael Schultz. VHS, Beta, LV $79.95 WAR ☾

Kuffs Slater stars as George Kuffs, a young guy who reluctantly joins his brother's highly respected private security team in this original action comedy. After his brother is gunned down in the line of duty, George finds himself the new owner of the business. Out to avenge his brother, George pursues a crooked art dealer as he battles crime on the streets of San Francisco. Worthwhile premise is hampered by a predictable plot and mediocre acting.
1991 (PG-13) 102m/C Christian Slater, Tony Goldwyn, Milla Jovovich, Bruce Boxleitner, Troy Evans, George De La Pena, Leon Rippy; *W/Dir:* Bruce A. Evans. VHS, Beta, LV MCA ☾☾

Kung Fu A fugitive Buddhist quasi-Asian martial arts master accused of murder in his native land, roams across the Old West fighting injustice. Served as a pilot for the successful television series and was reincarnated 14 years later in the sequel, "Kung Fu: The Movie."
1972 75m/C Keith Carradine, David Carradine, Barry Sullivan, Keye Luke; *Dir:* Jerry Thorpe. VHS, Beta $24.98 WAR ☾☾½

Kung Fu Avengers On his way to a karate show at Madison Square Garden, our hero Li confronts the abominable Mr. Chin and his deadly henchman. Another footstomper.
1985 90m/C Bruce Li, Carl Scott, Jim James, Ku Feng. VHS, Beta $59.95 PSM ☾

Kung Fu from Beyond the Grave The ultimate warrior/zombies are summoned from the cold earth for evil purposes. Not for the brain-active.
19?? 90m/C Billy Chong, Sung Kam Shing, Lo Lieh, Hui Piu Wan. VHS, Beta $29.95 OCE ☾

Kung Fu Commandos Five martial arts masters must face an evil warlord's army to rescue a captured agent.
1980 (R) 90m/C John Lui, Shangkuan Lung. VHS, Beta $59.95 PSM, GEM ☾

Kung Fu of Eight Drunkards Ching dynasty warriors battle it out using a karate technique that gives them the appearance of being drunk.
19?? 188m/C VHS, Beta BFV ☾

Kung Fu Genius Nasty assassin-types are trying to annex his home town and that is making the kung fu genius angry.
1969 88m/C VHS, Beta $39.95 TWE ☾

Kung Fu Hero Martial arts heroes try to help overthrow the "Evil Empire," by infiltrating the Forbidden City.
1989 96m/C VHS, Beta $24.95 JCI ☾

Kung Fu Hero 2 Chinese martial artists risk it all to take on the forces of evil.
1990 96m/C Wang Fu; *Dir:* Yu Lianqi. VHS, Beta JCI ☾

The Kung Fu Kid Thanks to his martial arts mastery, a young man is able to rise above his poor upbringing and become an undercover cop.
19?? 90m/C Weiman Chan, Nora Miao. VHS, Beta $49.95 OCE ☾

The Kung Fu Kid from Peking In 1940s Shanghai, martial artists slug it out for the skull of the Peking man in order to prove the theory of evolution.
19?? 90m/C VHS, Beta $29.95 *OCE* ✗

Kung Fu Kids A group of abandoned orphans become kung fu street performers to earn money.
1984 85m/C VHS, Beta $39.95 *OCE, TWE* ✗

Kung Fu Massacre Farmers hire hard-handed mercenaries to fight evil war lords.
1982 (R) 90m/C Yangzi, Pi Li Lee. VHS, Beta $12.95 *IVE* ✗

Kung Fu Powerhouse No one dares to get in the way of a powerful kung fu drug lord. If they do, they die!
19?? 90m/C Kam Kang, Yasuka Kurata, Lee Ka Ting. VHS, Beta $49.95 *OCE* ✗

Kung Fu Rebels Rebels are fearsome enough, but when they're kung fu masters, the establishment has real trouble on its hands.
198? 90m/C Wang Chung, Wen Ching Lung, Chang Mei. VHS, Beta $49.95 *REG* ✗ ½

Kung Fu for Sale A young kung fu enthusiast, along with his mentor, fight for the respect of his family.
1981 95m/C Chong Hua. VHS, Beta $54.95 *MAV* ✗

Kung Fu Shadow The forces of good and evil are at odds once again in this dubbed kung fu epic.
1985 90m/C Tien Peng, Chia Ling. VHS, Beta $14.95 *SVS* ✗

Kung Fu Strongman A peace-loving blacksmith is forced to crush, kill, and destroy when he is framed for a robbery.
19?? 90m/C Lei Chen, Lung Fei, Kam Kong. VHS, Beta $49.95 *OCE* ✗

Kung Fu Terminator A lady terminator proves that her fists are just as quick and deadly as any man's.
19?? 90m/C Wang Kuang Hsiung, Liu Hao Yi. VHS, Beta $49.95 *OCE* ✗

Kung Fu Vampire Buster Two Taoist kung fu priests take on coffins, cemeteries, and vampires. Good luck! The ultimate (tasteless) cross-genre flick.
19?? 90m/C Lu Fang, Wang Hsiao Feng, Chien Hsiao Hou. VHS, Beta $49.95 *OCE Woof!*

The Kung Fu Warrior A vacationing young Kung fu enthusiast gets caught up in a bizarre chain of events that force him to fight for his life.
198? 91m/C Chang Lei, Kuan Hai Shan. VHS, Beta $39.95 *TWE* ✗

Kung Fu Zombie A member of the undead rises and seeks vengeance on those responsible for his plight.
1982 (R) 90m/C Billy Chong, Chinag Tao. VHS, SVS *OCE* ✗

Kung Fu's Hero A detective must resort to killing with Kung Fu to stop a slave/drug ring.
19?? 90m/C VHS, Beta $49.95 *OCE* ✗

Kwaidan A haunting, stylized quartet of supernatural stories, each with a surprise ending. Adapted from the stories of Lafcadio Hearn, an American author who lived in Japan just before the turn of the century. The visual effects are splendid. Japanese dialogue with English subtitles.

1964 164m/C JP Michiyo Aratama, Rentaro Mikuni, Katsuo Makamura; *Dir:* Masaki Kobayashi. Cannes Film Festival '65: Special Jury Prize. VHS, Beta, LV $24.95 *NOS, HMV, SNC* ✗✗✗✗

La Amante In "The Lover," passionate love in the jet set leads to a game to conquer a lovely woman. In Spanish.
1971 90m/C SP Sarita Montiel, Patrick Bauchau. VHS, Beta *TSV* ✗ ½

L.A. Bad A Puerto Rican street thug learns he has cancer, and must come to terms with his own impending death.
1985 (R) 101m/C Esai Morales, Janice Rule, John Phillip Law, Chuck Bail, Carrie Snodgress. VHS, Beta $9.99 *STE, PSM* ✗

La Balance An underworld stool-pigeon is recruited by the Parisian police to blow the whistle on a murderous mob. Baye, as a prostitute in love with the pimp-stoolie, is a standout. Critically acclaimed in France; won many Caesar awards, including for best picture. In French with English subtitles.
1982 (R) 103m/C FR Philippe Leotard, Nathalie Baye, Bob Swaim; *Dir:* Reymond LePlont. VHS, Beta $79.95 *TAM, FOX* ✗✗✗

La Bamba A romantic biography of the late 1950s pop idol Ritchie Valens, concentrating on his stormy relationship with his half-brother, his love for his WASP girlfriend, and his tragic, sudden death in the famed plane crash that also took the lives of Buddy Holly and the Big Bopper. Soundtrack features Setzer, Huntsberry, Crenshaw, and Los Lobos as, respectively, Eddie Cochran, the Big Bopper, Buddy Holly, and a Mexican bordello band.
1987 (PG-13) 99m/C Lou Diamond Phillips, Esai Morales, Danielle von Zernaeck, Joe Pantoliano, Brian Setzer, Marshall Crenshaw, Howard Huntsberry, Rosana De Soto, Elizabeth Pena; *Dir:* Luis Valdez. VHS, Beta, LV, 8mm $14.95 *COL* ✗✗✗

La Bamba Party A Latin rock/salsa music-fest filmed live in San Antonio, Texas. Features performances by Lopez, Feliciano, Rodriguez and Martinez Herring.
198? 60m/C Trini Lopez, Jose Feliciano, Johnny Rodriguez, Laura Martinez Herring. VHS, Beta $24.95 *GHV* ✗✗✗

La Barbiec de Seville An early French musical version of the Rossini opera, without English subtitles.
1934 52m/B FR Josette Day, Andre Bauge, Charpin, Jean Galland. VHS, Beta *FCT* ✗✗✗

La Bella Lola A poor singer falls in love with a rich politician in nineteenth century society.
1987 113m/C MX Sara Montiel, Antonio Charriello, Frank Villard, Luisa Mattiol. VHS, Beta $53.95 *MAD* ✗✗

La Bete Humaine A dark, psychological melodrama about human passion and duplicity, as an unhinged trainman plots with a married woman to kill her husband. Wonderful performances and stunning photography. Based on the Emile Zola novel. In French with English subtitles. 1954 Hollywood remake, "Human Desire," was directed by Fritz Lang.
1938 90m/B FR Jean Gabin, Simone Simon, Julien Carette, Fernand Ledoux; *Dir:* Jean Renoir. VHS, Beta, 8mm $24.95 *NOS, HHT, CAB* ✗✗✗ ½

La Boca del Lobo Peru's civil strife inspired this sluggish drama about government troops occupying a jungle village. Tormented by unseen communist guerillas, the

stressed-out soldiers reach the breaking point.
1989 111m/C AR *Dir:* Francisco J. Lombardi. VHS $68.00 *CCN, APD* ✗✗

La Boum A teenager's adjustment to the changes brought about by a move to Paris is compounded by her parents' marital problems. Followed by "La Boum 2."
1981 90m/C FR Sophie Marceau, Claude Brasseur, Brigitte Fossey, Denise Grey, Bernard Giraudeau; *Dir:* Claude Pinoteau. VHS, Beta $59.95 *COL* ✗✗ ½

L.A. Bounty A beautiful cop and a bounty hunter go after the same murderer. Hauser steals the show as a psychopath who has completely blown a gasket.
1989 (R) 85m/C Sybil Danning, Wings Hauser, Henry Darrow, Lenore Kasdorf, Robert Hanley; *Dir:* Worth Keeter. VHS, Beta, LV $9.99 *CCB, IVE* ✗ ½

La Cage aux Folles Adaption of the popular French play. A gay nightclub owner and his transvestite lover play it straight when Tognazzi's son from a long-ago liaison brings his fiancee and her conservative parents home for dinner. Serrault received the French Oscar for his performance as the queen and the film got a nod from the Academy in the form of nominations for direction, screenplay, and costume design. Charming music and lots of fun. So successful, it was followed by two sequels, "La Cage Aux Folles II & III" and also a Broadway musical. In French with English subtitles.
1978 (R) 91m/C FR Ugo Tognazzi, Michel Serrault, Michel Galabru, Claire Maurier, Remy Laurent, Benny Luke; *Dir:* Edouard Molinaro. VHS, Beta, LV $19.98 *TAM, FOX, VYG* ✗✗✗ ½

La Cage aux Folles 2 Albin sets out to prove to his companion that he still has sex appeal, and gets mixed up in some espionage antics. This sequel to the highly successful "La Cage aux Folles" loses some steam, but is still worth seeing. Followed by "La Cage Aux Folles III: The Wedding."
1981 99m/C FR Ugo Tognazzi, Michel Serrault, Marcel Bozzuffi, Michel Galabru, Benny Luke; *Dir:* Edouard Molinaro. VHS, Beta $19.98 *FOX, FCT* ✗✗ ½

La Cage aux Folles 3: The Wedding The kooky gay duo must feign normalcy by marrying and fathering a child in order to collect a weighty inheritance. Final segment of the trilogy; inferior to the previous films.
1986 (PG-13) 88m/C FR Michel Serrault, Ugo Tognazzi, Michel Galabru, Benny Luke, Stephane Audran; *Dir:* Georges Lautner. VHS, Beta $29.95 *RCA* ✗ ½

La Campana del Infierno A lunatic strings his nubile young cousins in a slaughterhouse. In Spanish.
19?? 90m/C SP Renaud Verley, Viveca Lindfors. VHS, Beta $57.95 *UNI*

La Carcel de Laredo Two young lovers find they have made the biggest mistake of their lives when they have a baby out of wedlock.
1984 ?m/C MX Rodolfo de Anda, Eleazar Garcia; *Dir:* Rodolfo de Anda. VHS, Beta *WCV* ✗✗

La Casa del Amor A woman doctor visits a brothel in order to solve some of the sexual problems of her patients.
1987 90m/C MX Zulma Faiad, Elena Lucena, Enrique Liporace, Jorge Porcel. VHS, Beta $54.95 *MAD* ✗✗

La Casa de Madame Lulu A man who falls in love with a prostitute disrupts life in a brothel in this satire on the balance of power between men and women.
1965 90m/C *MX* Libertad Leblanc, Enzo Viena, Elena Lucena, Juan C. Altavista. **VHS, Beta** $39.95 *MAD* ♫♫

La Casita del Pecado "The Best Little Whorehouse" and it's beautiful occupants can fulfill every erotic fantasy...no matter how exotic!!
19?? 93m/C *MX* Polo Polo, Olivia Collins, Maria Cardinal, Alejandra Peniche. **VHS, Beta** $79.95 *MED* ♫♫

La Chamuscada A group of Mexican rebels fight against the army so they can live their lives freely.
1987 97m/C *MX* **VHS, Beta** $58.95 *MAD*

La Chevre A screwball French comedy about two policemen (Richard and Depardieu—picture Nick Nolte and Gene Wilder with accents) stumbling along the path of a missing heiress who suffers from chronic bad luck. Contains a hilarious scene with chairs used to test the luck of the investigative team; based on one partner's ability to sit on the only broken chair in a rather large collection, he is judged to be sufficiently jinxed to allow them to recreate the same outrageous misfortunes that befell the heiress in her plight. Also known as "The Goat." In French with English subtitles.
1981 91m/C *FR* Gerard Depardieu, Pierre Richard, Corynne Charbit, Michel Robin, Pedro Armendariz Jr.; *Dir:* Francis Veber. **VHS, Beta** $69.95 *FCT, TAM* ♫♫♫½

La Chienne This dark French film is a troubling tale of a bedraggled husband whose only excitement is his painting hobby until he becomes consumed by the ever-tempting prostitute Lulu. Director Renoir broke ground with his use of direct sound and Paris shooting locations, and the experiment was a hit. Film portrays marriage with acidity; Renoir's own marriage broke up as a result of this film's casting. Based on the novel by Georges de la Fouchardiere. Although produced in 1931, this film didn't reach American theaters until 1975. In French with English subtitles. Also known as "Isn't Life a Bitch?" and "The Bitch." Remade by Fritz Lang in 1945 as "Scarlet Street."
1931 93m/B *FR* Michel Simon, Janie Mareze, Georges Flament; *Dir:* Jean Renoir. **VHS, Beta, LV** $59.95 *INT, TAM* ♫♫♫♫

La Choca A gang of smugglers deep in the jungle turn on a member and slaughter him. The victim's wife and child must struggle to survive and avenge their loved one's fate.
1973 94m/C *MX* Pilar Pellicer, Gregorio Casal, Armando Silvestre; *Dir:* Emilio Fernandez. **VHS, Beta** *EAG* ♫

La Chute de la Maison Usher An expressionistic, abstracted adaptation of the Poe story "The Fall of the House of Usher" about an evil mansion and its ruined denizens. Less plot than the successful rendering of the story's mood through acclaimed director Bunuel's cinematic experiments. Silent.
1928 48m/B *FR* Marguerite Gance, Jean Debucourt, Charles Lamy; *Dir:* Jean Epstein. **VHS, Beta** $59.95 *FCT* ♫♫♫½

La Cicada (The Cricket) The sparks fly when a woman and her seventeen-year-old daughter become rivals for the affections of the same man.
1983 90m/C *IT* Clio Goldsmith, Virna Lisi, Anthony (Tony) Franciosa, Renato Salvatori; *Dir:* Alberto Lattuada. **VHS, Beta** $29.98 *SUE* ♫½

La Corona de un Campeon A boxer and a wrestler attempt to beat each others brains in.
1987 90m/C *MX* Andres Garcia, Rogelio Guerra, Norma Lazarendo, Crox Alvarado. **VHS, Beta** $53.95 *MAD* ♫♫

L.A. Crackdown A ruthless yet compassionate policewoman goes after crack dealers and pimps who are exploiting women. Pretty trashy fare; followed by "L.A. Crackdown II." Made for video.
1987 90m/C Pamela Dixon, Tricia Parks, Kita Harrison, Jeffrey Olson, Michael Coon; *Dir:* Joseph Merhi. **VHS, Beta** $79.95 *CLV* ♫

L.A. Crackdown 2 A woman cop goes undercover as a dancer in a dance hall to try to catch a psychotic killer who stalks hookers in this silly sequel to "L.A. Crackdown."
1988 90m/C Pamela Dixon, Anthony Gates, Joe Vance, Cynthia Miguel, Lisa Anderson, Bo Sabato; *Dir:* Joseph Merhi. **VHS, Beta** $79.95 *CLV, HHE* ♫

La Criada Malcriada A maid creates havoc for a Puerto Rican family.
196? 101m/B *SP* Velda Gonzalez, Jose Miguel Agrelot. **VHS, Beta** $49.95 *UNI* ♫

La Cucaracha A lavish production filled with Mexican songs and dances. The first three-strip, live-action Technicolor film ever made. Historically interesting.
1934 21m/C Steffi Duna. Academy Awards '34: Best Comedy Short. **VHS, Beta** $19.95 *VYY* ♫♫♫

La Dinastia de la Muerte When a man is killed in an automobile accident, his son starts an investigation when he suspects his father was really murdered. In Spanish.
19?? ?m/C *SP* Rodolfo de Anda, Mario Almada, Jorge Russek, Carmen Montejo. **VHS, Beta** *WCV* ♫♫♫

La Diosa Arrodillada (The Kneeling Goddess) A marriage ends in divorce because the husband is in love with another woman.
1947 89m/C *MX* Maria Felix, Arturo de Cordova, Rosario Granados; *Dir:* Roberto Gavaldon. **VHS, Beta** $39.95 *MAD* ♫

La Dolce Vita In this influential and popular work a successful, sensationalistic Italian journalist covers the show-biz life in Rome, and alternately covets and disdains its glitzy shallowness. The film follows his dealings with the "sweet life" over a pivotal week. A surreal, comic tableaux with award-winning costuming; one of Fellini's most acclaimed films. Music by Nino Rota. In Italian, subtitled in English.
1960 174m/B Marcello Mastroianni, Anita Ekberg, Anouk Aimee, Alain Cuny, Lex Barker, Yvonne Furneaux, Barbara Steele, Nadia Gray, Magali Noel, Walter Santesso, Jacques Sernas; *Dir:* Federico Fellini. Academy Awards '61: Best Costume Design; National Board of Review Awards '63: Best Foreign Language Film; New York Film Critics Awards '63: Best Foreign Language Film; Sight & Sound Survey '72: #4 of the Best Films of All Time. **VHS, LV** $24.98 *TAM, REP, APD* ♫♫♫♫

La Femme Nikita Besson—who directed "Subway" and "Le Dernier Combat"—wrote and directed this stylish French noir version of Pygmalion. Having killed a cop during a drugstore theft gone awry, young French sociopath Nikita (Parillaud) is reprieved from a death sentence in order to enroll in a government finishing school, of sorts. Trained in etiquette and assasination, she's released after three years, and starts a new life with a new beau, all the while carrying out agency-mandated assassinations. Parillaud is excellent as the once-amoral street urchin transformed into a woman of depth and sensitivity—a bitterly ironic moral evolution for a contract killer.
1991 (R) 117m/C *FR* Anne Parillaud, Jean-Hughes Anglade, Tcheky Karyo, Jeanne Moreau; *W/Dir:* Luc Besson. Cesar '91: Best Actress (Parillaud). **VHS** $89.95 *VMK, FCT* ♫♫♫

La Flor de la Mafia A tough Mafia don buys his way into legitimate business. In Spanish.
19?? 90m/C Zulma Faiad. **VHS, Beta** $57.95 *UNI*

La Forza del Destino The Metropolitan Opera Orchestra and Chorus spectacularly perform Verdi's moving tragedy.
1984 179m/C Leontyne Price. **VHS, LV** $39.95 *HMV, PIA, MVD*

La Fuerza Inutil A college professor tries to join the radical actions of his students.
1987 103m/C *MX* Rafael Baledon, Veronica Castro, Roberto Jordan, Silvia Mariscal. **VHS, Beta** $58.95 *MAD* ♫♫

La Fuga A dangerous criminal escapes from a penal colony and kidnaps a beautiful woman. Powerfully erotic and sexually tense.
19?? 90m/C Elenora Vallone, Rodrigo Obregon. **VHS, Beta** *NO*

La Fuga de Caro Rosa and Rolando attempt a daring escape in this sequel to "The Kidnapping of Lola." Action-packed!
19?? 103m/C *SP* Diana Ferreti, Alfredo Leal, Ana Louisa Peluffo, Frank Moro. **VHS, Beta** $72.95 *CVS*

La Fuga del Rojo An inmate is aided in his escape attempt so he can smuggle arms across the border.
1987 120m/C *MX* Mario Almada, Patricia Rivera, Carmen Del Valle, Noe Murayama. **VHS, Beta** $59.95 *MAD*

L.A. Gangs Rising The streets of Los Angeles run red with blood as a gang war explodes into bloody fury.
1989 87m/C David Kyle, Steve Bond, John Ashton; *Dir:* John Bushelman. **VHS, Beta** $9.95 *SIM*

La Generala A woman seeks vengeance for her brother's death after the Mexican Revolution.
1987 90m/C *MX* Maria Felix. **VHS, Beta** $59.95 *MAD* ♫♫

La Grande Bourgeoise A historical romance about the clash of aristocratic society and the new bourgeoisie in 1897 Italy. Great cast tries to save the true tale of murder and political intrigue from film's slow pacing, but in the end it's the movie's visual polish that shines forth. In Italian with subtitles.
1974 115m/C *IT* Catherine Deneuve, Giancarlo Giannini, Fernando Rey, Tina Aumont; *Dir:* Mauro Bolognini. **VHS, Beta** $69.95 *FCT* ♫♫½

La Guerra de los Sostenes Two bra factories in an obscure African nation are competing to win a contract with the military. They send sexy blondes to appease the colonel in control, but to their surprise the colonel is a woman.
19?? 102m/C *SP* Juan Carlos Calabro, Tristan, Mariquita Gallegos, Gogo Andreu. **VHS, Beta** $59.95 *UNI*

La Guerrillera de Villa (The Warrior of Villa) The story of the Warrior of Villa, who fought for love, happiness and ideals in old Mexico.
1977 100m/C Carmen Sevilla, Julio Aleman, Vicente Parra, Jose Elias Moreno. **VHS, Beta** MAD 𝄞

L.A. Heat When a vice cop's partner is killed, he seeks revenge on the murderers, only to find that his own department may have been involved. Followed by "L.A. Vice."
1988 90m/C Jim Brown, Lawrence Hilton-Jacobs; *Dir:* Joseph Merhi. **VHS, Beta** $69.95 PAR, HHE 𝄞 ½

La Hifa Sin Padre A country girl gains fame as a folk singer in Mexico City.
1987 90m/C MX Jaime Moreno, Mercedes Castro, Roberto "Flaco" Guzman, Dacia Gonzales. **VHS, Beta** $79.95 MAD 𝄞𝄞 ½

La Hija del General Moving story of a young girl becoming a woman in revolutionary Spain. She loses her mother and friends to the brutal government which her father heads.
1984 101m/C SP Jorge Russek, Juan Ferrara, Rocio Brambilia, Patricia Aspillaga. **VHS, Beta** $69.95 MED 𝄞

La Hora 24 A beggar with the power of premonition foresees dire consequences in the next 24 hours.
1987 102m/C SP Julio Aleman, Diana Ferreti, Jose Natera, Carlos Rotzinger; *Dir:* Fernando Duran. **VHS, Beta** $74.95 CCV 𝄞

La Isla Encantada (Enchanted Island) The new adventures of Robinson Crusoe and Man Friday, as they pursue wild beasts and cannibals and fight off pirates.
1984 90m/C **VHS, Beta** $57.95 UNI 𝄞 ½

La Isla de Tesoro de los Pinos Incredible adventures in the briny deep featuring a host of malevolent pirates.
19?? ?m/C SP **VHS, Beta** MAD 𝄞 ½

La Jetee/An Occurrence at Owl Creek Bridge Two of the greatest film shorts: first, a landmark sci-fi film, told almost completely through still photographs, about a time traveler from a post-World War III future who re-experiences a pivotal moment in his life as a child; second, adapted from the Ambrose Bierce story, about a soldier lynched during the civil war who appears to escape at the last moment. In French with English subtitles.
1964 60m/B FR *Dir:* Chris Marker, Robert Enrico. Academy Awards '63: Best Live Action Short Film ("Occurrence"); British Academy Awards '62: Best Short Film ("Occurrence"); Cannes Film Festival '62: Blue Ribbon ("Occurrence"). **VHS, Beta** $34.95 FCT 𝄞 ½

La Justicia Tiene Doce Anos A twelve-year-old girl decides to rid the world of evil.
1987 88m/C MX Joaquin Cordero, Iran Eroy, Jorge Mistral, Sergio Guzik. **VHS, Beta** $57.95 MAD 𝄞

L.A. Law The pilot episode of the acclaimed dramatic series, in which the staff of a Los Angeles law firm tries a variety of cases.
1986 97m/C Michael Tucker, Jill Eikenberry, Harry Hamlin, Richard Dysart, Jimmy Smits, Alan Rachins, Susan Ruttan, Susan Dey, Corbin Bernsen; *Dir:* Gregory Hoblit. **VHS, Beta** $79.98 FOX 𝄞𝄞𝄞

La Lectrice Artesian landscapes, Beethoven sonatas, and a Raymond Jean novel titled "La Lectrice" provide the backdrop for Constance (Miou-Miou) to read aloud to her boyfriend in bed, and from there she imagines herself the novel's heroine, who hires out her services as a reader to various odd characters. She reads from "L'Amant," "Alice," "War and Peace," and "Les Fleurs du Mal." Richly textured and not overly intellectual.
1988 (R) 98m/C FR Miou-Miou, Maria Casares, Christian Ruche; *Dir:* Michel DeVille. Montreal World Film Festival '88: Best Film. **VHS, Beta, LV** $79.98 ORI 𝄞𝄞

La Ley del Revolver (The Law of the Gun) This is an Old West adventure where the law of the gun speaks for and against justice. In Spanish.
1978 90m/C SP Michael Rivers, Angel Del Pozo, Lucia Gil Fernandez. **VHS, Beta** HHT 𝄞

La Mansion de la Locura A journalist goes into a horrifying castle owned by a madman.
1987 83m/C MX Claudio Brook, Arthur Dansel, Ellen Sherman, Martin La Salle. **VHS, Beta** $75.95 MAD 𝄞𝄞 ½

La Mansion de la Niebla Gore and madness run rampant in an old mansion. In Spanish.
197? 90m/C SP Evelyn Stewart. **VHS, Beta** $57.95 UNI 𝄞

La Marchanta A poor woman comes to Mexico City and becomes a street merchant.
1987 100m/C MX Lucha Villa, Resortes Cavillazo. **VHS, Beta** $59.95 MAD 𝄞𝄞

La Marseillaise Sweeping epic by Renoir made before he hit his stride with "Grand Illusion." It details the events of the French Revolution in the summer of 1789 using a cast of thousands. The opulent lifestyle of the French nobility is starkly contrasted with the peasant lifestyle of poverty and despair. The focus is on two new recruits who have joined the Marseilles division of the revolutionary army as they begin their long march to Paris, the heart of France. As they travel, they adopt a stirring and passionate song that embodies the spirit and ideals of the revolution known as "La Marseillaise," now France's national anthem. In French with English subtitles.
1937 130m/B FR Pierre Renior, Lisa Delamare, Louis Jouvet, Aime Clarimond, Andrex Andrisson, Paul Dullac; *W/Dir:* Jean Renoir. **VHS, Beta** $24.95 NOS, INT, HHT 𝄞𝄞𝄞 ½

La Mentira A man falls in love with a woman he had previously wanted to kill. Based on the novel by Caridad Bravo Adams.
1987 107m/C MX Enrique Lizalde, Julissa, Blanca Sanchez, David Estuardo. **VHS, Beta** $73.95 MAD 𝄞𝄞 ½

La Montana del Diablo (Devil's Mountain) Two men escape from slavery and then ask the superhero El Payo for help.
1987 89m/C MX Jorge Rivero, Ernesto Yanez, Carmen Vicarte, Miguel Garza. **VHS, Beta** $75.95 MAD 𝄞𝄞 ½

La Muerte del Che Guevara Che Guevara leads a group of rebels to Bolivia, where they hope to start a revolution. They meet, in a climactic battle, with the loyal forces.
197? 92m/C SP **VHS, Beta** MED 𝄞

La Muneca Perversa A beautiful woman kills her lovers compulsively. In Spanish.
19?? 90m/C SP Margo Lopez, Joaquin Cordero. **VHS, Beta** $59.95 UNI

La Noche de los Brujos Voodoo zombies kill, maim and cannibalize in the jungle. Plenty of blood and green skin. In Spanish.
19?? 90m/C SP Simon Andrew. **VHS, Beta** $49.95 UNI

La Notte Brava (Lusty Night in Rome) Italian drama focusing on the life of Rome's aimless youth in the late fifties.
1959 96m/B IT **VHS, Beta, 3/4U, Special order formats** CIG 𝄞𝄞

La Nueva Cigarra A brothel is quarantined with the bubonic plague and the customers are caught with their proverbial pants down. In Spanish.
1986 90m/C SP Toco Gomez, Haydee Balsa, Gogo Andreu, Olga Zubarry. **VHS, Beta** $57.95 UNI

La Nuit de Varennes This semi-historical romp is based on an actual chapter in French history when King Louis XVI and Marie Antoinette fled from revolutionary Paris to Varennes in 1971. On the way they meet an unlikely group of characters, including Cassanova and Thomas Paire. At times witty and charming, the melange of history and fiction is full of talk, sometimes profane, punctuated by sex and nudity. Director Scola's imagination stretches to light up this night. In French with English subtitles.
1982 (R) 133m/C FR IT Marcello Mastroianni, Harvey Keitel, Jean-Louis Barrault, Hanna Schygulla, Jean-Claude Brialy; *Dir:* Ettore Scola. **VHS, Beta, LV** $59.95 COL, IME, APD 𝄞𝄞𝄞

La Pachange (The Big Party) The tenants of a Mexican apartment enjoy wild nights of drinking and sex.
1987 85m/C MX Julissa, Claudia Islas, Gregorio Casal, Sergio Jimenez. **VHS, Beta** $58.95 MAD

La Passante An otherwise peace-loving man murders the Paraguayan ambassador to France. In somewhat clumsy flashback style the murderer's orphaned childhood and other memories provide motives for the murder. Though posed more slowly than typical intrigue plots, the movie blends love and the legacy of Nazism, passion and politics in an engaging way, spiced with nudity and some violence. In her last screen appearance Schneider movingly portrays dual roles. In French with English subtitles.
1983 106m/C FR GE Romy Schneider, Michel Piccoli, Helmut Griem, Gerard Klein, Matthieu Carriere, Maria Schell; *Dir:* Jacques Rouffio. **VHS, Beta, LV** $29.95 PAV 𝄞𝄞𝄞

La Pastorela An adaptation of the traditional Spanish Christmas play about the shepherds' journey to Bethlehem. Told through the eyes of a young Hispanic girl, using contemporary references to make this a modern-day story. Available in English and Spanish versions.
1992 80m/C Linda Ronstadt, Cheech Marin, Paul Rodriguez, Karla Montana, Don Novello, Robert Beltran, Freddy Fender, Falco Jimenez; *W/Dir:* Luis Valdez. **VHS** $19.98 BMG 𝄞𝄞𝄞

La Perichole Joan Sutherland and her puppet friends help tell this story, which is based on Merimee's "La Carrosse du Saint-Sacrament" and contains some fine soprano selections.

1973 30m/C Joan Sutherland. **VHS, Beta** BFA 🐾🐾🐾

La Petite Bande A gang of children escapes home and journeys to France, encountering Bedouins, Bavarians, circus performers, etc.
1984 (PG) 91m/C *Dir:* Michel DeVille. **VHS, Beta $59.95** COL 🐾🐾½

La Petite Sirene A young girl obsessed with the story of the Little Mermaid ingratiates herself to a 40-year-old garage mechanic. Seeing him as her prince charming, she persists in trying to make him a part of her life, in spite of his resistance. In French with English subtitles.
1980 104m/C *FR* Laura Alexis, Philippe Leotard, Evelyne Dress, Marie DuBois; *W/Dir:* Roger Andrieux. **VHS $59.95** WAC, FCT 🐾🐾½

La Primavera de los Escorpiones A strange man is rescued by a buxom woman, and starts a love triangle between her and her son. In Spanish.
19?? 90m/C Isela Vega. **VHS, Beta** UNI 🐾

La Puritaine Before his planned reconciliation with his daughter, the artistic manager of a theater has young actresses portray different sides of the daughter's personality. In French with English subtitles.
1986 90m/C Michel Piccoli, Sandrine Bonnaire; *Dir:* Lou Doillon. **VHS $49.95** FCT 🐾🐾½

La Rebelion de las Muertas In "The Rebellion of the Dead," two brothers get involved in a diabolical scheme to gain immortality and seek revenge. In Spanish.
1972 90m/C *SP* Paul Naschy, Rommy, Mirta Miller. **VHS, Beta** TSV 🐾

La Red del Egano (Web of Deception) Police search for the mastermind behind a series of jewel robberies. Dubbed into Spanish.
1971 90m/C *SP* Thomas Hunter, Marilia Branco, Gabriele Tinti. **VHS, Beta $39.98** JCI 🐾

La Ronde A classic comedy of manners and sharply witty tour-de-farce in which a group of people in 1900 Vienna keep changing romantic partners until things wind up where they started. Ophuls' swirling direction creates a fast-paced farce of desire and regret with wicked yet subtle style. Based on Arthur Schnitzler's play and remade as "Circle of Love." With English subtitles.
1951 97m/B *FR* Simone Signoret, Anton Walbrook, Simone Simon, Serge Reggiani, Daniel Gelin, Danielle Darrieux, Jean-Louis Barrault, Fernand Gravet, Odette Joyeux, Isa Miranda, Gerard Philippe; *Dir:* Max Ophuls. New York Film Festival '69: Best Film. **VHS, Beta $29.98** SUE, APD 🐾🐾🐾½

La Signora di Tutti An early Italian biography of Gaby Doriot, a movie star whose professional success is paired by personal misery. Prestigious director Ophuls, working in exile from his native Germany, was unable to complete many projects in the years leading up to and during WWII. This is one of the few. Watch for innovative camera work intended to underscore the film's mood. In Italian with subtitles.
1934 92m/B *IT* Isa Miranda; *Dir:* Max Ophuls. **VHS, Beta $59.95** CVC, FCT 🐾🐾🐾

La Sombra del Murcielago Monsters and Mexican wrestlers fight over a woman. In Spanish.
19?? 90m/C Blue Demon, Marta Romero. **VHS, Beta $59.95** UNI

L.A. Story Livin' ain't easy in the city of angels. Harris K. Telemacher (Martin), a weatherman in a city where the weather never changes, wrestles with the meaning of life and love while consorting with beautiful people, distancing from significant other Hennner, cavorting with valley girl Parker, and falling for newswoman Tennant (Martin's real life wife). Written by the comedian, the story's much more than a Martin vehicle, full of keen insights into the everyday problems and ironies of living in the Big Tangerine. (It's no wonder the script's full of so much thoughtful detail: Martin is said to have worked on it intermittently for seven years.) A charming, fault forgiving but not fault ignoring portrait that'll make you gush with that life isn't half bad feeling.
1991 (PG-13) 98m/C Steve Martin, Victoria Tennant, Richard E. Grant, Marilu Henner, Sarah Jessica Parker, Sam McMurray, Patrick Stewart, Iman, Kevin Pollack; *Dir:* Mick Jackson. **VHS, Beta $92.95** LIV, WAR 🐾🐾🐾

La Strada A simple-minded girl, played by Fellini's wife, Marsha, is sold to a brutal, coarse circus strong-man and she falls in love with him despite his abuse. They tour the countryside and eventually meet up with a gentle acrobat, played by Basehart, who alters their fate. This Fellini masterwork was the director's first internationally acclaimed film, and is, by turns, somber and amusing as it demonstrates the filmmaker's sensitivity to the underprivileged of the world and his belief in spiritual redemption. Features a haunting score by Nino Rota. Academy Award winner for Best Foreign Film. Subtitled in English.
1954 107m/B *IT* Giulietta Masina, Anthony Quinn, Richard Basehart, Aldo Silvani; *Dir:* Federico Fellini. Academy Awards '56: Best Foreign Language Film; National Board of Review Awards '56: 5 Best Foreign Films of the Year; New York Film Critics Awards '56: Best Foreign Film. **VHS, Beta, LV $29.98** SUE, IME, APD 🐾🐾🐾🐾

La Symphonie Pastorale A Swiss pastor takes in an orphan blind girl who grows up to be beautiful. The pastor then competes for her affections with his son. Quiet drama based on the Andre Gide novel, with breathtaking mountain scenery as the backdrop for this tragedy rife with symbolism.
1946 105m/B *FR* Pierre Blanchar, Michele Morgan, Jean Desailly, Line Noro, Andree Clement; *Dir:* Jean Delannoy. **VHS** FLI 🐾🐾🐾½

La Terra Trema The classic example of Italian neo-realism, about a poor Sicilian fisherman, his family and their village. A spare, slow-moving, profound and ultimately lyrical tragedy, this semi-documentary explores the economic exploitation of Sicily's fishermen. Filmed on location with the villagers playing themselves; highly acclaimed though not commercially successful. In Sicilian with English subtitles. Some radically cut versions may be available, but are to be avoided. Franco Zefferelli was one of the assistant directors.
1948 161m/B *IT* Antonio Pietrangeli; *Dir:* Luchino Visconti. Venice Film Festival '48: International Prize. **VHS, Beta $89.95** FCT 🐾🐾🐾½

La Tigeresa (The Tigress) A young girl, rape victim and witness to her father's murder, grows up to be a bitter, back-stabbing criminal.
1987 81m/C *SP* Perla Faith. **VHS, Beta $59.98** MAV 🐾

La Trampa Mortal A Spanish singing cowboy fights and wins the girl.

19?? 80m/C Luis Aguilar, Flor Silvestre. **VHS, Beta $49.95** UNI

La Traviata One of Verdi's most performed works is staged with the help of Joan Sutherland and her puppet friends. Based on the Dumas classic, "The Lady of the Camelias."
1973 30m/C Joan Sutherland. **VHS, Beta** BFA

La Traviata A film version of Giuseppe Verdi's opera classic, based on "La Dame aux Camelias" by Alexandre Dumas, tells the tale of a tragic romance in the 19th century. Subtitled.
1983 105m/C *IT* Teresa Stratas, Placido Domingo, Cornell MacNeil, Alan Monk, Axelle Gall, Pina Cei; *Dir:* Franco Zeffirelli. **VHS, Beta, LV $29.95** VTK, MCA, MVD 🐾🐾🐾🐾

La Traviata Giuseppe Verdi's opera, based on Dumas' "La Dame aux Camelias," is performed for British television at the Glyndebourne Festival Opera. Love, betrayal, death, and despair are wonderfully played out in period costume. Subtitled.
1988 135m/C *IT* Marie McLaughlin, John Gunther, Brent Ellis; *Dir:* Peter Hall. **VHS, Beta, LV $39.95** HMV, MVD 🐾🐾🐾🐾

La Trinchera A Spanish western centering around the Mexican Revolution.
1987 91m/C *MX* David Reynoso, Ignacio Lopez Tarso, Julio Aleman, Pilar Pellicer. **VHS, Beta $59.95** MAD

La Truite (The Trout) A young woman leaves her family's rural trout farm and a loveless marriage to seek her fortune in high finance and corporate mayhem. The complicated plot, full of intrigue and sexual encounters, sometimes lacks focus. Slickly filmed.
1983 (R) 80m/C *FR* Isabelle Huppert, Jean-Pierre Cassel, Daniel Olbrychski, Jeanne Moreau, Jacques Spiesser, Ruggero Raimondi, Alexis Smith, Craig Stevens; *Dir:* Joseph Losey. **VHS, Beta $59.95** RCA 🐾🐾🐾

La Venganza del Kung Fu A woman becomes a kung fu champ to avenge the death of her family.
1987 105m/C *MX* Charlie C.C. Lau, Wong Dong, Mao Ying, Ting Shan Si. **VHS, Beta $34.80** MAD 🐾🐾

La Venganza del Rojo A doctor helps the rebels in a revolution.
1987 90m/C *MX* Mario Almada, Ferdinando Almada, Carmen Del Valle, Noe Murayama, Jorge Patino. **VHS, Beta $59.95** MAD 🐾🐾

L.A. Vice A detective is transferred to the vice squad, where he must investigate a series of murders. Sequel to "L.A. Heat."
1989 90m/C Jim Brown, Lawrence Hilton-Jacobs, William Smith; *Dir:* Joseph Merhi. **VHS, Beta** PAR 🐾½

La Vida Sigue Igual An up-and-coming law student and soccer player becomes paralyzed in an auto accident and must contend with a change of attitude from his friends and lover.
19?? 102m/C *SP* Jean Harrington, Charo Lopez. **VHS, Beta** MED

La Vie Continue A middle-aged woman suddenly finds herself widowed after 20 years of marriage and tries to build a new life for herself and her children. Girardot's performance can't quite dispel melodramatic suds. Dubbed in English. Loosely and more effectively remade as "Men Don't Leave."

1982 93m/C *IT* Annie Girardot, Jean-Pierre Cassel, Michel Aumont, Pierre Dux, Guilia Salvatori, Emmanuel Goyet, Rivera Andres; *Dir:* Moshe Mizrahi. **VHS, Beta** $59.95 COL 🎬🎬½

La Violetera (The Violet) A poor flower vendor falls in love with a rich diplomat despite their opposite positions in society. **1971** 90m/C Sara Montiel, Raf Vallone, Frank Villard. **VHS, Beta** $39.95 MAD 🎬

La Virgen de Guadalupe A miracle occurs amidst fighting between the Aztecs and the Spaniards. **1987** 82m/C *SP* Fernando Allende, Valentin Trujillo, Angelica Chain. **VHS, Beta** $75.95 MAD 🎬🎬½

L.A. Woman Blond babes pose and pout, with their clothes and without. **1989** 60m/C Julie Magnum. **VHS, Beta** $19.95 AHV 🎬🎬½

Laboratory Things go awry when the earthling subjects of an alien experiment revolt against their captors. **1980** 93m/C Camille Mitchell, Corinne Michaels, Garnett Smith. **VHS, Beta** $49.95 UHV 🎬

Labyrinth Fascinating adventure directed by and written by Henson and Monty Python's Terry Jones. While baby-sitting her baby brother Toby, Connelly is so frustrated she asks the goblins to take him away. When the Goblin King, played by Bowie, comes to answer her idle wish, she must try to rescue Toby by solving the fantastic labyrinth. If she does not find him in time, Toby will become one of the goblins forever. During her journey, Connelly is befriended by all sorts of odd creatures created by Henson, and she grows up a little along the way. A wonderful movie for the whole family. Produced by George Lucas; songs written and sung by David Bowie. **1986** (PG) 90m/C David Bowie, Jennifer Connelly, Toby Froud; *Dir:* Jim Henson. **VHS, Beta, LV, 8mm** $14.98 SUE 🎬🎬🎬

Labyrinth of Passion A screwball farce, directed by Almodovar, featuring a host of strange characters running around Madrid in search of sex and laughter. They find plenty of both. An early film by an influential director. **1982** 100m/C *SP* Antonio Banderas, Imanol Arias, Celia Roth; *W/Dir:* Pedro Almodovar. **VHS, Beta, LV** $79.95 CCN, IME 🎬🎬½

The Lacemaker Huppert's first shot at stardom, on videodisc with digital sound and a letterboxed print. A young beautician and a university student fall in love on vacation but soon begin to realize the differences in their lives. Adapted from the novel by Pascal Laine. Original title "La Dentielliere." In French with English subtitles. **1977** 107m/C *FR SI GE* Isabelle Huppert, Yves Beneyton, Florence Giorgetti, Anna Marie Duringer; *Dir:* Claude Goretta. **LV** $79.95 HMV, CRC, VYG 🎬🎬🎬½

L'Addition A prison drama about the struggle between a soon-to-be-released prisoner and a sadistic guard. **1985** (R) 85m/C *FR* Richard Berry, Richard Bohringer, Victoria Abril; *Dir:* Denis Amar. **VHS, Beta** $69.95 MOR 🎬🎬½

Ladies of the Chorus Monroe stars as a burlesque chorus girl who shares the stage with her mom (Jergens) and a handful of other beauties. She meets and falls in love with a wealthy socialite (Brooks), but her mother completely disapproves of the relationship. Monroe later learns that her mother

was in love with the same man several years before. Although only her second film appearance, Monroe's talents are already quite apparent in this low-budget musical romance. **1949** 61m/B Adele Jergens, Marilyn Monroe, Rand Brooks, Nana Bryant, Steve Geray, Bill Edwards; *Dir:* Phil Karlson. **VHS** $19.95 COL 🎬🎬

Ladies Club A group of rape victims get together and begin to victimize rapists. **1986** (R) 86m/C Bruce Davison, Karen Austin, Diana Scarwid, Shera Danese, Beverly Todd; *Dir:* A.K. Allen. **VHS, Beta** $79.95 MED 🎬🎬½

Ladies of the Lotus Bloodthirsty Vancouver gangsters raid a modeling agency and sell beautiful young women into slavery. **1986** 120m/C Richard Dale, Angela Read, Patrick Bermel, Darcia Carnie; *Dir:* Lloyd A. Simandl, Douglas C. Nicolle. **VHS, Beta** $59.95 MAG 🎬½

The Ladies' Man Piecemeal Lewis farce, with Jerry playing a clutzy handyman working at a girls' boarding house. Some riotous routines balanced by slow pacing. **1961** 106m/C Jerry Lewis, Helen Traubel, Jack Kruschen, Doodles Weaver, Gloria Jean; *Dir:* Jerry Lewis. **VHS, Beta, LV** $14.95 PAR 🎬🎬

Ladies Night Out "Ladies Night Out" is an all-male burlesque, featuring the highly acclaimed Peter Adonis Traveling Fantasy Show. **1983** 80m/C *Performed by:* Peter Adonis Traveling Fantasy Show. **VHS, Beta** $29.98 LIV, VES 🎬🎬

Ladies on the Rocks Aspiring comediennes Micha and Laura pack up their van and take their bizarre cabaret act on the road through rural Denmark. They find their private lives disintegrating but turn each disaster into new material for their act. In Danish with English subtitles. **1983** 100m/C *DK* Helle Ryslinge, Anne Marie Helger, Flemming Quist Moller; *Dir:* Christian Braad Thomsen. Atlantic Film and Video Festival '91: Special Jury Prize. **VHS** $79.95 NYF 🎬🎬½

Lady Avenger A woman is bent on getting revenge on her brother's killers. **1989** (R) 90m/C Peggie Sanders. **VHS, Beta** $79.95 SOU 🎬

Lady Be Good Adapted from the 1924 Gershwin Broadway hit, the plot's been revamped, much to the critics' distaste ("Variety" called it "molasses paced"). Sothern and Young play a tunesmith duo who excel at musical harmony and marital strife. Applauded for its music (the critics loved Hammerstein's and the Gershwins' tunes), the show's Academy Award winning song was, ironically, written by Jerome Kern. And if that's not enough, levity man Skelton and hoofer Powell are thrown in for good measure. **1941** 111m/B Eleanor Powell, Ann Sothern, Robert Young, Lionel Barrymore, John Carroll, Red Skelton, Dan Dailey, Virginia O'Brien, Tom Conway, Phil Silvers, Doris Day; *Dir:* Norman Z. McLeod. Academy Awards '41: Best Song ("The Last Time I Saw Paris"). **VHS** $19.98 MGM, TTC, FCT 🎬🎬½

Lady Beware A psychotic doctor becomes obsessed with a beautiful store-window dresser who specializes in steamy, erotic fantasies. When he wages a campaign of terror against her, she realizes that she must stop him before he kills her. **1987** (R) 108m/C Diane Lane, Michael Woods, Cotter Smith; *Dir:* Karen Arthur. **VHS, Beta, LV** $14.95 IVE 🎬🎬

Lady of Burlesque Burlesque dancer is found dead, strangled with her own G-string. Clever and amusing film based on "The G-String Murders" by Gypsy Rose Lee. **1943** 91m/B Barbara Stanwyck, Michael O'Shea, Janis Carter, Pinky Lee; *Dir:* William A. Wellman. **VHS, Beta** $9.95 NEG, MRV, NOS 🎬🎬½

Lady on the Bus A sexually frustrated newlywed bride rides the city buses looking for men to satisfy her. Shot in Rio de Janeiro. **1978** (R) 102m/C *BR* Sonia Braga; *Dir:* Neville D'Almeida. **VHS, Beta, LV** $69.98 LIV, VES 🎬

Lady in a Cage A wealthy widow is trapped in her home elevator during a power failure and becomes desperate when hoodlums break in. Shocking violence ahead of its time probably plays better than when first released. Young Caan is a standout among the star-studded cast. **1964** 95m/B Olivia de Havilland, Ann Sothern, James Caan, Jennifer Billingsley, Jeff Corey, Scatman Crothers, Rafael Campos; *Dir:* Walter Grauman. **VHS, Beta** $49.95 PAR 🎬🎬🎬

The Lady of the Camellias Neumeier brings to film the ballet he created out of Dumas' "The Lady of the Camelias" and Abbe Prevost's "Manon Lescaut," set to the music of Chopin. **1987** 125m/C Marcia Haydee, Ivan Liska, Francois Klaus, Colleen Scott, Vladimir Klos, Lynne Charles, Jeffrey Kirk, Gigi Hyatt, Beatrice Cordua, Victor Hughes, William Parton, Richard Hoynes, John Neumeier; *Dir:* John Neumeier. **VHS, Beta** $39.95 MVD, KUL 🎬🎬🎬

Lady Caroline Lamb Lady Caroline is a passionate young lady in 19th-century England, who, although the wife of a member of Parliament, has an affair with Lord Byron and brings about her own downfall. Costume drama that never quite goes anywhere. **1973** 123m/C *GB IT* Sarah Miles, Richard Chamberlain, Jon Finch, Laurence Olivier, John Mills, Ralph Richardson; *Dir:* Robert Bolt. **Beta** $14.95 PSM 🎬🎬

Lady in Cement The second Tony Rome mystery, in which the seedy Miami dick finds a corpse with cement shoes while swimming. **1968** (R) 93m/C Frank Sinatra, Raquel Welch, Richard Conte, Martin Gabel, Lainie Kazan, Pat Henry, Steve Peck, Joe E. Lewis, Dan Blocker; *Dir:* Gordon Douglas. **VHS, Beta** $59.98 FOX 🎬½

Lady Chatterley's Lover An Englishwoman's husband is wounded in World War I and returns home paralyzed. In her quest for sexual fulfillment, she takes a new lover, the estate's earthy gamekeeper. Based on D.H. Lawrence's novel. In French with English subtitles. Cheaply remade in 1981 with softporn actress Sylvia Kristel. **1955** 102m/B *FR* Danielle Darrieux, Erno Crisa, Leo Genn; *Dir:* Marc Allegret. **VHS, Beta, LV** $24.95 NOS, MRV, HHT 🎬🎬

Lady Chatterley's Lover Remake of the 1955 film version of D.H. Lawrence's classic novel of an English lady who has an affair with the gamekeeper of her husband's estate. Basically soft-focus soft porn. **1981** 107m/C *GB FR* Sylvia Kristel, Nicholas Clay, Shane Briant; *Dir:* Just Jaeckin. **VHS, Beta, LV** $79.95 MGM 🎬½

Lady by Choice Lombard stars as a beautiful young fan dancer who is arrested for a lewd public performance. Taking the advice of her press agent, she hires an old bag lady to pose as her mother on Mother's Day. Robson portrays her "mother" and comes to think of Lombard as her own

daughter. Robson encourages Lombard to give up fan dancing and to strive for greater things in life. She also pushes her into romance with a wealthy young man (Pryor). Robson is excellent in her role as "mother" and Lombard shows great comic talent in this charming film.
1934 78m/C Carole Lombard, May Robson, Roger Pryor, Walter Connolly, Raymond Walburn, James Burke; *Dir:* David Burton. **VHS $19.95** COL ♨♨½

Lady Cocoa Routine story of a young woman who gets released from jail for twenty-four hours and sets out for Las Vegas to find the man who framed her.
1975 (R) 93m/C Lola Falana, Joe "Mean Joe" Greene, Gene Washington, Alex Dreier; *Dir:* Matt Cimber. **VHS, Beta $49.95** UNI ♨½

The Lady Confesses Average mystery that involves Hughes trying to clear her boyfriend of murder. Independently produced. "Leave it to Beaver" fans will want to watch for Hugh "Ward Cleaver" Beaumont.
1945 66m/B Mary Beth Hughes, Hugh Beaumont, Edmund McDonald, Claudia Drake, Emmett Vogan; *Dir:* Sam Newfield. **VHS $16.95** NOS, SNC ♨♨

Lady for a Day Delightful telling of the Damon Runyon story, "Madame La Gimp," about an apple peddler (Robson) down on her luck, who is transformed into a lady by a criminal with a heart. "Lady By Choice" is the sequel.
1933 96m/B May Robson, Warren William, Guy Kibbee, Glenda Farrell, Ned Sparks, Jean Parker, Walter Connolly; *Dir:* Frank Capra. **VHS, Beta, LV $29.95** CVC, VYG, CRC ♨♨♨½

Lady in the Death House A framed woman is set to walk the last mile, as a scientist struggles to find the real killer in time.
1944 57m/B Jean Parker, Lionel Atwill, Marcia Mae Jones. **VHS, Beta $24.95** VYY ♨♨

Lady in Distress A British drama about an unhappily married man who witnesses an apparent murder. He discovers, however, it was actually the prank of an illusionist and his flirtatious wife. He soon finds himself becoming increasingly involved with their lives. Alternate title: "A Window in London."
1939 59m/B GB Michael Redgrave, Paul Lukas, Sally Gray; *Dir:* Herbert Mason. **VHS, Beta $16.95** SNC, VYY, MLB ♨♨

The Lady with the Dog A bittersweet love story based on the Anton Chekhov story, about two married Russians who meet by chance in a park, fall in love, and realize they are fated to a haphazard, clandestine affair. In Russian with English titles.
1959 86m/B RU Iya Savvina, Alexei Batalov, Ala Chostakova, N. Alisova; *Dir:* Yosif Heifitz. Cannes Film Festival '60: Special Prize. **VHS, Beta $29.95** WST, FCT ♨♨♨

The Lady Eve Two con artists, out to trip up wealthy beer tycoon Fonda, instead find themselves tripped up when one falls in love with the prey. Ridiculous situations, but Sturges manages to keep them believable and funny. With a train scene that's every man's nightmare. Perhaps the best Sturges ever. Later remade as "The Birds and the Bees."
1941 93m/B Barbara Stanwyck, Henry Fonda, Charles Coburn, Eugene Pallette, William Demarest, Eric Blore, Melville Cooper; *Dir:* Preston Sturges. National Board of Review Awards '41: 10 Best Films of the Year. **VHS, Beta, LV $29.95** MCA, RKO ♨♨♨♨

Lady of the Evening A prostitute and a crook team up to seek revenge against the mob.
1979 (PG) 110m/C IT Sophia Loren, Marcello Mastroianni. **VHS, Beta** KOV ♨♨

Lady Frankenstein Frankenstein's lovely daughter graduates from medical school and returns home. When she sees what her father's been up to, she gets some ideas of her own. Good fun for fans of the genre.
1972 (R) 84m/C IT Joseph Cotten, Sara (Rosalba Neri) Bay, Mickey Hargitay, Paul Muller, Paul Whiteman; *Dir:* Mel Welles. **VHS, Beta $9.98** SNC, SUE ♨½

Lady Godiva Rides A bad, campy version of the story of Lady Godiva. Godiva comes to the United States with a bevy of scantily clad maidens and winds up in the Old West. When the town villain threatens to compromise her, Tom Jones comes to the rescue. This film of course features Godiva's naked ride on horseback through the town.
1968 88m/C Marsha Jordan, Forman Shane; *Dir:* A.C. Stephen. **VHS $19.95** VDM ♨½

Lady Grey A poor farmer's daughter rises to the top of the country music charts, but at a price.
1982 111m/C Ginger Alden, David Allen Coe; *Dir:* Worth Keeter. **VHS, Beta $59.99** HBO ♨

Lady of the House A true story about a madame who rose from operator of a brothel to become a political force in San Francisco.
1978 100m/C Dyan Cannon, Susan Tyrrell, Colleen Camp, Armand Assante; *Dir:* Ralph Nelson. **VHS, Beta $49.95** PSM ♨♨

Lady Ice Sutherland, as an insurance investigator on the trail of jewel thieves, follows them to Miami Beach and the Bahamas. After stealing a diamond he enters into partnership with a crook's daughter. Worth seeing for the cast.
1973 (PG) 93m/C Donald Sutherland, Jennifer O'Neill, Robert Duvall, Eric Braeden; *Dir:* Tom Gries. **VHS, Beta $49.95** WES ♨♨

Lady Jane An accurate account of the life of 15-year-old Lady Jane Grey, who secured the throne of England for 9 days in 1553 as a result of political maneuvering by noblemen and the Church-of-England. A wonderful film even for non-history buffs. Carter's first film.
1985 (PG-13) 140m/C GB Helena Bonham Carter, Cary Elwes, Sara Kestelman, Michael Hordern, Joss Ackland, Richard Johnson, Patrick Stewart; *Dir:* Trevor Nunn. **VHS, Beta, LV $79.95** PAR ♨♨♨

Lady Killers A series of British crime dramatizations, made for television, featuring the cases of Ruth Ellis and George Joseph Smith.
198? 60m/C *Dir:* Robert Lewis; *Nar:* Robert Morley. **Beta $29.99** PSM ♨♨♨½

Lady in the Lake Why is it everyone wants to get artsy with pulp fiction? Actor Montgomery directs himself in this Philip Marlowe go-round, using a subjective camera style to imitate Chandler's first person narrative (that means the only time we get to see Marlowe/Montgomery's mug is in a mirror. That pretty well says it all). Having decided to give up eyeing privately, Marlowe turns to the pen to tell the tangled tale of the lady in the lake: once upon a time, a detective was hired to find the wicked wife of a paying client...Some find the direction clever. Others consider MGM chief Lous B. Mayer,

who made sure this was Montgomery's last project with MGM, to be a wise man.
1946 103m/B Robert Montgomery, Lloyd Nolan, Audrey Totter, Tom Tully, Leon Ames, Jayne Meadows; *Dir:* Robert Montgomery. **VHS, Beta $19.98** MGM, FCT ♨♨½

Lady from Louisiana A lawyer in old New Orleans out to rid the city of corruption falls in love with the daughter of a big-time gambler. Great storm scene.
1942 84m/B John Wayne, Ona Munson, Dorothy Dandridge, Ray Middleton, Henry Stephenson, Helen Westley, Jack Pennick; *Dir:* Bernard Vorhaus. **VHS $14.98** REP ♨♨

Lady for a Night The female owner of a gambling riverboat does her best to break into high society, when murder threatens to spoil her plans. Cast and costumes burdened by pacing.
1942 88m/B John Wayne, Joan Blondell, Ray Middleton, Philip Merivale, Blanche Yurka, Edith Barrett, Leonid Kinskey, Montagu Love; *Dir:* Leigh Jason. **VHS $14.98** REP ♨♨

Lady from Nowhere A woman is the only witness to a gangland rub-out and is subsequently pursued by both the mob and the police. Unfortunately for her, the gangsters catch up with her first.
1936 60m/B Mary Astor, Charles Quigley, Thurston Hall, Victor Kilian, Spencer Charters; *Dir:* Gordon Wiles. **VHS, Beta $16.95** SNC ♨♨

The Lady in Question A Parisian shopkeeper, played by Aherne, sits on a jury that acquits Hayworth of murder. His interest doesn't end with the trial, however, and matters heat up when his son also becomes involved. Remake of a melodramatic French release.
1940 81m/B Rita Hayworth, Glenn Ford, Brian Aherne, Irene Rich, Lloyd Corrigan, George Coulouris, Evelyn Keyes, Curt Bois, Edward Norris; *Dir:* Charles Vidor. **VHS, Beta, LV $19.95** COL, CCB ♨½

Lady in Red A story of America in the '30s, and the progress through the underworld of the woman who was Dillinger's last lover. Also known as "Guns, Sin and Bathtub Gin."
1979 90m/C Pamela Sue Martin, Louise Fletcher, Robert Conrad, Christopher Lloyd; *Dir:* Lewis Teague. **VHS, Beta $69.98** LIV, VES ♨♨

The Lady Refuses A three-sided love affair leads to tragedy.
1931 70m/B Betty Compson, John Darrow, Margaret Livingston. **VHS, Beta $24.95** DVT ♨♨

Lady of the Rising Sun A female doctor in the Orient makes house calls at a brothel and finds reason to stay.
198? 91m/C Chai Lee, Ilona Staller. **VHS, Beta $29.95** MED Woof!

The Lady Says No A photographer must photograph an attractive author who has written a book uncomplimentary toward the male sex.
1951 80m/B Joan Caulfield, David Niven, James Robertson Justice, Francis Bavier, Henry Jones, Jeff York, Robert Williams; *Dir:* Frank Ross. **VHS $29.95** FCT ♨♨

Lady Scarface Murderous mobster queen breaks the law, bats men around, and takes the cops on a merry chase. Cheap, fast-paced action.
1941 66m/B Mildred Coles, Dennis O'Keefe, Frances Neal, Judith Anderson, Eric Blore; *Dir:* Frank Woodruff. **VHS, Beta $34.95** RKO ♨♨

Lady Scarface/The Threat A crime double feature: "Lady Scarface" (1941), a clever cop is on the trail of a dangerous gunwoman; and "The Threat" (1949), a ruthless killer escapes from prison intent on murdering those who convicted him.
1949 132m/B Dennis O'Keefe, Judith Anderson, Robert Knapp, Eric Blore, Charles McGraw, Michael O'Shea, Virginia Grey; *Dir:* Frank Woodruff. **VHS, Beta $34.95** *RKO*

The Lady from Shanghai An unsuspecting seaman becomes involved in a web of intrigue when a woman hires him to work on her husband's yacht. Hayworth (a one-time Mrs. Orson Welles), in her only role as a villainess, plays a manipulative, sensual schemer. Wonderful and innovative cinematic techniques from Welles, as usual, including a tense scene in a hall of mirrors. Filmed on a yacht belonging to Errol Flynn.
1948 87m/B Orson Welles, Rita Hayworth, Everett Sloane, Glenn Anders, Ted de Corsia, Erskine Sanford, Gus Schilling; *W/Dir:* Orson Welles. **VHS, Beta, LV $19.95** *COL, MLB* 𝄇𝄇𝄇 ½

Lady Sings the Blues Jazz artist Billie Holiday's life becomes a musical drama depicting her struggle against racism and drug addiction in her pursuit of fame and romance. What could be a typical price-of-fame story is saved by Ross' inspired performance as the tragic singer. Songs include "God Bless the Child" and "Lover Man." Oscar nominated for director.
1972 (R) 144m/C Diana Ross, Billy Dee Williams, Richard Pryor; *Dir:* Sidney J. Furie. **VHS, Beta, LV $29.95** *PAR* 𝄇𝄇 ½

Lady Stay Dead A gory Australian shocker about another faceless, knife-wielding psychotic chasing a girl.
1903 100m/C *AU* VHS, Beta $59.95 *VCD* 𝄇 ½

Lady Street Fighter A well-armed female combat fighter battles an organization of assassins to avenge her sister's murder.
1986 73m/C Renee Harmon, Joel McCrea Jr. **VHS, Beta $59.95** *UNI* 𝄇

Lady Takes a Chance A romantic comedy about a New York working girl with matrimonial ideas and a rope-shy rodeo rider who yearns for the wide open spaces. Fine fun on the range.
1943 86m/B John Wayne, Jean Arthur, Phil Silvers, Charles Winninger; *Dir:* William A. Seiter. **VHS, Beta $29.95** *VID* 𝄇𝄇𝄇

Lady Terminator A student on an anthropology expedition digs up more than she bargained for when she is possessed by an evil spirit and goes on a rampage.
1989 (R) 83m/C Barbara Constable, Christopher Hart, Joseph McGlynn, Claudia Rademaker; *Dir:* Jalil Jackson. **VHS, Beta** *SED* 𝄇 ½

Lady and the Tramp The animated Disney classic about two dogs who fall in love. Tramp is wild and carefree; Lady is a spoiled house pet who runs away from home after her owners have a baby. Songs by Peggy Lee and Sonny Burke include "He's a Tramp," "Bella Notte," and "The Siamese Cat Song." They don't make dog romances like this anymore.
1955 (G) 76m/C *Dir:* Hamilton Luske; *Voices:* Larry Roberts, Peggy Lee, Barbara Luddy, Stan Freberg, Alan Reed. **VHS, Beta, LV $29.95** *DIS, APD* 𝄇𝄇𝄇𝄇

The Lady Vanishes When a kindly old lady disappears from a fast-moving train, her young friend finds an imposter in her place and a spiraling mystery to solve. Hitchcock's first real winner, a smarmy, wit-drenched

British mystery that precipitated his move to Hollywood. Along with "39 Steps," considered an early Hitchcock classic. From the novel "The Wheel Spins," by Ethel Lina White. Special edition contains short subject on Hitchcock's cameos in his films. Remade in 1979.
1938 99m/B *GB* Margaret Lockwood, Paul Lukas, Michael Redgrave, May Whitty, Googie Withers, Basil Radford, Naunton Wayne, Cecil Parker; *Dir:* Alfred Hitchcock. **VHS, Beta, LV $16.95** *SNC, NOS, SUE* 𝄇𝄇𝄇𝄇

Lady Vanishes In this reworking of the '38 Alfred Hitchcock film, a woman on a Swiss bound train awakens from a nap to find that the elderly woman seated next to her was kidnapped. Falls short of the original movie due to the "screwball" nature of the main characters, Gould and Shepherd. George Axelrod wrote the script which is based on Ethel Lina White's novel "The Wheel Spins."
1979 95m/C *GB* Elliott Gould, Cybill Shepherd, Angela Lansbury, Herbert Lom; *Dir:* Anthony Page. **VHS, Beta, LV $9.95** *MED, MRV* 𝄇

The Lady in White Small-town ghost story about murder and revenge. Young Haas is accidentally locked in school one night and is visited by the ghost of a little girl; he is determined to find her murderer. Well-developed characters, interesting style, and suspenseful plot make for a sometimes slow but overall exceptional film.
1988 (PG-13) 92m/C Lukas Haas, Len Cariou, Alex Rocco, Katherine Helmond, Jason Presson, Renato Vanni, Angelo Bertolini, Jared Rushton; *Dir:* Frank Laloggia. **VHS, Beta, LV $19.95** *VIR* 𝄇𝄇𝄇

Lady Windermere's Fan The silent, Lubitsch adaptation of the Oscar Wilde tale concerning an upper-class couple's marriage being almost destroyed by suspected adultery.
1926 66m/B Ronald Colman, May McAvoy, Bert Lytell; *Dir:* Ernst Lubitsch. **VHS, Beta, 8mm $29.95** *VYY, GPV* 𝄇𝄇 ½

The Lady Without Camelias A young woman is discovered by a film producer who casts her in a successful movie and eventually marries her. Convinced that she can sustain a career as a serious actress, the producer is dismayed by the job offers that come her way, usually of a sexually exploitive nature. What follows is a tragic decline for both.
1953 106m/C *IT* VHS, Beta, 3/4U, Special order formats *CIG* 𝄇𝄇 ½

Lady from Yesterday An American couple's life is shattered by the appearance of the husband's Vietnamese mistress and her 10-year-old son. Made for television.
1985 87m/C Wayne Rogers, Bonnie Bedelia, Pat Hingle, Barrie Youngfellow, Blue Dedeort, Tina Chen; *Dir:* Robert Day. **VHS, Beta $79.98** *NSV* 𝄇

Ladybugs Hangdog salesman Dangerfield would like to move up his company's corporate ladder. But that's only going to be possible if the Ladybugs, a girls' soccer team sponsored by the company, turns into a group of winners. Dangerfield knows nothing about soccer and the girls are lousy but Dangerfield manages to persuade his secretary to serve as assistant coach and his fiancee's athletic son to masquerade as a girl and lead the team to victory. Obvious fluff.
1992 (PG-13) 91m/C Rodney Dangerfield, Jackee, Jonathan Brandis, Ilene Graff, Vinessa Shaw; *Dir:* Sidney J. Furie. **VHS, Beta** *PAR* 𝄇 ½

Ladyhawke In medieval times, a youthful pickpocket befriends a strange knight who is on a mysterious quest. This unlikely duo, accompanied by a watchful hawk, are enveloped in a magical adventure.
1985 (PG-13) 121m/C Matthew Broderick, Rutger Hauer, Michelle Pfeiffer, John Wood, Leo McKern; *Dir:* Richard Donner. **VHS, Beta, LV $19.98** *WAR* 𝄇𝄇 ½

Ladykillers A gang of bumbling bank robbers is foiled by a little old lady. Hilarious antics follow, especially on the part of Guinness, who plays the slightly demented-looking leader of the gang. Script by William Rose.
1955 87m/C *GB* Alec Guinness, Cecil Parker, Katie Johnson, Herbert Lom, Peter Sellers; *Dir:* Alexander MacKendrick. British Academy Awards '55: Best Actress (Johnson), Best Screenplay. **VHS, Beta $19.98** *HBO, FCT* 𝄇𝄇𝄇 ½

The Lady's Maid's Bell Based upon the story by Edith Wharton, this ghost story deals with the new maid in a family estate who discovers the ghost of another past maid.
1985 60m/C VHS, Beta $59.95 *PSM* 𝄇𝄇

Lagrimas de Amor A bank cashier, a night club singer, and her lover form an unusual love triangle.
1987 94m/C *MX* Ana Louisa Peluffo, Carlos Lopez Moctezuma, Sergio Bustamante. **VHS, Beta $24.95** *MAD* 𝄇𝄇 ½

Laguna Heat A tired LA cop investigates a series of murders that are related to an old friend of his father's. Based on a novel by T. Jefferson Parker. Made for television.
1987 110m/C Harry Hamlin, Jason Robards Jr., Catherine Hicks, Rip Torn, Anne Francis; *Dir:* Simon Langton. **VHS, Beta, LV $79.95** *LHV, WAR* 𝄇 ½

Lai Shi: China's Last Eunuch A starving Chinese child becomes a eunuch in order to garner wealth in the Emperor's palace. But he is castrated at a historically inopportune moment (as if there's a good time for such modification), just prior to the turbulent 1920s. In Cantonese with English subtitles.
1988 100m/C *Dir:* Chan Tsi-liang. **VHS $39.95** *FCT* 𝄇 ½

The Lair of the White Worm Scottish archaeologist uncovers a strange skull, and then a bizarre religion to go with it, and then a very big worm. An unusual look at the effects of Christianity and paganism on each other, colored with sexual innuendo and, of course, giant worms. A cross between a morality play and a horror film. Adapted from Bram Stoker's last writings, written while he was going mad from Bright's disease. Everything you'd expect from Russell.
1988 (R) 93m/C *GB* Amanda Donohoe, Sammi Davis, Catherine Oxenberg, Hugh Grant, Peter Capaldi, Stratford Johns, Paul Brooke; *Dir:* Ken Russell. **VHS, Beta, LV $14.98** *VES, LIV* 𝄇𝄇𝄇

The Lamb Intent on proving to his sweetheart that he's not the coward people say he is, a guy heads west to engage in various manly activities, including fisticuffs, karate and Indian kidnapping. Fairbanks' debut on film (he was already an established Broadway star) set the mold for the American leading man: moral, cheerful, physical, and not hard to look at. A popular guy with the public, Fairbanks formed his own film company the following year. Story's based on D.W. Griffith's book, "The Man and the Test."

1915 60m/B Douglas Fairbanks Sr., Seena Owen, Lillian Langdon, Monroe Salisbury, Kate Toncray, Alfred Paget, William E. Lowery; *Dir:* Christy Cabanne. **VHS, Beta** $24.95 *GPV, FCT* ♫♫½

Lambada By day, he's a high school teacher in Beverly Hills; by night, a Latin dirty dancer cum tutor of ghetto teens. The very first of several films based on the short-lived dance craze.
1989 (PG) 104m/C J. Eddie Peck, Melora Hardin, Adolfo "Shabba Doo" Quinones, Ricky Paull Goldin, Basil Hoffman, Dennis Burkley; *Dir:* Joel Silberg. **VHS, Beta, LV** $89.95 *WAR, OM* ♫

Lamp at Midnight This presentation from "George Schaefer's Showcase Theatre" deals with three critical periods in the life of Italian astronomer Galileo Galilei, from his invention of the telescope, through his appearance at the Holy Office of the Inquisition, to the publication of his "Dialogue on the Two Systems of the World."
1966 76m/C Melvyn Douglas, David Wayne, Michael Hordern, Hurd Hatfield, Kim Hunter; *Dir:* George Schaefer. **VHS, Beta** *FHS* ♫♫

Lana in Love A lonely woman places an ad in the personal section in hopes of finding her Mr. Right.
1992 90m/C Daphna Kastner, Clark Gregg. **VHS, Beta** *NVH* ♫♫

The Land Before Time Lushly animated children's film about five orphaned baby dinosaurs who band together and try to find the Great Valley, a paradise where they might live safely. Works same parental separation theme as Bluth's "American Tail." Charming, coy, and shamelessly tearjerking; producers included Steven Spielberg and George Lucas.
1988 (G) 70m/C *Dir:* Don Bluth; *Voices:* Pat Hingle, Helen Shaver, Gabriel Damon, Candice Houston, Burke Barnes, Judith Barsi, Will Ryan. **VHS, Beta, LV** $19.95 *MCA, FCT, APD* ♫♫♫

Land of Doom An amazon and a warrior struggle for survival in a post-holocaust fantasy setting.
1984 87m/C Deborah Rennard, Garrick Dowhen; *Dir:* Peter Maris. **VHS, Beta** $19.98 *LIV, LTG* ♫

The Land of Faraway A Swedish boy is whisked off to a magical land where he does battle with evil knights and flies on winged horses. Dubbed; based on a novel by Astrid Lindgren.
1987 (G) 95m/C Timothy Bottoms, Christian Bale, Susannah York, Christopher Lee, Nicholas Pickard; *Dir:* Vladimir Grammatikov. **VHS, Beta** $9.99 *STE, PSM* ♫½

Land of Fury British naval officer Hawkins steps on New Zealand's shore and into trouble when he accidentally walks on sacred Maori burial ground. Very British, very dated colonial saga, based on the novel "The Seekers" by John Guthrie.
1955 90m/C *GB* Jack Hawkins, Glynis Johns, Noel Purcell, Ian Fleming; *Dir:* Ken Annakin. **VHS** $19.95 *NOS* ♫♫

Land of Hunted Men The Range Busters are on the case as they track down the hideout of terrorizing gunmen.
1943 58m/B Ray Corrigan, Dennis Moore, Max Terhune. **VHS, Beta** $19.95 *NOS, DVT* ♫

Land of the Lawless A group of outlaws rule over a barren wasteland. What's the point?
1947 54m/B Rachel Brown, Raymond Hatton, Christine McIntyre, Tristram Coffin; *Dir:* Lambert Hillyer. **VHS** $19.95 *NOS, DVT* ♫

Land of the Minotaur A small village is the setting for horrifying ritual murders, demons, and disappearances of young terrorists. Score by Brian Eno. Fans of Pleasence and Cushing won't want to miss this.
1977 (PG) 88m/C *GB* Donald Pleasence, Peter Cushing, Luan Peters; *Dir:* Costa Carayiannis. **VHS, Beta** $29.95 *UHV, MRV* ♫½

Land of the Pharaohs An epic about the building of Egypt's Great Pyramid. Hawkins is the extremely talkative pharoah and Collins plays his sugary-sweet yet villainous wife. Sort of campy, but worth watching for the great surprise ending.
1955 106m/C Jack Hawkins, Joan Collins, James Robertson Justice, Dewey Martin, Alexis Minotis, Sydney Chaplin; *Dir:* Howard Hawks. **VHS, LV** *WAR* ♫♫½

Land Raiders Savalas plays a man who hates Apaches and wants their land, but is distracted when his brother arrives on the scene, igniting an old feud.
1969 101m/C Telly Savalas, George Maharis, Arlene Dahl, Janet Landgard; *Dir:* Nathan Juran. **VHS, Beta** *GKK* ♫♫

Land That Time Forgot A World War I veteran, a beautiful woman, and their German enemies are stranded in a land outside time filled with prehistoric creatures. Based on the 1918 novel by Edgar Rice Burroughs. Followed in 1977 by "The People that Time Forgot."
1975 (PG) 90m/C *GB* Doug McClure, John McEnery, Susan Penhaligon; *Dir:* Kevin Connor. **VHS, Beta** $59.98 *LIV, VES* ♫♫

Landlord Blues A trashy slumlord without morals or a conscience goes one step too far, and a tenant retaliates with munitions.
1987 96m/C Mark Boone, Raye Dowell, Richard Litt, Bill Rice, Mary Schultz, Gigi Williams; *Dir:* Jacob Burckhardt. **VHS, Beta** $79.95 *MNC* ♫♫

Lantern Hill During the Depression, a 12-year-old girl attempts to fan the embers between her estranged parents. Made for cable Wonderworks production, filmed partly on Canada's Prince Edward Island. Based on a story by Lucy Maud Montgomery, whose "Anne of Green Gables" is also part of the Wonderworks series.
1990 (G) 112m/C Sam Waterston, Colleen Dewhurst, Sarah Polley, Marion Bennett; *Dir:* Kevin Sullivan. **VHS** $29.95 *BVV, FCT* ♫♫½

Laramie Kid A vintage Tyler sagebrush saga.
1935 60m/B Tom Tyler, Alberta Vaughan, Al Ferguson, Murdock McQuarrie, George Cheseboro; *Dir:* Harry S. Webb. **VHS, Beta** $19.95 *NOS, DVT, RXM* ♫

Larks on a String Banned for 23 years, this wonderful film is Menzel's masterpiece, even better than his Oscar winning "Closely Watched Trains." Portrays the story of life in labor camps where men and women are re-educated at the whim of the government. No matter what the hardships, these people find humor, hope and love. Their individuality will not be lost, nor their humanity dissolved. Excellent performances, tellingly directed with a beautiful sense of composition and tone. Screenplay written by Menzel and Bohumil Hrabil, author of the short story on which it is based.
1968 96m/C *CZ* Vaclav Neckar, Jitka Zelenohorska, Jaroslav Satoransky, Rudolf Hrusinsky; *Dir:* Jiri Menzel. **VHS** $79.95 *FXL* ♫♫♫½

Larry Based on the Nevada State Hospital case history of a 26-year-old man, institutionalized since infancy, who is discovered to be of normal intelligence and must learn to live the life he has always been capable of living.
1974 80m/C Frederic Forrest, Tyne Daly, Katherine Helmond, Michael McGuire, Robert Walden; *Dir:* William A. Graham. **VHS, Beta** *LCA* ♫♫

Las Barras Bravas A Spanish drug-ring-in-the-big-city melodrama.
19?? 94m/C Mercedes Carreras, Tita Merello. **VHS, Beta** $59.95 *UNI* ♫½

Las Doce Patadas Mortales del Kung Fu Kung Fu masters square off in this martial arts epic.
1987 95m/C *MX* Liang Hsian Lung, Ku Feug, Han Kuo Tsai. **VHS, Beta** $34.80 *MAD* ♫♫

Las Munecas del King Kong A thief and rancher battle for vengeance, women and greed. In Spanish.
198? 82m/C *MX* Lyn May, Armando Silvestre, Diana Torres. **VHS, Beta** $77.95 *MAD* ♫

Las Vegas Hillbillys A pair of country-singing hillbillies inherit a saloon in Las Vegas and enjoy wine, moonshine, and song. These two make the Clampetts look like high society. Followed, believe it or not, by "Hillbillys in a Haunted House."
1966 85m/C Mamie Van Doren, Jayne Mansfield, Ferlin Husky, Sonny James; *Dir:* Arthur C. Pierce. **VHS, Beta** $29.95 *CHA* ♫

Las Vegas Lady Three shrewd casino hostesses plot a multi-million dollar heist in the nation's gambling capital.
1976 (PG) 90m/C Stella Stevens, Stuart Whitman, George de Cecenzo, Andrew Stevens; *Dir:* Noel Nosseck. **VHS, Beta** $49.95 *PSM* ♫

Las Vegas Serial Killer A serial killer, let out of prison on a technicality, starts killing again. Made for video by Camp.
1986 90m/C Pierre Agostino, Ron Jason, Tara MacGowan, Kathryn Downey. **VHS, Beta** $39.95 *CAM* Woof!

Las Vegas Story Gambling, colorful sights, and a murder provide the framework for this fictional, guided-tour of the city.
1952 88m/B Victor Mature, Jane Russell, Vincent Price, Hoagy Carmichael, Brad Dexter; *Dir:* Robert Stevenson. **VHS, Beta** *KOV* ♫

Las Vegas Strip War A Vegas hotel owner angers his rivals by building up a floundering casino. Hudson's last TV movie. Also known as "The Vegas Strip Wars."
1984 96m/C Rock Hudson, Sharon Stone, James Earl Jones, Pat Morita; *Dir:* George Englund. **VHS, Beta** $69.98 *LIV, LTG* ♫½

Las Vegas Weekend A computer nerd goes to Las Vegas and discovers fun.
1985 83m/C Barry Hickey, Macka Foley, Ray Dennis Steckler; *W/Dir:* Dale Trevillion. **VHS, Beta** $19.95 *STE, NWV* ♫

Laser Mission When it is discovered that the Soviets have laser weapon capabilities, an agent is given the task of destroying the weapon and kidnapping the scientist who developed it.
1990 83m/C Brandon Lee. **VHS** $79.98 *TTC* ♫♫½

Laser Moon A serial killer uses a surgical laser beam to kill his beautiful victims, and he strikes at every full moon. When he announces his next attack on a late-night radio talk show the police call in a beautiful rookie cop (Lords) to use as bait.

1992 90m/C Traci Lords, Crystal Shaw, Harrison Leduke, Bruce Carter; *Dir:* Bruce Carter. **VHS** $89.95 *HMD* ♫♫

Laserblast A standard wimp-gets-revenge story in which a frustrated young man finds a powerful and deadly laser which was left near his home by aliens; upon learning of its devastating capabilities, his personality changes and he seeks revenge against all who have taken advantage of him.
1978 (PG) 87m/C Kim Milford, Cheryl "Rainbeaux" Smith, Keenan Wynn, Roddy McDowall; *Dir:* Michael Raye. **VHS, Beta** $29.95 *MED* Woof!

The Lash of the Penitents An explosive documentary about Southern fundamentalist cults.
1939 70m/B **VHS, Beta** $16.95 *SNC, NOS, DVT*

Lassie, Come Home In first of the Lassie series, the famed collie is reluctantly sold and makes a treacherous cross-country journey back to her original family. Oscar-nominated for color cinematography.
1943 90m/C Roddy McDowall, Elizabeth Taylor, Donald Crisp, Edmund Gwenn, May Whitty, Nigel Bruce, Elsa Lanchester, J. Pat O'Malley, Lassie; *Dir:* Fred M. Wilcox. National Board of Review '43: 10 Best Films of the Year. **VHS, Beta, LV** $19.98 *MGM* ♫♫½

Lassie from Lancashire Not the famous Collie—but here, a Colleen. A pair of struggling lovebirds try to make it in show biz against all odds, including the girl's lunatic aunt, who locks her away before an audition.
1938 67m/B Marjorie Brown, Hal Thompson, Marjorie Sandford, Lassie, Mark Daly. **VHS, Beta** $29.95 *VYY* ♫

Lassie: The Miracle A TV-movie featuring the faithful collie saving new-born pups and befriending a mute boy.
1975 90m/C Lassie. **VHS, Beta** $29.95 *MGM* ♫

Lassie's Great Adventure Lassie and her master Timmy are swept away from home by a runaway balloon. After they land in the Canadian wilderness, they learn to rely on each other through peril and adventure.
1962 104m/C June Lockhart, Jon Provost, Hugh Reilly, Lassie; *Dir:* William Beaudine. **VHS, Beta** $29.99 *MGM* ♫♫½

Lassiter Tom Selleck plays a jewel thief who is asked to steal diamonds from the Nazis for the FBI. Supporting cast adds value.
1984 (R) 100m/C Tom Selleck, Lauren Hutton, Jane Seymour, Bob Hoskins, Ed Lauter; *Dir:* Roger Young. **VHS, Beta, LV** $19.98 *WAR* ♫♫½

The Last American Hero The true story of how former moonshine runner Junior Johnson became one of the fastest race car drivers in the history of the sport. Entertaining slice of life chronicling whiskey running and stock car racing, with Bridges superb in the lead. Based on articles, written by Tom Wolfe. Also known as "Hard Driver."
1973 (PG) 95m/C Jeff Bridges, Valerie Perrine, Gary Busey, Art Lund, Geraldine Fitzgerald, Ned Beatty; *Dir:* Lamont Johnson. National Board of Review Awards '73: 10 Best Films of the Year. **VHS, Beta** $59.98 *FOX* ♫♫♫

Last American Virgin Three school buddies must deal with a plethora of problems in their search for girls who are willing. Music by Blondie, The Cars, The Police, The Waitresses, Devo, U2, Human League, Quincy Jones.

1982 (R) 92m/C Lawrence Monoson, Diane Franklin, Steve Antin, Louisa Moritz; *Dir:* Boaz Davidson. **VHS, Beta** $79.95 *MGM* ♫

The Last Angry Man An old, idealistic Brooklyn doctor attracts a television producer wanting to make a documentary about his life and career, and the two conflict. Muni was Oscar-nominated for this, his last film.
1959 100m/B Paul Muni, David Wayne, Betsy Palmer, Luther Adler, Dan Tobin; *Dir:* Daniel Mann. **VHS, Beta** $69.95 *COL* ♫♫♫½

The Last Bastion A World War II drama emphasizing the political struggle between Churchill, MacArthur, Roosevelt, and Australia's John Curtin.
1984 160m/C *AU* Robert Vaughn, Timothy West. **VHS, Beta** $19.95 *ACA* ♫♫

The Last Boy Scout This formula thriller stars Willis as a private eye and Wayans as an ex-quarterback teaming up against a pro-football team owner who will stop at nothing to get political backing for a bill promoting legalized gambling on sports. Another variation of the violent buddy-picture.
1991 (R) 105m/C Bruce Willis, Damon Wayans, Halle Berry; *Dir:* Tony Scott. **VHS, LV** $94.99 *WAR* ♫♫

Last Bullet A young westerner chases after a band of outlaws to avenge the cold-blooded murder of his parents.
1950 55m/B James Ellison, Russell Hayden, John Cason. **VHS, Beta, LV** *WGE* ♫♫

Last Call Clearly, Joe Sixpack won't rent this bimbo fest for its subtle plot. Katt, a mafiosi cheated real estate guy, decides to even the score with the assistance of gal pal/playboy of the year Tweed, who's more than willing to compromise her position. A direct to video release, it also features the talents of playboy emerita Stevens.
1990 (R) 90m/C William Katt, Shannon Tweed, Joseph Campanella, Stella Stevens, Matt Roe; *Dir:* Jag Mundhra. **VHS, Beta** $89.95 *PAR* ♫

Last Challenge of the Dragon A son brings death to his family by humiliating the underworld martial arts king of the city.
1980 (R) 90m/C Bruce Lee. **VHS, Beta** $59.95 *GEM* ♫

The Last Chance A realistic look at the efforts of three WWII Allied officers to help a group of refugees escape across the Alps from Italy to Switzerland. The officers are played by former pilots who were shot down over Switzerland. In spite of this—or perhaps because of it—the acting is superb. Watch for inspirational scene of refugees singing.
1945 105m/B E.G. Morrison, John Hoy, Ray Reagan, Odeardo Mosini, Sigfrit Steiner, Emil Gerber; *Dir:* Leopold Lindtberg. **VHS** $19.95 *NOS* ♫♫♫

Last Chase A famed race car driver becomes a vocal dissenter against the sterile society that has emerged, in this drama set in the near future. Made for television.
1981 (PG) 106m/C *CA* Lee Majors, Burgess Meredith, Chris Makepeace, Alexandra Stewart; *Dir:* Martyn Burke. **VHS, Beta** $69.95 *VES* ♫

Last of the Clintons A range detective tracks an outlaw gang amid spur jinglin', sharp shootin' and cow punchin'.
1935 64m/B Harry Carey. **VHS, Beta, LV** $19.95 *NOS, WGE* ♫½

Last of the Comanches Cavalry and Indians fight for water when both are dying of thirst in the desert.

1952 85m/C Broderick Crawford, Barbara Hale, John Stewart, Lloyd Bridges, Mickey Shaughnessy, George Mathews; *Dir:* Andre de Toth. **VHS, Beta** $14.99 *GKK* ♫♫

Last Command Famous powerful silent film by Sternberg about an expatriate Czarist general forging out a pitiful living as a silent Hollywood extra, where he is hired by his former adversary to reenact the revolution he just left. Next to "The Last Laugh," this is considered Jannings' most acclaimed performance. Deeply ironic, visually compelling film with a new score by Gaylord Carter.
1928 88m/B Emil Jannings, William Powell, Evelyn Brent; *Dir:* Josef von Sternberg. Academy Awards '28: Best Actor (Jannings). **VHS, Beta** $14.95 *PAR, FCT, BTV* ♫♫♫

The Last Command Jim Bowie and his followers sacrifice their lives in defending the Alamo. A good cast holds its own against a mediocre script; battle scenes are terrific.
1955 110m/C Sterling Hayden, Richard Carlson, Ernest Borgnine, J. Carroll Naish, Virginia Grey; *Dir:* Frank Lloyd. **VHS** $14.95 *REP, PAR* ♫♫½

The Last Contract An artist who pays his bills by working as a hitman, kills the wrong man. As a result, he must fear for his life since a counterhit is inevitable.
1986 (R) 85m/C Jack Palance, Rod Steiger, Bo Svenson, Richard Roundtree, Ann Turkel; *Dir:* Allan A. Buckhantz. **VHS** $19.95 *ACA* ♫♫

Last Cry for Help A psychiatrist helps a seventeen-year-old high school coed who, fearing she has disappointed her parents, attempts suicide. Typical made-for-television melodrama that attempts to make a statement, but falls short. Definitely not the "Partridge Family."
1979 98m/C Linda Purl, Shirley Jones, Tony LoBianco, Murray Hamilton, Grant Goodeve; *Dir:* Hal Sitowitz. **VHS, Beta** $29.95 *UNI* ♫½

Last Dance Five scantily-clad exotic dancers are ready to compete for the title of Miss Dance-TV, but trouble ensues when the dancers start turning up dead.
1991 86m/C Cynthia Bassinet, Elaine Hendrix, Kurt T. Williams, Allison Rhea, Erica Ringston; *Dir:* Anthony Markes. **VHS, Beta** $79.95 *PSM* ♫½

Last Day of the War A German scientist is chased by Americans and Germans as WWII winds to its conclusion. Filmed in Spain.
1969 (PG) 96m/C *SP* George Maharis, Maria Perschy, James Philbrook, Gerard Herter; *Dir:* Juan Antonio Bardem. **VHS, Beta** $39.95 *IVE* ♫½

The Last Days of Dolwyn A man banned for thievery from his Welsh village returns bent upon revenge. His plans to buy the entire district when it is designated part of a water-reservoir project, but finds plans thwarted by a dowager and her stepson. Burton's first film was based on a true story.
1949 95m/B *GB* Edith Evans, Emlyn Williams, Richard Burton, Anthony James, Barbara Couper, Alan Aynesworth, Hugh Griffith, Roddy Hughes, Tom Jones; *W/Dir:* Emlyn Williams. **VHS** $29.95 *FCT* ♫♫½

Last Days of Frank & Jesse James A tired television-movie rehash of the well-known western legend in which the famous brothers try to be like others after their personal war against society ends.
1986 100m/C Johnny Cash, Kris Kristofferson, June Carter Cash, Willie Nelson, Margaret Gibson, Gail Youngs; *Dir:* William A. Graham. **VHS, Beta** $19.95 *STE, VMK* ♫½

The Last Days of Patton A made-for-television movie depicting the aging general's autumn years as a controversial ex-Nazi defender and desk-bound World War II historian. Scott reprises his feature-film role.
1986 146m/C George C. Scott, Ed Lauter, Eva Marie Saint, Richard Dysart, Murray Hamilton, Kathryn Leigh Scott; ***Dir:*** Delbert Mann. **VHS, Beta $89.98** *FOX* 🦴½

Last Days of Pompeii Vintage DeMille-style epic based on Lord Lytton's book, where the eruption of Vesuvius threatens noblemen and slaves alike. State-of-the-art special effects still look fantastic today. Remade in 1960; available colorized.
1935 96m/B Preston Foster, Alan Hale Jr., Basil Rathbone, John Wood, Louis Calhern, David Holt, Dorothy Wilson, Wryley Birch, Gloria Shea; ***Dir:*** Ernest B. Schoedsack. **VHS, Beta** *TTC* 🦴🦴½

The Last Days of Pompeii Superman Reeves plays a gladiator trying to clean up the doomed town of Pompeii in this remake of the 1935 classic. Some spectacular scenes, including the explosive climax when the mountain blows its top. Screenwriter Sergio Leone went on to do many "spaghetti westerns" and the epics "Once Upon a Time in America" and "Once Upon a Time in the West."
1960 105m/C *IT* Steve Reeves, Christine Kaufmann, Barbara Carroll, Anne Marie Baumann, Mimmo Palmara; ***Dir:*** Mario Bonnard. **VHS $9.95** *GKK* 🦴🦴

The Last Detail Two hard-boiled career petty officers (Nicholson and Young) are commissioned to transfer a young sailor facing an eight-year sentence for petty theft from one brig to another. In an act of compassion, they attempt to show the prisoner a final good time. Nicholson shines in both the complexity and completeness of his character. Adapted from a Daryl Ponicsan novel; Gilda Radner appears in cameo. Oscar nominations for Nicholson, Quaid, and Adapted Screenplay.
1973 (R) 104m/C Jack Nicholson, Randy Quaid, Otis Young, Carol Kane, Michael Moriarty, Nancy Allen, Gilda Radner; ***Dir:*** Hal Ashby. British Academy Awards '74: Best Actor (Nicholson); Cannes Film Festival '74: Best Actor (Nicholson); National Board of Review Awards '74: 10 Best Films of the Year; New York Film Critics Awards '74: Best Actor (Nicholson); National Society of Film Critics Awards '74: Best Actor (Nicholson). **VHS, Beta, LV $19.95** *COL* 🦴🦴🦴🦴

The Last Dragon It's time for a Kung Fu showdown on the streets of Harlem, for there is scarcely enough room for even one dragon. Also known as "Berry Gordy's Last Dragon."
1985 (PG-13) 108m/C Taimak, Vanity, Christopher Murney, Julius J. Carry III, Faith Prince; ***Dir:*** Michael Schultz. **VHS, Beta $29.95** *FOX* 🦴½

The Last Embrace A feverish thriller in the Hitchcock style dealing with an ex-secret serviceman who is convinced someone is trying to kill him.
1979 (R) 98m/C Roy Scheider, Janet Margolin, Christopher Walken, John Glover, Charles Napier, Mandy Patinkin; ***Dir:*** Jonathan Demme. **VHS, Beta $59.98** *FOX* 🦴🦴🦴

The Last Emperor Deeply ironic epic detailing life of Pu Yi, crowned at the age of three as the last emperor of China before the onset of communism. Follows Pu Yi from childhood to manhood (sequestered away in the Forbidden City) to fugitive to puppet-ruler to party proletariat. O'Toole portrays the sympathetic Scot tutor who educates the adult Pu Yi (Lone) in the ways of the western world after Pu Yi abdicates power in 1912. Shot on location inside the People's Republic of China with a cast of thousands; authentic costumes. Rich, visually stunning movie. Music by David Byrne, Ryuichi Sakamoto, and Cong Su; cinematography by Vittorio Storaro.
1987 (PG-13) 140m/C *IT* John Lone, Peter O'Toole, Joan Chen, Victor Wong, Ryuichi Sakamoto, Dennis Dun, Maggie Han; ***Dir:*** Bernardo Bertolucci. Academy Awards '87: Best Adapted Screenplay, Best Art Direction/Set Decoration, Best Cinematography, Best Costume Design, Best Director (Bertolucci), Best Film Editing, Best Picture, Best Sound, Best Original Score; Directors Guild of America Awards '87: Best Director (Bertolucci); Golden Globe Awards '88: Best Film—Drama; National Board of Review Awards '87: 10 Best Films of the Year. **VHS, Beta, LV, SVS, 8mm $29.98** *SUE, FCT, BTV* 🦴🦴🦴🦴

The Last of England A furious non-narrative film by avant-garde filmmaker Derek Jarman, depicting the modern British landscape as a funereal, waste-filled rubble-heap - as depleted morally as it is environmentally.
1987 87m/C *GB* Tilda Swinson, Spencer Leigh; ***Dir:*** Derek Jarman. Los Angeles Film Critics Association Awards '88: Best Independent Film. **VHS, Beta, LV $29.95** *MFV, IVA* 🦴🦴🦴½

Last Exit to Brooklyn Hubert Selby, Jr.'s shocking book comes to the screen in a vivid film. Leigh gives a stunning performance as a young Brooklyn girl caught between the Mafia, union men, and friends struggling for something better. Set in the 1950s. Fine supporting cast; excellent pacing.
1990 (R) 102m/C *GE* Jennifer Jason Leigh, Burt Young, Stephen Lang, Ricki Lake, Jerry Orbach, Maia Danzinger; ***Dir:*** Uli Edel. New York Film Critics Awards '89: Best Supporting Actress (Leigh). **VHS, LV $89.95** *COL, FCT* 🦴🦴🦴½

Last Fight A boxer risks his life and his girlfriend for a shot at the championship title. Watch and see if he could've been a contender.
1982 (R) 85m/C Fred Williamson, Willie Colon, Ruben Blades, Joe Spinell, Darlanne Fluegel; ***Dir:*** Fred Williamson. **VHS, Beta $69.99** *HBO* 🦴½

The Last of the Finest Overzealous anti-drug task force cops break the rules in trying to put dealer-drug lords in prison; ostensibly a parallel on the Iran-Contra affair.
1990 (R) 106m/C Brian Dennehy, Joe Pantoliano, Jeff Fahey, Bill Paxton, Deborra-Lee Furness, Guy Boyd, Henry Darrow, Lisa Jane Persky, Michael C. Gwynne; ***Dir:*** John MacKenzie. **VHS, Beta, LV $14.98** *ORI* 🦴½

Last Flight of Noah's Ark Adventure concerns a high-living pilot, a prim missionary and two stowaway orphans who must plot their way off a deserted island following the crash landing of their broken-down plane.
1980 (G) 97m/C Elliott Gould, Genevieve Bujold, Rick Schroder, Vincent Gardenia, Tammy Lauren; ***Dir:*** Charles Jarrott. **VHS, Beta $69.95** *DIS* 🦴

Last Flight Out: A True Story Hours before Saigon fell to the Vietcong in 1975, Vietnamese and Americans rushed to board the last commercial flight out of Vietnam. Lukewarm account of a true story.
1990 99m/C James Earl Jones, Richard Crenna, Haing S. Ngor, Eric Bogosian. **VHS, 8mm $39.95** *MNC* 🦴🦴

The Last Fling When Selleca has her mind set on a romantic fling, Ritter is headed for trouble. She wants love for one night, he wants it forever. Will they solve this difference of opinion? Will true love prevail? Will Ritter consent to being used as a one-night stand? Weakly plotted and poorly acted comedy.
1986 95m/C John Ritter, Connie Sellecca, Scott Bakula, Shannon Tweed, Paul Sand, John Bennett Perry; ***Dir:*** Corey Allen. **VHS $29.95** *ACA* 🦴½

Last Four Days A chronicle of the final days of Benito Mussolini.
1977 (PG) 91m/C Rod Steiger, Henry Fonda, Franco Nero; ***Dir:*** Carlo Lizzani. **VHS, Beta $69.98** *LIV, LTG* 🦴🦴

Last Frontier Serial of 12 chapters contains shades of spectacular figures in Western history: Custer, Hickok, and others in a background of grazing buffalo, boom towns, and covered wagon trails.
1932 216m/B Lon Chaney Jr., Yakima Canutt, Francis X. Bushman; ***Dir:*** Spencer Gordon Bennet, Thomas L. Story. **VHS, Beta $14.95** *VCN, VYY, DVT* 🦴🦴½

Last Full Measure A docudrama of the Battle of Gettysburg, made for video.
1985 30m/C Stacy Keach. **VHS, Beta $19.95** *BVG, FGF, AVL* 🦴🦴½

Last Game A college student is torn between his devotion to his blind father and going out for the college's football team.
1980 107m/C Howard Segal, Ed L. Grady, Terry Alden; ***Dir:*** Martin Beck. **VHS, Beta $59.99** *HBO* 🦴🦴

Last Gun A legendary gunman on the verge of retirement has to save his town from a reign of terror before turning his gun in.
1964 98m/C Cameron Mitchell. **VHS, Beta $29.98** *MAG* 🦴🦴

The Last of His Tribe In 1911 an anthropologist befriends an Indian and discovers that Ishi is the last surviving member of California's Yahi tribe. Ishi then becomes a media and scientific society darling, spending his remaining life in captivity to academia. A good portrayal of the Native American plight. Based on a true story.
1992 (PG-13) 90m/C Jon Voight, Graham Greene, Anne Archer, David Ogden Stiers; ***Dir:*** Harry Hook. **VHS $89.99** *WAR* 🦴🦴

Last Holiday A man who is told he has a few weeks to live decides to spend them in a posh resort where people assume he is important. Script by J. B. Priestley.
1950 89m/B *GB* Alec Guinness, Kay Walsh, Beatrice Campbell, Wilfrid Hyde-White, Bernard Lee; ***Dir:*** Henry Cass. **VHS, Beta** *LCA, RXM* 🦴🦴🦴

The Last Horror Film A beautiful queen of horror films is followed to the Cannes Film Festival by her number one fan who, unbeknownst to her, is slowly murdering members of her entourage in a deluded and vain attempt to capture her attentions.
1982 (R) 87m/C Caroline Munro, Joe Spinell, Judd Hamilton; ***Dir:*** David Winters. **VHS, Beta $59.95** *MED* 🦴

The Last Hour A Wall Street crook crosses a Mafia punk, so the latter holds the former's wife hostage in an unfinished skyscraper. Fortunately the lady's first husband is a gung-ho cop eager to commence the DIE HARD-esque action. Undistinguished fare that gives away the climax in an opening flash-forward.

1991 (R) 85m/C Michael Pare, Shannon Tweed, Bobby DiCicco, Robert Pucci; *Dir:* William Sachs. **VHS, Beta $19.95** *ACA* ✗

Last House on Dead End Street
Actors die for their art in this splatter flick about snuff films.
1977 (R) 90m/C Steven Morrison, Dennis Crawford, Lawrence Bornman, Janet Sorley; *Dir:* Victor Janos. **VHS, Beta $54.95** *SUN* ✗

Last House on the Left
Two girls are kidnapped from a rock concert by a gang of escaped convicts; the girls' parents exact bloody revenge when the guilty parties pay an intended housecall. Controversial and grim low-budget shocker; loosely based on Bergman's ''The Virgin Spring.''
1972 (R) 83m/C David Hess, Lucy Grantham, Sandra Cassel, Mark Sheffler, Fred J. Lincoln; *Dir:* Wes Craven. **VHS, Beta, LV $19.98** *LIV, VES, HHE* ✗

The Last Hunter
A soldier fights for his life behind enemy lines during the Vietnam War.
1980 (R) 97m/C *IT* Tisa Farrow, David Warbeck, Tony King, Bobby Rhodes, Margit Evelyn Newton, John Steiner, Alan Collins; *Dir:* Anthony M. Dawson. **VHS, Beta $69.98** *LIV, VES* ✗

Last Hurrah
An aging Irish-American mayor battles corruption and political backbiting in his effort to get re-elected for the last time. Semi-acclaimed heart warmer, based on the novel by Edwin O'Connor.
1958 121m/B Spencer Tracy, Basil Rathbone, John Carradine, Jeffrey Hunter, Dianne Foster, Pat O'Brien, Edward Brophy, James Gleason, Donald Crisp, Ricardo Cortez, Wallace Ford, Frank McHugh, Jane Darwell; *Dir:* John Ford. National Board of Review Awards '58: Best Actor (Tracy), Best Director; 10 Best Films of the Year '58: Best Actor. **VHS, Beta, LV $19.95** *COL* ✗✗½

Last Innocent Man
An attorney who quit law due to a guilty conscience is lured into defending a suspected murderer. Things become even more complicated when he begins having an affair with his client's seductive wife. A well-acted and suspense-filled flick. Made for cable television.
1987 113m/C Ed Harris, Roxanne Hart, Bruce McGill, Clarence Williams III, Rose Gregorio; *Dir:* Roger Spottiswoode. **VHS, Beta $79.95** *LHV, WAR* ✗✗½

The Last Laugh
An elderly man, who as the doorman of a great hotel was looked upon as a symbol of ''upper class,'' is demoted to washroom attendant due to his age. Important due to camera technique and consuming performance by Jannings.
1924 77m/B Emil Jannings, Maly Delshaft, Max Hiller; *Dir:* F. W. Murnau. Brussels World's Fair '58: #11 of the Best Films of All Time. **VHS, Beta, 3/4U $19.95** *NOS, MRV, IHF* ✗✗✗½

The Last Man on Earth
Price is the sole survivor of a plague which has turned the rest of the world into zombies, who constantly harass him. Uneven U.S./Italian production manages to convey a creepy atmosphere of dismay.
1964 86m/B *IT* Vincent Price, Franca Bettoya, Giacomo Rossi-Stuart, Emma Danieli; *Dir:* Ubaldo Ragona, Sidney Salkow. **VHS, Beta $16.95** *SNC, MRV* ✗✗

Last Man Standing
A cheap, gritty drama about bare-knuckle fist fighting. Wells stars as a down-on-his-luck prizefighter trying to find work that doesn't involve pugilism.
1987 (R) 89m/C William Sanderson, Vernon Wells, Franco Columbo; *Dir:* Damian Lee. **VHS, Beta $29.95** *ACA* ✗

The Last Married Couple in America
A couple fight to stay happily married amidst the rampant divorce epidemic engulfing their friends.
1980 (R) 103m/C George Segal, Natalie Wood, Richard Benjamin, Valerie Harper, Dom DeLuise, Priscilla Barnes; *Dir:* Gilbert Cates. **VHS, Beta $69.95** *MCA* ✗✗

Last Mercenary
An angry ex-soldier kills everyone who makes him mad.
1984 90m/C Tony Marsina, Malcolm Duff, Kitty Nichols, Louis Walser. **VHS, Beta $59.95** *MGL* Woof!

The Last Metro
Truffaut's alternately gripping and touching drama about a theater company in Nazi-occupied Paris, where the proprietress' husband is a Jew hiding in the cellar, and the leading man is a Resistance fighter.
1980 135m/C *FR* Catherine Deneuve, Gerard Depardieu, Heinz Bennent, Jean Poiret, Andrea Ferreol, Paulette Dubost, Sabine Haudepin; *Dir:* Francois Truffaut. New York Film Festival '80: #11 of the Best Films of All Time. **VHS, Beta $39.95** *HMV, FOX, APD* ✗✗✗

Last Mile
The staff of a prison prepares for the execution of a celebrated murderer.
1932 70m/B Preston Foster, Howard Phillips, George E. Stone; *Dir:* Sam Bischoff. **VHS, Beta $19.95** *NOS, MRV, KRT* ✗✗

Last of the Mohicans
Serial based on James Fenimore Cooper's novel of the life-and-death struggle of the Mohican Indians during the French and Indian War. Twelve chapters, 13 minutes each. Remade twice, as a movie in 1936 and as a television movie in 1977.
1932 230m/B Edwina Booth, Harry Carey, Hobart Bosworth, Junior Coghlan; *Dir:* Ford Beebe, B. Reeves Eason. **VHS, Beta $49.95** *NOS, VCN, VYY* ✗✗

Last of the Mohicans
James Fenimore Cooper's classic about the French and Indian War in colonial America is brought to the screen. Remake of the 1932 serial was remade itself in 1977 for television.
1936 91m/B Randolph Scott, Binnie Barnes, Bruce Cabot, Henry Wilcoxon, Heather Angel, Hugh Buckler; *Dir:* George B. Seitz. **VHS, Beta $12.95** *IVE, MED, CCB* ✗✗½

Last of the Mohicans
The classic novel by James Fenimore Cooper about the scout Hawkeye and his Mohican companions, Chingachgook and Uncas during the French and Indian war, comes to life in this made-for-television film. Remake of a 1932 serial and a 1936 film.
1985 (R) 97m/C Steve Forrest, Ned Romero, Andrew Prine, Don Shanks, Robert Tessier, Jane Actman; *Dir:* James L. Conway. **VHS, Beta $19.95** *STE, MAG, UHV* ✗✗

The Last Movie
A movie stunt man stays in a small Peruvian town after his filming stint is over. Hopper's confused, pretentious follow-up to ''Easy Rider,'' has a multitude of cameos but little else. Given an award by the Italians; nobody else understood it. Also known as ''Chinchero,'' the movie is Kristofferson's film debut; he also wrote the music.
1971 (R) 108m/C Dennis Hopper, Julie Adams, Peter Fonda, Kris Kristofferson, Sylvia Miles, John Phillip Law, Russ Tamblyn, Rod Cameron; *Dir:* Dennis Hopper. **VHS $79.95** *UHV* ✗½

The Last of Mrs. Cheyney
Remake of Norma Shearer's 1929 hit, based on the play by Frederick Lonsdale, about a sophisticated jewel thief who poses as a wealthy woman to get into parties hosted by London bluebloods. Dripping with charm, she works her way into Lord Drilling's mansion where she plans a huge heist. The film is handled well, and the cast gives solid performances throughout. This chic comedy of high society proved to be one of Crawford's most popular films of the 30's.
1937 98m/B Joan Crawford, Robert Montgomery, William Powell, Frank Morgan, Nigel Bruce, Jessie Ralph; *Dir:* Richard Boleslawski, George Fitzmaurice. **VHS $19.98** *MGM* ✗✗✗

Last of Mrs. Lincoln
Intimate portrayal of the famous first lady. Film focuses on Mary Todd Lincoln's life from the assassination of her husband, through her autumn years, and her untimely downfall.
1976 118m/C Julie Harris, Robby Benson, Patrick Duffy. **VHS, Beta $39.95** *USA* ✗✗½

Last Night at the Alamo
Patrons fight to stop the destruction of their Houston bar, the Alamo, which is about to be razed to make room for a modern skyscraper. Insightful comedy written by the author of ''The Texas Chainsaw Massacre.''
1983 80m/B Sonny Davis, Lou Perry, Steve Matilla, Tina Hubbard, Doris Hargrave; *Dir:* Eagle Pennell. **VHS $39.98** *CGI* ✗✗✗

Last of the One Night Stands
This award-winning documentary traces the career of the Lee Williams Band, which brought the sound of the swing era to rural America.
1983 28m/C *Dir:* Jeff Belker; *Nar:* Hugh Thomas. **VHS, Beta $39.95** *EVG* ✗✗✗

The Last Outlaw
Action-packed western with Cooper following his success of ''Wings'' with Paramount. Most excellent equine stunts.
1927 61m/B Gary Cooper, Jack Luden, Betty Jewel; *Dir:* Arthur Rosson. **VHS, Beta $19.95** *GPV, FCT* ✗✗

Last Outlaw
The last of the famous badmen of the old West is released from jail and returns home to find that times have changed. The action climaxes in an old-time blazing shoot-out.
1936 79m/B Harry Carey, Hoot Gibson, Henry B. Walthall, Tom Tyler, Russell Hopton, Alan Curtis, Harry Woods, Barbara Pepper; *Dir:* Christy Cabanne. **VHS, Beta $29.95** *VCN* ✗✗

The Last of Philip Banter
An alcoholic writer is terrified to learn events in his life exactly match a mysterious manuscript.
1987 (R) 105m/C Tony Curtis, Greg Henry, Irene Miracle, Scott Paulin, Kate Vernon; *Dir:* Herve Hachuel. **VHS $79.95** *REP* ✗½

The Last Picture Show
Working with Larry McMurtry's novel, McMurtry and Bogdanovich scripted this slice of life/nostalgic farewell to an innocent age that was Oscar nominated for best picture. Soap opera with heart and soul, breathing life into both the time and place. That place is Archer City, a backwater Texas town, and the time is the early fifties, when football jock buddies Bridges and Bottoms first cut their teeth in the Real World. Most of the story is played out at the local hangout run by Sam the Lion, an ex-cowboy both boys look up to. Bridges is hooked up with spoiled pretty girl Sheperd, while Bottoms, a sensitive guy, is having an affair with the coach's neglected wife (Leachman). Loss of innocence, disillusionment and confusion are played out against the backdrop of a town about to close its cinema. Shepherd's and Bottoms' film debut. Stunningly photographed in black and white (Bogdanovich claimed he didn't want to ''prettify''

the picture by shooting in color) by Robert Surtees. Available in a laserdisc special edition with 17 minutes restored. Followed by "Texasville."

1971 (R) 118m/B Jeff Bridges, Timothy Bottoms, Ben Johnson, Cloris Leachman, Cybill Shepherd, Ellen Burstyn, Eileen Brennan, Clu Gulager, Sharon Taggart, Randy Quaid; *Dir:* Peter Bogdanovich. Academy Awards '71: Best Supporting Actor (Johnson), Best Supporting Actress (Leachman). **VHS, LV, 8mm $19.95** COL, VYG, CRC 🐾🐾🐾🐾

Last Plane Out A Texas journalist sent out on assignment to Nicaragua falls in love with a Sandinista rebel. Based on producer Jack Cox's real-life experiences as a journalist during the last days of the Samosa regime. Low-budget propaganda.

1983 (R) 90m/C Jan-Michael Vincent, Lloyd Batista, Julie Carmen, Mary Crosby, David Huffman, William Windom; *Dir:* David Nelson. **VHS, Beta $19.95** STE, FOX 🐾½

The Last Polka SCTV vets Candy and Levy are Yosh and Stan Schmenge, polka kings interviewed for a "documentary" on their years in the spotlight and on the road. Hilarious spoof on Martin Scorsese's "The Last Waltz", about The Band's last concert. Several fellow Second City-ers keep the laughs coming. If you liked SCTV or "This is Spinal Tap", you'll like this. Made for cable (HBO).

1984 54m/C John Candy, Eugene Levy, Rick Moranis, Robin Duke, Catherine O'Hara; *Dir:* John Blanchard. **VHS, Beta $19.98** LIV, VES 🐾🐾🐾

Last of the Pony Riders When the telegraph lines linking the East and West Coasts are completed, Gene and the other Pony Express riders find themselves out of a job. Autry's final feature film.

1953 59m/B Gene Autry, Smiley Burnette, Kathleen Case, Buzzy Henry; *Dir:* George Archainbaud. **VHS, Beta $37.50** CCB 🐾½

The Last Porno Flick A pornographic movie script winds up in the hands of a couple of goofy cab drivers. The twosome sneak the movie's genre past the producers, families, and police.

1974 (PG) 90m/C Michael Pataki, Marianna Hill, Carmen Zapata, Mike Kellin, Colleen Camp, Tom Signorelli, Antony Carbone; *Dir:* Ray Marsh. **VHS $19.95** API 🐾

The Last Prostitute Two teenage boys search for a legendary prostitute to initiate them into manhood, only to discover that she has retired. They hire on as laborers on her horse farm, and one of them discovers the meaning of love. Made for cable TV.

1991 (PG-13) 93m/C Sonia Braga, Wil Wheaton, David Kaufman, Woody Watson, Dennis Letts, Cotter Smith; *Dir:* Lou Antonio. **VHS $89.95** MCA 🐾🐾

The Last Radio Station A mini-musical with a largely Motown soundtrack about a failing radio station supported by old rock songs: "Reflections," "Nowhere to Run," "Dancing in the Street" and others.

1986 60m/C VHS, Beta, LV $19.95 MCA, MVD 🐾🐾

Last of the Red Hot Lovers The unfunny film adaptation of Neil Simon's Broadway hit about a middle-aged man who decides to have a fling and uses his mother's apartment to seduce three very strange women.

1972 (PG) 98m/C Alan Arkin, Paula Prentiss, Sally Kellerman, Renee Taylor; *Dir:* Gene Saks. **VHS, Beta, LV $14.95** PAR 🐾½

The Last Remake of Beau Geste A slapstick parody of the familiar Foreign Legion story from the Mel Brooks-ish school of loud genre farce. Gary Cooper makes an appearance by way of inserted footage from the 1939 straight version.

1977 (PG) 85m/C Marty Feldman, Ann-Margret, Michael York, Peter Ustinov, James Earl Jones; *Dir:* Marty Feldman. **VHS, Beta $14.95** MCA 🐾🐾

Last Resort A married furniture executive unknowingly takes his family on vacation to a sex-saturated, Club Med-type holiday spot, and gets more than he anticipated.

1986 (R) 80m/C Charles Grodin, Jon Lovitz; *Dir:* Zane Buzby. **VHS, Beta, LV $79.98** LIV, VES 🐾🐾

The Last Reunion The only witness to a brutal killing of a Japanese official and his wife during WWII seeks revenge on the guilty American platoon, thirty-three years later. Violent.

1980 98m/C Cameron Mitchell, Leo Fong, Chanda Romero, Vic Silayan, Hal Bokar, Philip Baker Hall; *Dir:* Jay Wertz. **VHS $16.95** SNC 🐾½

Last Ride of the Dalton Gang A long-winded retelling of the wild adventures that made the Dalton gang legendary among outlaws.

1979 146m/C Larry Wilcox, Jack Palance, Randy Quaid, Cliff Potts, Dale Robertson, Don Collier; *Dir:* Dan Curtis. **VHS, Beta $19.98** WAR, OM 🐾🐾

The Last Riders Motorcycle centaur Estrada revs a few motors fleeing from cycle club cronies and crooked cops. Full throttle foolishness.

1990 (R) 90m/C Erik Estrada, William Smith, Armando Sylvester, Kathrin Lautner; *Dir:* Joseph Merhi. **VHS $79.95** PMH, HHE 🐾

Last Rites A priest at St. Patrick's Cathedral in New York allows a young Mexican woman to seek sanctuary from the Mob, who soon come after both of them.

1988 (R) 103m/C Tom Berenger, Daphne Zuniga, Chick Vennera, Dane Clark, Carlo Pacchi, Anne Twomey, Paul Dooley, Vassili Lambrinos; *Dir:* Donald P. Bellisario. **VHS, Beta, LV $19.98** FOX 🐾🐾

The Last Season Shoot 'em up involving a bunch of hunters. Their aimless destruction provokes the good guy to save the forest they are demolishing. A battle ensues.

1987 90m/C Christopher Gosch, Louise Dorsey, David Cox; *Dir:* Raja Zahr. **VHS** TPV Woof!

The Last of Sheila A movie producer invites six big-star friends for a cruise aboard his yacht the "Sheila." He then stages an elaborate "Whodunnit" parlor game to discover which one of them murdered his wife. Scripted by Stephen Sondheim and Anthony Perkins.

1973 (PG) 119m/C Richard Benjamin, James Coburn, James Mason, Dyan Cannon, Joan Hackett, Raquel Welch, Ian McShane; *Dir:* Herbert Ross. Edgar Allan Poe Awards '73: Best Screenplay. **VHS, Beta $64.95** WAR 🐾🐾½

The Last Slumber Party A slumber party is beset by a homicidal maniac. Heavy metal soundtrack.

1987 89m/C Jan Jensen, Nancy Meyer. **VHS, Beta $19.95** UHV Woof!

Last Song A singer's husband discovers a plot to cover up a fatal toxic-waste accident, and is killed because of it. It's up to her to warn the authorities before becoming the next victim.

1980 (PG) 96m/C Lynda Carter, Ronny Cox, Nicholas Pryor, Paul Rudd, Jenny O'Hara; *Dir:* Alan J. Levi. **VHS, Beta $59.98** FOX 🐾½

The Last Starfighter A young man who becomes an expert at a video game is recruited to fight in an inter-galactic war. Listless adventure which explains where all those video games come from. Watch for O'Herlihy disguised as a lizard.

1984 (PG) 100m/C Lance Guest, Robert Preston, Barbara Bosson, Dan O'Herlihy, Catherine Mary Stewart, Cameron Dye, Kimberly Ross, Wil Wheaton; *Dir:* Nick Castle. **VHS, Beta, LV $19.95** MCA 🐾🐾

Last Summer Three teenagers discover love, sex and friendship on the white sands of Fire Island, N.Y. The summer vacation fantasy world they create shatters when a sweet but homely female teenager joins their group. Based on a novel by Evan Hunter. Burns earned an Oscar nomination.

1969 (R) 97m/C Barbara Hershey, Richard Thomas, Bruce Davison, Cathy Burns; *Dir:* Frank Perry. **VHS, Beta $59.98** FOX 🐾🐾½

The Last Supper A repentant Cuban slave-owner decides to cleanse his soul and convert his slaves to Christianity by having twelve of them reenact the Last Supper. Based on a true story. In Spanish with English subtitles.

1976 110m/C SP Nelson Villagra, Silvano Rey, Lamberto Garcia, Jose Antonio Rodriguez, Samuel Claxton, Mario Balmaseda, Thomas Gutierrez Alea. **VHS, Beta $69.95** NYF, FCT 🐾🐾🐾

Last Tango in Paris Rated "X" upon release. Brando plays a middle-aged American who meets a French girl and tries to forget his wife's suicide with a short, extremely steamy affair. Bertolucci proves to be a master; Brando gives one of his best performances. Very controversial when made, still quite explicit. Visually stunning. Oscar nominations for actor Brando and director Bertolucci.

1973 129m/C FR IT Marlon Brando, Maria Schneider, Jean-Pierre Leaud; *Dir:* Bernardo Bertolucci. New York Film Critics Awards '73: Best Actor (Brando); National Society of Film Critics Awards '73: Best Actor (Brando); Harvard Lampoon Awards '72: Worst Film of the Year. **VHS, Beta, LV $19.98** MGM, FOX, VYG 🐾🐾½

The Last Temptation of Christ Scorsese's controversial adaptation of the Nikos Kazantzakis novel, portraying Christ in his last year as an ordinary Israelite tormented by divine doubt, human desires and the voice of God. The controversy engulfing the film, as it was heavily protested and widely banned, tended to divert attention from what is an exceptional statement of religious and artistic vision. Excellent score by Peter Gabriel.

1988 (R) 164m/C Willem Dafoe, Harvey Keitel, Barbara Hershey, Harry Dean Stanton, Andre Gregory, David Bowie, Verna Bloom, Juliette Caton, John Lurie, Roberts Blossom, Irvin Kershner, Barry Miller, Tomas Arana, Nehemiah Persoff, Paul Herman; *Dir:* Martin Scorsese. **VHS, Beta, LV $19.95** MCA 🐾🐾🐾½

Last Time I Saw Paris A successful writer reminisces about his love affair with a wealthy American girl in post-World War II Paris.

1954 116m/C Elizabeth Taylor, Van Johnson, Walter Pidgeon, Roger Moore, Donna Reed, Eva Gabor; *Dir:* Richard Brooks. **VHS, Beta $19.95** MGM, MRV, NOS 🐾🐾½

The Last Train Romance most hopeless in the heart of France during zee beeg war. French, very French.

1974 ?m/C Romy Schneider, Jean-Louis Trintignant. **VHS, Beta $59.99** WES 🐾🐾

Last Train from Gun Hill An all-star cast highlights this suspenseful story of a U.S. marshall determined to catch the man who raped and murdered his wife. Excellent action packed western.
1959 94m/C Kirk Douglas, Anthony Quinn, Carolyn Jones, Earl Holliman, Brad Dexter, Brian Hutton, Ziva Rodann; ***Dir:*** John Sturges. **VHS $19.95** *STE* 🦴🦴🦴

The Last Tycoon An adaptation of the unfinished F. Scott Fitzgerald novel about the life and times of a Hollywood movie executive of the 1920s. Confusing and slow moving despite a blockbuster conglomeration of talent. Joan Collins introduces the film.
1977 (PG) 123m/C Robert De Niro, Tony Curtis, Ingrid Boulting, Jack Nicholson, Jeanne Moreau, Peter Strauss, Robert Mitchum, Theresa Russell, Donald Pleasence, Ray Milland, Dana Andrews, John Carradine, Anjelica Huston; ***Dir:*** Elia Kazan. National Board of Review Awards '76: 10 Best Films of the Year. **VHS, Beta, LV $19.95** *PAR, PIA, FCT* 🦴🦴🦴

The Last Unicorn Peter Beagle's popular tale of a beautiful unicorn who goes in search of her lost, mythical "family." Music by Jimmy Webb.
1982 (G) 95m/C *Dir:* Jules Bass; ***Voices:*** Alan Arkin, Jeff Bridges, Tammy Grimes, Angela Lansbury, Mia Farrow, Robert Klein, Christopher Lee, Keenan Wynn. **VHS, Beta, LV $19.95** *JTC, FOX, KAR* 🦴🦴🦴

The Last Valley A scholar tries to protect a pristine 17th century Swiss valley, untouched by the Thirty Years War, from marauding soldiers. Historical action with an intellectual twist.
1971 (PG) 128m/C Michael Caine, Omar Sharif, Florinda Bolkan, Nigel Davenport, Per Oscarsson, Arthur O'Connell; ***Dir:*** James Clavell. **VHS** *ABC* 🦴🦴

The Last Waltz Martin Scorsese filmed this rock documentary featuring the farewell performance of The Band, joined by a host of musical guests that they have been associated with over the years. Songs include: "Up On Cripple Creek," "Don't Do It," "The Night They Drove Old Dixie Down," "Stage Fright" (The Band), "Helpless" (Young), "Coyote" (Mitchell), "Caravan" (Morrison), "Further On Up the Road" (Clapton), "Who Do You Love" (Hawkins), "Mannish Boy" (Waters), "Evangeline" (Harris), "Baby, Let Me Follow You Down" (Dylan).
1978 (PG) 117m/C The Band, Bob Dylan, Neil Young, Joni Mitchell, Van Morrison, Eric Clapton, Neil Diamond, Emmylou Harris, Muddy Waters, Ronnie Hawkins; ***Dir:*** Martin Scorsese. **VHS, Beta, LV $19.95** *MVD, MGM* 🦴🦴

Last War A nuclear war between the United States and Russia triggers Armageddon.
1968 79m/C Frankie Sakai, Nobuko Otawa, Akira Takarada, Yuriko Hoshi; ***Dir:*** Shue Matsubayashi. **VHS, Beta $19.95** *GEM* 🦴

Last of the Warrens A cowboy returns home after World War I to discover that an unscrupulous storekeeper has stolen his property.
1936 56m/C Bob Steele, Charles King. **VHS, Beta $19.95** *NOS, DVT, WGE* 🦴

Last Warrior A good-looking but exploitive "Hell in the Pacific" rip-off, as an American and a Japanese soldier battle it out alone on a remote island during World War II.
1989 (R) 94m/C Gary Graham, Cary-Hiroyuki Tagawa, Maria Holvoe; ***Dir:*** Martin Wragge. **VHS, Beta, LV $14.95** *SVS* 🦴½

The Last Wave An Australian attorney takes on a murder case involving an aborigine and he finds himself becoming distracted by apocalyptic visions concerning tidal waves and drownings that seem to foretell the future. Weir's masterful creation and communication of time and place are marred by a somewhat pat ending.
1977 (PG) 109m/C *AU* Richard Chamberlain, Olivia Hamnett, David Gulpilil, Frederick Parslow; ***Dir:*** Peter Weir. **VHS, Beta** *FCT* 🦴🦴🦴

Last of the Wild Horses Range war almost starts when ranch owner is accused of trying to force the small ranchers out of business.
1949 86m/B Mary Beth Hughes, James Ellison, Jane Frazee. **VHS, Beta** *WGE* 🦴

The Last Winter An American woman fights to find her Israeli husband who has disappeared in the 1973 Yom Kippur War. Trouble is, an Israeli woman thinks the man she's looking for is really her husband too.
1984 (R) 92m/C *IS* Kathleen Quinlan, Yona Elian, Zipora Peled, Michael Schnider; ***Dir:*** Riki Shelach. **VHS, Beta $79.95** *RCA* 🦴🦴

The Last Winters From Jean Charles Tacchella, writer-producer-director of "Cousin Cousine," comes his first short film about an old couple's bittersweet romance. In French with English subtitles.
1971 23m/C *FR* ***Dir:*** Jean-Charles Tacchella. **VHS, Beta $79.98** *INT, MTI* 🦴🦴½

Last Witness A man on the run from the government claims that he is innocent and was unjustly persecuted, but the government sees him as a dangerous threat. Nothing will stop the hunt for the escaped man, no matter what it costs.
1988 85m/C Jeff Henderson. **VHS, Beta $79.95** *TWE* 🦴

The Last Woman on Earth Two men vie for the affections of the sole surviving woman after a vague and unexplained disaster of vast proportions. Robert Towne, who appears herein under the pseudonym Edward Wain, wrote the script (his first screenwriting effort). You might want to watch this if it were the last movie on earth, although Corman fans will probably love it.
1961 71m/C Antony Carbone, Edward Wain, Betsy Jones-Moreland; ***Dir:*** Roger Corman. **VHS $16.95** *NOS, MRV, SNC Woof!*

Last Word A man fights to protect his home, family and neighbors from a corrupt real estate deal involving shady politicians, angry policemen, and a beautiful television reporter.
1980 (PG) 103m/C Richard Harris, Karen Black, Martin Landau, Dennis Christopher. **VHS, Beta $69.98** *SUE* 🦴½

Last Year at Marienbad A young man tries to lure a mysterious woman to run away with him from a hotel in France. Once a hit on the artsy circuit, it's most interesting for its beautiful photography. In French with English subtitles.
1961 93m/B *FR IT* Delphine Seyrig, Giorgia Albertazzi, Sacha Pitoeff, Luce Garcia-Ville; ***Dir:*** Alain Resnais. Venice Film Festival '61: Grand Prize. **VHS, Beta, 8mm $29.95** *NOS, MRV, VYY* 🦴🦴🦴

L'Atalante Vigo's great masterpiece, a slight story about a husband and wife quarreling, splitting, and reuniting around which has been fashioned one of the cinema's greatest poetic films. In French with English subtitles.
1934 82m/B *FR* Dita Parlo, Jean Daste, Michel Simon; ***Dir:*** Jean Vigo. Sight & Sound Survey '62: #10 of the Best Films of All Time. **VHS, Beta $24.95** *NOS, HHT, VYY* 🦴🦴🦴½

Late for Dinner In 1962 Willie, a young, married man, and his best friend Frank are framed by a sleazy land developer. On the run, the two decide to become guinea pigs in a cryonics experiment and wind up frozen for 29 years. It's now 1991 and Willie wants to find his family—only his wife is middle-aged and his daughter is all grown-up. Can he make a life with them again or has time indeed passed him by?
1991 (PG) 93m/C Brian Wimmer, Peter Berg, Marcia Gay Harden, Peter Gallagher; ***Dir:*** W.D. Richter. **VHS** *NLC* 🦴🦴

Late Extra A novice reporter attempts to track down a notorious bank robber and gets some help from a savvy female journalist. Mason's first role of substance.
1935 69m/B *GB* Virginia Cherrill, James Mason, Alastair Sim, Ian Colin, Clifford McLaglen, Cyril Cusack, David Horne, Antoinette Cellier, Donald Wolfit, Michael Wilding; ***Dir:*** Albert Parker. **VHS, Beta $16.95** *SNC* 🦴🦴

The Late Great Planet Earth Based on Hal Lindsey's best-seller, this documentary interprets Biblical prophecy to predict a possible disasterous fate for our planet.
198? 87m/C *Nar:* Orson Welles. **VHS $9.99** *VTR* 🦴🦴

The Late Show A veteran private detective finds his world turned upside down when his ex-partner comes to visit and winds up dead, and a flaky woman whose cat is missing decides to become his sidekick. Carney and Tomlin are fun to watch in this sleeper, a tribute to the classic detective film noirs.
1977 (PG) 93m/C Art Carney, Lily Tomlin, Bill Macy, Eugene Roche, Joanna Cassidy, John Considine; ***Dir:*** Robert Benton. Edgar Allan Poe Awards '77: Best Screenplay; National Board of Review Awards '77: 10 Best Films of the Year. **VHS, Beta, LV $64.95** *WAR* 🦴🦴🦴

Late Spring An exquisite Ozu masterpiece. A young woman lives with her widowed father for years. He decides to remarry so that she can begin life for herself. Highly acclaimed, in Japanese with English subtitles. Reworked in 1960 as "Late Autumn."
1949 107m/B *JP* Setsuko Hara, Chishu Ryu, Jun Usami, Haruko Sugimura; ***Dir:*** Yasujiro Ozu. **VHS, Beta $49.95** *FCT* 🦴🦴🦴🦴

Late Summer Blues Experience the life and death decisions of seven teenagers who have finished their finals and face now a more difficult test: serving in the Israeli Armed Forces in the Suez Canal. In Hebrew with English subtitles.
1987 101m/C *IS* Dor Zweigenbom, Shahar Segal, Yoav Zafir; ***Dir:*** Renen Schorr. **VHS $79.95** *HMV* 🦴🦴

Lathe of Heaven In a late twentieth century world suffocating from pollution, a young man visits a dream specialist. He then dreams of a world free from war, pestilence, and overpopulation. At times, these dreams have disastrous side effects. Based on futuristic novel by Ursula K. LeGuin; produced for PBS.
1980 120m/C Bruce Davison, Kevin Conway, Margaret Avery; ***Dir:*** David Loxton. **VHS, Beta** *WNE* 🦴🦴½

Latino The self-tortured adventures of a Chicano Green Beret who, while fighting a covert U.S. military action in war-torn Nicaragua, begins to rebel against the senselessness of the war.
1985 108m/C Robert Beltran, Annette Cardona, Tony Plana; ***Dir:*** Haskell Wexler. **VHS, Beta, LV $79.98** *FOX* ♂♂½

Lauderdale Two college schmoes hit the beach looking for beer and babes. Few laughs for the sober and mature. Surf's down in this woofer.
1989 (R) 91m/C Darrel Gilbeau, Michelle Kemp, Jeff Greenman, Lara Belmonte; ***Dir:*** Bill Milling. **VHS, Beta, LV $59.99** *PAR Woof!*

Laugh with Linder! Compilation of Linder shorts, including "Max Learns to Skate" and "Max and His Dog."
1913 70m/B Max Linder. **VHS, Beta $47.50** *GVV*

Laughfest Five classic slapstick shorts: "It's a Gift" (1923), with Snub Pollard, "Barney Oldfield's Race for a Life" (1913), "Kid Auto Races" and "Busy Day" (1914), both with Charlie Chaplin, and "Daredevil" (1923), starring Ben Turpin.
191? 59m/B Ben Turpin, Barney Oldfield, Mabel Normand, Charlie Chaplin, Snub Pollard. **VHS, Beta $19.98** *CCB*

Laughing at Life A mercenary leaves his family to fight in South America. Years later, when he is the leader of said country, he meets a man who turns out to be his son. Rehashed plot has been done better.
1933 72m/B Victor McLaglen, William Boyd, Lois Wilson, Henry B. Walthall, Regis Toomey; ***Dir:*** Ford Beebe. **VHS $24.95** *NOS, DVT* ♂½

Laughing Policeman Two antagonistic cops embark on a vengeful hunt for a mass murderer through the seamy underbelly of San Francisco. Adapted from the Swedish novel, written by Per Wahloo and Maj Sjowallo.
1974 (R) 111m/C Walter Matthau, Bruce Dern, Louis Gossett Jr.; ***Dir:*** Stuart Rosenberg. **VHS, Beta $59.98** *FOX* ♂♂½

Laughing Room Only A compilation of stand-up comic routines.
1986 52m/C Jeff Altman, Phyllis Diller, Mort Sahl, Harry Anderson, Henny Youngman. **VHS, Beta $24.95** *MON* ♂♂½

Laura A detective assigned to the murder investigation of the late Laura Hunt finds himself falling in love with her painted portrait. Superb collaboration by excellent cast and fine director. Superior suspense yarn, enhanced by a love story. Based on the novel by Vera Caspary. Rouben Mamoulian was the original director, then Preminger finished the film. Score by David Raskin. Oscar nominations: Best Supporting Actor (Webb), Black & White Interior Decoration.
1944 85m/B Gene Tierney, Dana Andrews, Clifton Webb, Lane Chandler, Vincent Price, Judith Anderson, Grant Mitchell, Dorothy Adams; ***Dir:*** Otto Preminger. Academy Awards '44: Best Black and White Cinematography. **VHS, Beta, LV $59.98** *FOX* ♂♂♂♂

Laurel & Hardy: Another Fine Mess The boys find themselves the owners of a magnificent Beverly Hills mansion. Yeah right. Lots of physical comedy and hilarious dialogue with a "special appearance" by Laurel in drag as a maid. Based on a sketch written by Laurel's father. Colorized.
1930 25m/C Stan Laurel, Oliver Hardy. **VHS $9.95** *CAF*

Laurel & Hardy: At Work Three short films collected here. Includes "Towed in a Hole," "Busy Bodies," and "The Music Box."
1932 70m/B Oliver Hardy, Stan Laurel. **VHS $14.98** *VTR, FCT*

Laurel & Hardy: Be Big Stan and Ollie ditch their wives to go to a swinging stag party at their club. Of course everything goes wrong in their quest for lascivious fun. Filled with tons of sight gags. Colorized.
192? 25m/C Stan Laurel, Oliver Hardy. **VHS $9.95** *CAF*

Laurel & Hardy: Below Zero It's the Great Depression of 1929, and Laurel and Hardy are about as depressed as you can get, working for pennies as street musicians. Their luck changes when they find a full wallet lying in the snow. But, as the saying goes, a fool and his money are soon parted. Colorized.
192? 25m/C Stan Laurel, Oliver Hardy. **VHS $9.95** *CAF*

Laurel & Hardy: Berth Marks Laurel & Hardy take a hilarious train trip to Pottsville for a vaudeville performance. Colorized.
1929 25m/C Stan Laurel, Oliver Hardy. **VHS $9.95** *CAF*

Laurel & Hardy: Blotto Laurel and Hardy want to get drunk, but they can't get any alcohol and their wives won't let them get away. When they finally do escape, they wind up at the swanky Rainbow Club, where they're drastically out of their element. Colorized.
1930 25m/C Stan Laurel, Oliver Hardy. **VHS $9.95** *CAF*

Laurel & Hardy: Brats Stan and Ollie play not only themselves but also their children when their wives go away and they have to babysit. The children, of course, wreak havoc wherever they go, playing on the hilariously oversized props and sets. One of the few films in which only Laurel and Hardy appear. Colorized.
192? 25m/C Stan Laurel, Oliver Hardy. **VHS $9.95** *CAF*

Laurel & Hardy: Chickens Come Home Ollie is running for mayor, but things go crazy when an ex-girlfriend shows up and blackmails him with a compromising photo. Stan, acting as Ollie's campaign manager, tries to help, but things just get worse. A remake of their own film "Love 'Em and Weep." Colorized.
192? 25m/C Stan Laurel, Oliver Hardy, James Finlayson. **VHS $9.95** *CAF*

Laurel & Hardy Comedy Classics Volume 1 A collection of four classic comedy shorts starring Laurel and Hardy: "The Music Box," which won an Academy Award for Best Short Subject, "Country Hospital," "The Live Ghost," and "Twice Two," all from 1932-33.
1933 84m/B Stan Laurel, Oliver Hardy, Mae Busch, May Wallace, Charlie Hall, Billy Gilbert. Academy Awards '31: Best Comedy Short ("The Music Box"). **VHS, Beta $9.95** *MED*

Laurel & Hardy Comedy Classics Volume 2 A collection of four Laurel and Hardy shorts from 1930 including "Blotto," "Towed in a Hole," "Brats," and "Hog Wild."
1930 75m/B Stan Laurel, Oliver Hardy, Billy Gilbert, Tiny Sandford, Anita Garvin. **VHS, Beta $9.95** *MED*

Laurel & Hardy Comedy Classics Volume 3 Laurel and Hardy star in four separate comedy shorts, including "Oliver the 8th," "Busy Bodies," "Their First Mistake," and "Dirty Work," all from 1933-34.
1934 75m/B Stan Laurel, Oliver Hardy, Billy Gilbert, Mae Busch, Tiny Sandford. **VHS, Beta $9.95** *MED*

Laurel & Hardy Comedy Classics Volume 4 Laurel and Hardy star in four of their comedy shorts from 1931-32. Included are "Another Fine Mess," "Come Clean," "Laughing Gravy," and "Any Old Part."
1932 75m/B Stan Laurel, Oliver Hardy, Jacqueline Wells, James Finlayson, Thelma Todd. **VHS, Beta $9.95** *MED*

Laurel & Hardy Comedy Classics Volume 5 Four classic comedy shorts starring Laurel and Hardy. Included are "Be Big," "The Perfect Day," "Night Owls," and "Help Mates," from 1929-31.
1931 84m/B Stan Laurel, Oliver Hardy, Edgar Kennedy, James Finlayson, Blanche Payson. **VHS, Beta $19.95** *MED*

Laurel & Hardy Comedy Classics Volume 6 Laurel and Hardy comedy shorts are presented in this package: "Our Wife," "The Fixer Uppers," "Them Thar Hills," and "Tit for Tat," from 1932-35.
1935 75m/B Stan Laurel, Oliver Hardy, Charlie Hall, Billy Gilbert, Ben Turpin, Mae Busch, James Finlayson. **VHS, Beta $19.95** *MED*

Laurel & Hardy Comedy Classics Volume 7 Four Laurel and Hardy two-reelers are combined on this tape: "Me and My Pal" (1933), "The Midnight Patrol" (1933), "Thicker than Water" (1935), and the classic "Below Zero" (1930).
193? 90m/B Stan Laurel, Oliver Hardy, Mae Busch, Daphne Pollard, James Finlayson. **VHS, Beta $19.95** *MED*

Laurel & Hardy Comedy Classics Volume 8 Compilation of Laurel and Hardy shorts includes: "Men O' War" (1929), "Scram" (1932), "Laurel and Hardy Murder Case" (1930) and "One Good Turn" (1931).
193? 90m/B Stan Laurel, Oliver Hardy, Lewis R. Foster, James W. Horne, James Parrott. **VHS, Beta $19.95** *MED*

Laurel & Hardy Comedy Classics Volume 9 Includes the shorts: "Beau Hunks" (1931), "Chickens Come Home" (1931), "Going Bye-Bye" (1934), and "Berth Marks" (1929).
193? 100m/B Stan Laurel, Oliver Hardy. **VHS, Beta $19.95** *MED*

Laurel & Hardy and the Family Four zany comedies seeing the boys in all kinds of trouble involving homelife. Includes "Brats," "Perfect Day," "Their First Mistake" and "Twice Two."
1933 85m/B Stan Laurel, Oliver Hardy. **VHS $14.98** *VTR, FCT*

Laurel & Hardy: Hog Wild Hilarity ensues when Laurel and Hardy try to fix the radio antenna on Hardy's roof. It may sound easy, but come on, this is Laurel and Hardy. Colorized.
192? 25m/C Stan Laurel, Oliver Hardy. **VHS $9.95** *CAF*

Laurel & Hardy: Laughing Gravy
The boys' pet dog, oddly named Laughing Gravy, gets them into trouble with their landlord who keeps trying to evict them. So funny even the mashed potatos will laugh. Colorized.
192? 25m/C Stan Laurel, Oliver Hardy. **VHS $9.95** *CAF*

Laurel & Hardy: Men O'War Laurel & Hardy star as two sailors on shore leave trying to impress the ladies. Things go from bad to worse when they take two ladies out on a date. Features the memorable soda fountain and canoe scenes. Colorized.
1929 25m/C Stan Laurel, Oliver Hardy, Gloria Greer, Anne Cornwall. **VHS $9.95** *CAF*

Laurel & Hardy: Night Owls When a cop needs to capture some crooks to save his job, he sets up a fake break-in and catches two bungling burglars—Laurel and Hardy. Lots of sight gags and slapstick. Colorized.
192? 25m/C Stan Laurel, Oliver Hardy, Edgar Kennedy. **VHS $9.95** *CAF*

Laurel & Hardy On the Lam Four Laurel & Hardy classics are collected, including "Scram," "Another Fine Mess," "One Good Turn," and "Going Bye-Bye."
1930 90m/B Oliver Hardy, Stan Laurel. **VHS $14.98** *VTR, FCT*

Laurel & Hardy: Perfect Day Laurel & Hardy plan a quiet picnic with their wives, but car trouble turns their day into a disaster. Colorized.
1929 25m/C Stan Laurel, Oliver Hardy, Edgar Kennedy. **VHS $9.95** *CAF*

Laurel & Hardy Special A unique program of Laurel and Hardy films including the first film they appeared in together—"Lucky Dog" (1918), "The Soilers," a wild Laurel spoof of The Spoilers, "Kill or Cure," "Near Dublin," "The Sleuths," "Laurel and Hardy's Trip to Britain," and "Stan Visits Ollie," a color home movie of their last visit together.
19?? 60m/B Stan Laurel, Oliver Hardy. **VHS, Beta $57.95** *GVV*

Laurel & Hardy Spooktacular Stan and Ollie scream through four spooky comedies, including "The Live Ghost," "The Laurel-Hardy Murder Case," "Oliver the Eighth" and "Dirty Work."
1934 95m/B **VHS $14.98** *VTR, FCT*

Laurel & Hardy: Stan "Helps" Ollie Four Stan and Ollie classics including "County Hospital," "Me and My Pal," "Hog Wild" and "Helpmates."
1933 85m/B Stan Laurel, Oliver Hardy. **VHS $14.98** *VTR, FCT*

Laurel & Hardy: The Hoose-Gow Laurel and Hardy are mistakenly picked up by the police and sent to prison to do hard labor. Hilarity ensues when they try to convince the warden that they're innocent. Colorized.
192? 25m/C **VHS $9.95** *CAF*

Laurel & Hardy Volume 1 Animated adventures of Laurel and Hardy.
196? 59m/C **VHS, Beta $57.95** *UNI*

Laurel & Hardy Volume 2 Animated short cartoons starring Laurel and Hardy.
196? 59m/C **VHS, Beta $57.95** *UNI*

Laurel & Hardy Volume 3 More animated adventures of Laurel and Hardy.
196? 59m/C **VHS, Beta $57.95** *UNI*

Laurel & Hardy Volume 4 Animated shorts starring cartoon characters of Laurel and Hardy.
196? 59m/C **VHS, Beta $57.95** *UNI*

The Lavender Hill Mob A prim and prissy bank clerk schemes to melt the bank's gold down and re-mold it into miniature Eiffel Tower paper-weights for later resale. The foolproof plan appears to succeed, but then develops a snag. Audrey Hepburn makes a cameo appearance. An excellent comedy that is still a delight to watch. Guinness received a Best Actor nomination. Also available with "The Man in the White Suit" on Laser Disc.
1951 78m/B *GB* Alec Guinness, Audrey Hepburn, Stanley Holloway, Sidney James, Alfie Bass, Marjorie Fielding, John Gregson; *Dir:* Charles Crichton. Academy Awards '52: Best Story & Screenplay. **VHS, Beta, LV $19.98** *HBO, FCT, RXM* ♫♫♫½

L'Avventura A stark, dry and minimalist exercise in narrative by Antonioni, dealing with the search for a girl on an Italian island by her lethargic socialite friends who eventually forget her in favor of their own preoccupations. A highly acclaimed, innovative film; somewhat less effective now, in the wake of many film treatments of angst and amorality. Subtitled in English. Laser edition features the original trailer, commentary and a collection of still photographs from Antonioni's work.
1960 145m/C *IT* Monica Vitti, Gabriel Ferzetti, Lea Massari, Dominique Blanchar, James Addams; *Dir:* Michelangelo Antonioni. Cannes Film Festival '60: Special Jury Prize; Sight & Sound Survey '62: #2 of the Best Films of All Time; Sight & Sound Survey '72: #5 of the Best Films of All Time; Sight & Sound Survey '82: #7 of the Best Films of All Time (tie). **VHS, Beta, LV $29.95** *CVC, VDM, VYG* ♫♫♫½

Law of Desire A wicked, Almodovarian attack-on-decency farce about a love-obsessed gay Madridian and his transsexual brother-sister, among others, getting hopelessly entangled in cross-gender relationships. Unlike the work of any other director; in Spanish with English subtitles.
1986 100m/C *SP* Carmen Maura, Eusebio Poncela, Antonio Banderas, Bibi Andersson; *Dir:* Pedro Almodovar. **VHS, Beta, LV $79.95** *CCN, TAM* ♫♫♫½

Law and Disorder Two average Joes, fed up with the rate of rising crime, start their own auxiliary police group. Alternately funny and serious with good performances from the leads.
1974 (R) 103m/C Carroll O'Connor, Ernest Borgnine, Ann Wedgeworth, Anita Dangler, Leslie Ackerman, Karen Black, Jack Kehoe; *Dir:* Ivan Passer. **VHS** *COL* ♫♫♫

Law of the Jungle A lady scientist and a fugitive team up to uncover a secret Nazi radio base in the jungle.
1942 61m/B Arlene Judge, John "Dusty" King, Mantan Moreland, Martin Wilkins, Arthur O'Connell; *Dir:* Jean Yarborough. **VHS, Beta $16.95** *SNC* ♫½

Law of the Land A made-for-television sagebrush epic with the author/artist of "Garfield" portraying an accused but innocent man searching for a murderer.
1976 100m/C Jim Davis, Don Johnson, Barbara Parkins, Charles Martin Smith. **VHS, Beta $49.95** *WOV* ♫

Law of the Lash Cowboys and rustlers battle it out.
1947 54m/B Lash LaRue, Al "Fuzzy" St. John. **VHS, Beta $9.99** *NOS, MRV, VCN* ♫

Law and Lawless A roving cowboy brings justice to a gang of cattle thieves in this film.
1933 58m/B Jack Hoxie. **VHS, Beta $19.95** *NOS, DVT* ♫

Law and Order Billy the Kid impersonates a Cavalry lieutenant in order to swindle his aunt out of her money. Also called "Billy the Kid's Law and Order."
1942 58m/B Buster Crabbe, Al "Fuzzy" St. John. **VHS, Beta $9.95** *VDM, MRV* ♫♫

The Law Rides Rush of gold claims causes an outbreak of murder and robbery.
1936 57m/B Bob Steele. **VHS, Beta $29.95** *DVT, WGE* ♫

The Law Rides Again Prairie lawmen bring a cheating Indian agent to justice.
1943 56m/B Ken Maynard, Hoot Gibson, Betty Miles, Jack LaRue; *Dir:* William Castle. **VHS $19.95** *NOS, MRV, DVT* ♫

Law of the Saddle The Lone Rider pits himself against a gang of outlaws.
1945 60m/B Bob Livingston. **VHS, Beta $29.95** *VCN, MLB* ♫

Law of the Sea A family on a sinking ship is rescued by a lecherous sea captain who makes advances on the wife and blinds the husband in a fight. Twenty years later, the husband meets him again, recognizes his laugh and exacts revenge.
1938 63m/B Priscilla Dean, Sally Blaine, Ralph Ince, Rex Bell, William Farnum. **VHS, Beta, LV $24.95** *VYY, WGE* ♫

Law of the Underworld An innocent couple is framed for the robbery of a jewelry store which resulted in the death of a clerk. When they are arrested, their fate lies in the hands of the gangster responsible for the crime. Will he let them fry in the electric chair, or will his conscience get the better of him?
1938 58m/B Chester Morris, Anne Shirley, Eduardo Ciannelli, Walter Abel, Richard Bond, Lee Patrick, Paul Guilfoyle, Frank M. Thomas, Eddie Acuff, Jack Arnold; *Dir:* Lew Landers. **VHS, Beta $16.95** *SNC* ♫½

Law West of Tombstone A former outlaw moves to a dangerous frontier town in order to restore the peace.
1938 73m/B Tim Holt, Harry Carey, Evelyn Brent; *Dir:* Glenn Tyron. **VHS, Beta** *MED* ♫♫½

Law of the Wild The search for a magnificent stallion that was hijacked by race racketeers before a big sweepstakes race is shown in this 12 chapter serial.
1934 156m/B Rex, Rin Tin Tin Jr, Bob Custer. **VHS, Beta $26.95** *SNC, NOS, VCN* ♫

Lawless Frontier In the early West, the Duke fights for law and order.
1935 53m/B John Wayne, George "Gabby" Hayes, Sheila Terry, Earl Dwire; *Dir:* Robert N. Bradbury. **VHS $12.98** *REP, NOS, VYY* ♫½

The Lawless Land Post-holocaust America is ruled by a tyrant. Two young lovers who can't take it anymore go on the lam to escape the despotic rule.
1988 (R) 81m/C Leon Berkeley, Xander Berkeley, Nick Corri, Amanda Peterson; *Dir:* Jon Hess. **VHS, Beta $79.95** *MGM* ♫♫

The Lawless Nineties Wayne plays a government agent sent to guarantee honest elections in the Wyoming territory. Hayes is exceptional as the newspaper editor who backs him up. Re-made in 1940 as "The Dark Command."

1936 55m/B John Wayne, Ann Rutherford, Lane Chandler, Harry Woods, George "Gabby" Hayes, Charles King, Sam Flint; *Dir:* Joseph Kane. **VHS** $12.98 *REP* ♂♂

Lawless Range John Wayne and the marshal's posse save the ranchers from trouble.
1935 56m/B John Wayne, Sheila Manners, Jack Curtis, Earl Dwire; *Dir:* Robert Bradbury. **VHS** $12.98 *NOS, REP, MRV* ♂

A Lawman is Born A tough-as-nails marshal goes gunning for a pack of no-good, land-stealing varmits in this installment of the Johnny Mack Brown movie serials. Be sure to count the number of bullets each cowboy can fire from a six-shooter before being forced to reload; you may be surprised.
1937 58m/B Johnny Mack Brown, Iris Meredith, Warner Richmond, Mary MacLaren, Dick Curtis, Earle Hodgins, Charles King, Frank LaRue; *Dir:* Sam Newfield. **VHS, Beta** $19.99 *NOS, MOV, MRV* ♂♂♂

Lawmen Government agents fight to enforce the law in the badlands.
1944 55m/B Johnny Mack Brown, Raymond Hatten. **VHS, Beta** $19.95 *NOS, DVT* ♂ ½

The Lawnmower Man Brosnan is a scientist who uses Fahey, a dim-witted gardener, as a guinea pig to test his experiments in "virtual reality," an artificial computer environment. With the use of drugs and high-tech equipment, Brosnan is able to increase Fahey's mental powers—but not necessarily for the better. An evil side of Fahey rears its ugly head and wreaks havoc. Fantastic special effects and a memorable "virtual reality" sex scene. Based on a short story by Stephen King. Available in an unrated version which contains 32 more minutes of footage.
1992 (R) 108m/C Jeff Fahey, Pierce Brosnan, Jenny Wright, Geoffrey Lewis; *Dir:* Brett Leonard. **VHS** *COL* ♂♂♂

Lawrence of Arabia Exceptional biography of T. E. Lawrence, a British military "observer" who strategically aids the Bedouins battle the Turks during World War 1. Lawrence, played masterfully by O'Toole in his first major film, is a hero consumed more by a need to reject British tradition than to save the Arab population. He takes on Arab costume and a larger-than-life persona. Stunning photography of the desert in all its harsh reality. Laser edition contains 20 minutes of restored footage and a short documentary about the making of the film. Oscar nominations for O'Toole, Sharif, and adapted screenplay, besides the ones it garnered. Available in letterboxed format.
1962 (PG) 221m/C *GB* Peter O'Toole, Omar Sharif, Anthony Quinn, Alec Guinness, Jack Hawkins, Claude Rains, Anthony Quayle, Arthur Kennedy, Jose Ferrer; *Dir:* David Lean. Academy Awards '62: Best Art Direction/Set Decoration (Color), Best Color Cinematography, Best Director (Lean), Best Film Editing, Best Picture, Best Sound, Best Original Score; British Academy Awards '62: Best Actor (O'Toole), Best Film, Best Screenplay; Directors Guild of America Awards '62: Best Director (Lean); Golden Globe Awards '63: Best Film—Drama; National Board of Review Awards '62: 10 Best Films of the Year, Best Director (Lean). **VHS, Beta, LV** $14.95 *COL, VYG, CRC* ♂♂♂♂

Lazarus Syndrome Astute doctor teams up with an ex-patient of the chief of surgery in an effort to expose the chief's unethical surgical procedures.
1979 90m/C Louis Gossett Jr., E.G. Marshall. **VHS, Beta** $29.95 *USA* ♂♂♂♂

Lazer Tag Academy: The Champion's Biggest Challenge A sequel to the first Lazer Tag movie, still essentially a promotion for the toy guns.
1990 110m/C **VHS, Beta** *JFK*

LBJ: A Biography Produced for the PBS series "The American Experience," this four-hour revisionist look at LBJ is one of the most critically acclaimed political documentaries of our time. Divided into four segments. "Beautiful Texas" chronicles Johnson's rise to power. "My Fellow Americans" focuses on his years as an unelected president. The Great Society and and the escalation of war in Vietnam are covered in "We Shall Overcome." "The Last Believer" highlights events leading to Johnson's withdrawal from politics.
1991 240m/C *W/Dir:* David Grubin. **VHS** $39.95 *PAV* ♂♂♂♂

LBJ: The Early Years The early years of President Lyndon Baines Johnson's political career are dramatized in this made-for-television movie. Winning performances hoist this above other biographies.
1988 144m/C Randy Quaid, Patti LuPone, Morgan Brittany, Pat Hingle, Kevin McCarthy, Barry Corbin, Charles Frank; *Dir:* Peter Werner. **VHS, Beta** $29.95 *FRH* ♂♂♂

Le Bal You won't find many films with the music of Paul McCartney and Chopin in the credits, and even fewer without dialogue. With only music and dancing the film uses a French dance hall to illustrate the changes in French society over a fifty-year period. Based on a French play, it won three French Academy Awards.
1982 112m/C *IT* *Dir:* Ettore Scola. **VHS** $79.99 *WAR, FCT, TAM* ♂♂♂

Le Beau Mariage An award-winning comedy from the great French director about a zealous woman trying to find a husband and the unsuspecting man she chooses to marry. In French with English subtitles.
1982 (R) 97m/C *FR* Beatrice Romand, Arielle Dombasle, Andre Dussollier; *Dir:* Eric Rohmer. Venice Film Festival '82: Best Actress (Romand). **VHS, Beta** $59.95 *TAM* ♂♂♂

Le Beau Serge A convalescing Frenchman returns to his home village and tries to help an old friend, who's become a hopeless drunkard, change his life. Chabrol's first film, and a major forerunner of the nouvelle vague. In French with English subtitles.
1958 97m/B *FR* Gerard Blain, Jean-Claude Brialy, Michele Meritz, Bernadette LaFont; *Dir:* Claude Chabrol. **VHS, Beta** $79.95 *FCT* ♂♂♂

Le Boucher In a provincial French town a sophisticated schoolmistress is courted by the shy local butcher—who turns out to be a sex murderer. A well-paced thriller that looks at sexual frustration. In French with English subtitles. Also known as "The Butcher."
1969 94m/C *FR* Stephane Audran, Jean Yanne, Antonio Passallia; *Dir:* Claude Chabrol. **VHS** $79.95 *CVC* ♂♂♂

Le Bourgeois Gentilhomme Translated "The Would-Be Gentleman." A comedy-ballet in five acts in prose, performed by the Comedie Francaise. Subtitled in English.
1958 97m/C *FR* *Dir:* Comedie Francaise. **VHS, Beta** $24.95 *VYY, APD, TAM* ♂♂♂

Le Cas du Dr. Laurent A country doctor in a small French town tries to introduce methods of natural childbirth to the native women, but meets opposition from the superstitious villagers. Fine performance by Gabin.

1957 88m/B *FR* Jean Gabin, Nicole Courcel; *Dir:* Jean-Paul LeChanois. **VHS, Beta** $29.95 *NOS, VYY, DVT* ♂♂♂

Le Cavaleur (Practice Makes Perfect) A light hearted comedy about a philandering concert pianist. Also known as "Practice Makes Perfect."
1978 90m/C *FR* Jean Rochefort, Lila Kedrova, Nicole Garcia, Annie Girardot, Danielle Darrieux; *Dir:* Philippe de Broca. **VHS, Beta** $39.95 *LUN* ♂♂

Le Chat A middle-aged couple's marriage dissolves into a hate-filled battle of wits, centering around the husband's love for their cat. In French with English subtitles.
1975 88m/C *FR* Jean Gabin, Simone Signoret, Annie Cordy, Jacques Rispal; *Dir:* Pierre Granier-Deferre. Berlin Film Festival '71: Best Actor (Gabin). **VHS, Beta, LV** $59.95 *FCT, LUM, PIA* ♂♂♂ ½

Le Complot (The Conspiracy) When de Gaulle announces his intention to abandon Algeria, several army officers, feeling that their service has been in vain, stage a coup. There's suspense a plenty as the Gaulists, the leftists and the police spy and are spied upon. In French with English subtitles.
1973 120m/C *FR* Michel Bouquet, Jean Rochefort; *Dir:* Rene Gainville. **VHS, LV** $79.95 *CVC, FCT, TAM* ♂♂♂

Le Corbeau A great, notorious drama about a small French village whose everyday serenity is ruptured by a series of poison pen letters that lead to suicide and despair. The film was made within Nazi-occupied France, sponsored by the Nazis, and has been subjected to much misdirected malice because of it. In French with English subtitles. Also known as "The Raven."
1943 92m/B *FR* Pierre Fresnay, Noel Roquevort, Ginette LeClerc; *Dir:* Henri-Georges Clouzot. **VHS, Beta** $24.95 *NOS, HHT, TAM* ♂♂♂

Le Crabe Tambour Rochefort plays a dying naval captain remembering his relationship with his first officer. Revealed via flashbacks, the recollections concern the adventures which transpired on a North Atlantic supply ship. Great performances from all and award-winning cinematography by Raoul Coutard. Winner of several French Cesars. A French script booklet is available.
1977 120m/C *FR* Jean Rochefort, Claude Rich, Jacques Dufilho, Jacques Perrin, Odile Versuis, Aurore Clement; *Dir:* Pierre Schoendoerffer. **VHS, LV** $59.95 *INT, IME, APD* ♂♂♂

Le Dernier Combat A stark film about life after a devastating nuclear war marks the directorial debut of Besson. The characters fight to survive in a now speechless world by staking territorial claims and forming new relationships with other survivors. An original and expressive film made without dialogue. Also known as "The Last Battle."
1984 (R) 93m/B *FR* Pierre Jolivet, Fritz Wepper, Jean Reno, Jean Bouise, Christiane Kruger; *Dir:* Luc Besson. **VHS, Beta** *COL, OM* ♂♂♂

Le Doulos Compelling story of an ex-convict and his buddy, a man who may be a police informant. Chronicles the efforts of the snitch (the "doulos" of the title) to bring the criminal element before the law. Melville blends in several plot twists and breathes a new-French life into the cliche-ridden genre.
1961 108m/B *FR* Serge Reggiani, Jean-Paul Belmondo, Michel Piccoli; *Dir:* Jean-Pierre Melville. **VHS, Beta** $59.95 *INT, TAM* ♂♂♂

Le Gentleman D'Epsom A French comedy about a breezy con man who scams everyone around him in order to keep up his upwardly mobile appearance and to put his bets down at the racetrack. Subtitled in English.
1962 83m/B *FR* Jean Gabin, Paul Frankeur. VHS, Beta *KRT* 🎞🎞

Le Grand Chemin Two young children wander about the French countryside learning about the various facets of life. Stars the director's son Antoine as one of the protaganists. Enjoyable look at chilhood wonder. Also known as "The Grand Highway."
1987 107m/C *FR* Vanessa Guedj, Antoine Hubert, Richard Bohringer, Anemone; *Dir:* Jean-Loup Hubert. VHS, Beta $29.95 *PAV, IME, FCT* 🎞🎞🎞½

Le Joli Mai A famous documentary that established Marker, depicting the everyday life of Paris denizens on the occasion of the end of the Algerian War, May 1962, when France was at peace for the first time since 1939. In French with English subtitles.
1962 180m/B *FR Dir:* Chris Marker. Cannes Film Festival '63: International Critics Award; New York Film Festival '63: International Critics Award; Venice Film Festival '63: Best First Film. VHS, Beta $29.95 *FCT* 🎞🎞🎞½

Le Jour Se Leve The dark, expressionist film about a sordid and destined murder/love triangle that starts with a police stand-off and evolves into a series of flashbacks. The film that put Carne and Gabin on the cinematic map. Highly acclaimed. French with English subtitles. Also known as "Daybreak." Remade in 1947 as "The Long Night."
1939 85m/B *FR* Jean Gabin, Jules Berry, Arletty, Jacqueline Laurent; *Dir:* Marcel Carne. Sight & Sound Survey '52: #8 of the Best Films of All Time. VHS, Beta $29.95 *HHT, VYY, APD* 🎞🎞🎞½

Le Magnifique Belmondo is a master spy and novelist who mixes fantasy with reality when he chases women and solves cases.
1976 84m/C *FR* Jean-Paul Belmondo, Jacqueline Bisset. VHS, Beta $59.95 *PSM, CCI* 🎞🎞

Le Mans The famous 24-hour Grand Prix race sets the stage for this tale of love and speed. McQueen (who did his own driving) is the leading race driver, a man who battles competition, fear of death by accident, and emotional involvement. Excellent documentary-style race footage almost makes up for weak plot and minimal acting.
1971 (G) 106m/C Steve McQueen, Elga Andersen, Ronald Leigh-Hunt, Luc Merenda; *Dir:* Lee H. Katzin. VHS $39.98 *FOX, CCI* 🎞🎞

Le Million A comedy/musical masterpiece of the early sound era which centers on an artist's adventures in searching for a winning lottery ticket throughout Paris. Highly acclaimed member of the Clair school of subtle French farce. In French with English subtitles.
1931 89m/B *FR* Annabella, Rene Lefevre, Paul Olivier, Louis Allibert; *Dir:* Rene Clair. National Board of Review Awards '31: 5 Best Foreign Films of the Year; Cinematheque Belgique '52: #4 of the Best Films of All Time (tie); Sight & Sound Survey '52: #10 of the Best Films of All Time (tie). VHS, Beta $29.95 *NOS, VYY, DVT* 🎞🎞🎞½

Le Petit Amour A popular French comedy about the fateful romance between a 40-year-old divorced woman and a 15-year-old boy obsessed with a kung fu video game.

Originally titled "Kung Fu Master." In French with English subtitles.
1987 (R) 80m/C *FR* Jane Birkin, Mathieu Demy, Charlotte Gainsbourg, Lou Doillon; *Dir:* Agnes Varda. VHS, Beta $79.95 *PSM* 🎞🎞🎞

Le Plaisir An anthology of three Guy de Maupassant stories about the search for pleasure, with English subtitles.
1952 97m/B *FR* Claude Dauphin, Simone Simon, Jean Gabin, Danielle Darrieux; *Dir:* Max Ophuls. VHS, Beta $24.95 *NOS, DVT* 🎞🎞½

Le Repos du Guerrier Star vehicle for Bardot in which she plays a respectable woman who abandons societal norms to pursue an unbalanced lover. Subtitled in English. Known in English as "Warrior's Rest."
1962 100m/C *FR* Brigitte Bardot, Robert Hossein, James Robertson Justice, Jean-Mark Bory; *Dir:* Roger Vadim. VHS, Beta $19.95 *KRT* 🎞🎞

Le Secret This classic French thriller with a shocking ending finds an escaped convict seeking shelter with a reclusive couple in the mountains. His tales of abuse and torture create tension for the pair as one believes the woeful story and the other doesn't. Dubbed in English.
1974 103m/C *FR* Jean-Louis Trintignant, Philippe Noiret, Jean-Francois Adam; *Dir:* Robert Enrico. VHS, Beta *CCI* 🎞🎞🎞

Le Sex Shop A man turns his little book store into a porn equipment palace, to make ends meet. When his relationship with his wife gets boring, they begin to use their erotic merchandise and adopt a swinging lifestyle.
1973 (R) 92m/C *FR* Claude Berri, Juliet Berto, Daniel Auteuil, Nathalie Delon; *Dir:* Claude Berri. VHS, Beta *COL* 🎞🎞

Leader of the Band An out-of-work musician tries to train the world's worst high school band. Music by Dick Hyman. Landesberg helps this attempted comedy along.
1987 (PG) 90m/C Steve Landesburg, Gailard Sartain, Mercedes Ruehl, James Martinez, Calvert Deforest; *Dir:* Nessa Hyams. VHS, Beta $14.95 *IVE* 🎞🎞

League of Gentlemen An ex-Army officer plots a daring bank robbery using specially skilled military personnel and irreproachable panache. Hilarious British humor fills the screen.
1960 115m/B *GB* Jack Hawkins, Nigel Patrick, Richard Attenborough, Roger Livesey, Bryan Forbes; *Dir:* Basil Dearden. VHS, Beta $39.95 *IND, FCT* 🎞🎞🎞½

Lean on Me The romanticized version of the career of Joe Clark, a tough New Jersey teacher who became the principal of the state's toughest, worst school and, through controversial hard-line tactics, turned it around.
1989 (PG-13) 109m/C Morgan Freeman, Robert Guillaume, Beverly Todd, Alan North, Lynne Thigpen; *Dir:* John G. Avildsen. VHS, Beta, LV, 8mm $19.98 *WAR, FCT* 🎞🎞🎞

Leap Year Made at the apex of the 300 pound comedian's career, "Leap Year" wasn't released in the States until the sixties. Tried for manslaughter in 1921 (with two hung juries), Fatty was forced off camera and his films were taken out of circulation (although he did return to cinema behind the camera using the pseudonym William Goodrich).
1921 60m/B Fatty Arbuckle. VHS, Beta $24.95 *GPV, FCT* 🎞🎞½

The Learning Tree A beautifully photographed adaptation of Parks' biographical novel about a fourteen-year-old black boy in the 1920s' South, living on the verge of manhood, maturity, love and wisdom.
1969 (PG) 107m/C Kyle Johnson, Alex Clarke, Estelle Evans, Dana Elcar; *Dir:* Gordon Parks. VHS, Beta $59.95 *WAR, KUI* 🎞🎞

Leather Boys A teen-aged girl marries a mechanic and then begins to look beyond the boundaries of marriage. Considered very controversial in the 60s, but less so now.
1963 103m/C *GB* Rita Tushingham, Dudley Sutton, Colin Campbell; *Dir:* Sidney J. Furie. VHS, Beta $59.95 *VID* 🎞🎞

Leatherface: The Texas Chainsaw Massacre 3 The human-skin-wearing cannibal is at it again in this, the second sequel to the Tobe Hooper protomess. This one sports a bit more humor and is worth seeing for that reason.
1989 (R) 81m/C Kate Hodge, William Butler, Ken Foree, Tom Hudson, R.A. Mihailoff; *Dir:* Jeff Burr. VHS, Beta, LV $19.95 *COL* 🎞

The Leatherneck Love unrequited in a faraway place starring soon to be cowboy (Hopalong Cassidy) matinee idol Boyd and Hale, the real life dad of Gilligan Island's skipper.
1928 65m/B William Boyd, Alan Hale Jr., Diane Ellis. VHS, Beta $24.95 *GPV, FCT* 🎞🎞

Leave 'Em Laughing Rooney plays Chicago clown Jack Thum in this based-on-real-life TV movie. Thum takes in orphans even though he cannot find steady work, and then discovers he has cancer.
1981 103m/C Mickey Rooney, Anne Jackson, Allen Goorwitz, Elisha Cook Jr., William Windom, Red Buttons, Michael Le Clair; *Dir:* Jackie Cooper. VHS, Beta $14.95 *FRH* 🎞🎞

Leave It to the Marines A jerky guy wanders into a recruiting office instead of a marriage license bureau and unwittingly signs on for military service.
1951 66m/B Sid Melton, Mara Lynn. VHS, Beta, LV *WGE* 🎞½

Leaves from Satan's Book Impressionistic episodes of Satan's fiddling with man through the ages, from Christ to the Russian Revolution. An early cinematic film by Dreyer, with ample indications of his later brilliance. Silent.
1919 165m/B *DK* Helge Milsen, Halvart Hoft, Jacob Texiere; *Dir:* Carl Theodor Dreyer. VHS, Beta $39.95 *FCT, GPV* 🎞½

Leaving Normal Darly, a fed-up waitress, and Marianne, an abused housewife, meet at a bus stop in Normal, Wyoming and decide to blow town. Together they travel across the American West, through Canada, and up to Alaska, where Lahti's ex-boyfriend has left her a house. Because both women have made bad choices all their lives, they decide to leave their futures to chance, and they end up finding their nirvana. Although the plot is similar to "Thelma & Louise," "Leaving Normal" doesn't even come close to the innovativeness of the former. This film is sappy and sentimental to the point of being annoying at times.
1992 (R) 110m/C Christine Lahti, Meg Tilly, Patrika Darbo, Lenny Von Dohlen, Maury Chaykin, Brett Cullen, James Gammon; *Dir:* Edward Zwick. VHS, LV *MCA* 🎞🎞

Led Zeppelin: The Song Remains the Same
Combination of fantasy sequences and concert footage from Led Zeppelin's 1973 Madison Square Garden concert.
1973 (PG) 136m/C John Paul Jones, Jimmy Page, Robert Plant, John Bonham; *Dir:* Peter Clifton. **VHS, Beta, LV $19.98** *MVD, WAR, VTK* 🎞🎞

Left for Dead
A millionaire is accused of brutally killing his wife, and discovers in his search for an alibi that he is the victim of a conspiracy.
1978 82m/C Elke Sommer, Donald Pilon; *Dir:* Murray Markowitz. **VHS, Beta $69.98** *LIV, VES* 🎞½

The Left Hand of God
After an American pilot escapes from a Chinese warlord in post-World War II, he disguises himself as a Catholic priest and takes refuge in a missionary hospital. Bogie is great as the flyboy/cleric.
1955 87m/C Humphrey Bogart, E.G. Marshall, Lee J. Cobb, Agnes Moorehead, Gene Tierney; *Dir:* Edward Dmytryk. **VHS, Beta $19.98** *FOX* 🎞🎞🎞

The Left-Handed Gun
An offbeat version of the exploits of Billy the Kid, which portrays him as a 19th-century Wild West juvenile delinquent. Newman's role, which he method-acted, was originally intended for James Dean. Based on a 1955 Philco teleplay by Gore Vidal.
1958 102m/B Paul Newman, Lita Milan, John Dehner; *Dir:* Arthur Penn. **VHS, Beta $19.98** *WAR, TLF* 🎞🎞🎞

The Legacy
An American couple become privy to the dark secrets of an English family gathering in a creepy mansion to inherit an eccentric millionaire's fortune. Death and demons abound.
1979 (R) 100m/C *GB* Katharine Ross, Sam Elliott, John Standing, Roger Daltrey, Ian Hogg; *Dir:* Richard Marquand. **VHS, Beta $19.95** *MCA* 🎞½

Legacy of Horror
A weekend at the family's island mansion with two unfriendly siblings sounds bad enough, but when terror, death and a few family skeletons pop out of the closets, things go from bad to weird. A remake of the director's own "The Ghastly Ones."
1978 (R) 83m/C Elaine Boies, Chris Broderick, Marilee Troncone, Jeannie Cusik; *Dir:* Andy Milligan. **VHS, Beta $59.95** *MPI* 🎞

A Legacy for Leonette
A girl is led into a web of murder and love in this torrid romance novel brought to video.
1985 90m/C Loyita Chapel, Michael Anderson Jr., Dinah Anne Rogers. **Beta $14.95** *PSM Woof!*

Legal Eagles
An assistant D.A. faces murder, mayhem, and romance while prosecuting a girl accused of stealing a portrait painted by her father. Redford sparkles in this otherwise convoluted tale from Reitman, while Hannah lacks depth as the daffy thief.
1986 (PG) 116m/C Robert Redford, Debra Winger, Darryl Hannah, Brian Dennehy, Terence Stamp, Steven Hill, David Clennon, Roscoe Lee Browne, Ivan Reitman; *Dir:* Ivan Reitman. **VHS, Beta, LV $14.95** *MCA* 🎞🎞½

Legal Tender
Rude talk-show host Downey as a corrupt S & L chief? Big, menacing Davi as a romantic hero? Give this often-inept thriller credit for creative casting, if little else. Roberts adds steamy sex appeal as a lovely saloon-keeper imperiled by deadly bank fraud.
1990 (R) 93m/C Robert Davi, Tanya Roberts, Morton Downey Jr.; *Dir:* Jag Mundhra. **VHS, Beta $89.95** *PSM* 🎞½

Legend
A colorful, unabashedly Tolkien-esque fantasy about the struggle to save an innocent waif from the Prince of Darkness. Set in a land packed with unicorns, magic swamps, bumbling dwarves and rainbows. Produced in Great Britain.
1986 (PG) 89m/C *GB* Tom Cruise, Mia Sara, Tim Curry, David Bennent, Billy Barty, Alice Playten; *Dir:* Ridley Scott. **VHS, Beta, LV $14.95** *MCA* 🎞🎞

Legend of Alfred Packer
The true story of how a guide taking five men searching for gold in Colorado managed to be the sole survivor of a blizzard.
1980 87m/C Patrick Dray, Ron Haines, Bob Damon, Dave Ellingson; *Dir:* Jim Roberson. **VHS, Beta $59.95** *MON* 🎞½

Legend of Big Foot
Facing a spree of killings, a local town mobilizes to stop the legendary beast.
1982 92m/C Stafford Morgan, Katherine Hopkins. **VHS, Beta $49.95** *AHV* 🎞

Legend of Billie Jean
Billie Jean believed in justice for all. When the law and its bureaucracy landed hard on her, she took her cause to the masses and inspired a generation.
1985 (PG-13) 92m/C Helen Slater, Peter Coyote, Keith Gordon, Christian Slater, Richard Bradford, Yeardley Smith, Dean Stockwell; *Dir:* Matthew Robbins. **VHS, Beta $79.98** *FOX* 🎞🎞

The Legend of Black Thunder Mountain
A children's adventure in the mold of "Grizzly Adams" and "The Wilderness Family."
1979 (G) 90m/C Holly Beeman, Steve Beeman, Ron Brown, F.A. Milovich; *Dir:* Tom Beeman. **VHS, Beta** *CON* 🎞½

The Legend of Blood Castle
Another version of the Countess of Bathory legend, in which an evil woman bathes in the blood of virgins in an attempt to keep the wrinkles away. Released hot on the heels of Hammer's "Countess Dracula." Also known as "The Female Butcher" and "Blood Ceremony."
1972 (R) 87m/C *IT SP* Lucia Bose, Ewa Aulin; *Dir:* Jorge Grau. **VHS, Beta $9.95** *NO* 🎞

Legend of Boggy Creek
A dramatized version of various Arkansas Bigfoot sightings.
1975 (G) 87m/C David Hess, Lucy Grantham, Sandra Cassel; *Dir:* Charles B. Pierce. **VHS, Beta $59.98** *LIV, MRV, LTG* 🎞½

Legend of the Dinosaurs
Slavering, teeth-gnashing dinosaurs are discovered on Mt. Fuji.
1983 92m/C **VHS, Beta $39.95** *CEL Woof!*

Legend of Earl Durand
In Wyoming 1939, a man spends his life searching for freedom and justice. His Robin Hood actions cause a manhunt of massive proportions.
1974 90m/C Martin Sheen, Peter Haskell, Keenan Wynn, Slim Pickens, Anthony Caruso; *Dir:* John D. Patterson. **VHS** *QUE* 🎞½

Legend of Eight Samurai
An ancient Japanese princess hires a group of eight samurai to destroy the witch who reigns over her clan.
1984 130m/C **VHS, Beta $9.99** *STE, PSM* 🎞

The Legend of Frank Woods
Gunslinger returns to the land of the free after an extended holiday south of the border. Mistaken for an expected preacher in a small town, he poses as the padre and signs on the dotted line to take out a new lease on life.
1977 88m/C Troy Donahue, Brad Steward, Kitty Vallacher, Michael Christian; *Dir:* Deno Paoli. **VHS $9.95** *SIM* 🎞🎞

Legend of Frenchie King
When prospectors discover oil on disputed land, two families feud over their conflicting claims to the property. Bardot heads a band of female outlaws.
1971 96m/C Brigitte Bardot, Claudia Cardinale, Michael J. Pollard, Micheline Presle; *Dir:* Christian Jaque. **VHS, Beta** *PGN* 🎞½

Legend of Hell House
A multi-millionaire hires a team of scientists and mediums to investigate his newly acquired haunted mansion. It seems that the creepy house has been the site of a number of deaths and may hold clues to the afterlife. A suspenseful, scary screamfest.
1973 (PG) 94m/C Roddy McDowall, Pamela Franklin, Clive Revill, Gayle Hunnicutt, Clive Revill; *Dir:* John Hough. **VHS, Beta, LV $19.98** *FOX* 🎞🎞🎞

The Legend of Hiawatha
Hiawatha must confront a demon who casts a plague on his people. This animated program is based on Henry Wadsworth Longfellow's poem.
1982 35m/C **VHS, Beta $29.95** *FHE* 🎞🎞🎞

The Legend of Hillbilly John
Hillbilly John holds off the devil with a strum of his six-string. While demons plague the residents of rural America, the hayseed messiah wanders about saving the day.
1973 (G) 86m/C Severn Darden, Denver Pyle, Susan Strasberg, Hedge Capers; *Dir:* John Newland. **VHS, Beta $19.95** *STE, NWV* 🎞

The Legend of Jedediah Carver
A rancher battles desert elements and hostile Indians in a desperate bid for survival.
197? 90m/C DeWitt Lee, Joshua Hoffman, Val Chapman, Richard Montgomery, David Terril; *Dir:* DeWitt Lee. **VHS $39.95** *XVC* 🎞

Legend of Lobo
The story of Lobo, a crafty wolf who seeks to free his mate from the clutches of greedy hunters. A Disney wildlife adventure.
1962 67m/C *Nar:* Rex Allen. **VHS, Beta $69.95** *DIS* 🎞½

Legend of the Lone Ranger
The fabled Lone Ranger (whose voice is dubbed throughout the entire movie) and the story of his first meeting with his Indian companion, Tonto, are brought to life in this weak and vapid version of the famous legend. The narration by Merle Haggard leaves something to be desired as do most of the performances.
1981 (PG) 98m/C Klinton Spilsbury, Michael Horse, Jason Robards Jr., Richard Farnsworth, Christopher Lloyd, Matt Clark; *Dir:* William A. Fraker. **VHS, Beta $29.98** *FOX* 🎞

Legend of the Lost
Two men vie for desert treasure and desert women. Interesting only because of Wayne, but certainly not one of his more memorable films.
1957 109m/C John Wayne, Sophia Loren, Rossano Brazzi, Kurt Kasznar, Sonia Moser; *Dir:* Henry Hathaway. **VHS $19.95** *MGM* 🎞🎞

L

Legend of the Northwest The loyalty of a dog is evidenced in the fierce revenge he has for the drunken hunter who shot and killed his master.
1978 (G) 83m/C Denver Pyle. **VHS, Beta** $19.95 *STE, GEM, HHE* ♂♂½

The Legend of Robin Hood Robin Hood of Sherwood Forest returns in this animated version of his adventures.
1951 47m/C **VHS, Beta** $19.95 *MGM, KUI* ♂♂½

Legend of the Sea Wolf Adaptation of Jack London's dramatic adventure novel, "The Sea Wolf." A slave driving ship captain rescues a shipwreck victim and sets to abusing him along with the rest of his crew. A rebellion ensues. Superb musical score benefits the strong acting and good pace. Also known as "Wolf Larsen."
1958 83m/B Barry Sullivan, Peter Graves, Gita Hall, Thayer David; *Dir:* Harmon Jones. **VHS** *API* ♂♂½

The Legend of Sea Wolf A sadistic sea captain forcefully rules his crew in this weak version of Jack London's novel. Also known as "Wolf Larsen."
1975 (PG) 90m/C Chuck Connors, Barbara Bach, Joseph Palmer. **VHS, Beta** $9.95 *NO* ♂♂

The Legend of Sleepy Hollow The story of Ichabod Crane and the legendary ride of the headless horseman by Washington Irving; narrated by Crosby. Also includes two classic short cartoons, "Lonesome Ghosts" (1932) with Mickey Mouse and "Trick or Treat" (1952) with Donald Duck.
1949 45m/C *Dir:* Jack Kinney, Clyde Geronimi, James Algar; *Nar:* Bing Crosby. **VHS, Beta** $12.99 *DIS* ♂♂

The Legend of Sleepy Hollow Washington Irving's classic tale of the Headless Horseman of Sleepy Hollow is brought to life on the screen.
1979 (G) 100m/C Jeff Goldblum, Dick Butkus, Paul Sand, Meg Foster, James Griffith, John S. White. **VHS, Beta, LV** $19.95 *STE, UHV, LME* ♂♂½

The Legend of Suram Fortress From old Russian folktales comes this story of a fearless young man and his devotion to his village.
1985 76m/C *RU Dir:* Sergei Paradjanov. **VHS** *TPV* ♂♂

Legend of Valentino Made-for-television docudrama traces the legendary exploits of one of the silver screen's greatest lovers, Rudolph Valentino. Typical TV-movie fare that lacks the power of the legend's life.
1975 96m/C Franco Nero, Suzanne Pleshette, Lesley Ann Warren, Yvette Mimieux, Judd Hirsch, Milton Berle, Harold J. Stone; *Dir:* Melville Shavelson. **VHS, Beta** $49.95 *PSM, MRV* ♂

The Legend of Valentino This is a biographical documentary of perhaps the world's greatest lover beginning with his immigration to America and ending at his unexpected, sudden death.
1983 71m/B **VHS, Beta** $24.98 *SUE*

Legend of Walks Far Woman A made-for-television film about a proud Sioux woman fighting for survival and her tribe during the American-Indian Wars. Welch is miscast. From Colin Stuart's novel.
1982 120m/C Raquel Welch, Nick Mancuso, Bradford Dillman. **VHS, Beta** $59.95 *VCL* ♂

Legend of the Werewolf A child who once ran with the wolves has forgotten his past, except when the moon is full.
1975 (R) 90m/C *GB* Peter Cushing, Hugh Griffith, Ron Moody, David Rintoul, Lynn Dalby; *Dir:* Freddie Francis. **VHS, Beta** *VCL, MRV* ♂♂

The Legend of the Wolf Woman The beautiful Daniella assumes the personality of the legendary wolfwoman, leaving a trail of gruesome killings across the countryside. Genre fans will find this one surprisingly entertaining. Also on video as "Werewolf Woman."
1977 (R) 84m/C *IT* Anne Borel, Fred Stafford, Tino Carey, Elliot Zamuto, Ollie Reynolds, Andrea Scott, Karen Carter; *Dir:* Raphael D. Silver. **VHS, Beta** $19.95 *VIP, UHV* ♂

The Legend of Young Robin Hood The early life of the robber of the rich is depicted in this movie. He learns to use a longbow, and forms his convictions as he and other Saxons struggle at their integration into the Norman culture.
197? (G) 60m/C **VHS, Beta** $19.95 *GEM* ♂♂

Legendary Ladies of Rock & Roll The young and old veterans of girl-group rock such as Brenda Lee, Mary Wells, Belinda Carlisle, Martha Reeves and Grace Slick perform live at the Latin Quarter in New York. Songs include "Tonight's the Night," "You Don't Own Me," "Soldier Boy," "I'm Sorry," "Be My Baby," "Heat Wave," "Will You Still Love Me Tomorrow," "Band of Gold," and more.
1986 57m/C Brenda Lee, Lesley Gore, Mary Wells, Belinda Carlisle, Martha Reeves, Ronnie Spector, Grace Slick, Clarence Clemmons, Shirley Alston Reeves. **VHS, Beta, LV** $19.95 *MVD, HBO* ♂♂

Legends of Rock & Roll A made-for-cable television performance of the great blues/rock vets, taped in Rome's Palaeur Arena. Songs include: "I Feel Good," "I'm a Man," "The Fat Man" and "Whole Lotta Shakin Going On."
1989 55m/C James Brown, Ray Charles, Bo Diddley, Fats Domino, B.B. King, Jerry Lee Lewis, Kareem Abdul-Jabbar. **VHS, Beta, LV** $19.95 *MVD, HBO* ♂♂

Legion of Iron Adventures in a computer-run, neo-Roman civilization where men and women battle for supremacy.
1990 (R) 85m/C Kevin T. Walsh, Erica Nann, Regie De Morton, Camille Carrigan; *Dir:* Yakov Bentsvi. **VHS, Beta, LV** $79.95 *COL* ♂

Legion of the Lawless A group of outlaws band together in order to spread terror and confusion among the populace.
1940 59m/B George O'Brien, Virginia Vale. **VHS, Beta** *MED* ♂

Legion of Missing Men Professional soldiers of fortune, the French Foreign Legion, fight the evil sheik Ahmed in the Sahara.
1937 62m/B Ralph Forbes, Ben Alexander, Hala Linda. **VHS, Beta** $29.95 *VYY* ♂½

Legs Backstage story of three girls who are competing for a job with Radio City Music Hall's Rockettes. Made for television.
1983 91m/C Gwen Verdon, John Heard, Sheree North, Shanna Reed, Maureen Teefy; *Dir:* Jerrold Freedman. **VHS, Beta** $59.95 *PSM* ♂♂

The Lemon Drop Kid Second version of the Damon Runyon chestnut about a racetrack bookie who must recover the gangster's money he lost on a bet. As the fast-talking bookie, Hope sparkles.

1951 91m/B Bob Hope, Lloyd Nolan, Fred Clark, Marilyn Maxwell, Jane Darwell, Andrea King, William Frawley, Jay C. Flippen, Harry Bellaver; *Dir:* Sidney Lanfield. **VHS, Beta, LV** $14.95 *COL* ♂♂½

The Lemon Sisters Three women, friends and performance partners since childhood, struggle to buy their own club. They juggle the men in their lives with less success. Great actresses like these should have done more with this interesting premise, and the excellent male cast has much more potential.
1990 (PG-13) 93m/C Diane Keaton, Carol Kane, Kathryn Grody, Elliott Gould, Ruben Blades, Aidan Quinn; *Dir:* Joyce Chopra. **VHS, LV** $92.99 *HBO, FCT* ♂♂

Lemora, Lady Dracula A pretty young church singer is drawn into the lair of the evil Lady Dracula, whose desires include her body as well as her blood. Horror fans will enjoy some excellent atmosphere, particularly in a scence where the girl's church bus is attacked by zombie-like creatures; Smith remains a '70s "B" movie favorite. Perhaps a double feature with "Lady Frankenstein...?"
1973 (PG) 80m/C Leslie Gilb, Cheryl "Rainbeaux" Smith, William Whitton, Steve Johnson, Monty Pyke, Maxine Ballantyne, Parker West, Richard Blackburn; *Dir:* Richard Blackburn. **VHS** $18.00 *FRG* ♂

Lena's Holiday Fluffy comedy about a winsome East German girl visiting L.A. for the first time, and the culture-shock that ensues. Script and performances are surprisingly sharp, but the story bogs down with a silly thriller plot and pointless celebrity cameos.
1991 (PG-13) 97m/C Felicity Waterman, Chris Lemmon, Pat Morita, Susan Anton, Michael Sarrazin, Nick Mancuso, Bill Dana, Liz Torres; *Dir:* Michael Keusch. **VHS, Beta, LV** $89.95 *PSM* ♂♂

Leningrad Cowboys Go America An outlandish Finnish band, the Leningrad Cowboys pack up their electric accordians and their one dead member and go on tour in America. With matching front-swept pompadours and pointy shoes, the group's appearance is the picture's best joke—and one of its only jokes, as the road trip plods in deadpan fashion that may bore viewers. It's best appreciated by fans of comic minimalist Jim Jarmusch (who has a guest role). English and Finnish dialogue.
1989 (PG-13) 78m/C *FI* Jim Jarmusch, Matti Pellonpaa, Kari Vaananen, Nicky Tesco; *Dir:* Aki Kaurismaki. **VHS** $79.98 *ORI, FCT* ♂♂½

Lenny Smokey nightclubs, drug abuse and obscenities abound in Hoffman's portrayal of the controversial comedian Lenny Bruce, whose use of street language led to his eventual blacklisting. Perrine is a gem as his stripper wife. Adapted from the Julian Barry play and filmed in black and white, this is a visually compelling piece that sharply divided critics upon release. Oscar nominations for Best Picture, actor Hoffman, actress Perrine, director Fosse, Adapted Screenplay, and Cinematography.
1974 (R) 111m/B Dustin Hoffman, Valerie Perrine, Jan Miner, Stanley Beck; *Dir:* Bob Fosse. Cannes Film Festival '75: Best Actress (Perrine); National Board of Review Awards '74: 10 Best Films of the Year, Best Supporting Actress (Perrine); Harvard Lampoon Awards '74: Worst Film of the Year. **VHS, Beta, LV** $19.98 *MGM, FOX* ♂♂♂

Lenny Bruce An uncensored San Francisco nightclub performance of Bruce's offbeat, bewitching, and often unprecedented humor.
1967 60m/B Lenny Bruce. **VHS, Beta $24.95** VYY ♂♂♂

Lenny Bruce Performance Film A videotape of one of Lenny Bruce's last nightclub appearances at Basin Street West in San Francisco. Also included is "Thank You Masked Man," a color cartoon parody of the Lone Ranger legend, with Lenny Bruce providing all the character voices.
1968 72m/B Lenny Bruce. **VHS, Beta $59.95** VES ♂♂♂

Leonard Part 6 A former secret agent comes out of retirement to save the world from a crime queen using animals to kill agents. In the meantime, he tries to patch up his collapsing personal life. Wooden and disappointing; produced and co-written by Cosby, who himself panned the film.
1987 (PG) 83m/C Bill Cosby, Gloria Foster, Tom Courtenay, Joe Don Baker; **Dir:** Paul Weiland. **VHS, Beta, LV $89.95** COL ♂

Leonor Liv Ullman demonstrates her versatility as an actress in this movie, in which she plays the mistress of the Devil.
1975 90m/C FR Liv Ullman; **Dir:** Juan Bunuel. **VHS, Beta** FOX ♂ 1/2

Leopard Fist Ninja A masked ninja whacks away at American hoodlums in a fit of Maoist, anti-capitalistic fury. Dubbed.
198? 88m/C Willie Freeman, Jack Lam, Chuck Horry. **VHS, Beta $54.95** MAV ♂ 1/2

The Leopard Man An escaped leopard terrorizes a small town in New Mexico. After a search, the big cat is found dead, but the killings continue. Minor but effective Jacques Tourneur creepie. Based on Cornell Woolrich's novel "Black Alibi." Another Val Lewton Horror production.
1943 66m/B Jean Brooks, Isabel Jewell, James Bell, Margaret Landry, Dennis O'Keefe, Margo, Rita Corday; **Dir:** Jacques Tourneur. **VHS, Beta, LV $19.95** MED ♂♂ 1/2

Leopard in the Snow The romance between a race car driver allegedly killed in a crash and a young woman is the premise of this film.
1978 (PG) 89m/C Keir Dullea, Susan Penhaligon, Kenneth More, Billie Whitelaw; **Dir:** Gerry O'Hara. **VHS, Beta $59.98** SUE ♂♂

The Leopard Woman Long before the cat lady cut her first claws, there was the leopard woman. Vamp Glaum finds much trouble to meow about on the Equator.
1920 66m/B Louise Glaum. **VHS, Beta $19.95** GPV, FCT ♂♂

Lepke The life and fast times of Louis "Lepke" Buchalter from his days in reform school to his days as head of Murder, Inc. and his execution in 1944.
1975 (R) 110m/C Tony Curtis, Milton Berle, Gianni Russo, Vic Tayback, Michael Callan; **Dir:** Menahem Golan. **VHS, Beta $59.95** WAR ♂♂ 1/2

Les Assassins de L'Ordre A righteous French judge faces the case of an innocent man killed by police brutality. Subtitled in English. Known in English as "Law Breakers."
1971 100m/C FR Jacques Brel, Catherine Rouvel, Michael Lonsdale, Charles Denner; **Dir:** Marcel Came. **VHS, Beta $29.95** TAM ♂♂

Les Biches An exquisite film that became a landmark in film history with its theme of bisexuality and upper class decadence. A rich socialite picks up a young artist drawing on the streets of Paris, seduces her, and then takes her to St. Tropez. Conflict arises when a suave architect shows up and threatens to come between the two lovers. In French with English subtitles.
1968 (R) 95m/C FR Stephane Audran, Jean-Louis Trintignant, Jacqueline Sassard; **Dir:** Claude Chabrol. **VHS $79.95** CVC ♂♂♂

Les Bons Debarras In French with English subtitles, this Canadian film follows a lonely 13-year-old's effort to win her mother's exclusive love.
1981 114m/C CA Marie Tifo, Charlotte Laurier, German Houde; **Dir:** Francis Mankiewicz. Genie Awards '81: Best Actress (Tifo), Best Cinematography, Best Director (Mankiewicz), Best Picture, Best Screenplay, Best Supporting Actor (Houde). **VHS, Beta** IFE ♂♂ 1/2

Les Carabiniers A cynical, grim anti-war tract, detailing the pathetic adventures of two young bums lured into enlisting with promises of rape, looting, torture and battle without justification. Controversial in its day, and typically elliptical and non-narrative. In French with English subtitles.
1963 80m/B GB IT FR Anna Karina, Genevieva Galea, Marino Mase; **Dir:** Jean-Luc Godard. **VHS, Beta $29.95** FCT ♂♂♂

Les Choses de la Vie A car crash makes a man re-evaluate his life, goals, and loves. In the end he must choose the course of his future and decide between his wife and mistress. In French with English subtitles.
1970 90m/C FR Romy Schneider, Michel Piccoli, Lea Massari; **Dir:** Claude Sautet. **VHS, Beta, LV $79.95** CVC ♂♂♂

Les Comperes A woman suckers two former lovers into finding her wayward son by secretly telling each ex he is the natural father of the punk. Depardieu is a streetwise journalist who teams up with a suicidal hypochondriac/wimp (Richard) to find the little brat. Humorous story full of bumbling misadventures. In French with English subtitles.
1983 92m/C FR Pierre Richard, Gerard Depardieu, Anny Duperey, Michel Aumont; **W/Dir:** Francis Veber. **VHS, Beta, LV $59.99** MED, FCT, TAM ♂♂♂

Les Contes D'Hoffman A poet relates the stories of his last three romantic encounters that were thwarted by his rivals.
1984 135m/C Placido Domingo, Ileana Cotrubas, Agnes Baltsa. **VHS, Beta $39.95** HBO ♂♂♂

Les Enfants Terrible The classic, lyrical treatment of adolescent deviance adapted by Cocteau from his own play, wherein a brother and sister born into extreme wealth eventually enter into casual crime, self-destruction, and incest. In French with English subtitles.
1950 105m/B FR Edouard Dermithe, Nicole Stephane, Jean Cocteau; **Dir:** Jean-Pierre Melville. New York Film Festival '74: Best Supporting Actor. **VHS, Beta $59.98** FCT ♂♂♂

Les Girls When one member of a performing troupe writes her memoirs, the other girls sue for libel. Told through a series of flashbacks, this story traces the girls' recollections of their relationships to American dancer Kelly. Cole Porter wrote the score for this enjoyable "Rashomon"-styled musical. Laserdisc features include the original trailer and letterboxed screen.

1957 114m/C Gene Kelly, Mitzi Gaynor, Kay Kendall, Taina Elg, Henry Daniell, Patrick Macnee; **Dir:** George Cukor. Academy Awards '57: Best Costume Design; Golden Globe Awards '58: Best Film—Musical/Comedy. **VHS, Beta, LV $19.98** MGM ♂♂ 1/2

Les Grandes Gueules From the director of the classic "An Occurrence at Owl Creek Bridge," comes this comedy-drama about a sawmill owner dealing with two ex-convicts hired as lumberjacks. In French with English subtitles. Also known as "Jailbirds' Vacation."
1965 125m/C FR Lino Ventura, Bourvil, Marie DuBois; **Dir:** Robert Enrico. **VHS, Beta $19.95** KRT ♂♂ 1/2

Les Miserables Victor Hugo's classic novel about small-time criminal Jean Valjean and 18th-century France. After facing poverty and prison, escape and torture, Valjean is redeemed by the kindness of a bishop. As he tries to mend his ways, he is continually hounded by the policeman Javert, who is determined to lock him away. The final act is set during a student uprising in the 1730s. This version is the best of many, finely detailed and well-paced with excellent cinematography by Gregg Toland. Received Oscar nominations for Best Picture, Best Cinematography, Best Assistant Director and Best Film Editing.
1935 108m/B Fredric March, Charles Laughton, Cedric Hardwicke, Rochelle Hudson, John Beal, Frances Drake, Florence Eldridge, John Carradine; **Dir:** Richard Boleslawski. National Board of Review Awards '35: 10 Best Films of the Year. **VHS, Beta, LV $59.98** FOX, APD ♂♂♂♂

Les Miserables Hugo's classic novel done up Italian style with lavish sets and spectacle. Dubbed in English.
1952 119m/B IT Gino Cervi, Valentina Cortese; **Dir:** Riccardo Freda. **VHS $29.95** FCT ♂♂ 1/2

Les Miserables An epic French adaptation of the Victor Hugo standard about Valjean, Javert and injustice. Although this doesn't reach the level of the 1935 classic, it is still worth watching. Dubbed in English.
1957 210m/C FR GE Jean Gabin, Daniele Delorme, Bernard Blier, Bourvil, Gianni Esposito, Serge Reggiani; **Dir:** Jean-Paul LeChanois. **VHS, Beta $29.95** FCT ♂♂♂

Les Miserables An excellent made-for-television version of the Victor Hugo classic about the criminal Valjean and the policeman Javert playing cat-and-mouse in 18th-century France. Dauphin's last film role.
1978 150m/C Richard Jordan, Anthony Perkins, John Gielgud, Cyril Cusack, Flora Robson, Celia Johnson, Claude Dauphin; **Dir:** Glenn Jordan. **VHS, Beta, LV $59.98** FOX ♂♂♂ 1/2

Les Miserables Victor Hugo's classic novel comes to life in this beautifully animated family feature.
1979 70m/C **VHS, Beta $29.95** FHE, APD ♂♂

Les Mistons This study of male adolescence finds five teen-age boys worshiping a beautiful girl from afar, following her everywhere, spoiling her dates and finally reaching maturity in light of their mistakes.
1957 18m/C FR **Dir:** Francois Truffaut. **VHS, Beta** TEX ♂♂♂

Les Patterson Saves the World An obnoxious ambassador to an anonymous Middle Eastern country teams up with the singularly distastefully named Dr. Herpes to stop a new killer disease about to engulf the world.

1990 (R) 105m/C Barry Humphries, Pamela Stephenson, Thaao Penghlis, Andrew Clarke, Joan Rivers; *Dir:* George Miller. **VHS $14.95** *HMD, SOU* 🎬½

Les Rendez-vous D'Anna An independent woman travels through Europe and comes face to face with its post-war modernism.
1978 120m/C *FR BE* Aurore Clement, Helmut Griem, Magali Noel, Hanns Zieschler, Lea Massari, Jean-Pierre Cassel; *W/Dir:* Chantal Akerman. **VHS $79.95** *WAC* 🎬½

Les Visiteurs du Soir A beautiful, charming fairy tale about the devil's intrepid interference with a particular love affair in 15th-century France, which he cannot squelch. Purportedly a parable about Hitler's invasion of France. Interestingly, this was released after the Nazi occupation of France, so one assumes that the Germans didn't make the connection. In French with English subtitles. Also known as "Devil's Envoys."
1942 110m/B *FR* Arletty, Jules Berry, Marie Dea, Alain Cuny, Fernand Ledoux, Marcel Herrand; *Dir:* Marcel Carne. **VHS, Beta $59.95** *FCT* 🎬🎬½

Less Than Zero An adaptation of Bret Easton Ellis' popular, controversial novel about a group of affluent, drug-abusing youth in Los Angeles. Although it tries, the film fails to inspire any sort of sympathy for this self-absorbed and hedonistic group. Mirrors the shallowness of the characters although Downey manages to rise above this somewhat. Music by the Bangles, David Lee Roth, Poison, Roy Orbison, Aerosmith and more.
1987 (R) 98m/C Andrew McCarthy, Jami Gertz, Robert Downey Jr., James Spader; *Dir:* Marek Kanievska. **VHS, Beta, LV $89.98** *FOX* 🎬

Lesson in Love A gynecologist's philandering is discovered by his wife, who then starts an affair of her own. Change of pace for Bergman as he does this modest comedy. Early Bergman.
1954 95m/B *SW* Gunnar Bjornstrand, Eva Dahlbeck, Harriet Andersson; *Dir:* Ingmar Bergman. **VHS, Beta $29.98** *SUE* 🎬🎬½

Let 'Em Have It Exciting action saga loosely based on the newly formed FBI and its bouts with John Dillinger. Car chases and tommy guns abound in this fairly ordinary film.
1935 90m/B Richard Arlen, Virginia Bruce, Bruce Cabot, Harvey Stephens, Eric Linden, Joyce Compton, Gordon Jones; *Dir:* Sam Wood. **VHS $19.95** *NOS, IVE, MGM* 🎬🎬½

Let 'er Go Gallegher A youngster sees a murder, tells his newspaper buddy about it, and the guy writes an article about the crime. It's a big hit, the guy's ego starts to overinflate, and his life starts to head downwind until he decides its time for him and the little guy to rope themselves a miscreant.
1928 57m/B Junior Coghlan, Elinor Fair, Wade Boteler; *Dir:* Elmer Clifton. **VHS, Beta $19.95** *GPV, FCT* 🎬

Let Him Have It A controversial film about youth in postwar England and a miscarriage of British justice. On November 2, 1952 two Londoners, Christopher Craig, 16, and Derek Bentley, 19, climbed onto the roof of a candy warehouse in an apparent burglary attempt. When police arrived Bentley was captured. Craig, who had a gun, shot and wounded one policeman and killed another. According to testimony Bentley shouted to his friend "Let him have it" but did he mean

shoot or give the officer the gun? With those words Bentley, whose IQ was 66, was sentenced to death by the British courts. The uproar over the sentence at the time was reignited by the film and has led to a request for a reexamination of evidence and sentencing by the British Home Office.
1991 (R) 115m/C *GB* Christopher Eccleston, Paul Reynolds, Tom Bell, Eileen Atkins, Clare Holman, Michael Elphick, Mark McGann, Tom Courtenay; *Dir:* Peter Medak. **VHS $89.95** *COL* 🎬🎬½

Let It Be Documentary look at a Beatles recording session, giving glimpses of the conflicts which led to their breakup. Features appearances by Yoko Ono and Billy Preston.
1970 80m/C John Lennon, Paul McCartney, George Harrison, Ringo Starr, Billy Preston, Yoko Ono; *Dir:* Michael Lindsay-Hogg. Academy Awards '70: Best Original Score. **VHS, Beta, LV** *FOX* 🎬🎬

Let It Ride Dreyfuss is a small-time gambler who finally hits it big at the track. Some funny moments but generally a lame script cripples the cast, although the horses seem unaffected. Garr is okay as his wife who slips further into alcoholism with each race.
1989 (PG-13) 91m/C Richard Dreyfuss, Teri Garr, David Johansen, Jennifer Tilly, Allen Garfield, Ed Walsh, Michelle Phillips, Mary Woronov, Robbie Coltrane, Richard Edson, Cynthia Nixon; *Dir:* Joe Pytka. **VHS, Beta, LV, 8mm $14.95** *PAR* 🎬🎬

Let It Rock A maniacal promoter makes superstars out of his new band by staging publicity stunts. Not one of Hopper's better efforts. It was filmed in Germany, but Roger Corman bought the U.S. rights and added clips from other movies before it was released here.
1986 75m/C *GE* Dennis Hopper, David Hess; *Dir:* Roland Klick. **VHS, Beta $79.95** *MED* 🎬

Let Sleeping Minnows Lie This is an animated sequel to "Gilligan's Island," depicting the castaways' efforts to build a ship.
1982 23m/C **VHS, Beta $14.95** *MGM* 🎬

Let There Be Light Moving documentary of shell-shocked soldiers in an army hospital was shelved by the War Department because of its revealing content.
1946 60m/B *Dir:* John Huston; *Nar:* Walter Huston. **VHS, Beta, 3/4U $24.95** *NOS, HHT, FST*

Lethal Obsession Two cops battle street crime and a powerful drug ring. Music by Tony Carey. Filmed in Germany.
1987 (R) 100m/C *GE* Michael York, Elliott Gould, Peter Maffay, Armin Mueller-Stahl; *Dir:* Peter Patzak. **VHS, Beta $29.95** *VMK* 🎬½

Lethal Pursuit A popular singer who returns to the town where she grew up is tormented by her ex-boyfriend, a criminal.
1988 90m/C Mitzi Donahue, Blake Gahner, John Wildman; *Dir:* Donald M. Jones. **VHS, Beta $79.95** *SOU* 🎬

Lethal Weapon In Los Angeles, a cop nearing retirement (Glover) unwillingly begins work with a new partner (Gibson), a suicidal, semi-crazed risk-taker who seems determined to get the duo killed. Both Vietnam vets, the pair uncover a vicious heroin smuggling ring run by ruthless ex-Special Forces personnel. Packed with plenty of action, violence, and humorous undertones. Eric Clapton's contributions to the musical score are an added bonus. Gibson and Glover work

well together and give this movie extra punch. Followed by (so far) two sequels.
1987 (R) 110m/C Mel Gibson, Danny Glover, Gary Busey, Mitchell Ryan, Tom Atkins, Darlene Love, Traci Wolfe; *Dir:* Richard Donner. **VHS, Beta, LV, 8mm $19.98** *WAR* 🎬🎬🎬

Lethal Weapon 2 This sequel to the popular cop adventure finds Gibson and Glover taking on a variety of blond South African "diplomats" who try to use their diplomatic immunity status to thwart the duo's efforts to crack their smuggling ring. Gibson finally finds romance, and viewers learn the truth about his late wife's accident. Also features the introduction of obnoxious, fast-talking con artist Leo ("OK, OK") Getz, adeptly played by Pesci, who becomes a third wheel to the crime-fighting team. Music by Eric Clapton, Michael Kamen, and David Sanborn. Followed by a second sequel.
1989 (R) 114m/C Mel Gibson, Danny Glover, Joe Pesci, Joss Ackland, Derrick O'Connor, Patsy Kensit, Darlene Love, Traci Wolfe; *Dir:* Richard Donner. **VHS, Beta, LV, 8mm $24.98** *WAR, BUR* 🎬🎬🎬

Lethal Woman A group of beautiful women, all once victimized by rapes, is recruited to an island by the "Lethal Woman," who plots revenge on any man who dares to vacation on the island.
1988 96m/C Shannon Tweed, Merete Van Kamp. **VHS, Beta $29.98** *VID* 🎬½

Let's Dance A young widow tries to protect her son from his wealthy, paternal grandmother, while Fred dances his way into her heart. More obscure Astaire vehicle, but as charming as the rest. Needless to say, great dancing.
1950 112m/C Betty Hutton, Fred Astaire, Roland Young, Ruth Warrick, Sheppard Strudwick, Lucile Watson, Barton MacLane, Gregory Moffett, Melville Cooper; *Dir:* Norman Z. McLeod. **VHS, Beta, LV $14.95** *PAR* 🎬🎬½

Let's Do It! Soft-core sex antics abound, with a special appearance by Playmate Amanda Cleveland.
1984 (R) 60m/C Greg Bradford, Brit Helfer, Amanda Cleveland. **VHS, Beta $39.95** *AHV* 🎬

Let's Do It Again An Atlanta milkman and his pal, a factory worker, milk two big-time gamblers out of a large sum of money in order to build a meeting hall for their fraternal lodge. A lesser sequel to "Uptown Saturday Night."
1975 (PG) 113m/C Sidney Poitier, Bill Cosby, John Amos, Jimmie Walker, Ossie Davis, Denise Nicholas, Calvin Lockhart; *Dir:* Sidney Poitier. **VHS, Beta $19.98** *WAR* 🎬½

Let's Get Harry American mercenaries mix it up with a Columbian cocaine franchise. Never released theatrically. Also known as "The Rescue."
1987 (R) 107m/C Robert Duvall, Gary Busey, Michael Schoeffling, Tom Wilson, Glenn Frey, Rick Rossovich, Ben Johnson, Matt Clark, Mark Harmon, Gregory Sierra, Elpidia Carrillo; *Dir:* Alan Smithee. **VHS, Beta, LV $14.99** *HBO* 🎬½

Let's Get Lost Absorbing documentary of the prominent jazz trumpeter (and occasional vocalist) who succumbed, like many of his peers, to drug addiction. Downbeat, but rarely less than worthy.
1989 125m/C Chet Baker; *Dir:* Bruce Weber. **VHS $89.95** *MVD, COL* 🎬🎬🎬½

Let's Get Married A young doctor's weakness under pressure brings him to the brink of babbling idiocy when his wife goes into early labor.

1963 90m/B Anthony Newley, Ann Aubrey. **VHS, Beta** *MON* 🎬

Let's Get Tough The East Side kids, unable to enlist because of their age, harass local Oriental shopkeepers and eventually get involved with a murder.
1942 52m/B Leo Gorcey, Bobby Jordan, Huntz Hall, Gabriel Dell, Sammy Morrison; *Dir:* Wallace Fox. **VHS, Beta, 8mm** $24.95 *NOS, VYY, BUR* 🎬

Let's Go! The son of a cement company president sets out to prove his worth to the firm by landing a contract with the town of Hillsboro. For some unknown reason, he sets off on an amazing set of stunts. Silent with original organ music.
1923 79m/B Richard Talmadge, Hal Clements, Eileen Percy, Tully Marshall; *Dir:* W.K. Howard. **VHS, Beta, 8mm** $29.99 *VYY, GPV* 🎬 ½

Let's Go Collegiate Two college rowers promise their girlfriends that they'll win the big race, but their plans are almost frustrated when their best oarsman is drafted. Their only hope lies in replacing him with a truck driver.
1941 62m/B Frankie Darro, Marcia Mae Jones, Jackie Moran, Keye Luke, Mantan Moreland, Gale Storm; *Dir:* Jean Yarborough. **VHS** $19.95 *NOS* 🎬 ½

Let's Have an Irish Party First Irish production to be put on the market, and it features some of Ireland's top stars.
1983 60m/C Paddy Noonan, Carmel Quinn, Richie O'Shea, John Scott Trotter, Johnny Hanley, Kenneth McLeod, Anna McGoldrick, Barley Bree. **VHS, Beta** *RGO, PPI* 🎬 ½

Let's Make it Legal Amusing comedy of a married couple who decide to get a divorce after 20 years of marriage. Colbert stars as the woman who decides to leave her husband, Carey, because he's a chronic gambler. They part as friends, but soon Colbert's old flame, Scott, is back in town and things get rather complicated. A solid cast serves as the main strength in this film.
1951 77m/B Claudette Colbert, MacDonald Carey, Zachary Scott, Barbara Bates, Robert Wagner, Marilyn Monroe; *Dir:* Richard Sale. **VHS** $14.98 *FXV* 🎬 ½

Let's Make Love An urbane millionaire discovers he is to be parodied in an off-Broadway play, and vows to stop it—until he falls in love with the star of the show. He then ends up acting in the play. Marilyn sings "My Heart Belongs to Daddy" and "Incurably Romantic."
1960 118m/C Yves Montand, Marilyn Monroe, Tony Randall, Frankie Vaughan, Bing Crosby, Gene Kelly, Milton Berle; *Dir:* George Cukor. **VHS, Beta** $14.98 *FOX* 🎬🎬 ½

Let's Make Up When it comes to giving her heart, Neagle can't decide betwen to dashing men. Also known as "Lilacs in the Spring."
1955 94m/C *GB* Errol Flynn, Anna Neagle, David Farrar, Kathleen Harrison, Peter Graves; *Dir:* Herbert Wilcox. **VHS** *MLB, MRV* 🎬🎬

Let's Scare Jessica to Death A young woman who was recently released from a mental hospital is subjected to unspeakable happenings at a country home with her friends. She encounters murder, vampires, and corpses coming out of nowhere. A supernatural thriller with quite a few genuine scares.
1971 (PG) 89m/C Zohra Lampert, Barton Heyman, Kevin J. O'Connor, Gretchen Corbett, Alan Manson; *Dir:* John Hancock. **VHS, Beta** $49.95 *PAR* 🎬🎬

Let's Sing Again Eight-year-old singing sensation Bobby Breen made his debut in this dusty musical vehicle, as a runaway orphan who becomes the pal of a washed-up opera star in a traveling show.
1936 70m/B Bobby Breen, Henry Armetta, George Houston, Vivienne Osborne, Grant Withers, Inez Courtney, Lucien Littlefield; *Dir:* Kurt Neumann. **VHS** $15.95 *NOS, LOO, DVT* 🎬🎬

The Letter When a man is shot and killed on a Malaysian plantation, the woman who committed the murder pleads self-defense. Her husband and his lawyer attempt to free her, but find more than they expected in this tightly-paced film noir. Based on the novel by W. Somerset Maugham. Davis emulated the originator of her role, Jeanne Eagels, in her mannerisms and line readings, although Eagels later went mad from drug abuse and overwork. Oscar nominations: Best Picture, Best Supporting Actor (Stephenson), Best Director, Black & White Cinematography, Original Score, and Film Editing.
1940 96m/B Bette Davis, Herbert Marshall, James Stephenson, Gale Sondergaard, Bruce Lester, Cecil Kellaway, Victor Sen Yung; *Dir:* William Wyler. **VHS, Beta, LV** $19.95 *FOX, MGM* 🎬🎬🎬

Letter to Brezhnev Acclaimed independent film about two working-class Liverpool girls who fall in love with two furloughed Russian sailors. When the men eventually return to the USSR, one of the girls contacts the Russian Secretary General in an effort to rejoin her lover. Amusing tale of different relationships.
1986 (R) 94m/C *GB* Margi Clarke, Alexandra Pigg, Peter Firth, Ken Campbell, Tracy Lea; *Dir:* Chris Bernard. **VHS, Beta, LV** $19.98 *LHV, IME, WAR* 🎬🎬🎬

Letter of Introduction A struggling young actress learns that her father is really a well known screen star and agrees not to reveal the news to the public.
1938 104m/B Adolphe Menjou, Edgar Bergen, George Murphy, Eve Arden, Ann Sheridan; *Dir:* John M. Stahl. **VHS, Beta** $19.95 *NOS, MRV, HHT* 🎬🎬 ½

A Letter to Three Wives Three women taking a Hudson River boat trip receive a letter from a mysterious woman. The letter tells them that the writer has eloped with one of their husbands but does not specify his identity. The women spend the rest of the trip agonizing over which of them has lost their husband and why this mystery woman has decided to put them through the emotional wringer. Oscar nominated for Best Picture. Remade for television in 1985.
1949 103m/B Jeanne Crain, Linda Darnell, Ann Sothern, Kirk Douglas, Paul Douglas, Thelma Ritter; *Dir:* Joseph L. Mankiewicz. Academy Awards '49: Best Director (Mankiewicz), Best Screenplay; Directors Guild of America Awards '48: Best Director (Mankiewicz). **VHS, Beta, LV** $39.98 *FOX* 🎬🎬🎬

Letter from an Unknown Woman A woman falls in love with a concert pianist on the eve of his departure. He promises to return but never does. The story is told in flashbacks as the pianist reads a letter from the woman. A great romantic melodrama.
1948 90m/B Joan Fontaine, Louis Jourdan, Mady Christians, Marcel Journet, Art Smith; *Dir:* Max Ophuls. **VHS** $19.95 *REP* 🎬🎬🎬

Letters from My Windmill A series of three short stories: "The Three Low Masses," "The Elixir of Father Gaucher," and "The Secret of Master Cornille" from

respected director Pagnol. The unique format only enhances this film. In French with English subtitles.
1954 116m/B *FR* Henri Velbert, Yvonne Gamy, Robert Vattier, Roger Crouzet; *Dir:* Marcel Pagnol. **VHS** $29.95 *NOS, DVT, APD* 🎬🎬🎬 ½

Letters from the Park In Cuba, 1913, two taciturn would be lovers hire a poet, Victor La Place, to write love letters to each other. Normal proboscis aside, the poet follows the fate of Cyrano, and finds himself the unhappy hypotenuse in a triangular romance. Made for Spanish television; in Spanish with English subtitles.
1988 85m/C *SP* Victor Laplace, Ivonne Lopez, Miguel Paneque; *Dir:* Thomas Gutierrez Alea. **VHS** $79.95 *FXL, FCT, APD* 🎬🎬🎬

Letters to an Unknown Lover During a unspecified war, a man and woman engage in sexual and emotional combat.
1984 101m/C *GB FR* Cherie Lunghi, Mathilda May, Yves Beneyton, Ralph Bates; *Dir:* Peter Duffell. **VHS, Beta** $29.95 *ACA* 🎬🎬 ½

Letting the Birds Go Free When a stranger is caught stealing from a farmer's barn, the farmer and his son decide to let the villain work off his crime. Some very interesting relationships develop from this arrangement.
1986 60m/C Lionel Jeffries. **VHS** $11.95 *PSM* 🎬🎬 ½

Letting Go Made for television comedy-drama about a broken-hearted career woman and a young widower who fall in love.
1985 (PG) 94m/C John Ritter, Sharon Gless, Joe Cortese; *Dir:* Jack Bender. **VHS, Beta** $79.95 *AIP* 🎬🎬

Leviathan A motley crew of ocean-floor miners are trapped when they are accidentally exposed to a failed Soviet experiment that turns humans into insatiable, regenerating fish-creatures.
1989 (R) 98m/C Peter Weller, Ernie Hudson, Hector Elizondo, Amanda Pays, Richard Crenna, Daniel Stern, Lisa Eilbacher, Michael Carmine, Meg Foster; *Dir:* George P. Cosmatos. **VHS, Beta, LV** $19.98 *MGM* 🎬🎬

Lianna Acclaimed screenwriter/director John Sayles wrote and directed this story of a woman's romantic involvement with another woman. Chronicles an unhappy homemaker's awakening to the feelings of love that she develops for a female professor. Sayles makes a cameo as a family friend.
1983 (R) 110m/C John DeVries, Linda Griffiths, Jane Halleren, Jo Henderson, Jessica Wright MacDonald; *Dir:* John Sayles. **VHS, Beta** $69.95 *VES* 🎬🎬🎬

Liar's Moon A local boy woos and weds the town's wealthiest young lady, only to be trapped in family intrigue.
1982 (PG) 106m/C Cindy Fisher, Matt Dillon, Christopher Connelly, Susan Tyrrell; *Dir:* David Fisher. **VHS, Beta** $29.98 *VES* 🎬🎬

Libeled Lady A fast, complicated screwball masterwork, dealing with a newspaper editor's efforts to get something on a bratty heiress with the help of his own fiancee and a reporter he recently fired. One of the era's funniest Hollywood films. Remade in 1946 as "Easy to Wed." Oscar nominated for Best Picture.
1936 98m/B Myrna Loy, Spencer Tracy, Jean Harlow, William Powell, Walter Connolly, Charley Grapewin, Cora Witherspoon, E.E. Clive, Charles Trowbridge, Dennis O'Keefe, Hattie McDaniel; *Dir:* Jack Conway. **VHS, Beta, LV** $29.95 *MGM* 🎬🎬🎬 ½

Liberation of L.B. Jones A wealthy black undertaker wants a divorce from his wife, who is having an affair with a white policeman. Wyler's final film.
1970 (R) 101m/C Lee J. Cobb, Lola Falana, Barbara Hershey, Anthony Zerbe, Roscoe Lee Browne; *Dir:* William Wyler. VHS, Beta $59.95 COL *ℐℐ*

Liberators Kinski stars as a criminal soldier battling with the American authorities, German troops and his fugitive partner.
1977 91m/C Klaus Kinski. VHS, Beta $9.95 UNI *ℐ*

Liberty & Bash Two boyhood friends who served together in Vietnam reunite to rid their neighborhood of drug pushers and save the life of a friend.
1990 (R) 92m/C Miles O'Keefe, Lou Ferrigno, Mitzi Kapture, Richard Eden, Cheryl Paris, Gary Conway; *Dir:* Myrl A. Schreibman. VHS, Beta $89.95 FRH *ℐ*

License to Drive When teen Haim fails the road test for his all-important driver's license, he steals the family car for a hot date with the girl of his dreams. The evening starts out quietly enough, but things soon go awry. If Haim survives the weekend, he'll definitely be able to pass his driving test on Monday morning. Semi-funny in a demolition derby sort of way.
1988 (PG-13) 90m/C Corey Feldman, Corey Haim, Carol Kane, Richard Masur; *Dir:* Greg Beeman. VHS, Beta, LV $19.98 FOX *ℐ ½*

License to Kill A young girl is killed by a drunk driver, devastating both families. A made-for-television movie with a strong message against drinking and driving.
1984 96m/C James Farentino, Don Murray, Penny Fuller, Millie Perkins, Donald Moffatt, Denzel Washington, Ari Meyers; *Dir:* Judd Taylor. VHS, Beta, LV $59.95 GEM *ℐℐ*

License to Kill The 18th Bond film, and Dalton's second, in which drug lords try to kill 007's best friend and former CIA agent. Disobeying orders for the first time and operating without his infamous "license to kill," Bond goes after the fiends. Fine outing for Dalton (and Bond, too).
1989 (PG-13) 133m/C *GB* Timothy Dalton, Carey Lowell, Robert Davi, Frank McRae, Talisa Soto, David Hedison, Anthony Zerbe, Everett McGill, Wayne Newton, Benicoi Del Toro, Desmond Llewelyn, Priscilla Barnes; *Dir:* John Glen. VHS, Beta, LV $19.98 FOX *ℐℐℐ*

The Lickerish Quartet It's hilarious, realistic family fun as a couple watches a porno flick with their young son, and then act out their own zany sexual fantasies!!!
1970 (R) 90m/C Silvana Venturelli, Frank Wolff; *Dir:* Radley Metzger. VHS, Beta MAG *ℐ ½*

The Lie An womanizing Italian journalist jumps from sack to sack in this nonsensical bed-hopping drama.
1984 97m/C *IT* Ben Cross, Stefania Sandrelli, Amanda Sandrelli, Leslie Lyon, Claudia Cavalcanti; *Dir:* Giovanni Soldati. VHS, Beta, LV $79.98 VES, IME *ℐ*

Liebestraum An architectural expert, his old college friend and the friend's wife form a dangerous triangle of passion and lust that strangely duplicates a situation that led to a double murder forty years earlier.
1991 (R) 109m/C Kevin Anderson, Bill Pullman, Pamela Gidley, Kim Novak; *W/Dir:* Mike Figgis. VHS, LV $89.99 MGM *ℐℐ*

Lies Complicated mystery about a murder/thriller movie plot becoming reality as it is being filmed, revolving around a scam to collect an inheritance from a rich guy in a mental hospital.
1983 (PG) 93m/C *GB* Ann Dusenberry, Gail Strickland, Bruce Davison, Clu Gulager, Bert Remsen, Dick Miller; *Dir:* Ken Wheat, Jim Wheat. VHS, Beta $79.98 FOX *ℐℐ ½*

Lies My Father Told Me Simple drama about growing up in the 1920s in a Jewish ghetto. The story revolves around a young boy's relationship with his immigrant grandfather. Quiet and moving.
1975 (PG) 102m/C *CA* Yossi Yadin, Len Birman, Marilyn Lightstone, Jeffery Lynas; *Dir:* Jan Kadar. VHS COL *ℐℐℐ*

Lies of the Twins A beautiful model becomes involved with her therapist and his identical, and irresponsible, twin brother.
1991 93m/C Isabella Rossellini, Aidan Quinn, Iman, John Pleshette; *Dir:* Tim Hunter. VHS $79.95 MCA *ℐℐ*

The Life and Adventures of Santa Claus Animated story of how Santa became the symbol of Christmas. Based on the novel by L. Frank Baum.
1985 49m/C *Voices:* Alfred Drake, Earle Hyman. VHS, Beta $12.95 WAR *ℐℐ*

The Life and Assassination of the Kingfish Recommended by the NEA, this story of Louisiana politician Huey Long is riveting. Asner gives a profound performance.
1976 96m/C Ed Asner, Nicholas Pryor, Diane Kagan; *Dir:* Robert Collins. VHS, Beta $59.95 USA *ℐℐ*

Life is Beautiful A politically neutral man is arrested and tortured in pre-revolutionary Lisbon, and forced to make, and act on, a political commitment. Dubbed.
1979 102m/C *PT* Giancarlo Giannini, Ornella Muti; *Dir:* Grigorij Ciukhrai. VHS, Beta $19.95 STE, NWV *ℐℐ ½*

Life Begins for Andy Hardy Andy gets a job in New York before entering college and finds the working world to be a sobering experience. Surprisingly downbeat and hard-hitting for an Andy Hardy film. Garland's last appearance in the series.
1941 100m/B Mickey Rooney, Judy Garland, Lewis Stone, Ann Rutherford, Fay Holden, Gene Reynolds, Ralph Byrd; *Dir:* George B. Seitz. VHS, Beta $19.95 MGM *ℐℐℐ*

Life in Camelot: The Kennedy Years The JFK story told from his perspective.
1988 53m/C VHS, Beta $19.95 HBO *ℐℐℐ*

The Life of Christ A series of 16 small films on three tapes, following Christ from birth to crucifixion.
197? 90m/C VHS, Beta $39.95 UHV, VGD *ℐℐℐ*

Life is a Circus, Charlie Brown/ You're the Greatest, Charlie Brown Two Peanuts favorites appear on this tape. In the Emmy award-winning "Life Is a Circus, Charlie Brown," Snoopy falls in love with Fifi, a French poodle circus performer, and winds up as a member of the traveling circus. In "You're the Greatest, Charlie Brown," our hero has a chance to win a decathlon in the Junior Olympics. With Peppermint Patty coaching, anything can happen!
1980 60m/C *Dir:* Phil Roman. VHS, Beta $14.95 SHV, VTR *ℐℐℐ*

Life & Death of Colonel Blimp Chronicles the life of a British soldier who survives three wars (Boer, World War I, and World War II), falls in love with three women (all portrayed by Kerr), and dances a fine waltz. Fine direction and performance abound.
1943 115m/C *GB* Roger Livesey, Deborah Kerr, Anton Walbrook, Ursula Jeans, Albert Lieven; *Dir:* Michael Powell, Emeric Pressburger. VHS, Beta, LV $19.98 VID, VYG *ℐℐℐ*

The Life of Emile Zola Writer Emile Zola intervenes in the case of Alfred Dreyfus who was sent to Devil's Island for a crime he did not commit. Well-crafted production featuring a handsome performance from Muni. Heavily Oscar nominated, including best actor (Muni), best director, and best score.
1937 117m/B Paul Muni, Gale Sondergaard, Gloria Holden, Joseph Schildkraut; *Dir:* William Dieterle. Academy Awards '37: Best Picture, Best Screenplay, Best Supporting Actor (Schildkraut). VHS, Beta $19.98 MGM, BTV *ℐℐℐ ½*

Life with Father Based on the writings of the late Clarence Day, Jr., this is the story of his childhood in New York City during the 1880s. A delightful saga about a stern but loving father and his relationship with his knowing wife and four red-headed sons. Oscar nominations for best actor (Powell), color cinematography and color art and set direction.
1947 118m/C William Powell, Irene Dunne, Elizabeth Taylor, Edmund Gwenn, ZaSu Pitts, Jimmy Lydon, Martin Milner; *Dir:* Michael Curtiz. New York Film Critics Awards '47: Best Actor (Powell). VHS, Beta $9.95 NEG, MRV, NOS *ℐℐℐ ½*

The Life & Loves of a Male Stripper A shot-on-video feature about a male stripper who aspires to be a songwriter. Produced by Chippendales alumni.
1987 82m/C Rafael, Carl Fuerst, Dennis Landry, Jeffrey Allen, Troy. Beta MFI *Woof!*

The Life and Loves of Mozart Despite the title, this is really about the great composer's later life at the time of the premiere of "Die Zauberfloete" ("The Magic Flute"). This is only partly saved by a stellar performance from Werner, as well as the music. Other than that, it gets bogged down in titillation about W.A.M.'s romantic life. In German with English subtitles.
1959 87m/C Oskar Werner, Johanna Matz, Angelika Hauff; *Dir:* Karl Hartl; *Voices:* Anton Dermota. VHS $24.95 NOS *ℐℐ ½*

Life with Luigi This tape features two episodes of the early sitcom with two different actors in the lead role.
1953 30m/B J. Carroll Naish, Vito Scotti. VHS, Beta $24.95 NOS, DVT *Woof!*

Life with Mickey A collection of Mickey Mouse cartoons, including "Alpine Climbers," "Mickey's Polo Team," "Mickey's Circus," and "Shanghaied," an early black and white starring Minnie Mouse.
1951 51m/C *Dir:* Walt Disney. VHS, Beta, LV DIS *Woof!*

Life and Nothing But Two young women search for their lovers at the end of World War I. They're helped by a French officer brutalized by the war and driven to find all of France's casualties. Romantic, evocative, and saddening. Stars Philippe Noiret who played Alfredo in "Cinema Paridiso." In French with English subtitles. Winner of of over 12 International Film awards. Also received a 1990 Cesar (French equivalent of an Oscar) for Best Actor.

1989 (PG) 135m/C *FR* Philippe Noiret, Sabine Azema, Francoise Perrot; *W/Dir:* Bertrand Tavernier. British Academy Awards '90: Best Foreign Film; Los Angeles Film Critics Association Awards '90: Best Foreign Language Film. **VHS, Beta** $79.98 *ORI, FCT* ✯✯✯ ½

Life of O'Haru A near masterpiece rivaled only by "Ugetsu" in the Mizoguchi canon, this film details the slow and agonizing moral decline of a woman in feudal Japan, from wife to concubine to prostitute. A scathing portrait of social pre-destination based on a novel by Ibara Saikaku.
1952 136m/B *JP* Kinuyo Tanaka, Toshiro Mifune; *Dir:* Kenji Mizoguchi. Venice Film Festival '52: International Prize (Mizoguchi). **VHS, Beta** $39.95 *VYY, APD* ✯✯✯ ½

Life Stinks So does the film. A grasping tycoon bets he can spend a month living on the street without money, resulting in cheap laughs, heavy-handed sentiment and one musical number. Those expecting the innovative, hilarious Brooks of "Young Frankenstein" or "Blazing Saddles" will be very disappointed—these jokes are stale and the timing is tedious. Those looking for a Chaplinesque tale for modern times should stick with Chaplin.
1991 (PG) 93m/C Mel Brooks, Jeffrey Tambor, Lesley Ann Warren, Stuart Pankin, Howard Morris, Teddy Wilson, Rudy DeLuca; *W/Dir:* Mel Brooks. **VHS, Beta, LV** $92.99 *MGM* ✯

Life is Sweet The consuming passions of food and drink focus the lives of an oddball English working-class family beset by hopeless dreams and passions. Mother is always fixing family meals—in between helping her friend open a gourmet restaurant which features such revolting dishes as pork cyst and prune quiche. Dad is a chef who buys a snack truck and dreams of life on the road. Natalie and Nicola, the grown twins, eat their meals in front of the television but Nicola is also a bulemic who binges and purges on choclate bars behind her bedroom door. An affectionate, if sometime unattractive, look at a chaotic family.
1991 (R) 103m/C *GB* Alison Steadman, Jane Horrocks, Jim Broadbent, Claire Skinner, Timothy Spall, Stephen Rea; *W/Dir:* Mike Leigh. Los Angeles Film Critics Association Awards '91: Best Supporting Actress (Horrocks); National Society of Film Critics Awards '91: Best Actress (Steadman), Best Picture, Best Supporting Actress (Horrocks). **VHS, LV** $89.98 *REP* ✯✯✯ ½

Life & Times of Captain Lou Albano The long and illustrious career of wrestling's most notorious manager.
1985 97m/C Captain Lou Albano, Mr. Fuji, Mr. Saito. **VHS, Beta** $59.95 *CSM* ✯✯✯ ½

The Life & Times of the Chocolate Killer Police turn against the hand that's helped them save property and lives, by framing a good samaritan with deeds done by the "Chocolate Killer."
1988 75m/C Michael Adrian, Rod Browning, Tabi Cooper. **VHS** $79.95 *VMK* ✯ ½

Life & Times of Grizzly Adams Lightweight family adventure film based on the rugged life of legendary frontiersman, Grizzly Adams, that served as the launching pad for the TV series. Grizzly is mistakenly chased for a crime he didn't commit and along the way befriends a big bear.
1974 (G) 93m/C Dan Haggerty, Denver Pyle, Lisa Jones, Marjorie Harper, Don Shanks; *Dir:* Richard Friedenberg. **VHS, Beta** $19.95 *UHV* ✯

Life & Times of Judge Roy Bean Based on the life of the famed Texas "hanging judge," the film features Newman as the legendary Bean who dispenses frontier justice in the days of the Wild West. Filled with gallows humor. Gardner sparkles as actress Lily Langtry.
1972 (PG) 124m/C Paul Newman, Stacy Keach, Ava Gardner, Jacqueline Bisset, Anthony Perkins, Roddy McDowall, Victoria Principal; *Dir:* John Huston. **VHS, Beta** $19.98 *WAR* ✯✯ ½

Life of Verdi Epic mini-series biography of the famous composer, with many excerpts of his music sung by Luciano Pavarotti, Renata Tebaldi, and Maria Callas.
1984 600m/C *IT* Ronald Pickup, Carla Fracci. **VHS, Beta** $124.95 *HMV, KUL, MVD* ✯✯ ½

Lifeboat When a German U-boat sinks a freighter during World War II, the eight survivors seek refuge in a tiny lifeboat. Tension peaks after the drifting passengers take in a stranded Nazi. Hitchcock saw a great challenge in having the entire story take place in a lifeboat and pulled it off with his usual flourish, earning another best director nomination. In 1989, the film "Dead Calm" replicated the technique. From a story by John Steinbeck. Bankhead shines.
1944 96m/B Tallulah Bankhead, John Hodiak, William Bendix, Canada Lee, Walter Slezak, Hume Cronyn, Henry Hull, Mary Anderson; *Dir:* Alfred Hitchcock. National Board of Review Awards '44: 10 Best Films of the Year; New York Film Critics Awards '44: Best Actress (Bankhead). **VHS, Beta, LV** $19.98 *FOX* ✯✯✯ ½

Lifeforce A beautiful female vampire from outer space drains Londoners and before long the city is filled with disintegrating zombies in this hi-tech thriller. Sex was never stranger.
1985 (R) 100m/C Steve Railsback, Peter Firth, Frank Finlay, Patrick Stewart, Michael Gothard, Nicholas Ball, Aubrey Morris, Nancy Paul, Mathilda May; *Dir:* Tobe Hooper. **VHS, Beta, LV** $9.99 *FHE, LIV, VES* ✯✯ ½

Lifeguard The lifeguard lives by the credo that work is for people who cannot surf. Aging, he questions whether he should give up the beach life and start selling Porsches. Made for television.
1975 (PG) 96m/C Sam Elliott, Anne Archer, Stephen Young, Parker Stevenson, Kathleen Quinlan; *Dir:* Daniel Petrie. **VHS, Beta** $14.95 *PAR* ✯✯

Lifepod A group of intergalactic travelers is forced to evacuate a luxury space liner when a mad computer sabotages the ship.
1980 94m/C Joe Penny, Jordan Michaels, Kristine DeBell. **VHS, Beta** $49.95 *UHV* ✯

Life's a Beach Surfing stunts are performed to a rock music soundtrack.
1989 60m/C **VHS, Beta** $19.95 *AHV*

Lifespan A young American scientist visiting Amsterdam discovers experiments involving a drug that halts aging.
1975 85m/C Klaus Kinski, Hiram Keller, Tina Aumont; *Dir:* Alexander Whitelaw. **VHS, Beta** $69.95 *VES* ✯

The Lifetaker A woman lures an unsuspecting young man into her home where she seduces him. Violence and sex ensue.
1989 97m/C Lea Dregorn, Peter Duncan, Terence Morgan; *Dir:* Michael Papas. **VHS** $79.98 *SVS* Woof!

The Lift Unsuspecting passengers meet an unfortunate fate when they take a ride in a demonic elevator in a highrise. In this film the last stop isn't ladies' lingerie, but death. In Dutch; dubbed into English.
1985 (R) 95m/C *NL* Huub Stapel, Willeke Van Ammelrooy; *Dir:* Dick Maas. **VHS, Beta, LV** $59.95 *MED* ✯✯

The Light Ahead Lovers Fishke and Hodel dream of escaping the poverty and prejudices of their shtetl for the possibilities of big city life in Odessa. They're aided in their quest by enlightened bookseller Mendele who turns the town's superstitions to their advantage. Based on the stories of social satirist Mendele Mokher Seforim. In Yiddish with English subtitles.
1939 94m/B David Opatoshu, Isadore Cashier, Helen Beverly; *Dir:* Edgar G. Ulmer. **VHS** $72.00 *NCJ* ✯✯ ½

Light of Day A rock-'n-roll semi-musical about a working class brother and sister who escape from their parents and aimless lives through their bar band. Script tends to fall flat although both Fox and Jett are believable. Title song written by Bruce Springsteen.
1987 (PG-13) 90m/C Michael J. Fox, Joan Jett, Gena Rowlands, Jason Miller, Michael McKean; *Dir:* Paul Schrader. **VHS, Beta, LV** $29.98 *LIV, VES* ✯✯

Light at the Edge of the World A band of pirates torments a lighthouse keeper near Cape Horn after he sees a shipwreck they caused.
1971 (PG) 126m/C Kirk Douglas, Yul Brynner, Samantha Eggar; *Dir:* Kevin Billington. **VHS, Beta** $69.95 *MED* ✯✯

Light in the Forest Disney adaptation of the classic Conrad Richter novel about a young man, kidnapped by Indians when he was young, who is forcibly returned to his white family.
1958 92m/C James MacArthur, Fess Parker, Carol Lynley, Wendell Corey, Joanne Dru, Jessica Tandy, Joseph Calleia, John McIntire; *Dir:* Herschel Daugherty. **VHS, Beta** $69.95 *DIS* ✯✯ ½

The Light in the Jungle The biography of Nobel Peace Prize winner Dr. Albert Schweitzer. He established a hospital in Africa and had to overcome many obstacles, including tribal superstitions and European bureaucracy to make it successful and bring health care to the area.
1991 (PG) 91m/C Malcolm McDowell, Susan Strasberg, Andrew Davis; *Dir:* Gray Hof-meyr. **VHS, LV** $89.98 *LIV* ✯✯ ½

The Light of Western Stars A proper Eastern woman goes West and falls in love with a drunken lout.
1940 67m/B Victor Jory, Jo Ann Sayers, Russell Hayden, Morris Ankrum; *Dir:* Lesley Selander. **VHS** $19.95 *NOS, DVT* ✯

Light Years Garish animated fantasy epic about an idyllic land suddenly beset by evil mutations and death rays. Re-edited and dubbed from the original French version created by Rene Laloux and written by Raphael Cruzel; script by Isaac Asimov.
1988 (PG) 83m/C *FR* *Dir:* Harvey Weinstein, Harvey Weinstein; *Voices:* Glenn Close, Jennifer Grey, Christopher Plummer, Penn Jillette, John Shea, David Johansen, Bridget Fonda, Paul Schaffer. **VHS, Beta, LV** $29.95 *VMK* ✯ ½

Lightblast A San Francisco policeman tries to stop a deadly explosive-wielding mad scientist from blowing the city to kingdom come.

1985 89m/C Erik Estrada, Michael Pritchard; *Dir:* Enzo G. Castellari. VHS, Beta $69.98 *LIV, LTG* ⫝̸

The Lighthorseman A compelling World War I drama following several battle-hardened men of the Australian Right-Horse mounted infantry stationed in the Middle Eastern desert and the new recruit who joins their ranks and craves acceptance. Superbly filmed, particularly the final battle scene which pits the Aussies against the Turks for control of the wells at the Beersheba. Fine performance by Andrews in this epic which contains the essential elements of a good war movie—horses, guns, and more horses. Originally filmed at 140 minutes.
1987 (PG-13) 110m/C Jon Blake, Peter Phelps, Tony Bonner, Bill Kerr, Nick Wateres, John Walton, Tim McKenzie, Sigrid Thornton, Anthony Andrews; *Dir:* Simon Wincer. VHS, Beta $79.95 *WAR* ⫝̸⫝̸⫝̸

Lightnin' Carson Rides Again A tough frontier lawman tracks down his payroll-carrying nephew who's been accused of murder and thievery.
1938 58m/B Colonel Tom McCoy, Joan Barclay, Ted Adams, Forrest Taylor; *Dir:* Sam Newfield. VHS, Beta, 8mm $19.95 *NOS, VYY* ⫝̸½

Lightning Bill Bill champions good in battling rustlers.
1935 46m/B Buffalo Bill Jr. VHS, Beta, LV *WGE* ⫝̸

Lightning Bill Crandall Gunman heads south for the quiet life. Unfortunately, Arizona proves the scene of a fierce battle between various factions of cattlemen. The gunman aids the good guys, and tries to win the daughter's heart.
1937 60m/B Bob Steele, Loie January, Charles King, Frank LaRue, Ernie Adams, Earl Dwire, Dave O'Brien; *Dir:* Sam Newfield. VHS $19.95 *NOS, DVT* ⫝̸½

Lightning Kung Fu Three brothers senselessly battle entire armies with their feet.
197? 94m/C John Li, Hung Kam-Bo, Cheung Ying. VHS, Beta $54.95 *MAV* ⫝̸

Lightning Over Water An odd, touching documentary co-arranged by Ray and Wenders, all about Ray's last months dying of cancer in his New York loft, his editing of his final film "We Can't Go Home Again," his film corpus and career in general.
1980 91m/C Nicholas Ray, Wim Wenders, Ronee Blakley, Tom Farrel, Gerry Bamman; *Dir:* Wim Wenders. VHS, Beta, LV $59.95 *PAV, IME*

Lightning Range The Good Guy beats the bad guys, gets the girl, feeds his horse, gets to shoot a little and then takes a vacation.
1933 50m/B Buddy Roosevelt. VHS, Beta $19.95 *NOS, WGE* ⫝̸

Lightning Strikes West A U.S. Marshal trails an escaped convict, eventually catches him, and brings him in to finish paying his debt to society.
1940 57m/B Ken Maynard. VHS, Beta $19.95 *VCN, UHV* ⫝̸

Lightning: The White Stallion An old gambler and his two young friends enter a horse race in order to win their beloved white stallion back from thieves.
1986 (PG) 93m/C Mickey Rooney, Susan George, Isabel Lorca; *Dir:* William A. Levey. VHS, Beta $19.98 *MED* ⫝̸½

Lightning Warrior Western suspense about pioneer life and the unraveling of a baffling mystery. Twelve chapters, 13 minutes each.
1931 156m/B George Brent, Frankie Darro, Rin Tin Tin Jr; *Dir:* Armand Schaefer, Ben Kline. VHS, Beta $27.95 *VCN* ⫝̸½

Lights, Camera, Action, Love A romance novel on tape, in which a young actress finds the path to stardom littered with pain, pornography and seediness.
1985 90m/C VHS, Beta $14.95 *PSM* ⫝̸

Lights! Camera! Murder! When a twelve-year-old boy witnesses a filmed murder, he becomes the next target.
1989 89m/C John Barrett; *Dir:* Frans Nel. VHS $79.95 *ATL* ⫝̸⫝̸

Lights of Old Santa Fe A cowboy rescues a beautiful rodeo owner from bankruptcy. The original, unedited version of the film.
1947 78m/B Roy Rogers, Dale Evans, George "Gabby" Hayes, Bob Nolan, The Pioneers. VHS, Beta $19.95 *NOS, CPB, DVT* ⫝̸½

The Lightship On a stationary lightship off the Carolina coast, the crew rescues three men from a disabled boat, only to find they are murderous criminals. Duvall as a flamboyant homosexual psychopath is memorable, but the tale is pretentious, overdone, and hackneyed. Based on Siegfried Lenz's story.
1986 (PG-13) 87m/C Robert Duvall, Klaus Maria Brandauer, Tom Bower, William Forsythe, Arliss Howard; *Dir:* Jerzy Skolimowski. VHS, Beta $79.95 *FOX* ⫝̸⫝̸

Like Father, Like Son First and worst of a barrage of interchangeable switcheroo movies that came out in '87-'88. Moore is in top form, but the plot is contrived.
1987 (PG-13) 101m/C Dudley Moore, Kirk Cameron, Catherine Hicks, Margaret Colin, Sean Astin; *Dir:* Rod Daniel. VHS, Beta, LV $19.95 *COL* ⫝̸

Li'l Abner Al Capp's famed comic strip comes somewhat to life in this low-budget comedy featuring all of the Dogpatch favorites.
1940 78m/C Granville Owen, Martha Driscoll, Buster Keaton; *Dir:* Albert Rogell. VHS, Beta $24.95 *NOS, MRV, DVT* ⫝̸½

Li'l Abner High color Dogpatch drama adapted from the Broadway play (with most of the original cast) based on the Al Capp comic strip. When Abner's berg is considered as a site for atomic bomb testing, the natives have to come up with a reason why they should be allowed to exist. A Jean De Paul-Johnny Mercer musical; choreography by Michael Kidd and Dee Dee Wood. Oscar nominated for the musical direction by Joseph Lilley and Nelson Riddle.
1959 114m/C Peter Palmer, Leslie Parrish, Stubby Kaye, Julie Newmar, Howard St. John, Stella Stevens, Billie Hayes, Joe E. Marks; *Dir:* Melvin Frank. VHS, Beta $14.95 *PAR, FCT* ⫝̸⫝̸½

Lili Delightful musical romance about a 16-year-old orphan who joins a traveling carnival and falls in love with a crippled, embittered puppeteer. Heartwarming and charming, if occasionally cloying. Leslie Caron sings the film's song hit, "Hi-Lili, Hi-Lo." Nominated for six Oscars: Best Actress (Caron), Best Color Art Direction and Set Design, Best Screenplay (Helen Deutsch) and Color Cinematography, plus its winning score.
1953 81m/C Leslie Caron, Jean-Pierre Aumont, Mel Ferrer, Kurt Kasznar, Zsa Zsa Gabor; *Dir:* Charles Walters. Academy Awards '53: Best Musical Score; British Academy Awards '53: Best Actress (Caron); National Board of Review Awards '53: 10 Best Films of the Year. VHS, Beta $19.98 *MGM* ⫝̸⫝̸⫝̸

Lilies of the Field Five East German nuns enlist the aid of a free-spirited U.S. Army veteran. They persuade him to build a chapel for them and teach them English. Poitier is excellent as the itinerant laborer, holding the saccharine to an acceptable level, bringing honesty and strength to his role. Also-nominated Skala had been struggling to make ends meet in a variety of day jobs until this opportunity. Poitier was the first African-American man to win an Oscar, and the first African-American person nominated since Hattie MacDaniel in 1939. Also Oscar nominated for Cinematography. Followed by "Christmas Lilies of the Field," (1979).
1963 94m/B Sidney Poitier, Lilia Skala, Lisa Mann, Isa Crino, Stanley Adams; *Dir:* Ralph Nelson. Academy Awards '63: Best Actor (Poitier); Berlin Film Festival '63: Best Actor (Poitier); National Board of Review Awards '63: 10 Best Films of the Year. VHS, Beta $29.98 *MGM, FOX, BTV* ⫝̸⫝̸⫝̸

Liliom Boyer goes to heaven and is put on trial to see if he is deserving of his wings. Lang's first film after leaving Nazi Germany is filled with wonderful ethereal imagery, surprising coming from the man responsible for such grim visions as "Metropolis."
1935 85m/B *FR* Charles Boyer, Madeleine Ozeray, Florelle, Roland Toutain; *Dir:* Fritz Lang. VHS, Beta $16.95 *SNC, DVT* ⫝̸⫝̸⫝̸

Lilith Therapist-in-training Beatty falls in love with beautiful mental patient Seberg and approaches madness himself. A look at the doctor-patient relationship among the mentally ill and at the nature of madness and love. Doesn't always satisfy, but intrigues. Rossen's swan song.
1964 114m/B Warren Beatty, Jean Seberg, Peter Fonda, Gene Hackman, Kim Hunter; *Dir:* Robert Rossen. New York Film Festival '64: 10 Best Films of the Year. VHS, Beta $59.95 *COL, MLB* ⫝̸⫝̸⫝̸

Lily was Here After her fiancee is senselessly murdered, a young woman named Lily is forced to turn to a life of crime in order to survive. A series of petty thefts evolves into a huge crime wave and Lily soon finds herself the object of a massive manhunt. Lily must make the ultimate choice between freedom and motherhood in this shocking thriller. Soundtrack by Dave Stewart features hit instrumental theme "Lily was Here" by saxophonist Candy Dulfur.
1989 (R) 110m/C Marion Van Thijn, Thom Hoffman, Adrian Brine, Dennis Rudge; *W/Dir:* Ben Verbong. VHS $79.95 *SVT* ⫝̸½

Lily of Killarney A musical-comedy-romance about a British lord who arranges to pay his debts via horse-races, arranged marriages and inheritances. Originally released in Great Britain as "The Bride of the Lake."
1934 82m/B *GB* Gina Malo, John Garrick, Stanley Holloway; *Dir:* Maurice Elvey. VHS, Beta, 8mm $29.95 *VYY* ⫝̸⫝̸

Lily in Love An aging stage star disguises himself as a suave Italian to star in his playwright wife's new play, and woos her to test her fidelity. Charming, warm, and sophisticated. Loosely based on Molnar's "The Guardsman."

1985 (PG-13) 100m/C *HU* Maggie Smith, Christopher Plummer, Elke Sommer, Adolph Green; *Dir:* Karoly Makk. **VHS, Beta, LV $79.98** *LIV, VES* 🐾🐾½

Lily Tomlin Lily portrays her hilarious characters in this award winning special and gets a little help from her friends Alda and Pryor.
198? 60m/C Lily Tomlin, Alan Alda, Richard Pryor. **VHS, Beta $39.95** *LHV* 🐾🐾½

Limelight A nearly washed-up music hall comedian is stimulated by a young ballerina to a final hour of glory. A subtle if self-indulgent portrait of Chaplin's own life, featuring an historic pairing of Chaplin and Keaton.
1952 120m/B Charlie Chaplin, Claire Bloom, Buster Keaton, Nigel Bruce; *Dir:* Charlie Chaplin. National Board of Review Awards '52: 10 Best Films of the Year. **VHS, Beta $19.98** *FOX* 🐾🐾🐾

Limit Up An ambitious Chicago Trade Exchange employee makes a deal with the devil to corner the market in soybeans. Turgid attempt at supernatural comedy, featuring Charles as God. Catch Sally Kellerman in a cameo.
1989 (PG-13) 88m/C Nancy Allen, Dean Stockwell, Brad Hall, Danitra Vance, Ray Charles, Luana Anders; *Dir:* Richard Martini. **VHS, Beta, LV $89.95** *VIR* 🐾🐾

The Limping Man Bridges returns to post-World War II London to renew a wartime romance. On the way, he gets caught up in solving a murder. Unexceptional of-its-era thriller.
1953 76m/B *GB* Lloyd Bridges, Moira Lister, Leslie Phillips, Helene Cordet, Alan Wheatley; *Dir:* Charles De Latour. **VHS, Beta $16.95** *SNC, RXM* 🐾🐾

Linchamiento (Lynching) Cattle ranchers and farmers battle it out again. Dubbed into Spanish.
1968 90m/C *IT* Glenn Saxon, Gordon Mitchell, Christine Schmidt, King MacQueen. **VHS, Beta $59.95** *JCI* 🐾

Lincoln County Incident A satiric western starring secondary-school children.
1980 47m/C **VHS, Beta $14.98** *VID* 🐾½

The Lindbergh Kidnapping Case The famous Lindbergh baby kidnapping in 1932 and the trial and execution of Bruno Hauptmann, convincingly portrayed by Hopkins. DeYoung as Lindbergh is blah, but the script is quite good. Made for TV.
1976 150m/C Anthony Hopkins, Joseph Cotten, Cliff DeYoung, Walter Pidgeon, Dean Jagger, Martin Balsam, Laurence Luckinbill, Tony Roberts; *Dir:* Buzz Kulik. **VHS, Beta $69.95** *COL* 🐾🐾🐾

The Line Not-so-hot anti-war drama about a sit-down strike at a military installation by Vietnam veteran prison inmates. Leans heavily on recycled footage from director Siegel's own ''Parade'' (1971).
1980 (R) 94m/C Russ Thacker, David Doyle, Erik Estrada; *Dir:* Robert Siegel. **VHS, Beta** *USA* 🐾½

The Linguini Incident An inept escape artist, a pathological liar, a lingerie designer, a deaf restaurant hostess who throws out one-liners in sign language, and two sinister, yet chic, restaurant owners get together in this comedy about magic and adventure.
1992 (R) 99m/C Rosanna Arquette, David Bowie, Eszter Balint, Marlee Matlin, Buck Henry, Andre Gregory; *W/Dir:* Richard Shepard. **VHS, LV $89.95** *IME* 🐾🐾½

Link A primatologist and his nubile assistant find their experiment has gone—you guessed it—awry, and their hairy charges are running—yep, that's right—amok. Run for your life!
1986 (R) 103m/C *GB* Elizabeth Shue, Terence Stamp; *Dir:* Richard Franklin. **VHS, Beta $19.99** *HBO* Woof!

Lion of Africa A down-to-earth woman doctor and an abrasive diamond dealer share a truck ride across Kenya. Filmed on location in East Africa. Fine lead performances in otherwise nothing-special made-for-cable adventure.
1987 115m/C Brian Dennehy, Brooke Adams, Joseph Shiloal; *Dir:* Kevin Connor. **VHS, Beta $79.95** *LHV, WAR* 🐾🐾

Lion of the Desert Bedouin horse militias face-off against Mussolini's armored terror in this epic historical drama. Omar Muktar (Quinn as the ''Desert Lion'') and his Libyan guerrilla patriots kept the Italian troops of Mussolini (Steiger) at bay for 20 years. Outstanding performances enhanced by the desert backdrop. A British-Libyan co-production.
1981 (PG) 162m/C *GB* Anthony Quinn, Oliver Reed, Irene Papas, Rod Steiger, Raf Vallone, John Gielgud; *Dir:* Moustapha Akkad. **VHS, Beta $19.95** *USA* 🐾🐾🐾

The Lion Has Wings The story of how Britain's Air Defense was set up to meet the challenge of Hitler's Luftwaffe during their ''finest hour.'' Dated, now-quaint but stirring wartime period piece. ''Docudrama'' style was original at the time.
1940 75m/B Merle Oberon, Ralph Richardson, Flora Robson, June Duprez; *Dir:* Michael Powell, Brian Desmond Hurst. **VHS, Beta, LV** *SUE, RXM* 🐾🐾½

Lion Man Raised by wild animals, the son of King Solomon returns to his father's kingdom and roars his way to the throne.
1979 91m/C Steve Arkin. **VHS, Beta $19.99** *BFV* 🐾

The Lion of Thebes A muscleman unhesitatingly jumps into the thick of things when Helen of Troy is kidnapped. A superior sword and sandal entry.
1964 ?m/C *IT* Mark Forest, Yvonne Furneaux. **VHS, Beta $16.95** *SNC* 🐾🐾🐾

The Lion of Venice A kid and an old Venetian gondolier see a stone lion move and uncover a legend of secret treasure.
1982 73m/C **VHS, Beta $29.95** *GEM* 🐾

The Lion in Winter Medieval monarch Henry II and his wife, Eleanor of Aquitane, match wits over the succession to the English throne and much else in this fast-paced film version of James Goldman's play. The family, including three grown sons, are visiting royalty are united for the Christmas holidays fraught with tension, rapidly shifting allegiances, and layers of psychological manipulation. Superb dialogue and perfectly realized characterizations. O'Toole and Hepburn are triumphant. Shot on location, this literate costume drama surprised the experts with its box-office success. Oscar nominations for Best Picture, actor O'Toole, director Harvey, and Costume Design.
1968 (PG) 134m/C Peter O'Toole, Katharine Hepburn, Jane Merrow, Nigel Terry, Timothy Dalton, Anthony Hopkins, John Castle; *Dir:* Anthony Harvey. Academy Awards '68: Best Actress (Hepburn), Best Adapted Screenplay, Best Musical Score; Golden Globe Awards '69: Best Film—Drama; National Board of Review Awards '68: 10 Best Films of the Year; New

York Film Critics Awards '68: Best Film. **VHS, Beta, LV, Special order formats $14.98** *SUE, BTV* 🐾🐾🐾🐾

The Lion, the Witch & the Wardrobe Four children stumble through an old wardrobe closet in an ancient country house and into the fantasy land of Narnia. Adapted from ''The Chronicles of Narnia'' by C. S. Lewis.
1979 174m/C *Dir:* Bill Melendez. **VHS, Beta $29.98** *LIV, VES, PME* 🐾🐾🐾🐾

Lionheart A romantic portrayal of the famous English King Richard the Lionheart's early years. Meant for kids, but no Ninja turtles herein—and this is just as silly, and slow to boot.
1987 (PG) 105m/C Eric Stoltz, Talia Shire, Nicola Cowper, Dexter Fletcher, Nicholas Clay, Deborah Barrymore, Gabriel Byrne; *Dir:* Franklin J. Schaffner. **VHS, Beta, LV $19.98** *WAR* 🐾½

Lionheart Van Damme deserts the foreign legion and hits the streets when he learns his brother has been hassled. Many fights ensue, until you fall asleep. Also called ''A.W.O.L.'' and ''Wrong Bet.''
1990 (R) 105m/C Jean-Claude Van Damme, Harrison Page, Deborah Rennard, Lisa Pelikan, Brian Thompson; *Dir:* Sheldon Lettich. **VHS, Beta, LV $19.98** *MCA, PIA* 🐾½

Lion's Den A night club performer and a detective head west to fight crime in this film.
1936 59m/B Tim McCoy, Dave O'Brien. **VHS, Beta $19.95** *NOS, DVT* 🐾½

The Lion's Share A gang of bank robbers have their loot stolen and go after the guy who ripped them off.
1979 105m/C *SP* Julio de Grazia, Luisina Brando, Fernanda Mistral, Ulises Dumont, Julio Chavez. **VHS, Beta $59.95** *MHV* 🐾

Lip Service Satirical comedy-drama about the TV news industry. An ambitious young newscaster befriends a veteran reporter. He then manipulates his way to replace him on the reporter's morning program. Dooley as the veteran and Dunne as the upstart are fun to watch in this well-done cable rip-off of ''Broadcast News.''
1988 67m/C Griffin Dunne, Paul Dooley; *Dir:* W.H. Macy. **VHS, Beta $79.95** *PSM* 🐾🐾🐾

Lipstick Fashion model Margaux seeks revenge on the man who brutally attacked and raped her, after he preys on her kid sister (real-life sis Mariel, in her debut). Exquisitely exploitative excuse for entertainment.
1976 (R) 90m/C Margaux Hemingway, Anne Bancroft, Perry King, Chris Sarandon, Mariel Hemingway; *Dir:* Lamont Johnson. **VHS, Beta, LV $59.95** *PAR* 🐾

Liquid Dreams In this fast-paced, futuristic thriller, Daly goes undercover as an erotic dancer in a glitzy strip joint to try and solve her sister's murder. She finds that the owner and clientele deal not only in sexual thrills, but also in a strange brain-sucking ritual that provides the ultimate rush. Also available in an unrated version.
1992 (R) 92m/C Richard Steinmetz, Candice Daly, Barry Dennen, Juan Fernandez, Tracy Walter; *Dir:* Mark Manos. **VHS, Beta $89.95** *ACA* 🐾🐾

Liquid Sky An androgynous bisexual model living in Manhattan is the primary attraction for a UFO, which lands atop her penthouse in search of the chemical nourishment that her sexual encounters provide. Low-budget, highly creative film may not be for everyone, but the audience for which it

was made will love it. Look for Carlisle also playing a gay male.
1983 (R) 112m/C Anne Carlisle, Paula Sheppard, Bob Brady; *Dir:* Slava Tsukerman. **VHS, Beta, LV $59.95** *MED* 🎬🎬🎬

Lisa A young girl develops a crush on the new guy in town and arranges a meeting with him in which she pretends to be her mother. Little does she realize he's a psychotic serial killer. Lock your doors! Don't let anyone in if they have this video!
1990 (PG-13) 95m/C Staci Keanan, Cheryl Ladd, D.W. Moffett, Tanya Fenmore, Jeffrey Tambor, Julie Cobb; *Dir:* Gary Sherman. **VHS $19.98** *FOX* 🎬

Lisa and the Devil An unfortunate outing for Savalas and Sommer, about devil worship. Poor writing and directing leave little room for redemption. Just like on the telly, Telly's sucking on a sucker. Alternately titled "The House of Exorcism."
1975 (R) 93m/C *IT* Telly Savalas, Elke Sommer, Sylva Koscina, Robert Alda; *Dir:* Mario Bava. **VHS $19.95** *VDM, MPI* Woof!

Lisbon First film directed by Milland, shot in Portugal, details the adventures of a sea captain entangled in international espionage and crime while attempting to rescue damsel O'Hara's husband from communist doings. A familiar plot told with less-than-average panache.
1956 90m/C Ray Milland, Claude Rains, Maureen O'Hara, Francis Lederer, Percy Marmont; *Dir:* Ray Milland. **VHS, Beta $59.98** *NO* 🎬

The List of Adrian Messenger A crafty murderer resorts to a variety of disguises to eliminate potential heirs to a family fortune. Solid Huston-directed thriller with a twist: you won't recognize any of the name stars.
1963 98m/B Kirk Douglas, George C. Scott, Robert Mitchum, Dana Wynter, Burt Lancaster, Frank Sinatra; *Dir:* John Huston. **VHS, Beta $59.95** *MCA* 🎬🎬🎬

Listen to Me A small-town college debate team heads for the big time when they go to a national debate tournament. The usual mutual-distaste-turns-to-romance thing. Cheesy-as-all-get-out climactic abortion debate, in front of supposed real-life Supreme Court justices—yeah, right. And Kirk Cameron: get an accent, will ya?
1989 (PG-13) 109m/C Kirk Cameron, Jami Gertz, Roy Scheider, Amanda Peterson, Tim Quill, Christopher Atkins; *Dir:* Douglas Day Stewart. **VHS, Beta, LV $19.95** *COL* 🎬½

Listen to Your Heart Office romance in the '80s! Sounds like a mediocre, forgettable made-for-TV comedy-drama—which is exactly what it is. If Jackson was the brainy one on "Charlie's Angels," why wasn't she smart enough to avoid this one?
1983 90m/C Tim Matheson, Kate Jackson; *Dir:* Don Taylor. **VHS, Beta $59.98** *FOX* 🎬🎬

Lisztomania Ken Russell's excessive vision of what it must have been like to be classical composer/musician Franz Liszt, who is depicted as the first pop star. Rock opera in the tradition of "Tommy" with none of the sense or music.
1975 (R) 106m/C *GB* Roger Daltrey, Sara Kestelman, Paul Nicholas, Fiona Lewis, Ringo Starr; *Dir:* Ken Russell. **VHS, Beta, LV $19.95** *WAR* 🎬½

The Little American World War I drama sees a young woman sending her lover off to war. Silent.

1917 80m/B Mary Pickford, Jack Holt, Raymond Hatton, Walter Long. **VHS, Beta $32.95** *GPV, DNB* 🎬🎬½

Little Annie Rooney A policeman's tomboy daughter spends her time mothering her father and brother while getting into mischief with street punks. Minor melodrama. Silent.
1925 60m/B Mary Pickford, William Haines, Walter James, Gordon Griffith, Vola Vale; *Dir:* William Beaudine. **VHS, Beta, LV $19.95** *NOS, CCB, IME* 🎬🎬

Little Ballerina A young dancer struggles against misfortune and jealousy to succeed in the world of ballet, under the auspices of Fonteyn.
1947 62m/B Margot Fonteyn, Anthony Newley, Martita Hunt, Yvonne Marsh. **VHS, Beta $19.95** *VYY* 🎬🎬

Little Big Horn Low-budget depiction of Custer et al. at Little Big Horn actually has its gripping moments; solid acting all around helps.
1951 88m/B Marie Windsor, John Ireland, Lloyd Bridges, Reed Hadley, Hugh O'Brian, Jim Davis; *Dir:* Charles Marquis Warren. **VHS, Beta $12.95** *AVD, MRV, WGE* 🎬🎬½

Little Big Man Based on Thomas Berger's picaresque novel, this is the story of 121-year-old Jack Crabb and his quixotic life as gunslinger, charlatan, Indian, ally to George Custer, and the only white survivor of Little Big Horn. Told mainly through flashbacks. Hoffman provides a classic portrayal of Crabb, as fact and myth are jumbled and reshaped. George received an Oscar nomination.
1970 (PG) 135m/C Dustin Hoffman, Faye Dunaway, Chief Dan George, Richard Mulligan, Martin Balsam, Jeff Corey, Aimee Eccles; *Dir:* Arthur Penn. **VHS, Beta, LV $19.98** *FOX* 🎬🎬🎬½

Little Big Master A crafty knight-errant who heads a gang of beggars finds it difficult adjusting to military life.
197? 90m/C Huang I Lung, Man Li Peng. **VHS, Beta $39.95** *UNI* 🎬

Little Boy Lost The true story of the disappearance of a young boy in Australia.
1978 (G) 92m/C John Hargreaves, Tony Barry, Lorna Lesley; *Dir:* Alan Spires. **VHS, Beta $59.98** *MAG* 🎬🎬

Little Caesar A small-time hood rises to become a gangland czar, but his downfall is as rapid as his rise. Still thrilling. The role of Rico made Robinson a star and typecast him as a crook for all time.
1930 80m/B Edward G. Robinson, Glenda Farrell, Sidney Blackmer, Douglas Fairbanks Jr.; *Dir:* Mervyn LeRoy. **VHS, Beta $19.98** *FOX, FCT* 🎬🎬

Little Church Around the Corner Silent small town melodrama about a preacher who falls in love with a mine owner's daughter, but finds he isn't dad's favorite fella when he confronts him about poor mining conditions. When the mine caves in, the preacher man's caught between a rock and hard place when his sweetie's family needs protection from an angry mob.
1923 70m/B Kenneth Harlan, Hobart Bosworth, Walter Long, Pauline Starke, Alec B. Francis, Margaret Seddon, George Cooper; *Dir:* William A. Seiter. **VHS, Beta $19.95** *GPV* 🎬🎬½

The Little Colonel After the Civil War, an embittered Southern patriarch turns his back on his family, until his dimple-cheeked granddaughter softens his heart. Hokey and heartwarming. Shirley's first teaming with Bill

"Bojangles" Robinson features the famous dance scene. Adapted by William Conselman from the Annie Fellows Johnston best-seller.
1935 80m/B Shirley Temple, Lionel Barrymore, Evelyn Venable, John Lodge, Hattie McDaniel, Bill Robinson, Sidney Blackmer; *Dir:* David Butler. **VHS, Beta $19.98** *FOX* 🎬🎬🎬

Little Darlings Distasteful premise has summer campers Kristy and Tatum in a race to lose their virginity. Kristy is better (at acting, that is); but who cares? And just who is meant to be the market for this movie, anyway?
1980 (R) 95m/C Tatum O'Neal, Kristy McNichol, Matt Dillon, Armand Assante, Margaret Blye; *Dir:* Ronald F. Maxwell. **VHS, Beta, LV $14.95** *PAR* 🎬½

Little Dorrit, Film 1: Nobody's Fault The mammoth version of the Dickens tome, about a father and daughter trapped interminably in the dreaded Marshalsea debtors' prison, and the good samaritan who works to free them. Told in two parts (on four tapes), "Nobody's Fault," and "Little Dorrit's Story."
1988 369m/C *GB* Alec Guinness, Derek Jacobi, Cyril Cusack, Sarah Pickering, Joan Greenwood, Max Wall, Amelda Brown, Daniel Chatto, Miriam Margolyes, Bill Fraser, Roshan Seth, Michael Elphick; *Dir:* Christine Edzard. Los Angeles Film Critics Association Awards '88: Best Film, Best Supporting Actor (Guinness). **VHS, Beta, LV $89.95** *WAR* 🎬🎬🎬

Little Dorrit, Film 2: Little Dorrit's Story Second half of the monumental adaptation of Dicken's most popular novel during his lifetime.
1988 369m/C *GB* Alec Guinness, Derek Jacobi, Cyril Cusack, Sarah Pickering, Joan Greenwood, Max Wall, Amelda Brown, Daniel Chatto, Miriam Margolyes, Bill Fraser, Roshan Seth, Michael Elphick; *Dir:* Christine Edzard. Los Angeles Film Critics Association Awards '88: Best Film, Best Supporting Actor (Guinness). **VHS, Beta, LV $89.95** *WAR* 🎬🎬🎬

Little Dragons A grandfather and two young karate students rescue a family held captive by a backwoods gang.
1980 90m/C Ann Sothern, Joe Spinell, Charles Lane, Chris Petersen, Pat Petersen, Sally Boyden, Rick Lenz, Sharon Weber, Tony Bill; *Dir:* Curtis Hanson. **VHS, Beta $19.95** *AHV* 🎬

The Little Drummer Boy The classic story of "The Little Drummer Boy" is brought to life with colorful animation in this delightful film.
1968 30m/C *Dir:* Takeya Nakamura; *Voices:* Teddy Eccles, Jose Ferrer, Paul Frees; *Nar:* Greer Garson. **VHS, Beta $14.95** *FHE* 🎬

The Little Drummer Girl An Israeli counterintelligence agent recruits an actress sympathetic to the Palestinian cause to trap a fanatical terrorist leader. Solid performances from Keaton as the actress and Kinski as the Israli counterintelligence office sustain interest through a puzzling, sometimes boring and frustrating, cinematic maze of espionage. Keaton is at or near her very best. Based on the bestselling novel by John Le Carre.
1984 (R) 130m/C Diane Keaton, Klaus Kinski, Yorgo Voyagis, Sami Frey, Michael Cristofer, Anna Massey, Thorley Walters; *Dir:* George Roy Hill. **VHS, Beta, LV $19.98** *WAR* 🎬🎬🎬

The Little Foxes A vicious southern woman will destroy everyone around her to satisfy her desire for wealth and power. Filled with corrupt characters who commit numerous revolting deeds. The vicious matriarch is a part made to fit for Davis, and she makes

the most of it. Script by Lillian Hellman from her own play. Oscar nominations: Best Picture, Best Actress (Davis), Best Supporting Actress (Wright and Collinge), Best Director, Best Screenplay, Black & White Interior Decoration, and Best Score.
1941 116m/B Bette Davis, Herbert Marshall, Dan Duryea, Teresa Wright, Charles Dingle, Richard Carlson; *Dir:* William Wyler. VHS, Beta, LV $14.98 SUE ✓✓✓½

The Little Girl Who Lives Down the Lane Engrossing, offbeat thriller about a strange 13-year-old girl whose father is never home, and who hides something - we won't say what - in her basement. Very young Foster is excellent, as are her supporters. Then there's a child molester who knows what she's hiding. . .
1976 (PG) 90m/C CA Jodie Foster, Martin Sheen, Alexis Smith, Scott Jacoby; *Dir:* Nicolas Gessner. VHS, Beta, LV $29.98 LIV, VES ✓✓½

Little Girl...Big Tease A 16-year-old society girl is kidnapped and exploited by a group which includes her teacher.
1975 (R) 86m/C Jody Ray, Rebecca Brooke; *Dir:* Roberto Mitrotti. VHS, Beta PGN Woof!

Little Gloria...Happy at Last Miniseries based on the best-selling book about the custody battle over child heiress Gloria Vanderbilt, and her tumultuous youth. Lansberry leads a fine cast as the poor little rich girl's aunt Gertrude, the matriarchal family power broker.
1984 180m/C Bette Davis, Angela Lansbury, Christopher Plummer, Maureen Stapleton, Martin Balsam, Barnard Hughes, John Hillerman; *Dir:* Waris Hussein. VHS, Beta $79.95 PSM, PAR ✓✓½

Little Heroes An impoverished little girl gets through hard times with the aid of her loyal dog (the Hound can relate). A low-budget family tearjerker that nonetheless works, it claims to be based on a true story.
1991 78m/C Raeanin Simpson, Katherine Willis, Keith Christensen, Hoover the Dog; *W/Dir:* Craig Clyde. VHS $79.95 HMD ✓✓½

Little Heroes of Shaolin Temple The "Prince of Blood" attacks the Shaolin temple, but the Master's Ninjas await him.
1972 90m/C Cheng Tain Syh, Hw-Luen, Li Shuen Hwa, Tang Chen Da; *Hosted:* Sho Kosugi. VHS, Beta $29.95 TWE ✓

Little Johnny Jones George M. Cohan's classic musical is featured in a 1980 revival. Songs include "Yankee Doodle Dandy" and "Give My Regards to Broadway."
1980 93m/C VHS, Beta $59.95 WAR, OM

Little Ladies of the Night A former pimp tries to save Purl and other teenagers from the world of prostitution. Exploitation posing as "significant drama". Also titled "Diamond Alley."
1977 96m/C Linda Purl, David Soul, Clifton Davis, Carolyn Jones, Louis Gossett Jr.; *Dir:* Marvin J. Chomsky. VHS, Beta $49.95 PSM ✓

Little Laura & Big John The true-life exploits of the small-time Ashley Gang in the Florida everglades around the turn of the century. Fabian's comeback vehicle, for what it's worth.
1973 82m/C Fabian Forte, Karen Black; *Dir:* Luke Moberly. VHS, Beta $14.98 VID ✓✓

Little Lips A young French girl is awakened sexually.
197? 77m/C Katya Berger. VHS, Beta $49.95 TWE ✓

Little Lord Fauntleroy The vintage Hollywood version of the Frances Hodgson Burnett story of a fatherless American boy who discovers he's heir to a British dukedom. Also available in computer-colorized version. Well cast, charming, remade for TV in 1980. Smith is loveable as the noble tyke's crusty old guardian.
1936 102m/B Freddie Bartholomew, C. Aubrey Smith, Mickey Rooney, Dolores Costello, Jessie Ralph, Guy Kibbee; *Dir:* John Cromwell. VHS, Beta $19.95 NOS, MRV, AMV ✓✓✓

Little Lord Fauntleroy A poor boy in New York suddenly finds himself the heir to his grandfather's estate in England. Lavish remake of th 1936 classic, adapted from Frances Hodgson Burnett's novel. Well-deserved Emmy winner for photography; Guiness is his usual old-pro self. Made for TV.
1980 98m/C Rick Schroder, Alec Guinness, Victoria Tennant, Eric Porter, Colin Blakely, Connie Booth, Rachel Kempson; *Dir:* Jack Gold. VHS, Beta $14.95 FHE, STE ✓✓½

Little Lulu The popular comic book heroine gets into mischief with her pals, Tubby, Iggie, Wilbur and others. Includes "Little Angel" and "Operation Babysitting." Available in English and Spanish versions.
194? 48m/C VHS, Beta $29.95 MED ✓✓½

Little Man Tate A seven-year-old genius is the prize in a tug of war between his mother, who wants him to lead a normal life, and a domineering school director who loves him for his intellect. An acclaimed directorial debut for Foster, with overtones of her own extraordinary life as a child prodigy.
1991 (PG) 99m/C Jodie Foster, Dianne Wiest, Harry Connick Jr., Adam Hann-Byrd; *Dir:* Jodie Foster. VHS $92.98 ORI, CCB ✓✓✓

The Little Match Girl A musical version of the Hans Christian Andersen classic about a girl whose dying grandmother tells her of the magic in the matches she sells.
1984 54m/C Monica McSwain, Nancy Duncan, Matt McKim, Don Hays; *Dir:* Mark Hoeger. VHS, Beta $14.98 SUE ✓✓

Little Match Girl The Hans Christian Andersen Yuletide classic about an orphan selling magical matches this time set in the 1920s, with the littlest "Cosby" kid in the title role. Ain't she cute?
1987 90m/C GB John Rhys-Davies, Rue McClanahan, Roger Daltrey, Twiggy, Natalie Morse; *Dir:* Michael Lindsay-Hogg. VHS, Beta $19.95 ACA ✓✓

Little Men A modern version of the classic juvenile story by Louisa May Alcott is too cute for words.
1940 86m/B Jack Oakie, Jimmy Lydon, Kay Francis, George Bancroft; *Dir:* Norman Z. McLeod. VHS, Beta $19.95 NOS, MRV, QNE ✓½

The Little Mermaid An animated version of Hans Christian Andersen's tale about a little mermaid who rescues a prince whose boat has capsized. She immediately falls in love and wishes that she could become a human girl. Not to be confused with the 1989 Disney version.
1978 (G) 71m/C *Dir:* Tim Reid. VHS, Beta $19.95 STE, GEM ✓✓½

The Little Mermaid From "Faerie Tale Theatre" comes an adaptation of the Hans Christian Andersen tale of a little mermaid who makes a big sacrifice to win the prince she loves.

1984 60m/C Karen Black, Brian Dennehy, Helen Mirren, Pam Dawber, Treat Williams; *Dir:* Robert Iscove. VHS, Beta $14.98 KUI, FOX, FCT ✓✓½

The Little Mermaid A headstrong teenage mermaid falls in love with a human prince and longs to be human too. She makes a pact with the evil Sea Witch to trade her voice for a pair of legs; based on the famous Hans Christian Andersen tale. Charming family musical, which harks back to the classic days of Disney animation. Songs include "Under the Sea" and "Kiss the Girl."
1989 (G) 82m/C *Dir:* John Musker, Ron Clements; *Voices:* Jodi Benson, Christopher Daniel Barnes, Pat Carroll, Rene Auberjonois, Samuel E. Wright, Buddy Hackett, Jason Marin, Edie McClurg, Kenneth Mars, Nancy Cartwright. Academy Awards '89: Best Song ("Under the Sea"), Best Musical Score. VHS, Beta, LV, 8mm $14.98 DIS, BVV, APD ✓✓✓✓

Little Minister An adaptation of the James Barrie novel about a prissy Scottish pastor who falls in love with a free-spirited gypsy...he thinks she is, in fact, the local earl's daughter, played to perfection by the young Hepburn.
1934 110m/B Katharine Hepburn, John Beal, Alan Hale Jr., Donald Crisp; *Dir:* Richard Wallace. VHS, Beta $19.98 RKO ✓✓✓

Little Miss Broadway Orphan Temple brings the residents of a theatrical boarding house together in hopes of getting them into show business. She and Durante give worthwhile performances. Songs include "Be Optimistic," "Hop Skip and Jump," and "If All the World Were Paper." Also available in computer-colorized version.
1938 70m/B Shirley Temple, George Murphy, Jimmy Durante, Phyllis Brooks, Edna May Oliver, George Barbier, Donald Meek, Jane Darwell; *Dir:* Irving Cummings. VHS, Beta $19.98 FOX ✓✓

Little Miss Innocence A recording executive tries to survive the amorous advances of the two attractive female hitchhikers that he drove home.
1973 79m/C John Alderman, Sandra Dempsey, Judy Medford; *Dir:* Chris Warfield. VHS, Beta $24.95 AHV ✓

Little Miss Marker Heartwarming story starring Temple, who is the IOU for a gambling debt and steals her way into everyone's heart. Remade three times as "Sorrowful Jones," as "40 Pounds of Trouble," and in 1980 with the original title.
1934 88m/B Adolphe Menjou, Shirley Temple, Dorothy Dell, Charles Bickford, Lynne Overman; *Dir:* Alexander Hall. VHS $49.99 MCA ✓✓

Little Miss Marker Mediocre remake of the often retold story of a bookie who accepts a little girl as a security marker for a $10 bet. Disappointing performance from Curtis adds to an already dull film.
1980 (PG) 103m/C Walter Matthau, Julie Andrews, Tony Curtis, Bob Newhart, Lee Grant, Sara Stimson, Brian Dennehy; *Dir:* Walter Berstein. VHS, Beta $49.95 MCA ✓½

Little Monsters A young boy (Savage) discovers a monster (Mandel) under his bed and eventually befriends him. The pair embark on adventures that land them in trouble. Hardworking, talented cast hurdles the weak script, but can't save the film.
1989 (PG) 86m/C Fred Savage, Howie Mandel, Margaret Whitton, Ben Savage, Daniel Stern, Rick Ducommun; *Dir:* Richard Alan Greenberg. VHS, Beta $19.99 MGM ✓✓

Little Moon & Jud McGraw A wronged Indian woman and a framed cowpoke take their revenge on a small, corrupt town, razing it overnight. Already-weak plot is disabled by too many flashbacks.
1978 92m/C James Caan, Sammy Davis Jr., Stefanie Powers, Aldo Ray; *Dir:* Bernard Girard. **VHS, Beta** $59.95 *PSM* ⚁ 1/2

Little Murders Black comedy set in New York City. A woman convinces a passive photographer to marry her. Gardenia gives an excellent performance as the woman's father. A shadow of crime, depression, and strife seem to hangs over the funny parts of this film; more often depressing than anything else. Adapted by Jules Feiffer from his own play.
1971 (PG) 108m/C Elliott Gould, Marcia Rodd, Vincent Gardenia, Elizabeth Wilson, Jon Korkes, Donald Sutherland, Alan Arkin, Lou Jacobi; *Dir:* Alan Arkin. **VHS, Beta** $59.98 *FOX* ⚂⚂⚂

A Little Night Music Adapted from the Broadway play, and based loosely on Bergman's ''Smiles of a Summer Night,'' this film centers around four interwoven, contemporary love stories. Taylor's pathetic rendition of ''Send in the Clowns'' should be banned. Filmed on location in Austria.
1977 (PG) 110m/C Elizabeth Taylor, Diana Rigg, Hermione Gingold, Len Cariou, Lesley-Anne Down; *Dir:* Harold Prince. Academy Awards '77: Best Adapted Score. **VHS, Beta, LV** $9.98 *SUE* ⚁ 1/2

Little Nikita A California boy (Phoenix) discovers that his parents are actually Soviet spies planted as American citizens for eventual call to duty. Poitier's performance as the FBI agent tracking the spies is about the only spark in this somewhat incoherent thriller.
1988 (PG) 98m/C River Phoenix, Sidney Poitier, Richard Bradford, Richard Lynch, Caroline Kava, Lucy Deakins; *Dir:* Richard Benjamin. **VHS, Beta, LV** $14.95 *COL* ⚁⚁

Little Noises Glover stars as an artist who seeks not only fame and fortune, but also the love of his best friend (O'Neal). He finally creates a piece that his agent loves, but problems arise.
1991 91m/C Crispin Glover, Tatum O'Neal, Rik Mayall, Tate Donovan, John C. McGinley; *W/Dir:* Jane Spencer. **VHS, Beta** $89.95 *PSM* ⚁⚁

Little Orphan Annie The unjustly forgotten first sound adaptation of the comic strip. Max Steiner's fun score adds a lot, as does the good cast.
1932 60m/B May Robson, Buster Phelps, Mitzie Green, Edgar Kennedy; *Dir:* John S. Robertson. **VHS, Beta** $29.98 *CCB* ⚁⚁ 1/2

Little Prince Disappointing adaptation of the classic children's story by Antoine de Saint-Exupery, about a little boy from asteroid B-612. Lousy Lerner and Loewe score underscores a general lack of magic or spontaneity.
1974 (G) 88m/C *GB* Richard Kiley, Bob Fosse, Steven Warner, Gene Wilder; *Dir:* Stanley Donen. **VHS, Beta, LV** $14.95 *KUI, PAR* ⚁⚁

The Little Princess Based on the Frances Hodgson Burnett children's classic; perhaps the best of the moppet's films. Shirley is a young schoolgirl in Victorian London sent to a harsh boarding school when her Army officer father is posted abroad. When her father is declared missing, the penniless girl must work as a servant at the school to pay her keep, all the while haunting the hospitals, never believing her father has died. A classic tearjerker.

1939 91m/B Shirley Temple, Richard Greene, Anita Louise, Ian Hunter, Cesar Romero, Arthur Treacher, Sybil Jason, Miles Mander, Marcia Mae Jones, E.E. Clive; *Dir:* Walter Lang. **VHS, Beta, LV** $9.95 *NEG, MRV, NOS* ⚂⚂⚂ 1/2

A Little Princess A three-cassette adaptation of Frances Hodgson Burnett's book. A kind rich girl is reduced to poverty in Victorian England. Originally aired on PBS as part of the WonderWorks Family Movie series.
1987 180m/C Amelia Shankley, Nigel Havers, Maureen Lipman; *Dir:* Carol Wiseman. **VHS** $79.95 *WNE, PME* ⚂⚂⚂ 1/2

Little Rascals A collection of early silent escapades of the Little Rascals circa 1924.
1924 60m/B Farina Hoskins, Joe Cobb. **VHS, Beta** $14.98 *VDM*

Little Rascals, Book 1 Three ''Our Gang'' shorts: ''Railroadin''' (1929), in which the gang takes off on a runaway train; ''A Lad and a Lamp'' (1932), wherein they find an Aladdin's lamp; and ''Beginner's Luck'' (1935), in which Spanky wins a dress for a young actress.
193? 59m/B Farina Hoskins, Joe Cobb, Matthew ''Stymie'' Beard, George ''Spanky'' McFarland, Carl ''Alfalfa'' Switzer, Mary Ann Jackson. **VHS, Beta** $29.98 *CCB*

Little Rascals, Book 2 A second package of ''Our Gang'' two-reelers: ''Bear Shooters'' (1930), in which the Gang goes hunting but runs into some bootleggers; ''Forgotten Babies'' (1933), has Spanky baby-sitting for the gang's brothers and sisters; and ''Teacher's Beau'' (1935), where the gang cooks up a scheme to chase their teacher's fiance away.
193? 56m/B Matthew ''Stymie'' Beard, Wheezer Hutchins, George ''Spanky'' McFarland, Carl ''Alfalfa'' Switzer. **VHS, Beta** $29.98 *CCB*

Little Rascals, Book 3 Three more ''Our Gang'' installments are packaged on this tape: ''Dogs Is Dogs''(1931), ''Anniversary Trouble'' (1935) and ''Three Men in a Tub'' (1938).
1938 54m/B Wheezer Hutchins, Dorothy de Borba, Matthew ''Stymie'' Beard, George ''Spanky'' McFarland, Carl ''Alfalfa'' Switzer, Darla Hood. **VHS, Beta** $29.98 *CCB*

Little Rascals, Book 4 Another package of Our Gang favorites, including ''Helping Grandma'' (1931), ''Little Papa'' (1935) and ''Bear Facts'' (1938).
193? 52m/B Matthew ''Stymie'' Beard, George ''Spanky'' McFarland, Carl ''Alfalfa'' Switzer, Darla Hood. **VHS, Beta** $29.98 *CCB*

Little Rascals, Book 5 The comic adventures of the Little Rascals continue in this package of three original shorts: ''Readin' and Writin','' (1932), ''Sprucin' Up'' (1935) and ''Reunion in Rhythm'' (1937).
193? 49m/B Breezy Brisbane, Matthew ''Stymie'' Beard, George ''Spanky'' McFarland, Carl ''Alfalfa'' Switzer, Darla Hood. **VHS, Beta** $29.98 *CCB*

Little Rascals, Book 6 Three more classic ''Our Gang'' comedy shorts are packaged on this tape: ''Free Eats'' (1932), ''Arbor Day'' (1936) and ''Mail and Female'' (1937).
193? 48m/B Matthew ''Stymie'' Beard, Breezy Brisbane, George ''Spanky'' McFarland, Carl ''Alfalfa'' Switzer, Darla Hood, Dorothy DeBorba, Billy Gilbert. **VHS, Beta** $29.98 *CCB*

Little Rascals, Book 7 The ''Our Gang'' kids serve up another portion of comedy in these three shorts: ''Hook and Ladder'' (1932), ''The Lucky Corner'' (1936) and ''Feed 'Em and Weep'' (1938).
193? 47m/B George ''Spanky'' McFarland, Breezy Brisbane, Dickie Moore, Matthew ''Stymie'' Beard, Scotty Beckett, Carl ''Alfalfa'' Switzer, Darla Hood. **VHS, Beta** $29.98 *CCB*

Little Rascals, Book 8 More fun and nuttiness with ''Our Gang'' in three original short comedies, ''Bouncing Babies'' (1929), ''Two Too Young'' (1936), and ''The Awful Tooth'' (1938).
193? 42m/B Mary Ann Jackson, Darla Hood, George ''Spanky'' McFarland, Carl ''Alfalfa'' Switzer. **VHS, Beta** $29.98 *CCB*

Little Rascals, Book 9 The Little Rascals scamper into more mischief in these three original shorts: ''Little Daddy'' (1931), ''Spooky Hooky'' (1936), and ''Hide and Shriek'' (1938).
193? 48m/B Darla Hood, George ''Spanky'' McFarland, Carl ''Alfalfa'' Switzer, Matthew ''Stymie'' Beard, Mary Ann Jackson, Jackie Cooper. **VHS, Beta** $29.98 *CCB*

Little Rascals, Book 10 The tenth compilation of original ''Our Gang'' comedies includes ''The First Seven Years'' (1929), ''Bored of Education'' (1936), and ''Rushin' Ballet'' (1937).
193? 41m/B Jackie Cooper, Mary Ann Jackson, George ''Spanky'' McFarland, Carl ''Alfalfa'' Switzer, Darla Hood. Academy Awards '36: Best Comedy Short (''Bored of Education''). **VHS, Beta** $29.98 *CCB*

Little Rascals, Book 11 The Little Rascals go dramatic in these three shorts, all with a ''putting-on-a-show'' theme: ''Pay As You Exit'' (1936), ''Three Smart Boys'' (1937) and ''Our Gang Follies of 1938.''
1938 42m/B George ''Spanky'' McFarland, Carl ''Alfalfa'' Switzer, Darla Hood. **VHS, Beta** $29.98 *CCB*

Little Rascals, Book 12 The Little Rascals explore a haunted house, take a train ride and play football in these three original shorts: ''Moan and Groan, Inc.'' (1929), ''Choo-Choo!'' (1932) and ''The Pigskin Palooka'' (1937).
1937 53m/B Jackie Cooper, Farina Hoskins, Mary Ann Jackson, George ''Spanky'' McFarland, Carl ''Alfalfa'' Switzer, Darla Hood. **VHS, Beta** $29.98 *CCB*

Little Rascals, Book 13 The Our Gang kids return in three more original shorts: ''Shivering Shakespeare'' (1930), ''The First Round-Up'' (1934) and ''Fishy Tales'' (1937).
1937 48m/B Jackie Cooper, George ''Spanky'' McFarland, Edgar Kennedy, Carl ''Alfalfa'' Switzer. **VHS, Beta** $29.98 *CCB*

Little Rascals, Book 14 Package of shorts leads off with Our Gang's first talkie, ''Small Talk'' (1929), a three-reeler. The other two entries are ''Little Sinner'' (1935) and ''Hearts Are Thumps'' (1937).
193? 52m/B Darla Hood, Mary Ann Jackson, George ''Spanky'' McFarland, Carl ''Alfalfa'' Switzer. **VHS, Beta** $29.98 *CCB*

Little Rascals, Book 15 Those rascally imps return once again in three more original shorts: ''When the Wind Blows'' (1930), ''For Pete's Sake'' (1934) and ''Glove Taps'' (1937).
193? 50m/B Edgar Kennedy, Tommy ''Butch'' Bond, George ''Spanky'' McFarland, Wheezer Hutchins. **VHS, Beta** $29.98 *CCB*

Little Rascals, Book 16 Three more charming episodes featuring the original brat pack. The urchins of yesteryear star in "Love Business," "Hi Neighbor," and "Came the Brawn."
193? 30m/B Matthew "Stymie" Beard, Carl "Alfalfa" Switzer, George "Spanky" McFarland, Darla Hood; *Dir:* Fred Wolf. **VHS, Beta $19.95** *REP, CCB*

Little Rascals, Book 17 Three classic shorts, "Boxing Gloves," "The Kid from Borneo," and "Roamin' Holiday."
1931 46m/B Jean Darling, George "Spanky" McFarland, Darla Hood. **VHS, Beta $29.98** *CCB*

Little Rascals, Book 18 Here are two early "Our Gang" comedies from the silent era: "Saturday's Lesson" (1929) and "Wiggle Your Ears" (1929).
1929 41m/B Farina Hoskins, Wheezer Hutchins, Mary Ann Jackson, Joe Cobb. **VHS, Beta $29.98** *CCB*

Little Rascals: Choo Choo/Fishy Tales Spanky, Darla, Buckwheat, Stymie, and Alfalfa in two uncut installments of the groundbreaking children's comedy.
1932 30m/B George "Spanky" McFarland, Darla Hood, Carl "Alfalfa" Switzer, Matthew "Stymie" Beard. **VHS $9.98** *REP, CCB*

Little Rascals Christmas Special Spanky and the Little Rascals attempt to raise enough money to buy a winter coat for Spanky's mom and learn the true meaning of Christmas along the way.
1979 60m/C Darla Hood, Matthew "Stymie" Beard; *Dir:* Fred Wolf, Charles Swenson. **VHS, Beta $29.95** *FHE*

Little Rascals Collector's Edition, Vol. 1-6 A collection put together by Spotlight containing six individual two-reelers. Twelve episodes total featuring the rambunctious bunch of child stars.
19?? 180m/B George "Spanky" McFarland, Darla Hood, Carl "Alfalfa" Switzer, Matthew "Stymie" Beard. **VHS, Beta $89.95** *REP*

Little Rascals Comedy Classics, Vol. 1 Vivacious collection of six of the gang's original episodes. This presentation contains "Railroadin'," "A Lad and a Lamp," "Beginner's Luck," "Bear Shooters," "Forgotten Babies," and "Teacher's Beau."
19?? 112m/B George "Spanky" McFarland, Darla Hood, Carl "Alfalfa" Switzer, Matthew "Stymie" Beard. **VHS, Beta $19.95** *REP, CCB*

Little Rascals Comedy Classics, Vol. 2 Second volume of the mischievous gang's uncut, original shorts. The six episodes on this installment are "Pay as You Exit," "Three Smart Boys," "Our Gang Follies of 1938," "The First Round-Up," "Shivering Shakespeare," and "Fishy Tales."
19?? 90m/B George "Spanky" McFarland, Darla Hood, Carl "Alfalfa" Switzer, Matthew "Stymie" Beard. **VHS $19.95** *REP, CCB*

Little Rascals: Fish Hooky/ Spooky Hooky Two uncut classics featuring those mischievous little pranksters.
1933 30m/B Carl "Alfalfa" Switzer, Darla Hood, George "Spanky" McFarland, Matthew "Stymie" Beard. **VHS $9.98** *REP, CCB*

Little Rascals: Honkey Donkey/ Sprucin' Up The gang's comic capers continue in "Honky Donkey" and "Sprucin' Up."

1934 30m/B George "Spanky" McFarland, Darla Hood, Carl "Alfalfa" Switzer, Matthew "Stymie" Beard. **VHS $9.95** *REP, CCB*

Little Rascals: Little Sinner/Two Too Young The gang stars in "Little Sinner" and "Two Too Young."
1935 30m/B George "Spanky" McFarland, Darla Hood, Carl "Alfalfa" Switzer, Matthew "Stymie" Beard. **VHS $9.98** *REP, CCB*

Little Rascals: Mush and Milk/ Three Men in a Tub Two uncensored features starring those lovable urchins.
1933 30m/B George "Spanky" McFarland, Carl "Alfalfa" Switzer, Darla Hood, Matthew "Stymie" Beard. **VHS $9.98** *REP, CCB*

Little Rascals on Parade A collection of six vintage "Our Gang" shorts from the '30s: "Free Eats," "Arbor Day," "Mail and Female," "Hook and Ladder," "The Lucky Corner" and "Feed 'Em and Weep."
1937 60m/B George "Spanky" McFarland, Carl "Alfalfa" Switzer, Darla Hood, Matthew "Stymie" Beard. **VHS $19.95** *REP, CCB*

Little Rascals: Pups is Pups/ Three Smart Boys Two of the most hilarious features from the Little Rascals.
1930 30m/B George "Spanky" McFarland, Carl "Alfalfa" Switzer, Darla Hood, Matthew "Stymie" Beard. **VHS $9.98** *REP, CCB*

Little Rascals: Readin' and Writin'/Mail and Female Two uncut features starring those delightful rapscallions.
1931 30m/B George "Spanky" McFarland, Carl "Alfalfa" Switzer, Darla Hood, Matthew "Stymie" Beard. **VHS $9.98** *REP, CCB*

Little Rascals: Reunion in Rhythm/Mike Fright Join the gang in the outrageous episodes "Reunion in Rhythm" and "Mike Fright."
1937 30m/B George "Spanky" McFarland, Darla Hood, Carl "Alfalfa" Switzer, Matthew "Stymie" Beard. **VHS $9.98** *REP, CCB*

Little Rascals: Spanky/Feed 'Em and Weep Two hilarious features with the antics of those crazy kids.
1932 30m/B George "Spanky" McFarland, Carl "Alfalfa" Switzer, Darla Hood, Matthew "Stymie" Beard. **VHS $9.98** *REP, CCB*

Little Rascals: The Pinch Singer/Framing Youth Mischievous youngsters run amok in "The Pinch Singer" and "Framing Youth."
1936 30m/B George "Spanky" McFarland, Darla Hood, Carl "Alfalfa" Switzer, Matthew "Stymie" Beard. **VHS $9.98** *REP, CCB*

Little Rascals Two Reelers, Vol. 1 Wacky youngsters two reeler containing "Lazy Days," and "Spooky Hooky."
1929 30m/B George "Spanky" McFarland, Carl "Alfalfa" Switzer, Darla Hood, Matthew "Stymie" Beard. **VHS $14.95** *REP*

Little Rascals Two Reelers, Vol. 2 Comic two-reeler from those mischievous youngsters containing "Bedtime Worries," and "Glove Taps."
1933 30m/B George "Spanky" McFarland, Carl "Alfalfa" Switzer, Darla Hood, Matthew "Stymie" Beard. **VHS $14.95** *REP*

Little Rascals Two Reelers, Vol. 3 Those troublesome youngsters are back in a two-reeler containing "Second Childhood" and "Hide and Shriek."

1936 30m/B George "Spanky" McFarland, Darla Hood, Carl "Alfalfa" Switzer, Matthew "Stymie" Beard. **VHS $14.95** *REP*

Little Rascals Two Reelers, Vol. 4 Two reeler containing "When the Wind Blows," and "Hearts are Thumps."
1930 30m/B George "Spanky" McFarland, Darla Hood, Carl "Alfalfa" Switzer, Matthew "Stymie" Beard. **VHS $14.95** *REP*

Little Rascals Two Reelers, Vol. 5 These two episodes of the gang feature "Hi Neighbor," and "Came the Brawn."
1934 30m/B George "Spanky" McFarland, Darla Hood, Carl "Alfalfa" Switzer, Matthew "Stymie" Beard. **VHS $14.95** *REP*

Little Rascals Two Reelers, Vol. 6 The gang frolics in this two-reeler containing "Love Business," and "Pigskin Palooka."
1936 30m/B George "Spanky" McFarland, Darla Hood, Carl "Alfalfa" Switzer, Matthew "Stymie" Beard. **VHS $14.95** *REP*

Little Rascals, Volume 1 Here is a collection of three uncut "Little Rascals" short subjects. In "Boxing Gloves" (1929), Chubby and Joe decide two settle their differences over who should have Jean Darling's love by staging a winner-take-all boxing match. In "The Kid from Borneo" (1933), Spanky thinks he's met a crazy relative when he confuses an untamed human from Borneo for his uncle from the circus. In "Roamin' Holiday" (1936), Alfalfa and Spanky lead the gang on a journey fit for a hobo.
1929 50m/B Darla Hood, George "Spanky" McFarland, Carl "Alfalfa" Switzer, Matthew "Stymie" Beard. **VHS $14.95** *REP*

Little Rascals, Volume 2 Another uncut compilation of three classic episodes featuring the goofy gang. In "The First Seven Years" (1930), Speck and Jackie battle for the attention of a beautiful Mary Ann. In the award-winning "Bored of Education" (1936), Spanky and Alfalfa devise a plan to play hooky from school by developing very acute tooth problems. In "Rushin' Ballet" (1937), Alfalfa and Spanky establish the Secret Revengers Club and liberate Buckwheat and Porky from the bonds of tyranny.
1930 50m/B George "Spanky" McFarland, Carl "Alfalfa" Switzer. Academy Awards '36: Best Comedy Short ("Bored of Education"). **VHS $9.95** *REP, CCB*

Little Red Riding Hood From "Faerie Tale Theatre" comes the retelling of the story about a girl (Mary Steenburgen) off to give her grandmother a picnic basket, only to get stopped by a wicked wolf McDowell. Not particularly faithful, but fun and scary.
1983 60m/C Mary Steenburgen, Malcolm McDowell; *Dir:* Graeme Clifford. **VHS, Beta, LV $14.98** *FOX, FCT* ⅊⅊⅊

Little Red Schoolhouse A hard-nosed schoolteacher hunts for a truant lad and both of them land in jail.
1936 64m/C Frank Coghlan Jr., Dickie Moore, Ann Doran. **VHS, Beta $19.95** *NOS, KRT, FCT* ⅊⅊

Little River Band Selections from this Australian rock'n'roll band's six LP's, such as "It's a Long Way There," "Mistress of Mine," and "Just Say That You Love Me," are featured.
1982 75m/C VHS, Beta, LV $19.99 *HBO, PIA* ⅊⅊

A Little Romance An American girl living in Paris falls in love with a French boy; eventually they run away, to seal their love with a kiss beneath a bridge. Olivier gives a

wonderful, if not hammy, performance as the old pickpocket who encourages her. Gentle, agile comedy based on the novel by Patrick Cauvin.
1979 (PG) 110m/C Laurence Olivier, Diane Lane, Thelonious Bernard, Sally Kellerman, Broderick Crawford; *Dir:* George Roy Hill. Academy Awards '79: Best Original Score. **VHS, Beta, LV $19.98** *WAR, PIA* ♫♫♫

A Little Sex Capshaw, in her screen debut, finds herself the wife of womanizer Matheson. Harmless, but pointless, with a TV-style plot.
1982 (R) 94m/C Tim Matheson, Kate Capshaw, Edward Herrmann, Wallace Shawn, John Glover; *Dir:* Bruce Paltrow. **VHS, Beta $59.95** *MCA* ♫♫

Little Shop of Horrors The landmark cheapie classic, which Roger Corman reputedly filmed in three days, about a nebbish working in a city florist shop who unknowingly cultivates an intelligent plant that demands human meat for sustenance. Notable for then-unknown Nicholson's appearance as a masochistic dental patient. Hilarious, unpretentious farce—if you liked this one, check out Corman's "Bucket of Blood" for more of the same. Inspired a musical of the same name, remade as a movie again in 1986. Available colorized.
1960 70m/B Jackie Joseph, Jonathan Haze, Mel Welles, Jack Nicholson, Dick Miller; *Dir:* Roger Corman. **VHS, Beta, LV $9.95** *NEG, MRV, NOS* ♫♫♫½

Little Shop of Horrors During a solar eclipse, Seymour buys an unusual plant and takes it back to the flower shop where he works. The plant, Audrey 2, becomes a town attraction as it grows at an unusual rate, but Seymour learns that he must feed Audrey fresh human blood to keep her growing. Soon, Audrey is giving the orders ("Feed me") and timid Seymour must find "deserving" victims. Steve Martin's cameo as the masochistic dentist is alone worth the price. Song, dance, gore, and more prevail in this outrageous musical/comedy. Four Top Levi Stubbs is the commending voice of Audrey 2. Based on the off-broadway play, which was based on Roger Corman's 1960 horror spoof.
1986 (PG-13) 94m/C Rick Moranis, Ellen Greene, Vincent Gardenia, Steve Martin, James Belushi, Christopher Guest, Bill Murray, John Candy; *Dir:* Frank Oz. **VHS, Beta, LV $19.98** *WAR* ♫♫♫

Little Sister Prankster Silverman, on a dare, dresses up as a girl and joins a sorority. Problems arise when he falls in love with his "big sister" (Milano) in the sorority. What will happen when she finds out the truth?
1992 (PG-13) 94m/C Jonathan Silverman, Alyssa Milano. **VHS $89.98** *LIV* ♫♫

Little Sweetheart A reworking of "The Bad Seed," with a nine-year-old girl engaging in murder, burglary and blackmail, ruining the adults around her. Based on "The Naughty Girls" by Arthur Wise.
1990 (R) 93m/C John Hurt, Karen Yahng, Barbara Bosson, John McMartin, Cassie Barasch; *Dir:* Anthony Simmons. **VHS, Beta, LV, 8mm $79.98** *ORI, SUE* ♫½

The Little Theatre of Jean Renoir A farewell by director Jean Renoir featuring three short films. In the first, Renoir's humanist beliefs are apparent in "The Last Christmas Dinner," a Hans Christian Andersen-inspired story. Next, "The Electric Floor Waxer" is a comic opera. The third piece is called "A Tribute to Tolerance."

Slight but important late statement by a great director.
1971 100m/C Jean Renoir, Fernand Sardou, Jean Carmet, Francoise Arnoul, Jeanne Moreau; *Dir:* Jean Renoir; *W/Dir:* Jean Renoir. **VHS, Beta, LV $49.95** *INT, TEX, IME* ♫♫♫

The Little Thief Touted as Francois Truffaut's final legacy, this trite mini-drama is actually based on a story he co-wrote with Claude de Givray about a post-World War II adolescent girl who reacts to the world around her by stealing and petty crime. Screenplay by Annie Miller. In French, with subtitles. Truffaut's hand is evidently absent, but this film is a testament to his abruptly and sadly truncated career. Director Miller was Truffaut's longtime assistant.
1989 (PG-13) 108m/C *FR* Charlotte Gainsbourg, Simon de la Brosse, Didier Bezace, Raoul Billerey, Nathalie Cardone; *Dir:* Claude Miller. **VHS, Beta, LV $89.99** *HBO* ♫♫♫

Little Tough Guys The Dead End Kids come to the rescue of Halop, a young tough guy gone bad to avenge his father's unjust imprisonment. One of the early movies for the kids who were later transformed into the East Side Kids and eventually into the Bowery Boys.
1938 84m/B Helen Parrish, Billy Halop, Leo Gorcey, Marjorie Main, Gabriel Dell, Huntz Hall; *Dir:* Harold Young. **VHS, Beta $19.95** *NOS, MRV, VYY* ♫½

Little Treasure A stripper heads for Mexico to search for her long-lost father, but ends up looking for treasure with an American guy. Disappointing effort of director Sharp.
1985 (R) 95m/C Burt Lancaster, Margot Kidder, Ted Danson, Joseph Hacker, Malena Doria; *W/Dir:* Alan Sharp. **VHS, Beta, LV $79.95** *COL* ♫½

The Little Valentino One day in the aimless life of a young punk who gets by on his wits and petty theft. Jeles' directorial debut. In Hungarian with English subtitles.
1979 102m/B *HU Dir:* Andras Jeles. **VHS $59.95** *FCT* ♫

Little Vegas Looks like easy street for a young man who inherits a bundle from his girlfriend, but he's forced to wake up and smell the cappucino when her family and the other residents of the tiny desert berg think there's a gigolo in the woodpile. Much strife with siblings as he wages a battle for the bucks with the son while wooing the woman's daughter, with a measure of mob inflicted plot twists. Mildly amusing, cut from vein of you can't get rich quick and get away with it.
1990 90m/C Michael Nouri, Jerry Stiller, John Sayles, Anthony Denison, Catherine O'Hara, Bruce McGill, Anne Francis, Bob Goldthwait; *W/Dir:* Perry Lang. **VHS, LV $89.95** *COL* ♫♫½

Little Vera Extremely well-done Soviet film chronicles the life of a young working-class woman who loves rock music and who has been profoundly affected by Western Civilization. Post-glasnost Soviet production gives Westerners a glimpse into the Russian way of life. A box-office bonanza back home. In Russian with English subtitles.
1988 130m/C *RU* Natalia Negoda, Andrei Sokolov, Yuri Nazarov, Ludmila Zaisova, Alexander Niegreva; *Dir:* Vassili Pitchul. **VHS, Beta, LV $34.95** *NVH, FCT, LUM* ♫♫♫

Little White Lies Jillian and Matheson meet and fall in love during an exotic vacation. But both are traveling under assumed identities—she's a policewoman tracking a

jewel thief and he's an incognito doctor. When they reunite stateside they try to stick to their ruses in humdrum comedic fashion. Made for network TV.
1989 88m/C Ann Jillian, Tim Matheson; *Dir:* Anson Williams. **VHS $79.95** *RHI* ♫♫

Little Women Louisa May Alcott's Civil War story of the four March sisters—Jo, Beth, Amy, and Meg—who share their loves, their joys, and their sorrows. Everything about this classic film is wonderful, from the lavish period costumes to the excellent script by Victor Heerman and Sarah Mason, and particularly the captivating performances by the cast. A must-see for fans of Alcott and Hepburn, and others will find it enjoyable. Remade in 1949 and again in 1978 for television.
1933 107m/B Katharine Hepburn, Joan Bennett, Paul Lukas, Edna May Oliver, Frances Dee, Spring Byington; *Dir:* George Cukor. Academy Awards '33: Best Adapted Screenplay; Venice Film Festival '34: Best Actress (Hepburn). **VHS, Beta, 8mm $24.95** *KUI, MGM* ♫♫♫♫

Little Women Stylized color remake of the George Cukor 1933 classic. Top-notch if too obvious cast portrays Louisa May Alcott's story of teenage girls growing up against the backdrop of the Civil War. Remade again, for television in 1978.
1949 121m/C June Allyson, Peter Lawford, Margaret O'Brien, Elizabeth Taylor, Janet Leigh, Mary Astor; *Dir:* Mervyn LeRoy. Academy Awards '49: Best Art Direction/Set Decoration (Color). **VHS $19.98** *MGM* ♫♫♫

Little Women The third screen version of Louisa May Alcott's classic story. Lackluster compared to the previous attempts, particularly the sterling 1933 film, but still worthwhile. During the Civil War, four sisters share their lives as they grow up and find romance. Garson's television debut. Followed by a television series.
1978 200m/C Meredith Baxter Birney, Susan Dey, Ann Dusenberry, Eve Plumb, Dorothy McGuire, Robert Young, Greer Garson, Cliff Potts, William Shatner; *Dir:* David Lowell Rich. **VHS** *NO* ♫♫

Little World of Don Camillo A French-made farce based on the beloved novels of Giovanni Guareschi. Earthy French priest Don Camillo clashes repeatedly with his friendly enemy, the communist mayor of a tiny Italian village. The hero talks directly to God, whose voice is provided by Orson Welles, also narrator of the English version. Charming, good-natured approach.
1951 106m/B *FR IT* Fernandel, Gino Cervi, Sylvia, Vera Talqui; *Dir:* Julien Duvivier; *Nar:* Orson Welles. **VHS, Beta, 8mm $29.95** *VYY, FCT* ♫♫♫

The Littlest Angel Musical about a shepherd boy who dies falling off a cliff and wants to become an angel. He learns a valuable lesson in the spirit of giving. Made for TV.
1969 77m/C Johnny Whitaker, Fred Gwynne, E.G. Marshall, Cab Calloway, Connie Stevens, Tony Randall. **VHS, Beta $14.98** *SUE, NEG, KAR* ♫♫

Littlest Horse Thieves Good Disney film about three children and their efforts to save some ponies who work in mines. The children take it upon themselves to see that the animals escape the abuse and neglect they are put through. Filmed on location in England with excellent photography. Also known as "Escape From the Dark."
1976 (G) 109m/C Alastair Sim, Peter Barkworth; *Dir:* Charles Jarrott. **VHS, Beta $69.95** *DIS* ♫♫½

Littlest Outlaw A Mexican peasant boy steals a beautiful stallion to save it from being destroyed. Together, they ride off on a series of adventures. Disney movie filmed on location in Mexico.
1954 73m/C Pedro Armendariz, Joseph Calleia, Andres Velasquez; ***Dir:*** Roberto Gavaldon. **VHS, Beta $69.95** *DIS* 🦴🦴½

Littlest Rebel Temple stars in this well-done piece set during the Civil War in the Old South. She befriends a Union officer while protecting her father at the same time. She even goes to Washington to talk with President Lincoln. Nice dance sequences by Temple and Robinson. Available in computer-colored version.
1935 70m/B Shirley Temple, John Boles, Jack Holt, Bill Robinson, Karen Morley, Willie Best; ***Dir:*** David Butler. **VHS, Beta $19.98** *FOX* 🦴🦴½

Littlest Warrior Zooshio, the littlest warrior, is forced to leave his beloved forest and experiences many adventures before he is reunited with his family. Animated.
1975 70m/C VHS, Beta $24.95 *FHE* 🦴🦴½

Live and Let Die Agent 007 is out to thwart the villainous Dr. Kananga, a black mastermind who plans to control western powers with voodoo and hard drugs. Moore's first appearance as Bond. Can't we have the real Bond back? Title song by Paul McCartney.
1973 (PG) 121m/C *GB* Roger Moore, Jane Seymour, Yaphet Kotto, Clifton James, Julius W. Harris, Geoffrey Holder, David Hedison, Gloria Hendry, Bernard Lee, Lois Maxwell, Madeleine Smith, Roy Stewart; ***Dir:*** Guy Hamilton. **VHS, Beta, LV $19.95** *MGM, FOX, TLF* 🦴🦴

Live a Little, Love a Little Itinerant photographer Elvis juggles two different jobs by running around a lot. Sexually more frank than earlier King vehicles.
1968 90m/C Elvis Presley, Michele Carey, Rudy Vallee, Don Porter, Dick Sargent, Sterling Holloway, Eddie Hodges; ***Dir:*** Norman Taurog. **VHS, Beta $19.95** *MGM* 🦴½

Live a Little, Steal a Lot Surprisingly well-done film of two Florida beach bums who engineer the heist of the Star of India. Nice chase scenes add to an already fine action film. Also called "Murph the Surf"; later retitled "You Can't Steal Love."
1975 (PG) 101m/C Robert Conrad, Don Stroud, Donna Mills, Robyn Millan, Burt Young; ***Dir:*** Marvin J. Chomsky. **VHS $19.98** *WAR, OM* 🦴🦴½

Live & Red Hot Top reggae stars perform live.
1987 60m/C Peter Metro, Yellowman, Conroy Smith, Tonto Metro, Little John, Macca P, Joe Lick Shot, Daddy Life, Danny Dread. **VHS, Beta** *KRV*

Live from Washington it's Dennis Miller Dennis Miller's live stand-up routine, filmed live in Washington, D.C. , with a guest cameo by Edwin Newman.
1988 60m/C Dennis Miller, Edwin Newman. **VHS, Beta $14.98** *LIV, VES* 🦴🦴½

The Live Wire Stunt-ridden action fare featuring Talmadge at his most daring. This adventure is complete with desert island, lost treasure, stowaways, evil-doers, and a race against the clock.
1934 57m/B Richard Talmadge, George Walsh, Charles French, Alberta Vaughn; ***Dir:*** Harry S. Webb. **VHS, Beta $24.95** *NOS, DVT, GPV* 🦴🦴

The Lives of a Bengal Lancer One of Hollywood's greatest rousing adventures. Three British Lancers, of varying experience, encounter a vicious revolution against colonial rule. Rivaled only by "Gunga Din" as the ultimate 1930s swashbuckler. Swell plot; lotsa action; great comraderie. Based on the novel by Major Francis Yeats-Brown. Received Oscar nominations for Best Picture, Best Director, Best Screenplay, Best Interior Decoration, Best Sound Recording, Best Film Editing and Best Assistant Director.
1935 110m/B Gary Cooper, Franchot Tone, Richard Cromwell, Guy Standing; ***Dir:*** Henry Hathaway. National Board of Review Awards '35: 10 Best Films of the Year. **VHS, Beta $29.95** *MCA* 🦴🦴🦴🦴

Livin' Large An African-American delivery boy gets the break of his life when a nearby newscaster is shot dead. Grabbing the microphone and continuing the story, he soon finds himself hired by an Atlanta news station as an anchorman, fulfilling a life-long dream. But problems arise when he finds himself losing touch with his friends, his old neighborhood and his roots. Comedy "deals" with the compelling issue of blacks finding success in a white world by trivializing the issue at every turn and resorting to racial stereotypes.
1991 (R) 96m/C Terrence "T.C." Carson, Lisa Arrindell, Blanche Baker, Nathaniel "Afrika" Hall, Julia Campbell; ***Dir:*** Michael Schultz. **VHS $92.99** *HBO* 🦴🦴

Livin' the Life A young Englishman entertains a series of troublesome Mitty-esque fantasies.
1984 83m/C Rupert Everett, Christina Raines, Catherine Rabett; ***Dir:*** Francis Megahy. **VHS, Beta $69.98** *LIV, VES* 🦴½

The Living Coffin Woman has alarm rigged on her coffin in case she's buried alive. And they called her paranoid. Loosely based on Poe's "Premature Burial."
1958 72m/C *MX* Gaston Santos, Maria Duval, Pedro d'Auillon; ***Dir:*** Fernando Mendez. **VHS $19.98** *SNC* 🦴½

The Living Daylights The 17th Bond film. After being used as a pawn in a fake Russian defector plot, our intrepid spy tracks down an international arms and opium smuggling ring. Fine debut by Dalton as 007 in a rousing, refreshing cosmopolitan shoot-em-up. Let's be frank: we were all getting a little fatigued of Roger Moore.
1987 (PG) 130m/C Timothy Dalton, Maryam D'Abo, Jeroen Krabbe, John Rhys-Davies, Robert Brown, Joe Don Baker, Desmond Llewelyn; ***Dir:*** John Glen. **VHS, Beta, LV $19.98** *FOX, TLF* 🦴🦴🦴

The Living Dead An English film about a mad, re-animating scientist.
1936 76m/B *GB* Gerald de Maurier, George Curzan. **VHS, Beta $24.95** *NOS, SNC, DVT* 🦴🦴

Living Desert The life cycle of an American desert is shown through the seasons in this trendsetting Disney documentary feature.
1953 (G) 69m/C ***Dir:*** James Algar. Academy Awards '53: Best Feature Documentary. **VHS, Beta $69.95** *DIS* 🦴🦴

Living to Die Vegas gumshoe is assigned to investigate a blackmailed official, and discovers a woman believed to be dead is alive and beautiful and living incognito. This mystifies him.
1991 (R) 84m/C Wings Hauser, Darcy Demoss, Asher Brauner, Arnold Vosloo, Jim Williams; ***Dir:*** Wings Hauser. **VHS $79.95** *PMH, HHE* 🦴½

Living Free Sequel to "Born Free," based on the nonfictional books by Joy Adamson. Recounts the travails of Elsa the lioness, who is now dying, with three young cubs that need care. Nice and pleasant, but could you pick up the pace?
1972 (G) 91m/C Susan Hampshire, Nigel Davenport; ***Dir:*** Jack Couffer. **VHS, Beta, LV $9.95** *COL, GKK* 🦴🦴½

The Living Head Archeologists discover the ancient sepulcher of the great Aztec warrior, Acatl. Ignoring a curse, they steal his severed head and incur the fury of Xitsliapoli. Dubbed in English.
1959 75m/B *MX* Mauricio Garces, Ana Louisa Peluffo, German Robles, Abel Salazar; ***Dir:*** Chano Urueto. **VHS, Beta $29.95** *SNC, VYY* 🦴½

Living on Tokyo Time Interesting but often dull, low budget independent comedy about a young Japanese woman who hitches up with a boorish Japanese-American man in order to stay in the United States. An Asian view of Asian-America, filmed on location in San Francisco.
1987 83m/C Minako Ohashi, Ken Nakagawa, Kate Connell; ***Dir:*** Steven Okazaki. **VHS, Beta $19.98** *CHA* 🦴🦴½

The Lizzies of Mack Sennett Three quick-paced, hilarious Sennett "Tin Lizzie" comedies: "Lizzies of the Field," "Leading Lizzie Astray" and "Love, Speed and Thrills." Plenty of signature slapstick from the master.
1915 51m/C Billy Bevan, Fatty Arbuckle, Mack Swain, Chester Conklin; ***Dir:*** Mack Sennett. **VHS, Beta $24.95** *VYY*

Loaded Guns An airline stewardess-cum-double agent must totally immobilize a top drug trafficking ring. Pathetic spy tale with plenty of unintended laughs.
1975 90m/C Ursula Andress, Woody Strode; ***Dir:*** Fernando Di Leo. **VHS, Beta $59.95** *IVE, MON* *Woof!*

Loaded Pistols Story set in the old West starring Gene Autry.
1948 80m/B Gene Autry; ***Dir:*** John English. **VHS, Beta $19.95** *NOS, VCN, DVT* 🦴

Loan Shark An ex-convict gets a job at a tire company, working undercover to expose a loan shark ring. Hackneyed plot moves quickly.
1952 79m/B George Raft, John Hoyt; ***Dir:*** Seymour Friedman. **VHS, Beta, LV** *WGE, BUR* 🦴🦴

Lobster for Breakfast An Italian gag-fest about a toilet salesman infiltrating his friend's confused marital existence just to make a sale. With English subtitles; also available in a dubbed version.
1982 (R) 96m/C *IT* Enrico Montesano, Claude Brasseur, Claudine Auger; ***Dir:*** Giorgio Capitani. **VHS, Beta $39.98** *SUE* 🦴🦴½

Lobster Man from Mars When rich movie producer (Curtis) learns from his accountant that he must produce a flop or be taken to the cleaners by the IRS, he buys a homemade horror movie from a young film-maker. The film is an ultra-low budget production featuring a kooky lobster man and a screaming damsel. The premise peters out about halfway through, but there are enough yuks to keep you going.
1989 (PG) 84m/C Tony Curtis, Deborah Foreman, Patrick Macnee. **VHS, Beta, LV $14.95** *IVE* 🦴🦴

Local Badman A dim-witted cowhand saves the day in this rip-roaring tale of the old west.

1932 60m/B Hoot Gibson. VHS, Beta $19.95 *NOS, UHV, RXM* 🐾

Local Hero Riegert is a yuppie representative of a huge oil company who endeavors to buy a sleepy Scottish fishing village for excavation, and finds himself hypnotized by the place and its crusty denizens. Back in Texas at company headquarters, tycoon Lancaster deals with a psycho therapist and gazes at the stars looking for clues. A low-key, charmingly offbeat Scottish comedy with its own sense of logic and quiet humor, poetic landscapes, and unique characters, epitomizing Forsyth's original style.
1983 (PG) 112m/C *GB* Peter Riegert, Denis Lawson, Burt Lancaster, Fulton McKay, Jenny Seagrove, Peter Capaldi; *Dir:* Bill Forsyth. VHS, Beta, LV $19.98 *WAR* 🐾🐾🐾🐾

The Loch Ness Horror The famed monster surfaces and chomps on the local poachers—and with good reason. Would that it had munched this movie's producer in, say, 1981.
1982 93m/C Barry Buchanan, Miki McKenzie, Sandy Kenyon; *Dir:* Larry Buchanan. VHS, Beta $39.95 *MON* Woof!

Lock and Load Why are all the soldiers in the 82nd Airborne stealing vast sums, then committing suicide? Find out why, if you can find this movie in the video store.
1990 (R) ?m/C Jack Vogel; *Dir:* David A. Prior. VHS $79.95 *AIP* 🐾

Lock Up Peaceful con Stallone, with only six months to go, is harassed and tortured by vicious prison warden Sutherland in retribution for an unexplained past conflict. Lackluster and moronic; semi-color; surely Sutherland can find better roles.
1989 (R) 115m/C Sylvester Stallone, Donald Sutherland, Sonny Landham, John Amos, Darlanne Fluegel, Frank McRae; *Dir:* John Flynn. VHS, Beta $14.95 *IVE, LIV* 🐾 ½

Lockdown Charged with his partner's murder, a detective is forced to ponder the perennial puzzle of life. Not a pretty sight.
1990 90m/C Joe Estevez, Mike Farrell, Richard Lynch. VHS $89.95 *VMK* 🐾 ½

The Lodger A mysterious lodger is thought to be a rampaging mass murderer of young women. First Hitchcock film to explore the themes and ideas that would become his trademarks. Silent. Climactic chase is memorable. Also titled "The Case for Jonathan Drew"; remade three times. Look closely for the Master in his first cameo.
1926 91m/B Ivor Novello, Marie Ault, Arthur Chesney, Malcolm Keen; *Dir:* Alfred Hitchcock. VHS, Beta $16.95 *NOS, GPV, VYY* 🐾🐾🐾

Logan's Run In the 23rd century, a hedonistic society exists in a huge bubble and people are only allowed to live to the age of 30. Intriguing concepts and great futuristic sets prevail here. Based on the novel by William Nolan and George Clayton.
1976 120m/C Michael York, Jenny Agutter, Richard Jordan, Roscoe Lee Browne, Farrah Fawcett, Peter Ustinov, Camilla Carr, Ann Ford; *Dir:* Michael Anderson Sr. Academy Awards '76: Best Visual Effects. VHS, Beta, LV $59.95 *MGM* 🐾🐾 ½

Lois Gibbs and the Love Canal Mason's television movie debut has her portraying Gibbs, a housewife turned activist, fighting the authorities over chemical-dumping in the Love Canal area of Niagara Falls, New York. The script doesn't convey the seriousness of developments in that region.

1982 100m/C Marsha Mason, Robert Gunton, Penny Fuller, Roberta Maxwell, Jeremy Licht, Louise Latham; *Dir:* Glenn Jordan. VHS $59.95 *TRY* 🐾🐾

Lola A wonderful tale of a nightclub dancer and her amorous adventures. Innovative French film with English subtitles that marked the beginning of French New Wave.
1961 90m/B *FR* Anouk Aimee, Marc Michel, Elina Labourdette, Jacques Harden; *Dir:* Jacques Demy. VHS $59.95 *INT* 🐾🐾🐾

Lola A teenaged girl links up romantically with a considerably older writer of pornographic books. One would expect more from such a good cast, but the film never follows through. Also known as "The Statutory Affair."
1969 (PG) 98m/C *GB IT* Charles Bronson, Michael Craig, Honor Blackman, Lionel Jeffries, Robert Morley, Jack Hawkins, Orson Bean, Kay Medford, Paul Ford; *Dir:* Richard Donner. VHS $59.95 *FCT, MRV* 🐾 ½

Lola Montes Ophuls' final masterpiece recounts the life and sins of the famous courtesan, mistress of Franz Liszt and the King of Bavaria. Ignored upon release, but hailed later by the French as a cinematic landmark. Presented in its original wide-screen format; in French with English subtitles. Also available in a 110 minute version.
1955 140m/C *FR* Martine Carol, Peter Ustinov, Anton Walbrook, Ivan Desny, Oskar Werner; *Dir:* Max Ophuls. New York Film Festival '68: Best Visual Effects. Beta, LV $69.95 *CRC* 🐾🐾🐾🐾

Lolita A middle-aged professor is consumed by his lust for a teen-age nymphet in this strange film considered daring in its time. Based on Vladimir Nabokov's novel. Watch for Winters' terrific portrayal as Lolita's sex-starved mother.
1962 152m/B *GB* James Mason, Shelley Winters, Peter Sellers, Sue Lyon; *Dir:* Stanley Kubrick. VHS, Beta, LV $19.98 *MGM, PIA, CRC* 🐾🐾🐾

London Melody Neagle is a Cockney flower girl who is taken in by a kind-hearted Italian diplomat who finances her musical training.
1937 71m/B *GB* Anna Neagle, Tullio Carminati, Robert Douglas, Horace Hodgers; *Dir:* Herbert Wilcox. VHS, Beta, 8mm $29.95 *VYY* 🐾 ½

The Lone Avenger Average oater centers around a bank panic, with two-gun Maynard clearing up the trouble.
1933 60m/B Ken Maynard, Muriel Gordon, James Marcus. VHS, Beta $19.95 *NOS, VCN, CAB* 🐾

Lone Bandit Vintage western starring Lane Chandler.
1933 57m/B Lane Chandler, Doris Brook, Wally Wales; *Dir:* J.P. McGowan. VHS, Beta $27.95 *VCN* 🐾 ½

The Lone Defender A dozen episodes of the western serial starring the canine crusader, Rin Tin Tin.
1932 234m/B Rin Tin Tin, Walter Miller, June Marlowe, Buzz Barton, Josef Swickard, Frank Lanning, Bob Kortman; *Dir:* Richard Thorpe. VHS, Beta $19.95 *VDM, GPV, MLB* 🐾 ½

The Lone Ranger Western serial about the masked man and his faithful Indian sidekick. From a long-sought print found in Mexico, this program is burdened by a noisy sound track, two completely missing chapters, an abridged episode #15, and Spanish subtitles.

1938 234m/B Lee Powell; *Dir:* William Witney, John English. VHS, Beta $49.95 *NOS, VYY, VCN*

Lone Ranger Tonto and that strange masked man must prevent a war between ranchers and Indians in the first of the "Lone Ranger" series. "Hi-ho Silver!"
1956 87m/C Clayton Moore, Jay Silverheels, Lyle Bettger, Bonita Granville; *Dir:* Stuart Heisler. VHS, Beta $29.95 *MGM, BUR*

Lone Ranger Two volumes of three cartoons each depict the adventures of the famous lawman.
1980 60m/C VHS, Beta $29.95 *FHE*

Lone Ranger and the Lost City of Gold Three Indian men are found dead, and the Lone Ranger and Tonto set out to discover the killers.
1958 80m/C Clayton Moore, Jay Silverheels, Douglas Kennedy; *Dir:* Lesley Selander. VHS $14.95 *WKV*

The Lone Rider in Cheyenne One of the last in the "Lone Rider" series. The Lone Rider is out to clear the name of an innocent man accused of murder.
1942 59m/B George Houston, Al "Fuzzy" St. John; *Dir:* Sigmund Neufeld. VHS $19.95 *NOS, DVT* 🐾 ½

The Lone Rider in Ghost Town Outlaws "haunt" a ghost town to protect their hidden mine and the Lone Rider and his pal, Fuzzy, are out to solve the mystery. Installment in the short-lived "Lone Rider" series.
1941 64m/B George Houston, Al "Fuzzy" St. John, Budd Buster; *Dir:* Sigmund Neufeld. VHS $19.95 *NOS, DVT* 🐾 ½

The Lone Runner An adventurer rescues a beautiful heiress from Arab kidnappers in this dud.
1988 (PG) 90m/C Miles O'Keefe, Ronald Lacey, Michael J. Aronin, John Steiner, Al Yamanouchi; *Dir:* Ruggero Deodato. VHS, Beta, LV $79.95 *MED* Woof!

Lone Star Law Men A U.S. marshal has his hands full in a border town overrun with bandits. Keene stars as the deputy marshall who goes undercover and joins the gang in an attempt to foil their crooked plans.
1942 58m/B Tom Keene, Frank Yaconelli, Sugar Dawn, Betty Miles; *Dir:* Robert Tansey. VHS $19.95 *NOS, DVT* 🐾

The Lone Star Ranger Former outlaw O'Brien is trying to go straight by helping to bring a gang of rustlers to justice. Interesting plot twist in the link between his girlfriend and the head of the gang.
1939 64m/B George O'Brien, Sue Carol, Walter McGrail, Warren Hymer, Russell Simpson; *Dir:* A.E. Erickson. VHS $19.95 *DVT* 🐾 ½

The Lone Wolf A boy learns kindness by befriending an old military dog which villagers think is mad and responsible for killing their sheep. After a brush with death, the boy convinces the villagers of the dog's good qualities and wins the admiration of his friends.
1972 45m/C VHS, Beta $39.98 *SUE* 🐾 ½

Lone Wolf A werewolf terrorizes a high school, and a few computerniks track it down.
1988 97m/C Dyann Brown, Kevin Hart, Jamie Newcomb, Ann Douglas; *Dir:* John Callas. VHS, Beta $79.95 *PSM* 🐾

Lone Wolf McQuade Martial arts action abounds in this modern-day Western which pits a Texas Ranger against a band of gun-running mercenaries. One of the mercenaries, however, happens to share the hero's love interest.
1983 107m/C Chuck Norris, Leon Isaac Kennedy, David Carradine, L.Q. Jones, Barbara Carrera; *Dir:* Steve Carver. **VHS, Beta, LV $19.98** VES ✗

The Loneliest Runner Based on the true story of an Olympic track star who, as a teenager, suffered humiliation as a bed-wetter. The real-life experiences of producer-director Landon are presented in a touching and sensitive manner.
1976 74m/C Michael Landon, Lance Kerwin, DeAnn Mears, Brian Keith, Melissa Sue Anderson; *Dir:* Michael Landon. **VHS, Beta $19.98** WAR, OM ✗✗

Lonely Boy/Satan's Choice This tape contains two cinema-verite documentaries from the National Film Board of Canada. "Lonely Boy" (1962) follows the early career of pop singer Paul Anka; and "Satan's Choice" (1966) provides an inside look at the members of a motorcycle gang.
19?? 55m/B VHS, Beta $29.95 VYY ✗✗

Lonely are the Brave A free-spirited cowboy out of sync with the modern age tries to rescue a buddy from a local jail, and in his eventual escape is tracked relentlessly by modern law enforcement. A compelling, sorrowful essay on civilized progress and exploitation of nature.
1962 107m/B Kirk Douglas, Walter Matthau, Gena Rowlands, Carroll O'Connor, George Kennedy; *Dir:* David Miller. **VHS, Beta $19.95** MCA ✗✗✗

The Lonely Guy Romantic comedy with Martin as a jilted writer who writes a best-selling book about being a lonely guy and finds stardom does have its rewards. Based on "The Lonely Guy's Book of Life" by Bruce Jay Friedman.
1984 (R) 91m/C Steve Martin, Charles Grodin, Judith Ivey, Steve Lawrence, Robyn Douglass, Merv Griffin, Dr. Joyce Brothers; *Dir:* Arthur Hiller. **VHS, Beta, LV $19.95** MCA ✗✗½

Lonely Hearts An endearing Australian romantic comedy about a piano tuner, who at 50 finds himself alone after years of caring for his mother, and a sexually insecure spinster, whom he meets through a dating service. Wonderful performances and a good script make this a delightful film that touches the human heart.
1982 (R) 95m/C AU Wendy Hughes, Norman Kaye, Jon Finlayson, Julia Blake, Jonathon Hardy; *W/Dir:* Paul Cox. Australian Film Institute '82: Best Film. **VHS, Beta $24.98** SUE ✗✗✗

The Lonely Lady A young writer comes to Hollywood with dreams of success, gets involved with the seamy side of movie-making, and is driven to a nervous breakdown. Pia needs acting lessons, yet it probably wouldn't have helped this trash. Adapted from the novel by Harold Robbins.
1983 92m/C Pia Zadora, Lloyd Bochner, Bibi Besch, Joseph Cali, Ray Liotta; *Dir:* Peter Sasdy. **VHS, Beta, LV $59.95** MCA Woof!

Lonely Man A gunfighter tries to end his career, but is urged into one last battle. Strong performances and tight direction make up for weak plot.
1957 87m/B Jack Palance, Anthony Perkins, Neville Brand, Elaine Aiken; *Dir:* Henry Levin. **VHS, Beta $19.95** KRT ✗✗½

The Lonely Passion of Judith Hearne A self-effacing Dublin spinster meets a man who gives her his attention, but she must overcome her own self-doubt and crisis of faith. Adapted from Brian Moore's 1955 novel. Excellent performances from both Hoskins and Smith.
1987 (R) 116m/C GB Maggie Smith, Bob Hoskins, Wendy Hiller, Marie Kean, Ian McNeice, Alan Devlin, Rudi Davies, Prunella Scales; *Dir:* Jack Clayton. **VHS $79.98** CAN ✗✗✗

Lonely Trail/Three Texas Steers Double your pleasure with two early Duke operas. In "Lonely Trail," Wayne dukes it out with dastardly carpetbaggers after the Civil War, while "Three Texas Steers," part of the "Three Mesquiteers" series, has Wayne and company spare a gal rancher/major babe from certain peril.
1939 112m/B John Wayne, George "Gabby" Hayes, Ray Corrigan, Raymond Hatton; *Dir:* George Sherman. **VHS $14.98** REP

Lonely Wives A lawyer hires an entertainer to serve as his double because of his marital problems.
1931 86m/B Edward Everett Horton, Patsy Ruth Miller, Laura La Plante, Esther Ralston; *Dir:* Russell Mack. **VHS, Beta $19.95** NOS, VYY ✗✗

Lonelyhearts Clift plays a reporter who is assigned the lovelorn column of his paper and gets too immersed in the problems of his readers. Given the superior cast and excellent material, this is a somewhat disappointing adaptation of the brilliant Nathanael West novel "Miss Lonelyhearts." Film debuts of both Stapleton and director Donehue. Stapleton received an Academy Award nomination for her portrayal of a lonely sex-starved woman.
1958 101m/B Montgomery Clift, Robert Ryan, Myrna Loy, Dolores Hart, Maureen Stapleton, Frank Maxwell, Jackie Coogan, Mike Kellin; *Dir:* Vincent J. Donehue. **VHS $19.98** MGM, SUE ✗✗½

Loners Three teenagers run from the Southwest police after they are accused of murdering a highway patrolman.
1972 (R) 80m/C Dean Stockwell, Gloria Grahame, Scott Brady, Alex Dreier, Pat Stich; *Dir:* Sutton Roley. **VHS, Beta $69.95** VID ✗✗

Lonesome Dove A classic western saga with Duvall and Jones outstanding as the two aging ex-Texas Rangers who decide to leave their quiet lives for a last adventure—a cattle drive from Texas to Montana. Along the way they encounter a new love (Lane), a lost love (Huston), and a savage renegade Indian (well-played by Forrest). Based on Larry McMurtry's Pulitzer Prize winning novel, this handsome television miniseries is a finely detailed evocation of the Old West, with a wonderful cast and an equally fine production. Available in a four volume boxed set.
1989 480m/C Robert Duvall, Tommy Lee Jones, Anjelica Huston, Danny Glover, Diane Lane, Rick Schroder, Robert Urich, D.B. Sweeney; *Dir:* Simon Windsor. **VHS $99.95** CAF ✗✗✗½

Lonesome Trail Routine Western that differs from the average oater by having the good guys use bows and arrows to fight their battles.
1955 73m/B Jimmy Wakely, Lee White, Lorraine Miller, John James; *Dir:* Richard Bartlett. **VHS, Beta** WGE ✗

Long Ago Tomorrow A paralyzed athlete enters a church-run home for the disabled rather than return to his family as the object of their pity. A love affair with a

woman, who shares the same disability, helps the athlete to adapt. Original British title: "The Raging Moon."
1971 (PG) 90m/C GB Malcolm McDowell, Nanette Newman, Bernard Lee, Georgia Brown, Gerald Sim; *Dir:* Bryan Forbes. **VHS, Beta $59.95** COL ✗✗

The Long Dark Hall Courtroom drama in which an innocent man is brought to trial when his showgirl mistress is found dead.
1951 86m/B GB Rex Harrison, Lilli Palmer, Denis O'Dea, Raymond Huntley, Patricia Wayne, Anthony Dawson; *Dir:* Anthony Bushell, Reginald Beck. **VHS, Beta $16.95** SNC ✗✗½

Long Day's Journey into Night A brooding, devastating film based on Eugene O'Neill's most powerful and autobiographical play. Depicts a day in the life of a family deteriorating under drug addiction, alcoholism and imminent death. Hepburn's Oscar-nominated performance is outstanding. Music by Andre Previn. In 1988, the Broadway version was taped and released on video.
1962 174m/B Ralph Richardson, Katharine Hepburn, Dean Stockwell, Jason Robards Jr.; *Dir:* Sidney Lumet. Cannes Film Festival '62: Best Acting (collective); National Board of Review Awards '62: 10 Best Films of the Year, Best Supporting Actor (Robards). **VHS, LV $39.98** KUI, REP ✗✗✗✗

Long Day's Journey into Night A taped version of the Broadway production of the epic Eugene O'Neill play about a Southern family deteriorating under the weight of terminal illness, alcoholism and drug abuse. In 1962 a movie adaptation of the play was released with outstanding performances from its cast.
1988 169m/C Jack Lemmon, Bethel Leslie, Peter Gallagher, Kevin Spacey, Jodie Lynne McLintock; *Dir:* Jonathan Miller. **VHS, Beta $59.98** LIV, VES ✗✗

The Long Days of Summer Set in pre-World War II America, this film portrays a Jewish attorney's struggle against the prejudices of the New England town where he lives. Sequel to "When Every Day Is the Fourth of July."
1980 105m/C Dean Jones, Joan Hackett, Louanne, Donald Moffatt, Andrew Duggan, Michael McGuire; *Dir:* Dan Curtis. **VHS $49.95** IVE ✗✗

Long Gone In the fifties, an over-the-hill minor-league player/manager is given a last lease on life and the pennant with two talented rookies and a sexy baseball groupie. Made for cable television.
1987 113m/C William L. Petersen, Henry Gibson, Katy Boyer, Virginia Madsen; *Dir:* Martin Davidson. **VHS, Beta, LV $19.98** LHV, WAR ✗✗½

The Long Good Friday Set in London's dockland, this is the story of an underworld king out to beat his rivals at their own game. One of the best of the crime genre, with an exquisitely charismatic performance by Hoskins.
1979 109m/C GB Bob Hoskins, Helen Mirren, Dave King, Bryan Marshall, George Coulouris, Pierce Brosnan, Derek Thompson, Eddie Constantine; *Dir:* John MacKenzie. Edgar Allan Poe Awards '82: Best Screenplay. **VHS, Beta $19.99** HBO ✗✗✗½

The Long Goodbye Raymond Chandler's penultimate novel's got that certain Altman feeling. Which is to say that some of the changes to the pulp auteur's story have pushed purist noses out of joint: Gould is cast as an insouciant anti-Marlowe, the film noir atmosphere has been transmuted into a

Hollywoodesque film neon, genre jibing abounds, and the ending has been rewritten. But the updating, tongue in cheeking and revamping all serve a purpose, which is to make Marlowe a viable character in a contemporary world. Adapted by Leigh Brackett (who co-scripted Hawks' "Big Sleep") and handsomely photographed by Vilmos Zsigmond, with a clever score by John Williams. And don't miss bulky boy Arnold in cameo (his second film appearance).
1973 (R) 112m/C Elliott Gould, Helen Mirren, Eddie Constantine, Dave King, Mark Rydell, Henry Gibson, Sterling Hayden; *Dir:* Robert Altman. **VHS, Beta, LV $19.98** MGM, FCT ⅋⅋⅋

The Long Gray Line Power gives an outstanding performance as Marty Maher, a humble Irish immigrant who became an institution at West Point. This is the inspiring story of his rise from an unruly cadet to one of the academy's most beloved instructors. O'Hara does a fine job of playing his wife, who like her husband, adopts the young cadets as her own. Director Ford gracefully captures the spirit and honor associated with West Point in this affectionate drama.
1955 138m/C Tyrone Power Jr., Maureen O'Hara, Robert Francis, Donald Crisp, Ward Bond, Betsy Palmer, Phil Carey, John Ford. **VHS $19.95** COL ⅋⅋⅋

The Long Haul A truck driver becomes involved with crooks as his marriage sours.
1957 88m/C Victor Mature, Diana Dors, Patrick Allen, G.M. Anderson; *Dir:* Ken Hughes. GKK ⅋⅋

The Long, Hot Summer A tense, well-played adaptation of the William Faulkner story about a wanderer latching himself onto a tyrannical Mississippi family. The first on-screen pairing of Newman and Woodward, and one of the best; music by Alex North. Remade for television in 1986.
1958 118m/C Paul Newman, Orson Welles, Joanne Woodward, Lee Remick, Anthony (Tony) Franciosa, Angela Lansbury, Richard Anderson; *Dir:* Martin Ritt. **VHS, Beta $19.98** FOX ⅋⅋⅋½

The Long, Hot Summer A made-for-television version of the William Faulkner story, "The Hamlet," about a drifter taken under a Southern patriarch's wing. He's bribed into courting the man's unmarried daughter. Wonderful performances from the entire cast, especially Ivey and surprisingly, Johnson. Remake of the 1958 film with Paul Newman and Joanne Woodward that is on par with the original.
1986 172m/C Don Johnson, Cybill Shepherd, Judith Ivey, Jason Robards Jr., Ava Gardner, William Russ, Wings Hauser, William Forsythe, Albert Hall; *Dir:* Stuart Cooper. **VHS, Beta $59.98** FOX ⅋⅋½

Long Island Four Four Nazi saboteurs are smitten by the decadent nightlife of the Big Apple in 1942. The late Anders Grafstrom's Super-8 film, based on a true story, was an underground hit.
1980 100m/C *Dir:* Anders Grafstrom. **VHS $59.95** KIV, MWF ⅋⅋½

Long John Silver Famed pirate John Silver plans a return trip to Treasure Island to search for the elusive treasure; unofficial sequel to "Treasure Island" by Disney.
1953 106m/C AU Robert Newton, Connie Gilchrist, Kit Taylor, Grant Taylor, Rod Taylor; *Dir:* Byron Haskin. **VHS, Beta $29.95** GEM, MRV, CAB ⅋½

Long Journey Back The biographical story of a young woman's rehabilitation after her injury in a school bus accident. Produced for television.
1978 100m/C Mike Connors, Cloris Leachman, Stephanie Zimbalist; *Dir:* Mel Damski. **VHS $39.95** IVE ⅋⅋

The Long, Long Trailer A couple on their honeymoon find that trailer life is more than they bargained for. Lots of fun with charming direction from Minelli, and Ball's incredible slapstick style.
1954 97m/C Desi Arnaz Sr., Lucille Ball, Marjorie Main, Keenan Wynn; *Dir:* Vincente Minnelli. **VHS, LV $19.98** BTV, MGM, FCT ⅋⅋½

The Long Riders Excellent mythic western in which the Jesse James and Cole Younger gangs raid banks, trains, and stagecoaches in post-Civil War Missouri. Stylish, meticulous and a violent look back, with one of the better slow-motion shootouts in Hollywood history. Notable for the portrayal of four sets of brothers by four Hollywood brother sets. Complimented by excellent Ry Cooder score.
1980 (R) 100m/C Stacy Keach, James Keach, Randy Quaid, Dennis Quaid, David Carradine, Keith Carradine, Robert Carradine, Christopher Guest, Nicholas Guest, Pamela Reed, Savannah Smith, James Whitmore Jr., Harry Carey Jr.; *Dir:* Walter Hill. **VHS, Beta, LV $19.95** MGM ⅋⅋⅋

Long Shadows An analysis of how the resonating effects of the Civil War can still be felt on society, via interviews with a number of noted writers, historians, civil rights activists and politicians.
1986 88m/C Robert Penn Warren, Studs Terkel, Jimmy Carter, Robert Coles, Tom Wicker. **VHS, Beta, 3/4U** JAF ⅋½

Long Shot Two foosball enthusiasts work their way through local tournaments to make enough money to make it to the World Championships in Tahoe.
1981 100m/C Ian Giatti. **VHS, Beta $19.99** HBO ⅋½

Long Time Gone An over-the-hill detective tries to solve a murder while dealing with his bratty, alienated 11-year-old son. Made for television.
1986 97m/C Paul LeMat, Wil Wheaton, Ann Dusenberry, Barbara Stock; *Dir:* Robert Butler. **VHS, Beta $79.95** WES ⅋

The Long Voyage Home A talented cast performs this must-see screen adaptation of Eugene O'Neill's play about crew members aboard a merchant steamer in 1939. Wayne plays a young lad from Sweden who is trying to get home and stay out of trouble as he and the other seaman get shore leave. Nominated for six Oscars: Best Picture, Best Screenplay, Black & White Cinematography, Special Effects, Original Score, and Film Editing.
1940 105m/B John Wayne, Thomas Mitchell, Ian Hunter, Barry Fitzgerald, Mildred Natwick, John Qualen; *Dir:* John Ford. National Board of Review Awards '40: 10 Best Films of the Year; New York Film Critics Awards '40: Best Director. **VHS, Beta, LV $19.98** WAR, LTG ⅋⅋⅋½

The Long Walk Home In Montgomery Alabama, in the mid 1950s, sometime after Rosa Parks refused to sit in the black-designated back of the bus, Martin Luther King Jr. led a bus boycott. Spacek is the affluent white wife of a narrow-minded businessman while Goldberg is her struggling black maid. When Spacek discovers that Goldberg is supporting the boycott by walk-

ing the nine-mile trek to work, she sympathizes with the woman's familial responsibilities and tries to help. Hubby becomes unhappy camper when given the news and the plot marches inevitably toward a white-on-white showdown on racism while more quietly exploring gender equality between the women. Outstanding performances by Spacek and Goldberg, and a great fifties feel.
1989 (PG) 95m/C Sissy Spacek, Whoopi Goldberg, Dwight Schultz, Ving Rhames, Dylan Baker; *Dir:* Richard Pearce. **VHS, LV $92.95** LIV, IME, FCT ⅋⅋⅋

A Long Way Home A television movie about a grown man searching for his long-lost siblings, after the three of them were given up for adoption after birth.
1981 97m/C Timothy Hutton, Brenda Vaccaro, Rosanna Arquette, Paul Regina, George Dzundza, John Lehne, Bonnie Bartlett; *Dir:* Robert Markowitz. **VHS, Beta $59.95** IVE ⅋⅋

The Long Weekend When a vacationing couple goes camping and carelessly destroys a forest, Mother Nature turns her powers of retribution on them.
1986 95m/C **VHS, Beta $59.95** TWE ⅋

The Longest Day The complete story of the D-Day landings at Normandy on June 6, 1944, as seen through the eyes of American, French, British, and German participants. Exhaustively accurate details and extremely talented cast make this one of the all-time great Hollywood epic productions. The first of the big-budget, all-star war productions; based on the book by Cornelius Ryan. Three directors share credit. Oscar nomination for Best Picture, Art Direction and Set Decoration and Film Editing.
1962 179m/C John Wayne, Richard Burton, Red Buttons, Robert Mitchum, Henry Fonda, Robert Ryan, Paul Anka, Mel Ferrer, Edmond O'Brien, Fabian, Sean Connery, Roddy McDowall, Arletty, Curt Jurgens, Rod Steiger, Jean-Louis Barrault, Peter Lawford, Robert Wagner, Sal Mineo, Leo Genn, Richard Beymer, Jeffrey Hunter, Stuart Whitman, Eddie Albert, Tom Tryon, Alexander Knox, Ray Darton, Kenneth More, Richard Todd, Gert Frobe, Christopher Lee; *Dir:* Ken Annakin. Academy Awards '62: Best Black and White Cinematography, Best Special Effects; National Board of Review Awards '62: Best Film. **VHS, Beta, LV $29.97** FOX, TLF ⅋⅋⅋

The Longest Drive Two brothers comb the wildest parts of the West for their sister, whom they believe is living with Indians. A made for television movie originally titled "The Quest," intended to be a series which never materialized. Highlights include colorful performances from the veteran actors and a unique horse/camel race.
1976 92m/C Kurt Russell, Tim Matheson, Brian Keith, Keenan Wynn, Neville Brand, Cameron Mitchell, Morgan Woodward, Iron Eyes Cody, Luke Askew; *Dir:* Lee H. Katzin. **VHS $89.95** VMK ⅋⅋

The Longest Yard A one-time pro football quarterback, now an inmate, organizes his fellow convicts into a football team to play against the prison guards. Filmed on location at Georgia State Prison.
1974 (R) 121m/C Burt Reynolds, Eddie Albert, Bernadette Peters, Ed Lauter, Richard Kiel; *Dir:* Robert Aldrich. Golden Globe Awards '75: Best Film—Musical/Comedy. **VHS, Beta, LV $14.95** PAR ⅋⅋½

The Longshot Four bumblers try to raise cash to put on a sure-bet racetrack tip in this sorry comedy. Mike Nichols is the executive producer.

1986 (PG-13) 89m/C Tim Conway, Harvey Korman, Jack Weston, Ted Wass, Jonathan Winters, Stella Stevens, Anne Meara; *Dir:* Paul Bartel. **VHS, Beta** $59.95 *HBO* Woof!

Longtime Companion Critically acclaimed film follows a group of gay men and their friends during the 1980s. The closely knit group monitors the progression of the AIDS virus from early news reports until it finally hits home and begins to take the lives of their loved ones. One of the first films to look at the situation in an intelligent and touching manner. Produced by the PBS ''American Playhouse'' company.
1990 (R) 100m/C Stephen Caffrey, Patrick Cassidy, Brian Cousins, Bruce Davison, John Dossett, Mark Lamos, Dermot Mulroney, Mary Louise Parker, Michael Schoeffling, Campbell Scott, Robert Joy; *Dir:* Norman Rene. Golden Globe Awards '90: Best Supporting Actor (Davison); New York Film Critics Awards '90: Best Supporting Actor (Davison); National Society of Film Critics Awards '90: Best Supporting Actor (Davison). **VHS** $89.95 *VMK, FCT* 🎞🎞🎞½

Look Back in Anger Based on John Osbourne's famous play, the first British ''angry young man'' film, in which a squalor-living lad takes out his anger on the world by seducing his friend's wife.
1958 99m/B *GB* Richard Burton, Claire Bloom, Mary Ure, Edith Evans, Gary Raymond; *Dir:* Tony Richardson. National Board of Review Awards '59: 5 Best Foreign Films of the Year. **VHS, Beta** $24.98 *SUE* 🎞🎞🎞½

Look Back in Anger A working-class man angered by society's hypocrisy lashes out at his upper-class wife, his mistress and the world. Inferior remake of 1958 film version starring Richard Burton, based on '50s stage hit.
1980 100m/C Malcolm McDowell, Lisa Barnes, Fran Brill, Raymond Hardie, Lindsay Anderson. **VHS, Beta** $19.98 *WAR, OM* 🎞🎞

Look Back in Anger There's something about John Osborne's play that brings out the angry young man in British leads. Richard Burton played Osborne's irascible guy in 1958, Malcom McDowell looked back angrily in '80, and now Branagh convincingly vents his spleen on wife and mistress in this made for British TV production. Director Jones earlier filmed ''84 Charing Cross Road'' and ''Jacknife.''
1989 114m/C *GB* Kenneth Branagh, Emma Thompson, Gerard Horan, Siobhan Redmond; *Dir:* David Jones. **VHS** $29.99 *HBO* 🎞🎞🎞

Look Who's Laughing Bergen's plane lands in a town conveniently populated by radio stars. Not much plot here, but it might be worth a look to fans of the stars including Jim and Marion Jordan, better known as Fibber McGee and Molly.
1941 79m/B Edgar Bergen, Jim Jordan, Marion Jordan, Lucille Ball, Harold Peary, Lee Bonnell; *Dir:* Allan Dwan. **VHS, Beta, LV** $19.98 *TTC, MLB* 🎞🎞

Look Who's Talking When Alley bears the child of a married, and quite fickle man, she sets her sights elsewhere in search of the perfect father; Travolta is the cabbie with more on his mind than driving Alley around and babysitting. All the while, the baby gives us his views via the voice of Bruce Willis. A very light comedy with laughs for the whole family.
1989 (PG-13) 90m/C John Travolta, Kirstie Alley, Olympia Dukakis, George Segal, Abe Vigoda, Bruce Willis; *Dir:* Amy Heckerling. **VHS, Beta, LV** $19.95 *COL, RDG* 🎞🎞🎞

Look Who's Talking, Too If Academy Awards for Stupidest Sequel and Lamest Dialogue existed, this diaper drama would have cleaned up. The second go-round throws the now married accountant-cabbie duo into a marital tailspin when Alley's babysitting brother moves in and the Saturday Night dancer moves out. Meanwhile, Willis cum baby smartasses incessantly. A once-clever gimmick now unencumbered by plot; not advised for linear thinkers. The voice of Arnold, though, is a guarantee you'll get one laugh for your rental.
1990 (PG-13) ?m/C Kirstie Alley, John Travolta, Olympia Dukakis, Elias Koteas; *Dir:* Amy Heckerling; *Voices:* Bruce Willis, Mel Brooks, Damon Wayans, Roseanne (Barr) Arnold. **VHS, LV, 8mm** $19.95 *COL, PIA* 🎞

Looker Stunning models are made even more beautiful by a plastic surgeon, but one by one they begin to die. Finney plays the Beverly Hills surgeon who decides to investigate when he starts losing all his clients.
1981 (PG) 94m/C Albert Finney, James Coburn, Susan Dey, Leigh Taylor-Young; *Dir:* Michael Crichton. **VHS, Beta, LV** $64.95 *WAR* 🎞½

Lookin' to Get Out Comedy about two gamblers running from their debts. They wind up at the MGM Grand in Las Vegas trying to get out of a mess.
1982 (R) 70m/C Ann-Margret, Jon Voight, Burt Young; *Dir:* Hal Ashby. **VHS, Beta** $59.98 *FOX* 🎞🎞½

The Looking Glass War A Polish defector is sent behind the Iron Curtain on a final mission to photograph a rocket in East Berlin. Adapted from John Le Carre's best-selling spy novel.
1969 (PG) 108m/C *GB* Christopher Jones, Ralph Richardson, Pia Degermark, Anthony Hopkins, Susan George; *Dir:* Frank Pierson. **VHS, Beta** $59.95 *COL, MLB* 🎞🎞½

Looking for Miracles Two brothers, separated during the Depression because of poverty, get a chance to cultivate brotherly love when they're reunited in 1935. A Wonderworks production based on the A.E. Hochner novel.
1990 (G) 104m/C Zachary Bennett, Greg Spottiswood, Joe Flaherty. **VHS** $29.95 *BVV, DIS, FCT* 🎞🎞½

Looking for Mr. Goodbar A young teacher, played by Keaton, seeks companionship and love by frequenting single's bars. Her imminent self-destruction is brought on by her aimless intake of drugs and alcohol. In need of a father figure, she makes herself available to numerous men and eventually regrets her hedonistic behavior. Based on the bestselling novel by Judith Rossner. Oscar nominations for supporting actress Weld, and cinematography.
1977 (R) 136m/C Diane Keaton, Tuesday Weld, Richard Gere, Tom Berenger, William Atherton, Richard Kiley; *Dir:* Richard Brooks. Harvard Lampoon Awards '77: Worst Film of the Year. **VHS, Beta, LV** $39.95 *PAR* 🎞½

Looney Looney Looney Bugs Bunny Movie A feature-length compilation of classic Warner Brothers cartoons tied together with new animation. Cartoon stars featured include Bugs Bunny, Elmer Fudd, Porky Pig, Yosemite Sam, Duffy Duck and Foghorn Leghorn.
1981 (G) 80m/C *Dir:* Friz Freleng, Friz Freleng, Chuck Jones, Bob Clampett; *Voices:* Mel Blanc, June Foray. **VHS, Beta** $19.98 *WAR, FCT, BUR* 🎞🎞🎞

Looney Tunes & Merrie Melodies 1 A collection of eight Warner Brothers Vitaphone cartoons dating from 1931-33, most with jazzy musical accompaniments. Titles include ''It's Got Me Again,'' ''You Don't Know What You're Doing,'' ''Moonlight for Two,'' ''Battling Bosko,'' ''Red-Headed Baby'' and ''Freddy the Freshman.''
1933 56m/B Bob Clampett, Chuck Jones. **VHS, Beta, 8mm** $39.95 *VYY*

Looney Tunes & Merrie Melodies 2 A second collection of seven Warner-Vitaphone cartoons from 1931-33, 1937, and 1941-43. Porky Pig, Daffy Duck, and Bugs Bunny are featured in the later World War II-oriented titles, ''Scrap Happy Daffy'' and ''Porky Pig's Feat.'' Earlier titles include ''One More Time,'' ''Smile, Darnya, Smile,'' and ''Yodeling Yokels.''
194? 51m/B *Dir:* Chuck Jones, Bob Clampett, Friz Freleng. **VHS, Beta** $24.95 *VYY*

Looney Tunes & Merrie Melodies 3 Cartoon classics from Warner Bros: ''A Corny Concerto'' (1943), with Porky and Bugs; ''Foney Fables'' (1942), a retelling of old fairy tales; ''The Wacky Wabbit'' (1942), featuring Bugs and Elmer Fudd; ''Have You Got Any Castles'' (1938); ''Fifth Column Mouse'' (1943); ''To Duck or Not to Duck'' (1943), with Elmer and Daffy; ''The Early Worm Gets the Bird'' (1940); and ''Daffy the Commando'' (1943), with Daffy Duck.
194? 60m/C *Dir:* Bob Clampett, Friz Freleng, Chuck Jones. **VHS, Beta** $24.95 *VYY*

Looney Tunes Video Show, Vol. 1 Seven Warner Brothers cartoon classics of the 1940s and 50s: Bugs Bunny and the Tasmanian Devil in ''Devil May Hare,'' Sylvester in ''Birds of a Father,'' Daffy Duck and Porky Pig in ''The Ducksters,'' the Road Runner and Wile E. Coyote in ''Zipping Along,'' Sylvester and Tweety in ''Room and Bird,'' Elmer Fudd in ''Ant Pasted,'' and Speedy Gonzales in ''Mexican Schmoes.''
195? 49m/C *Dir:* Chuck Jones, Bob Clampett, Friz Freleng. **VHS, Beta** $12.95 *WAR*

Looney Tunes Video Show, Vol. 2 More Warner Brothers cartoon favorites: Daffy Duck in ''Quackodile Tears,'' Porky Pig in ''An Egg Scramble,'' Sylvester and Speedy Gonzales in ''Cats and Bruises,'' Foghorn Leghorn in ''All Fowled Up,'' Bugs Bunny and Yosemite Sam in ''14 Carrot Rabbit,'' Professor Calvin Q. Calculus in ''The Hole Idea,'' and Pepe Le Pew in ''Two Scents Worth.''
195? 48m/C *Dir:* Chuck Jones, Friz Freleng, Bob Clampett. **VHS, Beta** $12.95 *WAR*

Looney Tunes Video Show, Vol. 3 Seven more Warner Brothers cartoon shorts: Daffy Duck and Speedy Gonzales in ''The Quacker Tracker,'' the Wolf and Sheepdog in ''Double or Mutton,'' Claude Cat and Bulldog in ''Feline Frameup,'' Bugs Bunny in ''Eight Ball Bunny,'' Foghorn Leghorn in ''A Featured Leghorn,'' Porky Pig and Sylvester in ''Scaredy Cat'' and Pepe Le Pew in ''Louvre, Come Back to Me.''
195? 48m/C *Dir:* Bob Clampett, Friz Freleng, Chuck Jones. **VHS, Beta** $12.95 *WAR*

Looney Tunes Video Show, Vol. 4 Another Warner Brothers cartoon assortment: Sylvester and Tweety in ''Ain't She Tweet,'' Daffy Duck and Speedy Gonzales in ''Astroduck,'' Bugs Bunny in ''Backwoods Bunny,'' Pepe Le Pew in ''Heaven Scent,''

Elmer Fudd in "Pests for Guests," Sylvester in "Lighthouse Mouse" and the Wolf and Sheepdog in "Don't Give Up the Sheep." **195? 47m/C** *Dir:* Chuck Jones, Bob Clampett. Friz Freleng. **VHS, Beta $12.95** *WAR, OM*

Looney Tunes Video Show, Vol. 5 An additional package of Warner Brothers cartoons: Sylvester and Tweety in "Tugboat Granny," Daffy Duck in "Stork Naked," the Road Runner and Wile E. Coyote in "Fastest with the Mostest," Bugs Bunny in "Forward March Hare," Foghorn Leghorn in "Feather Dusted," Daffy Duck and Porky Pig in "China Jones" and Pepe Le Pew in "Odor of the Day." **195? 51m/C** *Dir:* Chuck Jones, Bob Clampett, Friz Freleng. **VHS, Beta $12.95** *WAR, OM*

Looney Tunes Video Show, Vol. 6 More classic Warner Brothers cartoons: Foghorn Leghorn in "Feather Bluster," the Road Runner and Wile E. Coyote in "Lickety Splat," Bugs Bunny in "Bowery Bugs," Daffy Duck and Speedy Gonzales in "Daffy Rents," Porky Pig in "Dough for the Dodo," Sylvester and Elmer Fudd in "Heir Conditioned" and Pepe Le Pew in "Scent of the Matterhorn." **195? 49m/C** *Dir:* Chuck Jones, Friz Freleng, Bob Clampett. **VHS, Beta $12.95** *WAR, OM*

Looney Tunes Video Show, Vol. 7 Seven additional Warner Brothers cartoon classics: Bugs Bunny in "A-Lad-In His Lamp," the Road Runner and Wile E. Coyote in "Beep Beep," Yosemite Sam in "Honey's Money," Foghorn Leghorn in "Weasel Stop," Daffy Duck and Elmer Fudd in "Don't Ax Me," Sylvester and Tweety in "Muzzle Tough" and Foghorn Leghorn in "The Egg-Cited Rooster." **195? 48m/C** *Dir:* Chuck Jones, Bob Clampett. **VHS, Beta $12.95** *WAR, OM*

Loophole An out-of-work architect, hard pressed for money, joins forces with an elite team of expert criminals, in a scheme to make off with millions from the most established holding bank's vault. Also known as "Break In." **1983 105m/C** *GB* Albert Finney, Martin Sheen, Susannah York, Robert Morley, Colin Blakely, Jonathan Pryce; *Dir:* John Quested. **VHS, Beta $59.95** *MED, HHE, MTX* 🎞️🎞️

Loose Cannons Yet another mismatched-cop-partner comedy, wherein a mystery is ostensibly solved by a veteran cop and a schizophrenic detective. **1990 95m/C** Gene Hackman, Dan Aykroyd, Dom DeLuise, Ronny Cox, Nancy Travis, David Alan Grier; *Dir:* Bob Clark. **VHS, Beta, LV $19.95** *COL* 🎞️

Loose Connections A feminist driving to a convention in Europe advertises for a travelmate, and gets a hopeless chauvinist who is masquerading as a gay man in this offbeat cult comedy. **1987 90m/C** *GB* Lindsay Duncan, Stephen Rea, Robbie Coltrane; *Dir:* Richard Eyre. **VHS, Beta $79.95** *PAV* 🎞️🎞️

Loose in New York Much to her surprise, a cynical socialite begins to fall for her computer-arranged mate. **198? 91m/C** Rita Tushingham, Aldo Maccione; *Dir:* Gian Polidoro. **VHS, Beta** *WES* 🎞️🎞️

Loose Screws Four perverted teenagers are sent to a restrictive academy where they continue their lewd ways in this stupid sequel to "Screwballs."

1985 (R) 75m/C *CA* Bryan Genesse, Karen Wood, Alan Deveau, Jason Warren; *Dir:* Rafal Zielinski. **VHS, Beta $79.98** *LIV, LTG* 🎞️

Loose Shoes A collection of vignettes that satirize movie trailers, teasers and special announcements. Murray's film debut. **1980 (R) 84m/C** Bill Murray, Howard Hesseman, Jaye P. Morgan, Buddy Hackett, Misty Rowe, Susan Tyrrell; *Dir:* Ira Miller. **VHS, Beta $39.98** *FOX* 🎞️🎞️

Loot...Give Me Money, Honey! Black comedy about a motley crew of greed-driven golddiggers who chase after a heisted fortune in jewels, which is hidden in a coffin belonging to one of the thieves' mother. From Joe Orton's play. **1970 101m/C** *GB* Richard Attenborough, Lee Remick, Hywel Bennett, Milo O'Shea, Roy Holder; *Dir:* Silvio Narizzano. **VHS, Beta $59.95** *WES* 🎞️🎞️½

Lord of the Flies Proper English schoolboys stranded on a desert island during a nuclear war are transformed into savages. A study in greed, power, and the innate animalistic/survivalistic instincts of human nature. Based on William Golding's novel, which he described as a journey to the darkness of the human heart. **1963 91m/B** *GB* James Aubrey, Tom Chapin, Hugh Edwards, Roger Elwin, Tom Gamen; *Dir:* Peter Brook. National Board of Review Awards '63: 10 Best Films of the Year. **VHS, Beta, LV $37.95** *KOV* 🎞️🎞️🎞️½

Lord of the Flies Inferior second filming of the famed William Golding novel about schoolboys marooned on a desert island who gradually degenerate into savages. Lushly photographed, yet redundant and poorly-acted. **1990 (R) 90m/C** Balthazar Getty, Danuel Pipoly, Chris Furrh, Badgett Dale, Edward Taft, Andrew Taft; *Dir:* Harry Hook. **VHS, Beta, LV, 8mm $19.95** *COL, SUE* 🎞️🎞️

Lord Jim A ship officer (O'Toole) commits an act of cowardice that results in his dismissal and disgrace, which leads him to the Far East in search of self-respect. Excellent supporting cast. Based on Joseph Conrad's novel. **1965 154m/C** Peter O'Toole, James Mason, Curt Jurgens, Eli Wallach, Jack Hawkins, Paul Lukas, Akim Tamiroff, Daliah Lavi, Andrew Keir, Jack MacGowran, Walter Gotell; *Dir:* Richard Brooks; *W/Dir:* Richard Brooks. **VHS, Beta $19.95** *KUI, COL* 🎞️🎞️🎞️

Lord of the Rings An animated interpretation of Tolkien's classic tale of the hobbits, wizards, elves, and dwarfs who inhabit Middle Earth. Animator Ralph Bakshi used live motion animation to give his characters more life-like and human motion. Features the voice of Christopher Guard. Well done in spite of the difficulty of adapting from Tolkien's highly detailed and lengthy works. **1978 (PG) 128m/C** *Dir:* Ralph Bakshi; *Voices:* Christopher Guard, John Hurt. **VHS, Beta $19.98** *HBO, FCT* 🎞️🎞️

Lords of the Deep A Roger Corman cheapie about underwater technicians trapped on the ocean floor with a race of aliens. A film rushed out to capitalize on the undersea sci-fi subgenre highlighted by "The Abyss." **1989 (PG-13) 95m/C** Bradford Dillman, Priscilla Barnes, Melody Ryane, Eb Lottimer, Daryl Haney; *Dir:* Mary Ann Fisher. **VHS, Beta $79.95** *MGM Woof!*

Lords of Discipline A military academy cadet is given the unenviable task of protecting a black freshman from racist factions at a southern school circa 1964. Based on Pat Conroy's autobiographical novel. **1983 (R) 103m/C** David Keith, Robert Prosky, Barbara Babcock, Judge Reinhold, G.D. Spradlin, Rick Rossovich, Michael Biehn; *Dir:* Franc Roddam. **VHS, Beta, LV $39.95** *PAR* 🎞️🎞️½

The Lords of Flatbush Four street toughs battle against their own maturation and responsibilities in 1950s Brooklyn. Winkler introduces the leather-clad hood he's made a career of and Stallone introduces a character not unlike Rocky. Interesting slice of life. **1974 (PG) 88m/C** Sylvester Stallone, Perry King, Henry Winkler, Susan Blakely, Armand Assante, Paul Mace; *Dir:* Stephen Verona, Martin Davidson. **VHS, Beta, LV $14.95** *COL* 🎞️🎞️½

Lords of Magick Two warriors chase an evil sorcerer and the princess he's kidnapped across time to the 20th century. **1988 (PG-13) 98m/C** Jarrett Parker, Matt Gauthier, Brendan Dillon Jr.; *Dir:* David Marsh. **VHS, Beta $79.95** *PSM* 🎞️

Los 3 Reyes Magos An animated retelling of the legend of the three wise men who brought gifts to Jesus Christ on the Epiphany. **198? 86m/C VHS, Beta $49.95** *UNI* 🎞️

Los Amores de Marieta Love and jealousy abound amidst the Mexican Revolution. **1987 100m/C** *MX* Elvira Quintana, Joaquin Cordero, Gina Romand. **VHS, Beta $59.95** *MAD* 🎞️🎞️

Los Apuros de Dos Gallos (Troubles of Two Roosters) Four Mexican singing aces spend some time at a ranch serenading the ranch hands. **1969 102m/C** *MX* Miguel Aceves Mejia, Marco Antonio Muniz, Lilian de Celis. **VHS, Beta $69.95** *MAD* 🎞️

Los Asesinos Two rival bandits arrive in a lawless town and fight for control. Dialogue in Spanish. **1968 95m/C** *MX* Nick Adams, Regina Torne, Pedro Armendariz Jr., Elsa Cardenas. **VHS, Beta $29.95** *UNI* 🎞️

Los Caciques A comic-book hero comes alive and stops an evil empire of crime. **1987 96m/C** *MX* Jorge Rivero, Pedro Armendariz, Carmen Vicarte, Mario Alberto Rodriquez. **VHS, Beta $79.95** *MAD* 🎞️🎞️½

Los Cacos A group of eleven friends try to steal a large sum of money but they run into problems. **1987 80m/C** *MX* Silvia Pinal, Milton Rodriguez. **VHS, Beta $59.95** *MAD* 🎞️🎞️

Los Caifanes A group of men live life on the edge, challenging death at every minute. **1987 95m/C** *MX* Julissa, Enrique Alvarez Felix, Sergio Jimenez, Oscar Chavez. **VHS, Beta $55.95** *MAD* 🎞️🎞️

Los Chantas A small band of inept petty thieves attempt one last big heist. In Spanish. **198? 125m/C** *SP* Norberto Aroldi, Elsa Daniel, Angel Magana. **VHS, Beta $59.95** *UNI* 🎞️

Los Chicos Crecen Spanish film exploring the feelings of attachment versus loyalty. A man is asked by an old friend to assume responsibility for his family when his friend's bigamy secret is threatened.

19?? 90m/C Luis Sandrini, Eduardo Rudy, Susana Campos. **VHS, Beta** $59.95 *UNI*

Los Cuatro Budas de Kriminal In "The Four Buddhas of Kriminal," there is doublecross and murder as Kriminal attempts to get the pieces of a map which are hidden in four statuettes. In Spanish.
1972 88m/C Glen Saxson, Helga Line. **VHS, Beta** *TSV*

Los Dos Hermanos A woman causes two brothers to hate each other.
1987 92m/C *MX* Jorge Rivero, Gregorio Casal, Nadia Milton, Sergio Jimenez. **VHS, Beta** $75.95 *MAD* ♂♂½

Los Drogadictos A woman avenges her sister's death from a drug overdose by sabotaging a powerful drug ring.
198? 105m/C *SP* Mercedes Carreras, Graciela Alfano, Juan Jose Camero. **VHS, Beta** $57.95 *UNI* ♂

Los Hermanos Centella Two brothers from Northern Mexico change from bandits to people who protect the oppressed.
1965 105m/C *MX* Dacia Gonzales, Jaime Fernandez, Dagoberto Rodriguez, Guillermo Rivas. **VHS, Beta** $39.95 *MAD* ♂♂

Los Hermanos Diablo Jose, Julio, and Juan are the Diablo brothers, and they are determined to keep outlaws and Indians off their ranch.
1965 90m/B *MX* Mauricio Garces, Abel Salazar, Rafael Baledon. **VHS, Beta** $39.95 *MAD* ♂♂

Los Hijos de Lopez A powerful Spanish dynasty's trio of brothers compete for wealth and power.
19?? 87m/C *SP* Alberto Martin, Jorge Barreiro, Dorys Del Valle. **VHS, Beta** $59.95 *UNI*

Los Hombres Piensan Solo en Solo Two ne'er-do-wells get the experience of their lives when they baby-sit a "little sister."
1976 90m/C *SP* Alberto Olmedo, Susana Gimenez, Jorge Porcel. **VHS, Beta** $49.95 *MED*

Los Jinetes de la Bruja (The Horseman...) A witch and her horsemen avenge the death of innocent puppeteers on a Frenchman's ranch.
1976 93m/C Blanca Sanchez, Kitty De Hoyos, Ferdinando Almada, Mario Almada. **VHS, Beta** $69.95 *MAD* ♂

Los Meses y los Dias A teenage girl runs away from home and finds that her experiences are forcing her into womanhood.
1970 100m/C *MX* Maritza Olivares, Blanca Pastor; *Dir:* Alberto Bojorquez. **VHS, Beta** $59.95 *VLA* ♂

Los Muchachos de Antes No Usaban Arsenico Spanish murder mystery.
198? 90m/C *SP* Narciso Ibanez Menta, Barbara Mujica. **VHS, Beta** $59.95 *UNI* ♂

Los Olvidados From surrealist Luis Bunuel, a powerful story of the poverty and violence of young people's lives in Mexico's slums. In Spanish with English subtitles. Also known as "The Young and The Damned."
1950 81m/B *MX* Alfonso Mejia, Roberto Cobo; *Dir:* Luis Bunuel. Cannes Film Festival '51: Best Director (Bunuel). **VHS, Beta, 8mm** $68.00 *NOS, HHT, DVT* ♂♂♂½

Los Sheriffs de la Frontera Two lawmen have to rid a town of mean bad guys.

1947 90m/B *MX* Fernando Casanova, Juan Gallardo. **VHS, Beta** $39.95 *MAD* ♂♂

Los Tres Amores de Losa A girl becomes friends with a famous actress in Madrid and then falls in love with her cousin.
1947 90m/C *MX* Lola Flores, Agustin Lara, Luis Aguilar, Abel Salazar. **VHS, Beta** $39.95 *MAD* ♂♂

Los Tres Calaveras (The Three Skeletons) The story of the Skeleton trio from its beginnings to the top of the international music charts.
1964 105m/B *MX* Javier Solis, Lucha Villa, Joaquin Cordero, Manuel Lopez Ochoa, Ruben Zepeda Novelo; *Dir:* Fernando Cortes. **VHS, Beta** $44.95 *MAD* ♂

Los Vampiros de Coyacan Two wrestlers are summoned to help a girl who has fallen under the spell of a vampire.
1987 87m/C *MX* German Robles, Mil Mascaras, Superzan, Sasha Montenegro. **VHS, Beta** $69.95 *MAD* ♂

The Losers Four motorcyclists are hired by the U.S. Army to rescue a presidential advisor who is being held captive by Asian bad guys.
1970 (R) 96m/C William Smith, Bernie Hamilton, Adam Roarke, Houston Savage, Brad Johnson, Vernon Wells; *Dir:* Jack Starrett. **VHS, Beta** $29.95 *ACA* Woof!

Losin' It Four teens travel across the Mexican border to Tijuana on a journey to lose their virginity. Cruise meets a married woman who says she is in town for a divorce, while the others become caught up in frenzied undertakings of their own.
1982 (R) 104m/C Tom Cruise, John Stockwell, Shelley Long, Jackie Earle Haley; *Dir:* Curtis Hanson. **VHS, Beta, LV** $9.98 *SUE* ♂♂

Lost A young girl runs away into the wilderness because of the resentment she feels toward her new stepfather.
1983 92m/C Sandra Dee, Don Stewart, Ken Curtis, Jack Elam, Sheila Newhouse. **VHS, Beta** $49.95 *PSM* ♂½

Lost When their boat capsizes in the Pacific, three sailors desperately cling to life, drifting aimlessly for 74 days. Based on a true story, this film is adult-fare.
1986 93m/C *CA* Michael Hogan, Helen Shaver, Ken Welsh; *Dir:* Peter Rowe. **VHS, Beta** $79.98 *CGH, PEV, HHE* ♂♂

Lost in America After deciding that he can't "find himself" at his current job, advertising executive David Howard and his wife sell everything they own and buy a Winnebago to travel across the country. This Albert Brooks comedy is a must-see for everyone who thinks that there is more in life than pushing papers at your desk and sitting on "Mercedes leather." Available in widescreen on laserdisc.
1985 (R) 91m/C Albert Brooks, Julie Hagerty, Michael Greene, Tom Tarpey, Garry Marshall; *Dir:* Albert Brooks. **VHS, Beta, LV** $19.98 *WAR, LDC* ♂♂♂

Lost Angels A glossy "Rebel Without a Cause" 80s reprise providing a no-holds-barred portrait of life in the fast lane. A wealthy, disaffected San Fernando Valley youth immerses himself in sex, drugs and rock 'n' roll. Ultimately he is arrested and sent by his parents to a youth home, where a dedicated therapist assists his tortuous road back to reality.

1989 (R) 116m/C Donald Sutherland, Adam Horowitz, Amy Locane, Kevin Tighe, John C. McGinley; *Dir:* Hugh Hudson. **VHS, Beta, LV** $19.98 *ORI* ♂♂½

Lost in the Barrens Two young boys, one a Native American and the other a rich white boy, get lost in the wilderness. Out of necessity and common need they become close and form a lifelong friendship.
1991 95m/C Graham Greene, Nicholas Shields. **VHS** *CGV* ♂♂

The Lost Boys Santa Cruz seems like a dull town when Michael, his younger brother, and their divorced mom move into their eccentric grandfather's home. But when Michael falls for a pretty girl with some hard-living friends he takes on more than he imagines—these partying teens are actually a group of vampires. Some humor, some bloodletting violence, and an attractive cast help out this updated vampire tale. Rock-filled soundtrack.
1987 (R) 97m/C Jason Patric, Kiefer Sutherland, Corey Haim, Jami Gertz, Dianne Wiest, Corey Feldman, Barnard Hughes, Edward Herrmann; *Dir:* Joel Schumacher. **VHS, Beta, LV, 8mm** $19.98 *WAR* ♂♂

The Lost Capone Made for cable TV version of the story of Al Capone's youngest brother, a clean living small town sheriff who struggles with his sibling's reputation at every turn.
1990 (PG-13) 93m/C Ally Sheedy, Eric Roberts, Adrian Pasdar; *W/Dir:* John Gray. **VHS, Beta** $79.95 *TTC* ♂♂♂

The Lost City A feature version of the rollicking vintage movie serial about a lost jungle city, adventurers and mad scientists.
1934 74m/B William Boyd, Kane Richmond, George "Gabby" Hayes, Claudia Dell; *Dir:* Harry Revier. **VHS, Beta** $19.95 *NOS, SNC, DVT* ♂♂

Lost City of the Jungle 13-chapter serial focusing on a crazed Atwill, in his last screen role, believing that he can rule the world from the heart of a deep, dark jungle by utilizing a special mineral. The final Universal serial.
1945 169m/B Russell Hayden, Lionel Atwill, Jane Adams; *Dir:* Ray Taylor, Lewis D. Collins. **VHS, Beta** $49.95 *NOS, GPV, DVT* ♂♂

The Lost Command A French colonel, relieved of his command, endeavors to regain power by battling a powerful Arab terrorist with his own specially trained platoon of soldiers. Set in post-World-War-II North Africa. Based on "The Centurions" by Jean Larteguy.
1966 129m/C Anthony Quinn, Michele Morgan, George Segal, Alain Delon, Maurice Ronet, Claudia Cardinale; *Dir:* Mark Robson. **Beta** $69.95 *COL* ♂½

The Lost Continent An expedition searching for a lost rocket on a jungle island discovers dinosaurs and other extinct creatures.
1951 82m/B Cesar Romero, Hillary Brooke, Chick Chandler, John Hoyt, Acquanetta, Sid Melton, Whit Bissell, Hugh Beaumont; *Dir:* Sam Newfield. **VHS, Beta, LV** $19.98 *WGE, MRV, MLB* ♂½

Lost Diamond Bumbling spies search for a smuggled gem.
1974 83m/C Juan Ramon, Sonia Rivas, Ricardo Bauleo. **VHS, Beta** $39.95 *MED* ♂

The Lost Empire Three bountiful and powerful women team up to battle the evil Dr. Syn Do.

1983 (R) 86m/C Melanie Vincz, Raven De La Croix, Angela Aames, Paul Coufos, Robert Tessier, Angus Scrimm, Angelique Pettyjohn, Kenneth Tobey; *Dir:* Jim Wynorski. VHS, Beta $69.98 LIV, LTG *Ⅰ*

Lost and Found An American professor of English and an English film production secretary fall in love on a skiing vacation. Good cast but Segal and Jackson did romance better in "A Touch of Class."
1979 (PG) 104m/C George Segal, Glenda Jackson, Maureen Stapleton, Hollis McLaren, John Cunningham, Paul Sorvino, John Candy, Martin Short; *Dir:* Melvin Frank. VHS, Beta $9.95 GKK *ⅠⅠ*½

Lost Honeymoon Monkeyshines abound as a soldier marries a girl while in a state of amnesia, and then wakes up to find twin daughters.
1947 70m/B Franchot Tone, Ann Richards; *Dir:* Leigh Jason. VHS, Beta $24.95 NOS, DVT *ⅠⅠ*

The Lost Honor of Katharina Blum A woman becomes involved with a man who's under police surveillance and finds her life open to public scrutiny and abuse from the media and the government. Based on Heinrich Boll's Nobel Prize-winning novel. In German with English subtitles. Remade for television as "The Lost Honor of Kathryn Beck."
1975 (R) 97m/C GE Angela Winkler, Mario Adorf, Dieter Laser, Juergen Prochnow; *Dir:* Volker Schlondorff, Margarethe von Trotta. New York Film Festival '75: Best Director. VHS, Beta $29.98 SUE, APD, GLV *ⅠⅠⅠ*

Lost Horizon A group of strangers fleeing revolution in China are lost in the Tibetan Himalayas and stumble across the valley of Shangri-La. The inhabitants of this Utopian community have lived for hundreds of years in kindness and peace but what will the intrusion of these strangers bring. The classic romantic role for Colman. Capra's directorial style meshed perfectly with the pacifist theme of James Hilton's classic novel, resulting in one of the most memorable films of the 1930s. This version restores more than 20 minutes of footage which had been cut from the movie through the years. Music score by Dimitri Tiomkin. Oscar nominated for Best Picture, Best Supporting Actor (H.B. Warner), Best Sound Recording, Best Assistant Director, and Best Score.
1937 132m/B Ronald Colman, Jane Wyatt, H.B. Warner, Sam Jaffe, Thomas Mitchell, Edward Everett Horton, Isabel Jewell, John Howard, Margo; *Dir:* Frank Capra. Academy Awards '37: Best Art Direction/Set Decoration, Best Film Editing. VHS, Beta, LV $19.95 COL, PIA, RDG *ⅠⅠⅠⅠ*

The Lost Idol/Shock Troop In "The Lost Idol," a lone soldier raids a sacred temple in order to retrieve a golden idol. "Shock Troop" tells of top secret mission to destroy a powerful computer that lies behind enemy lines. Also available on Laser Disc.
198? 219m/C Danny Aiello, Lyle Alzado, Erik Estrada. VHS $99.98 SHG *ⅠⅠⅠⅠ*

The Lost Jungle Exciting animal treasure hunt; danger and mystery. Serial in 12 chapters, 13 minutes each.
1934 156m/B Clyde Beatty, Cecilia Parker. VHS, Beta $19.95 NOS, VCN, VDM *Ⅰ*½

Lost Legacy: A Girl Called Hatter Fox Tradition and technology are at odds in the life of a young Indian girl. Strong cast makes this work. Made for television and better than average.

1977 100m/C Ronny Cox, Joanelle Romero, Conchata Ferrell; *Dir:* George Schaefer. VHS $59.95 GEM *ⅠⅠ*½

The Lost Missile A lost, alien missile circles the Earth, causing overheating and destruction on the planet's surface. A scientist works to find a way to save the planet before it explodes into a gigantic fireball. Director Burke's last film.
1958 70m/B Robert Loggia, Ellen Parker, Larry Kerr, Phillip Pine, Marilee Earle; *Dir:* William Berke. VHS $18.00 FRG *ⅠⅠ*

Lost Moment A publisher travels to Italy to search for a valuable collection of a celebrated author's love letters, but finds a neurotic woman in his way. Based on Henry James' "Aspern Papers."
1947 89m/B Robert Cummings, Agnes Moorehead, Susan Hayward; *Dir:* Martin Gabel. VHS $14.98 REP *ⅠⅠ*

The Lost One A German scientist's lover is suspected of selling his findings to England during World War II. Based on a true story. Lorre's only directorial outing.
1951 97m/B GE Peter Lorre, Karl John, Renate Mannhardt; *Dir:* Peter Lorre. VHS KIV *ⅠⅠ*

The Lost Patrol WWI British soldiers lost in the desert are shot down one by one by Arab marauders as Karloff portrays a religious soldier convinced he's going to die. The usual spiffy Ford exteriors peopled by great characters with a stirring score by Max Steiner. Based on the story, "Patrol" by Philip MacDonald.
1934 66m/B Victor McLaglen, Boris Karloff, Reginald Denny, Wallace Ford, Alan Hale Jr.; *Dir:* John Ford. National Board of Review Awards '34: 10 Best Films of the Year. VHS, Beta $14.98 CCB, MED, RXM *ⅠⅠⅠ*½

Lost Planet Airmen A feature-length condensation of the 12-part sci-fi serial "King of the Rocket Men." Rocket Man is pitted against the sinister Dr. Vulcan in this intergalactic battle of good and evil.
1949 65m/B Tristram Coffin, Mae Clarke, Dale Van Sickel; *Dir:* Fred Brannon. VHS, Beta AOV, DVT, RXM *Ⅰ*½

The Lost Platoon A troop of soldiers are transformed into vampires.
1989 (R) 120m/C David Parry, William Knight, Sean Heyman; *Dir:* David A. Prior. VHS, Beta $79.95 AIP *Ⅰ*½

The Lost Samurai Sword A group of characters searches for the legendary sword, which bestows awesome powers on the person who possesses it.
198? 90m/C VHS, Beta $49.95 OCE *Ⅰ*½

Lost in Space The complete pilot episode of the series, in which the Robinson family becomes stranded somewhere in the universe.
1965 52m/B Guy Williams, June Lockhart, Mark Goddard, Marta Kristen, Angela Cartwright, Billy Mumy, Jonathan Harris; *Dir:* Irwin Allen. VHS, Beta VYY

Lost Squadron A look at the dangers stuntmen go through in movie-making.
1932 79m/B Richard Dix, Erich von Stroheim; *Dir:* George Archainbaud. VHS, Beta KOV, RXM *ⅠⅠ*½

The Lost Stooges From the trio's one year at MGM, rare clips of them performing their famous slaptick gags. In black and white and color.

1933 68m/B Moe Howard, Curly Howard, Larry Fine, Clark Gable, Joan Crawford, Jimmy Durante, Robert Montgomery; *Nar:* Leonard Maltin. VHS, Beta $14.98 TTC

The Lost Tribe A man takes his wife on a trek through the jungle to find his lost brother.
1989 96m/C John Bach, Darien Teakle; *Dir:* John Laing. VHS, Beta MED *Ⅰ*

The Lost Weekend The heartrending Hollywood masterpiece about alcoholism, depicting a single weekend in the life of a writer, who cannot believe he's addicted. Except for its pat ending, it is an uncompromising, startlingly harsh treatment, with Milland giving one of the industry's bravest lead performances ever. Acclaimed then and now. Besides the Oscars it won, also nominated for Black & White Cinematography, Best Score of a Drama or Comedy, and Film Editing.
1945 100m/B Ray Milland, Jane Wyman, Phillip Terry, Howard da Silva, Doris Dowling, Frank Faylen; *Dir:* Billy Wilder. Academy Awards '45: Best Actor (Milland), Best Director (Wilder), Best Picture, Best Screenplay; Cannes Film Festival '46: Best Actor (Milland), Best Film (tie); National Board of Review Awards '45: 10 Best Films of the Year, Best Actor (Milland). VHS, Beta, LV $14.95 MCA, PAR, BTV *ⅠⅠⅠⅠ*

The Lost World A zoology professor leads a group on a South American expedition in search of the "lost world," where dinosaurs roam in this silent film. Based on a story by A. Conan Doyle.
1925 62m/B Wallace Beery, Lewis Stone, Bessie Love, Lloyd Hughes; *Dir:* Harry Hoyt. VHS, Beta, LV $19.95 NOS, MRV, VYY *ⅠⅠ*½

Lost Zeppelin A dirigible becomes lost in the wastes of Antarctica, forcing its passengers to combat the elements. Impressive special effects and miniatures for its time.
1929 73m/B Conway Tearle, Virginia Valli, Ricardo Cortez, Duke Martin, Kathryn McGuire; *Dir:* Edward Sloman. VHS, Beta $16.95 SNC *ⅠⅠ*

Lots of Luck A knee-slapping comedy about a family that wins the million-dollar lottery and sees that money doesn't solve all problems. Disney made for cable.
1985 88m/C Martin Mull, Annette Funicello, Fred Willard, Polly Holliday; *Dir:* Peter Baldwin. VHS, Beta $69.95 DIS *ⅠⅠ*

Lottery Bride In this charming musical, Jeanette MacDonald is the lottery bride who is won by the brother of the man she really loves. A fine outing for all involved, particularly the supporting cast.
1930 85m/C Jeanette MacDonald, Joe E. Brown, ZaSu Pitts, John Garrick, Carroll Nye; *Dir:* Paul Stein. VHS, Beta $34.98 CCB *ⅠⅠ*½

Lou Gehrig: King of Diamonds An account of the baseball career of Lou Gehrig, "Iron Man" of baseball, including his touching farewell at Yankee Stadium, on July 4th, 1939. Classic newsreel footage.
1954 15m/B VHS, Beta $49.95 TSF *ⅠⅠ*½

Lou Gehrig Story An early television drama based upon the great ballplayer's life.
1956 60m/B Wendell Corey, Jean Hagen, James Gregory, Harry Carey Jr. VHS, Beta $24.95 NOS, VYY, DVT *ⅠⅠ*½

Louie Bluie A documentary about the life and times of Armstrong, the last original black string musician, with views on his other talents, such as painting.
1985 60m/C Howard Armstrong, Ted Bogan. VHS, Beta $14.95 MVD, PAV *ⅠⅠ*½

Louisiana A belle of the Old South tries to get back the family plantation by romancing the new owner, even though she loves someone else. Oh, and then the Civil War breaks out. Made for television.
1987 130m/C Ian Charleson, Margot Kidder, Victor Lanoux, Len Cariou, Lloyd Bochner; *Dir:* Phillipe DeBroca. **VHS, Beta** $9.99 STE, PSM ⅃

Louisiana Story The final effort by the master filmmaker, depicting the effects of oil industrialization on the southern Bayou country as seen through the eyes of a young boy. One of Flaherty's greatest, widely considered a premiere achievement. Music is provided by composer Virgil Thomson.
1948 77m/B *Dir:* Robert Flaherty. British Academy Awards '48: Best Documentary; National Board of Review Awards '48: 10 Best Films of the Year; Sight & Sound Survey '52: #5 of the Best Films of All Time (tie). **VHS, Beta** $29.95 HMV, VYY, DVT ⅃⅃⅃

Loulou A woman leaves her middle-class husband for a leather-clad, uneducated jock who is more attentive. Romantic and erotic. In French with English subtitles.
1980 (R) 110m/C *FR* Isabelle Huppert, Gerard Depardieu, Guy Marchand; *Dir:* Maurice Pialat. **VHS, Beta** $79.95 NYF ⅃⅃⅃½

The Lovable Cheat Pretty lame adapatation of Balzac's "Mercadet Le Falseur," about a father who cons money from his friends in order to line up a marriage suitable for his daughter.
1949 75m/B Charlie Ruggles, Peggy Ann Garner, Richard Ney, Alan Mowbray, Iris Adrian, Ludwig Donath, Fritz Feld; *Dir:* Richard Oswald. **VHS** FCT ⅃⅃

Love Torocsik, touted as Hungary's leading actress in the 70's, plays a young woman whose husband has been imprisoned for political crimes. Living in a cramped apartment with her mother in law, she keeps the news from the aged and dying woman (Darvas) by reading letters she's fabricated to keep alive the woman's belief that her son is a successful movie director in America. Set in their shabby apartment, the story is punctuated by the older woman's dreamy remembrances of things past. Exceptional performances by both women, it was Darvas' final film. Based on two novellas by Tibor Dery, it won a special jury prize at the 1973 Cannes film festival and was listed among the year's best films by "Time" and "The New York Times." In Hungarian with English subtitles.
1972 100m/C *HU* Lili Darvas, Mari Torocsik, Ivan Darvas; *Dir:* Karoly Makk. Cannes Film Festival '73: Special Jury Prize. **VHS, LV** $59.95 CVC, IME ⅃⅃⅃½

Love Affair An excellent comedy about a shipboard romance which gets side-tracked by events back on shore. McCarey remade his own film in 1957 with Cary Grant and Deborah Kerr as "An Affair to Remember." Received Oscar nominations for Best Picture, Best Actress (Dunne), Best Supporting Actress (Ouspenskaya), Best Original Story, and Best Interior Decoration.
1939 87m/B Irene Dunne, Charles Boyer, Maria Ouspenskaya, Lee Bowman, Astrid Allwyn; *Dir:* Leo McCarey. **VHS** $19.95 NOS, MRV, TTC ⅃⅃⅃½

The Love Affair, or Case of the Missing Switchboard Operator Makavejev's second film, a dissertation on the relationship between sex and politics, involving an affair between a switchboard operator and a middle-aged ex-revolutionary, is told in the director's unique, farcically disjointed manner. Alternate titles include "Switchboard Operator" and "Case of the Missing Switchboard Operator" and "Ljubarni Slucaj."
1967 73m/B *YU* Eva Ras, Slobodan Aligrudic, Ruzica Sokic; *Dir:* Dusan Makavejev. **VHS, Beta** $59.95 FCT ⅃⅃⅃

Love Affair: The Eleanor & Lou Gehrig Story The true story, told from Mrs. Gehrig's point of view, of the love affair between baseball great Lou Gehrig and his wife Eleanor from his glory days as a New York Yankee, to his battle with an incurable disease. A television drama that is supported by Herrmann and Danner's convincing portrayals.
1977 96m/C Blythe Danner, Edward Herrmann, Patricia Neal, Ramon Bieri, Lainie Kazan; *Dir:* Fielder Cook. **VHS, Beta** $7.50 WOV ⅃⅃⅃

Love in the Afternoon A Parisian private eye's daughter (Hepburn) decides to investigate a philandering American millionaire (Cooper) and winds up falling in love with him. Cooper's a little old for the Casanova role but Hepburn is always enchanting.
1957 126m/B Gary Cooper, Audrey Hepburn, John McGiver, Maurice Chevalier; *Dir:* Billy Wilder. **VHS, Beta** $49.98 FOX, IME ⅃⅃½

Love Among the Ruins A made-for-television romance about an aging, wealthy widow who, after being scandalously sued for breach of promise by her very young lover, turns for aid to an old lawyer friend who has loved her silently for more than forty years.
1975 100m/C Laurence Olivier, Katharine Hepburn, Leigh Lawson, Colin Blakely; *Dir:* George Cukor. **VHS, Beta, LV** $59.98 FOX ⅃⅃⅃

Love and Anarchy An oppressed peasant vows to assassinate Mussolini after a close friend is murdered. Powerful drama about the rise of Italian facism. In Italian with English subtitles.
1973 108m/C *IT* Giancarlo Giannini, Mariangela Melato; *Dir:* Lina Wertmuller. **VHS, Beta** $59.95 COL, APD ⅃⅃⅃

Love Angels Italian prostitutes are stalked by a killer. Soft-core nudity; dubbed.
1987 80m/C *IT* Maria Fiore, Krista Nell, Andrea Scotti. **VHS, Beta** $29.95 MED *Woof!*

Love by Appointment An unlikely romantic comedy with a very unlikely cast has two businessmen meeting up with European prostitutes. Made for television.
1976 96m/C Ernest Borgnine, Robert Alda, Francoise Fabian, Corinne Clery. **VHS, Beta** $59.98 CHA ⅃

The Love Bug A race car driver (Jones) is followed home by Herbie, a white Volkswagen with a mind of its own. Eventually, Jones follows the "Love Bug" to a life of madcap fun. Followed by several sequels.
1968 (G) 110m/C Dean Jones, Michele Lee, Hope Lange, Robert Reed, Bert Convy; *Dir:* Robert Stevenson. **VHS, Beta, LV** $19.99 DIS ⅃⅃½

Love and Bullets An Arizona homicide detective is sent on a special assignment to Switzerland to bring a mobster's girlfriend back to the United States to testify against him in court.
1979 (PG) 95m/C *GB* Charles Bronson, Jill Ireland, Rod Steiger, Strother Martin, Bradford Dillman, Henry Silva, Michael Gazzo; *Dir:* Stuart Rosenberg. **VHS, Beta** $19.98 FOX ⅃½

Love Butcher A crippled old gardener kills his female employers with his garden tools and cleans up neatly afterward.
1982 (R) 84m/C Erik Stern, Kay Neer, Robin Sherwood; *Dir:* Mikel Angel, Don Jones. **VHS, Beta** $39.95 MON ⅃½

Love Camp A woman is invited to a swinger's holiday camp, frolics for a while, then is told she can never leave. Suspenseful hijinks ensue.
1976 100m/C Laura Gemser, Christian Anders, Gabriele Tinti. **VHS, Beta** $29.95 AHV ⅃

Love Child The story of a young woman in prison who becomes pregnant by a guard and fights to have and keep her baby.
1982 (R) 97m/C Amy Madigan, Beau Bridges, MacKenzie Phillips, Albert Salmi; *Dir:* Larry Peerce. **VHS, Beta** $49.95 WAR ⅃⅃

Love Circles Soft-core fluff about a pack of cigarettes that perpetually exchanges hands (revealing a handful of smutty encounters), and evidently goes unsmoked.
1985 (R) 84m/C John Sibbit, Marie France, Josephine Jacqueline Jones, Pierre Burton. **VHS, Beta** $79.95 MGM *Woof!*

Love in the City Five stories of life, love and tears in Rome. In Italian with English subtitles and narration.
1953 90m/B *IT* Ugo Tognazzi, Maresa Gallo, Caterina Rigogloso, Silvia Lillo; *Dir:* Michelangelo Antonioni, Federico Fellini, Dino Risi, Carlo Lizzani, Alberto Lattuada, Francesco Maselli, Cesare Zavattini. **VHS, Beta** $24.95 NOS, FCT, APD ⅃⅃⅃

Love Crimes A con man (Bergin) poses as a photographer who sexually intimidates women while playing on their erotic fantasies. Young is the Atlanta district attorney who sets out to nail him when none of his victims will testify against him, only she may be enjoying her undercover work more than she realizes. Implausible and sleazy. Also available in an unrated version.
1991 (R) 84m/C Sean Young, Patrick Bergin, Arnetia Walker, James Read; *Dir:* Lizzie Borden. **VHS, LV** $92.99 HBO ⅃

Love and Death In 1812 Russia, a condemned man reviews the follies of his life. Woody Allen's satire on "War and Peace," and every other major Russian novel.
1975 (PG) 89m/C Woody Allen, Diane Keaton, Georges Adel, Despo, Frank Adu; *Dir:* Woody Allen. **VHS, Beta, LV** $19.98 FOX ⅃⅃⅃

Love Desperados Cowboys and ranchers' wives mix it up in a lot of softcore hay.
1986 99m/C Virginia Gordon, James Arena. **VHS, Beta** $29.95 MED *Woof!*

Love on the Dole In a gloomy industrial section of England during the early '30s a family struggles to survive and maintain dignity. Grim Depression drama salvaged by great acting.
1941 89m/B *GB* Deborah Kerr, Clifford Evans, George Carney; *Dir:* John Baxter. **VHS, Beta** $19.95 NOS, VYY ⅃⅃⅃

Love, Drugs, and Violence Features ten shorts by independent filmmakers including the satire, "The Reagans Speak Out on Drugs," created from clips of the president and his wife, and edited to reverse their stand on drugs.
198? 60m/C **VHS** $29.95 PIC, FCT ⅃⅃⅃

Love 'Em and Leave 'Em A tale of life and love in the dizzying, roaring 20's starring the sexy and capricious Louise Brooks. Silent with orchestral score.
1927 70m/B Louise Brooks, Evelyn Brent, Lawrence Gray; *Dir:* Frank Tuttle. **VHS, Beta** $24.95 GPV ⅃⅃⅃

Love and Faith Two lovers are torn between their love for each other and their faiths during sixteenth-century Japan. English subtitles.
1978 154m/C *JP* Toshiro Mifune, Takashi Shimura, Yoshiko Nakana; *Dir:* Kei Kumai. **VHS, Beta** *VDA* ♫♫♫

Love Finds Andy Hardy Young Andy Hardy finds himself torn between three girls before returning to the girl next door. Garland's first appearance in the acclaimed Andy Hardy series features her singing "In Between" and "Meet the Best of my Heart." Also available with "Andy Hardy Meets Debutante" on Laser Disc.
1938 90m/B Mickey Rooney, Judy Garland, Lana Turner, Ann Rutherford, Fay Holden, Lewis Stone, Marie Blake, Cecilia Parker, Gene Reynolds; *Dir:* George B. Seitz. **VHS, Beta, LV $19.95** *MGM* ♫♫♫

Love at First Bite Intentionally campy spoof of the vampire film. Dracula is forced to leave his Transylvanian home as the Rumanian government has designated his castle a training center for young gymnasts. Once in New York, the Count takes in the night life and falls in love with a woman whose boyfriend embarks on a campaign to warn the city of Dracula's presence. Hamilton of the never-fading tan is appropriately tongue-in-cheek in a role which resurrected his career.
1979 (PG) 96m/C George Hamilton, Susan St. James, Richard Benjamin, Dick Shawn, Arte Johnson, Sherman Hemsley, Isabel Sanford; *Dir:* Stan Dragoti. **VHS, Beta, LV** *WAR, OM* ♫♫½

Love at First Sight A pre-"Saturday Night Live" hack job for Aykroyd, playing a blind man who falls in love with a girl he bumps into.
1976 86m/C Dan Aykroyd, Mary Ann MacDonald, George Murray, Barry Morse; *Dir:* Rex Bromfield. **VHS, Beta $19.95** *VIR* ♫

The Love Flower A man kills his second wife's lover and escapes with his daughter to a tropical island, pursued by a detective and a young adventurer. Interesting ending wraps things up nicely. Silent.
1920 70m/B Carol Dempster, Richard Barthelmess, George MacQuarrie, Anders Randolf, Florence Short; *W/Dir:* D.W. Griffith. **VHS, Beta $29.95** *DNB* ♫♫½

Love and the Frenchwoman A French tale tracing the nature of love through stages. Deals with a story about where babies come from, puppy love, saving sex for marriage, and the way some men treat women.
1960 135m/B *FR* Jean-Paul Belmondo, Pierre-Jean Vaillard, Marie-Jose Nat, Annie Girardot; *Dir:* Jean Delannoy. **VHS, Beta $29.95** *DVT, MRV* ♫♫♫

A Love in Germany A tragic love affair develops between a German shopkeeper's wife and a Polish prisoner-of-war in a small German village during World War II.
1984 (R) 110m/C *FR GE* Hanna Schygulla, Piotr Lysak, Elisabeth Trissenaar, Armin Mueller-Stahl; *Dir:* Andrzej Wajda. **VHS, Beta $59.95** *COL, APD, GLV* ♫♫½

The Love Goddesses A sixty-year examination of some of the most beautiful women on the silver screen, reflecting with extraordinary accuracy the customs, manners and mores of the times. Released theatrically in 1972.
1965 83m/B Marlene Dietrich, Greta Garbo, Jean Harlow, Gloria Swanson, Mae West, Betty Grable, Rita Hayworth, Elizabeth Taylor, Marilyn Monroe, Theda Bara, Claudette Colbert, Dorothy

Lamour; *Dir:* Saul J. Turell. **VHS, Beta, LV $14.95** *COL, SUE, VYG* ♫♫½

Love Happy A group of impoverished actors accidentally gain possession of valuable diamonds. Unfortunately for them, detective Groucho is assigned to recover them!
1950 85m/B Groucho Marx, Harpo Marx, Chico Marx, Vera-Ellen, Ilona Massey, Marion Hutton, Raymond Burr, Marilyn Monroe; *Dir:* David Miller. **VHS, LV $19.98** *REP* ♫♫½

Love and Hate Marital hell/murder mystery based on a true Canadian case. Joanne (Nelligan), the wife of wealthy rancher-politico Colin (Walsh), leaves her publicly charismatic and privately abusive husband with two thirds of their brood. A bitter battle for custody is waged, and Joanne is soon found savagely slain. The number one suspect: philandering ex-spouse Colin. Based on Maggie Siggins' "A Canadian Tragedy," the faux biography originally aired in two parts on Canadian TV as "Love and Hate: The Story of Colin and Joanne Thatcher."
1990 156m/C Kate Nelligan, Ken Walsh, Leon Pownall, John Colicos, Noam Zylberman, Victoria Snow, Cedric Smith; *Dir:* Francis Mankiewicz. **VHS, Beta, LV $99.95** *PAR* ♫♫½

Love Hurts A guy looking for romance finds his hands full with a number of beautiful women. Will he find the love he craves, or will the pain be too much to bear?
1991 (R) 110m/C Jeff Daniels, Judith Ivey, John Mahoney, Cynthia Sikes, Amy Wright; *Dir:* Bud Yorkin. **VHS $89.98** *LIV, VES* ♫♫

Love Kills Is the man a beautiful heiress falls in love with actually an assassin hired by her husband to kill her? Find out in this steamy suspenser.
1991 (PG-13) 92m/C Virginia Madsen, Lenny Von Dohlen, Erich Anderson, Kate Hodge, Jim Metzler; *Dir:* Brian Grant. **VHS, Beta** *PAR* ♫♫

Love at Large Hired by a beautiful woman, a private detective accidentally follows the wrong man and winds up being followed himself. He vies with a female detective in solving this case of mistaken identity.
1989 (R) 90m/C Tom Berenger, Elizabeth Perkins, Anne Archer, Ann Magnuson, Annette O'Toole, Kate Capshaw, Ted Levine, Kevin J. O'Connor, Ruby Dee, Neil Young; *Dir:* Alan Rudolph. **VHS, Beta, LV $89.98** *ORI, FCT* ♫♫♫

Love Laughs at Andy Hardy Andy Hardy, college boy, is in love and in trouble. Financial and romantic problems come to a head when Andy is paired with a six-foot tall blind date. One in the series.
1946 93m/B Mickey Rooney, Lewis Stone, Sara Haden, Lina Romay, Bonita Granville, Fay Holden; *Dir:* Willis Goldbeck. **VHS, Beta $19.95** *NOS, MRV, QNE* ♫♫

Love Leads the Way The true story of how Morris Frank established the seeing-eye dog system in the 1930s.
1984 99m/C Timothy Bottoms, Eva Marie Saint, Arthur Hill, Susan Dey, Ralph Bellamy, Ernest Borgnine, Patricia Neal; *Dir:* Delbert Mann. **VHS, Beta $69.95** *DIS* ♫♫

Love Lessons An Italian-made softcore film about a millionaire who hires a sultry art teacher to instruct his virginal son. Dubbed.
1985 80m/C *IT* Maria Louise Zetha, Yara Dawe, Ben Carra. **VHS, Beta $29.95** *MED* ♫♫

Love Letters A young disc jockey falls under the spell of a box of love letters that her mother left behind which detailed her double life. She, in turn, begins an affair with a married man. Thoughtful treatment of the

psychology of infidelity. Also known as "Passion Play."
1983 (R) 102m/C Jamie Lee Curtis, Amy Madigan, Bud Cort, Matt Clark, Bonnie Bartlett, Sally Kirkland, James Keach; *Dir:* Amy Jones. **VHS, Beta, LV $79.98** *LIV, VES, IME* ♫♫♫

The Love Machine A power-hungry newscaster climbs the corporate ladder by sleeping with many, including the president's wife. An adaptation of Jacqueline Susann's novel.
1971 (R) 108m/C John Phillip Law, Dyan Cannon, Robert Ryan, Jackie Cooper, David Hemmings, Jodi Wexler, William Roerick, Maureen Arthur, Shecky Greene, Clinton Greyn, Sharon Farrell, Alexandra Hay, Eve Bruce, Greg Mullavey, Edith Atwater, Gene Baylos, Claudia Jennings, Mary Collinson, Madeleine Collinson, Ann Ford, Gayle Hunnicutt; *Dir:* Jack Haley Jr. **VHS, Beta $69.95** *COL, MLB* ♫½

Love has Many Faces Judith Crist was too kind when she said this was for connoisseurs of truly awful movies. Playgal Turner marries beachboy Robertson and many faces come between them. Much melodrama. Filmed on location.
1965 104m/C Lana Turner, Cliff Robertson, Hugh O'Brian, Ruth Roman, Stefanie Powers, Virginia Grey, Ron Husmann; *Dir:* Alexander Singer. **VHS, Beta $9.95** *GKK* ♫½

Love is a Many-Splendored Thing A married American war correspondent and a beautiful Eurasian doctor fall in love in post-World War II Hong Kong. They struggle with racism and unhappiness, until he's sent to Korea to observe the Army's activities there. Based on the novel by Han Suyin. The extensive Los Angeles Oriental acting community got some work out if this film, although the leads are played by Caucasians. Oscar winning song was a very big popular hit. Other Oscar nominations include: Best Picture, Best Actress (Jones), and Sound Recording.
1955 102m/C William Holden, Jennifer Jones, Torin Thatcher, Isobel Elsom, Jorja Curtright, Virginia Gregg, Richard Loo; *Dir:* Henry King. Academy Awards '55: Best Costume Design (Color), Best Song ("Love Is a Many-Splendored Thing"), Best Musical Score. **VHS, Beta, LV $34.98** *FOX* ♫♫½

Love Me Deadly A young woman tries to get her husband interested in her new hobby-necrophilia.
1976 (R) 95m/C Mary Wilcox, Lyle Waggoner, Christopher Stone, Timothy Scott. **VHS, Beta $24.95** *GEM* ♫

Love Me or Leave Me A hard-hitting biography of '20s torch singer Ruth Etting and her rise and fall at the hand of her abusive, gangster husband, a part just made for Cagney. Day emotes and sings expressively in one of the best performances of her career. Songs include "Ten Cents a Dance," "Mean To Me," "I'll Never Stop Loving You" and "Shakin' The Blues Away." Oscar nominations for Best Actor (Cagney), Screenplay, Sound Recording, Score of a Musical Picture.
1955 122m/C Doris Day, James Cagney, Cameron Mitchell, Robert Keith, Tom Tully; *Dir:* Charles Vidor. Academy Awards '55: Best Story; Writers Guild of America Awards '55: Best Written Musical. **VHS, Beta, LV $19.95** *MGM* ♫♫♫

Love Me Tender A Civil War-torn family is divided by in-fighting between two brothers who both seek the affections of the same woman. Presley's first film. Songs include "Poor Boy," "We're Gonna Move," and the title tune.

1956 89m/B Elvis Presley, Richard Egan, Debra Paget, Neville Brand, Mildred Dunnock, James Drury, Barry Coe; *Dir:* Robert D. Webb. **VHS, Beta** $19.95 *MVD, FOX* 🎞🎞

Love or Money? A small comedy about a yuppie who must choose between a woman and a real estate deal.
1988 (PG-13) 90m/C Timothy Daly, Haviland Morris, Kevin McCarthy, Shelley Fabares, David Doyle, Allen Havey; *Dir:* Todd Hallowell. **VHS, Beta** $29.95 *HMD, HBO* 🎞½

Love & Murder Poorly developed tale of a photographer and his bevy of with beautiful models involved with murder and love, not necessarily in that order.
1991 (R) 87m/C Todd Waring, Kathleen Lasky, Ron White, Wayne Robeson; *Dir:* Steven Hilliard Stern. **VHS** $89.95 *CHV* 🎞½

Love Nest Lundigan stars as Jim Scott, the landlord of an apartment building brimming with wacky tenants, including Monroe, Paar, and Fay. He dreams of becoming a famous writer, but his time is always filled with fixing up the building and trying to pay the mortgage. When one of the tenants ends up in jail because he was living off wealthy widows, Scott's luck changes. This moderately funny film is a good look at the early careers of Monroe and Paar.
1951 84m/B William Lundigan, June Haver, Frank Fay, Marilyn Monroe, Jack Paar; *Dir:* Joseph M. Newman VHS $14.98 *FXV* 🎞🎞

Love Notes Three short vignettes of soft core fantasy, in the life of an ordinary guy.
1988 60m/C Jeff Daniels, Christine Veronica. **VHS, Beta** $29.95 *AHV* 🎞

Love with a Perfect Stranger This Harlequin Romance takes place in Italy where a young widow meets a dashing Englishman. He changes her life forever.
1986 102m/C Marilu Henner, Daniel Massey. **VHS, Beta** $19.95 *PAR* 🎞🎞

The Love Pill Mankind takes giant step forward when medical profession develops pill to create insatiable appetite for sex in women. Best taken with No Doze.
1971 82m/C Henry Woolf, Toni Sinclair, David Pugh, Melinda Schurcher; *Dir:* Kenneth Turner. **VHS** $29.95 *ACA* 🎞½

Love in the Present Tense A woman's romantic novel on video, dealing with an affair that reawakens the torpid, tragic life of a beautiful ex-model.
1985 90m/C **VHS, Beta** $14.95 *PSM* 🎞

Love with the Proper Stranger A quiet, gritty romance about an itinerant musician and a young working girl in Manhattan awkwardly living through the consequences of their one night stand. Moved Wood well into the realm of adult roles, after years playing teen-agers and innocents. Although the story is about an Italian neighborhood and family, an exceptional number of the cast members were Jewish, as was the screenwriter. Story does not conclude strongly, although this did not hamper the box office returns. Oscar nominations for Wood, screenplay and cinematography.
1963 102m/B Natalie Wood, Steve McQueen, Edie Adams, Herschel Bernardi, Tom Bosley, Harvey Lembeck, Penny Santon, Virginia Vincent, Nick Alexander, Augusta Ciolli; *Dir:* Robert Mulligan. **VHS, Beta, LV** $19.95 *PAR* 🎞🎞🎞

Love on the Run The further amorous adventures of Antoine Doinel, hero of "The 400 Blows," "Stolen Kisses," and "Bed and Board." This time out, the women from Doinel's past resurface to challenge his emotions.
1978 (PG) 91m/C *FR* Jean-Pierre Leaud, Marie-France Pisier, Claude Jade; *Dir:* Francois Truffaut. **VHS, Beta** $39.95 *HMV, WAR* 🎞🎞🎞

Love on the Run A television-movie in which a beautiful lawyer helps a wrongly-accused convict escape from prison, and then they both evade the law.
1985 102m/C Stephanie Zimbalist, Adam Baldwin, Constance McCashin, Howard Duff; *Dir:* Gus Trikonis. **VHS, Beta** $69.95 *COL* 🎞½

Love Scenes A director casts his wife in a sexy movie role, and finds she's getting along famously with her co-star.
1984 (R) 83m/C Tiffany Bolling, Britt Ekland. **VHS, Beta** $59.95 *MGM* 🎞

Love Songs A French film about a mother of two confronted with her husband's abandonment, and a subsequent romance with a younger man. Subtitled in English.
1985 (R) 107m/C *FR CA* Catherine Deneuve, Christopher Lambert; *Dir:* Elie Chouraqui. **VHS, Beta** $79.98 *LIV, VES* 🎞🎞

Love at Stake It's condo owners versus the witches in this charming parody that features a hilarious cameo by Dr. Joyce Brothers and a very sexy performance from Carrera.
1987 (R) 83m/C Patrick Cassidy, Kelly Preston, Bud Cort, Barbara Carrera, Stuart Pankin, Dave Thomas, Georgia Brown, Annie Golden; *Dir:* John Moffitt. **VHS, LV** $19.98 *SUE* 🎞🎞½

Love Story The famous line providing an out for every apologetic partner fed the enormous popular appeal of this melodrama. Although O'Neal and McGraw's acting drew mixed reviews, they garnered Academy Award nominations as did the film, the director, the music and the screenplay. He is the son of Boston's upper crust at Harvard; she's the daughter of a poor Italian on scholarship to study music at Radcliffe. They find happiness, but only for a time. Timeless story, simply told, with artful direction from Hiller pulling exceptional performances from the young duo (who have never done as well since). The end result is perhaps better than Segal's simplistic novel, which was produced after he sold the screenplay and became a best-seller before the picture's release - great publicity for any film.
1970 (PG) 100m/C Ryan O'Neal, Ali MacGraw, Ray Milland, John Marley, Tommy Lee Jones; *Dir:* Arthur Hiller. Academy Awards '70: Best Original Score; Golden Globe Awards '71: Best Film—Drama; National Board of Review Awards '70: 10 Best Films of the Year. **VHS, Beta, LV** $19.95 *PAR* 🎞🎞🎞

Love Strange Love A young boy develops a bizarre relationship with his mother who works in a luxurious bordello. Also available in an unedited 120-minute version.
1982 97m/C Vera Fischer, Mauro Mendonca. **VHS, Beta** $79.98 *LIV, VES* 🎞½

Love from a Stranger Thriller about a working woman who wins a lottery. Soon she is charmed by and marries a man whom she later suspects may be trying to kill her. Remade in 1947.
1937 90m/B *GB* Ann Harding, Basil Rathbone, Binnie Hale, Bruce Seton, Bryan Powley, Jean Cadell; *Dir:* Rowland V. Lee. **VHS** $19.95 *NOS, SNC, NWV* 🎞🎞

Love from a Stranger In this remake of the 1937 film, a young newlywed bride fears that the honeymoon is over when she suspects that her husband is a notorious killer and that she will be his next victim.

1947 81m/B Sylvia Sidney, John Hodiak, John Howard, Ann Richards, Isobel Elsom, Ernest Cosart; *Dir:* Richard Whorf. **VHS, Beta, LV** $19.95 *STE, NWV* 🎞🎞½

Love Streams A quirky character drama about a writer and his sister who struggle to find love despite their personal problems.
1984 (PG-13) 122m/C Gena Rowlands, John Cassavetes, Diahnne Abbott; *Dir:* John Cassavetes. Berlin Film Festival '84: Golden Bear (Best Film). **VHS, Beta** $79.95 *MGM* 🎞🎞½

The Love of Three Queens The loves, on and off stage, of a beautiful actress in a European traveling theatre group.
1954 80m/C Gerard Oury, Massimo Serato, Robert Beatty, Cathy O'Donnell, Terence Morgan, Hedy Lamarr; *Dir:* Marc Allegret. **VHS, Beta** $29.95 *PGN, FCT, VDM* 🎞½

Love Thrill Murders A film about a Mansonesque lunatic who is worshipped and obeyed by a mob of runaways and dropouts.
1971 (R) 89m/C Troy Donahue. **VHS, Beta** $69.98 *LIV, VES* Woof!

Love at the Top A ladies' foundation designer falls in love with his boss's son-in-law, jeopardizing her career.
1986 90m/C Louis Jourdan. **Beta** $14.95 *PSM* 🎞

Love Under Pressure Man and woman's marriage starts to evaporate when they realize their son is seriously disturbed.
198? 92m/C Karen Black, Keir Dullea. **VHS** $19.99 *ACA* 🎞🎞

Love Without Fear A dramatic, heart-rendering look at children with AIDS. Story focuses on the lives of three women from completely different backgrounds who adopt babies with AIDS and develop a special love for these children. Also shows their confrontations with a society which fears the disease and the women's strength to reach out and make a difference.
1990 ?m/C **VHS** *GHA* 🎞🎞

Love Without Pity Eric Rochant's theatrical debut is a modern romance, with more distance and alienation than joy. A lovely, successful grad student loses her heart to a jobless, unambitious layabout who wants total commitment from her. Untypical, sardonic love story may be too cynical for sentimental types. In French with English subtitles.
1991 95m/C *FR* Hippolyte Girardot, Mireille Perrier, Jean Marie Rollin; *W/Dir:* Eric Rochant. **VHS** *ORI* 🎞🎞½

The Loved One A famously outlandish, death-mocking farce based on Evelyn Waugh's satire about a particularly horrendous California funeral parlor/cemetery and how its denizens do business. A shrill, protracted spearing of American capitalism.
1965 118m/B Robert Morse, John Gielgud, Rod Steiger, Liberace, Anjanette Comer, Jonathan Winters, James Coburn, Dana Andrews, Milton Berle, Tab Hunter, Robert Morley, Lionel Stander, Margaret Leighton, Roddy McDowall, Bernie Kopell, Alan Napier; *Dir:* Tony Richardson. **VHS, Beta** $19.95 *MGM* 🎞🎞🎞

Loveless A menacing glance into the exploits of an outcast motorcycle gang. In the 50s, a group of bikers on their way to the Florida Cycle Races stop for lunch in a small-town diner. While repairs are being made on their motorcycles, they decide to take full advantage of their situation.
1983 (R) 85m/C Robert Gordon, Willem Dafoe, J. Don Ferguson; *Dir:* Kathryn Bigelow. **VHS, Beta** $59.95 *MED* 🎞🎞

Lovelines Two rock singers from rival high schools meet and fall in love during a panty raid. Laughs uncounted ensue.
1984 (R) 93m/C Greg Bradford, Michael Winslow, Mary Beth Evans, Don Michael Paul, Tammy Taylor, Stacey Toten, Miguel Ferrer, Shecky Greene, Aimee Eccles, Sherri Stoner; *Dir:* Rod Amateau. **VHS, Beta** $79.98 *FOX* 🎬

Lovely to Look At Three wanna-be Broadway producers (Skelton, Keel, Champion) go to gay Paree to peddle Skelton's half interest in Madame Roberta's, a chi chi dress shop. There, they meet the shop's other half interest, two sisters (Champion and Miller), and together they stage a fashion show to finance the floundering hospice of haute couture. Lavish production, light plot. Filmed in Technicolor based on Kern's 1933 Broadway hit (inspired by Alice Duer Miller's "Gowns by Roberta"). Vincent Minelli staged the fashion show, with gowns by Adrian (watch for cop-beater Zsa Zsa as a model).
Laserdisc format features the original motion picture trailer.
1952 105m/C Kathryn Grayson, Red Skelton, Howard Keel, Gower Champion, Marge Champion, Ann Miller, Zsa Zsa Gabor, Kurt Kasznar, Marcel Dalio, Diane Cassidy; *Dir:* Mervyn LeRoy. **VHS, LV** $19.98 *MGM, TTC, PIA* 🎬🎬🎬

Lovely...But Deadly A young girl wages a war against the drug dealers in her school after her brother dies of an overdose.
1982 (R) 95m/C Lucinda Dooling, John Randolph, Richard Herd, Susan Mechsner, Mel Novak; *Dir:* David Sheldon. **VHS, Beta** $69.98 *LIV, VES* Woof!

Lover Come Back More Day-Hudson antics in which an advertising executive falls in love with his competitor but that doesn't stop him from stealing her clients. Is there no shame?
1961 107m/C Rock Hudson, Doris Day, Tony Randall, Edie Adams, Joe Flynn, Ann B. Davis, Jack Oakie, Jack Albertson, Jack Kruschen, Howard St. John; *Dir:* Delbert Mann. **VHS, Beta** $14.95 *KRT* 🎬🎬½

Loverboy A college schnook takes a summer job as a Beverly Hills pizza delivery boy and is preyed upon by many rich and sex-hungry housewives.
1989 (PG-13) 105m/C Patrick Dempsey, Kate Jackson, Barbara Carrera, Kirstie Alley, Carrie Fisher, Robert Ginty; *Dir:* Joan Micklin Silver. **VHS, Beta, LV** $19.95 *COL* 🎬

The Lovers Chic tale of French adultery with Moreau starring as a provincial wife whose shallow life changes overnight when she meets a young man. Had a controversial American debut because of the film's tender eroticism and innocent view of adultery. In French with English subtitles.
1959 90m/B *FR* Jeanne Moreau, Alain Cuny, Jose-Luis De Villalonga, Jean-Mark Bory; *W/Dir:* Louis Malle. Venice Film Festival '59: Special Jury Prize. **VHS** $69.95 *NYF, JLT, FCT* 🎬🎬🎬

Lovers and Liars A romantic adventure in Rome turns into a symphony of zany mishaps when the man forgets to tell the woman that he is married!!
1981 (R) 93m/C *IT* Goldie Hawn, Giancarlo Giannini, Laura Betti; *Dir:* Mario Monicelli. **VHS, Beta** $24.98 *SUE* 🎬🎬

Lovers Like Us Two people each leave their spouses, meet one another, and fall in love.
1975 103m/C *FR* Catherine Deneuve, Yves Montand, Luigi Vannucchi, Tony Roberts, Dana Wynter; *Dir:* Jean-Paul Rappeneau. **VHS, Beta** *JLT* 🎬🎬½

Lovers and Other Strangers Two young people decide to marry after living together for a year and a half. Various tensions surface between them and among their families as the wedding day approaches. Good comedy features some charming performances. Keaton's first film. Castellano received a supporting actor nomination from the Motion Picture Academy.
1970 (R) 106m/C Gig Young, Bea Arthur, Bonnie Bedelia, Anne Jackson, Harry Guardino, Michael Brandon, Richard Castellano, Bob Dishy, Marian Hailey, Cloris Leachman, Anne Meara, Diane Keaton; *Dir:* Cy Howard. Academy Awards '70: Best Song ("For All We Know"). **VHS** *FOX* 🎬🎬🎬

Lovers of Their Time A romantic novel on tape, about forbidden love and other problems.
1985 60m/C Edward Fetherbridge, Cheryl Prime, Lynn Farleigh; *Dir:* Robert Knights. **Beta** $11.95 *PSM* 🎬

Loves of a Blonde Milos Forman discusses filmmaking in Czechoslovakia and his film "Loves of a Blonde."
1965 30m/B *CZ* **VHS, Beta** *CMR, NYS*

Loves of a Blonde A shy girl falls in love with a pianist when the reservist army comes to her small town. In Czechoslovakian with English subtitles. Touching look at the complications of love and our expectations.
1966 88m/B *CZ* Jana Brejchova, Josef Sebanek, Vladimir Pucholt; *Dir:* Milos Forman. New York Film Festival '66: Best Song. **VHS, Beta** $59.95 *RCA* 🎬🎬🎬

The Loves of Carmen Film version of the classic Prosper Merrimee novel about a tempestuous Spanish gypsy and the soldier who loves her. Hayworth is great to look at and the film's main selling point.
1948 98m/C Rita Hayworth, Glenn Ford, Ron Randell, Victor Jory, Arnold Moss, Luther Adler, Joseph Buloff; *Dir:* Charles Vidor. **VHS, Beta** $19.95 *COL* 🎬🎬½

The Loves of Edgar Allen Poe A bland biographical drama about young Poe's adoption, his treatment at the hands of his foster father, his rejection by a woman for a more well-off man, his marriage to his first cousin, and his early death from alcoholism.
1942 67m/B Linda Darnell, Sheppard Strudwick, Virginia Gilmore, Jane Darwell, Mary Howard; *Dir:* Harry Lachman. **VHS** *TCF* 🎬🎬

The Loves of Hercules The mythic mesomorph finds a mate with equiponderant chest measurements, and must save her from an evil queen. Somehow it eludes him that both queen and maiden are Miss Jayne in red and black wigs. Kudos for worst dubbing and special effects; a must see for connoisseurs of kitsch.
1960 94m/C *IT FR* Jayne Mansfield, Mickey Hargitay; *Dir:* Carlo L. Bragaglia. **VHS** $19.98 *SNC, VDM* Woof!

The Loves of Irina Franco's dreamy, sanguine, produced-on-a-dime tale of a vampiress. She cruises the Riviera, seducing and nibbling on a variety of men and women. Spanish release known as "Erotikill," not for most tastes.
198? 95m/C Lina Romay, Monica Swin, Jack Taylor, Alice Arno; *Dir:* Jess (Jesus) Franco. **VHS, Beta** $29.95 *MED* 🎬

Love's Savage Fury Two escapees from a Union prison camp seek out a hidden treasure that could determine the outcome of the Civil War. A bad made-for-television "Gone With the Wind" ripoff.

1979 100m/C Jennifer O'Neill, Perry King, Robert Reed, Raymond Burr, Connie Stevens, Ed Lauter; *Dir:* Joseph Hardy. **VHS, Beta** $49.95 *PSM* Woof!

Loves & Times of Scaramouche An eighteenth century rogue becomes involved in a plot to assassinate Napoleon and winds up seducing Josephine in the process. Dubbed.
1976 92m/C *IT* Michael Sarrazin, Ursula Andress, Aldo Maccione; *Dir:* Enzo G. Castellari. **VHS, Beta** $19.95 *SUE* 🎬½

Lovesick A very-married New York psychiatrist goes against his own best judgment when he falls in love with one of his patients.
1983 (PG) 98m/C Dudley Moore, Elizabeth McGovern, Alec Guinness, John Huston, Ron Silver; *Dir:* Marshall Brickman. **VHS, Beta, LV** $19.98 *WAR, PIA* 🎬🎬

Lovespell A retelling of the legend of Isolde and Tristan. Stilted direction and writing, but another chance to reminisce on Richard Burton. Also titled "Tristan and Isolde."
1979 91m/C Richard Burton, Kate Mulgrew, Nicholas Clay, Cyril Cusack, Geraldine Fitzgerald, Niall Toibin, Diana Van Der Vlis, Niall O'Brien; *Dir:* Tom Donovan. **VHS** $39.95 *CON* 🎬

Lovey: A Circle of Children 2 A teacher for special children takes in a terrified girl diagnosed as brain damaged or schizophrenic, and discovers love and intelligence in the girl as well as finding some important insights into her own life.
1982 120m/C Jane Alexander, Kris McKeon; *Dir:* Judd Taylor. **VHS, Beta** *TLF* 🎬

Loving The full-length, shot-on-video pilot for the daytime television soap opera with various guest stars. Exteriors shot at C.W. Post Center on Long Island.
1984 120m/C Lloyd Bridges, Geraldine Page. **VHS, Beta** $69.95 *SVS* 🎬🎬½

Loving Couples Two happily married couples meet at a weekend resort...and switch partners. Predictable but hilarious at moments.
1980 (PG) 120m/C Shirley MacLaine, James Coburn, Susan Sarandon, Stephen Collins, Sally Kellerman; *Dir:* Jack Smight. **VHS, Beta** $29.98 *VES* 🎬½

Loving You A small town boy with a musical style all his own becomes a big success thanks to the help of a female press agent. Features many early Elvis hits, including "Teddy Bear."
1957 101m/C Elvis Presley, Wendell Corey, Lizabeth Scott, Dolores Hart, James Gleason; *Dir:* Hal Kanter. **VHS, Beta** $19.98 *WAR, OM* 🎬🎬

Low Blow A kung fu expert tracks down a mind-bending cult.
1985 90m/C Cameron Mitchell, Leo Fong; *Dir:* Frank Harris. **VHS, Beta** $69.98 *LIV, VES* 🎬

The Lower Depths Renoir's adaptation of the Maxim Gorky play about a thief and a financially ruined baron learning about life from one another. In French with English subtitles; alternate title: "Les Bas Fonds."
1936 92m/B *FR* Jean Gabin, Louis Jouvet, Vladimir Sokoloff, Robert Le Vigan, Suzy Prim; *Dir:* Jean Renoir. National Board of Review Awards '37: 10 Best Foreign Films of the Year. **VHS, Beta** $39.95 *FCT, DVT* 🎬🎬🎬½

The Lower Depths Kurosawa sets the Maxim Gorky play in Edo during the final Tokugawa period, using Noh theatre elements in depicting the lowly denizens of a low-rent hovel. In Japanese with English titles.

1957 125m/B *JP* Toshiro Mifune, Isuzu Yamada, Ganjiro Nakamura, Kyoko Kagawa, Bokuzen Hidari; *Dir:* Akira Kurosawa. **VHS, Beta $24.95** *SVS, FCT* ✠✠✠½

Lower Level While working late, an attractive business woman becomes trapped in an office building by her psychotic secret admirer.

1991 (R) 88m/C David Bradley, Elizabeth Gracen, Jeff Yagher; *Dir:* Kristine Peterson. **VHS, LV $89.98** *REP* ✠

Loyalties A strong treatment of the issue of sexual abuse of girls by men, and of an upper-class woman's willingness to establish a true friendship with a woman of a so-called lower economic class.

1986 (R) 98m/C *CA* Ken Welsh, Tantoo Cardinal, Susan Wooldridge, Vera Martin, Christopher Barrington-Leigh; *Dir:* Anne Wheeler. Genie Awards '87: Best Costume Design. **VHS $9.95** *SIM, CGI* ✠✠

Loyola, the Soldier Saint A biography of the founder of the Jesuits from his years as a page in the Spanish court to his daring exploits on the battle field and, finally, his spiritual awakening at the University of Paris. Narrated by Father Alfred J. Barrett.

1952 93m/B Rafael Duran, Maria Rosa Jiminez; *Dir:* Jose Diaz Morales. **VHS** *IVY, KEP* ✠✠½

Lt. Robin Crusoe, U.S.N. A lighthearted navy pilot crash lands amusingly on a tropical island, falls hard for an island babe and schemes intensely against the local evil ruler. Lackluster Disney debacle.

1966 (G) 113m/C Dick Van Dyke, Nancy Kwan, Akim Tamiroff; *Dir:* Byron Paul. **VHS, Beta** *DIS, OM* ✠

Lucas A high school brain falls in love with the new girl in town, and tries to win her by trying out for the football team. Genuine film about the perils of coming of age. Thoughtful, non-condescending, and humorous.

1986 (PG-13) 100m/C Corey Haim, Kerri Green, Charlie Sheen, Winona Ryder, Courtney Thorne-Smith, Thomas E. Hodges; *Dir:* David Seltzer. **VHS, Beta $14.98** *FXV, FOX* ✠✠✠

Lucia di Lammermoor One of the most popular Bel Canto operas is based on Sir Walter Scott's "The Bride of the Lammermoor." Joan Sutherland and her puppet friends help tell the story.

1973 30m/C Joan Sutherland. **VHS, Beta $29.95** *MVD, VAI, BFA* ✠✠

The Lucifer Complex Nazi doctors are cloning exact duplicates of such world leaders as the Pope and the President of the United States on a remote South American island in the year 1996.

1978 91m/C Robert Vaughn, Merrie Lynn Ross, Keenan Wynn, Aldo Ray; *Dir:* David L. Hewitt, Kenneth Hartford. **VHS, Beta $59.95** *UHV* ✠

Lucky Devil Lucky devil wins car and motors around the country to the tune of organ music.

1925 63m/B Richard Dix, Esther Ralston, Edna May Oliver, Tom Findley; *Dir:* Frank Tuttle. **VHS, Beta $19.95** *GPV, FCT* ✠½

Lucky Jim A junior lecturer in history at a small university tries to get himself in good graces with the head of his department, but is doomed from the start by doing the wrong things at the worst possible times. Minus the social satire of the Kingsley Amis novel on which it was based; what's left is a cheerful comedy.

1958 91m/B *GB* Ian Carmichael, Terry-Thomas, Hugh Griffith; *Dir:* John Boulting. **VHS, Beta $59.99** *HBO* ✠✠½

Lucky Luciano A violent depiction of the final years of Lucky Luciano, gangster king-pin.

1974 (R) 108m/C *IT FR* Edmond O'Brien, Rod Steiger, Vincent Gardenia, Gian Marie Volonte; *Dir:* Francesco Rosi. **VHS, Beta $9.95** *COL, CHA* ✠✠

Lucky Luke: Daisy Town A full-length animated feature starring Lucky Luke, the all-American cowboy who saves the little community of Daisy Town from the hot-headed Dalton Brothers gang. A wacky western spoof.

1971 75m/C VHS, Beta, LV $49.95 *DIS, LDC* ✠✠

Lucky Luke: The Ballad of the Daltons Easygoing cowboy hero Lucky Luke gets involved in a wild feud with the bumbling Dalton Brothers gang in this animated feature.

1978 82m/C VHS, Beta, LV $49.95 *DIS, LDC* ✠✠

Lucky Me A group of theater entertainers are stranded in Miami and are forced to work in a hotel kitchen. They soon acquire the support of a wealthy oilman (Goodwin) who invests in their show, but not before his spoiled daughter tries to thwart all plans. Musical numbers include "Lucky Me," "Superstition Song," "I Speak to the Stars," "Take a Memo to the Moon," "Love You Dearly," "Bluebells on Broadway," "Parisian Pretties," "Wanna Sing Like an Angel," "High Hopes" and "Men."

1954 100m/C Doris Day, Robert Cummings, Phil Silvers, Eddie Foy Jr., Nancy Walker, Martha Hyer, Bill Goodwin, Marcel Dalio, James Burke, Jack Shea, William Bakewell, Charles Cane, Ray Teal, Tom Powers, Angie Dickinson, Dolores Dorn; *Dir:* Jack Donohue. **VHS $29.98** *WAR, CCB* ✠✠

Lucky Partners When an artist and an errand girl share a winning lottery ticket, funny complications arise as they embark on a fantasy honeymoon. Although Rogers' innocence and Coleman's savoir faire provide an interesting contrast, the script isn't equal to the status of its stars.

1940 101m/B Ronald Colman, Ginger Rogers, Jack Carson, Spring Byington, Harry Davenport; *Dir:* Lewis Milestone. **VHS, Beta $19.98** *RKO* ✠✠

Lucky Stiff A fat, unpopular dweeb has the shock of his life when a radiant woman falls in love with him, the strangeness of which becomes evident when he meets her very weird family.

1988 (PG) 93m/C Donna Dixon, Joe Alaskey; *Dir:* Anthony Perkins. **VHS, Beta, LV $89.95** *COL* ✠✠

Lucky Terror Western with a style consistent with Hoot Gibson, the star.

1936 60m/B Hoot Gibson, Kone Andre, Charles King; *Dir:* Alan James. **VHS, Beta $19.95** *NOS, GPV, VCN* ✠

Lucky Texan Wayne plays a tough easterner who goes West and finds himself involved in a range war.

1934 61m/B John Wayne, George "Gabby" Hayes, Yakima Canutt; *Dir:* Robert N. Bradbury. **VHS $19.95** *COL, REP, MRV* ✠½

Luggage of the Gods A lost tribe of cave people are confronted with civilization when suitcases fall from an airplane. Low-budget comedy reminiscent of "The Gods Must Be Crazy."

1987 (G) 78m/C Mark Stolzenberg, Gabriel Barre, Gwen Ellison; *Dir:* David Kendall. **VHS, Beta $29.95** *ACA* ✠✠

Lullaby of Broadway Many songs highlight this flimsy musical about a girl (Day) who ventures from England to the Big Apple in search of an acting career. She soon discovers that her mom has become a has-been actress now performing in a Greenwich Village dive. Day struggles to gain her own success while coping with her mother's downfall. Songs include "Lullaby of Broadway," "You're Getting to be a Habit with Me," "Just one of Those Things," "Somebody Loves Me," "I Love the Way You Say Goodnight," "Fine and Dandy," "Please Don't Talk About Me When I'm Gone," "A Shanty in Old Town," "We'd Like to Go on a Trip," "Zing! Went the Strings of My Heart" and "You're Dependable."

1951 93m/C Doris Day, Gene Nelson, S.Z. Sakall, Billy DeWolfe, Gladys George, Florence Bates; *Dir:* David Butler. **VHS $29.98** *WAR, CCB* ✠✠½

Lumiere An acclaimed drama about four actresses who, during the course of one night together, make pivotal decisions about their lives and relationships. Moreau's first directorial effort; dubbed.

1976 (R) 101m/C *FR* Jeanne Moreau, Lucia Bose, Francine Racette, Caroline Cartier, Keith Carradine; *Dir:* Jeanne Moreau. **VHS, Beta $29.98** *SUE* ✠✠½

Lunatics: A Love Story Hank (Raimi) is an ex-mental patient poet suffering from hallucinations and afraid to leave his Los Angeles apartment. For companionship he dials a party-line until, one day, he misdials and insteads gets the number of a local pay telephone, which is answered by a naive woman dumped by her boyfriend. When Nancy (Foreman) decides to visit Hank you learn she's as crazy as he is—so it only makes sense they would fall in love. A well-acted fable of how love fragilely connects even the most unlikely of people.

1992 (PG-13) ?m/C Theodore Raimi, Deborah Foreman, Bruce Campbell; *W/Dir:* Josh Beck. **VHS, LV $79.95** *SVT* ✠✠✠

Lunatics & Lovers A musician meets a bizarre nobleman who is in love with an imaginary woman.

1976 (PG) 92m/C *IT* Marcello Mastroianni, Lino Toffalo; *Dir:* Flavio Mogherini. **VHS, Beta $59.98** *FOX* ✠½

Lunch Wagon Two co-eds are given a restaurant to manage during summer vacation and wind up involved in a hilarious diamond chase and sex romp.

1981 (R) 88m/C Rick Podell, Candy Moore, Pamela Bryant, Rosanne Katon; *Dir:* Ernie Pintoff. **VHS, Beta $19.95** *MED* ✠✠

Lupo When threatened with the loss of his home and the separation of his family, an exuberant Greek man challenges the modern world.

1970 (G) 99m/C Yuda Barkan, Gabi Armoni, Esther Greenberg; *Dir:* Menahem Golan. **VHS, Beta $59.95** *MGM* ✠✠

Lure of the Islands Federal agents track wanted criminals to an island retreat in this highly predictable low-budget thriller.

1942 61m/B Margie Hart, Robert Lowery, Guinn Williams, Warren Hymer, Gale Storm; *Dir:* Jean Yarborough. **VHS, Beta $16.95** *NOS, SNC* ✠

Lure of the Sila Italian melodrama about a young peasant girl infiltrating a landowner's home in order to avenge the death of her brother and mother. Dubbed and more interesting than it sounds.
1949 72m/B IT Silvana Mangano, Amadeo Nazzari, Vittorio Gassman; *Dir:* Duilio Coletti. **VHS, Beta, 8mm $29.95** *VYY ✍✍½*

Lurkers An unappealing metaphysical morass about a woman who has been haunted throughout her life by...something.
1988 (R) 90m/C Christine Moore, Gary Warner, Marina Taylor, Carissa Channing, Tom Billett; *Dir:* Roberta Findlay. **VHS $79.95** *MED* Woof!

Lust in the Dust When part of a treasure map is found on the derriere of none other than Divine, the hunt is on for the other half. This comedy western travels to a sleepy town called Chile Verde (green chili for those who don't speak Spanish) and the utterly ridiculous turns comically corrupt. Deliciously distasteful fun. Features Divine singing a bawdy love song in his/her break from John Waters.
1985 (R) 85m/C Tab Hunter, Divine, Lainie Kazan, Geoffrey Lewis, Henry Silva, Cesar Romero, Gina Gallego, Courtney Gains, Woody Strode; *Dir:* Paul Bartel. **VHS, Beta, LV $14.95** *NWV, STE, HHE ✍✍✍*

Lust for Freedom An undercover cop decides to hit the road after she sees her partner gunned down. She winds up near the California-Mexico border, where she is wrongly imprisoned with a number of other young women in a white slavery business. This low-budget film includes lots of steamy women-behind-bars sex scenes.
19?? 92m/C Melanie Coll, William J. Kulzer; *Dir:* Eric Louzil. **VHS $89.95** *AIP ✍*

Lust for Gold Ford battles against greedy former lover Lupino and her husband for control of the Lost Dutchman gold mine.
1949 90m/B Ida Lupino, Glenn Ford, Gig Young, William Prince, Edgar Buchanan, Will Geer, Paul Ford, Jay Silverheels; *Dir:* Sylvan Simon. **VHS, Beta** *GKK ✍✍*

Lust for Life An absorbing, serious biography of Vincent Van Gogh, from his first paintings to his death. Remarkable for Douglas' furiously convincing, Oscar nominated portrayal. Featuring dozens of actual Van Gogh works from private collections. Music by Miklos Rozsa; based on an Irving Stone novel, produced by John Houseman. Also received Oscar nominations for Color Art Direction and Set Decoration.
1956 122m/C Kirk Douglas, Anthony Quinn, James Donald, Pamela Brown, Everett Sloane, Henry Daniell, Niall MacGinnis, Noel Purcell, Lionel Jeffries, Jill Bennett; *Dir:* Vincente Minnelli. Academy Awards '56: Best Supporting Actor (Quinn); National Board of Review Awards '56: 10 Best Films of the Year; New York Film Critics Awards '56: Best Actor (Douglas). **VHS, Beta, LV $39.95** *MGM, BTV ✍✍✍½*

Lust for a Vampire The sanguine tale of a deadly vampire who indiscriminately preys on pupils and teachers when she enrolls at a British finishing school. Moody and erotically charged, with an impressive ending. Quasi-sequel to "The Vampire Lovers."
1971 (R) 95m/C GB Ralph Bates, Barbara Jefford, Suzanna Leigh, Michael Johnson, Yutte Stensgaard, Pippa Steele; *Dir:* Jimmy Sangster. **VHS, Beta, LV $59.99** *HBO, MLB ✍✍*

The Lusty Men Two rival rodeo champions, both in love with the same woman, work the rodeo circuit until a tragic accident occurs. Mitchum turns in a fine performance as

the has-been rodeo star trying to make it big again.
1952 113m/B Robert Mitchum, Susan Hayward, Arthur Kennedy, Arthur Hunnicutt; *Dir:* Nicholas Ray. **VHS, Beta, LV $15.95** *UHV ✍✍✍*

Luther A well-acted characterization of Martin Luther's development from a young seminarian to his leadership of the Reformation Movement.
1974 (G) 112m/C Stacy Keach, Patrick Magee, Hugh Griffith, Robert Stephens, Alan Badel; *Dir:* Guy Green. **VHS $39.95** *MGS ✍✍½*

Luther the Geek Little Luther's visit to the circus is dramatically changed when he sees the geek, a sideshow freak. Since then Luther has taken to biting off chicken's heads and drinking their blood in his small Illinois town; the town will never be the same. (And neither will you if you watch this stupid film).
1990 90m/C Edward Terry, Joan Roth, J. Jerome Clarke, Tom Mills; *Dir:* Carlton J. Albright. **VHS $79.95** *QUE* Woof!

Luv Although the cast tries hard, they can't do much with this one. Three intellectuals, including one that's suicidal, discuss the trials and tribulations of their middle-class New York existence.
1967 (PG) 95m/C Jack Lemmon, Peter Falk, Elaine May; *Dir:* Clive Donner. **VHS, Beta, LV $9.95** *GKK ✍½*

Lydia Sentimental drama in which an elderly lady (Oberon) gets to relive her romantic past when she has a reunion with four of her lost loves. Well acted and directed, Oberon gives one of her best performances ever. Adapted from the highly regarded French film "Un Carnet de Bal."
1941 98m/B Merle Oberon, Joseph Cotten, Alan Marshal, George Reeves; *Dir:* Julien Duvivier. **VHS, Beta, LV** *SUE ✍✍✍*

Lying Lips An all-black detective mystery about a young nightclub singer framed for murder. Her boyfriend then turns detective to clear her name.
1939 60m/B Edna Mae Harris, Carmen Newsome, Earl Jones, Amanda Randolph; *Dir:* Oscar Micheaux. **VHS, Beta $39.95** *NOS, VCN, DVT ✍½*

M The great Lang dissection of criminal deviance, following the tortured last days of a child murderer, and the efforts of both the police and the underground to bring him to justice. Poetic, compassionate, and chilling. Inspired by real-life serial killer known as "Vampire of Dusseldorf," Lang also borrowed story elements from "Jack the Ripper"'s killing spree. Lorre's screen debut. Lang's personal favorite among his own films. In German with English subtitles. Remade in 1951.
1931 99m/C GE Peter Lorre, Ellen Widmann, Inge Landgut, Gustav Grundgens; *Dir:* Fritz Lang. National Board of Review Awards '33: 10 Best Foreign Films of the Year. **VHS, Beta, LV, 8mm, 3/4U $16.95** *SNC, NOS, HHT ✍✍✍✍*

Ma Barker's Killer Brood Biography of the infamous American criminal and her four sons, edited together from a television serial. The shoot-em-up scenes and Tuttle's performance keep the pace from slackening.
1960 82m/B Lurene Tuttle, Tristram Coffin, Paul Dubov, Nelson Leigh, Myrna Dell, Vic Lundin, Donald Spruance; *Dir:* Bill Karn. **VHS, Beta $16.95** *SNC, DVT ✍✍*

Mabel & Fatty Three silent shorts featuring the two stars, made with Mack Sennett: "He Did and He Didn't," "Mabel and Fatty Viewing The World's Fair at San Fran-

cisco" (with scenes of the actual 1914 World's Fair) and "Mabel's Blunder."
1916 61m/B Mabel Normand, Fatty Arbuckle, Al "Fuzzy" St. John. **VHS, Beta, 8mm $24.95** *VYY*

Mac and Me A lost E.T.-like alien stranded on Earth befriends a wheelchair-bound boy. Aimed at young kids, it's full of continual product plugs, most notably for McDonald's. Make the kids happy and stick to the real thing.
1988 (PG) 94m/C Christine Ebersole, Jonathan Ward, Katrina Caspary, Lauren Stanley, Jade Calegory; *Dir:* Stewart Raffill. **VHS, Beta, LV $19.98** *ORI ✍*

Macabre A beautiful woman, lustful and precocious, kills her husband and his twin brother.
1977 89m/C Larry Ward, Teresa Gimpera, Jack Stuart. **VHS, Beta $59.95** *MGL ✍✍*

Macao On the lam for a crime he didn't commit, an adventurer sails to the exotic Far East, meets a buxom cafe singer, and helps Interpol catch a notorious crime boss. A strong film noir entry. Russell sneers, Mitchum wise cracks. Director von Sternberg's last film for RKO.
1952 81m/B Robert Mitchum, Jane Russell, William Bendix, Gloria Grahame; *Dir:* Josef von Sternberg. **VHS, Beta, LV $19.95** *MED ✍✍✍*

Macario Traditional Mexican fable about a desperate farmer who makes a deal with the Devil for healing powers. His gift only leads to greater misfortunes, of course.
1958 91m/B MX Ignacio Lopez Tarso, Pina Pellicer, Enrique Lucero; *Dir:* Roberto Gavaldon. **VHS, Beta, LV $59.95** *CVC, MAD ✍*

Macaroni An uptight American businessman returns to Naples forty years after being stationed there in World War II. Comedic situations abound when he is reunited with his Italian war buddy, brother of his lover. Pleasant acting can't save irritating script.
1985 (PG) 104m/C IT Jack Lemmon, Marcello Mastroianni, Daria Nicolodi, Isa Danieli; *Dir:* Ettore Scola. **VHS, Beta, LV $79.95** *PAR ✍½*

MacArthur General Douglas MacArthur's life from Corregidor in 1942 to his dismissal a decade later in the midst of the Korean conflict. Episodic sage with forceful Peck but weak supporting characters. Fourteen minutes were cut from the original version; intended to be Peck's "Patton," it falls short of the mark.
1977 (PG) 130m/C Gregory Peck, Ivan Bonar, Ward Costello, Nicholas Coster, Dan O'Herlihy; *Dir:* Joseph Sargent. **VHS, Beta, LV $19.95** *MCA, TLF ✍✍*

MacArthur's Children A poignant evocation of life in a Japanese fishing village as it faces the end of World War II and American occupation. In Japanese with English subtitles. Not as good as some of the other Japanese films, but it is an interesting look at the changes in culture caused by their defeat in the war.
1985 115m/C JP Masako Natsume, Shima Iwashita, Hiromi Go, Takaya Yamamauchi, Shiori Sakura, Ken Watanabe, Juzo Itami, Yoshiyuka Omori; *Dir:* Masahiro Shinoda. **VHS, Beta, LV $29.95** *PAV, FCT ✍✍✍*

Macbeth Shakespeare's classic tragedy is performed in this film with a celebrated lead performance by Welles, who plays the tragic king as a demonic leader of a barbaric society. A low budget adaptation with cheap sets, a 3-week shooting schedule, lots of mood, and an attempt at Scottish accents.

After making this film, Welles took a ten-year break from Hollywood.
1948 111m/B Orson Welles, Jeanette Nolan, Dan O'Herlihy, Roddy McDowall, Robert Coote; *Dir:* Orson Welles. **VHS $19.98** *REP* 𝅝𝅝𝅝 ½

Macbeth Polanski's notorious adaptation of the Shakespearean classic, marked by realistic design, unflinching violence, and fatalistic atmosphere. Contains stunning fight scenes and fine acting. Polanski's first film following the grisly murder of his pregnant wife, actress Sharon Tate, was torn apart by critics. It is in fact a worthy continuation of his work in the horror genre. Very well made. First film made by Playboy Enterprises. Originally rated X.
1971 (R) 139m/C Jon Finch, Nicholas Selby, Martin Shaw, Francesca Annis, Terence Baylor; *Dir:* Roman Polanski. British Academy Awards '72: Best Costume Design; National Board of Review Awards '71: Best Film. **VHS, Beta, LV $19.95** *COL* 𝅝𝅝𝅝

Macbeth An extraordinary version of Shakespeare's "Macbeth" in which all the fire, ambition and doom of his text come brilliantly to life.
1976 137m/C Eric Porter, Janet Suzman. **3/4U, Special order formats $19.95** *TLF* 𝅝𝅝𝅝

Macbeth Another production of the classic tragedy by Shakespeare. Produced as part of HBO's Thames Collection.
1990 110m/C Michael Jayston, Leigh Hunt. **VHS, Beta $39.99** *HBO* 𝅝𝅝

Machine Gun Kelly Corman found Euro appeal with this thirties style gangster bio. Bronson, who was just gaining a reputation as an action lead, stars as criminal Kelly, who's convinced by his moll to give up bank robbery for kidnapping. Amsterdam, the wisecracking writer from "The Dick Van Dyke Show," is the fink who turns him in. Scripted by R. Wright Campbell.
1958 80m/B Charles Bronson, Susan Cabot, Morey Amsterdam, Barboura Morris, Frank De Kova, Jack Lambert, Wally Campo; *Dir:* Roger Corman. **VHS $9.95** *COL, FCT* 𝅝𝅝 ½

Macho Callahan A Civil War convict escapes from a frontier jail bent on tracking down the man who imprisoned him. (The escaped prisoner bit is a recurring theme in Janssen's oeuvre). Confused script even has a one-armed Carradine!
1970 (R) 99m/C David Janssen, Jean Seberg, David Carradine, Lee J. Cobb; *Dir:* Bernard L. Kowalski. **VHS, Beta $9.95** *COL, CHA* 𝅝𝅝

Macho Dancer A seamy look at Manila's gay underworld. A rural lad comes to city looking for work to support his family and winds up as a male prostitute. Features nudity and explicit sexual situations. In Tagalog with English subtitles.
1988 136m/C *PH Dir:* Lino Brocka. **VHS $79.95** *FCT* 𝅝𝅝

Macho y Hembra Three young people explore their sexuality away from the confines of society. Venezuelan version of "Jules et Jim."
1985 90m/C Orlando Ordaneta, Elva Escobar, Irene Arcila; *Dir:* Mauricio Walerstein. **VHS, Beta** *MAD* 𝅝𝅝𝅝

Machoman The matchless, peerless, Machoman knows all the best killing methods including: Lizard Fists, Crane Kung Fu, and the Drunken Fist.
197? 94m/C Araujo, Enyaw Liew, Jerry Rages. **VHS, Beta $54.95** *MAV* 𝅝

Maciste in Hell Inexplicably living in 17th century Scotland, Italian hero Maciste pursues witch into the depths of Hell. Also known as "Witch's Curse."
1960 78m/C *IT* Kirk Morris, Helene Chanel, Vira Silenti, Andrea Bosic, Angelo Zanolli, John Karlsen; *Dir:* Riccardo Freda. **VHS $19.98** *UNI, SNC* 𝅝 ½

Mack The Mack is a pimp who comes out of retirement to reclaim a piece of the action in Oakland, California. Violent blaxploitation flick was a box office dynamite at time of release. Early Pryor appearance.
1973 (R) 110m/C Max Julien, Richard Pryor, Don Gordon, Roger E. Mosley, Carol Speed; *Dir:* Michael Campus. **VHS, Beta $9.98** *CHA* 𝅝 ½

Mack & Carole Lombard began her screen career, in earnest, as a Mack Sennett comedienne. These three shorts represent her earliest screen work.
1928 60m/B Carole Lombard. **VHS, Beta $49.50** *GVV*

Mack the Knife Terrible adaptation of Brecht and Weill's "Threepenny Opera," detailing the adventures of a master thief in love with an innocent girl. Too much music, too much dancing, too much emoting.
1989 (PG-13) 121m/C Raul Julia, Roger Daltrey, Richard Harris, Julie Walters, Clive Revill, Erin Donovan, Rachel Robertson, Julia Migenes-Johnson; *Dir:* Menahem Golan. **VHS, Beta, LV $89.95** *COL* Woof!

Mack Sennett Comedies: Volume 1 Four Sennett films are included: "The Eyes Have It," "The Cannon Ball," "The Desperate Scoundrel," and "Pride of Pikeville." Ben Turpin and the Keystone Cops at their manic best.
1921 85m/B Ben Turpin, Chester Conklin, Ford Sterling, Minta Durfee; *Dir:* Mack Sennett. **VHS $29.98** *REP*

Mack Sennett Comedies: Volume 2 Four more Sennett shorts are featured: "Mabel and Fatty Adrift," "Mabel, Fatty, and the Law," "Fatty's Tin-Type Tangle," and "Our Congressman." The last film features Will Rogers in a lampoon of Capitol Hill.
1916 84m/B Fatty Arbuckle, Mabel Normand, Will Rogers, Madge Hunt; *Dir:* Mack Sennett. **VHS $29.98** *REP*

MacKenna's Gold Grim desperados trek through Apache territory to uncover legendary cache of gold in this somewhat inflated epic. Subdued stars Peck and Shariff vie for attention here with such overactors as Cobb, Meredith, and Wallach. Meanwhile, Newmar (Catwoman of TV's Batman) swims nude. A must for all earthquake buffs.
1969 (PG) 128m/C Gregory Peck, Omar Sharif, Telly Savalas, Julie Newmar, Edward G. Robinson, Keenan Wynn, Ted Cassidy, Eduardo Ciannelli, Eli Wallach, Raymond Massey, Lee J. Cobb, Burgess Meredith, Anthony Quayle; *Dir:* J. Lee Thompson. **VHS, Beta, LV $12.95** *GKK* 𝅝𝅝 ½

Mackintosh Man An intelligence agent must undo a communist who has infiltrated the free world's network in this solid but somewhat subdued thriller. Good cast keeps narrative rolling, but don't look here for nudity-profanity-violence fix.
1973 (PG) 100m/C Paul Newman, Dominique Sanda, James Mason, Ian Bannen, Nigel Patrick, Harry Andrews, Leo Genn, Peter Vaughan, Michael Hordern; *Dir:* John Huston. **VHS $19.98** *WAR* 𝅝𝅝 ½

Macon County Line A series of deadly mistakes and misfortunes lead to a sudden turn-around in the lives of three young people when they enter a small Georgia town and find themselves accused of brutally slaying the sheriff's wife. Sequelled by "Return to Macon County" in 1975, starring Don Johnson and Nick Nolte.
1974 (R) 89m/C Alan Vint, Jesse Vint, Cheryl Waters, Geoffrey Lewis, Joan Blackman, Max Baer; *Dir:* Richard Compton. **VHS, Beta $9.98** *SUE* 𝅝𝅝 ½

Macumba Love An author journeys to Brazil to prove a connection between mysterious murders and voodoo. A tedious suspense exploitation travelogue with generous servings of cheesecake and calypso music.
1960 86m/C Walter Reed, Ziva Rodann, William Wellman Jr., June Wilkinson, Ruth de Souza; *Dir:* Douglas Fowley. **VHS** *SMW* 𝅝

Mad About Money An early sound comedy about a bespectacled mild-mannered fellow who runs up against a Mexican spitfire.
1930 80m/B Lupe Velez, Harry Langdon. **VHS, Beta $24.95** *NOS, DVT* 𝅝𝅝

Mad About You Millionaire helps his daughter measure prospective suitors. Adam West is not the fourth man in the Montgomery Clift-James Dean-Marlon Brando chain of American acting greats.
1990 (PG) 92m/C Claudia Christian, Joseph Gian, Adam West, Shari Shattuck; *Dir:* Lorenzo Doumani. **VHS, Beta $29.95** *ACA* Woof!

The Mad Bomber Grim lawman Edwards tracks deranged bomber Connors, who is determined to blow up anyone who ever offended him. A must for all connoisseurs of acting that is simultaneously overblown and flat.
1972 80m/C Vince Edwards, Chuck Connors, Neville Brand; *Dir:* Bert I. Gordon. **VHS, Beta $24.95** *KOV, MRV, WES* Woof!

Mad Bull Sensitive wrestler finds meaning in life when he falls in love. If you ever cared about Karras, then you probably already saw this. If you never cared about Karras, then you probably never heard of this. If you ever cared about Anspach, rent "Five Easy Pieces" instead.
1977 96m/C Alex Karras, Susan Anspach, Nicholas Colosanto, Tracy Walter; *Dir:* Walter Doniger. **VHS, Beta $59.99** *HBO* 𝅝

The Mad Butcher Typically unhinged mental patient seeks teenage flesh for his various instruments of torture and death. Buono has never been more imposing.
1972 (R) 90m/C *IT* Victor Buono, Karin Field, Brad Harris, John Ireland; *Dir:* John Zuru. **VHS, Beta $39.98** *MAG, GHV, MRV* 𝅝

Mad Death The British Isles take extreme steps to quarantine themselves against rabies from the European continent, and this condensed TV miniseries preys on fears of the worst. An outbreak of the virus spreads from dogs and other pets to humans, and special commandoes are called in.
1985 120m/C *GB* **VHS $49.95** *FCT* 𝅝𝅝

Mad Doctor of Blood Island Dull band of travelers arrive on mysterious tropical island and encounter bloodthirsty creature. Warning: this film is not recognized for outstanding achievements in acting, dialogue, or cinematography.
1969 110m/C John Ashley; *Dir:* Gerardo (Gerry) De Leon. **VHS, Beta $39.98** *MAG* Woof!

Mad Dog An escaped convict seeks revenge on the cop responsible for his imprisonment. You can't say you've seen all of Helmut Berger's films if you haven't seen this one.
19?? 90m/C Helmut Berger, Marisa Mell; ***Dir:*** Sergio Grieco. **VHS, Beta $24.95** *VDC Woof!*

Mad Dog Morgan Hopper delivers as engaging outlaw roaming outlands of nineteenth-century Australia. Quirky and violent G'day man. Based on a true story and also known as just plain "Mad Dog."
1976 (R) 93m/C *AU* Dennis Hopper, David Gulpilil; ***Dir:*** Philippe Mora. **VHS, Beta $59.99** *HBO, SIM* 🦴🦴½

Mad Dogs & Englishmen This chronicle of Joe Cocker's 1971 American tour features such songs as "Delta Lady" and "Feelin' Alright."
1971 55m/C Joe Cocker, Leon Russell, Rita Coolidge; ***Dir:*** Pierre Adidge. **VHS, Beta $29.95** *COL* 🦴🦴½

Mad Max Set on the stark highways of the post-nuclear future, an ex-cop seeks personal revenge against a rovin' band of vicious outlaw bikers who killed his wife and child. Futuristic scenery and excellent stunt work make for an exceptionally entertaining action-packed adventure. Followed by "The Road Warrior" (also known as "Mad Max 2") in 1981 and "Mad Max Beyond Thunderdome" in 1985.
1980 (R) 93m/C Mel Gibson, Joanne Samuel, Hugh Keays-Byrne, Steve Bisley, Tim Burns, Roger Ward; ***Dir:*** George Miller. **VHS, Beta, LV $29.98** *LIV, VES* 🦴🦴🦴½

Mad Max Beyond Thunderdome Max drifts into evil town ruled by Turner and becomes gladiator, then gets dumped in desert and is rescued by band of feral orphans. Third in a bleak, extremely violent, often exhilirating series.
1985 (PG-13) 107m/C *AU* Mel Gibson, Tina Turner, Helen Buday, Frank Thring, Bruce Spence, Robert Grubb, Angelo Rossitto, Angry Anderson, George Spartels, Rod Zuanic; ***Dir:*** George Miller, George Ogilve. **VHS, Beta, LV, 8mm $19.98** *WAR* 🦴🦴½

Mad Miss Manton A socialite turns detective to solve murder. Pleasant comedy-mystery provides occasional laughs and suspense.
1938 80m/B Barbara Stanwyck, Henry Fonda, Hattie McDaniel, Sam Levene, Miles Mander; ***Dir:*** Leigh Jason. **VHS, Beta, LV $39.95** *RKO* 🦴🦴

Mad Mission 3 Chinese man vacationing in Paris becomes involved in plot to recover precious jewels stolen from England's royal crown. Has there already been "Mad Mission" and "Mad Mission 2"?
1984 81m/C Richard Kiel, Sam Kui, Karl Muka, Sylvia Chang, Tsuneharu Sugiyama. **VHS, Beta $59.99** *HBO* 🦴½

The Mad Monster This "Wolf Man"-inspired cheapie looks like a misty relic today. A mad scientist furthers the war effort by injecting wolf's blood into a handyman, who becomes hairy and anti-social.
1942 77m/B Johnny Downs, George Zucco, Anne Nagel, Sarah Padden, Glenn Strange, Gordon DeMain, Mae Busch; ***Dir:*** Sam Newfield. **VHS $15.95** *NOS, LOO, VCN* 🦴½

Mad Monster Party Frankenstein is getting older and wants to retire from the responsibilities of being senior monster, so he calls a convention of creepy creatures to decide who should take his place—The Wolfman, Dracula, the Mummy, the Creature, It, the Invisible Man, or Dr. Jekyll and Mr. Hyde. Animated using the process of "Animagic."
1968 94m/C ***Dir:*** Jules Bass; ***Voices:*** Boris Karloff, Ethel Ennis, Phyllis Diller. **VHS, Beta $14.95** *COL, SUE* 🦴🦴½

Mad Wax: The Surf Movie A flimsy plot about a window-cleaner who discovers a magical surfing wax serves as a framework for lots and lots of scenes of surfin' dudes.
1990 45m/C Richard Cram, Marvin Foster, Aaron Napolean, Tom Carroll, Bryce Andrews, Ross Clarke-Jones; ***Dir:*** Michael Hohensee. **VHS, Beta $14.95** *PAV, FCT* 🦴

Mad Wednesday An accountant loses his job after twenty years, gets drunk, wins a pile of money, and buys a circus. All this in one day. Then he uses a lion to intimidate investors into backing him. A sequel to "The Freshman" from 1925, this was originally released as "The Sin of Harold Diddlebock" in 1947. Not classic Sturges but occaisionally hilarious and worth a peek since this was Lloyd's last feature film. As usual, Lloyd, though in his fifties, did his own dangerous stunts.
1951 79m/B Harold Lloyd, Raymond Walburn, Edgar Kennedy, Franklin Pangborn, Margaret Hamilton, Lionel Stander, Rudy Vallee; ***W/Dir:*** Preston Sturges. New York Film Festival '79: Best Film. **VHS, Beta $24.95** *VYY, MRV, WFV* 🦴🦴🦴

The Mad Whirl The Roarin' Twenties are a mad mad whirl in this silent silent piece of period period.
1925 80m/B Myrtle Stedman, Barbara Bedford, Alec B. Francis, George Fawcett, Joseph Singleton. **VHS, Beta $24.95** *GPV, FCT, DNB* 🦴½

Mad Youth A teenage girl falls in love with a man whom her slutty, alcoholic mother had been chasing. Bottom of the barrel production provides many unintentional laughs. Extraneous South American dance sequences included as character development.
1940 61m/B Mary Ainslee, Betty Atkinson, Willy Castello, Betty Compson, Tommy Wonder; ***W/Dir:*** Willis Kent. **VHS, Beta $16.95** *SNC Woof!*

Madame Bovary A young adultress with delusions of romantic love finds only despair in this offbeat adaptation of Flaubert's masterpiece. In French with English subtitles.
1934 102m/B *FR* Pierre Renoir, Valentine Tessier, Max Dearly; ***Dir:*** Jean Renoir. **VHS, Beta $59.95** *HTV* 🦴🦴🦴

Madame Bovary Young adultress with delusions of romantic love finds only despair, even in this Hollywood version of Flaubert's classic. Mason/Flaubert is put on trial for indecency following publication of the novel, with the story told from the witness stand. While this device occasionally slows the narrative, astute direction helps the plot along. Minnelli's handling of the celebrated ball sequence is superb.
1949 114m/B Jennifer Jones, Van Heflin, Louis Jourdan, James Mason, Gene Lockhart, Gladys Cooper, George Zucco; ***Dir:*** Vincente Minnelli. **VHS, Beta $19.98** *CCB, MGM* 🦴🦴🦴

Madame Bovary Provincial 19th century France is the setting for the tragedy of a romantic woman. Emma Bovary is bored by her marriage to a country doctor and longs for passion and excitement. She allows herself to be seduced (and abandoned) by a local aristocrat and herself seduces a young banker. As her husband's modest medical practice is even further reduced by a botched operation, Emma finds herself under an increasing burden of debt as she struggles to pay for all the finery bought in her quest for luxury. Unwilling to have her husband discover the extent of her deceit, and realizing she will never find the passion she desires, Emma takes drastic, and tragic, measures. A sumptuous-looking film, with an extraordinary performance by Huppert. In French with English subtitles. Based on the novel by Gustave Flaubert.
1991 (PG-13) 130m/C *FR* Isabelle Huppert, Jean-Francois Balmer, Christophe MaLavoy, Jean Yanne; ***Dir:*** Claude Chabrol. **VHS, LV $89.98** *REP* 🦴🦴🦴½

Madame Curie The film biography of Madame Marie Curie, the woman who discovered Radium. A deft portrayal by Garson, who is reteamed with her "Mrs. Miniver" co-star, Pidgeon. Certainly better than most biographies from this time period and more truthful as well. Nominated for five Oscars including, Best Picture, Best Actor, Best Actress, Best Cinematography and Best Music, but won none.
1943 113m/B Greer Garson, Walter Pidgeon, Robert Walker, May Whitty, Henry Travers, C. Aubrey Smith, Albert Basserman, Victor Francen, Reginald Owen, Van Johnson; ***Dir:*** Mervyn LeRoy. **VHS $19.98** *MGM, CCB* 🦴🦴🦴

Madame in Manhattan Wayland Flowers and Madame escort you through an unforgettable tour of the Big Apple.
1984 60m/C VHS, Beta $39.95 *RKO*

Madame Rosa Bloated madame nearing old age tends prostitutes' offspring in this warmhearted work. Academy Award for best foreign-language film.
1977 (PG) 105m/C *FR IS* Simone Signoret, Claude Dauphin; ***Dir:*** Moshe Mizrahi. Academy Awards '77: Best Foreign Language Film. **VHS, Beta $69.95** *VES* 🦴🦴🦴

Madame Sin Aspiring world dominator enlists former CIA agent in scheme to obtain nuclear submarine. Davis is appealing as evil personified. Wagner is, well, Wagneresque. Davis' first television movie.
1971 91m/C Bette Davis, Robert Wagner; ***Dir:*** David Greene. **VHS, Beta $9.99** *CCB, FOX* 🦴🦴½

Madame Sousatzka Eccentric, extroverted piano teacher helps students develop spiritually as well as musically. When she engages a teenage Indian student, however, she finds herself considerably challenged. MacLaine is perfectly cast in this powerful, winning film.
1988 (PG-13) 113m/C Shirley MacLaine, Peggy Ashcroft, Shabana Azmi, Twiggy, Leigh Lawson, Geoffrey Bayldon; ***Dir:*** John Schlesinger. Venice Film Festival '88: Best Actress, Best Actress (MacLaine-tie). **VHS, Beta, LV $19.95** *MCA* 🦴🦴🦴

Madame X Turner is perfectly cast in this oft-filmed melodrama about a social outcast who is tried for murder and defended by her unknowing son, who believes her long dead. Bennett's last film.
1966 100m/C Lana Turner, John Forsythe, Ricardo Montalban, Burgess Meredith, Virginia Grey, Constance Bennett, Keir Dullea; ***Dir:*** David Lowell Rich. **VHS, Beta $14.95** *MCA* 🦴🦴

M.A.D.D.: Mothers Against Drunk Driving Emotionally gripping television movie recounts the story of Candy Lightner, who, after her daughter was killed in a drunk driving accident, founded M.A.D.D. and built it into a nationwide organization. Similar to "License to Kill," which dramatized the effect that a drunk driving accident can have on the families of those involved.

1983 97m/C Mariette Hartley, Paula Prentiss, Bert Remsen; *Dir:* William A. Graham. **VHS, Beta** $39.95 MCA ♫♫½

Made in Argentina An Argentinean couple living in New York decide to return for a visit. Having left for political reasons, they still harbor bitterness.
1986 90m/C *AR* Luis Brandoni, Marta Bianchi, Leonor Manso; *Dir:* Juan Jose Jusid. **VHS, Beta** $69.98 VES ♫

Made for Each Other Newlyweds must overcome meddlesome in-laws, poverty, and even serious illness in this classic melodrama. Dated but appealing.
1939 85m/B James Stewart, Carole Lombard, Charles Coburn, Lucile Watson; *Dir:* John Cromwell. New York Times Ten Best List '39: Best Film. **VHS, Beta** $9.95 NEG, MRV, NOS ♫♫♫

Made in Heaven Arid comedy in which British newlyweds try to sustain honeymoon for entire year. The bride, however, eventually suspects her husband of shenanigans with the maid. If this sounds good, then you may like it.
1952 90m/C *GB* Petula Clark, David Tomlinson, Sonja Ziemann, A.E. Matthews; *Dir:* John Paddy Carstairs. **VHS, Beta** $29.98 VID ♫♫

Made in Heaven Two souls in heaven fall in love and must find each other after being reborn on Earth if they are to remain eternal lovers. Contains ethereal interpretations of heaven and cameo appearances by famous actors and musicians, including Debra Winger, Neil Young, Tom Petty, Ellen Barkin, Ric Ocasek, Tim Robbins, and Gary Larsen.
1987 (PG) 102m/C Timothy Hutton, Kelly McGillis, Maureen Stapleton, Mare Winningham, Ann Wedgeworth, Don Murray, Amanda Plummer, Ellen Barkin, Timothy Daly, Neil Young, Tom Petty, Ric Ocasek, Tim Robbins, Debra Winger, Gary Larsen; *Dir:* Alan Rudolph. **VHS, Beta, LV** $19.98 LHV, WAR ♫♫½

Made for Love Couple journeys romantically in Egypt until disaster strikes: they fall into a tomb and can't get out.
1926 65m/B Leatrice Joy, Edmund Burns, Ethel Wales, Brandon Hurst, Frank Butler; *Dir:* Paul Sloane. **VHS, Beta** $19.95 GPV, FCT ♫½

Made in the USA Wayward dudes engage vixen while cruising Midwest. Trouble, however, awaits. Note: This is not Godard's mid-sixties classic.
1988 (R) 82m/C Adrian Pasdar, Lori Singer, Christopher Penn; *Dir:* Ken Friedman. **VHS, Beta, LV** $19.98 SUE ♫½

Madeleine Courtroom drama of woman charged with poisoning her French lover in 1850s Scotland. Directed by the same David Lean who made "Bridge on the River Kwai" and "Lawrence of Arabia." Here, though, he was merely trying to provide a vehicle for his wife, actress Todd. Intrigued? Also known as "The Strange Case of Madeleine."
1950 114m/B Ann Todd, Leslie Banks, Ivan Desny; *Dir:* David Lean. **VHS, Beta** $14.95 LCA ♫♫½

Mademoiselle Striptease Bardot stars as a young woman who finds a valuable stolen book. Her attempts to disassociate herself from the actual theft are supposed to be funny. Bardot is characteristically appealing. Also known as "Please! Mr. Balzac."
1957 100m/B Brigitte Bardot, Robert Hirsch, Daniel Gelin; *Dir:* Marc Allegret. **VHS, Beta** $29.95 STV, MRV ♫♫½

Madhouse A troubled horror film star tries to bring his "Dr. Death" character to television, but during production people begin dying in ways remarkably similar to the script. A strong genre cast should hold the fans' attention during this mild adaptation of Angus Hall's novel "Devilday."
1974 92m/C *GB* Vincent Price, Peter Cushing, Robert Quarry, Adrienne Corri, Natasha Pyne, Linda Hayden, Michael Parkinson; *Dir:* Jim Clark. **VHS, Beta** HBO ♫♫

Madhouse A woman has bizarre recollections of her twin sister whom she finally meets when she escapes from a mental hospital. Lots of violence and gore, has definite cult potential.
1987 90m/C Trish Everly, Michael MacRae, Dennis Robertson, Morgan Hart; *Dir:* Ovidio G. Assonitis. **VHS, Beta** $19.95 VCL ♫♫

Madhouse New homeowners find themselves unable to expel loathsome, boorish guests. Presumably, a comedy cashing in on two of sitcom's brightest lights.
1990 (PG-13) 90m/C John Larroquette, Kirstie Alley, Alison La Placa, John Diehl, Jessica Lundy, Bradley Gregg, Dennis Miller, Robert Ginty; *Dir:* Tom Ropelewski. **VHS, Beta, LV** $14.98 ORI ♫♫

Madhouse Mansion A horror actor revives his career and becomes embroiled in murder. Plenty of in-jokes and clips from older movies. Originally titled "Ghost Story."
1974 (PG) 86m/C *GB* Marianne Faithfull, Leigh Lawson, Anthony Bate, Larry Dann, Sally Grace, Penelope Keith, Vivian Mackerell, Murray Melvin, Barbara Shelley; *Dir:* Stephen Weeks. **VHS** $29.98 CGI ♫♫½

Madigan Realistic and exciting and among the best of the behind-the-scenes urban police thrillers. Hardened NYC detectives (Widmark and Guardino) lose their guns to a sadistic killer and are given 72 hours to track him down. Fonda is the police chief none to pleased with their performance. Adapted by Howard Rodman, Abraham Polonsky, and Harry Kleiner from Richard Dougherty's "The Commissioner."
1968 101m/C Richard Widmark, Henry Fonda, Inger Stevens, Harry Guardino, James Whitmore, Susan Clark, Michael Dunn, Don Stroud; *Dir:* Don Siegel. **VHS, Beta** $59.95 MCA ♫♫♫

Madigan's Millions Incompetent Treasury agent treks to Italy to recover funds swiped from deceased gangster. This is Hoffman's first film, and is to his career what "The Last Chalice" was to Paul Newman's and "Studs and Kitty" was Sylvester Stallone. Recommended only to the terminally foolhardy.
1967 (G) 89m/C Dustin Hoffman, Elsa Martinelli, Cesar Romero; *Dir:* Stanley Prager. **VHS, Beta** $59.95 WES Woof!

Madman Deranged Soviet Jew joins Israeli army to more effectively fulfill his desire to kill loathsome Soviets. Good date movie for those nights when you're home alone. Weaver's first starring role.
1979 (R) 95m/C *IS* Sigourney Weaver, Michael Beck, F. Murray Abraham; *Dir:* Dan Cohen. **VHS, Beta** $19.95 NEG ♫

Madman A buffoon prompts terror when he revives legend regarding ax murderer. Pairs well with above title!
1982 (R) 89m/C Alexis Dubin, Tony Fish. **VHS, Beta** $69.99 HBO, HHE ♫

Mado A middle-aged businessman's life is undone when he falls for a mysterious woman. Piccoli and Schneider are, as usual, convincing. Another of underrated director Sautet's effective, low-key works. In French with subtitles.
1976 130m/C Romy Schneider, Michel Piccoli, Charles Denner; *Dir:* Claude Sautet. **VHS, Beta** $59.95 FCT ♫♫♫

Madox-01 Animated Japanese feature aimed at mature audiences. When a bungling tank operater gets trapped in the army's newest tank, two rivals join forces to free him. Japanese with English subtitles.
1991 50m/C *JP* VHS $34.95 CPM ♫♫♫

Madron Road western filmed in the Israel desert, complete with menacing Apaches, wagon train massacre, a nun and a gunslinger.
1970 (PG) 93m/C Richard Boone, Leslie Caron, Paul Smith; *Dir:* Jerry Hopper. **VHS, Beta** $59.95 UHV, TAV ♫½

Mae West This made-for-television film details West's life from her humble beginnings to her racy film stardom. Jillian does a good job bringing the buxom legend back to life.
1984 97m/C Ann Jillian, James Brolin, Piper Laurie, Roddy McDowall; *Dir:* Lee Philips. **VHS, Beta** $59.95 LHV ♫♫½

Maedchen in Uniform A scandalous early German talkie about a rebellious schoolgirl who falls in love with a female teacher and commits suicide when the teacher is punished for the relationship. A controversial criticism of lesbianism and militarism that impelled the Nazis, rising to power two years later, to exile its director. It was also banned in the United States. Subtitled in English. Remade in 1965.
1931 90m/B *GE* Dorothea Wieck, Ellen Schwannecke, Hertha Thiele; *Dir:* Leontine Sagan. National Board of Review Awards '32: Ten Best Films of the Year. **VHS, Beta** $29.98 SUE, MRV ♫♫♫½

Mafia la Ley que non Perdona A Mafia kingpin struggles to maintain his power through violence. Dubbed into Spanish.
1979 90m/C Gordon Mitchell, Antonella Lualdi. **VHS, Beta** $59.95 JCI ♫½

Mafia Princess The daughter of a Mafia boss tries to discover her own identity despite her father's involvement in crime. Based on a best-selling autobiography by a real-life member of a Mob family, Antoinette Giancana. Made for television.
1986 100m/C Tony Curtis, Susan Lucci, Kathleen Widdoes, Chuck Shamata, Louie Dibianco; *Dir:* Robert Collins. **VHS** $69.95 VHV ♫♫

Mafia vs. Ninja The mob wants to control the city, but the Ninja has other ideas.
1984 90m/C Alexander Lou, Silvio Azzolini, Wang Hsia, Charlema Hsu, Eugene Trammel. **VHS, Beta** $59.95 WES ♫½

Magdalena Viraga Experimental filmmaker Nina Menkes uses surrealism to explore the inner torment of a California prostitute.
1986 90m/C *Dir:* Nina Menkes. **VHS** $49.95 FCT ♫½

Magdalene Somewhere in medieval Europe, a violent Baron seeks revenge on a beautiful whore that spurned him. Small-budget costumer.

1988 (PG) 89m/C Steve Bond, Nastassia Kinski, David Warner, Gunter Meisner, Ferdinand "Ferdy" Mayne, Anthony Quayle, Franco Nero, Janet Agren; *Dir:* Monica Teuber. **VHS, Beta, LV** $89.99 PAR, FCT *♂♂* ½

Magee and the Lady The crusty captain of a rusty ship must warm up to a spoiled debutante or lose his ship to a foreclosure firm. Could be better. Made for TV.
1978 (PG) 92m/C *AU* Tony LoBianco, Sally Kellerman; *Dir:* Gene Levitt. **VHS, Beta** $19.95 ACA *♂♂*

Magic Ventriloquist Hopkins and his dummy, an all-too-human counterpart, get involved with a beautiful but impressionable woman lost between reality and the irresistible world of illusion. Spine-chilling psychodrama with a less-than-believable premise. Screenplay by William Goldman ("the Princess Bride"), from his novel.
1978 (R) 106m/C Anthony Hopkins, Ann-Margret, Burgess Meredith, Ed Lauter; *Dir:* Richard Attenborough. **VHS, Beta, LV** $14.98 SUE *♂♂* ½

Magic Adventure Animated adventure about a brother and his sister who are in for the time of their lives when they meet the Wind Wizard who sends them on an unexpected journey.
198? 82m/C VHS, Beta $19.95 ASE *♂♂* ½

The Magic Christian A series of related skits about a rich man (Sellers) and his son (Starr) who try to prove that anyone can be bought. Raucous, now somewhat dated comedy; music by Badfinger, including Paul McCartney's "Come and Get It."
1969 (PG) 101m/C *GB* Peter Sellers, Ringo Starr, Isabel Jeans, Wilfrid Hyde-White, Graham Chapman, John Cleese, Peter Graves, John Lennon, Yoko Ono, Richard Attenborough, Leonard Frey, Laurence Harvey, Christopher Lee, Spike Milligan, Yul Brynner, Roman Polanski, Raquel Welch; *Dir:* Joseph McGrath. **VHS, Beta, LV** $59.98 REP, MRV, PIA *♂♂♂*

Magic Christmas Tree A seemingly nonsensical fantasy with witches, pies, giants, fire trucks and Santa Claus.
198? 70m/C Chris Kroesen, Charles Nix, Terry Bradshaw; *Dir:* Richard C. Parish. **VHS, Beta** $19.95 UHV *♂*

Magic of Derek Dingle Considered to be one of the best manipulators in the world, Derek Dingle performs six of his favorite tricks, then teaches how each is done. Most are done with cards or coins and all are fascinating.
1981 57m/C Derek Dingle. **VHS, Beta** HSR

The Magic Flute In Swedish with English subtitles, this is Bergman's acclaimed version of Mozart's famous comic opera, universally considered one of the greatest adaptations of opera to film ever made. Staged before a live audience for Swedish television.
1973 (G) 134m/C *SW* Josef Kostlinger, Irma Urrila, Hakan Hagegard, Elisabeth Erikson; *Dir:* Ingmar Bergman. National Board of Review Awards '75: Special Citation; National Society of Film Critics Awards '75: Special Award. **VHS, Beta, LV** $29.95 HMV, PAR, VTK *♂♂♂* ½

Magic of Lassie Stewart is engaging as the nice grandpa who refuses to sell his land to mean rich guy Roberts. Innocuous, pleasant remake of "Lassie Come Home." Stewart sings, as do Pat Boone and daughter Debby.

1978 (G) 100m/C James Stewart, Mickey Rooney, Stephanie Zimbalist, Alice Faye, Pernell Roberts, Lassie; *Dir:* Don Chaffey. **VHS, Beta** $19.98 MGM *♂♂*

The Magic Legend of the Golden Goose A retelling of the familiar brothers Grimm chestnut.
1965 72m/C VHS, Beta UHV *♂*

Magic on Love Island Romantic misadventures ensue when eight ladies go on vacation to Love Island, a tropical paradise.
1980 96m/C Adrienne Barbeau, Bill Daily, Howard Duff, Dody Goodman, Dominique Dunne, Lisa Hartman, Janice Paige; *Dir:* Earl Bellamy. **VHS, Beta** $59.98 MAG *♂*

The Magic Sword A family-oriented adventure film about a young knight who sets out to rescue a beautiful princess who is being held captive by an evil sorcerer and his dragon.
1962 80m/C Basil Rathbone, Estelle Winwood, Gary Lockwood; *Dir:* Bert I. Gordon. **VHS, Beta** $29.95 SNC, MRV, VYY *♂♂*

Magic Town An opinion pollster investigates a small town which exactly reflects the views of the entire nation, making his job a cinch. The publicity causes much ado in the town with ensuing laughs. Uneven but entertaining.
1947 103m/B Jane Wyman, James Stewart, Kent Smith, Regis Toomey, Donald Meek; *Dir:* William A. Wellman. **VHS** $19.98 REP *♂♂* ½

The Magic Voyage of Sinbad Sinbad embarks on a fantastic journey after promising the people of his Covasian home that he will find the elusive Phoenix, the bird of happiness. Also known as "Sadko," this production was not released in the U.S. until 1962 and was rewritten for the American screen by young Francis Ford Coppola.
1952 79m/C *RU* Sergey Stolyarov, Alla Larionova, Mark Troyanovsky; *Dir:* Alexander Ptushko. **VHS, Beta** $16.95 SNC *♂♂*

Magical Mystery Tour On the road with an oddball assortment of people, the Beatles experience a number of strange incidents around the English countryside. Originally made for British television. Songs include: "The Fool on the Hill," "Blue Jay Way," "Your Mother Should Know," and the title tune.
1967 55m/C *GB* John Lennon, George Harrison, Ringo Starr, Paul McCartney, Victor Spinetti. **VHS, Beta, LV** $29.95 MVD, MED, WFV *♂♂*

Magical Wonderland Join a handsome prince and his fair young maiden as they venture through such fairy tale classics as "Beauty and the Beast" and "Sleeping Beauty."
1985 90m/C VHS, Beta $39.95 UHV

The Magician A master magician in 19th century Sweden (von Sydow) wreaks ill in this darkly comical, supernatural parable. Dark, well photographed early Bergman effort.
1958 101m/B *SW* Max von Sydow, Ingrid Thulin, Gunnar Bjornstrand, Bibi Andersson, Naima Wifstrand; *Dir:* Ingmar Bergman. Venice Film Festival '59: Special Jury Prize, Cinema Nuova Prize. **VHS, Beta, LV** $29.95 SUE, MRV, HHT *♂♂♂*

The Magician of Lublin Based on an unusual story by Issac Bashevis Singer. Follows the exploits of a magician/con man in turn-of-the-century Poland whose personal flaws kill his career, until a chance emerges for one last trick. A good example of movie

not as good as book; this rendering is superficial, badly acted and unsatisfying. Made soon after Singer won the Nobel Prize for literature in 1978.
1979 (R) 105m/C Alan Arkin, Valerie Perrine, Louise Fletcher, Lou Jacobi, Shelley Winters; *Dir:* Menahem Golan. **VHS, Beta** $79.99 HBO *♂* ½

The Magnificent When the government is overthrown by a band of mercenaries led by Lord Lu, the deposed emperor disappears. Lots of head kicking and action shots.
19?? 80m/C Chen Sing, Bruce Lai, Col Len Rei Lee, Jane Jennings; *Dir:* Chen Shao Peng. **VHS, Beta** $59.95 SVS *♂*

Magnificent Adventurer Ostensibly a biography of Benvenuto Cellini, the Florentine sculptor, with a concentration on his love life and swordplay.
1976 94m/C Brett Halsey. **VHS, Beta** $59.95 UNI *♂*

The Magnificent Ambersons Welles' second film. A fascinating, inventive translation of the Booth Tarkington novel about a wealthy turn-of-the-century family collapsing under the changing currents of progress. Pure Welles, except the glaringly bad tacked-on ending that the studio shot (under the direction of the great Robert Wise and Fred Fleck), after taking the film from him. It seems they wanted the proverbial happy ending. Also available on laserdisc, with interviews and the original story boards. Oscar nominations: Best Picture, Best Supporting Actress (Morehead), Black & White Cinematography, Black & White Interior Decoration.
1942 88m/B Joseph Cotten, Anne Baxter, Tim Holt, Richard Bennett, Dolores Costello, Erskine Sanford, Ray Collins, Agnes Moorehead; *Dir:* Orson Welles. New York Film Critics Awards '42: Best Actress (Moorehead); Sight & Sound Survey '72: #9 of the Best Films of All Time; Sight & Sound Survey '82: #7 of the Best Films of All Time (tie). **VHS, Beta, LV** $19.95 RKO, VID, VYG *♂♂♂♂*

The Magnificent Dope Fonda is likeable, as usual, as the hapless yokel in the big city. A trip to New York to take a course on being successful is his prize for winning a "laziest man" contest. He falls for his teacher; she uses his crush to motivate him. Funny and entertaining.
1942 83m/B Henry Fonda, Lynn Bari, Don Ameche, Edward Everett Horton, Hobart Cavanaugh, Pierre Watkin; *Dir:* Walter Lang. **VHS** TCF *♂♂* ½

Magnificent Duo Chop-socky head-kicking between two hard-footed veterans.
197? 84m/C Master Lee, Chris Kim. **VHS, Beta** $39.95 UNI *♂*

The Magnificent Kick The story of a master Wong-Fai-Hung, inventor of the "Kick without Shadow," which was practiced by the late Bruce Lee.
1980 90m/C VHS, Beta $54.95 MAV *♂*

The Magnificent Matador The story of an aging matador who faces death in the bullring to win the love of a woman. Quinn is the bullfighter on the horns of a dilemma. Lots of bull in script carried by harried bull in ring.
1955 94m/C Anthony Quinn, Maureen O'Hara, Thomas Gomez; *Dir:* Budd Boetticher. **VHS, Beta** $29.95 VHE *♂♂* ½

Magnificent Obsession A drunken playboy (Hudson) kills a man and blinds his wife in an automobile accident. Plagued by guilt, he devotes his life to studying medicine

in order to restore the widow's sight. Well-acted melodrama lifted Hudson to stardom and gave Wyman a Best Actress nomination. Faithful to the 1935 original, based on a novel by Lloyd C. Douglas.
1954 108m/C Jane Wyman, Rock Hudson, Barbara Rush, Agnes Moorehead; **Dir:** Douglas Sirk. **VHS, Beta** $14.95 *MCA* 🎬🎬½

Magnificent Seven Western remake of Akira Kurosawa's classic "The Seven Samurai." Mexican villagers hire gunmen to protect them from the bandits who are destroying their town. Most of the actors were relative unknowns, though not for long. Sequelled by "Return of the Seven" in 1966, "Guns of the Magnificent Seven" in 1969, and "The Magnificent Seven Ride" in 1972. Excellent score.
1960 126m/C Yul Brynner, Steve McQueen, Robert Vaughn, James Coburn, Charles Bronson, Horst Buchholz, Eli Wallach, Brad Dexter; **Dir:** John Sturges. **VHS, Beta, LV** $19.95 *MGM, FOX, TLF* 🎬🎬🎬🎬

Magnum Force Eastwood's second "Dirty Harry" movie. Harry finds a trail leading from a series of gangland killings straight back to the P.D. Less gripping than "Dirty Harry" (1971); followed by "The Enforcer" (1976), "Sudden Impact" (1983), and "The Dead Pool" (1988). Holbrook is cast intriguingly against type, and the premise is good, but little else recommends this too-violent installment.
1973 (R) 124m/C Clint Eastwood, Hal Holbrook, Mitchell Ryan, David Soul, Robert Urich, Tim Matheson, Kip Niven; **Dir:** Ted Post. **VHS, Beta, LV** $19.98 *WAR, TLF* 🎬🎬

Magnum Killers A young man becomes involved in a web of deadly intrigue when he sets out to find who cheated him out of a small fortune during a card game.
1976 92m/C Sombat Methance, Prichela Lee. **VHS, Beta** $29.95 *UNI* 🎬

Magnum: Live The heavy metal group performs hits like "Les Mort Dansant," "Kingdom of Madness" and "How Far Jerusalem" at the Camden Palace in London.
1985 60m/C VHS, Beta $19.95 *MVD, SUE*

Magoo's Puddle Jumper Magoo backs off a pier in his "new" used car and points out all the underwater sights to his nephew Waldo. Part of the "Mister Magoo" series.
1956 7m/C Voices: Jim Backus. Academy Awards '56: Best Animated Short Film. **VHS, Beta** *CHF* 🎬

The Mahabharata Adapted from the myths and folklore of ancient India, the screen version of the original 9-hour stage play is the story of a devastating war between two powerful clans. Initially broadcast on public television, the 6-hour movie is divided into three 2-hour tapes, "The Game of Dice," "Exile in the Forest," and "The War."
1989 318m/C *GB FR* Robert Langton-Lloyd, Antonin Stahly-Vishwanadan, Bruce Myers; **Dir:** Peter Brook. **VHS, LV** $99.95 *BOK, BTV, RMF* 🎬

Mahler Strange Russell effort on the life of the great composer Gustav Mahler. Imperfect script is rescued by fine acting.
1974 (PG) 110m/C Robert Powell, Georgina Hale, Richard Morant, Lee Montague, Terry O'Quinn; **Dir:** Ken Russell. **VHS, Beta** *HBO, OM* 🎬🎬🎬

Mahogany A world-famous high fashion model and designer gets a career boost when she daringly appears in a dress of her own creation at a Roman fashion show. Predictable melodrama.
1976 (PG) 109m/C Diana Ross, Billy Dee Williams, Jean-Pierre Aumont, Anthony Perkins, Nina Foch; **Dir:** Berry Gordy. **VHS, Beta, LV** $14.95 *PAR, CCB* 🎬½

The Maid An offbeat comedy romance with Sheen as the house husband to Bisset's female executive character. Nice acting; charming, if unoriginal, premise.
1990 (PG) 91m/C Martin Sheen, Jacqueline Bisset, Jean-Pierre Cassel, James Faulkner; **Dir:** Ian Toynton. **VHS, Beta** $89.98 *MED* 🎬🎬½

Maid Marian & Her Merry Men: How the Band Got Together Monty-Pythonesque spoof of the Robin Hood myths; in this version, Maid Marian does all the work. Three episodes: "How the Band Got Together", "Robert the Incredible Chicken" and "A Game called John."
1991 74m/C *GB* **VHS, Beta** $14.98 *FOX, FCT* 🎬🎬½

Maid Marian & Her Merry Men: The Miracle of St. Charlene Revamping of the Robin Hood legend featuring Maid Marian as the star. From members of the Monty Python troupe.
1991 74m/C *GB* **VHS, Beta** $14.98 *FOX, FCT* 🎬🎬½

Maid to Order Rich girl Sheedy's fairy godmother puts her in her place by turning her into a maid for a snooty Malibu couple. Good-natured and well-acted if rather mindless Cinderella story.
1987 (PG) 92m/C Ally Sheedy, Beverly D'Angelo, Michael Ontkean, Dick Shawn, Tom Skerritt, Valerie Perrine; **Dir:** Amy Jones. **VHS, Beta, LV** $14.95 *IVE* 🎬½

Maid in Sweden A naive country girl visits Stockholm for the first time and has some bizarre encounters with men.
1983 90m/C *SW* Kristina Lindberg, Monika Ekman; **Dir:** Floch Johnson. **VHS, Beta** $42.95 *PGN* 🎬½

Maid's Night Out Pleasant, simple-minded mistaken-identity comedy. Fontaine is an heiress; rich-guy posing-as-milkman Lane thinks she is a maid. They fall in love, of course.
1938 64m/B Joan Fontaine, Allan Lane, Hedda Hopper, George Irving, William Brisbane, Billy Gilbert, Cecil Kellaway; **Dir:** Ben Holmes. **VHS** *RKO, MLB* 🎬🎬

Mail Early Norman McLaren superimposed non-abstract symbols on clear 35mm film, in this publicity clip for Canada Post.
1941 2m/C VHS, Beta *NFC* 🎬🎬

The Main Event Streisand plays a wacky—and bankrupt—executive who must depend on the career of a washed up boxer to rebuild her fortune. Desperate (and more than a little smitten), she badgers and bullies him back into the ring. Lame, derivative comedy; unsuccessful repairing of "What's Up, Doc?" stars. Streisand sings the title song.
1979 (PG) 105m/C Barbra Streisand, Ryan O'Neal; **Dir:** Howard Zieff. **VHS, Beta** $19.98 *WAR* 🎬🎬

Main Street to Broadway Superficial story of a struggling playwright, interesting mainly for abundant big-name cameos.

1953 102m/B Tom Morton, Mary Murphy, Rex Harrison, Helen Hayes, Mary Martin, Ethel Barrymore, Lionel Barrymore, Lilli Palmer, Henry Fonda, Tallulah Bankhead, Cornel Wilde, Agnes Moorehead, Joshua Logan; **Dir:** Tay Garnett. **VHS, Beta** $59.95 *MPI, MLB* 🎬🎬

Maitresse An examination of the sexual underworld in the same vein as "Blue Velvet" and "Crimes of Passion." Director Schroeder also created "Reversal of Fortune." In French with English subtitles.
1976 112m/C Gerard Depardieu; **Dir:** Barbet Schroeder. **VHS** $59.99 *WAR, BTV, FCT* 🎬🎬

Major Barbara A wealthy, idealistic girl joins the Salvation Army against her father's wishes. Based on the play by George Bernard Shaw. The excellent adaptation of the original and the cast make this film a winner. Deborah Kerr's film debut.
1941 90m/B *GB* Wendy Hiller, Rex Harrison, Robert Morley, Sybil Thorndike, Deborah Kerr; **Dir:** Gabriel Pascal. **VHS, Beta** $39.95 *LCA, FCT* 🎬🎬🎬½

Major Bowes' Original Amateur Hour Two episodes of Bowes' original contest radio show in behind-the-scenes films.
1939 20m/B VHS, Beta *VYY*

Major Dundee A Union army officer (Heston) chases Apaches into Mexico with a motley collection of prisoner volunteers. Too long and flawed; would have been better had Peckinpah been allowed to finish the project. Excellent cast.
1965 124m/C Charlton Heston, Richard Harris, James Coburn, Jim Hutton, Ben Johnson, Slim Pickens; **Dir:** Sam Peckinpah. **VHS, Beta** $12.95 *GKK* 🎬🎬½

Major League Comedy about a pathetic major league baseball team whose new owner schemes to lose the season and relocate the team to Miami. Predictable sports spoof scripted by David Ward ("The Sting"). Sheen is okay as the pitcher with control problems (both on and off the field), while Bernsen seems to be gazing affectionately at "L.A. Law" from a distance. Good for a few laughs, particularly those scenes involving Dennis Haysbert as a slugger with voodoo on his mind (and in his locker) and Snipes as a base stealer whose only problem is getting on base.
1989 (R) 107m/C Tom Berenger, Charlie Sheen, Corbin Bernsen, James Gammon, Margaret Whitton, Bob Uecker, Rene Russo, Wesley Snipes, Dennis Haysbert; **Dir:** David S. Ward. **VHS, Beta, LV, 8mm** $14.95 *PAR* 🎬🎬½

The Majorettes Eek! Someone is lurking around a high school murdering the majorettes with their own batons. A bare-bones plot lurks within this otherwise worthless pic.
1987 93m/C Kevin Kinklin, Terrie Godfrey, Mark V. Jevicky; **Dir:** John Russo. **VHS, Beta** $69.98 *LIV, VES* 🎬

Make Haste to Live Good, scary thriller. A woman survives attempted murder by her husband; moves far away to raise their infant daughter. He does time, then returns for revenge. McGuire is good as the terrorized wife.
1954 90m/B Dorothy McGuire, Stephen McNally, Edgar Buchanan, John Howard; **Dir:** William A. Seiter. **VHS** $19.98 *REP* 🎬🎬½

Make Me an Offer Somewhat slow-moving comedy centering on an antique dealer who attempts to buy an expensive vase from an old man. From the novel by Wolf Mankowitz.
1955 88m/C *GB* Peter Finch, Adrienne Corri, Rosalie Crutchley, Finlay Currie; *Dir:* Cyril Frankel. **VHS** $59.98 *SUE &&*

Make Me an Offer Made-for-TV drivel about selling real estate in Hollywood. Better than similar efforts.
1980 97m/C Susan Blakely, Stella Stevens, Patrick O'Neal; *Dir:* Jerry Paris. **VHS, Beta** $59.98 *CHA & 1/2*

Make a Million The Depression is played (successfully) for yuks in this tale of an economics professor fired from his post for advocating radical income redistribution. He doesn't get mad, he gets even: he makes a million by advertising for money.
1935 66m/B Charles Starrett, Pauline Brooks. **VHS, Beta** $29.95 *VYY &&*

Make Mine Mink Oft-hilarious British comedy about guests at an elegant but run-down mansion who become unlikely thieves, stealing furs for charity. Good cast headed by Terry-Thomas.
1960 100m/C *GB* Terry-Thomas, Billie Whitelaw, Hattie Jacques; *Dir:* Robert Asher. **VHS, Beta** $19.98 *VID &&&*

Make Room for Tomorrow Subtle, somewhat funny French comedy about generations in a family. A man in mid-life crisis has to cope with his father, son, and grandfather on the grandfather's 90th birthday.
1981 (R) 106m/C *FR* Victor Lanoux, Jane Birkin, George Wilson; *Dir:* Peter Kassovitz. **VHS, Beta** $29.95 *SUE &&*

Make-Up A doctor turned circus clown uses his medical skills when an elephant renders a society girl unconscious. After she awakens the two become involved. The clown's daughter objects, but is soon caught up in an accusation that she murdered the lion tamer. Her father the clown tries to save the day. Predictable doctor/clown relationship.
1937 72m/B Nils Asther, June Clyde, Judy Kelly, Kenne Duncan, John Turnbull; *Dir:* Alfred Zeisler. **VHS** $24.95 *NOS, DVT &&*

Make a Wish A noted composer goes stale in this colorful musical about backstage life. The comic relief provides the films best moments. Music by Oscar Strauss.
1937 80m/B Basil Rathbone, Leon Errol, Bobby Breen, Ralph Forbes; *Dir:* Kurt Neumann. **VHS, Beta** $29.95 *VYY &&*

Makin' It Three trampy barmaids rise to the top of the music business due to every talent except singing. Also known as "Running Hot."
19?? (R) 92m/C Rory Calhoun, Darrow Igus, Gwen Owens. **VHS, Beta** $19.95 *ACA &*

Making Contact A small boy's telekinetic powers enable him to bring to life his favorite toys. The ridicule he endures because of this leads him to set off on terrifying adventures with only his toys and a friend for company.
1986 (PG) 83m/C Joshua Morrell, Eve Kryll; *Dir:* Roland Emmerich. **VHS, Beta** $14.95 *NWV, STE & 1/2*

Making of "Gandhi" This documentary shows the struggles that director Richard Attenborough went through to bring Mahatma Gandhi's life story to the screen. Some black-and-white segments.

1984 51m/C **VHS, Beta** *DCL*

Making the Grade Jersey tough kid owes the mob; attends prep school in place of a rich kid who can't be bothered. Better than similar '80s teen flicks, but not much.
1984 (PG) 105m/C Judd Nelson, Joanna Lee, Dana Olsen, Ronald Lacey, Scott McGinnis, Gordon Jump, Carey Scott, Andrew Dice Clay; *Dir:* Dorian Walker. **VHS, Beta** $79.95 *MGM & 1/2*

Making of a Legend: "Gone With the Wind" An exhaustive tribute to the American classic on its 50th anniversary, featuring miles of film clips, wardrobe tests, interviews with surviving cast and crew members, and scads of production trivia.
1989 130m/C Clark Gable, Vivien Leigh, Carole Lombard, Laurence Olivier, David O. Selznick, Margaret Mitchell, Ann Rutherford, Butterfly McQueen, Evelyn Keyes, Leslie Howard, Olivia de Havilland, Thomas Mitchell, Ona Munson, Ward Bond, Hattie McDaniel, Victor Fleming, Sam Wood, George Cukor, Paulette Goddard, Lana Turner, Jean Arthur, Joan Bennett, Tallulah Bankhead, Melvyn Douglas, George Reeves, Fred Crane. **VHS, Beta, LV** $29.95 *MGM*

Making Love A closet homosexual risks his 8-year marriage by getting involved with a carefree writer. What could be a powerful subject gets only bland treatment.
1982 (R) 112m/C Kate Jackson, Harry Hamlin, Michael Ontkean, Wendy Hiller, Arthur Hill, Nancy Olson, Terry Kiser, Camilla Carr, Michael Dudikoff; *Dir:* Arthur Hiller. **VHS, Beta** $69.98 *FOX &&*

Making Mr. Right Under-rated satire about a high-powered marketing and image consultant who falls in love with the android that she's supposed to be promoting. Unbelievable comedy is shaky at times, but Magnuson & Malkovich create some magic.
1986 (PG-13) 95m/C John Malkovich, Ann Magnuson, Glenne Headly, Ben Masters, Laurie Metcalf, Polly Bergen; *Dir:* Susan Seidelman. **VHS, Beta, LV** $19.99 *HBO && 1/2*

The Making of the Stooges Documents the 50-year television history of the Three Stooges including rare footage of these slapstick stars.
1985 60m/B Moe Howard, Larry Fine, Curly Howard, Shemp Howard; *Nar:* Steve Allen. **VHS, Beta** $19.98 *LHV, WAR*

Malarek Tough Montreal journalist Victor Malarek exposed abuse in that city's teen detention center in his book "Hey, Malarek." Compelling lead performance from Koteas in decent though not great screen adaptation.
1989 (R) 95m/C Michael Sarrazin, Elias Koteas, Al Waxman, Kerrie Keane; *Dir:* Roger Cardinal. **VHS, Beta, LV** $14.95 *SVS &&*

Malcolm An offbeat comedy about a slightly retarded young man who is mechanically inclined and his unusual entry into a life of crime. Directorial debut for actress Tass, whose husband David Parker wrote the screenplay (and designed the Tinkertoys). Music score performed by The Penguin Cafe Orchestra.
1986 (PG-13) 86m/C *AU* Colin Friels, John Hargreaves, Lindy Davies; *Dir:* Nadia Tass. Australian Film Institute '86: Best Actor (Friels), Best Film. **VHS, Beta, LV** $79.98 *LIV, VES && 1/2*

Male and Female A group of British aristocrats is shipwrecked on an island and must allow their efficient butler (Meighan) to take command for their survival. Swanson is the spoiled rich girl who falls for her social inferior. Their rescue provides a return to the rigid British class system. Based on the play

"The Admirable Crichton" by James M. Barrie.
1919 ?m/B Gloria Swanson, Thomas Meighan, Lila Lee, Raymond Hatton, Bebe Daniels; *Dir:* Cecil B. DeMille. **VHS, Beta** *DNB &&&&*

Malibu Beach The California beach scene is the setting for this movie filled with bikini clad girls, tanned young men, and instant romances.
1978 (R) 93m/C Kim Lankford, James Daughton; *Dir:* Robert J. Rosenthal. **VHS, Beta** $9.99 *STE, PSM, UHV &*

Malibu Bikini Shop Two brothers manage a beachfront bikini shop in Malibu, and spend their time ogling the customers and arguing about running the "business." Just another excuse for parading babes in bathing suits across the naked screen.
1986 (R) 90m/C Michael David Wright, Bruce Greenwood, Barbara Horan, Debra Blee, Jay Robinson, Galyn Gorg, Ami Julius, Frank Nelson, Kathleen Freeman, Rita Jenrette; *W/Dir:* David Wechter. **VHS, Beta** $19.98 *FOX &*

Malibu Express Mystery/adventure plot about a P.I. is the excuse; babes in swimsuits is the reason for this waste of time.
1985 (R) 101m/C Darby Hinton, Sybil Danning, Art Metrano, Shelley Taylor Morgan, Niki Dantine, Barbara Edwards; *Dir:* Andy Sidaris. **VHS, Beta** $14.95 *MCA &*

Malibu High The accidental death of a young prostitute's client leads her to a new series of illegal activities of the "sex and hit" variety. Dark, sleazy, and antisocial; belies Beach Boys-esque title.
1979 (R) 92m/C Jill Lansing, Stuart Taylor; *Dir:* Irvin Berwick. **VHS, Beta** $19.95 *UHV &*

Malibu Hot Summer Three aspiring young women combine their talents in acting, music and athletics to find their careers as they spend a fun-filled summer in Malibu.
1981 89m/C Terry Congie, Leslie Brander, Roslyn Royce. **VHS, Beta** $49.95 *VHE &*

Malicious A housekeeper hired for a widower and his three sons becomes the object of lusty affection of all four men. As Papa makes plans to court and marry her, his fourteen-year-old son plots to have her as a companion on his road to sensual maturity. Dubbed in English.
1974 (R) 98m/C *IT* Laura Antonelli, Turi Ferro, Alessandro Momo, Tina Aumont; *Dir:* Salvatore Samperi. **VHS, Beta** $49.95 *PAR &&*

Malone A burnt-out secret agent stumbles into a real-estate swindle/murder plot in Oregon and sets out to stop it. Film tries hard but doesn't succeed in being believable.
1987 (R) 92m/C Burt Reynolds, Lauren Hutton, Cliff Robertson, Kenneth McMillan, Scott Wilson, Cynthia Gibb; *Dir:* Harley Cokliss. **VHS, Beta, LV** $19.98 *ORI & 1/2*

Malou A young French woman tries to uncover the truth about her late mother's life, sifting through wildly contradictory evidence in the desire to discover her own identity and avoid her mother's mistakes. A feminist story. In German with English subtitles.
1983 (R) 95m/C *GE* Grischa Huber, Ingrid Caven, Helmut Griem; *Dir:* Jeanine Meerapfel. Cannes Film Festival '83: International Film Critics' Prize. **VHS, Beta** $34.95 *SUE, GLV &&*

Malta Story A British World War II flier becomes involved with the defense of Malta.
1953 103m/B Alec Guinness, Jack Hawkins, Anthony Steel, Flora Robson, Muriel Pavlow; *Dir:* Brian Desmond Hurst. **VHS, Beta** $14.95 *SUE && 1/2*

The Maltese Falcon A good first screen version of the Dashiell Hammett story about private detective Sam Spade's search for the elusive Black Bird. Remade five years later as ''Satan Met a Lady'' with Bette Davis. Its most famous remake, however, occurred in 1941 with Humphrey Bogart in the lead. Also called ''Dangerous Female.''
1931 80m/B Bebe Daniels, Ricardo Cortez, Dudley Digges, Thelma Todd, Una Merkel, Dwight Frye, Robert Elliott; **Dir:** Roy Del Ruth. VHS, Beta *GPV* 🎬🎬🎬

The Maltese Falcon After the death of his partner, detective Sam Spade finds himself enmeshed in a complicated, intriguing search for a priceless statuette. ''It's the stuff dreams are made of,'' says Bogart of the Falcon. Excellent, fast-paced film noir with outstanding performances, great dialogue, and concentrated attention to details. Director Huston's first film and Greenstreet's talky debut. First of several films by Bogart and Astor; watch for cameo by Walter Huston. Based on the novel by Dashiell Hammett. Oscar nominations: Best Picture, Best Supporting Actor (Greenstreet), and Best Screenplay. Astor received an Academy nomination and Award for her other 1941 movie, ''The Great Lie.'' Also available colorized.
1941 101m/B Humphrey Bogart, Mary Astor, Peter Lorre, Sydney Greenstreet, Ward Bond, Barton MacLane, Gladys George, Lee Patrick, Elisha Cook Jr., Walter Huston; **Dir:** John Huston. National Board of Review Awards '41: Best Performances (Bogart and Astor). VHS, Beta, LV $19.95 *MGM, FOX, TLF* 🎬🎬🎬

Mama Dracula Fletcher stars in this poor satire of the horror genre. She's a vampire who needs the blood of virgins to stay young. Her son helps out—what good son wouldn't?
1980 90m/C *FR* Louise Fletcher, Bonnie Schneider, Maria Schneider, Marc-Henri Wajnberg, Alexander Wajnberg, Jess Hahn; **Dir:** Boris Szulzinger. VHS $79.95 *TWE* Woof!

Mama, There's a Man in Your Bed When a powerful executive is framed for insider trading, the only witness to the crime and his only hope is an earthy cleaning woman named Juliette. Together they plot revenge and takeovers in her tiny apartment, filled with children. Soon, he regains his former position of power but realizes his life isn't complete without his co-conspirator, Juliette. In French with English subtitles.
1989 (PG) 111m/C *FR* Daniel Auteuil, Firmine Richard, Pierre Vernier, Maxime LeRoux, Gilles Privat, Muriel Combeau, Catherine Salviat, Sambou Tati; **W/Dir:** Coline Serreau. VHS, Beta $89.99 *HBO, FCT* 🎬🎬🎬

Mama Turns 100 Spanish comedy with English subtitles depicts the Saura family celebrating their mother's 100th birthday at her estate. A very bizarre set of characters gathers at Mama's estate. One of the daughter's seduces her brother-in-law, another has a uniform fetish, a religious zealot is desperate to hang-glide, and several of them are trying to kill poor Mama. But of course she prevails, although sick and ailing. Slow-paced yet uproarious film full of inept characters who bumble along.
1979 (R) 100m/C *SP* Geraldine Chaplin, Fernando Gomez, Amparo Munoz; **Dir:** Carlos Savrat. VHS, Beta $59.95 *INT* 🎬🎬🎬

Mama's Dirty Girls A gangster mom's daughters take over where their mother left off and have the time of their lives.

1974 (R) 82m/C Gloria Grahame, Paul Lambert, Sondra Currie, Candice Rialson, Mary Stoddard; **Dir:** John Hayes. VHS, Beta $69.95 *TWE* 🎬🎬🎬

Mambo A poor young dancer inspires patronage and lust by dancing the Mambo. Although attracted to Gassman, she marries Rennie for his money. Technically weak but artistically interesting, though Rennie and Mangano have done their roles somewhat better elsewhere.
1955 94m/B Silvana Mangano, Michael Rennie, Vittorio Gassman, Shelley Winters, Katherine Dunham, Mary Clare, Eduardo Ciannelli; **Dir:** Robert Rossen. VHS $19.95 *NOS, HTV* 🎬🎬½

Mambo Kings Armand and Banderas play Cesar and Nestor Castillo, two brothers who flee Cuba and go to New York to hit it big playing mambo music. Just as their star begins to rise, their dreams are crushed when the quick-tempered Cesar (Assante) angers a mob boss. Soon, however, Desi Arnaz Sr., played by his son Desi Arnaz Jr., sees the brothers and asks them to appear on the ''I Love Lucy'' show. This funny and technical scene leads to the climactic confrontation between the brothers. Based on the Pulitzer prizewinning novel ''Mambo Kings Play Songs of Love'' by Oscar Hijuelos.
1992 (R) 100m/C Armand Assante, Antonio Banderas, Cathy Moriarty, Maruschka Detmers, Desi Arnaz Jr., Celia Cruz, Tito Puente, Talisa Soto; **Dir:** Arne Glimcher. VHS, LV, 8mm $94.99 *WAR* 🎬🎬🎬

Mambo Mouth The one-man show which enjoyed great success on Broadway and as an HBO special features Leguizamo creating biting caricatures of Latino men. His characterizations ring true, while expanding the audience's boundaries of perception regarding Latin males.
1992 90m/C John Leguizamo; **Dir:** Peter Askin. VHS $19.95 *PGV, IVA* 🎬🎬🎬

Mame In an adaptation of the Broadway musical ''Auntie Mame'' by Jerry Herman, Ball plays a dynamic woman who takes it upon herself to teach a group of eccentrics how to live life to the fullest. Arthur plays Mame's friend just as splendidly as she did in the Broadway version, but Ball is a lame Mame. The production as a whole misses its mark through its overambition. Ball's last feature film role.
1974 (PG) 132m/C Lucille Ball, Bea Arthur, Robert Preston, Joyce Van Patten, Bruce Davison; **Dir:** Gene Saks. VHS, Beta, LV $19.98 *WAR, PIA* 🎬½

Mamele (Little Mother) Picon prematurely becomes the ''mother'' of her seven siblings when their mother dies. Quintessential Yiddish musical with the usual shimmering performance from Picon. In Yiddish with English subtitles.
1938 95m/B *PL* Molly Picon, Max Bozyk, Edmund Zayenda; **Dir:** Joseph Green. VHS, Beta $79.95 *ERG* 🎬½

A Man About the House Two unmarried English sisters move into the Italian villa they have inherited. There, one marries the caretaker, who secretly plans to regain the property that once belonged to his family. When the newly married sister is found dead, her sibling sets out to solve the murder.
1947 83m/B *GB* Kieron Moore, Margaret Johnston, Dulcie Gray, Guy Middleton, Felix Aylmer; **Dir:** Leslie Arliss. VHS $16.95 *NOS, SNC* 🎬🎬½

Man from the Alamo A soldier sent from the Alamo during its last hours to get help is branded as a deserter, and struggles to clear his name. Well acted, this film will satisfy those with a taste for action.
1953 79m/C Glenn Ford, Julie Adams, Chill Wills, Victor Jory; **Dir:** Budd Boetticher. VHS, Beta $14.95 *KRT* 🎬🎬🎬

A Man for All Seasons Sterling, heavily Oscar-honored biographical drama concerning the life and subsequent martyrdom of 16th-century Chancellor of England, Sir Thomas More (Scofield). Story revolves around his personal conflict when King Henry VIII (Welles) seeks a divorce from his wife, Catherine of Aragon, so he can wed his mistress, Anne Boleyn—events that ultimately lead the King to bolt from the Pope and declare himself head of the Church of England. Screenplay by Robert Bolt. Remade for television in 1988 with Charlton Heston in the lead role.
1966 (G) 120m/C *GB* Paul Scofield, Robert Shaw, Orson Welles, Wendy Hiller, Susannah York; **Dir:** Fred Zinneman. Academy Awards '66: Best Actor (Scofield), Best Adapted Screenplay, Best Color Cinematography, Best Costume Design (Color), Best Director (Zinneman), Best Picture; British Academy Awards '67: Best Actor (Scofield), Best Art Direction/Set Decoration (Color), Best Color Cinematography, Best Film, Best Screenplay; Directors Guild of America Awards '66: Best Director (Zinneman); Golden Globe Awards '67: Best Film—Drama; National Board of Review Awards '66: Best Actor (Scofield), Best Director (Zinneman), Best Picture, Best Supporting Actor (Shaw). VHS, Beta, LV $19.95 *COL, BTV* 🎬🎬🎬🎬

A Man for All Seasons Fresh from the London stage, Heston directs and stars in this version of Robert Bolt's play depicting the conflict between Henry VIII and his chief advisor, Sir Thomas More. Strong supporting cast. Made for cable TV.
1988 150m/C Charlton Heston, Vanessa Redgrave, John Gielgud, Richard Johnson, Roy Kinnear, Martin Chamberlain; **Dir:** Charlton Heston. VHS, LV $79.98 *TTC* 🎬🎬🎬

A Man Alone A man, falsely accused of robbing a stagecoach, hides not alone but with the comely sheriff's daughter. Milland's debut behind the camera.
1955 96m/C Ray Milland, Ward Bond, Mary Murphy, Raymond Burr, Lee Van Cleef; **Dir:** Ray Milland. VHS $14.95 *REP* 🎬½

Man of Aran Celebrated account of a fisherman's struggle for survival on a barren island near the west coast of Ireland, featuring amateur actors. Three years in the making, it's the last word in man against nature cinema, and a visual marvel. A former explorer, Flaherty became an influential documentarian. Having first gained fame with ''Nanook of the North,'' he compiled an opus of documentaries made for commercial release.
1934 132m/B **Dir:** Robert Flaherty. VHS $29.95 *HMV* 🎬🎬🎬

The Man from Atlantis Patrick Duffy stars as the water-breathing alien who emerges from his undersea home, the Lost City of Atlantis. The made-for-TV movie led to a brief television series.
1977 60m/C Patrick Duffy, Belinda J. Montgomery, Victor Buono; **Dir:** Lee H. Katzin. VHS, Beta $19.95 *WOV* 🎬🎬

Man Bait A complex web of intrigue and mystery surrounds a blackmailed book dealer when he allows a blonde woman to catch his eye. A competently made film, it was released in England as ''The Last Page''.

1952 78m/B *GB* George Brent, Marguerite Chapman, Diana Dors; *Dir:* Terence Fisher. **VHS, Beta, LV** *WGE, MLB* ⅃⅃½

Man Beast The abominable snowman is sought and found in this grade-Z '50s monster movie.
1955 65m/B Rock Madison, Virginia Maynor, George Skaff, Lloyd Nelson, Tom Maruzzi; *Dir:* Jerry Warren. **VHS, Beta $19.95** *NOS, MRV, RHI* ⅃

The Man from Beyond Frozen alive, a man returns 100 years later to try and find his lost love. Silent.
1922 50m/B Harry Houdini, Arthur Maude; *Dir:* Burton King. **VHS, Beta $16.95** *SNC, VYY, DVT* ⅃⅃½

The Man with Bogart's Face Sacchi is no Bogart, but he does imitate him well. Bogart fans will enjoy this fond tribute to the late great actor, but the story uncertainly wavers between genuine detective story and detective spoof. Also known as "Sam Marlowe, Private Eye".
1980 (PG) 106m/C Robert Sacchi, Misty Rowe, Sybil Danning, Franco Nero, Herbert Lom, Victor Buono, Olivia Hussey; *Dir:* Robert Day. **VHS, Beta $59.98** *FOX* ⅃⅃

Man & Boy A black Civil War veteran, played by Bill Cosby, encounters bigotry and prejudice when he tries to set up a homestead in Arizona. An acceptable family film, some might be disappointed that Cosby is not playing this one for laughs. Also known as "Ride a Dark Horse".
1971 (G) 98m/C Bill Cosby, Gloria Foster, George Spell, Henry Silva, Yaphet Kotto; *Dir:* E.W. Swackhamer. **VHS, Beta $59.95** *COL* ⅃⅃

Man from Button Willow Classic animated adventure is the story of Justin Eagle, a man who leads a double life. He is a respected rancher and a shrewd secret agent for the government, but in 1869 he suddenly finds himself the guardian of a four-year-old Oriental girl, leading him into a whole new series of adventures. Strictly for younger audiences.
1965 (G) 79m/C *Voices:* Dale Robertson, Edgar Buchanan, Barbara Jean Wong, Howard Keel. **VHS, Beta $42.95** *PGN* ⅃⅃

Man from Cairo An American in Algiers is mistaken for a detective in search of gold lost during World War II and decides to play along.
1954 82m/B *IT* George Raft, Giana Maria Canale; *Dir:* Ray Enright. **VHS, Beta, LV** *WGE* ⅃⅃

A Man Called Adam A jazz musician is tortured by prejudice and the guilt created by his having accidentally killed his wife and baby years before. Davis is appropriately haunted, but the film is poorly produced.
1966 103m/B Sammy Davis Jr., Louis Armstrong, Ossie Davis, Cicely Tyson, Frank Sinatra Jr., Lola Falana, Mel Torme, Peter Lawford; *Dir:* Leo Penn. **VHS, Beta $59.98** *CHA* ⅃½

A Man Called Horse After a wealthy Britisher is captured and tortured by the Sioux Indians in the Dakotas, he abandons his formal ways and discovers his own strength. As he passes their torture tests, he is embraced by the tribe. In this very realistic and gripping portrayal of American Indian life, Harris provides a strong performance. Sequelled by "Return of a Man Called Horse" (1976) and "Triumphs of a Man Called Horse" (1983).

1970 (PG) 114m/C Richard Harris, Judith Anderson, Jean Gascon, Stanford Howard, Manu Tupou, Dub Taylor; *Dir:* Elliot Silverstein. **VHS, Beta $59.98** *FOX* ⅃⅃⅃

A Man Called Peter A biographical epic about Peter Marshall, a Scottish chaplain who served the U.S. Senate. Todd does his subject justice by sensitively showing all that was human in Marshall, and a talented supporting cast makes for a thoroughly watchable film.
1955 119m/C Richard Todd, Jean Peters, Marjorie Rambeau, Jill Esmond, Les Tremayne, Robert Burton; *Dir:* Henry Koster. **VHS, Beta $19.98** *FOX* ⅃⅃⅃

A Man Called Rage Rage is the only man capable of safely escorting a group of pioneers through a nuclear wasteland infested with mutants and cannibals.
1984 90m/C *IT* Stelio Candelli, Conrad Nichols; *Dir:* Anthony Richmond. **VHS, Beta $39.95** *IVE* Woof!

A Man Called Sarge Sophomoric comedy about a daffy WWII sergeant leading his squad against the Germans at Tobruk.
1990 (PG-13) 88m/C Gary Kroeger, Marc Singer, Gretchen German, Jennifer Runyon; *Dir:* Stuart Gillard. **VHS, Beta, LV $79.95** *WAR, OM* ⅃½

A Man Called Sledge Garner fans might be surprised to see the star play a villain in this violent story of a gang of outlaws who wind up fighting each other over a cache of gold. This is mainstream Western entertainment.
1971 93m/C James Garner, Dennis Weaver, Claude Akins, John Marley, Laura Antonelli; *Dir:* Vic Morrow. **VHS, Beta** *GKK* ⅃⅃

Man Called Tiger A Chinese martial arts expert infiltrates a ruthless Japanese gang in order to find the man who murdered his father.
1981 (R) 97m/C **VHS, Beta $9.98** *SUE* ⅃

Man from Cheyenne Cowboy comes home to find his town under a siege of terror from a lawless gang of cattle rustlers.
1942 54m/B Roy Rogers, Gale Storm, George "Gabby" Hayes, Sally Payne. **VHS, Beta $19.95** *NOS, DVT, RXM* ⅃

Man from Clover Grove Hilarity takes over a town when a nutty boy inventor puts the sheriff in a spin.
1978 (G) 96m/C Ron Masak, Cheryl Miller, Jed Allan, Rose Marie. **VHS, Beta $19.95** *MED* ⅃

Man from Colorado An odd Technicolor western about two Civil War vets at odds, one an honest marshall, the other a sadistic judge. Solid Western fare with a quirky performance by Ford.
1949 99m/C William Holden, Glenn Ford, Ellen Drew, Ray Collins, Edgar Buchanan, Jerome Cortland, James Millican, Jim Bannon; *Dir:* Henry Levin. **VHS, Beta $14.95** *COL* ⅃⅃½

Man with a Cross Archaeologists of the cinema may want to unearth this early Rossellini, made by the future father of neorealist cinema as a propaganda piece for the fascist war effort. A heroic Italian chaplain on the Russian front ministers to foe and friend alike, and even converts a few commies to Christ. Not as awful as it sounds—but never forget where this came from.
1943 88m/B *IT* *Dir:* Roberto Rossellini. **VHS $59.95** *FST* ⅃⅃

Man from Deep River A photographer is captured by a savage tribe in Thailand and forced to undergo a series of grueling initiation rites. Full of very violent and sickening tortures inflicted on both human and animal victims.
1977 (R) 90m/C *IT* Ivan Rassimov; *Dir:* Umberto Lenzi. **VHS, Beta $59.98** *PSM* Woof!

Man of Destiny Bonaparte (Keach) and a mysterious woman battle good-humoredly over a collection of love letters. From a Bernard Shaw story. Charming and well-acted.
1973 60m/C Stacy Keach, Samantha Eggar. **VHS, Beta $29.95** *USA* ⅃⅃

Man on the Eiffel Tower Laughton plays Inspector Maigret, the detective created by novelist Georges Simenon, in a highly suspenseful and cerebral mystery about a crazed killer who defies the police to discover his identity. This is the first film Meredith directed.
1948 82m/C Charles Laughton, Burgess Meredith, Franchot Tone, Patricia Roc; *Dir:* Burgess Meredith. **VHS, Beta $19.95** *NOS, MRV, GEM* ⅃⅃⅃

Man Escaped There's an excruciating realism about Bresson's account of a WWII Resistance fighter's escape just before he was to be executed by the Gestapo. Definitely not the story an American director would have told. It's the sounds and lingering camera shots, not the wham bam variety of action, that create and sustain the film's suspense. Bresson, who had been a Nazi prisoner, solicited the supervision of Andre Devigny, whose true story the film tells. Contributing to the realistic feel was the use of non professional actors. Voted one of the five best foreign language films of 1957 by the National Board; Bresson received "Best Director" award at 1957 Cannes Film Festival, not to mention that Truffaut lauded the film as the most crucial French film of the previous ten years. Also known as "Un Condamne a mort s'est echappe."
1957 102m/B Francois Leterrier. Cannes Film Festival '57: Best Director (Bresson). **VHS** *FST* ⅃⅃⅃½

Man of Evil The hard times of the illegitimate daughter of a member of the British Parliament in the early 1900s, told with an astonishing number of plot twists and a plodding melodramtic style. Based on the novel "Fanny by Gaslight" and also known by that title.
1948 108m/B *GB* Phyllis Calvert, James Mason, Wilfred Lawson, Stewart Granger, Margaretta Scott, Jean Kent, John Laurie, Stuart Lindsell, Nora Swinburne, Amy Veness, Ann Wilton, Helen Haye, Cathleen Nesbitt, Guy le Feuvre, John Turnbull, Peter Jones; *Dir:* Anthony Asquith. **VHS $29.95** *FCT* ⅃⅃

Man Facing Southeast The acclaimed Argentinean film about the sudden appearance of a strange man in an asylum who claims to be an extraterrestrial, and a psychologist's attempts to discover his true identity. The sense of mystery intensifies when the new patient indeed seems to have some remarkable powers. Although the pace at times lags, the story intriguingly keeps one guessing about the stranger right to the end. In Spanish, with English subtitles.
1986 (R) 105m/C *AR* Lorenzo Quinteros, Hugo Soto, Ines Vernengo; *Dir:* Eliseo Subiela. **VHS, Beta, LV $19.95** *STE, NWV, IME* ⅃⅃⅃½

Man on Fire Told via flashback, a cynical ex-CIA man is hired as a bodyguard for the daughter of a wealthy Italian couple, who is soon thereafter kidnapped by terrorists. He goes to her rescue with all the subtlety of a

wrecking ball. Decent cast goes down the tubes in this botched thriller.
1987 (R) 92m/C FR Scott Glenn, Brooke Adams, Danny Aiello, Joe Pesci, Paul Shenar, Jonathan Pryce, Jade Malle; *Dir:* Elie Chouraqui. **VHS, Beta, LV $89.98** LIV, VES *ℐ*

Man of Flowers Because of this puritan upbringing, a reclusive art collector has trouble coping with his feelings of sexuality. He pays a woman to disrobe in front of him, but is never able to bring himself to see her naked. A moody piece with overtones of black humor, this work has limited audience appeal.
1984 91m/C AU Norman Kaye, Alyson Best, Chris Haywood, Sarah Walker, Julia Blake, Bob Ellis, Werner Herzog; *Dir:* Paul Cox. Australian Film Institute '83: Best Actor (Kaye). **VHS, Beta, LV $69.98** LIV, VES *ℐℐ*½

Man of the Forest A cowboy goes to the aid of a damsel in distress and he winds up being framed for murder for his efforts. Based on Zane Grey's novel.
1933 59m/B Randolph Scott, Verna Hillie, Harry Carey, Noah Beery, Barton MacLane; *Dir:* Henry Hathaway. **VHS $19.95** NOS, DVT, MLB *ℐℐ*

Man Friday Stranded on a deserted island, a man forces a native from a neighboring island to be his slave. Based on the classic story "Robinson Crusoe" by Daniel Defoe, this adaptation charts the often-brutal treatment the native receives as his captor tries to civilize him. Through his intelligence, the enslaved man regains his freedom and returns home with his former captor, who then seeks acceptance from the native's tribe. A sometimes confusing story line and excessive blood and guts detract from this message-laden effort.
1976 (PG) 115m/C GB Peter O'Toole, Richard Roundtree, Peter Cellier, Christopher Cabot, Joel Fluellen; *Dir:* Jack Gold. **VHS** FOX *ℐℐ*

Man of the Frontier A vital irrigation project is being sabotaged, but Autry exposes the culprits.
1936 60m/B Gene Autry, Smiley Burnette, Frances Grant; *Dir:* B. Reeves Eason. **VHS, Beta $19.95** NOS, VDM, VCN *ℐ*½

The Man with the Golden Arm A gripping film version of the Nelson Algren junkie melodrama, about an ex-addict who returns to town only to get mixed up with drugs again. Considered controversial in its depiction of addiction when released. Sinatra's Oscar nominated performance is a stand-out. Also received an Oscar nomination for Black & White Art Direction and Set Decoration.
1955 119m/B Frank Sinatra, Kim Novak, Eleanor Parker, Arnold Stang, Darren McGavin, Robert Strauss, George Mathews, John Conte, Doro Merande; *Dir:* Otto Preminger. **VHS, Beta $9.95** NOS, MRV, NEG *ℐℐℐ*

The Man with the Golden Gun Roger Moore is the debonair secret agent 007 in this ninth James Bond flick. Assigned to recover a small piece of equipment which can be utilized to harness the sun's energy, Bond engages the usual bevy of villains and beauties.
1974 (PG) 125m/C GB Roger Moore, Christopher Lee, Britt Ekland, Maud Adams, Herve Villechaize, Clifton James, Soon-Teck Oh, Richard Loo, Marc Lawrence, Bernard Lee, Lois Maxwell, Desmond Llewelyn; *Dir:* Guy Hamilton. **VHS, Beta, LV $19.95** FOX *ℐℐ*½

The Man in the Gray Flannel Suit A very long and serious adaptation of the Sloan Wilson novel about a Madison Avenue advertising exec trying to balance his life between work and family. The Hollywood treatment falls short of the adaptation potential of the original story.
1956 152m/C Gregory Peck, Fredric March, Jennifer Jones, Ann Harding, Arthur O'Connell, Henry Daniell, Lee J. Cobb, Marisa Pavan, Gene Lockhart, Keenan Wynn, Gigi Perreau, Joseph Sweeney, Kenneth Tobey, DeForest Kelley; *Dir:* Nunnally Johnson. **VHS, Beta, LV $39.98** FOX *ℐℐ*

The Man in Grey In a story of romantic intrigue set in nineteenth-century England, a Marquis's wife is betrayed by her vile husband and the schoolmate she once befriended who has an affair with him. Stunning costumes and fine performances compensate for the overly extravagant production values in a work that helped bring stardom to Mason.
1945 116m/B GB James Mason, Margaret Lockwood, Stewart Granger, Phyllis Calvert; *Dir:* Leslie Arliss. **VHS, Beta $29.95** VID *ℐℐℐ*

The Man from Gun Town McCoy comes to the rescue of a woman who has been framed by an evil gang for the murder of her brother.
1936 58m/B Tim McCoy, Billie Seward, Rex Lease, Jack Clifford, Wheeler Oakman, Bob McKenzie; *Dir:* Ford Beebe. **VHS $19.95** NOS, DVT *ℐ*½

The Man from Hell Sheriff Russell goes undercover to expose the head of a gang of outlaws. Low-budget fare.
1934 55m/B Reb Russell, Fred Kohler Jr., Ann Darcy, George "Gabby" Hayes, Jack Rockwell, Charles French, Charles "Slim" Whitaker, Yakima Canutt; *Dir:* Lewis Collins. **VHS $19.95** DVT *ℐ*

Man from Hell's Edges An innocent cowpoke escapes from jail and brings the real baddy to justice.
1932 63m/B Bob Steele, Nancy Drexel, Julian Rivero, Robert E. Homans, George "Gabby" Hayes; *Dir:* Robert N. Bradbury. **VHS, Beta $19.95** NOS, DVT *ℐ*½

Man Inside An undercover agent infiltrates a powerful underworld narcotics ring and finds his honesty tested when $2 million is at stake.
1976 96m/C CA James Franciscus, Stefanie Powers, Jacques Godin; *Dir:* Gerald Mayer. **VHS, Beta $49.95** TWE *ℐ*½

The Man Inside Lukewarm Cold War saga based on the true story of Gunther Wallraff, a West German journalist who risked all to expose the corruption behind a large European newspaper.
1990 (PG) 93m/C NL GE Juergen Prochnow, Peter Coyote, Nathalie Baye, Dieter Laser, Monique Van De Ven, Sylvie Granotier; *W/Dir:* Bobby Roth. **VHS, LV $89.95** COL, PIA *ℐℐ*

Man in the Iron Mask Swashbuckling tale about twin brothers separated at birth. One turns out to be King Louis XIV of France, and the other a carefree wanderer and friend of the Three Musketeers. Their eventual clash leads to action-packed adventure and royal revenge. Remake of the "The Iron Mask" (1929) with Douglas Fairbanks. Remade for television in 1977 with Richard Chamberlain.
1939 110m/B Louis Hayward, Alan Hale Jr., Joan Bennett, Warren Williams, Joseph Schildkraut, Walter Kingsford, Marion Martin; *Dir:* James Whale. **VHS, Beta $9.99** CCB, MED *ℐℐℐ*

Man in the Iron Mask A tyrannical French king kidnaps his twin brother and imprisons him on a remote island. Chamberlain, the king of the miniseries, is excellent in a dual role in this big production swashbuckler. Adapted from the Dumas classic. Made for television.
1977 105m/C Richard Chamberlain, Patrick McGoohan, Louis Jourdan, Jenny Agutter, Ian Holm, Ralph Richardson; *Dir:* Mike Newell. **VHS, Beta $59.98** KUI, FOX *ℐℐℐ*

Man in the Iron Mask Animated version of the Alexander Dumas novel about two brothers separated at birth, one of whom grows up to become the dastardly king of France, the other a swashbuckling hero of the people.
19?? 110m/C Louis Hayward, Alan Hale Jr., Warren Williams, Joseph Schildkraut, Walter Kingsford, Marion Martin; *Dir:* James Whale. **VHS** CVA *ℐℐ*½

Man of La Mancha Arrested by the Inquisition and thrown into prison, Miguel de Cervantes relates the story of Don Quixote. Film not nearly as good as the Broadway musical it is based on.
1972 (PG) 129m/C Peter O'Toole, Sophia Loren, James Coco, Harry Andrews, John Castle, Brian Blessed; *Dir:* Arthur Hiller. National Board of Review Awards '72: 10 Best Films of the Year, Best Actor (O'Toole). **VHS, Beta, LV $19.99** FOX, FCT *ℐ*½

Man from Laramie A ranch baron who is going blind worries about which of his two sons he will leave the ranch to. Into this tension-filled familial atmosphere rides Stewart, a cow-herder obsessed with hunting down the men who sold guns to the Indians that killed his brother. Needless to say, the tension increases. Tough, surprisingly brutal western, the best of the classic Stewart-Mann films.
1955 104m/C James Stewart, Arthur Kennedy, Donald Crisp, Alex Nicol, Cathy O'Donnell, Aline MacMahon, Wallace Ford, Jack Elam; *Dir:* Anthony Mann. **VHS, Beta $19.95** COL *ℐℐℐ*½

Man of Legend An adventure-romance filmed in Morocco; a WWI German soldier flees to the Foreign Legion and fights with nomadic rebels, ultimately falling in love with their chief's beautiful daughter. An unoriginal desert saga.
1971 (PG) 95m/C IT SP Peter Strauss, Tina Aumont, Pier Paola Capponi; *Dir:* Sergio Grieco. **VHS, Beta $59.95** UHV *ℐℐ*

A Man Like Eva A weird, morbid homage to and portrait of Rainer Werner Fassbinder after his inevitable death, detailing his work-obsessed self-destruction. Mattes, one of Fassbinder's favorite actresses, plays him in drag, in an eerie gender-crossing transformation. In German with English subtitles.
1983 92m/C GE Eva Mattes, Lisa Kreuzer, Charles Regnier, Werner Stocker; *Dir:* Radu Gabrea. **VHS, Beta $79.95** FCT, APD, GLV *ℐℐℐ*

A Man in Love An international romantic melodrama set during the Italian filming of a biography of suicidal author Cesar Pavese. The self-important lead actor and a beautiful supporting actress (Coyote and Scacchi) become immersed in the roles and fall madly in love, oblivious to the fact that Coyote is married and Scacchi's engaged. The two make a steamy pair, to the detriment of friends, family, and the movie they're making. Kurys' first English-language film is visually appealing with a lush, romantic score, seamlessly weaving the story lines among vivid characters.

1987 (R) 110m/C *FR* Peter Coyote, Greta Scacchi, Jamie Lee Curtis, Peter Riegert, Jean Pigozzi, John Berry, Claudia Cardinale, Vincent Lindon; *Dir:* Diane Kurys. **VHS, Beta, LV** $14.95 *COL, SUE* ♫♫♫

A Man with a Maid A bizarre British exploitation pic mixes spookhouse cliches and sex, as a young man finds his new bachelor pad haunted by Jack the Ripper. Originally shot in 3-D and known variously as "The Groove Room" and "What the Swedish Butler Saw." It didn't help.
1973 83m/C Sue Longhurst, Martin Long, Diana Dors; *Dir:* Vernon Becker. **VHS** $19.95 *VDM* ♫

Man of Marble A satire on life in post-WWII Poland. A young filmmaker sets out to tell the story of a bricklayer who, because of his exceptional skill, once gained popularity with other workers. He became a champion for worker rights, only to then find himself being persecuted by the government. The conclusion was censored by the Polish government. Highly acclaimed and followed by "Man of Iron" in 1981.
1976 160m/C *PL* Krystyna Janda, Jerzy Radziwilowicz, Tadeusz Lomnicki, Jacek Lomnicki, Krystyna Zachwatowicz; *Dir:* Andrzej Wajda. Cannes Film Festival '78: International Critics Award. **VHS** $79.95 *NYF, FCT* ♫♫♫

The Man in the Mist Private investigators Tommy and Tuppence Beresford find themselves trying to solve a murder while on vacation in a provincial hotel. Made for British television and based upon the Agatha Christie story.
1983 51m/C *GB* Francesca Annis, James Warwick; *Dir:* PAV ♫♫♫ **VHS, Beta** $14.95 *PAV* ♫♫♫

The Man and the Monster When a concert pianist sells his soul to the devil, he fails to realize that part of the deal has him turning into a hideous beast every time he hears a certain piece of music. Maybe it was "Stairway to Heaven."
1965 78m/B *MX* Enrique Rambal, Abel Salazar, Martha Roth; *Dir:* Rafael Baledon. **VHS** $19.98 *SNC* ♫ 1/2

The Man with a Movie Camera A plotless, experimental view of Moscow through the creative eye of the cameraman Dziga Vertov, founder of the Kino Eye. The editing methods and camera techniques used in this silent film were very influential and still stand up to scrutiny today.
1929 69m/B *RU* *Dir:* Dziga Vertov. **VHS, Beta, 3/4U** $35.95 *IHF, MRV, DVT* ♫♫♫ 1/2

The Man from Music Mountain Worthless mining stock is sold in a desert mining town, but Gene and Smiley clear that up, with a little singing as well.
1938 54m/B Gene Autry, Smiley Burnette, Carol Hughes, Polly Jenkins. **VHS, Beta** $19.95 *NOS, VDM, HHT* ♫♫

Man is Not a Bird Follows the destructive love of a factory engineer and a hairdresser in a small Yugoslavian mining town. In Serbian with English subtitles.
1965 80m/B *YU* Eva Ras, Milena Dravic, Janez Urhovec; *Dir:* Dusan Makavejev. **VHS, Beta** $59.95 *FCT* ♫♫ 1/2

Man from Nowhere A henpecked man gets the break of his life when his domineering wife and mother-in-law believe he's dead.
19?? 98m/C *FR* Pierre Blanchar, Ginette LeClerc; *Dir:* Pierre Chenal. **VHS, Beta** $59.95 *FCT* ♫♫ 1/2

The Man with One Red Shoe Hanks is a lovable clod of a violinist who ensnares himself in a web of intrigue when CIA agents, both good and evil, mistake him for a contact by his wearing one red shoe. Sporadically funny remake of the French "The Tall Blond Man with One Black Shoe."
1985 (PG) 92m/C Tom Hanks, Dabney Coleman, Lori Singer, Carrie Fisher, James Belushi, Charles Durning, Edward Herrmann, Tommy Noonan, Gerrit Graham, David Lander, David Ogden Stiers; *Dir:* Stan Dragoti. **VHS, Beta** $14.98 *FXV, FOX* ♫♫

The Man Outside After a CIA agent is fired for allegedly assisting another agent in defecting to the East, he becomes involved in further intrigue. A Russian spy is looking to defect. In the process, the ex-agent is framed for murder. Straightforward espionage tale taken from Gene Stackleborg's novel "Double Agent."
1968 (R) 98m/C *GB* Van Heflin, Heidelinde Weis, Pinkas Braun, Peter Vaughan, Charles Gray, Ronnie Barker; *Dir:* Samuel Gallu. **VHS** $19.95 *VIR* ♫

Man Outside Logan is an ex-lawyer who takes to the arkansas outback after his wife dies. Anthropoligist/teacher Quinlan takes a shine to him. He seems like an okay guy, but bad guy Dillman has made it look like he's a child snatcher. Slick on the outside, but empty inside independent effort. Look for former members of The Band in supporting roles.
1988 (PG-13) 109m/C Robert F. Logan, Kathleen Quinlan, Bradford Dillman, Rick Danko, Levon Helm; *Dir:* Mark Stouffer. **VHS, LV** *VIR* ♫

Man in the Raincoat French film about a bumbling clarinet player who is erroneously tracked down as a murderer. Strenuous efforts to evoke laughter usually fail. Dubbed.
1957 97m/B *FR* Fernandel, John McGiver, Bernard Blier; *Dir:* Julien Duvivier. **VHS, Beta, 8mm** $29.95 *VYY* ♫♫

Man on the Run A robbery takes place in the store where an Army deserter is trying to sell his gun, and he ends up taking the rap. It's up to a lovely lady lawyer to prove his innocence. Efficiently told "B" crime drama.
1949 82m/B *GB* Derek Farr, Joan Hopkins, Edward Chapman, Laurence Harvey, H. Marion Crawford, Alfie Bass, John Bailey, John Stuart, Edward Underdown, Leslie Perrins, Kenneth More, Martin Miller, Eleanor Summerfield; *W/Dir:* Lawrence Huntington. **VHS, Beta** $16.95 *SNC* ♫♫

Man on the Run After an unwitting involvement with a small robbery, a teenager finds himself the object of a police manhunt for a murder suspect.
1974 (R) 90m/C Kyle Johnson, James B. Sikking, Terry Carter; *Dir:* Herbert L. Strock. **VHS, Beta** $59.95 *PSM* ♫ 1/2

The Man in the Santa Claus Suit A costume shop owner has an effect on three people who rent Santa Claus costumes from him. Astaire plays seven different roles in this made-for-television film. Average holiday feel-good movie.
1979 96m/C Fred Astaire, Gary Burghoff, John Byner, Nanette Fabray, Bert Convy; *Dir:* Corey Allen. **VHS, Beta** $69.95 *MED* ♫♫

Man in the Silk Hat A collection of the nearly-forgotten French comic's early silent comedy shorts, made in France before his resettlement in America.
1915 96m/B *FR* Max Linder; *Dir:* Maud Linder. **VHS, Beta** $59.95 *TAM* ♫♫

Man in the Silk Hat Gabriel-Maximilien Leuvielle, known in films as Max Linder, is now credited with developing the style of silent-movie slapstick comedy that Mack Sennett, Charlie Chaplin, and others became more famous for in their time. Here, Linder's daughter has done a fine job writing and directing a film full of historic footage of her father's work.
1983 99m/B Mack Sennett, Buster Keaton, Charlie Chaplin, Max Linder; *Dir:* Maud Linder. **VHS, Beta** $24.95 *XVC, MED* ♫♫♫

Man from Snowy River Stunning cinematography highlights this otherwise fairly ordinary adventure story set in 1880's Australia. A young man comes of age in the outback while taming a herd of wild horses. In a dual role, Douglas portrays battling brothers, one a rich landowner and the other a one-legged prospector. Based on the epic poem by A.B. "Banjo" Paterson and followed by "Return to Snowy River Part II." A big hit in Australia and not directed by "Mad Max's" Miller, but another Miller named George.
1982 (PG) 104m/C Kirk Douglas, Tom Burlinson, Sigrid Thornton, Terence Donovan, Tommy Dysart, Jack Thompson, Bruce Kerr; *Dir:* George Miller. **VHS, Beta, LV** $14.98 *FOX* ♫♫ 1/2

Man from Texas Singing and gun-slinging Tex defends a kid accused of horse thieving until the kid turns bad and Tex must bring him in.
1939 55m/B Tex Ritter, Hal Price, Charles B. Wood, Vic Demourelle Sr.; *Dir:* Al Herman. **VHS, Beta** $19.95 *NOS, VDM* ♫♫

The Man That Corrupted Hadleyburg Entertaining adaptation of a Mark Twain short story about a stranger's plot of revenge against the hypocritical residents of a small town. Originally made for television.
1980 40m/C Robert Preston, Fred Gwynne, Frances Sternhagen; *Dir:* Ralph Rosenblum; *Hosted:* Henry Fonda. **VHS, Beta** $24.95 *MON, KAR* ♫♫

The Man They Could Not Hang A good doctor tinkering with artificial hearts is caught by police while experimenting on a willing student. When the doctor is convicted and hanged for a murder, his assistant uses the heart to bring him back to life. No longer a nice guy, he vows revenge against the jurors that sentenced him. Karloff repeated the same story line in several films, and this one is representative of the type.
1939 70m/B Boris Karloff, Lorna Gray, Roger Pryor, Robert Wilcox; *Dir:* Nick Grinde. **VHS, Beta** $9.95 *GKK* ♫♫ 1/2

Man of a Thousand Faces A tasteful and touching portrayal of Lon Chaney, from his childhood with his deaf and mute parents to his success as a screen star. Recreates some of Chaney's most famous roles, including the Phantom of the Opera and Quasimodo in "Notre Dame." Cagney is magnificent as the long-suffering film star who was a genius with makeup and mime.
1957 122m/B James Cagney, Dorothy Malone, Jane Greer, Marjorie Rambeau, Jim Backus, Roger Smith, Robert Evans; *Dir:* Joseph Pevney. **VHS** $14.98 *MCA* ♫♫♫ 1/2

Man from Thunder River A group of cowboys uncover a plot to steal gold ore and wind up saving a young girl's life in the process. Standard western with lots of action.
1943 55m/B Bill Elliott, George "Gabby" Hayes, Anne Jeffreys, Ian Keith, John James; *Dir:* John English. **VHS, Beta** $29.95 *DVT* ♫♫

The Man with Two Brains Did you hear the one about the brilliant neuro-surgeon who falls in love with a woman in his laboratory? He only has two problems: dealing with his covetous wife and finding a body for his cerebral lover. Plenty of laughs in this spoof of mad scientist movies that is redeemed from potential idiocy by the cast's titillating performances. Listen closely and you'll recognize the voice of Spacek as the brain-in-the-jar of Martin's dreams.
1983 (R) 91m/C Steve Martin, Kathleen Turner, David Warner, Sissy Spacek; *W/Dir:* Carl Reiner. **VHS, Beta, LV** $34.98 *WAR* 🐾🐾🐾

Man with Two Heads Doctor Jekyll gets in touch with his innermost feelings when his muffed experiments turn him into, gasp, Mr. Blood. Another forlorn addition to the Dr. Jekyll retread collection.
1972 (R) 80m/C *W/Dir:* Andy Milligan. **VHS** *MID* 🐾

Man from Utah The Duke tangles with the crooked sponsor of some rodeo events who has killed several of the participants. Also available with "Blue Steel" on Laser Disc.
1934 55m/B John Wayne, George "Gabby" Hayes, Polly Ann Young; *Dir:* Robert Bradbury. **VHS, Beta** $8.95 *NEG, NOS, MRV* 🐾

Man of Violence A vulgar, tasteless man spends the worthless hours of his wasted life lurking about the more wretched entrance-ways of his native land.
1971 107m/C Michael Latimer, Luan Peters; *Dir:* Pete Walker. **VHS, Beta** $39.95 *MON* 🐾

Man of the West Bad guy turned good guy Cooper is asked by his neighborhood town fathers to take a tidy hunk of moolah to another city to recruit a school marm. Ambushed en route by his former partners in crime (who are led by his wacko uncle) he's forced to revert to his wanton ways in order to survive and to save innocent hostages. There's a raging debate among Cooper-philes whether this late effort has been unduly overlooked or duly ignored. A number of things conspire to give it a bad rap: Cooper does little but look mournful until the very end, there's no hiding the fact that he's older than Cobb, who plays his uncle, and the acting is in general more befitting of a B-grade slice and dicer. You be the judge. Scripted by Reginald Rose.
1958 100m/C Gary Cooper, Julie London, Lee J. Cobb, Arthur O'Connell, Jack Lord, John Dehner, Royal Dano, Guy Wilkerson, Emory Parnell; *Dir:* Anthony Mann. **VHS, Beta** $19.98 *MGM, FCT* 🐾🐾½

Man in the White Suit A humble laboratory assistant in a textile mill invents a white cloth that won't stain, tear, or wear out, and can't be dyed. The panicked garment industry sets out to destroy him and the fabric, resulting in some sublimely comic situations and a variety of inventive chases. Also available with "Lavender Hill Mob" on Laser Disc.
1951 82m/B *GB* Alec Guinness, Joan Greenwood; *Dir:* Alexander MacKendrick. National Board of Review Awards '52: 10 Best Films of the Year. **VHS, Beta** $19.98 *HBO* 🐾🐾🐾½

The Man Who Broke 1,000 Chains A made-for-television movie about a man who is committed to a chain gang after World War II for a petty crime, and his efforts, after escaping, in making a new life for himself. An unimaginative plot is occasionally highlighted by a good scene or two.
1987 113m/C Val Kilmer, Charles Durning, Sonia Braga; *Dir:* Daniel Mann. **VHS, Beta** $79.95 *LHV, WAR* 🐾🐾

The Man Who Came to Dinner Based on the Moss Hart-George S. Kaufman play, this comedy is about a bitter radio celebrity (Woolley) on a lecture tour (a character based on Alexander Woolcott). He breaks his hip and must stay in a quiet suburban home for the winter. While there, he occupies his time by barking orders, being obnoxious and generally just driving the other residents nuts. Woolley reprises his Broadway role in this film that succeeds at every turn, loaded with plenty of satiric jabs at the Algonquin Hotel Roundtable regulars.
1941 112m/B Monty Woolley, Bette Davis, Ann Sheridan, Jimmy Durante, Reginald Gardiner, Richard Travis, Billie Burke, Grant Mitchell, Mary Wickes, George Barbier, Ruth Vivian, Elisabeth Fraser; *Dir:* William Keighley. **VHS, Beta** $19.95 *MGM* 🐾🐾🐾½

The Man Who Could Work Miracles A mild-mannered draper's assistant becomes suddenly endowed with supernatural powers to perform any feat he wishes. Great special effects (for an early film) and fine performances result in a classic piece of science fiction.
1937 82m/B Ralph Richardson, Joan Gardner, Roland Young; *Dir:* Lothar Mendes. **VHS, Beta, LV** $14.98 *SUE* 🐾🐾🐾½

The Man Who Envied Women The non-narrative feminist story of a smug womanizer; the man "who knows almost too much about women."
1985 125m/C Bill Raymond, Larry Loonin, Trisha Brown; *W/Dir:* Yvonne Rainer. **VHS, Beta, 3/4U** *ICA* 🐾½

The Man Who Fell to Earth Entertaining and technically adept cult classic about a man from another planet (Bowie, in a bit of typecasting) who ventures to earth in hopes of finding water to save his family and drought-stricken planet. Instead he becomes a successful inventor and businessman, along the way discovering the human vices of booze, sex, and television. Also available in a restored version at 138 minutes. Remade for television in 1987 and based on Walter Tevis' novel.
1976 (R) 118m/C *GB* David Bowie, Candy Clark, Rip Torn, Buck Henry, Bernie Casey; *Dir:* Nicolas Roeg. **VHS, Beta, LV** $69.95 *COL* 🐾🐾🐾½

The Man Who Guards the Greenhouse Tracy must come to terms with her attraction for Jeff as well as once again trying to write a meaningful novel.
1988 150m/C Christopher Cazenove, Rebecca Dewey. **VHS** $79.95 *LHV* 🐾🐾🐾½

Man Who Had Power Over Women Disappointing sex farce about the exploits of a carnally insatiable (and married) talent executive (Taylor) who has an affair with every woman he meets and creates problems aplenty. Adapted from a novel by Gordon Williams.
1970 (R) 89m/C *GB* Rod Taylor, James Booth, Carol White, Penelope Horner, Clive Francis; *Dir:* John Krish. **VHS, Beta** $59.98 *SUE* 🐾🐾

The Man Who Haunted Himself While a man lies in critical condition on the operating table after a car accident, his alter-ego emerges and turns his ideal life into a nightmare until the man recovers and moves toward a fateful encounter. An expanded version of an episode of the television series "Alfred Hitchcock Presents," this was Moore's first movie after having starred in the TV series "The Saint," and it was Dearden's last film; he died in a car accident the following year. Appeals primarily to those fascinated by "Hitchcock" or "The Twilight Zone"—where mystery matters most. Filmed in London.
1970 94m/C *GB* Roger Moore, Hildegarde Neil, Olga Georges-Picot; *Dir:* Basil Dearden. **VHS, Beta** $59.99 *HBO* 🐾🐾

The Man Who Knew Too Much Hitchcock's first international success. A British family man on vacation in Switzerland is told of an assassination plot by a dying agent. His daughter is kidnapped to force his silence. In typical Hitchcock fashion, the innocent person becomes caught in a web of intrigue; the sticky situation culminates with surprising events during the famous shoot-out in the final scenes. Remade by Hitchcock in 1956.
1934 75m/B *GB* Leslie Banks, Edna Best, Peter Lorre, Nova Pilbeam, Pierre Fresnay; *Dir:* Alfred Hitchcock. **VHS, Beta** $16.95 *SNC, MRV, NOS* 🐾🐾🐾

The Man Who Knew Too Much Hitchcock's remake of his 1934 film, this time about an American doctor and his family vacationing in Marrakech. They become involved in a complicated international plot involving kidnapping and murder. While Doris tries to save the day by singing "Que Sera, Sera," Stewart tries to locate his abducted son. More lavish settings and forms of intrigue make this a less focused and, to some, inferior version.
1956 (PG) 120m/C James Stewart, Doris Day, Brenda de Banzie, Bernard Miles, Ralph Truman, Daniel Gelin, Alan Mowbray, Carolyn Jones, Hillary Brooke; *Dir:* Alfred Hitchcock. Academy Awards '56: Best Song ("Whatever Will Be, Will Be (Que Sera, Sera)"). **VHS, Beta, LV** $19.95 *MCA, HHE, BUR* 🐾🐾½

The Man Who Lived Again Boris strives to be a brain-switcher, and suspense builds around the question of whether or not he will change his mind. Shot in England with fine sets and a definite Anglo feel to the proceedings, with Karloff doing one of his better mad scientist routines.
1936 61m/B *GB* Boris Karloff, Anna Lee, John Loder, Frank Cellier, Lyn Harding, Cecil Parker; *Dir:* Robert Stevenson. **VHS, Beta, 8mm** $14.95 *SNC, VYY, MLB* 🐾½

Man Who Loved Cat Dancing Reynolds is an outlaw on the run after avenging his wife's murder and robbing a safe with pals Hopkins and Warden, and Miles has recently escaped from her abusive husband. It's love on the run as Burt and Sarah are pursued by bounty hunters and their tragic pasts—coming close to making us care, but close doesn't mean as much in movies as it does in dancing. Based on Marilyn Durham's novel.
1973 (PG) 114m/C Burt Reynolds, Sarah Miles, Jack Warden, Lee J. Cobb, Jay Silverheels, Robert Donner; *Dir:* Richard Sarafian. **VHS, Beta** $19.98 *MGM, FCT* 🐾🐾

The Man Who Loved Women An intelligent, sensitive bachelor writes his memoirs and recalls the many many many women he has loved. Truffaut couples sophistication and lightheartedness, the thrill of the chase and, when it leads to an accidental death, the wondering what-it's-all-about in the mourning after. Remade in 1983.
1977 (R) 119m/C *FR* Charles Denner, Brigitte Fossey, Leslie Caron; *Dir:* Francois Truffaut. National Board of Review Awards '77: 5 Best Foreign Films of the Year; New York Film

Festival '77: 5 Best Foreign Films of the Year. **VHS, Beta, LV $59.95** COL 🎬🎬🎬

The Man Who Loved Women A
remake of the 1977 French film, this is slower, tries to be funnier, and is less subtle than the original. Reynolds is a Los Angeles sculptor whose reputation as a playboy leads him to a psychoanalyst's couch, where a lot of talk slows the action—though Burt & Julie (the shrink) do share the couch.
1983 (R) 110m/C Burt Reynolds, Julie Andrews, Kim Basinger, Marilu Henner, Cynthia Sikes, Jennifer Edwards; *Dir:* Blake Edwards. **VHS, Beta $79.95** RCA 🎬🎬

The Man Who Never Was
Tense true story (with melodramatic embroidery) from WWII shows in step-by-step detail how Britain duped the Axis by letting them find an Allied corpse bearing phony invasion plans. Based on the book by the scheme's mastermind Ewen Montagu, played by Webb; Peter Sellers provides the voice of an offscreen Winston Churchill.
1955 102m/C *GB* Clifton Webb, Gloria Grahame, Robert Flemyng, Josephine Griffin, Stephen Boyd, Andre Morell, Laurence Naismith, Geoffrey Keen, Michael Hordern; *Dir:* Ronald Neame. **VHS, Beta $19.98** FOX, FCT 🎬🎬🎬

The Man Who Saw Tomorrow An
examination of the prophecies of Nostradamus and their reputed relevance today.
1981 (PG) 88m/C *Nar:* Orson Welles. **VHS, Beta $59.95** WAR, VTK, HHE 🎬🎬🎬

The Man Who Shot Liberty Valance
Tough cowboy Wayne and idealistic lawyer Stewart join forces against dreaded gunfighter Liberty Valance, played leatherly by Marvin. While Stewart rides to Senatorial success on his reputation as the man who shot the villain, he suffers moral crises about the act, but is toughened up by Wayne. Wayne's use of the word "pilgrim" became a standard for his impersonators. Strong character acting, great Western scenes, and value judgements to pander over make this last of Ford's black and white westerns among his best.
1962 119m/B James Stewart, John Wayne, Vera Miles, Lee Marvin, Edmond O'Brien, Andy Devine, Woody Strode; *Dir:* John Ford. **VHS, Beta, LV $14.95** PAR, BUR 🎬🎬🎬½

The Man Who Skied Down Everest
Documentary about a Japanese athlete who undertakes a most demanding downhill run, but the exciting run proves less interesting than the cost (human and financial) of getting there.
1975 (G) 86m/C Academy Awards '75: Best Feature Documentary. **VHS** SLC 🎬🎬½

The Man Who Wagged His Tail A
mean slumlord is turned into a dog as the result of a curse cast upon him. In order to return to his human form, he must be loved by someone. Despite his attempts to be loved, the dog alienates his only friend and must try to redeem himself. Played for fun, this is a mildly amusing fantasy filmed in Spain and Brooklyn. Not released in the U. S. until 1961.
1957 91m/B *IT SP* Peter Ustinov, Pablito Calvo, Aroldo Tieri, Silvia Marco; *Dir:* Ladislao Vajda. **VHS $16.95** SNC 🎬🎬½

The Man Who Wasn't There A
member of the State department receives a formula from a dying spy that can render him invisible, see? He has to use the formula to protect himself from the police and other spies, becoming a comic "Invisible Man," see? Generally chaotic tale that's bad, but

not so bad that it's worth seeing, though you might want to see what invisibility looks like in 3-D.
1983 (R) 111m/C Steve Guttenberg, Jeffrey Tambor, Art Hindle, Lisa Langlois; *Dir:* Bruce Malmuth. **VHS, Beta $59.95** PAR 🎬

The Man Who Would Be King A
great, grand, old-fashioned adventure based on the classic story by Rudyard Kipling about two mercenary soldiers who travel from India to Kafiristan in order to conquer it and set themselves up as kings. Splendid characterizations by Connery and Caine, and given royal directorial treatment by Huston—adventurous, majestic sweeps, well-developed characters, and strong conflicts between the Englishmen and the natives.
1975 (PG) 129m/C Sean Connery, Michael Caine, Christopher Plummer, Saeed Jaffrey; *Dir:* John Huston. **VHS, Beta, LV $19.98** FOX 🎬🎬🎬🎬

The Man Who Would Not Die
During his investigation of several deaths, a man discovers that all of the deceased were actually the same man. It seems the deaths were part of a intricate scheme to cover up a million-dollar heist. Not very suspenseful, a drawback in mystery movies, and two of the top three actors are killed early on. Also called "Target in the Sun." From Charles Williams' novel "The Sailcloth Shroud."
1975 (PG) 83m/C Dorothy Malone, Keenan Wynn, Aldo Ray; *Dir:* Robert Arkless. **VHS, Beta $59.95** WAR, OM 🎬🎬

Man Without a Star A
cowboy helps ranchers fight off a ruthless cattle owner from taking over their land. The conflict between freedom in the wild west and the need for order and settlements is powerfully internalized in Douglas, whose fight for justice will tame the cowboy code he lives by. You'll shed a tear for the fading frontier.
1955 89m/B Kirk Douglas, Jeanne Crain, Claire Trevor, William Campbell; *Dir:* King Vidor. **VHS, Beta $19.95** MCA 🎬🎬🎬

A Man and a Woman
When a man and a woman, both widowed, meet and become interested in one another, they experience difficulties in putting their past loves behind them. Intelligently handled emotional conflicts within a well-acted romantic drama, acclaimed for excellent visual detail. Remade in 1977 as "Another Man, Another Chance." Followed in 1986 with "A Man and A Woman: 20 Years Later." In addition to Oscar wins, Aimee received an Oscar nomination, as did the director. Dubbed.
1966 102m/C *FR* Anouk Aimee, Jean-Louis Trintignant, Pierre Barouh, Valerie Lagrange; *W/ Dir:* Claude Lelouch. Academy Awards '66: Best Foreign Language Film, Best Original Screenplay. **VHS, Beta $19.98** WAR, APD 🎬🎬🎬

A Man and a Woman: 20 Years Later
Slouchy sequel to the highly praised "Man and a Woman" that catches up with a couple after a long separation. The sad romantic complications of the original are more mundane in this sequel, which is burdened by a film-within-a-film script as well as shots from the original.
1986 (PG) 112m/C *FR* Jean-Louis Trintignant, Anouk Aimee, Richard Berry; *Dir:* Claude Lelouch. **VHS, Beta $19.98** WAR 🎬½

A Man, a Woman, & a Bank
Two con-men plan to rob a bank by posing as workers during the bank's construction. An advertising agency woman snaps their picture for a billboard to show how nice the builders have been, then becomes romantically involved with one of the would-be

thieves. Nice performances, but wacky touches aren't plentiful enough or well-timed to sustain comedy. Also called "A Very Big Weekend."
1979 (PG) 100m/C *CA* Donald Sutherland, Brooke Adams, Paul Mazursky; *Dir:* Noel Black. **VHS, Beta $9.98** SUE 🎬🎬½

Man, Woman & Child A
close, upscale California family is shocked when a child from the husband's long-ago affair with a Frenchwoman appears at their door. Pure sentimentalism, as the family emotionally confronts this unexpected development. Two hankies—one each for fine performances by Sheen and Danner. Based on a sentimental novel by Erich Segal of "Love Story" fame, who co-wrote the script.
1983 (PG) 99m/C Martin Sheen, Blythe Danner, Craig T. Nelson, David Hemmings; *Dir:* Dick Richards. **VHS, Beta $59.95** PAR 🎬🎬

A Man, a Woman and a Killer
Unusual film within a film shows the actual making of a movie while unfolding a story of a hired killer. A joint project from the directors of the well-received "Emerald Cities" and "Chan is Missing."
19?? 75m/C *Dir:* Rick Schmidt, Wayne Wang. **VHS $59.95** FCT 🎬🎬½

Manchurian Avenger
Desperados have wrested control of a Colorado gold rush town from Joe Kim's family, but he'll put an end to that, chop chop.
1984 (R) 87m/C Bobby Kim, Bill Wallace; *Dir:* Ed Warnick. **VHS, Beta $69.99** HBO 🎬

The Manchurian Candidate
Political thriller about an American Korean War vet who suspects that he and his platoon may have been brainwashed during the war, with his highly decorated, heroic friend programmed by commies to be an operational assassin. Loaded with shocks, conspiracy, inventive visual imagery, and bitter political satire of naivete and machinations of the left and right. Excellent performances by an all-star cast, with McGiver and Lansbury particularly frightening. Based on the Richard Condon novel. Featuring a special interview with Sinatra and Frankenheimer in which Sinatra is deified. Best Supporting Actress Oscar nomination for Lansbury.
1962 126m/B Frank Sinatra, Laurence Harvey, Angela Lansbury, Janet Leigh, James Gregory, Leslie Parrish, John McGiver, Henry Silva; *Dir:* John Frankenheimer. National Board of Review Awards '62: Best Supporting Actress (Lansbury); New York Film Festival '87: Best Supporting Actress. **VHS, Beta, LV $19.95** MGM 🎬🎬🎬🎬

The Mandarin Mystery
In the process of trying to retrieve a stolen Mandarin stamp, detective Ellery Queen uncovers a counterfeiting ring. Some fine performances, but a muddled script creates a mystery as to whether or not the action is played for laughs.
1937 65m/B Eddie Quillan, Charlotte Henry, Rita La Roy, Wade Boteler, Franklin Pangborn, George Irving, Kay Hughes, William Newell; *Dir:* Ralph Staub. **VHS, Beta $16.95** NOS, SNC 🎬½

Mandela A
gripping, powerful drama about human rights and dignity, tracing the real-life trials of Nelson and Winnie Mandela. The story focuses on the couple's early opposition to South African apartheid, as well as the events leading up to Nelson's life-imprisonment sentencing in 1964. Excellent, restrained performances from Glover and Woodard. Made for cable television.

1987 135m/C Danny Glover, Alfre Woodard, John Matshikiza, Warren Clarke, Allan Cordunner, Julian Glover; *Dir:* Philip Saville. **VHS, Beta** $79.99 *HBO, KUI* ♂♂♂

Mandinga A plantation owner develops an obsession for one of his female slaves. Made to cash in on the already exploitive "Mandingo," but two wrongs don't make a right. **1977** 100m/C Anthony Gismond; *Dir:* Mario Pinzauti. **VHS, Beta** $59.95 *PAR, TAV* ♂

Mandingo Portrays the brutal nature of slavery in the South, dealing with the tangled loves and hates of a family and their slaves. Heavyweight boxer Ken Norton made his screen debut in the title role. The characters are all emotionally charged-up, flay away at each other, and lose by unanimous decision. Followed by 1975's "Drum." Based on the novel by Kyle Onstott. **1975 (R)** 127m/C James Mason, Susan George, Perry King, Richard Ward, Ken Norton, Ben Masters; *Dir:* Richard Fleischer. **VHS, Beta, LV** $59.95 *PAR* ♂♂

Mandy Poignant drama of a deaf child and the family who must come to terms with her deafness. This basic plot is well-developed and performed, an intelligent treatment in all facets. Also known as "Crash of Silence." **1953** 93m/B *GB* Phyllis Calvert, Jack Hawkins, Mandy Miller; *Dir:* Alexander MacKendrick. **VHS, Beta** *LCA* ♂♂♂

Manfish Two men venture out in the boat, the Manfish, to hunt for sunken treasure in the Caribbean. Only one survives the trip, as his greed destroys the other. The scenes off the Jamaican coast are lovely, but the story fails to take hold, and though there is a star aboard in Chaney, you'll look astern and bow out with a sinking feeling. Derived from two Edgar Allan Poe stories, "The Gold Bug" and "The Tell-Tale Heart." **1956** 76m/C John Bromfield, Lon Chaney Jr., Victor Jory, Barbara Nichols; *Dir:* W. Lee Wilder. **VHS** $16.95 *NOS, SNC* ♂½

The Mango Tree A young man comes of age in a small Australian town during the 1920s. Everything is well-done, if not dramatic or fascinating. **1977** 93m/C *AU* Geraldine Fitzgerald, Robert Helpmann, Diane Craig, Gerald Kennedy, Christopher Pate; *Dir:* Kevin Dobson. **VHS, Beta** $69.95 *VID* ♂♂½

The Manhandlers After the uncle of a young woman is killed by the mob, she goes after them for revenge. The commercial about the soup with the same name was better, and more realistic. **1973 (R)** 87m/C Cara Burgess, Judy Brown, Vince Cannon, Rosalind Miles; *Dir:* Lee Madden. **VHS, Beta** $69.95 *TWE* ♂

Manhattan Allen plays a successful television writer yearning to be a serious writer. He struggles through a series of ill-fated romances, including one with a high school senior Hemingway, who is about a quarter century younger, and another with Keaton, who's also having an on-again, off-again affair with Murphy, Allen's best friend. A scathingly serious and comic view of modern relationships in urban America and of the modern intellectual neuroses. Shot in black-and-white to capture the mood of Manhattan and mated with an excellent Gershwin soundtrack, the video version preserves the widescreen effect that adds yet another impressive element to the overall production.

1979 (R) 96m/B Woody Allen, Diane Keaton, Meryl Streep, Mariel Hemingway, Michael Murphy, Wallace Shawn, Annie Byrne, Karen Allen; *Dir:* Woody Allen. British Academy Awards '79: Best Picture, Best Screenplay; Los Angeles Film Critics Association Awards '79: Best Supporting Actress; National Board of Review Awards '79: Best Picture, Best Supporting Actress; New York Film Critics Awards '79: Best Director; Harvard Lampoon Awards '79: Worst Film of the Year; National Society of Film Critics '79: Best Actress, Best Director. **VHS, Beta, LV** $19.98 *MGM, PIA* ♂♂♂♂

Manhattan Baby Unscary horror film about an archaeologist who digs up a relic that draws evil into the world and infects an American girl with powers that lead to many deaths. Advice to you that might have saved the archaeologist: don't dig this. **1982** 90m/C *IT* Christopher Connelly, Martha Taylor; *Dir:* Lucio Fulci. **VHS, Beta** $69.98 *LIV, LTG* ♂

Manhattan Melodrama Powell and Gable are best friends from childhood, growing up together in an orphanage. Their adult lives take different paths, however, as Powell becomes a respected prosecuting attorney while Gable becomes a notorious gambler/racketeer. Lovely Loy is Gable's girl who comes between the two. Eventually, Powell must prosecute his life-long friend for murder in order to win the governorship. This excellent drama offered Gable one of his toughest characters yet, while Loy is as always beautiful. Powell's character, however, is a bit unbelievable as his ethics seem to extend beyond love and friendship. This is the first film to team Powell and Loy, who would go on to make 13 more films together, including the "Thin Man" series. **1934** 93m/B Clark Gable, William Powell, Myrna Loy, Leo Carrillo, Nat Pendleton, George Sidney, Isabel Jewell, Muriel Evans, Claudelle Kaye, Frank Conroy, Jimmy Butler, Mickey Rooney, Edward Van Sloan; *Dir:* W.S. Van Dyke. **VHS, Beta** $19.98 *MGM, CCB* ♂♂♂½

Manhattan Merry-go-round One of the movies where a corrupt boss—in this case a record producer—threatens a bunch of good people as a pretense for a plot when the movie simply serves as a showcase for stars. Features many singing stars of the thirties ("where have you gone Cab Calloway?") plus Joltin' Joe, who ended up having a hit-streak in another genre. **1937** 89m/B Cab Calloway, Louis Prima, Ted Lewis, Ann Dvorak, Phil Regan, Kay Thompson, Gene Autry, Joe DiMaggio; *Dir:* Charles Riesner. **VHS, Beta** $19.95 *NOS, VCN, HHT* ♂♂

The Manhattan Project An exceptionally bright teenager decides to build a nuclear bomb for his project at the New York City science fair. He's out to prove how dangerously easy it is to build big bombs. When he steals plutonium from a local government installation, the feds attempt to nab the precocious youngster. Light moral overtones abound. Director Brickman co-wrote a film with a similar title—"Manhattan." **1986 (PG-13)** 117m/C John Lithgow, Christopher Collet, Cynthia Nixon, Jill Eikenberry, John Mahoney, Sully Boyer, Richard Council, Robert Schenkkan, Paul Austin; *Dir:* Marshall Brickman. **VHS, Beta, LV** $19.99 *HBO, LDC* ♂♂

Manhunt Frightened young man flees to Mexico believing he killed a classmate. Sky King seeks to tell him the truth—his classmate is alive. From the television series "Sky King." **195?** 25m/B Kirby Grant, Gloria Winters, Ewing Mitchell. **VHS, Beta** $59.95 *CCB Woof!*

Manhunt A man marked for execution by the mob launches his own assault on the organization's headquarters. Action and Italian food, but not much else. Also known as "The Italian Connection." Dubbed. **1973 (R)** 93m/C *IT* Henry Silva, Mario Adorf, Woody Strode. **VHS, Beta** $59.95 *MED* ♂♂

The Manhunt A framed cowhand escapes from prison to prove his innocence. Made-for-TV movie. **1986** 89m/C Ernest Borgnine, Bo Svenson, John Ethan Wayne; *Dir:* Larry Ludman. **VHS, Beta** $59.95 *MED* ♂♂

Manhunt in the African Jungle An American undercover agent battles Nazi forces in Africa. A serial in fifteen episodes. **1954** 240m/B Rod Cameron, Joan Marsh, Duncan Renaldo. **VHS** $29.98 *VCN, REP, MLB* ♂♂

Manhunt for Claude Dallas Mountain man uses game wardens for firing practice, ticks off local sheriff Torn, is tossed behind bars, checks out ahead of schedule and becomes a legend in his own time. Mediocre made for TV macho man melodrama. **1986** 93m/C Matt Salinger, Rip Torn, Claude Akins, Pat Hingle, Lois Nettleton, Beau Starr, Frederick Coffin; *Dir:* Jerry London. **VHS, Beta** $79.95 *PSM* ♂♂

Manhunt of Mystery Island Serial about the super-powered Captain Mephisto. Also called "Captain Mephisto and the Transformation Machine." **1945** 100m/B Linda Stirling, Roy Barcroft, Richard Bailey, Kenne Duncan; *Dir:* Spencer Gordon Bennet. **VHS, Beta** *MED, MLB Woof!*

Manhunter A made-for-television escapade about a World War I Marine who returns home from China in 1933 to track down a bunch of gangsters headed by his sister. **1974** 78m/C Ken Howard, Stefanie Powers, Gary Lockwood, Tim O'Connor, L.Q. Jones; *Dir:* Walter Grauman. **VHS, Beta, LV** $19.98 *WAR, TWE, PIA* ♂

Manhunter A mercenary undergoes the task of disengaging organized crime from high-level politics. Produced by Owensby. **1983** 92m/C Earl Owensby, Johnny Popwell, Doug Hale, Elizabeth Upton; *Dir:* Earl Owensby. **VHS, Beta** $49.95 *MED* ♂♂½

Manhunter A FBI forensic specialist who retired after a harrowing pursuit of a serial killer is called back to duty to find a psychotic family killer. Petersen's technique is to match the thought processes of serial killers and thus anticipate their moves. Intense thriller, based on the Thomas Harris novel "Red Dragon." Harris also wrote "The Silence of the Lambs", which won several Academy Awards in 1992. The character Hannibal ("The Cannibal") Lecter, focus of "Lambs" appears in this movie as well. Director Mann applies the slick techniques he introduced in the popular television series "Miami Vice," creating a quiet, moody intensity broken by sudden onslaughts of violence. Available in widescreen on laserdisc. **1986 (R)** 100m/C William L. Petersen, Kim Greist, Joan Allen, Brian Cox, Dennis Farina, Stephen Lang, Tommy Noonan; *Dir:* Michael Mann. **VHS, Beta, LV** $49.95 *WAR, LDC* ♂♂♂

Mania Fine adaptation of the Burke and Hare grave robbing legend. Cushing is the doctor who needs corpses and Pleasence and Rose provide them by any means. Highly atmospheric representation of dismal, 19th century Edinburgh. Very graphic for its time. Also known as "The Flesh and the Fiends." **1960 87m/B** *GB* Peter Cushing, June Laverick, Donald Pleasence, George Rose, Dermot Walsh, Renee Houston, Billie Whitelaw, John Cairney, Michael Balfour; *Dir:* John Gilling. **VHS $16.95** *SNC, DVT, MLB* 🎬🎬½

Mania Four suspense/murder stories set in a sleazy urban milieu. **198? 120m/C** Deborah Grover, Stephen B. Hunter, Lenore Zann, Wayne Robson. **Beta $79.95** *VHV* 🎬🎬

Maniac A scientist has designs on raising the dead and searches for victims on which to experiment. Bizarre "adults only" exploitation feature was considered very risque for its time, and includes eaten eyeballs, a cat fight with syringes and a rapist who thinks he's an orangutan. A must for genre aficionados. **1934 67m/B** Bill Woods, Horace Carpenter, Ted Edwards, Phyllis Diller, Thea Ramsey, Jennie Dark, Marcel Andre, Celia McGann; *Dir:* Dwain Esper. **VHS, Beta $14.95** *VDM, SNC, MRV* Woof!

Maniac An American artist living in France becomes involved with the daughter of a cafe owner, not suspecting that murder will follow. Seems that her old man is locked up in an insane asylum for torching the daughter's rapist several years earlier. **1963 86m/B** *GB* Kerwin Mathews, Nadia Gray, Donald Houston, Liliane Brousse; *Dir:* Michael Carreras. **VHS, Beta $49.95** *COL, MLB* 🎬🎬

Maniac A New York cop hunts down an arrow-shooting and obviously crazed Vietnam veteran who endeavors to hold an entire Arizona town for ransom. Which is entirely appropriate, since the cast is in it only for the money. Also known as "Ransom" and "Assault on Paradise." **1978 (PG) 87m/C** Oliver Reed, Deborah Raffin, Stuart Whitman, Jim Mitchum, Edward Brett; *Dir:* Richard Compton. **VHS, Beta $24.98** *SUE* 🎬

Maniac A psycho murderer slaughters and scalps his victims, adding the "trophies" to his collection. Carries a self-imposed equivalent "X" rating due to its highly graphic gore quotient. For extremely strong stomachs only. **1980 91m/C** Joe Spinell, Caroline Munro, Gail Lawrence, Kelly Piper, Tom Savini; *Dir:* William Lustig; *W/Dir:* William Lustig. **VHS, Beta $54.95** *MED* Woof!

Maniac Cop In New York city, a cop goes beyond the realm of sanity and turns vigilante. Low-budget slasher/thriller that too often sags. **1988 (R) 92m/C** Tom Atkins, Bruce Campbell, Laurene Landon, Richard Roundtree, William Smith, Robert Z'Dar, Sheree North; *Dir:* William Lustig. **VHS, Beta $19.98** *TWE, HHE* 🎬

Maniac Cop 2 Recently released from prison, a grossly disfigured policeman forms a one man vigilante squad, seeking revenge for his unjust incarceration. Blood and guts fly as any plot shortcomings are cleverly disguised by an array of violent video deaths. Sequel to "Maniac Cop." **1990 (R) 90m/C** Robert Davi, Claudia Christian, Michael Lerner, Bruce Campbell, Laurene Landon, Robert Z'Dar, Clarence Williams III, Leo Rossi; *Dir:* William Lustig. **VHS, LV $89.95** *LIV* 🎬

The Manions of America The long and sometimes interesting rags-to-riches tale of Rory O'Manion, a feisty Irish patriot who leaves his native land during the potato famine of 1845 to settle in America. Originally a television miniseries.

1981 290m/C Pierce Brosnan, Kate Mulgrew, Linda Purl, David Soul, Kathleen Beller, Simon MacCorkindale; *Dir:* Joseph Sargent, Charles S. Dubin. **VHS, Beta $69.95** *PSM, PAR* 🎬🎬½

Manipulator A deranged ex-movie make-up man (Rooney, playing to type) kidnaps a young actress and holds her prisoner in a deserted Hollywood sound stage. **1971 (R) 91m/C** Mickey Rooney, Luana Anders, Keenan Wynn; *Dir:* Yabo Yablonsky. **VHS, Beta $69.98** *LIV, VES* 🎬

The Manitou A San Francisco woman suffers from a rapidly growing neck tumor which eventually grows into a 400-year-old Indian witch doctor. (I hate when that happens.) Redeemed only by good special effects, especially those that kept Curtis, Strasberg, and Meredith from laughing. **1978 (PG) 104m/C** Susan Strasberg, Tony Curtis, Stella Stevens, Ann Sothern, Burgess Meredith, Michael Ansara; *Dir:* William Girdler. **VHS, Beta, LV $9.95** *COL, CHA* 🎬🎬

Mankillers A group of tough female convicts are enlisted to hunt down and rub out a psycho drug dealer, all the while displaying their feminine charms. **1987 90m/C** Edd Byrnes, Gail Fisher, Edy Williams, Lynda Aldon, William Zipp, Christopher Lunde, Susanne Tegman, Marilyn Stafford, Paul Bruno, Byron Clark; *Dir:* David A. Prior. **VHS, Beta $14.95** *SVS* 🎬

Mannequin Tracy and Crawford star in this romantic story of a poor girl who finds temporary happiness by marrying a wealthy man after ditching her con-artist husband. Somewhat predictable, the movie reads like a "People" magazine story on The Donald and Ivana. Written by Joseph L. Mankiewicz, the story is kept afloat by Tracy and Crawford (in their only film together) with an able assist from Curtis. **1937 95m/B** Joan Crawford, Spencer Tracy, Alan Curtis, Ralph Morgan, Leo Gorcey, Elisabeth Risdon, Paul Fix; *Dir:* Frank Borzage. **VHS $19.98** *MGM, MED* 🎬🎬½

Mannequin A young artist creates a store window display using various mannequins, one of which contains the spirit of an ancient Egyptian woman. She comes to life when he is around, and naturally none of his co-workers believe him. Very light comedy, featuring two pretty stars and music by Jefferson Starship. **1987 (PG) 90m/C** Andrew McCarthy, Kim Cattrall, Estelle Getty, James Spader, Meshach Taylor, Carole Davis, G.W. Bailey; *Dir:* Michael Gottlieb. **VHS, Beta, LV $9.99** *CCB, MED* 🎬🎬

Mannequin 2: On the Move Less of a sequel, more of a lame rehash proving that the first "Mannequin" could have been even dumber. At this rate part three will be off the scale. Now it's a lovesick Teutonic princess frozen for 1,000 years who revives in a department store. Taylor reprises his grotesque gay role. **1991 (PG) 95m/C** Kristy Swanson, William Ragsdale, Meshach Taylor, Terry Kiser, Stuart Pankin; *Dir:* Stewart Raffill. **VHS, LV $92.98** *LIV, PIA* Woof!

Mannikin A chilling supernatural short by Robert Bloch about a young singer who is possessed by a demon. **1977 28m/C** Ronee Blakley, Keir Dullea; *Dir:* Donald W. Thompson. **VHS, Beta** *LCA* Woof!

Manny's Orphans An out-of-work teacher takes on a lovable home for orphaned boys. Dull remake of "Bad News Bears" that's also known as "Here Come the Tigers."

1978 (PG) 92m/C Richard Lincoln, Malachy McCourt, Sel Skolnick; *Dir:* Sean S. Cunningham. **VHS, Beta $69.98** *LIV, VES* 🎬

Manon of the Spring In this excellent sequel to "Jean de Florette," the adult daughter of the dead hunchback, Jean, discovers who blocked up the spring on her father's land. She plots her revenge, which proves greater than she could ever imagine. Montand is astonishing. Based on a Marcel Pagnol novel. English subtitles. **1987 (PG) 113m/C** *FR* Yves Montand, Daniel Auteuil, Emmanuelle Beart, Hippolyte Girardot; *Dir:* Claude Berri. National Board of Review Awards '87: Best Foreign Film. **VHS, Beta, LV $19.98** *ORI, APD* 🎬🎬🎬

Manos, the Hands of Fate Family vacations in the lone star state and find themselves deep in the heart of a Satanic cult. More low concept fodder from erstwhile El Paso fertilizer salesman Warren. **1966 74m/C** Tom Nayman, Diane Mahree, Hal P. Warren, John Reynolds; *W/Dir:* Hal P. Warren. **VHS $19.98** *SNC* 🎬

Manos Torpes A man who is left to die in the desert is rescued by someone who teaches him how to avenge his would-be killers. **1965 93m/C** *MX* Peter Lee Lawrence, Alberto De Mendoza, Pilar Velasquez. **VHS, Beta $45.95** *MAD* 🎬🎬

Man's Best Friend Here is a collection of nine cartoons, featuring such lovable canines as Cuddles, Snoozer, Duffy Dog and Dizzy. **1964 51m/C VHS, Beta $14.95** *MCA* 🎬🎬

Man's Country A rip-roarin' western saga filled with the usual action and danger, as ranger Randall leads the fight against a band of nasties headed by Long, in a dual role as twin brothers. **1938 55m/B** Jack Randall, Ralph Peters, Marjorie Reynolds, Walter Long; *Dir:* Robert Hill. **VHS, Beta $19.95** *NOS, DVT* 🎬🎬

Man's Favorite Sport? A slapstick comedy about a renowned fishing expert author who actually hates fishing, but is forced to compete in a major tournament by a romantically inclined publicity agent. Very funny in spots. Music by Henry Mancini. **1963 121m/C** Rock Hudson, Paula Prentiss, Charlene Holt, Maria Perschy, John McGiver; *Dir:* Howard Hawks. **VHS, Beta $14.95** *MCA, CCB* 🎬🎬½

Man's Land A ranch needs savin', and Hoot's the guy to do it in this formulaic Gibson epic. **1932 65m/B** Hoot Gibson, Marion Shilling, Skeeter Bill Robbins, Alan Bridge; *Dir:* Phil Rosen. **VHS, Beta** *UHV* 🎬½

Mansfield Park BBC mini-series adaptation of the Jane Austen classic concerning love and family in 19th century England. **1985 261m/C** *GB* Anna Massey, Donald Pleasence, Bernard Hepton; *Dir:* David Giles. **VHS, Beta $29.98** *FOX* 🎬🎬½

Manson This documentary examines the life of convicted murderer Charles Manson and his "Family." **1985 90m/C VHS, Beta $59.95** *UHV* 🎬🎬

The Manster Another masterpiece from the director who brought us "Monster from Green Hell." Womanizing whiskey swilling American journalist receives mysterious injection from crazed scientist and sprouts unsightly hair and extra head. Although shot

in the land of the red sun, lips move in sync with dialogue.
1959 72m/B JP Peter Dyneley, Jane Hylton, Satoshi Nakamura, Terri Zimmern; **Dir:** Kenneth Crane. **VHS $19.98** SNC, MRV ✰✰

Mantis in Lace A go-go dancer slaughters men while tripping on LSD. Also known as ''Lila'' with cinematography by pre-''Easy Rider'' Laslo Kovacs. Watch it for the ''hep'' dialogue.
1968 (R) 68m/C Susan Stewart, Steve Vincent, M.K. Evans, Vic Lance, Pat Barrington, Janu Wine, Stuart Lancaster, John Carrol, Judith Crane, Cheryl Trepton; **Dir:** William Rotsler. **VHS** CIC ✰✰½

Mantrap Early silent success by Fleming, who later directed ''Gone with the Wind'' and ''The Wizard of Oz.'' Fabled flapper Bow tempts a lawyer on retreat in the woods.
1926 66m/B Clara Bow, Ernest Torrance, Percy Marmont, Eugene Pallette, Tom Kennedy; **Dir:** Victor Fleming. **VHS, Beta $24.95** GPV, FCT, DNB ✰✰½

Manu Dibango: King Makossa The Camarounian artist Manu Dibango performs in Brussels. This Afrojazz concert features the saxophonist's biggest hit, ''Soul Makossa.'' Recorded in Hi-Fi Stereo.
1981 55m/C **VHS, Beta $29.95** MVD, VWV, MBP ✰✰½

The Manxman Hitchcock's last silent film, a romantic melodrama about ambition and infidelity on the Isle of Man.
1929 129m/B GB Carl Brisson, Anny Ondra; **Dir:** Alfred Hitchcock. **VHS, Beta $49.95** VYY, MRV, MLB ✰✰½

Many Faces of Sherlock Holmes A documentary look at the various incarnations of the famous sleuth from A. Conan Doyle's stories to various film portrayals.
1986 58m/C Christopher Plummer, Basil Rathbone, Christopher Lee. **VHS, Beta $79.95** PSM ✰✰½

Many Wonder A wealthy woman is pursued by three men...at the same time! Will one of them turn out to be Mr. Right?
1989 40m/C **Dir:** Craig Lowy. **VHS $59.95** CIG ✰½

Maps to Stars' Homes Video A guided tour through Beverly Hills. See the mansions of your favorite movie and TV stars. Also, the actual souvenir map that is sold for $5.00 in Beverly Hills is included free, so that you can follow the route while viewing the tape.
1990 55m/C **VHS, Beta $19.95** SHO ✰½

Marat/Sade (Persecution and Assassination...) A theatrical production presented by patients at a mental institution is directed by the Marquis de Sade. Well directed and visually effective, featuring Glenda Jackson's first film performance. Based on a play by Peter Weiss. Full title is ''The Persecution and Assassination of Jean-Paul Marat as Performed by the Inmates of the Asylum of Charenton Under the Direction of the Marquis de Sade.'' Some theaters were forced to add extensions to the marquee when the film was first released.
1966 (R) 115m/C GB Patrick Magee, Clifford Rose, Glenda Jackson, Ian Richardson, Brenda Kempner, Ruth Baker, Michael Williams; **Dir:** Peter Brook. **VHS, LV $34.95** WBF, MGM, LUM ✰✰✰½

Marathon When a middle-aged jogger's ego gets a boost through the attention of a beautiful young woman, he takes up marathon running. Light comedy. Made for television.
1980 100m/C Bob Newhart, Leigh Taylor-Young, Herb Edelman, Dick Gautier, Anita Gillette, John Hillerman; **Dir:** Jackie Cooper. **VHS $59.95** IVE ✰✰

Marathon Man Nightmarish chase-thriller in which a graduate student becomes entangled in a plot involving a murderous Nazi fugitive. As student Hoffman is preparing for the Olympic marathon, he is reunited with his secret-agent brother, setting the intricate plot in motion. Courtesy of his brother, Hoffman becomes involved with Olivier, an old crazed Nazi seeking jewels taken from concentration camp victims. Non-stop action throughout, including a torture scene sure to set your teeth on edge. William Goldman adapted the screenplay from his novel.
1976 (R) 125m/C Dustin Hoffman, Laurence Olivier, Marthe Keller, Roy Scheider, William Devane, Fritz Weaver; **Dir:** John Schlesinger. **VHS, Beta, LV $24.95** PAR ✰✰✰

Marauder The prince of Venice leads a sea-faring onslaught against ransacking pirates and enemy fleets.
1965 90m/C IT Gordon Scott, Maria Canale, Franca Bettoya; **Dir:** Luigi Capuano. **VHS, Beta $29.95** FOR ✰✰

Marbella A luxury resort in Spain is the setting for a big caper heist of three million dollars from a monarchical tycoon.
1985 96m/C Rod Taylor, Britt Ekland; **Dir:** Miguel Hermoso. **VHS, Beta $29.98** LIV, LTG ✰½

Marc and Ann Two subjects who have appeared in other Blank films are given sole billing in this documentary. Musician Marc Savoy builds accordians, tells great stories, and celebrates his Cajun culture. His wife, Ann, is an author and the mother of their four children.
1991 27m/C **Dir:** Les Blank. **VHS, Special order formats $49.95** FFM ✰½

Marcados por el Destino Three Mexicans come into America looking for a better life, but they don't find it.
1987 95m/C MX Julio Aleman, Roberto ''Flaco'' Guzman, Alma Murillo. **VHS, Beta $77.95** MAD ✰✰½

March or Die Great potential, unrealized. An American joins the French Foreign Legion during World War I after his dismissal from West Point. Following the brutality of training, he is assigned to guard an archeological expedition in Morocco, where he pulls together a rag-tag outfit for the mission. Hackman proves once again the wide range of his acting abilities, surmounting the cliched and fairly sadistic plot. Shot on location in the Sahara Desert.
1977 (PG) 104m/C GB Candice Bergen, Gene Hackman, Terence Hill, Max von Sydow, Catherine Deneuve, Ian Holm; **Dir:** Dick Richards. **VHS** FOX ✰✰½

March of the Wooden Soldiers The classic Mother Goose tale about the secret life of Christmas toys, with Laurel and Hardy as Santa's helpers who must save Toyland from the wicked Barnaby. A Yuletide ''must see,'' originally titled ''Babes in Toyland.'' Also available in a colorized version.

1934 73m/B Stan Laurel, Oliver Hardy, Charlotte Henry, Henry Kleinbach (Brandon), Felix Knight, Jean Darling, Johnny Downs, Marie Wilson; **Dir:** Charles ''Buddy'' Rogers, Gus Meins. **VHS, LV $29.95** NOS, GKK, IND ✰✰✰

Marciano A made-for-television version of the great boxer's life. Average fight movie.
1979 97m/C Tony LoBianco, Vincent Gardenia; **Dir:** Bernard L. Kowalski. **VHS, Beta $9.98** SUE, CHA ✰✰

Marco Entertaining musical adventure of Marco Polo's life casts Arnaz as Marco Polo and Mostel as Kublai Khan. A couple of cut-ups, right? Songs include ''By Damn,'' ''Walls,'' ''A Family Man,'' and that old favorite, ''Spaghetti.'' One of the first films to combine animation with live action. Shot partially on location in the Orient.
1973 109m/C Desi Arnaz Jr., Zero Mostel, Jack Weston, Cie Cie Win; **Dir:** Seymour Robbie. **VHS, Beta $49.95** PSM ✰✰

Marco Polo, Jr. Marco Polo, Jr., the daring descendant of the legendary explorer, travels the world in search of his destiny in this song-filled, feature-length, but poorly-animated, fantasy.
1972 82m/C **Dir:** Eric Porter; **Voices:** Bobby Rydell. **VHS, Beta $24.95** FHE ✰

Mardi Gras Massacre An Aztec priest arrives in New Orleans during Mardi Gras to revive the blood ritual of human sacrifice to an Aztec god. A police detective relentlessly pursues him. Much gore and gut-slicing, with no redeeming social value. Also known as ''Crypt of Dark Secrets.''
1978 92m/C Curt Dawson, Gwen Arment, Wayne Mack, Laura Misch; **Dir:** Jack Weis. **VHS, Beta $49.95** VHE Woof!

Margin for Murder Mike Hammer investigates a mysterious accident that killed his best friend. Made for television; also known as ''Mickey Spillane's Margin for Murder.''
1981 98m/C Kevin Dobson, Cindy Pickett, Donna Dixon, Charles Hallahan; **Dir:** Daniel Haller. **VHS, Beta $59.95** PSM ✰✰

Maria A young man returns to his hometown to rekindle a romance with his childhood sweetheart. Also known as ''Marie.''
1968 89m/B Fernando Allende, Taryn Power, Alicia Caro, Jose Suarez. **VHS, Beta $67.45** MAD ✰✰

Maria Candelaria Society rejects a woman because people don't like her mother. (She posed nude for an artist, and the town responds by stoning her to death, proving that going against small-town morals can be lethal.) Eventually, time and circumstance push daughter onto mother's path in this tragic soaper. Del Rio portrays the scorned young woman, although she was 40 years old. Also known as ''Portrait of Maria'' and ''Xochimilco.''
1946 96m/B MX Delores Del Rio, Pedro Armendariz, Margarita Cortes; **Dir:** Emilio Fernandez. **VHS, Beta $44.95** MAD ✰✰

Maria Chapdelaine An early film from the renowned French director, in which a brutish trapper and a sophisticate battle for the love of a girl in the Canadian wilderness. English subtitles. Remade in 1984.
1934 75m/B FR Jean Gabin, Jean-Pierre Aumont, Madeleine Renaud; **Dir:** Julien Duvivier. **VHS, Beta $44.95** HHT ✰✰✰

Maria Chapdelaine Around the turn of the century, a young girl in the Northern Canadian wilderness endures a year of passion, doomed love and tragedy. Winner of a

Canadian Genie. In French with subtitles. Remake of a 1935 French Film.
1984 108m/C *FR CA* Nick Mancuso, Carole Laure, Claude Rich, Pierre Curzi; ***Dir:*** Gilles Carle. Genie Awards '84: Best Art Direction/Set Decoration, Best Cinematography, Best Costume Design, Best Musical Score. **VHS, Beta $59.95** *MED* 🦴🦴½

Maria and Mirabella Two little girls rescue their friend KiKi the Frog in this action adventure. Combines animation with live action techniques.
1990 60m/C VHS $9.95 *AIP* 🦴🦴½

Mariachi Mexican folklore and music are included in this Western.
1965 90m/C *MX* Antonio Badu, Beatriz Aguirre, Ferrusquilla. **VHS, Beta $54.95** *MAD* 🦴🦴

Marianela A disfigured peasant girl cares for a blind man, and the two fall in love. She doesn't tell him about her appearance until the day comes that he regains his sight. In Spanish.
197? 113m/C *SP* Pierre Orcel, Rocio Durcal; ***Dir:*** Angelino Fons. **VHS, Beta** *MED* 🦴🦴½

Maria's Day The chronicling of the downfall of a formerly aristocratic and wealthy Hungarian family in the years following the failed 1849 Revolution, victims each of plague, syphilis and their own folly. Acclaimed; in Hungarian with subtitles.
1984 113m/C *Dir:* Judit Elek. **VHS, Beta $59.95** *FCT* 🦴🦴🦴

Maria's Lovers The wife (Kinski) of an impotent WWII veteran (Savage) succumbs to the charms of a rakish lady-killer (Spano). Savage turns to the charms of an older woman, finds love again with Kinski, but was still impotent; she gets pregnant by a wandering minstrel (Carradine) and on we go to film climax. Offbeat and uneven, representing Russian director Konchalovsky's first American film and one of Kinski's better roles.
1984 (R) 103m/C Nastassia Kinski, John Savage, Robert Mitchum, Keith Carradine, Anita Morris, Bud Cort, Karen Yahng, Tracy Nelson, John Goodman, Vincent Spano; ***Dir:*** Andrei Konchalovsky. **VHS, Beta $79.95** *MGM* 🦴🦴½

Maricela Young Maricela Flores and her mother have come to the U.S. from El Salvador to escape the fighting. Living with a Southern Californian family, Maricela has a hard time adjusting to American life and in particular, dealing with prejudice. An entry in the PBS "Wonderworks" series.
1988 55m/C Linda Lavin, Carlina Cruz. **VHS $29.95** *PME* 🦴🦴½

Marie In this true story, a divorced (and battered) mother works her way through school and the system to become the first woman to head the Parole Board in Tennessee. Finding rampant corruption, she blows the whistle on her bosses, who put her life in jeopardy. Spacek gives a powerful performance, as does first-time actor Thompson, portraying himself as the abused woman's attorney. Based on the book by Peter Maas.
1985 (PG-13) 113m/C Sissy Spacek, Jeff Daniels, Keith Szarabajka, John Cullum, Morgan Freeman, Fred Dalton Thompson, Don Hood; ***Dir:*** Roger Donaldson. **VHS, Beta $79.95** *MGM* 🦴🦴🦴

Marie Antoinette An elephantine costume drama chronicling the French queen's life from princesshood to her final days before the Revolution. A Shearer festival all the way, and a late example of MGM's overstuffed period style and star-power. Overlong, but engrossing for the wrong reasons. Morley, in his first film, plays Louis XVI. Power's only MGM loan-out casts him as a Swedish count and Marie's romantic dalliance. Shearer received an Oscar nomination for Best Actress, Morley for Best Supporting Actor. Based on a book by Stephan Zweig, with script assistance from (among others) F. Scott Fitzgerald.
1938 160m/B Norma Shearer, Tyrone Power Sr., John Barrymore, Robert Morley, Gladys George, Anita Louise, Joseph Schildkraut, Henry Stephenson, Reginald Gardiner, Peter Bull, Albert Dekker, Joseph Calleia, George Zucco, Cora Witherspoon, Barry Fitzgerald, Mae Busch, Harry Davenport; ***Dir:*** W.S. Van Dyke. **VHS, Beta, LV $19.98** *MGM, PIA* 🦴🦴½

Marie Galante Fine spy drama has Tracy running into Gallien in Panama (years earlier she had been left there after a kidnapping). He finds her most helpful in his attempt to thwart the bombing of the Panama Canal. It's the performances that raise this otherwise standard thriller up a few notches.
1934 88m/B Spencer Tracy, Katie Galian, Ned Sparks, Helen Morgan, Sig Rumann, Leslie Fenton, Jay C. Flippen; ***Dir:*** Henry King. **VHS, Beta $16.95** *NOS, SNC* 🦴🦴🦴

Marihuana An unintentionally hilarious, "Reefer Madness"-type cautionary film about the exaggerated evils of pot smoking. A real dopey film favored by right-winger zealots.
1936 57m/B Marley Wood, Hugh MacArthur, Pat Carlyle; ***Dir:*** Dwain Esper. **VHS, Beta $16.95** *SNC, VYY* 🦴

Marilyn Rock Hudson narrates this documentary on the life of Marilyn Monroe.
1963 83m/B *Nar:* Rock Hudson. **VHS, EJ** *ILC* 🦴

Marilyn: The Untold Story Nominated for an Emmy, Hicks elevates what could easily have been a dull made for TV movie with her remarkable performance as Marilyn Monroe. Based on the book by Norman Mailer.
1980 156m/C Catherine Hicks, Richard Basehart, Frank Converse, John Ireland, Sheree North, Anne Ramsey, Viveca Lindfors, Jason Miller, Bill Vint; ***Dir:*** Jack Arnold, John Flynn, Lawrence Schiller. **VHS** *ABC* 🦴🦴🦴

Marine Raiders Marines train, then fight at Guadalcanal. A typical flag-waver, watchable but not particularly engaging.
1943 90m/B Pat O'Brien, Robert Ryan, Ruth Hussey, Frank McHugh, Barton MacLane; ***Dir:*** Harold Schuster. **VHS, Beta $19.98** *TTC* 🦴🦴

Mario Lanza: The American Caruso PBS documentary about the troubled life of the popular opera star, making extensive use of clips from his films.
1983 68m/C *Hosted:* Placido Domingo. **Beta $29.95** *KUL, CMR, FCT* 🦴🦴

Marius This is the first of Marcel Pagnol's trilogy ("Fanny" and "Cesar" followed), about the lives and adventures of the people of Provence, France. Marius is a young man who dreams of going away to sea. When he acts on those dreams, he leaves behind his girlfriend, Fanny. Realistic dialogue and vivid characterizations help create the classic fare here. Adapted by Pagnol from his play. The musical play and film "Fanny" (1961) was adapted from this trilogy.
1931 125m/B *FR* Raimu, Pierre Fresnay, Charpin, Orane Demazis; ***Dir:*** Alexander Korda. **VHS, Beta, LV $24.95** *NOS, MRV, INT* 🦴🦴🦴🦴

Marjoe Documentary follows the career of rock-style evangelist Marjoe Gortner, who spent 25 years of his life touring the country as a professional preacher. Marjoe later went on to become an actor and professional fundraiser.
1972 (PG) 88m/C Marjoe Gortner; ***Dir:*** Howard Smith, Sarah Kernochan. Academy Awards '72: Best Feature Documentary. **VHS, Beta $59.95** *COL* 🦴🦴🦴🦴

Marjorie Morningstar A temperate Hollywood adaptation of the Herman Wouk story of a young actress who fails to achieve stardom and settles on being a housewife. Wood slipped a bit in this story.
1958 123m/C Natalie Wood, Gene Kelly, Martin Balsam, Claire Trevor, Ed Wynn, Everett Sloane, Carolyn Jones; ***Dir:*** Irving Rapper. **VHS $14.98** *REP* 🦴🦴

The Mark Story of a convicted child molester who cannot escape his past upon his release from prison. Whitman's riveting portrayal of the convict deservedly earned him an Oscar nomination.
1961 127m/B *GB* Stuart Whitman, Maria Schell, Rod Steiger, Brenda de Banzie, Maurice Denham, Donald Wolfit, Paul Rogers, Donald Houston; ***Dir:*** Guy Green. **VHS, Beta** *VTR* 🦴🦴🦴½

Mark of the Beast Two students inadvertently videotape an assassination, and are then relentlessly pursued by the merciless killer.
19?? 90m/C James Gordon, Carolyn Guillet, David Smulker; ***Dir:*** Robert Stewart. **VHS** *NAV* 🦴

Mark of Cain Twin brothers, one normal, the other a raving lunatic with murderous tendencies, confuse the authorities who imprison the nice guy, allowing the nutcase to chase after his beautiful sister-in-law.
1984 90m/C Robin Ward, Wendy Crewson, August Schellenberg; ***Dir:*** Bruce Pittman. **VHS, Beta $69.98** *LIV, VES* 🦴🦴½

Mark of the Devil Witchcraft and romance don't mix as a Medieval witch hunter and a sexy girl accused of witchery discover. Notoriously graphic torture scenes add to the mayhem but the weird part is that this is based on true stories.
1969 (R) 96m/C *GB GE* Herbert Lom, Olivera Vuco, Udo Kier, Reggie Nalder, Herbert Fux, Michael Maien, Ingeborg Schoener, Johannes Buzalski, Gaby Fuchs; ***Dir:*** Michael Armstrong. **VHS, Beta $69.98** *VES, LIV, LTG Woof!*

Mark of the Devil 2 Sadistic witch-hunters torture satan's servants and torch sisters of mercy, whilst trying to horn in on a nobleman's fortune. It's just not as gross without the vomit bags.
1972 (R) 90m/C *GB GE* Erika Blanc, Anton Diffring, Reggie Nalder; ***Dir:*** Adrian Hoven. **VHS $29.98** *VDM* 🦴

The Mark of the Hawk A uniquely told story of African nations struggling to achieve racial equality after gaining independence. Songs include "This Man Is Mine," sung by Kitt. Also called "The Accused."
1957 83m/C *GB* Sidney Poitier, Eartha Kitt, Juano Hernandez, John McIntire; ***Dir:*** Michael Audley. **VHS, Special order formats** *AVD, MRV* 🦴🦴🦴

Mark of the Scorpion A gang member is placed in infinite peril when he steals the ruby ring from a dead princess.
1990 90m/C Mahamed Attifi. **VHS, Beta** *IMP* 🦴

Mark of the Spur A cowboy must find out who left a mysterious spur mark on his horse.

1932 60m/B Bob Custer, Lillian Rich, Franklin Farnum, Bud Osborne; **Dir:** J.P. McGowan. **VHS, Beta** $19.95 *NOS, DVT* 🐾½

Mark Twain and Me Robards portrays the aging and irascible Mark Twain, attended by a neglected but devoted daughter, and befriended by a young girl.
1991 93m/C Jason Robards Jr., Talia Shire, Amy Stewart. **VHS** $19.99 *DIS* 🐾½

Mark Twain's A Connecticut Yankee in King Arthur's Court A man from Connecticut falls asleep and finds himself in King Arthur's Court.
1978 60m/C Richard Basehart, Roscoe Lee Browne, Paul Rudd. **VHS, Beta, LV** $64.95 *MAS* 🐾🐾½

Mark of the Vampire A murder in a small town is solved through the use of vaudeville actors who pose as vampires. Great cast, surprise ending.
1935 61m/B Lionel Barrymore, Bela Lugosi, Elizabeth Allan; **Dir:** Tod Browning. **VHS, Beta** $19.95 *MGM, MLB* 🐾🐾🐾

Mark of Zorro Fairbanks plays a dual role as the hapless Don Diego and his dashing counterpart, Zorro, the hero of the oppressed. Silent film.
1920 80m/B Douglas Fairbanks Sr., Marguerite de la Motte, Noah Beery; **Dir:** Fred Niblo. **VHS, LV** $19.95 *REP, MRV, GPV* 🐾🐾🐾

Marked for Death Having killed a prostitute, DEA agent Seagal decides it's time to roll out the white picket fence in the 'burbs with the little woman and brood. Trouble is, a bunch of guys with dreadlocks don't approve of his early retirement, and the Jamaican gangsters plan to send him and his family to the great Rasta playground in the sky. Whereupon the Stevester kicks and punches and wags his ponytail. Much blood flows, mon. Tunes by Jimmy Cliff.
1990 (R) 93m/C Steven Seagal, Joanna Pacula, Basil Wallace, Keith David, Tom Wright, Danielle Harris, Arlen Dean Snyder, Teri Weigel; **Dir:** Dwight Little. **VHS, LV** $19.98 *FXV, CCB* 🐾½

Marked Money Amiable captain takes in boy who'll soon roll in the dough and swindlers show up on the scene. The captain's courageous girl and her pilot beau step in to right the rookery.
1928 61m/B Junior Coghlan, George Duryea, Tom Kennedy, Bert Woodruff, Virginia Bradford, Maurice Black, Jack Richardson; **Dir:** Spencer Gordon Bennet. **VHS, Beta** $19.95 *GPV, FCT* 🐾🐾

Marked for Murder Another in the Texas Ranger series. In this entry, the Rangers come to the rescue when a rancher war appears inevitable.
1945 58m/B Tex Ritter, Dave O'Brien, Guy Wilkerson; **W/Dir:** Elmer Clifton. **VHS, LV** $79.95 *VMK* 🐾½

Marked for Murder Two TV station employees are framed for murder after being sent out to find a missing videocassette. Renee Estevez is part of the Sheen/Estevez dynasty.
1989 (R) 88m/C Renee Estevez, Wings Hauser, Jim Mitchum, Ross Hagen, Ken Abraham; **Dir:** Rick Sloane. **VHS, Beta** $79.95 *VMK* 🐾½

Marked Trails One clue stamps out the guilty parties in this wild west saga.
1944 59m/B Hoot Gibson, Bob Steele. **VHS, Beta** $19.95 *NOS, DVT* 🐾

Marked Woman Gangster drama about crusading District Attorney who persuades a group of clipjoint hostesses to testify against their gangster boss. A gritty studio melodrama loosely based on a true story.
1937 97m/B Bette Davis, Humphrey Bogart, Eduardo Ciannelli, Isabel Jewell, Jane Bryan, Mayo Methot, Allen Jenkins, Lola Lane; **Dir:** Lloyd Bacon. Venice Film Festival '37: Best Actress (Davis). **VHS, Beta, LV** $19.95 *MGM* 🐾🐾🐾

Marlon Brando Documentary recounting the life and career of Hollywood's most rebellious superstar.
1985 120m/C Marlon Brando; **Dir:** Claudio Masenza. **VHS, Beta** *KRT* 🐾🐾🐾

Marlowe Updated telling by Stirling Silliphant of Chandler's ''The Little Sister'' sports retro guy Garner as Philip Marlowe, gumshoe. Hired by a mystery blonde to find her misplaced brother, rumpled sleuth Marlowe encounters kicking Bruce Lee in his first film. Slick looking, but the story's a bit slippery.
1969 (PG) 95m/C James Garner, Gayle Hunnicutt, Carroll O'Connor, Rita Moreno, Sharon Farrell, William Daniels, Bruce Lee; **Dir:** Paul Bogart. **VHS, Beta** $19.98 *MGM, FCT* 🐾🐾½

Marnie A lovely blonde with a mysterious past robs her employers and then changes her identity. When her current boss catches her in the act and forces her to marry him, he soon learns the puzzling aspects of Marnie's background. Criticized at the time of its release, the movie has since been accepted as a Hitchcock classic.
1964 130m/C Tippi Hedren, Sean Connery, Diane Baker, Bruce Dern; **Dir:** Alfred Hitchcock. **VHS, Beta, LV** $19.95 *MCA* 🐾🐾🐾½

Marooned Tense thriller casts Crenna, Hackman, and Franciscus as astronauts stranded in space after a retro-rocket misfires and their craft is unable to return to earth.
1969 (G) 134m/C Gregory Peck, David Janssen, Richard Crenna, James Franciscus, Gene Hackman, Lee Grant; **Dir:** John Sturges. Academy Awards '69: Best Visual Effects. **VHS, Beta, LV** $9.95 *GKK* 🐾🐾

The Marquis de Sade's Justine Two sisters who have been abused in a convent are turned out. They each adapt to worldly life in different ways in this classic tale stretching the boundaries of human depravity.
19?? 90m/C Koo Stark, Martin Potter. **VHS, Beta** $79.95 *JEF* 🐾🐾

Marriage is Alive and Well A wedding photographer reflects on the institution of marriage from his unique perspective. Made for television.
1980 100m/C Joe Namath, Jack Albertson, Melinda Dillon, Judd Hirsch, Susan Sullivan, Fred McCarren, Swoosie Kurtz; **Dir:** Russ Mayberry. **VHS** $59.95 *IVE* 🐾

The Marriage Circle A pivotal silent comedy depicting the infidelity of several married couples in Vienna. Remade as a musical, ''One Hour With You,'' in 1932. Silent.
1924 90m/B Florence Vidor, Monte Blue, Marie Prevost, Creighton Hale, Adolphe Menjou, Harry Myers, Dale Fuller; **Dir:** Ernst Lubitsch. **VHS** $19.95 *NOS, MOV, DNB* 🐾🐾

The Marriage of Maria Braun In post-WWII Germany, a young woman uses guile and sexuality to survive as the nation rebuilds itself into an industrial power. The first movie in Fassbinder's trilogy about

German women in Germany during the post war years, it is considered one of the director's finest films, and an indispensable example of the New German Cinema. In German with English subtitles.
1979 (R) 120m/C *GE* Hanna Schygulla, Klaus Lowitsch, Ivan Desny; **Dir:** Rainer Werner Fassbinder. Berlin Film Festival '79: Best Technical Team, Best Actress (Schygulla). **VHS, Beta** $79.95 *NYF, COL, GLV* 🐾🐾🐾

Married? Couple endures 365 days of marital bliss in order to inherit big bucks. Some do it for less.
1926 65m/B Owen Moore, Constance Bennett; **Dir:** George Terwilliger. **VHS, Beta** $24.95 *GPV, FCT* 🐾½

A Married Man Story focusing on a bored British lawyer who begins cheating on his wife. Amid the affair, someone gets murdered. Made for television.
1984 200m/C *GB* Anthony Hopkins, Ciaran Madden, Lise Hilboldt, Yvonne Coulette, John LeMesurier, Sophie Ashton; **Dir:** John Davies. **VHS, Beta** $69.95 *IVE* 🐾

Married to the Mob After the murder of her husband, an attractive Mafia widow tries to escape ''mob'' life, but ends up fighting off amorous advances from the current mob boss. A snappy script and a spray performance by Pfeiffer pepper this easily watched film. Continued Demme's streak of great films that continued with ''Silence of the Lambs.''
1988 (R) 102m/C Michelle Pfeiffer, Dean Stockwell, Alec Baldwin, Matthew Modine, Mercedes Ruehl, Anthony J. Nici, Joan Cusack, Ellen Foley, Chris Isaak, Jonathan Demme; **Dir:** Jonathan Demme. **VHS, Beta, LV** $19.98 *ORI* 🐾🐾🐾

Married Too Young High school honeys elope, despite disapproval from the parental units. When the boy groom trades his med school plans for a monkey wrench in order to support the little missus, he finds much trouble with the hot rod heavies.
1962 76m/B Harold Lloyd Jr., Jana Lund, Anthony Dexter, Marianna Hill, Trudy Marshall, Brian O'Hara, Nita Loveless; **Dir:** George Moskov. **VHS** $59.98 *HHT* 🐾

A Married Woman Dramatizes a day in the life of a woman who has both a husband and a lover. One of Godard's more mainstream efforts. French title is ''Une Femme Mariee.''
1965 94m/B *FR* Macha Meril, Philippe LeRoy, Bernard Noel; **Dir:** Jean-Luc Godard. **VHS, Beta, 8mm** $29.95 *NOS, MRV, DVT* 🐾🐾

Marry Me, Marry Me While in Paris, a Jewish encyclopedia salesman falls in love with a pregnant Belgian woman. A sensitive story about European Jewish families. Berri wrote, produced, directed, and starred in this romantic comedy.
1969 (R) 87m/C *FR* Elisabeth Wiener, Regine, Claude Berri, Louisa Colpeyn; **Dir:** Claude Berri. **VHS** *LHV* 🐾🐾🐾

The Marrying Man A young man meets his match when his buddies take him to Las Vegas for his bachelor party. He falls like a ton of bricks for the singer, not knowing she belongs to the local crime lord, who catches them together and forces them to marry. They immediately get a divorce, but can't forget each other. They eventually remarry, again and again and again. Silly story supposedly based on the story of Harry Karl (eventual husband of Debbie Reynolds) and Marie MacDonald. Although the off-stage romance between Basinger and Baldwin

created quite a stir, none of the chemistry is captured here. Ineffective, not funny, poorly paced, not well acted. A good example of what happens when egotistical stars get their way.
1991 (R) 116m/C Alec Baldwin, Kim Basinger, Robert Loggia, Armand Assante, Elizabeth Shue; *Dir:* Jerry Rees. **VHS, LV** $19.99 *TOU, HPH* ⚟½

Mars Needs Women When the Martian singles scene starts to drag, Mars boys cross the galaxy in search of fertile earth babes to help them repopulate the planet. Seems Batgirl Craig, the go-go dancing lady scientist, is at the top of their dance cards.
1966 80m/C Tommy Kirk, Yvonne Craig; *W/Dir:* Larry Buchanan. **VHS** $19.98 *SNC* ⚟

Marshal of Cedar Rock Marshal Rocky Lane sets a prisoner free, thinking the guy will lead him to a stash of stolen bank funds. Seems he miscalculates, but by way of consolation, he routs a rotten railroad agent who's rooking innocent people. Features the equine talent of Black Star.
1953 54m/B Allan "Rocky" Lane, Phyllis Coates, Roy Barcroft, William Henry, Robert Shayne, Eddy Waller; *Dir:* Harry Keller. **VHS** $19.99 *HEG, DVT, NOS* ⚟⚟

Marshal of Heldorado An outlaw-infested town needs an injection of law and order.
1950 53m/B James Ellison, Russell Hayden, Raymond Hatton, Al "Fuzzy" Knight; *Dir:* Thomas Carr. **VHS** $19.95 *NOS, DVT* ⚟

The Marshal's Daughter A father and daughter team up to outwit an outlaw. Features many cowboy songs, including the title track by Tex Ritter.
1953 71m/B Tex Ritter, Ken Murray, Laurie Anders, Preston Foster, Hoot Gibson; *Dir:* William Berke. **VHS** $19.95 *NOS, MOV* ⚟½

Martial Arts: The Chinese Masters This is a film featuring the arduous brick, log and slab splitting accomplishments of real Chinese martial arts experts.
198? 53m/C **VHS, Beta** $29.95 *PAV* ⚟½

Martial Law A film solely for martial arts fans. Two cops use their hands, feet, and other body parts to fight crime.
1990 (R) 90m/C Chad McQueen, Cynthia Rothrock, David Carradine, Andy McCutcheon; *Dir:* S.E. Cohen. **VHS** $89.98 *MED, FOX* ⚟½

Martial Law 2: Undercover Two cops, martial arts experts and part of an elite police force called Martial Law, go undercover to investigate the murder of a colleague. They uncover a fast-growing crime ring headed by a bad cop and a nightclub owner. The nightclub is host to the city's rich and powerful, who are treated to a bevy of beautiful women, protected by martial arts experts, and entertained by martial arts fights to the death. Lots of high-kicking action.
1991 (R) 92m/C Jeff Wincott, Cynthia Rothrock, Paul Johansson, Evan Lurie, L. Charles Taylor, Sherrie Rose, Billy Drago; *Dir:* Kurt Anderson. **VHS, LV** *MCA* ⚟½

Martial Monks of Shaolin Temple/Dragon Against Vampire Two full-length martial arts thrillers.
1983 171m/C **VHS, Beta** $7.00 *VTR* ⚟½

The Martian Chronicles: Part 1 Series episode "The Explorers." Adapted from Ray Bradbury's critically acclaimed novel. Futuristic explorations of the planet Mars. Strange fates of the discovery teams make everything more curious. Made for television.

1979 120m/C Rock Hudson, Bernie Casey, Nicholas Hammond, Darren McGavin; *Dir:* Michael Anderson Sr. **VHS, Beta** $14.95 *FRH* ⚟½

The Martian Chronicles: Part 2 Episode following the television movie. This part is called "The Settlers." The planet Mars meets with its first colonization and the settlers watch the Earth explode.
1979 97m/C Rock Hudson, Fritz Weaver, Roddy McDowall, Bernie Casey, Darren McGavin, Gayle Hunnicutt, Barry Morse, Bernadette Peters; *Dir:* Michael Anderson Sr. **VHS, Beta** $14.95 *FRH* ⚟½

The Martian Chronicles: Part 3 In the final chapter of this space saga, the Martian's secrets become known and will forever change man's destiny. Adapted from Ray Bradbury's classic novel; made-for-television.
1979 97m/C Rock Hudson, Bernadette Peters, Christopher Connelly, Fritz Weaver, Roddy McDowall, Bernie Casey, Nicholas Hammond, Darren McGavin, Gayle Hunnicutt, Barry Morse; *Dir:* Michael Anderson Sr. **VHS, Beta** $14.95 *FRH* ⚟⚟½

Martians Go Home! Joke-loving Martians come to earth and pester a nerdy composer.
1990 (PG-13) 89m/C Randy Quaid, Margaret Colin, Anita Morris, John Philbin, Ronny Cox, Gerrit Graham, Barry Sobel, Vic Dunlop; *Dir:* David Odell. **VHS, Beta, LV** $9.99 *CCB, IVE* ⚟½

Martin Martin is a charming young man, though slightly mad. He freely admits of the need to drink blood. Contemporary vampire has found a new abhorrent means of killing his victims.
1977 (R) 96m/C John Amplas, Lincoln Maazel; *Dir:* George A. Romero. **VHS, Beta, LV** $19.99 *HBO* ⚟⚟

Martin Luther French-made biography of the 16th century reformer who began the Protestant Reformation.
1953 105m/B Niall MacGinnis, John Ruddock, Pierre Leeavre, Guy Verney. **VHS, Beta** $19.95 *NOS, UHV, VGD* ⚟½

Martin Mull Live: From Ridgeville, Ohio Mull returns to his hometown and regales viewers with dry, sarcastic observations about small-town Middle American life.
1988 60m/C Teri Garr, Fred Willard, Jack Riley, Martin Mull. **VHS, Beta** $14.99 *HBO, TLF* ⚟½

Martin's Day An unusual friendship develops between an escaped convict and the young boy he kidnaps.
1985 (PG) 99m/C *CA* Richard Harris, Lindsay Wagner, James Coburn, Justin Henry, Karen Black, John Ireland; *Dir:* Alan Gibson. **VHS, Beta** $79.98 *FOX* ⚟⚟

Marty The empty life of a Bronx butcher is forever altered when he meets an equally lonely young woman. The original television version of the famous story, later filmed in 1955, starring Ernest Borgnine. Written by Paddy Chayefsky.
1953 58m/B Rod Steiger, Nancy Marchand; *Dir:* Delbert Mann. **VHS, Beta** $14.95 *WKV* ⚟

Marty Marty is a painfully shy bachelor who feels trapped in a pointless life of family squabbles. When he finds love, he also finds the strength to break out of what he feels is a meaningless existence. A sensitive and poignant film; written by Paddy Chayefsky, who went on to script new territory in "Altered States." Remake of a television version that originally aired in 1953. Notable for Borg-

nine's sensitive portrayal, one of his last quality jobs before sinking into the B-Movie sludge Pit.
1955 91m/B Ernest Borgnine, Betsy Blair, Joe De Santis, Ester Minciotti, Jerry Paris, Karen Steele; *Dir:* Delbert Mann. Academy Awards '55: Best Actor (Borgnine), Best Director (Mann), Best Picture, Best Screenplay; British Academy Awards '55: Best Actor (Borgnine), Best Actress (Blair); National Board of Review Awards '55: Best Actor (Borgnine), Best Film; New York Film Critics Awards '55: Best Actor (Borgnine), Best Picture. **VHS, Beta, LV** $19.98 *MGM, FOX, BTV* ⚟⚟⚟½

Marvelous Land of Oz L. Frank Baum's sequel to "The Wonderful Wizard of Oz" picks up the story of the Scarecrow and Tin Woodman after Dorothy returns home to Kansas. Original musical production was taped live especially for video. Don't confuse this yawner with the Judy Garland classic. Produced by the Minneapolis Children's Theatre Company and School.
1982 104m/C **VHS, Beta** $39.95 *MCA* ⚟½

Marvelous Stunts of Kung Fu A beautiful woman, a fortune teller and a martial arts hero team up to seek vengeance against the evil empire.
1983 86m/C Ling Yun, Wei Ping Line, Lung Fei. **VHS, Beta** $39.95 *TWE* ⚟

Marvin Gaye A look at the life and music of Motown superstar Marvin Gaye. Special appearance by Ashford and Simpson.
1989 60m/C *Hosted:* Smokey Robinson. **VHS, Beta, LV** $19.95 *MVD, FRH* ⚟

Marvin & Tige A deep friendship develops between an aging alcoholic and a streetwise 11-year-old boy after they meet one night in an Atlanta park.
1984 (PG) 104m/C John Cassavetes, Gibran Brown, Billy Dee Williams, Fay Hauser, Denise Nicholas-Hill; *Dir:* Eric Weston. **VHS, Beta** $24.98 *SUE* ⚟⚟½

Mary and Joseph: A Story of Faith A speculative look at the experiences and courtship of Mary and Joseph before the birth of Jesus. Made for television.
1979 100m/C Blanche Baker, Jeff East, Colleen Dewhurst, Stephen McHattie, Lloyd Bochner, Paul Hecht; *Dir:* Eric Till. **VHS, Beta** $59.95 *IVE* ⚟

Mary, Mary, Bloody Mary Young beautiful artist ravages Mexico with her penchant for drinking blood. Turns out she's a bisexual vampire. When even her friends become victims, her father steps in to end the bloodbath.
1976 (R) 85m/C Christina Ferrare, David Young, Helena Rojo, John Carradine; *Dir:* Carlos Lopez Moctezuma. **VHS, Beta** $59.98 *CGI* ⚟

Mary Poppins A magical English nanny arrives one day on the East Wind and takes over the household of a very proper London banker. She introduces her two charges to her friends and family, including Bert, the chimney sweep (Van Dyke), and eccentric Uncle Albert (Wynn). She also changes the lives of everyone in the family. From her they learn that life can always be happy and joyous if you take the proper perspective. Score includes "Supercalifragilistic-expialidocious," "Chim Chim Cheree," and "A Spoonful of Sugar." Film debut of Andrews. Based on the books by P.L. Travers. A Disney classic that hasn't lost any of its magic. Oscar nominations for Best Picture, Best Director, Adapted Screenplay, Color Cinematography, Art Direction, Set Decora-

tion, Sound and Costume Design, in addition to the awards it won.
1964 139m/C Ed Wynn, Hermione Baddeley, Julie Andrews, Dick Van Dyke, David Tomlinson, Glynis Johns; *Dir:* Robert Stevenson. Academy Awards '64: Best Actress (Andrews), Best Film Editing, Best Song ("Chim Chim Cher-ee"), Best Visual Effects, Best Musical Score. **VHS, Beta, LV $24.99** *DIS, APD, BTV* ✓✓✓ ½

Mary of Scotland The historical tragedy of Mary, Queen of Scots and her cousin, Queen Elizabeth I of England is enacted in this classic film. Traces Mary's claims to the throne of England which ultimately led to her execution. Based on the Maxwell Anderson play.
1936 123m/B Katharine Hepburn, Fredric March, Florence Eldridge, Douglas Walton, John Carradine, Robert Barrat, Gavin Muir, Ian Keith, Moroni Olson, William Stack, Alan Mowbray; *Dir:* John Ford. **VHS, Beta, LV** *RKO* ✓✓✓

Mary White The true story of Mary White, the sixteen-year-old daughter of a newspaper editor who rejects her life of wealth and sets out to find her own identity.
1977 102m/C Ed Flanders, Kathleen Beller, Tim Matheson, Donald Moffatt, Fionnula Flanagan; *Dir:* Judd Taylor. **VHS, Beta $29.95** *PAR* ✓✓

Masada Based on Ernest K. Gann's novel "The Antagonists," this dramatization recreates the first-century A.D. Roman siege of the fortress Masada, headquarters for a group of Jewish freedom fighters. Abridged from the original television presentation.
1981 131m/C Peter O'Toole, Peter Strauss, Barbara Carrera, Anthony Quayle, Giulia Pagano, David Warner; *Dir:* Boris Sagal. **VHS, Beta $59.95** *MCA* ✓✓✓

Mascara A group inspecting a transvestite's death are led into the seedy underground world of Belgian nightlife.
1987 (R) 99m/C *BE* Derek De Lint, Charlotte Rampling, Michael Sarrazin; *Dir:* Patrick Conrad. **VHS, Beta $79.95** *WAR* ✓✓

Masculine Feminine A young Parisian just out of the Army engages in some anarchistic activities when he has an affair with a radical woman singer. Hailed as one of the best French New Wave films. In French with English subtitles.
1966 103m/B *FR* Jean-Pierre Leaud, Chantal Goya, Marlene Jobert; *Dir:* Jean-Luc Godard. Berlin Film Festival '66: Best Actor (Leaud); New York Film Festival '66: Best Actor. **VHS, Beta $69.95** *NYF, DVT, VDM* ✓✓✓ ½

M*A*S*H Hilarious, irreverent, and well-cast black comedy about a group of surgeons and nurses at a Mobile Army Surgical Hospital in Korea. The horror of war is set in counterpoint to their need to create havoc with episodic late-night parties, practical jokes, and sexual antics. An all-out anti-war festival, highlighted by scenes that starkly uncover the chaos and irony of war. Altman establishes his influential style here, using the deliberate fragmentation of images and overlapping of sound and dialogue. Ring Lardner, Jr., won an Oscar for his screenplay, loosely adapted from the novel by the pseudonymous Richard Hooker. Subsequent hit TV series moved even further from the source novel. Oscar nominations for Best Picture, supporting actress Kellerman, director, and film editing.
1970 (R) 116m/C Donald Sutherland, Elliott Gould, Tom Skerritt, Sally Kellerman, JoAnn Pflug, Robert Duvall, Rene Auberjonois, Roger Bowen, Gary Burghoff, Fred Williamson, John Schuck, Bud Cort, G. Wood; *Dir:* Robert Altman. Academy Awards '70: Best Adapted Screenplay;

Golden Globe Awards '71: Best Film—Musical/Comedy. **VHS, Beta, LV $19.98** *FOX* ✓✓✓✓

M*A*S*H: Goodbye, Farewell & Amen The final two-hour special episode of the television series "M*A*S*H" follows Hawkeye, B.J., Colonel Potter, Charles, Margaret, Klinger, Father Mulcahy, and the rest of the men and women of the 4077th through the last days of the Korean War, the declaration of peace, the dismantling of the camp, and the fond and tearful farewells.
1983 120m/C Alan Alda, Mike Farrell, Harry Morgan, David Ogden Stiers, Loretta Swit, Jamie Farr, William Christopher. **VHS, Beta, LV** *FOX, OM* ✓✓✓ ½

The Mask A deservedly obscure gory horror film about a masked killer, filmed mostly in 3-D. With special 3-D packaging and limited edition 3-D glasses.
1961 85m/B *CA* Paul Stevens, Claudette Nevins, Bill Walker, Anne Collings, Martin Lavut; *Dir:* Julian Roffman. **VHS, Beta, LV $12.95** *RHI, MLB* ✓✓

Mask A dramatization of the true story of a young boy afflicted with craniodiaphyseal dysplasia (Elephantitis). The boy overcomes his appearance and revels in the joys of life in the California bikers' community. Well acted, particularly the performances of Stoltz and Cher. A touching film, well-directed by Bogdanovich, that only occasionally slips into maudlin territory.
1985 (PG-13) 120m/C Cher, Sam Elliott, Eric Stoltz, Estelle Getty, Richard Dysart, Laura Dern; *Dir:* Peter Bogdanovich. Cannes Film Festival '85: Best Actress (Cher). **VHS, Beta, LV $19.95** *MCA* ✓✓✓

Mask of the Dragon A soldier's friend and girlfriend track down his killer after he delivers a golden curio to a shop in Los Angeles.
1951 54m/B Richard Travis, Sheila Ryan, Richard Emory, Jack Reitzen. **VHS, Beta, LV** *WGE* ✓

The Masked Marvel The Masked Marvel saves America's war industries from sabotage. Serial in twelve episodes.
1943 195m/B William Forrest, Louise Currie, Johnny Arthur; *Dir:* Spencer Gordon Bennet. **VHS $29.98** *VCN, REP, MLB* ✓✓

Masks of Death New adventure for Sherlock Holmes, as he is pulled from retirement to find the murderer of three unidentified corpses found in London's East End.
1986 80m/C Peter Cushing, John Mills, Anne Baxter, Ray Milland; *Dir:* Roy Ward Baker. **VHS, Beta, LV $19.98** *LHV, IME, WAR* ✓✓✓

Mason of the Mounted Harmless enough western in which a Canadian Mountie tracks down a murderer in the U.S.
1932 58m/B Bill Cody, Nancy Drexel, Art Smith; *Dir:* Trem Carr. **VHS $19.95** *NOS, DVT* ✓✓

Masque of the Red Death An integral selection in the famous Edgar Allan Poe/Roger Corman canon, it deals with an evil prince who traffics with the devil and playfully murders any of his subjects not already dead of the plague. Photographed by Nicholas Roeg. Remade in 1989 with Corman as producer.
1965 88m/C *GB* Vincent Price, Hazel Court, Jane Asher, Patrick Magee; *Dir:* Roger Corman. **VHS, Beta $59.98** *LIV, LTG, MLB* ✓✓✓

Masque of the Red Death Roger Corman's second attempt at Edgar Allan Poe's horror tale pales compared to his Vincent Price version made 25 years earlier. Under-aged cast adds youth appeal but

subtracts credibility from the fable of a sadistic prince and his sycophants trying to ignore the plague outside castle walls. Only late in the plot does veteran actor Macnee add proper note of doom.
1989 (R) 90m/C Patrick Macnee, Jeffery Osterhage, Adrian Paul, Tracy Reiner, Maria Ford; *Dir:* Larry Brand. **VHS, Beta, LV $79.95** *MGM, IME* ✓✓

Masque of the Red Death Unrecognizable Poe mutation has guests invited to the mansion of a dying millionaire, only to be murdered by an unknown stalker. One scene features a pendulum and that's it for literary faithfulness.
1990 (R) 94m/C Frank Stallone, Brenda Vaccaro, Herbert Lom, Michelle McBride, Christine Lunde; *Dir:* Alan Birkinshaw. **VHS, LV** *COL, PIA* ✓

Masque of the Red Death/Premature Burial Double feature of horror classics adapted from Edgar Allan Poe's work. Price stars in "Masque of the Red Death" as a satan worshiper. Millard plays a man obsessed with the fear of catastrophe in "Premature Burial."
1965 175m/C Vincent Price, Hazel Court, Jane Asher, Patrick Magee, Nigel Green, Ray Milland; *Dir:* Roger Corman. **VHS, LV $44.95** *IME* ✓

Masquerade A lonely young heiress meets a handsome "nobody" with a mysterious background and it is love at first sight. The romance distresses everyone in the circle of the elite because they assume that he is after her money and not her love. At first it seems decidedly so, then definitely not, and then nothing is certain. A real romantic thriller, with wonderful scenes of the Hamptons.
1988 (R) 91m/C Rob Lowe, Meg Tilly, John Glover, Kim Cattrall, Doug Savant, Dana Delaney; *Dir:* Bob Swaim. **VHS, Beta, LV $19.98** *FOX* ✓✓✓

Mass Appeal An adaptation of the Bill C. Davis play about the ideological debate between a young seminarian and a complacent but successful parish pastor. Lemmon has had better roles and done better acting.
1984 (PG) 99m/C Jack Lemmon, Zeljko Ivanek, Charles Durning, Louise Latham, James Ray, Sharee Gregory, Talia Balsam; *Dir:* Glenn Jordan. **VHS, Beta, LV $19.95** *MCA* ✓✓

Massacre at Central High A new student takes matters into his own hands when gang members harass other students at a local high school. Other than some silly dialogue, this low-budget production is above average.
1976 (R) 85m/C Derrel Maury, Andrew Stevens, Kimberly Beck, Robert Carradine, Roy Underwood, Steve Bond; *W/Dir:* Renee Daalder. **VHS $79.95** *MPI* ✓✓ ½

Massacre in Dinosaur Valley A dashing young paleontologist and his fellow explorers go on a perilous journey down the Amazon in search of the Valley of the Dinosaur.
1985 98m/C Michael Sopkiw, Suzanne Carvall. **VHS, Beta $69.98** *LIV, LTG* ✓ ½

Massacre in Rome A priest opposes a Nazi colonel's plan to execute Italian civilians in retaliation for the deaths of 33 German soldiers. Strong drama based on a real event.
1973 (PG) 110m/C Richard Burton, Marcello Mastroianni, Leo McKern, John Steiner, Anthony Steel; *Dir:* George P. Cosmatos. **VHS, Beta $29.99** *HBO* ✓✓ ½

Massive Retaliation Hordes of pesky villagers seek refuge within the secluded safety of a family's country house as World War III approaches.
1985 90m/C Tom Bower, Peter Donat, Karlene Crockett, Jason Gedrick, Michael Pritchard; *Dir:* Thomas A. Cohen. **VHS, Beta, LV $19.98** *LIV, VES* ✴

The Master of Ballantrae Flynn plays James Durrisdear, the heir to a Scottish title, who gets involved in a rebellion with Bonnie Prince Charlie against the English crown. When the rebellion fails, Flynn heads for the West Indies where he and his partner amass quite a fortune through piracy. Flynn eventually returns to Scotland where he finds that his brother has taken over his title as well as his longtime love. Based on the novel by Robert Louis Stevenson. Flynn's riotous life had put him long past his peak swash-buckling days, as this film unfortunately demonstrates.
1953 89m/C Errol Flynn, Roger Livesey, Anthony Steel, Beatrice Campbell, Yvonne Furneaux, Jacques Berthier, Felix Aylmer, Mervyn Johns; *Dir:* William Keighley. **VHS, Beta, LV $59.99** *WAR* ✴✴½

Master Blaster Friendly game of survival with paintball guns goes awry when one of the contestants exchanges the play guns for deadly weapons.
1985 (R) 94m/C Jeff Moldovan, Donna Rosae, Joe Hess, Peter Lunblad; *Dir:* Glenn Wilder. **VHS** *PSM* ✴½

Master Class Behind the scenes look at martial arts stunt-filming hosted by one of the genres' stars.
1985 60m/C Sho Kosugi; *Dir:* David Hilmer. **VHS, Beta $39.95** *IVE* ✴½

Master of Death A young man trains for 18 years in the martial arts to avenge his parents' murder.
19?? 90m/C VHS, Beta $29.95 *OCE* ✴

Master of Dragonard Hill A low-rent swashbuckling romance-novel pastiche.
1989 92m/C Oliver Reed, Eartha Kitt. **VHS, Beta $79.95** *MED* ✴½

Master Harold and the Boys Stagey cable presentation of South African Athol Fugard's play about relationship between white man and two black servants. Occasionally provocative.
1984 90m/C Matthew Broderick; *Dir:* Michael Lindsay-Hogg. **VHS, Beta $59.95** *LHV, WAR* ✴✴✴

Master of the House Also known as "Thou Shalt Honour Thy Wife," this program is the story of a spoiled husband, a type extinct in this country but still in existence abroad. Silent with titles in English.
1925 118m/B *Dir:* Carl Theodor Dreyer. **VHS, Beta $49.95** *FCT* ✴✴✴

Master Key Features 13 complete chapters of this action-adventure serial.
1944 169m/B Jane Wiley, Milburn Stone, Lash LaRue, Dennis Moore. **VHS, Beta $49.95** *NOS, DVT, CPB* ✴½

Master Killer (The Thirty-Sixth Chamber) Master Killers are made, not born. Liu trains as he readies himself for an attack on the brutal Manchu assassins.
1984 (R) 109m/C Liu Chia Hui; *Dir:* Liu Chia Liang. **VHS, Beta $9.98** *SUE* ✴

The Master Killers Two brothers who hate each other's guts try to kill each other for their dead father's money.

1973 95m/C Donald O'Brien; *Dir:* George Bonge. **VHS, Beta $69.95** *LTG* ✴

Master of Kung Fu A Kung Fu instructor calls upon his former teacher to help him fend off loan sharks who want to take over his school.
1977 88m/C Yu Chan Yuan, Chi Hsiao Fu. **VHS, Beta $59.98** *FOX* ✴

Master Mind A renowned Japanese super sleuth attempts to solve the theft of a sophisticated midget android.
1973 86m/C Zero Mostel, Keiko Kishi, Bradford Dillman, Herbert Berghof, Frankie Sakai; *Dir:* Alex March. **VHS, Beta $9.95** *UNI* ✴✴

Master Ninja The continuing adventures of the Master and Max follow the pair across the US in their search for the Master's long-lost daughter. Each tape includes two programs.
1978 93m/C Lee Van Cleef, Sho Kosugi, Timothy Van Patten. **VHS, Beta $9.99** *QVD, TWE* ✴

Master Ninja 2 The Master and his assistant Max travel across the United States to search for his long lost daughter in these two episodes from the series.
1983 92m/C Sho Kosugi, Lee Van Cleef, Timothy Van Patten, David McCallum, Cotter Smith. **VHS, Beta $39.95** *TWE* ✴

Master Ninja 3 The Master and Max are back again in two new adventures. In "Fat Tuesday," the duo chop-sock a gun-running operation in New Orleans. In "Jog-gernaut," they help a mother and daughter get their produce to market.
1983 93m/C Sho Kosugi, Lee Van Cleef, Timothy Van Patten, Diana Muldaur, Mabel King. **VHS, Beta $39.98** *TWE* ✴

Master Ninja 4 The Master and Max travel to New York and Las Vegas seeking his long lost daughter in this collection of two episodes from "The Master" series.
1983 93m/C Lee Van Cleef, Timothy Van Patten, Sho Kosugi, George Maharis. **VHS, Beta $39.95** *TWE* ✴

Master Ninja 5 Two more episodes from the series: In "Kunoichi," a female martial arts assassin stalks the Master; "The Java Tiger" features a deadly race to find buried treasure.
1983 92m/C Sho Kosugi, Lee Van Cleef, Timothy Van Patten. **VHS, Beta $39.95** *TWE* ✴

Master Race Absorbing cautionary tale of a German officer who escapes retribution when the Nazis collapse, continuing to hold control over the inhabitants of a small town through intimidation.
1944 98m/B George Coulouris, Lloyd Bridges, Osa Massen, Nancy Gates, Stanley Ridges; *Dir:* Herbert Biberman. **VHS, Beta $29.95** *RKO* ✴✴½

Master Touch When a legendary safe-cracker is released from prison, he attempts one last heist at a Hamburg insurance company.
1974 (PG) 96m/C *IT GE* Kirk Douglas, Florinda Bolkan, Giuliano Gemma; *Dir:* Michele Lupo. **VHS, Beta $59.95** *PSM* ✴½

Master of the World Visionary tale of a fanatical 19th-century inventor who uses his wonderous flying fortress as an antiwar weapon.
1961 95m/C Vincent Price, Charles Bronson, Henry Hull; *Dir:* William Witney. **VHS, Beta $19.98** *WAR, OM* ✴✴½

Master of the World Animated version of the famed Jules Verne classic about a mad scientist who endeavors to take over the world. In HiFi.
1976 50m/C VHS, Beta $19.95 *MGM* ✴✴½

Masters of Comedy Here are some classic short subjects featuring Hope ("Bob's Busy Day"), Kaye ("Birth of a Star"), Laurel & Hardy ("A Day at the Studio"), and Fields ("The Barbershop" and "The Fatal Glass of Beer").
193? 90m/B Bob Hope, Danny Kaye, W.C. Fields, Stan Laurel, Oliver Hardy. **VHS, Beta $19.95** *NO* ✴✴½

The Masters of Menace The men in blue are mighty miffed when a bunch of bikers break parole to pay their last respects to a comrade in leather. Comic bits by Candy, Belushi, Aykroyd and Wendt provide little relief.
1990 (PG-13) 97m/C David Rasche, Catherine Bach, Dan Aykroyd, James Belushi, John Candy, George Wendt, Tino Insana; *Dir:* Daniel Raskov. **VHS, LV $89.95** *COL* ✴½

Master's Revenge When a girl is kidnapped by a violent motorcycle gang, martial artists are called in to rescue her. Originally released as, "Devil Rider."
1971 78m/C Sharon Mahon, Ridgely Abele, Johnny Pachivas. **VHS $19.95** *ACA* ✴½

Masters of Tap This history of tap dancing, presented by three masters, leads from taps heyday in the 1930s to the renewed interest today. Also shown are parts of basic tap lessons. Some segments in black and white.
1988 61m/C Charles "Honi" Coles, Chuck Green, Will Gaines. **VHS, Beta $39.95** *MVD, HMV* ✴½

Masters of the Universe A big-budget live-action version of the cartoon character's adventures, with He-Man battling Skeletor for the sake of the universe.
1987 (PG) 109m/C Dolph Lundgren, Frank Langella, Billy Barty, Courtney Cox, Meg Foster; *Dir:* Gary Goddard. **VHS, Beta, LV $19.98** *WAR* ✴✴

Masters of Venus Eight-part serial about spaceships, space maidens and the like.
1959 121m/B Ferdinand "Ferdy" Mayne. **VHS, Beta $24.95** *SNC, VDM* ✴½

Mata Hari During World War I, a lovely German spy steals secrets from the French through her involvement with two military officers. Lavish production and exquisite direction truly make this one of Garbo's best. Watch for her exotic pseudo-strip tease.
1932 90m/B Greta Garbo, Ramon Novarro, Lionel Barrymore, Lewis Stone, C. Henry Gordon, Karen Morley, Alec B. Francis; *Dir:* George Fitzmaurice. **VHS, Beta $19.98** *MGM, FCT* ✴✴✴

Mata Hari The racy story of World War I's most notorious spy, Mata Hari, who uses her seductive beauty to seduce the leaders of Europe.
1985 (R) 105m/C Sylvia Kristel, Christopher Cazenove, Oliver Tobias, Gaye Brown, Gottfried John; *Dir:* Curtis Cunningham. **VHS, Beta $79.95** *MGM* ✴

Matador Bizarre, entertaining black comedy about a retired matador who satiates his desire to kill by appearing in snuff films. Not for all tastes, but fine for those who like the outrageous. In Spanish with English subtitles.

1986 90m/C *SP* Assumpta Serna, Antonio Banderas, Nacho Martinez, Eva Cobo, Carmen Maura, Julieta Serrano, Chus Lampreave; *Dir:* Pedro Almodovar. VHS, Beta $79.95 CCN 🐾🐾½

The Matchmaker An adaptation of the Thornton Wilder play concerning two young men in search of romance in 1884 New York. Later adapted as "Hello Dolly." An amusing diversion.
1958 101m/B Shirley Booth, Anthony Perkins, Shirley MacLaine, Paul Ford, Robert Morse, Perry Wilson, Wallace Ford, Russell Collins, Rex Evans, Gavin Gordon, Torben Meyer; *Dir:* Joseph Anthony. VHS, Beta, LV $19.95 PAR, PIA, FCT 🐾🐾🐾

Matewan An acclaimed dramatization of the famous Matewan massacre in the 1920s, in which coal miners in West Virginia, reluctantly influenced by a young union organizer, rebelled against terrible working conditions. Complex and imbued with myth, the film is a gritty, moving, powerful drama with typically superb Sayles dialogue and Haskell Wexler's beautiful and poetic cinematography. Jones delivers an economical yet intense portrayal of the black leader of the miners. The director makes his usual on-screen appearance, this time as an establishment-backed reactionary minister. Partially based on the Sayles novel "Union Dues."
1987 (PG-13) 130m/C Chris Cooper, James Earl Jones, Mary McDonnell, William Oldham, Kevin Tighe, David Straithairn; *Dir:* John Sayles. VHS, Beta, LV $79.95 LHV, WAR 🐾🐾🐾½

Matilda An entrepreneur decides to manage a boxing kangaroo, which nearly succeeds in defeating the world heavyweight champion.
1978 (PG) 103m/C Elliott Gould, Robert Mitchum, Harry Guardino, Clive Revill; *Dir:* Daniel Mann. VHS, Beta $69.98 LIV, VES 🐾

Matinee at the Bijou, Vol. 1 This volume in a series of eight contains the cartoon "Little Red Hen," the shorts "Candy For Your Health," and "Fox Movietone Newsreel," and the feature film "Fatal Glass of Beer."
19?? 90m/B VHS $14.99 HEG

Matinee at the Bijou, Vol. 2 This volume in a series of eight contains the cartoon "A Waif's Welcome," the short "Dangerous Females," and the feature "Little Men."
19?? 90m/B VHS $14.99 HEG

Matinee at the Bijou, Vol. 3 This volume in a series of eight has the cartoon "Magic Mummy," short "Desert Demons," and feature "One Frightened Night."
19?? 90m/B VHS $14.99 HEG

Matinee at the Bijou, Vol. 4 This volume in a series of eight contains the cartoon "The Pincushion Man," the short "Famous People at Play," and the feature "Ladies Crave Excitement."
19?? 90m/B VHS $14.99 HEG

Matinee at the Bijou, Vol. 5 This volume in a series of eight contains the cartoon "A Little Bird Told Me;" short "Star Reporter;" and feature "I'd Give My Life."
19?? 90m/B VHS $14.99 HEG

Matinee at the Bijou, Vol. 6 This volume in a series of eight contains the cartoon "Song a Day," the short "Musical Charmers," and the feature "Sing While You're Able."
19?? 90m/B VHS $14.99 HEG

Matinee at the Bijou, Vol. 7 This volume in a series of eight contains the cartoon "Cobweb Hotel," the short "Hotel Anchovy," and the feature "Young Dynamite."
19?? 90m/B VHS $14.99 HEG

Matinee at the Bijou, Vol. 8 This volume in a series of eight contains the cartoon "On with the New," the short "The Truth about Taxes," and the feature "Remedy for Riches."
19?? 90m/B VHS $14.99 HEG

Mating Season A successful female attorney meets a good-natured laundromat owner at a bird-watching retreat. Made for television.
1981 96m/C Lucie Arnaz, Laurence Luckinbill; *Dir:* John Llewellyn Moxey. Paris International Art Film Festival '81: Grand Prix. VHS, Beta $14.95 LCA 🐾½

Matrimaniac A man goes to great lengths to marry a woman against her father's wishes. Silent with music score.
1916 48m/B Douglas Fairbanks Sr., Constance Talmadge. VHS, Beta $29.95 VYY, GPV, DNB 🐾🐾

Matrimonio a la Argentina Spanish infidelity farce.
198? 96m/C *SP* Rodolfo Beban, Jorge Barreiro, Mercedes Carreras. VHS, Beta $59.95 UNI 🐾

A Matter of Degrees Weeks before his graduation a beatnik-type college senior finally acquires a goal: romancing a mystery girl...or maybe saving the student radio station...or not. The hero's aimlessness permeates the script, which never goes anywhere in its exploration of campus ennui. Good photography and a great stratum of alternative music groups on the soundtrack: Dream Syndicate, Pere Ubu, Schooly D, Pixies, Poi Dog Pondering, Minutemen and Throwing Muses. John Doe, John F. Kennedy, Jr. and Fred Schneider and Kate Pierson of the B-52's make guest cameos.
1990 (R) 89m/C Arye Gross, Judith Hoag, Tom Sizemore, John Doe; *Dir:* W.T. Morgan. VHS, Beta $89.95 PSM 🐾🐾

A Matter of Dignity Young girl, forced into marriage with millionaire, eventually brings ruin to the entire family. Why? In Greek with English subtitles.
1957 104m/B *GR* Georges Pappas, Ellie Lambetti, Athena Michaelidou, Eleni Zafirou; *Dir:* Michael Cacoyannis. VHS, Beta $59.95 FCT 🐾🐾

Matter of Heart Via home movies, interviews and archival footage, a detailed, sympathetic portrait of the life, career and work of Carl G. Jung. Highly acclaimed.
1985 107m/C *Dir:* Mark Whitney. VHS, Beta, LV $79.95 KIV 🐾🐾

A Matter of Life and Death True story follows the real-life experiences of nurse Joy Ufema who has devoted her life to helping terminally ill patients. Exceptional work from Lavin. Made for television.
1981 98m/C Linda Lavin, Salome Jens, Gail Strickland, Gerald S. O'Loughlin, Ramon Bieri, Tyne Daly, Larry Breeding, John Bennett Perry; *Dir:* Russ Mayberry. VHS $59.95 USA 🐾🐾

A Matter of Love Two couples indulge in spouse swapping while vacationing at the beach.
1978 (R) 89m/C Michelle Harris, Mark Anderson, Christy Neal, Jeff Alin; *Dir:* Chuck Vincent. VHS, Beta $69.98 LIV, VES 🐾

A Matter of Principle The perennial Christmas special about a Scrooge-like character's yuletide change of heart. He decides it's time to shape up when he realizes he may lose his family permanently.
197? 60m/C Alan Arkin, Barbara Dana, Tony Arkin; *Dir:* Gwen Arner. VHS, Beta $19.95 ACA 🐾

A Matter of Time A maid is taught to enjoy life by an eccentric, flamboyant contessa, then finds the determination to become an aspiring actress. Often depressing and uneven, arguably Minnelli's worst directing job. Director Minnelli and Boyer's last film; first film for Bergman's daughter, Isabella Rosellini, in a small part as a nun.
1976 (PG) 97m/C Liza Minnelli, Ingrid Bergman, Charles Boyer, Tina Aumont, Spiros Andros, Anna Proclemer; *Dir:* Vincente Minnelli. VHS, Beta $69.95 VES 🐾

A Matter of WHO A detective for the World Health Organization, or WHO, investigates the disease related deaths of several oil men. Travelling to the Middle East, he uncovers a plot by an unscrupulous businessman to control the oil industry by killing off its most powerful members. Although intended as a comedy, the subject matter is too grim to be taken lightly.
1962 92m/B *GB* Terry-Thomas, Alex Nicol, Sonja Ziemann, Richard Briers, Clive Morton, Vincent Ball, Honor Blackman, Carol White, Martin Benson, Geoffrey Keen; *Dir:* Don Chaffey. VHS, Beta $16.95 SNC 🐾🐾

Matters of the Heart Young pianist searches for an opportunity to display his talents, much to the disapproval of his veteran father. He meets a successful but embittered musician who takes him under her wing and passion between the two soon flares. Will the young man find the success he craves and if so, at what cost? Made for cable TV and based on "The Country of the Heart" by Barbara Wershba.
1990 94m/C Jane Seymour, Christopher Gartin, James Stacy, Geoffrey Lewis, Nan Martin, Allen Rich, Clifford David, Katherine Cannon; *Dir:* Michael Rhodes. VHS $89.95 MCA 🐾🐾

Maurice Based on E.M. Forster's novel about a pair of Edwardian-era Cambridge undergraduates who fall in love, but must deny their attraction and abide by British society's strict norms regarding homosexuality. Maurice finds, however, that he cannot deny his nature, and must come to a decision regarding family, friends, and social structures. A beautiful and stately film of struggle and courage.
1987 (R) 139m/C James Wilby, Hugh Grant, Rupert Graves, Mark Tandy, Ben Kingsley, Denholm Elliott, Simon Callow, Judy Parfitt, Helena Bonham Carter; *Dir:* James Ivory. VHS, Beta, LV $79.95 LHV, IME, WAR 🐾🐾🐾

Mausoleum Only one man can save a woman from eternal damnation.
1983 (R) 96m/C Marjoe Gortner, Bobbie Breese, Norman Burton, LaWanda Page, Shari Mann; *Dir:* Michael Dugan. VHS, Beta, LV $14.98 SUE 🐾🐾

Maverick Queen Barbara Stanwyck, owner of a gambling casino and a member of the "Wild Bunch" outlaw gang, is torn between going straight for the love of a lawman or sticking with the criminals. Stanwyck is perfectly cast in this interesting Western.
1955 90m/C Barbara Stanwyck, Barry Sullivan, Wallace Ford, Scott Brady, Jim Davis, Mary Murphy; *Dir:* Joseph Kane. VHS $14.95 REP 🐾🐾½

Max Compilation of hilarious Linder shorts plus two features, "Seven Years Bad Luck" and "The Three Must-Get-Theres"
1922 90m/B Max Linder. **VHS, Beta $49.50** *GVV* 🦴🦴½

Max A night watchman watches an aspiring actress who is working on her performance in a Broadway play. He tells her she would be better off marrying a doctor.
1979 20m/C Jack Gilford, Lynn Lipton; ***Dir:*** Joseph Gilford, Jennifer Lax. **VHS, Beta** *CFV* 🦴🦴

Max Dugan Returns A Simon comedy about an ex-con trying to make up with his daughter by showering her with presents bought with stolen money. Sweet and light, with a good cast.
1983 (PG) 98m/C Jason Robards Jr., Marsha Mason, Donald Sutherland, Matthew Broderick, Kiefer Sutherland; ***Dir:*** Herbert Ross. **VHS, Beta, LV $59.98** *FOX* 🦴🦴

Max Fleischer's Documentary Features Two scientific films, "Evolution" and "Einstein's Theory of Relativity," from Fleischer's early studio were made to finance their cartoons. Features animated sequences.
1923 60m/B ***Dir:*** Max Fleischer. **VHS, Beta** *GVV* 🦴🦴

Max Fleischer's Popeye Cartoons Compilation of three classic color Popeye two-reelers from the late 30's, wherein Popeye meets Ali Baba, Aladdin and Sindbad the the Sailor.
1939 56m/C ***Dir:*** Max Fleischer. **VHS, Beta, LV $14.95** *REP* 🦴🦴

Max Headroom, the Original Story The computer-generated, British Channel-4 comedy personality stars in his first film, actually a shot-on-video feature about an innocent reporter who is transformed by an evil futuristic TV station into Headroom, the perfect (read: artificial) talk show host.
1986 60m/C *GB* Matt Frewer, Nickolas Grace, Hilary Tindall, Amanda Pays, Rocky Morton, Annabel Jankel. **VHS, Beta, LV $29.95** *LHV, WAR* 🦴🦴½

Max and Helen Uneven, but sensitive and at times highly moving made for TV story of two lovers who are victims of the holocaust. Max (Williams) survives both Nazi and Stalinist camps out of both love for his fiancee and guilt for having lived while she did not.
1990 94m/C *GB* Treat Williams, Alice Krige, Martin Landau, Jodhi May, John Phillips; ***Dir:*** Philip Saville. **VHS, Beta $9.98** *TTC* 🦴🦴½

Max Maven's Mindgames Magician Max Maven performs mindgames with the use of playing cards, signs and symbols, and magic with money in such settings as Las Vegas, a tropical jungle, the moon and an operating room.
1984 56m/C VHS, Beta $29.95 *MCA* 🦴🦴

Maxie Highly predictable, and forgettable, comedy where a ghost of a flamboyant flapper inhabits the body of a modern-day secretary, and her husband is both delighted and befuddled with the transformations in his spouse. Close is okay, but the film is pretty flaky.
1985 (PG) 98m/C Glenn Close, Ruth Gordon, Mandy Patinkin, Barnard Hughes, Valerie Curtin, Harry Hamlin; ***Dir:*** Paul Aaron. **VHS, Beta, LV $19.99** *HBO* 🦴🦴

Maxim Xul A professor of the occult is forced to tangle with a beast from Hell that possesses enormous strength and an insatiable appetite for human blood.
1991 90m/C Adam West, Jefferson Leinberger, Hal Strieb, Mary Schaeffer; ***Dir:*** Arthur Egeli. **VHS $79.98** *MAG* 🦴🦴

Maximum Breakout A beautiful and wealthy girl is kidnapped, her boyfriend left for dead. But he recovers and leads a posse of mercenaries to the rescue.
1991 93m/C Sydney Coale Phillips; ***Dir:*** Tracy Lynch Britton. **VHS $56.96** *AIP* 🦴½

Maximum Overdrive Based upon King's story "Trucks," recounts what happens when a meteor hits Earth and machines run by themselves, wanting only to kill people. Score by AC/DC.
1986 (R) 97m/C Emilio Estevez, Pat Hingle, Laura Harrington, Christopher Murvey, Yeardley Smith; ***Dir:*** Stephen King. **VHS, Beta, LV $19.98** *LHV, WAR* 🦴

Maximum Security A small-budget prison film detailing the tribulations of a model prisoner struggling to resist mental collapse.
1987 113m/C Geoffrey Lewis, Jean Smart, Robert Desiderio; ***Dir:*** Bill Duke. **VHS, LV $19.95** *STE, NWV* 🦴

Maximum Thrust A few white men confront a deadly Caribbean voodoo tribe.
1987 80m/C Rick Gianasi, Joe Derrig, Jennifer Kanter, Mizan Nunes. **Beta $69.95** *UCV Woof!*

Maxwell Street Blues Sketches the historical background of the Maxwell Street open-air market, where nearly every important blues musician in Chicago has performed.
1989 56m/C Blind Arvella Gray, Carrie Robinson, Jim Brewer, John Henry Davis, Playboy Venson, Floyd Jones, Robert Nighthawk. **VHS, Beta $29.95** *MVD, KIV Woof!*

May Fools Malle portrays individuals collectively experiencing personal upheaval against the backdrop of unrelated social upheaval. An upper crusty family gathers at a country estate for the funeral of the clan's matriarch, while the May of '68 Parisian riots unfold. Few among the family members mourn the woman's passing, save her son Milou (Piccoli) who leads a pastoral existence tending grapes on the estate. Milou's daughter (Miou-Miou), like the others, is more concerned with her personal gain, suggesting, to her father's horror, that they divide the estate in three. Touching, slow, keenly observed. Malle cowrote with Jean Claude Carriere; music by Stephane Grappelli. French with English subtitles.
1990 (R) 105m/C *FR* Michel Piccoli, Miou-Miou, Michael Duchaussoy; ***Dir:*** Louis Malle. **VHS $79.95** *ORI, FXL, FCT* 🦴🦴🦴

May Wine A sexy, romantic comedy starring "Twin Peaks" alumna, Boyle.
1990 (R) 85m/C Guy Marchand, Lara Flynn Boyle, Joanna Cassidy. **VHS $89.98** *MED* 🦴🦴

Maya A teacher of fashion in a high school becomes the object of a student's devotion and another teacher's insane jealousy.
1982 114m/C Berta Dominguez, Joseph D. Rosevich; ***Dir:*** Agust Agustsson. **VHS, Beta $39.95** *USA Woof!*

Mayalunta A young artist is emotionally tortured by a dying older couple.
1986 90m/C *SP* Federico Luppi, Miguel Angel Sola, Barbara Mujica. **VHS $59.98** *VES Woof!*

Mayerling Considered one of the greatest films about doomed love. Story of the tragic and hopeless affair between the Crown Prince Rudolf of Hapsburg and young Baroness Marie Vetsera. Heart wrenching and beautiful, with stupendous acting.
1936 95m/B Charles Boyer, Danielle Darrieux; ***Dir:*** Anatole Litvak. New York Film Critics Awards '37: Best Foreign Film. **VHS, Beta $29.95** *NOS, MRV, FCT* 🦴🦴🦴½

Mayflower Madam A made-for-TV movie about Sydney Biddle Barrows, a prominent New York socialite and madam of an exclusive escort service, whose clientele includes businessmen and dignitaries. Recounts her business dealings and court battles.
1987 (R) 93m/C Candice Bergen, Chris Sarandon, Chita Rivera; ***Dir:*** Lou Antonio. **VHS, Beta $19.95** *STE, VMK* 🦴½

Mayhem Two loners in Hollywood confront all types of urban low-life murderers, drug pushers, and child molesters.
1987 90m/C Raymond Martino, Pamela Dixon, Robert Gallo, Wendy MacDonald. **VHS, Beta** *CLV* 🦴

Maytime An opera singer is reunited with her true love after she left him to marry her teacher seven years earlier.
1937 132m/B Jeanette MacDonald, Nelson Eddy, John Barrymore, Tom Brown, Sig Rumann; ***Dir:*** Robert Z. Leonard. **VHS, Beta $19.95** *MGM* 🦴🦴½

Maytime in Mayfair A dress shop owner's rival is getting all the goodies first, for which he takes the heat from his partner. When he finds out how the rival is doing it, tho partnoro up and hcad for a vacation in the south of France. Overly simplistic but charming kitsch, helped along by Technicolor shots of the fashions and sets.
1952 94m/C Anna Neagle, Michael Wilding, Peter Graves, Nicholas Phipps, Tom Walls, Tom Walls Jr.; ***Dir:*** Herbert Wilcox. **VHS $19.95** *NOS* 🦴🦴

The Maze Made-for-British-television mystery about a young girl who meets the ghost of her mother's first love in her garden estate.
1985 60m/C *GB* Francesca Annis, James Bolam, Sky McCatskill. **VHS, Beta $59.95** *PSM* 🦴🦴🦴

Mazes and Monsters A group of university students becomes obsessed with playing a real-life version of the fantasy role-playing game, Dungeons and Dragons. Adapted for television from the book by Rona Jaffe and also known as "Rona Jaffe's Mazes and Monsters."
1982 100m/C *CA* Tom Hanks, Wendy Crewson, David Wallace, Chris Makepeace, Lloyd Bochner, Peter Donat, Murray Hamilton, Vera Miles, Louise Sorel, Susan Strasberg, Steven Hilliard Stern, Anne Francis; ***Dir:*** Steven Hilliard Stern. **VHS $19.98** *LHV, HHE* 🦴½

Mazurka English-dubbed Danish sex farce about a parochial school's teacher and his efforts to lose his virginity at the school dance.
1970 91m/C *DK* **VHS, Beta $29.95** *FCT* 🦴🦴½

McBain In this exciting fast-paced action-thriller, POW Robert McBain (Walken) leads a group of veterans into battle against the Columbian drug cartel. Stunning cinematography and superb performances combine to make this a riveting action film you won't soon forget.

1991 (R) 104m/C Christopher Walken, Maria Conchita Alonso, Michael Ironside, Steve James, Jay Patterson, Thomas G. Waites; **W/Dir:** James Glickenhaus. VHS *MCA* 🎥🎥½

McCabe & Mrs. Miller Altman's characteristically quirky take on the Western casts Beatty as a self-inflated entrepreneur who opens a brothel in the Great North. Christie is the madame who helps stabilize the haphazard operation. Unfortunately, success comes at a high price, and when gunmen arrive to enforce a business proposition, Beatty must become the man he has, presumably, merely pretended to be. A poetic, moving work, and a likely classic of the genre. Music by Leonard Cohen. Based on the novel by Edmund Naughton. Christie received an Oscar nomination.
1971 (R) 121m/C Warren Beatty, Julie Christie, William Devane, Keith Carradine, John Schuck, Rene Auberjonois, Shelley Duvall; **Dir:** Robert Altman. VHS, Beta, LV $19.98 *WAR, PIA* 🎥🎥🎥🎥

The McConnell Story True story of ace flyer McConnell, his heroism during WWII and the Korean conflict, and his postwar aviation pioneer efforts. Fine acting from Allyson and Ladd, with good support from Whitmore and Faylen.
1955 107m/C Alan Ladd, June Allyson, James Whitmore, Frank Faylen; **Dir:** Gordon Douglas. VHS, Beta, LV $59.95 *WAR, FCT* 🎥🎥

The McGuffin A film critic's inquisitiveness about the activities of his neighbors gets him caught up in murder and a host of other problems. A takeoff of Alfred Hitchcock's 1954 classic "Rear Window."
1985 95m/C *GB* Charles Dance, Ritza Brown, Francis Matthews, Brian Glover, Phyllis Logan, Jerry Stiller, Anna Massey; **Dir:** Colin Bucksey. VHS $79.95 *SVS* 🎥🎥

The McMasters Set shortly after the Civil War, the film tells the story of the prejudice faced by a black soldier who returns to the southern ranch on which he was raised. Once there, the rancher gives him half of the property, but the ex-soldier has difficulty finding men who will work for him. When a group of Native Americans assist him, a band of bigoted men do their best to stop it. The movie was released in two versions with different endings: in one, prejudice prevails; in the other, bigotry is defeated. Also called "The Blood Crowd" and "The McMasters...Tougher Than the West Itself."
1970 (PG) 89m/C Burl Ives, Brock Peters, David Carradine, Nancy Kwan, Jack Palance, Dane Clark, L.Q. Jones, Alan Vint, John Carradine; **Dir:** Alf Kjellin. VHS $19.95 *STE, XVC* 🎥🎥

McQ After several big dope dealers kill two police officers, a lieutenant resigns to track them down. Dirty Harry done with an aging Big Duke.
1974 (PG) 116m/C John Wayne, Eddie Albert, Diana Muldaur, Clu Gulager, Colleen Dewhurst, Al Lettieri, Julie Adams, David Huddleston; **Dir:** John Sturges. VHS, Beta, LV $19.98 *WAR, PIA, BUR* 🎥🎥½

McVicar A brutish and realistic depiction of crime and punishment based on the life of John McVicar, who plotted to escape from prison. From the book by McVicar.
1980 (R) 90m/C *GB* Roger Daltrey, Adam Faith, Cheryl Campbell; **Dir:** Tom Clegg. VHS, Beta, LV $29.98 *VES* 🎥🎥½

MD Geist An animated adventure tale featuring a specially engineered fighting machine named MD Geist who is exiled from his home planet. After awakening from an en-

forced sleep on an orbiting satellite, he must choose between saving his planet from a civil war or letting those who betrayed him die. In Japanese with English subtitles.
199? 45m/C *JP* VHS $34.95 *CPM* 🎥🎥

Me and Him An unsuspecting New Yorker finds himself jockeying for position with his own instincts when his libido decides it wants a life of its own.
1989 (R) 94m/C Griffin Dunne, Carey Lowell, Ellen Greene, Craig T. Nelson; **Dir:** Doris Dorrie; **Voices:** Mark Linn-Baker. VHS, LV $79.95 *COL* 🎥½

The Meal During a dinner party held by a wealthy woman, the rich and powerful guests divulge each others' secrets with reckless disregard for the consequences. Also titled "Deadly Encounter."
1975 (R) 90m/C Dina Merrill, Carl Betz, Leon Ames, Susan Logan, Vicki Powers, Steve Potter; **Dir:** R. John Hugh. VHS $19.95 *UHV, MOV* 🎥

Mean Dog Blues A musician is convicted of hit-and-run driving after hitching a ride with an inebriated politician.
1978 (PG) 108m/C George Kennedy, Kay Lenz, Scatman Crothers, Greg Henry, Gregory Sierra, Tina Louise, William Windom; **Dir:** Mel Stuart. VHS, Beta $69.98 *LIV, VES, LTG* 🎥🎥½

Mean Frank and Crazy Tony A mobster and the man who idolizes him attempt a prison breakout in this fun, action-packed production.
1975 (R) 92m/C *IT* Lee Van Cleef, Tony LoBianco, Jean Rochefort, Jess Hahn; **Dir:** Michele Lupo. VHS, Beta $39.95 *IVE* 🎥🎥

Mean Johnny Barrows When Johnny Barrows returns to his home town after being dishonorably discharged from the Army he is offered a job as a gang hitman.
1975 83m/C Fred Williamson, Roddy McDowall, Stuart Whitman, Luther Adler, Jenny Sherman, Elliott Gould; **Dir:** Fred Williamson. VHS, Beta $49.95 *UNI* 🎥½

Mean Machine A man and his beautiful cohort plot revenge on organized crime for the murder of his father.
1973 (R) 89m/C Chris Mitchum, Barbara Bouchet, Arthur Kennedy. VHS, Beta *MON* 🎥🎥

Mean Season A vicious mass murderer makes a Miami crime reporter his confidante in his quest for publicity during his killing spree. In time, the madman's intentions become clear as the tensions and headlines grow with each gruesome slaying. Then the reporter must come to terms with the idea that he is letting himself be used due to the success that the association is bringing him. Suspenseful story with a tense ending. Good performance from Russell as the reporter.
1985 (R) 106m/C Kurt Russell, Mariel Hemingway, Richard Jordan, Richard Masur, Andy Garcia, Joe Pantoliano, Richard Bradford, William Smith; **Dir:** Phillip Borsos. VHS, Beta, LV $14.99 *HBO* 🎥🎥

Mean Streets A grimy slice of street life in Little Italy among lower echelon Mafiosos, unbalanced punks and petty criminals. A riveting, free-form feature film, marking the formal debut by Scorsese (five years earlier he had completed a student film, "Who's That Knocking At My Door?"). Unorthodox camera movement and gritty performances by De Niro and Keitel, with underlying Catholic guilt providing the moral conflict. Excellent early '60s soundtrack.
1973 (R) 112m/C Harvey Keitel, Robert De Niro, David Proval, Amy Robinson, Richard Romanus, David Carradine, Robert Carradine, Cesare Danova; **Dir:** Martin Scorsese. New York Film

Festival '73: Best Supporting Actor; National Society of Film Critics '73: Best Supporting Actor (De Niro). VHS, Beta, LV $19.98 *WAR, PIA* 🎥🎥🎥🎥

Mean Streets of Kung Fu Ugliness of the nastiest kind brings battle and bloodshed to the boulevards of China.
1983 84m/C Barry Chan. VHS, Beta $59.95 *WES* 🎥½

The Meanest Men in the West Two criminal half-brothers battle the frontier law and each other in this rip-off of two episodes of "The Virginian" television series.
196? (PG) 92m/C Charles Bronson, Lee Marvin, Lee J. Cobb, James Drury, Albert Salmi, Charles Grodin; **Dir:** Samuel Fuller. VHS, Beta $19.95 *MCA* 🎥

Measure for Measure "Measure for Measure" is one of Shakespeare's most controversial comedies. It deals with the law of Vienna—a law where a sexual relationship between unmarried people is punishable by death. Part of the television series "The Shakespeare Plays."
1979 145m/C *GB* Kate Nelligan, Tim Pigott-Smith, Christopher Strauli. VHS, Beta *TLF* 🎥🎥½

Meatballs The Activities Director at a summer camp who is supposed to organize fun for everyone prefers his own style of "fun." If you enjoy watching Murray blow through a movie, you'll like this one even as it lapses into box office sentimentality.
1979 92m/C *CA* Bill Murray, Harvey Atkin, Kate Lynch; **Dir:** Ivan Reitman. Genie Awards '80: Best Actress (Lynch), Best Screenplay. VHS, Beta, LV $24.95 *PAR* 🎥🎥

Meatballs 2 The future of Camp Sasquatch is in danger unless the camp's best fighter can beat Camp Patton's champ in a boxing match. The saving grace of Bill Murray is absent in this one.
1984 (PG) 87m/C Archie Hahn, John Mengatti, Tammy Taylor, Kim Richards, Ralph Seymour, Richard Mulligan, Hamilton Camp, John Larroquette, Pee-Wee Herman, Misty Rowe, Elayne Boosler; **Dir:** Ken Wiederhorn. VHS, Beta, LV, 8mm $79.95 *COL* 🎥½

Meatballs 3 Second sequel to the teen-age sex/summer camp comedy. Enough is enough!
1987 (R) 95m/C Sally Kellerman, Shannon Tweed, George Buza, Isabelle Mejias, Al Waxman, Patrick Dempsey; **Dir:** George Mendeluk. VHS, Beta $19.95 *IVE* Woof!

Meatballs 4 Feldman stars as a water skier hired to serve as recreation director of Lakeside Water Ski Camp. His enemy is Monica (Douglas) of a nearby rival camp who wants to buy out Lakeside's owner and use the land for real estate development. Lame plot only serves to showcase some good water skiing stunts.
1992 (R) 87m/C Corey Feldman, Jack Nance, Sarah Douglas, Bojesse Christopher; **Dir:** Bob Logan. VHS $89.99 *HBO* 🎥½

The Meateater Disgusting horror flick of a disfigured hermit who inhabits a closed-up movie house. When it is reopened he turns into a stereotypical slasher. Vegetarians beware: references throughout about eating meat.
1979 84m/C Arch Jaboulian, Diane Davis, Emily Spendler; **Dir:** Derek Savage. VHS, Beta $24.95 *AHV* Woof!

The Mechanic Bronson stars as Arthur Bishop, a wealthy professional killer for a powerful organization. He has innumerable ways to kill.

1972 (R) 100m/C Charles Bronson, Jan-Michael Vincent, Keenan Wynn, Jill Ireland, Linda Ridgeway; *Dir:* Michael Winner. **VHS, Beta, LV** $19.95 *MGM, FOX* 🎬🎬

Mechanical Crabs A scientist develops the ultimate war weapon: giant mechanical crabs, strong enough to crush anything in their path and devour metal, grow and multiply. They eventually turn on the scientist and eliminate him.
1977 10m/C **VHS, Beta** *BFA* 🎬

The Medal of Honor Features 17 segments which highlight amazing and daring aerial feats, in tribute to the flying Medal of Honor recipients.
1985 100m/B **VHS, Beta** $29.95 *BVG, FGF, AVL* **Woof!**

Medal of Honor Rag The story of a Vietnam veteran who was shot and killed while robbing a supermarket in Detroit.
1983 54m/C **VHS, Beta** *FLI* **Woof!**

Medea Medea, who betrayed her homeland for Jason, is becoming wary of his boredom with her. The classic story of love and revenge ensues thereafter in unbridled fury and passion.
1959 107m/C Colleen Dewhurst, Zoe Caldwell, Judith Anderson; *Dir:* Jose Quintero. **VHS** *IVY, FHS* **Woof!**

Medea Cinema poet Pasolini directs opera diva Callas in this straightforward adaptation of Euripides's classic about a sorceress whose escapades range from assisting in the theft of the Golden Fleece to murdering her own children. Not Pasolini at his best, but still better than most if what you're looking for is something arty. In Italian with English subtitles.
1970 118m/C *IT* Maria Callas, Guiseppi Gentile, Laurent Terzieff; *Dir:* Pier Paolo Pasolini. **VHS, Beta** $49.95 *HMV, VAI, APD* 🎬🎬🎬

Medic & the Star & the Story One episode each: the pilot of "Medic," and "Rheingold Theatre."
1955 30m/B Richard Boone, Zachary Scott. **VHS, Beta** $24.95 *NOS, DVT* 🎬🎬🎬

Medicine Man Benny stars as a con-man who fronts a medicine show in this early talkie comedy.
1930 57m/B Jack Benny, Betty Bronson, E. Allen Warren, George E. Stone, Tom Dugan; *Dir:* Scott Pembroke. **VHS, Beta** $19.95 *NOS, DVT, VYY* 🎬½

Medicine Man Connery's usual commanding presence and the beautiful scenery are the only things to recommend in this lame effort. Dr. Robert Campbell (Connery) is a biochemist working in the Amazon rain forest on a cancer cure. Bracco plays Dr. Rae Crane, a fellow researcher sent by the institute sponsoring Campbell to see how things are going. Although Crane is uptight and Campbell is gruff, they are supposed to fall in love, but they're sorely lacking in chemistry. Oh, Campbell's cancer cure is made from a rare flower being eradicated by the destruction of the rain forest. This environmental-cause-of-the-week-meets-romance-of-opposites falls short of ever being truly entertaining.
1992 (PG-13) 105m/C Sean Connery, Lorraine Bracco; *Dir:* John McTiernan. **VHS, Beta** $94.95 *HPH* 🎬🎬

The Mediterranean in Flames In Nazi-occupied Greece, a small but brave resistance is formed to fight its captors. One of the women within the group is forced to seduce a Nazi officer in an attempt to learn enemy secrets.
1972 85m/C *GR* Costas Karras, Costas Precas; *Dir:* Dimis Dadiras. **VHS, Beta** $39.95 *USA* 🎬½

The Medium A phony medium is done in by her own trickery in this filmed version of the Menotti opera.
1951 80m/B Marie Powers, Anna Maria Alberghetti, Leo Coleman; *Dir:* Gian-Carlo Menotti; *W/Dir:* Gian-Carlo Menotti. **VHS, Beta** $49.95 *MVD, VAI* 🎬🎬

Medium Cool Commentary on life in the '60s focuses on a television news cameraman and his growing apathy with the events around him. Shot during the riots of '68. A frightening depiction of detachment in modern society.
1969 111m/C Robert Forster, Verna Bloom, Peter Bonerz, Marianna Hill; *Dir:* Haskell Wexler. **VHS, Beta** *PAR, OM* 🎬🎬🎬½

Medusa A bizarre series of events occur when an abandoned yacht containing two lifeless bodies is found on the Aegean Sea. Also released on video as, "Twisted."
1974 103m/C George Hamilton, Cameron Mitchell, Lucianna Paluzzi, Theodore Roubanis; *Dir:* Gordon Hessler. **VHS, Beta** $16.95 *SNC, PSM, SIM* 🎬½

Medusa Against the Son of Hercules This time the strongman takes on the evil Medusa and her deadly army of rock men. Also known as "Perseus the Invincible."
1962 ?m/C *IT* Richard Harrison. **VHS, Beta** $16.95 *SNC* 🎬

The Medusa Touch A man is struck over the head and is admitted to a hospital. Meanwhile, strange disasters befall the surrounding city. It seems that despite his unconscious state, the man is using his telekinetic powers to will things to happen....
1978 (R) 110m/C *GB* Richard Burton, Lino Ventura, Lee Remick, Harry Andrews, Alan Badel, Marie-Christine Barrault, Michael Hordern, Derek Jacobi, Jeremy Brett; *Dir:* Jack Gold. **VHS** *FOX* 🎬

Meet Dr. Christian The good old doctor settles some problems. Part of a series.
1939 72m/C Jean Hersholt, Robert Baldwin, Paul Harvey; *Dir:* Bernard Vorhaus. **VHS, Beta** $19.95 *NOS, DVT, VYY* 🎬½

Meet the Hollowheads Family situation comedy set in a futuristic society.
1989 (PG-13) 89m/C John Glover, Nancy Mette, Richard Portnow, Matt Shakman, Juliette Lewis, Anne Ramsey; *Dir:* Tom Burman. **VHS, LV** $39.95 *MED, IME* 🎬½

Meet John Doe A social commentary about an unemployed, down-and-out man selected to be the face of a political goodwill campaign. Honest and trusting, he eventually realizes that he is being used to further the careers of corrupt politicians. Available in colorized version.
1941 123m/B Gary Cooper, Barbara Stanwyck, Edward Arnold, James Gleason, Walter Brennan, Spring Byington, Gene Lockhart, Regis Toomey, Ann Doran; *Dir:* Frank Capra. **VHS, LV** $14.95 *NOS, MRV, PSM* 🎬🎬🎬

Meet Me in Las Vegas Compulsive gambler/cowboy Dailey betters his luck when he hooks up with hoofer girlfriend Charisse. Much frolicking and dancing and cameo appearances by a truckload of stars. A little more entertaining than a game of solitaire.
1956 112m/C Dan Dailey, Cyd Charisse, Agnes Moorehead, Lili Darvas, Jim Backus, Cara Williams, Betty Lynn, Oscar Karlweis, Liliane Montevecchi, Jerry Colonna, Frankie Laine, Debbie Reynolds, Frank Sinatra, Peter Lorre, Vic Damone; *Dir:* Roy Rowland. **VHS** $19.98 *MGM, TTC, FCT* 🎬🎬

Meet Me in St. Louis Wonderful music in this charming tale of a St. Louis family during the 1903 World's Fair. Garland sings the title song, "Trolley Song," and "Have Yourself a Merry Little Christmas," as well as "I Was Drunk Last Night" (with O'Brien). One of Garland's better musical performances.
1944 113m/C Judy Garland, Margaret O'Brien, Mary Astor, Lucille Bremer, Tom Drake, June Lockhart, Harry Davenport; *Dir:* Vincente Minnelli. National Board of Review Awards '44: 10 Best Films of the Year. **VHS, Beta, LV, 8mm** $19.95 *MGM, TLF* 🎬🎬🎬½

Meet the Navy Post-war musical revue about a pianist and a dancer.
1946 81m/B Joan Pratt, Margaret Hurst, Lionel Murton. **VHS, Beta** $29.95 *VYY* 🎬🎬

Meet the Parents A young man visits his fiance's parents and every nightmare, every tragedy that can possibly happen, does. From the minds of "National Lampoon," with an appearance by beyond-bizarre comedian Emo Phillips.
1991 ?m/C Greg Glionna, Jacqueline Cahill, Dick Galloway, Carol Wheeler, Mary Ruth Clarke, Emo Phillips; *Dir:* Greg Glionna. **VHS** $19.95 *JTC* 🎬🎬

Meet Sexton Blake A detective is hired to find a ring and some secret papers that were stolen from the corpse of a man killed in an air raid. He discovers that the papers contained the plans for a new metal alloy to be used in planes. Entertaining for its melodramatic elements.
1944 80m/B *GB* David Farrar, John Varley, Magda Kun, Gordon McLeod, Manning Whiley; *W/Dir:* John Harlow. **VHS, Beta** $16.95 *SNC* 🎬🎬

The Meeting: Eddie Finds a Home Ben Davidson finds an orphan pup rummaging through the trash and cautiously begins a relationship of trust.
1990 ?m/C **VHS** $14.95 *SPW* 🎬🎬

Meeting at Midnight Charlie Chan is invited to a seance to solve a perplexing mystery. Chan discovers that they use mechanical figures and from there on solving the mystery is easy.
1944 67m/B Sidney Toler, Joseph Crehan, Mantan Moreland, Frances Chan; *Dir:* Phil Rosen. **VHS, Beta** $19.95 *QNE, VCN, MRV* 🎬🎬½

Meeting Venus Backstage drama featuring Close as Karin Anderson, a world-famous Swedish opera diva. An international cast of characters adds to this sophisticated film of romance, rivalry and political confrontation set amidst a newly unified Europe.
1991 (PG-13) 121m/C Glenn Close, Niels Arestrup, Erland Josephson, Johanna Ter Steege, Maria De Medeiros; *Dir:* Istvan Szabo; *Voices:* Kiri Te Kanawa. **VHS, Beta, LV** $92.99 *WAR* 🎬🎬

Meetings with Remarkable Men An acclaimed, visually awesome film version of the memoir by Gurdjieff, about his wanderings through Asia and the Middle East searching for answers and developing his own spiritual code. Mildly entertaining but only of true interest to those familiar with the subject.

1979 108m/C Terence Stamp, Dragan Maksimovic, Mikica Dmitrijevic; *Dir:* Peter Brook. **VHS, Beta $69.95** *TAM* ♫♫½

Megaforce Futuristic thriller follows the adventures of the military task force, Megaforce, on its mission to save a small democratic nation from attack. **1982 (PG) 99m/C** Barry Bostwick, Persis Khambatta, Edward Mulhare, Henry Silva, Michael Beck, Ralph Wilcox; *Dir:* Hal Needham. **VHS, Beta $19.98** *FOX* ♫½

Megaville Set in the not too distant future, corrupt politician Travanti struggles to corner the market on the evil technology "Dream-a-Life," which aims to control the rebellious "riff-raff" of Megaville. He does all of his dealings while up against a hitman who has a defective memory chip in his brain. **1991 (R) 96m/C** Billy Zane, Daniel J. Travanti, J.C. Quinn, Grace Zabriskie; *W/Dir:* Peter Lehner. **VHS $89.98** *LIV* ♫♫

Mein Kampf The rise and fall of German fascism and its impact upon the world is examined in this documentary. **1960 120m/C** *Dir:* Erwin Leiser. International Film Festival '60: Golden Award. **VHS, Beta, LV $39.95** *SUE, MRV, FCT* ♫♫

Mel Brooks: An Audience Mel Brooks puts on a number of sketches, sings and tells jokes in this live comedy concert appearance. **1984 60m/C** Ronny Graham, Jonathan Pryce, Mel Brooks, Anne Bancroft. **Beta $11.95** *PSM* ♫♫

Melanie Drama tells the tale of one woman's courage, determination, and optimism, as she tries to regain her child from her ex-husband. **1982 (PG) 109m/C** *CA* Glynnis O'Connor, Paul Sorvino, Don Johnson, Burton Cummings; *Dir:* Rex Bromfield. Genie Awards '83: Best Screenplay. **VHS, Beta $69.95** *VES* ♫½

Melodie en Sous-Sol Two ex-convicts, one old and one young, risk their lives and freedom for one last major heist: a gambling casino. Subtitled in English. Also known as "Any Number Can Win." **1963 118m/B** *FR* Jean Gabin, Alain Delon, Viviane Romance; *Dir:* Henri Verneuil. **VHS, Beta $19.95** *KRT* ♫♫

Melody A sensitive study of a special friendship which enables two adolescents to survive in a regimented and impersonal world. Features music by the Bee Gees. **1971 (G) 106m/C** *GB* Jack Wild, Mark Lester, Colin Barrie, Roy Kinnear; *Dir:* Waris Hussein. **VHS, Beta $69.98** *SUE* ♫½

Melody Cruise Boring musical romantic comedy that can't stay with any one of those three things for more than a minute without careening into one of the other two. **1932 75m/B** Phil Harris, Charlie Ruggles, Greta Nissen, Helen Mack, Chick Chandler; *Dir:* Mark Sandrich. **VHS, Beta $19.98** *TTC* ♫♫

Melody in Love A young girl visits her cousin and her swinging friends on a tropical island. Dubbed. **1978 90m/C** Melody O'Bryan, Sasha Hehn; *Dir:* Hubert Frank. **VHS, Beta $59.98** *FOX* ♫

Melody Master Romanticized biography of composer Franz Schubert, chronicling his personal life and loves, along with performances of his compositions. If you just have to know what drove the composer of "Ave Maria" it may hold your interest, but otherwise not very compelling. Original title: "New Wine."

1941 80m/B Alan Curtis, Ilona Massey, Binnie Barnes, Albert Basserman, Billy Gilbert, Sterling Holloway. **VHS, Beta $19.95** *NOS, VYY, DVT* ♫♫

Melody Ranch Gene returns to his home town as an honored guest and appointed sheriff. But gangster MacLane is determined to drive him out of town. **1940 84m/C** Gene Autry, Jimmy Durante, George "Gabby" Hayes, George "Gabby" Hayes, Ann Miller, Barton MacLane, Joe Sawyer, Horace McMahon; *Dir:* Joseph Santley. **VHS, Beta $19.95** *CCB, VCN* ♫

Melody for Three In the tradition of matchmaker, Doctor Christian aids in the reuniting of music-teacher mother and great-conductor father, who have been divorced for years, in order to help the couple's son, a violin prodigy. **1941 69m/B** Jean Hersholt, Fay Wray; *Dir:* Erle C. Kenton. **VHS, Beta, 8mm $19.95** *NOS, VYY, DVT* ♫

Melody Trail Autry wins $1000 in a rodeo, loses the money to a gypsy, gets a job, falls for his employer's daughter...and in the end captures a kidnapper and cattle rustlers. **1935 60m/B** Gene Autry, Smiley Burnette, Ann Rutherford; *Dir:* Joseph Kane. **VHS, Beta $37.95** *CCB, VCN* ♫

Melon Crazy Comedian Gallagher describes his unusual fondness for watermelon in this concert performance. **1985 58m/C** Gallagher. **VHS, Beta $19.95** *PAR* ♫

Melvin and Howard Story of Melvin Dummar, who once gave Howard Hughes a ride. Dummar later claimed a share of Hughes's will. Significant for Demme's direction and fine acting from Steenburgen/Le-Mat, and Robards in a small role as Hughes. Offbeat and very funny. **1980 (R) 95m/C** Paul LeMat, Jason Robards Jr., Mary Steenburgen, Michael J. Pollard, Dabney Coleman, Elizabeth Cheshire, Pamela Reed, Cheryl "Rainbeaux" Smith; *Dir:* Jonathan Demme. Academy Awards '80: Best Original Screenplay, Best Supporting Actress (Steenburgen); National Board of Review Awards '80: 10 Best Films of the Year; New York Film Festival '80: Best Supporting Actress; National Society of Film Critics Awards '80: Best Picture, Best Screenplay, Best Supporting Actress (Steenburgen). **VHS, Beta, LV $14.95** *MCA, BTV* ♫♫♫½

Melvin Purvis: G-Man Mediocre action film has dedicated federal agent tracking killer Machine Gun Kelly across American Midwest during Depression. Also called "The Legend of Machine Gun Kelly." Made for television. **1974 78m/C** Dale Robertson, Harris Yulin, Margaret Blye, Matt Clark, Elliot Street, Dick Sargent, John Karlen, David Canary; *Dir:* Dan Curtis. **VHS $69.95** *HBO* ♫♫

The Member of the Wedding While struggling through her adolescence, a twelve-year-old girl growing up in 1945 Georgia seeks solace from her family's cook. Based on Carson McCullers play, this story of family, belonging, and growth is well acted and touching. **1952 90m/C** Ethel Waters, Julie Harris, Brandon de Wilde, Arthur Franz; *Dir:* Fred Zinneman. **VHS, Beta $49.95** *VHE* ♫♫♫

The Member of the Wedding TV rehash of Carson McCuller's play about a young girl who finds she's got some growing up to do when big brother marries. No match for Zinnemann's '52 version, or for Mann's

earlier "All Quiet on the Western Front" and "Marty." **1983 90m/C** Dana Hill, Pearl Bailey, Howard E. Rollins Jr.; *Dir:* Delbert Mann. **VHS $49.95** *VHE* ♫

Memoirs of an Invisible Man Nick Halloway, a slick and shallow stock analyst, is rendered invisible by a freak accident. When he is pursued by a CIA agent-hit man who wants to exploit him, Nick turns for help to Alice, a documentary filmmaker he has just met. Naturally, they fall in love along the way. Effective sight gags, hardworking cast can't overcome pitfalls in script, which indecisively meanders between comedy and thrills. **1992 (PG-13) 99m/C** Chevy Chase, Darryl Hannah, Sam Neill; *Dir:* John Carpenter. **VHS, LV $94.99** *WAR* ♫½

Memorial Valley Massacre When campers settle for a weekend in a new, unfinished campground, they're slaughtered in turn by a nutty hermit. **1988 (R) 93m/C** Cameron Mitchell, William Smith, John Kerry, Mark Mears, Lesa Lee, John Caso; *Dir:* Robert C. Hughes. **VHS, Beta, LV $19.98** *SUE* ♫

Memories of Hell Examines the special bonds of friendship that have been maintained among the New Mexicans who survived one of World War II's most brutal battles and subsequent prisoner experiences. **1987 57m/C VHS, Beta, 3/4U** *CEN* ♫

Memories of a Marriage A nostalgic look at a couple's many years of marriage, from the husband's point of view, as they face their most trying times. Touchingly reminiscent as it spans the youthful years of new love and marriage to the more quiet, solid years of matured love. Oscar nomination for Best Foreign Film. **1990 90m/C** *DK* Ghita Norby, Frits Helmuth, Rikke Bendsen, Henning Moritzen; *Dir:* Kaspar Rostrup. **VHS $79.95** *FXL* ♫♫

Memories of Me A distraught doctor travels to L.A. to see his ailing father and make up for lost time. Both the doctor and his father learn about themselves and each other. Story of child/parent relationships attempts to pull at the heartstrings but only gets as far as the liver. Co-written and co-produced by Crystal, this film never quite reaches the potential of its cast. **1988 (PG-13) 103m/C** Billy Crystal, Alan King, JoBeth Williams, David Ackroyd, Sean Connery, Janet Carroll; *Dir:* Henry Winkler. **VHS, Beta, LV $89.98** *FOX* ♫♫

Memories of Murder A woman suffering from amnesia is stalked by a merciless killer, apparently seeking revenge. Can she regain her memory and understand why she is being hunted, before it's too late? **1990 104m/C** Vanity, Nancy Allen, Robin Thomas; *Dir:* Robert Lewis. **VHS, Beta $79.98** *PSM* ♫♫

Memories of Underdevelopment An antirevolutionary intellectual takes a stand against the new government in Cuba. Spanish with English subtitles. **1968 97m/C** Sergio Corrieri, Daisy Granados; *Dir:* Thomas Gutierrez Alea. **VHS $69.95** *NYF, FCT* ♫♫½

Memory of Us Still-relevant story of a married, middle-aged woman who starts to question her happiness as a wife and mother. Written by its star Geer, whose father, Will, plays a bit part.

1974 (PG) 93m/C Ellen Geer, Jon Cypher, Barbara Colby, Peter Brown, Robert Hogan, Rose Marie, Will Geer; *Dir:* H. Kaye Dyal. **VHS** $19.95 *ACA* 🎞🎞

Memphis In 1957 three white drifters (two men and a woman) decide to secure some easy money by kidnapping the young grandson of the richest black businessman in Memphis. They think it's an easy job since, given the prejudice of the times, the police will do little or nothing but they reckon without the resources of the black community itself. Based on the novel "September, September" by Shelby Foote. Originally made-for-cable.
1991 (PG-13) 92m/C Cybill Shepherd, John Laughlin, J.E. Freeman, Richard Brooks, Moses Gunn. **VHS, Beta** $89.98 *TTC* 🎞🎞½

Memphis Belle Remake of classic World War II story of the last mission of a bomber crew with a cast of young up and coming actors. The boys of the Memphis Belle were the first group of B-25 crewmen to complete a 25 mission tour. Many of the 1940 film techniques and characters were used and still managed to inspire audiences. Fine ensemble acting and thrilling flight footage. Film debut of singer Connick.
1990 (PG-13) 107m/C Matthew Modine, Eric Stoltz, Billy Zane, John Lithgow, Harry Connick Jr., Sean Astin, D.B. Sweeney, Tate Donovan; *Dir:* Michael Caton-Jones. **VHS, Beta, LV, 8mm** $92.95 *WAR, MRV, FCT* 🎞🎞🎞

The Men A paraplegic World War II veteran sinks into depression until his former girlfriend manages to bring him out of it. Marlon Brando's first film. A thoughtful story that relies on subtle acting and direction.
1950 85m/B Marlon Brando, Teresa Wright, Everett Sloane, Jack Webb; *Dir:* Fred Zinnemann. National Board of Review Awards '50: 10 Best Films of the Year. **VHS** $19.98 *REP* 🎞🎞🎞½

Men... A funny and insightful satire about a man who discovers his loving wife has been having an affair with a young artist. In a unique course of revenge, the husband ingratiates himself with the artist and gradually turns him into a carbon copy of himself. In German; dubbed in English.
1985 99m/C GE Heiner Lauterbach, Uwe Ochsenknecht, Ulrike Kriener, Janna Marangosoff; *Dir:* Doris Dorrie. **VHS, Beta** $79.95 *VHV* 🎞🎞🎞½

The Men in the Blue Suits A number of short performances that include Eddie Grant, the Ramones, Sunsplash VI and more.
1984 60m/C **VHS, Beta** *IRP* 🎞🎞🎞½

Men of Boys Town Sequel to 1938's "Boys Town" has the same sentimentality, even more if that's possible. Father Flanagan's reformatory faces closure, while the kids reach out to an embittered new inmate. Worth seeing for the ace cast reprising their roles.
1941 106m/B Spencer Tracy, Mickey Rooney, Darryl Hickman, Henry O'Neill, Lee J. Cobb, Sidney Miller; *Dir:* Norman Taurog. **VHS** $19.98 *MGM, FCT* 🎞🎞½

Men Don't Leave A recent widow tries to raise her kids single-handedly, suffers big city life and regains her love life. Good performances by all, especially Cusack. By the director of "Risky Business."
1989 (PG-13) 115m/C Jessica Lange, Arliss Howard, Joan Cusack, Kathy Bates, Charlie Kormso, Corey Carrier, Chris O'Donnell; *Dir:* Paul Brickman. **VHS, Beta, LV, 8mm** $19.98 *WAR* 🎞🎞½

Men are not Gods Verbose film (precursor to "A Double Life") about an actor playing Othello who nearly kills his wife during Desdemona's death scene.
1937 90m/B GB Rex Harrison, Miriam Hopkins, Gertrude Lawrence; *Dir:* Walter Reisch. **VHS, Beta, LV** *SUE* 🎞🎞½

Men of Ireland A Dublin medical student cavorts with islanders off the Irish coast. Uninspired program with authentic native folk of the Blasket Islands, plus their songs and dances.
1938 62m/B IR Cecil Ford, Eileen Curran, Brian O'Sullivan, Gabriel Fallon; *Dir:* Dick Bird. **VHS, Beta, 8mm** $24.95 *VYY* 🎞🎞

Men in Love A young San Franciscan travels to Hawaii with the cremated remains of his lover, who has succumbed to AIDS. There he encounters a supportive group of men who nurse him back to sexual and spiritual wholeness.
1990 87m/C Doug Self, Joe Tolbe, Emerald Starr, Kutira Decosterd; *Dir:* Mark Huestis. **VHS** $59.95 *WBF, FCT* 🎞🎞

Men in the Park The men in the park are black pencil drawings that live in the hurried world of a busy, imaginary city.
1973 7m/C **VHS, Beta** *NFC* 🎞🎞

Men of Respect Shakespeare meets the mafia in this misbegotten gangster yarn. Turturro, a gangster MacBeth, is prodded by wife and psychic to butcher his way to the top o' the mob. Well acted but ill conceived, causing Bard's partial roll over in grave.
1991 (R) 113m/C John Turturro, Katherine Borowitz, Peter Boyle, Dennis Farina, Chris Stein, Steven Wright, Stanley Tucci; *W/Dir:* William Reilly. **VHS, LV, 8mm** $19.95 *COL, PIA* 🎞🎞

Men of Sherwood Forest Robin Hood and his band take it to the sheriff et al in this colorful Hammer version of the timeless legend.
1957 77m/C GB Don Taylor, Reginald Beckwith, Eileen Moore, Davis King-Wood, Patrick Holt, John Van Eyssen, Douglas Wilmer; *Dir:* Val Guest. **VHS** $16.95 *NOS, SNC, DVT* 🎞🎞½

Men of Steel Two Canadian military officers, one French and one English, must struggle to survive when their plane crashes in the wilderness. Besides contending with the surroundings, they must deal with one another's instilled bitterness and underlying prejudices.
1988 (PG) 91m/C Allan Royal, Robert Lalonde, David Ferry, Mavor Moore, Yvan Ponton; *Dir:* Donald Shebib. **VHS, Beta** *TWE* 🎞

Men of Two Worlds Classical musician raised in Africa is the toast of Europe until he finds out he's got a voodoo curse, and suddenly he's got the heebie jeebies too bad to play chopsticks. Also known as "Witch Doctor" and "Kisenga, Man of Africa."
1946 90m/C GB Robert Adams, Eric Portman, Orlando Martins, Phyllis Calvert, Arnold Marle, Cathleen Nesbitt, David Horne, Cyril Raymond; *Dir:* Thorold Dickinson. **VHS, Beta** $16.95 *SNC* 🎞🎞

Men in War Korean War drama about a small platoon trying to take an enemy hill by themselves. A worthwhile effort with some good action sequences.
1957 100m/B Robert Ryan, Robert Keith, Aldo Ray, Vic Morrow; *Dir:* Anthony Mann. **VHS, Beta** $19.95 *FGF, MRV, KOV* 🎞🎞½

The Men Who Tread on the Tiger's Tail Twelfth-century Japan is the setting for this struggle of power between two brothers, one a reigning shogun, the other on the run. English subtitles.
1945 60m/B JP *Dir:* Akira Kurosawa. **VHS, Beta, 3/4U** $29.95 *IHF* 🎞🎞½

Men at Work Garbage collectors Sheen and Estevez may not love their work, but at least it's consistent from day to day. That is, until they get wrapped up in a very dirty politically-motivated murder. And who will clean up the mess when the politicians are through trashing each other? A semi-thrilling semi-comedy that may leave you semi-satisfied.
1990 (PG-13) 98m/C Charlie Sheen, Emilio Estevez, Leslie Hope, Keith David; *W/Dir:* Emilio Estevez. **VHS, Beta, LV** $19.95 *COL, FOX* 🎞🎞

Menace on the Mountain Family-oriented drama about a father and son who battle carpetbagging Confederate deserters during the Civil War.
1970 (G) 89m/C Patricia Crowley, Albert Salmi, Charles Aidman. **VHS, Beta** $69.95 *DIS* 🎞🎞

The Men's Club Seven middle-aged buddies get together for a single night, and bare their respective souls and personal traumas, talking about women and eating and drinking. Banal. Based on novel by Leonard Michael.
1986 (R) 100m/C Harvey Keitel, Roy Scheider, Craig Wasson, Frank Langella, David Dukes, Richard Jordan, Treat Williams, Stockard Channing, Jennifer Jason Leigh, Ann Dusenberry, Cindy Pickett, Gwen Welles; *Dir:* Peter Medak. **VHS, Beta, LV** $19.95 *PAR* 🎞

Mephisto An egomaniacal actor, compellingly played by Brandauer, sides with the Nazis to further his career, with disastrous results. The first of three brilliant films by Szabo and Brandauer exploring the price of power and personal sublimation in German history. Critically hailed. In German with subtitles. From the novel by Klaus Mann, son of Thomas, who allegedly killed himself because he could not get the book published.
1981 144m/C GE Klaus Maria Brandauer, Krystyna Janda, Ildiko Bansagi, Karin Boyd, Rolf Hoppe, Christine Harbort, Gyorgy Cserhalmi, Christiane Graskoff, Peter Andorai, Ildiko Kishonti; *Dir:* Istvan Szabo. Academy Awards '81: Best Foreign Language Film. **VHS, Beta** $69.95 *HBO, GLV* 🎞🎞🎞½

The Mephisto Waltz A journalist gets more than a story when he is granted an interview with a dying pianist. It turns out that he is a satanist and the cult he is a part of wants the journalist. A chilling adaptation of the Fred Mustard Stewart novel that also features a haunting musical score by Jerry Goldsmith.
1971 (R) 108m/C Alan Alda, Jacqueline Bisset, Barbara Parkins, Curt Jurgens, Bradford Dillman, William Windom, Kathleen Widdoes; *Dir:* Paul Wendkos. **VHS** $59.98 *FOX* 🎞🎞🎞

The Mercenaries A romantic warrior fights to save the lives and honor of his people who are under attack from mercenaries murdering and raping their way across Europe.
1965 98m/C Debra Paget. **Beta** $39.95 *AIP* 🎞

The Mercenaries A tough guy given the assignment to kill Fidel Castro, encounters all sorts of adversity along the way. Also called "Cuba Crossing," "Kill Castro," "Assignment: Kill Castro," "Key West Crossing," "Sweet Dirty Tony."

1980 92m/C Stuart Whitman, Robert Vaughn, Caren Kaye, Raymond St. Jacques, Woody Strode, Sybil Danning, Albert Salmi, Michael Gazzo; *Dir:* Chuck Workman. **VHS, Beta** $59.95 *ACA* ✗

Mercenary Fighters An American soldier-of-fortune fights for both sides of an African revolution. Shot in South Africa. **1988 (R)** 90m/C Peter Fonda, Reb Brown, Ron O'Neal, Jim Mitchum; *Dir:* Riki Shelach. **VHS, Beta, LV** $79.95 *MED* ✗½

The Mercenary Game This tape offers a behind-the-scenes peek at a mercenary training camp in Georgia, plus interviews with some of the leaders and footage of an actual mercenary raid. **1981** 60m/C **VHS, Beta, 3/4U** $29.95 *IHF, MPI* ✗

The Merchant of Four Seasons Story focuses on the depression and unfulfilled dreams of an average street merchant. Direction from Fassbinder is slow, deliberate, and mesmerizing. **1971** 88m/C Hans Hirschmuller, Irm Hermann, Hanna Schygulla, Andrea Schober, Gusti Kreissl; *Dir:* Rainer Werner Fassbinder. **VHS, Beta** $79.95 *NYF* ✗✗✗

Merchants of War Sent out by the CIA on a mission, two best friends are soon the target of one of the most dangerous terrorists in the world. The fanatical Islamic terrorist soon casts his wrath upon the two men. **1990 (R)** 100m/C Asher Brauner, Jesse Vint, Bonnie Beck; *Dir:* Peter M. MacKenzie. **VHS** $89.95 *VMK* ✗

Mere Jeevan Saathi A bizarre, unclassifiable Indian film with music, fantasy, special effects, violence, and flamboyant mise-en-scene. In Hindi with English subtitles. English title: "My Life Partner." **1965** 126m/C *IN* Lata Mangeskar. **VHS, Beta** $29.95 *VYY* ✗✗½

Meridian: Kiss of the Beast A beautiful heiress is courted, kidnapped and loved by a demonic man-beast. **1990 (R)** 90m/C Sherilyn Fenn, Malcom Jamieson, Hilary Mason, Alex Daniels, Phil Fondacaro; *Dir:* Charles Band. **VHS, Beta, LV** $19.95 *PAR* ✗½

Merlin and the Sword Poor use of a good cast in this treatment of the legend of Merlin, Arthur and Excalibur. Also known as "Arthur the King." **1985** 94m/C Malcolm McDowell, Edward Woodward, Candice Bergen, Dyan Cannon; *Dir:* Clive Donner. **VHS, Beta** $69.98 *LIV, VES* ✗

Mermaids It's the halcyon Kennedy days, when fun foods were a major food group. Cher's Mrs. Flax, the toney tushed mother of two who hightails out of town every time a relationship threatens to turn serious. Having moved some eighteen times, her daughters, Charlotte, 15, and Kate, 8, are a little the worse for wear, psychologically speaking. One aspires to join the sisterly persuasion though not Catholic, and the other holds her breath under water. Now residing in Massachusetts after her latest fiasco de coeur, Mrs. Flax starts having those I got you babe feelings for Hoskins, a shoe store owning kinda guy. Amusing, well-acted multi-generational coming of ager based on a novel by Patty Dann. **1990 (PG-13)** 110m/C Cher, Winona Ryder, Bob Hoskins, Christina Ricci, Michael Schoeffling; *Dir:* Richard Benjamin. National Board of Review Awards '90: Best Supporting Actress (Ryder). **VHS, Beta, LV** $14.98 *ORI* ✗✗✗

The Mermaids of Tiburon A marine biologist and a criminal travel to a remote island off Mexico in search of elusive, expensive "fire pearls." There they encounter a kingdom of lovely mermaids who promptly liven things up. Filmed in "Aquascope" for your viewing pleasure. **1962** 77m/C Diane Webber, George Rowe, Timothy Carey, Jose Gonzalez-Gonzalez, John Mylong, Gil Baretto, Vicki Kantenwine, Nani Morrissey, Judy Edwards, Jean Carroll, Diana Cook, Karen Goodman, Nancy Burns; *W/Dir:* John Lamb. **VHS** $18.00 *FRG* ✗

Merry Christmas Two guys from the countryside head for Warsaw with a truck load of Christmas trees and encounter a variety of comedic situations. **19??** 60m/C *PL Dir:* Jerzy Sztwiernia. **VHS** $39.95 *FCT* ✗

Merry Christmas, Mr. Lawrence An often overlooked drama about a World War II Japanese POW camp. The film is a taut psychological drama about clashing cultures and physical and emotional survival. Focusing on the tensions between Bowie and a British POW and camp commander Sakamoto, who also composed the outstanding score. A haunting and intense film about the combat horrors of war. **1983 (R)** 124m/C *GB JP* David Bowie, Tom Conti, Ryuichi Sakamoto, Takeshi, Jack Thompson; *Dir:* Nagisa Oshima. **VHS, Beta** $59.95 *MCA* ✗✗✗

Merry Christmas to You A collection of cartoons, singalongs, and Lone Ranger and Lassie adventures that all carry a Christmas theme are featured on this tape. **1980** 80m/C **VHS, Beta** *MED* ✗✗✗

Merry Widow The first sound version of the famous Franz Lehar operetta, dealing with a playboy from a bankrupt kingdom who must woo and marry the land's wealthy widow or be tried for treason. A delightful musical comedy, with a sterling cast and patented Lubitschian gaiety. Songs include "Girls, Girls, Girls" and "Tonight Will Teach Me To Forget." Made as a silent in 1912 and 1925; remade in color in 1952. Also called "The Lady Dances." **1934** 99m/B Maurice Chevalier, Jeanette MacDonald, Edward Everett Horton, Una Merkel, George Barbier, Minna Gombell, Ruth Channing, Sterling Holloway, Henry Armetta, Barbara Leanard, Donald Meek, Akim Tamiroff, Herman Bing; *Dir:* Ernst Lubitsch. Academy Awards '34: Best Art Direction/Set Decoration. **VHS, Beta** $29.95 *MGM* ✗✗✗½

Mesa of Lost Women Mad scientist creates brave new race of vicious women with long fingernails. So bad it's a wanna-B. Addams Family buffs will spot the Fester in Coogan. **1952** 70m/B Jackie Coogan, Richard Travis, Allan Nixon, Mary Hill, Robert Knapp, Tandra Quinn, Lyle Talbot, Katherine Victor, Angelo Rossitto, Herbert Tevos; *Dir:* Ron Ormond. **VHS** $16.95 *NOS, MRV, SNC* Woof!

Mesmerized Based on the work by Jerzy Skolimowski, the film is a dramatization of the Victoria Thompson murder case in 1880s' New Zealand. A teenaged orphaned girl marries an older man and decides after years of abuse to kill him through hypnosis. An unengaging drama, though the lovely New Zealand landscape serves as a fitting contrast to the film's ominous tone. **1984** 90m/C *GB NZ AU* John Lithgow, Jodie Foster, Michael Murphy, Dan Shor, Harry Andrews; *Dir:* Michael Laughlin. **VHS, Beta** $89.98 *LIV, VES* ✗✗

Mesquite Buckaroo Two cowboys bet over who is the better bronco buster, but one of them gets kidnapped. **1939** 59m/B Bob Steele, Carolyn Curtis, Frank LaRue, Juanita Fletcher, Charles King; *Dir:* Harry S. Webb. **VHS, Beta** $19.95 *NOS, DVT* ✗

Messalina, Empress of Rome Ancient Rome is torn to bits in the zaniest manner possible in this Italian-French coproduction. **1980** 100m/C **VHS, Beta** *VCD* ✗

Messalina vs. the Son of Hercules A Roman slave leads a rebellion against the emperor Messalina. Italian; dubbed into English. **1964** 105m/C *IT* Richard Harrison, Marilu Tolo, Lisa Gastoni; *Dir:* Umberto Lenzi. **VHS, Beta** $16.95 *SNC, CHA* ✗

The Messenger A man right out of prison sets out to avenge his wife's murder by an Italian drug syndicate. **1987 (R)** 95m/C Fred Williamson, Sandy Cummings, Christopher Connelly, Cameron Mitchell. **VHS, Beta** $79.98 *ORI* ✗

Messenger of Death A tough detective investigates the slaughter of a Mormon family, and uncovers a conspiracy centering around oil-rich real estate. **1988 (R)** 90m/C Charles Bronson, Trish Van Devere, Laurence Luckinbill, Daniel Benzali, Marilyn Hassett, Jeff Corey, John Ireland, Penny Peyser, Gene Davis; *Dir:* J. Lee Thompson. **VHS, Beta, LV** $89.95 *MED* ✗✗½

The Messiah Rossellini's final film, a rarely-seen version of the life of Christ, completely passed over any notions of divinity to portray him as a morally perfect man. In Italian with subtitles. **1975** 145m/C *IT Dir:* Roberto Rossellini. **VHS, Beta** $39.95 *FCT, MVD* ✗✗✗½

Messiah of Evil A California coastal town is invaded by zombies. A confusing production from the writers of "American Graffiti." Originally titled "Dead People," and also known as "Return of the Living Dead" and "Revenge of the Screaming Dead." **1974 (R)** 90m/C Marianna Hill, Joy Bang, Royal Dano, Elisha Cook Jr., Michael Greer; *W/Dir:* Gloria Katz, Willard Huyck. **VHS, Beta** $24.95 *GEM, MRV* ✗

Metallica Alien warmongers endeavor to conquer Earth and scientists try to stop them. **1985** 90m/C Anthony Newcastle, Sharon Baker; *Dir:* Al Bradley. **VHS, Beta** $59.95 *MGL* ✗

Metalstorm It's the science fiction battle of the ages with giant cyclopses and intergalactic magicians on the desert planet of Lemuria. **1983 (PG)** 84m/C Jeffrey Byron, Mike Preston, Tim Thomerson, Kelly Preston, Richard Moll; *Dir:* Charles Band. **VHS, Beta, LV** $59.95 *MCA* ✗

Metamorphosis Novice scientist foolishly uses himself as the guinea pig for his antiaging experiments. He quickly loses control of the project. **1990 (R)** 90m/C Gene Le Brock, Catherine Baranov, Stephen Brown, Harry Cason, Jason Arnold; *Dir:* G.L. Eastman. **VHS, Beta** $89.95 *IMP* ✗

Meteor American and Soviet scientists attempt to save the Earth from a fast-approaching barrage of meteors from space in this disaster dud. Destruction ravages parts of Hong Kong and the Big Apple.

1979 (PG) 107m/C Sean Connery, Natalie Wood, Karl Malden, Brian Keith, Martin Landau, Trevor Howard, Henry Fonda, Joseph Campanella, Richard Dysart; *Dir:* Ronald Neame. VHS, Beta $19.98 *WAR, OM* 𝒥1/2

Meteor Monster A young boy is hit by a meteor and is somehow transformed into a raving, slime-covered maniac.
1957 73m/C Anne Gwynne, Stuart Wade, Gloria Castillo, Charles Courtney; *Dir:* Jacques Marquette. VHS, Beta $39.95 *MON, SNC* Woof!

Metropolitan The Izod set comes of age on Park Avenue during Christmas break. Tom Townsend (Clements), a member of the, gasp, middle class, finds himself drawn into a circle of self proclaimed urban haute bourgeoisie types. It seems they're embarrassingly short on male escorts, and, even without the compulsory navy overcoat, he's an adequate stand-in at the group's many deb party/group dates. Much juvenescent philosophizing and hormone quelling. Intelligently written and carefully made, it transcends the flirting with adulthood genre. Festival praised.
1990 (PG-13) 98m/C Carolyn Farina, Edward Clements, Taylor Nichols, Christopher Eigeman, Allison Rutledge-Parisi, Dylan Hundley, Isabel Gillies, Bryan Leder, Will Kempe, Elizabeth Thompson; *W/Dir:* Whit Stillman. New York Film Critics Awards '90: Best First Film by a Director (Whit Stillman). VHS, LV $19.95 *COL, PIA, FCT* 𝒥𝒥𝒥

Mexican Bus Ride A good-natured Bunuel effort about a newlywed peasant who travels to the big city to attend to his mother's will. While en route, he encounters a diversity of people on the bus and some temptation. In Spanish with English subtitles.
1951 86m/B *MX* Lilia Prado, Esteban Marquez, Carmelita Gonzalez; *Dir:* Luis Bunuel. Cannes Film Festival '51: Best Avant-Garde Film. VHS, Beta $39.95 *FCT, APD* 𝒥𝒥1/2

Mexican Hayride Bud heads a gang of swindlers and Lou is the fall guy. When things up north get too hot, the boys head south of the border to cool off and start a mining scam. Watch for the hilarious bull-fighting scene. Believe it or not, this is based on a Cole Porter musical—minus the music.
1948 77m/B Bud Abbott, Lou Costello, Virginia Grey, Luba Malina, John Hubbard, Pedro de Cordoba, Fritz Feld, Tom Powers, Pat Costello, Frank Fenton; *Dir:* Charles T. Barton. VHS $14.98 *MCA* 𝒥𝒥

Mexican Spitfire/Smartest Girl in Town A comedy double feature: In "Mexican Spitfire," a man impersonates an English lord in order to save a contract for the spitfire's husband. In "The Smartest Girl in Town," a photographer's model mistakes a millionaire for a fellow magazine model.
1940 125m/B Lupe Velez, Leon Errol, Donald Woods, Ann Sothern, Gene Raymond. VHS, Beta $34.95 *RKO* 𝒥𝒥

MGM: When the Lion Roars Three-part documentary covering the history and films of the Metro-Goldwyn-Mayer (MGM) studio from its beginnings in 1924 with the release of the studio's first film ("He Who Gets Slapped"), through its glory years, to the studio's gradual decline and ever-changing leadership. MGM brought out such films as "The Wizard of Oz," "Gone With the Wind," the Andy Hardy series, and the Thin Man films. Features clips of films and stars as well as footage of studio executives, including Louis B. Mayer and legendary studio chief and producer Irving Thalberg.

Tapes are available individually or as a three-tape set. Originally made for cable television.
1992 360m/C *Nar:* Patrick Stewart. VHS, LV *MGM* 𝒥𝒥

Miami Blues Cold-blooded killer plays cat-and-mouse with bleary cop while diddling unflappable prostitute. Violent and cynical, with appropriate performances from three leads.
1990 (R) 97m/C Fred Ward, Alec Baldwin, Jennifer Jason Leigh, Nora Dunn, Charles Napier, Jose Perez, Paul Gleason; *Dir:* George Armitage. New York Film Critics Awards '90: Best Supporting Actress (Leigh). VHS, Beta, LV $89.95 *ORI, IME* 𝒥𝒥1/2

Miami Cops When a cop's father is killed by a drug smuggler, he and his partner pursue the murderer over two continents. Italian-made adventure is a bit drawn-out.
1989 103m/C *IT* Richard Roundtree, Harrison Muller, Dawn Baker, Michael J. Aronin; *Dir:* Al Bradley. VHS, Beta $79.95 *CAN* 𝒥

Miami Horror A Florida scientist is experimenting with bacteria from space, trying to recreate a human only to have his efforts stolen by a crook. His intentions for the use of the experiment are not exactly for the furtherance of science.
1987 85m/C *IT* David Warbeck, Laura Trotter, Lawrence Loddi, John Ireland; *Dir:* Martin Herbert. VHS, Beta $14.95 *TTE* 𝒥

Miami Supercops Two goofy policemen strike out against such threats as a gang that hassles buses, the attempted kidnapping of an Orange Bowl quarterback, and a multi-million-dollar robbery.
1985 97m/C Terence Hill, Bud Spencer, Jackie Castellano, C.V. Wood Jr.; *Dir:* Bruno Corbucci. VHS, Beta $19.98 *WAR* 𝒥1/2

Miami Vendetta An L.A. vice cop risks it all to avenge his friend's death at the hands of Cuban drug smugglers.
1987 90m/C Sandy Brooke, Frank Gargani, Maarten Goslins, Barbara Pilavin; *Dir:* Steven Seemayer. VHS, Beta $59.98 *CON* 𝒥

Miami Vice Pilot for the popular television series paired Crockett and Tubbs for the first time on the trail of a killer in Miami's sleazy underground. Music by Jan Hammer and other pop notables.
1984 99m/C Don Johnson, Philip Michael Thomas, Saundra Santiago, Michael Talbott, John Diehl, Gregory Sierra; *Dir:* Thomas Carter. VHS, Beta, LV $14.95 *MCA* 𝒥𝒥1/2

Miami Vice 2: The Prodigal Son The second full-length television movie spawned by the popular series, following the laconic Floridians to the gritty streets of New York.
1985 99m/C Don Johnson, Philip Michael Thomas, Edward James Olmos, Penn Jillette. VHS, Beta, LV $14.95 *MCA* 𝒥𝒥

Michael Prophet Popular reggae stars headline.
1987 60m/C Courtney Melody, Frankie Jones, Khaki Rables, Michael Prophet, Andrew Be, Mello Tuffy, Joe Manniks, Johnny Bee & Tuffi, Super Chic. VHS, Beta *KRV* 𝒥𝒥

The Michigan Kid Two boys fight for the same girl in this silent love triangle set in Alaska. Film's most notable scene is of a raging forest fire, considered a classic even today.
1928 62m/B Conrad Nagel, Renee Adoree, Fred Esmelton, Virginia Grey, Adolph Milar, Lloyd Whitlock; *Dir:* Irwin Willat. VHS, Beta *GPV* 𝒥𝒥

Mickey Spoof on high society as a penniless young woman moves in with relatives and works as the family's maid. Silent film.
1917 105m/B Mabel Normand, Lew Cody, Minta Durfee; *Dir:* Mack Sennett. VHS, Beta $19.95 *NOS, VYY, DVT* 𝒥𝒥

Mickey A tomboy becomes a woman even as she plays matchmaker for her own father and sings a few songs. Not too exciting; you may fall asleep if you're slipped this "Mickey." Based on the novel "Clementine," by Peggy Goodin.
1948 87m/C Lois Butler, Bill Goodwin, Irene Hervey, John Sutton, Hattie McDaniel; *Dir:* Ralph Murphy. VHS $19.95 *NOS* 𝒥

Mickey & the Beanstalk Mickey, Donald and Goofy enact their version of "Jack and the Beanstalk" in this featurette which originally comprised one-half of the film "Fun and Fancy Free." Notable as the last film for which Disney provided the voice of Mickey Mouse.
1947 29m/C *Voices:* Walt Disney. VHS, Beta $12.99 *DIS* 𝒥

Mickey Commemorative Edition Mickey Mouse celebrates his 60th birthday in this happy compilation of memorable moments from his career, which includes the marvelous "Sorcerer's Apprentice" sequence from "Fantasia."
1988 29m/C VHS, Beta $12.99 *DIS* 𝒥

Mickey the Great Stitched together from several late 20's Mickey McGuire shorts, starring 10-year-old Rooney.
1939 70m/B Mickey Rooney. VHS, Beta $24.95 *NOS, DVT* 𝒥

Mickey Knows Best Three Mickey Mouse shorts from the late '30s: "Moving Day," "Mickey's Amateurs," and "Mickey's Elephant." Walt Disney's genius began with shorts like these.
1937 26m/C VHS, Beta $14.95 *DIS* 𝒥

Mickey Mouse Cartoon Cartoons featuring everybody's favorite mouse.
1989 30m/C VHS *SEC, APD* 𝒥

Mickey Rooney in the Classic Mickey McGuire Silly Comedies The classic boyhood series starring Rooney is also hosted by the actor. Three volumes available.
19?? 40m/C *Hosted:* Mickey Rooney. VHS $9.99 *SUM* 𝒥

Mickey's Christmas Carol Mickey Mouse returns along with all the other Disney characters in this adaptation of the Charles Dickens classic. Included in this video cassette is a documentary on how the featurette was made.
1983 (G) 25m/C VHS, Beta, LV, 3/4U, EJ, Special order formats $12.99 *DIS, MTI, DSN* 𝒥𝒥𝒥

Micki & Maude When a man longs for a baby, he finds that his wife, Micki, is too busy for motherhood. Out of frustration, he has an affair with Maude that leads to her pregnancy. Still shocked by the news of his upcoming fatherhood, the man learns that his wife is also expecting.
1984 (PG-13) 117m/C Dudley Moore, Amy Irving, Ann Reinking, Richard Mulligan, Wallace Shawn, George Gaynes, Andre the Giant; *Dir:* Blake Edwards. VHS, Beta, LV $79.95 *COL* 𝒥𝒥

Microwave Massacre Killer kitchen appliances strike again as late lounge comic Vernon murders nagging wife and 'waves her. Overcome by that Betty Crocker feeling, he goes on a microwave murdering/feeding

spree of the local ladies. Lots of Roger Corman copying.
1983 (R) 80m/C Jackie Vernon, Loren Schein, Al Troupe; *Dir:* Wayne Berwick. **VHS $59.95** *RHI, MID, FCT* ♫

Mid-Channel Young stars as a neglected married woman.
1920 70m/B Clara Kimball Young. **VHS, Beta $16.95** *NOS, MRV, GPV* ♫

Mid Knight Rider A penniless actor becomes a male prostitute at the service of bored, rich women. At an all-night orgy, he suddenly goes on a rampage, nearly killing one of his customers.
1984 76m/C Michael Christian, Keenan Wynn. **VHS, Beta $49.95** *TWE* ♫

The Midas Touch A flea market merchant in 1956 Budapest has the ability to turn items into gold. In Hungarian with English subtitles.
1989 100m/C *HU W/Dir:* Geza Beremenyi. **VHS, Beta $79.95** *EVD, FCT* ♫

Middle Age Crazy Story of a Texas building contractor who takes life pretty lightly until his own father dies. Then he becomes immersed in a mid-life crisis and has an affair with a Dallas Cowboys' cheerleader. Ann-Margret plays the victim wife of the middle aged swingin' guy.
1980 95m/C *CA* Bruce Dern, Ann-Margret, Graham Jarvis, Eric Christmas, Deborah Wakeham; *Dir:* John Trent. **VHS $59.98** *FOX* ♫½

Midnight A jury foreman's daughter is romantically involved with a gangster who is interested in a particular case before it appears in court. The foreman, who sentenced a girl to death, faces a dilemma when his daughter is arrested for the same crime. Also known as "Call It Murder." An early Bogart appearance in a supporting role led to a rerelease of the film as "Call it Murder" after Bogart made it big. Weak melodrama.
1934 74m/B Humphrey Bogart, Sidney Fox, O.P. Heggie, Henry Hull, Richard Whorf, Margaret Wycherly, Lynne Overman; *Dir:* Chester Erskine. **VHS, Beta $34.98** *CCB, CCI, DVT* ♫½

Midnight Russo, who cowrote the original "Night of the Living Dead," wrote and directed this film about a runaway girl who is driven out of her home by a lecherous stepfather and meets two young thieves and then a family of cultists. Russo adapted his own novel. He also attains some of "Night of the Living Dead's" low-budget ambience. Also known as "Backwoods Massacre."
1981 (R) 88m/C Lawrence Tierney, Melanie Verliin, John Hall, John Amplas; *Dir:* John Russo. **VHS, Beta $79.95** *VMK* ♫

Midnight Murder-thriller involving the vampirish hostess of a television horror movie showcase and a fanatical fan.
1989 (R) 90m/C Lynn Redgrave, Tony Curtis, Steve Parrish, Rita Gam, Gustav Vintas, Karen Witter, Frank Gorshin, Wolfman Jack; *Dir:* Norman Thaddeus Vane. **VHS, Beta $89.85** *SVS* ♫

Midnight Auto Supply Dealers in hot car parts, working out of the garage behind the local brothel, become persuaded to donate some of their profits to the cause of Mexican farm workers. Also called "Love and the Midnight Auto Supply."
1978 91m/C Michael Parks, Rory Calhoun, Scott Jacoby, Rod Cameron, Colleen Camp, Linda Cristal, John Ireland; *Dir:* James Polakof. **VHS $79.95** *MNC* ♫½

Midnight Cabaret A New York nightclub-based Satanic cult selects a child actress to bear Satan's child.
1990 (R) 94m/C VHS, Beta, LV $89.98 *WAR* Woof!

Midnight Cop A young woman gets tangled up in a web of murder, intrigue, prostitution and drugs and she enlists the aid of a cop to help get her out of it.
1988 (R) 100m/C Michael York, Morgan Fairchild, Frank Stallone, Armin Mueller-Stahl; *Dir:* Peter Patzak. **VHS, Beta, LV $29.95** *VMK, HHE* ♫

Midnight Cowboy Drama about the relationship between a naive Texan hustler and a seedy derelict, set in the underbelly of New York City. Graphic and emotional character study is brilliantly acted and engaging. Shocking and considered quite risque at the time of its release, this film now carries an "R" rating. It is the only film carrying an "X" rating to ever win the Best Picture Oscar. Voight, Hoffman, and Miles received Oscar nominations for their performances. From James Leo Herlihy's novel. Laserdisc version features: audio commentary by director John Schlesinger and producer Jereome Hellman, behind the scenes production photos, and original movie trailer.
1969 (R) 113m/C Dustin Hoffman, Jon Voight, Sylvia Miles, Brenda Vaccaro, John McGiver, Bob Balaban, Barnard Hughes; *Dir:* John Schlesinger. Academy Awards '69: Best Adapted Screenplay, Best Director (Schlesinger), Best Picture; British Academy Awards '69: Best Actor (Hoffman), Best Director (Schlesinger), Best Film, Best Film Editing, Best Screenplay; Directors Guild of America Awards '69: Best Director (Schlesinger); New York Film Critics Awards '69: Best Actor (Voight); National Society of Film Critics Awards '69: 10 Best Films of the Year, Best Actor (Voight). **VHS, Beta, LV $19.98** *MGM, VYG, CRC* ♫♫½

Midnight Crossing Two married couples are subjected to jealousy, betrayal and uncloseted-skeletons as their pleasure cruise on a yacht turns into a ruthless search for sunken treasure.
1987 (R) 96m/C Faye Dunaway, Daniel J. Travanti, Kim Cattrall, John Laughlin, Ned Beatty; *Dir:* Roger Holzberg. **VHS, Beta, LV $89.98** *LIV, VES* ♫

Midnight Dancer A young ballerina is forced to work as an erotic dancer.
1987 97m/C *AU* Deanne Jeffs, Mary Regan; *Dir:* Pamela Gibbons. **VHS, Beta $29.95** *ACA* ♫♫

Midnight Express A gripping and powerful film based on the true story of Billy Hayes. Davis plays Hayes as a young American in Turkey who is busted trying to smuggle hashish. He is sentenced to a brutal and nightmarish prison for life. After enduring tremendous mental and physical torture, he seeks escape along the "Midnight Express." Not always easy to watch, but the overal effect is riveting and unforgettable. Oscar nominations for Best Picture, supporting actor Hurt, director, and editing.
1978 (R) 120m/C *TU* John Hurt, Randy Quaid, Brad Davis, Paul Smith, Bo Hopkins, Oliver Stone; *Dir:* Alan Parker. Academy Awards '78: Best Adapted Screenplay, Best Musical Score; British Academy Awards '78: Best Director (Parker), Best Film Editing, Best Supporting Actor (Hurt); Los Angeles Film Critics Association Awards '78: Best Musical Score; National Board of Review Awards '78: 10 Best Films of the Year. **VHS, Beta, LV, 8mm $14.95** *COL* ♫♫♫

Midnight Faces Mysterious doings abound in a house in the Florida bayous, recently inherited by Bushman. Silent, with original organ music.
1926 72m/B Francis X. Bushman, Rocky Aoki, Jack Perrin, Kathryn McGuire. **VHS, Beta, 8mm $16.95** *SNC, VYY* ♫♫

Midnight Girl A fading opera impresario plots to steal a family's fortune. Lugosi before he became Dracula. Silent.
1925 84m/B Bela Lugosi, Lila Lee. **VHS, Beta, 8mm $19.95** *NOS, VYY, DVT* ♫½

The Midnight Hour A group of high schoolers stumbles upon a vintage curse that wakes up the dead. Made for television. More humor than horror.
1986 97m/C Shari Belafonte-Harper, LeVar Burton, Lee Montgomery, Dick Van Patten, Kevin McCarthy, Jonelle Allen, Peter DeLuise, Dedee Pfeiffer, Mark Blankfield; *Dir:* Jack Bender. **VHS $89.95** *VMK* ♫½

Midnight Lace Acceptable thriller about woman in London who is being harassed by telephone creep. Day is at her frantic best, and Harrrison is suitably charming. Handsome Gavin later became American ambassador to Mexico. Adapted from the play "Matilda Shouted Fire."
1960 108m/C Doris Day, Rex Harrison, John Gavin, Myrna Loy, Roddy McDowall; *Dir:* David Miller. **VHS, Beta $59.95** *MCA* ♫♫½

Midnight Limited A detective sets out to thwart the criminals who would rob the "Midnight Limited" train on its route from New York to Montreal.
1940 61m/B John "Dusty" King, Marjorie Reynolds, George Cleveland, Edward Keane, Pat Flaherty, Monty Collins, I. Stanford Jolley; *Dir:* Howard Bretherton. **VHS, Beta $16.95** *SNC* ♫

Midnight Madness Five teams of college students search the city of Los Angeles for clues leading to hundreds and thousands of dollars in buried treasure.
1980 (PG) 110m/C David Naughton, Stephen Furst, Debra Clinger, Eddie Deezen, Michael J. Fox, Maggie Roswell; *Dir:* David Wechter. **VHS, Beta** *BVV, OM* ♫

Midnight Movie Massacre Retro splatterama has really gross alien outside a movie theater in 1956, and the really weird movie patrons try to terminate it.
1988 86m/C Robert Clarke, Ann Robinson; *Dir:* Mark Stock. **VHS $89.95** *UHV* ♫♫½

Midnight Run An ex-cop, bounty hunter must bring in an ex-mob accountant who has embezzled big bucks from his former boss. After he catches up with the thief, the hunter finds that bringing his prisoner from New York to Los Angeles will be very trying, especially when it is apparent that the Mafia and FBI are out to stop them. The friction between the two leads—De Niro and Grodin—is fun to watch, while the action and comic moments are enjoyable.
1988 (R) 125m/C Robert De Niro, Charles Grodin, Yaphet Kotto, John Ashton, Dennis Farina, Joe Pantoliano; *Dir:* Martin Brest. **VHS, Beta, LV $19.95** *MCA* ♫♫♫

Midnight Warning A woman thinks she's going off the deep-end when her brother and all records of his existence disappear. Based on an urban legend that appeared somewhere around the time of the 1893 World's Fair. Remade in 1952 as "So Long at the Fair."
1932 63m/B William Boyd, Claudia Dell, Henry Hall, John Harron, Hooper Atchley; *Dir:* Spencer Gordon Bennet. **VHS $12.95** *SNC, NOS, DVT* ♫♫½

Midnight Warrior A reporter strikes it big when he investigates the underside of L.A. nightlife, but things go terribly wrong when he becomes wrapped up in the sleaze.
1989 90m/C Bernie Angel, Michelle Berger, Kevin Bernhardt, Lilly Melgar; *Dir:* Joseph Merhi. **VHS, Beta $29.95** *PAR* 🐾

Midnight at the Wax Museum A man attempts to spend an evening in a wax museum's chamber of horrors, only to find that he is the target of a murder plot. Also known as "Midnight at Madame Tussaud's."
1936 66m/B Lucille Lisle, James Carew, Charles Oliver, Kim Peacock; *Dir:* George Pearson. **VHS $24.95** *NOS, DVT, SNC* 🐾

Midnite Spares A young man's search for the men who kidnapped his father leads him into the world of car thieves and chopshops.
1985 90m/C Bruce Spence, Gia Carides, James Laurie. **VHS, Beta $19.95** *VCL* 🐾🐾

A Midsummer Night's Dream Famed Reinhardt version of the Shakespeare classic, featuring nearly every star on the Warner Brothers lot. The plot revolves around the amorous battle between the king and queen of a fairy kingdom, and the humans who are drawn into their sport. Features de Havilland's first film role. Classic credit line: Dialogue by William Shakespeare. Received Oscar nominations for Best Picture, Best Film Editing and Best Cinematography.
1935 117m/B James Cagney, Dick Powell, Joe E. Brown, Hugh Herbert, Olivia de Havilland, Ian Hunter, Mickey Rooney, Victor Jory, Arthur Treacher, Billy Barty; *Dir:* Max Reinhardt, William Dieterle. Academy Awards '35: Best Cinematography, Best Film Editing. **VHS, Beta $19.98** *FOX* 🐾🐾🐾

A Midsummer Night's Dream Shakespeare has created some of his most fanciful and unforgettable characters in this tale of devilish fairies, bewitched lovers and stolid workingmen-cum-actors.
1982 120m/C Helen Mirren, Peter McEnery, Brian Clover. **VHS, Beta $19.98** *FOX, TLF* 🐾🐾½

A Midsummer Night's Sex Comedy Allen's homage to Shakespeare, Renoir, Chekhov, Bergman, and who knows who else in an engaging ensemble piece about hijinx among friends and acquaintances gathered at a country house at the turn of the century. Standouts include Ferrer as pompous professor and Steenburgen as Allen's sexually repressed wife. Mia's first for the Woodman.
1982 (PG) 88m/C Woody Allen, Mia Farrow, Mary Steenburgen, Tony Roberts, Julie Hagerty, Jose Ferrer; *W/Dir:* Woody Allen. **VHS, Beta $19.98** *WAR* 🐾🐾🐾½

Midway The epic World War II battle of Midway, the turning point in the war, is retold through Allied and Japanese viewpoints by a big all-star cast saddled with dumpy dialogue and enough weaponry to seize Hollywood on any given Wednesday.
1976 (PG) 132m/C Charlton Heston, Henry Fonda, James Coburn, Glenn Ford, Hal Holbrook, Robert Mitchum, Cliff Robertson, Robert Wagner, Kevin Dobson, Christopher George, Toshiro Mifune, Tom Selleck; *Dir:* Jack Smight. **VHS, Beta, LV $19.95** *MCA, TLF* 🐾🐾

Mighty Joe Young Tongue-in-cheek King Kong variation features giant ape brought to civilization and exploited in a nightclub act, whereupon things get darned ugly. Bullied and given the key to the liquor cabinet, mild-mannered Joe goes on a drunk-

en rampage, but eventually redeems himself by rescuing orphans from a fire. Special effects (courtesy of Willis O'Brien and the great Ray Harryhausen) probably film's greatest asset. Also available colorized.
1949 94m/B Terry Moore, Ben Johnson, Robert Armstrong, Frank McHugh; *Dir:* Ernest B. Schoedsack. Academy Awards '49: Best Special Effects. **VHS, Beta, LV $19.95** *TTC, KOV, VID* 🐾🐾½

Mighty Jungle Lost in the Amazon jungle, a hunter must fight off killer iguanas, man-eating crocodiles and bloodthirsty natives who perform human sacrifices.
1964 90m/C Marshall Thompson, David Dalie. **VHS, Beta** *PGN* 🐾

The Mighty Pawns Inner-city kids turn to chess when their teacher inspires them to stay off the streets. Originally aired on PBS as part of the WonderWorks Family Movie television series.
1987 58m/C Paul Winfield, Alfonso Ribeiro, Terence Knox, Rosalind Cash, Teddy Wilson; *Dir:* Eric Laneuville. **VHS $29.95** *PME* 🐾🐾

The Mighty Quinn While investigating the local murder of a rich white guy the black Jamaican head of police becomes convinced that the prime suspect, an offbeat childhood friend, is innocent. As the police chief, Denzel is good in this off-beat comedy mystery.
1989 (R) 98m/C Denzel Washington, Robert Townsend, James Fox, Mimi Rogers, M. Emmet Walsh, Sheryl Lee Ralph, Esther Rolle; *Dir:* Carl Schenkel. **VHS, Beta, LV $19.98** *FOX* 🐾🐾½

Mignon Opera based on Goethe's novel, "Wilhelm Meisters Lehrjahre." Sutherland and her puppets prepare the viewer for the performance, which is accompanied by the London Symphony Orchestra.
1973 30m/C Joan Sutherland. **VHS, Beta** *BFA* 🐾🐾½

The Mikado Japanese emperor's son poses as a commoner to better woo his dream gal in this acceptable take on the classic Gilbert and Sullivan comic operetta. Features members of the D'Oyly Carte Company. Big tunes such as "I've Got a Little List" and "Three Little Maids."
1939 90m/C *GB* Kenny L. Baker, Sydney Granville, Jean Colin, Martyn Green; *Dir:* Victor Schertzinger. **VHS $19.95** *NOS, DVT, MLB* 🐾🐾½

Mike's Murder Disjointed drama about a shy bank teller who falls for a slick tennis player. When he is murdered, she investigates the circumstances, placing herself in dangerously close contact with his seedy, drug-involved buddies. The twists and confused plot leave the viewer bewildered.
1984 (R) 109m/C Debra Winger, Mark Keyloun, Paul Winfield, Darrell Larson, Dan Shor, William Ostrander; *Dir:* James Bridges. **VHS, Beta $79.95** *WAR* 🐾½

Mikey Mikey seems like such a sweet little boy—but these awful things keep happening all around him. At every foster home and every school people have such dreadful, and deadly, accidents. But innocent Mikey couldn't be to blame—or could he.
1992 (R) ?m/C Ashley Laurence, John Diehl, Lyman Ward, Josie Bisset; *Dir:* Dennis Dimster. **VHS $89.95** *IMP* 🐾½

Mikey & Nicky Quirky, uneven film about longtime friends dodging a hit man during one long night. Bears little evidence of time and money invested. Cassavetes and Falk, however, provide some salvation.
1976 (R) 105m/C John Cassavetes, Peter Falk, Ned Beatty, Oliver Clark, William Hickey; *W/Dir:* Elaine May. **VHS, Beta $69.95** *WAR, OM* 🐾🐾½

The Milagro Beanfield War Redford's endearing adaptation of John Nichols's novel about New Mexican townfolk opposing development. Seemingly simple tale provides plenty of insight into human spirit. Fine cast, with especially stellar turns from Blades, Braga, and Vennera. Oscar-winning score by Dave Grusin.
1988 (R) 118m/C Chick Vennera, John Heard, Ruben Blades, Sonia Braga, Daniel Stern, Julie Carmen, Christopher Walken, Richard Bradford, Carlos Riqueline, James Gammon, Melanie Griffith, Freddy Fender, M. Emmet Walsh; *Dir:* Robert Redford. **VHS, Beta, LV $19.95** *MCA* 🐾🐾🐾½

Mildred Pierce Gripping melodrama features Crawford as hard-working divorcee rivaling daughter for man's love. Things, one might say, eventually get ugly. Adaptation of James M. Cain novel is classic of its kind. Arden's performance earned her a nomination for Best Supporting Actress; also earned nominations for Best Picture, Best Supporting Actress (Blyth), Screenplay, Black & White Cinematography.
1945 113m/B Joan Crawford, Jack Carson, Zachary Scott, Eve Arden, Ann Blyth, Bruce Bennett; *Dir:* Michael Curtiz. Academy Awards '45: Best Actress (Crawford). **VHS, Beta, LV $19.98** *MGM, FOX, BTV* 🐾🐾🐾

Miles to Go A successful businesswoman tries to enjoy her last days after learning she has terminal cancer.
1986 88m/C Mimi Kuzyk, Tom Skerritt, Jill Clayburgh; *Dir:* David Greene. **VHS, Beta $79.95** *NSV* 🐾½

Miles to Go Before I Sleep A retired, lonely man and a delinquent girl distrust each other but ovontually thoy reach out in mutual need.
1974 78m/C Martin Balsam, MacKenzie Phillips, Kitty Winn, Elizabeth Wilson; *Dir:* Fielder Cook. **VHS, Beta** *LCA* 🐾🐾½

Miles from Home Times are tough for farmers Frank and Terry Roberts. The brothers are about to have another bad harvest and the bank is threatening to foreclose. In a symbolic last ditch effort to save their pride, they decide to burn the farm and leave. On their journey, they meet strangers who recognize the pair and help them escape the police. A melodrama with many members of Chicago's Steppenwolf Theater.
1988 (R) 113m/C Richard Gere, Kevin Anderson, John Malkovich, Brian Dennehy, Judith Ivey, Penelope Ann Miller, Laurie Metcalf, Laura San Giacomo, Daniel Roebuck, Helen Hunt; *Dir:* Gary Sinise. **VHS, Beta, LV $19.98** *WAR* 🐾🐾

Militant Eagle Fight of good vs. evil, complete with nobles, warriors and villains who fight to the death.
19?? 90m/C Choi Yue, Lu Ping, Pai Ying. **VHS, Beta $39.95** *MAV* 🐾

Milky Way Loopy comedy about milkman who finds unhappiness after accidently knocking out champion boxer. Adequate, but not equal to Lloyd's fine silent productions.
1936 89m/B Harold Lloyd, Adolphe Menjou, Verree Teasdale, Helen Mack, William Gargan; *Dir:* Leo McCarey. **VHS, Beta, LV $29.95** *NOS, MRV, VYY* 🐾🐾½

The Milky Way Wicked anti-clerical farce. Two bums team on religious pilgrimage and encounter seemingly all manner of strangeness and sacrilege in this typically peculiar Bunuel work. Perhaps the only film in which Jesus is encouraged to shave. In French with English subtitles.

1968 102m/C *FR* Laurent Terzieff, Paul Frankeur, Delphine Seyrig, Alain Cuny, Bernard Verley, Michel Piccoli, Edith Scon; *Dir:* Luis Bunuel. New York Film Festival '74: Best Actress. **VHS, Beta, LV** $24.95 *XVC, MED, FCT* ✍✍✍

The Mill on the Floss Based on George Eliot's classic novel, this film follows the course of an ill-fated romance and family hatred in rural England. Underwhelming, considering the source.
1937 77m/B *GB* James Mason, Geraldine Fitzgerald, Frank Lawton, Victoria Hopper, Fay Compton, Griffith Jones, Mary Clare; *Dir:* Tim Whelan. **VHS, Beta** $19.95 *NOS, MRV, KRT* ✍✍

Mill of the Stone Women Sculpture-studying art student encounters strange carousel with beautiful babes rather than horseys, and soon finds out that the statues contain shocking secrets. Filmed in Holland, it's offbeat and creepy.
1960 94m/C *FR IT* Pierre Brice, Scilla Gabel, Danny Carrel, Wolfgang Preiss, Herbert Boenne, Liana Orfei; *Dir:* Giorgio Ferroni. **VHS** $19.99 *PGN* ✍✍ ½

Millenium The Earth of the future is running out of time. The people are sterile and the air is terrible. To keep the planet viable, Ladd and company must go back in time and yank people off planes that are doomed to crash. Great special effects and well-thought out script make this a ball of fun.
1989 (PG-13) 108m/C Kris Kristofferson, Cheryl Ladd, Daniel J. Travanti, Lloyd Bochner; *Dir:* Michael Anderson Sr. **VHS, Beta, LV** $9.99 *CCB, IVE* ✍✍ ½

Miller's Crossing From the Coen brothers (makers of "Blood Simple" and "Raising Arizona") comes this extremely dark entry in the gangster movie sweepstakes of 1990. Jewish, Italian and Irish mobsters spin webs of deceit, protection and revenge over themselves and their families. Byrne is the protagonist, but no hero, being as deeply flawed as the men he battles. Harden stuns as the woman who sleeps with Byrne and his boss, Finney, in hopes of a better life and protection for her small-time crook brother. Visually exhilarating, excellently acted and perfectly paced.
1990 (R) 115m/C Albert Finney, Gabriel Byrne, Marcia Gay Harden, John Turturro, Jon Polito, J.E. Freeman; *W/Dir:* Joel Coen. **VHS, Beta, LV** $19.98 *FOX, FCT* ✍✍✍ ½

Millhouse: A White Comedy A satirical look at the life and political career of Richard M. Nixon.
1971 90m/B *Dir:* Emile DeAntonio. **VHS, Beta, 3/4U** $39.98 *IHF, NEW, MPI* ✍✍✍ ½

Millie Creaky melodrama about a divorcee who wants every man but one. He pursues her teenage daughter instead, leading to tragedy and courtroom hand-wringing. Twelvetrees still stands out in this hokum, based on a Donald Henderson novel considered daring in its day.
1931 85m/B Helen Twelvetrees, Robert Ames, Lilyan Tashman, Joan Blondell, Peter Halliday, James Hall, Anita Louise, Frank McHugh; *Dir:* John Francis Dillon. **VHS** $19.95 *NOS* ✍ ½

Million Dollar Duck A family duck is doused with radiation and begins to lay gold eggs. Okay Disney family fare, especially for youngsters.
1971 (G) 92m/C Dean Jones, Sandy Duncan, Joe Flynn, Tony Roberts; *Dir:* Vincent McEveety. **VHS, Beta** *DIS, OM* ✍ ½

Million Dollar Haul Tarzan sniffs out a ring of warehouse thieves.
1935 60m/B Tarzan the Wonder Dog, Reed Howes, Janet Chandler, William Farnum. **VHS, Beta, LV** *WGE* ✍

Million Dollar Kid When a group of thugs wreak havoc in the neighborhood, the East Side Kids try to help a wealthy man put a stop to it. They face an even greater dilemma when they discover that the man's son is part of the gang. Part of the "Bowery Boys" series.
1944 65m/B Leo Gorcey, Huntz Hall, Gabriel Dell, Louise Currie, Noah Beery, Iris Adrian, Mary Gordon; *Dir:* Wallace Fox. **VHS** $24.95 *NOS, DVT* ✍✍

Million Dollar Mermaid The prototypical Williams aquashow, with the requisite awesome Berkeley dance numbers. As a biography of swimmer Annette Kellerman it isn't much, but as an MGM extravaganza, it fits the bill.
1952 115m/C Esther Williams, Victor Mature, Walter Pidgeon, Jesse White, David Brian, Maria Tallchief, Howard Freeman, Busby Berkeley; *Dir:* Mervyn LeRoy. **VHS, Beta** $19.98 *MGM* ✍✍

Million Dollar Mystery A dying man's last words indicate that several million dollars have been stashed near a diner. Chaos breaks out as nearly everyone in town tries to dig up the loot.
1987 (PG) 95m/C Eddie Deezen, Penny Baker, Tom Bosley, Rich Hall, Wendy Sherman, Rick Overton, Mona Lyden; *Dir:* Richard Fleischer. **VHS, LV** $79.95 *HBO* ✍ ½

A Million to One An athlete trains for the Olympic decathlon, and catches the eye of an upper-class woman in the process. The lead actor was actually a shot-putter in the 1932 Olympic Games.
1937 60m/B Herman Brix, Joan Fontaine, Bruce Bennett, Monte Blue, Kenneth Harlan, Reed Howes; *Dir:* Lynn Shores. **VHS** *VYY* ✍ ½

Milo Milo Comedy about some very silly French people who propose to steal the Venus de Milo from the Louvre.
196? 110m/C Antonio Fargas, Joe Higgins. **VHS, Beta** $39.95 *VCD* ✍ ½

The Milpitas Monster A creature spawned in a Milpitas, California, waste dump terrorizes the town residents.
1975 (PG) 80m/C Doug Hagdahl, Scott A. Henderson, Scott Parker; *Dir:* Robert L. Burrill; *Nar:* Paul Frees. **VHS, Beta** $19.98 *UHV* ✍

Mimi Melodrama with a down-on-his-luck playwright falling for a poor lass who provides him with the needed inspiration to be a winner. The question is, will she die before he hits the big time? Loosely based on "La Vie de Boheme" by Henri Murger.
1935 98m/B *GB* Douglas Fairbanks Jr., Gertrude Lawrence, Diana Napier, Harold Warrender, Carol Goodner, Richard Bird, Austin Trevor, Lawrence Hanray, Paul Graetz, Martin Walker; *Dir:* Paul Stein. **VHS** $29.95 *FCT* ✍✍

Min & Bill Patchy early talkie about two houseboat dwellers fighting to preserve their waterfront lifestyle and keep their daughter from being taken to a "proper" home.
1930 66m/B Marie Dressler, Wallace Beery, Marjorie Rambeau, Dorothy Jordan; *Dir:* George Roy Hill. Academy Awards '31: Best Actress (Dressler). **VHS, Beta** $19.98 *MGM, BTV, FHE* ✍✍

Mind Games A young couple and their ten-year-old son pick up a hitcher in their mobile home, never suspecting that he's a deranged psychology student ready to pit

the family against each other as an experiment.
1989 (R) 93m/C Edward Albert, Shawn Weatherly, Matt Norero, Maxwell Caulfield; *Dir:* Bob Yari. **VHS, Beta** $79.98 *FOX, HHE* ✍ ½

Mind Snatchers An American G.I. becomes involved in U.S. Army experimental psychological brain operations when he is brought into a western European hospital for treatment. Also know as "The Happiness Cage."
1972 (PG) 94m/C Christopher Walken, Ronny Cox, Ralph Meeker, Joss Ackland; *Dir:* Bernard Girard. **VHS, Beta** $59.95 *PSM* ✍ ½

Mind Trap A beautiful movie star's family is killed due to her father's involvement with naval research. The government wants no part of her troubles so she is forced to take the law into her own hands.
1991 90m/C Dan Haggerty, Lyle Waggoner, Martha Kincare, Thomas Elliot, Samuel Steven; *Dir:* Eames Demetrios. **VHS** $48.00 *AMI* ✍✍ ½

Mind Warp A future society exercises mental control and torture over its citizens. Also known as "Grey Matter" and originally titled, "The Brain Machine."
1972 (R) 92m/C James Best, Barbara Burgess, Gil Peterson, Gerald McRaney, Marcus J. Grapes, Doug Collins; *Dir:* Joy Houck Jr. **VHS, Beta** $19.95 *ACA* ✍ ½

Mindfield An innocent man gets trapped by the CIA and is used by them in mind control experiments.
1989 (R) 91m/C *CA* Michael Ironside, Lisa Langlois, Christopher Plummer, Stefan Wodoslowsky, Sean McCann; *Dir:* Jean-Claude Lord. **VHS, Beta, LV** $89.98 *MAG* ✍ ½

Mindkiller A tongue-in-cheek gore-fest about a shy young guy who develops his brain in an effort to be socially accepted. Unfortunately, he overdoes it and his brain mutates, bursts from his head and runs around on its own. An aspiring cult classic.
1987 84m/C Joe McDonald, Christopher Wade, Shirley Ross, Kevin Hart; *Dir:* Michael Krueger. **VHS, Beta** $79.95 *PSM* ✍ ½

Mindwalk A feature-length intellectual workout; a physicist, poet and politician stroll the ancient grounds of Mont St. Michel monastery in France and discuss the need to change mankind's view of the universe. Uncinematic? Perhaps, but it entertainingly conveys the epochal ideas of scientist/author Fritjof Capra (whose book was adapted for the screen by younger brother Bernt).
1991 (PG) 111m/C Liv Ullman, Sam Waterston, John Heard; *Dir:* Bernt Capra. **VHS** *PAR* ✍✍✍

Mindwarp Because of ecological disaster, future residents of Earth live in a sterile place called "Inworld" and program their dreams through a computer system that hooks to the base of the skull. Judy (Alicia) seeks answers about her missing father by hooking into her mother's brain, accidentally killing her. Judy is cast out of "Inworld" to Earth's wasteland where she is captured by "Crawlers" but soon rescued by human survivor, Stover (Campbell). Together they fight the horrid environment, but soon they are both captured by the "Crawlers." Judy discovers the awful truth about her father in the "Crawlers" landfill world, and the truth may kill her.
1991 (R) 91m/C Marta Alicia, Bruce Campbell, Angus Scrimm, Elizabeth Kent, Mary Becker; *Dir:* Steve Barnett. **VHS, LV** *COL* ✍✍

Mine Own Executioner Determined but unstable psychologist in postwar London struggles to treat schizophrenic who suffered torture by Japanese while wartime prisoner. Strong, visually engrossing fare.
1947 102m/B *GB* Burgess Meredith, Kieron Moore, Dulcie Gray, Christine Norden, Barbara White, John Laurie, Michael Shepley; *Dir:* Anthony Kimmins. **VHS, Beta $19.95** *NOS, DVT, HHT* ♂♂♂

Mines of Kilimanjaro An archaeology student heads for Africa to investigate his professor's murder and comes up with a heap of trouble.
1987 (PG-13) 88m/C Christopher Connelly, Tobias Hoels, Gordon Mitchell, Elena Pompei; *Dir:* Mino Guerrini. **VHS, Beta $79.95** *IMP* ♂½

Ministry of Vengeance A psychotic murderer who hates grapes of any kind finds the woman of his dreams - she's a psychotic murderer who hates grapes! Together they find bliss - until the plumber discovers their secret!
1989 (R) 90m/C John Schneider, Ned Beatty, George Kennedy, Yaphet Kotto, James Tolken, Apollonia. **VHS, Beta, LV $89.95** *IVE* ♂

Minnesota Clay A blind gunman, who aims by sound and smell, is marked by two rival gangs and a tempestuous tramp.
1965 89m/C *FR IT* Cameron Mitchell, Diana Martin; *Dir:* Sergio Corbucci. **VHS, Beta, LV** *WGE* ♂½

A Minor Miracle This G-rated movie is tender family fare about a group of orphaned children who band together under the loving guidance of their guardian (Huston) to save St. Francis School for boys from the town planners. Warms the cockles.
1983 (G) 100m/C John Huston, Pele, Peter Fox. **VHS, Beta, LV $9.98** *SUE* ♂♂

Minsky's Follies Recreation of an old time burlesque revue complete with strippers.
1983 60m/C Phyllis Diller, Rip Taylor, Stubby Kaye. **VHS, Beta $39.95** *RKO* ♂♂

Minute Movie Masterpieces "Speed talkers" David Starns and Judith Silinsky host this program which condenses 30 classic films into 30 fast minutes. Includes "The Lady Vanishes," "It's a Wonderful Life," "The Third Man," "The Birth of a Nation," "Dr. Jekyll and Mr. Hyde," and 25 more.
1989 30m/C David Starns, Judith Silinsky. **VHS, Beta $14.95** *RHI* ♂♂

A Minute to Pray, A Second to Die A notorious, wanted-dead-or-alive gunman retreats to the amnesty of the New Mexico Territory, but finds he cannot shake his past. Also on video as "Outlaw Gun."
1967 100m/C *IT* Robert Ryan, Arthur Kennedy, Alex Cord; *Dir:* Franco Giraldi. **VHS, Beta $19.98** *FOX, TPV* ♂♂

The Miracle An innocent peasant woman is seduced by a shepherd and becomes convinced that her pregnancy will produce a second Christ. Controversial, compelling film derived from a story by Federico Fellini. In Italian with English subtitles. Also called "Ways of Love."
1948 43m/B *IT* Anna Magnani, Federico Fellini; *Dir:* Roberto Rossellini. **VHS, Beta $24.95** *NOS, FCT, APD* ♂♂♂

The Miracle Irish teens, whose strong friendship is based upon their equally unhappy home lives, find both tested when a secretive American woman turns up in town.

Excellent debuts from Byrne and Pilkington but tedious pacing in a dreamy script that tells too much too soon. Worth watching just for D'Angelo's smouldering rendition of "Stardust."
1991 (PG) 97m/C *GB* Beverly D'Angelo, Donal McCann, Niall Byrne, Lorraine Pilkington; *W/Dir:* Neil Jordan. **VHS, LV $89.98** *LIV* ♂♂½

Miracle on 34th Street The actual Kris Kringle is hired as Santa Claus for the Macy's Thanksgiving parade but finds difficulty is proving himself to the cynical parade sponsor. When the boss's daughter also refuses to acknowledge Kringle, he goes to extraordinary lengths to convince her. Holiday classic equal to "It's a Wonderful Life," with Gwenn and Wood particularly engaging. Oscar nomination for Best Picture. Also available colorized.
1947 97m/B Maureen O'Hara, John Payne, Edmund Gwenn, Natalie Wood, William Frawley, Porter Hall, Gene Lockhart, Thelma Ritter, Jack Albertson; *W/Dir:* George Seaton. Academy Awards '47: Best Screenplay, Best Story, Best Supporting Actor (Gwenn). **VHS, Beta, LV $14.98** *CCB, FOX, BTV* ♂♂♂♂

The Miracle of the Bells A miracle occurs after a dead movie star is buried in the cemetary of her modest hometown. Given the premise, the casting is all the more peculiar. Adapted by Ben Hecht from Russell Janney's novel.
1948 120m/B Fred MacMurray, Alida Valli, Frank Sinatra, Lee J. Cobb; *Dir:* Irving Pichel. **VHS, Beta $19.98** *REP* ♂♂

Miracle Down Under A family endures arduous times in 1890s Australia before a Christmas miracle changes their fortunes. Inoffensive Disney-produced drama. Australia has two directors named George Miller—one directs Mad Max movies; the other directs family fare such as "The Man from Snowy River." This Miller is the latter.
1987 101m/C *AU* Dee Wallace Stone, John Waters, Charles Tingwell, Bill Kerr, Andrew Ferguson; *Dir:* George Miller. **VHS, Beta $29.95** *BVV* ♂♂

Miracle of the Heart: A Boys Town Story A television movie based on the story of Boys Town, about an old priest sticking up for Boys Town's principles in the face of a younger priest with rigid ideas.
1986 100m/C Art Carney, Casey Siemaszko, Jack Bannon; *Dir:* George Stanford Brown. **VHS, Beta $69.95** *COL* ♂♂

Miracle on Ice Occasionally stirring television film recounts the surprise triumph of the American hockey team over the touted Soviet squad during the 1980 Winter Olympics at Lake Placid.
1981 150m/C Karl Malden, Steve Guttenberg, Andrew Stevens, Lucinda Dooling, Jessica Walter; *Dir:* David Hillard Stern. **VHS, Beta $59.95** *TRY* ♂♂

Miracle in Milan An innocent, child-like fantasy about heavenly intervention driving capitalists out of a Milanese ghetto and helping the poor to fly to a new Utopia. Happy mixture of whimsicality and neo-realism, co-written with Cesare Zavattini. In Italian with English subtitles.
1951 95m/B *IT* Francesco Golisano, Brunella Bova, Emma Gramatica, Paolo Stoppa; *Dir:* Vittorio DeSica. Cannes Film Festival '51: Best Film (tie); National Board of Review Awards '51: 5 Best Foreign Films of the Year; New York Film Critics Awards '51: Best Foreign Film. **VHS, Beta, LV $79.95** *HMV, FCT, VYG* ♂♂♂

Miracle Mile A riveting, apocalyptic thriller about a mild-mannered misfit who, while inadvertently standing on a street corner at 2 a.m., answers a ringing pay phone. The caller is a panicked missile-silo worker who announces that the bombs have been launched for an all-out nuclear war. With about an hour left before the end, he decides to head into the city and rendezvous with his new girlfriend. A surreal, wicked farce sadly overlooked in theatrical release. Music by Tangerine Dream.
1989 (R) 87m/C Anthony Edwards, Mare Winningham, John Agar, Denise Crosby, Lou Hancock; *W/Dir:* Steve DeJarnatt. **VHS, Beta, LV $89.99** *HBO* ♂♂♂

Miracle at Moreaux Three Jewish children fleeing from Nazis find sanctuary with a nun and her wards. Based on Clare Huchet Bishop's book "Twenty and Ten." Originally aired on PBS as part of the WonderWorks Family Movie series.
1986 58m/C Loretta Swit, Marsha Moreau, Robert Joy, Ken Pogue, Robert Kosoy, Talya Rubin; *Dir:* Paul Shapiro. **VHS $29.95** *PME, HMV* ♂♂

Miracle of Morgan's Creek The breakneck comedy that stands as Sturges' premier achievement. Details the misadventures of a wartime floozy who gets drunk, thinks she marries a soldier on leave, gets pregnant by him, forgets the whole thing, and then tries to evade scandal by getting a local schnook to marry her again. And she's expecting sextuplets. Hilarious, out-to-make-trouble farce that shouldn't have, by all rights, made it past the censors of the time. Sturges' most scathing assault on American values earned him an Oscar nomination for writing. Remade as "Rock-A-Bye Baby" in 1958.
1944 98m/B Eddie Bracken, Betty Hutton, Diana Lynn, Brian Donlevy, Akim Tamiroff, Porter Hall, Emory Parnell, Alan Bridge, Julius Tannen, Victor Potel, Almira Sessions, Chester Conklin, William Demarest; *W/Dir:* Preston Sturges. **VHS, Beta $14.95** *PAR, CCB* ♂♂♂♂

Miracle of Our Lady of Fatima Slick cold-war version of the supposedly true events surrounding the sighting of a holy vision by three children in Portugal during World War I. Max Steiner's score was Oscar nominated.
1952 102m/C Gilbert Roland, Susan Whitney, Sherry Jackson, Sammy Ogg, Angela Clark, Frank Silvera, Jay Novello; *Dir:* John Brahm. **VHS, Beta, LV $19.98** *WAR, IGP* ♂♂

Miracle Rider: Vol. 1 Old West guns-and-hero tale. In fifteen chapters.
1935 195m/B Tom Mix, Joan Gale, Charles Middleton; *Dir:* Armand Schaefer. **VHS, Beta $49.95** *NOS, MRV, VCN* ♂♂

Miracle Rider: Vol. 2 Serial covers excitement in the Old West.
1935 135m/C Tom Mix, Joan Gale, Charles Middleton; *Dir:* Armand Schaefer. **VHS, Beta $69.95** *VCN, MRV, VYY* ♂½

Miracle in Rome Man digs up dead daughter and finds her body in perfect condition despite 12-year interment. Attempting to convince the local clergy that she's miracle material, he finds them less than eager to elect her to sainthood. Made for Spanish TV. In Spanish with English subtitles.
1988 76m/C *SP* Frank Ramirez, Gerardo Arellano, Amalia Duque Garcia, Lisandro Duque, Daniel Priolett; *Dir:* Lisandro Duque Naranjo. **VHS $79.95** *FXL, FCT, APD* ♂♂½

Miracle of the White Stallions A disappointing Disney adventure about the director of a Viennese riding academy who guides his prized Lipizzan stallions to safety when the Nazis occupy Austria in World War II.
1963 92m/C Robert Taylor, Lilli Palmer, Eddie Albert, Curt Jurgens; *Dir:* Arthur Hiller. **VHS, Beta** $69.95 *DIS* ⅋½

The Miracle Worker Depicts the unconventional methods that teacher Anne Sullivan used to help the deaf and blind Helen Keller adjust to the world around her and shows the relationship that built between the two courageous women. An intense, moving experience, with both Duke and Bancroft delivering Oscar-winning performances. Oscar-nominated William Gibson adapted his own play for the screen. Oscar nominations for director Penn, stars Bancroft and Duke, adapted screenplay, and costume design.
1962 97m/B Anne Bancroft, Patty Duke, Victor Jory, Inga Swenson, Andrew Prine, Beah Richards; *Dir:* Arthur Penn. Academy Awards '62: Best Actress (Bancroft), Best Supporting Actress (Duke); National Board of Review Awards '62: 10 Best Films of the Year, Best Actress (Bancroft). **VHS, Beta** $19.98 *MGM, BTV, CCB* ⅋⅋⅋

The Miracle Worker Remade for television story of blind, deaf and mute Helen Keller and her teacher, Annie Sullivan, whose patience and perseverance finally enable Helen to learn to communicate with the world. Keller in the 1962 original, Duke plays the teacher here.
1979 98m/C Patty Duke, Melissa Gilbert; *Dir:* Paul Aaron. **VHS, Beta** $59.95 *WAR, OM* ⅋⅋⅋

Miracles Comedic misfire pairs a recently divorced couple as reluctant adventurers dodging jewel theives and peculiar tribes in South America. In a desperate move to create a plot, Mexican jewel thieves, strange tribal rites, and some outrageously unlikely coincidences reunite the family. If this sounds good, see "Romancing the Stone" instead.
1986 (PG) 90m/C Tom Conti, Teri Garr, Paul Rodriguez, Christopher Lloyd, Jorge Russek, Charles Rocket, Bob Nelson; *Dir:* Jim Kouf. **VHS, Beta, LV** $79.99 *HBO, IME* ⅋½

Mirage An amnesiac finds himself the target of a dangerous manhunt in New York City in this offbeat thriller. Peck is particularly sympathetic in the lead, and McCarthy, Matthau, and Kennedy all shine in supporting roles. Worth it just to hear thug Kennedy grunt, "I owe this man some pain!" Music score by Quincy Jones.
1966 108m/B Gregory Peck, Diane Baker, Walter Matthau, Jack Weston, Kevin McCarthy, Walter Abel, George Kennedy; *Dir:* Edward Dmytryk. **VHS, Beta** $14.95 *KRT, MLB* ⅋⅋

Mirele Efros A widowed but successful businesswoman finds herself at odds with her daughter-in-law in this film set in the turn of the century. In Yiddish with English subtitles.
1938 80m/B Berta Gersten, Michael Rosenberg, Ruth Elbaum, Albert Lipton, Sarah Krohner, Moishe Feder; *Dir:* Joseph Berne. **VHS, Beta** $79.95 *ERG* ⅋⅋½

The Mirror Wonderful child's view of life in Russia during WWII. Black and white flashbacks of important events in the country's history are interspersed with scenes of day-to-day family life. In Russian with English subtitles.
1976 106m/C *RU* Margarita Terekhova, Philip Yankovsky, Ignat Daniltsev, Oleg Yankovsky; *Dir:* Andrei Tarkovsky. **VHS** *TPV* ⅋⅋⅋

The Mirror Crack'd While filming a movie in the English countryside, an American actress is murdered and Miss Marple must discover who the killer is. Based on the substantially better Agatha Christie novel.
1980 (PG) 105m/C *GB* Elizabeth Taylor, Rock Hudson, Kim Novak, Tony Curtis, Edward Fox, Geraldine Chaplin, Charles Gray, Pierce Brosnan; *Dir:* Guy Hamilton. **VHS, Beta, LV** $19.99 *HBO* ⅋⅋

Mirror of Death A woman subject to physical abuse gets more than she bargains for when she seeks revenge by unleashing the gruesome Queen of Hell. Watch this if you dare!
1987 85m/C Julie Merrill, Kuri Browne, John Reno, Deryn Warren. **VHS, Beta** $14.95 *SVS* ⅋

Mirror Images Although sexy twins Kaitlin and Shauna may look alike, they couldn't have more diverse personalities. Yet the gorgeous twins find their lives thrown together in a mad tornado of passion and danger when they each encounter a handsome, mysterious stranger. Could it be that this guy has something more on his mind than love, something like... murder? Also available in an even steamier unrated version.
1991 (R) 94m/C Delia Sheppard, Jeff Conaway, Richard Arbolino, John O'Hurley, Korey Mall, Julie Strain, Nels Van Patten; *Dir:* Alexander Gregory Hippolyte. **VHS** $89.95 *ACA* ⅋½

Mirror, Mirror The prolific Ms. Black turns on the ol' black magic when her daughter's classmates decide it's open season for taunting shrinking violets. Thanks to a magical mirror on the wall, the cheerleading classmates are willed to the great pep rally in the sky.
1990 (R) 105m/C Karen Black, Rainbow Harvest, Kristin Datillo, Ricky Paull Goldin, Yvonne de Carlo, William Sanderson; *Dir:* Marina Sargenti. **VHS** $19.95 *ACA* ⅋

Mirrors A woman finds that her dreams lead to death and destruction in New Orleans.
1978 83m/C Kitty Wynn, Peter Donat, William Swetland, Mary-Robin Redd, William Burns; *Dir:* Noel Black. **VHS, Beta** $59.95 *MON* ⅋

Mirrors A Detroit-born ballerina goes to the Big Apple in search of fame and glory, leaving behind her journalist boyfriend. This upbeat love story features some spectacular dance routines and a behind-the-scenes look at a Broadway dancer's lifestyle.
1985 99m/C Timothy Daly, Marguerite Hickey, Antony Hamilton; *Dir:* Harry Winer. **VHS** $89.95 *VMK* ⅋⅋½

Misadventures of Merlin Jones A pair of college sweethearts become embroiled in a rollicking chimp-napping scandal. Disney pap done appropriately. A sequel, "The Monkey's Uncle," followed.
1963 (G) 90m/C Tommy Kirk, Annette Funicello, Leon Ames, Stuart Erwin, Connie Gilchrist; *Dir:* Robert Stevenson. **VHS, Beta** $69.95 *DIS* ⅋½

The Misadventures of Mr. Wilt Bumbling college lecturer has to convince inept inspector that he murdered a blowup doll, not his more missing wife. Based on Tom Sharpe's bestseller. "WILT" was the original British title.
1990 (R) 84m/C *GB* Griff Rhys Jones, Mel Smith, Alison Steadman, Diana Quick; *Dir:* Michael Tuchner. **VHS** *VMK* ⅋⅋

Mischief Alienated youths form a friendship during James Dean's heyday. Warning: this film offers a fairly convincing recreation of the 1950s.

1985 (R) 97m/C Doug McKeon, Catherine Mary Stewart, Kelly Preston, Chris Nash, D.W. Brown, Jami Gertz, Margaret Blye, Graham Jarvis, Terry O'Quinn; *Dir:* Mel Damski. **VHS, Beta, LV** $79.98 *FOX* ⅋⅋

Misery Caan plays an author who decides to chuck his lucrative but unfulfilling pulp novels and get serious with his work by finishing off his most popular character, Misery Chastain, in the final installment of a tired romance series. But on his way to the city in a heavy snowstorm, he crashes on a remote highway and is seriously hurt. Fate seems to intervene when Bates, his biggest fan ever, saves his life with rudimentary medical care. Trouble is, she's a Stephen King kinda gal, and she forces him to resurrect Misery from dead letterdom. Glib, brutal, and ever so balmy, Bates garnered an Oscar for her role. Screenplay adapted from the King story by William Goldman.
1990 (R) 107m/C James Caan, Kathy Bates, Lauren Bacall, Richard Farnsworth, Frances Sternhagen, Graham Jarvis; *Dir:* Rob Reiner. Academy Awards '90: Best Actress (Bates). **VHS, Beta, LV, SVS** $19.95 *COL, ORI, NLC* ⅋⅋⅋

The Misfit Brigade Variation on the "Dirty Dozen" finds German misfits recruited for warfare by desperate Nazis during WWII. This was also shown as "Wheels of Terror."
1987 (R) 99m/C Bruce Davison, Oliver Reed, David Carradine, David Patrick Kelly, D.W. Moffett, Keith Szarabajka; *Dir:* Gordon Hessler. **VHS, Beta** $19.98 *TWE* ⅋⅋

The Misfits A cynical floozy befriends grim cowboys in this downbeat drama. Compelling performances from leads Clift, Monroe (screenwriter Miller's wife), and Gable. Last film for the latter two performers, and nearly the end for Clift.
1961 124m/B Clark Gable, Marilyn Monroe, Montgomery Clift, Thelma Ritter, Eli Wallach, James Barton, Estelle Winwood; *Dir:* John Huston. **VHS, Beta, LV** $19.98 *FOX, FCT* ⅋⅋⅋

Misfits of Science Pilot for a failed television series about a group of teens with special powers who fight to save the world.
1985 96m/C Dean Paul Martin, Kevin Peter Hall, Mark Thomas Miller, Kenneth Mars, Courtney Cox; *Dir:* Philip DeGuere. **VHS, Beta** $39.95 *MCA* ⅋

Mishima: A Life in Four Chapters Somewhat detached account and indulgent portrayal of the narcissistic Japanese author (and actor, filmmaker, and militarist) alternates between stylized interpretations of his books and a straightforward account of his life. Culminates in a pseudo-military operation that, in turn, resulted in Mishima's gruesome suicide. U.S./Japanese production written by Paul and Leonard Schrader. Compelling score by Philip Glass, innovative design by Eiko Ishioka.
1985 (R) 121m/C Ken Ogata, Kenji Sawada, Yasosuke Bando; *Dir:* Paul Schrader; *Nar:* Roy Scheider. **VHS, Beta** $79.95 *WAR* ⅋⅋⅋

Mision Suicida Santo joins forces with a beautiful Interpol agent in his quest to break a spy ring.
1987 90m/C *MX* Santo, Lorena Velasquez, Elsa Cardenas, Dagoberto Rodriguez. **VHS, Beta** $49.95 *MAD* ⅋

Miss A & Miss M A young girl is shocked when she realizes that two teachers she has befriended also live together.
1986 60m/C Kika Markham, Jennifer Hilary. **VHS** $11.95 *PSM* ⅋⅋

Miss All-American Beauty A behind-the-scenes look at a smalltime Texas beauty pageant. Lane stars as a contestant who comes to realize her self-respect is more important than winning an award. Meadow's mean pageant director adds an ironic edge that the movie as a whole lacks. Made for television.
1982 96m/C Diane Lane, Cloris Leachman, David Dukes, Jayne Meadows, Alice Hirson, Brian Kerwin; *Dir:* Gus Trikonis. **VHS, Beta** **$39.95** *IVE ∄ ½*

Miss Annie Rooney Intended to bill the child star in a more mature role, Temple receives her first screen kiss in this story about a poor Irish girl who falls in love with a wealthy young man. The well-worn plot is weighed down by lifeless dialogue.
1942 (G) 86m/B Shirley Temple, Dickie Moore, William Gargan, Guy Kibbee, Peggy Ryan, June Lockhart; *Dir:* Edwin L. Marin. **VHS, Beta $12.95** *IVE, MED ∄*

Miss Firecracker Hunter, longing for love and self-respect, decides to change her promiscuous image by entering the local beauty pageant in her conservative southern hometown. The somewhat drippy premise is transformed by a super script and cast into an engaging and upbeat film. Henley's script was adapted from her own Off-Broadway play where Hunter created the role. Actress Lahti, wife of director Schlamme, makes a brief appearance.
1989 (PG-13) 102m/C Holly Hunter, Scott Glenn, Mary Steenburgen, Tim Robbins, Alfre Woodard, Trey Wilson, Bert Remsen, Ann Wedgeworth, Christine Lahti; *Dir:* Thomas Schlamme. **VHS, Beta, LV $89.99** *HBO ∄∄∄*

Miss Grant Takes Richmond Zipperhead secretary finds herself in hot water a la Lucy when she finds out the company she's been working for is really the front for a gambling getup. Only Lucy fans need apply.
1949 87m/B Lucille Ball, William Holden, Janis Carter, James Gleason, Gloria Henry, Frank McHugh, George Cleveland, Arthur Space, Willaim Wright, Jimmy Lloyd; *Dir:* Lloyd Bacon. **VHS $19.95** *COL, FCT ∄∄*

Miss Julie Melodramatic stew adapted from from the August Strindberg play about a confused noblewoman who disgraces herself when she allows a servant to seduce her. Swedish with English subtitles.
1950 90m/B *SW* Anita Bjork, Ulf Palme, Anders Henrikson; *Dir:* Alf Sjoberg. **VHS, Beta $19.95** *COL, SUE ∄*

Miss Mary A compassionate English governess in Buenos Aires conflicts with the honor-obsessed family she works for and Argentina's tumultuous history. Christie gives a wonderful performance. Some profanity and nudity.
1986 (R) 100m/C *AR* Julie Christie, Donald McIntire, Sofia Viruboff; *Dir:* Maria-Luisa Bemberg. **VHS, Beta $9.99** *STE, NWV ∄∄ ½*

Miss Melody Jones A beautiful young black woman's dreams of stardom unravel as she is forced to pander to prurient men by disrobing publicly for financial recompense.
1973 86m/C Philomena Nowlin, Ronald Warren, Jacqueline Dalya, Peter Jacob; *W/Dir:* Bill Brame. **VHS $39.95** *XVC ∄*

Miss Peach of the Kelly School The students from the famed comic strip celebrate the opening of school, Thanksgiving, Valentine's Day, and the annual picnic. Animated.
1982 115m/C VHS, Beta $19.98 *FOX ∄ ½*

Miss Right A young man determines to find the ideal woman for himself and casts aside his other romantic interests. Sputtering lightweight comedy.
1981 (R) 98m/C William Tepper, Karen Black, Virna Lisi, Margot Kidder, Marie-France Pisier; *Dir:* Paul Williams. **VHS, LV $14.95** *SVS, IME ∄*

Miss Sadie Thompson Based on the novel "Rain" by Somerset Maugham. Promiscuous tart Hayworth arrives on a Pacific island occupied by a unit of Marines and a sanctimonious preacher played by Ferrer (father of Miguel Ferrer of TV's "Twin Peaks" and "On the Air"). While Hayworth parties with the Marines and becomes involved with Ray's Sgt. O'Hara, Ferrer moralizes and insists she return to the mainland to face moral charges. The quasi-musical, with a scattering of dubbed songs by Hayworth, includes a memorable erotic dance scene complete with a tight dress and dripping sweat. Hayworth's strong performance carries the picture with Ferrer and Ray turning in cardboard versions of their Maugham characters. Originally filmed in 3-D.
1954 91m/C Rita Hayworth, Jose Ferrer, Aldo Ray, Charles Bronson; *Dir:* Curtis Bernhardt. **VHS, Beta, LV $19.95** *COL ∄∄∄*

Missile to the Moon The first expedition to the moon encounters not acres of dead rock but a race of gorgeous women in lingerie and high heels. A bad but entertaining remake of "Cat Women of the Moon," featuring a bevy of beauty contest winners from New Hampshire to Yugoslavia. Who says truth is stranger than fiction.
1959 78m/B Gary Clarke, Cathy Downs, K.T. Stevens, Laurie Mitchell, Michael Whalen, Nina Bara, Richard Travis, Tommy Cook, Marjorie Hellen; *Dir:* Richard Cunha. **VHS, Beta $9.95** *RHI, SNC, CNM ∄ ½*

Missiles of October Telling the story of the October 1962 Cuban Missile crisis, this made-for-TV drama keeps you on the edge of your seat while unfolding the sequence of events within the U.S. Government. Well written, with a strong cast including Devane, who turns in a convincing performance as-guess who-J.F.K.
1974 155m/C William Devane, Ralph Bellamy, Martin Sheen, Howard da Silva; *Dir:* Anthony Page. **VHS, Beta $59.95** *MPI, LCA ∄∄∄*

Missing At the height of a military coup in Chile (never named in the movie), a young American writer (Shea) disappears. Lemon, as Shea's father, is a right-wing Christian Scientist who has flown into the country to get to the bottom of his son's disappearance. Shea's wife, played by Spacek, has been living a bohemian lifestyle with her husband in Chile and is the political opposite of her father-in-law. While chafing and bickering, they search for information about the missing Shea, encountering obnoxious American diplomats and scary Chilean officials. Outstanding performances by Spacek and Lemon along with excellent writing and direction result in a gripping and thought-provoking thriller. Based on the book by Thomas Hauser from the true story of Charles Horman.
1982 (PG) 122m/C Jack Lemmon, Sissy Spacek, John Shea, Melanie Mayron, David Clennon, Charles Cioffi, Joe Regalbuto, Richard Venture, Janice Rule; *Dir:* Constantin Costa-Gavras. Academy Awards '82: Best Adapted Screenplay; Cannes Film Festival '82: Best Actor (Lemmon), Best Film. **VHS, Beta, LV $19.95** *MCA ∄∄∄ ½*

Missing in Action An army colonel returns to the Vietnam jungle to settle some old scores and rescue some POWs while on an MIA fact-finding mission. Box office smash.
1984 (R) 101m/C Chuck Norris, M. Emmet Walsh; *Dir:* Joseph Zito. **VHS, Beta, LV $14.98** *MGM, TLF, PIA ∄ ½*

Missing in Action 2: The Beginning Set in Vietnam, this prequel to the original "Missing in Action" provides some interesting background on Norris's rocky relationship with communism. Packed with violence, bloodshed and torture.
1985 (R) 96m/C Chuck Norris, Soon-Teck Oh, Cosie Costa, Steven Williams; *Dir:* Lance Hool. **VHS, Beta, LV $14.98** *MGM ∄*

The Missing Corpse When a newspaper publisher's worst rival is murdered, he fears that he will be implicated. He decides the only thing to do is hide the body to prevent its being discovered. Enjoyable, light comedy/mystery.
1945 62m/B J. Edward Bromberg, Eric Sinclair, Frank Jenks, Isabelle Randolph, Paul Guilfoyle, John Shay, Lorell Sheldon; *Dir:* Albert Herman. **VHS $16.95** *NOS, MRV, SNC ∄∄*

Missing Link A strange, meditative quasi-documentary depicting the singular adventures of a humanoid silently transversing the African wilderness and confronting various natural phenomena. Beautifully photographed. Make-up by Rick Baker.
1988 (PG) 92m/C Peter Elliott, Michael Gambon; *Dir:* David Hughes. **VHS, Beta $89.95** *MCA ∄∄ ½*

Missing Pieces A woman takes a job as a private investigator in order to find her journalist husband's killer. Effective made-for-television mystery taken from the novel "A Private Investigation" by Karl Alexander.
1983 96m/C Elizabeth Montgomery, Louanne, John Reilly, Ron Karabatsos, Robin Gammell, Julius W. Harris; *Dir:* Mike Hodges. **VHS** *IVE ∄∄ ½*

The Mission Sweeping, cinematically beautiful historical drama about an 18th-century Jesuit mission in the Brazilian jungle. The missionaries struggle against the legalized slave trade of Portugal and political factions within the church. Written by Robert Bolt of "A Man for All Seasons" fame, its visual intensity is marred by length and so much overt symbolism that an emotional coolness surfaces when the action slows. Nonetheless, epic in ambition and nearly in quality. Nominated for an Academy Award for best picture. Ennio Morricone's Oscar-nominated musical score is magnificent.
1986 (PG) 125m/C Robert De Niro, Jeremy Irons, Ray McAnally, Philip Bosco, Aidan Quinn, Liam Neeson, Cherie Lunghi, Rev. Daniel Berrigan; *Dir:* Roland Joffe. Academy Awards '86: Best Cinematography; British Academy Awards '86: Best Film Editing, Best Supporting Actor (McAnally), Best Musical Score; Cannes Film Festival '86: Best Film. **VHS, Beta, LV, 8mm $19.98** *WAR, PIA ∄∄∄*

Mission Batangas Unremarkable adventure-war-heist movie tells the story of a shallow American WWII pilot (Weaver) and a missionary nurse (Miles), who team up to steal the Philippine government's entire stock of gold bullion from the Japanese who captured it. Beautiful scenery, shot in the Philippines, adds something to an otherwise vacant effort.
1969 100m/C Dennis Weaver, Vera Miles, Keith Larsen; *Dir:* Keith Larsen. **VHS, Beta, LV $19.95** *STE, NWV ∄ ½*

Mission Corbari One heavily armed and ill-tempered hombre takes on the Nazis, with indifferent results.
197? 88m/C VHS, Beta WES 🎬

Mission to Death An adventure extravaganza in which a group of friends must risk their very lives.
1966 76m/C Jim Brewer, James E. McLarty, Jim Westerbrook, Robert Stolper, Dudley Hafner, Jerry Lasater. **VHS** LBD 🎬

Mission Galactica: The Cylon Attack The Battlestar Galactica is stranded in space without fuel and open to attack from the chrome-covered Cylons. Adama (Greene) is forced to stop Commander Cain's (Bridges) efforts to launch an attack against the Cylons, while countering the attacks of the Cylon leader. Standard made for television sci-fi flick.
1979 108m/C Lorne Greene, Lloyd Bridges, Richard Hatch, Dirk Benedict. **VHS, Beta, LV $19.95** MCA 🎬½

Mission to Glory This dozer tells the story of Father Francisco "Kino" Kin, a tough, seventeenth-century priest in California who took on the Apaches and murderous Conquistadors in defense of his people. Also known as "The Father Kino Story."
1980 (PG) 100m/C Richard Egan, John Ireland, Cesar Romero, Ricardo Montalban, Rory Calhoun, Michael Ansara, Keenan Wynn, Aldo Ray; **Dir:** Ken Kennedy. **VHS, Beta $29.95** WES, IGP 🎬🎬

Mission Inferno The ups and downs of a hot-blooded Korean War commando taking revenge, escaping prison, fighting and just plain making war.
1984 90m/C VHS, Beta $39.95 VCD 🎬

Mission to Kill An American demolitions expert, Ginty, joins a Latin American guerrilla force in battling tyrannical junta forces. This action-packed, fast-paced thriller is big on revenge and violence but not much else.
1985 (R) 97m/C Robert Ginty, Olivia D'Abo, Cameron Mitchell; **Dir:** David Winters. **VHS, Beta $79.95** MED 🎬½

Mission Kiss and Kill Enchanting title. A movie filled with martial arts, beautiful girls, and exotic settings that takes place in the early days of the Chinese Republic.
19?? 90m/C CH Lung Fei, Lu I Lung, Yuan Lung. **VHS, Beta $29.95** OCE 🎬

Mission Manila An ex-CIA operative and Manila-based drug addict is called by his ex-lover to return to Manila to help his brother, who has gotten in much the same trouble he had.
1987 (R) 98m/C Larry Wilcox. **VHS, Beta, LV $79.95** VIR 🎬

Mission Mars American astronauts McGavin and Adams, on a mission to the red planet, discover the bodies of two cosmonauts floating in space. After landing on the planet's surface, they find a third cosmonaut, this one in a state of suspended animation. While putting the viewer to sleep, they proceed to revive the third cosmonaut and have at it with the sinister alien force responsible for all the trouble.
1967 87m/C Darren McGavin, Nick Adams, George DeVries; **Dir:** Nicholas Webster. **VHS, Beta $29.95** UNI 🎬

Mission: Monte Carlo An episode from the television series "The Persuaders," wherein our two vacationing adventurers discover a girl's drowned body and try to solve the case.

1975 102m/C Roger Moore, Tony Curtis, Susan George, Annette Andre, Alfred Marks, Laurence Naismith; **Dir:** Basil Dearden, Roy Ward Baker. **VHS, Beta $59.98** FOX 🎬🎬

Mission in Morocco When an oilman scouting for oil fields in the Sahara is killed, his partner, Barker, hops a jet to Morocco to find the whereabouts of a microfilm that gives the location of a valuable oil deposit. The search for the secret microfilm is complicated by competing parties also interested in the black gold. Ho-hum action adventure.
1959 79m/C Lex Barker, Juli Reding. **VHS $19.98** REP 🎬

Mission Phantom A group of spies on a mission in Russia plans to steal some diamonds and help a woman get to the United States.
1979 90m/C Andrew Ray, Ingrid Sholder, Peter Martell; **Dir:** James Reed. **VHS, Beta $59.95** UHV 🎬

Mission Stardust An internationally produced but thoroughly unambitious adaptation of the once-popular Perry Rhodan sci-fi serial, in which Rhodan and his team bring ill aliens back to Earth and defend them against evil spies. Dubbed.
1968 90m/C GE IT SP Essy Persson, Gianni Rizzo, Lang Jeffries, Pinkas Braun; **Dir:** Primo Zeglio. **VHS, Beta $9.95** RHI 🎬½

The Missionary A mild-mannered English missionary returns to London from his work in Africa and is recruited into saving the souls of a group of prostitutes. Aspiring to gentle comedy status, it's often formulaic and flat, with Palin and Smith fighting gamely to stay above script level. Still, good for some laughs, particularly during the near classic walking butler sequence.
1982 (R) 86m/C GB Michael Palin, Maggie Smith, Trevor Howard, Denholm Elliott, Michael Hordern; **Dir:** Richard Loncraine. **VHS, Beta $19.95** PAR 🎬🎬½

Mississippi Burning The hard-edged social drama centers around the civil rights movement in Mississippi in 1964. When three activists turn up missing, FBI agents Hackman and Dafoe are sent to head up the investigation. Unfortunately, this is another example of a "serious" film about racial conflict in which white characters predominate and blacks provide background. Oscar-nominated for best picture and best actor (Hackman).
1988 (R) 101m/C Gene Hackman, Willem Dafoe, Frances McDormand, Brad Dourif, Lee Ermay, Gailard Sartain, Stephen Tobolowsky, Michael Rooker, Pruitt Taylor Vince, Badja Djola, Kevin Dunn, Frankie Faison, Tom Mason; **Dir:** Alan Parker. Academy Awards '88: Best Cinematography; Berlin Film Festival '88: Best Actor; National Board of Review Awards '88: Best Actor, Best Director, Best Picture, Best Supporting Actress. **VHS, Beta, LV $19.98** ORI 🎬🎬🎬

Mississippi Masala "Masala" is an Indian seasoning made of different-colored spices blended together, as this film is a blend of romance, comedy, and social conscience. An interracial romance sets off a cultural collision in a small Southern town. Mina is a young Indian woman whose family were exiled from Uganda in a purge of Asians by Idi Amin. They've settled, somewhat unwillingly, into running a motel in Greenwood, Mississippi. Demetrius is an ambitious black man with his own carpet-cleaning business. When Mina and Demetrius fall in love both families disapprove and it appears the lovers will be parted by the

racial tensions. Washington and Choudhury are engaging as the lovers with Seth, as Mina's unhappy father, especially watchable.
1992 (R) 118m/C Denzel Washington, Sarita Choudhury, Roshan Seth, Charles Dutton, Sharmila Tagore; **Dir:** Mira Nair. **VHS, LV** COL 🎬🎬🎬

Missouri Breaks Thomas McGuane wrote the screenplay for this offbeat tale of Montana ranchers and rustlers fighting over land and livestock in the 1880s. Promising combination of script, cast and director unfortunately yields rather disappointing results, though both Brando and Nicholson chew up scenery to their hearts' content.
1976 (PG) 126m/C Jack Nicholson, Marlon Brando, Randy Quaid, Kathleen Lloyd, Frederic Forrest, Harry Dean Stanton; **Dir:** Arthur Penn. **VHS, Beta $19.98** FOX 🎬🎬

Missouri Traveler An orphan boy struggles to get his own farm in Missouri. Family fare based on John Buress' novel.
1958 103m/C Lee Marvin, Gary Merrill, Brandon de Wilde, Paul Ford. **VHS, Beta** MRV 🎬🎬🎬

Missourians Small-town sheriff Hale helps a Polish rancher overcome ethnic prejudice among the townspeople. The situation worsens with the arrival in town of the rancher's outlaw brother. Strong direction and engaging storyline.
1950 60m/B Monte Hale, Paul Hurst. **VHS, Beta $29.95** DVT 🎬🎬½

Mistaken Identity An all-black mystery set against the background of a nightclub, and featuring a production number, "I'm a Bangi from Ubangi."
1941 60m/B Nellie Hill, George Oliver. **VHS, Beta $29.95** VCN, DVT, RXM 🎬

Mister Arkadin Screenwriter, director, star Welles, adapting from his own novel, gave this plot a dry run on radio during early 1950s. Welles examines the life of yet another ruthless millionaire, but this one can't seem to remember the sordid source of all that cash. Investigator Arden follows the intriguing and descending trail to a surprise ending. As in "Citizen Kane," oblique camera angles and layered dialogue prevail, but this time only serve to confuse the story. Strong cameo cast. Shot over two years around Europe, required seven years of post production before finding distribution in 1962. British alternate title: "Confidential Report."
1955 99m/B GB Orson Welles, Akim Tamiroff, Michael Redgrave, Patricia Medina, Mischa Auer; **Dir:** Orson Welles. **VHS, Beta $24.95** VDM 🎬🎬🎬

Mister Halpren & Mister Johnson Gleason attends the funeral of a longtime friend and one time flame, and becomes acquainted with the woman's husband (Olivier). Heavyweight cast wasted on flimsy script...and away we go!
1983 57m/C Laurence Olivier, Jackie Gleason; **Dir:** Alvin Rakoff. **VHS, Beta $39.95** USA 🎬🎬

Mister Johnson In 1923 Africa, an educated black man working for the British magistrate constantly finds himself in trouble, thanks to backfiring schemes. This highly enjoyable film from the director of "Driving Miss Daisy" suffers only from the underdevelopment of the intriguing lead character. Based on the novel by Joyce Cary.
1991 (PG-13) 105m/C Pierce Brosnan, Edward Woodward, Maynard Eziashi, Beatie Edney, Denis Quilley, Nick Reding; **Dir:** Bruce Beresford. **VHS $89.98** VES, LIV 🎬🎬🎬

Mister Mean A man who once worked for a mafia don is now given the task of rubbing him out. For Cosa Nostra diehards and gluttons for Roman scenery.
1977 98m/C *IT* Fred Williamson; ***Dir:*** Fred Williamson. **VHS, Beta $79.98** *MAG* 🦴

Mister Roberts The crew of a Navy cargo freighter in the South Pacific during World War II relieves the boredom of duty with a series of elaborate practical jokes, mostly at the expense of their long-suffering and slightly crazy captain, who then determines that he will get even. The ship's cargo officer, Mr. Roberts, longs to be transferred to a fighting vessel and see some action. Originally a hit Broadway play (based on the novel by Thomas Heggen) which also featured Fonda in the title role. Great performance from Lemmon as Ensign Pulver. Powell's last film. In addition to Lemmon's statue, Oscar-nominated for best picture and best sound. Sequelled in 1964 by "Ensign Pulver," and later a short-lived television series as well as a live television special. Newly transferred in 1988 from a pristine stereo print.
1955 120m/C Henry Fonda, James Cagney, Jack Lemmon, William Powell, Betsy Palmer, Ward Bond, Harry Carey Jr.; ***Dir:*** John Ford, Mervyn Le Roy. Academy Awards '55: Best Supporting Actor (Lemmon); National Board of Review Awards '55: 10 Best Films of the Year. **VHS, Beta $19.98** *WAR, BTV* 🦴🦴🦴🦴

Misterios de Ultratumba All the horrors that go on inside an insane asylum occur in this film, some unintentionally.
1947 90m/B *MX* Gaston Santos, Rafael Bertrand, Mapita Cortes; ***Dir:*** Fernando Mendez. **VHS, Beta $39.95** *MAD* 🦴🦴

Mistral's Daughter Frothy miniseries, based on the Judith Krantz novel, about a French artist and his relationship with three beautiful women. Formulaic TV melodrama offers sexual scenes and revealing glimpses. Produced, like other Krantz miniseries, by author's hubby, Steve.
1984 390m/C Stefanie Powers, Lee Remick, Stacy Keach, Robert Urich, Timothy Dalton; ***Dir:*** Douglas Hickox. **VHS, Beta $79.95** *IVE* 🦴½

Mistress Weepy made-for-television melodrama about a woman who makes a living as a mistress. When her lover dies, she must learn to stand on her own two feet.
1987 96m/C Victoria Principal, Alan Rachins, Don Murray; ***Dir:*** Michael Tuchner. **VHS $79.98** *REP* 🦴½

Mistress of the Apes A woman searches for her missing husband in Africa with a group of scientists and discovers a tribe of near-men, who may be the missing link in evolution (or perhaps, a professional football team). Into their little group she is accepted, becoming their queen. Buchanan earned his reputation as a maker of horrible films honestly. Includes the songs "Mistress of the Apes" and "Ape Lady."
1979 (R) 88m/C Jenny Neumann, Barbara Leigh, Garth Pillsbury, Walt Robin, Stuart Lancaster, Suzy Mandel; ***W/Dir:*** Larry Buchanan. **VHS, Beta $39.95** *MON Woof!*

Mistress Pamela When young Pamela goes to work in the household of handsome Lord Devonish, he sets about in wild pursuit of her virginity. Loosely based on the 1740 work "Pamela" by Samuel Richardson, considered the first modern English novel. Read the book unless you only have 95 minutes to spend finding out who gets what.
1974 (R) 95m/C Anne Michelle, Julian Barnes, Anna Quayle, Rosemary Dunham; ***Dir:*** Jim O'Connor. **VHS, Beta $39.95** *MON* 🦴½

Mistress (Wild Geese) A classic Japanese period piece about an innocent woman who believes she's married to a ruthless industrialist, only to find he is already married and she is but his mistress. Her love for a medical student unleashes tragedy. Subtly moving in a low-key way; starkly beautiful in black and white. In Japanese, with English subtitles.
1953 106m/B *JP* Hideko Takamine, Hiroshi Akutagawa; ***Dir:*** Shiro Toyoda. **VHS, Beta $79.95** *SUE* 🦴🦴🦴

Mistress of the World A scientist, aided by Swedish Intelligence agent Ventura, works to protect his gravity-altering invention from Chinese agents. Partly based on a German serial from the silent film era, but not up to director Dieterle's usual fare. A must-see for Ventura fans and Mabuse mavens.
1959 107m/C Martha Hyer, Micheline Presle, Gino Cervi, Lino Ventura, Sabu, Wolfgang Preiss; ***Dir:*** William Dieterle. **VHS, Beta** *VMK* 🦴🦴

Misty Ladd and Smith are two kids who capture a wild pony and teach it to run around in circles and step over things. Based on Marguerite Henry's bestselling "Misty of Chincoteague," and filmed on an island off the coast of Virginia.
1961 92m/C David Ladd, Arthur O'Connell, Pam Smith; ***Dir:*** James B. Clark. **VHS $14.95** *PAR, WAX, FCT* 🦴🦴½

Misunderstood A former black market merchant has to learn how to relate to his sons after his wife dies. The father, now a legitimate businessman, is more concerned with running his shipping business than growing closer to his boys. Fine acting can't overcome a transparent plot.
1984 (PG) 92m/C Gene Hackman, Susan Anspach, Henry Thomas, Rip Torn, Huckleberry Fox; ***Dir:*** Jerry Schatzberg. **VHS, Beta $79.95** *MGM* 🦴🦴½

Misunderstood A father learns his lonely son is full of love and sensitivity when a tragedy reveals that it is this son who takes the blame for all the wrongs his brother commits.
1987 101m/C Anthony Quayle, Stefano Colagrande, Simone Gianozzi, John Sharp; ***Dir:*** Luigi Comencini. **VHS, Beta $59.95** *IMP* 🦴🦴½

Mitchell Tough cop battles drug traffic and insipid script. Big screen release with that certain TV look.
1975 (R) 90m/C Joe Don Baker, Linda Evans, Martin Balsam, John Saxon, Merlin Olsen, Harold J. Stone; ***Dir:*** Andrew V. McLaglen. **VHS $9.95** *SIM* 🦴

Mixed Blood From the renowned underground film-maker; a dark comedy that examines the seedy drug subculture in New York. Violent, fast, and funny.
1984 (R) 98m/C Marilia Pera, Richard Vlacia, Linda Kerridge, Geraldine Smith, Angel David; ***Dir:*** Paul Morrissey. **VHS, Beta $59.95** *MED* 🦴🦴½

M'Lady's Court A lawyer searching for an heiress in a convent of lusty maidens has loads of fun. None too captivating.
198? 92m/C Sonia Jeanine, Maja Hoppe. **VHS, Beta $29.95** *MED* 🦴

Mo' Better Blues Not one of his more cohesive or compelling works, Lee's fourth feature is on the surface a backstage jazz biopic. But there hasn't been a Lee feature that wasn't vitally concerned with complicated racial issues and though subtle and indirect, this is no exception. Bleek Gilliam is a handsome, accomplished jazz trumpeter who divides his limited extra-curricular time between Clarke (newcomer Williams), an aspiring jazz vocalist, and Indigo (junior Lee sibling Joie), a down-to-earth school teacher. What's interesting is not so much the story of self-interested musician and ladies' man Gilliam, but the subtle racial issues his life draws into focus. The Branford Marsalis Quartet provides the music for Bleek's group, scored by Lee's dad Bill (on whose life the script is loosely based).
1990 (R) 129m/C Denzel Washington, Spike Lee, Joie Lee, Wesley Snipes, Cynda Williams, Giancarlo Esposito, Robin Harris, Bill Nunn, John Turturro, Dick Anthony Williams, Ruben Blades; ***W/Dir:*** Spike Lee. **VHS, Beta, LV $19.95** *MCA, FCT* 🦴🦴½

Moana, a Romance of the Golden Age An early look through American eyes at the society of the people of Samoa, in the Pacific Islands. Picturesque successor to Flaherty's "Nanook of the North."
1926 76m/B Ta'avale, Fa'amgase, Moana; ***Dir:*** Robert Flaherty. **VHS $29.95** *GPV, FCT, DNB* 🦴🦴🦴½

Mob Boss Gangster films spoof starring Hickey as the head of a successful California crime ring. When his beautiful wife conspires to have him killed, he is left just breaths away from his demise. He calls his nerdy son to his bedside to ask him to take over the family business. But this kid thinks that money laundering requires detergent! Take this one down for a walk by the river in its new cement shoes.
1990 (R) 93m/C Eddie Dezen, Morgan Fairchild, William Hickey, Stuart Whitman; ***Dir:*** Fred Olen Ray. **VHS $89.95** *VMK* 🦴

Mob Story Vernon stars as a big-wig mob boss on the run from the government and a few of his "closest" friends. He plans to disappear into Canada, but if his girl (Kidder) and best friend (Waxman) find him first he'll be traveling in a body-bag. Very silly, but fun.
1990 (R) 98m/C John Vernon, Margot Kidder, Al Waxman, Kate Vernon; ***Dir:*** Gabriel Markiw, Jancarlo Markiw. **VHS, Beta, LV $89.98** *SHG, IME* 🦴½

Mob War A bloody Mafia adventure in which a sausage factory plays a large role.
1978 102m/C VHS, Beta, LV $79.98 *VCD Woof!*

Mob War A war breaks out between the head of New York's underworld and a media genius. After deciding to become partners, the "family" tries to take over. For fun, count how many times the word "respect" pops up in the script. Tries hard but fails.
1988 (R) 96m/C John Christian, David Henry Keller, Jake La Motta, Johnny Stumper; ***Dir:*** J. Christian Ingvordsen. **VHS, Beta $79.98** *MCG* 🦴½

Mobsters It sounded like a great idea, casting the hottest young actors of the 1990s as youthful racketeers 'Lucky' Luciano, Meyer Lansky, Bugsy Siegel and Frank Costello. But it would take an FBI probe to straighten out the blood-choked plot, as the pals' loyalties get tested the hard way in a dismembered narrative. Sicker than the violence is a seeming endorsement of the glamorous hoods.
1991 (R) 104m/C Christian Slater, Patrick Dempsey, Richard Grieco, Costas Mandylor, Anthony Quinn, F. Murray Abraham, Lara Flynn Boyle, Michael Gambon, Christopher Penn; ***Dir:***

Michael Karbelnikoff. **VHS, Beta, LV** $94.95 *CCB, MCA* ✂️½

Moby Dick Ray Bradbury and Huston co-wrote the screenplay adapted from Herman Melville's high seas saga. Peck as Captain Ahab, obsessed with desire for revenge upon the great white whale, Moby Dick, isn't always believable. Those moments when it clicks, however, make the film more than worthwhile.
1956 116m/C Gregory Peck, Richard Basehart, Orson Welles, Leo Genn, Harry Andrews, Friedrich Ledebur; *Dir:* John Huston. National Board of Review Awards '56: 10 Best Films of the Year, Best Director. **VHS, Beta, LV** $19.95 *FOX, BMV* ✂️✂️✂️

Moby Dick An animated version of the Herman Melville classic.
1977 52m/C VHS, Beta $19.95 *MGM* ✂️✂️✂️

Moby Dick & the Mighty Mightor A collection of two animated adventures featuring Moby Dick, a super whale who uses his incredible speed to fight crime, and The Mighty Mightor, a young man who uses his unlimited physical powers to protect his Stone Age village.
1967 38m/C VHS, Beta $19.95 *HNB* ✂️✂️

Model Behavior Fresh from college, an aspiring photographer pines for a glamorous model.
1982 86m/C Richard Bekins, Bruce Lyons, Cindy Harrel; *Dir:* Bud Gardner. **VHS, Beta** $79.98 *LIV, LTG* ✂️½

Modern Girls Teen comedy about three bubble-headed LA rock groupies and their various wild adventures during a single night on the town. Surprising in that the three lead actresses waste their talents in this lesson in exploitation. Bruce Springsteen's younger sister portrays a drug user.
1986 (PG-13) 82m/C Cynthia Gibb, Daphne Zuniga, Virginia Madsen; *Dir:* Jerry Kramer. **VHS, Beta, LV** $19.95 *PAR* ✂️½

Modern Love An average slob realizes that marriage, fatherhood and in-law-ship isn't quite what he expected. Benson stars, directs and produces as well as co-starring with real-life wife DeVito. A plotless hodge-podge of bits that were more successful in "Look Who's Talking" and "Parenthood."
1990 (R) 89m/C Robby Benson, Karla DeVito, Rue McClanahan, Kaye Ballard, Frankie Valli, Cliff Bemis, Louise Lasser, Burt Reynolds, Lyric Benson; *Dir:* Robby Benson. **VHS, Beta** $79.95 *SVS* ✂️

Modern Problems A man involved in a nuclear accident discovers he has acquired telekinetic powers, which he uses to turn the tables on his professional and romantic rivals. A fine cast but an unsuccessful fission trip.
1981 (PG) 93m/C Chevy Chase, Patti D'Arbanville, Mary Kay Place, Brian Doyle-Murray, Nell Carter, Dabney Coleman; *Dir:* Ken Shapiro. **VHS, Beta** $59.98 *FOX* ✂️✂️

Modern Romance The romantic misadventures of a neurotic film editor who is hopelessly in love with his girlfriend but can't seem to maintain a normal relationship with her. Smart and hilarious at times and always simmering with anxiety. Offers an honest look at relationships as well as an accurate portrait of filmmaking.
1981 (R) 102m/C Albert Brooks, Kathryn Harrold, Bruno Kirby, George Kennedy, James L. Brooks, Bob Einstein; *Dir:* Albert Brooks. **VHS, Beta, LV** $14.95 *COL* ✂️✂️✂️

Modern Times This "mostly" silent film finds Chaplin playing a factory worker who goes crazy from his repetitious job on an assembly line and his boss' demands for greater speed and efficiency. Ultimately encompassing the tyranny of machine over man, this cinematic masterpiece has more relevance today than ever. Chaplin wrote the musical score which incorporates the tune "Smile."
1936 87m/B Charlie Chaplin, Paulette Goddard, Henry Bergman, Stanley Sandford, Gloria de Haven, Chester Conklin; *Dir:* Charlie Chaplin. National Board of Review Awards '36: 10 Best Films of the Year. **VHS, Beta** $19.98 *FOX* ✂️✂️✂️✂️

The Moderns One of the quirkier directors around, this time Rudolph tries a comedic period piece about the avant-garde art society of 1920s Paris. Fleshed out with some familiar characters (Ernest Hemingway, Gertrude Stein) and some strange art-world types. The tone is not consistently funny but, instead, romantic as the main characters clash over art and love.
1988 (R) 126m/C Keith Carradine, Linda Fiorentino, John Lone, Genevieve Bujold, Geraldine Chaplin, Wallace Shawn, Kevin J. O'Connor; *Dir:* Alan Rudolph. **VHS, Beta, LV, 8mm** $14.98 *SUE* ✂️✂️✂️

Mogambo Remake of "Red Dust," this is the steamy story of a love triangle between an African game hunter, a proper British lady, and an American showgirl in the jungles of Kenya. Oscar nominations for Best Actress (Gardner), and Best Supporting Actress (Kelly).
1953 115m/C Clark Gable, Ava Gardner, Grace Kelly; *Dir:* John Ford. National Board of Review Awards '53: 10 Best Films of the Year. **VHS, Beta** $19.98 *MGM, FCT, CCB* ✂️✂️✂️

Mohammed: Messenger of God Sprawling saga of the genesis of the religion of Islam, with Anthony Quinn portraying Mohammad's uncle, Hamza, an honored warrior. The story behind the movie might prove much more successful as a sequel than did the movie itself. The filming itself created a religious controversy. Also known as "The Message."
1977 (PG) 180m/C *GB* Damien Thomas, Anthony Quinn, Irene Papas, Michael Ansara, Johnny Sekka, Michael Forest, Neville Jason; *Dir:* Moustapha Akkad. **VHS** *USA* ✂️✂️

Mohawk A cowboy and his Indian maiden try to stop a war between Indian tribes and fanatical landowners.
1956 80m/C Rita Gam, Neville Brand, Scott Brady, Lori Nelson; *Dir:* Kurt Neumann. **VHS, Beta** $19.95 *NOS, MRV, GEM* ✂️½

The Moistro Meet the Cool Ruler Two popular reggae stars perform.
1987 60m/C Keeling Beckford, Gregory Issacs. **Beta** *KRV* ✂️½

Mole Men Against the Son of Hercules Italian muscleman Maciste battles the pale-skinned denizens of an underground city.
1961 ?m/C *IT* Mark Forest. **VHS, Beta** $16.95 *SNC* ✂️½

Moliendo Vidrio A performance by Puerto Rican vocal group Moliendo Vidrio filmed during their U.S. tour.
1983 30m/C Beta *NJN* ✂️½

Moliere Mikhail Bulgakov's comedy/drama of the life of the great French playwright.
1990 112m/C Anthony Sher. **VHS, Beta, LV** $39.98 *TTC* ✂️✂️✂️

Molly & Lawless John Average western, with Miles the wife of a sadistic sheriff who helps a prisoner escape the gallows so they can run away together. Enough cliches for the whole family to enjoy.
1972 90m/C Sam Elliott, Vera Miles; *Dir:* Gary Nelson. **VHS, Beta** *PGN* ✂️½

Molly Maguires Dramatization based on a true story, concerns a group of miners called the Molly Maguires who resort to using terrorist tactics in their fight for better working conditions during the Pennsylvania Irish coal mining rebellion in the 1870s. During their reign of terror, the Mollies are infiltrated by a Pinkerton detective who they mistakenly believe is a new recruit. It has its moments but never fully succeeds. Returned less than fifteen percent of its initial eleven million dollar investment.
1969 (PG) 123m/C Sean Connery, Richard Harris, Samantha Eggar, Frank Finlay; *Dir:* Martin Ritt. **VHS, Beta** $14.95 *PAR, CCB* ✂️✂️½

Mom When his mother is bitten by a flesh-eater, Clay Dwyer is at a loss as to what to do. How do you tell your own mother that she must be destroyed, lest she continue to devour human flesh? A campy horror/comedy.
1989 (R) 95m/C Mark Thomas Miller, Art Evans, Mary McDonough, Jeanne Bates, Brion James, Stella Stevens, Claudia Christian; *Dir:* Patrick Rand. **VHS, LV** $79.95 *COL, PIA* ✂️✂️½

Mom & Dad An innocent young girl's one night of passion leads to an unwanted pregnancy. Stock footage of childbirth and a lecture on the evils of syphilis concludes this campy schlock that features the national anthem. Banned or denied release for years, this movie is now a cult favorite for its time-capsule glimpse at conventional 1940s sexual attitudes.
1947 97m/B Hardie Albright, Sarah Blake, George Eldridge, June Carlson, Jimmy Clark, Bob Lowell; *Dir:* William Beaudine. **VHS, Beta** $19.95 *VDM* ✂️

Mom, the Wolfman and Me An 11-year-old girl arranges and manages the love affair of her mother, an ultra-liberated photographer, and an unemployed teacher, who is also the owner of an Irish wolfhound. Charming adaptation of a novel by Norma Klein.
1980 100m/C Patty Duke, David Birney, Danielle Brisbois, Keenan Wynn, Viveca Lindfors, John Lithgow; *Dir:* Edmond Levy. **VHS, Beta** *TLF* ✂️✂️½

Moment in Time A romantic novel on tape, about a guy, a girl, and their various troubles.
1984 60m/C Jean Simmons. **Beta** $11.95 *PSM* ✂️

Mommie Dearest Film based on Christina Crawford's memoirs of her incredibly abusive and violent childhood at the hands of her mother, actress Joan Crawford. The story is controversial and sometimes trashy, but fairly well done nevertheless.
1981 (PG) 129m/C Faye Dunaway, Diana Scarwid, Steve Forrest, Mara Hobel, Rutanya Alda, Harry Goz, Howard da Silva; *Dir:* Frank Perry. **VHS, Beta, LV** $59.95 *PAR* ✂️✂️

Mon Oncle Tati's celebrated comedy contrasts the simple life of Monsieur Hulot with the technologically complicated life of his family when he aids his nephew in war against his parents' ultramodern, push-button home. An easygoing, delightful comedy, this is the director's first piece in color. Also called "My Uncle." Sequel to "Mr. Hulot's

Holiday,'' followed by ''Playtime.'' In French with English subtitles.
1958 110m/C FR Jacques Tati, Jean-Pierre Zola, Adrienne Serrantie, Alain Bacourt; **Dir:** Jacques Tati. Academy Awards '58: Best Foreign Language Film; National Board of Review Awards '58: 5 Best Foreign Films of the Year; New York Film Critics Awards '58: Best Foreign Film. **VHS, Beta, LV, 8mm $24.95** NOS, MRV, SUE 𝒵𝒵𝒵𝒵

Mon Oncle Antoine A splendid tale of young Benoit, who learns about life from a surprisingly compassionate uncle, who works as everything from undertaker to grocer in the depressed area where they live.
1971 104m/C CA Jean Duceppe, Olivette Thibault; **Dir:** Claude Jutra. **VHS $79.95** HMV, FLI, APD 𝒵𝒵𝒵½

Mon Oncle D'Amerique Three French characters are followed as they try to find success of varying kinds in Paris, interspersed with ironic lectures by Prof. Henri Laborit about the biology that impels human behavior. An acclaimed, witty comedy by the former Nouvelle Vague filmmaker, dubbed into English. Oscar-nominated for screenwriting.
1980 (PG) 123m/C FR Gerard Depardieu, Nicole Garcia, Roger-Pierre, Marie DuBois; **Dir:** Alain Resnais. New York Film Critics Awards '80: Best Foreign Film. **VHS, Beta $29.98** SUE 𝒵𝒵𝒵½

Mona Lisa A small-time hood gets personally involved with the welfare and bad company of the high-priced whore he's been hired to chauffeur. Brilliantly filmed and acted; critically lauded.
1986 (R) 104m/C GB Bob Hoskins, Cathy Tyson, Michael Caine, Clarke Peters, Kate Hardie, Robbie Coltrane, Zoe Nathenson, Sammi Davis, Rod Bedall, Joe Brown, Pauline Melville; **Dir:** Neil Jordan. British Academy Awards '86: Best Actor (Hoskins); Cannes Film Festival '86: Best Actor (Hoskins); Los Angeles Film Critics Association Awards '86: Best Actor (Hoskins); New York Film Critics Awards '86: Best Actor (Hoskins); National Society of Film Critics Awards '86: Best Actor (Hoskins). **VHS, Beta, LV $19.99** HBO 𝒵𝒵𝒵½

Mona's Place Some really tired (and probably dirty) miners take a weekend off to go to the local brothel. Unfortunately, a rival gang decides to attack.
1989 (R) 80m/C James Whitworth, Shawn Devereaux, Sebastian Gregory, Donna Stanley, Paul Harper; **W/Dir:** John Hayes. **VHS $39.95** AVD 𝒵

Mondo Africana Supposedly serious documentary of Africa featuring hot coal walkers, natives passing spears through their cheeks, and other events of varying weirdness, plus a narrator who hams up his end of things.
19?? 80m/C Nar: Quentin Reynolds. **VHS $24.95** FCT 𝒵

Mondo Cane A documentary showcasing the eccentricities of human behavior around the world, including cannibalism, pig killing and more. Dubbed in English. Inspired a rash of ''shockumentaries'' over the next several years. The song ''More'' made its debut in this film.
1963 (R) 105m/C IT Dir: Gualtiero Jacopetti; **Nar:** Stefano Sibaldi. **VHS, Beta $16.95** SNC, VDC, VDM 𝒵𝒵

Mondo Cane 2 More documentary-like views of the oddities of mankind and ethnic rituals around the world. Also known as ''Mondo Pazzo.'' Enough, already.

1964 (R) 94m/C IT Dir: Gualtiero Jacopetti. **VHS, Beta $19.95** VDC, HHE 𝒵

Mondo Elvis (Rock 'n' Roll Disciples) This is a sordid look at Elvis' legacy and his aimless, worshipping fans.
1985 60m/C VHS, Beta RHI, MNP, HHE 𝒵

Mondo Lugosi: A Vampire's Scrapbook The classically terrible actor is profiled campily in this collection of interviews, rare appearances and his worst screen performances.
1986 60m/C VHS, Beta $29.95 RHI 𝒵

Mondo Magic A compilation of tribal rituals offering viewers a look on the darker side of matters magical.
1976 100m/C Dir: Melvin Ashford. **VHS, Beta $39.98** MAG 𝒵½

Mondo Trasho A major trasho film of the last day in the life of a most unfortunate woman, complete with sex and violence. First full length effort from cult filmmaker Waters, who was also writer, producer, and editor.
1970 95m/B Divine, Mary Vivian Pearce, John Leisenring, Mink Stole, David Lochary, Chris Atkinson; **Dir:** John Waters. **VHS, Beta $39.98** FLL Woof!

The Money The quest for money drives a young man to kidnap the child that his girlfriend is baby-sitting. Average but Workman's direction is right on the money.
1975 (R) 88m/C Laurence Luckinbill, Elizabeth Richards, Danny De Vito, Graham Beckel; **Dir:** Carl Workman. **VHS $59.95** GHV 𝒵𝒵

Money to Burn An aging high school counselor and two senior citizens develop a plan to steal fifty million dollars from the Federal Reserve Bank. A harmless bit of fun.
1983 90m/C Jack Kruschen, Meegan King, David Wallace, Phillip Pine; **Dir:** Virginia Stone. **VHS, Beta $42.95** PGN 𝒵

Money Hunt $100,000 cash is secured in a safe deposit box; the first person who can solve the puzzle from the hints given in this program will claim the prize. Magnum P.I.'s John Hillerman is the host.
1984 30m/C John Hillerman. **VHS, Beta $29.95** LHV 𝒵½

Money Madness Popular rock singer Eddie Money's rise to the top of the music business is chronicled in this program, in which he performs some of his songs.
1979 92m/C Eddie Money. **VHS, Beta $29.98** FOX 𝒵½

Money Movers Aussie thieves plan megabuck bank robbery but heist and plot go sour. An oft told tale. Features early appearance from burly boy Brown.
1978 94m/C AU Terence Donovan, Ed Devereaux, Tony Bonner, Lucky Grills, Charles Tingwell, Candy Raymond, Bryan Brown, Alan Cassell; **W/Dir:** Bruce Beresford. **VHS $79.95** IMP 𝒵𝒵

The Money Pit A young yuppie couple encounter sundry problems when they attempt to renovate their newly purchased, seemingly self-destructive, Long Island home. However, the collapse of their home leads directly to the collapse of their relationship as well. Somewhat modeled after ''Mr. Blandings Builds His Dream House.'' Hanks and Long fail to jell as partners and the many sight gags are on the predictable side. A Spielberg production.

1986 (PG) 91m/C Tom Hanks, Shelley Long, Alexander Godunov, Maureen Stapleton, Philip Bosco, Joe Mantegna, Josh Mostel; **Dir:** Richard Benjamin. **VHS, Beta, LV $14.95** MCA 𝒵𝒵

The Mongols Splashy Italian production has Genghis Khan's son repelling invading hordes while courting buxom princess. Pairing of Palance and Ekberg makes this one worthy of consideration.
1960 105m/C IT FR Jack Palance, Anita Ekberg, Antonella Lualdi, Franco Silva, Gianno Garko, Roldano Lupi, Gabriella Pallotti; **Dir:** Andre de Toth, Leopoldo Savona, Riccardo Freda. **VHS $16.95** SNC 𝒵𝒵½

Mongrel El-cheapo kennel horror about a man who dreams he's a wild murderous mutt. His relief upon waking up is short-lived, however, when he discovers the people he killed in his dreams are actually dead. Flick will make you want to gnaw on a bone.
1983 90m/C Aldo Ray, Terry Evans; **Dir:** Robert Burns. **VHS, Beta** PGN Woof!

Monika Two teenagers who run away together for the summer find the winter brings more responsibility than they can handle when the girl becomes pregnant and gives birth. Lesser, early Bergman, sensitively directed, but dull. Adapted by Bergman from a Per Anders Fogelstrom novel. In Swedish with English subtitles. Also known as ''Summer with Monika.''
1952 96m/B SW Harriet Anderson, Lars Ekborg, John Harryson, Georg Skarstedt, Dagmar Ebbesen, Ake Gronberg; **W/Dir:** Ingmar Bergman. **VHS $29.95** CVC 𝒵𝒵½

Monique A sophisticated career woman is about to unleash a terrifying secret on her new husband.
1983 96m/C Florence Giorgetti, John Ferris. **VHS, Beta $49.95** VCL 𝒵

Monkey Boy Decent sci-fi horror from Britain, about a cunning human-ape hybrid that escapes from a genetics lab after massacring the staff. Not excessively gruesome or vulgar, and some sympathy is aroused for the killer mutant. Based on the novel ''Chimera'' by scriptwriter Stephen Gallagher.
1990 104m/C John Lynch, Christine Kavanagh, Kenneth Cranham; **Dir:** Lawrence Gordon Clark. **VHS, Beta $79.95** PSM 𝒵𝒵½

Monkey Business Marx Brothers run amok as stowaways on ocean liner. Fast-paced comedy provides seemingly endless amount of gags, quips, and pratfalls, including the fab four imitating Maurice Chevalier at Immigration. This film, incidentally, was the group's first to be written—by noted humorist Perelman—directly for the screen.
1931 77m/B Groucho Marx, Harpo Marx, Chico Marx, Zeppo Marx, Thelma Todd, Ruth Hall, Harry Woods; **Dir:** Norman Z. McLeod. **VHS, Beta, LV $19.95** MCA 𝒵𝒵𝒵½

Monkey Business A scientist invents a fountain-of-youth potion, a lab chimpanzee mistakenly dumps it into a water cooler, and then grown-ups start turning into adolescents. Top-flight crew occasionally labors in this screwball comedy, though comic moments shine. Monroe is the secretary sans skills, while absent-minded Grant and sexy wife Rogers race hormonally as teens. Written by Ben Hecht, I.A.L. Diamond, and Charles Lederer.
1952 97m/B Cary Grant, Ginger Rogers, Charles Coburn, Marilyn Monroe, Hugh Marlowe, Larry Keating, George Winslow; **Dir:** Howard Hawks, Howard Hawks; **W/Dir:** Charles Lederer. **VHS, Beta, LV $14.98** FOX 𝒵𝒵𝒵

Monkey Grip An unmarried woman copes with parenthood and a drug-addicted boyfriend while working on the fringe of Australia's music business. Grim but provocative, based on Helen Graham's novel (and adapted by Graham and Cameron).
1982 101m/C *AU* Noni Hazlehurst, Colin Friels, Alice Garner, Harold Hopkins, Candy Raymond; *Dir:* Ken Cameron. Australian Film Institute '82: Best Actress (Hazlehurst). **VHS $24.98** *SUE* 🐾🐾🐾

Monkey Hustle Vintage blaxploitation has trouble hustling laffs. The Man plans a super freeway through the ghetto, and law abiding do gooders join forces with territorial lords of vice to fight the project. Shot in the Windy City.
1977 (PG) 90m/C Yaphet Kotto, Rudy Ray Moore, Rosalind Cash, Debbi Morgan, Thomas Carter; *Dir:* Arthur Marks. **VHS $59.95** *ORI, FCT* 🐾

Monkey Kung Fu A young man learns kung fu from a master and, amidst kicking enemies in the head, cavorts and bumbles comically. Dubbed.
198? 96m/C Chen Mu-Chuan, Sun Jung-Chi. **VHS, Beta $54.95** *MAV* 🐾

Monkey in the Master's Eye A martial arts adventure spiced with low comedy, in the story of a buffoonish slave rising to martial arts infamy.
1972 92m/C **VHS, Beta $39.95** *UNI* 🐾

Monkey Shines Based on the novel by Michael Stuart, this is a sick, scary yarn about a quadriplegic who is given a specially trained capuchin monkey as a helpmate. However, he soon finds that the beast is assuming and acting on his subconscious rages. Also known as "Monkey Shines: An Experiment in Fear."
1988 (R) 108m/C Jason Beghe, John Pankow, Melanie Parker, Christine Forrest, Stephen Root, Kate McNeil, Joyce Van Patten; *Dir:* George A. Romero. **VHS, Beta, LV $19.98** *ORI* 🐾🐾

Monkeys, Go Home! Dumb Disney yarn about young American who inherits a badly neglected French olive farm. When he brings in four chimpanzees to pick the olives, the local townspeople go on strike. People can be so sensitive. Based on "The Monkeys" by G. K. Wilkinson. Chevalier's last film appearance.
1966 89m/C Dean Jones, Yvette Mimieux, Maurice Chevalier, Clement Harari, Yvonne Constant; *Dir:* Andrew V. McLaglen. **VHS, Beta $69.95** *DIS* 🐾½

Monkey's Uncle A sequel to Disney's "The Misadventures of Merlin Jones" and featuring more bizarre antics and scientific hoopla, including chimps and a flying machine.
1965 90m/C Tommy Kirk, Annette Funicello, Leon Ames, Arthur O'Connell; *Dir:* Robert Stevenson. **VHS, Beta $69.95** *DIS* 🐾½

The Monolith Monsters A geologist investigates a meteor shower in Arizona and discovers strange crystals. The crystals attack humans and absorb their silicone, causing them to grow into monsters. Good "B" movie fun.
1957 76m/B Grant Williams, Lola Albright, Les Tremayne, Trevor Bardette; *Dir:* John Sherwood. **VHS $18.00** *FRG, MLB* 🐾🐾

Monsieur Beaucaire The Duke of Chartres ditches France posing as a barber, and once in Britain, becomes a lawman. Not a classic Valentino vehicle.

1924 100m/B Rudolph Valentino, Bebe Daniels, Lois Wilson, Doris Kenyon, Lowell Sherman, John Davidson; *Dir:* Sidney Olcott. **VHS, Beta $19.95** *GPV, DNB* 🐾½

Monsieur Hire The usual tale of sexual obsesssion and suspense. Mr. Hire spends much of his time trying to spy on his beautiful young neighbor woman, alternately alienated and engaged by her love affairs. The voyeur soon finds his secret desires have entangled him in a vicious intrigue. Political rally setpiece is brilliant. Excellent acting, intense pace, elegant photography. Based on "Les Fiancailles de M. Hire" by Georges Simenone and adapted by Leconte and Patrick Dewolf. In French with English subtitles.
1989 (PG-13) 81m/C *FR* Michel Blanc, Sandrine Bonnaire, Luc Thuillier, Eric Berenger, Andre Wilms; *Dir:* Patrice Leconte. **VHS, LV $39.95** *ORI, FXL, FCT* 🐾🐾🐾½

Monsieur Verdoux A thorough Chaplin effort, as he produced, directed, wrote, scored and starred. A prim and proper bank cashier in Paris marries and murders rich women in order to support his real wife and family. A mild scandal in its day, though second-thought pacifism and stale humor date it. A bomb upon release (leading Chaplin to shelve it for 17 years) and a cult item today, admired for both its flaws and complexity. Raye fearlessly chews scenery and croissants. Chaplin was Oscar-nominated for the screenplay, which was initially based upon a suggestion from Orson Welles.
1947 123m/B Charlie Chaplin, Martha Raye, Isobel Elsom, Mady Correll, Marilyn Nash, Irving Bacon, William Frawley, Allison Roddan, Robert Lewis; *W/Dir:* Charlie Chaplin. National Board of Review Awards '47: Best Picture. **VHS, Beta $19.98** *FOX* 🐾🐾🐾

Monsieur Vincent True story of 17th century figure St. Vincent de Paul, who forsakes worldly possessions and devotes himself to tending the less fortunate. Inspirational. In French with English subtitles.
1947 111m/B *FR* Pierre Fresnay, Lisa Delamare, Aime Clarimond; *Dir:* Maurice Cloche. Academy Awards '47: Best Foreign Language Film. **VHS, Beta $29.95** *NOS, MRV, DVT* 🐾🐾🐾

Monsignor Callow, ambitious priest befriends mobsters and even seduces a nun while managing Vatican's business affairs. No sparks generated by Reeve and Bujold (who appears nude in one scene), and no real conviction related by most other performers. Absurd, ludicrous melodrama best enjoyed as unintentional comedy. Based on Jack Alain Leger's book. Woof!
1982 (R) 121m/C Christopher Reeve, Fernando Rey, Genevieve Bujold, Jason Miller; *Dir:* Frank Perry. **VHS, Beta $59.98** *FOX* Woof!

Monster Boodthirsty alien indiscriminately preys on gaggle of teens in wilds of civilization. Performers struggle with dialogue and their own self-esteem.
1978 (R) 98m/C Jim Mitchum, Diane McBain, Roger Clark, John Carradine, Phil Carey, Anthony Eisley, Keenan Wynn; *Dir:* Herbert L. Strock. **VHS, Beta $24.95** *GHV, MRV* 🐾

The Monster that Challenged the World Huge, ancient eggs are discovered in the Salton Sea and eventually hatch into killer, crustaceous caterpillars. Superior monster action.
1957 83m/B Tim Holt, Audrey Dalton, Hans Conried, Harlen Ward, Casey Adams, Mimi Gibson, Gordon Jones; *Dir:* Arnold Laven. **VHS $18.00** *FRG, MLB* 🐾🐾½

Monster in the Closet A gory horror spoof about a rash of San Francisco murders that all take place inside closets. A news reporter and his scientist friend decide they will be the ones to protect California from the evil but shy creatures. From the makers of "The Toxic Avenger."
1986 (PG) 87m/C Donald Grant, Claude Akins, Denise DuBarry, Stella Stevens, Howard Duff, Henry Gibson, Jesse White, John Carradine; *Dir:* Bob Dahlin. **VHS, Beta $19.98** *LHV, WAR* 🐾🐾

The Monster Club Price and Carradine star in this music-horror compilation, featuring songs by Night, B.A. Robertson, The Pretty Things and The Viewers. Soundtrack music by John Williams, UB 40 and The Expressos.
1985 104m/C *GB* Vincent Price, Donald Pleasence, John Carradine, Stuart Whitman, Britt Ekland, Simon Ward, Patrick Magee; *Dir:* Roy Ward Baker. **VHS, Beta, LV $39.95** *IVE* 🐾🐾

The Monster Demolisher Nostradamus' descendent shares the family penchant for sanguine cocktails, and threatens a professor who protects himself with an electronic wonder of modern technology. Edited from a sombrero serial; if you make it through this one, look up its Mexican siblings, "Curse of Nostradamus" and "Genie of Darkness."
1960 ?m/B *MX* German Robles; *Dir:* Frederick Curiel. **VHS $19.98** *NOS, LOO* 🐾

Monster Dog A rock band is mauled, threatened, and drooled upon by an untrainable mutant canine. Rock star Cooper plays the leader of the band. No plot, but the German shepherd is worth watching.
1982 88m/C Alice Cooper, Victoria Vera; *Dir:* Clyde Anderson. **VHS, Beta $69.95** *TWE* Woof!

Monster a Go-Go! A team of go-go dancers battle a ten-foot monster from outer-space whose mass is due to a radiation mishap. He can't dance, either.
1965 70m/B Phil Morton, June Travis, Bill Rebane, Sheldon Seymour; *Dir:* Herschell Gordon Lewis. **VHS, Beta $39.95** *UHV* Woof!

Monster from Green Hell An experimental rocket containing radiation-contaminated wasps crashes in Africa, making giant killer wasps that run amok. Stinging big bug horror.
1958 71m/B Jim Davis, Robert Griffin, Barbara Turner, Eduardo Ciannelli; *Dir:* Kenneth Crane. **VHS, Beta $14.95** *NOS, MRV, SNC* 🐾

Monster High Bloodthirsty alien indiscriminately preys on gaggle of teens in wilds of civilization. Even the people who made this one may not have seen it all the way through.
1989 (R) 89m/C David Marriott, Dean Iandoli, Diana Frank, D.J. Kerzner; *Dir:* Rudiger Poe. **VHS, Beta, LV $89.95** *COL* 🐾

The Monster of London City During a stage play about Jack the Ripper, murders occur paralleling those in the production. The lead actor finds himself to be the prime suspect. Lightweight horror from Germany.
1964 87m/B *GE* Hansjorg Felmy, Marianne Koch, Dietmar Schoenherr, Hans Nielsen; *Dir:* Edwin Zbonek. **VHS, Beta $16.95** *SNC* 🐾

The Monster Maker Deranged scientist develops serum that inflates heads, feet, and hands. He recklessly inflicts others with this potion, then must contend with deformed victims while courting a comely gal.
1944 65m/B J. Carroll Naish, Ralph Morgan, Wanda McKay, Terry Frost; *Dir:* Sam Newfield. **VHS, Beta $19.95** *NOS, MRV, SNC* 🐾

Monster from the Ocean Floor An oceanographer in a deep-sea diving bell is threatened by a multi-tentacled creature. Roger Corman's first production.
1954 66m/C Anne Kimball, Stuart Wade; *Dir:* Wyott Ordung. **VHS, Beta $29.98** *VMK* ✂️

The Monster of Piedras Blancas During a seaside festival, two fisherman are killed by a bloodthirsty oceanic critter with no respect for holidays. The fishing village is determined to find it and kill it. Low-budget, with amateurish effects and poor acting.
1957 72m/B Les Tremayne, Jeanne Carmen, Forrest Lewis; *Dir:* Irvin Berwick. **VHS $14.98** *VDM, MRV, REP Woof!*

Monster from a Prehistoric Planet Researchers visiting a tropical island find a giant egg and take it back to Tokyo with them, dismaying the egg's giant monster parents. Dubbed.
1967 (PG) 90m/C *JP* **VHS, Beta $19.98** *ORI* ✂️½

The Monster Squad Youthful monster enthusiasts find their community inundated by Dracula, Frankenstein creature, Wolf Man, Mummy, and Gill Man(!?!), who are all searching for a life-sustaining amulet. Somewhat different, but still somewhat mediocre.
1987 (PG-13) 82m/C Andre Govan, Stephen Macht, Tommy Noonan, Duncan Regehr; *Dir:* Fred Dekker. **VHS, Beta, LV $19.98** *LIV, VES* ✂️½

The Monster and the Stripper A bayou creature is pursued by a gangster who foolishly hopes to realize considerable profits from the venture.
1973 (R) 90m/C **VHS, Beta $59.95** *NO* ✂️

The Monster Walks A whodunit thriller, complete with stormy nights, suspicious cripples, weird servants, a screaming gorilla, and a spooky house. Not unique but entertaining.
1932 60m/B Rex Lease, Vera Reynolds, Mischa Auer, Willie Best; *Dir:* Frank Strayer. **VHS, Beta $12.95** *SNC, NOS, CCB* ✂️✂️

Monsters, Madmen, Machines The 80-year history of science fiction films from ''Metropolis'' to ''Star Wars'' is presented in this retrospective.
1984 57m/C *Nar:* Gil Gerard. **VHS, Beta $39.95** *RKO* ✂️✂️

Monsters on the March Consists of 14 movie trailers, including frightening coming attractions for movies such as the ''The Return of the Fly'' with Vincent Price, ''Isle of the Dead'' with Boris Karloff, ''I Walked with a Zombie'' with Frances Dee, and others, dating as far back as 1932 and up to 1960.
1960 25m/B **VHS, Beta $29.95** *VYY* ✂️✂️

Montana Hoyce and Bess Guthrie are long-married and strong-willed ranchers who struggle against strip miners who want to take over their land. Made for cable television.
1990 (PG) 91m/C Gena Rowlands, Richard Crenna, Lea Thompson, Justin Deas, Elizabeth Berridge, Scott Coffey; *Dir:* William A. Graham. **VHS, Beta $89.95** *TTC* ✂️✂️½

Montana Belle A buxom bandit teams with the notorious Dalton Brothers before embarking on a life of reform. In the card-game of life, Russell is always holding a pair.
1952 81m/C Jane Russell, George Brent, Scott Brady, Forrest Tucker, Andy Devine, Jack Lambert, John Litel, Ray Teal; *Dir:* Allan Dwan. **VHS, Beta $19.98** *TTC* ✂️✂️

Monte Carlo A sexy Russian woman aids the Allies by relaying important messages during World War II. Fun, made for television production featuring Collins at her seductive best.
1986 200m/C Joan Collins, George Hamilton, Lisa Eilbacher, Lauren Hutton, Robert Carradine, Malcolm McDowell; *Dir:* Anthony Page. **VHS, Beta $29.95** *STE, NWV* ✂️✂️

Monte Carlo Nights Wrongly convicted murderer determines to prove his innocence even as he is tracked by police.
1934 60m/B Mary Brian, John Darrow, Kate Campbell, Robert Frazer, Astrid Allwyn, George ''Gabby'' Hayes, George Cleveland; *Dir:* William Nigh. **VHS, Beta $16.95** *SNC* ✂️½

Monte Walsh Aging cowboy sees declining of Old West, embarks on mission to avenge best friend's death. Subdued, moving western worthy of genre greats Marvin and Palance. Cinematographer Fraker proves himself a proficient director in this, his first venture. Based on Jack Schaefer's novel.
1970 (PG) 100m/C Lee Marvin, Jack Palance, Jeanne Moreau, Jim Davis, Mitchell Ryan; *Dir:* William A. Fraker. **VHS, Beta $59.98** *FOX* ✂️✂️✂️

Montenegro Offbeat, bawdy comedy details experiences of bored, possibly mad housewife who lands in a coarse, uninhibited ethnic community. To its credit, this film remains unpredictable to the end. And Anspach, an intriguing, resourceful—and attractive—actress, delivers what is perhaps her greatest performance.
1981 97m/C *SW* Susan Anspach, Erland Josephson; *W/Dir:* Dusan Makavejev. **VHS, Beta $69.99** *HBO* ✂️✂️✂️½

Monterey Pop This pre-Woodstock rock 'n roll festival in Monterey, California features landmark performances by some of the most popular sixties rockers. Compelling for the performances, and historically important as the first significant rock concert film. Appearances by Jefferson Airplane, Janis Joplin, Jimi Hendrix, Simon and Garfunkel, The Who, and Otis Redding.
1968 72m/C The Jefferson Airplane, Janis Joplin, Jimi Hendrix, Art Garfunkel, Paul Simon, Pete Townsend, Roger Daltrey, Keith Moon, John Entwhistle; *Dir:* James Desmond. **VHS, Beta, LV, 8mm $29.95** *SVS, VTK, VYG* ✂️✂️✂️½

Montgomery Clift Tribute includes clips from many of Clift's magnetic screen portrayals from 1948 to 1966.
1985 120m/C Montgomery Clift; *Dir:* Claudio Masenza. **VHS, Beta** *KRT* ✂️✂️✂️½

Month in the Country The reverently quiet story of two British WWI veterans, one an archeologist and the other a church painting restorer, who are working in a tiny village while trying to heal their emotional wounds. Based on the novel by J.L. Carr with a screenplay by Simon Frag.
1987 92m/C *GB* Colin Firth, Natasha Richardson, Kenneth Branagh, Patrick Malahide; *Dir:* Pat O'Connor. **VHS, Beta, LV $79.95** *LHV, WAR* ✂️✂️

Monty Python and the Holy Grail Britain's famed comedy band assaults the Arthurian legend in a cult classic replete with a Trojan rabbit and an utterly dismembered, but inevitably pugnacious, knight. Fans of manic comedy—and graphic violence—should get more than their fill here.
1975 (PG) 90m/C *GB* Graham Chapman, John Cleese, Terry Gilliam, Eric Idle, Terry Jones, Michael Palin, Carol Cleveland, Connie Booth, Neil Innes, Patsy Kensit; *Dir:* Terry Gilliam, Terry Jones. **VHS, Beta, LV, 8mm $19.95** *COL* ✂️✂️✂️½

Monty Python Live at the Hollywood Bowl The revered comedy group perform many of their most celebrated routines in this uproarious performance documentary. A must-see for all Python aficionados.
1982 78m/C *GB* Eric Idle, Michael Palin, John Cleese, Terry Gilliam, Terry Jones, Graham Chapman; *Dir:* Terry Hughes. **VHS, Beta, LV $69.95** *HBO, PAR* ✂️✂️✂️

Monty Python Meets Beyond the Fringe More fine British humor is featured in this compendium of skits, sketches, routines, and sequences. Those who know, will know. Those who don't probably won't enjoy it anyway.
19?? 85m/C *GB* Graham Chapman, John Cleese, Terry Gilliam, Eric Idle, Terry Jones, Michael Palin. **VHS** *UHV* ✂️✂️½

Monty Python's Life of Brian Often riotous spoof of Christianity tracks hapless peasant mistaken for the messiah in 32 A.D. Film reels from routine to routine, and only the most pious will remain unmoved by a chorus of crucifixion victims. Probably the group's most daring, controversial venture.
1979 (R) 94m/C *GB* Graham Chapman, John Cleese, Terry Gilliam, Eric Idle, Terry Jones, Michael Palin, George Harrison; *Dir:* Terry Jones. **VHS, Beta, LV $19.98** *PAR, PIA* ✂️✂️✂️½

Monty Python's Parrot Sketch not Included To celebrate their 20th anniversary, the crazy guys of Monty Python put together this collection of some of their funniest sketches. The parrot was pretty disappointed that he wasn't included.
1990 75m/C Graham Chapman, John Cleese, Terry Gilliam, Eric Idle, Terry Jones, Michael Palin. **VHS $19.95** *PAR* ✂️✂️✂️½

Monty Python's The Meaning of Life Funny, technically impressive film conducts various inquiries into the most profound questions confronting humanity. Notable among the sketches here are a live sex enactment performed before bored schoolboys, a student-faculty rugby game that turns quite violent, and an encounter between a physician and a reluctant, untimely organ donor. Another sketch provides a memorable portrait of a glutton prone to nausea. And at film's end, the meaning of life is actually revealed.
1983 (R) 107m/C *GB* Graham Chapman, John Cleese, Terry Gilliam, Eric Idle, Terry Jones, Michael Palin, Carol Cleveland; *Dir:* Terry Jones. Cannes Film Festival '83: Jury Prize. **VHS, Beta, LV $14.95** *MCA, CCB* ✂️✂️✂️

Moon 44 A space prison is overrun by thugs who terrorize their fellow inmates. Fine cast, taut pacing. Filmed in Germany.
1990 (R) 102m/C *GE* Malcolm McDowell, Lisa Eichhorn, Michael Pare, Stephen Geoffreys, Roscoe Lee Browne; *Dir:* Roland Emmerich. **VHS, LV $89.95** *LIV, IME, BTV* ✂️✂️½

Moon is Blue A young woman flaunts her virginity in this stilted adaptation of F. Hugh Herbert's play. Hard to believe that this film was once considered risque. Good performances, though, from Holden and Niven.
1953 100m/B William Holden, David Niven. Maggie McNamara; *Dir:* Otto Preminger. **Beta, LV $29.98** *FOX* ✂️✂️½

Moon in the Gutter In a ramshackle harbor town, a man searches despondently for the person who killed his sister years before. Various sexual liaisons and stevedore fights intermittently spice up the action. Nasty story that doesn't make much sense on film. In French with English subtitles.
1984 (R) 109m/C FR IT Gerard Depardieu, Nastassia Kinski, Victoria Abril, Vittorio Mezzogiorno, Dominique Pinon; **Dir:** Jean-Jacques Beineix. **VHS, Beta $59.95** COL ✗½

Moon Madness An astronomer takes a trip to the moon to find the Selenites, proprietors of the fountain of youth.
1983 82m/C VHS, Beta $69.95 VES ✗½

Moon Over Harlem A grade-C racially integrated musical directed by the questionable filmmaker Ulmer.
1939 75m/B Bud Harris, Cora Green, Alec Lovejoy, Sidney Bechet; **Dir:** Edgar G. Ulmer. **VHS, Beta $24.95** NOS, DVT ✗

Moon Over Miami Manhunting trio meet their match in this engaging musical. Songs include "Three Little Girls in Blue," "I've Got You All to Myself," and "Oh Me Oh Mi-Am-Mi." A remake of 1938's "Three Blind Mice," this was later remade in 1946 as "Three Little Girls in Blue."
1941 91m/C Don Ameche, Betty Grable, Robert Cummings, Carole Landis, Charlotte Greenwood, Jack Haley; **Dir:** Walter Lang. **VHS, Beta $19.95** FOX ✗✗½

Moon Over Parador An uneven comedy about a reluctant American actor who gets the role of his life when he gets the chance to pass himself off as the recently deceased dictator of a Latin American country. A political strongman wants to continue the charade until he can take over, but the actor begins to enjoy the benefits of dictatorship.
1988 (PG-13) 103m/C Richard Dreyfuss, Sonia Braga, Raul Julia, Jonathan Winters, Fernando Rey, Ed Asner, Dick Cavett, Michael Greene, Sammy Davis Jr., Polly Holliday, Charo, Marianne Sagebrecht, Dana Delaney; **Dir:** Paul Mazursky. **VHS, Beta, LV $19.95** MCA ✗✗½

Moon Pilot An astronaut on his way to the moon encounters a mysterious alien woman who claims to know his future.
1962 98m/C Tom Tryon, Brian Keith, Edmond O'Brien, Dany Saval, Tommy Kirk; **Dir:** James Neilson. **VHS, Beta $69.95** DIS ✗✗

Moon in Scorpio Three Vietnam-vet buddies go sailing and are attacked by the back-from-the-dead victims of a bloody war crime.
1986 90m/C John Phillip Law, Britt Ekland, William Smith; **Dir:** Gary Graver. **VHS, Beta $79.95** TWE ✗½

The Moon and Sixpence Stockbroker turns ambitious painter in this adaptation of Somerset Maugham's novel that was, in turn, inspired by the life of artist Paul Gaugin. Fine performance from Sanders. Filmed mainly in black and white, but uses color sparingly to great advantage. Compare this one to "Wolf at the Door," in which Gaugin is played by Donald Sutherland.
1942 89m/B George Sanders, Herbert Marshall, Steve Geray, Doris Dudley, Eric Blore, Elena Verdugo, Florence Bates, Albert Basserman, Heather Thatcher; **Dir:** Albert Lewin. **VHS** MGM, IVY ✗✗✗

Moon-Spinners Lightweight Disney drama featuring Hayley Mills as a young tourist traveling through Crete who meets up with a young man, accused of being a jewel thief, and the two work together to find the real jewel thieves. Watch for silent film star Pola Negri.
1964 118m/C Hayley Mills, Peter McEnery, Eli Wallach, Pola Negri; **Dir:** James Neilson. **VHS, Beta** DIS, OM ✗✗

The Moon Stallion A charming tale of mystery and fantasy for children and adults. Professor Purwell is researching King Arthur. His blind daughter soon becomes involved with supernatural events and a white horse that appears to lead her into adventures not of this age.
19?? 95m/C GB VHS $29.95 BFS ✗✗✗

Moon of the Wolf A small town in bayou country is terrorized by a modern-day werewolf that rips its victims to shreds. Made-for-television.
1972 74m/C David Janssen, Barbara Rush, Bradford Dillman, John Beradino; **Dir:** Daniel Petrie. **VHS, Beta $49.95** WOV, GKK ✗

Mooncussers A children's film detailing the exploits of a precocious 12-year-old determined to exact revenge upon a band of ruthless pirates.
1962 85m/C Kevin Corcoran, Rian Garrick, Oscar Homolka; **Dir:** James Neilson. **VHS, Beta $69.95** DIS ✗½

Moonfleet Follows the adventures of an 18th-century buccaneer who tries to swindle a young lad in his charge of a valuable diamond. From J. Meade Falkner's novel.
1955 89m/C Stewart Granger, Jon Whiteley, George Sanders, Viveca Lindfors, Joan Greenwood, Ian Wolfe; **Dir:** Fritz Lang. **VHS, LV $34.98** MGM ✗✗½

Moonlight Sonata Professional concert pianist Paderewski performs his way through a confusing soap-opera of a film. Includes performances of Franz Liszt's "Second Hungarian Rhapsody" and Frederic Chopin's "Polonaise."
1938 80m/B GB Ignace Jan Paderewski, Charles Farrell, Marie Tempest, Barbara Greene, Eric Portman; **Dir:** Lothar Mendes. **VHS $19.95** NOS, DVT ✗✗½

Moonlight Sword & Jade Lion A Kung Fu action film taking place in ancient China.
197? 94m/B Mao Yin, Wong Do, So Chan Ping. **VHS, Beta $29.95** OCE, MAV ✗

Moonlighting Compelling drama about Polish laborers illegally hired to renovate London flat. When their country falls under martial law, the foreman conceals the event and pushes workers to complete project. Unlikely casting of Irons as foreman is utterly successful.
1982 (PG) 97m/C GB PL Jeremy Irons, Eugene Lipinski, Jiri Stanislay, Eugeniusz Haczkiewicz; **Dir:** Jerzy Skolimowski. Cannes Film Festival '82: Best Screenplay. **VHS, Beta, LV $49.95** MCA ✗✗✗½

Moonlighting The television-pilot for the popular detective show, where Maddie and David, the daffy pair of impetuous private eyes, meet for the first time and solve an irrationally complex case.
1985 93m/C Cybill Shepherd, Bruce Willis, Allyce Beasley; **Dir:** Robert Butler. **VHS, Beta $19.95** WAR, OM ✗✗½

The Moonraker Compelling action scenes highlight this tale set at the end of the English Civil War, as the Royalists attempt to sneak the king out of England and into France.
1958 82m/C GB George Baker, Sylvia Syms, Peter Arne, Marius Goring, Clive Morton, Gary Raymond, Richard Leech, Patrick Troughton; **Dir:** David MacDonald. **VHS $19.95** NOS ✗✗½

Moonraker Uninspired Bond fare has 007 unraveling intergalactic hijinx. Bond is aided by a female CIA agent, assaulted by a giant with jaws of steel, and captured by Amazons when he sets out to protect the human race. Moore, Chiles, and Lonsdale all seem to be going through the motions only.
1979 (PG) 126m/C GB Roger Moore, Lois Chiles, Richard Kiel, Michael Lonsdale, Corinne Clery, Geoffrey Keen, Emily Bolton, Walter Gotell, Bernard Lee, Lois Maxwell, Desmond Llewelyn; **Dir:** Lewis Gilbert. **VHS, Beta, LV $19.95** MGM, TLF ✗✗

Moonrise A melodramatic tale about the son of an executed murderer who becomes the object of derision in the small town where he lives. One tormentor finally attacks him, and the young man must make a split-second decision that may affect his own mortality.
1948 90m/B Dane Clark, Gail Russell, Ethel Barrymore, Allyn Joslyn, Harry Morgan, Lloyd Bridges, Selena Royle, Rex Ingram, Harry Carey Jr.; **Dir:** Frank Borzage. **VHS $19.98** REP ✗✗

Moon's Our Home A fast-paced, breezy, screwball comedy about an actress and adventurer impulsively marrying and then bickering through the honeymoon. Silliness at its height. Fonda and Sullavan were both married and then divorced before filming this movie.
1936 80m/B Margaret Sullavan, Henry Fonda, Beulah Bondi, Charles Butterworth, Margaret Hamilton, Walter Brennan; **Dir:** William A. Seiter. **VHS, Beta $14.95** KRT ✗✗

Moonshine County Express Three sexy daughters of a hillbilly moonshiner set out to run the still and avenge their father's murder.
1977 (PG) 97m/C John Saxon, Susan Howard, William Conrad, Morgan Woodward, Claudia Jennings, Jeff Corey, Dub Taylor, Maureen McCormick, Albert Salmi, Candice Rialson; **Dir:** Gus Trikonis. **VHS, Beta $39.98** WAR ✗½

Moonstruck Winning romantic comedy about widow engaged to one man but falling in love with his younger brother in Little Italy. Excellent performances all around, with Cher particularly fetching as attractive, hapless widow. Unlikely casting of usually dominating Aiello, as unassuming mama's boy also works well, and Cage is at his best as a tormented one-handed opera lover/baker.
1987 (PG-13) 103m/C Cher, Nicolas Cage, Olympia Dukakis, Danny Aiello, Vincent Gardenia, Julie Bovasso, Louis Guss, Anita Gillette, Feodor Chaliapin, John Mahoney; **Dir:** Norman Jewison. Academy Awards '87: Best Actress (Cher), Best Original Screenplay, Best Supporting Actress (Dukakis); National Board of Review Awards '87: Best Supporting Actress (Dukakis). **VHS, Beta, LV $19.98** MGM, BTV ✗✗✗½

Morals for Women This soap-opera's so old it almost predates soap. A secretary becomes the Kept Woman of her boss in the Big Bad City. She escapes, but not for long.
1931 65m/B Bessie Love, Conway Tearle, John Holland, Natalie Moorhead, Emma Dunn, June Clyde; **Dir:** Mort Blumenstock. **VHS $15.95** NOS, LOO ✗½

Moran of the Lady Letty Seaweed saga of Ramon Laredo (Valentino), a high society guy who's kidnapped aboard a pirate barge. When the salty dogs rescue/capture Moran (a theme ahead of its time?) from a ship fuming with burning coal, Ramon is

smitten in a big way with the boyish girl and battles contagiously to prevent the pirates from selling her as a slave. Excellent fight scenes with much hotstepping over gang-planks and swinging from masts, with a 60-foot death dive from above. Little known Dalton was a celluloid fave in the 'teens and twenties, and Valentino is atypically cast as a man's man. The sheik himself is said not to have like this one because it undermined his image as a ladies' man.
1922 71m/B Dorothy Dalton, Rudolph Valentino, Charles Brinley, Walter Long, Emil Jorgenson; *Dir:* George Melford. **VHS, Beta $32.95** *GPV, FCT, DNB* ✂✂½

More Smells like teen angst in the sixties when a German college grad falls for American in Paris, to the tune of sex, drugs and Pink Floyd. Schroeder's first effort as director, it's definitely a sixties pic. In French with English subtitles.
1969 110m/C Mimsy Farmer, Klaus Grunberg, Heinz Engelmann, Michel Chanderli; *Dir:* Barbet Schroeder. **VHS, Beta $59.99** *WAR, BTV, FCT* ✂✂½

More Laughing Room Only A recording of live stand-up comedy routines featuring some of today's most hilarious performers.
19?? 60m/C Arsenio Hall, Shelley Berman, Foster Brooks, Tom Dreeson, George Gobel, George Miller. **VHS, Beta $24.95** *MON* ✂✂½

The More the Merrier Likeable romantic comedy in which working girl must share apartment with two bachelors in Washington, D.C., during WWII. Arthur is especially endearing as a young woman in male company. Oscar-nominated for Best Picture, Best Actress (Arthur), Best Director, Original Story, and Screenplay.
1943 104m/B Joel McCrea, Jean Arthur, Charles Coburn, Richard Gaines, Bruce Bennett, Ann Savage, Ann Doran, Frank Tully, Grady Sutton; *Dir:* George Stevens. Academy Awards '43: Best Supporting Actor (Coburn); New York Film Critics Awards '43: Best Director (Stevens). **VHS, Beta, LV $19.95** *COL, BTV* ✂✂✂

More! Police Squad More episodes from the seminal spoof of television cop shows, episodes entitled "Revenge and Remorse," "The Butler Did It" and "Testimony of Evil."
1982 75m/C Leslie Nielsen, Alan North, Rex Hamilton, Peter Lupus. **VHS, Beta, LV $19.95** *PAR* ✂✂✂

More Song City U.S.A. Features rock videos for the kids such as "Clean Up My Room Blues" and "La Bamba." Children will love it.
1989 30m/C VHS, Beta $14.95 *FHE* ✂✂✂

More Wild, Wild West Another feature-length continuation of the satirical television western series, with Winters taking on Conrad and Martin. Made for television.
1980 94m/C Robert Conrad, Ross Martin, Jonathan Winters, Victor Buono; *Dir:* Burt Kennedy. **VHS, Beta $59.98** *FOX* ✂✂

Morgan! Offbeat comedy in which deranged artist copes with divorce by donning ape suit. Some laughs ensue. Based on David Mercer's play. Oscar nominations for Redgrave and Best Costume Design.
1966 93m/B *GB* Vanessa Redgrave, David Warner, Robert Stephens, Irene Handl; *Dir:* Karel Reisz. **VHS, Beta, LV** *HBO* ✂✂½

Morgan the Pirate Steve Reeves is at his finest as Morgan, an escaped slave who becomes a notorious pirate in the Caribbean and is wrongly condemned to death in this brisk, sea-going adventure.
1960 93m/C *FR IT* Steve Reeves, Valerie Lagrange; *Dir:* Andre de Toth. **VHS, Beta, LV $39.98** *SUE* ✂✂

Morgan Stewart's Coming Home When Dad needs a good family image in his political race, he brings Morgan home from boarding school. Fortunately, Morgan can see how his parents are using him and he doesn't approve at all. He decides to turn his family's life upside down while pursuing the love of his life.
1987 (PG-13) 96m/C Jon Cryer, Lynn Redgrave, Nicholas Pryor, Viveka Davis, Paul Gleason, Andrew Duncan, Savely Kramorov, John Cullum, Robert Sedgwick, Waweru Njenga, Sudhir Rad; *Dir:* Alan Smithee. **VHS, Beta, LV $34.95** *HBO* ✂

Morituri Gripping wartime drama in which an Allied spy tries to persuade German gunboat captain to surrender his vessel. Brando is—no surprise—excellent. Also known as "Saboteur: Code Name Morituri."
1965 123m/B Marlon Brando, Yul Brynner, Trevor Howard, Janet Margolin, Wally Cox, William Redfield; *Dir:* Bernhard Wicki. **VHS, Beta $19.98** *FOX* ✂✂✂

Mormon Maid A young woman and her family, after being saved from an Indian attack by a Mormon group, come to live among them without converting to their beliefs. Silent.
1917 78m/B Noah Beery, Frank Borzage, Hobart Bosworth. **VHS, Beta $52.95** *GVV* ✂½

Morning After A predictable suspense-thriller about an alcoholic actress who wakes up one morning with a corpse in bed next to her, but cannot remember anything about the night before. She evades the police, accidentally meets up with an ex-cop, and works with him to unravel an increasingly complicated mystery.
1986 (R) 104m/C Jane Fonda, Jeff Bridges, Raul Julia; *Dir:* Sidney Lumet. **VHS, Beta, LV $19.98** *LHV, WAR* ✂✂

Morning Glory Small-town girl finds love and fame in the big city. Predictable fare nonetheless boasts fine performances from Hepburn and Fairbanks. Adapted from Zoe Atkins' stage play.
1933 74m/B Katharine Hepburn, Douglas Fairbanks Jr., Adolphe Menjou, Mary Duncan; *Dir:* Lowell Sherman. Academy Awards '33: Best Actress (Hepburn). **VHS, Beta, LV $19.98** *RKO, CCB, MED* ✂✂½

Morocco A foreign legion soldier falls for a world-weary chanteuse along the desert sands. Cooper has never been more earnest, and Dietrich has never been more blase and exotic. In her American film debut, Dietrich sings "What am I Bid?" A must for anyone drawn to improbable, gloriously well-done kitsch. Based on Benno Vigny's novel, "Amy Jolly."
1930 97m/B Marlene Dietrich, Gary Cooper, Adolphe Menjou, Ullrich Haupt, Francis McDonald, Eve Southern, Paul Porcasi; *Dir:* Josef von Sternberg. **VHS, Beta, LV $14.95** *MCA* ✂✂✂½

Moron Movies A collection of one-hundred-fifty short films that describe comedic uses for everyday products.
1985 60m/C VHS, Beta $39.95 *MPI* ✂✂✂½

Morons from Outer Space Slow-witted aliens from elsewhere in the universe crash onto Earth, but unlike other sci-fis, these morons become internationally famous, despite them acting like intergalactic stooges. Plenty of sight gags.
1985 (PG) 87m/C *GB* Griff Rhys Jones, Mel Smith, James B. Sikking, Dinsdale Landen, Jimmy Nail, Joanne Pearce, Paul Brown; *Dir:* Mike Hodges. **VHS** *HBO, OM* ✂✂

Mortal Passions Another "Double Indemnity" rip-off, with a scheming tramp manipulating everyone around her with betrayal, sex, and murder. Her dead husband's brother eventually ignites old flames to steer her away from his money.
1990 (R) 96m/C Zach Galligan, Krista Errickson, Michael Bowen, Luca Bercovici, Sheila Kelley, David Warner; *Dir:* Andrew Lane. **VHS, Beta $89.98** *FOX* ✂½

Mortal Sins A TV evangelist finds more than salvation at the altar, when bodies begin to turn up. The detective he hires is close-mouthed, but more than a little interested in the preacher's daughter. A tense psychosexual thriller.
1990 (R) 85m/C Brian Benben, Anthony LaPaglia, Debrah Farentino; *Dir:* Yuri Sivo. **VHS $29.95** *ACA* ✂

Mortal Thoughts Best friends find their relationship tested when the brutal husband of one of them is murdered. Moore and Headly are exceptional, capturing the perfect inflections and attitudes of the hard-working New Jersey beauticians sure of their friendship. Excellent pacing, fine supporting cast, with Keitel and Willis stand-outs.
1991 (R) 104m/C Demi Moore, Bruce Willis, Glenne Headly, Harvey Keitel, John Pankow, Billie Neal; *Dir:* Alan Rudolph. National Society of Film Critics Awards '91: Best Supporting Actor (Keitel). **VHS, Beta, LV, 8mm $19.95** *COL, LDC* ✂✂½

Mortuary A young woman's nightmares come startlingly close to reality.
1981 (R) 91m/C Christopher George, Lynda Day George, Paul Smith; *Dir:* Howard Avedis. **VHS, Beta $79.98** *LIV, VES* ✂

Mortuary Academy To win an inheritance two brothers must attend the family mortician school, a situation paving the way for aggressively tasteless jokes on necrophilia. An attempt to recapture the successful black humor of the earlier Bartel/Woronov teaming "Eating Raoul," this one's dead on arrival.
1991 (R) 86m/C Christopher Atkins, Perry Lang, Paul Bartel, Mary Woronov, Tracy Walter, Lynn Danielson, Cesar Romero, Wolfman Jack; *Dir:* Michael Schroeder. **VHS, LV $79.95** *COL, LDC* ✂

Mosby's Marauders A young boy joins a Confederate raiding company during the Civil War and learns about bravery, war, and love.
1966 79m/C Kurt Russell, James MacArthur, Jack Ging, Peggy Lipton, Nick Adams; *Dir:* Michael O'Herlihy. **VHS, Beta $69.95** *DIS* ✂✂

Moscow Does Not Believe in Tears Three Soviet women realize very different fates when they pursue their dreams in 1950's Moscow. Bittersweet, moving fare that seems a somewhat surprising production from pre-Glasnost USSR. Brief nudity and violence. Also known as "Moscow Distrusts Tears."

1980 115m/C *RU* Vera Alentova, Irina Muravyova, Raisa Ryazanova, Natalie Vavilova, Alexei Batalov; *Dir:* Vladimir Menshov. Academy Awards '80: Best Foreign Language Film. **VHS, Beta, LV** $59.95 COL, IME *♪♪♪*

Moscow on the Hudson
Good-natured comedy has Williams as Soviet defector trying to cope with new life of freedom in fast-paced, freewheeling melting pot of New York City. Williams is particularly winning as naive jazzman, though Alonso also scores as his Hispanic love interest. Be warned though, it's not just played for laughs.
1984 (R) 115m/C Robin Williams, Maria Conchita Alonso, Cleavant Derricks, Alejandro Rey, Elya Baskin; *Dir:* Paul Mazursky. **VHS, Beta, LV, 8mm** $79.95 COL *♪♪♪*

Moses
Lancaster portrays the plight of Moses, who struggled to free his people from tyranny.
1976 (PG) 141m/C Burt Lancaster, Anthony Quayle, Ingrid Thulin, Irene Papas, William Lancaster; *Dir:* Gianfranco DeBosio. **VHS, Beta** $19.98 FOX *♪ ½*

The Mosquito Coast
Ambitious adaptation of Paul Theroux's novel about an asocial inventor who transplants his family to a rainforest to realize his utopian dream. A nightmare ensues. Ford tries hard, but supporters Mirren and Phoenix are main appeal of only intermittently successful drama. Written by Paul Schrader.
1986 (PG) 119m/C Harrison Ford, Helen Mirren, River Phoenix, Andre Gregory, Martha Plimpton, Conrad Roberts, Butterfly McQueen, Adrian Steele, Hilary Gordon, Rebecca Gordon, Dick O'Neil; *Dir:* Peter Weir. **VHS, Beta, LV** $19.98 WAR *♪♪ ½*

The Most Dangerous Game
Shipwrecked McRae washed ashore the island of Banks's Count Zaroff, a deranged sportman with a flair for tracking humans. Guess who becomes the mad count's next target. Oft-told tale is compellingly related in this, the first of many using Richard Connell's famous short story. If deja vu sets in, don't worry. This production uses most of the scenery, staff, and cast from its studio cousin, "King Kong." Remade in 1945 as "A Game of Death" and in 1956 as "Run for the Sun."
1932 78m/B Joel McCrea, Fay Wray, Leslie Banks, Robert Armstrong; *Dir:* Ernest B. Schoedsack, Irving Pichel. **VHS, Beta** $16.95 SNC, NOS, MED *♪♪♪*

The Most Death-Defying Circus Acts of All Time
A compilation of Ringling Bros. and Barnum & Bailey 115th edition Circus' most hair-raising stunt acts.
1987 30m/C **VHS, Beta** $14.98 IVE *♪♪♪*

Most Wanted
A psychopath with a penchant for raping nuns and collecting crucifixes is tracked down by a special police unit in this made-for-television film. Became a TV series.
1976 78m/C Robert Stack, Shelly Novack, Leslie Charleson, Tom Selleck, Sheree North; *Dir:* Walter Grauman. **VHS, Beta** $49.95 WOV *♪ ½*

Motel Hell
A completely tongue-in-cheek gore-fest about a farmer who kidnaps tourists, buries them in his garden, and reaps a human harvest to grind into his distinctive brand of smoked, preservative-free sausage.
1980 (R) 102m/C Rory Calhoun, Nancy Parson, Paul Linke, Nina Axelrod, Wolfman Jack, Elaine Joyce, Dick Curtis; *Dir:* Kevin Connor. **VHS, Beta, LV** $19.98 MGM *♪ ½*

Mother
Pudovkin's innovative classic about a Russian family shattered by the uprising in 1905. A masterpiece of Russian cinema that established Pudovkin, and rivaled only Eisenstein for supremacy in montage, poetic imagery, and propagandistic ideals. Based on Maxim Gorky's great novel, it's one of cinematic history's seminal works. Striking cinematography, stunning use of montage make this one important. Silent with English subtitles.
1926 70m/B *RU* Vera Baranovskaya, Nikolai Batalov; *Dir:* Vsevolod Pudovkin. Brussels World's Fair '58: #8 of the Best Films of All Time. **VHS, Beta, 3/4U** $24.95 NOS, IHF, DVT *♪♪♪ ½*

Mother
A Japanese family is undone after the devastation of WWII. An uncharacteristically dramatic, and thus more accessible, film from one of the Japanese cinema's greatest masters. In Japanese with English subtitles. Japanese title: "Okasan."
1952 98m/B *JP* Kinuyo Tanaka, Kyoko Kagawa, Eiji Okada, Akihiko Katayama; *Dir:* Mikio Naruse. **VHS, Beta, LV** $49.00 SVS, FCT, IME *♪♪♪ ½*

Mother & Daughter: A Loving War
A made-for-television film about three women experiencing motherhood. Special appearance by Harry Chapin as himself.
1980 96m/C Tuesday Weld, Frances Sternhagen, Kathleen Beller, Jeanne Lang, Ed Winter; *Dir:* Burt Brinckerhoff. **VHS, Beta** $59.98 CHA *♪ ½*

Mother Doesn't Always Know Best
A strange relationship between a young count and his widowed mother initiates a strange series of events.
196? 95m/C Senta Berger; *Dir:* Alfredo Gianelli. **VHS, Beta** INV *♪ ½*

Mother Goose Rock 'n' Rhyme
Duvall follows the success of her "Faerie Tale Theatre" series with a new comedic fable, starring some of Hollywood's top names. Great fun for the whole family.
1990 96m/C Shelley Duvall, Teri Garr, Howie Mandel, Jean Stapleton, Ben Vereen, Bobby Brown, Art Garfunkel, Dan Gilroy, Deborah Harry, Cyndi Lauper, Little Richard, Paul Simon, Stray Cats, ZZ Top, Harry Anderson, Elayne Boosler, Woody Harrelson, Cheech Marin, Katey Sagal, Garry Shandling. **VHS, Beta** $79.98 MED *♪ ½*

Mother, Jugs and Speed
Black comedy about the day-to-day tragedies encountered by a group of ambulance drivers. Interesting mix of stars.
1976 (PG) 95m/C Bill Cosby, Raquel Welch, Harvey Keitel, Allen Garfield, Larry Hagman, Bruce Davison, Dick Butkus, L.Q. Jones, Toni Basil; *Dir:* Peter Yates. **VHS** $79.98 FOX *♪ ½*

Mother of Kings
Livin' ain't easy for a widowed charwoman thanks to WWII and Stalinism. Innovative use of newsreel scenes. In Polish with English subtitles.
1982 126m/C *PL* Magda Teresa Wojcik. **VHS** $49.99 FCT *♪♪*

Mother Kusters Goes to Heaven
Mrs. Kusters's husband is a frustrated factory worker who goes over the edge and kills the factory owner's son and himself. Left alone, she learns that everyone is using her husband's death to further their own needs, including her daughter, who uses the publicity to enhance her singing career. A statement that you should trust no one, not even your family and friends. This film was banned from the Berlin Film Festival because of its political overtones. In German with English subtitles.
1976 108m/C *GE* Brigitte Mira, Ingrid Caven, Armin Meier, Irm Hermann, Gottfried John, Karl-Heinz Bohm, Margit Carstensen; *W/Dir:* Rainer Werner Fassbinder. **VHS** $79.95 NYF *♪♪♪*

Mother Lode
The violent conflict between twin brothers (played by Heston) is intensified by greed, near madness, and the all-consuming lust for gold in this action-adventure film. Heston's son, Fraser Clarke Heston, wrote as well as produced the film.
1982 (PG) 101m/C Charlton Heston, Nick Mancuso; *Dir:* Charlton Heston. **VHS, Beta** $79.95 VES *♪ ½*

Mother's Day
Three former college roommates plan a reunion together in the wilderness. All goes well until they are dragged into an isolated house by two insane boys who terrorize people to please their mother. Excessive gore.
1980 98m/C Tiana Pierce, Nancy Hendrickson, Deborah Luee; *Dir:* Charles Kaufman. **VHS, Beta** $49.95 MED *♪ ½*

Mothra
A giant moth wreaks havoc on Tokyo.
1962 101m/C *JP* Yumi Ito, Frankie Sakai, Lee Kresel, Emi Ito; *Dir:* Inoshiro Honda. **VHS, Beta** $9.95 GKK *♪*

Motor Patrol
A cop poses as a racketeer to infiltrate a car-stealing ring.
1950 66m/B Don Castle, Jane Nigh, Charles Victor. **VHS, Beta, LV** WGE *♪ ½*

Motor Psycho
When a motorcycle gang rapes a woman, she and her husband pursue them into the desert to seek their brutal revenge.
1965 73m/B Haji, Alex Rocco, Stephen Oliver, Holle K. Winters, Joesph Cellini, Thomas Scott, Coleman Francis, Sharon Lee; *Dir:* Russ Meyer. **VHS** RMF *♪*

Motorcycle Gang
It's the summer of '57, and Randy plans to bring home the title at the Pacific Motorcycle Championships for his club, the "Skyriders." But first his old rival Nick, who's just out of jail, turns up and challenges Randy to an illegal street race. Will Randy succumb to peer pressure and race?
1957 78m/B Steven Terrell, John Ashley, Anne Neyland, Carl "Alfalfa" Switzer, Raymond Hatton; *Dir:* Edward L. Cahn. **VHS** $9.95 COL *♪♪*

Motown 25: Yesterday, Today, Forever
The television all-star salute to Berry Gordy that features a musical duel between the Four Tops and The Temptations and the reunion of The Jackson Five. The show where Michael Jackson lip-synched "Billy Jean" and moonwalked to the big, big time.
1983 130m/C The Commodores, Michael Jackson, Richard Pryor, Diana Ross, Marvin Gaye, Stevie Wonder, Adam Ant, Dick Clark. **VHS, Beta, LV, 8mm** $19.95 MVD, MGM *♪♪*

Motown's Mustang
Eleven old Motown hits are used as the framework for this new musical drama featuring cameos by various artists, about a '64 Mustang passing through various owners' hands and cultural changes.
1986 43m/C Clyde Jones, Christi Shay, Louis Carr Jr., Billy Preston, The Temptations. **VHS, Beta, LV** $19.95 MCA *♪*

Mouchette
A lonely 14-year-old French girl, daughter of a drunk father and dying mother, eventually finds spiritual release by committing suicide. Typically somber, spiritual fare from unique master filmmaker Bresson. Perhaps the most complete expression

of Bresson's austere, Catholic vision. In French with subtitles.
1960 80m/B *FR* Nadine Nortier, Maria Cardinal, Paul Hebert; ***Dir:*** Robert Bresson. **VHS, Beta $59.95** *FCT* 🦴🦴🦴½

Moulin Rouge Colorful, entertaining portrait of acclaimed Impressionist painter Toulouse-Lautrec, more famous for its production stories than on-screen drama, Ferrer delivers one of his most impressive performances as the physically stunted, cynical artist who basked in the seamy Montmartre nightlife. In addition to grabbing a number of Oscars, was academy nominated for: Best Picture, Best Actor (Ferrer), Supporting Actress (Marchand), Director, and Film Editing.
1952 119m/C Jose Ferrer, Zsa Zsa Gabor, Christopher Lee, Peter Cushing, Colette Marchand, Katherine Kath, Michael Balfour, Eric Pohlmann; ***Dir:*** John Huston. Academy Awards '52: Best Art Direction/Set Decoration (Color), Best Costume Design (Color); National Board of Review Awards '53: 5 Best Foreign Films of the Year; Venice Film Festival '53: Silver Prize. **VHS, Beta $19.95** *MGM* 🦴🦴🦴½

The Mountain A man and his shady younger brother set out to inspect a Paris-routed plane that crashed in the French Alps. After some harrowing experiences in climbing the peak, it becomes evident that one brother has designs to save whatever he can, while the other intends to loot it. Many real, as well as staged, climbing scenes. Based on Henri Troyat's novel.
1956 105m/C Spencer Tracy, Robert Wagner, Claire Trevor, William Demarest, Richard Arlen, E.G. Marshall; ***Dir:*** Edward Dmytryk. **VHS, Beta $19.95** *PAR* 🦴🦴

Mountain Charlie A mountain girl's life is destroyed by three drifters.
1980 96m/C Denise Neilson, Dick Robinson, Rick Guinn, Lynn Seus; ***Dir:*** George Stapleford. **VHS, Beta** *VDY* 🦴½

Mountain Family Robinson An urban family, seeking escape from the hassles of city life, moves to the Rockies, determined to get back to nature. They soon find that nature may be more harsh than rush-hour traffic and nasty bosses when a bear comes calling. More "Wilderness Family"-type adventures.
1979 (G) 102m/C Robert F. Logan, Susan Damante Shaw, Heather Rattray, Ham Larsen, William Bryant, George Flower; ***Dir:*** John Cotter. **VHS, Beta $9.98** *MED* 🦴🦴

Mountain Fury A spoof based on the novel by Alexander Dumas wherein two brothers separated at birth are reunited and battle against an evil French baron.
1990 (R) 90m/C VHS *RAE* 🦴🦴

Mountain Man This historically true drama concerns Palan Clark's successful fight to save a magnificent wilderness and its animals. He learns, from naturalist John Muir, the need to save the land from destruction and goes to Washington to win President Lincoln's support.
1977 96m/C Denver Pyle, Cheryl Miller, John Dehner, Ken Berry, Ford Rainey; ***Dir:*** David O'Malley. **VHS, Beta** *LME* 🦴🦴½

Mountain Men A sweeping adventure drama set in the American West of the 1880s.
1980 (R) 102m/C Charlton Heston, Brian Keith, John Glover, Seymour Cassel, Victor Jory; ***Dir:*** Richard Lang. **VHS, Beta, LV $9.95** *COL* 🦴

The Mountain Road Listless drama of an American squadron stationed in China during the last days of WWII. Only Stewart makes this drab film worthwile.
1960 102m/B James Stewart, Harry Morgan, Glenn Corbett, Mike Kellin; ***Dir:*** Daniel Mann. **VHS $14.99** *COL* 🦴½

Mountains of the Moon Sprawling adventure detailing the obsessive search for the source of the Nile conducted by famed Victorian rogue/explorer Sir Richard Burton and cohort John Hanning Speke in the late 1800s. Spectacular scenery and images. Director Rafelson, better known for overtly personal films such as "Five Easy Pieces" and "The King of Marvin Gardens," shows considerable skill with this epic. Cinematography by Roger Deakins. From William Harrison's novel "Burton and Speke."
1990 (R) 140m/C Patrick Bergin, Iain Glen, Fiona Shaw, Richard E. Grant, Peter Vaughan, Roger Rees, Bernard Hill, Anna Massey, Leslie Phillips; ***Dir:*** Bob Rafelson. **VHS, Beta, LV $89.95** *IVE* 🦴🦴🦴½

Mountaintop Motel Massacre A resort motel's hostess is a raving lunatic who regularly slaughters her guests.
1986 (R) 95m/C Bill Thurman, Anna Chappell, Will Mitchell; ***Dir:*** Jim McCullough. **VHS, Beta $19.95** *STE, NWV* *Woof!*

Mouse and His Child A gentle animated fantasy adventure about a toy wind-up mouse and his child who fall into the clutches of a villainous rat when they venture into the outside world.
1977 83m/C ***Dir:*** Fred Wolf; ***Voices:*** Peter Ustinov, Cloris Leachman, Andy Devine. **VHS, Beta $19.95** *COL* 🦴🦴

The Mouse That Roared With its wine export business going down the drain, a tiny, desperate country decides to declare war on the United States in hopes that the U.S., after its inevitable triumph, will revive the conquered nation. So off to New York go 20 chain-mail clad warriors armed with bow and arrow. Featured in three roles, Sellers is great as the duchess, less effective (though still funny) as the prime minister, and a military leader. A must for Sellers' fans; maintains a sharp satiric edge throughout. Based on the novel by Leonard Wibberley.
1959 83m/C *GB* Peter Sellers, Jean Seberg, Leo McKern; ***Dir:*** Jack Arnold. **VHS, Beta $59.95** *COL* 🦴🦴🦴

Movers and Shakers An irreverent spoof of Hollywood depicting a filmmaker's attempt to render a best-selling sex manual into a blockbuster film. Written and co-produced by Charles Grodin, the film fails to live up to its potential and wastes a star-studded cast.
1985 (PG) 100m/C Walter Matthau, Charles Grodin, Gilda Radner, Vincent Gardenia, Bill Macy, Tyne Daly, Steve Martin, Penny Marshall, Luana Anders; ***Dir:*** William Asher. **VHS, Beta $79.95** *MGM* 🦴½

The Movie House Massacre A psychopath runs rampant in a theatre killing and maiming moviegoers. No refunds are given.
1978 80m/C Mary Woronov. **VHS, Beta $24.95** *AHV, MRV* 🦴

Movie Maker A young filmmaker ineptly raises money for his next film. Also known as "Smart Alec."
1986 87m/C Zsa Zsa Gabor, Antony Alda, Orson Bean, Bill Henderson; ***Dir:*** Jim Wilson. **VHS, Beta $79.98** *LIV, LTG* 🦴

Movie, Movie Acceptable spoof of 1930s films features Scott in twin-bill of black and white "Dynamite Hands," which lampoons boxing dramas, and "Baxter's Beauties," a color send-up of Busby Berkeley musicals. There's even a parody of coming attractions. Wholesome, mildly entertaining.
1978 (PG) 107m/B Stanley Donen, George C. Scott, Trish Van Devere, Eli Wallach, Red Buttons, Barbara Harris, Barry Bostwick, Harry Hamlin, Art Carney; ***Dir:*** Stanley Donen. **VHS** *FOX* 🦴🦴½

Movie Museum Series A fascinating and entertaining review of the first 25 years of the motion picture art form. The set comes on five cassettes and totals ten hours.
1980 600m/B ***Nar:*** Paul Killiam. **VHS, Beta $199.98** *CCB* 🦴🦴½

A Movie Star's Daughter The daughter of a famous movie star moves to a new town and school. She is afraid her popularity is due to her famous father. She must make choices and learn the real meaning of friendship. Also available in a 33-minute edited version.
1979 46m/C Frank Converse, Trini Alvarado; ***Dir:*** Robert Fuest. **VHS, Beta, 3/4U $14.95** *LCA* 🦴🦴½

Movie Struck Originally released as "Pick a Star," this Hollywood behind-the-scenes story of a girl waiting for her big break features guest appearances by Laurel and Hardy.
1937 70m/B Stan Laurel, Oliver Hardy, Jack Haley, Patsy Kelly, Rosina Lawrence; ***Dir:*** Edward Sedgwick. **VHS, Beta $19.95** *NOS, AMV, CAB* 🦴🦴

Movie Stunt Man A retired Hollywood stunt man takes a job for a dead friend so that the friend's widow can receive the fee. Good stunt work taken mostly from old Richard Taldmadge movies almost makes up for poor production. "Hollywood Thrillmakers" and "Hollywood Stunt Man" are alternate titles.
1953 56m/B James Gleason, William Henry; ***Dir:*** Bernard B. Ray. **VHS $19.95** *NOS* 🦴

Moving An engineer must relocate his family from New Jersey to Idaho in order to get his dream job. Predictable calamities ensue. Not apt to move you.
1988 (R) 89m/C Richard Pryor, Randy Quaid, Dana Carvey, Dave Thomas, Rodney Dangerfield, Stacey Dash; ***Dir:*** Alan Metter. **VHS, Beta, LV, 8mm $19.98** *WAR* 🦴

Moving Out An adolescent migrant Italian boy finds it difficult to adjust to his new surroundings in Melbourne, Australia.
1983 91m/C Vince Colosimo, Sally Cooper, Maurice Devincentis, Tibor Gyapjas; ***Dir:*** Michael Pattinson. **VHS, Beta $69.95** *VID* 🦴

Moving Target A young woman witnesses the brutal murder of her boyfriend by mobsters and flees to Florida. Unbeknownst to her, the thugs are still after her. Although top-billed, Blair's role is actually a supporting one.
1989 (R) 85m/C *IT* Linda Blair, Ernest Borgnine, Stuart Whitman, Charles Pitt, Jainine Linde, Kurt Woodruff; ***Dir:*** Marius Mattei. **VHS, Beta $14.95** *HMD, SOU* 🦴½

Moving Targets A young girl and her mother find themselves tracked by a homicidal maniac.
1987 95m/C *AU* Michael Aitkens, Carmen Duncan, Annie Jones, Shane Briant; ***Dir:*** Chris Langman. **VHS, Beta $29.95** *ACA* 🦴🦴

Moving Violation Crooked cops chase two young drifters who have witnessed the local sheriff commit a murder.
1976 (PG) 91m/C Eddie Albert, Kay Lenz, Stephen McHattie, Will Geer, Loni Chapman; *Dir:* Charles S. Dubin. **VHS, Beta, LV $59.98** FOX 🐾½

Moving Violations This could be entitled "Adventures in Traffic Violations School." A wise-cracking tree planter is sent to traffic school after accumulating several moving violations issued to him by a morose traffic cop. Bill Murray's little brother in feature role.
1985 (PG-13) 90m/C John Murray, Jennifer Tilly, James Keach, Brian Backer, Sally Kellerman, Fred Willard, Clara Peller, Wendie Jo Sperber; *Dir:* Neal Israel. **VHS, Beta, LV $79.98** FOX 🐾

Mozart: A Childhood Chronicle An experimental, semi-narrative portrait of the immortal composer, from the ages of seven to twenty, filmed on authentic locations and with the original instruments of the era. Entire soundtrack is composed of Mozart's music. In German with English subtitles.
1976 224m/B GE *Dir:* Klaus Kirschner. **VHS, Beta, LV $99.95** KIV, IME, GLV 🐾🐾🐾

The Mozart Brothers An angst-ridden Swedish opera director decides to break all the rules while doing an innovative production of Mozart's "Don Giovanni." At first he horrifies his conservative opera company but as his vision is gradually realized their resistance breaks down. A wonderful look at behind-the-scenes chaos and not just for opera aficionados. In Swedish with English subtitles.
1986 111m/C SW Etienne Glaser, Philip Zanden, Henry Bronett; *Dir:* Suzanne Osten. **VHS $59.95** FRI 🐾🐾🐾

The Mozart Story After Mozart's death, music minister to the Emperor, Antonio Salieri, reflects on how his jealousy and hatred of the musical genius held the great composer back and contributed to his death. Originally filmed in Austria in 1937, scenes were added for the U.S. release in 1948.
1948 91m/B Hans Holt, Winnie Markus, Irene von Meyendorf, Rene Deltgen, Edward Vedder, Wilton Graff, Walther Jansson, Curt Jurgens, Paul Hoerbiger; *Dir:* Karl Hartl. **VHS, Beta $24.95** VYY, GLV 🐾🐾

Mr. Ace A congresswoman decides to run for governor without the approval of the loyal political kingpin, Mr. Ace.
1946 85m/B George Raft, Sylvia Sidney, Sara Haden, Stanley Ridges; *Dir:* Edwin L. Marin. **VHS, Beta, LV $14.95** SVS 🐾½

Mr. Bill Looks Back Featuring Sluggo's Greatest Hits Mr. Bill creator Walter Williams has filmed all new material never before televised to include with some previous footage.
1983 31m/C VHS, Beta $39.95 PAV 🐾

The Mr. Bill Show Excerpts from Saturday Night Live's Mr. Bill Show in a hilarious video album of Mr. Bill's best.
1979 26m/C VHS, Beta $39.95 IND 🐾

Mr. Billion Engaging chase adventure comedy about an Italian mechanic who stands to inherit a billion dollar fortune if he can travel from Italy to San Francisco in twenty days. Of course, things get in his way. Hill made his American debut in this film.

1977 (PG) 89m/C Jackie Gleason, Terence Hill, Valerie Perrine, Slim Pickens, Chill Wills; *Dir:* Jonathan Kaplan. **VHS, Beta $59.98** FOX 🐾🐾

Mr. Bill's Real Life Adventures A made-for-television live-action adventure for the "Saturday Night Live" clay star, portrayed by Scolari, who is victimized as usual by his nemesis, Mr. Sluggo.
1986 43m/C Peter Scolari, Valerie Mahaffey, Lenore Kasorf, Michael McManus; *Dir:* Jim Drake. **VHS, Beta $29.95** PAR 🐾🐾

Mr. Blandings Builds His Dream House Classic comedy features Grant as an adman who tires of city life and moves his family to the country, where his troubles really begin. Grant is at his funniest. Loy and Douglas provide strong backup. A must for all homeowners.
1948 93m/B Cary Grant, Myrna Loy, Melvyn Douglas, Lex Barker, Reginald Denny, Louise Beavers; *Dir:* H.C. Potter. **VHS, Beta, LV $19.95** TTC, MED, MLB 🐾🐾🐾

Mr. Charlie The famous band leader is featured in this Technicolor special, with his orchestra and various guests.
194? 60m/C Charlie Barnet, Anita Ortez, Eddie Jones, Clark Terry, Fabulous Flippers. **VHS, Beta $59.95** JEF 🐾🐾½

Mr. Corbett's Ghost A young man, displeased with his boss's managerial finesse, cuts a deal with a soul collector who pink slips the guy to the great unemployment line in the sky. Once a boss always a boss, and the guy's ghost stops in to say boo, while the soul collector is hot to get his hands on his part of the bargain. Huston's final film appearance.
1990 60m/C John Huston, Paul Scofield, Burgess Meredith. **VHS $24.95** MON, KAR 🐾🐾

Mr. Deeds Goes to Town Typical Capra fare offers Cooper as philanthropic fellow who inherits $20 million and promptly donates it to the needy. He also manages to find time to fall in love with a beautiful reporter. Arthur is the hard-edged reporter determined to fathom the good guy's motivation. Superior entertainment. Based on Clarence Budington Kelland's play "Opera Hut." Bagged an Oscar for Capra and was nominated for Best Picture, Best Actor (Cooper), Best Screenplay, and Best Sound Recording.
1936 118m/B Gary Cooper, Jean Arthur, Raymond Walburn, Walter Catlett, Lionel Stander, George Bancroft, H.B. Warner, Ruth Donnelly, Douglas Dumbrille, Margaret Seddon, Margaret McWade; *Dir:* Frank Capra. Academy Awards '36: Best Director (Capra); National Board of Review Awards '36: Best Picture; New York Film Critics Awards '36: Best Picture. **VHS, Beta, LV $19.95** COL, PIA, BTV 🐾🐾🐾½

Mr. Destiny Mid-level businessman Belushi has a mid-life crisis of sorts when his car dies. Wandering into an empty bar, he encounters bartender Caine who serves cocktails and acts omniscient before taking him on the ten-cent tour of life as it would've been if he hadn't struck out in a high school baseball game. Less than wonderful rehash of "It's a Wonderful Life."
1990 (PG-13) 110m/C James Belushi, Michael Caine, Linda Hamilton, Jon Lovitz, Bill McCutcheon, Hart Bochner, Rene Russo, Jay O. Sanders, Maury Chaykin, Pat Corley, Douglas Seale, Courtney Cox, Kathy Ireland; *Dir:* James Orr. **VHS, Beta $19.99** BVV, FCT 🐾🐾

Mr. Emmanuel An elderly Jewish widower leaves his English home in 1935 in quest of the mother of a German refugee boy, and is subjected to shocking treatment in Germany.

1944 97m/B Felix Aylmer, Greta Gynt, Walter Rilla, Jean Simmons. **VHS, Beta** LCA 🐾🐾½

Mr. Frost Goldblum feigns fascination in this lackluster tale of the devil incarnate, who is imprisoned after owning up to a series of grisly murders, and who hides his true identity until approached by a woman psychiatrist.
1989 (R) 92m/C Jeff Goldblum, Kathy Baker, Alan Bates, Roland Giraud, Jean-Pierre Cassel; *Dir:* Phillip Setbon. **VHS $39.95** SVS, FCT, IME 🐾

Mr. Hobbs Takes a Vacation Good-natured comedy in which beleaguered parents try to resolve family squabbles while the entire brood is on a seaside vacation. Stewart and O'Hara are especially fine and funny as the well-meaning parents.
1962 116m/C James Stewart, Maureen O'Hara, Fabian, John Saxon, Marie Wilson, John McGiver, Reginald Gardiner; *Dir:* Henry Koster. Berlin Film Festival '62: Best Actor (Stewart). **VHS, Beta $19.98** FOX 🐾🐾🐾

Mr. Horn Based on the William Goldman western novel, this slow-paced film about folk-hero Scott Tom Horn capturing the famous Apache warrior, Geronimo, is average at best.
1979 200m/C David Carradine, Richard Widmark, Karen Black, Jeremy Slate, Enrique Lucero, Jack Starrett; *Dir:* Jack Starrett. **VHS $59.95** IVE 🐾🐾

Mr. Hulot's Holiday Superior slapstick details the misadventures of a dullard's seaside holiday. Inventive French comedian Tati at his best. Light-hearted and natural, with magical mime sequences. Followed by "My Uncle." In French with English subtitles.
1953 86m/B FR Jacques Tati, Natalie Pascaud, Michelle Rolia; *Dir:* Jacques Tati. National Board of Review Awards '54: 10 Best Foreign Films of the Year. **VHS, Beta, LV, 8mm $29.98** NOS, MRV, VYY 🐾🐾½

Mr. Inside, Mr. Outside A television movie about two New York City cops, one undercover, one not, trying to infiltrate a tough diamond-smuggling ring.
1974 74m/C Tony LoBianco, Hal Linden; *Dir:* William A. Graham. **VHS, Beta $59.95** VMK 🐾½

Mr. Kingstreet's War An idealistic game warden and his wife defend the wildlife of Africa against the fighting Italian and British armies at the dawn of World War II. Also released on video as "Heroes Die Hard."
1971 92m/C John Saxon, Tippi Hedren, Rossano Brazzi, Brian O'Shaughnessy; *Dir:* Percival Rubens. **VHS, Beta** UHV, SIM 🐾🐾½

Mr. Klein Cleverly plotted script and fine direction in this dark and intense film about a French-Catholic merchant who buys valuables from Jews trying to escape Nazi occupied France in 1942, paying far less than what the treasures are worth. Ironically, he is later mistaken for a Jew by the same name who has been using this man's reputation as a cover.
1975 (PG) 122m/C Alain Delon, Jeanne Moreau; *Dir:* Joseph Losey. **VHS, Beta $59.95** COL 🐾🐾🐾

Mr. Love A meek gardener is perceived by neighbors as a fool until he dies and numerous women arrive for his funeral. Is it his cologne? The radishes?
1986 (PG-13) 91m/C GB Barry Jackson, Maurice Denham, Margaret Tyzack; *Dir:* Roy Battersby. **VHS, Beta $19.98** WAR 🐾🐾½

Mr. Lucky Likeable wartime drama about a gambler who hopes to swindle a philanthropic organization, but then falls in love and determines to help the group in a fundraising effort. Grant is, no surprise, excellent as the

seemingly cynical con artist who actually has a heart of gold. Cliched, but nonetheless worthwhile. Later developed into a television series.
1943 99m/B Cary Grant, Laraine Day, Charles Bickford, Gladys Cooper, Paul Stewart, Henry Stephenson, Florence Bates; *Dir:* H.C. Potter. **VHS, Beta, LV** $19.98 *MED, RXM 🐾🐾*

Mr. Magoo in the King's Service Mr. Magoo is off on the King's business in this wacky full-length cartoon adventure.
1966 92m/C *Voices:* Jim Backus. **VHS, Beta** $24.95 *PAR 🐾🐾*

Mr. Magoo: Man of Mystery In this episode, Mr. Magoo plays four legendary literary and comic strip heroes: Dr. Watson, Dr. Frankenstein, the Count of Monte Cristo and Dick Tracy.
1964 96m/C *Voices:* Jim Backus. **VHS, Beta** *PAR 🐾🐾*

Mr. Magoo in Sherwood Forest As Friar Tuck, the nearsighted Mr. Magoo involves Robin Hood and his Merry Men in a series of zany adventures.
1964 83m/C *Voices:* Jim Backus. **VHS, Beta, LV** $12.95 *PAR 🐾🐾*

Mr. Magoo's Christmas Carol Nearsighted Mr. Magoo, as Ebeneezer Scrooge, receives Christmas visits from three ghosts in this entertaining version of Dickens' classic tale.
1962 52m/C *Voices:* Jim Backus. **VHS, Beta, LV** $12.95 *PAR 🐾🐾*

Mr. Magoo's Storybook Mr. Magoo acts out all the parts in versions of three famous tales of literature: "Snow White and the Seven Dwarfs," "Don Quixote," and "A Midsummer Night's Dream."
1964 113m/C *Voices:* Jim Backus. **VHS, Beta** $14.95 *PAR 🐾🐾*

Mr. Majestyk When a Vietnam veteran's attempt to start an honest business is thwarted by Mafia hitmen and the police, he goes after them with a vengeance.
1974 (PG) 103m/C Charles Bronson, Al Lettieri, Linda Cristal, Lee Purcell; *Dir:* Richard Fleischer. **VHS, Beta** $19.95 *MGM 🐾🐾½*

Mr. Mike's Mondo Video A bizarre, outrageous comedy special declared too wild for television conceived by the "Saturday Night Live" alumnus Mr. Mike.
1979 (R) 75m/C Michael O'Donoghue, Dan Aykroyd, Jane Curtin, Carrie Fisher, Teri Garr, Joan Haskett, Deborah Harry, Margot Kidder, Bill Murray, Laraine Newman, Gilda Radner, Julius LaRosa, Paul Schaeffer, Sid Vicious. **VHS, Beta** $59.95 *PAV 🐾🐾*

Mr. Mom A tireless auto exec loses his job and stays home with the kids while his wife becomes the breadwinner. He's forced to cope with the rigors of housework and child care, resorting to drugs, alcohol and soap operas. Keaton's funny as homebound dad chased by killer appliances and poker buddy to the ladies in the neighborhood. Written by John Hughes and produced by Aaron Spelling.
1983 (PG) 92m/C Michael Keaton, Teri Garr, Christopher Lloyd, Martin Mull, Ann Jillian, Jeffrey Tambor, Edie McClurg, Valri Bromfield; *Dir:* Stan Dragoti. **VHS, Beta, LV** $29.98 *LIV, VES 🐾🐾½*

Mr. Moto's Last Warning One of the better in the series of Mr. Moto, the wily detective! Lorre is convincing in the title role and gets good support from character villains Carradine and Sanders in this story of sabot-

eurs converging on the Suez Canal plotting to blow up the French Fleet.
1939 71m/B Peter Lorre, George Sanders, Ricardo Cortez, John Carradine, Virginia Field, Richard Coote; *Dir:* Norman Foster. **VHS, Beta** $16.95 *SNC, NOS, CAB 🐾🐾🐾*

Mr. Moto's Last Warning/ Meeting at Midnight The first film finds Moto at the Suez Canal trying to save France's flotilla from a deadly conspiracy. The second feature is classic Charlie Chan fare. This time his driver, Birmingham, gets Chan caught in the midst of the occult.
1944 130m/B Sidney Toler, Peter Lorre, Mantan Moreland. **VHS, Beta** *AOV 🐾🐾*

Mr. & Mrs. Bridge Set in the thirties and forties in Kansas City, Ivory's adaptation of Evan S. Connell's novels painstakingly portrays an upper middle-class family struggling to survive within an emotional vacuum. Newman and Woodward, together for the first time in many years as Walter and Ivory Bridge, bring a wealth of experience and insight to their characterizations. Oscar nominated Woodward won the New York Critics' award for Best Actress, and, although Newman received no Oscar nomination, many consider this to be his best, most subtle and nuanced performance.
1991 (PG-13) 127m/C Joanne Woodward, Paul Newman, Kyra Sedgwick, Blythe Danner, Simon Callow, Diane Kagan, Robert Sean Leonard, Saundra McClain, Margaret Welsh, Austin Pendleton, Gale Garnett, Remak Ramsay; *Dir:* James Ivory. **VHS, LV** $92.99 *HBO, PIA, FCT 🐾🐾🐾*

Mr. & Mrs. Smith Hitchcock's only screwball comedy, an underrated, endearing farce about a bickering but happy modern couple who discover their marriage isn't legitimate and go through courtship all over again. Vintage of its kind, with inspired performances. Smartly written by Norman Krasna, with crackling dialogue.
1941 95m/B Carole Lombard, Robert Montgomery, Gene Raymond, Jack Carson, Lucile Watson; *Dir:* Alfred Hitchcock. **VHS, Beta, LV** $19.95 *MED 🐾🐾🐾½*

Mr. Nice Guy In the near future, a hired gun strives for national recognition as the best of his now legitimate profession. Comedy with few laughs.
1986 (PG-13) 92m/C Michael MacDonald, Jan Smithers, Joe Silver. **VHS, Beta** $19.95 *STE, NWV 🐾*

Mr. North Capracorn fable about a charming, bright Yale graduate who encounters admiration and disdain from upper-crust Rhode Island residents when news of his miraculous "cures" spreads. Marks the directorial debut of Danny Huston, son of John Huston, who co-wrote the script and served as executive producer before dying several weeks into shooting. Set in the 1920s and adapted from Thornton Wilder's "Theophilus North."
1988 (PG) 90m/C Anthony Edwards, Robert Mitchum, Lauren Bacall, Harry Dean Stanton, Anjelica Huston, Mary Stuart Masterson, Virginia Madsen, Tammy Grimes, David Warner, Hunter Carson; *Dir:* Danny Huston. **VHS, Beta, LV** $36.95 *VIR 🐾🐾½*

Mr. Peabody & the Mermaid Lightweight fish story about a middle-aged married man who hooks a beautiful mermaid while fishing in the Caribbean and eventually falls in love with her. Powell is smooth as always, though hampered by the unrestrained absurdity of it all. Scripted by Nun-

nally Johnson from the novel by Guy and Constance Jones. Also available colorized.
1948 89m/B William Powell, Ann Blyth; *Dir:* Irving Pichel. **VHS** $19.98 *REP 🐾🐾½*

Mr. Peek-A-Boo A French clerk discovers he has the ability to walk through walls. Although friends try to coax him into a life of crime, he instead comes to the aid of an English girl who is being blackmailed. Good-natured French comedy was shot with English dialogue for the American market.
1950 74m/B FR Joan Greenwood, Bourvil, Marcel Arnold, Roger Treville; *Dir:* Jean Boyer. **VHS, Beta** $16.95 *SNC 🐾🐾½*

Mr. Reeder in Room 13 Based on the mystery stories created by Edgar Wallace, Mr. Reeder (a cultured English gentleman who fights crime) enlists the aid of a young man to get evidence on a gang of counterfeiters.
1938 66m/B Gibb McLaughlin. **VHS, Beta** $29.95 *VYY 🐾🐾*

Mr. Robinson Crusoe Rollicking adventure in the South Seas as Fairbanks makes a bet that he can live on a desert island for a year without being left any refinements of civilization. Of course, a woman arrives. Written by Fairbanks, with a score by Alfred Newman.
1932 76m/B Douglas Fairbanks Sr., William Farnum, Maria Alba, Earle Brown; *Dir:* Edward Sutherland. **VHS, Beta, LV** $19.95 *NOS, GPV, CCB 🐾🐾½*

Mr. Rossi's Dreams Mr. Rossi gets to act out his fantasies of being Tarzan, Sherlock Holmes, and a famous movie star in this film.
1983 80m/C *Dir:* Bruno Bozzetto. **VHS, Beta** $39.95 *FHE Woof!*

Mr. Rossi's Vacation Mr. Rossi and his dog Harold go off in search of quiet on a let's get away-from-it-all vacation.
1983 82m/C IT *Dir:* Bruno Bozzetto. **VHS, Beta** $39.95 *FHE Woof!*

Mr. Scarface A young man searches for the man who murdered his father years earlier in a dark alley.
1977 (R) 85m/C Jack Palance, Edmund Purdom, Al Cliver, Harry Baer, Gisela Hahn; *Dir:* Fernando Di Leo. **VHS, Beta** $19.95 *WES 🐾½*

Mr. Skeffington A super-grade soap opera spanning 26 years in the life of a ravishing, spoiled New York socialite. The beauty with a fondness for bedrooms marries for convenience, abuses her husband, then enjoys a highly equitable divorce settlement. Years later when diptheria leaves her totally deformed and no man will have her, she is saved by her former husband. Davis and Rains received Oscar nominations. Based on the novel by "Elizabeth" (Mary Annette Beauchamp Russell) and adapted by "Casablanca's" Julius and Philip Epstein.
1944 147m/B Bette Davis, Claude Rains, Walter Abel, Richard Waring, George Coulouris, John Alexander; *Dir:* Vincent Sherman. **VHS, Beta, LV** $19.95 *MGM 🐾🐾🐾*

Mr. Skitch Weak Rogers offering. Couple loses their farm and begins cross country jaunt with mishaps at every turn. All ends well when their daughter meets an Army cadet, but few laughs and lots of loose ends.
1933 70m/B Will Rogers, ZaSu Pitts, Rochelle Hudson, Charles Starrett; *Dir:* James Cruze. **VHS** $19.98 *FOX, FCT 🐾🐾*

Mr. Smith Goes to Washington In another classic from Hollywood's golden year of 1939, Jimmy Stewart plays an idealistic and naive young man who is selected to fill in for an ailing Senator. Upon his arrival in the Capitol, he is inundated by a multitude of corrupt politicians. He takes a stand for his beliefs and tries to denounce many of those he feels are unfit for their positions, meeting with opposition from all sides. Great cast is highlighted by Stewart in one of his most endearing performances. A quintessential Capra tale sharply written by Sidney Buchman from Lewis Foster's story. Outstanding in every regard. In addition to its single Oscar win, nominated for best picture, script, score (Dimitri Tiomkin), actor (Stewart), and supporting actors (Rains and Carey).
1939 130m/B James Stewart, Jean Arthur, Edward Arnold, Claude Rains, Thomas Mitchell, Beulah Bondi, Eugene Pallette, Guy Kibbee, Harry Carey, H.B. Warner, Porter Hall, Jack Carson; *Dir:* Frank Capra. Academy Awards '39: Best Story; National Board of Review Awards '39: 10 Best Films of the Year; New York Film Critics Awards '39: Best Actor (Stewart). **VHS, Beta, LV $19.95** *COL, PIA* 🎬🎬🎬🎬

Mr. Superinvisible Searching to cure the common cold, a bumbling scientist invents a bizarre virus , then strives with his loyal (of course) sheepdog to keep it from falling into the wrong hands. The Disney-like plot, featuring enemy agents and invisibilty may appeal to youngsters.
1973 (G) 90m/C *IT GE SP* Dean Jones, Ingeborg Schoener, Gastone Moschin; *Dir:* Anthony M. Dawson. **VHS, Beta $9.95** *NO* 🎬🎬

Mr. Sycamore A sappy mail carrier escapes his badgering wife by sprouting into a tree. A potentially whimsical piece that wilts.
1974 90m/C Jason Robards Jr., Jean Simmons, Sandy Dennis; *Dir:* Pancho Kohner. **VHS, Beta $39.95** *WOV* 🎬½

Mr. Walkie Talkie Tired of a sargeant's jabber-jaws, a soldier requests to be relocated to the front lines, only to find his catty comrade has followed. Set during the Korean War and featuring some truly amusing escapades set between the lulls in the war.
1952 65m/B William Tracy, Joe Sawyer; *Dir:* Fred Guiol. **VHS, Beta, LV** *WGE* 🎬🎬

Mr. Winkle Goes to War A weak, nerdy former banker is drafted for service during World War II and proves himself a hero by bulldozing a Japanese foxhole. Bits of genuine war footage add a measure of realism to an otherwise banal flag-waving comedy that is based on a novel by Theodore Pratt.
1944 80m/B Edward G. Robinson, Ruth Warrick, Richard Lane, Robert Armstrong; *Dir:* Alfred E. Green. **VHS, Beta $59.95** *COL* 🎬🎬

Mr. Wise Guy The East Side Kids break out of reform school to clear one of the Kids' brother of a murder charge. Typical pre-Bowery Boys vehicle.
1942 70m/B Leo Gorcey, Huntz Hall, Billy Gilbert, Guinn Williams, Benny Rubin, Douglas Fowley, Ann Doran, Jack Mulhall, Warren Hymer, David Gorcey; *Dir:* William Nigh. **VHS, Beta $19.95** *NOS, QNE, DVT* 🎬🎬

Mr. Wong in Chinatown Third of the Mr. Wong series finds James Lee Wong investigating the murder of a wealthy Chinese woman. She had been helping to fund the purchase of airplanes to equip China in its 1930s' struggle with Japan.

1939 70m/B Boris Karloff, Grant Withers, William Royle, Marjorie Reynolds, Peter George Lynn, Lotus Long, Richard Loo; *Dir:* William Nigh. **VHS $16.95** *NOS, MRV, SNC* 🎬½

Mr. Wong, Detective The first in the Mr. Wong series, the cunning detective traps a killer who feigns guilt to throw suspicion away from himself. The plot is loaded with the usual twists and villains. However, Karloff is a standout.
1938 69m/B Boris Karloff, Grant Withers; *Dir:* William Nigh. **VHS, Beta $19.95** *NOS, MRV, HHT* 🎬🎬

Mrs. Amworth E.F. Benson's classic tale of the occult, in which a mysterious and deadly epidemic attacking a quiet English village is diagnosed to be the work of a vampire.
1977 29m/C Glynis Johns; *Dir:* Alvin Rakoff. **VHS, Beta $14.95** *LCA* 🎬🎬

Mrs. Brown, You've Got a Lovely Daughter The Herman's Hermits gang inherits a greyhound and attempts to make a racer out of him while singing their songs. For the Brit group's diehard fans only.
1968 (G) 95m/C Peter Noone, Stanley Holloway, Mona Washbourne; *Dir:* Saul Swimmer. **VHS, Beta $59.95** *MGM* 🎬½

Mrs. Miniver A moving tale of a courageous, gentle middle-class British family and its struggle to survive during World War II. A Classic that garnered seven Academy Awards (including Best Picture), it's recognized for contributions to the Allied effort. Contains one of the most powerful orations in the film history, delivered by Wilcoxon, who portrayed the vicar. Followed by "The Miniver Story." Adapted from Jan Struther's book.
1942 134m/B Greer Garson, Walter Pidgeon, Teresa Wright, May Whitty, Richard Ney, Henry Travers, Reginald Owen, Henry Wilcoxon, Helmut Dantine, Aubrey Mather, Rhys Williams, Tom Conway, Peter Lawford; *Dir:* William Wyler. Academy Awards '42: Best Actress (Garson), Best Black and White Cinematography, Best Director (Wyler), Best Picture, Best Screenplay, Best Supporting Actress (Wright). **VHS, Beta, LV $19.98** *MGM, PIA, BTV* 🎬🎬🎬🎬

Mrs. R's Daughter An outraged mother fights the judicial systems in order to bring her daughter's rapist to trial. Based on a true story and made for television.
1979 97m/C Cloris Leachman, Season Hubley, Donald Moffatt, John McIntire, Stephen Elliott, Ron Rifkin; *Dir:* Dan Curtis. **VHS, Beta $19.98** *WAR, OM* 🎬🎬

Mrs. Silly Romance novel on tape, following the exploits of an aging woman who in losing material wealth, gains contentment.
1985 60m/C Maggie Smith, James Villers, Cyril Luckham. **Beta $11.95** *PSM* 🎬

Mrs. Soffel Falling in love with a convicted murderer and helping him flee confinement occupies the time of the prison warden's wife. Effectively captures the 1901 setting, yet a dark pall fairly strangles any emotion. Based on a true story.
1985 (PG-13) 113m/C Diane Keaton, Mel Gibson, Matthew Modine, Edward Herrmann, Trini Alvarado, Terry O'Quinn; *Dir:* Gillian Armstrong. **VHS, Beta $19.99** *MGM* 🎬½

Mrs. Wiggs of the Cabbage Patch A warm, funny, and altogether overdone story of a mother whose husband abandons her, leaving the woman to raise their four children alone. It's Thanksgiving, so the rich visit the poor household to deliver a turkey dinner, and in the end all live happily

ever after. Not one of Fields' larger roles; he doesn't even make an appearance until midway through the schmaltz.
1934 80m/B Pauline Lord, ZaSu Pitts, W.C. Fields, Evelyn Venable, Donald Meek; *Dir:* Norman Taurog. **VHS $9.95** *GKK* 🎬🎬½

Ms. 45 Rough, bristling cult favorite about a mute girl who, in response to being raped and beaten twice in one night, goes on a man-killing murder spree. Wild ending.
1981 (R) 84m/C Zoe Tamerlis, Steve Singer, Jack Thibeau, Peter Yellen, Darlene Stuto, Editta Sherman, Albert Sinkys, Jimmy Laine; *Dir:* Abel Ferrara. **VHS, Beta, LV $49.95** *IVE* 🎬🎬🎬

Mugsy's Girls A messy comedy about six sorority women, educated in the refined skills of bar mud wrestling, who wallow in a Las Vegas championship tournament. Features pop singer Laura Branigan.
1985 (R) 87m/C Ruth Gordon, Laura Branigan, Eddie Deezen; *Dir:* Kevin Brodie. **VHS, Beta $79.98** *LIV, VES* 🎬

Multiple Maniacs Waters at his most perverse. Divine's travelling freak show, filled with disgusting side-show sex acts, is the vehicle used for robbing and killing hapless spectators. One of the best scenes of a sordid lot is Divine being raped by Lobstora, a 15-foot broiled lobster.
1970 90m/B Divine, David Lochary, Mary Vivian Pearce, Edith Massey, Mink Stole; *Dir:* John Waters. **VHS $39.98** *CGH, NSV* 🎬

The Mummy Eerie chills mark this classic horror tale of an Egyptian priest, buried alive nearly 4000 years earlier, who comes back to life after a 1921 archeological dig. Eight hours of extraordinary makeup transformed Karloff into the macabre mummy, who believes the soul of the long-deceased (and probably long-decayed) lover resides in the body of a young woman. Marked the directing debut of famed German cinematographer Freund.
1932 72m/B Boris Karloff, Zita Johann, David Manners, Edward Van Sloan; *Dir:* Karl Freund. **VHS, Beta, LV $14.95** *MCA, TLF* 🎬🎬🎬½

The Mummy A group of British archaeologists discover they have made a grave mistake when a mummy kills off those who have violated his princess' tomb. A summation of all the previous "mummy" films, this one has a more frightening mummy (6'4" Lee) who is on screen much of the time. Additionally, there is pathos in this monster, not merely murder and revenge. An effective remake of the 1932 classic.
1959 88m/C *GB* Peter Cushing, Christopher Lee, Felix Aylmer, Yvonne Furneaux; *Dir:* Terence Fisher. **VHS, Beta, LV $19.98** *WAR, LDC, MLB* 🎬🎬🎬

Mummy & Curse of the Jackal Upon opening the tomb of female mummy, a man is cursed to roam the streets of Las Vegas as a werejackal. Impossibly inept.
1967 86m/C Anthony Eisley, Martina Pons, John Carradine, Saul Goldsmith; *Dir:* Oliver Drake. **VHS $19.95** *ACA Woof!*

The Mummy's Hand Two archaeologists (Foran and Ford) searching for the tomb of ancient Egyptian Princess Ananka discover more than they bargained for when they find that an evil and deadly mummy is the tomb's guardian. From there the plot unravels, but the good cast and a mix of scares and comedy make this worth watching.
1940 70m/B Dick Foran, Wallace Ford, Peggy Moran, Cecil Kellaway, George Zucco, Tom Tyler, Eduardo Ciannelli, Charles Trowbridge; *Dir:* Christy Cabanne. **VHS $14.98** *MCA* 🎬🎬½

The Mummy's Revenge A fanatic revives a mummy with virgin blood. The first mummy movie to feature open gore.
1973 91m/C *SP* Paul Naschy, Jack Taylor, Maria Silva, Helga Line; *Dir:* Carlos Aured. **VHS, Beta $49.95** UNI ☺

Munchie A forgotten alien critter is discovered in a mine shaft by young Gage. Munchie turns out to be a good friend, protecting Gage from bullies and granting other wishes. Frequent sight gags help keep the film moving. Sequel to "Munchies" (1987).
1992 (PG) 80m/C Loni Anderson, Andrew Stevens, Arte Johnson, Jamie McEnnan; *Dir:* Jim Wynorski; *Voices:* Dom DeLuise. **VHS $89.98** NHO ☺½

Munchies "Gremlins" rip-off about tiny aliens who love beer and fast food and invade a small town. Lewd and ribald.
1987 (PG) 83m/C Harvey Korman, Charles Stratton, Nadine Van Der Velde; *Dir:* Bettina Hirsch. **VHS, Beta $14.98** MGM ☺

Munster's Revenge Television movie based on the continuing adventures of the 1960s comedy series characters. Herman, Lily and Grandpa have to contend with robot replicas of themselves that were created by a flaky scientist.
1981 96m/C Fred Gwynne, Yvonne de Carlo, Al Lewis, Jo McDonnel, Sid Caesar, Ezra Stone, Howard Morris, Bob Hastings, K.C. Martel; *Dir:* Don Weis. **VHS, Beta $39.95** MCA ☺☺

Muppet Movie Seeking fame and footlights, Kermit the Frog and his pal Fozzie Bear travel to Hollywood, and along the way are joined by sundry human and muppet characters, including the lovely Miss Piggy. A delightful cult favorite filled with entertaining cameos, memorable (though somewhat pedestrian) songs and crafty special effects. Kermit rides a bike and rows a boat! A success for the late Jim Henson.
1979 (G) 94m/C *GB* Edgar Bergen, Milton Berle, Mel Brooks, Madeline Kahn, Steve Martin, Carol Kane, Paul Williams, Charles Durning, Bob Hope, James Coburn, Dom DeLuise, Elliott Gould, Cloris Leachman, Telly Savalas, Orson Welles, Jim Henson's Muppets; *Dir:* James Frawley; *Voices:* Jim Henson, Frank Oz. **VHS, Beta, LV $19.98** FOX ☺☺☺½

Muppets Take Manhattan Following a smashing success with a college musical, the Muppets take their show and talents to Broadway, only to face misfortune in the form of an unscrupulous producer. A less imaginative script than the first two Muppet movies, yet an enjoyable experience with numerous major stars making cameo appearances.
1984 (G) 94m/C Dabney Coleman, James Coco, Art Carney, Joan Rivers, Gregory Hines, Linda Lavin, Liza Minnelli, Brooke Shields, John Landis, Jim Henson, Frank Oz; *Dir:* Frank Oz. **VHS, Beta, LV $14.98** FOX ☺☺☺

Murder Believing in a young woman's innocence, one jurist begins to organize the pieces of the crime in order to save her. Fine early effort by Hitchcock based on play "Enter Sir John," by Clemense Dane and Helen Simpson.
1930 92m/B *GB* Herbert Marshall, Nora Baring, Phyllis Konstam, Miles Mander; *Dir:* Alfred Hitchcock. **VHS, Beta $16.95** SNC, NOS, HHT ☺☺☺

Murder 101 Brosnan plays Charles Lattimore, an author and English professor who specializes in murder mysteries. He gives his students a rather unique assignment of planning the perfect murder. When a student is killed right before his eyes and a fellow

teacher turns up dead, Lattimore realizes he's being framed for murder. Surprising twists abound in this stylish thriller.
1991 (PG-13) 93m/C Pierce Brosnan, Dey Young, Raphael Sbarge, Kim Thomson; *Dir:* Bill Condon. **VHS $89.98** MCA ☺☺

Murder Ahoy Miss Marple looks perplexed when dead bodies surface on a naval cadet training ship. Dame Marge is the dottie detective in the final, and least appealing, of her four Agatha Christie films of the sixties (although it was released in the States prior to "Murder Most Foul").
1964 74m/B Margaret Rutherford, Lionel Jeffries, Charles Tingwell, William Mervyn, Francis Matthews; *Dir:* George Pollock. **VHS, Beta $19.98** MGM, FCT ☺☺½

A Murder is Announced From the critically acclaimed BBC television series. While on holiday, super-sleuth Marple encounters a murder that was advertised in the local newspaper one week prior to its occurrence. Based on the 1952 Agatha Christie novel, Hickson shines once again as Christie's detective extraordinaire.
1987 155m/C *GB* Joan Hickson; *Dir:* David Giles. **VHS, Beta $29.98** FOX ☺☺

Murder at the Baskervilles Sherlock Holmes is invited to visit Sir Henry Baskerville at his estate, but then finds that Baskerville's daughter's fiance is accused of stealing a race horse and murdering its keeper. Based on Sir Arthur Conan Doyle's story "Silver Blaze."
1937 67m/B *GB* Arthur Wontner, Ian Fleming, Lyn Harding; *Dir:* Thomas Bentley. **VHS, Beta $16.95** SNC, CCB, REP ☺☺

Murder on the Bayou (A Gathering of Old Men) Down on the L'siana bayou, a white guy who thinks civil rights are color coded is murdered, and an elderly black man is accused of the crime. Made for TV, well performed, engaging.
1991 (PG) 91m/C Louis Gossett Jr., Richard Widmark, Holly Hunter. **VHS $89.95** VMK ☺☺☺

Murder by the Book An episode of "Partners in Crime," wherein Agatha Christie is visited on the publication eve of her first Hercule Poirot novel by the detective himself, determined to stop the book. Originally called "Alter Ego," sharing the title with the Mel Arrighi novel upon which it was based.
1987 52m/C *GB* Peggy Ashcroft, Robert Hays, Catherine Mary Stewart, Celeste Holm, Fred Gwynne, Christopher Murney, Ian Holm; *Dir:* Mel Damski. **VHS, Beta $14.95** PAV ☺☺☺

Murder on the Campus A reporter investigates the apparant suicide of his brother at Cambridge University. He discovers that a number of other mysterious deaths have occurred as well. Interesting characterizations hampered by predictability.
1952 61m/B *GB* Terence Longden, Donald Gray, Diane Clare, Robertson Hare, Dermot Walsh; *W/ Dir:* Michael Winner. **VHS, Beta $16.95** NOS, MRV, SNC ☺☺

Murder in Coweta County Griffith and Cash are strong in this true-crime drama based on the book by Margaret Anne Barnes. Griffith is a Georgia businessman who thinks he's gotten away with murder; Cash is the lawman who tenaciously pursues him. Based on an actual 1948 case. Made for television.
1983 100m/C Johnny Cash, Andy Griffith, Earl Hindman, June Carter Cash, Cindi Knight, Ed Van Nuys; *Dir:* Gary Nelson. **VHS, Beta $39.95** IVE ☺☺½

Murder by Death Capote is an eccentric millionaire who invites the world's greatest detectives to dinner, offering one million dollars to the one who can solve the evening's murder. Entertaining and hammy spoof of Agatha Christie's "And Then There Were None" and the earlier "Ten Little Indians."
1976 (PG) 94m/C Peter Falk, Alec Guinness, David Niven, Maggie Smith, Peter Sellers, Eileen Brennan, Elsa Lanchester, Nancy Walker, Estelle Winwood, Truman Capote; *Dir:* Robert Moore. **VHS, Beta, LV $19.95** COL ☺☺½

Murder by Decree Realistic and convincing version of the Jack the Ripper story. Sherlock Holmes and Dr. Watson find a vast web of conspiracy when they investigate the murders of Whitechapel prostitutes. Based partially on facts, it's a highly detailed suspenser with interesting camera work and fine performances.
1979 (PG) 120m/C *CA* Christopher Plummer, James Mason, Donald Sutherland, Genevieve Bujold, Susan Clark, David Hemmings, Frank Finlay, John Gielgud, Anthony Quayle; *Dir:* Bob Clark. Genie Awards '80: Best Actor (Plummer), Best Director (Clark), Best Supporting Actress (Bujold), Best Musical Score. **VHS, Beta $19.98** SUE ☺☺☺

Murder in the Doll House A private detective must find out who is systematically killing off the family of a Japanese toy executive. With English subtitles.
1979 92m/C *JP* Yusaku Matsuda, Hiroko Shino, Yoko Nosaki. **VHS, Beta $59.95** VDA ☺

Murder Elite A murdering maniac terrorizes the rural English countryside, where MacGraw has returned after losing all her money in America. Poor acting and direction make this one a dust-gatherer.
1986 98m/C Ali MacGraw, Billie Whitelaw, Hywel Bennett; *Dir:* Claude Whatham. **VHS, Beta $69.98** LIV, VES ☺½

Murder on Flight 502 A crisis arises on a 747 flight from New York to London when a terrorist runs amuck. Big cast of TV stars and Stack as the pilot keep this stale television flick from getting lost in the ozone.
1975 97m/C Farrah Fawcett, Sonny Bono, Ralph Bellamy, Theodore Bikel, Dane Clark, Polly Bergen, Laraine Day, Fernando Lamas, George Maharis, Hugh O'Brian, Molly Picon, Walter Pidgeon, Robert Stack; *Dir:* George McCowan. **VHS, Beta $49.95** PSM ☺☺

Murder at the Gallop Snooping Miss Marple doesn't believe a filthy rich oldtimer died of natural causes, despite the dissenting police point of view. Wheedling her way into the police investigation, she discovers the secret of the Gallop club, a place where people bounce up and down on top of horses. Much mugging between Dame Margaret and Morley. Marple's assistant, Mr. Stringer, is the real life Mr. Dame Margaret. Based on Christy's Poirot mystery "After the Funeral."
1963 81m/B Margaret Rutherford, Robert Morley, Flora Robson, Charles Tingwell, Duncan Lamont; *Dir:* George Pollock. **VHS, Beta $19.98** MGM, FCT ☺☺☺½

Murder on Line One A London murderer films his crimes for his later viewing pleasure. The police arrest a suspect, but the killings continue. Both the police and viewer are soon aware that they've made a mistake.
1990 103m/C Emma Jacobs, Peter Blake, Simon Shepherd, Allan Surtees, Andrew Wilde, Dirkan Tulane, Neil Duncan, Brett Forrest; *Dir:* Anders Palm. **VHS, Beta $29.95** ACA, TTC ☺☺

M

Murder Mansion Some decent scares ensue when a group of travelers is stranded in an old haunted mansion.
1970 84m/C *SP* Evelyn Stewart, Analia Gade; *Dir:* Francisco Lara Polop. **VHS, Beta $59.98** UNI ♫ ½

The Murder of Mary Phagan Lemmon stars as John Slaton, governor of Georgia during one of America's most notorious miscarriages of justice. In 1913, timid, Jewish factory manager Leo Frank is accused of the brutal murder of a female worker. Prejudice and a power hungry prosecuting attorney conspire to seal the man's fate at the end of the hangman's noose. Sensing the injustice, Slaton re-opens the case, causing riots in Atlanta. Top notch television drama, featuring a superb recreation of turn of the century atmosphere and a compelling, true story which was not finally resolved until the 1970s.
1987 (PG) 251m/C Jack Lemmon, Peter Gallagher, Richard Jordan, Robert Prosky, Paul Dooley, Rebecca Miller, Kathryn Walker, Charles Dutton, Kevin Spacey, Wendy J. Cooke; *Dir:* Billy Hale. **VHS $79.98** ORI ♫♫♫ ½

Murder Masters of Kung Fu An evil student kills his kind master and then tries to seduce his beautiful, grieving daughter. But she's no sap and gives him more than he bargained for, including a healthy dose of vengeance.
1985 93m/C Yim Nam Hei, Tien Pang. **VHS, Beta $39.95** WWV ♫

Murder or Mercy Still timely story of a famous doctor accused of killing his terminally ill wife. Focuses on the morality of mercy killing. Made for television.
1974 78m/C Melvyn Douglas, Bradford Dillman, David Birney, Denver Pyle, Mildred Dunnock; *Dir:* Harvey Hart. **VHS $49.95** WOV ♫♫ ½

Murder at Midnight The killings begin with a game of charades in which the gun wasn't supposed to be loaded, and continue as members of English high society die one by one. "Blondie" still working on his change of pace.
1931 69m/B Alice White, Leslie Fenton, Aileen Pringle, Hale Hamilton, Robert Elliott, Clara Blandick, Brandon Hurst; *Dir:* Frank Strayer. **VHS, Beta, 8mm $12.95** SNC, NOS, DVT ♫♫

Murder on the Midnight Express Made-for-television mystery about spies, thieves, honeymooners and a corpse aboard an all-night train. All aboard.
1974 70m/C Judy Geeson, James Smilie, Charles Gray. **VHS, Beta $12.95** IVE ♫ ½

Murder of a Moderate Man An Interpol agent sets for the Alps to apprehend an assassin gunning down political leaders. A British-made effort based on a novel by John Howlett.
197? 165m/C *GB* Denis Quilley. **VHS $49.95** FCT ♫ ½

Murder Most Foul Erstwhile school marm Dame Margaret is excellent as the only jury member to believe in the accused's innocence. Posing as a wealthy actress to insinuate herself into the local acting troupe, she sniffs out the true culprit. Based on the Poirot mystery "Mr. McGinty's Dead."
1965 90m/B Margaret Rutherford, Ron Moody, Charles Tingwell, Megs Jenkins, Dennis Price, Ralph Michael; *Dir:* George Pollock. **VHS, Beta $19.98** MGM, FCT ♫♫♫

Murder Motel Inept made-for-television spine-tingler about a motel keeper who makes a practice of killing his customers. There's always a vacancy at the Murder Motel, at least until the fiancee of one of his victims decides to look into the matter.
1974 80m/C Robyn Millan, Derek Francis, Ralph Bates, Edward Judd. **VHS, Beta $29.95** IVE ♫

Murder with Music A musical drama featuring an all-black cast.
1945 60m/B Bob Howard, Noble Sissle, Nellie Hill; *Dir:* George P. Quigley. **VHS, Beta $29.95** NOS, DVT, VDM ♫♫

Murder My Sweet Down-on-his-luck private detective Philip Marlowe (Powell) searches for an ex-convict's missing girlfriend through a dark world of murder, mayhem and ever-twisting directions. Classic film noir screen version of Raymond Chandler's tense novel "Farewell, My Lovely," which employs flashback fashion using that crisp Chandler narrative. A breakthrough dramatically for singer Powell; Chandler's favorite version. Remade for television using the novel's title in 1975.
1944 95m/B Dick Powell, Claire Trevor, Mike Mazurki, Otto Kruger, Anne Shirley, Miles Mander; *Dir:* Edward Dmytryk. Edgar Allan Poe Awards '45: Best Screenplay. **VHS, Beta, LV $19.95** MED, KOV, CCB ♫♫♫ ½

Murder by Natural Causes A made-for-TV brain teaser in which a woman and her lover plot to kill her mind-reader husband. Lots of twists make this fun for viewers who enjoy a challenge.
1979 96m/C Hal Holbrook, Barry Bostwick, Katharine Ross, Richard Anderson; *Dir:* Robert Day. **VHS, Beta $59.95** LTG ♫♫♫

Murder by Night Amnesia victim Urich is found next to a dead body, the result of a gruesome murder. Urich is the only witness and thinks he may be the next victim. Others think he's the killer. Made for cable mystery with enough twists to make it worthwhile.
1989 (PG-13) 95m/C Robert Urich, Kay Lenz, Jim Metzler, Richard Monette, Michael Ironside; *Dir:* Paul Lynch. **VHS, Beta $79.95** MCA ♫♫ ½

Murder: No Apparent Motive The producer of "60 Minutes" examines the lives of serial murderers, including such notorious psychos as the Son of Sam, the Boston Strangler, Ted Bundy, and John Wayne Gaycey.
1984 72m/C John Brotherton, Karen Levine, Joan Ranquet; *Dir:* Imre Horvath. **VHS, Beta $69.98** LIV, VES ♫♫ ½

Murder by Numbers Murder in the art world stumps a detective. What might have been a good suspense story is ruined by bad editing and a lack of continuity, resulting in confusion rather than suspense.
1989 (PG-13) 91m/C Sam Behrens, Shari Belafonte-Harper, Ronee Blakley, Stanley Kamel, Jayne Meadows, Debra Sandlund, Dick Sargent, Cleavon Little; *W/Dir:* Paul Leder. **VHS, Beta, LV $89.98** MAG ♫ ½

Murder in the Old Red Barn Based on the real murder of an unassuming girl by a randy squire. Stiff, melodramatic performances are bad enough, but the play-style production, unfamiliar to modern viewers, is the last nail in this one's coffin. Also known as "Maria Marten."
1936 67m/B *GB* Tod Slaughter, Sophie Stewart, D.J. Williams, Eric Portman; *Dir:* Milton Rosmer. **VHS, Beta $16.95** NOS, MRV, SNC ♫

Murder Once Removed A private eye discovers that a respectable doctor has a bedside manner that women are dying for. Made for television.
1971 74m/C John Forsythe, Richard Kiley, Barbara Bain, Joseph Campanella; *Dir:* Charles S. Dubin. **VHS, Beta $59.95** LHV ♫ ½

Murder One Two half-brothers escape from a Maryland prison and go on a killing spree, dragging their younger brother with them. Low-budget and it shows. Based on a true story.
1988 (R) 83m/C Henry Thomas, James Wilder, Stephen Shellan, Errol Slue; *Dir:* Graeme Campbell. **VHS, Beta, LV $19.98** SUE ♫

Murder Ordained Yet another true-crime network miniseries, this time about a Kansas minister and her lover plotting the demise of their spouses. Good acting for this sort of thing. In two-cassette package.
1987 183m/C Keith Carradine, JoBeth Williams, Terry Kinney, Guy Boyd, Terence Knox, Darrell Larson, M. Emmet Walsh, Kathy Bates; *Dir:* Mike Robe. **VHS $79.95** FRH ♫♫ ½

Murder on the Orient Express An Agatha Christie mystery lavishly produced with an all-star cast. In 1934, a trainful of suspects and one murder victim make the trip from Istanbul to Calais especially interesting. Super-sleuth Hercule Poirot sets out to solve the mystery. An entertaining whodunit, ably supported by the remarkable cast. Followed by "Death on the Nile."
1974 (PG) 128m/C *GB* Albert Finney, Martin Balsam, Ingrid Bergman, Lauren Bacall, Sean Connery, Richard Widmark, Anthony Perkins, John Gielgud, Jacqueline Bisset, Jean-Pierre Cassel, Wendy Hiller, Rachel Roberts, Vanessa Redgrave, Michael York, Colin Blakely, George Coulouris; *Dir:* Sidney Lumet. Academy Awards '74: Best Supporting Actress (Bergman); British Academy Awards '74: Best Supporting Actor (Gielgud), Best Supporting Actress (Bergman), Best Original Score; National Board of Review Awards '74: 10 Best Films of the Year. **VHS, Beta, LV $69.95** PAR, BTV ♫♫♫

Murder Over New York Episode in the Charlie Chan mystery series. Chan visits New York City and becomes involved in an investigation at the airport. With the aid of his klutzy son, he sleuths his way through a slew of suspects until the mystery is solved. Standard fare for the Chan fan.
1940 65m/B Sidney Toler, Marjorie Weaver, Robert Lowery, Ricardo Cortez, Donald MacBride, Melville Cooper, Sen Yung; *Dir:* Harry Lachman. **VHS, Beta $19.98** FOX ♫♫

Murder by Phone A deranged technician has turned his phone into an instrument of electronic death. Chamberlain is the visiting professor trying to discover who's permanently disconnecting the numbers of his students. Good cast decides to test schlock meter. Originally titled as "Bells."
1982 (R) 79m/C *CA* Richard Chamberlain, John Houseman, Sara Botsford; *Dir:* Michael Anderson Sr. **VHS, Beta $39.98** WAR ♫♫

Murder Rap An aspiring musician/sound technologist becomes involved with a mysterious woman and her plot to kill her husband. Complications galore, especially for the viewer.
1987 90m/C John Hawkes, Seita Kathleen Feigny; *Dir:* Kliff Keuhl. **VHS, Beta $79.95** UHV ♫

Murder for Sale Secret Agent 117 stages an elaborate scam in order to infiltrate a ring of terrorists and criminals. Ask for Bond next time.

1970 90m/C John Gavin, Margaret Lee, Curt Jurgens. **VHS, Beta** $59.95 *MED ℤ*

Murder She Said Dame Margaret, playing the benign Miss M for the first time, witnesses a murder on board a train, but the authorities don't seem inclined to believe her. Posing as a maid at an estate near where she thought the body was dropped, she solves the murder and lands three more Miss Marple movies. Based on Christy's "4:50 From Paddington," it features Marple-to-be Hickson as the cook.
1962 87m/B Margaret Rutherford, Arthur Kennedy, Nadia Pavlova, James Robertson Justice, Thorley Walters, Charles Tingwell, Conrad Phillips; *Dir:* George Pollock. **VHS, Beta** $19.98 *MGM, FCT ℤℤ½*

Murder in Space Nine multinational astronauts are stranded aboard a space station when they discover one of them is a murderer. This creates anxiety, particularly since the killer's identity is unknown. Oatmeal salesman Brimley is the earth-bound mission control chief trying desperately to finger a spaceman while the bodies pile up. Made for television.
1985 95m/C Wilford Brimley, Martin Balsam, Michael Ironside; *Dir:* Steven Hilliard Stern. **VHS, Beta** $79.95 *VMK ℤ½*

Murder Story An aspiring mystery writer finds himself mixed up in a real murder and winds up involving his mentor as well. An overdone story, but a reasonably enjoyable film.
1989 (PG) 90m/C Christopher Lee, Bruce Boa; *Dir:* Eddie Arno, Markus Innocenti. **VHS, Beta** $29.95 *FHS, ACA ℤℤ½*

Murder by Television A low-budget murder mystery with Lugosi in a dual role as twins—one good, one evil. When a professor and the evil twin are murdered, the instrument of death suspected is...television. Still an exotic and misunderstood invention when this was made, TV invoked fear and suspicion in the general public. Historically interesting and Lugosi fans will appreciate seeing him in a non-vampiric role.
1935 55m/B Bela Lugosi, George Meeker; *Dir:* Clifford Sandforth. **VHS, Beta** $24.95 *NOS, SNC, VYY ℤ½*

Murder in Texas Made for television docudrama looks at the strange but true events surrounding the death of society woman Joan Robinson Hill, first wife of prominent plastic surgeon, Dr. John Hill, and daughter of wealthy oilman Ash Robinson. Well-crafted script and effective performances keep your interest. Griffith was Emmy nominated. From the book, "Prescription Murder," written by the doctor's second wife.
1981 200m/C Farrah Fawcett, Katharine Ross, Andy Griffith, Sam Elliott, Craig T. Nelson, Barbara Sammeth; *Dir:* William Hale. **VHS, Beta** $59.95 *VHE, HHE ℤℤℤ*

Murder: Ultimate Grounds for Divorce A quiet weekend of camping turns into a night of horror for two couples when one of them plans an elaborate murder scheme.
1985 90m/C Roger Daltrey, Toyah Wilcox, Leslie Ash, Terry Raven. **VHS, Beta** $59.95 *LHV, WAR ℤ*

Murder at the Vanities Vintage murder mystery set against a musical revue format, in which a tough detective must find a killer before the Earl Carroll-based cabaret ends and he or she will escape with the exiting crowd. Ellington plays "Cocktails for

Two," "Ebony Rhapsody," and "Lovely One." Also featured is a mind-boggling production number based on the song "Marijuana."
1934 91m/B Victor McLaglen, Kitty Carlisle, Jack Oakie, Duke Ellington, Carl Brisson; *Dir:* Mitchell Leisen. **VHS, Beta** $59.95 *MCA ℤℤ½*

Murder Weapon One by one, a group of men learn just how dangerous women can be. Sex is just the first on their list of weapons! Epic trash nonsense.
1990 (R) 90m/C Linnea Quigley. **VHS, LV** $79.95 *CGH ℤ*

Murder on the Yukon An installment in the "Renfrew of the Royal Mounted" series. Newill and O'Brien's vacation plans are ruined when they find a corpse in a canoe and must investigate. Lots of action, but a somewhat threadbare production. Based on "Renfrew Rides North" by Lauri York Erskine.
1940 57m/B James Newill, Dave O'Brien, Polly Ann Young; *Dir:* Louis Grasnier. **VHS, Beta** $12.95 *VDM ℤ½*

Murderer in the Hotel An inspector is lead to a remote hotel run by a peculiar couple while investigating the murder of a wealthy woman. In Spanish.
1983 99m/C *MX* Jose Alonso, Blanca Guerra; *Dir:* Luis Mandoki. **VHS, Beta** $72.95 *MHV ℤℤ*

Murderers Among Us: The Simon Wiesenthal Story Powerful re-enactment of concentration camp survivor Simon Wiesenthal's search for war criminals. Kingsley's gripping performance drives this made-for-cable film. Be prepared for nauseating death camp scenes.
1989 157m/C Ben Kingsley, Renee Soutendijk, Craig T. Nelson, Paul Freeman; *Dir:* Brian Gibson. **VHS** $89.99 *HBO, FCT ℤℤℤ*

Murderer's Keep The ingredients used in the Central Meat Market's hamburger is discovered by a young, deaf girl. Beware of filler, she learns, particularly if it's someone you know.
1988 89m/C Vic Tayback, Talia Shire, Robert Walden; *Dir:* Paulmichel Miekhe. **VHS, Beta** $24.95 *GHV Woof!*

Murderers' Row Daredevil bachelor and former counter-espionage agent Matt Helm is summoned from his life of leisure to insure the safety of an important scientist. Martin's attempt as a super-spy doesn't wash, and Margret is implausible as the kidnapped scientist's daughter. Unless you want to hear Martin sing "I'm Not the Marrying Kind," don't bother. Second in the "Matt Helm" series.
1966 108m/C Dean Martin, Ann-Margret, Karl Malden, Beverly Adams, James Gregory; *Dir:* Henry Levin. **VHS, Beta** $14.95 *COL ℤ½*

Murderer's Wife A teacher at a private school seeks to prevent one of her students from making the same mistakes she did in her youth. An episode from the 1950's television drama series, "Fireside Theatre."
19?? 24m/B Audrey Totter, John Howard, June Kenny, Michael Chapin. **VHS, Beta** $19.98 *CCB ℤ½*

Murderlust Employers in search of a security guard make a poor choice in offering the position to a sexually frustrated man whose hobbies include strangling prostitutes. Poor hiring decisions come into play later when he secures a job at an adolescent crisis center.

1986 90m/C Eli Rich, Rochelle Taylor, Dennis Gannon, Bonnie Schneider, Lisa Nichols; *Dir:* Donald M. Jones. **VHS, Beta** $79.95 *PSM Woof!*

Murderous Vision Television fixture Boxleitner is a bored detective in this outing. But one day, while tracking a young mother in the missing persons bureau, he stumbles upon the trail of a serial killer. Enter a beautiful psychic who wants to help, add a race to catch the murderer, and Bruce suddenly has his hands full. Made for television.
1991 (R) 93m/C Bruce Boxleitner, Laura Johnson, Robert Culp; *Dir:* Gary Sherman. **VHS** *PAR ℤℤ*

Murders at Lynch Cross This is a made-for-British-television mystery about a weird, isolated hotel on the Yorkshire moors.
1985 60m/C *GB* Jill Bennett, Joanna David, Barbara Jefferd, Sylvia Syms. **VHS, Beta** $59.95 *PSM ℤℤ*

Murders in the Rue Morgue Lugosi stars as a deranged scientist (what a stretch) who wants to find a female companion for his pet gorilla. He kidnaps a beautiful woman and prepares to make her the gorilla's bride. Very loosely based on the story by Edgar Allan Poe, which has been remade several times.
1932 61m/B Bela Lugosi, Sidney Fox, Leon Ames, Brandon Hurst, Arlene Francis; *Dir:* Robert Florey. **VHS** $14.98 *MRV ℤℤ½*

Murders in the Rue Morgue A young woman has frightening dreams inspired by a play that her father is producing in Paris at the turn of the century. After many people associated with the production become murder victims, the girl becomes involved with one of her father's former associates, a man who killed her mother years ago and then faked his own suicide. The fourth film based on Edgar Allan Poe's classic horror story.
1971 (PG) 87m/C Jason Robards Jr., Lilli Palmer, Herbert Lom, Michael Dunn, Christine Kaufman, Adolfo Celi; *Dir:* Gordon Hessler. **VHS, Beta** $59.95 *LTG ℤℤ½*

The Murders in the Rue Morgue The fifth filmed version of the Edgar Allan Poe story. Set in 19th-century Paris; actors in a mystery play find their roles coming to life. Scott is terrific, with good supporting help. Made for television.
1986 (PG) 92m/C George C. Scott, Rebecca DeMornay, Val Kilmer, Ian McShane, Neil Dickson; *Dir:* Jeannot Szwarc. **VHS** $89.95 *VMK ℤℤℤ*

Muriel A complex, mosaic drama about a middle-aged woman who meets an old lover at Boulogne, and her stepson who cannot forget the needless suffering he caused a young woman named Muriel while he was a soldier at war. Throughout, director Alain Resnais plumbs the essential meanings of memory, age, and the anxieties created from the tension between personal and public actions. Acclaimed; in French with subtitles. Also called "The Time of Return."
1963 115m/C *FR* Delphine Seyrig, Jean-Pierre Kerien, Nita Klein, Jean-Baptiste Thierree; *Dir:* Alain Resnais. New York Film Festival '63: Best Actress; Venice Film Festival '63: Best Actress (Seyrig). **VHS, Beta** $79.95 *FCT ℤℤℤ½*

Murmur of the Heart Honest treatment of a fourteen-year-old's coming of age. After his older brothers take him to a prostitute for his first sexual experience, he comes down with scarlet fever. He then travels to a health spa with his mom to recover. There they find that their mother-son bond is stron-

ger than most. Music by Charlie Parker is featured in the score.
1971 (R) 118m/C *FR* Benoit Ferreux, Daniel Gelin, Lea Massari, Corinne Kersten, Jacqueline Chauveau, Marc Wincourt, Michael Lonsdale; *Dir:* Louis Malle. **VHS, Beta, LV $79.98** *ORI* 𝄢𝄢𝄢½

Murph the Surf Fact-based, engrossing story of two beach bums turned burglars who grow bored with small-time robbery and plan a trip to New York City to steal the Star of Africa sapphire. Notable among the many action scenes is a boat chase through the inland waterways of Miami, Florida. Also known as "Live a Little, Steal a Lot" and "You Can't Steal Love."
1975 (PG) 102m/C Robert Conrad, Don Stroud, Donna Mills, Luther Adler, Robyn Millan, Paul Stewart; *Dir:* Marvin J. Chomsky. **VHS, Beta $19.98** *WAR, OM* 𝄢𝄢½

Murphy's Fault Dark comedy about a night watchman/writer plagued by a series of bad luck incidents.
1988 (PG-13) 94m/C Patrick Dollaghan, Anne Curry, Stack Pierce; *Dir:* Robert J. Smawley. **VHS, Beta $89.95** *VMK* 𝄢𝄢

Murphy's Law A hard-headed cop gets framed for his ex-wife's murder and embarks on a mission to find the actual killer. He is slowed down by a smart-mouthed prostitute who is handcuffed to him during his search. The casting female in the role of a psycho-killer is unique to the genre.
1986 (R) 101m/C Charles Bronson, Carrie Snodgress, Kathleen Wilhoite, Robert F. Lyons, Richard Romanus, Angel Tompkins, Bill Henderson, James Luisi, Janet MacLachlan, Lawrence Tierney; *Dir:* J. Lee Thompson. **VHS, Beta, LV $19.95** *MED* 𝄢

Murphy's Romance A young divorced mother with an urge to train horses pulls up the stakes and heads for Arizona with her son. There she meets a pharmacist who may be just what the doctor ordered to help her build a new life. James Garner received an Oscar nomination for his performance.
1985 (PG-13) 107m/C James Garner, Sally Field, Brian Kerwin, Corey Haim; *Dir:* Martin Ritt. **VHS, Beta, LV $14.95** *COL* 𝄢𝄢𝄢

Murphy's War In World War II, the Germans sink an English ship and gun down most of its crew. An Irishman, however, survives and returns to health with the help of a nurse. He then seeks revenge on those who killed his crewmates, even after he learns the war has ended. O'Toole is interesting as the revenge-minded seaman though saddled with a mediocre script.
1971 (PG) 106m/C *GB* Peter O'Toole, Sian Phillips, Philippe Noiret; *Dir:* Peter Yates. **VHS, Beta $9.99** *CCB, PAR, TAV* 𝄢𝄢

Murri Affair A man kills his brother-in-law, touching off a political trial. Points for style but not for pace. Also Known As: "La Grande Bourgoise."
1974 120m/C *IT* Catherine Deneuve, Giancarlo Giannini, Fernando Rey. **VHS, Beta $33.95** *KOV* 𝄢𝄢½

Murrow A cable television biography of the renowed chain-smoking journalist who changed broadcasting history by fearlessly voicing his liberal ideas on the air waves. His life was more interesting though Travanti does his best to breathe some life into the script.
1985 114m/C Daniel J. Travanti, Dabney Coleman, Edward Herrmann; *Dir:* Jack Gold. **VHS, Beta $69.98** *LIV, LTG* 𝄢𝄢½

Muscle Beach Party Frankie and Annette team up for another romp in the sand. Trouble begins when a new gym opens up on beachfront property and hardbodies muscle in on surfer turf. Renowned actor Peter Lorre breaks up a fight between the two groups in a cameo role, his last screen appearance before he died. Watch for "Little" Stevie Wonder before he became big.
1964 94m/C Frankie Avalon, Annette Funicello, Buddy Hackett, Lucianna Paluzzi, Don Rickles, John Ashley, Jody McCrea, Morey Amsterdam, Peter Lupus, Candy Johnson, Dan Haggerty; *Dir:* William Asher. **VHS, Beta, LV $29.95** *VTR* 𝄢𝄢½

Muscle Rock Madness Rock music and pro wrestling, with both the music and the fighting staged in a chintzy fashion.
1988 30m/C VHS, Beta $9.95 *SIM* 𝄢𝄢½

Muse Concert: No Nukes Concert film of performances held at New York's Madison Square Garden for the benefit of the anti-nuclear power movement. Features Jackson Browne, Crosby, Stills and Nash, James Taylor, Bruce Springsteen, the Doobie Brothers, Bonnie Raitt, Carly Simon, and John Hall. Excellent editing keeps the joint hopping, while Bruce steals the show.
1980 (PG) 103m/C Jackson Browne, David Crosby, Stephen Stills, Graham Nash, James Taylor, Bruce Springsteen, The Doobie Brothers, Bonnie Raitt, Carly Simon, John Hall; *Dir:* Julian Schlossberg. **VHS, Beta** *FOX* 𝄢𝄢𝄢

Music Box An attorney defends her father against accusations that he has committed inhumane Nazi war crimes. If she loses, her father faces deportation. As the case progresses, she must struggle to remain objective in the courtroom and come to terms with the possibility that her father is guilty. Lange's Oscar-nominated portrayal of an ethnic character is highly convincing.
1989 (R) 126m/C Jessica Lange, Frederic Forrest, Lukas Haas, Armin Mueller-Stahl, Michael Rooker, Donald Moffatt, Cheryl Lynn Bruce; *Dir:* Constantin Costa-Gavras. **VHS, Beta, LV $19.95** *IVE, JCF* 𝄢𝄢½

Music Box/Helpmates These two classic, award-winning comedy shorts by the dynamic duo have been color-enhanced by Colorization.
1932 50m/B Stan Laurel, Oliver Hardy. **VHS, Beta $19.95** *QNE* 𝄢𝄢½

The Music Man A con man in the guise of a traveling salesman gets off the train in River City, Iowa. After hearing about plans to build a pool hall, he argues that such an establishment would be the gateway to hell for the young, impressionable males of the town. He then convinces the River Cityians to look toward the future of the community and finance a wholesome children's marching band. Although the huckster plans to take their money and run before band instruments arrive, his feelings for the town librarian cause him to think twice about fleeing the Heartland. This isn't just a slice of Americana; it's a whole pie. The acting and singing are terrific "with a capital 'T' and that rhymes with 'P' and that stands for" Preston, who epitomizes the charismatic pitchman. Songs include "Trouble," "Gary, Indiana," "Marian the Librarian", and "76 Trombones." Received Academy Award nomination for Best Picture, Art Direction, and Set Decoration, Sound and Film Editing.
1962 151m/C Robert Preston, Shirley Jones, Buddy Hackett, Hermione Gingold, Paul Ford, Pert Kelton, Ron Howard; *Dir:* Morton DaCosta. Academy Awards '62: Best Adapted Score; Golden Globe Awards '63: Best Film—Musical/

Comedy. **VHS, Beta, LV $19.98** *WAR, LDC, TLF* 𝄢𝄢𝄢𝄢

The Music of Melissa Manchester Manchester performs her greatest hits live in concert. Songs include "Don't Cry Out Loud," "Midnight Blue," and "Come in from the Rain."
1980 60m/C VHS, Beta, LV *PIA* 𝄢𝄢𝄢𝄢

Music in My Heart Two taxi cabs crash and spur the love of Martin and Hayworth. She proceeds to break off her engagement to save Martin from deportation. Received an Oscar nomination for the song "It's a Blue World," but does not offer much more. Hayworth's last low-budget film before becoming a Hollywood goddess.
1940 69m/B Rita Hayworth, Tony Martin, Edith Fellows, Alan Mowbray, Eric Blore, George Tobias, Joseph Crehan, George Humbert, Phil Tead; *Dir:* Joseph Santley. **VHS $19.95** *COL, CCB* 𝄢½

Music Shoppe Four teenagers form a rock band with the assistance of the local music store proprietor.
1982 (G) 93m/C Gary Crosby, Nia Peeples, Benny Medina, Stephen Schwartz, David Jackson, Jesse White, Doug Kershaw, Giselle MacKenzie. **VHS, Beta** *GEM* 𝄢½

Music & the Spoken Word An inspirational live television message, featuring a sermon by Spence Kinar coupled with music by the famous Morman Tabernacle Choir.
1974 30m/C Spence Kinard, Mormon Tabernacle Choir. **VHS, Beta $19.95** *VYY* 𝄢½

The Music Teacher A Belgian costume drama dealing with a famed singer who retires to devote himself to teaching two students exclusively. In time, he sees his prized pupils lured into an international competition by an old rival of his. In French with yellow English subtitles. Although the singing is sharp, the story is flat. Oscar-nominated for Best Foreign Film.
1988 (PG) 95m/C *BE* Patrick Bauchau, Sylvie Fennec, Phillipe Volter, Jose Van Dam, Johan Leysen, Anne Roussel; *Dir:* Gerard Corbiau. **VHS, Beta $79.98** *ORI* 𝄢𝄢½

Musica Proibita A serious opera-spiced drama about a young romance crushed by adultery in the teenage couple's ancestry.
1943 93m/B Tito Gobbi, Maria Mercader. **VHS, Beta $29.98** *VYY, APD* 𝄢𝄢½

Musicals 2: Trailers on Tape Here is a collection of previews for such musical favorites as "The King of Jazz," "The Dolly Sisters," "The Red Shoes" and "Lillian Russell," with some segments in black and white.
1962 60m/C VHS, Beta $34.95 *SFR* 𝄢𝄢½

Musico, Poeta y Loco Wild and crazy antics abound in this film.
1947 90m/B *MX* Tin Tan. **VHS, Beta $48.95** *MAD* 𝄢𝄢

Musicourt This program presents a superstar musical jam featuring Carlos Santana, Joe Cocker and others.
1984 58m/C VHS, Beta $19.95 *PAV* 𝄢𝄢

Mussolini & I Docudramatization of the struggle for power between Italy's Benito Mussolini and his son-in-law, Galeazzo Ciano. Narrated by Sarandon in the role of Il Duce's daughter. Made for cable television. Also called "Mussolini: The Decline and Fall of Il Duce."

1985 200m/C Bob Hoskins, Anthony Hopkins, Susan Sarandon, Annie Girardot; *Dir:* Alberto Negrin. **VHS, Beta $24.98** *SUE* ♂♂½

Mutant Another argument for the proper disposal of toxic waste. Hazardous materials transform the people of a southern town into monsters. Also known as "Night Shadows."
1983 (R) 100m/C Wings Hauser, Bo Hopkins, Jennifer Warren, Lee Montgomery; *Dir:* John Cardos. **VHS, Beta, LV $19.98** *LIV, VES* ♂½

Mutant on the Bounty A musician, lost in space as a beam of light for 23 years, is transported aboard a spaceship. Once he materializes, it becomes apparent that he has undergone some incredible physical changes. As if that's not enough, the crew then must flee from a couple of thugs who are trying to steal a vial of serum they have on board. A space spoof that never gets into orbit.
1989 93m/C John Roarke, Deborah Benson, John Furey, Victoria Catlin, John Fleck, Kyle T. Heffner; *Dir:* Robert Torrance. **VHS $14.95** *HMD, SOU* ♂½

Mutant Hunt Made for video movie about a battle between robots gone haywire and an all-American hero. The detective cyborgs are set free while high on a sexual stimulant called Euphoron to wreak their technological havoc on a rampage in Manhattan.
1987 90m/C Rick Gianasi, Mary-Anne Fahey; *Dir:* Tim Kincaid. **VHS, Beta $69.95** *UHV Woof!*

Mutant Video Six comedic, award-winning short-subject films from Jim Belushi and Rob Riley. Outrageous humor.
1976 60m/B James Belushi, Rob Riley. **VHS, Beta $19.95** *IVE*

Mutants In Paradise Comedy about a human guinea-pig who always gets involved in some sort of mishap.
1988 78m/C Brad Greenquist, Robert Ingham, Anna Nicholas, Edith Massey. **VHS, Beta $59.95** *TWE Woof!*

Mutator A corporation makes a tiny genetic mistake and winds up creating a new life form...a demonic pussycat who preys on humans.
1990 (R) 91m/C Brion James, Carolyn Ann Clark, Milton Raphiel Murrill; *Dir:* John R. Bowey. **VHS, Beta $89.98** *PSM* ♂♂

The Muthers Women prisoners escape from their confinement in a South American jungle.
1976 (R) 101m/C Jeanne Bell, Rosanne Katon, Jayne Kennedy, Trina Parks; *Dir:* Cirio H. Santiago. **VHS, Beta $39.98** *CON* ♂♂

The Mutilator After accidentally bumping off his mother, a young man stalks five high school students wanting to cut them up into tiny bits. An unedited version is also available. Bloody boredom. Also known as "Fall Break."
1985 (R) 85m/C Jack Chatham, Trace Cooper, Frances Raines, Bill Hitchcock; *Dir:* Buddy Cooper. **VHS, Beta, LV $79.98** *LIV, VES Woof!*

Mutiny In the War of 1812, the crew of an American ship carrying $10 million in gold fight among themselves for a part of France's donation to the war effort. The beautiful photography doesn't make up for the predictability of the storyline.
1952 76m/C Mark Stevens, Gene Evans, Angela Lansbury, Patric Knowles; *Dir:* Edward Dmytryk. **VHS, Beta $19.95** *NOS, UHV* ♂♂

Mutiny in the Big House A man is sent to prison for writing a bad ten dollar check and must choose between seeking salvation with the prison chaplain or toughing it out with a hardened convict. Suprisingly good low-budget programmer, produced by former actor Grant Withers.
1939 83m/B Charles Bickford, Barton MacLane, Pat Moriarity, Dennis Moore, William Royle, Charles Foy, George Cleveland; *Dir:* William Nigh. **VHS, Beta $16.95** *SNC* ♂♂½

Mutiny on the Bounty Compelling adaptation of the true story of sadistic Captain Bligh, Fletcher Christian and their turbulent journey aboard the HMS Bounty. No gray here: Laughton's Bligh is truly a despicable character and extremely memorable in this MGM extravaganza. Remade twice, in 1962 and again in 1984 as "The Bounty." Much, much better than the 1962 remake. Received Oscar nominations for Best Actor (Gable, Laughton, and Tone), Best Screenplay, Best Score, Best Film Editing, and Best Director.
1935 132m/B Clark Gable, Franchot Tone, Charles Laughton, Donald Crisp, Dudley Digges, Spring Byington, Henry Stephenson, Eddie Quillan; *Dir:* Frank Lloyd. Academy Awards '35: Best Picture; National Board of Review Awards '35: 10 Best Films of the Year; New York Film Critics Awards '35: Best Actor (Laughton). **VHS, Beta, LV, 8mm $19.98** *MGM, BMV, BTV* ♂♂♂♂

Mutiny on the Bounty Based on the novel by Charles Nordhoff and James Norman Hall, this account of the 1789 mutiny led by Fletcher Christian against Captain Bligh of the Bounty is highlighted by lavish photography. Can't navigate the same perfect course set by the 1935 classic. Marlon Brando seems miscast, giving a choppy performance as the mutiny leader. Oscar nominated for Best Picture, Cinematography, Art Direction and Set Decoration, Film Editing and Score.
1962 177m/C Marlon Brando, Trevor Howard, Richard Harris, Hugh Griffith; *Dir:* Lewis Milestone. **VHS, Beta, LV $29.98** *MGM* ♂♂

The Mutiny of the Elsinore A writer on board an old sailing ship fights mutineers and wins back both the helm and the captain's daughter. Tedious British flotsam that made hash of an oft-filmed Jack London tale, and made news even in 1939 for its use of modern-day slang and clothing in the period setting.
1939 74m/B Paul Lukas, Lyn Harding, Kathleen Kelly, Clifford Evans, Ben Soutten, Jiro Soneya; *Dir:* Roy Lockwood. **VHS $19.95** *NOS* ♂

Mutual Respect A dying wealthy man sets out to find the son he never knew.
1977 (PG) 88m/C Lloyd Bridges, Beau Bridges. **VHS, Beta $9.95** *ACA* ♂½

My African Adventure Silly tale of an ambassador's son discovering a talking monkey in Africa and meeting a slew of opportunistic misfits in his travels. A comedy unfortunately short on laughs. Based on Tamar Burnstein's book.
1987 93m/C Dom DeLuise, Jimmie Walker; *Dir:* Sam Firstenberg. **VHS, Beta** *MED Woof!*

My American Cousin Award-winning Canadian coming-of-age comedy about a 12-year-old girl who falls for her rebellious, fun-loving 17-year-old American cousin over summer vacation. Followed by "American Boyfriends."
1985 (PG) 94m/C *CA* Margaret Langrick, John Wildman, Richard Donat, Jane Mortifee; *W/Dir:* Sandy Wilson. Genie Awards '86: Best Actor (Wildman), Best Actress (Langrick), Best Director

(Wilson), Best Picture, Best Screenplay. **VHS, Beta, LV $19.95** *MED* ♂♂♂

My Beautiful Laundrette Omar, the nephew of a Pakistani businessman, is given the opportunity to better himself by turning his uncle's run-down laundry into a profitable business. He reunites with Johnny, a childhood friend and a working-class street punk, and they go into the business together. They find themselves battling the prejudice of each other's families and friends in order to succeed. An intelligent look at the sexuality, race relations and economic problems of Thatcher's London.
1986 (R) 93m/C Daniel Day Lewis, Saeed Jaffrey, Roshan Seth, Gordon Warnecke; *Dir:* Stephen Frears. **VHS, Beta, LV $79.95** *LHV, WAR* ♂♂♂

My Best Friend Is a Vampire Another teenage vampire story, about trying to cope with certain changes that adolescence and bloodsucking bring. Good supporting cast.
1988 (PG) 90m/C Robert Sean Leonard, Evan Mirand, Cheryl Pollak, Rene Auberjonois, Cecilia Peck, Fannie Flagg, Kenneth Kimmins, David Warner, Paul Wilson; *Dir:* Jimmy Huston. **VHS $14.99** *HBO* ♂♂

My Best Friend's Girl Coy, sardonic French romance about a woman who slaloms between two buddies working at a ski resort. Fine performances from all. Never makes it to the top of the hill. Has English subtitles.
1984 99m/C *FR* Isabelle Huppert, Thierry Lhermitte, Coluche, Francoise Perrot; *Dir:* Bertrand Blier. **VHS, Beta $29.98** *SUE* ♂♂½

My Best Girl Mary attempts to bring her new sweetheart home for dinner, with disastrous results. A gentle satire on middle-American life in the 1920's. Organ score by Gaylord Carter.
1927 78m/B Mary Pickford, Charles "Buddy" Rogers; *Dir:* Sam Taylor. **VHS, Beta, LV $19.95** *CCB* ♂♂

My Bloody Valentine A psychotic coal miner visits the peaceful little town of Valentine Bluffs on the night of the yearly Valentine Ball. Many of the townspeople have their hearts removed with a pick axe and sent to the sheriff in candy boxes. The bloodiest scenes were cut out to avoid an X-rating.
1981 (R) 91m/C *CA* Paul Kelman, Lori Hallier; *Dir:* George Mihalka. **VHS, Beta, LV $19.95** *PAR* ♂

My Blue Heaven After agreeing to rat on the Mafia, Martin is dropped into suburbia as part of the witness protection program. Moranis plays the FBI agent assigned to help the former mobster become an upstanding citizen. Adjusting to life in the slow lane isn't easy for an ex con who has grown accustomed to the big time. Not the typical role for Martin, who plays a brunette with a New York accent and is handcuffed by bad writing.
1990 (PG-13) 96m/C Steve Martin, Rick Moranis, Joan Cusack, Melanie Mayron, Carol Kane, Bill Irwin, William Hickey, Daniel Stern; *Dir:* Herbert Ross. **VHS, Beta, LV $92.95** *WAR, BTV* ♂♂

My Bodyguard An undersized high school student fends off attacking bullies by hiring a king-sized, withdrawn lad as his bodyguard. Their "business" relationship, however, develops into true friendship. An adolescent coming of age with more intelligence and sensitivity than most of its ilk, and

a pack of up and coming stars as well as old stand-bys Houseman and Gordon.
1980 (PG) 96m/C Chris Makepeace, Adam Baldwin, Martin Mull, Ruth Gordon, Matt Dillon, John Houseman, Joan Cusack, Craig Richard Nelson; *Dir:* Tony Bill. National Board of Review Awards '80: 10 Best Films of the Year. **VHS, Beta $69.98** *FOX* 🐾🐾🐾

My Boys are Good Boys A young foursome steal from an armored car and have to face the consequences. Quirky with an implausible storyline.
1978 (PG) 90m/C Ralph Meeker, Ida Lupino, Lloyd Nolan, David Doyle; *Dir:* Bethel Buckalew. **VHS, Beta $59.98** *MAG* 🐾

My Breakfast with Blassie Andy Kaufman and professional wrestler Fred Blassie meet for breakfast and carry on a comical dialogue.
1983 60m/C Andy Kaufman, Freddie Blassie; *Dir:* Johnny Legend, Linda Lautrec. **VHS, Beta $39.95** *RHI* 🐾

My Brilliant Career A headstrong young woman spurns the social expectations of turn-of-the-century Australia and pursues opportunities to broaden her intellect and preserve her independence. Davis is wonderful as the energetic and charismatic community trendsetter. Based on an autobiographical novel which has been marvelously transferred to the screen. Armstrong deserves credit for her fine direction.
1979 (G) 101m/C Judy Davis, Sam Neill, Wendy Hughes; *Dir:* Gillian Armstrong. Australian Film Institute '79: Best Film; British Academy Awards '80: Best Actress (Davis). **VHS, Beta $69.95** *VES* 🐾🐾🐾½

My Brother has Bad Dreams Young man with sibling tosses and turns in bed.
1988 97m/C Nick Kleinholz, Marlena Lustic. **VHS $29.98** *UHV* 🐾

My Champion A chance meeting propels Mike Gorman and Miki Tsuwa into a relationship based on the strong bonds of love and athletic competition.
1984 101m/C Yoko Shimada, Chris Mitchum. **VHS, Beta $59.95** *MED* 🐾

My Chauffeur When a wise-cracking female is hired on as a chauffeur at an all-male chauffeur service, sparks fly. And when the owner takes a definite liking to her work, things take a turn for the worse. As these sort of sexploitation flicks go, this is one of the better ones.
1986 (R) 94m/C Deborah Foreman, Sam Jones, Howard Hesseman, E.G. Marshall; *Dir:* David Beaird. **VHS, Beta, LV $79.98** *LIV, VES* 🐾🐾½

My Cousin Vinny Vinny Gambini (Pesci), an East Coast lawyer who took the bar exam six times before he passed it, goes to Wahzoo City, Alabama to get his cousin and his cousin's friend off the hook when they are accused of killing a store clerk. His leather jackets, gold chains and Brooklyn accent, and his fiancee's (Tomei) penchant for big hair and bold clothing, don't go over very well with the conservative judge (Gwynne), and although this is the first case he's ever tried, Vinny becomes a force to contend with in the courtroom. Some hilarious scenes between Pesci and Gwynne and Pesci and Tomei make this film a winner.
1992 (R) 119m/C Joe Pesci, Ralph Macchio, Marisa Tomei, Mitchell Whitfield, Fred Gwynne, Lane Smith, Austin Pendleton; *Dir:* Jonathan Lynn. **VHS $94.98** *FXV* 🐾🐾½

My Darling Clementine One of the best Hollywood westerns ever made, this recounts the precise events leading up to and including the gunfight at the O.K. Corral. Ford allegedly knew Wyatt Earp and used his stories to recount the details vividly, though not always accurately. Remake of 1939's "Frontier Marshal."
1946 97m/B Henry Fonda, Victor Mature, Walter Brennan, Linda Darnell, Tim Holt, Ward Bond, John Ireland; *Dir:* John Ford. **VHS, Beta, LV $19.98** *FOX, TLF* 🐾🐾🐾½

My Dear Secretary After she marries her boss, a woman grows jealous of the secretary that replaces her. Comedic dialogue and antics result.
1949 94m/C Kirk Douglas, Laraine Day, Keenan Wynn, Rudy Vallee, Florence Bates, Alan Mowbray; *Dir:* Charles Martin; *W/Dir:* Charles Martin. **VHS, Beta $19.95** *NOS, MRV, KRT* 🐾🐾½

My Demon Lover This sex comedy is complicated by the hero's transformation into a demon whenever he is aroused. Saved from complete mediocrity by Family Ties's Valentine in a likable performance.
1987 (PG-13) 90m/C Scott Valentine, Michelle Little, Arnold Johnson; *Dir:* Charles Loventhal. **VHS, Beta, LV $14.95** *COL* 🐾🐾

My Dinner with Andre Two friends talk about their lives and philosophies for two hours over dinner one night. A wonderful exploration into storytelling, the conversation juxtaposes the experiences and philosophies of nerdish, bumbling Shawn and the globe-trotting spiritual pilgrimage of Gregory, in this sometimes poignant, sometimes comic little movie that starts you thinking.
1981 110m/C Andre Gregory, Wallace Shawn; *Dir:* Louis Malle. **VHS, Beta, LV $19.95** *PAV* 🐾🐾🐾½

My Dog Shep An orphan and his dog run away and are pursued diligently when it is discovered that he is a wealthy heir.
1946 71m/B Tom Neal, William Farnum, Lannie Rees, Flame; *Dir:* Ford Beebe. **VHS, Beta, LV $19.95** *NOS, WGE, VYY* 🐾½

My Dog, the Thief A helicopter weatherman is unaware that the lovable St. Bernard he has adopted is a kleptomaniac. When the dog steals a valuable necklace from a team of professional jewel thieves, the fun begins.
1969 88m/C Joe Flynn, Elsa Lanchester, Roger C. Carmel, Mickey Shaughnessy, Dwayne Hickman, Mary Ann Mobley; *Dir:* Robert Stevenson. **VHS, Beta $69.95** *DVT* 🐾

My Dream is Yours Day plays an up and coming radio star in this Warner Brothers musical comedy with a cameo from Bugs Bunny in a dream sequnce. This fresh, fun remake of "Twenty Million Sweethearts" is often underrated, but is well worth a look.
1949 101m/C Jack Carson, Doris Day, Lee Bowman, Adolphe Menjou, Eve Arden, S.Z. Sakall, Selena Royle, Edgar Kennedy, Bugs Bunny; *Dir:* Michael Curtiz. **VHS $29.98** *FOX, FCT, MLB* 🐾🐾½

My Fair Lady Colorful production of Lerner and Loewe's musical version of "Pygmalion," about an ill-mannered cockney girl who is plucked from her job as a flower girl by Professor Henry Higgins. Higgins makes a bet with a colleague that he can turn this rough diamond into a "lady." Winner of eight Academy Awards. Songs include "The Rain in Spain," "I Could Have Danced All Night," "Get Me to the Church on Time," and "Wouldn't It Be Lovely." Hepburn's singing voice is dubbed by Marni Nixon, who was

also responsible for the singing in "The King and I" and "West Side Story;" the dubbing may have undermined her chance at an Oscar nomination. Typecasting role for Harrison as the crusty, egocentric Higgins. A timeless classic. Other Oscar nominations for supporting actor Holloway, supporting actress Cooper, Adapted Screenplay, and Film Editing.
1964 (G) 170m/C Audrey Hepburn, Rex Harrison, Stanley Holloway, Wilfrid Hyde-White, Theodore Bikel, Mona Washbourne, Jeremy Brett, Robert Coote, Gladys Cooper; *Dir:* George Cukor. Academy Awards '64: Best Actor (Harrison), Best Adapted Score, Best Art Direction/Set Decoration (Color), Best Color Cinematography, Best Costume Design (Color), Best Director (Cukor), Best Picture; British Academy Awards '65: Best Film; Directors Guild of America Awards '64: Best Director (Cukor); Golden Globe Awards '65: Best Film—Musical/Comedy; National Board of Review Awards '64: 10 Best Films of the Year; New York Film Critics Awards '64: Best Actor (Harrison), Best Film. **VHS, Beta, LV $29.98** *FOX, FCT, BUR* 🐾🐾🐾½

My Father, My Rival A high school student's crush on his teacher leads to emotional complications when his widowed father becomes romantically involved with her.
1985 60m/C Lance Guest. **VHS, Beta $39.98** *SCL* 🐾

My Father's Glory Based on Marcel Pagnol's tales of his childhood, this is a sweet, beautiful memory of a young boy's favorite summer in the French countryside of the early 1900s. Not much happens, yet the film is such a perfect evocation of the milieu that one is carried swiftly into the dreams and thoughts of all the characters. One half of a duo, followed by "My Mother's Castle."
1991 (G) 110m/C Julien Ciamaca, Philippe Caubere, Nathalie Roussel, Therese Liotard, Didier Pain; *Dir:* Yves Robert. **VHS** *ORI* 🐾🐾🐾

My Father's House A magazine editor recounts his youth while recovering from a heart attack. He measures his own quality of life against the life that his father led and begins to question his own choices with regard to his family.
1975 96m/C Cliff Robertson, Robert Preston, Eileen Brennan, Rosemary Forsyth. **VHS, Beta $49.95** *WOV* 🐾½

My Father's Nurse While father convalesces upstairs, the family frolics about the house and plots his dispatch. When his new nurse secretly cures him, he takes to a wheelchair and mounts a comedic covert counteroffensive.
197? 97m/C **VHS, Beta $39.95** *VCD* 🐾

My Father's Wife A bored wife gets sexual attention from her maturing stepson. Soft-core.
198? 80m/C Carroll Baker. **VHS, Beta $29.95** *MED* 🐾

My Favorite Brunette Detective parody starring Hope as a photographer turned grumbling private eye. He becomes involved with a murder, a spy caper, and a dangerous brunette (Lamour).
1947 85m/B Bob Hope, Dorothy Lamour, Peter Lorre, Lon Chaney Jr., Alan Ladd, Reginald Denny, Bing Crosby; *Dir:* Elliott Nugent. **VHS, Beta $9.95** *NEG, MRV, NOS* 🐾🐾½

My Favorite Wife A handsome widower remarries only to discover that his first wife, who was shipwrecked seven years prior and presumed dead, has reappeared. Eventually, a judge must decide what to do in

this most unusual situation, completely lacking precedent. A farcical and hilarious story filled with a clever cast. The 1963 remake "Move Over Darling" lacks the style and wit of this presentation. Also available colorized.
1940 88m/B Irene Dunne, Cary Grant, Randolph Scott, Gail Patrick; **Dir:** Garson Kanin. **VHS, Beta, LV $19.98** MED, MLB ♫♫♫

My Favorite Year A young writer on a popular live television show in the 1950s is asked to keep a watchful eye on the week's guest star—his favorite swashbuckling movie hero. Through a series of misadventures, he discovers his matinee idol is actually a drunkard and womanizer who has trouble living up to his cinematic standards. Sterling performance from O'Toole, with memorable portrayal from Bologna as the show's host, King Kaiser (a take-off of Sid Caesar from "Your Show of Shows").
1982 (PG) 92m/C Peter O'Toole, Mark Linn-Baker, Joseph Bologna, Jessica Harper, Lainie Kazan, Bill Macy, Anne De Salvo, Lou Jacobi, Adolph Green, Cameron Mitchell, Gloria Stuart; **Dir:** Richard Benjamin. **VHS, Beta, LV $19.95** MGM ♫♫♫

My First Wife Strong drama about a self-involved man's devastation when his wife abruptly leaves him after ten years of marriage. Realistic and well-acted.
1984 (PG) 95m/C AU John Hargreaves, Wendy Hughes, Lucy Angwin, Anna Jemison, David Cameron; **Dir:** Paul Cox. Australian Film Institute '84: Best Actor (Hargreaves). **VHS $39.98** CON ♫♫ ½

My Forbidden Past Melodrama set in 1890 New Orleans centers on a young woman (Gardner) with an unsavory past who unexpectedly inherits a fortune and vows to break up the marriage of the man (Mitchum) she loves. His wife is murdered and he is charged with doing the deed until Gardner, exposing her past, wins his love and helps to extricate him.
1951 81m/B Ava Gardner, Melvyn Douglas, Robert Mitchum; **Dir:** Robert Stevenson. **VHS, Beta $19.95** MED ♫♫ ½

My Friend Flicka Boy makes friends with four-legged beast. Dad thinks the horse is full of wild oats, but young Roddie trains it to be the best gosh darned horse in pre-Disney family faredom. Based on Mary O'Hara book, followed by "Thunderhead, Son of Flicka," and TV series.
1943 89m/C Roddy McDowall, Preston Foster, Rita Johnson, James Bell, Jeff Corey; **Dir:** Harold Schuster. **VHS, Beta $14.98** FXV, FCT ♫♫♫

My Friend Liberty This is a humorous look at the Statue of Liberty on the occasion of its renovation, using clay animation.
1986 30m/C VHS, Beta $14.98 LHV ♫♫♫

My Geisha MacLaine and husband experience bad karma because she wants to be in his new film. In pancake makeup and funny shoes, she poses as a geisha girl and is cast as Madame Butterfly. Her husband, however, is one sharp cookie. Filmed in Japan.
1962 120m/C Shirley MacLaine, Yves Montand, Edward G. Robinson, Bob Cummings; **Dir:** Jack Cardiff. **VHS $14.95** PAR, WAX, FCT ♫♫ ½

My Girl Chlumsky is delightful in her film debut as an 11-year old tomboy who must come to grips with the realities of life. Culkin plays her best friend Thomas, who understands her better than anyone else, including her father, a mortician, and his girlfriend, the makeup artist at the funeral parlor. Some

reviewers questioned whether young children would be able to deal with some unhappy occurrences in the film, but most seemed to classify it as a movie the whole family would enjoy.
1991 (PG) 102m/C Dan Aykroyd, Jamie Lee Curtis, Macauley Culkin, Anna Chlumsky, Griffin Dunne; **Dir:** Howard Zieff. MTV Movie Awards '92: Best Kiss. **VHS, LV, 8mm** COL ♫♫ ½

My Girl Tisa Chronicles the experiences of a young immigrant woman who struggles to survive in the U.S. at the turn of the century and hopes to bring her father over from their home country. Based on a play by Lucille S. Prumbs and Sara B. Smith.
1948 95m/B Lilli Palmer, Sam Wanamaker, Akim Tamiroff, Alan Hale Jr., Hugo Haas, Gale Robbins, Stella Adler; **Dir:** Elliott Nugent. **VHS $19.98** REP ♫♫ ½

My Grandpa is a Vampire When 12 year-old Lonny and his pal visit Grandpa Cooger in New Zealand Lonny discovers a long-hidden family secret—Grandpa's a vampire! This doesn't stop either boy from joining with Grandpa in lots of scary adventures.
199? (PG) 90m/C Al Lewis, Justin Gocke, Milan Borich, Noel Appleby; **Dir:** David Blyth. **VHS $89.98** REP ♫ ½

My Heroes Have Always Been Cowboys An aging rodeo rider returns to his hometown to recuperate and finds himself forced to confront his past. His ex-girlfriend, his dad and his sister all expect something from him. He learns how to give it, and gains the strength of purpose to get back on the bull that stomped him. Excellent rodeo footage, solid performances, but the story has been around the barn too many times to hold much interest.
1991 (PG) 106m/C Scott Glenn, Kate Capshaw, Ben Johnson, Balthazar Getty, Mickey Rooney, Gary Busey, Tess Harper; **Dir:** Stuart Rosenberg. **VHS, Beta, LV $92.98** COL, FXV ♫♫ ½

My Lady of Whims Clara Bow's dad disapproves of debutante daughter's desire to hang out with the girls who just want to have fun.
1925 42m/B Clara Bow, Betty Baker, Carmelita Geraghty, Donald Keith, Lee Moran. **VHS, Beta $19.95** GPV, FCT ♫

My Left Foot A gritty, unsentimental drama based on the life and autobiography of cerebral-palsy victim Christy Brown. Considered an imbecile by everyone but his mother (Fricker, in a stunning award-winning performance) until he teaches himself to write. He survives his impoverished Irish roots to become a painter and writer using his left foot, the only appendage over which he has control. He also falls in love and finds some heartaches along the way. Day-Lewis is astounding; the supporting cast, especially Shaw and Cusack, match him measure for measure. Oscar nominated for Best Picture.
1989 (R) 103m/C IR Daniel Day Lewis, Brenda Fricker, Ray McAnally, Cyril Cusack, Fiona Shaw, Hugh O'Conor, Adrian Dunbar, Ruth McCabe, Alison Whelan; **Dir:** Jim Sheridan. Academy Awards '89: Best Actor (Day Lewis), Best Supporting Actress (Fricker); Los Angeles Film Critics Association Awards '89: Best Actor (Day Lewis), Best Supporting Actress (Fricker); New York Film Critics Awards '89: Best Actor (Day Lewis); National Society of Film Critics Awards '89: Best Actor (Day Lewis). **VHS, Beta, LV $89.99** HBO, BTV ♫♫♫♫

My Life As a Dog A troublesome boy is separated from his brother and is sent to live with relatives in the country when his mother is taken ill. Unhappy and confused, he struggles to understand sexuality and love and tries to find security and acceptance. Remarkable Swedish film available with English subtitles or dubbed.
1987 101m/C SW Anton Glanzelius, Tomas Van Bromssen, Anki Liden; **Dir:** Lasse Hallstrom. Seattle Int'l. Film Festival '87: Best Film. **VHS, Beta, LV $79.95** PAR ♫♫♫♫

My Life to Live A woman turns to prostitution in this probing examination of sexual, and social, relations. Idiosyncratic Godard has never been more starstruck than in this vehicle for the endearing Karina, his wife at the time. A classic. French title: Vivre Savie. Also known as "It's My Life." In French with English subtitles.
1962 85m/B FR Anna Karina, Sady Rebbot, Andre S. Labarthe, Guylaine Schlumberger; **Dir:** Jean-Luc Godard. Venice Film Festival '62: Special Jury Prize. **VHS, Beta $24.95** NOS, NYF, DVT ♫♫♫♫

My Little Chickadee Classic comedy about a gambler and a fallen woman who marry for convenience so they can respectably enter an unsuspecting town. Sparks fly in their adventures together. Fields and West are both at their best playing their larger-than-life selves.
1940 91m/B W.C. Fields, Mae West, Joseph Calleia, Dick Foran, Margaret Hamilton, Donald Meek; **Dir:** Eddie Cline. **VHS, Beta, LV $39.95** MCA, MLB ♫♫♫

My Little Girl A rich Philadelphia girl idealistically volunteers her time to help local institutionalized orphans, but is met only with opposition.
1987 (R) 118m/C Mary Stuart Masterson, James Earl Jones, Geraldine Page, Anne Meara, Peter Gallagher; **Dir:** Connie Kaiserman. **VHS, Beta, LV $9.99** STE, PSM ♫♫

My Little Pony: The Movie Animated tale about the siege of Ponyland by malevolent witches.
1986 87m/C Dir: Michael Joens; **Voices:** Danny DeVito, Cloris Leachman, Tony Randall, Madeline Kahn. **VHS, Beta $9.99** VTR ♫♫

My Love For Yours The romantic tale of a young man hoping to win the love of a beautiful but icy girl. A tad silly in parts, but the clever dialogue moves the story along. Also known as "Honeymoon in Bali."
1939 99m/B Fred MacMurray, Madeleine Carroll; **Dir:** Edward H. Griffith. **VHS, Beta $19.95** NOS, CAB ♫♫ ½

My Lucky Stars A cop goes undercover and unfortunately for the criminals he happens to be a skull-fracturing martial arts expert!!
198? 90m/C Jackie Chan, Richard Ng, Samo Hung. **VHS, Beta $59.98** CON ♫♫ ½

My Man Adam A dreamy, Mitty-esque high schooler falls in love with a girl (Daryl's auburn-haired sister, Page Hannah), and becomes ensnared in a real life crime, leaving his friend to bail him out. Typical boy-meets-girl, boy-gets-in-trouble yarn.
1986 (R) 84m/C Raphael Sbarge, Veronica Cartwright, Page Hannah, Larry B. Scott, Charlie Barnett, Arthur Pendleton, Dave Thomas; **Dir:** Roger L. Simon. **VHS, Beta $79.98** FOX ♫♫

My Man Godfrey A spoiled rich girl picks up someone she assumes is a bum as part of a scavenger hunt and decides to keep him on as her butler. In the process, he teaches her about life, money, and happiness. Top-notch screwball comedy defines the genre. Lombard is a stunner alongside the equally charming Powell. Oscar-nominated for Best Actor (Powell), Best Actress (Lombard), Best Supporting Actor (Auer), Best Supporting Actress (Brady), Best Director, and Best Screenplay. Remade in 1957 with June Allyson and David Niven. From the novel by Eric Hatch.

1936 95m/B William Powell, Carole Lombard, Gail Patrick, Alice Brady, Mischa Auer, Eugene Pallette; *Dir:* Gregory La Cava. **VHS, Beta $9.95** *NEG, MRV, NOS* 🎞🎞🎞🎞

My Michael Sensitive adaptation of Amos Oz's novel concerns young woman stifled by marriage and the conventions of bourgeois life. In Hebrew with some English dialogue.

1975 90m/C *IS* Efrat Lavie, Oded Kotler; *Dir:* Dan Wolman. **VHS, Beta $79.95** *ERG, FCT* 🎞🎞🎞

My Mom's a Werewolf An average suburban mother gets involved with a dashing stranger and soon, to her terror, begins to turn into a werewolf. Her daughter and companion must come up with a plan to regain dear, sweet mom.

1989 (PG) 90m/C Susan Blakely, John Saxon, John Schuck, Katrina Caspary, Ruth Buzzi, Marilyn McCoo, Marcia Wallace, Diana Barrows; *Dir:* Michael Fischa. **VHS, Beta, LV $79.95** *PSM* 🎞 ½

My Name Called Bruce When the unstoppable "Bruce" is teamed with a sexy secret agent, the criminals of Hong Kong issue a collective shudder.

198? 90m/C Bruce Le. **VHS, Beta $19.99** *BFV* 🎞 ½

My Name Is Ivan Tarkovsky's first feature film is a vivid, wrenching portrait of a young Soviet boy surviving as a spy behind enemy lines during WWII. Technically stunning, heralding the coming of modern cinema's greatest formalist. In Russian with English subtitles. Also titled "Ivan's Childhood" and "The Youngest Spy."

1962 84m/B *RU* Kolya Burlyayev, Valentin Zubkov, Ye Zharikov, S. Krylov; *Dir:* Andrei Tarkovsky. Venice Film Festival '62: Best Film. **VHS, Beta $79.95** *FXL, FCT* 🎞🎞🎞 ½

My Name Is Nobody Fast-paced spaghetti-western wherein a cocky, soft-hearted gunfighter is sent to kill the famous, retired outlaw he reveres, but instead they band together.

1974 (PG) 115m/C Henry Fonda, Terence Hill, R.G. Armstrong; *Dir:* Tonino Valerii. **VHS, Beta $14.95** *KRT, HHE* 🎞🎞 ½

My Neighborhood A debonair bar-owner leads a double life as a knife wielding psychopath.

1984 35m/C VHS, Beta *NEW* 🎞🎞 ½

My New Partner Amiable French comedy in which cynical veteran cop is saddled with straight arrow rookie partner. French Cesars for best film and best director, but remember: They like Jerry Lewis too. In French with English subtitles.

1984 (R) 106m/C *FR* Philippe Noiret, Thierry Lhermitte, Regine, Grace de Capitani, Claude Brosset, Julien Guiomar; *W/Dir:* Claude Zidi. **VHS, Beta, LV $29.95** *MED, IME* 🎞🎞 ½

My Night at Maud's Typically subtle Rohmer entry concerns quandary of upright fellow who finds himself drawn to comparatively carefree woman. Talky, somewhat arid film is one of director's Six Moral Tales. You'll either find it fascinating or wish you

were watching "Rocky XXIV" instead. Cinematography by Nestor Almendros; in French with English subtitles. Oscar-nominated for Best Foreign Film and Best Original Screenplay. Also called "My Night With Maud."

1969 111m/B *FR* Jean-Louis Trintignant, Francoise Fabian, Marie-Christine Barrault, Antoine Vitez; *Dir:* Eric Rohmer. National Board of Review Awards '70: 5 Best Foreign Films of the Year; New York Film Critics Awards '70: Best Screenwriting; National Society of Film Critics Awards '70: Best Cinematography, Best Screenplay. **VHS, Beta, LV $29.98** *MED, TAM, APD* 🎞🎞🎞

My Nights With Susan, Sandra, Olga, and Julie A few extremely strange young women open their farm to a weary adventurer on his way to France.

19?? 78m/C *NL* Willeke Van Ammelrooy; *Dir:* Pin de La Parra. **VHS, Beta** *MED* 🎞

My Old Man Plucky teen and her seedy horsetrainer father come together over important horse race. Oates makes this one worth watching on a slow evening. A Hemingway story made for television.

1979 102m/C Kristy McNichol, Warren Oates, Eileen Brennan; *Dir:* John Erman. **VHS, Beta $9.95** *CAF* 🎞🎞 ½

My Old Man's Place A veteran, with two war buddies and a girl, returns to his father's run-down farm, hoping to fix it up. Sexual tensions arise and violence erupts.

1971 (R) 93m/C Arthur Kennedy, Michael Moriarty, Mitchell Ryan, William Devane, Topo Swope; *Dir:* Edwin Sherin. **VHS, Beta $79.95** *PSM* 🎞 ½

My Other Husband A woman has two husbands and families, one in Paris and one in Trouville, who eventually meet each other. Seems silly but grows into a sensitive and sad portrait of married life. Subtitled in English.

1985 (PG-13) 110m/C *FR* Miou-Miou, Rachid Ferrache, Roger Hanin; *Dir:* Georges Lautner. **VHS, Beta $59.95** *RCA* 🎞🎞 ½

My Outlaw Brother A man travelling West to visit his brother in Mexico meets a Texas Ranger on the train. The man discovers that his brother is an outlaw, and teams up with the Ranger to capture him. Also known as "My Brother, the Outlaw."

1951 82m/B Mickey Rooney, Wanda Hendrix, Robert Preston, Robert Stack, Jose Torvay; *Dir:* Elliott Nugent. **VHS $19.98** *MOV, MRV* 🎞🎞

My Own Private Idaho Director van Sant of "Drugstore Cowboy" returns to the underworld to examine another group of outsiders, this time young, homosexual hustlers. On the streets of Seattle, narcoleptic hustler Mike meets slumming rich boy, Scott and together they begin a search for Mike's lost mother, which leads them back to Mike's home in Idaho and on a journey to Italy. Stunning visuals and an elliptical plot, as well as a terrific performance by River Phoenix, highlight this search for love, the meaning of life and power. Van Sant couples these activities with scenes from Shakespeare's "Henry IV" for a sometimes inscrutable, but always memorable film. Look for director William Richert's Falstaff role as an aging chickenhawk.

1991 (R) 105m/C River Phoenix, Keanu Reeves, James Russo, William Richert, Rodney Harvey, Michael Parker, Flea, Chiara Caselli, Udo Kier; *Dir:* Gus Van Sant. National Society of Film Critics Awards '91: Best Actor. **VHS $89.98** *COL* 🎞🎞🎞

My Pal Trigger Roy is unjustly imprisoned in this high adventure on the plains. Better than usual script and direction makes this is one of the more entertaining of the singing cowboy's films.

1946 79m/B Roy Rogers, George "Gabby" Hayes, Dale Evans; *Dir:* Frank McDonald. **VHS, Beta $19.95** *NOS, MRV, HHT* 🎞🎞

My Pleasure Is My Business Cinematic autobiography of Xaveria Hollander, the world's most renowned prostitute of the seventies, and the zany occurrences which abound in her profession.

1974 (R) 85m/C *CA* Xaviera Hollander, Henry Ramer, Colin Fox, Ken Lynch, Jayne Eastwood; *Dir:* Al Waxman. **VHS $59.95** *GRG* 🎞🎞

My Science Project Teenager Stockwell stumbles across a crystal sphere with a funky light. Unaware that it is an alien time-travel device, he takes it to school to use as a science project in a last-ditch effort to avoid failing his class. Chaos follows and Stockwell and his chums find themselves battling gladiators, mutants, and dinosaurs. Plenty of special effects and a likeable enough, dumb teenage flick.

1985 (PG) 94m/C John Stockwell, Danielle von Zerneack, Fisher Stevens, Raphael Sbarge, Richard Masur, Barry Corbin, Ann Wedgeworth, Dennis Hopper, Candace Silvers, Beau Dremann, Pat Simmons, Pamela Springsteen; *Dir:* Jonathan Betuel. **VHS, Beta, LV $79.95** *TOU* 🎞 ½

My Side of the Mountain A thirteen-year-old boy decides to emulate his idol, Henry David Thoreau, and gives up his home and his family to live in the Canadian mountains.

1969 (G) 100m/C *CA* Teddy Eccles, Theodore Bikel; *Dir:* James B. Clark. **VHS, Beta $29.95** *KUI, PAR* 🎞🎞 ½

My Sister, My Love Odd tale of two sisters' incestuous relationship and what happens when one of them takes another lover. The cage of the alternate titles refers to the place where their pet apes are kept and seems to symbolize the confining nature of their life together. Also called "The Mafu Cage" and "The Cage."

1978 (R) 102m/C Lee Grant, Carol Kane, Will Geer, James Olson; *Dir:* Karen Arthur. **VHS $59.95** *UHV, MAG* 🎞 ½

My Son, the Vampire Part of Britain's Old Mother Riley series in which Lucan plays the Irish housekeeper in drag. Lugosi is a crazed scientist who wants to take over the world with his giant robot. Originally titled "Old Mother Riley Meets the Vampire."

1952 72m/B *GB* Bela Lugosi, Arthur Lucan, Dora Bryan, Richard Wattis; *Dir:* John Gilling. **VHS $19.98** *NOS, SNC* 🎞 ½

My Song Goes Round the World Chronicles the romantic foibles of a singing trio and the one girl they decide they all love.

1934 68m/B Joseph Schmidt, John Loder, Charlotte Ander; *Dir:* Richard Oswald. **VHS $29.95** *NOS, DVT* 🎞 ½

My Stepmother Is an Alien When eccentric physicist Aykroyd sends a message beam to another galaxy on a stormy night, the last thing he expects is a visit from beautiful alien Basinger. Unfortunately, he does not realize that this gorgeous blonde is an alien and he continues to court her despite her rather odd habits. Only the daughter seems to notice the strange goings on, and her dad ignores her warnings, enabling Basinger's evil sidekick to continue in its plot to take over the Earth.

1988 (PG-13) 108m/C Dan Aykroyd, Kim Basinger, Jon Lovitz, Alyson Hannigan, Joseph Maher, Seth Green, Wesley Mann, Adrian Sparks, Juliette Lewis, Tanya Fenmore; *Dir:* Richard Benjamin. **VHS, Beta, LV $14.95** *COL* 𝕚𝕚

My Sweet Charlie Unwed, pregnant white woman hides out with black lawyer in backwater Texas. better than it sounds, with nice performances by Duke and Freeman. Made for television.
1970 97m/C Patty Duke, Al Freeman Jr., Ford Rainey, William Hardy, Chris Wilson, Archie Moore, Noble Willingham; *Dir:* Lamont Johnson. **VHS, Beta $39.95** *MCA* 𝕚𝕚½

My Therapist A sex therapist's boyfriend cannot bear the thought of her having intercourse with other men as part of her work. Soft core.
1984 81m/C Marilyn Chambers; *Dir:* Gary Legon. **VHS, Beta $59.95** *VCL* Woof!

My Tutor When a high school student is in danger of flunking French, his parents hire a private tutor to help him learn the lessons. It becomes clear that his studies will involve many more subjects, however.
1982 (R) 97m/C Caren Kaye, Matt Lattanzi, Kevin McCarthy, Clark Brandon, Bruce Bauer, Arlene Golonka, Crispin Glover, Shelley Taylor Morgan, Francesca "Kitten" Natavidad, Jewel Shepard, Marilyn Tokuda; *Dir:* George Bowers. **VHS, Beta, LV $14.95** *MCA* 𝕚

My Twentieth Century This quirky gem of a movie is a charming, sentimental black-and-white journey through the early 1900s. Twins Dora and Lili are separated in early childhood. They reunite as grown, very different women on the Orient Express after they both (unknowingly) have sex with the same man. Dora is a sex kitten, while Lili is a radical equipped with explosives. When the two sisters come together, they both lose their destructive, dependent selves (Dora on men, Lili on politics) and become independent women. Lots of sidelights and subplots that are sure to amuse the viewer.
1990 104m/B HU CA Dortha Segda, Oleg Jankowski, Peter Andorai, Gabor Mate; *W/Dir:* Ildiko Enyedi. Cannes Film Festival '90: Camera d'Or. **VHS $89.95** *FXL* 𝕚𝕚𝕚

My Wicked, Wicked Ways A low-budget television movie based on the autobiography of Errol Flynn.
1984 95m/C Duncan Regehr, Barbara Hershey, Hal Linden, Darren McGavin; *Dir:* Don Taylor. **VHS, Beta $59.98** *FOX* 𝕚½

My Wonderful Life Softcore gristle about a beautiful tramp climbing the social ladder via the boudoir.
1990 (R) 107m/C Pierre Cosso, Jean Rochefort, Massimo Venturiello, Carol Alt, Elliott Gould; *Dir:* Carlo Vanzina. **VHS, Beta, LV $79.95** *COL* 𝕚

Myra Breckinridge A tasteless version of the Gore Vidal novel. An alleged satire of a film critic who undergoes a sex change operation and then plots the destruction of the American male movie star stereotype. Should have more laughs, especially with West in the title role. Created an outcry from all sides, and hung out to dry by studio where it's reportedly still blowing in the wind.
1970 (R) 94m/C Mae West, John Huston, Raquel Welch, Rex Reed, Farrah Fawcett, Jim Backus, John Carradine, Andy Devine, Tom Selleck; *Dir:* Michael Sarne. **VHS, Beta** *FOX, MLB* Woof!

The Mysterians A race of gigantic scientific intellects from a doomed planet attempts to conquer Earth. They want to rebuild their race by reproducing with earth women. Earth fights back. From the director of "Godzilla." Dubbed in English from Japanese.
1958 85m/C JP Kenji Sahara, Yumi Shirakawa, Takashi Shimura; *Dir:* Inoshiro Honda. **VHS, Beta $29.95** *UHV, MLB* 𝕚𝕚½

Mysteries A rich tourist becomes obsessed by a beautiful local girl. As his obsession grows, his behavior becomes stranger. Interesting and well-acted. The film is an adaptation of the famous love story by Nobel-laureate Knut Hamsun. Suffers from poor dubbing.
1984 100m/C Rutger Hauer, Sylvia Kristel, David Rappaport, Rita Tushingham; *Dir:* Paul de Lussanet. **VHS, Beta $69.95** *WES* 𝕚½

Mysteries From Beyond Earth The bizarre world of psychic phenomena and the paranormal is explored in this compelling film. Among the many areas of investigation are UFO's, Kirlian Photography, psychic healing, and witchcraft.
1976 95m/C **VHS, Beta $29.95** *UHV* 𝕚𝕚

Mysterious Desperado A young man, about to inherit a large estate, is framed on a murder charge by land grabbers. Not very memorable.
1949 61m/B Tim Holt, Richard Martin. **VHS, Beta $39.95** *RKO* 𝕚½

Mysterious Desperado/Rider From Tucson A western double feature. In "Mysterious Desperado," a young man is framed for a crime he did not commit. In "Rider From Tucson," evil claim jumpers will resort to murder to gain control of a gold mine.
1949 121m/B Tim Holt, Richard Martin. **VHS, Beta $39.95** *RKO* 𝕚𝕚½

Mysterious Doctor Satan A mad satanic scientist builds an army of mechanical robots to rob and terrorize the nation. In 15 episodes.
1940 250m/B Eduardo Ciannelli, Robert Wilcox, Ella Neal; *Dir:* William Witney. **VHS $29.95** *REP, VCN, MLB* 𝕚𝕚

Mysterious Island Exhilirating sci-fi classic adapted from Jules Verne's novel about escaping Civil War soldiers who go up in Verne balloon and come down on a Pacific island populated by giant animals. They also encounter two shipwrecked English ladies, pirates, and Captain Nemo (and his sub). Top-rate special effects by master Ray Harryhausen and music by Bernard Herrmann.
1961 101m/C GB Michael Craig, Joan Greenwood, Michael Callan, Gary Merrill, Herbert Lom, Beth Rogan, Percy Herbert, Dan Jackson, Nigel Green; *Dir:* Cy Endfield. **VHS, Beta, LV $19.95** *COL, MLB* 𝕚𝕚𝕚½

Mysterious Island Animated version of the Jules Verne fantasy classic, about prisoners from the Civil War who escape. After a balloon journey they land on an island where they struggle to survive.
1975 49m/C **VHS, Beta $19.95** *MGM* 𝕚𝕚

Mysterious Island of Beautiful Women A male sextet is stranded on a South Sea island, where they must endure the trials of an angry tribe of conveniently bikini-clad women.
1979 100m/C Jamie Lyn Bauer, Jayne Kennedy, Kathryn Davis, Deborah Shelton, Susie Coelho, Peter Lawford, Steven Keats, Clint Walker; *Dir:* Joseph Pevney. **VHS $39.95** *IVE* 𝕚

Mysterious Jane Amid soft-focus nudity, a husband and his lover conspire to institutionalize his wife.
198? 90m/C Amber Lee, Sandy Carey. **VHS, Beta $29.95** *MED* 𝕚

The Mysterious Lady Pre-Ninotchka Garbo plays Russian spy who betrays her mother country because she does not want to be alone.
1928 99m/B Greta Garbo, Conrad Nagel, Gustav von Seyffertitz, Richard Alexander, Albert Pollet, Edward Connelly. **VHS $29.99** *MGM, FCT* 𝕚𝕚½

The Mysterious Mr. Wong The Thirteen Coins of Confucius put San Francisco's Chinatown in a state of terror until Mr. Wong arrives.
1935 56m/B Arlene Judge, Wallace Ford; *Dir:* William Nigh. **VHS, Beta $24.95** *NOS, MRV, SNC* 𝕚½

Mysterious Planet Humans fight off alien hordes in this space flick.
1984 90m/C **VHS, Beta $59.95** *VCD* 𝕚

The Mysterious Rider A cowboy tries to prevent unscrupulous homesteaders from cheating farmers out of their land.
1933 59m/B Kent Taylor, Lona Andre, Gail Patrick, Warren Hymer, Berton Churchill; *Dir:* Fred Allen. **VHS $12.95** *SNC, DVT* 𝕚𝕚½

The Mysterious Rider A man framed for murder must help innocent homesteaders from being cheated out of their land. Also known as "Mark of the Avenger."
1938 75m/B Douglas Dumbrille, Sidney Toler, Russell Hayden, Charlotte Field; *Dir:* Lesley Selander. **VHS $19.95** *NOS, DVT, BUR* 𝕚½

The Mysterious Stranger Printer's apprentice is given to bouts of daydreaming about a magic castle in Austria. Based on Twain tale.
1982 89m/C Chris Makepeace, Lance Kerwin, Fred Gwynne; *Dir:* Peter Hunt. **VHS $19.95** *MCA* 𝕚𝕚

Mysterious Two Two aliens visit the Earth in an effort to enlist converts to travel the universe with them.
1982 100m/C John Forsythe, Priscilla Pointer, Noah Beery, Vic Tayback, James Stephens, Karen Werner, Robert Englund, Robert Pine; *Dir:* Gary Sherman. **VHS, Beta $59.95** *IVE* 𝕚𝕚

Mystery of Alexina True story of Herculine Adelaide Barbin, a 19th-century French hermaphrodite, who, after growing up a woman, fell in love with another woman and was actually revealed to be a man. Subtitled. Written by Jean Gruault.
1986 86m/C FR Vuillemin, Valeri Stroh; *Dir:* Rene Feret. **VHS, Beta $29.98** *SUE* 𝕚𝕚½

Mystery at Castle House Three precocious children stumble onto a mystery when their friends disappear at a neighborhood castle.
1982 80m/C **VHS, Beta $14.98** *VID* 𝕚𝕚½

Mystery Date A sort of teen version of "After Hours," in which a shy college guy gets a date with the girl of his dreams, only to be mistaken for a master criminal and pursued by gangsters, police and a crazed florist. Not terrible, but if you're old enough to drive you're probably too old to watch with amusement.
1991 (PG-13) 98m/C Ethan Hawke, Teri Polo, Brian McNamara, Fisher Stevens, B.D. Wong; *Dir:* Jonathan Wacks. **VHS $92.98** *ORI* 𝕚𝕚

Mystery at Fire Island A clever young girl and her cousin encounter some strange people as they try to find out why their fisherman friend mysteriously vanished. 1981 52m/C Frank Converse, Barbara Byrne, Beth Ehlers, Eric Gurry. VHS, Beta $39.98 SCL *

Mystery of the Hooded Horseman Tex finds himself pitted against a very strange adversary. 1937 61m/B Tex Ritter, Iris Meredith, Charles King, Joseph Girard, Lafe McKee; **Dir:** Ray Taylor. VHS, Beta $19.95 NOS, VDM, DVT **

Mystery Island Beautifully filmed underwater scenes in this children's film about four youths who discover a deserted island which they name Mystery Island, and a retired pirate who lives there. When the children find counterfeit money and the bad guys return for it, the old pirate's clever plans keep the kids safe. 1981 75m/C VHS, Beta $14.98 VID **½

Mystery of the Leaping Fish/ Chess Fever Two silent shorts. One, filmed in 1916, is a surreal comedy-mystery, written by Todd Browning, about a doped-up detective named Coke Ennyday and his hallucinatory, sped-up adventures. The second, made in 1925 during the Moscow Chess Tournament, is a parody of professional chess playing and the fever with which it grips a country. Pudovkin's second directorial credit. 1925 64m/B Douglas Fairbanks Sr., Bessie Love, Alma Rubens; **Dir:** John Emerson, Vsevolod Pudovkin. VHS, Beta, 8mm $29.95 VYY, MRV **½

Mystery Liner Dead bodies are found aboard ocean liner and passengers are concerned about it. Slow-moving sea cruise. 1934 62m/B Noah Beery, Astrid Allwyn, Cornelius Keefe, Gustav von Seyffertitz, Edwin Maxwell, Booth Howard, George "Gabby" Hayes; **Dir:** William Nigh. VHS, Beta $16.95 NOS, SNC **½

Mystery Magician An anonymous magician demonstrates how many classic magic tricks are achieved. 1986 52m/C VHS, Beta $29.98 FOX **½

The Mystery Man A Chicago reporter goes on a drinking binge and ends up in St. Louis. There he stumbles upon a mystery which may involve the paper for which he works. He teams up with a beautiful woman to try to crack the case. 1935 65m/B Robert Armstrong, Maxine Doyle, Henry Kolker, Leroy Mason, James Burke; **Dir:** Ray McCarey. VHS, Beta $16.95 SNC **½

Mystery Mansion A fortune in gold and a hundred-year-old mystery lead three children into an exciting treasure hunt. Family fare. 1983 (PG) 95m/C Dallas McKennon, Greg Wynne, Jane Ferguson. VHS, Beta $19.95 MED *

The Mystery of the Mary Celeste Tale of terror based on the bizarre case of the "Marie Celeste," an American ship found adrift and derelict on the Atlantic Ocean on December 5, 1872. Also known as "The Phantom Ship." 1937 64m/B Bela Lugosi, Shirley Grey, Edmund Willard. VHS, Beta $24.95 CCB, MRV, VYY **

Mystery of the Million Dollar Hockey Puck Sinister diamond smugglers learn a lesson on ice from a pair of orphan lads. Features the National Hockey League's Montreal Canadiens. 198? 88m/C Michael MacDonald, Angele Knight, The Montreal Canadiens; **Dir:** Jean LaFleur, Peter Svatek. VHS, Beta $59.98 LTG **½

Mystery Mountain Twelve episodes depict the villain known as the "Rattler" attempting to stop the construction of a railroad over Mystery Mountain. 1934 156m/B Ken Maynard, Gene Autry, Smiley Burnette; **Dir:** Otto Brower, B. Reeves Eason. VHS, Beta $49.95 NOS, VCN, CAB **½

Mystery of Mr. Wong The largest star sapphire in the world, the "Eye of the Daughter of the Moon," is stolen from a museum in its home country of China. Mr. Wong becomes involved in trying to trace its trail and the perpetrator of the murders that follow in its wake. One in the series of detective films. 1939 67m/B Boris Karloff, Grant Withers, Dorothy Tree, Lotus Long; **Dir:** William Nigh. VHS $19.98 MOV, MRV, DVT **½

Mystery Plane An inventor devises a new bomb-dropping mechanism, and must evade secret agents on his way to deliver it to the government. 1939 60m/B John Trent, Marjorie Reynolds, Milburn Stone, Peter George Lynn, Polly Ann Young; **Dir:** George Waggner. VHS, Beta $19.95 NOS, DVT **½

Mystery Ranch A cowboy finds himself in strange predicaments out on the range. 1934 52m/B George O'Brien, Charles Middleton, Cecilia Parker; **Dir:** David Howard. VHS, Beta $19.95 NOS, DVT **½

Mystery of the Riverboat Louisiana and the Mississippi River are the background for these 13 episodes of mystery and murder. 1944 90m/B Robert Lowery, Eddie Quillan; **Dir:** Henry MacRae. VHS, Beta $29.95 UHV, MLB **½

Mystery Squadron Twelve chapters, 13 minutes each. Daredevil air action in flight against the masked pilots of the Black Ace. 1933 240m/B Bob Steele, Guinn Williams, Lucille Browne, Jack Mulhall, J. Carroll Naish, Jack Mower; **Dir:** Colbert Clark. VHS, Beta $49.95 NOS, VCN, DVT **½

Mystery in Swing An all-black mystery with music, about a trumpet player who has snake venom put on his mouthpiece. 1940 66m/B F.E. Miller, Monte Hawley, Marguerite Whitten, Tom Moore, Ceepee Johnson; **Dir:** Arthur Dreifuss. VHS, Beta $29.95 VYY **

Mystery Theatre This tape offers two separate Mark Saber crime dramas back to back: "The Case of the Chamber of Death" and "The Case of the Locked Room." 1951 53m/B Tom Conway, James Burke, Verne Smith. VHS, Beta $29.95 VYY Woof!

Mystery Train A run down hotel in Memphis is the scene for three vignettes concerning the visit of foreigners to the U.S. Themes of mythic Americana, Elvis, and life on the fringe pervade this hip and quirky film. The three vignettes all tie together in clever and funny overlaps. Waits fans should listen for his performance as a DJ. 1989 (R) 110m/C Masatoshi Nagase, Youki Kudoh, Jay Hawkins, Cinque Lee, Joe Strummer; **Dir:** Jim Jarmusch. **Voices:** Tom Waits. VHS $79.98 ORI ***

Mystery Trooper Adventures in the great wilderness full of prospecting, greed, and Indians all being observed by the Royal Mounties. Ten chapters of the serial at 20 minutes each. 1932 200m/B Robert Frazer, Buzz Barton. VHS $24.95 GPV, NOS, CAB **

Mystery of the Wax Museum Rarely seen, vintage horror classic about a wax-dummy maker who, after a disfiguring fire, resorts to murder and installs the wax-covered bodies of his victims in his museum. Famous for its pioneering use of two-strip Technicolor. Remade in 1953 in 3-D as "House of Wax." 1933 77m/C Lionel Atwill, Fay Wray, Glenda Farrell; **Dir:** Michael Curtiz. VHS, Beta, LV $19.98 MGM, MLB ***

Mystic Pizza Two sisters and their best friend, work at the Mystic Pizza Parlor in the seaside town of Mystic, Connecticut. One is uncertain about marriage and leaves her fiance at the altar, another falls for a rich preppie, and the third, the smart girl with college plans, falls for a married man. Throughout several months, they struggle with their hopes, their loves, the caste system which keeps townies and tourists separate, and their family rivalries. Charming performances featuring the screen debut of Roberts. 1988 (R) 101m/C Annabeth Gish, Julia Roberts, Lili Taylor, Vincent D'Onofrio, William R. Moses, Adam Storke; **Dir:** Daniel Petrie. VHS, Beta, LV $89.95 VIR ***

Nabonga The daughter of an embezzler, whose plane crashes in the jungle, befriends a gorilla who protects her. Soon a young man comes looking for the embezzler's cash, meets the woman and the ape, and together they go on a wonderful journey. Also known as "Gorilla." 1944 72m/B Buster Crabbe, Fifi D'Orsay, Barton MacLane, Julie London, Herbert Rawlinson; **Dir:** Sam Newfield. VHS $16.95 SNC, NOS, MOV **

Nadia Entertaining account of the life of Nadia Comaneci, the Romanian gymnast who earned six perfect tens with her stunning performance at the 1976 Olympic Games. Made for television. 1984 100m/C Talia Balsam, Jonathan Banks, Simone Blue, Johann Carlo, Carrie Snodgress; **Dir:** Alan Cooke. VHS, Beta $19.95 FHE **½

Nadine In Austin circa 1954, an almost divorced beautician witnesses a murder and goes undercover with her estranged husband to track down the killer, before he finds her. Plenty of low-key humor. Well-paced fun; Basinger is terrific. 1987 (PG) 83m/C Jeff Bridges, Kim Basinger, Rip Torn, Gwen Verdon, Glenne Headly, Jerry Stiller; **Dir:** Robert Benton. VHS, Beta, LV $19.98 FOX **½

Nail Gun Massacre A crazed killer with a penchant for nailing bodies to just about anything goes on a hammering spree. In horrific, vivid color. 1986 90m/C Rocky Patterson, Ron Queen, Beau Leland, Michelle Meyer; **Dir:** Bill Lesley. VHS, Beta $79.95 MAG Woof!

Nairobi Affair Savage and Heston fight poachers and each other in Kenya, when the father has an affair with his son's ex-wife. Mediocre, derivative script, but beautiful scenery. Made for television. 1988 (PG) 95m/C Charlton Heston, John Savage, Maud Adams, John Rhys-Davies, Connie Booth; **Dir:** Marvin J. Chomsky. VHS, Beta $79.95 VMK *½

Nais Vintage Pagnolian peasant drama. A hunchback sacrifices himself for the girl he loves. She escapes to another man's arms. Adapted from the Emile Zola novel "Nais Micoulin." In French with English subtitles. **1945 105m/B** *FR* Fernandel; *Dir:* Fernand Lauterier. **VHS, Beta, 8mm $29.95** *VYY, FCT, DVT* ✶✶✶

The Naked Angels Rape, mayhem, beatings, and road-hogging streetbikes highlight this biker film, acted, in part, by actual bikers, and in part by others who were actually (almost) actors. The gang war rages from Los Angeles to Las Vegas. Roger Corman was the executive producer. **1969 (R) 83m/C** Michael Greene, Richard Rust, Felicia Guy; *W/Dir:* Bruce Clark. **VHS, Beta $39.95** *CRV* ✶

Naked Cage Brutal women's prison film that is borderline soft-core, about an innocent country woman, a horseriding fanatic, who is framed for a robbery and sent to the slammer. There, she contends with Lesbian wardens and inmates. Disgusting. **1986 (R) 97m/C** Shari Shattuck, Angel Tompkins, Lucinda Crosby, Christina Whitaker; *Dir:* Paul Nicholas. **VHS, Beta $19.95** *COL* Woof!

The Naked Civil Servant Remarkable film, based on the life of flamboyant homosexual Quentin Crisp, who came out of the closet in his native England, long before his lifestyle would be tolerated by Britains. For mature audiences. **1975 80m/C** John Hurt; *Dir:* Jack Gold. British Academy Awards '80: Best Actor (Hurt). **VHS, Beta $19.99** *HBO* ✶✶✶½

The Naked and the Dead Sanitized adaptation of Norman Mailer's bestselling novel about army life in the Pacific during World War II. Resentment between officers and enlisted men proves to be almost as dangerous as the Japanese. Uneven acting and cliched characterizations make this an average outing. Don't expect to hear the profane language that sensationalized the book, because most of it was edited out of the screenplay. **1958 131m/C** Aldo Ray, Cliff Robertson, Joey Bishop, Raymond Massey, Lili St. Cyr, James Best; *Dir:* Raoul Walsh. **VHS, Beta $15.95** *UHV* ✶✶½

Naked Eyes This program presents the rock group Naked Eyes performing their hits "Always Something There to Remind Me," "Promises, Promises" and "When the Lights Go Out." **1983 14m/C** Naked Eyes. **VHS, Beta, LV $9.95** *SVS* ✶✶½

The Naked Face A psychiatrist tries to find out why his patients are being killed. He gets no help from police, who suspect he is the murderer. A dull affair featuring unusual casting of Moore, while Steiger's suspicious police captain character provides spark. **1984 (R) 105m/C** Roger Moore, Rod Steiger, Elliott Gould, Art Carney, Anne Archer, David Hedison; *Dir:* Bryan Forbes. **VHS, Beta $79.95** *MGM* ✶✶

The Naked Gershwin This is a unique program revolving around the joining of Gershwin's music and the words written for it by friends and admirers, studying its relation to jazz and classical modes. **1985 60m/C** **VHS, Beta** *SDS* ✶✶

The Naked Gun: From the Files of Police Squad More hysterical satire from the creators of "Airplane!" The short-lived television cop spoof "Police Squad"

moves to the big screen and has Lt. Drebin uncover a plot to assassinate Queen Elizabeth while she is visiting Los Angeles. Nearly nonstop gags and pratfalls provide lots of laughs. Nielsen is perfect as Drebin and the supporting cast is strong; cameos abound. **1988 (PG-13) 85m/C** Leslie Nielsen, Ricardo Montalban, Priscilla Presley, George Kennedy, O.J. Simpson, Nancy Marchand, John Houseman, Weird Al Yankovic, Reggie Jackson, Dr. Joyce Brothers; *Dir:* David Zucker. **VHS, Beta, LV, 8mm $14.95** *PAR* ✶✶✶

The Naked Gun 2 1/2: The Smell of Fear Lt. Drebin returns to rescue the world from a faulty energy policy devised by the White House and oil-lords. A notch down from the previous entry but still hilarious cop parody. Nielsen has this character down to a tee, and there's a laugh every minute. Laserdisc version is letterboxed and features dolby surround sound. **1991 (PG-13) 85m/C** Leslie Nielsen, Priscilla Presley, George Kennedy, O.J. Simpson, Robert Goulet; *W/Dir:* David Zucker. **VHS, Beta, LV, 8mm $92.95** *PAR, PIA* ✶✶½

Naked Hills Meandering tale about a man (Wayne) who, suffering from gold fever, searches for 40 years in 19th Century California and ends up out of luck, losing his wife and family. **1956 72m/C** David Wayne, Keenan Wynn, James Barton, Marcia Henderson, Jim Backus, Denver Pyle, Myrna Dell, Frank Fenton, Al "Fuzzy" Knight; *W/Dir:* Josef Shaftel. **VHS** *AVC* ✶½

Naked Is Better A couple of buffoons pose as hospital workers so they can play doctor with a hospitalized actress. **1973 93m/C** Alberto Olmedo, Jorge Porcel, Jorge Barreiro, Maria Casan. **VHS, Beta $39.95** *MED* ✶

Naked Jungle Suspenseful, well-done jungle adventure of a plantation owner (Heston) in South America and his North American mail-order bride (Parker), who do battle with a deadly, miles-long siege of red army ants. Realistic and worth watching. Produced by George Pal, shot by Ernest Laszlo and based on the story "Leiningen vs. the Ants" by Carl Stephenson. **1954 95m/C** Charlton Heston, Eleanor Parker, Abraham Sofaer, William Conrad; *Dir:* Byron Haskin. **VHS, Beta, LV $19.95** *PAR* ✶✶✶

Naked Kiss Fuller's most savage, hysterical film noir. A brutalized prostitute escapes her pimp and tries to enter respectable small-town society. She finds even more perversion and sickness there. **1964 92m/B** Constance Towers, Anthony Eisley, Michael Dante, Virginia Grey, Patsy Kelly, Betty Bronson; *Dir:* Samuel Fuller. **LV $79.95** *HMV, VYG* ✶✶½

Naked Lie A District Attorney and a judge engage in an ultra-steamy affair and inevitably clash when she is assigned as a prosecutor on the politically explosive case over which he is presiding. **1989 89m/C** Victoria Principal, James Farentino, Glenn Withrow, William Lucking, Dakin Matthews; *Dir:* Richard A. Colla. **VHS $79.95** *ING* ✶✶

Naked Lunch Whacked-out movie based on William S. Burroughs' autobiographical account of drug abuse, homosexuality, violence and weirdness set in the drug-inspired land called Interzone. Hallucinogenic images are carried to the extreme: typewriters metamorphose into beetles, bloblike creatures with sex organs scurry about, and characters mainline insecticide. Some of the characters

are clearly based on writers of the Beat generation, including Jane (Davis), Paul Bowles, Allen Ginsberg, and Jack Kerouac. **1991 (R) ?m/C** Peter Weller, Judy Davis, Ian Holm, Julian Sands, Roy Scheider; *W/Dir:* David Cronenberg. New York Film Critics Awards '91: Best Supporting Actress (Davis); National Society of Film Critics Awards '91: Best Director (Cronenberg), Best Screenwriting. **VHS $94.98** *TCF* ✶✶✶

Naked Massacre A maniac kills nurses, leaving no clues for the police to follow. **198? 90m/C** Matthieu Carriere, Carol Laurie; *Dir:* Denis Heroux. **VHS, Beta $19.98** *VDC* ✶

Naked Paradise A young woman moves in with her mother and step-sister and finds difficulty in adjusting. **1978 (R) 86m/C** Laura Gemser, Annie Belle. **VHS, Beta** *UHV* ✶

The Naked Prey Unnerving African adventure which contains some unforgettably brutal scenes. A safari guide (Wilde), leading a hunting party, must watch as an indigenous tribe murders all his companions, and according to their customs, allows him to be set free, sans clothes or weapons, to be hunted by the best of the tribe. **1965 96m/C** Cornel Wilde, Gertrude Van Der Berger, Ken Gampu; *Dir:* Cornel Wilde. **VHS, Beta, LV $19.95** *PAR* ✶✶✶

The Naked Spur A compulsive bounty hunter tracks down a vicious outlaw and his beautiful girlfriend. An exciting film from the Mann-Stewart team and considered one of their best, infusing the traditional western with psychological confusion. Wonderful use of Rockies locations. Received Oscar nomination for best screenplay. **1952 93m/C** James Stewart, Robert Ryan, Janet Leigh, Millard Mitchell; *Dir:* Anthony Mann. **VHS, Beta $19.98** *MGM, TLF* ✶✶✶½

Naked in the Sun The true story of events leading to the war between the Osceola and Seminole Indians and a slave trader. Somewhat slow. **1957 95m/C** James Craig, Lita Milan, Barton MacLane, Tony Hunter; *Dir:* R. John Hugh. **VHS $19.98** *REP* ✶✶

Naked Sun Beguiling mystery set during the Carnival at Rio chock full of suspense, romance, and just a touch of the supernatural. **19?? 90m/C** **VHS $39.95** *CON* ✶✶

The Naked Sword of Zorro Deadly Don Luis taxes the town by day and violates the women by night, but Zorro takes a stand and saves the town. When he's not dressed as Zorro, he is Don Diego, an inept fool, but when he puts that suit on, look out. The ladies can't keep their hands off of him. There are more lusty females than villains, but there's still lots of swordplay and action. **1969 94m/C** **VHS $19.95** *VDM* ✶

The Naked Truth A greedy publisher tries to get rich quick by publishing a scandal magazine about the "lurid" lives of prominent citizens. Well-drawn characters in an appealing offbeat comedy. Also released as "Your Past is Showing." **1958 92m/C** *GB* Peter Sellers, Terry-Thomas, Shirley Eaton, Dennis Price; *Dir:* Mario Zampi. **VHS, Beta $59.95** *SUE* ✶✶½

Naked Vengeance A woman's husband is murdered, she is beaten and raped, and she becomes a vengeful, vicious killing machine. Gratuitous violence, distasteful to the max.

1985 (R) 97m/C Deborah Tranelli, Kaz Garaz, Bill McLaughlin; *Dir:* Cirio H. Santiago. **VHS, Beta** $79.98 *LIV, LTG* Woof!

Naked Venus Ulmer's last film is a slow-paced tale about an American artist, married to a model who poses in the nude and belongs to a nudist camp, whose wealthy mother tries to tear the marriage apart. Discrimination and intolerance of nudist camps is explored.
1958 80m/B Patricia Conelle; *Dir:* Edgar G. Ulmer. **VHS, Beta** $19.95 *VDM* 🎬½

Naked Warriors Ancient Romans capture beautiful women from around the world and force them to compete in gladiatorial games. Originally titled "The Arena," this New World exploitation gem features a mostly Italian cast, including Bay, who starred in the previous year's "Lady Frankenstein."
1973 (R) 75m/C Margaret Markov, Pam Grier, Lucretia Love, Paul Muller, Daniel Vargas, Marie Louise, Mary Count, Sara (Rosalba Neri) Bay, Vic Karis, Sid Lawrence, Peter Cester, Anna Melita; *Dir:* Steve Carver. **VHS, Beta** $79.95 *MGM* 🎬🎬

Naked Youth Seedy drive-in cheapie about two punks breaking out of juvenile prison and heading south of the border on a trail filled with crime, drugs and loose women. Also called "Wild Youth."
1959 80m/B Robert Hutton, John Goddard, Carol Ohmart, Jan Brooks, Robert Arthur, Steve Rowland, Clancy Cooper; *Dir:* John F. Schreyer. **VHS, Beta** $9.95 *RHI, SNC* Woof!

Nam Angels Hell's Angels enter Southeast Asia and rescue POWs.
1988 (R) 91m/C Brad Johnson, Vernon Wells, Kevin Duffis. **VHS, Beta** $79.95 *MED* 🎬

A Name for Evil Having grown tired of city living, a couple moves to an old family estate out in the country near the Great Lakes. Strange sounds in the night are the first signs of the terror to come.
1970 (R) 74m/C Robert Culp, Samantha Eggar, Sheila Sullivan, Mike Lane; *Dir:* Bernard Girard. **VHS, Beta** *PGN* 🎬🎬

The Name of the Rose An exhaustive, off-center adaptation of the bestselling Umberto Eco novel about violent murders in a 14th-century Italian abbey. An English monk struggles against religious fervor in his quest to uncover the truth.
1986 (R) 128m/C Sean Connery, F. Murray Abraham, Christian Slater, Ron Perlman, William Hickey, Feodor Chaliapin, Elya Baskin, Michael Lonsdale; *Dir:* Jean-Jacques Annaud. British Academy Awards '87: Best Actor (Connery). **VHS, Beta, LV** $14.98 *SUE* 🎬🎬½

Nana French version of Emile Zola's novel about an actress-prostitute who seduces the high society of Paris in the late 1880s, and suffers a heart-breaking downfall and death. Film has three remakes of the original 1926 version.
1955 118m/C *FR* Charles Boyer, Martine Carol. **VHS, Beta** $24.95 *NOS, DVT* 🎬🎬🎬

Nana A lavishly photographed, graphic adaptation of the Emile Zola novel about a Parisian whore who acquires power by seducing the richest and politically formidable men. Dubbed.
1982 92m/C *IT* Katya Berger, Jean-Pierre Aumont; *Dir:* Dan Wolman. **VHS, Beta** $79.95 *FCT, MGM* 🎬🎬½

Nanami, First Love A controversial Japanese film about a man madly in love with a beautiful prostitute he cannot satisfy sexually. Explicit; in Japanese with English subtitles.
1968 104m/B *JP* *Dir:* Susumu Hani. **VHS, Beta, 8mm** $29.95 *VYY, FCT* 🎬🎬🎬

Nancy Astor A portrait of the beautiful, spirited Virginian who became a member of the British Parliament.
1982 55m/C *GB* Lisa Harrow, James Fox, Dan O'Herlihy, Sylvia Syms. **VHS, Beta** *TLF* 🎬🎬

Nancy Goes to Rio Actresses Powell and Sothern compete for the same part in a play and for the same man. Catch is, they're mother and daughter. Zany consequences ensue. Sothern's last film for MGM. Songs include "Time and Time Again," "Shine on Harvest Moon," "Cha Boom Pa Pa," "Yipsee-I-O," "Magic is in the Moonlight" and more.
1950 99m/C Ann Sothern, Jane Powell, Barry Sullivan, Carmen Miranda, Louis Calhern, Scotty Beckett, Hans Conried, Glenn Anders; *Dir:* Robert Z. Leonard. **VHS** $19.98 *MGM, MLB, CCB* 🎬🎬½

Napoleon A vivid, near-complete restoration of Gance's epic silent masterpiece about the famed conqueror's early years, from his youth through the Italian Campaign. An innovative, spectacular achievement with its use of multiple split screens, montage, color and triptychs. Produced by Kevin Brownlow with a score by Carmine Coppola and given a gala theatrical re-release in 1981. Remade in 1955.
1927 235m/B *FR* Albert Dieudonne, Antonin Artaud, Pierre Batcheff, Gina Manes, Armand Bernard, Harry Krimer, Albert Bras, Abel Gance, Georges Cahuzac; *Dir:* Abel Gance. **VHS, Beta, LV** $29.95 *MCA* 🎬🎬🎬🎬

Napoleon Depicts life story of Napoleon from his days as a soldier in the French army to his exile on the Island of Elba. Falls short of the fantastic 1927 silent classic as it attempts to delve into Napoleon the man, as opposed to his conquests. Some might find it too slow for their tastes.
1955 123m/C *FR* Raymond Pellegrin, Orson Welles, Maria Schell, Yves Montand, Erich von Stroheim, Jean Gabin, Jean-Pierre Aumont; *Dir:* Sacha Guitry. **VHS, Beta, LV** $19.95 *NOS, MRV, MPI* 🎬🎬

Napoleon and Samantha Disney adventure about Napoleon, an orphan (Whitaker), who is befriended by Danny, a college student (Douglas). After the death of his grandfather, Napoleon decides to take Major, his elderly pet lion, and follow Danny, who is herding goats for the summer, across the American northwest mountains and they are joined by Samantha (Foster in her film debut). Worth watching.
1972 91m/C Jodie Foster, Johnny Whitaker, Michael Douglas, Will Geer, Henry Jones; *Dir:* Bernard McEveety. **VHS, Beta** $69.95 *DIS* 🎬🎬🎬

Narda or The Summer Two young men decide they will share a girlfriend during a summer romance, only Narda has other plans. She decides that both men will belong to her rather than the other way around. In Spanish with English subtitles.
1976 96m/C *MX* Hector Bonilla, Enrique Alvarez Felix; *Dir:* Juan Guerreo. **VHS** $24.98 *TAM* 🎬🎬

Narrow Margin Well-made, harrowing action adventure about a cop who is in charge of transporting a gangster's widow to a trial where she is to testify. On the train he must try to keep her safe from the hit-men who would murder her. A real cat and mouse game—one of the best suspense movies of the '50s.
1952 71m/B Charles McGraw, Marie Windsor, Jacqueline White, Queenie Leonard, Gordon Gebert, Don Beddoe, Harry Harvey; *Dir:* Alfred E. Green. **VHS** $19.98 *TTC, MLB* 🎬🎬🎬

Narrow Margin Archer reluctantly agrees to testify against the mob after witnessing a murder, and Los Angeles D.A. Hackman is assigned to protect her on the train through the Rockies back to L.A. Bad idea, she finds out. No match for its '52 predecessor.
1990 (R) 99m/C Gene Hackman, Anne Archer, James B. Sikking, J.T. Walsh, M. Emmet Walsh; *Dir:* Peter Hyams. **VHS, LV** $19.95 *LIV* 🎬🎬

Narrow Trail A silent classic that still stands as one of the best Westerns ever made. A tough cowboy with a troubled past hopes a good woman can save him. But she's seen trouble, too. They meet up again in San Francisco's Barbary Coast. A horse race might hold the answer.
1917 56m/B William S. Hart, Sylvia Bremer, Milton Ross; *Dir:* William S. Hart. **VHS, Beta** $15.95 *NOS, GPV, CCB* 🎬🎬🎬½

Nashville Altman's stunning, brilliant film tapestry that follows the lives of 24 people during a political campaign/music festival in Nashville. Seemingly extemporaneous vignettes, actors playing themselves (Elliott Gould and Julie Christie), funny, touching, poignant character studies concerning affairs of the heart and despairs of the mind. Repeatedly blurs reality and fantasy. "I'm Easy" by Carradine won Oscar for best song.
1975 (R) 159m/C Keith Carradine, Lily Tomlin, Henry Gibson, Ronee Blakley, Keenan Wynn, David Arkin, Geraldine Chaplin, Lauren Hutton, Shelley Duvall, Barbara Harris, Allen Garfield, Karen Black, Christina Raines; *Dir:* Robert Altman. Academy Awards '75: Best Song ("I'm Easy"); British Academy Awards '75: Best Musical Score; National Board of Review Awards '75: Best Director (Altman), Best Film, Best Supporting Actress (Blakley); National Society of Film Critics Awards '75: Best Director (Altman), Best Film, Best Supporting Actor (Gibson), Best Supporting Actress (Tomlin). **VHS, Beta, LV** $79.95 *PAR* 🎬🎬🎬🎬

Nashville Beat Gang targets the citizens of Nashville, Tennessee, until two tough cops stomp the gang's delusion that Southern folk are easy marks.
1989 110m/C Kent McCord, Martin Milner; *Dir:* Bernard L. Kowalski. **VHS** $59.95 *NAC* 🎬🎬

Nashville Girl Innocent country girl who wants to make good in the country music scene, rises to the top of the Nashville success ladder, by compromising everything and hanging out with stereotyped no-goods in the show biz world. Also known as "Country Music Daughter," it's as good as a yawn.
1976 (R) 90m/C Monica Gayle, Johnny Rodriguez; *Dir:* Gus Trikonis. **VHS, Beta** $59.98 *CHA* 🎬

Nashville Goes International A series of concert clips from country music's superstars as they travel and perform all over the world.
1989 59m/C **VHS, Beta** $9.95 *MVD, CAF* 🎬

Nashville Story Backstage look at the Grand Ole Opry, featuring copious amounts of boot-stompin' music.
1986 70m/C Minnie Pearl, Earl Scruggs, Roy Acuff, Dolly Parton. **VHS, Beta** $39.95 *INV* 🎬

The Nasty Girl A bright young German model plans to enter a national essay contest on the topic of her hometown's history during the Third Reich. While researching the paper, she's harassed and even brutalized, but refuses to cease her sleuthing. Excellent performances, tight direction, and Oscar-nominated drama with comedic touches that charmingly imparts an important message. Based on a true story. In German with English subtitles.
1990 (PG-13) 93m/C *GE* Lena Stolze, Monika Baumgartner, Michael Gahr; **W/Dir:** Michael Verhoeven. VHS *HBO ♫♫♫½*

Nasty Habits Broad farce depicting a corrupt Philadelphia convent as a satiric parallel to Watergate, with nuns modeled on Nixon, Dean and Mitchell. Includes cameo appearances by various media personalities.
1977 (PG) 92m/C *GB* Glenda Jackson, Geraldine Page, Anne Jackson, Melina Mercouri, Sandy Dennis, Susan Penhaligon, Anne Meara, Edith Evans, Rip Torn, Eli Wallach, Jerry Stiller; **Dir:** Michael Lindsay-Hogg. VHS, Beta $69.95 *MED ♫♫*

Nasty Hero Hard-boiled ex-con comes to Miami to avenge the car thieves who framed him and got him sent to the slammer. This is car-chase scene heaven—little plot or drama.
1989 (PG-13) 79m/C Robert Sedgwick, Carlos Palomino, Scott Feraco, Mike Starr, Rosanna DaVon; **Dir:** Nick Barwood. VHS, Beta, LV $79.95 *COL ♫½*

Nasty Rabbit Ridiculous spoof about Soviet spies intent on releasing a rabbit with a hideous Communist disease into the U.S. Everyone gets in on the chase: Nazi forces, cowboys and indians, sideshow freaks, banditos - what is the point? Also known as "Spies-A-Go-Go."
1964 88m/C Arch Hall Jr., Micha Terr, Melissa Morgan, John Akana; **Dir:** James Landis. VHS, Beta $59.95 *RHI ♫*

Natas es Satan Adventures of a corrupt New York cop who moonlights as a psychopath.
19?? 90m/C *SP* Miguel A. Alvarez, Perla Faith. VHS, Beta *MAV*

Natas...The Reflection A reporter persists in validating an Indian demon myth.
1983 (PG) 90m/C Randy Mulkey, Pat Bolt, Craig Hensley, Kelli Kuhn. VHS, Beta $29.95 *IVE ♫*

Nate and Hayes Set during the mid-1800s in the South Pacific, the notorious real-life swashbuckler Captain "Bully" Hayes ("good pirate") helps young missionary Nate recapture his fiancee from a cutthroat gang of evil slave traders. Entertaining "jolly rogers" film.
1983 (PG) 100m/C *NZ* Tommy Lee Jones, Michael O'Keefe, Max Phipps, Jenny Seagrove; **Dir:** Ferdinand Fairfax. VHS, Beta, LV $14.95 *PAR ♫♫½*

Nathalie Comes of Age Soft-core Italian flick in which a young girl is lured into making a dirty movie by a lustful entrepreneur. Dubbed.
1986 93m/C *IT* Marcella Petri, Grazia de Giorgi. VHS, Beta $29.95 *MED Woof!*

A Nation Aflame Anti-Klan film is based on a story by Thomas L. Dixon, the author of "The Clansman," from which D.W. Griffith made "The Birth of a Nation." In this story, a con man gains control of the state's government through an ultra-nationalist secret society, and eventually pays for his corruption.
1937 70m/B Lila Lee, Noel Madison, Snub Pollard; **Dir:** Norma Trelvar. VHS, Beta $16.95 *SNC, NOS, GVV ♫½*

National Adultery The eternal triangle is sketched in this comedy for adults only, starring 1981 Miss Spain.
19?? 90m/C *IT* VHS, Beta *VCD ♫½*

National Lampoon's Animal House Classic Belushi vehicle running amuck. Set in 1962 and responsible for launching Otis Day and the Knights and defining cinematic food fights. Every college tradition from fraternity rush week to the homecoming pageant is irreverently and relentlessly mocked in this wild comedy about Delta House, a fraternity on the edge. Climaxes with the homecoming parade from hell. Sophomoric, but very funny, with a host of young stars who went on to more serious work. Remember: "Knowledge is good."
1978 (R) 109m/C John Belushi, Tim Matheson, John Vernon, Donald Sutherland, Peter Riegert, Stephen Furst, Bruce McGill, Mark Metcalf, Verna Bloom, Karen Allen, Tom Hulce, Mary Louise Weller, Kevin Bacon; **Dir:** John Landis. VHS, Beta, LV $14.95 *MCA ♫♫♫½*

National Lampoon's Christmas Vacation The third vacation for the Griswold family finds them hosting repulsive relatives for Yuletide. The sight gags, although predictable, are sometimes on the mark. Quaid is a standout as the slovenly cousin.
1989 (PG-13) 93m/C Chevy Chase, Beverly D'Angelo, Randy Quaid, Diane Ladd, John Randolph, E.G. Marshall, Doris Roberts, Julia Louis-Dreyfus, Mae Questel, William Hickey, Brian Doyle-Murray, Juliette Lewis, Johnny Galecki, Nicholas Guest, Miriam Flynn; **Dir:** Jeremiah S. Chechik. VHS, Beta, LV, 8mm $19.98 *WAR ♫♫½*

National Lampoon's Class of '86 This is a live stage comedy show written and sponsored by the famous adult humor magazine, featuring a new cast of comics.
1986 86m/C Rodger Bumpass, Veanne Cox, Annie Golden, John Michael Higgins, Tommy Koenig. VHS, Beta $29.95 *PAR ♫♫½*

National Lampoon's Class Reunion A class reunion with some very wacky guests and a decided lack of plot or purpose. Things go from bad to worse when a crazed killer decides to join in on the festivities. Very few laughs.
1982 (R) 85m/C Shelley Smith, Gerrit Graham, Michael Lerner; **Dir:** Michael Miller. VHS, Beta, LV $29.98 *LIV, VES Woof!*

National Lampoon's European Vacation Sappy sequel to "Vacation" that has witless Chase and his family bumbling around in the land "across the pond." The Griswolds nearly redefine the term "ugly American." Stonehenge will never be the same.
1985 (PG-13) 94m/C Chevy Chase, Beverly D'Angelo, Dana Hill, Jason Lively, Victor Lanoux, John Astin; **Dir:** Amy Heckerling. VHS, Beta, LV, 8mm $19.98 *WAR ♫½*

National Lampoon's Vacation The Clark Griswold (Chase) family of suburban Chicago embarks on a westward cross-country vacation via car to the renowned "Wally World." Ridiculous and hysterical misadventures, including a falling asleep at the wheel sequence and the untimely death of Aunt Edna.
1983 (R) 98m/C Chevy Chase, Beverly D'Angelo, Imogene Coca, Randy Quaid, Christie Brinkley, Stacy Keach, Anthony Michael Hall, John Candy, Eddie Bracken, Brian Doyle-Murray, Eugene Levy; **Dir:** Harold Ramis. VHS, Beta, LV $19.98 *WAR ♫♫♫*

National Velvet A young English girl wins a horse in a raffle and is determined to train it to compete in the famed Grand National race. Taylor, only 12 at the time, is superb in her first starring role. Rooney also gives a fine performance. Filmed with a loving eye on lushly decorated sets, this is a masterpiece version of the story of affection between a girl and her pet. Based on the novel by Enid Bagnold and followed by "International Velvet" in 1978. Additional Oscar action includes nominations for Best Director, Best Color Cinematography, and Color Interior Decoration.
1944 124m/C Elizabeth Taylor, Mickey Rooney, Arthur Treacher, Donald Crisp, Anne Revere, Angela Lansbury, Reginald Owen; **Dir:** Clarence Brown. Academy Awards '45: Best Film Editing, Best Supporting Actress (Revere). VHS, Beta, LV, 8mm $19.98 *KUI, MGM, TLF ♫♫♫♫*

Native Son A young black man from the ghettos of Chicago is hired as a chauffeur by an affluent white family. His job is to drive their head-strong daughter anywhere she wants to go. Unintentionally, he kills her, tries to hide, and is ultimately found guilty. Based on the classic novel by Richard Wright, who also stars. Remade in 1986. Unprofessional direction and low budget work against the strong story.
1951 91m/B Richard Wright, Jean Wallace, Nicholas Joy, Gloria Madison, Charles Cane; **Dir:** Pierre Chenal. VHS $24.95 *NOS, LTG ♫♫½*

Native Son This second film adaptation of the classic Richard Wright novel is chock full of stars and tells the story of a poor black man who accidentally kills a white woman and then hides the body. Changes in the script soft-soap some of the novel's disturbing truths and themes—so-so drama for those who have not read the book.
1986 (R) 111m/C Geraldine Page, Oprah Winfrey, Matt Dillon, John Karlen, Elizabeth McGovern, Akosua Busia, Carroll Baker, Victor Love; **Dir:** Jerrold Freedman. VHS, Beta, LV $39.95 *KUI, LIV, LTG ♫♫½*

Nativity Unmemorable, made-for-TV portrayal of the romance between Mary and Joseph (of the Bible story), Joseph's response to Mary's pregnancy, and the birthing of Jesus. Not exactly a Bible epic, the direction and acting are uninspired.
1978 97m/C John Shea, Madeleine Stowe, Jane Wyatt, John Rhys-Davies, Kate O'Mara; **Dir:** Bernard L. Kowalski. VHS, Beta $59.98 *FOX ♫♫*

The Natural A beautifully filmed movie about baseball as myth. A young man, whose gift for baseball sets him apart, finds that trouble dogs him, particularly with a woman (fine cameo by Hershey). In time, as an aging rookie, he must fight against his past to lead his team to the World Series, and win the woman who is meant for him. From the Bernard Malamud story with a musical score by Randy Newman.
1984 (PG) 134m/C Robert Redford, Glenn Close, Robert Duvall, Kim Basinger, Wilford Brimley, Barbara Hershey, Richard Farnsworth, Robert Prosky, Darren McGavin, Joe Don Baker, Michael Madsen; **Dir:** Barry Levinson. VHS, Beta, LV $19.95 *COL, PIA ♫♫♫*

Natural Enemies A successful publisher begins to consider suicide and murder as the cure for his family's increasing alienation and despair. A dark and depressing drama written by Kanew.
1979 (R) 100m/C Hal Holbrook, Louise Fletcher, Jose Ferrer, Viveca Lindfors; *Dir:* Jeff Kanew; *W/Dir:* Jeff Kanew. **VHS, Beta $59.95** COL 🎞🎞½

Natural States Mountains, forests, and coastal regions drift by to the relaxing strains of new age music by David Lanz and Paul Speer.
1988 45m/C *Dir:* Jan Nickman. **VHS, Beta $29.95** MIR, TTE, VTK 🎞🎞½

Natural States Desert Dream Deserts, caves, rain, and dripping stalactites drift by to the relaxing strains of electronic music by Tangerine Dream.
1988 50m/C Tangerine Dream. **VHS, Beta $29.95** MIR, TTE, VTK 🎞🎞½

Nature's Playmates A beautiful private eye tours Florida nudist camps in search of a missing man with a distinctive tattoo on his posterior. One of H.G. Lewis' obscure "nudie" flicks, sexually tame by modern standards, awful by any standards.
1962 56m/B Vicki Miles, Scott Osborne; *Dir:* Herschell Gordon Lewis. **VHS $19.95** VDM 🎞

Naughty Co-Eds College girls tell each other their own first love experiences. Low-brow fun.
198? (R) 90m/C **VHS, Beta $39.95** WES Woof!

Naughty Knights Soft-core romp in a medieval setting. Toss it in the moat. Originally titled, "Up the Chastity Belt."
1971 (PG) 94m/C Frankie Howerd, Graham Crowden, Bill Fraser, Roy Hudd, Hugh Paddick, Anna Quayle, Eartha Kitt, Dave King, Fred Emney; *Dir:* Bob Kellett. **VHS, Beta $19.95** ACA 🎞

Naughty Marietta A French princess switches identities with a mail-order bride to escape from her arranged marriage, and is captured by pirates. When she's saved by a dashing Indian scout, it's love at first sight. The songs by Victor Herbert include "Ah, Sweet Mystery of Life," "Tramp, Tramp, Tramp," "Italian Street Song," and "I'm Falling in Love With Someone." The first MacDonald-Eddy match-up and very popular in its day. Received Oscar nominations for Best Picture.
1935 106m/B Jeanette MacDonald, Nelson Eddy, Frank Morgan, Elsa Lanchester, Douglas Dumbrille, Cecilia Parker. Academy Awards '35: Best Sound. **VHS, Beta, LV $19.95** MGM 🎞🎞½

Naughty Negligee Nights Nearly naked nymphs nestle naughtily in nice nighties.
1988 60m/C **VHS, Beta $24.95** AHV 🎞🎞½

The Naughty Nineties Bud and Lou help a showboat owner fend off crooks in the 1890s. Usual slapstick shenanigans, but highlighted by verbal banter. Includes the first on-screen rendition of the classic "Who's on First?" routine.
1945 72m/B Bud Abbott, Lou Costello, Henry Travers, Alan Curtis, Joe Sawyer, Rita Johnson, Joe Kirk, Lois Collier; *Dir:* Jean Yarborough. **VHS, Beta $14.95** MCA, MLB 🎞🎞

Naughty Roommates A pair of seductive women share everything from clothing, to housework, to boyfriends and back.
19?? 90m/C **VHS $39.95** ACF 🎞🎞

Navajo Joe The sole survivor of a massacre single-handedly kills each person involved in the atrocity, and aids a terrorized, though unappreciative, town in the process. Low-budget Spanish-Italian western only worth watching because of Reynolds. Filmed in Spain.
1967 89m/C *IT SP* Burt Reynolds, Aldo Sambrel, Tanya Lopert, Fernando Rey; *Dir:* Sergio Corbucci. **VHS** TAV 🎞½

The Navigator Ever the quick thinker, Keaton actually bought a steamer headed for the scrap heap and used it to film an almost endless string of sight gags. Rejected by a socialite, millionaire Keaton finds himself alone with her on the abandoned boat. As he saves her from various and sundry perils, the gags and thrills abound—including one stunt that was inspired by a near-accident on the set. Too bad they don't make 'em like this anymore. Silent.
1924 60m/B Buster Keaton, Kathryn McGuire, Frederick Vroom, Noble Johnson, Clarence Burton, H. M. Clugston; *Dir:* Buster Keaton, Donald Crisp. **VHS $49.95** EJB 🎞🎞🎞

The Navigator A creative time-travel story of a 14th-century boy with visionary powers who leads the residents of his medieval English village away from a plague by burrowing through the earth's core and into late 20th-Century New Zealand. Quite original and refreshing.
1988 (PG) 92m/C *NZ* Hamish McFarlane, Bruce Lyons, Chris Haywood, Marshall Napier, Noel Appleby; *Dir:* Vincent Ward. Australian Film Institute '88: Best Film. **VHS $89.95** TRY 🎞🎞🎞

The Navy Comes Through Salty sailors aboard a mangy freighter beat the odds and sink Nazi warships and subs with the greatest of ease. Boy meets girl subplot is added for good measure. Morale booster includes newsreel footage.
1942 81m/B Pat O'Brien, George Murphy, Jane Wyatt, Jackie Cooper, Carl Esmond, Max Baer, Desi Arnaz Sr., Ray Collins, Lee Bonnell, Frank Jenks; *Dir:* Edward Sutherland. **VHS $19.98** TTC 🎞🎞

Navy SEALS A group of macho Navy commandos, whose regular work is to rescue hostages from Middle Eastern underground organizations, finds a stash of deadly weapons. They spend the balance of the movie attempting to destroy the arsenal. Sheen chews the scenery as a crazy member of the commando team. Lots of action and violence, but simplistic good guys - bad guys philosophy and plot weaknesses keep this from being more than below average.
1990 (R) 113m/C Charlie Sheen, Michael Biehn, Joanne Whalley-Kilmer, Rick Rossovich, Cyril O'Reilly, Bill Paxton, Dennis Haysbert, Paul Sanchez, Ron Joseph, Nicholas Kadi; *Dir:* Lewis Teague. **VHS, Beta, LV $14.98** ORI 🎞½

Navy vs. the Night Monsters When horrifying, acid-secreting plant monsters try to take over the world, the Navy must come to the rescue. Unfortunately the whole Defense Department couldn't save this one.
1966 87m/C Mamie Van Doren, Anthony Eisley, Pamela Mason, Bobby Van; *Dir:* Michael Hoey. **VHS, Beta $42.95** PGN 🎞

Navy Way Hurriedly produced war propaganda film in which a boxer gets inducted into the Navy just before his title shot.
1944 74m/B Robert Lowery, Jean Parker; *Dir:* William H. Pine, William C. Thomas. **VHS $24.95** NOS, DVT 🎞

Nazarin Bunuel's scathing indictment of Christianity finds its perfect vehicle in the adventures of a defrocked priest attempting to relive Christ's life. Gathering a group of disciples, he wanders into the Mexican desert as a cross between Christ and Don Quixote. Filmed in Mexico, this is Bunuel at his grimmest. English subtitles.
1958 92m/B *MX* Francisco Rabal, Rita Macedo, Margo Lopez, Rita McLedo, Ignacio Tarso, Jesus Fernandez; *Dir:* Luis Bunuel. Cannes Film Festival '59: Special Prize. **VHS, Beta, LV $69.95** MAD 🎞🎞🎞

Nea Comic drama about a young girl from a privileged home who writes a best selling pornographic novel anonymously. When her book writing talents are made public through betrayal, she gets even in a most unique way. Subtitled "A Young Emmanuelle." Kaplan also performs. Appealing and sophisticated.
1978 (R) 101m/C Sami Frey, Ann Zacharias, Heinz Bennent; *Dir:* Nelly Kaplan. **VHS, Beta $59.95** COL 🎞🎞½

Near Dark A farm boy falls in unwillingly with a family of thirsty, outlaw-fringe vampires who roam the West in a van. The first mainstream effort by Bigelow, and a rollicking, blood-saturated, slaughterhouse of a movie, with enough laughs and stunning imagery to revive the genre. Music by Tangerine Dream.
1987 (R) 95m/C Adrian Pasdar, Jenny Wright, Bill Paxton, Jenette Goldstein, Lance Henriksen, Tim Thomerson, Joshua Miller; *Dir:* Kathryn Bigelow. **VHS, Beta, LV $19.99** HBO, IME 🎞🎞🎞

Near Misses Missable farce with Reinhold as a bigamist in the foreign service. To make time for a mistress he swaps identities with a young Marine in Paris, but then the KGB kidnaps the imposter. Doors slam in hallways all over the place as the performers crank up the mixups to maximum frenzy, but with little effect since you really don't care much about these fools.
1991 93m/C Judge Reinhold, Casey Siemaszko; *Dir:* Baz Taylor. **VHS, Beta, LV $89.98** FXV 🎞½

'Neath the Arizona Skies Low budget oater epic about a cowhand who finds all the action he and his friends can handle as they try to rescue a young Indian girl (and oil heiress) who has been kidnapped. 'Nuff said.
1934 54m/B John Wayne, George "Gabby" Hayes, Sheila Terry; *Dir:* Henry Frazer. **VHS $12.98** REP, NOS, SVS 🎞🎞

'Neath Brooklyn Bridge The Boys from the Bowery get tangled up in crime when they try to help a young girl whose guardian was murdered.
1942 61m/B Leo Gorcey, Huntz Hall, Bobby Jordan, Sammy Morrison, Anne Gillis, Noah Beery Jr., Marc Lawrence, Gabriel Dell; *Dir:* Wallace Fox. **VHS $19.95** NOS, MRV, PME 🎞🎞

'Neath Canadian Skies Mounted lawman who says "eh" a lot tracks murdering gang of mine looters.
1946 41m/B *CA* Russell Hayden, Inez Cooper, Cliff Nazarro, Kermit Maynard, Jack Mulhall. **VHS, Beta, LV, 8mm $24.95** VYY 🎞

Necessary Parties Based on Barbara Dana's book that has a 15-year-old boy filing a lawsuit to stop his parents' divorce. Originally aired on PBS as part of the Wonder-Works Family Movie series.
1988 120m/C Alan Arkin, Mark Paul Gosselaar, Barbara Dana, Adam Arkin, Donald Moffatt, Julie Hagerty, Geoffrey Pierson, Taylor Fry; *Dir:* Gwen Arner. **VHS $29.95** PME, WNE 🎞

Necessary Roughness After losing their NCAA standing, the Texas State University Armadillos football team looks like it's headed for disaster. The once proud football factory is now composed of assorted goof-balls instead of stud players. But hope arrives in the form of a 34-year-old farmer with a golden arm, ready to recapture some lost dreams at quarterback. Can this unlikey team rise above itself and win a game, or will their hopes be squashed all over the field?
1991 (PG-13) 108m/C Scott Bakula, Robert Loggia, Harley Jane Kozak, Sinbad, Hector Elizondo, Kathy Ireland; ***Dir:*** Stan Dragoti. **VHS, Beta, LV** *PAR* ☠☠

Necromancer: Satan's Servant After a woman is brutally raped, she contacts a sorceress and makes a pact with the devil to insure her successful revenge. Plenty of graphic violence and nudity.
1988 (R) 90m/C Elizabeth Cayton, Russ Tamblyn, Rhonda Dorton; ***Dir:*** Dusty Nelson. **VHS, Beta, LV $79.98** *MCG, IME* ☠☠

Necropolis A witch burned at the stake 300 years ago is brought back to life as a motorcycle punkette in New York searching for a sacrificial virgin (in New York?). New wave horror should have been watchable.
1987 (R) 77m/C Leeanne Baker, Michael Conte, Jacquie Fritz, William Reed, Paul Ruben; ***Dir:*** Bruce Hickey. **VHS, Beta $79.98** *LIV, VES, LTG* ☠

Nefertiti, Queen of the Nile Woman, married against her will, turns into Nefertiti. For fans of Italo-Biblical epics only.
1964 97m/C *IT* Jeanne Crain, Vincent Price, Edmund Purdom, Amadeo Nazzari, Liana Orfei; ***Dir:*** Fernando Cerchio. **VHS** *LTG* ☠☠

Negatives A couple indulge in sexual fantasies to dangerous extremes, with the man first impersonating Dr. Crippen, the early 1900s murderer, and later a World War I flying ace. When a photographer friend joins the fun, the action careens off the deep end.
1968 90m/C *GB* Peter McEnery, Diane Cilento, Glenda Jackson, Maurice Denham, Norman Rossington; ***Dir:*** Peter Medak. **VHS $39.98** *CGI* ☠☠

The Neighborhood Thief Two thieves team up for the heist of their lives, but can they trust each other?
1983 84m/C *MX* Alfonso Zayas, Alberto Rojas, Angelica Chain. **VHS, Beta $74.95** *VVC* ☠☠

Neighbors The Keeses (Belushi and Walker) live in a quiet, middle-class suburban neighborhood where life is calm and sweet. But their new neighbors, Ramona and Vic (Aykroyd and Moriarty), prove to be loud, obnoxious, crazy, and freeloading. Will Earl and Enid Keese mind their manners as their neighborhood disintegrates? Some funny moments, but as the script fades, so do the laughs. Belushi's last waltz is based on Thomas Berger's novel.
1981 (R) 90m/C John Belushi, Dan Aykroyd, Kathryn Walker, Cathy Moriarty; ***Dir:*** John G. Avildsen. **VHS, Beta, LV $14.95** *GKK* ☠½

Nelson Mandela 70th Birthday Tribute In June of 1988 a concert was held to celebrate the birthday of the famous South African Marxist activist who had been in jail for almost twenty-five years. Songs include "Brothers in Arms" and "Wonderful Tonight" (Dire Straits and Clapton), "Let's Stay Together" (Al Green), and a performance of "There Must Be an Angel" and "You Have Placed a Chill in My Heart" by Annie Lennox and Eurythmics. Taped at London's Wembley Stadium.
1988 90m/C Sting, George Michael, Eurythmics, Al Green, Tracy Chapman, Peter Gabriel, Simple Minds, UB40, Steve Van Zandt, Whitney Houston, Dire Straits, Eric Clapton, Harry Belafonte, Whoopi Goldberg. **VHS, LV $14.98** *MVD, SMV* ☠½

Nelvanamation Four cosmic fantasies are featured in this program, including "Please Don't Eat the Planet," "A Cosmic Christmas," "The Devil and Daniel Mouse," and "Romie-O and Julie-8." The latter two are also available individually.
1980 100m/C VHS, Beta *WAR, OM* ☠½

Nelvanamation II This is a second anthology of fantasy cartoons for children of all ages, featuring "Take Me Out to the Ball Game," with music by Rick Danko, and "The Jack Rabbit Story" with music by John Sebastian.
1981 50m/C *Voices:* Phil Silvers, Garrett Morris. **VHS, Beta** *WAR, OM* ☠½

Neon City Title refers to a rumored city of refuge in the year 2053, where eight misfit adventurers in an armored transport seek safety from the Earth's toxic environment and "Mad Max" cliches.
1991 (R) 99m/C Michael Ironside, Vanity, Lyle Alzado, Valerie Wildman, Nick Klar, Juliet Landau, Arsenio "Sonny" Trinidad, Richard Sanders; ***Dir:*** Monte Markham. **VHS $89.95** *VMK* ☠☠

The Neon Empire Tepid tale of two gangsters battling inner mob opposition and backstabbing to build Las Vegas. Ostensibly based on the story of Bugsy Siegel and Meyer Lansky. Written by Pete Hamill. Made for cable.
1989 (R) 120m/C Martin Landau, Ray Sharkey, Gary Busey, Harry Guardino, Julie Carmen, Linda Fiorontino, Dylan McDermott; ***Dir:*** Larry Peerce. **VHS, Beta $39.95** *FRH* ☠½

Neon Maniacs An even half-dozen fetish-ridden zombies stalk the streets at night, tearing their victims into teensy weensy bits. Brave teenagers try to stop the killing. Brave viewers will stop the tape.
1986 (R) 90m/C Allan Hayes, Lelani Sarelle, Bo Sabato, Donna Locke, Victor Elliot Brandt; ***Dir:*** Joseph Mangine. **VHS, Beta $79.98** *LIV, VES, LTG* ***Woof!***

Neptune Factor Scientists board a special new deep-sea sub to search for their colleagues lost in an undersea earthquake. Diving ever deeper into the abyss that swallowed their friend's Ocean Lab II, they end up trapped themselves. The plot's many holes sink it. Also called "An Underwater Odyssey" and "The Neptune Disaster."
1973 (G) 94m/C Ben Gazzara, Yvette Mimieux, Walter Pidgeon, Ernest Borgnine; ***Dir:*** Daniel Petrie. **VHS, Beta $59.98** *FOX* ☠

Neptune's Daughter A swim-happy bathing suit designer finds romance and deep pools of studio water to dive into. Typically lavish and ridiculous Williams aquaparade. Songs include "Baby, It's Cold Outside" and "I Love Those Men."
1949 92m/C Esther Williams, Red Skelton, Ricardo Montalban, Betty Garrett, Keenan Wynn, Xavier Cugat, Ted de Corsia, Mike Mazurki, Mel Blanc, Juan Duvall, George Mann; ***Dir:*** Edward Buzzell. Academy Awards '49: Best Song ("Baby It's Cold Outside"). **VHS, Beta $19.95** *MGM* ☠☠

The Nervous Wreck Harrison Ford (not the one from "Indiana Jones") heads West, seeking a cure for what he believes is a terminal illness. Once there, he becomes romantically entangled with the sheriff's fiancee, uses a wrench to hold up a gas station, and gets involved in other shenanigans. And then he discovers he's going to live after all.
1926 70m/B Harrison Ford, Phyllis Haver, Chester Conklin, Mack Swain, Hobart Bosworth, Charles Gerrard; ***Dir:*** Scott Sidney. **VHS, Beta $27.95** *DNB* ☠☠☠

The Nest Tragic romance of an elderly man and a 13-year-old girl is helped along by superior script and directing.
1981 109m/C *SP* Ana Torrent, Hector Alterio, Patricia Adriani; ***Dir:*** Jaime de Arminan. **VHS, Beta $68.00** *SUE, APD* ☠☠½

The Nest A small island is overcome by giant cockroaches created by, you guessed it, a scientific experiment gone wrong. Special effects make it watchable.
1988 (R) 89m/C Robert Lansing, Lisa Langlois, Franc Luz, Terri Treas, Stephen Davies, Diana Bellamy, Nancy Morgan; ***Dir:*** Terence H. Winkless. **VHS, Beta $79.95** *MGM* ☠☠½

The Nesting Too-long tale of a neurotic author who rents a haunted Victorian manor and finds herself a pawn in a ghostly plan for revenge. Features Grahame's last performance.
1980 (R) 104m/C Robin Groves, John Carradine, Gloria Grahame, Christopher Loomis, Michael Lally; ***Dir:*** Armand Weston. **VHS, Beta $59.95** *WAR* ☠½

Netherworld A young man investigating his mysterious, dead father travels to his ancestral plantation in the bayou. To his horror, he discovers that his father was involved in the black arts. Now two beautiful young witches are after him. Can he possibly survive this madness?
1991 87m/C Michael C. Bendetti, Denise Gentile, Anjanette Comer, Holly Floria, Robert Burr; ***Dir:*** David Schmoeller. **VHS, LV** *PAR* ☠½

Network As timely now as it was then; a scathing indictment of the television industry and its propensity towards self-prostitution. A television newscaster's mental breakdown turns him into a celebrity when the network tries to profit from his illness. The individual characters are startlingly realistic and the acting is excellent. Written by Paddy Chayefsky; nominated for nine Academy Awards and winner of four.
1976 (R) 121m/C Faye Dunaway, Peter Finch, William Holden, Robert Duvall, Wesley Addy, Ned Beatty, Beatrice Straight; ***Dir:*** Sidney Lumet. Academy Awards '76: Best Actor (Finch), Best Actress (Dunaway), Best Original Screenplay, Best Supporting Actress (Straight); National Board of Review Awards '76: 10 Best Films of the Year; New York Film Critics Awards '76: Best Screenwriting. **VHS, Beta, LV $19.98** *MGM, BTV* ☠☠☠½

Neurotic Cabaret Would-be actress dances exotically and finds more exotic ways to roll in dough.
1990 90m/C Edwin Neal, Tammy Stone. **VHS $79.95** *AIP* ☠½

Neutron and the Black Mask Man wears mask with lightning bolts and gains comic book style powers to do deathly battle with a scientist with a neutron bomb up his sleeve. First Neutron feature.
1961 80m/B *MX* Wolf Ruvinskis, Julio Aleman, Armando Silvestre, Rosita Arenas, Claudio Brook; ***Dir:*** Frederick Curiel. **VHS $19.98** *SNC* ☠

Neutron vs. the Amazing Dr. Caronte Dr. Caronte from "Neutron and the Black Mask" just won't give up that neutron bomb and let the series die.

1961 80m/B *MX* Wolf Ruvinskis, Julio Aleman, Armando Silvestre, Rosita Arenas, Rodolfo Landa; *Dir:* Frederick Curiel. **VHS** $19.98 *SNC Ⅱ*

Neutron vs. the Death Robots Guy with a mask takes on army of killer robots and neutron bomb to protect the world from future sequels.
1962 80m/B *MX* Wolf Ruvinskis, Julio Aleman, Armando Silvestre, Rosita Arenas; *Dir:* Frederick Curiel. **VHS** $19.98 *SNC Ⅱ*

Neutron vs. the Maniac Man in mask hunts guys who kill people.
1962 80m/B *MX* Wolf Ruvinskis. **VHS** $19.98 *SNC Ⅱ*

Nevada City Roy outwits a financier who is trying to monopolize transportation in California.
1941 54m/B Roy Rogers, Trigger. **VHS, Beta** $27.95 *VCN Ⅱ*

Nevada Smith The half-breed Nevada Smith (previously introduced in Harold Robbins' story "The Carpetbaggers" and film of same name) seeks the outlaws who killed his parents. Standard western plot, characters. Later remade as a TV movie.
1966 135m/C Steve McQueen, Karl Malden, Brian Keith, Arthur Kennedy, Raf Vallone, Suzanne Pleshette; *Dir:* Henry Hathaway. **VHS, Beta, LV** $14.95 *PAR ⅡⅡ*

The Nevadan A marshal goes in search of an outlaw's gold cache, only to be opposed by a crooked rancher. Good scenery and action, but not one of Scott's best.
1950 81m/C Randolph Scott, Dorothy Malone, Forrest Tucker, Frank Faylen, George Macready, Charles Kemper, Jeff Corey, Tom Powers, Jock Mahoney; *Dir:* Gordon Douglas. **VHS, Beta** *GKK Ⅱ ½*

Never Cry Wolf A young biologist is sent to the Arctic to study the behavior and habitation of wolves, then becomes deeply involved with their sub-society. Based on Farley Mowat's book. Beautifully photographed.
1983 (PG) 105m/C Charles Martin Smith, Brian Dennehy, Samson Jorah; *Dir:* Carroll Ballard. **VHS, Beta, LV** $29.95 *DIS ⅡⅡⅡ ½*

Never a Dull Moment New York songwriter falls for rodeo Romeo, moves out West, and finds welcome wagon deficient. Mediocre songs; many dull moments.
1950 89m/B Irene Dunne, Fred MacMurray, William Demarest, Andy Devine, Gigi Perreau, Natalie Wood, Philip Ober, Jack Kirkwood, Ann Doran, Margaret Gibson; *Dir:* George Marshall. **VHS** $19.98 *TTC, FCT ⅡⅡ*

Never a Dull Moment Mobsters mistake an actor for an assassin in this gag-filled adventure. They threaten the thespian into thievery, before the trouble really starts when Ace, the actual assassin, arrives.
1968 (G) 90m/C Dick Van Dyke, Edward G. Robinson, Dorothy Provine, Henry Silva, Joanna Moore, Tony Bill, Slim Pickens, Jack Elam; *Dir:* Jerry Paris. **VHS, Beta, LV** $39.95 *DIS ⅡⅡ*

Never Forget Made-for-cable-TV true story of California resident Mel Mermelstein (played by Nimoy), a survivor of Hitler's death camps who accepted a pro-Nazi group's challenge to prove in court that the Holocaust of six million Jews really happened. A sincere, well-meaning courtroom drama that just can't surmount the uncinematic nature of the source material.
1991 94m/C Leonard Nimoy, Blythe Danner, Dabney Coleman; *Dir:* Joseph Sargent. **VHS, Beta** $89.98 *TTC ⅡⅡ*

Never Give a Sucker an Even Break An almost plotless comedy, based on an idea reputedly written on a napkin by Fields (he took screenplay credit as Otis Criblecoblis), and features Fields at his most unleashed. It's something of a cult favorite, but not for all tastes. Classic chase scene ends it. Fields' last role in a feature-length film. Also released as "What A Man."
1941 71m/B W.C. Fields, Gloria Jean, Franklin Pangborn, Leon Errol, Margaret Dumont; *Dir:* Eddie Cline. **VHS, Beta, LV** $14.95 *MCA, KRT ⅡⅡⅡ ½*

Never Let Go A man unwittingly tracks down the mastermind of a gang of racketeers. Sellers sheds his comedic image to play the ruthless and brutal gang boss, something he shouldn't have done.
1960 91m/C *GB* Peter Sellers, Richard Todd, Elizabeth Sellars, Carol White; *Dir:* John Guillermin. **VHS, Beta** *SUE Ⅱ*

Never Love a Stranger A young man becomes a numbers runner for a mobster and ultimately winds up heading his own racket. Later he finds himself in conflict with his old boss and the district attorney. No surprises here. Based on the Harold Robbins' novel.
1958 93m/B John Barrymore Jr., Steve McQueen, Lita Milan, Robert Bray; *Dir:* Robert Stevens. **VHS** $19.98 *REP ⅡⅡ*

Never Pick Up a Stranger When a local hooker is murdered, her cop boyfriend, who never gives up on a case, hunts her killer with a vengeance. Extremely graphic. Also on video as "Bloodrage."
1979 82m/C Ian Scott, Judith-Marie Bergan, James Johnston, Lawrence Tierney; *Dir:* Joseph Bigwood. **VHS, Beta** $49.95 *IVE, BFV Ⅰ ½*

Never Say Die Two innocents are thrown into a web of international intrigue and acquit themselves rather well. Lots of explosions and car crashes for a film shot in New Zealand.
1990 98m/C *NZ* George Wendt, Lisa Eilbacher; *Dir:* Geoff Murphy. **VHS, Beta** $29.95 *HBO, JTC, NVH Ⅱ ½*

Never Say Never Again James Bond matches wits with a charming but sinister tycoon who is holding the world nuclear hostage as part of a diabolical plot by SPECTRE. Connery's return to the world of Bond after 12 years is smooth in this remake of "Thunderball" hampered by an atrocious musical score. Carrera is stunning as Fatima Blush.
1983 (PG) 134m/C Sean Connery, Klaus Maria Brandauer, Max von Sydow, Barbara Carrera, Kim Basinger, Edward Fox, Bernie Casey, Pamela Salem, Rowan Atkinson, Valerie Leon, Prunella Gee, Saskia Cohen Tanugi; *Dir:* Irvin Kershner. **VHS, Beta, LV** $19.98 *WAR, TLF ⅡⅡ ½*

Never So Few A military commander and his outnumbered troops overcome incredible odds against the Japanese. There is a lot of focus on romance, but the script and acting make a strong impression nonetheless. Based on the novel by Tom T. Chamales.
1959 124m/C Frank Sinatra, Gina Lollobrigida, Peter Lawford, Steve McQueen, Richard Johnson, Paul Henreid, Charles Bronson; *Dir:* John Sturges. **VHS** $29.95 *MGM ⅡⅡ ½*

Never Steal Anything Small A tough union boss pushes everyone as he battles the mob for control of the waterfront. A strange musical-drama combination that will be of interest only to the most ardent Cagney and Jones fans. Based on the Max-well Anderson/Rouben Mamoulian play "The Devil's Hornpipe." Largely forgettable songs include "Helping Out Friends," "I'm Sorry, I Want a Ferrari," and the title song.
1959 94m/C James Cagney, Shirley Jones, Cara Williams; *Dir:* Charles Lederer. **VHS, Beta** $59.95 *MCA Ⅱ ½*

Never on Sunday An American intellectual tries to turn a Greek prostitute into a refined woman. Fine performances and exhilarating Greek photography. Fun all around. Oscar nominations for Mercouri, director Dassin, Story and Screenplay, and Costume Design.
1960 94m/B *GR* Melina Mercouri, Tito Vandis, Jules Dassin, Mitsos Liguisos; *Dir:* Jules Dassin. Academy Awards '60: Best Song ("Never on Sunday"). **VHS, Beta** $59.95 *MGM ⅡⅡⅡ*

Never Too Late to Mend Victorian British play about the abuses of the penal system which had, upon its initial London West End run, caused Queen Victoria to institute a sweeping program of prison reform. The movie caused considerably less stir.
1937 67m/B Tod Slaughter, Marjorie Taylor, Jack Livesey, Ian Colin; *Dir:* David MacDonald. **VHS, Beta, 8mm** $29.95 *SNC, MRV, VYY ⅡⅡ*

Never Too Young to Die A young man is drawn into a provocative espionage adventure by his late father's spy associates. Together they try to discover who killed the young man's father. High point may be Simmons as a crazed hermaphrodite plotting to poison L.A.
1986 (R) 97m/C John Stamos, Vanity, Gene Simmons, George Lazenby; *Dir:* Gil Bettman. **VHS, Beta, LV** $19.98 *CHA Ⅰ*

Never on Tuesday Two jerks from Ohio head to California and find themselves stuck midway in the desert with a beautiful girl, who has plans of her own. Charlie Sheen and Emilio Estevez make cameo appearances.
1988 (R) 90m/C Claudia Christian, Andrew Lauer, Peter Berg; *Dir:* Adam Rifkin. **VHS, Beta, LV** $14.95 *PAR Ⅰ ½*

Never Wave at a WAC A Washington socialite joins the Women's Army Corps hoping for a commission that never comes. She has to tough it out as an ordinary private. A reasonably fun ancestor of "Private Benjamin," with a cameo by Gen. Omar Bradley as himself.
1952 87m/B Rosalind Russell, Paul Douglas, Marie Wilson, William Ching, Arleen Whelan, Leif Erickson, Hillary Brooke, Regis Toomey, Omar Bradley; *Dir:* Norman Z. McLeod. **VHS** $19.95 *NOS, DVT ⅡⅡ ½*

Never Weaken/Why Worry? Silent comedy shorts from Harold. In "Never Weaken" (1921), Lloyd hustles customers. In "Why Worry?" (1923), he plays a wealthy young man who journeys to South America for his health and gets caught in the middle of a revolution.
1923 78m/B Harold Lloyd. **VHS, Beta** *TLF ⅡⅡⅡ*

NeverEnding Story A lonely young boy helps a warrior save the fantasy world in his book from destruction by the Nothing. A wonderful, intelligent family movie about imagination, with swell effects and a sweet but not overly sentimental script. The music was composed by Klaus Doldinger and Giorgio Moroder. Petersen's first English-language film, based on the novel by Michael Ende.

1984 (PG) 94m/C Barret Oliver, Noah Hathaway, Gerald McRaney, Moses Gunn, Tami Stronach, Patricia Hayes, Sydney Bromley; *Dir:* Wolfgang Petersen. **VHS, Beta, LV, 8mm** $19.98 *WAR, PIA, GLV* 🎬🎬 ½

The NeverEnding Story 2: Next Chapter Disappointing sequel to the first story that didn't end. Boy Oliver staves off life of couch potatodom reading books and escaping into fantasy land. Even the kids may wander away. But wait: contains the first Bugs Bunny theatrical cartoon in 26 years, "Box Office Bunny."
1991 (PG) 90m/C Jonathan Brandis, Kenny Morrison, Clarissa Burt, John Wesley Shipp, Martin Umbach; *Dir:* George Miller. **VHS, LV, 8mm** $19.95 *WAR, PIA, APD* 🎬🎬

Nevil Shute's The Far Country Made-for-television adaptation of a typical gothic romance by the author of "On the Beach."
1985 115m/C Michael York, Sigrid Thornton. **VHS, Beta** *QNE* 🎬 ½

The New Adventures of Pippi Longstocking Another musical rehashing of the Astrid Lindgren children's books about a precocious red-headed girl and her fantastic adventures with horses, criminals and pirates.
1988 (G) 100m/C Tami Erin, Eileen Brennan, Dennis Dugan, Dianne Hull, George Di Cenzo, John Schuck, Dick Van Patten; *Dir:* Ken Annakin. **VHS, Beta, LV** $19.95 *COL* 🎬 ½

The New Adventures of Tarzan: Vol. 1 Twelve episodes, each 22 minutes long, depict the adventures of Edgar Rice Burrough's tree-swinging character—Tarzan. Also known as "Tarzan and the Green Goddess."
1935 260m/B Herman Brix, Ula Holt, Frank Baker, Dale Walsh, Lewis Sargent; *Dir:* Edward Kull. **VHS, Beta** $26.95 *SNC, NOS, VCN*

The New Adventures of Tarzan: Vol. 2 Tarzan goes nuts in this spectacular adventure.
1935 25m/B Bruce Bennett, Ula Holt; *Dir:* Edward Kull. **VHS, Beta** *PFI, BUR*

The New Adventures of Tarzan: Vol. 3 More adventure with the vine swinger.
1935 25m/B Bruce Bennett, Ula Holt; *Dir:* Edward Kull. **VHS, Beta** *PFI, BUR*

The New Adventures of Tarzan: Vol. 4 The jungle man swings into action in this incredible adventure.
1935 25m/B Bruce Bennett, Ula Holt; *Dir:* Edward Kull. **VHS, Beta** *PFI, BUR*

New Centurions Rookies training for the Los Angeles Police Department get the inside info from retiring cops. Gritty and realistic drama based on the novel by former cop Joseph Wambaugh. Tends to be disjointed at times, but overall is a good adaptation of the bestseller. Scott, excellent as the retiring beat-walker, is supported well by other performers.
1972 (R) 103m/C George C. Scott, Stacy Keach, Jane Alexander, Scott Wilson, Erik Estrada; *Dir:* Richard Fleischer. **VHS, Beta, LV** $14.95 *COL, FCT* 🎬🎬 ½

New Faces Based on the hit Broadway revue. The plot revolves around a Broadway show that is about to be closed down, and the performers who fight to keep it open. Lots of hit songs, including "C'est Si Bon," "Monotonous," "Time for Tea," and "Uska-

dara." Mel Brooks is credited as a writer, under the name Melvin Brooks.
1954 98m/C Ronny Graham, Eartha Kitt, Robert Clary, Alice Ghostley, June Carroll, Carol Lawrence, Paul Lynde; *Dir:* Harry Horner. **VHS, Beta** $49.95 *MVD, NOS, FOX* 🎬🎬 ½

New Fist of Fury During World War II, a former pickpocket becomes a martial arts whiz with the assistance of his fiancee, and fights the entire Imperial Army.
1985 120m/C Jackie Chan; *Dir:* Jackie Chang. **VHS, Beta** $24.95 *ASE* 🎬

The New Gladiators In the future, criminals try to kill each other on television for public entertainment. Two such gladiators discover that the network's computer is using the games in order to take over mankind, and they attempt to stop it. Even if the special effects were any good, they couldn't save this one.
1987 90m/C *IT* Jared Martin, Fred Williamson, Eleanor Gold, Howard Ross, Claudio Cassinelli; *Dir:* Lucio Fulci. **VHS, Beta** $79.95 *MED* Woof!

The New Invisible Man A so-so Mexican adaptation of the popular H.G. Wells novel has a prisoner receiving a vanishing drug from his brother, who created it—perhaps to aid an escape attempt. Also called "H.G. Wells' New Invisible Man."
1958 95m/B *MX* Arturo de Cordova, Ana Louisa Peluffo, Jorge Mondragon; *Dir:* Alfredo B. Crevenna. **VHS** $16.95 *SNC* 🎬🎬

New Jack City Just say no ghetto-melodrama. Powerful performance by Snipes as wealthy Harlem drug lord sought by rebel cops Ice-T and Nelson. Music by Johnny Gill, 2 Live Crew, Ice-T and others. Available with Spanish subtitles.
1991 (R) 101m/C Wesley Snipes, Ice-T, Mario Van Peebles, Chris Rock, Judd Nelson, Tracy C. Johns, Allen Payne, Kim Park; *Dir:* Mario Van Peebles. **VHS, LV** $19.98 *WAR, PIA, CCB* 🎬🎬 ½

The New Kids An orphaned brother and sister find out the limitations of the good neighbor policy. A sadistic gang terrorizes them after their move to a relatives' home in Florida. They go after expected revenge.
1985 (R) 90m/C Shannon Presby, Lori Loughlin, James Spader, Eric Stoltz; *Dir:* Sean S. Cunningham. **VHS, Beta, LV** $79.95 *COL* 🎬🎬

A New Kind of Love Romantic fluff starring real-life couple Newman and Woodward who meet en route to Paris and end up falling in love. Newman plays a reporter and Woodward a fashion designer in this light comedy set amidst the sights of Paris.
1963 110m/C Paul Newman, Joanne Woodward, Thelma Ritter, Eva Gabor, Maurice Chevalier, George Tobias; *W/Dir:* Melville Shavelson. **VHS, Beta** $14.95 *PAR* 🎬🎬 ½

A New Leaf A playboy who has depleted his financial resources tries to win the hand of a clumsy heiress. May was the first woman to write, direct and star in a movie. She was unhappy with the cuts that were made by the studio, but that didn't seem to affect its impact with the public. Even with the cuts, the film is still funny and May's performance is worth watching.
1971 (G) 102m/C Walter Matthau, Elaine May, Jack Weston, George Rose, William Redfield, James Coco; *W/Dir:* Elaine May. **VHS** $14.95 *PAR, CCB* 🎬🎬 ½

A New Life An uptight stockbroker is abandoned by his wife. The New York singles scene beckons both of them to a new chance at life. Appealing performances can-

not completely mask the old story line, but it's definitely worth a look.
1988 (PG-13) 104m/C Alan Alda, Ann-Margret, Hal Linden, Veronica Hamel, John Shea, Mary Kay Place; *W/Dir:* Alan Alda. **VHS, Beta, LV** $89.95 *PAR* 🎬🎬 ½

New Look Erotic men's video magazine includes a centerfold, interviews with film-makers Francis Ford Coppola and Bob Rafelson, and a series of vignettes.
1981 44m/C Francis Ford Coppola, Bob Rafelson. **VHS, Beta, LV** *SUE* 🎬🎬 ½

New Mafia Boss Italian-made plodder has Savalas taking over a huge Mafia family and all hell breaking loose. Originally titled "Crime Boss."
1972 (PG) 90m/C *IT* Telly Savalas, Lee Van Cleef, Antonio Sabato, Paolo Tedesco; *Dir:* Alberto De Martino. **VHS, Beta** $59.95 *VCD* 🎬 ½

New Moon An adaptation of the operetta by Sigmund Romberg and Oscar Hammerstein II. A French heiress traveling on a boat that is captured by pirates falls in love with their leader. Features the hit song "Lover, Come Back to Me" and also "Stout Hearted Men," "One Kiss," and "Dance Your Cares Away." Remake of the 1930 film. Includes the 1935 Robert Benchley MGM short "How to Sleep."
1940 106m/B Jeanette MacDonald, Nelson Eddy, Buster Keaton, Joe Yule, Jack Perrin, Mary Boland; *Dir:* Robert Z. Leonard. **VHS, Beta** $19.95 *MGM* 🎬🎬 ½

New Orleans After Dark A pair of New Orleans detectives tour the city, crime-buster style, and bag a dope ring. Average crime tale, shot on location in the Big Easy.
1958 69m/B Stacy Harris, Louis Sirgo, Ellen Moore; *Dir:* John Sledge. **VHS** $15.95 *NOS, LOO* 🎬🎬

New Orleans: 'Til the Butcher Cuts Him Down A look at the history of New Orleans jazz through the eyes of trumpet great Kid Punch Miller.
1971 53m/C Kid Punch Miller, King Oliver, Kid Ory, Jelly Roll Morton, Louis Armstrong. **VHS, Beta** $29.95 *KIV*

New Pastures Three ex-convicts run into some humorous situations when they go back to their small hometown. A semi-acclaimed Czech comedy, with subtitles.
1962 92m/B *CZ Dir:* Vladimir Cech. **VHS, Beta, 3/4U** $39.95 *IHF* 🎬🎬 ½

New Star Over Hollywood An entertainment and variety special with a Christian format.
197? 60m/C Carol Lawrence, Tom Netherton, Susan Stafford, Debbie Boone, Ronald Reagan, Dean Jones, Pat Boone. **VHS, Beta** *TVS* 🎬🎬 ½

New Wave Comedy A compilation of cutting-edge stand-up comedy routines.
1985 60m/C John Kassir, Steve Sweeney, Patty Rosborough, Wayne Federman, Marc Weiner. **VHS, Beta** $19.98 *LIV, VES* 🎬🎬 ½

New Year's Day Jaglom continues his look at modern relationships in this story of a man reclaiming his house from three female tenants. Introspective character study lightened by humor and insight.
1989 (R) 90m/C Maggie Jakobson, Gwen Welles, Melanie Winter, Milos Forman, Michael Emil, Henry Jaglom; *W/Dir:* Henry Jaglom. **VHS, Beta** $79.98 *PAR, BTV, FCT* 🎬🎬🎬

New Year's Evil Every hour during New Year's Eve, a madman kills an unsuspecting victim. After each killing, the murderer informs a local disc jockey of his deed.

Little does the disc jockey know that she will soon be next on his list. No suspense, bad music makes this one a holiday wrecker. **1978 (R) 88m/C** Roz Kelly, Kip Niven, Chris Wallace, Louisa Moritz, Grant Kramer, Jed Mills; *Dir:* Emmett Alston. **VHS, Beta $19.98** *PGN, CAN* ♫

New York, New York A tragic romance evolves between a saxophonist and an aspiring singer/actress in this salute to the big-band era. A love of music isn't enough to hold them together through career rivalries and life on the road. Fine performances by De Niro and Minnelli and the supporting cast. Songs include "New York, New York," "It's a Wonderful World," "You Are My Lucky Star," and "The Man I Love." Re-released in 1981 with the "Happy Endings" number, which was cut from the original. Look for "Big Man" Clarence Clemons on sax. **1977 (PG) 163m/C** Robert De Niro, Liza Minnelli, Lionel Stander, Barry Primus, Mary Kay Place, Dick Miller, Diahnne Abbott; *Dir:* Martin Scorsese. **VHS, Beta, LV $29.98** *FOX* ♫♫♫

New York Nights The lives of nine New Yorkers intertwine in a treacherous game of passion and seduction. Liberal borrowing from plot of "La Ronde." **1984 (R) 104m/C** Corinne Alphen, George Auyer, Bobbi Burns, Peter Matthey, Cynthia Lee, Willem Dafoe; *Dir:* Simon Nuchtern. **VHS, Beta $79.98** *LIV, LTG* ♫

New York Ripper A New York cop tracks down a rampaging murderer in this dull, mindless slasher flick. **1982 88m/C** *IT* Fred Williamson; *Dir:* Lucio Fulci. **VHS, Beta $49.95** *VMK* Woof!

New York Stories Entertaining anthology of three separate short films by three esteemed directors, all set in New York. In "Life Lessons" by Scorsese, an impulsive artist tries to prevent his live-in girlfriend from leaving him. "Life Without Zoe" by Coppola involves a youngster's fantasy about a wealthy 12-year-old who lives mostly without her parents. Allen's "Oedipus Wrecks," generally considered the best short, is about a 50-year-old man who is tormented by the specter of his mother. **1989 (PG) 124m/C** Nick Nolte, Rosanna Arquette, Woody Allen, Mia Farrow, Mae Questel, Julie Kavner, Talia Shire, Giancarlo Giannini, Don Novello, Patrick O'Neal, Peter Gabriel, Paul Herman, Deborah Harry; *Dir:* Woody Allen, Martin Scorsese, Francis Ford Coppola. **VHS, Beta, LV $19.99** *TOU* ♫♫♫

New York's Finest Three prostitutes are determined to leave their calling and marry very eligible millionaires. But becoming society ladies is more difficult than they thought. Would-be sophisticated comedy misfires. **1987 (R) 80m/C** Ruth Collins, Jennifer Delora, Scott Thompson Baker, Heidi Paine, Jane Hamilton, Alan Naggar, John Altamura, Alan Fisler, Josey Duval; *Dir:* Chuck Vincent. **VHS, Beta $29.95** *ACA* ♫

The Newlydeads An uptight, conservative, honeymoon resort owner murders one of his guests and finds out that "she" is really a he. Fifteen years later, on his wedding night, all of his guests are violently murdered by the transvestite's vengeful ghost. Oddball twist to the usual slasher nonsense. **1987 84m/C** Scott Kaske, Jim Williams, Jean Levine, Roger Mathews, Jay Richardson; *Dir:* Joseph Merhi. **VHS, Beta $39.95** *CLV* ♫ 1/2

Newman's Law A city detective is implicated in a corruption investigation. While on suspension, he continues with his own investigation of a large drug ring and finds corruption is closer than he thought. Director Heffron's first feature film. **1974 (PG) 98m/C** George Peppard, Roger Robinson, Eugene Roche, Gordon Pinsent, Louis Zorich, Abe Vigoda; *Dir:* Richard T. Heffron. **VHS, Beta $59.95** *MCA* ♫♫

News at Eleven A fading news anchorman is pressured by his ambitious young boss to expose a touchy local sex scandal, forcing him to consider the public's right to know versus the rights of the individual. About average for made-for-television drama. **1986 95m/C** Martin Sheen, Peter Riegert, Barbara Babcock, Sheree J. Wilson, Sydney Penny, David S. Sheiner, Christopher Allport; *Dir:* Mike Robe. **VHS, Beta $9.98** *TTC* ♫♫

News from Home A plotless look at life and survival in New York City, with narration from a young girl's mother. **1976 90m/C** *W/Dir:* Chantal Akerman. **VHS $79.95** *WAC, FCT* ♫♫

Newsfront A story about two brothers, both newsreel filmmakers, and their differing approaches to life and their craft in the 1940s and '50s. Tribute to the days of newsreel film combines real stories and fictionalized accounts with color and black and white photography. Noyce's feature film debut. **1978 110m/C** *AU* Bill Hunter, Gerard Kennedy, Angela Punch-McGregor, Wendy Hughes, Chris Haywood, John Ewart, Bryan Brown; *Dir:* Phillip Noyce. Australian Film Institute '78: Best Actor (Hunter), Best Film. **VHS, LV $59.98** *SUE* ♫♫♫

Newsies An unfortunate attempt at an old-fashioned musical with a lot of cute kids and cardboard characters and settings. The plot, such as it is, concerns the 1899 New York newsboys strike against penny-pinching publisher Joseph Pulitzer. Bale plays the newsboys leader and at least shows some charisma in a strictly cartoon setting. The songs are mediocre but the dancing is lively. However, none of it moves the story along. Choreographer Ortega's feature-film directorial debut. **1992 (PG) 121m/C** Christian Bale, David Moscow, Max Casella, Bill Pullman, Ann-Margret, Robert Duvall; *Dir:* Kenny Ortega. **VHS** *DIS* ♫ 1/2

Next of Kin A daughter moves into her dead mother's retirement home and discovers an unspeakable evil that lurks there. Typical horror stuff has some very creepy moments; filmed in New Zealand. **1982 90m/C** *AU* Jackie Kerin, John Jarratt, Gerida Nicholson, Alex Scott; *Dir:* Tony Williams. **VHS, Beta, LV, 8mm $14.95** *VIR, VCL* ♫

Next of Kin A young boy experiencing familial difficulties discovers photographs of an Armenian family in his parent's bedroom. He convinces himself that they are his real family, and plunges into a bizarre fantasy world. **1984 72m/C** *CA* Patrick Tierney; *W/Dir:* Atom Egoyan. **VHS $59.95** *CVC, FCT* ♫

Next of Kin A Chicago cop returns to his Kentucky home to avenge his brother's brutal murder. Swayze's return to action films after his success in "Dirty Dancing" is unimpressive. **1989 (R) 108m/C** Patrick Swayze, Adam Baldwin, Bill Paxton, Helen Hunt, Andreas Katsulas, Ben Stiller, Michael J. Pollard, Liam Neeson; *Dir:* John Irvin. **VHS, Beta, LV, 8mm $19.98** *WAR* ♫♫

Next One A mysterious visitor from another time winds up on an isolated Greek island as the result of a magnetic storm. The local inhabitants are amazed when the visitor displays some Christ-like characteristics. **1984 105m/C** Keir Dullea, Adrienne Barbeau, Jeremy Licht, Peter Hobbs; *Dir:* Nico Mastorakis. **VHS, Beta $69.95** *VES* ♫ 1/2

Next Stop, Greenwich Village An affectionate, autobiographical look by Mazursky at a Brooklyn boy with acting aspirations, who moves to Greenwich Village in 1953. Good performances, especially by Winters as the overbearing mother. **1976 (R) 109m/C** Lenny Baker, Christopher Walken, Ellen Greene, Shelley Winters, Lou Jacobi, Mike Kellin; *W/Dir:* Paul Mazursky. **VHS $59.98** *FXV* ♫♫♫

Next Summer A large, character-studded French family pursues power, love and beauty. Excellent performances. In French with English subtitles. **1984 120m/C** *FR* Jean-Louis Trintignant, Claudia Cardinale, Fanny Ardant, Philippe Noiret, Marie Trintignant; *Dir:* Nadine Trintignant. **VHS, Beta $69.95** *CPT, FCT* ♫♫♫

Next Time I Marry Lucy fraternizes with ditch digger Ellison because she needs to hitch a Yankee in order to inherit $20 mill. Seems she really loves a foreigner, though. Not much to bobaloo about. Director Kanin's second effort. **1938 65m/B** Lucille Ball, James Ellison, Lee Bowman, Granville Bates, Mantan Moreland, Florence Lake; *Dir:* Garson Kanin. **VHS $19.98** *TTC, FCT, MLB* ♫ 1/2

Next Victim The unfaithful wife of an Austrian diplomat attempts to find out who has been slicing up beautiful jet-setters. Tepid mystery. **1971 (PG) 87m/C** George Hilton, Edwige French, Christina Airoldi, Ivan Rassimov; *Dir:* Sergio Martino. **VHS, Beta $19.95** *GEM, NEG* ♫

Next Victim A beautiful woman confined to a wheelchair is stalked by a lunatic killer. You've seen it before, and done better. **1974 80m/C** *GB* Carroll Baker, T.P. McKenna, Ronald Lacey, Maurice Kaufman; *Dir:* James Ormerod. **VHS, Beta $29.95** *IVE, THR* ♫

The Next Voice You Hear Lives are changed forever when a group of people hear the voice of God on the radio. Interesting premise presented seriously. **1950 82m/B** James Whitmore, Nancy Davis, Gary Gray, Lillian Bronson, Art Smith, Tom D'Andrea, Jeff Corey, George Chandler; *Dir:* William A. Wellman. **VHS, Beta $19.98** *CRR* ♫♫ 1/2

Next Year If All Goes Well Two young lovers struggle to overcome insecurities while establishing a relationship. Dubbed. **1983 95m/C** *FR* Isabelle Adjani, Thierry Lhermitte, Mariann Chazel; *Dir:* Jean-Loup Hubert. **VHS, Beta $19.95** *STE, HBO* ♫♫ 1/2

Ni Chana Ni Juana A girl gets herself into lots of wacky situations on her trip to the capitol in search of her family. **1987 102m/C** *MX* Maria Elena Velasco, Armando Calvo, Carmen Montejo. **VHS, Beta $79.95** *MAD* ♫♫ 1/2

Niagara During their honeymoon in Niagara Falls, a scheming wife (Monroe) plans to kill her crazed war-vet husband (Cotten). Little does she know that he is plotting to double-cross her. Steamy, quasi-Hitchcockian mystery ably directed, with interesting performances. Monroes sings "Kiss."

1952 89m/B Joseph Cotten, Jean Peters, Marilyn Monroe, Casey Adams, Don "The Dragon" Wilson; *Dir:* Henry Hathaway. **VHS, Beta, LV $14.98** FOX 🎬🎬½

Nice Girl Like Me
An orphaned young girl roams Europe, getting pregnant twice by two different men. She eventually finds love in the form of a kind caretaker. Harmless fluff. Based on Anne Piper's novel "Marry At Leisure."
1969 (PG) 91m/C GB Harry Andrews, Barbara Ferris, Gladys Cooper, Bill Hinnant, James Villiers; *Dir:* Desmond Davis. **VHS, Beta $59.98** CHA 🎬🎬

Nice Girls Don't Explode
A girl telekinetically starts fires when sexually aroused. This puts a bit of a damper on her relationship with O'Leary, much to her mother's delight. Matchless entertainment.
1987 (PG) 92m/C Barbara Harris, Wallace Shawn, Michelle Meyrink, William O'Leary; *Dir:* Chuck Martinez. **VHS, Beta $19.95** STE, NWV 🎬🎬

Nice Neighbor
An acclaimed Hungarian film about an outwardly self-sacrificing neighbor who cold-bloodedly manipulates his fellow rooming house tenants for his own ends. With English subtitles.
1979 90m/C HU Laszlo Szabo, Margit Dayka, Agi Margittay; *Dir:* Zsolt Kedzi-Kovacs. **VHS, Beta $59.95** FCT 🎬🎬🎬

Nicholas and Alexandra
Epic chronicling the final years of Tsar Nicholas II, Empress Alexandra and their children. Their lavish royal lifestyle gives way to imprisonment and eventual execution under the new Lenin government. Beautiful, but overlong costume epic that loses steam in the second half. Suzman earned an Oscar-nomination as Alexandra, while the film picked up nominations for Best Picture and Cinematography. Available in widescreen format on laserdisc.
1971 (PG) 183m/C Michael Jayston, Janet Suzman, Tom Baker, Laurence Olivier, Michael Redgrave, Harry Andrews, Jack Hawkins, Alexander Knox, Curt Jurgens; *Dir:* Franklin J. Schaffner. Academy Awards '71: Best Art Direction/Set Decoration, Best Costume Design; National Board of Review Awards '71: 10 Best Films of the Year. **VHS, Beta, LV $29.95** COL, LDC 🎬🎬½

Nicholas Nickleby
An ensemble cast works hard to bring to life Charles Dickens' novel about an impoverished family dependent on their wealthy but villainous uncle. Young Nicholas is an apprentice at a school for boys, and he and a student run away to a series of exciting adventures. An enjoyable film, though it is hard to tell the entire story in such a small amount of time.
1946 108m/B GB Cedric Hardwicke, Stanley Holloway, Derek Bond, Alfred Drayton, Sybil Thorndike, Sally Ann Howes, Bernard Miles, Mary Merrall, Cathleen Nesbitt; *Dir:* Alberto Cavalcanti. **VHS, Beta $19.98** HBO, MRV, PSM 🎬🎬🎬

Nicholas Nickleby
An animated version of the classic Charles Dickens novel.
1987 72m/C **VHS, Beta $19.98** LIV, CVL 🎬🎬🎬

Nick Knight
An L.A. cop on the night beat is really a good-guy vampire, who quaffs cattle blood as he tracks down another killer who's draining humans of their plasma. A gimmicky pilot for a would-be TV series that never bit.
1987 92m/C Rick Springfield, Michael Nader, Laura Johnson. **VHS $19.95** STE 🎬🎬

Nickel Mountain
Heartwarming story of a suicidal 40-year-old man who finds a new reason to live when he falls in love with a pregnant 16-year-old girl who works at his diner. Langenkamp particularly appealing. Based on a John Gardner novel.
1985 88m/C Michael Cole, Heather Langenkamp, Ed Lauter, Brian Kerwin, Patrick Cassidy, Grace Zabriskie, Don Beddoe; *Dir:* Drew Denbaum. **VHS, Beta $59.95** LHV 🎬🎬½

Nicole
The downfall of a wealthy woman who is able to buy everything she wants except love.
1972 (R) 91m/C Leslie Caron, Catherine Bach, Ramon Bieri; *Dir:* Itsvan Ventilla. **VHS, Beta** GHV 🎬

Nido de Aguilas (Eagle's Nest)
Two brothers stripped of their ranch search for the band of thieves who killed their parents.
1976 93m/C MX Mario Almada, Martha Elena Cervantes. **VHS, Beta $55.95** MAD 🎬

The Night After Halloween
A young woman gets the shock of her life when she discovers that her boyfriend is a crazed killer.
1983 (R) 90m/C IT Chantal Contouri, Robert Bruning, Sigrid Thornton. **VHS, Beta $59.95** MAG 🎬

Night After Night After Night
A killer prowls the streets of London for beautiful female throats to slash. The police are baffled, as usual, and the citizens are terrified. A Jack the Ripper rip-off.
1970 98m/C Jack May, Linda Marlowe; *Dir:* Lewis J. Force. **VHS, Beta** MED 🎬

Night Ambush
A Nazi general is kidnapped on Crete by British agents, who embark on a dangerous trip to the coast where a ship awaits to take them to Cairo. The general tries to thwart their plans, but they outwit him at every turn. Exciting actioner was originally titled "Ill Met By Moonlight." Based on a novel by W. Stanley Moss that details a similar real-life event.
1957 100m/B GB Dirk Bogarde, Marius Goring, David Oxley, Cyril Cusack, Christopher Lee; *Dir:* Michael Powell, Emeric Pressburger. **VHS $29.95** VID 🎬🎬½

Night Angel
A beautiful seductress lures men into a deadly trap from which only true love can free them.
1990 (R) 90m/C Isa Anderson, Linda Ashby, Debra Feuer, Helen Martin, Karen Black; *Dir:* Dominique Othenin-Girard. **VHS, Beta, LV $89.95** FRH 🎬🎬½

Night of the Assassin
A priest leaves his pulpit to become a terrorist. He plans a surprise for a U.N. secretary visiting Greece that would put the U.S. and Greek governments in the palm of his hand. Hard to believe.
1977 98m/C Klaus Kinski, Michael Craig, Eva Renzi; *Dir:* Robert McMahon. **VHS, Beta $69.98** LIV, LTG 🎬

Night Beast
An alien creature lands his spaceship near a small town and begins a bloody killing spree. Recycled plot.
1983 90m/C Tom Griffith, Dick Dyszel, Jaimie Zemarel. **VHS, Beta $42.95** PGN 🎬

Night Beat
A labored crime drama from postwar Britain. Two army pals pursue diverging careers. One becomes a cop, the other a crook. Want to bet that they meet again?
1948 95m/B GB Anne Crawford, Maxwell Reed, Ronald Howard, Christine Norden; *Dir:* Harold Huth. **VHS $15.95** LOO 🎬½

The Night Before
Snobby high school beauty Loughlin loses a bet and has to go to the prom with the school geek Reeves. On the way, they get lost on the wrong side of the tracks, and become involved with pimps, crime and the police. A drunken Reeves loses Loughlin as well as his father's car. Typical teen farce.
1988 (PG-13) 90m/C Keanu Reeves, Lori Loughlin, Trinidad Silva, Michael Greene, Theresa Saldana, Suzanne Snyder; *Dir:* Thom Eberhardt. **VHS, Beta $79.99** HBO 🎬🎬

Night of the Blood Beast
An astronaut comes back from space with an alien growing inside his body. The low budget defeats a valiant attempt at a story. Alternative title: "Creature from Galaxy 27."
1958 65m/B Michael Emmet, Angela Greene, John Baer, Ed Nelson; *Dir:* Bernard L. Kowalski. **VHS $16.95** SNC, MRV 🎬

Night of the Bloody Apes
When a doctor transplants an ape's heart into his dying son's body, the son goes berserk. Gory Mexican-made horror at its finest. Alternate title: "Gomar the Human Gorilla."
1968 (R) 84m/C MX Jose Elias Moreno, Carlos Lopez Moctezuma, Armando Silvestre, Norma Lazarendo, Augustin Martinez Solares, Gina Moret, Noelia Noel, Gerard Zepeda; *Dir:* Rene Cardona Sr. **VHS, Beta $59.95** MPI Woof!

Night of Bloody Horror
Tale of a former mental patient who is believed to be responsible for the brutal murders of his ex-girlfriends. A night of bloody horror indeed, as the gore is liberally spread.
1976 (R) 89m/C Gaye Yellen, Evelyn Hendricks, Gerald McRaney. **VHS, Beta** PGN 🎬½

Night of the Bloody Transplant
A lunatic scientist switches hearts from one person to another in this bloody gorefest. Footage of real surgery adds authenticity, but not believeability.
198? (R) 90m/C **VHS, Beta $24.95** UHV Woof!

Night Breed
A teenager flees a chaotic past to slowly become a member of a bizarre race of demons that live in a huge, abandoned Canadian graveyard; a place where every sin is forgiven. Based on Barker's novel "Cabal," and appropriately gross, nonsensical and strange. Good special effects almost save this one.
1990 (R) 102m/C CA Craig Sheffer, Anne Bobby, David Cronenberg, Charles Haid; *Dir:* Clive Barker. **VHS, Beta, LV $89.98** MED 🎬½

The Night Brings Charlie
A disfigured tree surgeon is the prime suspect in some grisly murders committed with a tree trimming saw. Obviously, everyone in the town fears him as a result. Not a sequel to "Roots."
1990 90m/C Kerry Knight, Joe Fishback, Aimee Tenalia, Monica Simmons; *Dir:* Tom Logan. **VHS $79.95** QUE 🎬½

Night Call Nurses
Three gorgeous nurses find danger and intrigue on the night shift at a psychiatric hospital. Third in the "nurse" quintet is back on target, shrugging off the previous film's attempts at "serious" social commentary. Miller provides comic relief. Preceded by "Private Duty Nurses" and followed by "The Young Nurses." Also on video as "Young L.A. Nurses 2."
1972 (R) 85m/C Patricia T. Byrne, Alana Collins, Mittie Lawrence, Clinton Kimbrough, Felton Perry, Stack Pierce, Richard Young, Dennis Dugan, Dick Miller; *Dir:* Jonathan Kaplan. **VHS, Beta $59.98** CHA 🎬🎬

Night Caller from Outer Space When a woman-hunting alien arrives in London, women begin to disappear. At first, incredibly, no one makes the connection, but then the horrible truth comes to light. Also called "Blood Beast From Outer Space."
1966 84m/B *GB* John Saxon, Maurice Denham, Patricia Haines, Alfred Burke, Jack Watson, Aubrey Morris; ***Dir:*** John Gilling. **VHS, Beta, LV $14.95** *SVS* ☆½

A Night in Casablanca Groucho, Harpo and Chico find themselves in the luxurious Hotel Casablanca, going after some leftover Nazis searching for treasure. One of the later Marx Brothers' films, but still loaded with the familiar wisecracks and mayhem.
1946 85m/B Groucho Marx, Harpo Marx, Chico Marx, Charles Drake, Dan Seymour, Sig Rumann; ***Dir:*** Archie Mayo. **VHS, Beta $49.95** *IND* ☆☆☆

Night Children Tough L.A. cop teams with a parole officer to battle a nasty gang leader. The battle is quite violent and senseless.
1989 (R) 90m/C David Carradine, Nancy Kwan, Griffin O'Neal, Tawny Fere; ***Dir:*** Norbert Meisel. **VHS, Beta $79.95** *VMK* ☆

The Night Club Swinging single stands to gain fortune if he weds a certain girl, and, as luck would have it, he falls for her. When the usual hearts and roses don't convince her he's earnest, he tries to prove his love is true by attempting suicide and hiring a hit man.
1925 62m/B Raymond Griffith, Vera Reynolds, Wallace Beery, Louise Fadenza; ***Dir:*** Frank Urson. **VHS, Beta $19.95** *GPV* ☆☆½

Night of the Cobra Woman A woman who can turn herself into a cobra needs constant sex and snake venom to keep her eternally young. Shot on location in the Philippines. One-time underground filmmaker Meyer co-wrote the script, for the Cotman factory.
1972 (R) 85m/C *PH* Joy Bang, Marlene Clark, Roger Garrett, Slash Marks, Vic Diaz; ***Dir:*** Andrew Meyer. **VHS, Beta $59.99** *EMB* ☆

Night of the Comet After surviving the explosion of a deadly comet, two California girls discover that they are the last people on Earth. When zombies begin to chase them, things begin to lose their charm. Cute and funny, but the script runs out before the movie does.
1984 (PG-13) 90m/C Catherine Mary Stewart, Kelli Maroney, Robert Beltran, Geoffrey Lewis, Mary Woronov, Sharon Farrell, Michael Bowen; ***W/Dir:*** Thom Eberhardt. **VHS, Beta $19.98** *FOX* ☆☆½

Night Creature A tough, Hemingway-type writer is determined to destroy the man-eating black leopard which nearly killed him once before. Filmed in Thailand. Also known as "Out of the Darkness" and "Fear."
1979 (PG) 83m/C Donald Pleasence, Nancy Kwan, Ross Hagen; ***Dir:*** Lee Madden. **VHS, Beta $19.95** *UHV* ☆

Night of the Creeps In 1958 an alien organism lands on earth and infects a person who is then frozen. Thirty years later he is accidentally unfrozen and starts spreading the infection throughout a college town. B-movie homage contains every horror cliche there is. Director Dekker's first film.
1986 (R) 89m/C Jason Lively, Jill Whitlow, Tom Atkins, Dick Miller, Steve Marshall; ***Dir:*** Fred Dekker. **VHS, Beta, LV $14.99** *HBO* ☆☆½

Night Cries After a woman's baby dies at birth, she has persistent dreams that he is alive and in trouble. No one will believe her when she tells them that something is wrong. Made for television.
1978 100m/C Susan St. James, Michael Parks, William Conrad; ***Dir:*** Richard Lang. **VHS, Beta** *WOV* ☆☆½

A Night for Crime A movie star disappears during the filming of a movie. When she turns up dead, a bumbling cop suspects everyone on the set.
1942 75m/B Glenda Farrell, Lyle Talbot, Lena Basquette, Donald Kirke, Ralph Sanford; ***Dir:*** Alexis Thurn-Taxis. **VHS $24.95** *NOS, DVT* ☆☆

Night Crossing The fact-based story of two East German families who launch a daring escape to the West in a homemade hot air balloon. Exciting action for the whole family, or even parts of it.
1981 (PG) 106m/C John Hurt, Jane Alexander, Glynnis O'Connor, Doug McKeon, Beau Bridges; ***Dir:*** Delbert Mann. **VHS, Beta $69.95** *DIS* ☆☆½

The Night Cry Silent canine melodrama has sheep-murdering dog sentenced to death row until he saves little girl from unbelievably big bird.
1926 65m/B Rin Tin Tin, John Harron, June Marlowe, Mary Louise Miller; ***Dir:*** Herman C. Raymaker. **VHS, Beta** *GPV* ☆☆

Night of the Cyclone A complex but not overly interesting thriller about a big city cop who goes to an island paradise in search of his missing daughter and finds murder.
1990 (R) 90m/C Kris Kristofferson, Jeffrey Meek, Marisa Berenson, Winston Ntshona, Gerrit Graham; ***Dir:*** David Irving. **VHS, LV $89.98** *REP* ☆☆

Night of Dark Shadows An abominable film which even fans of the TV show have a hard time liking. The last of the Collins family and his new bride are sent 150 years into the past. Also known as "Curse of Dark Shadows." Sequel to "House of Dark Shadows."
1971 (PG) 97m/C David Selby, Lara Parker, Kate Jackson, Grayson Hall, John Karlen, Nancy Barrett, James Storm; ***Dir:*** Dan Curtis. **VHS, Beta $19.98** *MGM Woof!*

Night and Day Sentimental musical about the life of bon-vivant composer-extraordinaire Cole Porter. His intensity and the motivations for his music are dramatized, but this film succeeds best as a fabulous showcase for Porter's songs. Includes "Begin the Beguine," "My Heart Belongs to Daddy," "I've Got You Under My Skin," "I Get a Kick Out of You," and many more.
1946 128m/C Cary Grant, Alexis Smith, Jane Wyman, Eve Arden, Mary Martin, Alan Hale Jr., Monty Woolley, Ginny Simms, Dorothy Malone; ***Dir:*** Michael Curtiz. **VHS, Beta $29.98** *FOX, FCT* ☆☆½

Night of the Death Cult After moving to a quiet seaside community, a young couple is plagued by cult practising human sacrifice in order to appease the Templars, a zombie-like pack of ancient clergymen who rise from the dead and torture the living. Last in a four film series about the Templars. Alternate title: "Night of the Seagulls."
1975 90m/C *SP* Victor Petit, Julie James, Maria Kosti, Sandra Mozarowsky; ***Dir:*** Armando de Ossorio. **VHS, Beta $69.95** *SVS* ☆☆

Night of the Demon Anthropology students are attacked by the legendary Bigfoot. Later they discover that he has raped and impregnated a young woman. Gore and sex prevail.
1980 97m/C Jay Allen, Michael J. Cutt, Bob Collins, Jodi Lazarus; ***Dir:*** James C. Watson. **VHS, Beta $49.95** *VHE Woof!*

Night of the Demons A gory, special-effects-laden horror farce about teenagers calling up demons in a haunted mortuary. On Halloween, of course.
1988 (R) 92m/C Linnea Quigley, Cathy Podewell, Alvin Alexis, William Gallo, Mimi Kinkade, Lance Fenton; ***Dir:*** Kevin S. Tenney. **VHS, LV $89.98** *REP* ☆☆

The Night Evelyn Came Out of the Grave A wealthy Italian playboy, obsessed with his dead, flame-haired wife, lures living redheads into his castle, where he tortures and kills them. Standard '70s Euro/horror/sex stuff enlivened by the presence of the incredible Erika Blanc as a stripper who works out of a coffin.
1971 (R) 99m/C *IT* Anthony Steffen, Marina Malfatti, Rod Murdock, Erika Blanc, Giacomo Rossi-Stuart, Umberto Raho; ***Dir:*** Emilio P. Miraglio. **VHS $18.00** *FRG* ☆½

Night of Evil A girl is released from reform school and promptly wins a beauty contest. Her bid to win the Miss America title is blown to smithereens however, when it is discovered that she has been secretly married all along. From there she resorts to working in strip joints and attempts to pull off an armed robbery. As exploitation fare goes, a winner.
1962 88m/B Lisa Gaye, William Campbell; ***Dir:*** Richard Galbreath; ***Nar:*** Earl Wilson. **VHS, Beta $16.95** *SNC* ☆½

Night Eyes Surveillance can be a dangerous profession, especially when your boss happens to be the jealous husband of a very sexy woman. Watching turns to yearning for this professional peeper, but he may just get more than he bargained for. Roberts is worth watching.
1990 (R) 95m/C Andrew Stevens, Tanya Roberts, Warwick Sims, Cooper Huckabee; ***Dir:*** Emilio P. Miraglio. **VHS, Beta, LV $89.99** *PAR* ☆

Night Flight During early history of aviation, one young pilot gets lost in a storm, begins questioning immortality, human values, and decision making. Based on Antoine de Saint Exupery's novel.
1979 22m/C Trevor Howard, Bo Svenson. **VHS, Beta** *LCA* ☆☆

Night Flight from Moscow A Soviet spy defects with a fistful of secret documents that implicate every free government. Then the CIA must decide if he's telling the truth. Complex espionage thriller. Also known as "The Serpent."
1973 (PG) 113m/C *FR IT GE* Yul Brynner, Henry Fonda, Dirk Bogarde, Virna Lisi, Philippe Noiret, Farley Granger, Robert Alda, Marie DuBois, Elga Andersen; ***Dir:*** Henri Verneuil. **VHS, Beta $59.98** *CHA* ☆☆½

Night of the Fox An American officer with top secret knowledge is captured by Germans on the brink of D-day. His home team plans to kill him if he cannot be rescued before spilling the beans. On-location filming adds much. Made for cable.
1990 (R) 95m/C Michael York, Deborah Raffin, George Peppard; ***Dir:*** Charles Jarrott. **VHS, LV $89.95** *VMK* ☆☆

Night Friend A priest sets out to help a young woman caught up in a world of drug abuse and prostitution. Exploitation morality flick.
1987 (R) 94m/C Art Carney, Chuck Shamata, Jayne Eastwood, Heather Kjollesdal; *Dir:* Peter Gerretsen. **VHS, Beta $79.95** *PSM* 🎬

A Night Full of Rain Italian communist tries unsuccessfully to seduce vacationing American feminist. They meet again in San Francisco and wedding bells chime. They argue. They argue more. They put all but rabid Wertmuller fans to sleep. The director's first English-language film, originally titled "End of the World (in Our Usual Bed in a Night Full of Rain)."
1978 (R) 104m/C *IT* Giancarlo Giannini, Candice Bergen, Annie Byrne, Flora Carabella; *W/Dir:* Lina Wertmuller. **VHS $59.98** *WAR, FCT* 🎬½

Night Gallery Rod Serling is the tour guide through an unusual art gallery consisting of portraits that reflect people's greed, desire and guilt. Made for television pilot for the series, which ran from 1969 to 1973. Three stories, including "Eyes," which saw novice Spielberg directing veteran Crawford.
1969 95m/C Joan Crawford, Roddy McDowall, Tom Bosley, Barry Sullivan, Ossie Davis, Sam Jaffe, Kate Greenfield; *Dir:* Steven Spielberg, Boris Sagal. **VHS, Beta $59.95** *MCA* 🎬½

Night Game A cop links a string of serial killings to the night games won by the Houston Astros. Brave cast tries hard, but is retired without a hit.
1989 (R) 95m/C Roy Scheider, Karen Yahng, Paul Gleason, Lane Smith, Carlin Glynn; *Dir:* Peter Masterson. **VHS, Beta, LV $89.99** *HBO* 🎬½

Night Games Raped as a child, California girl Pickett has trouble relating to her husband. To overcome her sexual anxieties, she engages in a number of bizarre fantasies.
1980 (R) 100m/C Cindy Pickett, Joanna Cassidy, Barry Primus, Gene Davis; *Dir:* Roger Vadim. **VHS, Beta, LV $69.98** *SUE* 🎬

Night of the Generals A Nazi intelligence officer is pursuing three Nazi generals who may be involved in the brutal murder of a Warsaw prostitute. Dark and sinister, may be too slow for some tastes. Based on Hans Helmut Kirst's novel.
1967 (R) 148m/C Peter O'Toole, Omar Sharif, Tom Courtenay, Joanna Pettet, Donald Pleasence, Christopher Plummer, Philippe Noiret, John Gregson; *Dir:* Anatole Litvak. **VHS, Beta, LV $9.95** *COL* 🎬🎬🎬

Night of the Ghouls The last in Edward Wood's celebrated series of inept so-called horror films, begun with "Bride of the Monster" and "Plan 9 from Outer Space." This one tells of a spiritualist who raises the dead. Unreleased for over 20 years because Wood couldn't pay the film lab. Not quite as classically bad as his other films, but still a laugh riot. Also known as "Revenge of the Dead."
1959 69m/B Paul Marco, Tor Johnson, Duke Moore, Kenne Duncan, John Carpenter, Criswell; *Dir:* Edward D. Wood Jr. **VHS, Beta $9.95** *RHI, SNC, MED* Woof!

The Night God Screamed A fanatical cult leader is convicted of murder and the cult goes wild. Also known as "Scream."
198? (PG) 85m/C Michael Sugich, Jeanne Crain, James B. Sikking; *Dir:* Lee Madden. **VHS, Beta $19.98** *TWE* Woof!

Night of the Grizzly An ex-lawman's peaceful life as a rancher is threatened when a killer grizzly bear goes on a murderous rampage terrorizing the residents of the Wyoming countryside.
1966 99m/C Clint Walker, Martha Hyer; *Dir:* Joseph Pevney. **VHS, Beta $29.95** *PAR* 🎬🎬

The Night Has Eyes Tense melodrama concerns a young teacher who disappears on the Yorkshire moors at the same spot where her girlfriend had vanished the previous year. Early film appearance for Mason. Also known as "Terror House."
1942 79m/B *GB* James Mason, Joyce Howard, Wilfred Lawson, Mary Clare; *Dir:* Leslie Arliss. **VHS, Beta $19.95** *NOS, SNC, DVT* 🎬🎬½

A Night in Heaven A college teacher gets involved with one of her students, who moonlights as a male stripper. Will he earn that extra credit he needs to pass? Uninspired, overly explicit. Look for Denny Terrio of "Dance Fever" fame.
1983 (R) 85m/C Christopher Atkins, Lesley Ann Warren, Carrie Snodgress, Andy Garcia; *Dir:* John G. Avildsen. **VHS, Beta $59.98** *FOX* Woof!

Night of Horror Zombies attack four young people stranded in the wilderness.
198? (R) 76m/C Steve Sandkuhler, Gae Schmitt, Rebecca Bach, Jeff Canfield. **VHS, Beta** *GHV, MRV* 🎬½

Night of the Howling Beast The selling point of this one is an epic, first-time battle between a werewolf and a Yeti. Unfortunately it lasts about fifteen seconds. Naschy's eighth stint as the wolfman. Also known as "The Werewolf and the Yeti."
1975 (R) 87m/C *SP* Paul Naschy, Grace Mills, Castillo Escalona, Silvia Solar, Gil Vidal; *Dir:* Miguel Iglesias Bonns. **VHS, Beta** *REP* 🎬

The Night of the Hunter The nightmarish story of a psychotic preacher who marries a lonely widow with two children in the hopes of finding the cache of money her thieving husband had stashed. A dark, terrifying tale, completely unique in Hollywood's history. Mitchum is terrific. Written by James Agee, from novel by Davis Grubb and, sadly, Laughton's only directorial effort.
1955 93m/B Robert Mitchum, Shelley Winters, Lillian Gish, Don Beddoe, Evelyn Varden, Peter Graves, James Gleason, Billy Chapin; *Dir:* Charles Laughton. **VHS, Beta, LV $19.98** *MGM, VYG* 🎬🎬🎬🎬

The Night of the Iguana An alcoholic ex-minister acts as a tour guide in Mexico, becoming involved with a spinster and a hotel owner. Based on Tennessee Williams' play. Excellent performances from Burton and Gardner. Hall received an Oscar nomination for her supporting actress performance. Other Oscar nominations for Art Direction, Set Decoration and Cinematography.
1964 125m/B Richard Burton, Deborah Kerr, Ava Gardner, Grayson Hall, Sue Lyon; *Dir:* John Huston. Academy Awards '64: Best Costume Design (B & W). **VHS, Beta $19.98** *MGM* 🎬🎬🎬

Night Is My Future A blind young man meets a girl who tries to bring him happiness. Usual somber Bergman, but unimportant story. In Swedish with English subtitles. Also released as "Music in Darkness."
1947 89m/B *SW* Mai Zetterling, Birger Malmsten, Naima Wifstrand, Olof Winnerstrand, Hilda Borgstrom; *Dir:* Ingmar Bergman. **VHS, Beta** *WFV, VYY, DVT* 🎬🎬½

Night of the Juggler An ex-cop encounters countless obstacles in trying to track down his daughter's kidnapper in New York City. Just a tad too complicated for some tastes.
1980 (R) 101m/C James Brolin, Cliff Gorman, Richard Castellano, Mandy Patinkin, Julie Carmen; *Dir:* Robert Butler. **VHS, Beta $49.95** *MED* 🎬🎬½

Night of the Kickfighters A mighty band of martial artists takes on a terrorist group that has in their possession a secret weapon that could destroy the world.
1991 87m/C Andy Bauman, Adam West, Marcia Karr; *Dir:* Buddy Reyes. **VHS $89.98** *AIP* 🎬

Night of the Laughing Dead Veteran British comedy cast perform unthinkable spoof. Man stands to inherit lotsa money as his family members kick the bucket one by one. Also known as "Crazy House" and "House in Nightmare Park."
1973 90m/C *GB* Ray Milland, Frankie Howerd, Rosalie Crutchley, Kenneth Griffith; *Dir:* Peter Sykes. **VHS, LV** *MGM* 🎬½

Night Life A teenager gets the all-out, high-stakes ride of his life when four cadavers are re-animated in his uncle's mortuary.
1990 (R) 92m/C Scott Grimes, John Astin, Cheryl Pollak, Alan Blumenfeld; *Dir:* David Acomba. **VHS, LV $79.95** *COL* 🎬🎬

A Night in the Life of Jimmy Reardon A high school Casanova watches his friends leave for expensive schools while he contemplates a trip to Hawaii with his rich girlfriend, a ruse to avoid the dull business school his father has picked out. Well photographed, but acting leaves something to be desired. Based on Richert's novel "Aren't You Even Going to Kiss Me Good-bye?"
1988 (R) 95m/C River Phoenix, Meredith Salenger, Matthew L. Perry, Louanne, Ione Skye, Ann Magnuson, Paul Koslo, Jane Hallaren, Jason Court; *Dir:* William Richert. **VHS, Beta, LV $19.98** *FOX* 🎬½

Night Life in Reno When a man leaves his wife, she sets out for Reno and falls in love with a married man. When her new love is killed by his jealous wife, the woman returns to her husband and they begin their marriage anew. Watchable melodrama.
1931 58m/B Jameson Thomas, Dorothy Christy, Virginia Valli, Carmelita Geraghty. **VHS, Beta, LV** *WGE* 🎬🎬

The Night the Lights Went Out in Georgia Loosely based on the popular hit song, the film follows a brother and sister as they try to cash in on the country music scene in Nashville. McNichol is engaging.
1981 (PG) 112m/C Kristy McNichol, Dennis Quaid, Mark Hamill, Don Stroud; *Dir:* Ronald F. Maxwell. **VHS, Beta $69.95** *SUE, TWE* 🎬🎬

Night of the Living Babes Two yuppie men go looking for fun at a brothel. They get their just deserts in the form of sex-seeking female zombies.
1987 60m/C Michelle McClellan, Connie Woods, Andrew Nichols, Louie Bonanno; *Dir:* Jon Valentine. **VHS $39.95** *MAG* Woof!

Night of the Living Dead Cult favorite is low budget but powerfully frightening. Space experiments set off a high level of radiation that makes the newly dead return to life, with a taste for human flesh. Handful of holdouts find shelter in a farmhouse. Claustrophobic, terrifying, gruesome, extreme, and yes, humorous. Followed by "Dawn of the Dead" (1979) and "Day of the Dead" (1985).

Romero's directorial debut. Available in a colorized version.
1968 90m/B Judith O'Dea, Duane Jones, Russell Streiner, Karl Hardman; *Dir:* George A. Romero. **VHS, LV $19.95** *REP, MRV, NOS* ♂♂♂½

Night of the Living Dead A bunch of people are trapped in a farmhouse and attacked by ghouls with eating disorders. Remake of the '68 classic substitutes high tech blood 'n' guts for bona fide frights.
1990 (R) 92m/C Tony Todd, Patricia Tallman, Tom Towles; *Dir:* Tom Savini. **VHS, LV $19.95** *COL* ♂♂

Night Master A handful of karate students practice their homework with much higher stakes away from the classroom. For them, deadly Ninja games are the only way to study.
1987 87m/C Tom Jennings, Nicole Kidman, Vince Martin; *Dir:* Mark Joffe. **VHS, Beta $79.95** *ACA* ♂

'night, Mother A depressed woman, living with her mother, announces one evening that she is going to kill herself. Her mother spends the evening reliving their lives and trying to talk her out of it, but the outcome seems inevitable. Well acted, though depressing. Based on Marsha Norman's Pulitzer Prize-winning novel.
1986 (PG-13) 97m/C Sissy Spacek, Anne Bancroft; *Dir:* Tom Moore. **VHS, Beta, LV $79.95** *MCA* ♂♂½

Night Moves While tracking down a missing teenager, a Hollywood detective uncovers a bizarre smuggling ring in the Florida Keys. Hackman is realistic as the detective whose own life is unraveling and in one of her early roles, Griffith plays the teenager. The dense, disjointed plot reaches a shocking end. Underrated when released and worth a view.
1975 (R) 100m/C Gene Hackman, Susan Clark, Jennifer Warren, Melanie Griffith, Harris Yulin, Edward Binns, Kenneth Mars, James Woods, Dennis Dugan, Max Gail; *Dir:* Arthur Penn. **VHS, Beta $19.98** *WAR* ♂♂♂½

Night of the Ninja Hop 'n' kick mystery about the search for a missing diamond. Investigator is led down a deadly trail as he tracks Ninja assassins who know where the gem is hidden.
19?? ?m/C VHS $59.99 *IMP* ♂½

Night Nurse A young nurse signs on to care for an aging, wheelchair-ridden opera star, only to find that the house is haunted.
1977 80m/C Davina Whitehouse, Kay Taylor, Gary Day, Kate Fitzpatrick; *Dir:* Igor Auzins. **VHS, Beta $42.95** *PGN* ♂

A Night at the Opera The Marx Brothers get mixed up with grand opera in this, their first MGM-produced epic. Jones, as an opera singer on the rise, sings "Alone" and "Cosi Cosa." Written by George S. Kaufman and Mollie Ryskind, it's blessed with a big budget that the brothers use to reach epic anarchic heights. One of the best Marx Brothers' films, it features scenes which were perfected and tested on live audiences before their inclusion, including the Groucho-Chico paper-tearing contract negotiation and the celebrated stateroom scene, in which the Marxists are joined in a small shipboard closet by two maids, the ship's engineer, his assistant, a manicurist, a young woman, a cleaning lady, and four food-laden waiters. The first effort without Zeppo. Also available on laserdisc with letterboxing, digital sound, original trailers, pro-

duction photos, memorabilia, and soundtrack commentary by film critic Leonard Maltin.
1935 87m/B Groucho Marx, Chico Marx, Harpo Marx, Allan Jones, Kitty Carlisle, Sig Rumann, Margaret Dumont, Walter Woolf King; *Dir:* Sam Wood. **VHS, Beta, LV $19.95** *MGM, VYG, CRC* ♂♂♂♂

Night Partners Bored housewives assist the police patrol the streets after the kids have gone to bed. Originally made for television with the intention of turning it into a series, but the script doesn't hold water.
1983 100m/C Yvette Mimieux, Diana Canova, Arlen Dean Snyder, M. Emmet Walsh, Patricia Davis, Larry Linville; *Dir:* Noel Nosseck. **VHS, Beta $39.95** *NWV* ♂½

Night Patrol The streets of Hollywood will never be the same after the night patrol runs amuck in the town. Crude imitation of "Police Academy."
1985 (R) 87m/C Linda Blair, Pat Paulsen, Jaye P. Morgan, Jack Riley, Murray Langston, Billy Barty, Pat Morita, Sydney Lassick, Andrew Dice Clay; *Dir:* Jackie Kong. **VHS, Beta, LV $9.95** *NWV, STE* Woof!

The Night Porter Max, an ex-SS concentration camp officer, unexpectedly meets his former lover-victim at the hotel where he works as the night porter. After they get reacquainted, the couple must hide from the porter's ex-Nazi friends who want the woman dead because they fear she will disclose their past. Sleazy, sado-masochistic.
1974 115m/C *IT* Dirk Bogarde, Charlotte Rampling, Phillippe LeRoy, Gabriel Ferzetti, Isa Miranda; *Dir:* Liliana Cavani. **VHS, Beta, LV $19.98** *SUE, GLV* ♂

A Night to Remember A murder-mystery writer's wife convinces him to move to a new apartment because she thinks the change might help him finish a novel he started long ago. When they find a dead body behind their new building, they try their hands at sleuthing. A clever and witty mystery, indeed, supported by likeable performances.
1942 91m/B Loretta Young, Brian Aherne, Sidney Toler, Gale Sondergaard, Willaim Wright, Donald MacBride, Blanche Yurka; *Dir:* Richard Wallace. **VHS $69.95** *COL* ♂♂

A Night to Remember Gripping tale of the voyage of the Titanic with an interesting account of action in the face of danger and courage amid despair. Large cast is effectively used. Adapted by Eric Ambler from the book by Walter Lord.
1958 119m/B Kenneth More, David McCallum, Anthony Bushell, Honor Blackman, Michael Goodlife, George Rose, Frank Lawton, Alec McCowen; *Dir:* Roy Ward Baker. National Board of Review Awards '58: 5 Best Foreign Films of the Year. **VHS, Beta, LV $19.95** *PAR* ♂♂♂

Night Rhythms Nick West is a radio talk show host with a sexy voice that causes his women listeners to reveal their most intimate fantasies and problems to him on the air. Only one listener goes too far, now it's murder, and Nick must work to prove his innocence in the crime if he intends to save his career—and his life. Also available in an unrated version.
199? (R) ?m/C Martin Hewitt, Delia Sheppard, David Carradine, Terry Tweed, Sam Jones, Deborah Driggs, Julie Strain; *Dir:* Alexander Gregory Hippolyte. **VHS $89.95** *IMP* ♂½

Night Rider A cowboy posing as a gunman ends a trail of murder and violence in a small western town.
1932 54m/B Harry Carey, George "Gabby" Hayes. **VHS, Beta $19.95** *NOS, DVT, WGE* ♂

Night Riders of Montana A state ranger helps a group of ranchers to fight off a band of rustlers.
1950 60m/B Allan Lane. **VHS, Beta $24.95** *DVT* ♂

Night Ripper A psychopathic killer stalks high fashion models and kills them by carving them up. Lots of violence.
1986 88m/C James Hansen, April Anne, Larry Thomas; *Dir:* Jeff Hathcock. **VHS, Beta** *MAG* ♂

Night School A police detective must find out who has been decapitating the women attending night school at Wendell College. Ward's first film. Also called "Terror Eyes."
1981 (R) 89m/C Edward Albert, Rachel Ward, Leonard Mann, Drew Snyder, Joseph R. Sicari; *Dir:* Ken Hughes. **VHS, Beta $19.98** *FOX* ♂

Night Screams Violent scare-monger about two escaped convicts who crash a high school house party. Kids and convicts start getting killed, one by one.
1987 85m/C Joe Manno, Ron Thomas, Randy Lundsford, Megan Wyss; *Dir:* Allen Plone. **VHS, Beta $79.95** *PSM* ♂

Night Shadow A woman, returning home after being away for years, picks up a hitchhiker. Soon after her arrival, terrible serial killings begin, and only she has the nerve to track down the killer.
1990 (R) 90m/C Brenda Vance, Dana Chan, Tom Boylan; *W/Dir:* Randolph Cohlan. **VHS** *QUE* ♂

Night of the Sharks Mercenaries in a downed plane go after a jewel-ridden shipwreck despite a plethora of sharks.
1989 87m/C Treat Williams, Christopher Connelly, Antonio Fargas, Janet Agren; *Dir:* Anthony Richmond. **VHS, Beta, LV $79.95** *IVE* Woof!

Night Shift Two morgue attendants, dull Winkler and manic Keaton, decide to spice up their latenight shift by running a call-girl service on the side. Keaton turns in a fine performance in his film debut, and Howard's sure-handed direction almost overcome the silly premise.
1982 (R) 106m/C Henry Winkler, Michael Keaton, Shelley Long, Kevin Costner; *Dir:* Ron Howard. **VHS, Beta, LV $19.98** *WAR* ♂♂

Night of the Shooting Stars Set in an Italian village during the last days of World War II, this film highlights the schism in the village between those who support the fascists and those who sympathize with the Allies. This division comes to a head in the stunning final scene. A poignant, deeply moving film.
1982 (R) 107m/C *IT* Omero Antonutti, Margarita Lozano; *Dir:* Paolo Taviani, Vittorio Taviani. Cannes Film Festival '83: Jury Prize; National Society of Film Critics Awards '83: Best Film. **VHS, Beta $29.95** *MGM, APD* ♂♂♂♂

Night Slasher A madman enjoys spilling the innards of London prostitutes with his dagger.
1984 87m/C Jack May, Linda Marlowe. **VHS, Beta $49.95** *UNI* ♂

Night of the Sorcerers An expedition to the Congo uncovers a bizarre tribe of vampire leopard women who lure young girls to their deaths.
1970 85m/C Jack Taylor, Simon Andrew, Kali Hansa; *Dir:* Armando de Ossorio. **VHS, Beta $49.95** *UNI* Woof!

N

Night Stage to Galveston Autry leads his Texas Rangers on a mission to uncover corruption in the Texas State Police during the turbulent post-Civil War days.
1952 61m/B Gene Autry, Pat Buttram, Virginia Huston, Thurston Hall; **Dir:** George Archainbaud. **VHS, Beta $29.98** *CCB* 🎬½

Night Stalker A horror film about two 12,000-year-old cannibals who terrorize the citizens of Los Angeles in their search for immortality. Lots of violence and sex.
1979 90m/C Aldo Ray. **VHS, Beta $29.98** *FOX, HHE* 🎬½

Night Stalker A bloodthirsty serial killer is tracked by a detective through the streets in New York. The usual rigermorale of fiztizuffe, gunplay, and car chases ensure. Impressive acting from Napier raises the film from run-of-the-mill.
1987 (R) 91m/C Charles Napier, John Goff, Robert Viharo, Robert Z'Dar, Joseph Gian, Gary Crosby, Joan Chen, Michelle Reese; **Dir:** Max Cleven. **VHS, Beta $79.98** *VES, LTG* 🎬🎬

The Night Stalker: Two Tales of Terror Two episodes from the television series "Kolchak: The Night Stalker" in which McGavin plays a nosy newspaper reporter who constantly stumbles upon occult phenomenon. Episodes are "Jack the Ripper" and "The Vampire."
1974 98m/C Darren McGavin, Simon Oakland. **VHS, Beta** *MCA* 🎬🎬

Night of the Strangler A love affair between a white society girl and a young black man causes a chain of events that end with brutal murders in New Orleans.
1973 (R) 88m/C Mickey Dolenz, James Ralston, Susan McCullough; **Dir:** Joy Houck Jr. **VHS, Beta $42.95** *PGN* 🎬🎬

Night Terror Everyone's after a hausfrau who saw a highway patrolman murdered on an expressway...including the psychotic murderer. Pretty standard fare. Made for television.
1976 73m/C Valerie Harper, Richard Romanus, Michael Tolan, Beatrice Manley, John Quade, Quinn Cummings, Nicholas Pryor; **Dir:** E.W. Swackhamer. **VHS, Beta $59.95** *WOV* 🎬

Night of Terror A bizarre family conducts brain experiments on themselves and then begins to kill each other. Bloodshed for the whole family.
1987 105m/C Renee Harmon, Henry Lewis. **VHS $49.95** *IME* 🎬

Night Terror A man wakes up from a nightmare only to find that his terrors are still present.
1989 90m/C Lloyd B. Mote, Jeff Keel, Guy Ecker, Jon Hoffman, Michael Coopet; **Dir:** Michael Weaver, Paul Howard. **VHS, Beta $79.95** *MAG* 🎬🎬

The Night They Raided Minsky's Chaotic but interesting period comedy about a young Amish girl who leaves her tyrannical father to come to New York City in the 1920s. She winds up at Minsky's Burlesque and accidentally invents the striptease. Lahr's last performance—he died during filming. Contains the songs "The Night They Raided Minsky's" and "Penny Arcade." Also called "The Night They Invented Striptease." Cowritten by Norman Lear.
1969 (PG) 97m/C Jason Robards Jr., Britt Ekland, Elliott Gould, Bert Lahr, Norman Wisdom, Denholm Elliott; **Dir:** William Friedkin. **VHS, Beta $19.98** *FOX* 🎬🎬🎬

The Night They Robbed Big Bertha's A bungling burglar attempts to knock off Big Bertha's massage parlor.
1983 83m/C Robert Nichols. **VHS, Beta $42.95** *PGN* 🎬

The Night They Saved Christmas A Christmas special in which Santa's North Pole headquarters are endangered by the progress of an expanding oil company. Will the attempts of three children be enough to save the day?
1987 94m/C Art Carney, Jaclyn Smith, Paul Williams, Paul LeMat; **Dir:** Jackie Cooper. **VHS, Beta $9.95** *CAF, STE, PSM* 🎬½

Night of a Thousand Cats A reclusive playboy cruises Mexico City in his helicopter, searching for beautiful women. It seems he needs their bodies for his cats who, for some reason, eat only human flesh. He keeps the heads, for some reason, for his private collection. A really odd '70s cannibal cat entry, but it moves along nicely and features some truly groovy fashions (floppy hats, translucent blouses, etc.). Also on video as "Blood Feast," not to be confused with the 1963 film of the same name. From the director of "Night of the Bloody Apes" and "Survive!"
1972 (R) 83m/C *MX* Anjanette Comer, Zulma Faiad, Hugo Stiglitz, Christa Linder, Teresa Velasquez, Barbara Ange; **Dir:** Rene Cardona Sr. **VHS, Beta $19.95** *PGN, ACA, JEF* 🎬

Night Tide Hopper in another off the wall character study, this time as a lonely sailor. He falls for a mermaid (Lawson) who works at the dock. She may be a descendent of the man-killing Sirens. Interesting and different little love story, sometimes advertised as horror, which it is not.
1963 84m/B Dennis Hopper, Gavin Muir, Luana Anders, Marjorie Eaton, Tom Dillon; **Dir:** Curtis Harrington. **VHS $16.95** *SNC, FRG* 🎬🎬½

Night Time in Nevada Young woman heads West to claim $50,000 trust left by dear old dad, but his crooked lawyer and associate have their hands in the pot. Cowboy Roy rides horse and flexes dimples while helping helpless girl.
1948 67m/B Roy Rogers, Adele Mara, Andy Devine, Grant Withers, Joseph Crehan; **Dir:** William Witney. **VHS, Beta, 8mm $19.95** *NOS, HEG* 🎬🎬½

The Night Train to Kathmandu A young girl accompanies her parents on a research expedition to Nepal, where her head is turned by an exotically handsome young Sherpa. Old story without much help from actors or director. Made for cable.
1988 102m/C Milla Jovovich, Pernell Roberts, Eddie Castrodad; **Dir:** Robert Wiemer. **VHS, Beta, LV $19.95** *PAR* 🎬🎬

Night Train to Munich There's Nazi intrigue galore aboard a big train when a scientist's daughter joins allied intelligence agents in retrieving some secret documents. From the book "Report on a Fugitive" by Gordon Wellesley. Also known as "Night Train."
1940 93m/B *GB* Margaret Lockwood, Rex Harrison, Paul Henreid; **Dir:** Carol Reed. **VHS, Beta $24.95** *NOS, DVT* 🎬🎬🎬

Night Train to Terror Strange things start happening on the train where a rock band makes its last appearance. Clips from other horror flicks were pieced together to make this film that's so bad it's almost good.

1984 (R) 98m/C John Phillip Law, Cameron Mitchell, Marc Lawrence, Charles Moll, Ferdinand "Ferdy" Mayne; **Dir:** Jay Schlossberg-Cohen. **VHS, Beta $79.95** *PSM* Woof!

Night Vision A young writer in the big city is given a video monitor by a street thief which is equipped with some remarkable features, including the ability to present scenes of future murders and demon worship.
1987 102m/C Ellie Martins, Stacy Carson, Shirley Ross, Tony Carpenter; **Dir:** Michael Krueger. **VHS, Beta $79.95** *PSM* 🎬

The Night Visitor A man in a prison for the criminally insane seeks violent vengeance on those he believes have set him up. Heavily detailed, slow moving. Ullman and von Sydow can do better.
1970 (PG) 106m/C *GB DK* Max von Sydow, Liv Ullman, Trevor Howard, Per Oscarsson, Andrew Keir; **Dir:** Laslo Benedek. **VHS, Beta $59.95** *UHV* 🎬🎬½

The Night Visitor A retired police detective teams up with a teenage peeping tom to disclose the identity of a satanic serial killer. The killer, of course, is one of the youth's teachers.
1989 (R) 95m/C Derek Rydell, Shannon Tweed, Elliott Gould, Allen Garfield, Michael J. Pollard, Richard Roundtree, Henry Gibson; **Dir:** Rupert Hitzig. **VHS, Beta $79.95** *MGM* 🎬

Night Warning A slasher gorefest redeemed by Tyrrell's go-for-broke performance. She plays a sexually repressed aunt who makes up for her problems by going on murderous rampages. Hide the kitchen knives!
1982 (R) 96m/C Bo Svenson, Jimmy McNichol, Susan Tyrrell, Julia Duffy; **Dir:** William Asher. **VHS, Beta $19.99** *HBO* 🎬🎬

Night of the Warrior Music-videos and martial arts don't mix...not here, anyway. A hunky exotic-dance-club owner pays his disco bills by fighting in illegal, underground blood matches, but not enough to make it exciting. Lamas stars with real-life wife Kinmont and mom Dahl.
1991 (R) 96m/C Lorenzo Lamas, Anthony Geary, Kathleen Kinmont, Arlene Dahl. **VHS $89.95** *VMK* 🎬

Night Wars Two ex-POWs who are plagued by their memories turn to dream therapy to relive their escape. Eventually, they rescue a buddy they left behind.
1988 90m/C Dan Haggerty, Brian O'Connor, Cameron Smith; **Dir:** David A. Prior. **VHS, Beta $14.95** *SVS* 🎬½

Night Watch A woman recovering from a nervous breakdown witnesses a murder, but no one will believe her.
1972 100m/C *GB* Elizabeth Taylor, Laurence Harvey, Billie Whitelaw; **Dir:** Brian G. Hutton. **VHS, Beta $9.95** *MED* 🎬½

Night of the Wilding Very, very loosely based on the story of the female jogger who was gang-raped and left for dead in New York's Central Park. Made for television, soon after the actual incident occurred.
1990 90m/C Erik Estrada, Kathrin Lautner; **Dir:** Joseph Merhi. **VHS, Beta, LV $89.95** *VIR* 🎬

Night of the Zombies WWII soldiers with eating disorders shuffle through 88 minutes of gratuitous gore, while pursued by porn star cum intelligence agent. Alternate titles: "Gamma 693," "Night of the Wermacht Zombies." From the director of "Bloodsucking Freaks."

1981 (R) 88m/C James Gillis, Ryan Hilliard, Samantha Grey, Joel M. Reed; **W/Dir:** Joel M. Reed. **VHS** $49.99 PSM ⚁

Night of the Zombies The staff of a scientific research center are killed and then resurrected as cannibals who prey on the living.
1983 101m/C *IT SP* Margit Evelyn Newton, Frank Garfield, Selan Karay; **Dir:** Bruno Mattei. **VHS, Beta** $49.95 VES, PSM *Woof!*

Night Zoo A confusing story about a father/son reconciliation and the lurid underworld of drugs and sex in Montreal. The graphic sex and violence undermines the sensitive aspects of the film.
1987 115m/C *CA* Gilles Maheu, Roger Le Bel, Lynne Adams; **Dir:** Jean-Claude Lauzon. Genie Awards '88: Best Actor (Le Bel), Best Art Direction/Set Decoration, Best Cinematography, Best Costume Design, Best Director (Lauzon), Best Picture, Best Screenplay, Best Supporting Actor (Houde), Best Musical Score. **VHS, Beta, LV** $9.99 STE, NWV, IME ⚁

Nightbreaker Revelation of the U.S. government's deliberate exposure of servicemen to atomic bomb tests and the resulting radiation, in order to observe the effects on humans. Sheen and his son, Estevez, portray a U.S. doctor in the 1980s and the 1950s, respectively.
1989 100m/C Martin Sheen, Emilio Estevez, Lea Thompson, Melinda Dillon, Nicholas Pryor; **Dir:** Peter Markle. **VHS** $9.98 TTC ⚁⚁⚁

The Nightcomers A pretend "prequel" to Henry James's "The Turn of the Screw," wherein an Irish gardener trysts with the nanny of two watchful children who believe that lovers unite in death. Don't be fooled: stick to the original.
1972 (R) 96m/C *GB* Marlon Brando, Stephanie Beacham, Thora Hird; **Dir:** Michael Winner. **VHS, Beta, LV, CDV** $14.95 COL, CHA ⚁

Nightfall Ray, accused of a crime he didn't commit, is forced to flee from both the law and the underworld. Classic example of film noir, brilliantly filmed by Tourneur.
1956 78m/B Aldo Ray, Brian Keith, Anne Bancroft, Jocelyn Brando, James Gregory, Frank Albertson; **Dir:** Jacques Tourneur. **VHS** GKK ⚁⚁⚁

Nightfall Television adaptation of the classic Isaac Asimov short story. A planet that has two suns (and therefore no night) experiences an eclipse and its inhabitants go mad.
1988 (PG-13) 87m/C David Birney, Sarah Douglas, Alexis Kanner, Andra Millian; **Dir:** Paul Mayersberg. **VHS, Beta** $79.95 MGM ⚁⚁

Nightflyers Aboard a weathered space freighter, the crew experience a series of deadly accidents caused by an unknown evil presence. From a novella by George R.R. Martin; T.C. Blake is better known as Robert Collector.
1987 (R) 88m/C Michael Praed, Michael Des Barres, Catherine Mary Stewart, John Standing, Lisa Blount; **Dir:** T.C. Blake. **VHS, Beta, Special order formats** $9.99 CCB, IVE ⚁½

Nightforce Five buddies plunge into Central American jungles to rescue a young girl held by terrorists.
1986 (R) 87m/C Linda Blair, James Van Patten, Chad McQueen, Richard Lynch, Cameron Mitchell; **Dir:** Lawrence Foldes. **VHS, Beta** $79.98 LTG ⚁

Nighthawks New York City cops scour Manhattan to hunt down an international terrorist on the loose. They race from disco to subway to an airborne tramway. Exciting and well paced. Hauer's American film debut.

1981 (R) 100m/C Sylvester Stallone, Billy Dee Williams, Rutger Hauer, Lindsay Wagner, Nigel Davenport, Persis Khambatta, Catherine Mary Stewart, Joe Spinell; **Dir:** Bruce Malmuth. **VHS, Beta, LV** $19.98 MCA ⚁⚁½

The Nightingale From "Faerie Tale Theatre" comes the story of an Emperor who discovers the value of true friendship and loyalty from his palace kitchen maid who gives him a nightingale.
1983 60m/C Mick Jagger, Barbara Hershey, Bud Cort, Mako; **Dir:** Ivan Passer. **VHS, Beta, LV** $14.98 KUI, FOX, FCT ⚁⚁½

Nightkill A simple love affair turns into a deadly game of cat and mouse when a bored wife plots to do away with her wealthy, powerful husband with the aid of her attractive lover.
1980 (R) 104m/C Jaclyn Smith, Mike Connors, James Franciscus, Robert Mitchum, Sybil Danning; **Dir:** Ted Post. **VHS, Beta** $24.98 SUE ⚁½

Nightlife A vampiress, who rises from the dead after a hundred years, spends her time looking for blood to drink. Made for cable.
1990 93m/C Ben Cross, Maryam D'Abo, Keith Szarabajka; **Dir:** Daniel Taplitz. **VHS, Beta, LV** $79.95 MCA ⚁⚁½

Nightmare Boring splatter picture has a young boy hacking his father and his mistress to pieces when he discovers them in bed. He grows up to be a psycho who continues along the same lines. Humorless and dreadful, the original ads claimed Tom Savini did the special effects. He had nothing to do with it. Also on video as "Blood Splash."
1982 (R) 97m/C Baird Stafford, Sharon Smith, C.J. Cooke, Mik Cribben, Kathleen Ferguson, Danny Ronan; **W/Dir:** Romano Scavolini. **VHS, Beta** $39.98 NSV ⚁

Nightmare on the 13th Floor A travel writer at an old hotel glimpses murder on a supposedly non-existent floor. Nasty devil-worshippers are afoot, but the chills in this mild made-for-cable throwaway seldom dip below room temperature.
1990 (PG-13) 85m/C Michael Greene, John Karlen, Louise Fletcher, Alan Fudge, James Brolin; **Dir:** Walter Grauman. **VHS, Beta** PAR ⚁

Nightmare at 43 Hillcrest A family's life becomes a living hell when the police mistakenly raid their house. Based on a true story.
1974 72m/C Jim Hutton, Mariette Hartley. **VHS, Beta** $49.95 USA ⚁

Nightmare in Badham County Two girls get arrested on trumped-up charges in a small backwoods town. They soon discover that the prison farm is a front for a slavery ring. Made for television.
1976 100m/C Deborah Raffin, Lynne Moody, Chuck Connors, Della Reese, Robert Reed, Ralph Bellamy, Tina Louise; **Dir:** John Llewellyn Moxey. **VHS, Beta** $19.95 VMK ⚁

Nightmare at Bittercreek Four babes and a tour guide are pursued by psycho gang in the Sierra mountains. Made for television nightmare in your living room.
1991 (PG-13) 92m/C Tom Skerritt, Joanna Cassidy, Lindsay Wagner, Constance McCashin, Janne Mortil; **Dir:** Tim Burstall. **VHS** $89.95 VMK ⚁⚁

Nightmare in Blood Vampires lurk in San Francisco and wreak havoc on the night life.

1975 90m/C Kerwin Mathews, Jerry Walter, Barrie Youngfellow; **Dir:** John Stanley. **VHS, Beta** $59.95 VCD ⚁½

Nightmare Castle Scientist murders his wife and her lover in a bizarre experiment. He uses their blood to rejuvenate an old servant, then seeks to marry his late wife's sister when he realizes that she has been left the inheritance. In time, the ghosts of the late lovers appear and seek revenge. Also called "Night of the Doomed," "Lovers from Beyond the Tomb," and "The Faceless Monsters."
1965 90m/B *IT* Barbara Steele, Paul Muller, Helga Line; **Dir:** Allan Grunewald. **VHS, Beta** $44.95 SNC, HHT ⚁

A Nightmare on Elm Street A feverish and genuinely frightening horror film about Freddy Krueger (Englund), a scarred maniac in a fedora and razor-fingered gloves who kills neighborhood teens in their dreams and, subsequently, in reality. Of the children-fight-back genre, in which the lead victim (Langenkamp) goes to great lengths of ingenuity to destroy Freddy. In the tradition of "Friday the 13th" and "Halloween," this movie spawned a "Freddy" phenomenon: four sequels (to date); a TV series ("Freddy's Nightmares," a horror anthology show hosted by Englund that had little to do with the plot of the original movie but capitalized on his character); and an army of razor-clawed trick or treaters at Halloween. Englund went on to direct a horror movie of his own ("976-EVIL").
1984 (R) 92m/C John Saxon, Heather Langenkamp, Ronee Blakley, Robert Englund, Amanda Wyss, Nick Corri, Johnny Depp, Charles Fleischer; **W/Dir:** Wes Craven. **VHS, Beta, LV** $24.95 MED, CVS, IME ⚁⚁½

A Nightmare on Elm Street 2: Freddy's Revenge Mediocre sequel to the popular horror film. Freddy, the dream-haunting psychopath with the ginsu knife hands, returns to possess a teenager's body in order to kill again. Nothing new here, however praise is due for the stunning high-tech dream sequence.
1985 (R) 87m/C Mark Patton, Hope Lange, Clu Gulager, Robert Englund, Kim Myers; **Dir:** Jack Sholder. **VHS, Beta, LV** $34.95 MED, CVS, IME ⚁½

A Nightmare on Elm Street 3: Dream Warriors Chapter three in this slice and dice series. Freddy Krueger is at it again, haunting the dreams of unsuspecting suburban teens. Langenkamp, the nightmare-freaked heroine from the first film, returns to counsel the latest victims of Freddy-infested dreams. Noted for the special effects wizardry but little else, Part Three was produced on a $4.5 million shoe string and took in more than $40 million, making it one of the most successful independently produced films in Hollywood. Followed by "Nightmare on Elmstreet 4: Dream Master."
1987 (R) 96m/C Patricia Arquette, Robert Englund, John Saxon, Craig Wasson, Heather Langenkamp, Dick Cavett, Priscilla Pointer, Larry Fishburne; **Dir:** Chuck Russell. **VHS, Beta, LV** $36.95 MED, CVS, IME ⚁⚁

A Nightmare on Elm Street 4: Dream Master Freddy Krueger is still preying on people in their dreams, but he may have met his match as he battles for supremacy with a telepathically talented girl. What Part 4 lacks in substance, it makes up for in visual verve, including scenes of a kid drowning in his waterbed, and a pizza covered with pepperoni-like faces of Freddy's

N

previous victims. This box office bonanza set a new record as the most successful opening weekend of any independently released film. Followed by ''A Nightmare on Elm Street: A Dream Child.''
1988 (R) 99m/C Robert Englund, Rodney Eastman, Danny Hassel, Andras Jones, Tuesday Knight, Lisa Wilcox, Ken Sagoes; *Dir:* Renny Harlin. **VHS, Beta, LV $36.95** *MED, CVS, IME* ♫♫ ½

A Nightmare on Elm Street 5: Dream Child The fifth installment of Freddy Krueger's never-ending adventures. Here, America's favorite knife-wielding burn victim, unable to best the Dream Master from the previous film, haunts the dreams of her unborn fetus. Gore fans may be disappointed to discover that much of the blood and guts ended up on the cutting room floor.
1989 (R) 90m/C Robert Englund, Lisa Wilcox, Kelly Jo Minter, Danny Hassel; *Dir:* Stephen Hopkins. **VHS, Beta, LV, Special order formats $89.95** *IME, MED* ♫♫

Nightmare Festival This is a compilation of science fiction and fantasy movie previews from an assortment of classic films.
195? 120m/C VHS, Beta $24.95 *DVT, MLB* ♫

Nightmare at Noon Watered-down thriller about a small desert town beset by violent terrorists (alias locals gone mad from a chemical experiment dumped into the water system). Only the sheriff's small staff is there to stop them.
1987 (R) 96m/C Wings Hauser, George Kennedy, Bo Hopkins, Brion James, Kimberly Beck, Kimberly Ross; *Dir:* Nico Mastorakis. **VHS $79.98** *REP* ♫

Nightmare in Red China A campy film examining life in communist China with an all-Indian cast.
1955 67m/B VHS, Beta *JLT Woof!*

Nightmare Sisters Three sorority sisters become possessed by a demon and then sexually ravage a nearby fraternity.
1987 (R) 83m/C Brinke Stevens, Michelle McClellan, Linnea Quigley; *Dir:* David DeCoteau. **VHS, Beta $79.95** *TWE Woof!*

Nightmare in Wax After he is disfigured in a fight with a studio boss, a former make-up man starts a wax museum. For fun, he injects movie stars with a formula that turns them into statues. Also called ''Crimes in the Wax Museum.''
1969 91m/C Cameron Mitchell, Anne Helm; *Dir:* Bud Townsend. **VHS, Beta $29.95** *UHV, MRV* ♫ ½

Nightmare Weekend A professor's evil assistant lures three young women into his lab and performs cruel and vicious experiments that transform the girls and their dates into crazed zombies.
1986 (R) 86m/C Dale Midkiff, Debbie Laster, Debra Hunter, Lori Lewis; *Dir:* Henry Sala. **VHS, Beta $79.98** *LIV, LTG* ♫

The Nightmare Years An American reporter in Nazi Germany dares to report the truth to an unbelieving world. Based on a true story. Two cassettes.
1989 237m/C Sam Waterston, Marthe Keller, Kurtwood Smith; *Dir:* Anthony Page. **VHS, Beta $89.98** *TTC* ♫

Nightmares A less-than-thrilling horror anthology featuring four tales in which common, everyday occurrences take on the ingredients of a nightmare. In the same vein as ''Twilight Zone'' and ''Creepshow.''

1983 (PG) 99m/C Christina Raines, Emilio Estevez, Moon Zappa, Lance Henriksen, Richard Masur, Veronica Cartwright; *Dir:* Joseph Sargent. **VHS, Beta $59.95** *MCA* ♫♫

Nights of Cabiria Fellini classic which details the personal decline of a naive prostitute who thinks she's found true love. Dubbed; also known as ''Cabiria.'' Basis for the musical ''Sweet Charity.''
1957 111m/B *IT* Giulietta Masina, Amadeo Nazzari; *Dir:* Federico Fellini. Academy Awards '57: Best Foreign Language Film; Cannes Film Festival '57: Best Actress (Masina). **VHS, Beta $29.95** *NOS, VYY, APD* ♫♫♫ ½

Nights and Days Adaptation of writer Maria Dabrowska's tale about a Polish family chronicles the persecution, expulsions, and land grabbing that occurred after the unsuccessful Uprising of 1864. In Polish with English subtitles.
1976 255m/C *PL* Jadwiga Baranska, Jerzy Binczycki; *Dir:* Jerzy Antczak. **VHS $69.95** *FCT* ♫ ½

Nights in White Satin A made-for-video Cinderella story about the growing love between a fashion photographer and a young model whose boyfriend has just been murdered.
1987 96m/C Kenneth Gilman, Priscilla Harris; *Dir:* Michael Bernard. **VHS, Beta $79.95** *PSM, PAR* ♫ ½

Nightstalker A brother and sister who were condemned to eternal death 12,000 years ago must eat virgins to keep their bodies from rotting. Trouble is, while they search for dinner, the movie decomposes. For mature audiences with no fantasies of celluloid nirvana.
1979 90m/C Aldo Ray. **VHS, Beta $39.95** *LIV, IVE Woof!*

Nightstick New York cop carries big stick.
1987 (R) 90m/C Leslie Nielsen, Bruce Fairbairn, Robert Vaughn, Kerrie Keane; *Dir:* Joseph L. Scanlan. **VHS, Beta $79.95** *COL* ♫♫

Nightwing A suspense drama about three people who risk their lives to exterminate a colony of plague-carrying vampire bats. From the novel by Martin Cruz Smith. OK viewing for the undiscriminating palate.
1979 (PG) 103m/C Nick Mancuso, David Warner, Kathryn Harrold, Strother Martin; *Dir:* Arthur Hiller. **VHS, Beta, LV $39.95** *COL* ♫ ½

Nightwish Students do more than homework when a professor leads them into their own horrifying dreams. Soon it becomes impossible to distinguish dreams from reality.
1989 (R) 96m/C Jack Starrett, Robert Tessier; *Dir:* Bruce Cook Jr. **VHS, LV $89.95** *VES* ♫ ½

Nijinsky An opulent biography of the famous ballet dancer. His exciting and innovative choreography gets little attention. The film concentrates on his infamous homosexual lifestyle and his relationship with impresario Sergei Diaghilev (Bates, in a tour-de-force performance). Lovely to look at, but slow and unconvincing.
1980 (R) 125m/C Alan Bates, George De La Pena, Leslie Browne, Alan Badel, Carla Fracci, Colin Blakely, Ronald Pickup, Ronald Lacey, Vernon Dobtcheff, Jeremy Irons, Frederick Jaeger, Janet Suzman, Sian Phillips; *Dir:* Herbert Ross. **VHS, Beta $29.95** *PAR* ♫♫

Nikki, the Wild Dog of the North When a Malemute pup is separated from his Canadian trapper master, he teams up with a bear cub for a series of adventures. Later the pup is reunited with his former master for still

more adventures. Adapted from a novel by James Oliver Curwood for Disney.
1961 (G) 73m/C *CA* Jean Coutu; *Dir:* Jack Couffer, Don Haldane. **VHS, Beta $69.95** *DIS* ♫ ½

Nine Ages of Nakedness The story of a man whose ancestors have been plagued by a strange problem: beautiful, naked women who create carnal chaos.
1969 88m/C Harrison Marks; *Nar:* Charles Gray. **VHS, Beta $19.95** *VDM, MED, MRV* ♫

Nine Days a Queen An historical drama based on the life of Lady Jane Grey, proclaimed Queen of England after the death of Henry VIII of England and summarily executed for treason by Mary Tudor after a nine-day reign. An obscure tragedy with good performances and absorbing storyline. Also called ''Lady Jane Grey.'' Remade as ''Lady Jane'' (1985).
1934 80m/B *GB* John Mills, Cedric Hardwicke, Nova Pilbeam, Sybil Thorndike, Leslie Perrins, Felix Aylmer, Miles Malleson; *Dir:* Robert Stevenson. **VHS, Beta $44.95** *HHT, DVT* ♫♫♫

Nine Deaths of the Ninja A faceless ninja warrior leads a team of commandos on a mission to rescue a group of political prisoners held captive in the Phillipine jungles. The plot is nothing to write home about, but the two main villains—a neurotic Nazi in a wheelchair and a black lesbian amazon—are priceless. Their scenery chewing antics are well worth the price of admission.
1985 (R) 93m/C Sho Kosugi, Brent Huff, Emelia Lesniak, Regina Richardson; *W/Dir:* Emmett Alston. **VHS, Beta $69.95** *MED* ♫

9 to 5 In this caricature of large corporations and women in the working world, Coleman plays the male chauvinist boss who calls the shots and keeps his employees, all female, under his thumb. Three of the office secretaries daydream of Coleman's disposal and rashly kidnap him after a silly set of occurrences threaten their jobs. While they have him under lock and key, the trio take office matters into their own hands and take a stab at running things their own way, with amusing results. The basis for a television series.
1980 (PG) 111m/C Jane Fonda, Lily Tomlin, Dolly Parton, Dabney Coleman, Sterling Hayden; *Dir:* Colin Higgins. **VHS, Beta, LV $19.98** *FOX* ♫♫ ½

Nine Lives of Elfego Baca Venerable Western hero Loggia faces a veritable army of gunfighters and bandits. Action-packed but not too violent; family fun.
1958 78m/C Robert Loggia, Robert Simon, Lisa Montell, Nestor Paiva; *Dir:* Norman Foster. **VHS, Beta $69.95** *DIS* ♫♫

Nine Lives of Fritz the Cat Fritz feels that life's too square in the 70's, so he takes off into some of his other lives for more adventure. Cleaner but still naughty sequel to the X-rated *Fritz the Cat*, featuring neither the original's writer/director Ralph Bakshi nor cartoonist Robert Crumb. Tame and lame. Animated.
1974 77m/C *Dir:* Robert Taylor. **VHS, LV $19.98** *VES* ♫♫

9 1/2 Ninjas A cautious and disciplined martial artist trains a young and flirtatious woman in the ways of the ninja. His life becomes exciting in more ways than one, when he realizes she's being followed by ninjas with more on their minds than her training—they want to assassinate her! Crazy mixture of sex, kung fu and humor make this film one surprise after another. **1990 (R) 88m/C** Michael Phenicie, Andee Gray, Tiny Lister; *Dir:* Aaron Worth. **VHS, Beta, LV $9.98** *REP* 🐾🐾½

9 1/2 Weeks A chance meeting between a Wall Street executive and an art gallery employee evolves into an experimental sexual relationship that borders sado-masochism. The video version is more explicit than the one viewed in theaters, but not by much. Strong characterizations by both actors prevent this from being strictly pornography. Well written, with strength of male and female personalities nicely balanced. Not for all tastes, and definitely not for children. **1986 (R) 114m/C** Mickey Rourke, Kim Basinger, Margaret Whitton, Karen Yahng, David Margulies; *Dir:* Adrian Lyne. **VHS, Beta, LV $19.95** MGM 🐾🐾½

976-EVIL Englund (the infamous Freddy from the Nightmare on Elm Street epics) directs this horror movie where a lonely teenager dials direct to demons from hell. **1988 (R) 102m/C** Stephen Geoffreys, Jim Metzler, Maria Rubell, Sandy Dennis, Robert Picardo, Leslie Deane, Pat O'Bryan; *Dir:* Robert Englund. **VHS, Beta, LV $14.95** COL 🐾🐾

976-EVIL 2: The Astral Factor Satan returns the call in this supernatural thriller that sequels the original film. **1991 (R) 93m/C** Patrick O'Bryan, Rene Assa, Debbie James; *Dir:* Jim Wynorski. **VHS $89.98** LIV, VES 🐾🐾½

984: Prisoner of the Future A shocking, futuristic tale of human self-destruction. **1984 70m/C** Don Francks, Stephen Markle. **VHS, Beta $59.95** VCL 🐾½

1900 Bertolucci's impassioned epic about two Italian families, one land-owning, the other, peasant. Shows the sweeping changes of the 20th century begun by the trauma of WWI and the onslaught of Italian socialism. Edited down from its original 360-minute length and dubbed in English from three other languages, the film suffers somewhat from editing and from its nebulous lack of commitment to any genre. Elegantly photographed by Vittorio Storaro. Also titled "Novecento." **1976 (R) 255m/C** FR IT GE Robert De Niro, Gerard Depardieu, Burt Lancaster, Donald Sutherland, Dominique Sanda, Sterling Hayden, Laura Betti, Francesca Bertini, Werner Bruhns, Stefania Casini, Anna Henkel, Alida Valli, Stefania Sandrelli; *Dir:* Bernardo Bertolucci. New York Film Festival '77: Best Actress. **VHS, Beta, LV $29.95** PAR 🐾🐾🐾

1918 An adaptation of the Horton Foote play about the effects of World War I and an influenza epidemic on a small Texas town. Slow-moving but satisfying. Score by Willie Nelson. Originally produced for PBS's "American Playhouse." **1985 89m/C** Matthew Broderick, Hallie Foote, William Converse Roberts, Rochelle Oliver, Michael Higgins, Horton Foote Jr.; *Dir:* Ken Harrison. **VHS, Beta $79.98** FOX 🐾🐾½

The 1930s: Music, Memories & Milestones The Big Band era gets into full swing, Fascism takes control in Europe, King Edward VIII abdicates, and other exciting memories from the '30s are relived through newsreel footage. **1989 60m/C** **VHS, Beta $19.95** KUL, WST 🐾🐾½

1931: Once Upon a Time in New York Prohibition-era gangsters war, beat each other up, make headlines and drink bathtub gin. **1972 90m/C** Richard Conte, Adolfo Celi, Lionel Stander, Irene Papas. **VHS, Beta $59.95** VCD 🐾🐾

1939, The Movies' Vintage Year: Trailers on Tape A compilation of coming attractions from such classics as "Gunga Din," "Idiot's Delight," "Juarez," "Dark Victory" and "Gone With The Wind." Some segments are in color. **1939 60m/B** Cary Grant, Clark Gable, Vivien Leigh. **VHS, Beta $34.95** SFR 🐾🐾

1941 Proved to be the most expensive comedy of all time with a budget exceeding $35 million. The depiction of Los Angeles in the chaotic days after the bombing of Pearl Harbor combines elements of fantasy and black humor. **1979 (PG) 120m/C** John Belushi, Dan Aykroyd, Patti LuPone, Ned Beatty, Slim Pickens, Murray Hamilton, Christopher Lee, Tim Matheson, Toshiro Mifune, Warren Oates, Robert Stack, Nancy Allen, Elisha Cook Jr., Lorraine Gary, Treat Williams, Mickey Rourke; *Dir:* Steven Spielberg. **VHS, Beta, LV $14.95** MCA, PIA, CCB 🐾🐾½

1969 Three teenage friends during the 1960s become radicalized by the return of one of their friends from Vietnam in a coffin. Critically lambasted directorial debut for "On Golden Pond" author Ernest Thompson. **1989 (R) 96m/C** Kiefer Sutherland, Robert Downey Jr., Winona Ryder, Bruce Dern, Joanna Cassidy, Mariette Hartley; *Dir:* Ernest Thompson. **VHS, Beta, LV $19.98** MED 🐾🐾🐾½

1984 A very fine adaptation of George Orwell's infamous novel, this version differs from the overly simplistic and cautionary 1954 film because of fine casting and production design. The illegal love affair of a government official becomes his attempt to defy the crushing inhumanity and lack of simple pleasures of an omniscient government. Filmed in London, it skillfully visualizes our time's most central prophetic nightmare. **1984 (R) 117m/C** GB John Hurt, Richard Burton, Suzanna Hamilton, Cyril Cusack, Gregory Fisher, Andrew Wilde, Rupert Baderman; *Dir:* Michael Radford. **VHS, Beta, LV $19.95** IVE 🐾🐾🐾

1990: The Bronx Warriors Good street gang members combat evil corporate powers in a semi-futuristic South Bronx. Lame copy of "Escape from New York." **1983 (R) 86m/C** IT Vic Morrow, Christopher Connelly, Fred Williamson; *Dir:* Enzo G. Castellari. **VHS, Beta $59.95** MED 🐾

90 Days Charming independently made Canadian comedy about two young men handling their respective romantic dilemmas—one awaiting an oriental fiancee he's never met, the other who is being kicked out of his house by his wife. **1986 99m/C** CA Stefan Wodoslowsky, Sam Grana, Christine Pak; *Dir:* Giles Walker. **VHS, Beta $79.98** FOX 🐾🐾🐾

92 in the Shade Based upon McGuane's novel, the film deals with a bored, wealthy rogue who becomes a fishing guide in the Florida Keys, and battles against the competition of two crusty, half-mad codgers. Sloppy, irreverent comedy as only a first-time writer-turned-director can fashion. **1976 (R) 91m/C** Peter Fonda, Warren Oates, Margot Kidder, Burgess Meredith, Harry Dean Stanton; *Dir:* Thomas McGuane. **VHS, Beta $59.98** FOX 🐾🐾🐾

99 & 44/100 Dead Frankenheimer falters with this silly gangster flick. Harris is hired to kill Dillman, by local godfather O'Brien. Originally written as a satirical look at gangster movies, but it doesn't stick to satire, and as a result is disappointing. **1974 (PG) 98m/C** Richard Harris, Chuck Connors, Edmond O'Brien, Bradford Dillman, Ann Turkel; *Dir:* John Frankenheimer. **VHS** FOX 🐾

99 Women A sympathetic prison warden attempts to investigate conditions at a women's prison camp. Thin and exploitative view of lesbianism behind bars that sensationalizes the subject. **1969 (R) 90m/C** GB SP GE IT Maria Schell, Herbert Lom, Mercedes McCambridge, Lucianna Paluzzi; *Dir:* Jess (Jesus) Franco. **VHS, Beta $39.98** NO 🐾½

Ninja 3: The Domination A Ninja master must remove the spirit of a deadly Ninja assassin from a young woman. **1984 (R) 92m/C** Lucinda Dickey, Sho Kosugi; *Dir:* Sam Firstenberg. **VHS, Beta $14.98** MGM 🐾

Ninja Academy Seven wimps, losers, and spoiled brats come to the Ninja Academy to learn the art. Will they make it? **1990 (R) 93m/C** Will Egan, Kelly Randall, Gerald Okomura, Michael David, Robert Factor, Jeff Robinson; *Dir:* Nico Mastorakis. **VHS $79.95** QUE 🐾½

Ninja in Action An American woman, abetted by her head-kicking lover, hits Hong Kong to find her father's killer. **1987? 90m/C** Stuart Steen, Christine O'Hara. **VHS, Beta $59.95** TWE 🐾

Ninja: American Warrior An evil ninja takes on the U.S. Drug Enforcement Agency when the authorities threaten to shut him down. **1990 90m/C** Joff Houston, John Wilford; *Dir:* Tommy Cheng. **VHS, Beta $59.95** IMP 🐾½

Ninja Avengers A man seeks revenge on the ninja priest who imprisoned him. **1988 90m/C** **VHS, Beta $59.95** IMP 🐾

Ninja the Battalion Agents from America, the Soviet Union, and China try to recover germ warfare secrets stolen by the Japanese secret service. **1990 90m/C** Roger Crawford, Sam Huxley, Alexander Lou, Dickson Warn; *Dir:* Victor Sears. **VHS, Beta $59.95** TWE 🐾

Ninja Blacklist The Iron Hand leads an unchallenged reign of terror—until two warriors vow to take him down! **1990 90m/C** Sonny Qui, Chan Wai-Man. **VHS, Beta $19.99** BFV 🐾½

Ninja Brothers of Blood A guy falls for a rival gang member's girl. Neither the gang nor the guy's former girlfriend take kindly to this! **1989 90m/C** Marcus Gibson, Fonda Lynn, Brian McClave, Jonathan Soper; *Dir:* Raymond Woo. **VHS, Beta $59.95** IMP 🐾

Ninja Champion Interpol agents trained in kick fighting track down a sleazy diamond smuggler. **1987? 90m/C** *Hosted:* Sho Kosugi. **VHS, Beta $39.95** TWE 🐾

Ninja Checkmate An evil ninja seems indestructible, but finally he meets his match. **1990 90m/C** Simon Lee, Jack Long, Jeanie Chang. **VHS, Beta $49.95** OCE 🐾

Ninja in the Claws of the C.I.A.
KGB ninjas from the Soviet Union take on ninjas from the CIA in an all-out, high-stakes battle of immense proportions.
1983 102m/C John Liu. VHS, Beta MAG *

Ninja Commandments Everything is at stake when an evil warrior takes on the greatest fighter in the land: the winner will rule the empire!
1987 90m/C Richard Harrison, Dave Wheeler; *Dir:* Joseph Lai. VHS, Beta $59.95 IMP *

Ninja Condors A young man grows up to avenge the murder of his father.
1987 85m/C Alexander Lou, Stuart Hugh; *Dir:* James Wu. VHS, Beta $59.95 TWE *

Ninja Connection Ninja terrorism is employed as a scare tactic to deter a group who wants to break up an international drug ring.
1990 90m/C Patricia Greenford, Jane Kingsly, Joe Nelson, Louis Roth, Henry Steele, Stuart Steen; *Dir:* York Lam. VHS, Beta $59.95 IMP *

Ninja in the Deadly Trap Chinese general Chieh Chikuang uses his martial arts skills to defeat the Japanese forces that loot and plunder China's coastal provinces.
1985 90m/C VHS, Beta $49.95 WES, OCE *

Ninja Death Squad A team specializing in political assassinations is hunted down by a special agent.
1987 89m/C Glen Carson, Patricia Goodman, Joff Houston, Billy Jones, Wallace Jones, John Wilford; *Dir:* Tommy Cheng. VHS, Beta $59.95 IMP *

Ninja Demon's Massacre A ninja named Willie responds to federal investigations into his spying business by wearing a black robe and slashing them with long swords.
198? 89m/C Ted Brooke, James Lear. VHS, Beta $59.95 TWE *

Ninja Destroyer Ninja warriors battle over an emerald mine.
197? (R) 92m/C VHS, Beta $19.98 TWE *

Ninja Dragon Two Ninjas are hired in a gang war and dazzle all those who watch by kicking their enemies to death.
198? 92m/C Richard Harrison. VHS, Beta $59.95 TWE *

The Ninja Empire The feds fail to find the killer of young prostitutes and as a last result they call on the one group they know they can trust—Ninjas.
1990 90m/C VHS, Beta $59.95 IMP *

Ninja Enforcer A ninja warrior destroys all his foes.
197? ?m/C Lin Chang, Kim Bhang, Cze Young. VHS $19.98 TAV *

Ninja Exterminators Another rib-kicking bonanza of pain, revenge and intrigue.
1981 85m/C Jacky Cheng, Ko Shou Liang. VHS, Beta $59.95 WES *

Ninja Fantasy Ninja drug smugglers and government officials battle over a large drug shipment.
1986 95m/C Adam Nell, Ian Frank, Jordan Heller, Ken Ashley, Jenny Mills, Jack Rodman. VHS, Beta $59.95 TWE *

Ninja Force of Assassins Two espionage buddies part ways in Southeast Asia over a Ninja crimelord war, and engage in much head-kicking.
197? 79m/C Jim Davis, Louis Hill; *Dir:* Victor Sears. VHS, Beta $59.95 TWE *

Ninja Holocaust A running argument about a secret Swiss bank account gives license for rampant head-kicking and rib-cracking.
198? 90m/C Bruce Chen, Jim Brooks, Martin Lee, Joe Resinick. VHS, Beta MGL *

Ninja Hunt This time, ninjas kick international terrorists in the head.
198? 92m/C VHS, Beta $59.95 TWE *

The Ninja Hunter When a Shaolin temple is destroyed, an angry abbot swears vengeance on the perpetrators. He immediately makes good on his threat.
1990 90m/C VHS, Beta $34.95 OCE *

Ninja Kill Assassinations and national coups are the aim of the Purple Ninjas. Master Gordon must stop them all with some inspired heel-to-skull fighting.
1987 92m/C Richard Harrison, Stuart Smith. VHS, Beta $59.95 TWE *

Ninja Knight Heaven's Hell This time the ninja is given all he can handle in the way of bone-bruising kicks and chops.
19?? ?m/C VHS $59.95 IMP *

Ninja of the Magnificence When the ninja master is killed, factions within his group battle for control.
1989 90m/C Sam Baker, Patrick Frbezar, Clive Hadwen, Tim Michael, Renato Sala; *Dir:* Charles Lee. VHS, Beta $59.95 IMP *½

Ninja Massacre Gangsters covet a secret manual of kung fu techniques. Heel-splintering battles break out as a result.
1984 87m/C Lo Lieh, Pai Ying, Lung Jun Ehr, Tiki Shirtee. VHS, Beta $59.95 GEM, MRV *

Ninja Masters of Death Terrorists reign supreme until the white ninja saves the city.
1985 90m/C Mick Jones, Chris Peterson, Daniel Wells, Richard Young; *Dir:* Bruce Lambert. VHS, Beta $59.95 IMP *

Ninja Mission A CIA agent and his group of ninja fighters embark on a hazardous mission to rescue two people from a Soviet prison. They use their fighting skills against Russian soldiers, in a climactic scene of martial artistry and mayhem.
1984 (R) 95m/C Christopher Kohlberg, Curt Brober, Hanna Pola; *Dir:* Mats Helge. VHS, Beta $59.95 MED *½

Ninja Nightmare A man seeks revenge for his family's death.
1988 (R) 98m/C Leo Fong. VHS, Beta $9.99 STE, PSM *

Ninja Operation: Licensed to Terminate Two warriors risk broken noses and twisted limbs to take on the Black Ninja empire.
1987 89m/C Richard Harrison, Paul Marshall, Jack McPeat, Grant Temple; *Dir:* Joseph Lai. VHS, Beta $59.95 IMP *

Ninja Phantom Heroes Two Vietnam vets are imprisoned for war crimes. One escapes and forms his own secret ninja society.
1987 90m/C Glen Carson, George Dickson, Allen Leung, Christine Wells; *Dir:* Bruce Lambert. VHS, Beta $59.95 TWE *

The Ninja Pirates In the Sung Dynasty, a group of bloodthirsty pirates are hunted down after stealing a golden treasure.
1990 90m/C Chi Lan, Lo Lieh, Carter Wong, Nancy Yen. VHS, Beta $49.95 OCE *½

Ninja Powerforce Two childhood friends end up as members of rival ninja gangs in the midst of a bloody war. As is usual in ninja films, lots of injuries are the result.
1990 90m/C Jonathan Bould, Richard Harrison; *Dir:* Joseph Lai. VHS, Beta $59.95 IMP *½

Ninja the Protector A ninja tracks down and kicks the ninja head of a counterfeiting ring.
198? 92m/C Richard Harrison. VHS, Beta $19.98 TWE *

Ninja Showdown A warrior must take on a number of vicious bandits in order to defend his small town. He vows revenge on anyone who threatens his people.
1990 92m/C Richard Harrison; *Dir:* Joseph Lai. VHS, Beta $59.95 TWE *

Ninja: Silent Assassin An agent working for Interpol goes after a drug ring on a personal vendetta, avenging the brutal murder of his wife.
1990 90m/C VHS, Beta $59.95 IMP *½

Ninja Squad Empires, families, and honor are at stake in a fierce battle of heel-to-head action.
198? 92m/C Dave Wheeler, Richard Harrison, Timalden. VHS, Beta $59.95 TWE *

Ninja Strike Force The Black Ninjas steal the powerful "spirit sword," and go on a bloody rampage.
1988 89m/C Richard Harrison, Gary Carter; *Dir:* Joseph Lai. VHS, Beta $59.95 IMP *½

Ninja Supremo A rebellious teenager studies Kung-fu, and then returns to his village a hero after he gains expertise in his chosen field.
1990 100m/C David Tao, Yueh Wah, Sun Yueh. VHS, Beta $49.95 OCE *½

Ninja Swords of Death A noble warrior challenges a vicious assassin who uses a spur-sword. He is nearly decapitated in the process.
1990 90m/C Chang Ying Chun, Chia Ling, Tien Peng, Kong Pun. VHS, Beta $29.95 OCE *½

Ninja Terminator Possession of all three pieces of the Golden Warrior turns the holder into the magically super-powered "Ninja Terminator." Three students battle for that honor.
1969 92m/C *Hosted:* Sho Kosugi. VHS $59.95 TWE, TAV *

Ninja Thunderbolt A sleazy woman lusts after power and money and somehow becomes involved with ninjas.
1985 92m/C Jackie Chan, Richard Harrison, Wang Tao; *Dir:* Godfrey Ho. VHS, Beta $59.95 TWE *

Ninja Turf Teenage kick artists rip off a Hollywood drug dealer and are pursued by a swarm of ninja hitmen.
1986 (R) 83m/C Jun Chong, Bill Wallace, Rosanna King, James Lew, Phillip Rhee. VHS, Beta, LV $79.95 COL *

Ninja in the U.S.A. Drug lords hire men who kick and make funny noises to protect them.
1988 ?m/C Alexander Lou, Yau Jin Tomas. VHS $29.95 OCE *

Ninja USA A Vietnam vet hires Ninjas to help him smuggle drugs, and they have a hard time adjusting to American ways, eventually turning on each other.
198? 80m/C VHS, Beta $59.95 TWE *

Ninja, the Violent Sorcerer A murderer, with the help of a vampire, does battle with the ghost of a dead gambling lord's wife and the gambling lord's alive brother.
1986 90m/C *Dir:* Bruce Lambert. **VHS, Beta** $59.95 *TWE* 🎞

Ninja vs. Bruce Lee Lee investigates a secret ninja sect involved in smuggling.
1990 90m/C Bruce Lee. **VHS, Beta** $49.95 *OCE* 🎞

Ninja vs. Ninja A scientist has found a cure for heroin addiction, but ninja drug lords won't let his discovery see the light of day!
1990 90m/C Hui Hsiao Chiang, Callan Leong. **VHS, Beta** $49.95 *OCE* 🎞½

Ninja vs. the Shaolin Savage fighting between priests and ninjas results in one of the bloodiest pages in history.
1984 90m/C Alexander Lou, John Wu; *Dir:* Jacky Hwong. **VHS, Beta** *OCE* 🎞½

Ninja Warriors An evil warrior plans to take over the world with the help of his brainwashed Ninja zombies.
1985 90m/C Ron Marchini, Paul Vance. **VHS, Beta** $9.99 *CCB, IVE* 🎞½

Ninja Wars One ninja takes on five ninja assassins who, along with an evil ninja sorcerer, have kidnapped his ninja girlfriend.
1984 95m/C **VHS, Beta** $79.95 *PSM* 🎞

The Ninja Wolves During the Ming Dynasty, Japanese ninjas stage a bloody invasion of China.
1990 90m/C Yue Hwa. **VHS, Beta** $49.95 *OCE* 🎞🎞

Ninja the Wonder Boy The story of a young boy's magical and humorous journey toward becoming a master ninja.
1985 89m/C **VHS, Beta** $24.95 *PAR* 🎞🎞

Ninja's Extreme Weapons The Blue Ninja takes it to the streets when his brother is killed by the underworld.
1990 90m/C James Gray, Donald Muir; *Dir:* Victor Sears. **VHS, Beta** $59.95 *TWE, HHE* 🎞🎞

Ninotchka Delightful romantic comedy; Greta Garbo talks and laughs. Garbo is a cold Russian agent sent to Paris to check up on her comrades, who are being seduced by capitalism. She inadvertently falls in love with a playboy, who melts her communist heart. Satirical, energetic and witty. Later a Broadway musical called ''Silk Stockings.''
1939 110m/B Greta Garbo, Melvyn Douglas, Ina Claire, Sig Rumann, Felix Bressart, Bela Lugosi; *Dir:* Ernst Lubitsch. National Board of Review Awards '39: 10 Best Films of the Year. **VHS, Beta, LV, 8mm** $24.95 *MGM* 🎞🎞🎞½

The Ninth Configuration Based on Blatty's novel ''Twinkle, Twinkle, Killer Kane'' (also the film's alternate title), this is a weird and surreal tale of a mock rebellion of high-ranking military men held in a secret base hospital for the mentally ill. Keach is good as the commander who is just as insane as the patients. Available in many different lengths, this is generally considered to be the best.
1979 (R) 115m/C Stacy Keach, Scott Wilson, Jason Miller, Ed Flanders, Neville Brand, Alejandro Rey, Robert Loggia; *Dir:* William Peter Blatty. **VHS, Beta, LV** $19.95 *STE, NWV* 🎞🎞🎞

Nite Song Two inner city teenagers battle drugs by finding strength in Christian beliefs.
1988 65m/C Bobby Smith, Thom Hoffman, Vicki Nuzum; *Dir:* Russell S. Doughten Jr. **VHS, Beta** *MIV* 🎞

Nitti: The Enforcer Made-for network television saga about Al Capone's brutal enforcer and right hand man, Frank Nitti. Diversified cast, (Moriarty in particular) do their best to keep things moving along, and the atmosphere is consistently and appropriately violent. Made to capitalize on the success of 1987's ''The Untouchables.''
1988 (PG-13) 94m/C Anthony LaPaglia, Vincent Guastaferro, Trini Alvarado, Michael Moriarty, Michael Russo, Louis Guss, Bruno Kirby; *Dir:* Michael Switzer. **VHS** $89.95 *ACA* 🎞🎞

No Big Deal Dillon is a streetwise teenager who makes friends at his new school. Blah promise; bad acting makes this no big deal.
1983 (PG-13) 86m/C Kevin Dillon, Sylvia Miles, Tammy Grimes, Jane Krakowski, Christopher Gartin, Mary Joan Negro; *Dir:* Robert Charlton. **VHS** $79.95 *BTV* 🎞½

No Comebacks Two British mysteries, made for television from stories by Frederick Forsyth: ''A Careful Man,'' about a dying millionaire who cheats his heirs, and ''Privilege,'' about a clever stamp dealer who exacts revenge upon a libelous gossip columnist.
1985 60m/C *GB* Milo O'Shea, Cyril Cusack, Dan O'Herlihy, Gayle Hunnicutt. **VHS, Beta** $59.95 *PSM* Woof!

No Dead Heroes Green Beret Vietnam war hero succumbs to Soviet scheming when they plant a computer chip in his brain. Unoriginal and unworthy.
1987 86m/C Max Thayer, John Dresden, Toni Nero; *Dir:* J.C. Miller. **VHS, Beta** $69.95 *SVS* 🎞

No Deposit, No Return Tedious, silly, pointless Disney action comedy. Rich brats persuade bumbling crooks to kidnap them, offer them for ransom to millionaire grandfather.
1976 (G) 115m/C David Niven, Don Knotts, Darren McGavin, Barbara Feldon, Charles Martin Smith; *Dir:* Norman Tokar. **VHS, Beta** $69.95 *DIS* 🎞🎞

No Drums, No Bugles A West Virginia farmer and conscientious objector leaves his family to live alone in a cave for three years during the Civil War. Bad direction spoils Sheen's good performance.
1971 (G) 85m/C Martin Sheen, Davey Davison, Denine Terry, Rod McCarey; *Dir:* Clyde Ware. **VHS, Beta** $39.95 *NEG* 🎞½

No Greater Love Capraesque Soviet war movie. A peasant woman mobilizes her village against the Nazis to avenge her family's death. Dubbed into English by the Soviets for Western circulation during World War II.
1943 74m/B *RU* Vera Maretskaya; *Dir:* Frederic Ermler. **VHS, Beta, 3/4U** $35.95 *FCT, IHF* 🎞🎞½

No Grumps Allowed! A package of episodes of the children's fantasy series.
1969 90m/C **VHS, Beta** $19.98 *SUE* Woof!

No Holds Barred Cheesy, campy remake of cheesy, campy 1952 wrestling movie. Hulk Hogan on the big screen, at last.
1989 (PG-13) 98m/C Hulk Hogan, Kurt Fuller, Joan Severance, Tiny Lister; *Dir:* Thomas J. Wright. **VHS, Beta, LV** $19.95 *COL* 🎞🎞

No, I Have No Regrets Documentary on the life and music of Edith Piaf, ''the little sparrow.''
1973 52m/C *Nar:* Louis Jourdan, Charles Aznavour, Bruno Coquatrix, Eddie Constantine, Edith Piaf. **VHS, Beta** $24.95 *NOS, DVT* 🎞½

No Love for Johnnie Well acted if unoriginal political drama. Peter Finch as a British M.P. who fails in both public and private life. Based on the novel by Wilfred Fienburgh.
1960 105m/B *GB* Peter Finch, Stanley Holloway, Donald Pleasence, Mary Peach, Mervyn Johns, Dennis Price, Oliver Reed, Billie Whitelaw; *Dir:* Ralph Thomas. **VHS, Beta** $39.95 *IND* 🎞🎞🎞

No Man of Her Own Gable and Lombard in their only screen pairing. A gambler marries a local girl on a bet and attempts to hide his secret life from her. Neither star's best film.
1932 81m/B Carole Lombard, Clark Gable, Grant Mitchell, Elizabeth Patterson; *Dir:* Wesley Ruggles. **VHS, Beta, LV** $14.95 *MCA, KRT* 🎞🎞½

No Man's Land Undercover cop Sweeney tails playboy Sheen but is seduced by wealth and glamour. Flashy surfaces, shiny cars, little substance.
1987 (R) 107m/C Charlie Sheen, D.B. Sweeney, Lara Harris, Randy Quaid; *Dir:* Peter Werner. **VHS, Beta, LV** $19.98 *ORI* 🎞🎞

No Man's Range Unremarkable, typical, predictable Western: good guys, bad guys, guns, bullets, horses.
1935 56m/B Bob Steele. **VHS, Beta** *WGE* 🎞

No Mercy A Chicago cop (Gere) plunges into the Cajun bayou in order to avenge the murder of his partner. He falls for a beautiful girl enslaved by the killer, but that doesn't stop him from using her to flush out the powerful swamp-inhabiting crime lord. Absurd story without much plot.
1986 (R) 108m/C Richard Gere, Kim Basinger, Jeroen Krabbe, George Dzundza, William Atherton, Ray Sharkey, Bruce McGill; *Dir:* Richard Pearce. **VHS, Beta, LV** $19.95 *COL* 🎞🎞

No One Cries Forever When a prostitute breaks away from a gangster-madam after finding love, she is tracked down and disfigured. Swedish; dubbed.
1985 96m/C Elke Sommer, Howard Carpendale, Zoli Marks; *Dir:* Jans Rautenbach. **Beta** $79.95 *VHV* Woof!

No Place to Hide A girl's father drowns and she blames herself. Her mother and psychologist try to convince the girl that her father's spirit isn't stalking her. Made for television.
1981 120m/C Keir Dullea, Mariette Hartley, Kathleen Beller, Arlen Dean Snyder, Gary Graham, John Llewellyn Moxey; *Dir:* John Llewellyn Moxey. **VHS, Beta** $59.95 *PSM* 🎞½

No Place to Run An elderly man fights for custody of his grandson, then kidnaps him and flees with him to Canada. Made for TV.
1972 78m/C Herschel Bernardi, Larry Hagman, Stefanie Powers, Neville Brand; *Dir:* Delbert Mann. **VHS, Beta** $59.95 *VMK* 🎞🎞

No Prince for My Cinderella A schizophrenic girl who has turned to prostitution is sought desperately by her social worker.
1978 100m/C Robert Reed. **VHS, Beta** $49.95 *WOV* 🎞🎞

No Problem A man is pursued, shot, and drops dead in the apartment of an unsuspecting man who doesn't know what to do with the body. Dubbed in English.
197? 94m/C *FR* Miou-Miou. **VHS, Beta** *CCI* 🎞🎞

No Regrets for Our Youth A feminist saga depicting the spiritual growth of a foolish Japanese girl during the tumultuous years of World War II. In Japanese with English subtitles.
1946 110m/B JP Setsuko Hara; **Dir:** Akira Kurosawa. **VHS $29.95** CVC ♂♂♂

No Retreat, No Surrender A young American kick-boxer battles a formidable Russian opponent and wins, quite improbably, after having been tutored by the ghost of Bruce Lee in an abandoned house. Notable as Van Damme's debut, but little else recommends this silly "Rocky" rehash. Followed by "No Retreat, No Surrender II."
1986 (PG) 85m/C Kurt McKinney, J.W. Fails, Jean-Claude Van Damme; **Dir:** Corey Yuen. **VHS, Beta, LV $14.95** NWV, STE ♂

No Retreat, No Surrender 2 With help from two karate experts, a man sets out to find his girlfriend who has been kidnapped by Soviets. Has little or nothing to do with the movie to which it is ostensibly a sequel. High level kick-boxing action sequences.
1989 (R) 92m/C Loren Avedon, Max Thayer, Cynthia Rothrock; **Dir:** Corey Yuen. **VHS, Beta $79.98** MCG ♂♂

No Retreat, No Surrender 3: Blood Brothers Sibling martial arts rivals decide to bond in a manly way when CIA agent dad is most heinously slain by terrorists. The answer to the obscure question: "Whatever became of Joseph Campanella?"
1991 (R) 97m/C Keith Vitali, Loren Avedon, Joseph Campanella. **VHS $89.95** IMP ♂

No Room to Run A concert promoter's life turns into a nightmare of deadly corporate intrigue when he arrives in Australia. Luckily, he finds time to fall in love. Made for television.
1978 101m/C Richard Benjamin, Paula Prentiss, Barry Sullivan; **Dir:** Robert Lewis. **VHS, Beta $9.95** ACA ♂♂1/2

No Safe Haven A government agent seeks revenge for his family's death.
1987 (R) 92m/C Wings Hauser, Marina Rice, Robert Tessier; **Dir:** Ronnie Rondell. **VHS, Beta $79.98** MCG ♂

No Secrets A young man on the run seeks refuge with three girls in an isolated house. His dread secret isn't so dread, leaving this mild teen-oriented thriller starved for lack of menace.
1991 (R) 92m/C Adam Coleman Howard, Amy Locane, Heather Fairfield, Traci Lind; **Dir:** Dezso Magyar. **VHS, LV $89.98** COL, PIA ♂1/2

No Small Affair A 16-year-old aspiring photographer becomes romantically involved with a sultry 22-year-old rock star. Musical score by Rupert Holmes.
1984 102m/C Jon Cryer, Demi Moore, George Wendt, Peter Frechette, Elizabeth Daily, Tim Robbins, Jennifer Tilly, Rick Ducommun, Ann Wedgeworth; **Dir:** Jerry Schatzberg. **VHS, Beta, LV $19.95** COL ♂♂

No Surrender An unpredictable darkly charming comedy about a Liverpool nightclub newly managed by Angelis. On New Year's Eve, a small drunken war is triggered when two groups of irate senior citizens are booked into the club and clash over their beliefs. The group is made up of Protestants and Catholics and the fight resembles the ongoing conflicts in modern day Northern Ireland, although most of the action takes place in the loo. Watch for Costello as an inept magician.
1986 (R) 100m/C GB Ray McAnally, Michael Angelis, Avis Bunnage, James Ellis, Tom Georgeson, Mark Mulholland, Joanne Whalley-Kilmer, Elvis Costello; **Dir:** Peter Smith. London Evening Standard Film Awards '86: Best Actor (McAnally). **VHS, Beta $79.95** PSM ♂♂♂

No Survivors, Please Aliens from Orion take over politicians in order to rule the Earth. Bizarre, obscure, based on true story.
1963 92m/B GE Maria Perschy, Uwe Friedrichsen, Robert Cunningham, Karen Blanguernon, Gustavo Rojo; **Dir:** Hans Albin, Peter Berneis. **VHS $19.98** MOV ♂1/2

No Time to Die In Indonesia, two corporate pawns and a beautiful reporter battle for the possession of a new laser cannon.
1978 87m/C Chris Mitchum, John Phillip Law, Grazyna Dylong. **VHS, Beta $59.95** TWE ♂

No Time for Romance A sprightly musical with an all black cast. First such film to be shot in color.
1948 ?m/C Bill Walker. **VHS $39.95** FCT ♂♂

No Time for Sergeants Hiliarious film version of the Broadway play by Ira Levin. Young Andy Griffith is excellent as the Georgia farm boy who gets drafted into the service and creates mayhem among his superiors and colleagues. Story also told on an earlier television special and later a series. Note Don Knotts and Benny Baker in small roles along with Jameel Farah who went on to star in TV's M*A*S*H after channging his name to Jamie Farr.
1958 119m/B Andy Griffith, Nick Adams, Murray Hamilton, Don Knotts, Jamie Farr, Myron McCormick; **Dir:** Mervyn LeRoy. **VHS, Beta, LV $19.98** WAR ♂♂

No Trace A writer who broadcasts his crime stories as part of a radio show is the victim of blackmail. It seems a former associate is aware that the stories are all based in fact. A Scotland Yard detective investigates when murder rears its ugly head.
1950 76m/B GB Hugh Sinclair, Dinah Sheridan, John Laurie, Barry Morse, Michael Brennan, Dora Bryan; **Dir:** John Gilling. **VHS, Beta $16.95** SNC ♂♂♂

No Way Back Fred Williamson portrays Jess Crowder, a man-for-hire who is expert with guns, fists, and martial arts. Williamson, who also wrote, directed, and produced, creates an angry look at the white establishment. This genre of supermacho black exploitation films was so popular in the 1970s it earned its own appellation, "blaxploitation."
1974 (R) 92m/C Fred Williamson, Charles Woolf, Tracy Reed, Virginia Gregg, Don Cornelius; **Dir:** Fred Williamson. **VHS, Beta $49.95** UNI ♂

No Way Out Career Navy man Costner is involved with a beautiful, sexy woman who is murdered. Turns out she was also the mistress of the Secretary of Defense, Costner's boss. Assigned to investigate the murder, he suddenly finds himself the chief suspect. A tight thriller based on 1948's "The Big Clock," with a new surprise ending.
1987 (R) 114m/C Kevin Costner, Sean Young, Gene Hackman, Will Patton, Howard Duff, George Dzundza, Iman, Chris D, Marshall Bell; **Dir:** Roger Donaldson. **VHS, Beta, LV $14.95** HBO ♂♂♂

No Way to Treat a Lady Steiger is a psychotic master of disguise who stalks and kills various women in this suspenseful cat-and-mouse game. Segal, as the detective assigned to the case, uncovers clues, falls in love, and discovers that his new girl may be the killer's next victim.

1968 108m/C Rod Steiger, Lee Remick, George Segal, Eileen Heckart, Murray Hamilton; **Dir:** Jack Smight. **VHS, Beta $49.95** PAR ♂♂♂

Noa at Seventeen The political/social turmoil of Israel in 1951 is allegorically expressed by the school vs. kibbutz debate within a young girl's middle-class family. In Hebrew with subtitles.
1982 86m/C IS Dalia Shimko, Idit Zur, Shmuel Shilo, Moshe Havazelet; **Dir:** Isaac Yeshurun. **VHS, Beta $79.95** ERG, FCT ♂♂♂

Noah: The Deluge Part of "The Greatest Heroes of the Bible" series, this television film depicts the great flood and Noah's construction of the Ark.
1979 49m/C Lew Ayres, Eve Plumb, Ed Lauter, Robert Emhardt, Rita Gam. **VHS, Beta $19.98** MAG ♂♂

The Noble Ninja In a fantastic prehistoric world, a ninja does battle with foes both real and supernatural.
1990 90m/C Tsui Fung, Lo Lieh. **VHS, Beta** OCE ♂♂

Nobody's Daughter The tragedy of an eight year-old orphan shuttled from one family to another, all of whom are only interested in the money the government will pay for her care. In Hungarian with English subtitles.
? 90m/C HU Zsuzsi Czinkoczi; **Dir:** Lazlo Ranody. **VHS $59.95** EVD ♂♂1/2

Nobody's Fool Another entry in the genre of quirky Americana, this romantic comedy concerns a hapless Midwestern waitress suffering from low-self esteem who falls in love with a traveling stage-hand. Written by Beth Henley.
1986 (PG-13) 107m/C Rosanna Arquette, Eric Roberts, Mare Winningham, Louise Fletcher, Jim Youngs; **Dir:** Evelyn Purcell. **VHS, Beta, LV $19.98** LHV, WAR ♂♂

Nobody's Perfect Where "Tootsie" and "Some Like it Hot" collide (or more likely crash and burn). A lovesick teenager masquerades as a girl and joins the tennis team to be near his dream girl. Takes its title from Joe E. Brown's famous last line in "Some Like it Hot."
1990 (PG-13) 90m/C Chad Lowe, Gail O'Grady, Patrick Breen, Kim Flowers, Robert Vaughn; **Dir:** Robert Kaylor. **VHS, Beta, LV $89.98** MED ♂

Nobody's Perfekt Supposed comedy about three psychiatric patients who decide to extort $650,000 from the city of Miami when their car is wrecked. Lacks laughs and generally considered a turkey.
1979 (PG) 95m/C Gabe Kaplan, Robert Klein, Alex Karras, Susan Clark; **Dir:** Peter Bonerz. **VHS, Beta $69.95** COL Woof!

Nocturna Hard times have fallen upon the house of Dracula. To help pay the taxes on the castle, the owners have converted it to the Hotel Transylvania. In order to increase business and the blood supply at the hotel, Nocturna books a disco group to entertain the guests. Hard times fell on this script, too. It's no wonder that director Tampa used the alias Harry Hurwitz for this film.
1979 (R) 82m/C Yvonne de Carlo, John Carradine, Tony Hamilton, Nai Bonet; **Dir:** Harry Tampa. **VHS, Beta, LV $44.95** MED Woof!

Nocturne A police lieutenant investigates the supposed suicide of a famous composer and uncovers dark secrets that suggest murder is afoot. An overlooked RKO production shines thanks to Raft's inimitable tough

guy performance and some offbeat direction. For film noir completists.
1946 88m/B George Raft, Lynn Bari, Virginia Huston, Joseph Pevney, Myrna Dell, Edward Ashley, Walter Sande, Mabel Paige; *Dir:* Edwin L. Marin. **VHS, Beta $19.98** *TTC* 🐾🐾½

Noel's Fantastic Trip Noel and his dog travel through outer space in a single prop airplane seeking adventure.
1984 69m/C VHS, Beta $19.95 *COL* 🐾🐾½

Nomad Riders A la ''Mad Max,'' one rugged man goes after the bikers who killed his wife and daughter.
1981 82m/C Wayne Chema, Richard Cluck, Ron Gregg; *Dir:* Frank Roach. **VHS, Beta $69.98** *LIV, VES* 🐾

Nomads A supernatural thriller set in Los Angeles about a French anthropologist who is mysteriously killed and the woman doctor who investigates and becomes the next target of a band of strange street people with nomadic spirits. Nomad notables include pop stars Adam Ant and Josie Cotton.
1986 (R) 91m/C Pierce Brosnan, Lesley-Anne Down, Adam Ant, Anna Maria Monticelli; *Dir:* John McTiernan. **VHS, Beta, LV $29.95** *IME, PAR* 🐾🐾½

Nomads of the North Vintage silent melodrama set in the North Woods with the requisite young girl beset by evil villians, a climatic forest fire, and a dashing Mountie who allows a man wrongly sought by the law to be reunited with the woman he secretly loves.
1920 109m/B Lon Chaney Sr., Lewis Stone, Betty Blythe. **VHS, Beta, 8mm $16.95** *SNC, GPV, VYY* 🐾½

Nomugi Pass A young woman working in a Japanese silk mill in the early 1900s must endure hardship and abuse. English subtitles.
1979 154m/C *JP* Shinobu Otake, Meiko Harada, Rentaro Mikuni, Takeo Jii. **VHS, Beta** *VDA* 🐾🐾½

Non-Stop New York Mystery tale with interesting twist. A wealthy woman can give an alibi for a murder suspect, but no one will listen, and she is subsequently framed. Pays homage to Hitchcock with its photography and humor. Quick and charming.
1937 71m/B Anna Lee, John Loder, Francis L. Sullivan, Frank Cellier; *Dir:* Robert Stevenson. **VHS $16.95** *NOS, SNC* 🐾🐾½

None But the Brave During World War II, an American bomber plane crash lands on an island already inhabited by stranded Japanese forces. After a skirmish, the two groups initiate a fragile truce, with the understanding that fighting will resume if one or the other sends for help. The Americans repair their radio unit and must decide on their next actions. Sinatra's directorial debut is a poor effort.
1965 105m/C Frank Sinatra, Clint Walker, Tommy Sands, Brad Dexter, Tony Bill, Tatsuya Mihashi, Takeshi Kato, Sammy Jackson; *Dir:* Frank Sinatra. **VHS, Beta, LV $59.95** *WAR, FCT* 🐾🐾

None But the Lonely Heart In the days before World War II, a Cockney drifter wanders the East End of London. When Grant's get-rich-quick schemes fail, his dying shopkeeper-mother tries to help and lands in prison. Interesting characterization of life in the slums. Odets not only directed, but wrote the screenplay.
1944 113m/B Cary Grant, Ethel Barrymore, Barry Fitzgerald, Jane Wyatt, Dan Duryea, George Coulouris, June Duprez; *Dir:* Clifford Odets. Academy Awards '44: Best Supporting

Actress (Barrymore); National Board of Review Awards '44: Best Film. **VHS, Beta $19.95** *MED, KOV, BTV* 🐾🐾🐾

Noon Sunday A cold war situation in the Pacific islands explodes into an orgy of death.
1971 (PG) 104m/C Mark Lenard, John Russell, Linda Avery, Keye Luke. **VHS, Beta $19.95** *ACA* 🐾

Noon Wine A dramatization of Katherine Anne Porter's classic story about a Swedish worker on a Texas farm who becomes the center of familial warfare. Made for PBS' American Playhouse.
1984 60m/C Fred Ward, Lise Hilboldt, Stellan Skarsgard, Jon Cryer; *Dir:* Michael Fields. **VHS, Beta $24.95** *MON, FCT, KAR* 🐾🐾🐾

Norma After a young woman's husband is killed at war, she starts sleeping with every man she can find. After a year of this, she decides maybe she has a problem and enlists the aid of a psychotherapist. Through hypnosis he finally unlocks her terrible sexual secret. Truly awful.
1989 80m/C Art Metrano, Mady Maguire, Chris Warfield, William Rotsler, George Flower; *Dir:* William Rotsler. **VHS $29.95** *AVD, HHE* 🐾

Norma Rae A poor, uneducated textile worker joins forces with a New York labor organizer to unionize the reluctant workers at a Southern mill. Field was a surprise with her fully developed character's strength, beauty, and humor; her Oscar was well-deserved. Ritt's direction is top-notch. Jennifer Warnes sings the theme song, ''It Goes Like It Goes,'' which also won an Oscar.
1979 (PG) 114m/C Sally Field, Ron Leibman, Beau Bridges, Pat Hingle; *Dir:* Martin Ritt. Academy Awards '79: Best Actress (Field), Best Song (''It Goes Like It Goes''); Los Angeles Film Critics Association Awards '79: Best Actress (Field); National Board of Review Awards '79: Best Actress (Field); New York Film Critics Awards '79: Best Actress (Field). **VHS, Beta, LV $19.98** *FOX, BTV* 🐾🐾🐾

Norman Conquests, Part 1: Table Manners Part one of playwright Alan Ayckbourn's comic trilogy of love unfulfilled.
1980 108m/C Tom Conti, Richard Briers, Penelope Keith. **VHS, Beta $59.99** *HBO* 🐾🐾🐾

Norman Conquests, Part 2: Living Together Part two concerns the happenings in the living room during Norman's disastrous weekend of unsuccessful seduction.
1980 93m/C Tom Conti. **VHS, Beta $59.99** *HBO* 🐾🐾🐾

Norman Conquests, Part 3: Round and Round the Garden Part three concerns Norman's furtive appearance in the garden, which suggests that the weekend is going to misfire.
1980 106m/C Tom Conti. **VHS, Beta $59.99** *HBO* 🐾🐾🐾

Norman Loves Rose A precocious 13-year-old and his married sister-in-law join forces in this substandard comedy. When Rose tries to help Norman by teaching him about sex, she learns a little too—she's pregnant.
1982 (R) 95m/C *AU* Carol Kane, Tony Owen, Warren Mitchell, Myra de Groot; *Dir:* Henri Safran. **VHS, Beta $69.98** *LIV, VES* 🐾½

Norman's Awesome Experience Three adolescents are transported back in time to the Roman Empire.

1988 (PG-13) 90m/C Tom McCamus, Laurie Paton, Jaques Lussier; *Dir:* Paul Donovan. **VHS, Beta $14.95** *SOU* Woof!

Norseman The leader of a band of Norsemen sets sail for the New World in search of his missing, royal father.
1978 (PG) 90m/C Lee Majors, Cornel Wilde, Mel Ferrer, Christopher Connelly; *Dir:* Charles B. Pierce. **VHS, Beta $69.98** *LIV, VES* 🐾

North to Alaska A gold prospector encounters many problems when he agrees to pick up his partner's fiancee in Seattle and bring her home to Nome, Alaska, in the 1890s. Slapstick at times, but great fun nonetheless. Loosely based on Laszlo Fodor's play ''The Birthday Gift.''
1960 120m/C John Wayne, Stewart Granger, Ernie Kovacs, Fabian, Capucine; *Dir:* Henry Hathaway. **VHS, Beta, LV $19.98** *FOX* 🐾🐾🐾

The North Avenue Irregulars A slapstick Disney comedy along the same lines as some of their earlier laugh-fests. A priest and three members of the local ladies club try to bust a crime syndicate. Though the premise is silly, there are still lots of laughs in this family film.
1979 (G) 99m/C Edward Herrmann, Barbara Harris, Susan Clark, Karen Valentine, Michael Constantine, Cloris Leachman, Melora Hardin, Alan Hale Jr., Ruth Buzzi, Patsy Kelly, Virginia Capers; *Dir:* Bruce Bilson. **VHS $69.95** *DIS* 🐾🐾½

North of the Border A cowboy in Canada gets involved with back-stabbing and treachery in and around Alberta.
1946 40m/B *CA* Russell Hayden, Lyle Talbot, Inez Cooper. **VHS, Beta, LV** *WGE* 🐾

North Dallas Forty Based on the novel by former Dallas Cowboy Peter Gent, the film focuses on the labor abuses in pro-football. One of the best football movies ever made, it contains searing commentary and very good acting, although the plot is sometimes dropped behind the line of scrimmage.
1979 (R) 119m/C Nick Nolte, Mac Davis, Charles Durning, Bo Svenson, Brian Dennehy, John Matuszak, Dayle Haddon, Steve Forrest, Dabney Coleman, G.D. Spradlin; *Dir:* Ted Kotcheff. National Board of Review Awards '79: 10 Best Films of the Year. **VHS, Beta, LV $14.95** *PAR* 🐾🐾🐾½

North of the Great Divide Standard Rogers programmer, with Roy as a half-breed mediator between salmon fisherman and Indians.
1950 67m/C Roy Rogers, Penny Edwards, Gordon Jones, Roy Barcroft, Jack Lambert, Douglas Evans, Noble Johnson; *Dir:* William Witney. **VHS $12.98** *REP* 🐾½

North by Northwest Grant is a self-assured Madison Avenue ad executive who inadvertently gets involved with international spies when they believe him to be someone else. His problems are compounded when he is framed for murder. The movie that has Grant and Saint dangling from the faces of Mount Rushmore and also contains the segment where a plane chases Grant through farm fields. Performances by the cast are exceptional, particularly Grant's. As with other Hitchcock productions, plenty of plot twists are mixed with tongue-in-cheek humor. Considered by many to be one of Hitchcock's greatest films. Also available on videodisc with letter boxing, digital soundtrack, special interview with Hitchcock, production and publicity photos, storyboards, and original movie trailer.

1959 136m/C Cary Grant, Eva Marie Saint, James Mason, Leo G. Carroll, Martin Landau, Jessie Royce Landis; *Dir:* Alfred Hitchcock. Edgar Allan Poe Awards '59: Best Screenplay. **VHS, Beta, LV** $19.95 *MGM, PIA, VYG* 🐾🐾🐾🐾

North Shore A young surfer from Arizona hits the beaches of Hawaii and discovers love, sex, and adventure. Only redeeming quality is the surfing scenes.
1987 (PG) 96m/C Matt Adler, Nia Peeples, John Philbin, Gregory Harrison, Christina Raines; *Dir:* Will Phelps. **VHS, Beta** $79.95 *MCA* 🐾🐾½

The North Star Gripping war tale of Nazis over-running an eastern Russian city, with courageous villagers fighting back. Also called "Armored Attack." Colorized version available.
1943 108m/B Dana Andrews, Walter Huston, Anne Baxter, Farley Granger, Walter Brennan, Erich von Stroheim, Jack Perrin, Dean Jagger; *Dir:* Lewis Milestone. **VHS** $19.95 *NOS, MRV, VYY* 🐾🐾🐾

Northeast of Seoul Three people will stop at nothing to steal a legendary jewel-encrusted sword out of Korea.
1972 (PG) 84m/C Anita Ekberg, John Ireland, Victor Buono; *Dir:* David Lowell Rich. **VHS, Beta** $59.95 *MGM* 🐾

Northern Lights A small, black-and-white independent drama depicting the struggle of a lowly farmer combating governmental forces in the 1915 heartland. Subtitled.
1979 85m/B Robert Behling, Susan Lynch, Joe Spano, Rob Nilsson, Henry Martinson, Marianne Astrom-DeFina, Ray Ness, Helen Ness; *Dir:* Rob Nilsson, John Hanson. **VHS, Beta** $19.95 *STE, NWV* 🐾🐾🐾

Northern Pursuit A Canadian Mountie disguises himself to infiltrate a Nazi spy ring in this exciting adventure film. Based on Leslie White's "Five Thousand Trojan Horses."
1943 94m/B Errol Flynn, Helmut Dantine, Julie Bishop, Gene Lockhart, Tom Tully; *Dir:* Raoul Walsh. **VHS, Beta** $19.98 *MLB* 🐾🐾½

Northville Cemetery Massacre A gang of bikers comes to town and all havoc breaks loose. The result is a horribly bloody war between the townsfolk and the gang. Also known as "The Northfield Cemetery Massacre."
1976 (R) 81m/C David Hyry, Craig Collicott, Jan Sisk. **VHS, Beta** *PGN* 🐾

Northwest Frontier A turn-of-the-century adventure set in India about a courageous attempt to save the country from rebellion, and the infant prince from assassination. Also called "Flame Over India."
1959 129m/C Lauren Bacall, Herbert Lom, Kenneth More; *Dir:* J. Lee Thompson. **VHS, Beta** $59.95 *SUE, DVT* 🐾🐾

Northwest Outpost Eddy is a California cavalry officer in this lightweight operetta. He helps a young woman who is trying to free her husband from jail, and after his death they are able to pursue their relationship. Eddy's last film.
1947 91m/B Nelson Eddy, Ilona Massey, Hugo Haas, Elsa Lanchester, Lenore Ulric; *Dir:* Allan Dwan. **VHS** $19.98 *REP* 🐾½

Northwest Passage The lavish first half of a projected two-film package based on Kenneth Roberts' popular novel, depicting the troop of Rogers' Rangers fighting the wilderness and hostile Indians. Beautifully produced; the second half was never made and the passage itself is never seen.

1940 126m/C Spencer Tracy, Robert Young, Ruth Hussey, Walter Brennan, Nat Pendleton, Robert Barratt, Lumsden Hare; *Dir:* King Vidor. **VHS, Beta** $19.98 *MGM* 🐾🐾🐾

Northwest Trail Mounted policeman covers the wilderness in search of a killer.
1946 75m/C John Litel, Bob Steele, Joan Woodbury. **VHS, Beta** $19.95 *NOS, WGE, HHT* 🐾

A Nos Amours Craving the attention she is denied at home, a young French girl searches for love and affection from numerous boyfriends in hopes of eradicating her unhappy home. Occasional lapses in quality and slow pacing hamper an otherwise excellent effort. The characterization of the girl Suzanne is especially memorable. In French with English subtitles.
1984 (R) 99m/C *FR* Sandrine Bonnaire, Dominique Besnehard, Maurice Pialat, Evelyne Ker; *Dir:* Maurice Pialat. **VHS, Beta, LV** $59.95 *COL, TAM* 🐾🐾🐾

Nosotros (We) A thief tries to straighten up his life so he can win the love of a pretty girl.
1977 87m/C *MX* Ricardo Montalban, Emilia Guiu, Esther Luquin, Carlos M. Baena. **VHS, Beta** $34.95 *MAD* 🐾

Nostradamus A documentary that investigates Nostradamus, the 16th century doctor who predicted the future with a frightening accuracy.
1988 102m/C **VHS, Beta, LV** $29.95 *AVD, WSH, HHE* 🐾

Not My Kid The 15-year-old daughter of a surgeon brings turmoil to her family when she becomes heavily involved in drugs. Producer Polson, along with Dr. Miller Newton, wrote the original book for this emotional story that is one of the better treatments of this important subject. Made for television.
1985 120m/C George Segal, Stockard Channing, Viveka Davis; *Dir:* Michael Tuchner. **VHS, Beta** $79.95 *SVS* 🐾🐾½

Not for Publication A woman working as both a tabloid reporter and a mayoral campaign worker uncovers governmental corruption with the help of a shy photographer and a midget. Meant to be on par with older screwball comedies but lacking the wit and subtlety.
1984 (R) 87m/C Nancy Allen, David Naughton, Richard Paul, Alice Ghostley, Laurence Luckinbill; *Dir:* Paul Bartel. **VHS, Beta** $69.99 *HBO* 🐾🐾

Not Quite Love Pre-pubescent ghetto waif and a self-appointed prophet seek love in uncaring city. Much music, singing and foot-stomping.
1988 30m/C *Dir:* David Scmidlapp. **VHS** $29.95 *FCT* 🐾🐾

Not Quite Paradise A young American medical student falls in love with a young Israeli girl on an Israeli kibbutz. Also called "Not Quite Jerusalem."
1986 (R) 106m/C Sam Robards, Joanna Pacula; *Dir:* Lewis Gilbert. **VHS, Beta** $19.95 *STE, NWV* 🐾

Not of This Earth In a remake of the 1957 Roger Corman quickie, an alien wearing sunglasses makes an unfriendly trip to Earth. In order to save his dying planet he needs major blood donations from unsuspecting Earthlings. Not a match for the original version, some may nevertheless want to see ex-porn star Lords in her role as the nurse.
1988 (R) 92m/C Traci Lords, Arthur Roberts, Lenny Juliano, Rebecca Perle; *Dir:* Jim Wynorski. **VHS, Beta** $14.98 *MGM* 🐾

Not Tonight Darling A bored suburban housewife becomes involved with a fast-talking businessman who leads her into a web of deceit and blackmail.
1972 (R) 70m/C Luan Peters, Vincent Ball, Jason Twelvetrees. **VHS, Beta** $59.95 *GEM* 🐾

Not Without My Daughter Overwrought drama shot in Israel about American Field, who travels with Arab husband and their daughter to his native Iran, where (he must've forgotten to tell her) she has no rights. He decides the family will stay, using beatings and confinement to persuade his uncooperative wife, but she risks all to escape with daughter. Based on the true story of Betty Mahmoody.
1990 (PG-13) 107m/C Sally Field, Alfred Molina; *Dir:* Brian Gilbert. **VHS, Beta, LV, 8mm** $92.99 *MGM, PIA, FCT* 🐾🐾½

Nothing But the Night Lee's company produced this convoluted story of orphans who are victims of a cult that uses them in their quest for immortality. Alternate titles: "The Devil's Undead" and "The Resurrection Syndicate."
1972 (PG) 90m/C *GB* Christopher Lee, Peter Cushing, Diana Dors, Georgia Brown, Keith Barron, Fulton Mackay, Gwyneth Strong; *Dir:* Peter Sasdy. **VHS** *SNC* 🐾½

Nothing But Trouble Yuppie couple out for weekend drive find themselves smoldering in small town hell thanks to a traffic ticket. Horror and humor mix like oil and water in Aykroyd's debut as director.
1991 (PG-13) 94m/C Dan Aykroyd, Demi Moore, Chevy Chase, John Candy, Taylor Negron, Bertila Demas, Valri Bromfield; *Dir:* Dan Aykroyd. **VHS, Beta, LV, 8mm** $92.99 *WAR, PIA* 🐾

Nothing in Common In his last film, Gleason plays the abrasive, diabetic father of immature advertising agency worker Hanks. After his parents separate, Hanks learns to be more responsible and loving in caring for his father. Comedy and drama are blended well here with the help of satirical pokes at the ad business and a fine performance by Hanks, but an unorganized, lengthy plot may lose some.
1986 (PG) 119m/C Tom Hanks, Jackie Gleason, Eva Marie Saint, Bess Armstrong, Hector Elizondo; *Dir:* Garry Marshall. **VHS, Beta, LV** $19.99 *HBO* 🐾🐾½

Nothing Personal In a mix of romantic comedy and environmental themes, Somers plays a lawyer trying to help a college professor's attempts to prevent the slaughter of seals. The combination is only somewhat successful, and the comic bit undermines the comparatively serious theme. Somers' first starring film role.
1980 (PG) 96m/C Donald Sutherland, Suzanne Somers, Dabney Coleman, John Dehner, Roscoe Lee Browne, Catherine O'Hara; *Dir:* George Bloomfield. **VHS, Beta, LV** $69.98 *LIV, VES* 🐾½

Nothing Sacred A slick, overzealous reporter takes advantage of a small-town girl's situation. As a publicity stunt, his newspaper brings her to the Big Apple to distract her from her supposedly imminent death in order to manipulate the public's sentiment as a means to sell more copy. Innocent young Lombard, however, is far from death's door, and deftly exploits her exploitation. Scripted by Ben Hecht, it is a scathing indictment of the mass media and the bovine mentality of the masses. The diametric opposite of Frank Capra's optimism, it is both hysterically funny and bitterly cynical, and boasts Lombard's finest performance as the small-town rube who orches-

trates the ruse. Remade in 1954 as "Living It Up." Musical score by Oscar Levant.
1937 75m/C Fredric March, Carole Lombard, Walter Connolly, Sig Rumann, Charles Winninger, Margaret Hamilton; *Dir:* William A. Wellman. **VHS, Beta, LV $19.95** NOS, MRV, VYY 🎞🎞½

Nothing Underneath An American guy goes to Rome to search for his model twin sister, who may be one of the victims of a series of scissor killings.
1985 95m/C Tom Schanley, Renee Simonsen, Nicola Perring, Donald Pleasence; *Dir:* Carlo Vanzina. **VHS, Beta $79.95** SVS 🎞½

Notorious Post World War II story of a beautiful playgirl sent by the U.S. government to marry a suspected spy living in Brazil. Grant is the agent assigned to watch her. Duplicity and guilt are important factors in this brooding, romantic spy thriller. Suspenseful throughout, with a surprise ending. The acting is excellent all around and Hitchcock makes certain that suspense in maintained throughout this classy and complex thriller. Oscar nominations for Best Supporting Actor (Rains), Original Screenplay. Laser edition contains original trailer, publicity photos, and additional footage.
1946 101m/B Cary Grant, Ingrid Bergman, Claude Rains, Louis Calhern, Madame Konstantin; *Dir:* Alfred Hitchcock. **VHS, Beta, LV $19.98** FOX, VYG, CRC 🎞🎞🎞

A Nous le Liberte Two tramps encounter industrialization and automation, making one into a wealthy leader, the other into a nature-loving iconoclast. A poignant, fantastical masterpiece by Clair, made before he migrated to Hollywood. Though the view of automation may be dated, it influenced such films as Chaplin's "Modern Times." In French with English subtitles.
1931 87m/B FR Henri Marchand, Raymond Cordy, Rolla France, Paul Olivier; *W/Dir:* Rene Clair. National Board of Review Awards '32: Best Foreign Film; Venice Film Festival '31: Most Amusing Film. **VHS, Beta, 8mm $49.95** TAM, NOS, MRV 🎞🎞🎞

Nous N'Irons Plus au Bois On the outskirts of a forest killed by the Germans, a small group of French resistance fighters capture a German soldier during World War II. He actually wants to join forces with them, however, after falling in love with a lovely young French girl. In French; subtitled in English.
1969 98m/C FR Marie-France Pisier, Siegfried Rauch, Richard Leduc; *Dir:* Georges Dumoulin. **VHS, Beta $19.95** INV 🎞🎞

Now and Forever A young wife's life is shattered when her unfaithful husband is wrongly accused and convicted of rape. After he is sent to prison, she begins drinking and taking drugs. From the novel by Danielle Steel, it will appeal most to those who like their romances a la Harlequin.
1982 (R) 93m/C AU Cheryl Ladd, Robert Coleby; *Dir:* Adrian Carr. **VHS, Beta, LV $59.95** MCA 🎞½

Now, Voyager Davis plays a lonely spinster who is transformed into a vibrant young woman by therapy. She comes out of her shell to have a romantic affair with a suave European (who turns out to be married) but still utters the famous phrase "Oh, Jerry, we have the stars. Let's not ask for the moon." Definitely melodramatic, but an involving story nonetheless. Based on a novel by Olive Higgins Prouty.

1942 117m/B Bette Davis, Gladys Cooper, Claude Rains, Paul Henreid, Bonita Granville; *Dir:* Irving Rapper. Academy Awards '42: Best Musical Score. **VHS, Beta, LV $19.95** MGM 🎞🎞½

Now You See Him, Now You Don't Light Disney comedy involving a gang of crooks who want to use a college student's invisibility formula to rob a local bank. Sequel to Disney's "The Computer Wore Tennis Shoes."
1972 (G) 85m/C Kurt Russell, Joe Flynn, Cesar Romero, Jim Backus; *Dir:* Robert Butler. **VHS, Beta $69.95** DIS 🎞🎞

Nowhere to Hide A widow whose Marine officer husband has been assassinated is chased by some bad guys who are after a helicopter part (say what?) and she must fight for survival for herself and her 6-year-old son.
1987 (R) 91m/C Amy Madigan, Daniel Hugh-Kelly, Michael Ironside; *Dir:* Mario Azzopardi. **VHS, Beta $79.95** LHV, HHE, WAR 🎞🎞

Nowhere to Run Slipshod version of a true story about a series of murders in Caddo, Texas in 1960. Carradine, parolled from prison, murders for revenge while six high school seniors get involved in the chase and in the seedy side of politics, police, and their own puberty.
1988 (R) 87m/C David Carradine, Jason Priestly, Kieran Mulroney, Henry Jones; *Dir:* Carl Franklin. **VHS, Beta $79.95** MGM 🎞

Nuclear Conspiracy A reporter disappears while investigating a huge nuclear waste shipment and his wife searches for him.
1985 115m/C Birgit Doll, Albert Fortell. **VHS, Beta $69.95** VMK 🎞½

The Nude Bomb Maxwell Smart (would you believe?) tries to save the world from a bomb that would only destroy clothing and leave everyone in the buff. Old hat lines. Also known as "The Return of Maxwell Smart," followed by the television movie "Get Smart, Again!"
1980 (PG) 94m/C Don Adams, Dana Elcar, Pamela Hensley, Sylvia Kristel, Norman Lloyd, Rhonda Fleming, Joey Forman; *Dir:* Clive Donner. **VHS, LV $79.95** MCA, PIA 🎞🎞½

Nude on the Moon Lunar expedition discovers moon inhabited by people who bare skin as hobby. Groovy theme song, "I'm Mooning Over You, My Little Moon Doll." Part of Joe Bob Brigg's "Sleaziest Movies in the History of the World" series.
1961 83m/C *Dir:* Doris Wishman. **VHS $19.98** SVI Woof!

Nudity Required Two pals use the casting couch, pretending to be Hollywood producers, to seduce young women. Silly comedy with gratuitous nudity (Ms. Newmar included). Hardly a laugh to be found.
1990 90m/C Julie Newmar, Troy Donahue, Brad Zutaut, Billy Frank, Ty Randolph, Alvin Silver, Eli Rich, Phil Hock; *Dir:* John Bown. **VHS** RAE 🎞

Nudo di Donna A witty comedy about a Venetian bookseller who is becoming tired of marital struggles with his sexy wife of 16 years. He wanders into a fashion photography shop, sees a backside nude photo of a model who looks like his wife, and he takes off in fiery pursuit to find the subject of the photo. Location Venetian photography is quite nice. With English subtitles.
1983 112m/C IT FR Nino Manfredi, Jean-Pierre Cassel, George Wilson, Elenora Giorgi; *Dir:* Nino Manfredi. **VHS, Beta $29.95** PAV, APD 🎞🎞🎞

Number 1 of the Secret Service A handsome secret agent must foil the plans of a millionaire industrialist to destroy the economy. Spy spoof that's marginally interesting.
1977 (PG) 87m/C GB Nicky Henson, Geoffrey Keen, Sue Lloyd, Aimi MacDonald, Richard Todd; *Dir:* Lindsay Shonteff. **VHS, Beta $79.95** MGM 🎞🎞

Number One With a Bullet Two unorthodox "odd couple" detectives are demoted after losing a key witness, but still set out to unearth a drug czar on their own. Carradine and Williams are better than this standard action material.
1987 (R) 103m/C Robert Carradine, Billy Dee Williams, Peter Graves, Valerie Bertinelli; *Dir:* Jack Smight. **VHS, Beta $79.95** MGM 🎞🎞

Number Seventeen A humorous early thriller by the Master Hitchcock, filmed before the likes of "The 39 Steps." An unsuspecting hobo accidentally discovers a jewel thieve's cache. The chase is on—superb final chase sequence involving a bus and a train. Based on the play by J. Jefferson Farjeon.
1932 64m/B Leon M. Lion, Anne Grey, John Stuart, Donald Calthrop; *Dir:* Alfred Hitchcock. **VHS, Beta, 8mm $16.95** SNC, NOS, HHT 🎞🎞🎞

The Nun A young woman, unable to meet financial obligations, is forced into a convent. Victimized by the mother superior and betrayed by a clergyman who befriends her, she escapes from the convent and eventually ends up in a bordello, despairing and suicidal. Not a cheery story. Banned in France for two years. In French with English subtitles.
1966 155m/C FR Anna Karina, Liselotte Pulver, Micheline Presle, Christine Lenier, Francine Berge, Francisco Rabal, Wolfgang Reichmann, Catherine Diamant, Yori Bertin; *Dir:* Jacques Rivette. **VHS $59.95** INT, FCT, TAM 🎞🎞½

Nuns on the Run Idle and Coltrane are two nonviolent members of a robbery gang who double-cross their boss during a hold-up and disguise themselves as nuns while on the run from both the Mob and the police. Catholic humor, slapstick, and much fun with habits dominate. Idle and Coltrane do their best to keep the so-so script moving with its one-joke premise.
1990 (PG-13) 95m/C Eric Idle, Robbie Coltrane, Janet Suzman, Camille Coduri, Robert Patterson, Tom Hickey, Doris Hare, Lila Kaye; *Dir:* Jonathan Lynn. **VHS, Beta $89.98** FOX 🎞🎞

The Nun's Story The melancholy tale of a young nun working in the Congo and Belgium during World War II, and struggling to reconcile her free spirit with the rigors of the order. Highly acclaimed; from the Kathryn Hulme novel.
1959 152m/C Audrey Hepburn, Peter Finch, Edith Evans, Peggy Ashcroft, Mildred Dunnock, Dean Jagger, Beatrice Straight, Colleen Dewhurst; *Dir:* Fred Zinneman. British Academy Awards '59: Best Actress (Hepburn); National Board of Review Awards '59: Best Director (Zinneman), Best Picture, Best Supporting Actress (Evans). **VHS, Beta, LV $19.98** WAR 🎞🎞🎞½

Nurse A recently widowed woman resumes her career as a nurse in a large urban hospital, after her son leaves for college. Based on Peggy Anderson's book. Made for television; became a weekly series.
1980 105m/C Michael Learned, Robert Reed, Antonio Fargas; *Dir:* David Lowell Rich. **VHS, Beta $59.95** IVE 🎞🎞½

Nurse on Call A bevy of trampy nurses cavort throughout a big city hospital. Softcore.
1988 80m/C Anne Tilson, Jennie Martinez, Christopher Floyd. **VHS, Beta $29.95** *MED* 🦴

Nurse Edith Cavell Fine performances in this true story of Britain's famous nurse who aided the Belgium underground during World War I, transporting wounded soldiers out of the German-occupied country. Decidedly opposes war and, ironically, was released just as World War II began to heat up in 1939.
1939 95m/B George Sanders, Edna May Oliver, ZaSu Pitts, Robert Coote, May Robson, Anna Neagle; ***Dir:*** Herbert Wilcox. **VHS, Beta $19.95** *NOS, CAB, KRT* 🦴🦴🦴

The Nut House This is an unsold television pilot for a live comedy series that features a collection of skits and blackouts.
1962 34m/B VHS, Beta $24.95 *VYY* 🦴🦴🦴

Nutcase A trio of young children in New Zealand attempt to thwart a group of terrorists who threaten to reactivate a large city's volcanoes unless they are given a large sum of money. Novel twist there.
1983 49m/C *NZ* Nevan Rowe, Ian Watkin, Michael Wilson; ***Dir:*** Roger Donaldson. **VHS, Beta $14.98** *VID* 🦴

Nutcracker, A Fantasy on Ice An ice-skating version of the famed story and Tchaikovsky ballet aired as a television special in 1983.
1983 85m/C Dorothy Hamill; ***Nar:*** Lorne Greene. **VHS, Beta $19.95** *VMK* 🦴

The Nutcracker Prince The classic children's Christmas tale comes alive in this feature-length animated special.
1991 75m/C ***Dir:*** Paul Schibli; ***Voices:*** Kiefer Sutherland, Megan Follows, Michael MacDonald, Phyllis Diller, Peter O'Toole. **VHS, LV, 8mm $19.98** *WAR* 🦴

Nutcracker Sweet Ridiculous drama about a beautiful and powerful socialite (Collins) who runs a renowned ballet company with an iron fist. A Russian ballerina defects and infiltrates Collins' company to try out her own treacherous motives.
1984 101m/C Joan Collins, Finola Hughes, Paul Nicholas; ***Dir:*** Anwar Kawadri. **VHS, Beta $69.98** *LIV, VES* 🦴

Nutcracker: The Motion Picture A lavish, stage-bound filmization of the Tchaikovsky ballet, designed by Maurice Sendak.
1986 82m/C Pacific Northwest Ballet; ***Dir:*** Carroll Ballard. **VHS, Beta $14.95** *PAR, KRT, KAR* 🦴

Nuts A high-priced prostitute attempts to prove her sanity when she's accused of manslaughter. Ashamed of her lifestyle, and afraid of her reasons for it, her parents attempt to institutionalize her. A filmed version of Tom Topor's play that manages to retain its mesmerizing qualities. Very fine performances, although Barbra goes over the top on several occasions. Score by Streisand.
1987 (R) 116m/C Barbra Streisand, Richard Dreyfuss, Maureen Stapleton, Karl Malden, James Whitmore, Robert Webber, Eli Wallach, Leslie Nielsen; ***Dir:*** Martin Ritt. **VHS, Beta, LV, 8mm $19.98** *WAR* 🦴🦴🦴

The Nutty Professor A mild-mannered chemistry professor creates a potion that turns him into a suave, debonair, playboy type with an irresistible attraction to women. Lewis has repeatedly denied the slick character is a Dean Martin parody, but the evidence is quite strong. Easily Lewis' best film.
1963 107m/C Jerry Lewis, Stella Stevens, Howard Morris, Kathleen Freeman; ***Dir:*** Jerry Lewis. **VHS, Beta, LV $14.95** *PAR* 🦴🦴🦴

Nyoka and the Tigermen The adventures of the jungle queen Nyoka and her rival Vultura in their search for the lost tablets of Hippocrates. In 15 episodes. Also known as "Nyoka and the Lost Secrets of Hippocrates."
1942 250m/B Kay Aldridge, Clayton Moore; ***Dir:*** William Witney. **VHS $29.98** *VCN, REP, MLB* 🦴🦴🦴

O Lucky Man Surreal, black comedy following the rise and fall and eventual rebirth of a modern British coffee salesman. Several actors play multiple roles with outstanding performances throughout. Price's excellent score combines with the hilarity for an extraordinary experience.
1973 (R) 178m/C *GB* Malcolm McDowell, Ralph Richardson, Rachel Roberts, Arthur Lowe, Alan Price, Helen Mirren, Mona Washbourne; ***Dir:*** Lindsay Anderson. National Board of Review Awards '73: 10 Best Films of the Year. **VHS, Beta $29.95** *WAR* 🦴🦴🦴🦴

O Pioneers! Lange plays Alexandra Bergson, an unmarried woman at the turn of the century, who inherits her family's Nebraska homestead because her father knows how much she loves the land. Although the family has prospered through Alexandra's smart investments, her brothers come to resent her influence. When her first love returns after fifteen years and the romance is rekindled, family resentments surface once again. Based on the novel by Willa Cather. A Hallmark Hall of Fame presentation.
1991 (PG) 99m/C Jessica Lange, David Strathairn, Tom Aldredge, Reed Edward Diamond, Anne Heche, Heather Graham, Josh Hamilton, Leigh Lawson; ***W/Dir:*** Glenn Jordan. **VHS, LV $89.98** *REP* 🦴🦴½

O Youth and Beauty Film dealing with one of author Cheever's favorite themes, the American male's fear of losing his identity. Frustrated by the onset of age, a man tries to recapture his college years and athletic ability by hurdling furniture. From the "Cheever Short Stories" series.
1979 60m/C Michael Murphy, Kathryn Walker. **VHS, Beta** *FLI* 🦴🦴

Oasis of the Zombies Group of European students sets out to find buried treasure in Saharan oasis but instead finds bevy of hungry Nazis with an eating disorder.
1982 75m/C *FR SP* Manuel Gelin, France Jordan, Jeff Montgomery, Miriam Landson, Eric Saint-Just, Caroline Audret, Henry Lambert; ***Dir:*** A.M. Frank. **VHS, Beta $59.95** *UHV* 🦴

Oath of Vengeance Billy the Kid rides the West once more.
1944 50m/B Buster Crabbe, Al "Fuzzy" St. John, Jack Ingram, Charles King; ***Dir:*** Sam Newfield. **VHS, Beta $27.95** *VCN* 🦴½

Oblomov A production of the classic Goncharov novel about a symbolically inert Russian aristocrat whose childhood friend helps him find a reason for action. Well made, with fine performances. Russian dialogue with English subtitles. Also known as "A Few Days in the Life of I.I. Oblomov."
1981 145m/C *RU* Oleg Tabakov, Elena Solovei; ***Dir:*** Nikita Mikhailkov. **VHS, Beta $59.95** *TPV, IFE* 🦴🦴🦴½

The Oblong Box Coffins, blood, and live corpses fill drawn-out and lifeless adaptation of Edgar Allan Poe story. English aristocrat Price attempts to hide his disfigured brother in an old tower. Brother predictably escapes and rampages through town before being killed.
1969 (PG) 91m/C *GB* Vincent Price, Christopher Lee, Alastair Williamson, Hilary Dwyer, Peter Arne, Harry Baird, Carl Rigg, Sally Geeson, Maxwell Shaw; ***Dir:*** Gordon Hessler. **VHS, Beta $69.99** *KUI, HBO* 🦴🦴

The Obsessed Uninspiring story with Farrar suspected of murdering his wife. Competent performances, but this mystery isn't developed with any sense of style and the killer's real identity will be obvious to anyone watching. Cliches and stereotyped situations don't add anything.
1951 77m/B *GB* David Farrar, Geraldine Fitzgerald, Roland Culver, Jean Cadell, Mary Merrall; ***Dir:*** Maurice Elvey. **VHS, LV $16.95** *SNC, NOS, MOV* 🦴🦴

Obsessed Semi-realistic tale of a woman's exhausting desire for revenge against hit-and-run driver who killed her son. Cast acceptable but lack of psychological foundation weakens plot.
1988 (PG-13) 100m/C Kerrie Keane, Alan Thicke, Colleen Dewhurst, Saul Rubinek, Daniel Pilon, Lynne Griffin; ***Dir:*** Robin Spry. Genie Awards '89: Best Supporting Actress (Dewhurst). **VHS, Beta, LV $39.95** *NSV* 🦴½

Obsession A rich, lonely businessman meets a mysterious young girl in Italy, the mirror image of his late wife who was killed by kidnappers. Intriguing suspense film that's not quite up to comparisons with Hitchcock thrillers. Music by Bernard Herrmann.
1976 (PG) 98m/C Cliff Robertson, Genevieve Bujold, John Lithgow; ***Dir:*** Brian De Palma. National Board of Review Awards '76: 10 Best Films of the Year. **VHS, Beta, LV $12.95** *COL, PIA, HHE* 🦴🦴½

Obsession: A Taste for Fear Diane's high fashion models are being killed and although there are plenty of suspects, she can't find the killer.
1989 (R) 90m/C Virginia Hey, Gerard Darmon, Carlo Mucari; ***Dir:*** Piccio Raffanini. **VHS, Beta $89.95** *IMP* 🦴½

Obsessive Love A lonely, mentally unstable typist (Mimieux) becomes obsessed with her soap opera hero and decides to go to Hollywood to seduce him. She transforms herself into a sleek and stunning Hollywood temptress to woo him, and at first he goes along with her—until he realizes that she is insane. An average made for television movie of the older woman/younger man genre.
1984 97m/C Yvette Mimieux, Simon MacCorkindale, Kin Shriner, Constance McCashin, Allan Miller, Lainie Kazan; ***Dir:*** Steven Hilliard Stern. **VHS $59.98** *LIV* 🦴🦴½

O.C. and Stiggs Teenage bomb adapted from short story in National Lampoon. Held three years before released. Two teens spend a summer harassing a neighbor and his family. Flimsy attempts at comedy fall flat.
1987 (R) 109m/C Daniel H. Jenkins, Neill Barry, Jane Curtin, Tina Louise, Jon Cryer, Dennis Hopper, Paul Dooley, Ray Walston, Louis Nye, Martin Mull, Melvin Van Peebles; ***Dir:*** Robert Altman. **VHS, Beta $79.98** *FOX* 🦴½

The Occult Experience An engrossing study into the world of devil-worshippers, shamans, and the like. This documentary actually records exorcisms and other rituals. Not for the faint hearted!
1984 87m/C VHS, LV $14.95 *COL, SVS* 🦴½

The Occultist Satan worshippers do the dance of death as they skin men alive for their evil purposes.

1989 82m/C Rick Gianasi. **VHS, Beta** $79.95 *UNI* Woof!

An Occurrence at Owl Creek Bridge A man is about to be hanged when the rope snaps, and he is able to swim to safety. He entertains thoughts of his family while making his way home. Suddenly, this new-found sweetness of life comes to a halt. Based on a short story by Ambrose Bierce. Telecast on "The Twilight Zone" in 1963. **1962** 27m/B *FR Dir:* Robert Enrico. Cannes Film Festival '62: Blue Ribbon. **VHS, Beta** $19.95 *FST, VYY* 🐾🐾🐾½

Ocean Drive Weekend Several college students congregate to "Ocean Drive" for weekend of beer, sex, and dancing. Stereotypical characters, lack of plot, and poor cover versions of 1960 classics combine for boring and inane comedy. **1985** (PG-13) 98m/C Robert Peacock, Charles Redmond, Tony Freeman; *Dir:* Bryan Jones. **VHS, Beta** $69.98 *LIV, VES.* Woof!

Ocean's 11 A gang of friends make plans to rob five Las Vegas casinos simultaneously. Part of the "A Night at the Movies" series, this tape simulates a 1960 movie evening, with a Bugs Bunny cartoon, "Person to Bunny," a newsreel and coming attractions for "The Sundowners" and "Sunrise at Campobello." **1960** 148m/C Frank Sinatra, Dean Martin, Sammy Davis Jr., Angie Dickinson, Peter Lawford; *Dir:* Lewis Milestone. **VHS, Beta, LV** $19.98 *WAR* 🐾🐾½

Oceans of Fire Average rehash of the tension-filled world of oil riggers. Lives of two ex-cons are threatened when they hire on as divers for world's deepest undersea oil rig. **1986** (PG) 100m/C Gregory Harrison, Billy Dee Williams, Cynthia Sikes, Lyle Alzado, Tony Burton, Ray Mancini, David Carradine, Ken Norton; *Dir:* Steve Carver. **VHS, Beta** $79.95 *VMK* 🐾

Octagon Norris protects a woman from threatening Ninja warriors in average kung-fu adventure. Enough action-packed violence for fans of genre. **1980** (R) 103m/C Chuck Norris, Karen Carlson, Lee Van Cleef, Kim Lankford, Art Hindle, Jack Carter; *Dir:* Eric Karson. **VHS, Beta, LV** $19.95 *MED* 🐾½

Octaman Comical thriller featuring non-threatening octopus-man discovered by scientists in Mexico. Rip-off of director Essex's own "Creature from the Black Lagoon." Actress Angeli died during filming. **1971** 79m/C Kerwin Mathews, Pier Angeli, Harry Guardino, David Essex, Jeff Morrow, Norman Fields; *Dir:* Harry Essex. **VHS, Beta** $19.95 *GEM, WES, PSM* 🐾½

Octavia A contemporary fable about a blind girl who befriends a convict and learns about life and love. **1982** (R) 93m/C Susan Curtis; *Dir:* David Beaird. **VHS, Beta** $59.98 *CHA* 🐾

October Silent masterpiece based on John Reed's "Ten Days That Shook The World." Eisenstein, commissioned by the Soviet government, spared no expense to chronicle the Bolshevik Revolution of 1917 (in a flattering Communist light of course). Later he was forced to cut his portrayal of Leon Trotsky, who was then an enemy of the state. Includes rare footage of the Czar's Winter Palace in Leningrad. Haunting score by Shostakovich. Also known as "Ten Days That Shook The World." **1927** 103m/B *RU* **VHS, LV** $29.95 *WST* 🐾🐾🐾½

October Man When a model is found murdered, a stranger with mental problems must prove his innocence to others and himself. Strong characters make good, suspensful mystery reminiscent of Hitchcock. **1948** 95m/B John Mills, Joan Greenwood, Edward Chapman, Joyce Carey, Kay Walsh, Felix Aylmer, Juliet Mills; *Dir:* Roy Ward Baker. **VHS, Beta** *LCA* 🐾🐾🐾

Octopussy The Bond saga continues as Agent 007 is on a mission to prevent a crazed Russian general from launching a nuclear attack against the NATO forces in Europe. Lots of special effects and gadgets keep weak plot moving. **1983** (PG) 130m/C *GB* Roger Moore, Maud Adams, Louis Jourdan, Kristina Wayborn, Kabir Bedi; *Dir:* John Glen. **VHS, Beta, LV** $19.95 *MGM, FOX, TLF* 🐾🐾

The Odd Angry Shot Australian soldiers fighting in Vietnam discover the conflict is not what they expected. An ironic perspective is given as the men struggle with their feelings about the war. More of an unremarkable drama with comic overtones than a combat film, it will appeal to those who prefer good directing over bloodshed. **1979** 90m/C *AU Dir:* Tom Jeffrey. **VHS, Beta** $69.98 *LIV, VES* 🐾🐾½

The Odd Couple Two divorced men with completely opposite personalities move in together. Lemmon's obsession with neatness drives slob Matthau up the wall, and their inability to see eye-to-eye results in many hysterical escapades. A Hollywood rarity, it is actually better in some ways than Neil Simon's original Broadway version. Basis for the hit television series. **1968** (G) 106m/C Jack Lemmon, Walter Matthau; *Dir:* Gene Saks. **VHS, Beta, LV** $14.95 *PAR* 🐾🐾🐾½

The Odd Job Insurance salesman Chapman, depressed after his wife leaves him, hires a hit man to kill him. Chapman then has a hard time shaking his stalker after deciding he wants to live. Monty Python fans will enjoy seeing ex-troupe member Chapman again, but the comic's heart isn't in this unoriginal story. **1978** 100m/C *GB* Graham Chapman; *Dir:* Peter Medak. **VHS, Beta** $69.95 *VES* 🐾🐾

Odd Jobs When five college friends look for jobs during summer break, they wind up running their own moving business with the help of the mob. Good comic talent, but a silly slapstick script results in only a passable diversion. **1985** (PG-13) 89m/C Paul Reiser, Scott McGinnis, Rick Overton, Robert Townsend; *Dir:* Mark Story. **VHS, Beta** $79.95 *HBO* 🐾🐾

Odd Man Out In an adaptation of F.L. Green's novel that was previously filmed as "The Last Man," as Irish revolutionary is injured during a robbery attempt. Suffering from gunshot wounds and closely pursued by the police, he must rely on the help of others who could betray him at any moment. A gripping tale of suspense and intrigue that will keep the proverbial seat's edge warm until the final credits. **1947** 111m/B *GB* James Mason, Robert Newton, Dan O'Herlihy, Kathleen Ryan, Cyril Cusack; *Dir:* Carol Reed. British Academy Awards '47: Best Film; National Board of Review Awards '47: 10 Best Films of the Year. **VHS, Beta, LV** $19.95 *PAR* 🐾🐾🐾½

Odd Obsession One of Ichikawa's first films, about an elderly Japanese gentleman with a young wife whose feelings of jealousy and impotence eventually wreaks havoc on the marriage. Available in Japanese with English subtitles. Also known as "The Key." **1960** 96m/C *JP* Machiko Kyo, Ganjiro Nakamura; *Dir:* Kon Ichikawa. **VHS, Beta** $19.95 *COL, SUE* 🐾🐾½

The Odd Squad Five GIs in World War II defend a bridge from the enemy, and manage in doing so to have loads of laughs. **1986** (R) 82m/C Johnny Dorelli, Vincent Gardenia; *Dir:* D.E.P. Clucher. **VHS, Beta** $9.99 *STE, PSM* 🐾

Oddball Hall Two aging jewel thieves hide out in the African wilds, masquerading as powerful "wizards." When they head back to civilization, an African tribe beseeches the phony sorcerers to cure their long drought. **1991** (PG) 87m/C Don Ameche, Burgess Meredith. **VHS** $89.98 *WAR* 🐾🐾

Oddballs In another attempt to capitalize on the success of Bill Murray's "Meatballs," this summer camp story follows three campers and their pathetic attempts to lose their virginity. Brooks has his moments, but there is little else to recommend here. **1984** (PG-13) 92m/C Foster Brooks, Jason Sorokin, Wally Wodchis, Konnie Krome; *Dir:* Miklos Lente. **VHS, Beta** $79.98 *LIV, LTG* Woof!

Odds and Evens Italian supercops pretend they don't have badges in order to clean up illegal betting and gambling rings in sunny Miami. Odds are this spaghetti cop-o-rama won't keep Morpheus at bay. **1978** (PG) 109m/C *IT* Terence Hill, Bud Spencer; *Dir:* Sergio Corbucci. **VHS, Beta** $79.95 *WAR* 🐾½

Ode to Billy Joe The hit 1967 Bobby Gentry song of the same title is expanded to tell why a young man jumped to his death off the Tallahatchie Bridge. The problems of growing up in the rural South and teenage romance do not match the appeal of the theme song. Benson and O'Connor, however, work well together. **1976** (PG) 106m/C Robby Benson, Glynnis O'Connor, Joan Hotchkis, Sandy McPeak, James Best; *Dir:* Max Baer. **VHS, Beta** $19.98 *CCB, WAR* 🐾🐾

The Odessa File During 1963, a German journalist attempts to track down some SS war criminals who have formed a secret organization called ODESSA. The story, from Frederick Forsyth's novel, drags in some places. The high point is the scene where bad guy Schell and reporter Voight finally confront each other. Musical score by Andrew Lloyd Webber. **1974** 128m/C *GB* Jon Voight, Mary Tamm, Maximilian Schell, Maria Schell, Derek Jacobi; *Dir:* Ronald Neame. **VHS, Beta, LV** $9.95 *GKK* 🐾🐾½

L'Odeur des Fauves One of a cynical reporter's scandalous stories gets an innocent man killed, compelling the reporter to somehow make restitution. Subtitled in English. **1966** 86m/C *FR* Maurice Ronet, Josephine Chaplin, Vittorio De Sica; *Dir:* Richard Balducci. **VHS, Beta** *KRT* 🐾

The Odyssey of the Pacific Three young Cambodian refugees encounter retired train engineer Rooney living in the woods where a railway station once thrived. Together they work to restore an old locomo-

tive. Originally called "The Emperor of Peru," the ordinary, inoffensive script will not endanger quality family time.
1982 82m/C *CA* Mickey Rooney, Monique Mercure; *Dir:* Fernando Arrabal. **VHS, Beta $39.95** *MCA* 🎞🎞½

Oedipus Rex Contained, and highly structured rendering of Sophocles' Greek tragedy, by the Stratford (Ontario) Festival Players, but this production belongs on the stage and not the screen. Douglas Rain went on to be the voice of the Hal 9000 computer in Stanley Kubrick's "2001: A Space Odyssey."
1957 87m/C *CA* Douglas Campbell, Douglas Rain, Eric House, Eleanor Stuart; *Dir:* Tyrone Guthrie. **VHS** *HMV, WBF* 🎞🎞½

Oedipus Rex A new twist on the famous tragedy as Pasolini gives the story a modern prologue and epilogue. The classic plot has Oedipus spiral downward into moral horror as he tries to avoid fulfilling the prophecy that he will murder his father and sleep with his mother. Cross-cultural curiousities include Japanese music and Lenin-inspired songs.
1967 110m/C *IT* Franco Citti, Silvana Mangano, Alida Valli; *Dir:* Pier Paolo Pasolini. **VHS $59.95** *FST* 🎞🎞½

Of Cooks and Kung-Fu A martial arts culinary adventure.
198? 90m/C Jacky Chen, Chia Kai, Lee Kuen. **VHS, Beta $57.95** *OCE, UNI* 🎞🎞

Of the Dead This documentary looks at such funeral practices as an on-camera cremation and embalming. Not for the very young or the squeamish.
1979 90m/C **VHS, Beta $59.95** *MPI* 🎞🎞

Of Human Bondage The first movie version of W. Somerset Maugham's classic novel in which a young, handicapped medical student falls in love with a crude cockney waitress, in a mutually destructive affair. Established Davis' role as the tough, domineering woman. Remade in 1946 and 1964.
1934 84m/B Bette Davis, Leslie Howard, Frances Dee, Reginald Owen, Reginald Denny, Alan Hale Jr.; *Dir:* John Cromwell. New York Film Festival '65: 10 Best Films of the Year. **VHS, Beta, LV $9.95** *NEG, MRV, NOS* 🎞🎞🎞

Of Human Bondage An essentially decent man falls fatally in love with an alluring but heartless waitress, who subtly destroys him. Based on the W. Somerset Maugham novel. Miscast and least interesting of the three film versions, although Novak gives a good performance.
1964 100m/B *GB* Kim Novak, Laurence Harvey, Robert Morley, Roger Livesey, Siobhan McKenna; *Dir:* Henry Hathaway, Ken Hughes. **VHS, Beta $24.95** *MGM* 🎞🎞

Of Mice and Men A powerful adaptation of the classic Steinbeck tragedy about the friendship between two itinerant Southern ranch hands during the Great Depression. Chaney is wonderful as the gentle giant and mentally retarded Lenny, cared for by migrant worker Meredith. They both get into an irreversible situation when a woman is accidentally killed. Music by Aaron Copland. Oscar nominations: Best Picture, Sound Recording, Best Original Score.
1939 107m/C Lon Chaney Jr., Burgess Meredith, Betty Field, Bob Steele, Noah Beery Jr., Charles Bickford; *Dir:* Lewis Milestone. **VHS, Beta $59.95** *PSM* 🎞🎞🎞🎞

Of Mice and Men Made-for-television remake of the classic Steinbeck tale, casting TV's Baretta as George and Quaid as Lenny. Worth watching.
1981 150m/C Robert Blake, Randy Quaid, Lew Ayres, Mitchell Ryan, Ted Neely, Cassie Yates, Pat Hingle, Whitman Mayo, Dennis Fimple, Pat Corley; *Dir:* Reza Badiyi. **VHS, Beta $14.99** *STE, PSM, KUI* 🎞🎞🎞

Of Unknown Origin Weller encounters a mutated rampaging rat in his New York townhouse while his family is away on vacation. His house becomes the battleground in a terror-tinged duel of survival. A well-done rat thriller not for those with a delicate stomach.
1983 (R) 88m/C *CA* Peter Weller, Jennifer Dale, Lawrence Dane, Ken Welsh, Louis Del Grande, Shannon Tweed; *Dir:* George P. Cosmatos. **VHS, Beta $19.98** *WAR* 🎞🎞

Off Beat A shy librarian unluckily wins a spot in a police benefit dance troupe, and then falls in love with a tough police woman. With a screenplay by playwright Mark Medoff, and a good supporting cast, "Off Beat" still manages to miss the mark.
1986 (PG) 92m/C Judge Reinhold, Meg Tilly, Cleavant Derricks, Fred Gwynne, John Turturro, Jacques D'Amboise, James Tolken, Joe Mantegna, Harvey Keitel, Amy Wright; *Dir:* Michael Dinner. **VHS, Beta, LV $79.95** *TOU* 🎞½

Off on a Comet An animated adaptation of the Jules Verne story about a group of people carried through the cosmos on a comet. In HiFi.
1979 52m/C **VHS, Beta $19.95** *MGM* 🎞½

Off Limits When a boxing manager is drafted into the Army, he freely breaks regulations in order to train a fighter as he sees fit. Hope and Rooney, while not in peak form, make a snappy duo in this amusing romp.
1953 89m/B Bob Hope, Mickey Rooney, Marilyn Maxwell, Marvin Miller; *Dir:* George Marshall. **VHS, Beta, LV $19.98** *KRT* 🎞🎞½

Off Limits A spree of murders involving Vietnamese hookers draws two cops from the Army's Criminal Investigation Department into the sleazy backstreets of 1968 Saigon. There is little mystery as to who the killer is in this tale that seems written more for the sake of foul language and gratuitous gunfights than actual plot.
1987 (R) 102m/C Willem Dafoe, Gregory Hines, Fred Ward, Scott Glenn, Amanda Pays, Keith David, David Alan Grier; *Dir:* Christopher Crowe. **VHS, Beta, LV $14.98** *FOX* 🎞½

Off the Mark Neely is a young athlete suffering from a childhood affliction that causes spasms in his legs. Facing the ultimate challenge, he must overcome his handicap and defeat a talented woman athlete and a Russian student in a triathalon competition. An unremarkable effort, it had a limited one week run in theaters.
1987 (R) 89m/C Mark Neely, Terry Farrell, Jon Cypher, Clarence Gilyard Jr.; *Dir:* Bill Berry. **VHS, Beta** *FRH* 🎞🎞

Off the Wall Two hitchhikers wind up in a southern maximum security prison after being accused of a crime they did not commit. What follows is a series of shenanigans as they try to escape. This supposed comedy falls flat in every way with the possible exception of Arquette's portrayal of the governor's dauther.
1982 (R) 86m/C Paul Sorvino, Rosanna Arquette, Patrick Cassidy, Billy Hufsey, Monte Markham, Mickey Gilley; *Dir:* Rick Friedberg. **VHS, Beta $29.98** *LIV, VES* 🎞

Off Your Rocker Representatives of a corporate conglomerate find that the residents of Flo Adler's Mapleview Nursing Home can still muster stiff resistance to a threatened takeover.
1980 99m/C Milton Berle, Red Buttons, Lou Jacobi, Dorothy Malone, Helen Shaver, Sharon Acker, Helen Hughes; *Dir:* Morley Markson, Larry Pall. **VHS, Beta $59.95** *QNE* 🎞

Offerings After a boy is tormented by a gang of children who cause him to fall down a well, his resulting brain injuries turn him into a psychopathic killer. Seeking revenge ten years later, he systematically murders his oppressors, offering bits of their anatomy to the one girl who treated him kindly. A typical slasher movie, the plot is only slightly better than the worst examples of the genre.
1989 (R) 96m/C G. Michael Smith, Loretta L. Bowman; *Dir:* Christopher Reynolds. **VHS, Beta $89.95** *SGE* 🎞

Office Romances A woman who has moved to London from the country is seduced by her employer. Because of her plain appearance, she is flattered by his attention even though he is married and has no real feelings for her. Competently made, nonetheless hard to watch because of the slow pace.
1986 60m/C *GB* Judy Parfitt, Ray Brooks, Suzanne Burden. **VHS, Beta $11.95** *PSM* 🎞🎞

An Officer and a Gentleman Young drifter Gere, who enters Navy Officer Candidate School because he doesn't know what else to do with his life, becomes a better person almost despite himself. Winger is the love interest who sets her sights on marrying Gere, and Gossett is the sergeant who whips him into shape. Strong performances by the whole cast made this a must-see in 1982 that is still appealing despite the standard Hollywood premise.
1982 (R) 126m/C Richard Gere, Louis Gossett Jr., David Keith, Lisa Eilbacher, Debra Winger, Robert Loggia, Lisa Blount; *Dir:* Taylor Hackford. Academy Awards '82: Best Song ("Up Where We Belong"), Best Supporting Actor (Gossett). **VHS, Beta, LV, 8mm $14.95** *PAR, FCT, BTV* 🎞🎞🎞

The Official Story A devastating drama about an Argentinian woman who realizes her young adopted daughter may be a child stolen from one of the thousands of citizens victimized by the country's repressive government. A powerful, important film. English subtitles or dubbed.
1985 (R) 112m/C *AR* Norma Aleandro, Hector Alterio, Chunchuna Villafane, Patricio Contreras; *Dir:* Luis Puenzo. Academy Awards '85: Best Foreign Language Film. **VHS, Beta, LV $19.95** *PAV, APD* 🎞🎞🎞½

Offnight When a man is put on the day shift for the first time in fifteen years his life turns to chaos.
1984 47m/C **VHS, Beta** *NEW* 🎞½

The Offspring In four stories of past evils Price reveals his hometown can force folks to kill. Not the usual Price material: dismemberment, cannibalism, and necrophilia clash with his presence. Strong yuk factor. Also known as "From a Whisper to a Scream."
1987 (R) 99m/C Vincent Price, Cameron Mitchell, Clu Gulager, Terry Kiser, Susan Tyrrell, Harry Caesar, Rosalind Cash, Martine Beswick, Angelo Rossitto, Lawrence Tierney; *Dir:* Jeff Burr. **VHS, Beta $14.95** *IVE* 🎞🎞

Oh Alfie In every man's life, there comes a time to settle down... but never when you're having as much fun as Alfie! Inadequate sequel to "Alfie" was originally titled "Alfie Darling."
1975 (R) 99m/C *GB* Joan Collins, Alan Price, Jill Townsend; *Dir:* Ken Hughes. VHS, Beta $59.95 *MON, HHE* &&

Oh, Bloody Life! Young actress faces deportation during the Stalin era in Hungary because of her aristocratic family ties. In Hungarian with English subtitles.
1988 115m/C *HU* Udvaros Dorottya, Szacsvay Laszlo, Kern Andras, Bezeredi Zoltan, Oze Lajos, Lukacs Margit; *W/Dir:* Peter Bacso. VHS $49.95 *EVD* &&½

Oh! Calcutta! Film version of the first nude musical to play on Broadway, which caused a sensation in the late 1960s. It's really a collection of skits, some of which were written by such notables as John Lennon, Sam Shepard, and Jules Feiffer. And it's not that funny or erotic.
1972 105m/C Bill Macy, Mark Dempsey, Raina Barrett, Samantha Harper, Patricia Hawkins, Mitchell McGuire; *Dir:* Guillaume Martin Aucion. VHS, Beta $29.95 *VID* &½

Oh Dad, Poor Dad (Momma's Hung You in the Closet & I'm Feeling So Sad) Cult fave black comedy about a bizarre family on a vacation in Jamaica. The domineering mother travels with a coffin containing the stuffed body of her late husband. Additional corpses abound. Based on the play by Arthur L. Kopit.
1967 86m/C Rosalind Russell, Robert Morse, Barbara Harris, Hugh Griffith, Lionel Jeffries, Jonathan Winters; *Dir:* Richard Quine. VHS, Beta $49.95 *PAR* &&

Oh, God! God, in the person of Burns, recruits Denver as his herald in his plan to save the world. Despite initial skepticism, Denver, in his film debut, keeps faith and is rewarded. Sincere performances and optimistic end make for satisfying story. Followed by "Oh God! Book 2" and "Oh God! You Devil."
1977 (PG) 104m/C George Burns, John Denver, Paul Sorvino, Ralph Bellamy, Teri Garr, William Daniels, Donald Pleasence, Barnard Hughes, Barry Sullivan, Dinah Shore, Jeff Corey, David Ogden Stiers; *Dir:* Carl Reiner. VHS, Beta, LV $19.98 *WAR* &&½

Oh, God! Book 2 Burns returns as the "Almighty One" in this strained sequel to "Oh God!" This time he enlists the help of a young girl to remind others of his existence. The slogan she concocts saves God's image, but not the movie. Followed by "Oh, God! You Devil."
1980 (PG) 94m/C George Burns, Suzanne Pleshette, David Birney, Louanne, Conrad Janis, Wilfrid Hyde-White, Hans Conried, Howard Duff; *Dir:* Gilbert Cates. VHS, Beta $19.98 *WAR* &

Oh, God! You Devil During his third trip to earth Burns plays both the Devil and God as he first takes a struggling musician's soul, then gives it back. A few zingers and light atmosphere save unoriginal plot. The second sequel to "Oh, God!"
1984 (PG) 96m/C George Burns, Ted Wass, Roxanne Hart, Ron Silver, Eugene Roche, Robert Desiderio; *Dir:* Paul Bogart. VHS, Beta, LV $19.98 *WAR* &&

Oh Happy Day A performance of gospel singers recorded live; sequel to "Gospel." Includes performances from Shirley Caesar, the Clark Sisters, the Mighty Clouds of Joy,

the Rev. James Cleveland, and Walter Hawkins.
1987 60m/C Shirley Caesar, Joy, Rev. James Cleveland, Walter Hawkins. VHS, Beta $29.95 *MVD, MON, VTK* &&

Oh, Heavenly Dog! A private eye returns from the dead as a dog to solve his own murder. Man's best friend and intelligent to boot. The famous dog Benji's third film, and his acting improves with each one. Adults may find this movie slow, but kids will probably love it.
1980 (PG) 104m/C Chevy Chase, Jane Seymour, Omar Sharif, Robert Morley; *Dir:* Joe Camp. VHS, Beta $19.98 *FOX* &&

Oh, Kojo! How Could You! Adaptation of Ashanti folk tale in which young Kojo is taught invaluable lesson by cunning Ananse.
19?? 18m/C VHS $35.00 *AFR* &&

Oh, Mr. Porter Billeted at an obscure railway post, the stationmaster hero renovates the place and hires a special train to transport the area soccer team. Trouble arises when the train is hijacked by gun smugglers.
1937 85m/B *GB* Will Hay, Moore Marriot, Graham Moffatt, Frederick Piper; *Dir:* Marcel Varnel. VHS $19.95 *NOS* &&

Oh Susannah Robbed and thrown from a train, Autry and the two drifters who rescue him follow the thieves to Mineral Springs. Their journey is plagued by too many songs and the bad guys are predictably punished in the end.
1938 59m/B Gene Autry, Booth Howard, Smiley Burnette, Frances Grant, Earl Hodgins, Donald Kirke, Clara Kimball Young; *Dir:* Joseph Cane. VHS, Beta $19.95 *NOS, VDM, HHT* &½

Oh! Those Heavenly Bodies: The Miss Aerobics USA Competition Various scantily clad women dance for the title of Miss Aerobics USA; the tape's viewers are supposed to vote for the winner.
1987 60m/C VHS, Beta $29.95 *DWE* &½

O'Hara's Wife A loving wife continues to care for her family even after her untimely death. Only her husband, however, can see her ghost in this made-for-television movie.
1982 87m/C Ed Asner, Mariette Hartley, Jodie Foster, Tom Bosley, Perry Lang, Ray Walston; *Dir:* William S. Bartman. VHS, Beta $69.95 *VES* &

Oil Seven men fight a raging oil fire that threatens to destroy an entire country. Unfortunately, the movie is unable to ignite any interest at all.
1978 (PG) 95m/C Ray Milland, Woody Strode, Stuart Whitman, Tony Kendall, William Berger; *Dir:* Mircea Dragan. VHS, Beta $19.95 *UHV* &

Oklahoma! Jones's film debut; a must-see for musical fans. Songs include "Oh, What a Beautiful Mornin'", "The Surrey with the Fringe on Top", and "Oklahoma!" A cowboy and country girl fall in love, but she is tormented by another unwelcomed suitor. At over two hours, cuteness wears thin for some. Actually filmed in Arizona. Adapted from Rodgers and Hammerstein's broadway hit with original score; choreography by Agnes de Mille.
1955 145m/C Gordon MacRae, Shirley Jones, Rod Steiger, Gloria Grahame, Eddie Albert, Charlotte Greenwood, James Whitmore, Gene Nelson, Barbara Lawrence, Jay C. Flippen; *Dir:* Fred Zinneman. Academy Awards '55: Best Sound, Best Musical Score. VHS, Beta, LV $19.98 *FOX, RDG* &&&½

Oklahoma Annie Storekeeper Canova joins the new sheriff in booting undesirables out of town, and tries to sing her way into his heart. Amazingly, it works—not for delicate ears.
1951 90m/C Judy Canova, Al "Fuzzy" Knight, Grant Withers, John Russell, Denver Pyle, Allen Jenkins, Almira Sessions; *Dir:* R.G. Springsteen. VHS, Beta *MLB* &

Oklahoma Badlands Lane and his horse outsmart a corrupt newspaper publisher trying to steal land from a female rancher. An early directorial work for veteran stuntman Canutt.
1948 59m/B Allan Lane, Mildred Coles; *Dir:* Yakima Canutt. VHS, Beta $29.95 *DVT* &½

Oklahoma Bound A feisty farmer tries to save his failing farm through a series of ostensibly comedic, unsuccessful schemes.
1981 92m/C F.E. Bowling, Dan Jones; *Dir:* Patrick C. Poole. VHS, Beta $19.95 *UHV* &

Oklahoma Crude Sadistic oil trust rep Palence battles man-hating Dunaway for her well. Drifter Scott helps her resist on the promise of shared profits. In this 1913 setting, Dunaway tells Scott she wishes she could avoid men altogether, but later settles for him.
1973 (PG) 108m/C George C. Scott, Faye Dunaway, John Mills, Jack Palance, Harvey Jason, Woodrow Parfrey; *Dir:* Stanley Kramer. VHS *FOX* &&½

Oklahoma Cyclone A group of bronco busters find action and adventure on the prairie. Second of eight talkies in Steele series, includes his first of few singing roles.
1930 64m/B Bob Steele, Al "Fuzzy" St. John, Nita Ray, Charles L. King; *Dir:* John P. McCarthy. VHS $19.95 *NOS, GPV, DVT* &½

Oklahoma Kid Offbeat, hilarious western with Bogie as the villain and Cagney seeking revenge for his father's wrongful death. Highlight is Cagney's rendition of "I Don't Want To Play In Your Yard", complete with six-shooter accompaniment.
1939 82m/B James Cagney, Humphrey Bogart, Rosemary Lane, Ward Bond, Donald Crisp, Charles Middleton, Harvey Stephens; *Dir:* Lloyd Bacon. VHS, Beta $59.95 *MGM* &&&

The Oklahoman A routine western with some trivia value. Town doc McCrea helps Indian Pate keep his land. Talbott plays Indian maiden in same year as her title role in Daughter of Jekyll. Continuity buffs will note Hale wears the same outfit in m ost scenes.
1956 80m/C Joel McCrea, Barbara Hale, Brad Dexter, Gloria Talbott, Verna Felton, Douglas Dick, Michael Pate; *Dir:* Francis D. Lyon. VHS, Beta $19.98 *FOX* &&

Olatunji and His Drums of Passion The famous African percussionist was taped live at the Oakland Coliseum on New Year's Eve, 1985. Hosted by Grateful Dead drummer Mickey Hart. In HiFi Stereo.
1985 51m/C VHS, Beta $29.95 *MVD, VAI* &&

Old Barn Dance Autry and his singing cowboys are selling horses until a crooked tractor company puts them out of business. They join a radio program, discovers that it is owned by the same company that put them out of the horse-selling business, and runs the crooks out of town.
1938 54m/B Gene Autry, Smiley Burnette, Roy Rogers; *Dir:* Joseph Kane. VHS, Beta $19.95 *NOS, VCN, DVT* &

Old Boyfriends Shire is weak as a psychologist searching for old boyfriends in order to analyze her past. Strange combination of Carradine and Belushi may draw curious fans. Screenplay by Leonard and Paul Schrader.
1979 (R) 103m/C Talia Shire, Richard Jordan, John Belushi, Keith Carradine, John Houseman, Buck Henry; *Dir:* Joan Tewkesbury. **VHS, Beta, LV** $9.98 SUE *⏃*

Old Corral Sheriff Autry protets his love interest, a girl fleeing the Mob.
1936 54m/B Gene Autry, Roy Rogers, Smiley Burnette; *Dir:* Joseph Kane. **VHS, Beta** $19.95 NOS, MRV, VYY *⏃⏃*

The Old Curiosity Shop Webster and Benson are an old gambler and his daughter in this well-made film adaptation of the Dickens tale. Their miserly landlord tries to ruin their lives by evicting them and forcing them into a life of poverty. Remade as "Mr. Quilp" in 1957.
1935 90m/B Ben Webster, Elaine Benson, Hay Petrie, Beatrix Thompson, Gibb McLaughlin, Reginald Purdell; *Dir:* Thomas Bentley. **VHS** $19.95 NOS *⏃⏃ ½*

The Old Curiosity Shop Flat musical version of the Charles Dickens story about an evil man who wants to take over a small antique shop run by an elderly man and his granddaughter. Originally called "Mr. Quilp."
1975 (G) 118m/C GB Anthony Newley, David Hemmings, David Warner, Jill Bennett, Peter Duncan, Michael Hordern; *Dir:* Michael Tuchner. **VHS, Beta** $39.98 SUE *⏃⏃*

Old Curiosity Shop An animated adaptation of the classic Dickens story about a young girl and her grandfather who are evicted from their curiosity shop by an evil man.
1984 (G) 72m/C **VHS, Beta** $19.98 LIV, CVL *⏃⏃*

Old Enough Slow-moving coming-of-age comedy on the rich kid-poor kid friendship theme. Marisa Silver's directing debut; she's also responsible for the screenplay.
1984 (PG) 91m/C Sarah Boyd, Rainbow Harvest, Neill Barry, Danny Aiello, Susan Kingsley, Roxanne Hart, Alyssa Milano, Fran Brill, Anne Pitoniak; *Dir:* Marisa Silver. **VHS, Beta** $59.95 MED *⏃⏃*

Old Explorers Two old friends refuse to let old age lessen their
1990 (PG) 91m/C Jose Ferrer, James Whitmore. **VHS** $89.95 ACA *⏃⏃ ½*

Old Gringo Adapted from Carlos Fuentes' novelization of writer Ambrose Bierce's mysterious disappearance in Mexico during the revolution of 1913. Features Fonda in the unlikely role of a virgin schoolteacher, Smits as her revolutionary lover, and Peck as her hero. Soggy acting by all but Peck, whose presence is wasted in a sketchy character. Technical problems and cheesy sets and costumes—look for the dusk backdrop in the dance scene and Smits' silly moustache. Better to read the book.
1989 (R) 119m/C Jane Fonda, Gregory Peck, Jimmy Smits, Patricio Contreras, Jenny Gago, Gabriela Roel, Sergio Calderon, Guillermo Rios, Anne Pitoniak, Pedro Armendariz Jr.; *Dir:* Luis Puenzo. **VHS, Beta, LV, 8mm** $19.95 COL *⏃⏃ ½*

Old Gun A grief-plagued doctor finds he must seek out and kill each Nazi involved in the slaughter of his wife and child to ease his pain. Excellent directing and a fully formed main character.
1976 141m/C Philippe Noiret, Romy Schneider, Jean Bouise; *Dir:* Robert Enrico. **VHS, Beta** $49.95 MED *⏃⏃*

Old Ironsides Silent, black-and-white version of the big budget/important director and stars action-adventure. Merchant marines aboard the famous Old Ironsides battle 19th century Barbary pirates in rousing action scenes. Home video version features an engaging organ score by Gaylord Carter. Based on the poem "Constitution" by Oliver Wendell Holmes.
1926 111m/B Esther Ralston, Wallace Beery, Boris Karloff, Charles Farrell, George Bancroft; *Dir:* James Cruze. **VHS, Beta** $29.95 PAR *⏃⏃⏃*

Old Maid After her beau is killed in the Civil War, a woman allows her cousin to raise her illegitimate daughter, and therein begins a years-long struggle over the girl's affection. High grade soaper based on Zoe Adkin's stage adaptation of Edith Wharton's novel.
1939 96m/B Bette Davis, Miriam Hopkins, George Brent, Donald Crisp, Jane Bryan, Louise Fazenda, Henry Stephenson; *Dir:* Edmund Goulding. **VHS, Beta** $19.95 MGM *⏃⏃⏃*

Old Mother Riley's Ghosts A group of spies "haunt" Mother Riley's castle home in a futile attempt to scare her out; she turns the tables. Some good scares and laughs. Part of the Old Mother Riley series. Look for similarities to later Monty Python films.
1941 82m/B GB Arthur Lucan, Kitty McShane, John Stuart; *Dir:* John Baxter. **VHS, Beta** $16.95 NOS, SNC *⏃*

Old Spanish Custom Only a Keaton fan could love this comedy, set in Spain and filmed in England. One of his rarest—and poorest—sound films. He plays a bumbling rich yachtsman smitten with Tovar, a Spanish maiden who uses him to make her lover jealous. Originally titled "The Invader."
1936 58m/B Buster Keaton, Lupita Tova, Lyn Harding, Esme Percy; *Dir:* Adrian Brunel. **VHS, Beta** $19.95 GPV, VDM, DNB *⏃*

Old Swimmin' Hole Small-town friends Moran and Jones try to bring their single parents together; his mother can't afford to finance his dream to be a doctor. Dull, melodramatic ode to heartland America, reminiscent of "Our Town."
1940 78m/B Marcia Mae Jones, Jackie Moran, Leatrice Joy, Charles D. Brown; *Dir:* Robert McGowan. **VHS, Beta** $29.95 VYY *⏃⏃*

Old Yeller Disney Studios' first and best boy-and-his-dog film. Fifteen-year-old Kirk is left in charge of the family farm while Parker is away. When his younger brother brings home a stray dog, Kirk is displeased but lets him stay. Yeller saves Kirk's life, but contracts rabies in the process. Kirk is forced to kill him. Keep tissue handy, especially for the kids. Stong acting, effective scenery—all good stuff. Based on the novel by Fred Gipson. Sequel "Savage Sam" released in 1963.
1957 (G) 84m/C Dorothy McGuire, Fess Parker, Tommy Kirk, Kevin Corcoran, Jeff York, Beverly Washburn, Chuck Connors; *Dir:* Robert Stevenson. **VHS, Beta, LV** $19.99 DIS *⏃⏃⏃ ½*

Oldest Living Graduate Fonda is memorable in his last stage role as the oldest graduate of a Texas military academy. He and his son clash when Fonda refuses to give up his land. Teleplay features strong performances from Leachman and other big-name cast members.
1982 75m/C Cloris Leachman, Henry Fonda, John Lithgow, Timothy Hutton, Harry Dean Stanton, David Ogden Stiers, George Grizzard, Penelope Milford. **VHS, Beta** $49.95 VHE *⏃⏃⏃ ½*

Oldest Profession A study of prostitution from prehistoric times to the future. Generally unexciting and unfunny. In six segments meant as vehicles for their directors.
1967 97m/C FR GE IT Raquel Welch, Jeanne Moreau, Elsa Martinelli, Michele Mercier; *Dir:* Jean-Luc Godard, Philippe de Broca, Claude Autant-Lara, Franco Indovina, Mauro Bolognini, Michael Pfleghar. **VHS, Beta** $19.95 STE, NWV, IME *⏃*

Oliver! Splendid big-budget musical adaptation of Dickens' "Oliver Twist." An innocent orphan is dragged into a life of crime when he is befriended by a gang of pickpockets. Songs include "Food, Glorious Food," "Where Is Love?" and "Consider Yourself." Oscar nominations for actor Moody, supporting actor Wild, Adapted Screenplay, Cinematography, Film Editing and Costume Design.
1968 (G) 145m/C GB Mark Lester, Jack Wild, Ron Moody, Shani Wallis, Oliver Reed, Hugh Griffith; *Dir:* Carol Reed. Academy Awards '68: Best Art Direction/Set Decoration, Best Director (Reed), Best Picture, Best Sound, Best Musical Score; Golden Globe Awards '69: Best Film—Musical/Comedy; National Board of Review Awards '68: 10 Best Films of the Year. **VHS, Beta, LV** $19.95 COL, BTV *⏃⏃⏃ ½*

Oliver Twist Silent version of the Dickens classic is a vehicle for young Jackie Coogan. As orphan Oliver Twist, he is subjected to many frightening incidents before finding love and someone to care for him. Remade numerous times, most notably in 1933 and 1948 and as the musical "Oliver!" in 1968.
1922 77m/B Jackie Coogan, Lon Chaney Sr., Gladys Brockwell, George Siegmann, Esther Ralston; *Dir:* Frank Lloyd. **VHS, Beta** $19.95 NOS, CCB, BUR *⏃⏃⏃*

Oliver Twist The first talking version of Dickens' classic about an ill-treated London orphan involved with youthful gang of pickpockets. Moore was too young—at five—to be very credible in the lead role. The 1948 version is much more believable.
1933 70m/B Dickie Moore, Irving Pichel, William Boyd, Barbara Kent; *Dir:* William J. Cowan. **VHS, Beta** $19.95 NOS, MRV, HHT *⏃⏃*

Oliver Twist Charles Dickens' immortal story of a workhouse orphan who is forced into a life of crime with a gang of pickpockets. The best of many film adaptations, with excellent portrayals by the cast.
1948 116m/B GB Robert Newton, John Howard Davies, Alec Guinness, Francis L. Sullivan, Anthony Newley, Kay Walsh, Diana Dors, Henry Stephenson; *Dir:* David Lean. **VHS, Beta, LV** $19.95 PAR, DVT, FHS *⏃⏃⏃⏃*

Oliver Twist Animated version of the classic tale by Charles Dickens.
197? 72m/C **VHS, Beta** $29.95 CVL *⏃⏃⏃⏃*

Oliver Twist Good made-for-television version of the classic Dicken's tale of a boy's rescue from a life of crime. Scott's Fagin is a treat, and period details are on the mark. Scripted by James Goldman.
1982 72m/C George C. Scott, Tim Curry, Michael Hordern, Timothy West, Lysette Anthony, Eileen Atkins, Cherie Lunghi; *Dir:* Clive Donner. **VHS, Beta** $19.98 VES *⏃⏃⏃*

Oliver Twist A British television mini-series adaptation of the Charles Dickens classic about an orphan boy plunging into the underworld of 19th-century London. On two tapes.
1985 333m/C GB Ben Rodska, Eric Porter, Frank Middlemass, Gillian Martell. **VHS, Beta** $29.98 FOX *⏃⏃⏃*

Oliver's Story A "not so equal" sequel to "Love Story", where widower O'Neal wallows in grief untils rich heiress Bergen comes along. He falls in love again, this time with money.
1978 (PG) 90m/C Ryan O'Neal, Candice Bergen, Ray Milland, Edward Binns, Nicola Pagetti, Charles Haid; *Dir:* John Korty. **VHS, Beta $19.95** *PAR* ✓

Olivia Olivia hosts this musical television special, featuring such hits as "Hopelessly Devoted to You", "Have You Never Been Mellow", and "Please, Mister, Please". Olivia's guests include Andy Gibb and ABBA, performing a medley of classic rock songs.
1980 60m/C Olivia Newton-John, Andy Gibb, Abba. **Beta, LV $19.95** *MCA* ✓

Olivia An abused housewife moonlights as a prostitute and begins killing her customers. She falls in love with an American businessman and flees to America when her husband finds out about the affair. Revenge and murder are the result.
1983 90m/C Suzanna Love, Robert Walker, Jeff Winchester; *Dir:* Ulli Lommel. **VHS, Beta $49.95** *VHE* ✓½

Olivia in Concert Filmed during Olivia's first live show in five years, the concert reflects the popular singer's transformation from a sweet, romantic songstress to a strong, aggressive charmer and entertainer.
1983 78m/C Olivia Newton-John, John Farrar; *Dir:* Brian Grant. **VHS, Beta, LV $19.95** *MCA, MVD* ✓½

Olivia: Physical Olivia Newton-John performs favorites, modern funk, and earthy ballads in this video album. Songs include "Magic", "Physical", "A Little More Love", "Make a Move On Me", and "Hopelessly Devoted".
1981 54m/C Olivia Newton-John. **VHS, Beta, LV $19.95** *MCA, MVD* ✓½

Olivia: Twist of Fate Six of Olivia Newton-John's music videos are combined on this tape, including four songs from the movie "Two of a Kind:" the title tune, "Livin' in Desperate Times," "Take a Chance," and "Twist of Fate," plus "Heart Attack" and "Tied Up."
1984 19m/C Olivia Newton-John, John Travolta. **VHS, Beta, LV $14.95** *MVD, MCA* ✓½

Olly Olly Oxen Free Junkyard owner Hepburn helps two boys fix up and fly a hot-air balloon, once piloted by McKenzie's grandfather, as a surprize for the man's birthday. Beautiful airborne scenes over California and a dramatic landing to the tune of the "1812 Overture," but not enough to make the whole film interesting. Also known as "The Great Balloon Adventure."
1978 89m/C Katharine Hepburn, Kevin McKenzie, Dennis Dimster, Peter Kilman; *Dir:* Richard A. Colla. Film Advisory Board '78: Award of Excellence. **VHS, Beta** *TLF* ✓✓

Olongape: The American Dream A young woman works in a cheap bar while waiting for the chance to come to America.
1989 97m/C Jacklyn Jose, Susan Africa, Chanda Romero, Joel Torre, Marilou Sadiua; *Dir:* Chito Rono. **VHS, Beta $59.95** *TWE* ✓½

Omar Khayyam In medieval Persia, Omar (Wilde) becomes involved in a romance with the Shah of Persia's fiancee, while trying to fight off a faction of assassins trying to overthrow the Shah. Although this film has a great cast, the script is silly and juvenile, defeating the cast's fine efforts.
1957 101m/C Cornel Wilde, Michael Rennie, Debra Paget, Raymond Massey, John Derek, Yma Sumac, Margaret Hayes, Joan Taylor, Sebastian Cabot; *Dir:* William Dieterle. **VHS** *PAR* ✓✓

Omega Cop In a post-apocalyptic future, there's only one cop left. He uses his martial arts skills and tons of guns attempting to rescue three women, but the violence doesn't cover the poor acting and shoddy production. Fans of TV's Batman might enjoy this for West's presence.
1990 (R) 89m/C Ron Marchini, Adam West, Stuart Whitman, Troy Donahue, Meg Thayer, Jennifer Jostyn, Chrysti Jimenez, D.W. Landingham, Chuck Katzakian; *Dir:* Paul Kyriazi. **VHS $14.95** *HMD, SOU* ✓

Omega Man In post-holocaust Los Angeles, Heston is immune to the nuclear effects and battles those who aren't—an army of vampires bent on destroying what's left of the world. Strong suspense with considerable violence, despite the PG rating. Based on the science fiction thriller "I Am Legend" by Richard Matheson.
1971 (PG) 98m/C Charlton Heston, Anthony Zerbe, Rosalind Cash, Paul Koslo; *Dir:* Boris Sagal. **VHS, Beta $19.98** *WAR* ✓✓½

Omega Syndrome Neo-Nazis kidnap Wahl's daughter. He and Vietnam buddy DiCenzo get her back. Ho hum. DiCenzo's is the best performance. Tolerable for vigilante film fans.
1987 (R) 90m/C Ken Wahl, Ron Kuhlman, George DiCenzo, Doug McClure; *Dir:* Joseph Manduke. **VHS, Beta $14.95** *NWV, STE* ✓½

The Omen A young boy who has been adopted into an American diplomat's family is later realized to be the son of Satan. Bizarre and inexplicable deaths seem to happen when he is around. The shock and gore that was prevalent in "The Exorcist" is replaced in this film with more suspense and believable effects. Well-done horror film that doesn't insult the viewer's intelligence. Followed by two sequels: "Damien—Omen II" and "The Final Conflict."
1976 (R) 111m/C Gregory Peck, Lee Remick, Billie Whitelaw, David Warner, Holly Palance, Richard Donner; *Dir:* Richard Donner. Academy Awards '76: Best Original Score. **VHS, Beta, LV $19.98** *FOX* ✓✓½

Omoo Omoo, the Shark God A sea captain is cursed when he removes the black pearls from a stone shark god in this extremely cheap adventure.
1949 58m/B Ron Randell, Devera Burton, Trevor Bardette, Pedro de Cordoba, Richard Benedict, Rudy Robles, Michael Whalen, George Meeker; *Dir:* Leo Leonard. **VHS $16.95** *SNC* ✓

On the Air Live with Captain Midnight A socially challenged teen finds hipness as a rebel DJ operating an illegal radio station from his van. Leading actor is the son of the director/writer/producer but nepotism does not a good film make. Alson known as "Captain Midnight."
197? 90m/C Tracy Sebastian; *Dir:* Ferd Sebastian. **VHS, Beta $39.95** *IND* ✓½

On Any Sunday This exhilarating documentary examines the sport of motorcycle racing with a focus on three men: Mert Lawwill, Malcolm Smith, and Steve McQueen. Available in a special "director's edition," featuring previously unseen footage of McQueen.
1971 (G) 89m/C Steve McQueen; *Dir:* Bruce Brown. **VHS, Beta $39.95** *PAV, MON* ✓✓✓

On Approval Hilarious British farce in which two women trade boyfriends. Lillie's performance provides plenty of laughs. Brook runs the show as leading man, co-author, director, and co-producer. Based on the play by Frederick Lonsdale.
1944 80m/B *GB* Clive Brook, Beatrice Lillie, Googie Withers, Roland Culver; *Dir:* Clive Brook. **VHS, Beta $19.95** *NOS, MRV, HHT* ✓✓✓

On the Beach A group of survivors attempt to live normal lives in post-apocalyptic Australia, waiting for the inevitable arrival of killer radiation. Astaire is strong in his first dramatic role. Though scientifically implausible, still a good anti-war vehicle. Based on the best-selling novel by Nevil Shute.
1959 135m/B Gregory Peck, Anthony Perkins, Donna Anderson, Ava Gardner, Fred Astaire; *Dir:* Stanley Kramer. National Board of Review Awards '59: 10 Best Films of the Year. **VHS, Beta $19.98** *FOX, FCT* ✓✓✓½

On the Block Beautiful babe has trouble walking street in Baltimore.
1991 (R) 96m/C Howard E. Rollins Jr., Marilyn Jones, Blaze Starr. **VHS $89.95** *VMK* ✓½

On the Bowery A classic documentary covering the lives and situations of the alcoholic homeless in New York City.
1956 65m/B *Dir:* Lionel Rogosin. **VHS, Beta $29.95** *MFV* ✓½

On a Clear Day You Can See Forever A psychiatric hypnotist helps a girl stop smoking and finds that in trances she remembers previous incarnations. He falls in love with one of the women she used to be. Alan Jay Lerner of "My Fair Lady" and "Camelot" wrote the lyrics and the book. Based on a musical by Lerner and Burton Lane.
1970 (G) 129m/C Barbra Streisand, Yves Montand, Bob Newhart, Jack Nicholson, Simon Oakland; *Dir:* Vincente Minnelli. **VHS, Beta, LV $14.95** *PAR* ✓✓

On the Comet Zeman's fourth fantasy based on Jules Verne stories, about a chunk of the Earth's crust suddenly becoming a comet and giving its passengers a ride through the galaxy. Complete with Zeman's signature animation fantasias. In English. Also called "Mr. Sverdac's Ark."
1968 76m/C *CZ* Emil Horvath Jr., Magda Vasarykova, Frantisek Filipovsky; *Dir:* Karel Zeman. **VHS, Beta $19.99** *FCT, MRV* ✓✓✓

On Dangerous Ground A world-weary detective is sent to the countryside to investigate a murder. He encounters the victim's revenge-hungry father and the blind sister of the murderer. In the hateful father the detective sees a reflection of the person he has become, in the blind woman, he learns the redeeming qualities of humanity and compassion. A well-acted example of film noir that features the composer Bernard Herrman's favorite score.
1951 82m/B Robert Ryan, Ida Lupino, Ward Bond, Ed Begley Sr., Cleo Moore, Charles Kemper; *Dir:* Nicholas Ray. **VHS, Beta, LV $39.95** *TTC* ✓✓✓

On the Edge A narcotics agent tracks down a wealthy Brazilian drug kingpin.
1982 82m/C Anthony Steffen. **VHS, Beta $59.95** *MGL* ✓

On the Edge A drama about the inevitable Rocky-esque triumph of Dern as an aging marathon runner. Simultaneous to his running endeavor, Dern is trying to make up for lost time with his father. Available in two versions, one rated, the other unrated with a

racy appearance by Pam Grier as the runner's interracial lover.
1986 (PG-13) 86m/C Bruce Dern, John Marley, Bill Bailey, Jim Haynie, Pam Grier; *Dir:* Rob Nilsson. **VHS, Beta $79.95** *LTG* 🎬🎬 ½

On the Edge: Survival of Dana Another entry from the world of quality made-for-TV films. A young girl moves with her family to a new town. When she falls in with the "bad" crowd, her ethical standards are challenged. A stinker that may appeal to those with campy tastes. Not for the discriminating palate.
1979 92m/C Melissa Sue Anderson, Robert Carradine, Marion Ross, Talia Balsam, Michael Pataki, Kevin Breslin, Judge Reinhold, Barbara Babcock; *Dir:* Jack Starrett. Harvard Lampoon Awards '79: Worst Film of the Year. **VHS, Beta** *GEM* Woof!

On Golden Pond Henry Fonda won his first, and long overdue, Oscar for his role as the curmudgeonly patriarch of the Thayer family. He and his wife have grudgingly agreed to look after a young boy while at their summer home in Maine. Through his gradually affectionate relationship with the boy, Fonda learns to allay his fears of mortality. He also gains an understanding of his semi-estranged daughter. Jane Fonda plays his daughter and Hepburn is his loving wife in this often funny adaptation of Ernest Thompson's 1978 play. Predictable but deeply moving. Henry Fonda's final screen appearance.
1981 (PG) 109m/C Henry Fonda, Jane Fonda, Katharine Hepburn, Dabney Coleman, Doug McKeon, William Lanteau; *Dir:* Mark Rydell. Academy Awards '81: Best Actor (Fonda), Best Actress (Hepburn), Best Adapted Screenplay; Golden Globe Awards '82: Best Film—Drama. **VHS, Beta, LV, SVS $19.95** *FOX, JTC, SUP* 🎬🎬🎬 ½

On Her Majesty's Secret Service In the sixth 007 adventure, Bond again confronts the infamous Blofeld, who is planning a germ-warfare assault on the entire world. Australian Lazenby took a crack at playing the super spy, with mixed results. Many feel this is the best-written of the Bond films and might have been the most famous, had Sean Connery continued with the series. Includes the song "We Have All the Time In the World," sung by Louis Armstrong.
1969 142m/C *GB* George Lazenby, Diana Rigg, Telly Savalas, Gabriel Ferzetti, Ilse Steppat, Bernard Lee, Lois Maxwell, Desmond Llewelyn, Catherine von Schell, Julie Ege, Joanna Lumley, Mona Chong, Anouska Hempel, Jenny Hanley; *Dir:* Peter Hunt. **VHS, Beta, LV $19.95** *MGM, FOX, TLF* 🎬🎬 ½

On an Island with You Williams plays a movie star who finds romance on location in Hawaii in this musical extravaganza. Contains many of Williams' famous water ballet scenes and a bevy of bathing beauties. Songs include "On an Island with You," "Dog Song," and "Wedding Samba."
1948 107m/C Esther Williams, Peter Lawford, Ricardo Montalban, Jimmy Durante, Cyd Charisse, Leon Ames; *Dir:* Richard Thorpe. **VHS** *MGM* 🎬🎬 ½

On the Line Two Mexico-US border guards, one hard-bitten, the other sympathetic to the illegal immigrants' plight, battle it out over a beautiful Mexican whore. Disjointed and confusing.
1983 95m/C *SP* David Carradine, Victoria Abril, Scott Wilson, Sam Jaffe, Jesse Vint; *Dir:* Jose Luis Borau. **VHS, Beta $19.98** *SUE* 🎬 ½

On the Nickel An ex-alcoholic returns to Fifth Street in Los Angeles to save his friend from a life of despair. Tom Waits' musical score enhances this sentimental skid row drama. Scripted, directed and produced by "The Waltons'" Waite.
1980 (R) 96m/C Ralph Waite, Donald Moffatt, Hal Williams, Jack Kehoe; *Dir:* Ralph Waite. **VHS, Beta $69.95** *VES* 🎬🎬

On the Night Stage Gruff bandit (legendary Hart) loses his girl-of-questionable-values to town preacher. One of the first feature length westerns.
1915 83m/B William S. Hart, Robert Edeson, Rhea Mitchell, Shorty Hamilton; *Dir:* Reginald Barker. **VHS, Beta, 8mm $29.95** *GPV, VYY* 🎬🎬

On the Old Spanish Trail Rogers becomes a singing cowboy with a traveling tent show in order to pay off a note signed by the Sons of the Pioneers.
1947 56m/B Roy Rogers, Jane Frazee, Andy Devine, Tito Guizar; *Dir:* William Witney. **VHS, Beta $9.95** *NOS, MRV, VCN* 🎬 ½

On the Right Track A young orphan living in Chicago's Union Station has the gift of being able to pick winning race horses. May be appealing to fans of Coleman and Different Strokes television show, but this film does not have enough momentum to keep most viewers from switching tracks.
1981 (PG) 98m/C Gary Coleman, Lisa Eilbacher, Michael Lembeck, Norman Fell, Maureen Stapleton, Herb Edelman; *Dir:* Lee Philips. **VHS, Beta $59.95** *FOX* 🎬 ½

On the Run An ex-con and his gal cut a law-defying swath through the Bayou.
1985 (R) 96m/C Jack Conrad, Rita George, Dub Taylor, David Huddleston. **VHS, Beta $59.95** *GEM* 🎬

On the Third Day A headmaster finds a mysterious stranger in his house who turns out to be his long-lost illegitimate son.
1983 101m/C Richard Marant, Catherine Schell, Paul Williamson. **VHS, Beta $59.95** *LHV* 🎬

On Top of Old Smoky Autry and the Cass County Boys are mistaken for Texas Rangers by a gang of land poachers.
1953 59m/B Gene Autry, Smiley Burnette, Gail Davis, Sheila Ryan; *Dir:* George Archainbaud. **VHS, Beta $29.98** *CCB* 🎬

On Top of the Whale Two scholars move onto the estate of strange but wealthy benefactor to conduct research on the few survivors of a disappearing clan of Indians. Two tribesmen speak bizarre dialect which utilizes only one word while others confound the researchers with stories of ghosts and graveyards. Dutch comedy subtitled in English.
1982 93m/C *NL* *Dir:* Raul Ruiz. **VHS $59.95** *KIV, FCT* 🎬🎬

On the Town Kelly's directorial debut features three sailors on a one-day leave search for romance in the Big Apple. Filmed on location in New York City, with uncompromisingly authentic flavor. Based on the successful Broadway musical, with a score composed by Leonard Bernstein, Betty Comden, and Adolph Green. Additional songs by Roger Edens. Songs include "New York, New York," "On the Town," and "That's All There Is, Folks."
1949 98m/C Gene Kelly, Frank Sinatra, Vera-Ellen, Ann Miller, Betty Garrett; *Dir:* Gene Kelly, Stanley Donen. Academy Awards '49: Best Musical Score. **VHS, Beta, LV $19.95** *MGM* 🎬🎬🎬 ½

On Valentine's Day Author Horton Foote based this story loosely on his parents' lives. A wealthy young Southern girl marries a poor but decent young man and finds herself ostracized from her family. A prequel to the same author's "1918." Produced with PBS for "American Playhouse." Also known as "Story of a Marriage."
1986 (PG) 106m/C Hallie Foote, Matthew Broderick, Michael Higgins, Steven Hill, William Converse-Roberts; *Dir:* Ken Harrison. **VHS, Beta $19.98** *LHV* 🎬🎬 ½

On the Waterfront A trend-setting, gritty portrait of New York dock workers embroiled in union violence. Cobb is the gangster union boss, Steiger his crooked lawyer, and Brando, Steiger's ex-fighter brother who "could've been a contender!" Intense performances and excellent direction stand up well today. The picture was a huge financial success and won eight Academy Awards, plus receiving nominations for Best Supporting Actor for Malden, Cobb and Steiger, and for Score of a Comedy or Drama.
1954 108m/B Marlon Brando, Rod Steiger, Eva Marie Saint, Lee J. Cobb, Karl Malden; *Dir:* Elia Kazan. Academy Awards '54: Best Actor (Brando), Best Art Direction/Set Decoration (B & W), Best Black and White Cinematography, Best Director (Kazan), Best Film Editing, Best Picture, Best Story & Screenplay, Best Supporting Actress (Saint); Directors Guild of America Awards '54: Best Director (Kazan); Golden Globe Awards '55: Best Film—Drama; National Board of Review Awards '54: Best Picture; New York Film Critics Awards '54: Best Actor (Brando), Best Director (Kazan), Best Picture. **VHS, Beta, LV $19.95** *COL, PIA, BTV* 🎬🎬🎬🎬

On Wings of Eagles During the 1979 Iranian revolution two American executives are imprisoned by Islamic radicals. Help arrives in the form of covert agents of their boss—Texas tycoon H. Ross Perot (played by Crenna). Don't expect strict historical veracity or insight from this network miniseries (based on the 'nonfiction novel' by Ken Follett), an okay but lengthy 'mission impossible.' It came out in a two-cassette set, just in time for Perot's 1992 presidential campaign.
1986 221m/C Burt Lancaster, Richard Crenna, Paul LeMat, Esai Morales, Constance Towers, Jim Metzler, James Sutorius, Lawrence Pressman, Karen Carlson, Cyril O'Reilly; *Dir:* Andrew V. McLaglen. **VHS $89.95** *WOV* 🎬🎬 ½

On the Yard An attempt to realistically portray prison life. Focuses on a murderer who runs afoul of the leader of the prisoners and the system. A fairly typical prison drama with above average performances from Heard and Kellin.
1979 (R) 102m/C John Heard, Thomas G. Waites, Mike Kellin, Joe Grifasi; *Dir:* Raphael D. Silver. **VHS, Beta $69.95** *MED* 🎬🎬

Once Around Daughter of close-knit wealthy Italian family falls for a boisterous, self-assured Lithuanian salesman. Conflict ensues. Lightweight story buoyed by solid performances.
1991 (R) 115m/C Richard Dreyfuss, Holly Hunter, Danny Aiello, Laura San Giacomo, Gena Rowlands; *Dir:* Lasse Hallstrom. **VHS, Beta, LV $92.95** *MCA, PIA, FCT* 🎬🎬 ½

Once Before I Die Army soldiers caught in the Philippines during World War II struggle to survive and elude the Japanese. A single woman traveling with them becomes the object of their spare time considerations. A brutal, odd, and gritty war drama. Director/

actor Derek was Andress' husband at the time.
1965 97m/C PH Ursula Andress, John Derek, Richard Jaeckel, Rod Lauren, Ron Ely; *Dir:* John Derek. **VHS** *MOV* ♂♂½

Once Bitten A centuries-old though still remarkably youthful vampiress comes to Los Angeles to stalk virgins. That may be the wrong city, but she needs their blood to retain her youthful countenance. Vampire comedy theme was more effectively explored in 1979's Love at First Bite. Once Bitten is notable, however, for the screen debut of comic Carrey who would later gain fame on the TV show In Living Color.
1985 (PG-13) 94m/C Lauren Hutton, Jim Carrey, Cleavon Little, Karen Kopins, Thomas Balltore, Skip Lackey; *Dir:* Howard Storm. **VHS, Beta, LV** $79.98 *LIV, VES* ♂♂

Once a Hero The dregs of society, the Cicero Gang, have taken Captain Justice's number one fan and his mother hostage. Is it really curtains for rosy cheeks, bubble gum, and apple pie? But wait, it's Captain Justice, the man, the myth, to the rescue!
1988 74m/C Jeff Lester, Robert Forster, Milo O'Shea; *Dir:* Claudia Weill. **VHS** $9.99 *NWV, STE* ♂♂

Once is Not Enough Limp trash-drama concerning the young daughter of a has-been movie producer who has a tempestuous affair with a writer who reminds her of her father. Based on the novel by Jacqueline Susann. Vaccaro earned an Oscar nomination.
1975 (R) 121m/C Kirk Douglas, Deborah Raffin, David Janssen, George Hamilton, Brenda Vaccaro, Alexis Smith, Melina Mercouri; *Dir:* Guy Green. **VHS, Beta** $59.95 *PAR* ♂½

Once in Paris... A bittersweet romance about a scriptwriter working in Paris, the chauffeur who befriends him, and the aristocratic British woman with whom the writer falls in love. Beautiful French scenery and the engaging performance of Lenoir as the chauffeur make this film a diverting piece of entertainment.
1979 (PG) 100m/C Wayne Rogers, Gayle Hunnicutt, Jack Lenoir; *Dir:* Frank D. Gilroy. **VHS, Beta** $9.95 *SUE, MED* ♂♂½

Once Upon a Brothers Grimm An original musical fantasy in which the Brothers Grimm meet a succession of their most famous storybook characters including Hansel and Gretel, the Gingerbread Lady, Little Red Riding Hood, and Rumpelstiltskin.
1977 102m/C Dean Jones, Paul Sand, Cleavon Little, Ruth Buzzi, Chita Rivera, Teri Garr. **VHS, Beta** $39.95 *UHV* ♂♂

Once Upon a Crime Extremely disappointing comedy featuring a high profile cast set in Europe. The plot centers around Young and Lewis finding a dachshund and travelling from Rome to Monte Carlo to return the stray and collect a $5,000 reward. Upon arrival in Monte Carlo, they find the dog's owner dead and they end up getting implicated for the murder. Other prime suspects include Belushi, Candy, Hamilton, and Shepherd. Weak script is made bearable only by the comic genius of Candy.
1991 (PG) 94m/C John Candy, James Belushi, Cybill Shepherd, Sean Young, Richard Lewis, Ornella Muti, Giancarlo Giannini, George Hamilton, Joss Ackland; *Dir:* Eugene Levy. **VHS** $94.99 *MGM* ♂½

Once Upon a Honeymoon Set in 1938, Grant is an American radio broadcaster reporting on the oncoming war. Rogers is the ex-showgirl who unknowingly marries a Nazi. In this strange, uneven comedy, Grant tries to get the goods on him and also rescue Rogers. The plot is basic and uneven with Grant attempting to expose the Nazi and save Rogers. However, the slower moments are offset by some fairly surreal pieces of comedy.
1942 116m/B Ginger Rogers, Cary Grant, Walter Slezak, Albert Dekker; *Dir:* Leo McCarey. **VHS, Beta, LV** $19.99 *CCB, TTC, FCT* ♂♂½

Once Upon a Midnight Scary Price narrates a collection of three tales of terror, "The Ghost Belonged to Me," "The Legend of Sleepy Hollow," and "The House with a Clock in Its Walls." Aimed at the kiddies.
1990 50m/C Rene Auberjonois, Severn Darden; *Dir:* Neil Cox; *Nar:* Vincent Price. **VHS** *GEM, MRV* ♂♂

Once Upon a Scoundrel A ruthless Mexican land baron arranges to have a young woman's fiancee thrown in jail so he can have her all to himself.
1973 (G) 90m/C Zero Mostel, Katy Jurado, Tito Vandis, Priscilla Garcia, A. Martinez; *Dir:* George Schaefer. **VHS, Beta** $9.99 *STE, PSM* ♂♂½

Once Upon a Time A pretty girl, her puppy, a charming prince, and a grinch combine for a magical, wondrous animated fantasy.
1976 (G) 83m/C VHS, Beta $29.95 *GEM* ♂♂½

Once Upon a Time Two rival tribes spoil the love of a prince and princess in ancient times. Animated.
1987 92m/C VHS, Beta $39.95 *MED* ♂♂½

Once Upon a Time in America The uncut original version of director Leone's saga of five young men growing up in Brooklyn during the '20s who become powerful mob figures. Also available in a 143-minute version. Told from the perspective of De Niro's Noodles as an old man looking back at 50 years of crime, love, and death, told with a sweeping and violent elegance.
1984 (R) 225m/C Robert De Niro, James Woods, Elizabeth McGovern, Treat Williams, Tuesday Weld, Burt Young, Joe Pesci, Danny Aiello, Darlanne Fluegel, Jennifer Connelly; *Dir:* Sergio Leone. **VHS, Beta, LV** $29.98 *WAR* ♂♂♂½

Once Upon a Time in Vietnam Also called "How Sleep the Brave," this politically irksome drama about Vietnam was filmed in Georgia.
198? 90m/C VHS, Beta *VCD* ♂

Once Upon a Time in the West The uncut version of Leone's sprawling epic about a band of ruthless gunmen who set out to murder a mysterious woman waiting for the railroad to come through. Filmed in John Ford's Monument Valley, it's a revisionist western with some of the longest opening credits in the history of the cinema. Fonda is cast against type as an extremely cold-blooded villain. Script was coauthored by Leone, Bernardo Bertolucci, and Dario Argento and the brilliant score contributed by Ennio Morricone.
1968 (PG) 165m/C IT Henry Fonda, Jason Robards Jr., Charles Bronson, Claudia Cardinale, Keenan Wynn, Lionel Stander, Woody Strode, Jack Elam; *Dir:* Sergio Leone. New York Film Festival '80: Best Picture. **VHS, Beta, LV** $24.95 *PAR* ♂♂♂½

One A.M. Charlie is a drunk who must first battle a flight of stairs in order to get to bed. Silent with music track.
1916 20m/B Charlie Chaplin. **VHS, Beta** *CAB, FST* ♂♂♂½

One Arabian Night When an exotic dancer in a traveling carnival troupe is kidnapped into a harem, a dwarf acts on his unrequited love and avenges her death by murdering the sheik who killed her. Silent. Secured a place in American filmmaking for director Lubitsch.
1921 85m/B GE Pola Negri, Ernst Lubitsch, Paul Wegener; *Dir:* Ernst Lubitsch. **VHS, Beta** $29.95 *FCT, GPV, GLV* ♂♂♂

The One Arm Swordsmen Many one-armed men are persecuted for the murder committed by one fugitive one-armed man, and they decide to group together to track him down themselves.
1972 109m/C VHS, Beta $39.95 *UNI* ♂

One Armed Executioner An Interpol agent seeks revenge on his wife's murderers.
1980 (R) 90m/C Franco Guerrero, Jody Kay; *Dir:* Bobby A. Auarez. **VHS, Beta** *PGN* ♂♂

One Away A gypsy escapes from a South African prison and the police are in hot pursuit.
1980 (PG) 83m/C Elke Sommer, Bradford Dillman, Dean Stockwell. **VHS, Beta** $39.95 *MON* ♂

One Body Too Many A mystery spoof about a wacky insurance salesman who's mistaken for a detective. The usual comedy of errors ensues. A contrived mish-mash—but it does have Lugosi going for it.
1944 75m/B Jack Haley, Jean Parker, Bela Lugosi, Lyle Talbot; *Dir:* Frank McDonald. **VHS, Beta** $19.95 *NOS, SNC, HHT* ♂♂

One Brief Summer Love triangles and intrigues abound on a spectacular country manor. An aging woman is mortified by her father's interest in a young seductress, who finds herself attracted to the old man, but not only him.
1970 (R) 86m/C Clifford Evans, Felicity Gibson, Jennifer Hilary, Jan Holden, Peter Egan; *Dir:* John MacKenzie. **VHS** $14.95 *LPC* Woof!

One Cooks, the Other Doesn't A most unlikely story of a man's ex-wife and his son moving back into his house, where he's now living with his new wife. Made-for-television.
1983 96m/C Suzanne Pleshette, Joseph Bologna, Rosanna Arquette, Oliver Clark; *Dir:* Richard Michaels. **VHS, Beta** $39.95 *USA, IVE* ♂♂

One Crazy Summer A group of wacky teens spend a fun-filled summer on Nantucket Island in New England. This follow-up to "Better Off Dead" is offbeat and fairly charming, led by Cusack's perplexed cartoonist and with comic moments delivered by Goldthwait.
1986 (PG) 94m/C John Cusack, Demi Moore, William Hickey, Curtis Armstrong, Bob Goldthwait, Mark Metcalf; *Dir:* Steve Holland. **VHS, Beta, LV** $19.98 *WAR* ♂♂½

One Dark Night Two high school girls plan an initiation rite for one of their friends who is determined to shed her "goody-goody" image. West is the caped crusader of TV series "Batman" fame.
1982 (R) 94m/C Meg Tilly, Adam West, David Mason Daniels, Robin Evans; *Dir:* Tom McLoughlin. **VHS, Beta** $69.95 *HBO* ♂♂½

One Day in the Life of Ivan Denisovich The film version of Nobel Prize-winner Alexander Solzhenitsyn's novel about a prisoner's experiences in a Soviet labor camp. A testament to human endurance. Photography be Sven Nykvist.
1971 105m/C *NO GB* Tom Courtenay, Alfred Burke, Espen Skjonberg, James Maxwell, Eric Thompson; *Dir:* Caspar Wrede. **VHS, LV** $79.95 *SVS* ✍✍✍

One Deadly Owner A made-for-television film about a possessed Rolls-Royce torturing its new owner. Quite a vehicle for the usually respectable set of wheels.
1974 80m/C Donna Mills, Jeremy Brett, Robert Morris, Laurence Payne. **VHS, Beta** $29.95 *IVE* ✍

One Deadly Summer (L'Ete Meurtrier) Revenge drama about a young girl who returns to her mother's home village to ruin three men who had assaulted her mother years before. Very well acted. Subtitled in English. Released in France as "L'Ete Meurtrier." From the novel by Sebastien Japrisot.
1983 (R) 134m/C *FR* Isabelle Adjani, Alain Souchon, Suzanne Flon; *Dir:* Jean Becker. Cesar Award '83: Best Actress (Adjani). **VHS, Beta** $59.95 *MCA* ✍✍✍

One Down, Two to Go! When the mob is discovered to be rigging a championship karate bout, two dynamic expert fighters join in a climactic battle against the hoods. Example of really bad "blaxploitation" that wastes talent, film, and the audience's time.
1982 (R) 84m/C Jim Brown, Fred Williamson, Jim Kelly, Richard Roundtree; *Dir:* Fred Williamson. **VHS, Beta** $59.95 *MED, AVD, HHE* Woof!

One-Eyed Jacks An often engaging, but lengthy, psychological western about an outlaw who seeks to settle the score with a former partner who became a sheriff. Great acting by all, particularly Brando, who triumphed both as star and director. Stanley Kubrick was the original director, but Brando took over mid-way through the filming. The photography is wonderful and reflects the effort that went into it.
1961 141m/C Marlon Brando, Karl Malden, Katy Jurado, Elisha Cook Jr., Slim Pickens, Ben Johnson, Pina Pellicer; *Dir:* Marlon Brando. **VHS, Beta, LV** $79.95 *PAR* ✍✍✍½

One-Eyed Soldiers Young woman, criminal and dwarf follow trail to mysterious key to unlock $15 million treasure. Much cheesy intrigue.
1967 83m/C *GB YU* Dale Robertson, Lucianna Paluzzi; *Dir:* Jean Christophe. **VHS** *TAF* ✍

One-Eyed Swordsman A handicapped samurai battles shogun warriors. With English subtitles.
1963 95m/C *JP* Tetsuro Tanba, Haruko Wanibuchi; *Dir:* Seiichiro Uchikawa. **VHS, Beta** *VDA* ✍

One Flew Over the Cuckoo's Nest Sweeping all major Academy Awards, this film is a touching, hilarious, dramatic, and completely effective adaptation of Ken Kesey's novel. Nicholson plays a two-bit crook, who, facing a jail sentence, feigns insanity to be sentenced to a cushy mental hospital. The hospital is anything but cushy, with a tyrannical head nurse out to squash any vestige of the patients' independence. Nicholson proves to be a crazed messiah and catalyst for these mentally troubled patients and a worthy adversary for the head nurse. A classic that performs superbly on numerous levels. Oscar nominations for supporting actor Dourif and Film Editing, in addition to awards it garnered.
1975 (R) 129m/C Jack Nicholson, Brad Dourif, Louise Fletcher, Will Sampson, William Redfield, Danny DeVito, Christopher Lloyd, Scatman Crothers; *Dir:* Vincent Schiavelli, Milos Forman. Academy Awards '75: Best Actor (Nicholson), Best Actress (Fletcher), Best Adapted Screenplay, Best Director (Forman), Best Picture; British Academy Awards '76: Best Actor (Nicholson), Best Actress (Fletcher), Best Director (Forman), Best Film, Best Film Editing, Best Supporting Actor (Dourif); Golden Globe Awards '76: Best Film—Drama; National Board of Review Awards '75: 10 Best Films of the Year, Best Actor (Nicholson); National Society of Film Critics Awards '75: Best Actor (Nicholson); American Film Institute's Survey '77: 6th Best American Film Ever Made. **VHS, Beta, LV** $19.99 *HBO, BTV* ✍✍✍✍

One Frightened Night An eccentric millionaire informs his family members that he is leaving each of them one million dollars...as long as his long-lost granddaughter doesn't reappear. Guess who comes to dinner.
1935 69m/B Mary Carlisle, Wallace Ford, Hedda Hopper, Charley Grapewin; *Dir:* Christy Cabanne. **VHS, Beta** $19.95 *NOS, KRT* ✍✍½

One Good Cop A noble, inconsistent attempt to do a police thriller with a human face, as a young officer and his wife adopt the three little daughters of his slain partner from the force. But it reverts to a routine action wrapup, with 'Batman' Keaton even donning a masked-avenger getup to get revenge.
1991 (R) 105m/C Michael Keaton, Rene Russo, Anthony LaPaglia, Kevin Conway, Rachel Ticotin, Grace Johnston, Blair Swanson, Rhea Silver-Smith, Tony Plana; *W/Dir:* Heywood Gould. **VHS, Beta, LV** $19.99 *TOU, HPH* ✍✍½

One from the Heart The film more notable for sinking Coppola's Zoetrope Studios than for its cinematic contest. Garr and Forrest are a jaded couple who seek romantic excitement with other people. An extravagant (thus Coppola's finance problems) fantasy Las Vegas set, pretty to look at but does little to enhance the weak plot. Score by Waits, a much-needed plus.
1982 (R) 100m/C Teri Garr, Frederic Forrest, Nastassia Kinski, Raul Julia, Lainie Kazan, Rebecca DeMornay, Harry Dean Stanton; *Dir:* Francis Ford Coppola. **VHS, Beta, LV** $79.95 *COL* ✍✍

One Last Run It's drama on the slopes as a variety of individuals confront their pasts/fears/personal demons via extreme skiing. Features the performances of champion skiers Franz Weber and Scot Schmidt.
1989 82m/C Russell Todd, Ashley Laurence, Craig Branham, Jimmy Aleck, Tracy Scoggins, Nels Van Patten, Chuck Connors; *W/Dir:* Peter Winograd, Glenn Gebhard. **VHS, Beta** $79.95 *PSM* ✍✍

One Last Time: The Beatles' Final Concert The rare concert footage of The Beatles' final appearance in San Francisco's Candlestick Park on August 29, 1966, with a retrospective view of the concert's background and aftermath.
1966 50m/C John Lennon, Paul McCartney, Ringo Starr, George Harrison; *Performed by:* The Beatles. **VHS, Beta** *TEL* ✍✍

One Little Indian An AWOL cavalry man and his Indian ward team up with a widow and her daughter in an attempt to cross the New Mexican desert. A tepid presentation from the usually high quality Disney studio.
1973 90m/C James Garner, Vera Miles, Jodie Foster, Clay O'Brien, Andrew Prine, Bernard McEveety; *Dir:* Bernard McEveety. **VHS, Beta** $69.95 *DIS* ✍✍

One Magic Christmas Disney feel-good film about a disillusioned working woman whose faith in Christmas is restored when her guardian angel descends to Earth and performs various miracles. Somewhat cliched and tiresome, but partially redeemed by Stanton and Steenburgen's presence.
1985 (G) 88m/C Mary Steenburgen, Harry Dean Stanton, Gary Basaraba, Michelle Meyrink, Arthur Hill; *Dir:* Phillip Borsos. **VHS, Beta, LV** $19.99 *DIS* ✍✍½

One Man Force L.A. narcotics cop Jake Swan goes on a vigilante spree. Huge in body and vengeful in spirit, he makes his partner's murderers pay!
1989 (R) 92m/C John Matuszak, Ronny Cox, Charles Napier, Sharon Farrell, Sam Jones, Chance Boyer, Richard Lynch, Stacey Q; *Dir:* Dale Trevillion. **VHS, Beta, LV** $29.95 *ACA* ✍✍

One Man Jury An LAPD lieutenant, wearied by an ineffective justice system, becomes a one-man vigilante avenger. Pale Dirty Harry rip-off with over acting and an overdose of violence.
1978 (R) 95m/C Jack Palance, Chris Mitchum, Joe Spinell, Pamela Shoop; *Dir:* Charles Martin. **VHS, Beta** $19.95 *UHV* ✍

One Man Out An ex-CIA agent diverts his psychological problems into his new job: assassin for a South American despot. His jaded view of life changes when he meets and falls in love with an American journalist. Trouble in paradise when he is ordered to kill her. One man out was not enough, they should have done away with the entire crew of this stinker.
1989 (R) 90m/C Stephen McHattie, Deborah Van Valkenburgh, Aharon Ipale, Ismael Carlo, Michael Champion, Dennis A. Pratt; *Dir:* Michael Kennedy. **VHS, Beta, LV** $14.95 *SVS* ✍

One Man's War A human-rights crusader in repressive Paraguay won't be silenced, even after government thugs torture and murder his son. He fights obsessively to bring the killers to justice. The true story of the Joel Filartiga family is heartfelt but ultimately a dramatic letdown; an epilogue proves that full story hasn't been told.
1990 (PG-13) 91m/C Anthony Hopkins, Norma Aleandro, Fernanda Torres, Ruben Blades; *Dir:* Sergio Toledo. **VHS** $89.99 *HBO* ✍✍½

One Million B. C. The strange saga of the struggle of primitive cavemen and their battle against dinosaurs and other monsters. Curiously told in flashbacks, this film provided stock footage for countless dinosaur movies that followed. Portions of film rumored to be directed by cinematic pioneer D. W. Griffith.
1940 80m/B Victor Mature, Carole Landis, Lon Chaney Jr.; *Dir:* Hal Roach, Hal Roach Jr. **VHS, Beta** $9.95 *MED* ✍✍½

One in a Million: The Ron LaFlore Story The true story of Detroit Tigers' star Ron LeFlore, who rose from the Detroit ghetto to the major leagues. Adaptation of LeFlore's autobiography, "Breakout." A well-acted and compelling drama with Burton in a standout performance.
1978 90m/C LeVar Burton, Madge Sinclair, Billy Martin, James Luisi; *Dir:* William A. Graham. **VHS, Beta** $59.95 *VCL* ✍✍½

One Minute Before Death A woman is buried in a catatonic state, and returns from the grave to exact retribution. Supposedly based loosely on Poe.
1988 87m/C Giselle MacKenzie. **VHS, Beta** TAV *⅟*

One Minute to Zero Korean War action film divides its time between an army romance and war action. The lukwarm melodrama features Mitchum as a colonel in charge of evacuating American civilians but who ends up bombing refugees.
1952 105m/B Robert Mitchum, Ann Blyth, William Talman, Charles McGraw, Margaret Sheridan, Richard Egan, Eduard Fran, Robert Osterloh, Robert Gist; *Dir:* Tay Garnett. **VHS, Beta** RKO *⅟⅟*

One More Chance While he was in jail, an ex-con's family moves away without leaving a forwarding address. When he's released he strikes up a friendship with a woman from the neighborhood who knows where they've gone.
1990 102m/C Kirstie Alley, John Lamotta, Logan Clarke, Michael Pataki, Hector Maisonette; *Dir:* Sam Firstenberg. **VHS, Beta** $59.95 WAR, OM *⅟ ½*

One More Saturday Night Franken and Davis, "Saturday Night Live" alumni, wrote this film about a small town going wild on the weekend. Dry, taxing, and unfunny. Averages about 1 laugh per 1/2 hour—at 1 1/2 hours, it's way too long.
1986 (R) 96m/C Al Franken, Tom Davis, Nan Woods, Dave Reynolds; *Dir:* Dennis Klein. **VHS, Beta, LV** $9.95 RCA *⅟*

One of My Wives Is Missing An ex-New York cop tries to solve the mysterious disappearance of a newlywed socialite. Things become strange when she reappears but is discovered as an imposter. Above average acting in a film adapted from the play, "The Trap for a Lonely Man."
1976 97m/C Jack Klugman, Elizabeth Ashley, James Franciscus; *Dir:* Glenn Jordan. **VHS, Beta** $59.95 LHV *⅟⅟ ½*

One Night of Love Moore's best quasi-operetta, about a young American diva rebelling in response to her demanding Italian teacher. "Pygmallion"-like story has her falling in love with her maestro. Oscar-nominated for best picture, director and actress. Despite being nearly 60 years old, this film is still enchanting and fresh.
1934 95m/B Grace Moore, Tullio Carminati, Lyle Talbot, Jane Darwell, Nydia Westman, Mona Barrie, Jessie Ralph, Luis Alberni; *Dir:* Victor Schertzinger. Academy Awards '34: Best Sound, Best Musical Score. **VHS, Beta, LV** $19.95 COL *⅟⅟⅟ ½*

One Night Only A gorgeous law student decides to make big money by hiring herself and her friends out as hookers to the school football team.
1984 87m/C Lenore Zann, Jeff Braunstein, Grant Alianak; *Dir:* Timothy Bond. **VHS, Beta** $59.98 FOX *⅟⅟⅟ ½*

One Night Stand A woman's chance encounter with a man in a singles bar leads to an evening of unexpected terror. Adapted from a play for Canadian television. Billed as a horror flick, it is really more of a drama.
1977 90m/C Chapelle Jaffe, Brent Carver. **VHS, Beta** $49.95 TWE *⅟*

One Night Stand Four young people attempt to amuse themselves at the empty Sydney Opera House on the New Year's Eve, that night World War III begins. An odd commentary on nuclear war that sees the bomb as the ultimate bad joke. Features an appearance by alternative rock group Midnight Oil.
1984 94m/C Tyler Coppin, Cassandra Delaney, Jay Hackett, Saskia Post; *Dir:* John Duigan. **VHS, Beta** $59.98 SUE *⅟⅟⅟*

One on One A Rocky-esque story about a high school basketball star from the country who accepts an athletic scholarship to a big city university. He encounters a demanding coach and intense competition. Light weight drama that is economically entertaining.
1977 (PG) 100m/C Robby Benson, Annette O'Toole, G.D. Spradlin, Gail Strickland, Melanie Griffith; *Dir:* Lamont Johnson. **VHS, Beta** $39.98 WAR *⅟⅟ ½*

The One and Only An egotistical young man is determined to make it in show business. Instead, he finds himself in the world of professional wrestling. Most of the humor comes from the wrestling scenes, with Winkler and Darby's love story only serving to dilute the film.
1978 (PG) 98m/C Henry Winkler, Kim Darby, Gene Saks, William Daniels, Harold Gould, Herve Villechaize; *Dir:* Carl Reiner. **VHS, Beta** $69.95 PAR *⅟⅟ ½*

One and Only, Genuine, Original Family Band A harmonious musical family becomes divided when various members take sides in the presidential battle between Benjamin Harrison and Grover Cleveland, a political era that has been since overlooked.
1968 (G) 110m/C Walter Brennan, Buddy Ebsen, Lesley Ann Warren, Kurt Russell, Goldie Hawn, Wally Cox, Richard Deacon, Janet Blair; *Dir:* Michael O'Herlihy. **VHS, Beta, LV** $9.99 DIS *⅟⅟⅟*

One of Our Aircraft Is Missing The crew of an R.A.F. bomber downed in the Netherlands, struggle to escape Nazi capture. A thoughtful study of wars and the men who fight them, with an entertaining melodramatic plot. Look for the British version, which runs 106 minutes. Some of the American prints only run 82 minutes.
1941 103m/B GB Godfrey Tearle, Eric Portman, Hugh Williams, Pamela Brown, Googie Withers, Peter Ustinov; *Dir:* Emeric Pressburger, Michael Powell. National Board of Review Awards '42: 10 Best Films of the Year. **VHS** $19.95 NOS, MRV, REP *⅟⅟⅟ ½*

One of Our Dinosaurs Is Missing An English nanny and her cohorts help British Intelligence retrieve a microfilm—concealing dinosaur fossil from the bad guys that have stolen it. Disney film was shot on location in England.
1975 (G) 101m/C Peter Ustinov, Helen Hayes, Derek Nimmo, Clive Revill, Robert Stevenson, Joan Sims; *Dir:* Robert Stevenson. **VHS, Beta** $69.95 DIS *⅟⅟*

One Plus One A dramatization, believe it or not, of the Kinsey sex survey of the 1950s. Participants in a sex lecture talk about and demonstrate various "risque" practices, such as as premarital sex and extramarital affairs. Also known as "Exploring the Kinsey Report." Despite this film, the sexual revolution went on as planned.
1961 114m/B CA Leo G. Carroll, Hilda Brawner, William Traylor, Kate Reid, Ernest Graves; *W/Dir:* Arch Oboler. **VHS, Beta** $16.95 SNC *⅟*

One-Punch O'Day Boxing boy tries to buy back hometown's oil livelihood by winning prizefight. Much patient waiting by his honey.
1926 60m/B Billy Sullivan, Jack Herrick, Charlotte Merriam; *Dir:* Harry Joe Brown. **VHS, Beta** $19.95 GPV, FCT *⅟*

One Rainy Afternoon A bit-player kisses the wrong girl in a Paris theater causing a massive uproar that brands him as a notorious romantic "monster." Patterned after a German film, the story lacks depth and zest.
1936 80m/B Francis Lederer, Ida Lupino, Hugh Herbert, Roland Young, Donald Meek; *Dir:* Rowland V. Lee. **VHS, Beta** $19.95 NOS, MRV, VYY *⅟⅟*

One Russian Summer During the Russian Revolution, a cripple seeks revenge on the corrupt land baron who murdered his parents. An unfortunate attempt to dramatize a novel by M. Lermontov.
1973 (R) 112m/C Oliver Reed, Claudia Cardinale, John McEnery, Carole Andre, Ray Lovelock; *Dir:* Antonio Calenda. **VHS, Beta** $59.95 UHV Woof!

One Shoe Makes It Murder A shady casino owner hires ex-cop Mitchum to investigate the disappearance of his unfaithful wife. Adapted from Eric Bercovici's novel, the story does little to enhance Mitchum's TV debut.
1982 100m/C Robert Mitchum, Angie Dickinson, Mel Ferrer, Howard Hesseman, Jose Perez; *Dir:* William Hale. **VHS, Beta** $89.95 IVE *⅟⅟*

One Sings, the Other Doesn't Seeking contentment, a conservative widow and a liberal extrovert help each other cope in a man's world. Endearing characters, but a superficial treatment of dared feminist issues. In French with English subtitles.
1977 105m/C FR BE Valerie Mairesse, Therese Liotard; *Dir:* Agnes Varda. New York Film Festival '77: 10 Best Films of the Year. **VHS, Beta** $59.95 COL *⅟⅟ ½*

One Step to Hell Good cop rescues gold-miner's widow from three escaped convicts. Beautiful African scenery helps save an otherwise mediocre film.
1968 90m/C Ty Hardin, Rossano Brazzi, Pier Angeli, George Sanders, Tab Hunter, Sandy Howard. **VHS, Beta** $39.95 NO *⅟ ½*

One Summer Love Man checks out from loony bin, searches for family ties and befriends beautiful woman who works in a moviehouse. Also known as "Dragonfly."
1976 95m/C Beau Bridges, Susan Sarandon, Mildred Dunnock, Ann Wedgeworth, Michael B. Miller, Linda Miller, James Noble, Frederick Coffin; *Dir:* Gilbert Cates. **VHS** $79.99 HBO *⅟ ½*

One That Got Away A loyal German Luftwaffe pilot captured by the British becomes obsessed with escape. Fast paced and exciting. Based on a true story.
1957 111m/B Hardy Kruger, Colin Gordon, Alec Gordon; *Dir:* Roy Ward Baker. **VHS, Beta** $39.95 IND *⅟⅟⅟*

One Third of a Nation Depression era film contrasts the conditions of slum life with those in high society. A young entrepreneur inherits a city block in ruins, only to fall in love with a young woman who lives there and help her crippled brother. Timely social criticism.
1939 79m/B Sylvia Sidney, Leif Erickson, Myron McCormick, Sidney Lumet; *Dir:* Dudley Murphy. **VHS** $19.95 NOS, PAR, MLB *⅟⅟ ½*

One Touch of Venus Love fills a department store when a window dresser kisses a statue of the goddess Venus—and she comes to life. Appealing adaptation of the Broadway musical, with songs by Kurt

Weill and Ogden Nash, including "Speak Low." Also available colorized.
1948 82m/B Ava Gardner, Robert Walker, Eve Arden, Dick Haymes, Olga San Juan, Tom Conway; *Dir:* William A. Seiter. **VHS $19.98** *REP, MLB* 𝄞𝄞½

One Trick Pony Once-popular rock singer/songwriter struggles to keep his head above water in a turbulent marriage and in the changing currents of popular taste. Simon wrote the autobiographical screenplay and score, but let's hope he is more sincere in real life. A good story.
1980 (R) 100m/C Paul Simon, Blair Brown, Rip Torn, Joan Hackett, Mare Winningham, Lou Reed, Harry Shearer, Allen Goorwitz, Daniel Stern; *Dir:* Robert M. Young. **VHS, Beta $24.98** *WAR* 𝄞𝄞½

One, Two, Three Cagney, an American Coca-Cola exec in Germany, zealously pursues any opportunity to run Coke's European operations. He does the job too well and loses his job to a Communist hippy turned slick capitalist. Fast-paced laughs, wonderful cinematography, and a fine score.
1961 110m/B James Cagney, Horst Buchholz, Arlene Francis; *Dir:* Billy Wilder. National Board of Review Awards '61: 10 Best Films of the Year. **VHS, Beta, LV $19.98** *MGM* 𝄞𝄞𝄞½

One Voice, One Guitar A performance piece by Uruguayan singer Marinez Patingui.
1984 30m/C Beta *NJN* 𝄞𝄞𝄞½

One Wild Moment Comic complications arise when a divorced man is seduced by his best friend's daughter while vacationing. Warm and charming. Later re-made as "Blame It on Rio." French with English subtitles.
1978 (R) 88m/C *FR* Jean-Pierre Marielle, Victor Lanoux, Agnes Soral, Christine Dejoux, Martine Sarcey; *Dir:* Claude Berri. **VHS, Beta $59.95** *COL* 𝄞𝄞½

One Wish Too Many A young boy finds a marble that grants wishes. He has the time of his life with his teachers and the school bullies, but runs into trouble when he creates a giant steam roller which overruns London. Winner of Best Children's Film, 1956 Venice Film Festival, but it is hard to say why.
1955 55m/B *GB* Anthony Richmond, Rosalind Gourgey, John Pike; *Dir:* John Durst. **VHS, Beta $16.95** *SNC* 𝄞𝄞½

One Woman or Two Paleontologist Depardieu is duped by beautiful ad-exec who plans to use his findings to push perfume. Remake of "Bringing Up Baby," this screwball comedy has a few screws loose. Dubbed.
1986 100m/C *FR* Gerard Depardieu, Sigourney Weaver, Dr. Ruth Westheimer; *Dir:* Daniel Vigne. **VHS, Beta $29.98** *LIV, VES* 𝄞𝄞½

100% Bonded A collection of theatrical trailers from the James Bond films starring Sean Connery, including "Thunderball," "Dr. No" and "From Russia with Love."
1987 50m/C Sean Connery. **VHS, Beta $34.95** *NO* 𝄞𝄞½

100 Rifles A Native American bank robber and a Black American lawman join up with a female Mexican revolutionary to help save the Mexican Indians from annihilation by a despotic military governor. Quite racy in its day, but pretty tame and overblown by today's standards.
1969 (R) 110m/C Jim Brown, Raquel Welch, Burt Reynolds, Fernando Lamas; *Dir:* Tom Gries. **VHS, Beta $19.98** *FOX, FXV* 𝄞

1001 Arabian Nights In this Arabian nightmare, the nearsighted Mr. Magoo is known as "Azziz" Magoo, lamp dealer and uncle of Aladdin.
1959 76m/C *Dir:* Jack Kinney; *Voices:* Jim Backus, Kathryn Grant, Hans Conried, Herschel Bernardi. **VHS, Beta $19.95** *COL* 𝄞𝄞½

101 Dalmations Disney classic and one of the highest-grossing animated films in the history of Hollywood centers around two dogowners, Roger and Anita, and their respective spotted pets, Pongo and Perdita. When Pongo and Perdita's puppies are kidnapped by Cruella de Vil, villainess extraordinaire, to make a simply fabulous spotted coat, the parents enlist the aid of various animals, including a dog named Colonel, a horse, a cat, and a goose to rescue the doomed pups. Imagine their surprise when they find not only their own puppies, but 84 more as well. You can expect a happy ending and lots of spots—6,469,952 to be exact. Songs include "Remember When," "Cruella de Vil," "Dalmatian Plantation," and "Kanine Krunchies Kommercial." Based on the children's book by Dodie Smith. Technically notable for the first time use of the Xerox process to transfer the animator's drawings onto celluloid, which made the film's opening sequence of dots evolving into 101 barking dogs possible.
1961 (G) 79m/C *Dir:* Wolfgang Reitherman, Hamilton Luske; *Voices:* Rod Taylor, Betty Lou Gerson, Lisa Davis, Ben Wright, Frederick Worlock, J. Pat O'Malley. **VHS, Beta $24.99** *DIS* 𝄞𝄞𝄞½

125 Rooms of Comfort A mental patient has bizarre fantasies that create havoc among those who stand in his way.
1983 82m/C VHS, Beta *TWE* 𝄞𝄞

Onibaba A brutal parable about a mother and her daughter-in-law in war-ravaged medieval Japan who subsist by murdering stray soldiers and selling their armor. One soldier beds the daughter, setting the mother-in-law on a vengeful tirade. Review of the film varied widely, hailed by some as a masterpiece and by others as below average; in Japanese with subtitles.
1964 103m/B *JP* Nobuko Otowa, Yitsuko Yoshimura, Kei Sato; *Dir:* Kaneto Shindo. **VHS, Beta $79.95** *TAM* 𝄞𝄞𝄞

The Onion Field True story about the mental breakdown of an ex-cop who witnessed his partner's murder. Haunted by the slow process of justice and his own feelings of insecurity, he is unable to get his life together. Based on the novel by Joseph Wambaugh, who also wrote the screenplay. Compelling script and excellent acting.
1979 (R) 126m/C John Savage, James Woods, Ronny Cox, Franklyn Seales, Ted Danson; *Dir:* Harold Becker. **VHS, Beta, LV $14.98** *SUE* 𝄞𝄞𝄞½

Only Angels Have Wings Melodramatic adventure about a broken-down Peruvian air mail service. Large cast adds to the love tension between Grant, a pilot, and Arthur, a showgirl at the saloon. First ever special effects Oscar nomination.
1939 121m/B Cary Grant, Thomas Mitchell, Richard Barthelmess, Jean Arthur, Noah Beery Jr., Rita Hayworth, Sig Rumann, John Carroll; *Dir:* Howard Hawks. **VHS, Beta $19.95** *COL* 𝄞𝄞𝄞𝄞

Only the Lonely Middle-aged cop, Candy, falls in love with a shy undertaker's assistant, Sheedy, and is torn between love and dear old Mom, O'Hara, in her first role in

years. Candy is an unlikely leading man and even the jokes are forced. But the restaurant scene makes the whole thing well worth seeing.
1991 (PG) 102m/C John Candy, Ally Sheedy, Maureen O'Hara, Anthony Quinn, Kevin Dunn, James Belushi; *W/Dir:* Chris Columbus. **VHS, Beta, LV $19.98** *FXV* 𝄞𝄞

Only Once in a Lifetime A love story about an Hispanic immigrant painter looking for success in America.
198? 90m/C Miguel Robelo, Estrellita Lopez, Sheree North. **VHS, Beta $49.95** *VHE* 𝄞

Only One Night Upon finding out that he's the illegitimate son of an aristocrat, a circus attendant joins high society and is promptly matched with a beautiful woman. In Swedish with subtitles.
1942 87m/B *SW* Ingrid Bergman, Edvin Adolphson, Alno Taube, Olof Sandborg, Erik Berglund, Marianne Lofgren, Magnus Kesster; *Dir:* Gustaf Molander. **VHS, Beta, LV $29.95** *WAC, CRO, LUM* 𝄞

Only Two Can Play Sellers is a hilarious Casanova librarian who puts the moves on a society lady to get a promotion. Funny, of course, and based on Kingsley Amis' novel "That Uncertain Feeling."
1962 106m/C *GB* Peter Sellers, Virginia Maskell, Mai Zetterling, Richard Attenborough; *Dir:* Sidney Gilliat. **VHS, Beta $69.95** *COL* 𝄞𝄞𝄞

Only the Valiant Action-packed story of a cavalry officer who struggles to win his troops respect while warding off angry Apaches. Fast-paced Western fun requires little thought.
1950 105m/B Gregory Peck, Ward Bond, Gig Young, Lon Chaney Jr., Barbara Payton, Neville Brand; *Dir:* Gordon Douglas. **VHS, LV $14.95** *REP* 𝄞𝄞

Only Way A semi-documentary account of the plight of the Jews in Denmark during the Nazi occupation. Despite German insistence, the Danes succeeded in saving most of their Jewish population from the concentration camps.
1970 (G) 86m/C Jane Seymour, Martin Potter, Benjamin Christiansen. **VHS, Beta $19.95** *UHV* 𝄞𝄞½

The Only Way Home Two bikers end up in big trouble when one of them kills a wealthy man, and they kidnap his wife. Filmed entirely in Oklahoma.
1972 (PG) 86m/C Bo Hopkins, Beth Brickell, Steve Sandor, G.D. Spradlin; *Dir:* G.D. Spradlin. **VHS, Beta $39.95** *WES* 𝄞𝄞

Only When I Laugh Neil Simon reworked his Broadway flop "The Gingerbread Lady" to produce this poignant comedy about the relationship between an aging alcoholic actress and her teenage daughters. Excellent acting earned Oscar nomination for Mason, Coco, and Hackett.
1981 (R) 120m/C Marsha Mason, Kristy McNichol, James Coco, Joan Hackett, David Dukes, Kevin Bacon, John Bennett Perry; *Dir:* Glenn Jordan. **VHS, Beta, LV $19.95** *COL* 𝄞𝄞𝄞

Only With Married Men Carne's hassle-free dating routine is disrupted when a sly bachelor pretends that he's married to get a date. A middle-aged persons answer to a teenage sex comedy. Pretty bad.
1974 74m/C David Birney, Judy Carne, Gavin McLeod, John Astin; *Dir:* Jerry Paris. **VHS, Beta $59.95** *LHV* 𝄞

Open All Hours A British television comedy about a small corner shop and its viciously greedy, opportunistic owner.

1983 85m/C *GB* Ronnie Barker, Lynda Baron, David Jason, Stephenie Cole. **VHS, Beta $14.98** *FOX* 🦴

Open City A leader in the Italian underground resists Nazi control of the city. A stunning film, making Rossellini's realistic style famous. Co-scripted by Fellini. Italian dialogue with English subtitles.
1945 103m/B *IT* Anna Magnani, Aldo Fabrizi; ***Dir:*** Roberto Rossellini. National Board of Review Awards '46: Best Foreign Film; New York Film Critics Awards '46: Best Foreign Film; New York Times Ten Best List '46: Best Film. **VHS, Beta, 8mm $59.95** *CVC, MRV, NOS* 🦴🦴🦴🦴

Open Doors (Porte Aperte) A bitter review of Fascist rule and justice. The Fascist regime promises security, safety, and tranquility. So, on the morning that a white collar criminal murders his former boss, murders the man who got his job, and then rapes and murders his wife, tensions rise and the societal structures are tested. The people rally for his death. A judge and a juror struggle to uphold justice rather than serve popular passions. Winner of 4 Donatello Awards, Italian Golden Globes for Best Film, Best Actor, and Best Screenplay, and many international film awards. In Italian with English subtitles.
1989 (R) 109m/C *IT* Gian Marie Volonte, Ennio Fantastichini, Lidia Alfonsi; ***Dir:*** Gianni Amelio. **VHS $79.98** *ORI, FCT* 🦴🦴🦴½

Open House Radio psychologist and beautiful real estate agent search for the killer of real estate agents and their clients. A mystery-thriller for the very patient.
1986 95m/C Joseph Bottoms, Adrienne Barbeau, Mary Stavin, Rudy Ramos; ***Dir:*** Jag Mundhra. **VHS, Beta $79.95** *PSM* 🦴

Opera do Malandro A lavish Brazilian take off of "The Threepenny Opera." Married hustler falls for a beautiful entrepreneur longing to get rich off of American goods. Based on Chico Buarque's musical play. In Portuguese, with subtitles.
1987 106m/C *BR* Edson Celulari, Claudia Ohana, Elba Ramalho, Ney Latorraca; ***Dir:*** Ruy Guerra. **VHS, Beta, LV $79.95** *VIR* 🦴

Operation Amsterdam Four agents have fourteen hours to snare $10 million in diamonds from under the noses of local Nazis. Based on a true story. Full of 1940s wartime action and suspense.
1960 103m/B Peter Finch, Eva Bartok, Tony Britton, Alexander Knox; ***Dir:*** Michael McCarthy. **VHS, Beta $59.95** *SUE* 🦴🦴

Operation C.I.A. A plot to assassinate the U.S. ambassador in Saigon inspires brave CIA agent Reynolds to wipe out the bad guys. Action-packed and somewhat exciting.
1965 90m/C Burt Reynolds, John Hayt, Kieu Chin, Danielle Aubry; ***Dir:*** Christian Nyby. **VHS, Beta $59.98** *FOX, DVT* 🦴🦴

Operation Cross Eagles Routine WWII military thriller with Conte and Calhoun on a mission to rescue an American general in exchange for their German prisoner. Their mission is complicated by a traitor in the group. Conte's only directorship.
1969 90m/C *YU* Richard Conte, Rory Calhoun, Aili King, Phil Brown; ***Dir:*** Richard Conte. **VHS $79.95** *SVS* 🦴🦴

Operation Dames A squadron of soldiers must go behind Korean enemy lines to locate a missing U.S.O. troupe and bring them to safety. Low budget and not very funny for a supposed comedy.
1959 73m/B Eve Meyer, Chuck Henderson, Don Devlin, Ed Craig, Cindy Girard; ***Dir:*** Louis Clyde Stouman. **VHS $9.95** *COL* 🦴½

Operation Haylift The true story of the U.S. Air Force's efforts to rescue starving cattle and sheep herds during Nevada's blizzards of 1949. The Air Force provided realism for the film with planes, equipment, and servicemen.
1950 73m/B Bill Williams, Tom Brown, Ann Rutherford, Jane Nigh; ***Dir:*** William Berke. **VHS, Beta, LV** *WGE* 🦴🦴

Operation Inchon A foreign film depicting the battle in Devil Ridge during the Korean War.
1981 90m/C VHS, Beta $39.95 *VCD Woof!*

Operation Julie A detective searches out a huge drug ring that manufactures LSD.
1985 100m/C Colin Blakely, Lesley Nightingale, Clare Powney; ***Dir:*** Bob Mahoney. **VHS, Beta $59.98** *LIV, LTG* 🦴½

Operation 'Nam A group of bored Vietnam vets return to 'Nam to rescue their leader from a POW camp. Nothing special, but features John Wayne's son.
1985 85m/C Oliver Tobias, Christopher Connelly, Manfred Lehman, John Steiner, Ethan Wayne, Donald Pleasence; ***Dir:*** Larry Ludman. **VHS $79.95** *IMP* 🦴🦴

Operation Orient A statue laden with drugs is stolen when an international drug smuggler tries to bring it to the U.S.
19?? 94m/C Gianni Gori, Gordon Mitchell; ***Dir:*** Elia Milonakos. **VHS, Beta $39.95** *LUN* 🦴🦴

Operation Petticoat Submarine captain Grant teams with wheeler-dealer Curtis to make his sub seaworthy. They're joined by a group of Navy women, and the gags begin. Great teamwork from Grant and Curtis keeps things rolling. Jokes may be considered sexist these days. Later remake and TV series couldn't hold a candle to the original.
1959 120m/C Cary Grant, Tony Curtis, Joan O'Brien, Dina Merrill, Gene Evans, Arthur O'Connell; ***Dir:*** Blake Edwards. **VHS, LV $19.98** *REP* 🦴🦴🦴½

Operation Thunderbolt Israeli-produced depiction of Israel's July 14, 1976 commando raid on Entebbe, Uganda to rescue the passengers of a hijacked plane. Better than the American versions "Raid on Entebbe" and "Victory at Entebbe" in its performances as well as the information provided, much of it unavailable to the American filmmakers.
1977 120m/C Yehoram Gaon, Assaf Dayan, Ori Levy, Klaus Kinski; ***Dir:*** Menahem Golan. **VHS, Beta $59.95** *MGM* 🦴🦴½

Operation War Zone A platoon of soldiers uncovers a plot by corrupt officers to continue the Vietnam war and sell weapons.
1989 86m/C Joe Spinell. **VHS, Beta** *AIP* 🦴

The Oppermann Family On two tapes, the long epic made for German television about a family trying to survive in Berlin during the rise of Hitler. In German with subtitles.
1982 238m/C *Dir:* Egon Monk. **VHS, Beta $99.95** *FCT, APD* 🦴🦴½

The Opponent A young boxer saves a young woman's life, not realizing that she has mob connections, thus embroiling him in a world of crime.
1989 (R) 102m/C *IT* Daniel Greene, Ernest Borgnine, Julian Gemma, Mary Stavin, Kelly Shaye Smith; ***Dir:*** Sergio Martino. **VHS, LV $89.95** *VMK* 🦴🦴

Opportunity Knocks Carvey's first feature film has him impersonating a friend of a rich suburbanite's family while hiding from a vengeful gangster. They buy it, and give him a job and the daughter. Not hilarious, but not a dud either.
1990 (PG-13) 105m/C Dana Carvey, Robert Loggia, Todd Graff, Milo O'Shea, Julia Campbell, James Tolken; ***Dir:*** Donald Petrie. **VHS, Beta, LV $19.95** *MCA* 🦴🦴½

Opposing Force The commander of an Air Force camp simulates prisoner-of-war conditions for realistic training, but he goes too far, creating all too real torture situations. He preys on the only female in the experiment, raping her as part of the training. A decent thriller. Also known as "Hellcamp."
1987 (R) 97m/C Tom Skerritt, Lisa Eichhorn, Anthony Zerbe, Richard Roundtree, Robert Wightman, John Considine, George Kee Chueng, Paul Joynt, Jay Louden, Ken Wright, Dan Hamilton; ***Dir:*** Eric Karson. **VHS, Beta $19.99** *HBO* 🦴🦴½

The Opposite Sex Bevy of women battle mediocre script in adaptation of 1939's "The Women." Laserdisc includes original trailer and letterboxed screen.
1956 115m/C June Allyson, Joan Collins, Dolores Gray, Ann Sheridan, Ann Miller, Leslie Nielsen, Agnes Moorehead, Joan Blondell; ***Dir:*** David Miller. **VHS, LV $19.98** *MGM, PIA, FCT* 🦴🦴

Options Nerdy Hollywood agent Salinger treks to Africa to "option" a princess's life story—hence the title. He gets mixed up with her kidnapping and with her. Misplaced cameos by Roberts and Anton drag down the overall comedy content, which isn't to high to begin with.
1988 (PG) 105m/C Matt Salinger, Joanna Pacula, John Kani, Susan Anton, James Keach, Eric Roberts; ***Dir:*** Camilo Vila. **VHS, LV $89.98** *LIV, VES* 🦴🦴

The Oracle A woman takes a new apartment only to find that the previous occupant's spirit is still a resident. The spirit tries to force her to take revenge on his murderers. Not bad for a low-budget thriller, but bad editing is a distraction.
1985 94m/C Caroline Capers Powers, Roger Neil; ***Dir:*** Roberta Findlay. **VHS, Beta $59.95** *IVE* 🦴

Orca Ridiculous premise has a killer whale chasing bounty hunter Harris to avenge the murder of its pregnant mate. Great for gore lovers, especially when the whale chomps Derek's leg off.
1977 (PG) 92m/C Richard Harris, Charlotte Rampling, Bo Derek, Keenan Wynn, Will Sampson; ***Dir:*** Michael Anderson Sr. **VHS, Beta, LV $79.95** *PAR* 🦴½

Orchestra Wives A drama bursting with wonderful Glenn Miller music. A woman marries a musician and goes on the road with the band and the other wives. Trouble springs up with the sultry singer who desperately wants the woman's new husband. The commotion spreads throughout the group. Score includes "Serenade in Blue," "At Last," and "I've Got a Gal in Kalamazoo."
1942 98m/B George Montgomery, Glenn Miller, Lynn Bari, Carole Landis, Jackie Gleason, Cesar Romero, Ann Rutherford, Virginia Gilmore, Mary Beth Hughes, Harry Morgan; ***Dir:*** Archie Mayo. **VHS, Beta $19.98** *FXV, FCT* 🦴🦴🦴

The Ordeal of Dr. Mudd His name was Mudd—a fitting moniker after he unwittingly aided President Lincoln's assassin. Dr. Mudd set John Wilkes Boothe's leg, broken during the assassination, and was jailed for conspiracy. He became a hero in prison for

his aid during yellow fever epidemics and was eventually released. A strong and intricate performance by Weaver keeps this made-for-television drama interesting. Dr. Mudd's descendants are still trying to completely clear his name of any wrongdoing in the Lincoln assassination.
1980 143m/C Dennis Weaver, Susan Sullivan, Richard Dysart, Michael McGuire, Nigel Davenport, Arthur Hill; *Dir:* Paul Wendkos. **VHS, Beta** $39.95 *IVE ♫♫♫*

Ordeal by Innocence Sutherland is an amateur sleuth in 1950s England convinced he has proof that a convicted murderer is innocent, but no one wishes to reopen the case. The big-name cast is essentially wasted. Based on an Agatha Christie story.
1984 (PG-13) 91m/C Donald Sutherland, Christopher Plummer, Faye Dunaway, Sarah Miles, Ian McShane, Diana Quick, Annette Crosbie, Michael Elphick; *Dir:* Desmond Davis. **VHS, Beta, LV** $79.95 *MGM ♫ ½*

Order of the Black Eagle A Bondish spy and his sidekick, Typhoon the Baboon, battle neo-Nazis planning to bring Hitler back to life in this silly tongue-in-cheek thriller.
1987 (R) 93m/C *GB* Ian Hunter, Charles K. Bibby, William T. Hicks, Jill Donnellan; *Dir:* Worth Keeter. **VHS, Beta** $79.95 *CEL ♫ ½*

Order of the Eagle An innocent scouting trip turns into a non-stop nightmare when an Eagle Scout uncovers some dangerous information.
1989 88m/C Frank Stallone. **VHS, Beta** *AIP ♫ ½*

Order to Kill A gambling boss puts out a contract on a hit man.
1973 110m/C Jose Ferrer, Helmut Berger, Sydne Rome, Kevin McCarthy; *Dir:* Jose Maesso. **VHS, Beta** $33.95 *KOV ♫ ½*

Ordet A man who believes he is Jesus Christ is ridiculed until he begins performing miracles, which result in the rebuilding of a broken family. A statement on the nature of religious faith vs. fanaticism by the profoundly religious Dreyer, and based on the play by Kaj Munk. Also known as "The Word."
1955 126m/B *DK* Henrik Malberg; *Dir:* Carl Theodor Dreyer. National Board of Review Awards '57: Best Foreign Film; Venice Film Festival '55: Best Film. **VHS, Beta** $24.95 *NOS, WFV ♫♫♫ ½*

Ordinary Heroes The story is familiar but the leads make it worthwhile. Anderson is strong as a blinded Vietnam vet readjusting to life at home. Bertinelli's portrayal of his former girlfriend is eloquent. Nice work on an overdone story.
1985 105m/C Richard Dean Anderson, Doris Roberts, Valerie Bertinelli; *Dir:* Peter H. Cooper. **VHS** $79.99 *HBO ♫♫ ½*

Ordinary People Powerful, well-acted story of a family's struggle to deal with one son's accidental death and the other's subsequent guilt-ridden suicide attempt. Features strong performances by all, but Moore is especially believable as the cold and rigid mother. McGovern's film debut as well as Redford's directorial debut. Based on the novel by Judith Guest.
1980 (R) 124m/C Mary Tyler Moore, Donald Sutherland, Timothy Hutton, Judd Hirsch, M. Emmet Walsh, Elizabeth McGovern, Adam Baldwin, Dinah Manoff, James B. Sikking; *Dir:* Robert Redford. Academy Awards '80: Best Adapted Screenplay, Best Director (Redford), Best Picture, Best Supporting Actor (Hutton); Golden Globe Awards '81: Best Film—Drama; Los Angeles Film Critics Association Awards '80: Best Supporting Actor (Hutton); National Board

of Review Awards '80: Best Director (Redford), Best Picture; New York Film Critics Awards '80: Best Picture. **VHS, Beta, LV** $14.95 *PAR, BTV ♫♫♫ ½*

Oregon The band, Oregon, performs at the 1987 Arts Festival in Germany.
1988 60m/C Ralph Towner, Trilok Gurtu, Glenn Moore, Paul McCandless. **VHS, Beta, LV** $24.95 *MVD, PRS ♫♫♫ ½*

Oregon Trail A western serial in 15 chapters, each 13 minutes long.
1939 195m/B Johnny Mack Brown, Al "Fuzzy" Knight; *Dir:* Ford Beebe. **VHS, Beta** $29.95 *VCN, DVT, MLB ♫♫♫ ½*

The Organization Poitier's third and final portrayal of Detective Virgil Tibbs, first seen in "In the Heat of the Night." This time around he battles a drug smuggling ring with a vigilante group. Good action scenes and a realistic ending.
1971 (PG) 108m/C Sidney Poitier, Barbara McNair, Sheree North, Raul Julia; *Dir:* Don Medford. **VHS, Beta** $19.98 *FOX ♫♫♫*

Orgy of the Dead A classic anti-canon film scripted by Wood from his own novel. Two innocent travelers are forced to watch an even dozen nude spirits dance in a cardboard graveyard. Hilariously bad.
1965 90m/C Criswell, Fawn Silver, William Bates, Pat Barringer; *Dir:* A.C. Stephen. Golden Turkey Awards '65: One of the worst films ever. **VHS, Beta** $19.95 *RHI Woof!*

Orgy of the Vampires Tourists wander into village during cocktail hour. Originally titled "Vampire's Night Orgy."
1973 (R) 86m/C *SP IT* Jack Taylor, Charo Soriano, Dianik Zurakowska, John Richard; *Dir:* Leon Klimovsky. **VHS** $19.99 *SNC ♫*

Oriane A troubled Venezuelan woman must leave France for her native soil when her aunt dies. Once there, in the place where she spent youthful summers, she begins to recall her incestuous past.
1985 88m/C Deniela Silverio, Doris Wells, Philippe Rouleau, David Crotto; *Dir:* Fina Torres. **VHS** *TAM ♫*

The Original Adventures of Betty Boop Three Boop classics: "Betty and Henry," "Betty and the Little King," and "Betty in Blunderland."
1934 38m/B **VHS, Beta** $29.95 *JEF ♫*

The Original Fabulous Adventures of Baron Munchausen An adventure fantasy that takes the hero all over, from the belly of a whale, eventually landing him on the surface of the moon. Also known as "The Fabulous Baron Munchausen."
1961 84m/C *CZ* Milos Kopecky, Jana Brejchova, Rudolph Jelinek, Jan Werich; *Dir:* Karel Zeman. **VHS** $29.95 *AVD ♫♫ ½*

The Original Flash Gordon Collection Four vintage Flash Gordon serials from the 1930s were recently re-edited into feature length films and released as this collection. Great stuff, including "Spaceship to the Unknown," "Deadly Ray From Mars," "Peril From Planet Mongo," and "Purple Death From Outer Space."
1991 366m/B Buster Crabbe. **VHS** $79.95 *QHV, FCT ♫♫ ½*

Original Intent A powerful drama focusing on one man's crusade to help the homeless. A successful lawyer, facing a mid-life crisis, jeopardizes both his family and career

when he decides to defend a homeless shelter from eviction proceedings.
1992 (PG) 97m/C Kris Kristofferson, Candy Clark, Jay Richardson, Vince Edwards, Cindy Pickett, Joe Campanella, Martin Sheen; *W/Dir:* Robert Marcarelli. **VHS, Beta** *PAR ♫♫*

Orion's Belt Three Norwegian adventurers must flee through the wilderness when they discover a secret Soviet base on the Europe/Russia/Greenland border.
1985 (PG) 87m/C **VHS, Beta** $19.95 *STE, NWV ♫♫*

Orlak, the Hell of Frankenstein The monster of Frankenstein is revived for the purposes of revenge. The monster is forced to wear a helmet, as his face melted during the obligatory electrification. In Spanish only, no English subtitles.
1960 103m/B *MX* Joaquin Cordero, Andres Soler, Rosa de Castilla, Irma Dorante, Armando Calvo, Pedro de Aguillon, David Reynoso; *Dir:* Rafael Baledon. **VHS, Beta** $16.95 *SNC ♫*

Ornette Coleman Trio A philosophical as well as musical account of Coleman's work on a project called "Who's Crazy" for the Living Theatre.
1966 26m/B Ornette Coleman, David Izenon, Charles Moffett. **VHS, Beta** $24.95 *MVD, KIV ♫*

The Orphan A young orphaned boy seeks revenge against his cruel aunt who is harassing him with sadistic discipline.
1979 80m/C Mark Evans, Joanna Miles, Peggy Feury; *Dir:* John Ballard. **VHS, Beta** $49.95 *PSM ♫ ½*

Orphan Boy of Vienna A homeless street urchin with a beautiful singing voice is accepted into the wonderful world of the Choir, but is later unjustly accused of stealing. Performances by the Vienna Boys' Choir redeem the melodramatic plot. In German with English subtitles.
1937 87m/B *GE* The Vienna Boys Choir; *Dir:* Max Neufeld. **VHS, Beta, 3/4U** $29.95 *VYY, IHF, GLV ♫ ½*

Orphan Train A woman realizes her New York soup kitchen can't do enough to help the neighborhood orphans, so she takes a group of children out West in hopes of finding families to adopt them. Their journey is chronicled by a newspaper photographer and a social worker. Based on the actual "orphan trains" of the mid- to late 1800s. From the novel by Dorothea G. Petrie. Made for television.
1979 150m/C Jill Eikenberry, Kevin Dobson, Glenn Close, Linda Manz; *Dir:* William A. Graham. **VHS** $14.95 *PSM, KAR, HHE ♫♫♫*

Orphans A sensitive edge-of-glasnost portrait of a young boy's discovery of love, friendship and literature. In Russian with English subtitles.
1983 97m/C *Dir:* Nikolai Gubenko. Cannes Film Festival '83: Special Jury Prize. **VHS, Beta** $79.95 *FCT ♫♫♫*

Orphans A gangster on the run is kidnapped by a tough New York orphan but soon takes control by befriending his abductor's maladjusted brother. Eventually each brother realizes his need for the older man, who has become a father figure to them. This very quirky film is salvaged by good performances. Based on the play by Lyle Kessler.
1987 (R) 116m/C Albert Finney, Matthew Modine, Kevin Anderson; *Dir:* Alan J. Pakula. **VHS, Beta, LV** $19.95 *LHV, WAR ♫♫♫*

Orphans of the North The story of Bedrock Brown's search for gold and his lost partner. Filmed on location in Alaska with non-professional actors. Impressive footage of America's "last frontier," including the flora and fauna.
1940 56m/C W/Dir: Norman Dawn; **Nar:** Norman Dawn. **VHS, Beta, 8mm $24.95** *VYY* ♂

Orphans of the Storm Two sisters are separated and raised in opposite worlds—one by thieves, the other by aristocrats. Gish's poignant search for her sister is hampered by the turbulent maelstrom preceding the French Revolution. Silent. Based on the French play "The Two Orphans."
1921 190m/B Lillian Gish, Dorothy Gish, Monte Blue, Joseph Schildkraut; **Dir:** D.W. Griffith. **VHS, LV $19.98** *NOS, MRV, CCB* ♂♂♂½

Orpheus Cocteau's fascinating, innovative retelling of the Orpheus legend in a modern, though slightly askew, Parisian setting. Classic visual effects and poetic imagery. In French with English subtitles.
1949 95m/C Jean Marais, Francois Perier, Maria Casares; **Dir:** Jean Cocteau. **VHS, Beta, 8mm $16.95** *SNC, NOS, SUE* ♂♂½

Orpheus Descending Lust and hatred in small Southern town. Confusing, poorly paced, but interesting for Anderson as Elvis-style drifter and Redgrave as woman addicted to love. Made-for-cable version of the 1989 Broadway revival of Tennessee Williams play.
1991 117m/C Vanessa Redgrave, Kevin Anderson, Anne Twomey, Miriam Margolyes, Brad Sullivan; **Dir:** Peter Hall. **VHS $79.95** *TTC, FCT* ♂♂½

Osa Yet another post-nuke flick with the usual devastated landscape, leather-clad survivors, and hokey dialog. It's sometimes funny, despite everyone's effort to make it dramatic. The supposed plot centers on one woman's efforts to break up a man's monopoly on clean water.
1985 94m/C Kelly Lynch, Daniel Grimm, Phillip Vincent, Etienne Chicot, John Forristal, Pete Walker, David Hausman, Bill Moseley; **Dir:** Oleg Egorov. **VHS** *NEG* ♂½

Osaka Elegy A study of Japanese cultural rules when society condemns a woman for behavior that is acceptable for a man. In Japanese with English subtitles. Also known as "Woman of Osaka."
1936 71m/B *JP* Isuzu Yamada; **Dir:** Kenji Mizoguchi. **VHS $59.95** *SVS* ♂♂½

Oscar Unless you enjoy razzing bad acting, this film is not for you. Meant as a comeuppance for Hollywood, by Hollywood, this story of a star climbing the ladder of success and squashing fingers on every rung is too schlocky to succeed. The climax, when the scumbag star thinks he's won an Oscar and stands up, only to find it's for someone else, was based on Frank Capra's embarrassing experience with the same situation.
1966 (R) 119m/C Stephen Boyd, Elke Sommer, Jill St. John, Tony Bennett, Milton Berle, Eleanor Parker, Joseph Cotten, Edie Adams, Ernest Borgnine, Ed Begley Sr., Walter Brennan, Broderick Crawford, James Dunn, Peter Lawford, Merle Oberon, Bob Hope, Frank Sinatra; **Dir:** Russel Rouse. **VHS, Beta $69.98** *SUE* ♂½

Oscar The improbable casting of Stallone in a 1930s style crime farce (an attempt to change his image) is hard to imagine, and harder to believe. Stallone has little to do as he plays the straight man in this often ridiculous story of a crime boss who swears he'll go straight. Cameos aplenty, with Cur-

ry's the most notable. Based on a French play by Claude Magnier.
1991 (PG) 109m/C Sylvester Stallone, Ornella Muti, Peter Riegert, Tim Curry; **Dir:** John Landis. **VHS, LV $19.99** *TOU, IME* ♂

Oscar's Greatest Moments: 1971-1991 Collection of highlights from the Oscar ceremonies, featuring footage from the past 21 years. Proceeds from video sales benefit the Center for Motion Picture Study.
1992 110m/C Dir: Jeff Margolis; **Hosted:** Karl Malden. **VHS, LV, 8mm $19.95** *COL, BTV* ♂♂

Ossessione An adaptation of "The Postman Always Rings Twice," transferred to Fascist Italy, where a straggler and an innkeeper's wife murder her husband. Visconti's first feature, the film that initiated Italian neo-realism, and not released in the U.S. until 1975 due to a copyright dispute. In Italian with English subtitles.
1942 135m/B *IT* Massimo Girotti, Clara Calamai, Juan deLanda, Elio Marcuzzo; **Dir:** Luchino Visconti. **VHS, Beta $79.95** *FCT* ♂♂♂½

The Osterman Weekend Peckinpah was said to have disliked the story and the script in this, his last film, which could account for the convoluted and confusing end result. Adding to the problem is the traditional difficulty of adapting Robert Ludlum's complex psychological thrillers for the screen. The result: cast members seem to not quite "get it" as they portray a group of friends, one of whom has been convinced by the CIA that the others are all Soviet spies.
1983 (R) 102m/C Burt Lancaster, Rutger Hauer, Craig T. Nelson, Dennis Hopper, John Hurt, Chris Sarandon, Meg Foster, Helen Shaver; **Dir:** Sam Peckinpah. **VHS, Beta, LV $79.99** *HBO* ♂♂

Otello A performance of the Verdi opera taped at the Arena di Verona in Rome.
1982 135m/C Kiri Te Kanawa, Vladimir Atlantov, Piero Cappucilli. **VHS, Beta, LV $39.95** *HMV, HBO, PIA* ♂♂♂

Otello An uncommon film treat for opera fans, with a stellar performance by Domingo as the trouble Moor who murders his wife in a fit of jealous rage and later finds she was never unfaithful. Be prepared, however, for changes from the Shakespeare and Verdi stories, necessitated by the film adaptation. Highly acclaimed and awarded; in Italian with English subtitles.
1986 (PG) 123m/C *IT* Placido Domingo, Katia Ricciarelli, Justino Diaz; **Dir:** Franco Zeffirelli. **VHS, Beta $79.95** *HMV, MED, MVD* ♂♂♂½

Othello A silent version of Shakespeare's tragedy, featuring Emil Jannings as the tragic Moor. Titles are in English; with musical score.
1922 81m/B *GE* Emil Jannings, Lya de Putti, Werner Krauss. **VHS, Beta $29.95** *NOS, VYY, DVT* ♂♂½

Othello Shakespeare's keen understanding of jealousy in love results in perhaps his greatest triumph as a stage play and his prime example of the tragic hero, Othello. Made for television.
1982 120m/C *GB* Anthony Hopkins, Bob Hoskins, Penelope Wilton. **VHS, Beta, LV** *TLF* ♂♂

The Other Eerie, effective thriller adapted by Tyron from his supernatural novel. Twin brothers represent good and evil in a 1930s Connecticut farm town beset with gruesome murders and accidents. A good scare, with music by Jerry Goldsmith.

1972 (PG) 100m/C Martin Udvarnoky, Chris Udvarnoky, Uta Hagen, Diana Muldaur, Christopher Connelly, Victor French, John Ritter; **Dir:** Robert Mulligan. **VHS, Beta $59.98** *FOX* ♂♂♂

Other Hell Schlocky Italian-made chiller has the devil inhabiting a convent where he does his damnest to upset the nuns. Gross, but not scary—lots of cliche dark-hallway scenes and fright music.
1985 (R) 88m/C *IT* Carlo De Meyo, Francesca Carmeno; **Dir:** Stefan Oblowsky. **VHS, Beta $69.98** *LIV, VES* ♂

Other People's Money DeVito is "Larry the Liquidator," a corporate raider with a heart of stone and a penchant for doughnuts. When he sets his sights on a post-smokestack era, family-owned cable company, he gets a taste of love for the first time in his life. He and Miller, the daughter of the company president and also its legal advisor, court one another while sparring over the fate of the company. Unbelievably clipped ending mars otherwise enterprising comedy about the triumph of greed in corporate America. Based on the off-Broadway play by Jerry Sterner.
1991 (R) 101m/C Danny DeVito, Penelope Ann Miller, Dean Jones, Gregory Peck, Piper Laurie; **Dir:** Norman Jewison. **VHS, LV, 8mm $94.99** *WAR, CCB* ♂♂½

The Other Side of Midnight The dreary, depressingly shallow life story of a poor French girl, dumped by an American GI, who sleeps her way to acting stardom and a profitable marriage, then seeks revenge for the jilt. Based on Sidney Sheldon's novel, it's not even titillating—just a real downer.
1977 (R) 160m/C Susan Sarandon, Marie-France Pisier, John Beck, Raf Vallone, Clu Gulager, Sorrell Booke; **Dir:** Charles Jarrott. **VHS, Beta $59.98** *FOX* Woof!

The Other Side of the Mountain Tear-jerking true story of Olympic hopeful skier Jill Kinmont, paralyzed in a fall. Bridges helps her pull her life together. A sequel followed two years later. Based on the book "A Long Way Up" by E. G. Valens.
1975 (PG) 102m/C Marilyn Hassett, Beau Bridges, Dabney Coleman, John Garfield, Griffin Dunne; **Dir:** Larry Peerce. **VHS, Beta $59.95** *MCA* ♂♂

The Other Side of the Mountain, Part 2 Quadriplegic Jill Kinmont, paralyzed in a skiing accident that killed her hopes for the Olympics, overcomes depression and the death of the man who helped her to recover. In this chapter, she falls in love again and finds happiness. More tears are jerked.
1978 (PG) 99m/C Marilyn Hassett, Timothy Bottoms; **Dir:** Larry Peerce. **VHS, Beta $59.95** *MCA* ♂♂

The Other Side of Nashville Live performances, interviews and backstage footage combine to form a picture of the Nashville music scene. Over 40 songs are heard in renditions by country music's biggest stars.
1984 118m/C Johnny Cash, Kris Kristofferson, Bob Dylan, Kenny Rogers, Willie Nelson, Hank Williams, Gina Manes, Emmylou Harris, Carl Perkins. **VHS, Beta, LV $69.95** *MVD, MGM* ♂♂

Otis Day and the Knights: Otis My Man! A concert film starring the band featured in "Animal House," doing old rock 'n roll hits from the '50's and '60's, including "Louie, Louie," "Shout" and "Give Me Some Lovin'." In HiFi Digital Stereo Surround.

1986 54m/C VHS, Beta $19.95 *MVD, MCA* 🎞🎞

Otis Redding: Live in Monterey, 1967
The unedited footage of Redding's performance from the original Monterey Pop Festival. Songs include "Shake," "I've Been Loving You Too Long," "Satisfaction" and "Try a Little Tenderness."
1967 30m/C Otis Redding; *Dir:* D.A. Pennebaker. VHS, Beta $19.95 *MVD, HBO* 🎞🎞

Otis Redding: Ready, Steady, Go!
Otis Redding brings down the house with his soulful renditions of "Satisfaction" and "Respect" in this compilation of his performances in the "Ready Steady Go!" series.
1985 25m/B Otis Redding, Eric Burdon, Chris Farlow. VHS, Beta $19.95 *SVS, VTK, MVD* 🎞🎞

Otto Messmer and Felix the Cat
The creator of the cartoon superstar, now 84, talks about how Felix developed his remarkable personality. Excerpts from five vintage shorts demonstrate the cartoon cat's expressive skills.
1978 25m/C VHS, Beta *BFA* 🎞🎞

Oubliette
A peripheral version of Francois Villon, in what is a recently discovered and restored film, the earliest extant Chaney film.
1914 35m/B Lon Chaney Sr. VHS, Beta $39.50 *GVV* 🎞🎞

Our Daily Bread
Vidor's sequel to the 1928 "The Crowd." A young couple inherit a farm during the Depression and succeed in managing the land. A near-classic, with several sequences highly influenced by directors Alexander Dovshenko and Sergei Eisenstein. Director Vidor also co-scripted this film, risking bankruptcy to finance it.
1934 80m/B Karen Morley, Tom Keene, John Qualen, Barbara Pepper, Addison Richards; *Dir:* King Vidor. VHS, Beta, LV $19.95 *NOS, SUE, KRT* 🎞🎞🎞

Our Dancing Daughters
Flapper (Crawford on the brink of stardom) falls hard for millionaire who's forced into arranged marriage, but obliging little missus kicks bucket so Crawford can step in.
1928 98m/B Joan Crawford, Johnny Mack Brown, Nils Asther, Dorothy Sebastian, Anita Page; *Dir:* Harry Beaumont. VHS, Beta $19.98 *MGM, FCT* 🎞🎞½

Our Family Business
Made for television and meant as a series pilot, this generally plodding "Godfather"-type story is saved by good performances by Wanamaker and Milland.
1981 74m/C Sam Wanamaker, Vera Miles, Ray Milland, Ted Danson, Chip Mayer; *Dir:* Robert Collins. VHS $59.95 *IVE* 🎞🎞

Our Gang
Two silent "Our Gang" comedies, including "Thundering Fleas" and "Shivering Spooks."
1926 57m/B Joe Cobb, Farina Hoskins, Mary Kornman, Mickey Daniels, Scooter Lowry. VHS, Beta, 8mm $24.95 *VYY*

Our Gang Comedies
The five shorts on this cassette show the gang at their comedic best. They include "The Big Premiere," "Bubbling Troubles," "Clown Princes," "Don't Lie" and "Farm Hands."
194? 53m/B Robert Blake, Billie "Buckwheat" Thomas, George "Spanky" McFarland, Darla Hood, Carl "Alfalfa" Switzer, Tommy "Butch" Bond, Eugene "Porky" Lee, Shirley "Muggsy" Coates, Joe "Corky" Geil, Billy "Froggy" Lauglin. VHS, Beta $29.95 *MGM, BUR*

Our Gang Comedy Festival
A 20-year span of comedy from the Little Rascals, including rare silent features.
1988 45m/B Joe Cobb, Farina Hoskins, Mary Kornman, Mickey Daniels, Scooter Lowry. VHS *KID*

Our Gang Comedy Festival 2
More riotous good times with The Little Rascals, including silent footage.
1989 45m/B Joe Cobb, Farina Hoskins, Mary Kornman, Mickey Daniels, Scooter Lowry. VHS *KID*

Our Hospitality
One of Keaton's finest silent films, with all the elements in place. William McKay (Keaton) travels to the American South on a quaint train (a near-exact replica of the Stephenson Rocket), to claim an inheritance as the last survivor of his family. En route, a young woman traveler informs him that her family has had a long, deadly feud with his, and that they intend to kill him. McKay resolves to get the inheritance, depending on the Southern rule of hospitality to guests to save his life until he can make his escape. Watch for the river scene where, during filming, Keaton's own life was really in danger.
1923 74m/B Buster Keaton, Natalie Talmadge, Joe Keaton, Buster Keaton Jr., Kitty Bradbury, Joe Roberts; *Dir:* Buster Keaton. VHS, LV $49.95 *HBO* 🎞🎞🎞🎞

Our Little Girl
A precocious little tyke tries to reunite her estranged parents by running away to their favorite vacation spot. Sure to please Shirley Temple fans, despite a lackluster script.
1935 63m/B Shirley Temple, Joel McCrea, Rosemary Ames, Lyle Talbot, Erin O'Brien-Moore; *Dir:* John S. Robertson. VHS, Beta $19.98 *FOX* 🎞🎞

Our Man Flint
James Bond clone Derek Flint uses gadgets and his ingenuity to save the world from an evil organization, GALAXY, that seeks world domination through control of the weather. The plot moves quickly around many bikini-clad women, but still strains for effect. Spawned one sequel: "In Like Flint."
1966 107m/C James Coburn, Lee J. Cobb, Gila Golan, Edward Mulhare, Benson Fong, Shelby Grant, Sigrid Valdis, Gianna Serra, James Brolin, Helen Funai, Michael St. Clair; *Dir:* Daniel Mann. VHS, Beta, LV $19.98 *FOX* 🎞🎞

Our Miss Brooks
Quietly pleasing version of the television series. Miss Brooks pursues the "mother's boy" biology professor. The father of the child she begins tutoring appears taken with her. The professor takes notice.
1956 85m/B Eve Arden, Gale Gordon, Nick Adams, Richard Crenna, Don Porter; *Dir:* Al Lewis. VHS *WBV, WAR* 🎞🎞½

Our Relations
Confusion reigns when Stan and Ollie meet the twin brothers they never knew they had, a pair of happy-go-lucky sailors. Laurel directs the pair through madcap encounters with their twins' wives and the local underworld. One of the pair's

best efforts, though not well-remembered. Based on a story by W.W. Jacobs.
1936 94m/B Stan Laurel, Oliver Hardy, Alan Hale Jr., Sidney Toler, James Finlayson, Daphne Pollard; *Dir:* Harry Lachman. VHS, Beta $19.95 *MED, CCB, MLB* 🎞🎞½

Our Town
Small-town New England life in Grover's Corners in the early 1900s is celebrated in this well-performed and directed adaptation of the Pulitzer Prize-winning play by Thornton Wilder. Film debut of Martha Scott. Oscar nominations: Best Picture, Best Actress (Scott), Sound Recording, Best Score.
1940 90m/B Martha Scott, William Holden, Thomas Mitchell, Fay Bainter, Guy Kibbee, Beulah Bondi, Frank Craven; *Dir:* Sam Wood. National Board of Review Awards '40: 10 Best Films of the Year. VHS, Beta $9.95 *NEG, MRV, NOS* 🎞🎞🎞

Our Town
Television version of Thornton Wilder's classic play about everyday life in Grovers Corners, a small New England town at the turn of the century. It hews more closely to the style of the stage version than the earlier film version.
1977 120m/C Ned Beatty, Sada Thompson, Ronny Cox, Glynnis O'Connor, Robby Benson, Hal Holbrook, John Houseman; *Dir:* Franklin Schaffer. VHS, Beta, LV $74.95 *VTK, MAS* 🎞🎞🎞

Out
A drifter's travels throughout the U.S. from the sixties to the eighties. Successfully manages to satirize just about every conceivable situation but keeps from posturing by not taking itself too seriously. Adapted from an experimental novel by Ronald Sukenick.
1988 88m/C Peter Coyote, Danny Glover, O-Lan Shepard, Jim Haynie, Scott Beach, Semu Haute; *Dir:* Eli Hollander. VHS $79.98 *CGI* 🎞🎞½

Out of Africa
An epic film of the years spent by Danish authoress Isak Dinesen (her true name is Karen Blixen) on a Kenya coffee plantation. She moved to Africa to marry, and later fell in love with Denys Finch-Hatten, a British adventurer. Based on several books, including biographies of the two lovers. Some critics loved the scenery and music; others despised the acting and the script. Definitely no for those who love action.
1985 (PG) 161m/C Meryl Streep, Robert Redford, Klaus Maria Brandauer, Michael Kitchen, Malick Bowens, Michael Gough, Suzanna Hamilton; *Dir:* Sydney Pollack. Academy Awards '85: Best Adapted Screenplay, Best Art Direction/Set Decoration, Best Cinematography, Best Director (Pollack), Best Picture, Best Sound, Best Original Score; Golden Globe Awards '86: Best Film—Drama. VHS, Beta, LV $14.95 *MCA, BTV* 🎞🎞🎞

Out on Bail
A law-abiding citizen witnesses a murder, only to discover the crooked town council is behind the slaying.
1989 (R) 102m/C Robert Ginty, Kathy Shower, Tom Badal, Sydney Lassick; *Dir:* Gordon Hessler. VHS, Beta $89.95 *TWE* 🎞½

Out of the Blue
There's trouble in paradise for a married man when a shady lady passes out in his apartment. Thinking she is dead, he tries to get rid of the body. The antics with his neighbor and wife provide plenty of laughs.
1947 86m/B George Brent, Virginia Mayo, Carole Landis, Turhan Bey, Ann Dvorak; *Dir:* Leigh Jason. VHS, Beta, LV $19.95 *STE, NWV* 🎞🎞½

Out of the Blue
A harsh, violent portrait of a shattered family. When an imprisoned father's return fails to reunite this Woodstock-generation family, the troubled teenage daughter takes matters into her own

hands. "Easy Rider" star Hopper seems to have reconsidered the effects of the 1960s.
1982 (R) 94m/C Dennis Hopper, Linda Manz, Raymond Burr; *Dir:* Dennis Hopper. **VHS, Beta $59.95** *MED ♂♂♂*

Out of the Body A man is possessed by a spirit that likes to kill young, beautiful women. This makes him (and the viewer) uncomfortable.
1988 (R) 91m/C Mark Hembrow, Tessa Humphries; *Dir:* Brian Trenchard-Smith. **VHS, LV $79.95** *SVS ♂*

Out of Bounds An Iowa farmboy picks up the wrong bag at the Los Angeles airport, and is plunged into a world of crime, drugs, and murder. Plenty of action, but the fast pace can't hide huge holes in the script or the silliness of Hall playing a tough kid on the run from the law.
1986 (R) 93m/C Anthony Michael Hall, Jenny Wright, Jeff Kober; *Dir:* Richard Tuggle. **VHS, Beta, LV $79.95** *COL ♂1/2*

Out Cold Black comedy follows the misadventures of a butcher who believes he has accidentally frozen his business partner; the iced man's girlfriend, who really killed him; and the detective who tries to solves the crime. Too many poor frozen body jokes may leave the viewer cold.
1989 (R) 91m/C John Lithgow, Teri Garr, Randy Quaid, Bruce McGill; *Dir:* Malcolm Mowbray. **VHS, Beta, LV $89.98** *HBO ♂1/2*

Out of Control A plane full of rich teenagers crash on a secluded island and must battle a gang of vicious smugglers to survive. A teen sex theme keeps working its way in, leaving the viewer as confused as the actors and actresses appear to be.
1985 (R) 78m/C Betsy Russell, Martin Hewitt, Claudia Udy, Andrew J. Lederer; *Dir:* Allan Holzman. **VHS, Beta, LV $9.95** *STE ♂*

Out of the Dark The female employees of a telephone-sex service are stalked by a killer wearing a clown mask in this tongue-in-cheek thriller. A few laughs amid the slaughter. Look for Divine.
1988 (R) 98m/C Cameron Dye, Divine, Karen Black, Bud Cort, Lynn Danielson, Geoffrey Lewis, Paul Bartel, Tracy Walter, Silvania Gallardo, Starr Andreeff, Lainie Kazan, Tab Hunter; *Dir:* Michael Schroeder. **VHS, Beta, LV $89.95** *COL ♂1/2*

Out of the Darkness A television movie about the personal life of the New York detective who hunted and arrested sexual killer Son of Sam. Sheen shines in a tight suspenser.
1985 (R) 96m/C Martin Sheen, Hector Elizondo, Matt Clark; *Dir:* Judd Taylor. **VHS, Beta $19.95** *STE, VMK ♂♂♂*

Out for Justice A psycho Brooklyn hood goes on a murder spree, and homeboy turned lone-wolf cop Seagal races other police and the mob to get at him. Bloodthirsty and profane, it does try to depict N.Y.C.'s Italian-American community—but 90 percent of them are dead by the end so what's the point? Better yet, why does it open with a quote from Arthur Miller? Better still, what's Daffy Duck doing on this tape peddling Warner Brothers T-shirts to kid viewers?!
1991 (R) 91m/C Steven Seagal, William Forsythe, Jerry Orbach. **VHS, Beta, LV, 8mm $94.99** *WAR, PIA ♂1/2*

The Out-of-Towners A pair of Ohio rubes travels to New York City and along the way everything that could go wrong does. Lemmon's performance is excellent and Neil

Simon's script is, as usual, both wholesome and funny.
1970 (G) 98m/C Jack Lemmon, Sandy Dennis, Anne Meara, Sandy Baron, Billy Dee Williams; *Dir:* Arthur Hiller. **VHS, Beta, LV $14.95** *PAR ♂♂♂*

Out of Order A German-made film about people stuck in an office building's malevolent, free-thinking elevator. Dubbed.
1984 87m/C *GE* Renee Soutendijk, Goetz George, Wolfgang Kieling, Hannes Jaenicke; *Dir:* Carl Schenkel. **VHS, Beta $69.98** *LIV, VES ♂1/2*

Out of the Past A private detective gets caught in a complex web of love, murder, and money in this film noir classic. The plot is torturous but clear thanks to fine directing. Mitchum became an overnight star after this film, which was overlooked but now considered one of the best in its genre. Remade in 1984 as "Against All Odds."
1947 97m/B Robert Mitchum, Kirk Douglas, Jane Greer, Rhonda Fleming, Steve Brodie, Dickie Moore; *Dir:* Jacques Tourneur. **VHS, Beta, LV $19.98** *MED ♂♂♂1/2*

Out of the Rain A small town becomes a hot bed of deceit and lies due to the influence of drugs. A man and and a woman engage in a passionate struggle to free themselves from the town's grip.
1990 (R) 91m/C Bridget Fonda, Michael O'Keefe, John E. O'Keefe, John Seitz, Georgine Hall, Al Shannon; *Dir:* Gary Winick. **VHS, Beta, LV $9.99** *CCB, LIV ♂♂1/2*

Out of Season Mother and teenage daughter compete for the attentions of the mother's mysterious ex-lover. We never know who wins the man in this enigmatic drama set in an English village, and hints of incest make the story even murkier. Originally released as "Winter Rates," and had only a short run in the U.S.
1975 (R) 90m/C *GB* Cliff Robertson, Vanessa Redgrave, Susan George; *Dir:* Alan Bridges. **VHS, Beta $19.95** *UHV ♂♂*

Out of Sight, Out of Her Mind After witnessing her daughter burned alive, Alice is released from a mental institution. She tries to start a new life, but her daughter won't let her—she keeps appearing, crying out for help.
1989 (R) 94m/C Susan Blakely, Eddie Albert, Wings Hauser. **VHS, Beta, LV** *PSM, PAR ♂♂*

Outcast A young tough heads west to seize the family ranch wrongfully held by a conniving uncle. Of course he finds love as well, and enjoys gun and fistfights in this exciting though unspectacular western.
1954 90m/C John Derek, Jim Davis, Joan Davis; *Dir:* William Witney. **VHS $19.98** *REP ♂♂*

The Outcast Story of three brothers who live by the ax, love from the heart, hate with passion, and die violently. Trouble is they have to recite from a script written in crayon.
1984 86m/C Ben Dekker, Sandra Prinsloo. **VHS $19.95** *REP ♂1/2*

The Outcasts A Taiwanese film about young gay boys in an urban jungle who are given shelter and protection by an aging photographer. In Chinese with English subtitles.
1986 102m/C *Dir:* Yu Kan-Ping. **VHS, Beta $79.95** *FCT ♂♂♂*

Outcasts of the City The sort of negligible film filler that TV killed off. An American officer in post-WWII Germany romances a German girl, then gets blamed for the death of her Boche beau.

1958 61m/B Osa Massen, Robert Hutton, Maria Palmer, Nestor Paiva, John Hamilton, George Neise, Norbert Schiller, George Sanders; *Dir:* Boris L. Petroff. **VHS $19.95** *NOS ♂♂*

The Outing A group of high school kids sneaks into a museum at night to party, and get hunted down by a 3,000-year-old genie-in-a-lamp. The genie idea is original enough to make things interesting. Mediocre special effects and run-of-the-mill acting.
1987 (R) 87m/C Deborah Winters, James Huston; *Dir:* Tom Daley. **VHS, Beta $14.95** *IVE ♂♂*

Outland On a volcanic moon of Jupiter, miners begin suffering from spells of insanity. A single federal marshal begins an investigation that threatens the colony's survival. No more or less than a western in space, and the science is rather poor. Might make an interesting double feature with "High Noon," though.
1981 (R) 109m/C Sean Connery, Peter Boyle, Frances Sternhagen, James B. Sikking; *Dir:* Peter Hyams. **VHS, Beta, LV $19.98** *WAR, PIA ♂♂*

The Outlaw Hughes's variation on the saga of Billy the Kid, which spends more time on Billy's relationship with girlfriend Rio than the climactic showdown with Pat Garrett. The famous Russell vehicle isn't as steamy as it must have seemed to viewers of the day, but the brouhaha around it served to keep it on the shelf for six years. Also available colorized.
1943 123m/B Jane Russell, Jack Beutel, Walter Huston, Thomas Mitchell; *Dir:* Howard Hughes. **VHS, Beta, LV $8.95** *NEG, MRV, NOS ♂♂*

The Outlaw Bikers - Gang Wars The original road warriors take it to the streets as rival biker gangs square off.
1970 90m/C Clancy Syrko, Des Roberts, King John III, Linda Jackson; *Dir:* Lawrence Merrick. **VHS, Beta $59.95** *PSM ♂♂*

Outlaw Blues An ex-convict becomes a national folk hero when he sets out to reclaim his stolen hit song about prison life. St. James is charming in her first major movie role. A grab bag of action, drama, and tongue-in-cheek humor.
1977 (PG) 101m/C Peter Fonda, Susan St. James, Johnny Crawford, Michael Lerner, James Callahan; *Dir:* Richard T. Heffron. **VHS, Beta $59.95** *WAR ♂♂*

Outlaw Express After the annexing of California, a U.S. marshall is sent to stop a gang of outlaws who have been raiding Spanish landowners.
1938 57m/B Bob Baker, Cecilia Callejo, Leroy Mason, Don Barclay, Carleton Young; *Dir:* George Waggner. **VHS, Beta $24.95** *DVT ♂*

Outlaw Force A Vietnam veteran country singer tracks down the vicious rednecks that kidnapped his daughter and raped and killed his wife. Heavener did all the work in this low-budget, low-quality take-off on "Death Wish."
1987 (R) 95m/C David Heavener, Frank Stallone, Paul Smith, Robert Bjorklund, Devin Dunsworth; *Dir:* David Heavener. **VHS, Beta $19.98** *TWE ♂*

Outlaw Fury Another Shamrock Ellison epic, this time dealing with a coward proving himself by killing the bad guys.
1950 55m/B James Ellison, Russell Hayden. **VHS, Beta, LV** *WGE ♂1/2*

Outlaw Gang A marshal investigates a rash of rancher killings, leading to a conflict between Indians and local land and water companies.

1949 59m/B Donald (Don "Red") Barry, Robert Lowery, Betty Adams. **VHS, Beta, LV** WGE ✂✂½

Outlaw of Gor Once again, the mild professor is transported to Gor, where he has new, improved, bloody adventures. Lots of sword-play, but no magic. Sequel to "Gor."
1987 (PG) 89m/C Rebecca Ferratti, Urbano Barberini, Jack Palance, Donna Denton; *Dir:* John Cardos. **VHS, Beta** $19.98 WAR ✂✂½

The Outlaw and His Wife An early silent film about a fugitive linking up with a beautiful woman in the mountains. Powerful drama gave the Swedish film industry its breakthrough after WW I.
1917 73m/B SW Victor Sjostrom; *Dir:* Victor Sjostrom. **VHS, Beta, LV** $29.95 KIV, FCT ✂✂½

Outlaw Josey Wales Eastwood plays a farmer with a motive for revenge—his family was killed and for years he was betrayed and hunted. His desire to play the lone killer is, however, tempered by his need for family and friends. He kills plenty, but in the end finds happiness. Considered one of the last great Westerns, with many superb performances. Eastwood took over directorial chores during filming from Philip Kaufman, who co-scripted. Adapted from "Gone To Texas" by Forest Carter.
1976 (PG) 135m/C Clint Eastwood, Chief Dan George, Sondra Locke, Matt Clark, John Vernon, Bill McKinney, Sam Bottoms; *Dir:* Clint Eastwood, Philip Kaufman. **VHS, Beta, LV** $19.98 WAR, TLF ✂✂✂½

Outlaw Justice Worn oater plot has our hero infiltrating a gang of outlaws to bring them to justice and save the heroine.
1932 56m/C Jack Hoxie, Dorothy Gulliver, Donald Keith, Kermit Maynard, Charles King, Tom London; *Dir:* Armand Schaefer. **VHS, Beta** GPV ✂✂

Outlaw of the Plains Another of Crabbe's cowboy pictures in which he played hero Billy Carson, now bailing his trouble-prone sidekick out of a land swindle. Very plain indeed.
1946 56m/B Buster Crabbe, Al "Fuzzy" St. John, Patti McCarthy, Charles King, Karl Hackett, John Cason, Bud Osborne, Budd Buster, Charles "Slim" Whitaker; *Dir:* Sam Newfield. **VHS** $19.95 NOS ✂

Outlaw Riders Three outlaw bikers take it on the lam after committing a series of disastrous bank robberies.
1972 (PG) 84m/C Sonny West, Darlene Duralia, Bambi Allen, Bill Bonner; *Dir:* Tony Houston. **VHS, Beta** $59.95 WES ✂

Outlaw Roundup The Texas Rangers battle an outlaw gang. Guess who wins.
1944 51m/B Tex O'Brien, Jim Newill, Guy Wilkerson, Helen Chapman, Jack Ingram, I. Stanford Jolley; *Dir:* Harry Fraser. **VHS, Beta** $19.95 NOS, MRV, VDM ✂

Outlaw Rule An action-packed western adventure pitting outlaws against the law of the West.
1936 61m/B Reb Russell, Betty Mack, Yakima Canutt, Jack Rockwell, John McGuire, Alan Bridge; *Dir:* S. Roy Luby. **VHS, Beta** $19.95 NOS, VCN ✂½

The Outlaw Tamer Western adventure starring Lane Chandler.
1933 56m/B Lane Chandler, J.P. McGowan, Janet Morgan, George "Gabby" Hayes; *Dir:* J.P. McGowan. **VHS, Beta** VCN ✂

Outlaw Trail The Trail Blazers put an end to a counterfeiter's evil ways.

1944 53m/B Hoot Gibson, Bob Steele, Chief Thundercloud, Jennifer Holt, Cy Kendall; *Dir:* Robert Tansey. **VHS, Beta** $19.95 NOS, DVT ✂½

Outlaw Women A western town is run by a woman who won't let male outlaws in—until one wins her heart.
1952 76m/C Marie Windsor, Jackie Coogan, Carla Balenda; *Dir:* Sam Newfield. **VHS, Beta** $59.95 MON ✂

Outlaws This program presents the Outlaws playing some of their superb compositions.
1982 81m/C The Outlaws. **VHS, Beta** $29.95 SVS ✂

Outlaw's Paradise Cowboy hero McCoy plays not only a lawman out to catch mail thieves, but also the chief outlaw he's hunting! That's one way to keep the budget low.
1939 62m/B Tim McCoy, Benny Corbett, Joan Barclay, Ted Adams, Forrest Taylor, Bob Terry; *Dir:* Sam Newfield. **VHS** $15.95 NOS, LOO, RXM ✂½

Outlaws of the Range Action western featuring outlaws who terrorize the countryside.
1936 60m/B Bill Cody, Catherine Cotter, William McCall, Gordon Griffith; *Dir:* Al Herman. **VHS, Beta** VCN ✂½

Outlaws of Sonora A man transports money to a neighboring town while his outlaw double tries to steal it from him.
1938 58m/B Bob Livingston, Ray Corrigan, Max Terhune, Jack Mulhall, Otis Harlan; *Dir:* George Sherman. **VHS** $19.95 NOS, DVT ✂½

Outpost in Morocco A desert soldier is sent to quiet the restless natives and falls for the rebel leader's daughter. The good guys win, but the love interest is sacrificed. Glory before love boys, and damn the story.
1949 92m/B George Raft, Marie Windsor, Akim Tamiroff, John Litel, Eduard Franz; *Dir:* Robert Florey. **VHS** $19.95 NOS, MRV, MGM ✂½

Outrage An upstanding citizen guns down his daughter's killer and turns himself in to police. His ambitious lawyer fights a open-and-shut homicide case with both fair and foul means. Not a shootout like the cassette box suggests, but an okay courtroom drama attacking excesses of the legal system. From a novel by Henry Denker, made for TV by producer Irwin Allen.
1986 96m/C Robert Preston, Beau Bridges, Anthony Newley, Mel Ferrer, Burgess Meredith, Linda Purl, William Allen Young; *Dir:* Walter Grauman. **VHS** $89.95 VMK ✂✂½

Outrageous! An offbeat, low-budget comedy about the strange relationship between a gay female impersonator and his pregnant schizophrenic friend. The pair end up in New York, where they feel right at home. Russell's impersonations of female film stars earned him the best actor prize at the Berlin Film Festival.
1977 (R) 100m/C CA Craig Russell, Hollis McLaren, Richard Easley, Allan Moyle, Helen Shaver; *Dir:* Richard Benner. **VHS, Beta** $64.95 COL ✂✂✂

Outrageous Fortune Two would-be actresses - one prim and innocent and one wildly trampy - chase after the same two-timing boyfriend and get involved in a dangerous CIA plot surrounding a deadly bacteria. Tired plot. Mediocre acting and formula jokes, but somehow still funny. Disney's first foray into comedy for grown-ups.

1987 (R) 112m/C Shelley Long, Bette Midler, George Carlin, Peter Coyote; *Dir:* Arthur Hiller. **VHS, Beta, LV** $19.95 TOU ✂✂

Outside Chance A soapy, sanitized remake of Mimieux's film, "Jackson County Jail," wherein an innocent woman is persecuted in a small Southern jail. Made for television.
1978 92m/C Yvette Mimieux, Royce D. Applegate; *Dir:* Michael Miller. **VHS, Beta** $59.98 CHA ✂

Outside Chance of Maximillian Glick A sentimental Canadian comedy about a boy's dreams and his tradition-bound Jewish family.
1988 (G) 94m/C CA Noam Zylberman, Fairuza Balk, Saul Rubinek; *Dir:* Allan Goldstein. **VHS, Beta** $14.95 HMD, SOU ✂✂

Outside the Law Lon Chaney plays dual roles of the underworld hood in "Black Mike Sylva," and a Chinese servant in "Ah Wing." Silent.
1921 77m/B Lon Chaney Sr., Priscilla Dean, Ralph Lewis, Wheeler Oakman; *Dir:* Tod Browning. **VHS, Beta** $16.95 SNC, CCB, VYY ✂✂

The Outsiders Based on the popular S.E. Hinton book, the story of a teen gang from the wrong side of the tracks and their conflicts with society and each other. Melodramatic and over-done, but teen-agers still love the story. Good soundtrack and cast ripples with up and coming stars. Followed by a television series. Coppola adapted another Hinton novel the same year, "Rumble Fish."
1983 (PG) 91m/C C. Thomas Howell, Matt Dillon, Ralph Macchio, Patrick Swayze, Diane Lane, Tom Cruise, Emilio Estevez, Rob Lowe, Tom Waits, Leif Garrett; *Dir:* Francis Ford Coppola. **VHS, Beta, LV** $19.98 WAR ✂✂½

Outtakes A piecemeal compilation of comedy skits, concentrating on sex and genre parody.
1985 (R) 85m/C Forrest Tucker, Bobbi Wexler, Joleen Lutz; *Dir:* Jack M. Sell. **VHS, Beta** $79.95 SVS, SLP ✂½

The Oval Portrait An adaptation of the Edgar Allen Poe suspense tale about a Civil War-era maiden, her illicit love for a Confederate soldier and her eventual death.
1988 89m/C Giselle MacKenzie, Wanda Hendrix, Barry Coe; *Dir:* Regelio A. Gonzalez Jr. **VHS, Beta** $19.95 TAV ✂½

Over the Brooklyn Bridge A young Jewish man (Gould) must give up his Catholic girlfriend (Hemingway) in order to get his uncle (Caesar) to lend him the money he needs to buy a restaurant in Manhattan. Nowhere near as funny as it should be, and potentially offensive to boot.
1983 (R) 100m/C Elliott Gould, Sid Caesar, Shelley Winters, Margaux Hemingway, Carol Kane, Burt Young; *Dir:* Menahem Golan. **VHS, Beta** $79.95 MGM ✂✂

Over the Edge The music of Cheap Trick, The Cars, and The Ramones highlights this realistic tale of alienated suburban youth on the rampage. Dillon makes his screen debut in this updated, well-done "Rebel Without a Cause." Shelved for several years, the movie was finally released after Dillon made it big. Sleeper with excellent direction and dialogue.

▲ 523 • VIDEOHOUND'S GOLDEN MOVIE RETRIEVER

1979 (PG) 91m/C Michael Kramer, Matt Dillon, Pamela Ludwig, Vincent Spano; *Dir:* Jonathan Kaplan. **VHS, Beta** $59.95 *WAR* 🐾🐾🐾½

Over Indulgence A young woman witnesses a murder in an affluent society in British East Africa.
1987 (R) 95m/C Denholm Elliott, Holly Aird, Michael Byrne, Kathryn Pogson; *Dir:* Ross Devenish. **VHS, Beta** $79.98 *MCG* 🐾🐾

Over the Summer Strong drama about a troubled teen who escapes the city for a summer with her grandparents. No cliches here—the grandfather lusts after the granddaughter and eventually kills himself, and problems remain at the end. Intelligent and entertaining, though marred by needless nude scenes.
1985 97m/C Laura Hunt, Johnson West, Catherine Williams, David Romero; *Dir:* Teresa Sparks. **VHS, Beta** $69.98 *LIV, VES* 🐾🐾½

Over the Top The film that started a nationwide arm-wrestling craze. A slow-witted trucker decides the only way he can retain custody of his estranged son, as well as win the boy's respect, is by winning a big arm-wrestling competition. Stallone is an expert at grinding these movies out by now, and the kid (General Hospital's Mikey) is all right, but the end result is boredom, as it should be with an arm-wrestling epic.
1986 (PG) 94m/C Sylvester Stallone, Susan Blakely, Robert Loggia, David Mendenhall; *Dir:* Menahem Golan. **VHS, Beta, LV** $19.98 *WAR* 🐾½

Overboard A wealthy, spoiled woman falls off her yacht and into the arms of a low-class carpenter who picks her up and convinces her she is in fact his wife, and mother to his four brats. Just when she learns to like her life, the tables are turned again. Even though it's all been done before, you can't help but laugh at the screwy gags.
1987 (PG) 112m/C Goldie Hawn, Kurt Russell, Katherine Helmond, Roddy McDowall, Edward Herrmann; *Dir:* Garry Marshall. **VHS, Beta, LV** $19.98 *FOX* 🐾🐾

The Overcoat An adaptation of Nikolai Gogol's classic story about the dehumanizing life endured in 20th-century bureaucracy. All a menial civil servant wants is a new overcoat. When he finally gets one, this treasured article not only keeps him warm but makes him feel self-satisfied as well. English subtitles.
1959 93m/B *RU* Rolan Rykov; *Dir:* Alexei Batalov. National Board of Review Awards '65: 5 Best Foreign Films of the Year. **VHS, Beta** $59.95 *HTV, INV* 🐾🐾🐾½

Overdrawn at the Memory Bank In a futuristic tyranny, a romantic rebel becomes somehow fused with the spirit of Humphrey Bogart in the milieu of "Casablanca," and lives out a cliched version of the film character's adventures.
1983 84m/C Raul Julia, Linda Griffiths; *Dir:* Douglas Williams. **VHS, Beta** $19.95 *STE, NWV* 🐾🐾

Overexposed All the wackos love the sultry soap star Oxenburg; one is a killer. Eighty minutes and a bunch of bodies later the killer is unmasked. Plenty of suspense and some original gore. Based on the true stories of fans who stalk celebrities.
1990 (R) 80m/C Catherine Oxenberg, David Naughton, Jennifer Edwards, Karen Black; *Dir:* Larry Brand. **VHS** $79.98 *MGM* 🐾🐾

Overkill A detective and the vengeance-minded brother of a dead Japanese-American fight back against the controlling Yakuza in Little Tokyo.
1986 (R) 81m/C Steve Rally, John Nishio, Laura Burkett, Roy Summerset; *Dir:* Ulli Lommel. **Beta** $79.95 *VHV* 🐾

Overland Mail Fifteen episodes of the vintage serial filled with western action.
1941 225m/B Lon Chaney Jr., Helen Parrish. **VHS, Beta** $49.95 *NOS, VCN, MLB* 🐾

Overlanders The Japanese may invade, but rather than kill 1,000 head of cattle, these Aussies drive the huge herd across the continent, facing danger along the way. Beautifully photographed, featuring the "Australian Gary Cooper" and a stampede scene to challenge "Dances with Wolves."
1946 91m/B *AU* Chips Rafferty, Daphne Campbell, Jean Blue, John Nugent Hayward; *Dir:* Harry Watt. **VHS, Beta** $19.95 *NOS, DVT* 🐾🐾🐾

The Overlanders A group of nine men, two women and a boy challenge the forbidding Canadian Northwest circa 1862.
1979 76m/C **VHS, Beta** $59.95 *TWE* 🐾🐾

Overnight Sensation A modern adaptation of the Somerset Maugham story about a traditional marriage being tested by the wife's sudden literary success.
1983 30m/C Louise Fletcher, Robert Loggia, Shari Belafonte-Harper. **VHS, Beta, Special order formats** $225.00 *PYR* 🐾🐾

Overseas: Three Women with Man Trouble Three beautiful sisters in French colonial Algeria surround themselves in a life of luxury to avoid the incredible social changes going on around them. Each sister has a different perspective on her life, and the perspectives are graphically revealed. Lush photography set in the Mediterranean. In French with English subtitles.
1990 96m/C *FR* Nicole Garcia, Marianne Basler, Philippe Galland, Pierre Doris, Brigitte Rouan; *Dir:* Brigitte Rouan. Cannes Film Festival '90: Best Film. **VHS** $89.95 *FXL, ART, BTV* 🐾🐾🐾

Overthrow A nation torn apart by civil war experiences extreme bloodshed.
1982 90m/C John Phillip Law, Lewis Van Bergen, Roger Wilson; *Dir:* Larry Ludman. **VHS, Beta** $79.95 *IMP, HHE* 🐾½

Overture An episode of "The Persuaders." Brett and Danny are tricked into helping a retired judge. This adventure leads them to the Mediterranean and a gorgeous gal.
1975 52m/C *GB* Roger Moore, Tony Curtis. **VHS** $39.95 *SVS* 🐾½

Overture to Glory (Der Vilner Shtot Khazn) A Jewish cantor longs for the world of opera. He leaves his wife and son to fulfill his passion, but eventually he loses his voice and humbly returns home on Yom Kippur, the Jewish Day of Atonement. He learns that his son has passed away. Grief-stricken, he goes to the synagogue. There, the cantor regains his voice as he performs Kol Nidre in a passionate and melodious rendering. Yiddish with English subtitles.
1940 85m/B Helen Beverly, Florence Weiss. **VHS, Beta** $79.95 *ERG* 🐾🐾½

The Owl and the Pussycat Nerdy author gets the neighborhood hooker evicted from her apartment. She returns the favor, and the pair hit the street—and the sack—in Buck Henry's hilarious adaptation of the

Broadway play. Streisand's first non-singing role.
1970 (PG) 96m/C Barbra Streisand, George Segal, Robert Klein, Allen Garfield; *Dir:* Herbert Ross. **VHS, Beta, LV** $9.95 *GKK* 🐾🐾🐾

The Ox-Bow Incident A popular rancher is murdered, and a mob of angry townspeople can't wait for the sheriff to find the killers. They hang the young man, despite the protests of Fonda, a cowboy with a conscience. Excellent study of mob mentality with strong individual characterizations. A brilliant western based on a true story by Walter Van Tilburg Clark. Also see "Twelve Angry Men"—less tragic but just as moving.
1943 75m/B Henry Fonda, Harry Morgan, Dana Andrews, Anthony Quinn, Frank Conroy, Harry Davenport, Jane Darwell, William Eythe, Mary Beth Hughes; *Dir:* William A. Wellman. National Board of Review Awards '43: Best Director, Best Film. **VHS, Beta, LV** $19.98 *FOX* 🐾🐾🐾🐾

The Ox-Bow Incident Usually called "Lynch Mob," this is a television play based upon the famous book and 1943 film about mob ethics.
1955 45m/B Cameron Mitchell, Robert Wagner, Raymond Burr. **VHS, Beta** $19.98 *NOS, DVT* 🐾🐾🐾

Oxford Blues An American finagles his way into England's Oxford University and onto the rowing team in pursuit of the girl of his dreams. Beautiful scenery, but the plot is wafer-thin. Remake of "Yank At Oxford."
1984 (PG-13) 98m/C Rob Lowe, Ally Sheedy, Amanda Pays, Julian Sands, Michael Gough, Gail Strickland; *Dir:* Robert Boris. **VHS, Beta, LV** $19.98 *FOX* 🐾🐾

Pace That Kills An anti-drug cautionary drama from the Roaring 20s. One of many now marketed on cassette as campy fun. Beware, though, a little bit of this stuff goes a long way. A young man goes to the big city to find his missing sister and winds up hooked on heroin.
1928 87m/B Owen Gorin, Virginia Roye, Florence Turner. **VHS, Beta, 8mm** $29.95 *VYY* 🐾

Pacific Heights Young San Francisco couple takes on mammoth mortgage assuming tenants will write their ticket to the American dream, but psychopathic tenant Keaton moves in downstairs and redecorates. He won't pay the rent and he won't leave. Creepy psycho-thriller has lapses but builds to effective climax.
1990 (R) 100m/C Melanie Griffith, Matthew Modine, Michael Keaton, Mako, Nobu McCarthy, Laurie Metcalf, Carl Lumbly, Dorian Harewood, Luca Bercovici, Tippi Hedren, Sheila McCarthy, Dan Hedaya; *Dir:* John Schlesinger. **VHS, LV** $19.98 *FOX* 🐾🐾🐾

Pacific Inferno During World War II, American POWs endeavor to break out of a Japanese prison camp in the Philippines. Their goal is to prevent their captors from retrieving millions of dollars worth of sunken U.S. gold. Made for cable television.
1985 90m/C Jim Brown, Richard Jaeckel, Tim Brown; *Dir:* Rolf Bayer. **VHS, Beta** $59.95 *VIR, VCL* 🐾🐾

The Pack A group of dogs become wild when left on a resort island. A marine biologist leads the humans who fight to keep the dogs from using vacationers as chew toys. Fine production valves keep it from going to the dogs. Made with the approval of the American Humane Society who helped with the treatment of stage hands. Also known as "The Long, Dark Night."

1977 (R) 99m/C Joe Don Baker, Hope Alexander-Willis, R.G. Armstrong, Richard B. Shull; *Dir:* Robert Clouse. **VHS, Beta $19.98** *WAR* 🎞🎞½

Pack Up Your Troubles
Laurel and Hardy make good on a promise to help find the grandfather of a girl whose father was killed in WWI. All they know is the grandfather's last name though—Smith. Wholesome R and R.
1932 68m/B Stan Laurel, Oliver Hardy, James Finlayson, Jacquie Lyn; *Dir:* George Marshall. **VHS, Beta $14.98** *MED, FCT, MLB* 🎞🎞½

The Package
An espionage thriller about an army sergeant who loses the prisoner he escorts into the U.S. When he tries to track him down, he uncovers a military plot to start World War III. Hackman is believable in his role as the sergeant.
1989 (R) 108m/C Gene Hackman, Tommy Lee Jones, Joanna Cassidy, Dennis Franz, Pam Grier, John Heard; *Dir:* Andrew Davis. **VHS, Beta, LV $19.98** *ORI* 🎞🎞½

Packin' It In
A Los Angeles family is in for a rude awakening when they leave the city for a quieter life in the Oregon woods. There they encounter a band of survivalists with semi-comedic results.
1983 92m/C Richard Benjamin, Paula Prentiss, Molly Ringwald, Tony Roberts, Andrea Marcovicci; *Dir:* Judd Taylor. **VHS, Beta $24.95** *VCL* 🎞🎞

Paco
A young, South American boy heads for the city, meets his uncle who he discovers is the leader of a gang of youthful thieves. Predictable and sluggish.
1975 (G) 89m/C Jose Ferrer, Panchito Gomez, Allen Garfield, Pernell Roberts, Andre Marquis; *Dir:* Robert Vincent O'Neill. **VHS, Beta** *PGN, GHV* 🎞🎞

Paddy
A young Irish lad realizes his sexual potential by seducing every woman he can find. Freudians will have a field day.
1970 (PG) 97m/C *IR* Des Cave, Milo O'Shea, Darbnia Molloy, Peggy Cass; *Dir:* Daniel Haller. **VHS, Beta $59.98** *CHA* 🎞🎞

Padre Padrone
The much acclaimed adaptation of the Gavino Ledda autobiography about his youth in agrarian Sardinia with a brutal, tyrannical father. Eventually he overcomes his handicaps, breaks the destructive emotional ties to his father and successfully attends college. Highly regarded although low budget; in an Italian dialect (Sardinian) with English subtitles.
1977 113m/C *IT* Omero Antonutti, Saverio Marconi, Marcella Michelangeli, Fabrizio Forte, Salverio Marioni; *Dir:* Paolo Taviani, Vittorio Taviani. Cannes Film Festival '77: International Critics Prize, Best Film; New York Film Festival '77: International Critics Prize. **VHS, Beta $59.95** *COL* 🎞🎞🎞

Pagan Love Song
A dull musical which finds Keel in Tahiti taking over his uncle's coconut plantation and falling in love with Williams. Surprise! Williams performs one of her famous water ballets, which is the only saving grace in this lifeless movie.
1950 76m/C Esther Williams, Howard Keel, Minna Gombell, Rita Moreno; *Dir:* Robert Alton. **VHS** *MGM* 🎞🎞

Paid to Kill
A failing businessman hires a thug to kill him in order to leave his family insurance money, but when the business picks up, he can't contact the thug to cancel the contract. The actors' lack of talent is matched only by the characters' lack of motivation.

1954 71m/B *GB* Dane Clark, Paul Carpenter, Thea Gregory; *Dir:* Montgomery Tully. **VHS, Beta, LV** *WGE* 🎞🎞

Pain in the A—
A hit man helps a suicidal shirt salesman solve his marital problems in this black comedy that was later Americanized as "Buddy, Buddy." Not for all tastes, but has acquired a reputation for its dark wit. With English subtitles.
1977 (PG) 90m/C *FR IT* Lino Ventura, Jacques Brel, Caroline Cellier; *Dir:* Edouard Molinaro. **VHS, Beta $59.95** *COL* 🎞🎞🎞

Paint it Black
A violent/steamy mystery and action flick from the director of "River's Edge," in which a young man from the silverspoon set develops a fascination for an artist.
1989 (R) 101m/C Sally Kirkland, Rick Rossovich, Martin Landau, Doug Savant, Peter Frechette, Julie Carmen, Jason Bernhard, Monique Van De Ven; *Dir:* Tim Hunter. **VHS, Beta, LV $89.98** *LIV, VES* 🎞🎞½

Paint Your Wagon
A big-budget western musical-comedy about a gold mining boom town, and two prospectors sharing the same Mormon wife complete with a classic Lerner and Lowe score. Marvin chews up the sagebrush and Eastwood attempts to sing, although Seberg was mercifully dubbed. Overlong, with patches of interest, pretty songs, and plenty of panoramic scenery. Adapted by Paddy Chayeusky from L&L play. Nelson Riddle was nominated for an Oscar for musical direction.
1970 (PG) 164m/C Lee Marvin, Clint Eastwood, Jean Seberg, Harve Presnell; *Dir:* Joshua Logan. **VHS, Beta, LV $29.95** *PAR* 🎞🎞½

Painted Desert
Gable's first film role of any consequence came in this early sound western. Gable plays a villain opposite "good guy" William Boyd.
1931 80m/B William Boyd, Helen Twelvetrees, George O'Brien, Clark Gable, William Farnum; *Dir:* Howard Higgin. **VHS, Beta $19.95** *NOS, MRV, HHT* 🎞🎞

The Painted Hills
Lassie outsmarts crooked miners and rescues her friends. Surprise, Surprise! Overly sentimental, but action packed and beautifully shot. Lassie's last outing with MGM.
1966 90m/C Paul Kelly, Bruce Cowling, Gary Gray, Art Smith, Ann Doran, Lassie; *Dir:* Harold F. Kress. **VHS, Beta $9.95** *LGC, BPG, AMV* 🎞🎞½

The Painted Stallion
This 12-chapter serial features a mysterious figure on a painted stallion who attempts to maintain peace between Mexicans and Indians.
1937 212m/B Ray Corrigan, Hoot Gibson, Duncan Renaldo, Leroy Mason, Yakima Canutt; *Dir:* William Witney, Ray Taylor. **VHS, LV $29.98** *NOS, VDM, REP* 🎞🎞½

The Painted Veil
Adaptation of a Somerset Maugham story. Garbo, once again the disillusioned wife turning to the affections of another man, is magnificent, almost eclipsing the weak script. Lost money at the box office, but for Garbo fans, an absolute must.
1934 83m/B Greta Garbo, Herbert Marshall, George Brent, Warner Oland, Jean Hersholt; *Dir:* Richard Boleslawski. **VHS, Beta $19.98** *MGM, FCT* 🎞🎞½

Painting with Light
Four short experimental films by four major American painters - Larry Rivers, David Hockney, Jennifer Bartlett and Howard Hodgkin - in which they endeavor to "paint" with the technology of video and the broadcasting Quantel "Paint Box."

1987 60m/C VHS, Beta, LV $9.95 *PAV, CRY* 🎞🎞½

Paisan
Six episodic tales of life in Italy, several featuring Allied soldiers and nurses during World War II. One of the stories tells of a man who tries to develop a relationship without being able to speak Italian. Another focuses on a young street robber who is confronted by one of his victims. Strong stories that covers a wide range of emotions. Screenplay by Rossellini and Federico Fellini. In Italian with English subtitles.
1946 90m/B *IT* Maria Michi, Carmela Sazio, Gar Moore, William Tubbs, Harriet White, Robert Van Loon, Dale Edmonds, Carla Pisacane, Dots Johnson; *Dir:* Roberto Rossellini. National Board of Review Awards '48: Best Director, Best Picture; New York Film Critics Awards '48: Best Foreign Film; Venice Film Festival '46: Special Mention. **VHS, Beta, 8mm $24.95** *NOS, HHT, CAB* 🎞🎞🎞

Pajama Game
A spritely musical about the striking workers of a pajama factory and their plucky negotiator, who falls in love with the boss. Based on the hit Broadway musical, which was based on Richard Bissell's book "Seven and a Half Cents" and adapted for the screen by Bisell abd George Abbott. Songs include "Hey, There," "Steam Heat," and "Hernando's Hideaway." Bob Fosse choreographed the dance numbers.
1957 101m/C Doris Day, John Raitt, Eddie Foy Jr., Reta Shaw, Carol Haney; *Dir:* Stanley Donen, George Abbott. **VHS, Beta, LV $19.98** *WAR* 🎞🎞🎞½

Pajama Party
Beach blanket bingo lost in space. Martian Kirk comes to Earth, scouting for a possible invasion. Ho falls into the lap of Annette and decides instead to save the planet. All the usual beach movie faces are present, including cameos by Avalon, Rickles, and Keaton.
1964 82m/C Tommy Kirk, Annette Funicello, Elsa Lanchester, Harvey Lembeck, Jesse White, Jody McCrea, Donna Loren, Susan Hart, Bobbi Shaw, Cheryl Sweeten, Luree Holmes, Candy Johnson, Buster Keaton, Dorothy Lamour, Toni Basil, Frankie Avalon, Don Rickles; *Dir:* Don Weis. **VHS, LV $19.95** *TRY* 🎞🎞

Pajama Tops
A stage production of the classic French bedroom farce, taped live at the Music Hall Theatre in Toronto, Canada. Includes all the trials of marriage including adultery and deceit.
1983 105m/C Robert Klein, Susan George, Pia Zadora. **VHS $39.95** *USA* 🎞🎞½

Pal Joey
Musical comedy about an opportunistic singer who courts a wealthy socialite in hopes that she will finance his nightclub. His play results in comedic complications. Stellar choreography, fine direction, and beautiful costumes complement performances headed by Hayworth and Sinatra. Oscar overlooked his pal Joey when awards were handed out. Songs include some of Rodgers' and Hart's best—"The Lady is a Tramp," "My Funny Valentine," and "Bewitched, Bothered and Bewildered." Based on John O'Hara's book and play.
1957 109m/C Frank Sinatra, Rita Hayworth, Kim Novak; *Dir:* George Sidney. **VHS, Beta, LV $19.95** *COL, PIA* 🎞🎞🎞

Pale Blood
A serial killer in Los Angeles is leaving his victims drained of blood. Could it be that a vampire is stalking the modern American metropolis, or is this merely the workings of a bloodthirsty psychopath?

1991 (R) 93m/C George Chakiris, Wings Hauser, Pamela Ludwig, Diana Frank, Darcy Demoss; *Dir:* VV Dachin Hsu. Academy of Science Fiction, Fantasy, & Horror Film '91: Golden Scroll. **VHS, LV $79.95** *SVS* 🐾½

Pale Rider A mysterious nameless stranger rides into a small California gold rush town to find himself in the middle of a feud between a mining syndicate and a group of independent prospectors. Christ-like Eastwood evokes comparisons to "Shane." A classical western theme treated well complemented by excellent photography and a rock-solid cast.
1985 (R) 116m/C Clint Eastwood, Michael Moriarty, Carrie Snodgress, Sydney Penny, Richard Dysart, Richard Kiel, Christopher Penn, John Russell, Charles Hallahan; *Dir:* Clint Eastwood. **VHS, Beta, LV, 8mm $19.98** *WAR, TLF* 🐾🐾🐾

The Paleface A cowardly dentist becomes a gunslinging hero when Calamity Jane starts aiming for him. A rip-roarin' good time as the conventions of the Old West are turned upside down. Includes the Oscar-winning song "Buttons and Bows." The 1952 sequel is "Son of Paleface." Remade in 1968 as "The Shakiest Gun in the West."
1948 91m/C Jane Russell, Bob Hope, Robert Armstrong, Iris Adrian, Robert Watson; *Dir:* Norman Z. MacLeod. Academy Awards '48: Best Song ("Buttons and Bows"). **VHS, Beta, LV $14.95** *MCA* 🐾🐾½

The Palermo Connection Not the action-thriller the cassette box claims, but a cynical study of a crusading N.Y.C. mayoral candidate who honeymoons in Mafia-haunted Sicily and learns that the War On Crime has already been lost. Gore Vidal helped script this remote adaptation of "To Forget Palermo" by Edmonde Charles-Roux.
1991 (R) 100m/C James Belushi, Mimi Rogers, Joss Ackland, Philippe Noiret, Vittorio Gassman, Caroline Rosi; *Dir:* Francesco Rosi. **VHS, LV $89.95** *LIV* 🐾🐾

Palm Beach The lives of two petty thieves, a runaway, and a private detective intertwine at Palm Beach Down Under. Best appreciated by fans of things Aussie.
1979 90m/C *AU* Bryan Brown, Nat Young, Ken Brown, Amanda Berry; *Dir:* Albie Thomas. **VHS, Beta $19.95** *AHV* 🐾🐾

The Palm Beach Story Colbert and McCrea play a married couple who have everything—except money. The wife decides to divorce her current husband, marry a rich man, and finance her former husband's ambitions. She decides the best place to husband-hunt is Palm Beach. A Sturges classic screwball comedy.
1942 88m/B Claudette Colbert, Joel McCrea, Mary Astor, Rudy Vallee, William Demarest, Franklin Pangborn; *Dir:* Preston Sturges. **VHS, Beta, LV $29.95** *MCA* 🐾🐾🐾

Palm Springs Weekend A busload of love-hungry kids head south and get involved in routine hijinks. Actually shot in Palm Springs with adove average performances by a handful of stars.
1963 100m/C Troy Donahue, Ty Hardin, Connie Stevens, Stefanie Powers, Robert Conrad, Jack Weston, Andrew Duggan; *Dir:* Norman Taurog. **VHS, Beta, LV $19.98** *WAR* 🐾🐾½

Palm Tree Beach A continuous picture of a calm tropical beach viewed from under a shady palm tree to create a relaxed background.
1978 30m/C VHS, Beta *PCC* 🐾🐾½

Palooka Based on the comic strip, this film portrays a fast-talking boxing manager and his goofy, lovable protege. The young scrapper fights James Cagney's little brother. Durante sings his classic tune, "Inka-Dinka-Doo" and packs a punch in the lead. A two-fisted comedy with a witty dialogue and fine direction. Not a part of the Palooka series of the 1940's.
1934 (G) 86m/B Jimmy Durante, Stuart Erwin, Lupe Velez, Robert Armstrong, Thelma Todd; *Dir:* Ben Stoloff. **VHS, Beta $19.95** *NOS, QNE, VYY* 🐾🐾½

Pals Old friends stumble across $3 million in cash and learn the predictable lesson that money can't buy happiness. A terribly trite TV-movie with an exceptional cast.
1987 90m/C Don Ameche, George C. Scott, Sylvia Sidney, Susan Rinell, James Greene; *Dir:* Lou Antonio. **VHS $89.95** *VMK* 🐾🐾

Pals of the Range A framed ranch owner breaks out of jail in order to catch the real cattle thieves. Ho hum in the Old West.
1935 55m/B Rex Lease. **VHS, Beta $19.95** *NOS, WGE* 🐾

Panama Hattie Screen adaptation of Cole Porter's delightful Broadway musical. Unfortunately, something (like a plot) was lost in transition. Southern runs a saloon for our boys down in Panama. Among the musical numbers and vaudevillian acts some spies show up. Several screenwriters and directors, including Vincente Minnelli, worked uncredited on this picture, to no avail. Tunes include "It Was Just One of Those Things," "Fresh as a Daisy," "I've Still Got My Health," "Let's Be Buddies" and more. Horne's second screen appearance.
1942 79m/B Ann Sothern, Dan Dailey, Red Skelton, Virginia O'Brien, Rags Ragland, Alan Mowbray, Ben Blue, Carl Esmond, Lena Horne, Roger Moore; *Dir:* Norman Z. McLeod. **$19.98** *MGM, CCB* 🐾🐾

Panama Lady Ball stars as the sexy, sultry "Panama Lady" in this old-fashioned romance. She gets involved in some shady business south of the border. A lackluster remake of "Panama Flo."
1939 65m/B Lucille Ball, Evelyn Brent. **VHS, Beta $19.95** *VID* 🐾½

Panama Menace An agent heads to Panama to thwart spies who are after a special paint that makes things invisible. Poor effort all around, in spite of Beaumont's presence. Also known as "South of Panama."
1941 68m/B Roger Pryor, Virginia Vale, Lionel Royce, Lucien Prival, Duncan Renaldo, Lester Dorr, Hugh Beaumont; *Dir:* Jean Yarborough. **VHS, Beta $16.95** *SNC* 🐾

Panama Patrol Just as they are about to be married, two Army officers are called to duty in Panama. It seems that the Chinese have infiltrated the area with spies and something's got to be done. Too much attention to minor parts of the plot slows the story occasionally, but performances and photography compensate for some of the sluggishness.
1939 67m/B Leon Ames, Charlotte Wynters, Weldon Heyburn, Adrienne Ames, Abner Biberman, Hugh McArthur, Donald (Don "Red") Barry; *Dir:* Charles Lamont. **VHS, Beta $16.95** *SNC* 🐾🐾½

Panamint's Bad Man Ballaw is a poor man's Gary Cooper fighting for justice against stagecoach robbers. Almost as much flavor as tumbleweed.

1938 59m/B Smith Ballew, Evelyn Daw, Noah Beery; *Dir:* Ray Taylor. **VHS, Beta $19.95** *NOS, VDM, DVT* 🐾

Pancho Villa Savalas has the lead in this fictional account of the famous Mexican, who leads his men in a raid on an American fort after being hoodwinked in an arms deal. Connors tries to hold the fort against him. The finale, in which two trains crash head on, is the most exciting event in the whole darn movie.
1972 (PG) 92m/C *SP* Telly Savalas, Clint Walker, Anne Francis, Chuck Connors, Angel Del Pozo, Luis Davila; *Dir:* Eugenio Martin. **VHS $42.95** *PGN* 🐾½

Pancho Villa Returns It's 1913, and noble Mexican General Pancho Villa leads his merry men against the assassins of President Madera. This Mexican production (filmed in English) paints a partisan portrait of title character as a good-hearted revolutionary folk hero, not the roving bandit later notorious north of the border.
1950 95m/B *MX* Leo Carrillo, Esther Fernandez, Jeanette Comber, Rodolfa Acosta; *W/Dir:* Miguel Contreras Torres. **VHS $15.95** *NOS, LOO* 🐾🐾

Panda and the Magic Serpent An animated film from Japan for children, featuring a woman/ enchantress who must search for a way to bring her lover back to life, with the help of a mob of animated animals.
1961 74m/C *JP Dir:* Robert Tafur; *Nar:* Marvin Miller. **VHS, Beta $24.95** *VYY* 🐾🐾

Panda's Adventures An animated tale of a Panda prince who is exiled from his kingdom when he fails a test of courage. Through his adventures, he learns that true heroism and real courage come from the heart.
1984 60m/C VHS, Beta $19.95 *COL* 🐾🐾

Pandemonium A spoof of teen slasher films involving murder at Bambi's Cheerleading School. Seems that nationwide, cheerleaders have been brutally eliminated from their squads, leaving Bambi's as the last resort for the terminally perky. Smothers is interesting and more intelligent than his TV persona (although not by much) as the Mountie hero, assisted by Paul Reubens in a pre-Pee-Wee role. Kane steals the show as Candy, a pleasant lass with hyperkinetic powers. Original title: "Thursday the 12th."
1982 (PG) 82m/C *CA* Tom Smothers, Carol Kane, Miles Chapin, Paul Reubens, Judge Reinhold, Tab Hunter, Marc McClure, Donald O'Connor, Eve Arden, Eileen Brennan, Edie McClurg; *Dir:* Alfred Sole. **VHS, Beta $79.95** *MGM* 🐾🐾½

Pandora's Box This silent classic marked the end of the German Expressionist era, and established Brooks as a major screen presence. She plays a tempestuous prostitute who is eventually killed by Jack the Ripper. Silent with orchestral score.
1928 110m/B *GE* Louise Brooks, Fritz Kortner, Francis Lederer; *Dir:* G. W. Pabst. **VHS, Beta $29.98** *SUE, MRV, VDM* 🐾🐾🐾🐾

Panic A scientist terrorizes a small town when he becomes hideously deformed by one of his bacteria experiments. He should have known better.
1983 90m/C David Warbeck, Janet Agren. **VHS, Beta $59.95** *MPI* 🐾½

Panic on the 5:22 Wealthy commuters are kidnapped and held hostage by terrorist hoodlums in a suburban train club car. Standard made for television vehicle.

1974 78m/C Lynda Day George, Laurence Luckinbill, Ina Balin, Bernie Casey; *Dir:* Harvey Hart. **VHS, Beta $14.95** *WOV* ✯✯

Panic Button Mel Brooks took the plot from this film and made "The Producers," which was much better. Italian gangsters produce a television show and stack the deck so that it will fail. Unbeknownst to them, the star has figured out what they are doing and works to make it a success. Shot in Italy, mainly in Venice and Rome.
1962 90m/B *IT* Maurice Chevalier, Eleanor Parker, Jayne Mansfield, Michael Connors, Akim Tamiroff; *Dir:* George Sherman. **VHS $29.95** *SOU* ✯½

Panic in Echo Park A doctor races against time to find the cause of an epidemic that is threatening the health of a city. Good performances compensate for a predictable plot. Made for television.
1977 78m/C Dorian Harewood, Robin Gammell, Catlin Adams, Ramon Bieri, Movita; *Dir:* John Llewellyn Moxey. **VHS $29.95** *USA* ✯✯

Panic in Needle Park Drugs become an obsession for a young girl who goes to New York for an abortion. Her new boyfriend is imprisoned for robbery in order to support both their habits. She resorts to prostitution to continue her drug habit, and trouble occurs when her boyfriend realizes she was instrumental in his being sent to jail. Strikes a vein in presenting an uncompromising look at drug use. May be too much of a depressant for some. Pacino's first starring role.
1975 (R) 90m/C Al Pacino, Kitty Winn, Alan Vint, Richard Bright, Kiel Martin, Warren Finnerty, Raul Julia, Paul Sorvino; *Dir:* Jerry Schatzberg. **VHS** *FOX* ✯✯✯

Panic Station Two guys get lonely at remote satellite relay station. Originally titled "The Plains of Heaven."
1982 90m/C *AU* Richard Moir, Reg Evans, Gerard Kennedy; *Dir:* Ian Pringle. **VHS $19.95** *ACA* ✯✯

Panic in the Streets The Black Death threatens New Orleans in this intense tale. When a body is found on the water front, a doctor (Widmark) is called upon for a diagnosis. The carrier proves to be deadly in more ways than one. Fine performances, taut direction (this was one of Kazan's favorite movies). Filmed on location in New Orleans.
1950 96m/B Richard Widmark, Jack Palance, Barbara Bel Geddes, Paul Douglas, Zero Mostel; *Dir:* Elia Kazan. Academy Awards '50: Best Story. **VHS $39.98** *FOX, FCT* ✯✯✯

Panic in the Year Zero! Milland and family leave Los Angeles for a fishing trip just as the city is hit by a nuclear bomb. Continuing out into the wilderness for safety, the family now must try to survive as their world crumbles around them. Generally considered the best of Milland's five directorial efforts.
1962 92m/B Ray Milland, Jean Hagen, Frankie Avalon, Mary Mitchell, Joan Freeman, Richard Garland, Rex Holman; *Dir:* Ray Milland. **VHS $18.00** *FRG* ✯✯½

Panique A study of mob psychology in slums of post-World War II Paris. Two lovers frame a stranger for murder. Dark and tautly paced thriller taken form the novel by Georges Simenon. In French with English subtitles.
1947 87m/B *FR* Michel Simon, Viviane Romance; *Dir:* Julien Duvivier. **VHS, Beta $24.95** *NOS, HHT, CAB* ✯✯✯

Pantaloons A man who fancies himself to be irresistible chases women for sport. A quaint and lively period film.

1957 93m/C *FR* Fernandel, Carmen Sevilla, Fernando Rey; *Dir:* John Berry. **VHS** *VYY* ✯✯

Panther Squad Litter of sex kittens led by Danning get into major scraps to save world.
1984 77m/C Sybil Danning. **VHS $79.95** *LTG* ✯

The Panther's Claw Murder befalls an opera troupe, but a sleuth with the memorable name of Thatcher Colt is on the case. A quick-moving mystery quickie that delivers on its own modest terms.
1942 72m/B Sidney Blackmer, Byron Foulger, Rick Vallin, Herbert Rawlinson; *Dir:* William Beaudine. **VHS $15.95** *NOS, LOO* ✯✯

Papageno Set to the Bird Catcher theme from Mozart's Magic Flute, Lotte Reiniger's film uses the tradition of Eastern Shadow theatre to tell how the bird man, Papageno, reaches his Papagena.
1960 11m/C **VHS, Beta** *TEX* ✯✯

Papa's Delicate Condition Based on the autobiographical writings of silent screen star Corinne Griffith. Gleason is Papa whose "delicate condition" is a result of his drinking. His antics provide a constant headache to his family. A paean to turn-of-the-century family life. No I.D.s required as the performances are enjoyable for the whole family. Features the Academy Award-winning song "Call Me Irresponsible."
1963 98m/C Jackie Gleason, Glynis Johns, Charlie Ruggles, Laurel Goodwin, Elisha Cook Jr., Murray Hamilton. Academy Awards '63: Best Song ("Call Me Irresponsible"). **VHS, Beta $14.95** *KRT* ✯✯✯

The Paper Chase Students at Harvard Law School suffer and struggle through their first year. A realistic, sometimes acidly humorous look at Ivy League ambitions, with Houseman stealing the show as the tough professor. Wonderful adaptation of the John Jay Osborn novel which later became the basis for the acclaimed television series. Oscar nominations for Adapted Screenplay and Sound, plus Houseman's win.
1973 (PG) 111m/C Timothy Bottoms, Lindsay Wagner, John Houseman, Graham Beckel, Edward Herrmann, James Naughton, Craig Richard Nelson, Bob Lydiard; *Dir:* James Bridges. Academy Awards '73: Best Supporting Actor (Houseman); National Board of Review Awards '73: Best Supporting Actor (Houseman). **VHS, Beta, LV $59.98** *FOX, BTV* ✯✯✯

Paper Lion A comedy "documentary" about bestselling writer George Plimpton's tryout game as quarterback with the Detroit Lions. Film debut of Alan Alda. Helped Karras make the transition from the gridiron to the silver screen. Moves into field goal range but doesn't quite score.
1968 107m/C Alan Alda, Lauren Hutton, Alex Karras, Ann Turkel, John Gordy, Roger Brown, "Sugar Ray" Robinson, Roy Scheider, David Doyle; *Dir:* Alex March. **VHS, Beta $14.95** *WKV* ✯✯½

Paper Man A group of college students create a fictitious person in a computer for a credit card scam. The scheme snowballs, resulting in murder and hints of possible artificial intelligence. But it's just standard network TV fare, with a creepy performance by Stockwell as a computer whiz.
1971 90m/C Dean Stockwell, Stefanie Powers, James Stacy, Elliot Street, Tina Chen, James Olson, Ross Elliott; *Dir:* Walter Grauman. **VHS $39.95** *XVC* ✯✯

Paper Mask When a promising young doctor is killed in an auto accident, an unscrupulous, psychotic porter assumes his identity. He uses said identity to, among other things, initiate an affair with a sultry co-worker. How long will this madman play his unholy game, and at what cost?
1991 (R) 105m/C *GB* Paul McGann, Amanda Donohoe, Frederick Treves, Barbara Leigh-Hunt, Jimmy Yuill; *Dir:* Christopher Morahan. **VHS $89.95** *ACA* ✯✯½

Paper Moon Award-winning story set in depression-era Kansas with Ryan O'Neal as a Bible-wielding con who meets up with a nine-year old orphan. During their travels together, he discovers that the orphan (his daughter, Tatum) is better at "his" game than he is. Irresistible chemistry between the O'Neals, leading to Tatum's Oscar win (she is the youngest actor ever to take home a statue). Cinematically picturesque and cynical enough to keep overt sentimentalism at bay. Based on Joe David Brown's novel, "Addie Pray."
1973 (PG) 102m/B Ryan O'Neal, Tatum O'Neal, Madeline Kahn, John Hillerman, Randy Quaid; *Dir:* Peter Bogdanovich. Academy Awards '73: Best Supporting Actress (O'Neal); National Board of Review Awards '73: 10 Best Films of the Year. **VHS, Beta, LV $49.95** *PAR, BTV* ✯✯✯½

Paper Tiger Niven plays an imaginative English tutor who fabricates fantastic yarns fictionalizing his past in order to impress his student, the son of the Japanese ambassador to a Southeast Asian country. Poorly executed karate, misdirection, and a simplistic storyline work against it.
1974 104m/C *GB* David Niven, Toshiro Mifune, Eiko Ando, Hardy Kruger; *Dir:* Ken Annakin. **VHS, Beta** *SUE, OM* ✯✯

A Paper Wedding A middle-aged literature professor with a dead-end career and equally dead-end romance with a married man is persuaded by her lawyer sister to marry a Chilean political refugee to avoid his deportation. Of course, they must fool an immigration official and their fake marriage does turn into romance but this is no light-hearted "Green Card." Bujold shines. In French and Spanish with English subtitles.
1989 90m/C *CA* Genevieve Bujold, Manuel Aranguiz; *Dir:* Michel Brault. **VHS $79.95** *CTL* ✯✯✯

Paperback Hero A hot-shot hockey player turns to crime when his team loses its financial backing. Choopy storyline could use a Zamboni, while Dullea should be penalized for occasionally losing his rustic accent. Still the supporting actors skate through their roles and, for periods, it works. A slapshot that ultimately clangs off the goal post.
1973 94m/C *CA* Keir Dullea, Elizabeth Ashley, John Beck, Dayle Haddon; *Dir:* Peter Pearson. **VHS $79.95** *CGH* ✯✯

Paperhouse An odd fantasy about a young girl plagued by recurring dreams that begin to influence real life. Very obtuse, British-minded film with "Twilight Zone" feel that manages to capture genuine dreaminess. Not for all audiences, but intriguing nonetheless.
1989 92m/C *GB* Glenne Headly, Ben Cross, Charlotte Burke, Elliott Spiers; *Dir:* Bernard Rose. **VHS, Beta, LV $89.98** *LIV, VES* ✯✯✯

Papillon McQueen is a criminal sent to Devil's Island in the 1930s determined to escape from the Lemote prison. Hoffman is the swindler he befriends. A series of escapes and recaptures follow. Box-office winner based on the autobiographical writings of

French thief Henri Charriere. Excellent portrayal of prison life and fine performances from the prisoners. Certain segments would have been better left on the cutting room floor. The film's title refers to the lead's butterfly tattoo.
1973 (PG) 150m/C Steve McQueen, Dustin Hoffman, Victor Jory, George Coulouris, Anthony Zerbe; ***Dir:*** Franklin J. Schaffner. **VHS, Beta, LV $19.98** *FOX* 🦴🦴🦴

Para Servir a Usted A waiter gets mixed in with the high society crowd that he works for.
1987 95m/C *MX* Enrique Rambal, Claudia Islas, Norma Lazarendo, Ofelia Guillman. **VHS, Beta $55.95** *MAD* 🦴🦴

Parade A series of vignettes in a circus, "Parade" is actually a play within a play, meshing the action with events offstage. The laser disc edition includes Rene Clement's 1936 "Soigne Ton Gauche," Tati's inspiration.
1974 85m/C *FR* Jacques Tati; ***W/Dir:*** Jacques Tati. **VHS, LV $59.95** *HMV, VYG, CRC* 🦴🦴🦴

The Paradine Case A passable Hitchcock romancer about a young lawyer who falls in love with the woman he's defending for murder, not knowing whether she is innocent or guilty. Script could be tighter and more cohesive. Seventy thousand dollars of the three million budget were spent recreating the original Bailey courtroom. Based on the novel by Robert Hichens.
1947 125m/B Gregory Peck, Alida Valli, Ann Todd, Louis Jourdan, Charles Laughton, Charles Coburn, Ethel Barrymore, Leo G. Carroll; ***Dir:*** Alfred Hitchcock. **VHS, Beta, LV $19.98** *FOX* 🦴🦴½

Paradise A young American boy and a beautiful English girl are the sole survivors of a caravan massacre in the Middle East during the 19th century. They discover a magnificent oasis and experience their sexual awakening. Do a double-take, its the Blue Lagoon all over, with a bit more nudity and a lot more sand.
1982 (R) 100m/C Phoebe Cates, Willie Aames, Richard Curnock, Tuvio Tavi; ***Dir:*** Stuart Gillard. **VHS, Beta, LV $24.98** *SUE* 🦴

Paradise A young boy is sent to the country to live with his pregnant mother's married friends (real life husband and wife Johnson and Griffith). From the outset it is clear that the couple are experiencing marital troubles, making the boy's assimilation all the more difficult. Help arrives in the form of a sprightly ten-year-old girl, with whom he forms a charming relationship. Eventually the cause of the couple's problems are made known, and the rest of the film sees them trying to work things out. Largely predictable, this remake of the French film "Le Grand Chemin" works thanks to the surprisingly good work of its ensemble cast, and the gorgeous scenery of South Carolina, where the movie was filmed.
1991 (PG-13) 112m/C Melanie Griffith, Don Johnson, Elijah Wood, Thora Birch, Sheila McCarthy, Eve Gordon, Louise Latham, Greg Travis, Sarah Trigger; ***Dir:*** Mary Agnes Donoghue. **VHS, Beta $92.95** *TOU* 🦴🦴🦴

Paradise Alley Rocky tires of boxing, decides to join the WWF. Three brothers brave the world of professional wrestling in an effort to strike it rich and move out of the seedy Hell's Kitchen neighborhood of New York, circa 1946. Stallone wrote, stars in, and makes his directorial debut in addition to singing the title song. He makes a few good moves as director, but is ultimately pinned to the canvas.
1978 (PG) 109m/C Sylvester Stallone, Anne Archer, Armand Assante, Lee Canalito, Kevin Conway; ***Dir:*** Sylvester Stallone; ***W/Dir:*** Sylvester Stallone. **VHS, Beta, LV $19.98** *MCA* 🦴🦴

Paradise Canyon Early Wayne "B" thriller, in which he plays an undercover government agent sent to track down a group of counterfeiters.
1935 59m/B John Wayne, Yakima Canutt, Marion Burns. **VHS $12.98** *REP, NOS, SVS* 🦴½

Paradise in Harlem An all-black musical in which a cabaret performer witnesses a gangland murder, sees his sick wife die, and is pressured into leaving town by the mob. Have a nice day.
1940 83m/B Frank Wilson, Mamie Smith, Edna Mae Harris, Juanita Hall; ***Dir:*** Joseph Seiden. **VHS, Beta $39.95** *NOS, VYY, VCN* 🦴½

Paradise, Hawaiian Style Out-of-work pilot returns to Hawaii, where he and a buddy start a charter service with two helicopters. Plenty of gals are wooed by "the Pelvis." Filmed four years after Elvis's first Pacific piece, "Blue Hawaii." Presley showing the first signs of slow-down, is made the unlikely singer of "Bill Bailey, Won't You Please Come Home?" No surprises here.
1966 91m/C Elvis Presley, Suzanna Leigh, James Shigeta, Donna Butterworth, Irene Tsu, Julie Parrish, Philip Ahn, Mary Treen, Marianna Hill, John Doucette, Grady Sutton; ***Dir:*** Michael Moore. **VHS, Beta $14.98** *FOX, MVD* 🦴½

Paradise Island Down-scale musical romance about an ingenue on her way to join her fiance in the South Seas, only to find that he has gambled away his money. The opportunistic saloon owner tries to put the moves on her. Will the two lovers be reunited?
1930 68m/B Kenneth Harlan, Marceline Day, Thomas Santschi, Paul Hurst, Victor Potel, Gladden James, Will Stanton; ***Dir:*** Bert Glennon. **VHS $19.95** *NOS* 🦴

Paradise Motel Teen exploitation film centered around a local motel predominantly used for one-night rendezvous. High school heroes and jocks battle for the affections of a beautiful classmate.
1984 (R) 87m/C Bob Basso, Gary Hershberger, Jonna Leigh Stack, Robert Krantz; ***Dir:*** Cary Medoway. **VHS, Beta $19.98** *FOX* 🦴½

Paradise Now A videotape showing the Living Theatre's performances of the famous theatre piece in Brussels and Berlin. In French, German and English, with English subtitles.
1970 105m/C VHS, Beta $29.95 *MFV* 🦴½

Paradisio A dying inventor wills a professor an authentic pair of X-ray specs. Spies come gunning for the professor, chasing him through Europe. Focus is more on the young women the prof ogles (sans clothing, due to his special specs) than on the meager plot. Much of this movie was originally in 3-D.
1961 (R) 82m/C Arthur Howard, Eva Waegner. **VHS, Beta $19.95** *STE, NWV, VDM* 🦴½

Parallax View A reporter tries to disprove a report which states that a presidential candidate's assassination was not a conspiracy. As he digs deeper and deeper, he uncovers more than he bargained for and becomes a pawn in the conspirators' further plans. Beatty is excellent and the conspiracy is never less than believable. A lesser-known, compelling political thriller that deserves to be more widely seen.
1974 (R) 102m/C Warren Beatty, Hume Cronyn, William Daniels, Paula Prentiss, Kenneth Mars, Bill McKinney, Anthony Zerbe, Walter McGinn; ***Dir:*** Alan J. Pakula. **VHS, Beta $14.95** *PAR* 🦴🦴🦴½

Parallel Corpse A mortuary attendant finds a murder victim hidden in a coffin and blackmails the remorseless killer.
1983 89m/C Buster Larsen, Jorgen Kiil, Agneta Ekmanner, Masja Dessau. **VHS, Beta $49.95** *MED* 🦴

Paralyzed An unconscious journalist, mistaken for dead in a hospital emergency room, flashes back to the investigation that embroiled him in his predicament.
197? 90m/C Ingrid Thulin, Jean Sorel, Mario Adorf; ***W/Dir:*** Aldo Lado. **VHS, Beta $59.95** *MPI* 🦴🦴

Paramedics A motley crew of cavorting paramedics battle their evil captain. Another parody of people in uniform. A comedy with little pulse.
1988 (PG-13) 91m/C George Newbern, Christopher McDonald, John P. Ryan, Ray Walston, Lawrence Hilton-Jacobs; ***Dir:*** Stuart Margolin. **VHS, Beta, LV $79.98** *LIV, VES* 🦴½

Paramount Comedy Theater, Vol. 1: Well Developed A series of stand-up routines by contemporary comedians.
1987 71m/C Howie Mandel, Bob Saget, Judy Carter, Philip Welford, Bruce Mahler. **VHS, Beta, LV $29.95** *PAR*

Paramount Comedy Theater, Vol. 2: Decent Exposures Mandel hosts this concert tape of four contemporary stand-up comics doing their thing.
1987 67m/C Howie Mandel, Doug Ferrari, Marsha Warfield, Paul Feig, Joe Alaskey. **VHS, Beta, LV $29.95** *PAR*

Paramount Comedy Theater, Vol. 3: Hanging Party Five young stand-up comedians do their stuff. Introductions by Mandel.
1987 68m/C Howie Mandel, Ritch Shyder, Susan Healy, Billy Riback, Mark McCollum. **VHS, Beta, LV $29.95** *PAR*

Paramount Comedy Theater, Vol. 4: Delivery Man A sequence of stand-up comic bits featuring a slew of young comedians.
1987 71m/C Jimmy Aleck, Maurice LaMarche, Rick Overton, Lou Dinos, Pat Hazell. **VHS, Beta, LV $29.95** *PAR*

Paramount Comedy Theater, Vol. 5: Cutting Up More stand-up routines from rising comedians and comediennes.
1987 60m/C Dale Gonyea, Lois Broomfield, Teddy Bergeron, Louise Durant. **VHS, Beta, LV $29.95** *PAR*

Paranoia A beautiful jet-set widow is trapped in her own Italian villa by a young couple who drug her to get her to perform in various sex orgies, which are probably the most interesting part of this muddled affair. Also on video as "A Quiet Place to Kill."
1969 94m/C *IT* Carroll Baker, Lou Castel, Colette Descombes; ***Dir:*** Umberto Lenzi. **VHS, Beta** *REP, UNI* 🦴

Parasite A small town is beset by giant parasites. An "Alien" ripoff originally filmed in 3-D, during that technique's brief return in the early '80s. Bad films like this killed it both the first and second times. An unpardonable mess that comes off as a stinky sixth-grade film project.

1981 (R) 90m/C Demi Moore, Gale Robbins, Luca Bercovici, James Davidson, Al Fann, Cherie Currie, Cheryl "Rainbeaux" Smith, Vivian Blaine; *Dir:* Charles Band. **VHS, Beta** $89.95 *PAR ⅃*

Pardon Mon Affaire When a middle-aged civil servant gets a look at a model in a parking garage, he decides it's time to cheat on his wife. Enjoyable French farce. Re-made as in the United States "The Woman in Red" and followed by "Pardon Mon Affair, Too!" Subtitled.
1977 (PG) 107m/C *FR* Jean Rochefort, Guy Bedos, Anny Duperey; *Dir:* Yves Robert. **VHS, Beta, LV** $29.98 *SUE ⅃⅃⅃*

Pardon Mon Affaire, Too! Pale sequel to the first popular French comedy, "Pardon Mon Affair." This time the four fantasy-minded buddies withstand the trials of marriage and middle-class life. Also called "We Will All Meet in Paradise." With English subtitles or dubbed.
1977 105m/C *FR* Jean Rochefort, Claude Brasseur, Guy Bedos; *Dir:* Yves Robert. **VHS, Beta** $29.98 *SUE ⅃⅃ ½*

Pardon My Sarong Bud and Lou star as Chicago bus drivers who end up shipwrecked on a South Pacific island when they get involved with notorious jewel thieves. The island natives think Lou is a god! Standard Abbott & Costello fare.
1942 83m/B Bud Abbott, Lou Costello, Virginia Bruce, Robert Paige, Lionel Atwill, Leif Erickson, William Demarest, Samuel S. Hinds; *Dir:* Erle C. Kenton. **VHS** $14.98 *MCA ⅃⅃⅃*

Pardon My Trunk Struggling against poverty on his teacher's salary, a man and his family receive a gift from a Hindu prince—an elephant. De Sica carries this silly premise beyond mere slapstick. Also called "Hello Elephant." Dubbed.
1952 85m/B *IT* Sabu, Vittorio De Sica; *Dir:* Gianni Franciolini. **VHS, Beta** $29.95 *VYY ⅃⅃*

Pardon Us Laurel and Hardy are thrown into prison for bootlegging. Plot meanders along aimlessly, but the duo have some inspired moments. The first Laurel and Hardy feature.
1931 78m/B Stan Laurel, Oliver Hardy, June Marlowe, James Finlayson; *Dir:* James Parrott. **VHS, Beta** $19.95 *MED, CCB, MLB ⅃⅃ ½*

The Parent Trap Mills plays a dual role in this heartwarming comedy as twin sisters who conspire to bring their divorced parents together again. Well-known Disney fluff. Followed by several made-for-television sequels featuring the now grown-up twins.
1961 127m/C Hayley Mills, Maureen O'Hara, Brian Keith, Charlie Ruggles, Una Merkel, Leo G. Carroll; *Dir:* David Swift. **VHS, Beta, LV** $19.99 *DIS ⅃⅃ ½*

Parenthood Four grown siblings and their parents struggle with various levels of parenthood. From the college drop-out, to the nervous single mother, to the yuppie couple raising an overachiever, every possibility is explored, including the perspective from the older generation, portrayed by Robards. Genuinely funny with dramatic moments that work most of the time, with an affecting performance from Martin and Wiest. Director Howard has four kids and was inspired to make this film when on a European jaunt with them.
1989 (PG-13) 124m/C Steve Martin, Mary Steenburgen, Dianne Wiest, Martha Plimpton, Keanu Reeves, Tom Hulce, Jason Robards Jr., Rick Moranis, Harley Jane Kozak, Leaf Phoenix; *Dir:* Ron Howard. **VHS, Beta, LV** $19.95 *MCA ⅃⅃⅃*

Parents Dark satire of middle class suburban life in the 50s, centering on a young boy who discovers that his parents aren't getting their meat from the local butcher. Gives new meaning to leftovers and boasts a very disturbing barbecue scene. Balaban's debut is a strikingly visual and creative gorefest with definite cult potential. The eerie score is by Angelo Badalamenti, who also composed the music for "Blue Velvet," "Wild at Heart," and the television series "Twin Peaks."
1989 (R) 81m/C Randy Quaid, Mary Beth Hurt, Sandy Dennis; *Dir:* Bob Balaban. **VHS, Beta, LV** $14.98 *LIV, VES, HHE ⅃⅃⅃*

Paris Belongs to Us An early entry in the French "new wave" of naturalistic cinema, this psychological mystery drama has a woman investigating a suicide linked to a possible worldwide conspiracy. Interesting, if somewhat dated by Cold War elements.
1960 120m/B *FR* Jean-Claude Brialy, Betty Schneider, Gianni Esposito, Francoise Prevost; *W/Dir:* Jacques Rivette. **VHS** $29.95 *VDM ⅃⅃ ½*

Paris Blues Two jazz musicians, one white, one black, strive for success in Paris and become involved with American tourists who want to take them back to the States. Score by Duke Ellington and an appearance by Armstrong make it a must-see for jazz fans.
1961 100m/B Paul Newman, Sidney Poitier, Joanne Woodward, Diahann Carroll, Louis Armstrong, Barbara Lange; *Dir:* Martin Ritt. **VHS, Beta** $19.98 *FOX ⅃⅃ ½*

Paris is Burning Livingston's documentary portrayal of New York City's transvestite balls where men dress up, dance, and compete in various categories. Filmed between 1985 and 1989, this is a compelling look at a subculture of primarily black and Hispanic men and the one place they can truly be themselves. Madonna noted this look and attitude (much watered down) in her song "Vogue."
1991 (R) 71m/C Dorian Corey, Pepper Labeija, Venus Xtravaganza, Octavia St. Laurant, Willi Ninja, Anji Xtravaganza, Freddie Pendavis, Junior Labeija; *Dir:* Jennie Livingston. Los Angeles Film Critics Association Awards '92: 10 Best Films of the Year; Sundance Festival '92: Grand Jury Award. **VHS** $89.95 *ACA ⅃⅃⅃*

Paris Express When a man steals money from his employer, he boards the Paris Express to escape from the police. A bit convoluted, but well acted. Also known as "The Man Who Watched Trains Go By." Based on the George Simenon novel.
1953 82m/C Claude Rains, Herbert Lom, Felix Aylmer, Marius Goring, Anouk Aimee, Marta Toren; *Dir:* Harold French. **VHS, Beta** $16.95 *SNC, NOS, BUR ⅃⅃ ½*

Paris Holiday An actor heading for Paris to buy a noted author's latest screenplay finds mystery and romance. Entertaining chase scenes as the characters try to find the elusive script.
1957 100m/C Bob Hope, Fernandel, Anita Ekberg, Martha Hyer, Preston Sturges; *Dir:* Gerd Oswald. **VHS, Beta** $9.95 *UNI ⅃⅃ ½*

The Paris Reunion Band Jazz greats reunite for a performance in Freiburg, Germany.
1988 60m/C Nat Adderly, Woody Shaw, Joe Henderson, Nathan Davis, Curtis Fuller, Jimmy Woode, Idries Muhammed, Walter Bishop Jr. **VHS, Beta** $24.95 *MVD, PRS ⅃⅃ ½*

Paris, Texas After four years a drifter returns to find his son is being raised by his brother because the boy's mother has also disappeared. He tries to reconnect with the boy. Written by Sam Shepard in his usual introspective style. Acclaimed by many film critics, but others found it to be too slow. Music by Ry Cooder.
1983 (PG) 145m/C *FR GE* Harry Dean Stanton, Nastassia Kinski, Dean Stockwell, Hunter Carson, Aurore Clement; *Dir:* Wim Wenders. Cannes Film Festival '83: Best Film. **VHS, Beta, LV** $19.98 *FOX ⅃⅃⅃ ½*

Paris Trout Believing himself above the law, southern Trout (Hopper) shoots the mother and sister of a young black man who reneged on his IOU. Lawyer Harris is forced to defend a man he knows deserves to be punished, and wife Hershey suffers long. Adapted by Pete Dexter from his National Book Award-winning novel. Made for cable.
1991 (R) 98m/C Dennis Hopper, Barbara Hershey, Ed Harris; *Dir:* Stephen Gyllenhaal. **VHS, Beta, LV** $89.98 *MED, FXV, FCT ⅃⅃ ½*

Paris When It Sizzles A screenwriter and his secretary fall in love while working on a film in Paris, confusing themselves with the script's characters. Star-studded cast deserves better than this lame script. Shot on location in Paris. Holden's drinking—he ran into a brick wall while under the influence—and some unresolved romantic tension between him and Hepburn affected shooting. Deitrich, Sinatra, Astaire, Ferrer, and Curtis show up for a party on the set.
1964 110m/C William Holden, Audrey Hepburn, Gregoire Aslan, Raymond Bussieres, Tony Curtis, Fred Astaire, Frank Sinatra, Noel Coward, Marlene Dietrich, Mel Ferrer; *Dir:* Richard Quine. **VHS, Beta** $29.95 *PAR ⅃⅃*

The Park is Mine A deranged and desperate Vietnam vet takes hostages in Central Park. Semi-infamous film, notable for being the first movie made for HBO on cable and for being filmed in Toronto, before the practice of filming in Canada to save money was really widespread.
1985 102m/C Tommy Lee Jones, Yaphet Kotto, Helen Shaver; *Dir:* Steven Hilliard Stern. **VHS, Beta** $59.98 *FOX ⅃⅃*

Parker An executive is kidnapped, then released, but does not know why. Unable to live with the mystery, he tracks his kidnappers down into a world of blackmail, drugs and intrigue.
1984 100m/C Bryan Brown, Kurt Raab. **VHS, Beta** $59.95 *LHV ⅃⅃*

Parlor, Bedroom and Bath A family tries to keep a young woman from seeing that her love interest is flirting with other prospects. Doesn't live up to the standard Keaton set in his silent films. Keaton spoke French and German for foreign versions. Also known as "Romeo in Pyjamas."
1931 75m/B Buster Keaton, Charlotte Greenwood, Cliff Edwards, Reginald Denny; *Dir:* Edward Sedgwick. **VHS, Beta** $19.95 *NOS, VYY, VDM ⅃⅃*

Parole, Inc FBI takes on the underground in this early crime film. Criminals on parole have not served their sentences, and the mob is responsible. A meagerly financed yawner.
1949 71m/B Michael O'Shea, Evelyn Ankers, Turhan Bey, Lyle Talbot; *Dir:* Alfred Zeisler. **VHS** $16.95 *NOS, SNC ⅃*

Paroled to Die A man frames a young rancher for a bank robbery and a murder he committed in this wild west saga. Predictable and bland.
1937 66m/B Bob Steele, Kathleen Elliott. **VHS, Beta $19.95** *NOS, DVT* ℤ½

Parting Glances Low-budget but acclaimed film shows the relationship between two gay men and how they deal with a close friend's discovery of his exposure to the AIDS virus. Touching and realistic portrayals make this a must see. In 1990 Sherwood died of AIDS without completing any other films.
1986 (R) 90m/C John Bolger, Richard Ganoung, Steve Buscemi; ***Dir:*** Bill Sherwood. **VHS, Beta $79.98** *FOX* ℤℤℤ

Partners A straight, macho cop must pose as the lover of a gay cop to investigate the murder of a gay man in Los Angeles' homosexual community. Sets out to parody "Cruising"; unfortunately, the only source of humor the makers could find was in ridiculous homosexual stereotypes that are somewhat offensive and often unfunny. Written by "La Cage aux Folles'" Francis Veber.
1982 (R) 93m/C Ryan O'Neal, John Hurt, Kenneth McMillan, Robyn Douglass, Jay Robinson, Rick Jason; ***Dir:*** James Burrows. **VHS, Beta $79.95** *PAR* ℤℤ

The Party Disaster-prone Indian actor wreaks considerable havoc at a posh Hollywood gathering. Laughs come quickly in this quirky Sellers vehicle.
1968 99m/C Peter Sellers, Claudine Longet, Marge Champion, Sharron Kimberly, Denny Miller, Gavin MacLeod, Carol Wayne; ***Dir:*** Blake Edwards. **VHS $19.95** *MGM* ℤℤℤ

The Party A bachelorette party is highlighted by male dancers and a 42-foot limo.
1988 60m/C VHS, Beta $39.95 *AHV* ℤℤℤ

Party Animal A college stud teaches a shy farm boy a thing or two about the carnal aspects of campus life.
1983 (R) 78m/C Timothy Carhart, Matthew Causey, Robin Harlan; ***Dir:*** David Beaird. **VHS, Beta $79.98** *LIV, LTG* ℤ

Party Camp A rowdy summer camp counselor endeavors to turn a tame backwoods camp into a non-stop party. Late entry in the "Meatballs"-spawned genre fails to do anything of even minor interest.
1987 (R) 96m/C Andrew Ross, Billy Jacoby, April Wayne, Kirk Cribb; ***Dir:*** Gary Graver. **VHS, Beta $79.98** *LIV, LTG* ℤ

Party Girl A crime drama involving an attorney representing a 1920's crime boss and his henchmen when they run afoul of the law. The lawyer falls in love with a nightclub dancer who successfully encourages him to leave the mob, but not before he is wounded in a gang war attack, arrested and forced to testify against the mob as a material witness. The mob then takes his girl friend hostage to prevent his testifying, leading to an exciting climax. Must-see viewing for Charisse's steamy dance numbers.
1958 99m/C Robert Taylor, Cyd Charisse, Lee J. Cobb, John Ireland, Kent Smith, Claire Kelly, Corey Allen; ***Dir:*** Nicholas Ray. **VHS, Beta, LV $19.98** *MGM* ℤℤℤ½

Party Girls Escort service girls run afoul of the law in this hokey melodrama.
1929 67m/B Douglas Fairbanks Jr., Jeanette Loff, Judith Barrie, Marie Prevost, John St. Polis, Lucien Prival; ***Dir:*** Victor Halperin. **VHS, Beta $16.95** *NOS, LOO, SNC* ℤ½

Party Incorporated Marilyn Chambers (of "Behind the Green Door" fame) stars as a young widow with a huge tax load who gives parties to pay it off. Also known as "Party Girls."
1989 (R) 80m/C Marilyn Chambers, Kurt Woodruff, Christine Veronica, Kimberly Taylor; ***Dir:*** Chuck Vincent. **VHS, LV $19.95** *STE, NWV, IME* ℤℤ

Party Line A veteran police captain and a district attorney team up to track down a pair of killers who find their victims through party lines.
1988 (R) 90m/C Richard Hatch, Shawn Weatherly, Richard Roundtree, Leif Garrett, Greta Blackburn; ***Dir:*** William Webb. **VHS, Beta $14.95** *SVS* ℤℤ

Party! Party! A bunch of kids go nuts and have a party when their parents leave.
1983 100m/C *GB* **VHS $29.95** *ACA* ℤℤ

Party Plane Softcore fun-fest about a plane full of oversexed stewardesses.
1990 81m/C Kent Stoddard, Karen Annarino, John Goff, Jill Johnson; ***Dir:*** Ed Hansen. **VHS, Beta, LV $79.98** *VES* ℤ

Pascali's Island A Turkish spy becomes involved with an adventurer's plot to steal rare artifacts, then finds himself ensnared in political and personal intrigue. Superior tragedy features excellent performances from Kingsley, Dance, and Mirren.
1988 (PG-13) 106m/C *GB* Ben Kingsley, Helen Mirren, Charles Dance, Sheila Allen, Vernon Dobtcheff; ***Dir:*** James Dearden. **VHS, Beta, LV $14.95** *IVE* ℤℤℤ½

Pass the Ammo Entertaining comedy about a young couple who attempt to steal back $50,000 of inheritance money that a televangelist swindled from the family. One of the more creative satires on this religious TV phenomenon.
1988 (R) 93m/C Bill Paxton, Tim Curry, Linda Kozlowski, Annie Potts, Anthony Geary, Dennis Burkley, Glenn Withrow; ***Dir:*** David Beaird. **VHS, Beta, LV $89.95** *IVE* ℤℤ½

Passage of the Dragon A village is caught in the middle of a martial arts war.
19?? 86m/C VHS $14.98 *NEG* ℤℤ½

A Passage to India An ambitious adaptation of E.M. Forster's complex novel about relations between Brits and Indians in the 1920s. Drama centers on a young British woman's accusations that an Indian doctor raped her while serving as a guide in some rather ominous caves. Film occasionally flags, but is usually compelling. Features particularly strong performances from Banerjee, Fox, and Davis.
1984 (PG) 163m/C *GB* Peggy Ashcroft, Alec Guinness, James Fox, Judy Davis, Victor Banerjee, Nigel Havers; ***Dir:*** David Lean. Academy Awards '84: Best Supporting Actress (Ashcroft), Best Original Score; New York Film Critics Awards '84: Best Film. **VHS, Beta, LV $19.95** *COL, BTV* ℤℤℤ

Passage to Marseilles Hollywood propaganda in which convicts escape from Devil's Island and help French freedom fighters combat Nazis. Routine but entertaining. What else could it be with Bogart, Raines, Greenstreet, and Lorre, who later made a pretty good film set in Casablanca?
1944 110m/B Humphrey Bogart, Claude Rains, Sydney Greenstreet, Peter Lorre, Helmut Dantine, George Tobias, John Loder, Eduardo Ciannelli, Michele Morgan; ***Dir:*** Michael Curtiz. **VHS, Beta** *FOX* ℤℤℤ

The Passenger A dissatisfied television reporter changes identities with a dead man while on assignment in Africa, then learns that he is posing as a gunrunner. Mysterious, elliptical production from Italian master Antonioni, who co-wrote. Nicholson is fine in the low-key role, and Schneider is surprisingly winning as the woman drawn to him. The object of much debate, hailed by some as quintessential cinema and by others as slow and unrewarding.
1975 (PG) 119m/C *IT* Jack Nicholson, Maria Schneider, Ian Hendry; ***Dir:*** Michelangelo Antonioni. National Board of Review Awards '75: 10 Best Films of the Year. **VHS, Beta $19.98** *WAR* ℤℤℤ½

The Passing Two men find themselves trapped in the darker vicissitudes of life. The two lead almost parallel lives until an extraordinary event unites them.
1988 96m/C James Plaster, Welton Benjamin Johnson, Lynn Dunn, Albert B. Smith; ***Dir:*** John Huckert. **VHS** *VSI* ℤℤ½

The Passing of Evil Bisset is a starstruck Canadian undone by the bright lights and big cities of America. By age 22, she's a burnt-out prostitute in Las Vegas. Cheerless but compelling. Also released on video under its original title, "The Grasshopper."
1970 (R) 96m/C Jacqueline Bisset, Jim Brown, Joseph Cotten, Corbett Monica, Ramon Bieri, Christopher Stone, Roger Garrett, Stanley Adams, Dick Richards, Tim O'Kelly, Ed Flanders; ***Dir:*** Jerry Paris. **VHS $29.98** *WAR, AVD* ℤℤ½

The Passing of the Third Floor Back Boarding house tenants improve their lives after being inspired by a mysterious stranger. They revert, though, when he leaves. Now you know. Based on a Victorian morality play.
1936 80m/B *GB* Conrad Veidt, Rene Ray, Frank Cellier, Anna Lee, John Turnbull, Cathleen Nesbitt; ***Dir:*** Berthold Viertel. **VHS, Beta $16.95** *SNC* ℤℤ

Passion Respectable silent version of "Madame DuBarry" is a relatively realistic costume drama, but it doesn't rate with the best German films of this period.
1919 135m/B *GE* Pola Negri, Emil Jannings, Harry Liedtke; ***Dir:*** Ernst Lubitsch. **VHS, Beta $44.99** *VYY, GLV* ℤℤ½

Passion When a rancher's young family falls victim to rampaging desperadoes, he enlists an outlaw's aid to avenge the murders of his loved ones.
1954 84m/C Raymond Burr, Cornel Wilde, Yvonne de Carlo, Lon Chaney Jr., John Qualen; ***Dir:*** Allan Dwan. **VHS, Beta $39.95** *BVV* ℤℤ

The Passion of Evelyn An Italian soft-core adventure about a seductive count and a group of willing women. Dubbed and retitled.
198? 80m/C *IT* Femi Benussi, Krista Nell. **VHS, Beta $29.95** *MED* ℤ

Passion Flower A tropical romantic melodrama made for network TV and set in Singapore, where an ambitious banker falls in love with the daughter of island's wealthiest smuggler. The mixture of high finance, low-dealing and lust only proves moderately passionate.
1986 (PG-13) 95m/C Bruce Boxleitner, Barbara Hershey, Nicol Williamson; ***Dir:*** Joseph Sargent. **VHS $89.95** *VMK* ℤℤ

Passion of Joan of Arc Masterful silent film relates the events of St. Joan's trial and subsequent martyrdom. Moving, eloquent work features legendary performance by Falconetti. Also rates among master Dreyer's finest efforts. A classic.
1928 114m/B *FR* Maria Falconetti, Eugena Sylvaw, Maurice Schultz, Antonin Artaud; *Dir:* Carl Theodor Dreyer. Sight & Sound Survey '52: #9 of the Best Films of All Time; Brussels World's Fair '58: #4 of the Best Films of All Time; Sight & Sound Survey '72: #7 of the Best Films of All Time. **VHS, Beta $29.95** *NOS, MRV, VYY ⅄⅄⅄⅄*

Passion for Life A new teacher uses revolutionary methods to engage students, but draws ire from staid parents in rural France. Worthwhile drama. In French with English subtitles.
1948 89m/B Bernard Blier, Julliette Faber, Edouard Delmont; *Dir:* Jean-Paul LeChanois. **VHS, Beta $59.95** *FCT ⅄⅄½*

Passion of Love Military captain becomes the obsession of his commander's mysterious cousin when he reports to an outpost far from home. Daring and fascinating. An unrelentingly passionate historical romance guaranteed to heat up the VCR. Received a special award at the Cannes Film Festival. Also known as "Passione d'Amore."
1982 117m/C *IT FR* Laura Antonelli, Bernard Giraudeau, Valeria D'Obici, Jean-Louis Trintignant; *Dir:* Ettore Scola. **VHS, Beta $69.95** *VES ⅄⅄⅄*

Passion for Power Two men involved in a drug-smuggling syndicate attempt to take over the business. Complications arise when thoy both fall for a beautiful, deceitful woman who turns them against one another.
1985 94m/C *SP* Hector Suarez, Sasha Montenegro, Manuel Capetillo, Alejandra Peniche. **VHS $72.95** *MHV ⅄⅄⅄*

Passionate Thief Two pickpockets and a bumbling actress plan to rip off the guests at a posh New Year's Eve party with slightly comic results. Not much to recommend in this slow moving flick.
1960 100m/C *IT* Anna Magnani, Ben Gazzara, Toto, Fred Clark; *Dir:* Mario Monicelli. **VHS, Beta $19.95** *CHA ⅄⅄½*

Passione d'Amore A foolhardy cavalry officer already has buxom Antonelli for a lover, but he nonetheless falls in love with a mysterious but homely recluse. Intriguing, often moving. Costume drama set in 19th-century Italy.
1982 117m/C *IT FR Dir:* Ettore Scola. **Beta, 3/4U, Special order formats** *CIG ⅄⅄⅄*

The Passover Plot A controversial look at the crucifixion of Christ which depicts him as a Zealot leader who, aided by his followers, faked his death and then "rose" to win new converts.
1975 (PG) 105m/C Harry Andrews, Hugh Griffith, Zalman King, Donald Pleasence, Scott Wilson; *Dir:* Michael Campus. **VHS, Beta $59.95** *CAN ⅄⅄*

Passport to Pimlico Farce about a London neighborhood's residents who discover an ancient charter proclaiming their right to form their own country within city limits. Passable comedy.
1949 81m/B *GB* Stanley Holloway, Margaret Rutherford, Hermione Baddeley, Naunton Wayne, Basil Radford; *Dir:* Henry Cornelius. **VHS, Beta $19.95** *NOS, HBO, PSM ⅄⅄½*

Pastime A bittersweet baseball elegy set in the minor leagues in 1957. A boyish 41-year-old pitcher can't face his impending retirement and pals around with the team pariah, a 17-year-old black rookie. Splendidly written and acted, it's a melancholy treat whether you're a fan of the game or not, and safe for family attendance. The only drawback is a grungy, low-budget look.
1991 (PG) 94m/C William Russ, Scott Plank, Glenn Plummer, Noble Willingham, Jeffrey Tambor, Deidre O'Connell; *Dir:* Robin B. Armstrong. **VHS, LV $89.95** *COL ⅄⅄⅄*

Pat Garrett & Billy the Kid Coburn is Garrett, one-time partner of Billy the Kid (Kristofferson), turned sheriff. He tracks down and eventually kills the outlaw. The uncut director's version which was released on video is a vast improvement over the theatrically released and television versions. Dylan's soundtrack includes the now famous "Knockin' on Heaven's Door."
1973 106m/C Kris Kristofferson, James Coburn, Bob Dylan, Richard Jaeckel, Katy Jurado, Chill Wills, Charles Martin Smith, Slim Pickens, Harry Dean Stanton; *Dir:* Sam Peckinpah. **VHS, Beta, LV $19.95** *MGM, PIA ⅄⅄⅄*

Pat and Mike War of the sexes rages in this comedy about a leathery sports promoter who futilely attempts to train a woman for athletic competition. Tracy and Hepburn have fine chemistry, but supporting players contribute too. Script was nominated for an Academy Award. Watch for the first on-screen appearance of Bronson (then Charles Buchinski) as a crook.
1952 95m/B Spencer Tracy, Katharine Hepburn, Aldo Ray, Jim Backus, William Ching, Sammy White, Phyllis Povah, Charles Bronson, Chuck Connors, Mae Clarke, Carl "Alfalfa" Switzer; *Dir:* George Cukor. **VHS, Beta, LV $29.95** *MGM ⅄⅄⅄*

A Patch of Blue A kind-hearted blind girl falls in love with a black man without acknowledging racial differences. Good performances from Hartman and Poitier are film's strongest assets. Winters snagged an Oscar, while nominations went to actress Hartman, black & white cinematography, art direction, set decoration, and score.
1965 108m/C Sidney Poitier, Elizabeth Hartman, Shelley Winters, Wallace Ford, Ivan Dixon, John Qualen, Elisabeth Fraser, Kelly Flynn; *Dir:* Guy Green. Academy Awards '65: Best Supporting Actress (Winters). **VHS, Beta $19.98** *MGM, BTV ⅄⅄½*

Patchwork Girl of Oz A very early silent film production by Baum that brought his own Oz books to the screen.
1914 80m/B *Dir:* L. Frank Baum. **VHS, Beta $19.95** *NOS, DVT, VYY ⅄⅄*

Paternity Routine comedy about middle-aged manager of Madison Square Gardens (Reynolds) who sets out to find a woman to bear his child, no strings attached. The predictability of the happy ending makes this a yawner. Steinberg's directorial debut.
1981 (PG) 94m/C Burt Reynolds, Beverly D'Angelo, Lauren Hutton, Norman Fell, Paul Dooley, Elizabeth Ashley; *Dir:* David Steinberg. **VHS, Beta, LV $39.95** *PAR ⅄⅄*

Pather Panchali Somber, moving story of a young Bengali boy growing up in impoverished India. Stunning debut from India's master filmmaker Ray, who continued the story in "Aparajito" and "World of Apu." A truly great work. In Bengali with English subtitles.

1954 112m/B *IN* Kanu Banerjee, Karuna Banerjee, Uma Das Gupta, Subir Banerji; *Dir:* Satyajit Ray. **VHS, Beta $24.95** *NOS, FCT, MRV ⅄⅄⅄⅄*

Pathfinder A young boy in Lapland of 1,000 years ago comes of age prematurely after he falls in with cutthroat nomads who already slaughtered his family and now want to wipe out the rest of the village. Gripping adventure in the ice and snow features stunning scenery. Oscar nominated for best foreign film. In Norwegian with English subtitles and based on an old Lapp fable.
1988 88m/C *NO* Mikkel Gaup, Nils Utsi, Svein Scharffenberg, Helgi Skulason, Sara Marit Gaup, Sverre Porsanger; *W/Dir:* Nils Gaup. **VHS $79.95** *FXL, FCT ⅄⅄⅄½*

Paths of Glory Classic anti-war drama set in WWI France. A vain, ambitious officer imposes unlikely battle strategy on his hapless troops, and when it fails, he demands that three soldiers be selected for execution as cowards. Menjou is excellent as the bloodless French officer, and Carey shines as the officer who knows about the whole disgraceful enterprise. Fabulous, wrenching fare from filmmaking great Kubrick, who co-wrote. Based on a true story from Humphrey Cobb's novel of the same name.
1957 86m/B Kirk Douglas, George Macready, Ralph Meeker, Adolphe Menjou, Richard Anderson, Wayne Morris, Timothy Carey, Susanne Christian; *Dir:* Stanley Kubrick. **VHS, Beta, LV $19.98** *MGM, FOX ⅄⅄⅄⅄*

Paths to Paradise Compson and Griffith share criminal past, reunite at gala event to snatch priceless necklace and head south of the border. World class chase scene.
1925 78m/B Raymond Griffith, Betty Compson, Thomas Santschi, Bert Woodruff, Fred Kelsey; *Dir:* Clarence Badger. **VHS, Beta $19.95** *GPV, DNB ⅄⅄⅄*

Patrick A coma patient suddenly develops strange powers and has a weird effect on the people he comes in contact with.
1978 (PG) 115m/C *AU* Robert Helpmann, Susan Penhaligon, Rod Mullinar; *Dir:* Richard Franklin. **VHS, Beta** *HAR ⅄⅄*

Patriot An action film about an ex-Navy commando who battles a band of nuclear-arms smuggling terrorists. Edited via George Lucas' electronic editor, Edit Droid.
1986 (R) 88m/C Jeff Conaway, Michael J. Pollard, Leslie Nielsen, Greg Henry, Simone Griffeth; *Dir:* Frank Harris. **VHS, Beta, LV $79.98** *LIV, VES ⅄½*

The Patriots A German prisoner works as a shoemaker in a small village during WWI. A lyrical drama in German and Russian with English titles.
1933 82m/B *RU Dir:* Boris Barnett. **VHS, Beta, 3/4U $29.95** *FCT, IHF ⅄⅄⅄*

The Patsy Shady producers attempt to transform a lowly bellboy into a comedy superstar. Not one of Lewis's better efforts; Lorre's last film.
1964 101m/C Jerry Lewis, Ina Balin, Everett Sloane, Phil Harris, Keenan Wynn, Peter Lorre, John Carradine, Hans Conried, Richard Deacon, Scatman Crothers, Del Moore, Neil Hamilton, Buddy Lester, Nancy Kulp, Norman Alden, Jack Albertson; *Dir:* Jerry Lewis. **VHS, Beta $59.95** *USA, IME ⅄⅄*

Patsy, Mi Amor A modern woman eventually chooses to give herself to an older man, and finds life with him powerful and adventurous.

1987 94m/C *MX* Julio Aleman, Ofelia Medina, Joaquin Cordero, Leticia Robles, Carlos Cortes, Julian Pastor. **VHS, Beta $55.95** *MAD* ♫♫

Patterns Realistic depiction of big business. Heflin starts work at a huge New York office that is under the ruthless supervision of Sloane. Serling's astute screenplay is adept at portraying ruthless, power-struggling executives and the sundry workings of a large corporation. Film has aged slightly, but it still has some edge to it. Originally intended for television.
1956 83m/B Van Heflin, Everett Sloane, Ed Begley Sr., Beatrice Straight, Elizabeth Wilson; *Dir:* Fielder Cook. **VHS $19.95** *NOS, MGM, MRV* ♫♫♫

Pattes Blanches The title translates as "white leggings" a garb worn by a saloonkeeper resentful of a rich man's advances on his girlfriend. Moody, sensual French melodrama exploring class, money and sex. pretty explicit for its time. In French with English subtitles.
1949 92m/B *FR* Suzy Delair, Fernand Ledoux, Paul Bernard, Michel Bouquet; *Dir:* Jean Gremillon. **VHS** *INT* ♫♫

Patti Rocks Offbeat, realistic independent effort concerns a foul chauvinist who enlists a friend to accompany him on a visit to a pregnant girlfriend who turns out to be less than the bimbo she's portrayed as en route. Mulkey shines as the sexist. Written by Mulkey, Morris, Landry, and Jenkins. Same characters featured earlier in "Loose Ends."
1988 (R) 86m/C Chris Mulkey, John Jenkins, Karen Landry, David L. Turk, Stephen Yoakam; *Dir:* David Burton Morris. **VHS, Beta, LV $19.95** *VIR* ♫♫

Patton Lengthy but stellar bio of the vain, temperamental American general who masterminded significant combat triumphs during WWII. "Old Blood and Guts," who considered himself an 18th-century commander living in the wrong era, produced victory after victory in North Africa and Europe, but not without a decided impact upon his troops. Scott is truly magnificent in the title role, and Malden shines in the supporting role of General Omar Bradley. Not a subtle film, but neither is its subject. Co-written by Francis Ford Coppola and Edmund H. North. Interesting match-up with the 1986 made-for-TV movie "The Last Days of Patton," also starring Scott. Oscar nominated for Original Score and Cinematography as well as the several awards it won.
1970 (PG) 171m/C George C. Scott, Karl Malden, Stephen Young, Michael Strong, Frank Latimore, James Edwards, Lawrence Dobkin, Michael Bates, Tim Considine; *Dir:* Franklin J. Schaffner. Academy Awards '70: Best Actor (Scott), Best Art Direction/Set Decoration, Best Director (Schaffner), Best Film Editing, Best Picture, Best Sound, Best Story & Screenplay; Directors Guild of America Awards '70: Best Director (Coppola); National Board of Review Awards '70: Best Actor (Scott), Best Film; New York Film Critics Awards '70: Best Actor (Scott); National Society of Film Critics Awards '70: Best Actor (Scott). **VHS, Beta, LV $29.98** *FOX, TLF, BTV* ♫♫♫½

Patty Hearst Less than fascinating, expressionistic portrait of Hearst from her kidnapping through her brainwashing and eventual criminal participation with the SLA. An enigmatic film that seems to only make Hearst's transformation into a Marxist terrorist all the more mysterious. Based on Hearst's book, "Every Secret Thing."
1988 (R) 108m/C Natasha Richardson, William Forsythe, Ving Rhames, Frances Fisher, Jodi Long, Dana Delany; *Dir:* Paul Schrader. **VHS, Beta, LV $19.98** *MED* ♫♫½

Paul Bartel's The Secret Cinema A woman cannot determine whether her life is real or a film by a maniacal director. Offbeat, creative, and a little disturbing. Followed by Bartel's short "Naughty Nurse."
1969 37m/B *Dir:* Paul Bartel. **VHS, Beta $29.95** *RHI* ♫♫½

Paul Reiser: Out on a Whim A combination of stand-up routines by Reiser and extended skits populated by guest stars.
1988 60m/C Paul Reiser, Elliott Gould, Carrie Fisher, Teri Garr, Brooke Adams, Michael J. Pollard, Desi Arnaz Jr., Carol Kane; *Dir:* Carl Gottlieb. **VHS, Beta $19.98** *LIV, VES* ♫♫½

Paul Robeson Major events in the life of the popular actor are recounted in this one-man performance. Originally staged by Charles Nelson Reilly.
197? 118m/C James Earl Jones; *Dir:* Lloyd Richards. **VHS, Beta $39.95** *ALL, KAR* ♫♫

Paul Rodriguez Live! I Need the Couch The urban-ethnic comic does his stand-up routine, taped live at The Armory for the Arts in Santa Fe. Includes a skit featuring Father Guido Sarducci.
1989 60m/C Paul Rodriguez, Don Novello. **VHS, Beta $59.99** *HBO, TLF* ♫

Paula A woman loves an older man, and encounters many problems. In Spanish.
19?? 90m/C Julissa, Abel Salazar. **VHS, Beta** *UNI* ♫

Pauline at the Beach A young woman accompanies her more experienced cousin to the French coast for a summer of sexual hijinks. Contemplative, not coarse, though the leads look great in—and out—of their swimsuits. Breezy, typically talky fare from small-film master Rohmer. Cinematography by Nester Almendro. With English subtitles.
1983 (R) 95m/C *FR* Amanda Langlet, Arielle Dombasle, Pascal Greggory, Rosette; *W/Dir:* Eric Rohmer. **VHS, Beta, LV $24.95** *XVC, MED* ♫♫♫½

Paul's Case In turn-of-the-century Pittsburgh, an ambitious young man enters the upper crust of New York society through deceiving pretenses. Based on a story by Willa Cather. Made for television.
1980 52m/C Eric Roberts, Michael Higgins, Lindsay Crouse; *Dir:* Lamont Johnson. **VHS, Beta $24.95** *MON, MTI, KAR* ♫½

Pawnbroker A Jewish pawnbroker in Harlem is haunted by his grueling experiences in a Nazi camp during the Holocaust. Powerful and well done. Probably Steiger's best performance (Oscar nominated). Adapted from a novel by Edward Lewis Wallant.
1965 120m/B Rod Steiger, Brock Peters, Geraldine Fitzgerald, Jaime Sanchez, Thelma Oliver; *Dir:* Sidney Lumet. Berlin Film Festival '65: Best Actor (Steiger); British Academy Awards '66: Best Actor (Steiger). **VHS, LV $19.98** *REP* ♫♫♫½

Pawnshop Charlie is employed as a pawnbroker's assistant. Silent with music track.
1916 20m/B Charlie Chaplin. **VHS, Beta** *CAB, FST* ♫♫♫½

Pay or Die A crime bosses' men turn on him and kidnap his daughter.
1983 (R) 92m/C Dick Adair, Johnny Wilson; *Dir:* Bobby Suarez. **VHS, Beta $29.95** *IVE* ♫♫½

Pay Off A lunatic psychiatrist seeks murderous revenge on his ex-patient/girlfriend and her family.
1989 87m/C Michael Fitzpatrick, Veronika Mattson, Margareta Krook; *Dir:* George Tirl. **VHS, Beta $59.95** *TWE* ♫

Payback A brawny young man uses firepower to avenge those who have wronged him.
1989 90m/C Jean Carol. **VHS, Beta $29.95** *CLV* ♫

Payback A convict sets out to avenge his brother's death, and creates all sorts of mayhem in the process. Produced by Bob "Newlywed Game" Eubanks and scored by Daryl "Captain and Tenille" Dragon.
1990 (R) 94m/C Corey Michael Eubanks, Michael Ironside, Teresa Blake, Bert Remsen, Vincent Van Patten, Don Swayze; *Dir:* Russell Solberg. **VHS $89.98** *REP* ♫

Payday Rip Torn stars as a declining country music star on tour in this portrayal of the seamy side of show business, from groupies to grimy motels. Don Carpenter's script and fine performances make this an engaging, if rather draining, drama not easily found on the big screen.
1972 98m/C Rip Torn, Anna Capri, Michael C. Gwynne, Jeff Morris; *Dir:* Daryl Duke. **VHS, Beta $59.99** *HBO* ♫♫½

The Payoff The old-fashioned newspaper-reporter-as-crime-fighter routine. When the city's special prosecutor is murdered a daring newshawk investigates and tracks the bad guys. But will he spell their names right?
1943 74m/B Lee Tracy, Tom Brown, Tina Thayer, Evelyn Brent, Jack La Rue, Ian Keith, John Maxwell; *Dir:* Arthur Dreifuss. **VHS $19.95** *NOS, LOO* ♫½

Payoff An ex-cop discovers the identity of gangsters who killed his parents. He traces them to a Lake Tahoe resort and plots revenge. Familiar crime story with a good cast.
1991 (R) 111m/C Keith Carradine, Kim Griest, Harry Dean Stanton, John Saxon, Jeff Corey; *Dir:* Stuart Cooper. **VHS, Beta $89.98** *FXV* ♫♫

Peacekillers A mean bunch of bikers visit a commune to kidnap a young woman. The gang has a big surprise in store for them when the girl escapes.
1971 86m/C Michael Ontkean, Clint Ritchie, Paul Krokop; *Dir:* Douglas Schwartz. **VHS, Beta $19.95** *NWV, STE* ♫♫

Peacemaker Two aliens masquerading as human cops stalk each other through a major city.
1990 (R) 90m/C Robert Forster, Lance Edwards, Hilary Shepard, Bert Remsen, Robert Davi; *Dir:* Kevin S. Tenney. **VHS, Beta, LV $89.95** *FRH* ♫

The Peacock Fan The Peacock Fan is protected by a deadly curse, with certain death to anyone who possesses it.
1929 50m/B Lucian Preval. **VHS, Beta $16.95** *SNC, VYY* ♫½

The Peanut Butter Solution An imaginative 11-year-old boy investigates a strange old house which is haunted by friendly ghosts who are in possession of a magic potion.
1985 (PG) 96m/C *Dir:* Michael Rubbo. **VHS, Beta $14.95** *NWV, STE, HHE* ♫♫½

The Pearl Based on a John Steinbeck story, a Mexican-fisherman, living in poverty, finds a magnificent pearl and he thinks this find will improve the lives of his family. He is

bewildered by what this pearl really brings - liars and thieves. Simple, yet larger-than-life-film, it is beautifully photographed and is a timeless picture capturing human nature.
1948 77m/B *MX* Pedro Armendariz, Maria Elena Marques, Fernando Wagner, Charles Rooner; *Dir:* Emilio Fernandez. **VHS $69.95** *FCT, MAS* 🎧🎧🎧

The Pearl of Death Holmes and Watson investigate the theft of a precious pearl.
1944 69m/B Basil Rathbone, Nigel Bruce, Dennis Hoey, Miles Mander, Rondo Hatton, Evelyn Ankers; *Dir:* Roy William Neill. **VHS, Beta $19.98** *FOX* 🎧🎧½

Pearl of the South Pacific A trio of adventurers destroy a quiet and peaceful island when they ransack it for pearl treasures.
1955 85m/C Dennis Morgan, Virginia Mayo, David Farrar; *Dir:* Allan Dwan. **VHS, Beta $39.95** *BVV* 🎧

The Peasants Poignant rural Polish saga pits father against son due to the passion they share for one woman. In two tapes, based on Wladyslaw Reymont's Nobel prize-winning novel. In Polish with English subtitles.
19?? 184m/C *PL* **VHS $69.99** *FCT* 🎧🎧½

Peck's Bad Boy Impudent, precocious brat causes his parents considerable grief. Someone should lock this kid in a room with W.C. Fields or the Alien. Silent film based on the stories by George Wilbur Peck.
1921 51m/B Jackie Coogan, Thomas Meighan, Raymond Hatton, Wheeler Oakman, Lillian Leighton; *Dir:* Sam Wood. **VHS, Beta, LV $19.95** *NOS, MRV, CCB* 🎧🎧

Peck's Bad Boy with the Circus The circus will never be the same after the mischievous youngster and his buddies get done with it. Gilbert and Kennedy provide the high points.
1938 67m/B Tommy Kelly, Anne Gillis, George "Spanky" McFarland, Edgar Kennedy, Billy Gilbert; *Dir:* Eddie Cline. **VHS, Beta $19.95** *NOS, VYY, DVT* 🎧🎧

Pecos Kid A child has his parents killed, and grows up embittered, vengeful and thirsty for blood.
1935 56m/B Fred Kohler Jr. **VHS, Beta, LV $19.95** *NOS, WGE* 🎧

Peddlin' in Society A fruit vendor makes it big on the black market and lives the good life, until poor investments force her back to her old means. Strong perfomances by Magnani and De Sica. In Italian with English subititles.
1947 85m/B *IT* Anna Magnani, Vittorio De Sica, Virgilio Riento, Laura Gore; *Dir:* Gennaro Righelli. **VHS** *DVT* 🎧

The Pedestrian A prominent German industrialist is exposed as a Nazi officer who supervised the wholesale devastation of a Greek village during WWII. Impressive debut for director Schell, who also appears in a supporting role. Foreign title: "Der Fussgaenger."
1974 (PG) 97m/C *GE SI* Maximilian Schell, Peggy Ashcroft, Lil Dagover, Francoise Rosay, Elisabeth Bergner; *Dir:* Maximilian Schell. National Board of Review Awards '74: 5 Best Foreign Films of the Year. **VHS, Beta $69.98** *SUE, GLV* 🎧🎧🎧½

The Pee-Wee Herman Show The original HBO special which introduced Pee-Wee to the world. Hilarious stuff, with help from Captain Carl (Hartman), Miss Yvonne, and other assorted pals. Warning: Not to be confused with Pee Wee's children's show.

1982 60m/C Pee-Wee Herman, Phil Hartman, John Paragon. **VHS $29.95** *HBO* 🎧🎧½

Pee-Wee's Big Adventure Zany, endearing comedy about an adult nerd's many adventures while attempting to recover his stolen bicycle. Chock full of classic sequences, including a barroom encounter between Pee-Wee and several ornery bikers. A colorful, exhilerating experience written by Reubens, Phil Hartman, and Michael Varhol.
1985 (PG) 92m/C Pee-Wee Herman, Elizabeth Daily, Mark Holton, Diane Salinger, Judd Omen; *Dir:* Tim Burton. **VHS, Beta, LV $19.98** *WAR* 🎧🎧🎧½

Pee-Wee's Playhouse Christmas Special Pee Wee does an extended Playhouse episode for Christmas, with a load of guest stars.
1988 49m/C Pee-Wee Herman, Whoopi Goldberg, Zsa Zsa Gabor, Grace Jones, Dinah Shore, Frankie Avalon, Annette Funicello, Joan Rivers, Oprah Winfrey, Charo, Earle Johnson, k.d. lang, Kareem Abdul-Jabbar, Del Rubio Triplets, Little Richard. **VHS, Beta, LV $9.99** *VTR* 🎧🎧½

Pee-Wee's Playhouse Festival of Fun Five fun-filled episodes of the playhouse!
1988 123m/C Pee-Wee Herman. **VHS, Beta $79.95** *MED* 🎧🎧½

Peeping Tom Controversial, unsettling thriller in which a psychopath lures women before his film camera, then records their deaths at his hand. Unnerving subject matter is rendered impressively by British master Powell. A classic of its kind, but definitely not for everyone. This is the original uncut version of the film, released in the U.S. in 1979 with the assistance of Martin Scorsese.
1963 88m/C Karl Boehm, Moira Shearer, Anna Massey, Maxine Audley, Esmond Knight, Shirley Anne Field, Brenda Bruce, Pamela Green, Jack Watson, Nigel Davenport, Susan Travers, Veronica Hurst; *Dir:* Michael Powell. New York Film Festival '79: 5 Best Foreign Films of the Year. **VHS, Beta $39.95** *AOV* 🎧🎧🎧½

Peggy Sue Got Married Uneven but entertaining comedy about an unhappily married woman seemingly unable to relive her life when she falls unconscious at a high school reunion and awakens to find herself back in school. Turner shines, but the film flags often, and Cage isn't around enough to elevate entire work. O'Connor scores, though, as a sensitive biker. Look for musician Marshall Crenshaw as part of the reunion band.
1986 (PG-13) 103m/C Kathleen Turner, Nicolas Cage, Catherine Hicks, Maureen O'Sullivan, John Carradine, Helen Hunt, Lisa Jane Persky, Barbara Harris, Joan Allen, Kevin J. O'Connor, Barry Miller; *Dir:* Francis Ford Coppola. **VHS, Beta, LV $19.98** *FOX* 🎧🎧

The Peking Blond French fried spyfilm boasts lousy acting unencumbered by plot. Amnesiac may or may not hold secrets coveted by Americans, Russians and Chinese. Originally released as "The Blonde from Peking."
1968 80m/C *FR* Mireille Darc, Claudio Brook, Edward G. Robinson, Pascale Roberts; *Dir:* Nicolas Gessner. **VHS $29.99** *AVD, MRV* 🎧½

Pelle the Conqueror Overpowering tale of a Swedish boy and his widower father who serve landowners in late 19th-century Denmark. Compassionate saga of human spirit contains numerous memorable sequences. Hvenegaard is wonderful as young Pelle, but Oscar-nominated von Sydow delivers what is probably his finest performance as sympathetic weakling. American distributors foolishly trimmed the film by some 20 minutes (140 minute version). In Swedish with English subtitles. From the novel by Martin Anderson Nexo.
1988 160m/C *SW DK* Max von Sydow, Pelle Hvenegaard, Erik Paaske, Bjorn Granath, Axel Strobye, Astrid Villaume, Troels Asmussen, John Wittig, Anne Lise Hirsch Bjerrum; *Dir:* Bille August. Academy Awards '88: Best Foreign Language Film; Cannes Film Festival '88: Best Film. **VHS, Beta, LV $19.98** *HBO* 🎧🎧🎧🎧

Penalty Phase An up-for-election judge must decide a murder case in which his future, as well as the defendant's, is in question. Made for television.
1986 94m/C Peter Strauss, Melissa Gilbert, Jonelle Allen; *Dir:* Tony Richardson. **VHS, Beta $19.95** *STE, NWV* 🎧🎧

Pendulum A police captain struggles to prove himself innocent of his wife's—and his wife's lover's—murders. Wow.
1969 (PG) 106m/C George Peppard, Jean Seberg, Richard Kiley, Madeline Sherwood, Charles McGraw, Marj Dusay; *Dir:* George Schaefer. **VHS, Beta, LV $59.95** *COL* 🎧🎧

The Penitent A remote village's annual reenactment of the crucifixion of Christ serves as the backdrop for Assante's affair with his friend's (Julia) wife. Quirky little tale that focuses a shade too much on the romantic problems of the trio rather than the intriguing religious practices going on around them.
1988 (PG-13) 94m/C Raul Julia, Armand Assante, Rona Freed, Julie Carmen; *Dir:* Cliff Osmond. **VHS, Beta** *IVE* 🎧🎧

Penitentiary A realistic story of a black fighter who survives his prison incarceration by winning bouts against the other prisoners. Well-made and executed, followed by two progressively worse sequels.
1979 (R) 99m/C Leon Isaac Kennedy, Jamaa Fanaka; *Dir:* Jamaa Fanaka. **VHS, Beta $59.95** *UNI* 🎧🎧½

Penitentiary 2 A welterweight fighter is after the man who murdered his girlfriend, who, luckily, is incarcerated in the same prison as Our Hero. Sometimes things just work out right.
1982 (R) 108m/C Leon Isaac Kennedy, Mr. T, Leif Erickson, Ernie Hudson, Glynn Turman; *Dir:* Jamaa Fanaka. **VHS, Beta $69.95** *MGM, HHE* 🎧

Penitentiary 3 Once again, Kennedy as the inmate boxer extraordinaire punches his way out of various prison battles. Another punch-drunk sequel.
1987 (R) 91m/C Leon Isaac Kennedy, Anthony Geary, Steve Antin, Ric Mancini, Kessler Raymond, Jim Bailey; *Dir:* Jamaa Fanaka. **VHS, Beta $19.98** *WAR* 🎧½

Penn and Teller Get Killed The comedy team with a cult following are pursued by an assassin through dozens of pratfalls in this dark comedy.
1990 (R) 91m/C Penn Jillette, Jilette Teller, Caitlin Clarke, Leonardo Cimino, David Patrick Kelly; *Dir:* Arthur Penn. **VHS, Beta, LV $89.95** *WAR* 🎧🎧

Penn and Teller Go Public The "magicomic" duo play it to the hilt in this stunning and disgusting performance.
1989 30m/C Penn Jillette, Jilette Teller. **VHS $59.98** *LIV, VES* 🎧🎧

Penn and Teller's Cruel Tricks for Dear Friends A uniquely designed instructional video by the popular comedy/ magic/ cruel joke performance team. Seven scams are demonstrated, each using specially designed clips on the tape itself, so the interactive viewer can humiliate unsuspecting friends.
1988 59m/C Penn Jillette, Jilette Teller. **VHS, Beta $20.00** LHV, WAR ♂♂

Pennies From Heaven Underrated, one-of-a-kind musical about a horny sheet-music salesman in Chicago and his escapades during the Depression. Extraordinary musical sequences have stars lip-synching to great effect. Martin is only somewhat acceptable as the hapless salesman, but Peters and Harper deliver powerful performances as the women whose lives he ruins. Walken brings down the house in a stunning song-and-dance sequence. Songs include "Yes, Yes," "I Want to Be Bad," "Let's Misbehave," and "Let's Face the Music and Dance." Adapted by Dennis Potter from his British television series.
1981 (R) 107m/C Steve Martin, Bernadette Peters, Christopher Walken, Jessica Harper, Vernel Bagneris; **Dir:** Herbert Ross. **VHS, Beta, LV $19.99** MGM, FCT ♂♂♂ ½

Penny Serenade Newlyweds adopt a child, but tragedy awaits. Simplistic story nonetheless proves to be a moving experience. They don't make 'em like this anymore, and no one plays Grant better than Grant, who was Oscar nominated. Dunne is adequate. Also available colorized.
1941 120m/B Cary Grant, Irene Dunne, Beulah Bondi, Edgar Buchanan; **Dir:** George Stevens. **VHS, LV $9.95** NEG, NOS, PSM ♂♂♂

Pennywhistle Blues A thief loses his stolen money and goes mad trying to locate it. Little does he know that it has been recovered by honest people who are putting it to good use. Perhaps that would have given him peace of mind? Also titled "The Magic Garden."
1952 63m/B Tommy Ramokgopa; **Dir:** Donald Swanson. **VHS, Beta $29.95** VYY ♂

Penthouse Love Stories An array of sketches bringing the pages of Penthouse Forum to soft-focused life. Stars the "Happy Hooker."
1986 60m/C Xaviera Hollander. **VHS, Beta $29.98** VES

Penthouse Video, Vol. 1: The Girls of Penthouse The first volume of this video magazine features of some of Penthouse's most popular centerfolds in intimate photo sessions.
1985 60m/C VHS, Beta, LV $29.98 VES

Penthouse on the Wild Side A compilation of sex-news items from "Penthouse," including Madonna's nude photos, behind-the-scenes on the set of "Caligula," and various Penthouse Pets displaying anatomical virtues.
1988 (R) 58m/C VHS, Beta $29.95 VES

The People A young teacher takes a job in a small town and finds out that her students have telepathic powers and other strange qualities. Adapted from a novel by Zenna Henderson. Good atmosphere, especially for a TV movie.
1971 74m/C Kim Darby, Dan O'Herlihy, Diane Varsi, William Shatner; **Dir:** John Korty. **VHS, Beta $59.95** PSM ♂♂ ½

People Are Funny Battling radio producers vie to land the big sponsor with an original radio idea. Comedy ensues when one of them comes up with a great idea—stolen from a local station.
1946 94m/B Jack Haley, Rudy Vallee, Ozzie Nelson, Art Linkletter, Helen Walker; **Dir:** Sam White. **VHS, Beta $29.95** VYY ♂♂

The People Next Door Seventies attempt at exposing the drug problems of suburban youth. Typical middle-class parents Wallach and Harris are horrified to discover that daughter Winters is strung out on LSD. They fight to expose the neighborhood pusher while struggling to understand their now-alien daughter. Based on J.P. Miller's television play.
1970 93m/C Deborah Winters, Eli Wallach, Julie Harris, Stephen McHattie, Hal Holbrook, Cloris Leachman, Nehemiah Persoff; **Dir:** David Greene. **VHS** TAV ♂♂

The People That Time Forgot Sequel to "The Land That Time Forgot," based on the Edgar Rice Burroughs novel. A rescue team returns to a world of prehistoric monsters to rescue a man left there after the first film.
1977 (PG) 90m/C GB Doug McClure, Patrick Wayne, Sarah Douglas, Dana Gillespie, Thorley Walters, Shane Rimmer; **Dir:** Kevin Connor. **VHS, Beta $9.98** SUE ♂♂ ½

The People Under the Stairs Adams is part of a scheme to rob a house in the slums owned by a mysterious couple (Robie and McGill, both of "Twin Peaks" fame). After his friends are killed off in a gruesome fashion, he discovers that the couple aren't the house's only strange inhabitants—homicidal creatures also lurk within.
1991 (R) 102m/C Everett McGill, Wendy Robie; **W/Dir:** Wes Craven. **VHS, Beta, LV $92.95** MCA ♂♂ ½

The People vs. Jean Harris Follows the trial of Jean Harris. Shortly before this film was released, she had been convicted of murder in the death of Dr. Herman Tarnower, the author of "The Scarsdale Diet." Harris was a head mistress in a private school all the while. Made for television. Burstyn was nominated for an Emmy Award for best actress.
1981 147m/C Ellen Burstyn, Martin Balsam, Richard Dysart, Peter Coyote, Priscilla Morrill, Sarah Marshall, Millie Slavin; **Dir:** George Schaefer. **VHS, Beta $59.95** USA ♂♂♂

People Who Own the Dark A small farming community becomes possessed by a supernatural light during an explosion. The people, who have all been blinded, discover that the world is ending.
1975 (R) 87m/C Paul Naschy, Tony Kendall, Maria Perschy, Terry Kemper, Tom Weyland, Anita Brock, Paul Mackey; **W/Dir:** Armando de Ossorio. **VHS, Beta $19.95** SUN, STC, MRV ♂♂

People's Choice A small town boy suffers laryngitis, becomes a radio personality via his new huskiness, and claims he's a notorious criminal.
1946 68m/B Drew Kennedy, Louise Arthur, George Meeker, Rex Lease, Fred Kelsey. **VHS, Beta, LV** WGE ♂♂

Pepe Le Pew's Skunk Tales Eight Le Pew shorts, including "Who Scent You?," "The Cat's Bah" and "Scent-imental Romeo," most by Academy Award-winning director Chuck Jones.
1953 56m/C **Dir:** Chuck Jones, Robert McKimson, Bob Clampett. **VHS, Beta $12.95** WAR ♂♂

Pepe le Moko An influential French film about a notorious gangster holed up in the Casbah, emerging at his own peril out of love for a beautiful woman. Stirring film established Gabin as a matinee idol. Cinematography is particularly fine too. Based upon the D'Ashelbe novel. The basis for both "Algiers," the popular Boyer-Lamarr melodrama, and the musical "The Casbah." In French with English subtitles.
1937 87m/B FR Jean Gabin, Mireille Balin, Gabriel Gabrio, Lucas Gridoux; **Dir:** Julien Duvivier. National Board of Review Awards '41: Best Foreign Film. **VHS, Beta, 8mm $29.95** NOS, VYY, HHT ♂♂♂

Pepper International intrigue and super spying abound in this fast-paced, sexy adventure featuring Pepper, a female secret agent in the James Bond tradition.
1982 88m/C VHS, Beta $69.95 VES ♂

Pepper and His Wacky Taxi A father of four buys a '59 Cadillac and starts a cab company.
1972 (G) 79m/C John Astin, Frank Sinatra Jr., Jackie Gayle, Alan Sherman; **Dir:** Alex Grasshof. **VHS, Beta $19.95** UNI ♂

Perfect A "Rolling Stone" reporter goes after the shallowness of the Los Angeles health club scene, and falls in love with the aerobics instructor he is going to write about. "Rolling Stone" publisher Wenner plays himself. As bad as it sounds.
1985 (R) 120m/C John Travolta, Jamie Lee Curtis, Carly Simon, Marilu Henner, Laraine Newman, Jann Wenner, Anne De Salvo; **Dir:** James Bridges. **VHS, Beta, LV $14.95** COL ♂ ½

Perfect Crime A Scotland Yard inspector must find out who has been killing off executives of a powerful world trust.
1978 90m/C Joseph Cotten, Anthony Steel, Janet Agren. **VHS, Beta $29.95** VID ♂♂

Perfect Furlough A corporal wins a week in Paris with a movie star, accompanied by a female Army psychologist. Pretty bizarre, but it works.
1959 93m/C Tony Curtis, Janet Leigh, Keenan Wynn, Linda Cristal, Elaine Strich; **Dir:** Blake Edwards. **VHS, Beta $14.95** KRT, MLB ♂♂ ½

Perfect Killer Van Cleef stars as a world weary Mafia hit-man who is double-crossed by his girl, set up by his best friend, and hunted by another hired assassin. Also on video as "Satanic Mechanic."
1977 (R) 85m/C Lee Van Cleef, Tita Barker, John Ireland, Robert Widmark; **Dir:** Marlon Sirko. **VHS, Beta $49.95** PSM, SIM ♂♂ ½

Perfect Match A timid young woman and an unambitious young man meet each other through the personal ads and then lie to each other about who they are and what they do for a living.
1988 (PG) 93m/C Marc McClure, Jennifer Edwards, Diane Stilwell, Rob Paulsen, Karen Witter; **Dir:** Mark Deimel. **VHS, Beta $79.98** MCG ♂♂

Perfect Strangers Thriller develops around a murder and the child who witnesses it. The killer attempts to kidnap the young boy, but problems arise when he falls in love with the lad's mother.
1984 (R) 90m/C Anne Carlisle, Brad Rijn, John Woehrle, Matthew Stockley, Ann Magnuson, Stephen Lack; **Dir:** Larry Cohen. **VHS, Beta $24.98** SUE ♂♂

Perfect Timing Soft-core film about a fledgling photographer who can't turn his orgy-like photo sessions into anything profitable.

1984 87m/C Stephen Markle, Michelle Scarabelli; *Dir:* Rene Bonniere. **VHS, Beta $69.98** *LIV, LTG* Woof!

Perfect Victims
Beautiful models are being hunted by a killer, and a police officer wants to catch the psychopath before he strikes again.
1987 (R) 100m/C Deborah Shelton, Clarence Williams III, Lyman Ward, Tom Dugan; *Dir:* Shuki Levy. **VHS, Beta $29.95** *ACA* ✍

The Perfect Weapon
Kenpo karate master Speakman severs family ties and wears funny belt in order to avenge underworld murder of his teacher.
1991 (R) 85m/C Jeff Sanders, Jeff Speakman; *Dir:* Mark DiSalle. **VHS, Beta, LV $19.95** *PAR, PIA* ✍✍

Perfect Witness
A restaurant owner witnesses a mob slaying and resists testifying against the culprit to save his family and himself. Made for cable television. Filmed in New York City.
1989 104m/C Brian Dennehy, Aidan Quinn, Stockard Channing; *Dir:* Robert Mandel. **VHS, Beta, LV $89.99** *HBO* ✍✍½

Perfectly Normal
Two friends end up in hilarious situations when one opens an eccentric restaurant with a mysterious, newly found fortune.
1991 (R) 106m/C Michael Riley, Robbie Coltrane, Ken Welsh, Eugene Lipinski; *Dir:* Yves Simoneau. **VHS, Beta $19.95** *ACA* ✍✍

Performance
Grim and unsettling psychological account of a criminal who hides out in a bizarre house occupied by a peculiar rock star and his two female companions. Entire cast scores high marks, with Pallenberg especially compelling as a somewhat mysterious and attractive housemate to mincing Jagger. A cult favorite, with music by Jack Nitzsche under the direction of Randy Newman.
1970 (R) 104m/C *GB* James Fox, Mick Jagger, Anita Pallenberg; *Dir:* Donald Cammell, Nicolas Roeg. **VHS, Beta, LV $19.98** *WAR* ✍✍✍½

Perfumed Nightmare
Unique, appealing fable about a Filipino youth who discovers the drawbacks of social and cultural progress when he moves to Paris. Director Tahimik also plays the lead role. Imaginative, yet remarkably economical, this film was produced for $10,000 and received the International Critics Award at the Berlin Film Festival. In Tagalog with English dialogue and subtitles.
1989 93m/C *PH* Kidlat Tahimik, Dolores Santamaria, Georgette Baudry, Katrin Muller, Harmut Lerch; *Dir:* Kidlat Tahimik. **VHS, Beta $59.95** *FFM* ✍✍✍

Peril
From the novel "Sur La Terre Comme Au Ciel" by Rene Belletto. Deals with a music teacher's infiltration into a wealthy family and the sexually motivated, back-stabbing murder plots that ensue. DeVille does poorly with avant-garde technique. In French; with English subtitles.
1985 (R) 100m/C *FR* Michel Piccoli, Christophe MaLavoy, Richard Bohringer, Nicole Garcia, Anais Jeanneret; *Dir:* Michel DeVille. **VHS, Beta $59.95** *RCA* ✍✍½

Perils of the Darkest Jungle
Tiger Woman battles money-mad oil profiteers to protect her tribe. Also known as "The Tiger Woman." Also released as a feature, "Jungle Gold." This twelve episode serial comes on two tapes.
1944 180m/B Linda Stirling, Allan Lane, Duncan Renaldo, George Lewis; *Dir:* Spencer Gordon Bennet. **VHS $29.95** *REP, MLB* ✍✍½

The Perils of Gwendoline
A young woman leaves a convent to search for her long-lost father, in this adaptation of a much funnier French comic strip of the same name. What ends up on the screen is merely a very silly rip-off of the successful "Raiders of the Lost Ark," and of primary interest for its numerous scenes of amply endowed Kitaen in the buff.
1984 (R) 88m/C *FR* Tawny Kitaen, Brent Huff, Zabou, Bernadette La Font, Jean Rougerie; *Dir:* Just Jaeckin. **VHS, Beta, LV $79.98** *LIV, VES* ✍✍½

Perils of Pauline
All twelve episodes of this classic melodrama/adventure serial in one package, featuring dastardly villains, cliff-hanging predicaments, worldwide chases, and that remarkable character, Pearl White.
1934 238m/B Evelyn Knapp, Robert Allen, James Durkin, Sonny Ray, Pat O'Malley; *Dir:* Ray Taylor. **VHS, Beta $29.95** *NOS, VYY, MLB* ✍✍½

The Perils of Pauline
A musical biography of Pearl White, the reigning belle of silent movie serials. Songs by Frank Loesser include "I Wish I Didn't Love You So," an Academy Award nominee. Look for lots of silent film stars.
1947 96m/C Betty Hutton, John Lund, Constance Collier, William Demarest, Billy DeWolfe; *Dir:* George Marshall. **VHS, Beta $19.95** *NEG, NOS, PSM* ✍✍

Periphery
Shot in Berlin, this fictional narrative tells the story of a group of refugees from World War II who build their own society in which they communicate telepathically. They are hunted down by the press and the police. The film raises questions about the inability of society to deal with alternative lifestyles.
1986 29m/C VHS, 3/4U *WIF* ✍✍½

Permanent Record
A hyper-sincere drama about a popular high schooler's suicide and the emotional reactions of those he has left behind. Reeves gives a great performance. Cinematography by Frederick Elmes.
1988 (PG-13) 92m/C Alan Boyce, Keanu Reeves, Michelle Meyrink, Jennifer Rubin, Pamela Gidley, Kathy Baker; *Dir:* Marisa Silver. **VHS, Beta, LV $89.95** *PAR* ✍✍½

Permanent Vacation
A young man disenchanted with New York City decides to escape to Europe to forget about his problems.
1984 80m/C John Lurie, Chris Parker. **VHS, Beta** *NEW* ✍✍½

Permission To Kill
Grim spy drama features Bogard as an agent determined to prevent the return of a third-world radical to a totalitarian state.
1975 (PG) 96m/C *GB* Dirk Bogarde, Ava Gardner, Timothy Dalton, Frederic Forrest; *Dir:* Cyril Frankel. **VHS, Beta $39.98** *SUE* ✍✍½

Perry Como Show
Perry sings his biggest hits, Fonda does a spoof on westerns, and Anne Francis and Robbie the Robot (from "Forbidden Planet") perform.
1956 60m/B Perry Como, Henry Fonda, Vera-Ellen, Paul Winchell, Jerry Mahoney, The Platters, Anne Francis, Robbie the Robot. **VHS, Beta $19.95** *VDM* ✍✍½

Perry Mason Returns
Perry Mason returns to solve another baffling mystery. This time he must help his longtime assistant, Della Street, when she is accused of killing her new employer.

1985 95m/C Raymond Burr, Barbara Hale, William Katt, Patrick O'Neal, Richard Anderson, Cassie Yates, Al Freeman Jr.; *Dir:* Ron Satlof. **VHS, Beta $29.98** *MCG* ✍✍✍

Perry Mason: The Case of the Lost Love
Super-lawyer Mason defends the husband of a former lover he hadn't seen in 30 years. The solution to the mystery may catch viewer off guard.
1987 98m/C Raymond Burr, Barbara Hale, William Katt, Jean Simmons, Gene Barry, Robert Walden, Stephen Elliott, Robert Mandan, David Ogden Stiers; *Dir:* Ron Satlof. **VHS $19.95** *STE* ✍✍½

Person to Person
Edward R. Murrow electronically visits Groucho and Harpo Marx at their homes in California in this collection of two episodes from the series.
1958 30m/B *Hosted:* Edward R. Murrow. **VHS, Beta $19.95** *VYY* ✍✍½

Persona
A famous actress turns mute and is treated by a talkative nurse at a secluded cottage. As their relationship turns increasingly tense, the women's personalitites begin to merge. Memorable, unnerving—and atypically avant garde—fare from cinema giant Bergman. First of several collaborations between the director and leading lady Ullman. Subtitled in English.
1948 100m/B *SW* Bibi Andersson, Liv Ullman, Gunnar Bjornstrand; *Dir:* Ingmar Bergman. National Board of Review Awards '67: 5 Best Foreign Films of the Year; National Society of Film Critics Awards '67: Best Actress (Andersson), Best Director, Best Film; Sight & Sound Survey '72: #6 of the Best Films of All Time. **VHS, Beta, 8mm $39.95** *HTV, VYY, VDM* ✍✍✍✍

Personal Best
Lesbian lovers compete while training for the 1990 Olympics. Provocative fare often goes where few films have gone before, but overly stylized direction occasionally overwhelms characterizations. It gleefully exploits' locker-room nudity, with Hemingway in her pre-implant days. Still, an ambitious, often accomplished production. Towne's directorial debut. Music by Jack Nitzsche.
1982 (R) 126m/C Mariel Hemingway, Scott Glenn, Patrice Donnelly; *Dir:* Robert Towne. **VHS, Beta $69.95** *WAR* ✍✍✍

Personal Exemptions
In this stock comedy a frenzied IRS auditor learns that her daughter is smuggling aliens into the country, her son is brokering with some businessmen she's targeted for tax fraud, and her husband is having an affair with a high-school girl.
1988 (PG) 100m/C Nanette Fabray, John Cotton; *Dir:* Peter Rowe. **VHS, Beta $14.95** *HMD, SOU* ✍½

Personal Services
A bawdy satire loosely based on the life and times of Britain's modern-day madam, Cynthia Payne, and her rise to fame in the world of prostitution. Written by David Leland. Payne's earlier years were featured in the film "Wish You Were Here."
1987 (R) 104m/C *GB* Julie Walters, Alec McCowen, Shirley Stelfox; *Dir:* Terry Jones. **VHS, Beta, LV $29.98** *LIV, VES* ✍✍½

The Personals
A recently divorced young man takes out a personal ad in a newspaper to find the woman of his dreams. But it isn't quite that simple in this independent comedy shot entirely in Minneapolis.
1983 (PG) 90m/C Bill Schoppert, Karen Landry, Paul Eiding, Michael Laskin, Vickie Dakil; *Dir:* Peter Markle. **VHS, Beta $29.98** *SUE, EMB* ✍✍½

Personals Chilling drama that casts O'Neill as an unnoticed librarian, until she begins to meet (and kill) the men who answer her ads. Zimbalist is a widow who is determined to learn the truth. Made for television.
1990 93m/C Stephanie Zimbalist, Jennifer O'Neill, Robin Thomas, Clark Johnson, Gina Gallego, Rosemary Dunsmore, Colm Feore; *Dir:* Steven Hilliard Stern. VHS, Beta, LV $79.95 PAR, PIA *𝕀𝕀*

The Persuaders The first two episodes of Moore and Curtis as international playboy crime fighters.
1971 52m/C *GB* Roger Moore, Tony Curtis, Alex Scott, Joan Collins, Robert Hutton; *Dir:* Basil Dearden. VHS, Beta $39.95 SVS *𝕀𝕀*

Pet Sematary A quirky adaptation of Stephen King's bestseller about a certain patch of woods in the Maine wilderness that rejuvenates the dead, and how a newly located college MD eventually uses it to restore his dead son. Mildly creepy.
1989 (R) 103m/C Dale Midkiff, Fred Gwynne, Denise Crosby, Stephen King; *Dir:* Mary Lambert. VHS, Beta, LV, 8mm $14.95 PAR *𝕀𝕀*

Pete Kelly's Blues A jazz musician in a Kansas City speakeasy is forced to stand up against a brutal racketeer. The melodramatic plot is brightened by a nonstop flow of jazz tunes sung by Lee and Fitzgerald and played by an all-star lineup that includes Dick Cathcart, Matty Matlock, Eddie Miller and George Van Eps. Lee received an Oscar nomination.
1955 96m/C Jack Webb, Janet Leigh, Edmond O'Brien, Lee Marvin, Martin Milner, Peggy Lee, Ella Fitzgerald, Jayne Mansfield; *Dir:* Jack Webb. VHS, Beta, LV $19.98 WAR, PIA *𝕀𝕀*

Pete 'n' Tillie Amiable comedy turns to less appealing melodrama as couple meets, marries, and endures tragedy. Film contributes little to director Ritt's hitt-and-miss reputation, but Matthau and Burnett shine in leads. Adapted by Oscar-nominated Julius J. Epstein from a Peter de Vries' story. Page received an Oscar nomination.
1972 (PG) 100m/C Walter Matthau, Carol Burnett, Geraldine Page, Barry Nelson, Rene Auberjonois, Lee Montgomery, Henry Jones, Kent Smith; *Dir:* Martin Ritt. British Academy Awards '73: Best Actor (Matthau). VHS, Beta $79.95 MCA *𝕀𝕀½*

Pete Townshend: White City The Who guitarist portrays a musician who helps bring a close friend and and his estranged wife back together. This semi-autobiographical film includes music from Townsend.
1985 60m/C Pete Townsend, Andrew Wilde, Frances Barber; *Dir:* Richard Lowenstein. VHS, Beta, LV $19.95 MVD, VES *𝕀𝕀½*

Peter the First: Part 1 The first half of Petrov's Soviet epic about the early years of Tsar Peter I's reign. Lavish, in Russian with English subtitles.
1937 95m/B *RU Dir:* Vladimir Petrov. VHS, Beta $29.95 FCT *𝕀𝕀*

Peter the First: Part 2 The second half of Petrov's epic, also titled "The Conquests of Peter the Great," covering the triumphs and final years of the famous Tsar. In Russian with English subtitles.
1938 104m/B *RU Dir:* Vladimir Petrov. VHS, Beta $29.95 FCT *𝕀𝕀*

Peter the Great Dry but eye-pleasing TV mini-series follows the life of Russia's colorful, very tall ruler from childhood on. Much of the interesting cast is wasted in tiny roles.
1986 371m/C Maximilian Schell, Laurence Olivier, Omar Sharif, Vanessa Redgrave, Ursula Andress. Emmy Awards '86: Best Mini-Series. VHS $29.95 STE *𝕀𝕀𝕀*

Peter Gunn Detective Peter Gunn returns, only to find himself being hunted by both the mob and the Feds. A TV movie reprise of the vintage series that has none of the original cast members.
1989 97m/C Peter Strauss, Barbara Williams, Jennifer Edwards, Charles Cioffi, Pearl Bailey, Peter Jurasik, David Rappaport; *Dir:* Blake Edwards. VHS $19.95 STE *𝕀𝕀*

Peter Lundy and the Medicine Hat Stallion A teenaged Pony Express rider must outrun the Indians and battle the elements in order to carry mail from the Nebraska Territory to the West Coast in this made-for-television film. Good family entertainment.
1977 85m/C Leif Garrett, Mitchell Ryan, Bibi Besch, John Quade, Milo O'Shea; *Dir:* Michael O'Herlihy. VHS, Beta $69.95 LIV, VES *𝕀𝕀½*

Peter Pan The Disney classic about a boy who never wants to grow up. Based on J. M. Barrie's book and play. Still stands head and shoulders above any recent competition in providing fun family entertainment. Terrific animation and lovely hummable music.
1953 (G) 76m/C *Dir:* Hamilton Luske; *Voices:* Bobby Driscoll, Kathryn Beaumont, Hans Conried, Heather Angel, Candy Candido; *Nar:* Tom Conway. VHS, Beta, LV $24.99 DIS, APD *𝕀𝕀𝕀*

Peter Pan A TV classic, this videotape of a performance of the 1954 Broadway musical features Mary Martin in one of her most famous incarnations, as the adolescent Peter Pan. Songs include "I'm Flying," "Neverland," and "I Won't Grow Up."
1960 100m/C Mary Martin, Cyril Ritchard, Sondra Lee, Heather Halliday, Luke Halpin; *Dir:* Vincent J. Donehue. VHS, Beta, LV $24.99 GKK, COL *𝕀𝕀𝕀½*

Peter and the Wolf The classic tale of the young woodsman is told Disney-style.
1946 30m/C VHS, Beta $12.99 DIS *𝕀𝕀𝕀½*

Peter and the Wolf and Other Tales A collection of live-action stories using dance, music and poetry in "Jabberwocky," "The Ugly Duckling" and the title story.
1984 82m/C VHS, Beta $19.98 CVL *𝕀𝕀𝕀½*

Pete's Dragon Elliot, an enormous, bumbling dragon with a penchant for clumsy heroics, becomes friends with poor orphan Pete. Combines brilliant animation with the talents of live actors for an interesting effect.
1977 (G) 128m/C Helen Reddy, Shelley Winters, Mickey Rooney, Jim Dale, Red Buttons, Sean Marshall, Jim Backus, Jeff Conaway; *Dir:* Don Chaffey; *Voices:* Charlie Callas. VHS, Beta, LV $24.99 DIS, FCT *𝕀𝕀½*

Petit Con A live-action version of the popular French comic strip wherein a young rebel drops out of society to live with aging hippies. With English subtitles.
1984 (R) 90m/C *FR* Guy Marchand, Michel Choupon, Caroline Cellier, Bernard Brieux, Souâd Amidou; *Dir:* Gerard Lauzier. VHS, Beta $59.95 LHV *𝕀𝕀½*

Petrified Forest Patrons and employees are held captive by a runaway gangster's band in a desert diner. Often gripping, with memorable performances from Davis, Howard, and Bogart. Based on the play by Robert Sherwood recreating their stage roles.
1936 83m/B Bette Davis, Leslie Howard, Humphrey Bogart, Dick Foran, Charley Grapewin, Porter Hall; *Dir:* Archie Mayo. VHS, Beta $19.95 FOX, MGM *𝕀𝕀𝕀*

Petulia Overlooked, offbeat drama about a flighty woman who spites her husband by dallying with a sensitive, recently divorced surgeon. Classic '60s document and cult favorite offers great performance from the appealing Christie, with Scott fine as the vulnerable surgeon. On-screen performances by the Grateful Dead and Big Brother. Among idiosyncratic director Lester's best. Photographed by Nicholas Roeg.
1968 (R) 105m/C George C. Scott, Richard Chamberlain, Julie Christie, Shirley Knight, Arthur Hill, Joseph Cotten, Pippa Scott, Richard Dysart; *Dir:* Richard Lester. VHS, Beta $19.98 WAR *𝕀𝕀𝕀𝕀*

Peyton Place Passion, scandal, and deception in a small New England town set the standard for passion, scandal and deception in soap operadom. Shot on location in Camden, Maine. Performances and themes now seem dated, but box office blockbuster garnered nine Academy Award nominations, including best picture, actress (Turner), supporting actor (Kennedy, Tamblyn), supporting actress (Lange, Varsi), director, adapted screenplay, and cinematography. Written by John Michael Hayes from Grace Metalious' popular novel and followed by "Return to Peyton Place."
1957 157m/C Lana Turner, Hope Lange, Lee Philips, Lloyd Nolan, Diane Varsi, Lorne Greene, Russ Tamblyn, Arthur Kennedy, Terry Moore, Barry Coe, David Nelson, Betty Field, Mildred Dunnock, Leon Ames; *Dir:* Mark Robson. VHS $39.98 FOX, FCT *𝕀𝕀𝕀*

Phantasm A small-budgeted, hallucinatory horror fantasy about two parentless brothers who discover weird goings-on at the local funeral parlor, including the infamous airborne, brain-chewing chrome ball. Creepy, unpredictable nightmare fashioned on a shoestring by young independent producer Coscarelli. Scenes were cut out of the original film to avoid "X" rating. Followed by "Phantasm II."
1979 (R) 90m/C Michael Baldwin, Bill Thornbury, Reggie Bannister, Kathy Lester, Terrie Kalbus, Ken Jones, Susan Harper, Lynn Eastman, David Arntzen, Angus Scrimm, Bill Cone; *Dir:* Don A. Coscarelli. VHS, Beta, LV $14.95 COL, SUE *𝕀𝕀½*

Phantasm II Teen psychic keeps flashing on villainous Tall Man. A rehash sequel to the original, cultishly idiosyncratic fantasy, wherein more victims are fed into the interdimensional abyss and Mike discovers that the horror is not all in his head. Occasional inspired gore; bloodier than its predecessor. More yuck for the buck.
1988 (R) 97m/C James Le Gros, Reggie Bannister, Angus Scrimm; *Dir:* Don A. Coscarelli. VHS, Beta, LV $14.98 MCA *𝕀½*

The Phantom of 42nd Street An actor and a cop team up to find the killer of a wealthy uncle.
1945 58m/B Dave O'Brien, Kay Aldridge, Alan Mowbray, Frank Jenks; *Dir:* Martin Mooney. VHS $16.95 SNC, NOS, MRV *𝕀𝕀*

The Phantom from 10,000 Leagues Slimy sea monster attacks swimmers and fishermen; investigating oceanographer pretends not to notice monster is hand puppet. Early AIP release, when still named American Releasing Company.

1956 80m/B Kent Taylor, Cathy Downs, Michael Whalen, Helene Stanton, Phillip Pine; *Dir:* Dan Milner. **VHS $19.98** *NOS, SNC, MLB* ♦

The Phantom Broadcast A popular radio crooner is murdered. The main suspect is his hunchbacked accompanist and manager, who had all along secretly pre-recorded his own velvet voice for the crooner to lip-sync along with. A tightly plotted minor thriller that holds up surprisingly well.
1933 63m/B Ralph Forbes, Gail Patrick, Vivienne Osborne, Guinn Williams, George "Gabby" Hayes; *Dir:* Phil Rosen. **VHS, Beta, 8mm $29.95** *NOS, VYY, FCT* ♦♦½

Phantom Brother A semi-spoof about a teenager left orphaned by a car crash that wiped out his entire family, and who is subsequently haunted by his dead brother. Even thinner than it sounds.
1988 92m/C Jon Hammer, Patrick Molloy, John Gigante, Mary Beth Pelshaw; *Dir:* William Szarka. **VHS, Beta $69.95** *SOU* ♦

The Phantom Bullet Hoot wants to holler at his father's murderers.
1926 60m/B Hoot Gibson. **VHS, Beta $16.95** *GPV* ♦

The Phantom Chariot A fantasy depicting the Swedish myth about how Death's coach must be driven by the last man to die each year. Silent.
1920 89m/B *SW Dir:* Victor Sjostrom. **VHS, Beta $16.95** *SNC, FCT, DNB* ♦♦½

Phantom of Chinatown Mr. Wong is called in to solve another murder in the final entry in the series. The first to have an Asian actor portray the lead, as Luke replaced Boris Karloff.
1940 61m/B Keye Luke, Lotus Long, Grant Withers, Paul McVey, Charles Miller; *Dir:* Phil Rosen. **VHS $16.95** *NOS, SNC* ♦

The Phantom Creeps Evil Dr. Zorka, armed with a meteorite chunk which can bring an army to a standstill, not to mention build a robot and a giant spider, provides the impetus for this enjoyable serial in 12 episodes.
1939 75m/B Bela Lugosi, Dorothy Arnold, Robert Kent, Regis Toomey. **VHS, Beta $19.95** *NOS, SNC, VCN* ♦♦½

Phantom of Death A brilliant pianist, stricken by a fatal disease that causes rapid aging, goes on a murderous rampage.
1987 (R) 95m/C *IT* Michael York, Donald Pleasence, Edwige Fenech; *Dir:* Ruggero Deodato. **VHS, Beta $79.95** *VMK* ♦½

The Phantom Empire Autry faces the futuristic "Thunder Riders" from the subterranean city of Murania, located 20,000 feet beneath his ranch. A complete serial in twelve episodes. If you only see one science fiction western in your life, this is the one.
1935 245m/B Gene Autry, Frankie Darro, Betsy King Ross, Smiley Burnette. **VHS, Beta $24.95** *NOS, SNC, VYY* ♦♦

Phantom Empire A woman who rules over a lost city takes a bunch of scientists prisoner and forces them to be slaves.
1987 (R) 85m/C Ross Hagen, Jeffrey Combs, Dawn Wildsmith, Robert Quarry, Susan Stokey, Michelle Bauer, Russ Tamblyn, Sybil Danning; *Dir:* Fred Olen Ray. **VHS, Beta $79.95** *PSM* ♦♦

Phantom Express Experienced old engineer is dismissed from his job because no one believes his explanation that a mysterious train caused the wreck of his own train. He aims to find justice.

1932 65m/B J. Farrell MacDonald, Sally Blane, William Collier Jr., Hobart Bosworth; *Dir:* Emory Johnson. **VHS, Beta $12.95** *SNC, NOS, IUF* ♦♦

Phantom Fiend A gentle musician becomes a suspect when a series of Jack-the-Ripper type murders terrorize London. Is he the nice guy known by his girlfriend and the people in his lodging house, or is the musician really JTR? Based on the novel "The Lodger" by Marie Belloc-Lowndes.
1935 70m/B *GB* Ivor Novello, Elizabeth Allan, A.W. Baskcomb, Jack Hawkins, Barbara Everest, Peter Gawthorne, P. Kynaston Reeves; *Dir:* Maurice Elvey. **VHS $29.95** *FCT* ♦½

The Phantom Flyer A stunt pilot helps out a cattle-owning frontier family bedeviled by rustlers. Silent.
1928 54m/B Al Wilson. **VHS, Beta, 8mm $24.95** *VYY, GPV, RXM* ♦½

The Phantom in the House A woman commits a murder and has her inventor husband take the rap. While he's in prison, she gets rich off of his work. Things get hairy when he is released fifteen years later and is introduced to his daughter as a friend of the family. Melodramatic film was originally made with and without sound.
1929 64m/B Ricardo Cortez, Nancy Welford, Henry B. Walthall, Grace Valentine; *Dir:* Phil Rosen. **VHS, Beta $16.95** *SNC* ♦♦

Phantom of Liberty Master surrealist Bunuel's episodic film wanders from character to character and from event to event. Animals wander through a man's bedroom, soldiers conduct military exercises in an inhabited area, a missing girl stands before her parents even as they futilely attempt to determine her whereabouts, and an assassin ic found guilty, then applauded and led to freedom. They don't get much more surreal than this. If you think you may like it, you'll probably love it. Bunuel, by the way, is among the firing squad vicitms in the film's opening enactment of Goya's May 3, 1808. In French with English subtitles.
1974 104m/C *FR* Adrianna Asti, Jean-Claude Brialy, Michel Piccoli, Adolfo Celi, Monica Vitti; *Dir:* Luis Bunuel. National Board of Review Awards '74: 5 Best Foreign Films of the Year. **VHS, Beta, LV $24.95** *XVC, TAM, APD* ♦♦♦½

Phantom of the Mall: Eric's Revenge A murderous spirit haunts the local mall. Gore flows like water.
1989 91m/C Morgan Fairchild, Kari Whitman, Jonathan Goldsmith, Derek Rydell; *Dir:* Richard S. Friedman. **VHS, Beta, LV $19.95** *FRH* ♦

Phantom of the Opera Deranged, disfigured music lover haunts the sewers of a Parisian opera house and kills to further the career of an unsuspecting young soprano. First of many film versions still packs a wollop, with fine playing from Lon Chaney, Sr. Silent with two-color Technicolor "Bal Masque" sequence.
1925 79m/B Lon Chaney Sr., Norman Kerry, Mary Philbin, Gibson Gowland; *Dir:* Rupert Julian. **VHS, Beta, LV, 8mm $9.95** *NEG, NOS, CCB* ♦♦♦

Phantom of the Opera Second Hollywood version—the first with talking—of Gastron Leroux novel about a madman who promotes fear and mayhem at a Paris opera house. Raines is good in lead; Eddy warbles away.
1943 92m/C Claude Rains, Nelson Eddy, Susanna Foster, Edgar Barrier, Leo Carrillo, Hume Cronyn, J. Edward Bromberg; *Dir:* Arthur Lubin. Academy Awards '43: Best Art Direction/Set Decoration (Color), Best Color

Cinematography, Best Interior Decoration. **VHS, Beta, LV $14.95** *MCA* ♦♦♦

Phantom of the Opera A gory, "Elm Street"-ish version of the Gaston Leroux classic, attempting to cash in on the success of the Broadway musical.
1989 (R) 93m/C Robert Englund, Jill Schoelen, Alex Hyde-White, Bill Nighy, Terence Harvey, Stephanie Lawrence; *Dir:* Dwight Little. **VHS, Beta, LV $14.95** *COL* ♦♦

Phantom of the Paradise A rock 'n roll parody of "Phantom of the Opera." Splashy, only occasionally horrific spoof in which cruel music executive Williams, much to his everlasting regret, swindles a songwriter. Violence ensues. Not for most, or even many, tastes. Graham steals the film as rocker Beef. A failure at the box office, and now a cult item (small enthusiastic cult with few outside interests) for its oddball humor and outrageous rock star parodies. Williams also wrote the turgid score.
1974 (PG) 92m/C Paul Williams, William Finley, Jessica Harper, Gerrit Graham; *Dir:* Brian DePalma. **VHS, Beta, LV $19.98** *FOX* ♦♦

Phantom Patrol A mystery writer is kidnapped and someone impersonates him. Mountie Maynard must come to the rescue.
1936 60m/B Kermit Maynard, Joan Barclay, Paul Fix, Julian Rivero, Eddie Phillips, Roger Williams; *Dir:* Charles Hutchinson. **VHS, Beta $24.95** *DVT, NOS* ♦½

The Phantom Planet An astronaut crash-lands on an asteroid and discovers a race of tiny people living their. Having breathed the atmosphere, he shrinks to there dimunitive size and aids them in their war against brutal invaders. Infamously peculiar.
1961 82m/B Dean Fredericks, Coleen Gray, Tony Dexter, Dolores Faith, Francis X. Bushman, Richard Kiel; *Dir:* William Marshall. **VHS, Beta $16.95** *NOS, SNC* ♦½

Phantom Rancher Roaring melodrama finds Maynard donning a mask to find the real Phantom who is causing havoc.
1939 61m/B Ken Maynard. **VHS, Beta** *UHV, VCN* ♦

Phantom Ranger A federal agent masquerades as a crook to catch a band of counterfeiters.
1938 54m/B Tim McCoy, Suzanne Kaaren, Karl Hackett, John St. Polis, John Merton, Harry Strang; *Dir:* Sam Newfield. **VHS $19.95** *NOS, DVT* ♦½

Phantom Rider Mystery fills the old West in this serial composed of fifteen chapters.
1937 152m/B Buck Jones, Maria Shelton. **VHS, Beta $69.95** *VCN, MLB* ♦½

Phantom of the Ritz Bizarre accidents begin to take place when Ed Blake and his girlfriend start renovating the old Ritz Theater. It seems the theater is inhabited by a rather angry ghost and somebody's got to deal with it before the Ritz can rock and roll. Featuring the fabulous sounds of the Fifties.
1988 (R) 89m/C Peter Bergman, Deborah Van Valkenburgh, The Coasters; *Dir:* Allen Plone. **VHS, Beta $79.95** *PSM* ♦♦

The Phantom Ship The true story of the "Marie Celeste," a ship found in 1872 off the coast of England with her sails set but minus any of her crew. Also known as "The Mystery of the Marie Celeste."
1937 61m/B Bela Lugosi, Shirley Grey. **VHS, Beta $19.95** *NOS, SNC, DVT* ♦♦

The Phantom of Soho A Scotland Yard detective investigates the murders of several prominent businessmen and is assisted by a beautiful mystery writer.
1964 92m/B *GE* Dieter Borsche, Barbara Rutting, Hans Sohnker, Peter Vogel, Helga Sommerfeld, Werner Peters; *Dir:* Franz Gottlieb. VHS, Beta $16.95 *SNC* ♫♫

Phantom from Space An invisible alien lands on Earth, begins killing people, and is pursued by a pair of scientists.
1953 72m/B Ted Cooper, Rudolph Anders, Noreen Nash, James Seay, Harry Landers; *Dir:* W. Lee Wilder. VHS, Beta $16.95 *NOS, SNC* ♫♫½

Phantom Thunderbolt A man who has falsely spread the rumor that he's a gunfighter is hired to chase away some good-for-nothings.
1933 62m/B Ken Maynard, Frances Lee, Frank Rice, William Gould, Bob Kortman; *Dir:* Alan James. VHS, Beta $19.95 *NOS, DVT, MTX* ♫♫½

Phantom Tollbooth A young boy drives his car into an animated world, where the numbers are at war with the letters and he has been chosen to save Rhyme and Reason, to bring stability back to the Land of Wisdom. Completely unique and typically Jonesian in its intellectual level and interests. Bright children will be interested, but this is really for adults who will understand the allegory. Based on Norman Justers' metaphorical novel.
1969 (G) 89m/C *Dir:* Chuck Jones; *Voices:* Mel Blanc, Hans Conried. VHS, Beta $19.95 *MGM* ♫♫♫

The Phantom Treehouse This is an animated children's tale of two kids plunging into a fantasy world through a mysterious swamp-based treehouse.
1984 76m/C VHS, Beta $14.98 *VID* ♫♫½

Phantom of the West Ten-episode serial about a rancher who becomes "The Phantom of the West" in order to smoke out his father's killer.
1931 166m/B William Desmond, Thomas Santschi, Tom Tyler; *Dir:* Ross Lederman. VHS, Beta $49.95 *NOS, VYY, VCN* ♫♫

Phar Lap Saga of a legendary Australian racehorse who rose from obscurity to win nearly 40 races in just three years before mysteriously dying in 1932. American version runs 10 minutes shorter than the Aussie one.
1984 (PG) 107m/C *AU* Ron Leibman, Tom Burlinson, Judy Morris, Celia de Burgh; *Dir:* Simon Wincer. VHS, Beta $19.98 *FOX* ♫♫½

Pharaoh An expensive Polish epic about the power plays of royalty in ancient Egypt. Dubbed.
1966 125m/B *PL* Jerzy Zelnick; *Dir:* Jerzy Kawalerowicz. VHS, Beta *FCT* ♫♫

Pharmacist A day in the life of a hapless druggist undone by disgruntled customers and robbers. Typical Fields effort.
1932 19m/B Elise Cavanna, Babe Kane, W.C. Fields, Grady Sutton. VHS, Beta $24.98 *CCB, RXM* ♫♫½

Phase 4 A tale of killer ants retaliating against the humans attempting to exterminate them.
1974 (PG) 84m/C Nigel Davenport, Michael Murphy, Lynne Frederick; *Dir:* Saul Bass. VHS, Beta, LV $39.95 *PAR* ♫♫

Phedre Jean Racine's adaptation of the Greek legend involving Phedre, Theseus and Hippolyte is presented in French with English subtitles. Bell is the only one worth watching in this weak and stagy picture.

1968 93m/C Marie Bell; *Dir:* Pierre Jourdan. VHS, Beta $29.95 *VYY, APD* ♫♫½

Phenomenal and the Treasure of Tutankamen Fiendish criminal genius steals priceless golden mask of King Tut. Superhero Phenomenal is less than. Dry spaghetti sci-fi.
197? (R) 90m/C *IT* Mauro Parenti, Lucretia Love. VHS $29.99 *UHV* ♫♫

Phfffft! Holliday and Lemmmon are bored couple who decide to divorce, date others, and take mambo lessons. By no stretch of plausibility, they constantly run into each other and compete in the same mambo contest.
1954 91m/B Judy Holliday, Jack Lemmon, Jack Carson, Kim Novak, Luella Gear, Merry Anders; *Dir:* Mark Robson. VHS *GKK* ♫♫½

The Philadelphia Experiment A WWII sailor falls through a hole in time and lands in the mid-1980s, whereupon he woos a gorgeous woman. Sufficient chemistry between Pare and Allen, but PG rating is an indication of the film's less-than-graphic love scenes.
1984 (PG) 101m/C Michael Pare, Nancy Allen, Eric Christmas, Bobby DiCicco; *Dir:* Stewart Raffill. VHS, Beta, LV $19.99 *HBO, STE* ♫♫½

Philadelphia Story A woman's plans to marry again go awry when her dashing ex-husband arrives on the scene. Matters are further complicated when a loopy reporter—assigned to spy on the nuptials—falls in love with the blushing bride. Classic comedy, with trio of Hepburn, Grant, and Stewart all serving aces. Based on the hit Broadway play by Philip Barry, and remade as the musical "High Society" in 1956 (stick to the original). Clutched two Oscars and was also nominated for Best Picture, Best Actress (Hepburn), Best Supporting Actress (Hussey), Best Director. Also available colorized.
1940 112m/B Katharine Hepburn, Cary Grant, James Stewart, Ruth Hussey, Roland Young; *Dir:* George Cukor. Academy Awards '40: Best Actor (Stewart), Best Screenplay. VHS, Beta, LV $19.98 *MGM, BTV* ♫♫♫♫

Philby, Burgess and MacLean: Spy Scandal of the Century The true story of the three infamous British officials who defected to the Soviet Union in 1951, after stealing some vital British secrets for the KGB.
1984 83m/C *GB* Derek Jacobi, Anthony Bate, Michael Culver; *Dir:* Gordon Flemyng. VHS, Beta $59.95 *MTI* ♫♫½

Phobia Patients at a hospital are mysteriously being murdered. Stupid and unpleasant story that lasts too long and probably should never have started.
1980 (R) 91m/C *CA* Paul Michael Glaser, Susan Hogan; *Dir:* John Huston. VHS, Beta $69.95 *SOU, PAR* Woof!

Phoenix An aging queen summons her marksman to find the Phoenix, a mythical bird that she believes will bring her eternal life. With English subtitles.
1978 137m/C *JP* Tatsuya Nakadai, Tomisaburo Wakayama, Raoru Yumi, Reiko Ohara. VHS, Beta $29.95 *VDA* ♫½

Phoenix the Ninja A young girl grows up to be a ninja warrior and battles everyone who would deign to impede her.
198? 92m/C Sho Kosugi. VHS, Beta $59.95 *TWE* ♫

Phoenix Team Two agents discover their mutual attraction can be deadly in the spy business.
1980 90m/C Don Francks, Elizabeth Shepherd. VHS, Beta $49.95 *TWE* ♫

Phoenix the Warrior Sometime in the future, female savages battle each other for control of the now ravaged earth. A newcomer seeks the tribe most worthy of receiving the last man on the planet, thereby continuing the human race.
1988 90m/C Persis Khambatta, James H. Emery, Peggy Sands, Kathleen Kinmont; *Dir:* Robert Hayes. VHS $79.95 *SVS* ♫

The Phone Call 900-number morality tale about high-powered executive who calls phone sex line and is connected with escaped homicidal maniac who takes a toll on him and his family.
1989 (R) 95m/C Michael Sarrazin, Linda Smith, Ron Lea, Lisa Jakub; *Dir:* Allan Goldstein. VHS $79.95 *MNC* ♫

Phone Call from a Stranger After a plane crash, a survivor visits the families of three of the victims whom he met during the flight. Written by Nunnally Johnson.
1952 96m/B Bette Davis, Gary Merrill, Michael Rennie, Shelley Winters, Hugh Beaumont, Keenan Wynn, Eve Arden, Craig Stevens; *Dir:* Jean Negulesco. VHS, Beta $19.98 *FOX* ♫♫½

Photographer A photographer turns into a murderer, showing the negative side of his personality.
1975 (PG) 94m/C Michael Callan; *Dir:* William B. Hillman. VHS, Beta $59.98 *CHA* ♫

Photonos This is a video album by the Emmy-winning group Emerald Web. In stereo VHS.
1982 58m/C VHS, Beta *VCD* ♫

Physical Evidence A lawyer finds herself falling for an ex-cop turned murder suspect while she tries to defend him for a crime he doesn't remember committing.
1989 (R) 99m/C Burt Reynolds, Theresa Russell, Ned Beatty, Kay Lenz, Ted McGinley; *Dir:* Michael Crichton. VHS, Beta, LV $89.98 *LIV, VES, HHE* ♫♫

P.I. Private Investigations Corrupt, drug-dealing cops stalk another cop who can finger them.
1987 (R) 91m/C Martin Balsam. VHS, Beta *FOX* ♫

A Piano for Mrs. Cimino Declared senile and incompetent, a widowed woman fights for control of her own life. Good script.
1982 100m/C Bette Davis, Keenan Wynn, Alexa Kenin, Penny Fuller, Christopher Guest, George Hearn; *Dir:* George Schaefer. VHS, Beta $59.95 *GEM, IGV* ♫♫♫

Piano Players Rarely Ever Play Together A soulful film featuring many celebrated New Orleans piano players, including Professor Longhair, Allen Toussaint, and "Tuts" Washington. The film explores the links between three generations of players, and examines how they have influenced one another. A 76-minute version is also available. Highly recommended, winner of numerous awards.
1989 60m/C Professor Longhair, "Tuts" Washington, Allen Toussaint. VHS, Beta $39.95 *SPI, MVD, FFM* ♫♫♫

Picasso Trigger An American spy tries to catch an elusive murderer. The sequel to "Hard Ticket to Hawaii" and "Malibu Express."

1989 (R) 99m/C Steve Bond, Dona Spier, John Aprea, Hope Marie Carlton, Guich Koock; *Dir:* Andy Sidaris. **VHS** $19.98 *WAR ♂♂*

The Pick-Up Artist The adventures of a compulsive Don Juan who finds he genuinely loves the daughter of an alcoholic who's in debt to the mob. Standard story with no surprises.
1987 (PG-13) 81m/C Robert Downey Jr., Molly Ringwald, Dennis Hopper, Harvey Keitel, Danny Aiello, Vanessa Williams, Robert Towne, Mildred Dunnock, Lorraine Bracco, Joe Spinell; *Dir:* James Toback. **VHS, Beta, LV** $19.98 *FOX ♂♂½*

Pick-Up Summer Two suburban boys cruise their town after school lets out, chasing a pair of voluptuous sisters.
1979 (R) 99m/C *CA* Michael Zelniker, Carl Marotte; *Dir:* George Mihalka. **VHS, Beta** $59.95 *NWV ♂*

Pickles This is a series of a dozen vignettes commenting pointedly on the many absurd predicaments we humans get ourselves into. Uses irony, metaphor, and satire. No narration.
1973 11m/C **VHS, Beta** *BFA ♂*

Pickpocket Slow moving, documentary-like account of a petty theif's existence is a moral tragedy inspired by "Crime and Punishment." Ending is particularly moving. Classic filmmaking from Bresson, France's master of austerity. In French with English subtitles.
1959 75m/B *FR* Martin LaSalle, Marika Green, Pierre Leymarie; *Dir:* Robert Bresson. **VHS, Beta** $59.95 *FCT ♂♂♂½*

Pickup on South Street Petty thief Widmark lifts woman's wallet only to find it contains top secret Communist micro-film for which pinko agents will stop at nothing to get back. Intriguing look at the politics of the day. Ritter was nominated for an Oscar. The creme of "B" movies.
1953 80m/B Richard Widmark, Jean Peters, Thelma Ritter, Murvyn Vye, Richard Kiley, Milburn Stone; *W/Dir:* Samuel Fuller. **VHS, Beta** $19.98 *FOX, FCT ♂♂♂*

The Pickwick Papers Comedy based on the Dickens classic wherein Mrs. Bardell sues the Pickwick Club for breach of promise.
1954 109m/B *GB* James Hayter, James Donald, Nigel Patrick, Hermione Gingold, Hermione Baddeley, Kathleen Harrison; *Dir:* Noel Langley. **VHS, Beta** $19.98 *UHV ♂♂½*

Picnic A wanderer arrives in a small town and immediately wins the love a his friend's girl. The other women in town seem interested too. Strong, romantic work, with Holden excelling in the lead. Novak provides a couple pointers too. Lavish Hollywood adaptation of the William Inge play, with music score by George Duning, including the popular "Moonglow/Theme from Picnic." Grabbed two Oscars and was also nominated for Best Picture, Best Director, supporting actor Connell, and Score of a Comedy or Drama.
1955 113m/C William Holden, Kim Novak, Rosalind Russell, Susan Strasberg, Arthur O'Connell, Cliff Robertson, Betty Field, Verna Felton, Reta Shaw, Nick Adams, Phyllis Newman, Raymond Bailey; *Dir:* Joshua Logan. Academy Awards '55: Best Art Direction/Set Decoration (Color), Best Film Editing; National Board of Review Awards '55: 10 Best Films of the Year. **VHS, Beta, LV** $19.95 *COL, PIA ♂♂♂½*

Picnic on the Grass A strange, whimsical fantasy heavily evocative of the director's Impressionist roots; a science-minded candidate for the president of Europe throws a picnic as public example of his earthiness, and falls in love with a peasant girl. In French with English subtitles.
1959 92m/C *FR* Paul Meurisse, Catherine Rouvel, Fernand Sardou, Jacqueline Morane, Jean-Pierre Granval; *Dir:* Jean Renoir. **VHS, Beta** $49.95 *FCT, INT ♂♂*

Picnic at Hanging Rock School outing into a mountainous region ends tragically when three girls disappear. Eerie film is strong on atmosphere, as befits mood master Weir. Lambert is extremely photogenic—and suitable subdued—as one of the girls to disappear. Otherwise beautifully photographed on location.
1975 (PG) 110m/C *AU* Margaret Nelson, Rachel Roberts, Dominic Guard, Helen Morse, Jacki Weaver; *Dir:* Peter Weir. **VHS, Beta** *VES, OM ♂♂♂*

Picture of Dorian Gray Hatfield plays the rake who stays young while his portrait ages in this adaptation of Oscar Wilde's classic novel. Lansbury steals this one. Additional Oscar nominations for Best Supporting Actress (Lansbury), Black & White Interior Decoration.
1945 111m/B Hurd Hatfield, George Sanders, Donna Reed, Angela Lansbury, Peter Lawford, Lowell Gilmore; *Dir:* Albert Lewin. Academy Awards '45: Best Black and White Cinematography. **VHS, Beta, LV** $24.95 *MGM, FHE ♂♂♂*

Picture of Dorian Gray Another version of Wilde's renowned novel about a man who retains his youthful visage while his portrait shows the physical ravages of aging. Davenport is particularly appealing in the lead. Made for television.
1974 130m/C Shane Briant, Nigel Davenport, Charles Aidman, Fionnula Flanagan, Linda Kelsey, Vanessa Howard; *Dir:* Glenn Jordan. **VHS, Beta** $29.95 *IVE ♂♂♂*

Picture Mommy Dead Well acted melodrama involving a scheming shrew who struggles to drive her mentally disturbed stepdaughter insane for the sake of cold, hard cash.
1966 85m/C Don Ameche, Zsa Zsa Gabor, Martha Hyer, Susan Gordon; *Dir:* Bert I. Gordon. **VHS, Beta** $9.98 *CHA ♂♂*

Piece of the Action An ex-cop beats two con men at their own game when he convinces them to work for a Chicago community center. Music by Curtis Mayfield.
1977 (PG) 135m/C Sidney Poitier, Bill Cosby, James Earl Jones, Denise Nicholas, Hope Clarke, Tracy Reed, Tito Vandis, Ja'net DuBois; *Dir:* Sidney Poitier. **VHS, Beta** $19.98 *WAR ♂♂½*

A Piece of Pleasure Marriage declines due to a domineering husband in this familiar domestic study from French master Chabrol. Good, but not among the director's best efforts. In French with English subtitles.
1970 100m/C *Dir:* Claude Chabrol. **VHS, Beta** *FCT ♂♂♂*

Pieces A chain-saw wielding madman roams a college campus in search of human parts for a ghastly jigsaw puzzle.
1983 (R) 90m/C *IT SP* Christopher George, Lynda Day George, Paul Smith; *Dir:* Juan Piquer Simon. **VHS, Beta** $79.98 *VES Woof!*

Pied Piper of Hamelin The evergreen classic of the magical piper who rids a village of rats and then disappears with the village children into a mountain when the townspeople fail to keep a promise.
1957 90m/C Van Johnson, Claude Rains, Jim Backus, Kay Starr, Lori Nelson; *Dir:* Bretaigne Windust. **VHS, Beta** $14.95 *KAR, NOS, MED ♂♂*

Pied Piper of Hamelin From Shelley Duvall's "Faerie Tale Theatre" comes the story of how a man with a magic flute charmed the rats out of Hamelin.
1984 60m/C Eric Idle; *Dir:* Nicholas Meyer. **VHS, Beta** $14.98 *KUI, FOX, FCT ♂♂*

Pier 23 A private eye is hired to bring a lawless ex-convict to a priest's custody, but gets an imposter instead. A web of intrigue follows.
1951 57m/B Hugh Beaumont, Ann Savage, David Bruce, Raymond Greenleaf. **VHS, Beta, LV** *WGE ♂♂*

Pierrot le Fou A woman fleeing a gangster joins a man leaving his wife in this stunning, occasionally confusing classic from iconoclast Godard. A hallmark in 1960s improvisational filmmaking, with rugged Belmondo and always-photogenic Karina effortlessly excelling in leads. In French with English subtitles.
1965 110m/C *IT FR* Samuel Fuller, Jean-Pierre Leaud, Jean-Paul Belmondo, Anna Karina, Dirk Sanders; *Dir:* Jean-Luc Godard. **VHS, LV** $59.95 *INT, FCT, IME ♂♂♂½*

Pigeon Feathers A young man realizes a greater understanding of life in this adaptation of a John Updike tale.
1988 45m/C Christopher Collet, Caroline McWilliams. **VHS** $24.95 *MON, KAR ♂♂½*

Pigs A young woman who has escaped from a mental hospital teams up with an evil old man to go on a murdering spree. They complement each other beautifully. She kills them and he disposes of the bodies by making pig slop out of them. Also on video as "Daddy's Deadly Darling" and "The Killers."
1973 90m/C Toni Lawrence, Marc Lawrence, Jesse Vint, Katharine Ross; *W/Dir:* Marc Lawrence. **VHS, Beta** $9.95 *PGN, MRV ♂½*

Pilgrim Farewell This program is a dramatic story about a dying woman who goes looking for her estranged teenage daughter.
1982 110m/C Elizabeth Huddle, Christopher Lloyd, Laurie Pranage, Lesley Paxton, Shelley Wyant, Robert Brown; *Dir:* Michael Roemer. **VHS, Beta** *FLI ♂♂½*

Pillow Talk Sex comedy in which a man woos a woman who loathes him. By the way, they share the same telephone party line. Narrative provides minimal indication of the film's strengths, which are many. Classic '50's comedy with masters Day and Hudson exhibiting considerable rapport, even when fighting. Lighthearted, constantly funny. Oscar nominations for actress Day, supporting actress Ritter, Art Direction, Set Decoration, Musical Score in addition to the award it won.
1959 102m/C Rock Hudson, Doris Day, Tony Randall, Thelma Ritter, Nick Adams, Lee Patrick; *Dir:* Michael Gordon. Academy Awards '59: Best Story & Screenplay. **VHS, Beta** $14.95 *MCA, CCB ♂♂♂½*

The Pilot This film examines the value of honesty with one's friends.

1982 (PG) 98m/C Cliff Robertson, Diane Baker, Frank Converse, Dana Andrews, Amber O'Shea, Edward Binns, Gordon MacRae; *Dir:* Cliff Robertson. VHS, Beta $42.95 PGN, HHE ♫♫

Pilot A young girl discovers the debilitating effects of dishonesty when she becomes a break-dancing champion.
1984 27m/C Kelly Jo Minter, Adolph "Oz" Alvarez, Katie Rich. VHS, Beta PAU ♫ ½

Pimpernel Smith Seemingly scatter-brained archaeology professor is actually a dashing agent rescuing refugees from evil Nazis in WWII France. Howard is well cast, but the film is somewhat predictable, and it's too long. Wonderful scene, though, in which a Nazi officer champions Shakespeare as Aryan.
1942 121m/B GB Leslie Howard, Mary Morris, Francis L. Sullivan, David Tomlinson; *Dir:* Leslie Howard. VHS, Beta $29.95 VYY ♫♫ ½

Pin A boy's imaginary friend assists in the slaying of various enemies. Horror effort could be worse.
1988 (R) 103m/C Cyndy Preston, David Hewlett, Terry O'Quinn; *Dir:* Sandor Stern. VHS, Beta, LV $19.95 STE, NWV, IME ♫♫ ½

Pin-Up Girl Grable plays a secretary who becomes an overnight sensation during World War II. Loosely based on her famous pinup poster that was so popular at the time, the movie didn't even come close to being as successful. The songs aren't particularly memorable, although Charlie Spivak and his Orchestra perform.
1944 83m/C Betty Grable, Martha Raye, Jon Harvey, Joe E. Brown, Eugene Pallette, Mantan Moreland; *Dir:* H. Bruce Humberstone. VHS, LV $19.98 FOX, FXV ♫

The Pinchcliffe Grand Prix A master inventor designs the ultimate race car.
1981 (G) 88m/C VHS, Beta $19.95 STE, GEM ♫

Ping Pong A Chinese patriarch drops dead in London's Chinatown, leaving a young law student to disentangle his will, and get mixed up with his cross-cultured family.
1987 (PG) 100m/C David Yip, Lucy Sheen, Robert Lee, Lam Fung, Victor Kan, Barbara Yo Ling, Ric Young; *Dir:* Po Chich Leong. VHS, Beta, LV $79.95 VIR ♫♫ ½

The Pink Angels Transvestite bikers wheel their way to Los Angeles, turning many heads as they go.
1971 (R) 88m/C John Alderman, Tom Basham, Henry Olek, Dan Haggerty, Michael Pataki; *Dir:* Lawrence Brown. VHS, Beta $49.95 PSM ♫ ½

Pink Cadillac A grizzled, middle-aged bondsman is on the road, tracking down bail-jumping crooks. He helps the wife and baby of his latest target escape from her husband's more evil associates. Eastwood's performance is good and fun to watch, in this otherwise lightweight film.
1989 (PG-13) 121m/C Clint Eastwood, Bernadette Peters, Timothy Carhart, Michael Des Barres, William Hickey; *Dir:* Buddy Van Horn. VHS, Beta, LV, 8mm $19.98 WAR ♫♫ ½

The Pink Chiquitas Sci-fi spoof about a detective battling a mob of meteorite-traveling Amazons.
1986 (PG-13) 86m/C Frank Stallone, Eartha Kitt, Bruce Pirrie, McKinlay Robinson, Elizabeth Edwards, Claudia Udy; *Dir:* Anthony Currie. VHS, Beta $9.99 STE, PSM ♫ ½

Pink at First Sight The Pink Panther returns in a new animated collection of his funniest adventures.
1984 49m/C VHS, Beta MGM ♫

Pink Flamingos Divine, the dainty 300-pound transvestite, faces the biggest challenge of his/her career when he/she competes for the title of World's Filthiest Person. Tasteless, crude, and hysterical film; this one earned Waters his title as "Prince of Puke." If there are any doubts about this honor—or Divine's rep—watch through to the end to catch Divine chewing real dog excrement, all the time wearing a you-know-what grin.
1973 (R) 95m/C Divine, David Lochary, Mary Vivian Pearce, Danny Mills, Mink Stole, Edith Massey; *Dir:* John Waters. VHS, Beta, LV HAR, OM ♫♫

Pink Floyd: The Wall Film version of Pink Floyd's 1979 LP, "The Wall." A surreal, impressionistic tour-de-force about a boy who grows up numb from society's pressures. The concept is bombastic and over-wrought, but Geldof manages to remain somewhat likeable as the cynical rock star and the Gerald Scarfe animation perfectly complements the film. Laserdisc edition includes the original theatrical trailer and the letterboxed format of the film.
1982 (R) 95m/C GB Bob Geldof, Christine Hargreaves, Bob Hoskins; *Dir:* Alan Parker. VHS, Beta, LV $19.95 MVD, MGM, PIA ♫♫ ½

Pink Motel This is the story of several people and one hilarious night at a pink stucco motel which caters to couples.
1982 (R) 90m/C Phyllis Diller, Slim Pickens; *Dir:* Mike MacFarland. VHS, Beta $19.99 HBO Woof!

Pink Nights A high school nebbish is suddenly pursued by three beautiful girls, and his life gets turned upside-down.
1987 (PG) 87m/C Shaun Allen, Kevin Anderson, Larry King, Johnathan Jamcovic Michaels; *Dir:* Philip Koch. VHS, Beta, LV $19.95 STE, NWV ♫

The Pink Panther A bumbling, disaster-prone inspector invades a Swiss ski resort and becomes obsessed with capturing a jewel thief hoping to lift the legendary "Pink Panther" diamond. Said thief is also the inspector's wife's lover, though the inspector doesn't know it. Slick slapstick succeeds on strength of Sellers' classic portrayal of Clouseau, who accidentally destroys everything in his path while speaking in a funny French accent. Written by Edwards and Maurice Richlin. Followed by "A Shot in the Dark," "Inspector Clouseau" (without Sellers), "The Return of the Pink Panther," "The Pink Panther Strikes Again," "Revenge of the Pink Panther," "Trail of the Pink Panther," and "Curse of the Pink Panther." Memorable Pink Panther theme supplied by Henry Mancini, who received an Oscar nomination.
1964 113m/C GB Peter Sellers, David Niven, Robert Wagner, Claudia Cardinale, Capucine, Brenda de Banzie; *Dir:* Blake Edwards. VHS, Beta, LV $19.98 MGM, FOX ♫♫♫

Pink Panther: Pink-a-Boo The rascally Pink Panther returns in this collection of nine classic cartoons.
1985 56m/C VHS, Beta MGM ♫♫♫

The Pink Panther Strikes Again Strong series entry has the incompetent inspector tracking his former boss, who has gone insane and has become preoccupied with destroying the entire world. A must for Sellers buffs and anyone who appreciates slapstick. Fifth Panther film, once again scored by Mancini and written by Edwards and Frank Waldman.

1976 (PG) 103m/C GB Peter Sellers, Herbert Lom, Lesley-Anne Down, Colin Blakely, Leonard Rossiter, Burt Kwouk; *Dir:* Blake Edwards. VHS, Beta, LV $19.95 FOX ♫♫♫

Pink Strings and Sealing Wax A brutish pub owner in Victorian England is poisoned by his abused wife. She tries to involve the son of a chemist with the idea of blackmailing the father. Fine period flavor.
1945 75m/B GB Mervyn Johns, Mary Merrall, Gordon Jackson, Googie Withers, Sally Ann Howes, Catherine Lacey, Garry Marsh; *Dir:* Robert Hamer. VHS, Beta $16.95 SNC ♫♫

Pinocchio Pinocchio is a little wooden puppet, made with love by the old woodcarver Geppetto, and brought to life by a good fairy. Except Pinocchio isn't content to be just a puppet—he wants to become a real boy. Lured off by a sly fox, Pinocchio undergoes a number of adventures as he tries to return safely home. Has some scary scenes, including Geppetto, Pinocchio, and their friend Jiminy Cricket getting swallowed by a whale, and Pleasure Island, where naughty boys are turned into donkeys. An example of animation at its best and a Disney classic that has held up over time. Features the Oscar-winning song "When You Wish Upon a Star," as well as "Got No Strings," "Hi-Diddle-Dee-Dee," and "Give a Little Whistle."
1940 (G) 87m/C *Dir:* Ben Sharpsteen; *Voices:* Dick Jones, Cliff Edwards, Evelyn Venable. Academy Awards '40: Best Song ("When You Wish Upon a Star"), Best Original Score. VHS, Beta, LV DIS, OM ♫♫♫♫

Pinocchio This is the classical story of Pinocchio in musical form.
1976 76m/C Danny Kaye, Sandy Duncan. VHS, Beta $14.98 UHV ♫

Pinocchio Pee Wee Herman is the puppet who wants to be a real little boy in this "Faerie Tale Theatre" adaptation of this children's classic.
1983 60m/C Pee-Wee Herman, James Coburn, Carl Reiner, Lainie Kazan; *Dir:* Peter Medak. VHS, Beta $14.98 KUI, FOX, FCT ♫ ½

Pinocchio in Outer Space Adventure featuring Pinocchio and his friends on a magical trip to Mars.
1964 71m/C *Dir:* Ray Goosens; *Voices:* Arnold Stang, Minerva Pious, Peter Lazer, Conrad Jameson. VHS, Beta $19.95 COL ♫ ½

Pinocchio's Storybook Adventures Pinocchio takes over for an ailing puppet master and puts on a show for a group of children.
1979 80m/C *Dir:* Ron Merk. VHS, Beta $9.99 VTR ♫ ½

Pinto Canyon An honest sheriff does away with a band of cattle rustlers.
1940 55m/B Bob Steele, Louise Stanley, Kenne Duncan, Ted Adams, Steve Clark, Budd Buster; *Dir:* Raymond Johnson. VHS $19.95 NOS, DVT ♫

Pinto Rustlers A cowboy left orphaned by a gang of rustlers, seeks revenge.
1936 52m/C Tom Tyler, Catherine Cotter. VHS, Beta $19.95 NOS, DVT ♫ ½

Pioneer Cinema (1895-1905) This is a compilation of some early film works, including a Lumiere program from 1895, Melies' "Trip to the Moon," and "El Spectro Rojo," a color short.
191? 25m/B VHS, Beta $59.95 HHT ♫ ½

Pioneer Marshal Lawman Hale disguises himself as a criminal to track down an embezzler.
1949 60m/B Monte Hale, Paul Hurst. **VHS, Beta** $29.95 *DVT* ✗

Pioneer Woman A family encounters hostility when they set up a frontier homestead in Nebraska in 1867. Told from the feminine perspective, the tale is strewn with hurdles, both personal and natural. Also known as "Pioneers."
1973 74m/C Joanna Pettet, William Shatner, David Janssen; *Dir:* Buzz Kulik. **VHS, Beta** $34.95 *WOV, SIM* ✗✗

The Pioneers Tex Ritter sets out to protect a wagon train.
1941 59m/B Tex Ritter. **VHS, Beta** *VCN* ✗

Pioneers of the Cinema, Vol. 2 A collection of early-century French film shorts: "Where ," "A Father's Honor," "The Runaway Horse," "Scenes of Convict Life," "Revolution in Odyssey," "Down in the Deep."
1901 60m/B VHS, Beta *FCT* ✗✗½

Pipe Dreams A couple tries to repair their broken marriage against the backdrop of the Alaskan pipeline's construction.
1976 (PG) 89m/C Gladys Knight, Barry Hankerson, Bruce French, Sally Kirkland, Sherry Bain; *Dir:* Stephen Verona. **VHS, Beta** $39.98 *SUE* ✗✗

Pippi Goes on Board Pippi's father arrives one day to take her sailing to Taka-Kuka, his island kingdom. She can't bear to leave her friends and jumps off the ship to return home. Based on the classic by Astrid Lindgren, and part of a series of movies including "Pippi in the South Seas" and "Pippi on the Run."
1975 (G) 83m/C Inger Nilsson; *Dir:* Olle Hellbron. **VHS, Beta** $19.95 *GEM* ✗✗

Pippi Longstocking Mischievous Pippi creates havoc in her town through the antics of her pets, a monkey and a horse. Based on the children's book by Astrid Lindgren.
1973 (G) 99m/C Inger Nilsson. **VHS, Beta** $19.95 *GEM* ✗½

Pippi in the South Seas Pippi, a fun-loving, independent, red haired little girl, and her two friends decide to rescue her father, who is being held captive by a band of pirates. Based on the classic by Astrid Lindgren.
1974 (G) 99m/C Inger Nilsson; *Dir:* Olle Hellbron. **VHS, Beta** $19.95 *GEM* ✗✗½

Pippin Video version of the stage musical about the adolescent son of Charlemagne finding true love. Adequate record of Bob Fosse's Broadway smash. Features Vereen recreating his original Tony Award-winning role.
1981 120m/C Ben Vereen, William Katt, Martha Raye, Chita Rivera. **VHS, Beta, LV** $39.95 *FHE, PIA* ✗✗✗

Piranha A rural Texas resort area is plagued by attacks from ferocious man-eating fish which a scientist created to be used as a secret weapon in the Vietnam War. Spoofy horror film features the now-obligatory Dante film in-jokes in the background.
1978 (R) 90m/C Bradford Dillman, Heather Menzies, Kevin McCarthy, Keenan Wynn, Barbara Steele, Dick Miller, Paul Bartel; *Dir:* Joe Dante. **VHS, Beta** $39.98 *WAR, OM* ✗✗½

Piranha 2: The Spawning A diving instructor and a biochemist seek to destroy mutations that are murdering tourists at a club.
1982 (R) 88m/C Steve Marachuk, Lance Henriksen, Ricky Paul; *Dir:* James Cameron. **VHS, Beta, LV** $29.98 *SUE* ✗

The Pirate A traveling actor poses as a legendary pilot to woo a lonely woman on a remote Caribbean island. Minnelli always scores with this type of fare, and both Garland and Kelly make the most of the Cole Porter score, which includes "Be a Clown" and "You Can Do No Wrong."
1948 102m/C Judy Garland, Gene Kelly, Walter Slezak, Gladys Cooper, George Zucco, Reginald Owen, Nicholas Brothers; *Dir:* Vincente Minnelli. **VHS, Beta, LV** $19.95 *MGM* ✗✗✗

Pirate Movie Gilbert and Sullivan's "The Pirates of Penzance" is combined with new pop songs in this tale of fantasy and romance. Feeble attempt to update a musical that was fine the way it was.
1982 (PG) 98m/C Kristy McNichol, Christopher Atkins, Ted Hamilton, Bill Kerr, Garry McDonald; *Dir:* Ken Annakin. **VHS, Beta** $59.95 *FOX* ✗✗½

Pirate Warrior Grade-B pirate flick about slavery and the evil Tortuga.
1964 86m/C Ricardo Montalban, Vincent Price, Liana Orfei; *Dir:* Mary Costa. **VHS, Beta** $29.95 *FOR* ✗✗

Pirates Blustery, confused effort at a big-budgeted retro-adventure, about a highly regarded pirate and his gains and losses on the high seas. Broad comedy, no story to speak of.
1986 (PG-13) 124m/C Walter Matthau, Cris Campion, Damien Thomas, Richard Pearson, Charlotte Lewis, Olu Jacobs, David Kelly, Roy Kinnear, Bill Fraser, Jose Santamaria; *Dir:* Roman Polanski. **VHS, Beta** $19.95 *IVE* ✗

Pirates of the Coast A Spanish naval commander teams up with a group of pirates to even the score with an evil governor during the 1500s.
1961 102m/C Lex Barker, Estella Blain, Livio Lorenzon, Liana Orfei; *Dir:* Domenico Paolella. **VHS, Beta** $29.95 *UNI* ✗½

Pirates of the High Seas Crabbe helps a friend save his shipping line from sabotage. A serial in fifteen chapters.
1950 ?m/C Buster Crabbe, Lois Hall, Tommy Farrell, Gene Roth, Tristram Coffin; *Dir:* Spencer Gordon Bennet, Thomas Carr. **VHS** *MLB* ✗½

The Pirates of Penzance Gilbert and Sullivan's comic operetta is the story of a band of fun-loving pirates, their reluctant young apprentice, the "very model of a modern major general," and his lovely daughters. An adaptation of Joseph Papp's award-winning Broadway play.
1983 112m/C Kevin Kline, Angela Lansbury, Linda Ronstadt, Rex Smith, George Rose; *Dir:* Wilford Leach. **VHS, Beta, LV** $29.95 *MCA, FOX* ✗✗

Pirates of the Seven Seas In the further tales of "Sandokan the Great," the pirate hero helps save the heroine's father from an evil English Imperialist.
1962 90m/C Steve Reeves, Jacqueline Sassard, Andrea Bosic; *Dir:* Umberto Lenzi. **VHS, Beta** $69.95 *FOR* ✗½

Pistol: The Birth of a Legend Biography of "Pistol" Pete Maravich, the basketball star who defied age limitations in the 1960s to play on his varsity team.

1990 (G) 104m/C Adam Guier, Nick Benedict, Boots Garland, Millie Perkins; *Dir:* Frank C. Schroeder. **VHS, Beta** $89.95 *SVS* ✗✗½

The Pit A 12-year-old autistic boy gets his chance for revenge. The townspeople who humiliate him are in for a surprise after he stumbles across a huge hole in the forest, at the bottom of which are strange and deadly creatures.
1981 (R) 96m/C Sammy Snyders, Sonja Smits, Jeannie Elias, Laura Hollingsworth; *Dir:* Lew Lehman. **VHS, Beta** $19.95 *STE, SUE* ✗½

The Pit and the Pendulum A woman and her lover plan to drive her brother mad, and he responds by locking them in his torture chamber, which was built by his loony dad, whom he now thinks he is. Standard Corman production only remotely derived from the classic Poe tale, with the cast chewing on Richard Matheson's loopy script. A landmark in Gothic horror.
1961 80m/C Vincent Price, John Kerr, Barbara Steele, Luana Anders; *Dir:* Roger Corman. **VHS, Beta** $19.98 *WAR, OM* ✗✗✗

Pixote Wrenching, documentary-like account of an orphan-boy's life on the streets in a Brazil metropolis. Graphic and depressing, it's not for all tastes but nonetheless masterfully done. In Portuguese with English subtitles.
1981 127m/C *BR* Fernando Ramos Da Silva, Marilia Pera, Jorge Juliao; *Dir:* Hector Babenco. **VHS, Beta** $59.95 *COL* ✗✗✗✗

P.K. and the Kid A runaway kid meets up with a factory worker on his way to an arm wrestling competition and they become friends.
1985 90m/C Molly Ringwald, Paul LeMat, Alex Rocco, John Madden, Esther Rolle; *Dir:* Lou Lombardo. **VHS, Beta, LV** $79.95 *LHV, WAR* ✗✗

A Place Called Today Tale of big city politics where violence and fear in the streets is at the heart of the campaign. Also called "City in Fear."
1972 105m/C Cheri Caffaro, J. Herbert Kerr Jr., Lana Wood, Richard Smedley, Tim Brown; *Dir:* Don Schain. **VHS, Beta** $39.95 *MON, UNI* ✗

Place in Hell The Japanese armed forces camouflage a Pacific island beach which they are holding and invite the American forces to invade it during World War II.
1965 106m/C Guy Madison, Helene Chanel, Monty Greenwood. **VHS, Beta** $49.95 *UNI* ✗½

A Place in the Sun Melodramatic adaptation of "An American Tragedy," Theodore Dreiser's realist classic about an ambitious laborer whose aspirations to the high life with a gorgeous debutante are threatened by his relatively dull lover's pregnancy. Clift is magnificent in the lead, and Taylor and Winters also shine in support. Burr, however, grossly overdoes his role of the vehement prosecutor. Still, not a bad effort from somewhat undisciplined director Stevens. Triple Oscar winner also nominated for Best Picture, Best Actor (Clift), and Best Actress (Winters).
1951 120m/B Montgomery Clift, Elizabeth Taylor, Shelley Winters, Raymond Burr, Anne Revere; *Dir:* George Stevens. Academy Awards '51: Best Black and White Cinematography, Best Costume Design (B & W), Best Director (Stevens), Best Film Editing, Best Screenplay, Best Musical Score; Directors Guild of America Awards '51: Best Director (Stevens); Golden Globe Awards '52: Best Film—Drama; National Board of Review Awards '51: Best Film. **VHS, Beta, LV** $19.95 *PAR, PIA, BTV* ✗✗✗

P

Place of Weeping Early entry in the anti-apartheid sweepstakes (made in South Africa, no less). A South African woman opposes her nation's racist policies. A reporter supports her.
1986 (PG) 88m/C *SA* James Whylie, Geini Mhlophe; *Dir:* Darrell Roodt. VHS, Beta $19.95 *STE, NWV* ♫♫½

Places in the Heart A young widow determines to make the best of a bad situation on a small farm in Depression-era Texas, fighting poverty, racism, and sexism while enduring back-breaking labor. Support group includes a blind veteran and a black drifter. Hokey but nonetheless moving film is improved significantly by strong performances by virtually everyone in the cast. In his debut, Malkovich shines through this stellar group. Snared double Oscars and was also nominated for Best Picture, Supporting Actor (Malkovich), Supporting Actress (Crouse), and Director. Effective dust-bowl photography by Nestor Almendros.
1984 (PG) 113m/C Sally Field, John Malkovich, Danny Glover, Ed Harris, Lindsay Crouse, Amy Madigan, Terry O'Quinn; *W/Dir:* Robert Benton. Academy Awards '84: Best Actress (Field), Best Original Screenplay. VHS, Beta, LV $19.98 *FOX, BTV* ♫♫♫

Placido: A Year in the Life of Placido Domingo Domingo is followed through his hectic training schedule, and includes excerpts from major operatic productions.
1984 105m/C Placido Domingo, Kiri Te Kanawa, Marilyn Zschau, Katia Ricciarelli. VHS, Beta, LV, 8mm $39.95 *MVD, HMV, KUL* ♫♫♫

The Plague Dogs Two dogs carrying a plague escape from a research center and are tracked down in this unlikely animated film. A bit ponderous. And yes, that is Hurt's voice. Written by Rosen from the novel by Richard Adams, author of "Watership Down."
1982 99m/C *W/Dir:* Martin Rosen; *Voices:* John Hurt, Christopher Benjamin, Judy Geeson, Barbara Leigh-Hunt, Patrick Stewart. VHS, Beta $9.98 *CHA* ♫♫½

Plain Clothes An undercover cop masquerades as a high school student to solve the murder of a teacher. He endures all the trials that made him hate high school the first time around.
1988 (PG) 98m/C Arliss Howard, George Wendt, Suzy Amis, Diane Ladd, Abe Vigoda, Robert Stack; *Dir:* Martha Coolidge. VHS, Beta $14.95 *PAR* ♫♫½

The Plainsman Western legends Wild Bill Hickock, Buffalo Bill, and Calamity Jane team up for adventure in this big, empty venture. Just about what you'd expect from splashy director DeMille.
1937 113m/B Gary Cooper, Jean Arthur, Charles Bickford, Anthony Quinn, George "Gabby" Hayes, Porter Hall; *Dir:* Cecil B. DeMille. VHS, Beta, LV $29.95 *MCA, MLB* ♫♫½

Plan 9 from Outer Space Two or three aliens in silk pajamas conspire to resurrect several slow-moving zombies from a cardboard graveyard in order to conquer the Earth. Spaceships that look suspiciously like paper plates blaze across the sky. A pitiful, inadvertently hilarious fright qualifying as perhaps the absolute stupidest film ever made. Unforgettable. Lugosi's actual screen time is under two minutes, as he died before the film was complete and, unlike the cemetary dwellers, failed to resurrect. Enjoy the taller and younger replacement (the chiropractor of Wood's wife) they found for

Lugosi, who remains hooded to protect his identity. Alternate title: "Grave Robbers From Outer Space."
1956 78m/B Bela Lugosi, Tor Johnson, Lyle Talbot, Vampira, Gregory Walcott; *Dir:* Edward D. Wood Jr.; *Nar:* Criswell. Golden Turkey Award '56: Worst Film of All Time. VHS, Beta, LV $19.95 *NOS, SNC, MED* Woof!

Planes, Trains & Automobiles A one-joke Hughes comedy saved by Martin and Candy. A straight-laced businessman (played straight by Martin) on the way home for Thanksgiving meets up with a oafish, bad-luck-ridden boor who turns his efforts to get home upside down. Martin and Candy both turn in fine performances, and effectively straddle a thin line between true pathos and hilarious buffoonery. Bacon and McClurg, both Hughes alumni, have small but funny cameos.
1987 (R) 93m/C Steve Martin, John Candy, Edie McClurg, Kevin Bacon, Michael McKean, William Windom, Laila Robins, Martin Ferrero, Charles Tyner; *Dir:* John Hughes. VHS, Beta, LV, 8mm $19.95 *PAR* ♫♫½

Planet of the Apes Astronauts crash land on a planet where apes are masters and humans are merely brute animals. Superior science fiction with sociological implications marred only by unnecessary humor. Heston delivers on of his more plausible performances. Superb ape makeup creates realistic pseudo-simians of McDowall, Hunter, Evans, Whitmore, and Daly. Adapted by Rod Serling and Michael Wilson from Pierre Boulle's novel, "Monkey Planet." Musical score by Jerry Goldsmith was Oscar nominated. Followed by four sequels and two television series.
1968 (G) 112m/C Charlton Heston, Roddy McDowall, Kim Hunter, Maurice Evans, Linda Harrison, James Whitmore, James Daly; *Dir:* Franklin J. Schaffner. Academy Awards '68: Best Makeup; National Board of Review Awards '68: 10 Best Films of the Year. VHS, Beta, LV $19.98 *FOX* ♫♫♫½

Planet of Blood Space opera about an alien vampire discovered on Mars by a rescue team. Also known as "Queen of Blood." If you've ever seen the Soviet film "Niebo Zowiet," don't be surprised if some scenes look familiar; the script was written around segments cut from that film.
1966 81m/C John Saxon, Basil Rathbone, Judi Meredith, Dennis Hopper, Florence Marly; *Dir:* Curtis Harrington. VHS $16.95 *SNC, STC, NOS* ♫♫

Planet Burg A classic Soviet sci-fi flick about a space exploration team landing on Venus. Their job becomes a rescue mission when one of the crew is stranded. Although there are some silly moments, some good plot twists and acting make up for them. In Russian with English subtitles.
1962 90m/C *RU* Vladimir Temelianov, Gennadi Vernov, Kyunna Ignatova; *Dir:* Pavel Klushantsev. VHS, Beta $16.95 *SNC* ♫♫½

Planet of the Dinosaurs Survivors from a ruined spaceship combat huge savage dinosaurs on a swampy uncharted planet.
1980 (PG) 85m/C James Whitworth; *Dir:* James K. Shea. VHS, Beta $39.95 *AHV* ♫½

Planet Earth In the year 2133, a man who has been in suspended animation for 154 years is revived to lead the troops against a violent group of women (and mutants, too!).

1974 78m/C John Saxon, Janet Margolin, Ted Cassidy, Diana Muldaur, Johana DeWinter, Christopher Cary; *Dir:* Marc Daniels. VHS $59.95 *UNI* ♫♫½

Planet on the Prowl A fiery planet causes earthly disasters, so a troop of wily astronauts try to destroy it with the latest technology. They fail, leading one sacrificial soul to do it himself. Originally titled, "War Between the Planets."
1965 80m/C *IT* Jack Stuart, Amber Collins, Peter Martell, John Bartha, Halina Zalewska, James Weaver; *Dir:* Anthony M. Dawson. VHS, Beta $39.95 *MON* ♫

Planet of the Vampires Astronauts search for missing comrades on a planet dominated by mind-bending forces. Acceptable atmospheric filmmaking from genre master Bava, but it's not among his more compelling ventures.
1965 86m/C *IT* Barry Sullivan, Norman Bengell, Angel Aranda, Evi Marandi; *Dir:* Mario Bava. VHS, Beta $14.99 *HBO* ♫♫½

Planets Against Us Science fiction tale of escaped alien humanoid robots who take refuge on Earth, but whose touch is fatal. Good special effects.
1961 85m/B *IT FR* Michel Lemoine, Maria Pia Luzi, Jany Clair; *Dir:* Romano Ferrara. VHS $16.95 *SNC* ♫½

Platinum Blonde Screwball comedy in which a newspaper journalist (Williams) marries a wealthy girl (Harlow) but finds that he doesn't like the restrictions and confinement of high society. Yearning for a creative outlet, he decides to write a play and hires a reporter (Young) to collaborate with him. The results are funny and surprising when Young shows up at the mansion flanked by a group of hard-drinking, fun-loving reporters.
1931 86m/B Loretta Young, Robert Williams, Jean Harlow, Louise Closser Hale; *Dir:* Frank Capra. VHS $19.95 *COL* ♫♫♫

Platoon A grunt's view of the Vietnam War is provided in all its horrific, inexplicable detail. Sheen is wooden in the lead, but both Dafoe and Berenger are resplendent as, respectively, good and bad soldiers. Strong, visceral filmmaking from fearless director Stone, who based the film on his own GI experiences. Highly acclaimed; considered by many to be the most realistic portrayal of the war on film. Awarded four Oscars and also nominated for Supporting Actor (both Berenger and Dafoe), Stone for screenplay, and Robert Richardson for photography.
1986 (R) 113m/C Charlie Sheen, Willem Dafoe, Tom Berenger, Francesco Quinn, Forest Whitaker, John C. McGinley, Kevin Dillon, Richard Edson, Reggie Johnson, Keith David, Johnny Depp; *W/Dir:* Oliver Stone. Academy Awards '86: Best Director (Stone), Best Film Editing, Best Picture, Best Sound; Directors Guild of America Awards '86: Best Director (Stone); Golden Globe Awards '87: Best Film—Drama; National Board of Review Awards '86: 10 Best Films of the Year. VHS, Beta, LV $19.98 *LIV, HBO, BTV* ♫♫♫

Platoon Leader A battle-drenched portrait of a West Point lieutenant in Vietnam. When he first arrives, he must win the trust of his men, who have been on their tours for a much longer time. As time goes on, he slowly becomes hardened to the realities of the brutal life in the field. Made on the heels of the widely acclaimed "Platoon," but it doesn't have the same power. Norris is martial arts king Chuck Norris' brother.

1987 (R) 97m/C Michael Dudikoff, Brian Libby, Robert F. Lyons, Rick Fitts, Jesse Dabson, William Smith, Michael De Lorenzo; *Dir:* Aaron Norris. **VHS, Beta, LV $19.98** *MED* 🎬🎬

Platoon the Warriors The underworld rages with violence when two kingpins, Rex and Bill, become bitter and declare war over a botched drug deal.
1988 90m/C David Coley, Dick Crown, James Miller, Don Richard, Alex Sylvian; *Dir:* Philip Ko. **VHS, Beta $49.95** *HFE* 🎬🎬

Platypus Cove An adopted boy tracks down the culprits responsible for the sabotage of his family's boat, a crime for which he was suspected. Australian.
1986 72m/C Paul Smith; *Dir:* Peter Maxwell. **VHS, Beta $59.98** *EMB, SUE* 🎬

Play Dead Poor Yvonne De Carlo plays a psychotic woman who trains a dog to rip people to shreds.
1981 89m/C Yvonne de Carlo, Stephanie Dunham, David Culliname, Glenn Kezer, Ron Jackson, Carolyn Greenwood; *Dir:* Peter Wittman. **VHS $19.95** *ACA* 🎬

Play It Again, Charlie Brown The Peanuts gang goes bananas when Lucy enlists Schroeder to play rock music at a PTA meeting.
1970 25m/C Charles M. Schulz. **VHS, Beta $14.95** *SHV, VTR* 🎬

Play It Again, Sam Allen is—no surprise—a nerd, and this time he's in love with his best friend's wife. Modest storyline provides a framework of endless gags, with Allen borrowing heavily from "Casablanca." Bogey even appears periodically to counsel Allen on the ways of wooing women. Superior comedy isn't hurt by Ross directing instead of Allen, who adapted the script from his own play.
1972 (PG) 85m/C Woody Allen, Diane Keaton, Tony Roberts, Susan Anspach, Jerry Lacy, Jennifer Salt, Joy Bang, Viva, Herbert Ross; *Dir:* Herbert Ross. **VHS, Beta, LV, 8mm $19.95** *PAR* 🎬🎬🎬½

Play Misty for Me A radio deejay obliges a psychotic woman's song request and suddenly finds himself the target of her obsessive behavior, which rapidly turns from seductive to murderous. Auspicious directorial debut for Eastwood, borrowing from the Siegel playbook (look for the director's cameo as a barkeep).
1971 (R) 102m/C Clint Eastwood, Jessica Walter, Donna Mills, Don Siegel, John Larch, Jack Ging; *Dir:* Clint Eastwood. **VHS, Beta, LV $19.95** *MCA, TLF* 🎬🎬🎬

Play Murder for Me Saxophonist performing in a seedy Buenos Aires nightclub has his world turned upside down when a former lover suddenly re-enters his life. She still has her eye on him, but she's now the woman of a notorious mobster. Soon the musician finds himself drawn into the usual deadly world of crime, deceit and unharnessed passion.
1991 (R) 80m/C Jack Wagner, Tracy Scoggins; *Dir:* Hector Olivera. **VHS $89.98** *NHO* 🎬🎬

Playboy Playmates at Play Semiclad Playmates engage in various recreational activities, including ballooning, boating, flying planes, etc.
1990 53m/C Karen Witter, Carol Ficatier, Venice Kong, Jolanda Egger. **VHS, Beta $29.99** *UND* 🎬🎬

Playboy: Playmates of the Year-The '80s Playboy Magazine presents "intimate video portraits" of the top Bunnies of the decade.
1989 60m/C Dorothy Stratten, Kathy Shower, India Allen. **VHS, Beta, LV** *UND* 🎬🎬

Playboy Video Centerfold: 35th Anniversary Playmate A video record of the cataclysmic nationwide search for 1988's Playmate of the Year, the magazine's 35th.
1988 60m/C Hugh Hefner. **VHS, Beta, LV $19.99** *UND* 🎬🎬

Playboy Video Centerfold Double Header Another soft-focused Playboy collection with two Playmate gatefolds and a special "Women of Russia" segment.
1990 50m/C Karen Foster, Deborah Driggs. **VHS, Beta $29.99** *HBO* 🎬

Playboy Video Magazine: Volumes 1-7 This series features in-depth looks at Playboy's former Playmates of the year along with varied comedy line-ups and celebrity interviews.
1982 85m/C VHS, Beta $12.98 *FOX* 🎬🎬

Playboy of the Western World An innkeeper's daughter is infatuated with a young man who says he murdered his father. Adapted from the classic play by John Millington Synge.
1962 96m/C *IR* Siobhan McKenna, Gary Raymond; *Dir:* Brian Desmond Hurst. **VHS, Beta $59.99** *HBO* 🎬🎬½

The Playboys In 1957, in a tiny Irish village, Tara Maguire causes a scandal by having a baby out of wedlock and refusing to name the father. Tara's beauty attracts lots of men—there's a former beau who kills himself (Tara blames it on bad luck and bad land), the obsessive, middle-aged Sergeant Hegarty, and the newest arrival, Tom Castle. Tom is an actor with a rag-tag theatrical troupe called "The Playboys." In a tent in the village commons the troupe put on snippets of Shakespeare and sing Irish ballads but their time is coming to an end. As the first television set makes its appearance in the village the troupe decides to stage their own hysterically hammy version of "Gone with the Wind" (the most recent movie they've seen) as competition. The Playboys final performance is also the breaking point for Hegarty, as his jealousy of Tom and Tara's growing love forces him to violence. A slow-moving and simple story with particularly good performances by Wright as the strong-willed Tara and Finney as Hegarty, clinging to a last chance at love and family. Directorial debut of Gillies Mackinnon. Filmed in the village of Redhills, Ireland, the hometown of co-writer Shane Connaughton.
1992 (PG-13) 114m/C Robin Wright, Albert Finney, Aidan Quinn, Milo O'Shea, Alan Devlin, Niamh Cusack, Adrian Dunbar; *Dir:* Gillies Mackinnon. **VHS $92.99** *HBO* 🎬🎬🎬

Playboy's 1988 Playmate Video Calendar A mistily-shot video survey of Playboy's dozen playmates for 1988 in various modes of undress.
1987 80m/C Kim Morris, Kymberly Paige, Devin Devasquez, Luann Lee, Julie Peterson. **VHS, Beta, LV $19.98** *LHV, WAR* 🎬🎬🎬

Playboy's 1989 Playmate Video Calendar 1988's Playmates cavort in various stages of decoll-tage.

1989 60m/C Rebecca Ferratti, Diana Lee, Eloise Broady, Carmen Berg, Kimberley Conrad, India Allen, Terry Doss, Sharry Koponski. **VHS, Beta, LV** *UND* 🎬🎬🎬

Playboy's 1990 Playmate Video Calendar Twelve women wearing skimpy little nothings lounge about and breathe deeply.
1989 60m/C VHS, Beta, LV $19.99 *UND* 🎬🎬🎬

Playboy's Bedtime Stories Several soft-core vignettes from the Playboy stable.
1987 60m/C VHS, Beta $59.95 *LHV, WAR* 🎬🎬

Playboy's Farmers' Daughters A slew of healthy mid-Western farm girls strip for the camera.
1986 60m/C Brenda Adamson, Christine Rude, Jackie Lorenz, Colleen Donovan. **VHS, Beta, LV $19.95** *LHV, WAR* 🎬🎬🎬

Playboy's Inside Out A collection of nine separate stories full of humor, drama, fantasy, and eroticism. These tales of passion and obsession are perfect for "couples" entertainment.
1992 90m/C VHS $79.95 *UND* 🎬🎬🎬

Playboy's Party Jokes A series of skits in the manner of the reknowned magazine's bawdy humor page.
1989 55m/C VHS, Beta, LV $19.99 *UND* 🎬🎬🎬

Playboy's Playmate of the Year Video Centerfold The annual/semi-annual series of tapes profiling the bodies of Playboy's pick-of-the-year models.
1987 30m/C VHS, Beta $19.95 *UND, LHV* 🎬🎬🎬

Playboy's Sexy Lingerie A series of nudie vignettes in various milieus highlighting lingerie.
1989 55m/C VHS, Beta, LV $19.99 *UND* 🎬🎬🎬

Playboy's Wet & Wild A variety of Playboy playmates are taped in and out of skimpy swimsuits in assorted tropical locales.
1989 55m/C VHS, Beta, LV $19.99 *UND* 🎬🎬🎬

Playboy's Wet & Wild 2 Young lovelies cavort in tight clothes in exotic locales. They get their clothes wet, so they have to take them off to dry.
1990 60m/C VHS, Beta $19.99 *UND* 🎬🎬🎬

Players A young tennis hustler touring Mexico hooks up with a beautiful and mysterious older woman. They seem to be from different worlds yet their love grows. She inspires him enough to enter Wimbledon. Several tennis pros appear, including Guillermo Vilas, John McEnroe, and Ilie Nastase.
1979 (PG) 120m/C Ali MacGraw, Dean Paul Martin, Maximilian Schell, Pancho Gonzales; *Dir:* Anthony Harvey. **VHS, Beta $49.95** *PAR* 🎬

Playful Little Audrey This is a compilation of the Little Audrey cartoons, featuring the mischievous miss and her wacky cartoon friends.
1961 60m/C VHS, Beta $19.95 *WOV* 🎬

Playgirl Killer After impulsively murdering a restless model, an artist continues to kill indiscriminately, keeping his spoils on ice. Sedaka croons between kills and luxuriates poolside—seemingly oblivious to the plot of the film—while a female decoy is sent to bait the killer for the police, hence the alternate title, "Decoy for Terror." For adult viewers.
1966 86m/C *CA* William Kerwin, Jean Christopher, Andree Champagne, Neil Sedaka; *Dir:* Erick Santamaria. **VHS, Beta $19.95** *STE, NWV* 🎬

Playing Away A team of Caribbean cricket players meet a team of stuffy Brits on the field in a clash of the cultures.
1987 100m/C *GB* Norman Beaton, Robert Urquhart; *Dir:* Horace Love. **VHS, Beta, LV $19.98** *CHA* ⅃

Playing the Field Romantic entanglements defy explanation.
19?? (R) 90m/C Joan Collins. **VHS $69.95** *FOR, LTG* ⅃½

Playing with Fire A young girl gets caught up in drugs, casual sex and murder.
197? 87m/C Karen Tungay, Danny Keogh, Sam Marais; *Dir:* Alan Nathanson. **VHS, Beta $19.95** *WES* ⅃

Playing for Keeps Three high school grads turn a dilapidating hotel into a rock 'n roll resort. Music by Pete Townshend, Peter Frampton, Phil Collins, and others.
1986 (PG-13) 103m/C Daniel Jordano, Matthew Penn, Leon Grant, Harold Gould, Jimmy Baio; *Dir:* Bob Weinstein. **VHS, Beta, LV $79.95** *MCA* ⅃

Playing for Time Compelling television drama based on actual experiences of a Holocaust prisoner who survives by leading an inmate orchestra. Strong playing from Redgrave and Mayron. Pro-Palestinian Redgrave's political beliefs made her a controversial candidate for the role of Jewish Fania Fenelon, but her stunning performance is on the mark. Script from Arthur Miller won an Emmy, while the film took top drama special award.
1980 148m/C Vanessa Redgrave, Jane Alexander, Maud Adams, Verna Bloom, Melanie Mayron; *Dir:* Daniel Mann. Emmy Awards '80: Outstanding Drama Special. **VHS, Beta $59.95** *VIR* ⅃⅃⅃

Playmate Playoffs The gloves come off in this grueling competition between Playmates through mud, tires and water.
1986 75m/C Kimberly Evenson, Kymberly Herrin, Hope Marie Carlton, Kimberly McArthur. **VHS, Beta $19.98** *LHV* ⅃⅃⅃

Playmate Review Ten playmates from Playboy magazine, including 1982 Playmate of the Year Shannon Tweed, are featured in this candid pictorial.
1983 90m/C **VHS, Beta, LV $59.98** *FOX* ⅃⅃⅃

Playmate of the Year Video Centerfold 1990 The 1990 edition of the annual skinfest and personality profile.
1990 45m/C **VHS, Beta $19.99** *HBO* ⅃⅃

Playmates Barrymore is practically wasted in his last film as a down-on-his-luck actor who agrees to turn bandleader Kyser into a Shakespearean actor. Bizarre comedy is funny at times, but leaves the audience wondering what Barrymore thought he was doing.
1941 96m/B Kay Kyser, John Barrymore, Ginny Simms, Lupe Velez, May Robson, Patsy Kelly, Peter Lind Hayes, George Cleveland; *Dir:* David Butler. **VHS, Beta $19.98** *TTC* ⅃

Playmates Two divorced buddies fall in love with each other's ex-wives in this made for television flick. Typical Alda vehicle isn't bad.
1972 78m/C Doug McClure, Alan Alda, Connie Stevens, Barbara Feldon, Eileen Brennan, Tiger Williams, Severn Darden; *Dir:* Theodore J. Flicker. **VHS, Beta $19.98** *VMK* ⅃

Playroom Archaeologist McDonald is completing his father's search for the tomb of a medieval boy-prince. But someone, or something, has been waiting for him for a long, long time. When he finally reaches the long-sought-after tomb, he discovers that it is a torture chamber where his worst nightmares become realities. The demonic prince who was buried in the tomb has picked Chris as his playmate, and the rest of the staff as his personal toys.
1990 (R) 87m/C Lisa Aliff, Aron Eisenberg, Christopher McDonald, James Purcell, Jamie Rose, Vincent Schiavelli; *Dir:* Manny Coto. **VHS, LV $89.98** *REP* ⅃⅃

Playtime Occasionnally enterprising comedy in which the bemused Frenchman Hulot tries in vain to maintain an appointment in an urban landscape of glass and steel. The theme of cold, unfeeling civilization is hardly unique, but the film is nonetheless enjoyable. The third in the Hulot trilogy, preceded by "Mr. Hulot's Holiday" and "Mon Oncle." In French with English subtitles.
1967 108m/C *FR* Jacques Tati, Barbara Dennek, Jacqueline Lecomte, Jack Gautier; *Dir:* Jacques Tati. **VHS, Beta $29.98** *SUE* ⅃⅃⅃

Plaza Suite Three alternating skits from Neil Simon's play about different couples staying at the New York hotel. Matthau shines in all three vignettes. Some of Simon's funnier stuff, with the first sketch being the best: Matthau and Stapleton are a couple celebrating their 24th anniversary. She's sentimental, while he's yearning for his mistress. Number two has producer Matthau putting the make on old flame Harris, while the finale has father Matthau coaxing his anxious daughter out of the bathroom on her wedding day.
1971 (PG) 114m/C Walter Matthau, Maureen Stapleton, Barbara Harris, Lee Grant, Louise Sorel; *Dir:* Arthur Hiller. **VHS, Beta, LV $14.95** *PAR* ⅃⅃⅃

Please Don't Eat the Daisies City couple and kids leave the Big Apple for the country and are traumatized by flora and fauna. Goofy sixties fluff taken from Jean Kerr's book and the basis for the eventual TV series.
1960 111m/C Doris Day, David Niven, Janice Paige, Spring Byington, Richard Haydn, Patsy Kelly, Jack Weston, Margaret Lindsay; *Dir:* Charles Walters. **VHS, Beta $19.98** *MGM, FCT* ⅃⅃½

Please Don't Eat My Mother A softcore remake of "Little Shop of Horrors" in which a lonely voyeur plays host to a human-eating plant. Also known as "Hungry Pets," and "Glump."
1972 95m/C Buck Kartalian, Renee Bond; *Dir:* Carl Monson. **VHS $19.98** *MOV, VDM* Woof!

Pleasure A saga of love, treachery, and melodrama in English society where self-indulgence held priority and good manners were the rule.
1931 53m/B Conway Tearle, Roscoe Karns, Carmen Myers, Lena Basquette. **VHS, Beta, LV $19.95** *NOS, WGE* ⅃⅃

A Pleasure Doing Business Three high school buddies, now in their 40s, are reunited at a stag party, where they decide to go into business as managers in "the oldest profession."
1979 (R) 86m/C Conrad Bain, John Byner, Alan Oppenheimer, Misty Rowe, Phyllis Diller, Tom Smothers; *Dir:* Steve Vagnino. **VHS, Beta $49.95** *VHE* ⅃

Pleasure Palace A gambling lady's-man (Sharif) meets his match when he helps a woman casino owner (Lange). Made for television.
1980 96m/C Omar Sharif, Victoria Principal, J.D. Cannon, Gerald S. O'Loughlin, Jose Ferrer, Hope Lange, Alan King; *Dir:* Walter Grauman. **VHS, Beta $29.95** *USA, IVE* ⅃⅃

Pleasure Resort Trashy resort where sexually inhibited people come to work it all out. Staffed by professional, skilled technicians ready to help their guests any way they know how.
1990 (R) 80m/C Annette Haven, Kay Parker. **VHS $29.95** *AVD* ⅃

Pleasure Unlimited A housewife experiments with bisexuality, group sex, and more in this softcore release.
1986 (X) 85m/C Angela Carnon, Terry Johnson, Harvey Shane. **VHS, Beta $29.95** *MED* ⅃

Pledge Night Tale of horror and revenge. Killed years before in a fraternity hazing, Sid returns for the brothers who did him in. Gory and violent.
1990 (R) 90m/C Will Kempe, Shannon McMahon, Todd Eastland, Paul Ziller. **VHS $89.95** *IMP* ⅃⅃

Plenty Difficult but worthwhile film with Streep in top form as a former member of the French Resistance, who upon returning to England finds life at home increasingly tedious and banal and begins to fear her finest hours may be behind her. Gielgud is flawless as the aging career diplomat. Adapted by David Hare from his play, an allegory to British decline.
1985 (R) 119m/C Meryl Streep, Tracy Ullman, Sting, John Gielgud, Charles Dance, Ian McKellan, Sam Neill; *Dir:* Fred Schepisi. **VHS, Beta, LV $19.99** *HBO* ⅃⅃½

The Plot Against Harry Jewish racketeer Harry Plotnik checks out of prison, and finds the outside world ain't what it used to be. Attempting to lead an honest life only makes matters worse. Completely overlooked when first released in 1969 (and quickly shelved) because it was considered to have no commercial potential, Roemer's crime comedy found an enthusiastic audience when it was rediscovered twenty years later.
1969 81m/B *FR* Martin Priest, Ben Lang, Maxine Woods, Henry Nemo; *W/Dir:* Michael Roemer. **VHS $79.95** *NYF, FCT* ⅃⅃⅃

The Ploughman's Lunch A BBC news reporter claws and lies his way to the top. He then discovers that he is the victim of a far more devious plan. Engrossing and well made, although some of the political views are simplistic.
1983 107m/C Jonathan Pryce, Charlie Dore, Tim Curry, Rosemary Harris, Frank Finlay, Bill Paterson; *Dir:* Richard Eyre. **VHS, Beta $59.98** *SUE* ⅃⅃⅃

Plow That Broke the Plains/ River Two classic documentaries: "The Plow That Broke the Plains" deals with the New Deal efforts to improve the lot of Oklahoma "Dust Bowl" farmers. "The River" is a poetic history of the Mississippi River and its ecological balance.
1937 60m/B *W/Dir:* Pare Lorentz. **VHS, Beta $24.95** *NOS, CCB, WFV* ⅃⅃⅃

Plumber A plumber who makes a house call extends his stay to psychologically torture the woman of the house. Originally made for Australian television.
1979 76m/C *AU* Judy Morris, Ivar Karts, Robert Coleby; *Dir:* Peter Weir. **VHS, Beta $59.95** *MED* ⅃⅃½

Plunder Road A pair of crooks cook up a plan to rob a train bound for the San Francisco Mint that's carrying a fortune in gold bullion. Top notch B-film noir.
1957 76m/B Gene Raymond, Jeanne Cooper, Wayne Morris, Elisha Cook Jr.; *Dir:* Hubert Cornfield. **VHS** $19.98 *REP* 🎬🎬½

Plunge Into Darkness A quiet weekend in the mountains turns into a nightmare for an ex-Olympic runner and his family.
1977 77m/C Bruce Barry, Olivia Hamnett, Ashley Greenville, Wallace Eaton, Tom Richards; *Dir:* Peter Maxwell. **VHS, Beta** *PGN* 🎬½

Plutonium Baby A mutated kid, whose mother was killed by the same radiation exposure that infected him, tracks down the guilty party in New York in this comic-book style film.
1987 85m/C Patrick Molloy, Danny Guerra; *Dir:* Ray Hirschman. **VHS, Beta** $79.95 *TWE* 🎬

Plutonium Incident A female technician at a nuclear power plant suspects the facility to be less safe than it appears, and faces a diabolical plan of harassment when she attempts to bring attention to the hazards.
1982 90m/C Janet Margolin, Powers Boothe, Bo Hopkins, Joseph Campanella; *Dir:* Richard Michaels. **VHS, Beta** *TLF* 🎬🎬

Pocatello Kid Maynard plays a dual role as the wrongly convicted Kid and the Kid's no-good sheriff brother.
1931 60m/B Ken Maynard. **VHS, Beta** $19.95 *NOS, VCN, VDM* 🎬

Pocket Money Down-on-their-luck cowpokes foolishly do business with crooked rancher in attempt to make comeback in faltering acting careers. Star-powered, moderately entertaining modern western-comedy based on the novel "Jim Kane" by J.K.S. Brown.
1972 (PG) 100m/C Paul Newman, Lee Marvin, Strother Martin, Christine Belford, Wayne Rogers, Hector Elizondo, Gregory Sierra; *Dir:* Stuart Rosenberg. **VHS, Beta** $59.95 *WAR, FCT* 🎬🎬

Pocketful of Miracles Capra's final film, a remake of his 1933 "Lady for a Day," is just as corny and sentimental but doesn't work quite as well. Davis is delightful as Apple Annie, a down-on-her-luck street vendor who will go to any extreme to hide her poverty from the well-married daughter she adores. Ford is terrific as the man who transforms Annie into a lady in time for her daughter's visit. Touching. Maybe too touching. Also marks Ann-Margret's film debut.
1961 136m/C Bette Davis, Glenn Ford, Peter Falk, Hope Lange, Arthur O'Connell, Ann-Margret, Thomas Mitchell, Jack Elam, David Brian, Mickey Shaughnessy; *Dir:* Edward Everett Horton, Frank Capra. **VHS, Beta, LV** $19.98 *MGM, FOX, CCB* 🎬🎬🎬

A Pocketful of Rye Based on Agatha Christie's novel featuring the sleuthing Miss Marple. She faces another murderous puzzle, this one based on an old nursery rhyme. Made-for-British television.
1987 101m/C *GB* Joan Hickson; *Dir:* Guy Slater. **VHS, Beta** $29.98 *FOX* 🎬🎬

Poco The story of Poco, a shaggy little dog who travels across the country to search for the young girl who owns him.
1977 88m/C Chill Wills, Michelle Ashburn, John Steadman. **VHS, Beta** $69.98 *LIV, VES, CVL* 🎬

Pogo for President: "I Go Pogo" Walt Kelly's Pogo Possum becomes an unlikely presidential candidate when Howland Owl proclaims him the winner of a presidential election.
1984 (PG) 120m/C *Voices:* Jonathan Winters, Vincent Price, Ruth Buzzi, Stan Freberg, Jimmy Breslin. **VHS, Beta** $49.95 *DIS* 🎬½

Poil de Carotte A semi-famous French melodrama about a young boy harassed by his overbearing mother to the point of disaster, redeemed finally by the love of his father. In French with English subtitles.
1931 90m/B *FR* Harry Baur, Robert Lynen, Catherine Fontenoy; *Dir:* Julien Duvivier. **VHS, Beta** $24.95 *NOS, FCT, MRV* 🎬🎬½

The Point Charming and sincere made-for-television animated feature about the rejection and isolation of a round-headed child in a world of pointy-headed people. Excellent score written and performed by Harry Nilsson.
1971 74m/C *Dir:* Fred Wolf; *Nar:* Ringo Starr. **VHS, Beta, LV** $19.98 *LIV, VES* 🎬🎬½

Point Blank Adapted from Stark's "The Hunter." The film's techniques are sometimes compared to those of Resnais and Godard. Double-crossed and believed dead, gangster Marvin returns to claim his share of the loot from the Organization. Hard-nosed examination of the depersonalization of a mechanized urban world.
1967 92m/C Lee Marvin, Angie Dickinson, Keenan Wynn, Carroll O'Connor, Lloyd Bochner, Michael Strong; *Dir:* John Boorman. **VHS, Beta** $79.95 *MGM* 🎬🎬½

Point Break If you can suspend your disbelief—and you'd need a crane—then this crime adventure is just dandy. Reeves plays a young undercover FBI sent to infiltrate a gang of bank-robbing surfer dudes. Swayze is the leader of the beach subculture, a thrillseeker who plays cat-and-mouse with the feds in a series of excellent action scenes.
1991 (R) 117m/C Patrick Swayze, Keanu Reeves, Gary Busey, Lori Petty, John C. McGinley, Chris Pederson, Bojesse Christopher, Julian Reyes, Daniel Beer; *Dir:* Kathryn Bigelow. MTV Movie Awards '92: Most Desirable Male (Reeves). **VHS** $19.98 *FXV* 🎬🎬½

Point of Terror A handsome rock singer seduces a record company executive's wife in order to further his career.
1971 (R) 88m/C Peter Carpenter, Dyanne Thorne, Lory Hansen, Leslie Simms; *Dir:* Alex Nicol. **VHS, Beta** $39.95 *UHV* 🎬

The Pointsman A quirky, small film based on the novel by Jean Paul Franssens. A beautiful Scandinavian woman accidentally gets off a train in Northern Scotland, where the only shelter is a small railway outpost, and the only company is a mysterious pointsman who lives by the comings and goings of the trains. The couple's increasing isolation as winter falls and their lack of a common language persists is portrayed with both humor and poignancy.
1988 (R) 95m/C Jim Van Der Woude, Stephane Excoffier, John Kraaykamp, Josse De Pauw, Ton Van Dort; *Dir:* Jos Sterling. **VHS, Beta, LV** $79.98 *LIV, VES* 🎬🎬½

Poison A controversial, compelling drama weaving the story of a seven year-old boy's murder of his father with two other tales of obsessive, fringe behavior. From the director of the underground hit "Superstar: The Karen Carpenter Story," which was shot using only a cast of "Barbie" dolls.

1991 (R) 85m/C Edith Meeks, Larry Maxwell, Susan Norman, Scott Renderer, James Lyons; *Dir:* Todd Haynes. Sundance Film Festival '91: Grand Prize. **VHS** $89.95 *FXL* 🎬🎬🎬

Poison Ivy A routine, made for television comedy about a chaotic and lusty summer camp.
1985 97m/C Michael J. Fox, Nancy McKeon, Robert Klein, Caren Caye; *Dir:* Larry Elikann. **VHS, Beta** $69.95 *RCA* 🎬🎬

Poker Alice A lively Western starring Elizabeth Taylor as a sometime gambler who wins a brothel in a poker game.
1987 100m/C Elizabeth Taylor, George Hamilton, Tom Skerritt, Richard Mulligan, David Wayne, Susan Tyrrell, Pat Corley; *Dir:* Arthur Seidelman. **VHS, Beta** $19.95 *STE, NWV* 🎬🎬

Police French police drama with the intense Depardieu as a cop hunting an Algerian drug boss. Matters grow complicated when he falls for the elusive crook's girlfriend. Sometimes gripping, but uneven; the actors were encouraged to improvise. Inspired by the novel "Bodies Are Dust" by P.J. Wolfson.
1985 113m/C *FR* Gerard Depardieu, Sophie Marceau, Sandrine Bonnaire; *Dir:* Maurice Pialat. Venice Film Festival '85: Best Actor (Depardieu). **VHS** $79.95 *NYF, FCT* 🎬🎬

Police Academy In an attempt to recruit more cops, a big-city police department does away with all its job standards. The producers probably didn't know that they were introducing bad comedy's answer to the "Friday the 13th" series, but it's hard to avoid heaping the sins of its successors on this film. Besides, it's just plain dumb.
1984 (R) 96m/C Steve Guttenberg, Kim Cattrall, Bubba Smith, George Gaynes, Michael Winslow, Leslie Easterbrook, Georgina Spelvin, Debralee Scott; *Dir:* Hugh Wilson. **VHS, Beta, LV, 8mm** $19.98 *WAR* 🎬½

Police Academy 2: Their First Assignment More predictable idiocy from the cop shop. This time they're determined to rid the precinct of some troublesome punks. No real story to speak of, just more high jinks in this mindless sequel.
1985 (PG-13) 87m/C Steve Guttenberg, Bubba Smith, Michael Winslow, Art Metrano, Colleen Camp, Howard Hesseman, David Graf, George Gaynes; *Dir:* Jerry Paris. **VHS, Beta, LV** $19.98 *WAR* 🎬

Police Academy 3: Back in Training In yet another sequel, the bumbling cops find their alma mater is threatened by a budget crunch and they must compete with a rival academy to see which school survives. The "return to school" plot allowed the filmmakers to add new characters to replace those who had some scruples about picking up yet another "Police Lobotomy" check. Followed by three more sequels.
1986 (PG) 84m/C Steve Guttenberg, Bubba Smith, David Graf, Michael Winslow, Marion Ramsey, Art Metrano, Bob Goldthwait, Leslie Easterbrook, Tim Kazurinsky, George Gaynes, Shawn Weatherly; *Dir:* Jerry Paris. **VHS, Beta, LV** $19.98 *WAR* 🎬

Police Academy 4: Citizens on Patrol The comic cop cutups from the first three films aid a citizen's patrol group in their unnamed, but still wacky, hometown. Moronic high jinks ensue. Fourth in the series of five (or is it six?) that began with "Police Academy".

1987 (PG) **88m/C** Steve Guttenberg, Bubba Smith, Michael Winslow, David Graf, Tim Kazurinsky, George Gaynes, Colleen Camp, Bob Goldthwait, Sharon Stone; *Dir:* Jim Drake. **VHS, Beta, LV** $19.98 *WAR ♂*

Police Academy 5: Assignment Miami Beach
The fourth sequel, wherein the misfits-with-badges go to Miami and bumble about in the usual manner. It's about time these cops were retired from the force. **1988** (PG) **89m/C** Bubba Smith, David Graf, Michael Winslow, Leslie Easterbrook, Rene Auberjonois, Marion Ramsey, George Gaynes, Janet Jones, George Gaynes, Matt McCoy; *Dir:* Alan Myerson. **VHS, Beta, LV** $19.98 *WAR Woof!*

Police Academy 6: City Under Siege
In what is hoped to be the last in a series of bad comedies, the distinguished graduates pursue three goofballs responsible for a crime wave. **1989** (PG) **85m/C** Bubba Smith, David Graf, Michael Winslow, Leslie Easterbrook, Marion Ramsey, Matt McCoy, Bruce Mahler, G.W. Bailey, George Gaynes; *Dir:* Peter Bonerz. **VHS, LV** $89.95 *WAR ♂*

Police Court
The son of a faded, alcoholic screen star tries to bring the old fellow back to prominence. **1937** **62m/B** Nat Barry, Henry B. Walthall, Leon Janney. **VHS, Beta, LV** *WGE ♂♂*

Police Squad! Help Wanted!
Three episodes from the brilliant, short-lived television spoof, with the patented pun-a-second style of humor typical of the makers of "Airplane!" and "The Naked Gun," which was based on this show. Episodes included: "A Substantial Gift," "Ring of Fear" and "Rendezvous at Big Gulch." See also, "More! Police Squad." **1982** **75m/C** Leslie Nielsen, Alan North, Ed Williams, William Duell; *W/Dir:* Jerry Zucker, David Zucker, Jim Abrahams. **VHS, Beta, LV** $19.95 *PAR ♂♂*

Policewoman Centerfold
Exploitative rendering based on the true story of a cop who posed for a pornographic magazine. Anderson is appealing in the lead. **1983** **100m/C** Melody Anderson, Ed Marinaro, Donna Pescow, Bert Remsen, David Spielberg; *Dir:* Reza Badiyi. **VHS, Beta** $39.95 *IVE ♂*

Policewomen
A female undercover agent must stop a ring of gold smugglers. **1973** (R) **99m/C** Sondra Currie, Tony Young, Phil Hoover, Elizabeth Stuart, Jeanne Bell; *Dir:* Lee Frost. **VHS, Beta** $59.95 *VID ♂ 1/2*

A Polish Vampire in Burbank
A shy vampire in Burbank tries again and again to find blood and love. Wacky. **1980** **84m/C** Mark Pirro, Lori Sutton, Eddie Deezen; *Dir:* Mark Pirro. **VHS, Beta** $9.95 *NO ♂♂*

Pollyanna
A young orphan girl is adopted by her cold, embittered aunt and does her best to bring joy and gladness to all the new people she meets. Silent with music score. **1920** **60m/B** Mary Pickford. **VHS, Beta, LV** $19.95 *CCB ♂♂ 1/2*

Pollyanna
Based on the Eleanor Porter story about an enchanting young girl whose contagious enthusiasm and zest for life touches the hearts of all she meets. Mills is perfect in the title role and was awarded a special Oscar for outstanding juvenile performance. A distinguished supporting cast is the icing on the cake in this delightful Disney confection. Original version was filmed in 1920 with Mary Pickford.

1960 **134m/C** Hayley Mills, Jane Wyman, Richard Egan, Karl Malden, Nancy Olson, Adolphe Menjou, Donald Crisp, Agnes Moorehead, Kevin Corcoran; *Dir:* David Swift. **VHS, Beta, LV** *DIS ♂♂♂ 1/2*

Poltergeist
He's listed only as co-writer and co-producer, but this production has Stephen Spielberg written all over it. A young family's home becomes a house of horrors when they are terrorized by menancing spirits who abduct their five-year-old daughter...through the TV screen! Roller-coaster thrills and chills, dazzling special effects, and perfectly timed humor highlight this stupendously scary ghost story. **1982** (PG) **114m/C** JoBeth Williams, Craig T. Nelson, Beatrice Straight, Heather O'Rourke, Zelda Rubinstein, Dominique Dunne, Oliver Robbins, Richard Lawson, James Karen; *Dir:* Tobe Hooper. **VHS, Beta, LV** $19.95 *MGM ♂♂♂♂*

Poltergeist 2: The Other Side
Adequate sequel to the Spielburg-produced venture into the supernatural, where demons follow the Freeling family in their efforts to recapture the clairvoyant young daughter Carol Anne. The film includes sojourns into Indian lore and a four-foot high agave worm designed by H.R. Giger. The movie was followed by "Poltergeist 3" in 1988. **1986** (PG-13) **92m/C** Craig T. Nelson, JoBeth Williams, Heather O'Rourke, Will Sampson, Julian Beck, Geraldine Fitzgerald, Oliver Robbins, Zelda Rubenstein; *Dir:* Brian Gibson. **VHS, Beta, LV** $19.95 *MGM, PIA ♂♂ 1/2*

Poltergeist 3
Wrestling with the supernatural has finally unnerved Carol Ann and she's sent to stay with her aunt and uncle in Chicago where she attends a school for gifted children with emotional disorders. Guess who follows her? Uninspired acting, threadbare premise, and one ghastly encounter too many. Oddly, O'Rourke died suddenly four months before the film's release. **1988** (PG-13) **97m/C** Tom Skerritt, Nancy Allen, Heather O'Rourke, Lara Flynn Boyle, Zelda Rubinstein; *Dir:* Gary Sherman. **VHS, Beta, LV** $19.95 *MGM, PIA ♂ 1/2*

Polyester
Amusing satire on middle-class life, described by John Waters as "'Father Knows Best' gone berserk." A forlorn housewife (Divine) pines away for the man of her dreams while the rest of her life is falling apart at the seams. Written, produced, and directed by Waters, the movie was filmed in "Odorama," a hilarious gimmick in which theatre goers were provided with scratch-n-sniff cards, containing specific scents corresponding to key scenes. Video watchers will have to use their imagination in experiencing smells ranging from air-freshener to intestinal gas. The first of John Waters' more mainstream films. Features songs by Bill Murray and Debbie Harry. **1981** (R) **86m/C** Divine, Tab Hunter, Edith Massey, Mink Stole, Stiv Bators, David Samson, Mary Garlington, Kenneth King, Joni-Ruth White; *Dir:* John Waters. **VHS, Beta** $19.99 *HBO ♂♂ 1/2*

Pom Pom Girls
High school seniors, intent on having one last fling before graduating, get involved in crazy antics, clumsy romances, and football rivalries. **1976** (R) **90m/C** Robert Carradine, Jennifer Ashley, Michael Mullins, Cheryl "Rainbeaux" Smith, Dianne Lee Hart, Lisa Reeves, Bill Adler; *Dir:* Joseph Ruben. **VHS, Beta** $9.99 *STE, UHV ♂*

Pony Express
Buffalo Bill Cody and Wild Bill Hickok join forces to extend the Pony Express mail route west to California through rain and sleet, snow and hail. Far from a factual account but good for extending the myth of the Old West. **1953** **101m/C** Charlton Heston, Rhonda Fleming, Jan Sterling, Forrest Tucker; *Dir:* Jerry Hopper. **VHS, Beta** $14.95 *PAR, KRT, HHE ♂♂♂*

Pony Express Rider
Young man with a mission joins up with the Pony Express hoping to bag the male responsible for killing his pa. The well-produced script boasts a bevy of veteran western character actors, all lending, solid, rugged performances. **1976** (G) **100m/C** Stewart Peterson, Henry Wilcoxon, Buck Taylor, Maureen McCormick, Joan Caulfield, Ken Curtis, Slim Pickens, Dub Taylor, Jack Elam; *Dir:* Robert Totten. **VHS** $29.95 *TAV, TPI ♂♂♂*

Pool Hustlers
An Italian romance about an amateur billiards whiz who finds love and successfully defeats the national champ. With subtitles. **1983** **101m/C** *IT* Francesco Nuti, Guiliana de Sio, Marcello Loti; *Dir:* Maurizio Ponzi. **VHS, Beta** $24.95 *XVC, MED ♂ 1/2*

Pool Sharks
W.C. Fields' first film features the comedian's antics while playing a pool game to win the love of a woman. Silent. **1915** **10m/B** W.C. Fields. **VHS, Beta** *TEX ♂♂ 1/2*

Poor Girl, A Ghost Story
A young girl takes a job as a governess at an English mansion and is besieged by all manner of strange goings-on. **1974** **52m/C** *AU* Lynn Miller, Angela Thorne; *Dir:* Michael Apted. **VHS, Beta** $59.95 *PSM ♂♂*

A Poor Little Rich Girl
Mary Pickford received raves in this film, in which she portrayed Gwendolyn, one of her most tender child performances. Organ score. **1917** **64m/B** Mary Pickford, Madeline Traverse, Charles Wellesley, Gladys Fairbanks; *Dir:* Maurice Tourneur. **VHS, Beta** $19.98 *CCB, DNB ♂♂ 1/2*

The Poor Little Rich Girl
A motherless rich girl wanders away from home and is "adopted" by a pair of struggling vaudevillians. With her help, they rise to the big time. Songs include "Oh My Goodness," "When I'm With You" and "But Definitely." Also available with "Heidi" on Laser Disc. **1936** **79m/B** Shirley Temple, Jack Haley, Alice Faye, Gloria Stuart, Michael Whelan, Sara Haden, Jane Darwell; *Dir:* Irving Cummings. **VHS, Beta** $19.95 *FOX ♂♂ 1/2*

Poor Pretty Eddie
A young black singer gets waylaid and taken in by a twisted white Southern clan. An incredibly sleazy movie which boasts Shelly Winters performing a strip act. Also on video as "Black Vengeance," "Redneck County" and "Heartbreak Motel." **1973** (R) **90m/C** Leslie Uggams, Shelley Winters, Michael Christian, Ted Cassidy, Slim Pickens, Dub Taylor; *Dir:* Richard Robinson. **Beta** *MFI, VID, MRV ♂ 1/2*

Poor White Trash
An architect arrives in bayou country with plans to design a new building. He meets with resistance from the locals but falls for the sensual daughter of one of his staunchest detractors. Originally titled "Bayou." **1957** **83m/B** Peter Graves, Lita Milan, Douglas Fowley, Timothy Carey, Jonathan Haze; *Dir:* Harold Daniels. **VHS, Beta** $21.98 *CGI ♂ 1/2*

Poor White Trash 2 A young couple, vacationing in Louisiana's bayou country, are introduced to an unusual brand of southern hospitality by the eponymous group of locals.
1975 (R) 90m/C Gene Ross, Ann Stafford, Norma Moore, Camilla Carr; **Dir:** S.F. Brownrigg. **VHS, Beta** $59.98 MAG ℐ

Pop Always Pays A father gets in a jam when he has to make good on a bet with his daughter's boyfriend.
1940 67m/C Walter Catlett, Dennis O'Keefe, Leon Errol, Adele Pearce; **Dir:** Leslie Goodwins. **VHS, Beta** RKO, MLB ℐℐ

Pop Goes the Cork Three films by the man Chaplin referred to as his "professor," Max Linder. "Be My Wife," "Seven Years Bad Luck," and "The Three Must-Get-Theres." Linder was the foremost film comedian in early twentieth-century France, and made these three films during a stay in America.
1922 87m/B Max Linder. **VHS, Beta** CCB ℐℐ

Popcorn A killer stalks a movie audience who is unaware that his crimes are paralleling those in the very film they are watching.
1989 (R) 93m/C Jill Schoelen, Tom Villard, Dee Wallace Stone, Derek Rydell, Elliott Hurst, Kelly Jo Minter, Malcolm Danare, Ray Walston, Tony Roberts; **Dir:** Mark Herrier. **VHS, LV, 8mm** $89.95 COL ℐℐ

The Pope of Greenwich Village A film about two Italian-American cousins who struggle to escape the trap of poverty in New York's Greenwich Village. When a small crime goes wrong in a big way, the two must learn about deception and loyalty. Mostly character study; Page is exceptional. Inferior re-run of the "Mean Streets" idea.
1984 (R) 122m/C Eric Roberts, Mickey Rourke, Darryl Hannah, Geraldine Page, Tony Musante, M. Emmet Walsh, Kenneth McMillan, Burt Young; **Dir:** Stuart Rosenberg. **VHS, Beta, LV** $19.98 CCB, MGM ℐℐ½

Pope John Paul II A made-for-television biography of the Pontiff, from childhood to world eminence.
1984 150m/C Albert Finney, Michael Crompton, Nigel Hawthorne, John McEnery, Brian Cox; **Dir:** Herbert Wise. **VHS, Beta** $19.98 PSM ℐℐ

The Pope Must Diet Coltrane stars as a misfit who accidentally becomes Pope Dave I. Living up to the benevolence suggested by his title, Pope Dave proposes to use Vatican money to create a children's fund. But there are those in the organization who have different, less noble plans for the cash, and soon Dave finds himself the target of a mob hit. Frantic comedy caused a stir with its original title, "The Pope Must Die," with many newspapers refusing to run ads for the film. This, coupled with lukewarm critical reviews, led to a very brief stint at the box office.
1991 (R) 87m/C GB Robbie Coltrane, Alex Rocco, Beverly D'Angelo, Herbert Lom, Paul Bartel, Salvatore Cascio, Balthazar Getty; **Dir:** Peter Richardson. **VHS, Beta, LV** $92.98 MED ℐ½

Popeye The cartoon sailor brought to life is on a search to find his long-lost father. Along the way, he meets Olive Oyl and adopts little Sweet Pea. Script by Jules Feiffer, music by Harry Nilsson. Williams accomplishes the near-impossible feat of physically resembling the title character, and the whole movie does accomplish the maker's stated goal of "looking like a comic strip," but it isn't anywhere near as funny as it should be.
1980 (PG) 114m/C Robin Williams, Shelley Duvall, Ray Walston, Paul Dooley, Bill Irwin, Paul Smith, Linda Hunt, Richard Libertini; **Dir:** Robert Altman. **VHS, Beta, LV** $14.95 PAR ℐℐ

Popeye Assorted Cartoons Cartoon antics featuring Popeye the spinach-loving sailor man.
1989 30m/C VHS SEC ℐℐ

Popeye Cartoons Spinach-gobbling sailor is back in three favorite adventures: "Popeye the Sailor Meets Sinbad the Sailor," "Aladdin and His Wonderful Lamp," and "Popeye the Sailor Meets Ali Baba's Forty Thieves."
1936 56m/C VHS, LV $29.99 REP ℐℐ

Popeye Color Festival Popeye meets up with Sinbad the Sailor, Ali Baba, and Aladdin in this collection of vintage cartoons from the '30s.
193? 56m/C VHS, Beta $19.95 NO ℐℐ

Popeye and Friends in Outer Space Popeye and Olive Oyl take their act into the ozone and beyond. As always, they get into big trouble with their arch enemy, Bluto. Luckily for Popeye and all concerned, cans of spinach are not as rare in outer space as Halley's Comet.
19?? 60m/C VHS, Beta $9.95 MED ℐℐ

Popeye and Friends in the South Seas Join Popeye, Olive Oyl, Wimpy and Sweet Pea as they embark on a series of comedic adventures.
1961 59m/C **Voices:** Jack Mercer, Jackson Beck, Mae Questel, Arnold Stang. **VHS, Beta** $29.95 MED ℐℐ

Popeye and Friends in the Wild West The famous spinach eating sailor is back in ten gallon hat and spurs for uproarious western adventures. Also included in this cartoon collection are Krazy Kat, Beetle Bailey and Snuffy Smith.
1984 60m/C VHS, Beta $29.95 MED ℐℐ

Popeye Parade Contains three early episodes of the Popeye animated series including "Popeye Meets Sinbad," "Popeye Meets Aladdin" and "Popeye Meets Ali Baba."
1989 60m/C Popeye. **VHS, Beta** $19.95 MVC ℐ

Popeye the Sailor Cartoon package includes "Popeye Meets Aladdin and His Wonderful Lamp," "Popeye Meets Ali Baba and His 40 Thieves," and "Popeye the Sailor Meets Sinbad the Sailor."
193? 54m/C VHS, Beta, 8mm VYY, WFV, BTV ℐℐ

Popeye: Travelin' on About Travel Popeye and Olive Oyl visit foreign and exotic lands in this hilarious program.
1984 60m/C VHS, Beta $29.95 MED ℐℐ

Popi Arkin is the heart and soul of this poignant charmer in his role as a Puerto Rican immigrant hell-bent on securing a better life outside the ghetto for his two sons. His zany efforts culminate in one outrageous scheme to set them adrift off the Florida coast in hopes they will be rescued and raised by a wealthy family. Far fetched, but ultimately heartwarming.
1969 (G) 115m/C Alan Arkin, Rita Moreno, Miguel Alejandro, Ruben Figueroa; **Dir:** Arthur Hiller. **VHS, Beta** $14.95 WKV ℐℐℐ

The Poppy Is Also a Flower A star-laden, antidrug drama produced by the United Nations. Filmed on location in Iran, Monaco, and Italy. Based on a drug trade thriller by Ian Fleming that explains how poppies, converted into heroin, are brought into the United States. Made for television. Also known as "Poppies Are Also Flowers" and "Opium Connection."
1966 (PG) 100m/C E.G. Marshall, Trevor Howard, Gilbert Roland, Eli Wallach, Marcello Mastroianni, Angie Dickinson, Rita Hayworth, Yul Brynner, Trini Lopez, Jeanne Moreau, Omar Sharif; **Dir:** Terence Young. **VHS, Beta** $9.95 SIM, EMB ℐ½

Pork Chop Hill A powerful, hard-hitting account of the last hours of the Korean War. Peck is totally believable as the man ordered to hold his ground against the hopeless onslaught of Chinese Communist hordes. A chilling, stark look in the face of a no-win situation. Top notch cast and masterful directing.
1959 97m/B Gregory Peck, Harry Guardino, Rip Torn, George Peppard, James Edwards, Bob Steele, Woody Strode, Robert Blake, Martin Landau, Norman Fell, Bert Remsen; **Dir:** Lewis Milestone. **VHS, Beta, LV** $59.95 MGM ℐℐℐ

Porky Pig Cartoon Festival Porky is highlighted in this retrospective of his early career.
1945 36m/C **Dir:** Robert McKimson, Friz Freleng. **VHS, Beta** $14.95 MGM ℐℐℐ

Porky Pig Cartoon Festival: Tom Turk and Daffy Duck Five vintage Porky shorts, including "Old Glory," "My Favorite Duck" and the title cartoon are featured.
1944 38m/C **Dir:** Robert McKimson, Friz Freleng. **VHS, Beta** $14.95 MGM ℐℐℐ

Porky Pig and Company Seven vintage 'toons featuring Porky, Daffy and Hubie & Bertie: "Old Glory," "Tom Turk and Daffy," "House Menace," "The Aristo Cat," "Wagon Wheels" and "Roughly Speaking."
1948 60m/C **Dir:** Friz Freleng, Friz Freleng, Chuck Jones, Robert McKimson. **VHS, Beta** $14.95 MGM ℐℐℐ

Porky Pig and Daffy Duck Cartoon Festival Eight classic Daffy and Porky shorts, including "Tick Tock Tuckered," "Duck Soup to Nuts," and "Baby Bottleneck."
1944 57m/C **Dir:** Chuck Jones, Bob Clampett, Robert McKimson. **VHS, Beta** $19.95 MGM ℐℐℐ

Porky Pig Tales Six fabulous cartoons featuring that bumbling yet charming porker. You'll l-l-l-love it!
1989 45m/C **Voices:** Mel Blanc. **VHS** $14.95 WAR ℐℐℐ

Porky Pig's Screwball Comedies This compilation of Porky Pig's classic cartoons includes "You Ought to Be in Pictures," "Boobs in the Woods," and "Wearing of the Grim."
1952 59m/C **Dir:** Chuck Jones, Robert McKimson, Friz Freleng. **VHS, Beta** $12.95 WAR ℐℐℐ

Porky's Set in South Florida in the early 1950s, this irreverent comedy follows the misadventures of six youths of Angel Beach High School. Their main interest is, of course, girls. Followed by two sequels: "Porky's II: The Next Day" (1983) and "Porky's Revenge" (1985).
1982 (R) 94m/C CA Dan Monahan, Wyatt Knight, Scott Colomby, Tony Ganios, Mark Herrier, Cyril O'Reilly, Roger Wilson, Bob Clark, Kim Cattrall, Kaki Hunter; **Dir:** Bob Clark. **VHS, Beta, LV** $14.98 FXV, FOX ℐ½

Porky's 2: The Next Day More tame tomfoolery about teenage sex drives, Shakespeare, fat high school teachers and streaking.
1983 (R) 100m/C *CA* Bill Wiley, Dan Monahan, Wyatt Knight, Cyril O'Reilly, Roger Wilson, Tony Ganios, Mark Herrier, Scott Colomby. **VHS, Beta, LV $29.98** *FOX* ✍✍½

Porky's Revenge The Angel Beach High School students are out to get revenge against Porky who orders the school basketball coach to throw the championship game. The second of the "Porky's" sequels.
1985 95m/C *CA* Dan Monahan, Wyatt Knight, Tony Ganios, Nancy Parsons, Chuck "Porky" Mitchell, Kaki Hunter, Kimberly Evenson, Scott Colomby; *Dir:* James Komack. **VHS, Beta, LV $29.98** *FOX* ✍

The Pornographers Bizzare, black comedy from Japan focuses on a part-time porno filmmaker lusting after the daughter of the widow he lives with and trying to cope with his family, the world, and himself. A perversely fascinating exploration of contemporary Japenses society and the many facets of sexual emotion.
1966 128m/B *JP* Shoichi Ozawa, Massaomi Konda, Sumiko Sakamota, Haruo Tanaka, Keiko Sagowa; *Dir:* Shohei Imamura. **VHS $29.95** *CVC* ✍✍✍

Porridge A British comedy inspired by the popular BBC situation comedy of the title, about a habitual criminal and convict who makes the most of his time in prison.
1991 105m/C *GB* Ronnie Barker; *Dir:* Dick Clement. **VHS $29.95** *BFS* ✍✍

Port Arthur A sweeping Japanese film describing the Russo-Japanese Wars in the early 1900's.
1983 140m/C *JP* **VHS, Beta $69.95** *VCD* ✍✍½

Port of Call Early Bergman drama about a seaman on the docks who falls for a troubled woman whose wild, unhappy past has earned her an unsavory reputation. The hopeful, upbeat tone seems incongruous with the grim harbor/slum setting. It's minor Bergman but the seeds of his trademark themes can be seen taking shape, making it a must-see for avid fans. English subtitles.
1948 100m/B *SW* Ivine-Christine Jonsson, Bengt Eklund, Erik Hall, Berta Hall, Mimi Nelson; *Dir:* Ingmar Bergman. **VHS, Beta $29.98** *SUE* ✍✍

The Port of Missing Girls A young woman implicated in a murder stows away on a freighter. There she becomes caught up in a waterfront world of pirates, smugglers and other assorted undesirables.
1938 56m/B Harry Carey, Judith Allen, Milburn Stone, Betty Compson; *Dir:* Karl Brown. **VHS, Beta, 8mm $24.95** *VYY* ✍

Port of New York A narcotics gang is smuggling large quantities of drugs into New York. A government agent poses as a gang member in order to infiltrate the mob and get the goods on them. Brynner's film debut.
1949 82m/B Scott Brady, Yul Brynner, K.T. Stevens; *Dir:* Laslo Benedek. **VHS, Beta $19.95** *NOS, VYY, DVT* ✍✍

Portfolio Real-life models star in this gritty drama about rising to the top of the fashion heap. Music by Eurythmics, Fun Boy Three and others.
1988 (R) 83m/C Paulina Porizkova, Julie Wolfe, Carol Alt, Kelly Emberg. **VHS, LV $79.95** *SVS* ✍✍

Portnoy's Complaint The screen adaptation of Philip Roth's novel follows the frustrating experiences of a sexually obsessed young man as he relates them to his psychiatrist.
1972 (R) 101m/C Richard Benjamin, Karen Black, Lee Grant, Jeannie Berlin, Jill Clayburgh; *Dir:* Ernest Lehman. **VHS, Beta $64.95** *WAR* ✍½

Portrait of the Artist as a Young Man Based on Joyce's autobiographical novel. Dramatizes the childhood, school days, adolescence, and early manhood of Stephen Dedalus. Filmed in Ireland.
1978 93m/C *IR* T.P. McKenna, John Gielgud. **VHS, Beta $29.95** *MFV, TEX* ✍✍½

Portrait of Grandpa Doc As a young artist prepares for his first one—man show, he struggles to complete a painting which he especially wants to have ready for the show-a portrait of his grandfather, who died seven years earlier.
1977 28m/C Melvyn Douglas, Bruce Davison, Barbara Rush, Anne Seymour, Keith Blanchard. **VHS, Beta** *BFA, CCF* ✍½

Portrait of a Hitman An aspiring painter leads a double life as a professional hitman. Ragged feature feels more abandoned than finished, as if most of the movie had been shot and then the money ran out. All leads turn in paycheck performances, no more.
1977 85m/C Jack Palance, Rod Steiger, Richard Roundtree, Bo Svenson, Ann Turkel; *Dir:* Allan A. Buckhantz. **VHS, Beta $49.95** *WES* ✍

Portrait of Jennie In this haunting, romantic fable, a struggling artist is inspired by and smitten with a strange and beautiful girl who he also suspects may be the spirit of a dead woman. A fine cast works wonders with what could have been a forgettable story. The last reel was tinted green in the original release with the last scene shot in technicolor. Oscar-winning special effects.
1948 86m/B Joseph Cotten, Jennifer Jones, Cecil Kellaway, Ethel Barrymore, David Wayne, Lillian Gish, Henry Hull, Florence Bates, Felix Bressart, Anne Francis; *Dir:* William Dieterle. Academy Awards '48: Best Special Effects; Venice Film Festival '49: Best Actor (Cotten). **VHS, Beta, LV $39.98** *FOX* ✍✍✍½

Portrait of a Lady A spirited young American woman is taken to England and given complete freedom to choose her own future and make her own choices.
19?? 240m/C *GB* Richard Chamberlain. **VHS $59.95** *BFS* ✍✍✍½

Portrait of a Rebel: Margaret Sanger An important docudrama about the struggle of Margaret Sanger to repeal the Comstock Act of 1912, which prohibited the distribution of birth control information.
1982 96m/C Bonnie Franklin, David Dukes, Milo O'Shea; *Dir:* Virgil W. Vogel. **VHS, Beta** *TLF* ✍

Portrait of a Showgirl An inexperienced showgirl learns the ropes of Las Vegas life from a veteran of the Vegas stages.
1982 100m/C Lesley Ann Warren, Rita Moreno, Tony Curtis, Dianne Kay, Howard Morris; *Dir:* Steven Hilliard Stern. **VHS, Beta $59.95** *PSM* ✍½

Portrait of a Stripper A widowed mother works part-time as a stripper to support her son. Trouble arises when her father-in-law attempts to prove that she is an unfit mother.

1979 100m/C Lesley Ann Warren, Edward Herrmann, Vic Tayback, Sheree North; *Dir:* John Alonzo. **VHS, Beta $69.95** *VES* ✍

Portrait of Teresa Havana housewife has to balance motherhood, textile job and cultural group activities without the cooperation of her husband. Vega skillfully portrays the lingering archaic attitudes and insulting assumptions that still confront post-revolution women. A fine eye for the revealing moments and movements of everyday life. Spanish with English subtitles.
1979 115m/C Daisy Granados, Aldolfo Llaurado, Alina Sanchez, Alberto Molina; *Dir:* Pastor Vega. **VHS $69.95** *NYF, FCT* ✍✍✍

Portrait in Terror A master thief and a deranged artist plan a heist of a Titian painting in an oddball suspense piece. Not a great success, but atmospheric and weird.
1962 81m/B Patrick Magee, William Campbell, Anna Pavane; *Dir:* Jack Hill. **VHS $20.00** *SMW* ✍½

Portrait of a White Marriage An extended made-for-cable-TV comedy special revamping certain old "Mary Hartman, Mary Hartman" and "Fernwood 2-Night" conventions; a moronic talk show host moves his cheap show to his hometown of Hawkins Falls in order to boost the ratings.
1988 81m/C Martin Mull, Mary Kay Place, Fred Willard, Michael McKean, Harry Shearer, Jack Riley, Conchata Ferrell; *Dir:* Harry Shearer. **VHS, Beta $39.95** *MCA* ✍✍½

Posed For Murder A young centerfold is stalked by a psycho who wants her all for himself.
1989 (R) 90m/C Charlotte J. Helmkamp, Carl Fury, Rick Gianasi, Michael Merrins; *Dir:* Brian Thomas Jones. **VHS, Beta $79.95** *ACA* ✍✍

The Poseidon Adventure The cruise ship Poseidon is on its last voyage from New York to Athens on New Year's Eve when it is capsized by a tidal wave. The ten survivors struggle to escape the water-logged tomb. Oscar-winning special effects, such as Shelley Winters floating in a boiler room. Oscar nominations for Winters, cinematography, art direction and set decoration, sound, original score, film editing and costume design. Created an entirely new genre of film making - the big cast disaster flick.
1972 (PG) 117m/C Gene Hackman, Ernest Borgnine, Shelley Winters, Red Buttons, Jack Albertson, Carol Lynley; *Dir:* Ronald Neame. Academy Awards '72: Best Song ("The Morning After"), Best Visual Effects. **VHS, Beta, LV $19.98** *FOX* ✍✍½

Positive I.D. A troubled housewife learns that the man who raped her years before is getting released on parole. She devises a second persona for herself with which to entrap him and get her revenge.
1987 (R) 96m/C Stephanie Rascoe, John Davies, Steve Fromholz; *Dir:* Andy Anderson. **VHS, Beta $79.95** *MCA* ✍✍

Posse There's a hidden agenda, fueled by political ambition, in a lawman's (Douglas) dauntless pursuit of an escaped bandit (Dern). An interesting contrast between the evil of corrupt politics and the honesty of traditional lawlessness. Well performed, well photographed, and almost insightful.
1975 (PG) 94m/C Kirk Douglas, Bruce Dern, James Stacy, Bo Hopkins, Luke Askew, David Canary, Alfonso Arau, Kate Woodville, Mark Roberts; *Dir:* Kirk Douglas. **VHS, Beta, LV $14.95** *PAR* ✍✍✍

Possessed A poor factory girl becomes a wealthy Park Avenue sophisticate when she falls in love with a rich lawyer who wants to be governor. Not to be confused with Crawford's 1947 movie of the same name, but worth a look. Gable and Crawford make a great couple!
1931 77m/B Joan Crawford, Clark Gable, Wallace Ford, Skeets Gallager, John Miljan; *Dir:* Clarence Brown. VHS, Beta $19.98 *MGM* 𝐼𝐼½

The Possessed Farentino is a priest who loses his faith, but regains it after an apparently fatal auto accident. Hackett is the headmistress of a private girls' school which is in need of an exorcist, since one of her student's appears to be possessed, and Farentino seems just the man for the job. Fairly tame, since originally made for TV.
1977 (PG-13) 75m/C James Farentino, Joan Hackett, Diana Scarwid, Claudette Nevins, Eugene Roche, Ann Dusenberry, Dinah Manoff, P.J. Soles; *Dir:* Jerry Thorpe. VHS, Beta $59.95 *UNI* 𝐼

Possession Returned from a long mission, a secret agent notices that his wife is acting very strangely. She's about to give birth to a manifestation of the evil within her! Gory, hysterical, over-intellectual and often unintelligible. Originally shown in a longer version in Europe.
1981 (R) 97m/C *FR GE* Isabelle Adjani, Sam Neill, Heinz Bennent, Margit Carstensen, Shaun Lawtor; *Dir:* Andrzej Zulawski. VHS, Beta $69.98 *LIV, VES* 𝐼

Possession of Joel Delaney Blend of occult horror and commentary on social mores works for the most part, but some viewers may be put off by the low production values and spottiness of the script. MacLaine is a wealthy divorcee who must deal with the mysterious transformations affecting her brother. Skeptical at first, she begins to suspect he is the victim of Caribbean voodoo.
1972 (R) 105m/C Shirley MacLaine, Perry King, Michael Hordern, David Elliot, Robert Burr; *Dir:* Waris Hussein. VHS $19.95 *PAR* 𝐼𝐼½

Possession: Until Death Do You Part When one man's attraction for a beautiful young woman becomes an obsession, a bizarre series of inexplicable events unravels a terrifying account of passion and revenge.
1990 93m/C Monica Marko, John R. Johnston, Sharlene Martin, Cat Williamson; *Dir:* Michael Mazo, Lloyd A. Simandl. VHS $79.98 *CGH* 𝐼

Postal Inspector An extraordinary tale of a postal inspector's life. Crime, romance, and natural disaster all overtake this civil servant. Catch the song "Let's Have Bluebirds On All Our Wallpaper."
1936 58m/B Carlos Cortez, Patricia Ellis, Michael Loring, Bela Lugosi, David Oliver, Wallis Clark; *Dir:* Otto Brower. VHS $16.95 *SNC* 𝐼𝐼

Postcards from the Edge Fisher's best-selling novel, tamed and tempered for the big screen in a tour-de-force of talent. Streep received an Oscar nomination for her role as a delightfully harried actress struggling with her career, her drug dependence, and her competitive, overwhelming show-biz mother. The script, also written by Fisher, is bitingly clever and filled with refreshingly witty dialogue. Lots of cameos by Hollywood's hippest.
1990 (R) 101m/C Meryl Streep, Shirley MacLaine, Dennis Quaid, Gene Hackman, Richard Dreyfuss, Rob Reiner, Mary Wickes, Conrad Bain, Annette Bening, Michael Ontkean;

Dir: Mike Nichols. VHS, Beta, LV, 8mm $19.95 *COL, FCT* 𝐼𝐼𝐼½

The Postman Always Rings Twice Even without the brutal sexuality of the James M. Cain novel, Garfield and Turner sizzle as the lust-laden lovers in this lurid tale of fatal attraction. Garfield steals the show as the streetwise drifter who blows into town and lights a fire in Turner. As their affair steams up the two conspire to do away with her husband and circumstances begin to spin out of control. Tense and compelling. A classic.
1946 113m/B Lana Turner, John Garfield, Cecil Kellaway, Hume Cronyn, Leon Ames, Audrey Totter, Alan Reed; *Dir:* Tay Garnett. VHS, Beta, LV $19.98 *MGM* 𝐼𝐼𝐼½

The Postman Always Rings Twice It must be true because he's ringing again in Mamet's version of James M. Cain's depression-era novel. This time Nicholson plays the drifter and Lange the amoral wife with an aged husband. Truer to the original story than the 1946 movie in its use of brutal sex scenes, it nevertheless lacks the power of the original. Nicholson works well in this time era and Lange adds depth and realism to the character of Cora. But in the end it remains dreary and easily forgettable.
1981 (R) 123m/C Jack Nicholson, Jessica Lange, John Colicos, Anjelica Huston, Michael Lerner, John P. Ryan, Christopher Lloyd; *Dir:* Bob Rafelson. VHS, Beta, LV $19.98 *FOX, PIA, WAR* 𝐼𝐼

Postmark for Danger Detectives do their best to smash a diamond smuggling ring that operates between Britain and the U.S. Along the way a number of people are killed.
1956 84m/B *GB* Terry Moore, Robert Beatty, William Sylvester, Josephine Griffin, Geoffrey Keen, Henry Oscar; *Dir:* Guy Green. VHS, Beta $16.95 *SNC* 𝐼𝐼½

Postmark for Danger/Quicksand A "film noire" double feature. In "Postmark for Danger," a young actress returns from the dead to find a criminal, while in "Quicksand," a mechanic becomes indebted to the mob when he borrows extra money for a date.
1956 155m/B *GB* Terry Moore, Robert Beatty, Mickey Rooney, Jeanne Cagney, Peter Lorre; *Dir:* Guy Green. VHS, Beta *SNC* 𝐼𝐼

Pot O' Gold Stewart plays a wealthy young man who signs on with a struggling band. He convinces his uncle, who has a radio program, to let the band perform during a radio giveaway show he has concocted. Slight comedy, Stewart notwithstanding.
1941 87m/B Paulette Goddard, James Stewart, Art Carney, Horace Heidt Band; *Dir:* George Marshall. VHS, Beta $19.95 *NOS, CCB, VYY* 𝐼𝐼

Pot Shots A candid look at the home-growing of marijuana.
1982 65m/C VHS, Beta $39.95 *VCD* 𝐼𝐼

Pound Puppies and the Legend of Big Paw The Pound Puppies, known for breaking dogs out of pounds and delivering them to safe and secure homes, are featured in their first full-length musical (and merchandising effort). Fifties music.
1988 76m/C *Dir:* Pierre de Celles; *Voices:* Nancy Cartwright, George Rose, B.J. Ward, Ruth Buzzi, Brennan Howard. VHS $19.95 *FHE* 𝐼𝐼

Pouvoir Intime What happens when robbers of an armored car lock the guard in the vehicle, and he decides to fight back? This film offers one possible scenario. Available in dubbed or subtitled versions.
1987 (PG-13) 86m/C *FR* Marie Tifo, Pierre Curzi, Yvan Ponton, Jaques Lussier; *Dir:* Yves Simoneau. VHS $79.98 *NSV* 𝐼𝐼½

P.O.W. Deathcamp A combat unit is captured and tortured by the Vietcong in this typical Vietnam story.
1989 92m/C Charles Black, Bill Balbridge, Rey Malonzo; *Dir:* Jett C. Espirtu. VHS $79.95 *ATL* 𝐼𝐼

The P.O.W. Escape The adventures of a surly American commander captured as a POW during the final days of the Vietnam War.
1986 (R) 90m/C David Carradine, Mako, Charles R. Floyd, Steve James; *Dir:* Gideon Amir. VHS, Beta, LV $19.95 *RCA* 𝐼½

Powaqqatsi: Life in Transformation Director Reggio's follow-up to "Koyaanisqatsi" doesn't pack the wallop of its predecessor. Still, the cinematography is magnificent and the music of Philip Glass is exquisitely hypnotic as we are taken on a spellbinding video collage of various third world countries and see the price they've paid in the name of progress. Part 2 of a planned trilogy.
1988 (G) 95m/C *Dir:* Godfrey Reggio. VHS, Beta $89.95 *CAN* 𝐼𝐼𝐼½

Powder Keg A railroad company hires a rowdy team of investigators to retrieve a hijacked train. Action never quits. The idea for the television pilot "The Bearcats" came from this movie.
1970 93m/C Rod Taylor, Dennis Cole, Michael Ansara, Fernando Lamas, Tisha Sterling; *Dir:* Douglas Heyes. VHS, Beta $9.95 *SIM, WOV* 𝐼½

Powdersmoke Range A crooked frontier politician plots to steal valuable ranch property. The new owners, however, have other ideas. The first of "The Three Mesquiteer" series.
1935 71m/B Harry Carey, Hoot Gibson, Tom Tyler, Guinn Williams, Bob Steele, Sam Hardy, Boots Mallory, Franklin Farnum, William Desmond, William Farnum, Buzz Barton, Wally Wales, Art Mix, Buffalo Bill Jr., Buddy Roosevelt; *Dir:* Wallace Fox. VHS, Beta *MED, RXM* 𝐼½

Power Two dam-building construction workers vie with each other for the local girls. Silent.
1928 60m/B William Boyd, Alan Hale Jr., Carole Lombard, Joan Bennett; *Dir:* Howard Hughes. VHS, Beta $24.95 *FCT, GPV, GVV* 𝐼𝐼½

Power A Jewish ghetto inhabitant in 18th-century Wurtemburg works his way out of the gutter and into some authority by pleasing the whims of an evil duke. Based on the novel by Leon Fuechtwangler. Also known as "Jew Suss."
1934 105m/B *GB* Conrad Veidt, Benita Hume, Frank Vosper, Cedric Hardwicke, Gerald de Maurier, Pamela Ostrer, Joan Maude, Paul Graetz, Mary Clare, Percy Parsons, Dennis Hoey, Gibb McLaughlin, Francis L. Sullivan; *Dir:* Lothar Mendes. VHS *DVT* 𝐼𝐼½

The Power An ancient clay idol that was created by the Aztecs and possesses incredible destructive power is unleashed on modern man.
1980 (R) 87m/C Warren Lincoln, Susan Stokey, Lisa Erickson, Jeffrey Obrow; *Dir:* Stephen Carpenter. VHS $69.98 *LIV, VES* 𝐼½

Power A study of corporate manipulations. Richard Gere plays a ruthless media consultant working for politicians. Fine cast can't find the energy needed to make this great, but it is still interesting. Lumet did better with same material in ''Network.''
1986 (R) 111m/C Richard Gere, Julie Christie, E.G. Marshall, Gene Hackman, Beatrice Straight, Kate Capshaw, Denzel Washington, Fritz Weaver, Michael Learned, E. Katherine Kerr, Polly Rowles, Matt Salinger; *Dir:* Sidney Lumet. VHS, Beta, LV $19.98 LHV, WAR ♪♪½

The Power of the Ninjitsu A young man inherits the leadership of a martial arts crime gang, the Scorpions, but the elder members muscle him out of his place. He returns for revenge and has the old ones running scared. Power to the Ninjitsu!
1988 90m/C Master Lee, Adam Frank, Peter Ujaer; *Dir:* Joseph Lai. VHS, Beta $59.95 NVH, IME ♪

Power of One A good cast is generally wasted in another liberal, white look at apartheid. Set in South Africa during the 1940s, P.K. is a white orphan of British descendent who is sent to a boarding school run by Afrikaaners (South Africans of German descent). Humiliated and bullied, particularly when England and Germany go to war, P.K. is befriended by a German pianist and a black boxing coach who teaches him to box and stand up for his rights. Which is what P.K. does as he grows up to be a first-rate boxer and a teacher. Preachy and filled with stereotypes. Based on the novel by Bruce Courtenay.
1992 (PG-13) 126m/C Stephen Dorff, Armin Mueller-Stahl, Morgan Freeman, John Gielgud, Maria Marais; *Dir:* John G. Avildsen. VHS, LV $94.99 WAR ♪♪

Power, Passion & Murder A young, glamorous movie star has an affair with a married man which begins the end of her career in 1930s' Hollywood.
1983 104m/C Michelle Pfeiffer, Darren McGavin, Stella Stevens; *Dir:* Paul Bogart, Leon Ichaso. VHS, Beta, LV $79.95 VMK, IME, HHE ♪♪

Power Play A young army colonel from a small European country joins forces with rebels to overthrow the government. After the coup, it is discovered that one of the rebels is a traitor. The film never builds the suspense it should. Also known as ''Operation Overthrow.''
1981 95m/C CA GB Peter O'Toole, David Hemmings, Donald Pleasence, Barry Morse; *Dir:* Martyn Burke. VHS, Beta $49.95 MED, SIM ♪½

Power of the Resurrection A dramatization of the last days of Christ, focusing on the reformation of Peter.
197? 60m/C VHS, Beta $19.95 UHV, VGD ♪

The Power Within An electrified stuntman finds he can send electrical shocks from his hands and becomes the victim of a kidnapping plot.
1979 90m/C Eric Braeden, David Hedison, Susan Howard, Art Hindle; *Dir:* John Llewellyn Moxey. VHS, Beta $59.95 PSM ♪½

Powerforce A CIA agent must use martial arts in order to save the free world.
1983 (R) 98m/C Bruce Lee, Bruce Baron. VHS, Beta $49.95 IND ♪

Powwow Highway Remarkably fine performances in this unusual, thought-provoking, poorly-titled foray into the plight of Native Americans. Farmer shines as the unassuming, amiable Cheyenne traveling to New Mexico in a beat-up Chevy with his Indian activist buddy, passionately portrayed by Martinez. On the journey they are constantly confronted with the tragedy of life on a reservation. A sobering look at government injustice and the lingering spirit of a people lost inside their homeland.
1989 (R) 105m/C Gary Farmer, A. Martinez, Amanda Wyss, Rene Handren-Seals, Graham Greene; *Dir:* Joanelle Romero, Jonathan Wacks. VHS, Beta $89.95 WAR, OM ♪♪♪½

The Practice of Love A woman journalist's investigation of a murder casts suspicion on her two male lovers. The film raises the question of whether love is even possible in a world dominated by men's struggles for power. In German, with English subtitles.
1984 90m/C GE *Dir:* Valie Export. VHS $59.95 FCT, MOV ♪♪♪

Prairie Badmen Outlaws are after a treasure map in the possession of the owner of a medicine show. It's up to our hero to stop them. One of the ''Billy Carson'' western series.
1946 55m/B Buster Crabbe. VHS, Beta VCN ♪

The Prairie Home Companion with Garrison Keillor A tape of the in-concert final performance of Keillor's famous radio show in St. Paul's World Theatre on June 13, 1987.
1987 114m/C Garrison Keillor, Leo Kottke, Chet Atkins. VHS, Beta $29.95 DIS ♪

The Prairie King Will grants three people the same gold mine but stipulates only one may own it. Silent quandary.
1927 58m/B Hoot Gibson, Barbara Worth, Charles Sellon, Albert Priscoe. VHS, Beta $19.95 GPV ♪½

Prairie Moon Autry becomes the guardian of three tough kids from Chicago after they inherit a ranch. The kids help Autry round up a gang of rustlers.
1938 58m/B Gene Autry, Smiley Burnette. VHS, Beta CCB, VCN, DVT ♪½

Prairie Pals Deputies go undercover to rescue a kidnapped scientist.
1942 60m/B William Boyd, Lee Powell. VHS, Beta VCN ♪

Prancer An eight-year-old girl whose mother has recently died thinks an injured reindeer she has found belongs to Santa. She lovingly nurses him back to health. Harmless family entertainment.
1989 (G) 102m/C Sam Elliott, Rebecca Harrell, Cloris Leachman, Rutanya Alda, John Joseph Duda, Abe Vigoda, Michael Constantine, Ariana Richards, Mark Rolston; *Dir:* John Hancock. VHS, Beta, LV, 8mm $19.95 COL, ORI, FCT ♪♪

Pray for Death When a mild-mannered Japanese family is victimized by a crime syndicate, a master ninja comes to the rescue.
1985 (R) 93m/C James Booth, Robert Ito, Sho Kosugi, Shane Kosugi, Kane Kosugi, Donna Kei Benz; *Dir:* Gordon Hessler. VHS, Beta $79.95 USA ♪

Pray TV A sly con man turns a failing television station into a profitable one when the station starts to broadcast around-the-clock religious programming. Made for television. Also known as ''KGOD.''
1980 (PG) 92m/C Dabney Coleman, Archie Hahn, Joyce Jameson, Nancy Morgan, Roger E. Mosley, Marcia Wallace; *Dir:* Rick Friedberg. VHS, Beta, LV $69.95 VES ♪♪

Pray TV A television movie expose of broadcast religion in which a young preacher chooses between orthodox religion and lots of money. Not as cutting an indictment as it could have been.
1982 100m/C John Ritter, Ned Beatty, Madolyn Smith, Richard Kiley, Louise Latham, Jonathan Prince, Michael Currie, Lois Areno; *Dir:* Robert Markowitz. VHS, Beta SUE ♪♪

Pray for the Wildcats Three advertising executives' promotional trip from (or to?) Hell. Anything to please a client.
1974 96m/C Andy Griffith, William Shatner, Angie Dickinson, Marjoe Gortner, Lorraine Gary; *Dir:* Robert Lewis. VHS, Beta $69.95 REP ♪½

Prayer for the Dying An IRA hitman longs to quit, but he has to complete one last assignment. The hit is witnessed by a priest who becomes an unwitting associate when the hitman hides out at the church. Fine performances from Bates and Rourke, though Hodges and Rourke were not satisfied with finished film.
1987 (R) 104m/C Mickey Rourke, Alan Bates, Bob Hoskins, Sammi Davis, Liam Neeson; *Dir:* Mike Hodges. VHS, Beta, LV $19.95 VIR ♪♪½

Prayer of the Rollerboys Violent, futuristic, funky action as Haim infiltrates a criminal gang of syncopated roller-blading neo-nazi youth with plans for nationwide domination. Though routinely plotted and predictable, it's got great skating stunts and a wry vision of tomorrow's shattered USA—broke, drug-soaked, homeless, foreign-owned; even sharper when you realize this is a Japanese-American co-production.
1991 (R) 94m/C Corey Haim, Patricia Arquette, Christopher Collet, Julius W. Harris, J.C. Quinn, Jake Dengel; *Dir:* Rick King. VHS $89.95 ACA ♪♪½

Prayer for World Peace Pope John Paul II's sermon marking the beginning of the Marian Year, in the Basilica of St. Mary Major, on June 6, 1987. Translated from Italian; available also in Spanish.
1987 60m/C IT VHS, Beta $29.95 IVE ♪♪½

Praying Mantis The professor's scheming nurse murdered his family and plans to marry him for his money. She enlists the aid of the professor's assistant in her evil plot.
1983 119m/C Jonathan Pryce, Cherie Lunghi, Ann Cropper, Carmen Du Sautoy, Pinkas Braun; *Dir:* Jack Gold. VHS, Beta $59.95 ONE ♪♪

Preacherman A phony preacher travels through the South, fleecing gullible congregations wherever he goes.
1983 (R) 90m/C Amos Huxley, Marian Brown, Adam Hesse. VHS, Beta PGN ♪½

Predator Schwarzenegger leads a team of CIA-hired mercenaries into the Central American jungles to rescue hostages. They encounter an alien force that begins to attack them one by one. Soon it's just Arnold and the Beast in this attention-grabbing, but sometimes silly, suspense film.
1987 (R) 107m/C Arnold Schwarzenegger, Jesse Ventura, Sonny Landham, Bill Duke, Elpidia Carrillo, Carl Weathers, R.G. Armstrong; *Dir:* John McTiernan. VHS, Beta, LV $19.98 FOX ♪♪½

Predator 2 Tough cop takes time away from battling drug dealers to deal with malicious extraterrestrial who exterminated Arnold's band of commandos in ''Predator.'' Miss-billed as sequel, its only resemblance to the original is that the predator has inexplicably returned (this time to the thick of L.A.).

Gorey action aplenty, little logic. Fine cast dominated by minority performers can't save this one. **1990 (R) 105m/C** Danny Glover, Gary Busey, Ruben Blades, Maria Conchita Alonso, Bill Paxton, Robert Davi, Adam Baldwin, Kent McCord, Morton Downey Jr., Calvin Lockhart, Teri Weigel; *Dir:* Stephen Hopkins. **VHS, Beta, LV $19.98** *FXV, CCB* 🐾🐾

Prehistoric Women
A tribe of prehistoric women look for husbands the old-fashioned way—they drag them back to their caves from the jungle. So bad it's almost good.
1950 74m/C Laurette Luez, Allan Nixon, Mara Lynn, Joan Shawlee, Judy Landon; *Dir:* Greg Tallas. **VHS, Beta $9.99** *FHE, NOS, RHI* Woof!

Premature Burial
A cataleptic Englishman's worst fears come true when he is buried alive. He escapes and seeks revenge on his doctor and his greedy wife. Based upon the story by Edgar Allen Poe.
1962 81m/C Ray Milland, Richard Ney, Hazel Court, Heather Angel; *Dir:* Roger Corman. **VHS, Beta $69.98** *LIV, VES, MLB* 🐾🐾

Premonition
Three drug-riddled '60s' college students experience similar premonitions of death, and subsequently either die or become tormented.
1971 (PG) 83m/C Carl Crow, Tim Ray, Winfrey Hester Hill, Victor Izay; *Dir:* Alan Rudolph. **VHS, Beta $39.95** *AHV* 🐾½

The Premonition
A parapsychologist searching for a missing child is drawn into a frightening maze of dream therapy and communication with the dead. Filmed on location in Mississippi.
1975 94m/C Richard Lynch, Sharon Farrell, Jeff Corey, Ellen Barber, Edward Bell, Danielle Brisebois; *Dir:* Robert Allen Schnitzer. **VHS, Beta $24.98** *SUE* 🐾½

Prep School
A very proper New England prep school is turned topsy-turvy by two rambunctious co-eds. They're determined to break every one of the school's cardinal rules.
1981 (PG-13) 97m/C Leslie Hope, Andrew Sabiston; *Dir:* Paul Almond. **VHS, Beta $79.95** *PSM* 🐾

Preppies
Yet another teen sex comedy, but this time Ivy Leaguer Drake must pass his exams to receive his fifty-million dollar inheritance. Lots of skirt chasing as his conniving cousin leads him astray.
1982 (R) 83m/C Dennis Drake, Peter Brady Reardon, Steven Holt, Nitchie Barrett, Cindy Manion, Katt Shea, Lynda Wiesmeier; *Dir:* Chuck Vincent. **VHS, Beta $29.98** *LIV, VES* 🐾½

Presenting Lily Mars
A small-town girl comes to New York to make it on Broadway. Based on the Booth Tarkington novel.
1943 105m/B Judy Garland, Van Heflin, Fay Bainter, Richard Carlson, Tommy Dorsey, Bob Crosby & His Orchestra; *Dir:* Norman Taurog. **VHS, Beta $24.95** *MGM* 🐾🐾

Presidential Blooper Reel
This collection of excerpts from television and motion pictures features Ronald Reagan, Bette Davis, Humphrey Bogart, James Cagney, Red Skelton, James Arness, and Edward G. Robinson.
1981 55m/B VHS, Beta $19.95 *NOS, HHT, DVT* 🐾🐾

The President's Analyst
A superbly written, brilliantly executed satire from the mind of Theodore J. Flicker, who wrote as well as directed. Coburn steals the show as a psychiatrist who has the dubious honor of being appointed "secret shrink" to the President of the U.S. Pressures of the job steadily increase his paranoia until he suspects he is being pursued by agents and counter agents alike. Is he losing his sanity or...? Vastly entertaining.
1967 104m/C James Coburn, Godfrey Cambridge, Severn Darden, Joan Delaney, Pat Harrington, Will Geer, William Daniels; *W/Dir:* Theodore J. Flicker. **VHS $14.95** *PAR* 🐾🐾🐾½

President's Mistress
A security agent searches for his sister's murderer, while trying to obscure the fact that she was having an affair with the President.
1978 97m/C Beau Bridges, Susan Blanchard, Larry Hagman, Joel Fabiani, Karen Grassle; *Dir:* John Llewellyn Moxey. **VHS, Beta $69.95** *LTG* 🐾½

The President's Mystery
A lawyer decides to turn his back on society by giving up his practice and his marriage. Eventually he meets and falls in love with another woman. The most interesting aspect of the film is that it is based on a story by Franklin D. Roosevelt, which was published in "Liberty" magazine. The story was supposedly better than its screen adaptation.
1936 80m/B Henry Wilcoxon, Betty Furness, Sidney Blackmer, Evelyn Brent; *Dir:* Phil Rosen. **VHS, Beta $16.95** *SNC* 🐾½

The President's Plane is Missing
When Air Force One disappears, the less than trustworthy Vice President takes control. Could there be a dire plot in the making? Based on a novel by Rod Serling's brother Robert J. Serling.
1971 100m/C Buddy Ebsen, Peter Graves, Arthur Kennedy, Rip Torn, Louise Sorel, Raymond Massey, James Wainwright, Mercedes McCambridge, Dabney Coleman, Joseph Campanella; *Dir:* Daryl Duke. **VHS $69.98** *REP* 🐾🐾

The Presidio
An easy going police detective must investigate a murder on a military base where he and the base commander have sparred before. The commander becomes downright nasty when his daughter shows an interest in the detective. Good action scenes in San Francisco almost covers up script weaknesses, but not quite.
1988 (R) 97m/C Sean Connery, Mark Harmon, Meg Ryan, Jack Warden, Mark Blum, Jenette Goldstein; *Dir:* Peter Hyams. **VHS, Beta, LV, 8mm $89.95** *PAR* 🐾🐾

Presumed Innocent
Assistant district attorney is the prime suspect when a former lover turns up brutally murdered. Cover ups surround him, the political climate changes, and friends and enemies switch sides. Slow-paced courtroom drama with excellent performances from Ford, Julia and Bedelia. Skillfully adapted by Pakula and Frank Pierson from the best-seller by Chicago attorney Scott Turow.
1990 (R) 127m/C Harrison Ford, Brian Dennehy, Bonnie Bedelia, Greta Scacchi, Raul Julia, Paul Winfield; *Dir:* Alan J. Pakula. **VHS, Beta, LV, 8mm $92.98** *WAR, FCT* 🐾🐾🐾

Pretty Baby
Shield's launching pad and Malle's first American film is a masterpiece of cinematography and style, nearly upstaged by the plodding storyline. Carradine manages to be effective but never succeeds at looking comfortable as the New Orleans photographer besotted with and subsequently married to an 11-year old prostitute (Shields). Low key, disturbingly intriguing story, beautifully photographed by Sven Nykist.

1978 (R) 109m/C Brooke Shields, Keith Carradine, Susan Sarandon, Barbara Steele, Diana Scarwid, Antonio Fargas; *Dir:* Louis Malle. National Board of Review Awards '78: 10 Best Films of the Year. **VHS, Beta, LV $69.95** *PAR* 🐾🐾🐾

Pretty in Pink
More teenage angst from the pen of John Hughes. Poor girl falls for a rich guy. Their families fret, their friends are distressed, and fate conspires against them. The pair plans to go to prom together, but peer pressure creates difficulties. If you can suspend your disbelief that a teenaged girl with her own car and answering machine is financially inferior, then you may very well be able to accept the entire premise. Slickly done and adequately, if not enthusiastically, acted. In 1987, Hughes remade this film and put himself in the director's chair with much of the same result. "Some Kind of Wonderful" is the same story with the rich/pauper characters reversed by gender.
1986 (PG-13) 96m/C Molly Ringwald, Andrew McCarthy, Jon Cryer, Harry Dean Stanton, James Spader, Annie Potts, Andrew Dice Clay, Margaret Colin, Alexa Kenin, Gina Gershon, Dweezil Zappa; *Dir:* Howard Deutch. **VHS, Beta, LV, 8mm $14.95** *PAR* 🐾🐾½

Pretty Poison
You won't need an antidote for this one. Original, absorbing screenplay, top-notch acting and on target direction combine to raise this low-budget, black comedy above the crowd. Perkins at his eerie best as a burned-out arsonist who cooks up a crazy scheme and enlists the aid of a hot-to-trot high schooler, only to discover too late she has some burning desires of her own. Weld is riveting as the turbulent teen.
1968 89m/C Anthony Perkins, Tuesday Weld, Beverly Garland, John Randolph, Dick O'Neil; *Dir:* Noel Black. **VHS** *ING, MLB* 🐾🐾🐾½

Pretty Smart
Two diametrically opposed sisters at a European finishing school team up against a drug-dealing, voyeuristic teacher. Pretty lame.
1987 (R) 84m/C Tricia Leigh Fisher, Patricia Arquette, Dennis Cole, Lisa Lorient; *Dir:* Dimitri Logothetis. **VHS, Beta $19.95** *STE, NWV* Woof!

Pretty Woman
An old story takes a fresh approach as a successful but stuffy business man hires a fun-loving, energetic young hooker to be his companion for a week. The film caused some controversy over its upbeat portrayal of prostitution, but its popularity at the box office catapulted Oscar-nominated actress Roberts to stardom.
1989 (R) 117m/C Richard Gere, Julia Roberts, Ralph Bellamy, Jason Alexander, Laura San Giacomo, Hector Elizondo; *Dir:* Garry Marshall. Golden Globe Awards '89: Best Actress (Roberts). **VHS, Beta, LV $19.99** *TOU* 🐾🐾🐾

Prettykill
A confusing storyline with numerous subplots involves detective with a paramour/prostitute (Hubley) attempting to stalk a mad killer while contending with her split personality.
1987 (R) 95m/C David Birney, Susannah York, Season Hubley, Yaphet Kotto, Suzanne Snyder, Germane Honde; *Dir:* George Kaczender. **VHS, Beta, LV $39.98** *WAR* Woof!

The Prey
A poorly-done horror movie about a predator who is looking for a mate in the Colorado Rockies and kills five campers in the process.
1980 (R) 80m/C Debbie Thurseon, Steve Bond, Lori Lethin, Jackie Coogan; *Dir:* Edwin Scott Brown. **VHS, Beta $19.99** *HBO, STE* 🐾

Prey of the Chameleon A female serial killer escapes from her asylum, ready to rip more men to shreds. However a tough lady cop has other ideas and goes all out to put an end to the madwoman's doings. Can she stop this fiend before the man she loves becomes the next victim?
1991 (R) 91m/C Daphne Zuniga, James Wilder, Alexandra Paul, Don Harvey; **Dir:** Fleming Fuller. **VHS, Beta $89.95** PSM 🎞️

Priceless Beauty Musician Lambert was only looking when he spotted the bottle which changed his life. Beautiful genie (Lane) lives inside, and is waiting just for him! Fun premise, good score, nice chemistry between the actors.
1990 (R) 94m/C Christopher Lambert, Diane Lane, Francesco Quinn; **Dir:** Charles Finch. **VHS, LV $89.98** REP 🎞️🎞️½

Prick Up Your Ears Film biography of popular subversive playwright Joe Orton depicts his rise to fame and his eventual murder at the hands of his homosexual lover in 1967. Acclaimed for its realistic and sometimes humorous portrayal of the relationship between two men in a society that regarded homosexuality as a crime, the film unfortunately pays scant attention to Orton's theatrical success. The occasional sluggishness of Alan Bennett's script detracts a bit from the three leads' outstanding performances.
1987 (R) 110m/C Gary Oldman, Alfred Molina, Vanessa Redgrave, Julie Walters, Lindsay Duncan, Wallace Shawn, James Grant, Frances Barber, Janet Dale, David Atkins; **Dir:** Stephen Frears. New York Film Critics Awards '87: Best Supporting Actress (Redgrave). **VHS, Beta, LV $19.95** VIR 🎞️🎞️🎞️

Pride of the Bowery The Dead End Kids versus a boxing hopeful in training camp.
1941 60m/B Leo Gorcey, David Grocey, Huntz Hall, Gabriel Dell, Billy Halop, Bobby Jordan; **Dir:** Joseph H. Lewis. **VHS, Beta $19.95** NOS, CAB, DVT 🎞️

Pride of the Clan Silent drama with Pickford and Moore as Scottish sweethearts battling a bit of adversity. When Pickford's father is lost at sea, she moves onto his fishing boat. She meets Moore, a fishing boy, and falls in love. But, he inherits a fortune and his parents forbid him to see Pickford. He goes off to live the good life but comes back to dramatically rescue Pickford.
1918 80m/B Mary Pickford, Matt Moore; **Dir:** Maurice Tourneur. **VHS, Beta $19.95** GPV, DNB 🎞️🎞️½

Pride of Jesse Hallum A made-for-television film about an illiterate man, played by country singer Cash, who learns, after much trial and tribulation, to read.
1981 105m/C Johnny Cash, Brenda Vaccaro, Eli Wallach; **Dir:** Gary Nelson. **VHS, Beta $59.98** FOX 🎞️½

Pride and the Passion A small group of resistance fighters battling for Spanish independence in 1810 must smuggle a 6-ton cannon across the rugged terrain of Spain. Miscasting, especially of Sinatra as a Spanish peasant, hurts this film.
1957 132m/C Cary Grant, Frank Sinatra, Sophia Loren; **Dir:** Stanley Kramer. **VHS, Beta $19.98** FOX 🎞️🎞️

Pride and Prejudice Classic adaptation of Austen's classic novel of manners as a young marriageable woman spurns the suitor her parents choose for her. Excellent cast vividly recreates 19th century England,

aided by the inspired set design that won the film an Oscar.
1940 114m/B Greer Garson, Laurence Olivier, Edmund Gwenn, Edna May Oliver, Mary Boland, Maureen O'Sullivan, Ann Rutherford, Frieda Inescort; **Dir:** Robert Z. Leonard. Academy Awards '40: Best Art Direction/Set Decoration (B & W). **VHS, Beta $24.95** MGM 🎞️🎞️½

Pride and Prejudice BBC mini-series adaptation of Jane Austen's novel about 19th century British mores and the attempts of five sisters to get married.
1985 226m/C GB Elizabeth Garvie, David Rintoul; **Dir:** Cyril Coke. **VHS, Beta $29.98** KUI, FOX 🎞️🎞️

Pride of St. Louis A romanticized and humorous portrait of famed baseball player-turned-commentator Dizzy Dean. Oscar-nominated for story.
1952 93m/B Dan Dailey, Joanne Dru, Richard Crenna, Richard Hayden, Hugh Sanders; **Dir:** Harmon Jones. **VHS, Beta $19.98** FOX 🎞️🎞️

The Pride of the Yankees Excellent portrait of baseball great Lou Gehrig. Beginning as he joined the Yankees in 1923, the film follows this great American through to his moving farewell speech as his career was tragically cut short by the disease that bears his name. Cooper is inspiring in the title role and was the recipient of a well-deserved Oscar nomination. Film also nominated for best picture, best actress (Wright), original story, screenplay, black & white cinematography, black & white interior decoration, sound recording, and score for a dramatic picture.
1942 128m/B Gary Cooper, Teresa Wright, Babe Ruth, Walter Brennan, Dan Duryea; **Dir:** Sam Wood. Academy Awards '42: Best Film Editing. **VHS, Beta, LV $19.98** FOX 🎞️🎞️🎞️½

Priest Live A complete concert by the heavy-metal outfit, featuring 19 of their hits, including "Love Bites." In digital Hi-Fi Stereo.
1987 95m/C **VHS, Beta $19.98** FOX 🎞️🎞️🎞️½

Priest of Love Arty account of the final years of the life of then-controversial author D.H. Lawrence, during which time he published "Lady Chatterly's Lover." A slow-moving but interesting portrayal of this complex man and his wife as they grapple with his immanent death from tuberculosis.
1981 (R) 125m/C GB Ian McKellan, Janet Suzman, John Gielgud, Helen Mirren, Jorge Rivero; **Dir:** Christopher Miles. **VHS, Beta $79.99** HBO 🎞️🎞️

Primal Impulse An astronaut, stranded on the moon because of a sinister experimental double-cross, unleashes a mental scream which possesses a young woman's mind back on earth.
1974 90m/C Klaus Kinski. **VHS, Beta $69.95** LTG 🎞️½

Primal Rage A student is bitten by an experimental monkey, and begins to manifest his primal urges physically. Special effects by Carlo Rimbaldi.
1990 (R) 92m/C Bo Svenson, Patrick Lowe, Mitch Watson, Cheryl Arutt, Sarah Buxton; **Dir:** Vittoria Rambaldi. **VHS, Beta $19.98** WAR 🎞️

Primal Scream The Year is 1993. Earth's fuel sources are rapidly decaying and the top secret project to mine a revolutionary new energy source is underway—independently managed by a corrupt corporation.
1987 95m/C Kenneth John McGregor, Sharon Mason, Julie Miller, Jon Maurice, Mickey Shaughnessy; **Dir:** William Murray. **VHS $79.98** MAG, NVH 🎞️½

Primary Motive A taut political thriller with Nelson starring as Andy Blumenthal, a press secretary who finds that an opposing candidate's campaign is covered by a web of lies. When Blumental exposes the lies, the candidate denies them and pulls even farther ahead in the polls. When the candidate's wife gives Blumenthal an extremely damaging piece of information, will the polls finally turn against him?
1992 (R) 93m/C Judd Nelson, Richard Jordan, Sally Kirkland, Justine Bateman, John Savage; **W/Dir:** Daniel Adams. **VHS, Beta, LV $89.98** FXV 🎞️🎞️

Primary Target A mercenary reunites his 'Nam guerilla unit to rescue a diplomat's wife kidnapped by a Laotian jungle lord.
1989 (R) 85m/C John Calvin, Miki Kim, Joey Aresco, Chip Lucio, John Ericson, Colleen Casey; **Dir:** Clark Henderson. **VHS, Beta $79.98** MGM 🎞️

Prime Cut Veritable orgy of drug trafficking, prostitution, extortion, loan sharking, fisticuffs and gangsters getting ground into mincemeat. Sleazy but well-made crime melodrama has its followers, but is best known as Spacek's film debut.
1972 (R) 86m/C Lee Marvin, Gene Hackman, Sissy Spacek, Angel Tompkins, Gregory Walcott; **Dir:** Michael Ritchie. **VHS, Beta $59.98** FOX 🎞️🎞️½

Prime Evil A brave and determined nun infiltrates a sect of devil worshiping monks in an attempt to end their demonic sacrifices. The question is, will this sister slide beneath Satan's cleaver?
1988 (R) 87m/C William Beckwith, Christine Moore; **Dir:** Roberta Findlay. **VHS, LV $19.95** STE 🎞️🎞️

The Prime of Miss Jean Brodie Oscar-winning performance by Smith as a forward-thinking teacher in a Scottish Girls' school during the 1920s. She captivates her impressionable young students with her fascist ideals and free-thinking attitudes in this adaptation of the play taken from Muriel Spark's novel. Musical score by songwriter Rod McKuen was nominated for an Oscar.
1969 (PG) 116m/C GB Maggie Smith, Pamela Franklin, Robert Stephens, Celia Johnson, Gordon Jackson, Jane Carr; **Dir:** Ronald Neame. Academy Awards '69: Best Actress (Smith). **VHS, Beta $79.98** FOX, BTV 🎞️🎞️🎞️

Prime Risk A young engineer who discovers an electronic method to break into automated teller machines finds that it leads to more trouble when she discovers that foreign agents are hot on her trail.
1984 (PG-13) 98m/C Toni Hudson, Lee Montgomery, Sam Bottoms; **Dir:** W. Farkas. **VHS, Beta $79.98** LIV, LTG 🎞️½

Prime Suspect An honest citizen becomes the prime suspect after the coverage of a murder by an over-ambitious television reporter.
1982 100m/C Mike Farrell, Teri Garr, Veronica Cartwright, Lane Smith, Barry Corbin, James Sloyan, Charles Aidman; **Dir:** Noel Black. **VHS, Beta $79.95** SVC 🎞️🎞️

Prime Suspect A young man escapes from a mental institution to clear his name after being wrongfully accused of murdering his girlfriend.
1988 89m/C Susan Strasberg, Frank Stallone, Billy Drago, Doug McClure; **Dir:** Mark Rutland. **VHS $79.95** SVS 🎞️🎞️

Prime Target Small-town cop John Bloodstone (Heavener) is recruited by the FBI to transfer a mafia boss (Curtis) from a safehouse to the courthouse. He has to keep the former mobster alive long enough to testify against the ''family.'' Their cross-country adventure is heightened by a murderous confrontation with evil forces that want them both dead.
1991 (R) 87m/C David Heavener, Tony Curtis, Isaac Hayes, Jenilee Harrison, Robert Reed, Andrew Robinson, Don Stroud; **W/Dir:** David Heavener. VHS $59.95 *HMD* 🐾🐾

Prime Time A satirical comedy about what would happen if the censors took a day off from American television.
1980 (R) 73m/C Warren Oates, David Spielberg. VHS, Beta *PGN* 🐾 ½

The Primitive Lover Early silent comedy about young wife who decides she deserves more than her marriage is giving her.
1916 67m/B Constance Talmadge, Kenneth Harlan; **Dir:** Sidney Franklin. VHS, Beta $19.95 *GPV* 🐾🐾

Primrose Path Melodramatic soaper with comedic touches about a wrong-side-of-the-tracks girl falling for and then losing an ambitious young go-getter running a hamburger stand. Rambeau was Oscar-nominated.
1940 93m/B Ginger Rogers, Joel McCrea, Marjorie Rambeau, Henry Travers, Miles Mander, Queenie Vasser, Joan Carroll, Vivienne Osborne; **Dir:** Gregory La Cava. VHS, Beta $19.98 *RKO* 🐾🐾 ½

Prince of Bel Air A pool-cleaning playboy who makes a habit of one night stands meets a woman and starts falling in love with her. Originally a made-for-television movie.
1987 (R) 95m/C Mark Harmon, Kirstie Alley, Robert Vaughn, Patrick Laborteaux, Deborah Harmon; **Dir:** Charles Braverman. VHS, Beta $89.95 *ACA* 🐾 ½

The Prince of Central Park Two young orphans are forced by circumstance to live in a tree in New York's Central Park until they are befriended by a lonely old woman. An above-average made for television adaptation of the novel by Evan H. Rhodes, the story was later used for a Broadway play.
1977 76m/C Ruth Gordon, T.J. Hargrave, Lisa Richard, Brooke Shields, Marc Vahanian, Dan Hedaya; **Dir:** Harvey Hart. VHS, Beta $59.95 *IVE* 🐾🐾🐾

Prince of the City Docu-drama of a police officer who becomes an informant in an effort to end corruption within his narcotics unit, but finds he must pay a heavy price. Based on a true story, the powerful script by Sidney Lumet and Jay Preston (from Robert Daley's book) carries the tension through what would otherwise be an overly long film. Excellent performances make this a riveting character study.
1981 (R) 167m/C Treat Williams, Jerry Orbach, Richard Foronjy, Don Billett, Kenny Marino, Lindsay Crouse, Lance Henriksen; **Dir:** Sidney Lumet. VHS, Beta, LV $29.98 *WAR* 🐾🐾🐾

Prince of Darkness University students release Satan, in the form of a mysterious chemical, unwittingly on the world. Written by Martin Quatermass (a pseudonym of Carpenter). Strong personnel does not save this dreary and slow-moving cliche plot.
1987 (R) 102m/C Donald Pleasence, Lisa Blount, Victor Wong, Jameson Parker, Dennis Dun; **Dir:** John Carpenter. VHS, Beta, LV $14.98 *MCA* 🐾 ½

Prince and the Great Race Three Australian children search the outback to find their kidnapped horse who is scheduled to run in the big New Year's Day Race.
1983 91m/C *AU* John Ewart, John Howard, Nicole Kidman. VHS, Beta $14.98 *LIV, LTG* 🐾🐾

Prince Igor This dramatization of the Alexander Borodin opera tells the story of a Russian prince of the 12th century who protected his people from invading barbarians. This Kirov Opera production was the inspiration for Broadway's ''Kismet.''
1969 110m/C *RU* VHS, LV $49.95 *MVD, KUL* 🐾🐾

Prince Jack Profiles the turbulent political career of President John F. Kennedy.
1983 100m/C Lloyd Nolan, Dana Andrews, Robert Guillaume, Cameron Mitchell; **Dir:** Bert Lovitt. VHS, Beta $69.98 *LIV, VCL* 🐾🐾

Prince Jammy Features some of reggae's top performers, including Bammy Man, Shabba Ranks, Major Worries, Little Risto, Admiral Bailey, Little Twitch and more.
1987 60m/C Admiral Bailey, Little Twitch, Major Worries, Little Risto, Shaba Ranks, Bammy Man, Frankie Paul, Don Angelo. VHS, Beta $29.95 *MVD, KRV* 🐾🐾

Prince and the Pauper Satisfying adaptation of the classic Mark Twain story of a young street urchin who trades places with the young king of England. Wonderful musical score by noted composer Erich Wolfgang Korngold who provided the music for many of Flynn's adventure films. Also available in a computer-colorized version.
1937 118m/B Errol Flynn, Claude Rains, Alan Hale Jr., Billy Mauch, Montagu Love, Henry Stephenson, Barton MacLane; **Dir:** William Keighley. VHS, Beta $19.95 *FOX, MLB, OOD* 🐾🐾🐾

Prince and the Pauper A prince and a poor young boy swap their clothes and identities, thus causing a lot of confusion for their families. Based on the story by Mark Twain.
1962 93m/C Guy Williams, Laurence Naismith, Donald Houston, Jane Asher, Walter Hudd. VHS, Beta $12.99 *DIS* 🐾🐾

The Prince and the Pauper An animated version of the famous Mark Twain tale about a prince in medieval England who changes places with a pauper to see how the other half lives and finds out more than he expected.
1971 47m/C VHS, Beta $19.98 *MGM, KUI, MRV* 🐾🐾

Prince and the Pauper Remake of the 1937 Errol Flynn film employing lavish sets and a tongue-in-cheek attitude among the all-star cast, who occasionally wander adrift when the director stops for tea. When an English prince and a pauper discover that they have identical appearances, they decide to trade places with each other. Alternate title:''Crossed Swords.'' From Mark Twain's classic.
1978 (PG) 113m/C *GB* Oliver Reed, Raquel Welch, Mark Lester, Ernest Borgnine, George C. Scott, Rex Harrison, Charlton Heston, Sybil Danning; **Dir:** Richard Fleischer. VHS, Beta $19.98 *MED* 🐾🐾 ½

Prince of Pennsylvania A mild comedy about a spaced-out youth who kidnaps his own father in hopes of nabbing a family inheritance.
1988 (R) 113m/C Keanu Reeves, Fred Ward, Amy Madigan, Bonnie Bedelia, Jeff Hayenga; **Dir:** Ron Nyswaner. VHS, Beta, LV $29.95 *COL* 🐾 ½

The Prince and Princess of Wales...Talking Personally An interview with the royal couple covering a variety of topics.
1986 45m/C VHS, Beta $29.98 *LIV* 🐾 ½

Prince and the Revolution: Live A concert film featuring the rock star's biggest hits. Includes Prince favorites ''Let's Go Crazy,'' ''Delirious,'' ''1999,'' ''Little Red Corvette,'' ''Take Me With U,'' ''Do Me, Baby,'' ''Irresistible Bitch,'' ''Possessed,'' ''How Come U Don't Call Me Anymore,'' ''Let's Pretend We're Married,'' ''God,'' ''Computer Blue,'' ''Darling Niki,'' ''The Beautiful Ones,'' ''When Doves Cry,'' ''I Would Die 4 U,'' ''Baby I'm a Star,'' and ''Purple Rain.''
1985 116m/C *Performed by:* Prince. VHS, Beta, LV $19.99 *MVD, WRV, PIA* 🐾 ½

The Prince and the Showgirl An American showgirl in 1910 London is wooed by the Prince of Carpathia. Part of the ''A Night at the Movies'' series, this tape simulates a 1957 movie evening, with a Sylvester the Cat cartoon, ''Greedy for Tweety,'' a newsreel and coming attractions for ''Spirit of St. Louis.''
1957 127m/C Laurence Olivier, Marilyn Monroe, Sybil Thorndike; **Dir:** Laurence Olivier. VHS, Beta $19.98 *WAR* 🐾🐾 ½

The Prince of Thieves Robin Hood helps Lady Marian extricate herself from a forced marriage in this adventure made with younger audiences in mind.
1948 72m/C Jon Hall, Patricia Morison, Adele Jergens, Alan Mowbray, Michael Duane; **Dir:** Howard Bretherton. VHS *COL* 🐾🐾

The Prince of Tides Conroy's sprawling southern-fried saga is neatly pared down to essentials in this tale of the dysfunctional Wingo family. Tom Wingo (a bravura performance by Nolte), a failed teacher with a shaky marriage, leaves his home in South Carolina for New York, in order to aid his fragile twin sister Savannah who has attempted suicide. Susan Lowenstein (Streisand), Savannah's psychiatrist, asks Tom's help in discovering his sister's problems, rooted in their gothic family past. There is Lila, the self-possessed mother aiming for a better life; Henry, the brutal shrimper father; and Luke, the older brother who died violently. Dark family tragedies are gradually revealed as Tom, who has also blocked out the memories, comes to grips with his own demons under Lowenstein's ministering aid. Streisand is restrained in both her own performance and in her direction although a subplot dealing with Lowenstein's bad marriage and rebellious son is a predictable distraction. The South Carolina low country, and even New York City, never looked better. Received seven Academy Award nominations.
1991 (R) 132m/C Barbra Streisand, Nick Nolte, Blythe Danner, Kate Nelligan, Jeroen Krabbe, Melinda Dillon, Jason Gould; **Dir:** Barbra Streisand. Golden Globe Awards '92: Best Actor (Nolte); Los Angeles Film Critics Association Awards '91: Best Actor (Nolte). VHS, Beta, LV, 8mm *COL* 🐾🐾🐾

Prince Valiant When his royal dad is exiled by an evil tyrant, brave young Wagner brushes aside bangs and journeys to Camelot to seek the help of King Arthur. Based on Harold Foster's classic comic strip.

P

1954 100m/C James Mason, Janet Leigh, Robert Wagner, Debra Paget, Sterling Hayden, Victor McLaglen, Donald Crisp, Brian Aherne, Barry Jones, Mary Philips; **Dir:** Henry Hathaway. **VHS, Beta $14.98** FXV, FCT 𝄞𝄞½

Princess Academy A self-respecting young debutante battles the ways of her elitist finishing school.
1987 (R) 91m/C FR Eva Gabor, Lu Leonard, Richard Paul, Lar Park Lincoln, Carole Davis; **Dir:** Bruce Block. **VHS, Beta $79.98** LIV, LTG 𝄞

The Princess Bride A modern update of the basic fairy tale crammed with all the cliches, this adventurously irreverent love story centers around a beautiful maiden and her young swain as they battle the evils of the mythical kingdom of Florin to be reunited with one another. Great dueling scenes and offbeat satire of the genre make this fun for adults as well as children. Based on William Goldman's cult novel, with music by Mark Knopfler (vocal by Willy DeVille).
1987 (PG) 98m/C Cary Elwes, Mandy Patinkin, Robin Wright, Wallace Shawn, Peter Falk, Andre the Giant, Chris Sarandon, Christopher Guest, Billy Crystal, Carol Kane, Fred Savage, Peter Cook, Mel Smith; **Dir:** Rob Reiner. **VHS, Beta, LV, SVS, 8mm $19.95** COL, SUE, SUP 𝄞𝄞𝄞½

Princess and the Call Girl A call girl asks her look-alike roommate to take her place in Monaco for a lavishly erotic weekend.
1984 90m/C Carol Levy, Shannah Hall, Victor Bevine; **Dir:** Radley Metzger. **VHS, Beta $59.95** MON 𝄞𝄞

Princess Cinderella A live-action dramatization of the famous fantasy. Dubbed in English.
193? 75m/B SP Silvana Jachino, Roberto Villa. **VHS, Beta $59.95** JEF 𝄞𝄞

Princess Daisy Made-for-television movie about a beautiful model who claws her way to the top of her profession while trying to find true love and avoid the clutches of her rotten half-brother. Adapted from Judith Krantz's glitzy best-selling novel.
1983 200m/C Merete Van Kamp, Lindsay Wagner, Claudia Cardinale, Stacy Keach, Ringo Starr, Barbara Bach; **Dir:** Waris Hussein. **VHS, Beta $69.95** FOX 𝄞

Princess in Exile Young people struggle with their life-threatening illnesses at a special summer camp. They find that love and friendship hold the key to dreams about the future. Excellent cast of newcomers. Based on a novel of the same name by Mark Schreiber.
1990 (PG-13) 103m/C Zachary Ansley, Nicholas Shields, Stacy Mistysyn, Alexander Chapman, Chuck Shamata; **Dir:** Giles Walker. Montreal World Film Festival '90: Ecumenical Prize (Princes in Exile). **VHS, Beta, LV $89.95** FRH, FCT 𝄞𝄞½

Princess and the Pea From "Faerie Tale Theatre" comes the story of a princess who tries to prove that she's a blueblood by feeling the bump of a tiny pea under the thickness of twenty mattresses.
1983 60m/C Liza Minnelli, Tom Conti, Tim Kazurinsky, Pat McCormick, Beatrice Straight; **Dir:** Tony Bill. **VHS, Beta, LV $14.98** KUI, FOX, FCT 𝄞𝄞𝄞

The Princess and the Pirate Hope at his craziest as a vaudevillian who falls for a beautiful princess while on the run from buccaneers on the Spanish Main. Look for Crosby in a closing cameo performance. Available in digitally remastered stereo with original movie trailer.

1944 94m/C Bob Hope, Walter Slezak, Walter Brennan, Virginia Mayo, Victor McLaglen, Bing Crosby; **Dir:** David Butler. **VHS, Beta, LV $19.98** HBO, SUE 𝄞𝄞𝄞

The Princess and the Swineherd A German adaptation of the Brothers Grimm fairy tale about a young suitor who tries to please a materialistic princess.
1960 82m/B GE **VHS, Beta $29.95** VYY 𝄞½

Princess Tam Tam Pleasing French adaptation of Shaw's "Pygmalion," as a beautiful native African woman is "westernized" by a handsome writer and then introduced to high society as an exotic princess. A musical notable for its spectacular choreography and on-location Tunisian scenery. Story by Pepito Abatino, who was then Baker's husband. In French with English subtitles.
1935 77m/B FR Josephine Baker, Albert Prejean, Germaine Aussey, Viviane Romance; **Dir:** Edmond T. Greville. **VHS, Beta, LV $29.95** MVD, RHP, KIV 𝄞

The Princess Who Never Laughed A stern king holds a laugh-off contest to make his morose daughter happy in this adaptation of the Brothers Grimm story from the "Faerie Tale Theatre" series.
1984 60m/C Ellen Barkin, Howard Hesseman, Howie Mandel, Mary Woronov. **VHS, Beta $14.98** KUI, FOX, FCT 𝄞𝄞½

The Principal A tough, down-on-his-luck high school teacher is hired as the principal of a relentlessly violent, uncontrollable high school. Naturally he whips it into shape.
1987 (R) 109m/C James Belushi, Louis Gossett Jr., Rae Dawn Chong; **Dir:** Christopher Cain. **VHS, Beta, LV $14.95** COL 𝄞𝄞½

Prison The zombified body of an unjustly executed inmate haunts Creedmore Prison, stalking the guard that killed him. His search is aided by the terrified inmates.
1988 (R) 102m/C Lane Smith, Chelsea Field, Viggo Mortensen, Lincoln Kilpatrick, Tom Everett, Tiny Lister; **Dir:** Renny Harlin. **VHS, Beta, LV $19.95** STE, NWV 𝄞½

Prison Break A convict plans a daring prison escape in order to clear his name for a murder he did not commit.
1938 72m/B Barton MacLane, Glenda Farrell, Ward Bond. **VHS, Beta $16.95** SNC, NOS, BUR 𝄞𝄞

Prison Stories: Women on the Inside Three short dramatizations of stories depicting the life of women inside prison walls. The first, "New Chicks," directed by Spheeris, tells the story of two lifelong friends and partners in crime. When one woman (Chong) becomes pregnant, she goes to the leader of the toughest prison gang for protection, risking her only true friendship. "Esperanza," directed by Deitch, tells the story of a woman on the inside trying to prevent her family from following the same destructive path. Finally, Silver's "Parole Board" features Davidovich as a murderess up for parole but reluctant to leave the security of the prison. A blunt and gritty portrayal of prison gangs, the lifestyle of female inamtes, and their fears of returning to life on the outside. Originally produced for cable.
1991 94m/C Rae Dawn Chong, Annabella Sciorra, Lolita Davidovich, Talisa Soto, Rachel Ticotin, Grace Zabriskie; **Dir:** Penelope Spheeris, Donna Deitch, Joan Micklin Silver. **VHS, Beta** PSM 𝄞𝄞𝄞

Prison Train Travelogue of a convicted murderer's cross-country journey to begin his prison sentence at Alcatraz. Also called "People's Enemy."
1938 84m/B Fred Keating, Linda (Dorothy Comingore) Winters, Clarence Muse, Faith Bacon, Alexander Leftwich, Nestor Paiva, Franklin Farnum; **Dir:** Gordon Wiles. **VHS** MOV 𝄞𝄞½

The Prisoner Gritty drama about a Cardinal imprisoned in a Soviet bloc country as his captors attempt to break his determiniation not to be used as a propaganda tool. Interactions between the prisoner and his interogator are riveting. Based on the real-life experiences of Cardinal Mindszenty, a Hungarian activist during and after WWII. the Nazis in World War II and later the communists.
1955 91m/B GB Alec Guinness, Jack Hawkins, Raymond Huntley, Wilfred Lawson; **Dir:** Peter Glenville. **VHS $69.95** COL 𝄞𝄞𝄞

Prisoner of Honor A made-for-cable-TV retelling of notorious Dreyfus Affair, in which a Jewish officer in the 19th-century French military was accused of treason based on little evidence and lots of bigotry. George Piquart (Dreyfuss), the anti-semitic counterintelligence head, grows to realizes Dreyfus' innocence and fights zealously for the truth. Russell's flamboyant direction takes the heroic tale into the realm of surreal; this may not be a thoroughly accurate account, but it's one of the more eye-filling.
1991 (PG) 90m/C Richard Dreyfuss, Oliver Reed, Peter Firth, Jeremy Kemp, Brian Blessed; **Dir:** Ken Russell. **VHS $89.99** HBO 𝄞𝄞𝄞

Prisoner in the Middle Janssen is the only man who can stop a nuclear warhead from falling into the hands of rival Middle East factions. Originally released as "Warhead."
1974 (PG) 87m/C David Janssen, Karin Dor, Christopher Stone, Turia Tan, David Semadar, Art Metrano; **Dir:** John O'Conner. **VHS, Beta, LV, 8mm $69.95** ACA, MTX 𝄞𝄞

Prisoner of Rio The made-for-cable-TV satire about scruple-less television evangelists and their nefarious schemes at acquiring their audience's money.
1989 90m/C Steven Berkoff, Paul Freeman; **Dir:** Lech Majewski. **VHS, Beta $89.95** MOV 𝄞𝄞

Prisoner of Second Avenue A New Yorker in his late forties faces the future, without a job or any confidence in his ability, with the help of his understanding wife. Based on the Broadway play by Neil Simon.
1974 (PG) 98m/C Jack Lemmon, Anne Bancroft, Gene Saks, Elizabeth Wilson, Sylvester Stallone, F. Murray Abraham; **Dir:** Melvin Frank. **VHS, Beta, LV $19.98** WAR 𝄞𝄞½

Prisoner of Zenda An excellent cast and splendid photography make this the definitive film adaptation of Anthony Hope's swashbuckling novel. A British commoner is forced to pose as his cousin, the kidnapped king of a small European country, to save the throne. Complications of the romantic sort ensue when he falls in love with the queen. Excellent acting, robust sword play, and beautifully designed costumes make this an enjoyable spectacle.
1937 101m/B Ronald Colman, Douglas Fairbanks Jr., Madeleine Carroll, David Niven, Raymond Massey, Mary Astor, C. Aubrey Smith, Montagu Love, Byron Foulger, Alexander D'Arcy; **Dir:** John Cromwell. **VHS, Beta $19.98** MGM 𝄞𝄞𝄞½

Prisoner of Zenda Less-inspired remake of the 1937 version of Anthony Hope's novel, of a man resembling the monarch of a small country who is forced to pose as King during the coronation ceremony, becomes enamored with the queen, and finds himself embroiled in a murder plot. Worth watching for the luxurious costumes and lavish sets. Cast as a Cardinal here, Stone starred in the 1922 version.
1952 101m/C Stewart Granger, Deborah Kerr, Louis Calhern, James Mason, Jane Greer, Lewis Stone; *Dir:* Richard Thorpe. **VHS, Beta $19.98** *MGM* 🎬🎬½

Prisoner of Zenda Flat comedic interpretation of Anthony Hope's swashbuckling tale of two identical men who switch places, only to find things complicated by a murder. Sellers stars in the double role of Prince Rudolph of Ruritania and Syd, the cockney cab driver who doubles for Rudolph when the Prince is imprisoned by his jealous brother Michael.
1979 (PG) 108m/C Peter Sellers, Jeremy Kemp, Lynne Frederick, Lionel Jeffries, Elke Sommer; *Dir:* Richard Quine. **VHS, Beta $39.95** *MCA* 🎬🎬

Prisoners of Inertia Two newlyweds travel to New York city and find themselves caught up in a whirlwind adventure in this well-acted but lazily scripted comedy drama.
1989 (R) 92m/C Amanda Plummer, Christopher Rich, John C. McGinley; *Dir:* Jay Noyles Seles. **VHS $79.95** *BTV* 🎬🎬½

Prisoners of the Lost Universe A talk-show hostess and her buddy are transported to a hostile universe by a renegade scientist. The two terrified humans search desperately for the dimensional door that is their only hope of escape. Made for television.
1984 94m/C Richard Hatch, Kay Lenz, John Saxon; *Dir:* Terry Marcel. **VHS, Beta $19.95** *VCL* 🎬½

Prisoners of the Sun Right after WWII an Australian captain fights to convict Japanese officers for atrocities against Allied POWs, but he's stonewalled by both the US military and still-defiant enemy prisoners. This fiery drama from Down Under packs a punch as it questions whether wartime justice even exists; similar in that way to Brown's earlier "Breaker Morant." Takei (Sulu of the original "Star Trek") makes an imposing Japanese admiral.
1991 (R) 109m/C Bryan Brown, George Takei, Terry Quinn, John Back, Toshi Shioya, Deborah Unger; *Dir:* Stephen Wallace. **VHS, Beta $92.98** *PAR* 🎬🎬🎬

Private Affairs The mistress of a well-known surgeon becomes jealous when the doctor decides to have a fling with a gorgeous young swim instructor. Italian-made cheapie.
1989 (R) 83m/C *IT* Giuliana de Sio, Kate Capshaw, David Naughton, Luca Barbareschi, Michele Placido; *Dir:* Francesco Massaro. **VHS, Beta $29.95** *ACA* 🎬

The Private Affairs of Bel Ami "This is the story of a scoundrel," proclaims the opening. Sanders is ideally cast as a suave cad who rises in 1880s Parisian society, largely through stategic seduction of prominent women. Moralistically toned down Old Hollywood toned down the talky adaptation of the Guy de Maupassant novel, but it's still drama of a high order.
1947 112m/B George Sanders, Angela Lansbury, Ann Dvorak, Frances Dee, John Carradine, Susan Douglas, Hugo Haas, Marie Wilson, Albert Basserman, Warren William, Katherine Emery,

Richard Fraser; *W/Dir:* Albert Lewin. **VHS $19.98** *REP, FCT* 🎬🎬🎬

Private Benjamin Lighthearted sitcom about a pampered New York Jewish princess who impulsively enlists in the U.S. Army after her husband dies on their wedding night. Hawn, who also produced, creates a character loveable even at her worst moments and brings a surprising amount of depth to this otherwise frivolous look at high society attitudes. The basis for the television series. Academy-award nominated for best screenplay, actress (Hawn), and supporting actress (Brennan).
1980 (R) 110m/C Goldie Hawn, Eileen Brennan, Albert Brooks, Robert Webber, Armand Assante, Barbara Barrie, Mary Kay Place, Sally Kirkland, Craig T. Nelson, Harry Dean Stanton, Sam Wanamaker; *Dir:* Howard Zieff. **VHS, Beta, LV $19.98** *WAR* 🎬🎬½

Private Buckaroo War-time entertainment in which Harry James and his Orchestra get drafted. They decide to put on a show for the soldiers and get help from the Andrews Sisters.
1942 70m/B The Andrews Sisters, Harry James, Joe E. Lewis, Dick Foran, Shemp Howard, Mary Wickes, Donald O'Connor; *Dir:* Eddie Cline. **VHS, Beta $19.95** *NOS, VYY, CAB* 🎬½

Private Collection Beautiful strippers from various night clubs are showcased.
1989 60m/C VHS, Beta $39.95 *CEL* 🎬½

Private Contentment This is a drama about a young soldier's experiences before he goes off to war in 1945.
1983 90m/C Trini Alvarado, Peter Gallagher, John McMartin, Kathryn Walker; *Dir:* Vivian Matalon. **VHS, Beta** *FLI* 🎬🎬

Private Conversations: On the Set of "Death of a Salesman" Documentary about the filming of the successful Broadway adaptation of Miller's "Death of a Salesman."
1985 82m/C Dustin Hoffman, Volker Schlondorff, Arthur Miller, John Malkovich. **VHS, Beta $19.98** *LHV, WAR* 🎬🎬

Private Duty Nurses Three nurses take on racism, war wounds and a menage-a-trios (between one nurse, a doctor, and a drug addict). Second in Roger Corman's "nurse" quintet takes itself too seriously to be entertaining, but the gals make good use of those exciting new inventions, waterbeds. Preceded by "The Student Nurses" and followed by "Night Call Nurses." Also on video as "Young L.A. Nurses 3."
1971 (R) 80m/C Katherine Cannon, Joyce Williams, Pegi Boucher, Joseph Kaufmann, Dennis Redfield, Herbert Jefferson Jr., Paul Hampton, Paul Gleason; *Dir:* George Armitage. **VHS, Beta $59.98** *CHA, SUE, TAV* 🎬½

Private Eyes Two bungling sleuths, engaged to investigate two deaths, are led on a merry chase through secret passages to a meeting with a ghostly adversary.
1980 (PG) 91m/C Don Knotts, Tim Conway, Trisha Noble, Bernard Fox; *Dir:* Lang Elliott. **VHS, Beta, LV $29.98** *VES* 🎬½

The Private Files of J. Edgar Hoover Scandal-mongering "biography" of J. Edgar Hoover's private, sex-filled life.
1977 (PG) 112m/C Broderick Crawford, Dan Dailey, Jose Ferrer, Rip Torn, Michael Parks, Raymond St. Jacques, Ronee Blakley; *Dir:* Larry Cohen. **VHS $69.99** *HBO* 🎬🎬½

A Private Function A ribald gag-fest dealing with Palin as a Yorkshireman who steals and fattens a wily contraband pig against the backdrop of post-World War II rationing. The satire ranges from biting to downright nasty, but Palin is always likeable in the center of it all.
1984 (PG) 96m/C *GB* Michael Palin, Maggie Smith, Denholm Elliott, Bill Paterson; *Dir:* Malcolm Mowbray. **VHS, Beta $19.95** *PAR, HBO* 🎬🎬½

Private Hell 36 Two detectives become guilt-ridden after keeping part of some stolen money recovered after a robbery. Co-produced and co-written by Lupino, who also sings.
1954 81m/B Ida Lupino, Howard Duff, Steve Cochran; *Dir:* Don Siegel. **VHS $19.95** *REP* 🎬🎬

The Private History of a Campaign That Failed Adaptation of the Mark Twain story about a cowardly troop of Confederate soldiers.
1987 89m/C Pat Hingle, Edward Herrmann. **VHS, Beta $19.95** *MCA* 🎬½

Private Investigations A made-for-video thriller about a nosey reporter who gets himself and his adult son in trouble while investigating drug-pushing cops.
1987 91m/C Ray Sharkey, Clayton Rohner, Talia Balsam, Anthony Zerbe, Paul LeMat; *Dir:* Nigel Dick. **VHS, Beta $79.98** *FOX* 🎬½

Private Lessons A teenage boy is left alone for the summer in the care of an alluring maid and a scheming chauffeur.
1981 (R) 83m/C Eric Brown, Sylvia Kristel, Howard Hesseman; *Dir:* Alan Myerson. **VHS, Beta, LV $24.95** *MCA* 🎬½

Private Lessons A boy can't concentrate on his piano lessons when he gets a beautiful new teacher. You won't be able to concentrate on anything but comedy.
19?? 90m/C Carroll Baker. **VHS, Beta $79.95** *JEF* 🎬½

Private Life Suddenly with time on his hands, a Soviet official scrutinizes his relationships. After a number of revelations, he is forced to make new choices based on what he has learned. Oscar nominee in 1983. In Russian with English subtitles.
1983 112m/C *RU* Mikhail Ulyanov, Ita Sanvina, Irina Gubanova; *Dir:* Edgar Ryazanov. **VHS $59.95** *TPV, FCT* 🎬🎬½

Private Life of Don Juan Appropriately slow-moving British costume drama set in 17th-century Spain finds an aging Don Juan struggling to maintain his usual antics in the pursuit of beautiful women. Furthermore, his reputation is being upstaged by a young imposter. Notable only as the last film appearance by Douglas Fairbanks Sr., and based on the play by Henri Bataille.
1934 87m/B *GB* Douglas Fairbanks Sr., Merle Oberon, Binnie Barnes, Melville Cooper, Joan Gardner, Benita Hume, Athene Seyler; *Dir:* Alexander Korda. **VHS, Beta $9.95** *NEG, NOS, GPV* 🎬🎬

The Private Life of Henry VIII Lavish historical spectacle lustily portraying the life and lovers of notorious British Monarch, King Henry VIII. A tour de force for Laughton as the robust 16th-century king, with outstanding performances by the entire cast.
1933 97m/B *GB* Charles Laughton, Binnie Barnes, Elsa Lanchester, Robert Donat, Merle Oberon; *Dir:* Alexander Korda. Academy Awards '33: Best Actor (Laughton); Film Daily Poll '33:

10 Best Films of the Year. **VHS, Beta, LV** $19.98 *HBO, NOS, PSM* ✂✂✂✂

The Private Life of Sherlock Holmes A unique perspective on the life of the famous detective reveals a complex character. Beautifully photographed, with fine performances by the supporting cast, the film boasts a haunting musical score but received suprisingly little recognition despite the high caliber direction by Wilder, who co-wrote with I.A.L. Diamond.
1970 (PG) 125m/C *GB* Robert Stephens, Colin Blakely, Genevieve Page, Irene Handl, Stanley Holloway, Christopher Lee, Clive Revill; *Dir:* Billy Wilder. **VHS** $59.98 *FOX* ✂✂✂¹⁄₂

The Private Lives of Elizabeth & Essex Cast reads like a Who's Who in Hollywood in this lavishly costumed dramatization of the love affair between Queen Elizabeth I and the Second Earl of Essex. Forced to choose between her Kingdom and her lover, Davis' monarch is the epitome of a regal women. Fobray made her first film appearance as an adult in the adaptation of Maxwell Anderson's "Elizabeth the Queen."
1939 106m/C Bette Davis, Errol Flynn, Vincent Price, Nanette Fabray, Olivia de Havilland, Alan Hale Jr., Donald Crisp, Leo G. Carroll; *Dir:* Michael Curtiz. **VHS, Beta** $19.98 *FOX* ✂✂✂

Private Manoeuvres A comely Swiss military adviser gives her all to uplift the morale of the men at Camp Samantha.
1983 79m/C Zachi Noy, Yoseph Shiloah, Dvora Bekon. **VHS, Beta** $59.95 *MGM* Woof!

Private Passions A sultry woman gives her teenaged cousin a lesson in love during his European vacation.
1985 86m/C Sybil Danning. **VHS, Beta** $59.95 *PSM* ✂

Private Popsicle A hilarious sex comedy starring Europe's popular Popsicle team. Dubbed in English.
1982 (R) 111m/C Zachi Noy, Jonathan Segall, Yftach Katzur. **VHS, Beta** $59.95 *MGM, GLV* ✂

Private Practices: The Story of a Sex Surrogate A portrait of America's best-known sex surrogate and her treatment of two men during a four-month period.
1986 74m/C Maureen Sullivan; *Dir:* Kirby Dick. **VHS, Beta** $59.98 *LIV, LTG* ✂

Private Resort A curious house detective and a bumbling thief interrupt the highjinks of two girl-crazy teens on a quest for fun at an expensive Miami hotel.
1984 (R) 82m/C Johnny Depp, Rob Morrow, Karyn O'Bryan, Emily Longstreth, Tony Azito, Hector Elizondo, Dody Goodman, Leslie Easterbrook, Andrew Dice Clay; *Dir:* George Bowers. **VHS, Beta, LV** $29.95 *COL* ✂

Private Road: No Trespassing A stock car racer and a top engineer compete over a military project, cars and a rich heiress.
1987 (R) 90m/C George Kennedy, Greg Evigan, Mitzi Kapture; *Dir:* Raphael Nossbaum. **VHS, Beta** $79.95 *TWE* ✂

Private School Two high school girls from the exclusive Cherryvale Academy for Women compete for the affections of a young man from nearby Freemount Academy for Men, while Cherryvale's headmistress is trying to raise funds to build a new wing. Banal teen sexploitation comedy.
1983 (R) 89m/C Phoebe Cates, Betsy Russell, Kathleen Wilhoite, Sylvia Kristel, Ray Walston, Matthew Modine, Michael Zorek, Fran Ryan, Jonathan Prince, Kari Lizer, Richard Stahl; *Dir:* Noel Black. **VHS, Beta, LV** $14.95 *MCA* Woof!

Private Snafu Cartoon Festival A collection of cartoons produced for the Armed Forces during World War II as humorous instructional films. The language is strictly G.I. Titles include: "Censored," "Gas," "Goldbrick," "Going Home," "Rumors," and "Spies." Also included are two more Warner Bros. World War II cartoons, "Tokio Jokio," and "Confusions of a Nutzy Spy."
194? 55m/B VHS, Beta $34.95 *HHT, DVT* Woof!

Private War Training for an elite force turns deadly when the commanding officer makes the rules. Now it's a private war between two brutal fighting machines. Screenplay by DePalma, from a Jan Guillou story.
1990 (R) 95m/C Martin Hewitt, Joe Dallesandro, Kimberly Beck; *Dir:* Frank De Palma. **VHS** $89.98 *REP* ✂¹⁄₂

Privates on Parade Film centering around the comic antics of an Army song-and-dance unit entertaining the troops in the Malayan jungle during the late '40s. Occasionally-inspired horseplay based on Peter Nichols play.
1984 (R) 107m/C *GB* John Cleese, Denis Quilley, Simon Jones, Joe Melia, Nicola Pagetti, Julian Sands; *Dir:* Michael Blakemore. **VHS, Beta** $19.95 *PAR, HBO* ✂✂¹⁄₂

Prix de Beaute A woman's boyfriend does not know that she has won a beauty contest. French with English subtitles. Also known as "Miss Europe."
1930 93m/B *FR* Louise Brooks, Jean Bradin, George Charlia, Gaston Jacquet; *Dir:* Augusto Genina. **VHS, Beta** $59.95 *INT* ✂¹⁄₂

The Prize Gripping spy story laced with laughs based on a novel by Irving Wallace (as adapted by Ernest Lehman). In Stockholm, writer accepts the Nobel prize for dubious reasons and then finds himself in the midst of political intrigue. Newman and Sommer turn in great performances in this action drama.
1963 136m/C Paul Newman, Edward G. Robinson, Elke Sommer, Leo G. Carroll, Diane Baker, Micheline Presle, Gerard Oury, Sergio Fantoni; *Dir:* Mark Robson. **VHS** $19.98 *MGM, REP, MLB* ✂✂✂

Prize Fighter A fight manager and his pugilistic protege unknowingly get involved with a powerful gangster, who convinces them to fight in a fixed championship match. Set in the 1930; good for the kids.
1979 (PG) 99m/C Tim Conway, Don Knotts; *Dir:* Michael Preece. **VHS, Beta** $9.95 *MED* ✂

Prize of Peril A French television game show rewards it winners with wealth and its losers with execution. Not always the best policy to learn what's behind Door #1.
1984 95m/C *FR* Michel Piccoli, Marie-France Pisier; *Dir:* Yves Boisset. **VHS, Beta** $79.98 *LIV, LTG* ✂✂

Prizzi's Honor Highly stylized, sometimes leaden black comedy about an aging and none-to-bright hit man from a New York mob family who breaks with family loyalties when he falls for an upwardly mobile tax consultant who's also a hired killer. Skirting caricature in every frame, Nicholson is excellent in his portrayal of the thick-skulled mobster. Angelica Huston, daughter of the director, won an Oscar for her performance in this cynical view of the Mafia, while Nicholson was nominated for best actor and Hickey (as the Don) for best supporting. Oscar nominations were also received for best picture, direction, adapted screenplay, and editing. Adapted by Richard Condon and Janet Roach from Condon's novel.

1985 (R) 130m/C Jack Nicholson, Kathleen Turner, Robert Loggia, John Randolph, Anjelica Huston, Lawrence Tierney, William Hickey, Lee Richardson; *Dir:* John Huston. Academy Awards '85: Best Supporting Actress (Huston); Golden Globe Awards '86: Best Film—Musical/Comedy; New York Film Critics Awards '85: Best Picture. **VHS, Beta, LV** $29.98 *LIV, VES, BTV* ✂✂✂

Probation A dashing young man in trouble with the law receives an unusual sentence; he must become a chauffeur for a spoiled society girl. Grable's first film role is a small one.
1932 60m/B Sally Blane, J. Farrell MacDonald, Eddie Phillips, Clara Kimball Young, Betty Grable; *Dir:* Richard Thorpe. **VHS, Beta** $16.95 *NOS, SNC* ✂

Probe A detective uses computer-age technology to apprehend criminals. Pilot for the television series, "Search."
1972 95m/C Hugh O'Brian, Elke Sommer, John Gielgud, Burgess Meredith, Angel Tompkins, Lilia Skala, Kent Smith, Alfred Ryder, Jaclyn Smith; *Dir:* Russ Mayberry. **VHS** $59.95 *UNI* ✂✂¹⁄₂

Problem Child Ritter decides to adopt Oliver out of the goodness of his heart, but it seems young Oliver's already got a father figure named Beelzebub. Potential for laughs is unmet.
1990 (PG) 81m/C John Ritter, Michael Oliver, Jack Warden, Amy Yasbeck, Gilbert Gottfried, Michael Richards, Peter Jurasik; *Dir:* Dennis Dugan. **VHS, LV** $19.95 *MCA* ✂¹⁄₂

Problem Child 2 Ritter and his nasty adopted son are back, but this time there's an equally malevolent little girl. They team up to prevent Ritter's upcoming marriage to a socialite. Low slapstick junk.
1991 (PG-13) 91m/C John Ritter, Michael Oliver, Laraine Newman, Amy Yasbeck, Jack Warden, Ivyann Schwan, Gilbert Gottfried, James Tolken, Charlene Tilton, Alan Blumenfeld; *Dir:* Brian Levant. **VHS, Beta, LV** *MCA, PIA* ✂

Problems, 1950s' Style Three TV panel discussion/interview programs from the 1950's: "Stand Up and Be Counted!" (1956), in which audience members and guests discuss their problems; "People in Conflict," (1959), where guests ask advice of the panel; and "The Verdict Is Yours," (1958), with an actual courtroom trial.
195? 115m/B Robert Russell, Jim McKay. **VHS, Beta** $24.95 *SHO* ✂¹⁄₂

The Prodigal Luke's New Testament Bible story of the son seduced by greed slickly transfered to the silver screen by MGM. A colorful cast is the main attraction.
1955 113m/C Lana Turner, Edmund Purdom, Louis Calhern, Audrey Dalton, Neville Brand, Walter Hampden, Taina Elg, Francis L. Sullivan, Joseph Wiseman, Sandy Descher, John Dehner, Cecil Kellaway, Henry Daniell, Paul Cavanagh, Tracey Roberts, Jay Novello, Dorothy Adams, Richard Devon; *Dir:* Richard Thorpe. **LV** $34.98 *MGM* ✂✂¹⁄₂

The Prodigal A born-again family drama in which a sundered family is brought together by the return of a once-estranged son.
1983 (PG) 109m/C John Hammond, Hope Lange, John Cullum, Morgan Brittany, Ian Bannen, Arliss Howard, Joey Travolta, Billy Graham. **VHS, Beta, LV** $59.95 *IVE* ✂✂

Prodigal Boxer: The Kick of Death A son must avenge his father's death by facing two kung fu masters.
1973 (R) 90m/C Meng Fei, Pa Hung, Suma Wah Lung. **VHS, Beta, LV** $19.95 *GEM, MRV* ✂

The Prodigal Planet A small group of believers continue their struggle against the world government UNTIE by disrupting their communication network. Sequel to "Thief in the Night," "A Distant Thunder," and "Image of the Beast" (see separate entries).
1988 67m/C William Wellman Jr., Linda Beatie, Cathy Wellman, Thom Rachford; *Dir:* Donald W. Thompson. VHS, Beta *MIV, HHE* 🎬

The Producers Considered one of Brooks' best films, this hilarious farce follows an attempted swindle by a con artist and his accountant who attempt to stage a Broadway flop only to have it backfire. Achieving cult status, the film is known for musical interlude "Springtime for Hitler." Wilder received an Oscar nomination for his role in this classic of zaniness. The phony play was later actually produced by Alan Johnson.
1968 88m/C Zero Mostel, Gene Wilder, Dick Shawn, Kenneth Mars, Estelle Winwood; *W/Dir:* Mel Brooks. Academy Awards '68: Best Story & Screenplay. VHS, Beta, LV, 8mm $14.98 *SUE, VYG, CRC* 🎬🎬🎬½

Professional Killers 1 A trio of hired assassins wanders about the countryside killing people during Japan's Feudal Era.
1973 87m/C Jiro Tamiya, Koji Takahashi. VHS, Beta *VDA* 🎬

Professionals Action and adventure count for more than a storyline in this exciting western about four mercenaries hired by a wealthy cattle baron to rescue his young wife from Mexican kidnappers. Breathtaking photography recreates turn-of-the-century Mexico in this adaptation of the Frank O'Rourke novel. Nominated for three Oscars: Best Director, Adapted Screenplay, and Color Cinematography.
1966 (PG) 117m/C Burt Lancaster, Lee Marvin, Claudia Cardinale, Jack Palance, Robert Ryan, Woody Strode, Ralph Bellamy; *Dir:* Richard Brooks. VHS, Beta, LV $69.95 *COL, PIA* 🎬🎬🎬½

Professionals When the daily rigors of cocaine smuggling begin to dampen their spirits, the "Professionals" plan an exciting plutonium heist.
1976 104m/C Gordon Jackson, Martin Shaw, Lewis Collins. VHS, Beta, LV $59.95 *TWE* 🎬🎬

Professionals 2 Two episodes of the British action show: "Operation Susie" and "Heroes."
1985 100m/C Gordon Jackson, Martin Shaw, Lewis Collins. VHS, Beta $39.95 *TWE* 🎬🎬

Programmed to Kill A beautiful terrorist is captured by the CIA and transformed into a buxom bionic assassin.
1986 (R) 91m/C Robert Ginty, Sandahl Bergman, James Booth, Louise Caire Clark; *Dir:* Allan Holzman. VHS, Beta $79.95 *MED* 🎬🎬

Project A-ko Set in the near future, this Japanese animated feature, which is intended for adults, concerns teenagers with strange powers, an alien spaceship, and lots of action. Seventeen year-old A-ko possesses superhuman strength and a ditzy sidekick, C-ko. B-ko, the spoiled and rich daughter of a business tycoon, decides to fight A-ko for C-ko's companionship. Meanwhile, an alien spaceship is headed toward Earth with unknown intentions. Somehow, everything ties together—watch and see how. In Japanese with English subtitles.
1986 86m/C VHS $39.95 *CPM* 🎬🎬

Project: Alien Science fiction fans in search of a good extraterrestrial flick should avoid this teaser, for it has nothing to do with aliens. As the film begins, Earth is allegedly being attacked by beings from space, and a vast array of scientists, militia and journalists track the aliens. What's actually happening revolves around the testing of deadly biological weapons. Shot in Yugoslavia.
1989 (R) 92m/C Michael Nouri, Darlanne Fluegel, Maxwell Caulfield, Charles Durning; *Dir:* Frank Shields. VHS $89.95 *VMK* 🎬🎬

Project: Eliminator A group of terrorists kidnap a designer of "smart" weapons, and it's up to a hard-hitting special forces unit to get him back. Filmed in New Mexico.
1991 (R) 89m/C David Carradine, Frank Zagarino, Drew Snyder, Hilary English, Vivian Schelling; *W/Dir:* H. Kaye Dyal. VHS $89.95 *SOU* 🎬½

Project: Kill! The head of a murder-for-hire squad suddenly disappears and his former assistant is hired to track him down dead or alive.
1977 94m/C Leslie Nielsen, Gary Lockwood, Nancy Kwan. VHS, Beta *PGN* 🎬

Project Moon Base Espionage runs rampant on a spaceship headed by a female officer. Eventually the ship is stranded on the moon. Actually filmed for a television series "Ring Around the Moon." A cold-war sexist relic.
1953 64m/C Donna Martell, Hayden Rourke; *Dir:* Richard Talmadge. VHS, Beta *VMK* 🎬½

Project: Nightmare Seems not all dreams are wish fulfillment.
19?? 75m/C Charles Miller, Elly Koslo, Lance Dickson; *Dir:* Don Jones. VHS $19.99 *ACA* 🎬½

Project: Shadowchaser Action-packed would-be thriller about a billion dollar android who escapes from a top secret government laboratory. Programmed with superhuman strength and no human emotions the android and five terrorists take over a hospital. Their hostages include the President's daughter and the terrorists demand a 50 million dollar ransom. With a four-hour deadline the FBI calls in the hospital architect to advise them—only they've got the wrong man—and the android's creator, who wants his creation back—no matter what the cost. A low budget Terminator clone.
1992 (R) 97m/C Martin Kove, Meg Foster, Frank Zagarino, Paul Koslo, Joss Ackland; *Dir:* John Eyres. VHS $89.95 *PSM* 🎬½

Project X A bemused Air Force pilot is assigned to a special project involving chimpanzees. He must decide where his duty lies when he realizes the semi-intelligent chimps are slated to die.
1987 (PG) 107m/C Matthew Broderick, Helen Hunt; *Dir:* Jonathan Kaplan. VHS, Beta, LV $19.98 *FOX* 🎬🎬½

Projectionist A must-see for movie buffs, Dangerfield made his screen debut in this story about a projectionist in a seedy movie house whose real-life existence begins to blur into the films he continuously watches. Made on a limited budget, this creative effort by Hurwitz was the first film to utilize the technique of superimposition.
1971 (PG) 84m/C Rodney Dangerfield, Chuck McCann, Ina Balin; *Dir:* Harry Hurwitz. VHS, Beta $69.98 *LIV, VES* 🎬🎬🎬

Prom Night A masked killer stalks four high school girls during their senior prom as revenge for a murder which occurred six years prior. Sequelled by "Hello Mary Lou: Prom Night II" and "Prom Night III: The Last Kiss."

1980 (R) 91m/C *CA* Jamie Lee Curtis, Leslie Nielsen; *Dir:* Paul Lynch. VHS, Beta, LV *MCA, OM* 🎬🎬

Prom Night 3: The Last Kiss The second sequel, in which the reappearing high school ghoul beguiles a lucky teenager.
1989 (R) 97m/C Tim Conlon, Cyndy Preston, Courtney Taylor; *Dir:* Ron Oliver, Peter Simpson. VHS, Beta, LV $14.95 *IVE* 🎬

Prom Night 4: Deliver Us From Evil Yet another gory entry in the Prom series (one would hope it will be the last). Another group of naive teens decide that they can have more fun at a private party than at the prom. They host the party in a summer home that was once a monastery, but the festive affair soon turns into a night of terror when an uninvited guest crashes the party. For true fans of slasher flicks.
1991 (R) 95m/C Nikki DeBoer, Alden Kane, Joy Tanner, Alle Ghadban; *Dir:* Clay Borris. VHS $89.98 *LIV* 🎬

Promise Her Anything A widow (Caron) with a baby decides to make her boss (Cummings), a child psychologist who hates kids, her new husband. So she stashes the kid with her upstairs neighbor (Beatty), a would-be filmmaker who earns his living making blue movies. But the filmmaker has romantic designs on the widow and decides the baby may be his way to make a good impression. Beatty isn't very believable in this type of light-romantic comedy and the farcical situations are forced.
1966 98m/C *GB* Warren Beatty, Leslie Caron, Robert Cummings, Hermione Gingold, Lionel Stander, Keenan Wynn, Cathleen Nesbitt; *Dir:* Arthur Hiller. VHS $14.95 *PAR* 🎬🎬

Promise to Murder An early murder-drama episode from the "Climax" television series.
1956 60m/B Peter Lorre, Louis Hayward, Ann Harding. VHS, Beta $24.95 *NOS, DVT* 🎬½

Promised Land Two high school friends from the rural northwestern U.S. come together several years after graduation under tragic circumstances. Writer Hoffman's semi-autobiographical, disillusioned look at the American Dream was re-discovered by movie goers due to its excellent dramatic performances, notable also as the first film produced by Robert Redford's Sundance Institute.
1988 (R) 110m/C Kiefer Sutherland, Meg Ryan, Tracy Pollan, Jason Gedrick, Googy Gress, Deborah Richter, Sandra Seacat, Jay Underwood; *Dir:* Michael Hoffman. VHS, Beta, LV $29.98 *LIV, VES* 🎬🎬

Promises in the Dark Drama focusing on the complex relationship between a woman doctor and her 17-year-old female patient who is terminally ill with cancer. Gerome Hellman's directorial debut.
1979 (PG) 118m/C Marsha Mason, Ned Beatty, Kathleen Beller, Susan Clark, Paul Clemens, Donald Moffatt, Michael Brandon; *Dir:* Gerome Hellman. VHS, Beta $59.95 *WAR* 🎬🎬

Promises! Promises! Having difficulty getting pregnant, a woman goes on a cruise with her husband. While on board, they meet another couple, all get drunk, and change partners. Of course, both women find themselves pregnant, leaving the paternity in doubt. Famous primarily as the movie Mansfield told Playboy magazine she appeared "completely nude" in. It's also about the only reason to watch it.

1963 90m/B Jayne Mansfield, Marie McDonald, Tommy Noonan, Fritz Feld, Claude Stroud. **VHS, Beta $24.95** *WGE, SIM* 🐾½

Promoter Horatio Alger comedy stars Guinness as an impoverished student who gives himself a surrepticious leg up in life by altering his school entrance exam scores. Outstanding performances enliven this subtle British comedy of morals.
1952 87m/B *GB* Alec Guinness, Glynis Johns, Petula Clark, Valerie Hobson, Michael Hordern;; **Dir:** Ronald Neame. **VHS, Beta $39.95** *IND* 🐾🐾🐾

Proof of the Man A Japanese co-production about a murder in Tokyo that takes on international importance.
1984 100m/C *JP* George Kennedy, Broderick Crawford, Toshiro Mifune. **VHS, Beta $69.95** *VCD* 🐾

Property A comedy about a group of leftover '60s radicals adjusting to 1970s corporate America.
1978 92m/C VHS, 3/4U *ICA* 🐾½

Prophecy A doctor and his wife travel to Maine to research the effects of pollution caused by the lumber industry. They encounter several terrifying freaks of nature and a series of bizarre human deaths. Laughable horror film.
1979 (PG) 102m/C Talia Shire, Robert Foxworth, Armand Assante, Victoria Racimo, Richard Dysart; **Dir:** John Frankenheimer. **VHS, Beta $19.95** *PAR* 🐾½

A Propos de Nice First film by French director Vigo, the silent film parodies French travelogues in a manner that indicates the director's later brilliance.
1929 25m/B *FR* **Dir:** Jean Vigo. **VHS, Beta** *FST, MRV* 🐾🐾🐾

Protector A semi-martial arts cops 'n' robbers epic about the cracking of a Hong Kong-New York heroin route.
1985 (R) 94m/C Jackie Chan, Danny Aiello; **Dir:** James Glickenhaus. **VHS, Beta, LV $79.95** *WAR* 🐾

Protocol A series of comic accidents lead a Washington cocktail waitress into the U.S. State Department's employ as a protocol official. Once there she is used as a pawn to make an arms deal with a Middle Eastern country. Typical Hawn comedy with an enjoyable ending.
1984 (PG) 100m/C Goldie Hawn, Chris Sarandon, Andre Gregory, Cliff DeYoung, Ed Begley Jr., Gail Strickland, Richard Romanus, Keith Szarabajka, James Staley, Kenneth Mars, Kenneth McMillan, Archie Hahn, Amanda Bearse; **Dir:** Herbert Ross. **VHS, Beta, LV $19.98** *WAR* 🐾🐾

Prototype Made-for-television revision of the Frankenstein legend has award-winning scientist Plummer as the creator of the first android. Fearful of the use the military branch of government has in mind for his creation, he attempts to steal back his discovery, in this suspenseful adventure.
1983 100m/C Christopher Plummer, David Morse, Frances Sternhagen, James Sutorius; **Dir:** David Greene. **VHS $14.95** *IVE* 🐾🐾🐾

The Proud and the Damned Five Civil-War-veteran mercenaries wander into a Latin American war and get manipulated by both sides.
1972 (PG) 95m/C Chuck Connors, Aron Kincaid, Cesar Romero; **Dir:** Ferde Grofe Jr. **VHS, Beta $19.95** *NOS, UHV, MRV* 🐾

Proud Rebel A character study of a stubborn widower who searches for a doctor to aid him in dealing with the problems of his mute son, and of the woman who helps to tame the boy. Ladd's real-life son makes his acting debut.
1958 99m/C Alan Ladd, Olivia de Havilland, Dean Jagger, Harry Dean Stanton; **Dir:** Michael Curtiz. **VHS, Beta $69.98** *SUE* 🐾🐾🐾

Providence An interesting score highlights this British fantasy drama of a dying writer envisioning his final novel as a fusion of the people from his past with the circumstances he encounters on a daily basis. The first English-language effort by French director Resnais. Score by Miklos Rozsa's.
1976 (R) 104m/C *GB* John Gielgud, Dirk Bogarde, Ellen Burstyn, David Warner, Elaine Stritch; **Dir:** Alain Resnais. New York Film Critics Awards '77: Best Actor (Gielgud). **VHS, Beta $59.95** *COL* 🐾🐾🐾

Provoked A young cop pursues a serial killer who appears to have supernaturally reappeared after being sent to the electric chair.
1989 90m/C Harold W. Jones; **Dir:** Rick Pamplin. **VHS, Beta $69.95** *RAE, HHE* 🐾

Prowler Heflin is a cop who responds to a prowler scare and finds the beautiful married Keyes worried about an intruder. He learns she will inherit a lot of money if anything happens to her husband so he seduces her and then "accidentally" kills her husband, supposedly mistaking him for a prowler. Keyes then marries Heflin but when she learns the truth her own life is in danger. Tidy thriller with Heflin doing a good job as the unpredictable villain.
1951 92m/B Van Heflin, Evelyn Keyes, Katherine Warren, John Maxwell; **Dir:** Joseph Zito. **VHS, Beta $49.95** *VHE* 🐾🐾½

Prudential Family Playhouse Presentation of Sinclair Lewis' "Dodsworth." Tells the story of a wealthy American couple who travel from their small town to Europe.
1950 53m/B Ruth Chatterton, Walter Abel, Cliff Hall, Eva Marie Saint. **VHS, Beta $24.95** *VYY* 🐾🐾½

Prurient Interest A video play revolving around a movie producer's battle with the Board of Censors for approval of his feature "Eat Me in St. Louis."
1974 20m/B VHS, Beta *EAI* 🐾

PSI Factor A civilian researcher working at the NASA space track station observes and records signals coming from planet Serius B.
1980 91m/C Peter Mark Richman, Gretchen Corbett, Tom Martin; **Dir:** Quentin Masters. **VHS $79.95** *MNC* 🐾🐾½

Psych-Out A deaf girl searches Haight-Ashbury for her runaway brother. She meets hippies and flower children during the Summer of Love. Psychedelic score with Strawberry Alarm Clock and The Seeds. Somewhere in the picture, Nicholson whangs on lead guitar.
1968 95m/C Jack Nicholson, Bruce Dern, Susan Strasberg, Dean Stockwell, Henry Jaglom; **Dir:** Richard Rush. **VHS, Beta, LV $19.99** *HBO* 🐾🐾½

Psychedelic Fever A swinging look at the free-love decade of the 1960s.
1990 (R) 80m/C Dir: William Rostler. **VHS $29.95** *AVD* 🐾

The Psychic An ad executive gains psychic powers after he falls off a ladder and tries to conquer the world. A later effort by gore-king Lewis, notable only for sex scenes inserted by Lewis to make the movie salable.
1968 90m/C Dick Genola, Robin Guest, Bobbi Spencer; **Dir:** Herschell Gordon Lewis. **VHS, Beta** *CAM* Woof!

The Psychic A psychic envisions her own death and attempts to alter the prediction.
1978 (R) 90m/C Jennifer O'Neill, Marc Porel, Evelyn Stewart, Gabriel Ferzetti, Gianno Garko; **Dir:** Lucio Fulci. **VHS, Beta $69.98** *LTG* 🐾

Psychic A college student with psychic powers believes he knows who the next victim of a demented serial killer will be. The problem is: no one will believe him. And the victim is the woman he loves.
1991 (R) 92m/C Michael Nouri, Catherine Mary Stewart, Zach Galligan. **VHS $89.95** *VMK* 🐾½

Psychic Killer A wrongfully committed asylum inmate acquires psychic powers and decides to use them in a deadly revenge. Good cast in cheapie horror flick.
1975 (PG) 89m/C Jim Hutton, Paul Burke, Julie Adams, Neville Brand, Aldo Ray, Rod Cameron, Della Reese; **Dir:** Ray Danton. **VHS, Beta $14.98** *SUE* 🐾½

Psycho Hitchcock counted on his directorial stature and broke all the rules in this story of violent murder, transvestism, and insanity. Based on Robert Bloch's novelization of an actual murder, Leigh plays a fleeing thief who stops at the secluded Bates Motel where she meets her death in Hitchcock's classic "shower scene." Shot on a limited budget in little more than a month, "Psycho" changed the Hollywood horror film forever. Oscar nominations for Leigh, Hitchcock, Cinematography, Art Direction, and Set Design. Music by Bernard Herrman. Followed by "Psycho II" (1983), "Psycho III" (1986), and a made-for-television movie.
1960 109m/B Anthony Perkins, Janet Leigh, Vera Miles, John Gavin, John McIntire, Martin Balsam, Simon Oakland, Ted Knight; **Dir:** Alfred Hitchcock. Edgar Allan Poe Awards '60: Best Screenplay. **VHS, Beta, LV $19.95** *MCA, TLF* 🐾🐾🐾🐾

Psycho 2 This sequel to the Hitchcock classic finds Norman Bates returning home after 22 years in an asylum to find himself haunted by "mother" and caught up in a series of murders. In a surprisingly good horror film, Perkins and Miles reprise their roles from the original "Psycho." Perkins went on to direct yet another sequel, "Psycho 3."
1983 (R) 113m/C Anthony Perkins, Vera Miles, Meg Tilly, Robert Loggia, Dennis Franz; **Dir:** Richard Franklin. **VHS, Beta, LV $19.95** *MCA* 🐾🐾

Psycho 3 The second sequel to Hitchcock's "Psycho" finds Norman Bates drawn into his past by "mother" and the appearance of a woman who reminds him of this original victim. Perkins made his directorial debut in this film that stretches the plausibility of the storyline to its limits, but the element of parody throughout makes this an entertaining film for "Psycho" fans.
1986 (R) 93m/C Anthony Perkins, Diana Scarwid, Jeff Fahey, Roberta Maxwell, Robert Alan Browne; **Dir:** Anthony Perkins. **VHS, Beta, LV $19.95** *MCA* 🐾🐾

Psycho 4: The Beginning At the behest of a radio talk show host, Norman Bates recounts his childhood and reveals the circumstances that aided in the development of his peculiar neuroses.
1991 (R) 96m/C Anthony Perkins, Henry Thomas, Olivia Hussey; *Dir:* Mick Garris. **VHS, LV** $79.95 *MCA, PIA Woof!*

Psycho Cop Six college undergrads on a weekend retreat are menaced and picked off one at a time by a crazed local cop.
1988 89m/C Bobby Ray Shafer; *Dir:* Wallace Potts. **VHS, Beta** $89.95 *SOU* ♫½

Psycho Girls A pair of homicidal sisters torture an innocent couple.
1986 (R) 92m/C John Haslett Cuff, Darlene Mignacco, Agi Gallus; *Dir:* Gerard Ciccioritti. **VHS, Beta** $79.95 *MGM Woof!*

Psycho from Texas A quiet Southern town is disrupted by the kidnapping of a wealthy oil man, followed by a string of meaningless murders.
1983 (R) 85m/C King John III, Candy Dee, Janel King. **VHS, Beta** *PGN* ♫

Psychomania Former war hero and painter is suspected of being the demented killer who is stalking girls on campus. To prove his innocence, he tracks the killer himself. Also called "Violent Midnight."
1963 95m/C Lee Philips, Sheppard Strudwick, Jean Hale, Dick Van Patten, Sylvia Miles, James Farentino; *Dir:* Richard Hilliard. **VHS, Beta** $24.95 *SNC, HHT* ♫

Psychomania A drama of the supernatural, the occult, and the violence which lies just beyond the conventions of society for a group of dead motorcyclists, the Living Dead, who all came back to life after committing suicide with the help of the devil.
1973 (R) 89m/C *GB* George Sanders, Beryl Reid, Nicky Henson, Mary Laroche, Patrick Holt; *Dir:* Don Sharp. **VHS, Beta** $29.95 *MED, KOV, WES* ♫♫

Psychopath A film about a modern-day, slightly unhinged, Robin Hood and his escapades, stealing from other thieves and giving to the victimized.
1968 90m/C *GB* Klaus Kinski, George Martin, Ingrid Schoeller; *Dir:* Guido Zurli. **VHS, Beta** $59.95 *LTG* ♫½

Psychopath A nutso children's television personality begins to murder abusive parents. Fairly gory, and definitely not for family viewing.
1973 (PG) 85m/C Tom Basham, Gene Carlson, Gretchen Kanne; *Dir:* Larry Brown. **VHS, Beta** $69.95 *MED* ♫

Psychos in Love A psychotic murderer who hates grapes of any kind finds the woman of his dreams - she's a psychotic murderer who hates grapes! Together they find bliss - until the plumber discovers their secret!
1987 90m/C Carmine Capobianco, Debi Thibeault, Frank Stewart; *Dir:* Gorman Bechard. **VHS, Beta** $69.95 *UHV* ♫

The Psychotronic Man An innocent man suddenly finds he possesses amazing and dangerous powers, enabling him to control outside events with a thought.
1991 88m/C Peter Spelson, Christopher Carbis, Curt Colbert, Robin Newton, Paul Marvel; *Dir:* Jack M. Sell. **VHS** *UNI* ♫½

PT 109 The World War II exploits of Lieutenant (j.g.) John F. Kennedy in the South Pacific. Part of the "A Night at the Movies" series, this tape simulates a 1963

movie evening, with a Foghorn Leghorn cartoon ("Banty Raids"), a newsreel on the JFK assassination, and coming attractions for "Critic's Choice" and "Four for Texas."
1963 159m/C Cliff Robertson, Ty Hardin, Ty Hardin, Robert Blake, Robert Culp, James Gregory; *Dir:* Leslie Martinson. **VHS, Beta, LV** $19.98 *WAR, PIA* ♫♫

Puberty Blues Two Australian girls become part of the local surfing scene in order to be accepted by the "in crowd" at their high school.
1981 (R) 86m/C *AU* Neil Schofield, Jad Capelja; *Dir:* Bruce Beresford. **VHS, Beta** $59.95 *MCA* ♫♫½

A Public Affair Exposes the evils and abuses of collection agencies without even a shred of humor. Certainly gets its point across, but will bore any audience to tears or spontaneous naps.
1962 71m/B Myron McCormick, Edward Binns, Harry Carey Jr., Rocky Aoki. **VHS, Beta** $29.99 *VYY* ♫½

Public Cowboy No. 1 Cattle thieves use a radio, airplanes, and refrigerator trucks in their updated rustling schemes.
1937 54m/B Gene Autry, William Farnum. **VHS, Beta** $19.95 *NOS, VCN, DVT* ♫

Public Enemy Cagney's acting career was launched by this story of two Irish boys growing up in a Chicago shantytown to become hoodlums during the prohibition era. Cagney and Woods hook up with molls Clarke and Blondell on their rise to the top. Considered the most realistic "gangster" film, Wellman's movie is also the most grimly brutal due to its release prior to Hollywood censorship. The scene where Cagney smashes a grapefruit in Clarke's face was credited with starting a trend in abusing film heroines.
1931 85m/B James Cagney, Edward Woods, Leslie Fenton, Joan Blondell, Mae Clarke, Jean Harlow; *Dir:* William A. Wellman. **VHS, Beta, LV** $19.95 *MGM, FOX* ♫♫½

Pucker Up and Bark Like a Dog A young artist and his inspirationally lovely actress girl-friend find themselves in strange and crazy situations. Good cast gives this potential.
1989 (R) 94m/C Lisa Zane, Jonathan Gries, Paul Bartel, Robert Culp; *Dir:* Paul S. Parco. **VHS** $79.95 *FRH, BTV* ♫

Pudd'nhead Wilson An adaptation of a Mark Twain story about a small-town lawyer who discovers the illicit exchange of a white infant for a light-skinned negro infant by a slave woman. Made for television as an "American Playhouse" presentation on PBS.
1987 87m/C Ken Howard; *Dir:* Alan Bridges. **VHS, Beta** $19.95 *MCA* ♫½

Pueblo Affair A dramatization of the capture of the American spy ship "Pueblo" by the North Koreans, during which time the crew was tortured and then imprisoned and intelligence documents confiscated.
1973 99m/C Hal Holbrook, Andrew Duggan, Richard Mulligan, George Grizzard, Gary Merrill, Mary Fickett; *Dir:* Anthony Page. Chicago Film Festival '73: First Prize; Emmy Awards '74: Best Lead Actor in a Drama for a Special Program (Holbrook). **VHS, Beta** *LME* ♫♫♫

Pulp Caine, playing a mystery writer, becomes a target for murder when he ghostwrites the memoirs of a movie gangster from the 1930s, played deftly by Rooney.

1972 (PG) 95m/C *GB* Michael Caine, Mickey Rooney, Lionel Stander, Lizabeth Scott, Nadia Cassini, Dennis Price, Al Lettieri; *Dir:* Mike Hodges. **VHS, Beta** $14.95 *WKV* ♫♫½

Pulse Electricity goes awry in this science-fiction thriller about appliances and other household devices that become supercharged and destroy property and their owners.
1988 (PG-13) 90m/C Cliff DeYoung, Roxanne Hart, Joey Lawrence, Charles Tyner, Dennis Redfield, Robert Romanus, Myron Healey; *Dir:* Paul Golding. **VHS, Beta, LV** $89.95 *COL* ♫♫

Pulsebeat A made-for-television film about the troubled life of a health club owner.
1985 92m/C Daniel Greene, Alice Moore, Lee Taylor Allen, Peter Lupus. **VHS, Beta** $69.98 *LIV, LTG Woof!*

Puma Man Puma Man is a super hero who must stop the evil Dr. Kobras from using an ancient mask in his attempt to become ruler of the world.
1980 100m/C Donald Pleasence, Walter George Alton, Sydne Rome, Miguel Angel Fuentes; *Dir:* Alvin J. Neitz, Alberto De Martino. **VHS, Beta** $59.95 *PSM Woof!*

Pump Up the Volume High school newcomer leads double life as Hard Harry, sarcastic host of an illegal radio broadcast and Jack Nicholson soundalike. Anonymously popular with his peers, he invites the wrath of the school principal due to his less than flattering comments about the school administration. Slater seems to enjoy himself as defiant deejay while youthful cast effectively supports.
1990 (R) 105m/C Christian Slater, Scott Paulin, Ellen Greene, Samantha Mathis, Cheryl Pollak, Annie Ross; *W/Dir:* Allan Moyle. **VHS, LV, 8mm** $19.95 *COL, FCT* ♫♫½

Pumpkinhead A farmer evokes a demon from the earth to avenge his son's death. When it continues on its murdering rampage, the farmer finds he no longer has any control over the vicious killer.
1988 (R) 89m/C Lance Henriksen, John DiAquino, Kerry Remsen, Matthew Hurley, Jeff East, Kimberly Ross; *Dir:* Stan Winston. **VHS, Beta, LV** $19.95 *MGM* ♫♫½

Punch the Clock Attractive female thief finds her bail bondsman very exciting; he's ready to help her any way he can.
1990 (R) 88m/C Michael Rogan, Chris Moore, James Lorinz; *Dir:* Eric L. Schlagman. **VHS** $79.98 *VID* ♫♫

Punchline An ostensible look into the life of stand-up comics, following the career ups-and-downs of an embittered professional funny-man and a frustrated housewife hitting the stage for the first time. Written by Seltzer. Some very funny and touching moments and some not-so-funny and touching as the movie descends into melodrama without a cause.
1988 (R) 100m/C Tom Hanks, Sally Field, John Goodman, Mark Rydell, Kim Greist, Barry Sobel, Paul Mazursky, Pam Matteson, George Michael McGrath, Taylor Negron, Damon Wayans; *Dir:* David Seltzer. **VHS, Beta, LV** $19.95 *COL* ♫♫

The Punisher Lundgren portrays Frank Castle, the Marvel Comics anti-hero known as the Punisher. When his family is killed by the mob Castle set his eyes on revenge. Filmed in Australia.
1990 (R) 92m/C Dolph Lundgren, Louis Gossett Jr., Jeroen Krabbe, Kim Miyori; *Dir:* Mark Goldblatt. **VHS, Beta, LV** $92.95 *LIV* ♫

Punk Rock Movie The musical explosion known as punk rock is documented here in its early days (circa 1977) at the Roxy in England. Includes a look at leading punk rock artists and at the entire counterculture that is punk.
1978 80m/C The Sex Pistols, The Clash, Siouxsie & the Banshees, Slits, Wayne County. **VHS, Beta $19.95** *MVD, SUN* ⬙

Punk Vacation A gang of motorcycle mamas terrorize a small town. The usual biker stuff.
1990 90m/C Stephen Falchi, Roxanne Rogers, Don Martin, Sandra Bogan; *Dir:* Stanley Lewis. **VHS, Beta, LV** *RAE* ⬙

Puppet on a Chain An American narcotics officer busts an Amsterdam drug ring and the leader's identity surprises him. Based on the Alistair MacLean novel.
1972 (PG) 97m/C *GB* Sven-Bertil Taube, Barbara Parkins, Alexander Knox, Patrick Allen, Geoffrey Reeve; *Dir:* Geoffrey Reeve. **VHS, Beta $59.95** *PSM* ⬙ ½

Puppet Master Four psychics are sent to investigate a puppet maker who they think may have discovered the secret of life. Before they know it, they are stalked by evil puppets. Great special effects, not-so-great script.
1989 (R) 90m/C Paul LeMat, Jimmie F. Scaggs, Irene Miracle, Robin Rates, Barbara Crampton, William Hickey; *Dir:* David Schmoeller. **VHS, Beta, LV $14.95** *PAR* ⬙⬙ ½

Puppet Master 2 Puppets on a rampage turn hotel into den of special effects. Less effective string pulling than in original. Long live Pinocchio.
1991 (R) 90m/C Elizabeth MacLellan, Collin Bernsen, Greg Webb, Charlie Spradling, Nita Talbot, Steve Welles, Jeff Weston; *Dir:* David Allen. **VHS, Beta, LV $14.95** *PAR* ⬙⬙

Puppet Master 3 Prequel about the origin of the whole gory Puppet Master shebang, set in Nazi Germany. The weapon-hungry Third Reich tries to wrest secrets of artificial life from sorceror Andre Toulon, who sics his deadly puppets on them. Toulon's a good guy here, one of many contradictions in the series. Fine cast, but strictly for the followers.
1991 (R) 86m/C Guy Rolfe, Ian Abercrombie, Sarah Douglas; *Dir:* David DeCoteau. **VHS, LV $14.95** *PAR, LDC* ⬙⬙

The Puppetoon Movie A compilation of George Pal's famous Puppetoon cartoons from the 1940s, marking his stature as an animation pioneer and innovator. Hosted by Gumby, Pokey, Speedy Alka Seltzer and Arnie the Dinosaur in newly directed scenes.
1987 (G) 80m/C *Dir:* Arnold Leibovit. **VHS, Beta, LV $19.95** *FHE, IVE* ⬙⬙

The Pure Hell of St. Trinian's In this sequel to "Blue Murder at St. Trinian's," a sheik who desires to fill out his harem tries recruiting at a rowdy girls' school. Although it isn't as funny as the first, due to the lack of Alistair Sim, it is still humorous. Followed by "The Great St. Trinian's Train Robbery." Based on the cartoon by Ronald Searle.
1961 94m/B *GB* Cecil Parker, Joyce Grenfell, George Cole, Thorley Walters; *Dir:* Frank Launder. **VHS** *FCT* ⬙⬙ ½

Pure Luck A who-asked-for-it remake of a 1981 Franco-Italian film called "La Chevre." The premise is the same: an accident-prone heiress disappears in Mexico, and her father tries to locate her using an equally clumsy accountant. The nebbish and

an attendant tough private eye stumble and bumble south of the border until the plot arbitrarily ends. Pure awful.
1991 (PG) 96m/C Martin Short, Danny Glover, Sheila Kelley, Scott Wilson, Sam Wanamaker, Harry Shearer; *Dir:* Nadia Tass. **VHS, Beta, LV $34.98** *MCA, PIA* ⬙ ½

Purgatory While travelling abroad, two innocent women find themselves unjustly jailed. During their confinement they are tortured and raped.
1989 90m/C Tanya Roberts, Julie Pop; *Dir:* Ami Artzi. **VHS, Beta, LV $79.95** *NSV* ⬙

Purlie Victorious Presentation of the award-winning Broadway hit, which takes a look at racial integration.
1963 93m/C Ossie Davis, Ruby Dee, Godfrey Cambridge, Alan Alda; *Dir:* Nicholas Webster. **VHS, Beta $69.95** *MAS* ⬙⬙

The Purple Heart An American Air Force crew is shot down over Tokyo and taken into brutal POW camps. Intense wartime melodramatics.
1944 99m/B Dana Andrews, Richard Conte, Farley Granger, Donald (Don "Red") Barry, Sam Levene, Kevin O'Shea, Tala Birell, Nestor Paiva, Benson Fong, Marshall Thompson, Richard Loo; *Dir:* Lewis Milestone. **VHS, Beta $19.98** *FOX* ⬙⬙ ½

Purple Hearts Ken Wahl stars as a Navy doctor who falls in love with nurse Cheryl Ladd against the backdrop of the Vietnam war.
1984 (R) 115m/C Cheryl Ladd, Ken Wahl, Stephen Lee, Annie McEnroe, Paul McCrane, Cyril O'Reilly; *Dir:* Sidney J. Furie. **VHS, Beta $19.98** *WAR* ⬙ ½

The Purple Monster Strikes A martian plots to conquer Earth to save his dying planet. Serial in fifteen episodes.
1945 188m/B Dennis Moore, Linda Stirling, Roy Barcroft; *Dir:* Spencer Gordon Bennet. **VHS $29.98** *REP, VCN, MLB* ⬙

Purple People Eater The alien of the title descends to earth to mix with young girls and rock 'n roll. Based on the song of the same name who's performer, Sheb Wooley, appears in the film. Harmless fun for the whole family.
1988 (PG) 91m/C Ned Beatty, Shelley Winters, Neil Patrick Harris, Kareem Abdul-Jabbar, Little Richard, Chubby Checker, Peggy Lipton; *Dir:* Linda Shayne. **VHS, Beta, LV $79.95** *MED* ⬙

Purple Rain A quasi-autobiographical video showcase for the pop-star Prince. Film tells the tale of his struggle for love, attention, acceptance, and popular artistic recognition in Minneapolis. Not a bad film, for such a monumentally egotistical movie.
1984 (R) 113m/C Prince, Apollonia, Morris Day, Olga Karlatos, Clarence Williams III; *Dir:* Albert Magnoli. Academy Awards '84: Best Original Score. **VHS, Beta, LV $19.98** *MVD, WAR* ⬙⬙ ½

Purple Rose of Cairo A diner waitress, disillusioned by the Depression and a lackluster life, escapes into a film playing at the local movie house where a blond film hero, tiring of the monotony of his role, makes a break from the celluloid to join her in the real world. The ensuing love story allows director-writer Allen to show his knowledge of old movies and provide his fans with a change of pace. Farrow's film sister is also her real-life sister Stephanie, who went on to appear in Allen's "Zelig."
1985 (PG) 82m/C Mia Farrow, Jeff Daniels, Danny Aiello, Dianne Wiest, Van Johnson, Zoe Caldwell, John Wood, Michael Tucker, Edward Herrmann, Milo O'Shea, Glenne Headly; *Dir:*

Woody Allen. British Academy Awards '85: Best Film; French Film Critics Union Awards '85: Best Foreign Film. **VHS, Beta, LV $19.98** *LIV, VES* ⬙⬙⬙

The Purple Taxi A romantic drama revolving around several wealthy foreigners who have taken refuge in beautiful southern Ireland. From Michel Deon's bestselling novel.
1977 (R) 93m/C *FR* Fred Astaire, Charlotte Rampling, Peter Ustinov, Philippe Noiret, Edward Albert; *Dir:* Yves Boisset. **VHS, Beta $59.95** *COL* ⬙ ½

Purple Vigilantes The Three Musketeers uncover a gang of vigilantes. Part of the series.
1938 54m/B Bob Livingston, Ray Corrigan, Max Terhune. **VHS, Beta** *VCN, DVT* ⬙

Pursued Excellent performances by Mitchum and Wright mark this suspenseful Western drama of a Spanish-American war veteran in search of his father's killer.
1947 101m/B Teresa Wright, Robert Mitchum, Judith Anderson, Dean Jagger, Alan Hale Jr., Harry Carey Jr., John Rodney; *Dir:* Raoul Walsh. **VHS $14.95** *REP* ⬙⬙⬙

Pursuit A terrorist threatens to release a toxic nerve gas throughout a city hosting a political convention. Tension mounts as federal agents must beat the extremists' countdown to zero. Based on the novel by Michael Crichton.
1972 73m/C Ben Gazzara, E.G. Marshall, Martin Sheen, Joseph Wiseman, William Windom; *Dir:* Michael Crichton. **VHS, Beta $69.95** *REP* ⬙⬙⬙

Pursuit A mercenary comes out of retirement to join a renegade team recovering stolen gold. But they turn on him, stealing the treasure and taking a hostage. Now he must track them down, or there's no movie.
1990 (R) 94m/C James Ryan, Andre Jacobs; *Dir:* John H. Parr. **VHS $89.99** *HBO* ⬙⬙

Pursuit to Algiers The modernized Holmes and Watson guard over the King of fictional Rovenia during a sea voyage, during which assassins close in.
1945 65m/B Basil Rathbone, Nigel Bruce, Martin Kosleck, Marjorie Riordan, Rosalind Iven, John Abbott; *Dir:* Roy William Neill. **VHS, Beta $19.98** *FOX* ⬙⬙ ½

The Pursuit of D. B. Cooper Based on an actual hijacking that occurred on Thanksgiving Eve in 1971 in which J.R. Meade (alias D.B. Cooper) parachuted out of an airliner with $200,000 of stolen money. Although his fate and whereabouts have remained a mystery, this story speculates about what may have happened.
1981 (PG) 100m/C Robert Duvall, Treat Williams, Kathryn Harrold, Ed Flanders; *Dir:* Roger Spottiswoode. **VHS, Beta, LV $29.98** *LIV, VES* ⬙⬙

Pursuit of the Graf Spee Three small British ships destroy the mighty Graf Spee, a World War II German battleship. Originally titled "The Battle of the River Plate."
1957 106m/C *GB* Anthony Quayle, Peter Finch, Ian Hunter; *Dir:* Michael Powell. **VHS, Beta $39.95** *IND* ⬙⬙ ½

The Pursuit of Happiness An idealistic man is convicted of manslaughter after accidentally running down a woman. Once in prison, he's faced with the decision to try to escape.

1970 (PG) 85m/C Michael Sarrazin, Arthur Hill, E.G. Marshall, Barbara Hershey, Robert Klein; *Dir:* Robert Mulligan. **VHS, Beta** $69.95 *RCA* 🎬🎬½

Pushed to the Limit Harry Lee, the most feared gangster in Chinatown, is about to meet his match in martial-arts queen Lesseos. She's out for revenge when she learns that Lee is responsible for her brother's murder but first she must get by Lee's lethal bodyguard, the equally skilled Inga. **199? 88m/C** Mimi Lesseos. **VHS** $89.95 *IMP* 🎬½

Puss 'n Boots From "Faerie Tale Theatre" comes the story of a cat who makes his poor master a rich land-owning nobleman.
1984 60m/C Ben Vereen, Gregory Hines; *Dir:* Robert Iscove. **VHS, Beta** $14.98 *KUI, FOX, FCT* 🎬½

Putney Swope Comedy about a token black ad man mistakenly elected Chairman of the board of a Madison Avenue ad agency who turns the company upside-down. A series of riotous spoofs on commercials is the highpoint in this funny, though somewhat dated look at big business.
1969 (R) 84m/B Arnold Johnson, Laura Greene, Stanley Gottlieb, Mel Brooks; *Dir:* Robert Downey. **VHS, Beta** $14.95 *COL, PTB* 🎬🎬

Puzzle Franciscus searches for the urn that contains the remains of Buddha but instead finds danger and intrigue. Made for television.
1978 90m/C *AU* James Franciscus, Wendy Hughes, Robert Helpmann, Peter Gwynne, Gerald Kennedy, Kerry McGuire; *Dir:* Gordon Hessler. **VHS** $19.95 *ACA* 🎬🎬½

Pygmalion Oscar-winning film adaptation of Shaw's play about a cockney flower-girl who is transformed into a "lady" under the guidance of a stuffy phonetics professor. Shaw himself aided in writing the script in this superbly acted comedy that would be adapted into the musical, My Fair Lady, thirty years later. Oscar nominations: Best Picture, Best Actor (Howard), and Best Actress (Hiller).
1938 96m/B *GB* Leslie Howard, Wendy Hiller, Wilfred Lawson, Marie Lohr; *Dir:* Leslie Howard, Anthony Asquith. Academy Awards '38: Best Adapted Screenplay; Venice Film Festival '38: Best Actor (Howard). **VHS, Beta, LV** $14.95 *COL, NOS, SUE* 🎬🎬🎬½

Python Wolf It's non-stop action in South Africa as a counter-terrorist group sets out to thwart a drug smuggler's plans to export plutonium. As usual, intense pacing from director Friedkin. Made for television; also known as "C.A.T. Squad: Python Wolf." Sequel to "C.A.T. Squad" (1986).
1988 (R) 100m/C Joe Cortese, Jack Youngblood, Steve James, Deborah Van Valkenburgh, Miguel Ferrer, Alan Scarfe; *Dir:* William Friedkin. **VHS, LV** $19.95 *STE, VMK* 🎬🎬

The Pyx Canadian suspense thriller about the murder of a prostitute with satanic overtones. Police sargeant investigates the crime and enters a world of devil worship and decadence. Based on the novel by John Buell, the film was also released as "The Hooker Cult Murders."
1973 (R) 111m/C *CA* Karen Black, Christopher Plummer, Donald Pilon; *Dir:* Harvey Hart. **VHS** $39.95 *PSM, TAV* 🎬🎬🎬

Q & A Semi-taut thriller with Nolte playing a high-strung, hair-trigger cop trying to hide a murder. The question is: Who's gonna get it in the end?

1990 (R) 132m/C Nick Nolte, Timothy Hutton, Armand Assante; *Dir:* Sidney Lumet. **VHS, Beta, LV** $19.98 *HBO* 🎬🎬

Q Ships Silent German drama about a submarine commander who has inner conflicts with his country's war effort.
1928 78m/B *GE* J.P. Kennedy, Roy Travers, Johnny Butt, Philip Hewland; *Dir:* Geoffrey Barkas. **VHS, Beta** $19.95 *NOS, DVT, GLV* 🎬🎬½

Q (The Winged Serpent) A cult of admirers surrounds this goony monster flick about dragonlike Aztec god Quetzlcoatl, summoned to modern Manhattan by gory human sacrifices, and hungry for rooftop sunbathers and construction teams. Direction and special effects are pretty ragged, but witty script helps the cast shine, especially Moriarty as a lowlife crook who's found the beast's hidden nest. Also called "Q: The Winged Serpent."
1982 (R) 92m/C Michael Moriarty, Candy Clark, David Carradine, Richard Roundtree; *Dir:* Larry Cohen. **VHS, Beta** $69.95 *MCA* 🎬🎬🎬

QB VII A knighted physician brings a suit for libel against a novelist for implicating him in war crimes. Hopkins as the purportedly wronged doctor and Gazzara as the writer are both superb. Ending is stunning. Adapted from the novel by Leon Uris. Over five hours long; made for TV. Available only as a three-cassette set.
1974 313m/C Anthony Hopkins, Ben Gazzara, Lee Remick, Leslie Caron, Juliet Mills, John Gielgud, Anthony Quayle; *Dir:* Tom Gries. **VHS, Beta, LV** $139.95 *COL* 🎬🎬🎬½

Quackser Fortune Has a Cousin in the Bronx An Irish fertilizer salesman meets an exchange student from the U.S., who finds herself attracted to this unlearned, but not unknowing, man. An original love story with drama and appeal.
1970 (R) 88m/C *IR* Gene Wilder, Margot Kidder; *Dir:* Waris Hussein. **VHS, Beta** $19.95 *UHV* 🎬🎬🎬

Quadrophenia Pete Townshend's excellent rock opera about an alienated youth looking for life's meaning in Britain's rock scene circa 1963. Music by The Who is powerful and apt. Fine performance by Sting in his acting debut.
1979 (R) 115m/C *GB* Phil Daniels, Mark Wingett, Philip Davis, Leslie Ash, Sting; *Dir:* Franc Roddam. **VHS, Beta, LV** $19.95 *MVD, COL* 🎬🎬🎬

Quality Street English lovers Tone and Hepburn are separated when he leaves to fight in the Napoleonic Wars. When Tone returns years later, he's forgotten his erstwhile heartthrob. He also fails to recognize the 16-year old coquette who's caught his eye is his old beloved in disguise. Cast works hard to overcome the absurd premise, based on the play by Sir James Barrie.
1937 84m/B Katharine Hepburn, Franchot Tone, Fay Bainter, Eric Blore, Cora Witherspoon, Estelle Winwood, Florence Lake, Bonita Granville; *Dir:* George Stevens. **VHS, LV** $19.98 *NOS, FCT* 🎬🎬

Quarantine Quarantine camps for carriers of a fatal virus are the serious measures taken by the nation. The solution is clear to a rebel and an inventor; all must be liberated or exterminated.
1989 (R) 92m/C Beatrice Boepple, Garwin Sanford, Jerry Wasserman, Charles Wilkinson. **VHS** $89.98 *REP* 🎬🎬½

Quarterback Princess Workaday made-for-TV telling of the real-life girl who goes out for football and becomes homecoming queen. Heartwarming, if you like that kinda stuff, but not exciting.
1985 96m/C Helen Hunt, Don Murray, John Stockwell, Daphne Zuniga; *Dir:* Noel Black. **VHS, Beta** $59.98 *FOX* 🎬🎬

Quartet A young French women is taken in by an English couple after her husband goes to prison. The husband seduces her, and she becomes trapped emotionally and socially. Superbly acted, claustrophobic drama based on a Jean Rhys novel.
1981 (R) 101m/C *GB FR* Isabelle Adjani, Alan Bates, Maggie Smith, Anthony Higgins; *Dir:* James Ivory. Cannes Film Festival '81 '81: Best Actress (Adjani). **VHS, Beta** $69.95 *WAR* 🎬🎬🎬

Quatermass 2 In the well-made sequel to "Quatermass Experiment," (also known as "The Creeping Unknown") Professor Quatermass battles blobs and brainwashed zombies to rescue government officials whose bodies have been invaded by aliens. Also released as "Enemy from Space." "Five Million Years to Earth" concluded the trilogy.
1957 84m/B *GB* Brian Donlevy, John Longden, Sidney James, Bryan Forbes, William Franklyn, Vera Day, John Van Eyssen, Michael Ripper, Michael Balfour; *Dir:* Val Guest. **VHS, Beta, LV** $54.95 *FCT, CTH, SNC* 🎬🎬🎬

Quatermass Conclusion An elderly British scientist comes out of retirement to stop an immobilizing death ray from outer space from destroying Earth. Edited version of a television miniseries continues earlier adventures of Quatermass on film and television.
1979 105m/C *GB* John Mills, Simon MacCorkindale, Barbara Kellerman, Margaret Tyzack; *Dir:* Piers Haggard. **VHS, Beta** $59.99 *HBO* 🎬🎬

The Quatermass Experiment Excellent British production about an astronaut who returns to Earth carrying an alien infestation that causes him to turn into a horrible monster. Competent acting and tense direction. Also known as "The Creeping Unknown." Followed by "Enemy From Space."
1956 78m/B *GB* Brian Donlevy, Margia Dean, Jack Warner, Richard Wordsworth; *Dir:* Val Guest. **VHS** $19.95 *NOS, SNC* 🎬🎬🎬

Quatorze Juliet A taxi driver and a flower girl meet on Bastille Day and fall in love. When he becomes involved with gangsters, she steers him back to virtue in this minor French relic from director Clair.
1932 85m/B *FR* Anabella Rigaud, Jorge Rigaud; *Dir:* Rene Clair. **VHS** $29.95 *VDM, FCT* 🎬🎬½

Que Viva Mexico Eisenstein's grand unfinished folly on the history of Mexico, reconstructed and released in 1979 by his protege, Grigori Alexandrov. Along with "Greed" and "Napoleon," it remains as one of cinema's greatest irrecoverable casualties. Beautiful photographs; pedantic director.
1932 85m/B *RU* Eduard Tisse; *Dir:* Sergei Eisenstein, Grigori Alexandrov. **VHS, Beta** $59.95 *IFE* 🎬🎬🎬

Queen of the Amazons A girl searches for her fiance and finds him reluctantly held captive by a tribe of women who rule the jungle.
1947 60m/B Robert Lowery, Patricia Morison, J. Edward Bromberg, John Miljan. **VHS, Beta, LV** $16.95 *SNC, NOS, WGE* 🎬½

Queen Boxer A female kick-boxer beats up those who reprehensibly enslaved her family.
1973 (R) 92m/C Judy Lee, Peter Yang Kwan. VHS, Beta $12.95 *IVE* 🐾

Queen Christina A stylish, resonant star vehicle for Garbo, portraying the 17th-century Swedish queen from ascension to the throne to her romance with a Spanish ambassador. Alternately hilarious and moving, it holds some of Garbo's greatest and most memorable moments. Gilbert's second to last film and his only successful outing after the coming of sound.
1933 101m/B Greta Garbo, John Gilbert, Lewis Stone, C. Aubrey Smith, Ian Keith, Reginald Owen, Elizabeth Young; *Dir:* Rouben Mamoulian. VHS, Beta, LV $19.98 *MGM* 🐾🐾🐾½

Queen for a Day Three differing vignettes, based on a rather notorious radio/TV game show of the time, in which 'deserving' (or sufficiently pathetic) working-class women were rewarded for their selflessness. Part one concerns a perfect suburban family whose son contracts polio. Part two has a teen spooking his immigrant parents by working as a carnival high diver to earn college cash. The finale is a Dorothy Parker farce about a homely but kindhearted nurse caring for the child of an unappreciative couple. So popular was this segment that the film was retitled ''Horsie,'' after the main character. Overall, not bad considering the source. special contest would have the film premiere in her home town.
1951 107m/B Phyllis Avery, Darren McGavin, Tristram Coffin, Adam Williams, Tracey Roberts, Jack Bailey, Jim Morgan, Fort Pearson; *Dir:* Arthur Lubin. VHS $19.95 *NOS* 🐾🐾½

Queen of Diamonds A woman pulls off the biggest diamond heist of all time.
198? 90m/C Claudia Cardinale, Stanley Baker, Henri Charriere. VHS, Beta $59.95 *MON* 🐾🐾

Queen of Hearts Excellent, original romantic comedy is a directorial triumph for Amiel in his first feature. An Italian couple defy both their families and marry for love. Four children later, we find them running a diner in England. Humorous, dramatic, sad—everything a move can and should be. Fine performances.
1989 (PG) 112m/C *GB* Anita Zagaria, Joseph Long, Eileen Way, Vittorio Duse, Vittorio Amandola, Ian Hawkes; *Dir:* Jon Amiel. VHS, Beta, LV $89.95 *VIR* 🐾🐾🐾½

Queen of the Jungle Re-edited adventure serial featuring a white woman cast off in a hot-air balloon and landing in Africa, where she is hailed as a Queen. High camp, starring ex-Our Ganger Kornman.
1935 85m/B Mary Kornman, Reed Howes; *Dir:* Robert Hill. VHS, Beta, 8mm $26.95 *SNC, NOS, MLB* 🐾

Queen Kelly The popularly known, slap-dash version of von Stroheim's famous final film, in which an orphan goes from royal marriage to white slavery to astounding wealth. Never really finished, the film is the edited first half of the intended project, prepared for European release after von Stroheim had been fired. Even so, a campy, extravagant and lusty melodrama. Silent.
1929 113m/B Gloria Swanson, Walter Byron, Seena Owen, Tully Marshall, Madame Sul Te Wan; *Dir:* Erich von Stroheim. VHS, Beta, LV, 8mm $19.95 *NOS, KIV, VYY* 🐾🐾🐾

Queen: Live in Rio Queen rocks Rio di Janeiro in this concert performance that features ''Bohemian Rhapsody'' and ''Tie Your Mother Down'' in VHS and Beta Hi-Fi Stereo.
1985 60m/C VHS, Beta, LV $29.95 *SVS, PIA, MVD* 🐾🐾🐾

The Queen of Mean Tabloid TV-movie also known as ''Leona Helmsley: The Queen of Mean,'' based on Ransdell Pierson's scandal-sheet bio of the hotel magnate and convicted tax cheat. See Leona connive. See Leona bitch. See Leona get hers. What's the point? If you ask that you're too bright to watch.
1990 94m/C Suzanne Pleshette, Lloyd Bridges, Bruce Weitz, Joe Regalbuto; *Dir:* Richard Michaels. VHS $64.95 *FRH* 🐾½

Queen of Outer Space The notorious male-chauvinist sci-fi cheapie starts out slow, but then the laughs keep coming as the cast plays the hyperdumb material straight. Space cadets crash on Venus, find it ruled by women—and the 'dolls' have wicked plans in store for mankind.
1958 80m/C Zsa Zsa Gabor, Eric Fleming, Laurie Mitchell, Paul Birch, Barbara Darrow, Dave Willcock, Lisa Davis, Marilyn Buferd, Marjorie Durant, Lynn Cartwright, Gerry Gaylor; *Dir:* Edward L. Bernds. VHS, Beta $14.98 *FXV* Woof!

Queen of the Road A feisty Aussie schoolteacher starts a new life as a tractor-trailer driver.
1984 96m/C *AU* Joanne Samuel, Amanda Muggleston; *Dir:* Bruce Best. VHS, Beta $69.98 *LIV, VES* 🐾🐾

The Queen of Spades Mystical drama about a Russian soldier who ruins his life searching for winning methods of card playing. Well-made version of the Pushkin story.
1949 95m/B *GB* Anton Walbrook, Edith Evans, Ronald Howard, Mary Jerrold, Yvonne Mitchell, Anthony Dawson; *Dir:* Thorold Dickinson. VHS $79.99 *HBO* 🐾🐾🐾

Queen of the Stardust Ballroom Well-made television drama about a lonely widow goes to a local dance hall, where she meets a man and begins an unconventional late love.
1975 98m/C Maureen Stapleton, Charles Durning, Michael Strong, Charlotte Rae, Sam O'Steen; *Dir:* Michael Brandon. Beta $11.95 *PSM* 🐾🐾🐾

Queenie Television miniseries Loosely based on the life of actress Merle Oberon, exploring her rise to stardom. Based on the best-selling novel by Michael Korda, Oberon's nephew by marriage. Many long, dry passages fit for leaving the room for snacks.
1987 233m/C Mia Sara, Kirk Douglas, Martin Balsam, Claire Bloom, Chaim Topol, Joel Grey, Sarah Miles, Joss Ackland; *Dir:* Larry Peerce. VHS $29.95 *STE, NWV* 🐾🐾½

Queens Logic Ensemble comedy in the tradition of ''The Big Chill'' featuring the trials and tribulations of the ''old-neighborhood'' gang, who gather again on hometurf and reminisce. Most of the film centers around Olin and girlfriend Webb and whether or not he will chicken out of their wedding. Not bad comedy, with Mantegna delivering most of the good lines.
1991 (R) 116m/C John Malkovich, Kevin Bacon, Jamie Lee Curtis, Linda Fiorentino, Joe Mantegna, Ken Olin, Tom Waits, Chloe Webb; *Dir:* Steve Rash. VHS, LV $92.98 *LIV* 🐾🐾½

Querelle Querelle, a handsome sailor, finds himself involved in a bewildering environment of murder, drug smuggling and homosexuality in the port of Brest. Highly-stylized and erotic sets, along with a dark look, give the film an interesting feel. Nice performances by Davis, Nero, and Moreau. Fassbinder's controversial last film, based on the story by Jean Genet. Strange, difficult narrative (such as it is), alternately boring and engrossing.
1983 (R) 106m/C *GE* Brad Davis, Jeanne Moreau, Franco Nero, Laurent Malet; *Dir:* Rainer Werner Fassbinder. VHS, Beta $59.95 *COL, APD, GLV* 🐾🐾½

The Quest Australian film never released theatrically in the United States. A young boy (Thomas, of ''E.T.'' fame) raised in the outback investigates a local Aboriginal superstition about a monster living in an ancient cemetery. Not bad, just dull.
1986 (PG) 94m/C *AU* Henry Thomas, Tony Barry, John Ewart, Rachel Friend, Tamsin West, Dennis Miller, Katya Manning; *Dir:* Brian Trenchard-Smith. VHS, Beta $9.98 *CHA* 🐾🐾

Quest for Fire An interesting story sans the usual dialogue thing. A group of men during the Ice Age must wander the land searching for fire after they lose theirs fending off an attack. During their quest, they encounter and battle various animals and tribesmen in order to survive. The special language they speak was developed by Anthony Burgess, while the primitive movements were Desmond ''The Naked Ape'' Morris. Perlman went on to become the Beast in TV's ''Beauty and the Beast''; Chong is the daughter of Tommy Chong of the comic duo Cheech and Chong.
1982 (R) 75m/C *FR CA* Everett McGill, Ron Perlman, Nameer El-Kadi, Rae Dawn Chong; *Dir:* Jean-Jacques Annaud. Academy Awards '82: Best Makeup; Genie Awards '83: Best Actress (Chong), Best Costume Design. VHS, Beta, LV $69.98 *FOX* 🐾🐾🐾

Quest for Love Quirky sci-fi story of a man who passes through a time warp and finds himself able to maintain two parallel lives. Based on John Wyndham's short story.
1971 90m/C *GB* Joan Collins, Tom Bell, Denholm Elliott, Laurence Naismith; *Dir:* Ralph Thomas. VHS, Beta $39.95 *IND* 🐾🐾½

Quest for the Mighty Sword A warrior battles dragons, demons and evil wizards.
1990 (PG-13) 94m/C Eric Allen Kramer, Margaret Lenzey, Donald O'Brien, Dina Marrone, Chris Murphy; *Dir:* David Hills. VHS, Beta, LV $79.95 *COL* 🐾

A Question of Guilt When her child turns up dead, a free-wheeling divorcee is the prime suspect. Made-for-television soaper.
1978 100m/C Tuesday Weld, Ron Leibman, Peter Masterson, Alex Rocco, Viveca Lindfors, Lana Wood; *Dir:* Robert Butler. VHS $39.95 *IVE* 🐾🐾

Question of Honor Under governmental pressures, an honest narcotics cop decides to inform on his department's corruption. Co-written and produced by ex-cop Sonny Grosso, on who ''The French Connection'' was based. Made for television and based on the book, ''Point Blank,'' by Grosso and Philip Rosenberg, with script by Budd Schulberg.
1980 134m/C Ben Gazzara, Paul Sorvino, Robert Vaughn, Tony Roberts, Danny Aiello, Anthony Zerbe; *Dir:* Judd Taylor. VHS, Beta $59.95 *VCL* 🐾🐾½

Q

A Question of Love Fine performances in this well-done television movie about two lesbians—one of whom is fighting her ex-husband for custody of their child. **1978 100m/C** Gena Rowlands, Jane Alexander, Ned Beatty, Clu Gulager, Bonnie Bedelia, James Sutorius, Jocelyn Brando; *Dir:* Jerry Thorpe. **VHS** $29.95 *USA* ⅃⅃⅃

Question of Silence Three women, strangers to each other, stand trial for a murdering a man. Intricately analyzed courtroom drama with much to say about male domination. Foreign title is "De Stilte Rond Christine M." Subtitled and dubbed versions available. **1983 92m/C** *NL* Cox Habrema, Nelly Frijda, Henriette Tol, Edda Barends; *Dir:* Morleen Gorris. **VHS, Beta** $59.95 *FRI, SUE* ⅃⅃⅃

Qui Etes Vous, Mr. Sorge? The story of German journalist Richard Sorge, who was hanged by the Japanese in 1944 for being a Soviet spy. Uninteresting docudrama which focuses on the opinions of various witnesses as to the accuracy of the espionage charge. In French with English subtitles. **1961 130m/B** *FR* Thomas Holtzman, Keiko Kishi; *Dir:* Andre Girard. **VHS, Beta** $19.95 *KRT* ⅃⅃

Quick Change Murray, Davis, and Quaid form bumbling trio of New York bank robbers who can't seem to exit the Big Apple with loot. Based on Jay Cronley's book, it's Murray's directing debut (with help from screenwriter Franklin). Engaging minor caper comedy displaying plenty of NYC dirty boulevards. **1990 (R) 89m/C** Bill Murray, Geena Davis, Randy Quaid, Jason Robards Jr., Bob Elliott, Victor Argo, Kathryn Grody; *Dir:* Bill Murray. **VHS, Beta, LV, 8mm** $92.98 *WAR* ⅃⅃½

The Quick and the Dead Based on a Louis L'Amour story, this made-for-cable western features a lone gunslinger who protects a defenseless settler's family from the lawless West, only to become a source of sexual tension for the wife. **1987 91m/C** Sam Elliott, Tom Conti, Kate Capshaw, Kenny Morrison, Matt Clark; *Dir:* Robert Day. **VHS, Beta** $79.95 *LHV, WAR* ⅃⅃

Quick, Let's Get Married Quick, let's not watch this movie. Even the people who made it must have hated it: they waited seven years to release it. Rogers runs a whorehouse; Eden is a gullible, pregnant prostitute; Gould in his big-screen debut(!) is a deaf-mute. A must-not see. Also known as "Seven Different Ways." **1971 96m/C** Ginger Rogers, Ray Milland, Barbara Eden, Carl Schell, Michael Ansara, Walter Abel, Scott Meyer, Cecil Kellaway, Elliott Gould; *Dir:* William Dieterle. **VHS, Beta, LV** *WGE* ⅃⅃

Quicker Than the Eye When a magician gets mixed up in an assassination plot, he must use his wit, cunning, and magic to get out of it. **1988 94m/C** Ben Gazzara, Mary Crosby, Catherine Jarrett, Ivan Desny, Eb Lottimer, Sophie Carle, Wolfram Berger, Dinah Hinz, Jean Yanne; *Dir:* Nicolas Gessner. **VHS, Beta** $79.95 *ACA* ⅃⅃

Quicksand Mechanic Rooney borrows $20 from his boss's cash register, intending to return it. One thing leads to another, the plot thickens, and it's downhill (for Rooney) from there. Good, tense suspense drama. **1950 79m/B** Mickey Rooney, Peter Lorre, Jeanne Cagney; *Dir:* Irving Pichel. **VHS, Beta** $14.95 *NOS, QNE, BUR* ⅃⅃½

Quicksilver A young stockbroker loses all, then quits his job to become a city bicycle messenger. Pointless, self-indulgent yuppie fantasy. Songs by Roger Daltrey and Ray Parker, Jr. **1986 (PG) 106m/C** Kevin Bacon, Jami Gertz; *Dir:* Tim Donnelly. **VHS, Beta, LV** $79.95 *COL* ⅃½

Quiet Cool When his former girlfriend's family is killed, a New York cop travels to a sleepy California town run by a pot-growing tycoon. He subsequently kills the bad guys in a Rambo-like fit of vengeance. Dopey exercise in sleepy-eyed bloodshed. **1986 (R) 80m/C** James Remar, Daphne Ashbrook, Adam Coleman Howard, Jared Martin, Fran Ryan; *Dir:* Clay Borris. **VHS, Beta, LV** $79.95 *COL* ⅃

Quiet Day In Belfast Tragedy occurs when northern Irish patriots and British soldiers clash in an Irish betting parlor. Kidder plays a duel role: an Irish woman in love with a British soldier and the woman's twin sister. Convincingly adapted from a Canadian stage play. **1974 92m/C** *CA* Barry Foster, Margot Kidder; *Dir:* Milad Bessada. **VHS, Beta** $59.95 *MED* ⅃⅃⅃

The Quiet Earth Serious science fiction film about a scientist who awakens to find himself seemingly the only human left on earth as the result of a misfired time/space government experiment. He later finds two other people, a girl and a Maori tribesman, and must try to repair the damage in order to save what's left of mankind. **1985 (R) 91m/C** *NZ* Bruno Lawrence, Alison Routledge, Peter Smith; *Dir:* Geoff Murphy. **VHS, Beta, LV** $19.98 *FOX* ⅃⅃

Quiet Fire A health-club owner tries to get the goods on the arms-dealing congressman who killed his best friend. Hilton-Jacobs looks properly pumped up since his sweathog days on TV's "Welcome Back Kotter." **1991 (R) 100m/C** Lawrence Hilton-Jacobs, Robert Z'Dar, Nadia Marie, Karen Black, Lance Lindsey; *Dir:* Lawrence Hilton-Jacobs. **VHS** $79.95 *PMH* ⅃½

A Quiet Hope Seven veterans of the Vietnam War discuss how they were affected by their experiences, and how they overcame the war's invisible wounds. **1990 58m/C VHS** $39.95 *LGC, BPG* ⅃½

The Quiet Man The classic incarnation of Hollywood Irishness, and one of Ford's best, and funniest films. Wayne is Sean Thornton, a weary American ex-boxer who returns to the Irish hamlet of his childhood and tries to take a spirited lass as his wife, despite the strenuous objections of her brawling brother (Malaglen). Thornton's aided by the leprechaun-like Fitzgerald and the local parish priest, Bond. A high-spirited and memorable film filled with Irish stew, wonderful banter, and shots of the lush countryside. Oscar nominations: Best Picture, Best Actor (McLaglen), Screenplay. Listen for the Scottish bagpipes at the start of the horse race, a slight geographic inaccuracy. **1952 129m/C** John Wayne, Maureen O'Hara, Barry Fitzgerald, Victor McLaglen, Arthur Shields, Jack MacGowran, Ward Bond, Mildred Natwick; *Dir:* John Ford. Academy Awards '52: Best Color Cinematography, Best Director (Ford); National Board of Review Awards '52: Best Film; Venice Film Festival '52: International Prize (Ford). **VHS, LV** $19.98 *REP, TLF, BUR* ⅃⅃⅃

Quiet Thunder A hard-drinking bush pilot in Africa and a beautiful senator's wife are thrown together on the run after both witness an assassination. Mindless, thoroughly derivative "adventure." **1987 (PG-13) 94m/C** Wayne Crawford, June Chadwick, Victor Steinbach; *Dir:* David Rice. **VHS, Beta** $79.95 *IVE* ⅃

Quigley Down Under A Western sharpshooter moves to Australia in search of employment. To his horror, he discovers that he has been hired to kill aborigines. Predictable action is somewhat redeemed by the terrific chemistry between Selleck and San Giacomo and the usual enjoyable theatrics from Rickman as the landowner heavy. **1990 (PG-13) 121m/C** *AU* Tom Selleck, Laura San Giacomo, Alan Rickman, Chris Haywood, Ron Haddrick, Tony Bonner, Roger Ward, Ben Mendelsohn; *Dir:* Simon Wincer. **VHS, Beta, LV, 8mm** $92.99 *MGM* ⅃⅃½

The Quiller Memorandum An American secret agent travels to Berlin to uncover a deadly neo-Nazi gang. Refreshingly different from other spy tales of its era. Good screenplay by Harold Pinter taken from Adam Hall's novel, "The Berlin Memorandum." **1966 103m/C** George Segal, Senta Berger, Alec Guinness, Max von Sydow, George Sanders; *Dir:* Michael Anderson Sr. **VHS, Beta** $59.98 *FOX* ⅃⅃⅃

Quilombo The title refers to a legendary settlement of runaway slaves in 17th-century Brazil; an epic chronicles its fortunes as leadership passes from a wise ruler to a more militant one who goes to war against the government. Stunning scenery, tribal images, and folk songs, but the numerous characters seldom come to life as personalities. One of Brazil's most expensive films, in Portuguese with English subtitles. **1984 114m/C** *BR* Vera Fischer, Antonio Pompeo, Zeze Motta, Toni Tornado; *W/Dir:* Carlos Diegues. **VHS** $79.95 *NYF, FCT* ⅃⅃½

Quincy Jones: A Celebration Jones conducts a full orchestra playing his most famous music, with special guest stars. Songs include "Just Once," "Baby Come To Me" and "One-Hundred Ways." **1983 60m/C** Quincy Jones, James Ingram, Ray Charles. **VHS, Beta** $24.95 *VHE, HHE* ⅃⅃½

Quintet Atypical Altman sci-fi effort. The stakes in "Quintet," a form of backgammon, are high—you bet your life. Set during the planet's final ice age. Newman and wife Fossey wander into a dying city and are invited to play the game, with Fossey losing quickly. Bizarre and pretentious, with heavy symbolic going. **1979 118m/C** Paul Newman, Bibi Andersson, Fernando Rey, Vittorio Gassman, David Langton, Nina Van Pallandt, Brigitte Fossey; *Dir:* Robert Altman. **VHS, Beta** $19.98 *FOX* ⅃⅃

Quo Vadis One of the first truly huge Italian silent epics. The initial adaptation of the Henryk Sienkiewicz novel about Nero and ancient Rome. Probably the cinema's first great financial success. **1912 45m/B** *Dir:* Enrico Guazzoni. **VHS, Beta** $89.95 *FCT* ⅃⅃½

Quo Vadis Larger-than-life production about Nero and the Christian persecution. Done on a giant scale: features exciting fighting scenes, romance, and fabulous costumes. Definitive version of the classic novel by Henryk Siekiewicz. Remade for television (Italian) in 1985. Oscar nominations: Best Picture, Best Supporting Actor (Ustinov),

Color Cinematography, Color Art Direction, Score of a Comedy or Drama, and Best Color Film Editing. Taylor has a cameo, Loren is an extra.
1951 171m/C Robert Taylor, Deborah Kerr, Peter Ustinov, Patricia Laffan, Finlay Currie, Abraham Sofaer, Marina Berti, Buddy Baer, Felix Aylmer, Nora Swinburne, Sophia Loren, Elizabeth Taylor; *Dir:* Mervyn LeRoy, *Nar:* Walter Pidgeon. **VHS, Beta $29.98** MGM, APD 𝄢𝄢𝄢

Quo Vadis Third screen version of Henryk Sienkiewicz's book. See the other two first. This one is slow and perfunctory. Brandauer is memorable as Nero; Quinn (in the lead) is the son of Anthony.
1985 (R) 122m/C *IT* Klaus Maria Brandauer, Frederic Forrest, Christina Raines, Maria Therese Relin, Francesco Quinn, Barbara de Rossi, Phillippe LeRoy, Max von Sydow; *Dir:* Franco Rossi. **VHS, Beta, LV $89.95** PSM, APD 𝄢𝄢½

Rabbit Test In Rivers' first directorial effort, a clumsy virginal guy becomes the world's first pregnant man. So irreverent it's almost never in good taste, and so poorly written it's almost never funny.
1978 (PG) 86m/C Billy Crystal, Roddy McDowall, Imogene Coca, Paul Lynde, Alex Rocco, George Gobel; *Dir:* Joan Rivers. **VHS, Beta, LV $9.95** COL, SUE 𝄢½

Rabid A young girl undergoes a radical plastic surgery technique and develops a strange and unexplained lesion in her armpit. She also finds she has an unusual craving for human blood.
1977 (R) 90m/C *CA* Marilyn Chambers, Frank Moore, Joe Silver; *Dir:* David Cronenberg. **VHS, Beta $29.98** WAR 𝄢½

Rabid Grannies Octogenarians rip flesh from young bodies with their gnashing teeth. Only for fans of gory Troma trash.
1989 (R) 89m/C **VHS, Beta, LV $79.95** MED Woof!

Race with the Devil Vacationers are terrorized by devil worshippers after they witness a sacrificial killing. Heavy on car chases; light on plot and redeeming qualities. Don't waste your time.
1975 (PG) 88m/C Peter Fonda, Warren Oates, Loretta Swit, Lara Parker; *Dir:* Jack Starrett. **VHS, Beta $14.98** FOX, FCT 𝄢½

Race for Glory A young motorcyclist builds a super-bike he hopes will help him win a world championship race. Will he win? Can you stand the suspense? Let's just say there's a happy ending.
1989 (R) 102m/C Alex MacArthur, Peter Berg, Pamela Ludwig; *Dir:* Rocky Lang. **VHS, Beta $89.99** HBO 𝄢½

Race for Life A race car driver attempts to make a comeback despite objections from his wife. He races around Europe (great scenery); she leaves him; he tries to win her back and salvage his career. Standard, un-gripping story. Car scenes are great.
1955 69m/B *GB* Richard Conte, Mari Aldon, George Coulouris; *Dir:* Terence Fisher. **VHS, Beta, LV** WGE 𝄢½

Race to the Yankee Zephyr Bad guy Peppard and good guys Pleasance and Wahl race to recover $50 million in gold from a downed World War II plane. Confused, ill plotted and predictable.
1981 108m/C *AU NZ* Ken Wahl, Lesley Ann Warren, Donald Pleasence, George Peppard, Bruno Lawrence, Grant Tilly; *Dir:* David Hemmings. **VHS** VES 𝄢𝄢

Race for Your Life, Charlie Brown Another in the popular series of "Peanuts" character films. This one features Charlie Brown, Snoopy, and all the gang spending an exciting summer in the American wilderness.
1977 (G) 76m/C **VHS, Beta, LV $14.95** PAR 𝄢𝄢½

The Racers Douglas brings power to the role of a man determined to advance to the winners' circle. Exciting European location photography, but not much plot.
1955 112m/C Gilbert Roland, Kirk Douglas, Lee J. Cobb, Cesar Romero, Bella Darvi; *Dir:* Henry Hathaway. **VHS $39.98** FOX, FCT 𝄢𝄢½

The Rachel Papers Based on the Martin Amis novel, this is the funny/sad tale of an Oxford youth who plots via his computer the seduction of a beautiful American girl. For anyone who has ever loved someone just out of their reach.
1989 (R) 92m/C Dexter Fletcher, Ione Skye, James Spader, Jonathan Pryce, Bill Paterson, Michael Gambon; *Dir:* Damien Harris. **VHS, Beta $79.98** FOX 𝄢𝄢𝄢

Rachel, Rachel Rachel teaches by day, wearing simple, practical dresses and her hair up. By night she caters to her domineering mother by preparing refreshments for her parties. This sexually repressed, spinster schoolteacher, however, gets one last chance at romance in her small Connecticut town. Woodward mixes just the right amounts of loneliness and sweetness in the leading role. A surprising award-winner that was an independent production of Newman. Based on Margaret Laurence's "A Jest of God." Oscar nominations for Woodward, Parsons, and adapted screenplay.
1968 (R) 102m/C Joanne Woodward, James Olson, Estelle Parsons, Geraldine Fitzgerald, Donald Moffatt; *Dir:* Paul Newman. National Board of Review Awards '68: 10 Best Films of the Year; New York Film Critics Awards '68: Best Actress (Woodward), Best Director. **VHS, Beta $19.98** WAR 𝄢𝄢𝄢½

Rachel River A divorced radio personality struggles to make something of her life in her small Minnesota town.
1989 (PG-13) 88m/C Pamela Reed, Craig T. Nelson, Viveca Lindfors, James Olson. **VHS $89.95** ACA 𝄢𝄢½

Rachel and the Stranger A God-fearing farmer declares his love for his wife when a handsome stranger (Mitchum) nearly woos her away. Well-cast, well-paced, charming Western comedy-drama.
1948 93m/B Loretta Young, Robert Mitchum, William Holden, Gary Gray; *Dir:* Norman Foster. **VHS, Beta $19.95** MED 𝄢𝄢

Rachel's Man A big-screen version of the Biblical love story of Jacob and Rachel.
1975 115m/C Mickey Rooney, Rita Tushingham, Leonard Whiting; *Dir:* Moshe Mizrahi. **VHS, Beta $59.95** LIV, LTG 𝄢

Racing with the Moon A sweet, nostalgic film about two buddies awaiting induction into the Marines in 1942. They have their last chance at summer romance. Benjamin makes the most of skillful young actors and conventional story. Great period detail. Written by Steven Kloves.
1984 (PG) 108m/C Sean Penn, Elizabeth McGovern, Nicolas Cage, John Karlen, Rutanya Alda, Max Showalter, Crispin Glover, Page Hannah, Michael Madsen, Carol Kane; *Dir:* Richard Benjamin. **VHS, Beta, LV $14.95** PAR 𝄢𝄢𝄢

The Racket Police captain Mitchum tries to break up mob racket of gangster Ryan. Internecine strife on both sides adds complexity. Mitchum and especially Ryan are super; fine, tense melodrama.
1951 88m/B Robert Ryan, Robert Mitchum, Lizabeth Scott, Ray Collins, William Conrad, Don Porter; *Dir:* John Cromwell. **VHS, Beta, LV $19.98** MED 𝄢𝄢𝄢

Racket Squad, Vol. 1 An episode from the popular television show "Racket Squad," entitled "Flying Fish." Details the life of a meek accountant who answers an ad which promises an escape from his humdrum lifestyle.
1952 55m/B Reed Hadley. **VHS, Beta $19.99** HEG, BUR 𝄢½

The Racketeer A racketeer falls in love with a pretty girl and attempts to win her over. As part of his plan, the gangster arranges to help the girl's boyfriend begin his musical career in exchange for the girl's promise to marry him.
1929 68m/B Carole Lombard, Robert Armstrong, Hedda Hopper; *Dir:* Howard Higgin. **VHS, Beta $16.95** SNC, NOS, CAB 𝄢𝄢½

Racketeers of the Range A cattleman fights a crooked attorney who wants to sell his client's stock to a large meat packing company.
1939 62m/B George O'Brien, Marjorie Reynolds, Chill Wills, Ray Whitley; *Dir:* Ross Lederman. **VHS, Beta** MED 𝄢½

Racquet Tennis pro Convy searches for true love and a tennis court of his own in Beverly Hills. Lame, sophomoric, and unfunny.
1979 87m/C Bert Convy, Edie Adams, Lynda Day George, Phil Silvers, Bobby Riggs, Bjorn Borg, Tanya Roberts; *Dir:* David Winters. **VHS, Beta $49.95** VHE 𝄢

Rad Teenage drama revolving around BMX racing and such dilemmas as: can the good guys beat the bad guys, who's got the fastest bike, must you cheat to win, and should our hero miss his SAT tests to compete in "the big race.." Directed by stunt-expert Needham.
1986 (PG) 94m/C Bill Allen, Bart Conner, Talia Shire, Jack Weston, Lori Loughlin; *Dir:* Hal Needham. **VHS, Beta, LV $9.98** SUE 𝄢½

Radar Men from the Moon Commando Cody, with his jet-pack, fights to defend the earth from invaders from the moon. Also released as a feature, "Retik, the Moon Menace." Twelve-episode serial on two tapes. Silly sci-fi.
1952 152m/B George Wallace, Aline Towne, Roy Barcroft, William Bakewell, Clayton Moore; *Dir:* Fred Brannon. **VHS, LV $49.95** NOS, REP, SNC 𝄢½

Radar Secret Service Two servicemen witness the hijacking of a truck loaded with nuclear material.
1950 59m/B John Howard, Adele Jergens, Tom Neal, Ralph Byrd. **VHS, Beta, LV** WGE 𝄢½

The Radical Spacek is convincing as a middle class student who becomes involved with a radical political group, eventually becoming a terrorist.
1975 100m/C Julie Kavner, Hector Elias, Jenny Sullivan, Sissy Spacek, Art Carney, Henry Winkler, Jane Wyatt. **VHS $9.95** SIM 𝄢½

Radio Days A lovely, unpretentious remembrance of the pre-television radio culture. Allen tells his story in a series of vignettes centering around his youth in

Brooklyn, his eccentric extended family, and the legends of radio they all followed. The ubiquitous Farrow is a young singer hoping to make it big.
1987 (PG) 96m/C Mia Farrow, Dianne Wiest, Julie Kavner, Michael Tucker, Wallace Shawn, Josh Mostel, Tony Roberts, Jeff Daniels, Kenneth Mars, Seth Green, William Magerman, Diane Keaton, Renee Lippin, Danny Aiello, Gina DeAngelis, Kitty Carlisle, Mercedes Ruehl; **W/Dir:** Woody Allen; **Nar:** Woody Allen. **VHS, Beta, LV** $19.95 *HBO* 🐾🐾🐾

Radio Flyer Its 1969 and Mike and Bobby have just moved to northern California with their divorced mom. Everything would seem to be idyllic if only their mother hadn't decided to marry a drunken child abuser who beats Bobby whenever the mood strikes him. Mike decides to help Bobby escape by turning their Radio Flyer wagon into a magic rocketship that will carry Bobby to safety—and lead to further tragedy. The film's version of childhood dreams is appealing but the child abuse angle (toned down though it was) is abruptly and unsatisfactorily handled.
1991 (PG-13) 114m/C Elijah Wood, Joseph Mazzello, Lorraine Bracco, Adam Baldwin, John Heard, Ben Johnson; **Dir:** Richard Donner. **VHS, LV** *COL* 🐾🐾

Radio Patrol Plenty of action and thrills abound in this complete collection of all the chapters of this vintage serial. Pinky Adams, radio cop, is assisted by his trusty canine partner, Irish (Silverwolf). A cop's best friend is his dog.
1937 235m/B Mickey Rentschler, Adrian Morris, Monte Montague, Silver Wolf, Grant Withers, Catherine Hughes; **Dir:** Ford Beebe. **VHS, Beta** $49.95 *NOS, VDM, MLB* 🐾🐾

Radio Ranch Gene Autry and friends are up against an underground world that is complete with robots and death rays. Movie version of the serial "Phantom Empire."
1935 80m/B Gene Autry, Frankie Darro, Betsy King Ross, Smiley Burnette; **Dir:** Otto Brewer. **VHS, Beta** $19.95 *NOS, VYY, DVT* 🐾🐾½

Radioactive Dreams Surreal, practically senseless fantasy wherein two men, trapped in a bomb shelter for 15 years with nothing to read but mystery novels, emerge as detectives into a post-holocaust world looking for adventure.
1986 (R) 94m/C John Stockwell, Michael Dudikoff, Lisa Blount, George Kennedy, Don Murray, Michelle Little; **Dir:** Albert Pyun. **VHS, Beta, LV** $79.98 *LIV, VES* 🐾🐾

Rafael en Raphael The great Spanish singer/actor Rafael is seen on tour in several countries. In Spanish.
1975 90m/C Rafael. **VHS, Beta** *TSV* 🐾🐾

Rafferty & the Gold Dust Twins Two women kidnap a motor vehicle inspector at gunpoint in Los Angeles and order him to drive to New Orleans. En route, they become pals. Wandering, pointless female buddy flick. Also called "Rafferty and the Highway Hustlers."
1975 (R) 91m/C Alan Arkin, Sally Kellerman, MacKenzie Phillips, Charles Martin Smith, Harry Dean Stanton; **Dir:** Dick Richards. **VHS, Beta** $19.98 *WAR* 🐾½

Rage Scott's directorial debut. A military helicopter sprays peace-loving rancher Scott and his son with nerve gas. The son's death is covered up; the rancher embarks on a vengeful rampage. Scott's sudden transformation to avenging killer is implausible and overwrought.

1972 (PG) 100m/C George C. Scott, Richard Basehart, Martin Sheen, Barnard Hughes, Kenneth Tobey, Ed Lauter, Nicholas Beauvy, Dabbs Greer; **Dir:** George C. Scott. **VHS, Beta** $59.95 *WAR* 🐾🐾

Rage A well-acted, well-written, nerve-wracking drama about a convicted rapist who finds help through difficult therapy. Focuses on how the sex-offender discovers the "whys" behind his assaults. Made for television.
1980 100m/C David Soul, James Whitmore, Yaphet Kotto, Caroline McWilliams, Vic Tayback, Sharon Farrell, Craig T. Nelson, Garry Walberg; **Dir:** William A. Graham. **VHS, Beta** $59.95 *WAR* 🐾🐾🐾

Rage of Angels Lengthy television miniseries adaptation of the Sidney Sheldon novel about an ambitious female attorney, her furs, and her men. Later sequelled by "Rage of Angels: The Story Continues" (1986).
1985 192m/C Jaclyn Smith, Ken Howard, Armand Assante, Ronald Hunter, Kevin Conway, George Coe, Deborah May; **Dir:** Buzz Kulik. **VHS, Beta** $19.95 *COL* 🐾

Rage at Dawn An outlaw gang is tracked by a special agent who must "bend" the rules a little in order to get the bad guys. Not surprisingly, he gets his girl as well. A solid standard of the genre, with some clever plot twists.
1955 87m/C Randolph Scott, Forrest Tucker, Mala Powers, J. Carroll Naish, Edgar Buchanan; **Dir:** Tim Whelan. **VHS** $13.95 *NOS, GHV, BUR* 🐾🐾½

A Rage in Harlem The crime novels of Chester A. Himes were translated into the best movies of the early '70s blaxploitation era. Now, a Himes story gets the big budget Hollywood treatment with juice and aplomb. A voluptuous lady crook enters Harlem circa 1950 with a trunkful of stolen gold sought by competing crooks, and the chase is on, with one virtuous soul (Whitaker) who only wants the girl. Great cast and characters, much humor, but unsparing in its violence. heart out.
1991 (R) 108m/C Forest Whitaker, Gregory Hines, Robin Givens, Zakes Mokae, Danny Glover; **Dir:** Bill Duke. **VHS, LV** $92.99 *HBO, PNR* 🐾🐾🐾

Rage of Honor A high-kicking undercover cop seeks vengeance on bad guys for his partner's murder. Standard of its type; why bother?
1987 (R) 92m/C Sho Kosugi, Lewis Van Bergen, Robin Evans, Gerry Gibson; **Dir:** Gordon Hessler. **VHS, Beta** $19.95 *MED* 🐾

Rage to Kill Linguistic retelling of the 1982 Grenada invasion, complete with helpless American med. students. Good-guy race car driver whips them into shape to fight the Soviet-supplied general (he's the bad guy). Guess who wins.
1988 (R) 94m/C Oliver Reed, James Ryan, Henry Cele, Cameron Mitchell; **Dir:** David Winters. **VHS, Beta** $29.95 *AIP* 🐾½

Rage of the Master During the Ching dynasty, good and bad guys bound about, kicking each other and themselves in the head and temples.
197? 100m/C Ng Ming Tsai, Tiger Yang. **VHS, Beta** $19.95 *UHV* 🐾

The Rage of Paris Scheming ex-actress and head waiter hope to gain by helping a beautiful French girl (Darrieux) snag a rich hubby. She comes to her senses, realizing

that love, true love, matters more than wealth. Well-acted, quaint comedy.
1938 78m/B Danielle Darrieux, Douglas Fairbanks Jr., Mischa Auer, Helen Broderick; **Dir:** Henry Koster. **VHS, Beta** $19.95 *NOS, DVT, CAB* 🐾🐾🐾

Rage of Wind A master boxer seeks revenge on a Japanese commandant who kidnaps his wife during World War II.
1982 97m/C Chen Hsing, Yasuka Kurata, Irene Ryder. **VHS, Beta** *WVE* 🐾

Raggamuffin Tek' Over Top reggae stars in performance.
1987 60m/C Ragamuffin, Take Over. **VHS, Beta** *KRV* 🐾

Raggedy Ann and Andy: A Musical Adventure Here are the fun-filled exploits of America's favorite fictional dolls, transformed into an enchanting animated musical. Includes 16 songs by Joe Raposo.
1977 (G) 87m/C Dir: Richard Williams; **Voices:** Didi Conn, Joe Silver. **VHS, Beta** $14.98 *FOX* 🐾

Raggedy Man Spacek in her signature role as a lonely small-town woman. Here she's raising two sons alone in a small Texas town during World War II. Spacek's strong acting carries a well-scripted story, unfortunately marred by an overwrought ending.
1981 (PG) 94m/C Sissy Spacek, Eric Roberts, Sam Shepard, Tracy Walter, William Sanderson, Henry Thomas; **Dir:** Jack Fisk. **VHS, Beta, LV** $59.95 *MCA* 🐾🐾🐾

The Raggedy Rawney A young Army deserter dresses in women's clothing and hides out as a mad woman with a band of gypsies. Good first directing effort by English actor Bob Hoskins. Unpretentious and engaging.
1990 (R) 102m/C Bob Hoskins, Dexter Fletcher, Zoe Nathenson, David Hill, Ian Dury, Zoe Wanamaker; **W/Dir:** Bob Hoskins. **VHS** $19.98 *CAN* 🐾½

Ragin' Cajun Retired kickboxer is forced into a death match to save his girlfriend.
1990 91m/C David Heavener, Charlene Tilton, Sam Bottoms, Samantha Eggar, Dr. Death; **Dir:** William B. Hillman. **VHS** $79.95 *AIP, HHE* 🐾

Raging Bull Scorsese's depressing but magnificent vision of the dying American Dream and suicidal macho codes in the form of the rise and fall of middleweight boxing champ Jake LaMotta, a brutish, dull-witted animal who can express himself only in the ring and through violence. A photographically expressive, brilliant drama, with easily the most intense and brutal boxing scenes ever filmed. De Niro provides a vintage performance, going from the young LaMotta to the aging has-been, and is ably accompanied by Moriarty as his girl and Pesci as his loyal, much beat-upon bro. Script by Paul Schrader and Mardik Martin.
1980 (R) 128m/B Robert De Niro, Cathy Moriarty, Joe Pesci, Frank Vincent; **Dir:** Martin Scorsese. Academy Awards '80: Best Actor (De Niro), Best Film Editing; Los Angeles Film Critics Association Awards '80: Best Actor (De Niro), Best Film; National Board of Review Awards '80: Best Actor (De Niro), Best Supporting Actor (Pesci). **VHS, Beta, LV, 8mm** $29.98 *FOX, MGM, VYG* 🐾🐾🐾🐾

Rags to Riches A wealthy Beverly Hills entrepreneur decides to improve his public image by adopting six orphan girls. TV pilot.
1987 96m/C Joseph Bologna, Tisha Campbell. **VHS, Beta** $19.95 *STE, NWV* 🐾

R

Ragtime The lives and passions of a middle-class family weave into the scandals and events of 1906 America. A small, unthinking act represents all the racist attacks on one man, who refuses to back down this time. Wonderful period detail. Music composed by Randy Newman. From the E.L. Doctorow novel, but not nearly as complex. Features Cagney's last film performance.
1981 (PG) 156m/C Howard E. Rollins Jr., Kenneth McMillan, Brad Dourif, Mary Steenburgen, James Olson, Elizabeth McGovern, Pat O'Brien, James Cagney, Debbie Allen, Jeff Daniels, Moses Gunn, Donald O'Connor, Mandy Patinkin, Norman Mailer; *Dir:* Milos Forman. **VHS, Beta, LV $29.95** PAR ♂♂♂

Raid on Entebbe Dramatization of the Israeli rescue of passengers held hostage by terrorists at Uganda's Entebbe Airport in 1976. A gripping actioner all the more compelling because true. Finch received an Emmy nomination in this, his last film. Made for television.
1977 (R) 113m/C Charles Bronson, Peter Finch, Horst Buchholz, John Saxon, Martin Balsam, Jack Warden, Yaphet Kotto, Sylvia Sydney; *Dir:* Irvin Kershner. **VHS, Beta $69.95** HBO ♂♂♂

Raid on Rommel A British soldier (Burton) poses as a Nazi and tries to infiltrate Rommel's team with his rag-tag brigade of misfits. Predictable drivel. Contains action footage from the 1967 film "Tobruk."
1971 (PG) 98m/C Richard Burton, John Colicos, Clinton Greyn, Wolfgang Preiss; *Dir:* Henry Hathaway. **VHS, Beta $19.95** MCA ♂ 1/2

Raiders in Action Tale following the activities of a band of allied commandos in the closing days of the war.
1971 89m/C Paul Smith. **VHS, Beta $59.95** TWE ♂

Raiders of Atlantis Battles break out when the lost continent surfaces in the Caribbean. The warriors in these apocalyptic frays deploy atomic arsenals.
197? 100m/C Christopher Connelly; *Dir:* Roger Franklin. **VHS, Beta $69.95** PSM ♂

Raiders of the Border A white-hatted cowboy must prevent the takeover of a trading post.
1944 55m/B Johnny Mack Brown, Raymond Hatton, Ellen Hall. **VHS, Beta $19.95** VDM ♂

Raiders of the Buddhist Kung Fu Conspiracy to restore Manchu dynasty results in deadly duel between Nat Kit of the Manchu dynasty and Wong Lung of the new Buddhist government.
1984 40m/C VHS, Beta $14.95 NEG ♂♂

Raiders of the Lost Ark Breathless neo-classic '30s-style adventure made the now-ubiquitous Ford a household name. Intrepid archeologist Indiana Jones in his first and best adventure battles mean Nazis, decodes hieroglyphics and hates snakes in his search for the biblical Ark of the Covenant. He stops in Kathmandu en route to Egypt and teams up with old flame Allen who says, "Indiana Jones. I always knew some day you'd come walkin' back into my life." "I'm making it up as I go along," he admits to her once when they're in a bind; the whole story is wonderfully, playfully picaresque. Every chase and stunt tops the last, though "Raiders" avoids the excess and gore of its sequel, "Indiana Jones and the Temple of Doom." Allen is fetching as the feisty gal. Of course there's romance, and of course the good guys win (sort of).
1981 (PG) 115m/C Harrison Ford, Karen Allen, Wolf Kahler, Paul Freeman, John Rhys-Davies, Denholm Elliott; *Dir:* Steven Spielberg. Academy Awards '81: Best Art Direction/Set Decoration, Best Film Editing, Best Sound, Best Visual Effects. **VHS, Beta, LV, 8mm $19.95** PAR ♂♂♂♂

Raiders of Red Gap A cattle company tries running homesteaders off their land to get control of it. The Lone Rider saves the day.
1943 56m/B Al "Fuzzy" St. John, Bob Livingston. **VHS, Beta $19.98** DVT, VYY, VCN ♂ 1/2

Raiders of the Sun After the Earth has been ruined in a biological disaster, a futuristic warrior arrives to help restore world peace and order.
1992 (R) 80m/C Richard Norton, Rick Dean, William Steis, Blake Boyd, Brigitta Stenberg; *Dir:* Cirio H. Santiago. **VHS $89.98** NHO ♂♂

Raiders of Sunset Pass Wartime western quickie makes an interesting novelty today. With most cowboys off fighting the Axis, the ladies form the Women's Army of the Plains to watch out for 4-F rustlers.
1943 57m/B Eddie Dew, Smiley Burnette, Jennifer Holt, Roy Barcroft, Mozelle Cravens, Beverly Aadland, Nancy Worth, Kenne Duncan, Jack Rockwell, Budd Buster, Jack Ingram; *Dir:* John English. **VHS $19.95** NOS ♂♂

Railroaded The police seek a demented criminal who kills his victims with perfumed-soaked bullets. Tense, excellent noir Anothony Mann crime drama.
1947 72m/B John Ireland, Sheila Ryan, Hugh Beaumont, Ed Kelly; *Dir:* Anthony Mann. **VHS, Beta, LV $19.95** STE, NOS, NWV ♂♂

Railroader Buster Keaton as the railroader in this slapstick short fumbles his way across Canada. As in the days of the silents, he speaks not a word.
1965 25m/C Buster Keaton. New York Film Festival '65: Best Visual Effects. **VHS, Beta** NFC ♂♂♂

Railway Children At the turn of the century in England, the father of three children is framed and sent to prison during Christmas. The trio and their mother must survive on a poverty-stricken farm near the railroad tracks. They eventually meet a new friend who helps them prove their father's innocence. Wonderfully directed by Jeffries. From the classic Edith Nesbitt children's novel.
1970 104m/C GB Jenny Agutter, William Mervyn, Bernard Cribbins, Dinah Sheridan, Iain Cuthbertson; *Dir:* Lionel Jeffries. **VHS, Beta $79.99** HBO ♂♂♂ 1/2

Rain Somerset Maugham's tale of a puritanical minister's attempt to reclaim a "lost woman" on the island of Pago Pago. Crawford and Huston work up some static. Remake of the 1928 silent film "Sadie Thompson." Remade again in 1953 as "Miss Sadie Thompson."
1932 92m/B Joan Crawford, Walter Huston, William Gargan, Guy Kibbee, Beulah Bondi, Walter Catlett; *Dir:* Lewis Milestone. **VHS, Beta $8.95** NEG, NOS, UHV ♂♂ 1/2

The Rain Killer Big city cop and fed don goloshes to track serial killer who slays rich ladies during heavy precipitation. Soggy story.
1990 (R) 94m/C Ray Sharkey, David Beecroft, Maria Ford, Woody Brown, Tania Coleridge; *Dir:* Ken Stein. **VHS, LV $89.95** COL ♂

Rain Man When his father dies, ambitious and self-centered Charlie Babbit finds he has an older autistic brother who's been institutionalized for years. Needing him to claim an inheritance, he liberates him from the institution and takes to the road, where both brothers undergo subtle changes. The Vegas montage is wonderful. Critically acclaimed drama and a labor of love for the entire cast. Cruise's best performance to date as he goes from cad to recognizing something wonderfully human in his brother and himself. Hoffman is exceptional.
1988 (R) 128m/C Dustin Hoffman, Tom Cruise, Valeria Golino, Jerry Molen, Jack Murdock, Michael D. Roberts, Ralph Seymour, Lucinda Jenney, Bonnie Hunt, Kim Robillard, Beth Grant; *Dir:* Barry Levinson. Academy Awards '88: Best Actor (Hoffman), Best Director (Levinson), Best Original Screenplay, Best Picture; Berlin Film Festival '88: Golden Bear, Honorary Golden Bear (Hoffman). **VHS, Beta, LV, 8mm $19.98** MGM, BTV, JCF ♂♂♂ 1/2

Rain People Pregnant housewife Knight takes to the road in desperation and boredom; along the way she meets retarded ex-football player Caan. Well directed by Coppola from his original script. Pensive drama.
1969 (R) 102m/C Shirley Knight, James Caan, Robert Duvall; *Dir:* Francis Ford Coppola. **VHS, Beta $59.95** WAR ♂♂♂

Rainbow Broadway's "Annie" is badly miscast as Judy Garland from her early years in vaudeville to her starring years at M.G.M. Based on the book by Christopher Finch. Made for television; directed by sometime Garland flame Jackie Cooper.
1978 100m/C Andrea McArdle, Jack Carter, Don Murray; *Dir:* Jackie Cooper. **VHS, Beta, LV $49.95** VHE ♂

The Rainbow Mature, literate rendering of the classic D.H. Lawrence novel about a young woman's sexual awakening. Beautiful cinematography. Companion/prequel to director Russell's earlier Lawrence adaptation, "Women in Love" (1969).
1989 (R) 104m/C GB Sammi Davis, Amanda Donohoe, Paul McGann, Christopher Gable, David Hemmings, Glenda Jackson, Kenneth Colley; *Dir:* Ken Russell. **VHS, LV $49.95** VES, IME ♂♂♂

Rainbow Bridge The adventures of a group of hippies searching for their consciousness in Hawaii. Features concert footage from Jimi Hendrix's final performance.
1971 (R) 74m/C *Dir:* Chuck Wein. **VHS, Beta $19.95** AHV, RHI ♂

Rainbow Drive Made-for-cable cop thriller promises great things but fails to deliver. Wellers is a good cop trapped in the political intrigues of Hollywood. He discovers five dead bodies; when the official count is four, he detects funny business. Music by Tangerine Dream.
1990 (R) 93m/C Peter Weller, Sela Ward, Bruce Weitz; *Dir:* Bobby Roth. **VHS $89.95** BTV, VMK ♂♂ 1/2

The Rainbow Gang A trio of unlikely prospectors heads into a legendary mine in search of riches and fame.
1973 90m/C Donald Pleasence, Don Calfa, Kate Reid. **VHS, Beta $19.95** STE, NWV ♂

Rainbow Goblins Story In a live concert at Budokan, Masayoshi Takanaka performs the music he composed to interpret a book called "Rainbow Goblins Story."
1981 60m/C Masayoshi Takanaka. **VHS, Beta, LV $19.95** PAR, PNR ♂

Rainbow Parade Animator/director Burt Gillette's "Rainbow Parade" contains some of the most subtly textured color work to be found in animation. Features cartoon characters "Felix," "Molly Moo Cow," and "the Toonerville Folks."
193? 50m/C VHS, Beta *CCB* 🐾

Rainbow Parade 2 A collection of three animated classics from Burt Gillette including "Molly Moo Cow and Robinson Crusoe," "Trolley Ahoy," and "Waifs Welcome."
1936 21m/C VHS, Beta *CCB* 🐾

Rainbow Ranch A Navy boxer returns home to his ranch to find murder and corruption. Needless to say, he rides off to seek revenge.
1933 54m/B Rex Bell, Cecilia Parker, Robert Kortman, Henry Hall, Gordon DeMain; *Dir:* Harry Fraser. VHS $19.95 *NOS, DVT* 🐾½

Rainbow: The Final Cut The veteran rock band performs their greatest hits.
1986 60m/C VHS, Beta $19.95 *MVD, COL* 🐾½

Rainbow's End Gibson must defend Gale, the female rancher, from an evil adversary who is trying to run Gale and her invalid father off of their land. Things get complicated when Gibson discovers that Stark works for his father!
1935 54m/B Hoot Gibson, June Gale, Oscar Apfel, Ada Ince, Charles Hill, Warner Richmond; *Dir:* Norman Spencer. VHS $19.95 *NOS, DVT* 🐾🐾

The Rainmaker The music and Hepburn were nominated for Academy Awards, although both she and Lancaster were a little long in the tooth for the parts. Reminiscent of "Elmer Gantry" in his masterful performance, Lancaster makes it all believable as a con man who comes to a small midwestern town and works miracles not only on the weather but on spinster Hepburn.
1956 121m/C Burt Lancaster, Katharine Hepburn, Wendell Corey, Lloyd Bridges, Earl Holliman, Cameron Prudhomme, Wallace Ford; *Dir:* Joseph Anthony. VHS, LV $14.95 *PAR, CCB* 🐾🐾🐾

Raintree County A lavish, somewhat overdone epic about two lovers caught up in the national turmoil of the Civil War. An Indiana teacher marries a southern belle just after the outbreak of war. The new wife battles mental illness. Producers had hoped this would be another "Gone With the Wind." Adapted from the novel. Oscar nominations for actress Taylor, Art Direction and Set Decoration, Score and Costume Design.
1957 175m/C Elizabeth Taylor, Montgomery Clift, Eva Marie Saint, Lee Marvin, Nigel Patrick, Rod Taylor, Agnes Moorehead, Walter Abel; *Dir:* Edward Dmytryk. Harvard Lampoon Awards '57: Worst Film of the Year. VHS, Beta, LV $29.95 *MGM, PIA* 🐾🐾½

Rainy Day A successful actress' return home for her father's funeral brings back unpleasant childhood memories that force her to face the truth about herself.
1979 35m/C Mariette Hartley. VHS, Beta *LCA* 🐾🐾

Raise the Titanic A disaster about a disaster. Horrible script cannot be redeemed by purported thrill of the ship's emergence from the deep after 70 years. It's a shame, because the free world's security hangs in the balance. Based on Clive Cussler's best seller.

1980 (PG) 112m/C Jason Robards Jr., Richard Jordan, Anne Archer, Alec Guinness, J.D. Cannon; *Dir:* Jerry Jameson. VHS, Beta, LV $29.98 *FOX* 🐾

A Raisin in the Sun Outstanding story of a black family trying to make a better life for themselves in an all-white neighborhood in Chicago. The characters are played realistically and make for a moving story. Each person struggles with doing what he must while still maintaining his dignity and sense of self. Based on the Broadway play by Lorraine Hansberry.
1961 128m/B Diana Sands, John Fiedler, Ivan Dixon, Louis Gossett Jr., Sidney Poitier, Claudia McNeil, Ruby Dee; *Dir:* Daniel Petrie. National Board of Review Awards '61: 10 Best Films of the Year, Best Supporting Actress (Dee). VHS, Beta, LV $14.95 *COL* 🐾🐾🐾🐾

A Raisin in the Sun An "American Playhouse," made-for-TV presentation of the Lorraine Hansberry play about a black family threatened with dissolution by the outside forces of racism and greed when they move to an all-white neighborhood in the 1950s.
1989 171m/C Danny Glover, Esther Rolle, Starletta DuPois; *Dir:* Bill Duke. VHS, Beta $69.95 *FRH* 🐾🐾🐾

Raising Arizona Hi's an ex-con and the world's worst hold-up man. Ed's a policewoman. They meet, fall in love, marry, and kidnap a baby (one of a family of quints). Why not? Ed's infertile and the family they took the baby from has "more than enough," so who will notice? But unfinished furniture tycoon Nathan Arizona wants his baby back, even if he has to hire an axe-murderer on a motorcycle to do it. A brilliant, original comedy written by Coen and brother Ethan. Narrated in notorious loopy deadpan style by Cage. Innovative camerawork by Barry Sonnenfeld. Wild, surreal and hilarious.
1987 (PG-13) 94m/C Nicolas Cage, Holly Hunter, John Goodman, William Forsythe, Randall "Tex" Cobb, Trey Wilson, M. Emmet Walsh, Frances McDormand; *W/Dir:* Joel Coen. VHS, Beta, LV $19.98 *FOX* 🐾🐾🐾½

Ramblin' Man Con artists Selleck and Reed alienate victims and viewers.
198? (R) 100m/C Tom Selleck, Jerry Reed, Morgan Fairchild, Barbara Mandrell, Claude Akins, Ray Stevens; *Dir:* Burt Kennedy. VHS *EDV, HHE, MTX* 🐾🐾

Rambo: First Blood, Part 2 If anyone can save Our Boys still held prisoner in Asia it's John Rambo. Stallone stars in the sequel to "First Blood" (1982); followed by "Rambo 3" (1988).
1985 (R) 93m/C Sylvester Stallone, Richard Crenna, Charles Napier, Steven Berkoff, Julia Nickson, Martin Kove; *Dir:* George P. Cosmatos. VHS, Beta, LV $14.95 *HBO, LIV* 🐾½

Rambo: Part 3 John Rambo, the famous Vietnam vet turned Buddhist monk, this time invades Afghanistan to rescue his mentor. Filmed in Israel. Typical Rambo-style, but not as action-packed as the first two. At the time, the most expensive film yet made, costing $58 million. Co-scripted by Stallone.
1988 (R) 102m/C Sylvester Stallone, Richard Crenna, Marc De Jonge, Kurtwood Smith, Spiros Focas; *Dir:* Peter McDonald. VHS, Beta, LV $14.95 *IVE, LIV* 🐾

Ramparts of Clay A young woman in Tunisia walks the line between her village's traditional way of life and the modern world just after her country's independence from France. Brilliantly shot on location; exquisitely poignant.

1971 (PG) 87m/C *FR Dir:* Jean-Louis Bertucelli. VHS, Beta $59.95 *COL* 🐾🐾🐾

Ramrod Lake is a tough ranch owner at odds with her father, who is being manipulated by a big-time cattleman into trying to put them out of business. She fights back, and McCrea is caught in the middle as the only good guy. Nothing special.
1947 94m/B Veronica Lake, Joel McCrea, Arleen Whelan, Don DeFore, Preston Foster, Charlie Ruggles, Donald Crisp, Lloyd Bridges; *Dir:* Andre de Toth. VHS $14.95 *REP* 🐾🐾

The Ramrodder An adult western spoof.
1969 92m/C VHS, Beta $19.95 *VDM* 🐾

Ran The culmination of Kurosawa's life's work stands as one of the most auspicious and awesome films ever made. He loosely adapted Shakespeare's "King Lear," with plot elements from "Macbeth," and has fashioned an epic, heartbreaking statement about honor, ambition and the futility of war. Unanimously acclaimed. Costumes alone took three years to create. In Japanese with English subtitles.
1985 (R) 160m/C *JP* Tatsuya Nakadai, Akira Terao, Jinpachi Nesu, Daisuke Oka, Meiko Harada, Hisashi Igawa; *Dir:* Akira Kurosawa. Academy Awards '85: Best Costume Design; New York Film Critics Awards '85: Best Foreign Film; National Society of Film Critics '85: Best Film. VHS, Beta, LV $29.98 *FOX, APD* 🐾🐾🐾🐾

Rana: The Legend of Shadow Lake Gold at the bottom of a lake is guarded by a frog-monster, but treasure hunters try to retrieve it anyway.
1975 96m/C Alan Ross, Karen McDiarmid, Jerry Gregoris. VHS, Beta $24.95 *AHV* 🐾

Ranch An executive who loses everything inherits a dilapidated ranch and renovates it into a health spa.
1988 (PG-13) 97m/C Andrew Stevens, Gary Fjellgaard, Lou Ann Schmidt, Elizabeth Keefe; *Dir:* Stella Stevens. VHS, Beta $14.95 *HMD, SOU* 🐾🐾

Rancho Notorious Kennedy, on the trail of his girlfriend's murderer, falls for dance hall girl Dietrich. Fine acting. A "period" sample of 50's westerns, but different. A must for Dietrich fans.
1952 89m/C Marlene Dietrich, Arthur Kennedy, Mel Ferrer, William Frawley, Jack Elam, George Reeves; *Dir:* Fritz Lang. VHS, Beta, LV $15.95 *UHV* 🐾🐾🐾

Random Harvest A masterful, tearjerking film based on the James Hilton novel. A shell-shocked World War I amnesiac meets and is made happy by a beautiful music hall dancer. He regains his memory and forgets about the dancer and their child. This is Garson's finest hour, and a shamelessly potent sobfest. Oscar nominations: Best Picture, Best Actor (Coleman), Best Supporting Actress (Peters), Best Director, Screenplay, Black & White Interior Decoration, and Best Score of a Dramatic Film.
1942 126m/B Greer Garson, Ronald Colman, Reginald Owen, Philip Dorn, Susan Peters, Henry Travers, Margaret Wycherly, Bramwell Fletcher; *Dir:* Mervyn LeRoy. VHS, Beta $19.98 *MGM* 🐾🐾🐾

Randy Rides Alone Very young Wayne is good in this slappable but pleasant B effort. Wayne single-handedly cleans up the territory and rids the land of a passel o'bad guys.
1934 53m/B John Wayne, Alberta Vaughan, George "Gabby" Hayes; *Dir:* Henry Frazer. VHS $12.98 *REP, NOS, VCN* 🐾🐾

Range Busters Range Busters are called in to find the identity of the phantom killer.
1940 55m/B Ray Corrigan, Max Terhune. **VHS, Beta $19.95** *NOS, VCN, DVT* 🎬

The Range Feud/Two Fisted Law Two early, short Wayne oaters which reveal the Duke developing the drawl, gait and pleasingly profane style that would carry him to fame. In "Feud" he is a young rancher accused of murdering a cattle rustler. "Law" finds him playing a character named Duke who loses his land and girl to a neighboring land grabber.
1931 112m/C Buck Jones, Tim McCoy, John Wayne, Susan Fleming, Alice Fay, Walter Brennan; *Dir:* Ross Lederman. **VHS $19.95** *COL* 🎬🎬

Range Law Another entry in the infamous Maynard series of horseplay, cliche, and repetitive plot elements.
1931 60m/B Ken Maynard. **VHS, Beta $19.95** *NOS, UHV, DVT* 🎬

Range Rider Eight episodes of the series are on four tapes, including "Greed Rides the Range," "Silver Blade" and "Fatal Bullet."
1952 60m/B Jock Mahoney. **VHS, Beta** *DVT* 🎬

Range Riders A gunman poses as a wimp and cleans out an outlaw gang.
1935 46m/B Buddy Roosevelt. **VHS, Beta** *WGE* 🎬

Rangeland Empire Shamrock and Lucky are at it again, implicated as being members of an outlaw gang until that gang attacks them.
1950 59m/B James Ellison, Russell Hayden, Stanley Price, John Cason. **VHS, Beta, LV** *WGE* 🎬

Rangeland Racket The Lone Rider (Houston) has been wrongly accused of a crime.
1941 60m/B George Houston, Hillary Brooke, Al "Fuzzy" St. John. **VHS, Beta** *VYY* 🎬½

Ranger and the Lady Roy Rogers, a Texas Ranger, romances the woman who is the leader of a wagon train.
1940 54m/B Roy Rogers, George "Gabby" Hayes; *Dir:* Joseph Kane. **VHS, Beta $19.95** *NOS, DVT, KRT* 🎬½

Ranger's Roundup The Rangers prove their courage by rounding up an outlaw gang.
1938 57m/B Fred Scott, Al "Fuzzy" St. John. **VHS, Beta $19.95** *NOS, DVT* 🎬

Rangers Take Over Gunlords are driven out by the Texas Rangers.
1943 62m/B Dave O'Brien, James Newill. **VHS, Beta $19.95** *NOS, VYY, DVT* 🎬

Rank & File: Live The punkers perform "Amanda Ruth," "The Wreck," "Lucky Day," and others at the Target studio.
1981 30m/C **VHS, Beta $34.95** *MVD, TGV* 🎬

Ransom A group of wealthy citizens decides to take on an assassin who stalks a resort town. Also known as "Maniac" and "Assault on Paradise."
1984 90m/C Oliver Reed, Deborah Raffin, Stuart Whitman, John Ireland; *Dir:* Richard Compton. **VHS, Beta $69.98** *LIV, VES* 🎬🎬

Ransom Money A kidnapping scheme involving millions of dollars, in and around the Grand Canyon, backfires.
1988 87m/C Broderick Crawford, Rachel Romen, Gordon Jump, Randy Whipple; *Dir:* DeWitt Lee. **VHS, Beta $9.95** *SIM, TAV* 🎬½

Ranson's Folly Barthelmess makes a wager that he can impersonate a famous outlaw well enough to rob a stage with only a pair of scissors. When the army paymaster is killed, guess who gets caught with his swash unbuckled?
1926 80m/B Richard Barthelmess, Dorothy Mackaill, Anders Randolf, Pat Hartigan, Brooks Benedict; *Dir:* Sidney Olcott. **VHS, Beta $27.95** *DNB* 🎬🎬½

Rap Master Ronnie Reagan-era social phenomena are targeted in a series of musical sketches/parodies. Based on an off-Broadway hit.
1988 60m/C Jim Morris, Carol Kane, Jon Cryer, Dick Smothers, Tom Smothers, Tom Smothers; *Dir:* Jay Dublin. **VHS, Beta $39.99** *HBO, TLF* 🎬½

Rape Two guys investigate a third pal's mysterious death. They discover that his supernatural-power-imbued girlfriend is at the root of the matter.
198? 90m/C Rick Joss, Gil Vidal. **VHS, Beta $59.95** *MGL* Woof!

Rape of Love The story begis with one of the most chilling rape scenes on film. The film attempts to analyze the emotional impact of rape on its victim. Well-acted and directed. French with English subtitles.
1979 117m/C *FR* Nathalie Nell, Alain Foures; *Dir:* Yannick Bellon. **VHS, Beta $59.95** *COL* 🎬🎬½

Rape and Marriage: The Rideout Case The true story of an Oregon wife who accused her husband of rape. Explores thoughtfully the legal and moral questions raised by the case. Interpreted well enough, though not superbly, by a decent cast. Rourke is too intense. Made for television.
1980 96m/C Mickey Rourke, Linda Hamilton, Rip Torn, Eugene Roche, Conchata Ferrell, Gail Strickland, Bonnie Bartlett, Alley Mills; *Dir:* Peter Levin. **VHS, Beta $49.95** *IVE* 🎬🎬½

Rape of the Sabines The story of Romulus, king of Rome, and how he led the Romans to capture the women of Sabina. The battles rage, the women plot, and Romulus fights and lusts. Dubbed in English.
1961 101m/C *IT FR* Roger Moore, Mylene Demongeot, Jean Marais; *Dir:* Richard Pottier. **VHS, Beta $29.95** *UNI* Woof!

Rapid Fire Cheap, made-for-video quickie about a good guy U.S. agent who battles terrorists. Easy to skip.
1989 90m/C Joe Spinell, Michael Wayne, Ron Waldron; *Dir:* David A. Prior. **VHS, Beta $79.95** *AIP* 🎬

Rappaccini's Daughter Adaptation of a Nathaniel Hawthorne short story about a doctor's daughter who is cursed with the ability to kill anything she touches. Originally made for television.
1980 59m/C Kathleen Beller, Kris Tabori, Michael Egan; *Dir:* Dezso Magyar; *Hosted by:* Henry Fonda. **VHS, Beta $24.95** *MON, MTI, KAR* 🎬½

Rappin' Ex-con Van Peebles gets into it with the landlord and a street gang leader. Forgettable action/music drivel.
1985 (PG) 92m/C Mario Van Peebles, Tasia Valenza, Harry Goz, Charles Flohe; *Dir:* Joel Silberg. **VHS, Beta $79.95** *MGM* 🎬

The Rapture A beautiful telephone operator engages in indiscrimate sexual adventures to relieve the boredom of her job and life. She becomes curious by, and eventually converted to, evangelical Christianity, which leads her to a contented marriage and the birth of a daughter. When her husband is tragically killed she becomes convinced that she and her child will be taken by God into heaven if she only waits for the proper sign.
1991 (R) 100m/C Mimi Rogers, David Duchovny, Patrick Bauchau, Will Patton; *W/Dir:* Michael Tolkin. **VHS, LV** *NLC* 🎬🎬

Rapunzel A beautiful young woman locked in a tall tower is saved by the handsome prince who climbs her golden tresses. Part of Shelly Duvall's "Faerie Tale Theatre" series.
1982 60m/C Shelley Duvall, Gena Rowlands, Jeff Bridges. **VHS, Beta, LV $14.95** *KUI, FOX, FCT* 🎬🎬

The Rare Breed Plodding but pleasant Western. A no-strings ranch hand (Stewart) agrees to escort a Hereford Bull to Texas, where the widow of an English breeder plans to crossbreed the bull with longhorn cattle. The widow (O'Hara) insists that she and her daughter accompany Stewart on the trip, which features every kind of western calamity imaginable. When all others believe the attempt to crossbreed has failed, Stewart sets out to prove them wrong.
1966 97m/C James Stewart, Maureen O'Hara, Brian Keith, Juliet Mills, Jack Elam, Ben Johnson; *Dir:* Andrew V. McLaglen. **VHS, Beta $19.95** *MCA* 🎬🎬½

A Rare Breed Real-life story of a kidnapped horse in Italy and a young girl's quest to retrieve it. Directed by David Nelson of television's "Ozzie and Harriet" fame, this movie is cute and old-fashioned.
1981 (PG) 94m/C George Kennedy, Forrest Tucker, Tracy Vaccaro, Tom Hallick, Don DeFore; *Dir:* David Nelson. **VHS, Beta $19.95** *STE, IVE* 🎬½

Rascal Dazzle Montage-like tribute to that best-loved group of children, the Our Gang kids. Narrated by Jerry Lewis. Includes scenes with Spanky, Alfalfa, Darla and the rest of the gang.
1981 100m/B *Nar:* Jerry Lewis. **VHS, Beta, LV $39.98** *SUE* 🎬

The Rascals An irrepressible youth at a rural Catholic boys' school comes of age.
1981 (R) 93m/C *FR* Thomas Chabrol; *Dir:* Bernard Revon. **VHS, Beta $29.98** *SUE* 🎬🎬½

Rashomon In 12th century Japan, two travelers attempt to discover the truth about an ambush-rape-murder. They get four completely different versions of the incident from the three people involved in the crime and the single witness. An insightful masterpiece that established Kurosawa and Japanese cinema as major artistic forces. Fine performances, particularly Mifune as the bandit. Visually beautiful and rhythmic. Remade as a western, "The Outrage," in 1964.
1951 83m/B *JP* Machiko Kyo, Toshiro Mifune, Massayura Mori; *Dir:* Akira Kurosawa. Academy Awards '51: Best Foreign Language Film; National Board of Review Awards '51: Best Director, Best Foreign Film; Venice Film Festival '51: Best Film (Lion of St. Mark); New York Times Ten Best List '51: Best Foreign Film. **VHS, Beta, LV $29.98** *SUE, CRC, VYG* 🎬🎬🎬🎬

Rasputin The long-censored and banned film of the story of the mad monk and his domination of the royal family before the Russian Revolution. Petrenko is superb. First released in the United States in 1988. In Russian with subtitles.
1985 104m/C *RU* Alexei Petrenko, Anatoly Romashin, Velta Linne, Alice Freindlikh; *Dir:* Elem Klimov. **VHS, Beta $79.95** *IFE, WBF* 🎬🎬🎬

Rat Pfink and Boo-Boo A parody on "Batman" in which a bumbling superhero and his sidekick race around saving people. Notoriously inept.
1966 72m/B Vin Saxon, Carolyn Brandt, Titus Moede, Mike Kannon; *Dir:* Ray Dennis Steckler. VHS, Beta **$29.95** *CAM*

The Rat Race A dancer and a musician venture to Manhattan to make it big, and end up sharing an apartment. Their relationship starts pleasantly and becomes romantic. Enjoyable farce. Well photographed and scripted, with the supporting characters stealing the show.
1960 105m/C Tony Curtis, Debbie Reynolds, Jack Oakie, Kay Medford, Don Rickles, Joe Bushkin; *Dir:* Robert Mulligan. VHS *PAR ⅃⅃⅃*

Ratboy An unscrupulous woman attempts to transform a boy with a rat's face into a celebrity, with tragic results. First directorial effort by Locke that gradually loses steam.
1986 (PG-13) 105m/C Sondra Locke, Sharon Baird, Robert Townsend; *Dir:* Sondra Locke. VHS, Beta **$79.95** *WAR ⅃⅃½*

Rate It X Sexism in a variety of national institutions, from pornography to consumerism, is revealed through the average man on the street via microphone interviews.
1986 (R) 95m/C *Dir:* Lucy Winer, Paula de Koenigsberg. VHS, Beta **$59.95** *IVE ⅃⅃½*

Ratings Game A bitter, out-of-work actor and a woman who works at the ratings service manage to mess up the television industry. Early directorial effort by DeVito is uneven but funny. Made for cable television. Also known as "The Mogul."
1984 102m/C Danny DeVito, Rhea Perlman, Gerrit Graham, Kevin McCarthy, Jayne Meadows, Steve Allen, Ronny Graham, George Wenolt; *Dir:* Danny DeVito. VHS, Beta, LV **$79.95** *PAR ⅃⅃*

Rats In 2225, the beleaguered survivors of a nuclear holocaust struggle with a mutant rodent problem.
1983 100m/C Richard Raymond, Richard Cross; *Dir:* Vincent Dawn. VHS, Beta **$69.95** *LTG ⅃⅃*

Rats are Coming! The Werewolves are Here! Blood-thirsty werewolves and carnivorous rats are featured in this thriller depicting the demented diversions of one delusion-ridden family.
1972 (R) 92m/C Hope Stansbury, Jackie Skarvellis; *Dir:* Andy Milligan. VHS, Beta *MID Woof!*

Rattle of a Simple Man A naive, chaste, middle-aged bachelor (Corbett) who lives in London must spend the night with a waitress (Cilento) to win a bet. She knows of the bet and kindly obliges. Enjoyable sex comedy.
1964 91m/B *GB* Harry H. Corbett, Diane Cilento, Michael Medwin, Thora Hird; *Dir:* Muriel Box. VHS, Beta **$59.99** *HBO ⅃⅃½*

Rattler Kid Cavalry sergeant wrongly accused of murdering his commanding officer escapes from prison and finds the real killer.
197? 86m/C Richard Wyler, Brad Harris. VHS, Beta **$49.95** *UNI ⅃*

Rattlers Made for television movie featuring poisonous rattlesnakes who attack at random.
1976 (PG) 82m/C Sam Chew, Elizabeth Chauvet, Dan Priest; *Dir:* John McCauley. VHS, Beta **$29.95** *USA ⅃*

The Raven A lunatic surgeon (Lugosi), who has a dungeon full of torture gadgets inspired by Edgar Allan Poe's stories, is begged by a man to save his daughter's life. The surgeon does, and then falls in love with the girl. But when she rejects his love (she's already engaged), he plans revenge in his chamber of horrors. Karloff plays the criminal who winds up ruining the mad doctor's plans. Lugosi is at his prime in this role, and any inconsistencies in the somewhat shaky script can be overlooked because of his chilling performance.
1935 62m/B Boris Karloff, Bela Lugosi, Irene Ware, Lester Matthews, Samuel S. Hinds; *Dir:* Lew Landers. VHS **$14.98** *MCA ⅃⅃⅃*

The Raven A small French town brimming with unique characters is turned on its ear when a poison pen writer begins revealing its secrets. Suspicion and paranoia run rampant. Suspenseful and excellently directed by Clouzot. In French with English subtitles.
1943 90m/B *FR* Pierre Fresnay, Pierre Larquey, Noel Roquevort, Antoine Belpetre; *Dir:* Henri-Georges Clouzot. VHS, Beta **$16.95** *SNC, MRV, INT ⅃⅃⅃*

The Raven This could-have-been monumental teaming of horror greats Karloff, Lorre and Price is more of a satire than a true horror film. One of the more enjoyable of the Corman/Poe adaptations, the movie takes only the title of the poem. As for the storyline: Price and Karloff play two rival sorcerers who battle for supremacy, with Lorre as the unfortunate associate turned into the bird of the title.
1963 86m/C Vincent Price, Boris Karloff, Peter Lorre, Jack Nicholson, Hazel Court; *Dir:* Roger Corman. VHS, Beta, LV *MLB ⅃⅃⅃*

Ravishing Idiot An unemployed bank clerk becomes mixed up with Soviet spies. Also known as "Agent 38-24-36," "Adorable Idiot," and "The Warm-Blooded Spy."
1964 99m/B *FR* Brigitte Bardot, Anthony Perkins; *Dir:* Edouard Molinaro. VHS, Beta *MON, OM ⅃⅃*

Raw Courage Three long-distance runners relax in New Mexico, but are kidnapped by vigilantes. Also called "Courage." Loosely based on a James Dickey novel.
1984 90m/C Ronny Cox, Lois Chiles, Art Hindle, Tim Maier, M. Emmet Walsh; *Dir:* Robert L. Rosen. VHS, Beta, LV **$9.95** *NWV, STE ⅃*

Raw Deal Don't ever give Schwarzenegger a raw deal! FBI agent Schwarzie infiltrates the mob and shoots lots of people. Action flick that
1986 (R) 106m/C Arnold Schwarzenegger, Kathryn Harrold, Darren McGavin, Sam Wanamaker, Paul Shenar, Steven Hill; *Dir:* John Irvin. VHS, Beta, LV **$14.95** *HBO, TLF ⅃½*

Raw Force Three karate buffs visit an island inhabited by a sect of cannibalistic monks who have the power to raise the dead. They fix the villains by kicking them all in the head. Also called "Shogun Island."
1981 (R) 90m/C *PH* Cameron Mitchell, Geoffrey Binney, John Dresden, John Locke, Ralph Lombardi; *Dir:* Edward Murphy. VHS, Beta **$49.95** *MED ⅃*

Raw Nerve A cast with possibilities can't overcome this poorly reasoned suspenser. A young man has visions of the local serial killer, but when he reports them to police he becomes the suspect. There are a few twists beyond that, none too thrilling.

1991 (R) 91m/C Glenn Ford, Traci Lords, Sandahl Bergman, Randall "Tex" Cobb, Ted Prior, Jan-Michael Vincent. VHS, Beta **$89.98** *AIP ⅃*

Raw Power Two thrill-seeking rich kids get their greedy hands on a super-car that defies any known power and gets them in loads of trouble. Some kids never learn.
19?? (R) 90m/C VHS *PMH ⅃*

Raw and Rough Pinchers Top reggae performers headline.
1987 60m/C Daddy Life, Danny Dread, Culture Martin, Little Rohan, Ram Stepper, Joe Lickshot. VHS, Beta *KRV ⅃*

Rawhead Rex An ancient demon is released from his underground prison in Ireland by a plowing farmer, and begins to decapitate and maim at will. Adapted by Clive Barker from his own short story. But Barker later disowned the film, so make your own judgment.
1987 (R) 89m/C David Dukes, Kelly Piper; *Dir:* George Pavlou. VHS, Beta, LV **$19.98** *LIV, VES ⅃½*

Rawhide Rancher's Protection Association forces landowners to knuckle under resulting in friction. Lou Gehrig plays a rancher in a western released the year before he died. Bandleader Ballew could be more gripping in the lead.
1938 60m/B Lou Gehrig, Smith Ballew, Evelyn Knapp; *Dir:* Ray Taylor. VHS, Beta **$19.98** *NOS, CCB, DVT ⅃⅃*

Rawhide Four escaped convicts hijack a stagecoach way station and hold hostages while waiting for a shipment of gold. A suspenseful B grade western with a bang-up ending. Well paced and cast with an abundance of good talent. Also called "Desperate Siege," a remake of "Show Them No Mercy" (1938).
1950 86m/B Tyrone Power Sr., Susan Hayward, Hugh Marlowe, Jack Elam, Dean Jagger, George Tobias, Edgar Buchanan, Jeff Corey; *Dir:* Henry Hathaway. VHS, Beta **$19.98** *FOX ⅃⅃⅃*

Rawhide Romance Bill finds justice and love in this chap-slappin' opus.
1934 47m/B Buffalo Bill Jr., Genee Boutell, Lafe McKee, Si Jenks; *Dir:* Victor Adamson. VHS, Beta, LV *WGE ⅃*

Ray Bradbury Theater, Volume 1 Two chilling tales from the pen of Ray Bradbury. A husband makes a duplicate of himself to solve his marital problems in "Marionettes," and a doting father is determined to protect his son from "The Playground."
1985 60m/C James Coco, William Shatner. VHS, Beta **$69.95** *BVV ⅃*

Ray Bradbury Theater, Volume 2 Another in the dramatic series of adaptations of Bradbury stories. Concerns a man who discovers a mysterious crowd that always assembles at traffic accidents.
1985 28m/C Nick Mancuso, R.H. Thompson. VHS, Beta **$39.95** *BVV ⅃⅃⅃*

Razorback A young American travels to the Australian outback in search of his missing journalist wife. During his quest, he encounters a giant killer pig that has been terrorizing the area. Firmly establishes that pigs, with little sense of natural timing, make lousy movie villains.
1984 (R) 95m/C *AU* Gregory Harrison, Bill Kerr, Arkie Whiteley, Judy Morris, Chris Haywood; *Dir:* Russell Mulcahy. VHS, Beta **$19.98** *WAR ⅃*

The Razor's Edge Adaptation of the Somerset Maugham novel. A rich young man spends his time between World Wars I and II searching for essential truth, eventually landing in India. A satisfying cinematic version of a difficult novel, supported by an excellent cast. Remade in 1984 with Bill Murray in the lead role. Oscar nominations for Best Picture, Best Supporting Actor (Webb), Black & White Interior Decoration.
1946 146m/B Tyrone Power Sr., Gene Tierney, Anne Baxter, Clifton Webb, Herbert Marshall, John Payne, Elsa Lanchester; **Dir:** Edmund Goulding. Academy Awards '46: Best Supporting Actress (Baxter). **VHS, Beta, LV** $59.98 FOX, BTV ✍✍✍

The Razor's Edge A beautifully filmed but idiosyncratic version of the Somerset Maugham novel. WWI-ravaged Larry Darrell combs the world in search of the meaning of life. Murray is a little too old and sardonic to seem really tortured. Russell is great as the loser he rescues. Subtly unique, with real shortcomings.
1984 (PG-13) 129m/C Bill Murray, Catherine Hicks, Theresa Russell, Denholm Elliott, James Keach, Peter Vaughan, Saeed Jaffrey, Brian Doyle-Murray; **Dir:** John Byrum. **VHS, Beta, LV** $79.95 COL ✍✍ 1/2

RCA's All-Star Country Music Fair A hoe-down recorded live at the 1982 Nashville Fan Fair, featuring a line-up of Nashville talent.
1982 82m/C Charlie Pride, Razzy Bailey, Sylvia, Earl Thomas Conley. **VHS, Beta** $9.98 COL ✍✍ 1/2

The Re-Animator Based on an H.P. Lovecraft story, this grisly film deals with a medical student who re-animates the dead. It has quickly turned into a black humor cult classic. Also available in an "R" rated version, and followed by "Bride of Reactor."
1985 86m/C Jeffrey Combs, Bruce Abbott, Barbara Crampton, David Gale, Robert Sampson; **Dir:** Stuart Gordon. **VHS, Beta, LV** $19.98 LIV, VES ✍✍✍

Reach for the Sky WW II flying ace loses both legs in an accident and learns to fly again. He becomes a hero during the Battle of Britain, then is shot down over France and held prisoner by the Germans. At war's end he returns to England to lead 3,000 planes over London in a victory flight. Inspirationaly true story.
1957 136m/B Kenneth More, Alexander Knox, Nigel Green; **Dir:** Lewis Gilbert. **VHS, Beta** LCA ✍✍ 1/2

Reaching For the Moon A day-dreaming department store employee finds himself in the unlikely position of ruling a kingdom. Vintage Fairbanks fun, based on a story by Anita Loos. Silent.
1917 91m/C Douglas Fairbanks Sr., Eileen Percy, Millard Webb; **Dir:** John Emerson. **VHS, Beta, 8mm** $29.95 VYY ✍✍ 1/2

Reaching for the Moon Fairbanks is a businessman who falls for liquor and Daniels on a transatlantic cruise. Dull and cruel; not as light-hearted as it intends. One Irving Berlin song sung by Crosby.
1931 62m/B Douglas Fairbanks Sr., Bebe Daniels, Bing Crosby, Edward Everett Horton; **Dir:** Edmund Goulding. **VHS, Beta, 8mm** $19.95 NOS, VYY, CAB ✍✍ 1/2

Reactor A low-budget sci-fi film about kidnapped scientists, alien ships and an activated nuclear reactor.

1985 90m/C Yanti Somer, Melissa Long, James R. Stuart, Robert Barnes, Nick Jordan. **VHS, Beta** $59.95 MGL ✍

Ready Steady Go, Vol. 1 First in a series of classic rock video collectibles from the 1960s. Contains 16 of the '60s top hits.
1983 60m/B Mick Jagger, Keith Richards, Charlie Watts, Ron Wood, Bill Wyman, Pete Townsend, Keith Moon, Roger Daltrey, John Entwhistle, John Lennon, Paul McCartney, Ringo Starr, George Harrison, The Animals, Gerry & the Pacemakers. **VHS, Beta, LV** $19.99 HBO, PIA ✍

Ready Steady Go, Vol. 2 This second volume highlights more original performances from the classic British rock television show of the sixties.
1985 60m/B Ringo Starr, George Harrison, Paul McCartney, John Lennon, Pete Townsend, Keith Moon, Roger Daltrey, John Entwhistle, Mick Jagger, Ron Wood, Bill Wyman, Charlie Watts, Keith Richards, The Beach Boys, Marvin Gaye, Gene Pitney, Jerry Lee Lewis, Rolling Stones. **VHS, Beta, LV** HBO, OM ✍

Ready Steady Go, Vol. 3 This volume features more clips from the British music series including the Beatles performing "She's A Woman" and The Rolling Stones singing "Little Red Rooster."
1984 57m/B Marvin Gaye, Martha & the Vandellas, Mick Jagger, Ron Wood, Keith Richards, Bill Wyman, Charlie Watts, John Lennon, Paul McCartney, Ringo Starr, George Harrison. **VHS, Beta, LV** HBO, OM ✍

A Real American Hero Made for TV story of Tennessee sheriff Buford Pusser, the subject of three "Walking Tall" films. Here, he battles moonshiners, with the usual violence. Also known as "Hard Stick."
1978 94m/C Brian Dennehy, Brian Kerwin, Forrest Tucker; **Dir:** Lou Antonio. **VHS, Beta** $9.99 STE, PSM ✍ 1/2

Real Bruce Lee Contains actual early films of Bruce Lee, once feared lost but recently discovered in the Chinese film archives, (the movies, not Bruce).
1979 (R) 108m/C Bruce Lee. **VHS, Beta** $24.95 GEM, SUN ✍ 1/2

The Real Buddy Holly Story Documentary serves as a tribute to the rock 'n' roll legend who met an untimely death in 1959.
1987 90m/C **Nar:** Paul McCartney. **VHS, Beta, LV** $19.95 SMV, IME, VTK ✍ 1/2

Real Bullets Stunt team performs the real thing when two members get imprisoned in a bad guy's desert castle.
1990 86m/C John Gazarian, Martin Landau; **Dir:** Lance Lindsay. **VHS, Beta, LV** $89.95 VMK ✍

Real Genius Brainy kids in California work with lasers on a class project that is actually intended for use as an offensive military weapon. When they learn of the scheme, they use their brilliant minds to mount an amusingly elaborate strategic defense initiative of their own. Eccentric characters and an intelligent script provide a bevy of laughs for the family.
1985 (PG) 108m/C Val Kilmer, Gabe Jarret, Jonathan Gries, Michelle Meyrink, William Atherton, Patti D'Arbanville, Severn Darden; **Dir:** Martha Coolidge. **VHS, Beta, LV** $14.95 COL ✍✍✍

The Real Glory After the U.S. capture of the Philippine islands during the Spanish-American war, an uprising of Moro tribesmen spreads terror. After most of the islands are evacuated only a small group of Army officers is left to lead the Filipino soldiers against

the rebels. Cooper plays the heroic doctor, who is not afraid to fight, especially when it comes to the rescue of the beseiged Army fort. Great action sequences and paticularly good performances by Cooper and Niven.
1939 95m/B Gary Cooper, Andrea Leeds, David Niven, Reginald Owen, Broderick Crawford; **Dir:** Henry Hathaway. **VHS** $19.98 MGM ✍✍✍

Real Life Writer/comedian Brooks's first feature sags in places, but holds its own as a vehicle for his peculiar talent. Brooks plays a pompous director whose ambition is to make a documentary of a "typical" American family. Good, intelligent comedy.
1979 (R) 99m/C Charles Grodin, Frances Lee McCain, Albert Brooks; **Dir:** Albert Brooks. **VHS, Beta** $49.95 PAR ✍✍✍

Real Men The junior Belushi brother is a spy forced to recruit ordinary guy Ritter to help him negotiate with aliens to save the world. Bizarre premise shows promise, but spy spoof doesn't fire on all comedic pistons. Nice try.
1987 (PG-13) 86m/C James Belushi, John Ritter, Barbara Barrie; **Dir:** Dennis Feldman. **VHS, Beta, LV** $79.98 FOX ✍ 1/2

The Real Patsy Cline The story of Patsy Cline's brilliant rise to stardom and her tragic death at the pinnacle of her career.
1990 60m/C Patsy Cline, Loretta Lynn, Carl Perkins, Dottie West. **VHS** $14.95 MVD, CAF ✍ 1/2

Really Weird Tales Several SCTV alumni highlight this three-story movie, which is science fiction with a satirical edge.
1986 85m/C John Candy, Martin Short, Joe Flaherty, Catherine O'Hara. **VHS, Beta** $7.00 VTR ✍✍ 1/2

Reap the Wild Wind DeMille epic about salvagers off the Georgia coast in the 1840s featuring Wayne as the captain and Milland and a giant squid as the villains. Good cast, lesser story, fine underwater photography.
1942 123m/C Ray Milland, John Wayne, Paulette Goddard, Raymond Massey, Robert Preston, Lynne Overman, Susan Hayward, Charles Bickford, Walter Hampden, Louise Beavers, Martha O'Driscoll, Elisabeth Risdon, Hedda Hopper, Raymond Hatton; **Dir:** Cecil B. DeMille. Academy Awards '42: Best Special Effects. **VHS, Beta** $29.95 MCA ✍✍ 1/2

Rear Window A newspaper photographer with a broken leg (Stewart) passes the time recuperating by observing his neighbors through the window. When he sees what he believes to be a murder, he decides to solve the crime himself. With help from his beautiful girlfriend and his nurse, he tries to catch the murderer without getting killed himself. Top-drawer Hitchcock blends exquisite suspense with occasional on-target laughs. Oscar nominations for Hitchcock's direction, Color Cinematography, and Sound Recording.
1954 112m/C James Stewart, Grace Kelly, Thelma Ritter, Wendell Corey, Raymond Burr; **Dir:** Alfred Hitchcock. Edgar Allan Poe Awards '54: Best Screenplay. **VHS, Beta, LV** $19.95 MCA, TLF ✍✍✍✍

Reason to Die Bounty hunter uses his own girlfriend as a decoy to trap a psychotic prostitute killer.
1990 (R) 86m/C Liam Cundill, Wings Hauser, Anneline Kriel, Arnold Vosloo; **Dir:** Tim Spring. **VHS, LV** $89.95 VMK ✍

A Reason to Live, A Reason to Die A bland Italian-French-German-Spanish western in which a group of condemned men attempt to take a Confederate fort. Also known as "Massacre at Fort Holman."
1973 90m/C James Coburn, Telly Savalas, Bud Spencer, Robert Burton; ***Dir:*** Tonino Valerii. **VHS, Beta $9.95** *NO* 🦴🦴

Reba McEntyre A collection of Reba's hits including, "Whoever's in New England," "Cathy's Clown," "The Last One to Know," and more.
1990 30m/C Reba McEntire. **VHS, LV $14.95** *PIA* 🦴🦴

Rebecca Based on Daphne du Maurier's best-selling novel about a young unsophisticated girl who marries a moody and prominent country gentleman haunted by the memory of his first wife. Fontaine and Olivier turn in fine performances as the unlikely couple. Suspenseful and surprising. Hitchcock's first American film and only Best Picture Oscar. Oscar nominations include: Best Actor (Olivier), Best Actress (Fontaine), Best Supporting Actress (Anderson), Best Director, Best Screenplay, Best Black & White Interior Decoration, Best Original Score, and Best Special Effects. Laserdisc features: rare screen tests of Vivien Leigh, Anne Baxter, Loretta Young, and Joan Fontaine, footage from Rebecca's winning night at the Academy Awards, original radio broadcasts of film by Orson Welles and David O. Selznick, and commentary of film with interview excerpts with Hitchcock.
1940 130m/B Joan Fontaine, Laurence Olivier, Judith Anderson, George Sanders, Nigel Bruce, Florence Bates, Gladys Cooper, Reginald Denny, Leo G. Carroll; ***Dir:*** Alfred Hitchcock. Academy Awards '40: Best Black and White Cinematography, Best Picture, National Board of Review Awards '40: 10 Best Films of the Year. **VHS, Beta, LV $19.98** *FOX, VYG, BTV* 🦴🦴🦴🦴

Rebecca of Sunnybrook Farm The original film version of the tale about an orphan who spreads sunshine and good cheer to all those around her. Silent with organ score by Gaylord Carter. Based on the popular novel by Kate Douglas Wiggin; remade with Shirley Temple in 1938.
1917 77m/B Mary Pickford, Eugene O'Brien, Marjorie Daw, Helen Jerome Eddy; ***Dir:*** Marshall Neilan. **VHS, Beta, LV $19.98** *CCB* 🦴🦴

Rebecca of Sunnybrook Farm Temple becomes a radio star over her aunt's objections in this bouncy musical that has nothing to do with the famous Kate Douglas Wiggin novel. Temple sings "Old Straw Hat," a medley of her hits, and dances the "Toy Trumpet" finale with Bill "Bojangles" Robinson.
1938 80m/B Shirley Temple, Randolph Scott, Jack Haley, Phyllis Brooks, Gloria Stuart, Slim Summerville, Bill Robinson, Helen Westley, William Demarest; ***Dir:*** Allan Dwan. **VHS, Beta, LV $19.98** *FOX, MLB* 🦴🦴½

Rebel A student radical must decide between his love for a country girl and his loyalty to an underground terrorist organization. One of Stallone's early vehicles.
1973 (PG) 80m/C Sylvester Stallone, Anthony Page, Rebecca Grimes. **VHS, Beta** *PGN* 🦴½

Rebel A U.S. Marine falls in love with a Sydney nightclub singer and goes AWOL during World War II. Stylish but empty and badly cast.
1986 (R) 93m/C *AU* Matt Dillon, Debbie Byrne, Bryan Brown, Bill Hunter, Ray Barrett; ***Dir:*** Michael Jenkins. **VHS, Beta, LV $79.98** *LIV, VES* 🦴½

Rebel High Hijinks occur at a high school where youngsters study guerilla warfare and teachers wear bullet-proof vests.
1988 (R) 92m/C Harvey Berger, Stu Trivax, Larry Gimple, Shawn Goldwater; ***Dir:*** Harry Jacobs. **VHS, Beta $29.95** *ACA* 🦴½

Rebel Love Love grows between a northern widow and a Confederate spy. In better hands it might have been a good story. Here, it's overwrought and embarrassing historical melodrama.
1985 84m/C Terence Knox, Jamie Rose, Fred Ryan; ***Dir:*** Milton Bagby Jr. **VHS, Beta $79.98** *LIV, VES* 🦴

Rebel Rousers A motorcycle gang wrecks havoc in a small town where a drag-race is being held to see who will get the pregnant girlfriend of Dern's high-school buddy as the prize. Young Nicholson in striped pants steals the show.
1969 81m/C Jack Nicholson, Cameron Mitchell, Diane Ladd, Bruce Dern, Harry Dean Stanton; ***Dir:*** Martin B. Cohen. **VHS, Beta $29.95** *MED, KOV* 🦴🦴

The Rebel Set The owner of a beat generation coffeehouse plans an armed robbery with the help of some buddies. Genuinely suspenseful, competently directed.
1959 72m/B Gregg Palmer, Kathleen Crowley, Edward Platt, John Lupton, Ned Glass, Don Sullivan, Vicki Dougan, I. Stanford Jolley; ***Dir:*** Gene Fowler. **VHS, Beta $16.95** *SNC* 🦴🦴½

Rebel Storm A group of freedom fighters team up to rescue America from the totalitarian rulers that are in charge in 2099 A.D.
1990 (R) 99m/C Zach Galligan, Wayne Crawford, June Chadwick, Rod McCary, John Rhys-Davies, Elizabeth Kiefer; ***Dir:*** Francis Schaeffer. **VHS, Beta, LV $29.95** *ACA* 🦴

Rebel Vixens Even though the war is over, ex-confederate soldiers continue looting and raping, until a brothel-full of whores concoct a plan that has a lot to do with softcore sex.
1985 94m/C Maria Lease, Roda Spain. **VHS, Beta $29.95** *MED Woof!*

Rebel Without a Cause James Dean's most memorable screen appearance. In the second of his three films, he plays a troubled teenager from the right side of the tracks. Dean's portrayal of Jim Stark, a teen alienated from both his parents and peers, is excellent. He befriends outcasts Wood and Mineo in a police station and together they find a commonground. Superb young stars carry this in-the-gut story of adolescence. All three leads met with real-life tragic ends.
1955 111m/C James Dean, Natalie Wood, Sal Mineo, Jim Backus, Nick Adams, Dennis Hopper, Ann Doran; ***Dir:*** Nicholas Ray. **VHS, Beta, LV $19.98** *WAR* 🦴🦴🦴🦴

The Rebellious Reign In an effort to resurrect the Ming Dynasty, a single soldier seeks to unseat the sovereign power, the Ching Dynasty.
19?? (R) 100m/C Jimmy Lee, Tsui Siu Keung. **VHS, Beta $49.95** *OCE* 🦴

The Rebels In the sequel to the mini-series "The Bastard," Philip Kent (Stevens) continues his Revolutionary War battle on the side of American independence. He is assisted by Southerner Judson Fletcher (Johnson) in thwarting an assassination attempt on George Washington. In addition to Tom Bosley's Ben Franklin, "The Rebels" offers Jim Backus as John Hancock. Followed by "The Seekers;" based on the novel by John Jakes.
1979 (PG) 190m/C Andrew Stevens, Don Johnson, Doug McClure, Jim Backus, Richard Basehart, Joan Blondell, Tom Bosley, Rory Calhoun, MacDonald Carey, Kim Cattrall, William Daniels, Anne Francis, Peter Graves, Pamela Hensley, Wilfrid Hyde-White, Nehemiah Persoff, William Smith, Forrest Tucker, Tanya Tucker, Robert Vaughn, Debi Richter; ***Dir:*** Russ Mayberry; ***Nar:*** William Conrad. **VHS $79.95** *MCA* 🦴🦴

Reborn A faith healer and a talent scout hire actors to be cured of fake ailments.
1978 91m/C Dennis Hopper, Michael Moriarty, Francisco Rabal, Antonella Murgia. **VHS, Beta $69.95** *VES* 🦴🦴

Reckless A sincere movie about a "good" girl who finds herself obsessed with a rebel from the wrong side of her small town. Differs from 1950s' wrong-side-of-the-track flicks only in updated sex and music.
1984 (R) 93m/C Aidan Quinn, Darryl Hannah, Kenneth McMillan, Cliff DeYoung, Lois Smith, Adam Baldwin, Dan Hedaya, Jennifer Grey, Pamela Springsteen; ***Dir:*** James Foley. **VHS, Beta, LV $24.95** *MGM* 🦴🦴½

Reckless Disregard A doctor whose career is ruined by a news report accusing him of involvement in a drug scam sues the newscaster. Made-for-TV saga based on Dan Rather's "60 Minutes" story and ensuing lawsuit.
1984 92m/C Leslie Nielsen, Tess Harper, Ronny Cox, Kate Lynch; ***Dir:*** Harvey Hart. **VHS, Beta $69.98** *LIV, VES* 🦴½

Reckless Moment A mother commits murder to save her daughter from an unsavory older man, and finds herself blackmailed. Gripping, intense thriller.
1949 82m/B James Mason, Joan Bennett, Geraldine Brooks; ***Dir:*** Max Ophuls. **VHS, Beta** *LCA* 🦴🦴🦴

The Reckless Way A young woman struggles for her big break into the movies.
1936 72m/B Marion Nixon, Kane Richmond, Inez Courtney, Malcolm McGregor, Harry Harvey, Arthur Howard; ***Dir:*** Raymond K. Johnson. **VHS, Beta** *GPV* 🦴½

Recruits Sophomoric gagfest about an inept Californian police force. Typical 80's-style mindless "comedy:" dumb gags ensue from perfunctory premise.
1986 (R) 81m/C Alan Deveau, Annie McAuley, Lolita David; ***Dir:*** Rafal Zielinski. **VHS, Beta, LV $79.98** *LIV, VES* 🦴

Rectangle & Rectangles The abstract nature of time and film itself is revealed in this computer-animated film.
1987 9m/C ***Dir:*** Rene Jodoin. **VHS, Beta** *NFC* 🦴

Red Alert Good, suspenseful, topical made-for-TV thriller about nuclear meltdown. Will Minneapolis be saved? Usually Barbeau's name in the credits is a red alert, but this one is better than most.
1977 95m/C William Devane, Ralph Waite, Michael Brandon, Adrienne Barbeau; ***Dir:*** Billy Hale. **VHS, Beta $19.95** *PAR* 🦴🦴½

The Red Badge of Courage John Huston's adaptation of the Stephen Crane Civil War novel is inspired, despite cutting-room hatchet job by the studio. A classic study of courage and cowardice. Sweeping battle scenes and intense personal drama.
1951 69m/B Audie Murphy, Bill Mauldin, Douglas Dick, Royal Dano, Andy Devine; ***Dir:*** John Huston. National Board of Review Awards '51: 10 Best Films of the Year. **VHS, Beta $24.95** *KUI, MGM* 🦴🦴🦴½

R

The Red Balloon The story of Pascal, a lonely French boy who befriends a wondrous red balloon which follows him everywhere. Lovely, finely done parable of childhood, imagination and friendship.
1956 34m/C FR Pascal Lamorisse; *Dir:* Albert Lamorisse. Academy Awards '56: Best Original Screenplay; National Board of Review Awards '57: 5 Best Foreign Films of the Year. **VHS, Beta, LV, 8mm $14.98** *KUI, HHT, DVT* 🎬🎬🎬½

Red Balloon/Occurrence at Owl Creek Bridge Two classic award-winning shorts have been combined on one tape: "The Red Balloon" (color, 1956) and "An Occurrence at Owl Creek Bridge." The former is a children's story about a balloon on the loose; the latter is a perfectly rendered adaptation of the Ambrose Bierce short story of the same title, a taut narration of a Civil War hanging.
1962 60m/C FR Pascal Lamorisse, Roger Jacquet; *Dir:* Robert Enrico, Albert Lamorisse. Academy Awards '56: Best Original Screenplay ("The Red Balloon"); Academy Awards '63: Best Live Action Short Film ("Occurence at Owl Creek Bridge"); British Academy Awards '62: Best Short Film ("Occurence at Owl Creek Bridge"); Cannes Film Festival '62: Blue Ribbon ("Occurence at Owl Creek Bridge"); National Board of Review Awards '57: 5 Best Foreign Films of the Year ("The Red Balloon"). **VHS, Beta** *SUE, VYY, WFV* 🎬🎬🎬½

Red Balloon/White Mane The two classic, award-winning children's films by the French master, released only on laserdisc in pristine transfers.
1956 72m/C *Dir:* Albert Lamorisse. **Beta, LV** *CRC* 🎬🎬🎬½

Red Barry A comic strip detective tries to track down criminals who threaten the world. Thirteen episodes on one videotape.
1938 ?m/B Buster Crabbe, Frances Robinson, Edna Sedgwick, Cyril Delevanti, Frank Lackteen; *Dir:* Ford Beebe, Alan James. **VHS $49.95** *NOS, MLB* 🎬🎬🎬½

Red Beard An uncharacteristic drama by Kurosawa, about a young doctor in feudal Japan awakening to life and love under the tutelage of a compassionate old physician. Highly acclaimed; in Japanese with English subtitles.
1965 185m/B JP Toshiro Mifune, Yuzo Kayama, Yoshio Tsuchiya, Reiko Dan; *Dir:* Akira Kurosawa. **VHS, Beta, LV $59.95** *TAM, VYG, CRC* 🎬🎬🎬

The Red and the Black A big-budgeted French adaptation of the classic Stendhal novel. In French with English subtitles.
1957 134m/B FR Danielle Darrieux, Gerard Philippe; *Dir:* Claude Autant-Lara. L'Academie Du Cinema '57: Best Actor (Phillipe), Best Actress (Darrieux), Best Film (Red and the Black). **VHS, Beta** *FCT, IVY, MRV* 🎬🎬🎬

Red Blooded American Girl A scientist develops a virus that turns people into vampires. Those infected hit the streets in search of blood.
1990 (R) 89m/C CA Christopher Plummer, Andrew Stevens, Heather Thomas, Kim Coates; *Dir:* David Blyth. **VHS, Beta $29.95** *PSM* 🎬

Red Dawn During World War III, Russian invaders overrun America's heartland and take over the country. Eight small-town teenagers, calling themselves the Wolverines, hide out in the rugged countryside and fight the Russians. Swayze and Grey met again in "Dirty Dancing."

1984 (PG-13) 114m/C Patrick Swayze, C. Thomas Howell, Harry Dean Stanton, Powers Boothe, Lea Thompson, Charlie Sheen, Ben Johnson, Jennifer Grey, Ron O'Neal, William Smith; *Dir:* John Milius. **VHS, Beta, LV $19.95** *MGM* 🎬🎬½

Red Desert The Pecos Kid is sent to track down the theft of a shipment of gold.
1950 60m/B Jack Holt, Donald (Don "Red") Barry, Tom Neal, Joseph Crehan, Tom London; *Dir:* Charles Marquis Warren. **VHS, Beta, LV $79.95** *WGE* 🎬

The Red Desert Antonioni's first color film, depicting an alienated Italian wife who searches for meaning in the industrial lunar landscape of northern Italy, to no avail. Highly acclaimed, and a masterpiece of visual form. In Italian with subtitles. Cinematography by Carlo di Palma.
1964 120m/C IT Monica Vitti, Richard Harris, Carlos Chionetti; *Dir:* Michelangelo Antonioni. Venice Film Festival '64: International Critics Award, Best Film. **VHS, Beta $79.95** *FCT* 🎬🎬🎬

Red Desert Penitentiary A spoof of movie-making featuring caricatures of industry-types.
1983 104m/C James Michael Taylor, Cathryn Bissell, Will Rose; *Dir:* George Sluizer. **VHS, Beta $79.98** *PEV* 🎬🎬🎬

Red Dragons of Shaolin Acrobats/kung-fu performers are caught up in a bloodbath. Warriors teach them how to spill a little blood of their own.
19?? 90m/C VHS $49.95 *OCE* 🎬

Red Dust The overseer of a rubber plantation in Indochina causes all kinds of trouble when he falls in love with a young engineer's wife. Filled with free-spirited humor and skillfully directed; remarkably original. Remade in 1940 as "Congo Maisie" and in 1954 as "Mogambo."
1932 83m/B Clark Gable, Jean Harlow, Mary Astor, Gene Raymond, Donald Crisp; *Dir:* Victor Fleming. **VHS, Beta $19.98** *MGM* 🎬🎬🎬

Red Flag: The Ultimate Game Pilots rival for success in battle simulation training with terrible consequences. Fine performances cover plot weaknesses. Well-written dialogue.
1981 100m/C Barry Bostwick, William Devane, Joan VanArk, Fred McCarren, Debra Feuer, George Coe, Linden Chiles, Arlen Dean Snyder; *Dir:* Don Taylor. **VHS, Beta $39.95** *IVE* 🎬🎬½

The Red Fury Young Indian boy struggles to overcome prejudice and caring for his horse shows him the way. Family stuff.
197? 105m/C William Jordan, Katherine Cannon. **VHS $59.95** *TPV, HHE* 🎬🎬

Red Grooms Avant-garde filmmaker Red Grooms discusses his work and screens two of his films.
196? 30m/B VHS, Beta *CMR, NYS* 🎬🎬

The Red Half-Breed A lonely half-breed Indian, wrongly accused of murder, discovers the real killer while running from the law.
1970 103m/C Daniel Pilon, Genevieve Deloir; *Dir:* Gilles Carle. **VHS, Beta $59.95** *NWV* 🎬½

Red Headed Stranger Willie Nelson vehicle based on his 1975 album. Nelson is a gun-totin' preacher who kills his wife and is led to salvation by a good farm woman. But honestly: Morgan Fairchild as Willie Nelson's wife? Please.
1987 (R) 108m/C Willie Nelson, Katharine Ross, Morgan Fairchild; *Dir:* Bill Wittliff. **VHS, Beta, LV $9.98** *CHA* 🎬½

Red Headed Woman Harlow plays an unscrupulous woman who charms her way into the lives of influential men. She causes her boss to divorce his wife and marry her while she sleeps with the chauffeur (Boyer) and others. Audiences loved the scandalous material, but the Hays Office objected to the fact that the immoral woman goes unpunished. Boyer's small but important role salvaged his floundering American film career. Harlow is sultry and funny.
1932 79m/B Jean Harlow, Chester Morris, Lewis Stone, Leila Hyams, Una Merkel, Henry Stephenson, May Robson, Charles Boyer, Harvey Clark; *Dir:* Jack Conway. **VHS, Beta $19.98** *MGM, FCT* 🎬🎬🎬

Red Heat Blair is a tourist in East Germany mistakenly arrested and sent to a rough women's prison. Her fiance tries to free her. Meanwhile, she has to deal with tough-lady fellow jail bird Kristel. Familiar and marginal.
1985 104m/C Linda Blair, Sylvia Kristel, Sue Keil, William Ostrander; *Dir:* Robert Collector. **VHS, Beta, LV $29.98** *LIV, VES* 🎬½

Red Heat Two cops—one from the Soviet Union, one from Chicago—team up to catch the Eastern Bloc's biggest drug czar. Lots of action, but at times it seems too similar to Hill's earlier hit "48 Hours." Film claims to be the first major U.S. production shot in Red Square, Moscow.
1988 (R) 106m/C Arnold Schwarzenegger, James Belushi, Peter Boyle, Ed O'Ross, Larry Fishburne, Gina Gevshon, Richard Bright; *Dir:* Walter Hill. **VHS, Beta, LV $14.95** *IVE, TLF* 🎬🎬

The Red House Robinson plays a crippled farmer who, after his daughter brings home a suitor, attempts to keep everyone from a mysterious red house located on his property. Madness and murder prevail. Strange film noir about tangled relationships and unsuccessful attempts to bury the horrid past. Based on the novel by George Agnew Chamberlain.
1947 100m/B Edward G. Robinson, Lon McCallister, Judith Anderson, Allene Roberts, Rory Calhoun, Ona Munson; *Dir:* Delmer Daves. **VHS, Beta $9.95** *NEG, NOS, SNC* 🎬🎬🎬

Red Kimono Exploitative silent melodrama about a young woman ditched by her husband. She becomes a prostitute, but is redeemed by true love. Interesting slice of its period.
1925 95m/B Tyrone Power Sr., Priscilla Boner, Nellie Bly Baker, Mary Carr; *Dir:* Walter Lang. **VHS, Beta $19.95** *NOS, VYY, DVT* 🎬🎬

Red King, White Knight An assassin tries to kill Mikhail Gorbachev during super-power peace talks. Ex-s pook Skerritt returns to duty to thwart him. Better than most "topical" thrill ers. Made for cable.
1989 (R) 106m/C Tom Skerritt, Max von Sydow, Helen Mirren; *Dir:* Geoff Murphy. **VHS, Beta $89.99** *HBO* 🎬🎬½

Red Kiss (Rouge Baiser) Paris, 1952, a 15-year-old French Stalinist falls in love with older, apolitical American photographer, and she's torn between love and politics. French with English subtitles.
1985 110m/C FR Charlotte Valandrey, Lambert Wilson, Marthe Keller, Gunter Lamprecht, Laurent Terzieff; *Dir:* Vera Belmont. **VHS $79.95** *FXL, FCT* 🎬🎬🎬

The Red Light Sting A Justice Department rookie (Bridges) and a call girl (Fawcett) reluctantly team up to con the Mafia in order to convict a San Francisco rackets czar. Goes down easy, but also easily forgotten or skipped.

1984 96m/C Farrah Fawcett, Beau Bridges, Harold Gould, Paul Burke, Sunny Johnson; *Dir:* Rod Holcomb. **VHS, Beta $39.95** *MCA* 🎬

Red Line 7000 High stakes auto racers drive fast cars and date women. Excellent racing footage in otherwise routine four-wheel fest.
1965 110m/C James Caan, Laura Devon, Gail Hire, Charlene Holt, John Crawford, Marianna Hill, George Takei; *Dir:* Howard Hawks. **VHS $39.95** *PAR, FCT* 🎬🎬

Red Lion A bumbling horse-tender in feudal Japan impersonates a military officer to impress his family, only to be swept into leading a liberating revolution. With English subtitles.
1969 115m/C *JP* Toshiro Mifune, Shima Iwashita; *Dir:* Kihachi Okamoto. **VHS, Beta** *VDA* 🎬🎬½

Red Nightmare Anti-communist propaganda film, produced for the Department of Defense by Warner Brothers, features the cast of "Dragnet." The story dramatizes the Red Menace, with communists conspiring to take over America, and shows what would happen to life in a small American town if the commies took over.
1962 30m/B Jack Webb, Jack Kelly, Jeanne Cooper, Peter Brown. **VHS, Beta, 3/4U $19.95** *VYY, IHF* 🎬

Red Nights A country boy goes to Hollywood seeking success and finds instead corruption, decadence, and dishonesty. Excellent soundtrack by Tangerine Dream.
1987 90m/C Christopher Parker, Jack Carter, Brian Matthews; *Dir:* Izhak Hanooka. **VHS, Beta $19.98** *TWE* 🎬

The Red Pony A young boy escapes from his family's fighting through his love for a pet pony. Based on the novel by John Steinbeck, with musical score by Aaron Copeland. Remade for television in 1976. Timeless classic that the whole family can enjoy.
1949 89m/C Myrna Loy, Robert Mitchum, Peter Miles, Louis Calhern, Sheppard Strudwick, Margaret Hamilton, Beau Bridges; *Dir:* Lewis Milestone. **VHS $19.98** *KUI, REP* 🎬🎬🎬

The Red Pony Excellent TV remake of a classic 1949 adaptation of Steinbeck. Fonda is superb as a troubled young boy's difficult father.
1976 101m/C Henry Fonda, Maureen O'Hara; *Dir:* Robert Totten. **VHS, Beta, LV $19.98** *BFA* 🎬🎬🎬

The Red Raven Kiss-Off A seedy Hollywood detective becomes the key suspect in a movie-making murder. Based on the 1930s' Dan Turner mystery stories. Made for video.
1990 (R) 93m/C Marc Singer, Tracy Scoggins, Nicholas Worth, Arte Johnson; *Dir:* Christopher Lewis. **VHS, Beta $19.95** *FRH* 🎬

Red Riding Hood Musical version of story of wolf who has little girl fetish.
1988 84m/C Isabella Rossellini, Craig T. Nelson, Rocco Sisto. **VHS, Beta $19.98** *CAN* 🎬🎬½

Red River The classic Hawks epic about a gruelling cattle drive and the battle of wills between father and son. Generally regarded as one of the best westerns ever made, and Wayne's shining moment as a reprehensible, obsessive man. Featuring Montgomery Clift in his first film. Restored version has eight minutes of previously edited material. Remade for television in 1988 with James Arness and Bruce Boxleitner.

1948 133m/B John Wayne, Montgomery Clift, Walter Brennan, Joanne Dru; *Dir:* Howard Hawks. **VHS, Beta, LV $34.95** *FOX, TLF, BUR* 🎬🎬🎬🎬

Red River Valley Autry and partner Burnette go undercover to find out who's plaguing a dam construction site with explosions. Not to be confused with a 1941 Roy Rogers pic bearing the same title.
1936 60m/B Gene Autry, Smiley Burnette, Frances Grant, Booth Howard, Sam Flint, George Cheseboro, Charles King, Eugene Jackson, Frank LaRue, Lloyd Ingraham, Champion; *Dir:* B. Reeves Eason. **VHS $19.95** *NOS* 🎬🎬

Red Rock Outlaw A sleazy character tries to kill his twin brother and takes his place as an honest rancher.
1947 56m/B Lee White. **VHS, Beta, LV** *WGE* 🎬

The Red Rope A Western hero cuts his honeymoon short to track down some villains.
1937 56m/B Bob Steele, Lois January, Forrest Taylor, Charles King, Karl Hackett, Bobby Nelson; *Dir:* S. Roy Luby. **VHS $19.95** *NOS, DVT* 🎬½

Red Scorpion A Soviet soldier journeys to Africa where he is to assassinate the leader of a rebel group. Will he succeed or switch allegiances? Poor acting and directing abound.
1989 (R) 102m/C Dolph Lundgren, M. Emmet Walsh, Al White, T.P. McKenna, Carmen Argenziano, Brion James, Regopstann; *Dir:* Joseph Zito. **VHS, LV $89.95** *SHP* 🎬

Red Shoe Diaries When a woman commits suicide her grieving lover discovers her diaries and finds out she led a secret erotic life, revolving around a shoe-salesman lover and a pair of sexy red shoes. Made for cable television; also available in an unrated version.
1992 (R) 105m/C David Duchovny, Billy Wirth, Brigitte Bako; *Dir:* Zalman King. **VHS, LV** *REP* 🎬🎬

The Red Shoes British classic about a young ballerina torn between love and success. Inspired by the Hans Christian Anderson fairy tale. Written by Powell and Pressburger. Noted for its lavish use of Technicolor. Oscar nominations for Best Picture, Story, and Film Editing.
1948 136m/C *GB* Anton Walbrook, Moira Shearer, Marius Goring, Leonide Massine, Robert Helpmann, Albert Basserman, Emeric Pressburger; *Dir:* Michael Powell. Academy Awards '48: Best Art Direction/Set Decoration (Color), Best Musical Score; National Board of Review Awards '48: 10 Best Films of the Year. **VHS, Beta, LV $19.95** *PAR, VTK* 🎬🎬🎬🎬

The Red Shoes The Children's Theatre Company and School of Minneapolis perform their own unique interpretation of the Hans Christian Anderson story. In VHS and Beta Hi-Fi Stereo.
1983 79m/C **VHS, Beta $39.95** *MCA* 🎬🎬🎬🎬

Red Signals Mystery criminal makes trains collide.
1927 70m/B Wallace MacDonald, Earl Williams, Eva Novak, J.P. McGowan, Frank Rice; *Dir:* J.P. McGowan. **VHS, Beta $24.95** *GPV, FCT* 🎬🎬

Red Skelton: A Comedy Scrapbook Highlighted by footage from Skelton's hilarious television escapades, featuring his zany characterizations of Freddy the Freeloader, Clem Kadiddlehopper and others.
198? 64m/B **VHS $9.99** *VTR* 🎬🎬

Red Skelton: King of Laughter Uncensored and unedited material from the Comedy King's shows.
1989 60m/C Red Skelton. **VHS $19.95** *HBO* 🎬🎬

Red Skelton's Funny Faces The super-comic takes it to the hilt and pulls out all the stops for this now legendary performance.
1979 60m/C Red Skelton, Marcel Marceau. **VHS, Beta $39.95** *IVE* 🎬🎬

Red Skelton's Funny Faces 3 Here he is again threatening to kill you with yuks in this made for cable feature.
1982 57m/C Red Skelton. **VHS, Beta $39.95** *USA* 🎬🎬

Red Snow High in the Cascade Mountains, Kyle Lewis is the new snowboard instructor for the Hurricane Ridge Ski Resort. The job is fine in the beginning, until Kyle learns of the tragic fate of the last instructor. Before he knows it, Kyle is the next target and is framed for two murders. He secures the help of his fellow snowboard buddies in hopes of solving the mystery and getting the girl, too. High speed ski scenes and dangerous stunts are the film's only redeeming qualities.
1991 86m/C Carlo Scandiuzzi, Scott Galloway, Darla Slavens, Mitch Cox, Tamar Tibbs, Brian Mahoney; *Dir:* Phillip J. Roth. **VHS $79.95** *VHE* 🎬

Red Sonja Two warriors join forces against an evil queen in this sword and sorcery saga. Big, beautiful bodies everywhere, with Bergman returning from her Conan adventures. Little humor, few special effects, and weak acting make this a poor outing.
1985 (PG-13) 89m/C Arnold Schwarzenegger, Brigitte Nielsen, Sandahl Bergman, Paul Smith; *Dir:* Richard Fleischer. **VHS, Beta, LV $19.98** *FOX* 🎬½

Red Sorghum A stunning visual achievement, this new wave Chinese film (and Yimou's directorial debut) succeeds on many levels—as an ode to the color red, as dark comedy, and as a sweeping epic with fairy tale overtones. A young woman traded for a donkey and betrothed to a leper in the rural China of the 1920s is reprieved when the leper is murdered, presumably by her lover. She marries her lover, and they take over the distillery owned by the leper. Further complications ensue when the Japanese invade. The sorghum plot nearby is a symbolic playing field in the movie's most stunning scenes. Here, people make love, murder, betray, and commit acts of bravery, all under the watchful eye of nature. In Mandarin with English subtitles.
1987 91m/C *CH* Gong Li, Jiang Wen, Ji Cun Hua; *Dir:* Zhang Yimou. **VHS $79.95** *NYF* 🎬🎬🎬½

The Red Stallion Good film for animal lovers, but it loses something when it comes to human relationships. A young boy raises his pony into an award-winning racehorse that saves the farm when it wins the big race. Good outdoor photography.
1947 82m/B Robert Paige, Noreen Nash, Ted Donaldson, Jane Darwell; *Dir:* Lesley Selander. **VHS $19.95** *NOS, MOV* 🎬½

Red Sun A gunfighter, a samurai, and a French bandit fight each other in various combinations in the 1860s. Ludicrous and boring.
1971 115m/C *FR IT SP* Charles Bronson, Toshiro Mifune, Alain Delon, Ursula Andress; *Dir:* Terence Young. **VHS, Beta $24.95** *GEM* 🎬🎬

Red Surf Action abounds in this surfer film. A couple of hard-nosed wave-riders get involved with big money drug gangs and face danger far greater than the tide.
1990 (R) 104m/C George Clooney, Doug Savant, Dedee Pfieffer, Gene Simmons, Rick Najera, Philip McKeon; *Dir:* H. Gordon Boos. VHS, Beta $29.95 ACA ♂♂

Red Tent A robust, sweeping man-versus-nature epic based on the true story of the Arctic stranding of Italian explorer Umberto Nobile's expedition in 1928. Music by Ennio Morricone. A Russian-Italian co-production.
1969 (G) 121m/C Sean Connery, Claudia Cardinale, Peter Finch, Hardy Kruger, Massimo Girotti, Luigi Vannucchi; *Dir:* Mikhail Kalatozov. Beta, LV, Q $19.95 PAR ♂♂♂

The Red and the White Epic war drama about the civil war between the Red Army and the non-communist Whites in Russia in 1918. Told from the perspective of Hungarians who fought alongside the Reds. Little dialogue. Subtitled.
1968 92m/B HU *Dir:* Miklos Jancso. VHS, LV $79.95 KIV ♂♂♂

Red, White & Busted Amnesty or prison? That is the scorching, virtually unanswerable question asked by indictment of America. We follow three friends during the Vietnam War and see how they were affected by the events, whether or not they went.
1975 85m/C Darrell Larson, Heather Menzies, Dennia Oliveri, John Bill. VHS, Beta, LV WGE ♂

Red Wind Kris Morrow is a psychotherapist whose latest patient, Lila, has a number of disturbing sadomasochistic fantasies. When Kris feels she's getting too close to the situation, she refers Lila to another therapist. But Lila calls and tells Kris she's already acted out one of her fantasies by killing her husband. In order to stop Lila from killing again, Kris goes on a dangerous search for her deadly patient.
1991 (R) 93m/C Lisa Hartman, Deanna Lund, Philip Casnoff, Christopher McDonald; *Dir:* Alan Metzger. VHS MCA ♂

Redd Foxx—Video in a Plain Brown Wrapper Foxx tackles all of his favorite subjects-sex, marriage, death, crime and more sex.
1983 60m/C VHS, Beta, LV $39.98 LIV, VES ♂♂

The Redeemer A group of adults at their high school reunion play the children's game of hide-and-seek, with deadly results.
1976 80m/C F.G. Finkbinder, Damien Knight, Jeanetta Arnetta; *Dir:* Constantine S. Gochis. VHS, Beta GHV ♂

Redneck Implausible, distasteful purported ''thriller'' about criminals on the run.
1973 92m/B IT GB Telly Savalas, Franco Nero, Mark Lester, Ely Galleani, Dulio Del Prete, Maria Michi; *Dir:* Silvio Narizzano. VHS, Beta $18.95 KOV ♂

Redneck Zombies A bunch of backwoods rednecks become zombies after chug-a-lugging some radioactive beer. Eating local tourists becomes a hard habit to break. Betcha can't have just one!
1988 (R) 83m/C Floyd Piranha, Lisa DeHaven, W.E. Benson, William W. Decker, James Housely, Zoofoot, Tyrone Taylor, Perieles Lewnes. VHS, Beta $79.95 TWE ♂

Reds The recreation of the life of author John Reed (''Ten Days that Shook the World''), his romance with Louise Bryant, his efforts to start an American Communist party, and his reporting of the Russian Revolution. A sweeping, melancholy epic using dozens of ''witnesses'' who reminisce about what they saw. Won several Oscars, including Best Director for Beatty.
1981 (PG) 195m/C Warren Beatty, Diane Keaton, Jack Nicholson, Edward Herrmann, Maureen Stapleton, Gene Hackman, Jerzy Kosinski, George Plimpton, Paul Sorvino, William Daniels, M. Emmet Walsh, Dolph Sweet, Josef Sommer; *Dir:* Warren Beatty. Academy Awards '81: Best Cinematography, Best Director (Beatty), Best Supporting Actress (Stapleton); New York Film Critics Awards '81: Best Film. VHS, Beta, LV $29.95 PAR, BTV ♂♂♂

Reefer Madness Considered serious at the time of its release (it was originally titled ''Tell Your Children''), this low-budget depiction of the horrors of marijuana has become an underground comedy favorite. Overwrought acting and lurid script contribute to the fun.
1938 (PG) 67m/B Dave O'Brien, Dorothy Short, Warren McCollum, Lillian Miles, Thelma White, Carleton Young; *Dir:* Louis Gasnier. VHS, Beta $16.95 SNC, NOS, HHT ♂♂♂

Reet, Petite and Gone All-Black musical featuring the music of neglected jive singer Louis Jordan, and his band, The Tympany Five. A girl's mother dies; sneaky lawyer tries to cheat her. Slick and enjoyable.
1947 75m/B Louis Jourdan, June Richmond; *Dir:* William Forest Crouch. VHS, Beta $19.95 MVD, NOS, VYY ♂♂♂

The Reflecting Skin In a 1950s prairie town a small boy sees insanity, child-murder and radiation sickness, leading him to fantasize that the tormented young widow next door is a vampire. The pretentious drama/ freak show rationalizes its ghastly events as symbolizing the hero's loss of youthful innocence. But the Hound knows the score; this is ''Faces of Death'' for the arts crowd, a grotesque menagerie that dares you to watch. The exploding-frog opener is already notorious. Beautiful photography, with vistas inspired by the painting of Andrew Wyeth.
1991 116m/C Viggo Mortensen, Lindsay Duncan, Jeremy Cooper; *W/Dir:* Philip Ridley. VHS, LV $89.98 INV ♂♂

A Reflection of Fear Lame psycho-thriller about a young girl's jealousy of her father's girlfriend, with family members dying of supernatural causes.
1972 (PG) 90m/C Robert Shaw, Sally Kellerman, Sondra Locke, Mary Ure; *Dir:* William A. Fraker. VHS, Beta $69.95 COL ♂♂

Reflections A continuous picture of a pond shimmering in the afternoon hours to create a relaxed background.
1978 30m/C VHS, Beta PCC ♂♂

Reflections: A Moment in Time Lesbian production chronicles the formation of a relationship between two women through a personals ad. Looks at issues of codependency.
19?? 30m/C VHS $34.95 NWV ♂♂♂

Reflections in a Golden Eye Huston's film adaptation of Carson McCullers's novel about repressed homosexuality, madness, and murder at a Southern Army base in 1948. Star-studded cast cannot consistently pull off convoluted lives of warped characters; not for everyone, though it holds some interest.
1967 109m/C Elizabeth Taylor, Marlon Brando, Brian Keith, Julie Harris; *Dir:* John Huston. VHS, Beta $59.95 WAR ♂♂½

Reflections of Murder Evil schoolteacher's wife and mistress decide to kill him off and then are plagued by his phantom. Well done made-for-television remake of the French ''Diabolique.'' Suspenseful, with a great ending.
1974 97m/C Tuesday Weld, Joan Hackett, Sam Waterston, Lucille Benson, R.G. Armstrong, Michael Lerner, Ed Bernard, Lance Kerwin; *Dir:* John Badham. VHS $14.98 REP ♂♂½

Reform School Girls Satiric raucous women's prison film, complete with tough lesbian wardens, brutal lesbian guards, sadistic lesbian inmates, and a single, newly convicted heterosexual heroine. Wendy O. Williams as a teenager? Come on. Overdone, over-campy, exploitative.
1986 (R) 94m/C Linda Carol, Wendy O. Williams, Pat Ast, Sybil Danning, Charlotte McGinnis, Sherri Stoner; *Dir:* Tom DeSimone. VHS, Beta, LV $19.95 STE, NWV ♂♂

Refuge of Fear Drama set in a post-nuclear-holocaust world.
198? 90m/C VHS, Beta $19.95 ASE ♂

Regarding Henry A cold-hearted lawyer gets shot in the head during a holdup, and survives with memory and personality erased. During recovery the new Henry displays compassion and conscience the old one never had. Though too calculated in its yuppie-bashing ironies, the picture works thanks to splendid acting and low-key, on-target direction.
1991 (PG-13) 107m/C Harrison Ford, Annette Bening, Bill Nunn, Mikki Allen; *Dir:* Mike Nichols. VHS, Beta, LV PAR ♂♂♂

Reggie Mixes In Wealthy man brawls in bar room and falls for barmaid. Taking a job to be near his beloved, he wipes out local thugs while tending to her welfare. Exhilarating fight scenes in which Fairbanks fists it out with real life boxers.
1916 58m/B Douglas Fairbanks Sr., Bessie Love, Joseph Singleton, William E. Lowery, Wilbur Higby, Frank Bennett; *Dir:* Christy Cabanne. VHS, Beta $24.95 GPV, FCT ♂♂♂

Regina A woman controls all the activities of her husband and son. At age 36, her son is ready to leave home and she refuses to let him. Fine performances in this strange and disturbing film.
197? 86m/C IT Anthony Quinn, Ava Gardner, Ray Sharkey, Anna Karina; *Dir:* Jean-Yves Prate. VHS, Beta $69.95 VMK ♂♂½

Reg'lar Fellers A gang of kids save the town and soften the heart of their grandmother, too. Production is sloppy.
1941 66m/B Billy Lee, Carl ''Alfalfa'' Switzer, Buddy Boles, Janet Dempsey, Sarah Padden, Roscoe Ates; *Dir:* Arthur Dreifuss. VHS $19.95 NOS, FMT ♂♂

Rehearsal There is some rare footage of the rehearsal sessions for ''The Bell Telephone Hour'' radio program at Studio 6B in Radio City.
1947 24m/B Ezio Pinza, Blanche Thebom, Donald Voorhees, Floyd Mack, Tom Shirley. VHS, Beta, 8mm VYY ♂♂

Rehearsal for Murder Movie star Redgrave is murdered on the night of her Broadway debut. Seems it might have been someone in the cast. Brought to you by the creative team behind ''Columbo.'' Challenging made-for-TV whodunit with a twist. Good cast.

1982 96m/C Robert Preston, Lynn Redgrave, Patrick Macnee, Lawrence Pressman, Madolyn Smith, Jeff Goldblum, William Daniels; *Dir:* David Greene. VHS, Beta $59.95 REP 𝔊𝔊½

Reign of Terror British-made adventure set during the French Revolution, with everyone after the 'Black Book' (pic's alternate title) that holds the names of those arch-fiend Robespierre plans to guillotine in his ruthless bid for power. Well-mounted, but historical personages and events are reduced to cartoon form.
1949 89m/B Robert Cummings, Arlene Dahl, Richard Hart, Arnold Moss, Richard Basehart; *Dir:* Anthony Mann. VHS $19.95 NOS, MRV 𝔊𝔊

Reilly: The Ace of Spies Exploits of the real-life spy Sydney Reilly who uncovers Russian secrets in 1901 and helps plot the Revolution. Made for British television.
1987 80m/C *GB* Sam Neill, Sebastian Shaw, Jeananne Crowley; *Dir:* Jim Goddard. VHS, Beta, LV $29.95 CCB, HBO 𝔊

Reinaldo Jorge: Parts 1 & 2 Reinaldo Jorge and his Orchestra plays their latest Salsa hits in this series of two untitled programs.
1983 30m/C Beta NJN 𝔊

The Reincarnate Cult guarantees lawyer will live forever if he can find a new body. Enter gullible sculptor; lawyer uses skill at persuasion.
1971 (PG) 89m/C *CA* Jack Creley, Jay Reynolds, Trudy Young, Terry Tweed; *Dir:* Don Haldane. VHS, Beta $59.98 MAG 𝔊

The Reincarnation of Golden Lotus In this highly erotic story of love and revenge, Joi Wong escapes China for decadent Hong Kong by marrying a wealthy but foolish man. She has numerous sadomasochistic affairs and begins having flashbacks, revealing her to be the reincarnation of Golden Lotus, a courtesan of ancient China. Directed by internationally acclaimed Clara Law. In Mandarin Chinese with English subtitles.
1989 99m/C *CH* Joi Wong, Eric Tsang, Lam Chen Yen; *Dir:* Clara Law. VHS $79.95 CVC 𝔊𝔊𝔊𝔊

Reincarnation of Peter Proud A college professor has nightmares which lead him to believe the spirit of a murdered man is now possessing him. Kinda scary, but predictable. Screenplay adapted by Max Ehrlich from his novel.
1975 (R) 105m/C Michael Sarrazin, Jennifer O'Neill, Margot Kidder, Cornelia Sharpe; *Dir:* J. Lee Thompson. VHS, Beta, LV $29.98 LIV, VES 𝔊𝔊

The Reivers Young boy and two adult pals journey from small-town Mississippi (circa 1905) to the big city of Memphis in a stolen car. Picaresque tale is delightful on-screen, as in William Faulkner's enjoyable last novel. Oscar nomination for Crosse.
1969 (PG) 107m/C Steve McQueen, Sharon Farrell, Will Geer, Michael Constantine, Rupert Crosse; *Dir:* Mark Rydell. VHS, Beta $59.98 FOX 𝔊𝔊𝔊½

Rejoice Highlights of the 1983 Tournament of Roses Parade in Pasadena with Grand Marshal Merlin Olsen are included on this video.
1983 45m/C VHS, Beta TVT 𝔊½

The Rejuvenator Also known as "Rejuvenatrix." An oft-told film story: an aging actress discovers a "youth" serum but finds out that the serum affects things other than her aging. Well-done.
1988 (R) 90m/C Marcus Powell, John MacKay, James Hogue, Vivian Lanko, Jessica Dublin; *Dir:* Brian Thomas Jones. VHS, LV $79.95 SVS 𝔊𝔊½

Relentless Twisted psycho Nelson, once rejected by the LAPD Academy on psychological grounds takes his revenge by murdering people and using his police training to cover his tracks. Good acting keeps sporadically powerful but cliched thriller afloat.
1989 (R) 92m/C Judd Nelson, Robert Loggia, Meg Foster, Leo Rossi, Pat O'Bryan, Mindy Seeger, Angel Tompkins, Ken Lerner, George Flower; *Dir:* William Lustig. VHS, Beta, LV $19.95 COL 𝔊𝔊

Relentless 2: Dead On Rossi, the detective from the first film, tracks yet another murderer whose occult-style mutilations mask an international political conspiracy. So-so slaughter with artsy camerawork. Honestly, how many "Relentless" fans can you name who've been waiting with anticipation?
1991 (R) 93m/C Leo Rossi, Ray Sharkey, Meg Foster, Miles O'Keefe; *Dir:* Michael Schroeder. VHS, LV $89.95 SVS, PIA 𝔊𝔊

The Reluctant Debutante Harrison and Kendall are the urbane parents of Dee who are trying to find a suitable British husband for their girl. It seems their choices just won't do however, and Dee falls for American bad boy musician Saxon. A very lightweight yet extremely enjoyable romantic comedy thanks to Harrison and in particular his real-life wife Kendall, who unfortunately died the following year.
1958 94m/C Rex Harrison, Kay Kendall, John Saxon, Sandra Dee, Angela Lansbury, Diane Clare; *Dir:* Vincente Minnelli. VHS, Beta $19.98 MGM 𝔊𝔊𝔊

The Reluctant Dragon This is the classic Disney short about a fun-loving dragon, based on the Kenneth Grahame story; accompanied also by the Disney short "Morris, the Midget Moose." Not to be confused with the full-length 1941 Disney feature film.
1941 28m/C *Dir:* Alfred Werker. VHS, Beta $12.99 DIS, KUI 𝔊𝔊𝔊

Rembrandt Necessarily very visual biography of the great Dutch painter, Rembrandt. Superb acting by Laughton.
1936 86m/B Charles Laughton, Elsa Lanchester, Gertrude Lawrence, Walter Hudd; *Dir:* Alexander Korda. National Board of Review Awards '36: 10 Best Foreign Films of the Year. VHS, Beta, LV $14.98 SUE 𝔊𝔊𝔊

Remedy for Riches The fourth "Dr. Christian" comedy, has the small-town doctor trying to uncover a real estate fraud before it bankrupts his community. Mildly funny.
1940 66m/B Jean Hersholt, Dorothy Lovett, Edgar Kennedy, Jed Prouty, Walter Catlett; *Dir:* Erle C. Kenton. VHS, Beta, 8mm $19.95 NOS, VYY 𝔊𝔊

Remember Me Just as a woman is starting to gain control of her life, her husband returns home from the mental hospital.
1985 95m/C Robert Grubb, Wendy Hughes, Richard Moir; *Dir:* Lex Marinos. VHS, Beta $59.95 MNC 𝔊

Remembering LIFE A look back on 49 years of LIFE magazine, focusing on the photography and editing that made the periodical so popular.
1985 60m/C *Dir:* David Hoffman; *Nar:* Walter Cronkite. VHS, Beta, 3/4U AHV, KAR, PBS 𝔊

Remembering Winsor McCay This is a profile of the American animation pioneer, Winsor McCay (1876-1934). Three of his classics are highlighted: "Gertie the Dinosaur" (1914), "The Sinking of the Lusitania" (1918), and the rare, hand-colored version of the "Little Nemo" (1911).
1978 20m/C VHS, Beta BFA 𝔊

Reminiscences of a Journey to Lithuania Filmmaker Jonas Mekas returns to his native Lithuania for the first time in twenty-five years in this documentary.
1984 102m/C VHS, Beta NEW 𝔊

Remo Williams: The Adventure Begins Adaptation of "The Destroyer" adventure novel series with a Bond-like hero who can walk on water and dodge bullets after being instructed by a Korean martial arts master. Funny and diverting, and Grey is excellent (if a bit over the top) as the wizened oriental. The title's assumption is that the adventure will continue.
1985 (PG-13) 121m/C Fred Ward, Joel Grey, Wilford Brimley, Kate Mulgrew; *Dir:* Guy Hamilton. VHS, Beta, LV $14.99 HBO 𝔊𝔊

Remote Control A circulating videotape of a 1950's sci-fi flick is turning people into murderous zombies, thanks to some entrepreneurial aliens. Silly, but performed by young actors with gusto.
1988 (R) 88m/C Kevin Dillon, Deborah Goodrich, Christopher Wynne, Jennifer Tilly; *Dir:* Jeff Lieberman. VHS, Beta $79.95 IVE, HHE 𝔊½

Rendez-Moi Ma Peau A modern-day witch transposes two people's personalities with comic results. In French with subtitles.
1981 82m/C *FR* *Dir:* Patrick Schulmann. VHS, Beta FCT 𝔊𝔊½

Rendez-vous A racy, dark film about the sensitive balance between sex and exploitation, the real world and the stage world. Some may be disturbed by chillingly explicit scenes. Impressive tour de force of imagination, direction, and cinematography.
1985 (R) 82m/C *FR* Juliette Binoche, Lambert Wilson, Wadeck Stanczak, Jean-Louis Trintignant; *Dir:* Andre Techine. VHS $79.95 CVC 𝔊𝔊𝔊

Renegade Girl A special agent is sent west to capture the female head of a band of outlaws.
1946 65m/B Allan Curtis, Ann Savage, Edward Brophy, Ray Corrigan. VHS, Beta, LV $19.95 NOS, RXM, WGE 𝔊

Renegade Monk An invincible warrior-monk dispenses his own brand of justice with his fists.
1982 (R) 90m/C Lui Chang Liang, Huang Hsing Shaw, Ko Shou Liang, Lang Shih Chia, Hsu Chung Hsing. VHS, Beta $59.95 GEM 𝔊½

Renegade Ninja A man whose father was murdered kicks everyone within reach in his quest for revenge.
1983 109m/C Kensaku Marita; *Dir:* Sadao Nakajima. VHS, Beta $79.95 PSM 𝔊

Renegade Ranger/Scarlet River A western double feature: A Texas Ranger and his singing sidekick attempt to help out a woman accused of murder in the "The Renegade Ranger," and a cowboy tries to free his sweetheart from a band of cattle rustlers in "Scarlet River." **1938 113m/B** Rita Hayworth, Tim Holt, George O'Brien, Ray Whiltey, Tom Keene, Lon Chaney Jr., Myrna Loy, Bruce Cabot, Betty Furness; *Dir:* David Howard. **VHS, Beta $34.95** RKO ✂½

Renegade Trail Hopalong Cassidy helps a marshall and a pretty widow round up ex-cons and rustlers. The King's Men, radio singers of the era, perform a fistful of songs. **1939 61m/B** William Boyd, George "Gabby" Hayes, Russell Hayden, Charlotte Wynters, Russell Hopton, Sonny Bupp, Jack Rockwell, Roy Barcroft, Eddie Dean; *Dir:* Lesley Selander. **VHS $19.95** NOS, BUR ✂✂

Renegades A young cop from Philadelphia and a Lakota Indian begrudgingly unite to track down a gang of ruthless crooks. Good action, and lots of it; not much story. **1989 (R) 105m/C** Kiefer Sutherland, Lou Diamond Phillips, Rob Knepper, Jami Gertz, Bill Smitrovich; *Dir:* Jack Sholder. **VHS, Beta, LV $19.95** MCA ✂✂

Renfrew on the Great White Trail A frustrated fur trader takes off after a pack of thieves. "On the Great White Trail" and "Renfrew of the Royal Mounted on the Great White Trail" are alternate titles. **1938 58m/B** James Newill, Terry Walker, Robert Frazer, Richard Alexander; *Dir:* Al Herman. **VHS $19.95** NOS, DVT, BUR ✂½

Renfrew of the Royal Mounted A Canadian Mountie goes after counterfeiters and sings a few songs in this mild adaptation of a popular kids' radio serial. Newill was a popular tenor of the day, but the Hound thinks Renfrew's dog 'Lightning' should have been the star. **1937 57m/B** Carol Hughes, James Newill, Kenneth Harlan; *Dir:* Al Herman. **VHS, Beta** DVT, GPV ✂✂

Reno and the Doc Enjoyable comedy about two endearingly cranky middle-aged men battling mid-life crises join together and test their mettle in the world of professional skiing. **1984 88m/C** CA Ken Walsh, Henry Ramer, Linda Griffiths; *Dir:* Charles Dennis. **VHS, Beta $19.95** STE, NWV ✂✂½

Renoir Shorts Two early silent Renoir fantasies: "The Little Matchgirl," a lyrical adaptation of the Hans Christian Anderson fairy tale, and "Charleston," in which a space-traveling African and Parisian girl dance comically. **1927 50m/B** Catherine Hessling; *Dir:* Jean Renoir. **VHS, Beta $24.95** NOS, FCT ✂✂✂

Rent-A-Cop A cop (Reynolds) is bounced from the force after he survives a drug-bust massacre under suspicious circumstances. He becomes a security guard and continues to track down the still-at-large killer with help from call girl Minnelli. **1988 (R) 95m/C** Burt Reynolds, Liza Minnelli, James Remar, Richard Masur, Bernie Casey, John Stanton, John P. Ryan, Dionne Warwick, Robby Benson; *Dir:* Jerry London. **VHS, Beta, LV $14.99** HBO ✂✂

Rentadick A precursor to the Monty Python masterpieces, written by Graham Chapman and John Cleese. Private eye spoof isn't as funny as later Python efforts; but it is an indication of what was yet to come and fans should enjoy it. **1972 94m/C** GB James Booth, Julie Ege, Ronald Fraser, Donald Sinden, Michael Bentine, Richard Briers, Spike Milligan; *Dir:* Jim Clark. **VHS** VIR ✂✂

Rented Lips Two documentary film makers sell out, taking over the production of a porno film at the request of a public television producer who promises to help their careers. Mull and Shawn are well supported by an impressive cast, but everyone's effort is wasted on a very unfunny attempt at comedy. **1988 (R) 82m/C** Martin Mull, Dick Shawn, Jennifer Tilly, Kenneth Mars, Edy Williams, Robert Downey Jr., June Lockhart, Shelley Berman, Mel Welles, Pat McCormick, Eileen Brennan; *Dir:* Robert Downey. **VHS, Beta, LV $9.99** CCB, IVE ✂

Repentance A popular, surreal satire of Soviet and Communist societies. A woman is arrested for repeatedly digging up a dead local despot, and put on trial. Wickedly funny and controversial; released under the auspices of glasnost. In Russian with English subtitles. **1987 (PG) 151m/C** RU Avtandil Makharadze, Zeinab Botsvadze, Ia Ninidze, Edisher Giorgobiani; *Dir:* Tenghiz Abuladze. Cannes Film Festival '87: Special Jury Prize; Chicago Film Festival '87: Special Jury Prize. **VHS, Beta, LV $79.95** MED ✂✂✂

Repo Jake Jake Baxter learns the myth of the carefree life of a repossession man in this action thriller. His illusion is destroyed by angry clients, pornography, a deadly underworld racetrack, and a sinister crime boss. **1990 90m/C** Dan Haggerty, Robert Axelrod. **VHS $79.95** PMH ✂½

Repo Man An inventive, perversely witty portrait of sick modern urbanity, following the adventures of a punk rocker taking a job as a car repossessor in a barren city. The landscape is filled with pointless violence, no-frills packaging, media hypnosis and aliens. Executive producer: none other than ex-Monkee Michael Nesmith. **1983 (R) 93m/C** Emilio Estevez, Harry Dean Stanton, Sy Richardson, Tracy Walter, Olivia Barash, Fox Harris, Jennifer Balgobin, Vonetta McGee, Angelique Pettyjohn; *Dir:* Alex Cox. **VHS, Beta, LV $19.95** MCA ✂✂✂½

Report to the Commissioner A rough, energetic crime-in-the-streets cop thriller. A young detective accidentally kills an attractive woman who turns out to have been an undercover cop, then is dragged into the bureaucratic cover-up. **1974 (R) 112m/C** Michael Moriarty, Yaphet Kotto, Susan Blakely, Hector Elizondo, Richard Gere, Tony King, Michael McGuire, Stephen Elliott; *Dir:* Milton Katselas. **VHS, Beta $29.95** MGM ✂✂✂

The Report on the Party and the Guests A controversial, widely banned political allegory, considered a masterpiece from the Czech new wave. A picnic/lawn party deteriorates into brutality, fascist intolerance and persecution. Many Czech film makers, some banned at the time, appear. In Czech with English titles. **1966 71m/B** CZ Jiri Nemec, Evald Schorm, Ivan Vyskocil, Jan Klusak, Zdena Skvorecka, Pavel Bosek; *Dir:* Jan Nemec. New York Film Festival '68: Special Jury Prize. **VHS, Beta $59.95** FCT ✂✂✂✂

Reporters An ironic look at Parisian photojournalists that features many of their notable subjects: Francois Mitterand, Jean Luc Godard, Gene Kelly and others. Conveys a sense of the entire human comedy as well as the work of photographers. **1981 101m/C** Catherine Deneuve, Richard Gere. Caeser '81: Best Documentary. **VHS, 3/4U, Special order formats** CIG ✂✂✂✂

Repossessed A corny, occasionally funny takeoff on the human-possessed-by-the-devil films with Blair, as usual, as the afflicted victim. Nielsen, in another typecast role, plays the goofy man of the cloth who must save the day. Numerous actors appear in brief comedy sequences which lampoon current celebrities and constitute the film's highlights. For fans of "Airplane" and "The Naked Gun"; not necessarily for others. **1990 (PG-13) 89m/C** Linda Blair, Ned Beatty, Leslie Nielsen, Anthony Starke, Jesse Ventura; *W/Dir:* Bob Logan. **VHS, LV $89.95** LIV ✂✂

Repulsion Character study of a young French girl who is repulsed and attracted by sex. Left alone when her sister goes on vacation, her facade of stability begins to crack, with violent and bizarre results. Polanski's first film in English and his first publicly accepted full-length feature. Suspenseful, disturbing and potent. **1965 105m/B** GB Catherine Deneuve, Yvonne Furneaux, Ian Hendry, John Fraser, Patrick Wymark, James Villers; *Dir:* Roman Polanski. **VHS, Beta $29.95** VDM, VCN, MRV ✂✂✂½

Requiem for Dominic A Communist-bloc country in the midst of revolution accuses a man of terrorist activities. Can his friend, exiled for years, help him discover the truth? Interesting and timely premise matched with outstanding ensemble acting make for a superior thriller. Amnesty International receives partial proceeds from all sales of this video. In German with English subtitles. **1991 (R) 88m/C** GE Felix Mitterer, Viktoria Schubert, August Schmolzer, Angelica Schutz; *Dir:* Robert Dornhelm. **VHS, LV $79.95** HMD ✂✂✂½

Requiem for a Heavyweight The original television version of the story about an American Indian heavyweight boxer played by Palance who risks blindness in order to help his manager pay off bookies. Highly acclaimed teleplay written by Rod Sterling for "Playhouse 90." **1956 90m/B** Jack Palance, Keenan Wynn, Ed Wynn, Kim Hunter, Ned Glass; *Dir:* Ralph Nelson. **VHS, Beta $14.95** WKV ✂✂✂

The Rescue An elite team of U.S. Navy Seals is captured after destroying a disabled submarine. When the U.S. government writes off the men, their children decide to mount a rescue. Puerile, empty-headed trash for teens. **1988 (PG) 97m/C** Marc Price, Charles Haid, Kevin Dillon, Christina Harnos, Edward Albert; *Dir:* Ferdinand Fairfax. **VHS, Beta, LV $89.95** TOU, HHE ✂½

Rescue Force A crack team of anti-terrorists, led by Lt. Col. Steel, takes on the kidnappers of an American ambassador. Action filled excitement. **1990 85m/C** Richard Harrison. **VHS** RAE ✂✂

The Rescuers Bernard and Miss Bianca, two mice who are members of the Rescue Aid Society, attempt to rescue a girl named Penny from the evil Madame Medusa. They are aided by comic sidekick, Orville the seagull. Very charming in the best Disney tradition. Based on the stories of margery Sharp. Followed by "The Rescuers Down Under." **1977 (G) 76m/C** *Dir:* Wolfgang Reitherman, John Lounsbery; *Voices:* Bob Newhart, Eva Gabor, Geraldine Page, Jim Jordan, Joe Flynn, Jeanette Nolan, Pat Buttram. **VHS $24.99** DIS ✂✂✂

The Rescuers Down Under 'Crocodile' Dundee goes Disney; the followup to "The Rescuers" places its characters in Australia with only mild results for a Magic Kingdom product. Heroic mice Bernard and Bianca protect a young boy and a rare golden eagle from a poacher. The great bird closely resembles the logo of Republic Pictures.
1990 (G) 77m/C Dir: Hendel Butoy, Mike Gabriel; **Voices:** Bob Newhart, Eva Gabor, John Candy, Tristan Rogers, George C. Scott. **VHS, Beta, LV** $29.99 DIS, IME, RDG ⅜⅜½

The Respectful Prostitute Sartre's tale of hooker who is cajoled into providing false testimony as a witness in the trial of a member of a prominent family. English language version of French drama.
1952 75m/C FR Ivan Desny, Barbara Laage, Walter Bryan; **Dir:** Marcel Pagliero, Charles Brabant. **VHS** $39.95 INF, FCT ⅜⅜½

Rest In Pieces A young woman inherits a Spanish mansion, and weird things begin to happen. It seems a cult of Satan worshippers is already living on the estate. As one might infer from the title, quite a bit of hacking and slashing ensues.
1987 (R) 90m/C Scott Thompson Baker, Lorin Jean, Dorothy Malone; **Dir:** Joseph Braunstein. **VHS, Beta** $14.95 IVE ⅜

Restless A restless homemaker begins an affair with a childhood friend. Director Cosmatos later oversaw "Rambo." Also known as "The Beloved."
1972 75m/C GR Raquel Welch, Richard Johnson, Jack Hawkins, Flora Robson; **Dir:** George P. Cosmatos. **VHS, Beta** $79.95 VMK ⅜

The Restless Breed Son is restless after dad, a secret service agent, is slain by gunrunners. Routine oater-revenge epic done in by listless script.
1958 81m/C Scott Brady, Anne Bancroft, Jay C. Flippen, Rhys Williams, Jim Davis; **Dir:** Allan Dwan. **VHS, LV** $19.95 FRH, IME ⅜

The Resurrected One of the best recent H.P. Lovecraft horror adaptations, a fairly faithful try at "The Case of Charles Dexter Ward." Ward learns he has a satanic ancestor who possessed the secret of resurrection and eternal life—but, as the warlock says, it is very messy. Surprisingly tasteful even with occasional gore and truly ghastly monsters; in fact this pic could have used a bit more intensity.
1991 (R) 108m/C John Terry, Jane Sibbet, Chris Sarandon; **Dir:** Dan O'Bannon. **VHS** $89.98 LIV ⅜⅜½

Resurrection After a near-fatal car accident, a woman finds she has the power to heal others by touch. She denies that God is responsible, much to her Bible Belt community's dismay. Acclaimed and well-acted.
1980 (PG) 103m/C Ellen Burstyn, Sam Shepard, Roberts Blossom, Eva Le Gallienne, Clifford David, Richard Farnsworth, Pamela Payton-Wright; **Dir:** Daniel Petrie. National Board of Review Awards '80: Best Supporting Actress (Le Gallienne); National Board of Review '80: 10 Best Films of the Year. **VHS, Beta** $59.95 MCA ⅜⅜⅜½

Resurrection at Masada A celebration of the fortieth anniversary of the founding of the nation of Israel.
1989 100m/C Hosted: Gregory Peck, Yves Montand. **VHS, Beta** $29.95 KUL ⅜⅜⅜½

Resurrection of Zachary Wheeler A presidential candidate, who narrowly escaped death in an auto crash, is brought to a mysterious clinic in New Mexico. A reporter sneaks into the clinic and discovers the horrors of cloning.
1971 (G) 100m/C Angie Dickinson, Bradford Dillman, Leslie Nielsen, Jack Carter, James Daly; **Dir:** Bob Wynn. **VHS, Beta** $29.95 UHV ⅜⅜½

Retreat, Hell! Korean War drama about the retreat from the Changjin Reservoir. Standard gun-ho military orientation with dismal results.
1952 95m/B Frank Lovejoy, Richard Carlson, Russ Tamblyn, Anita Louise; **Dir:** Joseph H. Lewis. **VHS** $19.98 REP ⅜⅜

Retribution An artist survives a suicide attempt, only to find himself possessed by the spirit of a criminal who was tortured to death. Seeing the hood's gruesome demise in his dreams, the survivor sets out to bring his murderers face-to-face with their maker.
1988 (R) 106m/C Dennis Lipscomb, Hoyt Axton, Leslie Wing, Suzanne Snyder; **Dir:** Guy Magar. **VHS, Beta, LV** $89.95 VIR ⅜

Retrievers A young man and a former CIA agent team up to expose the unsavory practices of the organization.
1982 90m/C Max Thayer, Roselyn Royce. **VHS, Beta** $69.98 VES ⅜

The Return Vincent and Shepherd meet as adults and discover that they had both, as children, been visited by aliens who had given technology to a cattle-mutilating prospector. A mess with no idea of what sort of film it wants to be. Also titled, "The Alien's Return."
1980 (PG) 91m/C Raymond Burr, Cybill Shepherd, Martin Landau, Jan-Michael Vincent, Neville Brand; **Dir:** Greydon Clark. **VHS, Beta** $19.99 HBO Woof!

Return A woman believes that the man she loves is the reincarnation of her dead grandfather. Interesting concept, but overall, a disappointing mystery.
1988 (R) 78m/C Frederic Forrest, Anne Francis, Karlene Crockett, John Walcutt, Lisa Richards; **Dir:** Andrew Silver. **VHS, Beta** $79.95 ACA ⅜⅜

Return to Africa Rebels threaten the familial paradise of the tightknit Mallory clan.
1989 95m/C Stan Brock, Anne Collings, David Tors, Ivan Tors, Peter Tors, Steven Tors; **Dir:** Leslie Martinson. **VHS, Beta** $59.95 PSM ⅜⅜

Return of the Aliens: The Deadly Spawn Aliens infect the earth and violently destroy humans. Extremely violent and gory. Watch out for those officious offspring.
1983 (R) 90m/C Charles George Hildebrandt; **Dir:** Douglas McKeown. **VHS** $39.98 CON ⅜

Return of the Ape Man Campy fun with Carradine and Lugosi. A mad scientist transplants Carradine's brain into the body of the "missing link." Hams galore.
1944 51m/B Bela Lugosi, John Carradine, George Zucco, Judith Gibson, Michael Ames, Frank Moran, Mary Currier; **Dir:** Phil Rosen. **VHS** $59.95 MED ⅜⅜

Return of the Bad Men Scott is a retired marshall who must fight against a gang of outlaws lead by the Sundance Kid. Sequel to "Badman's Territory."
1948 90m/C Randolph Scott, Robert Ryan, Anne Jeffreys, George "Gabby" Hayes, Jason Robards Jr., Jacqueline White; **Dir:** Ray Enright. **VHS, Beta** $19.98 RKO, CCB, MED ⅜⅜

Return to the Blue Lagoon Neither the acting nor the premise has improved with age. Another photogenic adolescent couple experiences puberty on that island; for continuity, the young woman is the daughter of the lovers in the first "Blue Lagoon." Breathtaking scenery—just turn down the sound.
1991 (PG-13) 102m/C Milla Jovovich, Brian Krause, Lisa Pelikan; **Dir:** William A. Graham. **VHS, LV** COL, PIA ⅜

Return to Boggy Creek The townspeople in a small fishing village learn from a photographer that a "killer" beast, whom they thought had disappeared, has returned and is living in Boggy Creek. Some curious children follow the shutterbug into the marsh, despite hurricane warnings, and the swamp monster reacts with unusual compassion. OK for the kiddies. Ficticious story unlike other "Boggy Creek" films, which are billed as semi-documentaries.
1977 (PG) 87m/C Dawn Wells, Dana Plato, Louise Belaire, John Hofeus; **Dir:** Tom Moore. **VHS, Beta** $19.98 FOX ⅜½

The Return of Boston Blackie The oft-filmed crook/detective tries to retrieve an heiress's stolen jewels. Silent.
1927 77m/B Raymond Glenn, Corliss Palmer, Strongheart, Rosemary Cooper, Coit Albertson; **Dir:** Harry Hoyt. **VHS, Beta, 8mm** $29.95 VYY ⅜½

Return of Bruce No one is quite sure if Bruce Le was ever here before, but he claims to be returning now. Slam-bang, chop-socky action.
1982 90m/C Bruce Le. **VHS, Beta** $29.99 BFV ⅜½

Return of Captain Invincible Arkin is a derelict super-hero persuaded to fight crime again in this unique spoof. One liners fly faster than a speeding bullet. Lee gives one of his best performances as the mad scientist. Musical numbers by Rocky Horror's O'Brien and Hartley sporadically interrupt. Made on a shoe-string budget, offbeat film is entertaining but uneven.
1983 (PG) 90m/C AU Alan Arkin, Christopher Lee, Kate Fitzpatrick, Bill Hunter, Graham Kennedy, Michael Pate, Hayes Gordon, Max Phipps, Noel Ferrier; **Dir:** Philippe Mora. **VHS, Beta, LV** $79.98 MAG ⅜⅜½

Return of Chandu This serial in 12 chapters features Bela Lugosi as Chandu, who exercises his magical powers to conquer a religious sect of cat worshippers inhabiting the island of Lemuria. In the process, he fights to save the Princess Nadji from being sacrificed by them.
1934 156m/B Bela Lugosi, Maria Alba, Clara Kimball Young; **Dir:** Ray Taylor. **VHS, Beta** $24.95 DVT, NOS, SNC ⅜⅜

Return of the Chinese Boxer Samurai and ninja warriors fight, usually to the death.
1974 93m/C Jimmy Wang Yu, Lung Fei, Cheung Ying Chen, Chin Kang. **VHS, Beta** $39.95 UNI ⅜

Return of the Dinosaurs A comet is about to collide with the Earth. Never fear! The dinosaur patrol will interfere (and ultimately save the planet).
1983 82m/C VHS, Beta $39.95 TWE ⅜

The Return of Dr. Mabuse The evil doctor is back, this time sending his entranced slaves to attack a nuclear power plant. Frobe, the inspector, and Barker, the FBI man, team up to thwart him. Fun for fans of serial detective stories.

1961 88m/B *GE FR IT* Gert Frobe, Lex Barker, Daliah Lavi, Wolfgang Preiss, Fausto Tozzi, Rudolph Forster; *Dir:* Harold Reinl. **VHS, Beta** $16.95 SNC ♫♫½

Return of the Dragon In Lee's last picture, a Chinese restaurant in Rome is menaced by gangsters who want to buy the property. On behalf of the owners, Lee duels an American karate champ in the Roman forum. The battle scenes between Lee and Norris are great and make this a must-see for martial arts fans.
1973 (R) 91m/C *CH* Bruce Lee, Nora Miao, Chuck Norris; *Dir:* Bruce Lee. **VHS, Beta, LV** $19.98 FOX, GEM ♫♫½

Return of the Dragon For vengeance's sake, a wronged karate master goes on a rampage of abundant whacking, slashing and head kicking. Not to be confused with the 1973 film of the same name that marked Bruce Lee's final appearance.
1984 110m/C Ramon Zanora, Lotis Key, Leila Hermosa. **VHS, Beta** $49.98 WES ♫

Return of Draw Egan Vintage silent western wherein Hart is an outlaw-turned-greenhorn sheriff of a lawless town. Music score.
1916 64m/B William S. Hart, Louise Glaum; *Dir:* William S. Hart. **VHS, Beta, 8mm** $29.95 GPV, VYY ♫½

Return to Earth Docudrama deals with the self-doubts faced by Apollo 11 astronaut Buzz Aldrin after his triumphant return to Earth.
1976 74m/C Cliff Robertson, Shirley Knight, Ralph Bellamy, Stefanie Powers, Charles Cioffi; *Dir:* Judd Taylor. **VHS, Beta** $19.98 TWE ♫♫

Return to Eden A young woman falls in love with her philosophy tutor, but he can't return her affections. An interesting comparison between the passion of the primitive and the tangle of modern-day romance.
1989 90m/C Sam Bottoms, Edward Binns, Renee Coleman; *Dir:* William Olsen. **VHS** $69.95 QUE ♫♫

Return Engagement A lonely middle-aged ancient history professor falls in love with one of her students.
1978 76m/C Elizabeth Taylor, Joseph Bottoms, Peter Donat, James Ray; *Dir:* Joseph Hardy. **VHS, Beta** $69.98 LTG ♫½

Return of the Evil Dead The sightless dead priests return to attack still more 1970s' Europeans in this second installment of the "blind dead" trilogy. Preceded by "Tombs of the Blind Dead" and followed by "Horror of the Zombies." Originally titled "Return of the Blind Dead." Not to be confused with Raimi's "Evil Dead" slasher flicks.
1975 85m/C *SP PT* Tony Kendall, Esther Roy, Frank Blake; *Dir:* Armando de Ossorio. **VHS** $59.95 JEF ♫

Return of the Family Man A mass murderer seems unstoppable as he continues to slash whole families at a time...the family man is redefined in this bloody gorefest.
1989 90m/C Ron Smerczak, Liam Cundill, Terence Reis, Michelle Constant; *Dir:* John Murlowski. **VHS** RAE ♫

Return to Fantasy Island The second pilot television movie for the popular series wherein three couples get their most cherished fantasy fulfilled by Mr. Roarke and Company.

1977 100m/C Ricardo Montalban, Herve Villechaize, Adrienne Barbeau, Pat Crowley, Joseph Campanella, Karen Valentine, Laraine Day, George Maharis, Horst Buchholz, France Nuyen, Joseph Cotten, Cameron Mitchell, George Chakiris; *Dir:* George McCowan. **VHS, Beta** $79.95 PSM ♫½

Return Fire A former employee of the government finds that he must now fight the system if he wants to save his son. Trivial action fare.
1991 (R) 97m/C Adam West, Ron Marchini. **VHS** $79.95 AIP ♫½

Return of Fist of Fury A kung-fu film featuring Bruce Le.
1977 88m/C Bruce Le. **VHS, Beta** $29.99 BFV ♫

Return of the Fly The son of the scientist who discovered how to move matter through space decides to continue his father's work, but does so against his uncle's wishes. He soon duplicates his dad's experiments with similar results. This sequel to "The Fly" doesn't buzz like the original. Followed by "Curse of the Fly."
1959 80m/B Vincent Price, Brett Halsey, John Sutton, Dan Seymour, David Frankham; *Dir:* Edward L. Bernds. **VHS, Beta, LV** $14.98 FOX, FCT ♫♫

Return of Frank Cannon Portly private eye Frank Cannon comes out of retirement to investigate the alleged suicide of an old friend. Made-for-television.
1980 96m/C William Conrad, Diana Muldaur, Joanna Pettet, Allison Argo, Arthur Hill, Ed Nelson; *Dir:* Corey Allen. **VHS, Beta** $39.95 WOV ♫♫

Return of Frank James One of Lang's lesser Hollywood works, this is nonetheless an entertaining sequel to 1939's "Jesse James." Brother Frank tries to go straight, but eventually has to hunt down the culprits who murdered his infamous outlaw sibling. Tierney's first film.
1940 92m/C Henry Fonda, Gene Tierney, Jackie Cooper, Henry Hull, John Carradine, Donald Meek, J. Edward Bromberg; *Dir:* Fritz Lang. **VHS, Beta, LV** $19.98 FOX ♫♫♫

The Return of Grey Wolf A wild dog must fight for survival in Northern Quebec and is befriended by a fur trapper. An intelligent hound, he repays the trapper's generosity by helping him put the moves on some fur thieves. Silent.
1922 62m/B Helen Lynch, Henry Pierce. **VHS, Beta, 8mm** VYY ♫♫

Return to Horror High A horror movie producer makes a film in an abandoned and haunted high school, where a series of murders occurred years earlier. As in most "return" flicks, history repeats itself.
1987 (R) 95m/C Alex Rocco, Vince Edwards, Philip McKeon, Brendan Hughes, Lori Lethin, Scott Jacoby; *Dir:* Bill Froehlich. **VHS, Beta** $19.95 STE, NWV ♫

Return of the Jedi Third film in George Lucas' popular space saga. Against seemingly fearsome odds, Luke Skywalker battles such worthies as Jabba the Hutt and heavy-breathing Darth Vader to save his comrades and triumph over the evil Galactic Empire. Han and Leia reaffirm their love and team up with C3PO, R2-D2, Chewbacca, Calrissian, and a bunch of furry Ewoks to aid in the annihilation of the Dark Side. The special effects are still spectacular, even the third time around. Sequel to "Star Wars" (1977) and "The Empire Strikes Back" (1980).

1983 (PG) 132m/C Mark Hamill, Carrie Fisher, Harrison Ford, Billy Dee Williams, David Prowse, James Earl Jones, Kenny Baker, Denis Lawson, Anthony Daniels, Peter Mayhew; *Dir:* Richard Marquand; *Voices:* Alec Guinness, Frank Oz. Academy Awards '83: Best Visual Effects. **VHS, Beta, LV** $19.98 FOX, BUR, RDG ♫♫♫½

Return of Jesse James Look-alike of dead outlaw Jesse James joins in with some former members of the James Gang, leading townsfolk to believe the notorious bank robber never died. It's up to brother Frank James, now an upstanding citizen, to set the record straight. A little slow moving at times, but still worth viewing.
1950 77m/B John Ireland, Ann Dvorak, Reed Hadley, Henry Hull, Hugh O'Brian, Tommy Noonan, Peter Marshall; *Dir:* Arthur Hilton. **VHS, Beta** WGE ♫♫

The Return of Josey Wales No match for the original featuring Clint Eastwood, the title character is played woodenly by Parks. This installment pitted Wales against a bumbling lawman. Sequel to 1976's "The Outlaw Josey Wales."
1986 (R) 90m/C Michael Parks, Rafael Campos, Bob Magruder, Paco Vela, Everett Sifuentes, Charlie McCoy; *Dir:* Michael Parks. **VHS** $79.95 MAG, HHE ♫

Return of the Killer Tomatoes The man-eating plant-life from 1980's "Attack of the Killer Tomatoes" is back, able to turn into people due to the slightly larger budget. Astin is mad as the scientist. Not as bad as "Attack," representing a small hurdle in the history of filmdom.
1988 (PG) 98m/C Anthony Starke, George Clooney, Karen Mistal, Steve Lunquist, John Astin, Charlie Jones, Rock Peace, Frank Davis, C.J. Dillon, Teri Weigel; *Dir:* John DeBello. **VHS, Beta, LV** $19.95 STE, NWV ♫

Return of the Lash Lash LaRue once again defends settlers' rights from ruthless outlaws.
1947 53m/B Lash LaRue, Al "Fuzzy" St. John, Mary Maynard, George Cheseboro; *Dir:* Ray Taylor. **VHS, Beta** $12.95 COL, SVS ♫

Return of the Living Dead Poisonous gas revives a cemetery and a morgue rendering an outrageous spoof on the living-dead sub-genre with fast-moving zombies, punk humor, and exaggerated gore. Its humor does not diminish the fear factor, however. Sequel follows. Directed by "Alien" writer O'Bannon.
1985 (R) 90m/C Clu Gulager, James Karen, Linnea Quigley, Don Calfa, Jewel Shepard; *Dir:* Dan O'Bannon. **VHS, Beta, LV** $14.95 HBO, HMD ♫♫½

Return of the Living Dead Part 2 An inevitable sequel to the original Dan O'Bannon satire about George Romero-esque brain-eating zombies attacking suburbia with zest and vigor.
1988 (R) 89m/C Dana Ashbrook, Marsha Dietlein, Philip Burns, James Karen, Thom Mathews; *Dir:* Ken Weiderhorn. **VHS, Beta, LV** $89.95 LHV, WAR ♫½

Return to Macon County Tale of three young, reckless youths who become involved with drag racing and a sadistic law enforcement officer. Nolte turns in a good performance in his first film, as does Johnson, despite a mediocre script. Sequel to 1974's "Macon County Line."
1975 90m/C Nick Nolte, Don Johnson, Robin Mattson; *Dir:* Richard Compton. **VHS, Beta** $69.98 LIV, VES ♫½

The Return of a Man Called Horse Sequel to "A Man Called Horse" tells the story of an English aristocrat who was captured and raised by Sioux Indians, and then returned to his native homeland. Contains more of the torture scenes for which this series is famous, but not much of Harris.
1976 (PG) 125m/C Richard Harris, Gale Sondergaard, Geoffrey Lewis; *Dir:* Irvin Kershner. VHS, Beta $59.98 FOX ♂♂

Return of the Man from U.N.C.L.E. Those dashing super agents Napoleon Solo and Illya Kuryakin come out of retirement to settle an old score with their nemesis THRUSH. Made-for-television.
1983 96m/C Robert Vaughn, David McCallum, Patrick Macnee, Gayle Hunnicutt, Geoffrey Lewis; *Dir:* Ray Austin. VHS, Beta $69.95 TWE ♂♂

Return of Martin Guerre In this medieval tale, a dissolute village husband disappears soon after his marriage. Years later, someone who appears to be Martin Guerre returns, allegedly from war, and appears much kinder and more educated. Starring in a love story of second chances, Depardieu does not disappoint, nor does the rest of the cast. In French with English subtitles. Based on an actual court case.
1983 111m/C FR Gerard Depardieu, Roger Planchon, Maurice Jacquemont, Barnard Pierre Donnadieu, Nathalie Baye; *Dir:* Daniel Vigne. VHS, Beta, LV $29.98 SUE, APD ♂♂♂½

Return to Mayberry Andy Taylor returns to Mayberry after 20 years to reobtain his old job as sheriff. Sixteen of the original actors reappeared for this nostalgia-fest. Made for television.
1985 95m/C Andy Griffith, Ron Howard, Don Knotts; *Dir:* Bob Sweeney. VHS, Beta $29.98 MCG ♂♂

Return of Our Gang Contains three vintage silent Pathe "Our Gang" shorts: "The School Play (Stage Fright)," "Summer Daze (The Cobbler)," and "Dog Days."
1925 57m/B Joe Cobb, Ernie Morrison, Mickey Daniels, Farina Hoskins, Mary Kornman, Jackie Condon, Jack Davis, Jannie Hoskins, Andy Samuels, Eugene Jackson, Pete the Pup. VHS, Beta, 8mm $24.95 VYY ♂½

Return to Oz Picking up where "The Wizard of Oz" left off, Auntie Em and Uncle Ed place Dorothy in the care of a therapist to cure her "delusions" of Oz. A natural disaster again lands her in the land of the yellow brick road, where the evil Nome King and Princess Mombi are spreading terror and squalor. Based on a later L. Frank Baum book. Enjoyable for the whole family although some scenes may frighten very small children.
1985 (PG) 109m/C Fairuza Balk, Piper Laurie, Matt Clark, Nicol Williamson, Jean Marsh; *Dir:* Walter Murch. VHS, Beta, LV $29.95 DIS ♂♂½

The Return of Peter Grimm Man returns from death to try to tidy up the mess he left behind. Enjoyable if trite fantasy based on a play by David Belasco.
1935 82m/B Lionel Barrymore, Helen Mack, Edward Ellis, Donald Meek, George Breakston, James Bush; *Dir:* George Nicholls. VHS $19.98 TTC, FCT ♂♂½

Return to Peyton Place A young writer publishes novel exposing town as virtual Peyton Place, and the townsfolk turn against her and her family. Sequel to the scandalously popular '50s original and inspiration for the just-as-popular soap opera. Astor is excellent as the evil matriarch.

1961 122m/C Carol Lynley, Jeff Chandler, Eleanor Parker, Mary Astor, Robert Sterling, Lucianna Paluzzi, Tuesday Weld, Brett Halsey, Bob Crane; *Dir:* Jose Ferrer. VHS $39.98 FOX, FCT ♂♂

Return of the Pink Panther Bumbling Inspector Clouseau is called upon to rescue the Pink Panther diamond stolen from a museum. Sellers manages to produce mayhem with a vacuum cleaner and other devices that, in his hands, become instruments of terror. Clever opening credits. Fourth installment in the Pink Panther series but the first with Sellers since 1964's "A Shot In The Dark."
1974 (G) 113m/C Peter Sellers, Christopher Plummer, Catherine Schell, Herbert Lom, Victor Spinetti; *Dir:* Blake Edwards. VHS, Beta, LV $19.95 FOX, JTC, FCT ♂♂½

Return of the Rebels Made for television perishable sports erstwhile genie Eden as a motorcycle matron whose campground is rid of riff-raff by the crow-lined participants in a 25-year reunion of a biker gang.
1981 100m/C Barbara Eden, Robert Mandan, Jamie Farr, Patrick Swayze, Don Murray, Christopher Connelly; *Dir:* Noel Nosseck. VHS $69.99 TRY ♂

Return of the Red Tiger Kung-Fu master Bruce Lee displays one of the most exciting styles of kung-fu: The Tiger.
1981 82m/C Bruce Lee. VHS, Beta $59.95 HAR ♂½

The Return of Ruben Blades Musical portrait of musician/actor Ruben Blades ("Milagro Bean War"; "Mo' Better Blues") contains an interview with Pete Hamill, a performance with his band Seis del Solar and Linda Ronstadt, and musical selections such as "Buscando America," "Tiburon," "Todos Vuelven," and "Silencios."
1987 82m/C Ruben Blades, Linda Ronstadt. VHS $29.95 SMV, MVD ♂♂

Return to Salem's Lot Enjoyable camp sequel to the Stephen King tale, this time involving a cynical scientist and his son returning to the town only to find it completely run by vampires.
1987 (R) 101m/C Michael Moriarty, Ricky Addison Reed, Samuel Fuller, Andrew Duggan, Evelyn Keyes, Jill Gatsby, June Havoc, Ronee Blakley, James Dixon, David Holbrook; *Dir:* Larry Cohen. VHS, Beta $79.95 WAR ♂♂

Return of the Scorpion Big trouble abounds as that crazy kung fu guy, the "Scorpion," comes back to town.
198? 90m/C Yang Pan Pan, Lam Men Wei, Philip Ko. VHS, Beta $49.95 REG ♂

Return of the Secaucus 7 Centers around a weekend reunion of seven friends who were activists during the Vietnam War in the turbulent '60s. Now turning 30, they evaluate their present lives and progress. Writer and director Sayles plays Howie in this excellent example of what a low-budget film can and should be. A less trendy predecessor of "The Big Chill" (1983) which, perhaps, was a few years ahead of its time.
1980 110m/C Mark Arnott, Gordon Clapp, Maggie Cousineau-Arndt, David Strathairn; *Dir:* John Sayles. Los Angeles Film Critics Association Awards '80: Best Screenplay (John Sayles). VHS, Beta $59.95 COL ♂♂♂½

Return of the Seven The first sequel to "The Magnificent Seven" features the group liberating a compatriot who is held hostage. Yawn.

1966 97m/C Yul Brynner, Warren Oates, Robert Fuller, Claude Akins; *Dir:* Burt Kennedy. VHS, Beta $19.98 MGM ♂♂

Return to Snowy River Continues the love story of the former ranch hand and the rancher's daughter in Australia's Victoria Alps that began in "The Man From Snowy River." Dennehy takes over from Kirk Douglas as the father who aims to keep the lovers apart. The photography of horses and the scenery is spectacular, making the whole film worthwhile.
1988 (PG) 99m/C AU Tom Burlinson, Sigrid Thornton, Brian Dennehy, Nicholas Eadie, Mark Hembrow, Bryan Marshall; *Dir:* Geoff Burrowes. VHS, Beta, LV $19.99 DIS ♂♂♂

Return of the Soldier A shell-shocked World War I veteran has no memory of his marriage, leaving his wife, his childhood flame, and an unrequited love to vie for his affections. Adapted from the novel by Rebecca West.
1984 101m/C GB Glenda Jackson, Julie Christie, Ann-Margret, Alan Bates, Ian Holm, Frank Finlay; *Dir:* Alan Bridges. VHS, Beta $79.99 HBO ♂♂♂

Return of Superfly Man with insect moniker pesters drug dealers and cops in Harlem. Long-awaited sequel to urban epic "Superfly," with music again by Curtis Mayfield.
1990 (R) 94m/C Margaret Avery, Nathan Purdee, Sam Jackson. VHS $92.95 VMK ♂

The Return of Swamp Thing The DC Comics creature rises again out of the muck to fight mutants and evil scientists. Tongue-in-cheek, and nothing at all like the literate, ecologically-oriented comic from which it was derived.
1989 (PG-13) 95m/C Louis Jourdan, Heather Locklear, Sarah Douglas, Dick Durock; *Dir:* Jim Wynorski. VHS, Beta, LV $14.95 COL ♂½

Return of the Tall Blond Man With One Black Shoe Sequel to 1972's "The Tall Blond Man With One Black Shoe." Again a "klutz" is mistaken for a master spy and becomes unknowingly involved in the world of international intrigue. French with English subtitles. Not quite as good as the original.
1974 84m/C FR Pierre Richard, Mireille Darc, Michael Duchaussoy, Jean Rochefort, Jean Carmet; *Dir:* Yves Robert. VHS, Beta $49.95 PSM, CCI ♂♂½

Return of the Tiger The Hovver Night Club in Bangkok fronts the operations of an international narcotics group headed by an American. When a rival Chinese gang tries to dominate the drug market, conflict and lots of kicking ensue.
1978 95m/C Bruce Li, Paul Smith, Chaing I, Angela Mao; *Dir:* Jimmy Shaw. VHS, Beta $29.95 MED ♂

Return to Treasure Island, Vol. 1 Series based on the Robert Louis Stevenson classic with the continuing adventures of Jim Hawkins and Long John Silver. Episodes: "The Map" and "The Mutiny."
1985 101m/C Brian Blessed, Kenneth Colley, Christopher Guard, Reiner Schone; *Dir:* Piers Haggard. VHS, Beta $49.95 DIS ♂

Return to Treasure Island, Vol. 2 Episodes include: "The Island of the Damned" and "Jamaica." Drifting aimlessly, Jim Hawkins and Long John Silver land on an island filled with dangers, including madmen and Spanish soldiers.

1985 108m/C Brian Blessed, Kenneth Colley, Christopher Guard, Reiner Schone; *Dir:* Piers Haggard. **VHS, Beta $49.95** *DIS* *✗*

Return to Treasure Island, Vol.
3 Episodes include: "Manhunt" and "The Crow's Nest." Framed for murder, Jim Hawkins goes on the lam and runs into his old pal Long John Silver.
1985 107m/C Brian Blessed, Kenneth Colley, Christopher Guard, Reiner Schone; *Dir:* Piers Haggard. **VHS, Beta $49.95** *DIS* *✗*

Return to Treasure Island, Vol.
4 Episodes include: "Fugitives" and "In Chains." Held prisoner, Jim Hawkins gets an unexpected cell-mate in his friend, Long John Silver.
1985 108m/C Brian Blessed, Kenneth Colley, Christopher Guard, Reiner Schone; *Dir:* Piers Haggard. **VHS, Beta $49.95** *DIS* *✗*

Return to Treasure Island, Vol.
5 Episodes include: "Spanish Gold" and "Treasure Island." In the final volume, Jim Hawkins and Long John Silver escape from prison and head for Treasure Island in search of gold.
1985 107m/C Brian Blessed, Kenneth Colley, Christopher Guard, Reiner Schone; *Dir:* Piers Haggard. **VHS, Beta $49.95** *DIS* *✗*

Return of the Vampire A Hungarian
vampire and his werewolf servant seek revenge on the family who drove a spike through his heart two decades earlier.
1943 69m/B Bela Lugosi, Nina Foch, Miles Mander, Matt Willis, Frieda Inescort, Roland Varno; *Dir:* Lew Landers, Kurt Neumann. **VHS, Beta, LV $9.95** *GKK, MLB* *✗✗*

Return of Video Yesterbloop A
compilation of on-set movie blunders and foul-ups, from 1941 to 1947.
1947 27m/B Gary Cooper, Edward G. Robinson, Humphrey Bogart, Bette Davis, James Cagney, Pat O'Brien, George Raft, Ronald Reagan, Eddie Albert, James Stewart, Claude Rains, Dick Foran, Errol Flynn, Eve Arden, Ann Sheridan, Lauren Bacall, Edward Arnold, Patricia Neal, Joan Crawford. **VHS, Beta, 8mm** *VYY* *✗✗*

Return of Wildfire Two sisters are left
to run the family horse ranch when their brother is killed. A drifter aids the sisters in bringing the murderer to justice.
1948 83m/B Richard Arlen, Patricia Morison, Mary Beth Hughes. **VHS, Beta** *WGE* *✗½*

Return from Witch Mountain A pair
of evil masterminds use a boy's supernatural powers to place Los Angeles in nuclear jeopardy. Sequel to Disney's ever-popular "Escape to Witch Mountain."
1978 (G) 93m/C Christopher Lee, Bette Davis, Ike Eisenmann, Kim Richards, Jack Soo; *Dir:* John Hough. **VHS, Beta** *✗½*

Reuben, Reuben Brilliant, but drunken
poet Conti turns himself around when he falls in love with earthy college girl McGillis in her film debut. The student's dog Reuben, unwittingly alters Conti's progress, however, in the film's startling conclusion. Nominated for best actor and best screenplay. Based on the writings of Peter DeVries.
1983 (R) 100m/C Tom Conti, Kelly McGillis, Roberts Blossom, E. Katherine Kerr, Cynthia Harris, Joel Fabiani, Kara Wilson, Lois Smith; *Dir:* Robert Ellis Miller. **VHS, Beta $59.98** *FOX* *✗✗✗*

Reunion Well-heeled young Jewish boy is
sent to the U.S. for safety in the 1930s, while his friend, an ambassador's son, remains in Germany. Fifty years later, he returns to Stuttgart on business and encounters his old

friend, a shock for both. Thoughtful and well-acted though occasionally plodding drama written by Harold Pinter and based on Fred Uhlman's novel.
1988 (PG-13) 120m/C Jason Robards Sr.; *Dir:* Jerry Schatzberg. **VHS, Beta, LV $29.95** *FRH, IME, FCT* *✗✗✗*

Reunion: 10th Annual Young
Comedians The mega-comedians return to their roots and give the most hilarious performances of their lives.
1987 60m/C Robin Williams, Steven Wright, Howie Mandel, Harry Anderson, Richard Belzer. **VHS, Beta $14.95** *HBO* *✗✗*

Reunion in France A flag-waving
woman's picture about a Parisian dress designer who sacrifices her lifestyle to help an American flier flee France after the Nazis invade. Dated patriotic flag-waver.
1942 104m/B Joan Crawford, John Wayne, Philip Dorn, Reginald Owen, Albert Basserman, John Carradine, Anne Ayars, J. Edward Bromberg, Henry Daniell, Moroni Olsen, Howard da Silva, Ava Gardner, John Considine; *Dir:* Jules Dassin. **VHS, Beta $19.98** *MGM* *✗✗*

Revealing of Elsie A young dancer
goes from bed to bed in her rise to the top. Previously titled "Strip for Action."
1986 70m/C Christa Free, Marianne Dupont. **VHS, Beta $29.95** *MED* *✗*

Revenge The parents of a girl who was
brutally killed take the law into their own hands in this bloody, vengeful thriller. Also known as "Terror under the House," "After Jenny Died," and "Inn of the Frightened People."
1971 89m/C *GB* Joan Collins, James Booth, Ray Barrett, Sinead Cusack, Kenneth Griffith; *Dir:* Sidney Hayers. **VHS, Beta $29.95** *SUE, AXV, MRV* *✗½*

Revenge Shot-on-video sequel to "Blood
Cult" about a devilworshippers lead by horror king Carradine.
1986 104m/C John Carradine, Patrick Wayne, Bennie Lee McGowan; *Dir:* Christopher Lewis. **VHS, Beta $19.95** *UHV* Woof!

Revenge Retired pilot Costner makes the
mistake of falling in love with another man's wife. Quinn gives a first-rate performance as the Mexican crime lord who punishes his spouse and her lover, beginning a cycle of vengeance. Sometimes contrived, but artfully photographed with tantalizing love scenes. Based on the Jim Harrison novel.
1990 (R) 123m/C Kevin Costner, Anthony Quinn, Madeleine Stowe, Sally Kirkland, Joe Santos, Miguel Ferrer, James Gammon, Tomas Milian; *Dir:* Tony Scott. **VHS, Beta, LV, 8mm $19.95** *COL* *✗✗*

Revenge of the Barbarians As the
barbarians descend on Rome, Olympus must decide whether to save the empire or the beautiful Gallo, the woman he loves.
1964 99m/C Robert Alda, Anthony Steele. **Beta $39.95** *AIP* *✗*

Revenge of the Barbarians A
Scandinavian adventure about a young Celt seeking revenge against the Viking hordes that killed his family.
1985 95m/C *Dir:* Hrafn Gunnlaugsson. **Beta $79.95** *VHV* *✗*

Revenge of the Cheerleaders The
cheerleaders pull out all the stops to save their school from a ruthless land developer. What spirit.

1976 (R) 86m/C Jerii Woods, Cheryl "Rainbeaux" Smith, Helen Lang, Patrice Rohmer, Susie Elene, Eddra Gale, William Bramley, Carl Ballantine, David Hasselhoff; *Dir:* Richard Lerner. **VHS, Beta $69.98** *LIV, LTG* *✗*

Revenge of the Dead A European
archeological team discovers the existence of a powerful force that allows the dead to return to life.
1984 100m/C *IT* Gabriele Lavia, Anne Canoras; *Dir:* Pupi Avati. **VHS, Beta $69.98** *LIV, LTG* Woof!

Revenge of Doctor X A man/beast
lies waiting for the pretty reporter who seeks her shipwrecked father in the jungles of an uncharted island.
198? 90m/C John Ashley, Ronald Remy, Angelique Pettyjohn. **VHS, Beta $49.95** *REG* Woof!

Revenge of Fist of Fury Kung fu
masters beat one another to a pulp in tribute to Bruce Lee.
197? 93m/C **VHS, Beta $39.95** *UNI* *✗*

Revenge in the House of Usher
House holds grudge.
1982 90m/C Howard Vernon, Dan Villers; *Dir:* Jess (Jesus) Franco. **VHS $29.99** *MOV* *✗✗*

Revenge of a Kabuki Actor When
Kabuki actor Kazuo Hasegawa's parents are viciously murdered, he seeks their killer with a vengeance. In Japanese with English subtitles.
1963 114m/C *JP Dir:* Kon Ichikawa. **VHS $59.95** *FST, FCT* *✗✗*

Revenge of the Living Zombies
Teens on a Halloween hayride run headlong into flesh-craving zombies. Available in a slightly edited version.
1988 (R) 85m/C Bill Hinzman, John Mowod, Leslie Ann Wick, Kevin Kindlan; *Dir:* Bill Hinzman. **VHS, Beta $79.98** *MAG* *✗*

Revenge of the Nerds When nerdy
college freshmen are victimized by jocks, frat boys and the school's beauties, they start their own fraternity and seek revenge. Carradine and Edwards team well as the geeks in this better than average teenage sex comedy. Guess who gets the girls? Sequel was much worse.
1984 (R) 90m/C Robert Carradine, Anthony Edwards, Timothy Busfield, Andrew Cassese, Curtis Armstrong, Larry B. Scott, Brian Tochi, Julia Montgomery, Michelle Meyrink, Ted McGinley, John Goodman, Bernie Casey; *Dir:* Jeff Kanew. **VHS, Beta, LV $14.98** *FOX* *✗✗½*

Revenge of the Nerds 2: Nerds
in Paradise The nerd clan from the first film (minus Edwards who only makes a brief appearance) travels to Fort Lauderdale for a fraternity conference. They fend off loads of bullies and jocks with raunchy humor. Boy can Booger belch. Not as good as the first Nerds movie; a few laughs nonetheless.
1987 (PG-13) 89m/C Robert Carradine, Curtis Armstrong, Timothy Busfield, Andrew Cassese, Ed Lauter, Larry B. Scott, Courtney Thorne-Smith, Anthony Edwards, James Hong; *Dir:* Joe Roth. **VHS, Beta, LV $89.98** *FOX* *✗*

Revenge of the Ninja A Ninja hopes
to escape his past in Los Angeles. A drug trafficker, also Ninja-trained, prevents him. The two polish off a slew of mobsters before their own inevitable showdown. Sequel to "Enter the Ninja," followed by "Ninja III."
1983 (R) 90m/C Sho Kosugi; *Dir:* Sam Firstenberg. **VHS, Beta $19.95** *MGM* *✗½*

Revenge of the Pink Panther Inspector Clouseau survives his own assassination attempt, but allows the world to think he is dead in order to pursue an investigation of the culprits in his own unique, bumbling way. The last "Pink Panther" film before Sellers died, and perhaps the least funny.
1978 (PG) 99m/C Peter Sellers, Herbert Lom, Dyan Cannon, Robert Webber, Burt Kwouk, Robert Loggia; **Dir:** Blake Edwards. **VHS, Beta, LV** $19.98 FOX, PIA, FCT 🐾🐾½

Revenge of the Radioactive Reporter Nuclear plant honchos don't like journalist who's about to hang their dirty laundry, so they toss him into vat of nuclear waste. He returns as a reporter with a cosmetic peel.
1991 90m/C David Scammell, Kathryn Boese, Randy Pearlstein, Derrick Strange; **Dir:** Craig Pryce. **VHS** $79.98 MAG 🐾🐾

The Revenge Rider Orphan McCoy goes after the villains who killed his parents. Good performance by McCoy in an otherwise average film.
1935 60m/B Tim McCoy, Robert Allen, Billie Seward, Edward Earle, Jack Clifford, Lafe McKee; **Dir:** David Selman. **VHS** $19.95 DVT 🐾🐾

Revenge of the Stepford Wives Made-for-television sequel to the Ira Levin fantasy about suburban husbands who create obedient robot replicas of their wives. In this installment, the wives are not robots, but are simply made docile by drugs. Strange casting pairs Johnson and Kavner as husband and stepford wife. Begat "The Stepford Children."
1980 95m/C Sharon Gless, Julie Kavner, Don Johnson, Audra Lindley, Mason Adams, Arthur Hill; **Dir:** Robert Fuest. **VHS, Beta** $9.98 SUE 🐾½

Revenge of the Teenage Vixens from Outer Space A low-budget film about three sex starved females from another planet who come to earth to find men. When the ones they meet do not live up to their expectations, the frustrated females turn the disappointing dudes are turned into vegetables.
1986 84m/C Lisa Schwedop, Howard Scott; **Dir:** Jeff Ferrell. **VHS, Beta** $39.98 CGI 🐾🐾

Revenge of TV Bloopers A group of stars including Sammy Davis Jr., E.G. Marshall, Roddy McDowall, Jack Palance, James Coburn, Suzanne Pleshette, Eddie Fisher, Zero Mostel, Goldie Hawn, Mamie Van Doren, and others are caught in the act of making blunders on the set. Also included are rare old television commercials featuring Dick Van Dyke, the Marx Brothers, the Three Stooges and others. Bloopers from such series as "The Defenders," "Ben Casey," "McHale's Navy," "Ozzie and Harriet," "Laugh-In," "The Nurses," "Burke's Law," and more are shown.
196? 50m/B Tim Conway, Bing Crosby, Jack Klugman, Don Rickles, Ozzie Nelson, Orson Welles, Dick Van Dyke, Moe Howard, Shemp Howard, Larry Fine, Groucho Marx, Zeppo Marx, Chico Marx, Harpo Marx, Sammy Davis Jr., E.G. Marshall, Roddy McDowall, Jack Palance, James Coburn, Suzanne Pleshette, Eddie Fisher, Zero Mostel, Goldie Hawn, Mamie Van Doren. **VHS, Beta** $19.95 NO 🐾

Revenge of the Virgins A tribe of topless 'Indian' women protect their sacred land from the white men. Exhumed exploitation for bad-film junkies, narrated by Kenne Duncan, an Ed Wood cohort.

1966 53m/B Jewell Morgan, Charles Veltman, Jodean Russo, Stanton Pritchard; **Dir:** Paul Perry Jr. **VHS** $19.95 NOS, SMW, VDM 🐾

Revenge of the Zombies In this sequel of sorts to "King of the Zombies," Carradine is creating an army of the undead for use by an evil foreign entity.
1943 61m/B John Carradine, Robert Lowery, Gale Storm, Veda Ann Borg, Mantan Moreland; **Dir:** Steve Sekely. **VHS** MLB 🐾½

The Revenger A framed man returns from prison and finds that a mobster has kidnapped his wife and wants $50,000 to give her back.
1990 91m/C Oliver Reed, Frank Zagarino; **Dir:** Cedric Sundstrom. **VHS, Beta** $79.95 AIP 🐾

Reversal of Fortune True tale of wealthy socialite Claus von Bulow (Irons) accused of deliberately giving his wife Sunny (Close) a near-lethal overdose of insulin. Comatose Close narrates history of the couple's courtship and married life, a saga of unhappiness and substance abuse, while Silver (as lawyer Dershowitz) et al prepare von Bulow's defense. An unflattering picture of the idle rich that never spells out what really happened. Irons is excellent as the eccentric and creepy defendant and richly deserved his Best Actor Oscar.
1990 (R) 112m/C Jeremy Irons, Glenn Close, Ron Silver, Annabella Sciorra, Uta Hagen, Fisher Stevens, Julie Hagerty; **Dir:** Barbet Schroeder. Academy Awards '90: Best Actor (Irons). **VHS, Beta, LV, 8mm** $19.98 WAR, PIA, FCT 🐾🐾🐾½

Revolt of the Barbarians Ancient Rome is plagued by barbarian invasion and piracy. A consul travels to Gaul to try to put an end to their misdeeds.
1964 ?m/C IT Roland Carey, Grazia Maria Spina, Mario Feliciani. **VHS, Beta** $16.95 SNC 🐾

Revolt of the Dragon A martial arts expert visits friends in his hometown and finds himself battling a ruthless gang that terrorizes the people.
1975 (R) 90m/C **VHS, Beta** $54.95 HHT 🐾

The Revolt of Job An elderly Jewish couple adopts an 8-year-old Gentile boy although it is illegal in World War II-torn Hungary and against the beliefs of the orthodox community. Moving story with fine performances.
1984 98m/C GE HU Fereno Zenthe, Hedi Tenessy; **Dir:** Imre Gyongyossy, Barna Kabay. **VHS, Beta** $79.95 MGM 🐾🐾🐾

The Revolt of Mother Mary E. Wilkins Freeman story of two hard-working farmers. Set in 1890 New England.
19?? 60m/C Amy Madigan, Jay O'Saunders. **VHS** $24.95 MON, KAR 🐾🐾🐾½

Revolt of the Zombies A mad scientist learns the secret of bringing the dead to life and musters the zombies into an unique, gruesome military unit during World War I. Strange early zombie flick.
1936 65m/B Dean Jagger, Roy D'Arcy, Dorothy Stone, George Cleveland, Robert Nolan; **Dir:** Victor Halperin. **VHS, Beta** $14.95 SNC, QNE, MLB Woof!

Revolution Documentary of hippie life in San Francisco during the turbulent '60s.
1969 90m/C Today Malone; **Dir:** Jack O'Connell. **VHS** WAR Woof!

Revolution The American Revolution is the setting for this failed epic centering on an illiterate trapper and his son who find themselves caught up in the fighting. Long and dull, which is unfortunate, because the story

had some potential. The cast is barely believable in their individual roles. Where did you get that accent, Al?
1985 (R) 125m/C GB Al Pacino, Donald Sutherland, Nastassia Kinski, Annie Lennox, Joan Plowright, Steven Berkoff, Dave King; **Dir:** Hugh Hudson. **VHS, Beta, LV** $19.98 WAR 🐾

Revolution! A Red Comedy A group of students are lured away from their revolutionary tendencies by materialistic indulgences. This satirical comedy is in color and black and white.
1991 84m/B Kimberly Flynn, Christopher Renstrom, Georg Osterman; **W/Dir:** Jeff Kahn. **VHS** $69.95 WAC 🐾🐾🐾

Rhapsody in Blue Standard Hollywood biography of the great composer features whitewashed and non-existent characters to deal with the spottier aspects of Gershwin's life. Still, the music is the main attraction and it doesn't disappoint.
1945 139m/B Robert Alda, Joan Leslie, Alexis Smith, Charles Coburn, Julie Bishop, Albert Basserman, Morris Carnovsky, Rosemary de Camp, Herbert Rudley, Charles Halton, Robert Shayne, Johnny Downs; **Dir:** Irving Rapper. **VHS, LV** $29.98 FOX, PIA, FCT 🐾🐾🐾

Rhinestone A country singer claims she can turn anyone, even a cabbie, into a singing sensation. Stuck with Stallone, Parton prepares her protege to sing at New York City's roughest country-western club, The Rhinestone. Only die-hard Dolly and Rocky fans need bother with this bunk. Some may enjoy watching the thick, New York accented Stallone learn how to properly pronounce dog ("dawg") in country lingo. Yee-haw.
1984 (PG) 111m/C Sylvester Stallone, Dolly Parton, Ron Leibman, Richard Farnsworth, Tim Thomerson; **Dir:** Bob Clark. **VHS, Beta, LV** $79.98 FOX 🐾

Rhino's Guide to Safe Sex A campy collection of public service films from the '40s and '50s about teenage sex and its many imagined hazards.
1986 60m/C **VHS, Beta** $29.95 RHI 🐾

Rhodes The story of Cecil Rhodes, a man who was instrumental in the Boer War in Africa and established the Rhodes Scholarships with $10 million, is told in this biography. Fine performance by Huston and Homolka. Alternate title: "Rhodes of Africa."
1936 91m/B GB Walter Huston, Oscar Homolka, Basil Sydney, Frank Cellier, Peggy Ashcroft, Renne De Vaux, Percy Parsons, Bernard Lee, Ndanisa Kumalo; **Dir:** Berthold Viertel. **VHS, CV** $19.95 NOS, SNC 🐾🐾½

The Ribald Tales of Robin Hood The Robin Hood legend is satirized in this comedy which concentrates on Prince John and his favorite pastimes: rape, pillage and plunder.
1980 (R) 83m/C **VHS, Beta** $49.95 MED 🐾🐾

Ricco A young man swears vengeance against the mobsters who killed his father. He kidnaps the daughter of a mafia kingpin and the battle begins. Originally titled "Summertime Killer."
1974 (PG) 90m/C FR IT SP Chris Mitchum, Karl Malden, Olivia Hussey, Raf Vallone, Claudine Auger, Gerard Tichy; **Dir:** Antonio Isasi. **VHS, Beta** $29.95 AVD 🐾🐾

Rich and Famous The story of the 25-year friendship of two women, through college, marriage, and success. George Cukor's last film.

1981 (R) 117m/C Jacqueline Bisset, Candice Bergen, David Selby, Hart Bochner, Meg Ryan, Steven Hill, Michael Brandon, Matt Lattanzi; *Dir:* George Cukor. VHS, Beta, LV $79.95 MGM ♂♂

Rich Girl A winsome, wealthy wench hides her millions by going to work in a rock'n'roll bar and falls in love with the local Bruce Springsteen wannabe. Despite the thick coating of music, sex, drugs and profanity, this consists of cliches that were old before talkies. Too earnestly-acted to be campy fun. Poor movie.
1991 (R) 96m/C Jill Schoelen, Don Michael Paul, Ron Karabatsos, Sean Kanan, Willie Dixon, Paul Gleason; *Dir:* Joel Bender. VHS, LV $89.95 COL, PIA ♂

Rich Hall's Vanishing America Funnyman Hall travels across the country looking for the mail-order company that promised to give him a genuine Wilt Chamberlain basketball hoop.
1986 50m/C Rich Hall, M. Emmet Walsh, Peter Isacksen, Harry Anderson, Wilt Chamberlain; *Dir:* Steve Rash. VHS, Beta $29.95 PAR ♂

Rich Kids A young girl trying to cope with the divorce of her parents is aided by a boyfriend whose parents have already split.
1979 (PG) 97m/C John Lithgow, Kathryn Walker, Trini Alvarado, Paul Dooley, David Selby, Jill Eikenberry, Olympia Dukakis; *Dir:* Robert M. Young. VHS, Beta $59.95 MGM ♂♂

Rich Little: One's a Crowd A free-form video in which the impressionist talks about his craft and satirizes many current celebrities.
1988 (PG) 86m/C Rich Little; *Dir:* Thomas E. Engel. VHS, Beta, LV $59.98 ORI ♂♂½

Rich Little's Christmas Carol Rich Little acts out this Dickens classic with his vast repertoire of characters.
1990 55m/C Rich Little. VHS $14.95 COL, SVS, SYB ♂♂½

Rich Little's Great Hollywood Trivia Game The multi-talented impressionist brings his Hollywood repertoire together for the showbiz trivia challenge of the year.
1984 60m/C Rich Little. VHS, Beta $59.98 LIV, VES ♂♂½

Rich and Strange Early Hitchcock movie is not in the same class as his later thrillers. A couple inherits a fortune and journeys around the world. Eventually the pair gets shipwrecked. Also known as "East of Shanghai."
1932 92m/B GB Henry Kendall, Joan Barry, Betty Amann, Percy Marmont, Elsie Randolph; *Dir:* Alfred Hitchcock. VHS, LV $19.95 NOS, SNC, MLB ♂♂½

Richard II In the 14th century, a rebellion against even the most corrupt monarch was viewed as an uprising against heaven itself. Shakespeare's "Richard II" is the story of just such a momentous rebellion. Part of the television series "The Shakespeare Plays."
1979 157m/C GB Derek Jacobi, Jon Finch, John Gielgud. VHS, Beta TLF ♂♂♂

Richard III This landmark film version of the Shakespearean play features an acclaimed performance by Laurence Olivier, who also directs. The plot follows the life of the mentally and physically twisted Richard of Gloucester and his schemes for the throne of England. Olivier received a Best Actor Oscar nomination.

1955 138m/C GB Laurence Olivier, Cedric Hardwicke, Ralph Richardson, John Gielgud, Stanley Baker, Michael Gough, Claire Bloom; *Dir:* Laurence Olivier. British Academy Awards '55: Best Actor (Olivier), Best Director (Olivier), Best Film. VHS, Beta $14.95 COL, SUE ♂♂♂½

Richard III Shakespeare's play about Richard's quest for the British throne is performed in this BBC Television production.
1982 120m/C GB Martin Shaw, Brian Protheroe. VHS, Beta $14.98 TLF ♂♂½

Richard Lester Interview with film director Richard Lester and a study of his work including "The Four Musketeers."
1975 30m/C VHS, Beta CMR, NYS ♂♂½

Richard Lewis: I'm Exhausted The neuroses-addled comic raves for a live audience at the Park West Theatre in Chicago.
1989 60m/C Richard Lewis. VHS, Beta $14.99 HBO ♂♂½

Richard Lewis: I'm in Pain Irreverence abounds in this stand-up comic's improvised act.
1985 51m/C Richard Lewis. VHS, Beta $39.95 PAR ♂♂½

Richard Petty Story The biography of race car driver Petty, played by himself, and his various achievements on the track.
1972 (G) 83m/C Richard Petty, Darren McGavin, Kathi Browne, Lynn Marta, Noah Beery Jr., L.Q. Jones. VHS, Beta $59.95 GEM ♂♂

Richard Pryor: Here and Now Filmed at the Saenger Theater in New Orleans, Pryor's sharp, witty commentaries are delivered with a unique piercing humor.
1983 (R) 75m/C Richard Pryor; *Dir:* Richard Pryor. VHS, Beta, LV $79.95 COL ♂

Richard Pryor: Live in Concert Richard Pryor demonstrates his unique brand of humor before an enthusiastic audience. Extremely vulgar.
1979 (R) 78m/C Richard Pryor; *Dir:* Jeff Margolis. VHS, Beta $19.95 VES ♂

Richard Pryor: Live and Smokin' Richard Pryor displays his comedic talents in this 1971 performance where he does his classic "The Wino and the Junkie" routine.
1971 47m/C Richard Pryor; *Dir:* Michael Blum. VHS, Beta, LV $19.98 LIV, VES ♂

Richard Pryor: Live on the Sunset Strip Filmed live at the Hollywood Palladium, this program captures Richard Pryor at his funniest, including his segment about "Pryor on Fire."
1982 (R) 82m/C Richard Pryor; *Dir:* Joe Layton. VHS, Beta, LV $14.95 COL ♂

Richard Thompson A concert performance from the highly regarded Scottish Celtic/folk electric guitarist.
1983 84m/C VHS, Beta, LV $29.95 SVS ♂

Richard's Things A man's wife and his girlfriend find love and comfort in each other after his death. Screenplay by Frederic Raphael, from his novel. Very slow and dreary, with a weak performance from Ullman.
1980 (R) 104m/C GB Liv Ullman, Amanda Redman; *Dir:* Anthony Harvey. VHS, Beta $69.98 SUE ♂♂

Richie Experiments with drugs lead a young man's family life down a dead end. Performances are good, script gets sappy. Originally titled "The Death of Richie." Made for TV, and based on a Thomas Thompson story.

1977 (R) 86m/C Ben Gazzara, Eileen Brennan, Robby Benson; *Dir:* Paul Wendkos. VHS, Beta $59.95 PSM ♂♂

Rick Derringer This program presents Rick Derringer performing several of his monster hit songs at the Ritz night club in New York City.
1982 58m/C Rick Derringer. VHS, Beta, LV $19.95 SVS ♂♂

Rickisha Kuri A rickshaw boy gets mad and kicks everyone within reach.
197? 98m/C VHS, Beta UNI ♂

Rick's: Your Place for Fantasy 45 minutes of strippers in a Texas bar.
1988 45m/C VHS, Beta $19.95 CEL ♂

Ricky 1 Mumbling pugilist spoof. Da Mob wants to make Ricky da fall guy, but uh, he's not as dumb as he looks.
1988 90m/C Michael Michaud, Maggie Hughes, James Herbert, Lane Montano; *Dir:* Bill Naud. VHS $69.99 TPV, HHE ♂½

Ricochet Rookie cop Washington causes a sensation when he single handedly captures notorious psychopath Lithgow. But this particular criminal is a twisted genius, and from his prison cell he comes up with a plan to destroy the young cop. Teaming up with old friend Ice-T, the rookie tries to outwit his evil arch-nemesis, but to do so may cost him his very life. A genuinely scary, tense and violent thriller.
1991 (R) 104m/C Denzel Washington, John Lithgow, Ice-T, Jesse Ventura, Kevin Pollak; *Dir:* Russell Mulcahy. VHS, LV, 8mm $94.99 HBO, CCB ♂♂♂

Riddance The second of Meszaros' trilogy, about a Hungarian couple, an upper-class male student and a factory-working young woman, and their difficulties in meshing their lives together. In Hungarian with English subtitles.
1973 84m/B HU *Dir:* Marta Meszaros. VHS, Beta $69.95 KIV ♂♂♂

The Riddle of the Sands Two English yachtsmen in 1903 inadvertently stumble upon a German plot to invade England by sea. British film based on the Erskine Childers spy novel.
1984 99m/C GB Michael York, Jenny Agutter, Simon MacCorkindale, Michael York, Alan Badel; *Dir:* Tony Maylam. VHS, Beta $39.95 BVG, VID, BMV ♂♂½

Ride 'Em Cowboy Two peanut and hotdog vendors travel west to try their hand as cowpokes. Usual Abbott and Costello fare with a twist—there are lots of great musical numbers, including a rousing version of "A Tisket, A Tasket" by Fitzgerald.
1942 86m/B Bud Abbott, Lou Costello, Anne Gwynne, Samuel S. Hinds, Dick Foran, Richard Lane, Johnny Mack Brown, Ella Fitzgerald, Douglas Dumbrille; *Dir:* Arthur Lubin. VHS $14.98 MCA ♂♂½

Ride 'Em Cowgirl Singing cowgirl, Page, enters a horse race to expose the crook who cheated her father.
1941 51m/B Dorothy Page, Vince Barnett, Milton Frome; *Dir:* Samuel Diege. VHS, 3/4U VCN ♂♂

Ride to Glory In 1886, a cavalry officer deserts his troops to seek revenge on the Apaches that brutally killed his family. Hearing this, a general offers to pardon him if he will lead a dangerous group of men on a mission against the Apaches. Fast-paced and very bloody. Also released on video as "The Deserter."

1971 90m/C Bekim Fehmiu, Richard Crenna, Chuck Connors, Ricardo Montalban, Ian Bannen, Slim Pickens,. Woody Strode, Patrick Wayne, John Huston; *Dir:* Burt Kennedy. **VHS $19.95** *SIM, PAR, AVD* 🐾

Ride the High Country The cult classic western about two old friends who have had careers on both sides of the law. One, Joel McCrea, is entrusted with a shipment of gold, and the other, Randolph Scott, rides along with him to steal the precious cargo. Although barely promoted by MGM, became a critics' favorite. Grimacing and long in the tooth, McCrea and Scott enact a fitting tribute and farewell to the myth of the grand ol' West. Wonderfully photographed by Lucien Ballard. Also known as "Guns in the Afternoon." The laserdisc edition carries the film in widescreen format along with the original movie trailer.
1962 93m/C Randolph Scott, Joel McCrea, Mariette Hartley, Edgar Buchanan, R.G. Armstrong, Ronald Starr, John Anderson, James Drury, L.Q. Jones, Warren Oates; *Dir:* Sam Peckinpah. **VHS, Beta, LV $19.98** *MGM, PIA* 🐾🐾🐾

Ride the Man Down Story of a murderous land war between neighboring land-owners tells the same old story: Greed knows no boundaries.
1953 90m/C Brian Donlevy, Chill Wills, Jack LaRue, Rod Cameron, Ella Raines; *Dir:* Joseph Kane. **VHS $19.98** *REP* 🐾 1/2

Ride in a Pink Car When a Vietnam vet returns home, he discovers that his welcoming committee is really a lynch mob.
1974 (R) 83m/C Glenn Corbett, Morgan Woodward, Ivy Jones; *Dir:* Robert Emery. **VHS, Beta $19.95** *ACA* 🐾

Ride, Ranger, Ride Autry is a Texas Ranger out to stop a group of Comanches from raiding a wagon train.
1936 56m/B Gene Autry, Smiley Burnette, Kay Hughes, Max Terhune, Monte Blue, Chief Thundercloud; *Dir:* Joseph Kane. **VHS, Beta $19.95** *NOS, MED, VYY* 🐾 1/2

Ride in the Whirlwind Three cowboys are mistaken for members of a gang by a posse. Screenplay for the offbeat western was by Nicholson.
1967 83m/C Jack Nicholson, Cameron Mitchell, Millie Perkins, Katherine Squire, Harry Dean Stanton; *Dir:* Monte Hellman. **Beta, LV, CDV $19.99** *MED, HHT, VYY* 🐾🐾

Ride a Wild Pony A poor Australian farmer's son is allowed to pick a horse of his own from a neighboring rancher's herd. After he trains and grows to love the pony, the rancher's daughter, a handicapped rich girl, decides to claim it for herself. Enjoyable Disney story is based on the tale "Sporting Proposition," by James Aldridge.
1975 (G) 86m/C John Meillon, Michael Craig, Robert Bettles, Eva Griffith, Graham Rouse; *Dir:* Don Chaffey. **VHS, Beta $69.95** *DIS* 🐾🐾🐾

Ride the Wind The "Bonanza" gang goes to the aid of the Pony Express.
1966 120m/C Michael Landon, Lorne Greene, Dan Blocker. **VHS, Beta $14.95** *REP* 🐾 1/2

Ride the Wind The price of competition unfolds in this story of the personal struggles of two boys who entangle their families. Finally, they find solace in the Christian faith.
1988 60m/C Kent Petersen, Marty Baldwin, Maribeth Murray; *Dir:* Dick Talarico. **VHS, Beta $19.95** *MIV* 🐾

Rider of the Law A government agent poses as a greenhorn easterner to nab a gang of outlaws.
1935 56m/B Bob Steele. **VHS, Beta** *WGE* 🐾

Rider on the Rain A young housewife is viciously raped by an escaped sex maniac. She kills him and disposes of his body, not knowing that he is being relentlessly pursued by mysterious American Bronson. Suspenseful, French-made thriller featuring one of Bronson's best performances.
1970 115m/C *FR* Marlene Jobert, Charles Bronson, Jill Ireland; *Dir:* Rene Clement. **VHS, Beta $29.95** *MON* 🐾🐾

Rider from Tucson Two rodeo riders head out to Colorado for a friend's wedding. They find out the bride has been kidnapped by a bunch of claim jumpers after control of a gold mine.
1950 60m/B Tim Holt, Richard Martin, Elaine Riley, Douglas Fowley; *Dir:* Lesley Selander. **VHS, Beta** *RKO* 🐾

Riders of Death Valley Splendid western action serial in 15 thirteen-minute episodes. Foran, Jones, and Carillo head a passel o'men watching for thieves and claim jumpers in a mining area.
1941 195m/B Dick Foran, Buck Jones, Leo Carillo, Lon Chaney Jr.; *Dir:* Ford Beebe, Ray Taylor. **VHS, Beta $49.95** *NOS, VCN, DVT* 🐾🐾🐾

Riders of the Desert Steele tracks down the loot from a stagecoach robbery.
1932 57m/B Bob Steele. **VHS, Beta** *VYY, VCN* 🐾

Riders of Destiny Wayne plays a government agent sent to find out who is stealing water from the local farmers. Very early Wayne has the Duke looking awfully young as an agent securing water rights for ranchers.
1933 59m/B John Wayne, George "Gabby" Hayes, Cecilia Parker, Forrest Taylor, Al "Fuzzy" St. John; *Dir:* Robert N. Bradbury. **VHS $12.98** *REP, NOS, SVS* 🐾🐾 1/2

Riders of the Range Sheep invade the cattle range and a range war flares up between those who would herd and the ranchers. Silent flick of interest because of the stars.
1924 62m/B Edmund Cobb, Helen Hayes, Dolly Dale; *Dir:* Otis Thayer. **VHS, Beta** *GPV* 🐾🐾

Riders of the Range A cowboy rides into town just in time to save a girl's brother from gamblers.
1949 60m/B Tim Holt, Richard Martin. **VHS, Beta** *RKO* 🐾

Riders of the Range/Storm Over Wyoming A double feature package of Tim Holt Westerns.
1950 120m/B Tim Holt, Richard Martin, Jacqueline Wells. **VHS, Beta $39.95** *RKO* 🐾 1/2

Riders of the Rockies An honest cowboy turns rustler in order to trap a border gang. One of Ritter's very best cowboy outings; good fun, and a very long fistfight with King.
1937 60m/B Tex Ritter, Yakima Canutt, Louise Stanley, Charles King, Snub Pollard; *Dir:* Robert N. Bradbury. **VHS, Beta $19.95** *NOS, UHV, VCN* 🐾🐾🐾

Riders to the Seas A mother envisions the death of her last son in the same manner her other sons died—by drowning.
1988 90m/C Geraldine Page, Amanda Plummer, Sachi Parker, Barry McGovern; *Dir:* Ronan O'Leary. **VHS, Beta $59.98** *DMD* 🐾 1/2

Riders of the Storm A motley crew of Vietnam vets runs a covert television broadcasting station from an in-flight B-29, jamming America's legitimate airwaves. Interesting premise with boring result. Originally titled "The American Way."
1988 (R) 92m/C Dennis Hopper, Michael J. Pollard, Eugene Lipinski, James Aubrey, Nigel Pegram; *Dir:* Maurice Phillips. **VHS, Beta, LV $19.98** *SUE* 🐾 1/2

Riders of the Timberline Unusual for a Hopalong Cassidy programmer in that it puts him in timber country. Yes, he's a lumberjack and he's okay, as he fights against land-grabbing crooks all day.
1941 59m/B William Boyd, Brad King, Andy Clyde, J. Farrell MacDonald, Elaine Stewart, Anna Q. Nilsson, Hal Taliaferro, Tom Tyler, Victor Jory; *Dir:* Lesley Selander. **VHS $19.95** *NOS* 🐾

Riders of the West An honest cowpoke fights off a band of rustlers.
1942 58m/B Buck Jones, Tim McCoy, Raymond Hatton, Sarah Padden, Harry Woods; *Dir:* Howard Bretherton. **VHS $19.95** *NOS, DVT* 🐾

Riders of the Whistling Pines Our ever-so-bland hero solves a murder perpetrated by a passel of lumber thieves and saves the day for yet another damsel in distress.
1949 70m/B Gene Autry, Patricia White, Jimmy Lloyd; *Dir:* John English. **VHS, Beta $19.95** *NOS, MED, DVT* 🐾🐾

Riders of the Whistling Skull The Three Mesquiteers are guiding an expedition to an ancient Indian city. A gang of rustlers looting the city are holding an archeology professor hostage, and the Mesquiteers must rescue him. Swell cast; nifty, creepy plot; solid-as-usual leads.
1937 58m/B Bob Livingston, Ray Corrigan, Max Terhune, Yakima Canutt, Roland Winters; *Dir:* Mack V. Wright. **VHS, Beta $12.95** *SNC, NOS, VCN* 🐾🐾 1/2

Ridin' the California Trail The Cisco Kid is wanted dead or alive in several states.
1947 40m/B Gilbert Roland. **VHS, Beta $24.95** *DVT* 🐾 1/2

Ridin' the Lone Trail Steele attempts to stop some criminals who've been harassing a stagecoach line. One of Republic's poorest westerns.
1937 56m/B Bob Steele, Claire Rochelle, Charles King, Ernie Adams, Lew Meehan, Julian Rivero; *Dir:* Sam Newfield. **VHS $19.95** *NOS, DVT* Woof!

Ridin' on a Rainbow A has-been performer on a steamboat decides to rob a bank in the hopes of starting a new life for himself and his daughter. The money he robs had just been deposited by some cattlemen, one of whom joins the steamboat's crew, wins the daughter's heart, and gets to the father. Slow and dull, with too much singing and too little plot and scenery.
1941 79m/B Gene Autry, Smiley Burnette, Mary Lee; *Dir:* Lew Landers. **VHS, Beta $39.98** *REP, CCB, VCN* 🐾 1/2

Ridin' Thru Another Tom Tyler western, in which the hero does battle with unscrupulous villains.
1935 60m/B Tom Tyler, Ruth Hiatt, Lafe McKee, Philo McCullough, Lew Meehan, Bud Osborne; *Dir:* Harry S. Webb. **VHS, Beta $16.95** *NOS, DVT* 🐾🐾

Ridin' the Trail Scott stars as a cowboy out to rid the countryside of bad guys in this western.

1940 60m/B Fred Scott, Iris Lancaster, Harry Harvey, Jack Ingram, John Ward; *Dir:* Raymond K. Johnson. **VHS, Beta $19.95** *NOS, DVT* ♫

Riding on Air Small-town newspaper editor Brown discovers that his bumbling methods get him the right results. Smugglers and crazy inventions in a ho-hum Brown entry.
1937 58m/B Joe E. Brown; *Dir:* Edward Sedgwick. **VHS, Beta $19.95** *NOS, KRT, FCT* ♫

The Riding Avenger To gain the love of a beautiful lady, a cowboy vows never to fight again, but is forced to break his promise when the town is terrorized.
1936 56m/B Hoot Gibson, Ruth Mix,. Buzz Barton. **VHS, Beta $19.95** *VCN* ♫½

Riding Bean Animated Japanese feature aimed at a mature audience. An adventure about a kidnapping blamed on an innocent courier. Japanese with English subtitles.
1991 46m/C *JP* **VHS $34.95** *CPM* ♫½

Riding the Edge Teenage motocross rider rescues his scientist dad from Middle Eastern terrorists. Thoroughly unoriginal.
1990 (R) 100m/C **VHS, LV $89.99** *HBO* ♫

Riding High It's a battle for supremacy between the kings of the motorcycle stunt world. Made for television movie's soundtrack includes songs from The Police, Dire Straits, Jerry Lee Lewis, and more.
1978 92m/C Murray Salem, Marella Oppenheim, Eddie Kidd, Irene Handl; *Dir:* Ross Cramer. **VHS, Beta $39.95** *IVE* ♫

Riding On Hayseed falls for the daughter of his father's arch rival. While his father battles it out over cattle, he pines for his enemy's daughter. Lots of action which is typical fare in mediocre western production.
1937 59m/B Tom Tyler, Germaine Greer, Rex Lease, John Elliott, Earl Dwire; *Dir:* Harry S. Webb. **VHS $19.95** *NOS, DVT* ♫♫

Riding Speed Our hero breaks up a gang of smugglers along the Mexican border.
1935 50m/B Buffalo Bill Jr. **VHS, Beta, LV** *WGE* ♫

Riding With Death A man who can become invisible, thanks to a military experiment, uses his power to uncover a dastardly plot by a mad scientist. So-so TV pilot.
1976 97m/C Ben Murphy, Andrew Prine, Katherine Crawford, Richard Dysart; *Dir:* Alan J. Levi. **VHS, Beta $39.95** *MCA* ♫

Riel The story of Louis Riel, a half-French, half-Indian visionary who challenged the Canadian army in his fight for equality and self-rule.
1979 150m/C Raymond Cloutier, Arthur Hill, William Shatner, Christopher Plummer, Leslie Nielsen; *Dir:* George Bloomfield. **VHS, Beta $49.95** *PSM* ♫

Riff-Raff Various dramatic and comic complications arise when a dying man gives Panama City con man O'Brien a map to oil deposits. Interesting and quick-moving melodrama.
1947 80m/B Pat O'Brien, Walter Slezak, Anne Jeffreys, Jason Robards Sr., Percy Kilbride, Jerome Cowan; *Dir:* Ted Tetzlaff. **VHS, Beta $19.98** *TTC* ♫♫½

Rififi Perhaps the greatest of all "heist" movies. Four jewel thieves pull off a daring caper, only to fall prey to mutual distrust. The long scene of the actual theft, completely in silence, will have your heart in your throat. In French with English subtitles.

1954 115m/B *FR* Jean Servais, Carl Mohner, Robert Manuel, Jules Dassin; *Dir:* Jules Dassin. National Board of Review Awards '56: 5 Best Foreign Films of the Year. **VHS, Beta $29.95** *NOS, VYY, HHT* ♫♫♫

The Right Hand Man A brother and sister must deal with their crusty patriarch's demise in the Victorian-era Australian outback. Disappointing and overwrought.
1987 (R) 101m/C *AU* Rupert Everett, Hugo Weaving, Arthur Dignam; *Dir:* Di Drew. **VHS, Beta $9.99** *STE, NWV* ♫½

The Right Stuff A rambunctious adaptation of Tom Wolfe's nonfiction book about the beginnings of the U.S. space program, from Chuck Yeager's breaking of the sound barrier to the last of the Mercury missions. Featuring an all-star cast and an ambitious script. Rowdy, imaginative and thrilling, though broadly painted and oddly uninvolving. Former astronaut John Glenn was running for president when this was out.
1983 (PG) 193m/C Ed Harris, Dennis Quaid, Sam Shepard, Scott Glenn, Fred Ward, Charles Frank, William Russ, Kathy Baker, Barbara Hershey, Levon Helm, David Clennon, Kim Stanley, Mary Jo Deschanel, Veronica Cartwright, Pamela Reed, Jeff Goldblum, Harry Shearer, Donald Moffatt, Scott Paulin, Lance Henriksen, Scott Wilson, John P. Ryan, Royal Dano; *Dir:* Philip Kaufman. Academy Awards '83: Best Film Editing, Best Sound, Best Original Score. **VHS, Beta, LV $29.98** *WAR* ♫♫♫

Right of Way Stewart plays an elderly man who makes a suicide pact with his wife (Davis) when he learns of her terminal illness. The two stars' first work together is disappointing and stodgy, with an abrupt ending that looks like bad editing by a frightened studio. Made for TV.
1984 102m/C Priscilla Morrill, Bette Davis, James Stewart, Melinda Dillon; *Dir:* George Schaefer. **VHS, Beta** *VCL* ♫♫½

Rigoletto Capturing Victor Hugo's middle period of writing, Verdi's "Rigoletto" is adapted from "The King Amuses Himself." Joan Sutherland and her puppet friends help tell the story.
1973 30m/C Joan Sutherland. **VHS, Beta $39.99** *BFA* ♫♫½

Rigoletto Produced by Jonathan Miller, the famous opera by Verdi is performed in English.
1982 140m/C John Rawnsley, Marie McLaughlin, Arthur Davies, Jean Rigby. **VHS $39.95** *HMV, HBO, MVD* ♫♫½

Rikisha-Man A tragic melodrama about a rickshaw puller who helps raise a young boy after his father has died, and loves the boy's mother from afar. Also called "Rickshaw Man" and "Muhomatsu no Issho." Inagaki's remake of his original 1943 version. In Japanese with English subtitles.
1958 105m/B *JP* Toshiro Mifune; *Dir:* Hiroshi Inagaki. Venice Film Festival '58: Grand Prize. **VHS, Beta $59.95** *FCT* ♫♫♫½

Rikky and Pete An engaging Australian comedy about a bizarre, precociously eccentric brother and sister. He's a Rube Goldberg-style mentor; she's a scientist-cum-country singer. They move together to the outback and meet a score of weird characters. A worthy follow-up to director Tass's "Malcolm."
1988 (R) 103m/C *AU* Nina Landis, Stephen Kearney, Tetchie Agbayani, Bruce Spence, Bruno Lawrence, Bill Hunter, Dorothy Allison, Don Reid; *Dir:* Nadia Tass, David Parker. **VHS, Beta, LV $79.98** *MGM, FOX* ♫♫½

Rikyu Set in 16th century Japan, Rikyu is a Buddhist priest who elevated the tea ceremony to an art form. To him it is a spiritual experience, to his master, Lord Hideyoshi Toilyotomi, the ruler of Japan, mastery of the ceremony is a matter of prestige. Conflict arises between Rikyu's ideal of profound simplicity, symbolized by the ceremony, and Toilyotomi's planned conquest of China, which Rikyu opposes. In Japanese with English subtitles.
1990 116m/C *JP* Rentaro Mikuni, Tsutomu Yamazaki; *Dir:* Hiroshi Teshigahara. **VHS $79.95** *CTL* ♫♫♫

Rim of the Canyon Autry has a dual role as a lawman tracking villains who were jailed by his father years before.
1949 70m/B Gene Autry, Nan Leslie, Thurston Hall, Clem Bevans, Walter Sande, Jock Mahoney, Alan Hale Jr., Denver Pyle; *Dir:* John English. **VHS, Beta $19.95** *NOS, DVT, BUR* ♫½

Rime of the Ancient Mariner A recitation of Samuel Coleridge's epic poem is coupled with an enactment of the sailor's fateful voyage. A combination of live action and animation is featured. The first part explores Coleridge the poet, while the second part features the poem itself.
198? 60m/C *Dir:* Raul DeSilva; *Nar:* Michael Redgrave. **VHS, Beta $59.95** *KUL, VTK* ♫½

Rimfire An undercover agent tracks down some stolen U.S. Army gold, aided by the ghost of a wrongly-hanged gambler. Fun B western.
1949 65m/B Mary Beth Hughes, Henry Hull, Al "Fuzzy" Knight, James Millican, Victor Kilian, Margia Dean, Jason Robards Sr., Reed Hadley; *Dir:* B. Reeves Eason. **VHS, Beta, LV** *WGE* ♫♫½

Rin Tin Tin, Hero of the West In one of his last adventures, the famous German Shepherd proves his courage and hyper-canine intelligence. Colorized.
1955 75m/C James Brown, Lee Aaker, Rin Tin Tin. **VHS, Beta $24.95** *MON* ♫♫

The Ring Very early Hitchcock about boxer who marries carnival girl, loses carnival girl, and wins carnival girl back again. Standard romantic punch and roses yarn interesting as measure of young director's developing authority.
1927 82m/B *GB* Carl Brisson, Lillian Hall-Davies, Ian Hunter, Gordon Harker; *Dir:* Alfred Hitchcock. **VHS $19.98** *VEC, MLB* ♫♫½

The Ring Poignant early fight movie that gives a lesson in prejudice against Chicanos. The main character feels that the only way to get respect from the world is to fight for it. Well-directed and perceptive.
1952 79m/B Gerald Mohr, Rita Moreno, Lalo Rios, Robert Arthur, Art Aragon, Jack Elam; *Dir:* Kurt Neumann. **VHS, CV** *MOV, DVT* ♫♫½

Ring of Bright Water Well-done story of a pet otter from autobiography of Gavin Maxwell. The film stars the couple that made the delightful "Born Free." Beautiful Scottish Highlands photography adds to this captivating and endearing tale of a civil servant who purchases an otter from a pet store and moves to the country.
1969 (G) 107m/C *GB* Bill Travers, Virginia McKenna, Peter Jeffrey, Archie Duncan; *Dir:* Charles Lamont. **VHS, Beta $14.98** *FOX, FCT* ♫♫♫

Ring of Death A tough cop investigating a routine case becomes a hunted man after someone kills the man he is following.

1972 93m/C Franco Nero, Florinda Bolkan, Adolfo Celi. **VHS, Beta $59.95** *GEM, MRV* 🎬🎬½

Ring of Fire Wilson is one of very few modern American kung-fu heroes of Asian descent, and he works messages against prejudice into this interracial Romeo-and-Juliet chopsocky tale. What foot through yonder window breaks?
1991 (R) 100m/C Don "The Dragon" Wilson, Maria Ford, Vince Murdocco, Dale Jacoby, Michael Delano, Eric Lee; *Dir:* Richard W. Munchkin. **VHS $89.95** *IMP* 🎬🎬½

Ring of Terror A college fraternity hazing goes deadly wrong in this cheaply made flop. Allegedly based on a true incident.
1962 72m/B George Mather, Esther Furst, Austin Green, Joseph Conway; *Dir:* Clark Paylow. **VHS $16.95** *NOS, SNC* Woof!

Rings Around the World A series of some of the world's most spectacular circus acts from around the world. Features appearances by Don Ameche, Gunther Gebel-Williams and La Mara.
1967 79m/C Don Ameche, Gunther Gebel-Williams, La Mara; *Dir:* Gilbert Cates. **VHS, Beta $69.95** *COL* Woof!

Ringside A concert pianist turns boxer to avenge his brother who was blinded in a fight.
1949 64m/B Donald (Don "Red") Barry, Sheila Ryan, Tom Brown. **VHS, Beta, LV $16.95** *SNC, WGE* 🎬½

The Rink Chaplin plays a waiter who spends his lunch hour on roller skates. Silent with musical soundtrack added.
1916 20m/B Charlie Chaplin; *Dir:* Charlie Chaplin. **VHS, Beta** *CAB, FST* 🎬🎬🎬

Rink/Immigrant Two Chaplin shorts: "The Rink" (1916), in which Charlie defends his girlfriend's honor at a local roller skating rink; "The Immigrant" (1917), features Charlie as a newcomer to America.
191? 56m/B Charlie Chaplin, Mack Swain, Edna Purviance. **VHS, Beta $29.95** *VYY* 🎬🎬🎬

Rio Bravo The sheriff of a Texas border town (Wayne) takes a brutal murderer into custody and faces a blockade of hired gunmen determined to keep his prisoner from being brought to justice. Long, but continually entertaining. Generally regarded as an American film classic, with an all-star cast, didn't impress the critics at the time. Sequel: "El Dorado." Semi-remake was called "Assault on Precinct 13."
1959 140m/C John Wayne, Dean Martin, Angie Dickinson, Ricky Nelson, Walter Brennan, Ward Bond, Claude Akins, Bob Steele, John Russell; *Dir:* Howard Hawks. **VHS, Beta $19.98** *WAR, TLF, BUR* 🎬🎬🎬½

Rio Conchos Nifty nonstop action in this western set in Texas after the Civil War. Three Army buddies search for 2,000 stolen rifles. Boone is understated, O'Brien is good, and Brown memorable in his debut.
1964 107m/C Richard Boone, Stuart Whitman, Edmond O'Brien, Anthony (Tony) Franciosa, Jim Brown; *Dir:* Gordon Douglas. **VHS, Beta $19.98** *FOX* 🎬🎬🎬

Rio Grande The last entry in Ford's cavalry trilogy following "Fort Apache" and "She Wore a Yellow Ribbon." A U.S. cavalry unit on the Mexican border conducts an unsuccessful campaign against marauding Indians. The commander (Wayne) faces an unhappy wife and is estranged from his son, a new recruit. Featuring an excellent Victor Young score and several songs by the Sons of the Pioneers.

1950 105m/B John Wayne, Maureen O'Hara, Ben Johnson, Claude Jarman Jr., Harry Carey Jr., Victor McLaglen, Chill Wills, J. Carroll Naish; *Dir:* John Ford. **VHS $19.98** *REP, TLF, BUR* 🎬🎬🎬

Rio Grande Raiders A Stagecoach driver discovers that his kid brother is working for a crooked rival stage company.
1946 54m/B Sunset Carson, Linda Stirling, Bob Steele. **VHS, Beta $29.95** *DVT, RXM* 🎬

Rio Hondo An ex-gunfighter comes back to his home town only to find it filled with outlaws.
1947 90m/B Carlos Cortez, Elsa Cardenas, Alfonso Mejias, Alfredo Leal. **VHS, Beta $39.95** *MAD* 🎬🎬

Rio Lobo Hawks's final film takes place after the Civil War, when Union Colonel Wayne goes to Rio Lobo to take revenge on two traitors. Disappointing. The Duke has to carry weak supporting performances on his brawny shoulders—and nearly does.
1970 (G) 114m/C John Wayne, Jorge Rivero, Jennifer O'Neill, Jack Elam, Chris Mitchum, David Huddleston, George Plimpton; *Dir:* Howard Hawks. **VHS, Beta, LV $19.98** *FOX, BUR* 🎬🎬½

Rio Rattler A vintage fast-shooting matinee western.
1935 60m/B Tom Tyler, Marion Shilling, Eddie Gribbon, William Gould, Tom London; *Dir:* Bernard B. Ray. **VHS, Beta $27.95** *VCN* 🎬½

Riot Men in cages throw tantrums until warden returns from vacation. Filmed in Arizona State Prison with real convicts as extras. Strong performances by fullback Brown and durable Hackman. Screenplay by James Poe based on story by ex-con Frank Elli. Too oft played theme song sung by Bill Medley of the Righteous Brothers.
1969 (R) 97m/C Gene Hackman, Jim Brown, Mike Kellin, Ben Carruthers, Frank Eyman; *Dir:* Buzz Kulik. **VHS** *PAR* 🎬🎬½

Riot in Cell Block 11 A convict leads four thousand prisoners in an uprising to improve prison conditions. Based on producer/ex-con Walter Wanger's own experience. Powerful and still timely. Filmed at Folsom Prison.
1954 80m/B Neville Brand, Leo Gordon, Emile Meyer, Frank Faylen; *Dir:* Don Siegel. **VHS, Beta $19.98** *REP* 🎬🎬🎬

Riot Squad An intern poses as a doctor in order to find out who was responsible for putting a mob hit on a police captain.
1941 55m/B Richard Cromwell, Rita Quigley, John Miljan, Mary Ruth, Herbert Rawlinson, Mary Gordon, Arthur Space; *Dir:* Edward Finney. **VHS, Beta $16.95** *SNC, RXM* 🎬🎬

Rip-Off Two Greek wanderers enter the American way of life via sex, drugs, syndicated crime and terrorism.
1977 78m/C Michael Benet, Michelle Simone, James Masters, Johnny Dark; *Dir:* Manolis Tsafos. **VHS, Beta $49.95** *UNI* 🎬

The Rip Off A colorful gang of hoodlums goes for the biggest heist of their lives—six million bucks in diamonds!
1978 (R) 99m/C Lee Van Cleef, Karen Black, Robert Alda, Edward Albert; *Dir:* Anthony M. Dawson. **VHS, Beta $49.95** *WOV* 🎬

Rip Roarin' Buckaroo A fighter sets out to avenge himself when he is framed in a dishonest fight.
1936 51m/B Tom Tyler, B.J. Quinn Jr., Beth Marion, Charles King; *Dir:* Robert Hill. **VHS, Beta $19.95** *NOS, DVT, RXM* 🎬

Rip van Winkle Sleepy Copola adaptation of the Washington Irving story about a man who falls asleep for twenty years after a group of ghosts get him drunk. From the "Faerie Tale Theatre" series.
1985 60m/C Harry Dean Stanton, Talia Shire; *Dir:* Francis Ford Coppola. **VHS, Beta $14.95** *KUI, FOX, FCT* 🎬🎬

Ripley's Believe It or Not! Each episode of this series outlines incredible facts based on the long-running newspaper feature. New volumes are being added to this on-going series on a regular basis. Volumes are available separately.
1985 80m/C **VHS, Beta, Special order formats $39.95** *IND* 🎬

The Ripper The spirit of Jack-the-Ripper possesses a university professor's body. Ultra-violent. Shot on videotape for the video market.
1986 90m/C Wade Tower, Tom Schreir, Mona Van Pernis, Andrea Adams; *Dir:* Christopher Lewis. **VHS, Beta $24.95** *NEG* 🎬

Ripping Yarns Three episodes of the popular British comedy series written by Monty Python's Terry Jones and Michael Palin: "Tomkinson's Schooldays," "Escape from Stalag Luft 112B," and "Golden Gordon." The subsequent episodes are released as "More Ripping Yarns" and "Even More Ripping Yarns."
1975 90m/C *GB* Michael Palin, Gwen Watford, Ian Ogilvy, Roy Kinnear, Bill Fraser; *Dir:* Terry Hughes, Jimmy Franklin, Alan Bell. **VHS, Beta $14.98** *FOX* 🎬

Rise of Catherine the Great Slow but lavish and engrossing British dramatization of the tortured and doomed love affair between Catherine, Empress of Russia, and her irrational, drunken husband Peter. Also released as "Catherine the Great."
1934 88m/B *GB* Douglas Fairbanks Jr., Elisabeth Bergner, Flora Robson; *Dir:* Paul Czinner. National Board of Review Awards '34: 5 Best Foreign Films of the Year. **VHS, Beta, LV $14.98** *SUE, KRT* 🎬🎬½

Rise and Fall of the Borscht Belt A look at how a new form of humor evolved from the stages at resorts in the Catskills, often known as the Borscht Belt. Provides a glimpse of an ironic and fascinating part of American Jewish life.
1982 80m/C **VHS $59.95** *ERG* 🎬🎬½

The Rise and Fall of Legs Diamond Not quite historically accurate but fast-paced and entertaining gangster bio of legendary booze trafficker Legs Diamond. Danton's debut.
1960 101m/B Ray Danton, Karen Steele, Jesse White, Simon Oakland, Robert Lowery, Elaine Stewart; *Dir:* Budd Boetticher. **VHS, LV $59.99** *WAR, PIA, FCT* 🎬🎬½

Rise and Fall of the Third Reich This is a pictorial record which tracks Adolf Hitler's path through the years between 1920 and 1945. Some segments are in black and white.
1968 120m/C **VHS, Beta $59.95** *MGM* 🎬🎬½

The Rise of Louis XIV A masterful docudrama detailing the life and court intrigues of Louis XIV of France. Successfully captures the attitudes and mores of the royalty at the time. One of a series of historical films directed by Rossellini. Made for French television.

1966 100m/C *FR* Jean-Marie Patte, Raymond Jourdan, Dominique Vincent, Silvagni, Pierre Barrat; *Dir:* Roberto Rossellini. **VHS** $59.95 *HTV* ♂♂♂

Rise & Rise of Daniel Rocket A exceptional young boy believes he can fly without the aid of a flying machine. From the "American Playhouse" series.
1986 90m/C Tom Hulce, Timothy Daly, Valerie Mahaffey; *Dir:* Emile Ardolino. **VHS, Beta** $59.95 *ACA, PBS* ♂♂♂

Rising Son A family man who loves his job faces trauma when his factory closes at the same time his son informs him he's quitting medical school. Solid family drama and parable of economic hard times of the early '80s, sustained by top-drawer performances by Dennehy, Damon, and Laurie. Well directed by Coles. Made for cable.
1990 92m/C Brian Dennehy, Matt Damon, Piper Laurie, Graham Beckel, Ving Rhames, Jane Adams, Richard Jenkins, Emily Longstreth; *Dir:* John Coles. **VHS, Beta** $79.98 *TTC* ♂♂♂

Risky Business Fortune-seeking mother disapproves of daughter's doctor fiance, and educates girl in marital woes. Video includes early Pollard short "Sold at Auction."
1928 104m/B Vera Reynolds, ZaSu Pitts, Ethey Clayton; *Dir:* Alan Hale Jr. **VHS, Beta, LV** $24.95 *GPV, FCT, WAR* ♂♂♂

Risky Business With his parents out of town and awaiting word from the college boards, a teenager becomes involved in unexpected ways with a quick-thinking prostitute, her pimp, and assorted others. Cruise is likeable, especially when dancing in his underwear. Funny, well-paced, stylish prototypical '80s teen flick reintroduced Ray-Bans as the sunglasses for the wannabe hip. What a party!
1983 (R) 99m/C Tom Cruise, Rebecca DeMornay, Curtis Armstrong, Bronson Pinchot, Joe Pantoliano, Kevin Anderson, Richard Masur; *Dir:* Paul Brickman. **VHS, Beta, LV, 8mm** $19.98 *WAR* ♂♂♂

Rita Hayworth: The Love Goddess Made-for-television drama details the tragic life of the beautiful film star. Lynda Carter, lovely though she is, plays Rita weakly and without depth.
1983 100m/C Lynda Carter, Michael Lerner, Alejandro Rey, John Considine; *Dir:* James Goldstone. **VHS, Beta** $39.95 *IVE* ♂ ½

Rita, Sue & Bob Too A middle-aged Englishman gets involved in a menage a trois with two promiscuous teenagers, until the whole town gets wind of it. A raunchy, amoral British comedy. Written by Andrea Dunbar.
1987 (R) 94m/C *GB* Michelle Holmes, George Costigan, Siobhan Finneran; *Dir:* Alan Clarke. **VHS, Beta, LV** $79.95 *LHV, WAR* ♂♂♂

Rituals A group of five calm, rational men suddenly turn desperate after a chain of nightmarish events on a camping trip. Yet another "Deliverance" rip off. See the real thing instead of this garbage.
1979 100m/C *CA* Hal Holbrook, Lawrence Dane; *Dir:* Peter Carter. **VHS, Beta** $24.98 *SUE* ♂ ½

The Ritz Weston tries to get away from his gangster brother-in-law by hiding out in a gay bathhouse in New York. Moreno plays a talentless singer Googie Gomez, who performs in the bathhouse while waiting for her big break. Moreno is great reprising her Tony-winning stage role, and Lester's direc-

tion is spiffy. Written by Terence McNally from his play.
1976 (R) 91m/C Rita Moreno, Jack Weston, Jerry Stiller, Kaye Ballard, Treat Williams, F. Murray Abraham; *Dir:* Richard Lester. **VHS, Beta, LV** $19.98 *WAR* ♂♂♂

Rivals Shallow, unconvincing drama about a stepfather challenged by his stepson, who wants to kill this new contender for his mother's love. Nice handling of cast credits, but other details—like acting, directing and photography—leave much to be desired.
1972 (R) 103m/C Robert Klein, Joan Hackett, Scott Jacoby; *Dir:* Krishna Shah. **VHS** $54.95 *TPI* ♂ ½

Rivals of the Dragon/ Magnificent Shaolin Fist Fighter Two action-packed oriental martial arts movies are crammed onto one tape, fairly bursting with head-kicking pleasure.
1979 176m/C **VHS, Beta** $7.00 *VTR* ♂ ½

Rivals of the Silver Fox Young kickers prove no match for their hoary, hard-chopping elder.
198? 90m/C Casanova Wong, Barry Lam, Chen Shao Ping, Lee Fat Yuen. **VHS, Beta** $49.95 *REG* ♂ ½

The River A massively lauded late film by Renoir about three British girls growing up in Bengal India, all developing crushes on a one-legged American vet. Lyrical and heart-warming, with hailed cinematography by Claude Renoir. Based on a novel by Rumer Godden; script by Godden and director Renoir. Satyajit Ray, who assisted Renoir, is one of India's greatest filmmakers.
1951 99m/C *FR* Patricia Walters, Adrienne Corri, Nora Swinburne, Radha, Arthur Shields, Thomas E. Breen; *Dir:* Jean Renoir. National Board of Review Awards '51: 5 Best Foreign Films of the Year; Venice Film Festival '51: International Prize. **VHS, Beta, LV** $79.95 *TAM, VYG, CRC* ♂♂♂♂

The River Farmers battle a river whose flood threatens their farm. Spacek, as always, is strong and believable as the wife and mother, but Gibson falters. Beautiful photography. The third in an onslaught of films in the early '80s that dramatized the plight of the small American farmer. "The River" isn't as strong as "Country" and "Places in the Heart" which managed to convey important messages less cloyingly.
1984 (PG) 124m/C Mel Gibson, Sissy Spacek, Scott Glenn, Billy "Green" Bush; *Dir:* Mark Rydell. **VHS, Beta, LV** $19.95 *MCA* ♂♂

River Beat A woman aboard an American freighter is unwittingly used as diamond smuggler, and is arrested when the ship docks in London. An inspector does his best to clear the hapless woman.
1954 70m/B Phyllis Kirk, John Bentley, Robert Ayres, Lenny White, Glyn Houston, Charles Lloyd Pack; *Dir:* Guy Green. **VHS, Beta** $16.95 *SNC* ♂♂

River of Death Absurd adventure, based on an Alistair McLean novel, about a white man entering the Amazon jungle world of a forgotten tribe in search of wealth, tripping over Neo-Nazi scientists and war criminals. Too complex to be harmlessly enjoyable, too mindless for the complexity to be worth unraveling.
1990 (R) 103m/C Michael Dudikoff, Robert Vaughn, Donald Pleasence, Herbert Lom, L.Q. Jones, Cynthia Erland, Sarah Maur-Thorp; *Dir:* Steve Carver. **VHS, Beta, LV** $89.95 *WAR, OM* ♂ ½

River of Diamonds A daring adventurer, with the standard beautiful woman at his side, battles evil curses and Nazis buried alive to find a fortune in diamonds. Yeah, sure. Another miserable "Indiana Jones" rip off.
1990 88m/C Dack Rambo, Angela O'Neill, Ferdinand "Ferdy" Mayne, Graham Clark, David Sherwood, Tony Caprari, Dominique Tyawa; *Dir:* Robert J. Smawley. **VHS, Beta** $79.98 *TTC* ♂

The River Niger Jones is riveting, and Tyson is good in an otherwise muddling adaptation of the Tony-award winning play about black ghetto life. Realistic emotions and believable characters.
1976 (R) 105m/C James Earl Jones, Cicely Tyson, Glynn Turman, Louis Gossett Jr., Roger E. Mosely, Jonelle Allen; *Dir:* Krishna Shah. **VHS** $39.95 *CON, PLV* ♂♂ ½

River of No Return During the gold rush, an itinerant farmer and his young son help a heart-of-gold saloon singer search for her estranged husband. Rather crummy script is helped by the mere presence of Mitchum and Monroe. Marilyn sings the title song, "Down in the Meadow," and "I'm Going to File My Claim."
1954 91m/C Robert Mitchum, Marilyn Monroe, Tommy Rettig, Rory Calhoun, Murvyn Vye, Douglas Spencer; *Dir:* Otto Preminger. **VHS, Beta, LV** $14.98 *FOX* ♂♂ ½

The River Rat Ex-con Jones is reunited with his daughter after spending thirteen years in prison. There's something in there about stashed loot that should have been jettisonned in favor of more getting-to-know-you father-daughter drama, which is good.
1984 (PG) 93m/C Tommy Lee Jones, Brian Dennehy, Martha Plimpton, Shawn Smith, Melissa Davis; *Dir:* Tom Rickman. **VHS, Beta, LV** $79.95 *PAR* ♂♂ ½

River of Unrest Remarkable visual expressiveness and emotional power mark this depiction of terrorism and open warfare during the Irish Rebellion. Based on a play, hence slow; but good acting pulls it up half a bone.
1937 69m/B John Lodge, John Loder, Antoinette Cellier. **VHS, Beta** $29.95 *VYY* ♂♂ ½

Riverbend A renegade black Army major escapes a rigged court martial only to be persecuted by racist whites in a small Southern town. Made for television.
1989 (R) 106m/C Steve James, Tony Frank, Julius Tennon, Margaret Avery; *Dir:* Sam Firstenberg. **VHS, Beta** $89.95 *PAR* ♂♂

River's Edge A drug-addled high school student strangles his girlfriend and casually displays the corpse to his apathetic group of friends, who leave the murder unreported for days. Harrowing and gripping; based on a true story. Aging biker Hopper is splendid.
1987 (R) 99m/C Keanu Reeves, Crispin Glover, Daniel Roebuck, Joshua Miller, Dennis Hopper, Ione Skye, Roxana Zal; *Dir:* Tim Hunter. **VHS, Beta, 8mm** $14.98 *SUE* ♂♂♂

Road to 1984 Biodrama about futuristic author George Orwell centers on his classic "1984."
197? 86m/C *GB* Edward Fox. **VHS** $19.99 *PAR* ♂♂ ½

Road Agent A crooked attorney tries to beat a young woman out of her inherited ranch.
1926 70m/B Al Hoxie. **VHS, Beta** $19.95 *NOS, DVT, MLB* ♂ ½

Road to Bali Sixth Bob-n-Bing road show, the only one in color, is a keeper. The boys are competing for the love of—that's right—Lamour. She must be some gal, cuz they chase her all the way to Bali, where they meet cannibals and other perils, including the actual Humphrey Bogart. Jones's debut, in a bit role.
1953 90m/C Bob Hope, Bing Crosby, Dorothy Lamour, Murvyn Vye, Ralph Moody, Jane Russell, Jerry Lewis, Dean Martin, Carolyn Jones; *Dir:* Hal Walker. **VHS, Beta $29.95** UNI, DVT, MLB *♫♫♫*

Road Games Trucker Keach is drawn into a web of intrigue surrounding a series of highway "Jack the Ripper"-style murders. Curtis is a hitchhiker. Nothing special. Director Franklin later helmed "Psycho II."
1981 (PG) 100m/C *AU* Stacy Keach, Jamie Lee Curtis; *Dir:* Richard Franklin. **VHS, Beta $9.95** COL, SUE, CHA *♫♫*

Road to Hollywood Compilation of early musical shorts featuring Bing.
1942 70m/B Bing Crosby. **VHS, Beta $29.95** DVT *♫♫*

The Road to Hong Kong Last of the Crosby/Hope team-ups shows some wear, but still manages charm and humor. Lamour appears only briefly in this twisted comedy of hustlers caught in international espionage and cosmic goings-on.
1962 91m/B Bob Hope, Bing Crosby, Joan Collins, Dorothy Lamour, Peter Sellers; *Dir:* Norman Panama. **VHS $19.98** MGM, BTV, FCT *♫♫½*

Road House Nightclub singer Lupino inspires noir feelings between jealous road house owner Widmark and his partner Wilde. Widmark sets up Wilde to take the fall for a faked robbery, convinces the law to release him into his custody, then dares him to escape.
1948 95m/B Richard Widmark, Ida Lupino, Cornel Wilde, Celeste Holm, O.Z. Whitehead; *Dir:* Jean Negulesco. **VHS, Beta $19.98** FOX, FCT *♫♫½*

Road House Bouncer Swayze is hired to do the impossible: clean up the toughest bar in Kansas City. When he lays down his rules he makes a lot of enemies, including ex-bar employees and local organized crime. Ample violence; an example of formula filmmaking at its most brain-numbing, with a rock soundtrack.
1989 (R) 115m/C Patrick Swayze, Sam Elliott, Kelly Lynch, Ben Gazzara, Kevin Tighe, Marshall Teague, Julie Michaels, Jeff Healey; *Dir:* Rowdy Herrington. **VHS, Beta, LV, 8mm $19.98** MGM *♫½*

Road Lawyers and Other Briefs Several recent comedic shorts. "Road Lawyers," the title short, is a hilarious look at lawyers in a post-apocalyptic world that won major awards at film festivals in New York, Houston, Australia, and Chicago. Also included are: "Escape from Heaven," about a nun with an attitude problem; "Radar Men on the Moon," a comic reworking of an old "Commander Cody" serial; and "Hairline."
1989 79m/C VHS $79.95 AIP *♫♫½*

Road to Lebanon An original musical comedy featuring many stars, spoofing the famous Hope-Crosby "Road" films.
1960 50m/B Danny Thomas, Bing Crosby, Bob Hope, Hugh Downs, Claudine Auger, Sheldon Leonard. **VHS, Beta $24.95** NOS, VYY, DVT *♫♫*

Road to Morocco The third in the road movie series finds Hope and Crosby in Morocco, stranded and broke. To get some money, Crosby sells Hope into slavery to be the princess's (Lamour) personal plaything. Feeling guilty, Crosby returns to the palace to rescue Hope, only to find that he and the princess are getting married because the royal astrologer said it was in the stars. Crosby then tries to woo Lamour and, when the astrologer discovers the stars were mistaken, those two decide to marry. Quinn, however, also wants her and hilarious scenes ensue when the boys rescue Lamour from him. One of the funniest in the series. Watch for the camel at the end.
1942 83m/C Bing Crosby, Bob Hope, Dorothy Lamour, Anthony Quinn, Dona Drake, Vladimir Sokoloff, Yvonne de Carlo; *Dir:* David Butler. **VHS $14.98** MCA *♫♫♫½*

Road to Nashville An agent travels to Nashville to enlist talent for a new musical. Why do so many of these semi-musicals insist on having a plot? Good music.
1967 (G) 110m/C Marty Robbins, Johnny Cash, Doodles Weaver, Connie Smith, Richard Arlen; *Dir:* Robert Patrick. **VHS, Beta $29.95** UHV *♫♫*

Road Racers The roar of the engine, fast cars, and family ties combine in this movie about the thrill of speed.
195? 73m/B *Dir:* Arthur Swerdloff. **VHS $9.95** COL *♫♫*

Road to Rio The wisecracking duo travel to Rio De Janeiro to prevent Spanish beauty Lamour (there she is again) from going through with an arranged marriage. Top-notch entry; fifth in the "Road" series. Written by Jack Rose and Edmund Beloin. Songs include "But Beautiful," and, "You Don't Have to Know the Language."
1947 100m/B Bing Crosby, Bob Hope, Dorothy Lamour, Gale Sondergaard, Frank Faylen, The Andrews Sisters, The Wiere Brothers; *Dir:* Norman Z. MacLeod. **VHS, Beta $14.95** COL *♫♫♫½*

The Road to Ruin The evils of smoking and drinking plummet a girl into a life of prostitution in this melodramatic, moralistic tale. An unintentional laugh-riot right up there with "Reefer Madness." Silent.
1928 45m/B Grant Withers, Helen Foster; *Dir:* Norton S. Parker. **VHS, Beta $16.95** SNC, NOS *♫♫*

Road to Ruin A rich playboy pretends to lose all his money to see if his girlfriend loves him for himself or for his wealth. But when his business partner embezzles his fortune, he must regain both his money and his girlfriend's trust.
1991 (R) 94m/C Peter Weller, Carey Lowell, Michael Duchaussoy; *Dir:* Charlotte Brandstrom. **VHS $89.95** LIV *♫♫*

The Road Runner vs. Wile E. Coyote: Classic Chase Wile E. Coyote engages in a battle of wits against the Road Runner in this collection of eight classic cartoons. Watching this entire tape straight through is almost a religious experience in the Zen of Futility.
1963 54m/C *Dir:* Chuck Jones. **VHS, Beta $12.95** WAR *♫♫♫*

Road to Salina A drifter is mistaken for (and might actually be) a desert-restaurant owner's long-lost son. He plays along, eventually seducing the family's daughter, who might be his sister.
1968 (R) 97m/C *FR* Robert Walker Jr., Rita Hayworth, Mimsy Farmer, Ed Begley Sr.; *Dir:* Georges Lautner. **VHS, Beta $59.95** CHA *♫♫♫*

Road to Singapore This is the movie that started it all. Crosby and Hope decide to swear off women and escape to Singapore to enjoy the free life. There they meet Lamour, a showgirl who is abused by Quinn. The boys rescue Lamour, but soon find they are both falling for her. She's in love with one of them, but won't reveal her feelings. Who will get the girl? Not as funny as some of the other road movies, but it's great for a first try. Songs include "Too Romantic," "Kaigoon," and "Sweet Potato Piper."
1940 84m/C Bing Crosby, Bob Hope, Dorothy Lamour, Charles Coburn, Judith Barrett, Anthony Quinn, Jerry Colonna, Johnny Arthur, Pierre Watkin; *Dir:* Victor Schertzinger. **VHS $14.98** MCA *♫♫½*

Road to Utopia Fourth of the "Road" films, wherein the boys wind up in Alaska posing as two famous escaped killers in order to locate a secret gold mine. One of the series' funniest and most spontaneous entries, abetted by Benchley's dry, upper-crust comments. Songs by Johnny Burke and James Van Heusen, including "Personality" and "Put It There, Pal."
1946 90m/B Bing Crosby, Bob Hope, Dorothy Lamour, Jack LaRue, Robert Benchley, Douglas Dumbrille, Hillary Brooke; *Dir:* Hal Walker. **VHS, Beta, LV $29.95** MCA *♫♫♫*

The Road Warrior The first sequel to "Mad Max" takes place after nuclear war has destroyed Australia. Max helps a colony of oil-drilling survivors defend themselves from the roving murderous outback gangs and escape to the coast. The climactic chase scene is among the most exciting ever filmed; this film virtually created the "action-adventure" picture of the 1980s.
1982 (R) 95m/C Mel Gibson, Bruce Spence, Emil Minty, Vernon Wells; *Dir:* George Miller. **VHS, Beta, LV $19.98** WAR, PIA *♫♫♫½*

The Road to Yesterday Two couples together on a crashing train are somehow thrown into the eighteenth century in roles parallel to their own lives. DeMille's first independent film; intriguing action/melodrama. Silent with musical soundtrack.
1925 136m/B Joseph Schildkraut, William Boyd, Jetta Goudal, Vera Reynolds, Iron Eyes Cody; *Dir:* Cecil B. DeMille. **VHS, Beta $49.95** VYY, GPV *♫♫♫*

Road to Yesterday/The Yankee Clipper Four train passengers are transported back to previous lives in "The Road to Yesterday." "The Yankee Clipper" races the British Lord of the Isles from China to New England to capture the tea trade. Both of these pictures are abridged versions. Silent.
1927 56m/B William Boyd, Elinor Fair, Joseph Schildkraut, Jetta Boudal, Vera Reynolds; *Dir:* Cecil B. DeMille. **VHS, Beta $29.95** CCB *♫♫½*

Road to Zanzibar After selling a fake diamond mine to a criminal, Crosby and Hope flee to Zanzibar, where they meet up with Lamour and Merkel. The guys put up the money for a safari, supposedly to look for Lamour's brother, but they soon discover that they too have been tricked. Deciding to head back to Zanzibar, Crosby and Hope find themselves surrounded by hungry cannibals. Will they survive, or will they be someone's dinner? Not as funny as the other road movies, but amusing nonetheless. Songs include "On the Road to Zanzibar," "It's Always You," and "Birds of a Feather."
1941 92m/C Bing Crosby, Bob Hope, Dorothy Lamour, Una Merkel, Eric Blore, Iris Adrian, Lionel Royce; *Dir:* Victor Schertzinger. **VHS $14.98** MCA *♫♫♫*

Roadhouse 66 Utterly unoriginal broke-down-in-a-hick-town nonsense. Snooty Reinhold and scruffy Dafoe go through the motions, and of course find a pair of female companions. Soundtrack features Los Lobos, the Pretenders, and Dave Edmunds. **1984 (R) 90m/C** Willem Dafoe, Judge Reinhold, Karen Lee, Kate Vernon, Stephen Elliott; *Dir:* John Mark Robinson. **VHS, Beta $79.98** *FOX 𝄫 ½*

Roadhouse Girl A sexy woman kindles a romance with one of her husband's employees and the result is murder. Reminiscent of "The Postman Always Rings Twice," but with poor production values and performances. Also known as "Marilyn." **1953 70m/B** *GB* Maxwell Reed, Sandra Dorne, Leslie Dwyer, Vida Hope, Ferdinand "Ferdy" Mayne; *W/Dir:* Wolf Rilla. **VHS, Beta $16.95** *SNC 𝄫*

Roadie Supposedly a look at the backstage world of rock n' roll, but the performances and direction leave a lot to be desired. Meat Loaf is a roadie who desperately wants to meet Alice Cooper, and spends the movie trying to do so. Features Art Carney, and musical names like Blondie, Roy Orbison, Hank Williams, Jr. and Don Cornelius (of "Soul Train" fame). **1980 (PG) 105m/C** Meatloaf, Kaki Hunter, Alan Carney, Art Carney, Gailard Sartain, Alice Cooper, Blondie, Roy Orbison, Hank Williams Jr., Ramblin' Jack Elliot; *Dir:* Alan Rudolph. **VHS $14.95** *MVD, WKV 𝄫 ½*

Roads to the South One-time Spanish revolutionary turned screenwriter returns to his homeland to fight fascism, but finds himself attracted to his son's beautiful girlfriend. Much soul searching. **1978 100m/C** *FR* Yves Montand, Miou-Miou; *Dir:* Joseph Losey. **VHS, LV $59.95** *CVC, IME, FCT, 𝄫𝄫 ½*

The Roamin' Cowboy In a town where the local ranchers are being threatened, a drifter comes to their rescue and wins both a girl and a job. **1937 56m/B** Fred Scott, Al "Fuzzy" St. John, Lois Wilde, Forrest Taylor, Roger Williams; *Dir:* Robert Hill. **VHS $19.95** *NOS, DVT 𝄫*

Roamin' Wild Exciting, action-packed early western with marshal Tyler saving a lady's stagecoach line from unsavory varmints. **1936 60m/B** Tom Tyler, Carol Wyndham, Al Ferguson, George Cheseboro, Max Davidson, Fred Parker; *Dir:* Bernard B. Ray. **VHS, Beta $19.95** *NOS, DVT 𝄫𝄫 ½*

Roanoak A dramatization in three parts of the mysterious fate of the early English settlement of Roanoak in North Carolina, which disappeared without a trace. Long, but intriguing. Made for PBS TV. **1986 180m/C** Will Sampson, Victoria Racimo, Porky White, Adrian Sparks, Patrick Kilpatrick; *Dir:* Jan Egleson. **Beta $150.00** *PBS 𝄫𝄫 ½*

Roarin' Lead The Three Musketeers stop a gang of rustlers, helping orphans' and cattlemen's associations. **1937 54m/B** Bob Livingston, Ray Corrigan, Max Terhune. **VHS, Beta $19.95** *VCN 𝄫 ½*

Roaring City A hard-boiled detective searches for the person behind the murders of gangsters and prize fighters. **1951 58m/B** Hugh Beaumont, Edward Brophy, Richard Travis; *Dir:* William Berke. **VHS, Beta, LV** *WGE 𝄫 ½*

Roaring Fire A martial arts thriller set against the backdrop of exotic Japan.

1982 95m/C Sonny Chiba. **VHS, Beta $69.99** *HBO 𝄫 ½*

Roaring Guns An unflappable McCoy rides in to thwart a despicable landowner trying to drive all the small ranchers out of business. Low-budget fun. **1936 66m/B** Tim McCoy, Rex Lease, Wheeler Oakman; *Dir:* Sam Newfield. **VHS, Beta $19.95** *NOS, VCN, DVT 𝄫𝄫 ½*

The Roaring Road Romantic comedy about a car salesman named Toodles, who drives for his company in a big race. Vintage racing footage; silent with original organ music. **1919 57m/B** Wallace Reid, Ann Little, Theodore Roberts; *Dir:* James Cruze. **VHS, Beta, 8mm $24.99** *VYY, GPV 𝄫𝄫 ½*

Roaring Roads An heir to a fortune acquires a lust for race car driving, and his family thinks, a reckless death wish. **1935 60m/B** Gertrude Messinger, David Sharpe. **VHS, Beta, LV** *WGE 𝄫*

Roaring Six Guns Maynard protects his property against his crooked partner and his fiancee's father. **1937 60m/B** Kermit Maynard, Mary Hayes, John Merton, Edward Cassidy. **VHS, Beta $19.95** *NOS, DVT, VCN 𝄫*

Roaring Speedboats Bakewell is a speedboat racer up against crooks who want to con him out of his money by tinkering with his invention. Originally titled "Mile a Minute Love." **1937 62m/B** William Bakewell, Arletta Duncan, Duncan Renaldo, Vivien Oakland, Wilfrid Lucas; *Dir:* Elmer Clifton. **VHS, Beta, 8mm $24.95** *VYY 𝄫*

The Roaring Twenties Newsreels picture the Ku Klux Klan, the automobile era, the first filmed total solar eclipse, the Charleston, Babe Ruth, Rudolph Valentino's death, Charles Lindbergh, and Black Tuesday when the stock market crashed. **192? 60m/B** Dan Kuchuck, Peggy Jacobsen; *Dir:* J. Christian Ingvordsen. **VHS, Beta $19.95** *CCB 𝄫*

The Roaring Twenties Three World War I buddies find their lives intersecting unexpectedly in Prohibition-era New York. Cagney becomes a bootlegger and vies with Bogart for status as crime boss. Lynn is the attorney working to prosecute them. Greatest of all gangster flicks established director Walsh's reputation. Cheesy script delivered with zest by top pros. **1939 106m/B** James Cagney, Humphrey Bogart, Jeffrey Lynn, Priscilla Lane, Gladys George, Frank McHugh; *Dir:* Raoul Walsh. National Board of Review Awards '39: 10 Best Films of the Year. **VHS, Beta, LV $19.98** *MGM, FOX, PIA 𝄫𝄫𝄫 ½*

Rob McConnell & the Boss Brass This program presents the jazz music of Rob McConnell featuring his hits "The Waltz I Blew for You" and "My Man Bill." **1983 25m/C** Rob McConnell. **VHS, Beta $19.95** *MVD, SVS 𝄫𝄫𝄫 ½*

Rob Roy—The Highland Rogue In the early 18th century, Scottish Highlander Rob Roy must battle against the King of England's secretary, who would undermine the MacGregor clan to enact his evil deeds. Dull Disney drama. **1953 84m/C** Richard Todd, Glynis Johns, James Robertson Justice, Michael Gough; *Dir:* Harold French. **VHS, Beta $69.95** *DIS 𝄫 ½*

Robbers of the Sacred Mountain TV rip-off of "Raiders of the Lost Ark", loosely based on Conan Doyle's story "Challenger's Gold." Adventurers seek meteorites in the Mexican jungle. Also called "Falcon's Gold"; billed as the first made-for-cable movie. **1983 95m/C** John Marley, Simon MacCorkindale, Louise Vallance, George Touliatos; *Dir:* Bob Schulz. **VHS, Beta $49.95** *PSM 𝄫𝄫*

Robbery Unoriginal but competent rendition of the hackneyed British Royal Mail robbery. Thieves plan and execute a heist of a late-night mail train carrying $10 million. **1967 113m/C** *GB* Stanley Baker, James Booth, Joanna Pettet, Frank Finlay, Barry Foster; *Dir:* Peter Yates. **VHS, Beta $9.98** *CHA 𝄫𝄫 ½*

Robbery A six-man platoon of Vietnam war veterans endeavors to assault and rob a mob of shiftless bookies. **1985 91m/C** John Sheerin, Tony Richards, Tim Hughes; *Dir:* Michael Thornhill. **VHS, Beta $69.98** *LIV, LTG 𝄫*

Robbery Under Arms From Australia in the 1800s, similar to the American wild west, comes a tale of love and robbery. Finch is the leader of a band of outlaws. Rather dull story redeemed by excellent photography of beautiful landscape. **1957 83m/C** *GB* Peter Finch, Ronald Lewis, Laurence Naismith, Maureen Swanson, David McCallum; *Dir:* Jack Lee. **VHS $79.95** *IMP 𝄫𝄫 ½*

Robby A young white boy and a young black boy shipwrecked on an island form a strong friendship that sees them through a series of adventures. **1968 60m/C** Warren Raum, Ryp Siani, John Garces; *Dir:* Ralph Blumke. **VHS, Beta** *AWF 𝄫𝄫*

Robby the Rascal The evil Mr. Bullion devises a plan to steal a cybot, Robby the Rascal, from his inventor Dr. Rumplechips. **1985 90m/C VHS, Beta $24.95** *PAR 𝄫𝄫*

The Robe This moving, religious portrait follows the career and religious awakening of drunken and dissolute Roman tribune Marcellus (Burton), after he wins the robe of the just-crucified Christ in a dice game. Mature plays Burton's slave surprisingly well, and reprised the role in the sequel, "Demetrius and the Gladiators"; Burton is wooden. "The Robe" was the first movie to be filmed in CinemaScope. Oscar nominations for Burton, and the color cinematography. Based on the novel by Lloyd C. Douglas. **1953 133m/C** Richard Burton, Jean Simmons, Victor Mature, Michael Rennie, Richard Boone, Dean Jagger, Jeff Morrow, Jay Robinson, Dawn Addams, Ernest Thesiger; *Dir:* Henry Koster. Academy Awards '53: Best Art Direction/Set Decoration (Color), Best Costume Design (Color); Golden Globe Awards '54: Best Film—Drama; National Board of Review Awards '53: 10 Best Films of the Year, Best Actress (Simmons). **VHS, Beta, LV $19.98** *FOX, IGP 𝄫𝄫 ½*

Robert A. Taft: Mr. Republican An account of the political career of Robert A. Taft, son of the U.S. President, who was noted as a great leader of the Republican Party. Classic newsreel footage. **1954 15m/B VHS, Beta $49.95** *TSF 𝄫𝄫 ½*

Robert Kennedy and His Times Another look at the Kennedy clan, this time from the Bobby angle. Good acting. Based on the book by Arthur Schlesinger. **1990 335m/C** Brad Davis, Veronica Cartwright, Cliff DeYoung, Ned Beatty, Beatrice Straight, Jack Warden; *Dir:* Marvin J. Chomsky. **VHS $59.95** *COL 𝄫𝄫 ½*

Robert Klein on Broadway Record of Klein's comedy stage show, filmed at Broadway's Nederlander Theater.
1986 60m/C Robert Klein. **VHS, Beta, LV** $19.98 *LIV, VES* ♫♫½

Robert Klein: Child of the '50s, Man of the '80s The hilarious Robert Klein performs some of his classic routines in this concert taped at New York University in Manhattan's Greenwich Village.
1984 60m/C Robert Klein. **VHS, Beta** $29.99 *HBO* ♫♫½

Robert et Robert Two 'ineligible' bachelors resort to using a computerized matrimonial agency to find the girls of their dreams. They become friends while they wait for their dates. Sensitive tale of loneliness and friendship. Warm and witty with fine performances throughout. French with English subtitles.
1979 95m/C *FR* Charles Denner, Jacques Villeret, Jean-Claude Brialy; *Dir:* Claude Lelouch. **VHS, Beta** $59.95 *COL* ♫♫♫

Roberta A football player inherits his aunt's Parisian dress shop and finds himself at odds with an incognito Russian princess. Dumb plot aside, this is one of the best Astaire-Rogers efforts, featuring Jerome Kern's "Smoke Gets in Your Eyes," "Yesterdays," "I Won't Dance," and "I'll Be Hard to Handle." A later remake was titled "Lovely to Look At."
1935 106m/B Fred Astaire, Ginger Rogers, Irene Dunne, Lucille Ball, Randolph Scott; *Dir:* William A. Seiter. **VHS, Beta** $29.95 *MGM* ♫♫♫

Robin Hood Extravagant production casts Fairbanks as eponymous gymnastic swashbuckler who departs for Crusades as Earl of Huntington and returns as the hooded one to save King Richard's throne from the sinister Sheriff of Nottingham. Best ever silent swashbuckling.
1922 110m/B Douglas Fairbanks Sr., Wallace Beery, Sam DeGrasse, Enid Bennett, Paul Dickey, William E. Lowery, Roy Coulson, Bill Bennett, Merrill McCormick, Wilson Benge, Willard Louis, Alan Hale Jr., Maine Geary, Lloyd Talman; *Dir:* Allan Dwan. **VHS, Beta** $19.95 *NOS, GPV, FCT* ♫♫♫½

Robin Hood This time the Sherwood Forest crew are portrayed by appropiate cartoon animals, hence, Robin is a fox, Little John a bear, etc. Good family fare, but not as memorable as other Disney features.
1973 83m/C *Dir:* Wolfgang Reitherman; *Voices:* Roger Miller, Brian Bedford, Monica Evans, Phil Harris, Andy Devine, Carol Shelley, Peter Ustinov, Terry-Thomas, Pat Buttram, George Lindsey, Ken Curtis. National Board of Review Awards '73: Special Citation. **VHS, Beta, LV** $24.99 *DIS, FCT, RDG* ♫♫♫

Robin Hood An animated version of the classic legend.
1985 60m/C **VHS, Beta** $19.95 *LTG* ♫♫♫

Robin Hood Dark version of the medieval tale. Bergin's Prince of Thieves is well developed and Thurman is a graceful Marion. Three studios announced plans to remake "Robin Hood" in 1990 and two were completed, including this one which was scaled down for cable television.
1991 116m/C Patrick Bergin, Uma Thurman, Juergen Prochnow, Edward Fox, Jeroen Krabbe; *Dir:* John Irvin. **VHS, Beta, LV** $89.98 *FXV, FCT* ♫♫♫

Robin Hood, the Legend: Volume 2 Another volume of four episodes from the vintage television series featuring Robin and the characters we know by heart, Friar Tuck, Maid Marian, Little John and many more.
1952 120m/B Richard Greene, Bernadette O'Farrell, Alexander Gauge. **VHS** $9.99 *VTR* ♫♫♫

Robin Hood of the Pecos Ex-Confederate soldier takes on northern post-war politicians and carpetbaggers. Decent Rogers/Hayes outing.
1941 56m/B Roy Rogers, George "Gabby" Hayes, Marjorie Reynolds, Jay Novello, Roscoe Ates; *Dir:* Joseph Kane. **VHS, Beta** $19.95 *NOS, DVT, VYY* ♫♫½

Robin Hood: Prince of Thieves The huge-budget ($50 million) rethinking of the Robin Hood legend is a surprisingly eccentric effort, both helped and hindered by its odd choices: an uninspiring yank Robin; American cast members with accents that come and go; a comic neurotic Sheriff of Nottingham; and an African co-hero who becomes the Merry Men's science officer, or something. The action, adventure and romance survives, however, and undiscriminating viewers should be satisfied. Still, when Bryan Adams and his band show up in Sherwood Forest crooning the pic's hit MTV love theme the Hound has had enough. Merry Men. Revisionist in its ideas about the times and people, critics generally disapproved of the changes in the story and Costner's performance, though their comments about his lack of an English accent seem nitpicky in light of their problems with scenic reality. Lots of fun for lovers of action, romance and fairy tales. Did big box office in summer 1991.
1991 (PG-13) 144m/C Kevin Costner, Morgan Freeman, Christian Slater, Alan Rickman, Mary Elizabeth Mastrantonio, Brian Blessed, Geraldine McEwan, Michael Wincott, Sean Connery, Harold Innocent, Jack Wild; *Dir:* Kevin Reynolds. MTV Movie Awards '92: Best Song ("(Everything I Do) I Do for You"). **VHS, LV** $24.99 *WAR, PIA, RDG* ♫♫½

Robin Hood of Texas Autry finds his calm life disturbed when he is falsely accused of robbing a bank. Pretty good plot for once, in Autry's last Republic western, and plenty of action.
1947 71m/B Gene Autry, The Cass County Boys, Lynn Roberts, Sterling Holloway, Adele Mara; *Dir:* Lesley Selander. **VHS, Beta** $29.95 *CCB, VCN* ♫♫½

Robin Hood: The Movie A colorized and re-mixed version of several episodes of the British television series adds up to a fine version of the classic tale. Excellent cast, clever editing and stunning colorization.
1991 90m/C *GB* Richard Greene, Bernadette O'Farrell, Alan Wheatley; *Dir:* Ralph Smart, Daniel Birt, Terence Fisher. **VHS** $14.99 *GKK* ♫♫½

Robin Hood...The Legend British television version of the swashbuckling hero's medieval exploits.
1983 115m/C *GB* Michael Praed, Anthony Valentine, Nickolas Grace; *Dir:* Ian Sharp. **VHS, Beta** $39.95 *FOX* ♫½

Robin Hood...The Legend: Herne's Son When Herne's son is killed by the Sheriff of Nottingham, he begs Robert of Hunnington (Connery) to continue his work. Connery refuses; the band dissolves; Maid Marion is kidnapped, and they reunite to save her. Good entry in the BBC series.

1985 101m/C *GB* Jason Connery, Oliver Cotton, Michael Craig, Nickolas Grace, George Baker; *Dir:* Robert M. Young. **VHS, Beta** $14.98 *FOX, FCT* ♫♫½

Robin Hood...The Legend: Robin Hood and the Sorcerer In the introduction to the series, Robin is informed by an enchanted forestman of his duties as Robin Hood.
1983 115m/C Michael Praed, Anthony Valentine, Nickolas Grace, Clive Mantle, Peter Williams; *Dir:* Ian Sharp. **VHS** $14.98 *FOX, FCT* ♫♫♫♫

Robin Hood...The Legend: The Swords of Wayland A wicked sorceress provides a supernatural threat to Robin and his merry men. Second, even better entry in the BBC series. Convoluted, breathless plot makes for swashbuckling fun and much derring-do.
1983 115m/C *GB* Michael Praed, Rula Lenska, Nickolas Grace; *Dir:* Robert M. Young. **VHS, Beta** $14.98 *FOX, FCT* ♫♫♫

Robin Hood...The Legend: The Time of the Wolf Robin Hood finds out that an old enemy is about to release his evil doings on the world and he must stop him.
1985 105m/C *GB* Jason Connery, Nickolas Grace, Richard O'Brien; *Dir:* Syd Roberson. **VHS, Beta** $14.98 *FOX, FCT* ♫♫♫

Robin and Marian After a separation of twenty years, Robin Hood is reunited with Maid Marian, who is now a nun. Their dormant feelings for each other are reawakened as Robin spirits her to Sherwood Forest. In case you wanted to see Robin Hood robbed of all magic, spontaneity, and fun. Connery is dull, dull, dull, working with an uninspired script.
1976 (PG) 106m/C *GB* Sean Connery, Audrey Hepburn, Robert Shaw, Richard Harris; *Dir:* Richard Lester. **VHS, Beta, LV** $19.95 *COL, FOX* ♫♫

Robin and the Seven Hoods Runyon-esque Rat Pack version of 1920s' Chicago, with Frank and the boys as do-good gangsters in their last go-round. Fun if not unforgettable. Songs include the Oscar-nominated "My Kind of Town (Chicago Is)."
1964 124m/C Frank Sinatra, Bing Crosby, Dean Martin, Sammy Davis Jr., Peter Falk, Barbara Rush, Allen Jenkins; *Dir:* Gordon Douglas. **VHS, Beta** $19.98 *WAR, PIA* ♫♫½

Robin! Tour de Face! Stand-up performance at New York's Metropolitan Opera House by the irrepressible comedian.
1987 60m/C Robin Williams. **VHS, Beta** *VES* ♫♫½

Robin Williams Live! Williams performs his popular stand-up act before a live audience.
1981 70m/C **VHS, Beta, LV** $14.98 *VES* ♫♫½

Robinson Crusoe A real oddity: a British silent film originally made in 1927, re-edited ten years later. Music and sound effects were added, along with a narration by children's radio personality "Uncle Don" Carney. Follows the plot of the famous adventure story by Daniel Defoe.
1936 34m/B *Nar:* Uncle Don Carney. **VHS, Beta** $11.98 *VYY* ♫♫

Robinson Crusoe The classic survival story by Daniel Defoe is retold with animation.
1972 48m/C **VHS, Beta** $19.95 *MGM* ♫♫

Robinson Crusoe An animated adaptation of the classic Daniel Defoe story about a man marooned on a small island.
1978 86m/C VHS, Beta $11.95 FHE ♂♂

Robinson Crusoe of Clipper Island Fourteen-episodes serial featuring the investigative expertise of Mala, a Polynesian in the employ of the U.S. Intelligence Service. Each episode runs 16 minutes. Shortened version titled "Robinson Crusoe of Mystery Island."
1936 256m/B Ray Mala, Rex, Buck, Kate Greenfield, John Ward, Tracy Lane, Robert Kortman, Herbert Rawlinson; **Dir:** Mack V. Wright. **VHS, LV $26.95** SNC, NOS, REP ♂♂

Robinson Crusoe of Mystery Island A Polynesian employed by the U.S. Intelligence Service investigates saboteurs on a mysterious island. Would-be cliffhanger is a yawner; where are the chills and thrills? Edited from the serial "Robinson Crusoe of Clipper Island."
1936 100m/B Mala, Rex, Buck, Herbert Rawlinson, Ray Taylor; **Dir:** Mack V. Wright. **VHS, Beta $27.95** VCN ♂♂

Robinson Crusoe & the Tiger A tiger tells the famous story of how Robinson Crusoe became stranded on a desert island.
1972 (G) 109m/C Hugo Stiglitz, Ahui; **Dir:** Rene Cardona Jr. **VHS, Beta $29.98** SUE ♂ ½

Robo-Chic A woman who is part cop, part machine, becomes a defender of justice.
1989 90m/C Kathy Shower, Jack Carter, Julie Newmar, Burt Ward, Lyle Waggoner. **VHS, Beta** AIP ♂ ½

RoboCop A nearly dead Detroit cop is used as the brain for a crime-fighting robot in this bleak vision of the future. Humor, satire, action, and violence keep this moving in spite of its underlying sadness. Slick animation techniques from Phil Tippet. Verhoeven's first American film.
1987 (R) 103m/C Peter Weller, Nancy Allen, Ronny Cox, Kurtwood Smith, Ray Wise, Miguel Ferrer; **Dir:** Paul Verhoeven. **VHS, Beta, LV, SVS $19.98** ORI, SUP ♂♂♂

RoboCop 2 Grimer, more violent sequel to the initial fascinating look at the future, where police departments are run by corporations hungry for profit at any cost. A new and highly addictive drug has made Detroit more dangerous than ever. Robocop is replaced by a stronger cyborg with the brain of a brutal criminal. When the cyborg goes berserk, Robocop battles it and the drug lords for control of the city. Dark humor and graphic savagery, with little of the tenderness and emotion of the original.
1990 (R) 117m/C Peter Weller, Nancy Allen, Belinda Bauer, Dan O'Herlihy, Tommy Noonan, Gabriel Damon, Galyn Gorg, Felton Perry, Patricia Charbonneau; **Dir:** Irvin Kershner. **VHS, Beta, LV $94.98** ORI ♂♂

Roboman American scientist gets in car crash in Soviet Union. When he returns to the States as a cyborg, his friends seem to notice a change. Also known as "Who?"
1975 (PG) 91m/C GB GE Elliott Gould, Trevor Howard, Joe Bova, Ed Grover, James Noble, John Lehne; **Dir:** Jack Gold. **VHS** ACE, MTX ♂♂ ½

Robot Holocaust Somewhere in the doomed future, after the Earth is laid waste by robots with superior intelligence, a human survivor dares to challenge them. If you're doomed to watch this, you just might survive; it's so-so, unsurprisingly sci-fi.
1987 79m/C Norris Culf, Nadine Hart, Joel von Ornsteiner, Jennifer Delora, Andrew Howarth, Angelika Jager, Rick Gianasi; **W/Dir:** Tim Kincaid. **VHS, Beta $69.95** UHV ♂♂

Robot Jox Two futuristic warriors battle to the finish in giant mechanical robots. Preteen fare.
1989 (PG) 84m/C Gary Graham, Anne-Marie Johnson, Paul Koslo, Robert Sampson, Danny Kamekona, Hilary Mason, Michael Alldredge; **Dir:** Stuart Gordon. **VHS, LV $14.95** COL, PIA ♂ ½

Robot Monster Ludicrous cheapie is widely considered one of the worst films of all time, as a single alien dressed in a moth-eaten gorilla suit and diving helmet conspires to take over the Earth from his station in a small, bubble-filled California cave. Available in original 3-D format.
1953 62m/B George Nader, Claudia Barrett, Gregory Moffett, Selena Royle; **Dir:** Phil Tucker. **VHS, Beta, LV $12.95** RHI, SVS, FCT ♂♂ ½

Robot Ninja A comic-book ninja comes to life to destroy criminals. Good genre cast.
1989 (R) 90m/C Michael Todd, Linnea Quigley, Burt Ward. **VHS $79.95** CIN, PHX ♂ ½

Robot Pilot Tucker stars as a WWII test pilot trying to promote a friend's invention. Along the way, he's involved in a subplot with enemy agents who steal a bomber. He also finds time to become romantically entangled with the boss' spoiled daughter. Originally titled "Emergency Landing."
1941 68m/B Forrest Tucker, Carol Hughes, Evelyn Brent, Emmett Vogan, William Halligan; **Dir:** William Beaudine. **VHS, Beta, 8mm $29.95** VYY, SNC ♂

The Robot vs. the Aztec Mummy This is a grade-Z Mexican horror film involving mummies and such.
1959 72m/B MX Ramon Gay, Rosita Arenas; **Dir:** Rafael Portillo. **VHS, Beta $19.98** SNC, AOV, MRV Woof!

Robotix This is a feature-length animated film starring a bunch of robots who struggle to save the universe from evil forces.
1986 90m/C VHS, Beta $69.98 LIV, VES Woof!

Robotman & Friends 2: I Want to be Your Robotman This is another entry in the children's animated series about robots who fight the denizens of evil.
1986 42m/C VHS, Beta $19.98 LIV, CVL Woof!

Rocco and His Brothers A modern classic from top director Visconti, about four brothers who move with their mother from the Italian countryside to Milan. Very long, sometimes ponderous, but engrossing, complex drama. Available shortened to 90 minutes, but the unedited version is much more rewarding.
1960 175m/B IT Alain Delon, Renato Salvatori, Annie Girardot, Katina Paxinou, Claudia Cardinale, Roger Hanin; **Dir:** Luchino Visconti. **VHS, LV $29.95** VDM, CVW ♂♂♂ ½

The Rock A continuous picture of a calm, shimmering ocean inlet featuring a large rock which has been shaped by wind and water to create a relaxed background.
1978 30m/C VHS, Beta PCC ♂♂♂ ½

Rock-a-Doodle A sun-raising rooster named Chanticleer is tricked into neglecting his duties by an evil barnyard owl. Humiliated, he leaves his home and winds up in a Las Vegas-like city as an Elvis-impersonating singer (complete with pompadour). He finds success and acquires a hound-dog pal, an unscrupulous manager, and a cutie-chicken girlfriend. Only back at the farm things aren't going well, so Chanticleer's friends decide to go to the city and persuade him to come home. Mildly amusing, but bland by today's standards of animation and the music is certainly nothing to crow about.
1992 (G) 77m/C **Dir:** Don Bluth; **Voices:** Glen Campbell, Christopher Plummer, Phil Harris, Sandy Duncan, Ellen Greene, Charles Nelson Reilly. **VHS, LV $24.98** HBO ♂♂

Rock Adventure Music by Baenzai and breathtaking visuals combine to express a mood of wild adventure.
1981 29m/C Beta, LV PNR ♂♂

Rock, Baby, Rock It Weak and silly rock-and-roll crime drama about teens trying to fight the Mafia in Dallas in 1957. Features performances by many regional bands including Kay Wheeler, the Cell Block Seven, Johnny Carroll, Preacher Smith and the Deacons, and the Five Stars. If you grew up in Dallas in the '50s, the musical groups may bring back some fond memories; otherwise, this obscure film probably isn't worth your time.
1957 84m/B Kay Wheeler, John Carroll; **Dir:** Murray Douglas Sporup. **VHS, Beta $59.95** RHI ♂

Rock House When an L.A. narcotics officer's wife is murdered by drug runners he goes on an obsessive revenge mission to destroy the city's cocaine operations.
1988 (R) 98m/C Joseph Jennings, Michael Robbin, Alan Shearer; **W/Dir:** Jack Vacek. **VHS $89.95** CHV, HMD ♂♂

Rock & the Money-Hungry Party Girls A struggling musician battles against Gabor-sister-worshipping bimbos in an effort to find unpublished songs by a '50s music legend.
1989 90m/C Paul Sercu, Adam Small, Judi Durand, Mary Baldwin, Debra Lamb; **Dir:** Kurt MacCarley. **VHS, Beta** CAM ♂

Rock 'n' Roll Heaven A Don Kirschner compilation of original performances by the pioneers of rock and roll and those who followed.
1984 60m/C Buddy Holly, Elvis Presley, Eddie Cochran, Bill Haley, Paul McCartney, John Lennon, George Harrison, Ringo Starr; **Dir:** Don Kirshner. **VHS, Beta $29.95** MGM

Rock 'n' Roll High School The music of the Ramones highlights this non-stop high-energy cult classic about a high school out to thwart the principal at every turn. If it had been made in 1957, it would have been the ultimate rock 'n' roll teen movie. As it is, its 1970s' milieu works against it, but the performances are perfect for the material and the Ramones are great. Songs include "Teenage Lobotomy," "Blitzkrieg Bop," "I Wanna Be Sedated" and the title track, among others.
1979 (PG) 94m/C The Ramones, P.J. Soles, Vincent Van Patten, Clint Howard, Dey Young, Mary Woronov, Alix Elias, Dick Miller, Paul Bartel; **Dir:** Allan Arkush. **VHS, Beta $89.95** WAR, OM ♂♂♂

Rock 'n' Roll High School Forever Jesse Davis and his band just want to rock 'n' roll, but the new principal doesn't share their enthusiasm. Way late, way lame sequel to "Rock 'n' Roll High School." Soundtrack includes music from The Divinyls, Dee Dee Ramone, Mojo Nixon, Will and the Bushmen, The Pursuit of Happiness and more.

1991 (PG-13) 94m/C Corey Feldman, Mary Woronov, Larry Linville, Evan Richards, Liane Curtis, Mojo Nixon, Sarah Buxton; *W/Dir:* Deborah Brock. **VHS, Beta, LV** $89.95 *LIV, IME* Woof!

Rock 'n' Roll History: American Rock, the '60s Videos of performances by some of the American groups formed in the wake of the '60s British Invasion. Featured appearances include the Byrds, Lovin' Spoonful, the Mamas and the Papas, and the Young Rascals.
1989 30m/C The Lovin' Spoonful, Jay & the Americans, The McCoys, The Byrds, The Mamas & the Papas, The Young Rascals, Sam the Sham & the Pharoahs. **VHS, Beta** $9.95 *MVD, GKK*

Rock 'n' Roll History: English Invasion, the '60s Performances from the best of the British Invasion.
1990 30m/C The Beatles, The Dave Clark Five, Gerry & the Pacemakers, The Zombies, The Animals, Manfred Mann, The Yardbirds. **VHS, Beta** $9.95 *MVD, GKK*

Rock 'n' Roll History: Girls of Rock 'n' Roll Performances from some of the best girl groups of the 50's and 60's including the Crystals, the Ronettes, the Dixie Cups and Diana Ross and the Supremes. Connie Francis and Mary Wells also make appearances.
1990 30m/C Connie Francis, The Crystals, Ronettes, The Supremes, Mary Wells, The Dixie Cups. **VHS, Beta** $9.95 *MVD, GKK*

Rock 'n' Roll History: Rock 'n' ... Heaven Performances from the late stars ... were influential in American music history.
1990 30m/C Elvis Presley, Jackie Wilson, Bobby Darin, Otis Redding, The Beatles, Marvin Gaye, Tammy Terrell, Buddy Holly. **VHS, Beta** $9.95 *MVD, GKK*

Rock 'n' Roll History: Rockabilly Rockers Performances from the greats who started it all in rock 'n' roll.
1990 30m/C Hank Williams, Carl Perkins, Elvis Presley, Sonny James, Buddy Holly, The Everly Brothers, Roy Orbison. **VHS, Beta** $9.95 *MVD, GKK*

Rock 'n' Roll History: Sixties Soul Great performances from artists of the sixties in both color and black & white.
1990 30m/C The Four Tops, The Temptations, Otis Redding, Joe Tex, Stevie Wonder, Jerry Butler. **VHS, Beta** $9.95 *MVD, GKK*

Rock 'n' Roll Nightmare A rock band is pursued by demons from another dimension. The usual lame garbage.
1985 (R) 89m/C *CA* Jon Mikl Thor, Paula Francescatto, Rusty Hamilton; *Dir:* John Fasano. **VHS, Beta** $29.95 *ACA* 🎞️½

Rock 'n' Roll: The Early Days Rare footage of the performers who started it all - including Chuck Berry, Jerry Lee Lewis, The Everly Brothers, Bill Haley & the Comets and more!
1984 45m/C **VHS, Beta, LV** $16.95 *MVD, PIA*

Rock, Pretty Baby Early rock-'n'-roll movie about a teenage band and high school riots. Black and white.
1956 89m/C Sal Mineo, John Saxon, Rod McKuen, Luanna Patten, Fay Wray, Edward Platt; *Dir:* Richard Bartlett. **VHS, Beta** $14.95 *KRT* 🎞️

Rock River Renegades The Range Busters capture a band of renegades and restore peace to the territory once more.
1942 59m/B Ray Corrigan, John "Dusty" King. **VHS, Beta** $19.95 *NOS, DVT* 🎞️

Rock, Rock, Rock A young girl tries to earn enough money for a prom gown after her father closes her charge account. Screen debut of Tuesday Weld. Ultra-low-budget, but a classic after a fashion. Includes classic musical numbers performed by Chuck Berry and other rock'n'roll pioneers.
1956 78m/B Alan Freed, Chuck Berry, Fats Domino, Tuesday Weld, Frankie Lymon & the Teenagers; *Dir:* Will Price. **VHS, Beta** $12.95 *MVD, MED, MRV* 🎞️🎞️½

Rock & Roll Call Compilation of performances by some of the most popular vocal artists of the '70s.
1984 60m/C Neil Diamond, John Denver, Helen Reddy, Wayne Newton, Diana Ross. **VHS, Beta** $12.95 *MVD, IVE*

Rock & Roll Review An all-star show filmed on stage at Harlem's Apollo Theater.
1955 60m/B Duke Ellington, Clovers, Nat King Cole, Joe Turner. **VHS, Beta** *GVV*

Rock & Rule Animated sword & sorcery epic with a rock soundtrack. Voices are provided by Deborah Harry, Cheap Trick, Lou Reed, and Iggy Pop.
1983 85m/C *Dir:* Clive A. Smith; *Voices:* Paul LeMat; Catherine O'Hara. **VHS, Beta, LV** $79.95 *MGM* 🎞️🎞️

Rockers '87 Famous Jamaican reggae stars perform.
1987 60m/C Early B, Johnny Osbourne, Tonto Irie, Echo Minott, Ardie Wallace. **VHS, Beta** *KRV*

Rocket Attack U.S.A. Antiquated, ridiculous tale of nuclear warfare about the time of Sputnik, with Russia's first strike blowing up New York City and environs. Mercifully short, with time for a little romance.
1958 70m/B Monica Davis, John McKay, Dan Kern; *Dir:* Barry Mahon. **VHS, Beta** $14.95 *SVS* Woof!

Rocket Gibraltar On the occasion of a crusty patriarch's birthday, a large family unites on his remote estate to carry out his special birthday wish. Fine performance by Lancaster. Supported by a solid cast, overcomes a slim story, Culkin, later of "Home Alone," is loveable as the precocious five-year-old. Picturesque Long Island scenery.
1988 (PG) 92m/C Burt Lancaster, Bill Pullman, John Glover, Suzy Amis, Macauley Culkin; *Dir:* Daniel Petrie. **VHS, Beta** $14.95 *COL* 🎞️🎞️🎞️

Rocket to the Moon An adaptation of the Clifford Odets story about the mundane life of a 30-year-old Manhattan dentist and how he comes to terms with his life. From the "American Playhouse" series.
1986 118m/C Judy Davis, John Malkovich, Eli Wallach; *Dir:* John Jacobs. **VHS, Beta** $59.95 *ACA, PBS* 🎞️🎞️

The Rocketeer Lightheaded fun. A stunt flyer in the 1930s finds a prototype jet backpack sought by Nazi spies. Donning a mask, he becomes a flying superhero. Breezy family entertainment with stupendous effects; even better if you know movie trivia, as it brims with Hollywood references, like a great villain (Dalton) clearly based on Errol Flynn.

1991 (PG) 109m/C Bill Campbell, Jennifer Connelly, Alan Arkin, Timothy Dalton, Paul Sorvino; *Dir:* Joe Johnston. **VHS, Beta** $94.95 *DIS, CCB* 🎞️🎞️🎞️

Rocketship X-M A lunar mission goes awry and the crew lands on Mars, where they discover ancient ruins. Well acted and nicely photographed. Contains footage, a tinted sequence and previews of coming attractions from classic science fiction files.
1950 77m/B Lloyd Bridges, Osa Massen, John Emery, Hugh O'Brian, Noah Beery Jr.; *Dir:* Kurt Neumann. **VHS, Beta, LV** $19.95 *MED* 🎞️🎞️½

Rockin' the '50s Performances of some great '50s tunes. Little Richard does "Tutti Frutti," the Platters do "The Pretender," Johnny Otis does the original version of "Willie and the Hand Jive," Pat Boone sings "Ain't That a Shame" and the Diamonds perform "The Stroll."
195? 30m/C **VHS, Beta** $9.95 *SIM, KAR*

Rockin' Road Trip A guy meets a girl in a Boston bar, and finds himself on a drunken, slapstick road trip down the Eastern seaboard with a rock band.
1985 (PG-13) 101m/C Garth McLean, Katherine Harrison, Margaret Currie, Steve Boles. **VHS, Beta** $79.98 *FOX* 🎞️

Rockin' Ronnie A rough, sophomoric, uproarious send-up of President Reagan, using news footage of him and of Queen Elizabeth, Ayatollah Khomeini, Jayne Mansfield, and other figures.
1987 45m/C **VHS, Beta** $19.98 *MPI* 🎞️🎞️½

The Rocking Horse Winner Poignant tale of a young boy who tries to keep his parents from splitting up by predicting horse race winners. His mother's greed spurs tragedy. Based on the short story by D.H. Lawrence.
1950 91m/B Kenneth More; *Dir:* Peter Meda. **VHS, Beta, 3/4U** $59.95 *LCA, FLI* 🎞️🎞️½

Rocktober Blood A convicted and executed rock star comes back from the grave and starts killing people all over again. Yet another gorefest.
1985 (R) 88m/C Donna Scoggins, Tray Loren, Nigel Benjamin, Beverly Sebastian; *Dir:* Ferd Sebastian. **VHS, Beta** $19.98 *LIV, VES* 🎞️½

Rockula Teen rock comedy about a young-yet-300-year-old vampire looking to lose his virginity. Recommended only for hard-core Diddley fans.
1990 (PG-13) 87m/C Dean Cameron, Bo Diddley, Tawny Fere, Susan Tyrrell, Thomas Dolby, Toni Basil; *Dir:* Luca Bercovici. **VHS, Beta, LV** $89.95 *WAR, OM* 🎞️½

Rocky Box office smash about a young man from the slums of Philadelphia who dreams of becoming a boxing champion. Stallone wrote the script and plays Rocky, the underdog hoping to win fame and self-respect. Rags-to-riches story seems to parallel Stallone's life; he had been previously virtually unknown before this movie. Intense portrayal of the American Dream; loses strength in the subsequent (and numerous) sequels. Powerful score by Bill Conti.
1976 (PG) 119m/C Sylvester Stallone, Talia Shire, Burgess Meredith, Burt Young, Carl Weathers; *Dir:* John G. Avildsen. Academy Awards '76: Best Director (Avildsen), Best Film Editing, Best Picture; Golden Globe Awards '77: Best Film—Drama; Los Angeles Film Critics Association Awards '76: Best Film; National Board of Review Awards '76: 10 Best Films of the Year, Best Supporting Actress (Shire). **VHS, Beta, LV, 8mm** $19.98 *MGM, FOX, BTV* 🎞️🎞️🎞️½

Rocky 2 Time-marking sequel to the box office smash finds Rocky frustrated by the commercialism which followed his match to Apollo, but considering a return bout. Meanwhile, his wife fights for her life. The overall effect is to prepare you for the next sequel.
1979 (PG) 119m/C Sylvester Stallone, Talia Shire, Burt Young, Burgess Meredith, Carl Weathers; *Dir:* Sylvester Stallone. VHS, Beta, LV $19.95 *FOX*

Rocky 3 Rocky is beaten by big, mean Clubber Lang (played to a tee by Mr. T). He realizes success has made him soft, and has to dig deep to find the motivation to stay on top. Amazingly, Stallone regains his underdog persona here, looking puny next to Mr. T, who is the best thing about the second-best "Rocky" flick.
1982 (PG) 100m/C Sylvester Stallone, Talia Shire, Burgess Meredith, Carl Weathers, Mr. T, Leif Erickson, Burt Young; *Dir:* Sylvester Stallone. VHS, Beta, LV $19.95 *FOX* 🎬🎬½

Rocky 4 Rocky travels to Russia to fight the Soviet champ who killed his friend during a bout. Will Rocky knock the Russkie out? Will Rocky get hammered on the head a great many times and sag around the ring? Will Rocky ever learn? Lundgren isn't nearly as much fun as some of Rocky's former opponents and Stallone overdones the hyper-patriotism as well as relying too heavily on uplifting footage from the earlier "Rocky" movies.
1985 (PG) 91m/C Sylvester Stallone, Talia Shire, Dolph Lundgren, Brigitte Nielsen, Michael Pataki, Burt Young, Carl Weathers; *Dir:* Sylvester Stallone. VHS, Beta, LV $19.95 *FOX* 🎬½

Rocky 5 Brain damaged and broke, Rocky finds himself back where he started; on the streets of Philadelphia. Boxing still very much in his blood, Rocky takes in a protege, training him in the style that made him a champ (take a lickin' and keep on tickin'). However an unscrupulous promoter has designs on the young fighter and seeks to wrest the lad from under the former champ's wing. This eventually leads to a showdown between Rocky and the young boxer in a brutal streetfight. Supposedly the last "Rocky" film, it's clear that the formula has run dry.
1990 (PG-13) 105m/C Sylvester Stallone, Talia Shire, Burt Young, Sage Stallone, Tom Morrison, Burgess Meredith; *Dir:* John G. Avildsen. VHS, Beta, LV, 8mm $92.99 *MGM* 🎬🎬

The Rocky Horror Picture Show When a young couple take refuge in a haunted castle, they find themselves the unwilling pawns in a warped scientist's experiment. Cult camp classic has been a midnight movie favorite for years and has developed an entire subculture built around audience participation. Songs include "The Time Warp," "Sweet Transvestite" (from Transsexual Transylvania, no less), and "Touch-a, Touch-a Touch Me." The tape includes a seven-minute short detailing the story behind the movie's popularity. May not be as much fun watching it on the little screen unless you bring the rice and squirt guns.
1975 (R) 105m/C Tim Curry, Susan Sarandon, Barry Bostwick, Meatloaf, Little Nell, Richard O'Brien; *Dir:* Jim Sharman. VHS $89.98 *FOX, FCT* 🎬🎬🎬

Rocky Jones, Space Ranger: Pirates of Prah Rocky Jones, sidekick Winky, and Professor Newton confront space pirates in this episode.
1953 75m/B Richard Crane, Sally Mansfield, Maurice Cass. VHS, Beta $29.95 *VYY* 🎬🎬🎬

Rocky Jones, Space Ranger: Renegade Satellite Nostalgia freaks may embrace this revived edition of the infamous early live-TV space cadet show. Rocky squares off against enemies of the United Solar System like Dr. Reno and Rudy DeMarco. Really cheap.
1954 69m/B Richard Crane, Sally Mansfield, Maurice Cass. VHS $19.95 *VDM* 🎬½

Rocky Jones, Space Ranger: The Cold Sun An adventure in the series "Rocky Jones, Space Ranger," as Rocky, Winky, and Professor Newton travel through space.
1953 75m/B Richard Crane, Sally Mansfield, Maurice Cass. VHS, Beta $29.95 *VYY* 🎬½

Rocky Jones, Space Ranger: Trial of Rocky Jones Rocky Jones, trouble-shooter for the Office of Space Affairs, travels through the universe in his spaceship "The Orbit Jet" along with his pal Winky and Professor Newton.
1953 75m/B Richard Crane, Sally Mansfield, Maurice Cass. VHS, Beta $29.95 *VYY* 🎬½

Rocky Jones, Space Ranger, Vol. 3: Blast Off Rocky, his sidekick Winky, and Professor Newton travel through another space adventure in their ship, "The Orbit Jet."
1953 75m/B Richard Crane, Sally Mansfield, Maurice Cass. VHS, Beta $24.95 *NOS, VYY* 🎬½

Rocky Jones, Space Ranger, Vol. 7: Forbidden Moon Here are three episodes of the early 1950's television series about Rocky Jones, chief of the Space Rangers, an organization established to protect the planets of a united solar system.
1953 75m/B Richard Crane, Sally Mansfield, Robert Lyden. VHS, Beta $19.95 *NOS, VDM, DVT* 🎬½

Rodan A gigantic prehistoric bird is disturbed from his slumber by H-bomb tests. He awakens to wreak havoc on civilization. Big bugs also run amok. From the director of "Godzilla" and a host of other nuclear monster movies.
1956 74m/C JP Kenji Sahara, Yumi Shirakawa; *Dir:* Inoshiro Honda. VHS, Beta, LV $14.98 *VES, MLB* 🎬🎬

Rodeo Champ A made-for-television drama for kids about a young girl determined to win a barrel-racing event with her horse at a local rodeo.
1985 60m/C VHS, Beta $14.95 *NWV* 🎬🎬

Rodeo Girl Ross is a restless housewife who joins the rodeo in this television drama based on the life of cowgirl Sue Pirtle. Particularly strong supporting cast of Hopkins, Clark, and Brimley.
1980 100m/C Katharine Ross, Bo Hopkins, Candy Clark, Jacqueline Brookes, Wilford Brimley, Parley Baer; *Dir:* Jackie Cooper. VHS, Beta $39.95 *IVE* 🎬🎬½

Rodgers & Hammerstein: The Sound of American Music The career of the famed song-writing team is examined, with many song sequences from their shows and films.
1985 92m/C *Hosted:* Mary Martin. Beta *CMR* 🎬🎬½

Rodney Dangerfield: Nothin' Goes Right A made-for-cable assortment of young stand-up comedians, introduced by Rodney. For mature audiences.
1988 83m/C Rodney Dangerfield, Andrew Dice Clay, Dom Irrera, Carol Leifer, Robert Schimmel, Barry Sobel, Lenny Clarke, Bill Hicks. VHS, Beta, LV $19.98 *ORI* 🎬🎬½

The Rodney King Case: What the Jury Saw in California vs. Powell A condensed version of the Los Angeles police brutality trial involving Rodney King, the results of which caused the 1992 Los Angeles riots. Includes explanations of the law and a review of prosecution and defense tactics. Originally aired on cable's Courtroom Television Network.
1992 116m/C *Hosted:* Fred Graham; *Nar:* Fred Graham. VHS $24.98 *MPI* 🎬🎬½

Rodrigo D: No Future Gritty, realistic portrayal of the youth killing culture of Medellin, Columbia. Minimal plot involves a young drummer and his drug-running friends. The actors used are actual untrained street kids, several of whom died after the film's completion.
1991 92m/C Ramiro Meneses, Carlos Mario Restrepo; *Dir:* Victor Gaviria. VHS $79.95 *KIV* 🎬🎬

Roe vs. Wade Hunter is excellent as the single woman from Texas who successfully challenged the nation's prohibitive abortion laws in a landmark Supreme Court case. Television drama based on the actual 1973 case of Norma McCorvey.
1989 92m/C Holly Hunter, Amy Madigan, Terry O'Quinn, Stephen Tobolowsky, Dion Anderson, Kathy Bates, James Gammon, Chris Mulkey; *Dir:* Gregory Hoblit. VHS, Beta, LV $14.95 *PAR* 🎬🎬🎬

Roger Corman: Hollywood's Wild Angel Entertaining documentary looks at the life and films of legendary B-movie producer/director Corman and features interviews with a host of admirers and fellow directors, including Scorsese, Dante, and Demme.
1978 58m/C David Carradine, Peter Fonda, Jonathan Demme, Ron Howard, Paul Bartel, Martin Scorsese, Joe Dante, Roger Corman; *Dir:* Christian Blackwood. VHS, Beta, 3/4U $29.95 *IHF, MPI* 🎬🎬🎬

Roger & Me The hilarious, controversial and lauded atypical semi-documentary. Details Moore's protracted efforts to meet General Motors president Roger Smith and confront him with the poverty and despair afflicting Flint, Michigan, after GM closed its plants there. Includes some emotionally grabbing scenes: a Flint family is evicted just before Christmas; a woman makes a living by selling rabbits for food or pets; and a then soon-to-be Miss America addresses the socioeconomic impact of GM's decision. One of the highest-grossing non-fiction films ever released, and Moore's first.
1989 (R) 91m/C Michael Moore, Roger Smith, Bob Eubanks, Pat Boone, Anita Bryant; *Dir:* Michael Moore. New York Film Critics Awards '89: Best Documentary. VHS, Beta, LV $19.95 *WAR* 🎬🎬🎬½

Roger Whittaker Concert performance by the familiar, goateed folk singer.
1982 53m/C VHS, Beta $29.95 *SVS, MVD* 🎬🎬🎬½

Rogue The Rogue is a ruthless man who can make beautiful women do anything that he desires; so, of course, he does.
1976 (R) 87m/C Milan Galvonic, Barbara Bouchet, Margaret Lee; *Dir:* Gregory Simpson. VHS, Beta $59.95 *MAG* 🎬

Rogue Lion A quasi-incompetent game warden more than meets his match when a voraciously carnivorous lion begins stalking an African community.
1975 89m/C VHS, Beta $59.95 UHV 🎬½

Rogue of the Range A lawman pretends that he is a bandit in order to track down an outlaw gang.
1936 60m/B Johnny Mack Brown. **VHS, Beta $19.95** NOS, DVT, WGE 🎬

Rogue of the Rio Grande A musical western in which the renowned bandit El Malo arrives in the peace-loving town of Sierra Blanca. One of Loy's earlier films.
1930 60m/B Myrna Loy, Jose Bohr, Raymond Hatton, Carmelita Geraghty. **VHS, Beta $27.95** VCN 🎬½

Rogue's Gallery A reporter and photographer covering the story of an inventor and his latest device are caught up in intrigue when he is found murdered.
1944 60m/B Frank Jenks, Robin Raymond, H.B. Warner, Ray Walker, Davison Clark, Robert E. Homans, Frank McGlynn, Pat Gleason, Edward Keane; **Dir:** Albert Herman. **VHS, Beta $16.95** SNC, RXM 🎬

Rogue's Tavern In this grade B curio, a blood-thirsty killer stalks the occupants of a country inn.
1936 70m/B Wallace Ford, Joan Woodbury; **Dir:** Bob Hill. **VHS, Beta $24.95** NOS, DVT 🎬🎬

Rogue's Yarn An investigator probes the death of a woman, ostensibly the result of a boating accident. Gradually he uncovers a plot involving the woman's husband and his mistress in this semi-entertaining British mystery.
1956 80m/B GB Nicole Maurey, Derek Bond, Elwyn Brook-Jones, Hugh Latimer; **Dir:** Vernon Sewell. **VHS, Beta $16.95** SNC 🎬🎬

Roland the Mighty A muscular, Herculeanesque hero takes on all comers in this sword and sandal epic set in the Asian steppes.
1958 ?m/C IT Rick Battaglia. **VHS, Beta $16.95** SNC 🎬½

Roland and Rattfink Five cartoons starring the good/evil cartoon characters, including "Hawks and Doves," "Sweet and Sourdough" and "Flying Feet." In HiFi.
1969 32m/C VHS, Beta $14.95 MGM 🎬½

Roll Along Cowboy A vintage sagebrush saga about a cowboy caught by love. Based on Zane Grey's "The Dude Ranger."
1937 55m/B Smith Ballew, Cecilia Parker, Buster Fite & his Saddle Tramps; **Dir:** Gus Meins. **VHS, Beta $19.95** NOS, CCB, VDM 🎬½

Roll on Texas Moon A cowpoke is hired to prevent a feud between the sheep men and the cattle ranchers.
1946 68m/B Roy Rogers, George "Gabby" Hayes, Dale Evans, Dennis Hoey; **Dir:** William Witney. **VHS $19.95** NOS, DVT, MTX 🎬

Roll of Thunder, Hear My Cry A black family struggles to survive in Depression-era Mississippi in this inspiring, if somewhat predictable, television production. Look for the always impressive Freeman in a supporting role. Set in the South of 1933 and based on the novels of Mildred Taylor.
1978 110m/C Claudia McNeil, Janet MacLachlan, Morgan Freeman; **Dir:** Jack Smight. **VHS, Beta $14.95** LCA 🎬½

Roll, Wagons, Roll Tex Ritter leads a wagon train to Oregon.
1939 52m/B Tex Ritter; **Dir:** Al Herman. **VHS, Beta $19.95** NOS, VCN 🎬

Roller Blade In a post-holocaust world a gang of Amazon-style nuns, who worship a "have a nice day" happy face, battle the forces of evil with martial arts and mysticism. Has to be seen to be believed. Extremely low-budget outing.
1985 88m/C Suzanne Solari, Jeff Hutchinson, Shaun Mitchelle; **Dir:** Donald G. Jackson. **VHS, Beta $19.95** NWV 🎬

Roller Blade Warriors: Taken By Force Roller babes battle evil mutant while balancing on big boots with small wheels.
1990 90m/C Kathleen Kinmont, Rory Calhoun, Abby Dalton, Elizabeth Kaitan. **VHS $89.99** RAE 🎬🎬

Rollerball Caan is utterly convincing in this futuristic tale in which a brutal sport assumes alarming importance to a sterile society. Flashy, violent, sometimes exhilirating.
1975 (R) 123m/C James Caan, John Houseman, Maud Adams, Moses Gunn; **Dir:** Norman Jewison. **VHS, Beta, LV $29.95** MGM 🎬🎬½

Rollercoaster A deranged extortionist threatens to sabotage an amusement park ride. Plucky Segal must stop him. Video renters will be spared the nauseating effects of film's original "Sensurround."
1977 (PG) 119m/C George Segal, Richard Widmark, Timothy Bottoms, Henry Fonda, Susan Strasberg, Harry Guardino; **Dir:** James Goldstone. **VHS, Beta $59.95** MCA 🎬🎬

Rollin' Plains A Texas Ranger tries to settle a feud between cattlemen and sheepmen over water rights.
1938 60m/B Tex Ritter; **Dir:** Al Herman. **VHS, Beta $19.95** NOS, VCN, UHV 🎬

Rolling Home Tame tale about aging cowpoke, his grandson, and a really swell horse. Saddle up and snooze.
1948 71m/B Jean Parker, Russell Hayden, Pamela Blake, Buss Henry, Raymond Hatton. **VHS, Beta, LV** WGE 🎬🎬

Rolling Stone Presents Twenty Years of Rock & Roll Retrospective of two decades of rocking and rolling featuring the likes of Madonna, Springsteen, U2, Blondie, The Beatles, Led Zeppelin, Sex Pistols, Iggy Pop, Elvis Costello, Jimi Hendrix, Jackson Browne, Buffalo Springfield, David Bowie and many more.
1987 97m/C Hosted: Dennis Hopper; **Performed by:** Madonna. **VHS, Beta $19.95** MVD, MGM

Rolling Thunder Vietnam vet turns vigilante when he arrives home from POW camp and sees his family slaughtered. Graphically violent, potentially cathartic. Typical of screenwriter Paul Schrader.
1977 (R) 99m/C William Devane, Tommy Lee Jones, Linda Haynes; **Dir:** John Flynn. **VHS, Beta $79.98** LIV, VES 🎬🎬½

Rolling Vengeance Enterprising fellow constructs powerful truck to better facilitate the extermination of gang members who earlier slaughtered his family. Keep on truckin'.
1987 (R) 90m/C Don Michael Paul, Ned Beatty, Lawrence Dane, Lisa Howard; **Dir:** Steven Hilliard Stern. **VHS, Beta, LV $19.98** CHA 🎬½

Rollover Turgid big-budget drama about Arab undermining of American economy. Kristofferson is lifeless, Fonda is humorless. Supporting players Sommer and Cronyn fare better. Not even a good one for those of us who don't understand economics and don't think that economists do either.
1981 (R) 117m/C Jane Fonda, Kris Kristofferson, Hume Cronyn, Bob Gunton, Josef Sommer, Martha Plimpton; **Dir:** Alan J. Pakula. **VHS, Beta $39.98** WAR 🎬½

Roman Holiday Hepburn's first starring role is a charmer as a princess bored with her official visit to Rome who slips away and plays at being an "average Jane." A reporter discovers her little charade and decides to cash in with an exclusive story. Before they know it, love calls. Screenplay by John Dighton from Ian McLellan Hunter's award-winning story. In addition to claiming a bunch of Oscars, nominated for Best Picture, Best Supporting Actor (Albert), Director, Screenplay, Black & White Cinematography, Black & White Art Direction and Scene Decoration, and Film Editing.
1953 118m/B Audrey Hepburn, Gregory Peck, Eddie Albert, Tullio Carminati; **Dir:** William Wyler. Academy Awards '53: Best Actress (Hepburn), Best Costume Design (B & W), Best Story; British Academy Awards '53: Best Actress (Hepburn); New York Film Critics Awards '53: Best Actress (Hepburn). **VHS, Beta, LV, 8mm $14.95** PAR, BTV 🎬🎬🎬½

Roman Scandals A penniless young man daydreams himself back to ancient Rome with uproarious results. Songs include "No More Love," "Keep Young and Beautiful," and "Build a Little Home." Lucille Ball appears as one of the Goldwyn Girls. Dance sequences were choreographed by Busby Berkeley.
1933 91m/B Eddie Cantor, Ruth Etting, Gloria Stuart, Edward Arnold, Lucille Ball, Busby Berkeley; **Dir:** Frank Tuttle. **VHS, Beta $19.98** SUE, HBO 🎬🎬½

Roman Spring of Mrs. Stone An aging actress determines to revive her career in Rome but finds romance with a gigolo instead in this adaptation of Tennessee Williams's novella. Leigh and Beatty are compelling, but Oscar-nominated Lenya nearly steals the show as Leigh's distinctly unappealing confidant.
1961 104m/C Warren Beatty, Vivien Leigh, Lotte Lenya, Bessie Love, Jill St. John; **Dir:** Jose Quintero. **VHS, Beta $19.98** WAR, FHE 🎬🎬🎬

Romance with a Double Bass When a musician and a princess take a skinny dip in the royal lake and get their clothes stolen, comic complications arise. Reminiscent of Cleese's sojourn as a Monty Python member, with surreal, frequently raunchy humor.
1974 40m/C GB John Cleese, Connie Booth, Graham Crowden, Desmond Jones, Freddie Jones, Andrew Sachs; **Dir:** Robert Young. **VHS, Beta, LV $19.95** PAV 🎬🎬🎬

Romance on the High Seas A woman is scheduled to take a cruise vacation but skips the boat when she believes her husband is cheating on her. Her husband believes she is taking the cruise to cheat on him and hires a private detective to follow her. Pleasant comedy features the film debut of Doris Day.
1948 99m/C Jack Carson, Janice Paige, Don DeFore, Doris Day, Oscar Levant, S.Z. Sakall, Eric Blore, Franklin Pangborn, Leslie Brooks, William Bakewell. **VHS, LV $29.98** FOX, FCT 🎬🎬🎬

Romance of a Horsethief Brynner leads this entertaining but slow-paced "Fiddler on the Roof" comedic romp without the music. In 1904 Poland, a Cossack captain takes horses for the Russo-Japanese war.

The residents of the town rise up, goaded on by Birkin. **1971 (PG) 100m/C** *YU* Yul Brynner, Eli Wallach, Jane Birkin, Oliver Tobias, Lainie Kazan, David Opatoshu; *Dir:* Abraham Polonsky. **VHS** *ACE, MOV* 🎞🎞½

Romance in Manhattan Recent immigrant to New York struggles to build new life, in spite of unemployment, language barriers and loneliness. He meets Broadway chorine Rogers and song and dance ensues. America, ain't it a great place? **1934 78m/B** Ginger Rogers, Francis Lederer, J. Farrell MacDonald; *Dir:* Pandro S. Berman. **VHS** **$19.98** *TTC, FCT* 🎞🎞

Romance on the Orient Express Sparks fly when former lovers meet again on the Orient Express. Don't expect any flesh, though, as film was made for television. Among film's supporting players is distinguished actor John Gielgud. Why? **1989 96m/C** Cheryl Ladd, Stuart Wilson, John Gielgud; *Dir:* Lawrence Gordon Clark. **VHS, Beta** **$9.98** *TTC* 🎞½

Romance on the Range Rogers sets out to trap a gang of fur thieves and finds romance. **1942 60m/B** Roy Rogers, George "Gabby" Hayes, Sally Payne. **VHS** **$19.95** *NOS, VCN, REP* 🎞

Romance Theatre Collection A lengthy series of one-hour filmed romance novels. **1986 60m/C Beta** **$11.95** *PSM* 🎞

Romancing the Stone Turner is an uptight author of romance novels who gets to live out her fantasies when she finds that her sister has been kidnapped in South America. She's helped and hindered by an American soldier of fortune (Douglas) along the way. Great chemistry between the appealing stars and loads of clever dialogue. Sequel is "The Jewel of the Nile." **1984 (PG) 106m/C** Michael Douglas, Kathleen Turner, Danny DeVito, Zack Norman, Alfonso Arau, Ron Silver; *Dir:* Robert Zemeckis. Golden Globe Awards '85: Best Film—Musical/Comedy. **VHS, Beta, LV** **$19.98** *FOX* 🎞🎞🎞

Romantic Comedy Writing duo never seem to synchronize their desires for each other in this dull comedy. Engaging stars don't inspire each other much, and supporting cast can't make up the difference. Adapted from the Bernard Slade play. **1983 (PG) 102m/C** Dudley Moore, Mary Steenburgen, Frances Sternhagen, Ron Leibman; *Dir:* Arthur Hiller. **VHS, Beta** **$29.98** *FOX* 🎞🎞

Romantic Encounters See what happens after people call a sexy singles hotline. **199? 90m/C VHS** **$49.99** *CEL* 🎞

Romantic Englishwoman Literate comedy-drama about the intertwined lives of several sophisticated and restrained Brits. Caine is a successful novelist with writer's block whose wife falls for another man while on a trip alone. He then invites his wife's lover to stay with them in order to generate ideas for his writing until his jealousy begins to surface. Scripted by Tom Stoppard and Thomas Wiseman. **1975 (R) 117m/C** *GB* Glenda Jackson, Michael Caine, Helmut Berger, Kate Nelligan; *Dir:* Joseph Losey. **VHS, Beta** *WAR, OM* 🎞🎞🎞

Rome '78 Acclaimed underground film centered around Caligula and the Queen of Sheba in ancient Rome.

1978 60m/C Eric Mitchell; *Dir:* James Nares. **VHS** **$59.95** *MWF* 🎞🎞½

Rome Adventure Spinster librarian Pleshette takes Roman vacation hoping to meet handsome prince. Donahue takes shine to her while girlfriend Dickinson is away, as does suave Roman Brazzi. Soapy, ill-paced romance. **1962 119m/C** Troy Donahue, Angie Dickinson, Suzanne Pleshette, Rossano Brazzi, Constance Ford, Al Hirt, Chad Everett; *W/Dir:* Delmer Daves. **VHS, LV** **$59.99** *PAR, PIA, FCT* 🎞🎞

Romeo and Juliet One of MGM producer Irving Thalberg's pet projects (and starring Thalberg's wife, Norma Shearer), this Shakespeare classic was given the spare-no-expense MGM treatment. Physically too old to portray teen-age lovers, both Howard and Shearer let their acting ability supply all necessary illusions. Also notable is Barrymore's over-the-top portrayal of Mercutio. **1936 126m/B** Leslie Howard, Norma Shearer, John Barrymore, Basil Rathbone, Edna May Oliver; *Dir:* George Cukor. **VHS** **$19.98** *MGM, CCB* 🎞🎞🎞½

Romeo and Juliet Unfulfilling adaptation of Shakespeare's timeless drama of young love cast against family antagonists. Peculiar supporting cast features both Cabot and ubiquitous master Gielgud. **1954 138m/C** *IT* Laurence Harvey, Susan Shantall, Aldo Zollo, Sebastian Cabot, Flora Robson, Mervyn Johns, Bill Travers, John Gielgud; *Dir:* Renato Castellani. National Board of Review Awards '54: Best Foreign Film of the Year, Best Director. **VHS, Beta** **$29.95** *SUE, LCA* 🎞🎞

Romeo and Juliet Young couple share love despite prohibitive conflict between their families in this adaptation of Shakespeare's classic play. Director Zeffirelli succeeds in casting relative novices Whiting and Hussey in the leads, but is somewhat less proficient in lending air of free-wheeling sixties appeal to entire enterprise. Kudos, however, to cinematographer Pasquale de Santis and composer Nina Rota. In addition to Oscars awarded, Oscar-nominated for Best Picture and Also available in a 45-minute edited version. **1968 (PG) 138m/C** *GB IT* Olivia Hussey, Leonard Whiting, Michael York, Milo O'Shea; *Dir:* Franco Zeffirelli; *Nar:* Laurence Olivier. Academy Awards '68: Best Cinematography, Best Costume Design; National Board of Review Awards '68: 10 Best Films of the Year, Best Director. **VHS, Beta, LV** **$19.95** *PAR* 🎞🎞🎞½

Romeo and Juliet (Gielgud) Entertaining enactment of 400-year-old Shakespearean classic about the passion and tragedy of a family feud and young star-crossed lovers. Part of the television series "The Shakespeare Plays." **1979 167m/C** *GB* John Gielgud, Rebecca Saire, Patrick Ryecart. **VHS, Beta** *TLF* 🎞🎞🎞

Romero Julia is riveting as the Salvadoran archbishop who championed his destitute congregation despite considerable political opposition. A stirring biography financed by the United States Roman Catholic Church, with a screenplay by John Sacret Young of television's "China Beach." **1989 102m/C** Raul Julia, Richard Jordan, Ana Alicia, Eddie Velez, Alejandro Bracho, Tony Plana, Lucy Reina, Harold Gould, Al Ruscio, Robert Viharo; *Dir:* John Duigan. **VHS, Beta, LV** **$29.95** *VMK* 🎞🎞½

Romola Silent adventure. After Powell and his father are attacked by pirates, Powell escapes. But instead of rescuing his father, he opts for a life of corruption. Dad eventually escapes and returns to exact vengeance. A rather expensive film in its day, troubled by modern-day sights in what was supposed to be old Florence. Gish's drowning scene had to be re-shot because she wouldn't sink. Colman arranged for his then-wife Raye to have a bit part, but they were divorced soon after the film was finished. **1925 120m/B** Lillian Gish, Dorothy Gish, William H. Powell, Ronald Colman, Charles Lane, Herbert Grimwood, Bonaventure Ibanez, Frank Puglia, Thelma Raye; *Dir:* Henry King. **VHS, Beta** **$32.95** *DNB* 🎞🎞½

Ron Reagan Is the President's Son/The New Homeowner's Guide to Happiness Two half-hour comedies, "The New Homeowner's Guide to Happiness," and "Ron Reagan is the President's Son." **1985 60m/C** Ron Reagan, Judge Reinhold, Demi Moore. **VHS** **$39.95** *HBO, TLF* 🎞🎞½

Rona Jaffe's Mazes & Monsters Geeky collegians get more than they bargain for when they become preoccupied with a fantasy game. Early Hanks appearance is among this television film's few assets. Also known as "Mazes and Monsters." **1982 103m/C** Tom Hanks, Chris Makepeace, Lloyd Bochner, Anne Francis, Susan Strasberg, Murray Hamilton, Vera Miles, Louise Sorel; *Dir:* Steven Hilliard Stern. **VHS, Beta** **$59.95** *LHV, WAR* 🎞½

Ronald Reagan & Richard Nixon on Camera Nixon's famous "Checkers" speech and Reagan's early political career highlights are combined on this tape. **195? 30m/B** Ronald Reagan. **VHS, Beta** **$24.95** *NOS, DVT* 🎞½

Ronnie Dearest: The Lost Episodes A satirical look at Reagan's acting career and political tenure. Hosted by a young Nancy Reagan. **1988 45m/C** Ronald Reagan, Nancy Davis, Jayne Mansfield, Mickey Rooney, Edward R. Murrow, James Cagney, Ava Gardner; *Dir:* Ira Gallen. **VHS, Beta** **$14.95** *NEG* 🎞½

The Roof Uninteresting outing about a young couple who married against their families' wishes and find that setting up a home of their own is more difficult than they thought. Italian dialogue, English subtitles. **1956 98m/B** *IT* Gabriella Pallotti, Giorgio Listuzzi; *Dir:* Vittorio DeSica. **VHS, Beta** **$24.95** *NOS, HHT, FCT* 🎞🎞

Rooftops Peculiar but predictable, centering on love between Hispanic girl and white youth who has mastered martial arts dancing. Director Wise is a long way from his earlier "West Side Story." See this one and be the only person you know who has. **1989 (R) 108m/C** Jason Gedrick, Troy Beyer, Eddie Velez, Tisha Campbell; *Dir:* Robert Wise. **VHS, Beta, LV** **$19.95** *IVE* 🎞½

The Rookie Routine cop drama has worldly veteran and wide-eyed newcomer team to crack stolen-car ring managed by Germans. Eastwood and Sheen are reliable, but Hispanics Julia and Braga, though miscast, nonetheless steal this one as the German villains. **1990 (R) 121m/C** Clint Eastwood, Charlie Sheen, Raul Julia, Sonia Braga, Lara Flynn Boyle, Pepe Serna, Marco Rodriguez, Tom Skerritt, Roberta Vasquez; *Dir:* Clint Eastwood. **VHS, Beta, LV, 8mm** **$19.98** *WAR, PIA* 🎞🎞

The Room Altman takes on Pinter in this brooding mini-play about two young strangers who try to rent the room already occupied by a disconnected woman and her semi-catatonic husband. Made for television. 1987 48m/C Julian Sands, Linda Hunt, Annie Lennox, David Hemblen, Donald Pleasence; *Dir:* Robert Altman. **VHS, Beta $79.95** *PSM* ⟨⟨

Room 43 Campy fable about a British cab driver who falls in love with a French girl, and stumbles onto a white-slavery ring when she falls victim to it. Also called "Passport to Shame." Features a young Caine and future novelist Collins in bit parts. 1958 93m/B *GB* Diana Dors, Herbert Lom, Eddie Constantine, Michael Caine, Jackie Collins; *Dir:* Alvin Rakoff. **VHS, Beta $19.95** *VDM* ⟨½

Room to Let An elderly woman and her daughter take in boarders in Victorian England. They rent a room to a man who claims to be a doctor, but the two women eventually become prisoners of fear, believing the man is actually Jack the Ripper. Will there be a vacancy soon? Somewhat disturbing and suspenseful. 1949 68m/B *GB* Jimmy Hanley, Valentine Dyall, Christine Silver, Merle Tottenham, Charles Hawtrey, Connie Smith, Laurence Naismith; *Dir:* Godfrey Grayson. **VHS, Beta $16.95** *NOS, SNC* ⟨⟨

Room Service Marx Brothers provide less than the usual mayhem here with Groucho as penniless theatrical producer determined to remain in his hotel room until he can secure funds for his next play. Ball doesn't help matters much either. Not bad, but certainly not up to the zany Marx clan's usual stuff. 1938 78m/B Groucho Marx, Harpo Marx, Chico Marx, Lucille Ball, Ann Miller, Frank Albertson, Donald MacBride; *Dir:* William A. Seiter. **VHS, Beta, LV $19.95** *RKO, VID, CCB* ⟨⟨½

Room at the Top Ambitious factory man forsakes true love and marries boss's daughter instead in this grim drama set in industrial northern England. Cast excels, with Harvey and Baddeley both securing Oscar nominations as the worker and his wife. Oscar-winning Signoret is also quite compelling as the abandoned woman. Adapted from John Braine's novel and followed by "Life at the Top" and "Man at the Top." Also nominated for Best Actor (Harvey), Supporting Actress (Baddeley), and Director. 1959 118m/B *GB* Laurence Harvey, Simone Signoret, Heather Sears, Hermione Baddeley, Anthony Elgar, Donald Wolfit; *Dir:* Jack Clayton. Academy Awards '59: Best Actress (Signoret), Best Adapted Screenplay. **VHS** *AVT* ⟨⟨½

Room with a View Engaging adaptation of E. M. Forster's novel of requited love. Carter stars as the feisty British idealist who rejects dashing Sands for supercilious Lewis, then repents and finds (presumably) eternal passion. A multi-Oscar nominee, with great music (courtesy of Puccini), great scenery (courtesy of Florence), and great performances (courtesy of practically everybody, but supporters Smith, Dench, Callow, and Elliott must be particularly distinguished). Truly romantic, and there's much humor too. 1986 117m/C Helena Bonham Carter, Julian Sands, Denholm Elliott, Maggie Smith, Judy Dench, Simon Callow, Daniel Day Lewis, Rupert Graves; *Dir:* James Ivory. Academy Awards '86: Best Adapted Screenplay, Best Art Direction/Set Decoration, Best Costume Design; British Academy Awards '86: Best Actress (Smith), Best Cinematography, Best Picture, Best Supporting Actress (Dench). **VHS, Beta, LV $19.98** *FOX* ⟨⟨⟨⟨

Roommate A valedictorian church-goer and a rebellious iconoclast room together at Northwestern University in 1952, with the expected humorous results. Public television presentation based on a John Updike story. 1984 96m/C Lance Guest, Barry Miller, Elaine Wilkes, Melissa Ford, David Bachman; *Dir:* Neil Cox. **VHS, Beta $69.98** *LIV, VES* ⟨⟨⟨⟨

Rooster Cogburn A Bible-thumping schoolmarm joins up with a hard-drinking, hard-fighting marshal in order to capture a gang of outlaws who killed her father. Tired sequel to "True Grit" but the chemistry between Wayne and Hepburn is right on target. 1975 (PG) 107m/C John Wayne, Katharine Hepburn, Richard Jordan, Anthony Zerbe, John McIntire, Strother Martin; *Dir:* Stuart Millar. **VHS, Beta, LV $19.95** *MCA, TLF, BUR* ⟨⟨½

Rooster—Spurs of Death! An expose on cock fighting, in which a young idealist tries to stop the thriving sport in a small southern town. 1983 (PG) 90m/C Vincent Van Patten, Ty Hardin, Kristine DeBell, Ruta Lee. **VHS, Beta $59.95** *WES* Woof!

Rootin' Tootin' Rhythm All's not quiet on the range, but Autry comes along to sing things back to normal. Goofy, clumsy, ubiquitous sidekick Burnette is a symbol for this whole effort. 1938 55m/B Gene Autry, Smiley Burnette, Monte Blue; *Dir:* Mack V. Wright. **VHS, Beta $19.95** *NOS, VYY, DVT* ⟨

Roots of Evil This erotic thriller follows a cop on the trail of a murderer who specializes in killing Hollywood's most beautiful prostitutes. Also available in unrated and Spanish-language versions. 1991 (R) ?m/C Alex Cord, Delia Sheppard, Charles Dierkop, Jillian Kesner; *Dir:* Gary Graver. **VHS $89.99** *CAN* ⟨½

Roots, Vol. 1 The complete version of Alex Haley's saga following a black man's search for his heritage, revealing an epic panorama of America's past. Available on six 90-minute tapes. 1977 90m/C Ed Asner, Lloyd Bridges, LeVar Burton, Chuck Connors, Lynda Day George, Lorne Greene, Burl Ives, O.J. Simpson, Cicely Tyson, Ben Vereen, Sandy Duncan; *Dir:* David Greene. **VHS, Beta $64.95** *WAR, KUI* ⟨⟨½

Rope In New York City, two gay college students murder a friend for kicks and store the body in a living room trunk. They further insult the dead by using the trunk as the buffet table and inviting his parents to the dinner party in his honor. Very dark humor is the theme in Hitchcock's first color film, which he innovatively shot in uncut ten-minute takes, with the illusion of a continuous scene maintained with tricky camera work. Based on the Patrick Hamilton play and on the Leopold-Laeb murder case; script by Arthur Laurents. 1948 (PG) 81m/C James Stewart, John Dall, Farley Granger, Cedric Hardwicke, Constance Collier; *Dir:* Alfred Hitchcock. **VHS, Beta, LV $19.95** *MCA* ⟨⟨⟨½

Rosalie A gridiron great from West Point falls for a mysterious beauty from Vassar. He soon learns that her father reigns over a tiny Balkan nation. Expensive, massive, imperfect musical; exotic romantic pairing fails to conceal plot's blah-ness. Oh well, the music's good. Cole Porter's score includes "I've Got a Strange New Rhythm in my Heart" and "In the Still of the Night."

1938 118m/B Nelson Eddy, Eleanor Powell, Frank Morgan, Ray Bolger, Ilona Massey, Reginald Owen, Edna May Oliver, Jerry Colonna; *Dir:* W.S. Van Dyke. **VHS, Beta $24.95** *MGM* ⟨⟨½

Rosalie Goes Shopping Satire about American consumerism hiding behind slapstick comedy, and it works, most of the time. Misplaced Bavarian (Sagebrecht) moves to Arkansas and begins spending wildly and acquiring "things." Twisty plot carried confidently by confident wackiness from Sagebrecht and supporters. 1989 (PG-13) 94m/C *GE* Marianne Sagebrecht, Brad Davis, Judge Reinhold, Willie Harlander, Alex Winter; *Dir:* Percy Adlon. **VHS, LV $89.95** *VMK* ⟨⟨

Rosanne Cash: Retrospective Country music star Rosanne Cash sings "Blue Moon With Heartache," "Runaway Train," "I Wonder," and other hits. 1989 40m/C Rosanne Cash. **VHS, LV $14.98** *MVD, FOX* ⟨⟨⟨

Rosary Murders Based on the William X. Kienzle mystery about a serial killer preying on nuns and priests. Father Koesler (Sutherland) must betray the confessional to stop him. Interesting, though sometimes muddled and too rarely tense. Screenplay co-written by Elmore Leonard and Fred Walton. Filmed on location in Detroit. 1987 (R) 105m/C Donald Sutherland, Charles Durning, Belinda Bauer, Josef Sommer; *Dir:* Fred Walton. **VHS, Beta, LV $19.95** *VIR* ⟨⟨

Rose A simple, wordless story about a youth and a maiden who meet and fall in love in ancient Greece. Unique animation techniques are used. 1979 10m/C *Dir:* Mark Rydell. **VHS, Beta $19.98** *BFA* ⟨⟨

The Rose Modeled after the life of Janis Joplin, Midler plays a young, talented and self-destructive blues/rock singer. Professional triumphs don't stop her lonely restlessness and confused love affairs. The best exhibition of the rock and roll world outside of documentaries. Midler's electrifying film debut features an incredible collection of songs, including Amanda McBroom's "The Rose." Oscar nominations for Midler, Forrest, and Film Editing. 1979 (R) 134m/C Bette Midler, Alan Bates, Frederic Forrest, Harry Dean Stanton, David Keith; *Dir:* Mark Rydell. **VHS, Beta, LV $19.98** *FOX* ⟨⟨

The Rose Garden In modern Germany, a Holocaust survivor is put on trial for assaulting an elderly man. The victim, it turns out, was a Nazi guilty of heinous war crimes. Ullmann and Schell lend substance, though treatment of a powerful theme is inadequate. 1989 (PG-13) 112m/C Liv Ullman, Maximilian Schell, Peter Fonda; *Dir:* Fons Rademakers. **VHS, Beta, LV $89.95** *CAN* ⟨⟨½

The Rose and the Jackal Fun, intriguing fictionalization about the founder of the Pinkerton detective agency. During the Civil War, bearded, Scottish Reeve tries to persuade a wealthy Southern lady to help save the Union. Made for cable television. 1990 94m/C Christopher Reeve, Madolyn Smith, Granville Van Dusen, Carrie Snodgress, Kevin McCarthy; *Dir:* Jack Gold. **VHS, Beta $79.98** *TTC* ⟨

Rose Marie An opera star falls in love with the mountie who captured her escaped convict brother. Classic MacDonald-Eddy operetta. The Rudolf Friml score features "Indi-

an Love Call," "The Song of the Mounties" and the title song also called "Indian Love Call." Remade in 1954.
1936 112m/B Jeanette MacDonald, Nelson Eddy, James Stewart, Allan Jones, David Niven, Reginald Owen; *Dir:* W.S. Van Dyke. **VHS, Beta, LV $19.95** MGM, PIA 🎬🎬🎬

Rose Marie Blyth is a lonely Canadian woman wooed by a mean spirited fur trapper and a gallant mountie. Unremarkable remake of the Eddy/MacDonald musical is saved by specatular technicolor and CinemaScope photography. Choreography by Busby Berkeley.
1954 (G) 115m/C Ann Blyth, Howard Keel, Fernando Lamas, Bert Lahr, Marjorie Main, Joan Taylor, Chief Yowlachie, Abel Fernandez, Al Ferguson, Dabbs Greer, Lumsden Hare; *Dir:* Mervyn LeRoy. **VHS, LV** MGM, MLB 🎬🎬🎬

The Rose Parade: Through the Years A survey of Rose Bowl Parades through the decades, featuring news footage of its various incarnations from the 1800's to today.
1988 60m/C Mary Pickford, Shirley Temple, Walt Disney, Herbert Hoover; *Hosted:* Bill Welsh. **Beta $19.95** RHI 🎬🎬🎬

Rose of Rio Grande A western musical about a Mexican gang that avenges the deaths of members of the upper strata.
1938 60m/B John Carroll, Movita, Antonio Moreno, Lena Basquette, Duncan Renaldo; *Dir:* William Nigh. **VHS, Beta** GPV 🎬

The Rose Tattoo Magnani, in her U.S. screen debut, is just right as a Southern widow who cherishes her husband's memory, but falls for virile trucker Lancaster. Williams wrote this play and screenplay specifically for Magnani, who was never as successful again. Interesting character studies, although Lancaster doesn't seem right as an Italian longshoreman. Oscar nominations for Best Picture, supporting actress Pavan, Costume Design and Film Editing.
1955 117m/B Anna Magnani, Burt Lancaster, Marisa Pavan, Ben Cooper, Virginia Grey, Jo Van Fleet; *Dir:* Daniel Mann. Academy Awards '55: Best Actress (Magnani), Best Art Direction/ Set Decoration (B & W), Best Black and White Cinematography. **VHS, Beta, LV $19.95** PAR, PIA, FCT 🎬🎬🎬½

The Roseanne Barr Show This video contains new material from the original Roseanne television series.
1989 60m/C Roseanne (Barr) Arnold, John Goodman. **VHS, Beta $14.99** HBO, TLF 🎬🎬🎬½

Rosebud Beach Hotel Wimpy loser tries his hand at managing a run-down hotel in order to please his demanding girlfriend. Camp is okay, but this soft-porn excuse is horrible in every other way. Originally called "The Big Lobby."
1985 (R) 82m/C Colleen Camp, Peter Scolari, Christopher Lee, Fran Drescher, Eddie Deezen, Chuck McCann, Hank Garrett, Hamilton Camp, Cherie Currie; *Dir:* Harry Hurwitz. **VHS, Beta, LV $79.98** LIV, VES 🎬

Roseland Three interlocking stories, set at New York's famous old Roseland Ballroom, about lonely people who live to dance. Not fully successful, but strong characters, especially in the second and third stories, make it worth watching, although it lacks energy.
1977 (PG) 103m/C Christopher Walken, Geraldine Chaplin, Joan Copeland, Teresa Wright, Lou Jacobi; *Dir:* James Ivory. New York Film Festival '77: Best Black and White Cinematography. **VHS, Beta $59.95** VES 🎬🎬🎬

Rosemary's Baby A young woman, innocent and religious, and her husband, ambitious and agnostic, move into a new apartment. Soon the woman is pregnant, but she begins to realize that she has fallen into a coven of witches and warlocks, and that they claim the child as the antichrist. Gripping and powerful, subtle yet utterly horrifying, with luminous performances by all. Polanski's first American film; from Levin's best-seller. Oscar nomination for Adapted Screenplay.
1968 (R) 134m/C Mia Farrow, John Cassavetes, Ruth Gordon, Maurice Evans, Patsy Kelly, Elisha Cook Jr., Charles Grodin, Sidney Blackmer, William Castle; *Dir:* Roman Polanski. Academy Awards '68: Best Supporting Actress (Gordon). **VHS, Beta, LV $14.95** PAR, MLB, BTV 🎬🎬🎬🎬

Rosencrantz & Guildenstern Are Dead Playwright Tom Stoppard adapted his own absurdist 1967 play to film—which makes as much sense as a "Swan Lake" ballet on radio. Two tragicomic minor characters in "Hamlet" squabble rhetorically and misperceive Shakespeare's plot tightening fatally around them. Uprooted from the stage environment, it's affected, arcane and pointless for all but fervent Stoppard fans and Bard completists.
1990 (PG) 118m/C Richard Dreyfuss, Gary Oldman, Tim Roth, Iain Glen; *Dir:* Tom Stoppard. Venice Film Festival '91: Best Picture. **VHS, Beta $92.95** BVV, FCT 🎬🎬½

Roses Bloom Twice When a middleaged widow decides to make up for some of the wild living that she missed during her marriage, she shocks her children.
1977 87m/C Glynis McNicoll, Michael Craig, Diane Craig, John Allen, Jennifer West; *Dir:* David Stevens. **VHS, Beta** PGN 🎬½

Rosie: The Rosemary Clooney Story Above average biography of Clooney's up-and-down career. Locke lip-synches Clooney's voice in songs. Orlando doesn't help matters portraying Jose Ferrer. Made for television.
1982 100m/C Sondra Locke, Tony Orlando, Penelope Milford, Katherine Helmond, Kevin McCarthy, John Karlen, Cheryl Anderson, Robert Ridgely, Joey Travolta; *Dir:* Jackie Cooper; *Performed by:* Rosemary Clooney. **VHS, Beta $39.95** USA, IVE 🎬🎬

R.O.T.O.R. R.O.T.O.R. (Robotic Officer of Tactical Operations Research) is a law enforcement robot that is supposed to stop criminals. But watch out! It might go beserk and wreck your town.
1988 90m/C Richard Gesswein, Margaret Trigg, Jayne Smith; *Dir:* Cullen Blaine. **VHS $79.95** IMP 🎬

Rough Cut American diamond thief Reynolds lives in London, pursued by an aging Scotland Yard detective who wants to end his career in a blaze of glory. A beautiful lady is the decoy. Good chemistry; lousy script.
1980 (PG) 112m/C Burt Reynolds, Lesley-Anne Down, David Niven, Timothy West, Joss Ackland, Patrick Magee; *Dir:* Don Siegel. **VHS, Beta, LV $59.95** PAR 🎬🎬

Rough Justice Spaghetti western about a lone gunfighter tracking down three killers. Kinski is a sexually aroused bad guy in this miserable excuse. What could he have been thinking?
1987 95m/C *IT* Klaus Kinski, Steven Tedd; *Dir:* Mario Costa. **VHS, Beta $59.95** TWE 🎬½

Rough Riders of Cheyenne Sunset Carson ends the feud between the Carsons and the Sterlings.

1945 54m/B Sunset Carson, Peggy Stewart. **VHS, Beta** MED 🎬

Rough Riders' Roundup Roy and the boys rid a mining town of its crooked manager. Less yodeling than later Roger's outings; also better all around.
1934 54m/B Roy Rogers, Raymond Hatton, Mary Hart, Dorothy Sebastian, Duncan Renaldo, George Montgomery; *Dir:* Joseph Kane. **VHS, Beta $9.95** NOS, VDM, DVT 🎬

Rough Riding Rangers Threatening letters plague a western ranch family.
1935 57m/B Rex Lease, Janet Chandler. **VHS, Beta $27.95** NCN, WGE 🎬

Roughnecks A team of Texas oil drillers attempt to dig the deepest oil well in history and chase women on the side. Too long, and not particularly action-packed or gripping.
1980 240m/C Sam Melville, Cathy Lee Crosby, Vera Miles, Harry Morgan, Steve Forrest, Stephen McHattie, Wilford Brimley; *Dir:* Bernard McEveety. **VHS, Beta $69.95** PSM 🎬½

Round Midnight An aging, alcoholic jazz saxophonist comes to Paris in the late 1950s seeking an escape from his self-destructive existence. A devoted young French fan spurs him to one last burst of creative brilliance. A moody, heartfelt homage to such expatriate bebop musicians as Bud Powell and Lester Young. Available in a Spanish-subtitled version.
1986 (R) 132m/C *FR* Dexter Gordon, Lonette McKee, Francois Cluzet, Martin Scorsese, Herbie Hancock, Sandra Reaves-Phillips; *Dir:* Bertrand Tavernier. Academy Awards '86: Best Original Score. **VHS, Beta, LV $19.98** MVD, WAR 🎬🎬🎬

Round Numbers Judith Schweitzer believes her husband "Big Al the Muffler King" is having an affair with the gorgeous Muffler Mate of the Month, Mitzi (Playboy Playmate Hope Marie Carlton). Determined to get revenge, Judith books herself into the same health spa that Mitzi attends and the result is pure comedy.
1991 (R) 98m/C Kate Mulgrew, Samantha Eggar, Marty Ingles, Hope Marie Carlton, Shani Wallis, Natalie Barish, Debra Christofferson; *W/ Dir:* Nancy Zala. **VHS $89.95** HMD 🎬½

Round Trip to Heaven Party guy Larry (Feldman) and his innocent cousin (Galligan) take off in a borrowed Rolls Royce to go meet Larry's dream centerfold in a super model contest. Unfortunately, the Rolls' trunk is loaded with stolen money and Stoneface (Sharkey), the felon who stashed the cash, wants the money back. With Stoneface in pursuit these guys are in for the ride of their lives. Also available with Spanish subtitles.
1992 (R) 97m/C Corey Feldman, Zach Galligan, Julie McCullough, Rowanne Brewer, Ray Sharkey; *Dir:* Alan Roberts. **VHS $89.95** PSM 🎬🎬

The Round Up Brutal tale spins a realistic political web of fear. Suspected Hungarian subversives, largely peasants and herdsmen, are thrown into an antiquated jail and tortured. A direct and shocking exploration of life under dehumanizing, totalitarian rule. In Hungarian with English subtitles.
1966 90m/C *HU Dir:* Miklos Jancso. **VHS $59.95** FST 🎬🎬½

Round-Up Time in Texas Autry delivers a herd of horses to his diamond-prospector brother in South Africa and has to deal with bandits after the jewels. Typical early Autry entry, with singin' on the side.
1937 54m/B Gene Autry, Smiley Burnette; *Dir:* Joseph Kane. **VHS, Beta $19.95** NOS, VCN, VYY 🎬🎬

Rounding Up the Law Good guy dupes bad guy while organ plays.
1922 62m/b Russell Gordon, Chet Ryan, Patricia Palmer, William McCall, Guinn Williams; *Dir:* Charles Seeling. **VHS, Beta, 8mm $24.95** VYY ♫♫½

Roustabout A roving, reckless drifter joins a carnival and romances the owner's daughter. Elvis sings ''Little Egypt,'' ''Poison Ivy League'' and others. Elvis rides on good support from Stanwyck et al. Welch has a bit part in her film debut; look for Terri Garr as a dancer.
1964 101m/C Elvis Presley, Barbara Stanwyck, Joan Freeman, Leif Erickson, Sue Ann Langdon, Pat Buttram, Joan Staley, Dabbs Greer, Steve Brodie, Jack Albertson, Marianna Hill, Beverly Adams, Billy Barty, Richard Kiel, Raquel Welch; *Dir:* John Rich. **VHS, Beta $14.98** FOX, MVD ♫♫½

The Rousters Descendants of Wyatt Earp run a carnival in a small Western town. Comic mis-adventures arise when a modern varmint named Clayton comes looking for action. Stephen J.Cannell (of ''The A-Team'' and ''Riptide'' fame) wrote and produced—with his usual extraordinary mix of the absurd and the dangerous. Unusual vehicle for Rogers, but she keeps her style and humor. Made for TV.
1990 (PG) 72m/C Jim Varney, Mimi Rogers, Chad Everett, Maxine Stuart, Hoyt Axton; *W/Dir:* E.W. Swackhamer. **VHS, LV $79.95** VMK ♫♫½

Rover Dangerfield Lovable Las Vegas hound Rover Dangerfield searches for respect but finds none. The results are hilarious. Quality animation from a project developed by Rodney Dangerfield and Harold Ramis.
1991 (G) 78m/C *Voices:* Rodney Dangerfield. **VHS, LV, 8mm $92.99** WAR, PIA ♫♫

Roxanne A modern comic retelling of ''Cyrano de Bergerac.'' The romantic triangle between a big-nosed, small-town fire chief, a shy fireman and the lovely astronomer they both love. Martin gives his most sensitive and believable performance. Don't miss the bar scene where he gets back at a heckler. A wonderful adaptation for the modern age, memorably written by Martin.
1987 (PG) 107m/C Steve Martin, Darryl Hannah, Rick Rossovich, Shelley Duvall, Michael J. Pollard, Fred Willard; *Dir:* Fred Schepisi. **VHS, Beta, LV, 8mm $14.95** COL ♫♫♫

Roy Acuff's Open House: Vol. 1 Veteran country music star Roy Acuff performs his big hits in this pair of episodes from the series.
1959 60m/B Roy Acuff. **VHS, Beta $24.95** NOS, DVT ♫♫♫

Roy Acuff's Open House: Vol. 2 Join Roy Acuff and his special guests as they get together for a good old-fashioned hoe-down.
1959 60m/B **VHS, Beta $24.95** NOS, DVT ♫♫♫

The Royal Bed King Eric's queen really wears the pants in the family, and when she leaves on a trip, he's left helpless and hapless. A revolution erupts, and his daughter announces she plans to marry a commoner. What's a king to do?
1931 74m/B Lowell Sherman, Nance O'Neill, Mary Astor, Anthony Bushell, Gilbert Emery, Robert Warwick, J. Carroll Naish; *Dir:* Lowell Sherman. **VHS, Beta $29.95** VYY ♫♫½

Royal Wedding Astaire and Powell play a brother-and-sister dance team who go to London during the royal wedding of Princess Elizabeth, and find their own romances. Remembered for the inspired songs and Astaire's incredible dancing on the ceiling and walls, this film was Alan Jay Lerner's first screenplay. The idea came from Adele Astaire's recent marriage to a British Lord.
1951 100m/C Fred Astaire, Jane Powell, Peter Lawford, Keenan Wynn; *Dir:* Stanley Donen. **VHS, Beta, LV $19.95** NOS, MGM, CAB ♫♫♫

R.P.M.* **(*Revolutions Per Minute)** Student activists force the university hierarchy to appoint their favorite radical professor president. Hip prof-turned-prez Quinn is between a rock and a hard place, and allows police to crack down. Could have been intriguing, but loses steam; in any case, the script and direction both stink.
1970 (R) 90m/C Anthony Quinn, Paul Winfield, Gary Lockwood, Ann-Margret; *Dir:* Stanley Kramer. **VHS, Beta $59.95** COL ♫½

R.S.V.P. A Hollywood party honoring a writer turns tragic when a body is found in the guest of honor's pool. Most interesting part of movie is watching the porn stars (who make up most of the movie's cast) try to act with their clothes on. They fail miserably.
1984 (R) 87m/C Harry Reems, Lynda Wiesmeier; *Dir:* Lem Amero. **VHS, Beta, LV $69.98** LIV, VES ♫½

Rub a Dub Dub A jazzy score backs this modernization of classic Mother Goose tales. Animated.
1983 60m/C **VHS, Beta $29.95** MED

Rub a Dub Dub—There's a Zoo in This Tub!, Vol. 2 Another volume of songs & rhymes animated for children.
1987 60m/C **VHS, Beta $29.95** MED

Rubber Rodeo This program presents the band with the unique punk-country sound. Songs included are: ''Anywhere with You,'' ''How the West Was Won,'' ''The Theme from Rubber Rodeo'' and many more.
1984 18m/C **VHS, Beta, LV $9.95** SVS

Rubin & Ed Ed is a real estate groupie, with an ex-wife from hell, who bribes the equally strange Rubin to attend a get rich quick seminar. In return, Rubin decides Ed will be the perfect traveling companion to take on the road in search of the perfect gravesite for Rubin's long-dead feline. But the fun really begins when these two dysfunctional buddies are stranded in Death Valley and begin suffering from sunstroke and hallucinations. This entire movie is a hallucination—a bad one.
1992 (PG-13) 82m/C Crispin Glover, Howard Hesseman, Karen Black; *W/Dir:* Trent Harris. **VHS, LV $89.95** COL ♫

Ruby A young woman, christened in blood and raised in sin, has a love affair with the supernatural and murders up a storm at a drive-in. Confused, uneven horror a step up from most similar flicks.
1977 (R) 85m/C Roger Davis, Janet Baldwin, Piper Laurie, Stuart Whitman; *Dir:* Curtis Harrington. **VHS, Beta $14.95** NEG ♫♫

Ruby Another in the line of Kennedy assassination/conspiracy films—this one told from the viewpoint of Jack Ruby, the killer of Lee Harvey Oswald. Ruby is a Dallas strip club owner who is a mob hanger-on and also serves as a minor informant to both the Dallas police and the FBI. The film proposes Ruby assassinated Oswald to expose a mob and CIA conspiracy to get Kennedy. Fenn is a Marilyn Monroe look-alike, working as a stripper in Ruby's club, who has a tryst with President Kennedy. Confusing plot, somewhat redeemed by the performances of Aiello and Fenn. The film combines actual footage of Ruby's shooting of Oswald with black and white filmed scenes.
1992 (R) 100m/C Danny Aiello, Sherilyn Fenn, Arliss Howard; *Dir:* John MacKenzie. **VHS, LV** COL ♫♫♫

Ruby Gentry White-trash girl Jones, cast aside by man-she-loves Heston, marries wealthy Malden to spite him, then seeks revenge. Classic Southern theme of comeuppance, good direction and acting lift it a notch.
1952 82m/B Charlton Heston, Jennifer Jones, Karl Malden; *Dir:* King Vidor. **VHS, Beta $59.98** FOX ♫♫½

Ruckus A Vietnam vet uses his training to defend himself when he runs into trouble in a small Alabama town. Less obnoxious than ''First Blood,'' (which came later), but basically the same movie.
1981 (PG) 91m/C Dirk Benedict, Linda Blair, Ben Johnson, Richard Farnsworth, Matt Clark; *Dir:* Max Kleven. **VHS, Beta** PGN ♫♫

Ruddigore The Lords of Ruddigore have been bound for centuries by a terribly inconvenient curse; they must commit a crime every day or die a horribly painful death. When the Lord of Ruddigore passes his mantle on to the new heir, the young heir loses both his good reputation and his very proper fiancee. A new version of Gilbert and Sullivan's opera.
197? 112m/C Vincent Price, Keith Michell, Sandra Dugdale. **VHS, Beta $49.98** FOX ♫♫

Rude Awakening A real estate agent is tortured by bizarre nightmares which lead to an unusual series of events. Part of the ''Thriller Video'' series.
1981 60m/C Denholm Elliott, James Laurenson, Pat Heywood; *Dir:* Peter Sasdy. **VHS, Beta $29.95** IVE ♫♫½

Rude Awakening Hippies Cheech and Roberts (where's Chong?) settled in a South American commune 20 years ago. Now they're back, and don't know what to make of old pals Carradine and Hagerty who have become - that's right - yuppies. Rip Van Winkle-ian farce draws a few yuks, but could have been much funnier. Points for trying.
1989 (R) 100m/C Eric Roberts, Cheech Marin, Julie Hagerty, Robert Carradine, Buck Henry, Louise Lasser, Cindy Williams, Andrea Martin, Cliff DeYoung; *Dir:* Aaron Russo, David Greenwalt. **VHS, Beta, LV $89.99** HBO ♫♫

Rudolph & Frosty's Christmas in July More animated adventures of Rudolph and friends. This time they are transported to a glittering, sunny seashore for a magical summer weekend that almost turns into tragedy.
1982 97m/C *Voices:* Red Buttons, Ethel Merman, Mickey Rooney, Jackie Vernon, Shelley Winters. **VHS, Beta $14.98** LIV, LTG, TLF ♫♫½

Rudolph the Red-Nosed Reindeer The durable Christmas story of the nasally enhanced reindeer who saved the holiday; colorful stop-motion animation and character voices bring it to life, with a memorable guest visit from an abominable snowman. Originally made for television.
1964 53m/C *Dir:* Larry Roemer; *Nar:* Burl Ives. **VHS, Beta, LV, 8mm $14.95** FHE ♫♫♫

Rudolph's Shiny New Year A puppet-animated Christmas special, showing Rudolph rescuing the Baby New Year from certain peril. Made for television.
1979 50m/C *Voices:* Red Skelton, Frank Gorshin. **VHS, Beta $14.98** *LIV, LTG* ✗✗✗

Rudyard Kipling's Classic Stories Three Kipling stories come to animated life: ''How the Elephant Got His Trunk,'' ''How the Whale Got His Throat'' and ''How the First Letter Was Written.''
1970 25m/C VHS, Beta $14.95 *NWV* ✗✗✗

The Rue Morgue Massacres A modern reprise of the Poe story with touches of Frankensteinia thrown in; plenty of gore.
1973 (R) 90m/C *SP* Paul Naschy, Rossana Yanni, Maria Perschy, Vic Winner, Mary Ellen Arpon; *Dir:* Javier Aguirre. **VHS, Beta $59.95** *ASE* ✗ 1/2

Ruggles Vintage television sitcom with Charlie, his wife and children. In this episode, the Ruggleses are presented with a pair of rabbits.
1951 24m/B Charlie Ruggles, Erin O'Brien-Moore. **VHS, Beta $19.95** *VYY, DVT* ✗✗

Ruggles of Red Gap Classic comedy about an uptight British butler who is ''won'' by a barbarous American rancher in a poker game. Laughton as the nonplussed manservant is hilarious; supporting cast excellent. Third and far superior filming of Harry Leon Wilson story was nominated for Best Picture. One of the all-time great comedies, the film was remade musically with Bob Hope and Lucille Ball as ''Fancy Pants'' (1950).
1935 90m/B Charles Laughton, Mary Boland, Charlie Ruggles, ZaSu Pitts, Roland Young, Leila Hyams, James Burke, Maude Eburne; *Dir:* Leo McCarey. National Board of Review Awards '35: 10 Best Films of the Year; New York Film Critics Awards '35: Best Actor (Laughton). **VHS, Beta $29.95** *MCA* ✗✗✗

Rulers of the City A young gangster avenges his father's death.
1971 91m/C Jack Palance, Edmund Purdom, Al Cliver, Harry Baer. **VHS, Beta $19.98** *VDC* ✗

The Rules of the Game Renoir's masterpiece, in which a group of French aristocrats, gathering for a weekend of decadence and self-indulgence just before World War II, becomes a metaphor for human folly under siege. The film was banned by the French government, pulled from distribution by the Nazis, and not restored to its original form until 1959, when it premiered at the Venice Film Festival. A great, subtle, ominous film landmark. Also known as ''La Regle du Jeu.'' Much copied and poorly remade in 1989 as ''Scenes from the Class Struggle in Beverly Hills.''
1939 110m/B *FR* Marcel Dalio, Nora Gregor, Jean Renoir, Mila Parely, Julien Carette, Gaston Modot, Roland Toutain; *Dir:* Jean Renoir. Sight & Sound Survey '52: #10 of the Best Films of All Time (tie); Sight & Sound Survey '62: #3 of the Best Films of All Time; Sight & Sound Survey '72: #2 of the Best Films of All Time; Sight & Sound Survey '82: #2 of the Best Films of All Times. **VHS, Beta, LV, 8mm, 3/4U $19.95** *NOS, SUE, HHT* ✗✗✗✗

The Ruling Class The classic cult satire features O'Toole as the unbalanced 14th Earl of Gurney, who believes that he is either Jesus Christ or Jack the Ripper. Tongue-in-cheek look, complete with dance and music, at eccentric upper-class Brits and their institutions. Uneven, chaotic, surreal and noteworthy. Oscar nomination for O'Toole.

1972 (PG) 154m/C *GB* Peter O'Toole, Alastair Sim, Arthur Lowe, Harry Andrews, Coral Browne, Nigel Green; *Dir:* Peter Medak. National Board of Review '72: Best Actor (O'Toole), 10 Best Films of the Year. **VHS, Beta, LV $19.98** *SUE* ✗✗✗

Rumble Fish A young street punk worships his gang-leading older brother, the only role model he's known. Crafted by Coppola into an important story of growing up on the wrong side of town, from the novel by S. E. Hinton. Atmospheric musical score by Stewart Copeland. Ambitious and experimental; in black and white.
1983 (R) 94m/B Matt Dillon, Mickey Rourke, Dennis Hopper, Diane Lane, Vincent Spano, Nicolas Cage, Diana Scarwid, Christopher Penn, Tom Waits; *Dir:* Francis Ford Coppola. **VHS, Beta, LV $59.95** *MCA* ✗✗✗

The Rumor Mill The careers and titanic rivalry of influential Hollywood gossip writer Hedda Hopper and Louella Parsons, played with verve by Taylor and Alexander. Fictionalized script by Jacqueline Feather and David Seidler based on the book ''Hedda and Louella,'' by George Eels. Made for TV; originally called ''Malice in Wonderland.''
1986 94m/C Elizabeth Taylor, Jane Alexander, Richard Dysart, Joyce Van Patten; *Dir:* Gus Trikonis. **VHS, Beta $79.95** *VMK* ✗✗ 1/2

A Rumor of War The first big Vietnam drama for TV is a triumph, portraying the real-life experience of author Philip Caputo (based on his bestseller), from naive youth to seasoned soldier to bitter murder suspect at court martial. Succeeds where Stone's later, much-ballyhooed ''Born on the Fourth of July'' fails. Adapted by John Sacret Young.
1980 195m/C Brad Davis, Keith Carradine, Michael O'Keefe, Stacy Keach, Steve Forrest, Richard Bradford, Brian Dennehy, John Freidrich, Perry Lang, Chris Mitchum, Dan Shor, Jeff Daniels; *Dir:* Richard T. Heffron. **VHS, Beta $14.95** *FRH, USA, IVE* ✗✗ 1/2

Rumpelstiltskin From ''Faerie Tale Theatre'' comes the story of a woman who can spin straw into gold (Duvall) and the strange man who saves her life (Villechaize). Enjoyable for the whole family.
1982 60m/C Bud Cort, Ned Beatty, Shelley Duvall, Herve Villechaize, Paul Dooley; *Dir:* David Irving. **VHS, Beta, LV $14.98** *KUI, FOX, FCT* ✗✗✗ 1/2

Rumpelstiltskin Musical retelling of the classic Brother Grimm fairy tale. Irving plays a young girl who says she can spin straw into gold. Barty is a dwarf who helps her. The catch: she must give up her first born to pay him. Lackluster direction and uninspired acting make this version a yawner. Irving's real-life mother (Pointer) and brother have roles too.
1986 (G) 84m/C Amy Irving, Billy Barty, Robert Symonds, Priscilla Pointer, Clive Revill, John Moulder-Brown; *Dir:* David Irving. **VHS, Beta, LV $19.98** *MED* ✗✗

Run, Angel, Run! An ex-biker is on the run from his former gang after he helps expose them in a magazine article.
1969 (R) 90m/C William Smith, Margaret Markov, Valerie Starrett; *Dir:* Jack Starrett. **VHS, Beta $29.98** *VID* ✗ 1/2

Run of the Arrow Rather than surrender to the North at the end of the Civil War, an ex-Confederate soldier (Steiger) joins the Sioux nation, engaged in war against the white man. Steiger's Irish brogue is weird and distracting, and the intriguing premise fails to satisfy, but the cast is competent and interesting.

1956 85m/C Rod Steiger, Brian Keith, Charles Bronson, Ralph Meeker, Tim McCoy; *Dir:* Samuel Fuller. **VHS, Beta, LV $54.95** *UHV* ✗✗

Run If You Can A woman sees a brutal murderer killing his victims on her television. Will she be next?
1987 (R) 92m/C Martin Landau, Yvette Nipar, Jerry Van Dyke; *Dir:* Virginia Lively Stone. **Beta $79.95** *ALL* ✗

Run, Rebecca, Run Interesting family adventure yarn for the family about a South American refugee who tries to stop a young girl from leaving from an Australian island where they are both stranded. She eventually befriends him, and attempts to get permission for him to enter the country.
1981 90m/C *AU* Simone Buchanan, Henri Szeps; *Dir:* Peter Maxwell. **VHS, Beta $14.98** *VID* ✗✗ 1/2

Run for the Roses ''National Velvet'' clone about a young Puerto Rican boy and a champion race horse in Kentucky. Cheery family fare.
1978 (PG) 93m/C Lisa Eilbacher, Vera Miles, Stuart Whitman, Sam Groom; *Dir:* Henry Levin. **VHS, Beta $59.98** *LIV, LTG* ✗ 1/2

Run Silent, Run Deep Submarine commander Gable battles his officers, especially the bitter Lancaster who vied for the same command, while stalking the Japanese destroyer that sunk his former command. Top-notch World War II sub action, scripted from Commander Edward L. Beach's novel.
1958 93m/B Burt Lancaster, Clark Gable, Jack Warden, Don Rickles, Brad Dexter; *Dir:* Robert Wise. **VHS, Beta $19.95** *FOX, FCT* ✗✗✗

Run, Stranger, Run The inhabitants of a seaside house in Nova Scotia are paralyzed with fear when a number of brutal murders occur. A couple of sicko slasher scenes, but hard to swallow. Originally titled, ''Happy Mother's Day, Love, George.''
1973 (PG) 110m/C Ron Howard, Patricia Neal, Cloris Leachman, Bobby Darin; *Dir:* Darren McGavin. **VHS $59.95** *COL* ✗✗

Run Virgin Run The village women find a beautiful virgin to marry the blacksmith so he will stay in town and they will still be able to enjoy him.
1990 (R) 90m/C Cheryl Ross, Kyle Coleman, Suzanne Baron. **VHS, Beta $39.95** *MON* ✗

Runaway A cop and his sidekick track down a group of killer robots wreaking havoc. Self-serious, sorry sci-fi. Features Simmons of the rock group KISS.
1984 (PG-13) 100m/C Tom Selleck, Cynthia Rhodes, Gene Simmons, Stan Shaw, Kirstie Alley; *Dir:* Michael Crichton. **VHS, Beta, LV $14.95** *COL* ✗ 1/2

Runaway A boy blames himself for his friend's accidental death and runs away to live in the subway tunnels of New York City. After four months of life underground, a disabled veteran and a waitress lend the boy a hand. Made for PBS and aired as part of the WonderWorks Family Movie television series.
1989 58m/C Charles Dutton, Jasmine Guy, Gavin Allen; *Dir:* Gilbert Moses. **VHS $29.95** *PME* ✗ 1/2

The Runaway Barge Tiresomely ordinary TV yarn about three adventurers earning their living on a riverboat. The usual cliches are present, including a gang of bad guys.
1975 78m/C Bo Hopkins, Tim Matheson, Jim Davis, Nick Nolte, Devon Ericson, Christina Hart, James Best; *Dir:* Boris Sagal. **VHS $29.95** *IVE* ✗✗

The Runaway Bus Filled with the requisite motley crew of passengers, including a gold-carrying thief and the ever-befuddled Margaret Rutherford, a bus gets lost between airports and ends up in a deserted village. Would-be wacky comedy.
1954 78m/B *GB* Frankie Howerd, Margaret Rutherford, Petula Clark, George Coulouris; *Dir:* Val Guest. **VHS, Beta** $29.95 *VYY, FCT* ✕✕

Runaway Island, Part 1 Episode: "Quest for Jamie McLeod." In 1830's Australia, two youths battle corruption when their father leaves for England.
19?? 96m/C Miles Buchanan, Simone Buchanan. **VHS, Beta** $59.95 *SVS* ✕✕

Runaway Island, Part 2 Episode: "The Exiles." When their father dies, the McLeod children realize their only hope of obtaining his property is to find his hidden will.
19?? 103m/C Miles Buchanan, Simone Buchanan. **VHS, Beta** $59.95 *SVS* ✕✕

Runaway Island, Part 3 Episode: "The Bushranger." Jamie and Jemma McLeod continue the search for their father's will, but are captured by a gang of terrorists.
19?? 101m/C Miles Buchanan, Simone Buchanan. **VHS, Beta** $59.95 *SVS* ✕✕

Runaway Island, Part 4 Episode: "Treasure of the Conquistadors." Having found the missing portion of the map that will lead them to their father's will, the McLeod children encounter a band of pirates, Aborigines and other dangers that might prevent them from staking their claim.
19?? 104m/C Miles Buchanan, Simone Buchanan. **VHS, Beta** $59.95 *SVS* ✕✕

Runaway Nightmare Two western worm ranchers are kidnapped by a female gang whose members torture and initiate them, then talk about their plan to steal from organized criminals. Just another typical worm rancher flick.
1984 104m/C Michael Cartel, Al Valletta; *Dir:* Michael Cartel. **VHS** *ASE* ✕

Runaway Train A tough jailbird and his sidekick break out of the hoosegow and find themselves trapped aboard a brakeless freight train heading for certain derailment in northwestern Canada. Harrowing existential action drama based on a screenplay by Akira Kurosawa. Voigt is Superb.
1985 (R) 112m/C Jon Voight, Eric Roberts, Rebecca DeMornay, John P. Ryan, T.K. Carter, Kenneth McMillan; *W/Dir:* Andrei Konchalovsky. **VHS, Beta, LV** $14.98 *MGM* ✕✕

Runaway Truck Sky King takes to the air with novice pilot. Student pilot is then forced to make crucial decision. From the television series "Sky King."
19?? 25m/B Kirby Grant. **VHS, Beta** *CCB* ✕½

The Runaways A boy runs away from an unhappy foster home situation and becomes friends with a leopard that has escaped from an animal park. Sentimental but well-done family fun.
1975 76m/C Dorothy McGuire, John Randolph, Neva Patterson, Josh Albee; *Dir:* Harry Harris. **VHS, Beta** $49.95 *USA* ✕✕

Runaways A corrupt politician's wife falls in love with the man who saved her when she attempted suicide. The two try to escape her husband's wrathful vengeance.
1984 95m/C Steve Oliver, Sondra Currie, John Russell. **VHS, Beta** $49.95 *REG* ✕

The Runestone After the eponymous viking relic is unearthed, an archaeologist's meddling frees a long-imprisoned wolf-beast that terrorizes New York. An adequate horror flick that taps the seldom-explored vein of Norse mythology. Based on a novel by Mark E. Rogers.
1991 (R) 105m/C Peter Riegert, Joan Severance, William Hickey, Tim Ryan, Chris Young, Alexander Godunov; *Dir:* Willard Carroll. **VHS, LV** $89.98 *LIV* ✕✕

The Runner Stumbles Dated melodrama about a priest who goes on trial for murdering the nun with whom he fell in love in a mining town in the 1920s. Based on a true incident, from the play by Milan Stiff. Van Dyke in an unwanted serious role is miserably grim and wooden.
1979 (R) 109m/C Dick Van Dyke, Kathleen Quinlan, Maureen Stapleton, Ray Bolger, Beau Bridges, Tammy Grimes; *Dir:* Stanley Kramer. **VHS, Beta** $59.98 *FOX* ✕½

The Runnin' Kind A yuppie expecting to inherit his father's Ohio law firm goes on a wild spree with an all-girl rock band. Sexy drummer Howard shows him the sights of certain parts of L.A. Original, fun comedy. Independent.
1989 (R) 89m/C David Packer, Pleasant Gehman, Brie Howard, Susan Strasberg; *Dir:* Max Tash. **VHS, Beta** $79.98 *FOX* ✕✕½

Running Against Time A teacher tries to change history (and erase his brother's Vietnam War death) by time-warping back to 1963 and preventing JFK's murder. The made-for-cable-TV movie takes an offhanded gee-whiz approach to a tantalizing premise, as temporal paradoxes multiply in "Back to the Future" style. No assassination-conspiracy theories, by the way, but Lyndon Johnson fans may not like his portrayal here. Based on the novel "A Time to Remember" by Stanley Shapiro.
1990 (PG) 93m/C Robert Hays, Catherine Hicks, Sam Wanamaker; *Dir:* Bruce Seth Green. **VHS** $79.95 *MCA* ✕✕½

Running Away Bittersweet story of wartime romance, with a twist of adventure. Strong cast creates good chemistry, but somehow the story is not engaging. Loren and Penny are mother and daughter, leaving Rome for the hills and encountering dangers.
1989 (PG-13) 101m/C Sophia Loren, Robert Loggia. **VHS** $89.95 *PSM, PAR, FCT* ✕✕

Running Blind Espionage/adventure thriller has a double agent and a beautiful woman on the run in Iceland.
197? 120m/C *GB* Stuart Wilson. **VHS** $49.95 *FCT* ✕✕

Running Brave The true story of Billy Mills, a South Dakota Sioux Indian who won the Gold Medal in the 10,000 meter run at the 1964 Tokyo Olympics. Not bad, but holey in the way of many of these plodding true-story flicks.
1983 (PG) 90m/C Robby Benson, Claudia Cron, Pat Hingle, Denis Lacroix; *Dir:* D.S. Everett. **VHS, Beta, LV** *DIS* ✕✕

Running on Empty Two 1960s radicals are still on the run in 1988 for a politically motivated Vietnam War-era crime. Though they have managed to stay one step ahead of the law, their son wants a "normal" life, even if it means never seeing his family again. Well-performed, quiet, plausible drama. Phoenix was Oscar-nominated.

1988 (PG-13) 116m/C Christine Lahti, River Phoenix, Judd Hirsch, Martha Plimpton, Jonas Arby, Ed Crowley, L.M. Kit Carson, Steven Hill, Augusta Dabney, David Margulies, Sidney Lumet; *Dir:* Sidney Lumet. National Board of Review Awards '88: Best Supporting Actor (Phoenix). **VHS, Beta, LV, 8mm** $19.95 *WAR* ✕✕✕

Running Hot Seventeen-year-old convicted murderer Stoltz gets love letter in prison from older woman Carrico. When he escapes, they flee the law together. Quick-paced and entertaining. Also known as "Highway to Hell" and "Lucky 13."
1983 (R) 88m/C Monica Carrico, Eric Stoltz, Stuart Margolin, Virgil Frye, Richard Bradford, Sorrels Pickard, Juliette Cummins; *Dir:* Mark Griffiths; *W/Dir:* Mark Griffiths. **VHS, Beta** $69.98 *LIV, VES, RAE* ✕✕½

The Running Kind A young man set to inherit a law firm drops it all to manage an all-girl rock band. His new career takes him into the L.A. punk-scene in this clever comedy.
1989 (R) 100m/C David Packer, Pleasant Gehman, Brie Howard, Susan Strasberg; *Dir:* Max Tash. **VHS, Beta** $79.98 *FOX* ✕✕½

The Running Man Special-effects-laden adaptation of the Stephen King novel about a futuristic television game show. Convicts are given a chance for pardon—all they have to do is survive an on going battle with specially-trained assassins in the bombed-out sections of Los Angeles. Sci-fi with an attitude. The muscle man as the accused good guy Danson as his "Family Feud" self adds the right extra towel.
1987 (R) 101m/C Arnold Schwarzenegger, Richard Dawson, Maria Conchita Alonso, Yaphet Kotto, Mick Fleetwood, Dweezil Zappa, Jesse Ventura; *Dir:* Paul Michael Glaser. **VHS, Beta, LV** $14.98 *LIV, VES* ✕✕½

Running Mates Occasionally intriguing drama of two teens who fall in love, but are kept apart by their fathers' local political rivalry. Could have been better, but cardboard characters abound.
1986 (PG) 90m/C Greg Webb, Barbara Howard, J. Don Ferguson, Clara Dunn; *Dir:* Thomas L. Neff. **VHS, Beta** $19.95 *STE, NWV* ✕✕

Running Out of Luck A light, song-packed story about a rock star very much like Jagger who is abandoned in South America and believed dead, until he returns to his world. Features songs from Jagger's solo album "She's the Boss".
1986 (R) 88m/C Mick Jagger, Dennis Hopper, Jerry Hall, Rae Dawn Chong; *Dir:* Julien Temple. **VHS, Beta, LV** $19.95 *MVD, FOX* ✕½

Running Scared Two young men, returning from military service as stowaways aboard an Army cargo plane, are caught by an intelligence agent and thought to be spies, because Reinhold has unknowingly photographed a secret US base wire his stolen Army-issue camera. Not-bad spy thriller with a twist - the "spies" are hapless nobodies, now on the lam.
1979 (PG) 92m/C Ken Wahl, Judge Reinhold, John Saxon, Annie McEnroe, Bradford Dillman, Pat Hingle; *Dir:* Paul Glickler. **VHS, Beta, LV** $59.99 *HBO* ✕✕

Running Scared Hip, unorthodox Chicago cops Hines and Crystal have to handle an important drug arrest before taking an extended vacation in Key West. Not as relentless as "48 Hrs." but more enjoyable in some ways. Ever wonder what it's like to ride in a car at high speed on the tracks? Find out here.

1986 (R) 107m/C Gregory Hines, Billy Crystal, Dan Hedaya, Jimmy Smits, Darlanne Fluegel, Joe Pantoliano, Steven Bauer; *Dir:* Peter Hyams. **VHS, Beta, LV $14.98** *MGM* 🎞🎞½

Running Wild Early silent comedy starring mustachioed Fields as an eternal coward who suddenly turns into a lion by hypnosis. Odd, often unfunny Fields entry sometimes hits the mark. He's awfully mean, though. Music score by Gaylord Carter. **1927 68m/B** W.C. Fields, Mary Brian, Claud Buchanan; *Dir:* Gregory La Cava. **VHS, Beta $29.95** *PAR* 🎞🎞

Running Wild A free-lance photographer becomes involved in a dispute to save a corral of wild mustang horses. The usual great Colorado scenery enhances a good, family-view story. **1973 (G) 102m/C** Lloyd Bridges, Dina Merrill, Pat Hingle, Gilbert Roland, Morgan Woodward. **VHS, Beta $19.95** *MED* 🎞🎞½

Rush A post-nuclear world war movie in which a rebel named Rush battles the rulers of a slave colony. **1984 83m/C** *IT* Conrad Nichols, Gordon Mitchell, Laura Trotter, Rita Furlan; *Dir:* Tonino Ricci. **VHS** *USA* 🎞

Rush The Texas drug culture, circa 1975, is portrayed in this bleak cop drama. Kristen Cates (Leigh) is a rookie narcotics officer who goes undercover with the experienced Raynor (Patric) to catch a big-time dealer, menacingly played by Allman, and their bosses don't much care how they do it. But when Cates falls in love with Raynor she is drawn ever deeper into the drug-addicted world they are supposd to destroy. Fine performances and a great blues score by Eric Clapton. The directorial debut of Lili Fini Zanuck. Based on ex-narcotics cop Kim Wozencraft's autobiographical novel. **1991 (R) 120m/C** Jason Patric, Jennifer Jason Leigh, Gregg Allman, Max Perlich; *Dir:* Lili Fini Zanuck. **VHS $94.99** *MGM* 🎞🎞🎞

Rush It Scriptwriters ponder possible plots over capucino: "I've got it! 'Bike Messengers in Love'!" "Nah. The title's too obvious. Why don't we call it 'Rush It'?" Much hard work and many months later, the result is forgettable. **1977 78m/C** Tom Berenger, Jill Eikenberry. **VHS, Beta $49.95** *UNI* 🎞½

Rush Week Dead coeds populate campus during frat week. Greg Allman has bit part and the Dickies perform two songs. **1988 (R) 93m/C** Dean Hamilton, Gregg Allman, Kathleen Kinmont, Roy Thinnes, Pamela Ludwig; *Dir:* Bob Bravler. **VHS, LV $79.95** *COL* 🎞

The Russia House Russian scientist Brandauer attempts to publish book debunking Soviet Union's claims of military superiority by passing it through ex-lover Pfeiffer to British editor Connery, apparently unaware of impending glasnost. Star-studded spy thriller directed by Aussie Schepisi aspires to heights it never quite reaches. Adapted by Tom Stoppard from John Le Carre's novel. Beautiful on-location photography. Stoppard's screenplay is very fine, actually making some aspects of the novel work better. **1990 (R) 122m/C** Sean Connery, Michelle Pfeiffer, Roy Scheider, James Fox, John Mahoney, Klaus Maria Brandauer, Ken Russell, J.T. Walsh, Michael Kitchen; *Dir:* Fred Schepisi. **VHS, LV, 8mm $94.99** *MGM, PIA, FCT* 🎞🎞½

Russian Roulette The Russian premier is visiting Vancouver in 1970, and the Mounties must prevent a dissident KGB terrorist from assassinating him. Un-thrilling spy yarn fails to deliver on intriguing premise. **1975 100m/C** George Segal, Christina Raines, Bo Brundin, Denholm Elliott, Louise Fletcher; *Dir:* Lou Lombardo. **VHS, Beta $59.98** *FOX* 🎞🎞

The Russian Terminator An FBI agent is up against a one-man "death squad." **1990 (R) 90m/C** Helena Michaelson, Frederick Offrein, Harley Melin, Tina Tjung; *W/Dir:* Mats Helge. **VHS $79.95** *XVC, PHX* 🎞

The Russians are Coming, the Russians are Coming Based on the comic novel "The Off-Islanders" by Nathaniel Benchley, this is the story of a Russian sub which accidentally runs aground off the New England coast. The residents falsely believe that the nine-man crew is the beginning of a Soviet invasion, though the men are only looking for assistance. A memorable set of silly events follows the landing, engineered by a gung-ho police chief and a town filled with overactive imaginations. Oscar-nominated for Best Picture, Best Actor (Arkin), and Film Editing. **1966 126m/C** Alan Arkin, Carl Reiner, Theodore Bikel, Eva Marie Saint, Brian Keith, Paul Ford, Jonathan Winters, Ben Blue, Tessie O'Shea, Doro Merande, John Phillip Law; *Dir:* Norman Jewison. Golden Globe Awards '67: Best Film—Musical/Comedy; National Board of Review Awards '66: 10 Best Films of the Year. **VHS, Beta $19.95** *FOX, MLB* 🎞🎞🎞

Russkies A jolly comedy about three adorable American kids who capture, and eventually grow to like, a stranded Russian sailor. Friendly and peace-loving, but dull. **1987 (PG) 98m/C** Leaf Phoenix, Whip Hubley, Peter Billingsley, Stefan DeSalle; *Dir:* Rick Rosenthal. **VHS, Beta, LV $19.98** *LHV, WAR* 🎞½

Rustler's Hideout One of numerous westerns starring Crabbe as cowboy do-gooder Billy Carson, here plodding through standard histrionics against rustlers, a card-sharp, and a business fraud. **1944 60m/C** Buster Crabbe, Al "Fuzzy" St. John, Charles King, John Merton, Lane Chandler, Hal Price, Edward Cassidy, Bud Osborne; *Dir:* Sam Newfield. **VHS, Beta** *GPV* 🎞

Rustler's Paradise Carey seeks revenge on the man who stole his wife and daughter. **1935 59m/B** Harry Carey, Gertrude Messinger, Edmund Cobb; *Dir:* Harry Fraser. **VHS, Beta, LV $19.95** *NOS, WGE* 🎞

Rustler's Rhapsody A singing cowboy rides into a small western town and encounters all kinds of desperados in this earnest would-be satire of '40s B-movie westerns. **1985 (PG) 89m/C** Tom Berenger, Patrick Wayne, G.W. Bailey, Andy Griffith, Marilu Henner; *Dir:* Hugh Wilson. **VHS, Beta, LV $14.95** *PAR* 🎞🎞

Rustler's Valley Hopalong Cassidy is a ranch foreman who tries to save his employer from the clutches of a crooked lawyer. Pretty fun ordinary range-ridin' saga. **1937 59m/B** William Boyd, George "Gabby" Hayes, Lee J. Cobb; *Dir:* Nate Watt. **VHS, Beta $24.95** *VYY* 🎞🎞

The Rutanga Tapes Soviet bloc country has representatives in Africa producing deadly chemicals used for war and a dissident has taped evidence of native deaths. Reporter babe Simpson wants an exclusive, Petersen wants the dissident, Libyan agents

want the tapes, and you'll want to fast forward to the end. **1990 (R) 88m/C** Henry Cele, Arnold Vosloo, Wilson Dunster, David Dukes, Susan Anspach; *Dir:* David Lister. **VHS $89.98** *SHG* 🎞

The Rutherford County Line A tough small-town sheriff searches for the killer of his deputies. **1987 98m/C** Earl Owensby; *Dir:* Thom McIntyre. **VHS, Beta $59.95** *MCA* 🎞

The Ruthless Four Quartet of unlikely mining partners match wills and wits and battle the elements. Not-so-hot spaghetti western, with good lead from Heflin. Much-retitled; also "Every Man for Himself," "Each One for Himself," and "Sam Cooper's Gold." **1970 97m/C** *IT* Van Heflin, Klaus Kinski, Gilbert Roland, George Hilton; *Dir:* Giorgio Capitani. **VHS, Beta $24.95** *MON* 🎞

Ruthless People DeVito and his mistress spend a romantic evening plotting his obnoxious wife's untimely demise. Before he can put his plan into action, he's delighted to discover she's been kidnapped by some very desperate people—who don't stand a chance with Bette. High farcical entertainment script by Dale Lanner; variation on the story "The Rousom of Red Chief," by O. Henry. **1986 (R) 93m/C** Bette Midler, Danny DeVito, Judge Reinhold, Helen Slater, Anita Morris, Bill Pullman; *Dir:* Jim Abrahams, David Zucker. **VHS, Beta, LV $19.95** *TOU* 🎞🎞

Ryan's Daughter An Irish woman marries a man she does not love and then falls for a shell-shocked British major who arrives during the 1916 Irish uprising to keep the peace. Miles is accused of betraying the local IRA gunrunners to her British lover. Tasteful melodrama with lots of pretty scenery that goes on a bit too long. Oscar nominations for sound, as well as awards it won. **1970 (PG) 194m/C** Sarah Miles, Robert Mitchum, John Mills, Trevor Howard, Christopher Jones, Leo McKern; *Dir:* David Lean. Academy Awards '70: Best Cinematography, Best Supporting Actor (Mills). **VHS, Beta, LV $29.98** *MGM, PIA, BTV* 🎞🎞½

Ryder P.I. P.I Ryder and his sidekick fight crime and solve weird cases in the big city. **1988 (PG-13) 92m/C** Bob Nelson. **VHS, Beta $79.98** *FOR* 🎞½

Sabaka Adventure set in India about a scary religious cult. Karloff and the cast of stalwart "B" movie performers have fun with this one, and you should, too. Also known as "The Hindu." **1955 81m/C** Boris Karloff, Reginald Denny, Victor Jory, Lisa Howard, Jeanne Bates, Jay Novello, June Foray; *Dir:* Frank Ferrin. **VHS $16.95** *SNC* 🎞

Sabotage Early Hitchcock thriller based on Conrad's "The Secret Agent". A woman who works at a movie theater (Sidney) suspects her quiet husband (Homolka) might be the terrorist planting bombs around London. Numerous sly touches of the Master's signature humor. **1936 81m/B** Oscar Homolka, Sylvia Sidney, John Loder; *Dir:* Alfred Hitchcock. **VHS, Beta, LV, 8mm $16.95** *SNC, NOS, HHT* 🎞🎞🎞

Saboteur A man wrongly accused of sabotaging an American munitions plant during World War II sets out to find the traitor who framed him. Hitchcock uses his locations, including Boulder Dam, Radio City Music

Hall, and the Statue of Liberty, to greatly intensify the action. Stunning resolution. **1942 108m/B** Priscilla Lane, Robert Cummings, Otto Kruger, Alan Baxter, Norman Lloyd; *Dir:* Alfred Hitchcock. **VHS, Beta, LV $19.95** MCA ♂♂♂

Sabre Jet Run-of-the-mill wartime drama of hubbies flying combat missions in Korea and long-suffering spouses on the home front. **1953 90m/C** Robert Stack, Amanda Blake, Louis King; *Dir:* Louis King. **VHS, Beta $39.95** VCD ♂♂

Sabrina Two wealthy brothers, one an aging businessman and the other a dissolute playboy, vie for the attention of their chauffeur's daughter, who has just returned from a French finishing school. Typically acerbic, in the Wilder manner, with Bogart and Holden cast interestingly against type. Written by Wilder from the play "Sabrina Fair" by Samuel Taylor. Oscar nominations for Best Actress (Hepburn), Director, Screenplay, Black & White Cinematography, Art Direction, and Set Design. **1954 113m/B** Audrey Hepburn, Humphrey Bogart, William Holden, Walter Hampden, Francis X. Bushman, John Williams, Martha Hyer, Marcel Dalio; *Dir:* Billy Wilder. Academy Awards '54: Best Costume Design (B & W); National Board of Review Awards '54: 10 Best Films of the Year, Best Supporting Actor (Williams); Directors Guild of America Awards '54: Outstanding Directorial Achievement. **VHS, Beta, LV $19.95** PAR ♂♂♂

Sabrina, the Teenage Witch What happens when a pretty teenage girl is also a witch? If it's Sabrina, you know you're in for plenty of amazing, amusing and action-packed adventures. Join her and her pals Archie, Betty, Veronica and Jughead for magical fun at Riverdale High. Nine volumes available. **1969 57m/C** *Voices:* Jane Webb. **VHS, Beta $39.98** SUE ♂♂♂

Sacco & Vanzetti Two Italian immigrants and acknowledged anarchists are caught amidst communist witch-hunts and judicial negligence when they are tried and executed for murder in 1920s America. Based on the true-life case, considered by some a flagrant miscarriage of justice and political martyrdom, by others, honest American judicial-system proceedings. Well-made and acted. Joan Baez sings the title song "The Ballad of Sacco and Vanzetti." **1971 (PG) 120m/C** Milo O'Shea, Gian Maria Volonte, Cyril Cusack, Geoffrey Keen; *Dir:* Giuliano Montaldo. **VHS, Beta $59.95** UHV ♂♂

The Sacketts Follows the adventures of three Tennessee brothers who migrate to the West after the Civil War. Based on two Louis L'Amour novels. Made for television. **1979 198m/C** Jeffery Osterhage, Tom Selleck, Sam Elliott, Glenn Ford, Ben Johnson, Mercedes McCambridge, Ruth Roman, Jack Elam, Gilbert Roland; *Dir:* Robert Totten. **VHS, Beta $14.99** WAR, VHE, HHE ♂♂½

Sacred Ground A trapper and his pregnant Apache wife unknowingly build shelter on the Paiute Indians' sacred burial ground. When the wife dies in childbirth, the pioneer is forced to kidnap a Paiute woman who has just buried her own deceased infant. Average western drama. **1983 (PG) 100m/C** Tim McIntire, Jack Elam, L.Q. Jones, Mindi Miller; *Dir:* Charles B. Pierce. **VHS, Beta $59.98** FOX ♂♂

Sacred Mountain/White Flame Two short films starring Riefenstahl, featuring mostly fluid dance sequences and Teutonic mountain boy/girl plots. Silent. **1931 60m/B** *GE* Leni Riefenstahl. **VHS, Beta $34.95** FCT, GLV, MRV ♂♂♂

The Sacred Music of Duke Ellington A performance of Ellington's three Sacred Concerts, taped at St. Paul's Cathedral in London. **1982 90m/C** Tony Bennett, Phyllis Hyman; *Hosted:* Rod Steiger; *Nar:* Douglas Fairbanks Jr. **VHS, Beta $29.95** MGM ♂♂♂

The Sacrifice Tarkovsky's enigmatic final film, released after his death, deals with a retired intellectual's spiritually symbolic efforts at self-sacrifice in order to save his family on the eve of a nuclear holocaust. Stunning cinematography by Sven Nykvist. Acclaimed, but sometimes slow going; not everyone will appreciate Tarkovsky's visionary spiritualism. **1986 (PG) 145m/C** *SW* Erland Josephson, Susan Fleetwood, Valerie Mairesse, Allan Edwall, Gudrun Gisladottir, Sven Wollter, Filippa Franzen; *Dir:* Andrei Tarkovsky. British Academy Awards '87: Best Foreign Film; Cannes Film Festival '86: Special Grand Jury Prize; New York Film Festival '87: Special Grand Jury Prize. **VHS, Beta, LV $29.95** PAV ♂♂♂

Sacrilege A nun and a nobleman engage in a hot love affair that fosters scandal, murder and revenge. Cynical use of religious theme masks otherwise ordinary sex/romance flick. Controversial Italian-made erotic drama. **1986 104m/C** *IT* Myriem Roussel, Alessandro Gassman; *Dir:* Luciano Odorisio. **VHS, Beta $70.05** ΓOM ♂½

The Sad Sack A bumbling hero with a photographic memory winds up in Morocco as a member of the French Foreign Legion. Although a success at the box office, Lewis's second movie (without partner Dean Martin) seems jerky and out of sorts today. Based on the comic strip character by George Baker, but not effectively. Hal David and Burt Bacharach wrote the music. **1957 98m/B** Jerry Lewis, Phyllis Kirk, David Wayne, Peter Lorre, Gene Evans, Mary Treen; *Dir:* George Marshall. **VHS** COL ♂♂

Saddle Buster A rodeo cowboy loses his nerve after being thrown by a tough horse, but gets it back again by the end of the hour. Actual rodeo footage is incorporated into this low-budget bronco saga. **1932 59m/B** Tom Keene, Helen Foster, Charles Quigley, Marie Quillen, Robert Frazer, Charles "Slim" Whitaker; *Dir:* Fred Allen. **VHS $19.95** NOS ♂♂

Saddle Mountain Round-Up The Range Busters are on the trail again as they smoke out a murderer. **1941 61m/B** Ray Corrigan, John "Dusty" King, Max Terhune. **VHS, Beta $19.95** NOS, DVT, BUR ♂♂

Sadie McKee A melodrama with Crawford as a maid searching for love in the big city. She falls for a self-destructive ne'er-do-well, marries an alcoholic millionaire, and eventually finds true love with the wealthy Tone (who would become Crawford's third husband). Professional acting and directing elevate the story. **1934 90m/B** Joan Crawford, Franchot Tone, Gene Raymond, Edward Arnold; *Dir:* Clarence Brown. **VHS $19.98** BTV ♂♂♂

Sadie Thompson Swanson plays a harlot with a heart of gold, bawdy and good-natured, in the South Seas. A zealot missionary (Barrymore) arrives and falls in love with her. The last eight minutes of footage have been recreated by using stills and the original title cards, to replace the last reel which had decomposed. Remade as "Rain," "Dirty Gertie From Harlem," and "Miss Sadie Thompson." Based on W. Somerset Maugham's "Rain." Oscar nominations for Swanson and George Barnes for the photography. **1928 97m/B** Gloria Swanson, Lionel Barrymore, Raoul Walsh, Blanche Frederici, Charles Lane, James Marcus; *Dir:* Raoul Walsh. **VHS, LV $29.95** KIV, MOV ♂♂♂

The Sadist Three teachers on their way to Dodger Stadium find themselves stranded at a roadside garage and terrorized by a snivelling lunatic. Tense and plausible. Cinematography by Vilmos Zsigmond. Alternately titled "The Profile of Terror." **1963 95m/B** Arch Hall Jr., Helen Hovey, Richard Alden, Marilyn Manning; *W/Dir:* James Landis. **Beta $29.95** RHI ♂♂½

Safari 3000 A "Playboy" writer is assigned to do a story on a three-day, 3,000 kilometer car race in Africa. Doesn't take itself too seriously, and neither should we. Dumb, dull, and disjointed. **1982 (PG) 91m/C** Stockard Channing, David Carradine, Christopher Lee; *Dir:* Harry Hurwitz. **VHS, Beta $59.95** MGM ♂½

Saga of Death Valley Rogers battles a band of outlaws and discovers that their leader is his own brother. Exciting and likeable early Rogers western. **1939 56m/B** Roy Rogers, George "Gabby" Hayes, Donald (Don "Red") Barry, Doris Day; *Dir:* Joseph Kane. **VHS, Beta $19.95** NOS, VDM, VCN ♂♂½

The Saga of the Draculas An aging vampire wishes to continue his bloodline and seeks to convert his niece's baby to his bloodsucking ways. Also on video as "Dracula Saga" and "Dracula: The Bloodline Continues..." Originally titled "The Saga of Dracula." **1972 (R) 90m/C** *SP* Narciso Ibanez Menta, Tina Sainz, Tony Isbert, Maria Koski, Cristina Suriani, Helga Line; *Dir:* Leon Klimovsky. **VHS $19.98** SNC, ASE, TSV ♂

Saga of the Vagabond Arranged by master director Akira Kurosawa. A band of ruthless outlaws careen across a civil war-torn countryside. With English subtitles. **1959 115m/C** *JP* Toshiro Mifune, Michiyo Aratama; *Dir:* Toshio Sugie. **VHS, Beta $59.95** VDA ♂♂½

Sagebrush Trail Wayne's second-ever western, complete with all sorts of tumbleweed and white-hat cliches. Our hero is wrongly accused of killin' a man, and busts out of the hoosegow to clear his name. Interesting plot twists makes for fun, compelling viewing. **1933 53m/B** John Wayne, Yakima Canutt, Wally Wales; *Dir:* Armand Schaefer. **VHS, LV $8.95** NEG, REP, NOS ♂♂♂

Sahara A British-American unit must fight the Germans for their survival in the Libyan desert during World War II. Plenty of action and suspense combined with good performances makes this one a step above the usual war movie. Naish received a Best Supporting Actor nomination.

1943 97m/B Humphrey Bogart, Dan Duryea, Bruce Bennett, Lloyd Bridges, Rex Ingram, J. Carroll Naish, Richard Nugent; *Dir:* Zoltan Korda. **VHS, Beta $9.95** *COL* 🎯🎯🎯

Sahara A disappointing "Perils of Pauline" type adventure. Shields disguises herself as a man (sure) to complete a race across the desert in honor of her late father, and a sheik captures her.
1983 (PG) 111m/C Brooke Shields, John Rhys-Davies, Lambert Wilson, John Mills, Horst Buchholz, Perry Lang, Steve Forrest; *Dir:* Andrew V. McLaglen. **VHS, Beta, LV $79.95** *MGM Woof!*

Sahara Cross (Extrana Aventura en el Sahara) Researchers in the Sahara are ambushed by saboteurs, and the chase is on.
19?? 95m/C VHS, Beta $59.95 *JCI* 🎯

Saigon Commandos Commandos, on the eve of the fall of Saigon in the 1970s, attempt to prevent heroin dealers from toppling the already-shaky government. So-so war/action flick.
1988 (R) 84m/C Richard Young, P.J. Soles, John Allen Nelson, Jimi B Jr.; *Dir:* Clark Henderson. **VHS, Beta, LV $79.95** *MED* 🎯

Saigon: Year of the Cat A made for British television drama about an American ambassador, a CIA operative and a British bank clerk who try to leave Saigon in 1974 before the Vietcong enter the city.
1987 106m/C *GB* Frederic Forrest, E.G. Marshall, Judy Dench; *Dir:* Stephen Frears. **VHS, Beta $89.99** *HBO* 🎯

Sailing Along British musical about a young starlet who sacrifices her career for true love.
1938 80m/B *GB* Jessie Mathews, Barry Mackay, Jack Whiting, Roland Young, Noel Madison, Athene Seyler, Patrick Barr; *Dir:* Sonnie Hale. **VHS $29.95** *FCT* 🎯🎯

Sailor-Made Man/Grandma's Boy Silent comedy shorts. In "A Sailor-Made Man" (1921) Lloyd joins the Navy. As shy "Grandma's Boy" (1922), he is given courage and good advice to win the hand of a girl and defeat a bully.
192? 83m/B Harold Lloyd. **VHS, Beta** *TLF* 🎯🎯🎯

The Sailor Who Fell from Grace with the Sea Perverse tale of a disillusioned sailor who rejects the sea for the love of a lonely young widow and her troubled son. Graphic sexual scenes. Based on the novel by Yukio Mishima, the film suffers from the transition of Japanese culture to an English setting.
1976 (R) 105m/C *GB* Sarah Miles, Kris Kristofferson, Jonathan Kahn, Margo Cunningham; *Dir:* Lewis John Carlino. **VHS, Beta, LV** *FOX, SUE* 🎯🎯

The Saint Moore is Simon Templar, Leslie Charteris' mysterious British detective/hero in this adventure tale that takes him to Naples. Revival of and improvement on earlier movie series, and based on the very popular British television show.
1968 98m/C *GB* Roger Moore, Ian Hendry. **VHS $59.98** *FOX* 🎯🎯

St. Benny the Dip Three con-men evade police by posing as clergymen, and end up going straight after a series of adventures in a skid row mission. Funny in parts but extremely predictable. Still, good for a few laughs.

1951 80m/B Dick Haymes, Nina Foch, Roland Young, Lionel Stander, Freddie Bartholomew; *Dir:* Edgar G. Ulmer. **VHS $19.95** *NOS, FCT* 🎯🎯½

St. Elmo's Fire Seven Georgetown graduates confront adult problems during their first post-graduate year. Reminiscent of "The Big Chill," but a weak story wastes lots of talent and time.
1985 (R) 110m/C Martin Balsam, Emilio Estevez, Rob Lowe, Andrew McCarthy, Demi Moore, Judd Nelson, Ally Sheedy, Mare Winningham, Jenny Wright, Andie MacDowell; *Dir:* Joel Schumacher. **VHS, Beta, LV $12.95** *COL* 🎯🎯½

St. Helen's, Killer Volcano A young man and an old man develop a deep friendship amid the devastation, fear, greed and panic surrounding the eruption of the Mt. St. Helen's volcano. Based on the true story of Harry Truman (!), who refused to leave his home. Pretty good.
1982 95m/C Art Carney, David Huffman, Cassie Yates, Bill McKinney, Ron O'Neal, Albert Salmi, Cesare Danova; *Dir:* Ernie Pintoff. **VHS, Beta $69.95** *VES* 🎯🎯½

St. Ives Former police reporter Bronson agrees to recover some stolen ledgers and finds himself dealing with betrayal and murder. Bisset is sultry and the tale is slickly told, but dumb. Co-stars Travanti, later of TV's "Hill Street Blues."
1976 (PG) 94m/C Charles Bronson, Jacqueline Bisset, John Houseman, Harry Guardino, Maximilian Schell, Harris Yulin, Elisha Cook Jr., Daniel J. Travanti; *Dir:* J. Lee Thompson. **VHS, Beta $59.95** *WAR* 🎯🎯

Saint Jack The story of a small-time pimp with big dreams working the pleasure palaces of late-night Singapore. Engrossing and pleasant. Based on Paul Theroux's novel.
1979 (R) 112m/C Ben Gazzara, Denholm Elliott, Joss Ackland, George Lazenby; *Dir:* Peter Bogdanovich. **VHS, Beta $69.98** *LIV, VES* 🎯🎯🎯

Saint Joan Film of the George Bernard Shaw play, adapted by Graham Greene, about the French Maid of Orleans at her trial. Otto Preminger went on a nationwide talent hunt for his leading actress and chose the inexperienced Seberg (her screen debut). A good thing for her career, but not for this illbegotten, overambitious opus. Seberg doesn't fit. Also available colorized.
1957 110m/B Jean Seberg, Anton Walbrook, Richard Widmark, John Gielgud, Harry Andrews, Felix Aylmer, Richard Todd; *Dir:* Otto Preminger. **VHS, Beta $29.95** *HRS, QNE* 🎯🎯

Saint in London The Saint investigates a gang trying to pass counterfeit banknotes. The third in "The Saint" series.
1939 72m/B George Sanders, Sally Gray; *Dir:* John Paddy Carstairs. **VHS, Beta** *MED* 🎯🎯

The Saint in New York The Saint, Simon Templar, turns Robin Hood to help the Civic Committee clean up a gang of desperados. First of "The Saint" series.
1938 71m/B Louis Hayward, Kay Sutton, Jack Carson; *Dir:* Ben Holmes. **VHS, Beta** *MED* 🎯🎯

The Saint Strikes Back Sanders debuts as the mysterious Simon Templar in this competent series. On a trip to San Francisco, the Saint aims to clear the name of a murdered man. The second in the series.
1939 67m/B George Sanders, Wendy Barrie, Jonathan Hale, Jerome Cowan, Neil Hamilton, Barry Fitzgerald, Edward Gargan, Robert Strange; *Dir:* John Farrow. **VHS $19.95** *MOV* 🎯🎯

Saint Strikes Back: Criminal Court A suspense double feature: The Saint helps a daughter clear the name of her murdered father, a San Francisco police commissioner in "The Saint Strikes Back," and a young lawyer becomes involved in murder in "Criminal Court."
1946 37m/B George Sanders, Wendy Barrie, Barry Fitzgerald, Tom Conway, Steve Brodie; *Dir:* John Farrow. **VHS, Beta $34.95** *RKO* 🎯🎯

The Saint Takes Over More adventures with the British mystery man; this time involving racetrack gambling. An original story that wasn't based on a Leslie Charteris novel as were the others.
1940 69m/B George Sanders, Jonathan Hale, Wendy Barrie, Paul Guilfoyle, Morgan Conway, Cy Kendall; *Dir:* Jack Hively. **VHS** *MOV* 🎯🎯

The St. Valentine's Day Massacre Corman's big studio debut recreates the events leading to one of the most violent gangland shootouts in modern history: the bloodbath between Chicago's Capone and Moran gangs on February 14, 1929. Uninspired and unremittingly violent. Watch for Jack Nicholson's bit part.
1967 100m/C Jason Robards Jr., Ralph Meeker, Jean Hale, Joseph Campanella, Bruce Dern, Clint Ritchie, Richard Bakalayan, George Segal; *Dir:* Roger Corman. **VHS, Beta $19.98** *FOX, FCT* 🎯🎯½

The Saint's Double Trouble Leslie Charteris' suave detective becomes embroiled in another mystery, this one involving jewel thieves. Sanders plays a dual role as the hero and the crook. The fourth in the series.
1940 68m/B George Sanders, Helene Whitney, Jonathan Hale, Bela Lugosi, Donald MacBride, John Hamilton; *Dir:* Jack Hively. **VHS** *MED* 🎯🎯

The Saint's Vacation/The Narrow Margin A mystery double feature. In "The Saint's Vacation," Simon Templar prevents a valuable secret from getting into the wrong hands, and in "The Narrow Margin," strange things happen to a detective when he takes a train ride with a witness to Chicago. Sinclair took over the sleuth's role from George Sanders in this series.
1952 148m/B Hugh Sinclair, Sally Gray, Charles McGraw, Marie Windsor, Jacqueline White. **VHS, Beta $34.95** *RKO* 🎯🎯

Sakharov The true story of the Russian physicist who designed the H-bomb and then rose to lead the Soviet dissident movement and win the Nobel Peace Prize. Robards is excellent. Made for cable; topical and powerful.
1984 120m/C Glenda Jackson, Jason Robards Jr., Michael Bryant; *Dir:* Jack Gold. **VHS, Beta $69.95** *KUI, PSM* 🎯🎯🎯

Sakura Killers Killer Ninjas steal a valuable gene-splicing formula and are hunted down by an American colonel.
1986 87m/C Chuck Connors, George Nichols, Mike Kelly, Brian Wong, Thomas Lung; *Dir:* Richard Ward. **VHS, Beta $59.98** *FOX* 🎯

Salaam Bombay! A gritty film about a child street beggar in the slums of Bombay trying to raise enough money to return to his mother's house in the country. The boy experiences every variety of gutter life imaginable, from humiliation to love. Director debut for Mira Nairi moving and searing. Filmed on location, with actual homeless children; Nair's first feature. In Hindi with subtitles. Nominated for a Best Foreign Film Oscar.

1988 114m/C *IN* Shafiq Syed, Hansa Vithal, Chanda Sharma, Nana Patekar, Aneeta Kanwar; *Dir:* Mira Nair. Cannes Film Festival '88: Camera d'Or. **VHS, Beta, LV** $79.95 *VIR* 🎞🎞🎞

The Salamander A French detective tries to find the assassin of the leaders of a neo-fascist underground movement in Italy and follows clues that lead back to World War II. Disappointing, overwrought adaptation of a Morris West novel.
1982 (R) 101m/C *GB* Franco Nero, Anthony Quinn, Martin Balsam, Sybil Danning, Christopher Lee, Cleavon Little, Paul Smith, Claudia Cardinale, Eli Wallach; *Dir:* Peter Zinner. **VHS, Beta** $59.95 *CHA* 🎞½

Salammbo An early silent epic in the grand Italian tradition. Loosely based on Flaubert's novel about ancient Carthage.
1914 49m/B *IT* Ernesto Pagani; *Dir:* Giovanni Pastrone. **VHS, Beta** $49.95 *FCT* 🎞🎞½

Salem's Lot: The Movie Based on Stephen King's novel about a sleepy New England village which is infiltrated by evil. A mysterious antiques dealer takes up residence in a forbidding hilltop house—and it becomes apparent that a vampire is on the loose. Generally creepy; Mason is good, but Soul only takes up space in the lead as a novelist returning home.
1979 (PG) 112m/C David Soul, James Mason, Lance Kerwin, Bonnie Bedelia, Lew Ayres, Bo Fanders, Elisha Cook Jr., Reggie Nalder, Fred Willard, Kenneth McMillan, Marie Windsor; *Dir:* Tobe Hooper. **VHS, Beta** $19.98 *WAR* 🎞🎞½

Sallah A North African Jew takes his family to Israel in 1949 in hopes of making his fortune. He finds himself in a transit camp and runs up against the local bureaucracy in a quest for permanent housing as well as the European work ethic. Amusing and enjoyable satire. In Hebrew with English subtitles.
1963 105m/B *IS* Chaim Topol, Geula Noni, Gila Almagor, Arik Einstein, Shraga Friedman, Esther Greenberg; *Dir:* Ephraim Kishon. **VHS, Beta** $79.99 *MOV, ERG, NCJ* 🎞🎞🎞

Sally of the Sawdust W. C. Fields' first silent feature; he is a carnival barker who adopts a young woman. Allows him to demonstrate his talent for juggling, conning customers, and car chasing. Musical score. Interesting movie caught director Griffith on the decline and Fields on the verge of stardom. Remade as a talkie in 1936 entitled "Poppy."
1925 124m/B W.C. Fields, Carol Dempster, Alfred Lunt; *Dir:* D.W. Griffith. **VHS, Beta** $19.95 *NOS, VYY, CCB* 🎞🎞½

Salo, or the 120 Days of Sodom Extremely graphic film follows sixteen children (eight boys and eight girls) who are kidnapped by a group of men in Fascist Italy. On reaching a secluded villa in the woods, the children are told to follow strict rules and then subjected to incredible acts of sadomasochism, rape, violence, and mutilation. This last film of Pasolini's was taken from a novel by the Marquis de Sade. Viewers are strongly recommended to use their utmost discretion when watching this controversial film.
1975 117m/C *IT* Giorgio Cataldi, Umberto P. Quinavalle, Paolo Bonacelli; *Dir:* Pier Paolo Pasolini. **VHS** *WBF, WBV, FCT* 🎞🎞

Salome Garish silent rendition of Oscar Wilde's scandalous tale.
1922 54m/B Alla Nazimova, Rose Dione, Mitchell Lewis, Earl Schenck; *Dir:* Charles Bryant. **VHS, Beta** $19.95 *GPV, FCT* 🎞🎞½

Salome An over-costumed version of the Biblical story about King Herod's lascivious stepdaughter who danced her way to stardom and tried to save the life of John the Baptist. Talented cast Can't overcome hokey script.
1953 103m/C Rita Hayworth, Stewart Granger, Charles Laughton, Judith Anderson, Cedric Hardwicke, Basil Sydney, Maurice Schwartz; *Dir:* William Dieterle. **VHS, Beta** $19.95 *COL, CCB* 🎞🎞

Salome An updated and graphic version of the Oscar Wilde historical fantasy about the temptress Salome who helped topple Herod's Biblical kingdom with her dance of the Seven Veils. Photographed by Pasqualino De Santis. Not really a remake of the 1953 film since this is set during the 1940s.
1985 (R) 100m/C Jo Ciampa, Tomas Milian, Pamela Salem; *Dir:* Claude D'Anna. **VHS, Beta** $79.95 *MGM* 🎞½

Salome/Queen Elizabeth Two famous silent shorts: "Salome," (1923), an atmospheric adaptation of Oscar Wilde's tale, with sets designed after Aubrey Beardsley; and "Queen Elizabeth," (1912), Bernhardt's famous, and famously stilted, singular film appearance.
1923 91m/B Alla Nazimova, Sarah Bernhardt, Lou Tellegen, Mitchell Lewis, Nigel de Brulier; *Dir:* Charles Bryant, Henri Desfontaines, Louis Mercanton. **VHS, Beta, 8mm** $29.95 *VYY* 🎞½

Salome, Where She Danced Set during the Franco-Prussian war. A European dancer helps an American reporter in a spy scheme. She eventually relocates to America and becomes the toast of San Francisc. Somehow, De Carlo managed to spring from this ridiculous camp outing to stardom.
1945 90m/C Yvonne de Carlo, Rod Cameron, David Bruce, Walter Slezak; *Dir:* Charles Lamont. **VHS, Beta** $19.95 *NOS, KRT, DVT* 🎞

Salome's Last Dance A theatrical, set-surreal adaptation of the Wilde story. Typically flamboyant Russell.
1988 (R) 113m/C Glenda Jackson, Stratford Johns, Nickolas Grace, Douglas Hodge, Imogen Millais Scott; *Dir:* Ken Russell. **VHS, Beta, LV** $89.98 *LIV, IME, VES* 🎞🎞

Salsa An auto repairman would rather dance in this "Dirty Dancing" clone. Rosa was formerly a member of the pop group Menudo.
1988 (PG) 97m/C Robby Rosa, Rodney Harvey, Magali Alvarado, Miranda Garrison, Moon Orona; *Dir:* Boaz Davidson. **VHS, Beta, LV** $19.98 *CAN* 🎞🎞

Salt of the Earth Finally available in this country after being suppressed for thirty years, this controversial film was made by a group of blacklisted filmmakers during the McCarthy era. It was deemed anti-American, communist propaganda. The story deals with the anti-Hispanic racial strife that occurs in a New Mexico zinc mine when union workers organize a strike.
1954 94m/B Rosaura Revueltas, Will Geer, David Wolfe; *Dir:* Herbert Biberman. **VHS, Beta, LV, 3/4U** $19.95 *IHF, NOS, VYG* 🎞🎞🎞½

Salty A lovable but mischievous sea lion manages to complicate two brothers' lives when they volunteer to help a friend renovate a Florida marina, which is threatened by a mortgage foreclosure. Straightforward and inoffensive tale.
1973 93m/C Clint Howard, Mark Slade, Nina Foch; *Dir:* Ricou Browning. **VHS, Beta** $59.95 *LTG* 🎞🎞

Salut l'Artiste Two bad actors try to make it in the industry. Delightfully funny with an excellent cast, especially Mastroianni. Snide and irreverant, yet loving and comic look at the world of acting.
1974 96m/C *FR* Marcello Mastroianni, Jean Rochefort, Francoise Fabian, Carla Gravina; *Dir:* Yves Robert. **VHS** $79.95 *FCT* 🎞🎞🎞

Salute to Chuck Jones Collection of eight Chuck Jones favorites including "Duck Dodgers in the 24 1/2th Century," "One Froggy Evening," "Rabbit Seasoning," and "Feed the Kitty."
1960 56m/C *Dir:* Chuck Jones; *Voices:* Mel Blanc. **VHS, Beta** $12.95 *WAR* 🎞🎞🎞

Salute to Friz Freleng Several classic Friz Freleng cartoons, including the Academy Award winners "Speedy Gonzales" (1955), "Birds Anonymous" (1957), and "Knighty Knight Bugs" (1958).
1958 57m/C *Dir:* Friz Freleng; *Voices:* Mel Blanc. **VHS, Beta** $12.95 *WAR* 🎞🎞🎞

Salute John Citizen The life of an average English family during the early days of World War II is depicted, focusing on the deprivation and horror of the Nazi blitzkrieg.
1942 74m/B Peggy Cummins, Stanley Holloway, Dinah Sheridan, Jimmy Hanley. **VHS, Beta** $19.98 *VYY* 🎞🎞🎞

Salute to the "King of the Cowboys," Roy Rogers Glimpses of Roy from song clips, home movies, trailers, out-takes, and other sources.
197? 60m/B Roy Rogers. **VHS, Beta** $29.95 *DVT* 🎞½

Salute to Mel Blanc This compilation pays tribute to the multi-talented Mel Blanc and features such memorable cartoons including "The Rabbit of Seville," "Little Boy Boo," "Robin Hood Daffy," and "Past Performances."
1958 58m/C *Dir:* Chuck Jones, Friz Freleng, Bob Clampett; *Voices:* Mel Blanc. **VHS, Beta** $12.95 *WAR* 🎞½

Salvador Photo journalist Richard Boyle's unflinching and sordid adventures in war-torn El Salvador. Boyle (Woods) must face the realities of social injustice. Belushi and Woods are hard to like, but excellent. Early critical success for director Stone.
1986 (R) 123m/C James Woods, James Belushi, John Savage, Michael Murphy, Elpidia Carrillo, Cynthia Gibb; *Dir:* Oliver Stone. **VHS, Beta, LV** $29.98 *LIV, IME, VES* 🎞🎞½

Salvation! A mean-tempered, timely satire about a money-hungry television evangelist beset by a family of fervent, equally money-hungry followers, who proceed to cheat, seduce and blackmail him. Occasionally quite funny, but sloppy and predictable. Underground director Beth B's first above-ground film.
1987 (R) 80m/C Stephen McHattie, Exene Cervenka, Dominique Davalos, Viggo Mortensen, Rockets Redglare, Billy Bastiani; *Dir:* Beth B. **Beta** $79.95 *VHV* 🎞🎞½

The Salzburg Connection A list of Nazi collaborators is discovered in Austria. A vacationing American lawyer is caught up in the battle for its possession. Filmed on location. Poor, cheesy adaptation of Helen MacInnes novel.
1972 (PG) 94m/C Barry Newman, Anna Karina, Klaus Maria Brandauer, Karen Jensen, Lee H. Katzin; *Dir:* Lee H. Katzin. **VHS, Beta** $59.98 *FOX* 🎞½

Sam Kinison Live! The often controversial, always LOUD comic performs his stand-up act at the Roxy Theatre in L.A. **1988 60m/C** Sam Kinison. **VHS, Beta $14.99** *HBO* ✗½

Samar A liberal penal colony commandant rebels against his superiors by leading his prisoners through the Philippine jungles to freedom. Original action drama with an interesting premise. **1962 (PG) 89m/C** George Montgomery, Gilbert Roland, Joan O'Brien, Ziva Rodann; *Dir:* George Montgomery. **VHS, Beta $19.95** *NOS, MON* ✗✗½

Samaritan: The Mitch Snyder Story Effective television drama based upon the true story of Vietnam vet Mitch Snyder (Sheen) who battles various government agencies and ultimately fasts to call national attention to his crusade against homelessness. Tyson appears as bag lady; Synder eventually committed suicide. **1986 90m/C** Martin Sheen, Roxanne Hart, Joe Seneca, Stan Shaw, Cicely Tyson; *Dir:* Richard T. Heffron. **VHS, Beta $14.95** *FRH* ✗✗✗

Same Time, Next Year A chance meeting between an accountant and a housewife results in a sometimes tragic, always sentimental 25-year affair in which they meet only one weekend each year. Well-cast leads carry warm, touching story based on the Broadway play by Bernard Slade. Oscar nominations for Burstyn, Slade, Robert Surtees for photography, and the song "The Last Time I Felt Like This." **1978 119m/C** Ellen Burstyn, Alan Alda; *Dir:* Robert Mulligan. **VHS, Beta, LV $79.95** *MCA* ✗✗✗

Sammy A small boy, crippled since birth, helps his family find faith through his gentle spirit. **1988 68m/C** Eric Buhr, Peter Hedges, Tom McDonald, Bill Cort, Carol Locatell; *Dir:* Russell S. Doughten Jr. **VHS, Beta** *BPG, MIV* ✗

Sammy Bluejay Features two escapades—"Brainy Bluejay" and "Sammy's Revenge," starring Sammy Bluejay, Peter Cottontail, and Reddy the Fox. **1983 60m/C VHS, Beta $29.95** *FHE* ✗

Sammy & Rosie Get Laid An unusual social satire about a sexually liberated couple living in London whose lives are thrown into turmoil by the arrival of the man's father - a controversial politician in India. Provides a confusing look at sexual and class collisions. From the makers of "My Beautiful Launderette." **1987 97m/C** *GB* Shashi Kapoor, Frances Barber, Claire Bloom, Ayub Khan Din, Roland Gift, Wendy Gazelle, Meera Syal; *Dir:* Stephen Frears. **VHS, Beta, LV $79.95** *LHV, WAR* ✗✗✗

Sammy, the Way-Out Seal Two young boys bring a mischievous seal to live in their beach house and try to keep it a secret from their parents. Disney in its early '60s phase. **1962 89m/C** Michael McGreevey, Billy Mumy, Patricia Barry, Robert Culp; *Dir:* Norman Tokar. **VHS, Beta $69.95** *DIS* ✗✗

Sam's Son Sentimental, autobiographical tale about how a young man's athletic prowess opens the door for an acting career in Hollywood. Good performance by Wallach as the boy's father. Self-indulgent, though entertaining. Written and directed by Landon, who also appears briefly. **1984 (PG) 107m/C** Michael Landon, Eli Wallach, Anne Jackson, Timothy Patrick Murphy, James Karen; *Dir:* Michael Landon. **VHS, Beta $59.95** *WOV* ✗✗

Samson Samson attempts to keep wits and hair about him. **1961 90m/C** *IT* Brad Harris, Brigitte Corey, Alan Steel, Serge Gainsbourg, Mara Berni; *Dir:* Gianfranco Parolini. **VHS $14.99** *SNC, MRV* ✗

Samson Against the Sheik Samson survives long enough to battle a sheik in the Middle Ages. **1962 ?m/C** *IT* Ed Fury. **VHS, Beta $16.95** *SNC* ✗

Samson and Delilah The biblical story of the vindictive Delilah, who after being rejected by the mighty Samson, robbed him of his strength by shearing his curls. Delivered in signature DeMille style. Wonderfully fun and engrossing. Mature is excellent. **1949 128m/C** Victor Mature, Hedy Lamarr, Angela Lansbury, George Sanders, Henry Wilcoxon, Olive Deering, Fay Holden; *Dir:* Cecil B. DeMille. Academy Awards '50: Best Art Direction/Set Decoration (Color), Best Costume Design (Color). **VHS, Beta, LV $29.95** *PAR, PIA* ✗✗✗

Samson and Delilah A made-for-television version of the Biblical romance, semi-based on DeMille's 1949 version. Original Samson (Mature) plays Samson's father. Too long but inoffensive; see the DeMille version instead. **1984 95m/C** Anthony Hamilton, Belinda Bauer, Max von Sydow, Jose Ferrer, Victor Mature, Maria Schell; *Dir:* Lee Philips. **VHS, Beta $9.99** *STE, PSM* ✗✗

Samson and His Mighty Challenge In this rarely seen muscle epic, Hercules, Maciste, Samson and Ursus all take part in a battle royale. **1964 ?m/C** *IT* Alan Steel. **VHS, Beta $16.95** *SNC* ✗

Samson vs. the Vampire Women Santo (here called Samson) the masked hero and athlete, battles the forces of darkness as a horde of female vampires attempt to make an unsuspecting girl their next queen. **1961 89m/B** *MX* Santo, Lorena Velasquez, Jaime Fernandez, Maria Duval; *Dir:* Alfonso Corona Blake. **VHS, Beta $19.98** *SNC, HHT* ✗✗½

Samson in the Wax Museum Masked Mexican wrestling hero Santo (here called Samson) does battle with a scientist who has discovered a way to make wax monsters come to life. **1963 92m/B** *MX* Santo, Claudio Brook, Rueben Rojo, Norma Mora, Roxana Bellini; *Dir:* Alfonso Corona Blake. **VHS $19.98** *SNC, HHT* ✗✗

The Samurai An unstoppable ninja warrior seeks revenge against a sadistic clan of martial arts experts for the murder of his fiance's father. He will settle for nothing less than the decapitation of those who have destroyed his life. **19?? 90m/C** Chung Wong, Tien Pang. **VHS $49.95** *OCE* ✗

Samurai 1: Musashi Miyamoto The first installment in the film version of Musashi Miyamoto's life, as he leaves his 17th century village as a warrior in a local civil war only to return beaten and disillusioned. Justly award-winning. Subtitled. **1955 92m/C** *JP* Toshiro Mifune; *Dir:* Hiroshi Inagaki. Academy Awards '55: Best Foreign Language Film. **VHS, Beta $29.98** *EMB, SUE* ✗✗✗½

Samurai 2: Duel at Ichijoji Temple Inagaki's second film depicting the life of Musashi Miyamoto, the 17th century warrior, who wandered the disheveled landscape of feudal Japan looking for glory and love. Subtitled. **1955 102m/C** *JP* Toshiro Mifune; *Dir:* Hiroshi Inagaki. **VHS, Beta $29.98** *EMB, SUE* ✗✗✗½

Samurai 3: Duel at Ganryu Island The final film of Inagaki's trilogy, in which Musashi Miyamoto confronts his life-long enemy in a climactic battle. Depicts Miyamoto's spiritual awakening and realization that love and hatred exist in all of us. Subtitled. **1956 102m/C** *JP* Toshiro Mifune, Koji Tsurata; *Dir:* Hiroshi Inagaki. **VHS, Beta $29.98** *SUE* ✗✗✗½

Samurai Blood, Samurai Guts The wooden hag offers The Book of 9 Yin to the man who will retrieve the Devil Knight's head. Yet the man who would do so will risk his very blood, guts, and brains. **19?? 90m/C** Kong Pun, Ng Pun, Lee Shue. **VHS, Beta $29.95** *OCE* ✗

Samurai Death Bells Gory Samurai sword fighting flick full of conspiracy and deceit. **19?? 90m/C** Chung Wong, Lee Le Le, Miao Ko Sou. **VHS, Beta $49.95** *OCE* ✗

Samurai Reincarnation After the Shogunate government kills 18,000 Christian rioters in the revolt of 1638, and publicly beheads the leader Shiro Amakusa, Shiro reincarnates during a monstrous thunderstorm. Consumed with hatred, he discards the teachings of Jesus Christ and seeks revenge. In Japanese with English subtitles. **1981 122m/C** *JP Dir:* Kinji Fukasaku. **VHS $59.95** *FCT* ✗

Samurai Sword of Justice A pair of gruelling Samurai Warriors seek and destroy their evil nemesis Chi. When they finally meet up with him, you can bet some guts and brains will be splattered about. **19?? 90m/C** Kam Hon, Wong Kuan Hung, Mo Si Shing, Lee Shue. **VHS, Beta $49.95** *OCE* ✗

San Antonio A bad girl working in a dance hall turns over a new leaf on meeting the good guy. Trite plot, but good production. **1945 105m/C** Errol Flynn, Alexis Smith, S.Z. Sakall, Victor Francen, Florence Bates, John Litel, Paul Kelly; *Dir:* David Butler. **VHS, Beta $19.98** *MGM, FCT, MLB* ✗✗

San Demetrio, London Set in 1940, the crew of a crippled tanker endures harrowing situations to reach port. A decent anti-war movie in its time, it's now showing its age. Based on a true story, the plot sags with unbelievable scenes and holes on the plot side. **1947 76m/B** Walter Fitzgerald, Mervyn Johns, Ralph Michael, Robert Beatty, Charles Victor, Frederick Piper; *Dir:* Charles Frend. **VHS $19.95** *NOS* ✗✗½

San Fernando Valley Also called "San Fernando". Another lawman-in-evil-town epic. Rogers received his first on-screen kiss in 1944's biggest western. Fairly typical, pleasant cowboy tale. **1944 54m/B** Roy Rogers, Dale Evans. **VHS, Beta $19.95** *NOS, DVT* ✗

San Francisco The San Francisco Earthquake of 1906 serves as the background for a romance between an opera singer and a Barbary Coast saloon owner.

Somewhat overdone but gripping tale of passion and adventure in the West. Wonderful special effects. Finale consists of historic earthquake footage. Script by Anita Loos. Also available colorized. Oscar nominated for Best Picture, Best Director, Best Actor (Tracy), Best Original Story, and Best Assistant Director.
1936 116m/B Jeanette MacDonald, Clark Gable, Spencer Tracy, Jack Holt, Jessie Ralph, Al Shear; ***Dir:*** W.S. Van Dyke. Academy Awards '36: Best Sound. **VHS, Beta, LV $19.98** *MGM* 🦴🦴🦴½

San Francisco de Asis A Mexican biography of St. Francis of Assisi, a wealthy young man who gave up his riches and became men and animals' protector.
19?? 118m/C Jose Luis Jimenez, Lucia de Phillips, Elena Dorgaz. **VHS, Beta $42.95** *MAD* 🦴½

Sanctuary of Fear A priest in New York City helps an actress subjected to a series of terrorizing incidents. Uninspired would-be TV pilot based on Chesterton's "Father Brown" series. Made for television; also called "Girl in the Park."
1979 98m/C George Hearn, Barnard Hughes, Kay Lenz; ***Dir:*** John Llewellyn Moxey. **VHS, Beta $29.98** *FOX* 🦴½

Sand and Blood A young matador who wishes to escape poverty and a young, cultivated doctor, who despises bullfighting, meet and eventually become friends. A woman enters the picture and completes the love triangle. Widely praised by critics. Interesting and moving.
1987 101m/C *FR* Sami Frey, Andre Dussollier, Clementine Celarie, Patrick Catalifo, Maria Casares, Catherine Rouvel; ***Dir:*** Jeanne Labrune. **VHS, Beta $29.95** *NYF, FCT* 🦴🦴½

The Sand Pebbles An American expatriate engineer, transferred to a gunboat on the Yangtze River in 1926, falls in love with a missionary teacher. As he becomes aware of the political climate of American imperialism, he finds himself at odds with his command structure; the treatment of this issue can be seen as commentary on the situation in Vietnam at the time of the film's release. Considered one of McQueen's best performances, blending action and romance. Received Oscar nominations for Best Picture, Best Supporting Actor, and Best Musical Score, Sound, and Film Editing.
1966 193m/C Steve McQueen, Richard Crenna, Richard Attenborough, Candice Bergen, Marayat Andriane, Gavin MacLeod, Larry Gates, Mako, Simon Oakland; ***Dir:*** Robert Wise. **VHS, Beta, LV $29.98** *FOX, FCT* 🦴🦴🦴½

Sandakan No. 8 A Japanese woman working as a journalist befriends an old woman who was sold into prostitution in Borneo in the early 1900s. Justly acclaimed feminist story dramatizes the role of women in Japanese society. Japanese with English subtitles. Also titled "Brothel 8" and "Sandakan House #8."
1974 121m/C *JP* Kinuyo Tanaka, Yoko Takakashi, Komake Kurihara, Eitaro Ozawa; ***Dir:*** Kei Kumai. Berlin Film Festival '75: Best Actress. **VHS, Beta $59.95** *FCT* 🦴🦴🦴

Sanders of the River A British officer in colonial Africa must work with the local chief to quell a rebellion. Tale of imperialism. Dated but still interesting and of value. Robeson is very good; superb location cinematography.
1935 80m/B *GB* Paul Robeson, Leslie Banks, Robert Cochran; ***Dir:*** Zoltan Korda. **VHS, Beta $19.95** *NOS, SUE, VYY* 🦴🦴🦴

The Sandpiper Free-spirited artist Taylor falls in love with Burton, the married headmaster of her son's boarding school. Best Song Oscar for title tune. Muddled melodrama offers little besides starpower. Filmed at Big Sur, California.
1965 117m/C Elizabeth Taylor, Richard Burton, Charles Bronson, Eva Marie Saint, Robert Webber, Morgan Mason; ***Dir:*** Vincente Minnelli. Academy Awards '65: Best Song ("The Shadow of Your Smile"); Harvard Lampoon Awards '65: Worst Film of the Year. **VHS, Beta $19.98** *MGM* 🦴🦴

Sands of Iwo Jima Wayne earned his first Oscar as a tough Marine sergeant, in one of his best roles. He trains a squad of rebellious recruits in New Zealand in 1943. Later they are responsible for the capture of Iwo Jima from the Japanese - one of the most difficult campaigns of the Pacific Theater. Includes striking real war footage. Oscars also for story, sound recording, and film editing.
1949 109m/B John Wayne, Forrest Tucker, John Agar, Richard Jaeckel; ***Dir:*** Allan Dwan. **VHS, LV $19.98** *REP, TLF, BUR* 🦴🦴🦴½

Sandstone A pair of film-makers visit a California hedonist retreat and study the effects of nudity and free sexual expression on the modern psyche. Pretty lame and prurient.
1974 (R) 76m/C ***Dir:*** Jonathan, Bonny Peters. **VHS, Beta $69.98** *VES* 🦴½

Sanjuro In this offbeat, satiric sequel to "Yojimbo," a talented but lazy samurai comes to the aid of a group of naive young warriors. The conventional ideas of good and evil are quickly tossed aside; much less earnest than other Kurosawa Samurai outings. In Japanese with English subtitles.
1962 96m/B *JP* Toshiro Mifune, Tatsuya Nakadai; ***Dir:*** Akira Kurosawa. **VHS, Beta, LV $29.95** *NOS, SUE, HHT* 🦴🦴🦴

Sanshiro Sugata Kurosawa's first film. A young man learns discipline in martial arts from a patient master. Climactic fight scene is early signature Kurosawa. In Japanese with English subtitles.
1943 82m/B *JP* Susumo Fusuita, Takashi Shimura; ***Dir:*** Akira Kurosawa. **VHS $59.95** *SVS, FCT* 🦴🦴🦴

Sansho the Bailiff A world masterpiece by Mizoguchi about feudal society in 11th century Japan. A woman and her children are sold into prostitution and slavery. As an adult, the son seeks to right the ills of his society. Powerful and tragic, and often more highly esteemed than "Ugetsu." In Japanese with English subtitles. Also known as "The Bailiff."
1954 132m/B *JP* Kinuyo Tanaka, Yoshiaki Hanayagi, Kyoko Kagawa, Eitaro Shindo, Ichiro Sugai; ***Dir:*** Kenji Mizoguchi. Venice Film Festival '54: Silver Lion. **VHS, Beta $39.95** *SVS, TAM* 🦴🦴🦴🦴

Santa A beautiful woman falls victim to corruption and narcotics. In Spanish.
19?? 90m/C Julissa. **VHS, Beta $59.95** *UNI Woof!*

Santa Claus Santa Claus teams up with Merlin the Magician to fend off an evil spirit who would ruin everyone's Christmas. A camp classic.
1959 94m/C *MX* Jose Elias Moreno, Cesareo Quezadas; ***Dir:*** Rene Cardona Sr.; ***Nar:*** Ken Smith. **VHS, Beta $16.95** *SNC, MLB* 🦴

Santa Claus A kiddie fantasy about Saint Nick and his Arctic Playworld.
1985 90m/C VHS, Beta, LV $49.95 *UHV* 🦴🦴

Santa Claus is Coming to Town Kris Kringle takes you along on his quest to give toys to the citizens of Sombertown. An animated classic.
1970 53m/C ***Dir:*** Arthur Rankin Jr., Jules Bass; ***Voices:*** Mickey Rooney, Keenan Wynn, Paul Frees; ***Nar:*** Fred Astaire. **VHS, Beta, LV $14.95** *FHE* 🦴🦴

Santa Claus Conquers the Martians A Martian spaceship comes to Earth and kidnaps Santa Claus and two children. Martian kids, it seems, are jealous that Earth tykes have Christmas. Features then-child star Pia Zadora.
1964 80m/C John Call, Pia Zadora, Leonard Hicks, Vincent Beck, Victor Stiles, Donna Conforti; ***Dir:*** Nicholas Webster. **VHS, Beta $19.95** *COL, SNC, SUE* 🦴

Santa Claus: The Movie A big-budgeted spectacle about an elf who falls prey to an evil toy maker and almost ruins Christmas and Santa Claus. Boring, 'tis-the-season fantasy-drama meant to warm our cockles.
1985 (PG) 112m/C Dudley Moore, John Lithgow, David Huddleston, Judy Cornwell, Burgess Meredith; ***Dir:*** Jeannot Szwarc. **VHS, Beta, LV $9.98** *MED* 🦴🦴

Santa Fe Bound Tyler must work to clear his name when he is falsely accused of having murdered a man.
1937 58m/B Tom Tyler, Jeanne Martel, Richard Cramer, Charles "Slim" Whitaker, Edward Cassidy, Lafe McKee, Dorothy Woods, Charles King; ***Dir:*** Harry S. Webb. **VHS $19.95** *DVT* 🦴🦴

Santa Fe Marshal The 27th Hopalong Cassidy movie, if anyone's counting. In the title job the Hopster goes undercover with a medicine show to smash a criminal gang. A litle different than usual, right up to the surprise identity of the crooked mastermind.
1940 66m/B William Boyd, Russell Hayden, Marjorie Rambeau, Bernadene Hayes, Earle Hodgins, Kenneth Harlan, Eddie Dean; ***Dir:*** Lesley Selander. **VHS $19.95** *NOS* 🦴🦴

Santa Fe Trail Historically inaccurate but entertaining tale about the pre-Civil War fight for "bloody Kansas." The action-adventure depicts future Civil War Generals J.E.B. Stuart (Flynn) and George Armstrong Custer (Reagan!) as they begin their military career (although Custer was really just a youth at this time). Good action scenes. Also available colorized.
1940 110m/B Errol Flynn, Olivia de Havilland, Ronald Reagan, Van Heflin, Raymond Massey, Alan Hale Jr.; ***Dir:*** Michael Curtiz. **VHS, Beta $9.95** *NEG, NOS, PSM* 🦴🦴½

Santa Fe Uprising Red Ryder has to save Little Beaver from kidnappers. Good first entry for Lane in the "Red Ryder" series.
1946 54m/B Allan "Rocky" Lane, Robert Blake. **VHS, Beta** *MED* 🦴🦴½

Santa Sangre A circus in Mexico City, a temple devoted to a saint without arms, and a son who faithfully dotes upon his mother. These are just a few of the things you will see in Alejandro Jodorowsky's fantastic film, "Santa Sangre." Young Fenix witnesses his father's cavorting and subsequent suicide following a spousal fight which leaves his mother armless. After years in an insane asylum, Fenix's mother calls on him to help her out in her nightclub act. Fenix acts as his mother's arms, plays the piano and carries out any wish she desires. Eventually Fenix is forced to kill off any woman who comes into

his mother's sight. Visually intoxicating film that may prove too graphic for some viewers. From the director who brought us "El Topo." Not as rigorous as other Jodorowsky outings. Also available in an NC-17 version.
1990 (R) 123m/C *IT MX* Axel Jodorowsky, Sabrina Dennison, Guy Stockwell, Blanca Guerra, Thelma Tixou, Adan Jodorowsky, Faviola Tapia, Jesus Juarez; *Dir:* Alejandro Jodorowsky. **VHS, LV $89.98** *REP, FCT* ♂♂♂

Santee A father-son relationship develops between a bounty hunter and the son of a man he killed. Good, but not great, Western.
1973 (PG) 93m/C Glenn Ford, Dana Wynter, Jay Silverheels, John Larch; *Dir:* Gary Nelson. **VHS, Beta $49.95** *PSM, HHE* ♂♂

The Saphead Keaton's first outing as the rich playboy has him playing the none-to-bright son of a Wall Street mogul. Some great moments provide a glimpse of cinematic greatness to come from the budding comedic genius. Based the play "The New Henrietta by Winchell Smith and Victor Mapes." Silent.
1921 70m/B William H. Crane, Buster Keaton, Carol Holloway, Edward Connelly, Irving Cummings; *Dir:* Herbert Blache. **VHS $59.95** *EJB* ♂♂½

Sapphire Two Scotland Yard detectives seek the killer of a beautiful black woman who was passing for white. Good mystery and topical social comment; remains interesting and engrossing. Superbly acting all around.
1959 92m/C *GB* Nigel Patrick, Yvonne Mitchell, Michael Craig, Paul Massie, Bernard Miles; *Dir:* Basil Dearden. **VHS, Beta $39.95** *IND* ♂♂♂

Saps at Sea A doctor advises Ollie to take a rest away from his job at a horn factory. He and Stan rent a boat, which they plan to keep tied to the dock, until an escaped criminal happens by and uses the boys (and their boat) for his getaway. Cramer is a good bad guy; Stan and Ollie cause too much trouble, as always.
1940 57m/B Stan Laurel, Oliver Hardy, James Finlayson, Ben Turpin, Richard Cramer; *Dir:* Gordon Douglas. **VHS, Beta, LV $12.98** *CCB, MED, CCB* ♂♂½

Sara Dane A strong young woman in the 18th century rises in the world through marriage and business acumen. Interesting but imperfect and slow.
1981 150m/C Harold Hopkins, Brenton Whittle, Barry Quin, Sean Scully; *Dir:* Rod Hardy. **VHS, Beta $79.95** *PSM* ♂♂

Sarah Bailey Portrait of the Harlan County folk artist who teaches residents of an Elderhostel to make corn-shuck flowers and dolls.
1984 28m/C VHS, Beta, 3/4U $150.00 *AAI, APL* ♂½

Sarah, Plain and Tall Close stars as a New England school teacher who travels to 1910 Kansas to care for the family of a widowed farmer who has advertised for a wife. Superior entertainment for the whole family. Adapted for television from Patricia MacLachlan's novel of the same name by MacLachlan and Carol Sobieski. Nominated for nine Emmy Awards.
1991 (G) 98m/C Glenn Close, Christopher Walken, Lexi Randall, Margaret Sophie Stein, Jon De Vries, Christopher Bell; *Dir:* Glenn Jordan. **VHS, LV $89.98** *REP, PIA* ♂♂♂

Sarah & the Squirrel A young girl learns to survive in a forest after she becomes separated from her family during a war.

1983 74m/C *Voices:* Mia Farrow. **VHS, Beta $29.98** *FOX* ♂♂

Sardinia Kidnapped A beautiful girl is trapped in a clash between two powerful families while on vacation in Sardinia, where the peasants' custom is to kidnap a member of a rival family and then exchange the victim for land. Former documentary maker Mingozzi creates a realistic and uncompromising look at culture and change. In Italian; dubbed.
1975 95m/C *IT* Charlotte Rampling, Franco Nero; *Dir:* Gianfranco Mingozzi. **VHS, Beta $24.95** *AHV* ♂♂½

Sgt. Pepper's Lonely Hearts Club Band "Rip-off" would be putting it kindly. A classic album by one of the greatest rock bands of all time deserves the respect of not having a star-studded extravaganza "filmization" made of it. Gratuitous, weird casting, bad acting give it a surreal feel. And the Bee Gees? Please. A nadir of seventies popular entertainment.
1978 (PG) 113m/C Peter Frampton, The Bee Gees, Steve Martin, Aerosmith, Earth Wind & Fire, George Burns, Donald Pleasence; *Dir:* Michael Schultz. Harvard Lampoon Awards '78: Worst Film of the Year. **VHS, Beta, LV $29.95** *MCA* ♂♂

Sartana's Here...Trade Your Pistol for a Coffin A soldier of fortune searches for a missing shipment of gold in the Old West.
1970 92m/C George Hilton, Charles Southwood, Erika Blanc, Linda Sini. **VHS, Beta $49.95** *UNI* ♂

S.A.S. San Salvador A CIA agent tries to prevent a psychotic from running amok in El Salvador.
1984 95m/C Miles O'Keefe, Dagmar Lassander, Catherine Jarrett. **VHS, Beta $69.98** *LIV, VES* ♂

Sasquatch Purported "documentary" about the mythical creature Bigfoot. Includes "real" pictures of the "actual" monster.
1976 94m/C *Dir:* Ed Ragozzini. **VHS, Beta $19.95** *UHV* Woof!

The Satanic Rites of Dracula Count Dracula is the leader of a satanic cult of prominent scientists and politicians who develop a gruesome plague virus capable of destroying the human race. Also available in an 84-minute version under the title "Count Dracula and His Vampire Bride." Preceded by "Dracula A.D. 1972" and followed by "The 7 Brothers Meet Dracula."
1973 88m/C *GB* Christopher Lee, Peter Cushing, Michael Coles, William Franklyn, Freddie Jones, Joanna Lumley, Richard Vernon, Patrick Barr, Barbara Yo Ling; *Dir:* Alan Gibson. **VHS $14.95** *LEG, MLB* ♂♂

Satanik A hideous old hag is turned into a lovely young woman by a stolen potion, but it doesn't last long and she becomes a monster.
1969 (R) 85m/C *IT SP* Magda Konopka, Julio Pena, Lugi Montini, Armando Calva, Umi Raho; *Dir:* Piero Vivarelli. **VHS $16.95** *SNC* ♂

Satan's Black Wedding The ghouls arrive at a Monterey monastery in their Sunday best to celebrate devildom's most diabolical ritual, the "Black Wedding."
1975 (R) 61m/C Greg Braddock, Ray Miles, Lisa Milano. **VHS, Beta $39.95** *WES* ♂

Satan's Blade The owner of an ancient talisman goes on a killing rampage at a remote mountain lodge.
1984 87m/C VHS, Beta $9.99 *STE, PSM* Woof!

Satan's Cheerleaders A demonic high school janitor traps a bevy of buxom cheerleaders at his Satanic altar for sacrificial purposes. The gals use all of their endowments to escape the clutches of the evil sheriff and his fat wife. Is it ever campy!
1977 (R) 92m/C John Carradine, John Ireland, Yvonne de Carlo, Kerry Sherman, Jacqulin Cole, Hilary Horan, Alisa Powell, Sherry Marks, Jack Kruschen, Sydney Chaplin; *Dir:* Greydon Clark. **VHS, Beta $24.95** *UHV* Woof!

Satan's Harvest An American detective inherits an estate in Johannesburg, only to find that the place is being used by drug smugglers. Sidesteps the political climate and shows the beauty of the land. And a good thing: the story is virtually nonexistent.
1965 88m/C George Montgomery, Tippi Hedren; *Dir:* Ferde Grofe Jr. **VHS, Beta $59.95** *MON* ♂½

Satan's Princess Satanic cult with a female leader runs amuck. Sloppy and forgettable.
1990 90m/C Robert Forster, Caren Kaye, Lydie Denier; *Dir:* Bert I. Gordon. **VHS, Beta $29.95** *PAR* ♂

Satan's Satellites Space zombies invade Earth. Ironically, one of the evil aliens is played by Leonard Nimoy. Originally released in serial form as "Zombies of the Stratosphere."
1958 70m/B Judd Holdren, John Crawford, Leonard Nimoy, Ray Boyle; *Dir:* Fred Brannon. **VHS, Beta** *AOV* ♂½

Satan's School for Girls When a young woman investigates the circumstances that caused her sister's suicide, it leads her to a satanic girl's academy. Dumb and puerile made-for-TV "horror."
1973 74m/C Pamela Franklin, Roy Thinnes, Kate Jackson, Lloyd Bochner, Jamie Smith-Jackson, Jo Van Fleet, Cheryl Ladd; *Dir:* David Lowell Rich. **VHS, Beta $19.99** *PSM, HFE* ♂½

Satan's Touch A simpleton worships the Devil and gets burned for his trouble.
1984 (R) 86m/C James Lawless, Shirley Venard. **VHS, Beta** *REG* Woof!

Satanwar A fictional look at various forms of Satan worship.
1979 95m/C Bart LaRue, Sally Schermerhorn, Jimmy Drankovitch. **VHS, Beta $59.95** *UHV* Woof!

Satisfaction An all-girl, high-school rock band play out the summer before college. The Keatons should have sent Bateman to her room for this stunt.
1988 (PG-13) 93m/C Justine Bateman, Trini Alvarado, Britta Phillips, Julia Roberts, Scott Coffey, Liam Neeson, Deborah Harry; *Dir:* Joan Freeman. **VHS, Beta, LV $19.98** *FOX* ♂

Saturday the 14th A parody of the popular axe-wielding-maniac genre, about a family inheriting a haunted mansion. Poorly made; not funny or scary. Followed by even worse sequel: "Saturday the 14th Strikes Back".
1981 (PG) 91m/C Richard Benjamin, Paula Prentiss, Severn Darden; *Dir:* Howard R. Cohen. **VHS, Beta $19.98** *SUE* ♂½

Saturday the 14th Strikes Back Pathetic sequel to the horror film spoof of the "Friday the 13th" series.
1988 91m/C Ray Walston, Avery Schreiber, Patty McCormack, Julianne McNamara, Jason Presson; *Dir:* Howard R. Cohen. **VHS, Beta $79.95** *MGM* ♂

Saturday Night Fever Travolta plays a Brooklyn teenager, bored with his daytime job, who becomes the nighttime king of the local disco. Based on a story published in "New York Magazine" by Nik Cohn. Acclaimed for its disco dance sequences, memorable soundtrack by the Bee Gees, and carefree yet bleak script; extremely dated, although it made its mark on society in its time. Followed by the sequel "Staying Alive." Oscar nomination for Travolta. Also available in a 112-minute "PG" rated version.
1977 (R) 118m/C John Travolta, Karen Gorney, Barry Miller, Donna Pescow; *Dir:* John Badham. National Board of Review Awards '77: 10 Best Films of the Year, Best Actor (Travolta). **VHS, Beta, LV** $29.95 *PAR* ✂✂½

Saturday Night Live: Carrie Fisher Carrie Fisher joins in a spoof of her Star Wars character in the sketch "New Kid on Earth," a beach party parody.
1978 67m/C Carrie Fisher, Dan Aykroyd, John Belushi, Laraine Newman, Jane Curtin, Garrett Morris, Bill Murray, Gilda Radner. **VHS, Beta** $14.98 *WAR, OM*

Saturday Night Live: Richard Pryor Richard Pryor appears in a samurai bellhop sketch, a spoof of "The Exorcist," and undergoes a peculiar personnel interview in this "SNL" episode. Other highlights include two songs by Gil Scott-Heron and a film by Albert Brooks.
1975 67m/C Richard Pryor, Gil Scott-Heron, John Belushi, Dan Aykroyd, Gilda Radner, Laraine Newman, Jane Curtin, Chevy Chase, Garrett Morris. **VHS, Beta** $14.98 *WAR, OM*

Saturday Night Live: Steve Martin, Vol. 1 Highlights from this 1978 show include the Czechoslovakian brothers, Steve Martin singing "King Tut," an appearance by the Blues Brothers and Steve and Gilda dancing frenetically to "Dancing in the Dark."
1978 110m/C Steve Martin, John Belushi, Dan Aykroyd, Laraine Newman, Gilda Radner, Bill Murray, Garrett Morris, Jane Curtin. **VHS, Beta** $14.98 *WAR, OM*

Saturday Night Live: Steve Martin, Vol. 2 One of the "wild and crazy guy's" hilarious SNL appearances.
1978 110m/C Steve Martin, Jane Curtin, Gilda Radner, John Belushi, Dan Aykroyd, Laraine Newman, Garrett Morris, Bill Murray. **VHS, Beta** $14.98 *WAR, OM*

Saturday Night Live, Vol. 1 Premiere episode of the late night classic, SNL. Another episode is also included.
197? 90m/C Dan Aykroyd, John Belushi, Jane Curtin, Garrett Morris, Chevy Chase, Gilda Radner, Laraine Newman. **VHS, Beta** *COL*

Saturday Night Serials A compilation of excerpts from four serials: "The Phantom Creeps," "The Phantom Empire," "Undersea Kingdom," and "Junior G-Men."
1940 90m/B Bela Lugosi, Gene Autry, Ray Corrigan, Lon Chaney Jr. **VHS, Beta** $29.98 *RHI* ✂✂½

Saturday Night Shockers Each volume in this series is comprised of two full-length horror flicks, featuring twisted twins, barbarous butchers and sinister snow creatures. Terror abounds.
1950 150m/B Daisy Hilton. **VHS, Beta** $49.95 *RHI* ✂✂½

Saturday Night Shockers, Vol. 2 Two horror film compilations, each with copious cartoons, film clips and trailers added. Volume 2 has "Human Gorilla" and "Manbeast" and some added clips and attractions.
1958 150m/B Tod Slaughter, Eric Portman, Richard Carlson. **VHS, Beta** $49.95 *RHI* ✂✂½

Saturday Night Shockers, Vol. 3 A double feature with "Murders in the Red Barn," and "Face at the Window," plus some added classic previews and shocking shorts.
19?? 150m/B **VHS** $39.95 *RHI* ✂✂½

Saturday Night Shockers, Vol. 4 Features "The Monsters of Piedras Blancas" and "Mesa of Lost Women."
19?? 145m/B **VHS** $39.95 *RHI* ✂✂½

Saturday Night Sleazies, Vol. 1 Two sleazy films. "College Girl Confidential" (1968) and "Suburban Confidential" (1966) are each directed by the man who helmed "Orgy of the Dead." Also includes sleazy short subjects.
1966 150m/C Marsha Jordan, Harvey Shane; *W/Dir:* A.C. Stephen. **VHS, Beta** $49.95 *RHI* ✂

Saturn 3 Two research scientists create a futuristic Garden of Eden in an isolated sector of our solar system, but love story turns to horror story when a killer robot arrives. Sporadically promising, but ultimately lame; dumb ending. For Farrah fans only.
1980 (R) 88m/C *GB* Farrah Fawcett, Kirk Douglas, Harvey Keitel, Ed Bishop; *Dir:* Stanley Donen. **VHS, Beta, LV** $19.98 *FOX* ✂½

Saul and David Beautifully filmed story of David's life with King Saul, the battle with Goliath, and the tragic end of Saul. From the "Bible" series.
197? 120m/C **VHS, Beta** $19.95 *UHV* ✂½

Sauteuse de L'ANGE (The Angel Jump Girl) A whimsical study of eating strawberries and high-diving by the innovative filmmaker Pascal Aubier. Also titled "The Angel Jump Girl."
1989 13m/C *FR Dir:* Pascal Aubier. **VHS, Beta** *FFM* ✂½

Savage! A savage foreign mess with ex-baseball star Iglehardt killing people with grenades and guns, abetted by an army full of murderous models. Also on video as "Black Valor."
1973 (R) 81m/C *MX* James Iglehardt, Lada Edmund, Carol Speed, Rossana Ortiz, Sally Jordan, Aura Aurea, Vic Diaz; *Dir:* Cirio H. Santiago. **VHS, Beta** $59.95 *CHA, SIM Woof!*

Savage Abduction Two girls visiting Los Angeles are kidnapped by a bizarre man.
1973 (R) 84m/C Tom Drake, Stephen Oliver, Sean Kenney; *Dir:* John Lawrence. **VHS, Beta** $24.95 *GHV* ✂

Savage Attraction The true story of a 16-year-old girl forced to marry an ex-Nazi, and her hellish life thereafter. Frustrating because it could have been much better. Boring and unoriginal.
1983 93m/C *AU GE* Kerry Mack, Ralph Schicha; *Dir:* Frank Shields. **VHS, Beta** $19.95 *STE, SUE* ✂½

Savage Beach A pair of well-endowed female federal agents battle assorted buccaneers on a remote Pacific isle over a rediscovered cache of gold from WWII. Exploitative, pornographic, and degrading to watch. Sequel to "Picasso Trigger."
1989 (R) 90m/C Dona Spier, Hope Marie Carlton, Bruce Penhall, Rodrigo Obregon, John Aprea, Teri Weigel, Lisa London; *Dir:* Andy Sidaris. **VHS, Beta, LV** $79.95 *COL* ✂

The Savage Bees A South American ship harbored at New Orleans during Mardi Gras unleashes killer bees into the celebratory crowds. Not bad, relatively exciting, TV thriller.
1976 90m/C Ben Johnson, Michael Parks, Paul Hecht, Horst Buchholz, Gretchen Corbett; *Dir:* Bruce Geller. **VHS, Beta** $49.95 *IVE* ✂✂

Savage Dawn In yet another desert town, yet another pair of combat-hardened vets are confronted by yet another vicious motorcycle gang. Haven't we seen this one before?
1984 (R) 102m/C George Kennedy, Karen Black, Richard Lynch, Lance Henriksen, William Forsythe; *Dir:* Simon Nuchtern. **VHS, Beta** $59.95 *MED* ✂½

Savage Drums Sabu returns to his South Seas island to help end tribal warfare there. Dated and dumb, but fun-to-watch melodrama/action.
1951 70m/B Sabu, Lita Baron; *Dir:* William Berke. **VHS, Beta, LV** *WGE* ✂½

Savage Fury Also called "Call of the Savage," this is a feature version of the popular movie serial.
1933 80m/B Noah Beery Jr., Dorothy Short. **VHS, Beta** $19.95 *NOS, DVT* ✂

The Savage Girl A hunter journeys through the darkest jungle in search of prey. Instead, he stumbles across a beautiful white jungle girl, falls in love with her and attempts to take her back with him.
1932 64m/B Rochelle Hudson, Walter Byron, Harry Myers, Ted Adams, Adolph Milar, Floyd Shackleford; *Dir:* Harry Fraser. **VHS** $16.95 *SNC, NOS, DVT* ✂

Savage Guns A gunslinger goes after his brother's murderer.
1968 85m/C **VHS, Beta** $39.95 *TWE* ✂✂½

Savage Hunger Ten plane crash survivors struggle to survive in the Baja desert.
1984 90m/C Chris Makepeace, Scott Hylands, Anne Lockhart; *Dir:* Sparky Green. **VHS, Beta** $69.95 *VES* ✂

Savage Intruder A wandering opportunist worms his way into the heart and home of a rich ex-movie star. She soon discovers that he's a sexual deviant who mutilates and murders for fun. Twisted doings with Hopkins in her final role.
1968 (R) 90m/C John Garfield, Miriam Hopkins, Gale Sondergaard, Florence Lake, Joe Besser, Minta Durfee; *Dir:* Donald Wolfe. **VHS** $59.95 *UNI* ✂

Savage Island Women's prison in the tropics sets the scene for the usual exploitative goings-on. Blair is actually in the film only for a few minutes. Chopped-up, even worse version of "Escape From Hell."
1985 (R) 74m/C *SP IT* Nicholas Beardsley, Linda Blair, Anthony Steffen, Ajita Wilson, Christina Lai, Leon Askin; *Dir:* Edward Muller. **VHS** $69.95 *FOR* ✂½

Savage Journey A wagon train of pioneers has difficulty on its westward journey. Simple story with no real appeal.
1983 (PG) 99m/C Richard Moll, Maurice Grandmaison, Faith Clift; *Dir:* Tom McGowan. **VHS, Beta** $79.95 *PSM, SIM* ✂

Savage Justice A young woman seeks revenge against leftist rebels (boo, hiss) in a Southeast Asian country who killed her parents and raped her. She hooks up with an ex-Green Beret, and they fight and love their way through the jungle. The usual nudity and violence; derivative and dumb.
1988 90m/C Julia Montgomery, Steven Memel. **Beta $39.95** NSV ♂

Savage is Loose A scientist (Scott), his wife, and their son struggle with uncontrolled primal instincts, when they spend twenty years on a deserted island. The son grows up without a mate. Pseudo-Freudian claptrap.
1974 (R) 114m/C George C. Scott, Trish Van Devere, John David Carson; **Dir:** George C. Scott. **VHS, Beta $29.95** VHE, MLB ♂

Savage Run Reynolds is Papago Indian framed and imprisoned for his brother's murder. Sprung from the pen, he heads back to the reservation to find the real killers and to avenge his brother's death. Convincing made for television drama originally titled "Run, Simon, Run." Stevens' last role before her suicide.
1970 73m/C Burt Reynolds, Inger Stevens, James Best, Rodolfa Acosta, Don Dubbins, Joyce Jameson, Barney Phillips, Eddie Little Sky; **Dir:** George McCowan. **VHS $9.95** SIM, IPI ♂♂♂

Savage Sam Intended as a sequel to "Old Yeller." Sam, the offspring of the heroic dog named Old Yeller, tracks down some children kidnapped by Indians. Fun, but not fully successful.
1963 103m/C Tommy Kirk, Kevin Corcoran, Brian Keith, Dewey Martin, Jeff York, Marta Kristen; **Dir:** Norman Tokar. **VHS, Beta $69.95** DIS ♂♂

The Savage Seven An underground classic where everyone loses. A motorcycle gang rides into an Indian town which is at the mercy of corrupt businessmen. The gang's leader becomes involved in the Indians' problems but the businessmen and the cops cause trouble for all concerned. Less than memorable. Produced by Dick Clark.
1968 96m/C Larry Bishop, Joanna Frank, Adam Roarke, Robert Walker Jr.; **Dir:** Richard Rush. **VHS $39.95** TRY ♂♂

Savage Streets Blair seeks commando-style revenge on the street gang that raped her deaf sister. Too-long rape scene betrays the exploitative intentions, though other bits are entertaining, in a trashy kind of way.
1983 (R) 93m/C Linda Blair, John Vernon, Sal Landi, Robert Dryer, Debra Blee, Linnea Quigley; **Dir:** Danny Steinmann. **VHS, Beta, LV $29.98** LIV, VES ♂ ½

Savage Weekend A killer behind a ghoulish mask stalks human prey in the boonies, of couse. Astoundingly, there is one interesting role—William Sanderson's looney. Otherwise, throw this one no bones.
1980 (R) 88m/C Christopher Allport. **VHS, Beta $42.95** PGN Woof!

Savages A group of savages descend on a palatial mansion, and after living there for some time, become refined ladies and gentlemen. The moral savages and "civilized" men are really the same.
1972 106m/C Lewis J. Stadlen, Anne Francine, Thayer David, Salome Jens, Susan Blakely, Kathleen Widdoes, Sam Waterston; **Dir:** James Ivory. **VHS $59.98** LTG ♂♂

Savages Andy Griffith as a demented nutcase stalking another man in the desert? That's right, Opie. "The Most Dangerous Game" remade-sort of-for TV.
1975 74m/C Andy Griffith, Sam Bottoms, Noah Beery Jr.; **Dir:** Lee H. Katzin. **VHS, Beta $49.95** PSM ♂♂ ½

Savannah Smiles Poor little rich girl Anderson runs away from home, into the clutches of two ham-handed crooks. She melts their hearts, and they change their ways. Decent, sentimental family drama.
1982 (PG) 104m/C Bridgette Andersen, Mark Miller, Donovan Scott, Michael Parks; **Dir:** Pierre DeMoro. **VHS, Beta, LV $14.98** SUE ♂♂ ½

Save the Lady Four children save an old ferry from the scrap heap.
1982 76m/C **VHS, Beta $14.98** VID Woof!

Save the Tiger A basically honest middle-aged man sees no way out of his failing business except arson. The insurance settlement will let him pay off his creditors, and save face. David, as the arsonist, and Gilford, as Lemmon's business partner, are superb. Lemmon is also excellent throughout his Oscar-winning performance. Gilford was Oscar-nominated, as was the Original Screenplay.
1973 (R) 100m/C Jack Lemmon, Jack Gilford, Laurie Heineman, Patricia Smith, Norman Burton, Thayer David; **Dir:** John G. Avildsen. Academy Awards '73: Best Actor (Lemmon). **VHS, Beta, LV $69.95** PAR, BTV ♂♂♂

Saving Grace A tale of the fictional youngest Pope of modern times, Pope Leo XIV (Conti). Beset by duty, he decides to shed his robes and get in touch with the world's peasantry. Improbable, but that's okay; the mortal sin is that it's improbable and slow and boring.
1986 (PG) 112m/C Tom Conti, Fernando Rey, Giancarlo Giannini, Erland Josephson, Donald Hewlett, Edward James Olmos; **Dir:** Robert M. Young. **VHS, Beta, LV, 8mm $14.98** SUE ♂♂

Sawdust & Tinsel Early Bergman film detailing the grisly, humiliating experiences of a traveling circus rolling across the barren Swedish countryside. Lonely parable of human relationships. Also entitled "The Naked Night." English subtitles.
1953 85m/B SW Harriet Andersson, Ake Gronberg; **Dir:** Ingmar Bergman. **VHS, Beta, LV $29.98** SUE ♂♂♂

Sawmill/The Dome Doctor Two silent comedies starring the inimitable klutz of the golden era, Larry Semon, whose name has been nearly forgotten today.
1922 71m/B Larry Semon, Dorothy Dwan. **VHS, Beta $29.95** VYY ♂♂ ½

Say Amen, Somebody A documentary look at the joyful world of gospel music. Two old-timers, Thomas A. Dorsey, the "Father of Gospel Music" and Sallie Martin, also known as Mother Smith, talk and sing about the gospel heritage as they experienced it.
1980 100m/C Willie May Ford Smith, Thomas A. Dorsey, Sallie Martin, Delois Barrett Cambell; **Dir:** George T. Nierenberg. **VHS, Beta, LV $14.95** MVD, PAV, VTK ♂♂♂

Say Anything A semi-mature, successful teen romance about an offbeat loner interested in the martial arts (Cusack, in a winning performance) who goes after the beautiful class brain of his high school. Things are complicated when her father is suspected of embezzling from the IRS. Joan Cusack, John's real-life sister, also plays his sister in the film.

Works well on the romantic level without getting too sticky.
1989 (PG-13) 100m/C John Cusack, Ione Skye, John Mahoney, Joan Cusack, Lili Taylor, Richard Portnow; **Dir:** Cameron Crowe. **VHS, Beta, LV $19.98** FOX ♂♂♂

Say Goodbye, Maggie Cole A widow goes to work as a doctor in a Chicago street clinic and becomes emotionally involved with a dying child. Susan Hayward's final role in moving made-for-TV drama.
1972 78m/C Susan Hayward, Darren McGavin, Michael Constantine, Nichelle Nichols, Dane Clark; **Dir:** Judd Taylor. **VHS, Beta $69.95** VMK ♂♂ ½

Say Hello to Yesterday A May-December romance that takes place entirely in one day. An unhappy housewife meets an exciting young traveler while both are in London. Cast and director try hard but fail to get the point across to the audience.
1971 (PG) 91m/C GB Jean Simmons, Leonard Whiting, Evelyn Laye; **Dir:** Alvin Rackoff. **VHS, Beta $59.95** PSM ♂ ½

Say Yes! A nutty millionaire (Winters) bets $250 million that his inept nephew cannot get married within 24 hours. Three times the length of usual wacky sitcom, alas.
1986 (PG-13) 87m/C Jonathan Winters, Art Hindle, Logan Ramsey, Lissa Layng; **Dir:** Larry Yust. **VHS, Beta, LV $79.95** RCA ♂ ½

Sayonara An Army major is assigned to a Japanese airbase during the Korean conflict at the behest of his future father-in-law. Dissatisfied with his impending marriage, he finds himself drawn to a Japanese dancer and becomes involved in the affairs of his buddy who, against official policy, marries a Japanese woman. Tragedy surrounds the themes of bigotry and interracial marriage. Received ten Oscar nominations, including Best Picture, Best Director, and Best Actor, Best Adapted Screenplay, Cinematography and Film Editing, and won four. Based on the novel by James Michener.
1957 147m/C Marlon Brando, James Garner, Ricardo Montalban, Patricia Owens, Red Buttons, Miyoshi Umeki, Martha Scott; **Dir:** Joshua Logan. Academy Awards '57: Best Art Direction/Set Decoration, Best Sound, Best Supporting Actor (Buttons), Best Supporting Actress (Umeki). **VHS, Beta $19.98** FOX, BTV ♂♂♂

The Scalawag Bunch Yet another retelling of Robin Hood and his Merry Men and their adventures in Sherwood Forest.
1975 100m/C IT Mark Damon, Luis Davila, Silvia Dionisio; **Dir:** George Ferron. **VHS, Beta $59.95** RAE, HHE ♂

Scalp Merchant A private investigator named Cliff Rowan hunts for a missing payroll. Someone would rather see him dead than successful.
1977 108m/C Cameron Mitchell, John Waters, Elizabeth Alexander, Margaret Nelson; **Dir:** Howard Rubie. **VHS, Beta** PGN ♂

Scalpel After his daughter's death, a plastic surgeon creates her image on another woman to get an inheritance. Violent and graphic, with a surprise ending. Not too bad; interesting premise. Also called "False Face".
1976 (R) 95m/C Robert Lansing, Judith Chapman, Arlen Dean Snyder; **Dir:** John Grissmer. **VHS, Beta $9.95** COL, CHA ♂

Scalphunters, The Semi-successful, semi-funny western about an itinerant trapper (Lancaster) who is forced by Indians to trade his pelts for an educated black slave

(Davis); many chases and brawls ensue. Good performances.
1968 102m/C Burt Lancaster, Ossie Davis, Telly Savalas, Shelley Winters, Nick Cravat, Dabney Coleman, Paul Picerni; *Dir:* Sydney Pollack. **VHS, Beta $19.98** MGM 🎬🎬½

Scalps Hunted Indian princess and man whose family was killed by Indians form uneasy alliance in old West.
1983 (R) 90m/C Karen Wood, Albert Farley, Benny Cardosa, Charlie Bravo, Vassili Garis; *Dir:* Werner Knox. **VHS, Beta $79.95** IMP, HHE 🎬🎬

Scandal A dramatization of Britain's Profumo government sex scandal of the 1960s. Hurt plays a society doctor who enjoys introducing pretty girls to his wealthy friends. One of the girls, Christine Keeler, takes as lovers both a Russian government official and a British Cabinet Minister. The resulting scandal eventually overturned an entire political party, and led to disgrace, prison, and death for some of those concerned. Also available in an unedited 115-minute version which contains more controversial language and nudity. Top-notch performances make either version well worth watching.
1989 (R) 105m/C GB John Hurt, Joanne Whalley-Kilmer, Ian McKellan, Bridget Fonda, Jeroen Krabbe, Britt Ekland, Roland Gift; *Dir:* Michael Caton-Jones. **VHS, Beta, LV $19.95** HBO 🎬🎬🎬

Scandal A college co-ed has an affair with her professor and eventually ends up ruining his life.
198? 95m/C Glen Saxson, Melissa Long, Patricia Reed. **VHS, Beta $59.95** MGL 🎬🎬

Scandal Man A French photographer gets in over his head when he snaps the daughter of an American KKK leader embracing a black man in a nightclub. In French with subtitles.
1967 86m/C Maurice Ronet; *Dir:* Richard Balducci. **VHS, Beta $19.95** FCT 🎬🎬½

Scandal in a Small Town A cocktail waitress in a small town has to defend herself and her daughter from the town's critical eye and hateful actions. Made for television. Self-righteous made-for-TV drama rehashes several earlier plots.
1988 90m/C Raquel Welch, Christa Denton, Peter Van Norden, Ronny Cox; *Dir:* Anthony Page. **VHS, LV $9.99** STE, VMK 🎬🎬

Scandalous A bumbling American television reporter becomes involved with a gang of British con artists. Starts well but peters out. Look for the scene where Gielgud attends a Bow Wow Wow concert.
1984 (PG) 93m/C Robert Hays, Pamela Stephenson, John Gielgud, Jim Dale, M. Emmet Walsh; *Dir:* Rob Cohen. **VHS, Beta, LV $29.98** LIV, VES 🎬½

Scandalous A duke and his uncle work together to find a secretary who disappeared after witnessing a murder.
1988 90m/C Albert Fortell, Lauren Hutton, Ursula Carven, Capucine; *Dir:* Robert W. Young. **VHS, Beta $9.99** STE, PSM 🎬

Scandalous John Comedy western about a last cattle drive devised by an aging cowboy (Keith) in order to save his ranch. Keith's shrewd acting as the ornery cattle man who won't sell to developers carries this one.
1971 (G) 113m/C Brian Keith, Alfonso Arau, Michele Carey, Rick Lenz; *Dir:* Robert Butler. **VHS, Beta $69.95** DIS 🎬🎬

Scanners "Scanners" are telepaths who can will people to explode. One scanner in particular harbors Hitlerian aspirations for his band of psychic gangsters. Gruesome but effective special effects.
1981 (R) 102m/C CA Stephen Lack, Jennifer O'Neill, Patrick McGoohan, Lawrence Dane; *Dir:* David Cronenberg. **VHS, Beta, LV, SVS $14.95** COL, SUE, SUP 🎬🎬½

Scanners 2: The New Order This not-bad sequel to the 1981 cult classic finds a power-mad police official out to build a psychic militia of captured Scanners. More story than last time, plus more and better special effects.
1991 (R) 104m/C CA David Hewlett, Deborah Raffin, Yvan Ponton, Isabelle Mejias, Valentin Trujillo, Tom Butler, Vlasta Vrana, Dorothee Berryman; *Dir:* Christopher Duguay. **VHS, Beta, LV $92.98** MED, IME 🎬🎬½

Scanners 3: The Takeover The third sequel in which Scanner Siblings battle for control of the world.
1992 (R) 101m/C VHS, LV $92.98 REP 🎬½

The Scar A cunning criminal robs the mob, then hides by "stealing" the identity of a lookalike psychologist. But he's overlooked one thing...or two. Farfetched film noir showcases a rare villainous role for Henreid (who also produced). Based on a novel by Murray Forbes, also known as "Hollow Triumph."
1948 83m/C Paul Henreid, Joan Bennett, Eduard Franz, Leslie Brooks, John Qualen, Mabel Paige, Herbert Rudley; *Dir:* Steve Sekely. **VHS $19.98** REP, FCT 🎬🎬½

Scaramouche Thrilling swashbuckler about a nobleman (Granger, very well cast) searching for his family during the French Revolution. To avenge the death of a friend, he joins a theater troupe where he learns swordplay and becomes the character "Scaramouche." Features a rousing six-and-a-half-minute sword battle.
1952 111m/C Stewart Granger, Eleanor Parker, Janet Leigh, Mel Ferrer, Henry Wilcoxon, Nina Foch, Richard Anderson, Robert Coote, Lewis Stone, Elisabeth Risdon, Howard Freeman; *Dir:* George Sidney. **VHS, LV $19.98** EMB, VYG, FCT 🎬🎬🎬½

Scarecrow Two homeless drifters (Hackman and Pacino) walk across America, heading toward a car wash business they never reach. An oft-neglected example of the early 70s extra-realistic subgenre initiated by "Midnight Cowboy." Engrossing until the end, when it falls flat. Filmed on location in Detroit.
1973 (R) 112m/C Gene Hackman, Al Pacino, Ann Wedgeworth, Eileen Brennan, Richard Lynch; *Dir:* Jerry Schatzberg. Cannes Film Festival '73: Best Film. **VHS, Beta $19.98** WAR 🎬🎬½

Scared to Death A woman dies of fright when shown the death mask of the man she framed. Clues to her devise are revealed along the way. Lugosi's only color film, but probably still not worth seeing.
1947 70m/C Bela Lugosi, George Zucco, Douglas Fowley, Nat Pendleton, Joyce Compton; *Dir:* Walt Mattox. **VHS, Beta $19.95** NOS, SNC, MWP 🎬½

Scared to Death A scientific experiment goes awry as a mutation begins killing off the residents of Los Angeles.
1980 (R) 93m/C John Stinson, Diana Davidson, David Moses, Kermit Eller; *Dir:* William Malone. **VHS, Beta $49.95** MED 🎬

Scared to Death/Dick Tracy's Dilemma On one tape, two 40's programmers-Lugosi's only color film, about a psychotic magician who tortures a young girl, and a standard black-and-white Tracy entry.
1947 125m/C Bela Lugosi, George Zucco, Ralph Byrd, Kay Christopher, Ian Keith. **VHS, Beta** AOV 🎬🎬

Scared Stiff Fleeing a murder charge, Martin and Lewis find gangsters and ghosts on a Caribbean island. Funny and scary, a good remake of "The Ghost Breakers," with cameos by Hope and Crosby.
1953 108m/B Dean Martin, Dean Martin, Jerry Lewis, Lizabeth Scott, Carmen Miranda, Dorothy Malone; *Dir:* George Marshall. **VHS $14.98** REP, MLB 🎬🎬🎬

Scared Stiff Three people move into a mansion once owned by a sadistic slave trader. Once there, they become unwilling victims of a voodoo curse. Bad special effects and script, not redeemed by decent but unoriginal story.
1987 (R) 85m/C Andrew Stevens, Mary Page Keller; *Dir:* Richard S. Friedman. **VHS, LV $14.95** REP, PIA 🎬

Scarface The violent rise and fall of a 1930s Chicago crime boss (based on the life of notorious gangster Al Capone). Release was held back by censors due to the amount of violence and its suggestion of incest between the title character and his sister. Producer Howard Hughes recut and filmed an alternate ending, without director Howard Hawks' approval, to pacify the censors, and both versions of the film were released at the same time. Almost too violent and intense at the time. Remains brilliant and impressive. Originally titled: "Scarface: The Shame of a Nation." Remade in 1983.
1931 93m/B Paul Muni, Ann Dvorak, Karen Morley, Osgood Perkins, George Raft, Boris Karloff; *Dir:* Howard Hawks. National Board of Review Awards '32: 10 Best Films of the Year. **VHS, Beta, LV $19.95** MCA 🎬🎬🎬

Scarface Al Pacino is a Cuban refugee who becomes powerful in the drug trade until the life gets the better of him. A remake of the 1932 classic gangster film of the same name, although the first film has more plot. Extremely violent, often unpleasant, but not easily forgotten.
1983 (R) 170m/C Al Pacino, Steven Bauer, Michelle Pfeiffer, Robert Loggia, F. Murray Abraham, Mary Elizabeth Mastrantonio, Harris Yulin, Paul Shenar, Oliver Stone; *Dir:* Brian DePalma. **VHS, Beta, LV $29.95** MCA 🎬🎬🎬

Scarface Mob This is the original television pilot film for the popular series "The Untouchables," about Eliot Ness's band of good guys battling Chicago's crime lord, Al Capone. Still strong stuff. Narrated by Walter Winchell.
1962 120m/B Robert Stack, Neville Brand, Barbara Nichols; *Dir:* Phil Karlson. **VHS, Beta** PAR 🎬🎬

The Scarlet & the Black A priest clandestinely works within the shield of the Vatican's diplomatic immunity to shelter allied soldiers from the Nazis in occupied Rome. His efforts put him at odds with the Pope and target him for Gestapo assassination. Swashbuckling adventure at its second-best. Made for television. Based on the nonfiction book "The Scarlet Pimpernel of the Vatican" by J. P. Gallagher.
1983 145m/C Gregory Peck, Christopher Plummer, John Gielgud, Raf Vallone; *Dir:* Jerry London. **VHS, Beta $59.98** FOX 🎬🎬½

The Scarlet Car Embezzlers knock off bank teller to cover themselves, but teller isn't dead and has a tale to tell.
1917 50m/B Lon Chaney Sr., Franklin Farnum, Edith Johnson, Sam DeGrasse; *Dir:* Joseph DeGrasse. **VHS, Beta $16.95** SNC, GPV, FCT �277

Scarlet Claw Holmes and Watson solve the bloody murder of an old lady in the creepy Canadian village of Le Mort Rouge. The best and most authentic of the Sherlock Holmes series.
1944 74m/B Basil Rathbone, Nigel Bruce, Miles Mander, Gerald Hamer, Kay Harding; *Dir:* Roy William Neill. **VHS, Beta $19.98** FOX �277

The Scarlet Letter Unlikely comic relief provides only measure of redemption for this poorly rendered version of Hawthorne's classic novel about sin and Hester Prynne. Actor Hale is Skipper from "Gilligan's Island".
1934 69m/B Colleen Moore, Hardie Albright, Henry B. Walthall, Alan Hale Jr., Cora Sue Collins, Betty Blythe; *Dir:* Robert G. Vignola. **VHS, Beta $19.95** NOS, KRT, VYY �277

The Scarlet Letter A studied, thoughtful, international production of the Nathaniel Hawthorne classic about a woman's adultery which incites puritanical violence and hysteria in colonial America. Fine modernization by Wenders, in German with English subtitles.
1973 90m/C GE SP Senta Berger, Lou Castel, Yella Rottlaender, William Layton, Yelena Samarina, Hans-Christian Blech; *Dir:* Wim Wenders. **VHS, Beta $59.95** FCT, GLV �277

The Scarlet Pimpernel "The Scarlet Pimpernel," supposed dandy of the English court, assumes a dual identity to outwit the French Republicans and aid innocent aristocrats during the French Revolution. Classic rendering of Baroness Orczy's novel, full of exploits, 18th century costumes, intrigue, damsels, etc. Produced by Alexander Korda, who fired the initial director, Rowland Brown. Remade for television in 1982. Available in colorized version.
1934 95m/B GB Leslie Howard, Joan Gardner, Merle Oberon, Raymond Massey, Anthony Bushell; *Dir:* Harold Young. **VHS, Beta $9.95** NEG, NOS, PSM �277½

The Scarlet Pimpernel A British television remake of the classic about a British dandy who saved French aristocrats from the Reign of Terror guillotines during the Revolution. Made-for-television version is almost as good as the original 1935 film, with beautiful costumes and sets and good performances from Seymour and Andrews.
1982 142m/C GB Anthony Andrews, Jane Seymour, Ian McKellan, James Villers, Eleanor David; *Dir:* Clive Donner. **VHS, Beta, LV $69.98** LIV, VES �277

Scarlet Spear Filmed in Nairobi National Park in Kenya. A great white hunter convinces a young African chief that the ancient practices of his people are uncivilized. Pathetically stated and culturally offensive by today's standards.
1954 78m/C John Bentley, Martha Hyer, Morasi; *Dir:* George Breakston, Roy Stahl. **VHS, Beta $39.95** MSP Woof!

Scarlet Street A mild-mannered, middle-aged cashier becomes an embezzler when he gets involved with a predatory, manipulating woman. Darle Lang remake of Jean Renoir's "La Chienne" (1931). Set apart from later attempts on the same theme by excellent direction by Lang and acting. Also available Colorized.
1945 95m/B Edward G. Robinson, Joan Bennett, Dan Duryea, Samuel S. Hinds; *Dir:* Fritz Lang. **VHS, Beta $9.95** NEG, NOS, HHT �277

Scarred An unwed teenage mother becomes a prostitute to support her baby.
1983 85m/C Jennifer Mayo, Jackie Berryman, David Dean; *W/Dir:* Rosemarie Turko. **VHS, Beta $69.98** LIV, VES �7½

The Scars of Dracula A young couple tangles with Dracula in their search for the man's missing brother. Gory, creepy, violent, sexy tale from the dark side. Don't see it late at night. Preceded by "Taste the Blood of Dracula" and followed by "Dracula A.D. 1972."
1970 (R) 96m/C GB Christopher Lee, Jenny Hanley, Dennis Waterman, Wendy Hamilton, Patrick Troughton, Michael Gwynn, Anouska Hempel, Michael Ripper; *Dir:* Roy Ward Baker. **VHS, Beta $59.99** HBO, MLB �27½

Scavenger Hunt The action begins when a deceased millionaire's will states that his 15 would-be heirs must compete in a scavenger hunt, and whoever collects all the items first wins the entire fortune. Big cast of mostly B-actors who should have wandered off the set instead.
1979 (PG) 117m/C Richard Benjamin, James Coco, Ruth Buzzi, Cloris Leachman, Cleavon Little, Roddy McDowall, Scatman Crothers, Tony Randall, Robert Morley, Richard Mulligan, Dirk Benedict, Willie Aames, Vincent Price; *Dir:* Michael Schultz. **VHS, Beta $59.98** FOX �27

Scavengers A man and a woman become involved in a spy plot and wind up in all sorts of danger. Pathetic excuse for a "thriller."
1987 (PG-13) 94m/C Kenneth Gilman, Brenda Bakke, Ken Gampu, Norman Anstey, Crispin De Nys; *Dir:* Duncan Maclachlan. **VHS, Beta, LV $29.95** ACA �27

Scenario du Film Passion A classically self-analyzing video piece by Godard, in which he recreates on tape the scenario for his film "Passion," resulting in a reflexive, experimental essay on the process of image-making and conceptualizing films, especially Godard's. In French with English subtitles.
1982 54m/C *Dir:* Jean-Luc Godard, Anne-Marie Mieville, Barnard Menoud. **VHS, Beta** EAI �277

Scene of the Crime Three short mysteries, which the audience is asked to solve: "The Newlywed Murder," "Medium Is the Murder" and "Vote for Murder." Narrated by Orson Welles. Originally made for network television.
1985 74m/C Markie Post, Alan Thicke, Ben Piazza; *Dir:* Andre Techine, Walter Grauman; *Hosted:* Orson Welles. **VHS, Beta $39.95** MCA �27½

Scene of the Crime A beautiful widow, trapped in a small French town, is sexually awakened by an escaped convict hiding out near her home. Acclaimed, but sometimes distracting camera technique and slow pace undermine the film. In French with English subtitles.
1987 90m/C FR Catherine Deneuve, Danielle Darrieux, Wadeck Stanczak, Victor Lanoux, Nicolas Giraudi; *Dir:* Andre Techine. **Beta $79.95** VHV �277

Scenes from the Class Struggle in Beverly Hills Social satire about the wealthy, their servants, and their hangers-on in two Hollywood households. Two chauffeurs spend their spare time lusting after their lovely employers, and bet to see who can accomplish reality first. All kinds of erotic capers result. Usually funny and "irreverent"; occasionally misfires or takes itself too seri-

ously. Remake of Renoir's "The Rules of the Game."
1989 (R) 103m/C Jacqueline Bisset, Ray Sharkey, Mary Woronov, Robert Beltran, Ed Begley Jr., Wallace Shawn, Paul Bartel, Paul Mazursky, Arnetia Walker; *Dir:* Paul Bartel. **VHS, Beta, LV $89.95** VIR �27½

Scenes from the Goldmine A young woman joins a rock band and falls in love with its lead singer. Realistic but bland; music not memorable (performed mostly by the actors themselves).
1987 (R) 99m/C Catherine Mary Stewart, Cameron Dye, Joe Pantoliano, John Ford Coley, Steve Railsback, Timothy B. Schmit, Jewel Shepard, Alex Rocco, Lee Ving; *Dir:* Marc Rocco. **VHS, Beta, LV $19.98** CHA �7½

Scenes from a Mall Conspicuous consumers spend sixteenth wedding anniversary waltzing in mall while marriage unravels with few laughs. Surprisingly superficial comedy given the depth of talent of Allen and Midler.
1990 (R) 87m/C Woody Allen, Bette Midler; *W/Dir:* Paul Mazursky. **VHS, Beta, LV $19.99** TOU, IME �27

Scenes from a Marriage Originally produced in six one-hour episodes for Swedish television, this bold and sensitive film excruciatingly portrays painful, unpleasant, the disintegration of a marriage. Ullman is superb. Realistic and disturbing. Dubbed.
1973 (PG) 168m/C SW Liv Ullman, Erland Josephson, Bibi Andersson; *Dir:* Ingmar Bergman. National Board of Review Awards '74: 5 Best Foreign Films of the Year; New York Film Critics Awards '74: Best Actress (Ullman), Best Screenwriting; National Society of Film Critics Awards '74: Best Actress (Ullman), Best Picture, Best Screenplay (Bergman), Best Supporting Actress (Andersson). **VHS, Beta $89.95** COL �27777

Scenes from a Murder A killer (Savalas) stalks a beautiful actress (Heywood). Dull excuse for a suspense flick.
1972 91m/C IT Telly Savalas, Anne Heywood, Giorgio Piazza; *Dir:* Alberto DeMartino. **VHS, Beta $69.98** LTG �7½

Scheherazade Like Scheherazade of the Arabian Nights, a young woman trapped in her apartment by a would-be rapist must use her wits to fend off her attacker. Theatrical production written by Marisha Chamberlain.
198? 60m/C CBS New Playwright's Award '89: Best Supporting Actress; Dramatists Guild Award '89: Best Supporting Actress. **VHS, 3/4U $100.00** KTC �7½

Scherzo One of McLaren's earliest films, an abstract meditation lost in 1939 and subsequently found and reconstructed in 1984.
1939 2m/C *Dir:* Norman McLaren. **VHS, Beta** NFC �7½

Schizo Devious intentions abound as a middle-aged man is overcome by weird scenes and revelations, caused by the impending wedding of the figure skater he adores. Tries to gripping, but succeeds only in being confusing.
1977 (R) 109m/C Lynne Frederick, John Layton; *Dir:* Pete Walker. **VHS, Beta $24.95** UHV, MED �7

Schizoid An advice-to-the-lovelorn columnist (Hill) receives a series of threatening letters causing her to wonder whether a psychiatrist (Kinski) is bumping off his own patients. Difficult to follow.
1980 (R) 91m/C Klaus Kinski, Marianna Hill, Craig Wasson; *Dir:* David Paulsen. **VHS, Beta $19.98** MCA �7

Schlock Accurately titled horror parody is first directorial effort for Landis, who does double duty as a missing link who kills people and falls in love with blind girl. Look for cameo appearance by Forrest J. Ackerman; ape-makeup by Rick Baker. Also known as "The Banana Monster."
1973 (PG) 78m/C John Landis, Saul Kahan, Joseph Piantadosi, Eliza Garrett; **W/Dir:** John Landis. **VHS** $69.95 *LTG* ✂✂

Schnelles Geld A mild-mannered guy gets involved with a prostitute, and is plunged into a world of crime and drugs. In German with English subtitles.
1984 90m/C *GE* **VHS, Beta** $39.95 *VCD, GLV*

School Daze Director/writer/star Lee's second outing is a rambunctious comedy (with a message, of course) set at an African-American college in the South. Skimpy plot revolves around the college's homecoming weekend and conflict among frats and sororities and African-Americans who would lose their racial identity and others who assert it. Entertaining and thought provoking. A glimpse at Lee's "promise," fulfilled in "Do the Right Thing."
1988 (R) 114m/C Spike Lee, Larry Fishburne, Giancarlo Esposito, Tisha Campbell, Ossie Davis, Joe Seneca, Art Evans, Ellen Holly, Branford Marsalis; **Dir:** Spike Lee. **VHS, Beta, LV** $14.95 *COL* ✂✂✂

School for Scandal British television adaptation of Richard Sheridan's play of the morals and manners of 18th century England.
1965 100m/B Joan Plowright, Felix Aylmer. **VHS, Beta** $29.95 *VYY* ✂½

School for Sex A young man starts a school to teach young girls how to marry well and divest their husbands of their money. Not satirical, but funny in places, relies heavily on slapstick.
1969 81m/C Derek Aylward, Rose Alba, Hugh Latimer, Cathy Howard. **VHS, Beta** $59.95 *MON* ✂½

School Spirit A hormonally motivated college student is killed during a date. He comes back as a ghost to haunt the campus, disrupt the stuffy president's affair, and fall in love. Forgettable, lame, low-grade teen sex flick.
1985 (R) 90m/C Tom Nolan, Elizabeth Foxx, Larry Linville; **Dir:** Allan Holleb. **VHS, Beta** $79.95 *MED* ✂

Science Crazed A scientist creates a monster who could conceivably destroy the world.
19?? 90m/C Cameron Klein; **Dir:** Ron Switzer. **VHS, Beta** *IEC* ✂½

Scissors Yes, this Hitchcock imitation needs trimming. An unstable young woman contends with a rapist, devious lookalikes, birds, and a prison-like apartment, not all of which are relevant to the plot. An unsharp stab at suspense from novelist/filmmaker DeFelitta.
1991 (R) 105m/C Sharon Stone, Steve Railsback, Michelle Phillips, Ronny Cox; **W/Dir:** Frank de Felitta. **VHS, Beta** *PAR* ✂½

Scorchers Back in the bayou, Splendid (Lloyd), a newlywed who won't sleep with her husband, and her cousin Talbot (Tilly), a preacher's daughter whose husband prefers the town whore Thais (Dunaway), find their lives intertwined because of their marital and sexual problems. A real "scorcher."
1992 (R) 81m/C Faye Dunaway, Denholm Elliott, James Earl Jones, Emily Lloyd, Jennifer Tilly, Leland Crooke, James Wilder, Anthony Geary; **W/Dir:** David Beaird. **VHS, Beta, LV** $89.98 *FXV* ✂✂

Scorchy Low-budget female cop action flick about a narcotics agent (Stevens) on the trail of a drug kingpin. Blood and guts, guns, and violence. Poor casting.
1976 (R) 100m/C Connie Stevens, William Smith, Cesare Danova, Marlene Schmidt; **Dir:** Hikmet Avedis. **VHS, Beta** $29.98 *LIV, LTG* ✂

Score A married couple introduces newlywed neighbors to the world of kinky sexual experiments. The two couples try all combinations. Director Metzger was well known in the '60s for soft-core porn.
1973 (R) 89m/C Claire Wilbur, Calvin Culver, Lynn Lowry; **Dir:** Radley Metzger. **VHS, Beta** $59.95 *MAG* ✂½

Scoring A battle between an all-women's basketball team and a men's team combining fun, sex, and athletic prowess. Ever so zany.
1980 90m/C Myra Taylor, Greg Perrie, Freya Crane, Pete Maravich. **VHS, Beta** $49.95 *VHE Woof!*

The Scorpio Factor Murder and mayhem follow microchip heist.
1990 87m/C Attila Bertalan, David Nerman, Wendy Dawn Wilson. **VHS** $69.98 *MAG* ✂✂

Scorpion A karate-master and anti-terrorist expert defuses a skyjacking and infiltrates international assassination conspiracies. Ex-real life karate champ Tulleners is a ho-hum hero, and the story is warmed over.
1986 (R) 98m/C Tonny Tulleners, Allen Williams, Don Murray; **Dir:** William Reed. **VHS, Beta, LV** $79.95 *COL, HHE* ✂

Scorpion with Two Tails In an underworld of terror, people die grotesque deaths that a woman dreams of.
1982 99m/C John Saxon, Van Johnson. **VHS** $59.98 *CGH* ✂

The Scorpion Woman A female Viennese judge becomes entwined in an unusual case which partially parallels her own life. Her new young lover turns out to be bisexual. Not gripping. In German with English subtitles.
1989 101m/C *GE* Angelica Domrose, Fritz Hammel; **Dir:** Susanne Zanke. **VHS, Beta** $59.98 *CRO* ✂✂

Scotland Yard Inspector An American newspaperman (Romero) in London looks for a killer. Nothing special.
1952 73m/B *GB* Cesar Romero, Bernadette O'Farrell; **Dir:** Sam Newfield. **VHS, Beta, LV** $24.95 *WGE, MLB* ✂✂

Scott of the Antarctic Drama of doomed British expedition of 1911 struggling to be the first group to reach the South Pole. Much of the stunning location filming was shot in the Swiss Alps. Story is authentic, but oddly uninvolving, as though seen from afar.
1948 111m/C *GB* John Mills, Christopher Lee, Kenneth More, Derek Bond; **Dir:** Charles Frend. **VHS, Beta** $24.95 *NOS, DVT, WES* ✂✂½

Scoumoune The local "fixer" and heavy gets into trouble and no one comes to help. A movie that searches for honesty. Dubbed in English.
1974 87m/C *FR* Jean-Paul Belmondo, Claudia Cardinale. **VHS, Beta** *CCI* ✂✂½

Scout's Honor An orphan (Coleman) is determined to become the best Cub Scout ever when he joins a troop led by an executive who dislikes children. Harmless, enjoyable family tale.
1980 96m/C Wilfrid Hyde-White, Pat O'Brien, Joanna Moore, Meeno Peluce, Jay North, Gary Coleman, Katherine Helmond, Harry Morgan; **Dir:** Henry Levin. **VHS, Beta** $59.95 *LTG* ✂✂

Scrambled Feet An uninhibited satire of the world of show business.
1983 100m/C Madeline Kahn. **VHS, Beta** $39.95 *RKO* ✂✂½

Scream Vacationers on a raft trip down the Rio Grande are terrorized by a mysterious murderer. Hopelessly dull.
1983 (R) 86m/C Pepper Martin, Hank Warden, Alvy Moore, Woody Strode, John Ethan Wayne; **Dir:** Byron Quisenberry. **VHS, Beta** $69.98 *LIV, VES* ✂

Scream, Baby, Scream An unsuccessful artist switches from sculpting clay to carving young models' faces into hideous deformed creatures. Also known as "Nightmare House."
1969 86m/C Ross Harris, Eugenie Wingate, Chris Martell, Suzanne Stuart, Larry Swanson, Brad Grinter; **Dir:** Joseph Adler. **VHS, Beta** $49.95 *REG, CAM* ✂

Scream Blacula Scream Blacula returns from his dusty undoing in the original movie to once again suck the blood out of greater Los Angeles. A voodoo priestess (Grier) is the only person with the power to stop him. A weak follow up to the great "Blacula," but worth a look for Marshall and Grier.
1973 (R) 96m/C William Marshall, Don Mitchell, Pam Grier, Michael Conrad, Richard Lawson, Lynne Moody, Janee Michelle, Barbara Rhoades, Bernie Hamilton; **Dir:** Bob Kelljan. **VHS** $59.98 *ORI* ✂✂

Scream Bloody Murder A young boy grinds his father to death with a tractor but mangles his own hand trying to jump off. After receiving a steel claw and being released from a mental institution he continues his murderous ways in and around his home town.
1972 (R) 90m/C Fred Holbert, Leigh Mitchell, Robert Knox, Suzette Hamilton; **Dir:** Robert Emery. **VHS, Beta** $19.95 *UHV Woof!*

Scream of the Demon Lover A young biochemist has a busy day as she works for a reclusive Baron. She fantasizes about him, tries to track down a murderer, and eventually discovers a mutant in the cellar. About as bad as they come, but short!
1971 (R) 75m/C Jennifer Hartley, Jeffrey Chase; **Dir:** J.L. Merino. **VHS, Beta** $59.98 *CHA Woof!*

Scream and Die A model is harassed by a murderous thief and is forced to run for her life.
1974 (R) 98m/C Andrea Allan, Karl Lanchbury, Maggie Walker, Judy Matheson; **Dir:** Joseph Larraz. **VHS, Beta** $69.98 *LIV, LTG Woof!*

Scream Dream A beautiful rock star uses her supernatural powers to control her fans, turning them into revenge-seeking monsters when she wants to. Heavy on the blood and skin; heavily exploitative.
1989 (R) 80m/C Melissa Moore, Carole Carr, Nikki Riggins, Jesse Ray; **W/Dir:** Donald Farmer. **VHS** $19.95 *AVP* ✂

Scream of Fear A wheelchair-bound young woman goes to visit her father and new stepmother only to find her father is away on business. But she believes she sees

her father's corpse. Is someone trying to drive her mad? Abounds in plot twists and mistaken identities. A truly spooky film, suspenseful, and well-made.
1961 81m/B *GB* Susan Strasberg, Ronald Lewis, Ann Todd, Christopher Lee; *Dir:* Seth Holt. **VHS, Beta** $69.95 *COL, MLB* 🎞🎞🎞

Scream Free! A gang of thoughtless hippie drug smugglers and abusers escape from the police, drink acid-spiked wine, shoot each other, and have love-ins.
1971 82m/C Russ Tamblyn, Richard Beymer, Casey Kasem, Lana Wood, Warren Finnerty. **VHS, Beta, LV** *WGE* 🎞

Scream for Help A young girl discovers that her cheating stepfather is plotting to murder her mother, but no one will believe her. Is she paranoid? So bad it's funny; otherwise, it's just bad, and far from suspenseful.
1986 (R) 95m/C Rachael Kelly, David Brooks, Marie Masters; *Dir:* Michael Winner. **VHS, Beta, LV** $19.95 *LHV, WAR* 🎞

Scream and Scream Again Price is a sinister doctor who tries to create a super race of people devoid of emotions. Cushing is the mastermind behind the plot. Lee is the agent investigating a series of murders. Three great horror stars, a psychadelic disco, great '60s fashions; it's all here.
1970 (PG) 95m/C *GB* Vincent Price, Christopher Lee, Peter Cushing, Judy Huxtable, Alfred Marks, Anthony Newlands, Uta Levka, Judi Bloom, Yutte Stensgaard; *Dir:* Gordon Hessler. **VHS, Beta** $69.98 *LIV, VES* 🎞🎞½

Screamer A woman wants revenge after she is viciously attacked and raped. Now every man she sees is her attacker as she becomes more and more unbalanced.
1974 71m/C Pamela Franklin, Donal McCann. **VHS, Beta** $29.95 *THR* 🎞

Screamers A mad scientist on a desert island gleefully turns escaped convicts into grotesque monstrosities. Gory and gratuitous.
1980 (R) 83m/C Richard Johnson, Joseph Cotten, Barbara Bach; *Dir:* Dan T. Miller, Sergio Martino. **VHS, Beta** $19.98 *SUE* 🎞

The Screaming Dead Monsters rise from the tomb to do battle with planet Earth and each other. Originally titled "Dracula vs. Frankenstein". Not to be confused with the Al Adamson epic of the same name.
1972 84m/C *SP* Dennis Price, Howard Vernon, Alberto Dalbes, Mary Francis, Genevieve Deloir, Josianne Gibert, Fernando Bilbao; *Dir:* Jess (Jesus) Franco. **VHS** $69.95 *UHV* 🎞

The Screaming Skull Man redecorates house with skulls in attempt to drive already anxious wife insane.
1958 68m/B William Hudson, Peggy Webber, Toni Johnson, Russ Conway, Alex Nicol; *Dir:* Alex Nicol. **VHS** $19.98 *NOS, SNC* 🎞½

Screams of a Winter Night Ghostly tale of an evil monster from the lake and the terror he causes.
1979 (PG) 92m/C Matt Borel, Gil Glasco. **VHS, Beta** $54.95 *UHV* 🎞

Screamtime Two fiendish friends filch a trilogy of horror tapes for home viewing. After the show, real scary things happen.
1983 (R) 89m/C Jean Anderson, Robin Baily, Dora Bryan, David Van Day; *Dir:* Al Beresford. **VHS, Beta** $69.98 *LIV, LTG* 🎞

Screen Test Oversexed silly teenagers arrange fake screen tests in order to meet girls.

1985 (R) 84m/C Michael Allan Bloom, Robert Bundy, Paul Lueken, David Simpatico; *Dir:* Sam Auster. **VHS, Beta, LV** $79.95 *COL Woof!*

Screwball Academy A beautiful female director makes a soft-core film on a secluded island, and gets trouble from thugs, evangelists, horny crew members and others. Former "SCTV" director John Blanchard understandably used a pseudonym to direct this lame effort.
1986 (R) 90m/C Colleen Camp, Ken Welsh, Christine Cattall; *Dir:* Reuben Rose. **VHS, Beta** $69.95 *TWE* 🎞½

Screwball Hotel A beauty pageant is organized to help fund a nearly bankrupt hotel. Three bell hops are the saviors of this madcap motor inn. Exploitative skin flick.
1988 (R) 101m/C Michael C. Bendetti, Andrew Zeller, Corinne Alphen, Kelly Monteith; *Dir:* Rafal Zielinski. **VHS, Beta** $89.95 *MCA* 🎞½

Screwballs A freewheeling group of high school boys stirs up trouble for their snooty and virginal homecoming queen. Another inept teen sex comedy with no subtlety whatsoever. Sequel: "Loose Screws."
1983 (R) 80m/C Peter Keleghan, Lynda Speciale; *Dir:* Rafal Zeilinski. **VHS, Beta** $19.98 *WAR* 🎞

Scrooge On Christmas Eve a miser changes his ways after receiving visits from the ghosts of Christmas past, present, and future. Based on the classic novel "A Christmas Carol" by Charles Dickens. Screenplay co-written by Hicks, who gives an interesting performance as Ebeneezer Scrooge.
1935 61m/B *GB* Seymour Hicks, Maurice Evans, Robert Cochran, Donald Calthrop, Mary Glynne, Oscar Asche; *Dir:* Henry Edwards. **VHS, Beta** $24.95 *VYY, CCB, DVT* 🎞🎞🎞

Scrooge Well-done musical version of Charles Dickens' classic "A Christmas Carol," about a miserly old man who is faced with ghosts on Christmas Eve. Received Academy Award nominations for the music and the song "Thank You Very Much" by Leslie Bricusse. Finney is memorable in the title role.
1970 (G) 86m/C *GB* Albert Finney, Alec Guinness, Edith Evans, Kenneth More; *Dir:* Ronald Neame. **VHS, Beta, LV** $19.98 *FOX* 🎞🎞½

Scrooged Somewhat disjointed big-budgeted version of the hallowed classic. A callous television executive staging "A Christmas Carol" is himself visited by the three ghosts and sees the light. Music by Danny Elfman. Kane is terrific as one of the ghosts. Film is heavy-handed, Murray too sardonic to be believable.
1988 (PG-13) 101m/C Bill Murray, Carol Kane, John Forsythe, David Johansen, Bob Goldthwait, Karen Allen, Michael J. Pollard, Brian Doyle-Murray, Alfre Woodard, John Glover, Robert Mitchum, Buddy Hackett, Robert Goulet, Jamie Farr, Mary Lou Retton, Lee Majors; *Dir:* Richard Donner. **VHS, Beta, LV, 8mm** $89.95 *PAR* 🎞🎞

Scrubbers A young girl is sent to reform school where she's forced to survive in a cruel and brutal environment. Low-budget "reform school" movie with no point, but lots of lesbianism.
1982 (R) 93m/C *GB* Amanda York, Chrissie Cotterill, Elizabeth Edmonds, Kate Ingram, Debbie Bishop, Dana Gillespie; *Dir:* Mai Zetterling. **VHS, Beta** $69.99 *HBO* 🎞

Scruffy An orphan and his dog survive with the help of two crooks who are reformed through their love for the little boy. Animated.

1980 72m/C *Voices:* Alan Young, June Foray, Hans Conried, Nancy McKeon. **VHS, Beta** $19.95 *WOV, GKK* 🎞

Scruggs Tribute to banjo virtuoso Earl Scruggs, featuring Scruggs in performance, along with Dylan, Baez, and others.
1970 87m/C Earl Scruggs, Bob Dylan, Joan Baez, Doc Watson, The Byrds. **VHS, Beta** *FOX* 🎞

Scruples Set in the glamorous, jet-setting world of Beverly Hills, this top-rated miniseries follows the career of Billy Ikehorn (Wagner) as she weds a wealthy industrialist and opens a clothing boutique named "Scruples." As she caters to high society's haute couture, she encounters unscrupulous individuals who threaten to dethrone her from her position of power and privilege. Based on the best-selling novel by Judith Krantz.
1980 279m/C Lindsay Wagner, Barry Bostwick, Kim Cattrall, Gavin MacLeod, Connie Stevens, Efrem Zimbalist Jr., Gene Tierney. **VHS, Beta** $79.99 *WAR* 🎞

Scuba Three daring couples dive for hidden treasure in beautiful waters of the Caribbean.
1972 (G) 83m/C *Dir:* Ambrose Gaines III; *Nar:* Lloyd Bridges. **VHS, Beta** $59.95 *PSM* 🎞

Scum Adapted from Roy Minton's acclaimed play, this British production looks at the struggle among three young men in a British Borstal (a prison for young convicts.) Portrays the physical, sexual, and psychological violence committed. Horrifying and powerful.
1979 96m/C *GB* Phil Daniels, Mick Ford, Ray Winstone; *Dir:* Alan Clarke. **VHS, Beta** $59.95 *PSM* 🎞🎞🎞

The Sea Around Us Science documentary on the history and life of the ocean; based on Rachel Carson's study.
1952 61m/C *Nar:* Don Forbes. Academy Awards '52: Best Feature Documentary. **VHS, Beta** $19.95 *CCB, MED* 🎞🎞🎞

Sea Chase An odd post-war sea adventure, wherein a renegade German freighter captain is pursued by British and German navies as he leaves Australia at the outbreak of World War II. A Prussian Wayne rivaled only by his infamous Genghis Khan in "The Conqueror" for strange character selection. Lana Turner is on board as Wayne's girlfriend.
1955 117m/C John Wayne, Lana Turner, Tab Hunter, James Arness, Lyle Bettger, David Farrar, Richard Davalos, Claude Akins, John Qualen; *Dir:* John Farrow. **VHS, Beta, LV** $59.95 *WAR* 🎞🎞½

Sea Devils Sentenced to prison for a crime he didn't commit, a man escapes and joins a boatload of treasure hunters, but mutiny is afoot. Don't get soaked by this cheapie.
1931 77m/B Walter Long, Edmund Burns, Henry Otto, James Donnelly; *Dir:* Joseph Levering. **VHS** $16.95 *SNC, NOS* 🎞

Sea Devils The tale of a sea captain and his lovely daughter who have differing opinions on whom she should marry. While her father would like her to marry a tame gentlemen under his command, she has her heart set on another beau. Sound predictable? It is, but fun.
1937 88m/B Victor McLaglen, Preston Foster, Ida Lupino, Donald Woods, Gordon Jones; *Dir:* Ben Stoloff. **VHS** $9.95 *MED* 🎞🎞

Sea Devils A smuggler and a beautiful spy come together during the Napoleonic Wars in this sea romance filled with intrigue and adventure.
1953 86m/C *GB* Rock Hudson, Yvonne de Carlo, Maxwell Reed; *Dir:* Raoul Walsh. **VHS, Beta** $9.95 *MED* 🎞🎞

Sea of Dreams Don't expect much in the way of a script from this soft-core production.
1990 80m/C Jon Rodgers, Chad Scott, Jacky Peel, Renee O'Neil. **VHS, Beta** $39.95 *AHV, HHE* 🎞

Sea Gypsies A sailing crew of five is shipwrecked off the Aleutian Islands. They must escape before winter or learn to survive. Passable family drama.
1978 (G) 101m/C Robert F. Logan, Mikki Jamison-Olsen, Heather Rattray, Cjon Damitri; *Dir:* Stewart Raffill. **VHS, Beta** $19.98 *WAR* 🎞🎞

The Sea Hawk An English privateer learns the Spanish are going to invade England with their Armada. After numerous adventures, he is able to aid his queen and help save his country, finding romance along the way. One of Flynn's swashbuckling best. Available colorized.
1940 128m/B Errol Flynn, Claude Rains, Donald Crisp, Alan Hale Jr., Flora Robson, Brenda Marshall, Henry Daniell, Gilbert Roland, James Stephenson, Una O'Connor; *Dir:* Michael Curtiz. **VHS, Beta, LV** $19.95 *MGM, TTC, FOX* 🎞🎞🎞½

Sea Hound A 15-part series set in the late 1940s that charts the voyage of a group of pirates searching for buried treasure. Everything goes well until Buster Crabbe emerges to wreck their hopes.
1947 ?m/B Buster Crabbe, Jimmy Lloyd, Pamela Blake, Ralph Hodges, Robert Barron; *Dir:* Walter B. Eason, Mack V. Wright. **VHS** *MLB* 🎞🎞

Sea Lion A vicious sea captain, embittered by a past romance, becomes sadistic and intolerable, until the truth emerges. Lots of action, but plodding, rehashed plot. Silent.
1921 50m/B Hobart Bosworth. **VHS, Beta** $29.95 *VYY* 🎞

Sea of Love A tough, tightly wound thriller about an alcoholic cop with a mid-life crisis. While following the track of a serial killer, he begins a torrid relationship with one of his suspects. Pacino doesn't stand a chance when Barkin heats up the screen. Music by Trevor Jones.
1989 (R) 113m/C Al Pacino, Ellen Barkin, John Goodman, Michael Rooker, William Hickey, Richard Jenkins; *Dir:* Harold Becker. **VHS, Beta, LV** $19.95 *MCA* 🎞🎞🎞½

Sea Prince and the Fire Child Japanese animated film follows two young lovers who set off on an adventure to escape the disapproval of their parents.
1982 70m/C *JP* **VHS, Beta** $19.95 *COL* 🎞🎞

Sea Racketeers Soggy old tribute to the Coast Guard inadvertently makes them look like a bunch of bumblers, as they belatedly uncover a fur-smuggling racket working under their noses from a floating nightclub.
1937 64m/B Weldon Heyburn, Jeanne Madden, Warren Hymer, Dorothy McNulty, J. Carroll Naish, Joyce Compton, Charles Trowbridge, Syd Saylor, Lane Chandler, Benny Burt; *Dir:* Hamilton McFadden. **VHS, Beta** $16.95 *SNC* 🎞½

The Sea Serpent A young sea captain and a crusty scientist unite to search out a giant sea monster awakened by atomic tests. Not one of Millan's better films.

1985 92m/C *SP* Timothy Bottoms, Ray Milland, Jared Martin; *Dir:* Gregory Greens. **VHS, Beta** $69.98 *LIV, LTG* 🎞½

The Sea Shall Not Have Them A British bomber crashes into the North Sea during World War II. Film tells of the survivors' rescue by the Air-Sea Rescue Unit. Decent drama/adventure.
1955 92m/B Michael Redgrave, Dirk Bogarde; *Dir:* Lewis Gilbert. **VHS** $39.98 *REP* 🎞🎞½

Sea Wolves True World War II story about a commando-style operation undertaken by a group of middle-aged, retired British calvarymen in India in 1943. Decent acting, though Peck's British accent fades in and out, with Moore as Bond.
1981 (PG) 120m/C *GB* Gregory Peck, Roger Moore, David Niven, Trevor Howard, Patrick Macnee; *Dir:* Andrew V. McLaglen. **VHS, Beta** $59.98 *FOX, WAR* 🎞🎞½

The Seagull A pensive, sensitive adaptation of the famed Chekhov play about the depressed denizens of an isolated country estate. In Russian with English subtitles.
1971 99m/B Alla Demidova, Lyudmila Savelyeva, Yuri Yakovlev; *Dir:* Yuri Karasik. **VHS, Beta** $29.95 *FCT, MRV* 🎞🎞🎞

Sealab 2020 Two animated episodes chronicling the underwater adventures a scientific expedition encounters in the year 2020.
1972 50m/C *Voices:* Ross Martin, Ann Jillian, Pamelyn Ferdin. **VHS, Beta** $19.95 *HNB* 🎞🎞🎞

Seance on a Wet Afternoon Dark, thoughtful film about a crazed pseudo-psychic who coerces her husband into a kidnapping so she can gain recognition by divining the child's whereabouts. Stanley's performance as the medium won her a well-deserved Academy Award nomination. Directed splendidly by Forbes, and superb acting from Stanley and Attenborough, who co-produced with Forbes.
1964 111m/B *GB* Kim Stanley, Richard Attenborough, Margaret Lacey, Maria Kazan, Mark Eden, Patrick Magee; *Dir:* Bryan Forbes. British Academy Awards '64: Best Actor (Attenborough); National Board of Review Awards '64: 10 Best Films of the Year, Best Actress (Stanley). **VHS, Beta** $29.95 *VID* 🎞🎞🎞½

The Search Clift, an American solider stationed in post-World War II Berlin, befriends a homeless nine-year-old amnesiac boy (Jandl) and tries to find his family. Meanwhile, his mother has been searching the Displaced Persons camps for her son. Although Clift wants to adopt the boy, he steps aside, and mother and son are finally reunited. "The Search" was shot on location in the American Occupied Zone of Germany. Clift and director Zinnemann received Oscar nominations and Jandl won a special juvenile Oscar in his first (and only) film role. This was also Clift's first screen appearance, although this movie was actually filmed after his debut in "Red River" it was released first.
1948 105m/B Montgomery Clift, Aline MacMahon, Ivan Jandl, Jarmila Novotna, Wendell Corey; *Dir:* Fred Zinneman. Academy Awards '48: Best Original Screenplay. **VHS** $19.98 *MGM* 🎞🎞🎞½

The Search for Bridey Murphy Long before channelers, Ramtha and New-Age profiteers, there was a Colorado housewife who under hypnosis described a previous life as an Irish girl. Her famous claim was never proven—but that didn't stop this well-acted but dull dramatization. Neither sensationalist, nor likely to persuade skeptics.

Based on the book by Morey Bernstein, hypnotist in the case.
1956 84m/B Teresa Wright, Louis Hayward, Nancy Gates, Kenneth Tobey, Richard Anderson; *W/Dir:* Noel Langley. **VHS** $14.95 *PAR, FCT* 🎞🎞

Search and Destroy A deadly vendetta is begun by a Vietnamese official during the war. It is continued in the U.S. as he hunts down the American soldiers he feels betrayed him. Ordinary, forgettable war/action flick. Also known as "Striking Back."
1981 (PG) 93m/C Perry King, Don Stroud, Park Jong Soo, George Kennedy, Tisa Farrow; *Dir:* William Fruet. **VHS, Beta, LV** $69.95 *VES* 🎞

Search and Destroy Sci-fi action flick about the capture of a secret biological warfare research station. Lotsa action, that's for sure - where's the plot?
1988 (R) 87m/C Stuart Garrison Day, Dan Kuchuck, Peggy Jacobsen; *Dir:* J. Christian Ingvordsen. **VHS, Beta** $19.95 *MCG* 🎞

The Searchers The classic Ford western, starring John Wayne as a hard-hearted frontiersman who spends years doggedly pursuing his niece, who was kidnapped by Indians. A simple western structure supports Ford's most moving, mysterious, complex film. Many feel this is the best western of all time.
1956 119m/C John Wayne, Jeffrey Hunter, Vera Miles, Natalie Wood, Ward Bond, John Qualen, Harry Carey Jr.; *Dir:* John Ford. Sight & Sound Survey '82: #10 of the Best Films of All Time (tie). **VHS, Beta, LV** $19.98 *WAR, PIA, TLF* 🎞🎞🎞🎞

Seaside Swingers A group of teenagers at a seaside resort work to win a talent competition on television. But where is the talent? Dumb teen romance comedy with accidental plot.
1965 94m/C *GB* Michael Sarne, Grazina Frame, John Leyton, Freddie & the Dreamers; *Dir:* James Hill. **VHS, Beta** $59.98 *CHA* 🎞½

Season for Assassins Police commissioner Castroni is the only man who can stop a wave of violence enveloping Rome.
1971 102m/C Joe Dallesandro, Martin Balsam. **VHS, Beta** $29.95 *IVE* 🎞½

Season of Fear A young man visits his estranged father's home. Abandoned as a child, he's embroiled in a murder scheme engineered by his father's beautiful young wife. Intriguing premise, forgettable execution.
1989 (R) 89m/C Michael Bowen, Clancy Brown, Clare Wren, Ray Wise, Michael J. Pollard; *Dir:* Douglas Campbell. **VHS, Beta** $79.98 *FOX* 🎞½

Season of the Witch A frustrated housewife becomes intrigued with a neighboring witch and begins to practice witchcraft herself, through murder and seduction. Meant to be suspenseful, thrilling, and topical, it is none of these. Poorly acted at best. Originally 130 minutes! Also titled "Hungry Wives" and "Jack's Wife."
1973 (R) 89m/C Jan White, Ray Laine, Bill Thunhurst, Joedida McClain, Virginia Greenwald; *Dir:* George A. Romero. **Beta** $79.95 *VHV* 🎞½

Sebastian A mathematics whiz is employed by the British to decipher codes. He falls in love with York, another code breaker, and becomes involved in international intrigue. A bit too fast-paced and busy, but enjoyable spy drama. Donald Sutherland has a small role.
1968 100m/C *GB* Dirk Bogarde, Susannah York, Lilli Palmer, John Gielgud, Margaret Johnston, Nigel Davenport; *Dir:* David Greene. **VHS** *MOV* 🎞🎞½

Sebastiane An audacious film version of the legend of St. Sebastian, packed with homoerotic imagery and ravishing visuals. Honest, faithful rendering of the Saint's life and refusal to obey Roman authorities. Jarman's first film, in Latin with English subtitles. Music by Brian Eno.
1979 90m/C Leonardo Treviglio, Barney James, Richard James, Neil Kennedy; **Dir:** Derek Jarman. **VHS, Beta $29.95** FCT, CNT 🎬🎬🎬

Second Best Secret Agent in the Whole Wide World A klutzy agent attempts to prevent a Swedish anti-gravity formula from falling into Russian hands. Surprisingly good early spoof of the James Bond genre. Originally titled ''Licensed to Kill.''
1965 93m/C GB Tom Adams, Veronica Hurst, Peter Bull, Karel Stepanek, John Arnatt; **Dir:** Lindsay Shonteff. **VHS, Beta $59.98** CHA 🎬🎬½

Second Chance A former prizefighter (Mitchum) travels to Mexico where he protects a gangster's moll (Darnell) targeted for murder. But Palance is on their tail. Originally released theatrically on a wide screen in 3-D. Good melodrama.
1953 82m/C Robert Mitchum, Linda Darnell, Jack Palance, Reginald Sheffield, Roy Roberts; **Dir:** Rudolph Mate. **VHS, Beta, LV $19.95** MED 🎬🎬½

Second Chance When a woman finds her husband is in love with another, her marital happiness is shattered. She leaves her betrayer and struggles to adapt to life as a single mother. Made for British television.
1980 270m/C GB Susannah York, Ralph Bates; **Dir:** Gerry Mill, Richard Handford. **VHS $79.95** SVS 🎬🎬½

Second Chorus Rivalry of two trumpet players for a girl and a job with Artie Shaw Orchestra. Music, dance, and romance. Nothing great, but pleasant.
1940 83m/B Fred Astaire, Paulette Goddard, Burgess Meredith, Artie Shaw, Charles Butterworth; **Dir:** H.C. Potter. **VHS, Beta $19.95** NOS, HHT, CAB 🎬🎬½

Second City Comedy Special Comedy special featuring the current Second City Comedy Troupe, hosted by alumni comics Candy and Willard, satirizing popular television shows.
1979 61m/C John Candy, Fred Willard. **VHS, Beta $19.98** WAR, OM 🎬🎬½

Second City Insanity The famed Second City improvisational troupe performs their unique brand of humor.
1981 60m/C Fred Willard, John Candy. **VHS, Beta** LHV 🎬🎬

Second Coming of Suzanne A young actress encounters a hypnotic film director. Her role: to star in a Crucifixion - which may be more real than she imagines. Winner of two international film festivals.
1980 90m/C Sondra Locke, Richard Dreyfuss, Gene Barry; **Dir:** Michael Barry. **VHS, Beta $19.95** GEM 🎬½

Second Sight A detective (Larroquette) and a goofy psychic (Pinchot) set out to find a kidnapped priest. Flick is sidetracked along the way by such peculiarities as a pixieish nun for romantic, not religious, intrigue. Remember this one? No? That's because you blinked when it was released to theaters. Eminently missable.

1989 (PG) 84m/C John Larroquette, Bronson Pinchot, Bess Armstrong, James Tolken, Christine Estabrook, Cornelia Guest, Stuart Pankin, William Prince, John Schuck; **Dir:** Joel Zwick. **VHS, Beta, LV $19.95** WAR 🎬

Second Thoughts A frustrated woman attorney divorces her stuffy husband and takes up with an aging hippie. She comes to regret her decision when it becomes obvious that her new man has his head firmly stuck in the 60s. A comedy-drama that lacks real laughs or drama.
1983 (PG) 109m/C Lucie Arnaz, Craig Wasson, Ken Howard, Joe Mantegna; **Dir:** Lawrence Turman. **VHS, Beta $59.99** HBO 🎬

Second Time Lucky The devil makes a bet with God that if the world began all over again Adam and Eve would repeat their mistake they made in the Garden of Eden. Bites off a lot, but has no teeth nor laughs.
1984 98m/C AU NZ Diane Franklin, Roger Wilson, Robert Morley, Jon Gadsby, Bill Ewens; **Dir:** Michael Anderson Sr. **VHS, Beta $14.95** HMD, LHV 🎬

Second Wind Wagner and Naughton are trying to settle into their new life together, but find they must make a variety of difficult choices.
1990 (PG) 93m/C Lindsay Wagner, James Naughton. **VHS $79.98** VID 🎬🎬

Second Woman An architect, suffering from blackouts and depression, believes himself responsible for the death of his fiancee. Well-done psycho-drama.
1951 91m/B Robert Young, Betsy Drake, John Sutton; **Dir:** James V. Kern. **VHS, Beta $14.95** NOS, QNE, UHV 🎬🎬½

Secret Admirer A teenager's unsigned love letter keeps falling into the wrong hands. Intends to be funny, and sometimes is; too often, though, it surrenders to obviousness and predictability.
1985 (R) 98m/C C. Thomas Howell, Cliff DeYoung, Kelly Preston, Dee Wallace Stone, Lori Loughlin, Fred Ward, Casey Siemaszko, Corey Haim, Leigh Taylor-Young; **Dir:** David Greenwalt. **VHS, Beta, LV $19.99** HBO 🎬🎬

The Secret Adversary Set in 1920s London. Ace detectives Tommy and Tuppence Beresford must find a secret treaty before it falls into the wrong hands. Pilot for the ''Partners in Crime'' series. Based on the characters created by Agatha Christie.
1983 120m/C GB Francesca Annis, James Warwick. **VHS, Beta $19.95** PAV 🎬🎬

The Secret Agent Presumed dead, a British intelligence agent (Gielgud) reappears and receives a new assignment. Using his faked death to his advantage, he easily journeys to Switzerland where he is to eliminate an enemy agent. Strange Hitchcockian melange of comedy and intrigue; atypical, but worthy offering from the Master.
1936 83m/B GB Madeleine Carroll, Peter Lorre, Robert Young, John Gielgud, Lilli Palmer; **Dir:** Alfred Hitchcock. **VHS, Beta, LV $16.95** SNC, NOS, HHT 🎬🎬🎬

Secret Agent 00 A master criminal plans to blackmail the western world and only special Secret Agent 00 can stop him. Bad spy genre ripoff, interesting only as the screen debut of Sean Connery's brother. Originally titled ''Operation Kid Brother.''
1990 (R) 97m/C Neil Connery, Daniela Bianchi, Adolfo Celi; **Dir:** Alberto De Martino. **VHS $49.95** AVD 🎬

Secret Agent Super Dragon The CIA calls in the agent known as ''Super Dragon'' when it is discovered that a Venezuelan drug czar plans to spike U.S. gum and candy with an LSD-like drug. Less than competent production offers some unintended laughter.
1966 95m/C FR IT GE Ray Danton, Marisa Mell, Margaret Lee, Jess Hahn, Carlo D'Angelo, Andriana Ambesi; **Dir:** Calvin Jackson Padget. **VHS, Beta $16.95** SNC 🎬

Secret Beyond the Door A wealthy heiress marries a widower and soon discovers that he murdered his first wife. Understandably, she wonders what plans he might have for her. Capably done chiller, but the plot is hackneyed pseudo-Hitchcock.
1948 99m/B Joan Bennett, Michael Redgrave, Barbara O'Neill, Anne Revere; **Dir:** Fritz Lang. **VHS $19.98** REP 🎬🎬

Secret Ceremony An aging prostitute and a young, aimless waif resemble each other's dead mother and dead daughter, respectively, and the relationship goes on, strangely, from there. This original big-screen version is very good Freudian psycho-drama; 101-minute TV version is unfortunately greatly diluted.
1969 109m/C Elizabeth Taylor, Robert Mitchum, Mia Farrow, Pamela Brown, Peggy Ashcroft; **Dir:** Joseph Losey. **VHS, Beta $14.95** KRT 🎬🎬½

The Secret Diary of Sigmund Freud Spoofy look at the early years of Sigmund Freud, the father of modern psychoanalysis. Freud (Cort) has an affair with lisping nurse Kane while tending to patient Shawn and answering to mama Baker, who's carrying on with mad doctor Kinski. Talented cast falls asleep on couch.
1984 129m/C Bud Cort, Carol Kane, Carroll Baker, Klaus Kinski, Marisa Berenson, Dick Shawn, Ferdinand ''Ferdy'' Mayne; **Dir:** Danford B. Greene. **VHS, Beta $79.98** FOX 🎬🎬

Secret of El Zorro Don Diego's friend, Don Ricardo, threatens to unmask Zorro's secret identity when he challenges the legendary swordsman to a duel.
1957 75m/B Guy Williams. **VHS, Beta $39.95** DIS 🎬½

Secret Executioners A singular, lone, almost defenseless guy battles the mob's assassins with his feet.
197? 90m/C Wong Chen-Li, Jim Norris, Peggy Min. **VHS, Beta $54.95** MAV 🎬

Secret Fantasy A musician overcomes his fears of inferiority as he makes his fantasies a reality by having other men admire his wife's beautiful body. A lesson in self-esteem.
1981 (R) 88m/C Laura Antonelli. **VHS, Beta $54.95** MED 🎬

Secret File of Hollywood A down-and-out private eye gets a job taking photos for a sleazy Hollywood scandal sheet and finds himself in the midst of a blackmail plot masterminded by his editor. An often childish potboiler, perhaps inspired by the once-feared Confidential Magazine. Also known as ''Secret File: Hollywood.''
1962 85m/B Robert Clarke, Francine York, Syd Mason, Maralou Gray, John Warburton; **Dir:** Ralph Cushman. **VHS $15.95** NOS, LOO 🎬½

The Secret Four Three young men seek to avenge the death of their friend at the hands of British traitors, and become involved in a plot to sabotage the Suez canal. Poorly adapted from the novel by Edgar

Wallace. Also known as ''The Four Just Men.''
1940 79m/B *GB* Hugh Sinclair, Griffith Jones, Francis L. Sullivan, Frank Lawton, Anna Lee, Alan Napier, Basil Sydney; *Dir:* Walter Forde. **VHS, Beta** $16.95 *SNC* 🎬🎬

Secret Games An unhappily married woman (Brin) is searching for relief from her restrictive marriage. At the ''Afternoon Demitasse,'' an exclusive brothel where women are paid for fulfilling their ultimate fantasies, she meets a man (Hewitt) who pushes her beyond her sexual limits and threatens to totally possess her.
199? (R) 90m/C Martin Hewitt, Michele Brin, Delia Sheppard, Billy Drago; *Dir:* Alexander Gregory Hippolyte. **VHS** $89.95 *IMP* 🎬½

The Secret Garden The well-loved Frances Hodgson Burnett story has been remade and even revamped as a Broadway musical, but this is considered the definitive version. An orphan girl arrives at her uncle's somber Victorian estate. She starts tending a long-forgotten garden, and with its resurrection joy returns to the household. O'Brien leads an outstanding cast in one of her final juvenile roles. In black and white, with Technicolor for later garden scenes.
1949 92m/B Margaret O'Brien, Herbert Marshall, Dean Stockwell, Gladys Cooper, Elsa Lanchester, Brian Roper; *Dir:* Fred M. Wilcox. **VHS** $19.98 *MGM* 🎬🎬🎬

The Secret Garden Based on the children's novel by Francis Hogdon Burnett. Made-for-British-TV fantasy about a spoiled girl transformed by a secret fairy garden.
1975 107m/C *GB* Sarah Hollis Andrews, David Patterson; *Dir:* Katrina Murray. **VHS, Beta** $19.98 *KUI, FOX* 🎬🎬

The Secret Garden Orphaned Mary Lennox is sent to live with her mysterious uncle after her parents die. Mary is wilful and spoiled and her uncle's house holds a number of secrets, including a crippled cousin. When Mary discovers a mysteriously abandoned locked garden she makes it her mission to restore the garden to life, a mission that changes Mary's character and brings her family together. Based on the children's classic by Frances Hodgson Burnett.
1984 107m/C *GB* Sarah Hollis Andrews, David Patterson; *Dir:* Katrina Murray. **VHS** *MOV* 🎬🎬½

The Secret Garden Mary Lennox is a lonely orphan who is sent to live with her unknown uncle in England after her parents deaths. Mary, who has grown up in India, is selfish, spoiled, and unhappy until she discovers two secrets on her uncle's estate. One is her crippled cousin, as demanding as Mary, and the other is a walled, neglected garden. With the aid of Dickon, the gardener's boy, Mary begins to transform the garden into a living area once again, and the garden works its wonders not only on Mary but on her family as well. A prologue and afterword to the story (showing Mary as an adult) are unnecessary additions but this is a class production with wonderful performances. Based on the children's classic by Frances Hodgson Burnett.
1987 (PG) 100m/C Gennie James, Barret Oliver, Jadrien Steele, Michael Hordern, Derek Jacobi, Billie Whitelaw, Lucy Gutteridge, Julian Glover, Colin Firth, Alan Grint. **VHS, LV** $29.98 *REP* 🎬🎬🎬

Secret Honor Idiosyncratic, single-set, one-man film version adapted by Donald Freed and Arnold Stone from their stage play about Richard Nixon coping with the death of his presidency on the night he's decided to blow his brains out. Made with students at the University of Michigan, and carried by the ranting and raving of Hall as a tragic Shakespearean Nixon with plenty of darkly humorous lines.
1985 90m/C Philip Baker Hall; *Dir:* Robert Altman. **VHS, Beta** $29.98 *LIV, VES* 🎬🎬🎬

Secret of the Ice Cave A motley band of explorers, hijackers and mercenaries go after a secret treasure in the remote mountains.
1989 (PG-13) 106m/C Michael Moriarty, Sally Kellerman, David Mendenhall, Virgil Frye, Gerald Anthony, Norbert Weisser; *Dir:* Radu Gabrea. **VHS, Beta** $79.95 *WAR, OM* 🎬

Secret Life of an American Wife A bored housewife (Jackson) tries to seduce a client of her husband's, with less than hilarious results. Too much dialogue, not enough genuine humor. Disappointing inversion of director Axelrod's ''The Seven Year Itch.''
1968 97m/C Walter Matthau, Anne Jackson, Patrick O'Neal, Edy Williams; *Dir:* George Axelrod. **VHS, Beta** $59.98 *FOX* 🎬🎬

The Secret Life of Sergei Eisenstein A detailed, comprehensive portrayal of the early Russian filmmaker. Material from Eisenstein's production notes and sketchbooks were used to fill out the picture, and scenes from his movies ''Strike,'' ''Battleship Potemkin,'' ''Alexander Nevsky,'' and ''Ivan the Terrible'' are included.
1985 60m/C Sergei Eisenstein; *Dir:* Gian Carlo Bertelli. **VHS** $29.95 *MFV, IVA* 🎬🎬

Secret Life of Walter Mitty An entertaining adaptation of the James Thurber short story about a meek man (Kaye) who lives an unusual secret fantasy life. Henpecked by his fiancee and mother, oppressed at his job, he imagines himself in the midst of various heroic fantasies. While Thurber always professed to hate Kaye's characterization and the movie, it scored at the box office and today stands as a comedic romp for Kaye. Available with digitally remastered stereo and original movie trailer.
1947 110m/C Danny Kaye, Virginia Mayo, Boris Karloff, Ann Rutherford, Fay Bainter, Florence Bates; *Dir:* Norman Z. MacLeod. **VHS, Beta, LV** $19.98 *HBO, SUE* 🎬🎬

Secret Lives of Waldo Kitty Volume 1 Animated adventure wherein Waldo Kitty imagines himself to be a variety of heroes such as Robin Cat and the Lone Kitty.
1975 48m/C **VHS, Beta** *SUE, OM* 🎬🎬🎬

The Secret of the Loch Scottish scientist claims to have seen Loch Ness Monster but everyone thinks his bagpipe's blown a reed.
1934 80m/B *GB* Seymour Hicks, Nancy O'Neil, Gibson Gowland, Frederick Peisley; *Dir:* Milton Rosmer. **VHS** $19.98 *NOS, SNC* 🎬½

The Secret of My Success Country bumpkin Fox goes to the Big Apple to make his mark. He becomes the corporate mailboy who rises meteorically to the top of his company (by impersonating an executive) in order to win the love of an icy woman executive. He spends his days running frantically between his real job in the mailroom and his fantasy position, with various sexual shenanigans with the boss's wife thrown in to keep the viewer alert. Fox is charismatic while working with a cliche-ridden script that ties up everything very neatly at the end.
Written by top guns Jim Cash and Jack Epps, Jr., with an assist by Carothers.
1987 (PG-13) 110m/C Michael J. Fox, Helen Slater, Richard Jordan, Margaret Whitton; *Dir:* Herbert Ross. **VHS, Beta, LV** $19.95 *MCA* 🎬🎬

The Secret of Navajo Cave Fair family adventure. Two young friends explore the mysterious Navajo cave, where one can be assured that a secret awaits. Also known as ''Legend of Cougar Canyon.''
1976 (G) 84m/C Holger Kasper, Steven Benally Jr., Johnny Guerro; *Dir:* James T. Flocker; *Nar:* Rex Allen. **VHS** $39.95 *XVC, UHV* 🎬🎬

Secret of NIMH Animated tale, produced by a staff of Disney-trained artists led by ''American Tail's'' Bluth; concerns a newly widowed mouse who discovers a secret agency of superintelligent rats (they've escaped from a science lab) who aid her in protecting her family. As is usually the case with Bluth films, the animation is superb while the socially aware plot struggles to keep pace. That aside, it's still an interesting treat for the youngsters. Adapted from Robert C. O'Brien's ''Mrs. Frisby and the Rats of N.I.M.H.''
1982 (G) 84m/C *Dir:* Don Bluth; *Voices:* John Carradine, Derek Jacobi, Dom DeLuise, Elizabeth Hartman, Peter Strauss, Aldo Ray, Edie McClurg, Wil Wheaton. **VHS, Beta, LV, 8mm** $19.98 *MGM* 🎬🎬🎬

Secret Obsession A love triangle among a father, his illegitimate son, and the woman they both love, set in North Africa in 1955. Sounds intriguing, but don't be fooled. Slow, dull, and poorly acted.
1988 (PG) 82m/C Julie Christie, Ben Gazzara, Patrick Bruel, Jean Carmet; *Dir:* Henri Vart. **VHS, Beta** $79.95 *VMK* 🎬

Secret Passions A young couple are haunted by a dark, romance-novel-type secret. Made for television.
1978 99m/C John James, Susan Lucci; *Dir:* Michael Pressman. **VHS, Beta** $79.95 *JTC* 🎬½

Secret Passions Produced for a public-access cable station in California, this program became the first soap opera aired in the U.S. to feature continous gay and lesbian themes. Tape includes the first two episodes, plus scenes never broadcast. Also includes an interview with the show's creator David Gadberry.
199? ?m/C **VHS** $39.95 *FCT* 🎬½

Secret of the Phantom Knight An animated film about a magical kingdom, princes and white horses. A Choppy and the Princess adventure.
197? 86m/C **VHS, Beta** $14.95 *VDC* 🎬

Secret Places On the brink of World War II, a German girl in an English school finds friendship with a popular classmate. Touching and involving, but somehow unsatisfying.
1985 (PG) 98m/C *GB* Maria Therese Relin, Tara MacGowan, Claudine Auger, Jenny Agutter; *Dir:* Zelda Barron. **VHS, Beta** $79.95 *FOX* 🎬🎬½

The Secret Policeman's Other Ball Engaging live performance by most of the Monty Python troupe and guest rock artists, staged for Amnesty International. Follows 1979's ''The Secret Policeman's Ball,'' and followed by 1987's ''The Secret Policeman's Third Ball.''
1982 (R) 101m/C *GB* John Cleese, Graham Chapman, Michael Palin, Terry Jones, Pete Townsend, Sting, Billy Connolly, Bob Geldof, Jeff Beck, Eric Clapton; *Dir:* Julien Temple. **VHS, Beta, LV** $14.95 *MGM* 🎬🎬🎬

Secret Policeman's Private Parts Various sketches and performances from the various Secret Policeman occasions, featuring classic Python sketches including ''I'm a Lumberjack and I'm OK.'' Also featured are performances by Phil Collins, Pete Townshend, Donovan, and Bob Geldof. Thank you very much.
1981 77m/C *GB* John Cleese, Michael Palin, Terry Jones, Pete Townsend, Julien Temple; *Dir:* Roger Graef. VHS, Beta, LV $59.95 MED ♪♪½

The Secret of the Snake and Crane A resistance group relies on ancient fighting techniques to battle the rule of the Ching Dynasty in this Kung Fu action film.
197? 90m/C VHS, Beta $54.95 MAV ♪½

A Secret Space Story of a young man, born of secular parents, searching for the meaning of his Jewish roots as he nears his Bar Mitzvah. He is helped along by a group of Jews who are also searching for this meaning.
1988 80m/C Robert Klein, Phyllis Newman, John Matthews, Sam Schacht, Virginia Graham; *Dir:* Roberta Hodes. VHS, Beta $79.95 ERG ♪♪

Secret Squirrel Here are eight episodes of adventure with the clever secret agent.
196? 53m/C VHS, Beta $19.95 HNB ♪♪½

Secret Squirrel's Undercover Capers The world's only animated undercover rodent, Secret Squirrel battles a motley crew of zany villians with his cohorts Winsome Witch and Squiddley Diddley.
1966 52m/C VHS, Beta $19.95 HNB ♪♪½

Secret of the Sword He-Man must rescue his sister She-Ra from the evil clutches of the oppressive ruler Hordak.
1985 100m/C VHS, Beta $29.95 COL ♪♪

Secret of Tai Chi A young man uses his wits and martial arts to battle an evil Chinese general.
19?? 90m/C VHS, Beta $49.95 OCE ♪

The Secret War of Harry Frigg Private Harry Frigg, a nonconformist World War II G.I., is promoted to general as part of a scheme to help five Allied generals escape from the custody of the Germans. Rare Newman bomb; dismal comedy.
1968 (R) 123m/C Paul Newman, Sylva Koscina, John Williams, Tom Bosley, Andrew Duggan; *Dir:* Jack Smight. VHS, Beta $49.95 MCA ♪♪½

Secret Weapon An Israeli technician flees his country with atomic secrets and a beautiful government agent is sent to induce him to return. Based on the true case of Mordecai Vanunu, but fictionalized. Slow-paced and not very suspenseful. Made for cable.
1990 95m/C Griffin Dunne, Karen Allen, Stuart Wilson, Jeroen Krabbe, Brian Cox, John Rhys-Davies, Ian Mitchell; *Dir:* Ian Sharp. VHS, Beta $19.98 TTC ♪♪

Secret Weapons Sexy Russian babes are KGB-trained to get any secret from any man. Strains credibility as well as patience. Also known as ''Secrets of the Red Bedroom.'' Made for TV.
1985 100m/C Linda Hamilton, Sally Kellerman, Hunt Block, Viveca Lindfors, Christopher Atkins, Geena Davis, James Franciscus; *Dir:* Don Taylor. VHS $69.95 SVS ♪½

The Secret World of Reptiles A series that presents rare living relics of primeval times and traces the history of the reptile kingdom.
1977 94m/C VHS, Beta $69.95 DIS ♪½

Secret of Yolanda A steamy romance about a young deaf-mute whose guardian and riding instructor both fall for her.
1982 (R) 90m/C Aviva Ger, Asher Zarfati, Shraga Harpaz. VHS, Beta $59.95 MGM ♪

Secrets A wife, her husband, and her daughter each have a sexual experience which they must keep secret. Bisset's nude love scene is an eye-opener. Fast forward through the rest of this dull, overwrought drama.
1971 (R) 86m/C Jacqueline Bisset, Per Oscarsson, Shirley Knight Hopkins, Robert Powell, Tarka Kings, Martin C. Thurley; *Dir:* Philip Saville. VHS, CV $59.95 PSM ♪½

Secrets An unhappy young bride starts having numerous affairs when her mother dies. Made for television. Hardly worth the effort.
1977 100m/C Susan Blakely, Roy Thinnes, Joanne Linville, John Randolph, Anthony Eisley, Andrew Stevens; *Dir:* Paul Wendkos. VHS MGM ♪

Secrets British drama about the confused life of an innocent schoolgirl who's the victim of a mess of authoritative misunderstandings. Written by Noella Smith; one of David Puttnam's ''First Love'' series.
1982 79m/C *GB* Helen Lindsay, Anna Campbell-Jones, Daisy Cockburn; *Dir:* Gavin Millar. VHS, Beta $59.95 MGM ♪

Secrets in the Dark A small-town group of womanizers get together and engage in soft-core ''fun.''
1983 84m/C Viktor Lange, Ulrike Butz, Logena Marks, Martha Meding. VHS, Beta $29.95 MED ♪

Secrets of Life Documentary captures the various ways that organisms on earth respond to the challenges of survival and reproduction. An installment of the Disney ''True-Life Adventures'' series.
1956 69m/C *Dir:* James Algar. VHS, Beta $69.95 DIS ♪♪½

Secrets of a Married Man A married man's philandering ways are his ruination when he falls hard for a beautiful prostitute. Shatner and Shepherd contain their laughter as they make their way through this hyper-earnest family drama.
1984 96m/C William Shatner, Cybill Shepherd, Michelle Phillips, Glynn Turman; *Dir:* William A. Graham. VHS, Beta $29.98 LIV, LTG ♪

Secrets of a Soul A visually impressive presentation of Freudian psychoanalytic theory, in which a professor, wanting a child and jealous of his wife's childhood sweetheart, moves toward madness (we did say Freudian) and is cured through dream interpretation. Great dream sequences bring out the arm-chair psychoanalyst. Silent.
1925 94m/B *GE* Werner Krauss, Ruth Weyher, Jack Trevor; *Dir:* G. W. Pabst. VHS, Beta $49.95 FCT ♪♪½

Secrets of Sweet Sixteen Gag-fest about budding sexuality in high school.
1974 (R) 84m/C Suzie Atkins. VHS, Beta $59.95 WES Woof!

Secrets of Three Hungry Wives When a multi-millionaire is killed, the women with whom he was having affairs are suspected. Tacky made-for-television flick.
1978 97m/C Jessica Walter, Gretchen Corbett, Eve Plumb, Heather MacRae, James Franciscus, Craig Stevens; *Dir:* Gordon Hessler. VHS UNR ♪

Secrets of Women A rare Bergman comedy about three sisters-in-law who tell about their affairs and marriages as they await their husbands at a lakeside resort. His first commercial success, though it waited nine years for release (1961).
1952 108m/B *SW* Anita Bjork, Karl Arne Homsten, Eva Dahlbeck, Maj-Britt Nilsson, Jarl Kulle; *W/Dir:* Ingmar Bergman. VHS, Beta $29.98 SUE ♪♪♪

The Secrets of Wu Sin A suicidal news writer in Chinatown is given a reason to live by a news editor who gives her a job. She starts investigating a ring smuggling Chinese workers, with the trail leading to, of course, Chinatown. Low-budget crime thriller.
1932 65m/B Lois Wilson, Grant Withers, Dorothy Revier, Robert Warwick, Toschia Mori; *Dir:* Richard Thorpe. VHS $12.95 SNC, NOS, DVT ♪♪

Seduced When a rich businessman turns up dead, the ambitious politician who was involved with his wife must find the killer. Not memorable or exceptional, but not boring or offensive either. Made for television.
1985 100m/C Gregory Harrison, Cybill Shepherd, Jose Ferrer, Adrienne Barbeau, Michael C. Gwynne, Karmin Murcelo, Paul Stewart; *Dir:* Jerrold Freedman. VHS $9.99 STE, PSM ♪♪

Seduced and Abandoned A lothario seduces his fiancee's young sister. When the girl becomes pregnant, he refuses to marry her. Family complications abound. A delightfully comic look at the Italian code of honor.
1964 118m/B *IT* Saro Urzi, Stefania Sandrelli, Aldo Paglisi; *Dir:* Pietro Germi. National Board of Review Awards '64: 5 Best Foreign Films of the Year; Italian Academy Awards '64: Best Director. VHS, Beta $24.95 NOS, HHT, FST ♪♪♪

The Seducer Things get dangerous when a playboy's goals turn from love to death.
1982 (R) 72m/C Christopher Mathews; *Dir:* Donovan Winters. VHS $12.95 AVD ♪♪

Seducers A wealthy, middle-aged man unsuspectingly allows two young women to use his telephone. A night of bizarre mayhem and brutal murder begins. They tease him, tear apart his house, and generally make him miserable. Why? Good question. Also known as ''Death Game.''
1977 (R) 90m/C Sondra Locke, Colleen Camp, Seymour Cassel; *Dir:* Peter S. Traynor. VHS, Beta $59.95 GEM ♪

The Seduction A superstar television anchorwoman is harassed by a psychotic male admirer. Usual run of the mill exploitive ''B'' thriller with no brains behind the camera.
1982 (R) 104m/C Morgan Fairchild, Michael Sarrazin, Vince Edwards, Andrew Stevens, Colleen Camp, Kevin Brophy; *Dir:* David Schmoeller. VHS, Beta $9.98 MED ♪

The Seduction of Joe Tynan Political drama about a young senator (Alda) torn between his family, his political career, and his mistress (Streep). Alda also wrote the thin screenplay, which reportedly is loosely based on the remarkable life of Ted Kennedy. Relatively shallow treatment of the meaty themes of power, hypocrisy, sex, and corruption in our nation's capital.
1979 (R) 107m/C Alan Alda, Meryl Streep, Melvyn Douglas, Barbara Harris, Rip Torn; *Dir:* Jerry Schatzberg. Los Angeles Film Critics Association Awards '79: Best Supporting Actor (Douglas), Best Supporting Actress (Streep); New York Film Critics Awards '79: Best Supporting Actress (Streep); National Society of Film Critics Awards '79: Best Supporting Actress (Streep). VHS, Beta, LV $69.95 MCA ♪♪½

Seduction of Mimi Comic farce of politics and seduction about a Sicilian laborer's escapades with the Communists and the local Mafia. Giannini is wonderful as the stubborn immigrant to the big city who finds himself in trouble. One of the funniest love scenes on film. Basis for the later movie "Which Way is Up?"
1974 (R) 92m/C *IT* Giancarlo Giannini, Mariangela Melato; *Dir:* Lina Wertmuller. **VHS, Beta** FOX, OM ♂♂♂

See How She Runs Running becomes a central and redefining experience for a 40-year-old divorced schoolteacher (Woodward). Her training culminates with a run in the Boston marathon. Her loved ones, though concerned for her health and sanity, cheer her on. Excellent made-for-TV drama. Screen daughter Newman is also Woodward's real-life daughter.
1978 92m/C Joanne Woodward, John Considine, Lissy Newman, Barbara Manning; *Dir:* Richard T. Heffron. **VHS, Beta $59.95** LTG ♂♂♂

See It Now A rather dated exploration of how automation changes the way America works and how computers revolutionize industry.
1957 82m/B Edward R. Murrow. **VHS, Beta $29.95** VYY ♂♂♂

See No Evil A blind girl gradually discovers the murdered bodies of her uncle's family. Trapped in the family mansion, she finds herself pursued by the killer. Chilling and well crafted. Also known as "Blind Terror."
1971 (PG) 90m/C Mia Farrow, Dorothy Allison, Robin Bailey; *Dir:* Richard Fleischer. **VHS, Beta, LV $39.95** COL ♂♂♂

See No Evil, Hear No Evil Another teaming for Pryor and Wilder, in which they portray a blind man and a deaf man both sought as murder suspects. Pryor and Wilder deserve better. Music by Stewart Copeland.
1989 105m/C Gene Wilder, Richard Pryor, Joan Severance, Anthony Zerbe; *Dir:* Arthur Hiller. **VHS, Beta, LV, 8mm $89.95** COL ♂♂

See You in the Morning Ill-conceived romantic comedy-drama about a divorced psychiatrist and a widow, both of whom had unhappy marriages, who meet and marry. They must cope with their respective children, family tragedies, and their own expectations, in order to make this second marriage work. Director Pakula also wrote the disappointing screenplay.
1989 (PG-13) 119m/C Jeff Bridges, Alice Krige, Farrah Fawcett, Drew Barrymore, Lukas Haas, Macaulay Culkin, David Dukes, Frances Sternhagen, Theodore Bikel, George Hearn, Linda Lavin; *Dir:* Alan J. Pakula. **VHS, Beta, LV $19.98** WAR ♂ 1/2

Seedpeople Great mindless horror flick has bloodthirsty plants inhabiting peaceful Comet Valley after their seeds fall from outer space and germinate. These "seedpeople" possess tremendous powers and soon have the rural residents transformed into zombies. Will anyone be able to stop the "Seedpeople" before they attempt to pollinate and rule the world? Also available with Spanish subtitles.
1992 87m/C Sam Hennings, Andrea Roth, Dane Witherspoon, David Dunard, Holly Fields, Bernard Kates; *Dir:* Peter Manoogian. **VHS, Beta $19.95** PAR ♂ 1/2

Seeds of Evil Warhol alumnus Dallesandro is a strange gardener who grows flowers that can kill. He can also turn himself into a tree and figures to seduce rich and bored

housewife Houghton (niece of Katherine Hepburn) after he finishes tending her garden. Strange, quirky horror flick. Also called "The Gardener."
1976 (R) 97m/C Katherine Houghton, Joe Dallesandro, Rita Gam; *Dir:* James H. Kay. **VHS, Beta $29.95** UNI ♂♂

The Seekers The story of the Kent family continues, as Abraham Kent, his wife and young son are forced to leave the safety of Boston for the rugged, untamed Pacific Northwest. TV movie based on the novel by John Jakes; preceded by "The Bastard" and "The Rebels."
1979 200m/C Randolph Mantooth, Edie Adams, Neville Brand, Delta Burke, John Carradine, George DeLoy, Julie Gregg, Roosevelt Grier, George Hamilton, Alex Hyde-White, Brian Keith, Ross Martin, Gary Merrill, Martin Milner, Vic Morrow, Timothy Patrick Murphy, Hugh O'Brian, Robert Reed, Allen Rich, Barbara Rush, Sarah Rush, Stuart Whitman, Ed Harris, Jeremy Licht, Eric Stoltz; *Dir:* Sidney Hayers. **VHS $79.95** MCA ♂♂ 1/2

Seems Like Old Times A sweet lawyer (Hawn) finds herself helping her ex-husband (Chase) when two robbers force him to hold up a bank. Grodin is the new spouse threatened by Chase's appearance and by the heavies he's trying to escape. Better-than-average Simon script. Funny, with appealing characters.
1980 (PG) 102m/C Goldie Hawn, Chevy Chase, Charles Grodin, Robert Guillaume, Harold Gould, George Grizzard, J.K. Carter; *Dir:* Jay Sandrich. **VHS, Beta, LV $19.95** COL ♂♂ 1/2

Seize the Day A man approaching middle age (Williams) feels that he is a failure. Brilliant performances by all, plus a number of equally fine actors in small roles. Based on the short novel by Saul Bellow.
1986 93m/C Robin Williams, Joseph Wiseman, Jerry Stiller, Glenne Headly, Tony Roberts; *Dir:* Fielder Cook. **VHS, Beta, LV $19.99** HBO ♂♂♂

Seizure Three demonic creatures from a writer's dreams come to life and terrorize him and his houseguests at a weekend party. Slick but disjointed; Stone's directorial debut. Interesting cast includes Tattoo from "Fantasy Island."
1974 (PG) 93m/C *CA* Jonathan Frid, Herve Villechaize, Christina Pickles; *Dir:* Oliver Stone. **VHS, Beta $9.99** STE, PSM ♂♂

Seizure: The Story of Kathy Morris The real-life anguish and struggle of a young music student who must have brain surgery and comes out of it unable to read or count. Workaday dramatization.
1982 104m/C Leonard Nimoy, Penelope Milford; *Dir:* Gerald I. Isenberg. **VHS, Beta** TLF ♂♂

Self-Defense When a city's police force goes on strike, the citizens must band together into vigilante troops.
1988 (R) 85m/C Tom Nardini, Brenda Bazinet. **VHS, Beta, LV $79.95** MED ♂ 1/2

Sell Out A former spy living in Jerusalem is called out of retirement; his protege, who defected to the Soviets, now wants out.
1976 (PG) 102m/C Richard Widmark, Oliver Reed, Gayle Hunnicutt, Sam Wanamaker; *Dir:* Peter Collinson. **VHS, Beta $59.95** MED ♂♂ 1/2

Semaforo en Rojo Five men who are planning a robbery run into trouble from some ex-lovers.
1947 95m/B *MX* Jose Galvez, Enrique Ponton, Jaime Velasquez, German Robles. **VHS, Beta $44.95** MAD ♂ 1/2

Semi-Tough Likeable, still current social satire involving a couple of pro-football buddies and their mutual interest in their team owner's daughter. Romantic comedy, satire of the sports world, zany highjinks—it's all here, though not enough of any of these. Pleasant and enjoyable. Based on the novel by Dan Jenkins.
1977 (R) 107m/C Burt Reynolds, Kris Kristofferson, Jill Clayburgh, Lotte Lenya, Robert Preston, Bert Convy, Richard Masur, Carl Weathers, Brian Dennehy, John Matuszak, Ron Silver; *Dir:* Michael Ritchie. **VHS, Beta, LV $19.98** MGM, FOX ♂♂ 1/2

The Senator was Indiscreet A farcical comedy concerning a slightly loony senator with presidential aspirations. He thinks he can win the nomination because he keeps a diary that would prove embarrassing to numerous colleagues. The diary then falls into the hands of a journalist. Powell is wonderful as the daffy politician. Playwright Kaufman's only directorial effort.
1947 75m/B William Powell, Ella Raines, Hans Conried; *Dir:* George S. Kaufman. **VHS $19.98** REP ♂♂♂

Send Me No Flowers Vintage Hudson-Day comedy. Hudson plays a hypochondriac who thinks his death is imment. He decides to provide for his family's future by finding his wife a rich husband. She thinks he's feeling guilty for having an affair.
1964 100m/C Rock Hudson, Doris Day, Tony Randall, Paul Lynde; *Dir:* Norman Jewison. **VHS, Beta $14.95** MCA, CCB ♂♂♂

The Sender An amnesiac young man is studied by a psychiatrist. She discovers that her patient is a "sender," who can transmit his nightmares to other patients at a hospital.
1982 (R) 92m/C *GB* Kathryn Harrold, Zeljko Ivanek, Shirley Knight, Paul Freeman, Sean Hewitt; *Dir:* Roger Christian. **VHS, Beta, LV $19.95** PAR ♂♂

Senior Trip New York City will never be the same after a bunch of rowdy Midwestern high school seniors tear up the town on their graduation trip. Made for television.
1981 96m/C Scott Baio, Mickey Rooney, Faye Grant, Vincent Spano, Jane Hoffman; *Dir:* Kenneth Johnson. **VHS, Beta $39.95** WOV ♂

Senior Week A batch of girl-hungry teens have a wacky time on their Florida vacation just before graduation.
1988 97m/C Michael St. Gerard, Devon Skye, Leesa Bryte; *Dir:* Stuart Goldman. **VHS, Beta $29.98** LIV, VES Woof!

The Seniors A group of college students decide to open a phony sex clinic, but the joke is on them when the clinic becomes a success. Better than it sounds. Pretty good acting and much harmless goofiness.
1978 (R) 87m/C Dennis Quaid, Priscilla Barnes, Jeffrey Byron, Gary Imhoff; *Dir:* Rod Amateau. **VHS, Beta $29.98** VES ♂♂

Senora Tentacion Musical melodrama about a composer who fights to leave his mother, sister, and girlfriend in order to flee with Hortensia, a famous singer.
1949 82m/B David Silva, Susana Guizar, Ninon Sevilla. **VHS, Beta $29.95** VYY, APD ♂♂

Sensations A woman and a man who aren't exceptionally fond of one another share a lottery ticket and win. Vincent is best known as a former director of porno flicks, and he manages to work in sexual overtones.

1987 (R) 91m/C Rebecca Lynn, Blake Bahner, Jennifer Delora, Rick Savage, Loretta Palma, Frank Stewart; *Dir:* Chuck Vincent. **VHS, Beta** $29.95 ACA 🎬🎬

Sensations of 1945 A press agent turns his firm over to one of his clients, a dancer with some wild promotional ideas. Brings together numerous variety and musical acts. W. C. Fields has a cameo in his last film role. Mostly unremarkable, occasionally fun musical.
1944 86m/B W.C. Fields, Eleanor Powell, Sophie Tucker, Cab Calloway, Woody Herman, C. Aubrey Smith, Lyle Talbot, Marie Blake, Eugene Pallette, Dennis O'Keefe; *Dir:* Andrew L. Stone. **VHS, Beta** $19.95 VDM 🎬🎬½

A Sense of Freedom A hopeless racketeering criminal is thrust from institution to institution until the Scottish authorities decide on an innovative style of reform. Based on Jimmy Boyle's autobiographical book. Excellent and disturbing.
1978 (R) 81m/C *GB* David Hayman, Alex Norton, Jake D'Arcy, Sean Scanlan, Fulton McKay; *Dir:* John MacKenzie. **VHS, Beta** $79.99 HBO 🎬🎬🎬

Sense & Sensibility BBC mini-series adaptation of the Jane Austen classic.
1985 174m/C *Dir:* Rodney Benett. **VHS, Beta** $29.98 FOX 🎬½

Senso Tragic story of romance and rebellion as Italian patriots battle the Austro-Hungarian empire for independence in 1866. An Italian noblewoman betrays her marriage, and almost her country, to be with a cynical Austrian soldier in this visually stunning piece of cinematography. Also released as "The Wanton Contessa."
1968 125m/C *IT* Alida Valli, Massimo Girotti, Heinz Moog, Farley Granger; *Dir:* Luchino Visconti. **VHS** $29.95 CEG, APD 🎬🎬🎬

The Sensual Man A hot-blooded Italian falls in love and gets married only to find out that his wife cannot consummate their marriage. Exploitative, misoginistic and unfunny.
1974 (R) 90m/C Giancarlo Giannini, Rossana Podesta, Lionel Stander; *Dir:* Marco Vicario. **VHS, Beta** $59.95 COL 🎬½

Sensual Partners Young actresses are lured to Tangier by a fake movie producer and sold as sex slaves. Emphasis on soft-focus, soft-core flesh.
1987 87m/C Gina Jansen, Eric Falk. **VHS, Beta** $29.95 MED *Woof!*

The Sensual Taboo A collection of footage of bizarre sexual practices from the fleshpots of Third World cities.
1985 99m/C VHS, Beta $29.95 MED *Woof!*

Sensuous Caterer Marc Stevens hosts a video Valentine's Day orgy and invites an uninhibited crowd of erotic stars and starlets to come and show their stuff.
1982 60m/C Marc Stevens. **VHS, Beta** HAR *Woof!*

The Sensuous Nurse Italian comedy about a beautiful nurse hired by the greedy, treacherous relatives of a weak-hearted count in hopes that her voluptuousness will give him a heart attack. It doesn't, and she falls in love. Mindless, but fun and sexy. Dubbed in English.
1976 (R) 79m/C *IT* Ursula Andress, Mario Pisu, Dulio Del Prete, Jack Palance; *Dir:* Nello Rosatti. **VHS, Beta** $59.98 FOX 🎬🎬

The Sensuous Teenager A teenage girl is overwhelmingly erotic and uncontrollably seductive, and she aims to please by releasing all the sensuality within her. Also known as "Libido" and "Forbidden Passions."
1970 (R) 80m/C Sandra Jullien, Janine Reynaud, Michel Lemoine; *Dir:* Max Pecas. **VHS, Beta** SUN *Woof!*

The Sentinel A model, who has moved into a New York City brownstone, encounters an aging priest and some unusual neighbors. When she investigates strange noises she finds out that the apartment building houses the doorway to hell - and that she's intended to be the next doorkeeper. Modest suspense film with a good cast and enough shock and special effects to keep the viewer interested.
1976 (R) 92m/C Chris Sarandon, Christina Raines, Ava Gardner, Jose Ferrer, Sylvia Miles, John Carradine, Burgess Meredith, Tom Berenger, Beverly D'Angelo, Jeff Goldblum; *Dir:* Michael Winner. **VHS, Beta** $14.98 MCA 🎬🎬

Separacion Matrimonial A woman decides to forgive her husband and stop divorce proceedings when his outside love affair ends. Soon, however, he is up to his old tricks. In Spanish.
197? 96m/C *SP* Jacqueline Andere, Ana Belen, Simon Andrew. **VHS, Beta** $29.95 MED 🎬½

Separate But Equal One of TV's greatest history lessons, a powerful dramatization of the 1954 Brown vs. The Board of Education case that wrung a landmark civil rights decision from the Supreme Court. Great care is taken to humanize all the participants, from the humblest schoolchild to NAACP lawyer Thurgood Marshall (Poitier). On two cassettes.
1990 (PG) 194m/C Sidney Poitier, Burt Lancaster, Richard Kiley; *W/Dir:* George Stevens. Emmy '91: Best Miniseries, Best Casting. **VHS, LV** $89.98 REP, PIA, FCT 🎬🎬🎬½

A Separate Peace Two young men in a New England prep school during World War II come to grips with the war, coming of age, and a tragic accident. Cheap rendition of a not-so-great novel by John Knowles.
1973 (PG) 104m/C John Heyl, Parker Stevenson, William Roerick; *Dir:* Larry Peerce. **VHS, Beta** $45.95 PAR, KUI 🎬🎬

Separate Tables Adaptation of the Terence Rattigan play about a varied cast of characters living out their personal dramas in a British seaside hotel. Guests include a matriarch and her shy daughter, a divorced couple, a spinster, and a presumed war hero. Their secrets and loves are examined in grand style. Fine acting all around. Received seven Academy Award nominations including Adapted Screenplay, Black & White Cinematography, Score of a Drama or Comedy, and Best Actress (Kerr).
1958 98m/B Burt Lancaster, David Niven, Rita Hayworth, Deborah Kerr, Wendy Hiller, Rod Taylor, Gladys Cooper, Felix Aylmer, Cathleen Nesbitt; *Dir:* Delbert Mann. Academy Awards '58: Best Actor (Niven), Best Supporting Actress (Hiller); New York Film Critics Awards '58: Best Actor (Niven). **VHS, Beta** $29.98 FOX, MGM, FCT 🎬🎬🎬

Separate Tables A remake of the 1958 film adapted from Terence Rattigan's play. Separated into two parts, "Table by the Window" and "Table Number Seven." The lives of the lonely inhabitants of a hotel are dramatized through wonderful acting by Christie and Bates. Made for cable television.

1983 (PG) 50m/C Julie Christie, Alan Bates, Claire Bloom, Irene Worth; *Dir:* John Schlesinger. **VHS, Beta** $59.95 MGM 🎬½

Separate Vacations A married man leaves his wife and kids to take a separate vacation, hoping he'll find exciting sex. Wife and kids end up at a ski resort, where romance comes her way. A lightweight, old story.
1986 (R) 91m/C David Naughton, Jennifer Dale, Mark Keyloun; *Dir:* Michael Anderson Sr. **VHS, Beta** $79.98 LIV, LTG 🎬½

Separate Ways An unhappily married couple, each having an affair, split up in order to deal with themselves and their marriage. Good cast; bad script wandering down Maudlin Lane.
1982 (R) 92m/C Karen Black, Tony LoBianco, David Naughton, Sybil Danning; *Dir:* Howard Avedis. **VHS, Beta, LV** $79.95 VES 🎬½

Sepia Cinderella A songwriter, who finds himself with a hit, abandons his current life and love to try high society. He finds out it's not what he wants after all.
1947 67m/B Sheila Guyse, Rubel Blakely, Freddie Bartholomew, The John Kirby Sextet, Sid Catlett, Deke Wilson. **VHS, Beta** $29.95 NOS, DVT 🎬🎬

September Woody does Bergman again with a shuttered, claustrophobic drama about six unhappy people trying to verbalize their feelings in a dark summer house in Vermont. Well-acted throughout and interesting at first, but the whining and angst attacks eventually give way to boredom. Of course, the Woodman went on to "Crimes and Misdemeanors," blending his dark and comedic sides masterfully, so the best way to look at this is as a training film. Nice music by Art Tatum.
1988 (PG-13) 82m/C Mia Farrow, Dianne Wiest, Denholm Elliott, Sam Waterston, Elaine Stritch, Jack Warden; *Dir:* Woody Allen. **VHS, Beta, LV** $89.98 ORI 🎬🎬½

September 30, 1955 A college undergrad (Thomas) in a small Arkansas town is devastated when he learns of the death of his idol, James Dean. (The title refers to the day Dean died.) Jimmy gathers a group of friends together for a vigil which turns into a drinking-binge, resulting in police chases and, finally, tragedy. Film debut of Quaid. Also known as "9/30/55."
1977 (PG) 107m/C Richard Thomas, Lisa Blount, Deborah Benson, Tom Hulce, Dennis Christopher, Dennis Quaid, Susan Tyrrell; *W/Dir:* James Bridges. **VHS** MCA 🎬🎬½

September 1939 A Canadian documentary about the German invasion of Poland that began World War II.
1961 60m/B *CA Nar:* Frank Willis. **VHS, Beta** $24.95 VYY 🎬🎬½

September Affair A married engineer and a classical pianist miss their plane from Naples. When the plane crashes, they're presumed dead, and they find themselves free to continue their illicit love affair. They find they cannot hide forever and must make peace with their pasts. Features the hit "September Song," recorded by Walter Huston.
1950 104m/B Joseph Cotten, Joan Fontaine, Francoise Rosay, Jessica Tandy, Robert Arthur; *Dir:* William Dieterle. **VHS, LV** $14.95 PAR 🎬½

September Gun A nun hires an aging gunfighter to escort her and a group of Apache children to a church school two hundred miles away. Preston is excellent as

the ornery varmint in this enjoyable, pleasant western comedy. Made for television.
1983 94m/C Robert Preston, Patty Duke, Christopher Lloyd, Geoffrey Lewis, Sally Kellerman; *Dir:* Don Taylor. **VHS, Beta $39.95** WOV ⅃⅃½

Sergeant Matlovich vs. the U.S. Air Force A serviceman fights to remain in the Air Force after admitting his homosexuality. Compelling theme (from a true story) and good lead performance from Dourif are not enough to carry a rather flat made-for-TV script.
1978 100m/C Brad Dourif, Marc Singer, Frank Converse, William Daniels, Stephen Elliott, David Spielberg, Rue McClanahan, Mitchell Ryan, David Ogden Stiers; *Dir:* Paul Leaf. **VHS $59.95** IVE ⅃⅃½

Sergeant Ryker Confident lead acting from Marvin as a U.S. serviceman accused of treason during the Korean War carries this otherwise blah army/courtroom drama. Remake of 1963 TV film "The Case Against Sergeant Ryker." Dillman is good as Marvin's energetic lawyer.
1968 85m/C Lee Marvin, Bradford Dillman, Vera Miles, Peter Graves, Lloyd Nolan, Murray Hamilton; *Dir:* Buzz Kulik. **VHS $9.95** CAR ⅃⅃½

Sergeant Sullivan Speaking A romance develops between a widow and a police sergeant as the two search for the widow's lost sons. From the ABC television series "Return Engagement."
1953 24m/B William Bendix, Joan Blondell, Sarah Selby. **VHS, Beta $19.98** CCB ⅃⅃½

Sergeant York Timely and enduring war movie based on the true story of Alvin York, the country boy from Tennessee drafted during World War I. At first a pacifist, Sergeant York (Cooper, well cast in an Oscar-winning role) finds justification for fighting and becomes one of the war's greatest heros. Gentle scenes of rural life contrast with horrific battlegrounds. York served as a consultant. Oscar nominations: Best Picture, Best Supporting Actor (Brennan - his fourth nomination), Supporting Actress (Wycherly), Best Director, Original Screenplay, Black & White Cinematography, Black & White Interior Decoration, Sound Recording, Musical Score for a Dramatic Film.
1941 134m/B Gary Cooper, Joan Leslie, Walter Brennan, Dickie Moore, Ward Bond, George Tobias, Noah Beery Jr., June Lockhart, Abe Finkel; *Dir:* Howard Hawks. Academy Awards '41: Best Actor (Cooper), Best Film Editing. **VHS, Beta $19.98** FOX, MGM, BTV ⅃⅃⅃⅃

Serial Fun spoof of hyper-trendiness—open marriage, health foods, fad religions, navel-gazing—in Marin County, the really cool place to live across the Golden Gate from San Francisco. Mull in his first lead is the oddly normal guy surrounded by fruits and nuts. Based on Cyra McFadden's novel.
1980 (R) 90m/C Martin Mull, Sally Kellerman, Tuesday Weld, Tom Smothers, Bill Macy, Peter Bonerz, Barbara Rhoades, Christopher Lee; *Dir:* Bill Persky. **VHS, Beta, LV $49.95** PAR ⅃⅃⅃

Sermon on the Mount Filmization of Jesus' famous sermon, here consisting of illustrative paintings and music by Bach, Beethoven, and Prokofiev.
1964 29m/B *Dir:* Andre Girard. **VHS, Beta $19.95** VYY ⅃⅃⅃

The Serpent of Death An archaeologist uncovers a statue and a curse which changes his life. Horror and adventure rolled into one.
1990 (R) 90m/C VHS, Beta, LV $79.95 PAR ⅃⅃

Serpent Island Two seamen fight over a Caribbean woman amid voodoo rituals, sea monsters and buried treasure.
1954 63m/C Sonny Tufts. **VHS, Beta $59.95** MED ⅃½

The Serpent and the Rainbow A good, interesting book (by Harvard ethnobotanist Wade Davis) offering serious speculation on the possible existence and origin of zombies in Haiti was hacked and slashed into a cheap Wes Craven-ized screen semblance of itself. And a shame it is: the result is racist, disrespectful, exploitative and superficial. From the director of "Nightmare on Elm Street."
1987 (R) 98m/C Bill Pullman, Cathy Tyson, Zakes Mokae, Paul Winfield, Conrad Roberts, Badja Djola, Theresa Merritt; *Dir:* Wes Craven. **VHS, Beta, LV $19.95** MCA ⅃½

The Serpent's Egg Big disappointment from the great Bergman in his second film in English. Ullman is not sultry, and Carradine is horrible. Big budget matched by big Bergman ego, making a big, bad parody of himself. The plot concerns a pair of Jewish trapeze artists surviving in Berlin during Hitler's rise by working in a grisly and mysterious medical clinic. Foreign title is "Das Schlangenei." Nylerist's customary superb cinematography is far from enough.
1978 (R) 119m/C GE David Carradine, Liv Ullman, Gert Frobe, James Whitmore; *Dir:* Ingmar Bergman. **VHS, Beta $79.98** LIV, LTG ⅃½

Serpico Based on Peter Maas's book about the true-life exploits of Frank Serpico, a New York undercover policeman who exposed corruption in the police department. Known as much for his nonconformism as for his honesty, the real Serpico eventually retired from the force and moved to Europe. South Bronx-raised Pacino gives the character reality and strength and received an Oscar nomination for his performance. Excellent New York location photography.
1973 (R) 130m/C Al Pacino, John Randolph, Jack Kehoe, Barbara Eda-Young, Cornelia Sharpe, F. Murray Abraham, Tony Roberts; *Dir:* Sidney Lumet. **VHS, Beta, LV $14.95** PAR ⅃⅃⅃

The Servant A dark, intriguing examination of British class hypocrisy and the master-servant relationship. A rich, bored playboy is ruined by his socially inferior but crafty and ambitious manservant. Playwright Harold Pinter wrote the adaptation of Robin Maugham's novel in his first collaboration with director Losey. The best kind of British societal navel-gazing.
1963 112m/B GB Dirk Bogarde, James Fox, Sarah Miles, Wendy Craig; *Dir:* Joseph Losey. British Academy Awards '63: Most Promising Newcomer (Fox), Best Actor (Bogarde), Best Black and White Cinematography. **VHS, Beta $59.95** HBO ⅃⅃⅃½

Servants of Twilight Religious zealots target a small boy for assassination because their cult leader says he's the anti-Christ, in an adaptation of the Dean Koontz novel. Packed with action, its spell on the viewer hinges on a cruel shock ending that undercuts what came before.
1991 (R) 95m/C Bruce Greenwood, Belinda Bauer, Grace Zabriskie, Richard Bradford, Jarrett Lennon, Carel Struycken, Jack Kehoe, Kelli Maroney; *Dir:* Jeffrey Obrow. **VHS $89.98** VMK ⅃⅃

Sesame Street Presents: Follow That Bird Television's Big Bird suffers an identity crisis, and leaves Sesame Street to join a family of real birds. He soon misses his home, and returns, in a danger-filled journey.
1985 (G) 92m/C Sandra Bernhard, John Candy, Chevy Chase, Joe Flaherty, Dave Thomas, Waylon Jennings, Jim Henson's Muppets; *Dir:* Ken Kwapis. **VHS, Beta, LV $19.98** WAR ⅃⅃

A Session with the Committee Collection of skits performed by a Los Angeles comedy troupe.
1969 90m/C Peter Bonerz, Barbara Bosson, Garry Goodrow, Carl Gottlieb, Jessica Myerson, Christopher Ross, Melvin Stewart, Melvin Sturdy; *Dir:* Jack Del. **VHS $59.95** PAV ⅃⅃

Sessions Mentally exhausted by her roles as sister, single parent, lover, exercise enthusiast, and high-priced prostitute, Leigh Churchill (Hamel) seeks professional counselling. Hamel's first role after "Hill Street Blues" is only average in this made-for-TV drama.
1983 96m/C Veronica Hamel, Jeffrey DeMunn, Jill Eikenberry, David Marshall Grant, George Coe, Henderson Forsythe, Deborah Hedwall; *Dir:* Richard Pearce. **VHS, Beta $59.95** VCL ⅃⅃

The Set-Up Excellent, original if somewhat overwrought morality tale about integrity set in the world of boxing. Filmed as a continuous narrative covering only 72 minutes in the life of an aging fighter (Ryan). Told to throw a fight, he must battle gangsters when he refuses. Powerful, with fine performances, especially from Ryan in the lead. Inspired by Joseph Moncure March's narrative poem.
1949 72m/B Robert Ryan, Audrey Totter, George Tobias, Alan Baxter, James Edwards, Wallace Ford; *Dir:* Robert Wise. **VHS, Beta, LV $19.95** CCB, MED ⅃⅃⅃½

Seven Freelance desperado Smith takes $7 million of your tax dollars at work from the feds to get rid of Hawaiian gangsters. Might be tongue in cheek, though sometimes it's hard to tell. Fun action pic unfortunately characterized by bad camera work and over-earnest direction. Features the original of a gag made famous in "Raiders of the Lost Ark."
1979 (R) 100m/C William Smith, Barbara Leigh, Guich Koock, Art Metrano, Martin Kove, Richard Le Pore, Susan Kiger; *Dir:* Andy Sidaris. **VHS, Beta $59.95** IVE ⅃⅃½

Seven Alone Family adventure tale (based on Monroe Morrow's book "On to Oregon," based in turn on a true story) about seven siblings who undertake a treacherous 2000-mile journey from Missouri to Oregon, after their parents die along the way. Inspiring and all that; good family movie.
1975 (G) 85m/C Dewey Martin, Aldo Ray, Anne Collins, Dean Smith, Stewart Peterson; *Dir:* Earl Bellamy. **VHS, Beta $59.98** LGC, CVL ⅃⅃½

Seven Beauties Very dard war comedy about a small-time Italian crook in Naples with seven ugly sisters to support. He survives a German prison camp and much else; unforgettably, he seduces the ugly commandant of his camp to save his own life. Good acting and tight direction. Giannini received a well-deserved Oscar nomination, as did Wertmuller (the first woman director nominee), and the Screenplay.
1976 116m/C IT Giancarlo Giannini, Fernando Rey, Shirley Stoler; *Dir:* Lina Wertmuller. **VHS, Beta, LV $59.95** COL, APD ⅃⅃⅃⅃

Seven Blows of the Dragon The Chinese novel, "All Men Are Brothers," is the basis for this spectacular martial arts epic.

1973 (R) 81m/C David Chiang. VHS, Beta
WAR &&½

Seven Brides for Seven Brothers

The eldest of seven fur-trapping brothers in the Oregon Territory brings home a wife. She begins to civilize the other six, who realize the merits of women and begin to look for romances of their own. Thrilling choreography by Michael Kidd - don't miss "The Barn Raising." Charming performances by Powell and Keel, both in lovely voice. Oscar-nominated for Best Picture, Screenplay, Color Cinematography, and Film Editing. Based on Stephen Vincent Benet's story. Thrills, chills, singin', dancin'—a classic Hollywood good time. **1954 103m/C** Howard Keel, Jane Powell, Russ Tamblyn, Julie Newmar, Jeff Richards, Tommy Rall, Virginia Gibson; *Dir:* Stanley Donen. Academy Awards '54: Best Musical Score; National Board of Review Awards '54: Special Citation for Choreography (Michael Kidd), 10 Best Films of the Year. **VHS, Beta, LV, 8mm $19.95** MGM &&&½

The 7 Brothers Meet Dracula

Seven siblings meet guy with dental anomaly. Also known as "Legend of the Seven Golden Vampires." **1973 110m/C** *GB* Peter Cushing, David Chang, Robin Stewart, Julie Ege; *Dir:* Roy Ward Baker. **VHS $19.98** AVD &½

Seven Chances

Silent classic that Keaton almost didn't make, believing instead that it should go to Harold Lloyd. Lawyer Keaton finds that he can inherit $7 million if he marries by 7:00 p.m. After his girlfriend turns down his botched proposal, chaos breaks loose when he advertises for someone—anyone—to marry him and make him rich. Suddenly he finds what seems to be hundreds of women willing to make the sacrifice, setting up one of the great film pursuits. Memorable boulder sequence was added after preview audience indicated climax was lacking that certain something. **1925 60m/B** Buster Keaton, T. Roy Barnes, Snitz Edwards, Ruth Dwyer, Frankie Raymond; *Dir:* Buster Keaton. **VHS $49.95** EJB &&&½

Seven Cities of Gold

An expedition of Spanish conquistadors and missionaries descend on 18th-century California, looking for secret Indian caches of gold. Semi-lavish costume epic. **1955 103m/C** Anthony Quinn, Michael Rennie, Richard Egan, Rita Moreno, Jeffrey Hunter, Eduardo Noriega, John Doucette; *Dir:* Robert D. Webb. **VHS, Beta $19.98** FOX &&

Seven Days Ashore/Hurry, Charlie, Hurry

A comedy double feature. In "Seven Days Ashore," three merchant marines on leave in San Francisco become romantically involved with three women. In "Hurry, Charlie, Hurry," a henpecked husband finds himself in a lot of trouble. **1944 141m/B** Gordon Oliver, Virginia Mayo, Dooley Wilson, Margaret Dumont, Leon Errol, Mildred Coles. **VHS, Beta $34.95** RKO &&

Seven Days' Leave

Radio star-crammed musical comedy about a soldier who must marry a certain already betrothed gal (Ball) within a week so he can inherit a fortune. Songs include "Can't Get Out of this Mood," "A Touch of Texas," and "Baby, You Speak My Language." Following the completion of the film, Mature joined the Coast Guard. Lively but not memorable. **1942 87m/B** Victor Mature, Lucille Ball, Harold Peary, Mary Cortes, Ginny Simms, Marcy McGuire, Peter Lind Hayes, Walter Reed, Wallace Ford, Arnold Stang, Buddy Clark,

Charles Victor; *Dir:* Tim Whelan. **VHS, Beta $19.98** TTC, NVH &&

Seven Days in May

Topical but still gripping Cold War nuclear-peril thriller. An American general plans a military takeover because he considers the president's pacifism traitorous. Screenplay by Rod Serling of "Twilight Zone" fame. Highly suspenseful, with a breathtaking climax. Houseman's film debut in a small role. O'Brien won a Best Supporting Actor Oscar nomination as did the Art Direction and Set Decoration. **1964 120m/B** Burt Lancaster, Kirk Douglas, Edmond O'Brien, Fredric March, Ava Gardner, John Houseman; *Dir:* John Frankenheimer. **VHS, Beta, LV $49.95** PAR &&&½

Seven Deadly Sins

A collection of tales concerning destructive emotions and the downfalls they cause. Contains shorts from some of Europe's finest directors (five French and two Italian). Stories and their directors are "Avarice and Anger" (de Filippo), "Sloth" (Dreville), "Lust" (Allegret), "Envy" (Rossellini), "Gluttony" (Rim), "Pride" (Autant-Lara) and "The Eighth Sin" (Lacombe). **1953 127m/B** *FR IT* Eduardo de Filippo, Isa Miranda, Jacqueline Plessis, Louis de Funes, Paolo Stoppa, Frank Villard, Jean Richard, Francoise Rosay, Jean Debucourt, Louis Seigner; *Dir:* Eduardo de Filippo, Jean Dreville, Yves Allegret, Roberto Rossellini, Carlo Rim, Claude Autant-Lara, Georges Lacombe. **VHS** IVY &&&

Seven Deaths in the Cat's Eye

A ravenous beast slaughters people in a small Scottish village. This flick reveals a little-known fact about felines. **1972 90m/C** *IT* Anton Diffring, Jane Birkin; *Dir:* Anthony M. Dawson. **VHS, Beta $79.95** PSM &&

Seven Doors to Death

A young architect tries to solve a crime to avoid being placed under suspicion. The suspects are six shop owners in this low-budget attempt at a murder mystery. **1944 70m/B** Chick Chandler, June Clyde, George Meeker, Gregory Gay, Edgar Dearing; *Dir:* Elmer Clifton. **VHS, Beta $14.95** UHV, MRV, SNC &

Seven Doors of Death

A young woman inherits a possessed hotel. Meanwhile, hellish zombies try to check out. Chilling Italian horror flick. Originally titled "The Beyond." **1982 (R) 80m/C** *IT* Katherine MacColl, David Warbeck, Farah Keller, Tony St. John; *Dir:* Louis Fuller. **VHS, Beta $39.95** IVE &&

Seven Faces of Dr. Lao

Dr. Lao is the proprietor of a magical circus that changes the lives of the residents of a small western town. Marvelous special effects and makeup (Randall plays seven characters) highlight this charming family film in the Pal tradition. Charles Finney adapted from his novel. **1963 101m/C** Tony Randall, Barbara Eden, Arthur O'Connell, Lee Patrick, Noah Beery Jr., John Qualen; *Dir:* George Pal. Academy Awards '64: Best Makeup. **VHS, Beta $59.95** MGM &&&

Seven Hours to Judgment

When the punks who murdered his wife go free, a psychotic man (Liebman) decides to take justice into his own hands. He kidnaps the wife of the judge in charge (Bridges, who also directed) and leads him on a wild goose chase. Liebman steals the film as the psycho. Ex-Springsteen spouse Phillips plays the judge's wife. Ambitious but credulity-stretching revenge/crime drama.

1988 (R) 90m/C Beau Bridges, Ron Leibman, Julianne Phillips, Al Freeman Jr., Reggie Johnson; *Dir:* Beau Bridges. **VHS, Beta, LV $19.95** MED &&

Seven Indignant

A group of elite fighters is China's only hope against the invading Japanese. **19?? ?m/C** Kong Hoi. **VHS, Beta** OCE &

The Seven Little Foys

Enjoyable musical biography of Eddie Foy (played ebulliantly by Hope) and his famed vaudevillian troupe. Cagney's appearance as George M. Cohan is brief, but long enough for a memorable dance duet with Hope. **1955 95m/C** Bob Hope, Milly Vitale, George Tobias, Angela Clarke, James Cagney; *Dir:* Melville Shavelson. **VHS, Beta $19.95** COL &&&

Seven Magnificent Gladiators

Seven gladiators team up to save a peaceful Roman village from total annihilation. Dismal effort at a remake of "The Seven Samurai"—see the original instead, please. Ferrigno was "The Incredible Hulk" on TV. **1984 (PG) 86m/C** Lou Ferrigno, Sybil Danning, Brad Harris, Dan Vadis, Carla Ferrigno; *Dir:* Bruno Mattei. **VHS, Beta $59.95** MGM &

Seven Miles from Alcatraz/Flight from Glory

Dramatic double feature: Two escaped convicts discover a hideout for Nazi spies in "Seven Miles From Alcatraz," and an evil man tries to recruit pilots to fly planes over the Andes Mountains in "Flight From Glory." **1943 127m/B** James Craig, Bonita Granville, Chester Morris, Van Heflin, Onslow Stevens; *Dir:* Edward Dmytryk. **VHS, Beta $34.95** RKO &&

Seven Minutes in Heaven

Sensitive love story about a 15-year-old girl who invites her platonic male friend to live in her house, and finds it disturbs her boyfriend, as these things will. Ever so tasteful (unlike most other teen comedies), with gentle comedy—but forgettable. **1986 (PG) 90m/C** Jennifer Connelly, Byron Thames, Maddie Corman; *Dir:* Linda Feferman. **VHS, Beta $19.98** WAR &&

The Seven Percent Solution

Dr. Watson persuades Sherlock Holmes to meet with Sigmund Freud to cure his cocaine addiction. Holmes and Freud then find themselves teaming up to solve a supposed kidnapping. Adapted by Nicholas Meyer from his own novel. One of the most charming Holmes films; well-cast, intriguing blend of mystery, drama, and fun. **1976 (PG) 113m/C** Alan Arkin, Nicol Williamson, Laurence Olivier, Robert Duvall, Vanessa Redgrave, Joel Grey, Samantha Eggar, Jeremy Kemp, Charles Gray, Regine; *Dir:* Herbert Ross. National Board of Review Awards '76: 10 Best Films of the Year. **VHS, Beta, LV $34.98** MCA &&&

Seven Samurai

Kurosawa's masterpiece, set in 16th-century Japan. A small farming village, beset by marauding bandits, hires seven professional soldiers to rid itself of the scourge. Wanna watch a samurai movie? This is the one. Sweeping, complex human drama with all the ingredients: action, suspense, comedy. Available in several versions of varying length, all long—and all too short. Splendid acting. The classic western "The Magnificent Seven" is the best known U.S. remake. In Japanese with English subtitles. **1954 204m/B** *JP* Toshiro Mifune, Takashi Shimura, Yoshio Inaba, Kuninori Kodo; *Dir:* Akira Kurosawa. Venice Film Festival '54: Silver Prize; Sight & Sound Survey '82: #3 of the Best Films

of All Time (tie). **VHS, Beta, LV, 3/4U $24.95**
NOS, SUE, HHT 🐾🐾🐾🐾

Seven Sinners A South Seas cabaret singer (Dietrich) attracts sailors like flies, resulting in bar brawls, romance, and intrigue. Manly sailor Wayne falls for her. A good-natured, standard Hollywood adventure. Well cast; performed and directed with gusto.
1940 83m/B Marlene Dietrich, John Wayne, Albert Dekker, Broderick Crawford, Mischa Auer, Billy Gilbert, Oscar Homolka; *Dir:* Tay Garnett. **VHS, Beta $16.95** *SNC, NOS, KRT* 🐾🐾🐾

Seven Sixgunners A western classic from the book by Nelson Nye.
1987 60m/C *Dir:* George Potter. **VHS $49.95** *AVD* 🐾🐾

Seven Thieves Charming performances and nice direction make this tale of the perfect crime especially watchable. Robinson, getting on in years, wants one last big heist. With the help of Collins and Steiger, he gets his chance. From the Max Catto novel "Lions at the Kill." Surprisingly witty and light-hearted for this subject matter, and good still comes out ahead of evil.
1960 102m/C Joan Collins, Edward G. Robinson, Eli Wallach, Rod Steiger; *Dir:* Henry Hathaway. **VHS $39.98** *FOX, FCT* 🐾🐾🐾

The Seven-Ups An elite group of New York City detectives seeks to avenge the killing of a colleague and to bust crooks whose felonies are punishable by jail terms of seven years or more. Unoriginal premise portends ill; plotless cop action flick full of car chases. Scheider tries hard. Directed by the producer of "The French Connection."
1973 (PG) 109m/C Roy Scheider, Tony LoBianco, Larry Haines, Jerry Leon; *Dir:* Phil D'Antoni. **VHS, Beta, LV $59.98** *FOX* 🐾½

The Seven Year Itch Classic, sexy Monroe comedy. Stunning blonde model (who else?) moves upstairs just as happily married guy Ewell's wife leaves for a long vacation. Understandably, he gets itchy. Monroe's famous blown skirt scene is here, as well as funny situations and appealing performances.
1955 105m/C Marilyn Monroe, Tom Ewell, Evelyn Keyes, Sonny Tufts, Victor Moore, Doro Merande, Robert Strauss, Oscar Homolka, Carolyn Jones; *Dir:* Billy Wilder. **VHS, Beta, LV $14.98** *FOX* 🐾🐾🐾

Seven Years Bad Luck French comic Linder has the proverbial seven years' bad luck all in one day (!) after he breaks a mirror. Original and quite funny, full of fetching sophisticated sight gags. Silent with music score.
1920 67m/B Max Linder. **VHS, Beta $29.95** *VYY* 🐾🐾🐾

1776 A musical comedy about America's first Continental Congress. The delegates battle the English and each other trying to establish a set of laws and the Declaration of Independence. Adapted from the Broadway hit with many members of the original cast. Available in widescreen format on laserdisc.
1972 (G) 141m/C William Daniels, Howard da Silva, Ken Howard, Donald Madden, Blythe Danner, Ronald Holgate, Virginia Vestoff, Stephen Nathan, Ralston Hill; *Dir:* Peter Hunt. National Board of Review Awards '72: 10 Best Films of the Year. **VHS, Beta, LV $19.95** *COL, PIA, FCT* 🐾🐾🐾

The Seventeenth Bride Set in a Czechoslovakian town; a strong-willed young woman is slowly destroyed by the insanity of war and racism.

1984 92m/C Lisa Hartman, Rosemary Leach; *Dir:* Israeli Nadav Levitan. **VHS $79.95** *SVS* 🐾🐾

Seventh Cavalry A somewhat different look at Custer's defeat at the Little Big Horn. A soldier who was branded a coward for not taking part in the festivities tries to assuage his guilt by heading up the burial detail. The muddled ending tries to tell us that the Indians were afraid of Custer's horse.
1956 75m/C Randolph Scott, Barbara Hale, Jay C. Flippen, Jeanette Nolan, Frank Faylen, Leo Gordon, Denver Pyle, Harry Carey Jr., Michael Pate, Donald Curtis, Frank Wilcox, Pat Hogan, Russell Hicks; *Dir:* Joseph H. Lewis. **VHS $19.95** *NOS* 🐾🐾½

The 7th Commandment Unlikely melodrama about a man afflicted with amnesia following an auto accident. He becomes a successful evangelist only to be blackmailed by an old girlfriend. Don't you just hate when that happens?
1961 82m/B Jonathan Kidd, Lynn Statten; *Dir:* Irvin Berwick. **VHS $15.95** *NOS, LOO* 🐾½

The Seventh Seal As the plague sweeps through Europe a weary knight convinces "death" to play one game of chess with him. If the knight wins, he and his wife will be spared. The game leads to a discussion of religion and the existence of God. Considered by some Bergman's masterpiece. Von Sydow is stunning as the knight. In Swedish with English subtitles. Also available on laserdisc with a filmography of Bergman's work and commentary by film historian Peter Cowie.
1956 96m/B *SW* Gunnar Bjornstrand, Max von Sydow, Bibi Andersson, Bengt Ekerot, Nils Poppe, Gunnel Lindblom; *Dir:* Ingmar Bergman. **VHS, Beta, LV, 8mm $19.95** *VDM, SUE, VYG* 🐾🐾🐾🐾

The Seventh Sign A pregnant woman realizes that the mysterious stranger boarding in her house and the bizarre events that accompany him are connected to Biblical prophesy and her unborn child. Tries hard, but it's difficult to get involved in the supernatural goings-on.
1988 (R) 105m/C Demi Moore, Juergen Prochnow, Michael Biehn, John Heard, Peter Friedman, Manny Jacobs, John Taylor, Lee Garlington, Akosua Busia; *Dir:* Carl Schultz. **VHS, Beta, LV $14.95** *COL* 🐾🐾

Seventh Veil A concert pianist loses the use of her hands in a fire, and with it her desire to live. Through the help of her friends and a hypnotizing doctor, she regains her love for life. Superb, dark psycho-drama. Todd as the pianist and Mason as her guardian are both unforgettable. Wonderful music and staging.
1946 91m/B *GB* James Mason, Ann Todd, Herbert Lom, Hugh McDermott, Albert Lieven; *Dir:* Compton Bennett. Academy Awards '46: Best Original Screenplay. **VHS, Beta $29.95** *VID* 🐾🐾🐾½

The Seventh Victim Another Val Lewton-produced low-budget exercise in shadowy suggestion, dealing with a woman searching for her lost sister, who'd gotten involved with Satanists. The Hays Office's squeamishness regarding subject matter makes the film's action a bit cloudy but it remains truly eerie.
1943 71m/B Kim Hunter, Tom Conway, Jean Brooks, Hugh Beaumont, Erford Gage, Isabel Jewell, Evelyn Brent; *Dir:* Mark Robson. **VHS, Beta, LV $14.95** *RKO* 🐾🐾🐾

The Seventh Voyage of Sinbad Sinbad seeks to restore his fiancee from the midget size to which an evil magician (Thatcher) has reduced her. Ray Harryhausen works his animation magic around a well-developed plot and engaging performances by the real actors. Bernard Herrmann's score is great, as is fun, fast-moving plot.
1958 (G) 94m/C Kerwin Mathews, Kathryn Grant, Torin Thatcher, Richard Eyer; *Dir:* Nathan Juran. **VHS, Beta, LV $14.95** *COL, MLB, CCB* 🐾🐾🐾

Severance An ex-pilot attempts to save his daughter from the world of drugs and violence.
1988 93m/C Lou Liotta, Lisa Nicole Wolpe, Linda Christian-Jones; *Dir:* David Max Steinberg. **VHS, Beta $79.95** *MNC* 🐾

The Severed Arm Trapped in a cave, five men cut off the arm of a companion in order to ward off starvation.
1973 (R) 89m/C Deborah Walley, Marvin Kaplan, Paul Carr, John Crawford, David Cannon; *Dir:* Thomas Alderman. **VHS, Beta $59.95** *GEM, MRV* 🐾

Severed Ties A brilliant genetic scientist experiments with human limb regeneration in order to re-grow his accidentally severed arm. But his strange mixture of lizard and serial-killer genes results in a repulsive repitilian limb with a nasty habit of slithering out of his shoulder. When his mother discovers his experiments she tries to sell his discovery to the Nazis but her son escapes into a bizarre subterranean society and vows revenge.
1992 (R) 95m/C Billy Morrissette, Elke Sommer, Oliver Reed, Garrett Morris; *Dir:* Damon Santostefano. **VHS $89.95** *COL* 🐾½

Sewing Woman This program presents the story of the life of a Chinese woman who uses her sewing machine to represent the passage of time.
1981 60m/C **VHS, Beta** *PIC* 🐾½

Sex Interesting dated relic. Morality play about a dance-hall girl who uses her charms to destroy a marriage. The businessman she lands shamelessly betrays her. What a title, especially for its time!
1920 87m/B Adrienne Renault, Louise Glaum, Irving Cummings, Peggy Pearce, Myrtle Steadman. **VHS, Beta $19.98** *CCB, GPV* 🐾🐾½

Sex Adventures of the Three Musketeers Soft-core spoof of the Dumas classic. Dubbed.
198? 79m/C Inga Steeger, Achim Hammer, Peter Graf, Jurg Coray. **VHS, Beta $29.95** *MED* Woof!

Sex Appeal Another smut comedy depicting the sex-gathering adventures of a horny young man.
1986 (R) 84m/C Brian Bonanno, Tally Brittany, Marcia Karr, Jerome Brenner, Marie Sawyer, Philip Campanaro, Jeff Eagle, Gloria Leonard, Molly Morgan; *Dir:* Chuck Vincent. **VHS, Beta $79.98** *LIV, VES* 🐾

Sex on the Brain A PhD student gets a little too involved with her research on prostitution.
19?? 90m/C Pilar Valesquez. **VHS, Beta $59.95** *MPI* 🐾🐾

Sex and Buttered Popcorn Stylish and entertaining documentary of the Hollywood exploitation moguls, and their films, from the 1920s through the 1950s. These films shamelessly catered to audience's prurient interests, but typically delivered saccharin-coated morality messages and stock footage of routine baby births or the horrible

effects of sexually transmitted diseases. Hosted by Ned Beatty, and featuring clips from the actual films, plus an interview with the widow of Kroger Babb, the most notorious of the exploiteers. A real hoot.
1991 (R) 70m/C David Friedman; *Dir:* Sam Harrison; *Hosted:* Ned Beatty. **VHS $29.95** *CPM, FCT* 🎞️🎞️

Sex and the College Girl Folks talk about the relationship problems they are having while vacationing in Puerto Rico. Tame talkfest providing Grodin with his first role. Included as part of Joe Bob Brigg's "Sleaziest Movies in the History of the World" series, probably just for the title.
1964 ?m/C Charles Grodin, Richard Arlen, Luana Anders; *Dir:* Joseph Adler. **VHS** *SVI* 🎞️

Sex in the Comics Some of your favorite comic-strip characters are portrayed in this collection of erotic skits.
19?? 70m/C VHS, Beta $19.99 *BFV* 🎞️

Sex, Drugs, and Rock-n-Roll Several girls frequent a rock club looking for all the action they can get their hands on.
1984 90m/C Jeanne Silver, Sharon Kane, Tish Ambrose, Josey Duval. **VHS, Beta $59.98** *VHM Woof!*

sex, lies and videotape The acclaimed, popular independent film by first-timer Soderbergh, detailing the complex relations among a childless married couple, the wife's adulterous sister, and a mysterious college friend of the husband's obsessed with videotaping women as they talk about their sex lives. Heavily awarded, including first prize at Cannes. Confidently uses much (too much?) dialogue and slow (too slow?) pace. Available on laserdisc with a deleted scene and interviews with the director; also includes two theatrical trailers and a short film by Soderbergh.
1989 (R) 101m/C James Spader, Andie MacDowell, Peter Gallagher, Laura San Giacomo, Ron Vawter, Steven Brill; *Dir:* Steven Soderberg. Cannes Film Festival '89: Best Actor (Spader), Best Film; Los Angeles Film Critics Association Awards '89: Best Actress (MacDowell). **VHS, Beta, LV, 8mm $19.95** *COL, VYG* 🎞️🎞️🎞️

The Sex Machine In the year 2037, a scientist finds two of the world's greatest lovers and unites them so he can transform their reciprocating motion into electricity.
1976 (R) 80m/C Agostina Belli. **VHS, Beta $19.95** *MED Woof!*

Sex Madness Then—deadly serious, now—campy educational melodrama about the "madness" of sex and the peril of syphilis. Quaint, to say the least, in view of all that has intervened—and all the onscreen sex—since 1937.
1937 50m/B VHS, Beta $19.95 *NOS, HHT, BUR Woof!*

Sex O'Clock News Fervently paced series of comedy skits. Made for television. Meant to be zany and irreverent; succeeds only in being dull and dorky.
1985 (R) 80m/C Doug Ballard, Lydia Mahan, Wayne Knight, Don Pardo, Kate Weiman, Joy Bond; *Dir:* Roman Vanderbes. **VHS, Beta $79.95** *PSM* 🎞️

Sex and the Office Girl An advertising agency turns into an after-hours pleasure dome of erotic complications.
197? 76m/C VHS, Beta $19.95 *MED* 🎞️

Sex and the Other Woman A series of vignettes depicting how women manipulate men into sexual submission.

1979 (R) 80m/C Richard Wattis, Maggie Wright, Felicity Devon, Jane Carder. **VHS, Beta $69.95** *VID* 🎞️

Sex Ray Inexplicable light ray zaps everyone's clothing and trepidations. A sex frolic ensues.
19?? 90m/C VHS $39.95 *ACF* 🎞️

Sex on the Run The love-starved wife of an oil-rich sheik, stimulated by the idea of having Casanova for her lover, teases her master into delivering him, but Casanova finds peace in the arms of three convent lovelies. Disgusting, amateurish and offensive. Yuk!
1978 (R) 88m/C Tony Curtis, Marisa Berenson, Britt Ekland; *Dir:* Francosis Legrand. **VHS, Beta, LV $69.95** *VES* 🎞️

Sex and the Single Girl Curtis is a reporter for a trashy magazine who intends to write an expose on "The International Institute of Advanced Marital and Pre-Marital Studies," an organization run by Wood. Curtis poses as a man having marital trouble, using his neighbor's name and marital problems, in order to get close to Wood. Things get sticky when Wood wants to meet Curtis' wife and three women show up claiming to be her. This confusing but amusing tale twists and turns until it reaches a happy ending. Loosely based on the book by Helen Gurley Brown.
1964 114m/C Tony Curtis, Natalie Wood, Henry Fonda, Lauren Bacall, Mel Ferrer, Fran Jeffries, Leslie Parrish, Edward Everett Horton, Larry Storch; *Dir:* Richard Quine. **VHS, LV** *WAR* 🎞️🎞️½

Sex and the Single Parent Saint James and Farrell, both divorced, both with kids, have a relationship. Complications set in—just how does one be a single parent and have a fulfilling sex life in this day and age? Some of these made-for-TV efforts are so formulaic!
1982 98m/C Susan St. James, Mike Farrell, Dori Brenner, Warren Berlinger, Julie Sommars, Barbara Rhoades. **VHS, Beta** *TLF* 🎞️

Sex with a Smile Five slapstick episodes by five different directors with lots of sexual satire pointed at religion and politics in Italy. Don't know much about Italy? Then you'll be left out of the joke here. And really, only one episode is funny anyway (the one with Feldman of "Young Frankenstein" fame).
1976 100m/C *IT* Marty Feldman, Edwige Fenech, Alex Marino, Enrico Monterrano, Giovanna Ralli; *Dir:* Sergio Martino. **VHS, Beta $59.95** *GEM* 🎞️

Sex with the Stars In order to get a feel for his newspaper column, a horoscope writer conducts interviews with a beautiful woman born under each astrological sign.
1981 90m/C Martin Burrows, Faith Davkin, Suzannah Willis, Jada; *Dir:* Anwar Kawadri. **VHS, Beta $19.98** *TWE* 🎞️

Sex Surrogate A psychologist/sex surrogate gets very involved in her work.
1988 81m/C Marilyn Chambers. **Beta $79.95** *NSV Woof!*

Sex Through a Window A TV reporter becomes an obsessive voyeur after filing a report on high tech surveillance equipment. Flimsy premise is a yucky, lame excuse to show skin. Originally titled "Extreme Close-Up."
1972 (R) 81m/C James McMullan, James A. Watson Jr., Kate Woodville, Bara Byrnes, Al Checco, Antony Carbone; *Dir:* Jeannot Szwarc. **VHS, Beta $69.98** *LIV, VES* 🎞️

Sexcetera A Playboy Video program featuring sex-related news items.
1984 63m/C VHS, Beta $59.98 *FOX* 🎞️

Sexo Y Crimen Two lovers find themselves inextricably tangled in a violent situation.
1987 90m/C *MX* Enrique Lizalde, Emily Cranz, Enrique Rambal, Yoland Ciani. **VHS, Beta $54.95** *MAD* 🎞️🎞️

Sexpot A voluptuous woman's rich husbands mysteriously keep dying, but not before they leave their fortunes to her.
1988 (R) 95m/C Ruth Collins, Troy Donahue, Joyce Lyons, Gregory Patrick, Frank Stewart, Jack Carter; *Dir:* Chuck Vincent. **VHS, Beta $29.95** *ACA* 🎞️

Sextette Lavish film about an elderly star who is constantly interrupted by former spouses and well-wishers while on a honeymoon with her sixth husband. Mae West unwisely came out of retirement for this last film role, based on her own play. Exquisitely embarrassing to watch. Interesting cast.
1978 (PG) 91m/C Mae West, Timothy Dalton, Ringo Starr, George Hamilton, Dom DeLuise, Tony Curtis, Alice Cooper, Keith Moon, George Raft, Rona Barrett, Walter Pidgeon; *Dir:* Ken Hughes. **VHS, Beta $9.95** *MED, MLB Woof!*

Sexton Blake and the Hooded Terror Sexton Blake a British private detective a la Holmes, is after "The Snake," a master criminal, and his gang The Hooded Terror. One of a series of melodramatic adventures interesting chiefly for Slaughter's performance as the gangleader.
1938 70m/B *GB* Tod Slaughter, Greta Gynt, George Culzon, George King. **VHS, Beta $29.95** *SNC, VYY, MRV* 🎞️🎞️

Sexual Response Eve, a radio talk show "sexologist" is ironically unhappy with her marriage, but she scoffs at her producer's suggestion of having an affair. That is until she meets a brash sculptor and they begin a torrid, and increasingly obsessive, romance. Her lover suggests that her husband have an "accident" with his hunting rifle, but Eve refuses and throws him out. The sculptor, however, breaks into the husband's loft to steal the rifle and finds some news clippings about the man's past that could ruin the lives of many. Also available in an unrated version.
1992 (R) 87m/C Shannon Tweed, Catherine Oxenberg, Vernon Wells, Emile Levisetti; *Dir:* Yaky Yosha. **VHS, LV $89.95** *COL* 🎞️🎞️

Sexy Cat A comic strip character comes to life to duplicate the murders committed in the strip.
196? 90m/C Monika Kolpek, Maria Villa, Vidal Molina, Marques DeToro. **VHS, Beta** *TSV* 🎞️🎞️

Shack Out on 101 A waitress (Moore) in an isolated cafe on a busy highway notices suspicious doings among her customers. What could they be up to? Communist subversion, of course. Of-its-era anti-pinko propoganda, but with a twist: Moore uncovers commie plots, pleases her customers and fends off unwelcome lecherous advances all in 80 minutes, on a single set!
1955 80m/B Lee Marvin, Terry Moore, Keenan Wynn, Frank Lovejoy, Whit Bissell; *Dir:* Edward Dein. **VHS $29.95** *REP* 🎞️🎞️🎞️

Shades of Darkness This is a collection of British dramatizations of famous suspense tales from the likes of Daphne Du Maurier, Edith Wharton and Somerset Maugham.
198? 60m/C *GB* Beta *PSM* 🎞️🎞️🎞️

Shades of Love: Champagne for Two I really love him—but I want a career too! What's a gal to do? Vintage eighties yuppie self-involvement brought to the big screen.
1987 80m/C Kirsten Bishop, Nicholas Campbell. **VHS, Beta $9.98** LHV, WAR ♫♫½

Shades of Love: Echoes in Crimson Woman must deal with new career at art gallery and former lover who wants her back.
1987 90m/C Greg Evigan, Patty Talbot. **VHS $9.98** LHV, WAR ♫♫

Shadey The premise for this oddball comedy came to someone in the wee hours: I've got it! Mild-mannered mechanic can film people's thoughts; wants to use his gift only for good, but needs money for a sex-change operation. Assorted bad guys want to use him for evil purposes. Strange, but not bad; Helmond's character is memorable.
1987 (PG-13) 90m/C GB Anthony Sher, Billie Whitelaw, Patrick Macnee, Katherine Helmond; **Dir:** Philip Saville. **VHS, Beta $79.98** FOX ♫♫½

The Shadow Scotland Yard pulls out all the stops to track down a murderous extortionist, enlisting the aid of a novelist in their efforts. Long on talk and short on action.
1936 63m/B GB Henry Kendall, Elizabeth Allan, Jeanne Stuart, Felix Aylmer, Cyril Raymond; **Dir:** George Cooper. **VHS, Beta $16.95** NOS, SNC ♫½

The Shadow Box Three terminally ill people at a California hospice confront their destinies. Pulitzer Prize-winning play by Michael Cristofer is actually improved by director Newman and a superb, well-chosen cast. Powerful.
1980 96m/C Joanne Woodward, Christopher Plummer, Robert Urich, Valerie Harper, Sylvia Sidney, Melinda Dillon, Ben Masters, John Considine, James Broderick; **Dir:** Paul Newman. **VHS, Beta** KAR ♫♫♫

Shadow of Chikara A Confederate Army Captain and an orphan girl encounter unexpected adventures as they search for a fortune in diamonds hidden in a river in northern Arkansas.
1977 (PG) 96m/C Joe Don Baker, Sondra Locke, Ted Neely, Slim Pickens. **VHS, Beta, LV $19.95** STE, NWV ♫

Shadow of Chinatown A mad scientist creates a wave of murder and terror in Chinatown.
1936 70m/B Bela Lugosi, Herman Brix, Joan Barclay. **VHS, Beta $27.95** NOS, SNC, VCN ♫½

Shadow Dancing The proprietor of the auspicious Beaumont Theater of Dance is haunted by one of his former dancers who died unexpectedly during a performance of "Medusa." Now, over fifty years later, Jessica, the pick for the lead in a new production, begins to take on her predecessor's mannerisms and spirit as she faces her impending doom. Surprisingly believable and suspenseful—and scary!
1988 (PG) 100m/C Nadine Van Der Velde, James Kee, John Colicos, Shirley Douglas, Christopher Plummer; **Dir:** John Furey. **VHS, Beta $89.95** SGE ♫♫♫

Shadow of a Doubt Uncle Charlie has come to visit his relatives in Santa Rosa. Although he is handsome and charming, his young niece slowly comes to realize he is a wanted mass murderer - and he comes to recognize her suspicions. Hitchcock's personal favorite movie; a quietly creepy venture into Middle American menace. Good performances, especially by Cronyn. Co-written by Thornton Wilder, and score by Dmitri Tiomkin.
1942 108m/B Teresa Wright, Joseph Cotten, Hume Cronyn, MacDonald Carey, Henry Travers, Wallace Ford; **Dir:** Alfred Hitchcock. **VHS, Beta, LV $19.95** MCA ♫♫♫½

Shadow of the Eagle The intrigue and mystery of carnival life. Twelve chapters, 13 minutes each.
1932 226m/B John Wayne, Dorothy Gulliver; **Dir:** Ford Beebe. **VHS, Beta $24.95** NOS, VYY, VCN ♫♫♫½

Shadow Killers There's no time to fear your own shadow, for there is a far more terrifying darkness stalking the city.
198? 90m/C Chen Ching, Chang Wu Lang, Chen Chiang. **VHS, Beta $59.95** REG ♫½

The Shadow Man A casino operator in London (Romero) is accused of killing a former girlfriend and works to clear himself. Family conventional action mystery: nothing to write home about.
1953 76m/B GB Cesar Romero, Simone Silva, Kay Kendall, John Penrose, Edward Underdown; **Dir:** Richard Vernon. **VHS, Beta, LV $59.95** WGE ♫♫

Shadow Man A mysterious thief, a master of disguise, sets out to steal the Treasure of the Knights Templar and nothing will stand in his way.
1975 (PG) 89m/C Josephine Chaplin, Gert Frobe, Gayle Hunnicutt. **VHS, Beta $59.95** MED ♫♫

Shadow Play A successful female playwright is haunted by visions of a past lover. Her life slowly falls apart, but not before this film does.
1986 (R) 98m/C Dee Wallace Stone, Cloris Leachman; **Dir:** Susan Shadburne. **VHS, Beta $19.95** STE, NWV ♫½

The Shadow Riders Two brothers who fought on opposite sides during the Civil War return home to find their brother's fiancee kidnapped by a renegade Confederate officer who plans to use her as ransom in a prisoner exchange, and they set out to rescue the woman. Made for television as "Louis L'Amour's 'The Shadow Riders'."
1982 (PG) 96m/C Tom Selleck, Sam Elliott, Ben Johnson, Katharine Ross, Jeffery Osterhage, Gene Evans, R.G. Armstrong, Marshall Teague, Dominique Dunne, Jeanetta Arnetta; **Dir:** Andrew V. McLaglen. **VHS $79.98** VMK ♫♫

The Shadow Strikes An adaptation of the radio program. The sleuth pursues a killer and a gangster. Meanwhile, the police try to pin a robbery on the "Shadow." Who did it? The Shadow knows!
1937 61m/B Rod LaRocque, Lynn Anders. **VHS, Beta $16.95** SNC, NOS, CAB ♫

Shadow on the Sun The life of Beryl Markham, a pioneer female aviator and adventurer of the 1930s. Way too long and boring made-for-British-TV answer to "Out of Africa." On two cassettes.
1988 192m/C GB Stefanie Powers, Claire Bloom, Frederic Forrest, James Fox, John Rubenstein, Jack Thompson; **Dir:** Tony Richardson. **VHS $29.95** STE, NWV ♫½

Shadow of the Thin Man In the fourth "Thin Man" film, following "Another Thin Man," Nick and Nora stumble onto a murder at the racetrack. The rapport between Powell and Loy is still going strong, providing us with some wonderful entertainment. Followed by "The Thin Man Goes Home."
1941 97m/B William Powell, Myrna Loy, Barry Nelson, Donna Reed, Sam Levene, Alan Baxter, Dickie Hall, Loring Smith, Joseph Anthony, Henry O'Neill; **Dir:** W.S. Van Dyke. **VHS, Beta $19.98** MGM ♫♫♫

Shadow World A small boy transforms himself into a superhero and battles alien evil. Animated.
1983 77m/C VHS, Beta $9.95 MED ♫

Shadows A Chinese laundryman (Chaney) lives with a group of his countrymen in a New England seacoast village. All is peaceful until the local minister decides to convert the "heathen" Chinese. Ludicrous but worth seeing for Chaney's fun performance. Silent with music score.
1922 70m/B Lon Chaney Sr. **VHS, Beta, 8mm $16.95** SNC, NOS, CCB ♫♫

Shadows of Death Crabbe is the hero who must track down a gang of outlaws who kill to grab the land along the path of a new railroad. One of the "Billy Carson" series.
1945 60m/B Buster Crabbe, Al "Fuzzy" St. John. **VHS, Beta $27.95** VCN, UHV ♫½

Shadows and Fog Offbeat, unpredictable Allen film that is little more than an exercise in expressionistic visual stylings. The action centers around the Kafkaesque story of a haunted, alienated clerk (Allen) who is awakened in the middle of the night to join a vigilante group searching the streets for a killer. Although reminiscent of a silent film, Di Palma's black and white cinematography is stunning. Several stars also appear in cameo roles throughout this extremely unfocused comedy.
1992 (PG-13) 85m/B Woody Allen, Michael Kirby, Mia Farrow, John Malkovich, Madonna, Donald Pleasence, Lily Tomlin, Jodie Foster, Kathy Bates, John Cusack, Kate Nelligan, Kenneth Mars; **W/Dir:** Woody Allen. **VHS** ORI ♫♫

Shadows of Forgotten Ancestors Set in rural Russia in the early 20th century. Brings to expressive life the story of a man whose entire life has been overtaken by tragedy. Folk drama about a peasant who falls in with the daughter of his father's killer, then marries another woman. Strange, resonant and powerful with distinctive camera work. In Ukrainian with English subtitles.
1964 99m/C RU Ivan Micholaichuk, Larisa Kadochnikova; **Dir:** Sergei Paradjanov. New York Film Festival '66: Best Actress. **VHS, Beta $29.95** CVC, TAM ♫♫♫

Shadows of the Orient Two detectives are out to smash a Chinese smuggling ring.
1937 70m/B Regis Toomey, Esther Ralston. **VHS $16.95** SNC, NOS, DVT ♫♫♫

Shadows Over Shanghai This adventure relic was set during the Japanese invasion of China prior to WWII and uses much newsreel footage of that infamy. As for the plot, it's a groaner about good guys and bad skulking around the eponymous port in search of a treasure.
1938 66m/B James Dunn, Ralph Morgan, Linda Gray, Robert Barrat, Paul Sutton, Edward Woods; **Dir:** Charles Lamont. **VHS $19.95** NOS, LOO ♫½

Shadows Run Black A police detective must save a college coed from the clutches of a maniac wielding a meat cleaver. Ordinary slash-em-up. Bet Costner's embarrassed now about this early role - like Stallone's porno role in "The Italian Stallion."

1984 89m/C William J. Kulzer, Elizabeth Trosper, Kevin Costner; *Dir:* Howard Heard. **VHS, Beta** $59.95 *LTG* 𝄪 ½

Shadows on the Stairs A creepy boarding house is the scene of a number of mystery-drenched murders. A bizarre assortment of suspects make things even crazier.
1941 63m/B Frieda Inescort, Paul Cavanagh, Heather Angel, Bruce Lester, Miles Mander, Lumsden Hare, Turhan Bey; *Dir:* Ross Lederman. **VHS, Beta** $16.95 *SNC* 𝄪𝄪

Shadows in the Storm A librarian/ poet retreats to a secluded forest after losing his job. There he meets an exotically beautiful young woman who becomes his lover. With her love, though, she brings blackmail, murder and mystery.
1988 (R) 90m/C Ned Beatty, Mia Sara, Michael Madsen; *Dir:* Terrell Tannen. **VHS** *VMK* 𝄪𝄪

Shadowzone As a result of NASA experiments in dream travel, an interdimensional monster invades our world in search of victims. Begins well, with slightly interesting premise, but degenerates into typical monster flick. Good special effects.
1990 (R) 88m/C Louise Fletcher, David Beecroft, James Hong, Shawn Weatherly, Lu Leonard; *Dir:* J.S. Cardone. **VHS, Beta, LV** $19.95 *PAR* 𝄪𝄪

Shaft A black private eye is hired to find a Harlem gangster's kidnapped daughter. Lotsa sex and violence; suspenseful and well directed. Great ending. Academy Award winning theme song by Isaac Hayes, the first music award from the Academy to an African American.
1971 (R) 98m/C Richard Roundtree, Moses Gunn, Charles Cioffi; *Dir:* Gordon Parks. Academy Awards '71: Best Song (Theme from "Shaft"). **VHS, Beta** $19.95 *MGM* 𝄪𝄪𝄪

Shag: The Movie Four beautiful Southern girls party at Myrtle Beach for their last week-end of fun together before one gets married, one moves out of town, and two go off to college. Fun music, cool dancing, and great '60s fashions.
1989 (PG) 96m/C Phoebe Cates, Annabeth Gish, Bridget Fonda, Page Hannah, Scott Coffey, Robert Rusler, Tyrone Power Jr., Carrie Hamilton; *Dir:* Zelda Barron. **VHS, Beta** $89.99 *HBO* 𝄪𝄪½

The Shaggy D.A. Wilby Daniels is getting a little worried about his canine alter ego as he is about to run for District Attorney. Fun sequel to "The Shaggy Dog."
1976 (G) 90m/C Dean Jones, Tim Conway, Suzanne Pleshette, Keenan Wynn; *Dir:* Robert Stevenson. **VHS, Beta** *DIS; OM* 𝄪𝄪

Shaggy Dog When young Wilby Daniels utters some magical words from the inscription of an ancient ring he turns into a shaggy dog, causing havoc for family and neighbors. Disney slapstick is on target at times, though it drags in places. Followed by "The Shaggy D.A." and "Return of the Shaggy Dog."
1959 (G) 101m/B Fred MacMurray, Jean Hagen, Tommy Kirk, Annette Funicello, Tim Considine, Kevin Corcoran; *Dir:* Charles T. Barton. **VHS, Beta, LV** $69.95 *DIS* 𝄪𝄪½

Shaka Zulu British miniseries depicting the career of Shaka, king of the Zulus (Cole). Set in the early 19th century during British ascendency in Africa. Good, absorbing cross-cultural action drama - though could have been directed better.
1983 300m/C *GB* Edward Fox, Robert Powell, Trevor Howard, Christopher Lee, Fiona Fullerton, Henry Cele; *Dir:* William C. Faure. **VHS, Beta** $14.99 *STE, PSM* 𝄪𝄪𝄪

Shake Hands with Murder Three people spend their time bailing criminals out of jail for profit. One is murdered, and the other two become implicated in comedy.
1944 63m/B Iris Adrian, Frank Jenks, Douglas Fowley, Jack Raymond, Claire Rochelle, Herbert Rawlinson, Forrest Taylor; *Dir:* Albert Herman. **VHS, Beta** $16.95 *SNC* 𝄪𝄪

Shakedown Power-packed action film. An overworked attorney and an undercover cop work together to stop corruption in the N.Y.P.D. Although lacking greatly in logic or plot, the sensational stunts make this an otherwise entertaining action flick.
1988 (R) 96m/C Sam Elliott, Peter Weller, Patricia Charbonneau, Antonio Fargas, Blanche Baker, Richard Brooks, Jude Ciccolella, George Loros, Tom Waits, Shirley Stoler, Rockets Redglare; *Dir:* James Glickenhaus. **VHS, Beta, LV** $19.98 *MCA* 𝄪𝄪½

Shaker Run A stunt car driver and his mechanic transport a mysterious package. They don't know what to do—they're carrying a deadly virus that every terrorist wants! Car chases galore, and not much else. However, if you like chase scenes...
1985 91m/C *NZ* Leif Garrett, Cliff Robertson, Lisa Harrow; *Dir:* Bruce Morrison. **VHS, Beta, LV** $19.98 *SUE* 𝄪½

Shakes the Clown The rise and fall of Shakes, an alcoholic clown wandering through the all-clown town of Palukaville, as he tries to deal with his life. Framed for the murder of his boss by his archrival, Binky, Shakes takes it on the lam in order to prove his innocence, aided by his waitress girlfriend Judy, who dreams of becoming a professional bowler. Meant as a satire of substance-abuse recovery programs and the supposed tragedies of a performer's life, the film is merely unpleasant and unamusing. Robin Williams has a brief role as a mime.
1992 (R) 83m/C Bob Goldthwait, Julie Brown, Paul Dooley, LaWanda Page, Florence Henderson, Tom Kenny, Robin Williams; *W/Dir:* Bob Goldthwait. **VHS, LV** *COL* 𝄪

Shakespeare Wallah Tender, plausible drama of romance and postcolonial relations in India, from half of the splendid Merchant Ivory. A troupe of traveling Shakespeareans quixotically tours India trying to make enough money to return to England. Wonderfully acted and exquisitely and sensitively directed by Ivory. Written by Ivory and novelist Ruth Prawer Jhabvala.
1965 120m/B Laura Liddell, Geoffrey Kendal, Felicity Kendal, Shashi Kapoor, Madhur Jaffrey; *Dir:* James Ivory. New York Film Festival '65: Best Song. **VHS, Beta** $29.98 *SUE* 𝄪𝄪𝄪½

The Shakiest Gun in the West Don Knotts isn't Bob Hope, but he's pretty funny in the same bumbling way. Remake of Hope's "Paleface" has Philadelphia dentist Knotts unwittingly taking on bad guys and sultry Rhoades.
1968 101m/C Don Knotts, Barbara Rhoades, Jackie Coogan, Donald (Don "Red") Barry, Ruth McDevitt; *Dir:* Alan Rafkin. **VHS** *MCA* 𝄪𝄪½

Shaking the Tree A group of young adults try to avoid their march to adulthood and all its problems by seeking romance, success, and adventure, with each other if necessary. Also available in a PG-13 rated version.
1991 (R) 97m/C Arye Gross, Gale Hansen, Doug Savant, Courtney Cox; *W/Dir:* Duane Clark. **VHS** $89.95 *ACA* 𝄪𝄪

Shakma A group of medical researchers involved with experiments on animal and human tendencies toward aggression take a night off to play a quiet game of "Dungeons and Dragons." The horror begins when their main experimental subject comes along and turns things nasty!
1989 101m/C Roddy McDowall, Christopher Atkins, Amanda Wyss, Ari Meyers; *Dir:* Hugh Parks. **VHS** $29.95 *QUE, HHE* 𝄪½

Shalako Connery/Bardot pairing promises something special, but fails to deliver. European aristocrats on a hunting trip in New Mexico, circa 1880, are menaced by Apaches. U.S. Army scout tries to save captured countess Bardot. Strange British attempt at a Euro-western. Poorly directed and pointless. Based on a Louis L'Amour story.
1968 113m/C Brigitte Bardot, Sean Connery, Stephen Boyd, Jack Blackman, Woody Strode, Alexander Knox; *Dir:* Edward Dmytryk. **VHS, Beta** $19.98 *FOX* 𝄪𝄪

Shall We Dance And shall we ever! Seventh Astaire-Rogers pairing has a famous ballet dancer and a musical-comedy star embark on a promotional romance and marriage, to boost their careers, only to find themselves truly falling in love. Score by George and Ira Gershwin includes such memorable songs as "They All Laughed," "Let's Call the Whole Thing Off" and "They Can't Take That Away from Me." Thin, lame plot - but that's okay. For fans of good singing and dancing, and especially of this immortal pair.
1937 116m/B Fred Astaire, Ginger Rogers, Edward Everett Horton, Eric Blore; *Dir:* Mark Sandrich. **VHS, Beta, LV** $14.98 *TTC, RKO, CCB* 𝄪𝄪𝄪

Shallow Grave Four coeds en route to Florida witness a murder in rural Georgia and then are pursued relentlessly by the killer, a local sheriff. Decent premise and talented cast could have offered something better, given a meatier script.
1987 (R) 90m/C Tony March, Lisa Stahl, Tom Law, Carol Cadby; *Dir:* Richard Styles. **VHS, Beta** $14.95 *PSM, PAR* 𝄪½

The Shaman Members of a family succumb to the hypnotic spell of the Shaman as he searches for the perfect one to inherit his powers.
1987 (R) 88m/C Michael Conforti, Elvind Harum, James Farkas, Lynn Weaver; *Dir:* Michael Yakub. **VHS, Beta** $79.95 *IMP* 𝄪½

Shame Strangely unsuccessful low-budget Corman effort, starring pre-"Star Trek" Shatner as a freelance bigot who travels around Missouri stirring up opposition to desegregation. Moralistic and topical but still powerful. Adapted from the equally excellent novel by Charles Beaumont. Uses location filming superbly to render a sense of everydayness and authenticity. Also known as "The Intruder" and "I Hate Your Guts."
1961 84m/B William Shatner, Frank Maxwell, Jeanne Cooper, Robert Emhardt, Leo Gordon; *Dir:* Roger Corman. **VHS, Beta, LV** $19.95 *NOS, DVT* 𝄪𝄪𝄪½

Shame Strange Australian revenge drama about a female lawyer/biker who rides into a small town and finds herself avenging the rape of a 16-year-old girl. Is it a genuinely feminist movie about justice and a strong woman taking charge? Or is it exploitative trash using a politically correct theme and "message" as a peg for much - very much - gratuitous violence? Hard to tell. Cultish for sure; made on a B budget.

1987 (R) 95m/C *AU* Deborra-Lee Furness, Tony Barry, Simone Buchanan, Gillian Jones; *Dir:* Steve Jodrell. **VHS, LV** $89.99 *REP* ✂

Shame of the Jungle Animated spoof on Tarzan featuring the voices of John Belushi, Bill Murray, and Johnny Weissmuller, Jr.
1990 (R) 73m/C *Dir:* Boris Szulzinger, Picha; *Voices:* Johnny Weissmuller Jr., John Belushi, Bill Murray, Brian Doyle-Murray, Christopher Guest, Andrew Duncan. **VHS** $79.95 *JEF* ✂✂

Shame, Shame on the Bixby Boys Somewhere in the Old West, outside a small town, lives a band of misfits who persist in rustling cattle. The Bixby Boys and their Pa have made a bad habit into a family tradition.
1982 90m/C Monte Markham, Sammy Jackson; *Dir:* Anthony Bowers. **VHS, Beta** *TLF* ✂

The Shaming A puritanical schoolteacher is raped by a janitor, then develops a voracious sexual appetite. Horrible script, offensive premise can't be saved by good cast. Also called "The Sin" and "Good Luck, Miss Wyckoff." Based on a novel by William Inge.
1971 (R) 90m/C Anne Heywood, Donald Pleasence, Robert Vaughn, Carolyn Jones; *Dir:* Marvin J. Chomsky. **VHS, Beta** $14.95 *ONE* ✂½

Shampoo A satire of morals (and lack thereof) set in Southern California, concerning a successful hairdresser (Beatty) and the many women in his life. A notable scene with Julie Christie is set at a 1968 presidential election-night gathering. Jack Warden received an Oscar nomination; Fisher's screen debut, only one year before "Star Wars" made her famous. Beatty and Robert Towne wrote the script. Has a healthy glow in places and a perky bounce, but too many split ends.
1975 (R) 112m/C Warren Beatty, Julie Christie, Goldie Hawn, Jack Warden, Lee Grant, Tony Bill, Carrie Fisher, William Castle, Howard Hesseman; *Dir:* Hal Ashby. Academy Awards '75: Best Supporting Actress (Grant). **VHS, Beta, LV** $14.95 *COL, VYG, CRC* ✂✂½

Shamus Private dick Reynolds investigates a smuggling ring, beds a sultry woman, gets in lotsa fights. Classic Burt vehicle meant as a send-up. Unoriginal but fun.
1973 (PG) 91m/C Burt Reynolds, Dyan Cannon, John P. Ryan; *Dir:* Buzz Kulik. **VHS, Beta, LV** $9.95 *GKK* ✂✂

Shane A retired gunfighter, now a drifter, comes to the aid of a homestead family threatened by a land baron and his hired gun. Ladd is the mystery man who becomes the idol of the family's young son. Classic, flawless western. Well-deserved Oscar for cinematography; nominations for screenplay (by Pulitzer Prize-winning western novelist A.B. Guthrie, Jr., from the novel by Jack Schaefer); director; picture; Palance; de Wilde. Long and stately - worth savoring.
1953 117m/C Alan Ladd, Jean Arthur, Van Heflin, Brandon de Wilde, Jack Palance, Ben Johnson, Elisha Cook Jr., Edgar Buchanan; *Dir:* George Stevens. Academy Awards '53: Best Color Cinematography; National Board of Review Awards '53: 10 Best Films of the Year, Best Director. **VHS, Beta, LV** $14.95 *PAR, TLF* ✂✂✂✂

The Shanghai Gesture A wildly baroque, subversive melodrama. An English financier tries to close a gambling den, only to be blackmailed by the female proprietor - who tells him not only is the man's daughter heavily indebted to her, but that she is the wife he abandoned long ago. Based on a notorious Broadway play. Von Sternberg had

to make numerous changes to the script in order to get it past the Hays censors; the director's final Hollywood work, and worthy of his oeuvre, but oddly unsatisfying.
1941 97m/B Walter Huston, Gene Tierney, Victor Mature, Ona Munson, Albert Basserman, Eric Blore, Maria Ouspenskaya, Phyllis Brooks, Mike Mazurki; *Dir:* Josef von Sternberg. **VHS, Beta, LV** $29.95 *MFV* ✂✂½

Shanghai Massacre It's a violent battle when two Chinese gangs vie for control of the underworld.
19?? 83m/C *Dir:* Lee Tso Nan. **VHS, Beta** $42.95 *PGN* ✂

Shanghai Surprise Tie salesman Penn and missionary Madonna (yeah, right) are better than you'd think, and the story (of opium smuggling in China in the '30s) is intrepid and wildly fun, but indifferently directed and unsure of itself. Executive producer George Harrison wrote the songs and has a cameo.
1986 (PG-13) 90m/C Sean Penn, Madonna, Paul Freeman, Richard Griffiths; *Dir:* Jim Goddard. **VHS, Beta, LV** $79.98 *LIV, VES* ✂½

The Shaolin: Blood Mission/Deadly Shaolin Long Fist Two full-length kung fu spectaculars with double the head-kicking and rib-crunching fun.
1975 172m/C **VHS, Beta** $7.00 *VTR* ✂½

The Shaolin Brothers The brothers earn their living by putting on street gymnastics exhibitions. Foolishly, a Shaolin rebel attempts to disrupt one of their performances. Thus begins the battle between the rebels and the brothers.
198? ?m/C **VHS** $29.95 *OCE* ✂½

Shaolin Challenges Ninja Ninja and shaolin masters kick it out over honor and their respective orders.
198? 107m/C Liu Chia Hui. **Beta** $69.95 *VHV* ✂

Shaolin Death Squad A ruthless Japanese premier murders a statesman as the first step in his plan to become Emperor. The murdered man's daughter swears vengeance and enlists the Shaolin Death Squad to help.
1983 (R) 90m/C **VHS, Beta** $16.50 *OCE, FOX* ✂

Shaolin Devil/Shaolin Angel To prevent more needless kick murders, two warriors learn the dreaded Dammoth Kung Fu and kick everyone anyway.
197? 87m/C **VHS, Beta** $29.95 *UNI* ✂

Shaolin Disciple Kung Fu action-adventure featuring real monks.
19?? 90m/C Ku Feng. **VHS** $49.95 *OCE* ✂

Shaolin Drunk Fighter When a nasty spate of intemperance plagues the Shaolin training grounds, the temple masters seek new cures to an old problem.
1969 95m/C *Hosted:* Sho Kosugi. **VHS, Beta** $29.95 *TWE* ✂

The Shaolin Drunk Monkey The dreaded Shaolin Drunk Monkey faces his enemies in a veritable array of fights and lethal kicks.
1985 85m/C **VHS, Beta** $14.95 *SVS* ✂

Shaolin Ex-Monk A renegade Shaolin Monk is pursued by a Kung Fu master and his disciple.
1982 94m/C John Liu, Ko Shou Liang, Huang Hsing Shaw. **VHS, Beta** *WVE* ✂

Shaolin Executioner Several generations of Shaolin defend the Shaolin Temple against all high-kicking odds.
198? 104m/C Chen Kuan-Tai, Lo Lieh; *Dir:* Lu Chia-Liang. **Beta** $69.95 *VHV* ✂

Shaolin Fox Conspiracy This kung-fu epic features unusual moves, including the Mantis Strike, the Dragon Claw and the Crescent Kick.
1985 75m/C **VHS, Beta** $14.95 *SVS* ✂

Shaolin Incredible Ten A Shaolin abbot and his daughters kick a great many Manchu warriors in their helmeted heads.
197? 87m/C Sue Lee, Elton Chong, Eagle Han. **VHS, Beta** $54.95 *MAV* ✂

The Shaolin Invincibles Shaolin buddies team up to fight the evil Emperor Yung Ching eventually coming to final blows with him in the Forbidden City Golden Palace.
198? ?m/C **VHS** *OCE* ✂½

Shaolin Kids Shaolin warriors discover a conspiracy and kick it to death.
197? 91m/C **VHS, Beta** $29.95 *UNI* ✂

Shaolin Kung Fu Mystagogue In this action-packed chop-fest, the mysterious "Floating in Blood" weapon enlivens the battle between the Ching Dynasty and Ming rebels.
1984 90m/C *Dir:* Chang Peng-I. **VHS, Beta** $49.95 *WES* ✂

The Shaolin Kung Fu Mystagogue/Iron Ox, the Tiger's Killer Double your pleasure, double your fun, with two martial arts flicks in one.
19?? 180m/C **VHS** $9.99 *VTR* ✂

Shaolin Red Master/Two Crippled Heroes Kung Fu is back, choppin' 'em down and spreadin' 'em around in this double feature, martial arts adventure.
19?? 180m/C **VHS** $9.99 *VTR* ✂

Shaolin Temple Strikes Back The protective inhabitants of the Shaolin temple defend themselves against invading Manchu Barbarians.
1972 90m/C Chen Chien Chang, Sho Kosugi, Chang Shan, Lung Shao Fei; *Dir:* Joseph Kuo. **VHS, Beta** $29.95 *TWE* ✂

Shaolin Thief A traitor gives the invading Japanese a priceless treasure and the loyal Chinese troops must fight to get it back.
19?? 90m/C **VHS** $29.95 *OCE* ✂

Shaolin Traitor Exotic weapons and dazzling martial-arts action are featured in this exciting tale of intrigue and mystery.
1982 (R) 99m/C Carter Wong, Shangkuan Ling-Feng, Lung Chun-Eng, Chang Yi, Lung Fei. **VHS, Beta** $59.95 *GEM* ✂½

Shaolin Vengence Framed for murder and deceived into accepting guilt for the crime, a man does penance in a Shaolin temple. When he learns of the deception, he uses his newly learned Shaolin boxing skills to relentlessly pound the creeps who set him up.
1982 ?m/C Chen Sing. **VHS** $24.50 *OCE* ✂½

Shaolin vs. Lama A group of Shaolin Kung Fu fighters participates in a series of battles against Tibetan Monks.
1984 90m/C **VHS, Beta** $39.95 *WES* ✂

Shaolin vs. Manchu A Manchu spy infiltrates a Shaolin temple, and causes much head-splitting on both sides.

198? 91m/C Romy Tso. **VHS, Beta** $54.95 MAV ✰

Shaolin Warrior A young man must battle an evil Tartar warlord who massacred his family.
19?? 90m/C Kam Kong, Chan Wai Man, Chen Sing. **VHS, Beta** $29.95 OCE ✰

Shaolin Wooden Men A young man vows revenge, and gets it, on the masked ninja who killed his father.
19?? 90m/C Jackie Chan; **Dir:** Lo Wei. **VHS, Beta** $24.95 ASE ✰

Shark! An American gun smuggler (Reynolds) stranded in a tiny seaport in Africa joins the crew of a marine biologist's boat. He soon discovers the boat's owner and his wife are trying to retrieve gold bullion that lies deep in shark-infested waters. Typical Reynoldsian action-infested dumbness. Like "Twilight Zone: The Movie", earned notoriety because of on-location tragedy: a stunt diver really was killed by a shark. Re-released as "Maneater." Edited without the consent of Fuller, who disowned it.
1968 (PG) 92m/C Burt Reynolds, Barry Sullivan, Arthur Kennedy; **Dir:** Samuel Fuller. **VHS, Beta** $24.95 REP, MRV ✰ 1/2

Shark Attack Rodney Fox, a businessman left scarred and bitter from an attack by a great white shark, recuperates and kills the shark to satisfy his thirst for revenge. Later, having overcome his fear of the "White Death," he realizes that sharks are people too, so to speak.
197? 30m/C Jean-Michel Cousteau; **Nar:** Allen Ludden. **VHS, Beta** $29.95 VVI ✰

Shark Hunter A shark hunter gets ensnared in the mob's net off the Mexican coast as they race for a cache of sunken millions. The usual garden-variety B-grade adventure.
1984 (PG) 95m/C Franco Nero, Jorge Luke, Mike Forrest. **VHS, Beta** $49.95 PSM, MRV ✰

Shark River A man attempts to get his brother, a Civil War veteran, accused of murder, through the Everglades and send him to safety in Cuba. Good photography of the great swamp; disappointing story.
1953 80m/C Steve Cochran, Carole Matthews, Warren Stevens; **Dir:** John Rawlins. **VHS, Beta** $39.98 MAG ✰ 1/2

Shark's Paradise A madman threatens to set sharks loose on a popular Australian surf coast if his demands are not met.
1986 93m/C **VHS, Beta** $59.95 WOV ✰

Sharks, Past & Present Jean-Michel Cousteau, son of famed oceanographer Jacques Cousteau, meets with famed shark photographer Ron Taylor in Cabo San Lucas, where the Pacific Ocean and the Sea of Cortez join, to study the shark in its natural habitat so as to understand modern day sharks.
197? 30m/C Jean-Michel Cousteau; **Nar:** Allen Ludden. **VHS, Beta** VVI ✰

Sharks' Treasure A band of escaped convicts commandeer a boat filled with gold. Old-fashioned sunken-treasure tale unexceptional except as a virtual one-man show by writer/actor/director/producer/fitness nut Wilde.
1975 (PG) 96m/C Cornel Wilde, Yaphet Kotto, John Nellson, Cliff Osmond, David Canary, David Gilliam; **Dir:** Cornel Wilde. **VHS, Beta** $59.95 MGM ✰ 1/2

Sharky's Machine A tough undercover cop (Reynolds) is hot on the trail of a crooked crime czar. Meanwhile he falls for a high-priced hooker. Well done but overdone action, with much violence. Based on the William Diehl novel.
1981 (R) 119m/C Burt Reynolds, Rachel Ward, Vittorio Gassman, Brian Keith, Charles Durning, Bernie Casey, Richard Libertini, Henry Silva, John Fiedler, Earl Holliman; **Dir:** Burt Reynolds. **VHS, Beta, LV** $19.98 WAR ✰✰ 1/2

Sharma & Beyond A teenage would-be science fiction writer falls in love with the daughter of a famous sci-fi author.
1984 85m/C Suzanne Burden, Robert Urquhart, Michael Maloney; **Dir:** Brian Gilbert. **VHS, Beta** $59.95 MGM ✰

Sharp Fists in Kung Fu The cruel son of a gentle man returns to terrorize his hometown. Strictly for fans of the genre.
19?? 90m/C **VHS, Beta** $49.95 OCE Woof!

Shattered Finch plays Harry, a man blamed for a failed marriage, held hostage in his home by his own paranoia, and driven to bouts of drinking. Slowly his tenuous grip on sanity slips and at any moment his fragile and crumbling life may be shattered. Also known as "Something to Hide." Good idea to hide from anyone suggesting to rent it - tired, overwrought domestic drama.
1972 (R) 100m/C GB Peter Finch, Shelley Winters, Colin Blakely; **Dir:** Alastair Reid. **VHS, Beta** $24.95 MED ✰ 1/2

Shattered An architect recovering from a serious automobile accident tries to regain the memory that he has lost. As he begins to put together the pieces of his life, some parts of the puzzle don't quite fit. For example, he recalls his now loving wife's affair as well as his own. He remembers that he hired a private detective to follow his wife and shockingly, he remembers that he once believed his wife had planned to kill him. Are these memories the real thing, or are they all part of some mad, recuperative nightmare? Who knows.
1991 (R) 98m/C Tom Berenger, Bob Hoskins, Greta Scacchi, Joanne Whalley-Kilmer, Corbin Bernsen; **W/Dir:** Wolfgang Petersen. **VHS, Beta, LV, 8mm** $94.99 MGM ✰✰

Shattered Innocence A journey through beauty, sensuality and innocence revolving around a 16-year-old ballet dancer's first stirrings of sexuality. Originally titled "Laura," and available on video under that name.
1979 (R) 95m/C Maud Adams, Dawn Dunlap. **VHS, Beta** $9.95 SIM, SUE ✰

Shattered Silence A woman is tormented by phone calls that seem to be coming from her dead son as her life is torn apart by divorce. Leaves unanswered the question: Will there be any suspense? Made for TV. Worthless and dumb. Originally titled "When Michael Calls."
1971 73m/C Elizabeth Ashley, Ben Gazzara, Michael Douglas, Karen Pearson; **Dir:** Philip Leacock. **VHS** $9.99 CCB, TAV ✰ 1/2

Shattered Spirits A quiet family man cracks and goes on a rampage through his normally peaceful suburban neighborhood.
1991 (PG-13) 93m/C Martin Sheen, Melinda Dillon, Matthew Laborteaux, Roxana Zal, Lukas Haas. **VHS** $89.98 VMK ✰ 1/2

Shattered Vows A young nun struggles with her love for a priest and her desire for a child. Bertinelli is convincing as the confused teen in otherwise ordinary TV drama. Adapt-

ed from the true story of Dr. Mary Gilligan Wong as told in her book, "Nun: A Memoir."
1984 95m/C Valerie Bertinelli, David Morse, Caroline McWilliams, Patricia Neal, Millie Perkins, Leslie Ackerman, Lisa Jane Persky; **Dir:** Jack Bender. **VHS, Beta** $14.95 FRH ✰✰

She H. Rider Haggard's famous story about the ageless Queen Ayesha, (Blythe), who renews her life force periodically by walking through a pillar of cold flame. Story and titles by Haggard. Blythe is a stirringly mean queen, and the story remains fresh and fun. Silent film with music score.
1921 69m/B Betty Blythe, Carlyle Blackwell, Mary Odette. **VHS, Beta** $19.95 NOS, VYY, DVT ✰✰ 1/2

S*H*E Omar Sharif makes a cameo appearance in this flashy made-for-TV spy thriller.The female agent S*H*E (Securities Hazards Expert) must save the world's oil; good locations, nifty plot and looker Sharpe as the she-Bond add up to a reasonably fun time. Written by Richard Maibaum, late of several Bond scripts.
1979 100m/C Omar Sharif, Cornelia Sharpe, Robert Lansing, William Taylor, Isabella Rye, Anita Ekberg; **Dir:** Robert Lewis. **VHS, Beta** $59.95 PSM ✰✰

She A beautiful female warrior rules over the men in a post-holocaust world. She is kidnapped by a wealthy merchant who uses her to fight evil mutants. Utter rubbish vaguely based on the Haggard novel.
1983 90m/C Sandahl Bergman, Harrison Muller, Quin Kessler, David Goss; **Dir:** Avi Nesher. **VHS, Beta** $29.98 LIV, LTG ✰

The She-Beast Burned at the stake in 18th-century Transylvania, a witch returns in the body of a beautiful young English woman on her honeymoon (in Transylvania?!), and once again wreaks death and destruction. Caution, Barbara Steele fans: she appears for all of fifteen minutes. Sporadically funny.
1965 74m/C IT YU Barbara Steele, Ian Ogilvy, Mel Wells, Lucretia Love; **Dir:** Michael Reeves. **VHS, Beta** $59.95 SNC, MPI ✰ 1/2

She Came to the Valley A tough pioneer woman becomes embroiled in political intrigue during the Spanish-American War. Based on Cleo Dawson's book. Also known as "Texas in Flames."
1977 (PG) 90m/C Ronee Blakley, Dean Stockwell, Scott Glenn, Freddy Fender. **VHS, Beta** $69.95 MED ✰

She Couldn't Say No A wealthy oil heiress (Simmons) with good intentions plans to give her money away to those friends who had helped her when she was struggling. Things don't go quite as planned; her philanthropy leads to unexpected mayhem in her Arkansas hometown. Passable, but not memorable comedy. Mitchum as a small-town doctor?
1952 88m/B Robert Mitchum, Jean Simmons, Arthur Hunnicutt, Edgar Buchanan, Wallace Ford; **Dir:** Lloyd Bacon. **VHS, Beta** CCB ✰✰

She Demons A mad ex-Nazi scientist transforms pretty girls into rubber-faced Frankensteins on a far-off island on which a pleasure craft becomes shipwrecked. Very low-budget and very stupid—in fact it's so bad that some people love it!
1958 68m/B Irish McCalla, Tod Griffin, Victor Sen Yung; **Dir:** Richard Cunha. **VHS, Beta** $9.95 RHI, SNC, MWP Woof!

She-Devil A comic-book version of the acidic Fay Weldon novel "The Life and Loves of a She-Devil"; a fat, dowdy suburban wife (Arnold) becomes a vengeful beast when a smarmy romance novelist steals her husband. Uneven comedic reworking of a distinctly unforgiving feminist fiction. Arnold is given too much to handle (her role requires an actual range of emotions); Streep's role is too slight, though she does great things with it.
1989 (PG-13) 100m/C Meryl Streep, Roseanne (Barr) Arnold, Ed Begley Jr., Linda Hunt, Elizabeth Peters, Bryan Larkin, A. Martinez, Sylvia Miles; *Dir:* Susan Seidelman. **VHS, Beta, LV** $14.98 *ORI* 🎞🎞½

She Devils in Chains A group of travelling female athletes are kidnapped by a groups of sadists who torture and beat them. In the end, however, the girls get their revenge, and it's bloody.
19?? (R) 82m/C Colleen Camp, Rosanne Katon, Sylvia Anderson, Ken Washington, Leo Martinez; *Dir:* Cirio H. Santiago. **VHS** $9.95 *SIM* 🎞🎞

She-Devils on Wheels Havoc erupts as an outlaw female motorcycle gang, known as "Maneaters on Motorbikes," terrorizes a town - especially the men. A really, really bad biker flick. Finely honed by Herschell Gordon Lewis.
1968 83m/C Betty Connell, Christie Wagner, Pat Poston, Nancy Lee Noble, Ruby Tuesday; *Dir:* Herschell Gordon Lewis. **VHS, Beta** $19.95 *SVI, WES* Woof!

She Done Him Wrong West stars as Lil, an 1890s saloon singer in the screen version of her Broadway hit "Diamond Lily", and imparts on the negligible plot. Her usual share of double entendres and racy comments. Her funniest film. She sings "Frankie and Johnnie" and "I Wonder Where My Easy Rider's Gone." Grant is likeable as the hapless sap she beds.
1933 65m/B Mae West, Cary Grant, Owen Moore, Noah Beery, Gilbert Roland, Louise Beavers; *Dir:* Lowell Sherman. National Board of Review Awards '33: 10 Best Films of the Year. **VHS, Beta** $29.95 *MCA, MLB* 🎞🎞🎞

She-Freak Remake of Tod Brownings' "Freaks." A cynical waitress burns everyone in a circus and gets mauled by the resident freaks. Pales beside its unacknowledged, classic original.
1967 87m/C Claire Brennan, Lynn Courtney, Bill McKinney, Lee Raymond, Madame Lee; *Dir:* Byron Mabe. **VHS, Beta** $39.98 *MAG* 🎞

She Gods of Shark Reef Typical no-budget Corman exploitation tale of two brothers shipwrecked on an island inhabited by beautiful pearl-diving women. Filmed in Hawaii.
1956 63m/C Bill Cord, Don Durant, Lisa Montell, Carol Lindsay, Jeanne Gerson; *Dir:* Roger Corman. **VHS** $16.95 *SNC, MRV* 🎞

She Goes to War King's first part-sound effort that takes place in the milieu of World War I.
1929 50m/B Eleanor Boardman, Alma Rubens, Al "Fuzzy" St. John; *Dir:* Henry King. **VHS, Beta, 8mm** $19.95 *NOS, GVV, DVT* 🎞🎞

She Likes You, Charlie Brown Romantic travails for Charlie Brown and the gang. Animated.
1986 41m/C **VHS, Beta** $14.95 *KRT* 🎞🎞

She Shall Have Music A magnate hires a dance band to broadcast from a cruise ship. Features music by Jack Hylton and his band, one of Britain's most popular orchestras of the 1930s.
1936 76m/B Jack Hylton, June Clyde, Claude Dampier, Bryan Lawrence, Gwen Farrar. **VHS, Beta** *MWP* 🎞½

She Waits When a newlywed couple moves into an old house, the bride becomes possessed by the spirit of her husband's first wife. Unoriginal and mediocre horror. Made for television.
1971 74m/C Dorothy McGuire, Patty Duke, David McCallum; *Dir:* Delbert Mann. **VHS, Beta** $59.95 *PSM* 🎞🎞

She Wore a Yellow Ribbon An undermanned cavalry outpost makes a desperate attempt to repel invading Indians. Wayne shines as an officer who shuns retirement in order to help his comrades. Excellent color photography by Winston C. Hoch. Still fun and compelling. The second chapter in director Ford's noted cavalry trilogy, preceded by "Fort Apache" and followed by "Rio Grande."
1949 93m/C John Wayne, Joanne Dru, John Agar, Ben Johnson, Harry Carey Jr., Victor McLaglen, Mildred Natwick, George O'Brien; *Dir:* John Ford. Academy Awards '49: Best Color Cinematography. **VHS, Beta, LV** $19.95 *TTC, RKO, VID* 🎞🎞🎞½

Sheba, Baby A female dick (Grier) heads to Louisville where someone is trying to threaten her rich father and his loan company. Oddly non-violent for action-flick vet Grier; poorly written and directed.
1975 (PG) 90m/C Pam Grier, Rudy Challenger, Austin Stoker, D'Urville Martin, Charles Kissinger; *Dir:* William Girdler. **VHS, Beta** $59.98 *ORI* 🎞🎞

Sheena A television sportscaster aids a jungle queen in defending her kingdom from being overthrown by an evil prince. Horrid bubble-gum "action" fantasy.
1984 (PG) 117m/C Tanya Roberts, Ted Wass, Donovan Scott, Elizabeth Toro; *Dir:* John Guillermin. **VHS, Beta, LV** $12.95 *GKK* 🎞

The Sheep Has Five Legs Quintuplet brothers return from around the world for a reunion in their small French village. Fernandel plays the father and all five sons; otherwise, comedy is only average-to-good. In French with English subtitles.
1954 96m/B **FR** Fernandel, Edouard Delmont, Louis de Funes, Paulette Dubost; *Dir:* Henri Verneuil. **VHS, Beta, 8mm** $24.95 *NOS, DVT, VYY* 🎞🎞½

Sheer Heaven Nine illustrious females slip in and out of their lingerie in this video.
1988 60m/C Ginger Lynn. **VHS, Beta** $14.95 *CEL* 🎞🎞½

Sheer Madness Focuses on the intense friendship between a college professor and a troubled artist, both women. Engrossing and subtle exposition of a relationship. Ambiguous ending underscores film's general excellence. Subtitled.
1984 105m/C **GE FR** Hanna Schygulla, Angela Winkler; *Dir:* Margarethe von Trotta. **VHS, Beta** *MGM* 🎞🎞🎞

The Sheik High camp Valentino has English woman fall hopelessly under the romantic spell of Arab sheik who flares his nostrils. Followed by "Son of the Sheik."
1921 80m/B Agnes Ayres, Rudolph Valentino, Adolphe Menjou, Walter Long, Lucien Littlefield, George Waggner, Patsy Ruth Miller; *Dir:* George Melford. **VHS, Beta** $19.95 *PAR, GPV* 🎞🎞🎞

She'll Be Wearing Pink Pajamas Eight women volunteer for a rugged survival course to test their mettle. The intense shared experience gives them all food for thought. Well acted from a pretty thin story.
1984 90m/C Julie Walters, Anthony Higgins; *Dir:* John Goldschmidt. **VHS, Beta** $59.95 *LHV* 🎞🎞

Shell Shock Based on personal experience during the 1973 Israeli war, this film explores the brutal consequences of war.
1989 90m/C **VHS, Beta** *JCI* 🎞🎞

Shell Shock Four GI's fight off the Germans and the longings for home during World War II.
198? 90m/C *Dir:* John Hayes. **VHS, Beta** $34.95 *KOV* 🎞½

Shelley Duvall's Bedtime Stories A series of stories based on popular children's books. Each tale is narrated by a celebrity. Two stories are included on each tape. Originally produced for cable television.
1992 25m/C *Hosted:* Shelley Duvall; *Nar:* Ringo Starr, Bette Midler, Jean Stapleton, Dudley Moore, Rick Moranis, Bonnie Raitt. **VHS** $12.98 *MCA* 🎞½

The Sheltering Sky American couple Winger and Malkovich flee the plasticity of their native land for a trip to the Sahara desert where they hope to renew their spirits and rekindle love. Accompanied by socialite acquaintance Scott with whom Winger soon has an affair, their personalities and belief systems deteriorate as they move through the grave poverty of North Africa in breathtaking heat. Based on the existential novel by American expatriate Paul Bowles who narrates and appears briefly in a bar scene. Overlong but visually stunning, with cinematography by Vittorio Storaro.
1990 (R) 139m/C Debra Winger, John Malkovich, Campbell Scott, Jill Bennett, Timothy Spall, Eric Vu-An; *Dir:* Bernardo Bertolucci. Golden Globe Awards '90: Best Musical Score; Los Angeles Film Critics Association Awards '90: Best Musical Score; New York Film Critics Awards '90: Best Cinematography. **VHS, Beta, LV** $19.98 *WAR, PIA, FCT* 🎞🎞🎞

Shenandoah A Virginia farmer (Stewart, in a top-notch performance) who has raised six sons and a daughter, tries to remain neutral during the Civil War. War takes its toll as the daughter marries a Confederate soldier and his sons become involved in the fighting. Screen debut for Ross.
1965 105m/C James Stewart, Doug McClure, Glenn Corbett, Patrick Wayne, Rosemary Forsythe, Katharine Ross, George Kennedy; *Dir:* Andrew V. McLaglen. **VHS, Beta, LV** $19.95 *MCA* 🎞🎞🎞

Shep Comes Home Shep the wonder canine accompanies an orphan through the Midwest where he nabs a band of bankrobbers. Sequel to "My Dog Shep."
1949 62m/B Robert Lowery, Sheldon Leonard, Flame. **VHS, Beta, LV** *WGE* 🎞

The Shepherd Two veteran pilots, still experiencing the pain of the Vietnam War, find personal salvation in Christian beliefs.
1988 79m/C Christopher Stone, Dee Wallace Stone, Pepper Martin, Robert Ayres, Patty Dunning; *Dir:* Donald W. Thompson. **VHS, Beta** $39.95 *MIV* 🎞

Sheriff of Tombstone Trouble begins when a "judge" takes a sharpshooter in a poker game.
1941 60m/B Roy Rogers. **VHS, Beta** $19.95 *NOS, DVT, RXM* 🎞

Sherlock Holmes Episode of the early British television series entitled "The Man Who Disappeared."
1951 27m/B *GB* John Longden, Campbell Singer. **VHS, Beta, 8mm $24.95** VYY ♫

Sherlock Holmes: A Study in Scarlet Holmes and Watson have only one clue to go on when they investigate the murder of an American tourist in London.
1984 49m/C *Voices:* Peter O'Toole. **VHS, Beta, LV $14.95** PAV ♫♫ 1/2

Sherlock Holmes Faces Death Dead bodies are accumulating in a mansion where the detecting duo are staying. Underground tunnels, life-size dress boards, and unanswered mysteries... Top-notch Holmes. Peter Lawford appears briefly as a sailor. Also available with "Hound of the Baskervilles" on Laser Disc.
1943 68m/B Basil Rathbone, Nigel Bruce, Hillary Brooke, Milburn Stone, Halliwell Hobbes, Arthur Margetson, Gavin Muir; *Dir:* Roy William Neill. **VHS, Beta, LV $19.98** FOX ♫♫♫

Sherlock Holmes and the Incident at Victoria Falls A substandard Holmes excursion brings the Baker Street sleuth out of retirement to transport the world's largest diamond from Africa to London. The resulting mystery involves Teddy Roosevelt, the inventor of radio, and poor plotting.
1991 120m/C *GB* Christopher Lee, Patrick Macnee, Jenny Seagrove; *Dir:* Bill Corcoran. **VHS $89.98** VES ♫ 1/2

Sherlock Holmes and the Secret Weapon Based on "The Dancing Men" by Sir Arthur Conan Doyle. Holmes battles the evil Moriarty in an effort to save the British war effort. Good Holmes mystery with gripping wartime setting. Hoey is fun as bumbling Inspector Lestrade. Available colorized.
1942 68m/B Basil Rathbone, Nigel Bruce, Karen Verne, William Post Jr., Dennis Hoey, Holmes Herbert, Mary Gordon, Henry Victor, Philip Van Zandt, George Eldridge, Leslie Denison, James Craven, Paul Fix, Hugh Herbert, Lionel Atwill; *Dir:* Roy William Neill. **VHS, LV $19.98** REP, MED, AOV ♫♫♫

Sherlock Holmes: The Baskerville Curse Animated feature film with Peter O'Toole as the voice of Sherlock Holmes.
1984 70m/C *Voices:* Peter O'Toole. **VHS, Beta $14.95** PAV ♫♫

Sherlock Holmes: The Sign of Four Sherlock Holmes must find the man who murdered Bartholomew Sholto with a poison dart in the neck.
1984 48m/C *Voices:* Peter O'Toole. **VHS, Beta $14.95** PAV ♫♫

Sherlock Holmes: The Silver Blaze Sherlock Holmes mystery about the theft of a racehorse known as "The Silver Blaze." Remade in 1977.
1941 70m/B Arthur Wontner, Ian Fleming, Lyn Harding. **VHS, Beta $19.95** NOS, KRT, MWP ♫ 1/2

Sherlock Holmes: The Silver Blaze Sherlock Holmes, with Dr. Watson, in one of the legendary detective's most famous adventures involving the theft of a racehorse and the mysterious murder of his trainer. Remake of a 1941 film.
1977 31m/C Christopher Plummer, Thorley Walters; *Dir:* John Davies. **VHS, Beta $24.98** LCA ♫♫

Sherlock Holmes: The Valley of Fear Sherlock Holmes and Dr. Watson investigate a man who faked his own death to escape retribution from an American secret society.
1984 49m/C *Voices:* Peter O'Toole. **VHS, Beta $14.95** PAV ♫♫ 1/2

Sherlock Holmes: The Voice of Terror Holmes and Watson try to decode German radio messages during World War II. First Rathbone-as-Holmes effort for Universal, and first in which we are asked to believe the Victorian sleuth and his hairdo could have lived in this century. Patriotic and all that - and lots of fun.
1942 65m/B Basil Rathbone, Nigel Bruce, Hillary Brooke, Reginald Denny, Evelyn Ankers, Montagu Love; *Dir:* John Rawlins. **VHS, Beta, LV $19.98** FOX ♫♫

Sherlock Holmes in Washington A top-secret agent is murdered; seems it's those blasted Nazis again! Holmes and Watson rush off to Washington, D.C. to solve the crime and to save some vitally important microfilm. Heavily flag-waving Rathbone-Bruce episode. Dr. Watson is dumbfounded by bubble gum.
1943 71m/B Basil Rathbone, Nigel Bruce, Henry Daniell, George Zucco, Marjorie Lord, John Archer; *Dir:* Roy William Neill. **VHS, Beta $19.98** FOX ♫♫ 1/2

Sherlock Hound: Tales of Mystery Sherlock, Watson and Moriarty are all dogs in this animated children's adaptation of the popular Arthur Conan Doyle stories. Well worth a howl.
1990 115m/C **VHS, Beta** JFK ♫♫ 1/2

Sherlock's Rivals & Where's My Wife Double feature includes a detective movie in which two used car dealers become amateur sleuths and a comedy in which a man has a hard time enjoying his weekend at a seashore resort.
1927 53m/B Milburn Morante, Monty Banks. **VHS, Beta $19.98** VYY ♫ 1/2

She's in the Army Now Farce about military life. Two bubble-headed females undergo basic training. Made-for-TV "Private Benjamin" ripoff was meant as a series pilot.
1981 97m/C Jamie Lee Curtis, Kathleen Quinlan, Melanie Griffith, Susan Blanchard, Julie Carmen, Janet MacLachlan; *Dir:* Hy Averback. **VHS, Beta $9.98** SUE, CHA ♫

She's Back Murderee Fisher pesters hubby Joy from beyond the grave to avenge her death at the hands of a motorcycle gang. Uneven, but occasionally funny.
1988 (R) 89m/C Carrie Fisher, Robert Joy; *Dir:* Tim Kincaid. **VHS, Beta, LV $79.98** VES, LIV ♫ 1/2

She's Dressed to Kill Beautiful models are turning up dead during a famous designer's comeback attempt at a mountain retreat. Who could be behind these grisly deeds? Suspenseful in a made-for-TV kind of way, but not memorable.
1979 100m/C Eleanor Parker, Jessica Walter, John Rubinstein, Connie Sellecca; *Dir:* Gus Trikonis. **VHS $29.95** IVE ♫♫

She's a Good Skate, Charlie Brown Snoopy becomes Peppermint Patty's ice-skating coach, and he helps her train for the local competition.
1980 30m/C *Dir:* Phil Roman. **VHS, Beta $14.95** WPB, VTR, HSE ♫♫

She's Gotta Have It Lee wrote, directed, edited, produced and starred in this romantic comedy about an independent-minded black girl in Brooklyn and the three men and one woman who compete for her attention. Full of rough edges, but vigorous, confident, and hip. Filmed entirely in black and white except for one memorable scene. Put Lee on the film-making map.
1986 (R) 84m/B Spike Lee, Tommy Redmond Hicks, Raye Dowell; *Dir:* Spike Lee. **VHS, Beta $19.95** FOX, FCT ♫♫♫

She's Having a Baby Newlyweds tread the marital waters with some difficulty, as the news of an impending baby further complicates their lives. Told from the viewpoint of the young husband, who's feeling trapped. Charming leads don't fully make up for retrograde, arguably sexist premise and dull resolution.
1988 (PG-13) 106m/C Kevin Bacon, Elizabeth McGovern, William Windom, Paul Gleason, Alec Baldwin, Cathryn Damon, Holland Taylor, James Ray, Isabel Lorca, Dennis Dugan, Edie McClurg; *Dir:* John Hughes. **VHS, Beta, LV, 8mm $19.95** PAR ♫♫♫

She's Out of Control Dad Danza goes nuts when teen daughter Dolenz (real-life daughter of Monkee Mickey Dolenz) takes the advice of Dad's girlfriend on how to attract boys. Formulaic plot could almost be an episode of "Who's the Boss?" Danza is appealing, but not enough to keep this one afloat.
1989 95m/C Tony Danza, Catherine Hicks, Wallace Shawn, Dick O'Neill; *Dir:* Stan Dragoti. **VHS, Beta, LV, 8mm $19.95** COL, LDC ♫ 1/2

Shin Heike Monogatari Mizoguchi's second to last film, in which a deposed Japanese emperor in 1137 endeavors to win back the throne from the current despot, who cannot handle the feudal lawlessness. Acclaimed; his second film in color. In Japanese with English subtitles. Also called "New Tales of the Taira Clan."
1955 106m/C *JP* Raizo Ichikawa, Ichijiro Oya, Michiyo Kogure, Eijiro Yanagi, Tatsuya Ishiguro, Yoshiko Kuga; *Dir:* Kenji Mizoguchi. New York Film Festival '64: Best Cinematography. **VHS, Beta $29.95** FCT ♫♫♫

Shinbone Alley Animated musical about Archy, a free-verse poet reincarnated as a cockroach, and Mehitabel, the alley cat with a zest for life. Based on the short stories by Don Marquis.
1970 (G) 83m/C *Dir:* John D. Wilson; *Voices:* Carol Channing, Eddie Bracken, John Carradine, Alan Reed. **VHS, Beta $14.95** SIM, GEM, KAR ♫♫♫♫

Shine on, Harvest Moon Roy Rogers brings a band of outlaws to justice and clears an old man suspected of being their accomplice. The title of the film has nothing to do with its storyline.
1938 60m/B Roy Rogers, Mary Hart, Stanley Andrews; *Dir:* Joseph Kane. **VHS, Beta $19.95** NOS, VYY, VDM ♫ 1/2

The Shining Very loose adaptation of the Stephen King horror novel about a writer and his family, snowbound in a huge hotel, who experience various hauntings caused by either the hotel itself or the writer's dementia. Technically stunning, and pretty dang scary, but too long, pretentious and implausible. Nicholson is excellent as the failed writer gone off the deep end.
1980 (R) 143m/C Jack Nicholson, Shelley Duvall, Scatman Crothers, Danny Lloyd, Joe Turkel, Barry Nelson; *Dir:* Stanley Kubrick. **VHS, Beta, LV $19.98** WAR, PIA ♫♫ 1/2

The Shining Hour A compelling melodrama in which Crawford portrays a New York night club dancer who is pursued by the rather conservative Douglas. His brother tries to persuade him from marrying Crawford, but then he soon finds himself attracted to her. A devastating fire wipes out all problems of family dissension in this intelligent soap opera.
1938 76m/B Joan Crawford, Margaret Sullavan, Melvyn Douglas, Robert Young, Fay Bainter, Allyn Joslyn; ***Dir:*** Frank Borzage. **VHS $19.98** *MGM* 🦴🦴½

A Shining Season Tearjerker about a potential-packed long-distance runner, John Baker, stricken with cancer. He spends his last year training a girls' track team to a championship. Earnest, well-performed and involving. Based on a true story. Made for TV.
1979 100m/C Timothy Bottoms, Rip Torn, Allyn Ann McLerie, Ed Begley Jr., Benjamin Bottoms, Mason Adams, Constance Forslund, Ellen Geer; ***Dir:*** Stuart Margolin. **VHS, Beta $59.95** *BTV* 🦴🦴🦴

Shining Star A recording company is run by the mob. Potentially interesting film is hampered by quality of sound and photography and general lack of purpose or direction. Music by Earth, Wind and Fire. Originally called "That's the Way of the World."
1975 (PG) 100m/C Harvey Keitel, Ed Nelson, Cynthia Bostick, Bert Parks; ***Dir:*** Sig Shore. **VHS $39.95** *IVE* 🦴½

Shining Through An old-fashioned blend of romance and derring-do with a noble hero and a spunky heroine threatened by Nazis in World War II Germany. Linda Voss (Griffith) gets a job as a legal secretary to a secretive lawyer (Douglas) she suspects may be a government agent. It turns out he works for the OSS and Linda (who speaks German) eventually gets work as a spy. Her job is to infiltrate the home of a Nazi officer (Neeson) and photograph some vital plans. But when her cover is blown, thanks to a double-crossing comrade, can her lawyer/lover rescue her in time? The action stops dead in a series of flash-forwards with Linda's retelling of her adventures on a talk show but everything (and everyone) looks good and the film tries hard for thrills. The film, however, in no way resembles the best-selling Susan Isaacs novel it was adapted from.
1991 (R) ?m/C Michael Douglas, Melanie Griffith, Liam Neeson, Joely Richardson, John Gielgud; ***W/Dir:*** David Seltzer. **VHS $94.98** *FXV* 🦴🦴½

Shinobi Ninja A ninja fights off enemy forces to save his homeland.
1984 103m/C Tadashi Yamashita, Eric Lee, Karen Lee Sheperd. **VHS, Beta $24.95** *AHV* 🦴

Shiokari Pass A man has a divine revelation at the pass. In Japanese with English subtitles.
1977 60m/C *JP* ***Dir:*** Novuru Nakamuru. **VHS $29.95** *REP, VBL* 🦴🦴

Ship of Fools A group of passengers sailing to Germany in the '30s find mutual needs and concerns, struggle with early evidence of Nazi racism, and discover love on their voyage. Twisted story and fine acting maintain interest. Appropriate tunes written by Ernest Gold. Based on the Katherine Ann Porter novel. Leigh's last film role; she died 2 years later. Kramer grapples with civil rights issues in much of his work. Oscar nominations for Best Picture, actress Signoret, supporting actor Dunn, Adapted Screenplay, and Costume Design.
1965 149m/B Vivien Leigh, Simone Signoret, Jose Ferrer, Lee Marvin, Oskar Werner, Michael Dunn, Elizabeth Ashley, George Segal, Jose Greco, Charles Korvin, Heinz Ruehmann; ***Dir:*** Stanley Kramer. Academy Awards '65: Best Art Direction/Set Decoration (B & W), Best Black and White Cinematography; National Board of Review Awards '65: 10 Best Films of the Year, Best Actor (Marvin). **VHS, Beta, LV $19.95** *COL* 🦴🦴🦴

Ships in the Night Couple find silent trouble on the high seas when they try to locate missing relative.
1928 63m/B Frank Moran, Jacqueline Logan, Jack Mower, Andy Clyde. **VHS, Beta $19.95** *GPV, FCT* 🦴🦴

Shipwreck Beach A continuous picture of a long beach shot highlighting cliff formations in the distance to create a relaxed background.
1978 30m/C VHS, Beta *PCC* 🦴🦴

Shipwreck Island Eight boys are shipwrecked on a deserted island, then make their way to another island where they must survive against wild animals and the elements for two years.
1961 93m/C VHS $19.95 *HTV, FCT* 🦴🦴

Shipwrecked Kiddie swashbuckler based on the 1873 popular novel "Haakon Haakonsen."
1990 (PG) 93m/C Gabriel Byrne, Stian Smestad, Louisa Haigh; ***Dir:*** Nils Gaup. **VHS, LV $19.99** *BVV* 🦴🦴½

Shirley Temple Festival A collection of four Shirley Temple short subjects from the 30s: "Dora's Dunkin Doughnuts," "Pardon My Pups," "War Babies" and "Kid N' Africa."
1933 55m/B Shirley Temple, Andy Clyde. **VHS, Beta $19.95** *WES, MRV* 🦴🦴½

Shirley Valentine A lively middle-aged English housewife gets a new lease on life when she travels to Greece without her husband. Collins reprises her London and Broadway stage triumph. The character frequently addresses the audience directly to explain her thoughts and feelings; her energy and spunk carry the day. Good script by Willy Russell from his play. From the people who brought us "Educating Rita."
1989 108m/C *GB* Pauline Collins, Tom Conti, Alison Steadman, Julia McKenzie; ***Dir:*** Lewis Gilbert. **VHS, LV, 8mm $14.95** *PAR, SOU* 🦴🦴🦴

Shirt Tales This is a series of cartoons featuring the animated crime-stoppers.
1985 45m/C VHS, Beta $19.95 *HNB* 🦴🦴🦴

Shiver, Gobble & Snore Series of animations based on stories by Mary Winn: "The Kings of Snark," "The Town That Had No Policemen," "The Fisherman Who Needed a Knife," and the title tale.
1970 45m/C VHS, Beta $19.95 *NWV* 🦴🦴🦴

The Shock Crippled low-life Chaney becomes restored spiritually by a small-town girl and rebels against his Chinese boss. This causes an unfortunate string of melodramatic tragedies, including the San Francisco earthquake of 1906. Silent. Odd, desultory tale with bad special effects is worth seeing for Chaney's good acting.
1923 96m/B Lon Chaney Sr., Virginia Valli; ***Dir:*** Lambert Hillyer. **VHS, Beta $16.95** *SNC, NOS, DNB* 🦴🦴½

Shock A psychiatrist is called on to treat a woman on the edge of a nervous breakdown. He then discovers she saw him murder his wife, and tries to keep her from remembering it. Interesting premise handled in trite B style. Price's first starring role.
1946 70m/B Vincent Price, Lynn Bari, Frank Latimore, Anabel Shaw; ***Dir:*** Alfred Werker. **VHS, Beta $16.95** *SNC, NOS, KRT* 🦴🦴

Shock Corridor A reporter, dreaming of a Pulitzer Prize, fakes mental illness and gets admitted to an asylum, where he hopes to investigate a murder. He is subjected to disturbing experiences, including shock therapy, but does manage to solve the murder. However, he suffers a mental breakdown in the process and is admitted for real. Disturbing and lurid.
1963 101m/B Peter Breck, Constance Towers, Gene Evans, Hari Rhodes, James Best, Philip Ahn; ***Dir:*** Samuel Fuller. **VHS, LV $79.95** *CRC, VYG* 🦴🦴🦴

Shock 'Em Dead A devil worshipper trades his life to Lucifer for a chance at rock and roll fame and beautiful Miss Lords. Sexy thriller, with a good share of violence and tension. Ironically, ex porn great Lords is one of the few starlets who doesn't disrobe in the film.
1990 (R) 94m/C Traci Lords, Aldo Ray, Troy Donahue, Stephen Quadros; ***Dir:*** Mark Freed. **VHS $19.95** *ACA* 🦴🦴

Shock! Shock! Shock! Another low-budget slasher flick, with a homicidal lunatic wielding a butcher knife. Space alien jewel thieves add an element, for what it's worth. Makes one ponder the cosmic question: What's the diff between a bad slasher flick and a spoof of a bad slasher flick? Whichever this one is, it's really bad!
1987 60m/B Brad Issac, Cyndy McCrossen, Allen Rickman, Brian Fuorry; ***Dir:*** Todd Rutt, Arn McConnell. **VHS, Beta $29.95** *RHI* 🦴

A Shock to the System Business exec. Caine is passed over for a long-deserved promotion in favor of a younger man. When he accidentally pushes a panhandler in front of a subway in a fit of rage, he realizes how easy murder is and thinks it may be the answer to all his problems. Tries to be a satire take on corporate greed, etc., but somehow loses steam. Excellent cast makes the difference; Caine adds class. Also available subtitled in Spanish.
1990 (R) 88m/C Michael Caine, Elizabeth McGovern, Peter Riegert, Swoosie Kurtz, Will Patton, Jenny Wright; ***Dir:*** Jan Egleson. **VHS, Beta, LV $89.99** *HBO* 🦴🦴🦴

Shock Treatment A mediocre semi-sequel to the cult classic, "The Rocky Horror Picture Show" (1975). Brad and Janet, now married and portrayed by different leads, find themselves trapped on a TV gameshow full of weirdos. Several original cast members do make an appearance.
1981 (PG) 94m/C Richard O'Brien, Jessica Harper, Cliff DeYoung, Patricia Quinn, Charles Gray, Ruby Wax; ***Dir:*** Jim Sharman. **VHS, Beta $19.98** *FOX* 🦴½

Shock Waves A group of mutant-underwater-zombie-Nazi-soldiers terrorizes stranded tourists staying at a deserted motel on a small island. Cushing is the mad scientist intent on recreating the Nazi glory days with the seaweed-attired zombies. Odd B-grade, more or less standard horror flick somehow rises (slightly) above badness. Also known as "Death Corps."
1977 (PG) 90m/C Peter Cushing, Brooke Adams, John Carradine, Luke Halprin; ***Dir:*** Ken Wiederhorn. **VHS, Beta $9.99** *STE, PSM, AVD* 🦴🦴

Shocker Another Craven gore-fest. A condemned serial killer is transformed into a menacing electrical force after being fried in the chair. Practically a remake of Craven's original "Nightmare on Elm Street." Great special effects, a few enjoyable weird and sick moments. What's Dr. Timothy Leary doing here?
1989 (R) 111m/C Michael Murphy, Peter Berg, Cami Cooper, Mitch Pileggi, Richard Price; *Dir:* Wes Craven, Timothy Leary. **VHS, Beta, LV** **$19.98** *MCA* 🐾🐾½

Shocking Asia For the eye-covering delight of Western audiences, Mr. Fox compiles gruesome Asian oddities.
1975 94m/C *GE Dir:* Emerson Fox. **VHS, Beta** **$39.98** *MAG, HHE* 🐾½

Shocking Asia 2 Second "Mondo Cane"-style documentary about various unpleasant Asian oddities, including voodoo, cannibalism, and other strange practices.
1976 90m/C *Dir:* Emerson Fox. **VHS, Beta** **$49.98** *VCD* 🐾½

Shockwave This is a tape of eight European heavy metal rock groups "performing" in concert, including Alaska, Black Lace, Venom and Faithful Breath. In HiFi Stereo.
1986 58m/C VHS, Beta $19.95 *MVD, LIV, VES* 🐾½

The Shoes of the Fisherman Morris West's interesting, speculative best seller about Russian Pope brought to the big screen at much expense, but with little care or thought. Siberian prison-camp vet Quinn, elected Pope, tries to arrest nuclear war. Director Anderson wasted the prodigious talents of Olivier, Gielgud, et al. Sloppy use of a good cast and a promising plot.
1968 (G) 160m/C Anthony Quinn, Leo McKern, Laurence Olivier, John Gielgud, Vittorio De Sica, Oskar Werner, David Janssen; *Dir:* Michael Anderson Sr. **VHS, Beta, LV $29.99** *MGM, PIA, FCT* 🐾½

Shoeshine Two shoeshine boys struggling to survive in post-war Italy become involved in the black market and are eventually caught and imprisoned. The prison scenes detail the sense of abandonment and tragedy that destroys the boys' friendship. A rich, sad achievement in neo-realistic drama. In Italian with English subtitles.
1946 90m/B *IT* Franco Interlenghi, Rinaldo Smordoni, Anniello Mele, Bruno Ortensi, Pacifico Astrologo; *Dir:* Vittorio DeSica. National Board of Review Awards '47: 10 Best Films of the Year. **VHS, Beta** *FCT* 🐾🐾🐾🐾

Shogun Complete version of the television mini-series that chronicles the saga of a shipwrecked English navigator who becomes the first Shogun, or Samurai warrior chief, from the Western world. Colorfully adapted from the James Clavell bestseller. Also released in a two-hour version, but the full-length version is infinitely better.
1980 550m/C Richard Chamberlain, Toshiro Mifune, Yoko Shimada, John Rhys-Davies, Damien Thomas; *Dir:* Jerry London; *Nar:* Orson Welles. **VHS, Beta, LV $29.95** *PAR* 🐾🐾🐾½

Shogun Assassin Story of a proud samurai named Lone Wolf who served his Shogun master well as the Official Decapitator, until the fateful day when the aging Shogun turned against him. Extremely violent, with record-breaking body counts. Edited from two other movies in a Japanese series called "Sword of Vengeance"; a tour de force of the cutting room. The samurai pushes his son's stroller through much of the film-sets it aside to hack and slash.

1980 (R) 89m/C *JP* Tomisaburo Wakayama; *Dir:* Kenji Misumi, Robert Houston. **VHS, Beta, LV** *MCA, OM* 🐾🐾½

Shogun's Ninja In 16th century Japan, an age-old rivalry between two ninja clans sparks a search for a dagger which will lead to one clan's hidden gold.
1983 115m/C Henry Sanada, Sue Shiomi, Sonny Chiba. **VHS, Beta $49.95** *MED* 🐾

Shoot Five hunting buddies fall prey to a group of crazed killers after one man is shot by accident. What's the message: Anti-gun? Anti-hunting? Silly, unbelievable, irresponsibly moralistic, and gratuitously violent.
1985 (R) 98m/C *CA* Ernest Borgnine, Cliff Robertson, Henry Silva, Helen Shaver; *Dir:* Harvey Hart. **VHS, Beta $59.98** *SUE* 🐾

Shoot It Black, Shoot It Blue A rogue cop shoots a black purse snatcher and thinks he has gotten away with it. Unknown to him, a witness has filmed the incident and turns the evidence over to a lawyer.
1974 93m/C Michael Moriarty. **VHS, Beta $69.99** *HBO* 🐾

Shoot to Kill B-grade noir-esque mystery about a crooked D.A. and a gangster who bite the dust. Suspenseful, dark, street-level crime drama uncovered by ambitious journalist Ward.
1947 64m/B Russell Wade, Susan Walters, Nestor Paiva, Edmund MacDonald, Vince Barnett; *Dir:* William Berke. **VHS, Beta, LV $19.99** *SNC, WGE* 🐾🐾½

Shoot to Kill A city cop (Poitier, better than ever after 10 years off the screen) and a mountain guide (Berenger) reluctantly join forces to capture a killer who is part of a hunting party traversing the Pacific Northwest and which is being led by the guide's unsuspecting girlfriend (Alley). Poitier may be a bit old for the role, but he carries the implausible plot on the strength of his performance. Good action.
1988 (R) 110m/C Sidney Poitier, Tom Berenger, Kirstie Alley, Clancy Brown, Richard Masur, Andrew Robinson; *Dir:* Roger Spottiswoode. **VHS $89.95** *DIS* 🐾🐾🐾

Shoot the Living, Pray for the Dead While travelling through Mexico, the leader of a band of killers promises his guide half of a share in stolen gold if he can lead them to it.
1973 90m/C Klaus Kinski. **VHS, Beta $19.95** *NEG* 🐾🐾

Shoot Loud, Louder, I Don't Understand! A sculptor who has a hard time separating reality from dreams thinks he witnessed a murder. Shenanigans follow. Meant to be a black comedy, but when not dull, it is confusing. Welch looks good, as usual, but doesn't show acting talent here. In Italian with English subtitles.
1966 101m/C *IT* Marcello Mastroianni, Raquel Welch; *Dir:* Eduardo DeFilippo. **VHS, Beta $59.95** *CHA* 🐾½

Shoot the Moon A successful writer, married and with four children, finds his life unrewarding and leaves his family to take up with a younger woman. The wife must learn to deal with her resentment, the fears of her children, and her own attempt at a new love. Fine acting but a worn-out story.
1982 (R) 124m/C Diane Keaton, Albert Finney, Karen Allen, Peter Weller, Dana Hill, Viveka Davis, Tracey Gold, Tina Yothers; *Dir:* Alan Parker. **VHS, Beta $19.99** *MGM* 🐾🐾½

Shoot the Piano Player A former concert pianist (Aznavour, spendidly cast) has changed his name and now works playing piano at a low-class Paris cafe. A convoluted plot ensues; he becomes involved with gangsters, though his girlfriend wants him to try a comeback. Lots of atmosphere, character development, humor and romance. A Truffaut masterpiece. In French, with English subtitles.
1962 84m/B *FR* Charles Aznavour, Marie DuBois, Nicole Berger; *Dir:* Francois Truffaut. **VHS, Beta, LV $24.95** *NOS, HHT, WFV* 🐾🐾🐾🐾

Shoot the Sun Down Four offbeat characters united in a search for gold turn against each other. "Gilligan's Island"-style assortment of characters (Indian; gunfighter; girl; sea captain) are cast away in a pointless plot from which they never escape. Too bad: talented cast could have done better.
1981 (PG) 102m/C Christopher Walken, Margot Kidder, Geoffrey Lewis, Bo Brundin, Sacheen Little Feather; *Dir:* David Leeds. **VHS, Beta $19.95** *VID* 🐾½

Shooters A wacky platoon of misfits is paired off against a group of vicious killers in a "war game" training session. Not completely unfunny or worthless.
1989 84m/C Ben Schick, Robin Sims, Aldo Ray. **VHS, Beta $79.95** *AIP* 🐾½

The Shooting A mysterious woman, bent on revenge, persuades a former bounty hunter and his partner to escort her across the desert, with tragic results. Offbeat, small film filled strong by performances of Nicholson and Oates. Filmed concurrently with "Ride in the Whirlwind," with the same cast and director. Bang-up surprise ending.
1966 82m/C Warren Oates, Millie Perkins, Jack Nicholson, Will Hutchins; *Dir:* Monte Hellman. **VHS, Beta** *MED, DVT, KOV* 🐾🐾🐾

Shooting Three boys run away when they think they have killed a man in a hunting accident.
1982 60m/C Lynn Redgrave, Lance Kerwin, Barry Primus. **VHS, Beta $39.98** *SCL* 🐾

The Shooting Party A group of English aristocrats assemble at a nobleman's house for a bird shoot on the eve of World War I. Splendid cast crowned by Mason, in his last role. Fascinating crucible class anxieties, rich wit social scheming, personality conflicts, and things left unsaid. Julian Bond wrote the script from Isabel Colegate's novel.
1985 97m/C *GB* James Mason, Dorothy Tutin, Edward Fox, John Gielgud, Robert Hardy, Cheryl Campbell, Judi Bowker; *Dir:* Alan Bridges. **VHS, Beta $69.99** *HBO* 🐾🐾🐾½

Shooting Stars Two TV actors portraying private eyes are pushed out of their jobs by a jealous co-star. They take to the streets as "real" crime-fighting dicks. Interesting premise goes nowhere. Made for TV.
1985 96m/C Billy Dee Williams, Parker Stevenson, Efrem Zimbalist Jr., Edie Adams; *Dir:* Richard Lang. **VHS, Beta $24.95** *AHV* 🐾

The Shootist Wayne, in a support last role, plays a legendary gunslinger afflicted with cancer, who seeks peace and solace in his final days. Town bad guys Boone and O'Brian aren't about to let him rest and are determined to gun him down to avenge past deeds. One of Wayne's best and most dignified performances about living up to a personal code of honor. Stewart and Bacall head excellent supporting cast. Based on Glendon Swarthout's novel.

1976 (PG) 100m/C John Wayne, Lauren Bacall, Ron Howard, James Stewart, Richard Boone, Hugh O'Brian, Bill McKinney, Harry Morgan, John Carradine, Sheree North, Scatman Crothers; *Dir:* Don Siegel. National Board of Review Awards '76: 10 Best Films of the Year. **VHS, Beta, LV $14.95** PAR, TLF, BUR 🎞🎞½

Shop Angel A department store dress designer encounters romance and scandal in this low-budget drama of intrigue and romance. She falls in love with the fiance of her boss's daughter, and schemes to blackmail him (the boss). Competently rendered if unexceptional tale.
1932 66m/B Marion Shilling, Holmes Herbert, Creighton Hale; *Dir:* Mason Hopper. **VHS, Beta, 8mm** VYY 🎞🎞

The Shop Around the Corner A low-key romantic classic in which Stewart and Sullavan are feuding clerks in a small Budapest shop, who unknowingly fall in love via a lonely hearts club. Charming portrayal of ordinary people in ordinary situations. Excellent script adapted by Simon Raphaelson, from the play "Parfumerie" by Nikolaus Laszlo. Later made into a musical called "In the Good Old Summertime" and, on Broadway, "She Loves Me."
1940 97m/B Margaret Sullavan, James Stewart, Frank Morgan, Joseph Schildkraut, Sara Haden, Felix Bressart; *Dir:* Ernst Lubitsch. **VHS, LV $19.95** MGM, PIA 🎞🎞🎞½

Shop on Main Street During World War II, a Slovak takes a job as an "Aryan comptroller" for a Jewish-owned button shop. The owner is an old deaf woman; they slowly build a friendship. Tragedy ensues when all of the town's Jews are to be deported. Sensitive and subtle. Surely among the most gutwrenching portrayals of human tragedy ever on screen. Exquisite plotting and direction. Subtitled in English. Well-deserved Oscar nomination for Kaminska. Also known as "The Shop on High Street."
1965 111m/B CZ Ida Kaminska, Josef Kroner, Hana Slivkoua, Elmar Klos; *Dir:* Jan Kadar, Elmar Klos. Academy Awards '65: Best Foreign Language Film; New York Film Critics Awards '66: Best Foreign Film; New York Film Festival '65: Best Film; New York Times Ten Best List '66: Best Film. **VHS, Beta, LV $19.98** COL 🎞🎞🎞🎞

Shore Leave Dressmaker Mackaill isn't getting any younger. Tough-guy sailor Bilge Smith (Barthelmess, in top form) meets her on shore leave; little does he realize her plans for him! She owns a dry-docked ship, you see, and it (and she) will be ready for him when he comes ashore next. Lovely, fun (if rather plodding) romantic comedy. Later made into musicals twice, as "Hit the Deck" and "Follow the Fleet."
1925 74m/B Richard Barthelmess, Dorothy Mackaill; *Dir:* John S. Robertson. **VHS** DNB 🎞🎞🎞

Short Circuit A newly-developed robot designed for the military is hit by lightning and begins to think it's alive. Sheedy and Guttenberg help it hide from the mean people at the weapons lab who want to take it home. Followed two years later, save Sheedy and Guttenberg, by the lame "Short Circuit 2."
1986 (PG) 98m/C Steve Guttenberg, Ally Sheedy, Austin Pendleton, Fisher Stevens, Brian McNamara; *Dir:* John Badham. **VHS, Beta, LV $19.98** FOX 🎞🎞½

Short Circuit 2 A sequel to the first adorable-robot-outwits-bad-guys tale. The robot, Number Five, makes his way through numerous plot turns without much human assistance or much purpose. Harmless (unless you have to spend time watching it), but pointless and juvenile. Very occasional genuinely funny moments.
1988 (PG) 95m/C Fisher Stevens, Cynthia Gibb, Michael McKean, Jack Weston, David Hemblen; *Dir:* Kenneth Johnson. **VHS, Beta, LV $19.95** COL 🎞½

Short Eyes When a child molester (Davison) enters prison, the inmates act out their own form of revenge against him. Filmed on location at New York City's Men's House of Detention, nicknamed "The Tombs." Script by Manuel Pinero from his excellent play; he also acts in the film. Top-notch performances and respectful direction from Young bring unsparingly realistic prison drama to the screen. Title is prison jargon for child molester. Also called "The Slammer."
1979 (R) 100m/C Bruce Davison, Miguel Pinero, Nathan George, Donald Blakely, Curtis Mayfield, Jose Perez, Shawn Elliott; *Dir:* Robert M. Young. New York Film Festival '77: Best Film. **VHS, Beta $69.98** LIV, LTG 🎞🎞🎞½

A Short Film Festival A festive collection of short animated films. Includes Oscar nominee "The Family That Lived Apart," and "Hot Stuff."
1990 57m/C VHS, Beta $29.95 WTA, FCT 🎞🎞½

Short Films of D.W. Griffith, Volume 1 Three early shorts by the great silent-film director. "The Battle" is a lavish Civil War drama. "The Female of the Species" stars pickford McDowell and Bernard as pioneers struggling to survive hardship and each other's company in the desert. "The New York Hat" is a comedy (written by Anita Loos, aged 14!) starring Pickford as a girl who endures the malice of her peers when she is given a nice hat. Landmark early films from a master director.
1912 59m/B Mary Pickford, Charles West, Blanche Sweet, Claire McDowell; *Dir:* D.W. Griffith. **VHS, Beta $24.95** VYY, DVT 🎞🎞🎞🎞

Short Films of D.W. Griffith, Volume 2 Five early silent shorts by Griffith, with original organ music and some familiar faces in the casts: "A Corner in Wheat," "The Revenue Man and his Girl," "The Musketeers of Pig Alley," "A Girl and her Trust" and "The Battle of Elderbush Gulch."
1909 112m/B Lillian Gish, Bobby Harron, Mae Marsh, Lionel Barrymore, Dorothy Gish, Henry B. Walthall, Kate Bruce, James Kirkwood, Harry Carey. **VHS, Beta, 8mm $29.99** VYY, DVT 🎞🎞½

Short Fuse A Washington, D.C. reporter uncovers the truth behind the rape and murder of a nurse. Interesting "go-go" music from the Washington D.C. ghetto aids an otherwise uninteresting film. Released as "Good to Go" for theaters.
1988 (R) 91m/C Art Garfunkel, Robert DoQui, Harris Yulin, Richard Brooks, Reginald Daughtry; *Dir:* Blaine Novak. **VHS, Beta $79.95** VMK 🎞½

Short Time Somewhere between "Tango & Cash" and "Airplane" is where you'll find "Short Time." Non-stop action/comedy stars Dabney Coleman and Teri Garr as partners in crime and humor. Coleman is a cop days from retirement; wrongly told he is dying, he tries hard to get killed so his family will be provided for. Coleman is wonderful. Garr, no longer the long-suffering wife of "Oh, God"

and "Close Encounters," is appealing here as the indulgent sidekick.
1990 (PG-13) 100m/C Dabney Coleman, Teri Garr, Matt Frewer, Barry Corbin, Joe Pantoliano, Xander Berkeley; *Dir:* Gregg Champion. **VHS, LV $14.95** IME, LIV 🎞🎞🎞

A Shot in the Dark A minister suspects murder when a despised miser is thought to have committed suicide. He takes it on himself to investigate in this routine mystery.
1933 53m/B GB Dorothy Boyd, O.B. Clarence, Jack Hawkins, Russell Thorndike, Michael Shepley; *Dir:* George Pearson. **VHS, Beta $16.95** SNC 🎞½

A Shot in the Dark When his son is murdered at a New England college, a distraught dad takes it on himself to investigate. Former movie cowboy Starrett plays the sleuthing pop in this undistinguished mystery.
1935 69m/B Charles Starrett, Robert Warwick, Edward Van Sloan, Marion Shilling, Doris Lloyd, Helen Jerome Eddy, James Bush; *Dir:* Charles Lamont. **VHS, Beta $16.95** NOS, SNC 🎞🎞

A Shot in the Dark Second and possibly the best in the classic "Inspector Clouseau-Pink Panther" series of comedies. The bumbling Inspector Clouseau (Sellers, of course) investigates the case of a parlor maid (Sommer) accused of murdering her lover. Clouseau's libido convinces him she's innocent,even though all the clues point to her. Classic gags, wonderful Henry Maucini music. After this film, Sellers as Clouseau disappears until 1975's "Return of the Pink Panther" (Alan Arkin played him in "Inspector Clouseau," made in 1968 by different folks).
1964 101m/C Peter Sellers, Elke Sommer, Herbert Lom, George Sanders, Bryan Forbes; *Dir:* Blake Edwards. **VHS, Beta, LV $19.98** FOX, FCT 🎞🎞🎞½

Shotgun A sheriff on the trail of a killer is accompanied by a girl he's saved from Indians. Average western cowritten by western actor Rory Calhoun, who had hoped to star in the film, but was turned down by the studio.
1955 81m/C Sterling Hayden, Zachary Scott, Yvonne de Carlo; *Dir:* Lesley Selander. **VHS $29.95** PAR 🎞🎞½

Shotgun The main character, "Shotgun," is out to avenge a prostitute's murder. Along the way, he has to deal with sex, drugs, and violence.
1989 90m/C Stuart Chapin, Katie Caple. **VHS $29.95** PMH 🎞

The Shout From a strange Robert Graves story, an even stranger film. A lunatic befriends a young couple, moves in with them, and gradually takes over their lives. The movie's title refers to the man's ability to kill by shouting, a power he learned from his Australian aboriginal past.
1978 (R) 88m/C GB Alan Bates, Susannah York, John Hurt, Tim Curry; *Dir:* Jerzy Skolimowski. Cannes Film Festival '78: Jury Prize; New York Film Festival '78: Jury Prize. **VHS, Beta $39.98** COL, VID 🎞🎞🎞

Shout Romance and rebellion set in a sleepy Texas town during the 1950s. Jesse Tucker's (Walters) rebellious ways land him in the Benedict Home for Boys and he seems lost until Jack Cabe (Travolta) enters town. Jack introduces Jesse and the gang to the exciting new sounds of rock 'n' roll. Been done before and better.

1991 (PG-13) 93m/C John Travolta, James Walters, Heather Graham, Richard Jordan, Linda Fiorentino; **Dir:** Jeffrey Hornaday. **VHS, Beta, LV** $34.98 *MCA* ⅃⅃

Shout at the Devil An Englishman and an Irish-American seek to blow up a German battleship out for repairs in East Africa in 1913. From the novel by Wilbur Smith. Based on an actual incident, but that doesn't make it more palatable. Marvin is ludicrous, Moore is wooden, the movie drags on - and on - with way too much action and too many things exploding.
1976 (PG) 128m/C *GB* Lee Marvin, Roger Moore, Barbara Parkins, Ian Holm; **Dir:** Peter Hunt. **VHS, Beta, LV** $29.98 *LIV, VES* ⅃ ½

Shout!: The Story of Johnny O'Keefe Compilation of performance clips documenting O'Keefe's heyday; songs include "Shout!," "Johnny B. Goode," "The Great Pretender," and many, many more.
1985 260m/C Helen Reddy. **VHS, Beta** $29.95 *FRH* ⅃ ½

Show Boat The second of three film versions of the Jerome Kern-Oscar Hammerstein musical, filmed previously in 1929, about a Mississippi showboat and the life and loves of its denizens. Wonderful romance, unforgettable music, including "I Have the Room above Her," a lovely original song and, of course "Old Man River," sung by Robeson. Director Whale also brought the world "Frankenstein." The laser edition includes a historical audio essay by Miles Kreuger, excerpts from the 1929 version, Ziegfeld's 1932 stage revival, "Life Aboard a Real Showboat" (a vintage short), radio broadcasts, and a 300-photo essay tracing the history of showboats. Screenplay by Hammerstein. As is this version was not the end-all, it was filmed yet again 15 years later.
1936 110m/B Irene Dunne, Allan Jones, Paul Robeson, Helen Morgan, Hattie McDaniel, Charles Winninger, Donald Cook; **Dir:** James Whale. **VHS, Beta, LV** $29.95 *CRC, MLB* ⅃⅃⅃⅃

Show Boat Third movie version of Jerome Kern and Oscar Hammerstein II's 1927 musical play about the life and loves of a Mississippi riverboat theater troupe. The famous score includes "Old Man River," "Make Believe," "Can't Help Lovin' That Man," "Why Do I Love You," and "Bill." Young girl raised as a showboat entertainer falls in love with a gambler. Heartache follows. Terrific musical numbers, with fun dance from Champion who went on to great fame as a choreographer. Grayson is somewhat vapid, but lovely to look at and hear. Gardner didn't want to do the part of Julie, although she eventually received fabulous reviews. Her singing was dubbed by Annette Warren. The 171-foot "Cotton Blossom" boat was built on the Tarzan lake on the MGM back lot at a cost of $126,468.00 - a lot of money in 1951. Warfield's film debut - his "Ole Man River" - was recorded in one take. Laserdisc version features the original motion picture trailer, excerpts from "Broadway," "Silent Film," and radio version of "Show Boat," a documentary featuring life on a real show boat, and a commentary on the performers, the film, and the history of show boats.
1951 (G) 115m/C Kathryn Grayson, Howard Keel, Ava Gardner, William Warfield, Joe E. Brown, Agnes Moorehead; **Dir:** George Sidney. **VHS, Beta, LV** $19.95 *MGM, VYG, TLF* ⅃ ½

Show Business Historically valuable film record of classic vaudeville acts, especially Cantor and Davis. A number of vaudevillians recreate their old acts for director Marin - unforgettable slapstick and songs. All this pegged on a plot that follows Cantor's rise to fame with the Ziegfield Follies.
1944 92m/B Eddie Cantor, Joan Davis, George Murphy; **Dir:** Edwin L. Marin. **VHS, Beta, LV** $19.98 *CCB, TTC* ⅃⅃⅃

A Show of Force Reporter Irving investigates the coverup of a murder with political ramifications. Brazilian director Barreto cast (surprise!) his girlfriend in the lead; she doesn't exactly carry the day. Phillips is good, but you'll end up feeling cheated if you expect to see much of highly-billed Duvall or Garcia. Based on a real incident of 1978, but hardly believable as political realism or even moralism.
1990 (R) 93m/C Erik Estrada, Amy Irving, Andy Garcia, Robert Duvall, Lou Diamond Phillips; **Dir:** Bruno Barreto. **VHS, Beta, LV** $19.95 *PAR* ⅃⅃

Show My People Two separate programs featuring instrumental and vocal pieces performed by Bob Jones University musical groups.
1989 58m/C **VHS, Beta** $39.95 *BUP* ⅃⅃

Show People A pretty girl from the boonies tries to make it big in Tinseltown. But as a slapstick star?! She wanted to be a leading lady! Enjoyable, fun silent comedy shows Davies's true talents. Interesting star cameos, including director Vidor at the end.
1928 82m/B Marion Davies, William Haines, Dell Henderson, Paul Ralli; **Dir:** King Vidor. **VHS** $29.95 *MGM* ⅃⅃⅃

Show Them No Mercy A couple and their baby out for a drive unwittingly stumble into a kidnapping that doesn't go as planned. Tense gangster drama reincarnated as the Western "Rawhide."
1935 76m/B Rochelle Hudson, Cesar Romero, Bruce Cabot, Edward Norris, Edward Brophy, Warren Hymer, Herbert Rawlinson; **Dir:** George Marshall. **VHS** $19.98 *FOX, FCT* ⅃⅃⅃

Showbiz Goes to War The activities of Hollywood's biggest stars during World War II are the subject of this feature. Pin-up girls, USO shows, propaganda films, patriotic musicals and war bond drives are a few of the subjects seen along the way.
1982 90m/C *Nar:* David Steinberg. **VHS, Beta** $12.95 *IVE* ⅃⅃⅃

The Showdown The title actually refers to a tricky poker game, the highlight of yet another Hopalong Cassidy versus hoss thieves quickie epic. Kermit Maynard is the brother of cowboy hero Ken Maynard.
1940 65m/B William Boyd, Russell Hayden, Britt Wood, Morris Ankrum, Jan Clayton, Roy Barcroft, Kermit Maynard; **Dir:** Howard Bretherton. **VHS** $19.95 *NOS, BUR* ⅃⅃

Showdown A woman comes between two men and sends them their separate ways. One becomes a sheriff and must pursue the other, who is now a criminal. Seaton's final film is mediocre.
1973 (PG) 99m/C Rock Hudson, Dean Martin, Susan Clark, Donald Moffatt, John McLiam; **Dir:** George Seaton. **VHS** *TAV* ⅃⅃ ½

Showdown at Boot Hill A U.S. Marshal (Bronson) kills a wanted murderer but cannot collect the reward because the townspeople will not identify the victim.
1958 76m/B Charles Bronson, Robert Hutton, John Carradine; **Dir:** Gene Fowler. **VHS** $19.98 *REP* ⅃⅃

Showdown at the Equator Chinese cops are trained to kill criminals with their feet, and they do.
1983 (R) 95m/C Lo Lieh, Bruce Liang, Nora Maio, Lee Kan Kwun. **VHS, Beta** $24.95 *GEM* ⅃

Showdown in Little Tokyo Lundgren stars as a martial arts master/L.A. cop who was raised in Japan, and has all the respect in the world for his "ancestors" and heritage. Brandon Lee (son of Bruce) is Lundgren's partner, and he's a bona fide American-made, pop-culture, mall junkie. Together, they go after a crack-smuggling gang of "yakuza" (Japanese thugs). Lots of high-kicking action and the unique angle on stereotypes make this a fun martial arts film.
1991 (R) 78m/C Dolph Lundgren, Brandon Lee, Tia Carrere, Cary-Hiroyuki Tagawa; **Dir:** Mark L. Lester. **VHS, Beta, LV, 8mm** $92.99 *WAR* ⅃⅃

Showdown at Williams Creek A graphic Canadian Western set in the old Montana territory, where an outcast settler goes on trial for killing an old man. Testimony recounts a shocking history of greed and betrayal. The dark side of the Gold Rush, generally well-acted. Inspired by an actual incident.
1991 (R) 97m/C *CA* Tom Burlinson, Donnelly Rhodes, Raymond Burr; **Dir:** Allen Kroeker. **VHS** $89.98 *REP* ⅃⅃ ½

Shredder Orpheus Rock star Orpheus skateboards through hell to stop deadly television transmissions and rescue his wife.
1989 93m/C Jesse Bernstein, Robert McGinley, Vera McCaughan, Megan Murphy, Carlo Scandiuzzi; **Dir:** Robert McGinley. **VHS, Beta** $79.95 *AIP* ⅃

Shriek of the Mutilated An anthropological expedition on a deserted island turns into a night of horror as a savage beast kills the members of the group one by one.
1974 (R) 85m/C Alan Brock, Jennifer Stock, Michael Harris; **Dir:** Michael Findlay. **VHS, Beta** $59.98 *LIV, LTG Woof!*

Shriek in the Night Two rival reporters (Rogers and Talbot, previously paired in "The Thirteenth Guest") turn detective to solve a string of apartment murders. Made on a proverbial shoestring, but not bad.
1933 66m/B Ginger Rogers, Lyle Talbot, Harvey Clark; **Dir:** Albert Ray. **VHS, Beta** $12.95 *SNC, NOS, KRT* ⅃⅃

The Shrieking Black magic women hang out with biker types in Nebraska in 1919. Originally titled "Hex."
1973 (PG) 93m/C Keith Carradine, Christina Raines, Gary Busey, Robert Walker, Dan Haggerty, John Carradine, Scott Glenn; **Dir:** Leo Garen. **VHS, Beta** $29.95 *PSM, IME* ⅃

The Shrimp on the Barbie When daddy refuses to bless her marriage to dim bulb boyfriend, Australian Samms hires L.A. low life Marin to pose as new beau. Another pseudonymous Smithee effort.
1990 (R) 86m/C Cheech Marin, Emma Samms, Vernon Wells, Bruce Spence, Carole Davis; **Dir:** Alan Smithee. **VHS, Beta** $89.98 *MED* ⅃ ½

Shy People An urbanized New York journalist and her spoiled daughter journey to the Louisiana bayous to visit long-lost relatives in order to produce an article for "Cosmopolitan." They find ignorance, madness and ancestral secrets and are forced to examine their motives, their relationships and issues brought to light in the watery, murky, fantastic land of the bayous. Well-acted melodrama with an outstanding performance by Hershey as the cajun ma-

triarch. Music by Tangerine Dream; cinematography by Chris Menges.
1987 (R) 119m/C Jill Clayburgh, Barbara Hershey, Martha Plimpton, Mare Winningham, Merritt Butrick, John Philbin; *Dir:* Andrei Konchalovsky. VHS, Beta, LV **$19.98** *WAR* 🎞🎞🎞

The Sibling Two sisters become involved in the accidental murder of a man who was a husband to one woman and lover to the other. Of course, one sister has just been released from the mental rehabilitation clinic. Originally titled "Psycho Sisters" and also known as "So Evil, My Sister."
1972 (PG) 85m/C Susan Strasberg, Faith Domergue, Sydney Chaplin, Steve Mitchell; *Dir:* Reginald LeBorg. VHS, Beta **$9.95** *SIM, PSM, IPI* 🎞🎞

Sibling Rivalry Repressed doctor's wife (redundant) Alley rolls in hay with soon to be stiff stranger upon advice of footloose sister. Stranger expires from heart attack in hay and Alley discovers that the corpse is her long-lost brother-in-law. Slap-stick cover-up ensues.
1990 (PG-13) 88m/C Kirstie Alley, Bill Pullman, Carrie Fisher, Sam Elliott, Jami Gertz, Ed O'Neill, Scott Bakula, Frances Sternhagen, Bill Macy; *Dir:* Carl Reiner. VHS, Beta, LV, 8mm **$19.95** *COL, SUE* 🎞🎞

The Sicilian Adapted from the Mario Puzo novel and based on the life of Salvatore Giuliano. Chronicles the exploits of the men who took on the government, the Catholic Church, and the Mafia in an effort to make Sicily secede from Italy and become its own nation in the 1940s. Pretentious, overdone, and confused. This long, uncut version was unseen in America, but hailed by European critics; the 115-minute, R-rated American release is also available, but not as good. See "Salvatore Giuliano" (Francesco Rosi, 1962) instead of either version.
1987 146m/C Christopher Lambert, John Turturro, Terence Stamp, Joss Ackland, Barbara Sukowa; *Dir:* Michael Cimino. VHS, Beta, LV **$14.98** *LIV, VES* 🎞½

Sicilian Connection A narcotics agent poses as a nightclub manager to bust open a drug-smuggling operation.
1974 100m/C *IT* Ben Gazzara, Silvia Monti, Fausto Tozzi. VHS, Beta **$79.95** *KOV, VCD* 🎞½

Sicilian Connection Two brothers in the mob serve out vendettas and generally create havoc wherever they go.
1985 (R) 117m/C Michele Placido, Mark Switzer, Simona Cavallari. VHS, Beta **$79.95** *MGM* 🎞

Sid & Nancy The tragic, brutal, true love story of The Sex Pistols' Sid Vicious and American groupie Nancy Spungen, from the director of "Repo Man." Remarkable lead performances in a very dark story that manages to be funny at times. Depressing but engrossing; no appreciation of punk music or sympathy for the self-destructive way of life is required. Oldman and Webb are superb. Music by Joe Strummer, the Pogues, and Pray for Rain.
1986 (R) 111m/C Gary Oldman, Chloe Webb, David Hayman, Debbie Bishop; *Dir:* Alex Cox. National Society of Film Critics Awards '86: Best Actress (Webb); Boston Society of Film Critics Awards '86: Best Actress (Webb). VHS, Beta, LV **$14.95** *MVD, SUE* 🎞🎞🎞½

Side By Side In the same vein as "Cocoon," three senior citizens who aren't ready to retire decide to start their own business and launch a sportswear company designed for seniors. A witty portrayal of graceful aging that is uneven at times.

1988 100m/C Milton Berle, Sid Caesar, Danny Thomas; *Dir:* Jack Bender. VHS, Beta **$79.95** *TRI* 🎞🎞

Side Out The first major film about volleyball!? What a claim! What a bore. Midwestern college guy spends summer in Southern Cal. working for slumlord uncle; instead enters "the ultimate" beach volleyball touring. Bogus. Don't see it, dude.
1990 (PG-13) 100m/C C. Thomas Howell, Peter Horton, Kathy Ireland, Sinjin Smith, Randy Stoklos, Courtney Thorne-Smith, Harley Jane Kozak, Christopher Rydell; *Dir:* Peter Israelson. VHS, Beta, LV **$89.95** *COL* 🎞½

Side Show A runaway teen joins the circus, witnesses a murder, and must use his wits to stay out of reach of the killer. Forced "suspense" and "drama" of the bad made-for-TV ilk.
1984 98m/C Lance Kerwin, Red Buttons, Anthony (Tony) Franciosa, Connie Stevens; *Dir:* William Conrad. VHS, Beta **$19.98** *TWE* 🎞

Sideburns The Pushkin Club, a group of reactionaries who try to remove the western influence from Russia, attack first a rock band, then innocent civilians as they act as a social cleaning service. Uses humor to try and warn people against the rising fascism in Russia because of the battle between conservative and reformist forces. In Russian with English subtitles.
1991 110m/C *RU Dir:* Yurii Mamin. VHS **$59.95** *FCT* 🎞🎞

Sidewalks of London Laughton's a sidewalk entertainer who takes in homeless waif Leigh and puts her in his act and in his heart. Harrison steals her away and before long she's a star in the music halls. Meanwhile, Laughton has fallen on hard times. Memorable performances. Released in Great Britain under the title "St. Martin's Lane."
1940 86m/B *GB* Charles Laughton, Vivien Leigh, Rex Harrison; *Dir:* Tim Whelan. VHS, Beta **$19.95** *NOS, HHT, CAB* 🎞🎞🎞

Sidewinder One Motocross racing is the setting for a romance between a racer and an heiress. Good racing footage, but where's the plot? If you like cars a whole lot .

1977 (PG) 97m/C Michael Parks, Marjoe Gortner, Susan Howard, Alex Cord, Charlotte Rae; *Dir:* Earl Bellamy. VHS, Beta **$59.98** *CHA* 🎞

Sidney Sheldon's Bloodline A wealthy businesswoman (Hepburn; didn't she have anything better to do?) finds she is marked for death by persons unknown. Exquisitely bad trash from another Sheldon bestseller.
1979 (R) 116m/C Audrey Hepburn, Ben Gazzara, James Mason, Michelle Phillips, Omar Sharif, Irene Papas, Romy Schneider, Gert Frobe, Maurice Ronet, Beatrice Straight; *Dir:* Terence Young. VHS, Beta *PAR* 🎞

The Siege of Firebase Gloria The story of the Marines who risked their lives defending an outpost against overwhelming odds during the 1968 Tet offensive in Vietnam. Purportedly patriotic war drama made by an Australian director; lead Hauser is a disgusting sadist, and plot is hopelessly hackneyed.
1989 95m/C Wings Hauser, Lee Ermey; *Dir:* Brian Trenchard-Smith. VHS, Beta, LV **$29.95** *FRH* 🎞½

Siegfried Half of Lang's epic masterpiece "The Niebelungen," based on German mythology. Title hero bathes in the blood of a dragon he has slain. he marries a princess,

but wicked Queen Brumhilde has him killed. Part two, in which Siegfried's widow marries Attila the Hun, is titled "Kriemhilde's Revenge." These dark, brooding, archetypal tours de force were patriotic tributes, and were loved by Hitler.
1924 100m/B *GE* Paul Richter, Margareta Schoen; *Dir:* Fritz Lang. VHS, LV **$16.95** *SNC, NOS, GPV* 🎞🎞🎞🎞

Siempre Hay una Primera Vez The different loves of two women are contrasted.
1987 94m/C *MX* Ana Martin, Ana Louisa Peluffo, Helena Rojo. VHS, Beta **$47.95** *MAD* 🎞🎞

Siesta Barkin is a professional stunt woman who leaves her current lover-manager, played by Sheen, and returns to visit her former lover and trainer, Byrne, on the eve of his marriage to another woman. Her trip, marked by flashbacks and flights of seemingly paranoid fantasy, leads to the discovery of murder, but she cannot remember who, when, why or where. The film distorts time, reality and perception in a sometimes fascinating, sometimes frustrating psychological mystery. Attractively filmed by video director Lambert with music by Miles Davis. Barkin and Byrne were married in real life a year after the film's release.
1987 (R) 97m/C Ellen Barkin, Gabriel Byrne, Jodie Foster, Julian Sands, Isabella Rossellini, Martin Sheen, Grace Jones, Alexei Sayle; *Dir:* Mary Lambert. VHS, Beta, LV **$89.95** *LHV, WAR* 🎞🎞½

Sifted Evidence A woman archeologist who becomes involved with a man promises to show her an obscure archeological site in Mexico.
1983 42m/C 3/4U, Special order formats *WIF* 🎞

Sign of Zorro The adventures of the masked swordsman as he champions the cause of the oppressed in early California. A full-length version of the popular late-50s Disney television series.
1960 89m/C Guy Williams, Henry Calvin, Gene Sheldon, Romney Brent, Britt Lomond, George Lewis, Lisa Gaye; *Dir:* Norman Foster, Lewis R. Foster. VHS, Beta **$69.95** *DIS* 🎞½

Signal 7 An improvised, neo-verite document of a night in the lives of two San Francisco taxi drivers. Nilsson's first major release and a notable example of his unique scriptless, tape-to-film narrative technique.
1983 89m/C Bill Ackridge, Dan Leegant; *Dir:* Rob Nilsson. VHS, Beta **$39.98** *WAR* 🎞🎞

Signals Through the Flames A film by Sheldon Rochlin and Maxine Harris outlining the tumultuous history of the famous experimental group, The Living Theatre.
1983 97m/C Julian Beck, Judith Malina, Maxine Harris, Living Theatre Company; *Dir:* Sheldon Rochlin. VHS, Beta **$29.95** *MFV, IVA* 🎞🎞🎞

Signs of Life A boat-building company in Maine closes its doors after centuries in business; the employees and families whose lives have been defined by it for generations learn to cope. Wonderful performances compensate only partly for a week script. An episode on PBS's "American Playhouse."
1989 (PG-13) 95m/C Beau Bridges, Arthur Kennedy, Vincent D'Onofrio, Kevin J. O'Connor, Will Patton, Kate Reid; *Dir:* John David Coles. VHS, Beta, LV **$89.95** *IVE* 🎞🎞½

Silas Marner Superb made-for-British-television adaptation of the George Eliot classic about an itinerant weaver subjected to criminal accusation, poverty and exile. Wonderful detail and splendid acting.

1985 92m/C *GB* Ben Kingsley, Jenny Agutter, Patrick Ryecart, Patsy Kensit; *Dir:* Giles Foster. **VHS, Beta $29.98** *KUI, FOX* ♫♫♫

The Silence A brutal, enigmatic allegory about two sisters, one a nymphomaniac, the other a violently frustrated lesbian, traveling with the former's young son to an unnamed country beset by war. Surreal and dark; in Swedish with English subtitles or dubbed. Fascinating and memorable but frustrating and unsatisfying: What is it about? What is it an allegory of? Where is the narrative? The third in Bergman's crisis-of-faith trilogy following "Through a Glass Darkly" and "Winter Light."
1963 95m/C *SW* Ingrid Thulin, Gunnel Lindstrom, Birger Malmsten; *Dir:* Ingmar Bergman. **VHS, Beta $29.98** *SUE* ♫♫♫

Silence An autistic boy gets lost in the wilderness and faces an array of difficulties while his foster parents search for him. Also known as "Crazy Jack and the Boy."
1973 (G) 82m/C Will Geer, Ellen Geer, Richard Kelton, Ian Geer Flanders, Craig Kelly; *Dir:* John Korty. **VHS, Beta $29.98** *MAG* ♫♫

Silence of the Heart Mother copes with aftermath of suicide of teenage son following his recent divorce. Scripted by Phil Penningroth.
1984 100m/C Mariette Hartley, Dana Hill, Howard Hesseman, Chad Lowe, Charlie Sheen; *Dir:* Richard Michaels. **VHS $29.95** *PMH, HHE* ♫♫♫

The Silence of the Lambs Foster plays FBI cadet Clarice Starling, a woman with ambition, a cum laude degree in psychology and a traumatic childhood. When a serial killer begins his ugly rounds, the FBI wants psychological profiles from other serial killers. Starling's sent to collect a profile from one who's exceptionally clever - psychiatrist Hannibal Lector, a vicious killer prone to dining on his victims. He trades clues for intimate details of Starling's life and those clues aim her for a head-on collision with the mad man. Brilliant performances from Foster and Hopkins, finely detailed supporting characterizations, and elegant pacing from Demme. Some brutal visual effects. Excellent portrayals of women who refuse to be victims. Based on the Thomas Harris novel.
1991 (R) 118m/C Jodie Foster, Anthony Hopkins, Scott Glenn, Ted Levine, Brooke Smith, Charles Napier, Roger Corman, Anthony Heald, Diane Baker, Chris Isaak; *Dir:* Jonathan Demme. Academy Awards '91: Best Actor (Hopkins), Best Actress (Foster), Best Adapted Screenplay, Best Director (Demme), Best Picture; Chicago Film Critics Awards '91: Best Actor (Hopkins), Best Actress (Foster), Best Director (Demme), Best Picture, Best Screenplay; Directors Guild of America Awards '91: Best Director (Demme); Golden Globe Awards '92: Best Actress (Foster); National Board of Review Awards '91: 10 Best Films of the Year, Best Director (Demme), Best Picture; New York Film Critics Awards '91: Best Actor (Hopkins), Best Actress (Foster), Best Director (Demme), Best Picture. **VHS, Beta, LV $19.98** *CCB, ORI, FCT* ♫♫♫♫

Silence Like Glass Diagnosed with life-threatening illness, two young women struggle to overcome their anger at their fate. They find friendship and together, search for reasons to live. Fine performances, with pacing that keeps the melodrama to a minimum.
1990 (R) 102m/C Jami Gertz, Martha Plimpton, George Peppard, Rip Torn, James Remar; *Dir:* Carl Schenkel. **VHS, Beta $89.98** *MED* ♫♫½

Silence of the North A widow (Burstyn) with three children struggles to survive under rugged pioneer conditions on the Canadian frontier. The scenery is, not surprisingly, stunning. Based on a true but generic story.
1981 (PG) 94m/C *CA* Ellen Burstyn, Tom Skerritt; *Dir:* Allan Winton King. Genie Awards '82: Best Cinematography. **VHS, Beta $59.95** *MCA* ♫♫½

Silent Assassins Fists and bullets fly when a scientist is kidnapped in order to gain secrets to biological warfare. Chong and Rhee, real-life owners of a martial arts studio, produced and choreographed the film. Blair is here, but doesn't figure much. A notch up from most similar martial arts pics.
1988 92m/C Sam Jones, Linda Blair, Jun Chong, Phillip Rhee; *Dir:* Scott Thomas, Lee Doo-yong. **VHS $79.98** *MCG* ♫♫

Silent Code A Mountie is framed for the murder of a young miner.
1935 55m/B Tom Tyler. **VHS, Beta** *WGE* ♫

The Silent Enemy The true-life exploits of British frogmen battling Italian foes during World War II. Suspenseful and engrossing; good performances and good rendering of underwater action. Video release snips twenty minutes from the original and adds color.
1958 91m/C *GB* Laurence Harvey, John Clements, Michael Craig, Dawn Addams, Sidney James, Alec McCowen, Nigel Stock; *Dir:* William Fairchild. **VHS, Beta $7.00** *VTR* ♫♫½

Silent Killers Much maiming and slaughtering, little sound.
19?? 90m/C Stephen Soul, Daniel Garfield. **VHS** *AVD* ♫½

Silent Laugh Makers, No. 1 Four hilarious short movies from the silent era: "Looking for Sally," about an immigrant in search of his best girlfriend; "Lucky Dog," with the young Laurel and Hardy; "His Night Out," about a wild evening with Charlie Chaplin and Ben Turpin; and "Hop-a-Long." Non-stop side-splitting fun.
192? 50m/B Charley Chase, Stan Laurel, Oliver Hardy, Charlie Chaplin, Ben Turpin, Arthur Lake. **VHS, Beta $29.95** *VYY* ♫♫♫

Silent Laugh Makers, No. 2 Several short funny films of the silent picture era: "Fluttering Hearts," "Long Live the King," about a king who just can't get along with his nobles, starring Chase and Hardy; and "A Sea Dog Tale." All are unremittingly hilarious.
192? 50m/B Charley Chase, Oliver Hardy, Billy Bevan. **VHS, Beta $29.95** *VYY* ♫♫♫

Silent Laugh Makers, No. 3 Four comedy shorts: "Picking Peaches" (1924), directed by Frank Capra, Langdon's first for Mack Sennett; "All Tied Up" (1925), featuring "A Ton of Fun"-four rotund comedians; "Don't Shove" (1919), featuring Harold Lloyd wooing Bebe Daniels; and "Some Baby" (1922), with Snub Pollard, written by Hal Roach.
192? 55m/B Harry Langdon, Harold Lloyd, Bebe Daniels, Snub Pollard. **VHS, Beta $24.95** *VYY* ♫♫♫

Silent Madness A psychiatrist must stop a deranged killer, escaped from an asylum, from slaughtering helpless college coeds in a sorority house. Meanwhile, hospital execs send orderlies to kill the patient, to conceal their mistake. Ludicrous.
1984 (R) 93m/C Belinda J. Montgomery, Viveca Lindfors, Sydney Lassick; *Dir:* Simon Nuchtern. **VHS, Beta $59.95** *MED* ♫

Silent Movie A has-been movie director (Brooks) is determined to make a comeback and save his studio from being taken over by a conglomerate. Hilarious at times but uneven. An original idea; not as successful as it could have been. Has music and sound effects, but only one word of spoken dialogue by famous mime Marceau.
1976 (PG) 88m/C Mel Brooks, Marty Feldman, Dom DeLuise, Burt Reynolds, Anne Bancroft, James Caan, Liza Minnelli, Paul Newman, Sid Caesar, Bernadette Peters, Harry Ritz, Marcel Marceau; *Dir:* Mel Brooks. National Board of Review Awards '76: 10 Best Films of the Year. **VHS, Beta, LV $19.98** *FOX* ♫♫½

Silent Movies: In Color Five films produced in France between 1904 and 1914 in which color tints were added one frame at a time by hand. Films include: "The Nobleman's Dog," "Bob's Electric Theatre," "A Slave's Love," "A New Way of Traveling," and "The Life of Our Savior."
1914 52m/C **VHS, Beta $19.98** *VYY* ♫♫½

Silent Movies...Just For the Fun of It! Three short silent films: "The Iron Mule" (1925), "For Sadie's Sake" (1926), and "Hazel from Hollywood" (1918).
1926 56m/B Jimmie Adams, Dorothy Devore, Al "Fuzzy" St. John. **VHS, Beta, 8mm $24.95** *VYY* ♫♫½

The Silent Mr. Sherlock Holmes These two silent shorts are a rare treat for Holmes fans. "The Copper Beeches" was produced with the "personal supervision of Arthur Conan Doyle." In "The Man with the Twisted Lip," Holmes is hired to find out whether a banker was murdered.
1912 68m/B *GB* Ellie Norwood. **VHS** *VYY* ♫♫

Silent Night, Bloody Night An escaped lunatic terrorizes a small New England town, particularly a mansion that was once an insane asylum. Not great, but well done by director Gershuny, with some nail-biting suspense and slick scene changes.
1973 (R) 83m/C Patrick O'Neal, John Carradine, Walter Abel, Mary Woronov, Astrid Heeren, Candy Darling; *Dir:* Theodore Gershuny. **VHS $42.95** *PGN* ♫♫

Silent Night, Deadly Night A psycho ax-murders people while dressed as jolly old St. Nick. Violent and disturbing, to say the least. Caused quite a controversy when it was released to theaters. Santa gimmick sold some tickets at the time, but resist the urge to rent it: it's completely devoid of worth, whatever the killer's outfit. As if one were not enough, we've been blessed with four sequels.
1984 (R) 92m/C Lilyan Chauvan, Gilmer McCormick, Toni Nero; *Dir:* Charles E. Sellier. **VHS, Beta, LV $14.95** *IVE, IME* ♫

Silent Night, Deadly Night 2 The psychotic little brother of the psychotic, Santa Claus-dressed killer from the first film exacts revenge, covering the same bloody ground as before. Almost half this sequel consists of scenes lifted whole from the original.
1987 (R) 88m/C Eric Freeman, James Newman, Elizabeth Cayton, Jean Miller; *Dir:* Lee Harry. **VHS, Beta, LV $14.95** *IVE* ♫

Silent Night, Deadly Night 3: Better Watch Out! The now grown-up psycho goes up against a young blind woman. Santa is no longer the bad guy, thank goodness. The least bad of the lot, with black humor—though not enough to make it worth seeing.

1989 (R) **90m/C** Richard Beymer, Bill Moseley, Samantha Scully, Eric Dare, Laura Herring, Robert Culp; *Dir:* Monte Hellman. **VHS, Beta, LV** **$14.95** IVE *ℐ* ½

Silent Night, Deadly Night 4: Initiation A secret L.A. cult of she-demons use the slasher Ricky for their own ends—making for more mayhem and horror. Has virtually nothing to do with its three "sequels"—which is not to say it's very good.
1990 (R) **90m/C** Maud Adams, Allyce Beasley, Clint Howard, Reggie Bannister; *Dir:* Brian Yuzna. **VHS, LV $14.95** MOV *ℐ*

Silent Night, Deadly Night 5: The Toymaker A young boy's Christmas is overrun by murderous Santas and viscious stuffed animals. Definitely for fans of the genre only.
1991 (R) **90m/C** Mickey Rooney, William Thorne, Jane Higginson. **VHS $89.95** LIV *ℐℐ*

Silent Night, Lonely Night Two lonely middle-aged people (Bridges and Jones) begin an affair at Christmas to stem the pain of their separately disintegrating marriages. Poignant but not successfully credible. Based on the play by Robert Anderson. Made for television.
1969 **98m/C** Lloyd Bridges, Shirley Jones, Jeff Bridges, Cloris Leachman, Carrie Snodgress, Lynn Carlin; *Dir:* Daniel Petrie. **VHS, Beta** **$39.95** MCA *ℐℐ*

The Silent One Featuring underwater photography by Ron and Valerie Taylor, this is the odd story of a mysterious Polynesian boy who has a nautical relationship with a sea turtle.
1986 **96m/C** NZ Telo Malese, George Henare; *Dir:* Yvonne Mackay. **VHS, Beta $69.95** VID *ℐ* ½

The Silent Partner A bank teller (Gould) foils a robbery, but manages to take some money for himself. The unbalanced robber (Plummer) knows it and wants the money. Good script well directed by Duke, with emphasis on suspense and detail. Unexpectedly violent at times. Early, non-comedic role for big guy Candy.
1978 **103m/C** CA Elliott Gould, Christopher Plummer, Susannah York, John Candy; *Dir:* Daryl Duke. **VHS, Beta, LV $29.98** VES *ℐℐℐ*

The Silent Passenger Amateur sleuth Lord Peter Wimsey makes cinematic debut investigating murder and blackmail on the British railway. Dorothy Sayer's character later inspired BBC mystery series.
1935 **75m/B** GB John Loder, Peter Haddon, Mary Newland, Austin Trevor, Donald Wolfit, Leslie Perrins, Aubrey Mather, Ralph Truman; *Dir:* Reginald Denham. **VHS, Beta $16.95** NOS, SNC *ℐℐ* ½

Silent Rage The sheriff of a small Texas town (Norris) must destroy a killer (Libby) who has been made indestructible through genetic engineering. Chuck Norris meets Frankenstein, sort of. Nice try, but thoroughly stupid and boring.
1982 (R) **100m/C** Chuck Norris, Ron Silver, Steven Keats, Toni Kalem, Brian Libby; *Dir:* Michael Miller. **VHS, Beta, LV, 8mm $9.95** COL, PIA *ℐ* ½

Silent Raiders A commando unit is sent to France in 1943 to knock out a Nazi communications center.
1954 **72m/B** Richard Bartlett, Earle Lyon, Jeanette Bordeau. **VHS, Beta $39.95** SVS *ℐ*

Silent Rebellion A Greek immigrant (Savalas) and his son return to Greece, where they reunite with mother and brother and re-discover their heritage and each other. Cross-cultural misunderstandings abound; could have been interesting, but the story is just too generic and uninspired.
1982 **90m/C** Telly Savalas, Keith Gordon, Michael Constantine, Yula Gavala. **VHS, Beta** VCL *ℐ* ½

Silent Running Members of a space station orbiting Saturn care for the last vegetation of a nuclear-devastated earth. When orders come to destroy the vegetation, Dern takes matters into his own hands. Speculative sci-fi at its best. Trumbull's directorial debut; he created special effects for "2001" and "Close Encounters." Screenplay written by Michael Cimino ("The Deerhunter"), Steven Bochco ("Hill Street Blues") and Deric Washburn. Peter Schickele did the strange music.
1971 (G) **90m/C** Bruce Dern, Cliff Potts, Ron Rifkin; *Dir:* Douglas Trumbull. **VHS, Beta, LV $19.95** MCA *ℐℐℐ*

Silent Scream College kids take up residence with the owners of an eerie mansion complete with obligatory murders. Obvious to the point of being gratuitous—and just plain uninteresting. Sure, there are some scary parts.
1980 (R) **87m/C** Rebecca Balding, Cameron Mitchell, Avery Schreiber, Barbara Steele, Steve Doubet, Brad Reardon, Yvonne de Carlo; *Dir:* Denny Harris. **VHS, Beta $29.95** MED *ℐ* ½

Silent Scream A former Nazi concentration camp commandant collects unusual animals and humans as a hobby.
1984 **60m/C** GB Peter Cushing, Brian Cox, Elaine Donnelly; *Dir:* Alan Gibson. **VHS, Beta** **$29.95** IVE *ℐ*

Silent Valley Adventure of the Old West with American cowboy star Tom Tyler.
1935 **60m/B** Tom Tyler. **VHS, Beta $19.95** NOS, VCN, DVT *ℐ*

Silent Victory: The Kitty O'Neil Story Genuinely stirring real-life overcoming-adversity story of the deaf woman who became a top stunt woman in Hollywood and holder of the land speed record for women. Channing is very good as O'Neil. Made for television.
1979 **96m/C** Stockard Channing, Brian Dennehy, Colleen Dewhurst, Edward Albert, James Farentino; *Dir:* Lou Antonio. **Beta $69.95** VID *ℐℐℐ*

Silent Witness A woman (Bertinelli) witnesses her brother-in-law and his friend rape a young woman. She must decide whether to testify against them or keep the family secret. Exploitative and weakly plotted. Rip-off of the much-discussed Massachusetts barroom rape case. Made for television.
1985 (R) **97m/C** Valerie Bertinelli, John Savage, Chris Nash, Melissa Leo, Pat Corley, Steven Williams, Jacqueline Brookes, Alex MacArthur, Katie McCombs; *Dir:* Michael Miller. **VHS, Beta** **$79.95** ACA *ℐ*

Silhouette Detained in a small Texas town, a woman witnesses the murder of a local girl—but in silhouette, so the killer's identity takes a feature-length running time to resolve. A fair but contrived made-for-cable-TV thriller that the leading lady co-produced.
1991 **89m/C** Faye Dunaway, David Rasche, John Terry, Carlos Gomez, Ron Campbell, Margaret Blye, Talisa Soto, Ritch Brinkley; *Dir:* Carl Schenkel. **VHS $79.95** MCA *ℐℐ*

Silk A beautiful and bloodthirsty detective cuts a swath through local heroin rings.
1986 (R) **84m/C** Cec Verrell, Bill McLaughlin, Fred Bailey; *Dir:* Cirio H. Santiago. **VHS, Beta** **$79.95** MGM *ℐ*

Silk 2 A foreign-made sequel, in which a beautiful cop stops crime in Honolulu. Ludicrous plot, bad acting, lots of skin and violence—what more could you want? Dreadful.
1989 (R) **85m/C** Monique Gabrielle, Peter Nelson, Jan Merlin, Maria Clair; *Dir:* Cirio H. Santiago. **VHS, Beta $44.98** MGM *ℐ*

The Silk Road Set in the 11th-century, the ancient trading route leads to Dun Huang, a desert city that is the last Chinese outpost on the road. A young scholar, Zhao, travels in a caravan which is attacked by Chinese mercenaries. The leader of the group takes a liking to the young man and makes him his protege so that Zhao may witness history unfolding. When Zhao's new home of Dun Huang is attacked by a neighboring nation, Zhao is there to conveniently hide the city's treasures in some nearby caves. (This is based on the Thousand Buddha Caves, where a treasure of Buddhist icons and scrolls were discovered in 1900.) There's lots of pageantry and some impressive battle scenes but the film is too stately to be interesting. Adapted from a novel by Yasushi Inoue.
1992 (PG-13) **99m/C** CH Koichi Sato, Toshiyuki Nishida, Anna Nakagawa, Tsunehiko Watase; *Dir:* Junya Sato. **VHS, LV $89.95** VMK *ℐℐ* ½

Silk Stockings Splendid musical comedy adaptation of "Ninotchka," with Astaire as a charming American movie man, and Charisse as the cold Soviet official whose commie heart he melts. Music and lyrics by Cole Porter, including "All of You" and "Stereophonic Sound," highlight this film adapted from George S. Kaufman's hit Broadway play. Director Mamoulian's last film.
1957 **117m/C** Fred Astaire, Cyd Charisse, Janice Paige, Peter Lorre, George Tobias; *Dir:* Rouben Mamoulian. **VHS, Beta, LV $29.95** MGM, TTC *ℐℐℐ*

Silkwood The story of Karen Silkwood, who died in a 1974 car crash under suspicious circumstances. She was a nuclear plant worker and activist who was investigating shoddy practices at the plant. Streep acts up a storm, disappearing completely into her character. Cher surprises with her fine portrayal of a lesbian co-worker, and Russell is also good. Nichols has a tough time since we already know the ending, but he brings topnotch performances from his excellent cast.
1983 (R) **131m/C** Meryl Streep, Kurt Russell, Cher, Diana Scarwid, Bruce McGill, Fred Ward, David Straithairn, Ron Silver, Josef Sommer, Craig T. Nelson; *Dir:* Mike Nichols. **VHS, Beta, LV $9.98** SUE *ℐℐℐ*

Silver Bandit A bookkeeper is sent to investigate a mine theft. Along the way he sings country and western songs.
1947 **54m/B** Spade Cooley. **VHS, Beta, LV** **$19.95** NOS, WGE *ℐ*

Silver Bears Michael Caine is an unlikely mobster in this comedy. He's sent to Switzerland to buy a bank for laundering purposes and ends up being swindled himself. So-so entry adapted from the novel by Paul Erdman.

1978 (PG) 114m/C Michael Caine, Cybill Shepherd, Louis Jourdan, Stephane Audran, David Warner, Tom Smothers, Martin Balsam, Jay Leno, Charles Gray, Joss Ackland; *Dir:* Ivan Passer. **VHS, Beta** *USA* ♪♪

Silver Bullet Adapted from Stephen King's "Cycle of the Werewolf," about a town whose inhabitants are being brutally murdered. It finally dawns on them the culprit is a werewolf. Action moves along at a good clip and the film has its share of suspense. Also known as "Stephen King's Silver Bullet."
1985 (R) 95m/C Corey Haim, Gary Busey, Megan Follows, Everett McGill; *Dir:* Daniel Attias. **VHS, LV** $14.95 *PAR* ♪♪

Silver Chalice Paul Newman's career somehow survived his movie debut in this bloated, turgid Biblical epic about the momentous events that befall a young Greek sculptor who fashions the cup that will be used at the Last Supper. Newman later took out an ad in Variety to apologize for the film, in which Lorne Greene also made his debut. Based on the novel by Thomas Costain with a music score by Franz Waxman.
1954 135m/C Paul Newman, Virginia Mayo, Pier Angeli, Jack Palance, Natalie Wood, Joseph Wiseman, Lorne Greene, E.G. Marshall; *Dir:* Victor Saville. **VHS, Beta** $19.98 *WAR* ♪½

Silver City A saga depicting the plight of Polish refugees in Australia in 1949. A pair of lovers find each other in a crowded refugee camp. He's married to her friend. Too-complex, slow but earnest effort.
1984 (PG) 110m/C *AU* Gosia Dobrowolska, Ivar Kants; *Dir:* Sophia Turkiewicz. **VHS, Beta** $59.95 *LHV* ♪♪

Silver City Kid A foreman for a silver mine exposes a corrupt local banker as a thief.
1945 56m/B Allan Lane. **VHS, Beta** $24.95 *DVT* ♪

Silver Dream Racer An English grease monkey wants to win the World Motorcycle Championship title away from an American biker. He's also after someone else's girlfriend. Boring, stilted, unoriginal "Big Race" flick. Essex also wrote and performed the music.
1983 (PG) 103m/C Beau Bridges, David Essex, Christina Raines, Diane Keen, Harry H. Corbett; *Dir:* David Wickes. **VHS, Beta** $69.98 *LIV, VES* ♪½

Silver Hermit from Shaolin Temple/The Devil's Assignment Two kung fu thrillers on one action-filled tape.
1981 175m/C **VHS, Beta** $7.00 *VTR* ♪½

Silver Lode A man accused of murder on his wedding day attempts to clear his name while the law launches an intensive manhunt for him. Ordinary story improved by good, energetic cast.
1954 92m/C John Payne, Dan Duryea, Lizabeth Scott, Stuart Whitman; *Dir:* Allan Dwan. **VHS, Beta** $39.95 *BVV* ♪♪½

Silver Queen A woman finds out her father has gambled away a silver mine and left her with debts. She opens a saloon, only to have her fiance use the money to hunt for more silver.
1942 81m/B Priscilla Lane, George Brent, Bruce Cabot, Eugene Pallette; *Dir:* Lloyd Bacon. **VHS, Beta** $39.95 *IND* ♪♪

Silver Spurs A villain tries to get some oil-rich property by murdering the ranch's owner. Ranch foreman Rogers saves the day.
1943 60m/B Roy Rogers, Jerome Cowan, John Carradine, Phyllis Brooks, Smiley Burnette, Joyce Compton; *Dir:* Joseph Kane. **VHS, Beta** $19.95 *NOS, VCN, DVT* ♪

Silver Stallion Thunder the Wonder Horse plays the title character, who fights rattlesnakes, wild dogs and hoss thieves. Too bad he doesn't give acting lessons to the humans in the this frail family western with nice scenery and action scenes.
1941 59m/C David Sharpe, Carol Hughes, Leroy Mason, Walter Long; *Dir:* Edward Finney. **VHS, Beta** *GPV* ♪½

Silver Star A man elected sheriff of a western town turns down the job because he is a pacifist. He changes his mind when his defeated opponent hires a trio of killers to come after him.
1955 73m/B Jimmy Wakely, Edgar Buchanan, Marie Windsor. **VHS, Beta** *WGE* ♪

The Silver Streak The sickly son of a diesel train designer needs an iron lung—pronto. A rival's super-fast locomotive is the only hope for the boy. Murders, runaway engines, and a crew that would rather walk enliven this race against time.
1934 72m/B Sally Blane, Charles Starett, Arthur Lake, Edgar Kennedy, William Farnum; *Dir:* Thomas Atkins. **VHS, Beta** $17.95 *CCB* ♪♪

The Silver Streak Pooped exec Wilder rides a train from L.A. to Chicago, planning to enjoy a leisurely, relaxing trip. Instead he becomes involved with murder, intrigue, and a beautiful woman. Energetic Hitchcock parody action comedy written by Colin Higgins; successful first pairing of Wilder and Pryor.
1976 (PG) 113m/C Gene Wilder, Richard Pryor, Jill Clayburgh, Patrick McGoohan, Ned Beatty, Ray Walston, Richard Kiel, Scatman Crothers; *Dir:* Arthur Hiller. **VHS, Beta, LV** $59.98 *FOX* ♪♪♪

Silverado Affectionate pastiche of western cliches has everything a viewer could ask for--except Indians. The straightforward plot finds four virtuous cowboys rising up against a crooked lawman in a blaze of six guns. No subtlety from the first big Western in quite a while, but plenty of fun and laughs. Special laser disc edition features a wide screen film-to-tape transfer monitored by the photography director. Also included are set photos, release trailers, and other publicity hoohah as well as a special time-lapse sequence of the set construction, and interviews with the stars and director Lawrence Kasdan. Letterboxed laserdisc version is available with dolby surround sound.
1985 (PG-13) 132m/C Kevin Kline, Scott Glenn, Kevin Costner, Danny Glover, Brian Dennehy, Linda Hunt, John Cleese, Jeff Goldblum, Rosanna Arquette, Jeff Fahey; *Dir:* Lawrence Kasdan. **VHS, Beta, LV** $14.95 *COL, VYG, CRC* ♪♪♪

Simba A young Englishman arrives at his brother's Kenyan farm to find him murdered in a local skirmish between the Mau Maus and white settlers. Well made, thoughtful look at colonialism and racial animosity and violence.
1955 98m/C *GB* Dirk Bogarde, Donald Sinden, Virginia McKenna, Orlando Martins; *Dir:* Brian Desmond Hurst. **VHS, Beta** $39.95 *IND* ♪♪½

Simon A group of bored demented scientists brainwash a college professor into believing he is an alien from a distant galaxy, whereupon he begins trying to correct the evil in America. Screwball comedy, or semi-serious satire of some kind? Hard to tell. Some terrific set pieces but the movie as a whole doesn't quite hold together. Directorial debut of Marshall Brickman, who previously worked as a scriptwriter with Woody Allen ("Sleeper," etc.).
1980 (PG) 97m/C Alan Arkin, Madeline Kahn, Fred Gwynne, Adolph Green, Wallace Shawn, Austin Pendleton; *Dir:* Marshall Brickman. **VHS, Beta** $19.98 *WAR* ♪♪½

Simon Blanco A hero of the Mexican Revolution is tricked by men who want to kill him and steal his wife.
1987 145m/C *MX* Antonio Aguilar, Jacqueline Andere, Mario Almada, Valentin Trujillo. **VHS, Beta** $59.95 *MAD* ♪♪

Simon Boccanegra The stunning opera by Verdi about a complex man governed by his love and devotion for both his family and his country.
1984 153m/C Sherrill Mines, Anna Tomowa-Sintow. **VHS, Beta** $39.95 *HMV, PAR* ♪♪

Simon Bolivar The title character leads the Venezuelan revolution in 1817.
1969 110m/C Maximilian Schell, Rosanna Schiaffino. **VHS, Beta** $49.95 *UNI* ♪♪½

Simon of the Desert Not Bunuel's very best, but worthy of the master satirist. An ascetic stands on a pillar in the desert for several decades—closer to God, farther from temptation. Pinal is a gorgeous devil that tempts Simon. Hilarious, irreverent, sophisticated. What's with the weird ending, though? In Spanish with English subtitles.
1966 43m/B *Dir:* Luis Bunuel. New York Film Festival '66: 10 Best Films of the Year. **VHS, Beta** $68.00 *NOS, HHT, DVT* ♪♪♪½

Simon, King of the Witches An L.A. warlock who lives in a sewer drain finds himself the center of attention when his spells actually work. This interesting hippie/witchcraft entry bogs down now and then but Prine's performance is droll and lively.
1971 (R) 90m/C Andrew Prine, Brenda Scott, George Paulsin, Norman Burton, Ultra Violet; *Dir:* Bruce Kessler. **VHS, Beta** $59.95 *UNI* ♪♪½

Simple Justice Mindless, justice-in-own-hands anti-liberal hogwash. Overwrought, smug, and violent story of a young couple beaten by robbers who remain at large.
1989 (R) 91m/C Cesar Romero, John Spencer, Doris Roberts, Candy McClain; *Dir:* Deborah Del Prete. **VHS** $59.95 *JTC* ♪½

Simple Story A woman faces her fortieth birthday with increasing uneasiness, though her life seems perfect from the outside. She evaluates her chances at love, child-bearing, and friendship, after having an abortion and breaking up with her lover. Schneider's performance is brilliant in this gentle, quiet drama. French with English subtitles.
1980 110m/C *FR* Romy Schneider, Bruno Cremer, Claude Brasseur; *Dir:* Claude Sautet. **VHS, Beta** $59.95 *COL, FRI* ♪♪♪

The Simple Truth A woman's romance novel is dramatized, dealing with a receptionist who has a fling with the company president.
1985 90m/C **Beta** $14.95 *PSM* *Woof!*

The Simpsons Christmas Special America's favorite cartoon family celebrates Christmas as only they can; Bart gets a tattoo, Homer works as a mall Santa and the Christmas spirit arrives in the unlikely pack-

age of a greyhound named "Santa's Little Helper."
1989 30m/C Voices: Dan Castellaneta, Julie Kavner, Harry Shearer, Maggie Roswell, Nancy Cartwright, Yeardley Smith. **VHS $9.98** *FXV* Woof!

Sin of Adam & Eve The story of Adam and Eve in the Garden of Eden and their fall from grace. Voice-over narrator is only sound in this strange dialogue-less story of the Adam and Eve, how they got kicked out of the Garden of Eden, and how they lose and then find each other outside. Ancient plot, to say the least.
1972 (R) 72m/C Candy Wilson, Jorge Rivero; **Dir:** Michael Zachary. **VHS, Beta $19.95** *VDM, UHV* ✂½

The Sin of Harold Diddlebock A man gets fired from his job, stumbles around drunk, and wins a fortune gambling. He then buys a circus and uses a lion to frighten investors into backing him. Inventive comedy, but missing the spark and timing of "The Freshman" (1925), of which it is a sequel. Edited to 79 minutes and re-released as "Mad Wednesday" in 1951. The final feature film for Lloyd, made it at the urging of director Sturges, who did all his own stunts.
1947 89m/B Harold Lloyd, Margaret Hamilton, Frances Ramsden, Edgar Kennedy, Lionel Stander, Rudy Vallee, Franklin Pangborn; **Dir:** Preston Sturges. **VHS, Beta $14.95** *QNE, NOS, KRT* ✂✂½

The Sin of Madelon Claudet Hayes plays common thief who works her way into upper crust of Parisian society only to tumble back into the street, all in the name of making a better life for her illegitimate son. Very sudsy stuff, with an outstanding performance by Hayes.
1931 74m/B Helen Hayes, Lewis Stone, Neil Hamilton, Robert Young, Cliff Edwards, Jean Hersholt, Marie Prevost, Karen Morley, Charles Winninger, Alan Hale Jr.; **Dir:** Edgar Selwyn. Academy Awards '32: Best Actress (Hayes). **VHS $29.98** *MGM, FCT* ✂✂✂

Sin Retorno Crime, torture and martial arts on the streets of New York. In Spanish.
19?? 90m/C Fred Williamson. **VHS, Beta $57.95** *UNI*

Sin Salida A man is chased through Mexico City by the Mafia.
1987 97m/C *MX* Jorge Rivero, Nadia Milton, Mario Almada, Daniela Rosen. **VHS, Beta $58.95** *MAD*

Sinai Commandos The story of a group of commandos in the Israeli Six-Day War in 1967 who are assigned to destroy important Arab radar installations. Includes actual combat footage.
1968 99m/C Robert Fuller; **Dir:** Raphael Nussbaum. **VHS $29.95** *ATL, FCT* ✂✂½

Sinbad A sardonic portrait of an aging hedonist as he tries to hold onto the pleasures of drink, sex and gluttony. In Magyar (Hungarian), with English subtitles.
1971 98m/C *HU* **Dir:** Zoltan Huszarik. **VHS, Beta $69.95** *KIV, FCT* ✂✂✂

Sinbad and the Eye of the Tiger The swashbuckling adventures of Sinbad the Sailor as he encounters the creations of Ray Harryhausen's special effects magic. Don't see this one for the plot, which almost doesn't exist. Otherwise, mildly fun.
1977 (G) 113m/C *GB* Patrick Wayne, Jane Seymour, Taryn Power, Margaret Whiting; **Dir:** Sam Wanamaker. **VHS, Beta, LV $14.95** *COL, CCB* ✂✂

Sinbad the Sailor Fairbanks fits well in his luminent father's swashbuckling shoes, as he searches for the treasure of Alexander the Great. Self-mocking but hamhanded, and confusing if you seek the hidden plot. Still, it's all in fun, and it is fun.
1947 117m/C Douglas Fairbanks Jr., Maureen O'Hara, Anthony Quinn, Walter Slezak, George Tobias, Jane Greer, Mike Mazurki, Sheldon Leonard; **Dir:** Richard Wallace. **VHS, Beta, LV $24.95** *RKO, MED, VID* ✂✂✂

Sinbad of the Seven Seas Italian muscle epic based on the ancient legends. Ferrigno isn't green, but he's still a hulk, and he still can't act. It's poorly dubbed, which makes little difference; it would be stupid regardless. Italian muscle epic based on the ancient legends.
1989 90m/C *IT* Lou Ferrigno, John Steiner, Leo Gullotta, Teagan Clive; **Dir:** Enzo G. Castellari. **VHS, Beta, LV $89.95** *WAR, OM* ✂✂½

Since You Went Away An American family copes with the tragedy, heartache and shortages of wartime in classic mega-tribute to the home front. Be warned: very long and bring your hankies. Colbert is superb, as is the photography. John Derek unobtrusively made his film debut, as an extra. Script by producer David O. Selznick. Oscar-nominated for Best Picture, Best Actress (Colbert), Best Supporting Actor (Woolley), Best Supporting Actress (Jones), Black & White Interior Decoration, Black & White Cinematography, and Film Editing.
1944 172m/B Claudette Colbert, Jennifer Jones, Shirley Temple, Joseph Cotten, Agnes Moorehead, Monty Woolley, Guy Madison, Lionel Barrymore, Robert Walker, Hattie McDaniel, Keenan Wynn, Craig Stevens, Albert Basserman, Alla Nazimova, Lloyd Corrigan, Terry Moore, Florence Bates, Ruth Roman, Andrew V. McLaglen, Dorothy Dandridge, Rhonda Fleming; **Dir:** John Cromwell. Academy Awards '44: Best Musical Score. **VHS, Beta, LV $39.98** *FOX* ✂✂✂½

Sincerely Charlotte Caroline Huppert directs her sister in this film about a beautiful singer framed for her boyfriend's murder. She enlists an old lover to help her flee across the countryside. Love story/thriller is pleasingly odd and absorbing. In French, subtitled in English.
1986 92m/C *FR* Isabelle Huppert, Niels Arestrup, Christine Pascal; **Dir:** Caroline Huppert. **VHS, Beta, LV $79.95** *COL* ✂✂✂

Sincerely Yours Liberace wisely stayed away from acting after inauspiciously debuting in this horrible, maudlin remake of "The Man Who Played God." He plays a pianist who loses his hearing and decides to become a philanthropist to help those less fortunate than himself. Laughably cheesy. Script by Irving Wallace. Thirty-one musical numbers, including Liberace's inimitable arrangement of "Chopsticks."
1956 116m/C Liberace, Joanne Dru, Dorothy Malone, William Demarest; **Dir:** Gordon Douglas. **VHS, Beta, LV $19.98** *WAR* Woof!

The Sinful Bed A bed tells of its witnessed liasons over the years.
1978 84m/C **VHS, Beta $19.95** *ACA, HHE* Woof!

Sinful Life A strained, offbeat, B-grade comedy about an odd, infantile mother fighting to keep her unusual child from being taken away. Definitely not a must-see; can be irritating and obnoxious, depending on viewer and mood. Morris is at her oddball comedic best as the mother and former show dancer. Adult Tefkin plays her little girl.

Based on the play "Just Like the Pom Pom Girls."
1989 (R) 112m/C Anita Morris, Rick Overton, Dennis Christopher, Blair Tefkin, Mark Rolston, Cynthia Szigeti; **Dir:** William Schreiner. **VHS, Beta, LV $89.95** *COL* ✂✂

Sing The students in the real-life Brooklyn public school "Sing" program endure the trials of adolescence while putting together a musical revue. Hurtles over the edge into cheesiness from the start, with way too much (cheesy) music. From the creator of "Fame" and "Footloose."
1989 (PG-13) 111m/C Lorraine Bracco, Peter Dobson, Jessica Steen, Louise Lasser, George DiCenzo, Patti LaBelle; **Dir:** Richard Baskin. **VHS, Beta, LV $89.95** *COL* ✂✂

Sing, Cowboy, Sing Decent adventure story about a wagon train as it makes its way west and runs into a band of outlaws who wants to steal its supplies. Lots of singing and little bit of fighting.
1937 60m/B Tex Ritter. **VHS, Beta $19.95** *NOS, DVT* ✂✂½

Sing Sing Nights A world-famous journalist is killed with three bullets. Three men have confessed, but only one could have actually done it. A professor sets out to solve the case.
1935 60m/B Conway Tearle, Mary Doran, Hardie Albright, Boots Mallory, Ferdinand Gottschalk, Berton Churchill, Jameson Thomas; **Dir:** Lewis Collins. **VHS $16.95** *SNC* ✂✂½

Sing Your Worries Away Songwriter Ebsen inherits 3 million dollars and finds the money causes him nothing but trouble. Band of crooks cause the requisite wacky complications. Fun, harmless musical comedy.
1942 71m/B Buddy Ebsen, Bert Lahr, June Havoc, Patsy Kelly, Margaret Dumont; **Dir:** Edward Sutherland. **VHS, Beta** *CCB* ✂✂

Singin' in the Rain One of the all-time great movie musicals—an affectionate spoof of the turmoil that afflicted the motion picture industry in the late 1920s during the changeover from silent films to sound. Co-director Kelly and Hagen lead a glorious cast. Songs include the classic title tune sung (of course) by Kelly, "Make 'Em Laugh," "All I Do Is Dream of You," and "You Are My Lucky Star." Music and lyrics by Arthur Freed and Nacio Herb Brown. Hagen received a Best Supporting Actress Oscar nomination, as did the score. Served as basis of story by Betty Comden and Adolph Green. Also available on laserdisc with the original trailer, outtakes, behind the scenes footage, and commentary by film historian Ronald Haver. Later a Broadway musical.
1952 103m/C Gene Kelly, Donald O'Connor, Jean Hagen, Debbie Reynolds, Rita Moreno, King Donovan, Millard Mitchell, Cyd Charisse; **Dir:** Gene Kelly, Stanley Donen. National Board of Review Awards '52: 10 Best Films of the Year; American Film Institute's Survey '77: 7th Best American Film Ever Made; Sight & Sound Survey '82: #3 of the Best Films of All Time (tie). **VHS, Beta, LV $19.95** *MGM, PIA, VYG* ✂✂✂✂

The Singing Blacksmith (Yankl der Shmid) A relic of American Yiddish cinema, adapting popular 1909 play by David Pinski. A married blacksmith is wooed by another woman and falls victim to alcoholism. Overlong, but Oysher and his rich baritone voice still shine. Yiddish with English subtitles.
1938 95m/B *PL* Miriam Riselle, Florence Weiss, Moishe Oysher; **Dir:** Edgar G. Ulmer. **VHS $79.95** *ERG* ✂✂½

Singing the Blues in Red Dark, pessimistic character study of a protest singer/songwriter who leaves his native East Germany to escape repression and finds his loyalties divided. A moving personal story, as well as insightful, subtle social commentary. In English and German with subtitles. Cinematography by Chris Menges.
1987 110m/C *GE* Gerulf Pannach, Fabienne Babe, Cristine Rose, Trevor Griffiths; *Dir:* Ken Loach. **VHS, Beta $79.95** *JCI* 𝗜𝗜𝗜

Singing Buckaroo Dastardly bandits try to pilfer money from pretty blonde damsel and have to battle with yodelin' cowboy. B-level western good guy Scott belts out a few numbers and saves the day. Innocuously pleasant.
1937 58m/B Fred Scott, Victoria Vinton, Cliff Nazarro; *Dir:* Tom Gibson. **VHS, Beta $19.95** *NOS, VYY* 𝗜𝗜

Single Bars, Single Women Utterly formulaic made-for-TV comedy-drama based on Dolly Parton's song is not entirely without merit. Lonely people gather in a local pickup joint and share their miseries and hopes.
1984 96m/C Shelley Hack, Christine Lahti, Tony Danza, Mare Winningham, Keith Gordon, Paul Michael Glaser; *Dir:* Harry Winer. **VHS, Beta $79.95** *PSM, PAR* 𝗜½

Single Fighter A martial arts expert in China seeks out traitors who collaborated with the Japanese during World War II.
1978 (R) 79m/C Tu Chiang; *Dir:* Lee Chin-Chuan. **VHS $49.95** *PGN* 𝗜

Single Room Furnished The fall of a buxom blonde from uncorrupted innocence through pregnancies to desperate prostitution. Fails to demonstrate any range of talent in Mansfield, who died before it was completed. Exploitative and pathetic.
1968 93m/C Jayne Mansfield, Dorothy Keller; *Dir:* Matt Cimber, Matteo Ottaviano. **VHS, Beta $29.95** *UHV* 𝗜½

The Single Standard San Francisco deb Garbo flings with artsy Asther and finds out the Hayes Code is just around the corner.
1929 93m/B Greta Garbo, Nils Asther, Johnny Mack Brown, Dorothy Sebastian, Lane Chandler, Zeffie Tilbury; *Dir:* John S. Robertson. **VHS $29.99** *MGM, FCT* 𝗜𝗜

Singleton's Pluck Touching British comedy about a determined farmer who must walk his 500 geese 100 miles to market because of a strike. He becomes a celebrity when the TV stations start covering his odyssey. Also known as ''Laughterhouse.''
1984 89m/C *GB* Ian Holm, Penelope Wilton, Bill Owen, Richard Hope; *Dir:* Richard Eyre. **VHS, Beta $19.95** *STE, NWV* 𝗜𝗜

The Sinister Invasion A turn-of-the-century scientist (Karloff) discovers a death ray. Aliens, who would like a closer peek at what makes it work, use a sex-fiend's body to do so. One of Karloff's last four films, made simultaneously in Mexico. The great horror master's final offering. Excruciating. Also known as ''Alien Terror'' and ''The Incredible Invasion.''
1968 95m/C *MX* Boris Karloff, Enrique Guzman, Jack Hill; *Dir:* Juan Ibanez. **VHS, Beta $49.95** *SNC, UNI* 𝗜

Sinister Journey Hopalong Cassidy helps clear a young man wrongly accused of murder on a west-bound train. Not the best Hopalong effort.

1948 58m/B William Boyd, Andy Clyde, Rand Brooks, Elaine Riley; *Dir:* George Archainbaud. **VHS, Beta $39.95** *BVV* 𝗜

The Sinister Urge Vice cops Duncan and Moore search for the murderer of three women. Seems the disturbed slayer is unbalanced because he's been looking at pictures of naked ladies. Was this meant to be taken seriously at the time? The last film by camp director Ed Wood, maker of the infamous ''Plan 9 From Outer Space.'' A must see for Wood fans. Black and White.
1960 82m/B Kenne Duncan, Duke Moore, Jean Fontaine; *Dir:* Edward D. Wood Jr. **VHS, Beta $59.98** *AOV* Woof!

Sink the Bismarck The British navy sets out to locate and sink the infamous German battleship during World War II. Good special effects with the battle sequences in this drama based on real incidents. One of the better of the plethora of World War II movies, with stirring naval battles and stylish documentary-style direction.
1960 97m/B Kenneth More, Dana Wynter, Karel Stepanek, Carl Mohner, Laurence Naismith, Geoffrey Keen, Michael Hordern, Maurice Denham, Esmond Knight; *Dir:* Lewis Gilbert. **VHS, Beta $19.98** *FOX* 𝗜𝗜𝗜

Sinners Outrageous portrait of an Italian family in the Big Apple trying to come to terms with the violence that surrounds their neighborhood.
1989 90m/C Joey Travolta, Robert Gallo, Joe Palese, Lou Calvelli, Angie Daglas, Sabrina Ferrand; *Dir:* Charles Kanganis. **VHS, Beta $29.95** *PMH* 𝗜

Sinner's Blood Bikers terrorize and torture a small town.
1970 (R) 81m/C Stephen Jacques, Crusty Beal, Nancy Sheldon, Parker Herriott, Julie Connors; *Dir:* Neil Douglas. **VHS, Beta $24.95** *GHV* Woof!

Sinners in Paradise An assortment of trouble-plagued characters alternately hide and face up to their mysterious pasts when their plane crashes on a deserted island. Interesting ''crucible'' premise runs out of steam, but plucky, resourceful cast and competent direction keep it going.
1938 64m/B Madge Evans, John Boles, Bruce Cabot, Marion Martin, Gene Lockhart, Dwight Frye, Charlotte Wynters, Nana Bryant, Milburn Stone, Donald (Don ''Red'') Barry, Morgan Conway; *Dir:* James Whale. **VHS, Beta $16.95** *SNC, MLB* 𝗜𝗜

Sins On her way up the ladder of success in the fashion industry, Helene has stepped on a few toes. Those rivals and her ever-increasing acquisition of power and money make this film an exciting drama.
1985 336m/C Joan Collins, Timothy Dalton, Catherine Mary Stewart, Gene Kelly, James Farentino; *Dir:* Douglas Hickox. **VHS, Beta $29.95** *STE, NWV, HHE* 𝗜

Sins of Dorian Gray A modernized adaptation of ''The Portrait of Dorian Grey'' by Oscar Wilde, with Dorian as a beautiful woman who remains young for 30 years, while a video screen test ages, like the original character's mirror image. Might have been intriguing and stylish; instead, only disappointing and incompetent. Made for television.
1982 (PG) 95m/C Anthony Perkins, Joseph Bottoms, Belinda Bauer; *Dir:* Tony Maylam. **VHS, Beta $59.98** *FOX* 𝗜

Sins of Jezebel A Biblical epic about Jezebel, who worships an evil god. She marries the king of Israel and brings the kingdom nothing but trouble.

1954 74m/C Paulette Goddard, George Nader, John Hoyt, Eduard Franz; *Dir:* Reginald LeBorg. **VHS, Beta, LV** *WGE* 𝗜

Sins of Rome Spartacus risks it all in a bold attempt to free his fellow slaves. Not nearly the equal of Kirk Douglas's ''Spartacus,'' but much better than later Italian adventure epics.
1954 75m/B *IT* Ludmila Tcherina, Massimo Girotti, Giana Maria Canale, Yves Vincent; *Dir:* Riccardo Freda. **VHS $16.95** *SNC* 𝗜

Sioux City Sue Talent scouts looking to cast a western musical find Autry, then trick him into being the voice of a singing donkey in an animated production. But, the yodelin' cowboy belts out a number or two, and the poobahs give him the lead. Singin' and fancy ridin' abound in Autry's first post-WW II role.
1946 69m/B Gene Autry, Lynne Roberts, Sterling Holloway. **VHS, Beta $29.98** *CCB, VCN* 𝗜𝗜

Sir Arthur Conan Doyle An intimate portrait of the man who created Sherlock Holmes.
1927 11m/B VHS, Beta *CCB* 𝗜𝗜

Sirocco An American gun-runner (Bogart) stuck in Syria in 1925 matches wits with a French intelligence officer amid civil war and intrigue. About the underbelly of human affair.
1951 111m/B Humphrey Bogart, Lee J. Cobb, Zero Mostel, Everett Sloane, Gerald Mohr; *Dir:* Curtis Bernhardt. **VHS, Beta, LV $9.95** *COL* 𝗜𝗜½

Sister Dora An idealistic young woman joins a nursing sisterhood in 19th century England but she questions her vocation when she falls in love. Adapted from a popular romance novel.
1977 147m/C *GB* Dorothy Tutin, James Grout, Peter Cellier; *Dir:* Mark Miller. **VHS $59.95** *SVS* 𝗜𝗜½

The Sister-in-Law Shady dealings, seduction, and adultery run rampant. Savage (who also wrote and sings the folk score) reluctantly agrees to deliver a package across the Canadian border for his brother. Intriguing.
1974 (R) 80m/C John Savage, Anne Saxon, W.G. McMillan, Meredith Baer; *Dir:* Joseph Rubin. **VHS, Beta $79.95** *PSM* 𝗜𝗜

Sister Kenny Follows the story of a legendary Australian nurse crusading for the treatment of infantile paralysis. Stirring, well-made screen biography. Co-scripted by Mary McCarthy. Based on Elizabeth Kenny's memoir, ''And They Shall Walk.'' Russell received a highly deserved Best Actress nomination.
1946 116m/B Rosalind Russell, Dean Jagger, Alexander Knox, Philip Merivale, Beulah Bondi; *Dir:* Dudley Nichols. **VHS, Beta, LV $19.98** *NOS, CCB, TTC* 𝗜𝗜

Sister, Sister A Congressional aide on vacation in Louisiana takes a room in an old mansion. He gradually discovers the secret of the house and its resident sisters. Dark Southern gothicism, full of plot surprises, with a twisted ending.
1987 (R) 91m/C Eric Stoltz, Judith Ivey, Jennifer Jason Leigh, Dennis Lipscomb, Anne Pitoniak, Natalija Nogulich; *Dir:* Bill Condon. **VHS, Beta, LV $19.95** *STE, NWV* 𝗜𝗜

Sister Street Fighters Boy meets girl, boy and girl terrorize the neighborhood. Typical martial arts fest.
1976 (R) 82m/C Sonny Chiba; *Dir:* Kazuhiko Yamaguchi. **VHS** *UHV* 𝗜

The Sisterhood A pair of amazons fight for women's rights in a post-nuclear future. Cheap feminist theme laid over warmed-over sci-fi plot.
1988 (R) 75m/C Rebecca Holden, Chuck Wagner, Lynn-Holly Johnson, Barbara Hooper; ***Dir:*** Cirio H. Santiago. **VHS, Beta, LV $79.95** *MED* 🦴

Sisters Siamese twins are separated surgically, but one doesn't survive the operation. The remaining sister is scarred physically and mentally with her personality split into bad and good. The bad side commits a murder, witnessed (or was it?) by an investigative reporter. And then things really get crazy. DePalma's first ode to Hitchcock, with great music by Hitchcock's favorite composer, Bernard Herrmann. Scary and suspenseful.
1973 (R) 93m/C Margot Kidder, Charles Durning, Barnard Hughes; ***Dir:*** Brian DePalma. **VHS, Beta $29.98** *WAR, OM* 🦴🦴🦴

Sisters of Death Five members of a sorority gather together for a reunion in a remote California town. Little do they know, a psychopath is stalking them, one by one. Could it have something to do with the terrible secret they each keep? Good, cheap thrills and some Bicentennial fashions to boot.
1976 (PG) 87m/C Arthur Franz, Claudia Jennings, Cheri Howell, Sherry Boucher, Paul Carr; ***Dir:*** Joseph Mazzuca. **VHS, Beta $29.95** *UHV, VCI* 🦴🦴

Sisters of Gion Mizoguchi's most famous and arguably best pre-war film. Two geisha sisters in Tokyo's red-light district reflect the tension in Japanese culture by waging a quiet battle of tradition vs. progressiveness. Highly acclaimed. In Japanese with English subtitles.
1936 69m/B *JP* Isuzu Yamada, Yoko Umemura; ***Dir:*** Kenji Mizoguchi. **VHS, Beta** *FCT* 🦴🦴🦴½

Sisters of Satan The nuns of the infamous convent St. Archangelo choose to survive by trading God for the devil.
1975 (R) 91m/C *MX* Claudio Brook, David Silva, Tina Romero, Susana Kamini; ***Dir:*** Juan Monteczuma. **VHS, Beta $19.95** *ACA, MPI* 🦴🦴½

Sitting Ducks Mild-mannered accountant and lecherous pal rip off and then attempt to outrun the mob—all while swapping songs and confessions. Emil and Norman make it up as they go along; their hilarious repartee is largely improvised. They pick up a gorgeous lady (Townsend) and go about their way.
1980 (R) 88m/C Michael Emil, Zack Norman, Patrice Townsend, Richard Romanus, Irene Forrest, Henry Jaglom; ***Dir:*** Henry Jaglom. **VHS, Beta $19.95** *MED* 🦴🦴🦴

The Situation Two years of life and death in El Salvador's civil war. Also the tale of an American film crew who spent the time with a Salvadoran family covering the events. Many events in the movie "Salvador" were inspired by incidents portrayed here.
1987 91m/C Richard Boyle. **Beta** *VDE* 🦴🦴🦴

Six Fois Deux/Sur et Sous la Communication A six-part experimental series made for French television, in which Godard and Mieville analyze modern culture and socioeconomics via a free-form essay format, combining interviews, screened text, self-referential devices and news footage. In French with English voice-over.
1976 100m/C *FR* ***Dir:*** Jean-Luc Godard, Anne-Marie Mieville. **VHS, Beta** *EAI* 🦴🦴🦴

Six Gun Gospel Hatton masquerades as a preacher to investigate some shady dealings. Laughs come when the women of the congregation prevail upon him to sing, and he croons a ballad about Jesse James!
1943 59m/B Johnny Mack Brown, Raymond Hatton, Inna Gest, Eddie Dew, Roy Barcroft, Kenneth MacDonald, Bud Osborne; ***Dir:*** Lambert Hillyer. **VHS $19.95** *DVT* 🦴🦴

Six Gun Rhythm A football player goes to Texas to avenge his father's murder.
1939 59m/B Tex Fletcher. **VHS, Beta $19.95** *NOS, VDM* 🦴

Six Pack Rogers, in his film debut, stars as Brewster Baker, a former stock car driver. He returns to the racing circuit with the help of six larcenous orphans adept at mechanics. Not dreadful as an actor, but Rogers is a great singer.
1982 (PG) 108m/C Kenny Rogers, Diane Lane, Erin Gray, Barry Corbin, Anthony Michael Hall; ***Dir:*** Daniel Petrie. **VHS, Beta $59.98** *FOX* 🦴½

Six Shootin' Sheriff A cowboy framed for bank robbery is released and seeks to rid his town of all evil. And he gets the girl! Reynolds later played Peg on "Life of Riley" on TV.
1938 59m/B Ken Maynard, Marjorie Reynolds, Walter Long; ***Dir:*** Harry Fraser. **VHS, Beta $47.95** *UHV, VYY, VCN* 🦴🦴

Six Weeks A young girl dying of leukemia brings together her work-driven mother and an aspiring married politician. Manipulative hanky-wringer has good acting from both Moores but oddly little substance.
1982 (PG) 107m/C Dudley Moore, Mary Tyler Moore, Katherine Healy; ***Dir:*** Tony Bill. **VHS, Beta, LV $12.95** *COL* 🦴🦴

Sixteen A naive country lass is attracted to the glitter and hum of the outside world. Her determination and optimism help her triumph. Originally called "Like a Crow on a June Bug."
1972 (R) 84m/C Mercedes McCambridge, Parley Baer, Ford Rainey, Beverly Powers, John Lozier, Simone Griffeth, Maddie Norman; ***Dir:*** Lawrence Dobkin. **VHS, Beta $19.95** *STE, NWV* 🦴🦴

Sixteen Candles Hughes gathers his stable of young stars again for one of his best films. Every girl's sixteenth birthday is supposed to be special, but in the rush of her sister's wedding nobody remembers Ringwald's. After that, everything seems to go downhill in hilarious fashion. Ringwald and Hall are especially charming in this humorous look at teenage traumas. Title song performed by The Stray Cats.
1984 (PG) 93m/C Molly Ringwald, Justin Henry, Michael Schoeffling, Haviland Morris, Gedde Watanabe, Anthony Michael Hall, Paul Dooley, Carlin Glynn, Blanche Baker, Edward Andrews, Carole Cook, Max Showalter, Liane Curtis, John Cusack, Joan Cusack, Brian Doyle-Murray, Jami Gertz, Cinnamon Idles, Zelda Rubenstein; ***W/Dir:*** John Hughes. **VHS, Beta, LV $19.95** *MCA* 🦴🦴🦴

Sixteen Fathoms Deep A sponge fisherman risks it all, including his love, when a mean businessman threatens his operation. Good underwater photography, but otherwise undistinguished.
1934 57m/B Sally O'Neil, George Regas, Maurice Black, Russell Simpson; ***Dir:*** Armand Schaefer. **VHS $16.95** *SNC, NOS, DVT* 🦴🦴

'68 A Hungarian family struggles with the generation gap in 1968 America.
1987 (R) 99m/C Eric Larson, Terra Vandergaw, Neil Young; ***Dir:*** Steven Kovacks. **VHS, Beta, LV $19.95** *STE, NWV* 🦴

Sizzle When her boyfriend is murdered by the mob, nightclub singer Anderson, newly arrived from Hicksville in Roaring Twenties, Chicago, stops at nothing to get revenge. Bad-guy gangster Forsythe falls for her charms. Fun but unsophisticated. Made for TV.
1981 100m/C Loni Anderson, John Forsythe, Leslie Uggams, Roy Thinnes, Richard Lynch, Michael Goodwin; ***Dir:*** Don Medford. **VHS, Beta $59.95** *PSM* 🦴🦴

Sizzle Beach U.S.A. Three young women who want a shot at becoming famous travel to Los Angeles and play on the beach with little budget and no particular purpose. Also available in an unrated version. Although released in 1986, this was made in 1974. Kevin Costner's film debut.
1986 (R) 90m/C Terry Congie, Leslie Brander, Roselyn Royce, Kevin Costner; ***Dir:*** Richard Brander. **VHS, Beta, LV $19.95** *STE, VMK* 🦴

Skag Disabled following a stroke, a Pittsburgh steel worker confronts a change in family roles and long-ignored family problems during his convalescence. Malden is exceptional and believable as a Joe Lundbucket in this well-written, well-cast TV outing, pilot for the NBC series.
1979 145m/C Karl Malden, Piper Laurie, Craig Wasson, Leslie Ackerman, Peter Gallagher, Kathryn Holcomb, Powers Boothe; ***Dir:*** Frank Perry. **VHS, Beta $49.95** *IVE, IND* 🦴🦴🦴

Skateboard A down-and-out Hollywood agent creates a pro skateboarding team and enters them in a race worth $20,000. Quickie premise executed lamely.
1977 (PG) 97m/C Allen Garfield, Kathleen Lloyd, Chad McQueen, Leif Garrett, Richard Van Der Wyk, Tony Alva, Antony Carbone. **VHS, Beta $79.95** *VHV* 🦴

Skateboard Madness This program offers a tour of skating spots and performing skateboarding celebrities.
1980 (PG) 92m/C Stacy Peralta, Kent Senatore; ***Dir:*** Julian Pene Jr. **VHS, Beta $39.95** *IVE, MON* 🦴

Skeezer A dog becomes a key factor in a sympathetic doctor's efforts to communicate with emotionally unstable children. Based on reality, and a good cast makes it believable and moving. Quality family fare.
1982 100m/C Karen Valentine, Dee Wallace Stone, Tom Atkins, Mariclare Costello, Leighton Greer, Justine Lord; ***Dir:*** Peter Hunt. **VHS $39.95** *IVE* 🦴🦴½

Skeleton Coast Borgnine plays a retired U.S. Marine colonel who organizes a Magnificent Seven-like group of tough mercenaries to go into eastern Africa to save hostages held by Angolan terrorists. Cliches abound, including a token large-breasted woman, Mulford, getting her t-shirt ripped open. No plot and a bad script. Also called "Fair Trade."
1989 (R) 94m/C Ernest Borgnine, Robert Vaughn, Oliver Reed, Herbert Lom, Daniel Greene, Nancy Mulford, Leon Isaac Kennedy; ***Dir:*** John Cardos. **VHS, LV $19.98** *SUE* 🦴

Sketches of a Strangler A psychotic art student sketches, then murders prostitutes. Interesting lead character but hardly chilling—just horrible to watch. Made for TV.
1978 91m/C Allen Garfield, Meredith McRae; ***Dir:*** Paul Leder. **VHS, Beta $59.95** *MED* 🦴½

Ski Bum A ski bum discovers corruption and violence at a Colorado Rockies resort. Eminently forgettable trash. Scantily based on a novel by Romain Gary.

1975 **(R)** 94m/C Charlotte Rampling, Zalman King, Dimitra Arliss, Anna Karina; *Dir:* Bruce Clark. **VHS, Beta** $59.98 *CHA* ⅃

Ski Patrol Wacky ski groupies try to stop an evil developer. Good ski action in a surprisingly plotful effort from the crazy crew that brought the world "Police Academy."
1989 **(PG)** 85m/C Roger Rose, Yvette Nipar, T.K. Carter, Leslie Jordan, Ray Walston, Martin Mull; *Dir:* Richard Correll. **VHS, Beta, LV** $89.95 *COL* ⅃

Ski School Rival ski instructors compete for jobs and babes. Brow lowering.
1991 **(R)** 89m/C Ava Fabian, Dean Cameron, Tom Breznahan, Stuart Fratkin; *Dir:* Damian Lee. **VHS, LV** $89.99 *HBO* ⅃

Ski Troop Attack During World War II, an American ski patrol is sent behind Nazi lines to blow up a German railway bridge. Unexceptional low-budget Corman outing; not among the greatest of war movies.
1960 61m/B Michael Forest, Frank Wolff, Sheila Carol. **VHS, Beta** $16.95 *SNC, NOS, DVT* ⅃⅃

Skier's Dream This spectacular action-packed film focuses on a young executive in search of the ultimate run. Shot on the most exotic ski locations in the world; covers freestyle skiing, powder skiing, snowboarding, extreme skiing, cliff jumping, paragliding, and wave riding.
1988 75m/C John Eaves, Ian Boyd. **VHS** $79.95 *SYB, RAP* ⅃

Skin Deep A boyish Don Juan tries everything to win back his ex-wife. Ritter whines about his mid-life crisis and seduces women; this substitutes for plot. A few funny slapstick scenes still don't make this worth watching.
1989 **(R)** 102m/C John Ritter, Vincent Gardenia, Julianne Phillips, Alyson Reed, Nina Foch, Chelsea Field, Denise Crosby; *Dir:* Blake Edwards. **VHS, Beta, LV** $9.99 *CCB, MED* ⅃½

Skin Game Two British families feud over land rights. Not thrilling; not characterisitic of working with Hitchcock. Way too much talking in excruciating, drawn-out scenes.
1931 87m/B *GB* Edmund Gwenn, Helen Hayes, Frank Lawton; *Dir:* Alfred Hitchcock. **VHS, Beta** $16.95 *SNC, NOS, MLB* ⅃½

Skin Game A fast talking con-artist (Garner) and his black partner (Gossett) travel throughout the antebellum South setting up scams—Gossett is sold to a new owner by Garner, who helps him escape. Garner and Gossett make a splendid comedy team in this different kind of buddy flick. All is well until Asner turns the tables on them. Finely acted comedy-drama.
1971 **(PG)** 102m/C James Garner, Louis Gossett Jr., Susan Clark, Ed Asner, Andrew Duggan; *Dir:* Paul Bogart. **VHS, Beta** $19.98 *WAR* ⅃⅃⅃

Skinheads: The Second Coming of Hate Neo-fascists run rampant. Exploitative effort from schlock-doyen Clark.
1988 93m/C Chuck Connors, Barbara Bain, Brian Brophy, Jason Culp, Elizabeth Sagal; *Dir:* Greydon Clark. **VHS** $79.95 *NSV* ⅃

Skinned Alive A woman and her children travel cross-country to sell leather goods. When a detective discovers where the leather comes from he's hot on their trail.
1989 90m/C Mary Jackson, Scott Spiegel; *W/ Dir:* Jon Killough. **VHS, Beta** *CIN* ⅃

Skinner's Dress Suit Meek, henpecked office clerk tells domineering wife he got a raise so she'll get off his back. She quickly insinuates them into upper crusty

social circle, where the fib pays off big. Remake of the 1917 version based on Henry Irving Dodge's novel.
1926 79m/B Reginald Denny, Laura LaPlante, Arthur Lake, Hedda Hopper; *Dir:* William A. Seiter. **VHS, Beta** $19.95 *GPV* ⅃⅃½

Skirts Ahoy! Williams, Evans, and Blaine are three WAVES who have their eyes set on three handsome men. To get them, of course, they must sing and dance a lot, and Williams must perform one of her famous water ballets. Songs include "Hold Me Close to You," "What Makes a WAVE?," and "I Get a Funny Feeling."
1952 109m/C Esther Williams, Joan Evans, Vivian Blaine, Barry Sullivan, Keefe Brasselle, Billy Eckstine, Debbie Reynolds; *Dir:* Sidney Lanfield. **VHS** *MGM* ⅃⅃

The Skull Horror abounds when Cushing gets his hands on the skull of the Marquis de Sade that has mysterious, murderous powers. Based on a story by Robert Bloch.
1965 83m/C Peter Cushing, Patrick Wymark, Christopher Lee, Nigel Green, Jill Bennett, Michael Gough, George Coulouris, Patrick Magee; *Dir:* Freddie Francis. **VHS** *PAR* ⅃⅃

Skull: A Night of Terror A cop vows not to use guns after a tragic accident, then single-handed and unarmed, takes on terrorists who kidnap his family.
1988 **(R)** 80m/C Nadia Capone, Robert Bideman, Robbie Fox, Paul Sanders; *Dir:* Robert Bergman. **VHS, Beta** $79.95 *ACA* ⅃⅃

Skull & Crown Rin Tin Tin, Jr. helps the hero break up a group of smugglers. A favorite of Tin fans everywhere.
1938 58m/B Rin Tin Tin Jr, Regis Toomey, Jack Mulhall. **VHS, Beta** $24.95 *NOS, DVT* ⅃⅃

Skullduggery When a peaceful race of blond ape-people are discovered in New Guinea, a courtroom battle ensues to prevent their slaughter at the hands of developers. Interesting premise sadly flops.
1970 **(PG)** 105m/C Burt Reynolds, Susan Clark, Roger C. Carmel, Chips Rafferty, Edward Fox, Wilfrid Hyde-White, Rhys Williams; *Dir:* Gordon Douglas. **VHS, Beta** *MED* ⅃½

Skullduggery Costume-store minion Haverstock carries a curse that makes him kill and mutilate. The usual unspeakable horror ensues for a group of medieval-game players.
1979 **(PG)** 95m/C Thom Haverstock, Wendy Crewson, David Calderisi; *Dir:* Ota Richter. **VHS, Beta** $59.95 *MED* Woof!

The Sky Above, the Mud Below An acclaimed documentary following a band of explorers transversing New Guinea in 1959, and their confrontations with native rituals, heretofore unknown cannibal tribes and physical calamities. French title: "Le Ciel et la Boue." In HiFi.
1961 90m/C *Dir:* Pierre-Dominique Gaisseau. Academy Awards '61: Best Feature Documentary. **VHS, Beta** *LHV, WAR* Woof!

Sky Bandits Mounties uncover the mystery of a disappearing plane carrying gold from a Yukon mine. Last of the "Renfrew of the Mounties" pictures.
1940 56m/B James Newill, Dave O'Brien, Louise Stanley. **VHS, Beta, 8mm** $14.95 *NOS, SNC, VCN* ⅃

The Sky is Gray A young Louisiana black boy journeys with his mother to the dentist, and on the way is confronted with prejudice and poverty. Set in the 1940s. Based on a story by Ernest Gaines. Made for

television, with an introduction by Henry Fonda.
1980 46m/C Olivia Cole, James Bond III, Margaret Avery, Cleavon Little; *Dir:* Stan Lathan. **VHS, Beta** $24.95 *MON, KAR* ⅃⅃⅃

Sky Hei$t Criminals plan robbery of police helicopter carrying fortune in gold bullion.
1975 96m/C Stefanie Powers, Joseph Campanella, Don Meredith, Larry Wilcox, Frank Gorshin, Shelley Fabares, Ken Swofford, Ray Vitte, Nancy Belle Fuller, Suzanne Somers; *Dir:* Lee H. Katzin. **VHS** $29.95 *UNI* ⅃⅃

Sky High A spy at an Air Force base is trying to get hold of the plans to a secret plane. A GI is recruited to catch him.
1951 60m/B Sid Melton, Mara Lynn, Douglas Evans. **VHS, Beta, LV** $79.95 *WGE* ⅃

Sky High Three college students become immersed in international intrigue when the C.I.A. and the K.G.B. pursue them through Greece looking for a secret Soviet tape.
1984 103m/C Daniel Hirsch, Clayton Norcross, Frank Schultz, Lauren Taylor; *Dir:* Nico Mastorakis. **VHS, Beta** $79.98 *LIV, LTG* ⅃

Sky Liner An FBI agent is trailing a spy who has taken secret documents aboard a west-bound flight.
1949 62m/B Richard Travis, Pamela Blake. **VHS, Beta, LV** *WGE* ⅃

The Sky Pilot Young clergyman sets up a parish on North Pacific coast and finds he must prove himself to the cowboys. He saves young woman from stampede, but her father blames him for her maimed legs and sets the preacher's church on fire, providing girl with excuse to overcome handicap.
1921 63m/B John Bowers, Colleen Moore, David Butler, Donald Ian Macdonald; *Dir:* King Vidor. **VHS, Beta** $29.95 *GPV, FCT* ⅃⅃½

Sky Pirates Space epic deals about pieces of ancient stone left by prehistoric extraterrestrials, now lost in a time warp. Bad stunts, bad script and notably bad acting make it a woofer in any era.
1987 88m/C *AU* John Hargreaves, Max Phipps, Meredith Phillips; *Dir:* Colin Eggleston. **VHS, Beta** $79.98 *FOX* Woof!

Sky Riders Hang-gliders risk it all to take on a group of political kidnappers. Fine hang-gliding footage and glorious Greek locations make up for garden-variety plot.
1976 **(PG)** 93m/C James Coburn, Susannah York, Robert Culp, Charles Aznavour, Harry Andrews, John Beck; *Dir:* Douglas Hickox. **VHS** *FOX* ⅃⅃½

Skyline A Spanish photographer comes to New York City to work for a magazine. Quietly funny, with a startling ending. Partly in Spanish with English subtitles.
1984 84m/C *SP* Antonio Resines, Beatriz Perez-Porro, Jaime Nos, Roy Hoffman; *Dir:* Fernando Colombo. **VHS, Beta, LV** $29.95 *PAV* ⅃⅃⅃

Skyrider Astronauts, space capsules, witches on broomsticks, flying toasters, and chandeliers make up this animated fantasy of space travel.
1978 7m/C **VHS, Beta** *BFA* ⅃⅃⅃

The Sky's the Limit Astaire spends his leave in Manhattan and falls in love with fetching journalist Leslie. He's in civvies, so little does she know he's a war hero. Nothing-special semi-musical, with Fred-Ginger spark missing. Songs include "One For My Baby" and "My Shining Hour."

1943 89m/B Fred Astaire, Joan Leslie, Robert Benchley, Robert Ryan, Elizabeth Patterson; *Dir:* Edward H. Griffith. **VHS, Beta $19.95** *RKO, VYY, KOV* ✓✓½

Slacker Filmed on location in Austin, Texas this is a definition of a new generation. "Slackers" are so overwhelmed by the world and it's demands that they've retreated into lives of minimal expectations. The film is set as a series of improvisational stories of people living on the fringes of the working world and their reactions (or lack thereof) to the life swirling around them. The first feature of writer/director Linklater who had a budget of $23,000 and a cast primarily of nonprofessional actors to work with.
1991 (R) 97m/C Richard Linklater; *W/Dir:* Richard Linklater. **VHS** *ORI* ✓✓✓

Slamdance A struggling cartoonist is framed for the murder of a beautiful young woman while being victimized by the real killer. A complicated murder mystery with a punk beat and visual flash. But the tale is unoriginal - and where's the slamdancing?
1987 (R) 99m/C Tom Hulce, Virginia Madsen, Mary Elizabeth Mastrantonio, Harry Dean Stanton, Adam Ant, John Doe; *Dir:* Wayne Wang. **VHS, Beta $19.98** *FOX* ✓½

Slammer Girls In this sex-drenched, unfunny spoof on women's prison films, the inmates of Loch Ness Penitentiary try to break out using their sexual wiles. The actresses are pseudonymous porn stars.
1987 (R) 82m/C Tally Brittany, Jane Hamilton, Jeff Eagle, Devon Jenkin; *Dir:* Chuck Vincent. **VHS, Beta, LV $79.98** *LIV, LTG* ✓

The Slap French teens learn about love the French way.
1976 (PG) 103m/C *FR* Isabelle Adjani, Lino Ventura; *Dir:* Claude Pinoteau. **VHS, CDV** *SUE* ✓

Slap Shot A profane satire of the world of professional hockey. Over-the-hill player-coach Newman gathers an odd-ball mixture of has-beens and young players and reluctantly initiates them, using violence on the ice to make his team win. The strip-tease on ice needs to be seen to be believed. Charming in its own bone-crunching way.
1977 (R) 123m/C Paul Newman, Michael Ontkean, Jennifer Warren, Lindsay Crouse, Jerry Houser, Melinda Dillon, Strother Martin; *Dir:* George Roy Hill. **VHS, Beta, LV $19.95** *MCA* ✓✓✓

Slapstick An anthology of great moments from Mack Sennett comedy shorts, featuring many great stars in some of their earliest film appearances.
193? 76m/B Charlie Chaplin, Harold Lloyd, Charley Chase, Buster Keaton, Ben Turpin, Stan Laurel. **VHS, Beta $39.95** *REP* ✓

Slapstick of Another Kind Lewis and Kahn play dual roles as an alien brother and sister and their adoptive Earth parents, who are being pursued by U.S. agents. Rather miserable rendition of a Kurt Vonnegut novel.
1984 (PG) 85m/C Jerry Lewis, Madeline Kahn, Marty Feldman, Jim Backus, Pat Morita, Samuel Fuller, Orson Welles; *Dir:* Steven Paul. **VHS, Beta, LV $79.98** *LIV, VES* ✓½

Slash Each of four people caught in a political takeover harbors a secret that could harm one of the others. Extremely violent.
1987 90m/C Romano Kristoff, Michael Monty, Gwen Hung; *Dir:* John Gale. **VHS, Beta $59.95** *KOV* ✓½

Slashdance A really pathetic thriller follows a lady cop undercover in a chorus line to find out who's been murdering the dancers. The acting is on the level of pro wrestling. Don't be fooled by the naked babes on the cassette box - there's no nudity.
1989 83m/C Cindy Maranne, James Carroll Jordan, Queen Kong, Joel von Ornsteiner, Jay Richardson; *W/Dir:* James Shyman. **VHS, Beta $9.95** *SIM* Woof!

Slashed Dreams Hippie couple travels to California wilderness in search of friend. A pair of woodsmen assault them and rape the woman, which really messes with their heads. Retitled and packaged as slasher movie, it was originally titled "Sunburst." Mountain men Keach and Pritchard wrote the screenplay.
1974 (R) 74m/C Peter Hooten, Katherine Baumann, Ric Carrott, Anne Lockhart, Robert Englund, Rudy Vallee, James Keach, David Pritchard, Peter Brown; *Dir:* James Polakof. **VHS $19.95** *ACA* ✓

The Slasher A policeman must find the madman who has been killing off unfaithful married women. Miserably gory, pointless slasher flicker (hence the title) interesting only as an early example of its kind.
1974 88m/C Farley Granger, Sylva Koscina, Susan Scott. **VHS, Beta $59.95** *MON* Woof!

Slate, Wyn & Me Two bank-robbing, kidnapping brothers flee cross-country from the law, and fall in love with the girl they've snatched. Pointless, and pointlessly violent and vulgar.
1987 (R) 90m/C Sigrid Thornton, Simon Burke, Martin Sacks; *Dir:* Don McLennan. **VHS, Beta $19.98** *SUE* ✓

Slaughter After his parents are murdered, a former Green Beret goes after their killers. Plenty of brutality. Followed by "Slaughter's Big Ripoff."
1972 (R) 92m/C Jim Brown, Stella Stevens, Rip Torn, Cameron Mitchell, Don Gordon, Marlene Clark; *Dir:* Jack Starrett. **VHS, Beta, LV $59.98** *FOX* ✓

Slaughter High A high school nerd is accidentally disfigured by a back-firing prank. Five years later, he returns to exact bloody revenge. Available in an unrated version.
1986 (R) 90m/C Caroline Munro, Simon Scuddamore, Kelly Baker; *Dir:* George Dugdale. **VHS, Beta $79.98** *LIV, VES* ✓

Slaughter Hotel An asylum already inhabited by extremely bizarre characters is plagued by a series of gruesome murders. Lots of skin and lots of blood; Neri shines as a nymphomaniacal lesbian nurse. Also known as "Asylum Erotica."
1971 (R) 72m/C Klaus Kinski, Rosalba (Sara Bay) Neri, Margaret Lee, John Ely; *Dir:* Fernando Di Leo. **VHS, Beta** *MPI, GRG* ✓

Slaughter in San Francisco A Chinese-American cop leads a one-man fight against corruption in the department in Daly City, near San Francisco. Norris is the bad guy; made in 1973 but not released until he had an established following. Plenty of kicking and karate chops.
1981 (R) 92m/C *CH* Chuck Norris, Don Wong; *Dir:* William Lowe. **VHS, Beta $9.95** *SUE* ✓

Slaughter Trail Rancher-turned-stage robber Young shoots two Indians and cavalry colonel Donlevy has to deal with the fallout. Peace returns when Young dies. Ordinary western with balladeer device helping the plot along, or trying to.

1951 78m/C Brian Donlevy, Gig Young, Virginia Grey, Andy Devine, Robert Hutton; *Dir:* Irving Allen. **VHS, Beta $19.98** *TTC* ✓✓

Slaughter of the Vampires Newlyweds meet a bloodthirsty vampire in a Viennese chalet. Real, real bad horror flick, poorly dubbed. Also know as "Curse of the Blood-Ghouls."
1962 81m/C *IT* Walter Brandi, Dieter Eppler, Graziella Granta; *W/Dir:* Robert Mauri. **VHS, Beta $39.95** *SNC, MON* Woof!

Slaughterday An innocent woman is caught in the middle of an ex-con's elaborate scheme to commit the largest heist of his life.
1977 (R) 87m/C Rita Tushingham, William Berger, Frederick Jaeger, Michael Hauserman, Gordon Mitchell; *Dir:* Peter Patzak. **VHS, Beta** *GHV* ✓✓

Slaughterhouse A rotund, pig-loving country boy kills, maims, and eats numerous victims. Features the requisite dumb teens and plenty of blood.
1987 (R) 87m/C Joe Barton, Sherry Bendorf, Don Barrett; *Dir:* Rick Roessler. **VHS, Beta $14.98** *CHA* Woof!

Slaughterhouse Five A suburban optometrist becomes "unstuck" in time and flits randomly through the experiences of his life, from the Dresden bombing to an extra-terrestrial zoo. Noticed at Cannes but not at theatres; ambitious failure to adapt Kurt Vonnegut's odd novel.
1972 (R) 104m/C Michael Sacks, Valerie Perrine, Ron Leibman, Eugene Roche, Perry King; *Dir:* George Roy Hill. Cannes Film Festival '72: Jury Prize. **VHS, Beta, LV $19.95** *KUI, MCA* ✓✓

Slaughterhouse Rock A rock star goes on a cannibalistic rampage after being possessed by an evil spirit. Music group Devo did the score for this flick.
1988 (R) 90m/C Nicholas Celozzi, Donna Denton, Toni Basil, Hope Marie Carlton; *Dir:* Dimitri Logothetis. **VHS, Beta $14.95** *SVS* Woof!

Slaughter's Big Ripoff Slaughter is back battling the Mob with guns, planes and martial arts. This undistinguished sequel features McMahon as a mob boss.
1973 (R) 92m/C Jim Brown, Brock Peters, Don Stroud, Ed McMahon, Art Metrano, Gloria Hendry; *Dir:* Gordon Douglas. **VHS, Beta $59.98** *FOX* ✓✓

Slave of the Cannibal God Beautiful Andress is captured by "native" cannibals; Keach must save her. Pathetically unsuspenseful garbage. Also called "Mountain of the Cannibal Gods."
1979 (R) 86m/C *IT* Stacy Keach, Ursula Andress, Claudio Cassinelli; *Dir:* Sergio Martino. **VHS, Beta $49.95** *VES* Woof!

Slave Girls from Beyond Infinity In this B-movie spoof, two beautiful intergalactic slave girls escape their penal colony, land on a mysterious planet, and meet a cannibalistic despot. Fun spoof of '50s "B" sci-fi movies.
1987 80m/C Elizabeth Cayton, Cindy Beal, Brinke Stevens; *Dir:* Ken Dixon. **Beta $79.95** *UCV* ✓✓

A Slave of Love Poignant love story set in the Crimea as the Bolshevik Revolution rages around a film crew attempting to complete a project. Interesting as ideological cinema but also enjoyable romantic drama.
1978 94m/C *RU* Elena Solovei, Rodion Nakhapetov, Alexander Kalyagin; *Dir:* Nikita Mikhailkov. National Board of Review '78: 5 Best Foreign Films of the Year. **VHS, Beta $29.95** *COL* ✓✓✓

Slave Trade in the World Today Slavery practices in the modern world are shown, including actual scenes of a sheik's harem, slave auctions and the kidnapping and selling of young women.
1964 87m/C *Dir:* Roberto Malenotti. VHS, Beta $19.95 VDC ♂

Slavers Detailed depiction of the 19th century African slave trade with a little romance thrown in.
1977 (R) 102m/C GE Trevor Howard, Britt Ekland, Ron Ely, Cameron Mitchell, Ray Milland; *Dir:* Jurgen Goslar. VHS, Beta $69.95 VCD, TAV ♂♂

Slaves of Love Two men, stranded on a desert island, discover it is inhabited by sex-starved females.
197? 86m/C VHS, Beta $59.95 MED ♂

Slaves of New York Greenwich Village artists worry about life and love in the '80s and being artistic enough for New York. Adapted from the stories of Tama Janowitz, who also wrote the screenplay and appears as Abby. Disastrous adaptation of a popular novel.
1989 (R) 115m/C Bernadette Peters, Chris Sarandon, Mary Beth Hurt, Madeleine Potter, Adam Coleman Howard, Nick Corri, Mercedes Ruehl, Joe Leeway, Charles McCaughan, John Harkins, Anna Katarina, Tama Janowitz; *Dir:* James Ivory. VHS, Beta, LV $89.95 COL ♂ 1/2

The Slayer It's movies like this that give getting back to nature a bad name. That horrible monster is after those nice young people again! This time it's on an island off the coast of Georgia.
1982 (R) 95m/C *Dir:* J.S. Cardone. VHS, Beta NSV Woof!

The Slayer/The Scalps Double feature. "Scalps" is a slasher set in a Native American burial ground. "The Slayer" is about vacationing couples who run into more than they bargained for on their island retreat.
1982 (R) 150m/C *Dir:* J.S. Cardone. VHS, Beta $49.98 CGI ♂

Slayground A man, distraught at the accidental death of his daughter, hires a hitman to exact revenge. Excruciating adaptation of the novel by Richard Stark (Donald E. Westlake).
1984 (R) 85m/C GB Peter Coyote, Mel Smith, Billie Whitelaw, Philip Sayer, Kelli Maroney; *Dir:* Terry Bedford. VHS, Beta $69.95 HBO ♂ 1/2

Sleazemania This compilation of trailers 'n bits reviews some of the sleazier, sicker, and sexier selections of celluloid from the past 50 years.
1985 60m/C VHS, Beta $19.95 RHI ♂ 1/2

Sleazemania III: The Good, the Bad & the Sleazy Another compilation of the cheapest, funniest and scuriest moments, from grade-Z horror films, including "Mondo Psycho" and "Smut Peddler Meets Emmanuel."
1986 90m/B VHS, Beta $39.95 RHI ♂ 1/2

Sleazemania Strikes Back A scintillating array of sleazy scenes from the world's worst movies.
1985 60m/C VHS, Beta $39.95 RHI ♂ 1/2

The Sleazy Uncle A successful but bored businessman is energized by an encounter with his vice-ridden uncle, a mooching, womanizing beatnik poet who knows how to enjoy life. An energetic but rather

---

formula comedy, in Italian with English subtitles.
1990 105m/C IT Giancarlo Giannini, Vittorio Gassman; *Dir:* Franco Brusati. VHS, Beta NVH ♂♂

Sledgehammer A madman is wreaking havoc on a small town, annihilating young women with a sledgehammer.
1983 87m/C Ted Prior, Doug Matley, Steven Wright. VHS, Beta $9.95 WES Woof!

Sleep of Death A young English man's pursuit of a blue-blooded woman seems to set off a series of bizarre and mysterious murders. A gothic thriller based on Sheridan Le Fanu's short story.
1979 90m/C Brendan Price, Marilyn Tolo, Patrick Magee, Curt Jurgens, Per Oscarsson; *Dir:* Calvin Floyd. VHS, Beta $59.95 PSM ♂ 1/2

Sleepaway Camp A crazed killer hacks away at the inhabitants of a peaceful summer camp in this run-of-the-mill slasher.
1983 (R) 88m/C Mike Kellin, Jonathan Tiersten, Felissa Rose, Christopher Collet. VHS, Beta $9.95 MED ♂

Sleepaway Camp 2: Unhappy Campers A beautiful camp counselor is actually a blood-thirsty, murdering madwoman. Sequel to the 1983 slasher, "Sleepaway Camp."
1988 82m/C Pamela Springsteen, Renee Estevez, Walter Gotell, Brian Patrick Clarke; *Dir:* Michael A. Simpson. VHS, Beta $14.98 SUE ♂ 1/2

Sleepaway Camp 3: Teenage Wasteland This second sequel is as bad as the first two movies. Another disturbed camper hacks up another bevy of teenagers. Better luck at the Motel Six.
1989 (R) 80m/C Pamela Springsteen, Tracy Griffith, Michael J. Pollard; *Dir:* Michael A. Simpson. VHS, Beta, LV $19.98 SUE Woof!

Sleeper Hapless nerd Allen is revived two hundred years later after an operation gone bad. Keaton portrays Allen's love interest in a futuristic land of robots and giant vegetables. He learns of the hitherto unknown health benefits of hot fudge sundaes; discovers the truth about the nation's dictator, known as The Leader; and gets involved with revolutionaries seeking to overthrow the government. Hilarious, fast-moving comedy, full of slapstick and satire. Don't miss the "orgasmatron."
1973 (PG) 88m/C Woody Allen, Diane Keaton, John Beck, Howard Cosell; *Dir:* Woody Allen. National Board of Review Awards '73: Special Citation (Allen). VHS, Beta, LV $19.98 MGM, FOX ♂♂♂ 1/2

Sleeping Beauty Classic Walt Disney version of the famous fairy tale is set to the music of Tchaikovsky's ballet. Lavishly produced. With the voices of Mary Costa, Bill Shirley, and Vera Vague.
1959 (G) 75m/C *Dir:* Clyde Geronomi, Eric Larson, Wolfgang Reitherman, Les Clark. VHS, Beta, LV DIS, OM ♂♂♂

Sleeping Beauty Peters is the princess put to sleep by evil queen Kellerman. Handsome prince Reeve comes to the rescue in this classy retelling of the Sleeping Beauty legend. Combines live action and animation. From Shelley Duvall's "Faerie Tale Theatre."
1983 60m/C Christopher Reeve, Bernadette Peters, Beverly D'Angelo. VHS, Beta, LV $14.95 FOX, FCT, WKV ♂♂♂

The Sleeping Car A vindictive ghost haunts a group of people living in an old railway car. Fairly fun, fairly competent horror.
1990 87m/C David Naughton, Judie Aronson, Kevin McCarthy, Jeff Conaway, Dani Minnick, John Carl Buechler; *Dir:* Douglas Curtis. VHS, LV $89.95 VMK ♂♂

Sleeping Car to Trieste Conspirators in the theft of a diplomat's diary ride the Orient Express, as a wily detective on board tries to piece together murder clues. Based on the story "Rome Express" by Clifford Grey.
1945 95m/B Jean Kent, Albert Lieven, David Tomlinson, Finlay Currie; *Dir:* John Paddy Carstairs. VHS, Beta LCA ♂♂

Sleeping Dogs A man in near-future New Zealand finds it hard to remain neutral when he is caught between a repressive government and a violent resistance movement. The first New Zealand film ever to open in the U.S., and a fine debut for director Donaldson.
1982 107m/C NZ Sam Neill, Ian Mune, Nevan Rowe, Dona Akersten, Warren Oates; *Dir:* Roger Donaldson. VHS, Beta $69.95 VID ♂♂ 1/2

Sleeping with the Enemy Roberts escapes from abusive husband by faking death, flees to Iowa, falls for drama professor, and, gasp, is found by psycho husband. Occasionally chilling but oft predictable thriller based on novel by Nancy Price.
1991 (R) 99m/C Julia Roberts, Kevin Anderson, Patrick Bergin, Elizabeth Lawrence, Kyle Secor, Claudette Nevins; *Dir:* Joseph Ruben. VHS, Beta, LV $19.98 FXV, IME, FCT ♂♂ 1/2

Sleeping Fist A young man learns the sleeping fist style of Kung fu in order to save the girl he loves from a ruthless gang of marauders.
198? 85m/C Yuan Hsiao-Tien, Liang Chia-Yen. VHS, Beta $39.95 TWE ♂

The Sleeping Tiger A thief breaks into the home of a psychiatrist, who captures him. In exchange for his freedom, the thief agrees to become a guinea pig for the doctor's rehabilitation theories with ultimately tragic results. Director Losey was originally compelled to release the film under a pseudonym, "Victor Hanbury," because he had been blacklisted by Hollywood during the 1950s red scare. First pairing of Losey and Bogarde, who collaborated on several later films, including "Modesty Blaise" (1966) and "Accident" (1967).
1954 89m/B Alexis Smith, Alexander Knox, Dirk Bogarde, Hugh Griffith; *Dir:* Joseph Losey. VHS $16.95 NOS, SNC ♂♂ 1/2

Sleepwalk Odd independent effort about a bored New York desk worker who is given an ancient text of Chinese fairy tales to translate, and finds they're manifesting themselves in dangerous ways in real life. Cinematography by Jim Jarmusch and Franz Prinzi.
1988 (R) 78m/C Suzanne Fletcher, Ann Magnuson; *Dir:* Sara Driver. VHS, Beta, LV $19.98 SUE ♂♂

The Slender Thread Based on a true story, Poitier plays a college student who volunteers at a crisis center and must keep would-be suicide Bancroft on the phone until the police can find her. Filmed on location in Seattle.
1965 98m/C Sidney Poitier, Anne Bancroft, Telly Savalas, Steven Hill; *Dir:* Sydney Pollack. VHS $14.95 PAR ♂♂♂

Sleuth A mystery novelist and his wife's lover face off in ever shifting, elaborate, and diabolical plots against each other, complete with red herrings, traps, and tricks. Playful, cerebral mystery thriller from top director Mankiewicz. Oscar nominations for Caine and Olivier, who also scripted "Frenzy" for Hitchcock, from his play.
1972 (PG) 138m/C Laurence Olivier, Michael Caine; *Dir:* Joseph L. Mankiewicz. Edgar Allan Poe Awards '72: Best Screenplay. **VHS, Beta, LV** $19.95 MED ⅋⅋⅋½

Slightly Honorable A lawyer gets involved in political scandals and becomes a murder suspect. Snappy comedy-drama with good performances but too many subplots. Based on the novel "Send Another Coffin" by F. G. Presnell.
1940 75m/B Pat O'Brien, Broderick Crawford, Edward Arnold, Eve Arden, Evelyn Keyes, Phyllis Brooks; *Dir:* Tay Garnett. **VHS, Beta** $19.95 NOS, VYY, KRT ⅋⅋½

A Slightly Pregnant Man Comic complications abound as a construction worker becomes the world's first pregnant man. Surprise ending saves this simple and not very exciting film.
1979 (PG) 92m/C Catherine Deneuve, Marcello Mastroianni; *Dir:* Jacques Demy. **VHS, Beta** $19.98 VID ⅋⅋

Slightly Scarlet Small-time hood Payne carries out an assignment from boss DeCorsia to smear a law-and-order politico running for mayor. He falls in love with the candidate's secretary, tries to go straight, and ends up running the mob when DeCorsia flees town. A spiffy, low-budget noir crime drama.
1956 99m/C Rhonda Fleming, Arlene Dahl, John Payne, Kent Taylor, Ted DeCorsia; *Dir:* Allan Dwan. **VHS, Beta** $39.95 BVV ⅋⅋½

Slime City The widow of an alchemist poisons her tenants so they will join her husband in the hereafter. Very low-budget; occasionally funny.
1989 90m/C Robert C. Sabin, Mary Huner, T.J. Merrick, Dick Biel; *Dir:* Gregory Lamberson. **VHS, Beta** CAM ⅋

The Slime People Huge prehistoric monsters are awakened from long hibernation by atomic testing in Los Angeles. They take over the city, creating the fog they need to live. Thank goodness for scientist Burton, who saves the day. Filmed in a butcher shop in Los Angeles.
1963 76m/B Robert Hutton, Robert Burton, Susan Hart, William Boyce, Les Tremayne; *Dir:* Robert Hutton. **VHS, Beta** $19.95 NOS, GEM ⅋

Slipping into Darkness Three spoiled, rich college girls are held responsible for a retarded local boy's death. Honest and overwrought.
1988 (R) 86m/C Belle Mitchell, Laslo Papas, Rigg Kennedy, Beverly Ross, T.J. McFadden, Michelle Johnson, John DiAquino; *Dir:* Eleanor Gaver. **VHS, Beta, LV** $24.95 GHV, HHE ⅋½

Slipstream "Turgid" and "gloomy" are the words for this romantic drama set in remote Alberta. The minor plot concerns a young woman who teaches a lonely disc jockey a lesson about life and love. This film feels like a long Canadian winter.
1972 93m/C Julie Askew, Patti Oatman, Eli Rill. **VHS, Beta** $19.95 TWE ⅋⅋

Slipstream A sci-fi adventure set on a damaged Earthscape where people seek to escape a giant jetstream. While tracking down a bounty hunter gone bad, a futuristic cop follows his quarry into the dangerous river of wind. Ambitious "Blade Runner" clone with big names (Hamill is good) was never released theatrically in the US. Musical score by Elmer Bernstein.
1989 (PG-13) 92m/C GB Mark Hamill, Bill Paxton, Bob Peck, Eleanor David, Kitty Aldridge, Robbie Coltrane, Ben Kingsley, F. Murray Abraham; *Dir:* Steven Lisberger. **VHS, Beta, LV** $79.95 VIR ⅋⅋

Slither Caan and Boyle become wrapped up in a scheme to recover $300,000 in cash, stolen seven years previously. Along the way they pick up speed freak Kellerman, who assists them in a variety of loony ways. Frantic chase scenes are the highlight.
1973 (PG) 97m/C James Caan, Peter Boyle, Sally Kellerman, Louise Lasser, Allen Garfield, Richard B. Shull, Alex Rocco; *Dir:* Howard Zieff. **VHS, Beta** $19.98 MGM ⅋⅋⅋

Slithis Slithis, the nuclear-waste creature, is the menace of Venice, California. Keep your pets (and yourself) indoors!
1978 (PG) 86m/C Alan Blanchard, Judy Motulsky; *W/Dir:* Stephen Traxler. **VHS, Beta** $19.95 MED Woof!

Sloane Typical sleazy sex and violence flick. For the record: organized crime baddies kidnap good guy Resnick's girlfriend and he seeks revenge.
1984 95m/C Robert Resnick, Debra Blee, Paul Aragon; *Dir:* Dan Rosenthal. **VHS, Beta** $69.98 LIV, VES Woof!

Slow Burn An ex-reporter tries some detective work in the seamy side of Palm Springs, and becomes embroiled in the standard drug, kidnapping, murder routine. Based on the novel "Castles Burning" by Arthur Lyons. Made for cable television.
1986 92m/C Eric Roberts, Beverly D'Angelo, Dan Hedaya, Dennis Lipscomb; *Dir:* Matthew Chapman. **VHS, Beta** $59.95 MCA ⅋⅋

Slow Burn The Mafia and Chinese Triads are having troubles and a tired cop attempts to reach a peaceful solution.
1990 (R) 90m/C William Smith, Anthony James, Ivan Rogers. **VHS** $79.95 XVC, PHX ⅋

Slow Men Working Side-splitting stand-up comedy from the fresh new talent of Steve Geyer.
1990 60m/C Steve Geyer. **VHS** $14.95 SPW ⅋

Slow Moves Slow story of slow people who meet slowly, have sluggish affair, and find tragedy eventually.
1984 93m/C Roxanne Rogers; *Dir:* Jon Jost. **VHS, Beta** $69.95 FCT ⅋½

The Slugger's Wife The marriage between an Atlanta Braves outfielder and a rock singer suddenly turns sour when their individual careers force them to make some tough choices.
1985 (PG-13) 105m/C Michael O'Keefe, Rebecca DeMornay, Martin Ritt, Randy Quaid, Loudon Wainwright III, Rene Lefevre; *Dir:* Hal Ashby. **VHS, Beta, LV** $79.95 COL ⅋½

Slugs A health inspector discovers that spilled toxic waste is being helpfully cleaned up by the slug population, saving Uncle Sam countless dollars. But wait! The slugs are mutating into blood-thirsty man-eaters. Is this the answer to military cut-backs?
1987 (R) 90m/C Michael Garfield, Kim Terry, Philip Machale; *Dir:* J.P. Simon. **VHS, Beta, LV** $19.95 STE, NWV Woof!

Slumber Party '57 At a slumber party, six girls get together and exchange stories of how they lost their virginity. Lots of great music by the Platters, Big Bopper, Jerry Lee Lewis, the Crewcuts, and Paul and Paula.
1976 (R) 83m/C Noelle North, Bridget Hollman, Debra Winger, Mary Ann Appleseth, Cheryl "Rainbeaux" Smith, Janet Wood, R.L. Armstrong, Rafael Campos, Larry Gelman, Will Hutchins, Joyce Jillson, Victor Rogers, Joe E. Ross, Bill Thurman; *Dir:* William A. Levey. **VHS, Beta, LV** $69.98 VES Woof!

The Slumber Party Massacre A psychotic killer with a power drill terrorizes a high school girls' slumber party. Contrived and forced, but not always unfunny.
1982 (R) 84m/C Michele Michaels, Robin Stille, Andre Honore, Michael Villela, Debra Deliso, Gina Mari, Brinke Stevens; *Dir:* Amy Jones, Aaron Lipstadt. **VHS, Beta** $14.98 SUE ⅋½

Slumber Party Massacre 2 Drowsy babes in lingerie are drilled to death by a perverse madman. Another disappointing sequel.
1987 75m/C Crystal Bernard, Kimberly McArthur, Juliette Cummins, Patrick Lowe; *Dir:* Deborah Brock. **VHS, Beta, LV** $14.98 SUE Woof!

Slumber Party Massacre 3 Parents: Don't let your daughters have any slumber parties! Yes, it's a drill—for the third time.
1990 (R) 76m/C Keely Christian, Brittain Frye, M.K. Harris, David Greenle, Hope Marie Carlton, Maria Ford; *Dir:* Sally Mattison. **VHS, Beta** $79.98 MGM Woof!

The Small Back Room A crippled World War II munitions expert leads a tormented existence and laments government bureaucracy. Powerfully presented adult storyline. Also known as "Hour of Glory."
1949 106m/B Emeric Pressburger, David Farrar, Jack Hawkins, Cyril Cusack, Kathleen Byron, Anthony Bushell, Michael Gough, Robert Morley; *Dir:* Michael Powell. **VHS** $39.95 NOS, LOO, FCT ⅋⅋⅋

Small Change Pudgy, timid Desmouceaux and scruffy, neglected Goldman lead a whole pack of heartwarming tykes realistically and tenderly portrayed, testament to the great director's belief in childhood as a "state of grace." Criticized for sentimentality, "Small Change" followed Truffaut's gloomy "The Story of Adele H." Steven Spielberg suggested the English translation of "L'Argent de Poche." In French, with subtitles.
1976 (PG) 106m/C FR Geory Desmouceaux, Philippe Goldman, Jean-Francois Stevenin, Chantal Mercier, Francis Devlaeminck; *Dir:* Francois Truffaut. National Board of Review Awards '76: 5 Best Foreign Films of the Year; New York Film Festival '76: 5 Best Foreign Films of the Year. **VHS, Beta** $19.99 WAR, FCT ⅋⅋⅋⅋

A Small Circle of Friends Three Harvard students struggle through their shifting relationships during their college years in the 1960s.
1980 (R) 112m/C Brad Davis, Jameson Parker, Karen Allen, Shelley Long; *Dir:* Rob Cohen. **VHS, Beta** $59.98 FOX ⅋½

Small Hotel After discovering that he is to be replaced as headwaiter by a young woman, Harker plots to save his job through espionage, blackmail and cajolery. Average comedy, adapted from a popular British stage play as a star vehicle for Harker.

1957 59m/B Gordon Harker, Marie Lohr, John Loder, Irene Handl, Janet Munro, Billie Whitelaw, Francis Matthews, Dora Bryan; *Dir:* David MacDonald. **VHS** $19.95 *NOS* 𝆑𝆑

A Small Killing An undercover cop and a college professor pose as a wino and a bag lady, when trying to put the tabs on a druglord. Made-for-TV mystery/romance based on "The Rag Bag Clan" by Richard Barth.
1981 100m/C Ed Asner, Jean Simmons, Sylvia Sidney, Andrew Prine, Pat O'Malley, Anne Ramsey; *Dir:* Steven Hilliard Stern. **VHS, Beta** $59.95 *IVE* 𝆑𝆑

Small Sacrifices The true story of Diane Downs, an Oregon woman who may have murdered her own three children in 1983. Fine performance from Fawcett, who seems to like made-for-TV tales of true-life domestic violence as an antidote to her earlier ditzy persona as a Charlie's Angel. Gripping; superb script by Joyce Eliason.
1989 159m/C Farrah Fawcett, Ryan O'Neal, John Shea; *Dir:* David Greene. **VHS, Beta, LV** $19.95 *FRH* 𝆑𝆑½

Small Town Boy The cast does what they can with the brief, hackneyed story of a guy whose personality changes for the worse when he finds a 'fortune' (one thousand pre-inflationary dollars).
1937 61m/B Stuart Erwin, Joyce Compton, Jed Prouty, Clara Blandick, James "Doc" Blakely; *W/ Dir:* Glenn Tryon. **VHS** $19.95 *NOS, LOO* 𝆑𝆑

Small Town Girl Typical romantic musical of the era, as a city slicker picked up for speeding in a hick town is pursued by the sheriff's daughter. Several Busby Berkeley blockbuster musical numbers are shoehorned incongruously into the rural doings.
1953 93m/C Jane Powell, Farley Granger, Bobby Van, Ann Miller, Billie Burke, Robert Keith, S.Z. Sakall, Fay Wray, Nat King Cole, Chill Wills; *Dir:* Leslie Kardos. **VHS, Beta** $19.95 *MGM* 𝆑𝆑½

A Small Town in Texas An ex-con returns home looking for the sheriff who framed him on a drug charge, and who has stolen his woman. Not-too-violent, predictable revenge flick.
1976 (PG) 96m/C Timothy Bottoms, Susan George, Bo Hopkins; *Dir:* Jack Starrett. **VHS, Beta** $69.98 *VES, LIV* 𝆑𝆑

The Smallest Show on Earth A couple inherit not only an old movie house, but the three people who work there as well. Very funny and charming, with a wonderful cast. Sellers is delightful as the soused projectionist. Also titled "Big Time Operators."
1957 80m/B *GB* Bill Travers, Virginia McKenna, Margaret Rutherford, Peter Sellers, Bernard Miles, Leslie Phillips; *Dir:* Basil Dearden. **VHS** $19.95 *NOS, FCT* 𝆑𝆑𝆑

Smart-Aleck Kill An episode of the cable-television-produced Phillip Marlowe series involving the drug-related murder of a Hollywood actor in the 20s.
1985 53m/C Powers Boothe, Michael J. Shannon, Liza Ross; *Dir:* Peter Hunt. **VHS, Beta** $39.98 *FOX* 𝆑

Smart Money A wrongly-jailed man enlists the help of his oddball pals to get revenge on the real thief who committed the computer fraud that put him behind bars. They get even while padding their pockets.
1988 88m/C Spencer Leigh, Alexandra Pigg, Ken Campbell. **VHS** $79.95 *SVS* 𝆑𝆑½

Smash Palace A compelling drama of a marriage jeopardized by his obsession with building a race car and her need for love and affection. Melodramatic, but worth watching. Robson as their young daughter is wonderful.
1982 100m/C *NZ* Bruno Lawrence, Anna Jemison, Greer Robson, Keith Aberdein; *Dir:* Roger Donaldson. **VHS, Beta, LV** $69.98 *LIV, VES* 𝆑𝆑½

Smash-Up: The Story of a Woman A famous nightclub singer gives up her career for marriage and a family, only to become depressed when her husband's career soars. She turns to alcohol and her life falls apart. When her husband sues for divorce and custody of their child, she fights to recover from alcoholism. The original sceenplay by John Howard Lawson was nominated for an Oscar, as was Hayward, in her first major role.
1947 103m/B Susan Hayward, Lee Bowman, Marsha Hunt, Eddie Albert; *Dir:* Stuart Heisler. **VHS, Beta** $19.95 *NOS, PSM, DVT* 𝆑𝆑𝆑

Smile Barbed, merciless send-up of small-town America focusing on a group of naive California girls who compete for the "Young American Miss" crown amid rampant commercialism, exploitation and pure middle-class idiocy. Hilarious neglected '70s-style satire. Early role for Griffith.
1975 (PG) 113m/C Bruce Dern, Barbara Feldon, Michael Kidd, Nicholas Pryor, Geoffrey Lewis, Colleen Camp, Joan Prather, Annette O'Toole, Melanie Griffith, Denise Nickerson; *Dir:* Michael Ritchie. New York Film Festival '75: Best Short Documentary. **VHS, Beta** $19.95 *MGM* 𝆑𝆑𝆑

Smile, Jenny, You're Dead Janssen discovers that close friend's son-in-law has been murdered, and falls in love with the daughter, the main suspect. Pilot for Janssen's "Harry-O" detective series.
1974 100m/C David Janssen, John Anderson, Howard da Silva, Martin Gabel, Clu Gulager, Zalman King, Jodie Foster, Barbara Leigh; *Dir:* Jerry Thorpe. **VHS** $59.95 *UNI* 𝆑𝆑

Smiles of a Summer Night The best known of Bergman's rare comedies; sharp satire about eight Swedish aristocrats who become romantically and comically intertwined over a single weekend. Inspired Sondheim's successful Broadway musical "A Little Night Music," and Woody Allen's "A Midsummer Night's Sex Comedy." Subtitled.
1955 108m/B *SW* Gunnar Bjornstrand, Harriet Andersson, Ulla Jacobsson, Eva Dahlbeck, Jarl Kulle, Margit Carlquist; *Dir:* Ingmar Bergman. **VHS, Beta, LV** $29.95 *SUE, VYG, DVT* 𝆑𝆑𝆑½

Smilin' Through Third filming, second with sound, first in color, of a popular melodrama. An embittered man whose wife was murdered on their wedding day raises an orphaned niece, only to have her fall in love with the son of her aunt's murderer. Made MacDonald into a singing star. Songs include title tune and "A Little Love, a Little Kiss."
1941 101m/C Jeanette MacDonald, Brian Aherne, Gene Raymond, Ian Hunter, Frances Robinson; *Dir:* Frank Borzage. **VHS, Beta** $24.95 *MGM* 𝆑𝆑½

Smiling Madame Beudet/The Seashell and the Clergyman Two famous shorts by Dulac; the second scripted by Antonin Artaud. Both are pioneering examples of surrealistic film. Silent.
1928 62m/B *FR* Alix Allin; *Dir:* Germaine Dulac. **VHS, Beta** $19.95 *FCT* 𝆑𝆑𝆑½

Smith! Naive but well intentioned look at present-day treatment of Native Americans. Rancher Ford becomes embroiled in the trial of murder suspect Ramirez. Based on Paul St. Pierre's novel "Breaking Smith's Quarter Horse."
1969 (G) 101m/C Glenn Ford, Frank Ramirez, Keenan Wynn; *Dir:* Michael O'Herlihy. **VHS, Beta** $69.95 *DIS* 𝆑𝆑½

Smithereens A working-class girl leaves home for New York's music scene. Rugged, hip character study. Director Seidleman's first feature.
1984 (R) 90m/C Susan Berman, Brad Rinn, Richard Hell; *Dir:* Susan Seidelman. **VHS, Beta, LV** $59.95 *MED, AVD, FCT* 𝆑𝆑½

Smoke A young boy nurses a lost German sheperd back to health with the help of his new stepfather, whom he learns to trust. Runs away with the dog when the original owners show up. Vintage live-action Disney starring Opie/Riche (and later successful director) Howard.
1970 89m/C Earl Holliman, Ron Howard. **VHS, Beta** $69.95 *DIS* 𝆑𝆑½

Smoke Screen Cigarette advertising is contrasted with reality.
1988 5m/C **VHS, Beta, 3/4U** *PYR* 𝆑𝆑½

Smoke in the Wind Subpar western about men accused of complicity with the Union in postbellum Arkansas. Minor role for Brennan was his last; also director Kane's last film.
1975 (PG) 93m/C John Ashley, Walter Brennan, John Russell, Myron Healy; *Dir:* Joseph Kane. **VHS, Beta** $49.95 *PSM* 𝆑½

Smokescreen A young ad exec meets the girl of his dreams, and finds himself working for her gangster boyfriend. Competently made and acted, but phoney ending wrecks it.
1990 (R) 91m/C Kim Cattrall, Dean Stockwell, Matt Craven, Kim Coates, Brian George, Michael Hogan; *Dir:* Martin Lavut. **VHS, LV** $89.95 *COL* 𝆑𝆑½

Smokey and the Bandit If you don't know how to end a movie, you call for a car chase. The first and best of the horrible series about bootlegger Reynolds is one long car chase. Reynolds makes a wager that he can have a truck load of Coors beer—once unavailable east of Texas—delivered to Atlanta from Texas in 28 hours. Gleason is the "smokey" who tries to stop him. Field is the hitchhiker Reynolds picks up along the way. Great stunts; director Needham was a top stunt man.
1977 (PG) 96m/C Burt Reynolds, Sally Field, Jackie Gleason, Jerry Reed, Mike Henry, Paul Williams, Pat McCormick; *Dir:* Hal Needham. **VHS, Beta, LV** $19.95 *MCA* 𝆑½

Smokey and the Bandit, Part 2 This pathetic sequel to "Smokey and the Bandit" grossed $40 million. The Bandit is hired to transport a pregnant elephant from Miami to the Republican convention in Dallas. Sheriff Buford T. Justice and family are in hot pursuit.
1980 (PG) 101m/C Burt Reynolds, Sally Field, Jackie Gleason, Jerry Reed, Dom DeLuise, Pat McCormick, Paul Williams; *Dir:* Hal Needham. **VHS, Beta, LV** $19.95 *MCA* 𝆑

Smokey and the Bandit, Part 3 You thought the second one was bad? Another mega car chase, this time sans Reynolds and director Needham.

1983 (PG) 88m/C Burt Reynolds, Jackie Gleason, Jerry Reed, Paul Williams, Pat McCormick, Mike Henry, Colleen Camp; *Dir:* Dick Lowry. **VHS, Beta, LV** $59.95 *MCA* 🐾

Smokey Bites the Dust Car-smashing gag-fest about a sheriff's daughter kidnapped by her smitten beau. Near-plotless and literally unoriginal: lifted footage from several other Corman-produced flicks, a technique that can aptly be called garbage picking.
1981 (PG) 87m/C Janet Julian, Jimmy McNichol, Patrick Campbell, Kari Lizer, John Blythe Barrymore; *Dir:* Charles B. Griffith. **VHS, Beta** $19.98 *CHA* 🐾

Smokey & the Hotwire Gang A convoy of truckers try to track down a beautiful woman driving a stolen car.
1979 85m/C James Keach, Stanley Livingston, Tony Lorea, Alvy Moore, George Barris. **VHS, Beta** $49.95 *WES* 🐾½

Smokey & the Judge A police officer has his hands full with a trio of lovely ladies.
1980 (PG) 90m/C Gene Price, Wayde Preston, Juanita Curiel. **VHS, Beta** *SUN* 🐾

Smokey Smith Ubiquitous Hayes and steely Steele hunt down the slayers of an elderly couple. Ordinary oater directed by Steele's father, Bradbury.
1936 59m/B Bob Steele; *Dir:* Robert N. Bradbury. **VHS, Beta** $27.95 *VCN, WGE* 🐾½

A Smoky Mountain Christmas Dolly gets away from it all in a secluded cabin that has been appropriated by a gang of orphans. Innocuous seasonal country fun. Dolly sings a half dozen songs and co-wrote the script. Winkler's TV directing debut.
1986 94m/C Dolly Parton, Bo Hopkins, Dan Hedaya, Gennie James, David Ackroyd, Rene Auberjonois, John Ritter, Anita Morris, Lee Majors; *Dir:* Henry Winkler. **VHS, Beta** $79.98 *FOX* 🐾🐾

Smooth Talk An innocent, flirtatious teenager catches the eye of a shady character, played by Williams. Disturbing and thought-provoking film that caused some controversy when it opened. Dern gives a brilliant performance as the shy, sheltered girl. Based on the Joyce Carol Oates story of the same name.
1985 (PG-13) 92m/C Laura Dern, Treat Williams, Mary Kay Place, Levon Helm; *Dir:* Joyce Chopra. **VHS, Beta, LV** $29.98 *LIV, VES* 🐾🐾🐾

Smooth Talker A serial killer stalks the women of a 976 party line. Bad acting, bad dialog, not a scare to be found.
1990 (R) 89m/C Joe Guzaldo, Peter Crombie, Stuart Whitman, Burt Ward, Sydney Lassick, Blair Weickgenant; *Dir:* Tom Milo. **VHS, Beta** $89.95 *ACA* Woof!

Smothering Dreams A video piece expressing the horrors of war with a succession of suggestive images.
1983 30m/C Dan Reeves. **VHS, Beta** $59.95 *SVA, FCT* Woof!

Smouldering Fires A tough businesswoman falls in love with an ambitious young employee, who is fifteen years her junior. After they marry, problems arise in the form of the wife's attractive younger sister. Surprisingly subtle melodrama, if that's not an oxymoron. Silent.
1925 100m/B Pauline Frederick, Laura La Plante, Tully Marshall; *Dir:* Clarence Brown. **VHS, Beta** $29.95 *VYY, DNB* 🐾🐾½

Smugglers Opportunistic smugglers take advantage of the Russian Revolution to sack the land and make a bundle. Original title: "Lover of the Great Bear."
1975 110m/C *IT* Senta Berger, Giuliano Gemma; *Dir:* Valentino Orsini. **VHS, Beta** $59.95 *VCD* 🐾

Smugglers' Cove Gorcey wrongly believes he has inherited a mansion. He and the Bowery Boys move in, only to stumble across a smuggling ring. Plenty of slapstick; the boys at their best.
1948 66m/B Leo Gorcey, Huntz Hall, Gabriel Dell; *Dir:* William Beaudine. **VHS** $14.95 *VID* 🐾🐾🐾

Smuggler's Cove A young boy doggedly pursues a mysterious trawler captain despite warnings from his parents.
1983 75m/C *AU* **VHS, Beta** $14.98 *VID* 🐾½

Snake and Crane Arts of Shaolin There's plenty of action and battles going on in this martial arts film.
1980 93m/C **VHS, Beta** $24.95 *ASE* 🐾

Snake Dancer A softcore epic concerning the gyrations of a particular dancer.
19?? 87m/C Glenda Kemp, Peter Elliott, Christine Basson; *Dir:* Dirk DeVilliers. **VHS** $29.95 *ACA* 🐾

Snake in the Eagle's Shadow 2 Three champions of self-defense mount a major offensive and win battle after battle.
1983 89m/C Wang Tao, Cheng Shing, Lung Fei; *Dir:* Chang Shinn I. **VHS, Beta** $39.95 *TWE* 🐾

Snake Fist of the Buddhist Dragon In medieval China, a peaceful Buddhist goes mad and hits everyone around him in the head.
19?? 88m/C Martial Lee, Lee King, Debbie Lam **VHS, Beta** $49.95 *REG* 🐾

Snake Fist Dynamo A kick-throwin' fight takes place over some valuable jewels.
1984 86m/C Eric Yee, Nancy Leung, Dick Lee. **VHS, Beta** $49.95 *REG* 🐾

Snake Fist Fighter In one of the highest-grossing martial arts movies of all time, Jackie Chan uses the deadly "snake fist" technique against the bad guys.
1981 (R) 83m/C Jackie Chan. **VHS, Beta** *PLV* 🐾½

The Snake Hunter Strangler A young girl is rescued after fifteen years from an evil cult of snake people.
1966 65m/C Guy Madison, Ivan Desny; *Dir:* Luigi Capuano. **VHS, Beta** *LTG* 🐾

Snake in the Monkey's Shadow A martial arts expert demonstrates the deadliest fighting techniques.
1982 (R) 85m/C **VHS, Beta** $39.95 *TWE* 🐾

The Snake People A police captain investigates a small island littered with LSD-experimenting scientists, snake-worshippers, and voodoo. Perhaps you've heard of this as "Isle of the Snake People," "Cult of the Dead" or "La Muerte Viviente." One of the infamous quartet of Karloff's final films, all made in Mexico.
1968 90m/C *MX* Boris Karloff, Julissa, Carlos East; *Dir:* Enrique Vergara. **VHS, Beta** $49.95 *SNC, MPI, UNI* 🐾

The Snake Woman A herpetologist working in a small English village during the 1890s is conducting strange experiments to try to cure his wife's madness. He tries injecting his pregnant wife with snake venom and she gives birth to a cold-blooded daughter, who, when she grows up, has the ability to change herself into a deadly snake. She promptly begins killing the local male populace until Scotland Yard is called in to investigate. The curvy Travers is appropriately snakey but this movie is dull.
1961 68m/C *GB* John P. McCarthy, Susan Travers, Arnold Marle; *Dir:* Sidney J. Furie. **VHS** *CNM* 🐾

SnakeEater A television heart throb takes to the silver screen while trying to track down the swine who killed his parents and kidnapped his sister.
1989 (R) 89m/C Lorenzo Lamas, Larry Csonka, Ron Palillo; *Dir:* George Erschbamer. **VHS, Beta** $89.95 *MED* 🐾

SnakeEater 3...His Law Tough cop Jack Kelly (Lamas), a.k.a. SnakeEater, is out to avenge a woman's beating by a gang of thug bikers. Lots of action and violence; based on W. Glenn Duncan's novel "Rafferty's Rules."
1992 (R) 109m/C Lorenzo Lamas, Minor Mustain, Tracy Cook, Holly Chester, Scott "Bam Bam" Bigelow; *Dir:* George Erschbamer. **VHS, Beta** *PAR* 🐾🐾

Snapshot Typical grade 2 sex comedy; a free-lance photographer ogles lots of variously dishabilled young women through his lens.
1977 (R) 84m/C Jim Henshaw, Susan Petrie. **VHS, Beta** $19.95 *UHV* 🐾

Snatched The wives of three wealthy men are held for ransom, but one man doesn't want to pay, while one of the victims needs insulin. Standard made for television fare.
1972 73m/C Howard Duff, Leslie Nielsen, Sheree North, Barbara Parkins, Robert Reed, John Saxon, Tisha Sterling; *Dir:* Sutton Roley. **VHS** $79.95 *REP* 🐾

Sneakers A pair of old, worn-out sneakers is joined by pairs of moccasins, boots, and sandals for a merry dance around the living room floor. The program demonstrates that by means of single-frame shooting, an animated film can be made using everyday objects.
1978 5m/C **VHS, Beta** *BFA* 🐾🐾

Sneakin' and Peekin' Three short documentaries by the irreverent cinema-verite documentarian: "I Was a Teenage Contestant at Mother's Wet T-Shirt Contest," "Hot and Nasty" and the title film, all of which deal with the comically seamy side of American life.
1976 60m/C *Dir:* Tom Palazzolo. **VHS, Beta** $39.95 *FCT* 🐾🐾

Sniper Suspenseful television drama based on true story of Charles Whitman, who shot at University of Texas students from Texas Tower on a summer day in 1966. Disney alumnus Russell breaks type as mass killer, with fine support by Yniguez as a police officer on the scene and Beatty as the passerby who lends hand. Finely crafted recreation of disturbing, true event. Originally broadcast as "The Deadly Tower."
1975 85m/C Kurt Russell, Richard Yniguez, John Forsythe, Ned Beatty, Pernell Roberts, Clifton James, Paul Carr, Alan Vint, Pepe Serna; *Dir:* Jerry Jameson. **VHS** $39.95 *XVC* 🐾🐾🐾

The Sniper A law-abiding citizen becomes a vigilante killer after witnessing a violent crime.
1987 90m/C Claudia Cardinale. **VHS, Beta** $59.95 *MGL* 🐾½

Sno-Line A New York gangster moves to Texas and begins to wipe out the competition on his way to building his drug and gambling dynasty. Bo-ring excuse for entertainment. No hope, no redemption, no titillation: just ultra- "realistic" depiction of organized crime, poorly acted.
1985 (R) 89m/C Vince Edwards, Paul Smith, June Wilkinson; *Dir:* Douglas F. Oneans. **VHS, Beta $69.98** *LIV, LTG* 🐾

Snoopy, Come Home Snoopy leaves Charlie Brown to visit his former owner Lila in the hospital and returns with her to her apartment house. From Charles Schultz's popular comic strip "Peanuts."
1972 (G) 80m/C *Dir:* Bill Melendez. **VHS, Beta, LV $14.98** *FOX* 🐾🐾🐾

Snoopy's Getting Married, Charlie Brown An animated Peanuts adventure wherein Snoopy prepares to get married.
1985 30m/C VHS, Beta $14.95 *KRT* 🐾🐾🐾

Snoopy's Getting Married, Charlie Brown/You're In Love, Charlie Brown Two episodes of the famed Peanuts gang.
1985 55m/C VHS $14.95 *KRT* 🐾🐾🐾

Snow Country Romantic drama of a unsuccessful, married artist from Tokyo who travels to a mountain resort to find himself. He requests a Geisha for the evening and, when none are available, the beautiful adopted daughter of the local music teacher is sent to him instead. Based on the novel by Nobel Prize winner Yasunari Kawabata. In Japanese with English subtitles. Remade in 1965.
1957 104m/B *JP* Ryo Ikebe, Keiko Kishi, Akira Kubo; *Dir:* Shiro Toyoda. **VHS, Beta $29.95** *CVC* 🐾🐾½

The Snow Creature A troop of explorers bring back a snow creature from the Himalayas; it escapes in the United States and terrorizes all in its path. Horribly bad monster flick (the first about a snow monster), by Billy Wilder's brother. Once in a while, the monster suit doesn't look fake. And wouldn't the creature either melt, or at least collapse from heat exhaustion, with all that running around L.A.?
1954 72m/B Paul Langton, Leslie Denison; *Dir:* W. Lee Wilder. **VHS, Beta $19.95** *NOS, SNC, UHV* Woof!

Snow Kill Five city-dwellers travel into the mountains for a fun survival training weekend, but the fun turns deadly when they accidentally come across a murderous drug gang. Pseudo - '30s suspense adventure fails to thrill or surprise. Made for cable.
1990 90m/C Terence Knox, Patti D'Arbanville, David Dukes; *Dir:* Thomas J. Wright. **VHS, Beta $29.95** *PAR, TTC* 🐾½

Snow Queen Music by Glazunov, Mussorgsky and Rimsky-Korsakov combines with Jean Pierre Bonnefoux's ice choreography in this setting of the Hans Christian Andersen tale.
1982 85m/C Janet Lynn, John Curry, Sandra Bezic, Jo Jo Starbuck, Dorothy Hamill, Toller Cranston; *Dir:* John Thompson. **VHS, Beta $19.98** *BVV* 🐾🐾

Snow Queen From "Fairie Tale Theatre" comes this adaptation of the Hans Christian Andersen tale. A boy and girl who grow up together are separated by evil when the boy is held captive in the icy palace of the Snow Queen. The girl sets out to rescue him.

1983 60m/C Lauren Hutton, Linda Manz, David Hemmings, Melissa Gilbert, Lee Remick, Lance Kerwin; *Dir:* Peter Medak. **VHS, Beta $14.98** *KUI, FOX, FCT* 🐾🐾

Snow Queen An animated version of the Hans Christian Andersen story about a young girl who battles the evil Snow Queen for possession of her young boy friend.
1990 60m/C VHS $9.95 *AIP* 🐾🐾

Snow: The Movie Skiing, growing up, romance, and good times are what it's all about in this film from "down under."
1983 83m/C David Argue, Lance Curtis; *Dir:* Robert Gibson. **VHS, Beta $59.95** *VCD* 🐾

Snow Treasure With the help of an underground agent, Norwegian children smuggle gold out of the country right under the noses of the Nazis. Based on the novel by Marie McSwigan.
1967 (G) 96m/C James Franciscus, Paul Anstad; *Dir:* Irving Jacoby. **VHS, Beta $29.95** *IVE* 🐾

Snow White and the Seven Dwarfs From cable television's "Faerie Tale Theatre" comes the story of a princess who befriends seven little men to protect her from the jealous evil queen.
1983 60m/C Elizabeth McGovern, Rex Smith, Vincent Price, Vanessa Redgrave; *Dir:* Peter Medak. **VHS, Beta, LV $14.98** *KUI, FOX, FCT* 🐾

Snow White and the Three Stooges The Stooges fill in for the Seven Dwarfs when they go off prospecting in King Solomon's mines. Alas, see any Stooge movie but this one.
1961 107m/C Moe Howard, Curly Howard, Larry Fine, Carol Heiss, Patricia Medina; *Dir:* Walter Lang. **VHS, Beta $14.98** *FOX, FCT* 🐾

Snowball Express When a New York City accountant inherits a hotel in the Rocky Mountains, he decides to move his family west to attempt to make a go of the defunct ski resort, only to find that the place is falling apart. Run-of-the-mill Disney comedy, based on the novel "Chateau Bon Vivant" by Frankie and John O'Rear.
1972 (G) 120m/C Dean Jones, Nancy Olson, Harry Morgan, Keenan Wynn; *Dir:* Norman Tokar. **VHS, Beta $69.95** *DIS* 🐾🐾

Snowballin' A ski instructor gives private lessons to a bunch of snow bunnies.
1985 90m/C Seline Lomez, Daniel Pilon. **VHS, Beta $59.95** *JEF* 🐾

Snowballing Part of the polluted wave of teen-sex flicks of the 1980s, this is very mild for the genre but still no prize. Lusty high schoolers at a skiing competition look for action on and off the slopes, end up exposing a resort fraud.
1985 (PG) 96m/C Mary McDonough, Bob Hastings. **VHS, Beta $79.95** *PSM* 🐾

Snowbeast The residents of a ski resort are being terrorized by a half-human, half-animal beast leaving a path of dead bodies in its wake. "Jaws" hits the slopes. Not scary or even funny. Made for TV.
1977 96m/C Bo Svenson, Yvette Mimieux, Sylvia Sidney, Clint Walker, Robert F. Logan; *Dir:* Herb Wallerstein. **VHS, Beta $9.95** *WOV* 🐾½

Snowblind Two ski gondolas derail, placing the passengers in jeopardy. Among the passengers is a mobster being pursued by an assassin. Originally titled "Ski Lift to Death."

1978 98m/C Deborah Raffin, Charles Frank, Howard Duff, Don Galloway, Gail Strickland, Don Johnson, Veronica Hamel, Clu Gulager, Lisa Reeves, Suzy Chaffee; *Dir:* William Wiard. **VHS $9.95** *SIM* 🐾🐾

The Snows of Kilimanjaro Called by Hemingway "The Snows of Zanuck," in reference to the great producer, this film is actually an artful pastiche of several Hemingway short stories and novels. The main story, "The Snows of Kilimanjaro," acts as a framing device, in which the life of a successful writer is seen through his fevered flashbacks as he and his rich wife, while on safari, await a doctor to save his gangrenous leg. Superbly scripted by Casey Robinson.
1952 117m/C Gregory Peck, Susan Hayward, Ava Gardner, Hildegarde Neff, Leo G. Carroll, Torin Thatcher, Ava Norring, Helene Stanley, Marcel Dalio, Vincente Gomez, Richard Allen, Leonard Carey; *Dir:* Henry King. **Beta $7.50** *NOS, GKK, BUR* 🐾🐾🐾

Snuff-Bottle Connection A Russian envoy is spying for the government while delivering a gift to a Ching dynasty emperor.
1982 96m/C Hwang Jang Lee, John Liu, Roy Horan. **VHS, Beta** *WVE* 🐾

So Dear to My Heart A farm boy and his misfit black sheep wreak havoc at the county fair. Several sequences combine live action with animation. Heartwarming and charming; straightforward and likeable but never sentimental. Wonderful, vintage Disney. Features the hit song, "Lavender Blue (Dilly, Dilly)."
1949 82m/C Bobby Driscoll, Burl Ives, Beulah Bondi, Harry Carey, Luanna Patten; *Dir:* Harold Schuster. **VHS, Beta $69.95** *DIS* 🐾🐾🐾½

So Ends Our Night A German who scorns Nazi ideology flees Austria and meets a young Jewish couple seeking asylum. Fine adaptation of Erich Maria Remarque's novel "Flotsam," with splendid performances from Sullavan (on loan from Universal) and young Ford.
1941 117m/B Fredric March, Margaret Sullavan, Frances Dee, Glenn Ford, Anna Sten, Erich von Stroheim; *Dir:* John Cromwell. **VHS** *IVY* 🐾🐾🐾

So Fine Absent-minded English professor O'Neal tries to rescue his father's clothing business from going bottom-up. He accidentally invents peek-a-boo bottomed jeans, which become an immediate hit and make him rich. Comedy smorgasbord, setting out from sometime novelist Bergman, hits and misses. O'Neal is memorable, as is the ubiquitous Warden and his sidekick.
1981 (R) 91m/C Ryan O'Neal, Jack Warden, Mariangela Melato, Richard Kiel; *Dir:* Andrew Bergman. **VHS, Beta $19.98** *WAR* 🐾🐾

So This is Paris Roguish pre-Hayes Code dancing duo seeks spice through alternative lovemates. Classic sophisticated Lubitsch, with outstanding camerawork and a bit of jazz.
1926 68m/B Monte Blue, Patsy Ruth Miller, Lilyan Tashman, Andre de Beranger, Myrna Loy; *Dir:* Ernst Lubitsch. **VHS, Beta $29.95** *GPV, FCT* 🐾🐾🐾

So This is Washington The comedy team go to Washington with wacky inventions to help the war effort. Turns out there's too entirely too much nonsense going around, and the boys give 'em a piece of their mind. Fun, featherweight wartime comedy.
1942 70m/B Lum & Abner, Alan Mowbray. **VHS, Beta $24.95** *NOS, DVT* 🐾🐾½

Soap Opera Scandals Heroes and heroines of television's daytime soap operas are seen in outtakes as lines are blown, props fall, and the dialogue turns risque. **197? 50m/C** Ruth Warrick. **VHS, Beta $19.95** NO ⅋⅋½

Soapdish The back stage lives of a daytime soap opera and its cast. When the soap's ratings fall, a character written out of series via decapitation is brought back to give things a lift. While the writer struggles to make the reincarnation believable, the cast juggles old and new romances, and professional jealousies abound. Some really funny moments as film actors spoof the genre that gave many of them a start. **1991 (PG-13) 96m/C** Sally Field, Kevin Kline, Cathy Moriarty, Robert Downey Jr., Whoopi Goldberg; **Dir:** Michael Hoffman. **VHS, Beta, LV, 8mm** PAR, PIA ⅋⅋⅋

S.O.B. Blake Edwards' bitter farce about Hollywood and the film industry wheelers and dealers who inhabit it. When a multi-million dollar picture bombs at the box office, the director turns suicidal, until he envisions re-shooting it with a steamy, X-rated scene starring his wife, a star with a goody-two-shoes image. Edwards used his real-life wife, Julie ("Mary Poppins") Andrews, for the scene in which she bared her breasts. Oft-inspired, but oft-terrible. Zestfully vengeful. William Holden's last film. **1981 (R) 121m/C** William Holden, Robert Preston, Richard Mulligan, Julie Andrews, Robert Webber, Shelley Winters, Robert Vaughn, Larry Hagman, Stuart Margolin, Loretta Swit, Craig Stevens, Larry Storch, Jennifer Edwards, Robert Loggia, Rosanna Arquette, Marisa Berenson; **Dir:** Blake Edwards. **VHS, Beta, LV $19.98** FOX, WAR ⅋⅋½

Social Error An heiress mistakes her guards for a group of kidnappers. **1935 60m/B** Gertrude Messinger, David Sharpe. **VHS, Beta, LV $19.95** NOS, WGE ⅋⅋

The Social Secretary Driven from the rat race by lecherous men, Talmadge gets new job as rich woman's secretary, where she down-dresses to avoid further lewd encounters. She falls for her boss's son, but isn't considered marriage material until she proves herself. **1916 56m/B** Norma Talmadge, Kate Lester, Helen Weir, Gladden James, Herbert Frank; **Dir:** John Emerson. **VHS, Beta $19.95** GPV, FCT ⅋⅋½

Society A teenager wonders if his visions are real or hallucinations when he believes his family, and everyone else around him, are flesh-eating predators. The puzzle isn't much but wait for the special-effects ladened ending to get your fill of gore. **1992 (R) 99m/C** Billy Warlock, Devin Devasquez, Evan Richards; **Dir:** Brian Yuzna. **VHS $89.98** REP Woof!

Sodom and Gomorrah The Italian-made, internationally produced epic about Lot, the Hebrews and the destruction of the two sinful biblical cities. Moderately entertaining - but ponderous, to say the least, and very long. **1962 154m/C** IT Stewart Granger, Stanley Baker, Pier Angeli, Anouk Aimee, Rossana Podesta; **Dir:** Robert Aldrich. **VHS, Beta, LV $19.98** FOX, IME ⅋⅋

Soft and Hard Made for English television, this video by Godard and Mieville intends to address the role of media and television in everyday life via a seemingly informal look at the couple's daily existence

and the propensity for image production in it. In French with English subtitles. **1985 48m/C** FR **Dir:** Anne-Marie Mieville, Jean-Luc Godard. **VHS, Beta** EAI ⅋⅋½

The Soft Skin A classic portrayal of marital infidelity by the master director. A writer and lecturer has an affair with a stewardess. After the affair ends, his wife confronts him, with tragic results. Cliche plot is forgivable; acted and directed to perfection. "400 Blows" star Jean-Pierre Leaud served here as an apprentice director. In French with English subtitles. **1964 118m/B** FR Jean Desailly, Nelly Benedetti, Francoise Dorleac; **Dir:** Francois Truffaut. **VHS, Beta $39.95** HMV, FOX ⅋⅋⅋

Soggy Bottom U.S.A. A sheriff has his hands full trying to keep the law enforced in a small Southern town. Quite-good cast keeps the plot from disappearing altogether. **1984 (PG) 90m/C** Don Johnson, Ben Johnson, Dub Taylor, Ann Wedgeworth, Lois Nettleton, Anthony Zerbe; **Dir:** Theodore J. Flicker. **VHS, Beta $9.99** STE, PSM, HHE ⅋⅋

Sois Belle et Tais-Toi A police detective tracks jewel thieves and falls in love with a young waif. Several interesting and later famous actors have small parts. Light-hearted adventure. Subtitled in English. **1958 110m/B** FR Henri Vidal, Mylene Demongeot, Jean-Paul Belmondo, Alain Delon; **Dir:** Marc Allegret. **VHS, Beta $19.95** KRT ⅋⅋½

Solarbabies Rollerskating youths in a drought-stricken future vie for a mysterious power that will replenish the Earth's water. Shades of every sci-fi movie you've ever seen, from "Mad Max" to "Ice Pirates." Pathetic. **1986 (PG-13) 95m/C** Richard Jordan, Sarah Douglas, Charles Durning, Lukas Haas, Jami Gertz, Jason Patric; **Dir:** Alan Johnson. **VHS, Beta, LV $34.95** MGM ⅋

Solaris With this the USSR tried to eclipse "2001: A Space Odyssey" in terms of cerebral science-fiction. Some critics thought they succeeded. You may disagree now that the lumbering effort is available on tape. Adapted from a Stanislaw Lem novel, it depicts a dilapidated space lab orbiting the planet Solaris, whose ocean, a vast fluid 'brain,' materializes the stir-crazy cosmonauts' obsessions—usually morose ex-girl-friends. Talk, talk, talk (in Russian with English subtitles), minimal special effects. In a two-cassette package, with a letterbox format preserving Tarkovsky's widescreen compositions. **1972 167m/C** Donatas Banionis, Natalya Bondarchuk; **Dir:** Andrei Tarkovsky. **VHS $79.95** FXL, FCT ⅋⅋

Solarman A teenager is transformed into a powerful superhero who fights outerspace bad guys. Includes an interview with Marvel Comics' Stan Lee and a short feature on a cartoon artist. **1990 38m/C** VHS **$14.95** CAF, VTK ⅋⅋

Soldat Duroc...Ca Va Etre Ta Fete! Days after the 1944 liberation of Paris, a French soldier crosses back over the French lines to see his lover, and is caught as a German sympathizer by American troops. English title: "Soldier Duroc." **196? 105m/C** FR Pierre Tornade, Robert Webber, Michel Galabru, Roger Carel; **Dir:** INV ⅋⅋

The Soldier The Russians are holding the world at ransom with a pile of stolen plutonium, and a soldier finds himself in the position to carry out an unauthorized and dangerous plan to preserve the balance of world power. Wildly implausible, gratuitously violent spy trash. **1982 (R) 90m/C** Ken Wahl, Klaus Kinski; **Dir:** James Glickenhaus. **VHS, Beta $12.98** SUE ⅋

Soldier Blue Two survivors of an Indian attack make their way back to an army outpost. The cavalry then seeks revenge on the Cheyenne tribe accused of the attack. Gratuitously violent Vietnam-era western hits hard on racial themes. Based on the novel "Arrow in the Sun" by Theodore V. Olsen on the Sand Creek Indian massacre. **1970 (R) 109m/C** Candice Bergen, Peter Strauss, Donald Pleasence, Dana Elcar, Jorge Rivero; **Dir:** Ralph Nelson. **VHS, Beta $14.95** COL, SUE ⅋⅋

Soldier of Fortune A woman (Hayward) enlists mercenaries to help find her lost husband in Red China. Late Gable vehicle with the formula beginning to feel the post-war strain and the star looking a trifle long in the tooth. Still, a fun adventure from a top star. Hayward had been among the myriad starlets vying for the Scarlett O'Hara roles nearly two decades earlier. **1955 96m/C** Clark Gable, Susan Hayward, Gene Barry, Alexander D'Arcy, Michael Rennie, Tom Tully, Anna Sten, Russell Collins, Leo Gordon, Jack Kruschen, Robert Quarry; **Dir:** Edward Dmytryk. **VHS, Beta $39.98** FOX ⅋⅋

Soldier in Love Teleplay from "George Schaefer's Showcase Theatre" is based on the life of John Churchill, first Duke of Marlborough, and his wife Sarah. The ambitious young couple held the favor of Queen Anne for a time, until political intrigue brought about their downfall. **1967 76m/C** Jean Simmons, Claire Bloom, Keith Michell; **Dir:** George Schaefer. **VHS, Beta** FHS ⅋⅋

Soldier of the Night Toy-store worker by day, vigilante killer by night. Israeli thriller set in Tel Aviv lacks vim and is badly dubbed. **1984 89m/C Dir:** Dan Wolman. **VHS, Beta $59.95** MGM ⅋½

Soldier of Orange The lives of six Dutch students are forever changed by the World War II invasion of Holland by the Nazis. Based on the true-life exploits of Dutch resistance leader Erik Hazelhoff. Exciting and suspenseful; cerebral; carefully made and well acted. Made Rutger Hauer an international star. **1978 144m/C** NL Rutger Hauer, Jeroen Krabbe, Edward Fox, Susan Penhaligon; **Dir:** Paul Verhoeven. Los Angeles Film Critics Association Awards '79: Best Foreign Film. **VHS, Beta, LV $59.95** MED ⅋⅋

Soldier in the Rain An unusual friendship develops between career sergeant Gleason and wheeler-dealer McQueen. Gleason is in good form, but McQueen is listless, and the story (set in a Southern army camp) is unsatisfying. Weld is the comely teen who makes for an intriguing love triangle. **1963 88m/B** Steve McQueen, Jackie Gleason, Tuesday Weld, Tony Bill, Tom Poston, Ed Nelson. **VHS, Beta $59.98** FOX ⅋⅋½

Soldier's Fortune A travelling gun for hire leaves the combat of Central America for the jungles of Los Angeles, only to find out that his daughter had been taken hostage. That's the last thing these kidnappers will

ever do when they are found by this tough guy looking for his child.
1991 (R) 96m/C Gil Gerard, Charles Napier, Dan Haggerty, P.J. Soles, Barbara Bingham, Janus Blythe; *Dir:* Arthur N. Mele. VHS **$89.98** *REP* ✂✂½

Soldier's Home Ernest Hemingway's story of a soldier who returns from World War I and finds he can no longer fit into the life he left, but expertly glumly adapted, and introduced by Henry Fonda. From the "American Short Story" series.
1977 42m/C Richard Backus, Nancy Marchand, Robert McIlwaine, Lisa Essary, Mark LaMura. VHS, Beta **$24.95** *MON, MTI, KAR* ✂✂✂

Soldier's Revenge A Vietnam vet has a hard time adjusting to life at home.
1977? 92m/C John Savage. VHS, Beta **$19.98** *TWE* ✂✂

A Soldier's Story A black army attorney is sent to a Southern base to investigate the murder of an unpopular sergeant. Features World War II, Louisiana, jazz and blues, and racism in and outside the corps. From the Pulitzer-prize winning play by Charles Fuller, with most of the Broadway cast. Fine performances from Washington and Caesar.
1984 (PG) 101m/C Howard E. Rollins Jr., Adolph Caesar, Denzel Washington, Patti LaBelle, Robert Townsend, Scott Paulin, Wings Hauser, Art Evans, Larry Riley, David Alan Grier; *Dir:* Norman Jewison. Edgar Allan Poe Awards '84: Best Screenplay. VHS, Beta, LV **$14.95** *COL* ✂✂✂

Soldier's Tale Igor Stravinsky's "The Soldier's Tale" is animated by New Yorker cartoonist R.O. Bleechman with the voices provided by Max Von Sydow and Andre Gregory.
1984 60m/C *Voices:* Max von Sydow, Andre Gregory. VHS, Beta, LV **$39.95** *MGM* ✂✂✂

A Soldier's Tale During WWII, a menage-a-trois develops that can only lead to tragedy, yet the participants find themselves unable to resist.
1991 (R) 96m/C Gabriel Byrne, Marianne Basler, Judge Reinhold, Paul Wyett; *Dir:* Larry Parr. VHS **$89.98** *REP* ✂✂

Sole Survivor A group of zombies are searching for a beautiful advertising executive who was the sole survivor of a plane crash.
1984 (R) 85m/C Anita Skinner, Kurt Johnson, Caren Larkey; *Dir:* Thom Eberhardt. VHS, Beta **$79.98** *LIV, VES* ✂✂½

The Solitary Man Glum look at a family breakup, from the husband's point of view. Holliman is good, but the exercise at large is so regularly blues-inducing, and not especially inspiring or artful. Made for TV.
1982 96m/C Earl Holliman, Carrie Snodgress, Lara Parker, Lane Smith, Nicholas Coster, Michelle Pfeiffer; *Dir:* John Llewellyn Moxey. VHS, Beta *TLF* ✂

Solo Young hitchhiker Peers enters the lives of fire patrol pilot Gil and his teenage son. A plotless friendship develops, headed nowhere—like this movie. Great New Zealand scenery might be worth the price of rental.
1977 (PG) 90m/C Martyn Sanderson, Lisa Peers; *Dir:* Tony Williams. VHS, Beta **$69.98** *DAY, VES, HHE* ✂✂

Solo Voyage: The Revenge A Russian "Rambo" goes on a murderous rampage, the likes of which humanity has not yet seen.

1990 91m/C *RU* Mikhail Nojkine, Alexandre Fatiouchine; *Dir:* Mikhail Toumanichulli. VHS **$69.95** *WRD* ✂✂

Solomon and Sheba King Solomon's brother and the Egyptian Pharaoh send the Queen of Sheba to Israel to seduce King Solomon so they may gain his throne. Tyrone Power had filmed most of the lead role in this silly, overwrought epic when he died of a heart attack. The role was reshot, with Brynner (with a full head of hair) replacing him. Director Vidor's unfortunate last film. Shot on location in Spain.
1959 139m/C Yul Brynner, Gina Lollobrigida, Marisa Pavan, George Sanders, Alejandro Rey; *Dir:* King Vidor. VHS, Beta **$59.98** *FOX* ✂✂

The Sombrero Kid The Kid unwittingly gets involved in a murder and is nearly hanged before his friends help him discover the real culprits. Plenty of action in this straight forward, fun western.
1942 54m/B Donald (Don "Red") Barry, Lynn Merrick, Rand Brooks; *Dir:* George Sherman. VHS, Beta **$19.95** *VDM* ✂✂

Some Call It Loving Bad, pretentious modern version of Sleeping Beauty, set in L.A. King buys the sleeping girl (Farrow) from a carnival show, but she doesn't live up to his ideal.
1973 (R) 103m/C Zalman King, Carol White, Tisa Farrow, Richard Pryor, Pat Priest; *Dir:* James B. Harris. VHS **$39.95** *IVE, MON* ✂

Some Came Running James Jones's follow-up novel to "From Here to Eternity" does not translate nearly as well to the screen. Overlong and with little plot, the action centers around a would-be writer, his floozy girl friend, and the holier-than-thou characters which populate the town in which he grew up and to which he has now returned. Strong performances by all. Oscar nominations for MacLaine, Kennedy (his 5th nomination), Hyer, and Costume Design.
1958 136m/C Frank Sinatra, Dean Martin, Shirley MacLaine, Martha Hyer, Arthur Kennedy, Nancy Gates; *Dir:* Vincente Minnelli. VHS, Beta, LV **$19.98** *MGM* ✂✂✂

Some Girls A man goes to Quebec to see his college girlfriend who informs him that she is not in love with him anymore. But she has two sisters ready to comfort him! Strange black comedy; Gregory as the girl's father elevates so-so story.
1988 (R) 104m/C Patrick Dempsey, Andre Gregory, Lila Kedrova, Florinda Bolkan, Jennifer Connelly; *Dir:* Michael Hoffman. VHS, Beta **$89.95** *MGM* ✂✂½

Some Kind of Hero A Vietnam prisoner of war returns home to a changed world. Pryor tries hard, but can't get above this poorly written, unevenly directed film.
1982 (R) 97m/C Richard Pryor, Margot Kidder, Ray Sharkey, Ronny Cox, Lynne Moody, Olivia Cole; *Dir:* Michael Pressman. VHS, Beta, LV **$14.95** *PAR* ✂✂

Some Kind of Wonderful A high-school tomboy has a crush on a guy who also happens to be her best friend. Her feelings go unrequited as he falls for a rich girl with snobbish friends. In the end, true love wins out. Deutch also directed (and John Hughes also produced) the teen flick "Pretty in Pink," which had much the same plot, with the rich/outcast characters reversed by gender. OK, but completely predictable.

1987 (PG-13) 93m/C Eric Stoltz, Lea Thompson, Mary Stuart Masterson, Craig Sheffer, John Ashton, Elias Koteas, Molly Hagan; *Dir:* Howard Deutch. VHS, LV **$19.95** *PAR* ✂✂

Some Like It Hot Two unemployed musicians witness the St. Valentine's Day massacre in Chicago. They disguise themselves as women and join an all-girl band headed for Miami to escape the gangsters' retaliation. Flawless cast includes a fetching Monroe at her best; hilarious script. Curtis does his Cary Grant impression. Classic scenes between Lemmon in drag and Joe E. Brown as a smitten suitor. Brown also has the film's famous closing punchline. Monroe sings "I Wanna Be Loved By You," "Running Wild," and "I'm Through With Love." One of the very funniest movies of all time.
1959 120m/B Marilyn Monroe, Tony Curtis, Jack Lemmon, George Raft, Pat O'Brien, Nehemiah Persoff, Edward G. Robinson; *Dir:* Billy Wilder. Academy Awards '59: Best Costume Design (B & W); Golden Globe Awards '60: Best Film—Musical/Comedy; National Board of Review Awards '59: 10 Best Films of the Year. VHS, Beta, LV **$19.95** *FOX, MGM, VYG* ✂✂✂✂

Somebody Has to Shoot the Picture Photographer Scheider is hired by a convicted man to take a picture of his execution. Hours before the event, Scheider uncovers evidence that leads him to believe the man is innocent. He then embarks in a race against time to save him. Adapted by Doug Magee from his book "Slow Coming Dark." Unpretentious, tough drama. Made for cable.
1990 (R) 104m/C Roy Scheider, Bonnie Bedelia, Robert Carradine, Andre Braugher, Arliss Howard; *Dir:* Frank Pierson. VHS, Beta, LV **$79.98** *MCA* ✂✂✂

Somebody Up There Likes Me Story of Rocky Graziano's gritty battle from his poor, street-wise childhood to his prison term and his eventual success as the middle-weight boxing champion of the world. Adapted from Graziano's autobiography. Superior performance by Newman (in his third-ever screen role, after the miserable "The Silver Chalice" and forgettable "The Rack"); screen debuts for McQueen and Loggia.
1956 113m/B Paul Newman, Pier Angeli, Everett Sloane, Eileen Heckart, Sal Mineo, Robert Loggia, Steve McQueen; *Dir:* Robert Wise. Academy Awards '56: Best Art Direction/Set Decoration (B & W), Best Black and White Cinematography. VHS, Beta **$19.98** *MGM* ✂✂✂

Someone Behind the Door Evil brain surgeon Perkins implants murderous suggestions into psychopathic amnesia victim Bronson's mind, then instructs him to kill his (the surgeon's) wife and her lover. Also titled "Two Minds for Murder."
1971 (PG) 97m/C Charles Bronson, Anthony Perkins, Jill Ireland, Henri Garcin; *Dir:* Nicolas Gessner. VHS, Beta, LV **$39.95** *UNI* ✂½

Someone at the Door Reporter Medwin arranges to be suspected of killing his sister, thus providing him with a good story and advancing his career. His plan almost backfires, when he is almost executed. Remake of a 1936 film of the same name.
1950 65m/B *GB* Michael Medwin, Garry Marsh, Yvonne Owen, Hugh Latimer; *Dir:* Francis Searle. VHS, Beta **$16.95** *SNC* ✂✂

Someone I Touched Way-overdone melodrama starring Leachman as a finally-pregnant woman who learns her husband and a teenager he slept with have venereal disease.

1975 74m/C Cloris Leachman, James Olson, Glynnis O'Connor, Andy Robinson, Allyn Ann McLerie; **Dir:** Lou Antonio. **VHS, Beta $34.95** *WOV* *♫♫1/2*

Someone to Love Rootless filmmaker gathers all his single friends together and interviews them about their failed love lives. Welles's last film as an actor. Wildly uneven, interesting experiment.
1987 (R) 110m/C Henry Jaglom, Orson Welles, Sally Kellerman, Andrea Marcovicci, Michael Emil, Oja Kodar, Stephen Bishop, Ronee Blakley, Kathryn Harrold, Monte Hellman; **Dir:** Henry Jaglom. **VHS, Beta, LV $79.95** *PAR* *♫♫*

Someone to Watch Over Me After witnessing the murder of a close friend, a beautiful and very wealthy woman must be protected from the killer. The working-class New York detective assigned the duty is more than taken with her, despite the fact that he has both a wife and son at home. A highly watchable, stylish romantic crime thriller. Music by Michael Kamen.
1987 (R) 106m/C Tom Berenger, Mimi Rogers, Lorraine Bracco, Jerry Orbach, Andreas Katsulas, Tony DiBenedetto, James Moriarty, John Rubinstein; **Dir:** Ridley Scott. **VHS, Beta, LV $19.95** *COL* *♫♫♫*

Something in Common A woman discovers her 22-year-old son is having an affair with a woman her own age. Nicely done comedy; Burstyn and Weld both make good use of a solid script by Susan Rice. Made for TV.
1986 94m/C Ellen Burstyn, Tuesday Weld, Don Murray, Patrick Cassidy, Eli Wallach; **Dir:** Glenn Jordan. **VHS, Beta $19.95** *STE, NWV* *♫♫1/2*

Something for Everyone A corrupt footman uses sex and murder in an attempt to take over the estate of an aristocratic Bavarian family in post-World War II Germany. He nearly succeeds in marrying the countess (Lansbury), but someone has been keeping tabs on him. Funny, ambitious black comedy. From the novel "The Code" by Harry Kressing.
1970 (R) 112m/C Angela Lansbury, Michael York, Anthony Corlan, Heidelinde Weis; **Dir:** Harold Prince. **VHS, Beta $59.98** *FOX* *♫♫1/2*

Something Short of Paradise The owner of a Manhattan movie theater has an on again-off again romance with a magazine writer. Would-be Alleuesque romantic comedy is too talky, pretentious.
1979 (PG) 87m/C David Steinberg, Susan Sarandon, Jean-Pierre Aumont, Marilyn Sokol; **Dir:** David Helpern. **VHS, Beta $29.98** *LIV, VES* *♫1/2*

Something to Sing About Musical melodrama about a New York bandleader's attempt to make it big in Hollywood. Allows Cagney the opportunity to demonstrate his dancing talents; he also sings "Any Old Love," and Daw (at the heighth of her career) belts out "Right or Wrong," "Loving You," and "Out of the Blue." C. Bakaleinikoff's musical score was nominated for an Oscar. Rereleased as "Battling Hoofer" in 1947. Also available colorized. Frawley was Fred on "I Love Lucy."
1936 84m/B James Cagney, William Frawley, Evelyn Daw, Gene Lockhart; **Dir:** Victor Schertzinger. **VHS, Beta, LV $19.95** *NOS, VHE, CAB* *♫♫*

Something Special A fourteen-year-old girl has her wish come true when she is turned into a boy. Charming, original fantasy-comedy is fun for young and old. Also known as "Willy Milly" and "I Was a Teenage Boy."

1986 (PG-13) 93m/C Pamela Segall, Patty Duke, Eric Gurry, John Glover, Seth Green; **Dir:** Paul Schneider. **VHS, LV $29.95** *MAG* *♫♫1/2*

Something of Value Good ensemble performances in a serious colonial story about the Mau Mau rebellion in Kenya. Hudson and Poitier are torn between their friendship and their opposing loyalties. Drama solidly grounded in fact.
1957 113m/B Rock Hudson, Sidney Poitier, Wendy Hiller, Dana Wynter, Juano Hernandez; **Dir:** Richard Brooks. **VHS, Beta $24.95** *MGM* *♫♫♫*

Something Weird McCabe is disfigured horribly in an electrical accident. A seemingly beautiful witch fixes his face, on condition that he be her lover. The accident also gave him ESP—and it gets cheesier from there.
1968 80m/C Tony McCabe, Elizabeth Lee; **Dir:** Herschell Gordon Lewis. **VHS, Beta $29.95** *VMK, VDM, TWE* Woof!

Something Wicked This Way Comes Two young boys discover the evil secret of a mysterious traveling carnival that visits their town. Ray Bradbury wrote the screenplay for this much-anticipated, expensive adaptation of his own novel. Good special effects, but disappointing. Clayton earlier directed "The Great Gatsby."
1983 (PG) 94m/C Jason Robards Jr., Jonathan Pryce, Diane Ladd, Pam Grier, Richard Davalos, James Stacy; **Dir:** Jack Clayton. **VHS, Beta, LV $19.99** *DIS* *♫♫*

Something Wild Mild-mannered business exec Daniels is picked up by an impossibly free-living vamp with a Louise Brooks hairdo, and taken for the ride of his up-till-then staid life, eventually leading to explosive violence. A sharp-edged comedy with numerous changes of pace. Too-happy ending wrecks it, but it's great until then. Look for cameos from filmmakers John Waters and John Sayles.
1986 (R) 113m/C Jeff Daniels, Melanie Griffith, Ray Liotta, Margaret Colin, Tracy Walter, Dana Peru, Jack Gilpin, Su Tissue, John Sayles, John Waters, Kenneth Utt, Sister Carol East; **Dir:** Jonathan Demme. Edgar Allan Poe Awards '86: Best Screenplay. **VHS, Beta, LV $19.99** *HBO* *♫♫*

Sometimes Aunt Martha Does Dreadful Things Another deranged-slasher movie, this one featuring a female psychopath.
1988 95m/C VHS, Beta $24.95 *AHV* *♫*

Sometimes a Great Notion Trouble erupts in a small Oregon town when a family of loggers decide to honor a contract when the other loggers go on strike. Newman's second stint in the director's chair; Fonda's first role as an old man. Jaeckel received an Oscar nomination for his role. Based on the novel by Ken Kesey.
1971 (PG) 115m/C Paul Newman, Henry Fonda, Lee Remick, Richard Jaeckel, Michael Sarrazin; **Dir:** Paul Newman. **VHS, Beta $59.95** *MCA* *♫♫♫*

Sometimes It Works, Sometimes It Doesn't A film in two parts, the first being interviews with Merce Cunningham and John Case about their work together, the second being the filmdance "Channels/Inserts," in its entirety.
1983 63m/C VHS, Beta *CUN* *♫♫1/2*

Somewhere in Time Playwright Reeve (in his first post-Clark Kent role) falls in love with a beautiful woman in an old portrait. Through self-hypnosis he goes back in time to 1912 to discover what their relationship

might have been. The film made a star of the Grand Hotel, located on Mackinac Island in Michigan, where it was shot. Drippy rip-off of the brilliant novel "Time and Again" by Jack Finney. Reeve is horrible; Seymour is underused. All in all, rather wretched.
1980 (PG) 103m/C Christopher Reeve, Jane Seymour, Christopher Plummer, Teresa Wright; **Dir:** Jeannot Szwarc. **VHS, Beta, LV $19.95** *MCA* *♫♫*

Somewhere Tomorrow A lonely, fatherless teenage girl is befriended by the ghost of a young man killed in a plane crash. Charming and moving, if not perfect.
1985 (PG) 91m/C Sarah Jessica Parker, Nancy Addison, Tom Shea; **Dir:** Robert Wiemer. **VHS, Beta $59.95** *MED* *♫♫1/2*

Somos Novios (In Love and Engaged) A publicity agent and his friend, along with a salesclerk in a record store, find themselves at the top of the music charts.
1968 100m/C MX Angelica Maria, Armando Manzanero, Palito Ortega, Olga Zubarry, Raul Ross; **Dir:** Enrique Carreras. **VHS, Beta $69.95** *MAD* *♫*

Son of Blob A scientist brings home a piece of frozen blob from the North Pole; his wife accidentally revives the dormant gray mass. It begins a rampage of terror by digesting nearly everyone within its reach. A host of recognizable faces make for fun viewing. Post-Jeannie, pre-Dallas Hagman directed this exercise in zaniness; originally titled "Beware! The Blob."
1972 (PG) 87m/C Robert Walker Jr., Godfrey Cambridge, Carol Lynley, Shelley Berman, Larry Hagman, Burgess Meredith, Gerrit Graham, Dick Van Patten; **Dir:** Larry Hagman. **VHS, Beta $19.95** *GEM* *♫♫*

Son of Captain Blood The son of the famous pirate meets up with his father's enemies on the high seas. The son of famous actor Errol Flynn—Sean—plays the son of the character the elder Flynn played in "Captain Blood." Let's just say the gimmick didn't work.
1962 90m/C IT SP Sean Flynn, Ann Todd; **Dir:** Tulio Demicheli. **VHS, Beta $59.95** *PSM, MLB* *♫1/2*

Son of Dracula In this late-coming sequel to the Universal classic, a stranger named Alucard is invited to America by a Southern belle obsessed with eternal life. It is actually Dracula himself, not his son, who wreaks havoc in this spine-tingling chiller.
1943 80m/B Lon Chaney Jr., Evelyn Ankers, Frank Craven, Robert Paige, Louise Allbritton, J. Edward Bromberg, Samuel S. Hinds; **Dir:** Robert Siodmak. **VHS, Beta $14.95** *MCA* *♫♫♫*

Son of Flubber Sequel to "The Absent Minded Professor" finds Fred MacMurray still toying with his prodigious invention, Flubber, now in the form of Flubbergas, which causes those who inhale it to float away. Disney's first-ever sequel is high family wackiness.
1963 96m/C Fred MacMurray, Nancy Olson, Tommy Kirk, Leon Ames, Joanna Moore, Keenan Wynn, Charlie Ruggles, Paul Lynde; **Dir:** Robert Stevenson. **VHS, Beta** *DIS* *♫♫1/2*

Son of Frankenstein The second sequel (after "The Bride of Frankenstein") to the 1931 version of the horror classic. The good doctor's skeptical son returns to the family manse and becomes obsessed with his father's work and with reviving the creature. Full of memorable characters and brooding ambience. Karloff's last appearance as the monster.

1939 99m/C Basil Rathbone, Bela Lugosi, Boris Karloff, Lionel Atwill, Josephine Hutchinson, Donnie Dunagan, Emma Dunn, Edgar Norton, Lawrence Grant; *Dir:* Rowland V. Lee. **VHS, Beta, LV** $14.98 *MCA* ✓✓✓

Son of God's Country Utterly hackneyed, near-comic attempt at a western. Good-guy U.S. Marshall poses as a crook to get the lowdown on the varmints.
1948 60m/B Monte Hale, Pamela Blake, Adrian Booth. **VHS, Beta** $29.95 *DVT* ✓✓

Son of Godzilla Dad and junior protect beauty Maeda from giant spiders on a remote island ruled by a mad scientist. Fun monster flick with decent special effects.
1966 86m/C *JP* Tadao Takashima, Akira Kubo; *Dir:* Jun Fukuda. **VHS, Beta** $19.95 *PSM, HHT, DVT* ✓✓½

Son of a Gun A loveable but ornery cowboy gets banished from the county for disturbing the peace. However, he wins the favor of the townspeople when he stands up to a gang of gambling swindlers. Silent.
1919 68m/B Billy Anderson. **VHS, Beta** $19.95 *NOS, DVT, RXM* ✓½

Son of Hercules vs. Venus Mortals battle the gods in this low budget action-adventure.
19?? 93m/C **VHS, Beta** $19.99 *BFV* ✓

Son of Hollywood Bloopers More goofs and blunders of Hollywood stars.
194? 35m/B John Garfield, Errol Flynn, George Raft, Joan Blondell. **VHS, Beta** $19.95 *NO* ✓

Son of Ingagi Lonely ape-man kidnaps woman in search of romance. Early all-black horror film stars Williams of "Amos 'n' Andy."
1940 70m/B Zack Williams, Laura Bowman, Alfred Grant, Spencer Williams; *Dir:* Richard C. Kahn. **VHS** $19.98 *NOS, CHN, SNC* ✓½

Son of Kong King Kong's descendant is discovered on an island amid prehistoric creatures in this often humorous sequel to RKO's immensely popular "King Kong." Hoping to capitalize on the enormous success of its predecessor, director Schoedsack quickly threw this together. As a result, its success at the box office did not match the original's, and didn't deserve to, but it's fun. Nifty special effects from Willie O'Brien, the man who brought them to us the first time!
1933 70m/B Robert Armstrong, Helen Mack; *Dir:* Ernest B. Schoedsack. **VHS, Beta, LV** $19.98 *NOS, MED, FCT* ✓✓½

Son of Monsters on the March A sequel package to "Monsters on the March," this compilation offers trailers to ten horror and sci-fi classics, including "The Rocky Horror Picture Show," "Planet of the Apes," "The Fearless Vampire Killers," "This Island Earth" and "I Was a Teenage Frankenstein."
1977 27m/C **VHS, Beta** $19.95 *VYY* ✓✓½

The Son of Monte Cristo Illegitimate offspring of the great swashbuckler with Robert Donat proves they made pathetic, pointless sequels even back then.
1940 102m/B Louis Hayward, Joan Bennett, George Sanders, Florence Bates, Montagu Love, Ralph Byrd, Clayton Moore; *Dir:* Rowland V. Lee. **VHS, Beta, LV** $19.95 *DVT, KRT, CAB* ✓✓½

Son of the Morning Star Lavish made for TV retelling of Custer's famed butt kicking. Complex characterizations and unusual points of view, based on the book by Evan S. Connell.
1991 (PG-13) 186m/C Gary Cole, Rosanna Arquette, Dean Stockwell, Rodney Grant; *Dir:* Mike Robe. **VHS, LV** $89.98 *REP, PIA, FCT* ✓✓✓

Son of Paleface Hilarious sequel to the original Hope gag-fest, with the Harvard-educated son of the original character (again played by Hope) heading west to claim an inheritance. Hope runs away with every cowboy cliche and even manages to wind up with the girl. Songs include "Buttons and Bows" (reprised from the original), "There's a Cloud in My Valley of Sunshine," and "Four-legged Friend."
1952 95m/C Bob Hope, Jane Russell, Roy Rogers, Douglass Dambrille, Iron Eyes Cody, Bill Williams, Harry von Zell, Trigger; *Dir:* Frank Tashlin. **VHS, Beta, LV** $19.95 *COL* ✓✓✓½

Son of Samson Man with large pectoral muscles puts an end to the evil Queen of Egypt's reign of terror
1962 89m/C *FR IT YU* Mark Forest, Chelo Alonso, Angelo Zanolli, Vira Silenti, Frederica Ranchi; *Dir:* Carlo Campogalliani. **VHS** $19.98 *SNC* ✓½

Son of the Sheik A desert sheik abducts an unfaithful dancing girl. Valentino's well-done, slightly self-mocking last film. Silent. Sequel to "The Sheik" (1921).
1926 62m/B Rudolph Valentino, Vilma Banky, Agnes Ayres; *Dir:* George Fitzmaurice. **VHS, Beta, LV** $19.95 *NOS, CCB, VCN* ✓✓✓

Son of Sinbad Sinbad, captured by the Khalif of Baghdad, must bring him the secret of Greek fire to gain his freedom and free the city from the forces of mighty Tamarlane. About three dozen nubile young women wear very little and gambol with out hero herein; much-anticipated skinfest caused a scandal, nurtured by Howard Hughes for profit.
1955 88m/C Dale Robertson, Sally Forrest, Vincent Price, Lili St. Cyr, Mari Blanchard; *Dir:* Ted Tetzlaff. **VHS, Beta** $15.95 *UHV* ✓½

Son of Zorro Zorro takes the law into his own hands to protect ranchers from bandits. A serial in thirteen chapters.
1947 164m/B George Turner, Peggy Stewart, Roy Barcroft, Edward Cassidy. **VHS** $27.95 *REP, VCN, MLB* ✓½

Sonar no Cuesta nada Joven (Dreams are Free) A bumbling fool helps three girls in outwitting gangsters.
197? 88m/C *SP* Pedro Aviles Herrera, Perla Mar, Eduardo Davidson. **VHS, Beta** *MAV* ✓

Song of Arizona Rogers and the Sons of the Pioneers arrive to aid their pal Hayes against a gang of bank robbers in this modern-day oater. Eight songs perpetually punctuate the perfunctory plot, particularly "Will Ya Be My Darling," "Half-a-Chance Ranch," and the title tune.
1946 54m/B Roy Rogers, Dale Evans, George "Gabby" Hayes, Lyle Talbot, Bob Nolan, The Pioneers; *Dir:* Frank McDonald. **VHS, Beta** $19.95 *NOS, VCN, CPB* ✓✓

The Song of Bernadette Depicts the true story of a peasant girl who sees a vision of the Virgin Mary in a grotto at Lourdes in 1858. The girl is directed to dig at the grotto for water that will heal those who believe in its powers, much to the astonishment and concern of the townspeople. Based on Franz Werfel's novel. Directed with tenderness and carefully cast, and appealing to religious and sentimental susceptibilities, it was a box-office smash. Oscar-nominated for Best Supporting Actress (Revere), Best Director,
Screenplay, Sound Editing and Film editing, plus the five Oscars it won.
1943 156m/B Charles Bickford, Lee J. Cobb, Jennifer Jones, Vincent Price, Anne Revere, Gladys Cooper; *Dir:* Henry King. Academy Awards '43: Best Actress (Jones), Best Art Direction/Set Decoration (B & W), Best Black and White Cinematography, Best Interior Decoration, Best Musical Score; Golden Globe Awards '44: Best Film—Drama. **VHS, Beta, LV** $19.98 *FOX, IGP, BTV* ✓✓✓

Song of Ceylon A pictorial documentary in four sections: "The Buddha," "The Virgin Island," "The Voices of Commerce," and "The Apparel of a God."
1934 40m/B **VHS, Beta** $24.95 *NOS, WFV, FST* ✓✓✓

Song of Freedom The story of a black dockworker whose gift of voice is discovered by an opera impresario. After realizing a career as a concert performer, Robeson ventures to Africa to investigate his ancestry and finds he has royal roots. Robeson turns in a fine performance in what is otherwise an average film.
1938 70m/B *GB* Paul Robeson, Elizabeth Welch, George Mozart; *Dir:* J. Elder Wills. **VHS, Beta** $29.95 *NOS, HHT, VYY* ✓✓✓

Song of the Gringo Tex Ritter stars in his first singing western as a sheriff going after a gang of claim jumpers.
1936 57m/B Tex Ritter, Monte Blue, Joan Woodbury, Al "Fuzzy" Knight, Richard Adams, Warner Richmond; *Dir:* John P. McCarthy. **VHS, Beta** $19.95 *NOS, UHV, VCN* ✓

A Song for Ireland Popular Irish singers perform as the viewer is taken on a tour of Ireland.
1988 52m/C Phil Coulter, Mary Black, Furey Brothers; *Hosted:* Bryan Murray. **VHS, Beta** *RGO* ✓✓

Song of the Islands Landing families in Hawaii feud over a beach, while the son of one (Mature) and the daughter of the other (Grable) fall in love. Plenty of songs, and dancing to display Grable's legs. Of course happiness and concord reign at the end. Songs include "O'Brien Has Gone Hawaiian," "Sing Me a Song of the Islands," and "What's Buzzin', Cousin?"
1942 75m/C Betty Grable, Victor Mature, Jack Oakie, Thomas Mitchell, Billy Gilbert; *Dir:* Walter Lang. **VHS, Beta, LV** $19.98 *FOX, MLB* ✓✓½

Song of Nevada Rogers woos the high-society daughter (Evans, of course) of a rancher (Hall). Songs include "It's Love, Love, Love," "Hi Ho Little Dogies," and "A Cowboy Has to Yodel in the Morning." Quintesssential Roy-n-Dale.
1944 60m/B Roy Rogers, Dale Evans, Mary Lee; *Dir:* Joseph Kane. **VHS, Beta** $9.95 *NOS, HHT, DVT* ✓✓

Song of Norway A dramatization of the early life of the beloved Norwegian Romantic composer Edvard Grieg. Filmed against the beautiful mountains, waterfalls, and forests of Norway and based on the popular '40s stage production. The scenery is the best thing because Grieg's life was dull—even with cinematic liberties taken.
1970 (G) 143m/C Toralv Maurstad, Florence Henderson, Edward G. Robinson, Christina Schollin, Frank Porretta, Oscar Homolka, Robert Morley, Harry Secombe; *Dir:* Andrew L. Stone. **VHS, Beta** $59.98 *FOX* ✓✓

Song O' My Heart An early musical starring popular Irish tenor McCormack as a singer forced to abandon his career when he marries a woman he does not love. A tour de force for the lead in his movie debut. 11 songs in all, including "Little Boy Blue," "Paddy Me Lad" and "I Hear You Calling Me."

1929 91m/B John McCormack, Maureen O'Sullivan, John Garrick, J.M. Kerrigan, Tommy Clifford, Alice Joyce; *Dir:* Frank Borzage. **VHS, Beta $39.95** *VAI, FCT* 🎬🎬🎬

Song of Old Wyoming Three of those songs are presented in the course of an ordinary western about a spunky widow driving badmen out of Wyoming territory. La Rue plays the villainous Cheyanne Kid; this was before he got his own series as the heroic "Lash" La Rue.

1945 65m/C Eddie Dean, Sarah Padden, Al "Lash" LaRue, Jennifer Holt, Emmett Lynn, John Carpenter, Ian Keith, Bob Barron; *Dir:* Robert Tansey. **VHS $19.95** *NOS* 🎬½

Song to Remember With music performed by Jose Iturbi, this film depicts the last years of the great pianist and composer Frederic Chopin, including his affair with famous author George Sand, the most renowned French woman of her day. Typically mangled film biography still received Oscar nominations for Best Actor (Wilde), Sound Recording, Film Editing, and Best Score of A Drama or Comedy.

1945 112m/C Cornel Wilde, Paul Muni, Merle Oberon, Nina Foch, George Coulouris; *Dir:* Charles Vidor. **Beta, LV $19.95** *COL, RDG* 🎬🎬🎬

Song of Survival A documentary recounting the trials of 30 Dutch, British and Australian women who were captured by the Japanese in 1942 on the island of Sumatra.

1986 60m/C VHS, Beta *WMB* 🎬🎬🎬

Song of Texas When a young woman journeys west to visit her father at "his" ranch, Roy and his pals must keep the father's secret that he is just another hired hand. Whenever possible, everyone breaks into song.

1943 54m/B Roy Rogers, Harry Shannon, Pat Brady, Barton MacLane, Arlene Judge; *Dir:* Joseph Kane. **VHS, Beta $19.95** *NOS, VYY, VCN* 🎬½

Song of the Thin Man The sixth and final "Thin Man" mystery. This time Nick and Nora Charles (Powell and Loy) investigate the murder of a bandleader. Somewhat more sophisticated than its predecessor, due in part to its setting in the jazz music world. Sequel to "The Thin Man Goes Home."

1947 86m/B William Powell, Myrna Loy, Keenan Wynn, Dean Stockwell, Philip Reed, Patricia Morison, Gloria Grahame, Jayne Meadows, Don Taylor, Leon Ames, Ralph Morgan, Warner Anderson; *Dir:* Edward Buzzell. **VHS, Beta $19.98** *MGM* 🎬🎬½

Song of the Trail Our hero saves an old man hornswoggled at cards. Nothing special about this Maynard outing.

1936 65m/B Kermit Maynard, George "Gabby" Hayes, Al "Fuzzy" Knight, Wheeler Oakman, Evelyn Brent, Andrea Leeds; *Dir:* Russell Hopton. **VHS, Beta $19.95** *NOS, VDM, VCN* 🎬½

Song Without End This musical biography of 19th century Hungarian pianist/composer Franz Liszt is given the Hollywood treatment. The lavish production emphasizes Liszt's scandalous exploits with married women and his life among the royal courts of Europe rather than his musical talents. Features music from serveral composers including the "Mephisto Waltz," "Spozalizio," and "Sonata in B Minor." This film marked the debut of Capucine. Director Vidor died during

filming and Cukor stepped in, so there is a noticeable change in style. Although there is much to criticize in the story, the music is beautiful.

1960 130m/C Dirk Bogarde, Capucine, Genevieve Page, Patricia Morison, Ivan Desny, Martita Hunt, Lou Jacobi; *Dir:* Charles Vidor, George Cukor. Academy Awards '60: Best Musical Score; Golden Globe Awards '61: Best Film—Musical/Comedy. **VHS $19.95** *COL* 🎬🎬½

Songs and Bullets The Hound loves the zen-like simplicity of that title. The songs total five in number, the bullets somewhat more as hero cowboys battle rustlers in routine fashion. One of a handful of films (westerns and comedies) produced by funnyman Stan Laurel.

1938 57m/B Fred Scott, Al "Fuzzy" St. John, Alice Ardell, Charles King, Karl Hackett, Frank LaRue, Budd Buster; *Dir:* Sam Newfield. **VHS $19.95** *NOS* 🎬½

Songwriter A high-falutin' look at the lives and music of two popular country singers with, aptly, plenty of country tunes written and performed by the stars. Singer-businessman Nelson needs Kristofferson's help keeping a greedy investor at bay. Never mind the plot; plenty of good music.

1984 (R) 94m/C Willie Nelson, Kris Kristofferson, Rip Torn, Melinda Dillon, Lesley Ann Warren; *Dir:* Alan Rudolph. **VHS, Beta, LV $12.95** *COL* 🎬🎬½

Sonny Boy A transvestite and a psychopath adopt a young boy, training him to do their bidding, which includes murder.

1987 (R) 96m/C *IT* David Carradine, Paul Smith, Brad Dourif, Conrad Janis, Sydney Lassick, Savina Gersak, Alexandra Powers, Steve Carlisle, Michael Griffin; *Dir:* Robert Martin Carroll. **VHS, Beta, LV $89.98** *TWE, MED* Woof!

Sonny and Jed An escaped convict and a free-spirited woman travel across Mexico pillaging freely, followed determinedly by shiny-headed lawman Savalas. Lame rip-off of the Bonnie and Clyde legend.

1973 (R) 85m/C *IT* Tomas Milian, Telly Savalas, Susan George, Rosanna Janni, Laura Betti; *Dir:* Sergio Corbucci. **VHS, Beta $9.95** *NO* 🎬

Sonora Stagecoach Okay sagebrush saga with an interesting lineup; two cowboys, a heroic Indian and a tough gal pilot escort a murder suspect in the title conveyance and manage to clear him when the real bad guys attack.

1944 61m/B Hoot Gibson, Bob Steele, Chief Thunder Cloud, Rocky Camron, Betty Miles, Glenn Strange, George Eldridge, Karl Hackett; *Dir:* Robert Tansey. **VHS $19.95** *NOS* 🎬🎬

Sons A Vietnam veteran's son joins the military, causing strains within the family.

1991 ?m/C VHS $79.95 *YES* 🎬🎬

Sons of the Desert Laurel and Hardy in their best-written film. The boys try to fool their wives by pretending to go to Hawaii to cure Ollie of a bad cold when in fact, they are attending their lodge convention in Chicago. Also includes a 1935 Thelma Todd/Patsy Kelly short, "Top Flat."

1933 73m/B Stan Laurel, Oliver Hardy, Mae Busch, Charley Chase; *Dir:* William A. Seiter. **VHS, Beta $19.95** *MED, MLB* 🎬🎬🎬

Sons of Hercules in the Land of Darkness Hercules' sons kill dragons, fight titans, and generally flex up a storm.

1955 74m/B Dan Vadis. **VHS, Beta $29.95** *VYY* 🎬

Sons of Katie Elder After their mother's death, four brothers are reunited. Wayne is a gunman; Anderson is a college graduate; silent Holliman is a killer; and Martin is a gambler. When they learn that her death might have been murder, they come together to devise a way to seek revenge on the killer. The town bullies complicate matters; the sheriff tells them to lay off. Especially strong screen presence by Wayne, in his first role following cancer surgery. One of the Duke's most popular movies of the '60s.

1965 122m/C John Wayne, Dean Martin, Earl Holliman, Michael Anderson Jr., Martha Hyer; *Dir:* Henry Hathaway. **VHS, Beta, LV $14.95** *PAR, TLF, BUR* 🎬🎬🎬

Sons of the Pioneers Small town sheriff Hayes hires cowboy Roy to rid town of outlaws.

1942 61m/B Roy Rogers, George "Gabby" Hayes, Maris Wrixon, Forrest Taylor; *Dir:* Joseph Kane. **VHS $12.98** *REP* 🎬

Sooner or Later A 13 year-old girl passes herself off as a sixteen with a local rock idol and must decide whether to go all the way.

1978 100m/C Rex Smith, Judd Hirsch, Denise Miller, Morey Amsterdam, Lynn Redgrave; *Dir:* Bruce Hart. **VHS, Beta $19.95** *STE, NWV* 🎬½

Sophia Loren: Her Own Story The life story of Sophia Loren, from a spindly child growing up in working-class Naples, to a world-renowned movie star, beauty queen and mother. Loren plays herself and her mother in this overlong account of her less-than-fascinating life story. Scripted by Joanna Crawford from A.E. Hotchner's biography. Made or TV.

1980 150m/C Sophia Loren, Armand Assante, Ed Flanders, John Gavin; *Dir:* Mel Stuart. **VHS, Beta $69.95** *HBO* 🎬½

Sophie's Choice A haunting modern tragedy about Sophie Zawistowska, a beautiful Polish Auschwitz survivor settled in Brooklyn after World War II. She has intense relationships with a schizophrenic genius and an aspiring Southern writer. An artful, immaculately performed and resonant drama, with an astonishing, commanding performance by the versatile Streep; a chilling portrayal of the banality of evil. From the best-selling, autobiographical novel by William Styron.

1982 (R) 157m/C Meryl Streep, Kevin Kline, Peter MacNicol, Rita Karin, Stephen D. Newman, Josh Mostel; *Dir:* Alan J. Pakula. Academy Awards '82: Best Actress (Streep); National Board of Review Awards '82: Best Actress (Streep); New York Film Critics Awards '82: Best Actress (Streep); National Society of Film Critics Awards '82: Best Actress (Streep); Los Angeles Film Critics Awards '82: Best Actress (Streep). **VHS, Beta, LV $19.98** *FOX* 🎬🎬🎬½

Sophisticated Gents Nine boyhood friends of a black athletic club reunite after 25 years to honor their old coach and see how each of their lives has been affected by being black men in American society. Based on the novel "The Junior Bachelor Society" by John A. Williams. Originally made-for-television.

1981 200m/C Paul Winfield, Roosevelt Grier, Bernie Casey, Raymond St. Jacques, Thalmus Rasulala, Dick Anthony Williams, Ron O'Neal, Rosalind Cash, Denise Nicholas, Alfre Woodard, Melvin Van Peebles; *Dir:* Harry Falk. **VHS $89.95** *XVC* 🎬🎬🎬

Sophisticated Ladies A revival of the classic Ellington show with over 30 Ellington originals performed by his original orchestra, including "Solitude," "Take the 'A' Train,"

"Mood Indigo," "It Don't Mean a Thing," and more.
1984 108m/C Duke Ellington. **VHS, Beta, LV** $59.95 *JTC, FCT, MVD* ♫♫♫

Sorcerer To put out an oil fire, four men on the run in South America agree to try to buy their freedom by driving trucks loaded with nitroglycerin over dangerous terrain—with many natural and man-made obstacles to get in their way. Remake of "The Wages of Fear" is nowhere as good as the classic original, but has exciting moments. Puzzlingly retitled, which may have contributed to the box-office failure, and the near demise of Friedlein's directing career.
1977 (PG) 121m/C Roy Scheider, Bruno Cremer, Francisco Rabal, Soudad Amidou, Ramon Bieri; *Dir:* William Friedkin. **VHS, LV** $79.95 *MCA* ♫♫½

Sorcerer's Apprentice The classic fairy tale of a young boy who is hired by an evil sorcerer and eventually challenges him to a duel of magic.
1980 27m/C VHS, Beta $95.00 *PYR* ♫♫½

Sorceress The Harris girls are sisters who use their powers of sorcery and fighting skills to battle demons, dragons, and evil. Yeah, sure. Really, they're just a pair of babes parading for the camera.
1982 (R) 83m/C Leigh Anne Harris, Lynette Harris, Bob Nelson, David Milbern, Bruno Rey, Anna De Sade; *Dir:* Brian Stuart. **VHS, Beta** $19.99 *HBO* ♫

Sorceress A friar in medieval Europe feels insecure in his religious beliefs after encountering a woman who heals through ancient practices. Historically authentic and interestingly moody. Written, produced, and directed by two women: an art history professor and a collaborator of Francois Truffaut's. Available either in English or in French with English subtitles.
1988 98m/C Tcheky Karyo, Christine Boisson, Jean Carmet; *Dir:* Suzanne Schiffman. **VHS** $29.95 *MFV, IVA* ♫♫♫

Sorority Babes in the Slimeball Bowl-A-Rama An ancient gremlin-type creature is released from a bowling alley, and the great-looking sorority babes have to battle it at the mall, with the help of a wacky crew of nerds. Horrible horror spoof shows plenty of skin.
1987 (R) 80m/C Linnea Quigley, Brinke Stevens, Andras Jones, John Wildman; *Dir:* David DeCoteau. **Beta** $69.95 *UCV* ♫♫

Sorority House Massacre A knife-wielding maniac stalks a sorority girl while her more elite sisters are away for the weekend. Yawn—haven't we seen this one before?
1986 (R) 74m/C Angela O'Neill, Wendy Martel, Pamela Ross, Nicole Rio; *Dir:* Carol Frank. **VHS, Beta** $79.95 *WAR* Woof!

Sorority House Massacre 2 This time around three lingerie-clad lovelies are subjected to the terrors of a killer their first night in their new sorority house. College just keeps getting tougher all the time.
1992 80m/C Melissa Moore, Robyn Harris, Stacia Zhivago, Dana Bentley; *Dir:* Jim Wynorski. **VHS** $89.98 *NHO* Woof!

Sorority House Scandals These strippers have nothing whatsoever to do with a sorority, unless you believe the on-stage announcer, who claims they're pledges. The only pledge these girls have taken is to wear as little as possible as often as possible.
1986 60m/C *Dir:* Tony Christopher. **VHS, Beta** $19.95 *AHV* Woof!

The Sorrow and the Pity A classic documentary depicting the life of a small French town and its Resistance during the Nazi occupation. Lengthy, but totally compelling. A great documentary that brings home the atrocities of war. In French with English narration.
1971 265m/B *FR Dir:* Marcel Ophuls. National Board of Review Awards '72: Best Foreign Film; New York Film Critics Awards '72: Special Citation; New York Film Festival '71: Best Foreign Film; National Society of Film Critics '71: Special Prize. **VHS, Beta** *COL, OM* ♫♫♫½

Sorrowful Jones A "Little Miss Marker" remake, in which bookie Hope inherits a little girl as collateral for an unpaid bet. Good for a few yuks, but the original is much better.
1949 88m/B Bob Hope, Lucille Ball, William Demarest, Bruce Cabot, Thomas Gomez, Mary Jane Saunders; *Dir:* Sidney Lanfield. **VHS, Beta** $29.95 *MCA* ♫♫

Sorrows of Gin An eight-year-old girl searches for a sense of family amid her parents' lives. Her efforts to deal with alienation and emotional isolation lead her to run away from home. From a short story by John Cheever.
1979 60m/C Edward Herrmann, Sigourney Weaver, Mara Hobel, Eileen Heckart, Rachel Roberts. **VHS, Beta** *FLI* ♫½

The Sorrows of Satan Dempster and Cortez star as writers in love with each other, but Cortez can't take his poverty and the rejection of publishers. As he is about to end it all, the devil appears to him in disguise and makes an offer he can hardly refuse. He is introduced to London society and winds up marrying a fortune hunter. When he discovers this he goes back to Dempster, who still loves him, and vows to give up everything for her. Menjou then reveals himself to be the devil but Dempster's faith is so strong that the evil is vanquished.
1926 111m/B Adolphe Menjou, Carol Dempster, Ricardo Cortez, Lya de Putti; *Dir:* D.W. Griffith. **VHS, Beta** $32.95 *GPV, DNB* ♫♫♫

Sorry, Wrong Number A wealthy, bedridden wife overhears two men plotting a murder on a crossed telephone line, and begins to suspect that one of the voices is her husband's. A classic tale of paranoia and suspense. Based on a radio drama by Louise Fletcher, who also wrote the screenplay. Stanwyck received an Academy Award nomination for her excellent portrayal as the bedridden woman. Remade for television in 1989.
1948 89m/B Barbara Stanwyck, Burt Lancaster, Ann Richards, Wendell Corey, Harold Vermilyea, Ed Begley Sr.; *Dir:* Anatole Litvak. **VHS, Beta, LV** $19.95 *PAR* ♫♫♫½

Sorry, Wrong Number Cable remake of the 1948 Barbara Stanwyck thriller about a bed-ridden woman trying to get help after she realizes she's being stalked by a killer. Offers a new twist (drug dealing), but doesn't even approach the Stanwyck original. From the story by Lucille Fletcher.
1989 90m/C Loni Anderson, Hal Holbrook, Patrick Macnee, Miguel Fernandez, Carl Weintraub; *Dir:* Tony Wharmby. **VHS, Beta, LV** $79.95 *PAR* ♫½

S.O.S. Coast Guard A fiendish scientist creates a disintegrating gas and the U.S. Coast Guard must stop him from turning it over to unfriendly foreigners. Loads of action in this 12-part serial.
1937 224m/B Ralph Byrd, Bela Lugosi, Maxine Doyle. **VHS, LV** $29.98 *NOS, SNC, REP* ♫♫½

S.O.S. Titanic The story of the Titanic disaster, recounted in flashback, in docu-drama style. James Costigan's teleplay of the familiar story focuses on the courage that accompanied the horror and tragedy. Thoroughly professional, absorbing made-for-TV drama.
1979 98m/C David Janssen, Cloris Leachman, Susan St. James, David Warner, Ian Holm, Helen Mirren, Harry Andrews; *Dir:* Billy Hale. **VHS, Beta** $59.99 *HBO* ♫♫½

Sotto, Sotto A minor Wertmuller comedy about an Italian businessman who is constantly suspicious of his wife's fidelity. As it happens, she's attracted to another woman (de Santis). Massimo Wertmuller is Lina's nephew. Entertaining. Subtitled in English.
1985 (R) 104m/C *IT* Enrico Motesano, Veronica Lario, Massimo Wertmuller, Luisa de Santis; *Dir:* Lina Wertmuller. **VHS, Beta** $69.95 *COL, APD* ♫♫½

Soul of the Beast An elephant repeatedly saves Bellamy from the villainous Beery. Silent with original organ score.
1923 77m/B Madge Bellamy, Cullen Landis, Noah Beery. **VHS, Beta, 8mm** $29.99 *VYY* ♫½

Soul-Fire South Pacific high seas induce Cupid's arrow. Silent.
1925 100m/B Richard Barthelmess, Bessie Love, Walter Long, Arthur Metcalfe; *Dir:* John S. Robertson. **VHS, Beta** $24.95 *GPV, FCT, DNB* ♫♫

Soul Hustler A con man becomes rich and famous as a tent-show evangelist. Pathetic drivel.
1976 (PG) 81m/C Fabian, Casey Kasem, Larry Bishop, Nai Bonet; *Dir:* Burt-Topper. **VHS, Beta** $39.95 *IVE, MON* Woof!

Soul Man Denied the funds he expected for his Harvard tuition, a young white student (Howell) masquerades as a black in order to get a minority scholarship. As a black student at Harvard, Howell learns about racism and bigotry. Pleasant lightweight comedy with romance thrown in (Chong is the black girl he falls for), and with pretensions to social satire that it never achieves.
1986 (PG-13) 101m/C C. Thomas Howell, Rae Dawn Chong, James Earl Jones, Leslie Nielsen, Arye Gross; *Dir:* Steve Miner. **VHS, Beta, LV** $9.95 *NWV, STE* ♫♫

Soul Patrol A black newspaper reporter clashes with the all-white police department in a racist city.
1980 (R) 90m/C Nigel Davenport, Ken Gampu, Peter Dyneley; *Dir:* Christopher Rowley. **VHS, Beta** *TAV, SUN* ♫

Soul of Samurai Duo searches for the essence of kung fu but finds terror instead.
19?? 90m/C Kong Pun. **VHS, Beta** $29.95 *OCE* ♫

Soul Vengeance Black man is jailed and brutalized for crime he didn't commit and wants revenge when he's released. Many afros and platform shoes. Vintage blaxploitation, originally titled "Welcome Home Brother Charles."
1975 91m/C Marlo Monte, Reatha Grey, Stan Kamber, Tiffany Peters, Ven Bigelow, Jake Carter; *W/Dir:* Jamaa Fanaka. **VHS** $39.95 *XVC* ♫

Soultaker The title spirit is after a young couple's souls, and they have just an hour to reunite with their bodies after a car crash. Meanwhile, they're in limbo (literally) between heaven and earth. The ending's OK, if you can make it that far.

1990 (R) 94m/C Joe Estevez, Vivian Schilling, Gregg Thomsen, David Shark, Jean Reiner, Chuck Williams, Robert Z'Dar; *Dir:* Michael Rissi. **VHS** $89.95 *AIP* 🐾½

Sound of Horror The Hound usually admires creative efforts to keep budgets down, but this is too much (or too little). A dinosaur egg hatches, and out lashes an invisible predator. Yes, you'll have to use your imagination as archaeologists are slashed to bits by the no-show terror. 1964 85m/B SP James Philbrook, Arturo Fernandez, Soledad Miranda, Ingrid Pitt; *Dir:* Jose Antonio Nieves-Conde. **VHS** $15.95 *NOS, LOO* 🐾½

Sound of Love The mutual attraction between a deaf female hustler and a deaf race car driver leads to a deep relationship where they both learn about their own fears and desires. 1977 74m/C Celia de Burgh, John Jarratt. **VHS, Beta** $49.98 *SUE* 🐾

Sound of Murder A successful children's book writer wants to exact revenge on his adulterous wife, but is himself murdered before he can "execute" his plan. 1982 114m/C Michael Moriarty, Joanna Miles. **VHS, Beta** $59.95 *IVE, USA* 🐾

The Sound of Music The classic film version of the Rogers and Hammerstein musical based on the true story of the singing von Trapp family of Austria and their escape from the Nazis just before World War II. Beautiful Salzburg, Austria location photography and an excellent cast. Memorable songs include "Climb Every Mountain," "My Favorite Things," "Do-Re-Mi," and the title tune. Andrews, fresh from her Oscar for "Mary Poppins," is effervescent, in beautiful voice, but occasionally too good to be true. Not Rodgers & Hammerstein's most innovative or lyrical score, but lovely to hear and see. Plummer's singing was dubbed by Bill Lee. Marni Nixon, behind-the-scenes songstress for "West Side Story" and "My Fair Lady," makes her on-screen debut as one of the nuns. Oscar nominations for Andrews, Wood, Art Direction and Set Decoration. 1965 174m/C Julie Andrews, Christopher Plummer, Eleanor Parker, Peggy Wood, Charmian Carr, Heather Menzies, Marni Nixon; *Dir:* Robert Wise. Academy Awards '65: Best Adapted Score, Best Director (Wise), Best Film Editing, Best Picture, Best Sound; Directors Guild of America Awards '65: Best Director (Wise); Golden Globe Awards '66: Best Film—Musical/Comedy; National Board of Review Awards '65: 10 Best Films of the Year. **VHS, Beta, LV** $24.98 *FOX, BTV, RDG* 🐾🐾🐾🐾

Sounder The struggles of a family of black sharecroppers in rural Louisiana during the Depression. When the father is sentenced to jail for stealing in order to feed his family, they must pull together even more, and one son finds education to be his way out of poverty. Tyson brings strength and style to her role, with fine help from Winfield. Moving and well made, with little sentimentality and superb acting from a great cast. Adapted from the novel by William Armstrong. Score was written by Taj Mahal. Oscar nominations for Best Picture, actor Winfield, and actress Tyson. Director Ritt works well in this Southern-struggles genre; he went on to direct "Norma Rae" in 1979. 1972 (G) 105m/C Paul Winfield, Cicely Tyson, Kevin Hooks, Taj Mahal, Carmin Mathews, Carmen Best, Janet MacLachlan; *Dir:* Martin Ritt. National Board of Review Awards '72: 10 Best Films of the Year; National Society of Film

Critics Awards '72: Best Actress (Tyson). **VHS, Beta, LV** $14.95 *KUI, PAR, PTB* 🐾🐾🐾🐾

Sounds of Silence A young deaf boy fears that a killer arsonist may be at large—but can he prove it? One of a number of obscure oddities featuring former youth idol Donahue. 1991 108m/C Troy Donahue, Peter Nelson. **VHS, Beta** $79.98 *MAG* 🐾🐾

Soup for One Writer-director Kaufer's first film is a solid entry in the Woody Allen-esque Manhattan's angst genre. Young Jewish guy Rubinek pines for the woman of his dreams, finds her, is rejected, persists, gets her, wonders if she's what he wants after all. If you like Woody Allen, this is worth seeing. 1982 (R) 84m/C Saul Rubinek, Marcia Strassman, Gerrit Graham, Richard Libertini, Andrea Martin; *Dir:* Jonathan Kaufer. **VHS, Beta** $19.98 *WAR* 🐾🐾½

Sourdough Fur trapper Perry escapes the hustle and bustle of modern life by fleeing to the Alaskan wilderness. Near-plotless travelog depends heavily on scenery—and there's plenty of that, for sure. 1977 94m/C Gil Perry; *Nar:* Gene Evans. **VHS, Beta** *IGV* 🐾🐾

South of the Border Autry and Burnette are government agents sent to Mexico to investigate a possible revolution instigated by foreign agents. Propaganda-heavy singing western appeared just before the U.S. entered World War II; did well at the box office and boosted Autry's career. 1939 70m/B Gene Autry, Smiley Burnette. **VHS, Beta** $19.88 *CCB, VCN* 🐾🐾

South Bronx Heroes A police officer helps two children when they discover their foster home is the headquarters for a pornography ring. Also known as "The Runaways" and "Revenge of the Innocents." 1985 (R) 105m/C Brendan Ward, Mario Van Peebles, Megan Van Peebles, Melissa Esposito, Martin Zurla, Jordan Abeles; *Dir:* William Szarka. **VHS, Beta** $69.98 *CON, NSV* 🐾½

South of Hell Mountain On the run from a gold mine robbery where they left twenty dead, the McHenry brothers meet a mother-daughter team that slows them down. 1970 (R) 87m/C Ann Stewart, Sam Hall, Nicol Britton; *Dir:* William Sachs, Louis Leahman. **VHS, Beta** $59.95 *MGM* 🐾½

South of Monterey Western adventure with the Cisco kid. 1947 63m/B Gilbert Roland, Martin Garralaga, Frank Yaconelli, Marjorie Riordan; *Dir:* William Nigh. **VHS, Beta** $29.95 *VCN* 🐾½

South Pacific A young American Navy nurse and a Frenchman fall in love during World War II. Expensive production included much location shooting in Hawaii. Based on Rodgers and Hammerstein's musical; not as good as the play, but pretty darn good still. The play in turn was based on James Michener's novel "Tales of the South Pacific." Songs include "Bali Ha'i," "Some Enchanted Evening,"" I'm Gonna Wash That Man Right Out of My Hair," "There is Nothing Like a Dame," and "I'm in Love With a Wonderful Guy." 1958 167m/C Mitzi Gaynor, Rossano Brazzi, Ray Walston, France Nuyen, Beverly Aadland; *Dir:* Joshua Logan. Academy Awards '58: Best Sound; Harvard Lampoon Awards '58: Worst Film of the Year. **VHS, Beta, LV** $19.98 *FOX, RDG* 🐾🐾🐾½

South of Pago Pago The unsuspecting natives of a tropical isle are exploited by a gang of pirates searching for a seabed of rare pearls. A notch up from most similar adventure outings, with great violent action and excellent cinematography. Leading lady Farmer was thought to be a rising star, but retired at 29 in 1942 because of alcoholism and spent some time in mental institutions. "Frances" (1982), with Jessica Lange, is based on her life. 1940 98m/B Victor McLaglen, Jon Hall, Frances Farmer, Gene Lockhart; *Dir:* Alfred E. Green. **VHS, Beta** $12.95 *IVE, MED* 🐾🐾½

South of Reno A man living a secluded life in the desert dreams of moving to Reno and learns his wife is cheating on him. Odd, minimal plot. Impressive first feature for Polish diretor Rezyka, who cut his teeth on music videos. 1987 (R) 98m/C Jeffery Osterhage, Lisa Blount, Joe Phelan, Lewis Van Bergen, Julia Montgomery, Brandis Kemp, Danitza Kingsley, Mary Grace Canfield, Bert Remsen; *Dir:* Mark Rezyka. **VHS, LV** $79.98 *REP* 🐾🐾½

South of the Rio Grande The Cisco Kid rides again as he comes to the aid of a rancher whose horses were stolen. Garden-variety law-and-order western. 1945 60m/B Duncan Renaldo; *Dir:* Lambert Hillyer. **VHS, Beta** $12.95 *UHV* 🐾½

South of St. Louis A peaceful cattle rancher turns renegade gunrunner during the Civil War when his stock is destroyed by Union guerrillas. Overplotted but exciting and action-packed western, with Smith dazzling in a plethora of costumes. 1948 88m/C Joel McCrea, Zachary Scott, Victor Jory, Douglas Kennedy, Alexis Smith, Alan Hale Jr., Dorothy Malone; *Dir:* Ray Enright. **VHS** $14.95 *REP* 🐾🐾½

South of Santa Fe Rogers and friends take on the mob—yes, the mob, armed with machine guns and airplanes—after they are falsely implicated in a crime. In lulls between the rootin' tootin' gunfights, Rogers belts out "We're Headin' for the Home Corral," "Down the Trail," and "Open Range Ahead." 1942 60m/B Roy Rogers, George "Gabby" Hayes, Linda Hayes, Paul Fix, Judy Clark; *Dir:* Joseph Kane. **VHS, Beta** $19.95 *NOS, DVT, CAB* 🐾🐾

South Seas Massacre A cop and a prisoner, shackled together, try to survive modern-day pirates, natives, and each other. They may make it through, but viewers may wish they didn't. 19?? 60m/C Troy Donahue; *Dir:* Pablo Santiago. **VHS, Beta** $59.95 *PSM* Woof!

Southern Comfort A group of National Guardsmen are on weekend maneuvers in the swamps of Louisiana. They run afoul of some of the local Cajuns, and are marked for death in this exciting and disturbing thriller. Booth is excellent in a rare exploration of a little-understood way of life. Lots of blood. If you belong to the National Guard, this could make you queasy. 1981 (R) 106m/C Powers Boothe, Keith Carradine, Fred Ward, Franklyn Seales, Brion James, T.K. Carter, Peter Coyote; *Dir:* Walter Hill. **VHS, Beta, LV** $29.95 *SUE, HHE* 🐾🐾🐾

A Southern Yankee Skelton plays a bumbling bellboy who ends up as a Union spy during the Civil War. Enjoyable comedy, thanks largely to the off-screen input of Buster Keaton.

1948 90m/B Red Skelton, Brian Donlevy, Arlene Dahl, George Coulouris, Lloyd Gough, John Ireland, Minor Watson, Charles Dingle, Art Baker, Reed Hadley, Arthur Space, Addison Richards, Joyce Compton, Paul Harvey, Jeff Corey; **Dir:** Edward Sedgwick. **VHS, Beta $19.98** *MGM ☆☆½*

The Southerner A man used to working for others is given some land by an uncle and decides to pack up his family and try farming for himself. They find hardships as they struggle to support themselves. A superb, naturalistic celebration of a family's fight to survive amid all the elements. From the story "Hold Autumn in Your Hand," by George Sessions Perry. Renoir received a Best Director Oscar nomination. The great novelist William Faulkner had an uncredited hand in the script. He thought Renoir the best contemporary director, and later said "The Southerner" gave him more pleasure than any of his other Hollywood work (though this is faint praise; Faulkner is said to have hated Hollywood).
1945 91m/B Zachary Scott, Betty Field, Beulah Bondi, Norman Lloyd, Bunny Sunshine, Jay Gilpin, Estelle Taylor, Blanche Yurka, Percy Kilbride, J. Carroll Naish; **Dir:** Jean Renoir. National Board of Review Awards '45: Best Director; Venice Film Festival '46: Best Film. **VHS, Beta, 8mm $19.95** *NOS, PSM, VYY ☆☆☆☆*

Southward Ho! In the first western to team Rogers with sidekick Gabby Hayes, they portray ranchers investigating bloodshed in their valley. Roy meanwhile yodelin' four tunes. Fun outing.
1939 56m/B Roy Rogers, George "Gabby" Hayes. **VHS, Beta $19.95** *NOS, VCN, CAB ☆☆*

Souvenir A German soldier returns to France after World War II to find the woman he left behind and the daughter he never saw. Might have been good, but too melodramatic and overwrought, with lukewarm acting.
1988 (R) 93m/C Christopher Plummer, Catherine Hicks, Christopher Cazenove, Michael Lonsdale; **Dir:** Geoffrey Reeve. **VHS, Beta, LV** *PAR, SOU ☆☆*

Soviet Spy A French docu-drama written by Hans Otto Meissner, Chief of Protocol at the German Embassy in Tokyo during WWII, about the journalist Richard Sorge, a German hanged in 1944 for supplying information to the Soviets. In French with subtitles.
1961 130m/B *FR* **VHS, Beta $19.95** *FCT ☆☆*

Soy un Golfo A lazy bum is inspired to do something with his life by a pretty girl who he later finds out is his brother's fiancee.
1965 90m/B *MX* Adalberto Martinez, Luz Maria Aguilar, Che Reyes. **VHS, Beta $39.95** *MAD ☆☆*

Soylent Green In the 21st Century, hard-boiled police detective Heston investigates a murder and discovers what soylent green-the people's principal food-is made of. Robinson's final film is a disappointing end to a great career. Its view of the future and of human nature is relentlessly dark. Don't watch it with kids.
1973 (PG) 95m/C Charlton Heston, Leigh Taylor-Young, Chuck Connors, Joseph Cotten, Edward G. Robinson, Brock Peters; **Dir:** Richard Fleischer. **VHS, Beta, LV $59.95** *MGM ☆☆½*

Space Movie Consists entirely of footage from NASA and the U.S. National Archives which tells the story of America's space effort (specifically the Apollo 11 moon flight), The footage, most of which has not been seen before, is mixed with music and a little narration, showing the beauty of space.

1980 (G) 78m/C **Dir:** Tony Palmer. **VHS, Beta $49.95** *WAR, OM ☆☆½*

Space Mutiny A spaceship, the Southern Sun, falls under the attack of the mutinous Kalgan. To keep from being sold into slavery, a small band of loyal passengers strike back. Will they be successful in thwarting the attack? Who cares?
1988 (PG) 93m/C Reb Brown, James Ryan, John Phillip Law, Cameron Mitchell; **Dir:** David Winters. **VHS, Beta $79.95** *AIP ☆*

Space Ninja: Sword of the Space Ark A young fighter pilot returns home to find the planet under siege and his family dead. He vows to seek revenge and sets out to destroy the evil emperor who is wreaking such havoc.
1981 75m/C **Dir:** Bunker Jenkins. **VHS, Beta $39.95** *IVE ☆½*

Space Rage A criminal sentenced to life on a prison planet leads a revolt of the inmates. Lame, formula western set in outer space.
1986 (R) 78m/C Michael Pare, Richard Farnsworth, John Laughlin, Lee Purcell; **Dir:** Conrad Palmisano. **VHS, Beta $19.98** *LIV, LTG ☆*

Space Raiders A plucky 10-year-old blasts off into a futuristic world of intergalactic desperados, crafty alien mercenaries, starship battles and cliff-hanging dangers. Recycled special effects (from producer Corman's movie factory) and plot (lifted near-whole from "Star Wars").
1983 (PG) 84m/C Vince Edwards, David Mendenhall; **Dir:** Howard R. Cohen. **VHS, Beta $39.98** *WAR ☆☆*

Space Riders Action-adventure about the world's championship motorcycle race, with a rock soundtrack featuring Duran Duran, Simple Minds and Melba Moore.
1983 (PG) 93m/C Barry Sheene, Gavan O'Herlihy, Toshiya Ito, Stephanie McLean, Sayo Inaba; **Dir:** Joe Massot. **VHS, Beta $79.99** *HBO Woof!*

Space Soldiers Conquer the Universe From the "Flash Gordon" serial. The evil Emperor Ming introduces a horrible plague from outer space called "The Purple Death." Dr. Zarkov, Dale, and Flash Gordon travel from the frozen wastes of Frigia to the palaces of Mongo and must take risk after risk. Twelve chapters at 20 minutes each.
1940 240m/B Buster Crabbe, Charles Middleton, Carol Hughes. **VHS, Beta $49.95** *NOS, CAB ☆☆*

Spaceballs A humorous Brooks parody of recent science fiction pictures, mostly notably "Star Wars," with references to "Alien," the "Star Trek series," and "The Planet of the Apes." Disappointingly tame and tentative, but chuckle-laden enough for Brooks fans. The great man himself appears in two roles, including puny wise man/wise guy Yogurt.
1987 (PG) 96m/C Mel Brooks, Rick Moranis, John Candy, Bill Pullman, Daphne Zuniga, Dick Van Patten, John Hurt, George Wyner, Joan Rivers, Lorene Yarnell, Sal Viscuso, Stephen Tobolowsky, Dom DeLuise; **Dir:** Mel Brooks. **VHS, Beta, LV $19.95** *MGM ☆☆½*

SpaceCamp Gang o'teens and their instructor at the US Space Camp are accidentally launched on a space shuttle, and then must find a way to return to Earth. Hokey plot, subpar special effects; why bother? Well, it is "inspirational".

1986 (PG) 115m/C Kate Capshaw, Tate Donovan, Leaf Phoenix, Kelly Preston, Larry B. Scott, Tom Skerritt, Lea Thompson, Terry O'Quinn; **Dir:** Harry Winer. **VHS, Beta, LV $29.98** *LIV, VES, IME ☆☆*

Spaced Invaders Five ultra-cool aliens crash-land in a small midwestern town at Halloween. Local denizens mistake them for trick-or-treaters. Poorly made and a waste of time.
1990 (PG) 102m/C Douglas Barr, Royal Dano, Ariana Richards, Kevin Thompson, Jimmy Briscoe, Tony Cox, Debbie Lee Carrington, Tommy Madden; **Dir:** Patrick Read Johnson. **VHS, Beta, LV $19.99** *BVV ☆½*

Spaced Out A naughty sci-fi sex comedy that parodies everything from "Star Wars" to "2001," but not very well. The sultry female aliens are visually pleasing, though. Watch this one with the sound turned off. Also known as "Outer Reach".
1980 (R) 85m/C Barry Stokes, Glory Annen; **Dir:** Norman J. Warren. **VHS, Beta $19.99** *HBO Woof!*

Spacehunter: Adventures in the Forbidden Zone Galactic bounty hunter agrees to rescue three damsels held captive by a cyborg. Strauss ain't no Harrison Ford. Filmed in 3-D, but who cares?
1983 (PG) 90m/C *CA* Peter Strauss, Molly Ringwald, Michael Ironside, Ernie Hudson, Andrea Marcovicci; **Dir:** Lamont Johnson. **VHS, Beta, LV $12.95** *COL ☆½*

Spaceship Misguided attempt to spoof creature-features. Mad scientist tries to protect kindly monster from crazed crew. Not very funny, with the exception of the song-and-dance routine by the monster. Also known as "The Creature Wasn't Nice."
1981 (PG) 88m/C Cindy Williams, Bruce Kimmel, Leslie Nielsen, Gerrit Graham, Patrick Macnee, Ron Kurowski; **Dir:** Bruce Kimmel. **VHS $69.98** *LIV ☆½*

Spaceways Scientist Duff is beset by all sorts of troubles: his experimental rockets explode, his wife has an affair with an ambitious scientist, and when they disappear together, he's accused of killing them and placing their bodies in the exploded rocket. All in all, a pretty bad (and long) day for our hero. Why should we suffer through it with him?
1953 76m/B Howard Duff, Eva Bartok, Cecile Cheyreau, Andrew Osborn; **Dir:** Terence Fisher. **VHS, Beta, LV** *WGE, MLB ☆½*

Spaghetti House A handful of waiters are held hostage by ruthless killers in this innocuous but worthless Italian comedy. English subtitles.
1982 103m/C *IT* Nino Manfredi, Rita Tushingham; **Dir:** Giullo Paradisi. **VHS, Beta $29.95** *AXV ☆½*

Spaghetti Western A farmer and a publisher take on the vile oil baron who is taking over the town. A parody—or is it?—of Eastwoodian B-grade spaghetti westerns. No director credited.
196? 90m/C Franco Nero, Martin Balsam, Sterling Hayden. **VHS, Beta $19.98** *VDC ☆½*

Spalding Gray: Terrors of Pleasure The ultra-cool Gray tells the story of one man's dream to own land in his own inimitable way. HBO comedy special.
1988 60m/C Spalding Gray; **Dir:** Thomas Schlamme. **VHS, Beta $39.99** *HBO ☆☆☆*

Spangles Silent circus drama.

1926 58m/B Marion Nixon, Hobart Bosworth, Pat O'Malley, Gladys Brookwell; *Dir:* Frank O'Connor. **VHS, Beta $14.95** *GPV* ♫

The Spaniard's Curse When a man is convicted for a murder he did not commit, he puts a curse on the judge and jury responsible. Mysteriously, the marked people begin dying. Is he responsible? Then he dies and the mystery thickens. Intriguing but sloppy murder mystery.
1958 80m/B *GB* Tony Wright, Lee Patterson, Michael Hordern, Ralph Truman, Henry Oscar; *Dir:* Ralph Kemplen. **VHS, Beta $16.95** *SNC* ♫♫½

The Spanish Gardener A boy in a prominent family spends more time with the gentle gardner than with the domineering father, so dad arranges for his rival to be framed and sent to prison. An affecting British adaptation of the A.J. Cronin novel, turned into a showcase for Bogarde in the title role.
1957 95m/C Dirk Bogarde, Jon Whiteley, Michael Hordern, Cyril Cusack, Maureen Swanson, Lyndon Brook, Josephine Griffin, Bernard Lee; *Dir:* Philip Leacock. **VHS $19.95** *NOS* ♫♫♫

The Spanish Main Typical, gusto-laden swashbuckler, RKO's first in Technicolor. Evil Spanish governor Slezak captures Dutch crew led by Henreid. They escape and kidnap his fiancee (O'Hara) off a ship coming from Mexico. Henreid forces her to marry him, but his crew uses the might of the armada and returns O'Hara to Slezak behind Henreid's back. Wow!
1945 100m/C Paul Henreid, Maureen O'Hara, Walter Slezak, Binnie Barnes, John Emery, Barton MacLane; *Dir:* Frank Borzage. **LV $39.95** *IME* ♫♫½

Spare Parts Guests at a remote hotel discover that it is run by black marketeers who kill guests and sell the body parts. Of course, they don't bother signing out.
1985 108m/C Judith Speidel, Wolf Roth. **VHS, Beta $69.95** *VMK* **Woof!**

Sparkle The saga of three singing sisters struggling to rise to the top of the charts in the 1950s. Sound familiar? Well done but cliche fictional version of the Supremes' career. McKee shines. Alcohol, drugs, and mobsters get in the way. Curtis Mayfield wrote the excellent musical score.
1976 (PG) 98m/C Irene Cara, Lonette McKee, Dwan Smith, Philip Michael Thomas, Mary Alice, Dorian Harewood, Tony King; *Dir:* Sam O'Steen. **VHS, Beta $39.98** *WAR* ♫♫½

Sparrows Set to William Perry's score, a woman cares for nine orphan children on an impoverished farm, warding off starvation and kidnappers. Silent melodrama features notable performance by Pickford.
1926 81m/B Mary Pickford, Roy Stewart, Gustav von Seyffertitz; *Dir:* William Beaudine. **VHS, Beta, LV $19.95** *NOS, CCB, DVT* ♫♫♫

Spartacus The true story of a gladiator who leads other slaves in a rebellion against the power of Rome in 73 B.C. The rebellion is put down and the rebels are crucified. Douglas, whose political leanings are amply on display herein, also served as executive producer, surrounding himself with the best talent available. Magnificent climactic battle scene features 8,000 real, live Spanish soldiers to stunning effect. Additional Oscar nominations for Score and Film Editing. A newer version featuring Kubrick's "director's cut" is also available, featuring a restored, controversial homoerotic bath scene with

Olivier and Curtis. A box-office triumph that gave Kubrick much-desired financial independence.
1960 (PG-13) 196m/C Kirk Douglas, Laurence Olivier, Jean Simmons, Tony Curtis, Charles Laughton, Herbert Lom, Nina Foch, Woody Strode, Peter Ustinov, John Gavin, John Ireland; *Dir:* Stanley Kubrick. Academy Awards '60: Best Art Direction/Set Decoration (Color), Best Color Cinematography, Best Costume Design (Color), Best Supporting Actor (Ustinov). **VHS, Beta, LV $14.95** *MCA, APD, BTV* ♫♫♫♫

Spasms Reed is the unfortunate big game hunter who encounters The Demon Serpent, known as N'Gana Simbu, the deadliest snake in the world. Lame-brained horror fantasy; the producers should have released the nifty Tangerine Dream soundtrack and chucked the movie.
1982 (R) 92m/C *CA* Peter Fonda, Oliver Reed; *Dir:* William Fruet. **VHS, Beta $19.99** *HBO* ♫

The Spawning The sequel to "Piranha." At Club Elysium the main course is the vacationers themselves.
1982 (R) 88m/C Steve Marachuk; *Dir:* James Cameron. **VHS, Beta, LV $29.98** *EMB* ♫

Speak of the Devil A New Orleans evangelist who seduces his followers moves to Los Angeles and strikes a bargain with Satan. Exploitative, unthinking spoof of Swaggart/Bakker scandals of recent years.
1990 99m/C Robert Elarton, Jean Miller, Bernice Tamara Goor, Louise Sherill, Walter Kay, Shawn Patrick Greenfield. **VHS $79.95** *AIP* ♫

Speaking Parts VCR-obsessed laundry worker and another woman battle for the attention of bit-part actor McManus who works in ritzy hotel. Scripted by Canadian director Egoyan, it's a telling picture of the inextricable nature of modern technology.
1989 92m/C *CA* Michael McManus, Arsinee Khanjian, David Hemblen; *Dir:* Atom Egoyan. **VHS $79.95** *FXL, FCT* ♫♫♫

Special Bulletin A pacifistic terrorist threatens to blow up Charleston, South Carolina with a nuclear warhead. Done quite well in docu-drama style as a TV news bulletin. Interesting examination of the media's treatment of dramatic events written by director Zwick (of "thirtysomething")and Marshall Herskovitz. Top-notch made-for-TV fare.
1983 105m/C Ed Flanders, Christopher Allport, Kathryn Walker, Roxanne Hart; *Dir:* Edward Zwick. **VHS, Beta $59.95** *KRT, LHV, WAR* ♫♫♫

A Special Day The day of a huge rally celebrating Hitler's visit to Rome in 1939 serves as the backdrop for an affair between weary housewife Loren and lonely, unhappy homosexual radio announcer Mastroianni. Good performances from two thorough pros make a depressing film well worth watching. Italian with English subtitles. Oscar nomination for Mastroianni.
1977 105m/C *IT* Sophia Loren, Marcello Mastroianni; *Dir:* Ettore Scola. National Board of Review Awards '77: 5 Best Foreign Films of the Year. **VHS, Beta, LV $59.95** *COL, APD* ♫♫♫

Special Delivery Three unemployed Vietnam veterans decide to rob a bank but their getaway plans go awry. Svenson, the only robber who escapes, stashes the cash in a mailbox only to have it discovered by nutty artist, Shepherd, and crooked barkeep Gwynne.
1976 (PG) 99m/C Bo Svenson, Cybill Shepherd, Vic Tayback, Michael C. Gwynne, Tom Atkins, Sorrell Booke, Deidre Hall, Gerrit Graham, Jeff Goldblum; *Dir:* Paul Wendkos. **VHS, Beta $69.98** *LIV, VES* ♫♫½

Special Effects A desperate movie director murders a young actress, then makes a movie about her death. Solid, creepy premise sinks in the mire of flawed execution; a good film about Hollywood ego trips and obsession is lurking inside overdone script.
1985 (R) 103m/C Zoe Tamerlis, Eric Bogosian, Brad Rijn, Bill Oland, Richard Greene; *W/Dir:* Larry Cohen. **VHS, Beta $24.98** *SUE* ♫♫

Special Forces Eight specially trained soldiers drop behind enemy lines to rescue prisoners of war in World War II.
198? 90m/C Peter Lee Lawrence, Guy Madison, Erika Blanc, Tony Norton. **VHS, Beta** *MGL* ♫

The Specialist Lawyer West thinks he has seduced stunning Capri, but she has lured him into her clutches—she's been hired to kill him. Campy crud. It's old home week for '60s TV alums: West was Batman; bailiff Moore was Mr. Kimbell on Green Acres.
1975 (R) 93m/C Adam West, John Anderson, Anna Capri, Alvy Moore; *Dir:* Hikmet Avedis. **VHS, Beta $19.95** *VCI, VID* ♫½

The Speckled Band Set in 1930, this early talkie is a Sherlock Holmes adventure wherein the great detective must solve the mysterious death of a young woman. Massey makes his screen debut as Holmes, making the sleuth cynical, unhappy and pessimistic. Interesting prototypical Holmes case, faithful to the like-titled Conan Doyle story.
1931 84m/B Raymond Massey, Lyn Harding, Athole Stewart, Angela Baddeley, Nancy Price; *Dir:* Jack Raymond. **VHS, Beta, 8mm $16.95** *SNC, NOS, VYY* ♫♫♫

Specters Archaeologists excavating the Roman catacombs break open the gates of hell. Hellishly confused plot is unoriginal, to boot; decent production values hardly compensate.
1987 95m/C Donald Pleasence, John Pepper, Erna Schurer; *Dir:* Marcello Avallone. **VHS, Beta $79.95** *IMP* ♫

The Spectre of Edgar Allen Poe The horridly fictionalized writer of horror fiction visits the asylum where his love Lenore is being held, and discovers murder and torture. Based very loosely on Poe's own torments. Dorky, rip-off horror.
1973 87m/C Cesar Romero, Robert Walker Jr., Tom Drake; *Dir:* Mohy Quandour. **VHS, Beta $49.95** *UNI* ♫½

Spectre of the Rose Strange film set in the world of ballet. An impressario contends with an over-the-hill ballerina and a young male dancer with a deadly affinity for knives. Written, directed, and produced by Hecht.
1946 90m/B Judith Anderson, Michael Chekhov, Ivan Kirov, Viola Essen, Lionel Stander; *W/Dir:* Ben Hecht. **VHS $19.98** *REP* ♫♫½

The Speed Spook Silent ghost races car and steals documents.
1924 85m/B Johnny Hines, Warner Richmond; *Dir:* Charles Hines. **VHS, Beta $19.95** *GPV* ♫

Speed Zone Comic celebrities take over a high speed auto race when a redneck cop locks up the real drivers. Unfunny sequel to "Cannonball Run."
1988 (PG) 96m/C Melody Anderson, Peter Boyle, Tim Matheson, Donna Dixon, John Candy, Eugene Levy, Joe Flaherty, Matt Frewer, Shari Belafonte-Harper, Tom Smothers, Dick Smothers, Brooke Shields, Lee Van Cleef, Jamie Farr, John Schneider, Michael Spinks; *Dir:* Jim Drake. **VHS, Beta, LV $89.95** *MED* ♫

Speeding Up Time An angry man wants to ice the dudes who torched his mother's tenement.
1971 (R) 90m/C Winston Thrash, Pamela Donegan. VHS, Beta $59.98 MAG *

Speedtrap Typical chase scenes and cross-gender sparring characterize this tale of a private detective and a police officer pursuing car thieves. Good cast and fun chemistry between the leads; weak, typical script.
1978 (PG) 101m/C Joe Don Baker, Tyne Daly, Richard Jaeckel, Robert Loggia, Morgan Woodward, Timothy Carey; **Dir:** Earl Bellamy. VHS, Beta $69.95 MED **

Speedway Elvis the stock car driver finds himself being chased by Nancy the IRS agent during an important race. Will Sinatra keep to the business at hand? Or will the King melt her heart? Some cameos by real-life auto racers. Songs include "Speedway," "He's Your Uncle, Not Your Dad," and "There Ain't Nothing Like a Song." Movie number 27 for Elvis.
1968 90m/C Elvis Presley, Nancy Sinatra, Bill Bixby, Gale Gordon, William Schallert, Carl Ballantine, Ross Hagen; **Dir:** Norman Taurog. VHS, Beta $19.95 MGM **

Speedy Lloyd comes to the rescue when the last horse car in New York City, operated by his fiance's grandfather, is stolen by a gang. Thoroughly phoney, fun pursuit/action comedy shot on location. Look for a brief appearance by Babe Ruth.
1928 72m/B Harold Lloyd, Bert Woodruff, Ann Christy; **Dir:** Ted Wilde. VHS, Beta TLF ***

The Spell An obese fifteen-year-old girl has the power to inflict illness and death on the people she hates. Necessarily mean-spirited, if we're being asked to sympathize with the main character. Therein lies the rub. Made for TV.
1977 86m/C Lee Grant, James Olson, Susan Myers, Barbara Bostock, Lelia Goldoni, Helen Hunt; **Dir:** Lee Philips. VHS, Beta $14.99 GKK, WOV **

The Spellbinder L.A. lawyer Daly falls in love with a woman he saves from an attacker, then discovers she's a fugitive from a satanic cult that wants her back. Unoriginal, but slickly made.
1988 (R) 96m/C Timothy Daly, Kelly Preston, Rick Rossovich, Audra Lindley; **Dir:** Janet Greek. VHS, Beta $19.98 FOX *½

Spellbound Broken-hearted over dead girlfriend, young college student attempts to contact her through spiritualism, succeeds, and suffers nervous breakdown. Originally titled "The Spell of Amy Nugent."
1941 75m/B GB Derek Farr, Vera Lindsay, Frederick Leister, Hay Petrie, Felix Aylmer; **Dir:** John Harlow. VHS $19.98 SNC **

Spellbound Peck plays an amnesia victim accused of murder. Bergman plays the psychiatrist who uncovers his past through Freudian analysis and ends up falling in love with him. One of Hitchcock's finest films of the 1940s, with a riveting dream sequence designed by Salvador Dali. Full of classic Hitchcock plot twists and Freudian imagery. Oscar-nominated for Best Picture, Best Director, Best Supporting Actor (Chekov), Best Cinematography, and Best Special Effects. Based on Francis Bleeding's novel "The House of Dr. Edwardes."
1945 111m/B Ingrid Bergman, Gregory Peck, Leo G. Carroll, Michael Chekhov, Wallace Ford, Rhonda Fleming, Regis Toomey; **Dir:** Alfred Hitchcock. Academy Awards '45: Best Musical Score. VHS, Beta, LV $19.98 FOX, IME, TLF ***½

Spellcaster An evil wizard invites a DJ and a band of rock 'n roll fanatics to his thousand-year-old Italian castle for an evil and bloodcurdling treasure hunt.
1991 (R) 83m/C Richard Blade, Gail O'Grady, Harold Pruett, Bunty Bailey, Traci Lin, Adam Ant; **Dir:** Rafal Zielinski. VHS, LV $79.95 COL *

Spetters Four Dutch teenagers follow the motorcycle racing circuit and motocross champ Hauer. Misdirected youth film with a spicy performance from Soutendijk. Plenty of violence, sex, and gripping photography. Verhoeven went on to direct "Robocop" and "Total Recall."
1980 (R) 108m/C NL Rutger Hauer, Renee Soutendijk; **Dir:** Paul Verhoeven. VHS, Beta $29.98 SUE **½

The Sphinx A murderer causes havoc for his deaf-mute twin (Atwill, good in a dual role) when he frames him for his own crimes. Remade as "The Phantom Killer."
1933 63m/B Lionel Atwill, Theodore Newton, Sheila Terry, Paul Hurst, Luis Alberni, George "Gabby" Hayes; **Dir:** Phil Rosen. VHS $16.95 SNC, DNB **½

Sphinx A woman archaeologist searches for hidden riches in the tomb of an Egyptian king. The scenery is impressive, but otherwise, don't bother. Based on the novel by Robin Cooke.
1981 (PG) 117m/C Lesley-Anne Down, Frank Langella, John Gielgud, Maurice Ronet, John Rhys-Davies; **Dir:** Franklin J. Schaffner. VHS, Beta $19.98 WAR Woof!

Spices Poor woman Sonbai revolts against the sexist mores of rural colonial India in the 1940s. Features actress Patil in her final role. In Hindi with English subtitles.
1986 98m/C IN Smita Patil; **Dir:** Ketan Mehta. VHS $29.95 MFV, FCT, IVA ***

Spider Baby A tasteless horror-comedy about a chauffeur who takes care of a psychotic family. Also known as "The Liver Eaters." Theme song sung by Lon Chaney.
1964 86m/B Lon Chaney Jr., Mantan Moreland, Carol Ohmart, Sid Haig; **Dir:** Jack Hill. VHS, Beta $19.98 VEC, SNC, AOV *

Spider-Woman A collection of Spider-Woman's most daring escapades.
1982 100m/C VHS, Beta MCA *

Spiderman: The Deadly Dust A live-action episode of Spider-Man, as he attempts to prevent a city-destroying plutonium accident.
1978 93m/C Nicholas Hammond, Robert F. Simon, Chip Fields; **Dir:** Ron Satlof. VHS, Beta FOX *

Spiders Directed by Fritz Lang, this is one of his earliest surviving films, and predates Indiana Jones by almost 60 years. In these first two chapters ("The Golden Lake" and "The Diamond Ship") of an unfinished 4-part thriller, Carl deVogt battles with the evil Spider cult for a mystically powerful Incan diamond. Restored version has original color-tinted scenes. Silent with organ score by Gaylord Carter.
1918 137m/B GE Lil Dagover; **Dir:** Fritz Lang. VHS, Beta, LV $29.95 KIV, IME, GLV ***

The Spider's Stratagem Thirty years after his father's murder by the facists, a young man returns to a small Italian town to learn why his father was killed. The locals resist his efforts, and he is trapped in a mysterious web where history and lies exert a stranglehold on the truth. Intriguing, high literate thriller; outrageously lovely color photography. Based on a short story by Jorge Luis Borges. Italian with English subtitles.
1970 97m/C IT Giulio Brogi, Alida Valli; **Dir:** Bernardo Bertolucci. VHS $59.95 NYF, FCT, APD ***

Spiderwoman A modernized Holmes and Watson adventure as the duo track down a woman responsible for a series of murders. His adversary uses poisonous spiders to do her work. Zestful, superior Holmes. Also known as "Sherlock Holmes and the Spider Woman."
1944 62m/B Basil Rathbone, Nigel Bruce, Gale Sondergaard, Dennis Hoey; **Dir:** Roy William Neill. VHS, Beta $19.98 FOX ***

Spies A sly criminal poses as a famous banker to steal government information and create chaos in the world in this silent Lang masterpiece. Excellent entertainment, tight plotting and pacing, fine performances. Absolutely relentless intrigue and tension. Originally released as "Spione" in Germany.
1928 90m/B GE Rudolf Klein-Rogge, Lupo Pick, Fritz Rasp, Gerda Maurus, Willy Fritsch; **Dir:** Fritz Lang. New York Film Festival '78: Best Musical Score. VHS, Beta, LV $29.95 KIV, NOS, VYY ***½

Spies, Lies and Naked Thighs A pair of crackpot CIA agents find themselves in a whirlpool of comedic madness when they are assigned to track down a deadly assassin. The problem? The assassin happens to be the ex-wife of one of the guys. Can he go another round with her and this time come out on top? Who cares?
1991 90m/C Harry Anderson, Ed Begley Jr., Rachel Ticotin, Linda Purl, Wendy Crewson; **Dir:** James Frawley. VHS, LV $79.95 COL *½

Spies Like Us Chase and Aykroyd meet while taking the CIA entry exam. Caught cheating on the test, they seem the perfect pair for a special mission. Pursued by the Soviet government, they nearly start World War III. Silly, fun homage to the Bing Crosby-Bob Hope "Road" movies that doesn't capture those classics' quota of guffaws, but comes moderately close. Look for several cameos by film directors.
1985 (PG) 103m/C Chevy Chase, Dan Aykroyd, Steve Forrest, Bruce Davison, William Prince, Bernie Casey, Tom Hatton, Michael Apted, Frank Oz, Constantin Costa-Gavras, Terry Gilliam, Ray Harryhausen, Donna Dixon; **Dir:** John Landis. VHS, Beta, LV $19.98 WAR **½

Spike of Bensonhurst A Brooklyn street kid dreams of becoming a championship boxer. He tries to gain the mob's help by courting the local Don's daughter. Hip-but-stereotyped Morrissey effort is helped by great cast.
1988 (R) 101m/C Sasha Mitchell, Ernest Borgnine, Maria Pitillo, Talisa Soto, Sylvia Miles; **Dir:** Paul Morrissey. VHS, Beta, LV $89.95 VIR **½

The Spike Jones Story Musical satirist Spike Jones and his City Slickers are shown in compilation clips taken primarily from early 1950s telecasts, as well as in comments from Jones' family and surviving band members.
195? 60m/C Performed by: Spike Jones. VHS $19.95 WST **½

Spinout A pouty traveling singer decides to drive an experimental race car in a rally. Usual Elvis fare with the King being pursued by an assortment of beauties. Songs include "Stop, Look, Listen," "Never Say Yes,"

''Beach Shack,'' ''Adam and Evil,'' and ''I'll Be Back.'' Alternately titled ''California Holiday.''

1966 93m/C Elvis Presley, Shelley Fabares, Carl Betz, Diane McBain, Cecil Kellaway, Jack Mullaney, Deborah Walley, Una Merkel, Warren Berlinger, Will Hutchins, Dodie Marshall; *Dir:* Norman Taurog. **VHS, Beta $19.95** MGM 𝄞𝄞

The Spiral Staircase A mute servant, working in a creepy Gothic mansion, may be the next victim of a murderer preying on women afflicted with deformities, especially when the next murder occurs in the mansion itself. Great performance by McGuire as the terrified victim. Remade for television in 1975.

1946 83m/B Dorothy McGuire, George Brent, Ethel Barrymore, Kent Smith, Rhonda Fleming, Gordon Oliver, Elsa Lanchester, Sara Allgood; *Dir:* Robert Siodmak. **VHS, LV $39.95** FOX 𝄞𝄞𝄞

Spiral Staircase Mild TV remake of the 1946 classic about a mute servant who is menaced by a psychopathic killer. Why didn't they just show the original?

1975 99m/C GB Jacqueline Bisset, Christopher Plummer, John Phillip Law, Mildred Dunnock, Sam Wanamaker, Gayle Hunnicutt; *Dir:* Peter Collinson. **VHS, Beta $59.95** WAR 𝄞½

Spirit of '76 In the 22nd century, the Earth faces certain disaster as a magnetic storm wipes out all of American culture. Now time travellers must return to 1776 to reacquire the Constitution to fix things up. But when their computer goes on the blink, the do-gooders land, not in 1776, but in 1976 at the beginning of disco fever!

1991 (PG-13) 82m/C David Cassidy, Olivia D'Abo, Leif Garrett, Geoff Hoyle, Jeff McDonald, Steve McDonald, Liam O'Brien, Barbara Dain, Julie Brown, Thomas Chong, Devo, Iron Eyes Cody, Kipper Kids, Don Novello, Carl Reiner, Rob Reiner, Moon Zappa; *W/Dir:* Lucas Reiner. **VHS, LV $89.95** SVS 𝄞½

Spirit of the Beehive An acclaimed and haunting film about a young Spanish girl enthralled by the 1931 ''Frankenstein,'' embarking on a journey to find the creature in the Spanish countryside. One of the best films about children's inner life; in Spanish with subtitles.

1973 95m/C SP Fernando Gomez, Teresa Gimpera, Ana Torrent; *Dir:* Victor Erice. Venice Film Festival '73: Prix L'Age d'Or; Chicago Film Festival '73: Silver Hugo. **VHS, Beta, LV $29.95** CVC, TAM 𝄞𝄞𝄞

Spirit of the Eagle Man and young son wander in mountains and make friends with feathered creature. Then boy is kidnapped, creating problems for dad. Somnolent family fare.

1990 (PG) 93m/C Dan Haggerty, Bill Smith, Don Shanks, Jeri Arrendondo, Trever Yarrish; *W/Dir:* Boon Collins. **VHS, LV $89.98** SHG, IME 𝄞𝄞

Spirit of St. Louis A lavish Hollywood biography of famous aviator Charles Lindbergh and his historic transatlantic flight from New York to Paris in 1927, based on his autobiography. Intelligent; Stewart shines as the intrepid airman. Inexplicably, it flopped at the box office. Musical score by Franz Waxman.

1957 137m/C James Stewart, Patricia Smith, Murray Hamilton, Marc Connelly; *Dir:* Billy Wilder. National Board of Review '57: 10 Best Films of the Year. **VHS, Beta $19.98** WAR 𝄞𝄞𝄞

Spirit of the West Rodeo star Gibson and his ranch foreman brother work together to save a fair damsel's land from a gang of rustlers. Routine western of its era.

1932 61m/B Hoot Gibson, Hooper Atchley; *Dir:* Otto Brower. **VHS, Beta $19.95** NOS, DVT, UHV 𝄞½

The Spirit of West Point The true story of West Point's two All-American football greats, Doc Blanchard and Glenn Davis (who play themselves), while on leave after graduation from the title school.

1947 77m/B Felix ''Doc'' Blanchard, Glenn Davis, Tom Harmon, Alan Hale Jr., Anne Nagel, Robert Shayne; *Dir:* Ralph Murphy. **VHS, Beta, LV $19.95** STE, NWV 𝄞½

The Spirit of Youth Joe Louis supports his family with menial jobs until he shows his knack as a fighter. When he's knocked down, his gal appears at ringside to inspire him, and does.

1937 70m/B Joe Louis, Mantan Moreland. **VHS, Beta $29.95** NOS, VYY, VCN 𝄞𝄞

Spiritism Mother uses one of her three wishes to bring back her dead son. Adapted from the ''Monkey's Paw.''

1961 85m/B MX Nora Veyran, Jose Luis Jimenez, Jorge Mondragon, Rene Cardona Jr.; *Dir:* Benito Alazraki. **VHS $19.98** SNC 𝄞𝄞

Spirits A priest tormented by his lust-filled dreams breaks his vows of chastity with a woman who turns out to be a murderer. Bad luck.

1990 (R) 88m/C Erik Estrada, Carol Lynley, Robert Quarry, Brinke Stevens, Oliver Darrow, Kathrin Lautner; *Dir:* Fred Olen Ray. **VHS $89.98** VMK 𝄞½

Spirits of the Dead Three Edgar Allan Poe stories adapted for the screen and directed by three of Europe's finest. ''Metzengerstein,'' directed by Roger Vadim stars the Fonda siblings in a tale of incestuous lust. ''William Wilson'' finds Louis Malle directing Delon and Bardot in the story of a vicious Austrian army officer haunted by a murder victim. Finally, Fellini directs ''Never Bet the Devil Your Head'' or ''Toby Dammitt'' in which Stamp plays a drunken British film star who has a gruesome date with destiny. Although Fellini's segment is generally considered the best (and was released on its own) all three provide an interesting, atmospheric vision of Poe.

1968 (R) 117m/C IT FR Jane Fonda, Peter Fonda, Carla Marlier, Francoise Prevost, James Robertson Justice, Brigitte Bardot, Alain Delon, Katia Christine, Terence Stamp, Salvo Randone; *Dir:* Roger Vadim, Louis Malle, Federico Fellini; *Nar:* Vincent Price, Clement Biddle Wood. **VHS $18.00** FRG, MLB 𝄞𝄞𝄞

Spiritual Kung Fu A young man masters the art of spiritual kung fu and uses it to take on a roving band of warriors.

1978 97m/C Jackie Chan. **VHS, Beta $24.95** ASE 𝄞

Spite Marriage When Sebastian's lover dumps her like yesterday's garbage, she marries Keaton out of spite. Much postnuptual levity follows. Keaton's final silent.

1929 82m/B Buster Keaton, Dorothy Sebastian, Edward Earle, Leila Hyams, William Bechtel, Hank Mann; *Dir:* Edward Sedgwick. **VHS, Beta, LV $29.99** MGM, FCT 𝄞𝄞𝄞

Spitfire Sentimental comedy drama starring Hepburn as a hillbilly faith healer from the Ozarks who falls into a love triangle. One in a string of early box-office flops for Hepburn. She wanted to play a role other than patrician Eastern Seaboard, and did, but audiences didn't buy it.

1934 90m/B Katharine Hepburn, Robert Young, Ralph Bellamy, Sidney Toler; *Dir:* John Cromwell. **VHS, Beta, 8mm $44.95** VYY 𝄞𝄞𝄞

Spitfire True story of Reginald J. Mitchell, who designed ''The Spitfire'' fighter plane, which greatly assisted the Allies during World War II. Howard's last film. Also known as ''The First of the Few.'' Heavily propagandist but enjoyable and uncomplicated biography. Splendid score by composer William Walton.

1942 88m/B GB Leslie Howard, David Niven, Rosamund John, George King, Jon Stafford, Adrian Brunel; *Dir:* Leslie Howard. **VHS, Beta $9.95** NEG, NOS, HHT 𝄞𝄞𝄞

Spitfire: The First of the Few Critically-acclaimed biography of R.J. Mitchell, the designer of Spitfire—military plane. Focuses on the struggle and sacrifice the patriotic hero made in his life to provide a plane that would end the invasions of the Third Reich. Wonderful script and musical score.

1943 90m/B Leslie Howard, David Niven, Rosamund John, Roland Culver; *Dir:* Leslie Howard. **VHS, Beta $29.95** VYY 𝄞𝄞𝄞½

Spittin' Image After falling from her mean-spirited father's wagon, a young girl befriends a kind mountain man. She learns the ways of the wilderness and tobacco spitting.

1983 92m/C Sunshine Parker, Trudi Cooper, Sharon Barr, Karen Barr; *Dir:* Russell Kern. **VHS, Beta $59.95** PSM 𝄞

Splash A beautiful mermaid ventures into New York City in search of a man she's rescued twice when he's fallen overboard. Now it's her turn to fall—in love. Charming performances by Hanks and Hannah. Well-paced direction from Howard, with just enough slapstick. Don't miss the lobster scene.

1984 109m/C Tom Hanks, Darryl Hannah, Eugene Levy, John Candy, Dody Goodman, Shecky Greene, Richard B. Shull, Bobby DiCicco, Howard Morris; *Dir:* Ron Howard. **VHS, Beta, LV, 8mm $79.95** TOU, BVV 𝄞𝄞½

Splatter University A deranged killer escapes from an asylum and begins to slaughter and mutilate comely coeds at a local college. Abysmally motiveless killing and gratuitous sex. Also available in a 78-minute ''R'' rated version.

1984 79m/C Francine Forbes, Dick Biel, Cathy Lacommaro, Ric Randing, Dan Eaton; *Dir:* Richard W. Haines. **VHS, Beta $69.98** LIV, VES Woof!

Splendor in the Grass A drama set in rural Kansas in 1925, concerning a teenage couple who try to keep their love on a strictly intellectual plane and the sexual and family pressures that tear them apart. After suffering a mental breakdown and being institutionalized, the girl returns years later in order to settle her life. Film debuts of Warren Beatty, Sandy Dennis, and Phyllis Diller. William Inge wrote the screenplay specifically with Beatty in mind, after the actor appeared in one of Inge's stage plays. Wood won an Oscar nomination. Filmed not in Kansas, but on Staten Island and in upstate New York.

1961 124m/C Natalie Wood, Warren Beatty, Audrey Christie, Barbara Loden, Zohra Lampert, Phyllis Diller, Sandy Dennis; *Dir:* Elia Kazan. Academy Awards '61: Best Story & Screenplay. **VHS, Beta, LV $19.98** WAR, PIA 𝄞𝄞𝄞

Split Sixteen-year-old Susy James tells her own story about running away from home to live the life of a punk-type. Facts and fiction are mixed, with an original score by Jill Kroesen.

1981 20m/C VHS, 3/4U, Special order formats CST 𝄞𝄞𝄞

Split Humanoids from another dimension manipulate earth activity.
1990 85m/C John Flynn, Timothy Dwight, Chris Shaw, Joan Bechtel; *Dir:* Chris Shaw. VHS $79.95 *AIP* ℐ

Split Decisions An Irish family of boxers, dad and his two sons, slug it out emotionally and physically, as they come to terms with career choices and each other. Good scenes in the ring but the drama leans toward melodrama. Decent family drama, but somewhat of a "Rocky" rip-off.
1988 (R) 95m/C Gene Hackman, Craig Sheffer, Jeff Fahey, Jennifer Beals, John McLiam, Eddie Velez, Carmine Caridi, James Tolken; *Dir:* David Drury. VHS, LV, 8mm $19.95 *WAR* ℐℐ

Split Image An all-American boy comes under the spell of a cult. His parents then hire a deprogrammer to bring the boy back to reality. Exceptional performances by Dennehy and Woods. This is a worthwhile film that spent far too little time in theatres.
1982 (R) 113m/C Michael O'Keefe, Karen Allen, Peter Fonda, Brian Dennehy, James Woods; *Dir:* Ted Kotcheff. VHS, Beta $24.98 *SUE* ℐℐℐ

Split Second An escaped prisoner holds hostages in a Nevada atomic bomb testing area. Powell's directorial debut, and a script by William Bowers and Irving Wallace. McNally's excellent performance as the kidnapper, in addition to strong supporting performances, enhance a solid plot.
1953 85m/B Paul Kelly, Richard Egan, Jan Sterling, Alexis Smith, Stephen McNally; *Dir:* Dick Powell. VHS, Beta $39.95 *RKO* ℐℐ½

Split Second Hauer stars as a futuristic cop tracking down a vicious serial killer in London in the year 2008. The monster rips out the hearts of his victims and then eats them in what appears to be a satanic ritual in this blood-soaked thriller. Hauer gives a somewhat listless performance and overall, the action is quite dull. The music soundtrack also manages to annoy with the Moody Blues song "Nights in White Satin" playing at the most inappropriate times.
1992 (R) 91m/C *GB* Rutger Hauer, Kim Cattrall, Neil Duncan, Michael J. Pollard, Alun Armstrong, Peter Postlethwaite, Ian Dury, Roberta Eaton; *Dir:* Tony Maylam. VHS $92.99 *HBO* ℐ½

Splitz An all-girl rock band agrees to help out a sorority house by participating in a series of sporting events.
1984 (PG-13) 89m/C Robin Johnson, Patti Lee, Shirley Stoler, Raymond Serra; *Dir:* Domonic Paris. VHS, Beta $79.98 *LIV, VES Woof!*

The Spoilers Virtually unstaged account of gold hunting in the Alaskan wilderness enhanced by gritty film quality. Was a gold mine when released.
1914 110m/B William Farnum, Thomas Santschi, Kathlyn Williams, Bessie Eyton, Frank Clark, Wheeler Oakman; *Dir:* Colin Campbell. VHS, Beta $19.95 *GPV* ℐℐ½

The Spoilers Two adventurers in the Yukon are swindled out of their gold mine and set out to even the score. A trademark scene of all versions of the movie (and there are many) is the climactic fistfight, in this case between hero Wayne and bad-guy Scott. One of the better films adapted from the novel by Rex Beach. William Farnum, who starred in both the 1914 and the 1930 versions, has a small part. The film earned an Oscar nomination for set design.
1942 84m/B John Wayne, Randolph Scott, Marlene Dietrich, Margaret Lindsay, Harry Carey, Richard Barthelmess; *Dir:* Ray Enright. VHS, Beta, LV $19.95 *MCA, MLB* ℐℐℐ

Spontaneous Combustion A grisly horror film detailing the travails of a hapless guy who has the power to inflict the title phenomenon on other people.
1989 (R) 97m/C Brad Dourif, Jon Cypher, Melinda Dillon, Cynthia Bain, William Prince; *Dir:* Tobe Hooper. VHS, Beta, LV $89.98 *IVE* ℐℐ

Spook Busters The Bowery Boys take jobs as exterminators, only to find themselves assigned the unenviable task of ridding a haunted house of ghosts. To make matters worse, the resident mad scientist wants to transplant Sach's brain into a gorilla. Essential viewing for anyone who thought "Ghostbusters" was an original story.
1946 68m/B Leo Gorcey, Huntz Hall, Douglas Dumbrille, Bobby Jordan, Gabriel Dell, Billy Benedict, David Gorcey, Bernard Gorcey, Tanis Chandler, Maurice Cass, Charles Middleton; *Dir:* William Beaudine. VHS $14.98 *WAR* ℐℐ½

The Spook Who Sat By the Door A black CIA agent organizes an army of inner-city youths and launches a revolution. Based on the novel by Sam Greenlee.
1973 (PG) 95m/C Lawrence Cook, Paula Kelly, J.A. Preston; *Dir:* Ivan Dixon. VHS, Beta $69.95 *VCD* ℐ½

Spookies An old master sorcerer who lives in a run-down haunted house sacrifices humans to give eternal life to his unconscious bride. He needs only a few more vitims when a group of teenagers come along to explore the house.
1985 (R) 85m/C Felix Ward, Dan Scott, Maria Pechukas, Brendan Faulkner, Eugine Joseph; *Dir:* Thomas Doran. VHS, Beta, LV $14.95 *SVS Woof!*

Spooks Run Wild The East Side Kids seek refuge in a spooky mansion owned by the eerie Lugosi. A fun horror-comedy with the Kids' antics playing off Lugosi's scariness quite well. Co-scripter Carl Foreman later co-wrote "High-Noon," "The Bridge on the River Kwai," and "The Guns of Navarone."
1941 64m/B Huntz Hall, Leo Gorcey, Bobby Jordan, Sammy Morrison, Dave O'Brien, Dennis Moore, Bela Lugosi; *Dir:* Phil Rosen. VHS, Beta $19.95 *NOS, SNC, VYY* ℐℐ½

Spoon River Anthology This film is based on Edgar Lee Master's poetry in which we can sense the complexity of American life as it entered the twentieth century. Masters probes human strengths and weaknesses.
1976 21m/C VHS, Beta *BFA* ℐℐ½

Sport Goofy Six sport-themed cartoons starring that lovable mutt, Goofy.
1988 45m/C VHS, Beta *DIS* ℐℐ½

Sport Goofy's Vacation Six more hilarious cartoons starring Goofy.
1988 45m/C VHS, Beta *DIS* ℐℐ½

The Sporting Club A semi-gothic melodrama about a strict all men's club which, during a 100th anniversary party, reverts to savagery and primitive rites.
1972 (R) 104m/C Robert Fields, Margaret Blye, Nicholas Coster, Ralph Waite, Jack Warden, Linda Blair; *Dir:* Larry Peerce. VHS, Beta $24.98 *CHA Woof!*

Sports Illustrated's 25th Anniversary Swimsuit Video The 25th anniversary of the erstwhile skin issue of Sports Illustrated is honored by this making-of video.

1989 55m/C Christie Brinkley, Elle Macpherson, Stephanie Seymour, Cheryl Tiegs, Kathy Ireland. VHS, Beta $19.99 *ESP, HBO Woof!*

Spotlight Scandals A barber and a vaudevillian team up and endure the ups and downs of showbiz life in this low-budget musical from prolific B-movie director Beaudine.
1943 79m/B Billy Gilbert, Frank Fay, Bonnie Baker; *Dir:* William Beaudine. VHS $19.95 *NOS, LOO* ℐℐ

Spraggue A Boston professor and his eccentric aunt put the moves on a doctor who may have committed murder. Served as a TV pilot for a series that never materialized.
1984 78m/C Michael Nouri, Glynis Johns, James Cromwell, Mark Herrier, Patrick O'Neal, Andrea Marcovicci; *Dir:* Larry Elikann. VHS $59.95 *IVE* ℐℐ

The Spring Two archaeologists search Florida for the fountain of youth after they find new clues to its whereabouts. However, a greedy industrialist and an evil priest are both on their trail, fighting to keep the secret for themselves at any cost.
1989 110m/C Dack Rambo, Gedde Watanabe, Shari Shattuck, Steven Keats; *Dir:* John D. Patterson. VHS *QUE* ℐ½

Spring Break Four college students go to Fort Lauderdale on their spring vacation and have a wilder time that they bargained for.
1983 (R) 101m/C Perry Lang, David Knell, Steve Bassett, Paul Land, Jayne Modean, Corinne Alphen; *Dir:* Sean S. Cunningham. VHS, Beta, LV $14.95 *COL Woof!*

Spring Fever Heartaches of the junior tennis circuit are brought to the screen in this sports comedy. Previews promised a lot more than this film could ever hope to deliver. Even tennis fans will be disappointed.
1981 93m/C *CA* Susan Anton; *Dir:* Joseph L. Scanlan. VHS, Beta $69.98 *LIV, VES* ℐ½

Spring Symphony A moody, fairy-tale biography of composer Robert Schumann, concentrating on his rhapsodic love affair with pianist Claire Weick. Kinski is very good as the rebellious daughter, while the music is even better. Reasonably accurate in terms of history; dubbed.
1986 (PG-13) 102m/C *GE* Nastassia Kinski, Rolf Hoppe, Herbert Gronemeyer; *Dir:* Peter Schamoni. VHS, Beta $79.98 *LIV, LTG* ℐℐ½

Springhill The true story of a mining disaster in which seven miners fought to survive the elements while buried 13,000 feet underground.
197? 90m/C VHS, Beta $49.95 *TWE* ℐ½

Springtime in the Rockies Autry is a cattle-ranch foreman whose employer decides to try raising sheep. This doesn't set well with the other ranchers, but Autry manages to save the day and sing a little too.
1937 60m/B Gene Autry, Polly Rowles, Smiley Burnette; *Dir:* Joseph Kane. VHS, Beta $19.98 *NOS, VCN, DVT* ℐ½

Springtime in the Rockies A Broadway duo just can't get along despite being in love with each other. Top-notch musical, with a touch of romantic tension and comedy. Beautifully filmed in the Canadian Rockies.
1942 91m/C Betty Grable, John Payne, Carmen Miranda, Cesar Romero, Charlotte Greenwood, Edward Everett Horton, Jackie Gleason; *Dir:* Irving Cummings. VHS, LV $19.95 *FOX, MLB* ℐℐ½

Springtime in the Sierras Rogers and Devine band together to fight a gang of poachers who prey on the wildlife of a game preserve. Features a number of songs by Rogers and The Sons of the Pioneers.
1947 54m/B Roy Rogers, Andy Devine, Jane Frazee, Stephanie Bachelor; **Dir:** William Whitney. **VHS, Beta $19.95** *NOS, CAB, DVT* ✂✂½

The Sprinter A young homosexual, while trying to submerge his confusion in professional track and field, gets to know a homely shot-putter and is seduced by her. In German with English subtitles.
1984 90m/C VHS, Beta $39.95 *VCD, GLV* ✂✂½

Spurs A rodeo star takes on a whole gang of rustlers and their imposing machine gun. This transitional western was made both with and without sound.
1930 60m/B Hoot Gibson, Robert E. Homans; **Dir:** Hoot Gibson. **VHS $19.95** *NOS, DVT* ✂✂

Sputnik A Frenchman, amnesiac after a car crash, comes up against Russian scientists, space-bound dogs, and weightlessness. Pleasant and charming family fun though clearly dated. Another fine performance from Auer. This film is also titled "A Dog, a Mouse, and a Sputnik."
1961 80m/B *FR* Noelia Noel, Mischa Auer, Denise Gray; **Dir:** Jean Dreville. **VHS, Beta $39.95** *SVS* ✂✂

Spy A former spy who knows too much becomes the target of a renegade agency. Made for television.
1989 88m/C Ned Beatty, Tim Choate, Bruce Greenwood, Catherine Hicks, Jameson Parker, Michael Tucker; **Dir:** Philip Frank Messina. **VHS, Beta, LV $79.95** *PAR* ✂✂

Spy in Black A German submarine captain returns from duty at sea during World War I and is assigned to infiltrate one of the Orkney Islands and obtain confidential British information. Known in the U.S. as "U-Boat 29," this film is based on a J. Storer Clouston novel. This was the first teaming of director Powell and writer Pressburger, who followed with "Contraband" in 1940 .
1939 82m/B *GB* Conrad Veidt, Valerie Hobson, Sebastian Shaw, Marius Goring, June Duprez, Helen Hayes, Cyril Raymond, Hay Petrie; **Dir:** Michael Powell. **VHS, Beta, LV $19.95** *NOS, VYY, DVT* ✂✂

Spy of Napoleon A French aristocrat agrees to marry the illegitimate daughter of Emperor Napoleon III and finds himself working to uncover traitors to the throne. Based on Baroness Orczy's novel.
1936 77m/B Richard Barthelmess, Dolly Hass, Francis L. Sullivan. **VHS, Beta $29.95** *VYY* ✂✂

Spy Smasher Returns The exciting 12-part serial "Spy Smasher" was cut and edited into feature length and released under this title. The masked marvel and his twin brother battle enemy agents during World War II. This new version is just as exciting as the original cliffhanger.
1942 185m/B Kane Richmond, Marguerite Chapman, Sam Flint, Hans Schumm, Tristram Coffin; **Dir:** William Witney. **VHS $29.98** *REP, MLB* ✂✂

Spy Story A novel spy/counter-spy story written by Len Deighton.
197? 105m/C VHS, Beta $69.95 *VCD* ✂✂

The Spy Who Came in from the Cold The acclaimed adaptation of the John Le Carre novel about an aging British spy who attempts to infiltrate the East German agency. Prototypical Cold War thriller, with emphasis on de-glamorizing espionage. Gritty and superbly realistic with a documentary style which hampered it at the box office. Burton received a Best Actor nomination. Photographer Oswald Morris won Best Black-and-White Cinematographer honors from the British Film Academy.
1965 110m/B Richard Burton, Oskar Werner, Claire Bloom, Sam Wanamaker, Peter Van Eyck, Cyril Cusack, Rupert Davies, Michael Hordern; **Dir:** Martin Ritt. Edgar Allan Poe Awards '65: Best Screenplay. **VHS, Beta, LV $19.95** *PAR* ✂✂✂½

The Spy Who Loved Me James Bond teams up with female Russian Agent XXX to squash a villain's plan to use captured American and Russian atomic submarines in a plot to destroy the world. The villain's henchman, 7'2" Kiel, is the steel-toothed Jaws. Carly Simon sings the memorable, Marvin Hamlisch theme song, "Nobody Does It Better."
1977 (PG) 125m/C *GB* Roger Moore, Barbara Bach, Curt Jurgens, Richard Kiel, Caroline Munro, Walter Gotell, Geoffrey Keen, Valerie Leon, Bernard Lee, Lois Maxwell, Desmond Llewelyn; **Dir:** Lewis Gilbert. **VHS, Beta, LV $19.95** *MGM, FOX, TLF* ✂✂

Spy With a Cold Nose A sporadically funny spy spoof about a bugged dog passed between British and Soviet intelligence. Hard-working cast keeps it from collapsing.
1966 93m/C Laurence Harvey, Daliah Lavi, Lionel Jeffries; **Dir:** Daniel Petrie. **VHS, Beta $59.95** *CHA* ✂✂

Spymaker: The Secret Life of Ian Fleming Fictionalized account of the creator of the ultimate spy James Bond and his early days in the British Secret Service. Fans of the Bond movies will appreciate the numerous inside jokes. Connery is the son of Sean. Also known as "Spymaker." Made for British television.
1990 100m/C *GB* Jason Connery, Kristin Scott Thomas, Joss Ackland, Patricia Hodge, David Warner, Fiona Fullerton, Richard Johnson; **Dir:** Ferdinand Fairfax. **VHS** *TTC* ✂✂½

Spyro Gyra Live performances of some of Spyro Gyra's well-known hits are interspersed with interviews with band members.
1980 56m/C VHS, Beta $29.98 *WAR, OM* ✂½

S*P*Y*S An attempt to cash in on the success of "M*A*S*H," this unfunny spy spoof details the adventures of two bumbling CIA men who botch a Russian defection, and get both sides after them. Usually competent director Kershner had a bad day.
1974 (PG) 87m/C Donald Sutherland, Elliott Gould, Joss Ackland, Zouzou, Shane Rimmer, Vladek Sheybal, Nigel Hawthorne; **Dir:** Irvin Kershner. **VHS, Beta $59.98** *FOX* ✂

Squadron of Doom Government troubleshooter Ace Drummond battles enemy agents out to steal a mountain of jade.
1936 70m/B John "Dusty" King, Jean Rogers, Noah Beery Jr. **VHS, Beta $19.95** *NOS, VCN, DVT* ✂✂

Square Dance A Texas teenager leaves the farm where she's been raised by her grandfather to live in the city with her promiscuous mother (Alexander, cast against type) where she befriends a retarded young man (yes, it's Lowe, also cast against type). Too slow, but helped by good acting; TV title was "Home is Where the Heart Is."
1987 (PG-13) 118m/C Jane Alexander, Jason Robards Jr., Rob Lowe, Winona Ryder; **Dir:** Daniel Petrie. **VHS, Beta, LV $19.95** *PAV* ✂✂½

Square Dance Jubilee Television scouts hit Prairie City in search of cowboy stars, and stumble onto cattle rustlers. Twenty-one C & W tunes support a near-invisible plot, suitable for fridge runs.
1951 78m/B Mary Beth Hughes, Donald (Don "Red") Barry, Wally Vernon, John Eldridge; **Dir:** Paul Landres. **VHS, Beta, LV** *WGE* ✂½

Square Shoulders Adventures of young military school student with geometric anatomy (Coghlan).
1929 58m/B Junior Coghlan, Louis Wolheim, Anita Louis, Montague Shaw; **Dir:** E. Mason Hopper. **VHS, Beta $19.95** *GPV, FCT* ✂½

The Squeaker Complicated but interesting thriller about the underworld goings-on after a big-time diamond heist. German remake of an early (1930) British talkie, also done in 1937, based on the Edgar Wallace novel.
1965 95m/B *GE* Heinz Drache, Eddi Arent, Klaus Kinski, Barbara Rutting; **Dir:** Alfred Vohrer. **VHS** *SNC* ✂✂½

The Squeeze Scotland Yard detective Keach, fired for drunkenness, gets a chance to reinstate himself when his ex-wife is caught up in a brutal kidnapping scheme. Slim script gives good cast uphill work. Ordinary thriller. Available with Spanish subtitles.
1977 (R) 106m/C Stacy Keach, Carol White, David Hemmings, Edward Fox, Stephen Boyd; **Dir:** Michael Apted. **VHS, Beta, LV $19.98** *WAR* ✂✂

The Squeeze An aging safecracker is hired for a final job, but learns that his cohorts plan to kill him when the heist is finished. Blah revenge thriller. Dawson is a pseudonym for Antonio Margheriti.
1980 93m/C *IT* Lee Van Cleef, Karen Black, Edward Albert, Lionel Stander, Robert Alda; **Dir:** Anthony M. Dawson. **Beta $14.99** *MFI* ✂½

The Squeeze Keaton is wasted in this attempt at a comedy about a small-time con artist who discovers a Mafia plan to fix a lottery electromagnetically.
1987 (PG-13) 101m/C Michael Keaton, Rae Dawn Chong, John Davidson, Ric Abernathy, Bobby Bass, Joe Pantoliano, Meatloaf, Paul Herman; **Dir:** Roger Young. **VHS, Beta, LV $14.99** *HBO* ✂½

Squeeze Play A group of young women start a softball team and challenge their boyfriends' team to a game. It's a no-holds-barred competition, wet T-shirt contest and all. The usual cheap, plotless, offensive trash.
1979 92m/C Al Corley, Jenni Hetrick, Jim Metzler; **Dir:** Samuel Weil. **VHS, Beta, LV $19.99** *HBO, PIA* ✂

Squiddly Diddly Compilation of "Squiddly Diddly" cartoons, about a star-struck squid hoping to break into show business.
196? 55m/C VHS, Beta $19.95 *HNB* ✂

Squirm A storm disrupts a highly charged power cable, turning garden-variety worms into giant monsters that terrorize a small town in Georgia. The opening credits claim it's based on an actual 1975 incident.
1976 (R) 92m/C Don Scardino, Patricia Pearcy, Jean Sullivan; **Dir:** Jeff Lieberman. **VHS, Beta, LV $59.98** *LIV, VES* ✂½

Squizzy Taylor The rise and fall of real-life Australian mob boss Squizzy Taylor in the 1920s. Colorful period gangster drama loses steam about halfway through.

1984 82m/C *AU* Jacki Weaver, Alan Cassell, David Atkins; *Dir:* Kevin Dobson. **VHS, Beta** $69.95 *VID* 🐾🐾

S.S. Experiment Camp 2 Hitler's elite corps of sadists get their kicks torturing and maiming prisoners during the dark days of World War II.
1986 90m/C John Braun, Macha Magal. **VHS, Beta** $59.95 *MGL* 🐾

SS Girls After the assassination attempt of July 1944, Hitler does not trust the Wermacht and extends the power of the SS over Germany. General Berger entrusts Hans Schillemberg to recruit a specially chosen group of prostitutes who must test the fighting spirit and loyalty of the generals.
1978 82m/C Gabriele Carrara, Marina Daunia, Vassilli Karis, Macha Magal, Thomas Rudy, Lucic Bogoljub Benny, Ivano Staccioli. **VHS, Beta** *MED* 🐾

S.S. Love Camp No. 27 A beautiful young girl's life is torn to shreds when she is imprisoned in a Nazi brothel.
197? 90m/C **VHS, Beta** *VCD* 🐾

Stacey Ex-Playmate Randall stars as Stacey Hansen, a beautiful, implausibly talented private detective. Hyper-convoluted plot, too much sex and violence detract from a potentially not-wretched whodunnit. Also called "Stacey and Her Gangbusters."
1973 (R) 87m/C Anne Randall, Marjorie Bennett, Anitra Ford, Alan Landers, James Westmoreland, Christina Raines; *Dir:* Andy Sidaris. **VHS, Beta** $59.95 *GEM* 🐾½

Stacking In the 1950s, a Montana family is threatened with losing their farm. Typical, boring '80s farmland tragedy with coming of age tale. Originally titled "Season Of Dreams."
1987 (PG) 111m/C Christine Lahti, Megan Follows, Frederic Forrest, Peter Coyote, Jason Gedrick; *Dir:* Martin Rosen. **VHS, Beta, LV** $19.98 *CHA* 🐾🐾

Stacy's Knights A seemingly shy girl happens to have an uncanny knack for blackjack. With the odds against her and an unlikely group of "knights" to aid her, she sets up an implausible "sting" operation. Blah TV fodder for the big screen.
1983 (PG) 95m/C Kevin Costner, Andra Millian; *Dir:* Jim Wilson. **VHS, Beta** $29.98 *LIV, VES* 🐾🐾

Stag Party This is a video of a stag party, for viewing at stag parties.
1984 60m/C **VHS, Beta** $24.95 *AHV* 🐾🐾

Stage Door Based on the play by Edna Ferber and George Kaufman, an energetic ensemble peek at the women of the theater. A boarding house for potential actresses houses a wide variety of talents and dreams. Patrician Hepburn and wisecracking Rogers make a good team in a talent-packed ensemble. Realistic look at the sub-world of Broadway aspirations includes dialogue taken from idle chat among the actresses between takes. Based on the play by Edna Ferber and George S. Kaufman, who suggested in jest a title change to "Screen Door," since so much had been changed. Watch for young stars-to-be like Ball, Arden and Miller. Oscar nominated for Best Picture, Best Director, Best Supporting Actress (Leeds), and Best Screenplay.
1937 92m/B Katharine Hepburn, Ginger Rogers, Lucille Ball, Eve Arden, Andrea Leeds, Jack Carson, Adolphe Menjou, Gail Patrick; *Dir:* Gregory La Cava. National Board of Review Awards '37: 10 Best Films of the Year. **VHS, Beta** $19.98 *TTC, CCB, MED* 🐾🐾🐾½

Stage Door Canteen The Stage Door Canteens were operated by the American Theatre Wing during World War II for servicemen on leave. They were staffed by some of the biggest stars of the day, 65 of whom are featured here. The slight, hokey plot concerns three soldiers who fall for canteen workers while on furlough in New York City. Many musical numbers, cameos, and walk-ons by the plethora of stars.
1943 135m/B Tallulah Bankhead, Merle Oberon, Katharine Hepburn, Paul Muni, Ethel Waters, Judith Anderson, Ray Bolger, Helen Hayes, Harpo Marx, Gertrude Lawrence, Ethel Merman, Edgar Bergen, George Raft, Benny Goodman, Peggy Lee, Count Basie, Kay Kyser, Guy Lombardo, Xavier Cugat, Johnny Weissmuller; *Dir:* Frank Borzage. **VHS, Beta, LV** $9.95 *NEG, NOS, PSM* 🐾🐾½

Stage Fright Wyman will stop at nothing to clear her old boyfriend, who has been accused of murdering the husband of his mistress, an actress (Dietrich). Disguised as a maid, she falls in love with the investigating detective, and discovers her friend's guilt. Dietrich sings "The Laziest Gal in Town." The Master's last film made in England until "Frenzy" (1971).
1950 110m/B *GB* Jane Wyman, Marlene Dietrich, Alastair Sim, Sybil Thorndike, Michael Wilding, Kay Walsh; *Dir:* Alfred Hitchcock. National Board of Review Awards '50: 10 Best Films of the Year. **VHS, Beta, LV** $19.98 *WAR, LDC, MLB* 🐾🐾🐾

Stage Fright A bashful actress is transformed into a homicidal killer after a latent psychosis in her becomes active.
1983 82m/C Jenny Neumann, Gary Sweet. **VHS, Beta** $19.98 *VID* 🐾

Stage to Mesa City Marshall LaRue must find out who is behind the pesky attacks on the stage line to Mesa City. Cheseboro, as usual, is a good bad guy. A few chuckles between fist fights and shoot-em-ups.
1948 52m/B Lash LaRue, Al "Fuzzy" St. John, George Cheseboro, Jennifer Holt, Russell Arms; *Dir:* Ray Taylor. **VHS, Beta** $12.95 *COL, SVS* 🐾🐾

Stage Struck Strasberg reprises the role made famous by Katharine Hepburn as a determined, would-be actress in this mediocre remake of "Morning Glory" (1933). Christopher Plummer's screen debut shows little of his later talent.
1957 95m/B Henry Fonda, Susan Strasberg, Christopher Plummer, Herbert Marshall, Joan Greenwood; *Dir:* Sidney Lumet. **VHS, Beta, LV** $29.95 *UHV, IME* 🐾🐾½

Stagecoach A varied group of characters with nothing in common are stuck together inside a coach besieged by bandits and Indians. Considered structurally perfect, with excellent direction by Ford, it's the film that made Wayne a star as the Ringo Kid, an outlaw looking to avenge the murder of his brother and father. The first pairing of Ford and Wayne changed the course of the modern western. Stunning photography by Bert Glennon and Ray Binger captured the mythical air of Monument Valley, a site that Ford was often to revisit. Based on "Stage to Lordsburg" by Ernest Haycox. Remade miserably with Ann-Margret, Red Buttons, and Bing Crosby as the drunken doctor, in 1966 and again—why?—as a TV movie in 1986. Many Academy Awards plus the following nominations: Best Picture, Best Director, Best Interior Decoration, Best Film Editing.

1939 100m/B John Wayne, Claire Trevor, Thomas Mitchell, George Bancroft, John Carradine, Andy Devine, Donald Meek; *Dir:* John Ford. Academy Awards '39: Best Supporting Actor (Mitchell), Best Musical Score; National Board of Review Awards '39: 10 Best Films of the Year; New York Film Critics Awards '39: Best Director. **VHS, Beta, LV** $19.98 *WAR, VES, PIA* 🐾🐾🐾🐾

Stagecoach A forgettable made-for-television remake of the classic 1939 western about a motley crew of characters in a cross-country coach beset by thieves and Indians.
1986 95m/C Willie Nelson, Waylon Jennings, Johnny Cash, Kris Kristofferson, John Schneider, Elizabeth Ashley, Mary Crosby, Anthony Newley, Anthony (Tony) Franciosa; *Dir:* Ted Post. **VHS, Beta** $19.95 *STE, VMK* 🐾

Stagecoach to Denver Lane as Red Ryder investigates a stagecoach wreck and uncovers a land-grabbing plot led by—who else?—the pillar of the community (Barcroft). Well-directed "Red Ryder" episode.
1946 53m/B Allan Lane, Roy Barcroft, Robert Blake, Peggy Stewart, Martha Wentworth; *Dir:* R.G. Springsteen. **VHS, Beta** $19.95 *NOS, CPB, CAB* 🐾½

Stagefright A maniacal serial killer tries to cover his trail by joining the cast of a play about mass murder. The other actors soon have more to worry about than remembering their lines. Typical low-grade horror.
1987 95m/C David Brandon, Barbara Cupisti, Robert Gligorov; *Dir:* Michele (Michael) Soavi. **VHS, Beta** $79.95 *IMP* 🐾

Stakeout A sometimes violent comedy-thriller about a pair of detectives who stake out a beautiful woman's apartment, hoping for a clue to the whereabouts of her psycho boyfriend who has broken out of prison. One of them (Dreyfuss) then begins to fall in love with her. Natural charm among Estevez, Dreyfuss, and Stowe that adds to the proceedings, which are palpably implausible and silly.
1987 (R) 117m/C Richard Dreyfuss, Emilio Estevez, Madeleine Stowe, Aidan Quinn, Forest Whitaker; *Dir:* John Badham. Edgar Allan Poe Awards '87: Best Screenplay. **VHS, Beta, LV, 8mm** $89.95 *TOU* 🐾🐾½

Stalag 17 A group of American G.I.s in a German POW camp during World War II suspects the opportunistic Holden of being the spy in their midst. One of the very best American movies of the 1950's, adapted from the play by Donald Bevan and Edmund Trzcinski. Wilder, so good at comedy, proved himself equally adept at drama, and brought a top-drawer performance out of Holden. Superb photography from Laszlo and Wayman's wonderful score add much.
1953 120m/B William Holden, Don Taylor, Peter Graves, Otto Preminger, Harvey Lembeck, Robert Strauss, Sig Rumann; *Dir:* Billy Wilder. Academy Awards '53: Best Actor (Holden); National Board of Review Awards '53: 10 Best Films of the Year. **VHS, Beta, LV** $14.95 *PAR, TLF, BTV* 🐾🐾🐾🐾

Stalk the Wild Child Well-intentioned but unconvincing American remake of Truffaut's "The Wild Child," about a child psychologist's efforts to rehabilitate a boy raised by wolves. TV pilot inspired "Lucan," a short-lived series.
1976 78m/C David Janssen, Trish Van Devere, Benjamin Bottoms, Joseph Bottoms; *Dir:* William Hale. **VHS** *WOV* 🐾½

Stalking Danger A secret government group must terminate an assassination plot by a terrorist organization.

1988 (PG) 97m/C Joe Cortese, Steve James, Patricia Charbonneau, Jack Youngblood; *Dir:* William Friedkin. **VHS, Beta, LV $19.95** STE, VMK ✂✂½

The Stalking Moon Indian scout Peck, ready to retire, meets a woman and her half-breed son who have been captives of the Apaches for 10 years. He agrees to help them escape but learns that the woman's Indian husband is hunting them down. Skeletal plot with little meat on it; great scenery but you wouldn't know it.
1969 (G) 109m/C Gregory Peck, Eva Marie Saint, Robert Forster, Noland Clay; *Dir:* Robert Mulligan. **VHS** ACE, WAR ✂✂½

Stallion A city slicker brings his reputation as a womanizer to a small town.
1988 90m/C Alberto Rojas, Fernando Lujan, Leticia Perdigon, Charly Valentino, Luis Aguilar, Hilda Aguirre. **VHS, Beta $79.95** MED ✂

Stamp Day for Superman Based on the "Superman" television series, this program was made to promote defense savings stamps for kids. A "Superman" cartoon is also included.
1953 25m/B George Reeves, Jack Larson, Robert Shayne. **VHS, Beta $19.95** VDM ✂

Stamp of a Killer A tough cop tries to decipher the pattern of a rampaging serial killer. Made for television.
1987 (PG) 95m/C Jimmy Smits, Judith Light, Audra Lindley, Michael Parks, Billy Sullivan, Rhea Perlman; *Dir:* Larry Elikann. **VHS $19.95** STE, NWV ✂

Stamping Ground A post-Woodstock concert festival in Holland, featuring a host of legendary rock acts.
1971 90m/C Pink Floyd, The Jefferson Airplane, The Byrds, Carlos Santana, T. Rex, Canned Heat, Al Stewart. **VHS, Beta $12.95** MVD, JLT ✂

Stan Without Ollie, Volume 1 Three silent comedy shorts featuring Laurel on his own, years before teaming with Ollie: "Kill or Cure," "Oranges & Lemons," and "The Noon Whistle."
1923 53m/B Stan Laurel. **VHS, Beta $24.95** VYY ✂

Stan Without Ollie, Volume 2 Three silent shorts from Laurel before he met the big guy.
1923 55m/B Stan Laurel. **VHS, Beta $24.95** VYY ✂

Stan Without Ollie, Volume 3 Three early Laurel shorts: "White Wings," "Man About Town," and "Smithy."
1923 51m/B Stan Laurel, James Finlayson. **VHS, Beta $24.95** VYY ✂

Stan Without Ollie, Volume 4 Three classic pre-Hardy shorts featuring Laurel: "Half a Man," "Yes, Yes, Nanette," and "Save the Ship."
1925 64m/B Stan Laurel, Blanche Payson, James Finlayson. **VHS, Beta, 8mm $29.95** VYY ✂

Stand Alone World War II vet Durning battles local dope dealers in his New York neighborhood. Self-serious anti-drug flick miscasts Durning (overweight and looking silly in fight scenes) and Grier.
1985 (R) 94m/C Charles Durning, Pam Grier, James Keach; *Dir:* Alan Beattie. **VHS, Beta $19.95** NWV, STE ✂

Stand By Me A sentimental, observant adaptation of the Stephen King novella "The Body." Four 12-year-olds trek into the Oregon wilderness to find the body of a missing boy, learning about death and personal courage. Told as a reminiscence by narrator "author" Dreyfus with solid performances from all four child actors. Too much gratuitous obscenity, but a very good, gratifying film from can't-miss director Reiner.
1986 (R) 87m/C River Phoenix, Wil Wheaton, Jerry O'Connell, Corey Feldman, Kiefer Sutherland, Richard Dreyfuss, Casey Siemaszko, John Cusack; *Dir:* Rob Reiner. **VHS, Beta, LV, 8mm $14.95** COL ✂✂✂

Stand and Deliver War vet joins Greek ranks for the smell of gunpowder in the morning and meets woman of his dreams. Kidnapped by infamous outlaws, he pledges allegiance to them before turning them over to the authorities, and he's free to live happily ever after with heartthrob.
1928 57m/B Rod La Rocque, Lupe Velez, Warner Oland, James Dime, Frank Lanning, Donald Crisp; *Dir:* Donald Crisp. **VHS, Beta $14.95** GPV ✂✂

Stand and Deliver A tough teacher inspires students in an East L.A. barrio to take the Advanced Placement Test in calculus. A superb, inspirational true story, with a wonderful performance from Olmos.
1987 (PG) 105m/C Edward James Olmos, Lou Diamond Phillips, Rosana De Soto, Andy Garcia; *Dir:* Ramon Menendez. **VHS, LV, 8mm $19.98** WAR ✂✂✂

Stand-In When a Hollywood studio is threatened with bankruptcy, the bank sends in timid efficiency expert Howard to save it. Satire of studio executives and big-budget movie making. Bogart is interestingly cast and effective in his first comedy role, playing a drunken producer in love with star Shelton.
1937 91m/B Humphrey Bogart, Joan Blondell, Leslie Howard, Alan Mowbray, Jack Carson; *Dir:* Tay Garnett. **VHS, Beta $24.95** MON ✂✂✂

Stand-In A strange comedy-action film that takes a behind-the-scenes look at sleaze and organized crime.
1985 87m/C Danny Glover; *Dir:* Robert Zagone. **VHS, Beta $59.98** MAG ✂

Stand Off Two young hoodlums with guns take eighteen girls hostage in a dormitory. Hungarian with subtitles.
1989 97m/C HU Ary Beri, Gabor Svidrony, Zbigniew Zapassiewicz, Istvan Szabo; *Dir:* Gyula Gazdag. **VHS $79.95** EVD, FCT ✂½

Stand Up and Cheer The new federal Secretary of Entertainment organizes a huge show to raise the country's depressed spirits. Near-invisible plot, fantastic premise are an excuse for lots of imagery, dancing, and comedy, including four-year-old Temple singing "Baby Take a Bow." Also available colorized.
1934 80m/B Shirley Temple, Warner Baxter, Madge Evans, Nigel Bruce, Stepin Fetchit, Frank Melton, Lila Lee, James Dunn, John Boles; *Dir:* Hamilton McFadden. **VHS, Beta $19.98** FOX ✂✂½

Stand Up Reagan A review of Reagan's in-office humor, jokes and moments of whimsy, presumably all unintentional.
1989 40m/C Ronald Reagan. **VHS, Beta $19.95** JTC ✂✂½

Standing Tall A small-time cattle rancher takes on a high-class land baron in this average, made-for-TV western set during the Depression. Playing a "half-breed," Forster carries an otherwise mediocre effort.
1978 100m/C Robert Forster, Linda Evans, Will Sampson, L.Q. Jones; *Dir:* Harvey Hart. **VHS** WOV ✂✂

Stanley Seminole Vietnam veteran Robinson uses rattlesnakes as his personal weapon of revenge against most of mankind. Thoroughly wretched effort in the gross-pets vein of "Willard" and "Ben."
1972 (PG) 108m/C Chris Robinson, Alex Rocco, Susan Carroll; *Dir:* William Grefe. **VHS, Beta $14.98** VID ✂½

Stanley An animated featurette about a duckling named Stanley who's searching for his identity.
1983 30m/C **VHS, Beta $19.95** ISS ✂½

Stanley and Iris Blue collar recent widow Fonda meets co-worker De Niro, whose illiteracy she helps remedy. Romance follows, inevitably but excruciatingly. Leads' strong presence helps along a very slow, underdeveloped plot.
1990 (PG-13) 107m/C Jane Fonda, Robert De Niro, Swoosie Kurtz, Martha Plimpton, Harley Cross, Jamey Sheridan, Feodor Chaliapin, Zohra Lampert; *Dir:* Martin Ritt. **VHS, Beta, LV $19.99** MGM, FCT ✂✂½

Stanley and Livingstone The classic Hollywood kitsch version of the Victorian legend-based-on-fact. American journalist Tracy sets out into darkest Africa to locate a long lost Brisith explorer. Lavish, dramatically solid fictionalized history. (The real Stanley did not become a missionary—but hey, this is the movies). Tracy is excellent, as usual, and low-key.
1939 101m/B Spencer Tracy, Cedric Hardwicke, Nancy Kelly, Walter Brennan, Richard Greene, Charles Coburn, Henry Hull, Henry Travers, Miles Mander, Holmes Herbert, Paul Stanton, Brandon Hurst, Joseph Crehan, Russell Hicks; *Dir:* Henry King. **VHS, Beta $39.98** FOX ✂✂✂

Star 80 Based on the true-life tragedy of Playmate of the Year Dorothy Stratten and her manager-husband Paul Snider as they battle for control of her body, her mind and her money, with gruesome results. Roberts is overpowering as the vile Snider, but the movie is generally unpleasant. Fosse's last film.
1983 (R) 104m/C Mariel Hemingway, Eric Roberts, Cliff Robertson, David Clennon, Josh Mostel, Roger Rees, Carroll Baker; *Dir:* Bob Fosse. **VHS, Beta, LV $79.95** WAR ✂✂½

Star Bloopers A collection of great movie moments that never made it to the screen, including missed cues, flubbed deliveries, malfunctioning props, and some rare Ronald Reagan footage.
1979 47m/B Ronald Reagan. **VHS, Beta $59.99** HBO ✂✂½

A Star is Born A movie star declining in popularity marries a shy girl and helps her become a star. Her fame eclipses his and tragic consequences follow. Shows Hollywood-behind-the-scenes machinations. Stunning ending is based on the real-life tragedy of silent film star Wallace Reid, who died of a morphine overdose in 1923 at age 31. Remade twice, in 1954 and 1976. Oscar nominated for Best Picture, Best Actress (Gaynor), Best Actor (March), Best Director, and Best Screenplay (co-written by Dorothy Parker).
1937 111m/C Janet Gaynor, Fredric March, Adolphe Menjou, May Robson, Andy Devine, Lionel Stander, Franklin Pangborn; *W/Dir:* William A. Wellman. Academy Awards '37: Best Story; National Board of Review Awards '37: 10 Best Films of the Year. **VHS, Beta, LV $9.95** NEG, NOS, VYY ✂✂✂½

A Star is Born Aging actor helps a young actress to fame. She becomes his wife, but alcoholism and failure are too much for him. She honors his memory. Remake of the 1937 classic was Garland's triumph, a superb and varied performance. Newly restored version reinstates over 20 minutes of long-missing footage, including three Garland musical numbers, "Here's What I'm Here For," "Shampoo Commercial," and "Lose That Long Face." Oscar nominations for Best Actor (Mason), Color Art Direction and Set Design, and Color Costume Design. **1954 (PG) 175m/C** Judy Garland, James Mason, Jack Carson, Tommy Noonan, Charles Bickford, Emerson Treacy; *Dir:* George Cukor. National Board of Review Awards '54: 10 Best Films of the Year. **VHS, Beta, LV $19.95** *WAR* ⬛⬛⬛½

A Star is Born The tragic story of one rock star (the relentlessly un-hip Streisand) on her way to the top and another (Kristofferson) whose career is in decline. A miserable update of the 1937 and 1954 classics. Streisand's showcase of her hit song "Evergreen." **1976 (R) 140m/C** Barbra Streisand, Kris Kristofferson, Paul Mazursky, Gary Busey, Sally Kirkland, Oliver Clark; *Dir:* Frank Pierson. Academy Awards '76: Best Song ("Evergreen"); Golden Globe Awards '77: Best Film—Musical/Comedy. **VHS, Beta $19.98** *WAR* ⬛½

The Star Chamber A conscientious judge (Douglas) sees criminals freed on legal technicalities and wonders if he should take justice into his own hands. He finds a secret society that administers justice extra-legally. Implausible yet predictable. **1983 (R) 109m/C** Michael Douglas, Hal Holbrook, Yaphet Kotto, Sharon Gless; *Dir:* Peter Hyams. **VHS, Beta, LV $19.98** *FOX* ⬛⬛

Star Crash A trio of adventurers square off against interstellar evil by using their wits and technological wizardry. Semi-funny; otherwise cheesy sci-fi. **1978 (PG) 92m/C** *IT* Caroline Munro, Marjoe Gortner, Christopher Plummer, David Hasselhoff, Robert Tessier, Joe Spinell, Nadia Cassini, Judd Hamilton; *Dir:* Lewis (Luigi Cozzi) Coates. **VHS, Beta, LV $19.98** *CHA* ⬛½

Star Crystal Aboard a spaceship, an indestructible alien hunts down the human crew. Cheap imitation of "Alien." **1985 (R) 93m/C** C. Jutson Campbell, Faye Bolt, John W. Smith; *Dir:* Lance Lindsay. **VHS, Beta $14.95** *NWV, STE, HHE* ⬛

Star of Midnight A lawyer/detective becomes involved in the disappearance of the leading lady in a Broadway show and the murder of a columnist. Powell and Rogers take on characters similar to Nick and Nora Charles, but the pizazz of the "Thin Man" series is missing. **1935 90m/B** Ginger Rogers, William Powell, Paul Kelly; *Dir:* Stephen Roberts. **VHS, Beta $39.95** *RKO* ⬛⬛

The Star Packer Wayne puts on a marshal's badge, straightens out a gang of crooks, and still finds time for romance. Implausible and kind of dull early Wayne vehicle. **1934 53m/B** John Wayne, George "Gabby" Hayes, Earl Dwire, Yakima Canutt; *Dir:* Robert N. Bradbury. **VHS $19.95** *COL, REP, NOS* ⬛⬛

Star Quest A genetically engineered woman fights for freedom from her corporate creators. Mediocre effects and some pointless action; done much better by Ridley Scott in "Blade Runner." Also called "Beyond the Rising Moon." **1989 90m/C** Tracy Davis, Hans Bachman, Michael Mack; *Dir:* Phillip Cook. **VHS, Beta, LV $79.98** *VID* ⬛

Star Reporter A son takes over his father's newspaper when dad is killed by gangsters. The son uses the paper in an attempt to destroy the underworld, but stumbles on some startling personal information that could drastically backfire on him. Generally interesting plot, but has some implausibility and too-broad characterization. **1939 62m/B** Warren Hull, Marsha Hunt, Morgan Wallace, Clay Clement, Wallis Clark, Virginia Howell, Paul Fix, Joseph Crehan; *Dir:* Howard Bretherton. **VHS, Beta $16.95** *SNC* ⬛½

Star Shorts A compilation of dark-humored blackout sketches. **198? 51m/C** Bill Murray, Griffin Dunne, Brian Doyle-Murray, Dom DeLuise, Ellen Foley, Bob Goldthwait. **VHS, Beta $59.95** *MPI* ⬛½

Star Slammer A beautiful woman is unjustly sentenced to a brutal intergalactic prison ship. She leads the convicts to escape amid zany situations. Unevenly funny sci-fi comedy. **1987 (R) 85m/C** Ross Hagen, John Carradine, Sandy Brooke, Aldo Ray; *Dir:* Fred Olen Ray. **VHS, Beta $79.95** *VMK* ⬛½

Star Spangled Girl One of Neil Simon's lesser plays, one of his least movies. A pert, patriotic young lady captures the hearts of two left-wing alternative-newspaperguys next door; their political conflicts never rise above bland sitcom level. **1971 (G) 94m/C** Sandy Duncan, Tony Roberts, Todd Susman, Elizabeth Allan; *Dir:* Jerry Paris. **VHS $14.95** *PAR* ⬛½

Star Trek: The Motion Picture The Enterprise fights a strange alien force that threatens Earth in this first film adaptation of the famous television series. Very slow moving; followed by numerous sequels. Twelve additional minutes of previously unseen footage have been added to this home video version of the theatrical feature. The laserdisc edition offers the film in widescreen format. **1980 (G) 143m/C** William Shatner, Leonard Nimoy, DeForest Kelley, James Doohan, Stephen Collins, Persis Khambatta, Nichelle Nichols, Walter Koenig, George Takei; *Dir:* Robert Wise. **VHS, Beta, LV $14.95** *PAR, PIA* ⬛⬛½

Star Trek 2: The Wrath of Khan Picking up from the 1967 Star Trek episode "Space Seed," Admiral James T. Kirk and the crew of the Enterprise must battle Khan, an old foe out for revenge. Warm and comradly in the nostalgic mode of its successors. Introduced Kirk's former lover and unknown son to the series plot, as well as Mr. Spock's "death," which led to the next sequel (1984's "The Search for Spock"). Can be seen in widescreen format on laserdisc. **1982 (PG) 113m/C** William Shatner, Leonard Nimoy, Ricardo Montalban, DeForest Kelley, Nichelle Nichols, James Doohan, George Takei, Walter Koenig, Kirstie Alley, Merritt Butrick, Paul Winfield; *Dir:* Nicholas Meyer. **VHS, Beta, LV, SVS, 8mm $14.95** *PAR, PIA* ⬛⬛⬛

Star Trek 3: The Search for Spock Captain Kirk hijacks the USS Enterprise and commands the aging crew to go on a mission to the Genesis Planet to discover whether Mr. Spock still lives (supposedly he died in the last movie). Klingons threaten, as usual. Somewhat slow and humorless, but intriguing. Third in the series of six (so far) Star Trek movies. The laserdisc edition carries the film in widescreen format. **1984 (PG) 105m/C** William Shatner, Leonard Nimoy, DeForest Kelley, James Doohan, George Takei, Walter Koenig, Mark Lenard, Robin Curtis, Merritt Butrick, Christopher Lloyd, Judith Anderson, John Larroquette, James B. Sikking, Nichelle Nichols, Cathie Shirriff, Miguel Ferrer, Grace Lee Whitney; *Dir:* Leonard Nimoy. **VHS, Beta, LV, 8mm $14.95** *PAR, PIA* ⬛⬛½

Star Trek 4: The Voyage Home Kirk and the gang go back in time (to the 1980s, conveniently) to save the Earth of the future from destruction. Filled with hilarious moments and exhilarating action; great special effects enhance the timely conservation theme. Watch for the stunning going-back-in-time sequence. Spock is particularly funny as he tries to fit in and learn the 80s lingo! Can be seen in widescreen format on laserdisc. Also available as part of Paramount's "director's series," in which Nimoy discusses various special effects aspects in the making of the film. The best in the six-part (so far) series. **1986 (PG) 119m/C** William Shatner, DeForest Kelley, Catherine Hicks, James Doohan, Nichelle Nichols, George Takei, Walter Koenig, Mark Lenard; *Dir:* Leonard Nimoy. **VHS, Beta, LV, 8mm $14.95** *PAR, PIA* ⬛⬛⬛

Star Trek 5: The Final Frontier A renegade Vulcan kidnaps the Enterprise and takes it on a journey to the mythic center of the universe. Shatner's big-action directorial debut (he also co-wrote the script) is a poor follow-up to the Nimoy-directed fourth entry in the series. Heavy-handed and pretentiously pseudo-theological. Available in widescreen format on laserdisc. **1989 (PG) 107m/C** William Shatner, Leonard Nimoy, DeForest Kelley, James Doohan, Laurence Luckinbill, Walter Koenig, George Takei, Nichelle Nichols, David Warner; *Dir:* William Shatner. **VHS, Beta, LV, 8mm $14.95** *PAR, PIA* ⬛½

Star Trek 6: The Undiscovered Country The final chapter in the long running Star Trek series is finally here. The Federation and the Klingon Empire are preparing a much-needed peace summit but Captain Kirk has his doubts about the true intentions of the Federation's longtime enemies. When a Klingon ship is attacked, Kirk and the crew of the Enterprise, who are accused of the misdeed, must try to find the real perpetrator. Has an exciting, climactic ending. As is typical of the series, the film highlights current events—glasnost—in its plotlines. Meyer also directed Star Trek movies 2 ("The Wrath of Khan") and 4 ("The Voyage Home"). **1991 (PG) 110m/C** William Shatner, Leonard Nimoy, DeForest Kelley, James Doohan, George Takei, Walter Koenig, Nichelle Nichols, Christopher Plummer, Kim Cattrall, Iman, David Warner; *Dir:* Nicholas Meyer. **VHS, Beta** *PAR* ⬛⬛½

Star Trek 25th Anniversary Special Shatner and Nimoy are joined by other members of the original series as well as cast from "Star Trek: The Next Generation" in this tribute to the enduring popularity of the television/film series. Includes commentary by "Star Trek" creator Gene Roddenberry, highlights from the films and both television series, bloopers, and a behind-the-scenes look at "Star Trek VI: The Undiscovered Country." **1991 100m/C** Gene Roddenberry; *Hosted:* William Shatner, Leonard Nimoy. **VHS, Beta $19.95** *PAR* ⬛⬛½

Star Trek Bloopers The television cast of ''Star Trek'' appears in these flubs and goofs that were edited out of the actual episodes. The blooper reels from all three seasons are included on this tape.
1969 25m/C William Shatner, Leonard Nimoy, DeForest Kelley, Nichelle Nichols, James Doohan, George Takei, Walter Koenig, Frank Gorshin. **VHS, Beta $24.95** *MOV* ✂✂

Star Wars The first of Lucas's ''Star Wars'' trilogy and one of the biggest box-office hits of all time. A young hero, a captured princess, a hot-shot pilot, cute robots, a vile villain, and a heroic and mysterious Jedi knight blend together with marvelous special effects to make you care about the rebel forces engaged in a life-or-death struggle with the tyrant leaders of the Galactic Empire. Musical score composed by John Williams. Followed by ''The Empire Strikes Back'' (1980) and ''Return of the Jedi'' (1983). Oscar nominations for Best Picture, Supporting Actor Guinness, director, and Original Screenplay plus awards it won.
1977 (PG) 121m/C Mark Hamill, Carrie Fisher, Harrison Ford, Alec Guinness, Peter Cushing, Kenny Baker, James Earl Jones, David Prowse; **Dir:** George Lucas. Academy Awards '77: Best Art Direction/Set Decoration, Best Costume Design, Best Film Editing, Best Sound, Best Visual Effects, Best Original Score; Los Angeles Film Critics Association Awards: Best Original Score; Los Angeles Film Critics Association Awards '77: Best Film; National Board of Review Awards '77: 10 Best Films of the Year; American Film Institute's Survey '77: 8th Best American Film Ever Made. **VHS, Beta, LV $19.98** *FOX, BUR, RDG* ✂✂✂✂

Starbird and Sweet William On a solo plane flight, a young Native American crashes in the wilderness. He must fight for survival in the harsh woods with his only friend, a bear cub. Good family fare.
1973 (G) 95m/C A. Martinez, Louise Fitch, Dan Haggerty, Skip Homeier; **Dir:** Jack B. Hively. **VHS $79.95** *UHV* ✂✂

Starbirds Animated sci-fi epic featuring warriors, robots and battles for the supremacy of Earth and the universe.
1982 75m/C VHS, Beta $9.95 *MED* ✂ ½

Starchaser: The Legend of Orin Animated fantasy about a boy who must save the world of the future from malevolent hordes.
1985 (PG) 107m/C Dir: Steven Hahn. **VHS, Beta, LV $14.95** *PAR* ✂✂ ½

Stardust Memories Allen's ''8 1/2.'' A comic filmmaker is plagued with creative blocks, relationships, modern fears and fanatical fans. The last film in Allen's varying self-analysis, with explicit references to Fellini and Antonioni.
1980 (PG) 88m/B Woody Allen, Charlotte Rampling, Jessica Harper, Marie-Christine Barrault, Tony Roberts, Helen Hanft, Cynthia Gibb, Amy Wright; Daniel Stern; **Dir:** Woody Allen. **VHS, Beta, LV $19.98** *FOX* ✂✂ ½

The Starfighters Air Force Lieutenant Dornan must prove his courage to his disapproving war hero /Congressman father. Dornan flies his F-104 through a dangerous storm, thereby earning his father's respect, which leads them to form a bond.
1963 84m/C Robert Dornan, Richard Jordahl, Shirley Olmstead. **VHS, Beta $59.95** *WES* ✂ ½

Starflight One A space shuttle is called on to save the world's first hypersonic airliner trapped in an orbit above earth. Also called ''Starflight: The Plane That Couldn't Land.''

The film features good special effects, but also a predictable ''rescue-mission'' plot.
1983 155m/C Ray Milland, Lee Majors, Hal Linden, Lauren Hutton, Robert Webber, Terry Kiser; **Dir:** Jerry Jameson. **VHS, Beta $69.98** *LIV, VES* ✂✂

Starhops A trio of buxom young women attract business to their drive-in restaurant by wearing skimpy outfits. Surprisingly inoffensive for an ''R'' rating.
1978 (R) 92m/C Dorothy Buhrman, Sterling Frazier, Jillian Kesner, Peter Paul Liapis, Paul Ryan, Anthony Mannino, Dick Miller; **Dir:** Barbara Peeters. **VHS $19.95** *VCI* ✂ ½

Stark A Wichita detective journeys to Las Vegas in an effort to locate his missing sister. During his search, he finds himself taking on the mob. A television pilot for a series that never happened, and Ernest Tidyman's last screenplay. A sequal, ''Stark: Mirror Image,'' was made in 1986.
1985 94m/C Nicolas Surovy, Dennis Hopper, Marilu Henner, Seth Jaffe; **Dir:** Rod Holcomb. **VHS, Beta $59.98** *FOX* ✂ ½

Starlet Soft-core romp wherein a young actress sleeps her way to the top of the Hollywood ladder.
1987 80m/C Shari Mann, Stuart Lancaster, John Alderman. **VHS, Beta $29.95** *MED* Woof!

Starlight Hotel An unhappy teenage girl sets off to find her father, with the help of a shellshocked veteran wrongly accused of a crime. Set in New Zealand during the Depression. Lovely scenery and good acting highlight nice story of odd friendship.
1990 (PG) 90m/C NZ Greer Robson, Peter Phelps, Marshall Napier, Mr. Wizard; **Dir:** Sam Pillsbury. **VHS, LV $79.95** *HEP, VIK, PIA* ✂✂ ½

Starman An alien from an advanced civilization lands in Wisconsin. He hides beneath the guise of a grieving young widow's recently deceased husband. He then makes her drive him across country to rendezvous with his spacecraft so he can return home. Well-acted, interesting twist on the ''Stranger in a Strange Land'' theme. Bridges is fun as the likeable starman; Allen is lovely and earthy in her worthy follow-up to ''Raiders of the Lost Ark.'' Available in widescreen format on laserdisc.
1984 (PG) 115m/C Jeff Bridges, Karen Allen, Charles Martin Smith, Richard Jaeckel; **Dir:** John Carpenter. **VHS, Beta, LV $12.95** *COL, PIA* ✂✂✂

Starring the Barkleys Two more episodes from the cartoon series where Arnie Barkley puts his paw in his mouth once again.
1973 44m/C VHS, Beta $39.95 *TWE* ✂✂✂

Starring Bugs Bunny! Bug Bunny teams up with Yosemite Sam and others in some of these great hits. Friz Freleng's ''Buccaneer Bunny'' and ''Hare Force,'' as well as Chuck Jones' ''Hare Tonic.''
1948 54m/C dir: Friz Freleng, Chuck Jones, Robert McKimson. **VHS $14.95** *MGM* ✂✂✂ ½

Starring Tom and Jerry Maniacal entertainment from the legendary cat and mouse team.
1988 60m/C Tom & Jerry. **VHS, Beta $14.95** *MGM* ✂✂✂

Stars and Bars A stuffy English art expert travels from New York to Georgia to price a Renoir, and happens on a bizarre backwoods family marginally run by Stanton. Uses stereotypes in search of humor; teeters on the edge of black comedy without often

succeeding at being funny. Based on the novel by William Boyd.
1988 (R) 95m/C Daniel Day Lewis, Harry Dean Stanton, John Cusack, Joan Cusack, Spalding Gray, Will Patton, Martha Plimpton, Glenne Headly, Laurie Metcalf, Maury Chaykin; **Dir:** Pat O'Connor. **VHS, Beta, LV $79.95** *COL* ✂

Stars of the Century A never-before-seen peek into the intimate lives of our favorite stars. Guests include Cyd Charisse, Sammy Davis, Jr., Clint Eastwood, Sly Stallone and Zsa Zsa Gabor.
1988 87m/C Cyd Charisse, Sammy Davis Jr., Clint Eastwood, Zsa Zsa Gabor, Sylvester Stallone, Elizabeth Taylor, Madonna. **VHS $19.95** *AVD, KAR* ✂

The Stars Look Down A mine owner forces miners to work in unsafe conditions in a Welsh town and disaster strikes. Redgrave is a miner's son running for office, hoping to improve conditions, and to escape the hard life. Forceful, well-directed effort suffered at the box office, in competition with John Ford's similar classic ''How Green Was My Valley.'' From the novel by A.J. Cronin.
1939 96m/B GB Michael Redgrave, Margaret Lockwood, Emlyn Williams, Cecil Parker; **Dir:** Carol Reed. National Board of Review Awards '41: 10 Best Films of the Year. **VHS, Beta $19.95** *NOS, HHT, CAB* ✂✂✂

Stars on Parade/Boogie Woogie Dream Pair of all-black musicals which shows off a host of talent from the 1940s.
1946 55m/B Milton Wood, Jane Cooley, Francine Everett, Bob Howard, Eddie Smith, Phil Moore, Lena Horne, Teddy Wilson. **VHS, Beta $24.95** *VYY* ✂✂✂

Stars and Stripes Forever Sumptious, Hollywoodized bio of composer John Phillip Sousa, based on his memoir ''Marching Along,'' but more concerned with the romantic endeavors of young protege Wagner. Accuracy aside, it's solid entertainment even if you're not mad about march music. Those performed include ''Washington Post,'' ''Dixie,'' ''Battle Hymn of the Republic,'' and ''Hail to the Chief.''
1952 89m/C Clifton Webb, Debra Paget, Robert Wagner, Ruth Hussey, Finlay Currie; **Dir:** Henry Koster. **VHS $19.98** *FXV, FCT* ✂

Starship British sci-fi seems to be an oxymoron. To wit, this lame ''Star Wars'' ripoff is about human slaves on a planet run by evil robots.
1987 (PG) 91m/C GB John Tarrant, Cassandra Webb, Donough Rees, Deep Roy, Ralph Cotterill; **Dir:** Roger Christian. **VHS, Beta $29.98** *MAG, HHE* ✂

Starstruck Fun-loving folly about a teen who tries to help his talented cousin make it as a singer. Playfully tweeks Hollywood musicals. Enjoyable and fun.
1982 (PG) 95m/C AU Jo Kennedy, Ross O'Donovan, Pat Evison; **Dir:** Gillian Armstrong. **VHS $69.98** *SUE* ✂✂ ½

Start the Revolution Without Me Hilarious, Moliere-esque farce about two sets of identical twins (Wilder and Sutherland) separated at birth, who meet 30 years later, just before the French Revolution. About as hammy as they come; Wilder is unforgettable. Neglected when released, but now deservedly a cult favorite.
1970 (PG) 91m/C Gene Wilder, Donald Sutherland, Orson Welles, Hugh Griffith, Jack MacGowran, Billie Whitelaw, Victor Spinetti, Ewa Aulin; **Dir:** Bud Yorkin. **VHS, Beta $19.98** *WAR* ✂✂

Starting Over His life racked by divorce, Phil Potter learns what it's like to be single, self-sufficient, and lonely once again. When a blind date grows into a serious affair, the romance is temporarily halted by his hang-up for his ex-wife. Enjoyable love-triangle comedy loses direction after a while, but Reynolds is subtle and charming, and Bergen good as his ex, a very bad songwriter. Oscar nominations for Clayburgh and Bergen. Based on a novel by Dan Wakefield.
1979 (R) 106m/C Burt Reynolds, Jill Clayburgh, Candice Bergen, Frances Sternhagen, Austin Pendleton, Mary Kay Place, Kevin Bacon, Daniel Stern; *Dir:* Alan J. Pakula. **VHS, Beta, LV $59.95** PAR ♫♫♫

State Department File 649 Insipid spy drama set in northern China. Mongolian rebels hold U.S. agent Lundigan captive; he hopes to capture a Chinese warlord.
1949 87m/C Virginia Bruce, William Lundigan; *Dir:* Peter Stewart. **VHS, Beta $16.95** SNC, NOS, DVT ♫½

State Fair The second version of the glossy slice of Americana about a family at the Iowa State Fair, featuring plenty of great songs by Rodgers and Hammerstein. Screenplay by Hammerstein from the 1933 screen version of Phil Strong's novel. "It Might as Well Be Spring" won the Academy Award as Best Song. Other songs include "It's a Grand Night for Singing," "That's For Me," and "Isn't It Kinda Fun?" Remade again in 1962.
1945 100m/C Charles Winninger, Jeanne Crain, Dana Andrews, Vivian Blaine, Dick Haymes, Fay Bainter, Frank McHugh, Percy Kilbride, Donald Meek, William Marshall, Harry Morgan; *Dir:* Walter Lang. Academy Awards '45: Best Song ("It Might as Well Be Spring"). **VHS, Beta $19.98** FOX ♫♫♫

State Fair The third film version of the story of a farm family who travel to their yearly state fair and experience life. The original songs are still there, but otherwise this is a letdown. Texas setting required dropping the song "All I Owe Iowa." Songs by Rodgers and Hammerstein include "It's a Grand Night for Singing" and "It Might As Well Be Spring."
1962 118m/C Pat Boone, Ann-Margret, Bobby Darin, Tom Ewell, Alice Faye, Pamela Tiffin, Wally Cox; *Dir:* Jose Ferrer. **VHS, Beta $19.98** FOX ♫♫

State of Grace Irish hood Penn returns to old NYC neighborhood as undercover cop and becomes involved with an Irish Westies mob in a fight for survival as urban renewal encroaches on their Hell's Kitchen turf. Shrinking client base for shakedown schemes and protection rackets forces them to become contract killers for the Italian mafia, which mixes as well as Bushmills and chianti. Fine performances, with Penn tense but restrained, gang honcho Harris intense, and Oldham chewing up gritty urban scenery as psycho brother of Harris, but the story is long and meandering. Well-choreographed violence.
1990 (R) 134m/C Sean Penn, Ed Harris, Gary Oldman, Robin Wright, John Turturro, Burgess Meredith; *Dir:* Phil Joanou. **VHS, LV $14.98** ORI ♫♫♫

State of Siege Third pairing of Montand and Costa-Gavras, about the real-life death of USAID employee Daniel Mitrione, suspected to be involved in torture and murder in Uruguay in the '60s. Quietly suspenseful, with snazzy editing; conspiracy-theory premise is similar to Stone's "JFK," and similarly disturbing, whether you believe it or not.

1973 119m/C *FR* Yves Montand, Renato Salvatori, O.E. Hasse, Jacques Perrin; *Dir:* Constantin Costa-Gavras. **VHS, Beta, LV $59.95** COL ♫♫♫

State of Things Ostensibly a mystery involving a film crew trying to remake a B-movie, "The Most Dangerous Man On Earth," but also an in-depth look at the process of filmmaking, a scathing look at nuclear warfare and an homage to Roger Corman and Hollywood at large. An enigmatic, complex film from Wenders prior to his American years. In German with English subtitles; some dialogue in English.
1982 120m/B *GE* Patrick Bachau, Allen Garfield, Isabelle Weingarten, Viva, Samuel Fuller, Paul Getty III, Roger Corman; *Dir:* Wim Wenders. Venice Film Festival '82: Best Film. **VHS, Beta, LV $59.95** PAV, GLV ♫♫♫

State of the Union Tracy is a liberal multimillionaire seeking the Republican presidential nomination. His estranged wife (Hepburn) is asked to return so they can masquerade as a loving couple for the sake of his political career. Hepburn tries to help Tracy, as the backstage political machinations erode his personal convictions. From a highly successful, topical Broadway play whose writers changed dialogue constantly to reflect the news. Capra and his partners at Liberty Pictures originally hoped to cast Gary Cooper and Claudette Colbert. Hepburn and Menjou were at odds politically (over communist witch hunts in Hollywood at the time) and on the set, but are fine together onscreen.
1948 124m/B Spencer Tracy, Katharine Hepburn, Angela Lansbury, Van Johnson, Adolphe Menjou, Lewis Stone, Howard Smith; *Dir:* Frank Capra. **VHS, Beta, LV $39.95** MCA ♫♫♫½

Stateline Motel Cheap Italian ripoff of "The Postman Always Rings Twice," with a surprise ending but little of great interest. Also called "Last Chance for a Born Loser."
1975 (R) 86m/C *IT* Eli Wallach, Ursula Andress, Fabio Testi, Barbara Bach; *Dir:* Maurizio Lucidi. **VHS, Beta $24.95** NO ♫½

State's Attorney A burned-out attorney finds himself falling in love with the least likely person, a prostitute he's defending. Contrived, improbable story succumbs to star-quality lead acting from Barrymore. Remade in 1937 with Lee Tracy as "Criminal Lawyer."
1931 79m/B John Barrymore, Jill Esmond, William Boyd, Helen Twelvetrees; *Dir:* George Archainbaud. **VHS, Beta $19.98** TTC ♫♫

Static A strange, disquieting independent film about an eccentric youth who claims to have built a machine through which one can see heaven. Uneven, with some dull stretches. Co-written by Romanek and Gordon.
1987 (PG-13) 89m/C Keith Gordon, Amanda Plummer, Bob Gunton; *Dir:* Mark Romanek. **VHS, Beta, LV $19.98** MCG ♫♫½

Station A Japanese detective's love affair is corrupted by his ruthless search for a bloodthirsty cop killer. With English subtitles.
1981 130m/C *JP* Ken Takakura, Chieko Baisho; *Dir:* Yasuo Furuhata. **VHS, Beta** VDA ♫♫½

Station West A disguised Army officer is sent to uncover a mystery of hijackers and murderers. Along the way, he meets up with and falls for a beautiful woman who may be involved in the treachery. Good, solid western with a fine cast and plenty of fun action shot against the requisite beautiful western landscapes. Based on the Luke Short story. Also available colorized.

1948 92m/B Dick Powell, Jane Greer, Agnes Moorehead, Burl Ives, Tom Powers, Raymond Burr; *Dir:* Sidney Lanfield. **VHS, Beta $19.95** RKO, CCB ♫♫♫

The Statue A famed sculptress unveils a nude rendering of her Nobel Prize-winning husband. Much hilarity in the British vein is meant to ensue, but doesn't. Niven is OK, but this empty, plotless farce should have been much shorter.
1971 (R) 84m/C *GB* David Niven, Virna Lisi, Robert Vaughn, Ann Bell, John Cleese; *Dir:* Rod Amateau. **VHS, Beta $59.95** PSM ♫

Stavisky Sumptiously lensed story of Serge Stavisky, a con-artist and bon-vivant whose machinations almost brought down the French government when his corruption was exposed in 1934. Belmondo makes as charismatic an antihero as you could find, and Stephen Sondheim's musical score complements the visuals.
1974 117m/C Jean-Paul Belmondo, Anny Duperey, Charles Boyer, Francois Perier, Gerard Depardieu; *Dir:* Alain Resnais. **VHS** INT ♫♫♫

Stay As You Are Aging French architect Mastroianni has designs on a teenager who may be his illegitimate daughter from an earlier affair.
1978 95m/C *FR* Marcello Mastroianni, Nastassia Kinski, Francisco Rabal; *Dir:* Alberto Lattuada. **VHS, Beta $19.98** WAR, OM ♫½

Stay Awake A demon stalks, haunts and tortures eight young girls sleeping together at a secluded Catholic school. Title might be addressed to the viewer, who will be tempted to snooze.
1987 (R) 90m/C *SA* Shirley Jane Harris, Tanya Gordon, Jayne Hutton, Heath Porter; *Dir:* John Bernard. **VHS, Beta, LV $19.98** SUE ♫

Stay Away, Joe The King is a singing half-breed rodeo star who returns to his reservation where he finds love and trouble. Utterly cliche, embarrassing and stupid even by Elvis-movie standards.
1968 98m/C Elvis Presley, Burgess Meredith, Joan Blondell, Thomas Gomez, L.Q. Jones, Katy Jurado, Henry Jones; *Dir:* Peter Tewkesbury. **VHS, Beta $19.95** MGM ♫

Stay Hungry A wealthy southerner (Bridges) is involved in a real estate deal which depends on the sale of a gym where a number of body builders hang out. He becomes immersed in their world and finds himself in love with the working-class Fields. Big Arnold's first acting role. Offbeat and occasionally uneven comedy-drama based on a novel by Charles Gaines is a sleeper.
1976 (R) 102m/C Jeff Bridges, Sally Field, Arnold Schwarzenegger, Robert Englund, Scatman Crothers; *Dir:* Bob Rafelson. **VHS, Beta, LV** FOX ♫♫♫

Stay Tuned for Murder A sexy news reporter gets caught in a web of financial wrongdoings and corruption. The financiers, lawyers, and other sinister types she's investigating decide to give her a story she will never forget—if she ever gets the chance to tell it!
1988 92m/C Terry Reeves Wolf, Christopher Ginnaven; *Dir:* Gary Jones. **VHS, Beta $79.95** IME ♫½

Staying Alive "Saturday Night Fever" was the ultimate cheesy 70's musical, hence likeable in a dorky way. This pathetic sequel (directed by Stallone from a Rocky-esque script about beating the odds, etc.) is utterly predictable and forgettable. Music mostly by

Frank Stallone, the great heir to Rogers and Hammerstein.
1983 (PG) 96m/C John Travolta, Cynthia Rhodes, Finola Hughes; *Dir:* Sylvester Stallone. **VHS, Beta, LV $19.95** PAR 🎬½

Staying On Based on the Paul Scott novel, this English drama follows the life of a post-colonial British colonel and his wife who chose to remain in India.
1980 87m/C *GB* Trevor Howard, Celia Johnson; *Dir:* Irene Shubik. **VHS, Beta $19.95** MTI 🎬🎬½

Staying Together Three midwestern brothers go into a panic when their father decides to sell the restaurant they've worked at all their adult lives. Somehow they manage the transition and learn about life. Sloppy comedy-drama with way too many unresolved subplots.
1989 (R) 91m/C Dermot Mulroney, Tim Quill, Sean Astin, Stockard Channing, Melinda Dillon, Daphne Zuniga; *Dir:* Lee Grant. **VHS, Beta, LV $89.99** HBO 🎬½

Steagle The threat of the Cuban missile crisis lets a fantasizing professor loose to enact his wildest dreams. Odd Mitty-esque fantasy comedy might have worked, but ends up confusing and frustrating.
1971 (R) 101m/C Richard Benjamin, Cloris Leachman; *Dir:* Paul Sylbert. **VHS, Beta $9.98** CHA 🎬

Steal the Sky Israeli agent Hemingway seduces Iraqi pilot Cross and persuades him to steal a Soviet MIG jet and defect to Israel. Unlikely plot but swell flying scenes. Hemingway is poorly cast, as she often is. Made for cable television.
1988 110m/C Mariel Hemingway, Ben Cross, Etta Ankri; *Dir:* John Hancock. **VHS, Beta, LV $89.99** HBO 🎬🎬

Stealing Heaven Based on the lives of Abelard and Heloise. In 12th-century Paris, a theologian and teacher falls in love with a beautiful young woman, and the two must defend their bond from all comers, including her righteous uncle. Steamy; trimmed to 108 minutes after threatened X rating. Attractive and cerebral. Available in a 115 minute-uncut, unrated version.
1988 (R) 108m/C *GB YU* Derek De Lint, Kim Thomson, Denholm Elliott; *Dir:* Clive Donner. **VHS, Beta, LV $89.95** VIR 🎬🎬½

Stealing Home A washed-up baseball player learns that his former babysitter (who was also his first love and inspiration), has committed suicide. Their bittersweet relationship is told through flashbacks. Foster's superb performance steals the show from Harmon.
1988 (PG-13) 98m/C Mark Harmon, Jodie Foster, William McNamara, Blair Brown, Harold Ramis; *Dir:* Steven Kampmann, Will Aldis. **VHS, LV $19.98** WAR 🎬🎬

Steamboat Bill, Jr. City-educated student returns to his small home-town and his father's Mississippi river boat, where he's an embarrassment to dad. But bond they do, to ward off the owner of a rival riverboat, whose daughter Keaton falls for. Engaging look at small-town life and the usual wonderful Keaton antics, including braving the big tornado.
1928 75m/B Buster Keaton, Ernest Torrence, Marion Byron, Tom Lewis; *Dir:* Charles Riesner. **VHS, Beta, LV $19.95** NOS, CCB, DVT 🎬🎬🎬½

Steaming Six women get together in a London bathhouse and review their various troubles with men. Unfortunate adaptation of the good play by Nell Dunn. Where are the laughs - and what happened to the dramatic

tension? The last film of Dors and director Losey.
1986 (R) 102m/C *GB* Vanessa Redgrave, Sarah Miles, Diana Dors; *Dir:* Joseph Losey. **VHS, Beta $19.95** STE, NWV 🎬🎬

Steel Construction workers on a mammoth skyscraper face insurmountable odds and strong opposition. Will they finish their building? Worth watching, if you don't have anything else to do. Also called "Look Down and Die" and "Men of Steel."
1980 (R) 100m/C *Dir:* Steve Carver. **VHS, Beta, LV $69.95** VES 🎬

Steel Arena Several real life stunt-car drivers appear in this action-packed film, crammed with spins, jumps, explosions and world-record-breaking, life risking stunts.
1972 (PG) 99m/C Dusty Russell, Gene Drew, Buddy Love; *Dir:* Mark L. Lester. **VHS, Beta $69.98** LIV, VES 🎬

The Steel Claw One-handed ex-Marine Montgomery organizes guerilla forces against the Japanese in the Philippines in World War II. Good location shooting and plenty of action.
1961 95m/C George Montgomery, Charito Luna, Mario Barri; *Dir:* George Montgomery. **VHS, Beta $59.95** MON 🎬🎬

Steel Cowboy With his marriage, sanity and livelihood all on the line, an independent trucker agrees to haul a hot herd of stolen steers. Passable made-for-TV macho adventure.
1978 100m/C James Brolin, Rip Torn, Jennifer Warren, Strother Martin, Melanie Griffith, Lou Frizzell; *Dir:* Harvey Laidman. **VHS, Beta** VCL 🎬🎬

Steel Dawn Another "Mad Max" clone: A leather-clad warrior wields his sword over lots of presumably post-apocalyptic desert terrain. Swayze stars with his real-life wife Niemi in unfortunate follow-up to "Dirty Dancing."
1987 (R) 90m/C Patrick Swayze, Lisa Niemi, Christopher Neame, Brett Hool, Brion James, Anthony Zerbe; *Dir:* Lance Hool. **VHS, Beta, LV $14.98** LIV, VES 🎬½

Steel Fisted Dragon A son seeks revenge on the gang who murdered his mother and burned her house to the ground.
1982 (R) 85m/C Stephen Lee, Johnny Kong Kong, Peter Chan. **VHS, Beta $69.99** HBO 🎬

The Steel Helmet Hurriedly made Korean War drama stands as a top-notch war film. Brooding and dark, GIs don't save the world for democracy or rescue POWs; they simply do their best to survive a horrifying situation. Pointless death, confused loyalties and cynicism abound in writer-director Fuller's scathing comment on the madness of war.
1951 84m/B Gene Evans, Robert Hutton, Steve Brodie, William Chun, James Edwards, Richard Loo, Sid Melton; *W/Dir:* Samuel Fuller. **VHS, Beta, LV $14.95** AVD, WGE, RXM 🎬🎬🎬½

Steel and Lace Davison is the scientist brother of a classical pianist who commits suicide after being raped. He revives her as a cyborg, which promptly sets out to enact its revenge. Depressing, morbid sci-fi.
1990 (R) 92m/C Bruce Davison, Clare Wren, Stacy Haiduk, David Naughton, David Lander; *Dir:* Ernest Farino. **VHS, Beta, LV $89.95** FRH 🎬½

Steel Magnolias Julia Roberts plays a young woman stricken with severe diabetes who chooses to live her life to the fullest despite her bad health. Much of the action centers around a Louisiana beauty shop

where the women get together to discuss the goings-on of their lives. Screenplay by R. Harling, based on his partially autobiographical play. Sweet, poignant, and often hilarious, yet just as often overwrought. MacLaine is funny as a bitter divorcee; Parton is sexy and fun as the hairdresser. But Field and Roberts go off the deep end and make it all entirely too weepy.
1989 (PG) 118m/C Sally Field, Dolly Parton, Shirley MacLaine, Darryl Hannah, Olympia Dukakis, Julia Roberts, Tom Skerritt, Sam Shepard, Dylan McDermott, Kevin J. O'Connor, Bil McCutcheon, Ann Wedgeworth; *Dir:* Herbert Ross. **VHS, Beta, LV, 8mm $19.95** COL 🎬🎬🎬

Steele Justice A tough 'Nam vet takes on the whole Vietnamese Mafia in Southern California after his friend is murdered. One of those head-scratchers: Should I laugh, or be offended?
1987 (R) 96m/C Martin Kove, Sela Ward, Ronny Cox, Bernie Casey, Joseph Campanella, Sarah Douglas; *Dir:* Robert Boris. **VHS, Beta, LV $19.95** PAR 🎬

Steele's Law A loner cop is forced to take the law into his own hands in order to track down an insane international assassin.
1991 90m/C Fred Williamson, Bo Svenson, Doran Inghram, Phyllis Cicero; *Dir:* Fred Williamson, Fred Williamson. **VHS $89.95** ACA 🎬🎬

Steelyard Blues A motley-crew comedy about a wacky gang that tries to steal an abandoned WW II airplane. Zany pranks abound. Technically flawed direction from Myerson mars a potentially hilarious story.
1973 (PG) 93m/C Jane Fonda, Donald Sutherland, Peter Boyle, Howard Hesseman, John Savage; *Dir:* Alan Myerson. **VHS, Beta $19.98** WAR 🎬🎬½

Stella In this anachronistic remake update of "Stella Dallas," Midler plays a single mother who sacrifices everything to give her daughter a better life than hers. Barbara Stanwyck did it much better in 1937. Based on Olive Higgins Prouty's novel.
1989 (PG-13) 109m/C Bette Midler, John Goodman, Stephen Collins, Eileen Brennan, Ben Stiller, Trini Alvarado, Marsha Mason; *Dir:* John Erman. **VHS, Beta, LV $12.95** TOU 🎬½

Stella Dallas An uneducated woman lets go of the daughter she loves when she realizes her ex-husband can give the girl more advantages. What could be sentimental turns out believable and worthwhile under Vidor's steady hand. Stanwyck never makes a wrong step. From a novel by Olive Higgins Prouty. Remade in 1989 as "Stella," starring Bette Midler. Oscar nominated for Best Actress (Stanwyck), and Best Supporting Actress (Shirley).
1937 106m/B Barbara Stanwyck, Anne Shirley, John Boles, Alan Hale Jr., Marjorie Main; *Dir:* King Vidor. **VHS, Beta, LV $14.98** SUE 🎬🎬🎬

Stella Maris Pickford plays tattered waif who's a slave to an alcoholic who's married to the man she loves. But he's in love with Pickford playing a sheltered rich invalid. Brimming with the tragic consequences of love.
1918 70m/B Mary Pickford, Conway Tearle, Camille Ankewich, Ida Waterman; *Dir:* Marshall Neilan. **VHS, Beta $24.95** GPV, FCT 🎬🎬🎬

Step Lively A young playwright tries to recover the money he lent to a fast-talking Broadway producer and is forced to take the leading role in his play. Lively musical remake of "Room Service" (1938) was Sinatra's first

starring role. Look for Dorothy Malone as a brunette switchboard operator.
1944 88m/B Frank Sinatra, Gloria de Haven, George Murphy, Walter Slezak, Adolphe Menjou, Anne Jeffreys; *Dir:* Tim Whelan. **VHS, Beta, LV $19.98** *RKO, CCB* 𝄞𝄞½

Stepfather Creepy thriller about a seemingly ordinary stepfather who is actually a homicidal maniac searching for the "perfect family." An independently produced sleeper tightly directed by Ruben and well written by Donald E. Westlake. Followed by the inferior "Stepfather II."
1987 (R) 89m/C Terry O'Quinn, Shelley Hack, Jill Schoelen; *Dir:* Joseph Ruben. **VHS, Beta, LV $14.95** *COL, SUE* 𝄞𝄞𝄞

Stepfather 2: Make Room for Daddy A poor sequel to the suspenseful sleeper, wherein the psychotic family man escapes from an asylum and woos another suburban family, murdering anyone who may suspect his true identity.
1989 (R) 93m/C Terry O'Quinn, Meg Foster, Caroline Williams, Jonathan Brandis; *Dir:* Jeff Burr. **VHS, Beta, LV $89.99** *HBO* 𝄞½

Stephen Crane's Three Miraculous Soldiers This film immerses us in Stephen Crane's Civil War story, so we can see how irony in war emerges from the girl's encounter with both Confederate and Union soldiers.
1976 18m/C VHS, Beta *BFA* 𝄞½

Stephen King's Golden Years Stephen King creates a chilling vision of scientific progress gone awry in this shocking techno-thriller. After being accidentally exposed to exotic chemicals in a lab explosion, an aging janitor undergoes an extraordinary transformation and the government will sacrifice anything to learn more about it. Originally a made-for-television miniseries.
1991 232m/C Keith Szarabajka, Frances Sternhagen, Ed Lauter, R.D. Call. **VHS $89.95** *WOV* 𝄞½

Stephen King's Nightshift Collection Two short stories from the mind of Stephen King. The first, "Disciples of the Crow," tells the story of a community of children who worship crows and kill grownups. In "The Night Waiter," a waiter must deal with spooky Room 321.
198? 40m/C *Dir:* Jack Garrett, John Woodward. **VHS, Beta $14.95** *SIM, KAR* 𝄞𝄞

The Stepmother Yet another Hitchcock ripoff story involving an evil stepmother. Rey is passable, but there's not much else to recommend this dredge.
1971 100m/C Alejandro Rey, John Anderson, Katherine Justice, John Garfield, Marlene Schmidt, Claudia Jennings, Larry Linville; *Dir:* Hikmet Avedis. **VHS $19.95** *ACA* 𝄞½

Steppenwolf Static, enigmatic film version of the famous Herman Hesse novel about a brooding writer searching for meaning and self-worth. Interesting to watch, but the offbeat novel doesn't translate to the screen; leaves you flat.
1974 (PG) 105m/C *SI* Max von Sydow, Dominique Sanda; *Dir:* Fred Haines. **VHS, Beta, LV $59.95** *VMK, GLV* 𝄞½

Stepping Out Minnelli stars as a would-be Broadway dancer who gives tap dancing lessons in an old church to an assortment of offbeat and interesting characters. When the troupe is asked to perform for a local charity, they make the most of their opportunity. A warm and touching ensemble piece that avoids over-sentimentalizing and utilizes its cast to best advantage.
1991 (PG) 113m/C Liza Minnelli, Shelley Winters, Bill Irwin, Ellen Greene, Julie Walters, Robyn Stevan, Jane Krakowski, Sheila McCarthy, Andrea Martin, Carol Woods, Nora Dunn; *Dir:* Lewis Gilbert. **VHS, Beta, LV** *PAR* 𝄞𝄞𝄞

Stepsisters Murderous double-crosses occur among a pilot, his wife, and her sister.
1985 75m/C Hal Fletcher, Sharyn Talbert, Bond Gideon. **VHS, Beta $14.95** *REG* Woof!

The Sterile Cuckoo An aggressive co-ed pursues a shy freshman who seems to embody her romantic ideal. Minnelli's Oscar-nominated performance is outstanding; Burton as the naive young man is also fine. Alan Pakula's splendid first directing job.
1969 (PG) 108m/C Liza Minnelli, Wendell Burton, Tim McIntire; *Dir:* Alan J. Pakula. **VHS, Beta, LV $49.95** *PAR* 𝄞𝄞𝄞

Steve Martin Live Martin shows veteran comedians his famous Universal Amphitheatre performance and his short "The Absent-Minded Waiter."
1985 58m/C Steve Martin, Teri Garr, Buck Henry, David Letterman, Paul Simon. **VHS, Beta, LV $14.98** *LIV, VES* 𝄞𝄞½

Steven Wright Live This is a straight stand-up gig for deadpan comedian Wright.
1985 53m/C VHS, Beta, LV $19.98 *LIV, VES* 𝄞𝄞½

Stevie Jackson brings her flawless skill to the role of British poet Stevie Smith. Excellent performance from Washbourne as Stevie's spinster aunt. Wooden, lifeless screen rendition for the Hugh Whitemore stage play is helped greatly by good performances, but is too talky and, frankly, rather dull and self-absorbed.
1978 102m/C *GB* Glenda Jackson, Mona Washbourne, Alec McCowen, Trevor Howard; *Dir:* Robert Enders. Montreal World Film Festival '78: Best Actress (Jackson). **VHS $29.95** *AVD, UAV, SUE* 𝄞𝄞

Stewardess School Airline spoof aims at wackiness, but misses the runway. Generic title betrays probable badness which turns out to be all too real. An utter woofer.
1986 (R) 84m/C Sandahl Bergman, Wendie Jo Sperber, Judy Landers, Julia Montgomery, Corinne Bohrer; *Dir:* Ken Blancato. **VHS, Beta, LV $9.95** *RCA* Woof!

The Stewardesses The notorious 3-D spectacle of 56 stewardi and their lusty, druggy, rude activities during an 18-hour layover. Legendary sexploitation pasted together from peep-show loops by a San Francisco porn entrepeneur.
1969 90m/C VHS $19.95 *VDM* 𝄞

Stick Ex-con Stick (Reynolds, directing himself) wants to start a new life for himself in Miami. Lots of drug dealers and guns don't help the interest level in this dull underworld tale. Based upon the Elmore Leonard novel.
1985 (R) 109m/C Burt Reynolds, Candice Bergen, George Segal, Charles Durning, Dar Robinson; *Dir:* Burt Reynolds. **VHS, Beta, LV $19.95** *MCA* 𝄞½

The Stick-Up In 1935, a young American traveling in Great Britain is introduced to some rather illegal fun. Rather sorry attempt at romance/comedy/adventure. Apt alternate title is "Mud."
1977 101m/C David Soul, Pamela McMyler; *Dir:* Jeffrey Bloom. **VHS, Beta $79.95** *CGH, PEV, HHE* 𝄞

Stickfighter The oppressive reign of an evil Spanish ruler of the Philippines is challenged by the world's best stickfighter.
1989 (PG) 102m/C *PH* Dean Stockwell, Nancy Kwan, Alejandro Rey, Roland Dantes; *Dir:* Luis Nepomuceno. **VHS $59.95** *PSM* 𝄞

Sticks of Death When a man is left near death by gangsters, his grandfather teaches him the ancient martial art of the sticks of death.
1984 90m/C VHS, Beta *VCL* 𝄞

Sticky Fingers Two female musicians, asked to watch nearly a million bucks in drug money, go on a mega shopping spree. Completely incredible, unlikeable and mean-spirited attempt at zany comedy. Co-written by Mayron and Adams.
1988 (PG-13) 89m/C Melanie Mayron, Helen Slater, Eileen Brennan, Carol Kane, Christopher Guest, Danitra Vance, Gwen Welles, Stephen McHattie; *Dir:* Catlin Adams. **VHS, Beta, LV $19.95** *MED* 𝄞𝄞

Stigma "Miami Vice" star Thomas (then 23; later to restore his middle name, Michael) is a young doctor who treats a syphilis epidemic in a small town. He's indistinguishable, but better than anything else here. Ever seen close-ups of advanced syphilis? Here's your chance—but it's not pretty.
1973 93m/C Philip Michael Thomas, Harlan Cary Poe; *Dir:* David E. Durston. **Beta $79.95** *VHV* 𝄞½

Stiletto Good cast (including Raul Julia in a cameo) is wasted in this mediocre depiction of the trouble encountered by a contract Mafia assassin when he decides to change careers. Based on the usual pulp cheese by Harold Robbins.
1969 (R) 101m/C Alex Cord, Britt Ekland, Patrick O'Neal, Joseph Wiseman, Barbara McNair, Roy Scheider, M. Emmet Walsh; *Dir:* Bernard L. Kowalski. **VHS, Beta $59.98** *SUE* 𝄞½

Still of the Night A Hitchcock-style thriller about a psychiatrist infatuated with a mysterious woman who may or may not be a killer.
1982 (PG) 91m/C Meryl Streep, Roy Scheider, Jessica Tandy, Joe Grifasi, Sara Botsford, Josef Sommer; *Dir:* Robert Benton. **VHS, Beta, LV $29.95** *MGM* 𝄞𝄞

The Stilts From modern Spanish cinema's preeminent director, this is a study of sexual dynamics revolving around a doomed love triangle. Aging professor Gomez wants young Del Sol to commit to him, but she won't; she has another, younger lover. Overwrought at times, but well acted. In Spanish with English subtitles.
1984 95m/C *SP* Laura Del Sol, Francisco Rabal, Fernando Gomez; *Dir:* Carlos Savrat. **VHS, Beta $24.95** *XVC, MED* 𝄞𝄞𝄞

The Sting Newman and Redford together again in this sparkling story of a pair of con artists in 1930s Chicago. They set out to fleece a big-time racketeer, pitting brain against brawn and pistol. Very inventive, excellent acting, Scott Joplin's wonderful ragtime music adapted by Marvin Hamlisch. The same directorial and acting team from "Butch Cassidy and the Sundance Kid" triumphs again. Oscar nominations for Redford, cinematography, and sound, plus the awards it garnered.
1973 (PG) 129m/C Paul Newman, Robert Redford, Robert Shaw, Charles Durning, Eileen Brennan, Harold Gould, Ray Walston; *Dir:* George Roy Hill. Academy Awards '73: Best Adapted Score, Best Art Direction/Set Decoration, Best Costume Design, Best Director

S

(Hill), Best Film Editing, Best Picture, Best Story & Screenplay; Directors Guild of America Awards '73: Best Director (Hill). **VHS, Beta, LV $14.95** *MCA, BTV* ⅃⅃⅃ ½

The Sting 2 A complicated comic plot concludes with the final con game, involving a fixed boxing match where the stakes top a million dollars and the payoff could be murder. A lame sequel to "The Sting" (1973). **1983 (PG) 102m/C** Jackie Gleason, Mac Davis, Teri Garr, Karl Malden, Oliver Reed; *Dir:* Jeremy Paul Kagan. **VHS, Beta, LV $69.95** *MCA* ⅃ ½

Sting of the Dragon Masters Martial arts adventure featuring Bruce Lee's female Kung-Fu counterpart, Angela Mao. **1974 (R) 96m/C** Angela Mao, Thoon Rhee, Carter Huang; *Dir:* Huang Feng. **VHS, Beta $19.95** *UHV* ⅃

Sting of the West A journeyman con artist swindles his way across the Wild West. **1976 (PG) 90m/C** Jack Palance, Timothy Brent, Lionel Stander. **VHS, Beta $59.95** *MPI* ⅃

Stingiest Man in Town Dickens' tale inspires celebrity voices to break into frequent musical numbers. **1978 50m/C Voices:** Tom Bosley, Walter Matthau, Paul Frees, Theodore Bikel, Robert Morse, Dennis Day. **VHS $12.95** *WAR* ⅃

Stingray Two guys buy a Corvette, not knowing it's loaded with stolen heroin. Gangsters with an interest in the dope come after them, and the chase is on. Very violent and not all that funny. **1978 (PG) 105m/C** Chris Mitchum, Sherry Jackson, Les Lannom; *Dir:* Richard Taylor. **VHS, Beta $59.95** *SUE* ⅃ ½

Stir Crazy Two down-on-their-luck losers find themselves convicted of a robbery they didn't commit and sentenced to 120 years behind bars with a mean assortment of inmates. **1980 (R) 104m/C** Richard Pryor, Gene Wilder, Nicholas Coster, Lee Purcell, Craig T. Nelson, JoBeth Williams; *Dir:* Sidney Poitier. **VHS, LV $89.95** *COL* ⅃⅃ ½

Stitches Adolescent comedy about med students playing practical jokes on the dean using laboratory specimens. So bad the director (actually Rod Holcomb) allegedly did not want his name associated with it. **1985 (R) 92m/C** Eddie Albert, Parker Stevenson, Geoffrey Lewis, Brian Tochi; *Dir:* Alan Smithee. **VHS, Beta $19.95** *MED* ⅃

Stocks and Blondes The world of industrial finance is used as a background for some raunchy sex humor. **1985 (R) 79m/C VHS, Beta $69.98** *LIV, VES* Woof!

A Stolen Face Creepy, implausible drama of a plastic surgeon, spurned by a beautiful concert pianist, who transforms a female convict to look just like her. The convict runs away, but perhaps there's hope in the future with the pianist. **1952 71m/B** *GB* Paul Henreid, Lizabeth Scott, Andre Morell, Susan Stephen, John Wood; *Dir:* Terence Fisher. **VHS, Beta, LV** *WGE, MLB* ⅃⅃

Stolen Kisses Sequel to "The 400 Blows," the story of Antoine Doinel: his unsuccessful career prospects as a detective in Paris, and his initially awkward but finally successful adventures with women. Made during Truffaut's involvement in a political crisis involving the sack of Cinematique Francais director Henri Lauglois. Truffaut dedicated the film to Lauglois and the Cinematique, but it is a thoroughly apolitical,

small-scale, charming (some say too charming) romantic comedy, Truffaut-style. Followed by "Bed and Board." French title: "Baisers Voles." **1969 90m/C** *FR* Jean-Pierre Leaud, Delphine Seyrig; *Dir:* Francois Truffaut. National Board of Review Awards '69: 5 Best Foreign Films of the Year; National Society of Film Critics Awards '69: Best Director. **VHS, Beta $59.95** *COL* ⅃⅃⅃ ½

A Stolen Life Remake of 1939 film of the same title starring Elizabeth Berguer. Oddly, Davis chose this as her first and last producing effort. Implausible tale of an evil twin (Davis) who takes her sister's (Davis) place so she can have the man they both love. Davis pulls it off as both twins; Ford is good as the hapless hubby. **1946 107m/B** Bette Davis, Glenn Ford, Dane Clark, Walter Brennan, Charlie Ruggles, Bruce Bennett, Esther Dale, Peggy Knudsen; *Dir:* Curtis Bernhardt. **VHS, Beta $19.95** *MGM* ⅃⅃

Stolen Painting A man's wife is kidnapped by an evil gang while having her portrait painted. Her husband rescues her, and they, in turn, are chased by the kidnappers. **1979 11m/C VHS, Beta** *BFA* ⅃ ½

The Stone Boy A boy accidentally kills his older brother on their family's Montana farm. The family is torn apart by its sadness and guilt. Sensitive look at variety of reactions during a crisis, with an excellent cast led by Duvall's crystal-clear performance. **1984 (PG) 93m/C** Glenn Close, Robert Duvall, Jason Presson, Frederic Forrest, Wilford Brimley, Linda Hamilton; *Dir:* Christopher Cain. **VHS, Beta $59.98** *FOX* ⅃⅃⅃ ½

Stone Cold Flamboyant footballer Bosworth made his acting debut in this sensitive human document, playing the usual musclebound, one-punk-army terminator cop, out to infiltrate a sadistic band of fascist biker barbarians engaged in drug running and priest shooting. Profane, lewd, gory, self-deifying; a crash course (accent on crashes) in everything despicable about modern action pics. **1991 (R) 91m/C** Brian Bosworth, Lance Henriksen, William Forsythe, Arabella Holzbog, Sam McMurray; *Dir:* Craig R. Baxley. **VHS, Beta, LV, 8mm $64.99** *COL, PIA* Woof!

Stone Cold Dead Rugged cop Crenna battles crime lord Williams (Paul Williams? Yeah, sure) over a prostitution ring. Meanwhile, a sniper starts killing hookers. Unoriginal and thoroughly dull would-be thriller. **1980 (R) 100m/C** Richard Crenna, Paul Williams, Linda Sorensen, Belinda J. Montgomery; *Dir:* George Menduluk. **VHS, Beta $59.95** *MED* ⅃ ½

Stone Fox Heartwarming family drama in which a young man must win a dogsled race to save the family farm. Based on John Reynolds Gardiner's popular children's book. **1987 96m/C** Buddy Ebsen, Joey Cramer, Belinda J. Montgomery, Gordon Tootoosis; *Dir:* Harvey Hart. **VHS $69.95** *WOV* ⅃⅃

The Stone Killer Bronson stars as a tough plainclothes cop in this action-packed drama about a Mafia plot to use Vietnam vets in a mass killing. Violent but tense and action-packed revenge adventure set in the underworlds of New York and Los Angeles. **1973 (R) 95m/C** Charles Bronson, Martin Balsam, Norman Fell, Ralph Waite, John Ritter; *Dir:* Michael Winner. **VHS, Beta, LV $59.95** *COL* ⅃⅃ ½

Stoner A martial arts adventure featuring a showdown between special agents and a depraved crime lord. **1980 88m/C VHS, Beta $54.95** *HAR* ⅃

Stones of Death An aboriginal curse is invoked when a subdivision is built too close to an ancient burial site. Teenagers begin having all-too-real nightmares about death. **1988 (R) 90m/C** Tom Jennings, Natalie McCurray, Zoe Carides, Eric Oldfield; *Dir:* James Bagle. **VHS, Beta $79.95** *SVS* ⅃

Stoogemania A nerd becomes so obsessed with the Three Stooges that they begin to take over his life and ruin it. Harmless except as a waste of time. Includes actual Stooge footage including some colorized—but see an old Stooges movie instead. **1985 95m/C** Josh Mostel, Melanie Chartoff, Sid Caesar; *Dir:* Chuck Workman. **VHS, Beta $19.95** *PAR* ⅃

The Stoogephile Trivia Movie A program of trivia points, quizzes and clips about Larry, Moe, Curly and Shemp for Stooges fanatics. **1986 55m/C VHS, Beta $19.95** *MPI* ⅃

Stooges Shorts Festival This collection of shorts includes "Disorder in the Court" (1936), "Sing a Song of Six Pants" (1947), and "Malice in the Palace" (1949). **194? 55m/B VHS, Beta $34.95** *HHT* ⅃

The Stooges Story "The Stooges Anthology" presents a history of the comedy team, from their vaudevillian beginnings to their television resurgence in the 60s. "Disorder in the Court" finds the trio making a mockery of the United States justice system, and in "Brideless Groom," Shemp must find a bride within forty-eight hours in order to collect an inheritance. **1990 90m/B** Moe Howard, Larry Fine, Curly Howard, Shemp Howard. **VHS $9.95** *SIM* ⅃⅃⅃

Stop Making Sense The Talking Heads perform eighteen of their best songs in this concert filmed in Los Angeles. Considered by many to be the best concert movie ever made. The band plays with incredible energy and imagination, and Demme's direction and camera work is appropriately frenzied and original. Features such Talking Heads songs as "Burning Down the House," "Psycho Killer," and "Once in a Lifetime." Band member Tina Weymouth's Tom Tom Club also performs for the audience. **1984 99m/C** David Byrne, Tina Weymouth, Chris Franz, Jerry Harrison; *Dir:* Jonathan Demme. **VHS, Beta, LV $19.95** *MVD, COL* ⅃⅃⅃ ½

Stop! or My Mom Will Shoot Getty is an overbearing mother paying a visit to her cop son (Stallone) in Los Angeles. When mom witnesses a crime she has to stay in town longer than intended, which gives her time to meddle in her son's work and romantic lives. If Stallone wants to change his image this so-called comedy isn't the way to do it—because the joke is only on him. **1991 (PG-13) 87m/C** Sylvester Stallone, Estelle Getty, JoBeth Williams, Roger Rees; *Dir:* Roger Spottiswoode. **VHS, Beta, LV** *MCA* Woof!

Stop That Cab Sloppy crooks accidentally leave precious jewels in the back seat of a taxi, and pursue and torture the cabby who found and hid them. Rather mean-spirited, uninteresting comedy. **1951 56m/B** Sid Melton, Iris Adrian, Tom Neal. **VHS, Beta, LV** *WGE* ⅃ ½

Stop That Train Escaped convict plants explosives aboard train with railway president on it. With ten minutes to spare, Sky King tries to make a rescue. From the television series "Sky King."
19?? 25m/B Kirby Grant, Gloria Winters, Perry Kellman, Edward Foster, William Hale. **VHS, Beta** CCB 𝄞½

Stopover Tokyo An American intelligence agent uncovers a plot to assassinate the American ambassador while on leave in Japan. Nice location shooting and scenery; limp story and characters. Based on a novel by John P. Marquand.
1957 100m/C Robert Wagner, Joan Collins, Edmond O'Brien, Ken Scott; **Dir:** Richard L. Breen. **VHS, Beta $59.98** FOX 𝄞𝄞

Stories from Ireland Kelly tells tales of comedy and myth.
1988 60m/C Eamon Kelly. **VHS, Beta** RGO 𝄞𝄞

Stork Club Includes two full episodes from the early television series that feature host and owner Billingsley interviewing guests from his table at the posh Stork Club in New York City.
1952 29m/B Sherman Billingsley, Peter Donald, Dorothy Kilgallen. **VHS, Beta $29.95** VYY 𝄞𝄞

Storm Uneven thriller about college students on a camping trip who find themselves fighting for their lives when they are confronted by killer thieves. Somewhat contrived, but interesting ending to this low-budget outing.
1987 (PG-13) 99m/C CA David Palfy, Stan Kane, Harry Freedman, Lawrence Elion, Tom Schioler; **Dir:** David Winning. **VHS $17.95** WAR 𝄞½

Storm Over Asia A Mongolian trapper is discovered to be descended from Genghis Khan and made puppet emperor of a Soviet province. Beautiful and evocative. Silent masterpiece. Also titled "The Heir to Genghis Khan."
1928 70m/B RU I. Inkizhinov, Valeri Inkizhinov, A. Christiakov; **Dir:** Vsevolod Pudovkin. National Board of Review Awards '30: 5 Best Foreign Films of the Year. **VHS, Beta, 3/4U $35.95** IHF, FST 𝄞𝄞𝄞𝄞

Storm Over Wyoming The old sheep-and-cattle battle again; Holt and Martin find themselves haplessly embroiled in a Wyoming range war. Action-packed but thin, ordinary western.
1950 60m/B Tim Holt, Richard Martin. **VHS, Beta $39.95** RKO 𝄞𝄞

Storm in a Teacup A reporter starts a campaign to save a sheepdog that the town magistrate has ordered killed because the owner, an old woman, is unable to pay the license tax. As the dog's fate hangs in the balance, this often humorous film provides an interesting look at British society of the 1930s.
1937 80m/B GB Vivien Leigh, Rex Harrison, Cecil Parker, Sara Allgood; **Dir:** Victor Saville. **VHS, Beta $19.95** NOS, KRT, DVT 𝄞𝄞½

Stormquest Deep in the jungle, a band of women warriors live without men. When it is discovered that one of the group has a male lover, she is sentenced to death. Her man tries to rescue her, and it turns into a war of the sexes—literally. In Spanish with subtitles.
1987 90m/C SP Kai Baker, Brent Huff; **Dir:** Alex Sessa. **VHS $79.95** MED Woof!

Stormy Monday An American developer conspires to strike it rich in Newcastle, England when real estate by resorting to violence and political manipulations. Sting plays the jazz club owner who opposes him. Slow plot, but acted and directed well; interesting photography.
1988 (R) 108m/C Melanie Griffith, Tommy Lee Jones, Sting, Sean Bean; **Dir:** Mike Figgis. **VHS, Beta, LV, 8mm** PAR, OM 𝄞𝄞½

Stormy Trails Bell battles bellicose bad guys who are after his property. At least all the gunfights should keep a viewer awake!
1936 59m/B Rex Bell. **VHS, Beta $27.95** VCN 𝄞𝄞

Stormy Waters A romantic French drama about a sea captain falling in love with a woman he rescues from a storm. He goes back to his wife, however, when she becomes critically ill. Director Gremillon insisted on realistic footage of storms at sea; production was delayed because of the difficulties of the German occupation. Dubbed into English; original French title: "Remorques."
1941 75m/B FR Jean Gabin, Michele Morgan, Madeleine Renaud; **Dir:** Jean Gremillon. **VHS, Beta $29.95** VYY 𝄞𝄞½

Stormy Weather In this cavalcade of black entertainment the plot takes a back seat to the nearly non-stop array of musical numbers, showcasing this stellar cast at their performing peak. Lena Horne sings "Stormy Weather," "No Two Ways About Love," "Diga Do," and "I Can't Give You Anything But Love;" Bill Robinson sings his own "Ain't Misbehavin'" and is joined by Ada Brown for "That Ain't Right;" and Cab Calloway struts through "Geechee Joe" and is joined by The Nicholas Brothers for "The Jumpin' Jive."
1943 77m/B Lena Horne, Bill Robinson, Fats Waller, Dooley Wilson, Cab Calloway, Nicholas Brothers, Katherine Dunham & Her Dance Troupe; **Dir:** Andrew L. Stone. **VHS, Beta, LV $59.98** FOX, PTB 𝄞𝄞½

Story of Adele H. The story of Adele Hugo, daughter of Victor Hugo, whose love for an English soldier leads to obsession and finally to madness after he rejects her. Sensitive and gentle unfolding of characters and story. Beautiful photography. In French with English subtitles. Adjani received an Academy Award nomination for her leading role.
1975 (PG) 97m/C FR Isabelle Adjani, Bruce Robinson; **Dir:** Francois Truffaut. National Board of Review Awards '75: Best Actress (Adjani), Best Foreign Film; New York Film Critics Awards '75: Best Actress (Adjani), Best Screenwriting; New York Film Festival '75: Best Screenwriting; National Society of Film Critics Awards '75: Best Actress (Adjani). **VHS, Beta** FCT 𝄞𝄞𝄞

The Story of Boys & Girls Two very different families come together for the wedding feast of their children, during which we come to know and care for everyone present. Fine ensemble of actors, well directed, with a vivid evocation of 1930's Italy.
1991 92m/C IT Lucrezia Lante della Rovere, Massimo Bonetti, Davide Bechini; **W/Dir:** Pupi Avati. **VHS $89.95** FXL 𝄞𝄞𝄞

Story of a Cowboy Angel On the Christmas Mountain Ranch, a cowboy angel descends to Earth to bestow various beneficences.
1981 90m/C Slim Pickens. **VHS, Beta $69.98** LIV, VES 𝄞

Story of the Dragon How Bruce Lee developed the martial arts technique of Jeet Kune Do.
1982 93m/C Bruce Li, Hwang Jang Lee, Carter Wong, Roy Horan. **VHS, Beta** WVE 𝄞𝄞

The Story of Drunken Master A drunken kung fu master? That's the premise for this film.
19?? 90m/C Yang Pan Pan, Chi Hsiao Fu. **VHS $49.95** OCE 𝄞

The Story of Elvis Presley Provides a glimpse of the Presley legend through clips of his life from 1955 to 1977.
1977 53m/B Elvis Presley. **VHS $9.95** MVD, VTR, BUR 𝄞

The Story of Fausta A cleaning woman in the slums of Rio entices all the money and gifts she can out of an elderly widower. Decidedly grim comedy-drama, well-done but unsparing in its depiction of overwhelming greed. In Portugese with English subtitles.
1992 (R) 90m/C BR Betty Faria, Daniel Filho, Brandao Filho; **Dir:** Bruno Barreto. **VHS $79.95** FXL 𝄞𝄞½

The Story of the Late Chrysanthemum Classic drama about the son of a Kabuki actor who falls in love with a servant girl against his father's wishes. Their doomed affair is the center of the plot. Acclaimed and sensitive drama, in Japanese with English subtitles.
1939 115m/B JP Shotaro Hanayagi, Kakuo Mori, Kokichi Takada, Gonjuro Kawarazaki, Yoko Umemura; **Dir:** Kenji Mizoguchi. **VHS, Beta $79.95** FCT 𝄞𝄞𝄞½

The Story of Louis Pasteur A formulaic Hollywood biopic raised a notch or two by Muni's superb portrayal of the famous scientist and his career leading up to his most famous discoveries. Acclaimed in its time; excellent despite low budget. Oscar nominated for Best Picture.
1936 85m/B Paul Muni, Josephine Hutchinson, Anita Louise, Fritz Leiber, Donald Woods, Porter Hall, Akim Tamiroff, Walter Kingsford; **Dir:** William Dieterle. Academy Awards '36: Best Actor (Muni), Best Screenplay, Best Story; National Board of Review Awards '36: 10 Best Films of the Year. **VHS, Beta $19.98** MGM, BTV 𝄞𝄞𝄞

The Story of a Love Story Bates plays a writer whose imagination gets the better of him when he attempts an extra-marital affair. Was he with her or not? Unusual concept doesn't stand up in the long run. Fine cast is under-utilized. Also known as "Impossible Object."
1973 110m/C FR Alan Bates, Dominique Sanda, Evans Evans, Lea Massari, Michel Auclair, Laurence De Monaghan; **Dir:** John Frankenheimer. **VHS $59.99** USA 𝄞𝄞½

The Story of O A young woman's love for one man moves her to surrender herself to many men, in order to please him. Soft-core porn with bondage and S&M beautified by camera work. Based on the classic Freudian-erotic novel by Pauline Reage.
1975 97m/C Corinne Clery, Anthony Steel; **Dir:** Just Jaeckin. **VHS, Beta, LV $39.95** IND, HHE 𝄞𝄞

The Story of O, Part 2 A sort-of sequel to the erotic classic, in which the somewhat soiled vixen takes over an American conglomerate by seducing everyone in it.
1987 107m/C Sandra Wey, Carole James. **Beta, LV $69.98** LIV, VES, HHE 𝄞

Story of Robin Hood & His Merrie Men Well-made swashbuckler based on the English legend was Disney's second live action feature. Almost, but not quite, as memorable as the 1938 Micheal

Curtiz "Adventures of Robin Hood." Curtiz's version had Errol Flynn, after all.
1952 83m/C Richard Todd, Joan Rice, Peter Finch, Martita Hunt; *Dir:* Ken Annakin. **VHS, Beta** $19.99 *DIS ♂♂♂*

The Story of Ruth Biblical saga of adventures of Ruth as she denounces her pagan gods and flees to Israel. Typically "epic" with overwrought performances. Alternately not too bad to downright boring.
1960 132m/C Elana Eden, Viveca Lindfors, Peggy Wood, Tom Tryon, Stuart Whitman, Jeff Morrow, Thayer David, Eduard Franz; *Dir:* Henry Koster. **VHS, Beta** $19.98 *FOX, IGP ♂♂*

Story of the Silent Serials Girls in Danger These two selections capture the best of more than fourteen silent serials. Featured are Swanson tied to the railroad tracks by Beery, Marsh threatened with death (or worse) in caveman times, and other classic cliffhanger situations. Narration and musical score.
19?? 54m/B Gloria Swanson, Mae Marsh, Wallace Beery. **VHS, Beta** $29.98 *CCB ♂♂½*

Story in the Temple Red Lily Insurrectionists capture the Prince and Princess of the Sung Dynasty and only one man can save them in this Kung Fu bonanza.
197? 88m/C **VHS, Beta** $24.50 *OCE, MAV ♂*

The Story of a Three Day Pass A black American GI falls in love with a French girl he meets in peacetime Paris. With English subtitles. Made on a low budget and flawed, but poignant and impressive.
1968 87m/B *FR* Harry Barrd, Nicole Berger, Pierre Doris; *Dir:* Melvin Van Peebles. **VHS, Beta** $24.95 *DVT ♂♂½*

The Story of Vernon and Irene Castle In this, their last film together for RKO, Astaire and Rogers portray the internationally successful ballroom dancers who achieved popularity in the early 1900s. Irene Castle served as technical advisor for the film and exasperated everyone on the set by insisting that Rogers be a brunette. Still fun, vintage Fred and Ginger. The score includes "Oh, You Beautiful Doll," "By the Light of the Silvery Moon," and "Glow, Little Glow Worm."
1939 93m/B Fred Astaire, Ginger Rogers, Edna May Oliver, Lew Fields, Jack Perrin, Walter Brennan; *Dir:* H.C. Potter. **VHS, Beta, LV** $14.98 *TTC, MED ♂♂♂*

Story of William S. Hart/Sad Clowns Features William S. Hart as a cowboy in "Hell's Hinges," and Chaplin in "Tumbleweeds" (1925). From "The History of the Motion Picture" series.
192? 50m/B William S. Hart, Charlie Chaplin, Buster Keaton, Harry Langdon. **VHS, Beta** $29.98 *CCB ♂♂♂*

Story of Women Riveting factual account of woman (Huppert) who was guillotined for performing abortions in Nazi-occupied France. Huppert won best actress at Venice Film Festival. In French with English subtitles.
1988 110m/C *FR* Isabelle Huppert, Francois Cluzet, Marie Trintignant; *Dir:* Claude Chabrol. **VHS** $79.95 *NYF ♂♂♂*

Storytime Classics A collection of popular fairy tales narrated by Katharine Hepburn.
1983 78m/C *Nar:* Katharine Hepburn. **VHS, Beta** *VES ♂♂♂*

Storyville Love and music overcome prostitution and poverty in this turn-of-the-century New Orleans jazz drama, but tragedy prevails.
1974 (PG) 96m/C Tim Rooney, Jeannie Wilson, Butch Benit, Wayne Mack, Bond Gideon, Oley Sassone; *Dir:* Jack Weis. **VHS, Beta** *VHE ♂*

Stowaway After her missionary parents are killed in a Chinese revolution, Shirley stows away on a San Francisco-bound liner and plays cupid to a bickering couple who adopt her. Alice Faye sings "Goodnight My Love" and Shirley sings "You Gotta S-M-I-L-E to be H-A-P-P-Y" and "That's What I Want for Christmas."
1936 86m/B Shirley Temple, Robert Young, Alice Faye, Eugene Pallette, Helen Westley, Arthur Treacher, Astrid Allwyn; *Dir:* William A. Seiter. **VHS, Beta** $19.98 *FOX ♂♂♂*

Straight to Hell A wildly senseless, anachronistic western spoof about a motley, inept gang of frontier thieves. An overplayed, indiscriminating punk spaghetti oat-opera.
1987 (R) 86m/C Dennis Hopper, Joe Strummer, Elvis Costello, Grace Jones, Jim Jarmusch, Dick Rude, Courtney Love, Sy Richardson; *Dir:* Alex Cox. **VHS, Beta** *FOX, OM ♂½*

Straight Out of Brooklyn A bleak, nearly hopeless look at a struggling black family in a Brooklyn housing project. The son seeks escape through crime, his father in booze. An up-close and raw look at part of society seldom shown in mainstream film, its undeniable power is sapped by ragged production values and a loose narrative prone to melodrama. Rich (seen in a supporting role) was only 19 years old when he completed this, funded partly by PBS-TV's "American Playhouse."
1991 (R) 91m/C George T. Odis, Ann D. Sanders, Lawrence Gilliard, Mark Malone Jr., Reana E. Drummond, Barbara Sanon; *W/Dir:* Matty Rich. Sundance Film Festival '91: Special Jury Prize. **VHS** $92.99 *HBO ♂♂*

Straight Shooter An early sagebrush programmer.
1940 60m/B Tim McCoy. **VHS, Beta** $19.95 *NOS, DVT ♂*

Straight Shootin' Ford's first major effort launched both Carey and Gibson to national fame. Prototypical western; great action, great scenery. Silent.
1917 53m/B Harry Carey, Hoot Gibson, Mollie Malone; *Dir:* John Ford. **VHS, Beta** $27.95 *GPV, VCN, DVT ♂♂½*

Straight Talk Shirlee (Parton) is a down-home girl and former dance teacher from Arkansas who heads for Chicago to start life anew. She finds a job as a receptionist at WNDY radio, but she is mistaken for the new radio psychologist. Her homespun advice ("Get off the cross. Somebody needs the wood") becomes hugely popular and soon "Dr." Shirlee is the toast of the town. An investigative reporter (Woods), however, is looking for a scandal by trying to prove that Dr. Shirlee holds no degrees, but he soon falls in love with her. Parton's advice is the funniest part of this flimsy movie, but she is helped immensely by Dunne and Orbach. Woods, however, is not in his element in a romantic comedy, and holds the movie down.
1992 (PG) 88m/C Dolly Parton, James Woods, Griffin Dunne, Philip Bosco, Jerry Orbach, Spalding Gray, Michael Madsen, Deidre O'Connell, John Sayles, Teri Hatcher; *Dir:* Barnet Kellman. **VHS, Beta** $94.95 *HPH ♂♂½*

Straight Time Ex-con Hoffman hits the streets for the first time in six years and finds himself again falling into a life of crime. Well-told, sobering story flopped at the box office and has never received the recognition it deserved. Convincing, realistic portrayal of a criminal. Hoffman was the original director, but gave the reins to Grosbard. Based on the novel "No Beast So Fierce" by Edward Bunker.
1978 (R) 114m/C Dustin Hoffman, Harry Dean Stanton, Gary Busey, Theresa Russell, M. Emmet Walsh; *Dir:* Ulu Grosbard. **VHS, Beta** $19.98 *WAR ♂♂♂*

Straight Up A musical f/x-ridden attempts to warn MTV-era children about drugs, but visual gimmicks often overwhelm the message. Gossett sings nicely as a cosmic sage of abstinence, who cautions a boy against substance abuse. The smartest segment lambasts cigarette and alcohol ads.
1990 75m/C Chad Allen, Louis Gossett Jr. **VHS, Beta** *RHI ♂½*

Strait-Jacket After a woman is released from an insane asylum where she was sent 20 years earlier for axing her husband and his mistress, mysterious axe murders begin to occur in the neighborhood, and she is the prime suspect. Moderately creepy grade B+ horror lifted somewhat by Crawford's good portrayal. Written by Robert Block ("Psycho"). Never one to miss a gimmick, director William Castle arranged for the distribution of "bloody axes" to all theatre patrons attending the movie.
1964 89m/B Joan Crawford, Leif Erickson, Diane Baker, George Kennedy; *Dir:* William Castle. **VHS, Beta** $59.95 *COL ♂½*

Stranded A group of aliens escaping interplanetary persecution land on Earth and enlist the aid of an Earth family. Solid characters make it more than sci-fi; sort of a parable of intolerance, human (and alien) goodness, etc.
1987 (PG-13) 80m/C Maureen O'Sullivan, Ione Skye, Cameron Dye; *Dir:* Tex Fuller. **VHS, Beta, LV** $79.95 *COL ♂♂*

The Strange Affair of Uncle Harry Small-town gothic with fine acting by Sanders, an aging bachelor who plots murder when his romance is threatened by a jealous sister. A title card asks you not to reveal the 'surprise' ending—a hackneyed twist that appeased the censors but made producer Joan Harris resign in protest. Based on a play by Robert Job. Tape suffers from poor film-video transfer.
1945 80m/B George Sanders, Geraldine Fitzgerald, Ella Raines, Sara Allgood, Moyna MacGill, Samuel S. Hinds, Harry von Zell; *Dir:* Robert Siodmak. **VHS, LV** $19.98 *REP, PIA, FCT ♂♂½*

Strange Awakening Traveling in France while recuperating from amnesia, Barker is trapped in a plot of fraud and theft. Confused and contrived.
1958 75m/B *GB* Lex Barker, Carole Mathews, Nora Swinburne, Richard Molinos, Peter Dyneley; *Dir:* Montgomery Tully. **VHS** $16.95 *SNC ♂*

Strange Behavior In a small Midwestern town, the police chief follows the clues from a series of murders to the experimental lab of the local college. Seems there's a mad scientist involved... Grisley and creepy, but unduly ballyhooed when it appeared. Shot on location in New Zealand. Terrific alternate title: "Dead Kids."

1981 (R) 105m/C Michael Murphy, Louise Fletcher, Dan Shor, Fiona Lewis, Arthur Dignam, Marc McClure, Scott Brady; *Dir:* Michael Laughlin. **Beta $59.95** COL *ℐℐ*

Strange Brew The screen debut of the SCTV alumni's characters Doug & Bob MacKenzie, the Great White North duo. They do battle with a powerful, megalomaniacal brew master over - what else? - a case of beer. Dumb, but what did you expect? Watch it, or be a hoser.
1983 (PG) 91m/C Rick Moranis, Dave Thomas, Max von Sydow, Paul Dooley; *Dir:* Rick Moranis, Dave Thomas. **VHS, Beta $19.98** MGM *ℐℐ ½*

Strange Cargo Convicts escaping from Devil's Island are mystically entranced by a Christ-like fugitive en route to freedom. An odd, pretentious Hollywood fable waiting for a cult following. Gable and Crawford's eighth and final pairing. Adapted by Anita Loos from the book "Not Too Narrow...Not Too Deep" by Richard Sale.
1940 105m/B Clark Gable, Joan Crawford, Ian Hunter, Peter Lorre, Paul Lukas, Albert Dekker, J. Edward Bromberg, Eduardo Ciannelli, Frederick Worlock; *Dir:* Frank Borzage. **VHS, Beta $19.98** MGM *ℐℐ ½*

Strange Case of Dr. Jekyll & Mr. Hyde An adaptation of the classic Robert Louis Stevenson book about a scientist who conducts experiments on himself to separate good from evil. Palance is oddly but appealingly cast; Jarrott's bad direction wrecks it. Made for TV.
1968 128m/C Jack Palance, Leo Genn, Oscar Homolka, Billie Whitelaw, Denholm Elliott; *Dir:* Charles Jarrott. **VHS, Beta $49.95** IVE *ℐℐ*

Strange Case of Dr. Jekyll & Mr. Hyde The Robert Louis Stevenson classic, with Hyde portrayed as an icy, well-dressed sociopath. In Shelley Duvall's "Nightmare Classics" series. More psychological than special effect-y. Made for cable.
1989 60m/C Anthony Andrews, Laura Dern, George Murdock, Nicholas Guest; *Dir:* Michael Lindsay-Hogg. **VHS, Beta $59.95** WAR, OM *ℐℐ ½*

A Strange and Deadly Occurence Strange things start to happen to a family when they move to a house in a remote area. Made-for-television.
1974 74m/C Robert Stack, Vera Miles, L.Q. Jones, Herb Edelman; *Dir:* John Llewellyn Moxey. **VHS, Beta $34.95** WOV *ℐ ½*

Strange Fruit Based on Lillian Smith's novel, this program tells the story of a black painter in Georgia, 1948, who faces racism. At first avoiding voter registration, he becomes involved and is killed. His death serves as an inspiration to his community.
1979 33m/C **VHS, Beta** LCA *ℐ ½*

Strange Illusion Unbalanced teen Lydon believes his mother, about to remarry, was responsible for his father's death. He feigns insanity in a plan to catch her, but is sent to an asylum, where he nearly goes insane for real. Slow and implausible, but creepy enough to hold your interest.
1945 87m/B Jimmy Lydon, Warren William, Sally Eilers, Regis Toomey, Charles Arnt, George Reed; *Dir:* Edgar G. Ulmer. **VHS, Beta $16.95** SNC *ℐℐ ½*

Strange Interlude Slow PBS production of Eugene O'Neill's famous drama covering two decades of the lives and loves of an upper-class family. Excellent performances help to make up for talkiness. But why all these upper-class family PBS dramas? On two cassettes.

1990 190m/C Glenda Jackson, Jose Ferrer, David Dukes, Ken Howard, Edward Petherbridge; *Dir:* Herbert Wise. **VHS $29.95** FRH *ℐℐ ½*

Strange Invaders Body-snatchers-from-space sci-fi with an attitude - fun spoof of '50s alien flicks. Space folks had taken over a midwestern town in the '50s, assuming the locals' appearance and attire before returning to their ship. Seems one of them married an earthling - but divorced and moved with her half-breed daughter to New York City. So the hicksters from space arrive in Gotham wearing overalls...
1983 (PG) 94m/C Paul LeMat, Nancy Allen, Diana Scarwid, Michael Lerner, Louise Fletcher, Wallace Shawn, Fiona Lewis, Kenneth Tobey, June Lockhart, Charles Lane, Dey Young, Mark Goddard; *Dir:* Michael Laughlin. **VHS, Beta, LV $79.95** VES *ℐℐℐ*

The Strange Love of Martha Ivers Douglas is good in his screen debut as the wimpy spouse of unscrupulous Stanwyck. Stanwyck shines as the woman who must stay with Douglas because of a crime she committed long ago... Tough, dark melodrama; classic film noir.
1946 117m/B Barbara Stanwyck, Van Heflin, Kirk Douglas, Lizabeth Scott, Judith Anderson; *Dir:* Lewis Milestone. **VHS, Beta $19.95** NOS, KRT, DVT *ℐℐℐ*

Strange New World Three astronauts awake from 188 years in the fridge to find cloning has arrived. Made for TV as a pilot for a hoped-for series that might have been even worse.
1975 100m/C John Saxon, Kathleen Miller, Keene Curtis, Martine Beswick, James Olson, Catherine Bach, Richard Farnsworth, Ford Rainey; *Dir:* Robert Butler. **VHS $59.95** UNI *ℐ ½*

Strange Shadows in an Empty Room Sleuth Whitman wants answers about the murder of his kid sis. He beats lots of people up, and the viewer leaves the empty living room to look at the inside of the fridge, which is more interesting.
1976 (R) 97m/C Stuart Whitman, John Saxon, Martin Landau, Tisa Farrow, Carole Laure, Gayle Hunnicutt; *Dir:* Martin Herbert. **VHS, Beta $69.95** VES *ℐ*

Strange Tales A collection of fantasy shorts, including "Twilight Journey," "The Crystal Quest" and "The Visitant."
198? 60m/C **VHS, Beta** VDC

Strange Tales/Ray Bradbury Theatre An anthology of thrillers based on Bradbury stories: "Banshee," "The Screaming Woman," and "The Town Where No One Got Off."
1986 (R) 86m/C Jeff Goldblum, Peter O'Toole, Drew Barrymore, Charles Martin Smith. **VHS, Beta $19.99** HBO *ℐℐ ½*

The Strange Woman Uneventful Hollywood costume drama. Lamarr stalks man after man, but never creates much excitement in spite of Ulmer's fancy camera work and intense pace.
1946 100m/B Hedy Lamarr, George Sanders, Louis Hayward, Gene Lockhart, Hillary Brooke, June Storey; *Dir:* Edgar G. Ulmer. **VHS $29.95** NOS, FCT *ℐℐ*

The Strangeness Miners in search of gold release a "strange" creature from far beneath the earth—eek! Made on a negative budget - or so one assumes.
1985 90m/C Dan Lunham, Terri Berland; *Dir:* David Michael Hillman. **VHS, Beta $49.95** TWE **Woof!**

The Stranger Notably conventional for Welles, but swell entertainment nonetheless. War crimes tribunal sets Nazi thug Shayne free hoping he'll lead them to his superior, Welles. Robinson trails Shayne through Europe and South America to a small town in Connecticut. Tight suspense made on a tight budget saved Welles's directorial career.
1946 95m/B Edward G. Robinson, Loretta Young, Martha Wentworth, Konstantin Shayne, Richard Long, Orson Welles; *Dir:* Orson Welles. **VHS, Beta $9.95** NEG, NOS, PSM *ℐℐℐ ½*

The Stranger Corbett crash-lands on a planet an awful lot like Earth - and must stay on the run. Uneven fugitive thriller in sci-fi drag. Made for TV.
1973 100m/C Cameron Mitchell, Glenn Corbett, Sharon Acker, Lew Ayres, George Coulouris, Dean Jagger; *Dir:* Lee H. Katzin. **VHS** KOV *ℐ ½*

The Stranger Amnesiac car-wreck victim Bedelia begins regaining her memory, and realizes she witnessed several grisly murders. Is her shrink (Riegert) helping her remember, or keeping something from her? Good, neglected thriller.
1987 (R) 93m/C AR Bonnie Bedelia, Peter Riegert, Barry Primus, David Spielberg, Julio de Grazia, Celia Roth; *Dir:* Adolfo Aristarain. **VHS, Beta, 8mm $79.95** COL *ℐℐℐ*

The Stranger and the Gunfighter The baddest spaghetti western of all time - or is it a kung fu movie? Alcoholic cowpoke Van Cleef and kung fu master Lieh pair up to find hidden treasure. There is a map, but parts are printed on the assorted fannies of various women.
1976 106m/C Patty Shepard, Lee Van Cleef, Lo Lieh; *Dir:* Anthony M. Dawson. **VHS, Beta, LV $59.95** COL *ℐℐ*

A Stranger in My Forest Two children discover Christian love and joy after losing their parents in an automobile accident.
1988 75m/C Eddie Moran, Trent Dolan, Susan Backlinie, George Jones; *Dir:* Donald W. Thompson. **VHS, Beta** BPG, MIV *ℐ*

Stranger than Paradise Regular guy, nerdy sidekick and Hungarian girl cousin traipse around America being bored and having fun and adventure. New York, then - on to Cleveland! The thinking person's mindless flick. Inventive, independent comedy made on quite a low budget was acclaimed at Cannes.
1984 (R) 90m/B GE Richard Edson, Eszter Balint, John Lurie; *Dir:* Jim Jarmusch. Cannes Film Festival '84: Camera d'Or; National Society of Film Critics '84: Best Picture. **VHS, Beta, LV $79.98** FOX *ℐℐℐ*

Stranger in Paso Bravo A distraught drifter returns to a strange Italian town to avenge the murders of his wife and daughter years before. Dubbed.
1973 92m/C IT Anthony Steffen, Giulia Rubini, Eduardo Fajardo, Andriana Ambesi. **VHS, Beta $57.95** UNI *ℐ*

Stranger on the Third Floor Reporter McGuire's testimony helped convict cabbie Cook of murder, but he begins to have doubts. Odd, shadowy psycho-thriller considered by some the first film noir. The reporter comes to be suspected of the crime...and his fiancee pounds the pavement to prove him innocent. Average acting, but great camera work gives the whole a deliciously menacing feel.
1940 64m/B Peter Lorre, John McGuire, Elisha Cook Jr., Margaret Tallichet; *Dir:* Boris Ingster. **VHS, Beta, LV $19.95** CCB, MED *ℐℐℐ*

Stranger in Town A young journalist looks into the murder of an American composer in a sleepy English village. He discovers blackmail and intrigue. So-so thriller. Based on the novel "The Uninvited" by Frank Chittenden.
1957 73m/B GB Alex Nicol, Ann Page, Mary Laura Wood, Mona Washbourne, Charles Lloyd Pack; **Dir:** George Pollock. **VHS, Beta $16.95** SNC ✂✂

The Stranger from Venus "The Day the Earth Stood Still" warmed over. Venusian Dantine tells earth lady Neal he's worried about the future of her planet. Real low budget. Includes previews of coming attractions from classic science fiction films.
1954 78m/B Patricia Neal, Helmut Dantine, Derek Bond. **VHS, Beta $19.95** MED, MLB Woof!

A Stranger is Watching Rapist-murderer Torn holds his victim's 10-year-old daughter hostage, along with a New York television anchorwoman. Complicated and distasteful. Made for TV.
1982 (R) 92m/C Rip Torn, Kate Mulgrew, James Naughton; **Dir:** Sean S. Cunningham. **VHS, Beta $79.95** MGM ✂½

Stranger Who Looks Like Me Adoptees Baxter and Bridges set out to find their real parents. Not-bad TV fare. Blake is Baxter's real-life mom. Look for a young Patrick Duffy.
1974 74m/C Meredith Baxter Birney, Beau Bridges, Whitney Blake; **Dir:** Larry Peerce. **VHS, Beta $29.95** WOV ✂✂

The Stranger Within Made-for-TV "Rosemary's Baby" rip-off. Eden is in a family way, though hubby Grizzard is impotent. The stranger within begins commanding her to do its bidding, just as if she were a genie.
1974 74m/C Barbara Eden, George Grizzard, Joyce Van Patten, Nehemiah Persoff; **Dir:** Lee Philips. **VHS, Beta $49.95** IVE, IVE ✂✂½

Strangers in the City Puerto Ricans newly arrived in New York try to make their way. Serious but slightly overwrought immigrant-family melodrama.
1962 80m/B Robert Gentile, Camilo Delgado, Rosita De Triana; **Dir:** Rick Carrier. **VHS, Beta $59.98** CHA ✂✂

Strangers of the Evening In its day this dark comedy/mystery caught flak for its gruesomeness, dealing with a mixup at the undertaker's and the wrong body buried - possibly alive. Good photography and acting make the humor work, intentionally or not. Based on a novel "The Illustrious Corpse," by Tiffany Thayer.
1932 70m/B ZaSu Pitts, Eugene Pallette, Lucien Littlefield, Tully Marshall, Miriam Seegar, Theodore von Eltz; **Dir:** H. Bruce Humberstone. **VHS $12.95** SNC, NOS ✂✂

Stranger's Gold A mysterious gunslinger eliminates corruption in a western mining town.
1971 90m/C John Garko, Antonio Vilar, Daniela Giordano. **VHS, Beta $19.95** GEM ✂

Strangers in Good Company A loving metaphor to growing older. Director Scott uses non-actors for every role in this quiet little film about a bus-load of elderly women lost in the Canadian wilderness. They wait for rescue without hystrionics, using the opportunity instead to get to know each other and nature. Beautifully made, intelligent, uncommon and worthwhile.

1991 (PG) 101m/C Alice Diabo, Mary Meigs, Cissy Meddings, Beth Webber, Winifred Holden, Constance Garneau, Catherine Roche, Michelle Sweeney; **Dir:** Cynthia Scott. **VHS, Beta $94.95** TOU ✂✂✂

Strangers Kiss Circa 1955: Hollywood director encourages his two leads to have an off-screen romance to bring reality to his film. Conflict arises when the leading lady's boyfriend, the film's financier, gets wind of the scheme. Intriguing, fun, slightly off-center. Based on Stanley Kubrick's film "Killer Kiss."
1983 (R) 93m/C Peter Coyote, Victoria Tennant, Blaine Novak, Dan Shor, Richard Romanus, Linda Kerridge, Carlos Palomino; **Dir:** Matthew Chapman. **VHS, Beta $69.99** HBO ✂✂½

Strangers in Paradise A scientist who had cryogenically frozen himself to escape the Nazis is thawed out in the present, and his powers are used by a delinquent-obsessed sociopath.
1986 81m/C W/Dir: Ulli Lommel. **VHS, Beta $69.98** LIV, VES ✂½

Strangers: The Story of a Mother and Daughter Rowlands is the long-estranged daughter of Davis, who won an Emmy for her portrayal of the embittered widow. The great actress truly is at her recent best in this made-for-TV tearjerker, and Rowlands keeps pace.
1979 88m/C Bette Davis, Gena Rowlands, Ford Rainey, Donald Moffatt; **Dir:** Milton Katselas. **VHS, Beta $69.95** LTG, TLF ✂✂✂

Strangers on a Train Long before there was "Throw Momma from the Train," there was this Hitchcock super-thriller about two passengers who accidentally meet and plan to "trade" murders. Amoral Walker wants the exchange and the money he'll inherit by his father's death; Granger would love to end his stifling marriage and wed Roman, a senator's daughter, but finds the idea ultimately sickening. What happens is pure Hitchcock. Screenplay co-written by murder-mystery great Raymond Chandler. Patricia Hitchcock, the director's only child, plays Roman's sister. Deservedly Oscar-nominated for Best Cinematography; the concluding "carousel" scene is a masterpiece.
1951 101m/B Farley Granger, Robert Walker, Ruth Roman, Leo G. Carroll, Patricia Hitchcock; **Dir:** Alfred Hitchcock. National Board of Review Awards '51: 10 Best Films of the Year. **VHS, Beta, LV $19.98** WAR, MLB ✂✂✂

Strangers When We Meet A married architect and his equally married neighbor begin an affair. Their lives become a series of deceptions. Lavish but uninvolving soaper. Written by Evan Hunter and based on his novel.
1960 117m/B Kirk Douglas, Kim Novak, Ernie Kovacs, Walter Matthau, Barbara Rush, Virginia Bruce, Kent Smith; **Dir:** Richard Quine. **VHS, Beta $19.95** COL ✂✂½

The Strangler A confused, mother-fixated psychopath strangles young women. Made at the time the Boston Strangler was terrorizing Beantown. The film slayer strangles ten lasses before his love of dolls gives him away.
1964 89m/B Victor Buono, David McLean, Diane Sayer, Ellen Corby, Jeanne Bates; **Dir:** Burt Topper. **VHS, Beta $14.98** FXV, FOX, FCT ✂✂½

Strangler of the Swamp The ghost of an innocent man wrongly hanged for murder returns to terrorize the village where he was lynched. Decidedly B-grade horror features an out-of-focus ghost and pre-An-

drews Edwards as one of his would-be victims.
1946 60m/B Rosemary La Planche, Blake Edwards, Charles Middleton, Robert Barratt; **Dir:** Frank Wisbar. **VHS, Beta $14.95** COL, SVS ✂½

Strapless Just-turned-40 American doctor Brown lives and works in London. She has just ended a long-term romance, and takes up with suave foreigner Ganz. Her young sister Fonda, arrives for a visit. Good, interesting if sometimes plodding story of adult relationships.
1990 99m/C Blair Brown, Bridget Fonda, Bruno Ganz, Alan Howard, Michael Gough, Hugh Laurie, Suzanne Burden, Camille Coduri; **Dir:** David Hare. **VHS, Beta, LV $79.95** COL ✂✂½

Strategic Air Command A classic post-World War II chunk of Air Force patriotism. Veteran third baseman Stewart is recalled to flight duty at the hint of a nuclear war. He's already put in his time in the Big One and thinks he's being singled out now, but he answers his Uncle Sam's call. Allyson plays Stewart's wife for the third time.
1955 114m/C James Stewart, June Allyson, Frank Lovejoy, Barry Sullivan; **Dir:** Anthony Mann. **VHS, Beta, LV $14.95** PAR ✂✂½

Straw Dogs An American mathematician, disturbed by the predominance of violence in American society, moves with his wife to an isolated Cornish village. He finds that primitive savagery exists beneath the most peaceful surface. After his wife is raped, Hoffman's character seeks revenge. Hoffman is good, a little too wimpy at times. A violent, frightening film reaction to the violence of the 1960s.
1972 (R) 118m/C GB Dustin Hoffman, Susan George, Peter Vaughan, T.P. McKenna, David Warner; **Dir:** Sam Peckinpah. **VHS, Beta, LV, CDV $59.98** FOX ✂✂✂

Strawberry Blonde A romantic comedy set in the 1890s, with Cagney as a would-be dentist infatuated with money-grubbing Hayworth (the strawberry blonde of the title), who wonders years later whether he married the right woman (chestnut brunette de Havilland). Attractive period piece remade from 1933's "One Sunday Afternoon," and revived yet again in 1948 by Raoul Walsh.
1941 100m/B James Cagney, Olivia de Havilland, Rita Hayworth, Alan Hale Jr., George Tobias, Jack Carson, Una O'Connor, George Reeves; **Dir:** Raoul Walsh. **VHS, Beta $24.95** MGM ✂✂✂

Strawberry Roan Maynard plays a stubborn rodeo cowboy, out to prove he ain't whipped yet. One of Maynard's most popular films. Gene Autry bought the story to help him out in later years; the 1948 Autry "remake" uses only the title.
1933 59m/B Ken Maynard, Ruth Hall; **Dir:** Ken Maynard. **VHS $19.95** NOS, DVT ✂✂

The Strawberry Statement Dated message film about a campus radical who persuades a college student to take part in the student strikes on campus during the '60s. Ambitious anti-violence message is lost in too many subplots. Soundtrack features songs by Crosby, Stills, Nash and Young and John Lennon.
1970 (R) 109m/C Kim Darby, Bruce Davison, Bud Cort, James Coco, Kristina Holland, Bob Balaban, David Dukes, Jeannie Berlin; **Dir:** Stuart Hagmann. **VHS, Beta $69.95** MGM ✂✂

Stray Dog A tense, early genre piece by Kurosawa, about a police detective who has his revolver picked from his pocket on a bus, and realizes soon after it's being used in a series of murders. Technically somewhat flawed, but tense and intense. Mifune's

pursuit of the criminal with his gun becomes metaphorically compelling. In Japanese with English subtitles.
1949 122m/B *JP* Toshiro Mifune, Takashi Shimura; *Dir:* Akira Kurosawa. **VHS, Beta $24.95** *SVS, TAM* ♂♂♂

Streamers Six young soldiers in a claustrophobic army barracks tensely await the orders that will send them to Vietnam. Written by David Rabe from his play. Well acted but downbeat and drawn out.
1983 (R) 118m/C Matthew Modine, Michael Wright, Mitchell Lichenstein, George Dzundza; *Dir:* Robert Altman. **VHS, Beta, LV $69.95** *MED, IME* ♂♂

Street A dark classic of German Kammerspiel film, notable for its expressionist treatment of Paris street life. Silent.
1923 87m/B *GE* **W/Dir:** Karl Grune. **VHS, Beta $29.95** *VYY* ♂♂½

Street Asylum Real-life Watergate spook Liddy is pathetically bad as an evil genius who cooks up a scheme to rid the streets of deadbeats and scumbags by implanting cops with a gizmo that makes them kill. Also available in a 94-minute unrated version.
1990 (R) 94m/C Wings Hauser, Alex Cord, Roberta Vasquez, G. Gordon Liddy, Marie Chambers, Sy Richardson, Jesse Doran, Jesse Aragon, Brion James; *Dir:* Gregory Brown. **VHS, Beta, LV $89.98** *MAG* ♂

Street Crimes A street-wise cop convinces gang members to put down their weapons and settle their grudges in the boxing ring. But when a gang leader starts shooting down the police and civilians, the cop and his young partner must work together to keep the neighborhood safe.
1992 (R) ?m/C Dennis Farina, Max Gail, Mike Worth; *W/Dir:* Stephen Smoke. **VHS $89.95** *PMH* ♂♂

Street of Crocodiles The decrepit caretaker of an equally decrepit museum is transformed into another being by a Kinetoscope machine. An avant-garde animated classic by the Quay Brothers.
1986 21m/C VHS, 3/4U *ICA* ♂♂½

Street Fighter Local hoodlum engages in fast-paced marital arts action.
1975 (R) 77m/C Sonny Chiba. **VHS, Beta** *FOX* ♂

The Street Fighter's Last Revenge The theft of a formula for synthetic heroin brings martial artist Sonny Chiba to the rescue. Lots of action, not much acting.
1979 (R) 80m/C Sonny Chiba, Sue Shiomi; *Dir:* Sakae Ozawa. **VHS $49.98** *UHV* ♂

Street Fighters Part 2/Masters of Tiger Crane Martial arts action abounds in these two movies on one tape.
1985 176m/C VHS, Beta $7.00 *VTR* ♂½

Street Girls A father enters the world of urban drugs and prostitution to find his runaway daughter.
1975 (R) 77m/C Carol Case, Christine Souder, Paul Pompian; *Dir:* Michael Miller. **VHS, Beta $59.98** *CHA* ♂

Street Hawk A government agent operates an experimental, crime-fighting motorcycle in battling against a high-powered drug smuggling ring.
1984 60m/C Rex Smith, Jayne Modean, Richard Venture, Christopher Lloyd. **VHS, Beta $39.95** *MCA* ♂

Street Hero A young thug is torn between familial common sense and Mafia connections.
1984 102m/C Vince Colosimo, Sigrid Thornton, Sandy Gore. **VHS, Beta $69.98** *LIV, VES Woof!*

Street Hunter Former cop turned freelance fighter quit the force when his integrity was questioned. But he still helps out now and then. Scumbag drug dealers line up a thug to waste him; and epic kung fu battle ensues.
1990 (R) 120m/C Steve James, Reb Brown, John Leguizamo, Valarie Pettiford, Frank Vincent, Richie Havens; *Dir:* John A. Gallagher. **VHS, LV $79.95** *COL Woof!*

Street Justice Over-complex tale of a CIA agent on the lam. Returning from years of Russian imprisonment, Ontkean finds his wife remarried, his daughter grown and both battling the corrupt town government. Good performances can't help the confusion of too many characters too flatly developed.
1989 (R) 94m/C Michael Ontkean, Joanna Kerns, Catherine Bach, J.D. Cannon, Jeanette Nolan, Richard Cox, William Windom; *Dir:* Richard Sarafian. **VHS, Beta $79.95** *WAR* ♂½

Street Law A vivid and violent study of one man's frustrated war on crime.
1979 (R) 77m/C Franco Nero, Barbara Bach, Reno Palmer. **VHS, Beta $14.98** *VID Woof!*

Street Music The elderly people living in an old hotel join forces with a young couple to organize a protest that may save their building. Well-directed urban comedy-drama; good location shooting in San Francisco's Tenderloin.
1981 88m/C Larry Breeding, Elizabeth Daily, Ned Glass; *Dir:* Jenny Bowen. **VHS, Beta $69.98** *LIV, VES* ♂♂½

Street People Gratuitous car chases and violence do not a movie make, as in this case in point. Utter woofer has Brit Moore cast as a mafiosa. Yeah, right.
1976 (R) 92m/C Roger Moore, Stacy Keach, Ivo Gassani, Ettore Manni; *Dir:* Maurizio Lucidi. **VHS, Beta $29.98** *LIV, VES Woof!*

Street Scene Life in a grimy New York tenement district, circa 1937. Audiences nationwide ate it up when Elmer Rice adapted his own Pulitzer Prize-winning play and top helmsman Vidor gave it direction. Fine score by Alfred Newman.
1931 80m/B Sylvia Sydney, William Collier Jr., Estelle Taylor, Beulah Bondi, David Landau; *Dir:* King Vidor. **VHS, Beta, 8mm $19.95** *NOS, HHT, CAB* ♂♂♂

Street of Shame A sensitive portrayal of the abused lives of six Tokyo prostitutes. Typically sensitive to the roles and needs of women and critical of the society that exploits them, Mizoguchi creates a quiet, incisive coda to his life's work in world cinema. Kyo is splendid as a hardened hooker and has a memorable scene with her father. Kogure is also good. The great director's last finished work was instrumental in the outlawing of prostitution in Japan. Written by Masaichi Narusawa. English subtitles.
1956 88m/B *JP* Machiko Kyo, Aiko Mimasu; *Dir:* Kenji Mizoguchi. **VHS, Beta $29.98** *SUE* ♂♂♂½

Street Smart Reeve was blah as Superman (let's be frank), and he's blah here as a desperate New York freelance writer who fakes a dramatic story about prostitution. When his deception return to haunt him, he's in trouble with pimps and murderers, as well as the D.A. Freeman and Baker are both superb. Based on screenwriter David Freeman's own experience with "New York" magazine.
1987 (R) 97m/C Christopher Reeve, Morgan Freeman, Kathy Baker, Mimi Rogers, Andre Gregory; *Dir:* Jerry Schatzberg. **VHS, Beta, LV $19.95** *MED, IME* ♂♂½

Street Soldiers Martial artists are out to take back the streets from the punks and scum who now rule them.
1991 (R) 98m/C Jun Chong, Jeff Rector, David Homb, Jonathan Gorman, Joon Kim, Katherine Armstrong, Joel Weiss; *Dir:* Lee Harry. **VHS, Beta $19.95** *ACA* ♂♂½

Street Trash In Brooklyn, a strange poisonous liquor is being sold cheap to bums, making them melt and explode. A gross, cheap, tongue-in-cheek shocker from the makers of "The Toxic Avenger."
1987 91m/C R.L. Ryan, Vic Noto, Miriam Zucker, Bill Chepil, Mike Lackey; *Dir:* Jim Muro. **VHS, Beta $79.98** *LIV, LTG Woof!*

Street War A cop pursues the Mob in the name of revenge as well as duty.
197? 90m/C James Mason, Cyril Cusack, Raymond Pellegrin. **VHS, Beta $59.95** *MGL* ♂

Street Warriors Ostensibly "serious" look at juvenile crime. Actually miserable, exploitative, offensive junk.
1987 105m/C Christa Leem, Nadia Windell, Victor Petit, Francisco (Frank) Brana; *Dir:* Jose Antonio De La Loma. **VHS, Beta $34.95** *ASE Woof!*

Street Warriors, Part 2 Even worse than its predecessor, and that's saying something. Horrible dubbing from Spanish would be laughable, if the rapes weren't so graphic.
1987 105m/C Angel Fernandez, Paul Ramirez, Teresa Giminez, Veronica Miriel, Conrad Tortosa; *Dir:* Jose Antonio De La Loma. **VHS, Beta $34.95** *ASE Woof!*

The Street With No Name In follow-up to "Kiss of Death," Widmark confirms rep as one disturbed guy playing psychotic career criminal whose life is grisly trail of murder and brutality.
1948 93m/B Mark Stevens, Richard Widmark, Lloyd Nolan, Barbara Lawrence, Ed Begley Sr., Donald Buka, Joseph Pevney; *Dir:* William Keighley. **VHS $19.98** *FOX, FCT* ♂♂♂

A Streetcar Named Desire Powerful film version of Tennessee Williams' play about a neurotic southern belle with a hidden past who comes to visit her sister and is abused and driven mad by her brutal brother-in-law. Grim New Orleans setting for terrific performances by all, with Malden, Leigh, and Hunter winning Oscars, and Brando making highest impact on audiences. Brando disliked the role, despite the great impact it had on his career. Great jazz score by Alex North. Oscar nominations: Best Picture, Best Actor (Brando, who lost to Bogart in "The African Queen" - a tough choice if ever there was one), Best Director, Screenplay, Black & White Cinematography, Score of a Comedy or Drama and Black & White Film Editing. Remade for television in 1984.

1951 122m/B Vivien Leigh, Marlon Brando, Kim Hunter, Karl Malden; *Dir:* Elia Kazan. Academy Awards '51: Best Actress (Leigh), Best Art Direction/Set Decoration (B & W), Best Supporting Actor (Malden), Best Supporting Actress (Hunter); National Board of Review Awards '51: 10 Best Films of the Year; New York Film Critics Awards '51: Best Actress (Leigh), Best Director (Kazan), Best Picture. **VHS, Beta, LV, CDV** $19.98 *RCA, FOX, WAR* 🐾🐾🐾🐾

Streetfight Semi-animated racist exploitation from the creator of "Fritz the Cat." Features some superb animation. Originally called "Coonskin."
1975 (R) 89m/C Philip Michael Thomas, Scatman Crothers; *Dir:* Ralph Bakshi. **VHS, Beta** $29.95 *ACA* 🐾 ½

Streets Applegate (of Fox TV's "Married...With Children") is believable as an illiterate runaway teen. Good drama set in Venice, Calif. about life on the streets is flawed by near-gratuitous pairing with story of a crazy prostitute-killing cop.
1990 (R) 90m/C Christina Applegate, David Mendenhall, Eb Lottimer; *Dir:* Katt Shea Ruben. **VHS, Beta** $89.99 *MGM* 🐾🐾

Streets of Fire A soldier of fortune rescues his ex-girlfriend, now a famous rock singer, after she's been kidnapped by a malicious motorcycle gang. Violently energetic in its insistent barrage of imagery. Director Hill's never-never land establishes a retro-futuristic feel, set to the music of Ry Cooder and beautifully photographed, however vacuous the ending may be.
1984 93m/C Michael Pare, Diane Lane, Rick Moranis, Amy Madigan, Willem Dafoe, Deborah Van Valkenburgh, Richard Lawson, Rick Rossovich, Bill Paxton, Lee Ving, Stoney Jackson, Robert Townsend, Grand Bush, Mykel T. Williamson, Elizabeth Daily, Lynne Thigpen, Marine Jahan, Ed Begley Jr., John Dennis Johnston, Olivia Brown; *Dir:* Walter Hill. **VHS, Beta, LV** $19.95 *MCA* 🐾🐾 ½

Streets of Gold Brandauer is good as a Soviet boxer who defected and now washes dishes in Brooklyn. He trains a pair of street youngsters in the hope of beating his old coach in the Olympics. "Rocky"-like in the boxing, and in the rags-to-riches optimism.
1986 (R) 94m/C Klaus Maria Brandauer, Adrian Pasdar, Wesley Snipes; *Dir:* Joe Roth. **VHS, Beta, LV** $79.98 *LIV, VES* 🐾🐾 ½

The Streets of L.A. A frustrated middle-aged woman teaches some punks from an L.A. barrio a thing or two about pride and motivation when she demands reimbursement for the tires they slashed. Without thoroughly professional Woodward to carry it, there would not be much of a movie. Woodward's character is resourceful, and so is the actress. Made for TV.
1979 94m/C Joanne Woodward, Robert Webber, Michael C. Gwynne; *Dir:* Jerrold Freedman. **VHS, Beta** $59.95 *LTG* 🐾🐾 ½

The Streets of San Francisco The pilot that spawned the popular TV series. A streetwise old cop and his young college-boy partner (who else but Malden and Douglas as Stone and Keller) investigate the murder of a young woman. Adapted from "Poor, Poor Ophelia" by Carolyn Weston.
1972 120m/C Karl Malden, Robert Wagner, Michael Douglas, Andrew Duggan, Tom Bosley, Kim Darby, Mako; *Dir:* Walter Grauman. **VHS, Beta** $14.99 *GKK* 🐾🐾🐾

Streetwalkin' Life in the Big Apple isn't always rosy for a brother and sister who must contend with prostitutes, drug dealers, and tough cops. She turns to prostitution to get by. Miserable, exploitative trash.
1985 (R) 86m/C Melissa Leo, Dale Midkiff, Leon Robinson, Julie Newmar, Randall Batinkoff, Annie Golden, Antonio Fargas; *Dir:* Joan Freeman. **VHS, Beta, LV** $79.98 *LIV, VES* 🐾

Streetwise Eye-opening and candid documentary inspired by a "Life" magazine article. Frank and honest look at homeless children making a living on the streets. Seattle provides the backdrop to the story as the filmmakers hang around with the kids, earning their trust, and receive the truth from them. Cruel, funny, enraging, but always sobering look into a population of homeless in America. A deservedly acclaimed, terrifying vision of urban America.
1985 92m/C *Dir:* Martin Bell. **VHS, Beta, LV** $19.95 *STE, NWV* 🐾🐾🐾 ½

Strictly Business An upwardly mobile black broker has his career aspirations in order until he meets a beautiful club promoter who finds him square and boring. Wanting to impress her he asks the advice of a young man who works in the mail room for the proper way to dress and talk. A low-rent Pygmalion story.
1991 (PG-13) 83m/C Halle Berry, Tommy Davidson, Joseph C. Phillips; *Dir:* Kevin Hooks. **VHS** $94.99 *WAR* 🐾 ½

Strictly G.I. A collection of war-time patriotic shorts, including "All Star Bond Rally." The support and perception of World War II is historically intriguing.
1944 45m/B Bing Crosby, Frank Sinatra, Betty Grable, Judy Garland, Harpo Marx, Harry James, Bob Hope, Jerry Colonna. **VHS, Beta** $19.95 *NOS, DVT* 🐾 ½

Strike Eisenstein's debut, and a silent classic. Stirring look at a 1912 clash between striking factory workers and Czarist troops.
1924 78m/B *RU Dir:* Sergei Eisenstein. **VHS, Beta, 3/4U** $24.95 *NOS, KIV, IHF* 🐾🐾🐾 ½

Strike Back A convict escapes from prison to find the girl he loves, and he then is hounded by the police and the mob.
1980 89m/C Dave Balko, Brigette Wollner; *Dir:* Carl Schenkel. **VHS, Beta** $69.98 *LIV, VES* 🐾🐾

Strike Commando Cheap, exploitative ripoff of the expensive, exploitative megaaction thriller "Rambo." Sleepwalking actors; horrible, overwrought script.
1987 (R) 92m/C Reb Brown, Christopher Connelly, Locs Kamme; *Dir:* Vincent Dawn. **VHS, Beta** $9.99 *CCB, IVE* Woof!

Strike Force A New York City cop, a Federal agent and a state trooper work together to battle a large drug ring. Failed made-for-TV pilot. Early Gere appearance; not related to later Robert Stack film with the same title. Called "Crack" on video.
1975 74m/C Cliff Gorman, Richard Gere, Donald Blakely; *Dir:* Barry Shear. **VHS, Beta** $24.95 *VMK* 🐾

Strike Force The jurors on an embezzlement case turn up bloody and headless around town, so a special criminal strike force is brought in to solve the murders. Why do the murders happen only on Tuesdays? Made for TV.
1981 90m/C Robert Stack, Dorian Harewood, Herb Edelman; *Dir:* Richard Lang. **VHS, Beta** $24.95 *AHV* 🐾🐾

Strike It Rich A honeymooning couple find themselves resorting to the Monte Carlo gambling tables in order to raise money for the ride home. Lightweight fluff unfortunately adapted from Graham Greene's short novel "Loser Takes All." Only for hard-core Gielgudites.
1990 (PG) 86m/C Molly Ringwald, Robert Lindsay, John Gielgud, Max Wall, Simon de la Brosse; *Dir:* James Scott. **VHS, Beta, LV** $89.99 *HBO* 🐾 ½

Strike of Thunderkick Tiger Two gang members kill their leader, and then have a go at each other.
1987? 84m/C Casanova Wong, Charles Han, Lisa Lee. **VHS, Beta** $54.95 *MAV* 🐾

Strike Up the Band A high school band turns to hot swing music and enters a national radio contest. Rooney and Garland display their usual charm in this high-energy stroll down memory lane. Songs include "I Ain't Got Nobody," "Our Love Affair," "Drummer Boy" and "Do The Conga!"
1940 120m/B Judy Garland, Mickey Rooney, Paul Whiteman, William Tracy, June Preisser; *Dir:* Busby Berkeley. Academy Awards '40: Best Sound. **VHS, Beta, LV** $24.95 *MGM* 🐾🐾 ½

Strikeback A young boy comes painfully of age in West Berlin.
1984 88m/C **VHS, Beta** $69.98 *VES* 🐾🐾

Striker Mercenary John "Striker" Slade is sent to Nicaragua to free a captured American journalist, not knowing his superiors have betrayed him to the Contras and the Sandanistas. He battles it out with both groups while seeking revenge on the businessmen who set him up.
1988 (R) 90m/C Frank Zagarino, Melanie Rogers, Paul Werner; *Dir:* Stephen M. Andrews. **VHS, Beta** $39.95 *AIP* 🐾🐾

Striker's Mountain Story of a sports enthusiast who creates heliskiing and builds it into a business. Great ski scenes, but skip the rest.
198? 99m/C Bruce Greenwood, Mimi Kuzyk, Leslie Nielsen, August Schellenberg; *Dir:* Allen Simmonds. **VHS** $9.95 *SIM, CGH, PEV* 🐾 ½

Striking Back Fed up citizens decline to turn other cheek.
1981 91m/C Perry King, Tisa Farrow, Don Stroud, George Kennedy, Park Jong Soo; *Dir:* William Fruet. **VHS** *NO* 🐾 ½

Stripes Feeling like losers and looking to straighten out their lives, two friends enlist in the Army under the mistaken impression that military life is something like a summer camp for grownups. A box-office success (despite a poorly written and directed script) due in large part to Murray's charm and his verbal and sometimes physical sparring with Oates, who is good as the gruff-tempered platoon sergeant. Often humorous, it includes performances by various "Second City" players, including fat slob Candy, whom Murray turns into a "lean, mean fighting machine."
1981 (R) 105m/C Bill Murray, Harold Ramis, P.J. Soles, Warren Oates, John Candy, John Larroquette, Judge Reinhold, Sean Young, Dave Thomas, Joe Flaherty; *Dir:* Ivan Reitman. **VHS, Beta, LV** $29.95 *COL* 🐾🐾 ½

Stripped to Kill A female cop goes undercover to lure a psycho killing strippers. Not-bad entry from Corman and cohorts. Followed by - you guessed it - "Stripped to Kill II."
1987 (R) 88m/C Kay Lenz, Greg Evigan, Norman Fell; *Dir:* Katt Shea Ruben. **VHS, Beta** $79.95 *MGM* ♫♫

Stripped to Kill II: Live Girls A young woman with extra-sensory powers dreams of murders which she discovers are all too real. A weak sequel, despite the interesting premise.
1989 83m/C Maria Ford, Eb Lottimer, Karen Mayo Chandler, Marjean Holden, Birke Tan, Debra Lamb; *Dir:* Katt Shea Ruben. **VHS, Beta** $79.95 *MGM* ♫

The Stripper A middle-aged stripper in a traveling show returns to the small Kansas town of her youth. Left stranded, she moves in with an old friend whose teen-age son promptly falls in love with her. Based on the play "A Loss of Roses" by William Inge.
1963 95m/B Joanne Woodward, Richard Beymer, Claire Trevor, Carol Lynley, Robert Webber, Gypsy Rose Lee; *Dir:* Franklin J. Schaffner. **VHS, Beta** $39.95 *FOX* ♫♫½

Stripper A good under-the-covers glimpse of the life and work of several real-life strippers, culminating in the First Annual Stripper's Convention in 1984.
1986 (R) 90m/C Janette Boyd, Sara Costa, Kimberly Holcomb; *Dir:* Jerome Gary. **VHS, Beta** $79.98 *FOX* ♫♫

Stroke of Midnight Souped-up Cinderella with Lowe as egotistic high fashion designer, Grey as struggling shoe designer and a pumpkin-colored VW Bug.
1990 (PG) 102m/C Rob Lowe, Jennifer Grey, Andrea Ferreol; *Dir:* Tom Clegg. **VHS, Beta, LV** $89.98 *MED* ♫½

Stroker Ace A flamboyant stock car driver tries to break an iron-clad promotional contract signed with a greedy fried-chicken magnate. Off duty, he ogles blondes as dopey as he is. One of Reynolds's very worst flicks - and that's saying something.
1983 (PG) 96m/C Burt Reynolds, Ned Beatty, Jim Nabors, Parker Stevenson, Loni Anderson, Bubba Smith; *Dir:* Hal Needham. **VHS, Beta** $19.98 *WAR* Woof!

Stromboli Bergman is a Czech refugee who marries an Italian fisherman in order to escape from her displaced persons camp. He brings her to his home island of Stromboli where she finds the life bleak and isolated and her marriage a trial. The melodrama was even greater off-screen than on; this is the movie which introduced Bergman and Rossellini and began their (then) scandalous affair.
1950 81m/B *IT* Ingrid Bergman, Mario Vitale, Renzo Cesana; *Dir:* Roberto Rossellini. **VHS, Beta** $19.95 *UHV* ♫♫½

Strong Man A World War I veteran, passing himself off as an unlikely circus strongman, searches an American city for the girl whose letters gave him hope during the war. Perhaps Langdon's best full-length film. With a new score by Carl Davis.
1926 78m/B Harry Langdon, Gertrude Astor, Tay Garnett; *Dir:* Frank Capra. **VHS, Beta, LV** $39.95 *HBO, IME* ♫♫♫

Strong Medicine Strange, avant-garde film about a strange, off-written young woman. Cleverly directed experiment that's not for everyone.

1984 84m/C Carol Kane, Raul Julia, Wallace Shawn, Kate Manheim, David Warrilow; *Dir:* Guy Green. **VHS, Beta** *NEW* ♫♫½

Stroszek Three German misfits-a singer, a prostitute and an old man-tour the U.S. in search of their dreams. Touching, hilarious comedy-drama with a difference and an attitude. One of Herzog's easiest and also best films.
1977 108m/C *GE* Eva Mattes, Bruno S, Clemens Scheitz; *Dir:* Werner Herzog. **VHS** $79.95 *NYF, FCT* ♫♫♫

Struggle Through Death Two young men escape the clutches of evil Ching Kue and attempt to free their fellow prisoners so they can overthrow the despot.
197? 93m/C **VHS, Beta** $54.95 *MAV* ♫

Stryker The ever-lovin' nuclear holocaust has occurred, and good guys and bad guys battle it out for scarce water. If you liked "Mad Max," go see it again; don't watch this miserable effort.
1983 (R) 86m/C *PH* Steve Sandor, Andria Fabio; *Dir:* Cirio H. Santiago. **VHS, Beta** $19.98 *SUE, STE* ♫

Stuck on You A couple engaged in a palimony suit takes their case to a judge to work out their differences. Wing-clipped angel Gabriel (Corey) comes to earth to help them patch it up.
1984 (R) 90m/C Prof. Irwin Corey, Virginia Penta, Mark Mikulski; *Dir:* Samuel Wail, Michael Herz. **VHS, Beta** $9.98 *SUE* Woof!

Stuckey's Last Stand A group of camp counselors prepare to take 22 children on a nature hike they will never forget. Mindless teen fodder.
1978 (PG) 95m/C Whit Reichert, Tom Murray, Rich Casentino. **VHS, Beta** $59.95 *LTG* ♫

The Stud The owner of a fashionable "after hours" dance spot hires a young, handsome stud to manage the club and attend to her personal needs. Pathetic resemblance of entertainment, adapted from a novel by Collins's sister Jackie.
1978 (R) 90m/C *GB* Joan Collins, Oliver Tobias; *Dir:* Quentin Masters. **VHS, Beta** $19.99 *HBO* Woof!

Student Affairs A bunch of high-school misfits vie for roles in a movie about a bunch of high school misfits.
1988 (R) 92m/C Louie Bonanno, Jim Abele, Alan Fisler, Deborah Blaisdell; *Dir:* Chuck Vincent. **VHS, Beta** $79.98 *LIV, VES* Woof!

Student Bodies A "Halloween" - style spoof of high-school horror films, except it's not funny. And what's so funny about bloody murder anyway? When the on-the-set problems and strife arose (which they did), why didn't everyone just cut their losses and go home?
1981 (R) 86m/C Kristen Riter, Matthew Goldsby, Richard Brando, Joe Flood; *Dir:* Mickey Rose. **VHS, Beta, LV** $72.95 *PAR* ♫

The Student Body Every boy on campus hopes to win the heart (and everything that goes with it) of the lustiest girl on campus.
1976 (R) 85m/C Jillian Kesner, Janice Heiden. **VHS, Beta** *PLV* ♫

The Student Body/Jailbait Babysitter Double cassette containing "The Student Body" and "Jailbait Babysitter." In our first feature, three girls recently released from the local reform school, participate in a psychological experiment. In the

second feature, a 17-year-old discovers her sexual awakening can be fun.
197? (R) 150m/C **VHS, Beta** $49.98 *CGI* ♫

Student Confidential A Troma-produced spoof of seedy high school youth movies, new and old, involving four students who are led into the world of adult vices by a mysterious millionaire. Badly made and dull. Douglas and Jackson both have brothers named Michael.
1987 (R) 99m/C Eric Douglas, Marlon Jackson, Susan Scott, Ronee Blakley, Elizabeth Singer; *Dir:* Richard Horian. **VHS, Beta** $79.95 *MED* ♫

The Student Nurses The adventures, amourous and otherwise, of four last-year nursing students. Followed by four sequels, starting with "Private Duty Nurses." Better than average exploitation fare. First release from Roger Corman's New World studios; it goes down hill from there.
1970 89m/C Elaine Giftos, Karen Carlson, Brioni Farrell, Barbara Leigh, Reni Santoni, Richard Rust, Lawrence Casey, Darrell Larson, Paul Camen, Richard Stahl, Scottie MacGregor, Pepe Serna; *Dir:* Stephanie Rothman. **VHS, Beta** $59.98 *CHA* ♫♫

Student of Prague Early silent classic based on the German Faust legend. A student makes a pact with the devil to win a beautiful woman. Poor fellow . . . With either German or English title cards. Music track.
1913 60m/B *GE* Lothar Koemer, Grete Berger, Paul Wegener; *Dir:* Paul Wegener, Stellan Rye. **VHS, Beta** $39.95 *FST, DVT, GLV* ♫♫♫

The Student Prince in Old Heidelberg Prince Karl Heinrich silently falls in love with lowly barmaid.
1927 102m/B Ramon Novarro, Norma Shearer; *Dir:* Ernst Lubitsch. **VHS** $29.98 *MGM, FCT* ♫♫½

The Student Teachers Yet another soft-core Corman product. Three student teachers sleep around, on screen.
1973 79m/C Susan Damante Shaw, Brooke Mills, Bob Harris, John Cramer, Chuck Norris; *Dir:* Jonathan Kaplan. **VHS, Beta** $59.98 *SUE* ♫½

Studs Lonigan Drifting, too-artsy rendering a James T. Farrell's trilogy about an Irish drifter growing up in Chicago in the twenties. Good period detail, but oddly off-kilter and implausible as history.
1960 96m/B Christopher Knight, Frank Gorshin, Jack Nicholson; *Dir:* Irving Lerner. **VHS, Beta** $24.95 *MGM* ♫♫

A Study in Scarlet Owen played Watson the previous year in "Sherlock Holmes"; here he's not perfect as Holmes. The plot is not the same as in Conan Doyle story of the same title. Owen wrote much of the script for this one.
1933 77m/B Reginald Owen, Alan Mowbray, Anna May Wong, June Clyde, Alan Dinehart; *Dir:* Edwin L. Marin. **VHS, Beta** $16.95 *SNC, NOS, CAB* ♫♫

A Study in Terror A well-appointed Sherlock Holmes thriller. The second Holmes film in color. Premise has a young, athletic Holmes in pursuit of a well-educated Jack the Ripper in the atmosphere of London in the 1880s. US title was "Fog."

1966 94m/C *GE GB* John Neville, Donald Houston, Judy Dench, Anthony Quayle, Robert Morley, Frank Finlay, Cecil Parker; *Dir:* James Hill. VHS, Beta, LV $69.95 *COL* 𝕀𝕀𝕀

The Stuff Surreal horror semi-spoof about an ice cream mogul and a hamburger king who discover that the new, fast-selling confection in town zombifies its partakers. Forced, lame satire from producer/director/writer Cohen.
1985 (R) 93m/C Michael Moriarty, Andrea Marcovicci, Garrett Morris, Paul Sorvino, Danny Aiello, Brooke Adams; *Dir:* Larry Cohen. VHS, Beta $9.95 *NWV, STE* 𝕀½

Stuff Stephanie in the Incinerator A Troma gagfest about wealthy cretins who torture and kill young women. "Funny" title betrays utter, exploitive mindlessness.
1989 (PG-13) 97m/C Catherine Dee, William Dame, M.R. Murphy, Dennis Cunningham; *Dir:* Don Nardo. VHS, Beta, LV $79.95 *IVE* 𝕀

The Stunt Man A marvelous and unique exercise in meta-cinematic manipulation. O'Toole, in one of his very best roles, is a power-crazed movie director; Railsback is a fugitive sheltered by him from sherrif Rocco. When a stunt man is killed in an accident, O'Toole prevails on Railsback to replace him. Railsback wonders if O'Toole wants him to die . . . A labor of love for director-producer Rush, who spent nine years working on it, and waited two years to see it released, by Fox.
1981 (R) 129m/C Peter O'Toole, Steve Railsback, Barbara Hershey, Allen Garfield, Alex Rocco; *Dir:* Richard Rush. Montreal World Film Festival '80: Best Film; National Board of Review Awards '80: 10 Best Films of the Year; National Society of Film Critics Awards '80: Best Actor (O'Toole). VHS, Beta, LV $29.98 *FOX* 𝕀𝕀𝕀𝕀

Stunt Pilot A young man takes a job at a film studio as a stunt pilot and finds mayhem and romance on and off the ground. One in the "Tailspin Tommy" series. Well-paced and fun to watch.
1939 61m/C John Trent, Marjorie Reynolds, Milburn Stone, Jason Robards Sr., Pat O'Malley; *Dir:* George Waggner. VHS $19.95 *NOS, MOV* 𝕀𝕀

Stunt Rock A feature film combining rock music, magic, and some of the most incredible stunts ever attempted.
1980 (PG) 90m/C VHS, Beta $39.95 *MON* 𝕀𝕀

Stunts See "The Stunt Man" instead. OK script hides near-invisible plot; stunt man engage in derring-do. retitled "Who is Killing the Stuntmen?"
1977 (PG) 90m/C Robert Forster, Fiona Lewis, Joanna Cassidy, Darrell Fetty, Bruce Glover, James Luisi; *Dir:* Mark L. Lester. VHS, Beta $69.99 *HBO* 𝕀𝕀

Stuntwoman Welch plays a stuntwoman whose death-defying job interferes with her love life. What ever shall she do?
1981 95m/C Raquel Welch, Jean-Paul Belmondo; *Dir:* Claude Zidi. VHS $14.95 *LPC* 𝕀𝕀½

Submarine Seahawk When a group of Japanese warships mysteriously disappear, allied intelligence suspect a big battle is brewing. An inexperienced commander who has been tracking the movement of the Japanese ships is put in charge of the submarine Seahawk, even though the crew lacks faith in his command abilities. His own men may prove a bigger obstacle than any of the Japanese. Lots of action and stock World War II footage.

1959 83m/B John Bentley, Brett Halsey, Wayne Heffley, Steve Mitchell, Henry McCann, Frank Gerstle; *Dir:* Spencer Gordon Bennett. VHS $9.95 *COL* 𝕀

Submission A pharmacist's sensuality is reawakened when she has a provocative affair with a clerk in her shop.
1977 (R) 107m/C *IT* Franco Nero, Lisa Gastoni; *Dir:* Salvatore Samperi. VHS, Beta $59.95 *PSM* 𝕀½

Subspecies New improved vampire demons descend on earth. The first full-length feature film shot on location in Transylvania. For horror buffs only.
1991 90m/C Laura Tate, Michael Watson, Anders Hove, Michelle McBride, Irina Movila, Angus Scrimm; *Dir:* Ted Nicolaou. VHS, LV, 8mm $19.95 *PAR, PIA, BTV* 𝕀½

Substitute Teacher Every guy on campus is falling for the science department's latest experiment - the most beautiful teacher on campus.
19?? 90m/C Dayle Haddon. VHS, Beta *PLV* 𝕀½

Subterfuge When a special American security agent goes to England for a "vacation," his presence causes speculation and poses several serious questions for both British Intelligence and the underworld. Nothing special, with the occasional suspenseful moment.
1968 89m/C *GB* Joan Collins, Gene Barry, Richard Todd; *Dir:* Peter Graham Scott. VHS, Beta *REP, AVD* 𝕀

Suburban Commando A goofy, muscular alien mistakenly lands on Earth while on vacation. He does his best to remain inconspicuous, resulting in numerous hilarious situations. Eventually he is forced to confront his arch, interstellar nemesis in order to defend the family who befriended him. A harmless, sometimes cute vehicle for wrestler Hogan, which will certainly entertain his younger fans.
1991 (PG) 88m/C Hulk Hogan, Christopher Lloyd, Shelley Duvall, Larry Miller, William Ball, JoAnn Dearing, Jack Elam, Roy Dotrice, Christopher Neame, Tony Longo; *Dir:* Burt Kennedy. VHS *SVS, NLC* 𝕀𝕀

Suburban Roulette "Adults only" feature from splattermaster Lewis caters to the prurient. Groovy themes like wife swapping and other very daring subjects. Totally sixties. Presented as part of Joe Bob Brigg's "Sleaziest Movies in the History of the World" series.
1967 91m/C *Dir:* Herschell Gordon Lewis. VHS $19.98 *SVI* 𝕀

Suburbia When a group of punk rockers move into a condemned suburban development, they become the targets of a vigilante group. Low budget, needless violent remake of anyone of many '50s rebellion flicks that tries to have a "message." Also known as "The Wild Side."
1983 (R) 99m/C Chris Pederson, Bill Coyne, Jennifer Clay, Timothy Eric O'Brien, Andrew Pece, Don Allen; *Dir:* Penelope Spheeris. VHS, Beta, LV $29.98 *LIV, VES* 𝕀

Subway "Surreal," MTV-esque vision of French fringe life from the director of "Le Dernier Combat." A spike-haired renegade escapes the law by plunging into the Parisian subway system. Once there, he encounters a bizarre subculture living under the city. Plenty of angry-youth attitude, but where's the point? And frankly, too much bad New Wave music.

1985 (R) 103m/C *FR* Christopher Lambert, Isabelle Adjani; *Dir:* Luc Besson. VHS, Beta, LV $79.98 *FOX* 𝕀½

Subway to the Stars Young saxophonist Fontes searches the back streets of Rio for his missing girlfriend. Realistic and sometimes depressing, if aimless at times. Overall, an intriguing "underworld" flick from Brazil.
1987 103m/C *BR* Guilherme Fontes, Milton Goncalves, Taumaturgo Ferreira, Ze Trindade, Ana Beatriz Wiltgen; *Dir:* Carlos Diegues. VHS $19.95 *STE, NWV* 𝕀𝕀½

Success Is the Best Revenge A Polish film director living in London struggles with his rebellious teenage son. Both struggle with questions of exile and identity. Uneven but worthwhile and based on Skolimowski's own experiences. His better earlier film is "Moonlighting."
1985 95m/C *GB* Michael York, Michel Piccoli, Anouk Aimee, John Hurt, Jane Asher, Joanna Szerbic, Michael Lyndon, George Skolimowski; *Dir:* Jerzy Skolimowski. VHS, LV $79.98 *MAG, IME* 𝕀𝕀½

Sucker Money Newspaper reporter works to expose phony spiritualist. Good title; little else. Boring.
1934 59m/B Mischa Auer, Phyliss Barrington, Earl McCarthy, Ralph Lewis, Mae Busch; *Dir:* Dorothy Reid. VHS, Beta, LV $16.95 *NOS, SNC, WGE* 𝕀½

Sudden Death Shamrock and Lucky are back as cowboys posing as hired gunmen in order to foil a wealthy, homesteader-hating land baron.
1950 55m/B James Ellison, Russell Hayden, Stanley Price, Raymond Hatton, Betty Adams. VHS, Beta, LV $79.98 *WGE* 𝕀

Sudden Death Two professional violence merchants put themselves up for hire.
1977 (R) 84m/C Robert Conrad, Felton Perry, Don Stroud, Bill Raymond, Ron Vawter, Harry Roskolenko; *Dir:* Richard Foreman. VHS, Beta $49.95 *MED* 𝕀

Sudden Death A beautiful businesswoman decides to kill every rapist she can find after she herself has been attacked. Exploitative in its own fashion; fueled and felled by rage.
1985 (R) 95m/C Denise Coward, Frank Runyeon, Jamie Tirelli; *Dir:* Sig Shore. VHS, Beta $29.98 *LIV, VES* 𝕀

Sudden Impact Eastwood directs himself in this formula thriller, the fourth "Dirty Harry" entry. This time Dirty Harry Callahan tracks down a revenge-obsessed murderess and finds he has more in common with her than he expected. Meanwhile, local mobsters come gunning for him. This is the one where he says, "Go ahead. Make my day." Followed by "The Dead Pool."
1983 (R) 117m/C Clint Eastwood, Sondra Locke, Pat Hingle, Bradford Dillman; *Dir:* Clint Eastwood. VHS, Beta, LV, 8mm $19.98 *WAR, PIA, TLF* 𝕀𝕀½

Sudden Terror The murder of an African dignitary is witnessed by a young boy who has trouble convincing his parents and the police about the incident. Suspenseful but less than original. Shot on location in Malta. Also known as "Eyewitness."
1970 (PG) 95m/C *GB* Mark Lester, Lionel Jeffries, Susan George, Tony Bonner; *Dir:* John Hough. VHS, Beta *FOX* 𝕀𝕀

Suddenly Crazed gunman Sinatra (following his Oscar role in "From Here to Eternity") holds a family hostage in the hick town of Suddenly, Calif., as part of a plot to kill the president, who is passing through town. Tense thriller well displays Sinatra's acting talent, as does "The Manchurian Candidate." Both sadly are hard to find (though the latter was re-released a few years ago), because Sinatra forced United Artists to take them out of distribution after hearing that Kennedy assassin Lee Harvey Oswald had watched "Suddenly" only days before Nov. 22, 1963. Really, Ol' Blue Eyes should have stuck with making top-notch thrillers like this one, instead of degenerating into the world's greatest lounge singer.
1954 75m/B Frank Sinatra, Sterling Hayden, James Gleason, Nancy Gates, Paul Frees; *Dir:* Lewis Allen. **VHS, Beta $16.95** *SNC, NOS, JLT* ♫♫♫½

Suddenly, Last Summer A brain surgeon is summoned to the mansion of a rich New Orleans matron who wishes him to perform a lobotomy on her niece, who has supposedly suffered a mental breakdown. The doctor seeks to learn the truth about the young woman and her family. Based on the play by Tennessee Williams. Although softpedaled for the censors, this movie's themes of homosexuality, insanity and murder, its characterizations of evil and its unusual settings, presage many movies of the next two decades. Extremely fine performances from Hepburn and Taylor both received Oscar nominations. Clift, having never completely recovered from his automobile accident two years before and struggling with his own mental health, does not come across with the strength of purpose really necessary in his character. Still, fine viewing.
1959 114m/B Elizabeth Taylor, Katharine Hepburn, Montgomery Clift, Mercedes McCambridge, Albert Dekker; *Dir:* Joseph L. Mankiewicz. National Board of Review Awards '59: 10 Best Films of the Year. **VHS, Beta, LV $59.95** *COL* ♫♫♫½

Suds Pickford is tragic laundress with major crush on a guy who left his shirt at the laundry.
1920 75m/B Mary Pickford, William Austin, Harold Goodwin, Rose Dione, Theodore Roberts; *Dir:* Jack Dillon. **VHS, Beta, LV $19.95** *GPV* ♫♫½

Suffering Bastards Buddy and Al have never worked a day in their lives—Mom's nightclub always gave them something to do and a regular paycheck. But when Mom's swindled out of the family business, they decide it's time to take charge. Soon they're caught between bullets, babes, and belly laughs in this action-filled comedy.
1990 95m/C Eric Bogosian, John C. McGinley, David Warshofsky. **VHS $89.95** *AIP* ♫♫½

Sugar Cane Alley After the loss of his parents, an 11-year-old orphan boy goes to work with his grandmother on a sugar plantation. She realizes that her young ward's only hope is an education. Set in Martinique in the 1930s among black workers. Poignant and memorable.
1983 (PG) 106m/C *FR* Garry Cadenat, Darling Legitimus, Douta Seck; *Dir:* Euzhan Palcy. **VHS, Beta, LV $29.95** *MED* ♫♫♫

Sugar Cookies An erotic horror story in which young women are the pawns as a satanic satyr and an impassioned lesbian play out a bizarre game of vengeance, love, and death.

1977 (R) 89m/C Mary Woronov, Lynn Lowry, Monique Van Vooren; *Dir:* Michael Herz. **VHS, Beta $29.98** *VID* Woof!

Sugarbaby An acclaimed German film about a fat mortuary attendant living in Munich who transforms herself (but doesn't lose weight) in order to seduce a young, handsome subway conductor. Touching and warm film that introduced Sagebrecht to American audiences. A bit too cutting-edge cinematographically. In German with English subtitles. Remade as "Babycakes." Sagebrecht and director Adlon team up again for "Rosalie Goes Shopping" and "Bagdad Cafe."
1985 (R) 103m/C *GE* Marianne Sagebrecht, Eisi Gulp, Toni Berger, Will Spendler, Manuela Denz; *Dir:* Percy Adlon. **VHS, Beta $29.95** *LHV, GLV, APD* ♫♫♫

The Sugarland Express To save her son from adoption, a young woman helps her husband break out of prison. In their flight to freedom, they hijack a police car, holding the policeman hostage. Speilberg's first feature film is a moving portrait of a couple's desperation. Based on a true story, adapted by Hal Barwood and Matthew Robbins.
1974 (PG) 109m/C Goldie Hawn, Ben Johnson, Michael Sacks, William Atherton; *Dir:* Steven Spielberg. **VHS, Beta $59.95** *MCA* ♫♫♫

The Suicide Club A wealthy, depressed young woman joins a club filled with affluent people who play mysterious games. She finds that winning their games means more than she bargained for. Hemingway looks great in her off-kilter way, but really, there's no point, and the on-screen decadence gets tiresome.
1988 (R) 92m/C Mariel Hemingway, Robert Joy, Lenny Henry, Madeleine Potter, Michael O'Donoghue, Anne Carlisle; *Dir:* James Bruce. **VHS, Beta $29.95** *ACA* ♫½

Suicide Cult A CIA employee predicts the second coming of Christ, and a battle ensues between the forces of good and evil for possession of the child.
1977 82m/C **VHS, Beta $39.98** *NSV* Woof!

Suicide Patrol A small band of saboteurs patrols a German-controlled Mediterranean island during World War II.
197? 90m/C Gordon Mitchell, Pierre Richard, Max Dean. **VHS, Beta $57.95** *UNI* ♫

The Suitors When a group of Iranians decide to sacrifice a sheep in their apartment, the New York police send in SWAT team assuming they're terrorists. First time effort by Ebrahimian, in Farsi with English subtitles.
1988 106m/C *Dir:* Ghasem Ebrahimian. **VHS $59.95** *FCT* ♫

Sullivan's Travels Sturges' masterpiece is a sardonic, whip-quick romp about a Hollywood director tired of making comedies who decides to make a serious, socially responsible film and hits the road masquerading as a hobo in order to know hardship and poverty. Beautifully sustained, inspired satire that mercilessly mocked the ambitions of Depression-era social cinema. Gets a little over-dark near the end before the happy ending; Sturges insisted on 20-year-old Lake as The Girl; her pregnancy forced him to rewrite scenes and design new costumes for her. As ever, she is stunning.
1941 90m/B Joel McCrea, Veronica Lake, William Demarest, Robert Warwick, Franklin Pangborn, Porter Hall, Eric Blore, Byron Foulger, Robert Greig, Torben Meyer, Jimmy Conlin, Margaret Hayes, Chester Conklin; *Dir:* Preston Sturges. National Board of Review Awards '42:

10 Best Films of the Year. **VHS, Beta, LV $29.95** *MCA* ♫♫♫♫

Summer The fifth and among the best of Rhomer's "Comedies and Proverbs" series. A romantic but glum young French girl finds herself stuck in Paris during the tourist season searching for true romantic love. Original French title: "Le Rayon Vert." In French with English subtitles. Takes time and patience to seize the viewer; moving ending makes it all worthwile.
1986 (R) 90m/C *FR* Marie Riviere, Lisa Heredia, Beatrice Romand, Eric Hamm, Rosette, Isabelle Riviere; *Dir:* Eric Rohmer. Venice Film Festival '86: Golden Lion. **VHS, Beta, LV $29.95** *PAV, FCT* ♫♫♫½

Summer of '42 Touching if sentimental story about a 15-year-old boy's sexual coming of age during his summer vacation on an island off New England. While his friends are fumbling with girls their own age, he falls in love with a beautiful 22-year-old woman whose husband is off fighting in the war.
1971 (R) 102m/C Jennifer O'Neill, Gary Grimes, Jerry Houser, Oliver Conant; *Dir:* Robert Mulligan. Academy Awards '71: Best Original Score. **VHS, Beta, LV $19.98** *WAR, PNR* ♫♫♫

Summer Affair Two young lovers run away from their bickering parents and get shipwrecked on a desert island.
1979 90m/C *IT* Ornella Muti, Les Rannow; *Dir:* George S. Casorati. **VHS, Beta $39.95** *AHV* ♫½

Summer Camp A ten-year reunion at a summer camp is the inane excuse for a bizarre weekend of co-ed football, midnight panty-raids, coupling couples and a wild disco party.
1978 (R) 85m/C John McLaughlin, Matt Michaels, Colleen O'Neil. **VHS, Beta $54.95** *MED* ♫

Summer Camp Nightmare See "Lord of the Flies" instead. A young fascist incites children at two summer camps to overthrow and imprison the adults. Barbarism rears its ugly head. Self-serious and anticlimatic. Based on the novel "The Butterfly Revolution" by William Butler.
1986 (PG-13) 89m/C Chuck Connors, Charles Stratton, Harold Pruett, Tom Fridley, Adam Carl; *Dir:* Bert L. Dragin. **VHS, Beta, LV $19.95** *SUE* ♫½

Summer City Fun-and-sun surfing movie complete with romance, hot rods, murder and great shots of the sea features Gibson in his debut. Ever wonder what Australian surf bums do on weekends? The same as their California counterparts - except they talk funny. Also called "Coast of Terror" on video.
1977 83m/C *AU* Mel Gibson, Phillip Avalon, John Jarratt, Christopher Fraser. **VHS, Beta $59.95** *VCD* ♫½

Summer Dog A family falls in love with an abandoned dog they adopt and their landlord tries to evict them.
1978 (G) 88m/C James Congdon, Elizabeth Eisenman; *Dir:* John Clayton. **VHS, Beta $19.95** *STE, VES* ♫½

Summer Fantasy Mindless TV surf saga about a curvy teen who would rather watch the beach hunks then go to medical school.
1984 96m/C Julianne Phillips, Ted Shackleford, Michael Gross, Dorothy Lyman; *Dir:* Noel Nosseck. **VHS, Beta $79.95** *SVS* ♫

Summer of Fear A happy young woman must overcome evil forces when her cousin comes to live with her family. Based on a novel by Lois Duncan. Also known as "Stranger in Our House." Pretty scary if ordinary, dumb made-for-TV horror from later-famous Craven.
1978 94m/C Linda Blair; *Dir:* Wes Craven. **VHS, Beta $59.99** HBO ♫♫

Summer Heat When a young man spends his summer with his hypnotic aunt, we are taken on a no-holds-barred tour through the steamy world of the rich and the restless.
1973 71m/C Bob Garry, Nicole Avril, Pat Pascal. **VHS, Beta $19.95** VID ♫♫

Summer Heat In rural, mid-Depression North Carolina, a young, lonely wife and mother is seduced by a drifter and together they plot murder. Her husband neglects her, and you should neglect to see this utter yawner. Based on a Louise Shivers novel.
1987 (R) 80m/C Lori Singer, Anthony Edwards, Bruce Abbott, Kathy Bates; *Dir:* Michie Gleason. **VHS, Beta, LV $19.95** PAR ♫ ½

Summer Holiday Rooney comes of age with a vengeance during summer vacation in musical rendition of "Ah, Wilderness." Undistinguished musical numbers, inferior to the original, but jazzy Technicolor cinematography. Ended up in the red by over $1.5 million, lotsa money back then.
1948 92m/C Mickey Rooney, Gloria DeHaven, Walter Huston, Frank Morgan, Jackie "Butch" Jenkins, Marilyn Maxwell, Agnes Moorehead, Selena Royle, Anne Francis; *Dir:* Rouben Mamoulian. **VHS $19.98** MGM, TTC, FCT ♫♫ ½

Summer Interlude A ballerina recalls a romantic summer spent with an innocent boy, who was later tragically killed. Contains many earmarks and visual ideas of later Bergman masterpieces. A journalist who falls for her tries to help her overcome her tragic obsession. In Swedish with English subtitles. Alternate title: "Illicit Interlude." Originally called "Summerplay."
1950 95m/B *SW* Maj-Britt Nilsson, Birger Malmsten, Alf Kjellin; *Dir:* Ingmar Bergman. **VHS, Beta $19.98** SUE ♫♫♫ ½

Summer Job A group of college kids work and play at a decadent resort. In the "alleged comedy" genre, and among the very worst.
1988 (R) 92m/C Sherrie Rose, Fred Boudin, Dave Clouse, Kirk Earhardt; *Dir:* Paul Madden. **VHS, Beta, LV $14.95** SVS Woof!

Summer Lovers A young couple go to the exotic Greek island of Santorini to spend the summer and meet up with a fun-loving woman. They spend the rest of the summer becoming friends and trying to come to terms with their feelings for each other (the three of them, that is). From the director of "Grease" and "Blue Lagoon."
1982 (R) 98m/C Peter Gallagher, Darryl Hannah, Valerie Quennessen; *Dir:* Randal Kleiser. **VHS, Beta, LV $9.95** COL, SUE ♫ ½

Summer Magic An impecunious recent widow is forced to leave Boston and settle her family in a small town in Maine. Typical, forgettable Disney drama; early Mills vehicle. A remake of "Mother Carey's Chickens."
1963 116m/C Hayley Mills, Burl Ives, Dorothy McGuire, Deborah Walley, Una Merkel, Eddie Hodges; *Dir:* James Neilson. **VHS, Beta $69.95** DIS ♫♫

The Summer of Miss Forbes A governess rules over two young boys with an iron hand by day and cavorts nude and drunk by night. The boys fantasize about her murder, but fate has different plans.
1988 85m/C *SP* Hanna Schygulla, Alexis Castanares, Victor Cesar Villalobos, Guadalupe Sandoval, Fernando Balzaretti, Yuriria Munguia; *Dir:* Jaime Humberto Hermosillo. **VHS $79.95** FXL, FCT ♫♫

Summer of My German Soldier A young Jewish girl befriends an escaped German prisoner of war in a small town in Georgia during World War II. Occasionally sentimental but more often genuinely moving made-for-TV drama. Rolle won a deserved Emmy as the housekeeper. Adapted from Bette Green's novel.
1978 98m/C Kristy McNichol, Esther Rolle, Bruce Davison; *Dir:* Michael Tuchner. **VHS, Beta $14.95** MTI, KUI ♫♫♫

Summer Night A female tycoon who is concerned for the environment decides to strike out at the terrorists who are destroying it. This caper turns into one big comedy. In Italian with English subtitles.
1987 (R) 94m/C *IT* Mariangela Melato, Michele Placido, Roberto Herlitzka, Massimo Wertmuller; *Dir:* Lina Wertmuller. **VHS, Beta $79.95** IVE ♫♫

Summer Night Fever A pair of young males go girl crazy near the end of their summer vacation.
1979 101m/C Stephanie Hillel, Olivia Pascal, Betty Verges; *Dir:* Siggi Gotz. **VHS, Beta $19.98** TWE ♫

Summer Night With Greek Profile, Almond Eyes . . A wealthy woman tycoon hires an ex-CIA man to kidnap a high-priced, professional terrorist and hold him for ransom. An ironic battle of the sexes results. In Italian with English subtitles. Classic Wertmuller plotline.
1987 (R) 94m/C *IT* Mariangela Melato, Michele Placido, Roberto Herlitzka; *Dir:* Lina Wertmuller. **VHS, Beta $19.95** IVE ♫♫

A Summer Place Melodrama about summer liaisons amid the young and middle-aged rich on an island off the coast of Maine. Too slick; romantic drama is little more than skin-deep, and dialogue is excruciating. Donahue's first starring role. Based on the novel by Sloan Wilson. Featuring "Theme from a Summer Place," which was a number-one hit in 1959.
1959 130m/C Troy Donahue, Richard Egan, Sandra Dee, Dorothy McGuire, Arthur Kennedy, Constance Ford, Beulah Bondi; *Dir:* Delmer Daves. **VHS, Beta, LV $19.98** WAR ♫♫

A Summer to Remember A five-year old boy spends a summer with his stepfather on a Soviet collective farm. The two become deeply attached. Scenes of collective life will be of more interest to Westerners than the near-sentimental story. Star Bondarchuk later directed the epic, award-winning "War and Peace."
1961 80m/B *RU* Borya Barkhazov, Sergei Bondarchuk, Irana Skobtseva. National Board of Review Awards '61: 5 Best Foreign Films of the Year. **VHS, Beta $39.95** INV ♫♫♫

A Summer to Remember A deaf boy (played by Gerlis, himself deaf since birth) develops a friendship with an orangutan through sign language. Bad guys abduct the friendly ape - but all is right in the end. Nice, innocuous family viewing. Made for TV.

1984 (PG) 93m/C Tess Harper, James Farentino, Burt Young, Louise Fletcher, Sean Gerlis, Bridgette Anderson; *Dir:* Robert Lewis. **VHS, Beta $39.95** MCA ♫♫ ½

Summer Rental That John Candy just can't win, can he? Here, as a hopeless, harried air traffic controller, he tries to have a few days to relax in sunny Florida. Enter mean rich guy Crenna. Candy can add something hefty to the limpest of plots, and does so here. Watch the first hour for yuks, then rewind.
1985 (PG) 87m/C John Candy, Rip Torn, Richard Crenna, Karen Austen, Kerri Green, John Larroquette, Pierrino Mascarino; *Dir:* Carl Reiner. **VHS, Beta, LV, 8mm $19.95** PAR ♫♫

A Summer in St. Tropez Photographer David Hamilton's erotic and lyrical study of a household of girls in their first stages of womanhood living outside of society in the south of France.
1981 60m/C VHS, Beta $19.99 HBO ♫♫

Summer School A teenaged boy's girl-friend will stop at nothing to prevent him from going out with the pretty new girl in town.
1977 (R) 80m/C John McLaughlin, Steve Rose, Phoebe Schmidt. **VHS, Beta $19.95** AHV ♫

Summer School A high-school teacher's vacation plans are ruined when he gets stuck teaching remedial English in summer school. It seems all these students are interested in is re-enacting scenes from "The Texas Chainsaw Massacre." Actually, one of the better films of this genre.
1987 (PG-13) 98m/C Mark Harmon, Kirstie Alley, Nels Van Patten, Courtney Thorne-Smith, Lucy Lee Flippin, Shawnee Smith; *Dir:* Carl Reiner. **VHS, Beta, LV $14.95** PAR ♫♫ ½

Summer School Teachers Three sultry femmes bounce about Los Angeles high school and make the collective student body happy. Typical Corman doings; sequel to "The Students Teachers."
1975 (R) 87m/C Candice Rialson, Pat Anderson, Rhonda Leigh Hopkins, Christopher Wales; *Dir:* Barbara Peeters. **VHS, Beta $59.98** CHA Woof!

Summer of Secrets A young couple get more than they bargained for when they make love in a demented doctor's beach house.
1976 (PG) 100m/C VHS, Beta $14.98 VID ♫

Summer and Smoke Repressed, unhappy Page falls for handsome doctor Harvey. A tour de force for Page, but that's all; adapted clumsily from the overwrought Tennessee Williams play. Oscar nominations for Page, Merkel, the score, set decoration and art direction.
1961 118m/C Laurence Harvey, Geraldine Page, Rita Moreno, Una Merkel, John McIntire, Thomas Gomez, Pamela Tiffin, Lee Patrick, Casey Adams, Earl Holliman, Harry Shannon, Pattee Chapman; *Dir:* Peter Glenville. **VHS, Beta, LV $19.95** PAR, PIA, FCT ♫♫ ½

Summer Solstice Fonda and Loy are splendid as a couple married half a century who revisit the beach where they first met. Fonda especially shines, as a crusty old artist. Yes, it is an awful lot like "On Golden Pond." Made for TV.
1981 75m/C Henry Fonda, Myrna Loy, Lindsay Crouse, Stephen Collins; *Dir:* Ralph Rosenblum. **VHS, Beta $59.99** HBO, MTT ♫♫ ½

Summer Stock Garland plays a farm owner whose sister arrives with a summer stock troupe, led by Kelly, to rehearse a show in the family barn. Garland agrees, if the troupe will help her with the farm's

harvest. When her sister decamps for New York, leaving the leading lady role open, guess who steps into the breach. Slim plot papered over with many fun song-and-dance numbers. Garland sings "Get Happy." Also features Garland's first MGM short, "Every Sunday," made in 1936 with Deanna Durbin.
1950 109m/C Judy Garland, Gene Kelly, Gloria de Haven, Carleton Carpenter, Eddie Bracken, Phil Silvers, Hans Conried, Marjorie Main, Ray Collins, Deanna Durbin; *Dir:* Charles Walters. **VHS, Beta, LV $29.95** MGM, TTC ✍✍✍

A Summer Story Superbly acted, typically British period drama about beautiful farm girl Stubbs and city lawyer Wilby, who fall in love. But can they overcome difference of social class?
1988 (PG-13) 97m/C *GB* James Wilby, Imogen Stubbs, Susannah York, Sophie Ward; *Dir:* Piers Haggard. **VHS, Beta $79.98** MED ✍✍✍

Summer Switch A boy and his father magically switch bodies and learn about each other. Originally made for television.
1983 46m/C Robert Klein. **VHS, Beta $19.95** NWV ✍

Summer Vacation: 1999 Four boys from the land of the rising sun experience love and other requisite traumas. In Japanese with English subtitles.
1988 90m/C *JP* Eri Miyagima, Miyuki Nakano, Tomoko Otakra, Rie Mizuhara; *Dir:* Shusuke Kaneko. **VHS, Beta $69.95** NYF, FCT ✍✍

Summer Wishes, Winter Dreams Woodward and Balsam are a materially prosperous middle-aged couple with little but tedium in their lives - a tedium accurately replicated, in what feels like real life, in this slow, dull film. The two leads and Sidney - Oscar-nominated in her first screen role in 17 years - are all excellent, but the story of regret and present unhappiness wears the viewer down.
1973 (PG) 95m/C Joanne Woodward, Martin Balsam, Sylvia Sidney, Dori Brenner, Ron Richards; *Dir:* Gilbert Cates. National Board of Review Awards '73: Best Supporting Actress (Sydney); New York Film Critics Awards '73: Best Actress (Woodward). **VHS, Beta $59.95** COL ✍✍

Summerdog Harmless family tale about vacationers who rescue a cheerful, lovable mutt named Hobo. And a good thing they found him: He saves them from many perils. Thanks, Hobo!
1978 (G) 90m/C James Congdon, Elizabeth Eisenman, Oliver Zabriskie, Tavia Zabriskie; *Dir:* John Clayton. **VHS, Beta $29.95** GEM, LTG ✍✍

Summer's Children A man who suffers from amnesia as a result of a car crash becomes the target of a mysterious killer.
1984 90m/C Tom Haoff, Paully Jardine, Kate Lynch. **VHS, Beta $49.95** TWE ✍

Summer's Games Buxom girls cavort on the beach, competing in various games designed specifically to expose as much flesh as possible-mud wrestling, bikini boxing, wet T-shirting and more.
1987 60m/C VHS, Beta, LV $39.98 MAG ✍½

Summertime Spinster Hepburn vacations in Venice and falls in love with Brazzi. She is hurt when inadvertently she learns he is married, but her life has been so bleak she is not about to end her one great romance. Moving, funny, richly photographed in a beautiful Old World city. From Arthur Laurents' play "The Time of the Cuckoo." Hepburn and director Lean, who co-wrote the script were deservedly Oscar-nominated.

1955 98m/C Katharine Hepburn, Rossano Brazzi; *Dir:* David Lean. National Board of Review Awards '55: 10 Best Films of the Year; New York Film Critics Awards '55: Best Director. **VHS, Beta, LV $14.98** SUE, VYG, CRC ✍✍✍½

Summertree Michael Douglas stars as a young musician in the 1960s trying to avoid the draft and the wrath of his parents. Contrived and heavy-handed. Michael's father, actor Kirk Douglas, produced the movie. Adapted from the play by Ron Cower.
1971 (PG) 88m/C Michael Douglas, Jack Warden, Brenda Vaccaro, Barbara Bel Geddes, Kirk Callaway, Bill Vint; *Dir:* Anthony Newley. **VHS, Beta** WAR, OM ✍✍

The Sun Shines Bright Heavily stereotyped, contrived tale of a Southern judge with a heart of gold who does so many good deeds (defending a black man accused of rape; helping a desperate prostitute) that he jeopardizes his re-election. Set during Reconstruction. An unfortunate remake of Ford's own 1934 "Judge Priest," starring Will Rogers.
1953 92m/B Charles Winninger, Arleen Whelan, John Russell, Stepin Fetchit, Milburn Stone, Russell Simpson; *Dir:* John Ford. **VHS $19.98** REP ✍✍

Sun Valley Serenade Wartime musical fluff about a band that that adopts a Norwegian refugee waif as a publicity stunt. She turns out to be a full-grown man-chaser who stirs up things at a ski resort. Fun, but it ends abruptly—because, they say, Henie fell during the huge skating finale, and Darryl Zanuck wouldn't greenlight a reshoot. One of only two feature appearances by Glenn Miller and his Orchestra (the other was "Orchestra Wives"), on video with a soundtrack restored from original dual-track recordings.
1941 86m/C Sonja Henie, John Payne, Glenn Miller, Milton Berle, Lynn Bari, Joan Davis, Dorothy Dandridge; *Dir:* H. Bruce Humberstone. **VHS $19.98** FXV, FCT, MLB ✍✍½

Sunburn Insurance investigator Grodin hires model Farrah to pretend to be his wife to get the scoop on a suicide/murder case of a rich guy in Acapulco. Made-for-TV drivel.
1979 (PG) 110m/C Farrah Fawcett, Charles Grodin, Joan Collins, Art Carney, William Daniels; *Dir:* Richard Sarafian. **VHS, Beta $49.95** PAR ✍½

Sundae in New York New York Mayor Ed Koch sings "New York, New York" in this award winning animated short.
1982 4m/C Academy Awards '83: Best Animated Short Film. **VHS, Beta** DCL ✍

Sundance Surf dudes get together on the beaches of Hawaii and California for some romance and, of course, plenty of wave riding.
1972 87m/C VHS, Beta $59.95 SVN ✍

Sundance and the Kid Two brothers try to collect an inheritance against all odds. Originally known as "Sundance Cassidy and Butch the Kid."
1976 (PG) 84m/C John Wade, Karen Blake; *Dir:* Arthur Pitt. **VHS, Beta $39.95** MON ✍✍

Sunday, Bloody Sunday Adult drama centers around the intertwined love affairs of the homosexual Finch, the heterosexual Jackson, and self-centered bisexual artist Head, desired by both. Fully-drawn characters brought to life by excellent acting make this difficult story well worth watching-though Head's central character is sadly rather dull. Powerful, sincere and sensitive. Schlesinger, Jackson, Finch, and screenwri-

ter Penelope Gilliatt were nominated for Oscars.
1971 (R) 110m/C *GB* Glenda Jackson, Peter Finch, Murray Head, Daniel Day Lewis; *Dir:* John Schlesinger. British Academy Awards '71: Best Actor (Finch), Best Actress (Jackson), Best Director (Schlesinger). **VHS $59.98** FOX ✍✍✍

A Sunday in the Country A lush, distinctively French affirmation of nature and family life. This character study with a minimal plot takes place during a single summer day in 1910 France. An elderly impressionist painter-patriarch is visited at his country home by his family. Highly acclaimed, though the pace may be too slow for some. Beautiful location photography. In French with English subtitles.
1984 (G) 94m/C *FR* Louis Ducreux, Sabine Azema, Michel Aumont; *Dir:* Bertrand Tavernier. Cannes Film Festival '84: Best Director (Tavernier). **VHS, Beta $79.95** MGM ✍✍✍½

Sunday Daughters A teenage girl kept in a detention home constantly tries to escape and find some kind of familial love. Stronge performances mix with keen direction. In Hungarian with English subtitles.
1980 100m/C *HU* Julianna Nyako; *Dir:* Janosz Rozsa. **VHS, Beta $59.95** FCT ✍✍✍

Sunday in the Park with George Taped theatrical performance of the Tony, Grammy and Pulitzer Prize-winning musical play, which is based upon impressionist Georges Seurat's painting "A Sunday Afternoon on the Island of Grande Jatte." The celebrated music score was composed by Stephen Sondheim.
1986 120m/C Mandy Patinkin, Bernadette Peters, Barbara Byrne, Charles Kimbrough; *Dir:* James Lapine. **VHS, Beta, LV $19.98** LHV, WAR ✍✍✍

Sunday Too Far Away The rivalries between Australian sheep shearers and graziers leads to an ugly strike at a remote outback area in 1955. An uncomplicated story and a joy to watch.
1974 100m/C *AU Dir:* Ken Hannam, Ken Hannam. Australian Film Institute '74: Best Actor (Thompson), Best Film. **VHS, Beta $59.98** SUE ✍✍✍

Sundays & Cybele A ragged war veteran and an orphaned girl develop a strong emotional relationship, which is frowned upon by the townspeople. Warm and touching.
1962 110m/B *FR* Hardy Kruger, Nicole Courcel; *Dir:* Serge Bourguignon. Academy Awards '62: Best Foreign Language Film; National Board of Review Awards '62: Best Foreign Film. **VHS, Beta $24.95** NOS, HHT, DVT ✍✍✍½

Sundown In Africa at the beginning of World War II, a local girl aids the British against a German plot to run guns to the natives and start a rebellion. Engaging performances and efficient direction keep it above the usual cliches. Also available in a colorized version.
1941 91m/B Gene Tierney, Bruce Cabot, George Sanders, Harry Carey, Cedric Hardwicke, Joseph Calleia, Dorothy Dandridge, Reginald Gardiner; *Dir:* Henry Hathaway. **VHS, Beta $19.95** NOS, VYY, DVT ✍✍½

Sundown An ambitious shot at a vampire western fails because it drains almost all vampire lore and winds up resembling a standard oater. Carradine plays a reformed vampire king (guess who) running a desert clinic that weans bloodsuckers away from preying on humans. But undead renegades attack using sixguns and wooden bullets. The climax may outrage horror purists.

1991 (R) 104m/C David Carradine, Bruce Campbell, Deborah Foreman, Maxwell Caulfield, Morgan Brittany; *Dir:* Anthony Hickox. **VHS** $89.98 VES *🎞½*

Sundown Fury A young cowboy fights against a group of bandits to hold a telegraph office.
1942 56m/B Donald (Don "Red") Barry. **VHS, Beta $27.95** VCN *🎞*

Sundown Kid A Pinkerton agent goes undercover to break up a counterfeiting operation. Routine western.
1943 55m/B Donald (Don "Red") Barry. **VHS, Beta** DVT *🎞½*

Sundown Riders Cowhands are victimized by noose-happy outlaws. Filmed in just over a week on a $30,000 budget and originally shot on 16mm to make it accessible to hospitals and schools with basic viewing equipment. The lesson here: it takes time and money to make a good western.
1948 56m/B Russell Wade, Andy Clyde, Jay Kirby; *Dir:* Lambert Hillyer. **VHS, Beta, LV** WGE *🎞*

Sundown Saunders A standard oater about a cowboy whose prize for winning a horse race is a ranch. Trouble is, an outlaw wants the homestead as well.
1936 64m/B Bob Steele. **VHS, Beta $19.95** NOS, WGE *🎞*

The Sundowners Slow, beautiful, and often moving epic drama about a family of Irish sheepherders in Australia during the 1920s who must continually uproot themselves and migrate. They struggle to save enough money to buy their own farm and wind up training a horse they hope will be a money-winner in racing. Well-acted by all, with Johns and Ustinov providing some humorous moments. Oscar nominations for Best Picture, actress Kerr, supporting actress Johns, director, and Adapted Screenplay. Adapted from the novel by Jon Cleary, with music by Dimitri Tiomkin. Filmed in Australia and London studios.
1960 133m/C Deborah Kerr, Robert Mitchum, Peter Ustinov, Glynis Johns, Dina Merrill, Chips Rafferty, Michael Anderson Jr., Lola Brooks, Wylie Watson; *Dir:* Fred Zinneman. National Board of Review Awards '60: 10 Best Films of the Year, Best Actor (Mitchum). **VHS, Beta, LV** $19.98 WAR, PIA, CCB *🎞🎞½*

Sunny Another glossed-over, love-conquers-all musical with Neagle as a circus star who falls for a wealthy car maker's son. Dad and crew disapprove, putting a damper on the romance. In spite of the weak storyline and flat direction, Kerns' music and Bolger's dancing make it enjoyable.
1941 98m/B Anna Neagle, Ray Bolger, John Carroll, Edward Everett Horton, Frieda Inescort, Helen Westley, Benny Rubin, Richard Lane, Martha Tilton; *Dir:* Herbert Wilcox. **VHS $19.95** NOS *🎞🎞*

Sunny Side Up As stress of factory work takes toll on woman's friend, she makes heroic effort to help her to the country to recover. Silent.
1928 66m/B Vera Reynolds, ZaSu Pitts. **VHS, Beta $19.95** GPV *🎞🎞½*

Sunny Skies Bargain-basement retread has Lease donating a pint of blood
1930 75m/B Benny Rubin, Marceline Day, Rex Lease, Marjorie Kane, Wesley Barry; *Dir:* Norman Taurog. **VHS $19.95** NOS *🎞*

Sunnyside Travolta plays a street kid trying to bring an end to the local gang warfare so he can move to Manhattan. Features a host of pop tunes in an attempt to ride the coattails of brother John's successful "Saturday Night Fever." Laughably bad.
1979 (R) 100m/C Joey Travolta, John Lansing, Stacey Pickren, Andrew Rubin, Michael Tucci, Talia Balsam, Joan Darling; *Dir:* Timothy Galfas. **VHS $19.98** AIP, FOX Woof!

Sunrise Magnificent silent story of a simple country boy who, prodded by an alluring city woman, tries to murder his wife. Production values wear their age well. Gaynor won an Oscar for her stunning performance. Remade as "The Journey to Tilsit."
1927 110m/B George O'Brien, Janet Gaynor, Bodil Rosing, Margaret Livingston, J. Farrell MacDonald; *Dir:* F. W. Murnau. Academy Awards '28: Best Actress (Gaynor), Best Cinematography. **VHS** VTR, GPV *🎞🎞🎞🎞*

Sunrise at Campobello A successful adaptation (with Bellamy and Donehue repeating their roles) that chronicles Franklin D. Roosevelt's battle to conquer polio and ultimately receive the Democratic presidential nomination in 1924. Garson's excellent performance as Eleanor Roosevelt was nominated for an Oscar. Screenplay by Dore Schary, who adapted from his own play (and also produced).
1960 143m/C Ralph Bellamy, Greer Garson, Hume Cronyn, Jean Hagen, Jack Perrin, Lyle Talbot; *Dir:* Vincent J. Donehue. National Board of Review Awards '60: 10 Best Films of the Year, Best Actress (Garson). **VHS, Beta $59.95** WAR *🎞🎞🎞*

Sunset Edwards wanders the range in this soft-centered farce about a couple of Western legends out to solve a mystery. On the backlots of Hollywood, silent screen star Tom Mix (Willis) meets aging marshall Wyatt Earp (Garner) and participates in a time-warp western circa 1927. They encounter a series of misadventures while trying to finger a murderer. Garner ambles enjoyably, lifting him a level above the rest of the cast.
1988 (R) 101m/C Bruce Willis, James Garner, Mariel Hemingway, Darren McGavin, Jennifer Edwards, Malcolm McDowell, Kathleen Quinlan, M. Emmet Walsh, Patricia Hodge, Richard Bradford, Joe Dallesandro, Dermot Mulroney; *Dir:* Blake Edwards. **VHS, Beta, LV $14.95** COL *🎞🎞*

Sunset Boulevard Famed tale of Norma Desmond (Swanson), aging silent film queen, who refuses to accept that stardom has ended for her and hires a young down-on-his-luck screenwriter (Holden) to help engineer her movie comeback. The screenwriter, who becomes the actress' kept man, assumes he can manipulate her, but finds out otherwise. Reality was almost too close for comfort, as Swanson, von Stroheim (as her major domo), and others very nearly play themselves. A darkly humorous look at the legacy and loss of fame with witty dialog, stellar performances, and some now-classic scenes.
1950 100m/B Gloria Swanson, William Holden, Erich von Stroheim, Nancy Olson, Buster Keaton, Jack Webb, Cecil B. DeMille, Fred Clark; *Dir:* Billy Wilder. Academy Awards '50: Best Art Direction/Set Decoration (B & W), Best Story & Screenplay, Best Musical Score; Golden Globe

Awards '51: Best Film—Drama. **VHS, Beta, LV** $14.95 PAR *🎞🎞🎞½*

Sunset on the Desert Rogers gets to wear a white hat and a black hat as both the leader of a gang of outlaws and the hero who brings them to justice. A poor man's Western omelette.
1942 53m/B Roy Rogers, George "Gabby" Hayes, Trigger, Lynne Carver, Frank M. Thomas, Bob Nolan, Beryl Wallace, Glenn Strange, Douglas Fowley, Roy Bancroft, Pat Brady; *Dir:* Joseph Kane. **VHS, Beta $24.95** VYY *🎞½*

Sunset in El Dorado Roy and Dale thwart a villainous scheme to defraud farmers of their land.
194? 56m/B Roy Rogers, Dale Evans. **VHS, Beta $27.95** VCN *🎞½*

Sunset Limousine An out-of-work stand-up comic gets thrown out by his girlfriend, then takes a job as a chauffeur. Standard vehicle with occasional bursts of speed.
1983 92m/C John Ritter, Martin Short, Susan Dey, Paul Reiser, Audrie Neehan, Lainie Kazan; *Dir:* Terry Hughes. **VHS, Beta $59.95** IVE *🎞🎞½*

Sunset Range The "World's All-Around Champion Cowboy" of 1912 stars in this chaps-slappin', bit-champin', dust-raisin' saga of the plains.
1935 59m/B Hoot Gibson. **VHS, Beta $19.95** NOS, VCN, UHV *🎞*

Sunset Serenade Rogers outwits a murderous duo who plan to eliminate the new heir to a ranch.
1942 60m/B Roy Rogers, Trigger, George "Gabby" Hayes, Helen Parrish; *Dir:* Joseph Kane. **VHS, Beta $19.95** NOS, MED *🎞½*

Sunset Strip A photographer investigates a friend's murder, and enters the seamy world of drugs and rock and roll in L.A.
1985 83m/C Tom Elpin, Cheri Cameron Newell, John Mayall. **VHS, Beta $69.98** LIV, VES *🎞*

Sunset Strip A young dancer finds a job in a strip club and competes against the other women there to find the man of her dreams. The women take their jobs very seriously—even attending ballet classes to improve their performances. However, the viewer probably won't take this movie very seriously since it is just another excuse to show women in as little clothing as possible.
1991 (R) 95m/C Jeff Conaway, Michelle Foreman, Shelley Michelle; *Dir:* Paul G. Volk. **VHS $79.95** PMH Woof!

Sunset Trail Routine Maynard sagebrush saga with the hero protecting a woman rancher from outlaws.
1932 60m/B Ken Maynard. **VHS, Beta $19.95** NOS, UHV, DVT *🎞*

The Sunshine Boys Two veteran vaudeville partners, who have shared a love-hate relationship for decades, reunite for a television special. Adapted by Neil Simon from his play. Matthau was a replacement for Jack Benny, who died before the start of filming. Burns, for his first starring role since "Honolulu" in 1939, won an Oscar.
1975 (PG) 109m/C George Burns, Walter Matthau, Richard Benjamin, Lee Meredith, F. Murray Abraham, Carol Arthur, Howard Hesseman; *Dir:* Herbert Ross. Academy Awards '75: Best Supporting Actor (Burns); Golden Globe Awards '76: Best Film—Musical/Comedy. **VHS, Beta, LV $19.95** MGM, BTV *🎞🎞🎞*

Sunshine Run Two escaped slaves and a young widow search for Spanish treasure in the Everglades. Also on video as "Black Rage."
1979 (PG) 102m/C Chris Robinson, Ted Cassidy, David Legge, Phyllis Robinson; **Dir:** Chris Robinson. **VHS, Beta $79.95** *DMD, TAV* 🎬

The Super Pesci stars as a slumlord who faces a prison sentence thanks to his terminal neglect. The option given to him is to live in his own rat hole until he provides reasonable living conditions. This he does, and predictably learns a thing or two about his own greed and the people who suffer as a result of it. Pesci as always gives an animated performance but poor scripting laden with stereotypes and cliches successfully restricts effort.
1991 (R) 86m/C Joe Pesci, Vincent Gardenia, Madolyn Smith, Ruben Blades; **Dir:** Rod Daniel. **VHS $94.98** *FXV, CCB* 🎬½

Super Bitch Nasty, cold, and uncaring woman uses men to keep up her expensive habits, then ruthlessly tosses them aside when she is done. She purposely entangles them in her drug trade and thinks nothing of their deaths. Proving once again that you get what you pay for. Also known as "Mafia Junction."
1977 90m/C Stephanie Beacham, Patricia Hayes, Gareth Thomas; **Dir:** Massimo Dellamano. **VHS $79.95** *ATL* 🎬½

Super Bloopers #1 Outtakes in this collection come from Monogram Pictures, Abbott and Costello films movies, television shows, and the "Star Trek" series. Also featured are rare television commercials, trailers, and other short films. Some black and white segments.
197? 60m/C Bud Abbott, Lou Costello, William Shatner, Leonard Nimoy. **VHS, Beta $19.95** *NO* 🎬½

Super Bloopers #2 Bloopers from "Gunsmoke," CBS television shows, the network news, the "Star Trek" series, plus a network savings bond drive with the stars of "Mission Impossible," "Mannix," and "The Odd Couple."
197? 60m/C James Arness, William Shatner, Jack Klugman, Tony Randall, Peter Graves. **VHS, Beta $19.95** *NO* 🎬½

Super Brother The feared and respected leader of an oppressed people is imprisoned, much to everyone's regret.
1990 (R) 90m/C Woody Strode. **VHS, Beta** *ELE* 🎬

Super Dragons/Mafia vs. Ninja The baddest of the bad and the best of the good battle for ultimate control in two, full-length martial arts films.
19?? 180m/C Alexander Lou, Pad Yku La, Wang Hsia. **VHS $9.99** *VTR* 🎬

Super Dynamo An orphaned child is raised by monks, who teach him kung fu as part of his spiritual training and discipline. Once grown, the young man fights to stem the tide of smuggling rings and crime societies in Asia.
19?? 90m/C VHS $19.95 *OCE* 🎬

Super Fuzz A rookie policeman develops super powers after being accidentally exposed to radiation. Somewhat ineptly, he uses his abilities to combat crime. Somewhat inepty acted, written, and directed, as well. Also known as "Supersnooper."
1981 (PG) 97m/C *IT* Terence Hill, Joanne Dru, Ernest Borgnine; **Dir:** Sergio Corbucci. **VHS, Beta $69.98** *SUE* 🎬½

Super Gang Abounding fists and feet punctuate this tale of colliding gangs.
1985 75m/C Bruce Lee. **VHS, Beta $59.95** *SVS* 🎬

Super Kung Fu Kid A kung fu hero is pursued by countless villains out to murder him but no one knows why.
19?? 90m/C Nan Kung Fan, Chung Lick, Yeung Si, Kong Yeh. **VHS $29.95** *OCE* 🎬

Super Models A bevy of former centerfolds from major men's magazines stretch and bend for the camera, wearing little or nothing.
1984 30m/C VHS, Beta $14.95 *AHV* 🎬

Super Ninja A modern day ninja must battle modern-day bad guys.
19?? 90m/C Yu Jin Boa, Yi Dao Dao, Alexander Lou, Lou Mei, Wang Qiang, Yang Song, Yau Jin Tomas. **VHS $59.95** *OCE* 🎬

Super Ninjas A ninja battles evil hordes of like-trained ninja warriors, and wins over and over again.
198? 104m/C Cheng Tien Chi, Lung Tien Hsiang, Chen Pei Hsi. **VHS, Beta $69.95** *VHV* 🎬

Super Seal An injured seal pup disrupts a family's normal existence after the young daughter adopts him.
1977 (G) 95m/C Foster Brooks, Sterling Holloway, Sarah Brown; **Dir:** Michael Dugan. **VHS, Beta $39.95** *UHV* 🎬

Super Soul Brother A lazy no-goodnik accidentally ingests a strange potion which turns him into a master thief.
198? 80m/C Wildman Steve. **VHS $39.95** *XVC* 🎬

Super Weapon Masters of the martial arts demonstrate karate, kung fu, ju jitsu, tae kuan do, tai chi and aikido in this documentary-format program.
1975 (PG) 86m/C Ron Van Cliff, Frank Ruiz, Pete Siringano, Byong Hoong Park. **VHS, Beta $54.95** *SUN*

Superargo A wrestler becomes a superhero with psychic powers and a bulletproof leotard. He fights against a madman who is turning athletes into robots. Successful Italian hero who wouldn't last two minutes in the ring with Batman.
1967 95m/C *IT SP* Guy Madison, Ken Wood, Liz Barrett, Diana Loris; **Dir:** Paul Maxwell. **VHS, Beta $19.98** *SNC, MAD* 🎬🎬

Superboy Join Superboy and his dog Krypto as they fight crime in this collection of eight animated adventures.
1966 60m/C VHS, Beta $12.95 *WAR* 🎬🎬

Superbug Super Agent Dodo, the wonder car puts the brakes on crime in this silly action-adventure tale. A lemon in a lot full of Herbies.
1976 90m/C Mark Robert, Heidi Hansen, George Goodman. **VHS, Beta $59.95** *JEF* 🎬½

Supercarrier Comprised of two episodes of the television series which centered around the drama of life on a military aircraft carrier.
1988 90m/C Robert Hooks, Richard Jaeckel; **Dir:** William A. Graham. **VHS** *FRH* 🎬½

Superchick Mild-mannered stewardess by day, sexy blonde with karate blackbelt by night. In addition to stopping a skyjacking she regularly makes love to men around the world. A superbomb that never gets off the ground.
1971 (R) 94m/C Joyce Jillson, Louis Quinn, Thomas Reardon; **Dir:** Ed Forsyth. **VHS, Beta $19.95** *PSM* Woof!

Superdad A middle-aged parent is determined to bridge the generation gap by trying his hand at various teenage activities. Disney family fare that's about as complicated as a television commercial. The adolescents are two-dimensional throwbacks to the fun-loving fifties.
1973 (G) 94m/C Bob Crane, Kurt Russell, Joe Flynn, Barbara Rush, Kathleen Cody, Dick Van Patten; **Dir:** Vincent McEveety. **VHS, Beta** *DIS* 🎬

Superdrumming A group of jazz drummers play a few sessions in a West German cathedral.
1988 55m/C Louis Bellson, Gerry Brown, Nippy Noya, Ian Paice, Simon Phillips, Cozy Powell, Pete York, Brian Auger, Wolfgang Schmid, Gerd Wilden Jr., Peter Wolpl. **VHS, Beta $19.95** *MVD, PRS* 🎬

Superfly Controversial upon release, pioneering blaxploitation has Harlem dope dealer finding trouble with gangs and police as he attempts to establish retirement fund from one last deal. Excellent period tunes by Curtis Mayfield. Two lesser sequels.
1972 (R) 98m/C Ron O'Neal, Carl Lee, Sheila Frazier, Julius W. Harris; **Dir:** Gordon Parks Jr. **VHS, Beta $19.98** *WAR* 🎬🎬½

Supergirl Big-budget bomb in which Slater made her debut and nearly killed her career, with the help of Kryptonite. Unexciting and unsophisticated story of a young woman, cousin to Superman, with super powers, based on the comic book series. She's in pursuit of a magic paperweight, but an evil sorceress wants it too. Dunaway is a terrifically vile villainess with awesome black magic powers. Slater is great to look at, but is much better in almost any other film.
1984 (PG) 114m/C *GB* Faye Dunaway, Helen Slater, Peter O'Toole, Mia Farrow, Brenda Vaccaro, Peter Cook, Simon Ward, Marc McLure, Hart Bochner, Maureen Teefy, David Healy, Matt Frewer; **Dir:** Jeannot Szwarc. **VHS, Beta, LV $19.95** *IVE* 🎬½

Supergrass A low-brow British farce about a nebbish who poses as a drug smuggler to impress his girlfriend, and is then mistaken for a real one by the authorities.
1987 (R) 105m/C *GB* Adrian Edmondson, Peter Richardson, Nigel Planer, Jennifer Saunders, Ronald Allen, Dawn French; **Dir:** Peter Richardson. **VHS, Beta $19.98** *CHA* 🎬

Superman: The Movie The DC Comics legend comes alive in this wonderfully entertaining saga of Superman's life from a baby on the doomed planet Krypton to Earth's own Man of Steel. Hackman and Beatty pair marvelously as super criminal Lex Luthor and his bumbling sidekick. Award winning special effects and a script (written by Mario Puzo, Robert Benton, and David Newman) that doesn't take itself too seriously make this great fun. Followed by three sequels.
1978 (PG) 144m/C Christopher Reeve, Margot Kidder, Marlon Brando, Gene Hackman, Glenn Ford, Susannah York, Ned Beatty, Valerie Perrine, Jackie Cooper, Marc McClure, Trevor Howard, Sarah Douglas, Terence Stamp, Jack O'Halloran, Phyllis Thaxter; **Dir:** Richard Donner. Academy Awards '78: Best Visual Effects; British Academy Awards '78: Best Special Effects; National Board of Review Awards '78: 10 Best Films of the Year. **VHS, Beta, LV $19.95** *WAR* 🎬🎬🎬½

Superman 2 The sequel to "the movie" about the Man of Steel. This time, he has his hands full with three super-powered villains from his home planet of Krypton. The romance between reporter Lois Lane and our superhero is made believable and the story-

line has more pace to it than the original. A sequel that often equals the first film—leave it to Superman to pull off the impossible.
1980 (PG) 127m/C Christopher Reeve, Margot Kidder, Gene Hackman, Ned Beatty, Jackie Cooper, Sarah Douglas, Jack O'Halloran, Susannah York, Marc McClure, Terence Stamp, Valerie Perrine, E.G. Marshall; **Dir:** Richard Lester. VHS, Beta, LV $19.98 WAR, PIA *III*

Superman 3 A villainous businessman tries to conquer Superman via the expertise of a bumbling computer expert and the judicious use of Red Kryptonite. Superman allows his darker side to manifest itself but is defeated more by poor direction and boring physical comedy than by any villainous element. Notable is the absence of Lois Lane as a main character.
1983 (PG) 123m/C Christopher Reeve, Richard Pryor, Annette O'Toole, Jackie Cooper, Margot Kidder, Marc McClure, Annie Ross, Robert Vaughn; **Dir:** Richard Lester. VHS, Beta, LV $19.98 WAR *II 1/2*

Superman 4: The Quest for Peace The third sequel, in which the Man of Steel endeavors to rid the world of nuclear weapons, thereby pitting himself against nuclear-entrepreneur Lex Luthor and his superpowered creation, Nuclear Man. Special effects are dime-store quality and it appears that someone may have walked off with parts of the plot. Reeve deserves credit for remaining true to character through four films.
1987 (PG) 90m/C Christopher Reeve, Gene Hackman, Jon Cryer, Marc McClure, Margot Kidder, Mariel Hemingway, Sam Wanamaker; **Dir:** Sidney J. Furie; **Voices:** Susannah York. VHS, Beta, LV $19.98 WAR, APD *II*

Superman: The Cartoons Eight of the comic book hero's adventures from 1942 are featured along with a 1953 black and white short called "Stamp Day for Superman," starring the cast of the television show.
1942 89m/C Voices: Bud Collyer, Joan Alexander. VHS, Beta $14.98 AOV *II*

Superman: The Complete Cartoon Collection This tape features all 17 Superman cartoons made by the Max Fleischer Studios from 1941 to 1943. Several are in black & white.
1943 141m/C VHS, Beta $19.95 VDM *II*

Superman: The Serial: Vol. 1 Superman was first seen in live action in this 15-part theatrical serial that has been rarely shown since its original release.
1948 248m/B Kirk Alyn, Noel Neill, Pierre Watkin, Tommy Bond, Thomas Carr; **Dir:** Spencer Gordon Bennet. VHS, Beta $29.95 WAR *II*

The Supernaturals Confederate Civil War-era ghosts, bent on avenging their deaths, haunt a wooded area in which modern Yankee army maneuvers are practiced. Antebellum boredom.
1986 (R) 85m/C Maxwell Caulfield, LeVar Burton, Nichelle Nichols; **Dir:** Armand Mastroianni. VHS, Beta $19.98 SUE *II 1/2*

Supersonic Man Incoherent shoestring-budget Superman spoof with a masked hero fighting to save the world from the evil intentions of a mad scientist.
1978 (PG) 85m/C SP Michael Coby, Cameron Mitchell, Diana Polakov; **Dir:** Piquer Simon. VHS, Beta $59.95 UHV Woof!

Superstar: The Life and Times of Andy Warhol Even if you didn't grok Andy's 'pop' artwork and self-created celebrity persona, this ironic, kinetic, oft-rollicking documentary paints a vivid picture of the wild era he inspired and exploited. Interviewees range from Warhol cohorts like Dennis Hopper to proud executives at the Campbell Soup plant. One highlight: a Warhol guest shot on "The Love Boat."
1990 87m/C Chicago Film Festival '90: Silver Plaque. VHS $89.98 VES *III*

Superstition A reverend and his family move into a vacant house despite warnings about a curse from the townsfolk. Some people never learn.
1982 85m/C James Houghton, Albert Salmi, Lynn Carlin; James Robertson Justice. VHS, Beta $79.98 LIV, LTG *II 1/2*

Supervixens True to Meyer's low-rent exploitation film canon, this wild tale is filled with characteristic Amazons, sex and violence. A gas station attendant is framed for the grisly murder of his girlfriend and hustles out of town, meeting a succession of well-endowed women during his travels. As if it needed further problems, it's hampered by a tasteless storyline and incoherent writing.
1975 105m/C Shari Eubank, Charles Napier; **Dir:** Russ Meyer. VHS $59.99 RMF *II*

Support Your Local Sheriff Amiable, irreverent western spoof with more than its fair share of laughs. When a stranger stumbles into a gold rush town, he winds up becoming sheriff. Garner is perfect as the deadpan sheriff, particularly in the scene where he convinces Dern to remain in jail, in spite of the lack of bars. Contains many cameos by veteran western actors as it neatly subverts every western cliche it encounters. Written and produced by William Bowers. Followed by "Support Your Local Gunfighter."
1969 (G) 92m/C James Garner, Joan Hackett, Walter Brennan, Bruce Dern, Jack Elam, Harry Morgan; **Dir:** Burt Kennedy. National Board of Review Awards '69: 10 Best Films of the Year. VHS, Beta $19.98 FOX, MGM *III 1/2*

Suppose They Gave a War And Nobody Came? A small Southern town battles with a local army base in this entertaining but wandering satire. The different acting styles used as the producers wavered on making this a comedy or drama were more at war with one another than the characters involved. Also known as "War Games."
1970 (G) 113m/C Tony Curtis, Brian Keith, Ernest Borgnine, Suzanne Pleshette, Ivan Dixon, Bradford Dillman, Don Ameche; **Dir:** Hy Averback. VHS $59.98 FOX *II 1/2*

Surabaya Conspiracy Mystery and intrigue surround a quest for gold in Africa.
1975 90m/C Michael Rennie, Richard Jaeckel, Barbara Bouchet, Mike Preston; **Dir:** Roy Davis. VHS, Beta $59.95 MON *II 1/2*

The Sure Thing College students who don't like each other end up travelling to California together, and of course, falling in love. Charming performances make up for predictability. Can't-miss director (and ex-Meathead) Reiner's second direct hit at the box office.
1985 (PG-13) 94m/C John Cusack, Daphne Zuniga, Nicolette Sheridan, Viveca Lindfors, Anthony Edwards, Tim Robbins, Boyd Gaines; **Dir:** Rob Reiner. VHS, Beta, LV, 8mm $14.95 COL, SUE *III*

Surf A made-for-video look of the Pacific Northwest coastline, designed to soothe the viewer's frayed nerves.
1985 58m/C VHS, Beta $29.95 NOR

Surf 2 A most excellent group of surfers get sick from drinking tainted Buzz Cola concocted by demented chemist Deezen in most heinous effort to obliterate surfer population from Southern California. Bogus, dude. Not a sequel, for sure, but title is the sort of in-joke this exercise in mental meltdown perpetuates as comedy. Ha.
1984 (R) 91m/C Morgan Paull, Cleavon Little, Lyle Waggoner, Ruth Buzzi, Linda Kerridge, Carol Wayne, Eddie Deezen, Eric Stoltz, Lucinda Dooling, Brandis Kemp, Terry Kiser; **Dir:** Randall Badat. VHS, Beta $59.95 MVD, KOV, MED Woof!

Surf Nazis Must Die A piece of deliberate camp in the Troma mold, about a group of psychotic neo-Nazi surfers taking over the beaches of California in the wake of a devastating earthquake. Tongue-in-cheek, tasteless and cheap, but intentionally so.
1987 83m/C Barry Brenner, Gail Neely, Michael Sonye, Dawn Wildsmith, Tom Shell, Bobbie Bresee; **Dir:** Peter George. VHS, Beta $19.98 MED *I*

Surf Party Romance among the sands of Malibu with Vinton as the owner of a surf shop. Risque in 1964; utterly campy today. Songs include, "If I Were an Artist," "That's What Love Is," and "Pearly Shells."
1964 68m/C Bobby Vinton, Jackie De Shannon, Patricia Morrow, Kenny Miller; **Dir:** Maury Dexter. VHS $39.95 MOV *II 1/2*

Surfacing A girl braves the hostile northern wilderness to search for her missing father. This pseudo-psychological, sex-driven suspense flick makes little sense.
1984 90m/C Joseph Bottoms, Kathleen Beller, R.H. Thompson, Margaret Dragu; **Dir:** Claude Jutra. VHS, Beta $69.95 VCL Woof!

Surfing Beach Party This program recaptures the beach-blanket fun of the '50s and '60s California rock 'n' roll. Feature songs include "Surfin' Safari," "Fun, Fun, Fun," "Barbara Anne" and many more.
1984 56m/C VHS, Beta $9.95 MSM Woof!

Surprise Attack The Spanish Civil War forms the backdrop for this action opus.
197? 124m/C Simon Andrew, Danny Martin, Patty Shepherd; **Dir:** Joseph Loman. VHS, Beta $19.95 NEG *I*

Surrender A struggling woman artist and a divorced author fall in love but won't admit it for fear of being hurt again. Jumbled plot, unfortunate casting. Available in a Spanish-subtitled version.
1987 (PG) 95m/C Michael Caine, Sally Field, Steve Guttenberg, Peter Boyle, Jackie Cooper, Julie Kavner, Louise Lasser, Iman; **Dir:** Jerry Belson. VHS, Beta, LV $19.98 WAR *II 1/2*

Surrogate A young couple fights to keep their marriage afloat by turning to a sex surrogate. Strong cast helps carry odd, confusing script, with plenty of (you guessed it) sex, and a mystery to solve.
1988 100m/C CA Art Hindle, Shannon Tweed, Carole Laure, Michael Ironside, Marilyn Lightstone; **Dir:** Don Carmody. VHS, Beta MED *II*

Survival After their plane is forced down in the wilderness, a family finds the strength for survival in their Christian faith.
1988 73m/C Robby Sella, Terry Griffin, Pearl Braaten; **Dir:** Donald W. Thompson. VHS, Beta MIV *II 1/2*

Survival of a Dragon A medieval Chinese warrior is transported into the future, where he goes on a kicking rampage.
197? 88m/C Squall Hung, Jaguar Lee. **VHS, Beta** $49.95 *REG* ✗

Survival Earth After civilization collapses, humans begin reverting back to a primitive way of life. A young couple and a soldier of fortune fight against it, hoping for a better existence.
19?? 90m/C VHS *NAM* ✗

Survival Game A young combat expert (Chuck Norris's son) gets involved with some ex-hippies in search of a $2 million cache. High-kickin' action.
1987 91m/C Mike Norris, Deborah Goodrich, Seymour Cassel; *Dir:* Herb Freed. **VHS, Beta** $79.95 *MED* ✗

Survival Quest Students in a Rocky Mountain survival course cross paths with a band of bloodthirsty mercenaries-in-training. A battle to the death ensues; you'll wish they'd all put each other out of their misery a lot sooner.
1989 (R) 90m/C Lance Henriksen, Dermot Mulroney, Mark Rolston, Steve Antin, Paul Provenza, Ben Hammer, Traci Lin, Catherine Keener, Reggie Bannister. **VHS, Beta** $89.98 *FOX* ✗ ½

Survival Run Six California teenagers become stranded in the Mexican desert. They witness Graves and Millard making a shady deal; the chase is on. The story is lame, the gory parts aren't gory - so why bother?
1980 (R) 90m/C Peter Graves, Ray Milland, Vincent Van Patten; *Dir:* Larry Spiegel. **VHS, Beta** $54.95 *MED* ✗

Survival of Spaceship Earth Man, his world and the protection of its ecology are discussed in this beautiful and fascinating film. Winner of numerous international awards due to the strength of its production values and writing. Nominated for 10 Emmy awards, winner of 2 Emmys.
1990 63m/C *W/Dir:* Dirk Summers; *Hosted:* Hugh Downs; *Nar:* Raymond Burr. **VHS, Special order formats** $24.95 *CPM, FCT* ✗

Survival Zone Nuclear holocaust survivors battle a violent band of marauding motorcyclists on the barren ranches of the 21st century. Advice to director Rubens and cohorts: Next time, find a plot that's not growing mold.
1984 (R) 90m/C Gary Lockwood, Morgan Stevens, Camilla Sparv; *Dir:* Percival Rubens. **VHS, Beta** $9.99 *PSM, STE Woof!*

The Survivalist An arms-ready survivalist nut defends his family and supplies against a panicked society awaiting nuclear war. Poorly acted, little tension, and ineffective plotting make boredom seem a more critical issue than survival.
1987 (R) 96m/C Steve Railsback, Susan Blakely, Marjoe Gortner, David Wayne, Cliff DeYoung; *Dir:* Sig Shore. **VHS, Beta, LV** $79.98 *VES* ✗

Survivor A jetliner crashes, leaving but one survivor, the pilot, who is then plagued by visions, tragedies, and ghosts of dead passengers. Viewers also suffer.
1980 91m/C *AU* Robert Powell, Jenny Agutter, Joseph Cotten, Angela Punch McGregor; *Dir:* David Hemmings. **VHS, Beta** $19.95 *LHV, WAR* ✗

Survivor A lone warrior on a post-nuclear holocaust wasteland battles a megalomaniac ruler. Moll as the evil villain is watchable, but overall the film is a wasteland, too.

1987 92m/C Chip Mayer, Richard Moll, Sue Kiel; *Dir:* Michael Shackleton. **VHS, Beta, LV** $19.98 *LIV, VES* ✗✗

Survivors Two unemployed men find themselves the target of hit man, whom they have identified in a robbery attempt. One of the men goes gun-crazy protecting himself. Uneven comedy with wild Williams and laid-back Matthau.
1983 (R) 102m/C Robin Williams, Walter Matthau, Jerry Reed, John Goodman; *Dir:* Michael Ritchie. **VHS, Beta, LV** $12.95 *COL* ✗✗

Susan and God A selfish socialite returns from Europe and starts practicing a new religion, much to the dismay of her friends and family. Despite her preaching, her own domestic life is falling apart and she realizes that her preoccupation and selfish ways have caused a strain on her marriage and her relationship with her daughter. An excellent script and a fine performance from Crawford make this a highly satisfying film.
1940 115m/B Joan Crawford, Fredric March, Ruth Hussey, John Carroll, Rita Hayworth, Nigel Bruce, Bruce Cabot, Rose Hobart, Rita Quigley, Marjorie Main, Gloria de Haven; *Dir:* George Cukor. **VHS** $19.98 *MGM* ✗✗✗

Susan Lenox: Her Fall and Rise Garbo, the daughter of an abusive farmer, falls into the arms of the handsome Gable to escape an arranged marriage. Both stars are miscast, but the well-paced direction keeps the melodrama moving.
1931 84m/B Greta Garbo, Clark Gable, Jean Hersholt, John Miljan, Alan Hale Jr., Hale Hamilton; *Dir:* Robert Z. Leonard. **VHS, Beta** $19.98 *MGM, FCT* ✗✗ ½

Susan Slept Here While researching a movie on juvenile delinquents, a Hollywood script writer (Powell) is given custody of a spunky 18-year-old delinquent girl (Reynolds) during the Christmas holidays. Cute sex comedy. Based on the play "Susan," by Alex Gottlieb and Steve Fisher.
1954 98m/C Dick Powell, Debbie Reynolds, Anne Francis; *Dir:* Frank Tashlin. **VHS, Beta, LV** $15.95 *UHV* ✗✗ ½

Susana Minor though still interesting Bunuel, in which a sexy delinquent young girl is rescued from vagrancy by a Spanish family, and how she subsequently undermines the family's structure through sexual allurement and intimidation. In Spanish with English subtitles.
1951 87m/B *MX* Rosita Quintana, Fernando Soler, Victor Manuel Mendoza, Matilde Palou; *Dir:* Luis Bunuel. **VHS, Beta** $24.95 *XVC, TAM, APD* ✗✗✗

Susanna Pass Oil deposits underneath a fish hatchery lake attract bad guys. They set off explosions to destroy the fishery, but game warden Rogers investigates. Minor outing for Roy.
1949 67m/C Roy Rogers, Dale Evans, Estelita Rodriguez, Martin Garralaga, Robert Emmett Keane, Lucien Littlefield, Douglas Fowley; *Dir:* William Witney. **VHS** $12.98 *REP* ✗

Susannah of the Mounties An adorable young girl is left orphaned after a wagon train massacre and is adopted by a Mountie. An Indian squabble gives Shirley a chance to play little peacemaker and teach Scott how to tap dance, too. Could she be any cuter? Available colorized.
1939 78m/B Shirley Temple, Randolph Scott, Margaret Lockwood, J. Farrell MacDonald, Moroni Olsen, Victor Jory; *Dir:* William A. Seiter. **VHS, Beta** $19.98 *FOX* ✗✗ ½

Suspect An overworked Washington, DC, public defender (Cher) is assigned to a controversial murder case in which her client is a deaf-mute, skid-row bum. A cynical lobbyist on the jury illegally researches the case himself. They work together to uncover a far-reaching conspiracy. Unrealistic plot, helped along by good performances and tight direction.
1987 (R) 101m/C Dennis Quaid, Cher, Liam Neeson, E. Katherine Kerr, Joe Mantegna; *Dir:* Peter Yates. **VHS, Beta, LV, 8mm** $14.95 *COL* ✗✗ ½

Suspended Alibi British crime-drama about an adulterous man unjustly accused when his alibi is murdered. Overlook the convenient coincidence. Also known as "Suspected Alibi."
1956 64m/B *GB* Patrick Holt, Honor Blackman, Andrew Keir, Valentine Dyall; *Dir:* Alfred Shaughnessy. **VHS** $16.95 *NOS, SNC* ✗✗ ½

Suspense A collection of two episodes from the series: "F.O.B. Vienna;" and "All Hallows Eve" in which a man who murders his pawnbroker is plagued by a guilty conscience.
1953 54m/B Walter Matthau, Jayne Meadows, Franchot Tone, Romney Brent. **VHS, Beta** $24.95 *VYY* ✗ ½

Suspicion Alfred Hitchcock's suspense thriller about a woman who gradually realizes she is married to a killer and may be next on his list. An excellent production unravels at the end due to RKO's insistence that Grant retain his "attractive" image. This forced the writers to leave his guilt or innocence undetermined. Oscar nominations: Best Picture, Best Score for a Dramatic Film, and Fontaine's. Available colorized.
1941 99m/B Cary Grant, Joan Fontaine, Cedric Hardwicke, Nigel Bruce, May Whitty, Leo G. Carroll, Heather Angel; *Dir:* Alfred Hitchcock. Academy Awards '41: Best Actress (Fontaine); Film Daily Poll '42: 10 Best Films of the Year. **VHS, Beta, LV** $19.98 *RKO, VID, MED* ✗✗✗ ½

Suspicion Remake of the chilling Hitchcock tale. Newlywed bride is consumed with fears about her wealthy new husband. The suspense builds as she comes to suspect that she is married to a cold-blooded killer. Curtin in way over her head.
1987 97m/C Jane Curtin, Anthony Andrews; *Dir:* Andrew Grieve. **VHS, Beta** $79.95 *MED* ✗✗

Suspiria An American dancer enters a weird European ballet academy and finds they teach more than movement as bodies begin piling up. Sometimes weak plot is aided by great-but-gory special effects, fine photography, good music, and a chilling opening sequence. Also available in unrated version.
1977 (R) 92m/C *IT* Jessica Harper, Joan Bennett, Alida Valli, Udo Kier; *Dir:* Dario Argento. **VHS, Beta, LV** $89.98 *MAG, HHE* ✗✗✗

Suzanne Set in 1950s Quebec, the story of a good woman whose criminal boyfriend leaves her pregnant when he is sent to prison. Ten years later, he returns, threatening her happy marriage. Confused script can't be saved by valiant cast.
1980 (R) 90m/C Sondra Locke, Richard Dreyfuss, Gene Barry. **VHS, Beta** $69.98 *VES* ✗ ½

Svengali A music teacher uses his hypnotic abilities to manipulate one of his singing students and make her a star. Soon the young woman is singing for sell-out crowds, but only if her teacher is present. Barrymore in a hypnotic performance. Adapted from

"Trilby" by George Du Maurier. Remade in 1955 and 1983.
1931 76m/B John Barrymore, Marian Marsh, Donald Crisp; *Dir:* Archie Mayo. **VHS, Beta $19.95** *NOS, GPV, SNC* ⅃⅃⅃

Svengali Flamboyant but faded music star O'Toole mentors Foster, a young pop singer looking for stardom. Boring, made for television remake of the Barrymore classic.
1983 96m/C Peter O'Toole, Jodie Foster, Elizabeth Ashley, Larry Joshua, Pamela Blair, Barbara Byrne, Holly Hunter; *Dir:* Anthony Harvey. **VHS $39.95** *USA* ⅃⅃½

Swamp Fire Mississippi river-boat captain Weismuller, wrestles aligators and battles with Crabbe for the woman he loves in this turgid drama. Look for Janssen in an early screen performance.
1946 69m/B Johnny Weissmuller, Virginia Grey, Buster Crabbe, Carol Thurston, Edwin Maxwell, Pedro de Cordoba, Pierre Watkin, David Janssen; *Dir:* William H. Pine. **VHS, Beta $19.95** *NOS, SNC* ⅃½

Swamp of the Lost Monster An incredible Mexican horror/western/musical that features a mouse-like monster that'll scare the cheese out of you.
1965 88m/B *MX* Gaston Santos, Sarah Cabrera, Manuel Donde; *Dir:* Rafael Baledon. **VHS, Beta $24.95** *NOS, SNC, GHV* ⅃

Swamp Thing Camp drama of scientist turned into half-vegetable, half-man swamp creature. Vegetarian nightmare or ecology propoganda? You be the judge.
1982 (PG) 91m/C Adrienne Barbeau, Louis Jourdan, Ray Wise, Adam West; *Dir:* Wes Craven. **VHS, Beta, LV $14.95** *COL, SUE* ⅃½

Swamp Women Four escaped women convicts, known as the "Nardo Gang," chase after a stash of diamonds in this super cheap, super bad action adventure from cult director Roger Corman. Also known as "Swamp Diamonds" and "Cruel Swamp."
1955 73m/C Michael Connors, Marie Windsor, Beverly Garland, Carole Matthews, Susan Cummings; *Dir:* Roger Corman. **VHS $16.95** *SNC* ⅃

The Swan Princess' arranged marriage with a prince is complicated by a poor man also in love with her. Classic romantic comedy of the silent era. Remade in 1956. Based on a play by Ferenc Molnar.
1925 50m/B Adolphe Menjou, Ricardo Cortez, Frances Howard; *Dir:* Dimitri Buchowetzki. **VHS, Beta $19.95** *NOS, DVT* ⅃⅃½

The Swan A twist on the Cinderella story with Kelly (in her last film before her marriage) as the charming beauty waiting for her prince. Both Guinness, as the crown prince, and Jourdan, as her poor tutor, want her hand (and the rest of her). Attractive cast, but story gets slow from time to time. Remake of a 1925 silent film.
1956 112m/C Grace Kelly, Louis Jourdan, Alec Guinness, Jessie Royce Landis, Brian Aherne, Estelle Winwood; *Dir:* Charles Vidor. **VHS $19.98** *MGM* ⅃⅃⅃

Swan Lake Prince searches for a bride. The swan that bears the golden crown possesses magical powers that hold the key.
1982 75m/C VHS, Beta $9.95 *MED* ⅃⅃

Swann in Love A handsome, wealthy French aristocrat makes a fool of himself over a beautiful courtesan who cares nothing for him. Elegant production lacks spark. Based upon a section of Marcel Proust's "Remembrance of Things Past."

1984 110m/C *FR GE* Jeremy Irons, Ornella Muti, Alain Delon, Fanny Ardant, Marie-Christine Barrault; *Dir:* Volker Schlondorff. **VHS, Beta $24.95** *XVC, MED* ⅃⅃½

Swans A brother and sister battle an evil spell that keeps them apart and controls their little village. From the Hans Christian Anderson story.
1990 60m/C VHS $9.95 *AIP* ⅃⅃½

The Swap Ex-con searches for his brother's killer. Muddled story uses clips from early De Niro film "Sam's Song."
1971 (R) 120m/C Robert De Niro, Jered Mickey, Jennifer Warren, Terrayne Crawford, Martin Kelley; *Dir:* Jordan Leondopoulos. **VHS $39.98** *CAN* ⅃

Swap Meet Wacky teen sex comedy about shenanigans at a small town swap meet. Strictly bargain basement.
1979 (R) 86m/C Ruth Cox, Debi Richter, Danny Goldman, Cheryl Rixon, Jonathan Gries; *Dir:* Brice Mack. **VHS** *UHV* ⅃

The Swarm A scientist fends off a swarm of killer bees when they attack metro Houston. No sting to this one. The bees are really just black spots painted on the film!
1978 (PG) 116m/C Michael Caine, Katharine Ross, Richard Widmark, Lee Grant, Richard Chamberlain, Olivia de Havilland, Henry Fonda, Fred MacMurray, Patty Duke, Ben Johnson, Jose Ferrer, Slim Pickens, Bradford Dillman, Cameron Mitchell; *Dir:* Irwin Allen. **VHS, Beta $59.95** *WAR* ⅃

Swashbuckler Jaunty pirate returns from sea to find his friends held captive by dastardly dictator for their political views. He rescues them, and helps them overthrow the erstwhile despot.
1976 (PG) 101m/C Robert Shaw, James Earl Jones, Peter Boyle, Genevieve Bujold, Beau Bridges, Geoffrey Holder; *Dir:* James Goldstone. **VHS, LV $79.95** *MCA, PIA* ⅃

Swashbuckler Eighteenth-century pirate fracas that can't compete with Errol Flynn. Also known as "The Scarlet Buccaneer."
1984 100m/C *FR* Jean-Paul Belmondo, Marlene Jobert, Laura Antonelli, Michel Auclair, Julien Guiomar; *Dir:* Jean-Paul Rappeneau. **VHS** *MCA* ⅃½

Sweater Girls Sex comedy about two girls starting their own club called the "Sweater Girls." Terrible teen exploitaiton film.
1978 (R) 90m/C Charlene Tilton, Harry Moses, Meegan King, Noelle North, Kate Sarchet, Carol Seflinger, Tamara Barkley, Julie Parsons; *Dir:* Don Jones. **VHS, Beta** *WPM* Woof!

Swedenhielms A poor Swedish scientist struggles to support his family while hoping to win a Nobel prize. Fine performances by Ekman and the 20-year-old Bergman (in her third film). In Swedish with English subtitles.
1935 92m/B *SW* Gosta Ekman, Karin Swanstrom, Bjorn Berglund, Hakan Westergren, Tutta Rolf, Ingrid Bergman, Sigurd Wallen; *Dir:* Gustaf Molander. **VHS, Beta, LV $29.95** *WAC, CRO, LUM* ⅃½

Swedish Wildcats Three eccentric ladies run a brothel specializing in live theatrical productions.
19?? 90m/C Diana Dors. **VHS, Beta $59.95** *JEF* ⅃⅃½

Sweeney Todd: The Demon Barber of Fleet Street Barber with eating disorder catches up customers and serves tasty meat pies.
1936 68m/B Tod Slaughter; *Dir:* George King. **VHS, Beta, 8mm $14.95** *VYY* ⅃⅃

Sweeney Todd: The Demon Barber of Fleet Street A filmed performance of the Tony-winning Broadway musical by Stephen Sondheim, directed by Harold Prince. Creepy thriller about a demon barber, his razor, and his wife's meatpies.
1984 139m/C *GB* Angela Lansbury, George Hearn; *Dir:* Harold Prince. **VHS, Beta, LV $39.95** *MVD, RKO, VTK* ⅃⅃

Sweet Adeline Smalltime hayseed makes it big in the music business while brother steals his girl back home. Not easily put off, young songster perseveres.
1926 60m/B Charles Ray, Gertrude Olmsted, Jack Clifford, Ida Lewis; *Dir:* Jerome Storm. **VHS, Beta $19.95** *GPV, FCT* ⅃

Sweet Beat A young woman gets a chance to become a singer when she reaches the finals of a British beauty contest. Pedestrian story enlivened by music from popular British bands of the time. Also known as "The Amorous Sex."
1962 66m/B Julie Amber, Sheldon Lawrence, Irv Bauer, Billy Myles. **VHS, Beta $19.95** *NOS, INC* ⅃

Sweet Bird of Youth An acclaimed adaptation of the Tennessee Williams play about Chance Wayne, a handsome drifter who travels with an aging movie queen to his small Florida hometown, hoping she'll get him started in a movie career. However, coming home turns into a big mistake, as the town boss wants revenge on Chance for seducing his daughter, Heavenly. Williams' original stage ending was cleaned up, providing a conventional "happy" movie ending for the censors. Page, Begley and Knight were nominated for Oscars.
1962 120m/C Paul Newman, Geraldine Page, Ed Begley Sr., Mildred Dunnock, Rip Torn, Shirley Knight, Madeline Sherwood; *Dir:* Richard Brooks. Academy Awards '62: Best Supporting Actor (Begley). **VHS, Beta $19.95** *MGM, BTV* ⅃⅃⅃

Sweet Charity An ever-optimistic dime-a-dance girl has a hard time finding a classy guy to marry. MacLaine is appealing in this big-budget version of the popular Broadway musical by Neil Simon (derived from Fellini's "Notti di Cabiria"), featuring the songs "Big Spender," "If They Could See Me Now" and "Where Am I Going." Bob Fosse's debut as film director. Bud Cort plays a flower child.
1968 148m/C Shirley MacLaine, Chita Rivera, John McMartin, Paula Kelly, Sammy Davis Jr., Ricardo Montalban; *Dir:* Bob Fosse. **VHS, Beta, LV $19.95** *MCA* ⅃⅃½

Sweet Country During the overthrow of the Allende government in Chile, the lives of a Chilean family and an American couple intertwine. Heavy-going propaganda is a cross between tragedy and unintentional comedy. Made by the director of "Zorba the Greek."
1987 (R) 147m/C Jane Alexander, John Cullum, Carole Laure, Franco Nero, Joanna Pettet, Randy Quaid, Irene Papas, Jean-Pierre Aumont, Pierre Vaneck, Katia Dandoulaki; *Dir:* Michael Cacoyannis. **VHS, Beta, LV $79.98** *MAG* ⅃½

Sweet Country Road A rock singer journeys to Nashville to try to cross over into country music.
1981 95m/C Buddy Knox, Kary Lynn, Gordy Trapp, Johnny Paycheck, Jeanne Pruett. **VHS, Beta $19.95** *STE, NWV* ⅃½

Sweet Creek County War Retired sheriff must leave his quiet ranch to battle a greedy businessman.

1982 90m/C Richard Egan, Albert Salmi, Nita Talbot, Slim Pickens. **VHS, Beta** *PGN* 🐾

Sweet Dirty Tony Tale about renegade CIA agent plotting to assassinate Fidel Castro. Key West scenery is pretty. Also known as "Assignment Kill Castro," "Cuba Crossing," and "The Mercenaries."
1981 90m/C Sybil Danning, Michael Gazzo, Caren Kaye, Raymond St. Jacques, Woody Strode, Robert Vaughn, Stuart Whitman; *Dir:* Chuck Workman. **VHS, Beta** $79.95 *JEF* 🐾

Sweet Dreams Biography of country singer Patsy Cline, focusing mostly on her turbulent marriage. Her quick rise to stardom ended in an early death. Fine performances throughout. Cline's original recordings are used.
1985 (PG-13) 115m/C Jessica Lange, Ed Harris, Ann Wedgeworth, David Clennon; *Dir:* Karel Reisz. **VHS, Beta, LV** $19.99 *HBO* 🐾🐾½

Sweet Ecstasy Wealthy shenanigans on the French Riviera. Pezy is the upright young man who can't sleep with anyone unless he's in love, much to Sommer's dismay.
1962 75m/B *FR* Elke Sommer, Pierre Brice, Christian Pezy, Claire Maurier; *Dir:* Max Pecas. **VHS** $29.95 *NO* 🐾½

Sweet 15 A young Hispanic girl learns that there is more to growing up than parties when she learns her father is an illegal alien. Originally aired on PBS as part of the WonderWorks Family Movie television series.
1990 120m/C Karla Montana, Panchito Gomez, Tony Plana, Jenny Gago, Susan Ruttan; *Dir:* Victoria Hochberg. **VHS** $29.95 *PME* 🐾½

Sweet Georgia Soft-core star Jordan is a lusty wife whose husband is an alcoholic, so she has a roll in the hay with almost every other cast member. Lots of action, too, including a pitchfork fight. Ride 'em cowboy.
1972 80m/C Marsha Jordan. **VHS** $19.95 *VDM* 🐾

Sweet Hearts Dance Parallel love stories follow two long-time friends, one just falling in love, the other struggling to keep his marriage together. Charming performances from all, but a slow pace undermines the film. Written by Ernest Thompson.
1988 (R) 95m/C Don Johnson, Jeff Daniels, Susan Sarandon, Elizabeth Perkins, Justin Henry; *Dir:* Robert Greenwald. **VHS, Beta, LV** $89.95 *COL* 🐾🐾

Sweet Hostage An escaped mental patient kidnaps an uneducated farm girl and holds her captive in a remote cabin. Blair and Sheen turn in good performances. Adaptation of Nathaniel Benchley's "Welcome to Xanadu." Made-for-television.
1975 93m/C Linda Blair, Martin Sheen, Jeanne Cooper, Lee DeBroux, Dehl Berti, Bert Remsen; *Dir:* Lee Philips. **VHS** *IMP* 🐾🐾½

Sweet Liberty Alda's hometown is overwhelmed by Hollywood chaos during the filming of a movie version of his novel about the American Revolution. Pleasant but predictable.
1986 (PG) 107m/C Alan Alda, Michael Caine, Michelle Pfeiffer, Bob Hoskins, Lillian Gish; *Dir:* Alan Alda. **VHS, Beta, LV** $19.95 *MCA* 🐾🐾

Sweet Lies An insurance investigator tracking a scam artist in Paris is preyed upon by a group of single women betting one another that any man can be seduced. Slow comedy.
1988 (R) 86m/C Treat Williams, Joanna Pacula, Julianne Phillips, Laura Manszky, Norbert Weisser, Marilyn Dodd-Frank; *Dir:* Nathalie Delon. **VHS, Beta** $79.98 *FOX* 🐾½

Sweet Lorraine A bittersweet, nostalgic comedy about the staff and clientele of a deteriorating Catskills hotel on the eve of its closing.
1987 (PG-13) 91m/C Maureen Stapleton, Lee Richardson, Trini Alvarado, Freddie Roman, John Bedford Lloyd, Giancarlo Esposito; *Dir:* Steve Gomer. **VHS, Beta** $79.98 *PAR* 🐾🐾🐾

Sweet Love, Bitter The downfall of a black jazz saxophonist in the 1950's. The character is loosely based on the life of legendary musician Charlie "Bird" Parker. Alto sax man Charles McPherson dubs Dick Gregory's solos.
1967 92m/C Dick Gregory, Don Murray, Diane Varsi. **VHS** $59.95 *RHP* 🐾🐾½

Sweet Movie Collection of two cult classics. In the first, a young woman's new husband sterilizes her instead of consummating their marriage. In the second, a young revolutionary has a sexual encounter that leads to murder.
197? 120m/C *Dir:* Dusan Makavejev. **VHS, Beta** $79.95 *FCT* 🐾🐾½

Sweet Perfection Jackson plans a big promotion for her beauty contest. Who will be voted the "Perfect Woman?" Tired premise, but it has its moments.
1990 (R) 90m/C Stoney Jackson, Anthony Norman McKay, Catero Colbert, Liza Crusat, Reggie Theus, Tatiana Tumbtzen. **VHS** $59.95 *XVC, AVD, HHE* 🐾🐾

Sweet Poison Bobby Stiles is a ruthless criminal who has just broken out of prison. He wants to get home and settle the score with his brother, who actually committed the crime Bobby was imprisoned for. Henry and Charlene Odell are traveling cross-country when Bobby crosses their path and forces them to drive him home. Bobby's not too busy eluding the police to notice that his captives marriage is a little shaky—and that Charlene is attractive and sexy. As Bobby and Charlene get closer husband Henry just becomes another problem.
1991 101m/C Steven Bauer, Patricia Healy, Edward Herrmann; *Dir:* Brian Grant. **VHS** $79.98 *MCA* 🐾🐾

Sweet Revenge TV journalist kidnapped by a white slavery ring in the Asian jungles. After escaping, she returns for vegeance. Meek drama.
1987 (R) 79m/C Ted Shackleford, Nancy Allen, Martin Landau; *Dir:* Mark Sobel. **VHS, Beta** $79.95 *MED* 🐾

Sweet Revenge When a judge tells a divorced couple that the woman must pay the man alimony, she hires an actress to marry her ex. The tables are turned in this way-out marriage flick, reminiscent of the screwball comedies of the 1940s.
1990 89m/C Rosanna Arquette, Carrie Fisher, John Sessions; *Dir:* Charlotte Brandstrom. **VHS, Beta** $79.98 *TTC* 🐾🐾

Sweet Sea A fantastical children's animated adventure about good and evil battling it out under the sea.
1986 30m/C **VHS, Beta** $19.98 *LIV, CVL* 🐾🐾

Sweet 16 Sixteen-year-old Melissa is beautiful, mysterious, and promiscuous, but she can't understand why all her boyfriends end up dead. No real puzzle.
1981 (R) 90m/C Susan Strasberg, Bo Hopkins, Don Stroud, Dana Kimmell, Patrick Macnee, Larry Storch; *Dir:* Jim Sotos. **VHS, Beta** $69.95 *VES* 🐾

Sweet Smell of Success Evil New York City gossip columnist J.J. Hunsecker (Lancaster) and sleazy press agent (Curtis) cook up smear campaign to ruin career of jazz musician in love with J.J.'s sister. Engrossing performances, great dialogue. Music by Oscar winner Elmer Bernstein.
1957 96m/C Burt Lancaster, Tony Curtis, Martin Milner, Barbara Nichols, Sam Levene; *Dir:* Alexander MacKendrick. **VHS, LV** $19.98 *MGM* 🐾🐾🐾

Sweet Smell of Woman An acclaimed dark comedy that may be an acquired taste for some. A blinded military officer and his valet take a sensual tour of Italy, the sightless man seducing beautiful women on the way. But at the end of the journey awaits a shock, and the real point of the tale, based on a novel by Giovanni Arpino. Also known as "Scent of a Woman." In Italian with English subtitles.
1975 103m/C *IT* Vittorio Gassman; *Dir:* Dino Risi. **VHS** $59.95 *FST* 🐾🐾½

Sweet Spirits Soft-sell European fluff about a modern mannequin having a raucous affair with an artist.
198? 87m/C Erika Blanc, Farley Granger. **VHS, Beta** $29.95 *MED Woof!*

Sweet Sugar Slave girls try to escape from a Costa Rican sugar cane plantation and its cruel owner. Needless to say, vulgar exploitation runs hither and yon in this film. Also known as "Chaingang Girls."
1972 (R) 90m/C Phyllis E. Davis, Ella Edwards, Pamela Collins, Cliff Osmond, Timothy Brown; *Dir:* Michel Levesque. **VHS** $49.98 *CON* 🐾

Sweet Sweetback's Baadasssss Song A black pimp kills two policemen who beat up a black militant. He uses his streetwise survival skills to elude his pursuers and escape to Mexico. A thriller, but racist, sexist, and violent.
1971 97m/C Melvin Van Peebles, Simon Chuckster, John Amos; *Dir:* Melvin Van Peebles. **VHS, Beta** $59.98 *SUN, MAG* 🐾🐾🐾

Sweet Talker Following his release from prison, a charming con man shows up in a small coastal village, thinking the townsfolk will be ripe for the picking. What he doesn't know is they can do some sweet talking of their own, and he soon finds himself caring for a pretty widow and her son. Enjoyable light comedy works thanks to likable leads.
1991 (PG) 91m/C *AU* Bryan Brown, Karen Allen, Chris Haywood, Bill Kerr, Bruce Spence, Bruce Myles, Paul Chubb, Peter Hehir, Justin Rosniak; *Dir:* Michael Jenkins. **VHS, LV** $89.98 *LIV* 🐾🐾½

Sweet Trash A vice cop, having an affair with a 17-year-old prostitute, is blackmailed by an underworld figure. Lives up to its name.
1989 (R) 80m/C Sebastian Gregory, Sharon Matt, Luke Perry, Bonnie Clark, Gene Blackey; *Dir:* John Hayes. **VHS** $29.95 *AVD Woof!*

Sweet William A philandering and seemingly irresistible young man finds that one sensitive woman hasn't the patience or time for his escapades. Adult comedy concerned with sex without displaying any on the screen.
1979 (R) 88m/C *GB* Sam Waterston, Jenny Agutter, Anna Massey, Arthur Lowe; *Dir:* Claude Whatham. **VHS, Beta** $59.95 *PSM* 🐾🐾🐾

Sweet Young Thing Young students at a French girls' school receive all sorts of attention from the male staff.
19?? 83m/C Jean Tolzac, Bernard Musson. **VHS, Beta** $39.95 *LUN* 🐾🐾🐾

Sweetheart! Biography of Hal Banks, convicted for strong-arming shipping unions in an effort to destroy them.
19?? 115m/C Maury Chaykin. **VHS** $69.95 *VSV* ♟♟½

Sweethearts MacDonald and Eddy star as married stage actors trying to get some time off from their hectic schedule in this show-within-a-show. Trouble ensues when their conniving producer begs, pleads and tricks them into staying. Lots of well-staged musical numbers.
1938 114m/C Jeanette MacDonald, Nelson Eddy, Frank Morgan, Florence Rice, Ray Bolger, Mischa Auer; *Dir:* W.S. Van Dyke. **VHS** $29.95 *MGM* ♟♟♟

Sweetie Bizarre, expressive Australian tragicomedy about a pair of sisters—one a withdrawn, paranoid Plain Jane, the other a dangerously extroverted, overweight sociopath who re-enters her family's life and turns it upside down. Campion's first feature.
1989 (R) 97m/C *AU* Genevieve Lemon, Karen Colston, Tom Lycos, Jon Darling, Dorothy Barry, Michael Lake, Andre Pataczek; *Dir:* Jane Campion. **VHS, Beta, LV** $89.95 *IVE* ♟♟♟

Sweetwater An independent woman and a feisty young orphan find each other amid society's rejections.
1984 30m/C **VHS, Beta** *LCA* ♟♟

Swept Away... A rich and beautiful Milanese woman is shipwrecked on a desolate island with a swarthy Sicilian deck hand, who also happens to be a dedicated communist. Isolated, the two switch roles, with the wealthy woman dominated by the crude proletarian. Sexy and provocative. Scripted by Wortmuller.
1975 (R) 116m/C *IT* Giancarlo Giannini, Mariangela Melato; *Dir:* Lina Wertmuller. National Board of Review Awards '75: 5 Best Foreign Films of the Year. **VHS, Beta, LV** $59.95 *COL, APD* ♟♟♟

Swift Justice Mayor of a small town thinks he's gotten away with murder after leaving a young girl for dead in a junk yard. However, the mayor and his violent pals are soon hunted by an ex-Green Beret with a penchant for justice.
1988 90m/C Jon Greene, Cindy Rome, Cameron Mitchell, Aldo Ray, Chuck "Porky" Mitchell, Wilson Dunster, Ted Leplat; *Dir:* Harry Hope. **VHS, Beta** $79.95 *TWE* ♟♟

Swifty A cowpoke is accused of murder but escapes to track down the real culprit and win the hand of the girl he loves.
1936 60m/B Hoot Gibson. **VHS, Beta** $27.95 *VCN* ♟

Swim Team New coach for a terminally inept school swimming team whips them into shape. Doesn't hold much water.
1979 81m/C Stephen Furst, James Daughton, Jenny Neumann, Kim Day, Buster Crabbe; *Dir:* James Polakof. **VHS, Beta** $9.99 *STE, PSM* ♟

Swimmer A lonely suburbanite swims an existential swath through the pools of his neighborhood landscape in an effort at self-discovery. A surreal, strangely compelling work based on a story by John Cheever.
1968 (PG) 94m/C Burt Lancaster, Janice Rule, Janet Landgard; *Dir:* Frank Perry. **VHS, Beta, LV** $9.95 *GKK* ♟♟♟½

Swimming to Cambodia Gray tells the story of his bit part in "The Killing Fields," filmed in Cambodia, and makes ironic observations about modern life. It works. Music by Laurie Anderson.

1987 87m/C Spalding Gray; *Dir:* Jonathan Demme. **VHS, Beta, LV** $79.95 *LHV, WAR* ♟♟♟

Swimming Pool Two men and two women spend a weekend in a villa on the French Riviera, manipulating each other, playing sexual games and changing partners. Their escapades end in murder. Romantic melodrama is sensuous if slow-moving. French film dubbed in English.
1970 (PG) 85m/C *FR* Romy Schneider, Alain Delon, Maurice Ronet, Jane Birkin; *Dir:* Jacques Deray. Int'l. Rio de Janeiro Film Festival '70: Best Director. **VHS, Beta** $59.95 *CHA* ♟♟½

Swimsuit: The Movie A young ad executive decides to revitalize a swimsuit company's failing business by sponsoring a contest for the perfect swimsuit model. Made-for-television fluff.
1989 100m/C William Katt, Catherine Oxenberg, Cyd Charisse, Nia Peeples, Tom Villard, Cheryl Pollak, Billy Warlock, Jack Wagner; *Dir:* Chris Thomson. **VHS** $79.95 *AIP* ♟½

The Swindle Three Italian conmen pull capers in Rome trying to make a better life for themselves. Dark overtones permeate one of Fellini's lesser efforts. Good cast can't bring up the level of this film. Also known as "Il Bidone."
1955 92m/C *IT* Broderick Crawford, Giulietta Masina, Richard Basehart, Franco Fabrizi; *Dir:* Federico Fellini. **VHS** *IME* ♟♟

Swing An all-black musical revue.
1947 80m/B **VHS, Beta** $29.95 *NOS, DVT* ♟♟

Swing High, Swing Low A trumpet player fights the bottle and the dice to become a hit in the jazz world and marry the woman he loves. Solid drama. From the stage play "Burlesque." Made first as "The Dance of Life," then as "When My Baby Smiles At Me."
1937 95m/B Carole Lombard, Fred MacMurray, Charles Butterworth, Dorothy Lamour; *Dir:* Mitchell Leisen. **VHS, Beta** $19.95 *NOS, HHT, DVT* ♟♟½

Swing It, Professor A stodgy music professor refuses to recognize jazz as a valid musical form and subsequently loses his job. During his unemployment, he wanders into a nightclub and discovers jazz isn't so bad after all. Nice camera work.
1937 62m/B Pinky Tomlin, Paula Stone, Mary Kornman, Milburn Stone, Pat Gleason; *Dir:* Maurice Conn. **VHS** $24.95 *NOS, DVT* ♟♟½

Swing It, Sailor! Two footloose sailors go after the same woman while on shore leave. Decent screwball comedy.
1937 61m/B Wallace Ford, Ray Mayer, Isabel Jewell, Mary Treen; *Dir:* Raymond Connon. **VHS, Beta** $24.95 *NOS, VYY* ♟♟

Swing Parade of 1946 One of the multitude of 1940s musicals. Young songwriter falls in love with club owner. Highlights include an appearance by the Three Stooges and Louis Jordan performing "Caledonia."
1946 74m/B Gale Storm, Phil Regan, Moe Howard, Edward Brophy, Connee Boswell; *Dir:* Phil Karlson. **VHS** *MNE* ♟♟

Swing Shift When Hawn takes a job at an aircraft plant after her husband goes off to war, she learns more than riveting. Lahti steals the film as her friend and co-worker. A detailed reminiscence of the American home front during World War II, that never seems to gel. Produced by Hawn.

1984 (PG) 100m/C Goldie Hawn, Kurt Russell, Ed Harris, Christine Lahti, Holly Hunter, Chris Lemmon, Belinda Carlisle, Fred Ward, Roger Corman, Lisa Pelikan; *Dir:* Jonathan Demme. **VHS, Beta, LV** $19.98 *WAR* ♟♟½

Swing Time Astaire, a dancer who can't resist gambling, is engaged to marry another woman, until he meets Ginger. One of the team's best efforts. The score by Jerome Kern and Dorothy Fields includes "Pick Yourself Up," "Never Gonna Dance," "The Way You Look Tonight" (which won an Oscar), and "A Fine Romance." Laserdisc version includes: production photos and stills, commentary on Fred Astaire and musical films, and excerpts from "Hooray for Love" featuring Bill "Bojangles" Robinson and Fats Waller.
1936 103m/B Fred Astaire, Ginger Rogers, Helen Broderick, Betty Furness, Eric Blore, Victor Moore; *Dir:* George Stevens. Academy Awards '36: Best Song ("The Way You Look Tonight"). **VHS, Beta, LV** $14.98 *TTC, RKO, MED* ♟♟♟

A Swingin' Summer A bunch of groovy guys and gals go-go to the beach for a vacation of fun and sun and end up starting a rock 'n' roll concert series. Everybody gets part of the action...even the bespectacled Raquel Welch, who gets a chance to sing in her film debut. Performances from the Righteous Brothers and rock semi-legends Donnie Brooks, Gary Lewis and the Playboys, and the Rip Chords.
1965 81m/C James Stacy, William Wellman Jr., Quinn O'Hara, Martin West, Mary Mitchell, Allan Jones, Raquel Welch; *Dir:* Robert Sparr. **VHS** $14.95 *AVD* ♟½

The Swinging Cheerleaders A group of amorous cheerleaders turn on the entire campus in this typical sexploiter.
1974 (PG) 90m/C Cheryl "Rainbeaux" Smith, Colleen Camp, Jo Johnston; *Dir:* Jack Hill. **VHS, Beta** $29.95 *MON* ♟½

Swinging Ski Girls A group of free-loving university co-eds spend a wild weekend at a ski lodge.
197? 85m/C Cindy Wilton, Dick Cassidy. **VHS, Beta** $44.95 *MED* ♟

Swinging Sorority Girls An intimate glimpse behind the closed doors of a sorority house during a wild homecoming weekend.
197? 73m/C Susie Carlson, Anne Marlie. **VHS, Beta** *MED* ♟

Swinging Wives Ribald humor for the discriminating viewer.
1979 (R) 84m/C Gale Mayberrie, Ron James. **VHS, Beta** $39.95 *WES* ♟

Swiss Conspiracy Against the opulent background of the world's richest financial capital and playground of the wealthy, one man battles to stop a daring and sophisticated blackmailer preying on the secret bank account set. Fast-paced, if sometimes confusing.
1977 (PG) 92m/C *GE* David Janssen, Senta Berger; *Dir:* Jack Arnold. **VHS, Beta** $49.95 *UHV* ♟♟

Swiss Family Robinson A family, seeking to escape Napoleon's war in Europe, sets sail for New Guinea, but shipwrecks on a deserted tropical island. There they build an idyllic life, only to be confronted by a band of pirates. Lots of adventure for family viewing. Filmed on location on the island of Tobago. Based on the novel by Johann Wyss.

1960 126m/C John Mills, Dorothy McGuire, James MacArthur, Tommy Kirk, Janet Munro, Sessue Hayakawa; *Dir:* Ken Annakin. **VHS, Beta, LV** $19.99 DIS ♂♂♂

The Swiss Family Robinson An animated version of the classic adventure epic.
1972 49m/C **VHS, Beta** $19.95 MGM ♂♂♂

Swiss Miss Stan and Ollie are mousetrap salesmen on the job in Switzerland. Highlights include Stan's tuba serenade and the gorilla-on-the-bridge episode. Also included on this tape is a 1935 Thelma Todd/Patsy Kelly short, "Hot Money."
1938 97m/B Stan Laurel, Oliver Hardy, Della Lind, Walter Woolf King, Eric Blore, Thelma Todd, Patsy Kelly; *Dir:* John Blystone. **VHS, Beta** $14.98 CCB, MED, FCT ♂♂½

Switch An episode of the television series "The Persuaders," in which Sinclair and Wilde are assigned to protect an ex-union-boss congressional witness.
1977 102m/C Roger Moore, Tony Curtis; *Dir:* Val Guest, Roy Ward Baker. **VHS, Beta** $59.98 FOX ♂♂

The Switch An escaped convict takes the place of a talented conman and chaos ensues.
1989 (R) ?m/C Anthony Quinn, Adriano Celentano, Capucine, Corinne Clery. **VHS** $12.95 AVD ♂♂

Switch A chauvinist louse, slain by the girlfriends he misused, is sent back to Earth as an alluring female to learn the other side's point of view. The plot may lack urgency, but this is a sparkling adult comedy that scores as it pursues the gimmicky concept to a logical, outrageous and touching conclusion. Barkin's act as swaggering male stuck in a woman's body is a masterwork of physical humor.
1991 (R) 104m/C Ellen Barkin, Jimmy Smits, JoBeth Williams, Lorraine Bracco, Perry King; *W/ Dir:* Blake Edwards. **VHS, LV** $92.99 HBO, PIA, WAR ♂♂♂

Switchblade Sisters A crime gang of female ex-cons wreaks psuedo-feminist havoc. Kitty Bruce is Lenny Bruce's daughter. Also known as "The Jezebels" and "Playgirl Gang."
1975 90m/C Robbie Lee, Joanne Nail, Monica Gayle, Kitty Bruce; *Dir:* Jack Hill. **VHS, Beta** $59.95 MON, HHE Woof!

Switching Channels A modernized remake of "His Girl Friday," and therefore the 4th version of "The Front Page." Beautiful television anchorwoman (Turner) wants to marry handsome tycoon (Reeve) but her scheming ex-husband (Reynolds) gets in the way. Weak performances from everyone but Beatty and lackluster direction render this oft-told story less funny than usual.
1988 (PG) 108m/C Burt Reynolds, Kathleen Turner, Christopher Reeve, Ned Beatty, Henry Gibson; *Dir:* Ted Kotcheff. **VHS, Beta, LV** $19.95 COL ♂♂½

Sword & the Cross A grade-B adventure about the life of Mary Magdelene.
1958 93m/B Yvonne de Carlo, George Mistral; *Dir:* Carlo L. Bragaglia. **VHS, Beta** $29.95 FOR ♂½

Sword of Doom A rousing samurai epic detailing the training of an impulsive, blood-lusting warrior by an elder expert. Subtitled in English.
1967 120m/B JP Tatsuya Nakadai, Toshiro Mifune; *Dir:* Kihachi Okamoto. **VHS, Beta** $29.98 SUE ♂♂♂

Sword & the Dragon A young warrior must battle any number of giant mythical creatures.
1956 81m/C RU Boris Andreyer, Andrei Abrikosov; *Dir:* Alexander Ptushko. **VHS, Beta** $59.95 UHV, SNC ♂♂

Sword of Fury I How Musashi Miyamoto became Japan's greatest samurai. With English subtitles. Followed by a sequel.
1973 90m/C JP Hideki Takahashi, Jiro Tamiya. **VHS, Beta** VDA ♂½

Sword of Fury II Two samurai fight to the death to determine who will become the premier swordsman of all Japan. With English subtitles.
1973 77m/C JP Hideki Takahashi, Jiro Tamiya. **VHS, Beta** VDA ♂½

Sword of Gideon An action packed and suspenseful made for television film about an elite commando group who set out to avenge the Munich Olympic killings of 1972. Adapted from "Vengeance" by George Jonas.
1986 148m/C Steven Bauer, Michael York, Rod Steiger, Colleen Dewhurst, Robert Joy, Laurent Malet, Lino Ventura, Leslie Hope; *Dir:* Michael Anderson Sr. **VHS** $79.99 HBO ♂♂½

Sword of Heaven Forged from a meteorite in the fifteenth century, a sword holds magical powers. When it is stolen, a young warrior pursues the villains.
1981 87m/C JP Tadashi Yamashita. **VHS, Beta** $69.95 TWE ♂

Sword of Lancelot A costume version of the Arthur-Lancelot-Guinevere triangle, with plenty of swordplay and a sincere respect for the old legend. Originally titled "Lancelot and Guinevere."
1963 115m/C GB Cornel Wilde, Jean Wallace, Brian Aherne, George Baker; *Dir:* Cornel Wilde. **VHS, Beta** $59.95 MCA ♂♂½

Sword of Monte Cristo A masked female avenger, seeking the liberty of the French people, discovers a valuable sword that holds the key to a vast fortune. But the sword is also desired by an evil government minister.
1951 80m/C George Montgomery, Paula Corday; *Dir:* Maurice Geraghty. **VHS, Beta** $29.95 VHE ♂½

Sword & the Rose Mary Tudor, sister of King Henry VIII, shuns the advances of a nobleman for the love of a commoner. Johns is an obstinate princess; Justice, a fine king. Based on the book "When Knighthood was in Flower."
1953 91m/C Richard Todd, Glynis Johns, Michael Gough, Jane Barrett, James Robertson Justice; *Dir:* Ken Annakin. **VHS, Beta** $69.95 DIS, HHE ♂♂½

Sword of Sherwood Forest The Earl of Newark plots the murder of the Archbishop of Canterbury in yet another rendition of the Robin Hood legend. Peter Cushing is truly evil as the villain.
1960 80m/C GB Richard Greene, Peter Cushing, Niall MacGinnis, Richard Pasco, Jack Gwillim, Sarah Branch, Nigel Green, Oliver Reed; *Dir:* Terence Fisher. **VHS** COL, MLB ♂♂½

Sword & the Sorcerer A young prince strives to regain control of his kingdom, now ruled by an evil knight and a powerful magician. Mediocre script and acting is enhanced by decent special effects.
1982 (R) 100m/C Lee Horsely, Kathleen Beller, George Maharis, Simon MacCorkindale, Richard Lynch, Richard Moll, Robert Tessier, Nina Van Pallandt, Anna Bjorn, Jeff Corey; *Dir:* Albert Pyun. **VHS, Beta, LV** $24.95 MCA ♂½

Sword in the Stone The Disney version of the first volume of T. H. White's "The Once and Future King" wherein King Arthur, as a boy, is instructed in the ways of the world by Merlin and Archimedes the owl. Although not in the Disney masterpiece fold, boasts the usual superior animation and a gripping mythological tale.
1963 (G) 79m/C *Dir:* Wolfgang Reitherman, Wolfgang Reitherman; *Voices:* Ricky Sorenson, Sebastian Cabot, Karl Swenson, Junius Matthews, Alan Napier, Norman Alden, Martha Wentworth, Barbara Jo Allen. **VHS, Beta, LV** $24.99 DIS, KUI ♂♂♂

Sword of the Valiant The Green Knight arrives in Camelot to challenge Gawain. Remake of "Gawain And The Green Knight." Connery adds zest to a minor epic.
1983 (PG) 102m/C GB Sean Connery, Miles O'Keefe, Cyrielle Claire, Leigh Lawson, Trevor Howard, Peter Cushing, Wilfrid Brambell, Lila Kedrova, John Rhys-Davies; *Dir:* Stephen Weeks. **VHS, Beta** $69.95 MGM ♂♂

Sword of Venus Uninspired production finds the son of the Count of Monte Cristo the victim in a scheme to deprive him of his wealth.
1953 73m/B Robert Clarke, Catherine McLeod, Dan O'Herlihy, William Schallert; *Dir:* Harold Daniels. **VHS** $29.95 FCT ♂

Swords of Death Final film by Uchida, released posthumously, deals with the adventurous period exploits of the irrepressible Miyamoto Musashi. With English subtitles.
1971 76m/C JP Kinnosuke Nakamura, Rentaro Mikuni; *Dir:* Tomu Uchida. **VHS, Beta** VDA ♂♂

Sybil Fact-based story of a woman who developed 16 distinct personalities, and the supportive psychiatrist who helped her put the pieces of her ego together. Excellent made-for-television production with Field's Emmy-winning performance.
1976 122m/C Sally Field, Joanne Woodward, Brad Davis, Martine Bartlett, Jane Hoffman; *Dir:* Daniel Petrie. Emmy Awards '77: Outstanding Special (Drama). **VHS, Beta** $19.98 FOX ♂♂♂

Sykes Eric Sykes and the denizens of Sebastopol Terrace in three episodes from long-running BBC television show. Peter Sellers makes a guest appearance.
19?? 88m/C GB Peter Sellers. **VHS** $29.95 BFS ♂♂

Sylvester A sixteen-year-old girl and a cranky stockyard boss team up to train a battered horse named Sylvester for the National Equestrian trials. Nice riding sequences and good performances can't overcome familiar plot.
1985 (PG) 104m/C Melissa Gilbert, Richard Farnsworth, Michael Schoeffling, Constance Towers; *Dir:* Tim Hunter. **VHS, Beta, LV** $79.95 COL ♂♂

Sylvester & Tweety's Crazy Capers Sylvester keeps on the tail of Tweety Pie in this collection of eight classic cartoons that include "Tweet and Lovely," "Tree for Two," and "Hyde and Go Tweet."
1961 54m/C *Dir:* Friz Freleng, Chuck Jones; *Voices:* Mel Blanc. **VHS, Beta** $12.95 WAR ♂♂

Sylvia Biography of author/educator Sylvia Ashton-Warner, and her efforts to teach Maori children in New Zealand to read via unorthodox methods. David is a marvel. Based on Ashton-Warner's books "Teacher" and "I Passed This Way."
1986 (PG) 99m/C NZ Eleanor David, Tom Wilkinson, Nigel Terry, Mary Regan; *Dir:* Michael Firth. **VHS, Beta** $79.98 FOX ♂♂½

Sylvia and the Phantom A charming French comedy about a lonely young girl who is befriended by the ghost of a man killed decades earlier in a duel over the girl's grandmother. Based on the play by Alfred Adam. Subtitled in English.
1945 97m/B *FR* Odette Joyeux, Jacques Tati, Francois Perier; *Dir:* Claude Autant-Lara. **VHS, Beta** $29.98 *SUE* ✂✂✂

Sylvia Scarlett An odd British comedy about a woman who masquerades as a boy while on the run with her father, who causes all kinds of instinctive confusion in the men she meets.
1935 94m/B Katharine Hepburn, Cary Grant, Brian Aherne, Edmund Gwenn, Natalie Paley, Dennis Moore, Lennox Pawle, Daisy Belmore, Nola Luxford; *Dir:* George Cukor. **VHS, Beta, LV** $19.95 *RKO, FCT* ✂✂ ½

Sympathy for the Devil Rolling Stones provide music for confused, revolutionary documentary. Episodic jumble is odd, sometimes fascinating. Also known as "One Plus One."
1970 110m/C *FR* Mick Jagger, Rolling Stones; *Dir:* Jean-Luc Godard. **VHS, Beta** *FOX, OM* ✂✂

Symphony of Living A heart-warming drama about a man's bitter betrayal by his family and the hardship and tragedy he experiences.
1935 73m/B **VHS, Beta** $29.95 *VYY Woof!*

Symphony in the Mountains An early German sound romance, where an Austrian singing teacher falls in love with the sister of one of his student's. Featuring choral work by the Vienna Boys' Choir. In German with English subtitles.
1935 75m/B *GE Voices:* The Vienna Boys Choir. **VHS, Beta, 3/4U** $35.95 *FCT, IHF, GLV* ✂✂ ½

Symphony of Swing Four classic swing bands are featured in five film shorts: "Symphony of Swing," "Artie Shaw's Class in Swing" (both 1939), "Artistry in Rhythm" (1945), "Charlie Barnet and His Orchestra" and "Buddy Rich and His Orchestra" (both 1948).
1986 57m/B Artie Shaw, Stan Kenton, Charlie Barnet, Buddy Rich, Helen Forrest, Tony Pastor, June Christy. **VHS, Beta** $19.95 *MVD, AFE, AMV* ✂✂ ½

Syngenor Syngenor stands for Synthesized Genetic Organism, to differentiate it from all other organisms. Created by science, it escapes, and a crack team of scientists and gung-ho military types are mobilized to track it down.
1990 (R) 98m/C Starr Andreeff, Michael Laurence, David Gale, Charles Lucia, Riva Spier, Jeff Doucette, Bill Gratton, Lewis Arquette, Jon Korkes, Melanie Shatner; *Dir:* George Elanjian Jr. **VHS** $14.95 *HMD, SOU* ✂✂ ½

T-Bird Gang A high school student goes undercover to infiltrate a teen gang. When his cover is blown, things get hairy. Laughable sleazebag production.
1959 75m/B Ed Nelson, John Brinkley, Pat George, Beach Dickerson, Tony Miller; *Dir:* Richard Harbinger. **VHS** $19.98 *SNC* ✂✂ ½

T-Men Two agents of the Treasury Department infiltrate a counterfeiting gang. Filmed in semi-documentary style, exciting tale serves also as an effective commentary on the similarities between the agents and those they pursue. Mann and cinematographer John Alton do especially fine work here.

1947 96m/B Alfred Ryder, Dennis O'Keefe, June Lockhart, Mary Meade, Wallace Ford, Charles McGraw; *Dir:* Anthony Mann. **VHS, Beta, LV** $29.95 *MED* ✂✂✂

Table for Five A divorced father takes his children on a Mediterranean cruise and while sailing, he learns that his ex-wife has died. The father and his ex-wife's husband struggle over who should raise the children. Sentimental and well-acted.
1983 (PG) 120m/C Jon Voight, Millie Perkins, Richard Crenna, Robbie Kiger, Roxana Zal, Son Hoang Bui, Marie-Christine Barrault, Kevin Costner; *Dir:* Robert Lieberman. **VHS, Beta, LV** $59.98 *FOX* ✂✂ ½

Table Settings A taped performance of James Lapines' comedy about the lives of three generations of a Jewish family.
1984 90m/C Robert Klein, Stockard Channing, Dinah Manoff, Eileen Heckart. **VHS, Beta** $39.95 *RKO* ✂✂

Tabloid! Unorthodox woman prints and sells racy trash for enquiring minds.
19?? ?m/C Scott Davis, Glen Cobum; *Dir:* Bret McCormick, Matt Shaffen. **VHS** *TPV, HHE* ✂ ½

Tabu Fascinating docudrama about a young pearl diver's ill-fated romance. The gods have declared the young woman he desires "taboo" to all men. Filmed on location in Tahiti. Authored and produced by Murnau and Flaherty. Flaherty left in mid-production due to artistic differences with Murnau, who was later killed in an auto accident one week before the premiere of film. Oscar-winning cinematography by Floyd Crosby.
1931 81m/B *Dir:* Robert Flaherty, F. W. Murnau. Academy Awards '31: Best Cinematography; National Board of Review Awards '31: 10 Best Films of the Year. **VHS, Beta** $39.95 *MIL, GVV* ✂✂✂ ½

Tacos al Carbon A singer falls in love with a beautiful woman.
1987 103m/C *MX* Vicente Fernandez, Nadia Milton, Sonia Amelio, Ana Martin, Sergio Ramos. **VHS, Beta** $59.95 *MAD* ✂✂

Taffin A local bill collector battles a group of corrupt businessmen who want to replace the local soccer field with a hazardous chemical plant. Pretty dull fare, overall, with Brosnan much less charismatic than usual.
1988 (R) 96m/C Pierce Brosnan, Ray McAnally, Alison Doody; *Dir:* Francis Megahy. **VHS, Beta, LV** $79.95 *MGM* ✂ ½

Tag: The Assassination Game An exciting new game has been discovered by a group of university students. Players act as spies and assassins, "killing" each other to make points. But a school newspaper sportswriter finds the game has become more real than make-believe.
1982 (PG) 92m/C Robert Carradine, Linda Hamilton, Michael Winslow, Kristine DeBell, Perry Lang; *Dir:* Nick Castle. **VHS, Beta** $59.98 *SUE* ✂ ½

Tagget A former CIA operative who has erased his nightmarish stint in Vietnam gradually begins to recall his past. When pieces of the puzzle begin to come together, he remembers a few things that the government would prefer he forget. Soon, CIA men are on his trail and their mission is to see that he forgets his past - permanently. Made for cable television.
1990 (PG-13) 89m/C Daniel J. Travanti, William Sadler, Roxanne Hart, Stephen Tobolowsky, Peter Michael Goetz, Sarah Douglas, Noel Harrison; *Dir:* Richard T. Heffron. **VHS** $79.95 *MCA* ✂

Tailspin: Behind the Korean Airline Tragedy Made-for-TV docudrama on the 1983 downing of Korean Aire flight 007 in Soviet airspace. Purports to show the U.S. government's reaction to what became a major international incident.
1989 82m/C Michael Moriarty, Michael Murphy, Chris Sarandon, Harris Yulin, Gavan O'Herlihy, Ed O'Ross; *Dir:* David Darlow. **VHS, Beta, LV** *PSM, PAR* ✂✂

Tainted Small-town woman tries to endure after being raped. Both her husband and the attacker are killed, and she is forced to conceal their deaths. Poorly scripted.
1988 90m/C Shari Shattuck, Park Overall, Gene Tootle, Magilla Schaus, Blaque Fowler; *Dir:* Orestes Matacena. **VHS, Beta** *SOU* ✂ ½

Tainted Image Artist woman starts to lose sanity with help of boyfriend, psycho neighbor and rash of bizarre deaths. She turns to canvas for solace and decides she needs secret ingredient that can't be found in art supply store to complete masterpiece.
1991 95m/C Tom Saunders, Sandra Frances, Annetta Arpin, Ken La Mothe, Heidi Emerich, Steve Kornacki. **VHS** $79.95 *HHT, IPI, PHX* ✂✂

Taipan A 19th century Scottish trader and his beautiful Chinese mistress are the main characters in this confusing attempt to dramatize the story of Hong Kong's development into a thriving trading port. Too many subplots and characters are introduced in a short time to do justice to James Clavell's novel of the same name, the basis for the movie. The first American production completely filmed in China.
1986 (R) 130m/C Joan Chen, John Stanton, Kyra Sedgwick, Tim Guinee, Bryan Brown; *Dir:* Daryl Duke. **VHS, Beta** $79.98 *VES* ✂ ½

Takanaka: World As Masayoshi Takanaka performs his greatest hits, vivid visuals bring alive his daydreams.
1981 60m/C *JP* Masayoshi Takanaka. **VHS, Beta, LV** $19.95 *PAR, PNR*

The Take Recovering alcoholic cop faces wrath of drug lord and marriage on the rocks.
1990 (R) 91m/C Ray Sharkey, R. Lee Ermey, Lisa Hartman, Larry Manetti, Joe Lala; *Dir:* Leon Ichaso. **VHS** $79.95 *MCA* ✂

Take Down Hermann is charming as a high school English teacher turned reluctant wrestling coach.
1979 96m/C Lorenzo Lamas, Kathleen Lloyd, Maureen McCormack, Edward Herrmann; *Dir:* Keith Merrill. **VHS, Beta** $29.95 *UNI* ✂✂

Take a Hard Ride Dull spaghetti western about a cowboy transporting money to Mexico, evading bandits and bounty hunters.
1975 (R) 103m/C Jim Brown, Lee Van Cleef, Fred Williamson, Jim Kelly, Barry Sullivan; *Dir:* Anthony M. Dawson. **VHS, Beta** $19.98 *FOX* ✂ ½

Take It Big Uninspired B-musical starring the Tin Man from the "Wizard of Oz." Haley is an impoverished actor who inherits a ranch and saves it by—surprise—putting on a show. Songs include "Take It Big," "Love and Learn," "Sunday, Monday, and Always," and "I'm a Big Success with You."
1944 75m/B Jack Haley, Harriet Nelson, Mary Beth Hughes, Arlene Judge, Nils T. Granlund, Al "Fuzzy" Knight, Ozzie Nelson; *Dir:* Frank McDonald. **VHS, Beta** $19.95 *NOS, VYY, DVT* ✂ ½

Take Me Back to Oklahoma Musical western with comedy.
1940 64m/B Tex Ritter, Terry Walker, Karl Hackett, George Eldridge; *Dir:* Al Herman. **VHS, Beta** $19.95 *NOS, VCN, VYY* ✂

Take Me Out to the Ball Game Williams manages a baseball team, locks horns with players Sinatra and Kelly, and wins them over with song. Naturally, there's a water ballet scene. Contrived and forced, but enjoyable.
1949 93m/C Frank Sinatra, Gene Kelly, Esther Williams, Jules Munshin, Betty Garrett, Edward Arnold, Tom Dugan, Richard Lane; *Dir:* Busby Berkeley. **VHS, Beta, LV $19.95** MGM *♪♪½*

Take the Money and Run Allen's directing debut; he also co-wrote and starred. "Documentary" follows a timid, would-be bank robber who can't get his career off the ground and keeps landing in jail. Little plot, but who cares? Nonstop one-liners and slapstick.
1969 (PG) 85m/C Woody Allen, Janet Margolin, Marcel Hillaire, Louise Lasser; *Dir:* Woody Allen. **VHS, Beta, LV $19.98** FOX *♪♪♪*

Take This Job & Shove It The Johnny Paycheck song inspired this story of a hot-shot efficiency expert who returns to his hometown to streamline the local brewery. Encounters with old pals inspire self-questioning. Alternately inspired and hackneyed. Cameos by Paycheck and other country stars.
1981 (PG) 100m/C Robert Hays, Art Carney, Barbara Hershey, David Keith, Martin Mull, Eddie Albert, Penelope Milford; *Dir:* Gus Trikonis. **VHS, Beta $14.95** COL, SUE *♪♪*

Take Two A woman has an affair with her husband's twin brother. Interesting metaphysically: Are you my husband or my lover? But is an adultery/murder drama with Frank Stallone worth 101 minutes of your time?
1987 (R) 101m/C Grant Goodeve, Robin Mattson, Frank Stallone, Warren Berlinger, Darwyn Swalve, Nita Talbot; *Dir:* Peter Rowe. **VHS, Beta, LV $29.95** ACA *♪*

Take Your Best Shot Standard TV fare about an out-of-work actor struggling with his low self-esteem and his failing marriage. Made-for-television comedy.
1982 96m/C Robert Urich, Meredith Baxter Birney, Jeffrey Tambor, Jack Bannon, Claudette Nevins; *Dir:* David Greene. **VHS, Beta $39.95** IVE *♪½*

Takin' It All Off In this sequel to "Takin' It Off," Kitten does it again with unbelievable bumps and grinds.
1987 90m/C Francesca "Kitten" Natavidad; *Dir:* Ed Hansen. **VHS $69.98** LIV, VES *♪½*

Takin' It Off An exotic dancer's agent advises her to trim down her well-endowed figure in order to become a serious actress.
1984 90m/C Francesca "Kitten" Natavidad, Adam Hadum, Ashley St. John, Angelique Pettyjohn. **VHS, Beta, LV $69.98** LIV, VES Woof!

The Taking of Beverly Hills Deranged billionaire Bat Masterson designs a bogus toxic spill that leaves the wealth of Beverly Hills his for the taking. Football hero Boomer Hayes teams up with renegade cop Ed Kelvin to try and stop Masterson and save the city of Beverly Hills.
1991 (R) 96m/C Ken Wahl, Matt Frewer, Robert Davi, Harley Jane Kozak. **VHS** COL *♪½*

Taking Care of Business Too familiar tale of switched identity has crazed Cubs fan Belushi find Grodin's Filofax, allowing him to pose as businessman. Old jokes and a story full of holes.
1990 (R) 108m/C James Belushi, Charles Grodin, Anne DeSalvo, Loryn Locklin, Veronica Hamel, Hector Elizondo, Mako, Gates McFadden, Stephen Elliott; *Dir:* Arthur Hiller. **VHS, Beta, LV $19.99** BVV *♪½*

Taking Care of Terrific A young babysitter takes the naive boy she is caring for out into the streets to visit with a street musician. The trio plan an evening to help the homeless in this entry in the "Wonderworks" series.
1987 55m/C Melvin Van Peebles, Joanne Vannicola. **VHS $29.95** PME *♪½*

The Taking of Flight 847: The Uli Derickson Story Wagner plays the stewardess who acted as a go-between for the passengers during a 1985 terrorist hijacking, saving all but one life. True-story drama is suspenseful and entertaining, nominated for several Emmys. Made-for-television drama.
1988 100m/C Laurie Walters, Sandy McPeak, Ray Wise, Leslie Easterbrook, Lindsay Wagner, Eli Danker; *Dir:* Paul Wendkos. **VHS, LV** IME *♪♪♪*

Taking My Turn A taped performance of the off Broadway musical revue about growing old gracefully.
1984 87m/C **VHS, Beta $39.95** PAV *♪♪♪*

The Taking of Pelham One Two Three A hijack team, lead by the ruthless Shaw, seizes a New York City subway car and holds the 17 passengers for one million dollars ransom. Fine pacing keeps things on the edge. New cinematic techniques used by cameraman Owen Roizman defines shadowy areas like never before.
1974 (R) 102m/C Robert Shaw, Walter Matthau, Martin Balsam, Hector Elizondo, James Broderick; *Dir:* Joseph Sargent. **VHS, Beta $29.95** MGM, FOX *♪♪♪*

Taking it to the Street It's an all-out high-stakes battle between good and evil.
19?? 90m/C **VHS** AVD *♪♪♪*

Tale of the Frog and the Prince Amphibious prince befriends self-centered princess with aversion to osculation. Costumes by Maxfield Parrish.
1982 55m/C Teri Garr, Robin Williams; *Dir:* Eric Idle. **VHS $14.98** FOX *♪♪½*

The Tale of the Frog Prince Superb edition of Shelley Duvall's "Faerie Tale Theatre" finds Williams the victim of an angry fairy godmother's spell. Garr to the rescue as the self-centered princess who saves him with a kiss. Directed and written by Idle of Monty Python fame.
1983 60m/C Robin Williams, Teri Garr; *W/Dir:* Eric Idle. **VHS, Beta, LV $14.98** FOX *♪♪♪½*

The Tale of Ruby Rose Living in the mountains of Tasmania, Ruby Rose, her husband Henry, and their adopted son Gem, struggle to survive the '30s depression by hunting and trapping. Having spent all her life in the mountains, Ruby Rose has a very superstitious view of life and a terror of nightfall. But after an argument with her husband, Ruby Rose sets out on a cross-country journey to find her grandmother and discover the secrets of her past.
1987 (PG) 101m/C AU Melita Jurisic, Chris Haywood; *W/Dir:* Roger Sholes. **VHS $89.95** HMD *♪♪♪*

A Tale of Two Chipmunks Three Chip 'n' Dale Disney cartoons: "Chicken in the Rough," "The Lone Chipmunks" and "Chips Ahoy."
1953 (G) 24m/C **VHS, Beta, LV $14.95** DIS *♪♪♪*

A Tale of Two Cities Lavish production of the Dickens' classic set during the French Revolution, about two men who bear a remarkable resemblance to each other, both in love with the same girl. Carefree lawyer Sydney Carton (Colman) roused to responsibility, makes the ultimate sacrifice. A memorable Madame DeFarge from stage star Blanche Yurka in her film debut, with assistance from other Dickens' film stars Rathbone, Oliver and Walthall. Oscar-nominated for Best Picture and Best Film Editing.
1935 128m/B Ronald Colman, Elizabeth Allan, Edna May Oliver, Reginald Owen, Isabel Jewell, Walter Catlett, H.B. Warner, Donald Woods, Basil Rathbone; *Dir:* Jack Conway. **VHS, Beta $24.95** MGM, KUI *♪♪♪♪*

A Tale of Two Cities A well-done British version of the Dickens' classic about a lawyer who sacrifices himself to save another man from the guillotine during the French Reign of Terror. The sixth remake of the tale.
1958 117m/B GB Dirk Bogarde, Dorothy Tutin, Christopher Lee, Donald Pleasence, Ian Bannen, Cecil Parker; *Dir:* Ralph Thomas. **VHS, Beta $19.98** SUE, LCA, FHS *♪♪♪*

A Tale of Two Cities A cartoon adaptation of the Dickens classic. A man sacrifices his own life for his friend's to ensure the happiness of the woman they both love.
1984 72m/C **VHS, Beta $29.98** CVL *♪♪♪*

Tale of Two Cities/In the Switch Tower A package of two medium-length silent films, the latter featuring a rare acting appearance by director Borzage.
1915 70m/B Walter Edwards, Maurice Costello, Florence Turner, Frank Borzage, Norma Talmadge, Leo Delaney. **VHS, Beta, 8mm $29.95** VYY *♪½*

Tale of Two Sisters Sheen narrates sensual drama about two beautiful, ambitious sisters.
1989 90m/C Valerie Breiman, Claudia Christian, Sydney Lassick, Jeff Conaway; *Nar:* Charlie Sheen. **VHS $89.99** RAE *♪♪½*

Talent for the Game A slight but handsomely produced baseball pleasantry about a talent scout who recruits a phenomenal young pitcher, then sees the kid exploited by the team owner. More like an anecdote than a story, with an ending that aims a little too hard to please. Good for sports fans and family audiences.
1991 (PG) 91m/C Edward James Olmos, Lorraine Bracco, Jeff Corbett; *Dir:* Robert M. Young. **VHS, Beta, LV $19.95** PAR, LDC *♪♪½*

Tales of Beatrix Potter The Royal Ballet Company of England performs in this adaptation of the adventures of Beatrix Potter's colorful and memorable creatures. Beautifully done. Also known as "Peter Rabbit and Tales of Beatrix Potter."
1971 (G) 90m/C GB *Dir:* Reginald Mills. **VHS $29.95** HMV *♪♪♪½*

Tales from the Crypt A collection of five scary stories from the classic EC comics that bear the movie's title. Richardson tells the future to each of five people gathered in a cave, each tale involving misfortune and, of course, gore.
1972 (PG) 92m/C GB Ralph Richardson, Joan Collins, Peter Cushing, Richard Greene, Ian Hendry; *Dir:* Freddie Francis. **VHS, Beta $59.95** PSM *♪♪♪*

Tales from the Crypt Three contemporary tales of the macabre, "The Man Who Was Death," "'Twas the Night Before," and "Dig That Cat...He's Real Gone," linked together by a special-effects host. Ry Cooder provides the musical score for the second episode. Based on stories published by William M. Gaines in EC Comics; made-for-cable-television.
1989 81m/C Bill Sadler, Mary Ellen Trainer, Larry Drake, Joe Pantoliano, Robert Wuhl, Gustav Vintas; *Dir:* Walter Hill, Robert Zemeckis, Richard Donner. **VHS, Beta, LV $89.99** *HBO ♫♫♫*

Tales from the Darkside: The Movie Three short stories in the tradition of the ghoulish TV show are brought to the big screen, with mixed results. The plot centers around a boy who is being held captive by a cannibal. In order to prolong his life, he tells her horror stories. The tales were written by Sir Arthur Conan Doyle, Stephen King, and Michael McDowell.
1990 (R) 93m/C Deborah Harry, Christian Slater, David Johansen, William Hickey, James Remar, Rae Dawn Chong, Julianne Moore, Robert Klein; *Dir:* John Harrison. **VHS, Beta, LV $14.95** *PAR ♫ ½*

Tales of the Klondike: In a Far Country An adaptation of Jack London's story exploring the fears, dangers, and joys experienced by prospectors in the harsh wilderness of northwest Canadian gold country.
1981 60m/C Robert Carradine, Scott Hylands; *Dir:* Janine Manatis; *Nar:* Orson Welles. **VHS, Beta $19.95** *AHV ♫♫*

Tales of the Klondike: Race for Number One Two men compete in a highly contested dog-sled race—the goal of which is to be the first to reach a tract of gold-laden land. Adapted from a story by Jack London.
1981 60m/C David Ferry, John Ireland; *Dir:* David Cobham; *Nar:* Orson Welles. **VHS, Beta $19.95** *AHV ♫♫*

Tales of the Klondike: The Scorn of Women An available man is enthusiastically courted by every single woman in Dawson City, Alaska. From a story by Jack London.
1981 60m/C Tom Butler, Eva Gabor; *Dir:* Claude Fournier; *Nar:* Orson Welles. **VHS, Beta $19.95** *AHV ♫♫*

Tales of the Klondike: The Unexpected Five gold prospectors grow rich, but lose all that is most valuable through overwhelming greed. Adapted from a story by Jack London.
1983 60m/C John Candy, Cherie Lunghi; *Dir:* Peter Pearson; *Nar:* Orson Welles. **VHS, Beta $19.95** *AHV ♫♫*

Tales of Ordinary Madness Gazzara as a poet who drinks and sleeps with assorted women. Based on the stories of Charles Bukowski. Pretentious and dull.
1983 107m/C Ben Gazzara, Ornella Muti, Susan Tyrrell, Tanya Lopert, Roy Brocksmith, Katya Berger; *Dir:* Marco Ferreri. **VHS, Beta $69.98** *LIV, VES ♫*

Tales of Robin Hood A low-budget depiction of the Robin Hood legend.
1952 59m/B Robert Clark, Mary Hatcher; *Dir:* James Tinling. **VHS, Beta, LV** *WGE ♫½*

Tales from the Snow Zone "Tales from the Snow Zone" is the latest feature length ski film from RAP producers Jon Long and James Angrove. Filmed entirely in the 1991 ski season, this film captures the essence of the sport and lifestyle. "Tales" covers true extreme skiing, radical free skiing, moguls, downhill ski racing, telemarking, snowboarding, and monoskiing. Filmed in Europe and North America, "Tales" features skiers from the United States, Canada, France, Sweden, and Australia.
1991 70m/C VHS *RAP ♫½*

Tales of Terror Three tales of terror based on stories by Edgar Allan Poe: "Morella," "The Black Cat," and the "The Case of M. Valdemar." Price stars in all three segments and is excellent as the bitter and resentful husband in "Morella."
1962 90m/C Vincent Price, Peter Lorre, Basil Rathbone, Debra Paget, Joyce Jameson; *Dir:* Roger Corman. **VHS, Beta** *MLB ♫♫♫*

Tales of Washington Irving Animated adaptations of Irving's famous tales including "Legend of Sleepy Hollow" and "Rip Van Winkle."
1970 48m/C VHS, Beta $19.95 *MGM ♫♫♫*

Talion A western revenge movie that features two disabled men, a rancher and a bounty hunter, who team up to find the gang that slaughtered the rancher's family. Originally released as "An Eye for an Eye."
1966 (PG) 92m/C Robert Lansing, Slim Pickens, Patrick Wayne; *Dir:* Michael Moore. **VHS, Beta $9.95** *COL, CHA ♫♫*

Talk to Me A successful New York accountant checks into the Hollins Communication Institute to cure his stuttering. While there, he meets and falls in love with a squirrel-hunting stuttering woman from Arkansas.
1982 90m/C Austin Pendleton, Michael Murphy, Louise Fletcher, Brian Backer, Clifton James. **VHS, Beta $59.98** *FOX ♫½*

Talk Radio Riveting Oliver Stone adaptation of Bogosian's one-man play. An acidic talk radio host confronts America's evil side and his own past over the airwaves. The main character is loosely based on the life of Alan Berg, a Denver talk show host who was murdered by white supremacists.
1988 (R) 100m/C Eric Bogosian, Alec Baldwin, Ellen Greene, John Pankow, John C. McGinley, Michael Wincott, Leslie Hope; *Dir:* Oliver Stone. **VHS, Beta, LV $19.95** *MCA ♫♫♫*

Talk of the Town A brilliantly cast, strange mixture of screwball comedy and lynch-mob melodramatics co-written by Irwin Shaw. An accused arsonist, a Supreme Court judge and the girl they both love try to clear the former's name and evade the cops. Oscar nominations: Best Picture, Black & White Cinematography, Black & White Interior Decoration, Score of a Dramatic Picture and Best Film Editing.
1942 118m/B Ronald Colman, Cary Grant, Jean Arthur, Edgar Buchanan, Glenda Farrell, Rex Ingram, Emma Dunn; *Dir:* George Stevens. **VHS, Beta, LV $69.95** *COL ♫♫♫*

Talkin' Dirty After Dark Sexy comedy starring Lawrence as a suave comedian who will do anything to get a late-night spot at Dukie's comedy club.
1991 (R) 89m/C Martin Lawrence, Jedda Jones, Phyllis Stickney, Darryl Sivad; *W/Dir:* Topper Carew. **VHS $89.95** *NLC ♫♫*

The Talking Parcel An animated film about a parrot sweeping a young girl off to a fantasy world. Based on the book by Gerald Durrell.
1984 40m/C VHS, Beta $12.99 *HBO ♫♫*

Talking Walls A college sociology student uses high-tech surveillance equipment to spy on the clients of a seedy motel, in order to complete his thesis on human sexuality.
1985 85m/C Stephen Shellan, Marie Laurin, Barry Primus, Sally Kirkland; *Dir:* Stephen Verona. **VHS, Beta $19.95** *STE, NWV ♫½*

Talking in Your Sleep The video clip "Talking in Your Sleep" features The Romantics and a cast of 100 models dressed, as the song implies, in a variety of sleep wear from robes to negligees.
1984 4m/C Romantics. **VHS, Beta** *FOX*

The Tall Blond Man with One Black Shoe A violinist is completely unaware that rival spies mistakenly think he is also a spy, and that he is the center of a plot to booby-trap an overly ambitious agent at the French Secret Service. A sequel followed called "Return of the Tall Blond Man with One Black Shoe" which was followed by a disappointing American remake, "The Man with One Red Shoe." Dubbed in English.
1972 (PG) 90m/C *FR* Pierre Richard, Bernard Blier, Jean Rochefort, Mireille Darc, Jean Carmet; *Dir:* Yves Robert. **VHS, Beta, 8mm $59.95** *CVC, COL, VYY ♫♫♫*

Tall, Dark & Handsome A parade of male strip-acts, featuring the Chippendale dancers.
1987 70m/C Maureen Murphy, Judy Landers. **VHS, Beta, LV $39.95** *CEL*

The Tall Guy Goldblum is a too-tall actor who tries the scene in London and lands the lead in a musical version of 'The Elephant Man', becoming an overnight success. Not consistently funny, but good British comedy, including an interesting sex scene.
1989 (R) 92m/C Jeff Goldblum, Emma Thompson, Rowan Atkinson, Geraldine James; *Dir:* Mel Smith. **VHS, LV $89.95** *COL, FCT ♫♫♫*

Tall Lie A student quits a fraternity hazing exercise. He's pursued by a psychotic bunch of his fraternity brothers and is killed in a smash-up. A local doctor goes onto an anti-hazing campaign in the aftermath. Also known as "For Men Only."
1953 93m/B Paul Henreid, Robert Sherman; *Dir:* Paul Henreid. **VHS, Beta, LV** *WGE ♫♫*

The Tall Men Standard western features frontier hands on a rough cattle drive confronting Indians, outlaws, and the wilderness while vying with each other for the love of Russell.
1955 122m/C Clark Gable, Jane Russell, Robert Ryan, Cameron Mitchell, Mae Marsh; *Dir:* Raoul Walsh. **VHS, Beta $19.98** *FOX ♫♫*

Tall in the Saddle A misogynist foreman (Wayne) finds himself accused of murder. Meanwhile, he falls in love with his female boss's niece. An inoffensive and memorable western.
1944 79m/B John Wayne, Ella Raines, George "Gabby" Hayes, Ward Bond; *Dir:* Edwin L. Marin. **VHS, Beta, LV $19.95** *MED ♫♫½*

Tall Story Perkins is a star basketball player who must pass a crucial test in order to continue to play the game. In addition to the pressure of the exam, he is being pressured by gamblers to throw a game against the Russians. Fonda makes her screen debut as a cheerleader who is so awe-struck by Perkins that she takes the same classes just to be close to him. Based on the play by Howard Lindsay and Russel Crouse, and the

novel "The Homecoming Game" by Howard Nemerov.
1960 91m/C Anthony Perkins, Jane Fonda, Ray Walston, Marc Connelly, Anne Jackson, Tom Laughlin; **Dir:** Joshua Logan. **VHS, LV $19.98** WEA *♫♫*

The Tall T A veteran rancher (Pat) stumbles onto big trouble when by chance he catches a stage coach that is eventually overrun by an evil band of cut-throats. After killing everyone on the coach except Pat and the daughter of a wealthy copper mine owner (Doretta), the renegades decide to leave the seen with the unfortunate two as hostages. Pat and Doretta, although in dire straights, find time to fall in love and devise an intricate plan for their escape. Regarded in certain western-lover circles as a cult classic.
1957 77m/B Randolph Scott, Richard Boone, Maureen O'Sullivan, Arthur Hunnicutt, Skip Homeier, Henry Silva, Robert Burton, Robert Anderson; **Dir:** Budd Boetticher. **VHS** GKK *♫♫*

Tall Tales & Legends: Annie Oakley This biography of the acclaimed shooting star covers her career from age 15 to retirement, including some actual silent footage of Miss Oakley filmed by Thomas Edison in 1923. Made for television.
1985 52m/C Jamie Lee Curtis, Cliff DeYoung, Brian Dennehy; **Dir:** Michael Lindsay-Hogg. **VHS, Beta $19.98** FOX *♫♫½*

Tall Tales & Legends: Casey at the Bat An installment of Shelley Duvall's made-for-cable-television "Faerie Tale Theatre," in which the immortal poem is brought to life.
1985 52m/C Bill Macy, Hamilton Camp, Elliott Gould, Carol Kane, Howard Cosell; **Dir:** David Steinberg. **VHS, Beta $19.98** FOX *♫½*

Tall Tales & Legends: Johnny Appleseed True story of the legendary American who spent his life planting apple trees across the country.
1986 60m/C Martin Short, Rob Reiner, Molly Ringwald. **VHS $19.98** FOX *♫½*

Tall, Tan and Terrific Things really go crazy when the owner of a nightclub in Harlem is accused of murder.
1946 40m/C Mantan Moreland, Monte Hawley, Francine Everett, Dots Johnson; **Dir:** Bud Pollard. **VHS, Beta $24.95** NOS, HVL, BUR *♫½*

Tall Texan A motley crew seeks gold in an Indian burial ground in the desert. Flawed but interesting and suspenseful. Greed, lust, Indians and desert heat.
1953 82m/B Lloyd Bridges, Lee J. Cobb, Marie Windsor, George Steele; **Dir:** Elmo Williams. **VHS, Beta, LV** WGE, RXM *♫♫½*

Talmage Farlow A portrait of jazz guitarist Tal Farlow, who is also shown in performance with bassist Red Mitchell and pianist Tommy Flanagan.
1981 58m/C Tal Farlow, Tommy Flanagan. **VHS, Beta $29.95** MVD, RHP, FST

The Tamarind Seed Andrews and Sharif are star-crossed lovers kept apart by Cold War espionage. Dated, dull and desultory.
1974 123m/C Julie Andrews, Omar Sharif, Anthony Quayle, Dan O'Herlihy, Sylvia Syms; **Dir:** Blake Edwards. **VHS, Beta $24.98** SUE *♫½*

The Tamer of Wild Horses A parody of the mechanized world.
1967 8m/C VHS, Beta TEX *♫½*

The T.A.M.I. Show The Santa Monica Civic Auditorium was the site, the Teenage Awards Music International was the show featuring Rock and R&B performances by such greats as Smokey Robinson and the Miracles, Jan and Dean, Marvin Gaye, Chuck Berry, the Supremes and the Rolling Stones. Young Teri Garr is a go-go dancer.
1964 100m/B Dir: Steve Binder; **Performed by:** Rolling Stones, James Brown, Chuck Berry, Marvin Gaye, The Supremes, Jan & Dean, Gerry & the Pacemakers, Smokey Robinson & the Miracles, Lesley Gore. **VHS** MED

Taming of the Shrew Historically interesting early talkie featuring sole (if perhaps unfortunate) pairing of Pickford and Fairbanks. Features the legendary credit line "By William Shakespeare, with additional dialogue by Sam Taylor." Re-edited in 1966; 1967 Zeffirelli remake featured Taylor and Burton.
1929 66m/B Mary Pickford, Douglas Fairbanks Sr., Edwin Maxwell, Joseph Cawthorn; **Dir:** Sam Taylor. **VHS, Beta, LV $19.95** NOS, CCB *♫♫½*

Taming of the Shrew A lavish screen version of the classic Shakespearean comedy. Burton and Taylor are violently physical and perfectly cast as the battling Katherine and Petruchio. At the time the film was made, Burton and Taylor were having their own marital problems, which not only added an inner fire to their performances, but sent the interested moviegoers to the theatres in droves.
1967 122m/C IT Elizabeth Taylor, Richard Burton, Michael York, Michael Hordern, Cyril Cusack; **Dir:** Franco Zeffirelli. National Board of Review Awards '67: 10 Best Films of the Year. **VHS, Beta, LV $19.95** COL *♫♫♫½*

Taming of the Shrew The classic Shakespearean comedy about the battle of the sexes where Petruchio attempts to tame his fiery, free-spirited wife, Kate.
1981 152m/C Dir: Peter Dews. **VHS, Beta $59.95** SUE *♫♫*

Tammy and the Bachelor A backwoods Southern girl becomes involved with a romantic pilot and his snobbish family. They don't quite know what to make of her but she wins them over with her down-home philosophy. Features the hit tune, "Tammy." Charming performance by Reynolds.
1957 89m/C Debbie Reynolds, Leslie Nielsen, Walter Brennan, Fay Wray, Sidney Blackmer, Mildred Natwick, Louise Beavers; **Dir:** Joseph Pevney. **VHS, Beta $59.95** MCA *♫♫½*

Tammy and the Doctor Sandra Dee reprises Debbie Reynolds's backwoods gal ("Tammy and the Bachelor"). Dee becomes a nurse's aide and is wooed by doctor Fonda, in his film debut.
1963 88m/C Sandra Dee, Peter Fonda, MacDonald Carey; **Dir:** Harry Keller. **VHS, Beta $59.95** MCA *♫♫*

Tampopo A hilarious, episodic Japanese comedy. A widowed noodle shop owner is coached by a ten-gallon-hatted stranger in how to make the perfect noodle. Popular, free-form hit that established Itami in the West. In Japanese with English subtitles.
1986 114m/C JP Ken Watanabe, Tsutomu Yamazaki, Nobuko Miyamoto, Koji Yakusho, Rikiya Yasuoka, Kinzo Sakura, Shoji Otake; **Dir:** Juzo Itami. **VHS, LV $79.95** REP *♫♫♫*

Tangier The disappearance of a British intelligence agent in Gibraltar sets off a chain-reaction of violent incidents.
198? 82m/C Ronny Cox. **VHS** IND *♫*

Tango An early melodrama follows the ups and down of a chorus girl and the guy who sticks with her through thick and thin and poorly-filmed dance routines. Based on a novel by Vida Hurst.
1936 66m/B Marion Nixon, Chick Chandler, Matty Kemp, Marie Prevost, Warren Hymer, Herman Bing, Franklin Pangborn, George Meeker; **Dir:** Phil Rosen. **VHS $15.95** NOS, LOO *♫½*

Tango The Grand Theatre Company of Geneva's ballet ensemble tangos the night away.
1983 57m/C Oscar Araiz. **VHS, Beta $39.95** VVV, KAR, MBP *♫½*

Tango Bar A romantic triangle between a tango dancer who leaves Argentina during a time of political unrest, his former partner, and his wife, who both stay behind. They are reunited after years of separation. Gives the viewer a detailed vision of the historical and cultural significance of the tango. Charming, interesting and very watchable. In Spanish with subtitles.
1988 90m/C AR Raul Julia, Valeria Lynch, Ruben Juarez; **Dir:** Marcos Zurinaga. **VHS, Beta $79.95** WAR *♫♫½*

Tango and Cash Stallone and Russell are L.A. cops with something in common: they both think they are the best in the city. Forced to work together to beat drug lord Palance, they flex their muscles a lot. Directing completed by Albert Magnoli, after Andrei Konchalovsky left in a huff!
1989 (R) 104m/C Sylvester Stallone, Kurt Russell, Jack Palance, Brion James, Teri Hatcher, Michael J. Pollard, James Hong, Marc Alaimo, Robert Z'Dar; **Dir:** Andrei Konchalovsky. **VHS, Beta, LV, 8mm $24.98** WAR *♫♫*

Tank A retired Army officer's son is thrown into prison on a trumped-up charge by a small town sheriff. His father comes to the rescue with his restored Sherman tank. A trite and unrealistic portrayal of good versus bad.
1983 (PG) 113m/C James Garner, Shirley Jones, C. Thomas Howell, Mark Herrier, Sandy Ward, Jenilee Harrison, Dorian Harewood, G.D. Spradlin; **Dir:** Marvin J. Chomsky. **VHS, Beta, LV $19.95** MCA *♫*

Tanner '88 Murphy is excellent as a presidential candidate. Story by Gary Trudeau of Doonesbury fame. A precise political satire made for cable.
1988 120m/C Michael Murphy, Pamela Reed, Cynthia Nixon; **Dir:** Robert Altman. **VHS $79.99** HBO *♫*

Tanya's Island Abused girlfriend fantasizes about life on deserted island and romance with an ape. Vanity billed as D.D. Winters.
1981 (R) 100m/C Vanity, Richard Sargent, Mariette Levesque, Don McCleod; **Dir:** Alfred Sole. **VHS, Beta** NO *♫♫½*

Taoism Drunkard It's alcoholism at it's funniest as a juice-guzzling martial artist goes nuts.
19?? 100m/C Yuen Tat Chor, Yuen Cheung Yan, Chu Hai Ling. **VHS, Beta $49.95** OCE *♫♫½*

Tap The son of a famous tap dancer, a dancer himself, decides to try to get away from his former life of crime by helping an aging hoofer revitalize the art of tap dancing. Fun to watch for the wonderful dancing scenes. Davis' last big screen appearance. Hines is sincere in this old-fashioned story,

and dances up a storm. Captures some never-before filmed old hoofers.
1989 (PG-13) 106m/C Gregory Hines, Sammy Davis Jr., Suzanne Douglas, Joe Morton; *Dir:* Nick Castle. VHS, Beta, LV $19.95 COL 𝄞𝄞½

Tapeheads Silly, sophomonic, sexy comedy starring Cusack and Robbins as young wannabe rock-video producers who strike it big. Many big-name cameos; good soundtrack; laughs.
1989 (R) 93m/C John Cusack, Tim Robbins, Mary Crosby, Connie Stevens, Susan Tyrrell, Lyle Alzado, Don Cornelius, Katy Boyer, Doug McClure, Clu Gulager, Jessica Walter, Stiv Bators, Sam Moore, Junior Walker; *Dir:* Bill Fishman. VHS, Beta, LV $19.95 PAV 𝄞𝄞½

Taps Military academy students led by Hutton are so true to their school they lay siege to it to keep it from being closed. An antiwar morality play about excesses of zeal and patriotism. Predictable but impressive.
1981 (PG) 126m/C Timothy Hutton, George C. Scott, Ronny Cox, Sean Penn, Tom Cruise; *Dir:* Harold Becker. VHS, Beta $59.98 FOX 𝄞𝄞½

Tarantulas: The Deadly Cargo Eek! Hairy spiders terrizing our sleepy town and destroying our orange crop! Made-for-TV cheapie not creepy; will make you sleepy.
1977 100m/C Claude Akins, Charles Frank, Deborah Winters, Pat Hingle, Sandy McPeak, Bert Remsen, Howard Hesseman, Tom Atkins, Charles Siebert; *Dir:* Stuart Hagmann. VHS $49.95 IVE 𝄞

Taras Bulba Well-photographed costume epic on the 16th century Polish revolution. Brynner as the fabled Cossack; Curtis plays his vengeful son. Good score. Shot on location in Argentina. Based on the novel by Nicolai Gogol.
1962 122m/C Tony Curtis, Yul Brynner, Christine Kaufmann, Sam Wanamaker, George Macready, Vladimir Sokoloff, Perry Lopez; *Dir:* J. Lee Thompson. VHS $24.95 MGM 𝄞𝄞½

Target Normal dad/hubby Hackman slips into a figurative phone booth and emerges as former CIA when his better half is kidnapped in Paris. Good action scenes, but poorly scripted and too long. Matt Dillon is the teenaged son.
1985 (R) 117m/C Gene Hackman, Matt Dillon, Gayle Hunnicutt, Josef Sommer; *Dir:* Arthur Penn. VHS, Beta, LV $19.98 FOX 𝄞𝄞

Target Eagle Spanish police chief von Sydow hires mercenary Rivero to infiltrate drug ring. Good performances; ho-hum story.
1984 99m/C SP MX Max von Sydow, George Peppard, Maud Adams, Chuck Connors, George Rivero, Susana Dosamantes; *Dir:* D.J. Anthony Loma. VHS, Beta $24.95 VCL 𝄞𝄞

Target: Favorite Son An attractive, ammunition-laden woman sharp-shoots her way to power in this made for TV movie.
1987 (R) 115m/C Linda Kozlowski, Harry Hamlin, Robert Loggia, Ronny Cox, James Whitmore. VHS $89.95 VMK 𝄞

Target for Today: The 8th Air Force Story This program chronicles a day in the life of WW II's largest bomber squadron. The film goes from the planning session to the heart of Germany for a major air strike.
1943 90m/B VHS, Beta, 3/4U $29.95 VCD, ARP, IHF 𝄞𝄞

Targets Bogdanovich's suspenseful directorial debut. An aging horror film star plans his retirement, convinced that real life is too scary for his films to have an audience. A

mad sniper at a drive-in movie seems to prove he's right. Some prints still have anti-gun prologue, which was added after Robert Kennedy's assassination.
1969 (PG) 90m/C Boris Karloff, James Brown, Tim O'Kelly, James Brown; *Dir:* Peter Bogdanovich. VHS, Beta $44.95 PAR 𝄞𝄞𝄞

Tarka the Otter The story of Tarka the Otter who encounters many dangers as he pursues his favorite eel meals.
1987 91m/C *Nar:* Peter Ustinov. VHS, Beta $39.95 TWE

Taro, the Dragon Boy A young boy searches for his mother who has been turned into a dragon.
1985 75m/C VHS, Beta $19.95 COL 𝄞

Tartuffe Popular French leading man Depardieu makes directorial debut with Moliere classic about religious hypocrisy.
1984 140m/C FR Gerard Depardieu, Francois Perier, Elisabeth Depardieu; *Dir:* Gerard Depardieu. VHS, LV $79.95 CVC, FCT, APD 𝄞𝄞½

Tartuffe Moliere's famous comedy. Sher's interpretation of Tartuffe is brilliant.
1990 110m/C Anthony Sher, Nigel Hawthorne, Alison Steadman; *Dir:* Bill Alexander. VHS, Beta $39.95 TTC 𝄞𝄞½

Tarzan, the Ape Man The Tarzan movie; the first Tarzan talkie; the original of the long series starring Weissmuller. Dubiously faithful to the Edgar Rice Burroughs story, but recent attempts to remake, update or improve it (notably the pretentious 1984 Greystoke) have failed to near the original's entertainment value or even its technical quality. O'Sullivan as Jane and Weissmuller bring style and wit to their classic roles.
1932 99m/B Johnny Weissmuller, Maureen O'Sullivan, Neil Hamilton; *Dir:* W.S. Van Dyke. VHS, Beta, LV $19.98 MGM, FCT 𝄞𝄞𝄞

Tarzan, the Ape Man Plodding, perverted excuse to see Derek nude. So bad not even hard-core Tarzan fans should bother.
1981 (R) 112m/C Bo Derek, Richard Harris, John Phillip Law, Miles O'Keefe; *Dir:* John Derek. VHS, Beta, LV $79.95 MGM 𝄞

Tarzan of the Apes Brawny Lincoln is the original screen Tarzan. Silent film well done and more faithful to the book than later versions.
1917 63m/B Elmo Lincoln, Enid Markey; *Dir:* Scott Sidney. VHS, Beta $19.95 NOS, VYY, WFV 𝄞𝄞

Tarzan Escapes Jane is tricked by evil hunters into abandoning her fairy tale life with Tarzan, so the Ape Man sets out to reunite with his one true love. The third entry in MGM's Weissmuller/O'Sullivan series is still among the better Tarzan movies thanks to the leads, but the Hays Office made sure Jane was wearing a lot more clothes this time around.
1936 95m/B Johnny Weissmuller, Maureen O'Sullivan, John Buckler, Benita Hume, William Henry; *Dir:* Richard Thorpe. VHS $19.98 FOX, FCT 𝄞𝄞𝄞

Tarzan the Fearless Tarzan (Crabbe) helps a young girl find her missing father.
1933 84m/B Buster Crabbe, Jacqueline Wells, E. Allen Warren, Eddie Woods, Philo McCullough, Matthew Betz; *Dir:* Robert Hill. VHS, Beta $16.95 SNC, NOS, CAB 𝄞½

Tarzan and the Green Goddess Tarzan searches for a statue that could prove dangerous if in the wrong hands.

1938 72m/B Herman Brix, Ula Holt, Frank Baker; *Dir:* Edward Kull. VHS, Beta $19.95 HHT 𝄞½

Tarzan and His Mate Second entry in the lavishly produced MGM Tarzan series. Weissmuller and O'Sullivan cohabit in unmarried bliss before the Hays Code moved them to a tree house with twin beds. Many angry elephants, nasty white hunters, and hungry lions. Laserdisc includes the original trailer.
1934 93m/B Johnny Weissmuller, Maureen O'Sullivan, Neil Hamilton, Paul Cavanagh; *Dir:* Jack Conway. VHS, LV $19.98 MGM, FCT 𝄞𝄞𝄞

Tarzan the Tiger Series of 15 chapters is loosely based on the Edgar Rice Burroughs novel entitled "Tarzan and the Jewels of Opar." These chapters were filmed as silent pieces, and were later released with a musical score and synchronized sound effects. Here Merrill is the first to sound the cry of the bull ape, Tarzan's trademark.
1929 ?m/B Frank Merrill, Natalie Kingston, Lilian Worth, Al Ferguson; *Dir:* Henry McRae. VHS $49.95 NOS, FCT, MOV 𝄞𝄞

Tarzan and the Trappers Bad-guy trappers and would-be treasure seekers wish they hadn't messed with the ape man.
1958 74m/B Gordon Scott, Eve Brent, Ricky Sorenson, Maurice Marsac, Cheetah; *Dir:* H. Bruce Humberstone. VHS, Beta $16.95 SNC, NOS, VYY 𝄞

Tarzan's Revenge Morris is a better Olympic runner than actor; Holm is horrible. Only for serious Tarzan fans.
1938 70m/B Glenn Morris, Eleanor Holm, Hedda Hopper; *Dir:* Ross Lederman. VHS, Beta $16.95 3NC, NO3, VYY 𝄞

Taste of Death A youth discovers his courage within when he stands up to the bandits invading his village.
1977 90m/C John Ireland. VHS, Beta WES 𝄞

A Taste for Flesh and Blood A monster from outer space comes to earth and is delighted with the easy pickings for his insatiable appetite. A brave boy and girl, and a NASA commander must join forces to take the alien out. This one is billed as a campy salute to '50s B movies.
1990 (R) 84m/C Rubin Santiago, Lori Karz, Tim Ferrante. VHS $59.95 LGC 𝄞

A Taste of Hell Two American soldiers fight a guerilla outfit in the Pacific during World War II. A low-budget film with few redeeming characteristics.
1973 (PG) 90m/C William Smith, John Garwood. VHS, Beta $19.95 NWV, STE Woof!

A Taste of Honey A plain working-class girl falls in love with a black sailor. When she becomes pregnant, a homosexual friend helps her out. A moving film with strong powerful performances. Based on the London and Broadway hit play by Shelagh Delaney.
1961 100m/B GB Rita Tushingham, Robert Stephens, Dora Bryan, Murray Melvin, Paul Danquah; *Dir:* Tony Richardson. VHS CGI 𝄞𝄞𝄞

Tatie Danielle Pitch-black comedy from the director of "Life Is a Long Quiet River" sees a bitter, elderly widow moving in with her young nephew and making his life a hell. In French with yellow English subtitles.
1991 114m/C FR Tsilla Chelton, Catherine Jacob, Isabelle Nanty, Neige Dolsky, Eric Prat, Laurence Fevrier; *Dir:* Etienne Chatiliez. VHS, LV $89.98 LIV 𝄞𝄞𝄞

The Tattered Web A police sergeant confronts and accidently kills the woman with whom his son-in-law has been cheating. He's then assigned the case. When he tries to frame a wino for the murder, things get more and more messy. Fine performances aid this made-for-television story.
1971 74m/C Lloyd Bridges, Frank Converse, Broderick Crawford, Murray Hamilton, Sallie Shockley; *Dir:* Paul Wendkos. **VHS, Beta** $59.95 *LHV* ✂✂

Tattoo A model becomes the object of obsession for a crazy tattoo artist. He relentlessly pursues her, leaving behind all his other more illustrated clients. Starts out fine, but deteriorates rapidly. Controversial love scene between Dern and Adams.
1981 (R) 103m/C Bruce Dern, Maud Adams, Leonard Frey, Rikke Borge, John Getz; *Dir:* Bob Brooks. **VHS, Beta** $19.98 *FOX* ✂✂

Tattoo Connection A gang of lethal Oriental diamond thieves is pursued with whip-snapping flamboyance by a taciturn kung-fu investigator.
1978 (R) 95m/C Jim Kelly; *Dir:* Luk Pak Sang. **VHS, Beta** $9.98 *SUE* ✂

Tattooed Dragon Even the dumbest of grisly gangs should know better than to toy with the Tattooed Dragon; but they don't.
1982 (R) 91m/C Jimmy Wang Yu; *Dir:* Lo Wei. **VHS, Beta** $9.98 *SUE* ✂

Tattooed Hit Man Japanese gangster violence in a revenge-doesn't-pay action story.
1976 81m/C *JP* Mayumi Nagisa, Bunta Sugawara, Tsunehiko Watase. **VHS, Beta** $59.95 *MED* ✂

Tax Season A small-time businessman buys a Hollywood tax service sight-unseen, and finds it employing a hooker, a bookie and a variety of other unconventional types.
1990 (PG-13) 93m/C Fritz Bronner, James Hong, Jana Grant, Toru Tanaka, Dorie Krum, Kathryn Knotts, Arte Johnson, Rob Slyker; *Dir:* Tom Law. **VHS, Beta** $29.95 *PAR* ✂✂

Taxi Driver A psychotic New York City taxi driver tries to save a child prostitute and becomes infatuated with an educated political campaigner. He goes on a violent rampage when his dreams don't work out. Repellant, frightening vision of alienation and urban catharsis. On-target performances from Foster and De Niro. Oscar nominations for Best Picture, De Niro, Foster, and Score. The laserdisc version is slightly longer, with commentary by Scorsese and screenwriter Paul Schrader, storyboards, complete screenplay, and Scorsese's production photos.
1976 (R) 112m/C Robert De Niro, Jodie Foster, Harvey Keitel, Cybill Shepherd, Peter Boyle, Albert Brooks; *Dir:* Martin Scorsese. Cannes Film Festival '76: Best Film; Los Angeles Film Critics Association Awards '76: Best Score, Best Actor (De Niro); National Society of Film Critics Awards '76: Best Actor (De Niro), Best Director, Best Supporting Actress (Foster). **VHS, Beta, LV, 8mm** $14.95 *COL, VYG* ✂✂✂

Taxi zum Klo An autobiographical semi-documentary about the filmmaker's aimless existence and attempts at homosexual affairs after being fired from his job as a teacher. Explicit. In German with English subtitles. Sordid sexual situations.
1981 98m/C *GE* Bernd Broaderup, Frank Ripploh; *Dir:* Frank Ripploh. New York Film Festival '81: Best Supporting Actress. **VHS, Beta** $79.95 *FCT, GLV* ✂✂✂

The Taxi Mystery Guy saves Broadway starlet from band of ne'er-do-wells thanks to deserted taxi, then tries to figure out who she is, unaware that her sinister understudy plans to do her in.
1926 50m/B Edith Roberts, Robert Agnew, Virginia Pearson, Phillips Smalley; *Dir:* Fred Windermere. **VHS, Beta** $16.95 *GPV* ✂

A Taxing Woman Satiric Japanese comedy about a woman tax collector in pursuit of a crafty millionaire tax cheater. Followed by the equally hilarious "A Taxing Woman's Return." In Japanese with English subtitles.
1987 127m/C *JP* Nobuko Miyamoto, Tsutomu Yamazaki; *Dir:* Juzo Itami. Japanese Academy Award '88: Best Picture. **VHS, Beta, LV** $39.95 *LUM, TAM, FXL* ✂✂✂½

A Taxing Woman Returns Funny, sophisticated, action-packed. Miyamoto returns as the dedicated tax investigator to fight industrialists, politicians, and other big swindlers, who have contrived to inflate Tokyo's real estate values. Sequel to "A Taxing Woman."
1988 127m/C *JP* Nobuko Miyamoto, Rentaro Mikuni, Masahiko Tsugawa, Tetsuro Tamba, Toru Masuoka, Takeya Nakamura, Hosei Komatsu, Mihoko Shibata; *Dir:* Juzo Itami. **VHS, LV** $79.95 *NYF, CVW* ✂✂✂

Tchao Pantin An acclaimed French film noir about an ex-cop being drawn into drug smuggling underground via a young Arab. English subtitles. Finely drawn characters skillfully portrayed make this a winner.
1985 100m/C *FR* Coluche, Richard Anconina, Agnes Sorel; *Dir:* Claude Berri. French Academy Awards '85: Best Picture, Best Actor (Coluche), Best Supporting Actor (Anconina), Best Screenplay, Best Female Newcomer (Sorel). **VHS, Beta** $29.98 *SUE* ✂✂✂

Tea For Three Sexual trio develops and seeks therapy after a fledgling doctor has been having sexual relationships with two women, and the females finally meet.
1984 (R) 89m/C Iris Berben, Mascha Gonska, Heinz Marecek; *Dir:* Gerhard Janda. **VHS, Beta** $39.95 *LUN* ✂

Tea and Sympathy A young prep school student, confused about his sexuality, has an affair with an older woman, his teacher's wife. The three leads recreate their Broadway roles in this tame adaption of the Robert Anderson play which dealt more openly with the story's homosexual elements.
1956 122m/C Deborah Kerr, John Kerr, Leif Erickson, Edward Andrews, Darryl Hickman, Norma Crane, Dean Jones; *Dir:* Vincente Minnelli. **VHS, LV** $19.98 *MGM, FCT* ✂

Tea for Two This take-off on the Broadway play "No, No, Nanette" features Day as an actress who takes a bet that she can answer "no" to every question for twenty-four hours (life was less complex back then). If she can, she gets to finance and star in her own Broadway musical. Songs include "The West Point Story," "On Moonlight Bay" and "By the Light of the Silvery Moon."
1950 98m/C Doris Day, Gordon MacRae, Gene Nelson, Patrice Wymore, Eve Arden, Billy DeWolfe, S.Z. Sakall, Bill Goodwin, Bill Gibson, Virginia Gibson, Crauford Kent, Harry Harvey; *Dir:* David Butler. **VHS** $29.98 *WAR, CCB* ✂✂

The Teacher TV's "Dennis the Menace" (North) has an affair with his high school teacher; the pair is menaced by a deranged killer. Cheap but enjoyable.
1974 (R) 97m/C Angel Tompkins, Jay North, Anthony James; *Dir:* Hikmet Avedis. **VHS, Beta** $29.98 *VID* ✂✂

Teachers A lawsuit is brought against a high school for awarding a diploma to an illiterate student. Comedy-drama starts slowly and seems to condemn the school system, never picking up strength or resolving any issues, though Nolte is fairly intense. Shot in Columbus, Ohio.
1984 (R) 106m/C Nick Nolte, JoBeth Williams, Lee Grant, Judd Hirsch, Ralph Macchio, Richard Mulligan, Royal Dano, Morgan Freeman, Laura Dern, Crispin Glover, Madeline Sherwood, Zohra Lampert; *Dir:* Arthur Hiller. **VHS, Beta, LV** *FOX, OM* ✂✂

Teacher's Pet This charming comedy features a cynical newspaper editor who enrolls in a night college journalism course, believing that it won't teach anything of substance. He later becomes enamored of his instructor. The well written script and Young's superb performance as Gable's rival for Day both earned Oscar nominations.
1958 120m/B Clark Gable, Doris Day, Mamie Van Doren, Gig Young, Nick Adams; *Dir:* George Seaton. **VHS, Beta** $14.95 *PAR, KRT* ✂✂✂

The Teahouse of the August Moon An adaptation of the John Patrick play. Post-war American troops are assigned to bring civilization to a small village in Okinawa and instead fall for Okinawan culture and romance. Lively cast keeps it generally working, but assessments of Brando's comedic performance vary widely.
1956 123m/C Glenn Ford, Marlon Brando, Eddie Albert, Paul Ford, Machiko Kyo, Henry Morgan; *Dir:* Daniel Mann. **VHS, Beta** $19.98 *MGM* ✂✂½

Tearaway Australian story of a street punk with an alcoholic father. Story has potential, but ends up going nowhere.
1987 (R) 100m/C *NZ AU* Matthew Hunter, Mark Pilisi, Peter Bland, Kim Willoughby, Rebecca Saunders; *Dir:* Bruce Morrison. **VHS, Beta** $79.95 *LHV, WAR* ✂½

Teddy at the Throttle Teddy, the Great Dane, must rescue Gloria Swanson from a villain who has tied up her boyfriend. Silent.
1916 20m/B Bobby Vernon, Gloria Swanson, Wallace Beery; *Dir:* Clarence Badger. **VHS, Beta** *CCB* ✂✂

Teddy at the Throttle/Speeding Along A silent comedy about stealing fortunes from innocent heroines and another about racing cars in a farm area.
1917 54m/B Gloria Swanson, Wallace Beery, Bobby Vernon. **VHS, Beta** $29.95 *VYY* ✂✂

Tee Vee Treasures A series of nostalgic compilations of commercials, comedy shows, shorts and public service announcements from the '50s and '60s.
1986 120m/B VHS, **Beta** $39.95 *RHI* ✂✂

Teen Alien On Halloween, a young boy and his friends run into a hostile alien.
1988 (PG) 88m/C Vern Adix, Michael Dunn. **VHS, Beta** $79.95 *PSM* ✂

Teen Lust Two girls look for an exciting summer job. They get all the action they can handle when they work as undercover prostitutes for the local police department.
1978 (R) 90m/C Kirsten Baker, Perry Lang, Richard Singer. **VHS, Beta** $69.98 *LIV, LTG* ✂

Teen Mothers A bittersweet story of two unwed teenagers forced to leave their child and leave home for the tenements of New York.

T

1980 (R) 91m/C VHS, Beta *PGN* 🎞️

Teen Vamp Dorky high schooler is bitten by a prostitute; becomes a cool vampire. Mindless ''comedy.''
1988 87m/C Clu Gulager, Karen Carlson, Angela Brown; *Dir:* Samuel Bradford. VHS, Beta $19.95 *STE, NWV* 🎞️

Teen Witch A demure high schooler uses black magic to woo the most popular guy in school.
1989 (PG-13) 94m/C Robin Lively, Zelda Rubinstein, Dan Gauthier, Joshua Miller, Dick Sargent; *Dir:* Dorian Walker. VHS, Beta $89.95 *MED* 🎞️🎞️

Teen Wolf A nice, average teenage basketball player begins to show werewolf tendencies which suddenly make him popular at school when he leads the team to victory. The underlying message is to be yourself, regardless of how much hair you have on your body. Lighthearted comedy carried by the Fox charm; followed by subpar ''Teen Wolf Too.''
1985 (PG) 92m/C Michael J. Fox, James Hampton, Scott Paulin, Susan Ursitti; *Dir:* Rod Daniel. VHS, Beta, LV, 8mm $14.95 *PAR* 🎞️🎞️½

Teen Wolf Too The sequel to ''Teen Wolf'' in which the Teen Wolf's cousin goes to college on a boxing scholarship. More evidence that the sequel is rarely as good as the original.
1987 (PG) 95m/C Jason Bateman, Kim Darby, John Astin, Paul Sand; *Dir:* Christopher Leitch. VHS, Beta $14.95 *PAR* 🎞️🎞️

Teenage Bad Girl A woman who edits a magazine for teenagers can't control her own daughter. The kid does time in the pen after staying out all night and generally running around with the wrong crowd. Syms first lead performance is memorable in an otherwise routine teensploitation flick.
1959 100m/B Anna Neagle, Sylvia Syms, Norman Wooland, Wilfrid Hyde-White, Kenneth Haigh, Julia Lockwood; *Dir:* Herbert Wilcox. VHS, Beta $16.95 *SNC* 🎞️🎞️½

Teenage Caveman A teenage boy living in a post-apocalypse yet prehistoric world journeys across the river, even though he was warned against it, and finds an old man who owns a book about past civilizations in the 20th Century. Schlocky, and one of the better bad films around. The dinosaur shots were picked up from the film ''One Million B.C.'' Also known as ''Prehistoric World'' and ''Out of the Darkness.''
1958 66m/B Robert Vaughn, Darrah Marshall, Leslie Bradley, Frank DeKova; *Dir:* Roger Corman. VHS $9.95 *COL* 🎞️🎞️

Teenage Confidential A compilation of scenes from all those greasy 1950s' teen-rebellion movies, along with newsreel footage of teenage crime.
195? 60m/C VHS, Beta $9.95 *RHI* 🎞️🎞️

Teenage Devil Dolls A cheap exploitation teenage flick wherein an innocent girl is turned on to reefer and goofballs, and eventually ends up on the street.
1953 70m/C Barbara Marks, Bramlef L. Price Jr.; *Dir:* B. Lawrence Price. VHS, Beta $9.95 *RHI* 🎞️

Teenage Frankenstein A descendent of Dr. Frankenstein makes his own scramble-faced monster out of a teenage boy. Lurid '50s drive-in fare (companion to ''I Was a Teenage Werewolf'') is campy fun in small doses only.

1958 72m/B Whit Bissell, Phyllis Coates, Robert Burton, Gary Conway, George Lynn, John Cliff; *Dir:* Herbert L. Strock. VHS $29.95 *COL* 🎞️🎞️

Teenage Mother A Swedish sex education teacher comes to town and poisons the minds of the local high school kids. Naturally one becomes pregnant and has to try to trick the father into marrying her. More awful stuff from the director of ''Girl On a Chain Gang.''
1967 78m/C Arlene Sue Farber, Frederick Riccio, Julie Ange, Howard Le May, George Peters; *Dir:* Jerry Gross. VHS, Beta $16.95 *SNC* Woof!

Teenage Mutant Ninja Turtles: The Movie Four sewer-dwelling turtles that have turned into warrior ninja mutants due to radiation exposure take it upon themselves to rid the city of crime and pizza. Aided by a television reporter and their ninja master, Splinter the Rat, the turtles encounter several obstacles, including the evil warlord Shredder. A most excellent live-action version of the popular comic book characters which will hold the kids' interest. Much head-kicking and rib-crunching action as Leonardo, Donatello, Raphael, and Michelangelo fight for the rights of pre-adolescents everywhere. Combines real actors with Jim Henson creatures.
1990 (PG) 95m/C Judith Hoag, Elias Koteas; *Dir:* Steven Barron; *Voices:* Robbie Rist, Corey Feldman. VHS, Beta, LV, 8mm $24.99 *FHE, LIV* 🎞️🎞️½

Teenage Mutant Ninja Turtles 2: The Secret of the Ooze Amphibious pizza-devouring mutants search for the toxic waste that turned them into marketable martial artist ecologically correct kid idols. Same formula as the first go-round with some new characters tossed in. Animatronic characters from the laboratory of Jim Henson, first screen appearance by rapper Vanilla Ice. Marked end of pre-teen turtle craze.
1991 (PG) 88m/C Francois Chau, David Warner, Paige Turco, Ernie Reyes Jr., Vanilla Ice; *Dir:* Michael Pressman. VHS, Beta $22.98 *COL, NLC* 🎞️🎞️

Teenage Seductress A young girl connives her way into the life of a writer and finds ways to distract him from his work.
19?? 84m/C Sondra Currie, Chris Warfield. VHS, Beta *MAV* Woof!

Teenage Strangler A homicidal teen terrorizes his school. Cheap film features drag races, rumbles, dances, babes, rock 'n' roll and more. One for the ''bad enough to fun'' category.
1964 61m/C Bill A. Bloom, Jo Canterbury, John Ensign; *Dir:* Bill Posner. VHS, Beta $16.95 *SNC, SMW* Woof!

Teenage Zombies A mad scientist kidnaps teenagers and uses her secret chemical formula to turn them into zombies as part of her plan to enslave the world. Not as good as it sounds.
1958 71m/B Don Sullivan, Katherine Victor; *Dir:* Jerry Warren. VHS, Beta $19.95 *NOS, SNC, VYY* Woof!

Teenager Three desperate, reckless individuals, all caught up in making a low budget movie, interact and eventually self-destruct.
1974 (R) 91m/C Andrea Cagan, Reid Smith, Susan Bernard; *Dir:* Gerald Seth Sindell. VHS, Beta $19.95 *STE, NWV* 🎞️

Teenagers from Outer Space A low-budget sci-fi film in which extraterrestrial youngsters come to conquer the earth and find food for their ''monstrous pets'' (they're really lobster shadows.) So cheaply made

and melodramatic that it just may be good for a laugh.
1959 86m/B Tom Graeff, Dawn Anderson, Harvey B. Dunn, Bryant Grant, Tom Lockyear; *Dir:* Tom Graeff. VHS $16.95 *SNC* Woof!

Tekkaman, the Space Knight A Japanese-produced giant-robot-saves-the-universe cartoon series.
1986 30m/C VHS, Beta $19.95 *WES* 🎞️½

Tekkaman, the Space Knight: Vol. 2 Human space pilot Barry Gallagher dons a suit of space armor made from ''tekka,'' an indestructible alloy, and fights the evil robots that have invaded the world of 2037. Although this film was produced in the United States, it employs Japanese animation techniques.
199? 86m/C VHS $19.95 *LDV* 🎞️½

Telefon Suspenseful, slick spy tale. Soviet agents battle a lunatic comrade who tries to use hypnotized Americans to commit sabotage. Daly, later of TV's ''Cagney and Lacey,'' is memorable.
1977 (PG) 102m/C Charles Bronson, Lee Remick, Donald Pleasence, Tyne Daly, Patrick Magee, Sheree North; *Dir:* Don Siegel. VHS, Beta $19.95 *MGM* 🎞️🎞️

The Telephone A neurotic out-of-work actress begins a comedic chain of events via a few prank phone calls. Goldberg has managed to salvage some bad material with the strength of her comedic performances - but this isn't the case here. In fact, she sued to prevent this version from ever seeing the light of day. Too bad she didn't have a better lawyer. Torn's directorial debut.
1987 (R) 96m/C Whoopi Goldberg, Elliott Gould, John Heard, Amy Wright; *Dir:* Rip Torn. VHS, Beta, LV $19.95 *STE, NWV* Woof!

Television's Golden Age of Comedy ''You Bet Your Life,'' with Groucho Marx, Amos and Andy's ''Rare Coin,'' and an episode of ''The Jack Benny Show'' are featured in this vintage package of television comedy.
195? 75m/B Groucho Marx, Bob Hope, Dean Martin, Jerry Lewis, Jack Benny. VHS, Beta $24.95 *VYY* 🎞️

Tell Me a Riddle A dying woman attempts to reconcile with her family in this poignant drama about an elderly couple rediscovering their mutual love after 47 years of marriage. Fine acting; Grant's directorial debut.
1980 (PG) 94m/C Melvyn Douglas, Lila Kedrova, Brooke Adams, Peter Coyote; *Dir:* Lee Grant. VHS, Beta $19.95 *MED* 🎞️🎞️½

Tell Me That You Love Me A tear-jerking portrait of a disintegrating modern marriage.
1984 (R) 88m/C Nick Mancuso, Barbara Williams, Belinda J. Montgomery; *Dir:* Tzipi Trope. VHS, Beta $59.98 *LIV, VES* 🎞️½

Tell Me Where It Hurts A middle-aged housewife changes her life when she forms a women's consciousness-raising group with her friends. The script won an Emmy. Stapleton's performance is redeeming in this otherwise slow movie.
1974 78m/C Maureen Stapleton, Paul Sorvino; *Dir:* Paul Bogart. VHS, Beta *LCA* 🎞️🎞️

The Tell-Tale Heart Daydreaming author fantasizes about falling for major babe and killing his best friend for her. Adapted Poe tale.

T

1962 78m/B *GB* Laurence Payne, Adrienne Corri, Dermot Walsh, Selma Vaz Dias, John Scott, John Martin, Annette Carell, David Lander; *Dir:* Ernest Morris. **VHS** *$19.98 NOS, SNC, LOO ♂♂½*

Tell Them Willie Boy Is Here Western drama set in 1909 California about a Paiute Indian named Willie Boy and his white bride. They become the objects of a manhunt (led by the reluctant local sheriff, Redford) after Willie kills his wife's father in self-defense. This was once-blacklisted Polonsky's first film in 21 years.
1969 (PG) 98m/C Robert Redford, Katharine Ross, Robert Blake, Susan Clark, Barry Sullivan, John Vernon, Charles McGraw; *Dir:* Abraham Polonsky. **VHS, Beta** *$39.95 MCA ♂♂♂*

Tempest A Russian peasant soldier rises through the ranks to become an officer, only to be undone by his love for the daughter of his commanding officer. Silent with musical score.
1928 105m/B John Barrymore, Louis Wolheim; *Dir:* Sam Taylor. **VHS, Beta** *$9.95 NOS, CCB, VYY ♂♂♂*

Tempest Shakespeare's classic tale of the fantasy world of spirits, sorcerers, maidens, monsters, and scheming noblemen is brought to life in George Schaefer's ethereal production.
1963 76m/C Maurice Evans, Richard Burton, Roddy McDowall, Lee Remick, Tom Poston; *Dir:* George Schaefer. **VHS, Beta** *FHS ♂♂½*

The Tempest The lust for power is ultimately overcome by the need for love in this play, one of Shakespeare's most dramatic and emotional romances. Part of the "Shakespeare Plays" series.
1980 150m/C *GB* **VHS, Beta** *TLF ♂♂♂*

Tempest A New York architect, fed up with city living, chucks it all and brings his daughter to live with him on a barren Greek island. Based on Shakespeare's play of the same name. Well-written, thoughtfully acted, and beautifully filmed.
1982 (PG) 140m/C John Cassavetes, Gena Rowlands, Susan Sarandon, Vittorio Gassman, Molly Ringwald, Paul Stewart, Sam Robards, Raul Julia; *Dir:* Paul Mazursky. **VHS, Beta, LV** *COL ♂♂♂*

Tempest/Eagle A Russian soldier during the Revolution is betrayed and humiliated by the woman he loves in "The Tempest." In "The Eagle," an outcast guardsman becomes the Russian version of Robin Hood.
1927 56m/B John Barrymore, Camilla Horn, Rudolph Valentino, Louise Dressler; *Dir:* Sam Taylor, Clarence Brown. **VHS, Beta** *$29.98 CCB ♂♂♂*

Temporada Salvaje (Savage Season) Three criminals try to steal a fortune in platinum.
1987 90m/C *MX* Armando Silvestre, Isela Vega, Ron Harper, Diane McBain. **VHS, Beta** *$77.95 MAD ♂♂½*

The Tempter Also known as "L'Anticristo" or "The Antichrist." Mirrors the gruesomeness of "The Exorcist," with story following an invalid's adventures with witchcraft, bestiality, and Satanism.
1974 (R) 96m/C *IT* Mel Ferrer, Carla Gravina, Arthur Kennedy, Alida Valli, Anita Strinberg, George Coulouris; *Dir:* Alberto De Martino. **VHS, Beta** *$24.98 SUE ♂♂*

The Temptress A doctor attempting to find a cure for polio is coerced by an irresistible woman into murder and blackmail. Well-told story, if somewhat grim.

1949 85m/B *GB* Joan Maude, Arnold Bell, Don Stannard, Shirley Quentin, John Stuart, Ferdinand "Ferdy" Mayne; *Dir:* Oswald Mitchell. **VHS, Beta** *$16.95 SNC ♂♂½*

10 A successful songwriter who has everything finds his life is incomplete without the woman of his dreams, the 10 on his girl-watching scale. His pursuit brings surprising results. Music by Henry Mancini; also popularizes Ravel's "Bolero."
1979 (R) 121m/C Dudley Moore, Julie Andrews, Bo Derek, Dee Wallace Stone, Brian Dennehy, Robert Webber; *Dir:* Blake Edwards. **VHS, Beta, LV** *$19.98 WAR, FCT ♂♂½*

Ten Brothers of Shao-lin A martial arts adventure from the days of warlords and warriors.
198? 90m/C Chia Ling, Wang Tao, Chang Yi. **VHS, Beta** *$57.95 UNI ♂*

Ten Commandments The silent epic that established DeMille as a popular directorial force and which he remade 35 years later as an even bigger epic with sound. Follows Moses' adventures in Egypt, plus a modern story of brotherly love and corruption. Features a new musical score by Gaylord Carter. Remade in 1956.
1923 146m/B Theodore Roberts, Richard Dix, Rod La Rocque, Edythe Chapman; *Dir:* Cecil B. DeMille. **VHS, Beta** *$19.95 PAR ♂♂½*

The Ten Commandments DeMille's remake of his 1923 silent classic is a lavish Biblical epic that tells the life story of Moses, who turned his back on a privileged life to lead his people to freedom outside of Egypt. Exceptional cast. Parting of Red Sea rivals any modern special effects. Available in widescreen on laserdisc. A 35th Anniversary Collector's Edition is available in widescreen format in Dolby Surround stereo, and 1,000 copies of an Autographed Limited Edition that includes an engraved bronze plaque and an imprinted card written and personally signed by Charlton Heston are also available.
1956 (G) 219m/C Charlton Heston, Yul Brynner, Anne Baxter, Yvonne de Carlo, Nina Foch, John Derek, H.B. Warner, Henry Wilcoxon, Judith Anderson, John Carradine, Douglas Dumbrille, Cedric Hardwicke, Martha Scott, Vincent Price, Debra Paget; *Dir:* Cecil B. DeMille. Academy Awards '56: Best Special Effects; Harvard Lampoon Awards '56: Worst Film of the Year. **VHS, Beta, LV** *$35.00 PAR, LDC, IGP ♂♂♂*

Ten Days That Shook the World/October Silent Russian epic details the events that culminated in the Russian Revolution of October 1917, using actual locations and many actual participants. Based on the book by American author John Reed by the same name. Some of Eisenstein's most striking work. Alternate title, "October."
1927 104m/B Nikandrov, N. Popov, Boris Livanov; *Dir:* Sergei Eisenstein, Grigori Alexandrov. **VHS, Beta** *$29.95 VYY, IHF, WFV ♂♂♂½*

Ten Days Wonder Mystery/drama focuses on the patriarch of a wealthy family (Welles), his young wife, and his adopted son (Perkins), who is having an affair with his stepmother. It also turns out that Perkins is certifiable and is trying to break all of the Ten Commandments, which he succeeds in doing before killing himself (gruesomely). Based on the mystery novel by Ellery Queen.
1972 101m/C *FR* Orson Welles, Anthony Perkins, Marlene Jobert, Michel Piccoli, Guido Alberti; *Dir:* Claude Chabrol. **VHS** *MOV ♂♂½*

Ten Little Indians Ten people are gathered in an isolated inn under mysterious circumstances. One by one they are murdered, each according to a verse from a children's nursery rhyme. British adaptation of the novel and stage play by Agatha Christie.
1975 (PG) 98m/C *GB* Herbert Lom, Richard Attenborough, Oliver Reed, Elke Sommer, Charles Aznavour, Stephane Audran, Gert Frobe, Adolfo Celi, Orson Welles; *Dir:* Peter Collinson. **VHS, Beta, LV** *$59.98 CHA ♂♂½*

Ten Little Indians A group of prize-winning vacationers find themselves embarking on an African adventure. They all wind up at the same camp, and realize that they are being murdered one by one. Based on the work of Agatha Christie.
1989 (PG) 100m/C Donald Pleasence, Brenda Vaccaro, Frank Stallone, Herbert Lom, Sarah Maur-Thorp; *Dir:* Alan Birkinshaw. **VHS, Beta, LV** *$89.95 CAN ♂♂*

Ten to Midnight Vigilante Bronson is on the prowl again, this time as a police officer after a kinky serial murderer. The psychotic killer stalks his daughter and Dad's gonna stop him at any cost.
1983 (R) 101m/C Charles Bronson, Wilford Brimley, Lisa Eilbacher, Andrew Stevens; *Dir:* J. Lee Thompson. **VHS, Beta, LV** *$14.98 MGM ♂♂*

The Ten Million Dollar Getaway True story of the 1978 Lufthansa robbery at Kennedy Airport in which the criminals were forced to leave ten of their twenty-million-dollar heist behind when a pick-up van didn't show. Standard made-for-cable fare produced to cash in on the same scenario played out in "Goodfellas."
1991 (PG-13) 93m/C Jon Mahoney, Karen Yahng, Tony LoBianco, Gerry Bamman, Joseph Carberry, Terrence Mann, Kenneth John McGregor, Christopher Murney; *Dir:* James A. Contner. **VHS, Beta** *PAR ♂♂*

Ten Nights in a Bar Room The evils of alcohol are dramatized when they ruin a man's life. One of many propaganda films about alcohol and the problems that it caused that were made during the prohibition era. Silent, with a musical score. Remade in 1931.
1913 68m/B Robert Lawrence, Marie Trado. **VHS, Beta** *$29.95 VCN ♂♂*

Ten Nights in a Bar-Room A turn-of-the-century mill owner succumbs to alcoholism and ruins his life and family. Later, he sees the error of his ways and gives up alcohol. This was one of many films made during this era that chronicled the evils of alcohol and fanned the flames of the prohibition debate. Remake of the 1913 silent film.
1931 60m/B William Farnum, Thomas Santschi, John Darrow, Robert Frazer. **VHS, Beta** *$24.95 VYY ♂♂*

Ten North Frederick A shrewish wife (Fitzgerald) forces her gentle lawyer husband (Cooper) into cut-throat politics. He consoles himself in the arms of a much younger woman. Good performances, hokey script, based on the John O'Hara novel.
1958 102m/B Gary Cooper, Diane Varsi, Suzy Parker, Geraldine Fitzgerald, Tom Tully, Stuart Whitman; *Dir:* Philip Dunne. **VHS** *TCF ♂♂½*

10 Rillington Place A grimy, upsetting British film about the famed serial killer John Christie and the man wrongly hanged for his crimes, the incident that led to the end of capital punishment in England. Impeccably acted. One of Hurt's earliest films.

1971 (PG) 111m/C *GB* Richard Attenborough, John Hurt, Judy Geeson, Gabrielle Daye, Andre Morell; *Dir:* Richard Fleischer. **VHS, Beta, LV** $59.95 *COL* 🎬

Ten Speed Two advertising executives compete in a 400-mile bicycle race from San Francisco to Malibu.
1976 89m/C William Woodbridge, Patricia Hume, David Clover. **VHS, Beta** $69.98 *LIV, LTG* 🎬🎬

Ten Tigers of Shaolin Kicks and chops are dished out freely in this martial arts epic.
197? 90m/C VHS, Beta $29.95 *UNI* 🎬

10 Violent Women Ten women who take part in a million dollar jewelry heist are tossed into a women's prison where brutal lesbian guards subject them to degradation and brutality.
1979 (R) 97m/C Sherri Vernon, Dixie Lauren, Sally Gamble; *Dir:* Ted V. Mikels. **VHS, Beta** $49.95 *WES* Woof!

Ten Wanted Men A successful cattle baron is confronted by a pistol-wielding landowner determined to ruin him. Standard fare, with better than average performances from Scott and Boone.
1954 80m/C Randolph Scott, Jocelyn Brando, Richard Boone, Skip Homeier, Leo Gordon, Jack Perrin, Donna Martell; *Dir:* H. Bruce Humberstone. **VHS, Beta, LV** $69.95 *COL* 🎬

Ten Who Dared Ten Civil War heroes brave the Colorado River in an effort to chart its course. Although it's based on an actual historic event, the film is poorly paced and lacks suspense.
1960 92m/C Brian Keith, John Beal, James Drury; *Dir:* William Beaudine. **VHS, Beta** $69.95 *DIS* 🎬🎬

Ten Years After: Going Home The Alvin Lee-led sixties band reunites at London's Marquee Club in 1983, performing their landmark hits "Love Like a Man," "Help Me," "Slow Blues," "I'm Going Home," "Found Love" and more.
1983 55m/C VHS, Beta $19.95 *MVD, PMV* 🎬🎬

The Tenant Polanski co-wrote the screenplay, directed, and starred in this disturbing film about a hapless office worker who moves into a spooky Paris apartment house. Once lodged inside its walls, he becomes obsessed with the previous occupant—a suicide—and the belief that his neighbors are trying to kill him. Based on a novel by Roland Topor.
1976 (R) 126m/C *FR* Roman Polanski, Isabelle Adjani, Melvyn Douglas, Jo Van Fleet, Bernard Fresson, Shelley Winters; *Dir:* Roman Polanski. **VHS, Beta** $49.95 *PAR* 🎬🎬🎬½

The Tender Age A troubled 18 year-old girl gets out of a detention home and is put on probation. She captivates her idealistic probation officer, whose interest may extend beyond professional concern. Fine idea, but under-played. Also released as "The Little Sister."
1984 103m/C John Savage, Tracy Pollan, Roxanne Hart, Richard Jenkins; *W/Dir:* Jan Egleson. **VHS, Beta** $79.98 *LIV, LTG* 🎬½

Tender Comrade Flag-waving violin-accompanied tearjerker about five women who live together to make ends meet while their men do manly things during WWII. Intended to puff your heart up with patriotic gusto. Ironically, thanks to the girls' communal living arrangement, director Dmytryk was later accused of un-American activities.
1943 101m/B Ginger Rogers, Robert Ryan, Ruth Hussey, Patricia Collinge, Mady Christians, Kim Hunter, Jane Darwell, Mary Forbes, Richard Martin; *Dir:* Edward Dmytryk **VHS** $19.98 *TTC, FCT* 🎬🎬

Tender Is the Night A television-special adaptation of the F. Scott Fitzgerald classic, set in the roaring twenties, in which a psychiatrist marries one of his patients and then heads down the road to ruin, driven by his marriage and the times in which he lives. An entry from the "Front Row Center" series.
1955 60m/B Mercedes McCambridge, James Daly. **VHS, Beta** $24.95 *NOS, DVT* 🎬🎬½

Tender Loving Care Three nurses dispense hefty doses of T.L.C. in their hospital, and the patients aren't the only ones on the receiving end.
1973 72m/C Donna Desmond, Leah Simon, Anita King; *Dir:* Don Edmonds. **VHS, Beta** $39.98 *SUE* 🎬

Tender Mercies A down-and-out country western singer finds his life redeemed by the love of a good woman. Aided by Horton Foote's script, Duvall, Harper and Barkin keep this from being simplistic and sentimental. Duvall wrote as well as performed the songs in his Oscar-winning performance. Wonderful, life-affirming flick.
1983 (PG) 93m/C Robert Duvall, Tess Harper, Betty Buckley, Ellen Barkin, Wilford Brimley; *Dir:* Bruce Beresford. Academy Awards '83: Best Actor (Duvall), Best Original Screenplay. **VHS, Beta, LV** $19.99 *HBO, BTV* 🎬🎬🎬

The Tender Trap Charlie Reader (Sinatra) is a bachelor not content with the many women in his life. He meets the innocent Julie Gillis (Reynolds) and falls head over heels for her. He then torments himself over a marriage proposal, unwilling to let go of his freedom.
1955 116m/C Frank Sinatra, Debbie Reynolds, Celeste Holm, David Wayne, Carolyn Jones, Lola Albright, Tom Helmore, Howard St. John, Willard Sage, James Drury, Benny Rubin, Frank Sully, David White; *Dir:* Charles Walters. **LV** $39.98 *MGM* 🎬🎬½

The Tender Warrior A beautifully photographed animal adventure with Haggerty as the woodsman.
1971 (G) 85m/C Dan Haggerty, Charles Lee, Liston Elkins; *Dir:* Stewart Raffill. **VHS, Beta** $59.95 *VCD* 🎬🎬

The Tender Years Sentimental drama of a minister trying to outlaw dog fighting, spurred on by his son's fondness for a particular dog.
1947 81m/B Reb Brown, Richard Lyon, Noreen Nash, Charles Drake, Josephine Hutchinson; *Dir:* Harold Schuster. **VHS, Beta** $12.95 *TCF* 🎬🎬

Tendres Cousines Comedy follows the exploits of two cousins coming of age in the French countryside. Along with the rest of their relatives and friends, they become entwined in a web of unrequited love and intricate relationships.
1983 (R) 90m/C *FR* **VHS, Beta** $59.98 *VES* 🎬🎬

Tenement Unspeakably violent story of urban low-lifes, frightened slum-dwellers, and revenge.
1985 94m/C VHS, Beta $59.95 *USA* 🎬

Tennessee Stallion A low-class horse-trainer breaks into the world of aristocratic thoroughbred racing.
1978 87m/C Audrey Landers, Judy Landers, James Van Patten. **VHS, Beta** $69.98 *LIV, VES* 🎬

Tennessee's Partner Enjoyable, unexceptional buddy Western. Reagan intervenes in an argument and becomes Payne's pal.
1955 87m/C Ronald Reagan, Rhonda Fleming, John Payne; *Dir:* Allan Dwan. **VHS, Beta** $39.95 *WGE, BVV* 🎬🎬

Tension at Table Rock Accused of cowardice, a lone gunman must prove he killed in self-defense. Well-cast serious Western.
1956 93m/C Richard Egan, Dorothy Malone, Cameron Mitchell; *Dir:* Charles Marquis. **VHS, Beta** *UHV* 🎬🎬

Tentacles A cheesy monster disaster movie starring John Huston and Henry Fonda. Good, if you've got nothing better to do.
1977 (PG) 90m/C *IT* John Huston, Shelley Winters, Bo Hopkins, Henry Fonda, Cesare Danova; *Dir:* Oliver Hellman. **VHS, Beta** $59.98 *LIV, VES* 🎬½

Tenth Month Carol Burnett plays a divorced, middle-aged woman who becomes pregnant by a married man. She refuses help, choosing instead to live alone and keep an ever hopeful vigil for the birth of her child. Sappy but touching made-for-TV drama.
1979 123m/C Carol Burnett, Keith Mitchell, Dina Merrill; *Dir:* Joan Tewkesbury. **VHS, Beta** $59.95 *TLF, LTG* 🎬🎬

10th Victim Sci-fi cult film set in the 21st century has Mastroianni and Andress pursuing one another in a futuristic society where legalized murder is used as the means of population control. Intriguing movie where Andress kills with a double-barreled bra, the characters hang out at the Club Masoch, and comic books are considered literature. Based on "The Seventh Victim" by Robert Sheckley.
1965 92m/C *IT* Ursula Andress, Marcello Mastroianni, Elsa Martinelli, Salvo Randone, Massimo Serato; *Dir:* Elio Petri. **VHS, Beta** $69.98 *SUE* 🎬🎬🎬

Teorema A scathing condemnation of bourgeois complacency. Stamp, either a devil or a god, mysteriously appears and enters into the life of a well-to-do Milanese family and raises each member's spirituality by sleeping with them. Ultimately, the experience leads to tragedy.
1968 98m/C *IT* Terence Stamp, Silvana Mangano, Massimo Girotti, Anna Wiazemsky, Laura Betti, Andres Jose Cruz; *Dir:* Pier Paolo Pasolini. **VHS** $39.95 *CGI* 🎬🎬

Tequila Sunrise Writer-director Robert Towne's twisting film about two lifelong friends and a beautiful woman. Gibson is a (supposedly) retired drug dealer afraid of losing custody of his son to his nagging ex-wife. Russell is the cop and old friend who's trying to get the lowdown on a drug shipment coming in from Mexico. Pfeiffer runs the poshest restaurant on the coast and is actively pursued by both men. Questions cloud the plot: Is Gibson involved in the upcoming drug deal? Will Russell catch the notorious Mexican drug kingpin? Will beautiful Pfeiffer ever make up her mind about which man she loves? Loaded with double-crosses, intrigue and surprises around every corner. Steamy love scene between Pfeiffer and Gibson.
1988 (R) 116m/C Mel Gibson, Kurt Russell, Michelle Pfeiffer, Raul Julia, Arliss Howard, Arye Gross, J.T. Walsh, Ann Magnuson; *Dir:* Robert Towne. **VHS, Beta, LV, 8mm** $89.95 *WAR* 🎬🎬½

Teresa Venerdi An unusual Italian comedy about a man and his financial difficulties. His mistress, his fiance and a young girl make his troubles more confusing. Unusually sweet and humorous DeSica.
1941 90m/B IT Dir: Vittorio DeSica. **VHS, Beta, 3/4U, Special order formats $49.95** CIG ♪♪♪

Tereza In politically chaotic Czechoslovakia, a female police detective tries to solve an intricate murder mystery. In Czech with English subtitles.
1961 91m/B CZ Jirina Svorcova; **Dir:** Pavel Blumenfeld. **VHS, Beta, 3/4U $39.95** FCT, IHF ♪♪♪

Terminal Bliss Teen heartthrob Perry stars as a spoiled rich druggie in this ''Less Than Zero'' knock-off. Perry and childhood pal Owen grow up sharing everything—lacrosse, girlfriends, and drugs. However, their friendship is really put to the test when a beautiful new girl (Chandler) moves to town just before graduation. Release was delayed, but it will now try to capitalize on Perry's popularity.
1991 (R) 94m/C Luke Perry, Timothy Owen, Estee Chandler; **Dir:** Jordan Allen. **VHS, LV $89.99** WAR ♪ ½

Terminal Choice Staff at a hospital conduct a betting pool on patients' life expectancies. Someone decides he should win more often and uses a computer to help hedge his bets. Wasted cast on this unsuspenseful and bloody film.
1985 (R) 98m/C Joe Spano, Diane Venora, David McCallum, Ellen Barkin; **Dir:** Sheldon Larry. **VHS, Beta $79.95** VES ♪ ½

Terminal Entry Teen computer geeks accidentally find a terrorist online network, and inadvertently begin transmitting instructions to destroy U.S. targets. The premise is intriguing but poorly done.
1987 (R) 95m/C Edward Albert, Yaphet Kotto, Kabir Bedi; **Dir:** John Kincade. **VHS, Beta $79.95** CEL ♪ ½

Terminal Exposure Predictable but likeable comedy-mystery. A pair of Venice, CA, beach denizens capture a murder on film and try to solve it.
1989 105m/C Steve Donmyer, John Vernon, Ted Lange, Joe Phelan, Hope Marie Carlton, Mark Hennessy, Scott King; **Dir:** Nico Mastorakis. **VHS, Beta, LV $79.98** LIV, VES ♪♪

Terminal Force Stupid, poorly executed story of kidnapping and the mob. So bad it never saw the inside of a theater; went straight to video.
1988 (R) 83m/C Troy Donahue, Richard Harrison, Dawn Wildsmith; **Dir:** Fred Olen Ray. **VHS, LV $19.95** STE, NWV ♪

Terminal Island Tough southern California penal colony is crowded with inmates from death row. When prison becomes coed, violence breaks out. Exploitative and unappealing.
1973 (R) 88m/C Phyllis E. Davis, Tom Selleck, Don Marshall, Ena Hartman, Marta Kristen; **Dir:** Stephanie Rothman. **VHS, Beta $29.95** UHV ♪ ½

The Terminal Man A slick, visually compelling adaptation of the Michael Crichton novel. A scientist plagued by violent mental disorders has a computer-controlled regulator implanted in his brain. The computer malfunctions and he starts a murdering spree. Futuristic vision of man-machine symbiosis gone awry. Well acted, but still falls short of the novel.
1974 (R) 107m/C George Segal, Joan Hackett, Jill Clayburgh, Richard Dysart, James B. Sikking, Norman Burton; **Dir:** Mike Hodges. **VHS, Beta $19.98** WAR ♪♪ ½

The Terminator A futuristic cyborg is sent to present-day Earth. His job: kill the woman who will conceive the child destined to become the great liberator and arch-enemy of the Earth's future rulers. The cyborg is also pursued by another futuristic visitor, who falls in love with the intended victim. Cameron's pacing is just right in this exhilarating, explosive thriller which displays Arnie as one cold-blooded villain.
1984 (R) 108m/C Arnold Schwarzenegger, Michael Biehn, Linda Hamilton, Paul Winfield, Lance Henriksen, Bill Paxton, Rick Rossovich, Dick Miller; **Dir:** James Cameron. **VHS, Beta, LV, SVS $19.95** HBO, HMD ♪♪♪

Terminator 2: Judgement Day Arnold returns as the lethal android from the future, now programmed not to destroy but protect a boy who will be mankind's post-nuke resistance leader. But the T-1000, a shape-changing, ultimate killing machine, is also on the boy's trail, bent on terminating him. Twice the mayhem, five times the special effects, ten times the budget of the first ''Terminator,'' but without Arnold it'd be half the movie. The word hasn't been invented to describe the special effects, particularly THE scariest nuclear holocaust scene yet. While it was a worldwide megahit, a hundred-million dollar budget nearly ruined the studio; Arnold accepted his $12 million in the form of a jet. Laserdisc features include Pan and Scan, Widescreen and a ''Making of T-2'' short.
1991 (R) 139m/C Arnold Schwarzenegger, Linda Hamilton, Edward Furlong, Robert Patrick, Earl Boen, Joe Morton; **Dir:** James Cameron. Academy Awards '91: Best Makeup, Best Sound, Best Sound Effects Editing, Best Visual Effects; MTV Movie Awards '92: Best Movie, Best Male Performance (Schwarzenegger), Best Female Performance (Hamilton), Breakthrough Performance (Furlong), Most Desirable Female (Hamilton), Best Action Sequence. **VHS, LV, 8mm** LIV, PIA ♪♪♪ ½

Termini Station Dewhurst gives an excellent performance as the alcoholic matriarch of a small-town Canadian family. She drinks all day and dreams of traveling to Italy, while her children work and believe there is something more to life than their mundane existence. When they finally do leave, they discover an intriguing new world. Dewhurst and Follows also appear together in ''Anne of Green Gables'' and ''Anne of Avonlea.''
1991 105m/C Colleen Dewhurst, Megan Follows, Gordon Clapp, Hannah Lee; **Dir:** Allen King. **VHS $79.95** MNC ♪♪ ½

Terms of Endearment A weeper following the changing relationship between a young woman and her mother, over a thirty-year period. Beginning as a comedy, turning serious as the years go by, this was Brooks' debut as screenwriter and director. Superb supporting cast headed by Nicholson's slyly charming neighbor/astronaut, with stunning performances by Winger and MacLaine as the two women who often know and love each other too well. Adapted from Larry McMurtry's novel.
1983 (PG) 132m/C Shirley MacLaine, Jack Nicholson, Debra Winger, John Lithgow, Jeff Daniels, Danny DeVito; **Dir:** James L. Brooks. Academy Awards '83: Best Actress (MacLaine), Best Adapted Screenplay, Best Director (Brooks), Best Picture, Best Supporting Actor (Nicholson); Golden Globe Awards '84: Best Film—Drama;

New York Film Critics Awards '83: Best Film. **VHS, Beta, LV $19.95** PAR, BTV ♪♪♪

Terraces TV series pilot about high-rise tenants whose balconies are their only common ground. Good performances.
1977 90m/C Lloyd Bochner, Eliza Garrett, Julie Newmar, Tim Thomerson; **Dir:** Lila Garrett. **VHS, Beta** WOV ♪♪ ½

Terronauts When man begins space exploration, Earth is attacked by aliens. The defenders are taken to an out-dated fortress where they learn that their forebears were similarly attacked. Juvenile, lackluster, contrived, and really dumb.
1967 77m/C GB Simon Oates, Zena Marshall, Charles Hawtrey, Stanley Meadows; **Dir:** Montgomery Tully. **VHS, Beta $59.98** CHA Woof!

The Terror A lieutenant in Napoleon's army chases a lovely maiden and finds himself trapped in a creepy castle by a mad baron. Movie legend has it Corman directed the movie in 3 days as the sets (from his previous movie ''The Raven'') were being torn down around them.
1963 81m/C Boris Karloff, Jack Nicholson, Sandra Knight; **Dir:** Roger Corman. **VHS, Beta $19.95** NOS, SNC, PSM ♪♪

The Terror After 100 years, a man reveals in a film that his family killed a witch. Friends who see the film are attacked by supernatural forces. Originally double-billed with ''Dracula's Dog.''
1979 (R) 86m/C John Nolan, Carolyn Courage, James Aubrey; **Dir:** Norman J. Warren. **VHS, Beta $29.95** UHV, VCD, HHE ♪

Terror on the 40th Floor Seven people make an attempt to escape from the fortieth floor of an inflamed skyscraper. Poorly done re-hash of ''Towering Inferno.'' Uninspired.
1974 98m/C John Forsythe, Anjanette Comer, Don Meredith, Joseph Campanella; **Dir:** Jerry Jameson. **VHS, Beta $49.95** PSM ♪

Terror in the Aisles Pleasence and Allen take you on a terrifying journey through some of the scariest moments in horror film history.
1984 (R) 84m/C Donald Pleasence, Nancy Allen; **Dir:** Andrew J. Kuehn. **VHS, Beta $14.98** MCA ♪

Terror on Alcatraz The only prisoner to successfully escape from Alcatraz Island returns to retrieve the safety deposit box key he needs to get his stolen loot. And anyone who gets in his way is in for trouble.
1986 96m/C Veronica Porsche Ali, Sandy Brooke, Aldo Ray, Scott Ryder; **Dir:** Philip Marcus. **VHS $79.95** TWE, HHE ♪

Terror Beach A doctor and his bride move to a coastal village, only to find it enmeshed in witchcraft and devil worship.
1987 90m/C Victor Petit, Julie James, Mary Costi, Sandra Mozar. **VHS, Beta $59.95** MGL ♪

Terror Beneath the Sea A mad scientist wants to rule the world with his cyborgs. American and Japanese scientists unite to fight him. Fine special effects, especially the transformation from human to monster.
1966 85m/C JP Sonny Chiba, Peggy Neal, Franz Gruber, Gunther Braun, Andrew Hughes, Mike Daneen; **Dir:** Hajime Sato. **VHS, Beta $39.95** DVT ♪♪ ½

Terror in Beverly Hills When the President's daughter is kidnapped, it's up to an ex-marine to save her. The trouble is, the head terrorist hates the marine's guts as he blames him for the deaths of his wife and

children. So incredibly bad it's campy and fun.
1990 88m/C Frank Stallone, Cameron Mitchell, William Smith; *W/Dir:* John Myhers. **VHS, Beta** $79.95 *AIP Woof!*

Terror of the Bloodhunters A French author is sentenced to Devil's Island for a crime he didn't commit. The warden's daughter takes a liking to him and arranges for he and a friend to escape, but numerous perils await them in the jungle.
1962 60m/B Robert Clarke, Steve Conte, Dorothy Haney; *W/Dir:* Jerry Warren. **VHS, Beta** $16.95 *SNC* ✗

Terror Creatures from the Grave Husband summons medieval plague victims to rise from the grave to drop in on his unfaithful wife. Should've been better.
1966 85m/C *IT* Barbara Steele, Riccardo Garrone, Walter Brandi; *Dir:* Ralph Zucker. **VHS** $19.98 *SNC* ✗

Terror Eyes Spoof-thriller revolves around an "agent from Hell," sent to Earth to recruit writers for a horror film.
1987 (PG-13) 90m/C Daniel Roebuck, Vivian Schilling, Dan Bell, Lance August. **VHS, Beta** $39.95 *AIP* ✗ ½

Terror in the Haunted House Newlyweds move into an old house. The bride remembers it from her nightmares. First release featured the first use of Psychorama, a technique in which scary words or advertising messages were flashed on the screen for a fraction of a second - just long enough to cause subliminal response. The technique was banned later in the year.
1958 90m/C Gerald Mohr, Cathy O'Donnell, William Ching; *Dir:* Harold Daniels. **VHS, Beta** $9.95 *RHI, MLB, CCB* ✗ ½

Terror in the Jungle Plane crashes in Peruvian wilds and young boy survivor meets Jivaro Indians who think he's a god. Much struggling to survive and battling with horrible script.
1968 (PG) 95m/C Jimmy Angle, Robert Burns, Fawn Silver; *Dir:* Tom De Simone. **VHS** $19.95 *ACA* ✗

Terror at London Bridge London Bridge is transported brick by brick to Arizona. It carries its history with it, including the havoc-wreaking spirit of Jack the Ripper. Also released as "Bridge Across Time" and "Arizona Ripper." Campy made-for-TV effort.
1985 96m/C David Hasselhoff, Stephanie Kramer, Adrienne Barbeau, Randolph Mantooth; *Dir:* E.W. Swackhamer. **VHS, Beta** $29.95 *FRH* ✗✗

Terror is a Man A mad scientist attempts to turn a panther into a man on a secluded island. Early Filipino horror attempt inspired by H.G. Wells' "The Island of Doctor Moreau." Also released as "Blood Creature."
1959 89m/B *PH* Francis Lederer, Greta Thyssen, Richard Derr, Oscar Keesee; *Dir:* Gerardo (Gerry) De Leon. **VHS** $16.95 *SNC* ✗✗

Terror of Mechagodzilla It's monster vs. machine in the heavyweight battle of the universe as a huge mechanical Godzilla built by aliens is pitted against the real thing. The last Godzilla movie made until "Godzilla 1985."
1978 (G) 79m/C *JP* Katsuhiko Sasakai, Tomoko Ai. **VHS, Beta** $19.95 *PAR* ✗ ½

Terror by Night Holmes and Watson attempt to solve the murder of the owner of a gigantic, beautiful jewel. Their investigation must be completed before their train arrives at its destination, where the murderer can escape.
1946 60m/B Basil Rathbone, Nigel Bruce; *Dir:* Roy William Neill. **VHS, LV** $9.95 *NEG, REP, FOX* ✗✗ ½

Terror by Night/Meeting at Midnight In "Terror by Night" (1946), Sherlock Holmes and Dr. Watson board a train to protect a fabulous diamond. In "Meeting at Midnight" (1944), Charlie Chan becomes involved in magic and murder.
1946 122m/B Basil Rathbone, Nigel Bruce, Sidney Toler; *Dir:* Roy William Neill. **VHS, Beta** *MED* ✗✗

Terror at the Opera A bizarre staging of Verdi's "Macbeth" is plagued by depraved gore murders. But the show must go on, as one character chirps in badly-dubbed English. Italian horrormeister Argento's ever-fluid camera achieves spectacular shots, but the lurid, ludicrous script make this one for connoisseurs only. The operatic scenes employ the voice of Maria Callas. Available in an edited "R" rated version.
1988 107m/C Christina Marsillach, Ian Charleson, Urbano Barberini, William McNamara, Antonella Vitale, Barbara Cupisti, Coralina Cataldi Tassoni, Daria Nicolodi; *W/Dir:* Dario Argento. **VHS** $89.98 *SOU, FCT* ✗✗

Terror Out of the Sky A scientist must disguise himself as one of the insects to divert the attention of a horde of killing bees from a busload of elementary children. Sub-par sequel to "The Savage Bees."
1978 95m/C Efrem Zimbalist Jr., Dan Haggerty; *Dir:* Lee H. Katzin. **VHS, Beta** $49.95 *IVE* ✗ ½

Terror at Red Wolf Inn Young woman wins a vacation; finds she's been invited for dinner, so to speak. Not campy enough to overcome stupidity. Also known as "Club Dead" and "Terror House."
1972 (R) 90m/C Linda Gillin, Arthur Space, John Neilson, Mary Jackson, Janet Wood, Margaret Avery; *Dir:* Bud Townsend. **VHS, Beta** $29.95 *ACA* ✗✗

Terror of Rome Against the Son of Hercules Muscle-bound gladiators battle it out in this low-budget Italian action flick. Dubbed.
1964 100m/C *FR IT* Mark Forest, Marilu Tolo; *Dir:* Mario Ciano. **VHS, Beta** $16.95 *SNC, CHA* ✗

Terror Squad An unsuspecting high school population is put under siege by Libyan terrorists. The students get a hands-on lesson in revolution and guerrilla war tactics.
1987 95m/C Kerry Brennan, Bill Calvert, Chuck Connors, Greer Brodie; *Dir:* Peter Maris. **VHS** $79.98 *MCG* ✗✗

Terror of the Steppes Yet another sword and sandal extravaganza in which a musclebound hero conquers all.
1964 ?m/C *IT* Kirk Morris. **VHS, Beta** $16.95 *SNC* ✗

Terror Street An Air Force pilot's wife is murdered. He has 36 hours to clear his name and find the killer.
1954 84m/B Dan Duryea, Elsy Albiin, Ann Gudrun, Eric Pohlmann. **VHS, Beta, LV** *WGE* ✗

Terror in the Swamp When not hanging out in the local murky waters, a swamp creature terrorizes the residents of a small town.

1985 (PG) 89m/C Billy Holliday; *Dir:* Joe Catalanotto. **VHS, Beta, LV** $9.95 *NWV, STE* ✗

Terror on Tape Clip collection of horror films' most violent and graphic moments.
1985 90m/C *Hosted:* Cameron Mitchell. **VHS, Beta** $29.98 *CGI* ✗

Terror at Tenkiller Two girls vacationing in the mountains are seemingly surrounded by a rash of mysterious murders. Made for video.
1986 87m/C Mike Wiles, Stacey Logan. **VHS, Beta** $19.95 *UHV* ✗

Terror of Tiny Town Main characteristic of this musical/western is its entire midget cast, all members of Jed Buell's Midgets. Otherwise the plot is fairly average and features a bad guy and the good guy who finally teaches him a lesson. Newfield plays up the cast's short stature—they walk under saloon doors and ride Shetland ponies.
1933 65m/B Bill Curtis, Yvonne Moray; *Dir:* Sam Newfield. **VHS, Beta** $16.95 *SNC, NOS, DVT* ✗✗

Terror on Tour The Clowns, a rock group on their way up, center their stage performance around sadistic, mutilating theatrics. When real murders begin, they become prime suspects. Exploitive, bloody, disgusting.
1983 90m/C Dave Galluzzo, Richard Styles, Rick Pemberton; *Dir:* Don Edmonds. **VHS, Beta** $54.95 *MED* ✗

Terror in Toyland Once again a knife-wielding lunatic dresses as Santa Claus to strike terror and death into the hearts of children. Also known as "Christmas Evil."
1985 90m/C Brandon Maggart, Jeffrey DeMunn; *Dir:* Lewis Jackson. **VHS, Beta** $19.95 *ACA* ✗ ½

Terror Trail As usual, Mix uses guns and fists to bring bad guys to justice, and gets the girl in the bargain.
1933 62m/B Tom Mix, Naomi Judge, Arthur Rankin, Raymond Hatton, Francis McDonald, Robert Kortman, John St. Polis, Lafe McKee, Buffalo Bill Jr.; *Dir:* Armand Schaefer. **VHS** $19.98 *DVT* ✗ ½

Terror Train A masquerade party is held on a chartered train. But someone has more than mask-wearing in mind as a series of dead bodies begin to appear. Copperfield provides magic, Curtis provides screams in this semi-scary slasher movie.
1980 (R) 97m/C Jamie Lee Curtis, Ben Johnson, Hart Bochner, David Copperfield, Vanity; *Dir:* Roger Spottiswoode. **VHS, Beta, LV** $79.98 *FOX* ✗✗

Terror in the Wax Museum The owner of a wax museum is killed while mulling over the sale of the museum. The new owner has little better luck as the bodies continue to pile up. Production uses every out-of-work horror film actor of the time. The most suspenseful part of this low-budget flick is waiting to see how long the "wax dummies" can hold their breath.
1973 (PG) 94m/C Ray Milland, Broderick Crawford, Elsa Lanchester, Louis Hayward, Maurice Evans, John Carradine; *Dir:* Georg Fenady. **VHS, Beta** $69.98 *LIV, LTG Woof!*

The Terror Within Reptilian mutants hit the streets searching for human women to breed with. Have the Teenage Mutant Ninja Turtles grown up?
1988 (R) 90m/C George Kennedy, Andrew Stevens, Starr Andreeff, Terri Treas; *Dir:* Thierry Notz. **VHS, Beta** $79.95 *MGM* ✗

The Terror Within 2 In a world destroyed by biological warfare, a warrior and the woman he rescued traverse the badlands occupied by hideous mutants. What they don't know is that the real terror comes from within. How can they possibly survive? Not without decent dialogue, that's for sure.
1991 (R) 90m/C Andrew Stevens, Stella Stevens, Chick Vennera, R. Lee Ermey; *Dir:* Andrew Stevens. **VHS $89.98** VES 🎬½

Terrorgram Not a very special delivery. An offscreen James Earl Jones introduces three lesser tales of terror. Best seg: a sleazy filmmaker gets trapped in the world of his own exploitation scripts. The other two are bloody but negligible revenge-from-the-grave yarns. All start off with the receipt of a sinister package, hence a postal motif.
1990 (R) 88m/C Michael Hartson, J.T. Wallace; *Nar:* James Earl Jones. **VHS $79.95** MNC 🎬

The Terrorists Just as later he was a Scottish Lithuanian sub-commander in "The Hunt for Red October" (1990), here Connery is a Scottish Norwegian security chief. Good cinematography and premise are wasted; generic title betrays sloppy execution.
1974 (PG) 89m/C Sean Connery, Ian McShane, John Quentin; *Dir:* Caspar Wrede. **VHS, Beta $59.98** FOX 🎬½

Terrorists An Army investigator discovers a plot to assassinate the President. It's up to him to move quickly enough to stop it. Brutal action.
1983 61m/C Marland Proctor, Irmgard Millard. **VHS, Beta $59.95** WES 🎬½

The Terrornauts Astronomer inadvertently contacts alien race and his entire building is transported across the galaxy. Supposed nuclear precautionary.
1967 75m/C *GB* Simon Oates, Zena Marshall, Charles Hawtrey, Patricia Hayes; *Dir:* Montgomery Tully. **VHS** CHA 🎬½

Terrorvision OUR TV GAVE BIRTH TO SPACE ALIENS! It's amazing what modern technology can do: Suburban family buys a fancy satellite dish; bad black comedy and gory special effects result.
1986 (R) 84m/C Gerrit Graham, Mary Woronov, Diane Franklin, Bert Remsen, Alejandro Rey; *Dir:* Ted Nicolaou. **VHS, Beta $79.98** LIV, LTG Woof!

The Terry Fox Story In the spring of 1980, a young man who had lost his right leg to cancer dipped his artificial limb into the Atlantic Ocean and set off on a fund-raising "Marathon of Hope" across Canada, drawing national attention. Inspiring made-for-TV true story is well-scripted and avoids corniness. Good acting from real-life amputee Fryer and, as usual, from Duvall.
1983 96m/C Robert Duvall, Chris Makepeace, Eric Fryer, Rosalind Chao; *Dir:* Ralph L. Thomas. Genie Awards '84: Best Actor (Fryer), Best Picture, Best Supporting Actor (Zelniker). **VHS, Beta $69.95** VES 🎬🎬🎬

Tess Sumptuous adaptation of a Thomas Hardy novel. Kinski is wonderful as an innocent poor girl who marries into the aristocracy, where she is sexually and socially out of her depth. Polanski's direction is faithful and artful. Nearly three hours long, but worth every minute. Multiple Oscar winner.
1980 (PG) 170m/C *GB FR* Nastassia Kinski, Peter Firth, Leigh Lawson, John Collin; *Dir:* Roman Polanski. Academy Awards '80: Best Art Direction/Set Decoration, Best Cinematography, Best Costume Design; National Board of Review Awards '80: 10 Best Films of the Year; New York Film Critics Awards '80: Best Cinematography; Los Angeles Film Critics

Association '79: Best Director, Best Cinematography. **VHS, Beta, LV $14.95** COL 🎬🎬🎬

Test of Donald Norton Good early Western. A young man raised by Indians finds work with the Hudson's Bay Company and does well, but is haunted by suspicions that he is a half-breed.
1926 99m/B George Walsh, Tyrone Power Sr., Eugenia Gilbert. **VHS, Beta, 8mm** VYY 🎬🎬½

A Test of Love Incorrectly diagnosed as retarded, a young woman with cerebral palsy struggles to adjust to life outside an institution. Predictable but touching and well-acted.
1984 (PG) 93m/C Angela Punch McGregor, Drew Forsythe, Tina Arhondis; *Dir:* Gil Brealey. **VHS, Beta $69.95** MCA 🎬🎬

Test Pilot Gable and Spencer star as daring test pilot and devoted mechanic respectively. When Gable has to land his experimental craft in a Kansas cornfield, he meets and falls in love with farm girl Loy. The two marry and raise a family, all the while she worries over his dangerous profession. When the Air Force asks him to test their new B-17 bomber, she refuses to watch, thinking the test will end in tragedy. Superb aviation drama featuring excellent cast.
1939 118m/B Clark Gable, Myrna Loy, Spencer Tracy, Lionel Barrymore, Samuel S. Hinds; *Dir:* Victor Fleming. **VHS, Beta $19.98** MGM, TLF, CCB 🎬🎬🎬½

Test Tube Babies A married couple's morals begin to deteriorate as they mourn the fact they can't have a child. The day is saved, however, when they learn about the new artificial insemination process. Amusing, campy propaganda in the vein of "Reefer Madness."
1948 83m/B Dorothy Dube, Timothy Farrell, William Thompson; *Dir:* W. Merle Connell. **VHS, Beta $19.95** GVV Woof!

Testament Well-made and thought-provoking story of the residents of a small California town struggling to survive after a nuclear bombing. Focuses on one family who tries to accept the reality of post-holocaust life. We see the devastation but it never sinks into sensationalism. An exceptional performance from Alexander.
1983 (PG) 90m/C Jane Alexander, William Devane, Ross Harris, Roxana Zal, Kevin Costner, Rebecca DeMornay; *Dir:* Lynne Littman. **VHS, Beta, LV $14.95** PAR 🎬🎬🎬

The Testament of Dr. Cordelier A strange, experimental fantasy about a Jekyll and Hyde-type lunatic stalking the streets and alleys of Paris. Originally conceived as a television play, Renoir attempted to create a new mise-en-scene, using multiple cameras covering the sequences as they were performed whole. In French with English subtitles.
1959 95m/C *FR* Jean-Louis Barrault; *Dir:* Jean Renoir. **VHS, Beta $29.95** FCT 🎬🎬🎬

The Testament of Orpheus Superb, personal surrealism; writer-director Cocteau's last film. Hallucinogenic, autobiographical dream-journey through time. Difficult to follow, but rewarding final installment in a trilogy including "The Blood of the Poet" and "Orpheus."
1959 80m/B *FR* Jean Cocteau, Edouard Dermithe, Maria Casares, Francois Perier, Yul Brynner, Jean-Pierre Leaud, Daniel Gelin, Jean Marais, Pablo Picasso, Charles Aznavour; *Dir:* Jean Cocteau. **VHS, Beta $29.95** NOS, FCT 🎬🎬🎬½

Tevye (Teyve der Milkhiker) Sholom Aleichem's story of Jewish family life, intermarriage, and turmoil in Poland. When daughter of Tevye the dairyman seeks to marry a Ukrainian peasant, Tevye must come to terms with his love for his daughter and his faith and loyalty to tradition. Also the basis for the musical "Fiddler on the Roof." In Yiddish with English subtitles.
1939 96m/B Maurice Schwartz, Miriam Riselle; *Dir:* Maurice Schwartz. **VHS, Beta $89.95** ERG, NCJ 🎬🎬🎬½

Tex Fatherless brothers in Oklahoma come of age. Dillon is excellent. Based on the novel by S. E. Hinton.
1982 (PG) 103m/C Matt Dillon, Jim Metzler, Meg Tilly, Bill McKinney, Frances Lee McCain, Ben Johnson, Emilio Estevez; *Dir:* Tim Hunter. **VHS, Beta, LV** DIS, OM 🎬🎬½

Tex Avery's Screwball Classics, Vol. 1 Collection of eight classic cartoons compiled by renowned MGM director Avery.
194? 60m/C *Dir:* Tex Avery. **VHS, Beta, LV $14.95** MGM

Tex Avery's Screwball Classics, Vol. 2 More Averian cartoon classics, bearing the master's unmistakable stamp: "One Ham's Family," "Happy Go Nutty," "Slap Happy Lion," "Wild and Wolfy," "Ventriloquist Cat," "Big Heel Watha," "Northwest Hounded Police" and the semi-rare, long-banned "Red Hot Riding Hood."
194? 60m/C *Dir:* Tex Avery. **VHS, Beta $14.95** MGM

Tex Avery's Screwball Classics, Vol. 3 Hilarious cartoons from the man responsible for Bugs Bunny and other animated nuts.
194? 60m/C VHS, Beta $12.98 MGM, FCT

Tex Rides With the Boy Scouts Scouts aid Ritter in capturing gold bandits with a little bit of singing thrown in for good measure.
1937 60m/B Tex Ritter. **VHS, Beta $19.95** NOS, DVT 🎬½

Texas Two friends wander through the West after the Civil War getting into scrapes with the law and eventually drifting apart. Ford takes a job on a cattle ranch run by Trevor and discovers Holden has joined a gang of rustlers aiming to steal her herd. The two men vie for Trevor's affections. Although friends, a professional rivalry also existed between the two leading actors and they competed against each other, doing their own stunts during filming. Well-acted, funny, and enthusiastic Western.
1941 94m/B William Holden, Glenn Ford, Claire Trevor, Edgar Buchanan, George Bancroft; *Dir:* George Marshall. **VHS, Beta $9.95** COL 🎬🎬🎬

Texas to Bataan The Range Busters ship horses to the Philippines and encounter enemy spies. Part of the "Range Busters" series.
1942 56m/B Range Busters, John "Dusty" King, David Sharpe, Max Terhune; *Dir:* Robert Tansey. **VHS, Beta $24.95** VYY 🎬

Texas Carnival Williams and Skelton star as carnival performers who operate the dunk tank. When Skelton is mistaken for an oil tycoon, he lives high on the hog until the mistake is discovered. Believe it or not, this musical has only one water ballet sequence. Songs include "It's Dynamite," "Whoa! Emma," and "Young Folks Should Get Married."

1951 77m/C Esther Williams, Red Skelton, Howard Keel, Ann Miller, Paula Raymond, Keenan Wynn, Tom Tully; *Dir:* Charles Walters. VHS *MGM* 🎞🎞

Texas Chainsaw Massacre The movie that put the "power" in power tools. An idyllic summer afternoon drive becomes a nightmare for a group of young people pursued by a chainsaw-wielding maniac. Made with tongue firmly in cheek, this is nevertheless a mesmerizing saga of gore, flesh, mayhem, and violence.
1974 (R) 86m/C Marilyn Burns, Allen Danzinger, Paul A. Partain, William Vail, Teri McMinn, Edwin Neal, Jim Siedow, Gunnar Hansen, John Dugan, Jerry Lorenz; *W/Dir:* Tobe Hooper; *Nar:* John Larroquette. VHS, Beta, LV $39.95 *MED* 🎞🎞½

Texas Chainsaw Massacre Part 2 A tasteless, magnified sequel to the notorious blood-bucket extravaganza, about a certain family in southern Texas who kill and eat passing travelers. Followed by "Leatherface: Texas Chainsaw Massacre III."
1986 (R) 90m/C Dennis Hopper, Caroline Williams, Bill Johnson, Jim Siedow; *Dir:* Tobe Hooper. VHS, Beta, LV $19.95 *MED* Woof!

Texas Cyclone A stranger rides into a town that turns out to be incredibly friendly. Later he realizes the entire town thinks he is a prominent citizen who disappeared years ago.
1932 63m/B Tim McCoy, Wheeler Oakman, Shirley Grey, Walter Brennan, John Wayne, Wallace MacDonald, Vernon Dent, Mary Gordon; *Dir:* Irving Briskin. VHS $19.95 *NOS, DVT* 🎞🎞

Texas Detour On a trip across country, a trio of young Californians have their van stolen. They decide to take the law into their own hands when the redneck sheriff gives them no help. Dull, undistinguished "action" flick.
1977 90m/C Cameron Mitchell, Priscilla Barnes, Patrick Wayne, Mitch Vogel, Lindsay Bloom, R.G. Armstrong; *Dir:* Hikmet Avedis. VHS, Beta $59.95 *PSM* 🎞

Texas Gunfighter Your basic western from the Maynard series, with everything a western could have.
1932 60m/B Ken Maynard. VHS, Beta $19.95 *NOS, UHV, DVT* 🎞🎞

Texas Guns A gritty western about an old-time gunman (Nelson) and his quest for one last robbery, killing those who stand in his way and some just for fun. Also watch for Cassidy's comeback.
1990 (PG) 96m/C Willie Nelson, Richard Widmark, Shaun Cassidy, Angie Dickinson, Kevin McCarthy, Royal Dano, Chuck Connors, Ken Curtis, Dub Taylor. VHS $89.95 *VMK* 🎞🎞

Texas John Slaughter: Geronimo's Revenge An Indian-loving rancher frets and fights when Geronimo attacks innocent settlers. Compiled fom Walt Disney television episodes.
1960 77m/C Tom Tryon, Darryl Hickman, Betty Lynn; *Dir:* Harry Keller. VHS, Beta $69.95 *DIS* 🎞🎞

Texas John Slaughter: Stampede at Bitter Creek Good guy Texas John Slaughter meets a variety of threatening obstacles when he tries to move his cattle herd into New Mexico. Originally from the "Walt Disney Presents" television series.
1962 90m/C Tom Tryon; *Dir:* Harry Keller. VHS, Beta $69.95 *DIS* 🎞🎞

Texas Justice The "Lone Rider" cavorts around in the desert, dispensing justice in the guise of a monk.
1942 60m/B George Huston, Al "Fuzzy" St. John, Dennis Moore, Wanda McKay, Claire Rochelle; *Dir:* Sam Newfield. VHS, Beta $19.95 *NOS, DVT* 🎞½

Texas Kid A typical western. Hero helps a vengeance-minded kid catch the robbergang who robbed and killed his father. Yee-Haw.
1943 53m/B Johnny Mack Brown, Kermit Maynard. VHS, Beta $19.95 *VDM* 🎞½

Texas Lady When a woman wins $50,000 gambling, she buys a Texas newspaper on the stipulation that she can edit it in this standard fare western. Good vehicle for Colbert to look beautiful, but the plot and script are 'average and uninspired.
1956 86m/C Claudette Colbert, Barry Sullivan, Gregory Walcott; *Dir:* Tim Whelan. VHS $19.98 *REP* 🎞🎞

Texas Layover Impossible-to-describe western/exploitation entry from trashmeister Adamson involving multiple stewardesses and the remnants of the Ritz Brothers.
1975 88m/C Yvonne de Carlo, Bob Livingston, Donald (Don "Red") Barry, Connie Hoffman, Regina Carrol, Jimmy Ritz, Harry Ritz; *Dir:* Al Adamson. VHS *TAF* Woof!

Texas Legionnaires Roy comes back to his home town and gets caught between feuding cattle and sheep herders. Also known as "Man from Music Mountain."
1943 71m/B Roy Rogers, Ruth Terry, Paul Kelly, Anne Gillis, George Cleveland, Pat Brady; *Dir:* Joseph Kane. VHS $19.95 *NOS, DVT, BUR* 🎞🎞

Texas Lightning Innocuous B-grade touching movie about a father and son and their family woes.
1981 (R) 93m/C Cameron Mitchell, Channing Mitchell, Maureen McCormick, Peter Jason. VHS, Beta $49.95 *MED* 🎞½

Texas Masquerade Standard good cowboys versus land-grabbers action filler. William Boyd plays Hopalong Cassidy. Andy Clyde plays California Carson. Jimmy Rogers plays...Jimmy Rogers.
1944 59m/B William Boyd, Andy Clyde, Jimmy Rogers, Mady Correll, Don Costello, Russell Simpson, Nelson Leigh, Francis McDonald, J. Farrell MacDonald; *Dir:* George Archainbaud. VHS $19.95 *NOS* 🎞🎞

Texas Pioneers A frontier scout is sent to an outpost town besieged by crooks.
1932 54m/B Bill Cody. VHS, Beta $19.95 *VDM* 🎞½

The Texas Rangers A prisoner is given the opportunity to capture the notorious Sam Bass gang in exchange for freedom. Good B Western; lotsa shootin'.
1951 68m/C George Montgomery, Gale Storm, Jerome Courtland, Noah Beery Jr.; *Dir:* Phil Karlson. VHS, Beta $19.95 *GKK* 🎞🎞

Texas Terror Young Wayne resigns his badge when he mistakenly believes he has shot and killed his friend. He later rescues his friend's sister, learns the truth, and (naturally) finds true love.
1935 50m/B John Wayne, George "Gabby" Hayes, Lucille Browne. VHS $8.95 *NEG, REP, NOS* 🎞🎞

Texas Trouble Western adventure starring the American cowboy Bob Steele.
193? 0m/B Bob Steele. VHS, Beta $27.95 *VCN* 🎞

Texas Wildcats A cowboy helps make sure that sinister adversaries are brought to justice on the American Frontier.
194? 60m/B Tim McCoy. VHS, Beta $19.95 *NOS, DVT* 🎞½

Texasville Sequel to "The Last Picture Show" finds the characters still struggling after 30 years, with financial woes from the energy crisis, mental illness inspired by the Korean War, and various personal tragedies. Lacks the melancholy sensitivity of its predecessor, but has some of the wit and wisdom that comes with age. Based again on a Larry McMurtry novel. Not well received during its theatrical release.
1990 (R) 120m/C Timothy Bottoms, Jeff Bridges, Annie Potts, Cloris Leachman, Eileen Brennan, Randy Quaid, Cybill Shepherd, William McNamara; *Dir:* Peter Bogdanovich. VHS, LV $19.95 *COL, ORI, VMK* 🎞🎞

T.G.I.S.(Thank Goodness It's Shabbat) Designed to introduce the viewer to Jewish holidays; a young man recounts tales to his wife of his childhood visits with his aunt and uncle on Shabbat.
19?? 28m/C Theodore Bikel. VHS, Beta $34.95 *ERG*

Thank God It's Friday Episodic, desultory, boring disco-dancing vehicle won an Oscar for "Last Dance" by Donna Summer. Also featuring the Commodores.
1978 (PG) 100m/C Valerie Landsburg, Teri Nunn, Chick Vennera, Jeff Goldblum, Debra Winger; *Dir:* Robert Klane. Academy Awards '78: Best Song ("Last Dance"). VHS, Beta $12.95 *COL* 🎞

Thank You Mr. President The wit and humor of John F. Kennedy are captured in this documentary that features excerpts of his press conferences.
1984 55m/C *Nar:* E.G. Marshall. VHS, Beta $29.95 *WOV* 🎞

Thank Your Lucky Stars A lavish, slap-dash wartime musical that emptied out the Warner's lot for an array of uncharacteristic celebrity turns, including Davis' rendition of "They're Either Too Young or Too Old." In her movie debut, Shore sings "How Sweet You Are."
1943 127m/B Eddie Cantor, Dinah Shore, Joan Leslie, Errol Flynn, Bette Davis, Edward Everett Horton, Humphrey Bogart, John Garfield, Alan Hale Jr., Ann Sheridan, Ida Lupino, Jack Carson, Dennis Morgan, Olivia de Havilland; *Dir:* David Butler. VHS, Beta, LV $29.95 *MGM* 🎞🎞🎞

Thanksgiving Story When John-Boy's college career is jeopardized by a freak accident, the Walton family gathers together for comfort. Very sentimental family-fare, originally shown as a holiday special from the popular television series "The Waltons."
1973 95m/C Richard Thomas, Ralph Waite, Michael Learned, Ellen Corby, Will Geer; *Dir:* Philip Leacock. VHS, Beta $59.95 *LHV, WAR* 🎞🎞½

Tharus Son of Attila A couple of sword-wielding barbarians battle each other and an evil emperor.
197? 89m/C Jerome Courtland, Rik von Nutter, Lisa Gastoni; *Dir:* Robert Montero. VHS, Beta $59.95 *TWE* 🎞

That Brennan Girl A maudlin soap opera about how a girl's upbringing by an inconsiderate mother makes her a devious little wench. Fortunately her second husband had a better mom, and redemption is at hand.

1946 95m/B James Dunn, Mona Freeman, William Marshall, June Duprez, Frank Jenks, Charles Arnt; *Dir:* Alfred Santell. **VHS $19.95** NOS 𝄢½

That Certain Thing Very early silent Capra comedy about a bachelor with a silver spoon in his mouth who loses his inheritance when he marries for love. It's got that certain Capra screwball feeling.
1928 65m/B Viola Dana, Ralph Graves, Burr McIntosh, Aggie Herring, Syd Crossley; *Dir:* Frank Capra. **VHS, Beta $19.95** GPV 𝄢𝄢½

That Championship Season Long-dormant animosities surface at the reunion of a championship basketball team. Unfortunate remake of Miller's Pulitzer-prize winning play, with none of the fire which made it a Broadway hit.
1982 (R) 110m/C Martin Sheen, Bruce Dern, Stacy Keach, Robert Mitchum, Paul Sorvino, Jason Miller; *Dir:* Jason Miller. **VHS, Beta $24.95** MGM 𝄢𝄢

That Cold Day in the Park Early Altman. An unhappy woman takes in a homeless young man from the park near her home. The woman (Dennis) is obsessive and odd; so is the film. Dennis is excellent. Reminiscent of "The Collector."
1969 (R) 91m/C Sandy Dennis, Michael Burns, Suzanne Benton, Michael Murphy; *Dir:* Robert Altman. **VHS, Beta $14.98** REP, FCT 𝄢𝄢½

That Darn Cat Vintage Disney comedy about a Siamese cat that helps FBI Agent Jones thwart kidnappers. Could be shorter, but suspenseful and funny with characteristic Disney slapstick. Based on the book "Undercover Cat" by The Gordons.
1965 (G) 115m/C Hayley Mills, Dean Jones, Dorothy Provine, Neville Brand, Elsa Lanchester, Frank Gorshin; *Dir:* Robert Stevenson. **VHS, Beta** DIS, OM 𝄢𝄢½

That Darn Sorceress A child witch has grown up, and so has her power. To keep her disciples, she must perform increasingly horrible feats.
1988 89m/C Pauline Adams, Betty Page; *Dir:* Whitney Bain. **VHS, Beta $89.95** MRC 𝄢

That Forsyte Woman Based on the novel "A Man of Property" by John Galsworthy, a married Victorian woman falls for an architect engaged to be wed. Remade later, and better for BBC-television.
1949 112m/C Errol Flynn, Greer Garson, Walter Pidgeon, Robert Young, Janet Leigh, Harry Davenport, Stanley Logan, Lumsden Hare, Aubrey Mather, Matt Moore; *Dir:* Compton Bennett. **VHS $19.98** MGM, CCB 𝄢𝄢

That Gang of Mine "East Side Kids" episode about gang member Mugg's ambition to be a jockey. Good racing scenes.
1940 62m/B Bobby Jordan, Leo Gorcey, Clarence Muse, Dave O'Brien; *Dir:* Joseph H. Lewis. **VHS, Beta $24.95** NOS, DVT, BUR 𝄢𝄢

That Girl from Paris An opera singer from Paris stows away on an ocean liner to be near the swing bandleader she has fallen in love with. An entertaining film with many songs, including: "Seal It with a Kiss," "Moon Face," "The Call to Arms," and "Love and Learn." Remade as "Four Jacks and a Jill."
1936 105m/B Lily Pons, Gene Raymond, Jack Oakie, Herman Bing, Lucille Ball, Mischa Auer, Frank Jenks; *Dir:* Leigh Jason. **VHS, Beta $19.98** TTC 𝄢𝄢½

That Hamilton Woman Screen biography of the tragic 18th-century love affair between British naval hero Lord Nelson and Lady Hamilton. Korda exaggerated the film's historical distortions in order to pass the censor's production code about adultery. Winston Churchill's favorite film which paralleled Britain's heroic struggles in World War II.
1941 125m/B Laurence Olivier, Vivien Leigh, Gladys Cooper, Alan Mowbray, Sara Allgood, Henry Wilcoxon; *Dir:* Alexander Korda. Academy Awards '41: Best Sound. **VHS, Beta, LV $19.98** HBO, SUE 𝄢𝄢𝄢

That Long Night in '43 A gripping story of fascist terror during World War II.
1960 110m/B *Dir:* Florestano Vancini. **VHS, Beta, 3/4U, Special order formats** CIG 𝄢𝄢½

That Lucky Touch York is a reporter covering NATO games; Moore is Bond warmed over as an arms dealer. Sexual and professional tension becomes unlikely romance. Slight and dull. Cobb's last film.
1975 92m/C Roger Moore, Susannah York, Shelley Winters, Lee J. Cobb, Jean-Pierre Cassel, Raf Vallone, Sydne Rome; *Dir:* Christopher Miles. **VHS, Beta $69.95** TWE 𝄢𝄢

That Man Is Pregnant A NYPD detective gets the incredible news that he is pregnant.
1980 (PG) 80m/C **VHS, Beta $39.95** IND 𝄢

That Naughty Girl The bored daughter of a nightclub owner decides to experience all life has to offer.
1958 77m/C *FR* Brigitte Bardot, Jean Bretonniere, Francoise Fabian; *Dir:* Michel Boisrond. **VHS, Beta $69.95** JEF 𝄢𝄢

That Obscure Object of Desire Bunuel's last film, a comic nightmare of sexual frustration. A rich Spaniard obtains a beautiful girlfriend who, while changing physical identities, refuses to sleep with him. Based on a novel by Pierre Louys, which has been used as the premise for several other films, including "The Devil is a Woman," "La Femme et le Pantin," and "The Female."
1977 (R) 100m/C *SP* Fernando Rey, Carol Bouquet, Angela Molina; *Dir:* Luis Bunuel. Los Angeles Film Critics Association Awards '77: Best Foreign Film; National Board of Review Awards '77: Best Director, Best Foreign Film; New York Film Festival '77: Best Foreign Film; National Society of Film Critics Awards '77: Best Foreign Film. **VHS, Beta, LV $29.98** SUE, VYG, APD 𝄢𝄢𝄢½

That Sinking Feeling A group of bored Scot teenagers decide to steal 90 sinks from a plumber's warehouse. One of Forsythe's early films, it was well received and is genuinely funny as the boys try to get rid of the sinks and turn a profit.
1979 (PG) 82m/C *GB* Robert Buchanan, John Hughes, Billy Greenlees, Alan Love; *Dir:* Bill Forsyth. **VHS, Beta $59.98** SUE 𝄢𝄢𝄢

That Summer of White Roses Compelling action drama about a lifeguard working at a resort in Nazi-occupied Yugoslavia and his love for a woman resistance fighter and her child.
1990 (R) 98m/C Tom Conti, Susan George, Rod Steiger; *Dir:* Rajko Grlic. **VHS, Beta, LV $89.98** MED, IME 𝄢𝄢½

That Touch of Mink In New York City, an unemployed secretary finds herself involved with a business tycoon. On a trip to Bermuda, both parties get an education as they play their game of "cat and mouse." Enjoyable romantic comedy.

1962 99m/C Cary Grant, Doris Day, Gig Young, Audrey Meadows, John Astin, Dick Sargent; *Dir:* Delbert Mann. Golden Globe Awards '63: Best Film—Musical/Comedy. **VHS $19.98** REP, MLB 𝄢𝄢½

That Uncertain Feeling Light comedy about a couple's marital problems increasing when she develops the hiccups and a friendship with a flaky piano player. A remake of the earlier Lubitsch silent film, "Kiss Me Again." Available colorized.
1941 86m/B Merle Oberon, Melvyn Douglas, Burgess Meredith, Alan Mowbray, Eve Arden, Sig Rumann, Harry Davenport; *Dir:* Ernst Lubitsch. **VHS, Beta $14.95** NOS, VYY, DVT 𝄢𝄢½

That Was Then...This Is Now Lame adaptation of S. E. Hinton adolescent novel (script by Estevez) about a surly kid who is attached to his adoptive brother and becomes jealous when the brother gets a girlfriend.
1985 (R) 102m/C Emilio Estevez, Craig Sheffer, Kim Delaney, Jill Schoelen, Barbara Babcock, Frank Howard, Larry B. Scott, Morgan Freeman; *Dir:* Christopher Cain. **VHS, Beta, LV $79.95** PAR 𝄢𝄢

That'll Be the Day The early rock 'n' roll of the 1950s is the only outlet for a frustrated young working-class Brit. Prequel to "Stardust." Good, meticulous realism; engrossing story.
1973 86m/C Ringo Starr, Keith Moon, David Essex, Rosemary Leach; *Dir:* Claude Whatham. **VHS, Beta, LV $14.95** HBO 𝄢𝄢½

That's Action! Collection of film stunts; narrated by Culp.
1990 78m/C David Carradine, Robert Ginty, Oliver Reed, Reb Brown; *Nar:* Robert Culp. **VHS $79.95** AIP

That's Adequate Mock documentary about a fictional film studio. Lampoons just about every movie made in the last 60 years. Premise might have been adequate for a sketch, but hardly for a feature-length film. Zaniness gets old quickly.
1990 (R) 82m/C Tony Randall, Robert Downey, Rocky Aoki, Bruce Willis, Robert Townsend, James Coco, Jerry Stiller, Peter Riegert, Susan Dey, Richard Lewis, Robert Vaughn, Renee Taylor, Stuart Pankin, Brother Theodore, Anne Bloom, Chuck McCann, Anne Meara; *Dir:* Harry Hurwitz. **VHS $14.95** HMD, SOU 𝄢𝄢

That's Dancing! This anthology features some of film's finest moments in dance from classical ballet to break-dancing.
1985 104m/C Fred Astaire, Ginger Rogers, Ruby Keeler, Cyd Charisse, Gene Kelly, Shirley MacLaine, Liza Minnelli, Sammy Davis Jr., Mikhail Baryshnikov, Ray Bolger, Jennifer Beals, Dean Martin; *Dir:* Jack Haley Jr. **VHS, Beta, LV $29.95** MGM 𝄢𝄢

That's Entertainment A compilation of scenes from the classic MGM musicals beginning with "The Broadway Melody" (1929) and ending with "Gigi" (1958). Great fun, especially for movie buffs.
1974 (G) 132m/C Judy Garland, Fred Astaire, Frank Sinatra, Gene Kelly, Esther Williams, Bing Crosby; *Dir:* Jack Haley Jr. **VHS, Beta, LV $19.95** MGM, PIA 𝄢𝄢𝄢½

That's Entertainment, Part 2 A cavalcade of great musical and comedy sequences from MGM movies of the past. Also stars Jeanette MacDonald, Nelson Eddy, the Marx Brothers, Laurel and Hardy, Jack Buchanan, Ann Miller, Mickey Rooney, Louis Armstrong, Oscar Levant, Cyd Charisse, Elizabeth Taylor, Maurice Chevalier, Bing Crosby, Jimmy Durante, Clark Gable,

and the Barrymores. Not as unified as its predecessor, but priceless nonetheless.
1976 (G) 133m/C Fred Astaire, Gene Kelly. VHS, Beta, LV $19.95 *MGM* ℐℐ½

That's Life! A lackluster semi-home movie starring Edwards' family and friends. A single weekend in the lives of a writer who's turning sixty, his singer wife who has been diagnosed with cancer, and their family.
1986 (PG-13) 102m/C Jack Lemmon, Julie Andrews, Sally Kellerman, Chris Lemmon, Emma Walton, Rob Knepper, Robert Loggia, Jennifer Edwards; *Dir:* Blake Edwards. VHS, Beta, LV $9.99 *CCB, LIV, VES* ℐℐ

That's My Baby! A man wants his girlfriend to have their baby, but the only thing she wants is a career.
1988 (PG-13) 97m/C Sonja Smits, Timothy Webber; *Dir:* Edie Yolles, John Bradshaw. VHS, Beta $79.98 *TWE* ℐ

That's My Hero! An animated film about Waldo Kitty, a cartoon cat who fantasizes about heroic adventures much like Thurber's Mitty.
1975 72m/C VHS, Beta *SUE, OM* ℐ½

That's Singing A tribute to the American musical theatre featuring performances of memorable songs from twenty Broadway shows.
1984 111m/C *Hosted:* Tom Bosley, Nell Carter, Barry Bostwick, Robert Morse, Debbie Reynolds, Diahann Carroll, Chita Rivera, Ethel Merman, Ray Walston. VHS, Beta, LV $19.98 *LHV, VTK, WAR*

Theatre of Blood Critics are unkind to Shakespearian ham Price, and he eliminates them by various Bard-inspired methods with the assistance of his lovely daughter. Top drawer comedy noir.
1973 (R) 104m/C *GB* Vincent Price, Diana Rigg, Ian Hendry, Robert Morley, Dennis Price, Diana Dors, Milo O'Shea; *Dir:* Douglas Hickox. VHS, LV $59.95 *MGM* ℐℐℐ

Theatre of Death Horror master Lee is back, this time as a theatre director in Paris. Meanwhile, police are baffled by a series of mysterious murders, each bearing a trace of vampirism. Well-plotted suspense. A racy voodoo dance sequence, often cut, is available on some video versions. Also known as "Blood Fiend."
1967 90m/C Christopher Lee, Lelia Goldoni, Julian Glover, Evelyn Laye, Jenny Till, Ivor Dean; *Dir:* Samuel Gallu. VHS, Beta $54.95 *UHV, SNC, VCI* ℐℐ½

Their Only Chance Steve Hoddy, a man with an amazing talent for communicating with animals, is seen as he helps wounded beasts get back into the wilderness.
1978 (G) 84m/C Jock Mahoney, Steve Hoddy; *Dir:* David Siddon. VHS, Beta $9.95 *NO* ℐℐ½

Thelma & Louise Hailed as the first "feminist-buddy" movie, Sarandon and Davis bust out as best friends hittin' the open road and heading directly into one of the better movies of the year. Davis plays the ditzy Thelma, a housewife rebelling against her dominating, unfaithful, abusive husband (who, rather than being disturbing, provides some of the best comic relief in the film.) Sarandon is Louise, a hardened and world-weary waitress in the midst of an unsatisfactory relationship. These two hit the road for a respite from their mundane lives, only to find violence and a part of themselves they never knew existed. Outstanding performances from Davis and especially Sarandon. Laserdisc version features a letterboxed screen

and Dolby surround sound. Director Scott's fine eye for set details is omniscient, although at times the film appears to be a travelogue made for MTV.
1991 (R) 130m/C Susan Sarandon, Geena Davis, Harvey Keitel, Christopher McDonald, Michael Madsen, Brad Pitt, Timothy Carhart; *Dir:* Ridley Scott. Academy Awards '91: Best Original Screenplay; National Board of Review Awards '91: 10 Best Films of the Year, Best Actress (Sarandon, Davis). VHS, Beta, LV, 8mm $94.99 *MGM, PIA* ℐℐℐ

Them! A group of mutated giant ants wreak havoc on a southwestern town. The first of the big-bug movies, and far surpassing the rest, this is a classic fun flick. See how many names you can spot among the supporting cast. Academy Award nomination for Ralph Ayres for special effects.
1954 93m/B James Whitmore, Edmund Gwenn, Fess Parker, James Arness, Onslow Stevens, Jack Perrin; *Dir:* Gordon Douglas. VHS, Beta, LV $19.98 *WAR, LDC* ℐℐℐ

There was a Crooked Man An Arizona town gets some new ideas about law and order when an incorruptible and innovative warden takes over the town's prison. The warden finds he's got his hands full with one inmate determined to escape. Offbeat western black comedy supported by an excellent and entertaining production, with fine acting all around.
1970 (R) 123m/C Kirk Douglas, Henry Fonda, Warren Oates, Hume Cronyn, Burgess Meredith, John Randolph, Arthur O'Connell, Alan Hale Jr., Lee Grant; *Dir:* Joseph L. Mankiewicz. VHS, Beta $19.98 *WAR* ℐℐℐℐ

There's a Girl in My Soup Sellers is a gourmet who moonlights as a self-styled Casanova. Early post "Laugh-In" Hawn is the young girl who takes refuge at his London love nest when her boyfriend dumps her. Lust ensues. Has its funny parts, but not a highlight of anyone's career. Based on the hit play.
1970 (R) 95m/C *GB* Peter Sellers, Goldie Hawn, Diana Dors; *Dir:* Roy Boulting. VHS, Beta $9.95 *GKK* ℐℐ

There's Naked Bodies on My T.V.! A sexy spoof of TV shows, "Happy Daze," "Bernie Milner," and "Don't Come Back Kotler."
197? 79m/C VHS, Beta *MED* ℐ½

There's No Business Like Show Business A top husband and wife vaudevillian act make it a family affair. Showcases 24 songs by Irving Berlin including "There's No Business Like Show Business," "A Pretty Girl is Like a Melody," and Monroe's sultry rendition of "Heat Wave." Filmed in CinemaScope, allowing for full and lavish musical numbers. Good performances and Berlin's music make this an enjoyable film.
1954 117m/C Ethel Merman, Donald O'Connor, Marilyn Monroe, Dan Dailey, Johnny Ray, Mitzi Gaynor, Frank McHugh, Hugh O'Brian; *Dir:* Walter Lang. VHS, Beta, LV $19.98 *FOX, FCT* ℐℐℐ

There's No Time for Love, Charlie Brown An animated adventure with the Peanuts gang as they travel to a museum on a class trip.
1987 30m/C VHS, Beta $14.95 *KRT*

Therese Stylish biography of a young French nun and her devotion to Christ bordering on romantic love. Explores convent life. Real-life Therese died of TB and was made a saint. A directorial tour de force for Cavalier.

1986 90m/C *FR* Catherine Mouchet, Aurore Prieto, Sylvie Habault, Ghislaine Mona; *Dir:* Alain Cavalier. VHS, Beta *MED, APD* ℐℐℐ

Therese & Isabelle Story of growing love and physical attraction between two French schoolgirls. On holiday together, they confront their mutual desires. Richly photographed soft porn. Not vulgar, but certainly for adults.
1968 102m/C Essy Persson, Anna Gael, Barbara Laage, Anne Vernon; *Dir:* Radley Metzger. VHS, Beta $39.95 *MON* ℐℐ½

These Girls Won't Talk Three female stars of early silent pictures are featured separately in: "Her Bridal Nightmare," "Campus Carmen," and "As Luck Would Have It."
192? 50m/B Colleen Moore, Carole Lombard, Betty Compson. VHS, Beta $29.95 *VYY* ℐℐ

These Three A teenaged girl ruins the lives of two teachers by telling a malicious lie. Excellent cast. Script by Lillian Hellman, based on her play "The Children's Hour." Remade by the same director as "The Children's Hour" in 1961.
1936 92m/B Miriam Hopkins, Merle Oberon, Joel McCrea, Bonita Granville, Marcia Mae Jones, Walter Brennan, Margaret Hamilton; *Dir:* William Wyler. VHS, Beta, LV $19.98 *SUE* ℐℐℐ½

They Beings from beneath the earth's crust take their first trip to the surface.
198? 88m/C Paul Dentzer, Debbie Pick, Nick Holt. VHS, Beta $49.95 *REG* ℐ

They All Laughed Three detectives become romantically involved with the women they were hired to investigate. Essentially light-hearted fluff with little or no script. Further undermined by the real-life murder of Stratten before the film's release.
1981 (PG) 115m/C Ben Gazzara, John Ritter, Audrey Hepburn, Colleen Camp, Patti Hansen, Dorothy Stratten, Elizabeth Pena; *Dir:* Peter Bogdanovich. VHS, Beta, LV $19.98 *VES* ℐℐ

They Call Me Bruce A bumbling Bruce Lee look-alike meets a karate-chopping Mafia moll in this farce. Silly premise creates lots of laughs. Also known as "A Fist Full of Chopsticks" and followed by a sequel, "They Still Call Me Bruce."
1982 88m/C Johnny Yune, Margaux Hemingway; *Dir:* Elliot Hong. VHS, Beta $29.98 *VES* ℐℐ

They Call Me Mr. Tibbs! Lieutenant Virgil Tibbs (Poitier) investigates the murder of a prostitute. The prime suspect is his friend, the Reverend Logan Sharpe (Landau). He is torn between his duty as a policeman, his concern for the reverend, and the turmoil of his domestic life. Less tense, less compelling sequel to "In the Heat of the Night."
1970 (PG) 108m/C Sidney Poitier, Barbara McNair, Martin Landau, Juano Hernandez, Anthony Zerbe, Ed Asner, Norma Crane, Jeff Corey; *Dir:* Gordon Douglas. VHS, Beta $19.98 *FOX* ℐℐ

They Call Me Trinity Lazy drifter-gunslinger and his outlaw brother join forces with Mormon farmers to rout bullying outlaws. Spoofs every western cliche with relentless comedy, parodying "The Magnificent Seven" and gibing the spaghetti western. Followed by "Trinity is Still My Name." Dubbed.
1972 (G) 110m/C *IT* Terence Hill, Bud Spencer, Farley Granger, Steffen Zacharias; *Dir:* E.B. Clucher. VHS, Beta $14.95 *CCB, SUE, HHE* ℐℐ½

They Came from Beyond Space
Aliens invade the earth and possess the brains of humans. They only want a few slaves to help them repair their ship which crashed on the moon. The only person able to stop them is a scientist with a steel plate in his head. Silly and forgettable.
1967 86m/C Robert Hutton, Michael Gough; *Dir:* Freddie Francis. VHS, Beta $59.98 SUE *Ä*

They Came to Cordura Mexico, 1916: a woman and six American soldiers—five heroes and one who has been branded a coward—begin a journey to the military headquarters in Cordura. Slow journey brings out personalities. Look for ''Bewitched'' hubby Dick York. Based on Glendon Swarthout's best-seller.
1959 123m/C Gary Cooper, Rita Hayworth, Van Heflin, Tab Hunter, Dick York, Richard Conte; *Dir:* Robert Rossen. VHS, Beta, LV $14.95 COL *ÄÄ½*

They Came from Within The occupants of a high-rise building go on a sex and violence spree when stricken by an aphrodisiac parasite. Queasy, sleazy and weird. Originally released as ''Shivers'' and also known as ''The Parasite Murders.'' First major film by Cronenberg.
1975 (R) 87m/C *CA* Paul Hampton, Joe Silver, Lynn Lowry, Barbara Steele; *Dir:* David Cronenberg. VHS, Beta $69.98 LIV, VES *ÄÄ½*

They Died with Their Boots On The Battle of Little Big Horn is recreated Hollywood style. Takes liberties with historical fact, but still an exciting portrayal of General Custer's last stand. The movie also marks the last time de Havilland worked with Flynn. Nice cameos from the supporting cast. Also available colorized.
1941 141m/B Errol Flynn, Sydney Greenstreet, Anthony Quinn, Hattie McDaniel, Arthur Kennedy, Gene Lockhart, Regis Toomey, Olivia de Havilland, Charley Grapewin; *Dir:* Raoul Walsh. VHS, Beta $19.95 FOX, MLB *ÄÄÄ*

They Drive by Night Two truck-driving brothers break away from a large company and begin independent operations. After an accident to Bogart, Raft is forced to go back to the company where Lupino, the boss's wife, becomes obsessed with him and kills her husband to gain the company and win Raft. When he rejects her, she accuses him of the murder. Well-plotted film with great dialogue. Excellent cast gives it their all.
1940 97m/B Humphrey Bogart, Ann Sheridan, George Raft, Ida Lupino, Alan Hale Jr., Gale Page, Roscoe Karns; *Dir:* Raoul Walsh. VHS, Beta $19.98 FOX, FCT *ÄÄÄ½*

They were Expendable Two American captains pit their PT boats against the Japanese fleet. Based on the true story of a PT boat squadron based in the Philippines during the early days of World War II. One of the best (and most underrated) World War II films. Also available in a colorized version.
1945 135m/B Robert Montgomery, John Wayne, Donna Reed, Jack Holt, Ward Bond, Cameron Mitchell, Leon Ames, Marshall Thompson; *Dir:* John Ford. VHS, Beta $19.95 MGM, BUR *ÄÄÄ½*

They Got Me Covered Two World War II-era journalists get involved in a comic web of murder, kidnapping and romance. Hope is very funny and carries the rest of the cast. Look for Doris Day in a bit part and listen for Bing Crosby singing whenever Hope opens his cigarette case.

1943 95m/B Bob Hope, Dorothy Lamour, Otto Preminger, Eduardo Ciannelli, Donald Meek, Walter Catleh; *Dir:* David Butler. VHS, Beta, LV $19.98 SUE *ÄÄ*

They Knew What They Wanted A lonely San Francisco waitress begins a correspondence romance with a grape grower and agrees to marry him after he sends her a photo which shows him as being young and handsome. She arrives to discover Laughton had sent her a picture of another man. An accident, an affair, and a pregnancy provide further complications before a satisfactory ending is suggested. Flawed adaptation of Sidney Howard's play still has strong performances from Lombard and Laughton. Gargen received a Best Supporting Actor Oscar nomination.
1940 96m/B Charles Laughton, Carole Lombard, Harry Carey, Karl Malden, William Gargan; *Dir:* Garson Kanin. VHS, Beta $34.95 RKO *ÄÄÄ*

They Live A semi-serious science-fiction spoof about a drifter who discovers an alien conspiracy. They're taking over the country under the guise of Reaganism, capitalism and yuppiedom. Screenplay written by Carpenter under a pseudonym. Starts out fun, deteriorates into cliches and bad special effects makeup.
1988 (R) 88m/C Roddy Piper, Keith David, Meg Foster, George Flower, Peter Jason, Raymond St. Jacques, John Lawrence, Sy Richardson; *Dir:* John Carpenter. VHS, Beta, LV $19.95 MCA *ÄÄ½*

They Live by Night A young fugitive sinks deeper into a life of crime thanks to his association with two hardened criminals. Classic film noir was Ray's first attempt at directing. Based on Edward Anderson's novel ''Thieves Like Us,'' under which title it was remade in 1974. Compelling and suspensful.
1949 95m/B Cathy O'Donnell, Farley Granger, Howard da Silva, Jay C. Flippen, Helen Craig; *Dir:* Nicholas Ray. VHS, Beta, LV $59.95 HHT *ÄÄÄ½*

They Made Me a Criminal A champion prizefighter, believing he murdered a man in a drunken brawl, runs away. He finds refuge with the Dead End Kids. Remake of ''The Life of Jimmy Dolan.'' Berkeley was best-known for directing and choreographing musicals and surprisingly did very well with this movie.
1939 92m/B John Garfield, Ann Sheridan, Claude Rains, Leo Gorcey; *Dir:* Busby Berkeley. VHS, Beta, LV $9.95 NEG, NOS, PSM *ÄÄ½*

They Meet Again The final segment of the Dr. Christian series. The country doctor sets out to prove the innocence of a man who has been wrongly accused of stealing money. Slow moving and not particularly exciting.
1941 68m/B Jean Hersholt, Dorothy Lovett; *Dir:* Erle C. Kenton. VHS, Beta, 8mm $19.95 NOS, VYY, KRT *ÄÄ*

They Might Be Giants Woodward is a woman shrink named Watson who treats retired judge Scott for delusions that he is Sherlock Holmes. Scott's brother wants him committed so the family loot will come to him. Very funny in places; Woodward and Scott highlight a solid cast.
1971 (G) 98m/C George C. Scott, Joanne Woodward, Jack Gilford, Eugene Roche, Kitty Winn, F. Murray Abraham, M. Emmet Walsh; *Dir:* Anthony Harvey. VHS, Beta $12.95 MCA *ÄÄÄ*

They Never Come Back A boxer is thrown in jail and then stages a comeback.

1932 62m/B Regis Toomey, Dorothy Sebastian. VHS, Beta, LV $16.95 SNC, WGE *Ä½*

They Paid with Bullets: Chicago 1929 The story of one man's rise to power, and eventual downfall as a Mafia consigliatore.
1977 88m/C Peter Lee Lawrence, Ingrid Schoeller. VHS, Beta $19.95 NEG *Ä½*

They Saved Hitler's Brain Fanatical survivors of the Nazi holocaust gave eternal life to the brain of their leader in the last hours of the war. Now it's on a Caribbean island giving orders again. One of the truly great ''bad'' movies. Shot in pieces in the U.S., the Philippines, and elsewhere, with chunks of other films stuck in to hold the ''story'' together.
1964 91m/B Walter Stocker, Audrey Caire; *Dir:* David Bradley. VHS, Beta, 8mm $29.95 VYY, UHV Woof!

They Shall Have Music A poverty-stricken child hears a concert by violinist Heifetz and decides to become a musician. He joins Brennan's nearly bankrupt music school and persuades Heifetz to play at a benefit concert. Hokey but enjoyable Goldwyn effort to bring classical music to the big screen.
1939 101m/B Walter Brennan, Joel McCrea, Gene Reynolds, Jascha Heifetz, Marjorie Main, Porter Hall, Andrea Leeds, Terence Kilburn; *Dir:* Archie Mayo. VHS, Beta $19.98 SUE *ÄÄ*

They Shoot Horses, Don't They? Powerful period piece depicting the desperation of the Depression Era. Contestants enter a dance marathon in the hopes of winning a cash prize, not realizing that they will be driven to exhaustion. Fascinating and tragic. Oscar nominations for actress Fonda, supporting actress York, director, Adapted Screenplay, Score of a Musical Picture, and Costume Design.
1969 (R) 121m/C Jane Fonda, Michael Sarrazin, Susannah York, Gig Young, Red Buttons, Bonnie Bedelia, Bruce Dern, Allyn Ann McLerie, Chaing I; *Dir:* Sydney Pollack. Academy Awards '69: Best Supporting Actor (Young). VHS, LV ABC, OM *ÄÄ½*

They were So Young European models become the pawns of wealthy Brazilian magnates when they are sent to South America. They are threatened with death if they do not cooperate. The scenes were shot in Italy and Berlin. Well acted and believable, although rather grim story.
1955 78m/B *GE* Scott Brady, Johanna Matz, Raymond Burr; *Dir:* Kurt Neumann. VHS, Beta, LV WGE *ÄÄ½*

They Still Call Me Bruce A Korean searches for the American who saved his life when he was young and instead becomes big brother to an orphan. Supposedly a sequel to ''They Call Me Bruce,'' but has little in common with it. Poorly done and not particularly entertaining.
1986 (PG) 91m/C Johnny Yune, Robert Guillaume, Pat Paulsen; *Dir:* Johnny Yune. VHS, Beta $14.95 NWV, STE *Ä*

They were Ten Ten Soviet Jews make Palestine their home in the 1800s despite Arab and Turkish persecution and pressures among themselves. In Hebrew with English subtitles.
1961 105m/B *IS* Ninette, Oded Teomi, Leo Filer, Yosef Safra; *Dir:* Baruch Dienar. VHS, Beta $79.95 ERG, FCT *ÄÄ½*

They Went That-a-Way & That-a-Way Two bumbling deputies pose as convicts in this madcap prison caper. Script by Conway. Terrible re-hash of Laurel-and-Hardy has nothing going for it. Lackluster.
1978 (PG) 96m/C Tim Conway, Richard Kiel; *Dir:* Edward Montagne. **VHS, Beta** $69.98 *SUE* Woof!

They Won't Believe Me A man plots to kill his wife, but before he does, she commits suicide. He ends up on trial for her "murder." Interesting acting by Young and Hayward against type. Surprising, ironic ending.
1947 95m/B Robert Young, Susan Hayward, Rita Johnson, Jane Greer; *Dir:* Irving Pichel. **VHS, Beta, LV** $19.98 *MED* 🐾🐾🐾

They Won't Forget When a young girl is murdered in a southern town, personal interests take precedence over justice. Turner is excellent in her first billed role, as are the other actors. Superbly scripted by Robert Rossen and Aben Kandel. Expert direction by LeRoy pulls it all together. Based on Ward Greene's "Death in the Deep South."
1937 95m/B Claude Rains, Otto Kruger, Lana Turner, Allyn Joslyn, Elisha Cook Jr., Edward Norris; *Dir:* Mervyn LeRoy. **VHS** *WAR, OM* 🐾🐾🐾½

They're Playing with Fire An English teacher seduces a student and gets him involved in a murder plot to gain an inheritance. Turns out someone else is beating them to the punch. Sleazy semi-pornographic slasher.
1984 (R) 96m/C Sybil Danning, Eric Brown, Andrew Prine, Paul Clemens, K.T. Stevens, Alvy Moore; *Dir:* Howard Avedis. **VHS, Beta** $19.95 *OTC, HBO* Woof!

The Thief An American commits treason and is overcome by guilt. Novel because there is not one word of dialogue, though the gimmmjck can't sustain the ordinary story. A product of the Communist scare of the 1950s, it tends to be pretentious and melodramatic.
1952 84m/C Ray Milland, Rita Gam, Martin Gabel, Harry Bronson, John McKutcheon; *Dir:* Russel Rouse. **VHS, Beta** $19.95 *UHV* 🐾🐾

Thief A successful businessman (Crenna) attempts to put his criminal past behind him. He's trapped when looking for a quick way to get some money to pay off a gambling debt. Made for television drama that's well-written and directed with tepid acting.
1971 74m/C Richard Crenna, Angie Dickinson, Cameron Mitchell, Hurd Hatfield, Robert Webber; *Dir:* William A. Graham. **VHS, Beta** $59.95 *LHV* 🐾🐾

Thief A big-time professional thief likes working solo, but decides to sign up with the mob in order to make one more big score and retire. He finds out it's not that easy. Taut and atmospheric thriller with score by Tangerine Dream. Director Michael Mann's feature film debut.
1981 (R) 126m/C James Caan, Tuesday Weld, Willie Nelson, James Belushi, Elizabeth Pena, Robert Prosky; *Dir:* Michael Mann. **VHS, Beta, LV** $19.95 *MGM, FOX* 🐾🐾🐾

Thief of Baghdad The classic silent crowd-pleaser, about a roguish thief who uses a genie's magic to outwit Baghdad's evil Caliph. With famous special effects in a newly struck print, and a new score by Carl Davis based on Rimsky-Korsakov's "Scheherezade." Remade many times.
1924 153m/B Douglas Fairbanks Sr., Snitz Edwards, Charles Belcher, Anna May Wong, Etta Lee, Brandon Hurst, Sojin, Julanne Johnston; *Dir:* Raoul Walsh. **VHS, Beta, LV** $19.98 *HBO, SNC, NOS* 🐾🐾🐾

Thief of Baghdad A wily young thief enlists the aid of a powerful genie to outwit the Grand Vizier of Baghdad. An Arabian Nights spectacular with lush photography, fine special effects, and striking score by Miklos Rozsa. Outstanding performance by Ingram as the genie.
1940 106m/C Sabu, Conrad Veidt, June Duprez, Rex Ingram, Tim Whelan, Michael Powell; *Dir:* Ludwig Berger. **VHS, Beta, LV** $14.98 *SUE, MLB* 🐾🐾🐾½

Thief of Baghdad An Arabian Nights fantasy about a thief in love with a Sultan's daughter who has been poisoned. He seeks out the magical blue rose which is the antidote. Not as lavish as the previous two productions by this title, nor as much fun.
1961 89m/C *IT* Georgia Moll, Steve Reeves; *Dir:* Arthur Lubin. **VHS, Beta** $59.98 *SUE* 🐾🐾

Thief of Baghdad A fantasy-adventure about a genie, a prince, beautiful maidens, a happy-go-lucky thief, and magic. Easy to take, but not an extraordinary version of this oft-told tale. Ustinov is fun as he tries to marry off his daughter. Made for television.
1978 (G) 101m/C Peter Ustinov, Roddy McDowall, Terence Stamp, Frank Finlay, Ian Holm; *Dir:* Clive Donner. **VHS, Beta** $29.95 *GEM* 🐾🐾½

Thief of Hearts A thief steals a woman's diary when he's ransacking her house. He then pursues the woman, using his secret knowledge. Slick, but too creepy. Re-edited with soft porn scenes in some video versions. Stewart's first film.
1984 (R) 101m/C Steven Bauer, Barbara Williams, John Getz, Christine Ebersole, George Wendt; *Dir:* Douglas Day Stewart. **VHS, Beta, LV** $79.95 *PAR* 🐾🐾

Thief Who Came to Dinner A computer analyst and a wealthy socialite team up to become jewel thieves and turn the tables on Houston's high society set. Comedy plot gets complex, though O'Neal and Bisset are likeable. Early Clayburgh role as O'Neal's ex. Based on a novel by Terence Lore Smith, music by Henry Mancini.
1973 (PG) 103m/C Ryan O'Neal, Jacqueline Bisset, Warren Oates, Jill Clayburgh, Ned Beatty, Gregory Sierra, Michael Murphy, Austin Pendleton, John Hillerman, Charles Cioffi; *Dir:* Bud Yorkin. **VHS, Beta** $59.95 *WAR* 🐾🐾

Thieves of Fortune Former Miss Universe stars as a contestant in a high stakes fortune hunt that spans half the globe. She meets with more than her share of action-packed encounters among the ruffians who compete for the $28 million purse. Implausible; horrible script; cliched.
1989 (R) 100m/C Michael Nouri, Lee Van Cleef, Shawn Weatherly. **VHS, Beta** $29.95 *ACA* Woof!

Thighs and Whispers: The History of Lingerie A brief history of lingerie, entertaining and informative.
1982 45m/C VHS, Beta $39.98 *LHV*

The Thin Blue Line The lavishly acclaimed docudrama about the 1977 shooting of a cop in Dallas County, and the incorrect conviction of Randall Adams for the crime. A riveting, spellbinding experience. Due to continued lobbying by Morris and the film's impact, Adams has now been freed. Music by Philip Glass.

1988 101m/C *Dir:* Errol Morris. Edgar Allan Poe Awards '88: Best Screenplay. **VHS, Beta, LV** $19.98 *HBO, CCB* 🐾🐾🐾½

The Thin Man Married sleuths Nick and Nora Charles investigate the mysterious disappearance of a wealthy inventor. Charming and sophisticated, this was the model for all husband-and-wife detective teams that followed. Don't miss Asta, their wire-hair terrier. Based on the novel by Dashiell Hammett. Its enormous popularity triggered five sequels, starting with "After the Thin Man."
1934 90m/B William Powell, Myrna Loy, Maureen O'Sullivan, Cesar Romero, Porter Hall, Nat Pendleton, Minna Gombell, Natalie Moorhead, Edward Ellis; *Dir:* W.S. Van Dyke. **VHS, Beta, LV** $19.95 *MGM* 🐾🐾🐾

The Thin Man Goes Home Married sleuths Nick and Nora Charles solve a mystery with Nick's disapproving parents looking on in the fifth film from the "Thin Man" series. Despite a three year gap and slightly less chemistry between Powell and Loy, audiences welcomed the skinny dude. Sequel to "Shadow of the Thin Man;" followed by "Song of the Thin Man."
1944 100m/B William Powell, Myrna Loy, Lucile Watson, Gloria DeHaven, Anne Revere, Harry Davenport, Helen Vinson, Lloyd Corrigan, Donald Meek, Edward Brophy; *Dir:* Richard Thorpe. **VHS, Beta** $19.98 *MGM* 🐾🐾½

The Thing One of the best of the Cold War allegories and a potent lesson to those who won't eat their vegetables. In the original version of this science fiction classic, an alien craft and creature (Arness in monster drag) are discovered by an Arctic research team. They take it back to the base where the critter is accidentally thawed. The creature then wreaks havoc, sucking the blood from sled dog and scientist alike. The scientists soon discover they have a giant seed-dispersing vegetable running amuck, little affected by missing body parts, bullets, or cold. In other words, Big Trouble. Excellent direction, assisted substantially by producer Hawks, and supported by strong performances. Available colorized; remade in 1982. Loosely based on "Who Goes There?" by John Campbell; adapted by Charles Lederer; score by Dimith Tiomkin.
1951 87m/B James Arness, Kenneth Tobey, Margaret Sheridan, Dewey Martin; *Dir:* Christian Nyby, Howard Hawks. **VHS, Beta, LV** $19.95 *MED, TTC, KOV* 🐾🐾🐾½

The Thing A team of scientists at a remote Antarctic outpost discover a buried spaceship with an unwelcome alien survivor still alive. Bombastic special effects overwhelm the suspense and the solid cast. Less a remake of the 1951 science fiction classic than a more faithful version of the original novel, since the seeds/spores take on human shapes.
1982 (R) 127m/C Kurt Russell, Wilford Brimley, T.K. Carter, Richard Masur, Keith David, Richard Dysart, David Clennon, Donald Moffatt, Thomas G. Waites, Charles Hallahan; *Dir:* John Carpenter. **VHS, Beta, LV** $19.95 *MCA* 🐾🐾½

Things A man creates a monster during his freakish experiments with artificial insemination. The monster returns to his house seeking revenge, and the man's visiting brother and a friend disappear, probably a clever move given their lackluster circumstances.
198? 90m/C Barry Gillis, Amber Lynn, Doug Bunston, Bruce Roach; *Dir:* Andrew Jordan. **VHS, Beta** *IEC* 🐾

Things Change An old Italian shoeshine agrees, for a fee, to be a fall guy for the Mafia. A lower-echelon mob hood, assigned to watch over him, decides to give the old guy a weekend of fun in Vegas before going to jail. Mamet's second film as director; he co-wrote the screenplay with Silverstein, best-known for his children's books ("Where the Sidewalk Ends," "The Giving Tree"). Combines charm and menace with terrific performances, especially from Ameche and Mantegna.
1988 (PG) 114m/C Joe Mantegna, Don Ameche, Robert Prosky, J.J. Johnston, Ricky Jay; *Dir:* David Mamet. Venice Film Festival '88: Best Actor (Ameche & Mantegna). **VHS, Beta, LV $14.95** *COL* 🐾🐾🐾

Things to Come Using technology, scientists aim to rebuild the world after a lengthy war, followed by a plague and other unfortunate events. Massey and Scott each play two roles, in different generations. Startling picture of the world to come, with fine sets and good acting. Based on an H.G. Wells story, "The Shape of Things to Come."
1936 92m/B *GB* Raymond Massey, Ralph Richardson, Cedric Hardwicke, Derrick DeMarney; *Dir:* William Cameron Menzies. **VHS, Beta $19.95** *NOS, PSM, SNC* 🐾🐾🐾½

Things Happen at Night Scientist and insurance investigator determine that friendly ghost has possessed a family's youngest daughter. Not much happens.
1948 79m/B *GB* Gordon Harker, Alfred Drayton, Robertson Hare, Olga Lindo, Wylie Watson; *Dir:* Francis Searle. **VHS, Beta $16.95** *NOS, SNC* 🐾

Things in Their Season Made for TV melodrama about the imminent death of the mother of a Wisconsin farm family who makes those around her realize the value of happiness. Neal is excellent, rest of cast better than average in a well-told and sincere film.
1974 79m/C Patricia Neal, Ed Flanders, Marc Singer, Meg Foster; *Dir:* James Goldstone. **VHS, Beta** *LCA* 🐾🐾½

Things We Did Last Summer Members of the original cast of "Saturday Night Live" are featured in this special program of skits that show how they spent their summer vacations. Aykroyd and Belushi perform live as the Blues Brothers.
1978 46m/C John Belushi, Dan Aykroyd, Bill Murray, Gilda Radner, Garrett Morris, Laraine Newman; *Dir:* Gary Weis. **VHS, Beta $19.95** *PAV* 🐾🐾½

Think Big Former professional wrestlers, the Pauls are truck drivers in this silly, enjoyable comedy. They pick up a brilliant teenager running from bad guys. Mayhem ensues.
1990 (PG-13) 86m/C Peter Paul, David Paul, Martin Mull, Ari Meyers, Richard Kiel, David Carradine, Richard Moll, Peter Lupus. **VHS, Beta, LV $89.98** *IVE* 🐾🐾½

Think Dirty An advertising executive develops a series of sexy commercials to sell cereal. At the same time his wife forms a "clean up television" group. Not very interesting or funny. Originally titled "Every Home Should Have One."
1970 (R) 93m/C *GB* Marty Feldman, Judy Cornwell, Shelley Berman; *Dir:* Jim Clark. **VHS, Beta $59.95** *COL* 🐾

Thinkin' Big Guys and gals cavort on the Texas coast, except for one lonely fat guy who tries to increase his sexual prowess. The usual teen sex frenzy.

1987 (R) 96m/C Bruce Anderson, Kenny Sargent, Randy Jandt, Nancy Buechler, Darla Ralston; *Dir:* S.F. Brownrigg. **VHS, Beta $9.99** *STE, PSM* 🐾

Third Degree Burn A made-for-cable-television mystery about a small-time detective hired by a businessman to follow a woman, only to become a suspect when the woman's real husband is killed.
1989 97m/C Treat Williams, Virginia Madsen, Richard Masur; *Dir:* Roger Spottiswoode. **VHS, Beta $14.95** *PAR* 🐾🐾

The Third Key A solid British crime drama follows a dogged detective as he probes serial safe-crackings, with a climax set at London's Festival Hall. Originally known as "The Long Arm," based on the book by Robert Barr, who co-scripted.
1957 96m/B Jack Hawkins, Dorothy Allison, Geoffrey Keen, Richard Leech, Ian Bannen; *Dir:* Charles Frend. **VHS $19.95** *NOS* 🐾🐾½

The Third Man An American writer of pulp westerns (Cotten) arrives in post-war Vienna to take a job with an old friend, but discovers he has been murdered. Or has he? Based on Graham Greene's mystery, this classic film noir thriller plays on national loyalties during the Cold War. Welles is top-notch as the manipulative Harry Lime, black-market drugdealer extraordinare. The underground sewer sequence is not to be missed. With a haunting (sometimes irritating) theme by Anton Karas on unaccompanied zither and Oscar-winning cinematography by Robert Krasker. Oscar nominations: Best Director, Film Editing. Special edition features trailer.
1949 104m/B Joseph Cotten, Orson Welles, Alida Valli, Trevor Howard, Bernard Lee, Wilfrid Hyde-White; *Dir:* Carol Reed. Academy Awards '50: Best Black and White Cinematography; British Academy Awards '49: Best Film; Cannes Film Festival '49: Best Film; Directors Guild of America Awards '49: Best Director (Reed); National Board of Review Awards '50: 5 Best Foreign Films of the Year. **VHS, Beta, LV, 8mm $16.95** *SNC, REP, MED* 🐾🐾🐾🐾

Third Man on the Mountain A family epic about mountain climbing, shot in Switzerland and based on James Ramsey Ullman's "Banner in the Sky." A young man is determined to climb the "Citadel" as his ancestors have. He finds there's more to climbing than he imagined. Look for Helen Hayes (MacArthur's mother) in a cameo. Standard Disney adventure drama.
1959 106m/C James MacArthur, Michael Rennie, Janet Munro, James Donald, Herbert Lom, Laurence Naismith, Helen Hayes; *Dir:* Ken Annakin. **VHS, Beta $69.95** *DIS* 🐾🐾½

Third Solution An Italian-made film about the discovery of a secret pact made between the Kremlin and the Vatican that may start World War III. Well-cast, but poorly conceived spook drama, with music by Vangelis.
1989 (R) 93m/C *IT* Treat Williams, F. Murray Abraham, Danny Aiello, Nigel Court, Rita Rusic, Rossano Brazzi; *Dir:* Pasquale Squitieri. **VHS, Beta, LV $79.95** *COL* 🐾🐾

Third Walker A desperate woman sacrifices her marriage to immerse herself in efforts to reunite her twin sons, who were inadvertently separated at birth. An original and intriguing directorial debut by McLuhan.
1979 85m/C William Shatner, Colleen Dewhurst; *Dir:* Marshall McLuhan. **VHS, Beta $59.95** *USA* 🐾🐾🐾

Thirst A girl is abducted by a secret society that wants her to become their new leader. There is just one catch: she has to learn to like the taste of human blood. Chilling but weakly plotted.
1987 (R) 96m/C David Hemmings, Henry Silva, Chantal Contouri; *Dir:* Rod Hardy. **VHS, Beta $79.95** *MED* 🐾🐾

The Thirsty Dead Maybe they could learn to like Gatorade. An eternally young jungle king is looking for a wife. She can be young forever too, if she's not above vampirism and sacrificing virgins. In spite of the jungle vampire slant, this Filipino-made movie is a woofer. Also released on video, as "Blood Hunt."
1974 90m/C *PH* John Considine, Jennifer Billingsley, Judith McConnell, Fredricka Meyers, Tani Phelps Guthrie; *Dir:* Terry Becker. **VHS, Beta $59.95** *KOV, WES, SIM* Woof!

13 Ghosts A dozen ghosts need another member to round out their ranks. They have four likely candidates to choose from when a family moves into the house inhabited by the ghoulish group. Originally viewed with "Illusion-O," a technology much like 3-D, which allowed the viewing of the ghosts only through a special pair of glasses.
1960 88m/C Charles Herbert, Jo Morrow, Martin Milner, Rosemary de Camp, Donald Woods, Margaret Hamilton; *Dir:* William Castle. **VHS, Beta $9.95** *GKK* 🐾🐾½

13 Rue Madeleine Cagney plays a World War II spy who infiltrates Gestapo headquarters in Paris in order to find the location of a German missile site. Actual OSS footage is used in this fast-paced early postwar espionage propaganda piece. Rex Harrison rejected the part taken by Cagney.
1946 95m/B James Cagney, Annabella, Richard Conte, Frank Latimore, Walter Abel, Sam Jaffe, Melville Cooper, E.G. Marshall, Karl Malden, Red Buttons; *Dir:* Henry Hathaway. **VHS, Beta $19.98** *FOX, FCT* 🐾🐾🐾

Thirteenth Day of Christmas A made-for-British-television yarn about a psychotic boy left alone by his parents on Christmas night.
1985 60m/C *GB* Patrick Allen, Elizabeth Spiggs. **VHS, Beta $59.95** *PSM* 🐾🐾🐾

The 13th Floor A young girl fuses with the spirit of a young boy her father ruthlessly killed years before, and together they wreak havoc.
1988 (R) 86m/C Lisa Hensley, Tim McKenzie, Miranda Otto; *Dir:* Chris Roach. **VHS, Beta $89.99** *PAR* 🐾½

13th Guest Two people try to solve a murder that occurred at a dinner party. An incredibly creaky early talkie melodrama that created some of the cliches of the genre.
1932 70m/B Ginger Rogers, Lyle Talbot, J. Farrell MacDonald, Paul Hurst; *Dir:* Albert Ray. **VHS, Beta $12.95** *SNC, NOS, VYY* 🐾🐾

The Thirteenth Man When a journalist investigating the murder of a district attorney is also killed, the paper's gossip columnist takes the case. Entertaining, fast-paced mystery.
1937 70m/B Weldon Heyburn, Inez Courtney, Selmer Jackson, Milburn Stone, Matty Fain; *Dir:* William Nigh. **VHS, Beta $16.95** *NOS, SNC, MLB* 🐾🐾½

The 13th Mission A low-budget overseas combat adventure, no better than the first twelve.

1991 95m/C Robert Marius, Jeff Griffith, Michael Monty, David Morisson, Paul Home, John Falch, Albert Bronski, Chantal Manz; *Dir:* Antonio Perez. VHS *SHE* ♫½

Thirteenth Reunion A newspaper-woman uncovers a bizarre secret society when she does a routine story on a health spa.
1981 60m/C Julia Foster, Dinah Sheridan, Richard Pearson; *Dir:* Peter Sasdy. VHS, Beta $29.95 *IVE* ♫

'38: Vienna Before the Fall Wartime Vienna explodes with love and politics. Academy Award Nominee for best foreign language film. In German with English subtitles.
1988 97m/C *GE* Tobias Engel, Sunnyi Melles; *Dir:* Wolfgang Gluck. VHS, LV $34.95 *LUM, CRO* ♫♫½

30-Foot Bride of Candy Rock A junk dealer invents a robot, catapults into space and causes his girlfriend to grow to 30 feet in height. Lightweight, whimsical fantasy was Costello's only solo starring film, and his last before his untimely death.
1959 73m/B Lou Costello, Dorothy Provine, Gale Gordon; *Dir:* Sidney Miller. VHS, Beta $9.95 *GKK* ♫

30 Is a Dangerous Age, Cynthia Dudley Moore stars as a nightclub pianist who spends a lot of his time daydreaming about being rich, famous and married to a beautiful woman. Moore determines to attain all these dreams before his thirtieth birthday, which is only six weeks away. Light-hearted fun. Moore wrote the score and co-wrote the screenplay.
1968 85m/C *GB* Dudley Moore, Suzy Kendall, Eddie Foy Jr.; *Dir:* Joseph McGrath. VHS, Beta $59.95 *RCA* ♫♫½

The 39 Steps The classic Hitchcock mistaken-man-caught-in-intrigue thriller, featuring some of his most often copied set-pieces and the surest visual flair of his pre-war British period. The laserdisc version includes a twenty-minute documentary tracing the director's British period. Remade twice, in 1959 and 1979.
1935 81m/B *GB* Robert Donat, Madeleine Carroll, Godfrey Tearle, Lucie Mannheim, Peggy Ashcroft; *Dir:* Alfred Hitchcock. VHS, Beta, LV, 8mm $8.95 *NEG, NOS, SUE* ♫♫♫♫

The Thirty-Nine Steps Hitchcock remake is a visually interesting, but mostly uninvolving, mystery. Powell is the man suspected of stealing plans to begin World War I. Above average, but not by much.
1979 98m/C *GB* Robert Powell, David Warner, Eric Porter, Karen Dotrice, John Mills, Andrew Keir; *Dir:* Don Sharp. VHS, Beta $14.95 *MED* ♫♫½

Thirty Seconds Over Tokyo Written by Dalton Trumbo and based on a true story. Dated but still interesting classic wartime flagwaver details the conception and execution of the first bombing raids on Tokyo by Lt. Col. James Doolittle and his men. Look for Blake Edwards, as well as Steve Brodie in his first screen appearance. Received an Oscar for Best Special Effects.
1944 138m/B Spencer Tracy, Van Johnson, Robert Walker, Robert Mitchum, Phyllis Thaxter, Scott McKay, Stephen McNally, Louis Jean Heydt, Leon Ames, Paul Langton; *Dir:* Mervyn LeRoy. Academy Awards '44: Best Special Effects. VHS, Beta $19.98 *MGM, FCT, TLF* ♫♫♫

36 Fillete Intellectually precocious 14-year old French girl (dress size 36 fillete) discovers her sexuality. Critically acclaimed.

1988 88m/C *FR* Delphine Zentout, Etienne Chicot, Oliver Parniere, Jean-Pierre Leaud; *Dir:* Catherine Breillat. VHS $79.95 *FXL* ♫♫♫

Thirty-Six Hours of Hell A troop of Marines battle Japanese forces in the South Pacific during World War II. Dubbed.
1977 95m/C Richard Harrison, Pamela Tudor; *Dir:* Roberto Marrtero. VHS, Beta $59.95 *UNI* ♫½

This is the Army A robust tribute to the American soldier of World War II based on the hit play by Irving Berlin. Also contains many of his songs, including "This Is the Army, Mr. Jones," "I Left My Heart at the Stage Door Canteen," "Oh, How I Hate to Get Up in the Morning," and "God Bless America" (sung by Kate Smith). Murphy, who later was a senator from California, played Reagan's father.
1943 105m/C George Murphy, Joan Leslie, Ronald Reagan, Alan Hale Jr., Kate Smith, George Tobias, Irving Berlin, Joe Louis; *Dir:* Michael Curtiz. Academy Awards '43: Best Musical Score. VHS, Beta $19.95 *NOS, DVT, CAB* ♫♫½

This is Elvis The life of Elvis, combining documentary footage with dramatizations of events in his life. Includes more than three dozen songs. Generally seen as an attempt to cash in on the myth, but was well-received by his fans.
1981 (PG) 144m/C Elvis Presley; *Dir:* Malcolm Leo. VHS, Beta $19.95 *MVD, WAR* ♫♫

This Gun for Hire In his first major film role, Ladd plays a hired gun seeking retribution from a client who betrays him. Preston is the cop pursuing him and hostage Lake. Ladd's performance as the cold-blooded killer with a soft spot for cats is stunning; his train-yard scene is an emotional powerhouse. Based on Graham Greene's novel "A Gun for Sale." Remade as "Short Cut to Hell." The first of several films using the Ladd-Lake team, but the only one in which Ladd played a villain.
1942 81m/B Alan Ladd, Veronica Lake, Robert Preston, Laird Cregar; *Dir:* Frank Tuttle. VHS, Beta, LV $39.95 *MCA* ♫♫♫

This Happy Breed A celebrated film version of Noel Coward's classic play depicts the changing fortunes of a large family in England between the world wars. Happiness, hardships, triumph and tragedy mix a series of memorable episodes, including some of the most cherished moments in popular British cinema, though its appeal is universal. Narrated by Laurence Olivier.
1947 114m/C *GB* Robert Newton, Celia Johnson, John Mills, Kay Walsh, Stanley Holloway; *W/Dir:* David Lean. VHS $39.95 *HMV, FCT* ♫♫♫½

This Happy Feeling Reynolds is a young woman who is romantically attracted to both Jurgens, a sophisticated retired actor, and Saxon, a younger suitor who is Jurgens' neighbor. Good acting all around. Charming telling of old tale.
1958 83m/C Curt Jurgens, Debbie Reynolds, John Saxon, Alexis Smith, Mary Astor, Estelle Winwood; *Dir:* Blake Edwards. VHS, Beta $14.95 *KRT, FCT* ♫♫½

This is a Hijack A gambler hijacks the plane carrying his wealthy boss to help pay his debts. What it lacks in suspense, it lacks in performance and direction as well.
1975 (PG) 90m/C Adam Roarke, Neville Brand, Jay Robinson, Lynn Borden, Dub Taylor. VHS, Beta $39.95 *MON* ♫

This Island Earth The planet Metaluna is in desperate need of uranium to power its defense against enemy invaders. A nuclear scientist and a nuclear fission expert from Earth are kidnapped to help out. The first serious movie about interplanetary escapades. Bud Westmore created pulsating cranium special effects make-up.
1955 86m/C Jeff Morrow, Faith Domergue, Rex Reason, Russell Johnson; *Dir:* Joseph M. Newman. VHS, Beta, LV $19.95 *MCA, MLB* ♫♫½

This Land is Mine A timid French schoolteacher gathers enough courage to defy the Nazis when they attempt to occupy his town. Laughton's characterization is effective as the meek fellow who discovers the hero within himself in this patriotic wartime flick.
1943 103m/B Charles Laughton, Maureen O'Hara, George Sanders, Walter Slezak, George Coulouris, Una O'Connor; *Dir:* Jean Renoir. Academy Awards '43: Best Sound. VHS, Beta, LV $19.98 *RKO* ♫♫

This Lightning Always Strikes Twice An elegant murder mystery made for British television.
1985 60m/C *GB* Claire Bloom, Trevor Howard, Charles Dance. VHS, Beta $59.95 *PSM* ♫♫♫

This Man Must Die A man searches relentlessly for the driver who killed his young son in a hit-and-run. When found, the driver engages him in a complex cat-and-mouse chase. Another stunning crime and punishment tale from Chabrol.
1970 (PG) 112m/C *FR* Michael Duchaussoy, Caroline Cellier, Jean Yanne; *Dir:* Claude Chabrol. National Board of Review Awards '70: 5 Best Foreign Films of the Year. VHS, Beta $59.95 *FOX* ♫♫♫½

This is My Life The story of a working mother torn between her skyrocketing career as a stand-up comic and her two daughters. Kavner plays the divorced mom who is determined to chuck her cosmetic sales job for comic success. Her career starts to really take off; she hires an agent and makes appearances on talk shows. As offers pour in, she eventually realizes that her girls are suffering as a result of her success. Good performances highlight an otherwise average drama.
1991 (PG-13) 105m/C Julie Kavner, Samantha Mathis, Carrie Fisher, Dan Aykroyd, Gaby Hoffman; *Dir:* Nora Ephron. VHS $94.98 *FXV* ♫♫

This is Not a Test When news comes of an impending nuclear attack, a state trooper at a roadblock offers sanctuary to passing travellers. The effectiveness of the film's social commentary is hindered by its small budget.
1962 72m/B Seamon Glass, Mary Morlass, Thayer Roberts, Aubrey Martin; *Dir:* Frederic Gadette. VHS, Beta $16.95 *SNC* ♫♫

This Nude World A campy "expose" of the secret world of nudist colonies, including a look at volleyball players in Camp Olympia, New York.
1932 55m/B VHS, Beta $24.95 *NOS, VDM, MLB* ♫♫

This Property is Condemned The lure of steamy sex fills a southern boardinghouse when a mother offers her daughter (Wood) to a wealthy tenant (Redford). But the daughter loves another man. Based on the Tennessee Williams play; the author was reportedly unhappy with the script and wanted his name removed from the film.

1966 109m/C Robert Redford, Natalie Wood, Charles Bronson, Kate Reid, Mary Badham, Jon Provost, Robert Blake; *Dir:* Sydney Pollack. **VHS, Beta, LV** $19.95 *PAR, PIA* 𝒥𝒥

This Special Friendship Powerful, if overdone French drama about an emotionally sensitive young man at a Catholic boarding school. His sexual relationship with another boy leads to tragedy. Based on the novel by Roger Peyrefitte.
1967 99m/B *FR* Francis Lacombrade, Didier Haudepin, Lucien Nat, Louis Seigner, Michel Bouquet, Francois Leccia; *Dir:* Jean Delannoy. **VHS, Beta** *AWF* 𝒥𝒥½

This is Spinal Tap Pseudo-rockumentary about heavy-metal band Spinal Tap, profiling their career from "England's loudest band" to an entry in the "where are they now file." Hilarious satire, featuring music performed by Guest, McKean, Shearer, and others. Included are Spinal Tap's music video "Hell Hole," and an ad for their greatest hits album, "Heavy Metal Memories." Features great cameos, particularly David Letterman's Paul Schaefer as a record promoter and Billy Crystal as a surly mime. First feature for Reiner (Meathead on "All in the Family").
1984 (R) 82m/C Michael McKean, Christopher Guest, Harry Shearer, Tony Hendra, Bruno Kirby, Rob Reiner, June Chadwick, Howard Hesseman, Billy Crystal, Dana Carvey, Ed Begley Jr., Patrick Macnee, Paul Schaffer, Anjelica Huston; *Dir:* Rob Reiner. **VHS, Beta, LV, 8mm** $14.95 *MVD, SUE* 𝒥𝒥𝒥½

This Sporting Life A gritty, depressive portrait of a former coal miner breaking into the violent world of professional rugby whose inability to handle social differences causes problems. He begins an affair with the widow from whom he rents a room but finds they are capable of only a physical attachment. One of the best of the British early 60s working-class angry young man melodramas. Written by David Storey. Harris and Roberts received Oscar nominations for their work.
1963 134m/B *GB* Richard Harris, Rachel Roberts, Alan Badel, William Hartnell, Colin Blakely, Vanda Godsell, Arthur Lowe; *Dir:* Lindsay Anderson. British Academy Awards '63: Best Actor (Harris); Cannes Film Festival '63: International Critics Prize, Best Actor (Harris). **VHS, Beta** $19.95 *PAR* 𝒥𝒥𝒥

This Stuff'll Kill Ya! A backwoods preacher who believes in free love and moonshining runs into trouble with the locals when a series of gruesome religious murders are committed. Southern drive-in material from one of the genre's masters, Holt's last film.
1971 (PG) 100m/C Jeffrey Allen, Tim Holt, Gloria King, Ray Sager, Eric Bradly, Terence McCarthy; *W/Dir:* Herschell Gordon Lewis. **VHS** $29.95 *TWE, FRG Woof!*

This Time I'll Make You Rich Two American hoodlums work out a drug-ring heist in the Far East.
1975 (PG) 97m/C Tony Sabato, Robin McDavid; *Dir:* Frank Kramer. **VHS, Beta** $59.98 *CHA* 𝒥

Thomas Crown Affair A multi-millionaire (McQueen) decides to plot and execute the perfect theft, a daring daylight robbery of a bank. Dunaway is the gorgeous and efficient insurance investigator determined to nab him. One of the best visual scenes is the chess match between the two as they begin to fall in love. Strong production with Oscar-winning theme "Windmills of Your Mind."

1968 102m/C Steve McQueen, Faye Dunaway, Jack Weston, Yaphet Kotto; *Dir:* Norman Jewison. Academy Awards '68: Best Song ("Windmills of Your Mind"). **VHS, Beta** $19.98 *FOX, FCT* 𝒥𝒥𝒥

Thomas Graal's Best Film The famous comical adventures of a film scriptwriter wooing his secretary, followed by "Thomas Graal's First Child." Silent.
1917 70m/B *SW* Victor Sjostrom, Karin Molander; *Dir:* Mauritz Stiller. **VHS, Beta** $49.95 *FCT* 𝒥𝒥𝒥

Thomas Graal's First Child The sequel to "Thomas Graal's Best Film." The comical screenwriter and actress couple attempt to raise a baby. Silent.
1918 70m/B *SW* Victor Sjostrom, Karin Molander; *Dir:* Mauritz Stiller. **VHS, Beta** $49.95 *FCT* 𝒥𝒥𝒥

Thompson's Last Run Boyhood friends grow up on either side of the law. They come face-to-face, though, when one is assigned to transport the other to prison. Made for television.
1990 95m/C Robert Mitchum, Wilford Brimley, Kathleen York, Susan Tyrrell; *Dir:* Jerrold Freedman. **VHS, Beta, LV** $79.95 *NSV* 𝒥½

Thor and the Amazon Women The mighty Thor leads his enslaved men on an attack against the evil Queen Nera and her female-dominated Amazon society.
1960 85m/C Joe Robinson, Susy Anderson. **VHS** $39.95 *AIP* 𝒥

Thor el Conquistador The mighty Thor risks it all in a savage prehistoric land to avenge the slaying of his parents.
1986 90m/C *MX* Maria Romano, Malisa Lang, Ralf Falcone, Christopher Holm. **VHS, Beta** $59.95 *MED* 𝒥

Thorn The Virgin Mary and Joseph try to raise Jesus Christ in the modern world.
1973 (R) 90m/C Bette Midler, John Bassberger. **VHS, Beta** $59.98 *MAG* 𝒥

The Thorn Birds Pioneer couple finds danger and romance in the Australian outback as they struggle to begin a dynasty. Charismatic priest sometimes hurts, sometimes helps them. One of Stanwyck's last performances; Ward's American television debut; Brown went on to make F/X and F/X2. Originally a ten-hour television epic based on Colleen McCullough novel. Emmy nominations for Actor Chamberlain, Supporting Actor(s) Brown and Plummer, Supporting Actress Laurie, direction, photography, music, costume design, and editing. On four cassettes.
1983 486m/C Rachel Ward, Richard Chamberlain, Jean Simmons, Ken Howard, Mare Winningham, Richard Kiley, Piper Laurie, Bryan Brown, Christopher Plummer, Barbara Stanwyck; *Dir:* Daryl Duke. Emmy Awards '83: Best Actress in a Mini-series (Barbara Stanwyck), Best Supporting Actress in a Mini-series (Jean Simmons), Best Supporting Actor in a Mini-series (Richard Kiley). **VHS, Beta** $199.92 *WAR, FCT* 𝒥𝒥𝒥

Thoroughly Modern Millie Andrews is a young woman who comes to New York in the early 1920s where she meets another newcomer, the innocent Moore. Andrews decides to upgrade her image to that of a "modern" woman, a flapper, and sets out to realize her ambition, to become a stenographer and marry the boss. Meanwhile, Moore has become an object of interest to Lillie, who just happens to run a white-slavery ring. Lots of frantic moments and big production numbers in this campy film. Channing and

Lillie are exceptional fun. Oscar nominations for Channing and the Adapted Score.
1967 (G) 138m/C Julie Andrews, Carol Channing, Mary Tyler Moore, John Gavin, Beatrice Lillie, James Fox, Pat Morita; *Dir:* George Roy Hill. Academy Awards '67: Best Original Score. **VHS, Beta, LV** $19.95 *MCA* 𝒥𝒥½

Thorpe's Gold A documentary about Olympic star Jim Thorpe, multi-award winner in the 1912 Olympics. Some black-and-white segments.
1984 75m/C **VHS, Beta** $29.95 *UHV* 𝒥𝒥½

Those Calloways A small-town family attempts to establish a sanctuary for the flocks of wild geese who fly over the woods of Swiftwater, Maine. Fine Disney family fare, with good cast. Based on Paul Annixter's novel "Swiftwater."
1965 131m/C Brian Keith, Vera Miles, Brandon de Wilde, Walter Brennan, Ed Wynn, John Qualen, Linda Evans; *Dir:* Norman Tokar. **VHS, Beta** $69.95 *DIS* 𝒥𝒥½

Those Crazy Americans A look at the various cultural and popular tastes of Americans in the first half of the 20th Century.
1967 54m/B *Dir:* Laurence E. Mascott; *Nar:* George Gobel. **VHS, Beta, LV** *WGE*

Those Daring Young Men in Their Jaunty Jalopies Daring young men in noisy slow cars trek 1500 miles across country in the 1920s and call it a race.
1969 (G) 125m/C *GB* Tony Curtis, Susan Hampshire, Terry-Thomas, Eric Sykes, Gert Frobe, Peter Cook, Dudley Moore, Jack Hawkins; *Dir:* Ken Annakin. **VHS** $39.95 *PAR, FCT* 𝒥𝒥

Those Endearing Young Charms Romance develops between a young Air Corps mechanic and a salesgirl. Complications arise when another man enters the scene. Light romantic comedy. Based on the play by Edward Chodorov.
1945 82m/B Robert Young, Laraine Day, Anne Jeffreys, Lawrence Tierney; *Dir:* Lewis Allen. **VHS, Beta** *CCB* 𝒥𝒥

Those Fantastic Flying Fools A mad race to be the first on the moon brings hilarious results. Loosely based on a Jules Verne story.
1967 95m/C *GB* Burl Ives, Troy Donahue, Gert Frobe, Terry-Thomas, Hermione Gingold, Daliah Lavi, Lionel Jeffries; *Dir:* Don Sharp. **VHS** $79.95 *HBO* 𝒥𝒥½

Those Glory, Glory Days An English woman, secure in her position as an outstanding sports journalist, reminisces about the days when she and her friends idolized the boys on the soccer team. Nothing extraordinary, but nicely made.
1983 92m/C *GB* Zoe Nathenson, Liz Campion, Cathy Murphy; *Dir:* Philip Seville. **VHS, Beta** $59.95 *MGM* 𝒥½

Those Krazy Klassic Kolor Kartoons: Volume 1 A compilation of classic cartoons such as "Felix the Cat" and "Old Mother Hubbard" that the whole family can enjoy.
1985 58m/C **VHS, Beta** $29.95 *MPI*

Those Lips, Those Eyes A pre-med student takes a job as a prop boy in a summer stock company and winds up falling in love with the company's lead dancer. Sub-plot about aging actor is more interesting, better played by Langella. O'Conner is appropriately lovely. Nicely made, charming sleeper of a film.

T

1980 (R) 106m/C Frank Langella, Tom Hulce, Glynnis O'Connor, Jerry Stiller, Kevin McCarthy; *Dir:* Michael Pressman. **VHS, Beta $59.95** *MGM* ✪✪½

Those Magnificent Men in Their Flying Machines In 1910, a wealthy British newspaper publisher is persuaded to sponsor an air race from London to Paris. Contestants come from all over the world and shenanigans, hijinks, double-crosses, and romance are found along the route. Skelton has fun in prologue, while Terry-Thomas is great as the villain. Fun from start to finish.
1965 138m/C *GB* Stuart Whitman, Sarah Miles, Robert Morley, Alberto Sordi, James Fox, Gert Frobe, Jean-Pierre Cassel, Flora Robson, Sam Wanamaker, Terry-Thomas, Irina Demick, Benny Hill, Gordon Jackson, Millicent Martin, Red Skelton; *Dir:* Ken Annakin. **VHS, Beta, LV $79.98** *FOX* ✪✪✪

Those Wild Bloopers All kinds of foulups and blunders from Bogart, Abbott and Costello and Bette Davis are in this program.
1984 65m/B VHS, Beta *AOV*

Thou Shalt Not Kill...Except A Vietnam vet seeks revenge on the violent cult who kidnapped his girlfriend.
1987 84m/C Brian Schulz, Robert Rickman, John Manfredi, Tim Quill, Cheryl Hansen, Sam Raimi. **VHS, Beta $9.99** *STE, PSM* ✪

Thoughts are Free A young soldier, wounded in World War II, escapes from the train that is transporting him and his comrades to a Soviet labor camp and rejoins his family. Under the repressive communist regime, he plots his family's escape to the West but his plans are thwarted when the Berlin Wall is erected, leaving him in the West and his wife and daughter in the East. As the years go by, they can get together only by outwitting the bureaucrats. The film's title is taken from an old German folk song "Die Gedanken sind frei," sung here by Heino.
1984 (PG) 93m/C Herbert Ludwig, Kathrin Kratzer; *Dir:* Josef Sommer. **VHS, Beta $29.95** *ZBI* ✪✪✪½

Thoughts by the Ocean A continuous picture of peaceful waves rolling on to the shore to create a relaxed background.
1978 30m/C VHS, Beta *PCC* ✪✪✪½

Thousand Clowns A nonconformist has resigned from his job as chief writer for an obnoxious kiddie show in order to enjoy life. But his independence comes under fire when he becomes guardian for his young nephew and social workers take a dim view of his lifestyle. Balsam won an Oscar for his role as Robards' agent brother. Adapted from Herb Gardner's Broadway comedy. Other Oscar nominations: Best Picture, Adapted Screenplay, and Adapted Musical Score.
1965 118m/B Jason Robards Jr., Barry Gordon, William Daniels, Barbara Harris, Gene Saks, Martin Balsam; *Dir:* Fred Coe. Academy Awards '65: Best Supporting Actor (Balsam); National Board of Review Awards '65: 10 Best Films of the Year. **VHS, Beta $19.98** *MGM, FOX, BTV* ✪✪✪

The Thousand Eyes of Dr. Mabuse Lang's last film is a return to his pre-war German character, the evil Dr. Mabuse. A series of strange murders occur in a Berlin hotel and police believe the killer thinks he's a reincarnation of the doctor. Disorienting chiller. Also known as "The Secret of Dr. Mabuse" and "The Diabolical Dr. Mabuse."
1960 103m/B *GE* Dawn Addams, Peter Van Eyck, Gert Frobe, Wolfgang Preiss; *Dir:* Fritz Lang. **VHS $39.95** *FCT* ✪✪½

Thousand Mile Escort Chinese peasants struggle to survive amid poverty and power-mad warlords. Dubbed.
19?? 86m/C *CH* Pak Ying-Mai Suet. **VHS, Beta $54.95** *MAV* Woof!

Thousand Pieces of Gold A young Chinese woman is sold by her father to a marriage broker, but instead of a respectable marriage she is shipped to America and expected to work as a prostitute in an Idaho mining town. Instead she works taking in laundry as she tries to make her way in a man's world, finding a sweet romance of opposites along the way. Based on a true story. Excellent performances by Chao and Cooper.
1991 (PG-13) 105m/C Rosalind Chao, Dennis Dun, Michael Paul Chan, Chris Cooper, Jimmie F. Skaggs, William Oldham, David Hayward, Beth Broderick; *Dir:* Nancy Kelly. **VHS, LV $89.95** *HMD* ✪✪✪

Thousands Cheer A flag-waving wartime musical about a tap-dancing Army private who falls in love with the colonel's daughter, culminating in an all-star (MGM) USO show. Songs include "Honeysuckle Rose," sung by Horne, and "The Joint Is Really Jumping Down at Carnegie Hall," sung by Garland.
1943 126m/C Gene Kelly, Kathryn Grayson, Judy Garland, Mickey Rooney, Mary Astor, John Boles, Lucille Ball, Eleanor Powell, Virginia O'Brien, Margaret O'Brien, Red Skelton, Lionel Barrymore, June Allyson, Frank Morgan, Kay Kyser, Bob Crosby, Lena Horne, Donna Reed; *Dir:* George Sidney. **VHS, Beta, LV $19.98** *MGM* ✪✪½

Thrashin' A new-to-L.A. teen must prove himself to a tough gang on skateboards by skateboarding a certain treacherous race. For skate-boarding fans only.
1986 (PG-13) 93m/C Josh Brolin, Pamela Gidley, Robert Rusler, Chuck McCann; *Dir:* David Winters. **VHS, Beta $24.95** *VIR* ✪

Threads The famed dramatic recreation of the effects of nuclear war on a British city and two of its families. A disturbing, uncompromising, and somewhat plausible drama. Made for British television.
1985 110m/C *GB* Karen Meagher, Rita May, David Brierly, Reece Dinsdale, Harry Beety; *Dir:* Mick Jackson. **VHS, Beta $59.95** *WES* ✪✪✪

Threat A killer escapes from jail and returns to settle the score with those who convicted him. Tense, fast-moving thriller.
1949 66m/B Michael O'Shea, Virginia Grey, Charles McGraw; *Dir:* Felix Feist. **VHS, Beta $34.95** *RKO* ✪✪½

Three Ages A parody of D.W. Griffith's 1916 film "Intolerance." Keaton's first feature film casts him in prehistoric days, ancient Rome, and in modern times. Silent with musical score.
1923 59m/B Wallace Beery, Oliver Hardy, Buster Keaton; *Dir:* Eddie Cline. **VHS, Beta $19.95** *NOS, VDM, DVT* ✪✪✪

Three Amigos Three out-of-work silent screen stars are asked to defend a Mexican town from bandits; they think it's a public appearance stint. Spoof of Three Stooges and Mexican bandito movies that at times falls short, given the enormous amount of comedic talent involved. Generally enjoyable with some very funny scenes. Co-written by Martin, songwriter Randy Newman (who also appears in the film), and former "Saturday

Night Live" producer Lorne Michaels. Short's first major film appearance.
1986 (PG) 105m/C Chevy Chase, Steve Martin, Martin Short, Joe Mantegna, Patrice Martinez, Jon Lovitz, Alfonso Arau, Randy Newman; *Dir:* John Landis. **VHS, Beta, LV $14.99** *HBO* ✪✪

Three in the Attic A college student juggles three girlfriends at the same time. When the girls find out they are being two-timed, they lock their boyfriend in an attic and exhaust him with forced sexual escapades. Trashy and foolish.
1968 (R) 92m/C Christopher Jones, Yvette Mimieux, John Beck; *Dir:* Richard Wilson. **VHS, Beta $9.98** *SUE, MLB* ✪½

Three Avengers Trouble abounds when two kung fu masters and an American Chinese boy open a martial arts school.
1980 (R) 93m/C VHS, Beta $9.98 *SUE* ✪

Three for Bedroom C A movie star and a scientist make romance on a train bound for Los Angeles. Swanson's follow-up to "Sunset Boulevard."
1952 74m/C Gloria Swanson, Fred Clark, Steve Brodie, Hans Conried, Margaret Dumont; *Dir:* Milton Bren. **VHS, Beta $12.95** *IVE* ✪✪

Three Broadway Girls Three gold-diggers go husband hunting. Well-paced, very funny telling of this old story. Also released as "The Greeks Had A Word For Them." Remade many times, including "How to Marry a Millionaire," "Three Blind Mice," "Moon Over Miami," and "Three Little Girls in Blue."
1932 78m/B Joan Blondell, Ina Claire, Madge Evans, David Manners, Lowell Sherman; *Dir:* Lowell Sherman. **VHS, Beta $19.95** *NOS, CAB, KRT* ✪✪✪

Three Brothers An acclaimed Italian film by veteran Rosi about three brothers summoned to their small Italian village by a telegram saying their mother is dead. Sensitive and compassionate. In Italian with English subtitles. Adapted from Platonov's "The Third Son."
1980 (PG) 113m/C *IT* Philippe Noiret, Charles Vanel, Michele Placido, Vittorio Mezzogiorno, Andrea Ferreol; *Dir:* Francesco Rosi. **VHS, Beta $29.98** *SUE, APD* ✪✪½

Three Bullets for a Long Gun Two prairie renegades battle bandits as they search for an inherited gold mine. Written by Van Der Wat.
1973 89m/C Keith Van Der Wat, Beau Brummel, Patrick Munhardt; *Dir:* Peter Henkel. **VHS, Beta $9.95** *SIM, AVD, CHA* ✪½

The Three Caballeros Donald Duck stars in this journey through Latin America. Full of music, variety, and live-action/animation segments. Songs include "The Three Caballeros" and "Baia." Stories include "Pablo the Penguin," "Little Gauchito," and adventures with Joe Carioca, who was first introduced in Disney's "Saludos Amigos." Today this film stands as one of the very best pieces of animation ever created. Great familyfare.
1945 71m/C *Voices:* Sterling Holloway, Aurora Miranda. **VHS, Beta, LV $24.99** *DIS, APD* ✪✪½

Three Came Home Colbert is an American married to a British administrator in the Far East during WWII. Conquering Japanese throw the whole family into a brutal POW concentration camp, and their confinement is recounted in harrowing and unsparing detail. Superior drama, also laudable for a fairly even-handed portrayal of the enemy

captors. Based on an autobiographical book by Agnes Newton-Keith.
1950 106m/B Claudette Colbert, Patric Knowles, Sessue Hayakawa; *Dir:* Jean Negulesco. **VHS $19.95** *NOS* 🐾🐾🐾

Three in the Cellar A disgruntled college student seeks revenge on the school president by seducing his wife, daughter and mistress. Good screenplay helps carry it off. Also known as "Up in the Cellar."
1970 (R) 92m/C Wes Stern, Joan Collins, Larry Hagman, Judy Pace; *Dir:* Theodore J. Flicker. **VHS $79.95** *HBO* 🐾🐾

Three Charlies and One Phoney! Three of Charlie Chaplin's best comedy shorts, plus one featuring his best-known imitator - Billy West. Included are "Recreation" ("Fun is Fun") ("Spring Fever"), "His Musical Career" ("The Piano Movers") ("Musical Tramps"), "The Bond" (featuring Charlie in a Liberty Bond appeal) and "His Day Out" with Billy West imitating Chaplin. All but "His Day Out" were written and directed by Chaplin.
1918 69m/B Charlie Chaplin, Mack Swain, Charley Chase, Edna Purviance, Sydney Chaplin, Albert Austin; *W/Dir:* Charlie Chaplin. **VHS, Beta, 8mm $29.95** *VYY* 🐾🐾🐾

Three Coins in the Fountain Three women throw money into fountain and get romantically involved with Italian men. Outstanding CinemaScope photography captures beauty of Italian setting. Sammy Cahn theme song sung by Frank Sinatra.
1954 102m/C Clifton Webb, Dorothy McGuire, Jean Peters, Louis Jourdan, Maggie McNamara, Rossano Brazzi; *Dir:* Jean Negulesco. **VHS, Beta $59.98** *FOX* 🐾🐾½

Three Day Weekend Two female college students leave their urban campus for a rural respite, and end up in the same neck of the woods as a reclusive college drop-out who conveniently (for who?) hasn't seen females in a long time.
19?? 78m/C Dan Diego, Jody Lee Olhava, Blake Parrish; *Dir:* Emmett Alston. **VHS, Beta** *ACA* 🐾🐾½

Three Days in Beirut A soldier of fortune and a beautiful interpreter join forces against a backdrop of Middle Eastern politics, espionage, and violence.
1983 (PG) 94m/C Diana Sands, Calvin Lockhart; *Dir:* Michael Schultz. **VHS, Beta $59.95** *PSM* 🐾

Three Days of the Condor CIA researcher Redford finds himself on the run from unknown killers when he is left the only survivor of the mass murder of his office staff. Good performance by Dunaway as the photographer who shelters him. Excellent suspense tale. Based on "Six Days of the Condor" by James Grady.
1975 (R) 118m/C Robert Redford, Faye Dunaway, Cliff Robertson, Max von Sydow, John Houseman; *Dir:* Sydney Pollack. Edgar Allan Poe Awards '75: Best Screenplay. **VHS, Beta, LV $24.95** *PAR, PIA* 🐾🐾🐾

Three Desperate Men Three brothers accused of murder become outlaws. Not innovative, but satisfactory tale of the Old West.
1950 71m/B Preston Foster, Virginia Grey, Jim Davis; *Dir:* Sam Newfield. **VHS, Beta** *WGE* 🐾🐾

The Three Faces of Eve An emotionally disturbed woman seeks the help of a psychiatrist, who discovers she has three distinct personalities. Powerful performances highlight this fascinating study based on a true story.

1957 91m/B Joanne Woodward, David Wayne, Lee J. Cobb, Nancy Kulp, Vince Edwards; *Dir:* Nunnally Johnson; *Nar:* Alistair Cooke. Academy Awards '57: Best Actress (Woodward). **VHS, LV $59.98** *FOX, TCF* 🐾🐾🐾

Three Faces West A dust bowl community is helped by a Viennese doctor who left Europe to avoid Nazi capture. He's aided by the Duke in this odd combination of Western frontier saga and anti-Nazi propaganda. Works only part of the time.
1940 79m/B John Wayne, Charles Coburn, Sigrid Gurie, Sonny Bupp, Russell Simpson; *Dir:* Bernard Vorhaus. **VHS $14.98** *REP* 🐾

3:15—The Moment of Truth A vicious high school gang is confronted by an angry ex-member in this so-so teen delinquent film.
1986 (R) 86m/C Adam Baldwin, Deborah Foreman, Rene Auberjonois, Danny De La Paz; *Dir:* Larry Gross. **VHS, Beta $19.95** *MED* 🐾🐾

Three Fugitives An ex-con holdup man determined to go straight is taken hostage by a bungling first-time bank robber, who is only attempting the holdup in order to support his withdrawn young daughter. Ex-con Nolte winds up on the lam with the would-be robber and his daughter. The comedy is fun, the sentimental moments too sweet and slow. Remake of French "Les Fugitifs."
1989 (PG-13) 96m/C Nick Nolte, Martin Short, James Earl Jones, Kenneth MacMillan, Sarah Rowland Doroff, Alan Ruck; *Dir:* Francis Veber. **VHS, Beta, LV $19.99** *TOU* 🐾🐾

Three Godfathers A sweet and sentimental western has three half-hearted outlaws on the run, taking with them an infant they find in the desert. Dedicated to western star and Ford alumni Harry Carey, Sr. (whose son is one of the outlaws in the film), who died of cancer the year before. Ford had first filmed the tale with Carey Sr. as "Marked Men" in 1919.
1948 106m/C John Wayne, Pedro Armendariz, Harry Carey Jr., Ward Bond, Mae Marsh, Jane Darwell, Ben Johnson, Mildred Natwick, Guy Kibbee; *Dir:* John Ford. **VHS, Beta $19.95** *MGM* 🐾🐾🐾

Three Guys Named Mike Wyman plays an overly enthusiastic airline stewardess who finds herself the object of affection from three guys named Mike, including an airline pilot, an advertising man, and a scientist. A cute comedy.
1951 89m/B Jane Wyman, Van Johnson, Barry Sullivan, Howard Keel, Phyllis Kirk, Jeff Donnell; *Dir:* Charles Walters. **VHS $19.95** *VDM* 🐾🐾

Three the Hard Way An insane white supremacist has a plan to eliminate blacks by contaminating water supplies. A big blaxploitation money maker and the first to team Brown, Williamson, and Kelly, who would go on to make several more pictures together.
1974 (R) 93m/C Jim Brown, Fred Williamson, Jim Kelly, Sheila Frazier, Jay Robinson, Alex Rocco; *Dir:* Gordon Parks. **VHS $39.95** *XVC* 🐾🐾½

Three by Hitchcock Condensed versions of three of Hitchcock's early films (1927-28). Consists of "The Ring," "Champagne," and "The Manxman," his last silent film.
1928 90m/B *Dir:* Alfred Hitchcock. **VHS, Beta $60.50** *GVV* 🐾🐾

The 317th Platoon An emotionally gripping war-drama of the French-Vietnemese War in which the 317th Platoon of the French Army, consisting of 4 French commanders and 41 Laotians, is ordered to

retreat to camp Dien Bien Phu. The men begin their march and slowly succomb to the elements and ambushes. Days later, when what is left of the platoon finally reaches Dien Bien Phu, the camp is in enemy hands and the remnants of the 317th Platoon are killed in cold blood. Director Shoendoerffer portrays this struggle, conflict, and the quiet haunting tension of war with great skill. In French with English subtitles.
1965 100m/B *FR* Jacques Perrin, Bruno Cremer, Pierre Fabre, Manuel Zarzo; *W/Dir:* Pierre Shoendoerffer. **VHS, LV $49.95** *INT, IME, FCT* 🐾🐾

357 Magnum Jonathan Hightower, the CIA's special investigator in the Far East, speaks softly and carries a very big gun.
1977 71m/C Marland T. Stewart, James Whitworth, Kathryn Hayes. **VHS, Beta $59.95** *WES* 🐾

Three Husbands A deceased playboy leaves letters to the title characters incriminating their wives in extramarital affairs. The men's reactions are pure farce—even though this same concept was played straight with a sex change in the earlier "A Letter to Three Wives."
1950 79m/B Eve Arden, Ruth Warrick, Vanessa Brown, Howard da Silva, Sheppard Strudwick, Jane Darwell; *Dir:* Irving Reis. **VHS $19.95** *NOS, VDM* 🐾🐾½

Three of a Kind A compilation of the best skits and satires from the British comedy series.
198? 84m/C *GB* Tracy Ullman, David Copperfield, Lenny Henry. **VHS, Beta $14.98** *FOX* 🐾🐾½

Three Kinds of Heat Three Interpol agents track down a warring Oriental crime lord. Alternately tongue-in-cheek and silly. Never released in theaters, it found a home on video. Stevens created "The Outer Limits" for television.
1987 (R) 88m/C Robert Ginty, Victoria Barrett, Shakti; *Dir:* Leslie Stevens. **VHS, Beta $19.98** *WAR* 🐾

Three Legionnaires Three soldiers cavort in a post-World War I Siberian tank town.
1937 67m/B Robert Armstrong, Lyle Talbot, Fifi D'Orsay, Anne Nagel, Donald Meek. **VHS, Beta, LV** *WGE* 🐾🐾½

The Three Little Pigs From "Faerie Tale Theatre" comes the story of three little pigs, the houses they lived in and the wolf that tried to do them in.
1984 60m/C Billy Crystal, Jeff Goldblum, Valerie Perrine; *Dir:* Howard Storm. **VHS, Beta $14.98** *KUI, FOX, FCT* 🐾🐾½

Three Little Words A musical biography of songwriting team Harry Ruby and Bert Kalmar, filled with Kalmar-Ruby numbers. June Haver (famous for her boop-boop-de-boops) dubbed "I Wanna Be Loved By You" for Reynolds. Other songs include "Nevertheless," "Who's Sorry Now," "You Are My Lucky Star," and the title song. A musical in the best MGM tradition.
1950 102m/C Fred Astaire, Red Skelton, Vera-Ellen, Arlene Dahl, Keenan Wynn, Gloria de Haven, Debbie Reynolds, Gale Robbins; *Dir:* Richard Thorpe. **VHS, Beta $29.95** *MGM* 🐾🐾½

Three Lives of Thomasina In turn-of-the-century Scotland, a veterinarian orders his daughter's beloved cat destroyed when the pet is diagnosed with tetanus. After the cat's death (with scenes of kitty heaven), a beautiful and mysterious healer from the woods is able to bring the animal back to life

T

and restore the animal to the little girl. Lovely Disney fairy tale with good performances by all.
1963 95m/C Patrick McGoohan, Susan Hampshire, Karen Dotrice, Matthew Garber; *Dir:* Don Chaffey. **VHS, Beta $19.99** DIS 🎬🎬🎬

Three on a Match An actress, a stenographer and a society woman get together for a tenth year reunion. They become embroiled in a world of crime when Blondell's gangster boyfriend takes a liking to Dvorak. She leaves her husband and takes up with the crook. The results are tragic. Bogart's first gangster role.
1932 64m/B Joan Blondell, Warren William, Ann Dvorak, Bette Davis, Lyle Talbot, Humphrey Bogart, Patricia Ellis, Grant Mitchell; *Dir:* Mervyn LeRoy. **VHS, Beta $19.98** MGM, FCT 🎬🎬🎬

Three on a Meathook When a young man and his father living on an isolated farm receive female visitors, bloodshed is quick to follow. Essentially a remake of ''Psycho,'' and very loosely based on the crimes of Ed Gein. Filmed in Louisville, Kentucky.
1972 85m/C Charles Kissinger, James Pickett, Sherry Steiner, Carolyn Thompson; *W/Dir:* William Girdler. **VHS, Beta $49.95** REG, TAV **Woof!**

Three Men and a Baby The arrival of a young baby forever changes the lives of three sworn bachelors living in New York. Well-paced, charming and fun, with good acting from all. A remake of the French movie ''Three Men and a Cradle.''
1987 (PG) 102m/C Tom Selleck, Steve Guttenberg, Ted Danson, Margaret Colin, Nancy Travis, Philip Bosco, Celeste Holm; *Dir:* Leonard Nimoy. **VHS, Beta, LV, 8mm $19.99** TOU, BVV 🎬🎬🎬

Three Men and a Cradle Remade in the U.S. in 1987 as ''Three Men and a Baby,'' this French film is about three bachelors living together who suddenly find themselves the guardians of a baby girl. After the initial shock of learning how to take care of a child, they fall in love with her and won't let her go. Fans of the American version may find themselves liking the original even more. In French with English subtitles.
1985 (PG-13) 100m/C FR Roland Giraud, Michel Boujenah, Andre Dussollier; *Dir:* Coline Serreau. **VHS, LV $79.98** LIV, MGM 🎬🎬🎬½

Three Men and a Little Lady In this sequel to ''Three Men and a Baby,'' the mother of the once abandoned infant decides that her child needs a father. Although she wants Danson, he doesn't get the message and so she chooses a snooty British director. The rest of the movie features various semi-comic attempts to rectify the situation.
1990 (PG) 100m/C Tom Selleck, Steve Guttenberg, Ted Danson, Nancy Travis, Robin Weisman, Christopher Cazenove, Leona Shaw; *Dir:* Emile Ardolino. **VHS, Beta $19.99** BVV 🎬🎬½

The Three Musketeers D'Artagnan swashbuckles silently amid stylish sets, scores of extras and exquisite costumes. Relatively faithful adaptation of Alexandre Dumas novel, slightly altered in favor of D'Artagnan's lover. Classic Fairbanks, who also produced.
1921 120m/B Douglas Fairbanks Sr., Leon Barry, George Siegmann, Eugene Pallette, Boyd Irwin, Thomas Holding, Sidney Franklin, Charles Stevens, Nigel de Brulier, Willis Robards; *Dir:* Fred Niblo. **VHS, Beta, LV $29.95** GPV, FCT 🎬🎬🎬½

Three Musketeers Modern adaptation of the classic tale by Alexander Dumas depicts the three friends as members of the Foreign Legion. Weakest of Wayne serials, in 12 parts.
1933 215m/B John Wayne, Raymond Hatton, Lon Chaney Jr. **VHS, Beta $24.95** NOS, VCN, VDM 🎬½

Three Musketeers Three swordsmen, loyal to each other, must battle the corrupt Cardinal Richelieu. Generally regarded as the least exciting adaptation of Alexander Dumas' classic tale. The first talking version.
1935 95m/B Walter Abel, Paul Lukas, Moroni Olsen, Ian Keith, Margot Grahame, Heather Angel, Onslow Stevens, Miles Mander; *Dir:* Rowland V. Lee. **VHS, Beta $29.95** RKO, IME 🎬🎬

Three Musketeers The three musketeers who are ''all for one and one for all'' join forces with D'Artagnan to battle the evil Cardinal Richelieu in this rambunctious adaptation of the classic tale by Alexander Dumas. Good performances by the cast, who combined drama and comedy well. Turner's first color film. Oscar nomination for Best Cinematography.
1948 126m/C Lana Turner, Gene Kelly, June Allyson, Gig Young, Angela Lansbury, Van Heflin, Keenan Wynn, Robert Coote, Reginald Owen, Frank Morgan, Vincent Price; *Dir:* George Sidney. **VHS, Beta $19.98** CCB, MGM 🎬🎬½

Three Musketeers Extravagant and funny version of the Dumas classic. Three swashbucklers and their country cohort, who wishes to join the Musketeers, set out to save the honor of the French Queen. To do so they must oppose the evil cardinal who has his eyes on the power behind the throne. A strong cast leads this winning combination of slapstick and high adventure.
1974 (PG) 105m/C Richard Chamberlain, Oliver Reed, Michael York, Raquel Welch, Frank Finlay, Christopher Lee, Faye Dunaway, Charlton Heston, Geraldine Chaplin, Simon Ward, Jean-Pierre Cassel; *Dir:* Richard Lester. **VHS, Beta** COL 🎬🎬🎬½

The Three Musketeers (D'Artagnan) One of too many silent movie adaptation of Alexandre Dumas' adventure, but this ones comes from the hand of famed movie pioneer Thomas Ince. D'Artagnan joins the title characters in saving the queen's jewels. Also available on video under its original title, ''D'Artagnan,'' with the original organ score and presented in ''accuspeed.''
1916 74m/B Orin Johnson, Dorothy Dalton, Louise Glaum, Walt Whitman; *Dir:* Thomas Ince. **VHS, Beta, 8mm $29.95** VYY, GPV, CVL 🎬🎬

Three Nuts in Search of a Bolt Three neurotics send a surrogate to a comely psychiatrist for help, but the shrink thinks the surrogate's a three-way multiple personality. Raunchy monkey-shines ensue. Includes Van Doren's infamous beer bath scene.
1964 (R) 78m/C Mamie Van Doren, Tommy Noonan, Paul Gilbert, Alvy Moore; *Dir:* Tommy Noonan. **VHS, Beta $9.95** WGE 🎬½

Three O'Clock High A nerdy high school journalist is assigned a profile of the new kid in school, who turns out to be the biggest bully, too. He approaches his task with great unease. Silly teen-age farce is given souped-up direction by Spielberg protege Joanou. Features decent work by the young cast.
1987 (PG) 97m/C Casey Siemaszko, Anne Ryan, Stacey Glick, Jonathan Wise, Richard Tyson; *Dir:* Phil Joanou. **VHS, Beta, LV $79.95** MCA, TOU 🎬🎬½

The Three Penny Opera Mack the Knife presides over an exciting world of thieves, murderers, beggars, prostitutes and corrupt officials in a seedy section of London. Based on the opera by Kurt Weill and Bertold Brecht, which was based on ''The Beggar's Opera'' by John Gay. Remake of ''The Threepenny Opera'' (1931).
1962 124m/C GE Curt Jurgens, Hildegarde Neff, Gert Frobe, June Ritchie, Lino Ventura, Sammy Davis Jr. **VHS, Beta, LV $29.98** MVD, VYY, VYG 🎬🎬🎬

Three for the Road A young aspiring political aide and his roommate must escort a Senator's ill-mannered and spoiled daughter to a reform institution. On the road, they run into more than a few obstacles. Good performances elevate a bland script. Sheen made this movie before his riveting performance as a young GI in ''Platoon.''
1987 (PG) 98m/C Charlie Sheen, Kerri Green, Alan Ruck, Sally Kellerman; *Dir:* B.W.L. Norton. **Beta, LV $19.95** VHV 🎬🎬

Three in the Saddle Ritter and the Texas Rangers fight for law and order.
1945 61m/B Tex Ritter, Dave O'Brien, Charles King; *Dir:* Harry Fraser. **VHS, Beta $19.95** NOS, VCN 🎬

Three Secrets A plane crashes and the only survivor is a five-year-old boy. Three women, each with a secret, seek to claim him as the child each gave up for adoption five years before. Tearjerker with good cast.
1950 98m/B Eleanor Parker, Patricia Neal, Ruth Roman, Frank Lovejoy, Leif Erickson; *Dir:* Robert Wise. **VHS $19.98** REP 🎬🎬½

Three Sovereigns for Sarah Witchhunters tortured and toasted Sarah's two sisters for practicing witchcraft in the past. Now, accused of witchery herself, she struggles to prove the family's innocence. Excellent, made-for-PBS. Fine performances from all, especially Redgrave.
1985 152m/C Vanessa Redgrave, Phyllis Thaxter, Patrick McGoohan; *Dir:* Philip Leacock. **VHS, Beta $79.95** PSM, PBS 🎬🎬½

Three Stooges Comedy Capers: Volume 1 Compilation includes: ''Disorder in the Court'' (1936), ''Malice in the Palace'' (1949), ''Sing a Song of Six Pants'' (1947), and ''The Brideless Groom'' (1947).
194? 80m/B Moe Howard, Larry Fine, Curly Howard, Shemp Howard. **VHS, Beta $9.95** NOS, MED, BUR

Three Stooges Comedy Classics The antics of the Stooges are featured in this collection of five original shorts, along with an early Ted Healy comedy that features all four Stooges: ''Hollywood on Parade,'' ''Disorder in the Court,'' ''Malice in the Palace,'' ''Brideless Groom,'' and ''Sing a Song of Six Pants.''
1949 79m/B Moe Howard, Larry Fine, Curly Howard, Shemp Howard, Ted Healy. **VHS $19.95** REP

Three Stooges: Cookoo Cavaliers and Other Nyuks The boys star in ''Cookoo Cavaliers,'' ''Busy Buddies,'' and ''Booby Dupes.''
1940 55m/B Moe Howard, Curly Howard, Larry Fine. **VHS $14.95** COL

Three Stooges Festival Three original shorts by the comedy trio, including: ''Disorder in the Court'' (1936-Curly), ''Sing a Song of Six Pants'' (1949-Shemp) and ''Malice in the Palace'' (1949-Shemp).

1949 60m/B Moe Howard, Larry Fine, Shemp Howard, Curly Howard. **VHS, Beta $29.95** *AOV, VYY*

Three Stooges: From Nurse to Worse and Other Nyuks The irrepressible Stooges star in "From Nurse to Worse," "Crash Goes the Hash," and "G.I. Wanna Go Home."
1945 55m/B Moe Howard, Larry Fine, Curly Howard. **VHS $14.95** *COL*

Three Stooges: Half-Wit's Holiday and Other Nyuks Includes, "Half-Wit's Holiday," "Horse Collars," and "How High Is Up?"
1935 55m/B Moe Howard, Curly Howard, Larry Fine. **VHS $14.95** *COL*

Three Stooges: Idiots Deluxe and Other Nyuks Curly, Larry and Moe star in, "Idiots Deluxe," "I Can Hardly Wait," and "Idle Roomers."
1940 55m/B Moe Howard, Curly Howard, Larry Fine. **VHS $14.95** *COL*

The Three Stooges Meet Hercules The Three Stooges are transported back to ancient Ithaca by a time machine with a young scientist and his girlfriend. When the girl is captured, they enlist the help of Hercules to rescue her.
1961 80m/B Moe Howard, Larry Fine, Joe DeRita, Vicki Trickett, Quinn Redeker; ***Dir:*** Edward L. Bernds. **VHS, Beta $9.95** *GKK* 🦴🦴½

Three Stooges: Nutty but Nice and Other Nyuks The boys star in "Nutty but Nice," "The Sitter-Downers," and "Slippery Silks."
1936 55m/B Moe Howard, Curly Howard, Larry Fine. **VHS $14.95** *COL*

Three Stooges: They Stooge to Conga and Other Nyuks World War II madness from the Stooges in "They Stooge to Conga," "You Natzy Spy!" and "I'll Never Heil Again."
1940 55m/B Moe Howard, Curly Howard, Larry Fine. **VHS $14.95** *COL*

Three Stops to Murder An FBI agent investigates the murder of a beautiful model in this early, unspectacular Hammer production.
1953 76m/B *GB* Tom Conway, Mila Parely, Naomi Chance, Eric Pohlmann, Andrew Osborn, Richard Wattis, Eileen Way, Delphi Lawrence; ***Dir:*** Terence Fisher. **VHS, Beta $16.95** *SNC, MLB* 🦴

Three Strange Loves Men and women struggle with loneliness, old age, and sterility. Sometimes disjointed, but for the most part, well-made. Finely acted. In Swedish with English subtitles.
1949 88m/B *SW* Eva Henning, Brigit Tengroth, Birger Malmsten; ***Dir:*** Ingmar Bergman. **VHS, Beta $29.98** *NOS, HHT, SUE* 🦴🦴🦴

3:10 to Yuma In order to collect $200 he desperately needs, a poor farmer has to hold a dangerous killer at bay while waiting to turn the outlaw over to the authorities arriving on the 3:10 train to Yuma. Stuck in a hotel room with the outlaw's gang gathering outside, the question arises as to who is the prisoner. Farmer Heflin is continually worked on by the outlaw Ford, in a movie with more than its share of great dialogue. Suspenseful, well-made action western adapted from an Elmore Leonard story.
1957 92m/B Glenn Ford, Van Heflin, Felicia Farr, Richard Jaeckel; ***Dir:*** Delmer Daves. **VHS, Beta, LV $9.95** *GKK* 🦴🦴🦴½

Three Warriors A young Native American boy is forced to leave the city and return to the reservation, where his contempt for the traditions of his ancestors slowly turns to appreciation and love.
1977 100m/C Charles White Eagle, Lois Red Elk, McKee "Kiko" Red Wing, Christopher Lloyd, Randy Quaid. **VHS, Beta $59.99** *HBO* 🦴½

Three Way Weekend Two young girls set off on a back-packing trip through the mountains of Southern California to enjoy camping and romance.
1981 (R) 78m/C Dan Diego, Jody Lee Olhava, Richard Blye, Blake Parrish; ***Dir:*** Emmett Alston. **VHS, Beta $19.95** *ACA* 🦴

The Three Weird Sisters Title characters maintain life of luxury through murder and inheritance. Dylan Thomas co-wrote script.
1948 82m/B *GB* Nancy Price, Mary Clare; ***Dir:*** Daniel Birt. **VHS $19.95** *VDM, SNC, VMK* 🦴🦴

Three Word Brand William S. Hart plays three roles in this film, a homesteader who is killed by Indians and his twin sons, who are separated after their father's death and reunited many years later. Silent with musical score.
1921 75m/B William S. Hart, Jayne Navak, S.J. Bingham; ***Dir:*** Lambert Hillyer. **VHS, Beta $29.95** *GPV, VYY* 🦴½

The Three Worlds of Gulliver A colorful family version of the Jonathan Swift classic about an Englishman who discovers a fantasy land of small and giant people. Visual effects by Ray Harryhausen. Music by Bernard Herrmann.
1959 100m/C Kerwin Mathews, Jo Morrow, Basil Sydney, Mary Ellis; ***Dir:*** Jack Sher. **VHS, Beta $14.95** *COL, MLB, CCB* 🦴🦴½

The Threepenny Opera A musical about the exploits of gangster Mack the Knife. Adapted from Bertolt Brecht's play and John Gay's "The Beggar's Opera." In German with English subtitles.
1931 107m/B Rudolph Forster, Lotte Lenya, Carola Neher, Reinhold Schunzel, Fritz Rasp, Valeska Gert; ***Dir:*** G. W. Pabst. **VHS $14.95** *VDM* 🦴🦴

Three's Trouble A middle-class family erupts when a young male babysitter enters to care for its three mischievous sons.
1985 93m/C *AU* John Waters, Jacki Weaver, Steven Vidler; ***Dir:*** Chris Thomson. **VHS, Beta $19.95** *STE, NWV* 🦴🦴

Threshold An internationally-acclaimed research biologist is frustrated by his inability to save a dying 20-year-old woman born with a defective heart. She becomes the recipient of the first artificial heart. Solid performances.
1983 (PG) 97m/C *CA* Donald Sutherland, Jeff Goldblum, John Marley, Mare Winningham; ***Dir:*** Richard Pearce. Genie Awards '83: Best Actor (Sutherland), Best Cinematography. **VHS, Beta $59.98** *FOX* 🦴🦴½

The Thrill of It All! An average housewife becomes a star of television commercials, to the dismay of her chauvinist husband, a gynecologist. Fast and funny, with numerous acidic jokes about television sponsors and programs. Written by Carl Reiner.
1963 108m/C James Garner, Doris Day, Arlene Francis, Edward Andrews, Carl Reiner, Elliott Reid, Reginald Owen, ZaSu Pitts; ***Dir:*** Norman Jewison. **VHS, Beta $59.95** *MCA* 🦴🦴½

The Thrill Killers Young murderous thugs rampage through Los Angeles suburbs, killing indiscriminately. Pretentious and exploitative. Sometimes shown with Hallucinogenic Hypo-Vision - a prologue announcing the special effect would be shown before the film, and on cue, hooded ushers would run through the theatre with cardboard axes. Weird.
1965 82m/C Cash Flagg, Liz Renay, Brick Bardo, Carolyn Brandt, Atlas King; ***Dir:*** Ray Dennis Steckler. **VHS, Beta $39.98** *CAM Woof!*

Thrill of a Romance Johnson, an Army hero, meets Williams, a swimming instructor of all things, at a resort in the Sierra Nevadas. The only problem is that she's married. The one highlight of this otherwise typical swim-romance vehicle is the song "I Should Care."
1945 105m/C Esther Williams, Van Johnson, Frances Gifford, Henry Travers, Spring Byington, Lauritz Melchior; ***Dir:*** Richard Thorpe. **VHS** *MGM* 🦴🦴

Thrilled to Death An utterly naive husband and wife get caught in the middle of a lethal game when they befriend a scheming couple.
1988 (R) 90m/C Blake Bahner, Rebecca Lynn, Richard Maris, Christine Moore; ***Dir:*** Chuck Vincent. **VHS $79.98** *REP* 🦴½

Thrillkill A young girl is enmeshed in a multi-million-dollar burglary scheme after her computer-whiz/embezzler sister disappears.
1984 88m/C Robin Ward, Gina Massey, Anthony Kramroither; ***Dir:*** Anthony D'Andrea. **VHS, Beta $9.95** *MED* 🦴

Throne of Blood Kurosawa's masterful adaptation of "Macbeth," transposes the story to medieval Japan and the world of the samurai. Mifune and Chiaki are warriors who have put down a rebellion and are to be rewarded by their overlord. On their way to his castle they meet a mysterious old woman who prophesizes that Mifune will soon rule but his reign will be short. She is dismissed as crazy but her prophesies come to pass. Mifune's ambitions soon overwhelm him and his paranoia causes him to have his old friend Chiaki killed. Chiaki's son then seeks revenge against the tyrant. So steeped in Japanese style that it bears little resemblance to the Shakespearean original, this film is an incredibly detailed vision in its own right. In Japanese with English subtitles.
1957 105m/B *JP* Toshiro Mifune, Isuzu Yamada, Takashi Shimura, Minoru Chiaki, Akira Kubo; ***Dir:*** Akira Kurosawa. **VHS, Beta, LV, 3/4U $35.95** *IHF, VYG, CRC* 🦴🦴🦴🦴

Throne of Fire A mighty hero battles the forces of evil to gain control of the powerful Throne of Fire. Dubbed.
1982 91m/C *IT* Peter McCoy, Sabrina Siami; ***Dir:*** Franco Prosperi. **VHS, Beta $59.95** *MGM Woof!*

Throne of Vengeance Prince Sandokan's bride, his kingdom, and even his life are in danger when the nefarious "Leopard" plans to take over.
19?? 86m/C VHS $29.95 *LTG* 🦴

Through the Breakers Heartbroken woman suffers silently on South Pacific isle.
1928 55m/B Margaret Livingston, Holmes Herbert, Clyde Cook, Natalie Joyce. **VHS, Beta $19.95** *GPV, FCT* 🦴½

Through a Glass Darkly Oppressive interactions within a family sharing a holiday on a secluded island: a woman recovering from schizophrenia, her husband, younger brother, and her psychologist father. One of

Bergman's most mysterious, upsetting and powerful films. In Swedish with English subtitles. Part of Bergman's Silence-of-God trilogy followed by "Winter Light" and "The Silence."
1961 91m/B *SW* Harriet Andersson, Max von Sydow, Gunnar Bjornstrand, Lars Passgard; *Dir:* Ingmar Bergman. Academy Awards '61: Best Foreign Language Film; National Board of Review Awards '62: 5 Best Foreign Films of the Year. **VHS, Beta $24.95** *NOS, SUE* ♪♪♪

Through Naked Eyes Made for television thriller about two people who spy on each other. They witness a murder and realize that someone else is watching them.
1987 91m/C David Soul, Pam Dawber; *Dir:* John Llewellyn Moxey. **VHS, Beta $79.95** *MED* ♪

Throw Momma from the Train DeVito plays a man, henpecked by his horrific mother, who tries to persuade his writing professor (Crystal) to exchange murders. DeVito will kill Crystal's ex-wife and Crystal will kill DeVito's mother. Only mama isn't going to be that easy to get rid of. Fast-paced and entertaining black comedy. Ramsey steals the film. Inspired by Hitchcock's "Strangers on a Train."
1987 (PG-13) 88m/C Danny DeVito, Billy Crystal, Anne Ramsey, Kate Mulgrew; *Dir:* Danny DeVito. **VHS, Beta, LV $19.98** *ORI, TOU* ♪♪♪

Thumb Tripping A dated sixties/hippie film. Two flower children hitchhike and encounter all kinds of other strange people in their travels on the far-out roads.
1972 (R) 94m/C Meg Foster, Michael Burns, Bruce Dern, Marianna Hill, Michael Conrad, Joyce Van Patten; *Dir:* Quentin Masters. **VHS, Beta $59.98** *CHA* ♪♪

Thumbelina From "Faerie Tale Theatre" comes the story of a tiny girl who gets kidnapped by a toad and a mole, but meets the man of her dreams in the nick of time. Adapted from the classic tale by Hans Christian Andersen.
1982 60m/C Carrie Fisher, William Katt, Burgess Meredith; *Dir:* Michael Lindsay-Hogg. **VHS, Beta, LV $14.98** *KUI, FOX, FCT* ♪♪♪

Thumbelina An animated version of the Hans Christian Andersen fairy tale about a little girl who's only as tall as a thumb.
1984 45m/C VHS, Beta $19.95 *COL* ♪♪♪

Thunder Alley In middle America, two friends and their rock group struggle for success. One of the friends dies after an accidental overdose and the other's grief propels him to stardom. Songs include "Thunder Alley," "Heart to Heart," "Sometimes in the Night," and "Can't Look Back."
1985 102m/C Roger Wilson, Leif Garrett, Jill Schoelen; *Dir:* J.S. Cardone. **VHS, Beta $79.95** *MGM* ♪½

Thunder Bay Stewart and Duryea are a pair of Louisiana wildcat oil drillers who believe there is oil at the bottom of the Gulf of Mexico off the coast of the town of Port Felicity. They decide to construct an oil platform which the shrimp fisherman of the town believe will interfere with their livelihoods. Tensions rise between the two groups and violence seems likely. Action packed, with timely storyline as modern oil drillers fight to drill in waters that have historically been off-limits.
1953 82m/C James Stewart, Joanne Dru, Dan Duryea, Gilbert Roland, Marcia Henderson; *Dir:* Anthony Mann. **VHS, Beta $19.95** *MCA* ♪♪♪

Thunder in the City Robinson is an intense American promoter in the sales game who is sent to London by his employers to learn a more subdued way of doing business. Instead, he meets a pair of down-on-their-luck artistocrats whose only asset is an apparently worthless Rhodesian mine. Robinson, however, comes up with a way to promote the mine and get enough capital together to make it profitable. Good satire with Robinson well cast. Robinson wasn't taken with the script in its original form and persuaded his friend, playwright Robert Sherwood, to re-write it.
1937 85m/B *GB* Edward G. Robinson, Nigel Bruce, Ralph Richardson, Constance Collier; *Dir:* Marion Gering. **VHS, Beta $19.95** *NOS, DVT, KRT* ♪♪♪

Thunder County Federal agents and drug runners are after a group of four convicts who have just escaped from prison.
1974 78m/C Mickey Rooney, Ted Cassidy, Chris Robinson. **VHS, Beta $49.95** *PSM* ♪½

Thunder in the Desert A cowhand joins an outlaw gang in order to catch his uncle's killer.
1938 56m/C Bob Steele, Louise Stanley, Don Barclay, Charles King, Lew Meehan, Budd Buster; *Dir:* Sam Newfield. **VHS, Beta $19.95** *NOS, DVT* ♪

Thunder in God's Country A dishonest gambler eyes a peaceful western town as the next Vegas-like gambling capital. Before this occurs he is exposed as a cheat.
1951 67m/B Rex Allen, Mary Ellen Kay. **VHS, Beta $24.95** *DVT* ♪

Thunder and Lightning A mismatched young couple chase a truckload of poisoned moonshine. Action-packed chases ensue. Nice chemistry between Carradine and Jackson.
1977 (PG) 94m/C David Carradine, Kate Jackson, Roger C. Carmel, Sterling Holloway; *Dir:* Corey Allen. **VHS, Beta $59.98** *FOX* ♪♪

Thunder Mountain Zane Grey's novel about two prospectors who are bushwhacked on their way to file a claim comes to life on the screen.
1935 56m/B George O'Brien. **VHS, Beta $24.95** *DVT, HHE, MLB* ♪

Thunder & Mud Female mud wrestlers battle over their favorite rock band. Hahn's attempt to cash in on her notoriety after the Bakker scandal.
1989 90m/C Jessica Hahn; *Dir:* Penelope Spheeris. **VHS, Beta, LV $59.95** *COL* ♪

Thunder Over Texas The cowboy hero gets embroiled in a railroad scandal that involves a kidnapping. Directed by Edgar G. Ulmer, who used the alias John Warner because he was moonlighting. His first western; previously he was best known for "The Black Cat."
1934 52m/B Marion Schilling, Helen Westcott, Victor Potel, Tiny Skelton; *Dir:* Edgar G. Ulmer. **VHS, Beta $19.95** *VDM* ♪

Thunder Pass Gold and greed are the motives for murder as two young brothers become separated when their wagon train is attacked and their family is slaughtered. One brother searches for the other and finds him several years later. Based on the novel "Arizona Ames" by Zane Grey. Also released as "Thunder Trail."
1937 58m/B Charles Bickford, Marsha Hunt, Gilbert Roland, Monte Blue; *Dir:* Charles T. Barton. **VHS $14.95** *NOS, VDM, DVT* ♪♪

Thunder in the Pines Two lumberjack buddies send for the same French girl to marry, unbeknownst to each other. When she arrives, the dismayed men agree to a contest to see who will marry her. Meanwhile the French girl falls for the local saloon owner and his jilted girlfriend vows to get even. She does, with the help of the lumberjacks.
1949 61m/B Denise Darcel, George Reeves, Ralph Byrd, Lyle Talbot, Michael Whalen, Roscoe Ates, Vince Barnett; *Dir:* Robert Edwards. **VHS, Beta, LV** *WGE* ♪

Thunder River Feud The Rangebusters help two ranchers avoid a range war. Part of "The Rangebusters" series.
1942 58m/B Ray Corrigan, John "Dusty" King, Max Terhune. **VHS, Beta $19.95** *NOS, VCN, DVT* ♪

Thunder Road Mitchum comes home to Tennessee from Korea and takes over the family moonshine business, fighting both mobsters and federal agents. An exciting chase between Mitchum and the feds ends the movie with the appropriate bang. Robert Mitchum not only produced, wrote and starred in this best of the moonshine-running films, but also wrote the theme song "Whippoorwill" (which later became a radio hit). Mitchum's son, James, made his film debut, and later starred in a similar movie "Moonrunners." A cult favorite, still shown in many drive-ins around the country.
1958 94m/B Robert Mitchum, Jacques Aubuchon, Gene Barry, Keely Smith, Trevor Bardette, Sandra Knight, Jim Mitchum, Betsy Holt, Frances Koon; *Dir:* Arthur Ripley. **VHS, Beta $19.95** *MGM* ♪♪♪

Thunder Run When a load of uranium needs to be transferred across Nevada, a retired trucker takes on the job. The trip is made more difficult since terrorists are trying to hijack the shipment to use for a bomb.
1986 (PG-13) 84m/C Forrest Tucker, John Ireland, John Shepherd, Jill Whitlow, Wally Ward, Cheryl Lynn; *Dir:* Gary Hudson. **VHS, Beta $79.95** *MED* ♪½

Thunder Score Evil crime boss and his ruthless ninja hitmen are out to steal a fortune in gold.
19?? 92m/C Jack Fox, Stuart Lyne. **VHS $39.95** *TWE* ♪½

Thunder Warrior A young Indian turns into a one-man army determined to punish the local authorities who are abusing his fellow tribe members.
1985 (R) 84m/C Raymond Harmstorf, Bo Svenson, Mark Gregory; *Dir:* Larry Ludman. **VHS, Beta $79.95** *TWE* ♪

Thunder Warrior 2 A tough Native American is provoked to violence by small-town prejudice.
1985 (PG-13) 114m/C *IT* Mark Gregory, Karen Reel, Bo Svenson; *Dir:* Larry Ludman. **VHS, Beta $79.95** *TWE* ♪

Thunder Warrior 3 The Indian Thunder has vowed to live in peace, but forswears the oath when his wife is kidnapped and family is terrorized. Revenge is the top priority as he torments the bad guys.
1988 90m/C Mark Gregory, John Phillip Law, Ingrid Lawrence, Werner Pochat, Horts Schon; *Dir:* Larry Ludman. **VHS $79.95** *IMP* ♪

Thunderball The fourth installment in Ian Fleming's James Bond series finds 007 on a mission to thwart SPECTRE, which has threatened to blow up Miami by atomic bomb if 100 million pounds in ransom is not paid. One of the more tedious Bond entries but a

big box-office success. Tom Jones sang the title song. Remade as "Never Say Never Again" in 1983 with Connery reprising his role as Bond after a 12-year absence.
1965 125m/C GB Sean Connery, Claudine Auger, Adolfo Celi, Lucianna Paluzzi, Rick Van Nutter, Martine Beswick, Molly Peters, Guy Doleman, Bernard Lee, Lois Maxwell, Desmond Llewelyn; *Dir:* Terence Young. Academy Awards '65: Best Visual Effects. **VHS, Beta, LV $19.95** MGM, TLF 🎬🎬½

Thunderbolt & Lightfoot Eastwood is an ex-thief on the run from his former partners (Kennedy and Lewis) who believe he's made off with the loot from their last job, the robbery of a government vault. He joins up with drifter Bridges, who helps him to escape. Later, Eastwood manages to convince Kennedy he doesn't know where the money is. Bridges then persuades the men that they should plan the same heist and rob the same government vault all over again, which they do. But their getaway doesn't go exactly as planned. All-around fine acting. Bridges received an Academy Award nomination for his supporting role (look for his scene dressed in drag). First film for director Cimino.
1974 (R) 115m/C Clint Eastwood, Jeff Bridges, George Kennedy, Geoffrey Lewis, Gary Busey; *Dir:* Michael Cimino. **VHS, Beta, LV $19.95** MGM, PNR 🎬🎬🎬

Thunderfist A police detective, who is also a Kung Fu expert, goes after thieves who have stolen a precious jade pagoda.
1973 (PG) 82m/C Alex Lung, Steve Yu. **VHS $59.95** ACA 🎬

Thunderground A beautiful con artist teams with a fighter and together they travel to New Orleans to arrange a bout with the king of bare-knuckled fighting.
1989 (R) 92m/C Paul Coufos, Margaret Langrick, Jesse Ventura, M. Emmet Walsh; *Dir:* David Mitchell. **VHS, LV $39.95** SHG, IME 🎬

Thunderheart A young FBI agent (Kilmer) is sent to an Oglala Sioux reservation to investigate a murder. He is himself part Sioux but Ray Levoi is so assimilated that he resents being chosen for the assignment because of his forgotten heritage. Aided by a veteran partner (Shepard), Ray learns most, professionally and personally, from a shrewd tribal police officer (well played by Greene). Set in the late 70's the film is loosely based on actual events plaguing the violence-torn Native American community at that time. Great cinematography by Roger Deakins. Filmed on the Pine Ridge Reservation located in South Dakota. Director Apted deals with similar incidents in his documentary "Incident at Oglala."
1992 (R) 118m/C Val Kilmer, Graham Greene, Sam Shepard, Sheila Tousey, Chief Ted Thin Elk, Fred Ward, John Trudell; *Dir:* Michael Apted. **VHS, LV $94.95** COL 🎬🎬🎬

Thundering Gunslingers A group of gunmen terrorizing townspeople finally meet their match in this Western.
1940 61m/B Buster Crabbe. **VHS, Beta $19.95** NOS, DVT 🎬

Thundering Mantis A rousing martial arts film with a shattering climax that features the thundering mantis style of Kung-fu.
1976 85m/C VHS, Beta $39.95 TWE 🎬

Thundering Ninja Three secret agents fight off an army of ninjas. Plenty of heel-to-head antics.

198? 89m/C Bernard Geurts, Joe Redner, Klaus Mutter, Ken Kerr, Jimmy Wang Yu; *Dir:* Joseph Kong. **VHS, Beta $59.95** TWE 🎬½

Thundering Trail Two cowboys encounter plenty of trouble before they take the President's newly appointed territorial governor to his office.
1951 55m/B Lash LaRue, Al "Fuzzy" St. John, Sally Anglim, Archie Twitchell, Ray Bennett. **VHS, Beta $19.95** NOS, DVT 🎬

Thundersquad Mercenaries help a South American rebel leader rescue his kidnapped son.
1985 90m/C Sal Borgese, Julia Fursich, Anthony Sabato. **VHS, Beta $59.95** MGL 🎬

Thursday's Child A child's success in films causes trouble for her family. Melodramatic and sappy, but Howes, as the kid, is excellent. One of Granger's early supporting roles.
1943 95m/B Stewart Granger, Sally Ann Howes; *Dir:* Rodney Ackland. **VHS, Beta $24.95** NOS, DVT 🎬🎬

Thursday's Game Two crisis-besieged businessmen meet every Thursday, using poker as a ruse to work on their business and marital problems. Wonderful cast; intelligently written and directed by James L. Brooks. Made for television.
1974 (PG) 99m/C Gene Wilder, Ellen Burstyn, Bob Newhart, Cloris Leachman, Nancy Walker, Valerie Harper, Rob Reiner; *W/Dir:* James L. Brooks. **VHS, Beta, LV $69.95** VMK 🎬🎬🎬

THX 1138 In the dehumanized world of the future, people live in underground cities run by computer, are force-fed drugs to keep them passive, and no longer have names - just letter/number combinations (Duvall is THX 1138). Emotion is also outlawed and when the computer-matched couple THX 1138 and LUH 3417 discover love, they must battle the computer system to escape. George Lucas' first film, which was inspired by a student film he did at USC. The laserdisc format carries the film in widescreen.
1971 (PG) 88m/C Robert Duvall, Donald Pleasence, Maggie McOmie; *Dir:* George Lucas. **VHS, Beta, LV $19.98** WAR 🎬🎬🎬

Tiara Tahiti Intrigue, double-cross, romance, and violence ensue when two old army acquaintances clash in Tahiti. Very fine performances from the two leading men.
1962 100m/C James Mason, John Mills, Claude Dauphin, Rosenda Monteros, Herbert Lom; *Dir:* Ted Kotcheff. **VHS, Beta $59.95** SUE 🎬🎬

Ticket to Heaven A young man, trying to deal with the painful breakup of a love affair, falls under the spell of a quasi-religious order. His friends and family, worried about the cult's influence, have him kidnapped in order to be de-programmed. Mancuso is excellent, as are his supporting players Rubinek and Thomson.
1981 (PG) 109m/C Nick Mancuso, Meg Foster, Kim Cattrall, Saul Rubinek, R.H. Thomson; *Dir:* Ralph L. Thomas. Genie Awards '82: Best Actor (Mancuso), Best Picture, Best Supporting Actor (Rubinek). **VHS, Beta $79.95** MGM 🎬🎬🎬

Ticket of Leave Man London's most dangerous killer fronts a charitable organization designed to help reform criminals. It actually steers them into a crime syndicate that cheats philanthropists out of their fortunes.
1937 71m/B Tod Slaughter, John Warwick, Marjorie Taylor. **VHS, Beta $29.95** SNC, VYY 🎬🎬

Ticket to Tomahawk A gunslinger is hired by a stagecoach company to sabotage a railroad. Look for Marilyn Monroe as a chorus girl in a musical number.
1950 90m/C Anne Baxter, Rory Calhoun, Walter Brennan, Charles Kemper, Connie Gilchrist, Arthur Hunnicutt, Victor Sen Yung; *Dir:* Richard Sale. **VHS** TCF 🎬🎬½

Tickle Me An unemployed rodeo star finds work at an all-girl health spa and dude ranch. He falls in love with a young lady who has a treasure map and keeps her safe from the evil men who want her fortune. Lots of musical numbers, including "It Feels So Right," "I'm Yours," "Slowly but Surely," and "I Feel that I've Known You Forever." For die-hard Elvis fans only.
1965 90m/C Elvis Presley, Julie Adams, Jack Mullaney; *Dir:* Norman Taurog. **VHS, Beta $14.98** FOX, MVD 🎬½

Tidal Wave Scientists discover that Japan is slowly sinking and order the island to be evacuated. Originally a popular Japanese production, the American version is a poorly dubbed, re-edited mess.
1975 (PG) 82m/C JP Lorne Greene, Keiju Kobayashi, Rhonda Leigh Hopkins, Hiroshi Fujioka, Shiro Moriana; *Dir:* Andrew Meyer. **VHS** NWV 🎬½

Tides of War In the twilight of the second world war, a German officer begins to question his role in a proposed missile attack on Washington in light of the S.S. brutality he witnesses.
1990 90m/C Yvette Heyden, Rodrigo Obregon, David Soul, Bo Svenson; *Dir:* Neil Rossati. **VHS $89.95** QUE 🎬½

Tidy Endings A man who died from AIDS leaves behind his male lover and his ex-wife. The two form a friendship as they both try to cope with the man's death.
1988 54m/C Harvey Fierstein, Stockard Channing. **VHS, Beta $79.95** PSM 🎬½

Tie Me Up! Tie Me Down! A young psychiatric patient kidnaps a former porno actress he has always had a crush on, and holds her captive, certain that he can convince her to love him. Black comedy features a fine cast and is well directed. At least one fairly explicit sex scene and the bondage theme caused the film to be X-rated, although it was released unrated by the distributor. In Spanish with English subtitles.
1990 105m/C SP Victoria Abril, Antonio Banderas, Loles Leon, Francisco Rabal, Julieta Serrano, Maria Barranco, Rossy De Palma; *Dir:* Pedro Almodovar. **VHS, LV $14.95** COL 🎬🎬½

Tiefland Melodrama based on the libretto for D'Abert's opera about a gypsy dancer and her loves. Unreleased until 1954, this film is the last of Riefenstahl's career; she subsequently turned to still photography. In German with English subtitles.
1944 98m/B GE Leni Riefenstahl, Franz Eichberger, Bernard Minetti, Maria Koppenhofer, Luise Rainer; *Dir:* Leni Riefenstahl. **VHS, Beta $24.95** NOS, FCT, GLV 🎬🎬½

Tiffany Jones Tiffany Jones is a secret agent who has information that presidents and revolutionaries are willing to kill for.
1975 (R) 90m/C Anouska Hempel, Ray Brooks, Eric Pohlmann, Martin Benson, Susan Sneers; *Dir:* Pete Walker. **VHS, Beta $39.95** MON 🎬

Tiger Features some popular reggae stars.
1987 60m/C Tiger, Joseph Cotten, Winston Reedy. **VHS, Beta** KRV

Tiger Bay A young Polish sailor, on leave in Cardiff, murders his unfaithful girlfriend. Lonely ten-year-old Gillie sees the crime and takes the murder weapon, thinking it will make her more popular with her peers. Confronted by a police detective, she convincingly lies but eventually the sailor finds Gillie and kidnaps her, hoping to keep her quiet until he can get aboard his ship. A delicate relationship evolves between the child and the killer as she tries to help him escape and the police close in. Marks Hayley Mills' first major role and one of her finest performances.
1959 107m/B *GB* John Mills, Horst Buchholz, Hayley Mills, Yvonne Mitchell; *Dir:* J. Lee Thompson. **VHS, Beta $19.95** *PAR* 🐾🐾🐾

Tiger Claws The pretty, petite Rothrock has been called the female Bruce Lee. Perhaps, but at least he got a good script every once in a while. This time Rothrock's a kung-fu cop investigating the strange ritual-murders of martial-arts champions.
1991 (R) 93m/C Cynthia Rothrock, Bolo Yeung, Jalal Merhi; *Dir:* Kelly Markin. **VHS, Beta, LV $89.98** *MCA* 🐾½

Tiger Fangs When Nazis threaten the Far East rubber production with man-eating tigers, real-life big game hunter Frank Buck comes to the rescue.
1943 57m/B Frank Buck, June Duprez, Duncan Renaldo; *Dir:* Jack Schwartz. **VHS $16.95** *SNC, NOS, DVT* 🐾½

Tiger Joe There's trouble in Southeast Asia, and the U.S. military is there to confuse things even more.
1985 96m/C Alan Collins, David Warbeck; *Dir:* Anthony M. Dawson. **VHS $69.98** *LIV, LTG* 🐾

Tiger and the Pussycat A successful businessman's infatuation with an attractive American art student leads him to marital and financial woes.
1967 (R) 110m/C Ann-Margret, Vittorio Gassman, Eleanor Parker; *Dir:* Dino Risi. **VHS, Beta $59.98** *SUE, HHE* 🐾🐾½

Tiger of the Seven Seas A female pirate takes over her father's command and embarks on adventures on the high seas. Generous portions of action, intrigue, and romance. Sequel to "Queen of the Pirates."
1962 90m/C *IT FR* Giana Maria Canale, Anthony Steel, Grazia Maria Spina, Ernesto Calindri; *Dir:* Luigi Capuano. **VHS, EJ** *EMB* 🐾🐾

Tiger Town A baseball player, ending an illustrious career with the Detroit Tigers, sees his chance of winning a pennant slipping away. A young fan, however, proves helpful in chasing that elusive championship. Made for the Disney cable station.
1983 (G) 76m/C Roy Scheider, Justin Henry, Ron McLarty, Bethany Carpenter, Noah Moazezi; *Dir:* Alan Shapiro. **VHS, Beta $69.95** *DIS* 🐾🐾½

Tiger vs. Dragon The Hungry Tiger, undercover agent, takes on Chinese boxing expert The Crazy Dragon.
198? 88m/C **VHS, Beta $59.95** *SVS* 🐾

A Tiger Walks A savage tiger escapes from a circus and local children start a nationwide campaign to save its life. Notable for its unflattering portrayal of America's heartland and small town dynamics. Radical departure from most Disney films of the period.
1964 88m/C Sabu, Pamela Franklin, Brian Keith, Vera Miles, Kevin Corcoran, Peter Brown, Una Merkel, Frank McHugh, Edward Andrews; *Dir:* Norman Tokar. **VHS, Beta $69.95** *DIS* 🐾🐾½

Tiger Warsaw A young man returns to the town where he once lived, before he shot his father. He hopes to sort out his life and repair family problems. Muddled and sappy. Also known as "The Tiger."
1987 (R) 92m/C Patrick Swayze, Barbara Williams, Piper Laurie, Bobby DiCicco, Kaye Ballard, Lee Richardson, Mary McDonnell; *Dir:* Amin Q. Chaudhri. **VHS, Beta, LV $14.95** *SVS* 🐾½

Tiger's Claw Shen roves the country honing his Kung Fu technique, obsessed by his fanatical desire to challenge Kuo, the finest warrior in China.
197? 90m/C Chin Long. **VHS, Beta $39.95** *MAV* 🐾½

Tigers Don't Cry An unusual friendship develops between a male nurse and the African dignitary he kidnaps.
1981 105m/C Anthony Quinn. **VHS, Beta $59.95** *MPI* 🐾½

Tigers in Lipstick Seven short vignettes featuring the beautiful actresses as aggressive women and the men that they influence. Also known as "Wild Beds."
1980 (R) 88m/C *IT* Laura Antonelli, Monica Vitti, Ursula Andress, Sylvia Kristel; *Dir:* Luigi Zampa. **VHS, Beta $39.95** *MON* 🐾½

A Tiger's Tale A middle-aged divorced nurse begins an affair with her daughter's ex-boyfriend.
1987 (R) 97m/C Ann-Margret, C. Thomas Howell, Charles Durning, Kelly Preston, Ann Wedgeworth, Tim Thomerson, Steven Kampmann, Traci Lin, Angel Tompkins; *Dir:* Peter Douglas. **VHS, Beta, LV** *KRT, OM* 🐾🐾½

Tigers at the Top More Kung-Fu brutality and its jinks.
1985 75m/C **VHS, Beta $59.95** *SVS* 🐾

Tigershark A martial arts expert goes to Southeast Asia to rescue his girlfriend, who has been kidnapped by communist forces and held captive for arms and ammunition. Strictly for fans of the martial arts.
1987 (R) 97m/C Mike Stone, John Quade, Pamela Bryant; *Dir:* Emmett Alston. **VHS, Beta $79.98** *DMD* 🐾

Tightrope Police inspector Wes Block pursues a killer of prostitutes in New Orleans' French Quarter. The film is notable both as a thriller and as a fascinating vehicle for Eastwood, who experiments with a disturbing portrait of a cop with some peculiarities of his own.
1984 (R) 115m/C Clint Eastwood, Genevieve Bujold, Dan Hedaya, Jennifer Beck, Alison Eastwood, Randi Brooks, Regina Richardson, Jamie Rose; *Dir:* Richard Tuggle. **VHS, Beta, LV $19.98** *WAR, TLF* 🐾🐾🐾

Till the Clouds Roll By An all-star, high-gloss musical biography of songwriter Jerome Kern, that, in typical Hollywood fashion, bears little resemblance to the composer's life. Filled with wonderful songs from his Broadway hit "Showboat," among others.
1946 137m/C Robert Walker, Van Heflin, Judy Garland, Frank Sinatra, Lucille Bremer, Kathryn Grayson, June Allyson, Dinah Shore, Lena Horne, Virginia O'Brien, Tony Martin; *Dir:* Richard Whorf. **VHS, Beta, LV $9.95** *NEG, NOS, MGM* 🐾🐾½

Till the Clouds Roll By/A Star Is Born Ostensibly concerns the life and career of songwriter Jerome Kern. Highlighted by stellar performances from a number of musical stars of the day. Maudlin and unconcerned with factual accuracy, the film none-

theless is an exuberant celebration of Kern's talent.
1946 249m/C Judy Garland, Frank Sinatra, Robert Walker, Fredric March, Janet Gaynor, Andy Devine, Lionel Stander, Franklin Pangborn, Van Heflin, Lucille Bremer, Kathryn Grayson, Lena Horne, Tony Martin, Angela Lansbury, Dinah Shore, Virginia O'Brien, Cyd Charisse, Adolphe Menjou; *Dir:* William A. Wellman, Richard Whorf. **VHS, Beta $29.95** *DVT* 🐾🐾🐾

Till Death Do Us Part Three married couples spend a weekend at a counseling retreat, unaware that the proprietor is a murderous maniac.
1972 (PG) 77m/C James Keach, Claude Jutra, Matt Craven; *Dir:* Timothy Brand. **VHS, Beta $69.95** *VES* 🐾🐾

Till Death Do Us Part An ex-cop turns into an evil killer who seduces unsuspecting women then murders them for their insurance money. Contains footage not seen in the TV movie. Based on a true story and the book by Vincent Bugliosi.
1992 (R) 93m/C Treat Williams, Arliss Howard, Rebecca Jenkins; *Dir:* Yves Simoneau. **VHS, Beta $89.95** *TRI* 🐾🐾

Till the End of Time Three GIs have trouble adjusting to civilian life in their home town after World War II. Fine performances and excellent pacing. Popular title song.
1946 105m/C Robert Mitchum, Guy Madison, Bill Williams, Dorothy McGuire, William Gargan, Tom Tully; *Dir:* Edward Dmytryk. **VHS, Beta, LV $19.98** *RKO* 🐾🐾½

Till Marriage Do Us Part An innocent couple discover on their wedding night that they are really brother and sister. Since they can't in good conscience get a divorce or consummate their marriage, they must seek sexual gratification elsewhere. Antonelli is deliciously tantalizing in this silly satire.
1974 (R) 97m/C *IT* Laura Antonelli, Alberto Lionello, Jean Rochefort, Michele Placido, Karin Schubert; *Dir:* Luigi Comencini. **VHS, Beta, LV $59.95** *VES, AXV* 🐾🐾½

Till There was You A struggling musician travels to a faraway tropical isle to solve the mystery of his brother's death.
1991 (R) 94m/C Mark Harmon, Deborah Unger, Jeroen Krabbe, Shane Briant; *Dir:* John Seale. **VHS $92.95** *MCA* 🐾

Tillie's Punctured Romance The silent comedy which established Chaplin and Dressler as comedians. Chaplin, out of his Tramp character, is the city slicker trying to put one over on farm girl Dressler, who pursues revenge. First feature length comedy film. Followed by "Tillie's Tomato Surprise" and "Tillie Wakes Up," which starred Dressler but not Chaplin.
1914 73m/B Charlie Chaplin, Marie Dressler, Mabel Normand, Mack Swain, Chester Conklin; *Dir:* Mack Sennett. **VHS, Beta, LV $12.95** *NOS, CCB, HHT* 🐾🐾🐾

Tilt Flimsy storyline concerns a young runaway's adventures with an aspiring rock star. Shields' pinball expertise eventually leads her to a match against pinball champ Durning.
1978 (PG) 111m/C Charles Durning, Ken Marshall, Brooke Shields, Geoffrey Lewis; *Dir:* Rudy Durand. **VHS $39.98** *CON* 🐾

Tim Follows the relationship between a handsome, mentally retarded young man and an attractive older businesswoman. Sappy storyline is redeemed by Gibson's fine debut performance. Based on Colleen McCullough's first novel.

1979 94m/C *AU* Mel Gibson, Piper Laurie, Peter Gwynne, Alwyn Kurtis, Pat Evison; *Dir:* Michael Pate. Australian Film Institute '79: Best Actor (Gibson). **VHS, Beta** *MED, OM* 🎬🎬½

Timber Queen A WWII flier comes home and aids his best friend's widow, who must make a quota of timber or lose her sawmill to gangsters. Old-fashioned.
1944 66m/B Richard Arlen, Mary Beth Hughes, June Havoc, Sheldon Leonard, George E. Stone, Dick Purcell; *Dir:* Frank McDonald. **VHS $19.95** *NOS, LOO* 🎬🎬

Timber Terrors A Mountie investigates his partner's murder.
1935 50m/B **VHS, Beta** *WGE* 🎬

Time After Time In Victorian London, circa 1893, H.G. Wells is experimenting with his time machine. He discovers the machine has been used by an associate, who turns out to be Jack the Ripper, to travel to San Francisco in 1979. Wells follows to stop any further murders and the ensuing battle of wits is both entertaining and imaginative. McDowell is charming as Wells and Steenburgen is equally fine as Well's modern American love interest. Notable score by Miklos Rosza.
1979 (PG) 112m/C Malcolm McDowell, David Warner, Mary Steenburgen, Patti D'Arbanville, Charles Cioffi; *Dir:* Nicholas Meyer. National Board of Review Awards '79: 10 Best Films of the Year. **VHS, Beta, LV $19.98** *WAR* 🎬🎬🎬

Time of the Angels Faith Hubley interprets three Latin American poems in colorfully symbolic animation. Included are "Time of the Angels" by Emperor Nezahualcotl of Texcoco, "Horses of the Conquerors" by Jose Santos Chocano, and "The Guardian Angel" by Gabriela Mistral. Each is spoken in both Spanish and English.
1986 10m/C **VHS, Beta, Special order formats $195.00** *PYR*

Time of the Apes A Japanese-made horror film about a woman and two kids thrust into an underground world ruled by intelligent gorillas.
1987 98m/C *JP* **VHS, Beta $39.95** *CEL* 🎬

Time Bandits A group of dwarves help a young boy to travel through time and space with the likes of Robin Hood, Napoleon, Agamemnon, and other time-warp playmates. Epic fantasy from Monty Python alumni.
1981 (PG) 110m/C *GB* John Cleese, Sean Connery, Shelley Duvall, Katherine Helmond, Ian Holm, Michael Palin, Ralph Richardson, Kenny Baker, Peter Vaughan, David Warner; *Dir:* Terry Gilliam. **VHS, Beta, LV $19.95** *PAR* 🎬🎬½

Time Burst—The Final Alliance A man fights for his soul to protect the secret of immortality at any cost.
1989 93m/C **VHS, Beta** *AIP* 🎬🎬½

Time of Destiny During World War II, an American soldier vows to kill his brother-in-law, whom he believes caused his father's death. Meanwhile, the two men have become good friends in the army, never realizing their connection. A strange, overblown family saga from the director of "El Norte." Music by Ennio Morricone. Beautiful photography, terrific editing.
1988 (PG-13) 118m/C William Hurt, Timothy Hutton, Melissa Leo, Stockard Channing, Megan Follows; *Dir:* Gregory Nava. **VHS, Beta, LV, 8mm $14.95** *COL, SUE* 🎬🎬

Time to Die A victim of heinous war crimes, obsessed with revenge, stalks his prey for a final confrontation. Also known as "Seven Graves for Rogan."
1983 (R) 89m/C Edward Albert, Rex Harrison, Rod Taylor, Raf Vallone; *Dir:* Matt Cimber. **VHS, Beta $59.95** *MED* 🎬

A Time to Die Lords plays a police photographer who, well, photographs the police. But her camera catches one in the act of murder. Routine crime thriller.
1991 (R) 93m/C Traci Lords, Jeff Conaway, Richard Roundtree, Bradford Bancroft, Nitchie Barrett; *W/Dir:* Charles Kanganis. **VHS $89.95** *PMH* 🎬🎬

Time Fighters A time machine has the ability to travel at the speed of light from prehistoric times to the distant future. This is a great help to the Time fighters.
1985 60m/C **VHS, Beta** *SUE* 🎬½

Time Flies When You're Alive A made-for-cable-television presentation of actor Linke's performance monologue detailing the traumas and experiences he and his family underwent as his wife Francesca slowly died of cancer.
1989 79m/C Paul Linke; *Dir:* Roger Spottiswoode. **VHS, Beta $79.99** *HBO* 🎬½

The Time Guardian Time-travelers of the future arrive in the Australian desert in 1988, intent on warning the local populace of the impending arrival of killer cyborgs from the 40th century. Muddled and confusing.
1987 (PG) 89m/C Tom Burlinson, Carrie Fisher, Dean Stockwell, Nikki Coghill; *Dir:* Brian Hannant. **VHS, Beta, LV, 8mm $14.95** *COL, SUE* 🎬½

Time of the Gypsies Acclaimed Yugoslavian saga about a homely, unlucky Gypsy boy who sets out to steal his way to a dowry large enough to marry the girl of his dreams. Beautifully filmed, magical, and very long. In the Gypsy language, Romany, it's the first feature film made in the Gypsy tongue.
1990 (R) 136m/C *YU* Davor Dujmovic, Sinolicka Trpkova, Ljubica Adzovic, Hunsija Hasimovic, Bora Todorovic; *Dir:* Emir Kusturica. Cannes Film Festival '90: Best Director (Kusturica). **VHS, Beta, LV $79.95** *COL* 🎬🎬🎬

Time of Indifference Based on the novel by Alberto Moravia. Tale of the disintegration of values in 1920s Italy is unable to take full advantage of its fine cast or potentially thought-provoking themes.
1964 84m/B *IT* Rod Steiger, Shelley Winters, Claudia Cardinale, Paulette Goddard, Tomas Milian; *Dir:* Francesco Maselli. **VHS** *GKK* 🎬🎬

Time to Kill A young soldier in Africa wanders away from his camp and meets a woman whom he rapes and kills. But when he returns to his outfit he finds he can't escape his tormenting conscience.
1989 (R) 110m/C Nicolas Cage, Giancarlo Giannini, Robert Liensol; *Dir:* Guiliano Montaldo. **VHS, LV $89.98** *REP* 🎬🎬

Time to Live Made-for-television adaptation of Mary Lou Weisman's "Intensive Care" about a family's adjustment to a young boy's fight with Muscular Dystrophy. Minnelli's first appearance in a television movie.
1985 97m/C Liza Minnelli, Jefferey De Munn, Swoosie Kurtz, Corey Haim, Scott Schwartz; *Dir:* Rick Wallace. **Beta $79.95** *VHV* 🎬½

Time Lock An expert safecracker races against time to save a child who is trapped inside a bank's pre-set time-locked vault. Canada is the setting, although the film was actually made in England. Based on a play by Arthur Hailey.
1957 73m/B *GB* Sean Connery, Robert Beatty, Lee Patterson, Betty McDowall, Vincent Winter; *Dir:* Gerald Thomas. **VHS, Beta $39.95** *MON* 🎬🎬

A Time to Love & a Time to Die Lush adaptation of Erich Maria Remarque novel about WWII German soldier who falls in love with a young girl during furlough. The two are married, only to be separated when he is forced to return to the Russian front. Sympathetic treatment of Germans opposed to Hitler's policies.
1958 133m/C John Gavin, Lilo Pulver, Jock Mahoney, Keenan Wynn, Klaus Kinski, Jim Hutton, Dana J. Hutton; *Dir:* Douglas Sirk. **VHS, Beta $14.95** *KRT* 🎬🎬½

The Time Machine English scientist living near the end of the 19th century invents time travel machine and uses it to travel into various periods of the future. Rollicking version of classic H.G. Wells' cautionary tale boasts Oscar-winning special effects. Remade in 1978.
1960 103m/C Rod Taylor, Yvette Mimieux, Whit Bissell, Sebastian Cabot, Alan Young; *Dir:* George Pal. Academy Awards '60: Best Special Effects. **VHS, Beta, LV $19.95** *MGM, MLB* 🎬🎬🎬

Time Machine Yet another adaptation of H.G. Wells' classic novel about a scientist who invents a machine that enables him to travel through time. Inferior remake of the 1960 version.
1978 (G) 99m/C John Beck, Priscilla Barnes, Andrew Duggan; *Dir:* Henning Schellerup. **VHS, Beta $19.95** *UHV, LME* 🎬🎬

Time for Miracles Dramatic biography of the first native-born American saint, Elizabeth Bayley Seton, who was canonized in 1975. Made-for-television.
1980 97m/C Kate Mulgrew, Lorne Greene, Rossano Brazzi, John Forsythe, Jean-Pierre Aumont, Milo O'Shea; *Dir:* Michael O'Herlihy. **VHS, Beta $9.98** *CHA* 🎬½

Time for Murder A series of British dramatizations of famous murder mystery stories.
198? 60m/C *GB* **Beta** *PSM* 🎬½

A Time to Remember Sticky with sentiment, this family film tells of a boy with a gift for singing, encouraged by a priest, discouraged by his father. At one point the kid loses his voice—he probably choked on the pathos.
1990 90m/C Donald O'Connor, Morgana King. **VHS $29.95** *IGP* 🎬½

Time for Revenge A demolitions worker attempts to blackmail his corrupt employer by claiming a workplace explosion made him speechless. He finds himself in a contest of silence and wits. English subtitles.
1982 112m/C *AR* Federico Luppi, Haydee Padilla, Julio de Grazia, Ulises Dumont; *Dir:* Adolfo Aristarain. Montreal World Film Festival '82: Best Film. **VHS, Beta $24.95** *XVC, MED* 🎬🎬½

Time Stands Still Two brothers experience troubled youth in Budapest. Artful, somberly executed, with American pop soundtrack. Subtitled in English.
1982 99m/C *HU* Istvan Znamenak, Henrik Pauer, Aniko Ivan, Sander Soth, Peter Galfy; *Dir:* Peter Gothar. **VHS, Beta $59.95** *COL* 🎬🎬🎬

Time of Tears A young boy, grieving after his grandfather's death, meets a lonely old man and they fill each other's emotional needs.
1987 (PG) 95m/C VHS, Beta $19.95 STE, NWV ⅋½

The Time of Their Lives A pair of Revolutionary War-era ghosts haunt a modern country estate. One of the best A & C comedies.
1946 82m/B Bud Abbott, Lou Costello, Marjorie Reynolds, Binnie Barnes, Gale Sondergaard, John Shelton; *Dir:* Charles T. Barton. **VHS, Beta $14.95** MCA ⅋⅋½

Time Trackers A Roger Corman cheapie about a race through time, from present-day New York to medieval England, to recover a time machine before it alters the course of history.
1988 (PG) 87m/C Kathleen Beller, Ned Beatty, Will Shriner; *Dir:* Howard R. Cohen. **VHS, Beta $79.95** MGM ⅋½

The Time Travelers Scientists discover and pass through a porthole leading to earth's post-armageddon future where they encounter unfriendly mutants. Frightened, they make a serious effort to return to the past. Remade as ''Journey to the Center of Time.''
1964 82m/C Preston Foster, Phil Carey, Merry Anders, John Hoyt, Joan Woodbury, Dolores Wells, Dennis Patrick; *Dir:* Ib Melchoir. **VHS, Beta $69.95** HBO ⅋⅋

Time Troopers In another post-nuke society, special police must execute anyone who has misused their allotted ''energy clips'' for illicit behavior. Dime-store budget sci-fi.
1989 (R) 90m/C Albert Fortell, Hannelore Elsner; *Dir:* L.E. Neiman. **VHS, Beta $59.95** PSM ⅋

Time Walker An archaeologist unearths King Tut's coffin in California. An alien living inside is unleashed and terrorizes the public.
1982 (PG) 86m/C Ben Murphy; *Dir:* Tom Kennedy. **VHS, Beta $19.98** CHA ⅋

Timebomb When someone attempts to kill Biehn, he turns to a beautiful psychiatrist for help. She triggers flashbacks that send her hunted patient on a dangerous, perhaps deadly, journey into his past.
1991 (R) 96m/C Michael Biehn, Patsy Kensit, Tracy Scoggins, Robert Culp, Richard Jordan, Raymond St. Jacques; *W/Dir:* Avi Nesher. **VHS, Beta, LV $89.98** MGM ⅋½

Timepiece This tape features one of Jim Henson's earlier films about an overworked, young executive.
1965 10m/C Jim Henson. **VHS, Beta** FLL ⅋⅋

Timerider A motorcyclist riding through the California desert is accidentally thrown back in time to 1877, the result of a scientific experiment that went awry. Co-written and co-produced by Michael Nesmith, best known for his days with the rock group ''The Monkees.''
1983 (PG) 93m/C Fred Ward, Belinda Bauer, Peter Coyote, Richard Masur, Ed Lauter, L.Q. Jones, Tracy Walter; *Dir:* William Dear. **VHS, Beta, LV $19.95** PAV ⅋½

Times to Come Taut thriller reminiscent of bleak futuristic movies like ''Clockwork Orange.'' Won three ACE awards for best director, best actor, and best supporting actor. In Spanish with English subtitles.
1981 98m/C AR Hugo Soto, Juan Leyrado, Charly Garcia; *Dir:* Gustavo Mosquera. **VHS $79.95** CCN, FCT ⅋⅋⅋

Times of Harvey Milk A powerful and moving documentary about the life and career of San Francisco supervisor and gay activist Harvey Milk. The film documents the assassination of Milk and Mayor George Moscone by Milk's fellow supervisor, Dan White. Highly acclaimed by both critics and audiences, the film gives an honest and direct look at the murder and people's reactions. News footage of the murders and White's trial is included.
1983 90m/C *Dir:* Robert Epstein. Academy Awards '83: Best Feature Documentary. **VHS, Beta, LV $59.95** PAV ⅋⅋⅋

Times Square A 13-year-old girl learns about life on her own when she teams up with a defiant, anti-social child of the streets. Unappealing and unrealistic, the film features a New Wave music score. Film critic Jacob Brackman wrote the script.
1980 (R) 111m/C Tim Curry, Trini Alvarado, Robin Johnson, Peter Coffield, Elizabeth Pena, Anna Maria Horsford; *Dir:* Allan Moyle. **VHS, Beta $59.99** HBO ⅋⅋⅋

Timestalkers A college professor's infatuation with a young woman is complicated by their pursuit of a criminal from the 26th century into the past. Mildly entertaining adventure. Tucker's last film. Made for television.
1987 100m/C William Devane, Lauren Hutton, Klaus Kinski, John Ratzenberger, Forrest Tucker, Gail Youngs; *Dir:* Michael Schultz. **VHS $29.95** FRH ⅋⅋½

Timothy and the Angel A young man with a deep desire to be a ranger must pass up the rival temptations of a baseball scholarship and a life of speed and drugs offered by a motorcyclist.
1979 29m/C VHS, Beta LCA ⅋½

The Tin Drum German child in the 1920s wills himself to stop growing in response to the increasing Nazi presence in Germany. He communicates his anger and fear by pounding on a tin drum. Memorable scenes, excellent cast. In German with English subtitles.
1979 (R) 141m/C GE David Bennent, Mario Adorf, Angela Winkler, Daniel Olbrychski; *Dir:* Rainer Werner Fassbinder, Volker Schlondorff. Academy Awards '79: Best Foreign Language Film; Cannes Film Festival '79: Best Film; Los Angeles Film Critics Association Awards '80: Best Foreign Film; National Board of Review Awards '80: Best Foreign Film. **VHS, Beta $59.95** WAR, APD, GLV ⅋⅋⅋⅋

Tin Man A garage mechanic born totally deaf designs and builds a computer that can both hear and speak for him. His world is complicated, however, when a young speech therapist introduces him to a world both of new and wonderful sounds and of unscrupulous and exploitative computer salesmen.
1983 95m/C Timothy Bottoms, Deana Jurgens, Troy Donahue; *Dir:* John G. Thomas. **VHS, Beta $59.95** MED ⅋½

Tin Men Set in Baltimore in the 1960s, this bitter comedy begins with two aluminum-siding salesmen colliding in a minor car accident. They play increasingly savage pranks on each other until one seduces the wife of the other, ruining his marriage. Like ''Diner,'' the movie is full of Levinson's idiosyncratic local Baltimore color.
1987 (R) 112m/C Richard Dreyfuss, Danny DeVito, Barbara Hershey, John Mahoney, Jackie Gayle; *Dir:* Barry Levinson. **VHS, Beta, LV $19.95** TOU ⅋⅋½

Tin Star Perkins is the young sheriff who persuades veteran bounty hunter Fonda to help him rid the town of outlaws. Excellent Mann western balances humor and suspense.
1957 93m/B Henry Fonda, Anthony Perkins, Betsy Palmer, Neville Brand, Lee Van Cleef, John McIntire, Michel Ray; *Dir:* Anthony Mann. **VHS, Beta $14.95** PAR, KRT ⅋⅋⅋

Tin Toy Computer-animation techniques highlight this humorous story of a wind-up toy's first encounter with a boisterous baby.
1988 5m/C Academy Awards '88: Best Animated Short Film. **VHS, Beta, 3/4U $75.00** DCL ⅋⅋⅋

Tintorera...Tiger Shark Three shark hunters attempt to discover why buxom swimmers are disappearing. Phony takeoff of ''Jaws.'' Also known as ''Tintorera'' and ''Tintorera... Bloody Waters.''
1978 (R) 91m/C GB MX Susan George, Fiona Lewis, Jennifer Ashley; *Dir:* Rene Cardona Jr. **VHS, Beta $54.95** MED Woof!

Tiny Toon Adventures: How I Spent My Vacation How the tiny toons of Acme Acres spend their summer vacation in this animated adventure that spoofs several popular films and amusements. Plucky Duck and Hampton Pig journey to ''Happy World Land'' (a takeoff on Walt Disney World), Babs and Buster Bunny's water adventure parodies ''Deliverance,'' there's a spoof of ''The Little Mermaid,'' and the Road Runner even makes a cameo appearance. Parents will be equally entertained by the level of humor and the fast-paced action. Based on the Steven Spielberg TV cartoon series, this is the first made-for-home-video animated feature ever released in the United States.
1991 80m/C VHS, Beta, LV, 8mm $19.95 WAR ⅋⅋⅋

The Tip-Off Lovable boxer (Armstrong) and vivacious girlfriend (Rogers) save less-than-bright friend from getting involved with mean mobster's girlfriend. Major chemistry between Rogers and Armstrong.
1931 75m/B Eddie Quillan, Robert Armstrong, Ginger Rogers, Joan Peers, Ernie Adams; *Dir:* Albert Rogell. **VHS $19.98** TTC, FCT ⅋⅋½

'Tis a Pity She's a Whore A brother and sister engage in an incestuous affair. She becomes pregnant and is married off. Her husband, however, becomes infuriated when he learns of his wife's pregnancy. Photography is by Vittorio Storaro in this highly stylized drama set in Renaissance Italy.
1973 (R) 102m/C IT Charlotte Rampling, Oliver Tobias, Fabio Testi, Antonio Falsi, Rick Battaglia, Angela Luce, Rino Imperio; *Dir:* Giuseppe Patroni Griffi. **VHS, Beta $19.95** ACA ⅋⅋½

Title Shot A crafty manager convinces a millionaire to bet heavily on the guaranteed loss of his heavyweight contender in an upcoming fight. He doesn't realize the boxer is too stupid to throw the fight.
1981 (R) 88m/C Tony Curtis, Richard Gabourie, Susan Hogan, Robert Delbert; *Dir:* Les Rose. **VHS, Beta $69.98** LIV, LTG ⅋

TNT Jackson A kung fu mama searches for her brother while everyone in Hong Kong tries to kick her out of town.
1975 (R) 73m/C Jeanne Bell; *Dir:* Cirio H. Santiago. **VHS, Beta $19.95** CHA, TAV Woof!

To All a Goodnight Five young girls and their boyfriends are planning an exciting Christmas holiday until a mad Santa Claus puts a damper on things. A total gore-fest, with few, if any, redeeming virtues.
1980 (R) 90m/C Jennifer Runyon, Forrest Swanson, Linda Gentile, William Lover; ***Dir:*** David Hess. **VHS, Beta $59.95** *MED Woof!*

To All My Friends on Shore A made-for-television story about a father dealing with his young son's sickle cell anemia. Fine performances, realistic script make this an excellent outing.
1971 74m/C Bill Cosby, Gloria Foster, Dennis Hines; ***Dir:*** Gilbert Cates. **VHS, Beta $14.95** *SAT* 🦴🦴🦴

To Be or Not to Be Sophisticated black comedy set in wartime Poland. Lombard and Benny are Maria and Josef Tura, the Barrymores of the Polish stage, who use the talents of their acting troupe against the invading Nazis. The opening sequence is regarded as a cinema classic. One of Benny's finest film performances, the movie marks Lombard's final screen appearance. Classic Lubitsch. Remade in 1983.
1942 102m/B Carole Lombard, Jack Benny, Robert Stack, Sig Rumann, Lionel Atwill, Felix Bressart, Helmut Dantine, Tom Dugan; ***Dir:*** Ernst Lubitsch. **VHS, Beta $59.95** *VES, WAR* 🦴🦴🦴½

To Be or Not to Be In this remake of the 1942 film, Bancroft and Brooks are actors in Poland during World War II. They accidently become involved with the Polish Resistance and work to thwart the Nazis. Lots of laughs, although at times there's a little too much slapstick.
1983 (PG) 108m/C Mel Brooks, Anne Bancroft, Charles Durning, Jose Ferrer, Tim Matheson, Christopher Lloyd; ***Dir:*** Alan Johnson. **VHS, Beta, LV $19.98** *FOX* 🦴🦴½

To Be a Rose Poignant rendering of the person behind the prostitute. In-depth story following one woman's evening at work, from her point of view. Wonderful sense of her defenses and her feelings.
1974 (R) 92m/C Rose Kilpatrick; ***Dir:*** William A. Levey. **VHS, Beta $69.95** *TWE* 🦴🦴

To the Camp & Back A romantic novel transposed to tape, with equal amounts of unfettered sentiment, fantasy, and style.
1985 60m/C Jan Niklas. **Beta $11.95** *PSM* 🦴½

To Catch a Killer A chilling performance by Dennehy highlights this true-crime tale of a detective's relentless pursuit of serial killer John Wayne Gacy, who preyed on young men and hid the bodies in his home. Made for television.
1992 95m/C Brian Dennehy, Michael Riley, Margot Kidder, Meg Foster; ***Dir:*** Eric Till. **VHS $89.95** *WOV* 🦴🦴🦴

To Catch a King Garr is a singer working in Wagner's nightclub in 1940s' Lisbon. The pair becomes involved in trying to foil the Nazi plot to kidnap the vacationing Duke and Duchess of Windsor. Average made-for-cable fare.
1984 120m/C Robert Wagner, Teri Garr, Horst Janson, Barbara Parkins, John Standing; ***Dir:*** Clive Donner. **VHS, Beta $49.95** *PSM* 🦴🦴

To Catch a Thief On the French Riviera, a reformed jewel thief falls for a wealthy American woman, who suspects he's up to his old tricks when a rash of jewel thefts occur. Oscar-winning photography by Robert Burks, a notable fireworks scene, and snappy dialogue. A change of pace for Hitchcock, this charming comedy-thriller proved to be as popular as his other efforts. Kelly met future husband Prince Ranier during shooting in Monaco.
1955 103m/C Cary Grant, Grace Kelly, Jessie Royce Landis, John Williams; ***Dir:*** Alfred Hitchcock. Academy Awards '55: Best Color Cinematography. **VHS, Beta, LV $19.95** *PAR, MLB* 🦴🦴🦴

To the Devil, A Daughter A nun is put under a spell by a priest who has been possessed by Satan. She is to bear his child. A writer on the occult intervenes. Based on the novel by Dennis Wheatley. One of Kinski's early films. Also on video as "Child of Satan."
1976 (R) 93m/C Richard Widmark, Christopher Lee, Nastassia Kinski, Honor Blackman, Denholm Elliott, Michael Goodliffe; ***Dir:*** Peter Sykes. **VHS, Beta $19.95** *NSV, MLB* 🦴🦴

To Die for A vampire stalks, woos, and snacks on a young real estate woman.
1989 (R) 99m/C Brendan Hughes, Scott Jacoby, Duane Jones, Steve Bond, Sydney Walsh, Amanda Wyss, Ava Fabian; ***Dir:*** Deran Sarafian. **VHS, Beta, LV $29.95** *ACA* 🦴½

To Die for 2 This far outclasses part one in terms of acting and production values, but the plot still isn't back from the grave. Strange, because all the vampires are; despite their fiery deaths last time they flock around the adoptive mother of a baby secretly sired by a bloodsucker.
1991 (R) 95m/C Rosalind Allen, Steve Bond, Scott Jacoby, Michael Praed, Jay Underwood, Amanda Wyss, Remy O'Neill; ***Dir:*** David F. Price. **VHS $89.95** *VMK* 🦴½

To Die Standing Two mavericks team up to do what the D.E.A. and cops simply cannot... capture a dangerous international drug lord. High action.
1991 (R) 87m/C Cliff De Young, Robert Beltran, Jamie Rose, Gerald Anthony, Orlando Sacha, Michael Ford; ***Dir:*** Louis Morneau. **VHS, LV $89.95** *COL, PIA* 🦴½

To Forget Venice Portrays the sensitive relationships among four homosexuals, and their shared fears of growing older. Dubbed in English. Oscar nomination for Best Foreign Film.
197? 90m/C *IT* Erland Josephson, Mariangela Melato, David Pontremoli, Elenora Giorgi; ***Dir:*** Franco Brusati. **VHS, Beta $59.95** *COL* 🦴🦴½

To Have & Have Not Martinique charter boat operator gets mixed up with beautiful woman and French resistance fighters during World War II. Top-notch production in every respect. Classic dialogue and fiery romantic bouts between Bogart and Bacall. Bacall's first film. Based on a story by Ernest Hemingway. Remade in 1950 as "The Breaking Point" and in 1958 as "The Gun Runners."
1944 100m/B Humphrey Bogart, Lauren Bacall, Walter Brennan, Hoagy Carmichael, Marcel Dalio, Dolores Moran, Sheldon Leonard, Dan Seymour; ***Dir:*** Howard Hawks. **VHS, Beta, LV $19.95** *MGM* 🦴🦴🦴½

To Hell and Back Adaptation of Audie Murphy's autobiography. Murphy plays himself, from his upbringing to World War II, where he was the most-decorated American soldier. Features realistic battle sequences punctuated with grand heroics.
1955 106m/C Audie Murphy, Marshall Thompson, Jack Kelly; ***Dir:*** Jesse Hibbs. **VHS, Beta $19.95** *MCA, TLF* 🦴🦴½

To Joy The rocky marriage of a violinist and his wife illuminate the problems of the young in Swedish society. Early Bergman. With English subtitles.
1950 90m/B *SW* ***Dir:*** Ingmar Bergman. **VHS, Beta, LV $29.98** *SUE* 🦴🦴🦴

To Kill a Clown A painter and his wife are trapped on an isolated island. A crazed Vietnam veteran terrorizes them. Some holes in the plot, but generally effective.
1972 (R) 82m/C Alan Alda, Blythe Danner, Heath Lamberts; ***Dir:*** George Bloomfield. **VHS, Beta $9.95** *MED* 🦴½

To Kill with Intrigue After his entire family is brutally slaughtered, a young lord learns new fighting techniques and embarks on a mission of vengeance.
1985 107m/C Jackie Chan. **VHS, Beta $24.95** *ASE* 🦴

To Kill a Mockingbird Faithful adaptation of powerful Harper Lee novel, both an evocative portrayal of childhood innocence and a denunciation of bigotry. Peck's performance as southern lawyer defending a black man accused of raping a white woman is flawless. Duvall debuted as the dim-witted Boo Radley. Lee based her characterization of "Dill" on Truman Capote, a childhood friend. With an Oscar-winning script by Horton Foote and a nominated score by Elmer Bernstein. Other Oscar nominations include supporting actress Badham, director, and Best Picture.
1962 129m/B Gregory Peck, Brock Peters, Phillip Alford, Mary Badham, Robert Duvall, Rosemary Murphy, William Windom, Alice Ghostley; ***Dir:*** Robert Mulligan. Academy Awards '62: Best Actor (Peck), Best Adapted Screenplay, Best Art Direction/Set Decoration (B & W). **VHS, Beta, LV $14.95** *KUI, MCA, BTV* 🦴🦴🦴🦴

To Kill a Priest Based on the true story of Father Jerzy Popieluszko, a young priest in 1984 Poland who defies his church and speaks out publicly on Solidarity. He is killed by the government as a result. Harris is good as the menacing police official.
1989 (R) 110m/C Christopher Lambert, Ed Harris, David Suchet, Tim Roth, Joanne Whalley-Kilmer, Peter Postlethwaite, Cherie Lunghi, Joss Ackland; ***Dir:*** Agnieszka Holland. **VHS, Beta, LV $89.95** *COL* 🦴🦴

To Kill a Stranger A beautiful pop singer is stranded in a storm, and then victimized by a mad rapist/murderer.
197? 100m/C Donald Pleasence, Dean Stockwell, Angelica Maria, Aldo Ray, Sergio Aragones. **VHS, Beta $24.95** *VIR, VCL* 🦴🦴

To the Last Man An early Scott sagebrush epic, about two feuding families. Temple is seen in a small role. Based on Zane Grey's novel of the same name.
1933 70m/B Randolph Scott, Esther Ralston, Jack LaRue, Noah Beery, Buster Crabbe, Gail Patrick, Barton MacLane, Al "Fuzzy" Knight, John Carradine, Jay Ward, Shirley Temple; ***Dir:*** Henry Hathaway. **VHS, Beta $9.95** *NOS, DVT, MLB* 🦴

To the Lighthouse A proper British holiday turns into a summer of disillusionment in this adaptation of the Virginia Woolf novel.
1983 115m/C Rosemary Harris, Michael Clough, Suzanne Bertish, Lynsey Baxter. **VHS, Beta $69.95** *MAG* 🦴🦴

To Live and Die in Hong Kong Two sailors arrive in Hong Kong for a little R & R, until they are mistaken for spies by both the Hong Kong police and the Chinese mob. Action-packed tale.
1989 98m/C Rowena Cortes, Lawrence Jan, Mike Kelly, Laurens C. Postma; **Dir:** Lau Shing Hon. **VHS** ELV 🎬½

To Live & Die in L.A. Fast-paced, morally ambivalent tale of cops and counterfeiters in L.A. After his partner is killed shortly before his retirement, a secret service agent sets out to track down his ruthless killer. Lots of violence; some nudity. Notable both for a riveting car chase and its dearth of sympathetic characters.
1985 (R) 114m/C William L. Petersen, Willem Dafoe, John Pankow, Dean Stockwell, Debra Feuer, John Turturro, Darlanne Fluegel, Robert Downey Jr.; **Dir:** William Friedkin. **VHS, Beta, LV** $9.99 LIV, VES 🎬🎬

To Love Again A middle-aged love story about a reclusive college professor and the campus handyman.
1980 96m/C Lynn Redgrave, Brian Dennehy, Conchata Ferrell; **Dir:** Joseph Hardy. **VHS, Beta** $9.98 CHA 🎬½

To Paris with Love A British man and his son fall in love with a shop girl and her boss while on vacation in Paris. Charming and humorous, with a witty performance from Guinness.
1955 75m/C Alec Guinness, Odile Versuis, Vernon Gray; **Dir:** Robert Hamer. **VHS, Beta** $19.98 VID 🎬🎬½

To Race the Wind A blind man wants to be treated like a normal person as he struggles through Harvard Law School. Based upon Harold Krents' autobiography "Butterflies Are Free." Made-for-television.
1980 97m/C Steve Guttenberg, Lisa Eilbacher, Randy Quaid, Barbara Barrie; **Dir:** Walter Grauman. **VHS, Beta** $39.95 UNI 🎬🎬

To See Such Fun Hilarious excerpts from 80 years of the greatest British movie comedies.
1981 90m/C GB Peter Sellers, Marty Feldman, Benny Hill, Eric Idle, Alec Guinness, Margaret Rutherford, Dirk Bogarde, Spike Milligan. **VHS, Beta** $59.95 PAV

To the Shores of Hell Leaving his sweetheart behind, Major Greg Donahue attacks Da Nang and plots to rescue his brother from the Viet Cong.
1965 82m/C Marshall Thompson, Kiva Lawrence, Richard Jordahl. **VHS, Beta** $42.95 PGN 🎬🎬

To the Shores of Tripoli Wartime propaganda in the guise of drama, in which a smarmy playboy is transformed into a Marine in boot camp. Oscar-nominated for cinematography.
1942 82m/C John Payne, Maureen O'Hara, Randolph Scott, Nancy Kelly, Henry Morgan, Maxie Rosenbloom, William Tracy, Minor Watson, Alan Hale Jr., Hugh Beaumont, Hillary Brooke; **Dir:** H. Bruce Humberstone. **VHS, Beta** $14.98 FOX, MLB 🎬🎬

To Sir, with Love Teacher in London's tough East End tosses books in the wastebasket and proceeds to teach his class about life. Skillful and warm performance by Poitier as idealistic teacher; supporting cast also performs nicely. Based on the novel by E.R. Braithwaite. LuLu's title song was a big hit in 1967-68.
1967 105m/C GB Sidney Poitier, Lulu, Judy Geeson, Christian Roberts, Suzy Kendall, Faith Brook; **Dir:** James Clavell. **VHS, Beta** $14.95 COL 🎬🎬½

To Sleep with Anger At first a comic, introspective look at a black middle-class family going about their business in the heart of Los Angeles. Sly charmer Glover shows up and enthralls the entire family with his slightly sinister storytelling and a gnawing doom gradually permeates the household. Insightful look into the conflicting values of Black America. Glover's best performance.
1990 (PG) 105m/C Danny Glover, Mary Alice, Paul Butler, Richard Brooks, Carl Lumbly, Vonetta McGee, Sheryl Lee Ralph; **W/Dir:** Charles Burnett. Los Angeles Film Critics Association Awards '90: Best Screenplay; National Society of Film Critics Awards '90: Best Screenplay; United States Film Festival '90: Special Jury Prize. **VHS, Beta** $39.95 SVS, FCT, IME 🎬🎬🎬½

To Subdue the Evil In a tale of brutality and betrayal, twin brothers oppose each other after being separated at birth.
19?? 90m/C Lung Chunem, Chen Hung Lieh, Tien Pang. **VHS** $49.95 OCE 🎬

Toast to Lenny A tribute to Lenny Bruce, taped at Hollywood's Troubadour nightclub, with Bill Cosby, Steve Allen, Hugh Hefner and Bruce's family.
1984 60m/C Bill Cosby, Steve Allen. **VHS, Beta, LV** $59.95 MED

The Toast of New Orleans A poor fisherman rises to stardom as an opera singer. Likable, though fluffy production features a plethora of musical numbers, including the Oscar-nominated "Be My Love."
1950 97m/C Kathryn Grayson, Mario Lanza, David Niven, Rita Moreno, J. Carroll Naish; **Dir:** Norman Taurog. **VHS, Beta** $19.98 MGM 🎬🎬½

Toast of New York Arnold plays Jim Fisk, a New England peddler who rises to become one of the first Wall Street giants of industry. Atypical Grant performance.
1937 109m/B Edward Arnold, Cary Grant, Frances Farmer, Jack Oakie, Donald Meek, Billy Gilbert; **Dir:** Rowland V. Lee. **VHS, Beta** $19.98 KOV, RKO 🎬🎬🎬

Toast of the Town Two original Ed Sullivan television programs, both featuring appearances by the cast of "I Love Lucy." The program of October 3, 1954, is a full one-hour tribute to "the Ricardos." The show of February 5, 1956 features Lucy and Desi promoting their latest picture, "Forever Darling," along with other guest stars. Original commercials and I.D.'s are included.
1956 120m/B Ed Sullivan, Lucille Ball, Desi Arnaz Sr., Vivian Vance, William Frawley, Orson Welles, The Ames Brothers, Richard Rodgers, Oscar Hammerstein. **VHS, Beta** $24.95 SHO, MLB 🎬🎬🎬

Tobor the Great Sentimental and poorly executed, this film tells the tale of a boy, his grandfather, and Tobor the robot. Villainous communists attempt to make evil use of Tobor, only to be thwarted in the end.
1954 77m/B Charles Drake, Billy Chapin, Karin Booth, Taylor Holmes, Joan Gerber, Steve Geray; **Dir:** Lee Sholem. **VHS, LV** $19.98 REP 🎬½

Tobruk American GI's endeavor to knock out the guns of Tobruk, to clear the way for a bombing attack on German fuel supply depots of North Africa in this World War II actioner.
1966 110m/C Rock Hudson, George Peppard, Guy Stockwell, Nigel Green; **Dir:** Arthur Hiller. **VHS, Beta** $19.95 MCA 🎬🎬

Toby & the Koala Bear A little boy and his pet koala bear leave a convict colony in Australia. They befriend an Aborigine boy who teaches them to survive in the woods.
1983 77m/C **VHS, Beta** $29.98 FOX 🎬🎬

Toby McTeague A story about an Alaskan family that breeds Siberian Huskies. When his father is injured, the youngest son tries to replace him in the regional dog-sled race.
1987 (PG) 94m/C Winston Rekert, Wannick Bisson, Timothy Webber; **Dir:** Jean-Claude Lord. **VHS, Beta, LV** $19.98 CHA 🎬🎬

Toby Tyler A boy runs off to join the circus, and teams up with a chimpanzee. A timeless and enjoyable Disney film that still appeals to youngsters. Good family-fare.
1959 93m/C Kevin Corcoran, Henry Calvin, Gene Sheldon, Bob Sweeney, Mr. Stubbs; **Dir:** Charles T. Barton. **VHS, Beta** $69.95 DIS 🎬🎬🎬

Today I Hang A man is framed for murder and sentenced to hang. His buddies on the outside do their best to prove his innocence. Meanwhile, in prison, the accused encounters any number of interesting characters.
1942 67m/B Walter Woolf King, Mona Barrie, William Farnum, Harry Woods, James Craven; **Dir:** Oliver Drake. **VHS, Beta** $16.95 SNC 🎬🎬

Today We Kill, Tomorrow We Die A rancher is unjustly sent to prison; when his sentence is over he hires a gang to relentlessly track down the culprit who framed him.
1971 (PG) 95m/C Montgomery Ford, Bud Spencer, William Berger, Tatsuya Nakadai, Wayde Preston; **Dir:** Tonino Cervi. **VHS, Beta** $59.95 PSM 🎬

The Todd Killings A psychotic young man commits a series of murders involving young women. Sleazy, forgettable picture wastes a talented cast. Also released as "A Dangerous Friend."
1971 (R) 93m/C Robert F. Lyons, Richard Thomas, Barbara Bel Geddes, Ed Asner, Sherry Miles, Gloria Grahame, Belinda J. Montgomery; **Dir:** Barry Shear. **VHS** $59.95 WAR 🎬½

Toga Party A fraternity house throws a wild toga party in this raunchy depiction of college life.
1979 (R) 82m/C Bobby H. Charles, Mary Mitchell. **VHS, Beta** $59.95 MON Woof!

Together A divorced woman and a male chauvinist test the limits of their sexual liberation.
1979 (R) 91m/C IT Maximilian Schell, Jacqueline Bisset, Terence Stamp, Monica Guerritore; **Dir:** Armenia Balducci. **VHS, Beta** $24.98 SUE 🎬½

Togetherness Two wealthy, good-for-nothing playboys chase after the same blonde, a Communist, in Greece.
1970 101m/C George Hamilton, Peter Lawford, Olinka Berova, John Banner, Jesse White; **Dir:** Arthur Marks. **VHS, Beta, LV** WGE 🎬½

Tokyo Joe A war hero/nightclub owner returns to Tokyo and becomes ensnared in blackmail and smuggling while searching for his missing wife and child. Slow-moving tale has never been considered one of Bogart's better movies.
1949 88m/B Humphrey Bogart, Florence Marly, Alexander Knox, Sessue Hayakawa, Jerome Courtland, Lore Lee Michel; **Dir:** Stuart Heisler. **VHS, Beta, LV** $69.95 COL 🎬½

Tokyo Olympiad A monumental sports documentary about the 1964 Olympic Games in its entirety. Never before available for home viewing. Letterboxed with digital sound. In Japanese with English subtitles.
1966 170m/C *JP Dir:* Kon Ichikawa. **VHS, Beta, LV, 3/4U** $99.95 *TPV, WFV, CRC* ✍ ½

Tokyo Pop A punk rocker travels to Japan to find stardom and experiences a series of misadventures. Enjoyable in its own lightweight way.
1988 (R) 99m/C *Carrie Hamilton, Yutaka Tadokoro, Tetsuro Tamba, Taiji Tonoyama, Masumi Harukawa; Dir:* Fran Rubel Kazui. **VHS, LV** $19.95 *WAR* ✍✍½

Tokyo Story Poignant story of elderly couple's journey to Tokyo where they receive little time and less respect from their grown children. Masterful cinematography, and sensitive treatment of universally appealing story. In Japanese with English subtitles.
1953 134m/B *JP Chishu Ryu, Chieko Higashiyama, So Yamamura, Haruko Sugimura, Setsuko Hara; Dir:* Yasujiro Ozu. **VHS** $69.95 *CEG, NYF, FCT* ✍✍✍✍

Tol'able David A simple tale of mountain folk, done in the tradition of Mark Twain stories. A family of hillbillies is embroiled in a feud with a clan of outlaws. When the community's mail is stolen by the troublesome ruffians, the family's youngest member, who harbors dreams of becoming a mail driver, comes to the rescue. Silent film.
1921 79m/B *Richard Barthelmess, Gladys Hulette, Ernest Torrance; Dir:* Henry King. **VHS, Beta** $19.95 *NOS, FST, VYY* ✍✍

Toll of the Desert A lawman learns that his father is a ruthless, back-stabbing renegade. Now he must try and catch him.
1935 58m/B *Roger Williams, Ted Adams, Edward Cassidy, Tom London, John Elliott, Earl Dwire, Betty Mack, Fred Kohler Jr.; Dir:* William Berke. **VHS, Beta, LV** $19.95 *NOS, DVT, WGE* ✍

Toll Gate Quick on the draw outlaw Hart risks capture to rescue drowning child, and consequently gets involved with the boy's mother. Silent classic.
1920 55m/B *William S. Hart, Anna Q. Nilsson, Jack Richardson, Joseph Singleton; Dir:* Lambert Hillyer. **VHS, Beta** $27.95 *GPV, VCN* ✍✍½

Tom Brown's School Days Depicts life among the boys in an English school during the Victorian era. Based on the classic novel by Thomas Hughes. Also released as "Adventures at Rugby." Remade in 1951.
1940 86m/B *GB Cedric Hardwicke, Jimmy Lydon, Freddie Bartholomew; Dir:* Robert Stevenson. **VHS, Beta** $19.95 *NOS, HHT, PSM* ✍✍½

Tom Brown's School Days Tom enrolls at Rugby and is beset by bullies in classic tale of English school life. British remake of the 1940 version. Based on the novel by Thomas Hughes.
1951 93m/B *GB Robert Newton, John Howard Davies, James Hayter; Dir:* Gordon Parry. **VHS, Beta** $19.95 *UHV* ✍✍✍

Tom Corbett, Space Cadet Volume 1 The adventures of Tom Corbett, a Space Cadet at the U.S. Space Academy, where men and women train to become agents to protect Earth and its neighbor planets.
19?? 90m/B *Frankie Thomas, Jan Merlin.* **VHS, Beta** $19.95 *MED* ✍ ½

Tom Corbett, Space Cadet Volume 2 This early '50's TV series follows the adventures of Tom Corbett, a space cadet at the U.S. Space Academy.
195? 90m/B **VHS, Beta** $19.95 *MED* ✍ ½

Tom, Dick, and Harry Dreamy girl is engaged to three men and unable to decide which to marry. It all depends on a kiss. Remade as "The Girl Most Likely."
1941 86m/B *Ginger Rogers, George Murphy, Burgess Meredith, Alan Marshal, Phil Silvers; Dir:* Garson Kanin. National Board of Review Awards '41: 10 Best Films of the Year. **VHS, Beta, LV** $19.98 *CCB, TTC, FCT* ✍✍½

Tom Edison: The Boy Who Lit Up the World This is the story of Tom Edison's life and how he changed the world.
1978 49m/C *David Huffman, Adam Arkin, Michael Callan; Dir:* Henning Schellerup. **VHS, Beta** $9.98 *VID, CVL* ✍✍

Tom Jones Bawdy comedy based on Henry Fielding's novel about rustic playboy's wild life in 18th-century England. Hilarious and clever with a grand performance by Finney. One of the sexiest eating scenes ever. Screenplay by John Osbourne. Redgrave's debut. Oscar nominations for Best Actor (Finney), Supporting Actor (Griffith), Supporting Actresses (Cilento, Evans and Redman), Art Direction, and Set Decoration. Theatrically released at 129 minutes the film was recut by the director, who decided it needed tightening before its 1992 re-release on video.
1963 121m/C *GB Albert Finney, Susannah York, Hugh Griffith, Edith Evans, Joan Greenwood, Diane Cilento, George Devine, David Tomlinson, Joyce Redman, Lynn Redgrave, Julian Glover; Dir:* Tony Richardson. Academy Awards '63: Best Adapted Screenplay, Best Director (Richardson), Best Picture, Best Original Score; Golden Globe Awards '64: Best Film—Musical/Comedy. **VHS, Beta, LV** $59.95 *HBO, FOX, PIA* ✍✍✍✍

Tom Mix Compilation 1 This compilation of six "Great American Cowboy" films from Tom Mix's early period includes: "Jimmy Hayes and Muriel," "The Stagecoach Driver and the Girl," "Sagebrush Tom," "An Arizona Wooing," "Roping a Bride" and "Local Color."
1916 60m/B *Tom Mix.* **VHS, Beta** *GVV* ✍✍✍✍

Tom Mix Compilation 2 Five films from Tom Mix's early period include "Never Again," "An Angelic Attitude," "A Western Masquerade," "A $5,000 Elopement," and "Local Color."
1916 60m/B *Tom Mix.* **VHS, Beta** *GVV* ✍✍✍✍

Tom Sawyer Winning adaptation of Mark Twain's oft told tale of boyhood in Hannibal, Missouri. Cast reprised their roles the following year in "Huckleberry Finn."
1930 86m/B *Jackie Coogan, Mitzie Green, Lucien Littlefield, Tully Marshall; Dir:* John Cromwell. **VHS, Beta** $19.98 *MGM* ✍✍✍

Tom Sawyer Fun-filled adaptation of the Twain classic helped along nicely by a competent cast. The songs slow things down a bit but all in all, superior family fare.
1973 (G) 100m/C *Johnny Whitaker, Celeste Holm, Warren Oates, Jeff East, Jodie Foster, Lucille Benson, Henry Jones; Dir:* Don Taylor. **VHS, Beta** $19.98 *MGM* ✍✍✍

tom thumb Diminutive boy saves village treasury from bad guys. Adapted from classic Grimm fairy tale. Special effects combine live actors, animation, and puppets.

1958 92m/C *GB Russ Tamblyn, Peter Sellers, Terry-Thomas; Dir:* George Pal. Academy Awards '58: Best Special Effects. **VHS, Beta** $39.95 *MGM* ✍✍✍

Tomahawk Territory A formula western wherein Buffalo Bill Cody battles the Sioux.
1952 63m/B *Clayton Moore.* **VHS, Beta, 8mm** *VYY* ✍ ½

Tomb Fortune-seekers disturb the slumber of a magical, sadistic princess much to their everlasting regret. Adapted from a Bram Stoker novel.
1986 106m/C *Cameron Mitchell, John Carradine, Sybil Danning; Dir:* Fred Olen Ray. **VHS, Beta** $19.98 *TWE* ✍✍

Tomb of Ligeia The ghost of a man's first wife expresses her displeasure when groom and new little missus return to manor. One of the better Corman adaptations of Poe. Written by Robert Towne. Also available with "The Conqueror Worm" on Laser Disc.
1964 82m/C *GB Vincent Price, Elizabeth Sheppard, John Westbrook, Oliver Johnston, Richard Johnson; Dir:* Roger Corman. **VHS, Beta, LV** $59.99 *HBO, MLB* ✍✍✍

Tomb of Torture A murdered countess is reincarnated in the body of a newspaperman's mistress. Together they investigate monster reports in a murder-ridden castle. Filmed in Sepiatone.
1965 88m/B *Annie Albert, Thony Maky, Mark Marian, Queen Elizabeth; Dir:* William Grace. **VHS, Beta** $39.99 *MSP* ✍

Tomb of the Undead The sadistic guards of a prison camp are rudely awakened by the prisoners they tortured to death when they return as flesh-eating zombies looking to settle the score.
1972 60m/C *GB Duncan McLeod, Lee Frost, John Dennis; Dir:* John Hayes. **VHS, Beta** $19.95 *NOS, INC* ✍ ½

The Tomboy A young woman behaves boyishly in Colonial America.
1924 64m/B *Dorothy Devore, Lotta Williams, Herbert Rawlinson, Harry Gibbon, Lee Moran; Dir:* David Kirkland. **VHS, Beta** $19.95 *GPV, FCT* ✍ ½

Tomboy A shy country boy and a not-so-shy city girl meet, fall in love, and overcome obstacles that stand in their path.
1940 70m/B *Jackie Moran, Marcia Mae Jones.* **VHS, Beta** $29.95 *VYY* ✍

Tomboy A pretty auto mechanic is determined to win not only the race but the love and respect of a superstar auto racer.
1985 (R) 91m/C *Betsy Russell, Eric Douglas, Jerry Dinome, Kristi Somers, Richard Erdman, Toby Iland; Dir:* Herb Freed. **VHS, Beta** $79.98 *LIV, VES* Woof!

Tomboy & the Champ Despite many obstacles, a music-loving calf becomes a prize winner with the help of its loving owner. Kids will love it, but adults will find this one a little sugary.
1958 82m/C *Candy Moore, Ben Johnson, Jesse White; Dir:* Francis D. Lyon. **VHS, Beta** $29.95 *UHV* ✍✍

Tombs of the Blind Dead Blinded by crows for using human sacrifice in the thirteenth century, zombies rise from the grave to wreak havoc upon 20th-century Spaniards. Atmospheric chiller was extremely popular in Europe and spawned three sequels. Originally titled "The Blind Dead."

The sequels, also on video, are "Return of the Evil Dead" and "Horror of the Zombies."
1972 (PG) 86m/C *SP PT* Caesar Burner, Lone Fleming, Helen Harp; *Dir:* Armando de Ossorio. **VHS, Beta $42.95** *PGN* 𝟸𝟸½

Tombstone Canyon Death rides the range until Maynard and his horse Tarzan put a halt to it.
1935 60m/B Ken Maynard, Sheldon Lewis, Cecilia Parker. **VHS, Beta $12.95** *SNC, NOS, RXM* 𝟸

Tombstone Terror A cowboy is beleaguered by a case of mistaken identity.
1934 55m/B Bob Steele. **VHS, Beta** *WGE* 𝟸

Tommy Peter Townsend's rock opera as visualized in the usual hyper-Russell style about the deaf, dumb, and blind boy who becomes a celebrity due to his amazing skill at the pinball machines. A parade of rock musicians perform throughout the affair, with varying degrees of success. Despite some good moments, the film ultimately falls prey to ill-conceived production concepts and miscasting. Ann-Margret earned an Oscar nomination.
1975 (PG) 108m/C Ann-Margret, Elton John, Oliver Reed, Tina Turner, Roger Daltrey, Eric Clapton, Keith Moon, Pete Townsend, Jack Nicholson, Robert Powell; *W/Dir:* Ken Russell. **VHS, Beta, LV $14.95** *COL, MVD* 𝟸𝟸

Tomorrow Powerful tale of the love of two lonely people. Outstanding performance by Duvall as lumber mill worker who falls for a pregnant woman. Based on neglected Faulkner story with screenplay by Horton Foote.
1983 102m/B Robert Duvall, Olga Bellin, Sudie Bond; *Dir:* Joseph Anthony. **VHS, Beta $24.95** *MON* 𝟸𝟸𝟸

Tomorrow is Forever Welles and Colbert marry shortly before he goes off to fight in World War I. Badly wounded and disfigured, he decides to stay in Europe while Colbert, believing Welles dead, eventually marries Brent. Fast forward to World War II when Brent, a chemical manufacturer, hires a new scientist to work for him in the war effort (guess who). This slow-moving melodrama wastes a good cast. Six-year-old Wood debuts as Colbert's daughter.
1946 105m/B Claudette Colbert, Orson Welles, George Brent, Lucile Watson, Richard Long, Natalie Wood; *Dir:* Irving Pichel. **VHS $19.98** *MGM* 𝟸𝟸½

Tomorrow Never Comes When he discovers his girlfriend has been unfaithful, a guy goes berserk and eventually finds himself in a stand-off with the police. Violent.
1977 109m/C Oliver Reed, Susan George, Raymond Burr, John Ireland, Stephen McHattie, Donald Pleasence; *Dir:* Peter Collinson. **VHS, Beta $79.95** *UNI, HHE* 𝟸½

Tomorrow at Seven A mystery writer is determined to discover the identity of the Ace of Spades, a killer who always warns his intended victim, then leaves an ace of spades on the corpse as his calling card.
1933 62m/B Chester Morris, Vivienne Osborne, Frank McHugh, Allen Jenkins, Henry Stephenson; *Dir:* Ray Enright. **VHS, Beta $12.95** *SNC, NOS, VYY* 𝟸𝟸½

Tomorrow's Child Made-for-television drama about in vitro fertilization (test-tube babies) and surrogate motherhood.
1982 100m/C Stephanie Zimbalist, William Atherton, Bruce Davison, Ed Flanders, Salome Jens, James Shigeta, Susan Oliver, Arthur Hill; *Dir:* Joseph Sargent. **VHS, Beta $59.98** *FOX* 𝟸𝟸

Tomorrow's Children An alarmist melodrama warning against the threat of government-induced female sterilization asks the question: if a woman's family is weird, should her tubes be tied?
1934 55m/B Sterling Holloway, Diana Sinclair, Sarah Padden, Don Douglas; *Dir:* Crane Wilbur. **VHS, Beta $16.95** *SNC, NOS, VYY* 𝟸

The Tongfather The Tongfather is a ruthless crime boss who leads a Chinese opium ring. An undercover agent sets out to bring him down.
1978 (R) 92m/C VHS $14.95 *KBV* 𝟸

Tongs: An American Nightmare Chinese-American street gangs in Chinatown battle over turf and drug shipments.
1988 (R) 80m/C Ian Anthony Leung, Christopher O'Conner, Simon Yam; *Dir:* Philip Chan. **VHS, Beta $29.95** *ACA* 𝟸

Toni An Italian worker falls for his landlady and they make plans to marry. A grim turn of events, however, brings tragic consequences. Based on the lives of several townsfolk in the village of Les Martiques, the film was shot in the town and members of the local populace were used as characters. In French with English subtitles.
1934 90m/B *FR* Charles Blavette, Jenny Helia, Edouard Delmont, Celia Montalvan; *Dir:* Jean Renoir. New York Film Festival '68: Best Special Effects. **VHS, Beta $24.95** *DVT* 𝟸𝟸𝟸½

Tonight and Every Night Cabaret singer and RAF pilot fall in love during World War II. Her music hall post puts her in the midst of Nazi bombing, and her dedication to her song and dance career puts a strain on the romance. Imaginative production numbers outweigh pedestrian storyline
1945 92m/C Rita Hayworth, Lee Bowman, Janet Blair, Marc Platt, Leslie Brooks; *Dir:* Victor Saville. **VHS, Beta $19.95** *COL* 𝟸𝟸½

Tonight for Sure Coppola's first film, produced as a student at UCLA. Two men, one who spies on women and the other who imagines nude women everywhere, plan an escapade. Music by Carmine Coppola.
1961 66m/B *Dir:* Francis Ford Coppola, Francis Ford Coppola. **VHS, Beta $29.95** *VYY* 𝟸½

Tonio Kroger A young writer travels through Europe in search of intellectual and sensual relationships and a home that will suit him. He must balance freedom and responsibility. Works best if one is familiar with the Thomas Mann novel on which this is based. German dialogue with English subtitles.
1965 92m/B *GE* Jean-Claude Brialy, Nadja Tiller, Gert Frobe; *Dir:* Rolf Thiele. **VHS, Beta $29.95** *VYY, APD, GLV* 𝟸𝟸

Tonka A children's story about a wild horse tamed by a young Indian, only to have it recruited for the Battle of Little Bighorn. Mineo is fine as the Indian brave determined to be reunited with his steed. The film also makes a laudable effort to portray the Indians as a dignified race. The movie, however, stumbles at its conclusion and is contrived throughout. Also released as "A Horse Named Comanche."
1958 97m/C Sal Mineo, Phil Carey, Jerome Courtland; *Dir:* Lewis R. Foster. **VHS, Beta $69.95** *DIS* 𝟸𝟸½

Tono Bicicleta Tono Bicicleta cavorts with his wife and girlfriend and yet persists in kidnapping more women.
19?? 116m/C Tommy Vegas, Alida Arizmendy. **VHS, Beta** *MAV* Woof!

The Tonto Kid Action-packed western featuring Rex Bell.
1935 56m/B Rex Bell, Ruth Mix, Buzz Barton, Joseph Girard, Jack Rockwell, Murdock McQuarrie; *Dir:* Harry Fraser. **VHS, Beta $19.95** *NOS, DVT, VDM* 𝟸

Tony Draws a Horse An eight-year old draws an anatomically correct horse on the door of his father's office, leading to a rift between the parents as they argue how to handle the heartbreak of precociousness. Somewhat uneven but engaging comedy, based on a play by Lesley Storm.
1951 90m/B Cecil Parker, Anne Crawford, Derek Bond, Barbara Murray, Mervyn Johns, Barbara Everest, David Hurst; *Dir:* John Paddy Carstairs. **VHS $19.95** *NOS* 𝟸𝟸½

Tony Powers This program presents Tony Powers performing his songs "Don't Nobody Move," "Midnite Trampoline" and "Odyssey."
1981 54m/C Tony Powers. **VHS, Beta $19.95** *SVS* 𝟸𝟸½

Tony Rome The Marvin H. Albert private eye agency investigates a millionaire's daughter's disappearance and her involvement in organized crime. Entertaining and complex plot, although the cast is a numbingly large one.
1967 110m/C Frank Sinatra, Jill St. John, Simon Oakland, Gena Rowlands, Richard Conte, Lloyd Bochner, Jeffrey Lynn, Sue Lyon; *Dir:* Gordon Douglas. **VHS, Beta $59.98** *FOX* 𝟸𝟸½

Too Beautiful for You A successful car salesman, married for years to an extraordinarily beautiful woman, finds himself head over heels for his frumpy secretary. Depardieu is regular guy who finds he's never believed in the love and fidelity of a woman he thinks is too beautiful for him. In French with English sub-titles.
1988 (R) 91m/C *FR* Gerard Depardieu, Josiane Balasko, Carol Bouquet, Roland Blanche; *W/Dir:* Bertrand Blier. **VHS $79.95** *ORI, FXL, FCT* 𝟸

Too Hot to Handle Two rival photographers searching for a beautiful lady-pilot's missing brother wind up in Brazil, where they encounter a dangerous tribe of voodoo types. Amusing, if exaggerated, picture of the lengths to which reporters will go for a story. Classic Gable.
1938 105m/B Clark Gable, Myrna Loy, Walter Pidgeon, Walter Connolly, Leo Carrillo, Virginia Weidler; *Dir:* Jack Conway. **VHS $19.98** *MGM, FCT* 𝟸𝟸𝟸

Too Hot to Handle A voluptuous lady contract killer fights against the mob with all the weapons at her disposal. Filmed on location in Manila.
1976 88m/C Cheri Caffaro, Aharon Ipale, Vic Diaz, Corinne Calvert; *Dir:* Don Schain. **VHS, Beta $39.98** *WAR* 𝟸

Too Late the Hero Unlikely band of allied soldiers battle Japanese force entrenched in the Pacific during WWII. Rousing adventure, fine cast. Also known as "Suicide Run."
1970 (PG) 133m/C Michael Caine, Cliff Robertson, Henry Fonda, Ian Bannen, Harry Andrews, Denholm Elliott; *Dir:* Robert Aldrich. **VHS, Beta $14.98** *FOX* 𝟸𝟸𝟸

Too Late for Tears An honest husband and his not so honest wife stumble on a load of mob-stolen money and become entangled in a web of deceit and murder as the wife resorts to increasingly desperate measures to keep her newfound fortune. Atmospheric and entertaining film noir albeit some-

times confusing. Also released as "Killer Bait."
1949 99m/B Lizabeth Scott, Don DeFore, Dan Duryea, Arthur Kennedy, Kristine Miller, Barry Kelley, Denver Pyle; *Dir:* Byron Haskin. **VHS, Beta $16.95** *SNC, VDM, NOS* 🎬🎬🎬

Too Many Crooks A British spoof of crime syndicate films. Crooks try to extort, but bungle the job. Terry-Thomas is fun, as always.
1959 85m/B *GB* Terry-Thomas, Brenda de Banzie, George Cole; *Dir:* Mario Zampi. **VHS, Beta $39.95** *IND* 🎬🎬

Too Many Girls Beautiful heiress goes to a small New Mexico college to escape from a cadre of gold-digging suitors. Passable adaptation of the successful Rodgers and Hart Broadway show, with many original cast members and the original stage director. Songs include "I Didn't Know What Time It Was," "Love Never Went to College" and "Spic and Spanish." Lucy and Desi met while making this film.
1940 85m/B Lucille Ball, Richard Carlson, Eddie Bracken, Ann Miller, Desi Arnaz Sr., Hal LeRoy, Libby Bennett, Frances Langford, Van Johnson; *Dir:* George Abbott. **VHS, Beta, LV $14.98** *RKO* 🎬🎬½

Too Much Sun A dying man can prevent his fortune from falling into the hands of a corrupt priest simply by having one of his two children produce an heir. The problem is, they're both gay!
1990 97m/C Robert Downey Jr., Ralph Macchio, Eric Idle, Andrea Martin, Laura Ernst, Jim Haynie; *Dir:* Robert Downey. **VHS, LV $89.95** *COL* 🎬½

Too Pretty to Be Honest Four French girls witness a robbery. When they discover the thief's identity they decide to rip him off. In French with subtitles.
1972 95m/C *FR* Jane Birkin, Bernadette LaFont; *Dir:* Richard Balducci. **VHS, Beta $39.95** *FCT* 🎬🎬

Too Scared to Scream A policeman and an undercover agent team up to solve a bizarre series of murders at a Manhattan apartment house.
1985 (R) 104m/C Mike Connors, Anne Archer, Leon Isaac Kennedy, John Heard, Ian McShane, Maureen O'Sullivan, Murray Hamilton; *Dir:* Tony LoBianco. **VHS, Beta $79.98** *LIV, VES* 🎬½

Too Shy to Try A incredibly shy man meets his dream girl, and enrolls in a psychology course to overcome his ineptitude. A ribald, slapstick French comedy.
1982 (PG) 89m/C *FR* Pierre Richard, Aldo Maccione; *Dir:* Pierre Richard. **VHS, Beta $29.98** *SUE* 🎬🎬

Too Wise Wives A story of would-be marital infidelity and misunderstandings. Silent.
1921 90m/C Louis Calhern, Claire Windsor; *Dir:* Lois Weber. **VHS, Beta $57.95** *GVV* 🎬🎬

The Toolbox Murders An unknown psychotic murderer brutally claims victims one at a time, leaving police mystified and townsfolk terrified. Sick and exploitative, with predictably poor production values.
1978 (R) 93m/C Cameron Mitchell, Pamelyn Ferdin, Wesley Eure, Nicholas Beauvy, Aneta Corseaut, Tim Donnelly, Evelyn Guerrero; *Dir:* Dennis Donnelly. **VHS, Beta $19.95** *UHV, VTR* 🎬

Tootsie A stubborn, unemployed actor disguises himself as a woman to secure a part on a soap opera. As his popularity on television mounts, his love life becomes increasingly soap operatic. Hoffman is delightful, as is the rest of the stellar cast.

Laserdisc version features audio commentary by director Sidney Pollack, behind the scenes footage and photographs, and complete coverage of Tootsie's production.
1982 (PG) 110m/C Dustin Hoffman, Jessica Lange, Teri Garr, Dabney Coleman, Bill Murray, Charles Durning, Geena Davis; *Dir:* Sydney Pollack. Academy Awards '82: Best Supporting Actress (Lange); Golden Globe Awards '83: Best Film—Musical/Comedy; National Society of Film Critics Awards '82: Best Film. **VHS, Beta, LV $29.95** *COL, VYG, CRC* 🎬🎬🎬

Top Gun Young Navy pilots compete against one another on the ground and in the air at the elite Fighter Weapons School. Cruise isn't bad as a maverick who comes of age in Ray Bans, but Edwards shines as his buddy. Awesome aerial photography and high-cal beefcake divert from the contrived plot and stock characters. The Navy subsequently noticed an increased interest in fighter pilots. Features Berlin's Oscar-winning song "Take My Breath Away."
1986 (PG) 109m/C Tom Cruise, Kelly McGillis, Val Kilmer, Tom Skerritt, Anthony Edwards, Meg Ryan, Rick Rossovich, Michael Ironside, Barry Tubb, Whip Hubley, John Stockwell, Tim Robbins; *Dir:* Tony Scott. Academy Awards '86: Best Song ("Take My Breath Away"). **VHS, Beta, LV, SVS, 8mm $14.95** *PAR* 🎬🎬½

Top Gun: The Real Story An insider's look at "Fightertown U.S.A.," where the Navy trains the real Top Gun pilots. Features an assignment for the graduates over hostile Libya.
19?? 35m/C **VHS $14.95** *VTK* 🎬🎬½

Top Hat Ginger isn't impressed by Fred's amourous attentions since she's mistaken him for a friend's other half. Many believe it to be the duo's best film together. Musical numbers by Irving Berlin include "Top Hat, White Tie and Tails," "Cheek to Cheek," and "Isn't This a Lovely Day." Received four Academy Award nominations for Best Picture, Art Direction, Choreography, and Song ("Cheek to Cheek"). A classic Hollywood musical. Look for a young Lucille Ball as a clerk in a flower shop.
1935 97m/B Fred Astaire, Ginger Rogers, Erik Rhodes, Helen Broderick, Edward Everett Horton, Eric Blore, Lucille Ball; *Dir:* Mark Sandrich. **VHS, Beta, LV, 8mm $14.98** *TTC, RKO, VID* 🎬🎬🎬🎬

Top of the Heap A cop is denied a promotion and gets angry. Deciding to chuck the laws of due process, he goes on a criminal-killing rampage.
197? (R) 91m/C Christopher St. John, Paula Kelly, Patrick MacVey; *Dir:* Christopher St. John. **VHS, Beta $59.95** *UNI* 🎬

Top Secret! Unlikely musical farce parodies spy movies and Elvis Presley films. Young American rock Nick Rivers star goes to Europe on goodwill tour and becomes involved with Nazis, the French Resistance, an American refugee, and more. Sophisticated it isn't. From the creators of 1980's "Airplane!"
1984 (PG) 90m/C Val Kilmer, Lucy Gutteridge, Christopher Villiers, Omar Sharif, Peter Cushing, Jeremy Kemp, Michael Gough; *Dir:* Jim Abrahams, Jerry Zucker, David Zucker. **VHS, Beta, LV $14.95** *PAR* 🎬🎬🎬

Topaz American CIA agent and French intelligence agent combine forces to find information about Russian espionage in Cuba. Cerebral and intriguing, but not classic Hitchcock. Based on the novel by Leon Uris. Laser disc video release includes two alternate endings.

1969 (PG) 126m/C John Forsythe, Frederick Stafford, Philippe Noiret, Karin Dor, Michel Piccoli; *Dir:* Alfred Hitchcock. National Board of Review Awards '69: 10 Best Films of the Year, Best Director, Best Supporting Actor (Noiret). **VHS, Beta, LV $19.95** *MCA* 🎬🎬🎬

Topaze American adaptation of the Marcel Pagnol play, about an innocent, doltish schoolmaster who becomes the unknowing front for a baron's illegal business scam. Remade in two French versions, in 1933, and in 1951 by Pagnol.
1933 127m/B John Barrymore, Myrna Loy, Jobyna Howland, Jackie Searl; *Dir:* Harry D'Abbadie D'Arrast; *Dir:* **VHS, Beta, LV $39.98** *FOX* 🎬🎬🎬

Topaze A shy, depressed teacher is fired from his job at a private school and takes a new job, unaware that his boss is being him in a business scam. From the French play by Marcel Pagnol. Remake of an American version (1933); Pagnol produced his own version in '51. French with English subtitles.
1933 92m/B *FR* Louis Jouvet, Edwige Feuillere, Marcel Vallee; *Dir:* Louis Gasnier. **VHS, Beta $59.95** *INT, FOX* 🎬🎬🎬

Topaze After losing his job, a pathetic school teacher gets involved with the mob and winds up a powerful businessman. Remake of two earlier versions, one French and one American. Based on the play by Director Pagnol.
1951 95m/B *FR* Fernandel, Helen Perdriere, Pierre Larquey; *Dir:* Marcel Pagnol. **VHS, Beta $29.95** *DVT, APD* 🎬🎬🎬½

Topaze: Mark of a Forgotten Master Excerpts from the 1933 American version of "Topaze." Also includes an interview with director Harry d'Abbadie d'Arrest.
1971 30m/C *Voices:* Harry D'Abbadie D'Arrast. **VHS, Beta** *CMR, NYS* 🎬🎬🎬½

Topkapi An international bevy of thieves can't resist the treasurers of the famed Topkapi Palace Museum, an impregnable fortress filled with wealth and splendor. Comic thriller based on Eric Ambler's "The Light of Day."
1964 122m/C Melina Mercouri, Maximilian Schell, Peter Ustinov, Robert Morley, Akim Tamiroff; *Dir:* Jules Dassin. Academy Awards '64: Best Supporting Actor (Ustinov); National Board of Review Awards '64: 10 Best Films of the Year. **VHS, Beta $59.98** *FOX* 🎬🎬🎬

Topless Dancing Texas Style Taped at Rick's Cabaret in Houston, a selection of strip-dance routines.
1988 90m/C **VHS, Beta $39.95** *CEL*

Topper George and Marion Kerby return as ghosts after a fatal car accident, determined to assist their pal Cosmo Topper. Immensely popular at the box office, followed by "Topper Takes a Trip" and "Topper Returns." The series uses trick photography and special effects to complement the comedic scripts. Inspired a television series and remade in 1979 as television movie. Young received a Best Actor Nomination. Also available colorized.
1937 97m/B Cary Grant, Roland Young, Constance Bennett, Billie Burke, Eugene Pallette; *Dir:* Norman Z. MacLeod. **VHS, Beta $9.95** *MED, CCB* 🎬🎬🎬

Topper Returns Cosmo Topper helps ghost find the man who mistakenly murdered her and warns her friend, the intended victim. Humorous conclusion to the trilogy preceded by "Topper" and "Topper Takes a Trip."

Followed by a television series. Also available colorized.
1941 87m/B Roland Young, Joan Blondell, Dennis O'Keefe, Carole Landis, Eddie Anderson, H.B. Warner, Billie Burke; *Dir:* Roy Del Ruth. VHS, Beta, LV $16.95 *SNC, NOS, CAB* 🎞🎞🎞

Topper Takes a Trip Cosmo Topper and his wife have falling out and ghosts Kerby help them get back together. The special effects sequences are especially funny. Followed by "Topper Returns." Also available colorized.
1939 85m/B Constance Bennett, Roland Young, Billie Burke, Franklin Pangborn, Alan Mowbray; *Dir:* Norman Z. MacLeod. **VHS, Beta $14.95** *MED* 🎞🎞🎞

Topsy Turvy A comedy about reckless romance and not much else.
1984 (R) 90m/C Lisbet Dahl, Ebbe Rode; *Dir:* Edward Fleming. **VHS, Beta $69.98** *LIV, VES* 🎞½

Tor A medieval barbarian endeavors to free an enslaved village from a band of murderous hoodlums.
1964 94m/C Joe Robinson, Bella Cortez, Harry Baird; *Dir:* Antonio Leonuiola. **VHS, Beta $59.95** *TWE* 🎞½

Tora! Tora! Tora! The story of events leading up to December 7, 1941, is retold by three directors from both Japanese and American viewpoints in this tense, large-scale production. Well-documented and realistic treatment of the Japanese attack on Pearl Harbor that brought the US into WWII; notable for its good photography but lacks a storyline equal to its epic intentions.
1970 (G) 144m/C Martin Balsam, Soh Yomamura, Jocoph Cotton, E.G. Marshall, Tatsuya Mihashi, Wesley Addy, Jason Robards Jr., James Whitmore, Leon Ames, George Macready; *Dir:* Richard Fleischer, Toshio Masuda, Kinji Fukasaku. Academy Awards '70: Best Visual Effects; National Board of Review Awards '70: 10 Best Films of the Year. **VHS, Beta, LV $14.98** *FOX, TLF* 🎞🎞

Torch Song In this muddled melodrama, a tough, demanding Broadway actress meets her match when she is offered true love by a blind pianist. Contains a couple of notoriously inept musical numbers. Made to show off Crawford's figure at age 50. Crawford's singing in the movie is dubbed. Rambeau received Best Supporting Actress nomination.
1953 90m/C Joan Crawford, Michael Wilding, Marjorie Rambeau, Gig Young, Harry Morgan, Dorothy Patrick, Benny Rubin, Nancy Gates; *Dir:* Charles Walters. **VHS, LV $19.98** *MGM* 🎞🎞

Torch Song Trilogy Adapted from Fierstein's hit Broadway play about a gay man who "just wants to be loved." Still effective, but the rewritten material loses something in the translation to the screen. Bancroft heads a strong cast with a finely shaded performance as Fierstein's mother.
1988 (R) 126m/C Anne Bancroft, Matthew Broderick, Harvey Fierstein, Brian Kerwin, Karen Yahng, Charles Pierce; *Dir:* Paul Bogart. **VHS, Beta, LV** *COL* 🎞🎞½

Torchlight A young couple's life begins to crumble when a wealthy art dealer teaches them how to free base cocaine. Heavy-handed treatment of a subject that's been tackled with more skill elsewhere.
1985 90m/C Pamela Sue Martin, Steve Railsback, Ian McShane, Al Corley, Rita Taggart; *Dir:* Thomas J. Wright. **VHS, Beta, LV $19.98** *SUE* 🎞

Torero (Bullfighter) A young man becomes a bullfighter to overcome his fear of bulls.
1956 89m/B *MX* Luis Procuna, Manolete, Dolores Del Rio, Carlos Arruza; *Dir:* Carlos Velo. **VHS, Beta $44.95** *MAD* 🎞

Torment Tragic triangle has young woman in love with fellow student and murdered by sadistic teacher. Atmospheric tale hailed by many as Sjoberg's finest film. Ingmar Bergman's first filmed script. In Swedish with English subtitles. Alternate title: "Hets."
1944 90m/B *SW Dir:* Alf Sjoberg. **VHS, Beta $69.95** *SUE* 🎞🎞🎞

Torment A mild-mannered fellow is secretly a mass murderer specializing in young lovers. He struggles to hide his secret as his daughter approaches dating age and a relentless detective gets closer to the truth. A low-budget slasher.
1985 (R) 85m/C Taylor Gilbert, William Witt, Eve Brenner; *Dir:* Samson & Hopkins, John Aslanian. **VHS, Beta $19.95** *STE, NWV* 🎞

Tormented A man pushes his mistress out of a lighthouse, killing her. Her ethereal body parts return to haunt him.
1960 75m/B Richard Carlson, Susan Gordon, Juli Reding; *Dir:* Bert I. Gordon. **VHS $16.95** *SNC* 🎞

The Tormentors Vaguely neo-Nazi gangs rape and murder a man's family, and he kills them all in revenge.
198? (R) 78m/C James Craig, Anthony Eisley, Chris Nole; *Dir:* Boris Eagle. **VHS, Beta $69.95** *LTG Woof!*

Torn Allegiance An Australian anti-war epic set in the Boer War about the effects of war on one family and their fight for survival on the veldt.
1984 102m/C *AU* **VHS, Beta $79.95** *PSM* 🎞

Torn Apart Traditional Middle East hatreds undermine the love affair between two young people. Excellent performances from the principals highlight this drama. Adapted from the Chayin Zeldis novel "A Forbidden Love."
1989 (R) 95m/C Adrian Pasdar, Cecilia Peck, Machram Huri, Arnon Zadok, Barry Primus; *Dir:* Jack Fisher. **VHS, Beta $89.95** *WAR, BTV* 🎞🎞

Torn Between Two Lovers A beautiful married woman has an affair with an architect while on a trip. Remick is enjoyable to watch, but it's a fairly predictable yarn. Made for television.
1979 100m/C Lee Remick, Joseph Bologna, George Peppard, Giorgio Tozzi; *Dir:* Delbert Mann. **VHS $59.95** *IVE* 🎞🎞

Torn Curtain American scientist poses as a defector to East Germany in order to uncover details of the Soviet missile program. He and his fiancee, who follows him behind the Iron Curtain, attempt to escape to freedom. Derivative and uninvolving.
1966 125m/C Paul Newman, Julie Andrews, Lila Kedrova, David Opatoshu; *Dir:* Alfred Hitchcock. **VHS, Beta, LV $19.95** *MCA* 🎞🎞½

Tornado An army sergeant revolts against his superiors and the enemy when his captain leaves him stranded in Vietnam.
1983 90m/C Timothy Brent, Tony Marsina, Alan Collins. **VHS, Beta $69.98** *LIV, LTG* 🎞

Torpedo Alley A guilt-ridden WWII Navy pilot feels responsible for the deaths of his crew. After the war, he joins a submarine crew and gradually learns to deal with his guilt. Average drama with plenty of submarine footage.
1953 84m/B Dorothy Malone, Mark Stevens, Charles Winninger, Bill Williams; *Dir:* Lew Landers. **VHS $19.98** *REP, AVC* 🎞🎞

Torpedo Attack Greek sailors, in an outdated submarine, bravely but hopelessly attack Mussolini's Italy during World War II.
1972 88m/C *GR* John Ferris, Sidney Kazan; *Dir:* George Law. **VHS, Beta $39.95** *IVE* 🎞

Torpedo Run Standard submarine melodramatics, about a U.S. sub that must torpedo a Japanese aircraft carrier which holds some of the crew's family members. Sometimes slow, generally worthwhile.
1958 98m/C Glenn Ford, Ernest Borgnine, Dean Jones, Diane Brewster, L.Q. Jones; *Dir:* Joseph Pevney. **VHS, Beta $19.98** *MGM* 🎞🎞

Torrents of Spring Based on an Ivan Turgenev story, this lavishly filmed and costumed drama concerns a young Russian aristocrat circa 1840 who is torn between two women. Predictably, one is a good-hearted innocent, the other a scheming seductress.
1990 (PG-13) 102m/C Timothy Hutton, Nastassia Kinski, Valeria Golino, William Forsythe, Urbano Barberini, Francesca De Sapio, Jacques Herlin; *Dir:* Jerzy Skolimowski. **VHS, Beta, LV $89.99** *HBO* 🎞🎞

Torso A crazed psychosexual killer stalks beautiful women and dismembers them. Fairly bloodless and uninteresting, in spite of lovely Suzy Kendall. Italian title translates: The Bodies Showed Signs of Carnal Violence - quite an understatement for people without arms and legs.
1973 (R) 91m/C *IT* Suzy Kendall, Tina Aumont, John Richardson; *Dir:* Sergio Martino. **VHS, Beta $79.95** *MPI, PSM Woof!*

Tortilla Flat Based on the John Steinbeck novel of the same name, two buddies, Tracy and Garfield, struggle to make their way on the wrong side of the tracks in California. Garfield gets a break by inheriting a couple of houses, Tracy schemes to rip-off a rich, but eccentric, dog owner. In the meantime, both fall for the same girl, Lamarr, in perhaps the finest role of her career. Great performances from all, especially Morgan, who was Oscar nominated.
1942 100m/B Spencer Tracy, Hedy Lamarr, John Garfield, Frank Morgan, Akim Tamiroff, Sheldon Leonard, John Qualen, Donald Meek, Connie Gilchrist; *Dir:* Victor Fleming. **VHS $19.98** *MGM, CCB* 🎞🎞🎞

Torture Chamber of Baron Blood Baron with gorgeous thirst for the red stuff is reanimated and lots of innocent bystanders meet a grisly fate in this spaghetti gorefest. Familiar story has that certain Bava feel. Originally known as "Baron Blood."
1972 (PG) 90m/C *IT* Joseph Cotten, Elke Sommer, Massimo Girotti, Rada Rassimov; *Dir:* Mario Bava. **VHS, Beta $19.99** *HBO* 🎞🎞½

Torture Chamber of Dr. Sadism Decapitated and drawn and quartered for sacrificing twelve virgins, the evil Count Regula is pieced together forty years later to continue his wicked ways. Great fun, terrific art direction. Very loosely based on Poe's "The Pit and the Pendulum." Also known as "Blood Demon." Beware heavily edited video version entitled, "Castle of the Walking Dead."
1969 120m/C *GE* Christopher Lee, Karin Dor, Lex Barker, Carl Lange, Vladimir Medar, Christiane Rucker, Dieter Eppler; *Dir:* Harold Reinl. **VHS, Beta $29.98** *MAG, REG* 🎞🎞½

Torture Dungeon Director Milligan tortures audience with more mindless medieval pain infliction from Staten Island.
1970 (R) 80m/C Jeremy Brooks, Susan Cassidy; *Dir:* Andy Milligan. VHS *MID* 🐾

Torture Garden A sinister man presides over an unusual sideshow where people can see what is in store if they allow the evil side of their personalities to take over. Written by Robert Bloch ("Psycho") and based on four of his short stories.
1967 93m/C *GB* Jack Palance, Burgess Meredith, Peter Cushing, Beverly Adams; *Dir:* Freddie Francis. VHS, Beta, LV $59.95 *COL* 🐾🐾

Torture Ship Director Halperin, who earlier made the minor horror classic "White Zombie," was all at sea in this becalmed thriller about a mad doctor who uses convicts for glandular experiment in his shipboard laboratory. Based on the short story "A Thousand Deaths" by Jack London.
1939 57m/B Lyle Talbot, Irving Pichel, Jacqueline Wells, Sheila Bromley; *Dir:* Victor Halperin. VHS $15.95 *NOS, LOO* 🐾1/2

The Torture of Silence A simple little drama about a doctor's neglected wife seeking out the doctor's best friend for love, remade by Gance again in 1932; both originally titled "Mater Dolorosa." Silent.
1917 55m/B *Dir:* Abel Gance. VHS, Beta $44.95 *FCT, DNB* 🐾🐾1/2

Torture Train Mayhem, murder and knife-wielding psychosis occurs on board a train.
1983 (R) 78m/C Patty Edwards, Kay Beal. VHS, Beta $59.95 *JEF* 🐾1/2

Tosca Puccini's lyric drama of idealistic young love receives a lush, passionate treatment in this Arena of Verona production.
1984 118m/C Eva Marton, Giacomo Aragall, Ingvar Wixell, Daniel Oren. VHS, Beta $39.95 *HMV, HBO* 🐾1/2

Tosca Franco Zeffirelli produced this breathtaking version of Puccini's opera for the PBS "Live From the Met" series. He framed the performance against vast, beautifully detailed sets. With English subtitles.
1985 127m/C Hildegard Behrens, Placido Domingo, Cornell MacNeil, Italo Tajo. Emmy Award '85: '85-Best Musical Special. VHS, Beta, LV $29.95 *HMV, PAR* 🐾1/2

Tosca The Australian Opera presents Eva Marton in Puccini's famous opera at the Sydney Opera House.
1989 123m/C Eva Marton. VHS, Beta $39.95 *MVD, HMV, KUL* 🐾1/2

Total Exposure Total idiocy involving bimbous babes and hormonally imbalanced men in blackmail and murder. 1990 Playboy Playmate Deborah Driggs proves there is life of sorts after playmatedom.
1991 (R) 96m/C Michael Nouri, Season Hubley, Christian Burgess, Robert Prentiss, Deborah Driggs, Jeff Conaway; *Dir:* John Quinn. VHS, LV $89.95 *REP, PIA* 🐾

Total Recall Mind-bending sci-fi movie set in the 21st century. Construction worker Schwarzenegger travels to Mars to discover his true identity after learning that his memories have been artificially altered. Intriguing plot and spectacular special effects. Laced with graphic violence. Based on Phillip K. Dick's "We Can Remember It For You Wholesale." Received 3 Academy Award nominations: Best Sound, Best Sound Effects Film Editing, and Best Visual Effects (Special Achievement Award).

1990 (R) 113m/C Arnold Schwarzenegger, Rachel Ticotin, Sharon Stone, Michael Ironside, Ronny Cox, Roy Brocksmith, Marshall Bell, Mel Johnson Jr.; *Dir:* Paul Verhoeven. Academy Awards '90: Best Visual Effects. VHS, Beta, LV, 8mm $19.95 *IME, IVE, LIV* 🐾🐾🐾

The Touch Straightforward story of a woman who is satisfied with her husband until the arrival of a stranger. The stranger soon has her yearning for a life she has never known. Bergman's first English film lacks the sophistication of his earlier work.
1971 (R) 112m/C *SW* Bibi Andersson, Elliott Gould, Max von Sydow; *Dir:* Ingmar Bergman. VHS, Beta *FOX* 🐾1/2

A Touch of Class Married American insurance adjustor working in London plans quick and uncommitted affair but finds his heart doesn't obey the rules. Jackson and Segal create sparkling record of a growing relationship. Oscar nominations for Best Picture and Original Score.
1973 (PG) 105m/C George Segal, Glenda Jackson, Paul Sorvino, Hildegarde Neil, Callan K., Mary Barclay, Cec Linder; *Dir:* Melvin Frank. Academy Awards '73: Best Actress (Jackson). VHS, Beta $9.99 *MED* 🐾🐾🐾

Touch of Evil Stark, perverse story of murder, kidnapping and police corruption in Mexican border town. Welles portrays a police chief who invents evidence to convict the guilty. Filled with innovative photography reminiscent of "Citizen Kane." (Filmed by Russell Metty). Score by Henry Mancini.
1958 108m/B Charlton Heston, Orson Welles, Janet Leigh, Joseph Calleia, Akim Tamiroff, Marlene Dietrich, Ray Collins, Dennis Weaver, Zsa Zsa Gabor, Joseph Cotten, Mercedes McCambridge; *Dir:* Orson Welles. VHS, Beta, LV $29.95 *MCA* 🐾🐾🐾🐾

Touch & Go Three beautiful women commit grand larceny in order to raise funds for underprivileged children.
1980 (PG) 92m/C *AU* Wendy Hughes; *Dir:* Peter Maxwell. VHS, Beta $29.95 *VID* 🐾🐾

Touch and Go Sentimental drama about a self-interested hockey pro who learns about love and giving through a young delinquent and his attractive mother.
1986 (R) 101m/C Michael Keaton, Maria Conchita Alonso, Ajay Naidu, Maria Tucci, Max Wright; *Dir:* Robert Mandel. VHS, Beta, LV $14.99 *HBO* 🐾1/2

A Touch of Magic in Close-Up The amazing Siroco, renowned master of "close-up" magic, mystifies, and then demonstrates the secrets behind many of his illusions.
1982 (G) 78m/C Siroco. VHS, Beta $29.95 *GEM* 🐾

Touch Me Not A neurotic secretary is used by an industrial spy to gain information on her boss. Weak thriller. Also known as "The Hunted."
1974 (PG) 84m/C *GB* Lee Remick, Michael Hinz, Ivan Desny, Ingrid Garbo; *Dir:* Douglas Fifthian. VHS $59.95 *JLT* 🐾🐾

The Touch of Satan Lost on the road, a young man meets a lovely young woman and is persuaded to stay at her nearby farmhouse. Things there are not as they seem, including the young woman—who turns out to be a very old witch. Also known as "The Touch of Melissa."
1970 90m/C Michael Berry, Emby Mallay, Lee Amber, Yvonne Wilson, Jeanne Gerson; *Dir:* Don Henderson. VHS, Beta $44.95 *KOV* 🐾

Touched Two young people struggle with the outside world after their escape from a mental institution. Although melodramatic and poorly scripted, the film is saved by fine performances by Hays and Beller.
1982 (R) 89m/C Robert Hays, Kathleen Beller, Ned Beatty, Gilbert Lewis; *Dir:* John Flynn. VHS, Beta $59.95 *MED* 🐾🐾

Touched by Love The true story of a handicapped child who begins to communicate when her teacher suggests she write to Elvis Presley. Sincere and well performed. Originally titled "To Elvis, With Love."
1980 (PG) 95m/C Deborah Raffin, Diane Lane, Christina Raines, Clu Gulager, John Amos; *Dir:* Gus Trikonis. VHS, Beta $59.95 *COL* 🐾🐾1/2

Tough Assignment A reporter in the West discovers and infiltrates a gang of organized rustlers who are forcing butchers to buy inferior meat.
1949 64m/B Donald (Don "Red") Barry, Marjorie Steele, Steve Brodie. VHS, Beta, LV *WGE* 🐾

Tough Enough A country-western singer and songwriter decides to finance his fledgling singing career by entering amateur boxing matches. Insipid and predictable, with fine cast wasted.
1987 (PG) 107m/C Dennis Quaid, Charlene Watkins, Warren Oates, Pam Grier, Stan Shaw, Bruce McGill, Wilford Brimley; *Dir:* Richard Fleischer. VHS, Beta $59.98 *FOX* 🐾🐾

Tough Guy Based on the play by Bruce Walker, a London hood carouses, mugs, breaks hearts and personifies his generation's anxiety.
1953 73m/B *GB* James Kenney, Hermione Gingold, Joan Collins; *Dir:* Lewis Gilbert. VHS, Beta $19.95 *UNI* 🐾

Tough Guy Two undercover policemen battle local gangsters in a bid to smash their crime ring.
197? 90m/C Chen Ying, Charlie Chiang; *Dir:* Chieng Hung. VHS, Beta $54.95 *MAV* 🐾

Tough Guys Two aging ex-cons, who staged America's last train robbery in 1961, try to come to terms with modern life after 30 years in prison. Amazed and hurt by the treatment of the elderly in the 1980s, frustrated with their inability to find something worthwhile to do, they begin to plan one last heist. Tailor-made for Lancaster and Douglas, who are wonderful. Script becomes cliched at end.
1986 (PG) 103m/C Burt Lancaster, Kirk Douglas, Charles Durning, Eli Wallach, Lyle Alzado, Dana Carvey; *Dir:* Jeff Kanew. VHS, Beta, LV $19.95 *TOU* 🐾🐾1/2

Tough Guys Don't Dance Mailer directed this self-satiric mystery thriller from his own novel. A writer may have committed murder—but he can't remember. So he searches for the truth among various friends, enemies, lovers, and cohorts.
1987 (R) 110m/C Ryan O'Neal, Isabella Rossellini, Wings Hauser, Debra Sandlund, John Bedford Lloyd, Lawrence Tierney, Clarence Williams III; *Dir:* Norman Mailer. VHS, Beta, LV $19.95 *MED, CVS* 🐾🐾1/2

Tough to Handle ...And no picnic to watch, either. A reporter cracks a criminal racket that rips off lottery sweepstakes winners. A laughably inept production.
1937 58m/B Frankie Darro, Kane Richmond, Phyllis Fraser, Harry Worth; *Dir:* S. Roy Luby. VHS $15.95 *NOS, LOO* 🐾1/2

Tough Kid A bad kid with gang connections attempts to shield his boxer brother from corruption at the hands of the same gang. Poor production all around.
1939 61m/B Frankie Darro, Dick Purcell, Judith Allen, Lillian Elliot; *Dir:* Howard Bretherton. **VHS, Beta $16.95** SNC ⊿

Tougher than Leather The rap triumvirate battles criminals, vigilante-style, in their home neighborhood.
1988 (R) 92m/C Run DMC, The Beastie Boys, Slick Rick, Richard Edson, Jenny Lumet; *Dir:* Rick Rubin. **VHS, Beta $89.95** COL ⊿

Toughlove The parents of a juvenile delinquent organize a tough parental philosophy and a support group for other parents having a difficult time with their teens. Made for television.
1985 100m/C Lee Remick, Bruce Dern, Piper Laurie, Louise Latham, Dana Elcar, Jason Patric, Eric Schiff, Dedee Pfeiffer; *Dir:* Glenn Jordan. **VHS, Beta $39.95** FRH ⊿⊿

Tour of Duty Follows the trials that the members of an American platoon face daily during the Vietnam war. The pilot movie for a television series.
1987 93m/C Terence Knox, Stephen Caffrey, Joshua Maurer, Ramon Franco; *Dir:* Bill L. Norton. **VHS, Beta, LV $59.95** NWV, HHE ⊿⊿½

Tourist Trap While traveling through the desert, a couple's car has a flat. A woman's voice lures the man into an abandoned gas station, where he discovers that the voice belongs to a mannequin. Or is it? Not very interesting or suspenseful.
1979 (PG) 90m/C Tanya Roberts, Chuck Connors, Robin Sherwood; *Dir:* David Schmoeller. **VHS, Beta $79.95** PAR ⊿

Tournament Renoir's second to last silent film, in which Protestants and Catholics joust out their differences in era of Catherine de Medici. French title: "Le Tournoi dans la Cite."
1929 90m/B *FR* Aldo Nadi; *Dir:* Jean Renoir. **VHS, Beta $29.95** FCT ⊿⊿½

Toute Une Nuit Assorted people stumble melodramatically in and out of one another's lives on steamy summer night in Brussels. Fine combination of avant-garde technique and narrative. Also known as "All Night Long." French with subtitles.
1982 90m/C *BE FR* Aurore Clement, Tcheky Karyo, Veronique Silver, Angelo Abazoglou, Natalia Ackerman; *Dir:* Chantal Akerman. **VHS $79.95** WAC, FCT ⊿⊿⊿

Tower of Evil Anthropologists and treasure hunters unite in search of Phoenician hoard on Snape Island. Suddenly, inanimate bodies materialize. Seems the island's single inhabitant liked it quiet. Cult favorite despite uneven performances. So bad it's not bad!
1972 (R) 86m/C *GB* Bryant Halliday, Jill Haworth; *Dir:* James O'Connolly. **VHS, Beta $59.95** MPI ⊿

The Tower of London Tells the story of Richard III (Rathbone), the English monarch who brutally executed the people who tried to get in his way of the throne. This melodrama was considered extremely graphic for its time, and some of the torture scenes had to be cut before it was released.
1939 93m/B Basil Rathbone, Boris Karloff, Barbara O'Neil, Ian Hunter, Vincent Price, Nan Grey, John Sutton, Leo G. Carroll, Miles Mander; *Dir:* Rowland V. Lee. **VHS $14.98** MCA ⊿⊿½

Tower of London A deranged lord (Price) murdering his way to the throne of England is eventually crowned Richard III. Sophisticated and well-made Poe-like thriller. More interesting as historic melodrama than as horror film. A remake of the 1939 version starring Basil Rathbone, in which Price played a supporting role.
1962 79m/B Vincent Price, Michael Pate, Joan Freeman, Robert Brown; *Dir:* Roger Corman. **VHS, Beta $9.95** WKV ⊿⊿

Tower of Screaming Virgins The Queen of France maintains a tower in which she has her lovers killed - after she makes love to them. An extremely loose adaptation of a Dumas novel; with English subtitles.
1968 89m/C *GE* Terry Torday, Jean Piat. Golden Turkey Award '68: So Bad It's Good. **VHS, Beta, 8mm $14.95** VYY, VDM, SNC ⊿

Tower of Terror A woman escapes from a German concentration camp and takes refuge in a lighthouse. The proprietor sees in her an uncanny resemblance to his late wife, whom he killed, and soon plots to do away with her. Poor acting hampers the story.
1942 62m/B *GB* Wilfred Lawson, Movita, Michael Rennie, Morland Graham, John Longden, George Woodbridge; *Dir:* Lawrence Huntington. **VHS, Beta $16.95** NOS, SNC ⊿

Towering Inferno Raging blaze engulfs the world's tallest skyscraper on the night of its glamorous dedication ceremonies. Allen had invented the disaster du jour genre two years earlier with "The Poseidon Adventure." Features a new but equally noteworthy cast.
1974 (PG) 165m/C Steve McQueen, Paul Newman, William Holden, Faye Dunaway, Fred Astaire, Jennifer Jones, Richard Chamberlain, Susan Blakely, O.J. Simpson, Robert Vaughn, Robert Wagner; *Dir:* John Guillermin, Irwin Allen. Academy Awards '74: Best Cinematography, Best Film Editing, Best Song ("We May Never Love Like this Again"). **VHS, Beta, LV $19.98** FOX ⊿⊿

Towers Open Fire & Other Films A collection of experimental short films that attempt to transfer prose and ideas onto visual forms: "The Cut-Ups," "Bill & Tony," "William Buys a Parrot," and the title film.
1972 35m/B William S. Burroughs, Ian Sommerville, Brion Gysin, Anthony Balch; *Dir:* Anthony Balch. **VHS, Beta $29.95** MFV, IVA ⊿⊿½

A Town Called Hell Two men hold an entire town hostage while looking for "Aguila," the Mexican revolutionary. Greed, evil, and violence take over. Shot in Spain. Also known as "A Town Called Bastard."
1972 95m/C *GB SP* Robert Shaw, Stella Stevens, Martin Landau, Telly Savalas, Fernando Rey; *Dir:* Robert Parrish. **VHS, Beta $24.95** KOV, WES, GEM ⊿

Town that Dreaded Sundown A mad killer is on the loose in a small Arkansas town. Based on a true story, this famous 1946 murder spree remains an unsolved mystery.
1976 (R) 90m/C Ben Johnson, Andrew Prine, Dawn Wells; *Dir:* Charles B. Pierce. **VHS, Beta $19.98** WAR, OM ⊿⊿½

A Town Like Alice An Australian miniseries based on a novel by Nevil Shute. The story follows women prisoners of war during World War II. One of the women is separated from her Australian lover. They are reunited decades later. Remake of a 1951 film of the same name that is also known as "Rape of Malaya."
1985 301m/C *AU* Bryan Brown, Helen Morse, Gordon Jackson; *Dir:* David Stevens. **VHS, Beta $29.95** STE, NWV, IME ⊿⊿½

Toxic Avenger Tongue-in-cheek, cult fave has 98-pound weakling fall into barrel of toxic waste to emerge as a lumbering, bloodthirsty hulk of sludge and mire. Two sequels followed: "Toxic Avenger Part II" and "Toxic Avenger Part III: The Last Temptation of Toxie."
1986 (R) 90m/C Mitchell Cohen, Andree Maranda, Jennifer Baptist; *Dir:* Michael Herz. **VHS, Beta, LV $29.98** VES, LIV, LTG ⊿½

Toxic Avenger, Part 2 Sequel to "Toxic Avenger." Hulky slimer targets Japanese corporations that built toxic chemical dump in Tromaville. Followed by "Toxic Avenger Part III: The Last Temptation of Toxie."
1989 (R) 90m/C Ron Fazio, Phoebe Legere, Rick Collins; *Dir:* Michael Herz, Lloyd Kaufman. **VHS, Beta, LV $89.95** WAR ⊿½

Toxic Avenger, Part 3: The Last Temptation of Toxie Unemployed superhero is tempted by Devil to become greedy capitalist. Also available in an unrated version.
1989 (R) 102m/C Ron Fazio, Phoebe Legere, Rick Collins, Lisa Gaye, Jessica Dublin, John Altamura; *Dir:* Michael Herz, Lloyd Kaufman. **VHS $89.98** VES, LIV ⊿½

Toxic Reasons Live performances by the energetic San Francisco punk band, playing such songs as "War Hero," "White Noise," and "Drunk and Disorderly."
198? 60m/C The Toxic Reasons. **VHS, Beta $29.95** MVD, TGV

Toxic Zombies Pot farmers become crazy cannibals when the marijuana is sprayed with an untested chemical.
1989 90m/C VHS, Beta $99.97 MON ⊿

The Toy A penniless reporter finds himself the new "toy" of the spoiled son of a multimillionaire oil man. Unfortunate casting of Pryor as the toy owned by Gleason, with heavy-handed lecturing about earning friends. A slow and terrible remake of the Pierre Richard comedy "Le Jouet."
1982 (PG) 99m/C Richard Pryor, Jackie Gleason, Ned Beatty, Wilfrid Hyde-White; *Dir:* Richard Donner. **VHS, Beta, LV, 8mm $9.95** COL ⊿

Toy Soldiers A group of college students are held for ransom in a war-torn Central American country. When they escape, they join forces with a seasoned mercenary who leads them as a vigilante force.
1984 (R) 85m/C Cleavon Little, Jason Miller, Tim Robbins, Tracy Scoggins; *Dir:* David Fisher. **VHS, Beta, LV $9.95** NWV, STE ⊿

Toy Soldiers This unlikely action tale stops short of being laughable but still has a fair share of silliness. South American narco-terrorists seize an exclusive boys' school. The mischievous students turn their talent for practical jokes to resistance-fighting. Jerry Orbach has an uncredited cameo. Based on a novel by William P. Kennedy.
1991 (R) 104m/C Sean Astin, Wil Wheaton, Keith Coogan, Andrew Divoff, Jerry Orbach, Denholm Elliott, Louis Gossett Jr.; *W/Dir:* Daniel Petrie. **VHS, Beta, LV, 8mm $92.95** COL, PIA ⊿⊿

Toys in the Attic A man returns to his home in New Orleans with his child bride, where they will live with his two impoverished, spinster sisters. Toned-down version of the Lillian Hellman play.

1963 90m/B Dean Martin, Geraldine Page, Yvette Mimieux, Wendy Hiller, Gene Tierney, Larry Gates, Nan Martin; *Dir:* George Roy Hill. **VHS** $9.95 *WKV* 🐾🐾½

Track 29 A confusing black comedy about a lonely woman, her husband who has a model train fetish, and a stranger who claims to be her son. Filmed in North Carolina. **1988** (R) 90m/C Theresa Russell, Gary Oldman, Christopher Lloyd, Colleen Camp, Sandra Bernhard, Seymour Cassel, Leon Rippy, Vance Colvig; *Dir:* Nicolas Roeg. **VHS** $89.95 *CAN* 🐾½

Track of the Moonbeast An American Indian uses mythology to capture the Moonbeast, a lizard-like creature that is roaming the deserts of New Mexico. **1976** 90m/C Chase Cordell, Donna Leigh Drake; *Dir:* Richard Ashe. **VHS, Beta** $49.95 *PSM* 🐾🐾

Track of the Vampire Half an hour of leftover footage from a Yugoslavian vampire movie stuck into the story of a California painter who kills his models. Gripping and atmospheric, but disjointed. Also called ''Blood Bath.'' **1966** 80m/B *YU* William Campbell, Jonathan Haze, Sid Haig, Marissa Mathes, Lori Saunders, Sandra Knight; *Dir:* Stephanie Rothman, Jack Hill. **VHS** *SNC, GNS* 🐾½

The Tracker A retired gunman/tracker must again take up arms. His college educated son joins him as he searches for a murdering religious fanatic. Mediocre. Made for cable television. **1988** 102m/C Kris Kristofferson, Scott Wilson, Mark Moses, David Huddleston; *Dir:* John Guillermin. **VHS, Beta** $89.99 *HBO* 🐾½

The Trackers A frontier man searches for the people who kidnapped his daughter and killed his son. He is joined by a cocky black tracker. Borgnine brings strength to his character as the rancher seeking revenge. Made for television. **1971** 73m/C Sammy Davis Jr., Ernest Borgnine, Julie Adams, Connie Kreski, Jim Davis; *Dir:* Earl Bellamy. **VHS, Beta** $19.98 *FOX* 🐾½

Tracks A Vietnam veteran accompanies the body of a buddy on a long train ride home. He starts suffering flashbacks from the war and begins to think some of the passengers are out to get him. **1976** (R) 90m/C Dennis Hopper, Dean Stockwell, Taryn Power; *Dir:* Henry Jaglom. **VHS, Beta** $39.95 *PAR, MON* 🐾½

Trading Hearts An over-the-hill baseball player falls an unsuccessful singer at a Florida training camp in this period piece set in the 1950s. **1987** (PG) 88m/C Beverly D'Angelo, Raul Julia, Jerry Lewis, Parris Buckner, Robert Gwaltney; *Dir:* Neil Leifer. **VHS, Beta** $14.95 *IVE* 🐾½

Trading Places Two elderly businessmen wager that environment is more important than heredity in creating a successful life, using a rich nephew and an unemployed street hustler as guinea pigs. Curtis is winning as a hooker-with-a-heart-of-gold who helps the hapless Aykroyd. Oft-told tale succeeds thanks to strong cast. Murphy's second screen appearance. **1983** (R) 118m/C Eddie Murphy, Dan Aykroyd, Jamie Lee Curtis, Ralph Bellamy, Don Ameche, Denholm Elliott, Paul Gleason, James Belushi; *Dir:* John Landis. **VHS, Beta, LV, 8mm** $14.95 *PAR* 🐾🐾½

Traenen in Florenz (Tears in Florence) A woman involved in a car accident winds up falling in love with the man who injured her, only to discover later that

they are actually brother and sister, long separated. In German with English subtitles. **1984** 90m/C *GE* Dir: Marianne Schafer. **VHS, Beta** $39.95 *VCD, APD, GLV* 🐾

Traffic in Souls A silent melodrama dealing with white slavery. Immigrants and gullible country girls are lured into brothels where they are trapped with no hope of escape until, in a hail of bullets, the police raid the dens of vice and rescue them. **1913** 74m/B Matt Moore, Jane Gail, William Welsh; *Dir:* George Tucker. **VHS, Beta** $24.95 *GPV* 🐾🐾🐾

Tragedy of Flight 103: The Inside Story Dramatization of the events surrounding the destruction of Pan Am flight 103 over Lockerbie, Scotland, due to a terrorist-planted bomb. **1991** (PG) 89m/C Ned Beatty, Peter Boyle, Vincent Gardenia, Timothy West, Michael Wincott; *Dir:* Leslie Woodhead. **VHS** $89.95 *MGM* 🐾🐾🐾

Tragedy of King Richard II Shakespeare's tale of the fight for the English crown between the current king, Richard II, and the usurper, Henry Bolingbroke, who seizes the throne. **1982** 180m/C David Birney, Paul Shenar. **VHS, Beta** $175.00 *KUL* 🐾🐾

Tragedy of a Ridiculous Man The missing son of an Italian cheese manufacturer may or may not have been kidnapped by terrorists. Lesser Bertolucci effort may or may not have a point. In Italian with English subtitles. **1981** (PG) 116m/C *IT* Ugo Tognazzi, Anouk Aimee, Victor Cavallo, Ricardo Tognazzi, Laura Morante; *Dir:* Bernardo Bertolucci. **VHS** $59.99 *WAR, FCT* 🐾🐾

Trail Beyond A cowboy and his sidekick go on the trek to the northwest to find a girl and a gold mine. **1934** 57m/B John Wayne, Noah Beery, Verna Hillie; *Dir:* Robert N. Bradbury. **VHS** $19.95 *COL, REP, NOS* 🐾½

Trail Drive Adventures of a cowboy during a big cattle drive. **1935** 63m/B Ken Maynard. **VHS, Beta** $19.95 *NOS, VCN, CAB* 🐾

Trail of the Hawk A saga of frontier justice and love on the plains. **1937** 50m/B Yancey Lane, Betty Jordan, Dick Jones; *Dir:* Edward Dmytryk. **VHS, Beta** $19.95 *NOS, DVT, RXM* 🐾

Trail of the Mounties A Mountie gallops smack into a murdering slew of fur thieves. **1947** 41m/B Donald (Don ''Red'') Barry, Robert Lowery, Tom Neal, Russell Hayden. **VHS, Beta, LV** *WGE* 🐾

Trail of the Pink Panther The sixth in Sellers' ''Pink Panther'' series. Inspector Clouseau disappears while he is searching for the diamond known as the Pink Panther. Notable because it was released after Sellers' death, using clips from previous movies to fill in gaps. Followed by ''Curse of the Pink Panther.'' **1982** (PG) 97m/C Peter Sellers, David Niven, Herbert Lom, Capucine, Burt Kwouk, Robert Wagner, Robert Loggia; *Dir:* Blake Edwards. **VHS, Beta, LV** *FOX* 🐾🐾

Trail Riders The Range Busters set a trap to capture a gang of outlaws who killed the son of the town marshal during a bank robbery. Part of the ''Range Busters'' series.

1942 55m/B John ''Dusty'' King, David Sharpe, Max Terhune, Evelyn Finley, Forrest Taylor, Charles King. **VHS, Beta** $19.95 *NOS, VYY, DVT* 🐾

Trail of Robin Hood Rogers sees to it that poor families get Christmas trees, in spite of the fact that a big business wants to raise the prices. Features a number of western stars, including Holt playing himself. **1950** 67m/C Roy Rogers, Penny Edwards, Gordon Jones, Jack Holt, Emory Parnell, Rex Allen, Allan ''Rocky'' Lane, Clifton Young, Monte Hale, Kermit Maynard, Tom Keene, Ray Corrigan, William Farnum; *Dir:* William Witney. **VHS** $12.98 *REP* 🐾🐾½

Trail of the Royal Mounted A western adventure serial made up of ten 15-minute episodes. **1934** 150m/B Robert Frazer. **VHS, Beta** $27.95 *VCN, CAB* 🐾

Trail of the Silver Spurs A lone man remaining in a ghost town discovers gold and becomes rich. Part of the ''Range Busters'' series. **1941** 57m/B Ray Corrigan, John ''Dusty'' King, Max Terhune, I. Stanford Jolley, Dorothy Short; *Dir:* S. Roy Luby. **VHS** $19.95 *NOS, DVT, BUR* 🐾½

Trail Street Scott is Bat Masterson, the western hero. He aids the struggle of Kansans as they conquer the land and local ranchers. **1947** 84m/B Randolph Scott, Robert Ryan, Anne Jeffreys, George ''Gabby'' Hayes, Madge Meredith, Jason Robards Jr.; *Dir:* Ray Enright. **VHS, Beta** $34.95 *CCB, MED* 🐾

Trail of Terror A G-Man poses as an escaped convict to get the goods on a gang. **1935** 60m/B Bob Steele. **VHS, Beta, LV** *WGE* 🐾

Trailing Double Trouble A corrupt attorney murders a rancher to pursue a business scheme. Three ranch hands kidnap the rancher's orphaned baby to keep her safe from the attorney. The second entry in the ''Range Busters'' series. **1940** 56m/B Ray Corrigan, John ''Dusty'' King, Max Terhune. **VHS, Beta** $19.95 *NOS, VDM* 🐾½

Trailing Trouble A cattle driver is robbed of his money when he sells cattle. He also finds time to vie for a young woman's heart. Basic low-budget oater. **1937** 60m/B Ken Maynard. **VHS, Beta** $19.95 *NOS, VYY, VCN* 🐾

Trails West Western adventure starring Steele. **194?** ?m/B Bob Steele. **VHS, Beta** $27.95 *VCN* 🐾

The Train During the German occupation of Paris in 1944, a German general (Scofield) is ordered to ransack the city of its art treasures and put them on a train bound for Germany. Word gets to the French Resistance who then persuade the train inspector (Lancaster) to sabotage the train. An elaborate ruse is planned, which works but ultimately causes the Nazi general to take hostages and demand the art shipment reach its destination. A battle of wills ensues between Scofield and Lancaster as each becomes increasingly obsessed in outwitting the other—though the cost to all turns out to be as irreplaceable as the art itself. Filmed on location in France. Frankenheimer used real locomotives, rather than models, throughout the film, even in the spectacular crash sequences.

1965 133m/C Burt Lancaster, Paul Scofield, Jeanne Moreau, Michel Simon, Suzanne Flon, Wolfgang Preiss, Albert Remy; *Dir:* John Frankenheimer. **VHS** $19.98 *MGM* 𝄞𝄞𝄞

Train of Events An interesting look at lives affected by a train wreck just outside London, taking the form of four short episodes. Three are frankly somber, depressing tales of the ill-fated train driver, a murderer and an escaped prisoner-of-war. The fourth segment is a refreshingly lighthearted piece about romantic jealousies in a traveling orchestra. A precursor to the later swarm of crass disaster movies, except here emphasis is on character over special effects.
1949 89m/B Jack Warner, Gladys Henson, Susan Shaw, Patric Doonan, Miles Malleson, Leslie Phillips, Joan Dowling, Laurence Payne, Valerie Hobson, John Clements, Peter Finch, Mary Morris, Laurence Naismith, Michael Hordern; *Dir:* Sidney Cole, Basil Dearden, Charles Crichton. **VHS** $29.95 *NOS, FCT* 𝄞𝄞½

Train to Hollywood Marilyn Monroe-obsessed young Polish girl daydreams incessantly as she works on dining car. Funny and surreal. Polish dialogue with English subtitles.
1986 96m/C *PL Dir:* Radoslaw Piwowarski. **VHS** $49.98 *FCT* 𝄞𝄞𝄞

The Train Killer A mad Hungarian is bent on destroying the Orient Express.
1983 90m/C Michael Sarrazin; *Dir:* Sandor Simo. **VHS, Beta** $69.98 *VES, AVD* 𝄞

Train Robbers A widow employs the services of three cowboys to help her recover some stolen gold in order to clear her late husband's name. At least that's what she says.
1973 (PG) 92m/C John Wayne, Ann-Margret, Rod Taylor, Ben Johnson, Christopher George, Ricardo Montalban, Bobby Vinton, Jerry Gatlin; *Dir:* Burt Kennedy. **VHS, Beta** $19.98 *WAR* 𝄞𝄞

Train Station Pickups Trashy film about young girls who hook at a local train depot, until something goes wrong.
197? 96m/C Marco Knoger, Katja Carrol, Ingeborg Steinbach, Benjamin Carwath; *Dir:* Walter Boos. **VHS** $29.95 *ACN* 𝄞

Train to Tombstone A train running from Albuquerque to Tombstone carries a motley mob of the usual characters.
1950 60m/B Donald (Don "Red") Barry, Judith Allen, Robert Lowery, Tom Neal; *Dir:* Charles F. Reisner. **VHS, Beta, LV** *WGE* 𝄞

Trained to Kill, U.S.A. A Vietnam veteran relives his war experiences when a gang of terrorists threaten his hometown.
1983 (R) 88m/C Stephen Sander, Heidi Vaughn, Rockne Tarkington. **VHS, Beta** *PGN* 𝄞½

The Traitor Undercover man joins a gang of bandits.
1936 57m/B Tim McCoy; *Dir:* Sam Newfield. **VHS, Beta** $19.95 *NOS, UHV, VCN* 𝄞½

Traitors of the Blue Castle Two samurai must preserve the honor of the emperor in nineteenth-century Japan.
1958 100m/C *JP* Kanjuro Aroshi, Ryuzaburo Nakamura. **VHS, Beta** $54.00 *VDA* 𝄞

Tramp at the Door Delightful drama about a wandering, magical man who enters the lives of an embittered family. He helps them sort the skeletons in their closets, and works wonders in their relationships.
1987 81m/C Ed McNamara, August Schellenberg, Monique Mercure, Eric Peterson; *Dir:* Allen Kroeker. **VHS, Beta** $79.98 *CGH* 𝄞𝄞½

Tramp & a Woman Two shorts made for the Essanay Company in 1915 which offer the Little Tramp wooing Purviance in typical Chaplin fashion. Silent with musical score.
1915 45m/B Charlie Chaplin, Edna Purviance. **VHS, Beta** $24.95 *VYY* 𝄞𝄞

Tramplers A rebel father and son split over the hanging of a Yankee during the Civil War. Italian made spaghetti western without much plot and lots of gratuitous gunplay.
1966 103m/C *IT* Joseph Cotten, Gordon Scott; *Dir:* Albert Band, Mario Soqui. **VHS, Beta** $9.95 *COL, SUE, HHE* 𝄞

Trancers A time-traveling cult from the future goes back in time to 1985 to meddle with fate. Only Jack Deth, defender of justice, can save mankind. Low-budget "Blade Runner." Also released as "Future Cop."
1985 (PG-13) 76m/C Tim Thomerson, Michael Stefoni, Helen Hunt, Art LaFleur, Telma Hopkins; *Dir:* Charles Band. **VHS, Beta, LV** $19.98 *LIV, VES* 𝄞𝄞

Trancers 2: The Return of Jack Deth Retro cop is back from the future again, but seems to have lost his wit and nerve in between sequels. Ward is miscast and the pacing undermines whatever suspense that might have been.
1991 (R) 85m/C Tim Thomerson, Helen Hunt, Megan Ward, Biff Manard, Martine Beswick, Jeffrey Combs, Barbara Crampton, Richard Lynch; *Dir:* Charles Band. **VHS, Beta, LV** $19.95 *PAR, PIA* 𝄞

Trancers 3: Deth Lives In the third film of the "Trancers" series, time-traveling cop Jack Deth fights the deadliest form of government-sponsored Trancer yet—it has a brain.
1992 83m/C Tim Thomerson, Melanie Smith, Andrew Robinson, Tony Pierce, Dawn Ann Billings, Helen Hunt, Megan Ward, Stephen Macht, Telma Hopkins; *W/Dir:* C. Courtney Joyner. **VHS, Beta** *PAR* 𝄞𝄞

Transatlantic Merry-Go-Round An early semi-musical mystery about an eclectic assortment of passengers aboard a luxury liner. Songs include "Rock and Roll" and "If I Had a Million Dollars."
1934 90m/B Jack Benny, Gene Raymond, Nancy Carroll, Boswell Sisters, Ralph Morgan, Sidney Blackmer; *Dir:* Ben Stoloff. **VHS, Beta** $12.95 *IVE* 𝄞𝄞

Transatlantic Tunnel An undersea tunnel from England to America is attempted, despite financial trickery and undersea disasters. Made with futuristic sets which were advanced for their time. Also released as "The Tunnel" in Great Britain.
1935 94m/B *GB* Richard Dix, Leslie Banks, Madge Evans, Helen Vinson, C. Aubrey Smith, George Arliss, Walter Huston; *Dir:* Maurice Elvey. **VHS, Beta** $44.95 *GPV, SNC, HHT* 𝄞𝄞½

Transformations Ain interplanetary pilot battles a deadly virus that threatens life throughout the universe. Obscure space jetsam.
198? 84m/C Rex Smith, Patrick Macnee. **VHS** $19.95 *STE* 𝄞½

Transformers: The Movie A full-length animated film featuring the universe-defending robots fighting the powers of evil. These robots began life as real toys, so there's some marketing going on.
1986 (G) 85m/C *Dir:* Nelson Shin; *Voices:* Orson Welles, Eric Idle, Judd Nelson, Leonard Nimoy, Robert Stack. **VHS, Beta** $19.95 *FHE* 𝄞

Transmutations A mad scientist harbors a bevy of mind-expanding drug-addicted, disfigured victims in his underground lab, who eventually rebel. Written by Clive Barker, also known as "Underworld."
1986 (R) 103m/C *GB* Denholm Elliott, Steven Berkoff, Miranda Richardson, Nicola Cowper; *Dir:* George Pavlou. **VHS, Beta, LV** $79.98 *LIV, VES* 𝄞½

Transport from Paradise Based on a novel by Arnold Lustig. Follows the stories of the Jewish prisoners of a ghetto prison camp. There aren't any gas chambers where they are, but in the end some are shipped off to Auschwitz, where they face death. In Czech with subtitles.
1965 93m/B *CZ Dir:* Zbynek Brynych. **VHS, Beta** $59.95 *FCT* 𝄞𝄞

Transylvania 6-5000 Agreeably stupid horror spoof about two klutzy reporters who stumble into modern-day Transylvania and encounter an array of comedic creatures. Shot in Yugoslavia.
1985 (PG) 93m/C Jeff Goldblum, Joseph Bologna, Ed Begley Jr., Carol Kane, John Byner, Geena Davis, Jeffrey Jones, Norman Fell; *Dir:* Rudy DeLuca. **VHS, Beta, LV** $19.95 *STE, NWV* 𝄞

Transylvania Twist Moronic comedy about vampires, teenage vampire hunters and half-naked babes.
1989 (PG-13) 90m/C Robert Vaughn, Teri Copley, Steve Altman, Ace Mask, Angus Scrimm, Jay Robinson, Brinke Stevens; *Dir:* Jim Wynorski. **VHS, Beta** $79.95 *MGM* 𝄞½

Trap In trying to escape justice, a ruthless crime syndicate boss holds a small desert town in a grip of fear.
1959 84m/C Richard Widmark, Tina Louise, Lee J. Cobb, Earl Holliman, Lorne Greene, Carl Benton Reid; *Dir:* Norman Panama. **VHS, Beta** $19.95 *KRT* 𝄞𝄞½

Trap A trapper buys a mute girl as his wife. Together they try to make a life for themselves in the Canadian wilderness. Fine telling of interesting western tale. Unusual and realistic.
1966 106m/C *GB CA* Rita Tushingham, Oliver Reed, Rex Sevenoaks; *Dir:* Sidney Hayers. **VHS, Beta** $59.95 *IND* 𝄞𝄞

Trap on Cougar Mountain A young boy begins a crusade to save his animal friends from the traps and bullets of hunters.
1972 (G) 97m/C Erik Larsen, Keith Larsen, Karen Steele; *Dir:* Keith Larsen. **VHS, Beta, LV** $9.95 *NWV, STE* 𝄞

Trap Them & Kill Them A group of Americans traveling through the Amazon jungle encounter a terrifying aborigine tribe.
1984 90m/C Gabriele Tinti, Susan Scott, Donald O'Brien. **VHS, Beta** $59.95 *TWE* 𝄞

Trapeze Lancaster is a former trapeze artist, now lame from a triple somersault accident. Curtis, the son of an old friend, wants Lancaster to teach him the routine. Lollobrigida, an aerial acrobat, is interested in both men. Exquisite European locations, fine camera work. The actors perform their own stunts.
1956 105m/C Burt Lancaster, Tony Curtis, Gina Lollobrigida, Katy Jurado; *Dir:* Carol Reed. **VHS, Beta** $59.98 *FOX* 𝄞𝄞𝄞

Trapped Semi-documentary crime drama shows in semi-documentary style how the feds hunt down a gang of counterfeiters by springing one of their comrades from prison. Well-paced, suspenseful and believable.

T

1949 78m/B Lloyd Bridges, Barbara Payton, John Hoyt, James Todd; *Dir:* Richard Fleischer. **VHS** $19.95 NOS, DVT &&½

Trapped A woman working late in her high-rise office is stalked by a killer and must rely on her ingenuity to outwit him. Made for cable television thriller.
1989 (R) 93m/C Kathleen Quinlan, Bruce Abbott, Katy Boyer, Ben Loggins; *Dir:* Fred Walton. **VHS, Beta** $79.95 MCA &½

Trapped by the Mormons Mormons seduce innocent young girls to add to their harems. An interesting piece of paranoid propaganda. Silent with original organ music. Also known as "The Mormon Peril."
1922 97m/B Evelyn Brent, Lewis Willoughby. **VHS, Beta, 8mm** $19.95 NOS, VYY, GPV &&

Trapped in Silence Young Sutherland, abused since childhood, stops speaking to protect himself mentally and physically. Psychologist Mason is determined to break down the walls to help him confront his pain. Interesting cameo by Silver as a gay counselor forced out of his job when his sexual preference is discovered. Mildly melodramatic made-for-TV message drama.
1990 94m/C Marsha Mason, Kiefer Sutherland, John Mahoney, Ron Silver, Amy Wright; *Dir:* Michael Tuchner. **VHS** $59.98 TTC &&

Trapper County War A city boy and a Vietnam vet band together to rescue a young woman held captive by a backwoods clan.
1989 (R) 98m/C Robert Estes, Bo Hopkins, Ernie Hudson, Betsy Russell, Don Swayze; *Dir:* Worth Keeter. **VHS, LV** $9.98 REP, LDC &

Trash Andy Warhol's profile of a depraved couple (Warhol-veteran Dallesandro and female impersonator Woodlawn) living in a lower east side basement and scouting the streets for food and drugs. Not for those easily offended by nymphomaniacs, junkies, lice, and the like; a must for fans of underground film and the cinema verite style.
1970 110m/C Joe Dallesandro, Holly Woodlawn, Jane Forth, Micael Sklar, Geri Miller, Bruce Pecheur; *Dir:* Paul Morrissey. **VHS, Beta** $29.95 MFV, PAR &&

Trauma A girl suffers amnesia after the trauma of witnessing her aunt's murder. She returns to the mansion years later to piece together what happened. A sometimes tedious psychological thriller.
1962 93m/C John Conte, Lynn Bari, Lorrie Richards, David Garner, Warren Kemmerling, William Bissell, Bond Blackman, William Justine; *Dir:* Robert M. Young. **VHS, Beta** VES &½

Traveling Companions The beautiful, young owner of a truck stop has all the local truckers vying for her via macho contests.
1986 80m/C Elizabeth Turner, Nikki Gentile. **VHS, Beta** $29.95 MED Woof!

Traveling Man A made-for-cable-television drama about a veteran traveling salesman who's assigned an eager young apprentice when his sales go down. Lithgow gives a great performance as the burnt-out traveling salesman.
1989 105m/C John Lithgow, Jonathan Silverman, Margaret Colin, John Glover; *Dir:* Irvin Kershner. **VHS, Beta** $89.99 HBO &½

Travelling North A belligerent retiree falls in love with a divorcee and they move together to a rustic retreat. They have an idyllic existence until the man discovers he has a serious heart condition. Adaptation of David Williamson's play has fine performances.

1987 (PG-13) 97m/C AU Leo McKern, Julia Blake; *Dir:* Carl Schultz. Australian Film Institute '87: Best Actor (McKern); Montreal World Film Festival '87: Best Actor (McKern). **VHS, Beta** $79.95 VIR &&

Travels with My Aunt A banker leading a mundane life is taken on a wild, whirlwind tour of Europe by an eccentric woman claiming to be his aunt. Based on Graham Greene's best-selling novel. Smith earned an Oscar nomination.
1972 (PG) 109m/C Maggie Smith, Alec McCowen, Louis Gossett Jr., Robert Stephens, Cindy Williams; *Dir:* George Cukor. Academy Awards '72: Best Costume Design. **VHS** MGM &&½

Traxx Satire about an ex-cop soldier of fortune. Not surprisingly, Stevens doesn't do much with this lame script.
1987 (R) 84m/C Shadoe Stevens, Priscilla Barnes, William E. Pugh, John Hancock, Robert Davi, Rick Overton; *Dir:* Jerome Gary. **VHS, Beta** $79.99 HBO &

Treasure of Arne A 16th century Scottish mercenary kills and loots the estate of rich, well-to-do property owner Sir Arne. Famous early Swedish silent that established Stiller as a director.
1919 78m/B SW *Dir:* Mauritz Stiller. **VHS, Beta** $49.95 FCT &&½

The Treasure of Bengal Sabu attempts to stop his chief from trading the village's precious gem (the largest ruby in the world) for firearms.
1953 72m/C Sabu. **VHS** $14.95 MAG &½

Treasure of Bruce Le Bruce Le beats up on Samurai to avenge a friend's death.
1976 80m/C Bruce Le. **VHS, Beta** $19.99 BFV &&

Treasure of Fear A bungling newspaper reporter gets involved with four jade chessmen once owned by Kubla Khan. A few laughs in this contrived haunted house story. Also known as "Scared Stiff."
1945 66m/B Jack Haley, Barton MacLane, Ann Savage; *Dir:* Frank McDonald. **VHS, Beta** $16.95 SNC, NOS, FCT &½

Treasure of the Four Crowns An aging history professor hires a team of tough commandos to recover four legendary crowns containing the source of mystical powers. The crowns are being held under heavy guard by a crazed cult leader.
1982 (PG) 97m/C Tony Anthony, Anna Obregon, Gene Quintano; *Dir:* Ferdinando Baldi. **VHS, Beta** $59.95 MGM &

Treasure of the Golden Condor A swashbuckler heads to the jungles of 18th-century Guatemala in search of ancient treasure. Contains interesting footage of Mayan ruins, but otherwise ordinary. A remake of "Son of Fury."
1953 93m/C Cornel Wilde, Connie Smith, Fay Wray, Anne Bancroft, Leo G. Carroll, Bobby Blake; *Dir:* Delmer Daves. **VHS** TCF &&½

Treasure Island Fleming's adaptation of Robert Louis Stevenson's eighteenth-century English pirate tale of Long John Silver is a classic. Beery is great as the pirate; Cooper has trouble playing the boy. Also available colorized. Multitudinous remakes.
1934 102m/B Wallace Beery, Jackie Cooper, Lionel Barrymore, Lewis Stone, Otto Kruger, Douglas Dumbrille, Chic Sale, Nigel Bruce; *Dir:* Victor Fleming. **VHS, Beta, LV** $19.98 MGM, TLF &&&½

Treasure Island Spine-tingling Robert Louis Stevenson tale of pirates and buried treasure, in which young cabin boy Jim Hawkins matches wits with Long John Silver. Some editions excise extra violence. Stevenson's ending is revised. Full Disney treatment, excellent casting.
1950 (PG) 96m/C Bobby Driscoll, Robert Newton, Basil Sydney, Walter Fitzgerald, Denis O'Dea, Ralph Truman, Finlay Currie; *Dir:* Byron Haskin. **VHS, Beta, LV** $19.99 DIS &&&½

Treasure Island Unexceptional British reheat of familiar pirate tale. Welles' interpretation of Long John Silver may be truer to Stevenson, but it's one heckuva blustering binge.
1972 (G) 94m/C GB Orson Welles, Kim Burfield, Walter Slezak, Lionel Stander; *Dir:* John Hough. **VHS** $79.95 BTV &&

Treasure Island Excellent made for cable version of the classic Robert Louis Stevenson pirate tale with Heston as Long John Silver. Written, produced and directed by Fraser Heston, son of Charlton.
1989 131m/C Charlton Heston, Christian Bale, Julian Glover, Richard Johnson, Oliver Reed, Christopher Lee, Clive Wood, Nicholas Amer, Michael Halsey; *W/Dir:* Fraser Heston. **VHS** $79.98 TTC &&½

The Treasure of Jamaica Reef Adventurers battle sharks and other nasty fish as they seek a sunken Spanish galleon and its cache of golden treasure.
1974 (PG) 96m/C Cheryl Ladd, Stephen Boyd, Roosevelt Grier, David Ladd, Darby Hinton; *Dir:* Virginia Lively Stone. **VHS, Beta** $39.95 WES &½

Treasure of the Lost Desert A Green Beret crushes a terrorist operation in the Middle East.
1983 93m/C Bruce Miller, Susan West, Larry Finch; *Dir:* Tony Zarindast. **VHS, Beta** $69.98 LIV, LTG &½

Treasure of Matecumbe A motley crew of adventurers led by a young boy search for buried treasure as they are pursued by Indians and other foes. Filmed in the Florida Key Islands.
1976 (G) 107m/C Billy Attmore, Robert Foxworth, Joan Hackett, Peter Ustinov, Vic Morrow; *Dir:* Vincent McEveety. **VHS, Beta** $69.95 DIS &&

Treasure of Monte Cristo A woman marries a seaman said to be an ancestor of the Count for his inheritance. Instead, she finds love and mystery.
1950 78m/B GB Adele Jergens, Steve Brodie, Glenn Langan; *Dir:* William Berke. **VHS, Beta, LV** WGE &&

Treasure of the Moon Goddess A dizzy nightclub singer is mistaken by Central American pirates and natives for the earthly manifestation of their Moon Goddess.
1988 (R) 90m/C Don Calfa, Joann Ayres, Asher Brauner; *Dir:* Joseph Louis Agraz. **VHS, Beta** $79.95 VMK Woof!

Treasure of Pancho Villa An American adventurer plots a gold heist to help Villa's revolution. He encounters every obstacle in the West while on his way to help the Mexican rebel.
1955 96m/C Rory Calhoun, Shelley Winters, Gilbert Roland; *Dir:* George Sherman. **VHS, Beta** $15.95 UHV &½

The Treasure Seekers Four rival divers set off on a perilous Caribbean expedition in search of the legendary treasure of Morgan the Pirate.

1979 88m/C Rod Taylor, Stuart Whitman, Elke Sommer, Keenan Wynn, Jeremy Kemp. **VHS, Beta** $59.95 *MGM* ✍

Treasure of the Sierra Madre
Three prospectors in search of gold in Mexico find suspicion, treachery and greed. Bogart is superbly believable as the paranoid, and ultimately homicidal, Fred C. Dobbs. Huston directed his father and wrote the screenplay, based on a B. Traven story. Oscar nomination for Best Picture.
1948 126m/B Humphrey Bogart, Walter Huston, Tim Holt, Bruce Bennett, Barton MacLane, Robert Blake; **W/Dir:** John Huston. Academy Awards '48: Best Director (Huston), Best Screenplay, Best Supporting Actor (Huston); National Board of Review Awards '48: 10 Best Films of the Year, Best Actor (Huston), Best Screenplay; New York Film Critics Awards '48: Best Director (Huston), Best Film; Venice Film Festival '48: Best Musical Score; New York Times Ten Best List '48: Best Film. **VHS, Beta, LV** $19.98 *MGM, FOX, TLF* ✍✍✍✍

Treasure of the Yankee Zephyr A
trio join in the quest for a plane that has been missing for 40 years... with a cargo of $50 million.
1983 (PG) 97m/C Ken Wahl, George Peppard, Donald Pleasence, Lesley Ann Warren; **Dir:** David Hemmings. **VHS, Beta** $69.95 *LIV, VES, SIM* ✍

A Tree Grows in Brooklyn Sensi-
tive young Irish lass growing up in turn-of-the-century Brooklyn tries to rise above her tenement existence. Based on the novel by Betty Smith. Kazan's directorial debut. Oscar-nominated Screenplay.
1945 128m/B Peggy Ann Garner, James Dunn, Dorothy McGuire, Joan Blondell, Lloyd Nolan, Ted Donaldson, James Gleason, John Alexander; **Dir:** Elia Kazan. Academy Awards '45. Best Supporting Actor (Dunn), Special Achievement Award (Garner). **VHS, Beta, LV** $19.98 *KUI, FOX, BTV* ✍✍✍½

The Tree of Wooden Clogs Epic
view of the lives of four peasant families working on an estate in turn-of-the-century Northern Italy. Slow moving and beautiful, from former industrial filmmaker Olmi. In Italian with English subtitles.
1978 185m/C *IT* Luigi Ornaghi, Francesca Moriggi, Omar Brignoli, Antonio Ferrari; **W/Dir:** Ermanno Olmi. Cannes Film Festival '78: Best Film. **VHS, Beta, LV** $79.95 *FXL, APD* ✍✍✍½

Tremors A tiny desert town is attacked by
giant man-eating worm-creatures. Bacon and Ward are the handymen trying to save the town. Amusing, with good special effects.
1989 (PG-13) 96m/C Kevin Bacon, Fred Ward, Finn Carter, Michael Gross, Reba McEntire, Bibi Besch; **Dir:** Ron Underwood. **VHS, Beta, LV** $19.95 *MCA* ✍✍½

Trenchcoat An aspiring mystery writer
travels to Malta to research her new novel and is drawn into a real-life conspiracy. Silly and contrived spoof of the detective genre.
1983 95m/C Margot Kidder, Robert Hays; **Dir:** Michael Tuchner. **VHS, Beta** *DIS, OM* ✍½

Trespasser A painter has an affair with a
young woman. He comes to regret destroying his family and their life together. Based on a D.H. Lawrence novel.
1985 90m/C Alan Bates, Dinah Stabb, Pauline Morgan, Margaret Whiting. **VHS, Beta** $59.98 *MAG* ✍✍

Trespasses A rape victim and a cattle-
man get involved and become the target of her husband's revenge. Murder and lust in a small Texas town in this tedious film.

1986 (R) 90m/C Lou Diamond Phillips, Robert Kuhn, Ben Johnson, Adam Roarke, Mary Pillot, Van Brooks; **Dir:** Lauren Bivens. **VHS, Beta** $29.95 *ACA, MTX* ✍

Triad Savages Triad war with the Mafia.
19?? 90m/C Chow Yung Fat, Ko Chun Hsiung, Lam Wai. **VHS** $49.95 *OCE* ✍

The Trial Expressionistic Welles adapta-
tion of classic Kafka novella about an innocent man accused, tried, and convicted of an unknown crime in an unnnamed exaggeratedly bureaucratic country. Another Welles project that met with constant disaster in production, with many lapses in continuity.
1963 118m/B *FR* Anthony Perkins, Jeanne Moreau, Orson Welles, Romy Schneider, Akim Tamiroff, Elsa Martinelli; **Dir:** Orson Welles. **VHS, Beta** $39.95 *CVC, NOS, HHT* ✍✍✍

Trial of Bernhard Goetz A re-enact-
ment of the celebrated vigilante's trial, taken from the actual court transcripts.
1988 135m/C **VHS, Beta** $59.95 *ALL* ✍✍

Trial of Billy Jack Billy Jack takes on
the feds and beats the hell out of a lot of people to prove that the world can live in peace. Awful, pretentious film.
1974 (PG) 175m/C Tom Laughlin, Delores Taylor, Victor Izay, Terasa Laughlin, William Wellman Jr.; **Dir:** Frank Laughlin. **VHS** *MOV* Woof!

Trial of the Catonsville Nine Riv-
eting political drama that focuses on the trial of nine anti-war activists, including Father Daniel Berrigan, during the Vietnam War days of the late 1960s. Imperfect, but involving.
1972 (PG) 85m/C Ed Flanders, Douglas Watson, William Schallert, Peter Strauss; **Dir:** Gordon Davidson. **VHS, Beta** $59.95 *COL* ✍✍½

Trial of the Incredible Hulk The
Hulk returns to battle organized crime and is aided by his blind superhero/lawyer friend Daredevil. Followed by "The Death of the Incredible Hulk."
1989 96m/C Bill Bixby, Lou Ferrigno, Rex Smith, John Rhys-Davies, Marta Du Bois, Nancy Everhard, Nicholas Hormann; **Dir:** Bill Bixby. **VHS** $19.95 *STE* ✍✍

The Trial of Lee Harvey Oswald
What would've happened if Jack Ruby had not shot Oswald. Oswald's trial and its likely results are painstakingly created. In 2 volumes. Made for television.
1977 192m/C Ben Gazzara, Lorne Greene, John Pleshette, Lawrence Pressman; **Dir:** Gordon Davidson. **VHS, Beta** $69.95 *WOV* ✍✍

The Triangle Factory Fire
Scandal Based on the true-life Triangle factory fire at the turn of the century. The fire killed 145 garment workers and drastically changed industrial fire and safety codes. Made for television.
1979 100m/C Tom Bosley, David Dukes, Tovah Feldshuh, Janet Margolin, Stephanie Zimbalist, Lauren Front, Stacey Nelkin, Ted Wass, Charlotte Rae, Milton Selzer, Valerie Landsburg; **Dir:** Mel Stuart. **VHS** $59.95 *IVE* ✍✍

Tribes A made-for-television movie about a
hippie drafted into the Marines. Despite the furious efforts of his drill sergeant, he refuses to conform to military life. Soon his bootcamp mates are trying his methods of defiance and survival. Well-made mix of comedy and social commentary. Released in Europe as "The Soldier Who Declared Peace."
1970 90m/C Jan-Michael Vincent, Darren McGavin, Earl Holliman; **Dir:** Joseph Sargent. **VHS, Beta** $59.98 *MGM, FOX* ✍✍½

Tribute A dying man is determined to
achieve a reconciliation with his estranged son. Adapted by Bernard Slade from his play.
1980 (PG) 125m/C *CA* Jack Lemmon, Robby Benson, Lee Remick, Kim Cattrall; **Dir:** Bob Clark. **VHS, Beta, LV** $19.98 *VES, FCT* ✍✍

A Tribute to the Boys An all-star
salute to the comedy of Laurel & Hardy featuring clips from more than 30 of their comedies, including scenes from their silent film days. Guests include Johnny Carson, Steve Allen, Walter Mathau, Rich Little, Henny Youngman, Robert Klein, Bronson Pinchot, and The Smothers Brothers.
1992 85m/C **Hosted:** Dom DeLuise. **VHS** $59.95 *CAF*

A Tribute to Lucy Rare footage, films
and promotional material from Lucy's heyday.
1990 83m/B Lucille Ball. **VHS, Beta** $9.95 *GKK*

Tribute to Ricky Nelson A tribute to
the late rock and roll singer featuring Orbison, Fogerty, Domino and Lewis. Includes concert footage, interviews and clips from the television show "The Adventures of Ozzie and Harriet."
1986 45m/C Ricky Nelson, Roy Orbison, John Fogerty, Fats Domino, Jerry Lee Lewis; **Dir:** Taylor Hackford. **VHS, Beta, LV** $19.95 *RHI*

Tricheurs Scam artists hit largest casino
in the world and look forward to golden years of retirement. From the director of "Reversal of Fortune." In French with English subtitles.
1984 95m/C **Dir:** Barbet Schroeder. **VHS, Beta** $59.99 *WAR, BTV, FCT* ✍✍

Trick or Treat A high school student is
helped to exact violent revenge against his bullying contemporaries. His helper is the spirit of a violent heavy metal rock star who he raises from the dead. Sometimes clever, not terribly scary.
1986 (R) 97m/C Tony Fields, Marc Price, Ozzy Osbourne, Gene Simmons; **Dir:** Charles Martin Smith. **VHS, Beta** $19.98 *LHV, AVD, WAR* ✍½

Trick or Treats A young boy's pranks
on his terrified baby-sitter backfire on Halloween night. His deranged father, escaped from an asylum, shows up to help him "scare" her. Giroux doesn't look very frightened in this extremely tedious film from Welles' best camera man.
1982 (R) 90m/C Carrie Snodgress, David Carradine, Jackie Giroux, Steve Railsback; **Dir:** Gary Graver. **VHS, Beta** $69.98 *LIV, VES* Woof!

Tricks of the Trade Goody two shoe
housewife and hooker accomplice search for husband's murderer. Racy made for television fodder.
1988 (R) 100m/C Cindy Williams, Markie Post, Chris Mulkey, James Whitmore Jr., Scott Paulin, John Ritter, Apollonia; **Dir:** Jack Bender. **VHS** $19.95 *ACA* ✍

Trident Force An international counter-
terrorism team is ordered to destroy the Palestinian Revolutionary Legion by any means possible.
1988 90m/C Anthony Alonzo, Mark Gil, Nanna Anderson, Steve Rogers, Eddie M. Gaerlan; **Dir:** Richard Smith. **VHS, Beta** $79.98 *DMD* ✍

Trifles Five people enter a farmhouse
kitchen in the bleak Midwest plains. A man was murdered upstairs. The solution lies in piecing together certain "trifles."
1979 21m/C **VHS, Beta** *BFA* ✍½

Trigger, Jr. Rogers and Evans team up to battle an extortion ring and a murderous stallion.
1950 68m/C Roy Rogers, Dale Evans, Pat Brady, Gordon Jones, Grant Withers; **Dir:** William Witney. **VHS** $12.98 REP ♫♫

Trigger Pals Jarrett stars in this basic oater about bringing a gang of rustlers to justice. Besides the usual gunplay, he also attempts to sing a few songs.
1939 58m/B Art Jarrett, Lee Powell, Al "Fuzzy" St. John, Ted Adams; **Dir:** Philip Krasne. **VHS, Beta** $27.95 VCN ♫

Trigger Trio A rancher, with diseased cattle, kills a ranch inspector. Three men are called in to try and crack the case. Part of the "Three Musketeers" series.
1937 54m/B Ray Corrigan, Max Terhune, Ralph Byrd, Robert Warwick, Cornelius Keefe; **Dir:** William Witney. **VHS, Beta** $19.95 VCN, DVT ♫

Trilby Original screen production of Du Maurier's classic tale of Svengali.
1917 59m/B Clara Kimball Young, Wilton Lackaye, Chester Barnet; **Dir:** Maurice Tourneur. **VHS, Beta** $19.95 GPV, FCT ♫♫♫

Trilogy of Terror Black shows her versatility as she plays a tempting seductress, a mousy schoolteacher and the terrified victim of an African Zuni fetish doll in three horror shorts penned by Richard Mathes. Made-for-television.
1975 78m/C Karen Black, Robert Burton, John Karlen, Gregory Harrison; **Dir:** Dan Curtis. **VHS, Beta** $59.95 MPI ♫♫½

Trinity is Still My Name Sequel to "They Call Me Trinity." Insouciant bumbling brothers Trinity and Bambino, oblivious to danger and hopeless odds, endure mishaps and adventures as they try to right wrongs. A funny parody of Western cliches that never becomes stale.
1975 (G) 117m/C IT Bud Spencer, Terence Hill, Harry Caray Jr.; **Dir:** E.B. Clucher. **VHS, Beta, LV** $9.99 CCB, SUE, HHE ♫♫½

Trio Sequel to "Quartet," featuring the W. Somerset Maugham stories "The Verger," "Mr. Knowall" and "Sanatorium Acclaimed." Cinematography by Geoffrey Unsworth and Reg Wyer.
1950 88m/B Jean Simmons, Michael Rennie, Bill Travers, James Hayter, Kathleen Harrison, Felix Aylmer, Nigel Patrick, Finlay Currie, John Laurie; **Dir:** Ken Annakin, Harold French. **VHS, Beta** $39.95 AXV ♫♫♫

The Trip A psychedelic journey to the world of inner consciousness, via drugs. A television director, unsure where his life is going, decides to try a "trip" to expand his understanding. Hopper is the drug salesman. Dern, the tour guide for Fonda's trip through sex, witches, torture chambers and more. Great period music. Screenplay by Jack Nicholson.
1967 85m/C Peter Fonda, Dennis Hopper, Susan Strasberg, Dick Miller, Peter Bogdanovich, Bruce Dern; **Dir:** Roger Corman. **VHS, Beta** $29.98 LIV, VES ♫♫

Trip to Bountiful An elderly widow, unhappy living in her son's fancy modern home, makes a pilgrimage back to her childhood home in Bountiful, Texas. Based on the Horton Foote play. Fine acting with Oscar-winning performance from Page.
1985 (PG) 102m/C Geraldine Page, Rebecca DeMornay, John Heard, Carlin Glynn, Richard Bradford; **Dir:** Peter Masterson. Academy Awards '85: Best Actress (Page). **VHS, Beta, LV, 8mm** $14.98 SUE, BTV ♫♫♫

Triple Cross Plummer is a British safe-cracker imprisoned on the channel islands at the outbreak of World War II. When the Germans move in, he offers to work for them if they set him free. The Germans buy his story and send him to England, but once there he offers his services to the British as a double agent. Based on the exploits of Eddie Chapman, adapted from his book, "The Eddie Chapman Story."
1967 91m/C FR GB Christopher Plummer, Romy Schneider, Trevor Howard, Gert Frobe, Yul Brynner, Claudine Auger, Georges Lycan, Jess Hahn, Howard Vernon; **Dir:** Terence Young. **VHS, Beta** $39.99 ACE, MOV, HHE ♫♫♫

Triple Echo A young woman falls for an army deserter who is hiding from the military police by dressing as a woman. Mistrust and deception get the best of some of the involved parties. Also known as "Soldier in Skirts."
1977 (R) 90m/C GB Glenda Jackson, Oliver Reed, Brian Deacon; **Dir:** Michael Apted. **VHS, Beta** $42.95 PGN ♫♫½

Triple Trouble/Easy Street Two of Chaplin's classic two-reel silent comedy shorts are shown on video.
1918 60m/B Charlie Chaplin, Edna Purviance, Eric Campbell. **VHS, Beta, 8mm** $24.95 VYY ♫♫½

TripleCross Three former cops strike it rich and although they love the money, they miss the game, so they get involved in a murder case. These wealthy detectives encounter high-level gangsters, a baseball scandal and a hit man in their race to solve the perfect crime.
1985 97m/C Ted Wass, Markie Post, Gary Swanson; **Dir:** David Greene. **VHS, Beta** $79.95 PSM ♫½

Tripods: The White Mountains Sci-fi thriller about the takeover of earth by alien tripods. The conquerors start controlling human minds, but not until after they are sixteen years old. Two boys seek to end the terror.
1984 150m/C John Shackley, Jim Baker, Cari Seel; **Dir:** Groham Theakston. **VHS, Beta, LV** $79.95 SVS, IME ♫½

Tripwire A vengeance-crazed government agent tracks down the terrorist who murdered his wife.
1989 (R) 120m/C Terence Knox, David Warner, Charlotte Lewis, Isabella Hoffman, Yaphet Kotto, Thomas Chong, Meg Foster; **Dir:** James Lemmo. **VHS, Beta, LV** $89.95 COL ♫½

Triumph of Sherlock Holmes Wontner and Fleming teamed for an early series of Holmesian romps. In one of their best outings, Sherlock Holmes comes out of retirement as a series of bizarre murders of Pennsylvania coal miners lure him back into action. From "The Valley of Fear" by Sir Arthur Conan Doyle.
1935 84m/B GB Arthur Wontner, Ian Fleming; **Dir:** Leslie Hiscott. **VHS, Beta** $16.95 SNC, NOS, CAB ♫♫½

Triumph of the Son of Hercules All other victories pale when compared to this, the Son of Hercules' greatest!
1963 ?m/C IT Kirk Morris, Cathia Caro. **VHS, Beta** $16.95 SNC ♫♫½

Triumph of the Spirit Gritty account of boxer Salamo Arouch's experiences in Auschwitz. The boxer became champion in matches between prisoners conducted for the amusement of Nazi officers. Filmed on location in Auschwitz.
1989 (R) 115m/C Willem Dafoe, Robert Loggia, Edward James Olmos, Wendy Gazelle; **Dir:** Robert M. Young. **VHS, Beta, LV** $89.95 COL ♫♫♫

Triumph of the Will Director Riefenstahl's formidable, stunning film, documenting Hitler and the Sixth Nazi Party Congress in 1934 in Nuernberg, Germany. The greatest and most artful propaganda piece ever produced. German dialogue. Includes English translation of the speeches.
1934 115m/B GE **Dir:** Leni Riefenstahl. **VHS, Beta, 3/4U** $24.95 NOS, SUE, VYY ♫♫♫

Troilus & Cressida Here love is mocked and heroism made absurd in a powerful play revealing the darker side of Shakespeare.
1982 190m/C Anton Lesser, Suzanne Burden. **VHS, Beta** TLF ♫½

Trojan Women Euripides' tragedy on the fate of the women of Troy after the Greeks storm the famous city. The play does not translate well to the screen, in spite of tour-de-force acting.
1971 (G) 105m/C Katharine Hepburn, Vanessa Redgrave, Irene Papas, Genevieve Bujold, Brian Blessed, Patrick Magee; **Dir:** Michael Cacoyannis. National Board of Review Awards '71: Best Actress (Papas). **VHS, Beta** $39.95 FHS ♫♫

Troll A malevolent troll haunts an apartment building and possesses a young girl in hopes of turning all humans into trolls. Sometimes imaginative, sometimes embarrassing.
1985 (PG-13) 86m/C Noah Hathaway, Gary Sandy, Anne Lockhart, Sonny Bono, Shelley Hack, June Lockhart, Michael Moriarty; **Dir:** John Carl Buechler. **VHS, Beta, LV** $79.98 LIV, VES ♫½

Troll 2 A young boy can only rely on his faith in himself when no one believes his warnings about an evil coming to destroy his family. Entering into a nightmare world, Joshua must battle witches' spells and the evil trolls who carry out their bidding.
1992 (PG-13) 95m/C Michael Stephenson, Connie McFarland; **Dir:** Drago Floyd. **VHS** $79.95 COL ♫½

Troma's War The survivors of an air crash find themselves amid a tropical terrorist-run civil war. Also available in a 105-minute, unrated version. Devotees of trash films shouldn't miss this one; Weil is an alias used by Michael Herz and Lloyd Kaufman.
1988 (R) 90m/C Ara Romanoff, Michael Ryder, Carolyn Beauchamp, Sean Bowen; **Dir:** Michael Herz, Samuel Weil. **VHS, Beta** $19.98 MED Woof!

Tromba, the Tiger Man A circus tiger tamer uses a special drug that hypnotizes the critters into obeying his every whim. When he tries the same drug on women, things get out of hand.
1952 62m/B GE Rene Deltgen, Angelika Hauff, Gustav Knuth, Hilde Weissner, Grethe Weiser, Gardy Granass, Adrian Hoven; **Dir:** Helmut Weiss. **VHS, Beta** $16.95 SNC ♫½

Tron A video game designer enters into his computer, where he battles the computer games he created and seeks revenge on other designers who have stolen his creations. Sounds better than it plays. Terrific special effects, with lots of computer-created graphics.
1982 (PG) 96m/C Jeff Bridges, Bruce Boxleitner, David Warner, Cindy Morgan, Barnard Hughes; **Dir:** Steven Lisberger. **VHS, Beta, LV** DIS, OM ♫♫

Troop Beverly Hills When spoiled housewife Long is chosen to lead her daughter's Girl Scout troop from posh Beverly Hills, everyone gets lessons deviating from the standard fare. From "camp-outs" at exclusive hotels to minks in the rain, the troop of unwilling, uncooperative little brats must eventually learn to work together to survive the wilderness and come out on top. Silly, light fun.
1989 (PG) 105m/C Shelley Long, Craig T. Nelson, Betty Thomas, Mary Gross, Stephanie Beacham; **Dir:** Jeff Kanew. **VHS, Beta, LV, 8mm** **$89.95** COL 🎞🎞½

Trop Jolie pour Etre Honette Four women sharing an apartment witness a robbery and begin to suspect their neighbor is the culprit. So they decide to steal the money from him. In French with English subtitles. English title: "Too Pretty to Be Honest."
196? 95m/C FR Jane Birkin, Serge Gainsbourg, Bernadette LaFont. **VHS, Beta** INV 🎞🎞

Tropic of Desire A gigolo is swept up into a number of meaningless affairs in Key West, Florida. Based on Darwin Porter's novel "Butterflies in Heat."
198? 87m/C Matt Collins, Eartha Kitt, Pat Carroll, Tom Ewell, Barbara Baxley. **VHS $69.95** SVS 🎞

Tropic Heat An exploitative voodoo/ancient taboo adventure with plenty to avoid.
197? 90m/C Tudi Wiggins, Christopher St. John, Greer St. John, Sabra Welles. **VHS, Beta $19.98** VDC Woof!

Tropical Snow A pair of lovers living poorly in Colombia decide to enter the drug trade in exchange for passage to the U.S. offered to them by drug kingpin Carradine. Unfortunately, they get more than they bargained for. Shot on location.
1988 (R) 87m/C David Carradine, Madeleine Stowe, Nick Corri, Argermiro Catiblanco; **Dir:** Ciro Duran. **VHS, Beta, LV $34.95** PAR 🎞

Trottie True Charming, though trivial, costume picture about a Gay '90s show girl who marries into nobility. Also known as "The Gay Lady."
1949 98m/C Jean Kent, James Donald, Hugh Sinclair, Lana Morris, Andrew Crawford; **Dir:** Brian Desmond Hurst. **VHS, Beta $19.95** UNI 🎞🎞½

Trouble Along the Way Wayne, in an uncharacteristically sentimental role, stars as a once big-time college football coach who takes a coaching job at a small, financially strapped Catholic college in an effort to retain custody of his young daughter. He uses some underhanded methods to recruit good players (with poor scholastic records), promising them they will get a share of parking fees and concessions. Once the rector of the school finds out, he fires Wayne, even though the team won its first game. Reed, the probation officer, thus won't file a favorable report for Wayne. At the custody trial, however, the rector shows up, says the whole fiasco was his fault, and offers Wayne his job back. Good family fare proves that Wayne can handle comedy as well as action.
1953 110m/B John Wayne, Donna Reed, Charles Coburn, Tom Tully, Sherry Jackson, Marie Windsor, Tom Helmore, Dabbs Greer, Leif Erickson, Douglas Spencer; **Dir:** Michael Curtiz. **VHS** WAR 🎞🎞½

The Trouble with Angels Two young girls turn a convent upside down with their endless practical jokes. Russell is everything a Mother Superior should be: understanding, wise, and beautiful.
1966 112m/C Hayley Mills, June Harding, Rosalind Russell, Gypsy Rose Lee, Binnie Barnes; **Dir:** Ida Lupino. **VHS, Beta, LV $59.95** COL 🎞🎞½

Trouble Busters Troubled locals worry about who to call when oil is discovered in their town, bringing evil profiteers with it. The Trouble Busters come into town to set things straight.
1933 51m/B Jack Hoxie, Lane Chandler; **Dir:** Lewis Collins. **VHS, Beta $19.95** NOS, DVT 🎞

The Trouble with Dick An ambitious young science fiction writer's personal troubles begin to appear in his writing. Although the box cover displays a "Festival Winner" announcement, don't be fooled! Story becomes tedious after first five minutes.
1988 (R) 86m/C Tom Villard, Susan Dey; **Dir:** Gary Walkow. **VHS, Beta $29.95** ACA 🎞

The Trouble with Girls A 1920's traveling show manager tries to solve a local murder. Not as much singing as his earlier films, but good attention is paid to details. This is definitely one of the better Elvis vehicles.
1969 99m/C Elvis Presley, Marlyn Mason, Nicole Jaffe, Sheree North, Edward Andrews, John Carradine, Vincent Price, Joyce Van Patten, Dabney Coleman, John Rubinstein, Anthony Teague; **Dir:** Peter Tewkesbury. **VHS, Beta $19.95** MGM 🎞🎞🎞

Trouble in the Glen A Scottish-American soldier returns to his ancestral home and becomes involved in a dispute between the town residents and a lord over a closed road. Uneven script undermines comic idea.
1954 91m/C Orson Welles, Victor McLaglen, Forrest Tucker, Margaret Lockwood; **Dir:** Herbert Wilcox. **VHS $19.98** REP 🎞🎞½

The Trouble with Harry When a little boy finds a dead body in a Vermont town, it causes all kinds of problems for the community. No one is sure who killed Harry and what to do with the body. MacLaine's film debut and Bernard Hermann's first musical score for Hitchcock.
1955 (PG) 90m/C John Forsythe, Shirley MacLaine, Edmund Gwenn, Jerry Mathers, Mildred Dunnock, Mildred Natwick, Royal Dano; **Dir:** Alfred Hitchcock. **VHS, Beta, LV $19.95** MCA 🎞🎞🎞

Trouble in Mind Stylized romance is set in the near future in a rundown diner. Kristofferson is an ex-cop who gets involved in the lives of a young couple looking for a better life. Look for Divine in one of her/his rare appearances outside of John Waters' works.
1986 (R) 111m/C Kris Kristofferson, Keith Carradine, Genevieve Bujold, Lori Singer, Divine, Joe Morton; **Dir:** Alan Rudolph. **VHS, Beta, LV, 8mm $19.98** CHA 🎞🎞🎞

Trouble in Paris Softcore trash about two wives going to Paris without their husbands and having sex all over the place.
1987 83m/C VHS, Beta $29.95 MED 🎞🎞🎞

The Trouble with Spies A bumbling British spy goes to Ibiza to locate a Russian agent, makes a million mistakes and wins anyway. A loser of a film, wasted a fine cast. Made in 1984; no one bothered to release it for 3 years.

1987 (PG-13) 91m/C Donald Sutherland, Malcolm Morley, Lucy Gutteridge, Ruth Gordon, Ned Beatty, Michael Hordern, Robert Morley; **Dir:** Burt Kennedy. **VHS, Beta $79.99** HBO 🎞

Trouble in Store A bumbling department store employee stumbles on a gangster's plot. Fun gags and a good cast.
1953 85m/B Margaret Rutherford, Norman Wisdom, Moira Lister, Megs Jenkins; **Dir:** John Paddy Carstairs. **VHS** REP 🎞🎞½

Trouble in Texas Outlaws go to a rodeo and try to steal the prize money. Tex wants to find out who killed his brother and in his spare time, sings and tries to stop the crooks. Notable because it is the last film Hayworth made under her original stage name (Rita Cansino).
1937 65m/B Tex Ritter, Rita Hayworth, Earl Dwire, Yakima Canutt; **Dir:** Robert N. Bradbury. **VHS, Beta $19.95** NOS, DVT, UHV 🎞🎞

Truck Stop This truck stop has all a trucker needs—including the owner, voluptuous Pamela.
19?? 80m/C Nikki Gentile, Elizabeth Turner; **Dir:** Jean Marie Pallardy. **VHS $39.95** LUN 🎞

Truck Stop Women Female truckers become involved in smuggling and prostitution at a truck stop. The mob wants a cut of the business and will do anything to get their way. Better than it sounds.
1974 (R) 88m/C Claudia Jennings, Lieux Dressler, John Martino, Dennis Fimple, Dolores Dorn, Gene Drew, Paul Carr, Jennifer Burton; **Dir:** Mark L. Lester. **VHS, Beta $59.95** VES 🎞🎞

Truck Turner He's a bounty hunter, he's black and he's up against a threadbare plot. Hayes methodically eliminates everyone involved with his partner's murder while providing groovy soundtrack. Quintessential blaxploitation.
1974 (R) 91m/C Isaac Hayes, Yaphet Kotto, Annazette Chase, Nichelle Nichols, Scatman Crothers, Dick Miller; **Dir:** Jonathan Kaplan. **VHS $59.95** ORI, FCT 🎞½

Trucker's Woman A man takes a job driving an 18-wheel truck in order to find the murderers of his father.
1983 (R) 90m/C Michael Hawkins, Mary Cannon; **Dir:** Will Zens. **VHS, Beta $42.95** PGN 🎞

True Believer Cynical lawyer, once a '60s radical, now defends rich drug dealers. Spurred on by a young protege, he takes on the hopeless case of imprisoned Asian-American accused of gang-slaying. Tense thriller with good cast.
1989 (R) 103m/C James Woods, Robert Downey Jr., Yuji Okumoto, Margaret Colin, Kurtwood Smith, Tom Bower, Miguel Fernandez, Charles Hallahan; **Dir:** Joseph Ruben. **VHS, Beta, LV $14.95** COL 🎞🎞½

True Blood A man returns to his home turf to save his brother from the same ruthless gang who set him up for a cop's murder.
1989 (R) 100m/C Chad Lowe, Jeff Fahey, Sherilyn Fenn; **Dir:** Peter Maris. **VHS, Beta, LV $29.95** FRH, NVH, IME 🎞🎞

True Colors A woman fights for her homeland of France in order to save the country from being taken over by Hitler.
1987 (PG-13) 160m/C Noni Hazelhurst, John Waters, Patrick Ryecart, Shane Briant, Alan Andrews; **Dir:** Pina Amenta. **VHS, Beta $29.95** ACA 🎞🎞½

True Colors Law school buddies take divergent paths in post grad real world. Straight and narrow Spader works for Justice Department while dropout Cusack manipulates friends and acquaintances to further his position as Senator's aid. Typecast, predictable and, moralizing.
1991 (R) 111m/C John Cusack, James Spader, Imogen Stubbs, Mandy Patinkin, Richard Widmark; ***Dir:*** Herbert Ross. **VHS, LV $19.95** *PAR, LDC* 🦴🦴

True Confessions Tale of corruption pits two brothers, one a priest and the other a detective, against each other. Nice 1940s period look, with excellent performances from De Niro and Duvall. Based on a true case from the Gregory Dunne novel, with screenplay by the author and Joan Didion.
1981 (R) 110m/C Robert De Niro, Robert Duvall, Kenneth McMillan, Charles Durning, Cyril Cusack, Ed Flanders, Burgess Meredith, Louisa Moritz; ***Dir:*** Ulu Grosbard. **VHS, Beta, LV $19.98** *MGM* 🦴🦴🦴

True Duke: A Tribute to John Wayne A collection of trailers from greater and lesser Wayne films, including "Haunted Gold," "Flying Tigers," "Fort Apache," "Red River" and "The Searchers." Partial black and white.
1987 60m/C VHS, Beta $34.95 *SFR* 🦴🦴🦴

The True Game of Death A story of the circumstances behind the death of superstar Bruce Lee.
197? 90m/C Bruce Lee, Shou Lung. **VHS, Beta $54.95** *MAV* 🦴🦴🦴

True Grit Hard-drinking U.S. Marshal Rooster Cogburn (Wayne) is hired by a young girl (Darby) to find her father's killer. Won Wayne his only Oscar. Based on the Charles Portis novel. Prompted sequel, "Rooster Cogburn."
1969 (G) 128m/C John Wayne, Glen Campbell, Kim Darby, Robert Duvall; ***Dir:*** Henry Hathaway. Academy Awards '69: Best Actor (Wayne); National Board of Review Awards '69: 10 Best Films of the Year. **VHS, Beta, LV $14.95** *PAR, HHE, TLF* 🦴🦴🦴

True Heart Susie A simple, moving story about a girl who is in love with a man who marries another girl from the city. Silent.
1919 87m/B Lillian Gish, Robert Harring; ***Dir:*** D.W. Griffith. **VHS, Beta $19.95** *NOS, VYY, DVT* 🦴🦴½

True Identity Henry, a comic superstar in England, stars as an innocent black man marked for death by the Mafia. He hides by disguising himself as white and gets a taste of life on the other side of the color line. The interesting premise was based on a classic Eddie Murphy sketch on "Saturday Night Live," but isn't treated with sufficient imagination or humor here.
1991 (R) 94m/C Lenny Henry, Frank Langella, Anne-Marie Johnson; ***Dir:*** Charles Lane. **VHS, Beta $94.95** *TOU* 🦴🦴

True Love Low-budget, savagely observed comedy follows the family and community events leading up to a Bronx Italian wedding. Authentic slice-of-life about two young people with very different ideas about what marriage and commitment mean. Acclaimed script and performances.
1989 (R) 104m/C Annabella Sciorra, Ron Eldard, Aida Turturro, Roger Rignack, Michael J. Wolfe, Star Jasper, Kelly Cinnante, Rick Shapiro, Suzanne Costallos, Vinny Pastore; ***Dir:*** Nancy Savoca. **VHS, Beta, LV $89.95** *MGM* 🦴🦴🦴

True Stories Quirky, amusing look at the eccentric denizens of a fictional, off-center Texas town celebrating its 150th anniversary. Notable are Kurtz as the Laziest Woman in America and Goodman as a blushing suitor. Directorial debut of Byrne. Co-written with playwright Beth Henley. Music by Byrne and the Talking Heads. Worth a look.
1986 (PG) 89m/C David Byrne, John Goodman, Swoosie Kurtz, Spalding Gray, Annie McEnroe, Pops Staples, Tito Larriva; ***Dir:*** David Byrne. **VHS, Beta, LV $19.98** *MVD, WAR* 🦴🦴🦴½

True West Filmed performance of the acclaimed Sam Shepard play about two mismatched brothers. Made for Television.
1986 110m/C John Malkovich, Gary Sinise; ***Dir:*** Allan Goldstein. **VHS, Beta $59.95** *ACA, VTK* 🦴🦴🦴

Truly, Madly, Deeply The recent death of her lover drives a young woman into despair and anger, until he turns up at her apartment one day. Tender and well written tale of love and the supernatural, with believable characters and plot-line. Playwright Minghella's directorial debut.
1991 (PG) 107m/C Juliet Stevenson, Alan Rickman, Bill Paterson, Michael Maloney; ***W/Dir:*** Anthony Minghella. **VHS** *TOU* 🦴🦴🦴

Truly Tasteless Jokes Video version of the popular best-selling book featuring appearances by top comedians.
1985 60m/C Marsha Warfield, Denny Johnson, Andrew Dice Clay. **VHS, Beta $59.98** *LIV, VES* 🦴🦴🦴

Truman Capote's "The Glass House" A middle-aged college professor, convicted on a manslaughter charge, is sent to a maximum security prison and must learn to deal with life inside. Based on Truman Capote story. Filmed at Utah State Prison, using many actual prisoners. Alternate title: "Truman Capote's Glass House."
1973 91m/C Vic Morrow, Clu Gulager, Billy Dee Williams, Alan Alda, Kris Tabori, Dean Jagger; ***Dir:*** Tom Gries. **VHS, Beta** *HHT, WES* 🦴🦴🦴

The Trumpet and I An eight-year-old boy inherits magical powers when he plays a "silver sounding" trumpet.
1985 53m/C VHS, Beta $29.95 *GEM* 🦴🦴½

Trumpet Kings More than twenty jazz trumpet greats are featured in this overview of the instrument's development in jazz history from Dixieland to swing to bop and the modern eras. Footage from the collection of jazz film historian David Chertok. Portions in black and white.
1985 72m/C Louis Armstrong, Roy Eldridge, Dizzy Gillespie, Miles Davis, Clark Terry, Red Nichols, Bunny Berigan, Cootie Williams; ***Hosted:*** Wynton Marsalis. **VHS, Beta, LV $29.95** *MVD, VAI, KAR* 🦴🦴½

Trumps Startling relationship between man and woman has dire consequences.
1988 (R) 90m/C Ellen Barkin, Kevin Bacon, Maria Tucci, Ron McLarty, Zvee Scooler. **VHS $29.95** *HHE* 🦴🦴

Trust A spoiled girl, tossed out by her family after becoming pregnant, forms an loving bond with a strange, possibly deranged guy from an abusive household. Similar in style and theme to Hartley's "The Unbelievable Truth" - a peculiar sardonic comedy/drama not for every taste.
1991 (R) 107m/C Adrienne Shelly, Martin Donovan, Merritt Nelson, Edie Falco, John McKay; ***W/Dir:*** Hal Hartley. Sundance Film Festival '91: Best Screenplay. **VHS, LV $89.98** *REP* 🦴🦴

Trust Me A cynical L.A. art dealer decides to promote a young artist's work and then kill him off to increase the value of his paintings. Satirical view of the Los Angeles art crowd.
1989 (R) 94m/C Adam Ant, Talia Balsam, David Packer, Barbara Bain, Joyce Van Patten, William DeAcutis; ***Dir:*** Bobby Houston. **VHS, Beta, LV $79.95** *VIR* 🦴🦴

Truth About Women An old aristocrat recounts his youthful adventures to his son-in-law. A pale British comedy of manners.
1958 107m/C *GB* Laurence Harvey, Julie Harris, Mai Zetterling, Eva Gabor, Wilfrid Hyde-White, Derek Farr; ***Dir:*** Muriel Box. **VHS, Beta $19.95** *NOS, DWE* 🦴🦴

Truth or Dare? The childhood game of truth or dare turns deadly in this violent thriller.
1986 87m/C John Brace, Mary Fanaro, Geoffrey Lewis Miller; ***Dir:*** Yale Wilson. **VHS, Beta $49.95** *PII, HHE* 🦴

Truth or Dare A quasi-concert-documentary—here is music superstar Madonna tarted up in fact, fiction and fantasy—exhibitionism to the nth power. Tacky, self-conscious and ultimately, if you are a Madonna fan, moving. On camera Madonna stings ex-boyfriend Warren Beatty, disses admirer Kevin Costner, quarrels with her father, reminisces and does sexy things with a bottle. Oh yes, she occasionally sings and dances. Both those who worship and dislike the Material Girl will find much to pick apart here. Available in Widescreen format on laserdisc. Overall, Madonna's best since "Desperately Seeking Susan."
1991 (R) 118m/C Madonna; ***Dir:*** Alek Keshishian. **VHS, LV $92.95** *LIV, FCT, PIA* 🦴🦴½

Truth or Die This trite, cliched TV-movie chronicles the life of convict Jerry Rosenberg, the first prisoner to earn a law degree from his cell. Still, an uncharacteristic role for sitcom guy Danza. Based on Rosenberg's book "Doing Life," the title under which this first aired.
1986 (PG-13) 96m/C Tony Danza, John DeVries, Lisa Langlois, Rocco Sisto; ***Dir:*** Gene Reynolds. **VHS** *VMK* 🦴🦴

Try and Get Me A small town is incited into a manhunt for kidnappers who murdered their victim. Suspenseful melodrama analyzes mob rule and the criminal mind. Also known as "Sound of Fury."
1950 91m/B Lloyd Bridges, Kathleen Ryan, Richard Carlson, Frank Lovejoy, Katherine Locke; ***Dir:*** Cy Endfield. **VHS $19.95** *REP* 🦴🦴

The Tsar's Bride A lovely young woman is chosen to be the bride of the horrible Tsar Glebov but when the Tsar's bodyguard falls for her, romantic difficulties ensue. A hybrid of the 1899 Rimsky-Korsakov opera and an 1849 play.
1966 95m/B *RU* Raisa Nedashkovskaya, Natalya Rudnaya, Otar Koberidze; ***Dir:*** Vladimir Gorikker. **VHS $49.95** *KUL, MVD* 🦴🦴🦴

Tucker and the Horse Thief A girl dresses and lives like a boy for safety's sake during the California Gold Rush.
1985 45m/C VHS, Beta $39.98 *SCL* 🦴🦴

Tucker: The Man and His Dream Portrait of Preston Tucker, entrepreneur and industrial idealist, who in 1946 tried to build the car of the future and was effectively run out of business by the powers-that-were. Ravishing, ultra-nostalgic lullaby to the Amer-

ican Dream. Watch for Jeff's dad, Lloyd, in a bit role. Musical score by Joe Jackson.
1988 (PG) 111m/C Jeff Bridges, Martin Landau, Dean Stockwell, Frederic Forrest, Mako, Joan Allen, Christian Slater, Lloyd Bridges; *Dir:* Francis Ford Coppola. **VHS, Beta, LV, 8mm $14.95** *PAR ♪♪♪*

Tuff Turf The new kid in town must adjust to a different social lifestyle and a new set of rules when his family is forced to move to a low-class section of Los Angeles. He makes enemies immediately when he courts the girlfriend of one of the local toughs. Fast-paced with a bright young cast. Music by Jim Carroll, Lene Lovich, and Southside Johnny.
1985 (R) 113m/C James Spader, Kim Richards, Paul Mones, Matt Clark, Olivia Barash, Robert Downey Jr., Catya Sassoon; *Dir:* Fritz Kiersch. **VHS, Beta, LV $9.95** *NWV, STE ♪♪*

Tulips A would-be suicide takes a contract out on himself, and then meets a woman who makes life worth living again. Together they attempt to evade the gangland hit man.
1981 (PG) 91m/C *CA* Gabe Kaplan, Bernadette Peters, Henry Gibson; *Dir:* Stan Ferris. **VHS, Beta $19.95** *SUE ♪♪½*

Tulsa High-spirited rancher's daughter begins crusade to save her father's oil empire when he's killed. Having become ruthless and determined to succeed at any cost, she eventually sees the error of her ways. Classic Hayward.
1949 96m/C Susan Hayward, Robert Preston, Chill Wills, Ed Begley Sr.; *Dir:* Stuart Heisler. **VHS, Beta $19.95** *NOS, QNE, MED ♪♪*

Tumbledown Ranch in Arizona After a rodeo accident, an unconscious college student dreams of adventure out west. Part of the "Range Busters" series.
1941 60m/B Ray Corrigan, John "Dusty" King, Max Terhune, Sheila Darcy, Marian Kerby; *Dir:* S. Roy Luby. **VHS $19.95** *NOS, DVT ♪*

Tumbleweed Trail An action-packed thundering western.
1942 57m/B William Boyd, Art Davis, Lee Powell, Jack Rockwell; *Dir:* Sam Newfield. **VHS, Beta $27.95** *VCN ♪*

Tumbleweeds William S. Hart's last western. Portrays the last great land rush in America, the opening of the Oklahoma Territory to homesteaders. Preceded by a sound prologue, made in 1939, in which Hart speaks for the only time on screen, to introduce the story. Silent, with musical score.
1925 114m/B William S. Hart, Lucien Littlefield, Barbara Bedford; *Dir:* King Baggott. **VHS, Beta, LV $44.95** *VYY, CCB, LDC ♪♪♪*

Tune in Tomorrow Lovesick young Keanu wants to woo divorced babe aunt Hershey and is counseled by wacky soap opera writer Falk. Seems even soap operas draw on real life, and his story's immortalized on the silver screen. Sometimes funny, sometimes not. Adapted from the novel "Aunt Julia and the Scriptwriter."
1990 (PG-13) 90m/C Barbara Hershey, Keanu Reeves, Peter Falk, Bill McCutcheon, Patricia Clarkson, Peter Gallagher, Dan Hedaya, Buck Henry, Hope Lange, John Larroquette, Elizabeth McGovern, Robert Sedgwick, Henry Gibson; *Dir:* Jon Amiel. **VHS, LV $92.98** *HBO, NVH ♪♪*

Tunes of Glory Guinness and Mills are wonderful in this well-made film about a brutal, sometimes lazy colonel and a disciplined and educated man moving up through the ranks of the British military. York's film debut. From the novel by James Kennaway, who adapted it for the screen. Laserdisc

version contains interview with director Neame.
1960 107m/C *GB* Alec Guinness, John Mills, Dennis Price, Kay Walsh, Susannah York; *Dir:* Ronald Neame. **VHS, Beta, LV $19.98** *SUE, VYG ♪♪♪½*

Tunes of Glory/Horse's Mouth Guinness double feature. First, "Tunes of Glory" looks at the relationship between good and bad soldiers vying for the same rank. "Horse's Mouth" takes a much lighter tone. Guinness plays a truly obsessed painter who creates his work on public property. Wildly humorous.
1960 305m/C Alec Guinness, Kay Walsh, Renee Houston, John Mills, Robert Coote; *Dir:* James Kennaway, Ronald Neame. **VHS, Beta $69.95** *CRC ♪♪♪♪*

The Tunnel Obsessive romance film with a trace of suspense. Artist sees the woman of his dreams at a showing of his work. Her marriage, and her trepidation, don't keep his obsession from consuming him.
1989 (R) 99m/C Jane Seymour, Peter Weller, Fernando Rey; *Dir:* Antonio Drove. **VHS, Beta, LV $79.98** *VES, IME ♪♪½*

Tunnelvision A spoof of television comprised of irreverent sketches. Fun to watch because of the appearances of several now popular stars.
1976 (R) 70m/C Chevy Chase, John Candy, Laraine Newman, Joe Flaherty, Howard Hesseman, Gerrit Graham, Al Franken, Tom Davis, Ron Silver; *Dir:* Neal Israel. **VHS, Beta $59.95** *HAR ♪♪*

Turf Boy Desperate for cash, a boy and his uncle try to get an old horse into prime condition for racing. A minor equine melodrama, just the right length for an old-time matinee double-feature. Also known as "Mr. Celebrity."
1942 68m/B Buzzy Henry, James Seay, Doris Day, William Halligan, Gavin Gordon; *Dir:* William Beaudine. **VHS $15.95** *NOS, LOO ♪♪*

Turk 182! The angry brother of a disabled fireman takes on City Hall in order to win back the pension that he deserves. He attacks through his graffiti art. Hutton is too heavy for the over-all comic feel of this mostly silly film.
1985 (PG-13) 96m/C Timothy Hutton, Robert Culp, Robert Urich, Kim Cattrall, Peter Boyle, Darren McGavin, Paul Sorvino; *Dir:* Bob Clark. **VHS, Beta $79.98** *FOX ♪♪*

Turkish Delights Free spirit sculptor falls in love with free spirit gal from bourgeois family and artsy soft porn ensues. That is until a brain tumor puts an end to their fun, and he's left to wallow in flashbacks and more artsy soft porn. Nominated for Best Foreign Film. Dubbed. Verhoeven later became big box office with "Total Recall."
1974 100m/C *NL Dir:* Paul Verhoeven. **VHS $39.95** *NSV ♪♪*

Turn of the Screw Supernatural powers vie with a young governess for control of the souls of the two children in her charge. Redgrave does well in this chilling film adapted from the Henry James story. Remade in 1989.
1974 120m/C Lynn Redgrave, Jasper Jacobs, Eva Griffith; *Dir:* Dan Curtis. **VHS, Beta, LV $29.95** *IVE ♪♪*

Turn of the Screw The young charges of an English nanny are sought for their souls by a dark, unimaginable evil in the mansion. Another version of the Henry James classic.
1989 60m/C Amy Irving, David Hemmings; *Dir:* Graeme Clifford. **VHS, Beta, LV $59.95** *CAN ♪♪*

Turnaround Unusual action/suspense film set in a small town. Relentless gang rampages the citizens. Finally, a man decides he's had enough, and turns on the heathens with his knowledge of magic.
1987 (R) 97m/C Doug McKeon, Tim Maier, Eddie Albert, Gayle Hunnicutt; *Dir:* Ola Solum. **VHS, Beta $79.98** *CGH ♪♪*

Turner and Hooch A dog witnesses a murder, and a fussy cop is partnered with the drooling mutt and a weak script in his search for the culprit. Drool as a joke will only go so far, but Hanks is his usual entertaining self.
1989 (PG) 99m/C Tom Hanks, Mare Winningham, Craig T. Nelson, Scott Paulin, J.C. Quinn; *Dir:* Roger Spottiswoode. **VHS, Beta, LV $19.99** *TOU ♪♪*

Turning Point A woman who gave up ballet for motherhood must come to terms as her daughter launches a ballet career and falls for the lead male dancer. The mother finds herself threatened by her daughter's affection toward an old friend who sacrificed a family life for the life of a ballerina. Baryshnikov's film debut. Melodramatic ending and problems due to ballet sequences.
1979 (PG) 119m/C Shirley MacLaine, Anne Bancroft, Tom Skerritt, Leslie Browne, Martha Scott, Marshall Thompson, Mikhail Baryshnikov; *Dir:* Herbert Ross. Golden Globe Awards '78: Best Film—Drama; National Board of Review Awards '77: Best Actress (Bancroft), Best Picture, Best Supporting Actor (Skerritt). **VHS, Beta $79.98** *FOX ♪♪½*

Turtle Diary Two lonely Londoners collaborate to free giant turtles from the city aquarium with the aid of a zoo-keeper, and the turtles' freedom somehow frees them as well. Jackson and Kingsley are stunning. Adapted by Harold Pinter from the Russell Hoban book.
1986 (PG) 90m/C *GB* Ben Kingsley, Glenda Jackson, Richard Johnson, Michael Gambon, Rosemary Leach, Jeroen Krabbe, Eleanor Bron; *Dir:* John Irvin. **VHS, Beta, LV $79.98** *LIV, VES ♪♪♪*

Turumba A quiet satire about a family of papier-mache animal makers in a Philippine village who get a gargantuan order in time for the Turumba holiday, and desperately try to fill it. In Tagalog with English subtitles.
1984 94m/C *PH Dir:* Kidlat Tahimik. **VHS, Beta $59.95** *FFM, FCT ♪♪♪*

Tusks A ruthless hunter tries to kill elephants.
1989 (R) 99m/C Andrew Stevens, John Rhys-Davies, Lucy Gutteridge, Julian Glover; *Dir:* Tara Hawkins Moore. **VHS, Beta, LV $89.98** *MAG ♪½*

Tut & Tuttle A young dabbler in magic is transported to ancient Egypt where he is able to demonstrate courage and wits against the evil Horemheb, who has kidnapped Prince Tut.
1982 97m/C Christopher Barnes, Eric Greene, Hans Conried, Vic Tayback. **VHS, Beta** *TLF ♪♪½*

The Tuttles of Tahiti The Tuttles lead a life of leisure in the South Seas. Laughton's child becomes enamoured of his rival's offspring, causing problems between the two families. More interesting than it sounds.
1942 91m/C Charles Laughton, Jon Hall, Peggy Drake, Victor Francen, Gene Reynolds, Florence Bates, Mala, Alma Ross, Curt Bois; *Dir:* Charles Vidor. **VHS, Beta, LV $19.95** *HHT, NVH ♪♪♪*

Tuxedo Warrior Set in South Africa, this film involves the adventures of a well-dressed mercenary who is caught between the police, diamond thieves and an old girlfriend's bank robbing.
1982 93m/C John Wyman, Carol Royle, Holly Palance; *Dir:* Andrew Sinclair. **VHS, Beta $19.95** *STE, NWV* ♫ ½

TV/ARM Video Magazine Eight different American punk bands are seen in action, including R. Kern and PGR.
1985 60m/C VHS, Beta $24.95 *MVD, TGV*

Tweety & Sylvester A compilation of pre-1948 Tweety and Sylvester cartoons.
1948 60m/C *Dir:* Robert McKimson, Friz Freleng, Chuck Jones. **VHS, Beta $14.95** *MGM* ♫♫♫♫

Twelfth Night With its practical jokes, lyrics, poetry, and haunting songs, "Twelfth Night" is the most subtle of Shakespeare's comedies. An aristocratic country house and its inhabitants set the stage for a portrayal of the varieties of love. Part of the television series "The Shakespeare Plays."
1980 124m/C *GB* Alec McCowen, Trevor Peacock, Felicity Kendall. **VHS, Beta, LV** *TLF* ♫♫♫

Twelve Angry Men Fonda sounds the voice of reason as a jury inclines toward a quick and dirty verdict against a boy on trial. Excellent ensemble work. Lumet's feature film debut, based on a television play by Reginald Rose. Received 3 Oscar nominations: Best Picture, Best Director, and Best Adapted Screenplay.
1957 95m/B Henry Fonda, Martin Balsam, Lee J. Cobb, E.G. Marshall, Jack Klugman, Robert Webber, Ed Begley Sr., John Fiedler, Jack Warden, George Voskovec, Edward Binns, Joseph Sweeney; *Dir:* Sidney Lumet. Berlin Film Festival '57: Catholic Film Office Award, Best Picture; British Academy Awards '57: Best Actor (Fonda); Edgar Allan Poe Awards '57: Best Screenplay; National Board of Review Awards '57: 10 Best Films of the Year. **VHS, Beta, LV $19.98** *MGM, FOX, VYG* ♫♫♫♫

Twelve Chairs Take-off on Russian folk-tale first filmed in Yugoslavia in 1927. A rich matron admits on her deathbed that she has hidden her jewels in the upholstery of one of twelve chairs that are no longer in her home. A Brooksian treasure hunt ensues.
1970 (PG) 94m/C Mel Brooks, Dom DeLuise, Frank Langella, Ron Moody, Bridget Brice; *Dir:* Mel Brooks. **VHS, Beta, LV $59.95** *MED* ♫♫♫

Twelve Months An animated tale of a young girl who learns by trial and error that kindness is always rewarded.
1985 90m/C VHS, Beta $19.95 *COL* ♫♫♫

Twelve O'Clock High Epic drama about the heroic 8th Air Force, with Peck as bomber-group commander sent to shape up a hard-luck group, forced to drive himself and his men to the breaking point. Compelling dramatization of the strain of military command. Includes impressive footage of actual WWII battles. Best-ever flying fortress movie.
1949 132m/B Gregory Peck, Hugh Marlowe, Gary Merrill, Millard Mitchell, Dean Jagger, Paul Stewart; *Dir:* Henry King. Academy Awards '49: Best Sound, Best Supporting Actor (Jagger). **VHS, Beta, LV $19.98** *FOX, FCT, BTV* ♫♫♫ ½

12 Plus 1 A man sells 13 chairs left to him by his aunt, only to find that one of them contained a hidden fortune. He chases across Europe in search of each of them. Some good cameos help move this re-make of a Russian folk-tale along, but it remains lightweight. Tate's last movie, before her

death at the hands of Manson and his "family."
1970 (R) 94m/C *FR IT* Orson Welles, Sharon Tate, Vittorio Gassman, Mylene Demongeot, Terry-Thomas, Tim Brooke, Vittorio De Sica; *Dir:* Nicolas Gessner. **VHS, Beta** *CON, LTG* ♫♫♫

Twentieth Century Maniacal Broadway director Barrymore transforms shop girl Lombard into a smashing success adored by public and press. Tired of Barrymore's manic-excessive ways, she heads for the Hollywood hills, pursued by the Profile in fine form. Scripted by Hecht and MacArthur.
1934 91m/B Carole Lombard, John Barrymore, Walter Connolly, Roscoe Karns, Edgar Kennedy, Ralph Forbes, Charles Lane, Etienne Girardot, Snow Flake; *Dir:* Howard Hawks. **Beta $19.95** *COL* ♫♫♫♫

21 Hours at Munich Made for TV treatment of the massacre of Israeli athletes by Arab terrorists at the 1972 Olympics. Well-done. The film was produced using the actual Munich locations.
1976 100m/C William Holden, Shirley Knight, William Knight, Franco Nero, Anthony Quayle, Noel Willman; *Dir:* William A. Graham. **VHS** *MOV* ♫♫♫

23 1/2 Hours Leave Amusing WWI service comedy as a barracks wiseacre bets that by the following morning he'll be breakfasting with a general. It's a song-filled remake of a 1919 film, whose star Douglas MacClean later produced this.
1937 73m/B James Ellison, Terry Walker, Morgan Hill, Arthur Lake, Paul Harvey; *Dir:* John Blystone. **VHS $15.95** *NOS, LOO* ♫♫

Twenty-Four Eyes Free-thinking teacher takes on a school of traditionally-taught students on the eve of Japan's invasion of Manchuria in the late 1920s. Much lauded in its native country; in Japanese with English subtitles.
1954 158m/B *JP* Hideko Takamine, Chishu Ryu, Toshiko Kobayashi; *Dir:* Keisuke Kinoshita. **VHS, Beta $29.95** *SVS, TAM* ♫♫♫

24 Hours to Midnight A widow uses her ninja skills to wipe out the crime bosses and henchmen who killed her husband.
1992 91m/C Cynthia Rothrock, Stack Pierce; *Dir:* Leo Fong. **VHS $89.95** *AIP* ♫ ½

24 Hours in a Woman's Life A girl in love with an irresponsible guy hears a cautionary tale from grandma Bergman, who once had a whirlwind romance with a young gambler and naively thought she could reform him. A static TV adaptation of an oft-filmed Stefan Zweig novel, unusually lavish for its time but on the dull side. Script by John Mortimer.
1961 90m/C Ingrid Bergman, Rip Torn, John Williams, Lili Darvas, John Orbach; *Dir:* Silvio Narizzano. **VHS $29.95** *FCT* ♫♫

25 Fireman's Street An evocative drama about the residents of an old house in Hungary which is about to be torn down, and their evening of remembrances of life before, during and after World War II. In Hungarian with English subtitles.
1973 97m/C Rita Bekes, Peter Muller, Lucyna Winnicka; *Dir:* Istvan Szabo. **VHS, Beta $69.95** *KIV* ♫♫♫

25x5: The Continuing Adventures of the Rolling Stones Rare video clips, interviews, and history of twenty-five years of the Stones. Ends with the "Steel Wheels" tour.

1990 130m/C Mick Jagger, Keith Richards, Bill Wyman, Charlie Watts, Brian Jones, Mick Taylor; *Dir:* Nigel Finch; *Performed by:* Rolling Stones. **VHS, Beta, LV $19.95** *MGM* ♫♫♫

The 27thth Day Temperate cold war allegory has aliens deliver five mysterious capsules to five Earthlings from different countries. Each capsule, if opened, is capable of decimating the population of the entire planet, but is rendered ineffective after 27 days or upon the death of the holder. Wild ending. Based on John Mantley novel.
1957 75m/B Gene Barry, Valerie French, George Voskovec, Arnold Moss, Stefan Schnabel, Ralph Clanton, Friedrich Ledebur, Mari Tsien. **VHS $14.98** *MOV, MLB* ♫♫ ½

29thth Street Compelling comedy-drama based on the true story of Frank Pesce, a New York actor who won six million dollars in that state's first lottery. Great performance from Aiello, as usual, but the direction falters in the dramatic sequences. Based on the book by Frank Pesce and James Franciscus.
1991 (R) 101m/C Danny Aiello, Anthony LaPaglia, Lainie Kazan, Frank Pesce, Donna Magnani, Rick Aiello; *W/Dir:* George Gallo. **VHS $94.98** *FXV, CCB* ♫♫ ½

Twenty Dollar Star Steamy thriller about an prominent actress who leads a double life as a cheap hooker, and consequent occupational hazards. The title may as well refer to the budget.
1991 (R) 92m/C Rebecca Holden, Bernie White, Eddie Barth, Marilyn Hassett, Dick Sargent; *Dir:* Paul Leder. **VHS** *PAR* ♫ ½

20 Million Miles to Earth A spaceship returning from an expedition to Venus crashes on Earth, releasing a fast-growing reptilian beast that rampages throughout Athens. Another entertaining example of stop-motion animation master Ray Harryhausen's work, offering a classic battle between the monster and an elephant.
1957 82m/B William Hopper, Joan Taylor, Frank Puglia, Joan Zaremba; *Dir:* Nathan Juran. **LV** *PIA* ♫♫ ½

20,000 Leagues Under the Sea Outstanding silent adaptation of Jules Verne's "20,000 Leagues Under the Sea" and "The Mysterious Island" filmed with a then revolutionary underwater camera. Much octopus fighting. Look for newly mastered edition.
1916 105m/B Matt Moore, Allen Holubar, June Gail, Matt Welsh, Chris Benton, Dan Hamlon; *W/Dir:* Stuart Paton. **VHS, Beta $19.98** *KIV, GPV, DVT* ♫♫♫

20,000 Leagues Under the Sea From a futuristic submarine, Captain Nemo wages war on the surface world. A shipwrecked scientist and sailor do their best to thwart Nemo's dastardly schemes. Buoyant Disney version of the Jules Verne fantasy.
1954 127m/C Kirk Douglas, James Mason, Peter Lorre, Paul Lukas, Robert J. Wilke, Carleton Young; *Dir:* Richard Fleischer. Academy Awards '54: Best Art Direction/Set Decoration (Color), Best Special Effects; National Board of Review Awards '54: 10 Best Films of the Year. **VHS, Beta, LV $19.99** *DIS* ♫♫♫ ½

Twenty-One Modern morality tale of a young English woman who moves to the U.S. to make a new start, but falls into the same old habits. Kensit is frank and charming, in spite of difficulties with her drug addict boyfriend and the illegal alien she marries. Director Boyd leaves it to the viewer to decide between right and wrong and where Kensit fits.

T

1991 (R) 92m/C Patsy Kensit, Jack Shepherd, Patrick Ryecart, Maynard Eziashi, Rufus Sewell, Sophie Thompson, Susan Woolridge; *Dir:* Don Boyd. VHS, LV $89.95 SVS ♫♫½

Twice Dead A family moves into a ramshackle mansion haunted by a stage actor's angry ghost. The ghost helps them battle some attacking delinquent boys. Uninteresting story, poorly done. Never released in theaters.
1988 94m/C Tom Breznahan, Jill Whitlow, Sam Melville, Brooke Bundy, Todd Bridges, Jonathan Chapin, Christopher Burgard; *Dir:* Bert L. Dragin. VHS, Beta, LV $19.98 SUE *Woof!*

Twice a Judas An amnesiac is swindled and has his family killed by a ruthless renegade. He wants to get even, even though he can't remember who anyone is.
196? 90m/C Klaus Kinski. VHS, Beta $49.95 UNI ♫

Twice in a Lifetime Middle-aged man takes stock of his life when he's attracted to Ann-Margret. Realizing he's married in name only, he moves in with his new love while his former wife and children struggle with shock, disbelief, and anger. Well-acted, realistic, and unsentimental.
1985 (R) 117m/C Gene Hackman, Ellen Burstyn, Amy Madigan, Ann-Margret, Brian Dennehy, Ally Sheedy; *Dir:* Bud Yorkin. VHS, Beta, LV $79.98 LIV, VES ♫♫♫

Twice-Told Tales Horror trilogy based loosely on three Nathaniel Hawthorne tales, "Dr. Heidegger's Experiment," "Rappaccini's Daughter," and "The House of Seven Gables." Price is great in all three of these well-told tales.
1963 120m/C Vincent Price, Sebastian Cabot, Brett Halsey; *Dir:* Sidney Salkow. VHS, Beta $19.95 MGM, MLB ♫♫♫

Twice a Woman A man and woman divorce, and then both fall in love with the same provocative young woman.
1979 90m/C Bibi Andersson, Anthony Perkins, Sandra Dumas. VHS, Beta $59.95 PSM ♫

Twigs A taped performance of George Furthis' Tony Award-winning drama about a mother and her three daughters who meet the day before Thanksgiving.
1984 138m/C Cloris Leachman. VHS, Beta $39.95 RKO ♫½

Twilight People When a mad scientist's creations turn on him for revenge, he runs for his life. Boring, gory, with bad make-up and even poorer acting. This one should fade away into the night.
1972 (PG) 84m/C PH John Ashley, Pat Woodell, Jan Merlin, Pam Grier; *Dir:* Eddie Romero. VHS, Beta $29.95 UHV *Woof!*

Twilight on the Rio Grande Autry runs into a female knife thrower and some jewel smugglers in this far-fetched western.
1941 54m/B Gene Autry, Sterling Holloway, Adele Mara, Bob Steele; *Dir:* Frank McDonald. VHS, Beta $27.95 VCN ♫

Twilight in the Sierras Lots of heroics and music in this action-packed western. An ex-outlaw goes straight and is kidnapped for his notorious criminal finesse. Rogers must come to rescue.
1950 67m/C Roy Rogers, Dale Evans, Estelita Rodriguez, Pat Brady, Russ Vincent, George Meeker; *Dir:* William Witney. VHS $12.98 REP ♫♫½

Twilight on the Trail Cattle rustlers disappear and Hopalong Cassidy and his sidekicks become "dude" detectives to solve the crime.
1941 54m/B William Boyd, Brad King, Andy Clyde, Jack Rockwell, Wanda McKay, Howard Bretherton. VHS, Beta GVV ♫♫

Twilight Zone: The Movie Four short horrific tales are anthologized in this film as a tribute to Rod Sterling and his popular television series. Three of the episodes, "Kick the Can," "It's a Good Life" and "Nightmare at 20,000 Feet," are based on original "Twilight Zone" scripts.
1983 (PG) 101m/C Dan Aykroyd, Albert Brooks, Vic Morrow, Kathleen Quinlan, John Lithgow, Billy Mumy, Scatman Crothers, Kevin McCarthy, Bill Quinn, Selma Diamond, Abbe Lane, John Larroquette, Jeremy Licht, Patricia Barry, William Shallert, Burgess Meredith, Cherie Currie; *Dir:* John Landis, Steven Spielberg, George Miller, Joe Dante. VHS, Beta, LV $19.98 WAR ♫♫½

Twilight's Last Gleaming A maverick general takes a SAC missile base hostage, threatening to start World War III if the U.S. government doesn't confess to its Vietnam policies and crimes. Gripping film, with good performances. Based on a novel by Walter Wager, "Viper Three."
1977 (R) 144m/C Burt Lancaster, Charles Durning, Richard Widmark, Melvyn Douglas, Joseph Cotten, Paul Winfield, Burt Young; *Dir:* Robert Aldrich. VHS, Beta, LV $59.98 FOX ♫♫½

Twin Dragon Encounter Two kick-boxing twins go on vacation with their girlfriends. When terrorized by slavering criminals, they place many an instep to many a temple.
198? (PG) 79m/C Michael McNamara; *Dir:* Paul Dunlop. VHS, Beta $79.95 VMK ♫

Twin Peaks Sketchy home video distillation of the eccentric TV series. It preserves much of the setup episode (the murder of teen-queen Laura Palmer and nest of small-town weirdness it reveals), then rushes piecemeal through the rest of the season. There's an altered quasi-solution to the enigma and a bizarre backward sequence not broadcast in the US. In this compressed form it's best for initiates of the TP cult; others won't be converted.
1990 113m/C Kyle MacLachlan, Michael Ontkean, Sherilyn Fenn, Lara Flynn Boyle, Joan Chen, Peggy Lipton, Piper Laurie, Michael Horse, Russ Tamblyn, Richard Beymer, Madchen Amick, Catherine Coulson, Warren Frost, Everett McGill, Jack Nance, Ray Wise, Eric Da Re, Harry Goaz, Sheryl Lee; *Dir:* David Lynch. VHS, LV $79.99 WAR, LDC, FCT ♫♫

Twins Black comedy pits one twin against another.
1980 50m/C Michael Smith; *Dir:* Charlie Ahearn. VHS $19.95 FCT ♫½

Twins A genetics experiment gone awry produces twins rather than a single child. One is a genetically-engineered superman, the other a short, lecherous petty criminal. Schwarzenegger learns he has a brother, and becomes determined to find him, despite their having been raised separately. The two meet and are immediately involved in a contraband scandal. Amusing pairing of Schwarzenegger and DeVito.
1988 (PG) 107m/C Arnold Schwarzenegger, Danny DeVito, Kelly Preston, Hugh O'Brian, Chloe Webb, Bonnie Bartlett; *Dir:* Ivan Reitman. VHS, Beta, LV $19.95 MCA ♫♫½

Twins of Evil Beautiful female twins fall victim to the local vampire, and their God-fearing uncle is out to save/destroy them. The Collinsons were featured in the October 1970 issue of Playboy as that magazine's first twin Playmates.
1971 86m/C GB Madeleine Collinson, Mary Collinson, Peter Cushing, Kathleen Bryon, Dennis Price, Damien Thomas, David Warbeck, Katya Wyeth, Maggie Wright, Luan Peters, Kristen Lindholm, Judy Matheson; *Dir:* John Hough. VHS, Beta $14.98 VID, MLB ♫½

Twinsanity Evil twins are charged with the murder of a man who attempted to blackmail them. Originally titled "Goodbye Gemini."
1970 (R) 91m/C GB Judy Geeson, Martin Potter, Alexis Kanner, Michael Redgrave, Freddie Jones, Peter Jeffrey; *Dir:* Alan Gibson. VHS, Beta $59.95 PSM ♫

Twirl Satirical look at the cutthroat world of baton twirling. Parents apply relentless pressure on their daughters to win a contest. Decent made for TV movie.
1981 100m/C Stella Stevens, Charles Haid, Lisa Whelchel, Erin Moran, Edd Byrnes, Sharon Spelman, Matthew Tobin, Donna McKechnie, Rosalind Chao, Heather Locklear, Debi Richter, Jamie Rose, Tracy Scoggins; *Dir:* Gus Trikonis. VHS, Beta $39.95 IVE ♫♫

Twist A tepid French drama chronicling the infidelities of various wealthy aristocrats; dubbed. The aristocrats are boring, as is the film. Also known as "Folies Bourgeoises." In French and English versions.
1976 105m/C FR Bruce Dern, Stephane Audran, Ann-Margret, Sydne Rome, Jean-Pierre Cassel, Curt Jurgens, Maria Schell, Charles Aznavour; *Dir:* Claude Chabrol. VHS, Beta $49.95 IVE ♫½

Twist & Shout Sequel to the popular Danish import "Zappa," in which two teenage lovers discover sex and rock and roll. Transcends the teens discovering sex genre. Danish with English subtitles. Profanity, nudity, and sex.
1984 107m/C DK Lars Simonsen, Adam Tonsberg, Ulrikke Juul Bondo, Camilla Soeberg; *Dir:* Bille August. VHS, Beta $79.99 HBO ♫♫♫

Twisted This curio from Slater's early career came out on tape to take advantage of his rising stardom. He's properly creepy as a brilliant but evil teen using electronics, psychology, and swordplay to terrorize victims. The suspense isn't too bad when direction doesn't go overboard, which unfortunately is half the time. Based on the play "Children! Children!" by Jack Horrigan.
1986 (R) 87m/C Lois Smith, Christian Slater, Tandy Cronyn, Brooke Tracy, Dina Merrill, Dan Ziskie, Karl Taylor, Noelle Parker, John Cunningham, J.C. Quinn; *Dir:* Adam Holender. VHS $79.95 SEL, HMD ♫♫

Twisted Brain An honor student develops a serum that makes him half man and half beast. Now he can exact revenge on all the jocks and cheerleaders who have humiliated him in the past. Cardi is okay, the rest of the cast is awful. Silly, horrific teen monster epic. Also known as "Horror High."
1974 (R) 89m/C Pat Cardi, Austin Stoker, Rosie Holotik, John Niland, Joyce Hash, Jeff Alexander, Joe "Mean Joe" Greene; *Dir:* Larry N. Stouffer. VHS, Beta $29.95 UHV ♫

Twisted Justice Set in 2020, David Heavener is a cop determined to stop a ruthless killer. Matters become complicated, however, when his gun is taken away. Now he's forced to rely on his cunning to outwit his sadistic opponent.

1989 90m/**C** David Heavener, Erik Estrada, Jim Brown, Shannon Tweed, James Van Patten, Don Stroud, Karen Black, Lori Warren; *W/Dir:* David Heavener. **VHS** *$79.95 KBE* ½

Twisted Nightmare The people responsible for a young retarded boy's death are hunted by a strange figure lurking in the shadows.
1987 (R) 95m/**C** Rhonda Gray, Cleve Hall, Brad Bartrum, Robert Padilla; *Dir:* Paul Hunt. **Beta** *TWE*

Twisted Obsession A man becomes passionately obsessed with a woman, and it leads to murder.
1990 (R) 109m/**C** Jeff Goldblum, Miranda Richardson, Anemone, Dexter Fletcher, Daniel Ceccaldi, Liza Walker, Jerome Natali, Arielle Dombasle; *Dir:* Fernando Trueba. **VHS, Beta, LV** *$89.95 IVE*

Twister An intriguing independent feature about a bizarre midwestern family, its feverish eccentricities and eventual collapse. Awaiting a cult following.
1989 (PG-13) 93m/**C** Dylan McDermott, Crispin Glover, Harry Dean Stanton, Suzy Amis, Jenny Wright, Lindsay Christman, Lois Chiles; *Dir:* Michael Almereyda. **VHS, Beta, LV** *$89.98 LIV, VES*

Twister Kicker When the woman he loves is in jeopardy, an invincible man battles the best of fighters from three continents. Despite all this action, it's pretty ho-hum.
1985 93m/**C** Wang Hoi, Yeung Oi Wa, Po Pei. **VHS** *$39.95 WES*

Twitch of the Death Nerve A bevy of vacationers are hacked up by a crazy with a sickle. A superior inspiration to the "Friday the 13th" series. Also released as "Last House on the Left Part 2," although there is no connection between the two films.
1972 87m/**C** *IT* Claudine Auger, Chris Avran; *Dir:* Mario Bava. **VHS, Beta** *$79.95 MPI*

Two Assassins in the Dark A martial arts adventure about a pair of kung-fu killers.
198? 90m/**C** Wang Tao, Chang Yi, Lung Chun-Eng. **VHS, Beta** *$57.95 UNI*

Two Can Play This is a taped production of Trevor Rhone's play about political upheaval in Jamaica in the 1970's.
1983 105m/**C** **VHS, Beta** *KET*

Two Champions of Shaolin Sure, there are two champions now, but after the battle royale, only one will be left standing!
19?? ?m/**C** **VHS** *$59.98 SOU*

Two Daughters Director Ray wrote the scripts for two stories based on the writings of Nobel Prize winner Rabindranath Tagore, one about the relationship between an orphan and a postman, the other about a young man who declines to marry the girl arranged to be his intended. English subtitles.
1961 114m/**B** *IN Dir:* Satyajit Ray, Satyajit Ray. Berlin Film Festival '61: D.O. Selznick Golden Laurel Award. **VHS, Beta** *$29.98 SUE*

Two Dragons Fight Against the Tiger Kung fu equals survival in this martial arts adventure. Several men strike a gold vein in a Chinese mine. They wind up fighting against corruption and greed.
19?? 90m/**C** Kao Tien Chu, Wang Luan Hsiung, Tse Lan, Huang I Lung. **VHS** *$49.95 OCE*

Two English Girls A pre-World War I French lad loves two English sisters, one an impassioned, reckless artist, the other a repressed spinster. Tenderly delineates the triangle's interrelating love and friendship over seven years. Based on the novel "Les Deux Anglaises et le Continent" by Henri-Pierre Roche.
1972 130m/**C** *FR* Jean-Pierre Leaud, Kika Markham, Stacey Tendeter, Sylvia Marriott, Marie Mansert, Philippe Leotard; *Dir:* Francois Truffaut. New York Film Festival '72: D.O. Selznick Golden Laurel Award. **VHS, Beta** *$39.95 HMV, FOX, APD* ½

Two Evil Eyes Horror kings Romero and Argento each co-direct an Edgar Allan Poe tale in this release, hence the title. Barbeau, the scheming younger wife of a millionaire, hypnotizes her husband with the help of a her lover in "The Facts in the Case of M. Valdemar." When hubby dies too soon to validate changes made in his will, the lovers decide to freeze him for two weeks in order that death can be recorded at the correct time. In "Black Cat" a crime photographer used to photographing gore adopts a feline friend and starts getting sick on the job.
1991 (R) 121m/**C** Adrienne Barbeau, Ramy Zada, Harvey Keitel, Madeleine Potter; *Dir:* George Romero, Dario Argento. **VHS, Beta, LV** *$92.98 FXV*

Two-Faced Woman Garbo, in her last film, attempts a ruse when she finds her husband may be interested in an old flame. Pretending to be her twin sister in an attempt to lure Douglas away from the other woman, she's fooled a little herself. Romantic comedy unfortunately miscasts Garbo as an Americanized ski bunny.
1941 90m/**B** Greta Garbo, Melvyn Douglas, Constance Bennett, Roland Young, Ruth Gordon, Robert Sterling, George Cleveland, Frances Carson; *Dir:* George Cukor. **VHS, Beta** *$19.98 MGM*

The Two Faces of Evil A family's vacation turns into a night of unbearable terror when they pick up a sinister hitchhiker.
1982 60m/**C** Anna Calder-Marshall, Gary Raynmond, Denholm Elliott, Pauline Delany, Philip Latham; *Dir:* Peter Sasdy, Alan Gibson. **VHS, Beta** *$29.95 IVE*

Two Fathers' Justice Two dads, one tough and one a wimp, seek revenge on the men who killed their kids. Made for TV.
1985 100m/**C** Robert Conrad, George Hamilton, Brooke Bundy, Catherine Corkill, Whitney Kershaw, Greg Terrell; *Dir:* Rod Holcomb. **VHS** *$19.98 MCG*

Two Fisted Justice A cowboy protects a town against a daring and deadly Indian attack. From the "Range Busters" series.
1931 45m/**B** Tom Tyler. **VHS, Beta** *$19.95 VDM*

Two-Fisted Justice The Range Busters do their part to maintain law and order in the old West.
1943 55m/**B** John "Dusty" King, David Sharpe, Max Terhune. **VHS, Beta** *$19.95 NOS, DVT*

Two by Forsyth Two short films based on stories by Frederick Forsyth, one dealing with a dying millionaire's clever efforts to prevent his fortune from being inherited by greedy relatives, the other about a mild-mannered stamp dealer who avenges himself on a nasty gossip columnist.
1986 60m/**C** Dan O'Herlihy, Cyril Cusack, Milo O'Shea, Shirley Anne Field. **VHS, Beta** *$59.95 PSM*

Two Girls & a Sailor Wartime musical revue loosely structured around a love triangle involving a sailor on leave. Vintage hokum with lots of songs. Allyson and DeHaven sing "Sweet and Lovely," Horne does "Paper Doll," Helen Forrest and the Harry James Band perform "In a moment of Madness," and Cugat and Durante do their usual specialties.
1944 124m/**B** June Allyson, Gloria de Haven, Van Johnson, Xavier Cugat, Jimmy Durante, Tom Drake, Lena Horne, Harry James, Gracie Allen, Virginia O'Brien, Jose Iturbi, Carlos Ramirez, Donald Meek, Ava Gardner, Buster Keaton; *Dir:* Richard Thorpe. **VHS, Beta, LV** *$19.95 MGM*

Two Graves to Kung-Fu A young kung-fu student is framed for murder. When the real murderers kill his teacher, he escapes from prison to seek revenge.
1982 (R) 95m/**C** Liu Chia-Yung, Shek Kin, Chen Hung-Lieh. **VHS, Beta** *$59.95 GEM*

The Two Great Cavaliers A Ming warrior has a hard time with the Manchurian army, fending it off as he must with only his bare instep, heel, and wrist.
1973 95m/**C** Chen Shing, Mao Ying; *Dir:* Yeung Ching Chen. **VHS, Beta** *$39.95 UNI*

Two Gun Man America's first singing cowboy, Ken Maynard, performs hair-raising stunts in this old oater. One of the first "talkies!"
1931 60m/**B** Ken Maynard, Lafe McKee, Charles King, Tom London, Lucille Powers; *Dir:* Phil Rosen. **VHS, Beta** *$19.95 NOS, VCN, DVT*

Two-Gun Man from Harlem Big city guy travels to the wild West in search of a little truth.
1938 60m/**B** Herbert Jeffries, Marguerite Whitten, Mantan Moreland, Matthew "Stymie" Beard; *Dir:* Richard C. Kahn. **VHS** *$24.95 NOS, TPV, BUR*

Two-Gun Troubador A masked singing cowboy struggles to uncovers his father's murderer.
1937 59m/**B** Fred Scott, Claire Rochelle, John Merton. **VHS, Beta** *$19.95 NOS, KRT*

200 Motels A rambling, non-narrative self-indulgent video album by and about Frank Zappa and the Mothers of Invention, as they document a long and especially grueling road tour. For fans only.
1971 (R) 99m/**C** Frank Zappa, Ringo Starr, The Mothers of Invention; *Dir:* Frank Zappa. **VHS, Beta** *$19.95 MVD, MGM* ½

2 Idiots in Hollywood Two unapologetically moronic dimwits make their mark on Tinseltown amid underwear jokes, exaggerated leering and contrived subplots.
1988 (R) 87m/**C** Jeff Doucette. **VHS, Beta, LV** *$19.95 STE, NWV*

The Two Jakes Ten years have passed and Jake Gittes is still a private investigator in this sequel to 1974's "Chinatown." When a murder occurs while he's digging up dirt on an affair between the wife of a real estate executive and the executive's partner, Jake must return to Chinatown to uncover the killer and face the painful memories buried there. Despite solid dialogue and effective performances, it's unreasonably difficult to follow if you haven't seen "Chinatown." Outstanding photography by Vilmos Zsigmond.
1990 (R) 137m/**C** Jack Nicholson, Harvey Keitel, Meg Tilly, Madeleine Stowe, Eli Wallach, Ruben Blades, Frederic Forrest, David Keith, Richard Farnsworth, Tracy Walter; *Dir:* Jack Nicholson,

Vilmos Zsigmond. **VHS, Beta, LV $19.95** *PAR, FCT* 🦴🦴

Two of a Kind In his television movie debut, Burns plays an elderly man whose mentally handicapped grandson helps him put the starch back in his shirt. Sensitively produced and performed. Made for television.
1982 102m/C George Burns, Robby Benson, Cliff Robertson, Barbara Barrie, Frances Lee McCain, Geri Jewell, Ronny Cox; ***Dir:*** Roger Young. **VHS, Beta $49.95** *FOX, USA, IVE* 🦴🦴½

Two of a Kind Angels make a bet with God—two selfish people will redeem themselves or God can blow up the Earth. Bad acting, awful direction, and the script needs divine intervention.
1983 (PG) 88m/C John Travolta, Olivia Newton-John, Charles Durning, Beatrice Straight, Scatman Crothers, Oliver Reed; ***Dir:*** John Herzfeld. **VHS, Beta, LV $19.98** *FOX* 🦴

Two Kinds of Love Adapted from Peggy Mann's novel "There Are Two Kinds of Terrible" this is the story of a young boy who learns to love his distant, workaholic father after his protective mother suddenly dies. Made for television.
1985 94m/C Rick Schroder, Lindsay Wagner, Peter Weller; ***Dir:*** Jack Bender. **VHS, Beta $59.98** *FOX* 🦴🦴

Two Lost Worlds A young hero battles monstrous dinosaurs, pirates, and more in this cheapy when he and his shipmates are shipwrecked on an uncharted island. Don't miss the footage from "Captain Fury," "One Billion B.C. ," and "Captain Caution" and Arness long before his Sheriff Dillon fame, and his "big" role in "The Thing."
1950 63m/B James Arness, Laura Elliott, Bill Kennedy; ***Dir:*** Norman Dawn. **VHS, Beta $39.95** *SVS Woof!*

Two Men & a Wardrobe Polanski's project while a student at the Polish Film Institute. Consideration of modern man's lack of privacy and made without dialogue. Two men emerge from the sea carrying nothing but a large piece of furniture. Absurd and funny.
1958 19m/B *Dir:* Roman Polanski. **VHS, Beta** *IHF, TEX Woof!*

Two Minute Warning A sniper plans to take out the president of the United States at an NFL playoff game in this boring, pointless, disaster film that goes on forever. Features all the cookie cutter characters inherent in the genre, but, until its too late, precious little of the mayhem. To get caught watching the TV version, which features more characters and an additional subplot, would be truly disastrous.
1976 (R) 116m/C Charlton Heston, John Cassavetes, Martin Balsam, Beau Bridges, Marilyn Hassett, David Janssen, Jack Klugman, Walter Pidgeon, Gena Rowlands; ***Dir:*** Larry Peerce. **VHS, LV $89.95** *MCA* 🦴

Two Moon Junction A soon-to-be-wed Southern debutante enters into a wild love affair with a rough-edged carnival worker. Poorly acted, wildly directed, with many unintentional laughs.
1988 (R) 104m/C Sherilyn Fenn, Richard Tyson, Louise Fletcher, Burl Ives, Kristy McNichol, Millie Perkins, Don Galloway, Herve Villechaize, Dabbs Greer, Jay Hawkins; ***Dir:*** Zalman King. **VHS, Beta, LV $14.95** *COL* 🦴½

The Two Mrs. Carrolls Bogart is a psycho-killer/artist who paints portraits of his wives as the Angel of Death—and then kills them. The married Bogart falls for Stanwyck, poisons his current wife, and the two marry. After a few years, Bogart falls for Smith and decides to rid himself of Stanwyck in the same manner as his first killing. Stanwyck, however, becomes increasingly suspicious and calls on an old beau for help. Melodrama cast Bogart against type (not always successfully). Filmed in 1945, but was unreleased until 1947.
1947 99m/B Humphrey Bogart, Barbara Stanwyck, Alexis Smith, Nigel Bruce, Pat O'Moore, Ann Carter; ***Dir:*** Peter Godfrey. **VHS $19.98** *MGM* 🦴🦴½

Two Mules for Sister Sara American mercenary in 19th century Mexico gets mixed up with a cigar-smoking nun and the two make plans to capture a French garrison. MacLaine and Eastwood are great together. Based on a Boetticher story, scripted by Maltz. Filmed by Gabriel Figueroa.
1970 (PG) 105m/C Clint Eastwood, Shirley MacLaine; ***Dir:*** Don Siegel. **VHS, Beta $19.95** *MCA* 🦴🦴🦴

Two Nights with Cleopatra A piece of Italian pizza involving Cleopatra's double (also played by Loren) falling in love with one of the guards. Notable only for the 19-year-old Sophia's brief nudity.
1954 77m/C *IT* Sophia Loren, Ettore Manni, Alberto Sordi, Paul Muller, Alberto Talegalli, Rolf Tasna, Gianni Cavalieri, Fernando Bruno, Riccardo Garrone, Carlo Dale; ***Dir:*** Mario Mattoli. **VHS, Beta $29.95** *VDM* 🦴½

Two Reelers: Comedy Classics 1 Includes a 1944 Edgar Kennedy short, "Feather Your Nest" (RKO), and two 1933 films, "How Comedies Are Born" and "Dog Blight."
1933 54m/B Edgar Kennedy, Harry Gribbon, Harry Sweet, Jack Norton, Maxine Jennings, Willie Best. **VHS, Beta $24.95** *VYY* 🦴½

Two Reelers: Comedy Classics 2 Three vintage shorts: "Chicken Feed" (1939) with Billy Gilbert, "Twin Husbands" (1946) with Leon Errol, and Edgar Kennedy in "False Roomers" (1938).
194? 53m/B Leon Errol, Dorothy Granger, Billy Gilbert, Edgar Kennedy. **VHS, Beta $24.95** *VYY* 🦴½

Two Reelers: Comedy Classics 3 More laughs, with two Edgar Kennedy shorts, "A Merchant of Menace" (1933), and "Social Terrors" (1946). Also, Jed Prouty is featured in "Coat Tales" (193).
194? 52m/B Edgar Kennedy, Jed Prouty. **VHS, Beta $24.95** *VYY* 🦴½

Two Reelers: Comedy Classics 4 "Bridal Bail" (1934), "No More West" (Educational-1934) with Bert Lahr singing and clowning, and "Bad Medicine" (1936) with crooner Gene Austin comprise this comedy package.
1936 56m/B Bert Lahr, Gene Austin, June Brewster, Carol Tevis, Grady Sutton. **VHS, Beta $24.95** *VYY* 🦴½

Two Reelers: Comedy Classics 5 A compilation of three vintage comedy shorts by popular stars of the 1930's: "Dear Deer" (1942) with Leon Errol, "Hotel Anchovy" (1934) with the Ritz Brothers and "Love Your Neighbor" (1930) with Charlotte Greenwood.
1938 55m/B Leon Errol, The Ritz Brothers, Charlotte Greenwood. **VHS, Beta $24.95** *VYY* 🦴🦴½

Two Reelers: Comedy Classics 6 Three vintage comedy shorts: "LePutt, The Specialist," "Dumb Luck," and "Our Gang Follies of 1938."
1935 59m/B Charles Sale, Goodman Ace. **VHS, Beta $24.95** *VYY* 🦴🦴½

Two Reelers: Comedy Classics 7 Three more comedy shorts from the '30s: "Dumb Dicks," "Love and Onions," and "Julius Sizzler."
1938 56m/B Benny Rubin, Harry Gribbon, Johnny Johnson, Pat Rooney Jr. **VHS, Beta $24.95** *VYY* 🦴🦴½

Two Reelers: Comedy Classics 8 Three comedy shorts from the zany 30's.
1933 57m/B Billy Bevan, Tom Patricola, Charles Judels, Dorothy Sebastian, William Farnum, Mack Swain. **VHS, Beta $24.95** *VYY* 🦴🦴½

Two Reelers: Comedy Classics 9 Three early comedy shorts featuring several stars-to-be. Includes "A Slip at the Switch" (directed by Mark Sandrich), "No Sleep on the Deep," and "Going Spanish."
1934 57m/B Bob Hope, Leah Ray, Betty Compson, Robert Warwick, Chic Sale, George Hay. **VHS, Beta, 8mm $24.95** *VYY* 🦴🦴½

Two for the Road After over a decade of marriage, Finney and Hepburn look back and find only fragments of their relationship. Mancini score adds poignancy to the couple's reflections on their stormy life. Touching and well-acted. Scipted by Frederic Raphael.
1967 112m/C Audrey Hepburn, Albert Finney, Eleanor Bron, William Daniels, Claude Dauphin, Nadia Gray, Jacqueline Bisset; ***Dir:*** Stanley Donen. **VHS, LV** *TCF, OM* 🦴🦴🦴½

Two Rode Together A Texas marshal and an army lieutenant negotiate with the Comanches for the return of captives, but complications ensue.
1961 109m/C James Stewart, Richard Widmark, Shirley Jones, Linda Cristal, Andy Devine, John McIntire; ***Dir:*** John Ford. **VHS, Beta, LV $14.95** *RCA, COL* 🦴🦴½

The Two Sisters Goldstein, in her only film role, plays an older sister who sacrifices everything for the younger. The younger thanks her by stealing her fiance. In Yiddish with English subtitles.
1937 70m/B Jenney Goldstein, Michael Rosenberg. **VHS $89.95** *ALD* 🦴🦴½

2 Sound Clash Dance Hall Features some top reggae performers, including Redman, Robert, Bailey, Teppa Lee, Squiddly Ranks and more.
1988 60m/C Admiral Bailey, Rappa Robert, Tippa Lee, John Wayne, Squidley. **VHS, Beta $19.95** *MVD, KRV* 🦴🦴½

Two to Tango A hired assassin goes to Buenos Aires to kill a crime boss, and falls in love with his tango-dancing mistress. Based on the novel by J.P. Feinman.
1988 (R) 87m/C *AR* Don Stroud, Adrienne Sachs; ***Dir:*** Hector Olivera. **VHS, Beta $79.95** *VIR* 🦴½

2000 Maniacs One of cult director Lewis' most enjoyably watchable films. A literal Civil War "ghost town" takes its revenge 100 years after being slaughtered by renegade Union soldiers by luring unwitting "yankee" tourists to their centennial festival. The hapless Notherners are then chopped, crushed, ripped apart etc. while the ghostly rebels party. Quite fun in a cartoonishly gruesome sort of way. Filmed in St. Cloud, FL.

1964 75m/C Thomas Wood, Connie Mason, Jeffrey Allen, Ben Moore, Gary Bakeman, Jerome Eden, Shelby Livingston, Michael Korb, Yvonne Gilbert, Mark Douglas, Linda Cochran, Vincent Santo, Andy Wilson; *W/Dir:* Herschell Gordon Lewis. VHS, Beta $29.95 *RHI* ♫♫½

2000 Year Old Man An animated version of the Brooks-Reiner classic comedy routine, featuring the reminiscences of the bemused patriarch of the past, the 2000 Year Old Man.
1982 (G) 25m/C *Voices:* Carl Reiner, Mel Brooks. VHS, Beta $19.95 *MED* ♫♫½

2001: A Space Odyssey A space voyage to Jupiter turns into chaos when a computer, HAL 9000, takes over. Seen by some as a mirror of man's historical use of machinery and by others as a grim vision of the future, the special effects and music are still stunning. Critically acclaimed and well accepted by the public. Screenplay written by Arthur C. Clarke. Followed by a sequel "2010: The Year We Make Contact." Oscar nominations for director Kubrick, Original Screenplay, Art Direction, and Set Decoration. The laser disc edition is presented in letterbox format and features a special supplementary section on the making of "2001," a montage of images from the film, production documents, memos and photos. Also included on the disc is a NASA film entitled "Art and Reality," which offers footage from the Voyager I and II flybys of Jupiter.
1968 139m/C *GB* Keir Dullea, Gary Lockwood, William Sylvester, Dan Richter; *Dir:* Stanley Kubrick; *Voices:* Douglas Raines. Academy Awards '68: Best Visual Effects; British Academy Awards '68: Best Art Direction/Set Decoration, Best Cinematography, Best Film Sound; National Board of Review Awards '68: 10 Best Films of the Year; American Film Institute's Survey '77: 9th Best American Film Ever Made. VHS, Beta, LV $29.95 *MGM, LDC, VYG* ♫♫♫♫

2010: The Year We Make Contact The film version of Arthur C. Clarke's novel, "The Sentinel" continues in this sequel to "2001: A Space Odyssey." Americans and Russians unite to investigate the abandoned starship Discovery's decaying orbit around Jupiter and try to determine why the HAL 9000 computer sabotaged its mission years before, while signs of cosmic change are detected on and around the giant planet.
1984 (PG) 116m/C Roy Scheider, John Lithgow, Helen Mirren, Bob Balaban, Keir Dullea, Madolyn Smith, Mary Jo Deschanel; *Dir:* Peter Hyams; *Voices:* Douglas Raines. VHS, Beta, LV $19.95 *MGM* ♫♫♫

2020 Texas Gladiators In the post-nuclear holocaust world, two groups, one good, one evil, battle for supremacy.
1985 (R) 91m/C *IT* Harrison Muller, Al Cliver, Daniel Stephen, Peter Hooten, Al Yamanouchi, Sabrina Santi; *Dir:* Kevin Mancuso. VHS, Beta $59.95 *MED* Woof!

2069: A Sex Odyssey A team of beautiful, sensuous astronauts are sent to Earth to obtain male sperm which they must bring back to Venus. Tongue-in-cheek soft core sci-fi spoof.
1978 (R) 73m/C Alena Penz, Nina Fredric, Gerti Sneider, Raul Retzer, Catherine Conti, Heidi Hammer, Michael Mein, Herb Heesel; *Dir:* George Keil. VHS, Beta $19.95 *ACA, HHE* ♫

Two Tickets to Broadway A small-town singer and a crooner arrange a hoax to get themselves on Bob Crosby's television show. Appealing but lightweight. Full of

songs by Jule Stein and Leo Robin, including "Let the Worry Bird Worry for You."
1951 106m/C Tony Martin, Janet Leigh, Gloria de Haven; *Dir:* James V. Kern; *Performed by:* Smith & Dale. VHS, Beta *KOV* ♫♫

Two Top Bananas A Las Vegas-style burlesque show, featuring song, dance, sketch comedy, and impersonations.
1987 45m/C Don Rickles, Don Adams, Carol Wayne, Sandy O'Hara, Murray Langston, Dorit Stevens. VHS, Beta $11.95 *PSM* ♫♫

Two of Us Young Jewish boy flees Nazi-occupied Paris to live in the country with an irritable, bigoted guardian. Sensitive, eloquent movie about racial prejudice and anti-Semitism. In French with English subtitles.
1968 86m/B *FR* Michel Simon, Alain Cohen, Luce Fabiole, Roger Carel, Paul Preboist, Charles Denner; *Dir:* Claude Berri. VHS $59.95 *CCI* ♫♫♫½

Two-Way Stretch Three prison inmates in a progressive jail plan to break out, pull a diamond heist, and break back in, all in the same night. Fast-paced slapstick farce.
1960 84m/B *GB* Peter Sellers, Wilfrid Hyde-White, Liz Fraser, David Lodge; *Dir:* Robert Day. VHS, Beta $19.98 *HBO* ♫♫½

Two Weeks in Another Town Douglas and Robinson are a couple of Hollywood has-beens who set out to make a comeback picture but meet with adversity at every turn. Extremely sappy melodrama is based on Irwin Shaw's trashy novel and represents one of director Minelli's poorer efforts. Laserdisc features a letterboxed screen and the original movie trailer.
1962 107m/C Kirk Douglas, Edward G. Robinson, Cyd Charisse, George Hamilton, Daliah Lavi, Claire Trevor, James Gregory, Rosanna Schiaffino, George Macready, Stefan Schnabel, Vito Scotti, Leslie Uggams; *Dir:* Vincente Minnelli. VHS, LV *MGM, PIA* ♫♫

Two Weeks to Live The comedy team inherits a railroad line, only to find it's a pile of junk.
1943 60m/B Lum & Abner, Franklin Pangborn. VHS, Beta $24.95 *NOS, DVT* ♫½

Two Weeks with Love Reynolds and family wear funny bathing suits in the Catskills in the early 1900s while the Debster sings songs and blushes into young adulthood. Songs include the ever-hummable "By the Light of the Silvery Moon."
1950 92m/C Debbie Reynolds, Jane Powell, Ricardo Montalban, Louis Calhern, Ann Harding, Phyllis Kirk, Carleton Carpenter, Clinton Sundberg, Gary Gray; *Dir:* Roy Rowland. VHS $19.98 *MGM, FCT* ♫♫

Two Women A mother and her 13-year-old daughter travel war-torn Italy during World War II and must survive lack of food, crazed monks and brutal soldiers. Tragic, moving, well-directed. Loren received well-deserved Oscar. Available dubbed or with English subtitles.
1961 99m/B *IT* Sophia Loren, Raf Vallone, Eleonora Brown, Jean-Paul Belmondo; *Dir:* Vittorio DeSica. Academy Awards '61: Best Actress (Loren); British Academy Awards '61: Best Actress (Loren); Cannes Film Festival '61: Best Actress (Loren); National Board of Review Awards '61: 5 Best Foreign Films of the Year; New York Film Critics Awards '61: Best Actress (Loren). VHS, Beta $19.98 *SUE, APD, BTV* ♫♫♫♫

Two Women in Gold Two sexy housewives seduce everyone around, until a bird salesman has a coronary in their living room. A laugh riot.

1970 (R) 90m/C Louise Turcot, Monique Mercure, Donald Pilon, Marcel Sabourin. VHS, Beta $59.95 *NWV* Woof!

Two Wondrous Tigers An awesome Kung-fu gang terrorizes a small town until a local dares to out-chop the bullies.
1979 87m/C John Chang; *Dir:* Wilson Tong. VHS, Beta $39.95 *TWE* ♫

The Two Worlds of Jenny Logan A woman travels back and forth in time whenever she dons a 19th century dress she finds in her old house. She is also able to fall in love twice, in different centuries. Made for television and adapted from "Second Sight," a novel by David Williams. Stylish and well-made.
1979 97m/C Lindsay Wagner, Marc Singer, Alan Feinstein, Linda Gray, Constance McCashin, Henry Wilcoxon, Irene Tedrow, Joan Darling, Allen Williams, Pat Corley, Gloria Stuart; *Dir:* Frank de Felitta. VHS, Beta $14.95 *FRH, IVE* ♫♫½

Two Wrongs Make a Right A quiet nightclub owner beats and clubs his way through his tough neighborhood to protect himself and his woman.
1989 85m/C Ivan Rogers, Ron Blackstone, Rich Komenich; *Dir:* Robert Brown. VHS, Beta $79.95 *UNI* ♫

Tycoon Wayne goes to Latin America to build a road for an American industrialist. When the industrialist insists on a shorter but more dangerous route, Wayne must satisfy his own sense of honor. Meanwhile, he's found romance with the industrialist's half-Spanish daughter. Long, but well-acted.
1947 129m/C John Wayne, Laraine Day, Cedric Hardwicke, Judith Anderson, James Gleason, Anthony Quinn; *Dir:* Richard Wallace. VHS, Beta, LV $19.95 *CCB, RKO, KOV* ♫♫½

Typhoon Treasure Whilst recovering his sunken treasure, the hero battles bad guys, savage natives and a crocodile, all of whom could probably have made a more sophisticated film.
1939 68m/B *AU* Campbell Copelin, Gwen Munro, Joe Valli, Douglas Herald; *Dir:* Noel Monkman. VHS $15.95 *NOS, LOO* ♫

Tyrant of Lydia Against the Son of Hercules Hercules' son takes on a wicked king in this startling adventure.
1963 ?m/C *IT* Gordon Scott, Massimo Serato. VHS, Beta $16.95 *SNC* ♫

Ub Iwerks Cartoonfest There are seven delightful cartoons by one of the pioneering geniuses of animation. Includes: "The Brave Tin Soldier," "Happy Days," "Fiddlesticks," "Jack and the Beanstalk," "The Headless Horseman," and "The Little Red Hen."
193? 57m/C VHS, Beta $19.98 *CCB* ♫

Ub Iwerks Cartoonfest 2 Here are more cartoons from the pen of the immortal Iwerks. Six color masterpieces of such great tales as "Tom Thumb," "Jack Frost," "Aladdin and the Wonderful Lamp," "Ali Baba," "Sinbad," and "Spooks."
193? 46m/C VHS, Beta $19.98 *CCB* ♫

Ub Iwerks Cartoonfest 3 A compilation of Iwerks classics: "Simple Simon" (1935), "Puss in Boots" (1934), "Dick Whittington's Cat" (1936), and "Don Quixote" (1934).
193? 30m/C VHS, Beta $19.98 *CCB* ♫

Ub Iwerks Cartoonfest 4 Another collection of enjoyable cartoons from the pen of Ub Iwerks. Included are ''The Valiant Tailor,'' ''Mary's Little Lamb,'' ''The Bremen-town Musicians,'' and ''Balloonland,'' all from 1934-35.
1935 30m/C VHS, Beta $19.98 *CCB* 🐾🐾½

Ub Iwerks Cartoonfest 5 Three more fabulous Cinecolor cartoons from the Ub Iwerks studio are combined on this tape: ''Queen of Hearts'' (1934), ''Old Mother Hubbard'' (1935) and ''Humpty Dumpty'' (1935), all from the Comicolor Cartoons series.
193? 23m/C VHS, Beta $19.98 *CCB* 🐾🐾½

Ubu and the Great Gidouille Animation from one of the world's most exciting animators, based on the play by Alfred Jarry. Ubu captures the Polish throne and spreads chaos across the land. When the people revolt, he flees to Paris and successfully spreads his message around the world. Strong black humor. In French with English subtitles.
1979 80m/C *FR Dir:* Jan Lenica. **VHS $49.95** *FCT* 🐾🐾½

L'Udienza A bizarre black comedy about a guy who does anything he can, at any criminal cost, to obtain an audience with the Pope. In Italian with subtitles.
1971 111m/C *IT* Claudia Cardinale, Ugo Tognazzi; *Dir:* Marco Ferreri. **VHS, Beta $49.95** *FCT* 🐾🐾½

U.F.O. Live The Misdemeanor tour of the famous music group is taped live, including ''Wreckless,'' ''Meanstreets,'' and ''The Chase.''
1986 60m/C VHS, Beta $19.95 *MVD, SUE* 🐾🐾½

UFO: Target Earth Scientist attempts to fish flying saucer out of lake and costs studio $70,000.
1974 (G) 80m/C Nick Plakias, Cynthia Cline, Phil Erickson; *Dir:* Michael de Gaetano. **VHS $14.98** *MOV* 🐾

Uforia Two rival evangelists meet up with a UFO-infatuated girl. Under her guidance, they wait for UFO encounters they intend to use to milk their revivalist audiences. Sometimes clumsy, but more often fun.
1981 92m/C Cindy Williams, Harry Dean Stanton, Fred Ward; *Dir:* John Binder. **VHS, Beta $59.95** *MCA* 🐾🐾

Ugetsu The classic film that established Mizoguchi's reputation outside of Japan. Two 16th-century Japanese peasants venture from their homes in pursuit of dreams, and encounter little more than their own hapless folly and a bit of the supernatural. A wonderful mix of comedy and action with nifty camera movement. Complete title is ''Ugetsu Monogatari.'' Based on the stories of Akinara Ueda. In Japanese with English subtitles.
1953 96m/B *JP* Machiko Kyo, Masayuki Mori, Kinuyo Tanaka, Sakae Ozawa; *Dir:* Kenji Mizoguchi. National Board of Review Awards '54: Special Citation (Kyo); Venice Film Festival '53: Silver Prize; Sight & Sound Survey '52: #5 of the Best Films of All Time; Sight & Sound Survey '72: #10 of the Best Films of All Time. **VHS, Beta, 8mm $29.98** *NOS, SUE, WFV* 🐾🐾🐾½

The Ugly American A naive American ambassador to a small, civil-war-torn Asian country fights a miniature Cold War against northern communist influence. Too preachy, and the ''Red Menace'' aspects are now very dated, but Brando's performance is worth

watching. Based on the William J. Lederer novel.
1963 120m/C Marlon Brando, Sandra Church, Eiji Okada, Pat Hingle, Arthur Hill; *Dir:* George Englund. **VHS, Beta $19.95** *MCA* 🐾🐾½

The Ugly Dachshund Jones and Pleshette are married dog lovers who raise Dachshunds. When Ruggles convinces them to take a Great Dane puppy, the fun begins! The Great Dane thinks he is a Dachshund because he has been raised with them—just imagine what happens when such a large dog acts as if he is small! Kids will love this wacky Disney film.
1965 93m/C Dean Jones, Suzanne Pleshette, Charlie Ruggles, Kelly Thordsen, Parley Baer; *Dir:* Norman Tokar. **VHS, Beta** *DIS, OM* 🐾🐾½

Ugly Little Boy A nurse is placed in charge of a child brought back through time from the Neanderthal age. Problems occur when the scientists involved ignore the fact that he is human.
1977 26m/C Kate Reid; *Dir:* Barry Morse, Don Thompson. **VHS, Beta** *LCA* 🐾½

UHF A loser is appointed manager of a bargain-basement UHF television station. He turns it around via bizarre programming ideas. Some fun parodies of TV enhance this minimal story. Developed solely as a vehicle for Yankovic.
1989 (PG-13) 97m/C Weird Al Yankovic, Kevin McCarthy, Victoria Jackson, Michael Richards, David Bowie, Anthony Geary; *Dir:* Jay Levey. **VHS, Beta, LV $19.98** *ORI* 🐾🐾

Ultimate Desires When a high class hooker is brutally killed, beautiful Scoggins puts on what there is of the victim's clothes in order to solve the mystery. But trouble lies in wait as she actually begins to enjoy her funky new double life. Can she solve the crime before becoming lost in her own hot fantasy world?
1991 (R) 93m/C Tracy Scoggins, Brion James, Marc Singer, Marc Baur, Marc Bennett, Suzy Joachim, Jason Scott, Frank Wilson; *Dir:* Lloyd A. Simandl. **VHS, Beta $89.95** *PSM* 🐾🐾

The Ultimate Imposter A secret agent acquires voluminous knowledge through a computer brain linkup, but can only retain the knowledge for 72 hours. In that timespan he has to rescue a defecting Russian from hordes of assassins. Made for television.
1979 97m/C Keith Andes, Erin Gray, Joseph Hacker, Macon McCalman; *Dir:* Paul Stanley. **VHS, Beta $39.95** *MCA* 🐾

The Ultimate Ninja Two ninjas get together and batter the evil enemy with foot and sword.
1985 92m/C Stuart Smith, Bruce Baron, Sorapong Chatri; *Dir:* Godfrey Ho. **VHS, Beta $59.95** *TWE* 🐾

The Ultimate Ninja Challenge A police officer, disheartened by the corruption of his fellow officers, is alone in his integrity, refusing to accept the ''fringe benefits'' of drug-busting. To combat the evil, he forms a sort of vigilante group of wrongly accused illegal aliens.
1989 90m/C VHS, Beta $69.95 *NVH* 🐾

Ultimate Sampler Video Over 50 gorgeous women dance erotically.
1989 90m/C *Hosted:* Ginger Lynn. **VHS, Beta $24.95** *CEL* 🐾

The Ultimate Stuntman: A Tribute to Dar Robinson Details the life of Robinson, who had never broken a bone in 19 years of stunts, and ironically lost his life in a freak car accident; presented with typical Hollywood hoopla and ringmastered by Chuck Norris.
1990 60m/C Chuck Norris. **VHS, Beta $19.98** *MPI* 🐾

The Ultimate Teacher Japanese animated adventure in which a new teacher is assigned to the most violent school in the world. This dude will do anything, including insulting the underwear of the students, to keep law and order.
1992 60m/C *JP* **VHS** *CPM* 🐾

The Ultimate Thrill A businessman's paranoia about his wife's affairs leads him to follow her to Colorado on a ski holiday and stalk her lovers. Well-filmed skiing scenes make this slightly out of the ordinary. Also known as ''The Ultimate Chase.''
1974 (PG) 84m/C Britt Ekland, Barry Brown, Michael Blodgett, John Davis Chandler, Eric Braeden; *Dir:* Robert Butler. **VHS, Beta $19.95** *GEM, WES* 🐾½

The Ultimate Warrior Yul Brynner must defend the plants and seeds of a pioneer scientist to help replenish the world's food supply in this thriller set in 2012.
1975 (R) 92m/C Yul Brynner, Max von Sydow, Joanna Miles, Richard Kelton, Lane Bradbury, William Smith; *Dir:* Robert Clouse. **VHS, Beta $59.95** *WAR* 🐾🐾½

Ultra Flash A fantasy that is performed to some of today's hottest dance music. In stereo
1983 60m/C VHS, Beta $39.98 *LIV, VCO* 🐾🐾½

Ulysses An Italian-made version of the epic poem by Homer, as the warrior returns to his homeland and the ever-faithful Penelope after the Trojan war. Ambitious effort provides for the ultimate mixed review. Douglas is good in the role and his stops along the ten-year way are well visualized. Sometimes sluggish with poor dubbing. Seven writers helped bring Homer to the big screen.
1955 104m/C *IT* Kirk Douglas, Silvana Mangano, Anthony Quinn, Rossana Podesta; *Dir:* Mario Camerini. **VHS, Beta $39.98** *WAR, OM* 🐾🐾½

Ulysses James Joyce's probably unfilmable novel was given a noble effort in this flawed film covering a day in the life of Leopold Bloom as he wanders through Dublin. Shot in Ireland with a primarily Irish cast.
1967 140m/B Milo O'Shea, Maurice Roeves, T.P. McKenna, Martin Dempsey, Sheila O'Sullivan, Barbara Jefford; *Dir:* Joseph Strick. **VHS, LV $29.95** *MFV* 🐾🐾½

Ulzana's Raid An aging scout and an idealistic Cavalry lieutenant lock horns on their way to battling a vicious Apache chieftain. A violent, gritty western that enjoyed critical re-evaluation years after its first release.
1972 (R) 103m/C Burt Lancaster, Bruce Davison, Richard Jaeckel, Lloyd Bochner, Jorge Luke; *Dir:* Robert Aldrich. **VHS, Beta, LV $19.95** *MCA* 🐾🐾🐾

Umberto D A government pensioner, living alone with his beloved dog, struggles to keep up a semblance of dignity on his inadequate pension. DeSica considered this his masterpiece. A sincere, tender treatment of the struggles involved with the inevitability of aging. In Italian with English subtitles. Laser edition features letterboxed print.

1955 89m/B *IT* Carlo Battista, Maria Pia Casilio, Lina Gennari; *Dir:* Vittorio DeSica. New York Film Critics Awards '55: Best Foreign Film (co-winner). **VHS, Beta, LV, 8mm $29.98** *NOS, SUE, HHT ♂♂♂♂*

Umbrellas of Cherbourg A bittersweet film operetta with no spoken dialog. Deneuve is the teenaged daughter of a widow who owns an umbrella shop. She and her equally young boyfriend are separated by his military duty in Algeria. Finding she is pregnant, the girl marries a wealthy man, and when her former lover returns, he too marries someone else. But when they meet once again will their love be rekindled? Lovely photography and an evocative score enhance the story. In French with subtitles; also available dubbed in English (not with the same effectiveness). Oscar nominee for cinematography and story.
1964 90m/C *FR* Catherine Deneuve, Nino Castelnuovo, Anne Vernon, Ellen Farner, Marc Michel, Mireille Perrey, Jean Champion, Alfred Wolff, Dorothee Blanck; *W/Dir:* Jacques Demy. Cannes Film Festival '64: Catholic Film Office Award, Best Film. **VHS, Beta $39.95** *IVE, APD ♂♂♂½*

Un Chien Andalou Masterful surrealist short features a host of classic sequences, including a razor across an eye, a severed hand lying in the street, ants crawling from a hole in a man's hand, priests, dead horses, and a piano dragged across a room. A classic. Score features both Wagner and tango.
1928 20m/B *FR* Pierre Batcheff, Simone Marevil, Jaime Miravilles; *Dir:* Luis Bunuel, Salvador Dali. **VHS, Beta $34.95** *HHT, TEX, APD ♂♂♂♂*

Un Elefante Color Ilusion A little boy visits the circus, makes friends with an elephant, and decides to bring it home with him.
19?? 85m/C *SP* Waldo Martinez, Pablo Luis Codevilla. **VHS, Beta $39.95** *MED ♂♂*

Un Novio Para Dos Hermanas (One Boyfriend for Two Sisters) Details what happens when two sisters fall in love with the same man and the conflicts that arise from it.
197? 100m/C *SP* Sara Garcia, Pili & Mili, Angel Garasa. **VHS, Beta $49.95** *MAD ♂½*

Un Singe en Hiver A young wanderer (Belmondo) befriends a crusty old alcoholic (Gabin) who has kept his vow to abstain since his village survived the German bombing of World War II. Moving tale of regret and the passage of time. Black and White. In French with English subtitles.
1962 105m/B Jean-Paul Belmondo, Jean Gabin, Suzanne Flon, Paul Frankeur; *Dir:* Henri Verneuil. **VHS, Beta $19.95** *KRT ♂♂½*

Una Vez en la Vida A young farm girl travels to the big city, where she becomes a nightclub dancer and gets involved with a shady character. In Spanish.
1949 75m/B *SP* Libertad Lamarque, Luis Aldas, Raimundo Pastore. **VHS, Beta $29.95** *VYY, APD ♂*

Una Viuda Descocada A beautiful woman kills her lovers comically. In Spanish.
19?? 90m/C Isabel Sarfi, Jose Marrone. **VHS, Beta $57.95** *UNI*

The Unapproachable A rendition of the Polish director's two one-act plays concerning a reclusive aging stage star being manipulated by her hangers-on.
1982 92m/C *PL* Leslie Caron, Daniel Webb, Leslie Malton; *Dir:* Krzysztof Zanussi. **VHS, Beta $59.95** *MGM ♂♂½*

The Unbearable Lightness of Being Tomas (Lewis), a young Czech doctor in the late 1960s, leads a sexually and emotionally carefree existence with a number of women, including the provocative artist, played by Olin. When he meets the fragile Binoche, he may be falling in love for the first time. On the eve of the 1968 Russian invasion of Czechoslovakia the two flee to Switzerland, but Binoche can't reconcile herself to exile and returns, followed by the reluctant doctor who has lost his position because of his new-found political idealism. The two find themselves leading an increasingly simple life, drawn ever closer together. The haunting ending caps off superb performances. Based on the novel by Milan Kundra.
1988 (R) 172m/C Daniel Day Lewis, Juliette Binoche, Lena Olin, Derek De Lint, Erland Josephson, Pavel Landovsky, Donald Moffatt, Daniel Olbrychski, Stellan Skarsgard; *W/Dir:* Philip Kaufman. National Society of Film Critics Awards '88: Best Director, Best Picture. **VHS, Beta, LV $89.98** *ORI ♂♂♂½*

The Unbeaten 28 A group of kung fu masters is forced to fight for their lives against evil criminals.
198? 90m/C Lisa Chang, Meng Fei, Lung Sikar; *Dir:* Joseph Kuo. **VHS, Beta $39.95** *MPI ♂*

The Unbelievable Truth Ex-con Robocop-to-be Burke meets armageddon-obsessed model and sparks fly until bizarre murder occurs. Quirky black comedy shot in less than two weeks.
1990 (R) 100m/C Adrienne Shelly, Robert Burke, Christopher Cooke, Julia Mueller; *Dir:* Hal Hartley. **VHS $89.95** *VMK ♂♂½*

The Unborn An infertile wife gets inseminated at unorthodox clinic. But once pregnant she suspects that her unborn baby is a monstrous being. Tasteless B-movie with an A-performance from Adams; if the rest had been up to her standard this could have been another "Stepford Wives." Instead it cops out with cheap gore.
1991 (R) 85m/C Brooke Adams, Jeff Hayenga, James Karen, K. Callon, Jane Cameron; *Dir:* Rodman Flender. **VHS, LV $89.95** *COL, LDC ♂½*

The Unbreakable Alibi A wealthy young man asks Tommy and Tuppence Beresford to help him prove that one of two alibis an Australian journalist created is false. Based on the mysteries by Agatha Christie. Made for television.
1983 51m/C *GB* Francesca Annis, James Warwick. **VHS, Beta $14.95** *PAV ♂♂*

The Uncanny A horror anthology. Three tales concerned with feline attempts to control the world. Waste of good cast on silly stories. Not released in the U.S.
1978 85m/C *GB* Peter Cushing, Ray Milland, Samantha Eggar, Donald Pleasence; *Dir:* Denis Heroux. **VHS, Beta $49.95** *MED ♂*

Uncensored This fast-paced comedy special features dozens of hilarious skits and sketches.
1985 60m/C Murray Langston, Cassandra Peterson, John Paragon, Ben Powers. **VHS, Beta $29.95** *WES ♂½*

The Unchastened Woman A married couple is torn apart by infidelity despite their impending parenthood, and wife Bara heads overseas to run wild and have the baby. She returns with a foreign flame, whom she dumps when erstwhile husband finds fatherly feelings in his heart. One of Bara's last films.
1925 52m/B Theda Bara, Wyndham Standing, Dale Fuller, John Miljan; *Dir:* James Young. **VHS, Beta $19.95** *GPV, FCT ♂♂*

Uncle Buck When Mom and Dad have to go away suddenly, the only babysitter they can find is good ol' Uncle Buck, a lovable lout who spends much of his time smoking cigars, trying to make up with his girlfriend, and enforcing the teenage daughter's chastity. More intelligent than the average slob/teen comedy with a heart, due in large part to Candy's dandy performance. Memorable pancake scene.
1989 (PG) 100m/C John Candy, Amy Madigan, Jean Kelly, Macauley Culkin, Jay Underwood; *Dir:* John Hughes. **VHS, Beta, LV $19.95** *MCA ♂♂♂*

Uncle Moses The clash of old-world values and new-world culture in a story of East European Jewish immigrants transplanted to turn-of-the-century New York's Lower East Side. Patriarch Uncle Moses struggles to keep traditional family values as romantic difficulties and labor union struggles intrude. Based on the novel by Sholem Asch. In Yiddish with English subtitles.
1932 87m/B Maurice Schwartz, Rubin Goldberg; *Dir:* Sidney Goldin, Aubrey Scotto. **VHS $72.00** *NCJ ♂½*

Uncle Sam Magoo Mr. Magoo provides a history lesson in his own unmatchable style.
1970 53m/C *Voices:* Jim Backus. **VHS, Beta $12.95** *PAR, FCT ♂♂½*

Uncle Tom's Cabin Satisfying version of Harriet Beecher Stowe's tale from the view of a founder of the underground railroad. Lucas was the first black actor to garner a lead role.
1914 54m/C Mary Eline, Irving Cummings, Sam Lucas; *Dir:* William Robert Daly. **VHS, Beta $19.95** *GPV, FCT, DVT ♂♂*

Uncle Tom's Cabin An ambitious adaptation of Harriet Beecher Stowe's stirring book about the Southern slavery and the famous underground railroad devised to aid those most determined to reach freedom. The film has the habit of drifting away from the facts. Though set in Kentucky, the picture was filmed in Yugoslavia with an international cast. Also known as "Onkel Toms Hutte," "La Case de L'Oncle Tom," "Cento Dollari D'Odio" and "Cica Tomina Koliba."
1969 (G) 120m/C *FR IT YU GE* Herbert Lom, John Kitzmiller, O.W. Fischer, Eleanora Rossi-Drago, Mylene Demongeot, Juliette Greco; *Dir:* Geza von Radvanyi; *Voices:* Ella Fitzgerald. **VHS, Beta $39.95** *XVC ♂½*

Uncommon Valor After useless appeals to the government for information on his son listed as "missing in action" in Vietnam, Colonel Rhodes takes matters into his own hands. Hackman is solid and believable as always, surrounded with good cast in generally well-paced film.
1983 (R) 105m/C Gene Hackman, Fred Ward, Reb Brown, Randall "Tex" Cobb, Robert Stack, Patrick Swayze, Harold Sylvester, Tim Thomerson; *Dir:* Ted Kotcheff. **VHS, Beta, LV, 8mm $14.95** *PAR ♂♂½*

The Undaunted Wu Dang Set in 19th-century China, this film follows the adventures of the beautiful young daughter of a Chinese martial arts master seeking to avenge the murders of her father and brother.

1989 95m/C VHS, Beta *JCI* ⊘

The Undead A prostitute is accidentally sent back to the Middle Ages as the result of a scientific experiment and finds herself condemned to die for witchcraft. Early Corman script filled with violence and heaving bosoms.
1957 75m/B Pamela Duncan, Richard Garland, Allison Hayes, Mel Welles; *Dir:* Roger Corman. VHS $19.95 *NOS, AIP, MLB* ⊘⊘⊘

Undeclared War Suspenseful espionage thriller that takes you behind the scenes of an international terrorist plot, which has been cleverly disguised as a bloody global revolution. Intense action heightened by conflicts between worldwide intelligence networks, the news media and terrorist organizations.
1991 (R) 103m/C Vernon Wells, David Hedison, Olivia Hussey, Peter Lapis; *Dir:* Ringo Lam. VHS $89.95 *IMP* ⊘⊘

The Undefeated A Confederate and Yankee find they must team up on the Rio Grande. They attempt to build new lives in the Spanish held Mexico territory, but are caught in the battle for Mexican independence. Standard fare made palatable only by Wayne and Hudson.
1969 (G) 119m/C John Wayne, Rock Hudson, Lee Meriwether, Merlin Olsen, Bruce Cabot, Ben Johnson, Jan-Michael Vincent, Harry Carey Jr., Antonio Aguilar, Roman Gabriel; *Dir:* Andrew V. McLaglen. VHS, Beta, LV $19.98 *FOX* ⊘⊘

Under the Biltmore Clock F. Scott Fitzgerald's story, "Myra Meets His Family," is the basis for this American Playhouse installment for PBS television. It is outwit or be outwitted in this tale of a fortune-hunting '20s flapper who is outsmarted by the man she has set her sights on when he hires actors to play his eccentric "family."
1985 70m/C Sean Young, Lenny Von Dohlen, Barnard Hughes; *Dir:* Neal Miller. VHS, Beta, LV, 8mm $19.98 *SUE* ⊘⊘

Under the Boardwalk A pair of star-crossed teenage lovers struggle through familial and societal differences in 1980s California. This is, like, bogus, ya know.
1989 (R) 102m/C Keith Coogan, Danielle von Zernaeck, Richard Joseph Paul, Hunter von Leer, Tracy Walter, Roxana Zal, Dick Miller, Sonny Bono, Corky Carroll; *Dir:* Fritz Kiersch. VHS, Beta, LV $19.95 *STE, NWV* ⊘ ½

Under California Stars A shady gang making a living rounding up wild horses decides they can make more money by capturing Rogers' horse, Trigger. Also known as "Under California Skies."
1948 71m/C Roy Rogers, Andy Devine, Jane Frazee, The Pioneers; *Dir:* William Witney. VHS, Beta $19.95 *NOS, VYY, DVT* ⊘ ½

Under Capricorn Bergman is an Irish lass who follows her convict husband Cotten out to 1830s Australia where he makes a fortune. She turns to drink, perhaps because of his neglect, and has her position usurped by a housekeeper with designs on her husband. When Bergman's cousin (Wilding) arrives, Cotten may have cause for his violent jealousy. There's a plot twist involving old family skeletons, but this is definitely lesser Hitchcock. Adapted from the Helen Simpson novel. Remade in 1982.
1949 117m/C GB Ingrid Bergman, Joseph Cotten, Michael Wilding, Margaret Leighton, Jack Watling, Cecil Parker, Denis O'Dea; *Dir:* Alfred Hitchcock. VHS, Beta, LV $14.98 *VES, VID* ⊘⊘ ½

Under Capricorn An Australian remake of the 1949 Hitchcock film about family secrets, an unhappy marriage, and violence set in 1830's Australia.
1982 120m/C AU Lisa Harrow, John Hallam, Peter Cousens, Julia Blake, Catherine Lynch; *Dir:* Rod Hardy. VHS, Beta $59.95 *PSM* ⊘⊘

Under the Cherry Moon Prince portrays a fictional musician of the 1940s who travels to the French Riviera, seeking love and money. Songs include "Under the Cherry Moon," "Kiss," "Anotherloverholeinyohead," "Sometimes It Snows in April," and "Mountains." A vanity flick down to its being filmed in black and white.
1986 (PG-13) 100m/B Prince, Jerome Benton, Francesca Annis, Kristin Scott Thomas; *Dir:* Prince. VHS, Beta, LV $19.98 *MVD, WAR* ⊘ ½

Under the Doctor A silly British comedy about a psychiatrist who fantasizes about his gorgeous female patients who only want to talk about their sexual problems.
1976 (R) 84m/C GB *Dir:* Gerry Poulson. VHS, Beta $29.95 *ACA* Woof!

Under the Earth An Argentine-Czech co-production, depicting the plight of a family of Polish Jews forced to live underground for two years to avoid the Nazis. Harrowing and critically well-received. In Spanish with subtitles.
1986 (R) 100m/C SP Victor Laplace; *Dir:* Beda Ocampo Feijoo, Juan Bautista Stagnaro. VHS, Beta, LV $19.95 *STE, NWV* ⊘⊘⊘

Under Fire Three foreign correspondents, old friends from the past working together, find themselves in Managua, witnessing the 1979 Nicaraguan revolution. In a job requiring objectivity, but a situation requiring taking sides, they battle with their ethics to do the job right. Fine performances, including Camp as a mercenary. Interesting view of American media and its political necessities.
1983 128m/C Gene Hackman, Nick Nolte, Joanna Cassidy, Ed Harris, Richard Masur, Hamilton Camp, Jean-Louis Trintignant; *Dir:* Roger Spottiswoode. VHS, Beta, LV $19.98 *VES* ⊘⊘⊘½

Under the Gun A cop recruits a lawyer to help him find the man who murdered his brother. Action drives this formulaic plot along.
1988 (R) 89m/C Sam Jones, Vanessa Williams, John Russell, Michael Halsey; *Dir:* James Sbardellati. VHS, LV $79.89 *IME* ⊘⊘

Under Milk Wood An adaptation of the Dylan Thomas play about the lives of the residents of a village in Wales. Burton and O'Toole are wonderful in this uneven, but sometimes engrossing film.
1973 (PG) 90m/C GB Richard Burton, Elizabeth Taylor, Peter O'Toole, Glynis Johns, Vivien Merchant; *Dir:* Andrew Sinclair. VHS, Beta $59.98 *FOX* ⊘⊘½

Under Montana Skies An early-talkie western features music and comedy in abundance, relegating the action to second place. A singing cowboy protects a travelling burlesque show from evildoers.
1930 55m/B Kenneth Harlen, Dorothy Gulliver, Lafe McKee, Ethel Wales, Slim Summerville; *Dir:* Dave Thorpe. VHS, Beta *GPV* ⊘½

Under Nevada Skies Fairly good matinee-era western centers on a missing map to a uranium deposit, climaxing with Rogers heading a posse of Indian allies to the rescue. The Sons of the Pioneers contribute songs.

1946 69m/B Roy Rogers, George "Gabby" Hayes, Dale Evans, Douglas Dumbrille, Tristram Coffin, Rudolph Anders, Iron Eyes Cody; *Dir:* Frank McDonald. VHS $19.95 *NOS* ⊘⊘½

Under the Rainbow Comic situations encountered by a talent scout and a secret service agent in a hotel full of "The Wizard of Oz" Munchkin midgets. International intrigue adds to the strange and mostly un-funny mix.
1981 (PG) 97m/C Chevy Chase, Carrie Fisher, Eve Arden, Joseph Maher, Robert Donner, Mako, Pat McCormick, Billy Barty, Zelda Rubinstein; *Dir:* Steve Rash. VHS, Beta $19.98 *WAR* ⊘

Under the Red Robe A French soldier of fortune is trapped into aiding Cardinal Richelieu in his persecution of the Huguenots and winds up falling in love with the sister of his intended victim. Good costume adventure-drama. Sjostrom's last film as a director.
1936 82m/B GB Raymond Massey, Conrad Veidt, Annabella, Romney Brent; *Dir:* Victor Sjostrom. VHS, Beta $19.95 *NOS, HHT, WFV* ⊘⊘½

Under the Roofs of Paris (Sous les Toits de Paris) An early French sound film about the lives of young Parisian lovers. A gentle, highly acclaimed melodrama. In French with English subtitles.
1929 95m/B FR Albert Prejean, Pola Illery, Gaston Modot; *Dir:* Rene Clair. National Board of Review Awards '31: 5 Best Foreign Films of the Year. VHS, Beta $29.95 *NOS, VYY, CAB* ⊘⊘⊘½

Under Satan's Sun A rural priest is tortured by what he sees as his sins and failings to his parishioners. He is further tempted from the straight path by a beautiful murderess, a worldly priest, and, perhaps by Satan in disguise. Stylized film is not easily accessible, but Depardieu's performance is worth the effort. Based on the Georges Bernanos book "Diary of a Country Priest." Also known as "Under the Sun of Satan."
1987 97m/C FR Gerard Depardieu, Sandrine Bonnaire, Maurice Pialat, Yann Dedet, Alain Artur; *Dir:* Maurice Pialat. VHS *FCT* ⊘⊘⊘

Under the Sheets Four zany medical students would rather undress the nurses than hit the books.
1986 87m/C SP Karin Schubert. VHS $39.95 *MHV* ⊘⊘⊘

Under Suspicion Christmas 1959 in Brighton, England finds seedy private eye Tony Aaron doing his usual divorce work. He and wife/partner Hazel stage phony adultery cases for those desperate to get around England's strict divorce laws. Hazel checks into a hotel with the male client and after a suitable time, Tony bursts in to snap incriminating photos. Only this time, Tony finds both his wife and their client, a famous artist, murdered. Suspects include Angeline, the artist's mistress and his jealous wife, Selina. Even Tony is a suspect and as he works to clear his name, he also falls into a heated, if mistrustful, affair with the sultry Angeline. Conventional melodrama well-acted by Neeson as the charmingly sleazy Tony; San Giacomo is, however, miscast as the femme fatale. Theatrical feature debut for director Moore.
1991 (R) 99m/C GB Liam Neeson, Laura San Giacomo, Alphonsia Emmanuel, Kenneth Cranham, Maggie O'Neill; *W/Dir:* Simon Moore. VHS, LV *COL* ⊘⊘½

Under the Volcano An alcoholic British ex-consul finds his life further deteriorating during the Mexican Day of the Dead in 1939. His half-brother and ex-wife try to save him from himself, but their affair has sent him ever deeper into his personal hell. Finney's performance is pathetic and haunting. Adapted from the novel by Malcolm Lowry.
1984 112m/C Albert Finney, Jacqueline Bisset, Anthony Andrews, Katy Jurado; *Dir:* John Huston. **VHS, Beta, LV** $79.95 *MCA* 🎗🎗🎗

Under Western Stars Roy's first starring vehicle. A newly elected congressman goes to Washington and fights battles for his constituents, caught in the middle of the Dust Bowl and drought.
1945 83m/B Roy Rogers, Trigger, Smiley Burnette; *Dir:* Jean Yarborough. **VHS, Beta** $19.95 *NOS, DVT, VCN* 🎗

The Underachievers A night school is the scene for typical libido-oriented gags, but this time instead of teenagers its adult education provicing the plot. Only a few real laughs in this one.
1988 (R) 90m/C Barbara Carrera, Edward Albert, Michael Pataki, Vic Tayback, Garrett Morris, Susan Tyrrell; *Dir:* Jackie Kong. **VHS, Beta** $79.98 *LIV, VES, LTG* 🎗

Undercover Routine cop action film in which an eastern detective finds himself undercover in a southern high school in search of a cop-murdering drug ring. Plagued with cliches. Director Stockwell is appropriately better known for his role as "Cougar" in "Top Gun."
1987 (R) 92m/C David Neidorf, Jennifer Jason Leigh, Barry Corbin, David Harris, Kathleen Wilhoite; *Dir:* John Stockwell. **VHS** $79.98 *WAR* 🎗½

Undercover Man A Wells Fargo agent goes undercover to expose a crooked sheriff.
1936 57m/B Johnny Mack Brown, Suzanne Karen, Ted Adams, Lloyd Ingraham, Horace Murphy; *Dir:* Albert Ray. **VHS** $19.95 *NOS, DVT, RXM* 🎗½

Undercover Vixens Undercover Israeli beauties battle and seduce enemy agents.
1987 78m/C Monica Gayle, Sherrie Land. **VHS, Beta** $29.95 *MED* Woof!

The Underdog After the bank forecloses, a farm family must move to the city. The family's young son has only his loyal dog to turn to for friendship, but the gallant animal proves his mettle by rounding up a gang of spies plotting some World War II sabotage. Believe it if you dare.
1943 65m/B Barton MacLane, Bobby Larson, Jane Wiley; *Dir:* William Nigh. **VHS, Beta, 8mm** *VYY* 🎗½

Undergrads A bright generational Disney comedy with Carney, estranged from his stick-in-the-mud son, deciding to attend college with his free-thinking grandson. Made for television.
1985 102m/C Art Carney, Chris Makepeace, Jackie Burroughs, Len Birman; *Dir:* Steven Hilliard Stern. **VHS, Beta** $59.95 *DIS* 🎗🎗🎗

Underground Aces A group of parking lot attendants transforms a sheik into an attendant in order to help him meet the girl of his dreams. A bad "Car Wash" spinoff.
1980 (PG) 93m/C Dirk Benedict, Melanie Griffith, Frank Gorshin, Jerry Orbach, Robert Hegyes, Audrey Landers; *Dir:* Robert Butler. **VHS, Beta** $69.98 *LIV, VES* Woof!

Underground Forces Volumes 1-8 The Sex Pistols, Siouxie and the Banshees, and the Screamers are some of the many punk acts featured in this series.
1984 60m/C **VHS, Beta** $39.95 *MVD, TGV*

Underground Rustlers The Range Busters are commissioned by the government to stop unscrupulous gold-marketeers.
1941 58m/B Ray Corrigan, Max Terhune, John "Dusty" King; *Dir:* S. Roy Luby. **VHS, Beta** $19.95 *VDM* 🎗

Underground Terror A gang of murderers, organized by a psychopath, rampages through the New York City subway system. One cop enthusiastically tries to snuff out each one.
1988 91m/C Doc Dougherty, Lenny Y. Loftin; *Dir:* James McCalmont. **VHS** $79.95 *SVS* 🎗

Underground U.S.A. A street hustler picks up a has-been underground movie star at the chic New York new wave disco, The Mudd Club.
1984 85m/C Patti Astor, Eric Mitchell; *Dir:* Eric Mitchell. **VHS, Beta** *NEW* 🎗

Undersea Kingdom Adventure beneath the ocean floor. In twelve chapters of thirteen minutes each; the first chapter runs twenty minutes. Later re-edited into one film, "Sharad of Atlantis."
1936 226m/B Ray Corrigan, Lon Chaney Jr.; *Dir:* B. Reeves Eason. **VHS, LV** $29.98 *NOS, SNC, VCN* 🎗½

Understudy: The Graveyard Shift 2 Vampire survives poorly directed cliche-riddled original to be cast as a vampire cast as a vampire in a horror movie within a horrible sequel
1988 (R) 88m/C Wendy Gazelle, Mark Soper, Silvio Oliviero; *Dir:* Gerard Ciccoritti. **VHS, Beta, LV** $79.95 *VIR* 🎗½

The Undertaker and His Pals Undertaker teams up with diner owners in murder scheme to improve mortician's business and expand restaurateurs' menu. Pretty violent stuff, with some good laughs and a campy flare, but not for every one's palate.
1967 70m/C Ray Dannis, Brad Fulton, Larrene Ott, Robert Lowery, Sally Frei; *Dir:* David C. Graham. **VHS** *MTH* 🎗🎗

Underwater A team of skin divers faces danger when they try to retrieve treasure from a Spanish galleon. The second film after "The Outlaw" masterminded by Howard Hughes, primarily to show off Russell's figure.
1955 99m/C Jane Russell, Richard Egan, Gilbert Roland, Jayne Mansfield; *Dir:* John Sturges. **VHS, Beta, LV** $19.98 *MED* 🎗🎗

Underworld Underground mutants kidnap a prostitute in order to get a life-sustaining drug. A gangland boss hires a gunman to save her. Based on a Clive Barker novel.
1985 100m/C Denholm Elliott, Steven Berkoff, Larry Lamb, Nicola Cowper; *Dir:* George Pavlou. **VHS** *VES, LIV* 🎗½

The Underworld Story Big city journalist at large moves to smalltown New England after losing job for unethical reporting and uncovers scheme to frame innocent man for murder. Solid performances.
1950 90m/B Dan Duryea, Herbert Marshall, Gale Storm, Howard da Silva, Michael O'Shea, Mary Anderson, Gar Moore, Melville Cooper, Frieda Inescort, Art Baker, Harry Shannon, Alan Hale Jr., Stephen Dunne, Roland Winters; *Dir:* Cy Endfield. **VHS** $19.98 *FOX, FCT* 🎗🎗½

Underworld, U.S.A. A man infiltrates a tough crime syndicate to avenge his father's murder, which winds up with him caught between the mob and the feds. A well-acted and directed look at the criminal underworld.
1960 99m/B Cliff Robertson, Dolores Dorn, Beatrice Kay, Robert Emhardt, Larry Gates, Paul Dubov; *W/Dir:* Samuel Fuller. **VHS, Beta** $69.95 *COL* 🎗🎗🎗

The Undying Monster Upper crusty Brit is werewolf with insatiable appetite. Familiar story with lots of atmosphere and style. Photography by Lucien Ballard.
1942 63m/B James Ellison, Heather Angel, John Howard, Bramwell Fletcher, Heather Thatcher, Aubrey Mather, Halliwell Hobbes, Heather Wilde; *Dir:* John Brahm. **VHS** $19.98 *MOV* 🎗🎗½

Une Jeunesse A middle-aged French couple reminisce about their meeting and courtship. In French with English subtitles.
1983 100m/C *FR* **VHS, Beta** $29.95 *FCT* 🎗🎗🎗

The Unearthly A mad scientist is trying to achieve immortality through his strange experiments, but all he winds up with is a basement full of mutants. When his two latest about-to-be victims fall in love, the doctor's mutant assistant decides enough is enough and things come to an unpleasant end. Carradine is typecast. Absurd, but fun.
1957 76m/B John Carradine, Tor Johnson, Allison Hayes, Myron Healy; *Dir:* Brooke L. Peters. **VHS, Beta** $9.95 *RHI, SNC, DVT* 🎗½

The Unearthly Stranger Earth scientist marries woman and decides she's from another planet. Not a nineties style gender drama. Seems she's part of an invading alien force, but really does love her earth man. Surprisingly good low-budget sci-fi.
1964 75m/B *GB* John Neville, Gabriella Licudi, Philip Stone, Patrick Newell, Jean Marsh, Warren Mitchell; *Dir:* John Krish. **VHS** $19.98 *MOV* 🎗🎗½

Uneasy Terms Rennie is a detective investigating the murder of one of his own clients. Evidence points to stepchildren and inheritance money but things may not be as they seem. Written for the screen by Peter Cheyney, adapted from his own novel.
1948 91m/B *GB* Michael Rennie, Moira Lister, Faith Brook, Joy Shelton, Patricia Goddard, Barry Jones; *Dir:* Vernon Sewell. **VHS, Beta** $16.95 *SNC* 🎗

Unexplained Laughter A cynical journalist with vacations in Wales with her timid vegetarian friend and stumble across a mystery in this darkly comic British TV-movie.
1989 85m/C *GB* Diana Rigg, Elaine Page, Jon Finch. **VHS** $49.95 *FCT* 🎗🎗

Unfaithfully Yours A conductor suspects his wife is cheating on him and considers his course of action. He imagines punishment scenarios while directing three classical works. Well-acted by all, but particularly by Harrison as the egotistical and jealous husband. Another of Sturges' comedic gems. Remade in 1984.
1948 105m/B Rex Harrison, Linda Darnell, Kurt Krueger, Rudy Vallee, Lionel Stander, Edgar Kennedy; *Dir:* Preston Sturges. **VHS, Beta, LV** $59.98 *FOX* 🎗🎗🎗½

Unfaithfully Yours A symphony conductor suspects his wife of fooling around with a musician; in retaliation, he plots an elaborate scheme to murder her with comic results. No match for the 1948 Preston Sturges film it's based on.

1984 (PG) 96m/C Dudley Moore, Nastassia Kinski, Armand Assante, Albert Brooks, Cassie Yates, Richard Libertini, Richard B. Shull; *Dir:* Howard Zieff. **VHS, Beta, LV** $79.98 *FOX* 𝄞𝄞

Unfinished Business A 17 year-old girl leads a troubled adolescent life with her divorced parents. She decides running away will ease her problems — not a smart move. Sequel to "Nobody Waved Goodbye."
1989 88m/C *CA* Isabelle Mejias, Peter Kastner, Leslie Toth, Peter Spence; *Dir:* Don Owen. **VHS, Beta** $79.95 *JEF* 𝄞𝄞½

An Unfinished Piece for a Player Piano A general's widow invites family and friends to a weekend house party in 1910 Russia. Romantic and familial entanglements begin to intrude in a lyrical adaptation of Chekov's play "Platonov."
1977 100m/C *RU* Alexander Kaliagin, Elena Solovei, Antonina Shuranova, Oleg Tabakov, Yuri Bogatyrev, Nikita Mikhailkov; *Dir:* Nikita Mikhailkov. **VHS, Beta** $69.95 *FCT* 𝄞𝄞𝄞½

The Unforgiven A western family is torn asunder when it is suspected that the eldest daughter is of Indian birth. Film takes place in 1850s' Texas. One of Huston's weakest ventures, but viewed in terms of 1950s' prejudices it has more resonance. Fine acting from all the cast, especially Gish. Watch for the stunning Indian attack scene.
1960 123m/C Burt Lancaster, Audrey Hepburn, Lillian Gish, Audie Murphy, John Saxon, Charles Bickford, Doug McClure, Joseph Wiseman, Albert Salmi; *Dir:* John Huston. **VHS, Beta** $59.95 *MGM* 𝄞𝄞𝄞

The Unholy A New Orleans priest battles a demon that's killing innocent parishioners. Confusing and heavy-handed.
1988 (R) 100m/C Ben Cross, Hal Holbrook, Trevor Howard, Ned Beatty, William Russ, Jim Carroll; *Dir:* Camilo Vila. **VHS, Beta, LV** $19.98 *LIV, VES, HHE* 𝄞½

Unholy Four An amnesiac returns home after three years to attempt to find out which of his three fishing buddies left him for dead. Confusing at times, but has some suspenseful moments. British title was "A Stranger Came Home."
1954 80m/B *GB* Paulette Goddard, Paul Carpenter, William Sylvester, Patrick Holt, Russell Napier; *Dir:* Terence Fisher. **VHS, Beta, LV** *WGE* 𝄞½

Unholy Rollers Jennings stars as a factory worker who makes it big as a tough, violent roller derby star. Typical "Babes-on-Wheels" film that promises nothing and delivers even less.
1972 (R) 88m/C Claudia Jennings, Louis Quinn, Betty A. Rees, Roberta Collins, Alan Vint, Candice Roman; *Dir:* Vernon Zimmerman. **VHS** $79.95 *HBO Woof!*

The Unholy Three Ventriloquist Chaney, working with other carnival cohorts, uses his talent to gain entrance to homes which he later robs. Things go awry when two of the gang strike out on their own and the victim is killed. When the wrong man is accused, his girl, one of Chaney's gang, begs Chaney to get him free. Which he does by using his vocal talents. Chaney decides being a criminal is just too hard and goes back to his ventriloquism.
1925 70m/B Lon Chaney Sr., Harry Earles, Victor McLaglen, Mae Busch, Matt Moore, Matthew Betz, William Humphreys; *Dir:* Tod Browning. **VHS** $49.95 *EJB* 𝄞𝄞𝄞½

Unholy Wife A young woman plans to murder her wealthy husband, but her plan goes awry when she accidentally kills someone else. Muddled and heavy-handed, with Steiger chewing scenery.
1957 94m/C Rod Steiger, Diana Dors, Tom Tryon, Marie Windsor, Beulah Bondi; *Dir:* John Farrow. **VHS, Beta** $15.95 *UHV, HHE* 𝄞½

Unicorn A small boy hears the legend that if one rubs the horn of a Unicorn his wishes will come true. Mistakenly he buys a one-horned goat and sets out to fulfill his dreams.
1983 29m/C *GB* Diana Dors, David Kossoff, Celia Johnson; *Dir:* Carol Reed. **VHS, Beta** $24.98 *SUE* 𝄞𝄞

Unidentified Flying Oddball An astronaut and his robotic buddy find their spaceship turning into a time machine that throws them back into Arthurian times and at the mercy of Merlin the magician. Futuristic version of Twain's "A Connecticut Yankee at King Arthur's Court."
1979 (G) 92m/C Dennis Dugan, Jim Dale, Ron Moody, Kenneth More, Rodney Bewes; *Dir:* Russ Mayberry. **VHS, Beta** $69.95 *DIS* 𝄞𝄞½

The Uninvited A brother and sister buy a house in England, only to find it is haunted. Doors open and close by themselves, strange scents fill the air, and they hear sobbing during the night. Soon they are visited by a woman with an odd link to the house—her mother is the spirit who haunts the house. Chilling and unforgettable, this is one of the first films to deal seriously with ghosts.
1944 99m/B Ray Milland, Ruth Hussey, Donald Crisp, Cornelia Otis Skinner, Gail Russell; *Dir:* Lewis Allen. **VHS** $14.98 *MCA* 𝄞𝄞𝄞

The Uninvited A mutant cat goes berserk onboard a luxury yacht, killing the passengers one by one with big, nasty, pointy teeth.
1988 89m/C George Kennedy, Alex Cord, Clu Gulager, Toni Hudson, Eric Larson, Shari Shattuck, Austin Stoker; *Dir:* Greydon Clark. **VHS, Beta, LV** $79.95 *NSV Woof!*

Union City Deborah "Blondie" Harry's husband gets a little edgy when someone steals the milk. Murder ensues, and they're on the run from the law. Intended as a film noir spoof, and not without some good moments.
1981 (R) 82m/C Deborah Harry, Everett McGill, Dennis Lipscomb, Pat Benatar; *Dir:* Mark Reichart. **VHS, Beta, LV** $59.95 *COL* 𝄞𝄞

Union Station Holden plays the chief of the railway police for Chicago's Union Station. He learns the station is to be used as a ransom drop in a kidnapping; but for all his security, the main thug gets away with the money, and the hunt is on. Good acting raises this film above the ordinary.
1950 80m/B William Holden, Barry Fitzgerald, Nancy Olson, Jan Sterling, Lee Marvin, Allene Roberts, Lyle Bettger; *Dir:* Rudolph Mate. **VHS, Beta** $19.95 *KRT* 𝄞𝄞

United Front: Outside in Sight A look at the United Front, a San Francisco jazz quartet.
1989 30m/C Anthony Brown, Mark Izu, Lewis Jordan, George Sams. **VHS, Beta** $19.95 *MVD, KIV*

Unkissed Bride A young newlywed couple are driven to distraction by the husband's inexplicable fainting spells and his strange obsession with Mother Goose. Also known as "Mother Goose A Go-Go."

1966 (PG) 82m/C Tommy Kirk, Anne Helm, Danica D'Hondt, Henny Youngman; *Dir:* Jack H. Harris. **VHS, Beta** $19.95 *STE, NWV* 𝄞

The Unknown Typically morbid Chaney fare has him working as a circus freak while trying to win the heart of his assistant. After drastic romancing he's still rejected by her, so he plots to kill the object of his intentions. Ghoulish as it is, the picture is really top-notch.
1927 60m/B Lon Chaney Sr., Norman Kerry, Joan Crawford, Nick De Ruiz, Frank Lanning, John St. Polis; *Dir:* Tod Browning. **VHS** $49.95 *EJB* 𝄞𝄞𝄞

The Unknown Comedy Show The Unknown Comic goes nuts and hosts this deliciously off-the-wall comedy special.
1987 56m/C Unknown Comic, Murray Langston, Johnny Dark. **VHS, Beta** $29.95 *IVE* 𝄞𝄞𝄞½

Unknown Island Scientists travel to a legendary island where dinosaurs supposedly still exist. Bogus dinosaurs and cliche script.
1948 76m/C Virginia Grey, Philip Reed, Richard Denning, Barton MacLane; *Dir:* Jack Bernhard. **VHS** $19.95 *NOS, MOV Woof!*

Unknown Powers Science and drama are combined to examine ESP and magic. Are they gifts or curses, and how are peoples' lives affected by them? The cast introduce various sections of this totally inept film.
1980 (PG) 97m/C Samantha Eggar, Jack Palance, Will Geer, Roscoe Lee Browne; *Dir:* Don Como. **VHS, Beta** $39.95 *GEM Woof!*

Unknown World A group of scientists tunnel to the center of the Earth to find a refuge from the dangers of the atomic world. Dig start winds down fast.
1951 73m/B Bruce Kellogg, Marilyn Nash, Victor Kilian, Jim Bannon; *Dir:* Terrell O. Morse. **VHS, Beta** $19.95 *NOS, PSM, SNC* 𝄞

An Unmarried Woman Suddenly divorced by her husband of 17 years, a woman deals with change. She enters the singles scene, copes with her daughter's growing sexuality, and encounters a new self-awareness. Mazursky puts real people on screen from start to finish. Oscar nominations for picture, actress, and original screenplay.
1978 (R) 124m/C Jill Clayburgh, Alan Bates, Cliff Gorman, Michael Murphy; *Dir:* Paul Mazursky. Cannes Film Festival '78: Best Actress (Clayburgh); Los Angeles Film Critics Association Awards '78: Best Screenplay; National Board of Review Awards '78: 10 Best Films of the Year; New York Film Critics Awards '78: Best Screenwriting; National Society of Film Critics Awards '78: Best Screenplay. **VHS, Beta** $69.98 *FOX* 𝄞𝄞𝄞

Unmasked Part 25 A second-generation serial killer takes up where dad left off. This one is intended to be a parody of slasher films, but it's still very bloody.
1988 85m/C *GB* Gregory Cox, Fiona Evans, Edward Brayshaw, Debbie Lee London; *Dir:* Anders Palm. **VHS, Beta** $79.95 *ACA* 𝄞½

Unmasking the Idol A suave hero with tongue firmly in check battles Ninja warlords. Sequel to "Order of the Black Eagle."
1986 90m/C *GB* Ian Hunter, William T. Hicks, Charles K. Bibby; *Dir:* Worth Keeter. **VHS, Beta** $79.95 *CEL* 𝄞

The Unnamable The adaptation of the H.P. Lovecraft story about a particular New England ancestral home haunted by a typically Lovecraftian bloodthirsty demon borne of a woman hundreds of years before.

College students, between trysts, investigate the myths about it. Uncut version, unseen in theaters, available only on video good for a few giggles and thrills.
1988 87m/C Charles King, Mark Kinsey Stephenson, Alexandra Durrell; *Dir:* Jean-Paul Ouellette. **VHS, Beta, LV $19.95** *STE, VMK* ⚔

Unnatural A mad scientist creates a souless child from the genes of a murderer and a prostitute. The child grows up to be the beautiful Neff, who makes a habit of seducing and destroying men. Dark, arresting film from a very popular German story.
1952 90m/B Hildegarde Neff, Erich von Stroheim, Karl Boehm, Harry Meyen, Harry Helm, Denise Vernanc, Julia Koschka; *Dir:* Arthur Maria Rabenalt. **VHS $19.98** *SNC* ⚔⚔½

Unnatural Causes A dying Vietnam vet believes that his illness is the result of exposure to Agent Orange. With the help a VA counselor, they lobby for national programs to assist other veterans who have been exposed to the chemical and together bring publicity to the issue. A made-for-television drama that is exceptionally well-acted.
1986 96m/C John Ritter, Patti LaBelle, Alfre Woodard, Lamont Johnson; *W/Dir:* John Sayles. **VHS, Beta, LV $79.95** *NSV* ⚔⚔

An Unremarkable Life Two aging sisters live symbiotically together, until one views the other's romantic attachment to a charming widower as destructive to her own life.
1989 (PG) 97m/C Shelley Winters, Patricia Neal, Mako, Rochelle Oliver, Charles Dutton, Lily Knight; *Dir:* Amin Q. Chaudhri. **VHS, Beta $14.95** *SVS, IME* ⚔⚔

Unsane A mystery novelist realizes that a series of bizarre murders strangely resembles the plot of his latest book. Bloody fun from Argento. Originally titled, "Tenebrae."
1982 91m/C IT Anthony (Tony) Franciosa, John Saxon, Daria Nicolodi, Giuliano Gemma, Christian Borromeo, Mirella D'Angelo, Veronica Lario; *Dir:* Dario Argento. **VHS $9.95** *MED* ⚔⚔½

The Unseen Three young women from a television station are covering a story in a remote area of California. Before nightfall, two are horribly killed, leaving the third to come face to face with the terror.
1980 91m/C Barbara Bach, Sydney Lassick, Stephen Furst; *Dir:* Peter Foleg. **VHS, Beta $19.98** *VID* ⚔½

Unsettled Land Young Israeli settlers try to survive the elements and Bedouin attackers in Palestine during the 1920s.
1988 (PG) 109m/C *IS* Kelly McGillis, John Shea, Arnon Zadok, Christine Boisson; *Dir:* Uri Barbash. **VHS, Beta, LV, 8mm $19.98** *SUE* ⚔⚔

Unsinkable Donald Duck with Huey, Dewey & Louie Three vintage Disney cartoons featuring Donald and his nephews: "Sea Scouts," "Donald's Day Off" and "Lion Around."
1945 (G) 25m/C **VHS, Beta, LV $14.95** *DIS* ⚔⚔

The Unsinkable Molly Brown A spunky backwoods girl is determined to break into the upper crust of Denver's high society and along the way survives the sinking of the Titanic. This energetic version of the Broadway musical contains many Meredith Wilson (Music Man) songs like "Belly Up to the Bar Boys," "He's My Friend," "I'll Never Say No," and "I Ain't Down Yet." Hokey, good-natured fun. Earned Reynolds an Oscar nomination for Best Actress (she lost to Julie Andrews in "Mary Poppins"), was also nominated for

Cinematography, Art Direction, Set Decoration, Sound, Adapted Musical Score and Costume Design.
1964 128m/C Debbie Reynolds, Harve Presnell, Ed Begley Sr., Martita Hunt, Hermione Baddeley; *Dir:* Charles Walters. **VHS, Beta, LV $19.95** *MGM* ⚔⚔⚔

The Unstoppable Man The son of an American businessman is kidnapped while they are in London. No pushover he, dad develops his own plan to destroy the criminals, eventually pursuing them with a flame thrower.
1959 68m/B *GB* Cameron Mitchell, Marius Goring, Harry H. Corbett, Lois Maxwell, Denis Gilmore; *Dir:* Terry Bishop. **VHS, Beta $16.95** *NOS, SNC* ⚔½

Unsuitable Job for a Woman Independent Cordelia Gray, following the death of her boss, takes over his detective agency and gets involved with murder. Based on the novel by P.D. James.
1982 94m/C *GB* Pippa Guard, Paul Freeman, Billie Whitelaw; *Dir:* Christopher Petit. **VHS, Beta $59.95** *MON* ⚔⚔½

The Untamable Unusual silent about volatile woman suffering from schizophrenia.
1923 65m/B Gladys Walton, Malcolm McGregor, John Seinpolis, Etta Lee; *Dir:* Herbert Blache. **VHS, Beta $19.95** *GPV, FCT, DNB* ⚔⚔

Until September An American tourist becomes stranded in Paris. She meets and falls in love with a married banker while she is stuck in her hotel. Routine romance.
1984 (R) 96m/C Karen Allen, Thierry Lhermitte, Christopher Cazenove, Johanna Pavlis; *Dir:* Richard Marquand. **VHS, Beta $79.95** *MGM* ⚔⚔

Until They Get Me Northern Mountie rides horse in silence as he tracks criminal to the edge of the earth.
1918 58m/B Pauline Starke, Joe King. **VHS, Beta $24.95** *GPV, FCT* ⚔½

The Untouchables A big-budget, fast-paced and exciting re-evaluation of the famous television series about the real-life battle between Treasury officer Eliot Ness and crime boss Al Capone in 1920s' Chicago. Costner shows us the change in Ness from naive idealism to steely conviction. Splendid performances by De Niro as Capone and Connery as the cop playing mentor to Ness. Imaginative and beautifully filmed with excellent special effects. Music by Ennio Morricone and script by David Mamet.
1987 (R) 119m/C Kevin Costner, Sean Connery, Robert De Niro, Andy Garcia, Charles Martin Smith, Billy Drago, Richard Bradford, Jack Kehoe; *Dir:* Brian DePalma. Academy Awards '87: Best Supporting Actor (Connery); British Academy Awards '87: Best Original Score; National Board of Review Awards '87: 10 Best Films of the Year, Best Supporting Actor (Connery). **VHS, Beta, LV, 8mm $14.95** *PAR, PIA, BTV* ⚔⚔⚔½

Up the Academy Four teenaged delinquents are sent to an academy for wayward boys where they encounter a sadistic headmaster and a gay dance instructor. Sometimes inventive, often tasteless fare from "Mad" magazine.
1980 (R) 88m/C Ron Leibman, Ralph Macchio, Barbara Bach, Tom Poston, Stacey Nelkin, Wendell Brown, Tom Citera; *Dir:* Robert Downey. **VHS, Beta $19.98** *WAR* ⚔

Up Against the Wall A black kid from the Chicago projects attends school in the affluent suburbs, but there too he must resist temptation, crime and violence. A well-intentioned but didactic cautionary drama, adapt-

ed from the book by African-American author/commentator Dr. Jawanza Kunjufu.
1991 (PG-13) 103m/C Marla Gibbs, Stoney Jackson, Catero Colbert, Ron O'Neal, Salli Richardson; *Dir:* Ron O'Neal. **VHS $79.98** *BMG* ⚔

Up in Arms Danny Kaye's first film presents a typical Kaye scenario: he plays a twitching hypochondriac who is drafted into the Army and sneaks his girlfriend aboard the troopship bound for the Pacific. Features an appearance by the Goldwyn Girls. Danny performs his famous "Lobby Number" and "Malady in 4-F." Based on the film "The Nervous Wreck."
1944 105m/C Danny Kaye, Dinah Shore, Constance Dowling, Dana Andrews, Margaret Dumont, Goldwyn Girls, Lyle Talbot, Louis Calhern; *Dir:* Elliott Nugent. **VHS, Beta $19.98** *SUE* ⚔⚔

Up the Creek Crazy antics abound in this tale of an old British destroyer, a black-market scheme run by the crew, and the new skipper whose hobby is rocket building. Followed by "Further Up the River."
1958 83m/B *GB* David Tomlinson, Wilfrid Hyde-White, Peter Sellers, Vera Day, Michael Goodliffe, Lionel Jeffries; *Dir:* Val Guest. **VHS, Beta $39.95** *MON* ⚔⚔

Up the Creek Four college losers enter a whitewater raft race to gain some respect for their school. The soundtrack features songs by Heart, Cheap Trick and The Beach Boys. Routine.
1984 (R) 95m/C Tim Matheson, Jennifer Runyon, Stephen Furst, John Hillerman, James B. Sikking, Julia Montgomery, Jeana Tomasina; *Dir:* Robert Butler. **VHS, Beta, LV $29.98** *LIV, VES* ⚔½

Up from the Depths Something from beneath the ocean is turning the paradise of Hawaii into a nightmare. Prehistoric fish are returning to the surface with one thing on their minds—lunch. A "Jaws" rip-off played for humor.
1979 (R) 85m/C Sam Bottoms, Suzanne Reed; *Dir:* Charles B. Griffith. **VHS, Beta $59.98** *LIV, VES* ⚔

Up the Down Staircase A naive, newly trained New York public school teacher is determined to teach the finer points of English literature to a group of poor students. She almost gives up until one student actually begins to learn. Good production and acting. Based on Bel Kaufman's novel.
1967 124m/C Sandy Dennis, Patrick Bedford, Eileen Heckart, Ruth White, Jean Stapleton, Sorrell Brooke; *Dir:* Robert Mulligan. **VHS** *WAR, OM* ⚔⚔⚔

Up Periscope Garner is a demolitions expert unwillingly assigned to a submarine commanded by O'Brien. His mission: to sneak onto a Japanese-held island and steal a top-secret code book. Trouble is, O'Brien may not wait for Garner to complete his mission before taking the sub back underwater. A routine submarine film.
1959 111m/C James Garner, Edmond O'Brien, Andra Martin, Alan Hale Jr., Carleton Carpenter, Frank Gifford, Richard Bakalayan; *Dir:* Gordon Douglas. **VHS** *WAR* ⚔⚔½

Up River Wealthy land baron rapes and murders the wife of a simple pioneer, who is beaten and whose homestead is burned by the baron as well. The pioneer lives to seek terrible vengeance.
1979 90m/C Jeff Corey, Morgan Stevens, Debbie AuLuce; *Dir:* Carl Kitt. **VHS $69.95** *ASE* ⚔

Up the Sandbox A bored housewife fantasizes about her life in order to avoid facing her mundane existence. Fine acting from Streisand, with real problems of young mothers accurately shown.
1972 (R) 98m/C Barbra Streisand, David Selby, Jane Hoffman, Barbara Rhoades; ***Dir:*** Irvin Kershner. **VHS, Beta $19.98** *WAR* 🦴🦴

Up Your Alley A female reporter, pursuing a story on homelessness, and a skidrow bum find romance. Langston, who produced and co-wrote the film, is better known as "The Unknown Comic."
1989 (R) 90m/C Linda Blair, Murray Langston, Ruth Buzzi, Johnny Dark, Bob Zany, Yakov Smirnoff; ***Dir:*** Bob Logan. **VHS, Beta $89.95** *IVE* 🦴🦴

Up Your Anchor Two hyperactive nerds enlist in the Navy, and pursue every sexual fantasy within their reach.
1985 89m/C *IS* Yftach Katzur, Zachi Nay; ***Dir:*** Dan Wolman. **VHS, Beta $79.95** *MGM Woof!*

Uphill All the Way A couple of card-cheatin' good old boys are pursued by posses and cavalry alike, and end up killing real outlaws. Made for television.
1985 91m/C Roy Clark, Mel Tillis, Glen Campbell, Trish Van Devere, Burl Ives, Burt Reynolds; ***Dir:*** Frank Q. Dobbs. **VHS, Beta $19.95** *NWV, STE* 🦴🦴

Upper Crust A tale of corruption and greed that takes place at the highest levels of American business and government.
1988 95m/C Frank Gorshin, Broderick Crawford, Nigel Davenport. **VHS, Beta** *DMD* 🦴

The Uprising A Nicaraguan drama filmed just months after the 1979 Sandinista revolution, wherein a young guard of Somoza retains his job for the money, while his father is active in the revolutionary forces. In Spanish with English subtitles.
1981 96m/C *SP* ***Dir:*** Peter Lilienthal. **VHS, Beta $69.95** *KIV* 🦴🦴🦴

Uptown Angel A young woman is determined to work her way out of her home on the wrong side of town. Black cast brings new life to old premise.
1990 (R) 90m/C Caron Tate, Cliff McMullen, Gloria Davis Hill, Tracy Hill; ***W/Dir:*** Joy Shannon. **VHS $59.95** *XVC* 🦴🦴

Uptown Comedy Express Arsenio and other comedians gather for this laugh fest.
1989 60m/C Arsenio Hall, Chris Rock, Barry Sobel, Robert Townsend, Marsha Warfield; ***Dir:*** Russ Petranto; ***Hosted:*** Ray Murphy. **VHS $59.99** *HBO, TLF* 🦴🦴

Uptown New York A young man pressured by his family marries a rich girl instead of the woman he loves, who in turn marries a man she does not love. Routine melodrama.
1932 81m/B Jack Oakie, Shirley Green, Leon Waycoff, Shirley Grey, George Cooper, Raymond Hatton; ***Dir:*** Victor Schertzinger. **VHS, Beta $19.95** *NOS, FCT, DVT* 🦴🦴

Uptown Saturday Night Two working men attempt to recover a stolen lottery ticket from the black underworld after being ripped off at an illegal gambling place. Good fun, with nice performances from both leads and from Belafonte doing a black "Godfather" parody of Brando. Followed by "Let's Do It Again."
1974 (PG) 104m/C Sidney Poitier, Bill Cosby, Harry Belafonte, Flip Wilson, Richard Pryor, Calvin Lockhart; ***Dir:*** Sidney Poitier. **VHS, Beta $19.98** *WAR* 🦴🦴½

The Uranium Conspiracy A secret agent and a mercenary soldier try to stop a shipment of uranium out of Zaire from falling into enemy hands.
1978 (PG) 100m/C Fabio Testi, Janet Agren, Assaf Dayan. **VHS, Beta $42.95** *PGN* 🦴½

Urban Cowboy A young Texas farmer comes to Houston to work in a refinery. After work he hangs out at Gilley's, a roadhouse bar. Here he and his friends, dressed in their cowboy gear, drink, fight, and prove their manhood by riding a mechanical bull. Film made Winger a star, was an up in Travolta's roller coaster career, and began the craze for country western apparel and dance and them there mechanical bulls. Ride 'em, cowboy!
1980 (PG) 135m/C John Travolta, Debra Winger, Scott Glenn, Madolyn Smith, Barry Corbin; ***Dir:*** James Bridges. **VHS, Beta, LV** *PAR* 🦴🦴½

Urban Warriors Savage barbarians roam and pillage a post-nuclear war Earth, but one invincible warrior ushers them out.
1975 90m/C Karl Landgren, Alex Vitale, Deborah Keith; ***Dir:*** Joseph Warren. **VHS, Beta $59.95** *CAN* 🦴

Urbanscape A collection of four award-winning cartoon shorts from the Hubley Studio, including "The Hole" (1962), "Of Men and Demons" (1969), "Harlem Wednesday" (1957), and "Urbanissimo" (1966).
1986 41m/C Academy Awards '62: Best Animated Short Film ("The Hole"). **VHS, Beta $49.95** *DVT* 🦴🦴🦴

Urge to Kill A man returning from a mental hospital after pleading temporary insanity to the murder of his girlfriend returns to her town and tries to uncover the truth, with the help of her younger sister. Made for television.
1984 96m/C Karl Malden, Holly Hunter, William Devane, Alex MacArthur; ***Dir:*** Mike Robe. **VHS, Beta $59.98** *FOX* 🦴

URGH! A Music War Thirty-seven once-familiar new wave bands from Devo to XTC got together for this historic concert. Acts include Dead Kennedys, Chelsea, Police, Pere Ubu, X, Echo, Kluas Naomi, OMD and lots more.
1981 124m/C The Go Go's, The Cramps, XTC, Devo, The Police, Steel Pulse. **VHS, Beta $29.95** *MVD, FOX*

Ursus in the Valley of the Lions Ursus attempts to rescue love from druids but is dismayed to find that she has taken up homicide in her free time. Originally titled, "The Mighty Ursus."
1962 92m/C *IT SP* Ed Fury, Luis Prendes, Moira Orfei, Cristina Gajoni, Maria Luisa Merlo; ***Dir:*** Carlo Campogalliani. **VHS $19.98** *SNC* 🦴½

Used Cars A car dealer is desperate to put his jalopy shop competitors out of business. The owners go to great lengths to stay afloat. Sometimes too obnoxious, but often funny.
1980 (R) 113m/C Kurt Russell, Jack Warden, Deborah Harmon, Gerrit Graham, Joe Flaherty, Michael McKean; ***Dir:*** Robert Zemeckis. **VHS, Beta, LV, 8mm $12.95** *COL* 🦴🦴🦴

Users A small-town girl who worked as a prostitute meets a faded film star and becomes involved in the movie business. Ode to decadent Hollywood. Based on the Joyce Haber novel. Made for television.
1978 125m/C Jaclyn Smith, Tony Curtis, John Forsythe, Red Buttons, George Hamilton; ***Dir:*** Joseph Hardy. **VHS, Beta $49.95** *PSM* 🦴½

Utah A musical comedy star who inherits a ranch wishes to sell it to finance one of her shows. She is persuaded not to by Rogers, the ranch foreman.
1945 54m/B Roy Rogers, Dale Evans, George "Gabby" Hayes; ***Dir:*** John English. **VHS, Beta $19.95** *NOS, VDM, VCN* 🦴🦴

Utah Trail Ritter plays a lawman hired to stop a gang of cattle rustlers. A cheapie oater made for Grand National, which went bankrupt soon after.
1938 57m/B Tex Ritter; ***Dir:*** Al Herman. **VHS, Beta $19.95** *NOS, VCN, DVT* 🦴½

Utilities A frustrated social worker enlists the help of his friends in his efforts to impress an attractive policewoman by taking on a large corporate utility.
1983 (PG) 94m/C *CA* Robert Hays, Brooke Adams, John Marley, Ben Gordon, Helen Burns; ***Dir:*** Harvey Hart. **VHS, Beta, LV $79.98** *LIV, VES* 🦴½

Utopia The comic duo inherit an island rich in uranium. L & H's final film is poorly scripted and directed. Although aged, the comics still deliver laughs.
1952 80m/B Stan Laurel, Oliver Hardy, Suzy Delair, Max Elloy; ***Dir:*** Leo Joannon. **VHS $19.95** *NOS, LOO, AMV* 🦴🦴½

The Utopia Sampler Utopia, featuring Todd Rundgren, performs three songs on this Video 45: "Hammer in My Heart," "You Make Me Crazy," and "Feet Don't Fail Me Now."
1983 11m/C VHS, Beta $9.95 *SVS*

Utu A Maori tribesman serving with the colonizing British army in 1870 explodes into ritual revenge when his home village is slaughtered. Filmed in New Zealand.
1983 (R) 104m/C *NZ* Anzac Wallace, Kelly Johnson, Tim Elliot, Bruno Lawrence; ***Dir:*** Geoff Murphy. **VHS, Beta $79.98** *FOX* 🦴🦴🦴

The Vagabond Chaplin portrays a pathetic fiddler making a scanty living. Silent with musical soundtrack added.
1916 20m/B Charlie Chaplin; ***Dir:*** Charlie Chaplin. **VHS, Beta** *CAB, FST* 🦴🦴🦴½

Vagabond Bleak, emotionally shattering, powerful and compelling, this film traces the peripatetic life of an amoral and selfish young French woman who has no regard for social rules and tremendous fear of responsibility in her drifting yet inexorable journey into death. Told via flashbacks from the moment when she is found dead, alone, and unaccounted for by the roadside, this film will not leave you unscathed. Written by New Wave director Varda. In French with English subtitles.
1985 105m/C *FR* Sandrine Bonnaire, Macha Meril, Stephane Freiss, Elaine Cortadellas, Marthe Jarnias, Yolanda Moreau; ***Dir:*** Agnes Varda. Venice Film Festival '85: Golden Lion. **VHS, Beta, LV $29.95** *PAV* 🦴🦴🦴½

Vagabond Lover The amusing tale of the loves, hopes, and dreams of an aspiring saxophone player. Rudy croons through his megaphone in his movie debut. Appealing, with Dressler sparkling in the role of the wealthy aunt.
1929 66m/B Rudy Vallee, Sally Blane, Marie Dressler; ***Dir:*** Marshall Neilan. **VHS, Beta $29.95** *VYY, RXM* 🦴🦴½

Valdez is Coming Though not a world-class western, this filmed-in-Spain saga does feature a probing script (based on an Elmore Leonard novel) on the nature of race relations. Fine performance by Lancaster as a Mexican-American who ignites the passions

of a town, and ultimately confronts the local land baron.
1971 (PG) 90m/C Burt Lancaster, Susan Clark, Jon Cypher, Barton Heyman, Richard Jordan, Frank Silvera, Hector Elizondo; ***Dir:*** Edwin Sherin. **VHS $59.95** *SHE* 🦴🦴½

Valentino Returns A kid in 1955 California buys a pink Cadillac he nicknames "Valentino Returns," thinking it will help him meet girls, as his parents undergo a stormy divorce. Weak script fades fast from fine premise, good start. Nice period feel.
1988 (R) 97m/C Frederic Forrest, Veronica Cartwright, Jenny Wright, Barry Tubb; ***Dir:*** Peter Hoffman. **VHS, Beta $29.95** *VMK* 🦴🦴

Valet Girls Two Hollywood parking services engage in comedic competitive warfare by offering "special service" to their customers.
1986 89m/C Meri D. Marshall, April Stewart, Mary Kohnert, Christopher Weeks; ***Dir:*** Rafal Zielinski. **VHS, Beta, LV $79.98** *LIV, VES* 🦴

The Valley Hipsters search for Utopia in New Guinea and find sex, spirits, and Pink Floyd. Dated groovy road tripper with beautiful photography. In French with English subtitles.
19?? 106m/C ***Dir:*** Barbet Schroeder. **VHS, Beta $59.99** *WAR, BTV, TAM* 🦴🦴

Valley of the Eagles When his wife and partner take off into the Lapland wilderness with his new invention, a scientist leads the police across the tundra in pursuit.
1951 83m/B *GB* Jack Warner, Nadia Gray, John McCallum, Martin Boddeg, Christopher Lee; ***Dir:*** Terence Young. **VHS, Beta $39.95** *MSP* 🦴½

Valley of Fire Autry is the mayor of a boom town that suffers from a lack of women. He decides to import a caravan of brides for the men in town but villains attempt to kidnap the women.
1951 63m/B Gene Autry, Gail Davis, Pat Buttram; ***Dir:*** John English. **VHS, Beta $29.98** *CCB* 🦴

Valley Girl Slight but surprisingly likeable teen romantic-comedy inspired by Frank Zappa novelty tune. Title stereotype falls for a leather-jacketed rebel. Really. It may look like a music video, but the story is straight from "Romeo and Juliet" via Southern California. Helped launch Cage's career. Music by Men at Work, Culture Club, and others.
1983 (R) 95m/C Nicolas Cage, Deborah Foreman, Colleen Camp, Frederic Forrest; ***Dir:*** Martha Coolidge. **VHS, Beta, LV $29.98** *VES* 🦴🦴½

The Valley of Gwangi One of the best prehistoric-monster-westerns out there. Cowboys discover a lost valley of dinosaurs and try to capture a vicious, carnivorous allosaurus. Bad move, kemosabe! The creatures move via the stop-motion model animation by f/x maestro Ray Harryhausen, here at his finest.
1969 (G) 95m/C James Franciscus, Gila Golan, Richard Carlson, Laurence Naismith, Freda Jackson; ***Dir:*** James O'Connolly. **VHS, Beta, LV $59.95** *WAR, PIA* 🦴🦴🦴

Valley of the Lawless A cowboy rides into trouble aplenty while out searching for treasure. A routine western for Brown fans only.
1936 59m/B Johnny Mack Brown; ***W/Dir:*** Robert N. Bradbury. **VHS, Beta, LV** *WGE* 🦴

Valley of the Sun A government agent tracks a crooked Indian liaison to prevent an Indian uprising in Arizona. Ball, well before her "Lucy" days, plays the restaurant owner both men romance. Better than average, with good cast, some laughs, and lots of excitement.
1942 84m/B Lucille Ball, James Craig, Cedric Hardwicke, Dean Jagger, Peter Whitney, Billy Gilbert, Tom Tyler, Antonio Moreno, George Cleveland, Hank Bell; ***Dir:*** George Marshall. **VHS, Beta** *RKO* 🦴🦴🦴

Valley of Terror Maynard is framed for cattle rustling. Can he prove his innocence?
1938 59m/B Kermit Maynard, Rocky the Horse, Harley Wood, John Merton, Jack Ingram, Dick Curtis, Roger Williams; ***Dir:*** Al Herman. **VHS, Beta $19.98** *NOS, CCB, DVT* 🦴½

Valley of Wanted Men Three cons escape during a jailbreak and one uses his wits and nerve to prove another was framed.
1935 56m/B Frankie Darro. **VHS, Beta, 8mm $24.95** *VYY* 🦴½

Valmont Another adaptation of the Choderlos de Laclos novel "Les Liaisons Dangereuses." Various members of the French aristocracy in 1782 mercilessly play each other for fools in a complex game of lust and deception. Firth and Bening are at first playfully sensual, then the stakes get too high. They share an interesting bathtub scene. Well-acted, the 1988 Frears version, "Dangerous Liaisons," is edgier. Seeing the two films together makes for interesting comparisons of characters and styles.
1989 (R) 137m/C Colin Firth, Meg Tilly, Annette Bening, Fairuza Balk, Sian Philips, Jeffrey Jones, Fabia Drake, Henry Thomas, Vincent Schiavelli, T.P. McKenna, Ian McNeice; ***Dir:*** Milos Forman. **VHS, Beta, LV $89.98** *ORI* 🦴🦴🦴

Vals The glibly hip lethargy which is characteristic of four southern California teenaged girls turns to socially-conscious resolve when a local orphanage is threatened.
1985 (R) 100m/C Jill Carroll, Elana Stratheros, Gina Calabrese, Michelle Laurita, Chuck Connors, Sonny Bono, John Carradine, Michael Leon; ***Dir:*** James Polakof. **VHS, Beta $69.98** *LIV, VES* 🦴

Vamp Two college freshmen encounter a slew of weird, semi-vampiric people in a seamy red-light district nightclub. Starts cute but goes kinky. Jones is great as the stripping vampire.
1986 (R) 93m/C Grace Jones, Chris Makepeace, Robert Rusler, Gedde Watanabe; ***Dir:*** Richard Wenk. **VHS, Beta, LV $19.95** *NWV, STE* 🦴½

Vamping A struggling musician plans to burglarize a wealthy widow's home. He finds himself attracted to his proposed victim. Nice atmosphere, not much else to recommend it.
1984 (PG) 110m/C Patrick Duffy, Catherine Hyland, Rod Arrants, Fred A. Keller; ***Dir:*** Frederick King Keller. **VHS, Beta $59.98** *FOX* 🦴

The Vampire Pale man with big teeth attempts to swindle a beautiful babe out of fortune. Followed by "The Vampire's Coffin."
1957 95m/B *MX* Abel Salazar, Ariadne Welter, German Robles, Carmen Montejo, Jose Luis Jimenez; ***Dir:*** Fernando Mendez. **VHS $19.98** *SNC* 🦴🦴

The Vampire Bat A mad scientist and a vampire bat and its supernatural demands set the stage for murders in a small town. Sets and actors borrowed from Universal Studios in this low-budget flick that looks and plays better than it should. Weird and very exploitative for 1932, now seems dated.
1932 69m/B Lionel Atwill, Fay Wray, Melvyn Douglas, Dwight Frye, Maude Eburne, George E. Stone; ***Dir:*** Frank Strayer. **VHS, Beta $16.95** *NOS, SNC, WFV* 🦴🦴½

Vampire Circus A circus appears in an isolated Serbian village in the 19th century but instead of bringing joy and happiness, this circus brings only death, mutilation and misery. It seems all the members are vampires who have the unique ability to transform themselves into animals. They intend to take revenge on the small town, whose inhabitants killed their evil ancestor 100 years previously. Excellent Hammer production.
1971 (R) 84m/C *GB* Adrienne Corri, Laurence Payne, Thorley Walters, John Moulder-Brown, Lynne Frederick, Elizabeth Seal, Anthony Corlan, Richard Owens, Domini Blythe, David Prowse; ***Dir:*** Robert Young. **VHS $18.00** *FRG, MLB* 🦴🦴½

Vampire Cop A vampire cop (not to be confused with zombie, maniac, midnight, psycho, future or Robo) teams up with a beautiful reporter to 'collar' a drug kingpin.
1990 89m/C Melissa Moore, Ed Cannon, Terence Jenkins. **VHS $79.95** *ATL* 🦴½

The Vampire Happening An actress discovers to her chagrin that her ancestors were vampires and so is she.
1971 (R) 90m/C *GE* Ferdinand "Ferdy" Mayne; ***Dir:*** Freddie Francis. **VHS, Beta $24.95** *NEG, VCI* 🦴½

The Vampire Hookers Man in make-up recruits bevy of beautiful bloodsuckers to lure warm blooded victims to his castle. High ham performance by Carradine. Alternate titles: "Cemetery Girls," "Sensuous Vampires," "Night of the Bloodsuckers."
1978 (R) 82m/C *PH* John Carradine, Bruce Fairbairn, Trey Wilson, Karen Stride, Lenka Novak, Katie Dolan, Lex Winter; ***Dir:*** Cirio H. Santiago. **VHS, Beta $39.98** *NSV* 🦴

The Vampire Lovers An angry father goes after a lesbian vampire who has ravished his daughter and other young girls in a peaceful European village. Innovative story was soon used in countless other vampire vehicles. Hammer Studio's first horror film with nudity, another addition to the genre which spread rapidly. Based on the story "Carmilla" by Sheridan Le Fanu. Followed by, "Lust for a Vampire."
1970 (R) 91m/C *GB* Ingrid Pitt, Pippa Steele, Madeleine Smith, Peter Cushing, George Cole, Dawn Addams, Kate O'Mara, Ferdinand "Ferdy" Mayne; ***Dir:*** Roy Ward Baker. **VHS, Beta $9.98** *SUE, MLB* 🦴🦴½

Vampire at Midnight A homicide detective stalks a rampaging vampire in Los Angeles. Well done and suspenseful, but falters at the end.
1988 93m/C Jason Williams, Gustav Vintas, Jeanie Moore, Christina Whitaker, Leslie Milne; ***Dir:*** Gregory McClatchy. **VHS, Beta $79.98** *FOX* 🦴½

Vampire Princess Miyu These videos contain two episodes each from the popular Japanese cartoon series.
1990 60m/C VHS, LV $39.95 *CPM* 🦴½

Vampire Raiders - Ninja Queen The white ninjas versus the black ninjas. The evil black ninjas are plotting to infiltrate the hotel industry. The white ninjas come to the rescue.
1989 90m/C Agnes Chan, Chris Peterson; ***Dir:*** Bruce Lambert. **VHS $59.95** *TWE Woof!*

The Vampire's Coffin Count Lavud is relieved when faithful servant removes stake implanted in his heart.
1958 86m/B *MX* Abel Salazar, Ariadne Welter, German Robles; ***Dir:*** Fernando Mendez. **VHS $19.98** *SNC* 🦴½

The Vampire's Ghost A 400-hundred-year-old vampire/zombie, doomed to walk the earth forever, heads the African underworld. He can't be killed and can even go out during the day if he wears sunglasses. His future's so bright he's just gotta wear shades.
1945 59m/B John Abbott, Peggy Stewart; *Dir:* Lesley Selander. VHS $24.95 *NOS, DVT ⚔*

Vampire's Kiss Cage makes this one worthwhile; his twisted transformation from pretentious post-val dude to psychotic yuppie from hell is inspired. If his demented torment of his secretary (Alonso) doesn't give you the creeps, his scene with the cockroach will. Cage fans will enjoy his facial aerobics; Beals fans will appreciate her extensive sucking scenes (she's the vamp of his dreams). More for psych majors than horror buffs.
1988 (R) 103m/C Nicolas Cage, Elizabeth Ashley, Jennifer Beals, Maria Conchita Alonso, Kasi Lemmons, Bob Lujan; *Dir:* Robert Bierman. VHS, Beta $89.99 *HBO ⚔⚔⚔*

Vampyr Dreyer's classic portrays a hazy, dreamlike world full of chilling visions from the point of view of a young man who believes himself surrounded by vampires and who dreams of his own burial in a most disturbing way. Evil lurks around every corner as camera angles, light and shadow sometimes overwhelm plot. A high point in horror films based on a collection of horror stories by Sheridan le Fanu. Music by Wolfgang Zeller. Subtitled in English.
1931 75m/B *GE* Julien West, Sybille Schmitz, Harriet Gerard; *Dir:* Carl Theodor Dreyer. VHS, Beta $29.95 *KIV, NOS, SNC ⚔⚔⚔⚔*

Vampyres Alluring female vampires coerce unsuspecting motorists to their castle for a good time, which ends in death. Anulka was the centerfold girl in Playboy's May 1973 issue. Originally titled "Vampyres, Daughters of Dracula." Also known as "Blood Hunger" and "Satan's Daughter's."
1974 (R) 90m/C *GB* Marianne Morris, Anulka, Murray Brown, Brian Deacon, Sally Faulkner; *Dir:* Joseph Larraz. VHS, Beta $79.95 *MAG ⚔⚔ ½*

The Van A recent high school graduate passes on the college scene so he can spend more time picking up girls in his van. A lame sex (and sexist) comedy.
1977 (R) 92m/C Stuart Getz, Deborah White, Danny DeVito, Harry Moses, Maurice Barkin; *Dir:* Sam Grossman. VHS, Beta $9.95 *NO ⚔*

Van Nuys Blvd. The popular boulevard is the scene where the cool southern California guys converge for cruising and girl watching, so naturally it's where a country hick comes to test his drag racing skills and check out the action.
1979 (R) 93m/C Bill Adler, Cynthia Wood, Dennis Bowen, Melissa Prophet; *Dir:* William Sachs. VHS, Beta $9.99 *STE, UHV ⚔ ½*

Vanessa The story of an innocent young girl's introduction to the erotic pleasures of the Orient.
1977 (X) 90m/C VHS, Beta $34.95 *VID ⚔*

Vanina Vanini An acclaimed Rossellini historical drama about the daughter of an Italian aristocrat in 1824 who nurses and falls in love with a wounded patriot hiding in her house. Based on a Stendhal short story. In Italian with subtitles.
1961 113m/B *IT* Sandra Milo, Laurent Terzieff; *Dir:* Roberto Rossellini. VHS, Beta $59.95 *FCT ⚔⚔⚔ ½*

The Vanishing When his wife suddenly disappears, a young husband finds himself becoming increasingly obsessed with finding her. Three years down the road, his world has become one big, mad nightmare. Then, just as suddenly, the answer confronts him, but the reality of it may be too horrible to face. Well-made dark thriller based on "The Golden Egg" by Time Krabbe; originally known as "Spoorloos." In French and Dutch with English subtitles.
1988 107m/C *NL FR* Barnard Pierre Donnadieu, Johanna Ter Steege, Gene Bervoets; *Dir:* George Sluizer. VHS $89.95 *FXL ⚔⚔⚔*

Vanishing Act While on his honeymoon, Harry Kenyon reports his wife missing to the local police. Within a short period of time his wife is found, but Harry says the women is an impostor. This thriller follows Harry's desperate attempt to get to the truth. Made for television.
1988 (PG) 94m/C Mike Farrell, Margot Kidder, Elliott Gould, Fred Gwynne, Graham Jarvis; *Dir:* David Greene. VHS, Beta $89.95 *VMK ⚔⚔ ½*

Vanishing American The mistreatment of the American Indian is depicted in this sweeping Western epic. Musical score.
1926 114m/B Richard Dix, Noah Beery; *Dir:* George B. Seitz. VHS, Beta $49.95 *GPV, FST, VYY ⚔⚔*

Vanishing Legion Western serial with outdoor action and gunplay. Twelve chapters, 13 minutes each.
1931 156m/B Frankie Darro, Rin Tin Tin Jr, Harry Carey; *Dir:* B. Reeves Eason. VHS, Beta $24.95 *GPV, NOS, VCN ⚔⚔*

Vanishing Point An ex racer makes a bet to deliver a souped-up car from Denver to San Francisco in 15 hours. Taking pep pills along the way, he eludes police, meets up with a number of characters, and finally crashes into a roadblock. Rock score helps attract this film's cult following.
1971 (PG) 98m/C Barry Newman, Cleavon Little, Gilda Texler, Dean Jagger, Paul Koslo, Robert Donner, Severn Darden, Victoria Medlin; *Dir:* Richard Sarafian. VHS, Beta, LV $59.98 *FOX ⚔⚔*

Vanishing Prairie This documentary examines the wonders of nature that abound in the American prairie. From the "True-Life Adventures" series.
1954 60m/C *Dir:* James Algar. Academy Awards '54: Best Feature Documentary. VHS, Beta $69.95 *DIS ⚔⚔⚔*

The Vanishing Westerner A cowboy, falsely accused of murder, works to vindicate himself and uncovers a series of robberies. This one has enough plot twists to take it out of the routine.
1950 60m/B Monte Hale, Arthur Space, Aline Towne, Paul Hurst, Roy Barcroft, Richard Anderson, William Phipps, Rand Brooks; *Dir:* Philip Ford. VHS, Beta $24.95 *DVT ⚔⚔*

Vanishing Wilderness This documentary looks at the wild animals roaming across North America.
1973 (G) 90m/C *Dir:* Arthur Dubs. VHS, Beta $9.98 *MED, VTK ⚔⚔*

Varan the Unbelievable A chemical experiment near a small island in the Japanese archipelago disturbs a prehistoric monster beneath the water. The awakened monster spreads terror on the island. Most difficult part of this movie is deciding what the rubber monster model is supposed to represent.

1961 70m/B *JP* Myron Healy, Tsuruko Kobayashi; *Dir:* Inoshiro Honda. VHS, Beta $19.95 *NOS, UHV ⚔*

Varieties on Parade A vaudeville show on film, featuring an array of second-rate comics and musical personalities. Sad to watch its demise.
1952 55m/B Jackie Coogan, Eddie Garr, Tom Neal, Iris Adrian. VHS, Beta, LV *WGE ⚔*

Variety Simple and tragic tale of a scheming young girl and the two men of whom she takes advantage. The European circus in all its beautiful sadness is the setting. Extraordinary cast and superb cinematography. Silent.
1925 104m/B *GE* Emil Jannings, Lya de Putti, Warwick Ward, Werner Krauss; *Dir:* E.A. Dupont. VHS, Beta $29.95 *NOS, VYY, DVT ⚔⚔⚔*

Variety A girl takes a job selling tickets in a Times Square porn theater. Her initial loathing for the genre gives way to a fascination with its clients, one of whom is a gangster, and she becomes drawn to the seamy world inside. An independently made feature.
1983 101m/C Sandy McLeod, Will Patton, Richard Davidson; *Dir:* Bette Gordon. VHS, Beta $69.95 *MED ⚔⚔ ½*

Variety Lights Fellini's first (albeit joint) directorial effort, wherein a young girl runs away with a travelling vaudeville troupe and soon becomes its main attraction as a dancer. Filled with Fellini's now-familiar delight in the bizarre and sawdust/tinsel entertainment. In Italian with English subtitles.
1951 93m/B *IT* Giulietta Masina, Peppino de Filippo, Carla Del Poggio, Folco Lulli; *Dir:* Federico Fellini, Alberto Lattuada. VHS, Beta $29.95 *CVC, TAM, APD ⚔⚔⚔ ½*

Vasectomy: A Delicate Matter A mother of eight issues a final decree to her husband about their sex life. He must get a vasectomy or there won't be any. As good as it sounds. Also known as "Vasectomy."
1986 (PG-13) 92m/C Paul Sorvino, Abe Vigoda, Cassandra Edwards, Lorne Greene, Ina Balin, June Wilkinson, William Marshall; *Dir:* Robert Burge. VHS, Beta *WOV Woof!*

Vatican Conspiracy The members of the Vatican's College of Cardinals do everything in their power to discredit a newly-appointed radical pontiff.
1981 90m/C Terence Stamp; *Dir:* Marcello Aliprandi. VHS, Beta $59.95 *VCL ⚔ ½*

Vault of Horror A collection of five terrifying tales based on original stories from the E.C. comic books of the 1950s. Stories include, "Midnight Mess," "Bargain in Death," "This Trick'll Kill You," "The Neat Job" and "Drawn and Quartered." Also known as "Tales from the Crypt II."
1973 (R) 86m/C *GB* Terry-Thomas, Curt Jurgens, Glynis Johns, Dawn Addams, Daniel Massey, Tom Baker, Michael Craig, Anna Massey, Denholm Elliott; *Dir:* Roy Ward Baker. VHS, Beta *MED ⚔⚔ ½*

Vegas A private detective, with Las Vegas beauties as assistants and pursuers, solves the murder of a teenage runaway girl. Pilot for television series. Script by Michael Mann who went on to do "Miami Vice."
1978 74m/C Robert Urich, June Allyson, Tony Curtis, Will Sampson, Greg Morris; *Dir:* Richard Lang. VHS, Beta $49.95 *PSM ⚔*

The Vegas Strip Wars Rival casino owners battle it out in the land of lady luck. Unmemorable except for Jones's Don King impersonation and the fact that it was Hudson's last TV movie.

1984 100m/C Rock Hudson, James Earl Jones, Pat Morita, Sharon Stone, Robert Costanzo; *Dir:* George Englund. **VHS** *LTG* ✛✛

Velnio Nuotaka (The Devil's Bride) A popular Lithuanian musical based on the play ''Baltargio Malunas'' by K. Boruta. English subtitles.
1975 78m/C *LI* **VHS, Beta** *IHF* ✛✛ ½

Velvet Smooth A protection agency's sultry boss, Velvet Smooth, gets involved solving the problems of a numbers racket.
1976 (R) 89m/C Johnnie Hill; *Dir:* Janace Fink. **VHS** $42.95 *PGN* ✛

Velvet Touch Well-engineered thriller about an actress who craftily murders her producer. A theater-loving police detective winds up accusing the wrong woman—sending the overwrought murderess into a moral tailspin. Things don't work out as you may expect. Fine acting.
1948 97m/B Rosalind Russell, Leo Genn, Claire Trevor, Sydney Greenstreet, Leon Ames, Frank McHugh, Walter Kingsford, Dan Tobin, Lex Barker, Nydia Westman; *Dir:* John Gage. **VHS, Beta** *RKO* ✛✛✛

The Velvet Vampire Yarnall is a sexy, sun-loving, dune buggy-riding vampiress who seduces a young, sexy, swinging, Southern California couple in her desert home. Lots of atmosphere to go along with the blood and nudity.
1971 (R) 82m/C Michael Blodgett, Sherry Miles, Celeste Yarnall, Gene Shane, Jerry Daniels, Sandy Ward, Paul Prokop, Chris Woodley, Robert Tessier; *Dir:* Stephanie Rothman. **VHS, Beta, LV** $9.95 *SIM, SUE, IPI* ✛✛

Vendetta A woman gets herself arrested and sent to the penitentiary in order to exact revenge there for her sister's death. Better than average sexploitation film works because of acting and pacing.
1985 89m/C Karen Chase, Sandy Martin, Durga McBroom, Kin Shriner; *Dir:* Bruce Logan. **VHS, Beta** $79.98 *LIV, VES* ✛½

Vendetta for the Saint A feature-length episode of the television series ''The Saint,'' in which Simon Templar pursues a Sicilian mobster on a personal vendetta.
1968 98m/C *GB* Roger Moore; *Dir:* James O'Connolly. **VHS, Beta** $59.98 *FOX* ✛

Vengeance A police officer resigns in shame after failing to thwart a holdup. He attempts to redeem himself by working on his own to infiltrate the same gang.
1937 61m/B Lyle Talbot, Wendy Barrie, Wally Albright, Marc Lawrence; *Dir:* Del Lord. **VHS** *MOV* ✛✛

Vengeance Four burglars take hostages after their robbery is bungled, but they won't get away—especially after the hostages die.
1980 92m/C Sally Lockett, Nicholas Jacquez, Bob Elliott; *Dir:* Bob Blizz. **VHS, Beta** $39.98 *MAG* ✛

Vengeance A low-budget film about rebels fighting off an authoritarian government.
1986 114m/C Jason Miller, Lea Massari; *Dir:* Antonio Isasi. **VHS, Beta** $69.95 *TWE* ✛

Vengeance A bereaved young man strikes out against the injustice of his parents' murder. His revenge against the killers takes him several steps beyond the law, and he's not so sure it was the right idea. Made for television. Also known as ''Vengeance: The Story of Tony Cimo.''
1989 (R) 90m/C Brad Davis, Roxanne Hart, Brad Dourif, William Conrad; *Dir:* Marc Daniels. **VHS, LV** $39.95 *IME* ✛✛

Vengeance Is Mine Told in flashbacks, the film focuses on the life of a habitual criminal whose life of deprivation leads to murder. Contains violence and nudity. Based on a true story. In Japanese with English subtitles.
1979 129m/C *JP* Ken Ogata; *Dir:* Shohei Imamura. **VHS, LV** $59.95 *SVS, FCT* ✛✛✛

Vengeance Is Mine A demented farmer captures three criminals and tortures them in horrifyingly sadistic ways. Gratuitously grisly.
1984 (R) 90m/C Ernest Borgnine, Michael J. Pollard, Hollis McLaren; *Dir:* John Trent. **VHS, Beta, LV** $59.95 *VCL, GEM, VYG* Woof!

Vengeance of the Snowmaid Some Kung Fu creepies raped her mom. Now nobody can get a sword in edgewise on the Snow Maid as she seeks revenge.
197? 82m/C Mo Ka Kei, Chen Chen. **VHS, Beta** $54.95 *MAV* ✛½

Vengeance Valley Lancaster and Walker are foster brothers with Walker being an envious weasel who always expects Lancaster to get him out of scrapes. Lancaster is even accused of a crime committed by Walker and must work to clear himself. Good cast is let down by uneven direction.
1951 83m/C Burt Lancaster, Joanne Dru, Robert Walker, Sally Forrest, John Ireland, Hugh O'Brian; *Dir:* Richard Thorpe. **VHS, Beta** $14.95 *NOS, QNE, DVT* ✛✛

Vengeance of the Zombies A madman seeks revenge by setting an army of walking corpses to stalk the streets of London.
1972 90m/C *SP* Paul Naschy; *Dir:* Leon Klimovsky. **VHS, Beta** $19.95 *SNC, ASE* ✛½

Venice Beach Confidential A semi-documentary profile of the eccentric California beach and its variety of strange denizens.
198? 90m/C **VHS, Beta** $29.95 *RHI* ✛½

Venom A deadly black mamba is loose in an elegant townhouse. The snake continually terrorizes an evil kidnapper, his accomplices and his kidnapped victim. Wasted big-name cast all have that far-away look in their eyes - like they wish they were anywhere else.
1982 (R) 92m/C *GB* Sterling Hayden, Klaus Kinski, Sarah Miles, Nicol Williamson, Cornelia Sharpe, Susan George, Michael Gough, Oliver Reed; *Dir:* Piers Haggard. **VHS, Beta** $59.95 *VES* ✛

Venus Against the Son of Hercules Our hero must use all his genetically procured musculature in his battle against the lovely but deadly Venus.
1962 ?m/C *IT* Roger Brown, Jackie Lane. **VHS, Beta** $16.95 *SNC* ✛

Venus on Fire A photography crew searches for the perfect woman for their suntan oil ads on a tropical island, and one magically appears. French softcore, dubbed in English.
1984 91m/C *FR* Odile Michel, Nadege Clair, Francois Blanchard. **VHS, Beta** $29.95 *MED* Woof!

Venus in Furs A jazz musician working in Rio de Janeiro becomes obsessed with a mysterious woman; she resembles a murder victim whose body he discovered months earlier. Weird mix of horror, sadism, black magic, and soft porn.
1970 (R) 90m/C *GB* James Darren, Klaus Kinski, Barbara McNair, Dennis Price, Maria Rohm; *Dir:* Jess (Jesus) Franco. **VHS, Beta** $39.98 *NO* ✛

Venus the Ninja A comely female Ninja battles an evil warlord.
198? 90m/C Pearl Cheung, Meng Fei; *Dir:* Sze Ma Peng. **VHS, Beta** $39.95 *TWE* ✛

Vera Cruz Two soldiers of fortune become involved in the Mexican Revolution of 1866, a stolen shipment of gold, divided loyalties, and gun battles. Less than innovative plot is made into an exciting action flick.
1953 94m/C Gary Cooper, Burt Lancaster, Denise Darcel, Cesar Romero, George Macready, Ernest Borgnine, Charles Bronson; *Dir:* Robert Aldrich. **VHS, Beta** $19.98 *MGM, FOX, FCT* ✛✛ ½

Verboten! In post-war occupied Berlin, an American G.I. falls in love with a German girl. Good direction maintains a steady pace.
1959 93m/B James Best, Susan Cummings, Tom Pittman, Paul Dubov; *Dir:* Samuel Fuller. **VHS** *IME* ✛✛ ½

Verdi The story of the Italian operatic composer and the loves of his life. Dubbed in English.
1953 80m/C *IT* Pierre Cressoy. **VHS** $29.95 *FCT* ✛✛

The Verdict Newman plays an alcoholic failed attorney reduced to ambulance chasing. A friend gives him a supposedly easy malpractice case which turns out to be a last chance at redeeming himself and his career. Screenplay by David Mamet, from the novel by Barry Reed. One of Newman's finest performances.
1982 (R) 122m/C Paul Newman, James Mason, Charlotte Rampling, Jack Warden, Milo O'Shea, Lindsay Crouse; *Dir:* Sidney Lumet. **VHS, Beta, LV** $19.98 *FOX* ✛✛✛

Verne Miller A film of the true story of Verne Miller, Al Capone's hit man. After rescuing a friend from the Feds, Miller is hunted down by both the cops and the mob.
1988 (R) 95m/C Scott Glenn, Barbara Stock, Thomas G. Waites, Lucinda Jenney, Sonny Carl Davis; *Dir:* Rob Hewitt. **VHS, Beta, LV** $19.98 *SUE* ✛✛

Vernon, Florida A quixotic, eccentric documentary by Morris dealing simply with the backwater inhabitants of the small Florida town of the title, each speaking openly about his or her slightly strange lifestyle, dreams and life experiences.
1982 72m/C **VHS, Beta** $79.95 *COL* ✛✛✛

The Vernonia Incident Urban guerillas invade a small town, killing the police chief. The townspeople gather up their shotguns and fight back. That's entertainment.
1989 95m/C David Jackson, Shawn Stevens, Floyd Ragner, Ed Justice, Robert Louis Jakson; *W/Dir:* Ray Etheridge. **VHS, Beta** $9.95 *SIM* ✛½

Veronico Cruz Despite its sincerity, Pereira's feature debut offers muddled response to the human waste incurred by the Falklands War. Set in a tiny remote village in the Argentinean mountains, the film's narrative is derived from the growing friendship between a shepherd boy and a teacher from the city. Well-meaning anti-war movie. Also called ''La Dueda Interna.'' In Spanish with English subtitles.
1987 96m/C *AR GB* Juan Jose Camero, Gonzalo Morales, Rene Olaguivel, Guillermo Delgado; *Dir:* Miguel Pereira. **VHS** $79.95 *CCN* ✛✛½

Veronika Voss Highlights the real life of fallen star Sybille Schmitz who finally took her own life out of despair. Played by Zech, Voss is expoited by her physician to turn over all of her personal belongings for morphine. A lover discovers the corruption and reveals it to the authorities. This causes great upheaval resulting in Voss' suicide. Highly metaphoric and experimental in its treatment of its subject. In German with English subtitles.

1982 105m/C *GE* Conny Froboess, Anna Marie Duringer, Volker Spengler; *Dir:* Rainer Werner Fassbinder. **VHS** $48.00 *APD, FCT 🐾🐾🐾*

Vertigo Hitchcock's romantic story of obsession, manipulation and fear. Stewart plays a detective forced to retire after his fear of heights causes the death of a fellow policeman and, perhaps, the death of a woman he'd been hired to follow. The appearance of her double (Novak), whom he compulsively transforms into the dead girl's image, leads to a mesmerizing cycle of madness and lies. Haunting music by Bernard Herrmann.

1958 (PG) 126m/C James Stewart, Kim Novak, Barbara Bel Geddes, Tom Helmore, Ellen Corby; *Dir:* Alfred Hitchcock. Sight & Sound Survey '82: #7 of the Best Films of All Time (tie). **VHS, Beta, LV** $19.95 *MCA, TLF 🐾🐾🐾🐾*

Very Close Quarters Thirty people face many trials and tribulations as they share an apartment in Moscow. Interesting cast is wasted in this mostly un-funny flick.

1984 97m/C Shelley Winters, Paul Sorvino, Theodore Bikel, Farley Granger; *Dir:* Vladmir Rif. **VHS, Beta** $69.98 *LIV, VES 🐾*

Very Curious Girl A peasant girl realizes she is being used by the male population of her village, and decides to charge them for sex, creating havoc. Also known as "La Fiancee du Pirate" or "Dirty Mary." In French with English subtitles.

1969 105m/C *FR* Bernadette La Font; *Dir:* Nelly Kaplan. **VHS, Beta** *SUE, OM 🐾🐾½*

The Very Edge A pregnant woman suffers a miscarriage after being brutally attacked. The repercussions nearly destroy her marriage until her attacker is caught. A disturbing psycho-drama.

1963 90m/C *GB* Anne Heywood, Richard Todd, Jeremy Brett; *Dir:* Cyril Frankel. **VHS, Beta** $59.95 *VID 🐾🐾½*

A Very Merry Cricket An adventurous cricket learns about the spirit of Christmas. Sequel to "The Cricket in Times Square."

1973 30m/C *Dir:* Chuck Jones. **VHS, Beta** $14.95 *FHE 🐾🐾½*

A Very Old Man with Enormous Wings An angel, battered during a hurricane, seeks refuge on a tiny Caribbean island. There, two men keep him and charge an increasingly curious world admission to view the creature. A colorful, musical expose of human failings based on a story by Gabriel Garcia Marquez. In Spanish with English subtitles.

1988 90m/C *SP IT* Daisy Granados, Asdrubal Melendez, Luis Alberto Ramirez, Fernando Birri; *Dir:* Fernando Birri. **VHS** $79.95 *FXL, FCT 🐾🐾*

A Very Private Affair A movie star finds that she has no privacy from the hordes of fans and paparazzi who flock to her side. The glare of publicity helps to destroy her relationship with a married director, and to fuel her desire for privacy, a desire that ends tragically. An appealing, better-than-average Bardot vehicle, it is nonetheless below-average (early) Malle. In French with English subtitles.

1962 95m/C *FR* Marcello Mastroianni, Brigitte Bardot; *Dir:* Louis Malle. **VHS, Beta** $59.95 *MGM 🐾🐾½*

V.I. Warshawski The filmmakers seem to think they can slum with the oldest cliches in detective shows just as long as the tough gumshoe is a woman. They're wrong. Turner is terrific as the leggy shamus of the title (from a popular series of books by Sara Paretsky), but the plot is nothing special, featuring stock characters in the killing of a pro athlete and a real-estate deal.

1991 (R) 89m/C Kathleen Turner, Jay O. Sanders, Angela Goethals, Charles Durning; *Dir:* Jeff Kanew. **VHS, Beta, LV** $19.99 *TOU, HPH 🐾🐾*

Vibes Two screwball psychics are sent on a wild goose chase through the Ecuadorean Andes in search of cosmic power. They fall in love. Flat offering from the usually successful team of writers, Lowell Ganz and Babaloo Mandel ("Splash" and "Night Shift"). Lauper's first starring role (and so far, her last).

1988 (PG) 99m/C Jeff Goldblum, Cyndi Lauper, Julian Sands, Googy Gress, Peter Falk, Elizabeth Pena; *Dir:* Ken Kwapis. **VHS, Beta** $89.95 *COL 🐾*

Vice Academy Two females join the Hollywood vice squad. Allen was a former porn queen.

1988 (R) 90m/C Linnea Quigley, Ginger Lynn Allen, Karen Russell, Jayne Hamil, Ken Abraham, Stephen Steward, Jeannie Carol; *Dir:* Rick Sloane. **VHS, Beta, LV** $79.95 *PSM 🐾½*

Vice Academy 2 Two vice cop babes try to stop a female crime boss from dumping aphrodisiacs in the city's water supply.

1990 (R) 90m/C Linnea Quigley, Ginger Lynn Allen, Jayne Hamil, Scott Layne, Jay Richardson, Joe Brewer, Marina Benvenga, Teagan Clive; *Dir:* Rick Sloane. **VHS, Beta** $29.95 *PSM 🐾*

Vice Academy 3 The worst of the series, and not just because cult actress Linnea Quigley is absent. Wit, pacing and even sets are nonexistent as the girls battle a toxic villainess called Malathion.

1991 (R) 88m/C Ginger Lynn Allen, Elizabeth Kaitan, Julia Parton, Jay Richardson, Johanna Grika, Steve Mateo; *W/Dir:* Rick Sloane. **VHS, Beta** $79.95 *PSM Woof!*

Vice Squad Lukas is forced to turn police informer to save his own neck, which almost ruins him. Just when he thinks he's free of the cops they come back threatening the woman he loves unless he plays stoolie again. Rather flat, but the different cop/criminal angle makes for some unusual twists.

1931 80m/B Paul Lukas, Kay Francis, Helen Johnson, William B. Davidson, Esther Howard; *Dir:* John Cromwell. **VHS, Beta, LV** $19.98 *SUE 🐾🐾*

Vice Squad A violent and twisted killer-pimp goes on a murderous rampage, and a hooker helps a vice squad plainclothesman trap him. Sleazy and disturbing, with little to recommend it.

1982 (R) 97m/C Wings Hauser, Season Hubley, Gary Swanson, Cheryl "Rainbeaux" Smith; *Dir:* Gary Sherman. **VHS, Beta, LV** $19.98 *SUE 🐾*

Vice Versa Another 80s comedy about a workaholic father and his 11-year-old son who switch bodies, with predictable slapstick results. Reinhold and Savage carry this, appearing to have a great time in spite of over-done story.

1988 (PG) 97m/C Judge Reinhold, Fred Savage, Swoosie Kurtz, David Proval, Corinne Bohrer, Jane Kaczmarek, William Prince, Gloria Gifford; *Dir:* Brian Gilbert. **VHS, Beta, LV** $14.95 *COL 🐾🐾🐾*

Vicious A bored young woman falls in with people your mother warned you about and soon their high school hi-jinks turn into murder. Graphic violence.

1988 88m/C Tamblyn Lord, Craig Pearce, Tiffany Dowe; *Dir:* Karl Zwicky. **VHS** $79.95 *SVS 🐾*

The Vicious Circle A prominent London physician becomes involved in murder and an international crime ring when he agrees to perform an errand for a friend. Good performances and tight pacing. Also known as "The Circle."

1957 84m/B *GB* John Mills, Wilfrid Hyde-White, Rene Ray, Lionel Jeffries, Noelle Middleton; *Dir:* Gerald Thomas. **VHS, Beta** *MON 🐾🐾½*

Victim A successful married English barrister (Bogarde) with a hidden history of homosexuality is threatened by blackmail after the death of his ex-lover. When the blackmailers, who are responsible for his lover's suicide, are caught, Bogarde decides to prosecute them himself, even though it means revealing his hidden past. One of the first films to deal straightforwardly with homosexuality. Fine performances.

1962 100m/C *GB* Dirk Bogarde, Sylvia Syms, Dennis Price, Peter McEnery; *Dir:* Basil Dearden. **VHS, Beta** $19.98 *IND, SUE 🐾🐾🐾½*

The Victim Chan Wing meets up with Master Leung and decides to become his follower. Intent on impressing the master, Chan proves an inept chopsocker.

19?? 90m/C Samo Hung. **VHS, Beta** $24.50 *OCE 🐾🐾*

Victims Four young women head out for a weekend in the country not realizing that a host of freaks await them.

19?? (R) 85m/C **VHS** $59.95 *NO 🐾½*

Victims of the Assassin Two lovers are sentenced to decapitation but they use their martial arts skills to escape and wreak vengeance on those who would destroy them.

1988 ?m/C **VHS** $19.95 *OCE 🐾½*

Victor/Victoria An unsuccessful actress in Depression-era Paris impersonates a man impersonating a woman and becomes a star. Luscious music and sets. Warren as femme fatale and Preston as Andrews' gay mentor are right on target; Garner is charming as the gangster who falls for the woman he thinks she is.

1982 (PG) 133m/C Julie Andrews, James Garner, Robert Preston, Lesley Ann Warren, Alex Karras, John Rhys-Davies; *Dir:* Blake Edwards. Academy Awards '82: Best Musical Score. **VHS, Beta, LV** $19.95 *MGM 🐾🐾🐾*

Victoria Regina In a succession of vignettes, the life of Queen Victoria is viewed, from her ascension to the throne of England in 1837 through the celebration of her Diamond Jubilee. A presentation from "George Schaefer's Showcase Theatre." Made for television.

1961 76m/C Julie Harris, James Donald, Felix Aylmer, Pamela Brown, Basil Rathbone; *Dir:* George Schaefer. Emmy Awards '61: Program of the Year; Outstanding Single Performance/ Actress (Harris) '61: Program of the Year; Outstanding Performance in Supporting Role/ Actress '61: Program of the Year. **VHS, Beta** *FHS 🐾🐾*

Victorian Fantasies A softcore comedy about Victorian adults unleashing wild sexual fantasies.

1986 90m/C Ollie Soltoft, Sue Longhurst, Diana Dors. **VHS, Beta** $29.95 *MED 🐾½*

Victory A soccer match between World War II American prisoners of war and a German team is set up so that the players can escape through the sewer tunnels of Paris. Of course they want to finish the game first. Pele is the only part worth watching. 1981 (PG) 116m/C Sylvester Stallone, Michael Caine, Max von Sydow, Pele, Carole Laure, Bobby Moore, Daniel Massey; *Dir:* John Huston. **VHS, Beta** $29.98 *FOX, WAR* 🎞️½

The Victory In 1967 Montreal, an American college exchange student falls in love with a French-Canadian co-ed. Can love survive their different cultures? 1988 (PG) 95m/C Vincent Van Patten, Cloris Leachman, Eddie Albert, Claire Pimpare, Nicholas Campbell, Jack Wetherall, Jacques Godin, Marthe Mercure; *Dir:* Larry Kent. **VHS, Beta** $9.99 *STE, PSM* 🎞️🎞️

Video Americana A collection of seven video shorts by various avant-garde artists depicting the absurdities of modern life. 1985 60m/C **VHS, Beta** *SVA* 🎞️🎞️

Video Baby The joys and frustrations of parenting a video child can be yours without the mess, expense or longterm responsibility. Watch baby laugh, cry and sleep on command. 1989 13m/C **VHS, Beta** $9.98 *LIV, CRP, VTK* 🎞️

The Video Dead Gore-farce in which murderous zombies emerge from a possessed television and wreak havoc. 1987 (R) 90m/C Roxanna Augesen, Rocky Duvall, Michael St. Michaels; *Dir:* Robert Scott. **VHS, Beta** $19.98 *SUE Woof!*

Video Murders A police detective tracks down a rapist/murderer who tapes all his own crimes. 1987 90m/C Eric Brown, Virginia Loridans, John Ferita. **VHS, Beta** $19.98 *TWE* 🎞️

Video Violence A gory spoof about a video store owner who discovers that his customers have grown bored with the usual Hollywood horror movies and decide to shoot some flicks of their own. Followed by a sequel. 1987 90m/C Art Neill, Jackie Neill, William Toddie, Bart Summer; *Dir:* Gary P. Cohen. **VHS, Beta** $49.95 *CAM* 🎞️

Video Violence Part 2...The Exploitation! Two sickos named Howard and Eli run a cable television network where talk show guests are spindled and mutilated. 1987 90m/C Uke, Bart Sumner, Lee Miller; *Dir:* Gary P. Cohen. **VHS, Beta** *CAM* 🎞️

Video Vixens A television executive shakes up a permissive society by producing the most erotic awards show ever seen on television. Basic soft-core nudie. 1984 85m/C Robyn Hilton, Sandy Dempsey. **VHS, Beta** $69.98 *LIV, VES Woof!*

Video Wars Video games explode randomly. A wicked computer whiz is behind it all in a blackmail scheme. 198? 90m/C George Diamond. **VHS** $19.99 *BFV Woof!*

Video Yesterbloop A collection of bloopers from television, including the "Steve Allen Show," "All My Children," "The Price Is Right," "Happy Days," "Mork and Mindy," and "One Day at a Time." Contains some nudity and strong language. 197? 81m/C Steve Allen, Ron Howard, Robin Williams, Henry Winkler. **VHS, Beta** $49.95 *VYY Woof!*

Videodrome Woods is a cable TV programmer with a secret yen for sex and violence, which he satisfies by watching a pirated TV show. "Videodrome" appears to show actual torture and murder, and also seems to control the thoughts of its viewers—turning them into human VCRs. Cronenburg's usual sick fantasies are definitely love 'em or leave 'em. Special effects by Rick Baker. 1983 87m/C CA James Woods, Deborah Harry, Sonja Smitts, Peter Dvorsky; *Dir:* David Cronenberg. Genie Awards '84: Best Director (Cronenberg). **VHS, Beta, LV** $19.95 *MCA* 🎞️🎞️

Vietnam: Chronicle of a War Drawing upon the resources of the CBS News archives, this CBS News Collectors Series program presents a retrospective portrait of American military involvement as witnessed by on-the-scene correspondents and camera crews. Some portions are in black-and-white. 1981 88m/C *Nar:* Walter Cronkite, Dan Rather, Morley Safer, Charles Collingwood, Charles Kuralt, Mike Wallace, Eric Sevareid. **VHS, Beta** $49.98 *FOX, KUI* 🎞️🎞️

Vietnam, Texas Vietnam vet leaves his past to become a priest. But he returns to violence when he discovers his Vietnamese daughter is in the hands of Houston's most relentless gangster. 1990 (R) 101m/C Robert Ginty, Haing S. Ngor, Tamlyn Tomita, Tim Thomerson; *Dir:* Robert Ginty. **VHS, Beta, LV** $89.95 *COL* 🎞️

A View to a Kill This James Bond mission takes him to the United States, where he must stop the evil Max Zorin from destroying California's Silicon Valley. Feeble and unexciting plot with unscary villain. Duran Duran performs the catchy title tune. Moore's last appearance as 007. 1985 (PG) 131m/C GB Roger Moore, Christopher Walken, Tanya Roberts, Grace Jones, Patrick Macnee, Lois Maxwell, Dolph Lundgren, Desmond Llewelyn; *Dir:* John Glen. **VHS, Beta, LV** $19.98 *FOX, TLF* 🎞️🎞️

Vigil A stark, dreamy parable about a young girl, living on a primitive farm in a remote New Zealand valley, who watches her family collapse after a stranger enters their territory. Visually ravishing and grim; Ward's first American import. Predecessor to "The Navigator." 1984 90m/C NZ Penelope Stewart, Bill Kerr, Fiona Kay, Gordon Shields, Frank Whitten; *Dir:* Vincent Ward. **VHS, Beta** $79.95 *PSM* 🎞️🎞️🎞️

Vigilante A frustrated ex-cop, tired of seeing criminals returned to the street, joins a vigilante squad dedicated to law and order. Often ridiculous and heavy handed. 1983 (R) 91m/C Carol Lynley, Robert Forster, Rutanya Alda; *Dir:* William Lustig. **VHS, Beta** $79.95 *LIV, VES Woof!*

The Vigilantes Are Coming "The Eagle" sets out to avenge his family and upsets a would-be dictator's plot to establish an empire in California. In twelve chapters; the first is 32 minutes, and additional chapters are 18 minutes each. 1936 230m/B Bob Livingston, Kay Hughes, Guinn Williams; *Dir:* Mack V. Wright, Ray Taylor. **VHS, Beta** $49.95 *NOS, VCN, MED* 🎞️🎞️

Vigilantes of Boom Town A championship prize fight is the cover for a bank robbery which must be foiled by Red Ryder.

1946 54m/B Allan "Rocky" Lane, Robert Blake, Peggy Stewart, Martha Wentworth, Roscoe Karns, Roy Barcroft, George Turner, John Dehner; *Dir:* R.G. Springsteen; *Nar:* Leroy Mason. **VHS, Beta** $9.99 *NOS, VCN, MED* 🎞️

Viking Massacre Brutal ax-bearing Vikings ruin the days of hundreds in this primitive story of courage and desperation. John Hold is the pseudonym for director Mario Bava. Also known as "Knives of the Avenger." 197? 90m/C IT Cameron Mitchell; *Dir:* John Hold. **VHS, Beta** *WES* 🎞️½

The Vikings A Viking king and his son kidnap a Welsh princess and hold her for ransom. Depicts the Vikings' invasion of England. Great location footage of both Norway and Brittany. Basic costume epic with good action scenes. Narrated by Welles. 1958 116m/C Kirk Douglas, Ernest Borgnine, Janet Leigh, Tony Curtis, James Donald, Alexander Knox; *Dir:* Richard Fleischer; *Nar:* Orson Welles. **VHS, Beta** $19.98 *MGM* 🎞️🎞️½

Villa Rides A flying gun-runner aids Francisco "Pancho" Villa's revolutionary Mexican campaign. Considering the talent involved, this one is a disappointment. Co-written by Sam Peckinpah. Check out Brynner's hair. 1968 125m/C Yul Brynner, Robert Mitchum, Charles Bronson, Herbert Lom, Jill Ireland, Robert Towne; *Dir:* Buzz Kulik. **VHS, Beta** $19.95 *KRT* 🎞️🎞️

Village of the Damned A group of unusual children are born in a small English village. They avoid their fathers and other men, except for the one who is their teacher. He discovers they are the vanguard of an alien invasion and leads the counter-attack. Exciting and bone-chilling low-budget thriller. From the novel, "The Midwich Cuckoos," by John Wyndham. The sequel, "The Children of the Damned" (1964) is even better. 1960 78m/B GB George Sanders, Barbara Shelley, Martin Stephens, Laurence Naismith, Michael C. Goetz; *Dir:* Wolf Rilla. **VHS, Beta** $59.95 *MGM, MLB* 🎞️🎞️🎞️

Village on Fire A young guy whose family was slaughtered inexplicably wants retribution, and is willing to kick anything to get it. 198? 86m/C **VHS, Beta** $59.95 *SVS* 🎞️

Village of the Giants A group of beer-guzzling teenagers become giants after eating a mysterious substance invented by a twelve-year-old genius. Fun to pick out all the soon-to-be stars. Totally silly premise with bad special effects and minimal plot follow-through. Based on an H.G. Wells story. 1965 82m/C Ron Howard, Johnny Crawford, Tommy Kirk, Beau Bridges, Freddy Cannon, Beau Brummel; *Dir:* Bert I. Gordon. **VHS, Beta** $59.98 *SUE, MLB* 🎞️

The Villain Still Pursued Her A poor hero and rich villain vie for the sweet heroine in this satire of old-fashioned temperance melodrama. Keaton manages to shine as the hero's sidekick. 1941 67m/B Anita Louise, Alan Mowbray, Buster Keaton, Hugh Herbert; *Dir:* Eddie Cline. **VHS, Beta** $19.95 *NOS, VYY, DVT* 🎞️🎞️

Vincent, Francois, Paul and the Others Three middle-aged Frenchmen rely on their friendships to endure a host of mid-life crises. In French with subtitles. 1976 113m/C FR Yves Montand, Gerard Depardieu, Michel Piccoli, Stephane Audran; *Dir:* Claude Sautet. **VHS, Beta** $79.95 *FCT* 🎞️🎞️🎞️

Vincent: The Life and Death of Vincent van Gogh Van Gogh's work and creativity is examined in a documentary manner through his life and his letters. Thoughtful and intriguing production. Narrated by John Hurt as van Gogh.
1987 99m/C *AU Dir:* Paul Cox; *Voices:* John Hurt. VHS, LV $89.95 FCT, PIA 🎞🎞🎞

Vincent & Theo The story of Impressionist painter Vincent van Gogh, and his brother Theo, a gallery owner who loved his brother's work, yet could not get the public to buy it. Increasing despair and mental illness traps both men, as each struggles to create beauty in a world where it has no value. Altman has created a stunning portrait of "the artist" and his needs. The exquisite cinematography will make you feel as if you stepped into van Gogh's work.
1990 (PG-13) 138m/C Tim Roth, Paul Rhys, Johanna Ter Steege, Wladimir Yordanoff; *Dir:* Robert Altman. VHS, Beta $19.95 HMD 🎞🎞🎞½

The Vindicator A scientist killed in a lab accident is transformed into a cyborg who runs amok and murders indiscriminately. A modernized Frankenstein, with interesting special effects and well drawn characters. Also known as "Frankenstein '88."
1985 (R) 92m/C *CA* Terri Austin, Richard Cox, David McIlwraith, Pam Grier; *Dir:* Jean-Claude Lord. VHS, Beta $79.98 FOX 🎞🎞

The Vineyard Hapless victims are lured to a Japanese madman's island, where he drinks their blood and maintains a questionable immortality.
1989 (R) 95m/C James Hong, Karen Witter, Michael Wong; *Dir:* James Hong, Bill Rice. VHS, Beta, LV $19.95 STE, NWV 🎞

Vintage Cartoon Holiday A tape of eight cartoons from the major animation studios of the 1930s and 40s that cannot be shown on TV because of extreme violence, racial slurs or ethnic stereotypes; from the Fleischer Brothers, Warners and Ub Iwerks.
194? 60m/C VHS, Beta, 3/4U $19.00 IHF 🎞

Violated A detective endeavors to implicate a local businessman in the rapes of two beautiful women. Exploitative and dreary.
1984 88m/C John Heard, J.C. Quinn, April Daisy White, Samantha Fox; *Dir:* Richard Cannistraro. VHS, Beta $69.98 LIV, VES Woof!

Violence at Noon Highly disturbing film in which two women protect a brutal sex murderer from the law. Living among a quiet community of intellectuals, this conspiracy ends in a shocking finale in Oshima's stylized masterpiece. In Japanese with English subtitles.
19?? 99m/B *JP* Saeda Kawaguchi, Akiko Koyama; *Dir:* Nagisa Oshima. VHS $79.95 KIV 🎞🎞🎞

Violent Blood Bath Justice takes on a terrifying new meaning. A grisly death romp occurs after one judge becomes a bit too interested by his own most despicable cases. He sheds his robe to try a life of crime on for size.
19?? 91m/C Fernando Rey. VHS $14.98 VID 🎞🎞🎞

Violent Breed A C.I.A. operative is sent on a mission to put a black marketeer out of business.
1983 91m/C *IT* Henry Silva, Harrison Muller, Woody Strode; *Dir:* Fernando Di Leo. VHS, Beta $59.95 MGM 🎞

Violent Ones Three men who are suspected of raping a young girl are threatened with lynching by an angry mob of townspeople. Badly acted, poorly directed, uneven and feeble.
1968 96m/C Fernando Lamas, David Carradine; *Dir:* Fernando Lamas. VHS, Beta REP Woof!

Violent Professionals A suspended cop runs into resistance inside and outside the force when he infiltrates the mob to get the goods on a crime boss.
1982 100m/C *IT* Richard Conte, Luc Merenda; *Dir:* Sergio Martino. VHS, Beta $42.95 PGN 🎞½

Violent Women Five female convicts escape and embark on a bloody journey through the countryside, pursued by the authorities. Shows women can be just as brutal as any man.
1959 61m/C Jennifer Slater, Jo Ann Kelly, Sandy Lyn, Eleanor Blair, Pati Magee; *Dir:* Barry Mahon. VHS, Beta $29.95 VMK 🎞

The Violent Years Spoiled high-school debutantes form a vicious all-girl gang and embark on a spree that includes murder, robbery, and male rape. Justice wins out in the end. Exploitive trash written by Wood, who directed the infamous "Plan 9 from Outer Space."
1956 60m/B Jean Moorehead, Barbara Weeks, Glenn Corbett; Franz Eichorn; *W/Dir:* Edward D. Wood Jr. VHS, Beta $16.95 SNC, RHI, AOV Woof!

Violent Zone Mercenaries go on a supposed rescue mission in the wilderness.
1989 92m/C John Douglas, Chad Hayward, Christopher Weeks; *Dir:* John Garwood. VHS $69.95 SGE 🎞

Violets Are Blue Two high-school sweethearts are reunited in their hometown years later and try to rekindle their romance—even though the man is married.
1986 (PG-13) 86m/C Kevin Kline, Sissy Spacek, Bonnie Bedelia; *Dir:* Jack Fisk. VHS, Beta, LV $79.95 COL 🎞🎞½

Violette Fascinating true-life account of a nineteen-year-old French girl in the 1930s who, bored with her life and wanting to be with her lover, decides to poison her parents so she can receive her inheritance. Her mother survives but her father dies, and the girl is sent to prison for murder. Extraordinary performance by Huppert and the film is visually stunning.
1978 (R) 122m/C *FR* Isabelle Huppert, Stephane Audran, Jean Carmet, Jean Francoise, Bernadette LaFont; *Dir:* Claude Chabrol. VHS NYF 🎞🎞🎞

Vip, My Brother Superman From the creator of "Allegro Non Troppo" comes this enticing yet amusing piece of animation by Bruno Bozzetto.
1990 ?m/C *Dir:* Bruno Bozzetto. VHS $59.95 EXP 🎞🎞🎞

Viper A woman battles a cryptic anti-terrorist band to avenge the murder of her husband.
1988 96m/C Linda Purl, Chris Robinson, James Tolken; *Dir:* Peter Maris. VHS $29.95 FRH 🎞

Virgin Among the Living Dead Young woman travels to remote castle when she hears of relative's death. Once there, she finds the residents a tad weird and has bad dreams in which zombies chase her. Bizarre even for Franco, who seems to have been going through a "Pasolini" phase while making this one.

1971 (R) 90m/C *SP* Christina von Blanc, Britt Nichols, Howard Vernon, Anne Libert, Rose Kiekens, Paul Muller; *Dir:* Jess (Jesus) Franco. VHS, Beta $29.99 LTG 🎞

The Virgin and the Gypsy An English girl brought up in a repressive household in 1920s England falls in love with a gypsy. Based on the novel by D.H. Lawrence. Directorial debut of Miles.
1970 92m/C *GB* Joanna Shimkus, Franco Nero, Honor Blackman, Mark Burns; *Dir:* Christopher Miles. VHS KIV 🎞🎞½

Virgin High Three young men sneak into an all-girls Catholic boarding school with hilarious consequences. Ward (TV's Robin, from "Batman") makes a special appearance in bondage in this sex farce.
1990 (R) 90m/C Burt Ward, Linnea Quigley, Tracy Dali, Richard Gabai, Catherine McGuiness, Chris Dempsey; *W/Dir:* Richard Gabai. VHS, LV $79.95 COL 🎞

The Virgin of Nuremberg A young woman enters her new husband's ancestral castle and is stalked by the specter of a legendary sadist. Also known as "Horror Castle."
1965 82m/C *IT* Rossana Podesta, George Riviere, Christopher Lee, Jim Dolen; *Dir:* Anthony Dawson. VHS, Beta $59.95 TTE 🎞½

The Virgin Queen Davis stars in this historical drama, which focuses on the stormy relationship between the aging Queen and Sir Walter Raleigh. Collins is the lady-in-waiting who is the secret object of Raleigh's true affections. Previously, Davis played Queen Elizabeth I in "Elizabeth and Essex." Davis holds things together.
1955 92m/C Bette Davis, Richard Todd, Joan Collins, Herbert Marshall, Dan O'Herlihy, Jay Robinson, Romney Brent; *Dir:* Henry Koster. VHS, Beta $19.98 FOX 🎞🎞🎞

Virgin Queen of St. Francis High Two high school foes make a bet that one of them can take the "virgin" title away from gorgeous Christensen by summer's end. She has to fight off their advances, but grows to like Straface.
1988 (PG) 89m/C *CA* Joseph R. Straface, Stacy Christensen, J.T. Wotton; *Dir:* Francesco Lucente. VHS, Beta, LV $79.95 MED 🎞

Virgin Sacrifice A great white hunter looking for zoo-bound jaguars confronts the virgin-sacrificing, Tiger God-revering natives of Guatemala.
1959 67m/C David Dalie, Antonio Gutierrez, Angelica Morales, Fernando Wagner; *Dir:* Fernando Wagner. VHS, Beta, LV WGE 🎞

The Virgin Soldiers A British comedy about greenhorn military recruits stationed in Singapore, innocent of women as well as battle, and their struggles to overcome both situations. A good cast raises this above the usual low-brow sex farce. Based on the novel by Leslie Thomas. Followed by "Stand Up Virgin Soldiers."
1969 (R) 96m/C *GB* Hywel Bennett, Nigel Davenport, Lynn Redgrave, Nigel Patrick, Rachel Kempson, Jack Shepherd, Tsai Chin; *Dir:* John Dexter. VHS, Beta $69.95 COL 🎞🎞½

The Virgin Spring Based on a medieval ballad and set in 14th-century Sweden. The rape and murder of a young innocent spurs her father to vengeance and he kills her attackers. Over the girl's dead body, the father questions how God could have let any of it happen, but he comes to find solace and forgiveness. Stunning Bergman compositions. In Swedish with English subtitles; also available in dubbed version.

1959 88m/B *SW* Max von Sydow, Brigitta Valberg, Gunnel Lindblom, Brigitta Pattersson, Axel Duborg; *Dir:* Ingmar Bergman. Academy Awards '60: Best Foreign Language Film. **VHS, Beta $29.98** SUE 𝅘𝅥𝅮𝅘𝅥𝅮𝅘𝅥𝅮½

The Virgin Witch Two beautiful sisters are sent to the British countryside, ostensibly for a modeling job. They soon find themselves in the midst of a witches' coven however, and discover one of them is to be sacrificed. The Michelles were "Playboy" magazine's first sister centerfolds.
1970 (R) 89m/C *GB* Anne Michelle, Vicki Michelle, Patricia Haines, Keith Buckley, James Chase, Neil Hallett; *Dir:* Ray Austin. **VHS, Beta $59.95** PSM 𝅘𝅥𝅮½

Virginia City Action-packed western drama set during the Civil War. Flynn is a Union soldier who escapes from a Confederate prison run by Scott, after learning of a gold shipment being sent by Southern sympathizers to aid the Confederacy. He ends up in Virginia City (where the gold-laden wagon train is to leave from) and falls for a dance-hall girl (Hopkins) who turns out to be a Southern spy working for Scott but who falls for Flynn anyway. Bogart is miscast as a half-breed outlaw who aids Scott but wants the gold for himself. Considered a follow-up to "Dodge City."
1940 121m/C Errol Flynn, Miriam Hopkins, Randolph Scott, Humphrey Bogart, Frank McHugh, Alan Hale Sr., Guinn Williams, Douglas Dumbrille; *Dir:* Michael Curtiz. **VHS $19.98** MGM 𝅘𝅥𝅮𝅘𝅥𝅮𝅘𝅥𝅮

The Virginia Hill Story Fictionalized biography of mobster Bugsy Siegel's girlfriend who, in the mid-fifties, was subpoenaed to appear before the Kefauver investigation on crime in the U.S. As the examining lawyer presents questions regarding her background and connections with the underworld, we see the story of her life. Made-for-television.
1976 90m/C Dyan Cannon, Harvey Keitel, Robby Benson, Allen Garfield, John Vernon; *Dir:* Joel Schumacher. **VHS, Beta** LME 𝅘𝅥𝅮𝅘𝅥𝅮

The Virginian A classic early-talkie western about a ranch-hand defeating the local bad guys. One line of dialogue has become, with modification, a standard western cliche: "If you want to call me that, smile." Based on the novel by Owen Wister. With this starring role, Cooper broke away from the juvenile lovers he had been playing to the laconic, rugged male leads he would be known for. Huston is perfectly cast as the outlaw leader.
1929 95m/B Gary Cooper, Walter Huston, Richard Arlen, Chester Conklin, Eugene Pallette; *Dir:* Victor Fleming. **VHS, Beta $14.95** KRT 𝅘𝅥𝅮𝅘𝅥𝅮𝅘𝅥𝅮

Virgins of Purity House A group of chaste maidens decide to go all out to change their lifestyles.
1979 90m/C Juan Carlos Dual, Jorge Martinez, Elena Sedova, Patricia Dal. **VHS, Beta $39.95** MED Woof!

Viridiana An innocent girl with strong ideas about goodness is sent to her worldly uncle's home for a final visit before she takes her vows as a nun. Her uncle has developed a sick obsession for her, but after drugging the girl he finds he cannot violate her purity. He tells her, however, she is no longer chaste so she will not join the church. After her uncle's suicide, Viridiana learns she and his illegitimate son Jorge have inherited her uncle's rundown estate. Viridiana opens the house to all sorts of beggars, who take shameless advantage, while Jorge works

slowly to restore the estate and improve the lives of those around him. Considered to be one of Bunuel's masterpieces and a bitter allegory of Spanish idealism versus pragmatism and the state of the world.
1961 90m/B *SP MX* Silvia Pinal, Francisco Rabal, Fernando Rey, Margarita Lozano, Victoria Zinny; *Dir:* Luis Bunuel. Cannes Film Festival '61: Grand Prize Co-Winner. **VHS, Beta $68.00** NOS, HHT, DVT 𝅘𝅥𝅮𝅘𝅥𝅮𝅘𝅥𝅮𝅘𝅥𝅮

Virtue's Revolt A small-town girl is corrupted by the sleazy world of showbiz. Silent.
1924 51m/B Edith Thornton, Crauford Kent. **VHS, Beta $19.95** NO 𝅘𝅥𝅮½

Virus After nuclear war and plague destroy civilization, a small group of people gather in Antarctica and struggle with determination to carry on life. A look at man's genius for self-destruction and his endless hope. Foreign title is "Fukkatsu no Hi."
1982 (PG) 102m/C *JP* George Kennedy, Sonny Chiba, Glenn Ford, Robert Vaughn, Stuart Gillard, Stephanie Faulkner, Ken Ogata, Bo Svenson, Olivia Hussey, Chuck Connors, Edward James Olmos; *Dir:* Kinji Fukasaku. **VHS, Beta $19.95** STE, MED 𝅘𝅥𝅮𝅘𝅥𝅮

The Vision A British suspenser about television evangelists' efforts to control their viewer's minds through worldwide satellite broadcasting.
1987 103m/C *GB* Lee Remick, Dirk Bogarde, Helena Bonham Carter, Eileen Atkins; *Dir:* Norman Stone. **VHS, Beta $79.95** SVS 𝅘𝅥𝅮𝅘𝅥𝅮

Vision Quest A high school student wants to win the Washington State wrestling championship and the affections of a beautiful older artist. He gives it his all as he trains for the meet and goes after his "vision-quest." A winning performance by Modine raises this above the usual teen coming-of-age movie. Madonna sings "Crazy for You" in a nightclub, a reason that the film is known as "Crazy for You" in England. Based on novel by Terry Davis.
1985 (R) 107m/C Matthew Modine, Linda Fiorentino, Ronny Cox, Roberts Blossom, Daphne Zuniga, Charles Hallahan, Michael Schoeffling; *Dir:* Harold Becker; *Performed by:* Madonna. **VHS, Beta, LV $19.98** WAR 𝅘𝅥𝅮𝅘𝅥𝅮½

Visions A man's ability to "see" murders before they happen leads police to suspect him of committing them, and he must clear himself.
1990 90m/C Joe Balogh, Alice Villarreal, Tom Taylor, A.R. Newman, J.R. Pella; *Dir:* Stephen E. Miller. **VHS $79.95** MNC 𝅘𝅥𝅮½

Visions of Evil A young woman, recently released from a mental institution, is plagued by a series of terrifying visions when she moves into a house where a brutal axe murder took place.
1973 85m/C Lori Sanders, Dean Jagger. **VHS, Beta $49.95** PSM 𝅘𝅥𝅮

Visitants Aliens descend irreverently on a small town in the 1950s, with unexpected comedic results.
1987 93m/C Marcus Vaughter, Johanna Grika, Joel Hile, Nicole Rio; *Dir:* Rick Sloane. **VHS, Beta $69.95** TWE 𝅘𝅥𝅮

Visiting Hours A psycho-killer slashes his female victims and photographs his handiwork. Grant is one of his victims who doesn't die, so the killer decides to visit the hospital and finish the job. Fairly graphic and generally unpleasant.
1982 (R) 101m/C *CA* Lee Grant, William Shatner, Linda Purl, Michael Ironside; *Dir:* Jean-Claude Lord. **VHS, Beta $59.98** FOX 𝅘𝅥𝅮

The Visitor Affluent handsome doctor and mate conspire with grisly devil worshippers to conceive devil child.
1980 (R) 90m/C Mel Ferrer, Glenn Ford, Lance Henriksen, John Huston, Shelley Winters, Joanne Nail, Sam Peckinpah; *Dir:* Giullo Paradisi. **VHS $29.95** UAV, HHE 𝅘𝅥𝅮½

Visitor from the Grave When an American heiress and her boyfriend dispose of a dead man's body, his spirit comes back to haunt them.
1981 60m/C Simon MacCorkindale, Kathryn Leigh Scott, Gareth Thomas, Mia Nadasi. **VHS, Beta $29.95** IVE 𝅘𝅥𝅮

The Visitors Ghosts come to stay at a young family's dream house in Sweden. Not exactly a novel treatment or a novel premise.
1989 (R) 102m/C *SW* Keith Berkeley, Lena Endre, John Force, John Olsen, Joanna Berg, Brent Landiss, Patrick Ersgard; *Dir:* Joakim Ersgard. **VHS, Beta, LV $79.95** VMK 𝅘𝅥𝅮½

Vital Signs Hackneyed drama about six medical students enduring the tribulations of their profession.
1990 102m/C Adrian Pasdar, Diane Lane, Jack Gwaltney, Laura San Giacomo, Jane Adams, Tim Ransom, Bradley Whitford, Lisa Jane Persky, William Devane, Norma Aleandro, Jimmy Smits, James Karen, Tim Hopkins; *Dir:* Marisa Silver. **VHS, Beta, LV $89.98** FOX 𝅘𝅥𝅮

Viva Chihuahua (Long Live Chihuahua) Supposed to be one of the greatest achievements of Tin Tan.
1947 95m/B *MX* German Valdes, Miguel Aceves Mejia, Norma Mora, Marcelo Chavez. **VHS, Beta $44.95** MAD 𝅘𝅥𝅮𝅘𝅥𝅮½

Viva Knievel Crooks plan to sabotage Knievel's daredevil jump in Mexico and then smuggle cocaine back into the States in his coffin. Unintentionally campy.
1977 (PG) 106m/C Evel Knievel, Gene Kelly, Lauren Hutton, Red Buttons, Leslie Nielsen, Cameron Mitchell, Marjoe Gortner, Albert Salmi, Dabney Coleman; *Dir:* Gordon Douglas. **VHS, Beta $59.95** WAR Woof!

Viva Las Vegas Race care driver Elvis needs money to compete against rival Danova in the upcoming Las Vegas Grand Prix. He takes a job in a casino and romances fellow employee Ann-Margret, who turns out to be his rival for the grand prize in the local talent competition. Good pairing between the two leads, and the King does particularly well with the title song.
1963 85m/C Elvis Presley, Ann-Margret, William Demarest, Jack Carter, Cesare Danova, Nicky Blair, Larry Kent; *Dir:* George Sidney. **VHS, Beta, LV $19.95** MGM 𝅘𝅥𝅮𝅘𝅥𝅮½

Viva Max A blundering modern-day Mexican general and his men recapture the Alamo, and an equally inept American force, headed by Winters, is sent to rout them out. Mostly works, with some very funny scenes. Ustinov is great.
1969 93m/C Peter Ustinov, Jonathan Winters, John Astin, Pamela Tiffin, Keenan Wynn; *Dir:* Jerry Paris. **VHS $14.98** REP 𝅘𝅥𝅮𝅘𝅥𝅮½

Viva Zapata! John Steinbeck's screenplay of the life of Mexican revolutionary Emiliano Zapata. Brando is powerful as he leads the peasant revolt in the early 1900s, only to be corrupted by power and greed. Quinn well deserved his Best Supporting Actor Oscar for his performance as Zapata's brother. Oscar nominations: Best Actor (Brando), Screenplay & Story, Black & White Art Direction. Based on the novel "Zapata the Unconquered" by Edgcumb Pinchon.

1952 112m/B Marlon Brando, Anthony Quinn, Jean Peters, Margo, Arnold Moss, Joseph Wiseman, Mildred Dunnock; *Dir:* Elia Kazan. Academy Awards '52: Best Supporting Actor (Quinn). VHS, Beta, LV $19.98 *FOX, BTV* ⁗

Vivacious Lady Romantic comedy about a mild-mannered college professor who marries a chorus girl. Problems arise when he must let his conservative family and his former fiancee in on the marriage news. Good performances. Appealing.
1938 90m/B Ginger Rogers, James Stewart, James Ellison, Beulah Bondi, Charles Coburn, Jack Carson, Franklin Pangborn; *Dir:* George Stevens. VHS, Beta $39.95 *CCB, MED* ⫶⫶⫶

Vivo o Muerto Action and adventure abound in this shoot 'em up Spanish film.
1987 100m/B *SP* Manuel Capetillo, Flor Silvestre, Aldo Monti, Wolf Rubinskis. VHS, Beta $44.95 *MAD*

Vogues of 1938 As a lark, a rich girl takes a job as a fashion model and incurs the displeasure of Baxter, the owner of the chic fashion house where she works. An early Technicolor fashion extravaganza. Songs include "That Old Feeling."
1937 110m/C Joan Bennett, Warner Baxter, Helen Vinson, Mischa Auer, Hedda Hopper, Penny Singleton, Alan Mowbray; *Dir:* Irving Cummings. VHS, Beta $29.95 *AXV* ⫶⫶

Voice of Hollywood Some of Hollywood's great stars are caught off guard in this collection of short subjects.
1929 60m/B Buster Keaton, Ken Maynard, Johnny Mack Brown. VHS, Beta $19.95 *VDM, DVT* ⫶½

Volpone A classic adaptation of Ben Jonson's famous tale. A greedy merchant pretends he is dying, leaving a fortune behind, in order see what his family will do to become his heir. In French with English subtitles.
1939 95m/B *FR* Harry Bauer, Louis Jouvet, Fernand Ledoux; *Dir:* Maurice Tourneur. VHS, Beta $24.95 *NOS, HHT, APD* ⫶⫶⫶

Voltus 5 Five children stand against an alien race invading earth. Animated.
1983 75m/C VHS, Beta $14.95 *MED* ⫶½

Volunteers An Ivy League playboy joins the newly formed Peace Corps to escape gambling debts and finds himself on a bridge-building mission in Thailand. Has its comedic moments, especially with Candy.
1985 (R) 107m/C Tom Hanks, John Candy, Rita Wilson, Tim Thomerson, Gedde Watanabe, George Plimpton, Ernest Harada; *Dir:* Nicholas Meyer. VHS, Beta, LV $19.99 *HBO* ⫶½

Volver, Volver, Volver! A beautiful woman is abused by the brother of her dead husband.
1987 90m/C Antonio Aguilar, Jorge Rivero, Claudia Islas, Consuelo Frank, Bruno Rey. VHS, Beta $59.95 *MAD* ⫶

Von Ryan's Express An American Air Force colonel leads a group of prisoners-of-war in taking control of a freight train in order to make their exciting escape from a World War II P.O.W. camp in Italy. Strong cast.
1965 117m/C Frank Sinatra, Trevor Howard, Brad Dexter, Raffaella Carra, Sergio Fantoni, John Leyton, Vito Scotti, Edward Mulhare, Adolfo Celi, James Brolin; *Dir:* Mark Robson. VHS, Beta $19.98 *FOX, FCT* ⫶⫶⫶

Voodoo Black Exorcist Some 3000 years ago, a black prince was buried alive for messin' with another man's woman. Now he's back...and he's mad—real mad. He's prepared to kill just about everyone. Can he

be stopped before all in the modern world are dead??
1989 ?m/C Aldo Sambrel, Tenyeka Stadle, Fernando Sancho. VHS $19.95 *VTP, HHE* ⫶

Voodoo Dawn Two New Yorkers travel to the Deep South to visit a friend who, it turns out, is the latest victim in a series of really gross voodoo murders. A beautiful girl is written into the plot, and the New York guys have an excuse to stay in voodooville, even though bimbolina' southern accent comes and goes for no discernable reason. Filmed near Charleston, South Carolina. Co-written by John Russo of "Night of the Living Dead" fame.
1989 (R) 83m/C Raymond St. Jacques, Theresa Merritt, Gina Gershon, Kirk Baily, Billy "Sly" Williams, J. Grant Albrecht, Tony Todd; *Dir:* Steven Fierberg. VHS $19.95 *ACA* ⫶⫶

Voodoo Woman An innocent girl is lured into the jungle by an evil scientist who is trying to create the perfect woman to commit murders. He turns the girl into an ugly monster in an attempt to get her to obey his telepathic commands. A campy classic that's as bad as it sounds.
1957 77m/B Marla English, Tom Conway, Michael Connors, Lance Fuller; *Dir:* Edward L. Cahn. VHS $9.95 *COL, PAR* ⫶½

Vortex Punk/film noir style in which a female private eye becomes immersed in corporate paranoia and political corruption.
1981 87m/C Lydia Lunch, James Russo, Bill Rice; *Dir:* Scott B, Beth B. VHS, 3/4U *ICA* ⫶½

Voulez-Vous Danser Avec Moi? A young bride endeavors to clear her husband of murder charges and find the real killer. Also known as "Come Dance with Me." In French with English subtitles.
1959 90m/C *FR* Brigitte Bardot, Henri Vidal; *Dir:* Michel Boisrond. VHS, Beta $19.95 *KRT* ⫶⫶

Voyage en Balloon This is the delightful story of a young boy who stowsaway on his grandfather's hot air balloon.
1959 82m/C *FR* Andre Gille, Maurice Baquet, Pascal Lamorisse; *Dir:* Albert Lamorisse. VHS, Beta $39.95 *SUE* ⫶⫶½

Voyage to the Bottom of the Sea The crew of an atomic submarine must destroy a deadly radiation belt which has set the polar ice cap ablaze. Fun stuff, with good special effects and photography. Later became a television show.
1961 106m/C Walter Pidgeon, Joan Fontaine, Barbara Eden, Peter Lorre, Robert Sterling, Michael Ansara, Frankie Avalon; *Dir:* Irwin Allen. VHS, Beta $14.98 *FOX, FCT* ⫶⫶⫶

Voyage of the Damned The story of one of the most tragic incidents of World War II. In 1939, 1,937 German-Jewish refugees fleeing Nazi Germany are bound for Cuba aboard the Hamburg-America liner S.S. St. Louis. They are refused permission to land in Cuba (and everywhere else) and must sail back to Germany and certain death. Based on the novel by Gordon Thomas and Max Morgan-Witts. Oscar nomination for Grant.
1976 (G) 134m/C Faye Dunaway, Max von Sydow, Oskar Werner, Malcolm McDowell, Orson Welles, James Mason, Lee Grant, Katharine Ross, Ben Gazzara, Lynne Frederick, Wendy Hiller, Jose Ferrer, Luther Adler, Sam Wanamaker, Denholm Elliott, Nehemiah Persoff, Julie Harris, Maria Schell, Janet Suzman; *Dir:* Stuart Rosenberg. VHS, Beta $79.98 *FOX* ⫶⫶⫶

Voyage of the Heart An aging fisherman meets up with a sexy college girl and the passion begins.

1990 88m/C Bill Aldrij, Dunja Djordjenic. VHS $79.95 *AIP* ⫶⫶

Voyage to Italy An unhappy couple, on the verge of divorce, inherit a house in Italy and make a trip to see it. In Italy, they dally with other people but realize their attachment to each other demands they attempt to salvage their marriage. Not the best of the Rossellini-Bergman collaborations. Poor editing and directing undermine weak script. Sanders and Bergman do their best.
1953 87m/B Ingrid Bergman, George Sanders; *Dir:* Roberto Rossellini. VHS, Beta $29.95 *NOS, FCT* ⫶⫶½

Voyage to the Planet of Prehistoric Women Astronauts journey to Venus, where they discover a race of gorgeous, sea-shell clad women, as well as a few monsters. The third film incorporating the Russian "Planeta Burg" footage. Directed by Bogdanovich under the pseudonym Derek Thomas.
1968 78m/C Mamie Van Doren, Mary Mark, Paige Lee; *Dir:* Peter Bogdanovich. VHS $16.95 *SNC* ⫶

Voyage to the Prehistoric Planet In the year 2020, an expedition to Venus is forced to deal with dinosaurs and other perils. In the making of this movie, Roger Corman edited in special effects and additional footage from a recently acquired Russian film, "Planeta Burg," and his own "Queen of Blood."
1965 80m/C Basil Rathbone, Faith Domergue, Marc Shannon, Christopher Brand; *Dir:* John Sebastian. VHS $16.95 *NOS, SNC* ⫶½

Voyage of the Rock Aliens A quasi-satiric space farce about competing alien rock stars.
1987 97m/C Pia Zadora, Tom Nolan, Craig Sheffer, Rhema, Ruth Gordon, Michael Berryman, Jermaine Jackson; *Dir:* James Fargo. VHS, Beta $79.95 *PSM* ⫶

A Voyage Round My Father John Mortimer's adaptation of his semi-autobiographical stage play. Olivier is the eccentric, opinionated blind barrister-father and Bates the exasperated son as both try to come to terms with their stormy family relationship. Well-acted and directed. Made for television.
1989 85m/C *GB* Laurence Olivier, Alan Bates, Jane Asher, Elizabeth Sellars; *Dir:* Alvin Rakoff. VHS, Beta $89.99 *HBO* ⫶⫶⫶

Voyage Surprise A broad French farce about the dotty organizer of an impromptu tour, where anything can happen. Funny, not necessarily Art. An early appearance by Martine Carol, France's greatest pre-Bardot sex symbol.
1946 108m/C *FR Dir:* Pierre Prevert. VHS $59.95 *FST* ⫶⫶½

The Voyage of Tanai A young Polynesian battles nature's snarls on his quest to save his village.
1975 (G) 90m/C *Dir:* John Latos. VHS, Beta $29.98 *MAG* ⫶⫶

The Voyage of the Yes Two teenagers, one white and one black, in a small sailboat hit rough weather and battle the elements while learning about themselves. Average made-for-television movie.
1972 (PG) 100m/C Desi Arnaz Jr., Mike Evans, Beverly Garland, Skip Homeier, Della Reese, Scoey Mitchell; *Dir:* Lee H. Katzin. VHS $59.95 *KOV* ⫶⫶

Voyager Restless, middle-aged engineer Walter Faber (Shepherd) tells his life story in a series of flashbacks. Twenty years before (the film starts in 1957) he abandons his pregnant girlfriend who promises to get an abortion. He laters hears she married, had a child, and divorced. While sailing to New York from France Walter falls in love with a young student, Sabeth (Delpy), and after an idyllic interlude Walter agrees to accompany her to Greece to visit her mother. Only then does Walter realize Sabeth's true identity, which leads to a tragic conclusion. Shepherd's is a glum, repressed performance (in keeping with the character) while Delpy personifies youthful sweetness. Based on the novel ''Homo Faber'' by Max Frisch.
1992 (PG-13) 110m/C Sam Shepard, Julie Delphy, Barbara Sukowa; *Dir:* Volker Schlondorff. VHS $89.95 ACA ✂✂✂

Voyager from the Unknown A ''time cop'' and an orphan travel through time to set the course of history straight. Two episodes from the ''Voyagers!'' series.
1983 91m/C Jon-Erik Hexum, Meeno Peluce, Ed Begley Jr., Faye Grant, Fionnula Flanagan. VHS, Beta $39.95 MCA ✂✂

Vulcan God of Fire A foreign film featuring various Steve Reeves style musclemen portraying the gods of Olympus.
19?? 76m/C Rob Flash, Gordon Mitchell, Bella Cortez. VHS, Beta $59.95 TWE Woof!

Vulture In an attempt to carry out a curse on the descendants of the man who killed his forefather, a scientist tries an atomic transmutation experiment and winds up combining himself with a bird.
1967 91m/C Robert Hutton, Akim Tamiroff, Broderick Crawford, Diane Clare, Philip Friend, Patrick Holt, Annette Carell; *Dir:* Lawrence Huntington. VHS, Beta $14.95 NEG ✂

Vultures A dying patriarch summons his predatory family to his home in order to straighten out the distribution of inheritance. One by one, they fall victim to a mysterious murderer.
1984 101m/C Yvonne de Carlo, Stuart Whitman, Jim Bailey, Meredith McRae, Aldo Ray; *Dir:* Paul Leder. VHS, Beta $79.95 PSM ✂

W A woman and her husband are terrorized and must find out why. A single letter "W" is found at the scene of the crimes. Also released as ''I Want Her Dead.'' Notable only as model Twiggy's first film.
1974 (PG) 95m/C Twiggy, Dirk Benedict, John Vernon, Eugene Roche; *Dir:* Richard Quine. VHS, Beta $69.98 LIV, VES ✂½

W Hour A tragic story of young Polish soldiers participating in the Warsaw uprisings of 1944.
19?? 85m/C PL VHS $39.99 FCT ✂✂½

Wackiest Ship in the Army A completely undisciplined warship crew must smuggle an Australian spy through Japanese waters during World War II. Odd, enjoyable mixture of action and laughs. Became a TV series. In the middle of the war effort, Nelson straps on a guitar and sings ''Do You Know What It Means to Miss New Orleans.''
1961 99m/C Jack Lemmon, Ricky Nelson, Chips Rafferty, John Lund, Mike Kellin, Patricia Driscoll; *Dir:* Richard Murphy. VHS, Beta, LV $14.95 RCA ✂✂

Wackiest Wagon Train in the West A hapless wagon master is saddled with a dummy assistant as they guide a party of five characters across the West. Based on the minor TV sitcom ''Dusty's Trail.'' Pro-

duced by the same folks who delivered the similarly premised TV series ''Gilligan's Island'' and ''The Brady Bunch.''
1977 (G) 86m/C Bob Denver, Forrest Tucker, Jeannine Riley. VHS, Beta $19.95 MED ✂

Wacko A group of nymphettes and tough guys get caught up in a wild Halloween-pumpkin-lawnmower murder. This spoof of the ''Halloween'' series works too hard for as few laughs as it gets.
1983 (PG) 84m/C Stella Stevens, George Kennedy, Joe Don Baker, Andrew Dice Clay; *Dir:* Greydon Clark. VHS, Beta, LV VES ✂

The Wacky World of Mother Goose All the familiar Mother Goose characters are brought together in a delightful animated tale of secret agents and sinister surprises.
1967 81m/C *Voices:* Margaret Rutherford. VHS, Beta $19.98 SUE ✂

The Wacky World of Wills & Burke Comedy based on the adventures of Wills and Burke, the two 19th-century explorers who led the first unsuccessful expedition across the outback—the Lewis and Clark of Australia, if you will. In real life they died during the adventure, a less than wacky finale. ''Burke and Wills,'' out the same year, was a serious treatment of the same story.
1985 102m/C AU Garry McDonald, Kim Gyngell, Peter Collingwood, Jonathon Hardy, Mark Little; *Dir:* Bob Weis. VHS, Beta $79.95 MNC ✂

Wages of Fear An American oil company controls a desolate Central American town. Its denizens desperately want out—so desperately that four are willing to try a suicide mission to deliver tow truckfuls of nitroglycerine to put out a well-fire raging 300 miles away. The company's cynical head has offered $2,000 to each man—enough to finance escape from the hell-hole they live in. Complex, multi-dimensional drama. Second half focuses on the truck rides, but the first half, concentrating on character development, is crucial to the film's greatness. This is the restored version. Remade by William Friedkin as ''Sorcerer'' in 1977. Based on a novel by Georges Arnaud. In French with English subtitles.
1955 138m/B FR Yves Montand, Charles Vanel, Peter Van Eyck; *W/Dir:* Henri-Georges Clouzot. Berlin, Film Festival '53: Top Audience Award; British Academy Awards '54: Best Picture; Cannes Film Festival '53: Best Actor (Vanel), Best Film. VHS, Beta, LV $24.95 HHT, SNC, VYY ✂✂✂✂

Wagner: The Movie Slightly half it's original length, the world's biggest biopic is still huge, sweeping, and long. Incoherent and over-acted. Wagner's music is conducted by Sir Georg Solti.
1985 300m/C GB HU Richard Burton, Vanessa Redgrave, John Gielgud, Franco Nero, Laurence Olivier; *Dir:* Tony Palmer. VHS, Beta $29.95 MVD, SUE, GLV ✂½

Wagon Master Two cowboys are persuaded to guide a group of Mormons, led by Bond, in their trek across the western frontier. They run into a variety of troubles, including a band of killers who joins the wagon train to escape a posse. Sensitively directed, realistic and worthwhile. Inspired the TV series ''Wagon Train.'' Also available colorized.
1950 85m/B Ben Johnson, Joanne Dru, Harry Carey Jr., Ward Bond, Jane Darwell, James Arness; *Dir:* John Ford. VHS, Beta, LV $19.95 MED, MED ✂✂✂

Wagon Trail The sheriff's son is blackmailed into helping a gang of robbers, with terrible consequences for himself and his father (Carey), who is fired from his job after the son breaks jail. The new sheriff is secretly chief of the outlaw band. Carey rides back into town and eventually virtue triumphs. Good, tense—but not too tense—western.
1935 59m/B Harry Carey; *Dir:* Harry Fraser. VHS, Beta $27.95 VCN, WGE ✂✂½

Wagon Wheels Recycling was in vogue in '34, when unused footage from ''Fighting Caravans'' (1931) with Gary Cooper was used to remake the very same plot. The result is biodegradable. Settlers heading to Oregon are ambushed by bad guy Blue, employed by fur traders. Blue sics Injuns on them too, but don't worry, they make it.
1934 54m/B Randolph Scott, Gail Patrick, Billy Lee; *Dir:* Charles T. Barton. VHS, Beta $9.95 NOS, DVT, BUR ✂✂

Waikiki Two private eyes set out to prove that their friend is not the ''cane field murderer'' who is terrorizing the Hawaiian island of Oahu. Harmless but pointless TV pilot.
1980 96m/C Dack Rambo, Steve Marachuk, Donna Mills, Cal Bellini, Darren McGavin; *Dir:* Ron Satlof. VHS, Beta $59.95 LHV ✂½

Wait Till Your Mother Gets Home ''Mr. Mom''-like zaniness abounds as a football coach care for the kids and does chores while his wife takes her first job in 15 years. Almost too darn cut, but well written. Made for television.
1983 97m/C Paul Michael Glaser, Dee Wallace Stone, Peggy McKay, David Doyle, Raymond Buktencia, James Gregory, Joey Lawrence, Lynne Moody; *Dir:* Bill Persky. VHS, Beta $9.99 STE, PSM ✂✂

Wait Until Dark A photographer unwittingly smuggles a drug-filled doll into New York, and his blind wife, alone in their apartment, is terrorized by murderous crooks in search of it. A compelling thriller based on the Broadway hit by Frederick Knott, who also wrote ''Dial M for Murder.'' The individual actors' performances were universally acclaimed in this spinetingler. Hepburn received an Oscar nomination. Music by Henry Mancini.
1967 105m/C Audrey Hepburn, Alan Arkin, Richard Crenna, Efrem Zimbalist Jr., Jack Weston; *Dir:* Terence Young. VHS, Beta $59.95 WAR ✂✂✂

Wait Until Spring, Bandini A flavorful immigrant tale about a transplanted Italian family weathering the winter in 1925 Colorado, as seen through the eyes of a young son. Alternately funny and moving, with one of Dunaway's scenery-chewing performances as a local temptation for the father. Based on the autobiographical novel by John Fante, co-produced by Francis Ford Coppola.
1990 (PG-13) 104m/C Joe Mantegna, Faye Dunaway, Burt Young; *Dir:* Dominique Deruddere. VHS $89.99 WAR ✂✂½

Waiting for the Light When a woman takes over a small-town diner, she gets the surprise of her life. Seems an angel has made his home there and now the townsfolk are flocking to see him. MacLaine is wonderful and Garr is fetching and likeable. Set during the Cuban missile crisis. Enjoyable, tame comedy.
1990 (PG) 94m/C Shirley MacLaine, Teri Garr, Vincent Schiavelli, John Bedford Lloyd; *W/Dir:* Christopher Monger. VHS, LV, 8mm COL, PIA ✂✂½

Waiting for the Moon Hunt as Alice B. Toklas is a relative treat in this ponderous, frustrating biopic about Toklas and Gertrude Stein, her lover. Made for "American Playhouse" on PBS.
1987 (PG) 88m/C Linda Hunt, Linda Bassett, Andrew McCarthy; *Dir:* Jill Godmilow. **VHS, Beta** $79.98 FOX ✦✦½

Waitress Three beautiful girls are waitresses in a crazy restaurant where the chef gets drunk, the kitchen explodes, and the customers riot. Awful premise, worse production.
1981 85m/C Jim Harris, Carol Drake, Carol Bever; *Dir:* Samuel Weil. **VHS, Beta** $19.99 HBO Woof!

Wake Island After Pearl Harbor, a small group of Marines face the onslaught of the Japanese fleet on a small Pacific Island. Although doomed, they hold their ground for 16 days. Exciting, realistic, and moving, this movie was nominated for several Academy Awards including Best Picture, Director, Supporting Actor, and Original Screenplay. The first film to capitalize on early "last stands" of WWII; also among the most popular war movies. Shown to soldiers in training camps with great morale-raising success.
1942 88m/B Robert Preston, Brian Donlevy, William Bendix, William Bendix, MacDonald Carey, Albert Dekker, Walter Abel, Rod Cameron; *Dir:* John Farrow. New York Film Critics Award '42: Best Director (Farrow). **VHS, Beta** $19.95 MCA, TLF ✦✦✦

Wake of the Red Witch Wayne captains the ship of the title and battles shipping tycoon Adler for a fortune in pearls and the love of a beautiful woman (Russell). Wayne shows impressive range in a non-gun-totin' role.
1949 106m/B John Wayne, Gail Russell, Gig Young, Luther Adler, Henry Daniell; *Dir:* Edward Ludwig. **VHS, Beta** $19.98 REP ✦✦½

Walk, Don't Run Romantic comedy involving a British businessman (Grant) unable to find a hotel room in Tokyo due to the crowds staying for the 1964 summer Olympic Games. He winds up renting a room from an Embassy secretary (Eggar) and then meets and invites Hutton, a member of the U.S. Olympic walking team, also without a place to stay, to share it with him. Grant then proceeds to play matchmaker, despite the fact that Eggar has a fiance. Grant's last film. Innocuous, unnecessary remake of "The More the Merrier."
1966 114m/C Cary Grant, Samantha Eggar, Jim Hutton, John Standing, Miiko Taka; *Dir:* Charles Walters. **VHS, Beta** $69.95 COL ✦✦½

Walk into Hell A mining engineer and his assistant searching for oil meet the primitive natives of New Guinea. Tedious at times and dated, but pleasant enough.
1957 91m/C AU Chips Rafferty, Francoise Christophe; *Dir:* Lee Robinson. **VHS, Beta** $59.95 CHA ✦✦

Walk Like a Man In a take-off of Tarzan movies, Mandel plays a man raised by wolves. Comic problems arise when he is found by his mother and the family attempts to civilize him. Juvenile script wastes fine cast.
1987 (PG) 86m/C Howie Mandel, Christopher Lloyd, Cloris Leachman, Colleen Camp, Amy Steel; *Dir:* Melvin Frank. **VHS, Beta** $79.95 MGM ✦½

Walk Softly, Stranger Two-bit crook is reformed by the love of an innocent peasant girl. If you can stand the cliche masquerading as a plot, the performances are good.
1950 81m/B Alida Valli, Joseph Cotten, Spring Byington, Paul Stewart, Jack Paar, Jeff Donnell, John McIntire; *Dir:* Robert Stevenson. **VHS, Beta** $19.98 TTC ✦✦

Walk in the Spring Rain The bored wife of a college professor follows him to rural Tennessee when he goes on sabbatical, where she meets the married Quinn and the two begin an affair. When Quinn's disturbed son learns of the affair, he attacks the woman, with tragic results. Fine cast should have had better effect on low-key script by Stirling Silliphant.
1970 (PG) 98m/C Ingrid Bergman, Anthony Quinn, Fritz Weaver, Katherine Crawford, Tom Fielding, Virginia Gregg; *Dir:* Guy Green. **VHS, Beta** $59.95 RCA ✦✦½

A Walk in the Sun The trials of a group of infantrymen in World War II from the time they land in Italy to the time they capture their objective, a farmhouse occupied by the Germans. Excellent ensemble acting shows well the variety of civilians who make up any fighting force and explores their fears, motivations, and weaknesses. Producer and director Milestone also made "All Quiet on the Western Front" and the Korean War masterpiece "Pork Chop Hill." Released in the final days of the war, almost concurrently with two other WWII films of the first echelon, "The Story of G.I. Joe" and "They Were Expendable."
1946 117m/B Dana Andrews, Richard Conte, John Ireland, Lloyd Bridges, Sterling Holloway; *Dir:* Lewis Milestone. **VHS, Beta, LV, 8mm** $19.95 LUM, NOS, DVT ✦✦✦½

Walk on the Wild Side In 1930s New Orleans, a man searches for his long-lost love, finds her working in a whorehouse and fights to save her from the lesbian madame Stanwyck. Melodrama, based only loosely on the Nelson Algren novel and adapted by cult novelist John Fante, with Edmund Morris. Much-troubled on the set, and it shows.
1962 114m/B Jane Fonda, Laurence Harvey, Barbara Stanwyck, Capucine, Anne Baxter; *Dir:* Edward Dmytryk. **VHS, Beta** $69.95 COL ✦✦

Walker Slapdash, tongue-in-cheek historical pastiche about the real-life American William Walker (played previously in "Burn!" by Marlon Brando), and how he led a revolution in Nicaragua in 1855 and became its self-declared president. A bitter, revisionist farce never for a moment attempting to be accurate. Matlin's unfortunate, though fortunately brief, follow-up to her Oscar-winning performance. Music by Joe Strummer.
1987 (R) 95m/C Ed Harris, Richard Masur, Peter Boyle, Rene Auberjonois, Marlee Matlin, Miguel Sandoval; *Dir:* Alex Cox. **VHS, Beta** $79.95 MCA ✦✦

Walking on Air Ray Bradbury's story comes to life as a crippled boy dreams of completing a real space walk. Originally aired on PBS as part of the WonderWorks Family Movie series.
1987 60m/C Lynn Redgrave, Jordan Marder, James Treuer, Katheryn Trainor; *Dir:* Ed Kaplan. **VHS** $29.95 PME ✦✦

Walking Back Troubled teens go for joyride and bash mirthmobile to smithereens. The body shop agrees to give them a different car, but they find themselves in the middle of a heist at a plant belonging to the father of one of the boys.
1926 53m/B Sue Carol, Richard Walling, Ivan Lebedeff, Robert Edeson, Florence Turner, Arthur Rankin Jr.; *Dir:* Rupert Julian. **VHS, Beta** $19.95 GPV, FCT ✦✦

Walking the Edge A widow hires a taxi driver to help her seek vengeance against the men who killed her husband and her son. Forster does what he can with the lousy story and script.
1983 (R) 94m/C Robert Forster, Nancy Kwan, Joe Spinell, Aarika Wells; *Dir:* Norbert Meisel. **VHS, Beta** $79.98 LIV, LTG ✦½

Walking Tall A Tennessee sheriff takes a stand against syndicate-run gambling and his wife is murdered in response. Ultra-violent crime saga wowed the movie going public and spawned several sequels and a TV series. Based on the true story of folk-hero Buford Pusser, admirably rendered by Baker.
1973 (R) 126m/C Joe Don Baker, Elizabeth Hartman, Noah Beery Jr., Gene Evans, Rosemary Murphy, Felton Perry; *Dir:* Phil Karlson. **VHS, Beta, LV** $29.98 LIV, LTG ✦✦½

Walking Tall: Part 2 Club-wielding Tennessee sheriff Buford Pusser, this time played less memorably by Svenson, attempts to find the man who killed his wife. Even more violent than the original.
1975 (PG) 109m/C Bo Svenson, Noah Beery Jr., Angel Tompkins, Richard Jaeckel; *Dir:* Earl Bellamy. **VHS, Beta, LV** $29.98 LIV, VES ✦✦

Walking Tall: The Final Chapter The final months in the life of Tennessee sheriff Buford Pusser and the mystery surrounding his death. It wasn't the final chapter. Still to come: a TV flick and series.
1977 (PG) 112m/C Bo Svenson, Forrest Tucker, Leif Garrett, Morgan Woodward; *Dir:* Jack Starrett. **VHS, Beta, LV** $29.98 LIV, LTG ✦½

Walking Through the Fire A young woman's real-life struggle with Hodgkin's disease. Her fight for her own and her unborn baby's survival is a stirring testament to the power of faith. Absorbing made for television drama.
1980 143m/C Bess Armstrong, Tom Mason, Bonnie Bedelia, Richard Masur, Swoosie Kurtz, J.D. Cannon, June Lockhart; *Dir:* Robert Day. **VHS, Beta** TLF ✦✦✦

Wall The last film by Guney, author of "Yol," about orphaned boys in prison in the Turkish capitol of Ankara trying to escape and/or rebel after ceaseless rapings, beatings and injustice. An acclaimed, disturbing film made from Guney's own experience. He died in 1984, three years after escaping from prison. Brutal and horrifying. In Turkish with English subtitles.
1983 117m/C TU Ayse Emel Mesci, Saban, Sisko; *W/Dir:* Yilmaz Guney. **VHS, Beta** $69.95 KIV ✦✦½

The Wall: Live in Berlin The largest and most ambitious live musical project ever staged. Waters, former leader of Pink Floyd, leads a cast of hundreds in the performance of "The Wall." The historic Berlin Wall collapse provided the backdrop for this event on July 21, 1990. Interviews and behind-the-scenes shots are included.
1990 120m/C Van Morrison, Joni Mitchell, Bryan Adams, Sinead O'Connor, Thomas Dolby, The Band, Marianne Faithfull, Cyndi Lauper, Albert Finney, Roger Waters. **VHS, LV** $19.95 MVD, PGV, BTV ✦✦✦

Wall Street Stone's energetic, high-minded big business treatise in which naive, neophyte stockbroker Sheen is seduced into insider trading by sleek entrepreneur Douglas, much to his blue-collar father's chagrin. A fast-moving drama of 80's-style materialism with a mesmerizing, award-winning performance by Douglas as greed personified. Expert direction by Stone, who co-wrote the not-very-subtle script. His father, to whom this film is dedicated, was a broker. Look for Stone in a cameo.
1987 (R) 126m/C Michael Douglas, Charlie Sheen, Martin Sheen, Darryl Hannah, Sean Young, James Spader, Hal Holbrook, Terence Stamp, Richard Dysart, John C. McGinley; *Dir:* Oliver Stone. Academy Awards '87: Best Actor (Douglas); National Board of Review Awards '87: Best Actor (Douglas). **VHS, Beta, LV** $19.98 *FOX, BTV ♫♫♫*

Wall Street Cowboy When his land is threatened, a cowboy fights big business in the big city.
1939 54m/B Roy Rogers. **VHS, Beta** $9.99 *NOS, DVT, BUR ♫*

Wallaby Jim of the Islands The singing captain of a pearl fishing boat must protect his treasure from marauding pirates. Pleasant enough musical adventure.
1937 61m/B George Houston, Douglas Walton; *Dir:* Bud Barsky. **VHS** $24.95 *NOS, DVT ♫♫*

The Walloping Kid A B-western with photography done in Monument Valley. Silent with original organ music.
1926 67m/B Kit Carson; *Dir:* Pauline Curley. **VHS, Beta, 8mm** $29.95 *VYY ♫½*

Walls of Glass An aging New York cabby tries to make it as an actor. Effective performances by all override the thin plot. Slow and uneven, but involving. Also called "Flanagan."
1985 (R) 85m/C Geraldine Page, Philip Bosco, William Hickey, Olympia Dukakis, Brian Bloom, Linda Thorson; *Dir:* Scott Goldstein. **VHS, Beta** $79.95 *MED ♫♫½*

Walpurgis Night Soapy Swedish drama about abortion was racy for its time. Office gal Bergman secretly loves her boss (Hanson). His wife refuses to have children and goes through an abortion. An unscrupulous fellow blackmails her with this knowledge. Bergman bides her time. Stay tuned... Interesting document on the mores of another time. In Swedish with English subtitles.
1941 82m/B SW Lars Hanson, Karin Carlsson, Victor Sjostrom, Ingrid Bergman, Erik Berglund, Sture Lagerwall, Georg Rydeberg, Georg Blickingberg; *Dir:* Gustaf Edgren. **VHS, Beta** $29.95 *WAC, CRO ♫♫♫*

A Walt Disney Christmas Six classic cartoons with a wintry theme are combined for this program: "Pluto's Christmas Tree" (1952), "On Ice," "Donald's Snowball Fight;" two Silly Symphonies from 1932-33: "Santa's Workshop" and "The Night Before Christmas;" and an excerpt from the 1948 feature "Melody Time," entitled "Once Upon a Wintertime."
1982 46m/C **VHS, Beta** $12.99 *DIS ♫♫♫*

The Waltons' Christmas Carol Heart-warming made for TV holiday special involving the Appalachian family on the farm for a yuletide celebration.
198? 94m/C Will Geer, Michael Learned, Richard Thomas, Judy Norton-Taylor, Ralph Waite. **VHS** $14.95 *VTK, WAR ♫♫♫*

Waltz Across Texas Young oil man Farnsworth and good-lookin' rock scientist Place at first don't take to each other, but there's something in the air, and romance blossoms. Not great, but enjoyable. Farnsworth and Place cowrote and coproduced, and cohabitate as real-life spouses offscreen.
1983 100m/C Mary Kay Place, Richard Farnsworth, Terry Jastrow, Anne Archer; *Dir:* Ernest Day. **VHS, Beta** $69.95 *VES ♫♫½*

Waltz King Typically hokey Disney biography of the young composer Johann Strauss during his Old Viennese heyday. Fine music, pretty German locations. Well-made family fare.
1963 94m/C Kerwin Mathews, Senta Berger, Brian Aherne; *Dir:* Steve Previn. **VHS, Beta** $69.95 *DIS ♫♫*

A Waltz Through the Hills Two orphans head into the Australian outback and experience adventures worthy of public television. Aired on PBS as part of the WonderWorks Family Movie television series.
1988 116m/C Tina Kemp, Andre Jansen, Ernie Dingo, Dan O'Herlihy; *Dir:* Frank Arnold. **VHS** $29.95 *PME, HMV ♫½*

Waltz of the Toreadors Retired general Sellers doesn't care for his wealthy, shrewish wife and tries to re-kindle a 17-year-old romance with a French woman. It doesn't work out (seems his illegitimate son also has a soft spot for the lady), but the general decides to keep his eyes open for other possibilities. Interestingly cast adaptation of Jean Anouilh's play. Sellers is hilarious, as usual.
1962 105m/C GB Peter Sellers, Dany Robin, Margaret Leighton, Cyril Cusack; *Dir:* John Guillermin. **VHS, Beta** $19.98 *VID ♫♫♫*

Wanda Nevada Gambler Fonda wins nubile young Shields in a poker game, so he drags her with him to the Grand Canyon to look for gold. Director/star Fonda sure picked a lemon for his only screen appearance with dad Henry (a grizzled old varmint appearing briefly). Shields is in her usual form—stellar for a shampoo commercial.
1979 (PG) 105m/C Peter Fonda, Brooke Shields, Henry Fonda, Fiona Lewis, Luke Askew, Ted Markland, Severn Darden, Paul Fix; *Dir:* Peter Fonda. **VHS, Beta** $14.95 *WKV ♫½*

Wanderers Richard Price's acclaimed novel (look for Price in a cameo) about youth gangs coming of age in the Bronx in 1963. The "Wanderers," named after the Dion song, are a gang of Italian-American teenagers about to graduate high school, who prowl the Bronx with the feeling that something is slipping away from them. Fascinating, funny and touching. Manz is unforgettable as a scrappy gal. A wonderful 60s soundtrack (Dion, the Four Seasons) colors this "coming of age the hard way" film.
1979 (R) 113m/C Ken Wahl, John Friedrich, Karen Allen, Linda Manz; *Dir:* Philip Kaufman. **VHS, Beta** $19.98 *WAR ♫♫♫½*

Wandering Jew An excellent print of the rare Austrian film version of the classic legend, one of at least three silent versions. A Jew is condemned to wander the earth for eternity.
1920 65m/B Rudolf Schildkraut, Joseph Schildkraut. **VHS, Beta** *GVV ♫♫½*

Wanna Be's This Japanese animated adventure finds a female wrestling tag-team looking for a way to beef up their act. The girls locate some high tech equipment for

sale at a ridiculously low price, but there's a catch. The wrestling team must take part in evil genetic experiments designed to create monsters.
1992 45m/C JP **VHS** *CPM ♫♫½*

The Wannsee Conference A startling, important film depicting, in real time, the conference held at the Wannsee on January 20, 1942, during which 14 members of the Nazi heirarchy decided in 85 minutes the means and logistics of effecting the Final Solution. Recreated from the original secretary's notes. Horrifying and chilling. Along with "Shoah," a must-see for understanding the Holocaust and the psychology of genocide. In German with English subtitles.
1984 87m/C GE Dietrich Mattausch, Gerd Brockmann, Friedrich Beckhaus; *Dir:* Heinz Schirk. **VHS, Beta, LV** $79.95 *PSM, APD, GLV ♫♫♫½*

Wanted: Babysitter A young student accidentally becomes involved by her roommate in a plot to kidnap the child she is babysitting. Schneider is truly awful; Italian comic Pazzetto is at sea in a bad role as her boyfriend. Also released as "The Babysitter."
1975 90m/C Robert Vaughn, Vic Morrow, Maria Schneider, Renato Pozzetto, Nadja Tiller, Carl Mohner, Sydne Rome; *Dir:* Rene Clement. **VHS, Beta** $24.95 *AHV Woof!*

Wanted Dead or Alive Ex-CIA agent Hauer is now a high-tech bounty hunter assigned to bring in an international terrorist. When the terrorist kills Hauer's friend and girlfriend, he forgets the $50,000 bonus for bringing him in alive. Official "sequel" to the Steve McQueen television series with Hauer as the McQueen character's great-grandson. The link is meaningless, and the plot is a thin excuse for much violence and anti-terrorist flag-waving.
1986 (R) 104m/C Rutger Hauer, Gene Simmons, Robert Guillaume, William Russ, Jerry Hardin, Mel Harris; *Dir:* Gary Sherman. **VHS, Beta, LV** $14.95 *NWV, STE ♫½*

Wanted: The Perfect Guy Danny and Melanie try to find Mister Right for their divorced Mom (Kahn). An Emmy Award winner about single parents and their children.
1990 45m/C Madeline Kahn, Melanie Mayron. **VHS** $99.95 *AIM ♫♫½*

The Wanton Contessa In 1866, as the Austrians prepare for war with Italy, a love affair arises between an Italian countess and an Austrian lieutenant. This deeply romantic saga dwells on the complexities in an affair that cross lines of war, nation, and social class. Lush photography. Originally released as "Senso." Dubbed.
1955 125m/C IT Alida Valli, Farley Granger, Massimo Girotti, Heinz Moog; *Dir:* Luchino Visconti. **VHS, Beta** $59.95 *VCD ♫♫♫*

War Between the Tates A college professor's affair with a female student is the basis for a tension-filled stand-off with his wife. Mediocre made-for-television adaptation of the Allison Lurie novel, somewhat redeemed by Ashley's performance.
1976 90m/C Richard Crenna, Elizabeth Ashley, Granville Van Dusen; *Dir:* Lee Philips. **VHS, Beta** *TLF ♫♫*

The War Boy Difficulties encountered by a young Canadian boy as he grows up in Central Europe during WWII. Not to be compared with "Empire of the Sun" or "Hope and Glory," both of which were released later; successful in its modest ambi-

tions. Beautiful performance from 12-year-old star Hopely.
1985 (R) 86m/C *CA* Helen Shaver, Ken Welsh, Jason Hopely; *Dir:* Allan Eastman. **VHS** $79.98 *CON, HHE* 🐾🐾½

War Brides Civil War brides confront life after the war. Low-budget drama.
198? 100m/C Elizabeth Richardson, Sharry Fleet, Sonja Smits. **VHS, Beta** $49.95 *TWE* 🐾

War Bus Commando A fully loaded bus is the only way out when soldiers are caught behind enemy lines in Afghanistan. A case of the spoof coming first: The plot is almost exactly the same as ''Stripes''—only this one isn't funny. In Hi-Fi.
1989 90m/C *IT* Savina Gersak, Mark Gregory, John Vernon; *Dir:* Frank Valenti. **VHS** $79.95 *TRY* 🐾

War Games A young computer whiz, thinking that he's sneaking an advance look at a new line of video games, breaks into the country's NORAD missile-defense system and challenges it to a game of Global Thermonuclear Warfare. The game might just turn out to be the real thing. Slick look at the possibilities of an accidental start to World War III. Entertaining and engrossing, but with a B-grade ending.
1983 (PG) 110m/C Matthew Broderick, Dabney Coleman, John Wood, Ally Sheedy; *Dir:* John Badham. **VHS, Beta, LV** $19.98 *FOX* 🐾🐾½

War of the Gargantuas Tokyo is once again the boxing ring for giant monsters. This time it's a good gargantua (half human, half monster) against a bad gargantua. This is a strange one.
1970 92m/C *JP* Russ Tamblyn, Kimi Mizuno, Kipp Hamilton, Yu Fujiki. **VHS** *PAR* 🐾

War in the Gulf - The Complete Story: Saddam The fourth in a series of six cassette tapes describing the Persian Gulf War.
1991 90m/C *Nar:* Bernard Shaw. **VHS** $19.98 *TTC* 🐾

War in the Gulf - The Complete Story: The Aftermath The sixth in a series of six cassette tapes describing the Persian Gulf conflict.
1991 90m/C *Nar:* Bernard Shaw. **VHS** $19.98 *TTC* 🐾

War in the Gulf - The Complete Story: The Air War The second in a series of six cassette tapes desciribing the Persian Gulf War.
1991 90m/C *Nar:* Bernard Shaw. **VHS** $19.98 *TTC* 🐾

War in the Gulf - The Complete Story: The Conflict Begins The first of a series of six cassettes describing the Persian Gulf War.
1991 90m/C *Nar:* Bernard Shaw. **VHS** $19.98 *TTC* 🐾

War in the Gulf - The Complete Story: The General - "Stormin' Norman" The fifth in a series of six cassettes describing the Persian Gulf conflict.
1991 90m/C *Nar:* Bernard Shaw. **VHS** $19.98 *TTC* 🐾

War in the Gulf - The Complete Story: The Ground War The third in a series of six cassette tapes describing the Persian Gulf war.
1991 90m/C *Nar:* Bernard Shaw. **VHS** $19.98 *TTC* 🐾

War Kill A man fights for his country in this violent war saga.
1968 99m/C Tom Drake, George Montgomery; *Dir:* Ferde Grofe Jr. **VHS, Beta** $39.95 *AHV* 🐾

The War Lord Set in the 11th century, Heston stars as Chrysagon, a Norman knight and war lord who commands a peasant village. While battling his enemies, he becomes enamored of a peasant girl named Bronwyn (Forsyth), who is unfortunately engaged to someone else. Pulling rank, Chrysagon uses an ancient law that allows noblemen the first night with a bride and the two fall in love. The two vow to never part, but that sets the stage for even more bloody battles. Fine acting and great production values make this a well-adapted version of the play ''The Lovers'' by Leslie Stevens.
1965 121m/C Charlton Heston, Richard Boone, Rosemary Forsyth, Guy Stockwell, Niall MacGinnis, Henry Wilcoxon, James Farentino, Maurice Evans; *Dir:* Franklin J. Schaffner. **VHS, Beta** $19.98 *MCA* 🐾🐾🐾

War & Love Two Jewish teenagers in Warsaw are torn apart by WWII and Nazi persecution. After the war they search for each other. Sincere, but not adequately developed, drama based on the book ''The Survivors'' by film's producer Jack Eisner.
1984 (PG-13) 112m/C Sebastian Keneas, Kyra Sedgwick; *Dir:* Moshe Mizrahi. **VHS, Beta** $79.95 *MGM* 🐾🐾

War Lover An American daredevil flying captain and his co-pilot find themselves vying for the affections of a woman during WWII in England. Seeks human frailty beneath surface heroism. McQueen is impressive, no thanks to mediocre script. Excellent aerial photography, and featuring one of only a very few serviceable WWII B-17s then remaining. Based on John Hersey's novel.
1962 105m/B *GB* Steve McQueen, Robert Wagner, Shirley Anne Field, Bill Edwards, Gary Cockrell; *Dir:* Philip Leacock. **VHS, Beta, LV** $9.95 *GKK* 🐾🐾

War Party During a 100-year commemoration of an Indian massacre in the Midwest, a murder occurs and a lynch mob chases a pack of young Blackfeet into the mountains before the inevitable showdown. A feeble attempt at portraying the unfair treatment of Native Americans. Un-subtle, Hollywood style, this time in favor of the Indians, and therefore un-serious. Too bad; the premise had potential.
1989 (R) 99m/C Kevin Dillon, Billy Wirth, Tim Sampson, M. Emmet Walsh; *Dir:* Franc Roddam. **VHS, Beta, LV** $89.99 *HBO* 🐾

War and Peace Lengthy adaptation of Tolstoy's great (and likewise lengthy) novel about three families caught up in Russia's Napoleonic Wars from 1805 to 1812; filmed in Rome. Bad casting and confused script (by six writers) are somewhat overcome by awesome battle scenes and Hepburn. Score by Nino Rota. Academy Award nominations for Best Director, Color Cinematography, and Costume Design. Remade in 1968.
1956 208m/C Audrey Hepburn, Mel Ferrer, Henry Fonda, Anita Ekberg, Vittorio Gassman, John Mills, Oscar Homolka, Herbert Lom, Helmut Dantine; *Dir:* King Vidor. National Board of Review Awards '56: 5 Best Foreign Films of the Year. **VHS, Beta, LV** $14.95 *PAR* 🐾🐾½

War and Peace The massive Russian production of Leo Tolstoy's masterpiece, adapting the classic tome practically scene by scene. All of the production took place in the Soviet Union. So painstaking that it took more than five years to finish, no other

adaptation can touch it. Hugely expensive ($100 million, claimed the Russians), wildly uneven production. Great scenes of battle and aristocratic life. Though this version is far from perfect, one asks: Is it humanly possible to do screen justice to such a novel? In Russian with English subtitles. On four tapes. (Beware the two-part, poorly dubbed version that was also released.)
1968 373m/C *RU* Lyudmila Savelyeva, Sergei Bondarchuk, Vyacheslav Tihonor, Hira Ivanov-Golarko, Irina Gubanova, Antonina Shuranova; *Dir:* Sergei Bondarchuk. Academy Awards '68: Best Foreign Language Film; National Board of Review Awards '68: Best Foreign Film; New York Film Critics Awards '68: Best Foreign Film. **VHS, Beta, LV** $99.95 *VTK, KUL* 🐾🐾🐾

War & Remembrance, Part 1 Tedious sequel to the epic television mini-series ''The Winds of War,'' based on the novel by Herman Wouk. Historical fiction is created around the events of World War II, including Nazi persecution and naval battles in the Pacific. Seven cassette series. In Hi-Fi stereo. Followed by Volume II of this series.
1988 120m/C Robert Mitchum, Jane Seymour, Hart Bochner, Victoria Tennant, Barry Bostwick, Polly Bergen, David Dukes, Michael Woods, Sharon Stone, Robert Morley, Sami Frey, Chaim Topol, John Rhys-Davies, Ian McShane, William Schallert, Jeremy Kemp, Steven Berkoff, Robert Hardy, Ralph Bellamy, John Gielgud; *Dir:* Dan Curtis. **VHS, Beta** $139.98 *MPI* 🐾½

War & Remembrance, Part 2 The final installments of the excruciatingly boring and long mini-series. Herman Wouk's historical fiction concerning the last years of World War II. Six cassettes, in Hi-Fi stereo.
1989 120m/C Robert Mitchum, Jane Seymour, Hart Bochner, Ralph Bellamy, Victoria Tennant, Barry Bostwick, John Gielgud, Polly Bergen, David Dukes, Michael Woods; *Dir:* Dan Curtis. **VHS, Beta** *MPI* 🐾½

War of the Robots A dying alien civilization kidnaps two brilliant Earth scientists in the hope that they'll help it survive.
1978 99m/C Antonio Sabato, Melissa Long, James R. Stuart. **VHS, Beta, LV** $59.95 *UHV* 🐾

The War of the Roses Acidic black comedy about a well-to-do suburban couple who can't agree on a property settlement in their divorce so they wage unreserved and ever-escalating combat on each other, using their palatial home as a battleground. Expertly and lovingly (if that's the word) directed by DeVito, who plays the lawyer. Turner and Douglas are splendid. Laserdisc version includes previously deleted scenes and commentary from DeVito.
1989 (R) 116m/C Michael Douglas, Kathleen Turner, Danny DeVito, Marianne Sagebrecht, Sean Astin, G.D. Spradlin, Peter Donat; *Dir:* Danny DeVito. **VHS, Beta, LV** $12.00 *FOX, IME* 🐾🐾🐾

War of the Shaolin Temple In 14th century China, a young boy is the only hope against a band of raiding Manchu warriors. Jammed with action.
19?? 90m/C Chan Kin Cheong. **VHS, Beta** *OCE* 🐾½

War in the Sky Director William Wyler's documentary account of the Army Air Corps in action. Extraordinary footage of the Thunderbolt, the P47 Fighter, and the B17 Bomber.
1982 90m/C *Dir:* William Wyler; *Hosted:* Lloyd Bridges; *Nar:* Peter Lawford, James Stewart. **VHS, Beta** $59.95 *GEM* 🐾🐾🐾

War in Space Powerful U.N. Space Bureau Starships and UFOs band together to battle alien invaders among the volcanoes and deserts of Venus.
1977 91m/C *JP* Kensaku Marita, Yuko Asano, Ryo Ikebe. **VHS, Beta** *VDA* 🦴

War Victims Drama about Japanese POW prisons in World War II.
1985 90m/C *IT* **VHS, Beta $59.95** *VCD* 🦴½

The War Wagon The Duke plans revenge on Cabot, the greedy mine owner who stole his gold claim and framed him for murder for which he spent years in prison. He assembles a gang to aid him, including a wise-cracking Indian (Keel) and the man sent by Cabot to kill him (Douglas). Wayne's plan is to steal the gold being shipped in Cabot's armor-plated stagecoach, the "war wagon." Well-written, good performances, lots of action.
1967 101m/C John Wayne, Kirk Douglas, Howard Keel, Robert Walker Jr., Keenan Wynn, Bruce Dern; ***Dir:*** Burt Kennedy. **VHS, Beta, LV $19.98** *MCA* 🦴🦴🦴

War of the Wildcats Fast-moving western with Wayne as a tough cowboy battling a powerful land baron. They fight over land, oil and a woman. Unusual because Wayne's character acts on behalf of the Indians to drill and transport oil. Also known as "In Old Oklahoma."
1943 102m/B John Wayne, Martha Scott, Albert Dekker, George "Gabby" Hayes, Sidney Blackmer; ***Dir:*** Albert Rogell. **VHS $14.98** *REP* 🦴🦴½

War of the Wizards Low-budget sci-fi thriller about an alien woman with supernatural powers who comes to take over the Earth. But, lucky for all earthlings, she's challenged by a hero-type just in the nick of time.
1981 (PG) 90m/C Richard Kiel; ***Dir:*** Richard Caan. **VHS, Beta** *PLV* 🦴½

The War of the Worlds H.G. Wells's classic novel of the invasion of Earth by Martians, updated to 1950s California, with spectacular special effects of destruction caused by the Martian war machines. Pretty scary and tense; based more on Orson Welles's radio broadcast than on the book. Still very popular; hit the top 20 in sales when rleased on video. Classic thriller later made into a TV series. Produced by George Pal, who brought the world much sci-fi, including "The Time Machine," "Destination Moon," and "When Worlds Collide," and who appears here as a street person.
1953 85m/C Gene Barry, Ann Robinson, Les Tremayne, Lewis Martin, Robert Cornthwaite, Sandro Giglio; ***Dir:*** Byron Haskin. Academy Awards '53: Best Special Effects. **VHS, Beta, LV $14.95** *PAR* 🦴🦴🦴½

War Years This episode of "Life Goes to the Movies" examines an era in which an event larger than any movie overshadowed the country. Hollywood did its bit to bolster morale while filling the need for symbolic heroes like John Wayne.
1977 35m/C ***Hosted:*** Henry Fonda, Shirley MacLaine, Liza Minnelli. **VHS, Beta $39.95** *TLF* 🦴🦴🦴½

Warbirds American pilots are off to quell a revolution in a Middle East country. Too little action fails to support mediocre plot and characters.
1988 (R) 88m/C Jim Eldert, Cully Holland, Bill Brinsfield; ***Dir:*** Ulli Lommel. **VHS, Beta $19.95** *STE, VMK* 🦴🦴

Warbus Marines lead a school bus load of Americans through war-torn Vietnam to safety. Decent action and surprisingly enjoyable characters. Otherwise mediocre.
1985 (R) 90m/C Daniel Stephen, Romano Kristoff, Urs Althaus; ***Dir:*** Ted Kaplan. **VHS, Beta $19.98** *SUE* 🦴🦴

WarCat The daughter of a green beret fights for her life against a blood-thirsty gang. Also known as "Angel of Vengeance."
1988 78m/C Macka Foley, Jannina Poynter, David O'Hara, Carl Erwin; ***Dir:*** Ted V. Mikels. **VHS $79.95** *TWE* 🦴

Wardogs (The Assassination Team) A man tries to rescue his brother from becoming one of an elite, brainwashed group of ex-Vietnam vets trained by the government as professional assassins. Might not have been quite so bad; promising inversion of many 'Nam cliches wrecked by unnecessary violence and bad acting.
1987 95m/C *SW* Tim Earle, Bill Redvers, Daniel Hubenbecker; ***Dir:*** Daniel Hubenbecker. **Beta $79.95** *VHV* 🦴½

Warkill WWII action set in the jungles of the Philippines. Journalist who idolizes violent, hard-edged colonel, loses his illusions, but gains a more genuine respect for his men. Not-bad war flick.
1965 99m/C George Montgomery, Tom Drake, Eddie Infante; ***W/Dir:*** Ferde Grofe Jr. **VHS, Beta $19.95** *AHV* 🦴🦴½

Warlock Claustrophic, resonant town-bound tale of a marshal (Fonda) and his adoring sidekick (Quinn) who clean up a town which then turns against him. Unusual story, fine performances carry this well beyond the run-of-the-mill cow flick. Fonda re-established himself at the box office as a western star, after "Stage Struck" and "Twelve Angry Men." Script by Robert Alan Arthur from the novel by Oakley Hall. "Bones" McCoy from "Star Treck" has a small part.
1959 122m/C Henry Fonda, Anthony Quinn, Richard Widmark, Dorothy Malone, Wallace Ford, Richard Arlen, Regis Toomey, DeForest Kelley; ***Dir:*** Edward Dmytryk. **VHS, Beta $19.98** *FOX* 🦴🦴🦴

Warlock It's 1691 and the most powerful warlock (Sands) in the New World is only hours away from execution. Luckily, his pal, Satan, whisks him (and witchhunter Grant, by mistake) three hundred years in the future to present-day Los Angeles, where he crash-lands in Singer's house. Surprisingly witty dialogue and neat plot twists outshine occasionally cheesy special effects.
1991 (R) 103m/C Richard E. Grant, Julian Sands, Lori Singer, Mary Woronov; ***Dir:*** Steve Miner. **VHS $92.95** *VMK* 🦴🦴½

Warlock Moon A young woman is lured to a secluded spa and falls prey to a coven of witches.
1973 75m/C *MX* Laurie Walters, Joe Spano; ***Dir:*** Bill Herbert. **VHS, Beta $49.95** *UNI* 🦴½

Warlords Lone soldier Carradine battles mutant hordes in a post-apocalyptic desert. Amateurish futuristic drivel.
1988 (R) 87m/C David Carradine, Sid Haig, Ross Hagen, Fox Harris, Robert Quarry, Victoria Sellers, Brinke Stevens, Dawn Wildsmith; ***Dir:*** Fred Olen Ray. **VHS, Beta $79.95** *VMK* 🦴½

Warlords of the 21st Century A gang of bandits who speed around the galaxy in an indestructible battle cruiser is challenged by a space lawman. Action pic without much action. Filmed in New Zealand. Originally titled "Battletruck."
1982 91m/C Michael Beck, Annie McEnroe, James Wainwright; ***Dir:*** Harley Cokliss. **VHS, Beta $19.98** *SUE* 🦴½

Warlords from Hell Two young men motorcycling through Mexico are captured by a bloodthirsty gang and forced to do hard labor in the marijuana fields. We can only hope that they just learn to say no.
1987 (R) 76m/C Brad Henson, Jeffrey Rice; ***Dir:*** Clark Henderson. **VHS, Beta** *WAR Woof!*

Warm Nights on a Slow-Moving Train A hot-blooded schoolteacher turns tricks for cash to support her brother on the Sunday-night train to Melbourne, until she meets a stranger who has a deadly masquerade of his own. Interestingly different; much sex, as you can imagine. Australian-made.
1987 (R) 90m/C *AU* Wendy Hughes, Colin Friels, Norman Kaye, John Clayton, Peter Whitford; ***Dir:*** Bob Ellis. **VHS, Beta, LV $79.95** *PSM, PAR* 🦴🦴

Warm Summer Rain A young couple meet under unusual circumstances and develop a relationship in a desert cabin as they reflect on their unfulfilled pasts. Lust, self-doubt and longing for something to hold onto draw them together as the days pass. Thoroughly self-indulgent romantic comedy tht's oddly self-pitying and poorly directed.
1989 (R) 85m/C Kelly Lynch, Barry Tubb; ***Dir:*** Joe Gayton. **VHS, Beta, LV $19.95** *COL* 🦴½

Warner Brothers Cartoon Festival 1 Includes "Corny Concerto," "Dover Boys," "Hamateur Night," "Wackiki Wabbit," "Tale of Two Kitties," "Daffy and the Dinosaur," and "Falling Hare."
194? 55m/C ***Dir:*** Chuck Jones, Bob Clampett, Friz Freleng. **VHS, Beta $34.95** *HHT, DVT* 🦴½

Warner Brothers Cartoon Festival 2 Includes "Wacky Wabbit," "Daffy the Commando," "Case of the Missing Hare," "Jungle Jitters," "All This and Rabbit Stew," "Have You Got Any Castiles," "Fresh Hare," and "Inki and the Mina Bird."
194? 55m/C ***Dir:*** Friz Freleng, Bob Clampett, Chuck Jones, Robert McKimson. **VHS, Beta $34.95** *HHT, DVT* 🦴½

Warner Brothers Cartoon Festival 3 Includes "Robin Hood Makes Good," "Flop Goes the Weasel," "Fony Fables," "Pigs in a Polka," "Fox Pop," "Fifth Column Mouse," and "Wabbit Who Came to Supper."
194? 55m/C ***Dir:*** Chuck Jones, Friz Freleng, Robert McKimson, Bob Clampett. **VHS, Beta $34.95** *HHT, DVT* 🦴½

Warner Brothers Cartoon Festival 4 Eight cartoons are included in this package, including "Presto Change-O," "Porky's Bare Facts," "Get Rich Quick, Porky," "Finn 'N Catty," "Sheepish Wolf," "Yankee Doodle Daffy," "Porky's Railroad," and "Bugs Bunny Bond Rally."
194? 55m/C ***Dir:*** Chuck Jones, Bob Clampett, Friz Freleng. **VHS, Beta $34.95** *HHT, DVT* 🦴½

Warner Brothers Cartoons Included in this collection are "The Wabbit Who Came to Supper" (1942), "A Tale of Two Kitties" (1948), "Case of the Missing Hare" (1942), "Hamateur Night" (1938), "Wackiki Wabbit" (1953), "Daffy Duck and the Dinosaur" (1939), and "Fresh Hare" (1942).
195? 54m/C ***Dir:*** Chuck Jones, Chuck Jones, Robert McKimson, Friz Freleng. **VHS, Beta $24.95** *VYY* 🦴½

The Warning A pair of proverbial honest cops investigate ties between the mob and the police department. Hard to follow.
1980 101m/C *IT* Martin Balsam, Giuliano Gemma, Giancarlo Zanetti; ***Dir:*** Damiano Damiani. **VHS, Beta $49.95** *MED* 🦴

Warning Shadows Classic example of interior German Expressionism. Brilliantly portrays a jealous husband's emotions and obsessions through shadows. Innocent events seem to reek of sin. A seldom-seen study of the oft-precarious distinction between love and obsession; directly influenced by the classic "The Cabinet of Dr. Caligari." Silent with German and English titles.
1923 93m/B *GE* Fritz Kortner; ***Dir:*** Arthur Robinson. **VHS, Beta, 8mm $19.95** *GPV, FCT, VPR* 🦴🦴🦴

Warning Sign A high-tech thriller in which a small town is terrorized by the accidental release of an experimental virus at a research facility. Shades of "The Andromeda Strains" and "The China Syndrome," though less originality and quality.
1985 (R) 99m/C Sam Waterston, Kathleen Quinlan, Yaphet Kotto, Richard Dysart, Rick Rossovich; ***Dir:*** Hal Barwood. **VHS, Beta $79.98** *FOX* 🦴½

Warning from Space Aliens visit Earth to warn of impending cosmic doom. When it becomes apparent that the one-eyed starfish look is off-putting, they assume human form. Japanese sci fi.
1956 87m/B *JP* Toyomi Karita, Keizo Kawasaki, Isao Yamagata, Shozo Nanbu, Buntaro Miake, Mieko Nagai, Kiyoko Hirai; ***Dir:*** Koji Shima. **VHS $19.98** *NOS, SMW, SNC* 🦴½

Warren Zevon Zevon performs his hit songs in concert.
1982 68m/C Warren Zevon. **VHS, Beta $29.95** *SVS* 🦴½

The Warrior A young warrior is trapped in a corrupt ring of competitive boxing and karate.
19?? 50m/C Dennis Brown, Cheng Tien Chi. **VHS, Beta $49.95** *OCE* 🦴

Warrior of the Lost World A warrior must destroy the evil Omega Force who tyrannically rules the world in the distant future, etcetera, etcetera. One-size-fits-all premise; horrible special effects; miserably directed. Only for really hard-core Pleasance fans.
1984 90m/C Robert Ginty, Persis Khambatta, Donald Pleasence; ***Dir:*** David Worth. **VHS, Beta $69.99** *HBO* ***Woof!***

Warrior Queen Long ago in ancient Rome, overly histrionic mayor Pleasance married porn queen with a wandering eye Fox, and much gratuitous sex and erupting volcanoes resulted. Even the score blows. Also available in an "Unrated" version. Brit pop star Fox is billed as Stasia Micula.
1987 (R) 69m/C Sybil Danning, Donald Pleasence, Richard Hill, Josephine Jacqueline Jones, Tally Chanel, Samantha Fox. **VHS, LV $79.98** *LIV, VES* 🦴

The Warrior & the Sorceress Mercenary Carradine offers his services to rival factions fighting for control of a water well in an impoverished desert village located on a planet with two suns. He attempts to aggravate the conflicts between the factions, playing the shifty go-between. The sorceress is topless throughout. "A Fistful of Dollars" goes to outer space. Inoffensive, except that better judgment is expected of Carradine.
1984 (R) 81m/C David Carradine, Luke Askew, Maria Socas, Harry Townes; ***Dir:*** John Broderick. **VHS, Beta, LV $29.98** *LIV, VES* 🦴½

Warrior Within Eight Kung Fu greats pay tribute to the memory of Bruce Lee by kicking and jumping about.
1980 85m/C Chuck Norris, Dan Inosanto, Mike Stone, Fumio Demura, Ron Taganashi. **VHS, Beta $9.95** *VDA* 🦴

The Warriors Swashbuckling adventure about Prince Edward's valiant rescue of Lady Joan and her children from the clutches of the evil Count De Ville. Flynn looks old and pudgy, but buckles his way gallantly through intrepid adventure. Also the last Flynn to see: fun, but more or less completely derivative and familiar. Filmed in England. Also known as "The Dark Avenger."
1955 85m/C *GB* Errol Flynn, Peter Finch, Joanne Dru, Yvonne Furneaux, Noel Willman, Michael Hordern; ***Dir:*** Henry Levin. **VHS, Beta $59.98** *FOX, MLB* 🦴🦴½

The Warriors An action story about a turf battle between New York City street gangs that rages from Coney Island to the Bronx. Silly plot works because of fine performances and direction, excellent use of action and color, and nonstop pace. Fight scenes are very carefully, even obviously, choreographed.
1979 (R) 94m/C Michael Beck, James Remar, Deborah Van Valkenburgh, Thomas G. Waites, David Patrick Kelly; ***Dir:*** Walter Hill. **VHS, Beta, LV $14.95** *PAR* 🦴🦴½

Warriors of the Apocalypse After a nuclear holocaust, heavily sworded and loin-clothed men search for the secret of eternal life and instead find an Amazon realm in the jungle. A terrible waste of post-apocalyptic scenery. Alternate titles: "Searchers of the Voodoo Mountain" and "Time Raiders."
1985 (R) 96m/C Michael James, Debrah Moore, Ken Metcalfe, Franco Guerrero; ***Dir:*** Bobby Suarez. **VHS, Beta $19.98** *LIV, LTG* ***Woof!***

Warriors of Fire A crime lord has a young girl killed, and her sister becomes a bloodthirsty ninja in order to wreak revenge.
198? 91m/C Glen Carson, Peter Davis, Joff Houston; ***Dir:*** Bruce Lambert. **VHS, Beta $59.95** *TWE* 🦴

Warriors of the Sung Dynasty Warriors given amnesty by the Emperor are tricked into a bloody confrontation. Adaption of a classic Chinese tale.
19?? 90m/C Mu Huan Hu. **VHS $34.95** *OCE* 🦴

Warriors of the Wasteland It's the year 2019, and the world has been devastated by a nuclear war. The few survivors try to reach a distant land which emits radio signals indicating the presence of human life. They are hindered by attacks from the fierce homosexual Templars, led by a self-proclaimed priest called One. Mindless and obviously made on the proverbial shoestring budget. Dubbed. Also known as "The New Barbarians."
1983 (R) 92m/C *IT* Fred Williamson, Timothy Brent, Anna Kanakis; ***Dir:*** Enzo G. Castellari. **VHS, Beta $69.99** *HBO, SIM* ***Woof!***

Warriors of the Wind An animated sword-and-sorcery tale about a magical land being fought over by the forces of good and evil.
1985 (PG) 85m/C *JP* ***Dir:*** Kazuo Komatsubara. **VHS, Beta $9.95** *NWV, STE* 🦴🦴🦴

Warsaw Ghetto This documentary takes a chilling look at the Nazi atrocities committed in Warsaw, Poland during World War II.
1967 51m/B VHS, Beta $29.95 *NOS, DVT, HHT* 🦴½

The Wash Quiet domestic drama about a couple who drift apart after 40 years of marriage. Set interestingly among Japanese-Americans in California; treats social and cultural factors with intelligence. Superb cast takes it a notch higher.
1988 94m/C Mako, Nobu McCarthy, Sab Shimono; ***Dir:*** Michael Toshiyuki Uno. **VHS, Beta, LV $79.95** *ACA* 🦴🦴🦴

Washington Affair A tale of intrigue in the nation's capital. Sullivan is a businessman who uses women and blackmail to capture government contracts. Selleck is the hapless bureaucrat he preys on. Remake of Soloff's own "Intimacy" was not released until "Magnum P.I." became a hit.
1977 (PG) 90m/C Tom Selleck, Carol Lynley, Barry Sullivan; ***Dir:*** Victor Stoloff. **VHS, Beta $69.98** *SUE* 🦴🦴

Washington Mistress Emminently ordinary made-for-TV grist about Capitol Hill aide Arnaz, politician Jordan, and their affair. Listless performances.
1981 96m/C Lucie Arnaz, Richard Jordan, Tony Bill; ***Dir:*** Peter Levin. **VHS** *USA* 🦴½

W.A.S.P. Filmed at the Lyceum Theater in London, this concert performance by the heavy-metal band features memorable tunes such as "On Your Knees" and "I Wanna Be Somebody."
1984 30m/C WASP. **VHS, Beta $14.95** *SVS* 🦴½

The Wasp Woman In her quest for eternal beauty, a woman uses a potion made from wasp enzymes. Naturally, she turns into a wasp monster at night. Good fun, courtesy of Corman.
1949 84m/B Susan Cabot, Anthony Eisley, Barboura Morris, Michael Marks, William Roerick, Frank Gerstle, Bruno Ve Sota, Frank Wolff; ***Dir:*** Roger Corman. **VHS $9.95** *NOS, RHI, SNC* 🦴½

Watch the Birdie It's Skelton three times over as cameraman, father, and grandfather! First, he accidentally films a scam that would send the lovely Miss Dahl filing for bankruptcy. Later a crazy chase scene unfolds, that will have viewers rolling, as Skelton the cameraman nabs the bad guys and turns them over to the cops. Light fun that will charm Skelton fans.
1950 70m/B Red Skelton, Arlene Dahl, Ann Miller, Leon Ames; ***Dir:*** Jack Donohue. **VHS, Beta $19.98** *MGM* 🦴🦴½

Watch Me When I Kill A young nightclub dancer stops by a drugstore seconds after the owner was killed. She doesn't see the killer's face, but his rasping voice remains to torment her and the viewer. Dubbed.
1981 (R) 95m/C *IT* Richard Stewart, Sylvia Kramer, Anthony Bido. **VHS, Beta $19.99** *HBO* 🦴

Watch Out, Crimson Bat! A blind swords-woman must deliver a sack containing a valuable secret which can win a war.
1969 87m/C Yoko Matsuyama, Goro Ibuki. **VHS, Beta** *VDA* 🦴

Watch on the Rhine A couple involved with the anti-Nazi underground has escaped the country, but is pursued and harassed by Nazi agents. Adapted by Lillian Hellman and Dashiell Hammett from Hellman's play. Performed on stage before the

U.S. entered the war, it was the first American play and movie to portray the ugliness of fascism as an ideology, as opposed to the more devious evil of its practical side. The Production Code at the time required that a killer always be punished; the murderer (whose screen motives had been noble) refused to film the offending scene, which explains the tacked-on ending. Superb drama from a pair of highly gifted writers and a great cast. Shumlin also directed the play. Oscar Nominations: Best Picture, Best Supporting Actress (Watson), Screenplay.
1943 114m/B Bette Davis, Paul Lukas, Donald Woods, Beulah Bondi, Geraldine Fitzgerald, George Coulouris, Henry Daniell; **Dir:** Herman Shumlin. Academy Awards '43: Best Actor (Lukas); National Board of Review Awards '43: 10 Best Films of the Year; New York Film Critics Awards '43: Best Actor (Lukas), Best Film. **VHS, Beta $19.98** FOX, MGM, FCT 🐾🐾🐾½

Watched A former U.S. attorney, gone underground, has a nervous breakdown and kills the narcotics agent who has him under surveillance. Self-important, too serious social/political conspiracy drama of the Watergate era. Keach and Yulin are good, but not good enough to save it. Music score written and performed by the jazz/fusion band Weather Report.
1973 95m/C Stacy Keach, Harris Yulin, Bridgit Polk; **Dir:** John Parsons. **VHS, Beta $69.98** LIV, VES, SIM 🐾🐾

Watcher in the Woods When an American family rents an English country house, the children are haunted by the spirit of a long-missing young girl. A very bland attempt at a ghost story.
1981 (PG) 83m/C Bette Davis, Carroll Baker, David McCallum, Ian Bannen, Lynn-Holly Johnson; **Dir:** John Hough. **VHS, Beta $19.99** DIS 🐾🐾

Watchers From the suspense novel by Dean R. Koontz, a secret experiment goes wrong, creating half-human monsters. A boy and his extremely intelligent dog are soon pursued. Another human experiment gone awry, which obviously had an effect on the film.
1988 (R) 99m/C Barbara Williams, Michael Ironside, Corey Haim; **Dir:** Jon Hess. **VHS, Beta, LV $14.95** IVE 🐾

Watchers 2 Sequel to "Watchers" follows the further adventures of a super-intelligent golden retriever who leads a Marine to an animal psychologist and then attempts to warn them both of a mutant killer. The dog says woof, but this movie doesn't quite. Fun for lovers of hounds or horror flicks.
1990 (R) 101m/C Marc Singer, Tracy Scoggins; **Dir:** Thierry Notz. **VHS, Beta, LV $14.95** IVE 🐾🐾

Water The resident governor of a Caribbean British colony juggles various predatory interests when a valuable mineral water resource is found. Cameos by Eric Clapton, George Harrison, Fred Gwynne, and Ringo Starr. The good cast is wasted in this all-too-silly effort.
1985 (PG-13) 89m/C GB Michael Caine, Brenda Vaccaro, Leonard Rossiter, Valerie Perrine, Jimmie Walker; **Dir:** Dick Clement. **VHS, Beta $29.95** PAR 🐾

Water Babies When a chimney sweep's 12-year-old apprentice is wrongly accused of stealing silver, the boy and his dog fall into a pond and eventually rescue some of the characters they find there. Combination of live-action and animated fairy-tale story set in nineteenth-century London. Boring, unless you're a young child with equivalent stan-

dards. Based on the book by Charles Kingsley. Also called "Slip Slide Adventures."
1979 93m/C GB James Mason, Billie Whitelaw, David Tomlinson, Paul Luty, Sammantha Coates; **Dir:** Lionel Jeffries. **VHS, Beta $29.98** SUE 🐾🐾

Water Rustlers An unscrupulous land baron builds a dam on his side of the Silver Creek with the intention of drying out the cattle pastures. Heroine Page, whose pop has been rubbed out by varmints, and good guy/lover O'Brien save her property and avenge her father. Second of three Page westerns.
1939 55m/B Dorothy Page, Dave O'Brien; **Dir:** Samuel Diege. **VHS, Beta $27.95** VCN, VDM 🐾½

Waterfront Nazis coerce German-Americans into helping them in World War II-era San Francisco. Credulity defying spy doings, but that's okay. Fairly entertaining wartime drama about paranoia on the home front.
1944 68m/B John Carradine, J. Carroll Naish, Terry Frost, Maris Wrixon, Edwin Maxwell; **Dir:** Steve Sekely. **VHS, Beta $16.95** SNC, UHV 🐾🐾

Waterfront Effective romantic drama set in Melbourne, Australia. Australian workers strike after taking forced pay cuts. As a result, Italian immigrants are hired as scabs to keep the docks going. Despite the tension, an Italian woman and an Australian man fall deeply in love, only to find that they must struggle to keep that love alive.
1983 294m/C AU Jack Thompson, Greta Scacchi, Frank Gallacher, Tony Rickards, Mark Little, Jay Mannering, Ray Barrett, Chris Haywood, Warren Mitchell, Noni Hazlehurst, John Karlsen, Elin Jenkins; **Dir:** Chris Thomson. **VHS, Beta $79.95** PSM, PAR 🐾🐾½

Waterhole #3 Three Confederate army buddies steal a fortune in gold bullion from the Union Army and hide it in a waterhole in the desert. One of the funnier entries in the Western comedy genre.
1967 (PG) 95m/C James Coburn, Carroll O'Connor, Margaret Blye, Claude Akins, Bruce Dern, Joan Blondell, James Whitmore; **Dir:** William A. Graham. **VHS $14.95** PAR 🐾🐾½

Waterloo Massive chronicle of Napoleon's European conquests and eventual defeat at the hands of Wellington. Filmed on location in Italy and the Ukraine, it bombed due largely to Steiger's bizarre rendition of Napoleon.
1971 (G) 122m/C IT RU Rod Steiger, Orson Welles, Virginia McKenna, Michael Wilding, Donal Donnelly, Christopher Plummer, Jack Hawkins, Dan O'Herlihy, Terence Alexander, Rupert Davies, Ivo Garrani, Gianno Garko; **Dir:** Sergei Bondarchuk. **VHS $14.95** PAR, FCT 🐾🐾

Waterloo Bridge In London during World War I, a soldier from an aristocratic family and a ballet dancer begin a tragic romance when they meet by chance on the foggy Waterloo Bridge. When he's listed as dead, her despair turns her to prostitution. But then he returns from POW camp, they once again meet by accident on Waterloo Bridge, and their romance resumes. Four-hanky drama with fine performances by Leigh (her first after "Gone with the Wind") and Taylor.
1940 109m/B Vivien Leigh, Robert Taylor, Lucile Watson, C. Aubrey Smith, Maria Ouspenskaya, Virginia Field; **Dir:** Mervyn LeRoy. **VHS, Beta, LV $19.98** MGM 🐾🐾🐾

Watermelon Man The tables are turned for a bigoted white guy when he wakes up one morning to discover he has become a black man. Broad comedy with not much place to go is still engaging. Cambridge takes

on both roles, appearing in unconvincing white makeup.
1970 (R) 97m/C Godfrey Cambridge, Erin Moran, Estelle Parsons, Howard Caine; **Dir:** Melvin Van Peebles. **VHS, Beta, LV $59.95** COL 🐾🐾

Watership Down Wonderfully animated story based on Richard Adams's allegorical novel about how a group of rabbits escape fear and overcome oppression while searching for a new and better home. It's really an adult theme with sufficient violence to the poor wittle wabbits to scare the kiddies.
1978 (PG) 92m/C GB **Dir:** Martin Rosen; **Voices:** Ralph Richardson, Zero Mostel, John Hurt, Denholm Elliott, Harry Andrews, Michael Hordern, Joss Ackland. **VHS, Beta, 8mm $19.98** WAR, VTK 🐾🐾½

Wavelength A rock star living in the Hollywood Hills with his girlfriend stumbles on an ultra-secret government project involving friendly aliens from outer space recovered from a recent UFO crash site. Fun, enthusiastically unoriginal cheap sci-fi. Soundtrack by Tangerine Dream.
1983 (PG) 87m/C Robert Carradine, Cherie Currie, Keenan Wynn; **W/Dir:** Mike Gray. **VHS, Beta, LV $9.98** SUE, STE 🐾🐾½

Waxwork A wax museum opens up, and it is soon evident that the dummies are not what they seem. Garbled nonthriller. Available in a 100-minute unrated version.
1988 (R) 97m/C Zach Galligan, Deborah Foreman, Michelle Johnson, Dana Ashbrook, Miles O'Keefe, Patrick Macnee, David Warner; **Dir:** Anthony Hickox. **VHS, Beta, LV $14.98** LIV, VES 🐾½

Waxwork 2: Lost in Time A young couple (Galligan and Schnarre) barely escape with their lives when the infamous waxworks museum burns down. A severed hand also gets loose and follows Schnarre home and murders her stepfather, leaving her to take the blame. In order to prove her innocence, the couple must travel through a bizarre time machine. Extraordinary special effects, strange plot twists and recreations of scenes from past horror movies make this a highly entertaining sequel.
1991 (R) 104m/C Zach Galligan, Alexander Godunov, Bruce Campbell, Michael Des Barres, Monika Schnarre; **Dir:** Anthony Hickox. **VHS $49.98** LIV 🐾🐾½

Waxworks A major achievement in German Expressionism, in which a poet imagines scenarios in a wax museum fairground that involve Jack the Ripper, Ivan the Terrible and Haroun al-Rashid. Influential and considered ahead of its time. Silent.
1924 63m/B GE William Dieterle, Emil Jannings, Conrad Veidt, Werner Krauss, John Gottowt, Olga Belajeff; **Dir:** Paul Leni. **VHS, Beta $16.95** SNC, NOS, LIV 🐾🐾🐾

Way Back Home Man wants to adopt a boy to release him from the clutches of his cruel guardian. Minor role for the then unknown Davis.
1932 81m/B Phillips Lord, Frank Albertson, Bette Davis; **Dir:** William A. Seiter. **VHS, Beta $19.98** TTC, FCT 🐾🐾

Way of the Black Dragon When slave-trading drug traffickers threaten the moral fiber of our nation, two commandoes join forces to halt their operation.
1981 88m/C Ron Van Cliff, Carter Wang, Charles Bonet; **Dir:** Chan Wui Ngai. **VHS, Beta $39.95** TWE 🐾

Way Down East Melodramatic silent drama of a country girl who is tricked into a fake marriage by a scheming playboy. The famous final scene of Gish adrift on the ice floes is in color. One of Griffith's last critical and popular successes. This tape includes the original Griffith-approved musical score. Remade in 1935 with Henry Fonda.
1920 107m/B Lillian Gish, Richard Barthelmess, Lowell Sherman, Creighton Hale; *Dir:* D.W. Griffith. **VHS, Beta $19.95** *NOS, KRT, VYY* 🎬🎬🎬

Way He Was A satirical re-enactment of the events that led up to the Watergate burglary and the coverup that followed.
1976 (R) 87m/C Steve Friedman, Al Lewis, Merrie Lynn Ross, Doodles Weaver. **VHS, Beta $54.95** *UHV* 🎬🎬

Way Out West The classic twosome journey way out west to deliver the deed to a gold mine to the daughter of their late prospector pal. The obligatory romance is missing, but you'll be laughing so hard you won't notice. One of Stan and Ollie's best. Academy Award nomination for the very nice score which includes the song "Trail of the Lonesome Pine." Also included on this tape is a 1932 Todd and Pitts short, "Red Noses." Also available colorized.
1937 86m/B Stan Laurel, Oliver Hardy, Rosina Lawrence, James Finlayson, Sharon Lynne, ZaSu Pitts, Thelma Todd; *Dir:* James W. Horne. **VHS, Beta $19.95** *CCB, MED, FRH* 🎬🎬🎬

The Way We Were Big box-office hit follows a love story between opposites from the 1930s to the 1950s. Streisand is a Jewish political radical who meets the handsome WASP Redford at college. They're immediately attracted to one another, but it takes years before they act on it and eventually marry. They move to Hollywood where Redford is a screenwriter and left-wing Streisand becomes involved in the Red scare and the blacklist, much to Redford's dismay. Though always in love, their differences are too great to keep them together. An old-fashioned and sweet romance, with much gloss. Marvin Hamlisch's score and title song (sung by Streisand) won Oscars. Adapted by Arthur Laurents from his novel. Oscar nominations for Streisand, Cinematography, Art Direction, Set Decoration, and Costume Design.
1973 (PG) 118m/C Barbra Streisand, Robert Redford, Bradford Dillman, Viveca Lindfors, Herb Edelman, Murray Hamilton, Patrick O'Neal, James Woods, Sally Kirkland; *Dir:* Sydney Pollack. Academy Awards '73: Best Song ("The Way We Were"), Best Original Score; National Board of Review Awards '73: 10 Best Films of the Year. **VHS, Beta, LV, 8mm $19.95** *COL, PIA* 🎬🎬🎬

Way of the West Them durn pesky sheepherders are up to it again! Okay, so the cow/sheep thing is a classic western plot, but here it's offered perfunctorily.
1935 52m/B Wally Wales, William Desmond, Art Mix, Jim Sheridan, Bobby Nelson; *Dir:* Robert Emmett. **VHS, Beta $19.95** *NOS, DVT, WGE* 🎬½

The Way West A wagon train heads to Oregon. A poor and muddled attempt at recreating the style of a John Ford western. What really galls is that it's based on the Pulitzer-winning novel by A.B. Guthrie, Jr. Boy is the book better than the movie. Field's first film.
1967 122m/C Kirk Douglas, Robert Mitchum, Richard Widmark, Lola Albright, Michael Witney, Stubby Kaye, Sally Field, Jack Elam; *Dir:* Andrew V. McLaglen. **VHS $14.95** *WKV* 🎬½

Wayne Murder Case A fast-moving, cleverly constructed murder mystery. A rich old man dies just as he is about to sign a new will. Though no one was standing near him, a knife is found stuck in his back.
1938 61m/B June Clyde, Regis Toomey, Jason Robards Sr. **VHS, Beta $12.95** *SNC, VYY, WGE* 🎬🎬🎬

Wayne's World Destined to become one of the top movies of all time—Not! This "Saturday Night Live" skit proved to be so popular that it got its own movie, not unlike the plot, which has slimy producer Lowe take the public access "Wayne's World" into the world of commercial television. The zany duo of Wayne and Garth are as much fun on the big screen as they are on SNL and there are many funny moments, several of which are destined to become comedy classics.
1992 (PG-13) 93m/C Mike Myers, Dana Carvey, Rob Lowe, Brian Doyle Murray, Lara Flynn Boyle, Tia Carrere; *Dir:* Penelope Spheeris. MTV Movie Awards '92: Best On-Screen Duo. **VHS $24.95** *PAR* 🎬🎬🎬

Ways of Kung Fu Reincarnation and multi-generation vendettas spice up this martial arts festival.
197? 95m/C **VHS, Beta $29.95** *UNI* 🎬

W.C. Fields Comedy Bag Compilation of three classic Fields shorts: "The Golf Specialist" (1930), "The Dentist" (1932) and "The Fatal Glass of Beer" (1933).
1933 56m/B W.C. Fields, Babe Kane, Dorothy Granger, George Chandler. **VHS, Beta $19.95** *KRT* 🎬

W.C. Fields: Flask of Fields Three classic Fields shorts: "The Golf Specialist" (1930), in which J. Effington Bellweather finds himself teaching a lovely young lady how to play the game; "The Fatal Glass of Beer" (1933); and "The Dentist" (1932), in which Fields tackles a room filled with patients.
1930 61m/B W.C. Fields, Babe Kane, Elise Cavanna, Bud Jamison, Rosemary Theby. **VHS, Beta, 8mm $24.95** *VYY* 🎬

W.C. Fields Rediscovered Goldman and Claude Fields, Jr. discuss the films of W.C. Fields.
196? 30m/B W.C. Fields; *Voices:* Albert Goldman, Claude Fields Jr. **VHS, Beta** *CMR, NYS* 🎬

We All Loved Each Other So Much Sensitive comedy follows three friends over 30 years, beginning at the end of World War II. All three have loved the same woman, an actress. Homage to friendship and to postwar Italian cinema. Includes a full-scale recreation of the fountain scene in "La Dolce Vita." In Italian with English subtitles.
1977 124m/C IT Vittorio Gassman, Nino Manfredi, Stefania Satta Flores, Stefania Sandrelli, Marcello Mastroianni, Federico Fellini; *Dir:* Ettore Scola. **VHS, Beta, LV $59.95** *COL, APD* 🎬🎬🎬½

We are the Children Cynical reporter covering the famine in Ethiopia meets an idealistic nurse who's been living there for years. Romance blossoms among the starving in made-for-TV drama co-produced by Danson.
1987 92m/C Ally Sheedy, Ted Danson, Judith Ivey, Zia Mohyeddi; *Dir:* Robert M. Young. **VHS, Beta $79.95** *PSM* 🎬🎬

We Dive at Dawn Interesting, tense British submarine drama. The "Sea Tiger" attempts to sink the German battleship "Brandenburg" off Denmark. Good cast.

Mills prepared for the role by riding an actual submarine, turning "a pale shade of pea-green" when it crash-dived.
1943 98m/B GB Eric Portman, John Mills; *Dir:* Anthony Asquith. **VHS, Beta $19.95** *NOS, WES, DVT* 🎬🎬🎬

We the Living The torpid long-lost and restored Italian version of Ayn Rand's unique political tome. Deals with a young Soviet woman in revolutionary Petrograd who is slowly ruined by the system and her affair with a romantic counter-revolutionary. Made under the Fascists' nose during World War II. A fascinating dialectic between utopian melodrama and Rand dogma. In Italian with subtitles.
1942 174m/B IT Alida Valli, Rossano Brazzi, Fosco Giachetti; *Dir:* Goffredo Alessandrini. **VHS, Beta, LV $49.95** *LUM, JCI* 🎬🎬½

We of the Never Never In turn-of-the-century Australia a city-bred woman marries a cattle rancher and moves from civilized Melbourne to the barren outback of the Northern Territory. Based on the autobiographical story written by Jeannie Gunn, the first white woman to travel in the aboriginal wilderness. She finds herself fighting for her own rights as well as for those of the aborigines in this sincere, well-done film.
1982 (G) 136m/C Angela Punch McGregor, Arthur Dignam, Tony Barry; *Dir:* Igor Auzins. **VHS, Beta $59.95** *COL* 🎬🎬🎬

We Think the World of You When bisexual Oldman goes to prison, his lover Bates, whose feelings are unresolved and complex, finds friendship with Oldman's dog. Odd, oft-bitter comedy-drama of love and loyalty characterized by excellent acting and respectful, gentle direction.
1988 (PG) 94m/C GB Alan Bates, Gary Oldman, Frances Barber, Liz Smith, Max Wall, Kerry Wise; *Dir:* Colin Gregg. **VHS, Beta, LV $19.98** *SUE* 🎬🎬🎬

We Will Not Enter the Forest Moral dilemmas afflict a group of French Resistance fighters during WWII when they must decide what to do with four German deserters they've captured. In French with subtitles.
1979 88m/C FR Marie-France Pisier; *Dir:* Georges Dumoulin. **VHS, Beta $39.95** *FCT* 🎬🎬🎬

Weapons of Death Kung Fu expert Lee splatters his enemies all over San Francisco's Chinatown.
1983 90m/C Eric Lee, Bob Ramos, Ralph Castellanos. **VHS, Beta** *PGN* 🎬

Weather in the Streets A young woman enters into an ill-fated love affair after spending a few moments with an aristocratic married man. Cliched drama set in England between the two world wars.
1984 (PG) 108m/C Michael York, Joanna Lumley, Lisa Eichhorn, Isabel Dean, Norman Pitt. **VHS, Beta $59.95** *MAG* 🎬½

Web of Deceit Defense lawyer Purl returns to her Southern hometown to defend an unjustly accused teenager. She falls in love with the prosecutor, but could he be the killer? Shades of "Jagged Edge." Good, suspenseful made-for-TV movie.
1990 (PG-13) 93m/C Linda Purl, James Read, Paul DeSouza, Larry Black, Len Birman, Barbara Rush; *W/Dir:* Sandor Stern. **VHS, Beta** *PAR* 🎬🎬½

Web of Deception A jewel thief disappears with five million dollars worth of jewels. When the case turns to murder, a private detective becomes the prime suspect.

19?? **(R)** 90m/C Thomas Hunter, Aelina Nathaniel, Gabriele Tinti. **VHS, Beta $39.95** MAG *ℐℐ*

Web of the Spider
Kinski as Poe is really the only good thing about this run-of-the-mill horror outing. A man stays the night in a spooky house to prove it's not haunted. Remake of "Castle of Blood." 1970 94m/C *GE FR IT* Anthony (Tony) Franciosa, Klaus Kinski, Michele Mercier, Peter Carsten, Karen Field; *Dir:* Anthony Dawson. **VHS $16.95** SNC *ℐℐ*

A Wedding
The occasion of a wedding leads to complications galore for the relatives and guests on the happy day. Silent film legend Gish's 100th screen role. Miserable, self-indulgent outing for Altman, who went through quite a dry spell after "Nashville." 1978 **(PG)** 125m/C Mia Farrow, Carol Burnett, Lillian Gish, Lauren Hutton, Viveca Lindfors, Geraldine Chaplin, Paul Dooley, Howard Duff, Dennis Christopher, Peggy Ann Garner, John Considine, Nina Van Pallandt, Dina Merrill, Pat McCormick, Vittorio Gassman, Desi Arnaz Jr.; *Dir:* Robert Altman. New York Film Festival '77: Best On-Screen Duo. **VHS, Beta $59.98** FOX *ℐℐ½*

Wedding Band
Episodic comedy about a rock 'n roll band that plays all sorts of weddings. Thin and lame but inoffensive. 1989 95m/C William Katt, Joyce Hyser, Tino Insana, Fran Drescher, David Rasche, Joe Flaherty, Tim Kazurinsky, James Belushi; *Dir:* Daniel Raskov. **VHS, Beta, LV $89.95** COL *ℐℐ*

Wedding in Blood
Two French lovers plot to kill their respective spouses, and proceed to do so amid calamity and much table-turning. Sharp social satire and suspenseful mystery. Based on a true story. English subtitles. 1974 **(PG)** 98m/C *FR IT* Claude Pieplu, Stephane Audran, Michel Piccoli; *Dir:* Claude Chabrol. **VHS, Beta $69.95** COL *ℐℐℐ*

A Wedding in Galilee
The elder of a Palestinian village is given permission to have a traditional wedding ceremony for his son if Israeli military officers can attend as the guests of honor. As the event approaches, conflicts arise among the villagers, the family, and the Israelis. Long but fascinating treatment of traditional culture in the modern world amidst political tension. In Arabic and Hebrew, with English subtitles. 1987 113m/C *BE FR* Ali M. El Aleili, Nazih Akleh, Anna Achdian; *Dir:* Michel Khleifi. **VHS $79.95** KIV *ℐℐℐ*

Wedding March
Von Stroheim's famous silent film about romantic couplings in pre-World War I Vienna. The director also stars as a prince preparing to marry for money, who falls in love with a beautiful, but poor woman. Romance and irony, with a memorable finale. Like many of his other great films, it was taken from him and re-edited. Initially the first of two halves; the second, "The Honeymoon," has been lost. Next to "Greed," his most successful film. Score by Gaylord Carter. 1928 113m/B Erich von Stroheim, Fay Wray, ZaSu Pitts, George Fawcett, Maude George, Matthew Betz; *Dir:* Erich von Stroheim. New York Film Festival '65: Best On-Screen Duo. **VHS, Beta $29.95** PAR *ℐℐℐℐ*

The Wedding Party
An apprehensive groom is overwhelmed by his too-eager bride and her inquisitive relatives at a prenuptial celebration. Hokey, dull, would-be comedy. First screen appearances of both Clayburgh and De Niro (spelled DeNero in the credits).

1969 90m/C Jill Clayburgh, Robert De Niro, William Finley; *Dir:* Brian DePalma, Cynthia Munroe, Wilford Leach. **VHS, Beta $14.98** VID *ℐℐ½*

Wedding Rehearsal
A young nobleman thwarts his grandmother's plans to marry him off by finding suitors for all the young ladies offered to him until he finally succumbs to the charms of his mother's secretary. Enjoyable, bun comedy. First featured roles for actresses Oberon, Barrie, and Gardner. 1932 84m/B Merle Oberon, Roland Young, John Loder, Wendy Barrie, Maurice Evans, Joan Gardner; *Dir:* Alexander Korda. **VHS, Beta, LV** SUE *ℐℐ½*

Wedding in White
A rape and out-of-wedlock pregnancy trouble a poor British clan living in Canada during World War II. Sad, glum way to spend an evening. Kane is good in a non-comic role. Her screen father care more for her honor than he does for her. 1972 **(R)** 103m/C *CA* Donald Pleasence, Carol Kane, Leo Phillips; *Dir:* William Fruet. **VHS, Beta $39.95** SUE *ℐℐ*

Wee Wendy
Wee Wendy and her people are aliens who land on earth and set up a village on an island in a lake, in this children's animated movie. 1989 100m/C **VHS, Beta** CEL *ℐℐ*

Wee Willie Winkie
A precocious little girl is taken in by a British regiment in India. Sugar-coated. If you're a cinematic diabetic, be warned. If you're a Temple fan, you've probably already seen it. If not, you're in for a treat. Inspired by the Rudyard Kipling story. 1937 99m/B Shirley Temple, Victor McLaglen, C. Aubrey Smith, June Lang, Michael Whalen, Cesar Romero, Constance Collier; *Dir:* John Ford. **VHS, Beta $19.98** FOX *ℐℐℐ*

Weeds
A highly fictionalized account of the career of Rick Cluchey, who, as a lifer in San Quentin federal prison, wrote a play. He was eventually paroled and went on to form a theater group made up of ex-cons. Original and often enjoyable with a tight ensemble performance. Filmed on location at Stateville Correctional Center in Illinois, with inmates serving as extras. Soundtrack by Angelo Badalamenti ("Blue Velvet," "Wild at Heart"). 1987 **(R)** 115m/C Nick Nolte, Rita Taggart, William Forsythe, Lane Smith, Joe Mantegna, Ernie Hudson, John Toles-Bey, Mark Rolston, J.J. Johnson, Anne Ramsey; *Dir:* John Hancock. **VHS, Beta $19.99** HBO *ℐℐℐ*

Weekend
A Parisian couple embark on a drive to the country. On the way they witness and are involved in horrifying highway wrecks. Leaving the road they find a different, equally grotesque kind of carnage. Godard's brilliant, surreal, hyper-paranoiac view of modern life was greatly influenced by the fact that his mother was killed in an auto accident in 1954 (he himself suffered a serious motorcycle mishap in 1975). In French with English subtitles. 1967 105m/C *FR IT* Mireille Darc, Jean Yanne, Jean-Pierre Kalfon, Valerie Lagrange, Jean-Pierre Leaud, Yves Beneyton; *W/Dir:* Jean-Luc Godard. **VHS $79.95** NYF, FCT *ℐℐℐ*

Weekend with the Babysitter
Sordid teen drama about a weekend babysitter who goes to a film director's house and babysits everyone but the kids, running into a heroin-smuggling ring along the way. Casting-couch story with a twist.

1970 **(R)** 93m/C Susan Roman, George E. Carey, James Almanzar, Luanne Roberts; *Dir:* Don Henderson. **VHS, Beta $59.95** PSM *Woof!*

Weekend at Bernie's
Two computer nerds discover embezzlement at their workplace after being invited to their boss's beach house for a weekend party. They find their host murdered. They endeavor to keep up appearances by (you guessed it) dressing and strategically posing the corpse during the party. Kiser as the dead man is memorable, and the two losers gamely keep the silliness flowing. Lots of fun. 1989 **(PG-13)** 101m/C Andrew McCarthy, Jonathan Silverman, Catherine Mary Stewart, Terry Kiser; *Dir:* Ted Kotcheff. **VHS, Beta $14.95** IVE, BUR *ℐℐ½*

Weekend Pass
Three rookie sailors who have just completed basic training are out on a weekend pass, determined to forget everything they have learned. You'll wish you had forgotten to watch this lame excuse of a film. 1984 **(R)** 92m/C D.W. Brown, Peter Ellenstein, Phil Hartman, Patrick Hauser, Chip McAllister; *Dir:* Lawrence Bassoff. **VHS, Beta, LV $79.98** LIV, VES *ℐ*

Weekend of Shadows
Confident, moralistic murder drama about a posse after an innocent immigrant they believe has killed a woman. Police sergeant Roberts is interested only in his reputation, not in justice. Posse member Waters begins to sympathize with the hunted man and defends his innocence. Intelligent, well-conceived, and professionally directed. 1977 94m/C John Waters, Melissa Jaffer, Graeme Blundell, Wyn Roberts, Barbara West, Graham Rouse; *Dir:* Tom Jelfry. **VHS, Beta $49.98** SUE *ℐℐℐ*

Weekend Warriors
Young film-studio employees evade the draft in 1961 by enlisting for National Guard weekend duty. Grade "C" dumb comedy from professional celebrity and first-time director Convy. 1986 **(R)** 88m/C Chris Lemmon, Lloyd Bridges, Daniel Greene, Tom Villard, Vic Tayback; *Dir:* Bert Convy. **VHS, Beta $79.98** LIV, LTG *ℐ½*

Weird Science
Hall is appealing, and Hughes can write dialogue for teens with the best of them. However, many of the jokes are in poor taste, and the movie seems to go on forever. Hall and his nerdy cohort Mitchell-Smith use a very special kind of software to create the ideal woman who wreaks zany havoc in their lives from the outset. 1985 **(PG-13)** 94m/C Kelly Le Brock, Anthony Michael Hall, Ilan Mitchell-Smith, Robert Downey Jr.; *W/Dir:* John Hughes. **VHS, Beta, LV $19.95** MCA *ℐℐ*

The Weirdo
"The Jerk" meets Freddie Krueger. This film is absolutely wretched, even by horror movie standards. 1989 91m/C Steve Burlington, Jessica Straus; *Dir:* Andy Milligan. **VHS, Beta** RAE *ℐ*

Welcome to 18
Three girls, just out of high school, take summer jobs at a dude ranch, work a local casino, flirt, tease and get in trouble. The movie isn't funny in the least, and is filled with utter nonsense. Hargitay is the daughter of actress Jayne Mansfield. 1987 **(PG-13)** 91m/C Courtney Thorne-Smith, Mariska Hargitay, Jo Ann Willette, Christine Kaufman; *Dir:* Terry Carr. **VHS, Beta $19.95** IVE *ℐ*

Welcome Back Mr. Fox
An award-winning science fiction short wherein an obnoxious movie producer cheats death by being cryogenically frozen, and then wakes

up years later no wiser for the experience. Suitably, he receives his comeuppance.
1983 21m/C E.D. Phillips, Gustav Vintas; *Dir:* Walter W. Pitt III. **VHS, Beta** *IFF* 🎞

Welcome to Blood City An anonymous totalitarian organization kidnaps a man and transports him electronically to a fantasy western town where the person who murders the most people becomes the town's ''kill master.'' Amateurish and low-budget.
1977 96m/C Jack Palance, Keir Dullea, Samantha Eggar, Barry Morse. **VHS, Beta** $69.98 *LIV, LTG* 🎞½

Welcome Home A missing-in-action Vietnam vet (Kristofferson) leaves his wife and children in Cambodia to return to America more than 15 years after he was reported dead. He finds his first wife remarried and discovers a teen-aged son he unknowingly fathered. This plotless, sporadically-moving film was Schaffner's last.
1989 (R) 92m/C Kris Kristofferson, JoBeth Williams, Sam Waterston, Brian Keith, Trey Wilson, Thomas Wilson Brown; *Dir:* Franklin J. Schaffner. **VHS, Beta, LV** $89.98 *VES, LIV* 🎞🎞

Welcome Home, Roxy Carmichael Ryder, as a young misfit, is the bright spot in this deadpan, would-be satire. Hollywood star Roxy Charmichael returns to her small Ohio hometown and begins fantasizing that she is really her mother. It is obvious why this deadpan, hard-to-follow movie was a box-office flop.
1990 (PG-13) 98m/C Winona Ryder, Jeff Daniels, Laila Robins, Dinah Manoff, Ava Fabian, Robbie Kiger, Sachi Parker; *Dir:* Jim Abrahams. **VHS, Beta, LV, 8mm** $19.95 *PAR* 🎞🎞

Welcome to L.A. Rudolph's ambitious dirootorial debut foouoo on the ocxual escapades of a group of Southern Californians. Based on the music suite ''City of the One Night Stands'' by Richard Baskin, which also (unfortunately) serves as the soundtrack. Randolph improved greatly after this initial effort.
1977 (R) 106m/C Sissy Spacek, Sally Kellerman, Keith Carradine, Geraldine Chaplin, Lauren Hutton, Viveca Lindfors, Harvey Keitel, John Considine; *Dir:* Alan Rudolph. **VHS, Beta** *FOX* 🎞🎞🎞

Welcome to Spring Break College co-eds are stalked by a killer on the beaches of Florida.
1988 (R) 92m/C John Saxon, Michael Parks. **VHS, Beta, LV** $9.99 *CCB, IVE* 🎞½

The Well A young black girl disappears and a white man (Morgan) is accused of kidnapping her. When it is discovered that the girl is trapped in a deep well, Morgan's expertise is needed to help free her.
1951 85m/B Richard Rober, Harry Morgan, Barry Kelley, Christine Larson; *Dir:* Leo Popkin, Russel Rouse. **VHS, Beta** *UHV* 🎞🎞🎞

Well-Digger's Daughter As her lover goes off to war, a well-digger's daughter discovers that she is pregnant causing both sets of parents to feud over who's to blame. This is the first film made in France after the end of World War II and marks the return of the French film industry. In French with English subtitles.
1946 142m/B *FR* Raimu, Josette Day, Fernandel, Charpin; *Dir:* Marcel Pagnol. **VHS, Beta** $29.95 *NOS, DVT, APD* 🎞🎞🎞

Wellington: The Duel Scandal A look at how the man who was Britain's greatest hero managed to garner the hatred of his countrymen in eighteen months.

19?? 60m/C *GB* **VHS** *HRS* 🎞🎞🎞

We're in the Legion Now Denny and Barnett star as a couple of American gangsters who join the French Foreign Legion to escape rival hoods from back home.
1937 56m/B Reginald Denny, Vince Barnett, Esther Ralston. **VHS, Beta, 8mm** $24.95 *VYY* 🎞½

We're No Angels Three escapees from Devil's Island hide out with the family of a kindly French storekeeper. One of Bogart's few comedies. From the French stage play of the same name. Claustrophobic and talky. Later remade with De Niro and Penn.
1955 103m/C Humphrey Bogart, Aldo Ray, Joan Bennett, Peter Ustinov, Basil Rathbone, Leo G. Carroll; *Dir:* Michael Curtiz. **VHS, Beta** $14.95 *PAR* 🎞🎞½

We're No Angels Two escaped cons disguise themselves as priests and get in the appropriate series of jams. De Niro and Penn play off each other well, turning in fine comic performances. Distantly related to the 1955 film of the same name and the David Mamet play.
1989 (R) 110m/C Robert De Niro, Sean Penn, Demi Moore, Hoyt Axton, Bruno Kirby, James Russo; *Dir:* Neil Jordan. **VHS, Beta, LV** $19.95 *PAR* 🎞🎞½

We're Not Married Five couples learn that they are not legally married when a judge realizes his license expired before he performed the ceremonies. The story revolves around this quintet of couples who now must cope with whether or not they really do want to be married. Although the episodes vary in quality, the Allen-Rogers sequence is excellent. Overall, the cast performs well in this lightweight comedy.
1952 85m/B Ginger Rogers, Fred Allen, Victor Moore, Marilyn Monroe, Paul Douglas, David Wayne, Eve Arden, Louis Calhern, Zsa Zsa Gabor, James Gleason, Jane Darwell, Eddie Bracken, Mitzi Gaynor; *Dir:* Edmund Goulding. **VHS** $14.98 *FXV* 🎞🎞🎞

Werewolf in a Girl's Dormitory Girls' school headmaster undergoes dental transformation at night. Atrocious dubbing, with equally atrocious theme song ''The Ghoul in School.''
1961 82m/B *IT* Barbara Lass, Carl Schell, Curt Lowens, Maurice Marsac; *Dir:* Richard Benson. **VHS** $19.98 *SMW, SNC* 🎞

Werewolf of London A scientist searching for a rare Tibetan flower is attacked by a werewolf. He scoffs at the legend, but once he's back in London, he goes on a murderous rampage every time the moon is full. Dated but worth watching as the first werewolf movie made.
1935 75m/B Henry Hull, Warner Oland, Valerie Hobson, Lester Matthews, Spring Byington; *Dir:* Stuart Walker. **VHS** $14.98 *MCA* 🎞🎞½

The Werewolf vs. the Vampire Woman Hirsute Spanish wolfman teams with two female students in search of witch's tomb. One is possessed by the witch, and eponymous title results. Also on video as ''Blood Moon.''
1970 (R) 82m/C *SP GE* Paul Naschy, Gaby Fuchs, Barbara Capell, Patty Sheppared, Valerie Samarine, Julio Pena, Andres Resino; *Dir:* Leon Klimovsky. **VHS** $19.98 *SNC, HEG* 🎞½

Werewolf of Washington A sorry combination of horror spoof and political satire as a White House press secretary turns into a werewolf.

1973 (PG) 90m/C Dean Stockwell, Biff Maguire, Clifton James; *Dir:* Milton Moses Ginseberg. **VHS, Beta** $39.95 *MON* 🎞½

Werewolves on Wheels A group of bikers are turned into werewolves due to a Satanic spell. A serious attempt at a biker/werewolf movie, however, too violent and grim, not at all funny, and painful to sit through. McGuire had a hit with ''Eve of Destruction.''
1971 (R) 85m/C Stephen Oliver, Severn Darden, Donna Anderson, Duece Barry, Billy Gray, Barry McGuire; *Dir:* Michel Levesque. **VHS, Beta** $59.95 *UNI* Woof!

West-Bound Limited A Romantic adventurer rescues a girl from certain death as a train is about to hit her, and the two fall in love. Lucky for him, she happens to be the boss' daughter. This film is a good example of classic silent melodrama with music score.
1923 70m/B Johnny Harron, Ella Hall, Claire McDowell. **VHS, Beta** $29.95 *VYY* 🎞🎞½

West of the Divide A young cowboy (Wayne) impersonates an outlaw in order to hunt down his father's killer and find his missing younger brother. He saves a proud rancher and his feisty daughter along the way. A remake of ''Partners of the Trail,'' and a solid early Wayne oater.
1934 53m/B John Wayne, George ''Gabby'' Hayes, Lloyd Whitlock, Yakima Canutt; *Dir:* Robert N. Bradbury. **VHS** $19.95 *COL, REP, NOS* 🎞🎞½

West to Glory Glory isn't necessarily exciting, as the hero helps a Mexican rancher save his gold and jewels from evildoers in this slow-paced oater.
1947 61m/B Eddie Dean, Roscoe Ates, Dolores Castle, Gregg Barton, Alox Montoya, Harry Vejar; *Dir:* Ray Taylor. **VHS** $19.95 *NOS* 🎞½

West of the Law A group of ranchers turn to lawmen for protection from a band of outlaws. Part of the Rough Riders series.
1942 60m/B Buck Jones, Tim McCoy; *Dir:* Howard Bretherton. **VHS, Beta** $19.95 *NOS, VCN, DVT* 🎞🎞

West of Nevada A rare film, featuring the former Lieutenant Governor of Nevada (and Clara Bow's husband) in a fast-moving adventure about a band of Indians who must protect their gold from thieves. Bell is the good guy who aids the Indians.
1936 59m/B Rex Bell, Joan Barclay, Al ''Fuzzy'' St. John, Steve Clark. **VHS, Beta, Special order formats** $52.95 *GVV* 🎞🎞

West of Pinto Basin The Range Busters are riding around checking the territory for badmen in this tale of the old west. The music score includes ''That Little Prairie Gal of Mine,'' ''Rhythm of the Saddle,'' and ''Ridin' the Trail Tonight.''
1940 60m/B Ray Corrigan, Max Terhune, John ''Dusty'' King, Jack Perrin. **VHS, Beta** $19.95 *NOS, DVT* 🎞

West Side Story Gang rivalry and ethnic tension on New York's West Side erupts in a ground-breaking musical that won ten Academy Awards. Loosely based on Shakespeare's ''Romeo and Juliet,'' the story follows the Jets and the Sharks as they fight for their own turf while Tony and Maria fight for love. Frenetic and brilliant choreography by co-director Robbins, who also directed the original Broadway show, and a high-caliber score by Leonard Bernstein and Stephen Sondheim. Songs include ''I Feel Pretty,'' ''Something's Coming,'' ''Cool,'' and ''Somewhere.'' Wood's voice was

dubbed by Marni Nixon (she is also heard in "The King and I" and "My Fair Lady") and Jimmy Bryant dubbed Beymer's. The laserdisc version includes the complete storyboards, production sketches, re-issue trailer, and an interview with Wise. It also utilizes letterboxing and digital stereo surround sound.
1961 151m/C Natalie Wood, Richard Beymer, Russ Tamblyn, Rita Moreno, George Chakiris, Simon Oakland, Ned Glass; *Dir:* Robert Wise, Jerome Robbins. Academy Awards '61: Best Art Direction/Set Decoration (Color); Best Color Cinematography, Best Costume Design, Best Director (Wise), Best Film Editing, Best Picture, Best Sound, Best Supporting Actor (Chakiris), Best Supporting Actress (Moreno), Best Musical Score; Golden Globe Awards '62: Best Film—Musical/Comedy. **VHS, Beta, LV** $19.98 *KUI, MGM, FOX* 𝕀𝕀𝕀½

West and Soda Bozzetto, renowned artist and creator of "Allegro Non Troppo," presents animated tale about pincushion named Johnny who is the fastest gun in the West.
1990 ?m/C *Dir:* Bruno Bozzetto. **VHS** $59.95 *EXP* 𝕀𝕀𝕀½

Western Double Feature 1 In "Utah" (1945, 55 minutes), Rogers foils a land swindle and helps a lady ranch owner; in "Man from Music Mountain" (1938, 60 minutes), Autry foils a worthless gold mining stock swindle.
1938 115m/B Roy Rogers, Dale Evans, George "Gabby" Hayes, Gene Autry, Smiley Burnette; *Dir:* Joseph Kane. **VHS, Beta** $59.95 *HHT* 𝕀

Western Double Feature 1: Wild Bill & Sunset Carson Double feature: in "Calling Wild Bill Elliot" (1943), Wild Bill comes to the aid of homesteaders; in "Santa Fe Saddlemates" (1945), Carson breaks up a diamond smuggling ring.
194? 120m/C Wild Bill Elliott, Sunset Carson; *Dir:* Spencer Gordon Bennet. **VHS, Beta** *MED* 𝕀

Western Double Feature 2 Wayne and Hayes are teamed together in two hourlong B-westerns: "Star Packer" (1934) and "West of the Divide" (1933).
1933 120m/B John Wayne, George "Gabby" Hayes; *Dir:* Robert N. Bradbury. **VHS, Beta** $59.95 *HHT* 𝕀𝕀

Western Double Feature 2: Wayne & Hale Double feature; in "Night Riders" (1939), Wayne battles injustice; in "Home on the Range" (1946, color), Hale protects a wild animal refuge.
198? 120m/B John Wayne, Monte Hale; *Dir:* George Sherman. **VHS, Beta** *MED* 𝕀𝕀

Western Double Feature 3 Two Wayne westerns are on this tape: "Helltown" (1938, 59 minutes), in which Wayne catches a cattle rustler; and "Winds of the Wasteland" (1936, 54 minutes), where he plays an out-of-work Pony Express rider.
1936 113m/B John Wayne, Johnny Mack Brown, Marsha Hunt, Phyllis Fraser; *Dir:* Charles T. Barton, Mack V. Wright. **VHS, Beta** $59.95 *HHT* 𝕀𝕀½

Western Double Feature 3: Rogers & Allen Double feature. In "Twilight in the Sierras" (1950), Rogers catches a gang of crooks; in "Under Mexicali Stars" (1950), Allen is a cowboy who uncovers a counterfeiting ring.
1950 134m/C Roy Rogers, Dale Evans, Rex Allen; *Dir:* William Witney. **VHS, Beta** *MED* 𝕀𝕀½

Western Double Feature 4 Rogers plays a dual role in a case of mistaken identity in "Billy the Kid Returns" (1938, 56 minutes); in "Round Up Time in Texas" (1937, 56 minutes), Autry is framed on a charge of diamond smuggling.
1937 112m/B Roy Rogers, Gene Autry, Smiley Burnette; *Dir:* Joseph Kane. **VHS, Beta** $59.95 *HHT* 𝕀½

Western Double Feature 4: Rogers & Evans Double feature. In "Don't Fence Me In" (1945), Rogers helps out a woman reporter; in "Sheriff of Wichita" (1948), a frontier investigator solves a crime.
194? 120m/B Roy Rogers, Dale Evans, Trigger, George "Gabby" Hayes, Allan "Rocky" Lane. **VHS, Beta** *MED* 𝕀½

Western Double Feature 5 John Wayne stars in both halves of this double feature: in "Desert Trail" (1935, 54 minutes), he plays a rodeo star; in "Man from Utah" (1934, 55 minutes), he brings a gang of murderers to justice.
1935 109m/B John Wayne, Polly Ann Young; *Dir:* Robert N. Bradbury. **VHS, Beta** $59.95 *HHT* 𝕀

Western Double Feature 5: Rogers & Autry Double feature. "The Big Show" (1937) features songs, action, and horses; in "Home in Oklahoma" (1946), Roger tracks down a murderer.
1937 108m/B Gene Autry, Roy Rogers, Smiley Burnette, The Pioneers; *Dir:* Max Wright. **VHS, Beta** *MED* 𝕀

Western Double Feature 6: Rogers & Trigger Double feature. "Trigger Jr." (1950) is the story of Trigger's colt; "Gangs of Sonora" (1941, black and white) deals with a fight to save a small frontier newspaper.
1950 120m/C Roy Rogers, Dale Evans, Trigger, Bob Livingston, Bob Steele; *Dir:* William Witney. **VHS, Beta** *MED* 𝕀

Western Double Feature 7: Wild Bill & Crabbe Double feature. "Phantom of the Plains" deals with a plan to save a woman from a gigolo; "Prairie Rustlers" tells of the problems of a man who is the identical double of Billy the Kid.
1945 120m/B Wild Bill Elliott, Buster Crabbe. **VHS, Beta** *MED* 𝕀

Western Double Feature 8: Hale & Carradine Double feature. In "Silver Spurs" (1943), ranchers wanting to get the right-of-way through oil land run into an ambitious resort owner; in "Out California Way" (1946), Hale is a young cowboy looking for work in Hollywood.
194? 121m/B Roy Rogers, Monte Hale, John Carradine. **VHS, Beta** *MED* 𝕀

Western Double Feature 9: LaRue & Beery Double feature. In "Cheyenne Takes Over" (1947), LaRue battles outlaws; in "Tulsa Kid" (1940), a young man must face a gunfighter.
194? 120m/B Lash LaRue, Donald (Don "Red") Barry, Noah Beery. **VHS, Beta** *MED* 𝕀

Western Double Feature 10: Roy, Dale, & Trigger Double feature; in "Bells of Coronado" (1950), Rogers must solve a murder and uranium ore theft; in "King of the Cowboys" (1943, black and white), Roy investigates a band of saboteurs.
198? 120m/C Roy Rogers, Dale Evans, Trigger; *Dir:* William Witney. **VHS, Beta** *MED* 𝕀

Western Double Feature 11 Roy and Trigger act as modern day Robin Hoods to end crooked dealing in the West; Roy hides behind a smile and a song to capture murderers of a deputy and a ranch owner.
1950 121m/B Roy Rogers, Penny Edwards, Gordon Jones, Purple Sage, George "Gabby" Hayes. **VHS, Beta** *MED* 𝕀

Western Double Feature 12 Red Ryder learns of a plot to scare ranchers into selling out before an oil strike is discovered; the Three Musketeers are sent to the Caribbean as U.S. envoys to sell Army horses.
1944 108m/B Bob Livingston, Wild Bill Elliott, Duncan Renaldo, Robert Blake. **VHS, Beta** *MED* 𝕀

Western Double Feature 13 Red Ryder foils outlaws trying to swindle the Duchess' stage line in a small isolated town in "Wagon Wheels Westward" (1945). "Rocky Mountain Rangers" (1940), features the Three Musketeers pursuing the deadly Barton Gang.
194? 108m/B Red Ryder, Bob Livingston, Robert Blake. **VHS, Beta** *MED* 𝕀

Western Double Feature 14 Outlaws, good guys, and romance on the old frontier. Amateur Robin Hoods steal back money wrongfully taken.
1952 123m/B Tim Holt, Jack Holt, Nan Leslie, Noreen Nash. **VHS, Beta** *MED* 𝕀

Western Double Feature 15 Rogers breaks up a cattle rustling ring; the good guys try to bring law and order to a vast Wyoming territory plagued by a crooked political boss.
1950 122m/B Roy Rogers, Trigger, Penny Edwards, Donald (Don "Red") Barry, Lynn Merrick. **VHS, Beta** *MED* 𝕀

Western Double Feature 16 Autry is sent to quell a revolution in Mexico; Rogers' horse Lady and Hayes' horse Golden Sovereign are the focal points of the second part of this double feature.
1946 108m/B Gene Autry, Smiley Burnette, Roy Rogers, George "Gabby" Hayes. **VHS, Beta** *MED* 𝕀

Western Double Feature 17 Double feature stars Rogers in "The Golden Stallion" (1949). Roy battles diamond smugglers who use wild horses to transport the goods. The second feature, "The Cherokee Flash" (1945), starring Carson, deals with a respectable citizen's outlaw past catching up to him.
194? 121m/C Roy Rogers, Sunset Carson; *Dir:* William Witney. **VHS, Beta** *MED* 𝕀

Western Double Feature 18 A Rogers double feature: in "Eyes of Texas" (1948), a westerner turns his ranch into a camp for war-orphaned boys; in "Helldorado" (1946), Roy travels to Las Vegas.
194? 108m/B Roy Rogers, Andy Devine, Dale Evans; *Dir:* William Witney. **VHS, Beta** *MED* 𝕀

Western Double Feature 19 Roy helps a young woman foil a plot by crooks to swindle her inheritance. Roy appears in his first starring role (1938) and fights an outlaw gang.
1945 108m/B Roy Rogers, Dale Evans, Pat Brady, Trigger. **VHS, Beta** *MED* 𝕀

Western Double Feature 20 "Night Time in Nevada" (1948), has Roy bringing a ruthless murderer to justice. "The Old Corral" (1937) features a clash between gangsters in limousines and deputies on horseback.

19?? 108m/B Roy Rogers, Gene Autry; *Dir:* Joseph Kane. **VHS, Beta** *MED ℐ*

Western Double Feature 21 Roy battles crooks in "Susanna Pass" (1949). In "Sheriff of Las Vegas" (1944), Red must find the murderer of Judge Blackwell.
194? 121m/C Roy Rogers, Red Ryder, Dale Evans, Robert Blake; *Dir:* William Witney. **VHS, Beta** *MED ℐ*

Western Double Feature 22 A Rogers double feature: "Under California Stars" (1948), Roy rounds up a gang of wild horse rustlers; "San Fernando Valley" (1944), Roy brings law and order to the valley.
194? 108m/B Roy Rogers; *Dir:* William Witney. **VHS, Beta** *MED ℐ*

Western Double Feature 23 In "Spoilers of the Plains" (1951), Roy finds cattle rustlers and sets out to capture the spoilers. "Lights of Old Santa Fe" (1947) features Roy as a cowboy who rescues a beautiful rodeo owner from bankruptcy.
19?? 122m/B Roy Rogers, Dale Evans. **VHS, Beta** *MED ℐ*

Western Double Feature 24 In "Cowboy and the Senorita" (1944), Roy solves the mystery of a missing girl. "Colorado Serenade" (1946) is a musical western in Cinecolor starring Dean.
194? 122m/C Roy Rogers, Dale Evans, Eddie Dean. **VHS, Beta** *MED ℐ*

Western Double Feature 25 In "Pals of the Golden West" (1951), Roy discovers how valuable Trigger is. "Springtime in the Sierras" (1947) features Roy raising and selling thoroughbreds.
19?? 122m/B Roy Rogers, Dale Evans, Andy Dovino, The Pioneers; *Dir:* William Witney. **VHS, Beta** *MED ℐ*

Western Frontier Ordinary cowpoke Maynard is called on to lead the fight against a band of outlaws led by the Indian-raised sister he never knew he had. Action-packed, fast, and fun. Maynard's first outing for Columbia.
1935 56m/B Ken Maynard, Lucille Browne, Nora Lane, Robert Henry, Frank Yaconelli; *Dir:* Al Herman. **VHS, Beta** $19.95 *NOS, DVT ℐℐ*

Western Justice Good Guy Steele brings western-style justice to a lawless town. Ordinary oater with horses, etc., and a few chuckles on the side.
1935 56m/B Bob Steele; *Dir:* Robert N. Bradbury. **VHS, Beta** $19.95 *NOS, WGE ℐℐ*

Western Mail Yet another western in which the hero goes undercover so he can bring the bad guys to justice. Yaconelli plays a guitar tune, accompanied by his monkey.
1942 54m/B Tom Keene, Frank Yaconelli, Leroy Mason, Jean Trent, Fred Kohler Jr.; *Dir:* Robert Tansey. **VHS** $19.95 *NOS, DVT ℐℐ*

Western Pacific Agent A Western Pacific agent chases an outlaw who has committed robbery and murder. The agent falls in love with the victim's sister and, of course, prevails. Rather violent.
1951 62m/B Kent Taylor, Sheila Ryan, Mickey Knox, Robert Lowery; *Dir:* Sam Newfield; *Nar:* Jason Robards Sr. **VHS, Beta, LV** *WGE ℐℐ*

Western Trails Forgettable western about the clean-up of a Wild West town terrorized by outlaws. Thin script, bad acting, and a little singin' and funnin' around.
1938 59m/B Bob Baker, Marjorie Reynolds, Robert Burns; *Dir:* George Waggner. **VHS, Beta** $19.95 *NOS, DVT ℐℐ½*

Western Union A lavish, vintage epic romantically detailing the political machinations, Indian warfare and frontier adventure that accompanied the construction of the Western Union telegraph line from Omaha, Nebraska, to Salt Lake City, Utah, during the Civil War. A thoroughly entertaining film in rich Technicolor. This was Lang's second western, following his personal favorite, "The Return of Frank James." Writer Carson utilized the title, but not the storyline, of a Zane Grey book. The German Lang showed himself a master of the most American of genres, yet made only one more western, "Rancho Notorious" (1952), another masterpiece.
1941 94m/C Randolph Scott, Robert Young, Dean Jagger, Slim Summerville, Virginia Gilmore, John Carradine, Chill Wills, Barton MacLane, Minor Watson, Charles Middleton, Irving Bacon; *Dir:* Fritz Lang. **VHS, Beta** $39.98 *FOX, MLB ℐℐℐ½*

The Westerner A sly, soft-spoken drifter champions Texas border homesteaders in a land war with the legendary Judge Roy Bean. "The Law West of the Pecos" sentences drifter Cooper to hang but Cooper breaks out of jail. He falls for damsel Davenport and stays in the area, advocating the rights of homesteaders to Brennan. Brennan's Bean is unforgettable and steals the show from Cooper. Film debuts of actors Tucker and Andrews. Amazing cinematography; Brennan's Oscar was his third, making him the first performer to pull a hat trick. Also received an Oscar nomination for Black & White Interior Decoration.
1940 100m/B Gary Cooper, Walter Brennan, Doris Davenport, Dana Andrews, Forrest Tucker; *Dir:* William Wyler. Academy Awards '40: Best Supporting Actor (Brennan). **VHS, Beta, LV** $19.98 *HBO, SUE, BTV ℐℐℐ½*

Westfront 1918 A dogmatic anti-war film by the German master (his first talkie), about German and French soldiers on the fields of World War I, dying together without victory. Stunning in its portrayal of war's futility, with excellent photography that achieves a palpable realism. In German, with English subtitles.
1930 90m/C *GE* Gustav Diesl, Fritz Kampers, Claus Clausen, Hans Joachim Moebis; *Dir:* G. W. Pabst. **VHS, Beta** $29.98 *SUE, GLV ℐℐℐ*

Westward Ho Wayne is determined to bring to justice his parents' slayer. Haven't we seen this one before? Seems the bad guys have corrupted his brother. Wayne's group of vigilantes is called "The Singing Riders," so naturally he sings—or is made to appear to. He looks and sounds ridiculous.
1935 55m/B John Wayne, Sheila Manners; *Dir:* Robert N. Bradbury. **VHS** $12.98 *NOS, VCN, REP ℐ½*

Westward Ho, the Wagons! The promised land lies in the west, but to get there these pioneers must pass unfriendly savages, thieves, villains and scoundrels galore. Suitable for family viewing, but why bother? Well, the cast does include four Mouseketeers.
1956 94m/B Fess Parker, Kathleen Crowley, Jeff York, Sebastian Cabot, George Reeves; *Dir:* William Beaudine. **VHS, Beta** $69.95 *DIS ℐ½*

Westworld Michael Crichton wrote this story of an adult vacation resort of the future which offers the opportunity to live in various fantasy worlds serviced by lifelike robots. Brolin and Benjamin are businessmen who choose a western fantasy world. When an electrical malfunction occurs, the robots be-

gin to go berserk. Brynner is perfect as a western gunslinging robot whose skills are all too real.
1973 (PG) 90m/C Yul Brynner, Richard Benjamin, James Brolin, Dick Van Patten, Majel Barrett; *Dir:* Michael Crichton. **VHS, Beta, LV** $19.95 *MGM ℐℐℐ*

Wet Gold A beautiful young woman and three men journey to retrieve a sunken treasure. Nice scenery. Proves that it is possible to make "The Treasure of the Sierra Madre" without making a classic. John Huston, where are you? Made for television.
1984 90m/C Brooke Shields, Brian Kerwin, Burgess Meredith, Tom Byrd; *Dir:* Dick Lowry. **VHS, Beta** $9.99 *CCB, IVE ℐ*

Wetherby Playwright David Hare's first directorial effort, which he also wrote, about a Yorkshire schoolteacher whose life is shattered when a young, brooding stranger comes uninvited to a dinner party, and then shoots himself in her living room. Compelling but oh, so dark. Richardson, who plays a young Redgrave, is actually Redgrave's daughter.
1985 (R) 104m/C *GB* Vanessa Redgrave, Ian Holm, Judy Dench, Joely Richardson, Tim McInnery, Suzanna Hamilton; *Dir:* David Hare. Berlin Film Festival '85: Golden Bear (Best Film). **VHS, Beta** $79.95 *MGM ℐℐℐ*

Whale for the Killing An ecologist stranded in a remote Alaskan fishing village battles to save a beached humpbacked whale from a malicious Russian fisherman. Platitudinous and self-congratulatory, if well-meaning. Made for television from a book by Farley Mowat.
1981 145m/C Richard Widmark, Peter Strauss, Dee Wallace Stone, Bruce McGill, Kathryn Walker; *Dir:* Richard T. Heffron. **VHS, Beta** $59.98 *FOX ℐℐ½*

Whale of a Tale A young boy trains a killer whale to appear in the big show in the main tank at "Marine Land."
1976 (G) 90m/C William Shatner, Marty Allen, Abby Dalton, Andy Devine, Nancy O'Conner. **VHS, Beta** $39.95 *UHV ℐ*

Whales of August Based on the David Berry play, the story of two elderly sisters—one caring, the other cantankerous, blind, and possibly senile—who decide during a summer in Maine whether or not they should give up their ancestral house and enter a nursing home. Gish and Davis are exquisite to watch, and the all-star supporting cast is superb—especially Price as a suave Russian. Lovingly directed by Anderson in his first U.S. outing.
1987 91m/C Lillian Gish, Bette Davis, Vincent Price, Ann Sothern, Mary Steenburgen, Harry Carey Jr.; *Dir:* Lindsay Anderson. National Board of Review Awards '87: Best Actress (Gish). **VHS, Beta, LV, 8mm** $14.98 *SUE ℐℐℐ½*

What About Bob? Bob, a ridiculously neurotic patient, follows his psychiatrist on vacation, turning his life upside down. The psychiatrist's family find Bob entertaining and endearing. Murray is at his comedic best; Dreyfuss's overly excitable characterization wears thin occasionally.
1991 (PG) 99m/C Richard Dreyfuss, Bill Murray, Julie Hagerty, Charlie Korsmo, Tom Aldredge, Roger Bowen, Fran Brill; *Dir:* Frank Oz. **VHS, Beta, LV** $19.99 *BVV ℐℐℐ*

What a Carve-Up! A group of relatives gather in an old, spooky mansion to hear the reading of a will. Tries too hard. Remake of "The Ghoul." Also known as "No Place Like Homicide!"

1962 87m/B GB Kenneth Connor, Sidney James, Shirley Eaton, Donald Pleasence, Dennis Price, Michael Gough; *Dir:* Pat Jackson. **VHS** $16.95 SNC ✍✍

What Comes Around A good-ole-boy drama about a doped-up country singer who is kidnapped by his brother for his own good. The singer's evil manager sends his stooges out to find him. Might have been funny, but isn't. But is it meant to be? Good Reed songs; mildly interesting plot.
1985 (PG) 92m/C Jerry Reed, Bo Hopkins, Arte Johnson, Barry Corbin; *Dir:* Jerry Reed. **VHS, Beta, LV** $19.98 CHA ✍✍

What Do You Say to a Naked Lady? Allen Funt takes his Candid Camera to the streets to find out how people react when confronted by nude members of the opposite sex. Sporadically hilarious.
1970 92m/C *Dir:* Allen Funt. **VHS, Beta** FOX ✍✍ ½

What Ever Happened to Baby Jane? Davis and Crawford portray aging sisters and former child stars living together in a decaying mansion. When the demented Jane (Davis) learns of her now-crippled sister's (Crawford) plan to institutionalize her, she tortures the wheelchair-bound sis. Davis plays her part to the hilt, unafraid of Aldrich's unsympathetic camera, and the viciousness of her character. She received her tenth Oscar nomination for her role. Other nominations for Buono, cinematography, sound and costume design.
1962 132m/B Bette Davis, Joan Crawford, Victor Buono, Anna Lee, B.D. Merrill; *Dir:* Robert Aldrich. **VHS, Beta, LV** $19.98 WAR ✍✍✍ ½

What Have I Done to Deserve This? A savage parody on Spanish mores, about a speed-addicted housewife who ends up selling her son and killing her husband with a ham bone. Black, black comedy, perverse and funny as only Almodovar can be. In Spanish with subtitles. Nudity and profanity.
1985 100m/C Carmen Maura; *Dir:* Pedro Almodovar. **VHS, Beta, LV** $79.95 TAM ✍✍✍

What the Peeper Saw The wife of a wealthy author finds her comfortable life turning into a terrifying nightmare when her young stepson starts acting funny. Juicy and terrifying. Also known as "Night Hair Child."
1972 97m/C Britt Ekland, Mark Lester, Hardy Kruger, Lilli Palmer. **VHS, Beta** $29.95 UHV ✍✍

What Price Glory? Remake of the 1926 silent classic about a pair of comradely rivals for the affections of women in World War I France. Strange to have Ford directing an offbeat comedy, but it works: masterful direction, good acting. Demarest broke both legs in a motorcycle accident during shooting.
1952 111m/C James Cagney, Dan Dailey, Corinne Calvert, William Demarest, James Gleason, Robert Wagner, Casey Adams, Craig Hill, Marisa Pavan; *Dir:* John Ford. **VHS, Beta** $19.98 FOX, FCT ✍✍✍

What Price Hollywood? Aspiring young starlet decides to crash film world by using an alcoholic director as her stepping stone. Bennett is lovely; Sherman is superb as an aging, dissolute man who watches her potential slip away. From a story by Adela Rogers St. John. Remade three times as "A Star is Born."
1932 88m/B Constance Bennett, Lowell Sherman, Neil Hamilton; *Dir:* George Cukor. **VHS, Beta, LV** $19.98 CCB, TTC, FCT ✍✍✍

What Waits Below A scientific expedition encounters a lost race living in caves in South America. Good cast, bad acting, and lousy script. Might have been much better.
1983 (PG) 88m/C Timothy Bottoms, Robert Powell, Lisa Blount; *Dir:* Don Sharp. **VHS, Beta** $69.98 LIV, LTG ✍ ½

What a Way to Die An evil assassin plots dispatching with a difference.
1970 (R) 87m/C William Berger, Anthony Baker, Helga Anders, Georgia Moll; *Dir:* Helmut Foernbacher. **VHS, Beta** $39.95 MON ✍ ½

What Would Your Mother Say? A hidden camera takes you inside actual casting sessions with more than 40 of Hollywood's erotic stars and starlets who didn't know they were being filmed.
1981 83m/C Bill Margold, Tiffany Clark, Maria Tortuga, Kevin Gibson, Tamara Webb, Mike Ranger, Monique Monge. **VHS, Beta** HAR Woof!

Whatever Happened to Aunt Alice? After murdering her husband to inherit his estate, a poor, eccentric widow develops an awful habit: she hires maids only to murder them and steal their savings. The only evidence is a growing number of trees by the drive. Sleuth Ruth Gordon (of "Harold and Maude" fame, acting here just as odd) takes the job in hopes of solving the mystery. Thoroughly amusing.
1969 (PG) 101m/C Geraldine Page, Ruth Gordon, Rosemary Forsyth, Robert Fuller, Mildred Dunnock; *Dir:* Lee H. Katzin. **VHS, Beta** $59.98 FOX, MLB ✍✍✍

Whatever Happened to Susan Jane? A bored housewife searches for an off-beat college friend, and enters into the seamy, strange world of San Francisco's lunatic fringe. Music by Tuxedo Moon, Indoor Life and other punk groups.
1984 90m/C **VHS, Beta** $39.95 MPI ✍ ½

What's New Pussycat? A young engaged man, reluctant to give up the girls who love him, seeks the aid of a married psychiatrist who turns out to have problems of his own. Allen's first feature as both actor and screenwriter. Burt Bacharach and Hal David wrote the music which included the Oscar-nominated title song sung by Tom Jones.
1965 108m/C Peter Sellers, Peter O'Toole, Romy Schneider, Paula Prentiss, Woody Allen, Ursula Andress, Capucine; *Dir:* Clive Donner. **VHS, Beta** $19.98 FOX, FCT ✍✍

What's Up, Doc? A shy musicologist from Iowa (Ryan) has traveled to San Francisco with his fiance (Kahn) for a convention. He meets the eccentric Streisand at his hotel and becomes involved in a chase to recover four identical flight bags containing top secret documents, a wealthy woman's jewels, the professor's musical rocks, and Streisand's clothing. Bogdanovich's homage to the screwball comedies of the '30s. Kahn's feature film debut.
1972 (G) 94m/C Barbra Streisand, Ryan O'Neal, Kenneth Mars, Austin Pendleton, Randy Quaid; *Dir:* Peter Bogdanovich. **VHS, Beta** $19.98 WAR ✍✍✍

What's Up Front Occasionally funny '60s gag-fest about an ogle-eyed bra salesman who decides to sell door to door in order to save a failing brassiere company. Filmed in "Girl-O-Rama" by the young Vilmos Zsigmond. Dated and weird, with camp value. No nudity. Costumes were done by Frederick's of Hollywood.
1963 90m/C Tommy Holden, Marilyn Manning, Carolyn Walker, William Watters; *Dir:* Bob Whealing. **VHS, Beta** $59.95 RHI ✍ ½

What's Up, Tiger Lily? This legitimate Japanese spy movie—"Kagi No Kag" (Key of Keys), a 1964 Bond imitation—was re-edited by Woody Allen, who added a new dialogue track, with hysterical results. Characters Terri and Suki Yaki are involved in an international plot to secure egg salad recipe; Allen's brand of Hollywood parody and clever wit maintain the joke. Allen's first adventure as auteur. Music by the Lovin' Spoonful, who make a brief appearance.
1966 90m/C JP Woody Allen, Tatsuya Mihashi, Mie Hama, Akiko Wakabayashi; *Dir:* Woody Allen, Senkichi Taniguhi. **VHS, Beta, LV** $29.98 VES, CAB ✍✍✍

Wheel of Fortune A shrewd small-town lawyer, working on a case in the big city, is forced to expose his girlfriend's father as a crooked politician. Strange to see the Duke here in the lead, but what the hey. Not great, but interesting. Also known as "A Man Betrayed."
1941 83m/B John Wayne, Frances Dee, Edward Ellis; *Dir:* John H. Auer. **VHS** $14.98 REP ✍✍

Wheels of Fire The earth is a wasteland controlled by sadistic highway hoodlums. When they kidnap the hero's sister, he fights back in a flame-throwing car. Horrible "Road Warrior" rip-off.
1984 (R) 81m/C Gary Watkins, Lynda Wiesmeier; *Dir:* Cirio H. Santiago. **VHS, Beta, LV** $79.98 LIV, VES ✍

Wheels of Terror Possessed black car stalks children in isolated village. Bus driver Cassidy gets behind the wheel of V-8 supercharged school bus to initiate most interminable chase scene in screen history.
1990 (R) 86m/C Joanna Cassidy, Marcie Leeds, Carlos Cervantes, Arlen Dean Snyder; *Dir:* Christopher Cain. **VHS, Beta** $79.95 PAR ✍

When Angels Fly In this emotional tale of love and murder, a young woman sets out to find the exact circumstances surrounding the mysterious death of her sister.
197? 96m/C Jennifer Dale, Robin Ward. **VHS, Beta** $49.95 UNI ✍

When the Bough Breaks A psychologist helps the police with the case of a murder-suicide witnessed by a child. He finds one sick secret society at work in this unkinder, ungentler network TV thriller, based on the novel by Jonathan Kellerman. Danson co-produced.
1986 100m/C Ted Danson, Richard Masur, Rachel Ticotin, David Huddleston, James Noble, Kim Miyori, Merritt Butrick; *Dir:* Waris Hussein. **VHS** WOV ✍✍

When the Clouds Roll By Psycho-satire pits demented doctor against Fairbanks in experiment to make him suicidal basket case. Fairbanks seems to contract ferocious nightmares, passionate superstitions, a spurning lover, and a warrant for his arrest, none of which suppresses his penchant for acrobatics.
1919 77m/B Douglas Fairbanks Sr., Herbert Grimwood, Kathleen Clifford, Frank Campeau, Ralph Lewis, Daisy Robinson, Albert MacQuarrie; *Dir:* Victor Fleming. **VHS, Beta** $24.95 GPV, FCT, DNB ✍✍✍ ½

When Comedy was King A compilation of the greatest silent screen comedians, including Keaton, Chaplin, Langdon, and Turpin.
1959 81m/C Buster Keaton, Charlie Chaplin, Harry Langdon, Ben Turpin; *Dir:* Robert Youngson. **VHS, Beta, LV** $14.98 CFV, VID ✍✍✍

When Dinosaurs Ruled the Earth When "One Million Years B.C." ruled the box office the Brits cranked out a few more lively prehistoric fantasies. A sexy cavegirl, exiled because of her blond hair, acquires a cavebeau and a dinosaur guardian. Stop-motion animation from Jim Danforth, story by J.G. Ballard. Under the name Angela Dorian, Vetri was Playmate of the Year in 1968 (A.D.).
1970 (G) 96m/C *GB* Victoria Vetri, Robin Hawdon, Patrick Allen, Drewe Henley, Sean Caffrey, Magda Konopka, Imogen Hassall, Patrick Holt, Jan Rossini; *W/Dir:* Val Guest. **VHS, Beta, LV $59.95** *WAR, PIA, FCT ♂♂*

When Every Day was the Fourth of July A nine-year-old girl asks her father, a lawyer, to defend a mute handyman accused of murder, knowing this will bring on him the contempt of the community. Well handled, but see "To Kill a Mockingbird" first. Based on producer/director Curtis' childhood. Sequel: "The Long Days of Summer."
1978 100m/C Katy Kurtzman, Dean Jones, Louise Sorel, Harris Yulin, Chris Peterson, Geoffrey Lewis, Scott Brady, Henry Wilcoxon, Michael Pataki; *Dir:* Dan Curtis. **VHS, Beta $49.95** *IVE ♂♂½*

When Father was Away on Business Set in 1950's Yugoslavia, a family must take care of itself when the father is sent to jail for philandering with a woman desired by a Communist Party official. The moving story of the family's day to day survival is seen largely through the eyes of the father's six-year-old son, who believes dad is "away on business." In Yugoslav with English subtitles.
1985 (R) 144m/C *YU* Moreno D'EBartolli, Miki Manojlovic, Mirjana Karanovic; *Dir:* Emir Kusturica. Cannes Film Festival '85: Best Film. **VHS $89.95** *FXL ♂♂♂½*

When Gangland Strikes A country lawyer, blackmailed by the mob, tries to protect his family without caving in. Unexceptional. Remake of "Main Street Lawyer."
1956 70m/B Raymond Greenleaf, Marjie Millar, Anthony Caruso, Jack Perrin, John Hudson; *Dir:* R.G. Springsteen. **VHS $14.98** *REP ♂♂*

When Harry Met Sally A romantic comedy about the long relationship between two adults who try throughout the changes in their lives and their mates to remain platonic friends—and what happens when they don't. Wry and enjoyable with wonderful performances by all concerned. Another directional direct hit for "Meathead" Reiner, and a tour de force of comic screenwriting for Ephron, with improvisational help from Crystal. Great score by Harry Connick Jr.
1989 96m/C Billy Crystal, Meg Ryan, Carrie Fisher, Bruno Kirby, Lisa Jane Persky, Steven Ford; *Dir:* Rob Reiner. **VHS, Beta, LV, SVS, 8mm $19.95** *COL, SUE, SUP ♂♂♂*

When Hell Was in Session In more than seven years as a prisoner of the Viet Cong, Holbrook is subjected to torture, starvation, and psychological warfare to break his will. Based on the true story of Navy Commander Jeremiah Denton. Painful and violent. Made for TV.
1982 98m/C Hal Holbrook, Eva Marie Saint, Ronny Cox; *Dir:* Paul Krasny. **VHS, Beta** *TLF ♂♂♂*

When Knights were Bold An English nobleman living in India inherits a castle in his native land. Returning home, he is knocked unconscious by a falling suit of armor while trying to impress a young lady and dreams himself back to medieval days.

Enjoyable comedy about that ubiquitous British class hierarchy.
1936 55m/B Fay Wray, Jack Buchanan, Martita Hunt. **VHS, Beta $24.95** *VYY ♂♂½*

When Ladies Meet Entertaining story of a love quadrangle that features several of MGM's top stars of the 40's. In this remake of the 1933 film, Crawford plays a novelist and a forerunner of the women's liberation movement. She falls in love with her publisher (Marshall), who just happens to be married to Garson. Meanwhile, Taylor is in love with Crawford and attempts to show her that he is a more suitable match for her than Marshall, but Crawford has yet to catch on. Lengthy dialogue on women's rights is badly outdated, but the real-life rivalry between Crawford and Garson adds a certain bite to their witty exchanges. Also known as "Strange Skirts."
1941 105m/B Joan Crawford, Robert Taylor, Greer Garson, Herbert Marshall, Spring Byington; *Dir:* Robert Z. Leonard; *W/Dir:* Anita Loos, S.K. Lauren. **VHS $19.98** *MGM ♂♂½*

When the Legends Die A Ute Indian strives to preserve his heritage in an often harsh modern world. Recorded in hi-fi.
1972 (PG) 105m/C Richard Widmark, Frederic Forrest; *Dir:* Stuart Millar. **VHS, Beta $19.98** *FOX ♂♂♂*

When Lightning Strikes Lightning the Wonder Dog prevents the owner of a rival lumber company from stealing his master's land. He gets to use his talents of running, swimming, barking and smoking cigars. Proves that real woofers were made even back when, though in 1934 they didn't feature scantily-clad babes. Part of Video Yesteryear's Golden Turkey series.
1934 51m/B Francis X. Bushman, Lightning the Wonder Dog. **VHS, Beta $24.95** *VYY Woof!*

When the Line Goes Through A drifter arrives in a small West Virginia town and changes the lives of two pretty sisters.
1971 90m/C Martin Sheen, Davey Davidson, Beverly Washburn. **VHS, Beta $39.95** *NEG ♂*

When a Man Rides Alone Tyler relieves trains of their gold shipments in order to reimburse swindled investors. Lovely Lacey frowns on his initiative, so he kidnaps her to keep her quiet. The two fall in love, the man who rides alone works beside his beloved.
1933 60m/B Tom Tyler, Adele Lacey, Alan Bridge, Bob Burns, Frank Ball, Alma Chester, Bud Osborne; *Dir:* J.P. McGowan. **VHS $19.99** *NOS, ASE ♂♂*

When Nature Calls A city family "gets back to nature" in this collection of mostly ineffective gags and poor satirical ideas. Probably your first and last chance to see Liddy and Mays onscreen together.
1985 (R) 76m/C Davie Orange, Barbara Marineau, Nicky Beim, Tina Marie Staiano, Willie Mays, G. Gordon Liddy; *Dir:* Charles Kaufman. **VHS, Beta, LV $59.95** *MED Woof!*

When the North Wind Blows An old, lone trapper hunts for and later befriends the majestic snow tiger of Siberia in the Alaskan wilderness. A good family film.
1974 (G) 113m/C Henry Brandon, Herbert Nelson, Dan Haggerty; *Dir:* Stewart Raffill. **VHS, Beta $54.95** *UHV, LME ♂♂*

When the Screaming Stops A hunter is hired to find out who has been cutting the hearts out of young women who reside in a small village near the Rhine River. Turns out it's a she-monster who rules a kingdom

beneath the river. Bad effects; gory gore; outlandish plot. All the ingredients, in other words.
1973 (R) 86m/C *SP* Tony Kendall, Helga Line, Sylvia Tortosa; *Dir:* Armando de Ossorio. **VHS, Beta $59.98** *LIV, LTG ♂*

When a Stranger Calls A babysitter is terrorized by threatening phone calls and soon realizes that the calls are coming from within the house. Story was expanded from director Walton's short film "The Sitter." Distasteful and unlikely, though the first half or so is tight and terrifying.
1979 (R) 97m/C Carol Kane, Charles Durning, Colleen Dewhurst, Rachel Roberts; *Dir:* Fred Walton. **VHS, Beta, LV $64.95** *COL ♂½*

When Taekwondo Strikes One brave Taekwondo master leads the Korean freedom fighters against the occupying army of World War II Japan.
1983 (R) 95m/C Jhoon Rhee, Ann Winton, Angela Mao, Huang Ing Sik; *Dir:* Raymond Chow. **VHS, Beta $9.98** *SUE ♂*

When Thief Meets Thief A cat burglar finds his world turned upside down when he begins to fall in love with one of his victims. Not much happening except a few nice stunts by Fairbanks.
1937 85m/B *GB* Douglas Fairbanks Jr., Valerie Hobson, Alan Hale Jr., Jack Melford, Leo Genn, Ian Fleming; *Dir:* Raoul Walsh. **VHS, Beta $16.95** *NOS, SNC ♂½*

When the Time Comes A woman dying of cancer wants to take her own life, but when her husband refuses to assist she seeks the help of a male friend. Excellent acting doesn't bring the needed depth to an important subject. Less a disease-of-the-week movie, more of an ethical dilemma-of-the-week, and just as trivial.
1991 (PG-13) 94m/C Bonnie Bedelia, Brad Davis, Terry O'Quinn, Karen Austin, Donald Moffatt, Wendy Schaal; *Dir:* John Erman. **VHS $79.98** *REP ♂♂*

When Time Ran Out A volcano erupts on a remote Polynesian island covered with expensive hotels and tourists with no way to escape. Contains scenes not seen in the theatrically released print of the film. A very good cast is wasted in this compilation of disaster-film cliches.
1980 (PG) 144m/C Paul Newman, Jacqueline Bisset, William Holden, Ernest Borgnine, Edward Albert, Barbara Carrera, Valentina Cortese, Burgess Meredith, Pat Morita, Red Buttons; *Dir:* James Goldstone. **VHS, Beta $19.98** *WAR ♂*

When the West Was Young An outdoor adventure about rounding up horses in the Old West. Also titled "Wild Horse Mesa." Based on a novel by Zane Grey.
1932 58m/B Randolph Scott, Sally Blaine, Guinn Williams; *Dir:* Henry Hathaway. **VHS, Beta $19.95** *NOS, DVT, VYY ♂½*

When the Whales Came A conservationist fable about two children and a grizzled old codger living on a remote British isle during World War I. They try to avert disaster as mysterious narwhal whales descend on the island. Moralistic but not involving; poorly acted; slumber-inducing.
1989 (PG) 100m/C *GB* Paul Scofield, Helen Mirren, David Threlfall, David Suchet, Jeremy Kemp, Max Rennie; *Dir:* Clive Rees. **VHS, Beta, LV $89.98** *FOX ♂♂½*

When the Wind Blows An animated feature about a retired British couple when their peaceful—and naive—life in the country is destroyed by nuclear war. Poignant, sad, and thought-provoking, and just a little

scary. Features the voices of Ashcroft and Mills; Roger Waters, David Bowie, Squeeze, Genesis, Hugh Cornell, and Paul Hardcastle all contribute to the soundtrack. Based on the novel by Raymond Briggs.
1986 80m/C GB *Voices:* Peggy Ashcroft, John Mills. **VHS, Beta, LV $79.95** *IVE 🐾🐾🐾*

When Wolves Cry An estranged father and son reunite on a Corsican vacation. Their new-found joy sours when the young son is diagnosed with a fatal illness.
1969 108m/C William Holden, Virna Lisi, Brook Fuller, Bourvil; *Dir:* Terence Young. **VHS, Beta $24.95** *GEM 🐾½*

When Women Had Tails A primitive comedy about prehistoric man's discovery of sex. And boy, did they ever discover it. Harmless (more or less), though not exactly cerebral. Followed by ''When Women Lost Their Tails.''
1970 (R) 99m/C *IT* Senta Berger, Frank Wolff; *Dir:* Pasquale Festa Campanile. **VHS, Beta $24.95** *NO 🐾½*

When Women Lost Their Tails Ostensible sequel to ''When Women Had Tails'' about prehistoric cavemen and their sexual habits.
1975 (R) 94m/C *IT* Senta Berger; *Dir:* Pasquale Festa Campanile. **VHS, Beta $59.95** *WES 🐾*

When Worlds Collide Another planet is found to be rushing inevitably towards earth, but before the collision a select group of people attempt to escape in a spaceship; others try to maneuver their way on board. Oscar-quality special effects and plot make up for cheesy acting and bad writing.
1951 (G) 81m/C Richard Derr, Barbara Rush, Larry Keating, Peter Hanson; *Dir:* Rudolph Mate. Academy Awards '51: Best Special Effects. **VHS, Beta, LV $49.95** *PAR 🐾🐾½*

When Your Lover Leaves Alleged comedy about a woman dumped by her married boyfriend who gets involved in a short-lived relationship with a neighbor. When that also doesn't work out, she decides she must first do something to please herself. Here endeth the lesson. Produced by Fonzie and Richie of ''Happy Days.''
1983 100m/C Valerie Perrine, Betty Thomas, David Ackroyd, Ed O'Neill, Dwight Schultz, Shannon Wilcox; *Dir:* Jeff Bleckner. **VHS $9.99** *NWV, STE 🐾*

When's Your Birthday? Brown stars in this comedy about a prize-fighter who is working his way through astrology school with his fighting skills. The stars, in turn, tell him when to fight. Lame zaniness (what other kind is there?) meant as a vehicle for Brown, though Kennedy is funnier. The opening sequence, an animated cartoon showing the influence of the moon over the planets, was filmed in Technicolor.
1937 77m/B Joe E. Brown, Marian Marsh, Edgar Kennedy; *Dir:* Harry Beaumont. **VHS, Beta $19.95** *NOS, VYY, JEF 🐾🐾½*

Where Are the Children? Based on Mary Higgins Clark's bestseller. A woman who was cleared of murdering the children from her first marriage, remarries. Then the children from her second marriage are kidnapped. Sustains suspense completely, until it falls apart.
1985 (R) 92m/C Jill Clayburgh, Max Gail, Barnard Hughes, Clifton James, Harley Cross, Elisabeth Harnois, Elizabeth Wilson, Frederic Forrest; *Dir:* Bruce Malmuth. **VHS, Beta $79.95** *COL 🐾½*

Where the Boys Are Four college girls go to Fort Lauderdale to have fun and meet boys during their Easter vacation. Features the film debuts of Francis, who had a hit single with the film's title song, and Prentiss. Head and shoulders above the ludicrous '84 remake.
1960 99m/C George Hamilton, Jim Hutton, Yvette Mimieux, Connie Francis, Paula Prentiss, Dolores Hart; *Dir:* Henry Levin. **VHS, Beta, LV $59.95** *MGM 🐾🐾½*

Where the Boys Are '84 Horrible remake of the 1960 comedy still features girls searching for boys during spring break in Fort Lauderdale. Telling about its era: charm gives way to prurience.
1984 (R) 95m/C Lisa Hartman, Wendy Schaal, Lorna Luft, Lynn-Holly Johnson; *Dir:* Hy Averback. **VHS, Beta $79.98** *FOX 🐾*

Where the Buffalo Roam Early starring role for Murray as the legendary ''gonzo'' journalist Hunter S. Thompson. Meandering satire based on Thompson's books ''Fear and Loathing in Las Vegas'' and ''Fear and Loathing on the Campaign Trail '72.'' Either confusing or offensively sloppy, depending on whether you've read Thompson. Music by Neil Young, thank goodness, or this might be a woof.
1980 (R) 98m/C Bill Murray, Peter Boyle; *Dir:* Art Linson. **VHS, Beta $19.95** *MCA 🐾*

Where the Bullets Fly Fast-paced Bond spoof. A British spy takes on the intelligence forces of several governments in his search for a new fuel elixir. Fun, if not scintillating.
1966 88m/C *GB* Tom Adams, Dawn Addams, Michael Ripper, Tim Barrett; *Dir:* John Gilling. **VHS** *FOX 🐾*

Where Did You Get That Woman? A poignant profile of a black woman who works as a washroom attendant. Her telling insights of the affluent people she serves are especially ironic, given the circumstances of her meager existence.
198? 28m/C VHS, Beta **$39.95** *MVD, TEX 🐾🐾*

Where the Eagle Flies Ersatz sixties rock'n'roll road movie shuffles together free spirited college coed, hobo with heart of gold, and rock and roller who's making a lane change out of the fast track. Originally titled ''Pickup on 101.''
1972 (PG) 93m/C Jack Albertson, Lesley Ann Warren, Martin Sheen, Michael Ontkean; *Dir:* John Florea. **VHS $39.99** *TPV 🐾🐾*

Where Eagles Dare During World War II, a small group of Allied commandos must rescue an American general held by the Nazis in a castle in the Bavarian Alps. Relentless plot twists and action keep you breathless. Well-made suspense/adventure. Alistair MacLean adapted his original screenplay into a successful novel.
1968 (PG) 158m/C Clint Eastwood, Richard Burton, Mary Ure, Michael Hordern, Anton Diffring, Ingrid Pitt, Patrick Wymark, Robert Beatty; *Dir:* Brian G. Hutton. **VHS, Beta, LV $19.95** *MGM, TLF 🐾🐾🐾*

Where the Girls Are A military training film warning against the dangers of VD.
1969 23m/B VHS, Beta **$19.95** *VYY 🐾🐾🐾*

Where the Green Ants Dream A mining excavation in the Outback is halted by Aborigines who declare ownership of the sacred place where the mythical green ants are buried. A minor entry in the Herzog vision of modern-versus-primal civilization. Too obvious and somehow unsure of itself artistically.
1984 99m/C *GE* Bruce Spence, Wandjuk Marika, Roy Marika, Ray Barrett, Norman Kaye, Colleen Clifford; *Dir:* Werner Herzog. **VHS, Beta $24.95** *XVC, MED 🐾🐾½*

Where Have All the People Gone? A nuclear explosion turns most of Earth's inhabitants to dust while the Anders family vacationed in a cave. The Anderses try to return home amid the devastation. Bad acting, bad script.
1974 74m/C Peter Graves, Kathleen Quinlan, Michael-James Wixted, George O'Hanlon Jr., Verna Bloom; *Dir:* John Llewellyn Moxey. **VHS, Beta $59.95** *LHV 🐾½*

Where the Heart Is Wealthy dad Coleman kicks his spoiled kids out on the streets to teach them the value of money. Meant as farce with a message. Flops in a big way; one senses it should have been much better.
1990 (R) 111m/C Dabney Coleman, Uma Thurman, Joanna Cassidy, Suzy Amis, Crispin Glover, Christopher Plummer; *Dir:* John Boorman. **VHS, LV $89.95** *TOU 🐾🐾*

Where the Hot Wind Blows The sexual scene in Sicily is sadly not succinctly summarized in this so-so cinematic soap. Cross-generational and cross-class couplings are constant. Think of it as ''The Last Picture Show'' in a similarly sorry setting, except boring.
1959 120m/C *FR IT* Gina Lollobrigida, Marcello Mastroianni, Yves Montand, Melina Mercouri, Paolo Stoppa; *Dir:* Jules Dassin. **VHS, Beta $39.95** *NOS, MON, FCT 🐾🐾*

Where Jesus Walked Travelogue of Jesus' odyssey through and around Palestine. Explore Capernaum, Jericho, Bethlehem, Nazareth, and Jerusalem.
1987 60m/C VHS **$29.95** *QHV 🐾🐾*

Where is My Child? Immigration to the New World brings only misfortune and betrayal in this tale of Eastern European Jews set between 1911 and 1937, as many experience the loss of family ties and religion. In Yiddish with English subtitles.
1937 92m/B Celia Adler, Anna Lillian, Morris Strassberg; *Dir:* Abraham Leff, Harry Lynn. VHS **$54.00** *NCJ 🐾🐾½*

Where the North Begins A Mountie ransacks the Canadian countryside in search of crime.
1947 41m/B Russell Hayden, Jennifer Holt. **VHS, Beta, LV** *WGE 🐾*

Where the Red Fern Grows A young boy in Dust Bowl-era Oklahoma learns maturity through his love and responsibility for two Redbone hounds. Well produced, but way hokey; only for small children.
1974 (G) 97m/C James Whitmore, Beverly Garland, Jack Ging, Loni Chapman, Stewart Peterson; *Dir:* Norman Tokar. **VHS, Beta $29.98** *LGC, VES, HHE 🐾🐾*

Where the River Runs Black An orphaned Indian child is raised in the Brazilian jungles by river dolphins. He is eventually befriended by a kindly priest who brings him into the modern world of violence and corruption. Slow pace is okay until the boy arrives at the orphanage, at which point the dolphin premise sadly falls by the wayside.
1986 (PG) 96m/C Charles Durning, Peter Horton, Ajay Naidu, Conchata Ferrell, Alessandro Rabelo, Castulo Guerra; *Dir:* Christopher Cain. **VHS, Beta $79.98** *FOX 🐾🐾*

Where the Sky Begins From the director of "Black Orpheus," this is a seven-part recreated history of turn-of-the-century aviation.
1984 420m/C *FR Dir:* Marcel Camus. **VHS, Beta** $19.00 *VCD ♫♫½*

Where the Spirit Lives Native children kidnapped by Canadian government agents are forced to live in dreadful boarding schools where they are abused emotionally and physically. St. John is a new arrival who refuses to put up with it and tries to escape. Engrossing and moving.
1989 (PG) 97m/C Michelle St. John; *Dir:* Bruce Pittman. **VHS** $79.95 *BTV, HHE ♫♫½*

Where Time Began The discovery of a strange manuscript of a scientist's journey to the center of the earth leads to the decision to recreate the dangerous mission. Based on Jules Verne's classic novel "Journey to the Center of the Earth," but not anywhere near as fun or stirring.
1977 (G) 87m/C Kenneth More, Pep Munne, Jack Taylor; *Dir:* Juan Piquer Simon. **VHS, Beta** $69.98 *SUE ♫♫*

Where Trails End Novel sci-fi western set during WWII with just about everything, including phosphorescent-clothed good guys, Nazis, scoundrels, literate horses, and Gallic sidekicks. Interestingly different.
1942 54m/B Tom Keene, Joan Curtis, Charles King; *Dir:* Robert Tansey. **VHS, Beta** $12.95 *VDM, VYY ♫♫*

Where's Piccone A woman attempts, with the assistance of a two-bit hustler, to locate her missing husband, only to discover that he has led a double life in the Neapolitan underworld. Giannini as the sleazy hubby is perfectly cast and hysterical. On-target social and political satire. In Italian with subtitles.
1984 110m/C *IT* Giancarlo Giannini, Lina Sastri, Aldo Guiffre, Clelia Rondinelli; *Dir:* Nanni Loy. **VHS, Beta** *WBF, FCT ♫♫♫*

Where's Poppa? A Jewish lawyer's senile mother constantly ruins his love life, and he considers various means of getting rid of her, including dressing up as an ape to scare her to death. Filled with outlandish and often tasteless humor, befitting its reign as a black-comedy cult classic. Adapted by Robert Klane from his novel.
1970 84m/C George Segal, Ruth Gordon, Ron Leibman, Vincent Gardenia, Rob Reiner, Trish Van Devere; *Dir:* Carl Reiner. **VHS, Beta, LV** $19.95 *FOX ♫♫♫*

Where's Willie? Willie is a very bright boy; perhaps a little too bright. When he reveals his latest invention to the folks in his small town, everyone is out to get him. Find out why in this hilarious comedy.
1977 (G) 91m/C Guy Madison, Henry Darrow, Kate Woodville, Marc Gilpin; *Dir:* John Florea. **VHS, Beta** *IGV ♫♫♫*

Which Way to the Front? Assorted Army rejects form a guerilla band and wage their own small-scale war during World War II.
1970 (G) 96m/C Jerry Lewis, Jan Murray, John Wood, Steve Franken; *Dir:* Jerry Lewis. **VHS, Beta** $19.98 *WAR ♫½*

Which Way Home A Red Cross nurse attempts to flee Cambodia with four young orphans. An Australian smuggler befriends and helps her. Way too long and wandering, though like many a TV movie, it means well.
1990 141m/C Cybill Shepherd, John Waters, Peta Toppano, John Ewart, Ruben Santiago-Hudson, Marc Gray; *Dir:* Carl Schultz. **VHS, Beta** $79.98 *TTC ♫♫*

Which Way Is Up? Pryor plays three roles in this story of an orange picker who accidentally becomes a union hero. He leaves his wife and family at home while he seeks work in Los Angeles. There he finds himself a new woman, starts a new family, and sells out to the capitalists. American version of the Italian comedy "The Seduction of Mimi" tries with mixed success for laughs. Pryor as a dirty old man is the high point.
1977 (R) 94m/C Richard Pryor, Lonette McKee, Margaret Avery, Morgan Woodward, Marilyn Coleman; *Dir:* Michael Schultz. **VHS, Beta** $79.95 *MCA ♫♫½*

Whiffs A gullible man plays guinea pig in an Army experiment on germ warfare which leaves him with the intellect of a chimpanzee. Naturally, he then devises a plan to use the volatile gas in a chain of bank robberies. The title is appropriate—this movie is a stinker.
1975 (PG) 91m/C Elliott Gould, Eddie Albert, Harry Guardino, Godfrey Cambridge, Jennifer O'Neill, Alan Manson; *Dir:* Ted Post. **VHS** $79.95 *MED ♫*

While the City Sleeps Three newspaper reporters vie to crack the case of a sex murderer known as "The Lipstick Killer" with the editorship of their paper the prize. Good thriller-plus with the emphasis on the reporters' ruthless methods of gaining information rather than on the killer's motivations. Lang's last big success. Based on "The Bloody Spur" by Charles Einstein.
1956 100m/B Dana Andrews, Rhonda Fleming, George Sanders, Howard Duff, Thomas Mitchell, Ida Lupino, Vincent Price, Mae Marsh; *Dir:* Fritz Lang. **VHS, Beta, LV** $15.95 *UHV ♫♫♫*

While I Live Woman meets female pianist with uncanny resemblance to sister who's been dead 25 years and rethinks her position on reincarnation.
1947 85m/B *GB* Tom Walls, Clifford Evans, Carole Raye, Patricia Burke, Sonia Dresdel, John Warwick; *Dir:* John Harlow. **VHS, Beta** $16.95 *SNC ♫½*

Whirlwind Horseman Substandard horse opera. Maynard's search for a gold prospector friend leads him to a rancher and his pretty daughter. He beats up some bad guys, etcetera.
1938 60m/B Ken Maynard, Joan Barclay, Bill Griffith; *Dir:* Bob Hill. **VHS, Beta** $19.95 *NOS, DVT, VCN ♫½*

Whiskey Galore During World War II, a whiskey-less Scottish island gets a lift when a ship, carrying 50,000 cases of spirits, wrecks off the coast. A full-scale rescue operation and the evasion of both local and British government authorities ensue. The classic Ealing studio comedy is based on the actual wreck of a cargo ship off the Isle of Eriskay in 1941. Also known as "Tight Little Island."
1948 81m/B *GB* Basil Radford, Joan Greenwood, Gordon Jackson, James Robertson Justice; *Dir:* Alexander MacKendrick. **VHS, Beta** $59.99 *HBO ♫♫♫½*

Whiskey Mountain Two couples go on a treasure hunt to Whiskey Mountain but instead find terror.
1977 (PG) 95m/C Christopher George, Preston Pierce, Linda Borgeson, Roberta Collins, Robert Leslie; *Dir:* William Grefe. **VHS, Beta** $19.99 *BFV ♫*

A Whisper to a Scream An actress, in researching a film part, takes a job at a telephone sex service, creating different personas as she talks to keep herself interested. Soon, women resembling the personas are being murdered. Great premise fails to deliver fully, though there is some suspense.
1988 (R) 96m/C *CA* Nadia Capone, Yaphet Kotto, Lawrence Bayne, Silvio Oliviero; *Dir:* Robert Bergman. **VHS, Beta, LV** $79.95 *VIR ♫♫*

Whispering City A female reporter receives an inside tip incriminating a prominent attorney in a murder committed several years earlier. She tries to get the evidence she needs before she becomes his latest victim. Highly suspenseful, thanks to competent scripting and directing.
1947 89m/B *CA* Helmut Dantine, Mary Anderson, Paul Lukas; *Dir:* Fedor Ozep. **VHS, Beta** $16.95 *NOS, SNC ♫♫♫*

Whispering Shadow Serial starring the master criminal known as the "faceless whisperer." Twelve chapters, 13 minutes each.
1933 156m/B Bela Lugosi, Robert Warwick; *Dir:* Al Herman, Colbert Clark. **VHS, Beta** $49.95 *NOS, SNC, VCN ♫♫*

Whispers Psycho Le Clerc repeatedly bothers writer Tennant even though she seems to have killed him. This dismays police guy Sarandon. Based on Dean Koontz novel.
1989 (R) 96m/C Victoria Tennant, Chris Sarandon, Jean LeClerc; *Dir:* Douglas Jackson. **VHS, LV** $89.95 *LIV ♫♫*

The Whistle Blower A young government worker in England with a high-security position mysteriously dies. His father, a former intelligence agent, begins investigating his son's death and discovers sinister Soviet-related conspiracies. A lucid, complex British espionage thriller. Adapted from the John Hale novel by Julian Bond.
1987 98m/C *GB* Michael Caine, Nigel Havers, John Gielgud, James Fox, Felicity Dean; *Dir:* Simon Langton. **VHS, Beta, LV** $19.98 *SUE ♫♫♫*

Whistle Down the Wind Three children of strict religious upbringing find a murderer hiding in their family's barn and believe him to be Jesus Christ. A well-done and hardly grim or dull allegory of childhood innocence based on a novel by Mills' mother, Mary Hayley Bell. For a film relying heavily on child characters, it's important to portray childhood well and realistically. That is done here, as in "To Kill a Mockingbird." Mills is perfect. The film is Forbes' directorial debut and Richard Attenborough's second production.
1962 98m/B *GB* Hayley Mills, Bernard Lee, Alan Bates; *Dir:* Bryan Forbes. National Board of Review Awards '62: 10 Best Films of the Year. **VHS, Beta** $19.98 *SUE ♫♫♫½*

Whistle Stop A small-town girl divides her attentions between low-life gambler Raft and a villainous nightclub owner McLaglen who plans a robbery-murder to get rid of any rivals. A forgettable gangster drama.
1946 85m/B George Raft, Ava Gardner, Victor McLaglen; *Dir:* Leonide Moguy. **VHS, Beta** $19.95 *NOS, DVT, FCT ♫♫*

Whistlin' Dan Yet another lead cowpoke sets out to avenge the murder of a close relative. Maynard whistles while he walks (and rides) woodenly through this one.
1932 60m/B Ken Maynard, Joyzelle Joyner; *Dir:* Phil Rosen. **VHS, Beta** $19.95 *NOS, UHV ♫♫*

Whistling Bullets Action galore in the story of an undercover Texas Ranger infiltrating a gang of thieves. The low budget doesn't detract from an exciting script and tight direction.
1936 58m/B Kermit Maynard, Jack Ingram; *Dir:* John English. **VHS, Beta $19.95** *NOS, DVT, RXM* ✷✷½

Whitcomb's War A small town becomes a battleground among a host of comic characters who are unaware who really is in charge.
1988 67m/C Patrick Pankhurst, Leon Charles, Bill Morey, Robert Denison, Garnett Smith; *Dir:* Russell S. Doughten Jr. **VHS, Beta** *MIV* ✷

White Apache Emotionally charged saga of a man barred from both Indian and white societies, focusing on the impact of prejudice toward a white man raised by Apaches.
1988 90m/C VHS, Beta $79.95 *IMP* ✷✷

The White Buffalo A strange, semi-surreal western parable about the last days of Wild Bill Hickok (Bronson) and his obsession with a mythical white buffalo that represents his fear of mortality. Something of a ''Moby Dick'' theme set in the Wild West. Clumsy but intriguing.
1977 (PG) 97m/C Charles Bronson, Jack Warden, Will Sampson, Kim Novak, Clint Walker, Stuart Whitman, John Carradine, Slim Pickens, Cara Williams, Douglas Fowley; *Dir:* J. Lee Thompson. **VHS, Beta $19.98** *MGM* ✷✷

The White Butterfly Killer A young girl who witnesses her grandfather's murder by a group of ruthless opium dealers gets revenge by using her deadly kung fu.
19?? 90m/C Au Fong, Chee Fong, Wah Hon, Kwok H. Ting. **VHS, Beta $49.95** *OCE* ✷

White Cannibal Queen A young girl is captured by cannibals in the Amazon jungle, and her father spends years tracking her down.
1974 90m/C VHS, Beta $39.95 *VCD* Woof!

White Christmas Two ex-army buddies become a popular comedy team and play at a financially unstable Vermont inn at Christmas for charity's sake. Many swell Irving Berlin songs rendered with zest, including the title tune, ''Snow,'' ''Sisters,'' ''Count Your Blessings (Instead of Sheep),'' ''Choreography'' and that all-time classic, ''What Do You Do With a General When He Stops Being a General?'' Paramount's first Vista Vision film. Oscar nomination for Berlin for ''Count Your Blessings.'' Presented in widescreen on laserdisc.
1954 120m/C Bing Crosby, Danny Kaye, Rosemary Clooney, Vera-Ellen, Dean Jagger; *Dir:* Michael Curtiz. **VHS, Beta, LV, 8mm $14.95** *PAR, LDC* ✷✷✷

White Comanche Half-breed twins battle themselves and each other to a rugged climax.
1967 90m/C William Shatner, Joseph Cotten, Perla Cristal, Rossana Yanni; *Dir:* Gilbert Kay. **VHS, Beta $19.95** *NOS, UHV, TAV* ✷½

The White Dawn Three whalers are stranded in the Arctic in the 1890s and are taken in by an Eskimo village. They teach the villagers about booze, gambling, and other modern amenities. Resentment grows until a clash ensues. All three leads are excellent, especially Oates. Much of the dialogue is in an Eskimo language, subtitled in English, as in ''Dances with Wolves.''

1975 (PG) 110m/C Warren Oates, Timothy Bottoms, Louis Gossett Jr., Simonie Kopapik, Joanasie Salomonie; *Dir:* Philip Kaufman. **VHS, Beta, LV $19.95** *PAR* ✷✷½

White Eagle A white man who believes himself to be an Indian gets a job as a Pony Express rider and gets mixed up with horse rustlers. When he finds out who he really is, he is free to marry the white woman he's had his eye on. Interesting plot twist caused by the narrow-mindedness of the Hays Office.
1932 64m/B Buck Jones, Barbara Weeks, Robert Ellis, Jason Robards Sr., Robert Elliott, Jim Thorpe, Ward Bond; *Dir:* Lambert Hillyer. **VHS $19.95** *DVT* ✷✷

White of the Eye A murdering lunatic is on the loose in Arizona, and an innocent man must find him to acquit himself of the murders. A tense, effective thriller; dazzling technique recalls the experimental films of the 1960s.
1988 (R) 111m/C David Keith, Cathy Moriarty, Art Evans, Alan Rosenberg, Michael Greene, Alberta Watson; *Dir:* Donald Campbell. **VHS, Beta $14.95** *PAR* ✷✷✷

White Fang Boy befriends canine with big teeth and both struggle to survive in third celluloid rendition of Alaska during the Gold Rush. Beautiful cinematography. Based on Jack London book.
1990 (PG) 103m/C Klaus Maria Brandauer, Ethan Hawke, Seymour Cassel, James Remar, Susan Hogan; *Dir:* Randal Kleiser. **VHS $19.99** *BVV, DIS* ✷✷✷

White Fang and the Hunter The adventures of a boy and his dog who survive an attack from wild wolves and then help to solve a murder mystery. Loosely based on the novel by Jack London.
1985 (G) 87m/C Pedro Sanchez, Robert Wood; *Dir:* Alfonso Brescia. **VHS, Beta $19.95** *STE, MAG, HHE* ✷✷

White Fire Brady, in search of his lost brother, gets involved in a smuggling ring. Seems the brother is falsely accused of murder and about to hang. Brady and bar singer Castle solve the case. Good photography; bad script. Also known as ''Three Steps to the Gallows.''
1953 82m/C *GB* Scott Brady, Mary Castle, Ferdinand ''Ferdy'' Mayne; *Dir:* John Gilling. **VHS, Beta $16.95** *NOS, SNC* ✷

White Fire Two jewel thieves will stop at nothing to own White Fire, a two hundred-carat diamond.
1983 90m/C Robert Ginty, Fred Williamson. **VHS, Beta $19.98** *TWE* ✷

White Fury Couples on a campout are victimized by sleazy hoodlums during a snowstorm.
1990 89m/C Deke Anderson, Sean Holton, Douglas Harter, Christine Shinn, Chastity Hammons; *Dir:* David A. Prior. **VHS $79.95** *AIP* ✷

White Ghost An Intelligence Officer who disappeared in the jungles of Asia 18 years ago is ready to come back to the United States, but not everyone wants to see him again.
1988 (R) 90m/C William Katt, Rosalind Chao, Martin Hewitt, Wayne Crawford, Reb Brown; *Dir:* B.J. Davis. **VHS, Beta $19.98** *TWE* ✷

White Gold Shepherd's son weds Mexican woman and his disgruntled father contrives apparent rendezvous between the wife and a ranch-hand, who's found dead in her bedroom. Seems no one believes the wife's story. Unprecedented, dark silent drama.

1928 73m/B Jetta Goudal, Kenneth Thomson, George Bancroft, George Nichols, Clyde Cook; *Dir:* William K. Howard. **VHS, Beta $19.95** *GPV, DNB* ✷✷✷

White Gorilla A great white hunter sets out to kill a gigantic white gorilla that is terrorizing the black natives. Imperialist trash. On any informed person's list of all-time worst movies.
1947 62m/B Ray Corrigan, Lorraine Miller. **VHS, Beta, LV** *WGE* Woof!

White Heat A classic gangster film with one of Cagney's best roles as a psychopathic robber/killer with a mother complex. The famous finale—perhaps Cagney's best-known scene—has Cagney trapped on top of a burning oil tank shouting ''Made it, Ma! Top of the world!'' before the tank explodes. Cagney's character is allegedly based on Arthur ''Doc'' Barker and his ''Ma,'' and his portrayal is breathtaking. Also available colorized.
1949 114m/B James Cagney, Virginia Mayo, Edmond O'Brien, Margaret Wycherly, Steve Cochran; *Dir:* Raoul Walsh. **VHS, Beta, LV $19.95** *MGM, FOX, PIA* ✷✷✷½

White Hot A young businessman loses his prestigious Wall Street job, and in order to live the good life, he resorts to selling drugs. Benson stars and directs.
1988 (R) 95m/C Robby Benson, Tawny Kitaen, Danny Aiello, Sally Kirkland, Judy Tenuta; *Dir:* Robby Benson. **VHS, Beta $89.95** *ACA* ✷½

White Hot: Mysterious Murder of Thelma Todd Thelma Todd was a Hollywood starlet found dead under strange circumstances in 1935. Buffs have sought a solution to the maybe-murder ever since; this treatment (based on the book ''Hot Toddy'' by Andy Edmunds) leaves too many loose ends for purists and isn't sufficiently gripping for the uninitiated.
1991 95m/C Loni Anderson, Robert Davi, Paul Dooley, Lawrence Pressman; *Dir:* Paul Wendkos. **VHS $89.98** *SHG* ✷✷

White Hunter, Black Heart Eastwood casts himself against type as Huston-esque director who is more interested in hunting large tusked creatures than shooting the film he came to Africa to produce. Based on Peter Viertel's 1953 account of his experiences working on James Agee's script for Huston's ''African Queen.'' Viertel scripted together with James Bridges and Burt Kennedy. Eastwood's Huston impression is a highlight, though the story occasionally wanders off with the elephants.
1990 (PG) 112m/C Clint Eastwood, Marisa Berenson, George Dzundza, Jeff Fahey, Timothy Spall, Charlotte Cornwell; *Dir:* Clint Eastwood. **VHS, Beta, LV, 8mm $92.98** *MGM, FCT, WAR* ✷✷✷

White Huntress Two brothers venturing into the jungle encounter a beautiful young woman, the daughter of a settlement leader. Meanwhile, a killer is on the loose.
1957 86m/C Robert Urquhart, John Bentley, Susan Stephen; *Dir:* George Breakston. **VHS, Beta $16.95** *SNC* ✷½

The White Legion Workers and engineers push their way through steaming jungles and reeking swamps while building the Panama Canal. Many fall victim to yellow fever. Physician Keith does some medical sleuthing and saves the day. Plodding, predictable drama.
1936 81m/B Ian Keith, Tala Birell, Snub Pollard; *Dir:* Karl Brown. **VHS, Beta $29.95** *VYY* ✷½

White Lie Len Madison, Jr. (Hines), a press secretary for the mayor of New York, receives an old photo of a lynched black man who he learns is his father, hung for raping a white woman. Madison returns to his Southern hometown to find out the truth behind his father's death, assisted by a white doctor (O'Toole) whose mother was a rape victim. They fall in love, adding to an already complicated situation, and incur the wrath of locals who are trying to keep the truth hidden. Based on the novel "Louisiana Black" by Samuel Charters.
1991 (PG-13) 93m/C Gregory Hines, Annette O'Toole, Bill Nunn; *Dir:* Bill Condon. **VHS $89.98** MCA 𝄫𝄫 ½

White Light "Flatliners" with a flatfoot; after a near-death experience a cop returns with memories of a beautiful woman. He tries to find out who she is/was in plodding fashion. The voyage to the afterlife is represented by a jog through a storm sewer.
1990 (R) 96m/C Martin Kove, Allison Hossack, Martha Henry, Heidi von Palleske, James Purcell, Bruce Boa; *Dir:* Al Waxman. **VHS $89.95** ACA 𝄫𝄫

White Lightning Good ol' boy Reynolds plays a moonshiner going after the crooked sheriff who murdered his brother. Good stunt-driving chases enliven the formula. The inferior sequel is "Gator."
1973 (PG) 101m/C Burt Reynolds, Ned Beatty, Bo Hopkins, Jennifer Billingsley, Louise Latham; *Dir:* Joseph Sargent. **VHS, Beta $29.95** MGM 𝄫𝄫 ½

White Line Fever A young trucker's search for a happy life with his childhood sweetheart is complicated by a corrupt group in control of the long-haul trucking business. Well-done action film of good triumphing over evil.
1975 (PG) 89m/C Jan-Michael Vincent, Kay Lenz, Slim Pickens, L.Q. Jones, John Garfield; *Dir:* Jonathan Kaplan. **VHS, Beta, LV $59.95** COL 𝄫𝄫 ½

White Mama A poor widow (Davis, in a splendid role) takes in a street-wise black kid (Harden) in return for protection from the neighborhood's dangers, and they discover friendship. Poignant drama, capably directed by Cooper, featuring sterling performances all around Made-for-TV drama at its best.
1980 96m/C Bette Davis, Ernest Harden, Eileen Heckart, Virginia Capers, Lurene Tuttle, Anne Ramsey; *Dir:* Jackie Cooper. **VHS, Beta $59.95** IVE 𝄫𝄫𝄫

White Mane The poignant, poetic story of a proud and fierce white stallion that continually eludes attempts by ranchers to capture it.
1952 38m/B Alain Emery, Frank Silvera; *Dir:* Albert Lamorisse. Cannes Film Festival '52: Grand Prize. **VHS, Beta, LV $14.98** SUE, TEX, VYG 𝄫𝄫𝄫

White Men Can't Jump Small-time con man Harrelson stands around looking like a big nerd until someone dares him to play basketball and he proves to be more adept than he looks. After he beats Snipes, they become friends and start hustling together. There's more than just basketballs flying around here—one of the strengths of the film lies in the fast dialogue between the characters. Snipes and Harrelson put in enjoyable performances as they trade insults and Perez and Ferrell are strong as the women in their lives.

1992 (R) 115m/C Wesley Snipes, Woody Harrelson, Rosie Perez, Tyra Ferrell, Cylk Cozart, Kadeem Hardison, Ernest Harden Jr.; *W/Dir:* Ron Shelton. **VHS $94.98** FXV 𝄫𝄫𝄫

White Mischief An alternately ghastly and hilarious indictment of the English upper class between the World Wars, and the decadence the British colonists perpetrated in Kenya, which came to world attention with the murder of the philandering Earl of Errol in 1941. Exquisitely directed and photographed. Acclaimed and grandly acted, especially by Scacchi; Howard's last appearance.
1988 (R) 108m/C Greta Scacchi, Charles Dance, Joss Ackland, Sylvia Miles, John Hurt, Hugh Grant, Geraldine Chaplin, Trevor Howard, Murray Head, Susan Fleetwood, Alan Dobie; *Dir:* Michael Radford. **VHS, Beta, LV, SVS, 8mm $14.98** SUE, SUP 𝄫𝄫𝄫

White Nights Based on a love story by Dostoyevsky. A young woman pines for the return of her sailor while a mild-mannered clerk is smitten by her. Both of their romantic fantasies are explored the dance-hall cadence is mixed with dreamy, fantastic flashbacks. Equally good Soviet version made in 1959.
1957 107m/C IT Maria Schell, Jean Marais, Marcello Mastroianni; *Dir:* Luchino Visconti. **VHS $59.95** FCT 𝄫𝄫𝄫

White Nights A Russian ballet dancer who defected to the U.S. (Baryshnikov) is a passenger on a jet that crashes in the Soviet Union. With the aid of a disillusioned expatriate tap dancer (Hines), he plots to escape again. The excellent dance sequences elevate the rather lame story.
1985 (PG-13) 135m/C Mikhail Baryshnikov, Gregory Hines, Isabella Rossellini, Helen Mirren, Jerzy Skolimowski, Geraldine Page; *Dir:* Taylor Hackford. Academy Awards '85: Best Song ("Say You, Say Me"). **VHS, Beta, LV $19.95** COL 𝄫𝄫 ½

The White Orchid Romantic triangle ventures into the wilds of Mexico in search of Toltec ruins. Exceptional sets.
1954 81m/C William Lundigan, Peggy Castle, Armando Silvestre, Rosenda Monteros; *Dir:* Reginald LeBorg. **VHS, Beta $16.95** SNC, DVT 𝄫𝄫

White Palace Successful widowed Jewish lawyer Spader is attracted to older, less educated hamburger waitress Sarandon, and ethnic/cultural strife ensues, as does hot sex. Adapted from the Glenn Savan novel, it starts out with promise but fizzles toward the end.
1990 (R) 103m/C Susan Sarandon, James Spader, Jason Alexander, Eileen Brennan, Griffin Dunne, Kathy Bates, Steven Hill, Rachel Levin; *Dir:* Luis Mandoki. **VHS, LV $19.95** MCA, FCT, CCB 𝄫𝄫 ½

White Phantom: Enemy of Darkness Ninja warriors in modern day society battling against other evil ninjas.
1987 89m/C Page Leong, Jay Roberts Jr., Bo Svenson; *Dir:* Dusty Nelson. **VHS, Beta $79.95** VMK 𝄫

White Pongo A policeman goes undercover with a group of British biologists to capture a mythic white gorilla believed to be the missing link. A camp jungle classic with silly, cheap special effects.
1945 73m/B Richard Fraser, Maris Wrixon, Lionel Royce; *Dir:* Sam Newfield. **VHS, Beta $16.95** SNC, NOS, HHT Woof!

The White Rose An aspiring minister travels to see the world he intends to save, and winds up falling from grace. Complicated menage a trois is finally sorted out at the end. Silent.
1923 120m/B Mae Marsh, Carol Dempster, Ivor Novello, Neil Hamilton, Lucille LaVerne; *Dir:* D.W. Griffith. **VHS, Beta $27.95** DNB 𝄫𝄫𝄫

The White Rose The true story of a group of dissident students in Munich in 1942, who put their lives in danger by distributing leaflets telling the truth of what was going on in the concentration camps. Hans Scholl was an Army officer who discovered the truth and told the others. In the end, most of the students were captured by the Gestapo and executed. "The White Rose" was the name of the group, but none survived to tell where the name came from. Verhoeven's film telling is engrossing and compelling, with excellent acting from the young cast. In German with English subtitles.
1983 108m/C GE Lena Stolze, Wulf Kessler, Oliver Siebert, Ulrich Tucker; *Dir:* Michael Verhoeven. **VHS, Beta $79.00** MGM, GLV 𝄫𝄫𝄫

White Sands When a dead man's body is found at a remote Indian reservation clutching a gun and a briefcase filled with $500,000, the local sheriff (Dafoe) takes his identity to see where the money leads. Using a phone number found on a piece of paper in the dead man's stomach, the sheriff follows clues until he finds himself mixed up with a rich woman (Mastrantonio) who sells black market weapons and uses the money to support "worthy" causes, and an FBI man (Jackson) who uses him as bait to lure a CIA turncoat/arms dealer (Rourke). Sound confusing? It is, and despite the strong cast and vivid scenery of the southwest United States, this film just doesn't cut it.
1992 (R) 101m/C Willem Dafoe, Mary Elizabeth Mastrantonio, Mickey Rourke, Mimi Rogers, Sam Jackson, M. Emmet Walsh, James Rebhorn, Maura Tierney, Beth Grant, Miguel Sandoval, John Lafayette; *Dir:* Roger Donaldson. **VHS, Beta, LV $94.99** WAR 𝄫𝄫 ½

The White Sheik Fellini's first solo effort. A newly wed bride meets her idol from the comic pages (made with photos, not cartoons; called fumetti) and runs off with him. She soon finds he's as ordinary as her husband. Brilliant satire in charming garb. Remade as "The World's Greatest Lover." Woody Allen's "The Purple Rose of Caine" is in a similar spirt. In Italian with subtitles.
1952 86m/B JP Alberto Sordi, Giulietta Masina, Brunella Bova, Leopoldo Trieste; *Dir:* Federico Fellini. **VHS, Beta $29.95** NOS, CVC, TAM 𝄫𝄫𝄫

The White Sin An innocent country girl hires on as maid of a rich woman and is seduced and abandoned by the woman's profligate son, but everything turns out all right by the end of the third handkerchief. Silent with original organ music.
1924 93m/B Madge Bellamy, John Bowers, Billy Bevan; *Dir:* William A. Seiter. **VHS, Beta, 8mm $29.99** VYY 𝄫𝄫

The White Sister Gish is an Italian aristocrat driven from her home by a conniving sister. When her true love (Colman) is reported killed she decides to become a nun and enters a convent. When her lover does return he tricks her into leaving the convent but before he can persuade her to renounce her vows Vesuvius erupts and he goes off to warn the villagers, dying for his efforts. Gish then re-dedicates herself to her faith. Colman's first leading role, which made him a romantic star. Filmed on location in Italy.

1923 108m/B Lillian Gish, Ronald Colman, Gail Kane, J. Barney Sherry, Charles Lane; *Dir:* Henry King. VHS, Beta $19.98 *DVT* 🐾🐾🐾½

White Slave
An Englishwoman is captured by bloodthirsty cannibals and, rather than being eaten, is tormented and made a slave.
1986 (R) 90m/C Elvire Avoray, Will Gonzales, Andrew Louis Coppola; *Dir:* Roy Garrett. VHS, Beta $69.95 *LTG* 🐾

The White Tower
Five men and a woman set out to scale the infamous White Tower in the Alps. Each person's true nature is revealed as he or she scales the peak, which has defied all previous attempts. Slightly overwrought, but exciting. Filmed in Technicolor on location in the Swiss Alps.
1950 98m/C Glenn Ford, Claude Rains, Cedric Hardwicke, Oscar Homolka, Lloyd Bridges; *Dir:* Ted Tetzlaff. VHS, Beta $19.98 *HHT* 🐾🐾🐾

White Water Summer
A group of young adventurers trek into the Sierras, and find themselves struggling against nature and each other to survive. Bacon is the rugged outdoorsman (yeah, sure) who shows them what's what. Chances are it was never at a theater near you, and with good reason.
1987 (PG) 90m/C Kevin Bacon, Sean Astin, Jonathan Ward, Matt Adler; *Dir:* Jeff Bleckner. VHS, Beta, LV $29.95 *COL* 🐾½

The White Zombie
Lugosi followed his success in "Dracula" with the title role in this low-budget horror classic about the leader of a band of zombies who wants to steal a beautiful young woman from her new husband. Set in Haiti; the first zombie movie. Rich and dark, though ludicrous. Based on the novel "The Magic Island" by William Seabrook.
1932 73m/B Bela Lugosi, Madge Bellamy, John Harron; *Dir:* Victor Halperin. VHS, Beta $19.95 *NOS, PSM, QNE* 🐾🐾🐾

WhiteForce
A secret agent tracks an enemy. His government tracks him. He runs around with a big machine gun and a couple of hand grenades.
1988 (R) 90m/C Sam Jones, Kimberly Pistone; *Dir:* Eddie Romero. Beta $79.95 *NSV* 🐾

Whitewater Sam
Whitewater Sam and his Siberian Husky, Sybar, embark on an exciting trip through the uncharted wilds of the Great Northwest. The likeable, intelligent hound steals the show. Good family adventure. Written, directed and coproduced by star Larsen.
1978 (G) 87m/C Keith Larsen. VHS, Beta *MON* 🐾🐾½

Who Am I This Time?
Two shy people can express their love for each other only through their roles in a local theater production of "A Streetcar Named Desire." Good cast responds well to competent direction from Demme; poignant, touching and memorable. Based on a story by Kurt Vonnegut, Jr. Made for television.
1982 60m/C Susan Sarandon, Christopher Walken, Robert Ridgely; *Dir:* Jonathan Demme. VHS, Beta, LV $29.95 *MED* 🐾🐾🐾

Who Done It?
Average Abbott and Costello comedy about two would-be radio mystery writers, working as soda-jerks, who play detective after the radio station's president is murdered.

1942 77m/B Bud Abbott, Lou Costello, William Gargan, Patric Knowles, Louise Allbritton, Don Porter, Jerome Cowan, William Bendix, Mary Wickes; *Dir:* Erle C. Kenton. VHS, Beta $14.95 *MCA* 🐾🐾

Who Framed Roger Rabbit?
Technically marvelous, cinematically hilarious, eye-popping combination of cartoon and live action create a Hollywood of the 1940s where cartoon characters are real and a repressed minority working in films. A 'toon-hating detective is hired to uncover the unfaithful wife of 2-D star Roger, and instead uncovers a conspiracy to wipe out all 'toons. Special appearances by many cartoon characters from the past. Coproduced by Touchstone (Disney) and Amblin (Spielberg). Adapted from "Who Censored Roger Rabbit?" by Gary K. Wold.
1988 (PG) 104m/C Bob Hoskins, Christopher Lloyd, Joanna Cassidy, Alan Tilvern, Stubby Kaye; *Dir:* Robert Zemeckis; *Voices:* Charles Fleischer, Mae Questel, Kathleen Turner, Amy Irving, Mel Blanc, June Foray, Frank Sinatra. Academy Awards '88: Best Film Editing, Best Visual Effects; National Board of Review Awards '88: 10 Best Films of the Year. VHS, Beta, LV $19.99 *TOU, BUR* 🐾🐾🐾½

Who Has Seen the Wind?
Two boys grow up in Saskatchewan during the Depression. So-so family viewing drama. Ferrer as a bootlegger steals the otherwise small-paced show.
1977 102m/C Jose Ferrer, Brian Painchaud, Charmion King, Helen Shaver. VHS, Beta *SUE* 🐾🐾

Who Is the Black Dahlia?
In L.A. in 1947, a detective tries to find out who murdered a star-struck 22-year-old woman (Arnaz) nicknamed "The Black Dahlia." Well scripted by Robert W. Lenski and superb cast. Based on a famous unsolved murder case, the case was later the base for "True Confessions." Made for TV.
1975 96m/C Efrem Zimbalist Jr., Lucie Arnaz, Ronny Cox, MacDonald Carey, Linden Chiles; *Dir:* Joseph Pevney. VHS, Beta $29.95 *WOV* 🐾🐾🐾

Who Killed Baby Azaria?
Parents are arrested for infanticide after they claim their baby was stolen by a wild dog while vacationing on a supposedly haunted Australian mountain, or was it an aboriginal spirit that absconded with the tyke? Less compelling than the American remake, "A Cry in the Dark." Made for Australian TV.
1983 96m/C *AU* Elaine Hudson, John Hamblin, Peter Carroll; *Dir:* Judy Rymer. VHS, Beta $59.95 *PSM* 🐾🐾

Who Killed Doc Robbin?
A group of youngsters try to clear their friend, Dan, the town handyman, when the sinister Dr. Robbin is murdered. The follow-up to "The Adventures of Curly and His Gang;" intended as part of a Roach kid-comedy series in the manner of "Our Gang." Fast pace and lots of slapstick give a first impression of spirited juvenile hijinks, but not much is genuinely funny. And the names of the two black kids—"Dis" and "Dat"—aren't funny either.
1948 50m/C Larry Olsen, Don Castle, Bernard Carr, Virginia Grey. VHS, Beta $16.95 *SNC, NOS, HHT* 🐾🐾

Who Killed Mary What's 'Er Name?
A diabetic ex-fighter tracks a prostitute's killer through Greenwich Village. Illogical, incredible and disjointed attempt at comedy/mystery. Odd, unsatisfying ending, with the boxer going into a coma. Good supporting cast.

1971 (PG) 90m/C Red Buttons, Sylvia Miles, Conrad Bain, Ron Carey, David Doyle, Sam Waterston; *Dir:* Ernie Pintoff. VHS, Beta $59.95 *GEM, PSM* 🐾🐾

Who is Killing the Great Chefs of Europe?
A fast-paced, lightly handled black comedy. When a gourmand, well-played by Morley, learns he must lose weight to live, a number of Europe's best chefs are murdered according to their cooking specialty. A witty, crisp, international mystery based on Ivan and Nan Lyon's novel.
1978 (PG) 112m/C George Segal, Jacqueline Bisset, Robert Morley, Jean-Pierre Cassel, Philippe Noiret, Jean Rochefort, Joss Ackland, Nigel Havers; *Dir:* Ted Kotcheff. VHS $19.98 *HBO, WAR* 🐾🐾🐾

Who Slew Auntie Roo?
An updated twist on the Hansel and Gretel fairy tale. Features Winters as an odd, reclusive widow mistaken for the fairy tale's children-eating witch by one of the orphans at her annual Christmas party—with dire consequences. Also known as "Whoever Slew Auntie Roo?"
1971 (PG) 90m/C Shelley Winters, Ralph Richardson, Mark Lester, Lionel Jeffries, Hugh Griffith; *Dir:* Curtis Harrington. VHS, Beta $69.95 *VES* 🐾🐾½

Whodunit
The bad horror flicks always seem to have (usually young) characters murdered one at a time, often on a remote island. This loser is no exception.
1982 (R) 82m/C Rick Bean, Gary Phillips; *Dir:* Bill Naud. VHS, Beta $69.98 *LIV, VES* 🐾

Whoever Says the Truth Shall Die
A film portrait of the great Italian filmmaker Pier Paolo Pasolini, his work in film and letters, his life as homosexual, anti-Fascist and Communist, until his assassination in 1975. A highly acclaimed, searingly controversial documentary featuring interviews and copious clips from Pasolini's films. In English, Italian and French with English subtitles.
1981 60m/C Bernardo Bertolucci, Laura Betti, Pier Paolo Pasolini; *Dir:* Philo Bregstein. VHS, Beta $59.95 *FCT* 🐾

The Whole Shootin' Match
Two thirty something Texas nobodies chase the American Dream via a sure-fire invention called the Kitchen Wizard, after failing in the small-animal biz and polyurethane. Offbeat, very low-budget feature shot interestingly in sepia tones. Sympathetic and human.
1979 108m/C Matthew L. Perry, Sonny Davis, Doris Hargrave; *Dir:* Eagle Pennell. VHS, 3/4U *ICA* 🐾🐾🐾

Who'll Save Our Children?
Two children are abandoned and left on the doorstep of a middle-aged, childless couple. Years later the natural parents, now reformed, attempt to reclaim their children. Well cast and topical. Made for TV.
1982 96m/C Shirley Jones, Len Cariou; *Dir:* George Schaefer. VHS, Beta *TLF* 🐾🐾🐾

Who'll Stop the Rain?
A temperamental Vietnam vet (Nolte) is enlisted in a smuggling scheme to transport a large amount of heroin into California. An excellent blend of drama, comedy, action, suspense and tragedy. Based on Robert Stones novel "Dog Soldiers" (also its alternate title). Outstanding period soundtrack by Creedence Clearwater Revival, including title song. Violent and compelling tale of late-'60s disillusionment.

1978 (R) 126m/C Nick Nolte, Tuesday Weld, Michael Moriarty, Anthony Zerbe, Richard Masur, Ray Sharkey, David Opatoshu, Charles Haid, Gail Strickland; *Dir:* Karel Reisz. **VHS, Beta, LV** $29.95 MGM ♫♫♫½

Wholly Moses! Set in biblical times this alleged comedy concerns a phony religious prophet who begins to believe that his mission is to lead the chosen. Horrible ripoff of "Life of Brian," released the previous year. What a cast—what a waste!
1980 (PG) 125m/C Dudley Moore, Laraine Newman, James Coco, Paul Sand, Dom DeLuise, Jack Gilford, John Houseman, Madeline Kahn, Richard Pryor, John Ritter; *Dir:* Gary Weis. **VHS, Beta, LV** $9.95 COL ♫½

Whoopee! Cantor stars in his first sound picture as a rich hypochondriac sent out west for his health, where he encounters rugged cowboys and Indians. Filmed in two-color Technicolor and based on Ziegfeld's 1928 Broadway production with the same cast. Dances supervised by Busby Berkeley; songs include "Makin' Whoopee," "My Baby Just Cares for Me," "Stetson," "I'll Still Belong to You" and "A Girlfriend of a Boyfriend of Mine."
1930 93m/C Eddie Cantor, Ethel Shutta, Paul Gregory, Eleanor Hunt, Betty Grable, George Olsen & His Orchestra; *Dir:* Thornton Freeland. **VHS, Beta** $19.98 SUE, HBO ♫♫

The Whoopee Boys A New York lowlife in Palm Beach tries to reform himself by saving a school for needy children so his rich girlfriend will marry him. Full of unfunny jokes and very bad taste, with no redeeming qualities whatever.
1986 (R) 89m/C Michael O'Keefe, Paul Rodriguez; *Dir:* John Byrum. **VHS, Beta, LV** $79.95 PAH ♫

Whoops Apocalypse From the British hit television series. Biting account of events leading up to World War III, full of rapid fire wit, one liners, and manic energy.
1983 137m/C John Cleese, John Barron, Richard Griffiths, Peter Jones, Bruce Montague, Barry Morse. **VHS, Beta** $59.95 PAV ♫

Whoops Apocalypse America's first female president Swit, recently ensconced after the untimely death of the ex-circus clown previous president, is forced to contend with hotheaded superpowers who want to kick off WWIII with a bang. Uneven black comedic satire of holocaust genre. Based on the BBC TV series.
1988 (R) 93m/C Loretta Swit, Peter Cook, Michael Richards. **VHS, Beta** $79.95 MGM ♫♫

Whore Ken Russell's gritty night in the life of cynical prostitute Theresa Russell is strong medicine, a powerful antidote to sappy Hollywood films that glorify streetwalking. Also available in the 92-minute original uncut version, an 85-minute NC-17 version, and an alternate R-Rated version titled "If You Can't Say It, Just See It."
1991 (R) 80m/C Theresa Russell, Antonio Vargas; *Dir:* Ken Russell. **VHS** $92.95 VMK ♫♫

Who's Afraid of Virginia Woolf? Nichols debuts as a director in this biting Edward Albee play. A teacher and his wife (Segal and Dennis) are invited to the home of a burned-out professor and his foul-mouthed, bitter, yet seductive wife (Burton and Taylor). The guests get more than dinner, as the evening deteriorates into brutal verbal battles between the hosts. Taylor and Dennis won Oscars; Burton's Oscar-nominated portrait of the tortured husband is magnificent. Richard and Liz's best film

together. Other Oscar nominations for supporting actor Segal, director Nichols, screenplay, sound, score, and film editing.
1966 127m/B Richard Burton, Elizabeth Taylor, George Segal, Sandy Dennis; *Dir:* Mike Nichols. Academy Awards '66: Best Actress (Taylor), Best Art Direction/Set Decoration (B & W), Best Black and White Cinematography, Best Costume Design (B & W), Best Supporting Actress (Dennis); National Board of Review Awards '66: 10 Best Films of the Year, Best Actress (Taylor). **VHS, Beta, LV** $19.98 WAR, PIA, BTV ♫♫♫♫

Who's Harry Crumb? Bumbling detective Candy can't even investigate a routine kidnapping! His incompetence is catching: the viewer can't detect a single genuinely funny moment. Candy is the only likeable thing in this all-around bad, mindless force.
1989 (PG-13) 95m/C John Candy, Jeffrey Jones, Annie Potts, Tim Thomerson, Barry Corbin, Shawnee Smith; *Dir:* Paul Flaherty. **VHS, Beta, LV** $14.95 COL ♫½

Who's Minding the Mint? A money checker at the U.S. Mint must replace $50,000 he accidentally destroyed. He enlists a retired money printer and an inept gang to infiltrate the mint and replace the lost cash, with predictable chaos resulting. Nonstop zaniness in this wonderful comedy that never got its due when released. Thieves who befriend Hutton include Denver of "Gilligan's Island" and Farr, later of "M*A*S*H."
1967 97m/C Jim Hutton, Dorothy Provine, Milton Berle, Joey Bishop, Bob Denver, Walter Brennan, Jamie Farr; *Dir:* Howard Morris. **VHS, Beta** $9.95 GKK ♫♫♫

Who's That Girl? A flighty, wrongly-convicted parolee kidnaps her uptight lawyer and they have wacky adventures as she goes in search of the crumb that landed her in the pokey. Plenty of Madonna tunes, if that's what you like; they briefly keep you from wondering why you're not laughing. Kind of a remake (a bad one) of "Bringing Up Baby" (1938).
1987 (PG) 94m/C Madonna, Griffin Dunne, John Mills, Haviland Morris; *Dir:* James Foley. **VHS, Beta, LV, 8mm** $19.98 WAR ♫

Who's That Knocking at My Door? Interesting debut for Scorsese, in which he exercises many of the themes and techniques that he polished for later films. An autobiographical drama about an Italian-American youth growing up in New York City, focusing on street life, Catholicism and adolescent bonding. Begun as a student film called "I Call First"; later developed into a feature. Keitel's film debut.
1968 90m/B Harvey Keitel, Zena Bethune; *Dir:* Martin Scorsese. **VHS, Beta** $59.99 DCL, FCT, WAR ♫♫♫

Whose Child Am I? A woman so desperately wants a child of her own that she doesn't mind who the father is. Ludicrous and vulgar.
1975 (R) 90m/C Kate O'Mara, Paul Freeman, Edward Judd, Felicity Devonshire; *Dir:* Lawrence Britten. **VHS** $59.95 GRG, ATL ♫

Whose Life Is It Anyway? Black humor abounds. A sculptor Dreyfuss is paralyzed from the neck down in an auto accident. What follows is his struggle to persuade the hospital authorities to let him die. Excellent cast headed impressively by Dreyfuss; Lahti as his doctor and hospital head Cassavetes also are superb. From Brian Clark's successful Broadway play, scripted by Clark and Reginald Rose.

1981 (R) 118m/C Richard Dreyfuss, John Cassavetes, Christine Lahti, Bob Balaban, Kenneth McMillan, Kaki Hunter, Thomas Carter; *Dir:* John Badham. **VHS, Beta** $79.95 MGM ♫♫♫

Why Me? Two jewel thieves unknowingly steal an enormously valuable ruby ring and are in over their heads trying to evade the cops and everyone else. In the same vein as the much-funnier "Nuns on the Run," this caper is done in by bad acting and too much ill-conceived slapstick. Adapted by Donald E. Westlake from his own novel.
1990 (R) 96m/C Christopher Lambert, Christopher Lloyd, Kim Greist, J.T. Walsh, Michael J. Pollard, Tony Plana, John Hancock, Lawrence Tierney; *Dir:* Gene Quintano. **VHS, LV** $89.95 COL ♫♫

Why Shoot the Teacher A young teacher sent to a one-room schoolhouse in a small prairie town in Saskatchewan during the Depression gets a cold reception on the cold prairie. One of the more popular films at the Canadian box office. Cort was Harold in "Harold and Maude." Based on the novel by Max Braithwaite.
1979 101m/C CA Bud Cort, Samantha Eggar, Chris Wiggins; *Dir:* Silvio Narizzano. **VHS, Beta** $69.98 SUE ♫♫♫

The Wicked Though packaged as serious horror, this is a cheap, tacky spoof with folks stranded at the country residence of a vampire family. Sole point of interest: these Transylvanian cliches take place in the Australian outback. Kids might actually like this (gore isn't severe), if they can surmount the thick Down Under accents.
1989 87m/C Brett Cumo, Richard Morgan, Angela Kennedy, Maggie Blinco, John Doyle; *Dir:* Colin Eggleston. **VHS** $79.95 SFI, HMD ♫

The Wicked Lady Posh but lame costume drama about a bored 17th-century noblewoman who takes to highway robbery to spice up her life. Credulity-stretcher extraordinaire. Based on "The Life and Death of Wicked Lady Skelton" by Magdalen King-Hall. Mason is cheeky as a fellow highwayman. Remade in 1983 with Faye Dunaway.
1945 103m/B GB James Mason, Margaret Lockwood, Patricia Roc, Michael Rennie, Felix Aylmer, Enid Stamp-Taylor, Griffith Jones; *Dir:* Louis Levy. **VHS, Beta** $29.95 AXV ♫♫

The Wicked Lady Remake of the 1945 costumer about a 17th-century noblewoman trying her hand at highway robbery. More noted for its low-cut costumes and racy humor than any talent involved.
1983 (R) 99m/C GB Faye Dunaway, John Gielgud, Denholm Elliott, Alan Bates, Glynis Barber, Oliver Tobias, Prunella Scales; *Dir:* Michael Winner. **VHS, Beta** $79.95 MGM ♫½

Wicked Stepmother A family discovers that an aged stepmother is actually a witch. Davis walked off the film shortly before she died, and was replaced by Carrera. Davis's move was wise; the result is dismal, and would have been had she stayed. As it is the newer wonders: How come the stepmother isn't Davis anymore? Davis's unfortunate last role.
1989 (PG-13) 90m/C Bette Davis, Barbara Carrera, Colleen Camp, Lionel Stander, David Rasche, Tom Bosley, Seymour Cassel, Evelyn Keyes, Richard Moll, Laurene Landon; *Dir:* Larry Cohen. **VHS, Beta** $79.95 MGM ♫

The Wicker Man The disappearance of a young girl leads a devoutly religious Scottish policeman to an island whose denizens practice bizarre pagen sexual rituals. An example of occult horror that achieves its

mounting terror without gratuitous gore. The first original screenplay by playwright Anthony Shaffer. Beware shortened versions that may still lurk out there; the 103-minute restored director's cut is definitive.
1975 103m/C Edward Woodward, Britt Ekland, Diane Cilento, Ingrid Pitt, Christopher Lee; **Dir:** Robin Hardy. **VHS, Beta** $19.95 *MED* 🐾🐾🐾 ½

The Widow A story about a woman slowly learning to deal with her own grief, her children's traumas and monetary worries after the death of her husband. Straightforward, passable made-for-TV domestic drama. Based on Lynn Caine's best-selling novel.
1976 99m/C Michael Learned, Farley Granger, Bradford Dillman, Robert Lansing; **Dir:** J. Lee Thompson. **VHS, Beta** *TLF, LME* 🐾🐾

Widow Couderc A provincial widow unknowingly includes an escaped murderer among her liaisons. Based on a novel by Georges Simenon. In French with English subtitles.
1974 92m/C *FR* Simone Signoret, Alain Delon; **Dir:** Pierre Granier-Deferre. **VHS, Beta** $59.00 *FCT* 🐾🐾🐾

Widow's Nest Three widowed sisters live in a bizarre fantasy world when they lock themselves in a dingy mansion in Cuba of the 1930s. Filmed in Spain.
1977 90m/C Patricia Neal, Susan Oliver, Lila Kedrova, Valentina Cortese; **Dir:** Tony Navarro. **VHS, Beta** $59.98 *MAG* 🐾🐾

Wifemistress In the early 1900s, an invalid wife resents her neglectful husband. When she goes into hiding because of a murder he didn't commit, she begins to drift into a world of sexual fantasies. Fine performances. Italian dialogue with English subtitles.
1979 (R) 101m/C *IT* Marcello Mastroianni, Laura Antonelli; **Dir:** Marco Vicario. **VHS, Beta** $59.95 *COL* 🐾🐾 ½

The Wilby Conspiracy A political activist and an Englishman on the wrong side of the law team up to escape the clutches of a prejudiced cop in apartheid Africa. The focus of the film is on the chase, not the political uprising taking place around it. Well-done chase film with fine performances throughout.
1975 (PG) 101m/C Sidney Poitier, Michael Caine, Nicol Williamson, Prunella Gee, Persis Khambatta, Saeed Jaffrey, Rutger Hauer; **Dir:** Ralph Nelson. **VHS** $49.95 *MGM* 🐾🐾🐾

The Wild Aces A relentless action flick with real documentary footage of the aerial war above Vietnam.
1985 58m/C **VHS, Beta** $24.95 *BVG, FGF, AVL* 🐾 ½

The Wild Angels Excessively violent film but B-movie classic about an outlaw biker gang and the local townspeople. Typical Corman fodder was one of AIP's most successful productions.
1966 (PG) 124m/C Peter Fonda, Nancy Sinatra, Bruce Dern, Diane Ladd, Michael J. Pollard, Gayle Hunnicutt; **Dir:** Roger Corman. **VHS, Beta** $9.98 *SUE* 🐾🐾 ½

The Wild Beasts Children and animals clash violently after drinking water tainted with PCP. Disgusting premise; unredeemably violent.
1985 92m/C John Aldrich, Lorraine DeSelle; **Dir:** Franco Prosperi. **VHS, Beta** $59.98 *LIV, LTG* Woof!

The Wild Bunch Acclaimed western about a group of aging outlaws on their final rampage, realizing time is passing them by. Highly influential in dialogue, editing style, and lyrical slow-motion photography of violence; Peckinpah's main claim to posterity. Holden and Ryan create especially memorable characters. Arguably the greatest western and one of the greatest American films of all times. Beware of shortened versions; after a pre-release showing to the East Coast critics, producer Feldman cut key scenes without Peckinpah's knowledge or consent.
1969 (R) 145m/C William Holden, Ernest Borgnine, Robert Ryan, Warren Oates, Strother Martin, L.Q. Jones, Albert Dekker, Bo Hopkins, Edmond O'Brien; **Dir:** Sam Peckinpah. **VHS, Beta, LV** $19.98 *WAR* 🐾🐾🐾

Wild Country Tame western. A U.S. Marshall races to stop the bad guy from killing the late sheriff's daughter and snatching her ranch.
1947 57m/B Eddie Dean, Roscoe Ates, Peggy Wynn, Douglas Fowley, I. Stanford Jolley, Bill Fawcett; **Dir:** Ray Taylor. **VHS** $19.95 *NOS* 🐾 ½

Wild in the Country A backwoods delinquent (Elvis) is aided in his literary aspirations by a woman psychiatrist. Pretty lame and dull, really; and Elvis as a writer? Actually, his performance is a high point. And of course he sings: title track, "In My Way," "I Slipped, I Stumbled, I Fell," and "Lonely Man."
1961 114m/C Elvis Presley, Hope Lange, Tuesday Weld, Millie Perkins, John Ireland, Gary Lockwood, Alan Napier; **Dir:** Philip Dunne. **VHS, Beta** $14.98 *FOX, MVD* 🐾 ½

The Wild Country Children's fare detailing the trials and tribulations of a Pittsburgh family moving into an inhospitable Wyoming ranch in the 1880s. Based on the novel "Little Britches" by Ralph Moody.
1971 (G) 92m/C Steve Forrest, Ron Howard, Clint Howard; **Dir:** Robert Totten. **VHS, Beta** $69.96 *DIS* 🐾🐾 ½

The Wild Duck A father discovers that his beloved daughter is illegitimate and turns against her. To regain his love, she plans to sacrifice her most prized possession. Irons overacts; Ullman almost disappears into the woodwork. Child star Jones is insufferable, and the whole family squeaks with "literary" pomposity. A horrid adaptation of the classic Ibsen play.
1984 (PG) 96m/C *AU* Jeremy Irons, Liv Ullman, Lucinda Jones, Arthur Dignam; **Dir:** Henri Safran. **VHS, Beta** $79.98 *LIV, VES* 🐾 ½

Wild Engine/Mistaken Orders The early screen daredevil Holmes stars in two railroad-and-betrayal epics.
1923 48m/B Helen Holmes, Jack Hoxie, Hoot Gibson. **VHS, Beta** $19.98 *CCB* 🐾🐾

The Wild and the Free Light-hearted comedy about research chimps. The lab-raised bunch become the instruments for radioactive experimenting. Scheming scientists return the chimps to their home to save them from their demise. Made for TV.
1980 100m/C Granville Van Dusen, Linda Gray, Bill Gribble; **Dir:** James Hill. **VHS, Beta** $39.95 *IVE* 🐾

Wild Hearts Can't Be Broken A rather shallow telling of the true-life story of Sonora Webster, a small town Georgia girl who decides to join the carnival in the early 1930s, hoping to become a stunt rider. She becomes a horse-diver (a Depression-era sideshow phenomena), is blinded in a diving accident, but returns to find romance and

ride again. Exciting real-life story has little screen tension, but this is a good movie for pre-teen and young teenage girls with its depiction of a feisty heroine, a sweet romance, and horses. Nice performance from Anwar, who has a fresh screen presence.
1991 (G) 89m/C Gabrielle Anwar, Cliff Robertson, Michael Schoeffling, Cliff Robertson, Kathleen York, Dylan Kussman; **Dir:** Steve Miner. **VHS, Beta** $19.99 *DIS* 🐾🐾🐾

Wild Horse Gibson is the lead (in his favorite talking role), but his magnificent mount Mut commands most of the attention. He's a heck of a horse better be, because ain't much else to this early western. Interesting historically as an example of stars' (Gibson's in this case) difficulty adjusting to talking pictures.
1931 68m/B Hoot Gibson, Stepin Fetchit, Edmund Cobb, Alberta Vaughn; **Dir:** Richard Thorpe, Sidney Algier. **VHS, Beta** $19.95 *NOS, GPV, VYY* 🐾 ½

Wild Horse Canyon Yakima Canutt, the man credited with creating the profession of stunt man, stars in this tale about a lady rancher who requires saving from her evil foreman. The climactic stampede scene gives Yakima a chance to demonstrate a high dive off a cliff and a somersault onto his horse. Silent, with musical score.
1925 68m/B Yakima Canutt, Edward Cecil, Helene Rosson, Jay Talbet. **VHS, Beta** $19.95 *NOS, GPV, VYY* 🐾🐾

Wild Horse Canyon One of those westerns designed to run as half of a double-feature, this one's shorter than most, but hardly notable. The hero goes after a rustler who shot his brother.
1939 57m/B Jack Randall, Dorothy Short, Frank Yaconelli, Dennis Moore, Warner Richmond, Charles King, Sherry Tansey; **Dir:** Robert Hill. **VHS** $19.95 *NOS* 🐾 ½

Wild Horse Hank A young woman risks everything to save a herd of wild mustangs from being slaughtered for use as dog food. Creena is good; otherwise, it's sentimental family fare. Based on the novel "The Wild Horse Killers" by Mel Ellis.
1979 94m/C *CA* Linda Blair, Richard Crenna, Michael Wincott, Al Waxman; **Dir:** Eric Till. **VHS, Beta** $69.98 *LIV, VES* 🐾🐾

Wild Horse Phantom A Wild West banker plans to fake a robbery in a diabolical deception to part honest ranchers from their lands. Surely this man's not a member of the F.D.I.C. Okay sagebrush saga.
1944 56m/B Buster Crabbe, Al "Fuzzy" St. John, Charles King, John Merton, Lane Chandler, Edward Cassidy, Bud Osborne; **Dir:** Sam Newfield. **VHS, Beta** *GPV* 🐾🐾

Wild Horse Rodeo When a gunman appears at a rodeo, trouble starts. Part of "The Three Mesquiteers" series. Ordinary action-packed western, the good guys win. Actor Dick Weston, who sings, later gained fame as singing cowboy Roy Rogers.
1937 53m/B Ray Corrigan. **VHS, Beta** $9.95 *NOS, DVT* 🐾 ½

Wild Horses A man attempts to make a living by capturing and selling wild horses. Adventure fans will love this film which deals with the common theme of man against nature.
1982 88m/C *NZ* Keith Aberdein, John Bach, Robyn Gibbs; **Dir:** Derek Morton. **VHS, Beta** $29.98 *VES* 🐾🐾

Wild Horses Ex-rodeo champ Rogers is hanherm' to escape his dull life. He joins a roundup of wild horses and with Dawber's help exposes a bureaucrat's scheme for them. Rogers is likable and sings, but the supporting cast carries this one. Made for TV.
1984 90m/C Kenny Rogers, Pam Dawber, Ben Johnson, Richard Masur, David Andrews, Karen Carlson; *Dir:* Dick Lowry. **VHS, Beta $29.95** *VID* 🦴🦴

The Wild Life A recent high school graduate takes on a wild and crazy wrestler as his roommate in a swinging singles apartment complex. The same writer and producer as "Fast Times at Ridgemont High" but that film's commercial success. Just as adolescent, though.
1984 (R) 96m/C Christopher Penn, Randy Quaid, Rick Moranis, Hart Bochner, Eric Stoltz, Jenny Wright, Lea Thompson, Sherilyn Fenn, Lee Ving, Ashley St. John, Francesca "Kitten" Natavidad; *Dir:* Art Linson. **VHS, Beta, LV $79.98** *MCA* 🦴

Wild Man A Las Vegas gambling magnate with supernatural powers decides to avenge the murder of his friend and sets out on an adventure to find the killer. Crummy acting in yet another "ex-agent called back for one more mission" yarn.
1989 106m/C Ginger Lynn Allen, Michelle Bauer, Don Scribner, Kathleen Middleton; *Dir:* Fred J. Lincoln. **VHS, Beta** *CEL* 🦴

Wild Mustang Sheriff's son (Gordon) joins some outlaws to help father (Carey) spring a trap. The plan is sidetracked, and Gordon is captured and nearly branded by the baddies. Don't worry, though; everything turns out all right. Good western.
1935 62m/B Harry Carey; *Dir:* Harry Fraser. **VHS, Beta $19.95** *NOS, VCN, DVT* 🦴🦴½

The Wild One The original biker flick: two motorcycle gangs descend upon a quiet midwestern town and each other. Brando is the leader of one, struggling against social prejudices and his own gang's lawlessness to find love and a normal life. The classic tribute to 1950s rebelliousness. Based vaguely on a real incident in California. Quaint after nearly 40 years, but still the touchstone for much that has come since, and still a central role in Brando's now-long career. Brando himself believes it failed to explore Motivations for youth gangs and violence, only depicting them. Banked in Britain until 1967.
1954 79m/B Marlon Brando, Lee Marvin, Mary Murphy, Robert Keith, Jerry Paris, Alvy Moore, Jay C. Flippen; *Dir:* Laslo Benedek. **VHS, Beta, LV $19.95** *COL* 🦴🦴🦴½

Wild Orchid One of the most controversial theatrical releases of 1990. Rourke is a mystery millionaire involved with two beautiful women in lovely Rio de Janeiro. Very explicit sex scenes. Bisset is strange as an international real estate developer with unusual sexual mores. Mostly boring and unbelievable. Rourke allegedly took his method-acting technique to the limit in the last love scene with Otis. Available in an unrated version as well. From the producers of "9 1/2 Weeks."
1990 (R) 117m/C Mickey Rourke, Jacqueline Bisset, Carre Otis, Assumpta Serna, Bruce Greenwood; *Dir:* Zalman King. **VHS, Beta, LV $19.95** *COL, FCT* 🦴

Wild Orchid 2: Two Shades of Blue When Blue's jazz-musician, heroin-addicted father is violently killed, she is thrust into the world of prostitution. Taken into Elle's brothel, Blue fantasizes about Josh, a boy she met before her father's death, while she is working. When Josh's father brings him to the brothel to lose his virginity, he chooses Blue, not realizing who she is because a geisha wig hides her blonde curls. When Blue runs away from the brothel to flee a violent gang rape by a Senator and his friends, she tries to lead a normal teenage life and enrolls in high school where she meets Josh again. Another soft-core effort from the director of "9 1/2 Weeks" and "Two Moon Junction." This unrealistic film is available in rated and unrated versions.
1992 (R) 107m/C Nina Siemaszko, Wendy Hughes, Brent Fraser, Robert Davi, Tom Skerritt, Joe Dallesandro, Christopher McDonald, Liane Curtis; *W/Dir:* Zalman King. **VHS** *COL* 🦴

Wild Orchids A husband suspects his wife of infidelity while they take a business cruise to Java. One of Garbo's earliest silent films. Garbo is lushly surrounded with great scenery, costumes and sets, making the best of her as she makes the best of a cheery tale.
1928 119m/B Greta Garbo, Lewis Stone, Nils Asther; *Dir:* Sidney Franklin. **VHS, Beta $29.99** *MGM, FCT* 🦴🦴½

The Wild Pair Two cops track down a cocaine smuggling ring and uncover a private army planning to conquer the United States. Lloyd Bridges is memorable as a rabid right-winger, and the wild pairing of Bubba and Beau inspires head-scatching. Otherwise, what we have here is a typical action pic, complete with car chases. Beau Bridges' inauspicious feature directing debut. Also called "Hollow Point."
1987 (R) 89m/C Beau Bridges, Bubba Smith, Lloyd Bridges, Gary Lockwood, Raymond St. Jacques; *Dir:* Beau Bridges. **VHS, Beta, LV $19.95** *MED, 3IM* 🦴½

The Wild Panther There is Kung Fu combat and more in this story of spies, guerrillas and stolen military secrets.
1984 90m/C VHS, Beta $39.95 *WES* 🦴½

Wild Party It's 1929, a year of much frivolity in Hollywood; and drinking, dancing, maneuvering and almost every sort of romance are the rule of the night at silent-film comic Jolly Grimm's sumptuous, star-studded party that climaxes with, among other things, a murder. Well performed but somehow hollow; ambitious Ivory effort unfortunately falls short. Based on Joseph Moncure March's poem, and loosely on the Fatty Arbuckle scandal.
1974 (R) 90m/C Raquel Welch, James Coco, Perry King, David Dukes, Royal Dano, Tiffany Bolling; *Dir:* James Ivory. **VHS, Beta $9.98** *SUE* 🦴🦴½

Wild Pony A young boy spurns his new stepfather, preferring to live with his pony instead. Reversal of the evil-stepmother-and-girl-loves-horse theme.
1983 87m/C Marilyn Lightstone, Art Hindle, Josh Byrne; *Dir:* Kevin Sullivan. **VHS, Beta $29.98** *LIV, VES* 🦴

Wild Rapture The hunting and dissecting of an elephant and gorilla, a look at a tribe that practices lip-splitting, and other gruesome events abound in this shock-fest. Banned for 18 years.
1950 68m/B VHS, Beta $24.95 *NOS* 🦴

Wild Rebels A two-faced member of a ruthless motorcycle gang informs the police of the gang's plans to rob a bank. Lots of bullet-flying action. Dumb, ordinary biker flick; star Pastrano was a former boxing champ in real life.
1971 (R) 90m/C Steve Alaimo, Willie Pastrano, John Vella, Bobbie Byers; *Dir:* William Grefe. **VHS, Beta $59.95** *ACA* 🦴

The Wild Ride Nicholson, in an early starring role, portrays a rebellious punk of the Beat generation who hotrods his way into trouble and tragedy. He kidnaps now-straight ex-buddy Bean's squeeze (Carter); kills a few cops; then is killed. Interesting only if you're interested in Nicholson.
1960 59m/B Jack Nicholson, Georgianna Carter, Robert Bean; *Dir:* Harvey Berman. **VHS, Beta $16.95** *SNC, NOS, VYY* 🦴½

Wild Ride Ruthless bandits on horseback rape and pillage their way through a series of pioneer towns.
19?? 86m/C VHS $39.95 *ACF* 🦴½

Wild Riders Two ruthless, amoral bikers molest, kidnap, rape, and beat two beautiful, naive young society ladies. Despicable story, with deplorable acting, but the ending, in which a husband slays a biker with a cello, surely is unique.
1971 (R) 91m/C Alex Rocco, Elizabeth Knowles, Sherry Bain, Arell Banton; *Dir:* Richard Kanter. **VHS, Beta $19.95** *ACA Woof!*

Wild Rovers An aging cowboy and his younger colleague turn to bank robbing and are pursued by a posse in a wacky comedy-adventure from director Edwards. Uneven script and too much referential baggage (faint shades of "Butch Cassidy") doom this valiant effort, with the two stars hanging right in there all the way.
1971 (PG) 138m/C William Holden, Ryan O'Neal, Karl Malden, Lynn Carlin, Tom Skerritt, Joe Don Baker, Rachel Roberts, Moses Gunn; *Dir:* Blake Edwards. **VHS, Beta $59.95** *MGM* 🦴🦴

Wild Strawberries Bergman's landmark film of fantasy, dreams and nightmares. An aging professor, on the road to accept an award, must come to terms with his anxieties and guilt. Brilliant performance by Sjostrom, Sweden's first film director and star. Excellent use of flashbacks and film editing. An intellectual and emotional masterpiece. In Swedish with English subtitles.
1957 90m/B *SW* Victor Sjostrom, Bibi Andersson, Max von Sydow; *Dir:* Ingmar Bergman. National Board of Review Awards '59: Best Foreign Film; Sight & Sound Survey '72: #11 of the Best Films of All Time. **VHS, Beta, LV, 8mm $29.95** *CVC, VYY, FOX* 🦴🦴🦴🦴

Wild in the Streets Satire set in the future, where a malcontent rock star becomes president after the voting age is lowered to 14. Adults over 30 are imprisoned and fed a daily dose of LSD. The president gets his comeuppance when challenged by even younger youths. Very groovy, very political in its own way, very funny.
1968 97m/C Shelley Winters, Chris Jones, Hal Holbrook, Richard Pryor, Diane Varsi, Millie Perkins, Ed Begley Sr.; *Dir:* Barry Shear. **VHS, Beta, LV $19.99** *HBO* 🦴🦴½

Wild Style Zoro, a mild-mannered Bronx teenager, spends his evenings spray-painting subway cars. Intended as a depiction of urban street life, including graffiti, breakdancing and rap music. Mildly interesting social comment, but uncommincing as cinema. Too "realistic," with too little vision.
1983 82m/C Lee George Quinones, Fredrick Braithwaite, Dondi White; *Dir:* Charlie Ahearn. **VHS, Beta $49.95** *WES* 🦴🦴

Wild Thing A modern variation on the Wild Child/Tarzan myth, as a feral kid stalks a crime-ridden ghetto. Is this the best scripter John Sayles, a usually imaginative writer and director, can do? of course, the Troggs' overplayed title hit is abused as the theme.
1987 (PG-13) 92m/C Kathleen Quinlan, Rob Knepper, Bryan Brown, Betty Buckley, Sam Neill, Rachel Ward; *Dir:* Max Reid, Ken Cameron. **VHS, Beta, LV** *KRT, OM* ✍️ ½

Wild Times Too-long but pleasant rendition of the life of High Candill, early Wild Wet showman. Based on the novel by Brian Garfield. Made for TV, originally in two parts.
1979 200m/C Sam Elliott, Trish Stewart, Ben Johnson, Dennis Hopper, Pat Hingle; *Dir:* Richard Compton. **VHS, Beta $69.95** *PSM* ✍️✍️

Wild Weed Funny smelling cigarettes ruin the lives of all who inhale. Viewers' advice: just say no. Leed's actual drug bust with Robert Mitchum got her the lead. Also known as "The Devil's Weed," "She Shoulda Said No," and "Marijuana the Devil's Weed."
1949 70m/B Lila Leeds, Alan Baxter, Lyle Talbot, Jack Elam, Alan Gorcey; *Dir:* Sam Newfield. **VHS** *SNC* ✍️

Wild West The hero and loyal companions (with names like Skinny, Stormy and Soapy) tackle desperadoes trying to stop telegraph line construction. More mild than wild, it was later trimmed and released under the title "Prairie Outlaws," but here you're getting the original.
1946 73m/C Eddie Dean, Roscoe Ates, Al "Lash" LaRue, Robert "Buzzy" Henry, Sarah Padden, Louise Currie, Warner Richmond, Chief Yowlachie, Bud Osborne; *Dir:* Robert Tansey. **VHS $19.95** *NOS* ✍️✍️

Wild West A compilation of scenes from dozens of western films.
1987 90m/C John Wayne, Clint Eastwood, Steve McQueen, Charles Bronson. **VHS, Beta $79.95** *TTE* ✍️

Wild Wheels A group of dune-buggy enthusiasts seek revenge against a gang of motorcyclists who have ravaged a small California beach town.
1975 (PG) 81m/C Casey Kasem, Dovie Beams, Terry Stafford, Robert Dix. **VHS, Beta $59.95** *GEM* Woof!

Wild, Wild Planet Alien beings from a distant planet are miniaturizing Earth's leaders in a bid to destroy our planet, and a dubbed, wooden hero comes to the rescue. A rarely seen Italian SF entry, great fun for genre fans. A must for "robot girls in skin tight leather outfits" completists.
1965 93m/C IT Tony Russell, Lisa Gastoni, Massimo Serato, Franco Nero; *Dir:* Antonio Margheriti. **VHS $20.00** *SMW* ✍️ ½

Wild, Wild West Revisited A feature-length reprise of the tongue-in-cheek western series. Irreverent and fun, in the spirit of its admirable predecessor-though it's probably just as well the proposed new series didn't see fruition. Made for television.
1979 95m/C Robert Conrad, Ross Martin, Harry Morgan, Rene Auberjonois; *Dir:* Burt Kennedy. **VHS, Beta $59.98** *FOX* ✍️✍️ ½

Wild, Wild World of Jayne Mansfield The life and times of Mansfield are explored in this unique documentary, which was unfinished at the time of her death. Jayne takes the viewer on a world-wide examination of sexual mores, from prostitution, nudist colonies, and the racy strip clubs of Europe, to the transvestite and topless bars of America. Perversely (but did

you expect otherwise?), footage of her fatal car crash was included.
1968 89m/C Jayne Mansfield. **VHS, Beta $19.99** *VDM, DVT, VVP* ✍️ ½

Wild Women Five female convicts are released during the Texas/Mexico dispute to help smuggle arms to American forces. Innocuous western, original in a made-for-TV kind of way.
1970 90m/C Hugh O'Brian, Anne Francis, Marilyn Maxwell, Marie Windsor; *Dir:* Don Taylor. **VHS, Beta $19.98** *FOX* ✍️

Wild Women of Wongo The denizens of a primitive isle, essentially beautiful women and ugly men, meet the natives of a neighboring island, handsome men and ugly women. Not quite bad enough to be true camp fun, but stupidly silly in a low-budget way.
1959 73m/C Pat Crowley, Ed Fury, Adrienne Barbeau, Jean Hawkshaw, Johnny Walsh; *Dir:* James L. Wolcott. **VHS, Beta $19.98** *DVT, MED* ✍️ ½

Wild & Wooly In 1903, a tough cowgirl and her friends must prevent the assassination of the President.
1978 120m/C Chris DeLisle, Susan Bigelow, Elyssa Davalos, Jessica Walter, Doug McClure, David Doyle, Ross Martin, Vic Morrow, Charles Seibert, Sherry Bain, Paul Burke; *Dir:* Philip Leacock. **VHS, Beta $59.95** *PSM* ✍️ ½

The Wild World of Batwoman A campy cult film in which Batwoman and a bevy of Bat Girls are pitted against an evil doctor in order to find the prototype of an atomic hearing aid/nuclear bomb. Also released theatrically as "She Was a Hippy Vampire."
1966 70m/C Katherine Victor, George Andre, Steve Brodie, Lloyd Nelson; *Dir:* Jerry Warren. **VHS, Beta $19.99** *RHI, SNC, DVT* ✍️✍️

Wild Zone An American ex-soldier combats mercenaries in Africa in order to rescue his kidnapped father.
1989 (R) 100m/C Edward Albert, Philip Brown, Carla Herd; *Dir:* Percival Rubens. **VHS, Beta, LV $79.95** *COL* ✍️

Wildcat Crabbe, famous as the serials' Flash Gordon, is a villain in this petrochemical adventure. The hero overextends his credit when he buys an oil well and must produce a gusher or else.
1942 73m/B Richard Arlen, Arlene Judge, Buster Crabbe, William Frawley, Arthur Hunnicutt, Elisha Cook Jr., Billy Benedict; *Dir:* Frank McDonald. **VHS $19.95** *NOS* ✍️✍️

Wildcats A naive female phys-ed teacher (Hawn) is saddled with the job of coach for a completely undisciplined, inner city, high school football team. Formulaic, connect-the-dots comedy; moderately funny, and of course Hawn triumphs in adversity, but at nearly two hours a very long sitcom episode.
1986 (R) 106m/C Goldie Hawn, James Keach, Swoosie Kurtz, Bruce McGill, M. Emmet Walsh; *Dir:* Michael Ritchie. **VHS, Beta, LV $19.98** *WAR* ✍️✍️

Wilderness Family, Part 2 Also known as the "Further Adventures of the Wilderness Family." Depicts the Robinson family who left civilization for the freedom of the Colorado wild. Predictable retread, but more good family adventure drama from the makers of "Adventures of the Wilderness Family."
1977 (G) 104m/C Heather Rattray, Ham Larsen, George Flower, Robert F. Logan, Susan D. Shaw; *Dir:* Frank Zuniga. **VHS, Beta $9.98** *MED* ✍️✍️ ½

Wildest Dreams Released from its bottle by a lovely nebbish, a genie puts a love spell on one beautiful girl after another in order to find him the right one, resulting in a bevy of women clamoring for our hero's attention.
1990 (R) 84m/C James Davies, Heidi Pane, Deborah Blaisdell, Ruth Collins, Jane Hamilton, Jill Johnson; *Dir:* Chuck Vincent. **VHS, Beta, LV $79.98** *LIV, VES* ✍️ ½

The Wildest Office Party A sex-filled tape for and about wild, rowdy office parties.
1987 60m/C **VHS, Beta $59.98** *LIV, VES* Woof!

Wildfire Two horse traders come to the aid of ranchers beset by horse thieves. Early color outing for Steele. Harmless, ordinary oater.
1945 60m/C Bob Steele, Sterling Holloway; *Dir:* Robert Tansey. **VHS, Beta, LV $19.95** *NOS, VDM* ✍️

Wildfire As teenagers, Frank and Kay run away to get married but Frank is sent to prison for robbing a bank and Kay winds up making a new life for herself. Released from prison after eight years, Frank discovers Kay is married with two children and doesn't want anything further to do with him. But when Frank violates his parole, Kay discovers she can't leave him again and the two go on the run together.
1988 (PG) 98m/C Steven Bauer, Linda Fiorentino, Will Patton, Marshall Bell; *Dir:* Zalman King. **VHS, LV** *MCA* ✍️✍️

Wilding A pair of cops are on the trail of a gang of kids who go "wilding," which involves sprees of killing, raping and looting. Exploitive fare based on horrible, recent events in New York. The filmmakers skirt controversy by making the criminals rich suburban kids.
1990 92m/C Wings Hauser, Joey Travolta, Karen Russell, Steven Cooke; *Dir:* Eric Louzil. **VHS, Beta $79.95** *AIP* ✍️ ½

Wildrose Eichhorn is memorable as a recent divorcee who must assert herself among her otherwise all-male co-workers at a Minnesota strip mine. She finds a new love, and tries to put her life back together. Filmed on location.
1985 96m/C Lisa Eichhorn, Tom Bower, James Cada; *Dir:* John Hanson. **VHS, Beta $69.98** *LIV, LTG* ✍️✍️✍️

Will: G. Gordon Liddy The life of the Watergate conspirator, based on his autobiography, made for TV. Producer/star Conrad reportedly objected strongly to cuts from the original three-hour length, apparently with cause: the first half is superficial pap. The portrayal of Liddy's time in prison, though, is fascinating, and Conrad is excellent.
1982 100m/C Robert Conrad, Katherine Cannon, Gary Bayer, Peter Ratray, James Rebhorn, Red West, Maurice Woods, Danny Lloyd; *Dir:* Robert Lieberman. **VHS $39.95** *IVE* ✍️✍️

Will Penny Just back from a cattle drive, a range-wandering loner looks for work in the wrong place and offends a family of outlaws who come after him. His escape from them leads to another kind of trap—one set by a love-hungry woman (Hackett, in a strong performance). Heston considers this film his personal best, and he's probably right. Superbly directed western, with excellent cinematography and professional, realistic portrayals, flopped in theaters, moviegoers preferring simultaneous Heston outing "Planet of the Apes." Photography by Lucien Ballard. Music by Elmer Bernstein.

1967 109m/C Charlton Heston, Joan Hackett, Donald Pleasence, Lee Majors, Bruce Dern, Anthony Zerbe, Ben Johnson, Clifton James; *Dir:* Tom Gries. **VHS, Beta, LV** $14.95 *PAR* 🐾🐾🐾½

Willa Raffin, a hash-joint waitress with two kids and one in the pipeline, is deserted by her husband and determines to get ahead by becoming a trucker. Okay made-for-TV drama.
1979 95m/C Cloris Leachman, Deborah Raffin, Clu Gulager, John Amos, Diane Ladd; *Dir:* Claudio Guzman, Joan Darling. **VHS, Beta** $39.95 *PSM* 🐾🐾

Willard Willard is a lonely, psychotic youngster who trains a group of rats, his only friends, to attack his enemies. Not as disgusting as it might have been (rated PG), but pretty weird and not redeemed by any sense of style or humor. Popular at the box office; followed by inferior "Ben." Based on by Stephen Gilbert's novel "Ratman's Notebooks."
1971 95m/C Bruce Davison, Ernest Borgnine, Elsa Lanchester, Sondra Locke; *Dir:* Daniel Mann. **VHS, Beta, LV** $59.95 *PSM, PAR* 🐾🐾

The Willies Three youngsters gross each other out with juvenile tales of horror and scariness while camping out in the backyard. Pointless but not without a few yuks (in both senses of the term).
1990 (PG-13) 120m/C James Karen, Sean Astin, Kathleen Freeman, Jeremy Miller; *Dir:* Brian Peck. **VHS, LV** $29.95 *PSM* 🐾🐾

Willow Blockbuster fantasy epic combines the story of Moses with "Snow White," dwarves and all. Willow is the little Nelwyn who finds the lost baby Daikini and is assigned the task of returning her safely to her people. Willow discovers that the girl is actually a sacred infant who is destined to overthrow the evil queen Bavmorda and rule the land. As you might expect from executive producer George Lucas, there is much action and plenty of clever, high-quality special effects. But the "Star Wars"-esque story (by Lucas) is strangely predictable, and a bit too action-packed. Not really for children.
1988 (PG) 118m/C Warwick Davis, Val Kilmer, Jean Marsh, Joanne Whalley-Kilmer, Billy Barty, Pat Roach, Ruth Greenfield; *Dir:* Ron Howard. **VHS, Beta, LV, 8mm** $14.95 *COL* 🐾🐾½

Willy Wonka & the Chocolate Factory When the last of five coveted "golden tickets" falls into the hands of sweet but very poor Charlie, he and his Grandpa Joe get a tour of the most wonderfully strange chocolate factory in the world. The owner is the most curious hermit ever to hit the big screen. He leads the five young "winners" on a thrilling and often dangerous tour of his fabulous factory. Adapted from Roald Dahl's "Charlie and the Chocolate Factory," with a score, including "Candy Man," by Anthony Newley-Leslie Brieusse. Without a doubt one of the best "kid's" movies ever made; a family classic worth watching again and again and...
1971 (G) 100m/C Gene Wilder, Jack Albertson, Denise Nickerson, Peter Ostrum, Roy Kinnear, Aubrey Woods, Michael Bollner, Ursula Reit, Leonard Stone, Dodo Denney; *Dir:* Mel Stuart. **VHS, Beta, LV** $19.98 *WAR, VTK, APD* 🐾🐾🐾½

Wilma Based on the true story of Wilma Rudolph, a young black woman who overcame childhood illness to win three gold medals at the 1960 Olympics. Plodding made-for-TV biography suffers from sub-par script and acting.

1977 100m/C Joe Seneca, Cicely Tyson, Shirley Jo Finney; *Dir:* Bud Greenspan. **VHS, Beta** $14.95 *COL* 🐾🐾

Wimps A collegiate wimp is subjected to a brutal fraternity initiation, but eventually gets the girl. Cheap ripoff of Cyrano de Bergerac, turned into a teen sex flick.
1987 (R) 94m/C Louie Bonanno, Jim Abele, Deborah Blaisdell; *Dir:* Chuck Vincent. **VHS, Beta** $79.98 *LIV, LTG* 🐾

Win, Place, or Steal Lame comedy set at the racetrack about three grown men and their adolescent schemes to win big. Also called "The Big Payoff" on video.
1972 (PG) 88m/C McLean Stevenson, Alex Karras, Dean Stockwell; *Dir:* Richard Bailey. **VHS, Beta** $69.98 *LIV, VES* 🐾½

Winchester '73 Superb acting and photography characterize this classic, landmark western. Simple plot—cowboy Stewart pursues his stolen state-of-the-art rifle as it changes hands—speeds along and carries the viewer with it, ending with an engrossing and unforgettable shoot-out. Almost single-handedly breathed new life into the whole genre. Laser videodisc version contains a special narration track provided by Stewart. Mann's and Stewart's first teaming.
1950 82m/C James Stewart, Shelley Winters, Stephen McNally, Dan Duryea, Millard Mitchell, John McIntire, Will Geer, Jay C. Flippen, Rock Hudson, Tony Curtis, Charles Drake; *Dir:* Anthony Mann. **VHS, Beta, LV** $19.95 *MCA* 🐾🐾🐾½

The Wind One of the last great silents still stands as a magnificent entertainment. Gish, in possibly her best role, is an innocent Easterner who finds herself married to a rough cowpoke and raped by a married man in a bleak frontier town. Director Sjostrom has a splendid feel for landscape, and the drama—climaxing with a tumultuous desert storm—is intense yet fully believable.
1928 74m/B Lillian Gish, Lars Hanson, Montagu Love, Dorothy Cumming, Edward Earle, William Orlamond; *Dir:* Victor Sjostrom. **VHS, Beta** $29.95 *MGM* 🐾🐾🐾

The Wind Foster is terrorized by Hauser while attempting to write her next thriller in Greece. Let's hope her book is more exciting than this movie. Shot on location in Greece.
1987 92m/C Meg Foster, Wings Hauser, Steve Railsback, David McCallum, Robert Morley; *Dir:* Nico Mastorakis. **VHS, Beta** $79.98 *LIV, VES, LTG* 🐾

The Wind and the Lion In turn-of-the-century Morocco, a sheik (Connery) kidnaps a feisty American woman (Bergen) and her children and holds her as a political hostage. President Teddy Roosevelt (Keith) sends in the Marines to free the captives, who are eventually released by their captor. Directed with venue and style by Milius. Highly entertaining, if heavily fictionalized. Based very loosely on a historical incident.
1975 (PG) 120m/C Sean Connery, Candice Bergen, Brian Keith, John Huston; *Dir:* John Milius. **VHS, Beta, LV** $59.95 *MGM, PIA* 🐾🐾🐾

The Wind That Walks Two girls trek into the New Jersey wilds after an unknown animal called by the Indians "the wind that walks."
1985 30m/C Beta *NJN* 🐾🐾🐾

The Wind in the Willows "The Wind in the Willows," originally a part of the 1950 Disney feature "Ichabod and Mr. Toad," is the tale of J. Thaddeus Toad, who has a strange mania for fast cars. Based on a story

by Kenneth Grahame. Also on this tape are two Disney cartoons with similar automotive themes, "Motor Mania" with Goofy and "Trailer Horn" with Donald Duck and Chip 'n' Dale.
1949 34m/C Eric Blore; *Dir:* Wolfgang Reitherman; *Nar:* Basil Rathbone. **VHS, Beta** $12.99 *DIS* 🐾🐾🐾

Windjammer Western star O'Brien took a break from the lone prairie to make this seafaring rescue drama. He is a deputy state's attorney who signs up for a yacht race in order to serve a subpoena. Bad guys arrive on the scene, and the initially moody yacht denizens rely on our trusty hero to save them. Decent adventure.
1931 75m/B George O'Brien, Constance Worth; *Dir:* Ewing Scott. **VHS** $24.95 *NOS, DVT* 🐾🐾

Windom's Way Finch, in a strong role in this intriguing political drama, is an idealistic doctor in a remote village in Malaysia. He juggles a failing marriage and a budding romance, then finds himself caught in a local labor dispute. He works for a peaceful solution, but is captured by an insurgent army. Disillusioned, he plans to leave the country, but his wife persuades him to stay and tend the wounded. Based on a novel by James Ramsey Ullman.
1957 90m/B GB Peter Finch, Mary Ure; *Dir:* Ronald Neame. **VHS, Beta** $19.98 *VID* 🐾🐾🐾

Window A little boy (Disney star Driscoll, intriguingly cast by director, Tetzlaff) has a reputation for telling lies, so no one believes him when he says he witnessed a murder...except the killers. Almost unbearably tense, claustrophobic thriller about the helplessness of childhood. Tetzlaff clearly learned more than a thing or two from the master, Hitchcock, for whom he photographed "Notorious." Based on "The Boy Who Cried Murder" by Cornell Woolrich, on whose writing "Rear Window" was also based. Driscoll was awarded a special miniature Oscar as Outstanding Juvenile for his performance.
1949 73m/B Bobby Driscoll, Barbara Hale, Arthur Kennedy, Ruth Roman; *Dir:* Ted Tetzlaff. Edgar Allan Poe Awards '49: Best Screenplay. **VHS, Beta, LV** $19.95 *MED* 🐾🐾🐾½

Window Shopping Multiple, confusing romances occur at the "Toison d' Or" shopping mall in Paris. A modern, French homage to the classic Hollywood musical. In French with English subtitles.
1986 96m/C FR Delphine Seyrig, Charles Denner, Fanny Cottencon, Miriam Boyer, Lio, Pascale Salkin, Jean-Francois Balmer; *Dir:* Chantal Akerman. **VHS** $79.95 *WAC, FCT* 🐾🐾🐾½

Windows A lonely lesbian (Ashley) becomes obsessed with her quiet neighbor (Shire) and concocts a plot to get close to her. Director Willis is known for his unbeatable cinematography, most notably for Woody Allen in "Manhattan." His debut in the chair is a miserable, offensive flop.
1980 (R) 93m/C Talia Shire, Joe Cortese, Elizabeth Ashley, Kay Medford, Linda Gillin; *Dir:* Gordon Willis. **VHS** *VID* 🐾

Windrider Simple love story made in Australia. A wind surfer and a rock star become lovers. Self-indulgent overwrought garbage with rare moments of promise.
1986 83m/C AU Tom Burlinson, Nicole Kidman, Charles Tingwell, Jill Perryman, Simon Chilvers; *Dir:* Vincent Monton. **VHS** $79.95 *MGM* 🐾½

Winds of Change A retelling of five ancient Greek myths, which features a disco-rock musical score. Written by Norman Corwin.
1979 90m/C Voices: Peter Ustinov. **VHS, Beta $19.95** COL ₰½

The Winds of Jarrah English woman runs from broken heart to Australia to become nanny for misogynist, and Harlequin code prevails. Based on Harlequin romance novel.
1983 78m/C Terence Donovan, Susan Lyons; **Dir:** Mark Egerton. **VHS, Beta $59.98** CGI ₰½

Winds of Kitty Hawk Visually interesting but talky account of the Wright brothers attempt to beat their rival Glenn Curtiss and his backer, phone man Alexander Graham Bell. Rule of thumb: Fast-forward through the parts when they are on the ground, or see Stacy Keach's public-TV version instead.
1978 96m/C Michael Moriarty, David Huffman, Kathryn Walker, Eugene Roche, John Randolph, Scott Hylands; **Dir:** E.W. Swackhamer. **VHS, Beta $14.95** FRH ₰

Winds of War On seven tapes, the excruciatingly long and dull television mini-series based on Herman Wouk's bestseller about World War II. The book was much better; Mitchum appears to be fighting sleep unsuccessfully for much of the show. A follow-up miniseries, "War and Remember-ance," was produced in 1988, in two volumes.
1983 900m/C Robert Mitchum, Ali MacGraw, Ralph Bellamy, Polly Bergen, Jan-Michael Vincent, David Dukes, John Houseman, Victoria Tennant, Peter Graves, Chaim Topol, Ben Murphy, Jeremy Kemp, Anton Diffring, Lawrence Pressman, Andrew Duggan, Barbara Steele; **Dir:** Dan Curtis. **VHS, Beta $139.95** PAR ₰₰

Winds of the Wasteland Would-be Pony Express contractors, Wayne and Chandler, race rivals to land government work. Competent western that put Wayne on the Hollywood map.
1936 54m/B John Wayne, Phyllis Fraser, Lane Chandler, Yakima Canutt; **Dir:** Mack V. Wright. **VHS $19.95** REP, NOS, VCN ₰₰½

Windwalker Howard, the only cast member who is not a Native American, is an aged chief who shares the memories of his life with his grandchildren. Filmed in the Crow and Cheyenne Indian languages and subtitled in English throughout. A Beautifully photographed, intelligent independent project.
1981 108m/C Trevor Howard, James Remar, Dusty Iron Wing McCrea; **Dir:** Keith Merrill. **VHS, Beta $59.98** FOX ₰₰₰

Windy City A group of seven childhood friends must come to terms with the harsh realities of their failed ambitions when they meet for a weekend in Chicago. Told from the point of view of one of the group, recounting the sorts of memories we all prefer to forget. Typical and overwrought '80s yuppie reunion/regret drama. Aren't these people too young to be unhappy?
1984 (R) 103m/C John Shea, Kate Capshaw, Josh Mostel, Jeffrey DeMunn, Lewis J. Stadlen, James Sutorius; **Dir:** Armyan Bernstein. **VHS, Beta, LV $79.98** FOX ₰₰

A Wing and a Prayer Better-than-average WWII Air Force action flick. Battles rage throughout the Pacific theater and the men aboard the aircraft carrier struggle to do their part to save the world for freedom. Fine

cast receives Hathaway's excellent unsentimental direction.
1944 97m/B Don Ameche, Dana Andrews, William Eythe, Charles Bickford, Cedric Hardwicke, Kevin O'Shea, Richard Jaeckel, Harry Morgan; **Dir:** Henry Hathaway. **VHS $14.98** FXV ₰₰₰

Wings The silent classic about friends, Rogers and Arlen, their adventures in the Air Corps during WWI, and their rivalry for the hand of a woman. Contains actual footage of combat flying from the Great War. Won the very first Best Picture Oscar. Too-thin plot hangs on (barely) to the stirring, intrepid dogfight scenes.
1927 139m/B Clara Bow, Charles "Buddy" Rogers, Richard Arlen, Gary Cooper; **Dir:** William A. Wellman. Academy Awards '27: Best Picture. **VHS, Beta, LV $14.95** BTV, PAR ₰₰½

Wings of Danger A pilot tries to save his friend from blackmailing gold smugglers. Also known as "Dead on Course."
1952 72m/B Zachary Scott, Kay Kendall, Robert Beatty, Naomi Chance. **VHS, Beta, LV** WGE, MLB, RXM ₰

Wings of Desire An ethereal, haunting modern fable about one of many angels observing human life in and above the broken existence of Berlin, and how he begins to long to experience life as humans do. A moving, unequivocable masterpiece, with as many beautiful things to say about spiritual need as about the schizophrenic emptiness of contemporary Germany; Wenders' magnum opus. In German with English subtitles, and with black-and-white sequences. Magnificent cinematography by Henri Alekan.
1988 (PG-13) 130m/C GE Bruno Ganz, Peter Falk, Solveig Dommartin, Otto Sander, Curt Bois; **W/Dir:** Wim Wenders. Los Angeles Film Critics Association Awards '88: Best Foreign Film. **VHS, Beta, LV $79.98** ORI, GLV ₰₰₰½

Wings of Eagles Hollywood biography of Frank 'Spig' Wead, a famous WWI aviation pioneer turned screenwriter. Veers wildly from comedy to stolid drama, though Bond's lampoon of Ford as "John Dodge" is justly famous.
1957 110m/C John Wayne, Ward Bond, Maureen O'Hara, Dan Dailey, Edmund Lowe, Ken Curtis, Kenneth Tobey, Sig Rumann; **Dir:** John Ford. **VHS, Beta $19.95** MGM ₰₰

Wings of War (La Patrulla Suicida) Ten American soldiers mount an assault against a well-fortressed German radio station. Dirty Dozen repries done less well.
19?? 95m/C SP **VHS, Beta $29.95** JCI ₰½

Winner Takes All A young college student during the Vietnam War comes to grips with maturity and growing pains.
1986 (R) 94m/C GB Jason Connery, Diane Cilento. **VHS, Beta $59.95** ACA ₰

Winners Take All A handful of friends compete in Supercross races. Spirited but thoroughly cliche sports flick.
1987 (PG-13) 103m/C Don Michael Paul, Kathleen York, Robert Krantz; **Dir:** Fritz Kiersch. **VHS, Beta, LV $19.98** SUE ₰₰

Winners of the West A landowner schemes to prevent a railroad from running through his property. The railroad's chief engineer leads the good guys in an attempt to prevent sabotage. A fun serial in thirteen chapters, full of shooting, blown-up bridges, locomotives afire, etc.; and of course, the requisite damsel in distress.

1940 169m/B Anne Nagel, Dick Foran, James Craig; **Dir:** Ford Beebe, Ray Taylor. **VHS, Beta $49.95** NOS, VCN, VYY ₰₰½

Winning A race car driver (Newman) will let nothing, including his new wife (Woodward), keep him from winning the Indianapolis 500. Newman does his own driving. Thomas' film debut.
1969 (PG) 123m/C Paul Newman, Joanne Woodward, Robert Wagner, Richard Thomas, Clu Gulager; **Dir:** James Goldstone. **VHS, Beta $49.95** MCA ₰₰₰

The Winning Team Reagan stars in this biography of baseball legend Grover Cleveland Alexander and Day plays his dedicated wife. Some controversial issues were glossed over in the film, such as Alexander's real-life drinking problem and the fact that he had epilepsy. Although this was one of Reagan's favorite roles, he was upset by the studio's avoidance of all-too-human problems. Nonetheless, it's an entertaining enough movie about a very talented ballplayer.
1952 98m/B Doris Day, Ronald Reagan, Frank Lovejoy, Eve Miller, James Millican, Russ Tamblyn; **Dir:** Lewis Seiler. **VHS, LV $19.98** WAR ₰₰½

Winning of the West Yodelin' ranger Autry vows to protect a crusading publisher from unscrupulous crooks, including his own dog-gone brothers. The brothers see the lights, and the two corral the varmints. Autry tunes include "Cowboy Blues," "Cowpoke Poking Along," and "Find Me My 45."
1953 57m/B Gene Autry, Smiley Burnette, Gail Davis, Richard Crane; **Dir:** George Archainbaud. **VHS, Beta $24.95** CCB ₰₰

The Winslow Boy A cadet at the Royal Naval College is wrongly accused of theft and expelled. Donat as the boy's lawyer leads a splendid cast. Despite consequences for his family, the boy's father (Hardwicke) sues the government and fights his son's battle, as the case makes the papers and he approaches bankruptcy. Plot would seem far-fetched if it weren't based on fact. Absorbing. Based on a play by Terrence Rattigan, who co-wrote the script.
1948 112m/B GB Robert Donat, Cedric Hardwicke, Margaret Leighton, Frank Lawton, Basil Radford, Kathleen Harrison; **Dir:** Anthony Asquith. **VHS, Beta $59.95** HBO ₰₰₰½

Winter Flight A small, gentle British drama about an RAF recruit and a barmaid who quickly become lovers, and then are confronted with her pregnancy.
1984 105m/C GB Reece Dinsdale, Nicola Cowper, Gary Olsen; **Dir:** Roy Battersby. **VHS, Beta $59.95** MGM ₰₰

Winter Kills Distinctive political black comedy suffers a hilariously paranoid version of American public life. Bridges, the younger brother of a U.S. president assassinated 15 years earlier, uncovers the plot. Huston is delightful as Bridges' horny old father. Uneven, but well worth seeing; flopped at the box office, but was re-released in 1983. Taylor was not billed.
1979 (R) 97m/C Jeff Bridges, John Huston, Anthony Perkins, Richard Boone, Sterling Hayden, Eli Wallach, Ralph Meeker, Belinda Bauer, Dorothy Malone, Toshiro Mifune, Elizabeth Taylor, Donald Moffatt, Tisa Farrow, Brad Dexter, Joe Spinell; **Dir:** William Richert. **VHS, Beta, LV $14.95** COL, SUE ₰₰₰

The Winter Light The second film in Bergman's famous trilogy on the silence of God, preceded by ''Through a Glass Darkly'' and followed by ''The Silence.'' Bleak and disturbing view of a tortured priest who searches for the faith and guidance he is unable to give his congregation. Hard to swallow for neophyte Bergmanites, but challenging, deeply serious and rewarding for those accustomed to the Swede's angst. Polished and personal.
1962 80m/B *SW* Gunnar Bjornstrand, Ingrid Thulin, Max von Sydow; *Dir:* Ingmar Bergman. National Board of Review Awards '63: 5 Best Foreign Films of the Year. **VHS, Beta $29.98** *SUE* 🎞🎞🎞½

The Winter of Our Dreams Down Under slice of seamy life as a heroin-addicted prostitute becomes involved with an unhappy bookshop owner. Davis shines in otherwise slow and confusing drama.
1982 89m/C *AU* Judy Davis, Bryan Brown, Cathy Downes, Baz Luhrmann, Peter Mochrie, Mervyn Drak, Margie McCrae, Marcie Deane-Johns; *Dir:* John Duigan. Australian Film Institute '81: Best Actress (Davis). **VHS, Beta $59.95** *IVE* 🎞🎞½

Winter People A clock-making widower and a woman living alone with her illegitimate son experience tough times together with feuding families and a Depression-era Appalachian community. Silly, rehashed premise doesn't deter McGillis, who gamely gives a fine performance. Based on the novel by John Ehle.
1989 (PG-13) 109m/C Kurt Russell, Kelly McGillis, Lloyd Bridges, Mitchell Ryan, Jeffrey Meek; *Dir:* Ted Kotcheff. **VHS, Beta, LV, 8mm $19.98** *SUE* 🎞🎞½

Winterset A son seeks to clear his father's name of a falsely accused crime 15 years after his electrocution. Powerful at the time, though time has lessened its impact. Loosely based on the trial of Sacco and Vanzetti and adapted from Maxwell Anderson's Broadway play, with stars in the same roles. Meredith's film debut.
1937 85m/B Burgess Meredith, Margo, John Carradine; *Dir:* Alfred Santell. **VHS, Beta $19.95** *NOS, HHT, CAB* 🎞🎞½

Wired A justly lambasted, unintentionally hilarious biography of comic genius and overdose victim John Belushi, very loosely based on Bob Woodward's bestselling muckraking book. The chronicle of addiction and tragedy is tried here as a weird sort of comedy; was it meant as a tribute to its off-kilter subject? If so, it misses by a mile. And, we're sorry, but Chiklis doesn't cut is as Belushi—who would?!
1989 (R) 112m/C Michael Chiklis, Ray Sharkey, Patti D'Arbanville, J.T. Walsh, Gary Groomes, Lucinda Jenney, Alex Rocco, Jere Bums, Billy Preston; *Dir:* Larry Peerce. **VHS, Beta, LV $89.95** *IVE* 🎞

Wired to Kill In a futuristic world, two teenagers seek justice for their parents' murder by building a remote-controlled erector set programmed for revenge. Laughably porous plot drops any pretence of credibility. Dizzyingly bad.
1986 (R) 96m/C Merritt Butrick, Emily Longstreth, Devin Hoelscher, Frank Collison; *Dir:* Franky Schaeffer. **VHS, Beta $79.98** *LIV, LTG* Woof!

Wisdom Unemployed young guy (Estevez) becomes a bank robber with Robin Hood aspirations, coming to the aid of American farmers by destroying mortgage records. Estevez became the youngest person to star in, write, and direct a major motion picture. And, my goodness, it shows.
1987 (R) 109m/C Emilio Estevez, Demi Moore, Tom Skerritt, Veronica Cartwright; *Dir:* Emilio Estevez. **VHS, Beta, LV $19.98** *WAR* 🎞½

Wise Blood Gothic drama about a drifter who searches for sin and becomes a preacher for a new religion, The Church Without Christ. Excellent cast in achingly realistic portrayal of ersatz religion, southern style. Many laughs are more painful than funny. Superb, very dark comedy from Huston. Adapted from the Flannery O'Connor novel.
1979 (PG) 106m/C Brad Dourif, John Huston, Ned Beatty, Amy Wright, Harry Dean Stanton; *Dir:* John Huston. New York Film Festival '79: Best Actress. **VHS, Beta $59.95** *MCA* 🎞🎞🎞½

Wise Guys Two small-time hoods decide to rip off the mob. When their boss figures out their plan he decides to set them up instead. Lame black comedy that has too few moments.
1986 (R) 100m/C Joe Piscopo, Danny DeVito, Ray Sharkey, Captain Lou Albano, Dan Hedaya, Julie Bovasso, Patti LuPone; *Dir:* Brian DePalma. **VHS, Beta $19.99** *FOX, FCT* 🎞½

Wish You Were Here Poignant yet funny slice of British postwar life roughly based on the childhood memoirs of famous madame, Cynthia Payne. A troubled and freedom-loving teenager expresses her rebellion in sexual experimentation. Mum is dead and Dad just doesn't understand, so what's a girl to do, but get the boys excited? Lloyd, in her first film, plays the main character with exceptional strength and feistiness. Payne's later life was dramatized in the Leland-scripted ''Personal Services.''
1987 92m/C *GB* Emily Lloyd, Tom Bell, Clare Clifford, Barbara Durkin, Geoffrey Hutchings, Charlotte Barker, Chloe Leland, Trudy Cavanagh, Jesse Birdsall, Geoffrey Durham, Pat Heywood; *Dir:* David Leland. **VHS, Beta, LV $19.95** *FRH* 🎞🎞🎞

The Wishing Ring Light and charming romance, beautifully filmed by Tourneur, based on Owen Davis' play. The son of an earl in Old England is expelled from college and told by his father he must earn half a crown on his own before the family will take him back. With the help of a minister's daughter, he sets things right.
1914 50m/B Vivian Martin, Alec B. Francis, Chester Barnett, Simeon Wiltsie, Walter Morton; *W/Dir:* Maurice Tourneur. **VHS, Beta $27.95** *DNB* 🎞🎞🎞

The Wishmaker A young man squanders the three wishes granted to him for helping a strange old woman and her timid daughter.
1985 54m/C VHS, Beta $29.95 *GEM* 🎞🎞🎞

The Wistful Widow of Wagon Gap Lou, a traveling salesman, accidentally kills a man, and according to the law of the west he has to take care of the dead man's widow and children—all seven of them. Because the family is so unsavory, Lou knows that no other man will kill him, so he allows himself to be appointed sheriff and clears the town of lowlifes. Usual Abbott & Costello fare is highlighted with their zany antics.
1947 78m/B Bud Abbott, Lou Costello, Marjorie Main, George Cleveland, Gordon Jones, William Ching, Peter Thompson, Glenn Strange, Olin Howland; *Dir:* Charles T. Barton. **VHS $14.98** *MCA* 🎞🎞

The Witch Historian hired by family to assemble the late father's works falls in love with beautiful daughter. Seems she's no angel.
1966 103m/B *IT* Rosanna Schiaffino, Richard Johnson, Sarah Ferrati, Gian Marie Volonte, Margherita Guzzinati; *Dir:* Damiano Damiani. **VHS $19.98** *SNC* 🎞½

Witch Who Came from the Sea A witch terrorizes all the ships at sea, but doesn't exactly haunt the viewer.
1976 (R) 98m/C Millie Perkins, Loni Chapman, Vanessa Brown, Peggy Feury, Rick Jason; *Dir:* Matt Cimber. **VHS, Beta $49.95** *UNI* 🎞½

A Witch Without a Broom Shoestring time travel fantasy. When an American professor catches the eye of a 15th century apprentice witch, the trip begins. Since the witch is only learning the ropes, they end up visiting a number of periods other than their own before the professor gets home.
1968 78m/C *SP* Jeffrey Hunter, Maria Perschy, Perla Cristal, Gustavo Rojo; *Dir:* Joe Lacy. **VHS** *IMA* 🎞½

Witchboard During a college party, a group of friends bring out the Ouija board and play with it for laughs. One of the girls discovers she can use the board to communicate with a small boy who died years before. In her effort to talk with him she unwittingly releases the evil spirit of an ax murderer who haunts and murders members of the group. An entertaining, relatively inoffensive member of its genre that displays some attention to characterization and plot.
1987 (R) 98m/C Todd Allen, Tawny Kitaen, Steven Nichols, Kathleen Wilhoite, Burke Byrnes, Rose Marie, James W. Quinn, Judy Tatum, Gloria Hayes, J.P. Luebsen, Susan Nickerson; *Dir:* Kevin S. Tenney. **VHS, Beta, LV $79.98** *MAG, HHE* 🎞🎞½

Witchcraft A young mother meets a couple killed three centuries ago for performing witchcraft. They want her baby, of course, to be the son of the devil. ''Rosemary's Baby'' rip-off, that is thoroughly predictable.
1988 (R) 90m/C Anat ''Topol'' Barzilai, Gary Sloan, Lee Kisman, Deborah Scott; *Dir:* Robert Spera. **VHS, Beta $29.95** *ACA* Woof!

Witchcraft 2: The Temptress A sensuous woman seduces an innocent young man into the rituals of witchcraft and the occult. Sequel to ''Witchcraft'' does not succeed where original failed, but does have a fair share of sex and violence.
1990 (R) 88m/C Charles Solomon, Mia Ruiz, Delia Sheppard; *Dir:* Mark Woods. **VHS, Beta $29.95** *ACA* 🎞

Witchcraft 3: The Kiss of Death Once a master of the occult, William Spanner now seeks only to live a normal life. His plans are changed, however, when a sensual creature from Hell is sent to seduce him. More sex and violence, with an emphasis on sex. The best of the ''Witchcraft'' trio, for those trying to plan a festive evening.
1990 85m/C Charles Solomon, Lisa Toothman, William L. Baker, Lena Hall; *Dir:* R.L. Tillmanns. **VHS $19.95** *ACA* 🎞🎞

Witchcraft 4: Virgin Heart Supernatural horror continues as attorney Will Spanner sinks even deeper into his enemy's satanic trap in this shocking sequel to the popular series. His only hope is to use his own black magic powers and to enlist the help of a seductive stripper (Penthouse Pet Strain).

1992 92m/C Charles Solomon, Julie Strain, Clive Pearson, Jason O'Gulihar, Lisa Jay Harrington, Barbara Dow; **Dir:** James Merendino. **VHS** $89.95 ACA ✍✍

Witchcraft Through the Ages
The demonic Swedish masterpiece in which witches and victims suffer against various historical backgrounds. Nightmarish and profane, especially the appearance of the Devil as played under much make-up by Christiansen himself. Silent. Also called "Haxan."
1922 74m/B SW Maren Pedersen, Clara Pontoppidan; **Dir:** Benjamin Christiansen; **Nar:** William S. Burroughs. **VHS, Beta** $29.98 GPV, MPI, WFV ✍✍✍

Witchdoctor in Tails
Mondo sleaze with footage of a slaughterhouse, firewalkers, and the rites and "operations" performed by witchdoctors.
1966 96m/B **Nar:** George Sanders. **VHS** $24.95 FCT ✍✍✍

Witchery
A photographer and his girlfriend vacation at a New England hotel where they discover a horrifying, satanic secret. One by one (as always, in this kind of bad flick, in the interest of "suspense"), people are killed off. Seems it's a witch, bent on revenge. Forget room service and bar the door.
1988 96m/C David Hasselhoff, Linda Blair, Catherine Hickland, Hildegarde Knef, Leslie Cumming, Bob Champagne, Richard Farnsworth, Michael Manches; **Dir:** Fabrizio Laurenti. **VHS, Beta** $79.95 VMK ✍

The Witches
Nine-year-old boy on vacation finds himself in the midst of a witch convention, and grand high witch Huston plans to turn all children into furry little creatures. The boy, with the help of his good witch grandmother, attempts to prevent the mass transmutation of children into mice. Top-notch fantasy probably too spooky for the training wheel set. Wonderful special effects; the final project of executive producer Jim Henson. Based on Roald Dahl's story.
1990 (PG) 92m/C Anjelica Huston, Mai Zetterling, Jasen Fisher, Rowan Atkinson, Charlie Potter, Bill Paterson, Brenda Blethyn, Jane Horrocks; **Dir:** Nicolas Roeg. Los Angeles Film Critics Association Awards '90: Best Actress (Huston); National Society of Film Critics Awards '90: Best Actress (Huston). **VHS, Beta, LV, 8mm** $92.95 WAR, FCT ✍✍✍½

Witches' Brew
Three young women try to use their undeveloped skills in witchcraft and black magic to help Garr's husband get a prestigious position at a university, with calamitous results. Oft-funny spoof is silly and oft-predictable. Turner's role is small as an older, experienced witch.
1979 (PG) 98m/C Teri Garr, Richard Benjamin, Lana Turner, Kathryn Leigh Scott; **Dir:** Richard Shorr, Herbert L. Strock. **VHS, Beta** $59.98 SUE ✍✍

Witches of Eastwick
"Mad Max" director Miller meets Hollywood in this unrestrained, vomit-filled treatment of John Updike's novel about three lonely small-town New England women and their sexual liberation. A strange, rich, overweight and balding, but nonetheless charming man knows their deepest desires and makes them come true with decadent excess. Raunchy fun, with Nicholson over-acting wildly as the Mephisto. Miller lends a bombastic violent edge to the effort, sometimes at the expense of the story. Filmed on location in Cohasset, Massachusetts.
1987 (R) 118m/C Jack Nicholson, Cher, Susan Sarandon, Michelle Pfeiffer, Veronica Cartwright, Richard Jenkins, Keith Joakum, Carel Struycken; **Dir:** George Miller. Los Angeles Film Critics

Association Awards '87: Best Actor (tie) (Nicholson); New York Film Critics Awards '87: Best Actor (Nicholson). **VHS, Beta, LV, 8mm** $19.98 WAR ✍✍½

Witches' Mountain
A troubled couple is captured by a coven of witches in the Pyrenees. Dull and pointless.
1971 98m/C MX Patty Shepard, John Caffari, Monica Randall; **Dir:** Raul Artigot. **VHS, Beta** $49.95 UNI ✍

Witchfire
After their pyschiatrist dies in an automobile accident, three maniacal women escape from an asylum and hide out in the woods. Winters then holds seances to contact the dead doctor, but instead captures a young, very much alive hunter. For dedicated fans of Winters.
1986 (R) 92m/C Shelley Winters, Gary Swanson, David Mendenhall, Corinne Chateau; **W/Dir:** Vincent J. Privitera. **VHS, Beta** $79.98 LIV, LTG ✍

The Witching
A poorly made story of man's continuing quest for supernatural power. Welles plays the high priest out to get victim Franklin. Also known as "Necromancy."
1972 90m/C Orson Welles, Pamela Franklin, Michael Ontkean, Lee Purcell; **Dir:** Bert I. Gordon. **VHS, Beta** $42.95 PGN ✍

Witching Time
Elvira presents this film, in which a young composer is visited by a horny 17th-century witch while his wife is away from home. When the wife returns, both women fight to possess him.
1984 60m/C Jon Finch, Prunella Gee, Patricia Quinn, Ian McCulloch; **Dir:** Don Leaver; **Hosted:** Elvira. **VHS, Beta** $29.95 IVE ✍½

The Witchmaker
A remote, crocodile-infested bayou in Louisiana is the scene of witchcraft and the occult as young girls are murdered in order for a group of witches to become youthful again. Also known as "Legend of Witch Hollow" and "Witchkill."
1969 101m/C John Lodge, Alvy Moore, Thordis Brandt, Anthony Eisley, Shelby Grant, Robyn Millan; **W/Dir:** William O. Brown. **VHS, Beta** CCI ✍

The Witch's Mirror
A sorceress plots to destroy the murderer of her god-daughter. The murderer, a surgeon, begins a project to restore the disfigured face and hands of his burned second wife, and he doesn't care how he gets the materials. For true fans of good bad horror flicks. Badly dubbed (which adds to the charm); dark (of course); and offbeat (naturally).
1960 75m/B MX Rosita Arenas, Armando Calvo, Isabela Corona, Dina De Marco; **Dir:** Chano Urueto. **VHS, Beta** $59.95 SNC, HHT ✍✍

WitchTrap
Lame sequel to "Witchboard." A mansion's new owner hires psychics to exorcise the disturbed ghost of his predecessor.
1989 (R) 87m/C James W. Quinn, Kathleen Bailey, Linnea Quigley; **Dir:** Kevin S. Tenney. **VHS, Beta, LV** $89.98 MAG, NVH, IME ✍

With Kit Carson Over the Great Divide
A large-scale silent western about a doctor's family gone asunder and then reunited during the Fremont expeditions out of St. Louis in the 1840s. Snow's final screen appearance. Recently refound. Fun cinematic history (in two senses of the phrase). Beautiful landscape.
1925 72m/B Roy Stewart, Henry B. Walthall, Marguerite Snow, Sheldon Lewis, Earl Metcalfe; **Dir:** Frank S. Mattison. **VHS, Beta, 8mm** $29.95 VYY ✍✍½

With Six You Get Eggroll
A widow with three sons and a widower with a daughter elope and then must deal with the antagonism of their children and even their dogs. Brady Bunch-esque family comedy means well, but doesn't cut it. Jamie Farr, William Christopher, and Vic Tayback have small parts. To date, Day's last big-screen appearance.
1968 (G) 95m/C Doris Day, Brian Keith, Pat Carroll, Alice Ghostley; **Dir:** Howard Morris. **VHS, Beta** $59.98 FOX ✍✍

Withnail and I
A biting and original black comedy about a pair of unemployed, nearly starving English actors during the late 1960s. They decide to retreat to a country house owned by Withnail's uncle, for a vacation and are beset by comic misadventures, particularly when the uncle, who is gay, starts to hit on his nephew's friend. Robinson, who scripted "The Killing Fields," makes his successful directorial debut, in addition to drafting the screenplay from his own novel. Co-produced by George Harrison and Richard Starkey (Ringo Starr).
1987 (R) 108m/C GB Richard E. Grant, Paul McGann, Richard Griffiths; **Dir:** Bruce Robinson. **VHS, Beta, LV** $79.95 MED ✍✍✍½

Without a Clue
Spoof of the Sherlock Holmes legend, in which Holmes is actually portrayed by a bumbling, skirt-chasing actor, and Watson is the sole crime-solving mastermind, hiring the actor to impersonate the character made famous by the doctor's published exploits. The leads have some fun and so do we; but laughs are widely spaced.
1988 (PG) 107m/C GB Michael Caine, Ben Kingsley, Jeffrey Jones, Lysette Anthony, Paul Freeman, Nigel Davenport, Peter Cook; **Dir:** Thom Eberhardt. **VHS, Beta, LV** $19.98 ORI ✍✍½

Without Honors
A man seeks to restore the reputation of his dead brother, accused of murder and theft. He joins the Texas Rangers and brings the real criminals to justice. Excellent location shooting lifts ordinary early western.
1932 62m/B Harry Carey, Mae Busch, Gibson Gowland, George "Gabby" Hayes; **Dir:** William Nigh. **VHS, Beta, LV** WGE ✍✍

Without Love
Ambitious young journalist experiences resentment from her peers because of her aggressive methods, and they decide to teach her a lesson by drawing her into a situation destined for disaster. In Polish with English subtitles.
1980 104m/C PL **Dir:** Barbara Sass. **VHS** $49.95 FCT ✍✍

Without Reservations
Hollywood-bound novelist Colbert encounters Marine flyer Wayne and his pal (DeFore) aboard a train. She decides he would be perfect for her newest movie. They both dislike her famous book and don't realize they're traveling with the renowned author. Midadventures and misunderstandings abound as this trio make their way to Tinseltown. Of course, Colbert and Wayne fall in love. Box-office success with a tired script and too few real laughs. The Duke is interesting but miscast. Louella Parsons, Cary Grant, and Jack Benny are seen in cameos.
1946 101m/B Claudette Colbert, John Wayne, Don DeFore, Frank Puglia, Phil Brown; **Dir:** Mervyn LeRoy. **VHS, Beta** $19.98 UHV, VCN, TTC ✍✍½

Without a Trace One morning, a six-year-old boy is sent off to school by his loving mother, never to return. The story of the mother's relentless search for her son. Cardboard characters, wildly unrealistic ending that is different from the real-life incident on which it's based. Scripted by Beth Gutcheon from her book "Still Missing." Jaffe's directorial debut.
1983 (PG) 119m/C Kate Nelligan, Judd Hirsch, Stockard Channing, David Dukes; *Dir:* Stanley R. Jaffe. VHS, Beta $59.98 *FOX* ♪♪

Without Warning Inexperienced director and actors pull out a minor success in this fairly ordinary murder mystery, about a serial killer of beautiful blondes.
1952 75m/C *FR* Maurice Ronet, Adam Williams, Gloria Franklin, Edward Binns; *Dir:* Arnold Laven. VHS *API* ♪♪½

Without Warning: The James Brady Story White House Press Secretary Brady took a bullet in the brain during the 1981 shooting of President Reagan, and made a slow, grueling recovery. This fine made-for-cable-TV film spares none of it, concentrating on the stricken man and his family, and opting for disease-of-the-week cliches (though the pic's politics won't please the gun-adorers). Based on Mollie Dickinson's biography "Thumbs Up," with a script by Oscar-winning
1991 (R) 120m/C Beau Bridges, Joan Allen, David Strathairn; *Dir:* Michael Toshiyuki Uno. VHS $89.98 *HBO* ♪♪♪

Without You I'm Nothing Comedienne Bernhard's stunning, intelligent, and hilarious one-person show explores three decades of American pop culture. Bernhard challenges audience expectations of societal norms by assuming sterotypical ideals from the 60's, 70's and 80's.
1990 (R) 90m/C Sandra Bernhard, John Doe, Lu Leonard, Cynthia Bailey; *Dir:* John Boskovich. VHS $89.98 *HBO* ♪♪♪

Witness A young Amish boy traveling from his father's funeral witnesses a murder in a Philadelphia bus station. The investigating detective (Ford, in one of his best roles) soon discovers the killing is part of a conspiracy involving corruption in his department. He follows the child and his young widowed mother to their home in the country. A thriller with a difference, about the encounter of alien worlds, with a poignant love story. McGillis, in her first major role, is luminous as the Amish mother, while Ford is believable as both a cop and a sensitive man. An artfully crafted drama, richly focusing on the often misunderstood Amish lifestyle.
1985 (R) 112m/C Harrison Ford, Kelly McGillis, Alexander Godunov, Lukas Haas, Josef Sommer, Danny Glover, Patti LuPone; *Dir:* Peter Weir. Academy Awards '85: Best Film Editing, Best Original Screenplay; Edgar Allan Poe Awards '85: Best Screenplay. VHS, Beta, LV, 8mm $14.95 *PAR* ♪♪♪½

Witness for the Prosecution An unemployed man is accused of murdering a wealthy widow whom he befriended. A straightforward court case becomes increasingly complicated in this energetic adaptation of an Agatha Christie stage play. Ailing defense attorney Laughton can't resist taking an intriguing murder case. Outstanding performances by Laughton, with excellent support by real-life wife, Lanchester, as his patient nurse. Power, as the alleged killer, and Dietrich as a tragic woman are top-notch (see if you can detect Dietrich in an unbilled second role). Scripted by Wilder and Harry Kurnitz. Oscar nominations for Best Picture, Best Actor (Laughton), Best Supporting Actress (Lanchester), Best Director, Sound, and Film Editing.
1957 114m/B Charles Laughton, Tyrone Power Sr., Marlene Dietrich, Elsa Lanchester, John Williams, Henry Daniell, Una O'Connor; *Dir:* Billy Wilder. VHS, Beta, LV $19.98 *MGM, FOX, MLB* ♪♪♪

Wives Under Suspicion While prosecuting a man who murdered his wife out of jealousy, a district attorney finds his own home life filled with similar tension. Director Whale's unnecessary remake of his own earlier film, "The Kiss Before the Mirror." He is best known for "Frankenstein."
1938 75m/B Warren William, Gail Patrick, Constance Moore, William Lundigan; *Dir:* James Whale. VHS, Beta $24.95 *NOS, DVT* ♪♪½

The Wiz A black version of the long-time favorite "The Wizard of Oz," based on the Broadway musical. Ross plays a Harlem schoolteacher who is whisked away to a fantasy version of New York City in a search for her identity. Some good character performances and musical numbers, but generally an overblown and garish effort with Ross too old for her role. Pryor is poorly cast, but Jackson is memorable as the Scarecrow. High-budget ($24 million) production with a ton of name stars, lost $11 million, and cooled studios on black films. Horne's number as the good witch is the best reason to sit through this one.
1978 (G) 133m/C Diana Ross, Michael Jackson, Nipsey Russell, Ted Ross, Mabel King, Thelma Carpenter, Richard Pryor, Lena Horne; *Dir:* Sidney Lumet. VHS, Beta, LV $29.95 *MCA* ♪½

The Wizard Facing the dissolution of his dysfunctional family, a youngster decides to take his autistic, video game-playing little brother across the country to a national video competition. Way too much plot in pretentious, blatantly commercial feature-length Nintendo ad, featuring the kid from "The Wonder Years." For teen video addicts only.
1989 (PG) 99m/C Fred Savage, Beau Bridges, Christian Slater, Luke Edwards, Jenny Lewis; *Dir:* Todd Holland. VHS, Beta, LV $19.95 *MCA* ♪½

The Wizard of Gore The prototypical Lewis splatter party, about a magician whose on-stage mutilations turn out to be messily real. High camp and barrels of bright movie blood.
1970 (R) 96m/C Ray Sager, Judy Cler, Wayne Ratay, Phil Lauenson, Jim Rau, John Elliott, Don Alexander, Monika Blackwell, Corinne Kirkin; *Dir:* Herschell Gordon Lewis. VHS, Beta $39.98 *NSV, RHI* ♪

The Wizard of Loneliness A disturbed young boy goes to live with his grandparents during WWII, and slowly discovers family secrets centering on his aunt. Excellent performances barely save an aimless plot with an overdone denouement. Based on the novel by John Nichols.
1988 (PG-13) 110m/C Lukas Haas, Lea Thompson, John Randolph, Lance Guest, Anne Pitoniak, Jeremiah Warner, Dylan Baker; *Dir:* Jenny Bowen. VHS, Beta, LV $89.95 *VIR* ♪♪½

The Wizard of Oz An early silent version of the L. Frank Baum fantasy, with notable plot departures from the book and later 1939 adaptation, starring long-forgotten comedian Semon as the Scarecrow, supported by a pre-Laurel Hardy as the Tin Woodman. With music score.
1925 93m/B Larry Semon, Dorothy Dwan, Bryant Washburn, Charles Murray, Oliver Hardy, Josef Swickard, Virginia Pearson; *Dir:* Larry Semon. VHS, Beta, 8mm $19.95 *NOS, VYY, MLB* ♪♪½

The Wizard of Oz From the book by L. Frank Baum, a fantasy about a Kansas farm girl (Garland, in her immortal role) who rides a tornado to a brightly colored world over the rainbow, full of talking scarecrows, munchkins and a wizard who bears a strange resemblance to a Kansas fortune-teller. She must outwit the Wicked Witch if she is ever to go home. Delightful performances from Lahr, Bolger and Hamilton. Director Fleming originally wanted Deanna Durbin or Shirley Temple for the role of Dorothy, but settled for Garland who made the song "Somewhere Over the Rainbow" her own. She received a special Academy Award for her performance. For the 50th anniversary of its release, "The Wizard of Oz" was restored and includes rare film clips of Bolger's "Scarecrow Dance" and the cut "Jitterbug" number, and shots by Buddy Ebsen as the Tin Man before he became ill and left the production. The laserdisc edition includes digital sound, commentary by film historian Ronald Haver, and test footage, trailers and stills, as well as Jerry Maren talking about his experiences as a Munchkin. Additional Oscar nominations for Best Picture, Best Interior Decoration, and Best Special Effects.
1939 101m/C Judy Garland, Margaret Hamilton, Ray Bolger, Jack Haley, Bert Lahr, Frank Morgan, Charley Grapewin, Clara Blandick, Mitchell Lewis, Billie Burke; *Dir:* Victor Fleming. Academy Awards '39: Best Song ("Over the Rainbow"), Best Original Score; American Film Institute's Survey '77: 10th Best American Film Ever Made. VHS, Beta, LV $19.98 *MGM, VYG, APD* ♪♪♪♪

The Wizard of Speed and Time A special-effects wizard gets the break of his life when he's hired by a movie studio, but there's more to this particular studio than he realizes. Jittlov, a real special-effects expert, plays himself in this personally financed production. Although brimming with inside jokes and references, this self-indulgence succeeds with its enthusiasm and ambition, despite being rather obviously self-produced. Unique special effects make it memorable.
1988 (PG) 95m/C Mike Jittlov, Richard Kaye, Page Moore, David Conrad, Steve Brodie, John Massari, Frank Laloggia, Philip Michael Thomas, Angelique Pettyjohn, Arnetia Walker, Paulette Breen; *Dir:* Mike Jittlov. VHS, LV $89.95 *SHG* ♪♪½

Wizards A good, bumbling sorcerer battles for the sake of a magic kingdom and its princess against his evil brother who uses Nazi propaganda films to inspire his army of mutants. Profane, crude, & typically Bakshian fantasy with great graphics. Animated.
1977 (PG) 81m/C *Dir:* Ralph Bakshi. VHS, Beta, LV $79.98 *FOX* ♪♪½

Wizards of the Lost Kingdom A boy magician, aided by various ogres and swordster Svenson battles an all-powerful wizard for control of his kingdom. Family fare a bit too clean and harmless.
1985 (PG) 76m/C Bo Svenson, Vidal Peterson, Thom Christopher; *Dir:* Hector Olivera. VHS, Beta $19.95 *MED* ♪½

Wizards of the Lost Kingdom 2 A boy wizard is charged with vanquishing the evil tyrants from three kingdoms. Barely a sequel; no plot continuation or cast from earlier kiddie sword epic.

1989 (PG) 80m/C David Carradine, Bobby Jacoby, Lana Clarkson, Mel Welles, Susan Lee Hoffman, Sid Haig; **Dir:** Charles B. Griffith. **VHS, Beta** $79.95 MED *

Wolf To everyone's lasting regret, an ex-Vietnam POW strives to rescue an American ambassador kidnapped by Central American rebels.
1986 95m/C J. Antonio Carreon, Ron Marchini; **Dir:** Charlie Ordonez. **VHS, Beta** $79.95 TWE *

Wolf Call Carroll plays the son of a miner who travels with his devoted hound to see his father's old mine and start it working again. Gangsters and a beautiful Indian maiden get in his way. Master and pooch both find romance, and both sing. From the Jack London novel.
1939 62m/B John Carroll, Movita, Wheeler Oakman, Peter George Lynn; **Dir:** George Waggner. **VHS, Beta, 8mm** $24.95 VYY, FCT *½

Wolf Dog A boy and his dog. Outdoor action and adventure. Twelve chapters, 13 minutes each.
1933 156m/B Rin Tin Tin Jr, Frankie Darro, Boots Mallory; **Dir:** Harry Fraser, Colbert Clark. **VHS, Beta** $26.95 SNC, VCN, MLB *½

The Wolf at the Door A well-appointed, sincere biography of impressionist Paul Gauguin in the middle period of life, during his transition from the petty demands of his Parisian life to the freedom of Tahiti. Not definitive or compelling, but serves the purpose of a biography—arousing interest in the subject's life and art. Screenplay by Christopher Hampton.
1987 (R) 90m/C DK FR Donald Sutherland, Jean Yanne, Sofie Graboel, Ghita Noerby, Max von Sydow, Merete Voldstedlund, Fanny Bastien; **Dir:** Henning Carlsen. **VHS, Beta** $79.98 FOX **½

Wolf Lake WWII veteran, whose son was killed in Vietnam, and a Vietnam army deserter clash with tragic consequences during their stay at a Canadian hunting lodge. Steiger tries hard, but overwrought revenge pic is basically unserious.
1979 90m/C Rod Steiger, David Hoffman, Robin Mattson, Jerry Hardin, Richard Herd, Paul Mantu; **Dir:** Burt Kennedy. **VHS, Beta** $9.99 STE, PSM *½

The Wolf Man Fun, absorbing classic horror with Chaney as a man bitten by werewolf Lugosi. His dad thinks he's gone nuts, his screaming gal pal just doesn't understand, and plants on the Universal lot have no roots. Ouspenskaya's finest hour as the prophetic gypsy woman. Ow-ooo! Chilling and thrilling.
1941 70m/B Lon Chaney Jr., Claude Rains, Maria Ouspenskaya, Ralph Bellamy, Bela Lugosi, Warren William, Patric Knowles, Evelyn Ankers; **Dir:** George Waggner. **VHS, Beta, LV** $14.95 MCA ***½

Wolfen Surrealistic menace darkens this original and underrated tale of super-intelligent wolf creatures terrorizing New York City. Police detective Finney tries to track down the beasts before they kill again. Notable special effects in this thriller, which covers environmental and Native American concerns while maintaining the tension. Feature film debuts of Hines and Venora. Based on the novel by Whitley Strieber. Available in widescreen on laserdisc.
1981 (R) 115m/C Albert Finney, Gregory Hines, Tommy Noonan, Diane Venora, Edward James Olmos; **Dir:** Michael Wadleigh. **VHS, Beta, LV** $19.98 WAR, LDC ***

Wolfen Ninja A girl is raised by wolves and is taught by them the secrets of Ninja fighting.
1987? 92m/C **Hosted:** Sho Kosugi. **VHS, Beta** $59.95 TWE *

Wolfheart's Revenge An honest cowboy, helped by Wolfheart the Wonder Dog, steps in when a sheep rancher is murdered. Silent film with original organ score.
1925 64m/B Guinn Williams. **VHS, Beta, 8mm** $29.95 VYY *

The Wolfman In 1910, a young man learns that his family is heir to the curse of the Werewolf. Same old stuff.
1982 91m/C Earl Owensby, Kristina Reynolds; **Dir:** Worth Keeter. **VHS, Beta** $19.99 HBO *

The Wolves Gosha shows the world of the yakuza (gangster) during the 1920s. Reminiscent of the samurai, the movie combines ancient Japanese culture with the rapidly changing world of 20th Century Japan. In Japanese with English subtitles.
1982 131m/C JP Tatsuya Nakadai; **Dir:** Hideo Gosha. **VHS** $79.95 WAC ***

A Woman of Affairs Garbo, who can't have the guy she really loves (Gilbert), marries scoundrel. When she finds out husband's true profession, he kills himself and Gilbert comes back. In order to preclude happy ending, she decides the affair wasn't meant to be and expires. Tragically. Silent.
1928 90m/B Greta Garbo, John Gilbert, Lewis Stone, Johnny Mack Brown, Douglas Fairbanks Jr., Hobart Bosworth, Dorothy Sebastian; **Dir:** Clarence Brown. **VHS** $29.98 MGM, FCT **

Woman in Brown It's Hungary, 1882, and five Jewish men are on trial for the murder of a man who actually committed suicide. Despite the prejudice and hatred from the locals, their lawyer believes in and fights for their innocence. Also known as "The Vicious Circle" and "The Circle."
1948 77m/B Conrad Nagel, Fritz Kortner, Reinhold Schnuzel, Philip Van Zandt, Eddie LeRoy, Edwin Maxwell; **Dir:** W. Lee Wilder. **VHS, Beta** $16.95 SNC **½

A Woman Called Golda Political drama following the life and career of Golda Meir, the Israeli Prime Minister and one of the most powerful political figures of the 20th century. Davis portrays the young Golda, Bergman taking over as she ages. Superior made for television bio epic, scripted by Harold Gast and Steven Gethers.
1982 192m/C Ingrid Bergman, Leonard Nimoy, Anne Jackson, Ned Beatty, Robert Loggia, Judy Davis; **Dir:** Alan Gibson. **VHS, Beta** $69.95 PAR ***

A Woman Called Moses The story of Harriet Ross Tubman, who bought her freedom from slavery, founded the underground railroad, and helped lead hundreds of slaves to freedom before the Civil War. Wonderful performance by Tyson but the made for television film is bogged down by a so-so script.
1978 200m/C Cicely Tyson, Dick Anthony Williams, Will Geer, Robert Hooks, Hari Rhodes, James Wainwright; **Dir:** Paul Wendkos. **VHS** $89.95 XVC **½

Woman Condemned A reporter tries to clear a woman's name of murder.
1933 58m/B Mischa Auer, Lola Lane, Claudia Dell, Richard Hemingway. **VHS, Beta, LV** WGE *½

A Woman of Distinction Slapstick comedy with potential to be ordinary is taken over the tope by a good, vivacious cast. Cameo by Ball. Russell, the stuffy Dean of a women's college, is driven to distraction by a reporter linking her romantically to Milland, a visiting professor.
1950 85m/C Ray Milland, Rosalind Russell, Edmund Gwenn, Francis Lederer, Lucille Ball; **Dir:** Edward Buzzell. **VHS, Beta, LV** $69.95 COL ***

Woman in the Dunes Splendid, resonant allegorical drama. A scientist studing insects in the Japanese sand dunes finds himself trapped with a woman in a hut at the bottom of a pit. Superbly directed and photographed (by Hiroshi Segawa). Scripted by Kobo Abe from his acclaimed novel, and Oscar-nominated for Best Foreign Film. In Japanese with English subtitles.
1964 123m/B JP Eiji Okada, Kyoko Kishida; **Dir:** Hiroshi Teshigahara. Cannes Film Festival '64: Special Jury Prize; New York Film Festival '64: Special Jury Prize. **VHS, Beta** $29.95 CVC, NOS, VDM ****

A Woman in Flames A bored middle-class housewife leaves her overbearing husband and becomes a high-priced prostitute. She has a passionless affair with an aging bisexual gigolo. Dark and dreary tale of human relationships and the role of sex in people's lives. Interesting but depressing.
1984 106m/C GE Gudrun Landgrebe, Robert Van Ackeren, Matthieu Carriere, Gabriele Lafari, Hanns Zischler; **Dir:** Robert Van Ackeren. **VHS, Beta** $39.98 CGI, GLV ***

Woman in Grey A man and a woman battle wits when they attempt to locate and unravel the Army Code while staying one step ahead of J. Haviland Hunter, a suave villain after the same fortune. Silent.
1919 205m/B Arline Pretty, Henry Sell. **VHS, Beta** $49.98 NOS, CCB, MLB **

A Woman, Her Men and Her Futon A beautiful woman tries to find her identity by having a number of lovers but none can satisfy her every need.
1992 (R) 90m/C Jennifer Rubin, Lance Edwards, Grant Show, Michael Ceveris, Delaune Michel, Robert Lipton; **W/Dir:** Mussef Sibay. **VHS** $89.98 REP **

Woman Hunt Men kidnap women, then hunt them in the jungle for fun. So deeply offensive, we would give it negative bones if we could. "Hee Haw" bimbo Todd plays a sadistic lesbian.
1972 81m/C John Ashley, Lisa Todd, Eddie Garcia, Laurie Rose; **Dir:** Eddie Romero. **VHS, Beta** $59.98 CHA Woof!

The Woman Hunter A wealthy woman recovering from a traffic accident in Mexico believes someone is after her for her jewels and possibly her life. Eden dressed well, if nothing else; suspense builds ploddingly to a "Yeah, sure" climax. Made for television.
1972 73m/C Barbara Eden, Robert Vaughn, Stuart Whitman, Sydney Chaplin, Larry Storch, Enrique Lucero; **Dir:** Bernard L. Kowalski. **VHS, Beta** $59.95 KOV, QHV *½

The Woman Inside Low-budget gender-bender depicts the troubled life of a Vietnam veteran who decides on a sex-change operation to satisfy his inner yearnings. Blondell's unfortunate last role as his/her aunt. Cheesy like Limburger; almost too weird and boring even for camp buffs.
1983 (R) 94m/C CA Gloria Manon, Dave Clark, Joan Blondell; **Dir:** Joseph Van Winkle. **VHS, Beta** $24.95 NO *½

Woman in the Moon Assorted people embark on a trip to the moon and discover water, and an atmosphere, as well as gold. Lang's last silent outing is nothing next to "Metropolis," with a rather lame plot (greedy trip bashers seek gold), but interesting as a vision of the future. Lang's predictions about space travel often hit the mark. Also called "By Rocket to the Moon" and "Girl in the Moon."
1929 115m/B *GE* Klaus Pohl, Willy Fritsch, Gustav von Wagenheim, Gerda Maurus; *Dir:* Fritz Lang. **VHS, Beta $49.95** *IHF, VYY, FST* ♫♫½

The Woman Next Door One of Truffaut's last films before his sudden death in 1984. The domestic drama involves a suburban husband who resumes an affair with a tempestuous now-married woman after she moves next door, with domestic complications all around. An insightful, humanistic paean to passion and fidelity by the great artist, though one of his lesser works. Supported by strong outings from the two leads. In French with English subtitles.
1981 (R) 106m/C *FR* Gerard Depardieu, Fanny Ardant, Michele Baumgartner, Veronique Silver, Roger Van Hool; *Dir:* Francois Truffaut. **VHS, Beta $39.95** *HMV, FOX* ♫♫♫

A Woman Obsessed Pathetic, over-cooked drama about a woman obsessed with her long-lost son. "Ruth Raymond" is really porn-flick vet Georgina Spelvin. Soon-to-be camp classic takes itself with utter seriousness.
1988 (R) 105m/C Ruth Raymond, Gregory Patrick, Troy Donahue, Linda Blair; *Dir:* Chuck Vincent. **VHS, Beta $29.95** *ACA* ♫

Woman of Paris/Sunnyside This double feature highlights the talents of Charlie Chaplin. "A Woman of Paris" is the tragic story of a French country girl who winds up a kept woman. Chaplin's only dramatic directional outing; he makes only a cameo appearance. Straightforward but hip story of sex and society. In "Sunnyside," an example of high Chaplin highjinks, the comic plays a hotel handyman who can't win.
1923 111m/B Charlie Chaplin, Adolphe Menjou, Edna Purviance, Carl Miller, Lydia Knott; *Dir:* Charlie Chaplin. **VHS, Beta $19.98** *FOX* ♫♫♫½

A Woman Rebels A young Victorian woman challenges Victorian society by fighting for women's rights. Excellent performance from Hepburn lifts what might have been a forgettable drama. Screen debut of Van Heflin. Based on Netta Syrett's "Portrait of a Rebel."
1936 88m/B Katharine Hepburn, Herbert Marshall, Elizabeth Allan, Donald Crisp, Van Heflin; *Dir:* Mark Sandrich. **VHS, Beta $34.98** *CCB* ♫♫½

The Woman in Red Executive Wilder's life unravels when he falls hard for stunning Le Brock. Inferior Hollywood-ized remake of the ebulliant "Pardon Mon Affaire." Somehow the French seem to do the sexual force thing with more verve, but this one has its moments, and Wilder is likeable. Music by Stevie Wonder.
1984 (PG-13) 87m/C Gene Wilder, Charles Grodin, Kelly Le Brock, Gilda Radner, Judith Ivey, Joseph Bologna; *Dir:* Gene Wilder. Academy Awards '84: Best Song ("I Just Called to Say I Love You"). **VHS, Beta, LV $79.98** *LIV, VES* ♫♫½

A Woman of Rome Standard Italian star vehicle, with Lollobrigida portraying a successful prostitute in Rome who decides to change her life. Dubbed.

1956 93m/B *IT* Gina Lollobrigida, Daniel Gelin, Franco Fabrizi, Raymond Pellegrin; *Dir:* Luigi Zampa. **VHS, Beta $24.95** *NOS, FCT* ♫♫

Woman in the Shadows A ex-con retreats to the woods for serenity and peace, but is assaulted by mysterious women, jealous lovers, and gun-slinging drunks, until he explodes. A man can only take so much. From a Dashiell Hammett story.
1934 70m/B Fay Wray, Ralph Bellamy, Melvyn Douglas, Roscoe Ates, Joe King. **VHS, Beta $29.95** *VYY* ♫♫

A Woman of Substance, Episode 1: Nest of Vipers The woman of the title, Emma Harte, rises from poverty to wealth and power through self-discipline. First episode in five three-cassette made-for-TV saga has our heroine working as a servant for the snooty Fairley family.
1984 100m/C Jenny Seagrove, Barry Bostwick, Deborah Kerr; *Dir:* Don Sharp. **VHS, Beta $69.98** *LTG, LIV* ♫♫½

A Woman of Substance, Episode 2: The Secret is Revealed Emma Harte struggles to rise from a life of poverty to wealth and social acceptance. Second miniseries episode has Emma fighting to get by in the big city. Fun TV drama.
1984 100m/C Jenny Seagrove, Barry Bostwick, Deborah Kerr; *Dir:* Don Sharp. **VHS, Beta $69.98** *LTG, LIV* ♫♫½

A Woman of Substance, Episode 3: Fighting for the Dream Third installment is pretty good romantic TV saga. Rags-to-riches Emma finds love and happiness—at last!
1984 100m/C Jenny Seagrove, Barry Bostwick, Deborah Kerr; *Dir:* Don Sharp. **VHS, Beta $69.98** *LTG, LIV* ♫♫½

Woman Times Seven Italian sexual comedy: seven sketches, each starring Mac-Laine with a different leading man. Stellar cast and good director should have delivered more, but what they have provided has its comedic moments.
1967 99m/C *IT FR* Shirley MacLaine, Peter Sellers, Rossano Brazzi, Vittorio Gassman, Lex Barker, Elsa Martinelli, Robert Morley, Alan Arkin, Michael Caine, Patrick Wymark, Anita Ekberg, Philippe Noiret; *Dir:* Vittorio DeSica. **VHS, Beta $59.98** *SUE* ♫♫½

The Woman of the Town Frontier marshall and newspaperman Bat Masterson is portrayed convincingly by Dekker as a very human hero who seeks justice when the woman he loves, a dance hall girl who works in the town for social causes, is killed by an unscrupulous rancher. Fictionalized but realistic western shows "heroes" as good, ordinary people.
1943 90m/B Claire Trevor, Albert Dekker, Barry Sullivan, Henry Hull, Porter Hall, Percy Kilbride; *Dir:* George Archainbaud. **VHS, Beta $19.95** *NOS, IND, FCT* ♫♫½

A Woman Under the Influence Strong performances highlight this overlong drama about a family's disintegration. Rowland is the lonely, middle-aged housewife who's having a breakdown and Falk is her blue-collar husband who can't handle what's going on. Received Academy Award nominations for Best Actress and Best Director.
1974 (R) 147m/C Gena Rowlands, Peter Falk, Matthew Cassel, Matthew Laborteaux, Christina Grisanti; *W/Dir:* John Cassavetes. **VHS, Beta $94.95** *TOU* ♫♫♫

The Woman Who Came Back A young woman who believes she suffers from a witch's curse returns to her small hometown with unhappy results. Good cast; bad script. Indifference inducing.
1945 69m/B Nancy Kelly, Otto Kruger, John Loder, Ruth Ford, Jeanne Gail; *Dir:* Walter Colmes. **VHS, Beta $14.95** *COL, SVS* ♫♫

The Woman Who Willed a Miracle Made for television drama has a devoted mother encouraging her mentally retarded son to pursue his interest in the piano.
1983 72m/C Cloris Leachman, James Noble; *Dir:* Sharron Miller. **VHS, Beta $59.95** *USA* ♫♫

A Woman Without Love A rarely seen film from Bunuel's Mexican period based on a classic Guy de Maupassant tale about a forbidden romance. Family tragedy results later when the husband "misbequeaths" his fortune. Minor but fascinating bug-the-bourgeoisie Bunuel. In Spanish with English subtitles.
1951 91m/C *MX* Rosario Granados, Julio Villareal, Tito Junco; *Dir:* Luis Bunuel. **VHS, Beta $24.95** *XVC, MED* ♫♫♫

A Woman is a Woman Godard's affectionate sendup of Hollywood musicals is a hilarious comedy about a nightclub dancer (Karina) who desperately wants a baby. When boyfriend Belmondo balks, she asks his best friend Brialy. Much ado is had, with the three leads all splendid. Great music by Legrand. Godard's first film shot in color and cinemascope. In French with English subtitles.
1960 88m/C *FR* Jean-Claude Brialy, Jean-Paul Belmondo, Anna Karina, Marie DuBois; *Dir:* Jean-Luc Godard. **VHS $59.95** *INT* ♫♫♫

Woman of the Year First classic Tracy/Hepburn pairing concerns the rocky marriage of a renowned political columnist and a lowly sportswriter. Baseball scene with Hepburn at her first game is delightful. Hilarious, rich entertainment that tries to answer the question "What really matters in life?" Tracy and Hepburn began a close friendship that paralleled their quarter-century celluloid partnership. Hepburn shepherded Kamin and Lardner's Oscar-winning script past studio chief Louis B. Mayer, wearing four-inch heels to press her demands. Mayer caved in; Tracy was freed from making "The Yearling," and the rest is history. Hepburn received an Oscar nomination for Best Actress.
1942 114m/B Spencer Tracy, Katharine Hepburn, Fay Bainter, Dan Tobin, Reginald Owen, Roscoe Karns, William Bendix, Minor Watson; *Dir:* George Stevens. Academy Awards '42: Best Original Screenplay. **VHS, Beta, LV $19.95** *MGM* ♫♫♫♫

A Woman's Face An unpleasant, bitter woman with a hideous scar on her face contemplates murder. After plastic surgery, she becomes a nicer person and doubts her plan. Lean, tight suspense with a bang-up finale. In Swedish with English subtitles. Remade in Hollywood in 1941. Originally titled "En Kvinnas Ansikte."
1938 100m/B *SW* Ingrid Bergman, Anders Henrikson, Karin Carlsson, Georg Rydeberg, Goran Bernhard, Tore Svennberg; *Dir:* Gustaf Molander. **VHS, Beta $29.95** *WAC, CRO* ♫♫♫

A Woman's Face A physically and emotionally scarred woman becomes part of a blackmail ring. Plastic surgery restores her looks and her attitude. Begins with a murder trial and told in flashbacks; tight, suspenseful

remake of the 1938 Swedish Ingrid Bergman vehicle. Climax will knock you out of your chair.
1941 107m/B Joan Crawford, Conrad Veidt, Melvyn Douglas, Osa Massen, Reginald Owen, Albert Basserman, Marjorie Main, Donald Meek, Charles Quigley, Henry Daniell, George Zucco, Robert Warwick; *Dir:* George Cukor. **VHS, Beta** **$19.98** *MGM ♂♂♂*

A Woman's Secret O'Hara admits to murder she didn't commit. Why? It's as indiscernible as why the plot should be muddled by flashbacks. RKO lost big bucks on this one.
1949 85m/B Maureen O'Hara, Melvyn Douglas, Gloria Grahame, Bill Williams, Victor Jory, Jay C. Flippen; *Dir:* Nicholas Ray. **VHS, Beta $19.98** *TTC, FCT ♂♂*

Wombling Free An English girl makes friends with a race of tiny, litter-hating furry creatures called Wombles. The creatures are visible only to the little girl because she is the only one who believes. Together they try to clean up a dirty world. British society is less cutting-edge about social roles, etc. than our enlightened one; be prepared for retrograde characterizations. Fun scene with Kelly/Astaire-style dance number.
1977 86m/C *GB* Bonnie Langford, David Tomlinson; *Dir:* Lionel Jeffries. **VHS, Beta $39.95** *COL ♂♂*

The Women A brilliant adaptation of the Clare Boothe Luce stage comedy about a group of women who destroy their best friends' reputations at various social gatherings. Crawford's portrayal of the nasty husband-stealer is classic, and the fashion-show scene in Technicolor is one not to miss. Hilarious bitchiness all around. Script co-written by Anita Loos. Remade semi-musically as "The Opposite Sex." Another in that long list of stellar 1939 pics.
1939 133m/B Norma Shearer, Joan Crawford, Rosalind Russell, Joan Fontaine, Mary Boland, Lucile Watson, Margaret Dumont, Paulette Goddard, Ruth Hussey, Marjorie Main; *Dir:* George Cukor. **VHS, Beta, LV $19.98** *MGM ♂♂♂½*

The Women Also known as "Les Femmes" and "The Vixen," this film depicts a sexy secretary's seduction of a world-weary writer.
1969 86m/C *FR* Brigitte Bardot, Maurice Ronet. **VHS, Beta, LV $59.95** *UNI ♂♂½*

The Women of Brewster Place Seven black women living in a tenement fight to gain control of their lives. (Men in general don't come out too well.) Excellent, complex script by Karen Hall gives each actress in a fine ensemble headed by Winfrey (in her TV dramatic debut) time in the limelight. Pilot for the series "Brewster Place." Based on the novel by Gloria Naylor. Winfrey was executive producer.
1989 180m/C Oprah Winfrey, Mary Alice, Olivia Cole, Robin Givens, Moses Gunn, Jackee, Paula Kelly, Lonette McKee, Paul Winfield, Cicely Tyson; *Dir:* Donna Deitch. **VHS, Beta $79.95** *JTC ♂♂♂*

Women in Cell Block 7 Another exploitation film of women in prison suffering from their jailers and each other. A double woof.
1977 (R) 100m/C *IT* Anita Strinberg, Eve Czemeys, Olga Bisera, Jane Avril, Valeria Fabrizi, Jenny Tamburi; *Dir:* Rino Di Silvestro. **VHS, Beta $29.95** *PGN Woof!*

Women in Fury A woman is sentenced to imprisonment in a mostly lesbian Brazilian jail, and subsequently leads a breakout.

1984 94m/C *BR* Suzanne Carvalno, Gloria Cristal, Zeni Pereira, Leonardo Jose. **VHS, Beta $69.98** *LIV, LTG Woof!*

Women in Love Drama of two steamy affairs, based on D.H. Lawrence's classic novel. Deservedly Oscar-winning performance by Jackson; controversial nude wrestling scene with Bates and Reed is hard to forget. Plot is dumped on in favor of atmosphere. Followed nearly two decades later (1989) by a "prequel": "The Rainbow," also from Lawrence, also directed by Russell and starring Jackson.
1970 (R) 129m/C *GB* Glenda Jackson, Jennie Linden, Alan Bates, Oliver Reed, Michael Gough, Eleanor Bron; *Dir:* Ken Russell. Academy Awards '70: Best Actress (Jackson); National Board of Review Awards '70: 10 Best Films of the Year, Best Actress (Jackson); National Society of Film Critics Awards '70: Best Actress (Jackson). **VHS, Beta, LV $19.98** *FOX, MGM, BTV ♂♂♂*

Women & Men: Stories of Seduction Three famous short stories are brought to the screen in this made-for-TV collection. Mary McCarthy's "The Man in the Brooks Brothers Shirt," Dorothy Parker's "Dusk Before Fireworks," and Hemingway's "Hills Like White Elephants" between them cover every aspect of male-female relationships. Since there are three casts and three directors, there is little to join the stories in style, calling attention to some flaws in pacing and acting ability; still, worth watching.
1990 90m/C James Woods, Melanie Griffith, Peter Weller, Elizabeth McGovern, Beau Bridges, Molly Ringwald; *Dir:* Ken Russell, Tony Richardson, Frederic Raphael. **VHS, LV $89.99** *HBO, FCT ♂♂½*

Women & Men: Stories of Seduction, Part 2 Made for cable extravaganza chronicling the relationships of three couples, each adapted from the short story of a renowned author. Irwin Shaw's "Return to Kansas City" tells of a young boxer who is prematurely pushed into a match by his ambitious wife. In Carson McCullers' "A Domestic Dilemma," a marriage begins to crumble thanks to an alcoholic wife. Finally, Henry Miller's "Mara" has an aging man and a young Parisian prostitute spending a revealing evening together.
1991 (R) 90m/C Matt Dillon, Kyra Sedgwick, Ray Liotta, Andie MacDowell, Scott Glenn, Juliette Binoche; *Dir:* Walter Bernstein, Kristi Zea, Mike Figgis. **VHS $89.99** *HBO ♂♂½*

Women of the Prehistoric Planet On a strange planet, the members of a space rescue mission face deadly perils. Typical bad sci-fi of its era, with horrid special effects, including "giant" lizards. Get this: there's one woman, and she's not of the planet in question. See if you can last long enough to catch the amazing plot twist at the end.
1966 87m/C Wendell Corey, Irene Tsu, Robert Ito; *Dir:* Arthur C. Pierce. **VHS, Beta** *PGN Woof!*

Women Unchained Five women escape from a maximum security prison and make a run for the border. Along the way, they shun civilized behavior.
197? (R) 82m/C Carolyn Judd, Teri Guzman, Darlene Mattingly, Angel Colbert, Bonita Kalem. **VHS, Beta $59.95** *GEM ♂*

Women of Valor During WWII, a band of American nurses stationed in the Philippines are captured by the Japanese and struggle to survive in a brutal POW camp. TV feature was made 40 years too late, adding up to a surreal experience.

1986 95m/C Susan Sarandon, Kristy McNichol, Alberta Watson, Valerie Mahaffey, Suzanne Lederer, Patrick Bishop, Terry O'Quinn, Neva Patterson; *Dir:* Buzz Kulik. **VHS, Beta $9.99** *STE, PSM ♂½*

Women on the Verge of a Nervous Breakdown An acclaimed, surreal and hilarious romp through the lives of a film dubber, her ex-lover, his ex-wife, his new lover, and their various children. They meet in a comedy of errors, missed phone calls, and rental notices, while discovering the truth and necessity of love. Fast paced and full of black humor. Introduced Almodovar to American audiences. Nominated for Best Foreign Film. In Spanish with English subtitles.
1988 (R) 88m/C *SP* Carmen Maura, Fernando Guillen, Julieta Serrano, Maria Barranco, Rossy De Palma; *W/Dir:* Pedro Almodovar. National Board of Review Awards '88: Best Foreign Film; New York Film Critics Awards '88: Best Foreign Film; Venice Film Festival '88: Best Screenplay. **VHS, Beta, LV $79.98** *ORI ♂♂♂½*

Women's Club Table-turning sex comedy about a wealthy matron who patronizes a talented (?) young movie writer, but really wants—you know what. She begins lending his services to friends—and does he ever get worn out! This excuse may be intended to be uproarious, but it's hard to tell. And all that sex gets very perfunctory after awhile.
1987 (R) 89m/C Michael Pare, Maud Adams, Eddie Velez; *Dir:* Sandra Weintraub. **VHS, Beta $79.98** *LIV, VES, LTG ♂½*

Women's Prison Escape Four tough broads blow the pen and are forced to hightail it through the Everglades. They can't decide which is worse, the snakes and 'gators, or the corrupt and sleazy life they left behind. Only cinematic pairing of Rooney and Lurch.
19?? 90m/C Ted Cassidy, Chris Robinson, Mickey Rooney; *Dir:* Chris Robinson. **VHS** *ALS ♂½*

Women's Prison Massacre Four male convicts temporarily detained at a woman's prison (why?) violently take hostages and generally make trouble for the authorities and the women. Horribly dubbed.
1985 89m/C Laura Gemser, Lorraine De Selle, Francoise Perrot; *Dir:* Gilbert Roussel. **VHS, Beta $69.98** *LIV, VES Woof!*

Wonder Man When a brash nightclub entertainer (Kaye) is killed by gangsters, his mild-mannered twin brother (Kaye) takes his place to smoke out the killers. One of Kaye's better early films. The film debuts of Vera-Ellen and Cochran. "So in Love" was an Oscar-nominated song. Look for Mrs. Howell of "Gilligan's Island."
1945 98m/C Danny Kaye, Virginia Mayo, Vera-Ellen, Steve Cochran, S.Z. Sakall, Otto Kruger; *Dir:* H. Bruce Humberstone. Academy Awards '45: Best Special Effects. **VHS, Beta $19.98** *HBO, SUE ♂♂½*

Wonderful World of the Brothers Grimm Big-budget fantasy based very loosely on the Grimm Brothers' lives and three of their stories: "The Dancing Princesses," "The Cobbler and the Elves," and "The Singing Bone." Good, fun Puppetoon scenes, but leaves much else to be desired. A megawatt cast and the ubiquitous hand of producer-director Pal are not enough to cover the flaws. The biographical parts are hokey, and the tales used are the wrong ones. Disappointing but historically interesting. The second-ever story film done in Cinerama.

1962 134m/C Laurence Harvey, Karl Boehm, Claire Bloom, Buddy Hackett, Terry-Thomas, Russ Tamblyn, Yvette Mimieux, Oscar Homolka, Walter Slezak, Beulah Bondi, Martita Hunt, Otto Kruger, Barbara Eden, Jim Backus, Arnold Stang; *Dir:* Henry Levin, George Pal. Academy Awards '62: Best Costume Design. **VHS, Beta, LV $19.98** MGM ♂♂½

Wonderland Campy thriller that follows two young gay boys in England. Charles recites lines in old movies while watching them and sidesteps his father's homophobic insults. His best friend, Michael, is restless and always looking for adventure. The two show up at a gay nightclub known as the "Fruit Machine" and are the only witnesses to the brutal murder of the club's transvestite-owner. The pair are soon taken in by a male opera singer they both find attractive and taken to a south seaside town. The murderer catches up with the boys and seeks to dispose of them before they can report him to the authorities. Fast pace, dreamy atmosphere make up for the sometimes confusing plot. Often overlooked British import also known as "The Fruit Machine." **1988** 103m/C *GB* Emile Charles, Tony Forsyth, Robert Stephens, Clare Higgins, Bruce Payne, Robbie Coltrane; *Dir:* Philip Saville. **VHS, LV $79.98** LIV, VES ♂♂♂

Wonderland Cove A seafaring adventurer adopts five orphan children and embarks on journeys to exotic locales. **1975** 78m/C Clu Gulager; *Dir:* Jerry Thorpe. **Beta $59.95** PSM ♂

Wonders of Aladdin Slapstick comedy based on the ancient rubbed-lamp-and-genie chestnut. Kids'll love it; adults will yawn. **1961** 93m/C *FR IT* Donald O'Connor, Vittorio De Sica; *Dir:* Henry Levin, Mario Bava. **VHS, Beta $59.95** CHA ♂½

Wonderwall: The Movie Bizarre characters trip over themselves in love while groovy music plays in background. **1969** 82m/C *GB* Jack MacGowran, Jane Birkin, Irene Handl; *Dir:* Joe Massot. **VHS, Beta $79.95** FCT, MVD ♂½

Wonsan Operation U.N. soldiers plunge into the muddy depths of enemy territory in search of secret information that would end the Korean War. Dubbed. **1978** 100m/C Lew Montania, Frederick Hill; *Dir:* Terrence Sul. **VHS, Beta** VDA ♂

Woodchipper Massacre Aunt Tess is frozen in the freezer, waiting to be turned into whatever the fleshy equivalent of woodchips is by her three unloving relations, and her totally evil son has just broken out of prison looking to retrieve his inheritance. You think you have problems? You'll have one less if you leave this one on the shelf. **1989** 90m/C Jon McBride, Patricia McBride; *Dir:* Jon McBride. **VHS, Beta** DMP Woof!

The Wooden Gun Conflicts arise between two teenage factions in 1950's Tel Aviv: native Israelis, and the children of Jews arrived from Europe since World War II. In Hebrew with English subtitles. **1979** 91m/C *IS Dir:* Han Moshenson. **VHS, Beta $79.95** ERG, FCT ♂♂½

Wooden Horse Lean thriller of British POWs escaping a Nazi camp through a tunnel beneath their exercise horse. Based on a true incident in 1943 and on "The Tunnel Escape" by Eric Williams; set the tone for all British prison-camp movies to follow: stiff upper lip and all that, and adoles-

cent-style trickery reminiscent of English boarding schools. **1950** 98m/C Anthony Steel, Leo Genn, David Tomlinson, Bryan Forbes, Peter Finch; *Dir:* Jack Lee. **VHS, Beta $59.99** HBO ♂♂♂

Woodstock Step into the way-back machine and return to the times of luv, peace and understanding. A powerful chronicle of the great 1969 Woodstock rock concert, celebrating the music and lifestyle of the late sixties. More than 400,000 spectators withstood lack of privacy, bathrooms, parking, and food while wallowing in the mud for four days to catch classic performances by a great number of popular rock performers and groups, including Joan Baez, Country Joe and the Fish, Joe Cocker, Crosby, Stills and Nash, Arlo Guthrie, Richie Havens, Jimi Hendrix, Jefferson Airplane, Santana, John Sebastian, Sly and the Family Stone, Ten Years After, and The Who. Martin Scorcese helped edit the documentary, which was trail-blazing in its use of split-screen montage. **1970** (R) 180m/C Canned Heat, Richie Havens, Roger Daltrey, Pete Townsend, Keith Moon, Joan Baez, Country Joe & the Fish, Jimi Hendrix, Carlos Santana, David Crosby, Stephen Stills; *Dir:* Michael Wadleigh. Academy Awards '70: Best Feature Documentary. **VHS, Beta, LV $29.95** MVD, WAR, VTK ♂♂♂♂

The Word Intriguing premise has an archeologist finding a perviously unknown text purporting to be written by the younger brother of Jesus. If authentic and publicized, it would wreak theological havoc. Janssen is off to Italy to check up; he finds murder and skullduggery. Well-acted, interesting story. **1978** 300m/C David Janssen, James Whitmore, Florinda Bolkan, Eddie Albert, Geraldine Chaplin, Hurd Hatfield, John Huston, Kate Mulgrew, Janioo Rulo, Nicol Williamson; *Dir:* Richard Long. **VHS, Beta, LV $24.95** FRH ♂♂½

Words by Heart Based on a book by Ouida Sebestyen, "Words by Heart" is the story of a black family's struggle to survive in an all-white Missouri town. Aired by PBS as part of the WonderWorks Family Movie television series. **1984** 116m/C Charlotte Rae, Robert Hooks, Alfre Woodard; *Dir:* Robert Thompson. **VHS $29.95** PME, FCT ♂♂½

Words & Music Plot based on the careers of Rodgers and Hart is little more than a peg on which to hang lots of classic songs, sung by a parade of MGM stars: "Johnny One Note," "The Lady Is a Tramp," "Blue Moon," etc. Also includes Kelly's dance recreation of "Slaughter on Tenth Avenue." Good advice: if no one's singing or dancing, fast forward. **1948** 122m/C Mickey Rooney, Tom Drake, Judy Garland, Gene Kelly, Lena Horne, Mel Torme, Cyd Charisse, Marshall Thompson, Janet Leigh, Betty Garrett, June Allyson, Perry Como, Vera-Ellen, Ann Sothern; *Dir:* Norman Taurog. **VHS, Beta $29.95** MGM ♂♂½

Work/Police Two Chaplin two-reelers. In "Work (The Paper Hanger)" (1915), a family hires Charlie and cohorts Armstrong and Purviance to re-paper a house. Little did they know! In "Police" (1916), Charlie plays an ex-con released into the cruel world. Unremittingly hilarious. Silent with music score. **1916** 54m/B Charlie Chaplin, Billy Armstrong, Charles Insley, Marta Golden, Edna Purviance; *Dir:* Charlie Chaplin. **VHS, Beta $39.95** VYY ♂♂♂½

Workin' for Peanuts A young man who sells beer at a baseball stadium falls in love with the stadium owner's daughter.

1985 52m/C Carl Marotte, Jessica Steer. **VHS, Beta $39.98** SCL ♂♂♂½

Working Girl Romantic comedy set in the Big Apple has secretary Griffith working her way to the top in spite of her manipulative boss (Weaver in a powerful parody). Griffith gets her chance to shine when Weaver breaks a leg. She meets Ford for business that turns to romance. A 1980s Cinderella story that's sexy, funny, and sharply written and directed. Nice work by Ford, but this is definitely Griffith's movie. And keep an eye on gal pal Cusak. Oscar nominated theme song by Carly Simon; script by Kevin Wade. **1988** (R) 115m/C Melanie Griffith, Harrison Ford, Sigourney Weaver, Joan Cusack, Alec Baldwin, Philip Bosco, Ricki Lake, Nora Dunn, Olympia Dukakis, Oliver Platt; *Dir:* Mike Nichols. Academy Awards '88: Best Song ("Let the River Run"); Golden Globe Awards '89: Best Film—Musical/Comedy. **VHS, Beta, LV $19.98** FOX ♂♂♂

Working Girls Three girls who share an apartment in Los Angeles are willing to do anything for money. And they do. For lack of anything else to recommend, watch for the striptease by Peterson, better known as Elvira on TV. **1975** (R) 80m/C Sarah Kennedy, Laurie Rose, Mark Thomas, Cassandra Peterson; *Dir:* Stephanie Rothman. **VHS, Beta $19.98** UHV ♂

Working Girls An acclaimed, controversial look by independent filmmaker Borden into lives of modern brothel prostitutes over the period of one day. The sex is realistically candid and perfunctory; the docudrama centers on a prostitute who is a Yale graduate and aspiring photographer living with a female lover. Compelling, touching, and lasting, with sexually candid language and scenery. **1987** 93m/C Amanda Goodwin, Louise Smith, Ellen McElduff; *Dir:* Lizzie Borden. **VHS, Beta $14.95** COL, CHA, HHE ♂♂♂

Working Winnie/Oh So Simple Two rare, second-echelon short silent comedies. **1922** 61m/B Ethlyn Gibson, Bobby Ray, Rinkydinks. **VHS, Beta, 8mm $24.95** VYY ♂♂♂

The World According to Garp Comedy turns to tragedy in this relatively faithful version of John Irving's popular (and highly symbolic) novel, adapted by Steve Tesich. Chronicles the life of T.S. Garp, a struggling everyman beset by the destructive forces of modern society. Nevertheless, Garp maintains his optimism even as his life unravels around him. At the core of the film is a subplot involving a group of extreme feminists inspired in part by Garp's mother, the author of "A Sexual Suspect." Close and Lithgow (as a giant transsexual) are spectacular, while Williams is low-key and tender as the beleagured Garp. Ultimately pointless, perhaps, but effectively and intelligently so. **1982** (R) 136m/C Robin Williams, Mary Beth Hurt, John Lithgow, Glenn Close, Hume Cronyn, Jessica Tandy, Swoosie Kurtz, Amanda Plummer; *Dir:* George Roy Hill. **VHS, Beta, LV $19.98** WAR, PIA ♂♂♂

The World Accuses A woman takes a job in a nursery, unaware that one child is her own, put up for adoption after her wealthy husband died. Then, her ex-lover shows up in her attic after escaping from prison and takes her hostage. Whew! Everything works out, except that the viewer is left confused and incredulous. **1935** 62m/B Vivian Tobin, Dickie Moore, Russell Hopton, Cora Sue Collins, Mary Carr; *Dir:* Charles Lamont. **VHS $24.95** NOS, DVT ♂½

A World Apart Cinematographer Menges' first directoral effort is a blistering, insightful drama told from the point of view of a 13-year-old white girl living in South Africa, oblivious to apartheid until her crusading journalist mother is arrested under the 90-Day Detention Act, under which she might remain in prison permanently. Political morality tale is also a look at the family-vs-cause choices activists must make. Heavily lauded, with good reason; the autobiographical script is based on Slovo's parents, persecuted South African journalists Joe Slovo and Ruth First.
1988 (PG) 114m/C *GB* Barbara Hershey, Jodhi May, Linda Mvusi, David Suchet, Jeroen Krabbe, Paul Freeman, Tim Roth; *Dir:* Chris Menges. Cannes Film Festival '88: Special Jury Prize, Best Actress (Hershey, May, Mvusi); National Society of Film Critics '88: 10 Best of the Year. **VHS, Beta, LV $9.98** *MED* ✴✴✴½

World of Apu Finale of director Ray's acclaimed Apu trilogy (following "Pather Panchali" and "Aparajito"). Aspiring writer Apu drops out of the university for want of money and takes up with an old chum. An odd circumstance leads him to marry his friend's cousin, whom he comes to love. She dies in childbirth (though her baby boy lives); Apu is deeply distraught, destroys the novel he was working on, and becomes a wanderer. His friend finds him five years later and helps him begin again with his young son. Wonderfully human, hopeful story told by a world-class director. From the novel "Aparajito" by B. Bandopadhaya. In Bengali with English subtitles. Also known as "Apu Sansat."
1959 103m/B *IN* Soumitra Chatterjee, Sharmila Tagore, Alok Charkravarty, Swapan Makerji; *Dir:* Satyajit Ray. National Board of Review Awards '60: Best Foreign Film. **VHS, Beta $29.95** *HHT, VYY, DVT* ✴✴✴✴

The World is Full of Married Men A philandering ad exec gets involved with a reckless model, inspiring his fed-up wife also to look for extramarital sex. Promiscuously raunchy melodrama was written by the queen of sleaze, Jackie Collins, based on her novel.
1980 (R) 106m/C *GB* Anthony (Tony) Franciosa, Carroll Baker, Sherrie Croon, Gareth Hunt, Paul Nicholas; *Dir:* Robert Young. **VHS, Beta $59.98** *FOX Woof!*

The World Gone Mad During Prohibition, a tough reporter discovers the district attorney is the intended victim of a murder plot involving crooked Wall Street types. Full circle: this interesting drama of white-collar crime is again topical, though dialogue heavy and desultory.
1933 70m/B Pat O'Brien, Louis Calhern, J. Carroll Naish; *Dir:* Christy Cabanne. **VHS, Beta $19.95** *NOS, CAB, DVT* ✴✴½

World Gone Wild Action yarn about a post-apocalyptic world of the future where an evil cult leader brainwashes his disciples. Together they battle a small band of eccentrics for the world's last water source. Stale rehash (with ineffective satiric elements) of the Mad Max genre, served with Ant for campy appeal.
1988 (R) 95m/C Bruce Dern, Michael Pare, Adam Ant, Catherine Mary Stewart, Rick Podell; *Dir:* Lee H. Katzin. **VHS, Beta, LV $89.95** *MED* ✴½

World of Henry Orient Charming, eccentric comedy about two 15-year-old girls who, madly in love with an egotistical concert pianist, pursue him all around New York City. Sellers is hilarious, Walker and Spaeth are adorable as his teen groupies; Bosley and

Lansbury are great as Walker's indulgent parents. For anyone who has ever been uncontrollably infatuated. Screenplay by the father/daughter team, Nora and Nunnally Johnson, based on Nora Johnson's novel. Music by Elmer Bernstein.
1964 106m/C Peter Sellers, Tippy Walker, Merrie Spaeth, Tom Bosley, Angela Lansbury, Paula Prentiss; *Dir:* George Roy Hill. National Board of Review Awards '64: 10 Best Films of the Year. **VHS, Beta $59.98** *FOX* ✴✴✴½

The World of Suzie Wong Asian prostitute Kwan plays cat-and-mouse with American painter Holden. She lies to him about her profession, her family, and herself. His association with her ruins relationships in his life. Why, then, does he not get a clue? Good question, not answered by this soap opera that would be a serious drama. Offensively sanitized picture of the world of prostitution in an Asian metropolis. On the other hand, it's all nicely shot, much of it on location in Hong Kong. Based on Paul Osborn's play which was taken from Richard Mason's novel.
1960 129m/C *GB* William Holden, Nancy Kwan, Sylvia Syms, Michael Wilding, Laurence Naismith, Jackie Chan; *Dir:* Richard Quine. **VHS, Beta, LV $19.95** *PAR* ✴✴

World of the Vampires An evil count commands his vampire legions to wreak vengeance upon the family whose ancestors condemned him. Fairly miserable Mexican horror entry. Spanish title: "El Mundo de los Vampiros."
1960 75m/B *MX* Mauricio Garces, Erna Martha Bauman, Silvia Fournier, Guillermo Murray; *Dir:* Alfonso Corona Blake. **VHS, Beta $54.95** *SNC, HHT, MAD* ✴

World War III How's that for a title? A Russian plot is afoot to seize and destroy the Alaskan pipeline. When the plot is discovered, negotiation is needed to prevent world war. Executive branch showdown ensues between U.S. prez Hudson and Soviet chief Keith. Director Boris Sagal was killed on location, whereupon Greene took over and, shooting was moved indoors with dramatic tension lost in the transition. Made for TV.
1986 186m/C Brian Keith, David Soul, Rock Hudson, Cathy Lee Crosby, Katherine Helmond; *Dir:* David Greene. **VHS, Beta $79.98** *FOX* ✴✴½

World's Greatest Athlete Lame Disney comedy about a Tarzan-like jungle-man (Vincent) recruited by an unsuccessful American college coach (Amos) and his bumbling assistant (Conway). Fun special effects, weak script add up to mediocre family fare. Cameo by Howard Cosell as—who else?—himself.
1973 (G) 89m/C Jan-Michael Vincent, Tim Conway, John Amos, Roscoe Lee Browne, Dayle Haddon; *Dir:* Robert Scheerer. **VHS, Beta** *DIS, OM* ✴✴

World's Greatest Lover Milwaukee baker Rudi Valentine, played oft-hilariously by Wilder, tries to make it big in 1920's Hollywood. He has a screen test as a Hollywood movie sheik, but his wife (Kane) leaves him for the real McCoy. Episodic and uneven, it's alternately uproarious, touching and downright raunchy.
1977 (PG) 89m/C Gene Wilder, Carol Kane, Dom DeLuise, Fritz Feld, Carl Ballantine, Michael Huddleston, Matt Collins, Ronny Graham; *W/Dir:* Gene Wilder. **VHS, LV $59.98** *FOX* ✴✴½

The World's Greatest Stunts Reeve hosts this look at dozens of stunt scenes from hit movies and the preparations that went into keeping the stunt performers alive through the stunts.
1990 60m/C *Nar:* Christopher Reeve. **VHS, Beta $19.98** *MPI, RDG* ✴✴½

The World's Oldest Living Bridesmaid Light romantic comedy about a successful woman attorney who just can't find Mr. Right. Brenda feels even worse when she attends yet another friend's wedding. However, when she hires a younger, male secretary romantic sparks fly but will she be able to admit her feelings or will her pride get in the way? An appealing, attractive cast helps out this made for television fare.
1992 ?m/C Donna Mills, Brian Wimmer, Beverly Garland, Art Hindle. **VHS $89.95** *AIP* ✴✴½

The Worm Eaters Mean developers want to take over a reclusive worm farmer's land. He unleashes his livestock on them. The bad guys turn into—eeck!—"worm people." A truck runs over our hero nearly 75 minutes too late to save the viewer.
1977 (PG) 75m/C Herb Robins, Barry Hostetler, Lindsay Armstrong Black; *W/Dir:* Herb Robins. **VHS, Beta $19.95** *WES Woof!*

The Worst of Hollywood: Vol. 1 Two really bad movies for your viewing displeasure. "Maniac" is a mad-scientist fest and "Protect Your Daughter" warns of the perils of the Flapper era.
1990 115m/B Bill Woods, Horace Carpenter, Phyllis Diller; *Dir:* Dwain Esper. **VHS, Beta $29.95** *INC Woof!*

The Worst of Hollywood: Vol. 2 A von Stroheim-fest of mediocrity. In "Fugitive Road" Erich plays an Austrian officer of the First World War. "The Crime of Dr. Crespi" has him portray a mad scientist.
1990 131m/B Erich von Stroheim, Dwight Frye; *Dir:* Erich von Stroheim. **VHS, Beta $29.95** *INC Woof!*

The Worst of Hollywood: Vol. 3 Two more awful, exploitive docudramas. "Nation Aflame" features many silent screen stars who teach the virtues of living in racial harmony. "Probation" boasts the first screen appearance of Grable, as a promiscuous woman who ends up in prison.
1990 140m/B Betty Grable. **VHS, Beta $29.95** *INC Woof!*

Worth Winning A notoriously eligible Philadelphia bachelor takes a bet to become engaged to three women within three months, and finds himself in hot water. A critically dead-in-the-water chucklefest.
1989 (PG-13) 103m/C Mark Harmon, Lesley Ann Warren, Madeleine Stowe, Maria Holvoe, Mark Blum, Andrea Martin, Alan Blumenfeld, Brad Hall, Tony Longo; *Dir:* Will MacKenzie. **VHS, Beta, LV $89.98** *FOX* ✴½

Would-Be Gentleman The Comedie Francaise troupe performs Moliere's famous farce-comedy. Valuable as a record of the famous troupe on stage. When they're long gone, our grandkids can enjoy this. We can too (if we can't afford the trip to France), though it's less a film than a filming of a play.
1958 93m/C *FR* Louis Seigner, Jean Meyer, Jean Piat. **VHS, Beta $24.95** *DVT* ✴✴✴

Woyzeck Chilling portrayal of a man plunging into insanity. Mired in the ranks of the German Army, Woyzeck is harassed by his superiors and tortured in scientific experiments, gradually devolving into homicidal

maniac. In German with subtitles. Based on Georg Buchner play.
1978 82m/C GE Klaus Kinski, Eva Mattes, Wolfgang Reichmann, Josef Bierbichler; **W/Dir:** Werner Herzog. Cannes Film Festival '78: Best Supporting Actress (Mattes). **VHS $59.95** NYF 𝄞𝄞𝄞

Woza Albert! An international hit play makes tries to transfer its appeal to film, not an easy feat. Set in South Africa and performed by black South Africans, (using their townships as a backdrop) it explores in stylized fashion events resulting from the second coming of Christ to a hypocritical nation.
1982 55m/C VHS, Special order formats $195.00 CAL, FLI 𝄞𝄞½

WR: Mysteries of the Organism Makavejev's breakthrough film, a surreal, essayist exploration of the conflict/union between sexuality and politics—namely, Wilheim Reich and Stalin. A raunchy, bitterly satiric non-narrative that established the rule-breaking Yugoslav internationally. With English subtitles.
1971 84m/C Milena Dravic, Jagoda Kaloper, Tuli Kupferberg, Jackie Curtis; **Dir:** Dusan Makavejev. New York Film Festival '71: Best Supporting Actress. **VHS, Beta $79.95** FCT 𝄞𝄞𝄞

The Wraith Drag-racing Arizona teens find themselves challenged by a mysterious, otherworldly stranger. Hot cars; cool music; little else to recommend it. Lousy script; ludicrous excuse for a premise. Most of the stars herein are related to somebody famous.
1987 (PG-13) 92m/C Charlie Sheen, Nick Cassavetes, Sherilyn Fenn, Randy Quaid, Matthew Barry, Clint Howard, Griffin O'Neal; **Dir:** Mike Marvin. **VHS, Beta, LV $14.98** LIV, LTG 𝄞½

The Wrestler All-star wrestling, which is fictional anyway, gets said treatment in the appropriate way. Honest promoter (yeah, sure) bumps heads with bad-guy crooks who want in on the action. Made for TV opportunity for Asner to slum.
1973 103m/C Ed Asner, Elaine Giftos, Verne Gagne, Harold Sakata. **VHS, Beta $19.95** NWV, STE 𝄞

Wrestling Racket Girls An alleged white slavery expose, replete with seamy material that is only slightly laughable by modern terms, reducing overall camp appeal and increasing rating on boredom meter.
1939 72m/B VHS, Beta $29.95 JLT Woof!

Wrestling Women vs. the Aztec Ape An unclassifiable epic wherein female wrestlers battle a mad doctor and his Aztec robot gorilla. Dubbed badly, of course, and also known as "Doctor of Doom." Much anticipated follow-up to "Wrestling Women vs. the Aztec Mummy."
1962 75m/C MX Elizabeth Campbell, Lorena Velasquez, Armando Silvestre, Roberto Canedo; **Dir:** Rene Cardona Jr. **VHS, Beta $19.95** RHI Woof!

Wrestling Women vs. the Aztec Mummy Women, broad of shoulder, wrestle an ancient Aztec wrestler who comes to life. Furnished with a new rock soundtrack.
1959 88m/C MX Lorena Velasquez, Armando Silvestre; **Dir:** Rene Cardona Sr. **VHS, Beta $54.95** SNC, RHI, HHT Woof!

Write to Kill A young mystery writer seeks revenge when his brother is murdered by a ring of counterfeiters.

1991 (R) 120m/C Scott Valentine, Chris Mulkey, Joan Severance, G.W. Bailey, Ray Wise; **Dir:** Rubin Preuss. **VHS, LV** COL, PIA 𝄞½

Written on the Wind Sirk's frenzied, melodrama-as-high-art dissection of both the American Dream and American movies follows a Texas oil family's self-destruction through wealth, greed and unbridled lust. Exaggerated depiction of and comment on American ambition and pretension, scripted by George Zuckerman from Robert Wilder's novel. Oscar-nominated for supporting actor Stack and Best Song.
1956 99m/C Lauren Bacall, Rock Hudson, Dorothy Malone, Robert Stack, Robert Keith, Grant Williams, Edward Platt, Harry Shannon; **Dir:** Douglas Sirk. Academy Awards '56: Best Supporting Actress (Malone). **VHS, Beta $14.95** MCA, BTV 𝄞𝄞𝄞½

The Wrong Arm of the Law Loopy gangster yarn about a trio of Aussie gangsters who arrive in London and upset the local crime balance when they dress up as cops and confiscate loot from apprehended robbers. General confusion erupts among the police, the local crooks, and the imposters. Riotous and hilarious, with Sellers leading a host of familiar faces.
1963 94m/B GB Peter Sellers, Lionel Jeffries, Nanette Newman, Bernard Cribbins, Dennis Price; **Dir:** Cliff Owen. **VHS, Beta $39.95** MON 𝄞𝄞𝄞

The Wrong Box Two elderly Victorian brothers try to kill each other so that one of them may collect the large inheritance left to them. Based on a Robert Louis Stevenson novel. Well-cast black comedy replete with sight gags, many of which flop.
1966 105m/C GB Peter Sellers, Dudley Moore, Peter Cook, Michael Caine, Ralph Richardson, John Mills; **Dir:** Bryan Forbes. **VHS, Beta $59.95** COL 𝄞𝄞𝄞

The Wrong Guys Five giants of stand-up comedy star as a group of men who reunite their old boy scout pack and go camping. A crazed convict mistakes them for FBI agents. It's supposed to get zany after that, but succeeds only in being clumsy and embarassing.
1988 (PG) 86m/C Richard Lewis, Richard Belzer, Louie Anderson, Tim Thomerson, Franklin Ajaye, John Goodman, Ernie Hudson, Timothy Van Patten; **Dir:** Danny Bilson. **VHS, Beta, LV $19.95** STE, NWV 𝄞

The Wrong Man Nightclub musician Fonda is falsely accused of a robbery and his life is destroyed. Taken almost entirely from the real-life case of mild-mannered bass player "Manny" Balestrero; probes his anguish at being wrongly accused; and showcases Miles (later to appear in "Psycho") and her character's agony. Harrowing, especially following more lighthearted Hitchcock fare such as "The Trouble with Harry." Part of the "A Night at the Movies" series, this tape simulates a 1956 movie evening with a color Bugs Bunny cartoon, "A Star Is Bored," a newsreel and coming attractions for "Toward the Unknown."
1956 126m/B Henry Fonda, Vera Miles, Anthony Quayle, Nehemiah Persoff; **Dir:** Alfred Hitchcock. **VHS, Beta, LV $19.98** WAR, MLB 𝄞𝄞𝄞½

The Wrong Move A loose adaptation of Goethe's "Sorrows of Young Werther" by screenwriter Peter Handke. Justly acclaimed and engrossing, though slow. Kinski's first film. A young poet, searching for life's meaning, wanders aimlessly through Germany. In German with English subtitles.

1978 103m/C GE Nastassia Kinski, Hanna Schygulla, Ruediger Vogler, Hans-Christian Blech; **Dir:** Wim Wenders. **VHS, Beta $59.95** FCT, GLV 𝄞𝄞½

Wrong is Right A black action comedy about international terrorism, news reporting and the CIA. Connery is terrific, as usual, as a TV reporter in a head-scratching attempt at satire of our TV-influenced society.
1982 (R) 117m/C Sean Connery, Katharine Ross, Robert Conrad, George Grizzard, Henry Silva, G.D. Spradlin, John Saxon, Leslie Nielsen, Robert Webber, Rosalind Cash, Hardy Kruger, Dean Stockwell, Ron Moody, Jennifer Jason Leigh; **Dir:** Richard Brooks. **VHS, Beta, LV $79.95** COL 𝄞𝄞

The Wrong Road A young couple embark on a robbery spree, and subsequently end up in prison. On their release, they search for the loot they had previously hidden, but find the man who was holding it for them has died. Will they ever recover their ill-gotten stash?
1937 62m/B Richard Cromwell, Helen Mack, Lionel Atwill, Horace McMahon, Russ Powell, Billy Bevan, Marjorie Main, Rex Evans; **Dir:** James Cruze. **VHS, Beta $16.95** SNC 𝄞½

Wuthering Heights The first screen adaptation of Emily Bronte's romantic novel about the doomed love between Heathcliff and Cathy on the Yorkshire moors. This film dynamically captures the madness and ferocity of the classic novel, remaining possibly the greatest romance ever filmed. Excellent performances under Wyler's sure direction, particularly Olivier's, which made him a star, and Oberon in her finest hour as the exquisito but selfish Cathy. Remade twice, in 1953 (by Luis Bunuel) and in 1970. Oscar nominations: Best Picture, Best Director, Best Actor (Olivier), Best Supporting Actress (Fitzgerald), Best Screenplay, Best Interior Decoration, and Best Original Score.
1939 104m/B Laurence Olivier, Merle Oberon, David Niven, Geraldine Fitzgerald, Flora Robson, Donald Crisp, Cecil Kellaway, Leo G. Carroll, Miles Mander, Hugh Williams; **Dir:** William Wyler. Academy Awards '39: Best Black and White Cinematography; National Board of Review Awards '39: 10 Best Films of the Year; New York Film Critics Awards '39: Best Picture. **VHS, Beta, LV $19.98** HBO, SUE 𝄞𝄞𝄞𝄞

Wuthering Heights Bunuel tackles the Bronte classic during his Mexican period, and comes up with a film containing little passion, but much of the customary Bunuel manic cynicism. The 1939 version is far superior, but fans of Bunuel will probably enjoy this one. Loosely adapted; in Spanish with English subtitles. Spanish title: "Abismos de Pasion." Remade in 1970.
1953 90m/B MX Irasema Dilian, Jorge Mistral, Lilia Prado, Ernesto Alonso; **Dir:** Luis Bunuel. **VHS, Beta $24.95** XVC, MED, APD 𝄞𝄞½

Wuthering Heights The third screening of the classic Emily Bronte romance about two doomed lovers. Fuest's version features excellent photography, and Calder-Marshall's and Dalton's performances are effective, but fail even to approach the intensity and pathos of the original (or of the book). Filmed on location in Yorkshire, England.
1970 (G) 105m/C GB Anna Calder-Marshall, Timothy Dalton, Harry Andrews, Pamela Brown, Judy Cornwell, James Cossins, Rosalie Crutchley, Hilary Dwyer, Hugh Griffith, Ian Ogilvy; **Dir:** Robert Fuest. **VHS, Beta $9.95** NEG, KAR 𝄞½

X Marks the Spot When a police sergeant is killed, his detective son investigates his murder. He's drawn into a crime ring which smuggles rubber (a scarce commodity during World War II). Quick pacing enlivens standard cliches.
1942 55m/B Damian O'Flynn, Helen Parrish, Dick Purcell, Jack LaRue, Neil Hamilton, Robert E. Homans, Anne Jeffreys, Dick Wessel, Vince Barnett; *Dir:* George Sherman. **VHS, Beta** $16.95 *SNC* 𝄍𝄍

X from Outer Space A space voyage to Mars brings back a giant rampaging creature. Badly dubbed.
1967 (PG) 85m/C *JP* Eiji Okada, Peggy Neal, Toshiya Wazaki, Itoko Harada, Shinichi Yanagisawa, Franz Gruber, Keisuke Sonoi, Mike Daning, Torahiko Hamada; *Dir:* Nazui Nihonmatsu. **VHS, Beta** $19.98 *ORI* 𝄍

X: The Man with X-Ray Eyes First-rate Corman has Milland gain power to see through solid materials. Predates Little Caesars campaign.
1963 79m/C Ray Milland, Diana Van Der Vlis, Harold J. Stone, John Hoyt, Don Rickles, Dick Miller, Jonathan Haze; *Dir:* Roger Corman. **VHS, Beta, LV** $59.95 *WAR* 𝄍𝄍𝄍

X: The Unheard Music A look at the recording industry, the L.A. underground and American culture in the '80s via the adventures of the new wave rock band "X."
1986 (R) 87m/C Exene Cervanka, John Doe, Billy Zoom, D.J. Bonebrake; *Dir:* W.T. Morgan. **VHS, Beta, LV** $79.95 *MVD, FOX* 𝄍𝄍𝄍

X, Y & Zee A brassy, harsh version of the menage a trois theme, wherein the vicious wife (Taylor) of a philanderer (Caine) decides to avenge herself by seducing his mistress. Written by Edna O'Brien and an embarrassment for everyone involved. Also known as "Zee & Co."
1972 (R) 110m/C Michael Caine, Elizabeth Taylor, Susannah York, Margaret Leighton; *Dir:* Brian G. Hutton. **VHS, Beta** $69.95 *COL* 𝄍½

Xanadu Dorky star-vehicle remake (of 1947's "Down to Earth") eminently of its era, which is now better forgotten. Newton-John is a muse who descends to Earth to help two friends open a roller disco. In the process she proves that as an actor, she's a singer. Kelly attempts to soft shoe some grace into the proceedings, though he seems mystified as anyone as to why he's in the movie. Don Bluth adds an animated sequence.
1980 (PG) 96m/C Olivia Newton-John, Michael Beck, Gene Kelly, Sandahl Bergman; *Dir:* Robert Greenwald. **VHS, Beta, LV** $79.95 *MCA, MLB* Woof!

Xtro An Englishman, abducted by aliens three years before, returns to his family with a deadly, transforming disease. Slime-bucket splatter flick notable for the scene where a woman gives birth to a full-grown man, and various sex slashings.
1983 (R) 80m/C *GB* Philip Sayer, Bernice Stegers, Danny Brainin, Simon Nash, Maryam D'Abo, David Cardy, Anna Wing, Peter Mandell, Robert Fyfe; *Dir:* Harry Bromley Davenport. **VHS, Beta** *HBO* Woof!

Xtro 2: The Second Encounter The "Xtro" sequel earns a bone just for having nothing to do with the gross-out original. Second encounter is a standard "Alien" clone, as a toothy monster from another dimension pursues scientists through an underground complex.
1991 (R) 92m/C Jan-Michael Vincent, Paul Koslo, Tara Buckman; *Dir:* Harry Bromley Davenport. **VHS, LV** $89.95 *COL, NLC* 𝄍

Y Donde Esta el Muerto? (And... Where Is the Dead) This mystery/comedy centers around a weird gathering of people who are trying to find out if a man is really dead.
1987 73m/C *MX* Yvonne Fraissinet, Ricardo Fernandez, Nilda Munoz. **VHS, Beta** $69.95 *MAD* 𝄍𝄍½

The Yakuza An ex-G.I. (Mitchum) returns to Japan to help an old army buddy find his kidnapped daughter. He learns the daughter has been kidnapped by the Japanese version of the Mafia (the Yakuza) and he must call on old acquaintances to help free her. A westernized oriental gangster drama with a nice blend of buddy moments, action, ancient ritual, and modern Japanese locations. Also known as "Brotherhood of the Yakuza."
1975 (R) 112m/C Robert Mitchum, Richard Jordan, Ken Takakura, Brian Keith, Herb Edelman; *Dir:* Sydney Pollack. **VHS, Beta, LV** $24.98 *WAR, PIA* 𝄍𝄍½

Yanco A young Mexican boy makes visits to an island where he plays his homemade violin. Unfortunately, no one can understand or tolerate his love for music, save for an elderly violinist, who gives him lessons on an instrument known as "Yanco." No dialogue. Not exactly riveting, but sensitively played.
1964 95m/B *MX* Ricardo Ancona, Jesus Medina, Maria Bustamante. **VHS, Beta** $19.95 *WFV, FST, DVT* 𝄍𝄍

A Yank in Australia Australian farce/war mystery/romance/documentary tells the story of two World War II newspaper reporters whose ship gets torpedoed, stranding them in the Australian outback, where they manage to foil a Japanese invasion.
1943 56m/B *AU* **VHS, Beta, 8mm** $24.95 *VYY* 𝄍𝄍𝄍

Yank in Libya No, not the American bombing raid of Khadafy, rather a low-budget WWII adventure pitting an American man and an English girl against Libyan Arabs and Nazis.
1942 65m/B Walter Woolf King, John Woodbury, H.B. Warner, Parkyakarkus. **VHS, Beta** *UHV* 𝄍½

A Yank in the R.A.F. Power's enthusiastic perf as a brash American pilot boosts this dated WWII adventure. He and his British allies seem more concerned over who gets showgirl Betty Grable than in the Nazis, but climactic air attacks retain excitement. Produced by Darryl F. Zanuck to drum up American support for embattled Britain and France.
1941 98m/B Tyrone Power Sr., Betty Grable, John Sutton, Reginald Gardiner; *Dir:* Henry King. **VHS, Beta** $19.98 *FOX, FCT* 𝄍𝄍½

Yankee Clipper Deceit, treachery, and romance are combined in this depiction of a fierce race from China to New England between the American ship Yankee Clipper and the English ship Lord of the Isles. Silent.
1927 68m/B William Boyd, Elinor Fair, Frank Coghlan Jr., John Miljan, Walter Long; *Dir:* Rupert Julian. **VHS, Beta** $29.95 *CCB, GPV* 𝄍𝄍½

Yankee Doodle A blacksmith and two Militiamen help Paul Revere complete his midnight ride.
198? 30m/C **VHS, Beta** $6.95 *STE, PSM* 𝄍𝄍½

Yankee Doodle in Berlin A spoof of World War I dramas, with the hero dressing up as a woman and seducing the Kaiser into giving up his war plans. Typical Sennett slapstick with a music and effects score added to the silent film.

1919 60m/B Ford Sterling, Ben Turpin, Marie Prevost, Bothwell Browne. **VHS, Beta** $29.95 *CCB* 𝄍𝄍½

Yankee Doodle Dandy Nostalgic view of the Golden Era of show business and the man who made it glitter—George M. Cohan. His early days, triumphs, songs, musicals and romances are brought to life by the inexhaustible Cagney in a rare and wonderful song-and-dance performance. Told in flashback, covering the Irishman's struggling days as a young song writer and performer to his salad days as the toast of Broadway. Cagney, never more charismatic, dances up a storm, reportedly inventing most of the steps on ths spot. Classic tunes include "Yankee Doodle Dandy," "Give My Regards to Broadway," "Over There," and "You're a Grand Old Flag." In addition to its three Oscars, nominated for Best Picture, Best Supporting Actor (Huston), Best Director, Original Story, and Film Editing. Available colorized.
1942 126m/B James Cagney, Joan Leslie, Walter Huston, Richard Whorf, Irene Manning, Rosemary DeCamp, Jeanne Cagney, S.Z. Sakall, Walter Catlett, Frances Langford, Eddie Foy Jr., Michael Curtiz; *Dir:* Michael Curtiz. Academy Awards '42: Best Actor (Cagney), Best Sound, Best Musical Score. **VHS, Beta, LV** $29.98 *MGM, FOX, FCT* 𝄍𝄍𝄍𝄍

Yanks An epic-scale but uneventful drama depicts the legions of American soldiers billeted in England during WWII, and their impact—mostly sexual—on the staid Britons. No big story, no big deal, despite a meticulous recreation of the era.
1979 139m/C Richard Gere, Vanessa Redgrave, William Devane, Lisa Eichhorn, Rachel Roberts, Chick Vennera, Arlen Dean Snyder, Annie Ross; *Dir:* John Schlesinger. **VHS, LV** $79.95 *MCA, PIA* 𝄍𝄍

Ye-Yo A soldier returning home finds his wife betraying him, kills her, and embarks on an unstoppable murdering spree.
197? 80m/C Ricardo Reyes, Eppy Dueno. **VHS, Beta** *MAV* 𝄍

Year of the Dragon Polish police Captain Stanley White of the NYPD vows to neutralize the crime lords running New York's Chinatown. Brilliant cinematography, well-done action scenes with maximum violence, a racist hero you don't want to root for, murky script, and semi-effective direction are the highlights of this tour through the black market's underbelly and hierarchy. Based on Robert Daley's novel and adapted by Oliver Stone and Cimino.
1985 (R) 136m/C Mickey Rourke, John Lone, Ariane, Leonard Termo, Raymond J. Barry; *Dir:* Michael Cimino. **VHS, Beta, LV** $19.95 *MGM* 𝄍𝄍½

Year of the Gun McCarthy is an American journalist in Rome who begins a novel based on the political instability around him, using the names of real people in his first draft. Soon, ambitious photojournalist Stone wants to collaborate with him, and the Red Brigade terrorist group wants to "remove" anyone associated with the book. Although failing on occassion to balance the thin line it establishes between reality and perception, "Gun" aspires to powerful drama, offering a realistic look at the lives and priorities of political terrorists. Love scenes between Stone and McCarthy are torrid.
1991 (R) 111m/C Andrew McCarthy, Sharon Stone, Valeria Golino, John Pankow; *Dir:* John Frankenheimer. **VHS, LV, 8mm** *COL* 𝄍𝄍½

The Year of Living Dangerously
Political thriller with a love interest features Gibson as an Australian journalist covering a political story in Indonesia, circa 1965. During the coup against President Sukarno, he becomes involved with a British attache (Weaver) at the height of the bloody fighting and rioting in Jakarta. Oscar-winning Hunt is excellent as male photographer Billy Swan, central to the action as the moral center. Rumored to be based on the activities of CNN's Peter Arnett, although the original source is a novel by C.J. Koch, who reportedly collaborated/battled with Weir on the screenplay. A fascinating, suspenseful film, set up brilliantly by Weir and shot on location in the Philippines (then moved to Sydney after cast and crew were threatened). The first Australian movie financed by a U.S. studio.
1982 (PG) 114m/C **AU** Mel Gibson, Sigourney Weaver, Linda Hunt, Michael Murphy; *Dir:* Peter Weir. Academy Awards '83: Best Supporting Actress (Hunt). **VHS, Beta, LV** $19.98 *MGM, BTV* 🎬🎬🎬½

The Year My Voice Broke Above average adolescent drama: a girl breaks a boy's heart by getting pregnant by a tougher, older boy; then leaves town. Blues-inducing, explicit, and not pleasant; but good acting from newcomers carries the day.
1987 (PG-13) 103m/C **AU** Noah Taylor, Loene Carmen, Ben Mendelsohn; *Dir:* John Duigan. Australian Film Institute '87: Best Film. **VHS, Beta, LV** $89.95 *IVE* 🎬🎬🎬

Year of the Quiet Sun A poignant, acclaimed love story about a Polish widow and an American soldier who find each other in the war-torn landscape of 1946 Europe. Beautifully rendered, with a confident sense of time and place, making this much more than a simple love story. In Polish with English subtitles.
1984 (PG) 106m/C **PL GE** Scott Wilson, Maja Komorowska; *Dir:* Krzysztof Zanussi. Venice Film Festival '84: Golden Lion (Best Film). **VHS, Beta, LV** $29.98 *KRT, WAR* 🎬🎬🎬½

Year Without Santa Claus A puppet-animated Christmas special. Really a standout in the children's Christmas parade. Great musical numbers and fun for adults, too. Watch for Heatmeiser and Coldmeiser - pure camp! Made for television.
196? 50m/C **Voices:** Mickey Rooney, Shirley Booth. **VHS, Beta** $14.95 *LIV, LTG* 🎬🎬🎬½

The Yearling This family classic is a tearjerking adaptation of the Marjorie Kinnan Rawlings novel about a young boy's love for a yearling fawn during the post civil-war era. His father's encouragement and his mother's bitterness play against the story of unqualified love amid poverty and the boy's coming of age. Wonderful footage of Florida. In addition to claiming three Oscars, nominated for Best Picture, Best Actor (Peck), Best Actress (Wyman), Best Director, and Film Editing, Jarman was awarded a special Oscar.
1946 128m/C Gregory Peck, Jane Wyman, Claude Jarman Jr., Chill Wills, Henry Travers, Jeff York, Forrest Tucker, June Lockhart, Margaret Wycherly; *Dir:* Clarence Brown. Academy Awards '46: Best Art Direction/Set Decoration (Color), Best Color Cinematography, Best Interior Decoration; New York Times Ten Best List '47: Best Film. **VHS, Beta, LV** $19.98 *KUI, MGM, TLF* 🎬🎬🎬½

Yellow Cargo Ingenious ''B'' movie about a pair of undercover agents who blow the lid off a smuggling scam. It seems that the smugglers have been masquerading as a

movie crew, and use disguised Chinese ''extras'' to transport their goods. The agents pose as actors to infiltrate the gang.
1936 70m/B Conrad Nagel, Eleanor Hunt, Vince Barnett, Jack LaRue, Claudia Dell; **W/Dir:** Crane Wilbur. **VHS, Beta** $16.95 *SNC* 🎬🎬

Yellow Emanuelle Exploitation film about a woman bound to an arranged marriage who falls in love with someone else.
197? 105m/C Chai Lee, Giuseppe Pambiere, Ilona Staller. **VHS, Beta** $29.95 *UNI* 🎬

Yellow Fever A 19th-century Cajun man breaks a yellow fever quarantine on a village in order to visit his family, with disastrous results. In Cajun French with subtitles.
1979 30m/C **FR** *Dir:* Glen Pitre. **VHS, Beta** *FCT* 🎬

Yellow Hair & the Fortress of Gold A part-Indian princess and her sidekick fight bad guys and seek gold. Well-meaning, self-conscious parody. Also known as ''Yellow Hair and the Pecos Kid.''
1984 (R) 102m/C **SP** Laurene Landon, Ken Robertson; *Dir:* Matt Cimber. **VHS, Beta** $79.98 *LIV, LTG* 🎬½

The Yellow Rose of Texas Rogers works as an undercover insurance agent (he's singing on a river boat) to clear the name of an old man falsely accused of a stagecoach robbery. The usual melodious Rogers lead.
1944 55m/B Roy Rogers, Dale Evans, Grant Withers, Harry Shannon; *Dir:* Joseph Kane. **VHS, Beta** $19.95 *NOS, VCN, CAB* 🎬🎬

Yellow Submarine The acclaimed animated fantasy based on a plethora of mid-career Beatles songs, sees the Fab Four battle the Blue Meanies for the sake of Sgt. Pepper, the Nowhere Man, Strawberry Fields, and Pepperland. The first full-length British animated feature in 14 years features a host of talented cartoonists. Fascinating LSD-esque animation and imagery. Incidental music by George Martin, with a script assist by Love Story's Erich Segal.
1968 (G) 87m/C **GB** *Dir:* Dick Emery, George Duning; **Voices:** Paul McCartney, John Lennon, George Harrison, Ringo Starr. **VHS, Beta, LV** $19.95 *MVD, MGM* 🎬🎬🎬½

Yellowbeard An alleged comedy with a great cast who wander about with little direction. Follows the efforts of an infamous pirate (Chapman) to locate a buried treasure using the map tattooed on the back of his son's head. The final role for Feldman who died during production. Written by Chapman, Cook, and Bernard McKenna.
1983 (PG) 97m/C Graham Chapman, Peter Boyle, Cheech Marin, Thomas Chong, Peter Cook, Marty Feldman, Martin Hewitt, Michael Hordern, Eric Idle, Madeline Kahn, James Mason, John Cleese, Susannah York, David Bowie; *Dir:* Mel Damski. **VHS, Beta, LV** $14.95 *VES* 🎬🎬

Yellowneck A handful of Confederate Army soldiers desert, hoping to cross the Florida Everglades and eventually reach Cuba. The swamp takes its toll, however, and one by one the men fall by the wayside. A sole survivor reaches the coast. Will the escape boat be waiting for him?
1955 83m/C **SP** Lin McCarthy, Stephen Courtleigh, Berry Kroeger, Harold Gordon, Bill Mason; *Dir:* R. John Hugh. **VHS** $29.95 *FCT* 🎬

Yellowstone An ex-con is murdered at Yellowstone National Park, at the site of a hidden cache of money, of which assorted folks are in search. Uncomplicated, well-meaning mystery.

1936 65m/B Andy Devine, Ralph Morgan, Judith Barrett; *Dir:* Arthur Lubin. **VHS, Beta, 8mm** $14.95 *NOS, DVT, WKV* 🎬🎬

Yentl The famous Barbra adaptation of Isaac Bashevis Singer's story set in 1900s Eastern Europe about a Jewish girl who masquerades as a boy in order to study the Talmud, and who becomes enmeshed in romantic miscues. Lushly photographed, with a repetitive score by Alan and Marilyn Bergman and Michel Legrand that nevertheless won an Oscar. Singer was reportedly appalled by the results of Streisand's hyper-controlled project.
1983 (PG) 134m/C Barbra Streisand, Mandy Patinkin, Amy Irving, Nehemiah Persoff, Steven Hill, Allan Corduner, Ruth Goring, David DeKeyser, Bernard Spear; *Dir:* Barbra Streisand. Academy Awards '83: Best Original Score; Golden Globe Awards '84: Best Film—Musical/Comedy. **VHS, Beta, LV** $39.98 *MGM, IME* 🎬🎬½

Yes, Giorgio Pavarotti in his big-screen debut. He has a big advantage over other non-actors in similar ventures (e.g., Hulk Hogan): He can sing—and how! Lame plot (famous opera star falls for lady doctor) is the merest excuse for the maestro to belt out ''If We Were in Love,'' ''I Left My Heart in San Francisco,'' and arias by Verdi, Donizetti, and Puccini.
1982 (PG) 111m/C Luciano Pavarotti, Kathryn Harrold, Eddie Albert, James Hong; *Dir:* Franklin J. Schaffner. **VHS, Beta, LV** $79.95 *MGM* 🎬½

Yes, Sir, Mr. Bones Entertaining but dated musical about a boy who wanders into a rest home and inspires the old folks there to reminisce about their days as riverboat minstrels. Songs include ''I Want to Be a Minstrel Man,'' ''Stay Out of the Kitchen,'' and ''Is Your Rent Paid up in Heaven.'' Written by Ormond.
1951 60m/B Pete Daily, Jimmy O'Brien, Sally Anglim, Cotton Watts, Chick Watts, Chet Davis, F.E. Miller, Scatman Crothers; *Dir:* Ron Ormond. **VHS, Beta** *WGE* 🎬½

The Yesterday Machine Camp sci-fi: A mad doctor tries to bring back Hitler. Don't worry, though: the good guys win. Predictable, dumb drivel.
1963 85m/B Tim Holt, James Britton, Jack Herman; *Dir:* Russ Marker. **VHS, Beta** $19.95 *SNC, VCD* Woof!

Yesterday, Today and Tomorrow Trilogy of comic sexual vignettes featuring Loren and her many charms. She plays a black marketeer, a wealthy matron, and a prostitute. Funny, and still rather racy. Loren at her best, in all senses; includes her famous striptease for Mastroianni.
1964 119m/C **IT FR** Sophia Loren, Marcello Mastroianni, Tony Pica, Giovanni Ridolfi; *Dir:* Vittorio De Sica. Academy Awards '64: Best Foreign Language Film. **VHS** $29.95 *JEF, HHE* 🎬🎬🎬

Yesterday's Hero A fading soccer star finds his career and love-life on the upswing. Poor production with equally lame pop music score. Written by Jackie Collins.
1979 95m/C **GB** Ian McShane, Suzanne Somers, Adam Faith, Paul Nicholas, Glynis Barber, Sandy Ratcliff; *Dir:* Neil Leifer. **VHS, Beta** $79.98 *CGH, HHE* 🎬½

Yesterday's Witness Documentary examining the history of the newsreel and also features interviews with Ed Herlihy, Lowell Thomas and Harry Von Zello.
1983 52m/C **VHS, Beta, LV** $24.95 *PAV* 🎬½

Yidl Mitn Fidl (Yiddle with His Fiddle) Disguised as a man, a young woman travels about the countryside with a group of musicians, revealing her female identity only after falling for a man. One of Picon's finest roles and Green's biggest successes. Vaudevillian fare in Yiddish with English subtitles. Songs include "Yiddle with His Fiddle" and "Arye with His Bass."
1936 92m/B PL Molly Picon, Simche Fostel, Max Bozyk, Leon Liebgold; **Dir:** Joseph Green. **VHS, Beta** $89.95 ERG, DVT 🎬🎬

Yin & Yang of Mr. Go A CIA operative must retrieve the stolen plans of an awesome weapons system. Set in Hong Kong, features a strong cast in a tangled tale of no particular merit. In addition to starring and directing, Meredith scripted this unworthy spoof of Oriental intrigue flicks.
1971 (R) 89m/C Jeff Bridges, James Mason, Broderick Crawford, Burgess Meredith; **Dir:** Burgess Meredith. **VHS, Beta** $9.95 NO 🎬

Yodelin' Kid From Pine Ridge Gene Autry tries to stop a war between cattlemen and woodsmen in Georgia, in this standard horse opera, notable only for its location.
1937 59m/B Gene Autry, Smiley Burnette, Betty Bronson, Tennessee Ramblers, Charles Middleton, Art Mix; **Dir:** Joseph Kane. **VHS, Beta** $19.95 NOS, VCN, DVT 🎬

Yog, Monster from Space When a spaceship crashes somewhere near Japan, the aliens in it create monsters out of ordinary critters in order to destroy all the cities. A promoter gets a gleam in his eye and sees the potential for a vacation spot featuring the viscious creatures. Standard Japanese monster flick utilizing the usual out-of-synch dub machine. Dubbed.
1971 (G) 105m/C JP Akira Kubo, Yoshio Tsuchiya; **Dir:** Inoshiro Honda. **VHS, Beta** $49.95 SNC 🎬

Yogi's First Christmas Yogi Bear joins his friend, Huckleberry Hound, and others for a musical celebration of Christmas at the Jellystone Lodge.
1980 100m/C Dir: Ray Patterson; **Voices:** Daws Butler, Don Messick, John Stephenson, Janet Waldo, Hal Smith. **VHS, Beta, LV** $19.95 HNB, IME 🎬

Yojimbo Two clans vying for political power bid on the services of a samurai (Mifune) who comes to town. The samurai sells his services to both parties, with devastating results for all. The original Japanese version is available with English subtitles or dubbed into English. Re-made by Sergio Leone as the western "A Fist Full of Dollars."
1962 110m/B JP Toshiro Mifune, Eijiro Tono, Suuzu Yamda; **Dir:** Akira Kurosawa. **VHS, Beta, LV, 8mm** $29.98 NOS, SUE, HHT 🎬🎬🎬🎬

Yol Five Turkish prisoners are granted temporary leave to visit their families. An acclaimed, heartfelt film, written by Yilmaz Guney while he himself was in prison. A potent protest against totalitarianism. In Turkish with English subtitles.
1982 (PG) 126m/C TU Tarik Akan, Serif Sezer; **Dir:** Serif Goren. Cannes Film Festival '82: Best Film. **VHS, Beta** $19.98 COL 🎬🎬🎬½

Yolanda & the Thief A charming, forgotten effort from the Arthur Freed unit about a con man who convinces a virginal South American heiress that he is her guardian angel. Songs by Freed and Harry Warren include "Yolanda," "Coffee Time" and a lengthy Dali-esque ballet built around "Will You Marry Me?"

1945 109m/C Fred Astaire, Lucille Bremer, Leon Ames, Mildred Natwick; **Dir:** Vincente Minnelli. **VHS, Beta** $29.95 MGM 🎬🎬½

Yongary, Monster of the Deep A giant burrowing creature is causing earthquakes and generally ruining scores of Japanese models. Dubbed.
1967 (PG) 79m/C Oh Young Il, Nam Chung-Im; **Dir:** Kim Ki-dak. **VHS, Beta** $19.98 ORI 🎬

Yor, the Hunter from the Future Lost in a time warp where the past and the future mysteriously collide, Yor sets out on a search for his real identity, with his only clue a golden medallion around his neck.
1983 (PG) 88m/C IT Reb Brown, Corinne Clery, John Steiner; **Dir:** Anthony M. Dawson. **VHS, Beta, LV** $79.95 COL 🎬

You Are Not Alone Boys in a small parochial school grow up and try to draw a defining line between friendship and love. In Danish with English subtitles.
1982 90m/C DK **VHS, Beta** FCT 🎬🎬

You Can't Cheat an Honest Man The owner of a misfit circus suffers a variety of headaches including the wisecracks of Charlie McCarthy. Contains Field's classic ping-pong battle and some of his other best work.
1939 79m/B W.C. Fields, Edgar Bergen, Constance Moore, Eddie Anderson, Mary Forbes, Thurston Hall; **Dir:** George Marshall. **VHS, Beta** $14.95 KRT 🎬🎬🎬

You Can't Fool Your Wife Ball in two roles gives zip to this otherwise ordinary marital comedy. Previously blah hubby has a fling; previously blah better half wins him back with glamour.
1940 69m/C Lucille Ball, James Ellison, Robert Coote, Emma Dunn; **Dir:** Ray McCarey. **VHS, Beta** $19.98 RKO 🎬🎬

You Can't Hurry Love A jilted-at-the-altar Ohio bachelor moves to Los Angeles and flounders in the city's fast-moving fast lane. A dull film with Fonda, daughter of "Easy Rider" Peter, playing a minor role.
1988 (R) 92m/C David Leisure, Scott McGinnis, Sally Kellerman, Kristy McNichol, Charles Grodin, Anthony Geary, Bridget Fonda, David Packer, Frank Bonner; **Dir:** Richard Martini. **VHS, Beta, LV** $79.98 LIV, VES 🎬

You Can't Take It with You The Capra version of the Kaufman-Hart play about an eccentric New York family and their non-conformist houseguests. Alice Sycamore (Arthur), the stable family member of an offbeat clan of free spirits, falls for Tony Kirby (Stewart), the down-to-earth son of a snooty, wealthy and not always quite honest family. Amidst the confusion over this love affair, the two families rediscover the simple joys of life. In addition to winning two Oscars, the film received nominations for Best Supporting Actress (Byington), Best Cinematography, Best Screenplay, Best Film Editing, and Best Sound Recording.
1938 127m/B James Stewart, Jean Arthur, Lionel Barrymore, Spring Byington, Edward Arnold, Mischa Auer, Donald Meek, Samuel S. Hinds, Ann Miller, H.B. Warner, Halliwell Hobbes, Dub Taylor, Mary Forbes, Eddie Anderson, Harry Davenport, Lillian Yarbo; **Dir:** Frank Capra. Academy Awards '38: Best Director (Capra), Best Picture. **VHS, Beta, LV** $19.95 COL, PIA, BTV 🎬🎬🎬½

You Can't Take It with You Taped performance of the Kaufman and Hart comedy about the strange pastimes of the Sycamore family who must behave themselves to

impress their daughter's boyfriend's stuffy family.
1984 116m/C Colleen Dewhurst, James Coco, Jason Robards Jr., Elizabeth Wilson, George Rose; **Dir:** Ellis Raab. **VHS, Beta, LV** $59.98 LIV, VES 🎬🎬🎬

You Light Up My Life Sappy sentimental story of a young woman trying to break into the music business as a singer. Debbie Boone's title song was a radio hit and the basic premise for the movie.
1977 (PG) 91m/C Didi Conn, Michael Zaslow, Melanie Mayron; **Dir:** Joseph Brooks. Academy Awards '79: Best Song ("You Light Up My Life"). **VHS, Beta** $19.95 COL 🎬½

You Must Remember This When Uncle Buddy (Guillaume) receives a mysterious trunk, Ella's curiosity gets the best of her. She opens the trunk to discover a number of old movies made by W.B. Jackson—Uncle Buddy. Ella takes the films to a movie archive to find out about her uncle's past as an independent black filmmaker. After researching the history of black cinema, Ella convinces her uncle to be proud of his contribution to the film world. Includes a viewers' guide. A WonderWorks production.
199? 110m/C Robert Guillaume, Tim Reid, Daphne Maxwell Reid. **VHS** $29.95 PME 🎬½

You are Not I After a multi-car accident a young woman escapes from psychiatric confinement and goes to her sister's house.
1984 50m/B **VHS, Beta** NEW 🎬½

You Only Live Once Ex-con Fonda wants to mend his ways and tries to cross into Canada with his girlfriend in tow. Impressively scripted, but a glum and dated Depression-era tale.
1937 86m/C Henry Fonda, Sylvia Sidney, Ward Bond, William Gargan, Barton MacLane, Margaret Hamilton; **Dir:** Fritz Lang. **VHS, Beta** $19.95 MON 🎬🎬½

You Only Live Twice 007 travels to Japan to take on arch-nemesis Blofeld, who has been capturing Russian and American spacecraft in an attempt to start World War III. Great location photography; theme sung by Nancy Sinatra. Implausible plot, however, is a handicap, even though this is Bond.
1967 115m/C GB Sean Connery, Mie Hama, Akiko Wakabayashi, Tetsuro Tamba, Karin Dor, Charles Gray, Donald Pleasence, Tsai Chin, Bernard Lee, Lois Maxwell, Desmond Llewelyn; **Dir:** Lewis Gilbert. **VHS, Beta, LV, CDV** $19.95 MGM, TLF 🎬🎬

You Talkin' to Me? Fledgling actor who idolizes De Niro moves to the West Coast for his big break. He fails, so he dyes his hair blond and digs the California lifestyle. Embarrassingly bad.
1987 (R) 97m/C Chris Winkler, Jim Youngs, Faith Ford; **Dir:** Charles Winkler. **VHS, Beta** $79.95 MGM 🎬

You Were Never Lovelier Charming tale of a father who creates a phony Romeo to try to interest his daughter in marriage. Astaire appears and woos Hayworth in the flesh. Splendid score by Jerome Kern and Johnny Mercer. The dancing, of course, is superb, and Hayworth is stunning.
1942 98m/B Fred Astaire, Rita Hayworth, Leslie Brooks, Xavier Cugat, Adolphe Menjou, Larry Parks; **Dir:** William A. Seiter. **VHS, Beta, LV** $19.95 COL 🎬🎬🎬½

You'll Find Out A comic mix of music and mystery as Kay Kyser and his Band, along with a debutante in distress, are terrorized by Lugosi, Karloff, and Lorre.

1940 97m/B Peter Lorre, Kay Kyser, Boris Karloff, Bela Lugosi, Dennis O'Keefe, Helen Parrish; *Dir:* David Butler. **VHS, Beta $59.95** *HHT, MLB* 🎬🎬

You'll Never Get Rich A Broadway dance director is drafted into the Army, where his romantic troubles cause him to wind up in the guardhouse more than once. He of course gets the girl. Great Cole Porter songs include "Since I Kissed My Baby Goodnight" and "The Astairable Rag." Exquisitely funny.
1941 88m/B Fred Astaire, Rita Hayworth, Robert Benchley; *Dir:* Sidney Lanfield. **VHS, Beta, LV $19.95** *COL* 🎬🎬🎬

Young Aphrodites Two young teenagers of a primitive Greek tribe discover sexuality. An acclaimed film retelling the myth of Daphnis and Chloe. Narrated in English.
1963 87m/B *GR Dir:* Nikos Koundouros. Berlin Film Festival '63: Best Director; International Film Critics Awards '63: Best Film. **VHS, Beta $29.98** *SUE* 🎬🎬½

Young Avengers/18 Jade Pearls Two full-length Kung Fu films packed with action, and an emphasis on self-defense.
19?? 180m/C **VHS $9.99** *VTR* 🎬🎬½

Young Bill Hickok A Rogers vehicle in the form of a very fictionalized biography of the famous gunfighter.
1940 54m/B Roy Rogers, George "Gabby" Hayes. **VHS, Beta $9.99** *NOS, DVT* 🎬🎬

Young Bing Crosby Crosby stars in short, delightful musical comedies by Mack Sennett made before Crosby entered feature-length musical comedy: "Crooner's Holiday," "Blue of the Night," and "Bing, Bing, Bing."
193? 39m/B Bing Crosby. **VHS, Beta $24.95** *VYY* 🎬🎬½

Young Blood A cowboy robs from the rich to help the poor. He's also interested in a foreign actress having trouble bonding with the townfolk. Lots of old fashioned western violence.
1933 61m/B Bob Steele, Charles King, Helen Foster; *Dir:* Phil Rosen. **VHS, Beta $19.95** *NOS, DVT* 🎬🎬

The Young Bruce Lee Bruce learns and uses the deadly skills that made him famous while only knee-high to a grasshopper.
19?? 90m/C Han Kwok Choi, Bruce Lee. **VHS, Beta $19.95** *OCE* 🎬🎬

Young Buffalo Bill Buffalo Bill battles the Spanish land-grant patrons.
1940 54m/B Roy Rogers, George "Gabby" Hayes, Pauline Moore; *Dir:* Joseph Kane. **VHS, Beta $19.95** *NOS, DVT* 🎬🎬

Young Caruso A dramatic biography of legendary tenor Enrico Caruso, following his life from childhood poverty in Naples to the beginning of his rise to fame. Dubbed in English.
1951 78m/B Gina Lollobrigida, Ermanno Randi, Mario del Monaco. **VHS, Beta $29.95** *MVD, NOS, VYY* 🎬½

Young Catherine Made for TV account of Russia's strongest female ruler, the girl who would be Catherine the Great. Starstudded cast and excellent production values. Script and strong cast make Ormond look like a lightweight. Filmed in Leningrad. Also available in 186-minute version.

1991 150m/C Vanessa Redgrave, Christopher Plummer, Marthe Keller, Franco Nero, Julia Ormond, Maximilian Schell, Reece Dinsdale, Mark Frankel; *Dir:* Michael Anderson Sr. **VHS, Beta $89.98** *TTC* 🎬🎬🎬

Young Charlie Chaplin A British-made TV biography of the legendary silent screen comic.
1988 160m/C Ian McShane, Twiggy. **VHS, Beta $39.99** *HBO* 🎬🎬

Young Doctors in Love Spoof of medical soap operas features a chaotic scenario at City Hospital, where the young men and women on the staff have better things to do than attend to their patients. Good cast keeps this one alive, though many laughs are forced. Includes cameos by real soap star, including then General Hospital star Moore. Director Marshall is from "Laverne and Shirley" fame.
1982 (R) 95m/C Dabney Coleman, Sean Young, Michael McKean, Harry Dean Stanton, Hector Elizondo, Patrick Macnee, Pamela Reed, Saul Rubinek, Demi Moore; *Dir:* Garry Marshall. **VHS, Beta, LV $29.95** *VES* 🎬🎬

Young Dragon The Young Dragon's girlfriend is kidnapped by the underworld society, and he sets out to rescue her single-handedly.
1977 (R) 90m/C Yang Sze, Lo Lai, Yung Man Chi, Wong Ching, Nora Miao. **VHS, Beta $54.95** *SUN* 🎬½

Young Dragons: The Kung Fu Kids A kung fu master teaches his three grandsons the secrets of martial arts, which comes in handy on a trip to the big city. Followed by a sequel.
1987 95m/C *Dir:* Chang Mei Jun. **VHS, Beta $59.95** *CAN* 🎬

Young Dragons: The Kung Fu Kids 2 The Kung Fu Kids must choose between their beloved grandfather and their long-missing kung fu grandmother.
1989 100m/C *Dir:* Chen Che Hwa. **VHS, Beta** *CAN* 🎬🎬

Young Eagles Serial with 12 chapters, 13 minutes each.
1934 156m/B Bobby Cox, Jim Vance, Carter Dixon. **VHS, Beta $49.95** *NOS, VCN, MLB* 🎬🎬

Young Einstein A goofy, irreverent Australian farce starring, directed, co-scripted and co-produced by Serious, depicting Einstein as a young Outback clod who splits beer atoms and invents rock and roll. Winner of several Aussie awards. Fun for the kids.
1989 (PG) 91m/C *AU* Yahoo Serious, Odile Le Clezio, John Howard, Pee Wee Wilson, Su Cruickshank; *Dir:* Yahoo Serious. **VHS, Beta, LV $19.95** *WAR* 🎬🎬

Young Frankenstein Young Dr. Frankenstein (Wilder), a brain surgeon, inherits the family castle back in Transylvania. He's skittish about the family business, but when he learns his grandfather's secrets, he becomes obsessed with making his own monster. Wilder and monster Boyle make a memorable song-and-dance team to Irving Berlin's "Puttin' on the Ritz," and Hackman's cameo as a blind man is inspired. Garr ("What knockers!" "Oh, sank you!") is adorable as a fraulein, and Leachman ("He's vass my-boyfriend!") is wonderfully scary. Wilder saves the creature in with a switcheroo, in which the doctor ends up with a certain monster-sized body part. Hilarious parody.

1974 (PG) 108m/B Peter Boyle, Gene Wilder, Marty Feldman, Madeline Kahn, Cloris Leachman, Teri Garr, Kenneth Mars, Gene Hackman; *Dir:* Mel Brooks. **VHS, Beta, LV $19.98** *FOX* 🎬🎬🎬

Young & Free Following the death of his parents, a young man must learn to face the perils of an unchartered wilderness alone. Ultimately he must choose between returning to civilization, or remain with his beloved wife and life in the wild.
1978 87m/C Erik Larsen; *Dir:* Keith Larsen. **VHS, Beta $59.95** *MON* 🎬½

The Young Graduates Hormonally imbalanced teens come of age in spite of meandering plot. Features "Breaking Away" star Christopher in big screen debut.
1971 (PG) 99m/C Patricia Wymer, Steven Stewart, Gary Rist, Bruce Kirby, Jennifer Ritt, Dennis Christopher; *Dir:* Robert Anderson. **VHS $19.95** *ACA* 🎬

The Young and the Guilty Two star-crossed teenagers are frustrated in their romance when their parents find one of their love letters and hit the ceiling. Told not to see each other again, the two take to sneaking around, which adds an edge to their relationship. Moving performances by the two young leads.
1958 65m/B Phyllis Calvert, Andrew Ray, Edward Chapman, Janet Munro, Campbell Singer; *Dir:* Peter Cotes. **VHS $19.95** *NOS* 🎬🎬

Young Guns A sophomoric Wild Bunch look-alike that ends up resembling a western version of the Bowery Boys. Provides a portrait of Billy the Kid and his gang as they move from prairie trash to demi-legends. Features several fine performances by a popular group of today's young stars.
1988 (R) 107m/C Emilio Estevez, Kiefer Sutherland, Lou Diamond Phillips, Charlie Sheen, Casey Siemaszko, Dermot Mulroney, Terence Stamp, Terry O'Quinn, Jack Palance; *Dir:* Christopher Cain. **VHS, Beta, LV $14.98** *LIV, VES* 🎬🎬½

Young Guns 2 Brat Pack vehicle neo-Western sequel about Billy the Kid (Estevez) and his gang. This time they're making a run for the border with the law on their trail. Best Song Oscar nomination for Jon Bon Jovi's "Blaze of Glory."
1990 (PG-13) 105m/C Emilio Estevez, Kiefer Sutherland, Lou Diamond Phillips, Christian Slater, William L. Petersen, Alan Ruck, R.D. Call, James Coburn, Balthazar Getty, Jack Kehoe, Rob Knepper, Jenny Wright, Tracy Walter, Ginger Lynn Allen, Jon Bon Jovi; *Dir:* Geoff Murphy. **VHS, LV $19.98** *FXV, BTV, CCB* 🎬🎬

The Young in Heart A lonely, old woman allows a family of con-artists into her life for companionship. Impressed by her sweet nature, the parasitic brood reforms. The cute comedy was a real crowd-pleaser in its day, especially after the bittersweet ending was replaced with a happier variety. Based on the novel "The Gay Banditti" by I.A.R. Wylie.
1938 90m/C Janet Gaynor, Douglas Fairbanks Jr., Paulette Goddard, Roland Young, Billie Burke, Minnie Dupree, Richard Carlson; *Dir:* Richard Wallace. **VHS $9.99** *VTR* 🎬🎬🎬

Young at Heart Fanny Hurst's lighthearted tale of a cynical hard-luck musician who finds happiness when he falls for a small-town girl. A remake of the 1938 "Four Daughters." Songs include the title tune, "You, My Love," and "Just One of Those Things."

1954 117m/C Frank Sinatra, Doris Day, Gig Young, Ethel Barrymore, Dorothy Malone, Robert Keith, Elisabeth Fraser, Alan Hale Jr.; *Dir:* Gordon Douglas. VHS $19.98 REP *ZZ* ½

Young Hero The "Godfather of Chaos" has a town squashed under his thumb in this martial arts thriller. Enter Chang Ching, the new sheriff, and goodbye corruption. Also known as "Deadly Strike."
19?? 92m/C Huand Lund. VHS, Beta $19.99 TWE *ZZ* ½

Young Hero of Shaolin When facing the fury of a thousand violent monks, one is forced to use the infamous "swastika" formation if one is to achieve victory.
19?? 90m/C Guo Liang, Shih Bao Hwa. VHS $49.95 OCE *Z*

Young Hero of the Shaolin Part 2 A mercenary and his gang commandeer a private yacht, kill the crew, and kidnap the owner's children.
19?? 90m/C Guo Liang. VHS, Beta $49.95 OCE *Z*

Young Heroes The youthful protagonist uses his mastery of the "Shadow Fist" against "The Godfather of Chaos."
1969 87m/C Hwang Cheng Li, Kwon Young Moon, Wang Chang, Haht Dang, Lo Chine Po; *Hosted:* Sho Kosugi. VHS, Beta $29.95 TWE *Z*

Young and Innocent Somewhat uneven thriller about a police constable's daughter who helps a fugitive prove he didn't strangle a film star. Released in the U.S. as "The Girl Was Young."
1937 80m/B *GB* Derrick DeMarney, Nova Pilbeam, Percy Marmont; *Dir:* Alfred Hitchcock. VHS, Beta, LV $16.95 SNC, NOS, VDM *ZZZ*

Young Lady Chatterly D.H. Lawrence meets "Debbie Does Houston." Cynthia Chatterly turns 20 and inherits Chatterly Estate, where servants are at her beck and call. Includes fifteen minutes of uncensored footage not found in the "Restricted" version.
1985 100m/C Harlee MacBride. VHS $39.99 LIV, LTG *Z*

Young Lady Chatterly 2 A poor sequel to the popular MacBride film, with only the name of the Lawrence classic. Chatterly inherits the family mansion and fools around with the servants and any one else who comes along. Unrated version with 13 minutes of deleted footage is also available.
1985 87m/C Sybil Danning, Adam West, Harlee MacBride; *Dir:* Alan Roberts. VHS, Beta, LV $79.98 LIV, LTG *Z* ½

The Young Lions A cynical World War II epic following the experiences of a young American officer and a disillusioned Nazi in the war's last days. Martin does fine in his first dramatic role. As the Nazi, Brando sensitively considers the belief that the Hitler would save Germany. A realistic anti-war film.
1958 167m/B Marlon Brando, Montgomery Clift, Dean Martin, Hope Lange, Barbara Rush, Lee Van Cleef, Maximilian Schell; *Dir:* Edward Dmytryk. VHS, Beta, LV $19.98 FOX *ZZZ*

Young Love, First Love A made-for-television movie about a young woman who must decide whether to have sex with her boyfriend. Sound familiar? Pretty blah, earnest story, though well cast.

1979 100m/C Valerie Bertinelli, Leslie Ackerman, Timothy Hutton, Arlen Dean Snyder, Fionnula Flanagan; *Dir:* Steven Hilliard Stern. VHS, Beta $59.95 IVE, USA *Z* ½

Young Love - Lemon Popsicle 7 Three hunks stalk the beaches in search of babes.
1987 (R) 91m/C Yftach Katzur, Zachi Noy, Jonathan Segall, Sonja Martin; *Dir:* Walter Bennett. VHS, Beta $79.95 WAR *Z*

Young Man with a Horn Dorothy Baker's novel, which was loosely based on the life of jazz immortal Bix Beiderbecke, was even more loosely adapted for this film, featuring Kirk as an angst-ridden trumpeter who can't seem to hit that mystical "high note." Songs include "With a Song in My Heart," "Too Marvelous for Words," and "The Very Thought of You."
1950 112m/B Kirk Douglas, Doris Day, Lauren Bacall, Hoagy Carmichael; *Dir:* Michael Curtiz. VHS, Beta, LV $19.98 WAR *ZZZ*

Young Mr. Lincoln A classy Hollywood biography of Lincoln in his younger years from log-cabin country boy to idealistic Springfield lawyer. Written by Lamar Trotti. A splendid drama, and one endlessly explicated as an American masterpiece by the French auteur critics in "Cahiers du Cinema."
1939 100m/B Henry Fonda, Alice Brady, Marjorie Weaver, Arleen Whelan, Eddie Collins, Ward Bond, Donald Meek, Richard Cromwell, Eddie Quillan; *Dir:* John Ford. National Board of Review Awards '39: 10 Best Films of the Year. VHS, Beta, LV $19.98 FOX *ZZZ* ½

The Young Nurses The fourth entry in the Roger Corman produced "nurses" series. Three sexy nurses uncover a drug ring run from their hospital, headed by none other than director Fuller. Also present is Moreland, in his last role. Sequel to "Night Call Nurses," followed by "Candy Stripe Nurses." Also on video as "Young L.A. Nurses 3."
1973 (R) 77m/C Jean Manson, Ashley Porter, Angela Gibbs, Zack Taylor, Dick Miller, Jack LaRue, William Joyce, Sally Kirkland, Allan Arbus, Mary Doyle, Don Keefer, Nan Martin, Mantan Moreland, Samuel Fuller; *Dir:* Clinton Kimbrough. VHS, Beta $59.98 CHA *ZZ*

Young Nurses in Love Low-budget sex farce in which a foreign spy poses as a nurse to steal sperm from a sperm bank donated by world leaders, celebrities and geniuses.
1989 (R) 82m/C Jeanne Marie, Alan Fisher, Barbra Robb, James Davies; *Dir:* Chuck Vincent. VHS, Beta, LV $79.98 LIV, VES *Z*

The Young Ones Three episodes from the hilarious, offbeat British television sitcom about four violently eccentric college roommates. Episodes included are "Oil," "Boring," and "Flood."
1988 96m/C *GB* Rik Mayall, Ade Edmondson, Nigel Planer, Christopher Ryan, Alexei Sayle; *Dir:* Geoffrey Posner. VHS, Beta $14.98 FOX *Z*

The Young Philadelphians Ambitious young lawyer Newman works hard at making an impression on the snobbish Philadelphia upper crust. As he schemes and scrambles, he woos debutante Rush and defends buddy Vaughn on a murder charge. Long, but worth it. Vaughn received a supporting actor Oscar nomination. Part of the "A Night at the Movies" series, this package simulates a 1959 movie evening with a Bugs Bunny cartoon, "People Are Bunny," a newsreel and coming attractions for "The Nun's Story" and "The Hanging Tree."

1959 136m/B Paul Newman, Barbara Rush, Alexis Smith, Billie Burke, Brian Keith, John Williams, Otto Kruger, Robert Vaughn; *Dir:* Vincent Sherman. VHS, Beta $69.95 WAR *ZZZ*

Young Sherlock Holmes Holmes and Watson meet as schoolboys. They work together on their first case, solving a series of bizarre murders gripping London. Watch through the credits for an interesting plot twist. Promising "what if" sleuth tale crashes and burns, becoming a typical high-tech Spielberg film. Second half bears too strong a resemblance to "Indiana Jones and the Temple of Doom."
1985 (PG-13) 109m/C Nicholas Rowe, Alan Cox, Sophie Ward, Freddie Jones, Michael Hordern; *Dir:* Barry Levinson. VHS, Beta, LV, 8mm $14.95 PAR *ZZ*

The Young Taoism Fighter The kung-fu film that tells it like it is—or was. Wacky shenanigans and zany pranks are the order of the day in a Taoist temple. Confucianism was never like this.
19?? 90m/C Yuen Tat Chor, Kwan Chung, Liu Hao Yi. VHS, Beta $49.95 OCE *ZZ*

The Young Teacher A novice teacher finds out that trying to apply her college education to a real life elementary school environment isn't as easy as it sounds.
1985 90m/C VHS, Beta $29.95 GEM *Z* ½

Young Tiger First a ninety-minute movie never before seen in the U.S. featuring an expert in martial arts who is accused of murder. Next is a twelve-minute documentary featuring Jackie Chan, kung-fu sensation, demonstrating his skills. What an evening.
1980 (R) 102m/C Jackie Chan. VHS, Beta $24.95 GEM *ZZ*

Young Warlord Arthur roams western England in 500 AD, leading a band of guerrilla cavalrymen. When the Saxons invade, Arthur unites the tribe, holds off the attack and becomes king.
1975 97m/C Oliver Tobias, Michael Gothard, Jack Watson, Brian Blessed; *Dir:* Peter Sasdy. VHS, Beta $24.95 WES *ZZ*

Young Warriors Frat boys turn vigilante to avenge a street gang murder. Weird mix of teen sex comedy, insufferable self-righteous preachiness, and violence.
1983 (R) 104m/C Ernest Borgnine, James Van Patten, Richard Roundtree, Lynda Day George, Dick Shawn; *Dir:* Lawrence Foldes. VHS, Beta $59.95 MGM *Z* ½

Young & Willing A group of struggling actors gets hold of a terrific play and tries various schemes to get it produced. Based on the Broadway Play "Out of the Frying Pan" by Francis Swann. Unexceptional but enjoyable comedy.
1942 83m/B William Holden, Susan Hayward, Eddie Bracken, Robert Benchley, Martha O'Driscoll; *Dir:* Edward H. Griffith. VHS, Beta $19.95 MON *ZZ* ½

Young Winston Based on Sir Winston Churchill's autobiography "My Early Life: A Roving Commission." Follows him through his school days, journalistic career in Africa, early military career, and his election to Parliament at the age of 26. Ward is tremendous as the prime minister-to-be.
1972 (PG) 145m/C *GB* Simon Ward, Robert Shaw, Anne Bancroft, John Mills, Jack Hawkins, Ian Holm, Anthony Hopkins, Patrick Magee, Edward Woodward, Jane Seymour; *Dir:* Richard Attenborough. VHS, Beta $19.95 COL *ZZZ*

Youngblood An underdog beats the seemingly insurmountable odds and becomes a hockey champion. Some enjoyable hockey scenes although the success storyline is predictable.
1986 (R) 111m/C Rob Lowe, Patrick Swayze, Cynthia Gibb, Ed Lauter, George Finn, Fionnula Flanagan, Keanu Reeves; *Dir:* Peter Markle. **VHS, Beta, LV $14.98** MGM 🐾🐾

Your Place or Mine Self-indulgent made for television yuppie pap about middle-aged singles trying to find the right mate.
1983 100m/C Bonnie Franklin, Robert Klein, Peter Bonerz, Tyne Daly, Penny Fuller; *Dir:* Robert Day. **VHS, Beta $39.95** IVE 🐾 ½

Your Ticket Is No Longer Valid Prurient excuse for a serious drama. Impotent failed businessman has disturbing erotic dreams about his girlfriend. Adapted from a novel by Romain Gary.
1984 (R) 96m/C CA Richard Harris, George Peppard, Jeanne Moreau; *Dir:* George Kaczender. **VHS, Beta $69.95** VES 🐾

You're a Big Boy Now Kastner, a virginal young man working in the New York Public Library, is told by his father to move out of his house and grow up. On his own, he soon becomes involved with man-hating actress Hartman and a discotheque dancer. A wild and weird comedy. Coppola's commercial directorial debut. Page received an Oscar nomination.
1966 96m/C Elizabeth Hartman, Geraldine Page, Peter Kastner, Julie Harris, Rip Torn, Michael Dunn, Tony Bill, Karen Black; *W/Dir:* Francis Ford Coppola. **VHS, Beta $19.98** WAR 🐾🐾🐾

You're a Good Man, Charlie Brown A filmed version of Charles Schulz's Broadway musical, based on his famous characters.
1987 60m/C VHS, Beta $14.95 KRT, KAR 🐾🐾🐾

You're Jinxed, Friend, You've Met Sacramento The tongue-in-cheek adventures of a cool-headed cowboy as he passes from town to colorful town. Great title, too.
1970 99m/C Ty Hardin, Christian Hay, Jenny Atkins; *Dir:* George Cristallini. **VHS, Beta $57.95** UNI 🐾🐾

Yours, Mine & Ours Bigger, better, big screen version of "The Brady Bunch." It's the story of a lovely lady (Ball) with eight kids who marries a a widower (Fonda) who has ten. Imagine the zany shenanigans! Family comedy manages to be both wholesome and funny. Based on a true story.
1968 111m/C Lucille Ball, Henry Fonda, Van Johnson, Tim Matheson, Tom Bosley, Tracy Nelson; *Dir:* Melville Shavelson. **VHS, Beta, LV $19.98** MGM 🐾🐾 ½

Youth on Parole Same old story: Boy gets paroled. Girl gets paroled. Parolee meets parolette and romance blossoms as they beat a further rap.
1937 60m/B Marian Marsh, Gordon Oliver. **VHS** LOO 🐾 ½

You've Got to Have Heart A young bridegroom's life gets complicated after he does not sexually satisfy his bride on their wedding night.
1977 (R) 98m/C Carroll Baker, Edwige French. **VHS, Beta $49.95** PSM 🐾

Yukon Flight A Renfrew of the Mounties adventure. The hero finds illegal gold mining operations and murder in the Yukon.
1940 57m/B James Newill, Dave O'Brien. **VHS, Beta $19.95** NOS, VCN 🐾🐾

The Yum-Yum Girls A pair of innocent girls arrive in New York City to pursue their dreams. Gives an inside looks at the fashion industry. Not as funny or as cute as it wants to be.
1978 (R) 89m/C Judy Landers, Tanya Roberts, Barbara Tully, Michelle Daw; *Dir:* Barry Rosen. **VHS, Beta** MCA 🐾 ½

Yum, Yum, Yum! Director Blank returns to Southern Louisiana for another look at Cajun and Creole cooking. Among the Louisiana cooks encountered are Chef Paul Prudhomme, Queen Ida and Marc Savoy.
1991 31m/C *W/Dir:* Les Blank. **VHS $49.95** FFM 🐾 ½

Yuma An old-style made for television Western about a sheriff (Walker) who rides into town, cleans it up, and saves his own reputation from a plot to discredit him. Dull in places, but action-packed ending saves the day.
1970 73m/C Clint Walker, Barry Sullivan, Edgar Buchanan; *Dir:* Ted Post. **VHS, Beta $19.98** FOX 🐾🐾

Yuri Nosenko, KGB Fact-based account of a KGB defector and the CIA agent who must determine if he's on the up-and-up. Made by for British television.
1986 85m/C GB Tommy Lee Jones, Oleg Rudnik, Josef Sommer, Ed Lauter, George Morfogen, Stephen D. Newman; *Dir:* Mick Jackson. **VHS $19.99** HBO 🐾🐾 ½

Z The assassination of a Greek nationalist in the 1960s and its aftermath are portrayed by the notorious political director as a gripping detective thriller. Excellent performances, adequate cinematic techniques, and important politics in this highly acclaimed film. Took two Oscars and was nominated for Best Director and Adapted Screenplay.
1969 128m/C FR Yves Montand, Jean-Louis Trintignant, Irene Papas, Charles Denner, Georges Geret; *Dir:* Constantin Costa-Gavras. Academy Awards '69: Best Film Editing, Best Foreign Language Film; British Academy Awards '69: Best Original Score; Cannes Film Festival '69: Best Actor (Trintignant); Edgar Allan Poe Awards '69: Best Screenplay; New York Film Critics Awards '69: Best Director (Costa-Gavras), Best Film; National Society of Film Critics Awards '69: Best Film. **VHS, Beta, LV $64.95** COL 🐾🐾🐾🐾

Zabriskie Point Antonioni's first U.S. feature. A desultory, surreal examination of the American way of life. Worthy but difficult. Climaxes with a stylized orgy in Death Valley.
1970 (R) 111m/C Mark Frechette, Daria Halprin, Paul Fix, Rod Taylor, Harrison Ford, G.D. Spradlin; *W/Dir:* Michelangelo Antonioni. **VHS, Beta, LV $59.95** MGM, MLB 🐾🐾🐾

Zachariah A semi-spoof 60s rock western, wherein two gunfighters given to pursuing wealth-laden bands of outlaws separate and experience quixotic journeys through the cliched landscape. Scripted by members of The Firesign Theater and featuring appearances by Country Joe and The Fish, The New York Rock Ensemble and The James Gang.
1970 (PG) 93m/C Don Johnson, John Rubinstein, Pat Quinn, Dick Van Patten, Country Joe & the Fish; *Dir:* George Englund. **VHS, Beta $19.98** FOX 🐾🐾 ½

Zalmen or the Madness of God Elie Wiesel's mystical story about a rabbi's struggle against religious persecution in post-Stalin Russia.
1975 120m/C Joseph Wiseman, The Arena Stage Company. **VHS, Beta** WNE 🐾🐾 ½

Zandalee The sexual adventures of a bored sexy young woman who has a fling with her husband's friend. Bad script, graphic sex. Available in an edited, "R" rated version and an unrated version.
1991 100m/C Nicolas Cage, Judge Reinhold, Erika Anderson, Viveca Lindfors, Aaron Neville, Joe Pantoliano; *Dir:* Sam Pillsbury. **VHS, LV $89.95** LIV Woof!

The Zany Adventures of Robin Hood A hackneyed spoof on the Robin Hood legend about a medieval man who robs from the rich to help the poor. Not nearly as funny as Mel Brooks' series that spoofed "Robin Hood" - in fact, it isn't very funny at all.
1984 95m/C George Segal, Morgan Fairchild, Roddy McDowall, Roy Kinnear, Janet Suzman, Tom Baker; *Dir:* Ray Austin. **VHS, Beta $79.95** MED 🐾

Zapped! A teenage genius discovers he possesses telekinetic powers, which he is most interested in using to remove girls' dresses. A teen boy's dream come true.
1982 (R) 98m/C Scott Baio, Willie Aames, Robert Mandan, Felice Schachter, Scatman Crothers, Roger Bowen, Marya Small, Greg Bradford, Hilary Beane, Sue Ann Langdon, Heather Thomas, Merritt Butrick, LaWanda Page, Rosanne Katon; *Dir:* Robert J. Rosenthal. **VHS, Beta, LV $14.98** SUE 🐾

Zapped Again The lame-brained sequel to 1982's "Zapped" about a high schooler who has telekinetic powers and lust on his mind.
1989 (R) 93m/C Todd Eric Andrews, Kelli Williams, Reed Rudy, Linda Blair, Karen Black, Lyle Alzado; *Dir:* Douglas Campbell. **VHS, Beta, LV, 8mm $19.98** SUE, ORI 🐾

Zardoz A surreal parable of the far future, when Earth society is broken into strict classes: a society of intellectuals, a society of savages, and an elite unit of killers who keep the order. Connery is the man who destroys the old order of things. Visually interesting but pretentious.
1973 105m/C Sean Connery, Charlotte Rampling, John Alderton, Sara Kestelman; *Dir:* John Boorman. **VHS, Beta, LV $19.98** FOX 🐾🐾

Zatoichi, the Blind Samurai The legendary blind swordsman of the 1600s slaughters most of the population of Japan with his tricky cane/sword.
19?? 90m/C JP Shintaro Katsu. **VHS, Beta $29.95** OCE 🐾🐾

Zatoichi vs. Yojimbo The legendary blind warrior-samurai, Zatoichi, wants to retire, but his village is being held captive by outlaws. He is forced to fight Yojimbo, the crude wandering samurai without a master, and the sparks really fly! This was a comic send-up of Akiro Kurosawa's "Yojimbo" with Mifune recreating his role here and playing it for laughs. Also known as Zatoichi Meets Yojimbo; Zato Ichi To Yojimbo. Subtitled.
1970 90m/C JP Toshiro Mifune, Shintaro Katsu, Osamu Takizawa; *Dir:* Kihachi Okamoto. **VHS** VDA 🐾🐾🐾 ½

Zazie dans le Metro One of Malle's early movies, this is one of the best of the French New Wave comedies. A young girl, wise beyond her years, visits her uncle in Paris. She wants to ride the subway, but the ensuing hilarious adventures keep her from her goal. Also called "Zazie in the Underground," and "Zazie in the Subway."
1961 92m/C FR Catherine Demonget, Philippe Noiret, Carla Marlier; *Dir:* Louis Malle. **VHS $69.95** NYF 🐾🐾🐾 ½

Zebra Force A group of army veterans embark on a personal battle against organized crime, using their military training with deadly precision. Typical bad action-adventure.
1976 (R) 81m/C Mike Lane, Richard X. Slattery, Rockne Tarkington, Glenn Wilder, Anthony Caruso; *Dir:* Joe Tornatore. **VHS, Beta $69.95** MED 🎬

A Zed & Two Noughts A serio-comic essay by the acclaimed British filmmaker. Twin zoologists, after their wives are killed in an accident, explore their notions of death by, among other things, filming the decay of animal bodies absconded from the zoo. Heavily textured and experimental; Greenaway's second feature film.
1988 115m/C GB Eric Deacon, Brian Deacon, Joss Ackland, Andrea Ferreol; *Dir:* Peter Greenaway. **VHS, Beta $29.95** PAV 🎬🎬🎬

Zelig Documentary spoof stars Allen as Leonard Zelig, the famous "Chameleon Man" of the 1920s, whose personality was so vague he would assume the characteristics of those with whom he came into contact, and who had a habit of showing up among celebrities and at historic events. Filmed in black-and-white; intersperses bits of newsreel and photographs with live action. Allen-style clever filmmaking at its best.
1983 (PG) 79m/B Woody Allen, Mia Farrow, Susan Sontag, Saul Bellow, Irving Howe; *Dir:* Woody Allen. **VHS, Beta, LV, 8mm $19.98** WAR 🎬🎬½

Zelly & Me A strange little drama about a young orphan living with her maniacally possessive grandmother, who forces the child into her own interior life through humiliation and isolation from anyone she cares for. Written by Rathborne, this is a well acted and interesting film that suffers from an overly introspective plot and confusing gaps in the narrative. Look for director Lynch on the other side of the camera.
1988 (PG) 87m/C Isabella Rossellini, Alexandra Johnes, David Lynch, Glynis Johns, Kaiulani Lee; *Dir:* Tina Rathborne. **VHS, Beta, LV $79.95** COL 🎬🎬

Zeppelin During World War I, the British enlist the aid of York as a double agent. His mission is to steal Germany's plans for a super-dirigible. Accompanying the Germans on the craft's maiden voyage, York discovers they are actually on a mission to steal British treasures. Although the script is poor, the battle scenes and the blimp itself are very impressive.
1971 (G) 101m/C Michael York, Elke Sommer, Peter Carsten, Marius Goring, Anton Diffring, Andrew Keir, Rupert Davies, Alexandra Stewart; *Dir:* Etienne Perier. **VHS, Beta, LV $59.95** WAR, FCT 🎬🎬½

Zero WWII story about the building of the zero fighter that was used to devastating effect in the invasion of Pearl Harbor.
1984 128m/C Yuzo Kayama, Tetsuro Tamba; *Dir:* Toshio Masuda. **VHS, Beta $39.95** SVS 🎬🎬½

Zero Boys Teenage survivalists in the Californian wilderness are stalked by a murderous lunatic. Exploitative and mean-spirited.
1986 (R) 89m/C Daniel Hirsch, Kelli Maroney, Nicole Rio, Joe Phelan; *Dir:* Nico Mastorakis. **VHS, Beta $19.98** LIV, LTG Woof!

Zero for Conduct/Taris/A Propos de Nice Three short films by the famed Jean Vigo. "Zero for Conduct" is a fascinating and fluid surrealist film about rebellion in a boy's prep school. "Taris" consists of underwater footage of the swimmer Jean Taris, and "A Propos de Nice" is a documentary that focuses on the great economic schism of the rich and the poor in the popular French vacation spot, Nice.
1991 75m/B FR *Dir:* Jean Renoir. **VHS $39.95** FST Woof!

Zero for Conduct (Zero de Conduit) Vigo's classic French fantasy about an outrageous rebellion of schoolboys against bureaucratic adults. More of a visual poem than a drama, it inspired Lindsay Anderson's "If....." One of only four films created by Vigo before his early death. Banned across Europe at release. French with English subtitles.
1933 49m/B FR Jean Daste; *Dir:* Jean Vigo. **VHS, Beta $24.95** NOS, VYY, HHT 🎬🎬🎬🎬

Zero Pilot A Japanese film about the heroic fighter pilots of World War II
1984 90m/C JP **VHS, Beta** VCD Woof!

Zero to Sixty A newly divorced man finds his car has been repossessed for nonpayment. Seeking out the manager of the finance company, he gets a job as a repo man with a sassy 16-year-old girl as his assistant.
1978 (PG) 96m/C Darren McGavin, Sylvia Miles, Denise Nickerson, Joan Collins; *Dir:* Don Weis. **VHS, Beta $69.98** SUE 🎬

Zertigo Diamond Caper Utilizing his heightened senses, a blind boy solves a diamond caper and proves his mother's innocence.
1982 50m/C Adam Rich, David Groh, Jane Elliot. **VHS, Beta $29.95** GEM 🎬

Ziegfeld Follies A lavish revue of musical numbers and comedy sketches featuring many MGM stars of the World War II era. Highlights include Astaire and Kelly's only duet, "The Babbitt and the Bromide," Horne singing "Love," Garland as "Madame Cremation," and an Astaire-Bremer ballet set to "Limehouse Blues."
1946 109m/C Fred Astaire, Judy Garland, Gene Kelly, Red Skelton, Fannie Brice, William Powell, Jimmy Durante, Edward Arnold, Lucille Bremer, Hume Cronyn, Victor Moore, Lena Horne, Lucille Ball, Esther Williams; *Dir:* Vincente Minnelli. **VHS, Beta, LV $19.95** MGM, MLB 🎬🎬½

Ziegfeld Girl Three star-struck girls are chosen for the Ziegfeld follies and move on to success and heartbreak. Lavish costumes and production numbers in the MGM style. Songs include "You Stepped Out of a Dream," "I'm Always Chasing Rainbows," and "Minnie from Trinidad."
1941 131m/B James Stewart, Judy Garland, Hedy Lamarr, Lana Turner, Tony Martin, Jackie Cooper, Ian Hunter, Charles Winninger, Al Shear, Edward Everett Horton, Philip Dorn, Paul Kelly, Eve Arden, Dan Dailey, Fay Holden, Felix Bressart, Mae Busch, Reed Hadley; *Dir:* Robert Z. Leonard. **VHS, Beta, LV $29.95** MGM 🎬🎬🎬

Zis Boom Bah A musical-comedy star buys a cafe for her college son. He and his friends transform the place into a restaurant-theater with predictable results.
1942 61m/B Peter Lind Hayes, Mary Healey, Grace Hayes, Huntz Hall, Benny Rubin. **VHS, Beta $19.95** NOS, VYY, DVT 🎬½

The Zodiac Killer Based on a true story, this tells the violent tale of the San Francisco murders that occurred in the late 1960s. Doesn't have the suspense it should.
1971 (R) 87m/C Tom Pittman, Hal Reed, Bob Jones, Ray Lynch; *Dir:* Tom Hanson. **VHS, Beta $59.95** ACA 🎬½

Zoltan...Hound of Dracula The vampire and his bloodthirsty dog go to Los Angeles to find the last of Count Dracula's living descendants. Campy and just original enough to make it almost worth watching. Also on video as "Dracula's Dog."
1977 85m/C Michael Pataki, Reggie Nalder, Jose Ferrer; *Dir:* Albert Band. **VHS, Beta $69.99** HBO, UHV 🎬

Zombie Italian-made white-men-in-the-Caribbean-with-flesh-eating-zombies cheapie. Also known as "Zombie Flesh Eaters."
1980 91m/C IT Tisa Farrow, Ian McCulloch, Richard Johnson, Al Cliver, Arnette Gay, Olga Karlatos, Stefania D'Amario; *Dir:* Lucio Fulci. **VHS, Beta, LV** VES 🎬

Zombie High Students at a secluded academy are being lobotomized by the school president. Mindless, in two senses of the word. Also known as "The School That Ate My Brain."
1987 (R) 91m/C Virginia Madsen, Richard Cox, Kay E. Kuter, James Wilder, Sherilyn Fenn, Paul Williams; *Dir:* Ron Link. **VHS, Beta $9.95** SIM, CON 🎬

Zombie Island Massacre In this trying film, tourists travel to see a voodoo ritual, and then are systematically butchered by the rite-inspired zombies. Featuring former congressional wife and Playboy Magazine model Jenrette.
1984 (R) 89m/C Rita Jenrette, David Broadnax; *Dir:* John N. Carter. **VHS, Beta $79.95** MED 🎬

Zombie Lake Killed in an ambush by villagers during WWII, a group of Nazi soldiers turned zombies reside in the town's lake, preying on unsuspecting swimmers, especially dog paddling nude young women. Laughable FX, almost bad enough to be good. From the team that produced "Oasis of the Zombies."
1984 90m/C FR SP Howard Vernon, Pierre Escourrou, Anouchka, Anthony Mayans, Nadine Pascale, Jean Rollin; *Dir:* J.A. Laser. **VHS $59.95** UHV 🎬

Zombie Nightmare A murdered teenager is revived by a voodoo queen and slaughters his punk-teen assailants. Cheap and stupid just about sums this one up. Music by Motorhead, Death Mask, Girlschool and Thor.
1986 (R) 89m/C Adam West, Jon Mikl Thor, Tia Carrere, Frank Dietz, Linda Singer, Mandn E. Turbride, Hamibh McEwen; *Dir:* John Bravman. **VHS, Beta $19.95** NWV, STE Woof!

Zombie vs. Ninja Guy who kicks and screams meets guy with bad complexion.
19?? ?m/C Wang Li, Keth Uh Land. **VHS $59.95** IVE Woof!

Zombies on Broadway Two press agents travel to the Caribbean in search of new talent, but find Lugosi performing experiments on people and turning them into sequel material. RKO hoped this would be equally successful follow-up to "I Walked with a Zombie."
1944 68m/B Wally Brown, Alan Carney, Bela Lugosi, Anne Jeffreys, Sheldon Leonard, Frank Jenks, Russell Hopton, Joseph Vitale, Ian Wolfe, Louis Jean Heydt, Darby Jones, Sir Lancelot; *Dir:* Gordon Douglas. **VHS $29.95** TTC, FCT, IME 🎬🎬½

Zombies of Moratau A diver and his girlfriend seek to retrieve an undersea treasure of diamonds protected by zombies. So boring you'll want to die.
1957 70m/B Gregg Palmer, Allison Hayes, Jeff Clark, Autumn Russel, Joel Ashley; **Dir:** Edward L. Cahn. **VHS, Beta $59.95** COL, MLB Woof!

Zombies of the Stratosphere A serial in twelve chapters in which a cosmic policeman fights Zombies attempting to blow the Earth out of orbit. Also available in a 93 minute, colorized version.
1952 152m/B Judd Holdren, Aline Towne, Leonard Nimoy; **Dir:** Fred Brannon. **VHS, LV $14.98** REP, VCN, PIA Woof!

Zombiethon A campy compilation of clips from cheapo zombie films, such as "Space Zombies," "Zombie Lake," and "The Invisible Dead."
1986 90m/C Dominique Singleton, Janelle Lewis, Tracy Burton. **VHS, Beta** FOR ♫½

Zone of the Dead A mortician is using his morgue for other than embalming—and it isn't pretty!
1978 81m/C John Ericson, Ivor Francis, Charles Aidman, Bernard Fox. **VHS, Beta $79.95** MNC ♫

Zone Troopers Five American G.I.'s in World War II-ravaged Europe stumble upon a wrecked alien spacecraft and enlist the extraterrestrial's help in fending off the Nazis.
1985 (PG) 86m/C Timothy Van Patten, Tim Thomerson, Art LaFleur, Biff Manard; **Dir:** Danny Bilson. **VHS, Beta $19.98** LIV, LTG ♫½

Zontar, the Thing from Venus Scientist is taken over by alien batlike thing from Venus, and attempts to take over the Earth. A parody of itself.
1966 68m/C John Agar, Anthony Huston; **Dir:** Larry Buchanan. **VHS $19.98** SNC, VDM ♫

The Zoo Gang A group of teens want to open a nightclub. Will they bring their dream to fruition, or will the mean rival gang foil their plans?
1985 (PG-13) 96m/C Jackie Earle Haley, Tiffany Helm, Ben Vereen, Jason Gedrick, Eric Gurry; **Dir:** John Watson, Pen Densham. **VHS, Beta $9.99** NWV, STE ♫½

Zoo Radio Run by a wacky assortment of characters, a zany underground radio station embarks on a battle against bureaucracy.
1990 90m/C **VHS $89.95** WAX ♫½

Zoot Suit Based on Luis Valdez' play, this murder mystery/musical is rooted in the historical Sleepy Lagoon murder in the 1940s. Valdez plays a Mexican-American accused of the crime. His friends (and defense lawyers) rally around him to fight this travesty of justice. Lots o'music and dancing.
1981 (R) 104m/C Edward James Olmos, Daniel Valdez, Tyne Daly, Charles Aidman, John Anderson; **W/Dir:** Luis Valdez. **VHS, LV $79.95** MCA, LDC, FCT ♫♫♫

Zora Is My Name! The funny, moving story of Zora Neal Hurston, a Black writer known for her stories and folklore of the rural South of the '30s and '40s. From PBS's American Playhouse.
1990 90m/C Ruby Dee, Louis Gossett Jr. **VHS $19.95** PBS, PAV, KAR ♫♫½

Zorba the Greek A young British writer (Bates) comes to Crete to find himself by working his father's mine. He meets Zorba, an itinerant Greek laborer (Quinn), and they take lodgings together with an aging courtesan, who Zorba soon romances. The writer, on the other hand, is attracted to a lovely

young widow. When she responds to him, the townsmen jealously attack her. Zorba teaches the young man the necessary response to life and its tragedies. Based on a novel by Nicolai Kazantzakis. Masterpiece performance from Quinn, who received a Best Actor Oscar nomination. Beautifully photographed, somewhat overlong. Film later written for stage production. Other Oscar nominations for Best Picture, Director, and Adapted Screenplay.
1964 142m/B Anthony Quinn, Alan Bates, Irene Papas, Lila Kedrova; **Dir:** Michael Cacoyannis. Academy Awards '64: Best Art Direction/Set Decoration (B & W), Best Black and White Cinematography, Best Supporting Actress (Kedrova). **VHS, Beta, LV $29.98** FOX ♫♫♫½

Zorro, the Gay Blade Hamilton portrays the swashbuckling crusader and his long-lost brother, Bunny Wigglesworth, in this spoof of the Zorro legend. The fashion-conscious hero looks his best in plum. Leibman is fun to watch.
1981 (PG) 96m/C George Hamilton, Lauren Hutton, Brenda Vaccaro, Ron Leibman; **Dir:** Peter Medak. **VHS, Beta $59.98** FOX ♫½

Zorro Rides Again Zorro risks his life to outwit enemy agents endeavoring to secure ancestor's property. In twelve chapters, the first runs 30 minutes, the rest 17.
1937 217m/B John Carroll, Helen Christian, Noah Beery, Duncan Renaldo; **Dir:** William Witney, John English. **VHS, Beta $27.95** NOS, VCN, MED ♫½

Zorro's Black Whip A young girl dons the mask of her murdered brother (Zorro) to fight outlaws in the old West. Serial in twelve episodes.
1944 182m/B George Lewis, Linda Stirling, Lucien Littlefield, Francis McDonald, Tom London; **Dir:** Spencer Gordon Bennet. **VHS, LV $29.98** NOS, GPV, MED ♫½

Zorro's Fighting Legion Zorro forms a legion to help the president of Mexico fight a band of outlaws endeavoring to steal gold shipments. A serial in 12 chapters.
1939 215m/B Reed Hadley, Sheila Darcy; **Dir:** William Witney, John English. **VHS, LV $29.98** NOS, VCN, VDM ♫½

Zotz! The holder of a magic coin can will people dead by uttering "zotz"; spies pursue the mild-mannered professor who possesses the talisman. Russell wrote the screenplay from a Walter Karig novel. Typical William Castle fare; his gimic in the theatrical release of the movie was to distribute plastic "zotz" coins to the theatre patrons.
1962 87m/B Tom Poston, Julia Meade, Jim Backus, Fred Clark, Cecil Kellaway; **W/Dir:** Ray Russell. **VHS, Beta $59.95** COL ♫♫

Zou Zou Lavish backstage musical/drama of a laundress who fills in for the leading lady on opening night and becomes a hit. Baker's talking picture debut. In French with English subtitles.
1934 92m/B FR Josephine Baker, Jean Gabin; **Dir:** Marc Allegret. **VHS, LV $29.95** MVD, KIV, APD ♫♫½

Zulu In 1879, a small group of British soldiers try to defend their African outpost from attack by thousands of Zulu warriors. Amazingly, the British win. Dated colonial epic based on an actual incident; battle scenes are magnificent. Prequel "Zulu Dawn" (1979) depicts British mishandling of the situation that led to the battle.
1964 139m/C Michael Caine, Jack Hawkins, Stanley Baker, Nigel Green; **Dir:** Cy Endfield. **VHS, Beta, LV $19.98** CHA, VYG ♫♫½

Zulu Dawn An historical epic about British troops fighting the Zulus at Ulandi in 1878. Shows the increasing tensions between the British colonial government and the Zulus. Stunning landscapes unfortunately don't translate to the small screen. Good but unoriginal colonial-style battle drama.
1979 (PG) 117m/C Burt Lancaster, Peter O'Toole, Denholm Elliott, Nigel Davenport, John Mills, Simon Ward, Bob Hoskins, Freddie Jones; **Dir:** Douglas Hickox. **VHS, Beta, LV $19.98** TWE ♫♫♫

Zvenigora Dovzhenko's first major film, and a lyrical revelation in the face of Soviet formality: a passionate, funny fantasy tableaux of 1,000 years of Ukrainian history, encompassing wild folk myths, poetic drama, propaganda and social satire. Silent.
1928 73m/B RU Mikola Nademsy, Alexander Podorozhny, Semyon Svashenko; **Dir:** Alexander Dovzhenko. **VHS, Beta $39.95** FCT ♫♫♫♫

CAPTIONED INDEX

The **Captioned Index** identifies 2,500 videos available with captioning for hearing impaired viewers provided by either the National Captioning Institute (Falls Church, VA) or the Caption Center (Los Angeles, CA). Titles are arranged alphabetically.

CAPTIONED INDEX

Human

The Human Shield
The Hunt for Red October
Hunter's Blood
Hurricane Smith
Husbands and Lovers
Hush, Hush, Sweet Charlotte
The Hustler
I Come in Peace
I Don't Buy Kisses Anymore
I Live with Me Dad
I Live My Life
I Love You to Death
I Posed for Playboy
I Wake Up Screaming
I Wanna Hold Your Hand
If Looks Could Kill
Illegally Yours
Illicit Behavior
Immediate Family
Immortal Sergeant
Impulse
In the Cold of the Night
The In Crowd
In Harm's Way
In the Mood
In the Navy
In Search of the Castaways
In Search of a Golden Sky
The Incident
An Inconvenient Woman
The Incredible Mr. Limpet
The Indian Runner
Indiana Jones and the Last
   Crusade
Indiana Jones and the Temple
   of Doom
Indio
Indiscreet
The Inner Circle
Inner Sanctum
Innerspace
An Innocent Man
Innocent Prey
Instant Justice
Instant Karma
Internal Affairs
Internecine Project
Intimate Power
Intimate Stranger
Into the Night
Into the Sun
Intruder
Invaders from Mars
Invasion of the Body
   Snatchers
Irma La Douce
Iron Eagle
Iron Warrior
Ironclads
Ironweed
Irreconcilable Differences
Is Paris Burning?
Ishtar
It Came Upon a Midnight
   Clear
It Happened One Night
It Started in Naples
It Takes Two
The Italian Job
It's Not Easy Bein' Me
It's a Wonderful Life
Ivory Hunters
Jack & the Beanstalk
Jacknife
Jack's Back
Jacob's Ladder
The Jagged Edge
Jake Speed
The James Dean 35th
   Anniversary Collection
Jane Eyre
The January Man
Jaws: The Revenge
The Jayhawkers
Jesse
The Jetsons: Movie

Jewel of the Nile
JFK
Jo Jo Dancer, Your Life is
   Calling
John Lennon: Imagine
Johnny Be Good
Johnny Dangerously
Johnny Handsome
The Josephine Baker Story
Joshua Then and Now
Journey to the Center of the
   Earth
Journey of Honor
Judgment
Juggernaut
Juice
Julia and Julia
Jumpin' Jack Flash
The Jungle Book
Jungle Fever
Just Around the Corner
Just One of the Guys
Just the Way You Are
Justine
K-9
K-9000
K2
The Karate Kid
The Karate Kid: Part 2
The Karate Kid: Part 3
Keep 'Em Flying
Keeper of the City
Keeping Track
Key Exchange
The Keys of the Kingdom
Kid
Kid Colter
Kidco
Kidnapped
Kill Me Again
Killer Fish
Killer Image
Killer Klowns from Outer
   Space
Killer Tomatoes Strike Back
The Killing Fields
Killing at Hell's Gate
Killing Hour
Killing in a Small Town
Kindergarten Cop
King David
The King and I
The King of the Kickboxers
King Kong
King Kong Lives
King of New York
King Ralph
Kings Row
Kismet
A Kiss Before Dying
Kiss of Death
Kiss Me Deadly
Kiss Shot
Kiss of the Spider Woman
The Kissing Place
The Kitchen Toto
The Klansman
Knights & Emeralds
Kramer vs. Kramer
The Krays
Krull
Krush Groove
Kuffs
La Balance
La Bamba
L.A. Law
L.A. Story
Labyrinth
Ladies of the Chorus
The Ladies' Man
Lady by Choice
Lady Jane
The Lady in White
Ladybugs
Ladyhawke
Lambada

The Land Before Time
Land of the Pharaohs
Lantern Hill
Laser Mission
The Last Boy Scout
The Last Days of Patton
Last Days of Pompeii
The Last Dragon
The Last Emperor
The Last of the Finest
The Last of His Tribe
The Last of Mrs. Cheyney
The Last Picture Show
Last Plane Out
Last Rites
Last Song
The Last Unicorn
The Last Waltz
The Late Show
Latino
Laughing Policeman
The Lawnmower Man
Lawrence of Arabia
Le Mans
Lean on Me
Leave 'Em Laughing
Leaving Normal
Legal Eagles
Legend
Legend of Billie Jean
Legend of Hell House
The Lemon Sisters
Lenny
Leonard Part 6
Less Than Zero
Let It Ride
Lethal Weapon
Lethal Weapon 2
Let's Dance
Let's Get Harry
Let's Make it Legal
Letter to Brezhnev
Leviathan
Liberty & Bash
License to Kill
Lies
Life Stinks
Life is Sweet
Lifeboat
The Lighthorseman
Lightning: The White Stallion
The Lightship
Li'l Abner
Lili
Lilies of the Field
Limit Up
The Linguini Incident
Link
The Lion, the Witch & the
   Wardrobe
Lionheart
Lisa
Listen to Me
Little Big Man
The Little Colonel
Little Dorrit, Film 1: Nobody's
   Fault
Little Dorrit, Film 2: Little
   Dorrit's Story
The Little Drummer Boy
The Little Drummer Girl
Little Lord Fauntleroy
Little Man Tate
Little Match Girl
The Little Mermaid
Little Miss Broadway
Little Monsters
Little Nikita
The Little Princess
Little Red Riding Hood
Little Shop of Horrors
Little Treasure
Little Women
Littlest Rebel
Livin' Large
The Living Daylights

Local Hero
Lock Up
Lonesome Dove
The Long Gray Line
The Long, Hot Summer
The Long, Long Trailer
The Long Walk Home
Look Who's Talking
Look Who's Talking, Too
Looker
Looking for Miracles
Loose Cannons
Lord of the Flies
Lost in America
Lost Angels
The Lost Boys
The Lost Capone
Love in the Afternoon
Love Among the Ruins
The Love Bug
Love and Bullets
Love Finds Andy Hardy
A Love in Germany
Love Hurts
Love Kills
Love at Large
Love is a Many-Splendored
   Thing
Love Me or Leave Me
Love or Money?
Love with the Proper Stranger
Lovelines
Loverboy
Lower Level
Lucas
Lust in the Dust
Mac and Me
Macaroni
Mad Max Beyond
   Thunderdome
Made for Each Other
Made in Heaven
Madhouse
The Magnificent Ambersons
The Maid
Major League
Malibu Bikini Shop
Malone
The Maltese Falcon
Mambo Kings
A Man for All Seasons
A Man Called Horse
A Man Called Peter
The Man in the Gray Flannel
   Suit
Man in the Iron Mask
The Man with One Red Shoe
Man of the West
The Man Who Loved Women
The Man Who Would Be King
A Man and a Woman
The Manchurian Candidate
Manhunter
Maniac Cop 2
Mannequin
Mansfield Park
Marie
Marked for Death
Marked Woman
Married to the Mob
The Marrying Man
Martial Law 2: Undercover
Martin's Day
Marty
Mary Poppins
Mascara
Mask
Masquerade
Mass Appeal
Masters of the Universe
The Matchmaker
Matewan
Matters of the Heart
Maurice
Max Headroom, the Original
   Story

Max and Helen
Max Maven's Mindgames
May Wine
McBain
The McConnell Story
McQ
Me and Him
Mean Streets
Meatballs
Meatballs 2
Meatballs 4
Meet the Hollowheads
Melody in Love
Memories of Me
Memphis
Memphis Belle
Men Don't Leave
Men of Respect
Men at Work
Meridian: Kiss of the Beast
Mermaids
Messenger of Death
Miami Blues
Micki & Maude
The Mighty Pawns
Mikey
Mildred Pierce
Miles from Home
Miller's Crossing
Million Dollar Mystery
Mindwarp
Miracle on 34th Street
Miracle Down Under
Miracle Mile
Mirrors
Mischief
Misery
Mishima: A Life in Four
   Chapters
The Mission
Mississippi Burning
Mississippi Masala
Mister Johnson
Misty
Mo' Better Blues
Mobsters
Moby Dick
Modern Love
Mom
The Money Pit
Monkey Shines
Monster in the Closet
Montana
Monte Carlo
Monte Walsh
Monty Python Live at the
   Hollywood Bowl
Monty Python's Life of Brian
Monty Python's Parrot Sketch
   not Included
Moon 44
Moon Over Miami
Moonlighting
Moonraker
Moonstruck
Morituri
Morning After
Mortal Passions
Mortal Thoughts
Moscow on the Hudson
The Mosquito Coast
Mother Goose Rock 'n'
   Rhyme
The Mountain
Mountains of the Moon
Moving
Moving Violations
Mr. Billion
Mr. Corbett's Ghost
Mr. Destiny
Mr. Frost
Mr. Hobbs Takes a Vacation
Mr. Love
Mr. Lucky
Mr. Mom
Mr. & Mrs. Bridge

Mr. North
Mrs. Soffel
The Mummy's Hand
Munchie
Muppets Take Manhattan
Murder 101
A Murder is Announced
Murder Ordained
Murder Over New York
Murderers Among Us: The
  Simon Wiesenthal Story
Murderous Vision
Murders in the Rue Morgue
Murphy's Law
Murphy's Romance
The Music Man
My Beautiful Laundrette
My Best Friend Is a Vampire
My Blue Heaven
My Cousin Vinny
My Darling Clementine
My Fair Lady
My Geisha
My Girl
My Grandpa is a Vampire
My Heroes Have Always Been
  Cowboys
My Little Pony: The Movie
My Man Adam
My Own Private Idaho
My Stepmother Is an Alien
Mysterious Island
Mystery Date
Mystery Magician
Mystic Pizza
Nadine
The Naked Gun: From the
  Files of Police Squad
The Naked Gun 2 1/2: The
  Smell of Fear
Naked Lunch
The Name of the Rose
Narrow Margin
National Lampoon's
  Christmas Vacation
National Lampoon's European
  Vacation
National Velvet
Native Son
Nativity
The Natural
Navy SEALS
Necessary Roughness
Neon City
Neptune Factor
Neptune's Daughter
Netherworld
Never Forget
Never Say Never Again
Never Too Young to Die
Never on Tuesday
The NeverEnding Story
The NeverEnding Story 2:
  Next Chapter
The New Adventures of Pippi
  Longstocking
The New Kids
A New Kind of Love
A New Life
New Year's Day
New York, New York
Next of Kin
Next Stop, Greenwich Village
The Night Before
Night Breed
Night of the Comet
Night of the Creeps
Night of the Cyclone
Night of Dark Shadows
A Night in Heaven
A Night in the Life of Jimmy
  Reardon
'night, Mother
Night Patrol
Night Rhythms

The Night They Raided
  Minsky's
The Night Train to Kathmandu
The Nightingale
Nightmare on the 13th Floor
A Nightmare on Elm Street
A Nightmare on Elm Street 2:
  Freddy's Revenge
A Nightmare on Elm Street 3:
  Dream Warriors
A Nightmare on Elm Street 4:
  Dream Master
A Nightmare on Elm Street 5:
  Dream Child
Nijinsky
9 1/2 Ninjas
976-EVIL
976-EVIL 2: The Astral Factor
1918
1969
90 Days
Ninja the Battalion
No Holds Barred
No Man's Land
No Retreat, No Surrender 3:
  Blood Brothers
No Small Affair
No Time for Sergeants
Nobody's Fool
Nobody's Perfect
None But the Brave
North to Alaska
Not Without My Daughter
Nothing But Trouble
Nothing in Common
Notorious
Nowhere to Hide
The Nutcracker Prince
Nuts
O Pioneers!
O.C. and Stiggs
The Odessa File
Off Limits
Oh, God! You Devil
Oh, Heavenly Dog!
Oklahoma!
Old Curiosity Shop
Old Gringo
Old Yeller
Oliver!
Oliver Twist
Omar Khayyam
On the Line
On the Waterfront
Once Around
Once Bitten
Once Upon a Crime
Once Upon a Time in America
One Crazy Summer
One Deadly Summer (L'Ete
  Meurtrier)
One Good Cop
One Magic Christmas
One Man's War
One More Saturday Night
Opera do Malandro
Operation C.I.A.
Operation Petticoat
Opportunity Knocks
Orchestra Wives
The Organization
Original Intent
Orphans
Orpheus Descending
Oscar
Oscar's Greatest Moments:
  1971-1991
Other People's Money
Our Little Girl
Out of Africa
Out of Bounds
Out of the Dark
Outland
Outlaw of Gor
Outrageous Fortune
Over the Top

Overboard
The Ox-Bow Incident
Oxford Blues
Pacific Heights
The Package
Pajama Game
Pale Rider
The Palermo Connection
Panic in the Streets
Papa's Delicate Condition
A Paper Wedding
The Paradine Case
Paradise
Paradise Motel
Parasite
The Parent Trap
Parenthood
Paris Blues
Paris, Texas
Paris Trout
The Park is Mine
Parting Glances
The Party
A Passage to India
Pat and Mike
Paths of Glory
Payback
Payoff
The Pearl of Death
Pee-Wee's Big Adventure
Pee-Wee's Playhouse Festival
  of Fun
Peggy Sue Got Married
Penalty Phase
Penitentiary 3
Penn and Teller Get Killed
The People Under the Stairs
Perfect
The Perfect Weapon
Permanent Record
Personals
Pet Sematary
Peter Pan
Pete's Dragon
Peyton Place
Phantom of the Opera
Phantom of the Ritz
Phar Lap
Phobia
Phone Call from a Stranger
Physical Evidence
The Pick-Up Artist
Pied Piper of Hamelin
Pin-Up Girl
Pink Cadillac
Pinocchio
Pistol: The Birth of a Legend
Places in the Heart
The Plague Dogs
Plain Clothes
Planes, Trains & Automobiles
Planet of the Apes
Platinum Blonde
Platoon
Playboy Video Magazine:
  Volumes 1-7
Playboy's 1988 Playmate
  Video Calendar
Playboy's Bedtime Stories
Playboy's Farmers' Daughters
Playboy's Playmate of the
  Year Video Centerfold
Playing for Keeps
Playmate Playoffs
Playroom
Pocketful of Miracles
A Pocketful of Rye
Point Blank
Point Break
Poker Alice
Police Academy
Police Academy 2: Their First
  Assignment
Police Academy 3: Back in
  Training

Police Academy 4: Citizens on
  Patrol
Police Academy 5:
  Assignment Miami Beach
Police Academy 6: City Under
  Siege
Pollyanna
Poltergeist 2: The Other Side
Poltergeist 3
A Poor Little Rich Girl
The Poor Little Rich Girl
The Pope Must Diet
Pork Chop Hill
Porky's Revenge
Portrait of Jennie
Posse
Possession of Joel Delaney
Postcards from the Edge
The Postman Always Rings
  Twice
Power
Power of One
Prancer
Prayer for the Dying
Predator
Predator 2
The Presidio
Presumed Innocent
Pretty in Pink
Pretty Woman
Prey of the Chameleon
Priceless Beauty
Pride of Jesse Hallum
Pride and Prejudice
The Pride of the Yankees
Prime Cut
The Prime of Miss Jean
  Brodie
Prime Target
The Prince of Tides
The Princess Bride
Princess in Exile
Princess and the Pea
The Princess Who Never
  Laughed
Prison
Prison Stories: Women on the
  Inside
Private Investigations
Prizzi's Honor
Problem Child
Problem Child 2
Programmed to Kill
Project: Shadowchaser
Project X
Prom Night 4: Deliver Us From
  Evil
Protector
Protocol
Psychic
Psycho 3
Pucker Up and Bark Like a
  Dog
Pulse
Pump Up the Volume
Pumpkinhead
Punchline
The Punisher
Puppet Master
Puppet Master 2
Puppet Master 3
Pure Luck
Purple Rain
Purple Rose of Cairo
Pursuit
Pursuit to Algiers
Puss 'n Boots
Q & A
Quarterback Princess
Queen: Live in Rio
The Queen of Mean
Queens Logic
The Quest
Quick Change
Quicksilver
The Quiet Earth

The Quiet Man
Quigley Down Under
The Quiller Memorandum
Quo Vadis
Race with the Devil
The Racers
The Rachel Papers
Racing with the Moon
Rad
Radio Flyer
Rafferty & the Gold Dust
  Twins
A Rage in Harlem
Raggedy Ann and Andy: A
  Musical Adventure
Raiders of the Lost Ark
Raiders of the Sun
Rain Man
The Rainmaker
A Raisin in the Sun
Raising Arizona
Rambo: First Blood, Part 2
Ran
Rap Master Ronnie
Rapunzel
Ratboy
Ratings Game
Raw Deal
The Razor's Edge
Real Genius
Real Men
Rebecca
Rebecca of Sunnybrook Farm
Rebel
Rebel Without a Cause
The Rebels
Red Headed Stranger
Red Line 7000
Red River
Red Shoe Diaries
The Red Shoes
Red Sonja
Red Wind
Reds
Regarding Henry
Relentless
Renegades
Rent-A-Cop
Repossessed
The Rescue
The Rescuers
The Rescuers Down Under
The Restless Breed
Return of the Bad Men
Return to the Blue Lagoon
Return of the Dragon
Return of Frank James
Return of the Jedi
Return of the Killer Tomatoes
Return of the Living Dead Part
  2
Return to Peyton Place
Return to Salem's Lot
Return of Superfly
The Return of Swamp Thing
Reuben, Reuben
Reunion
Revenge of the Nerds
Revenge of the Nerds 2:
  Nerds in Paradise
Reversal of Fortune
Revolution
Rhinestone
Rich Girl
Rich Little: One's a Crowd
Rich Little's Christmas Carol
Richard Lewis: I'm Exhausted
Richard Pryor: Here and Now
Ricochet
Riding the Edge
The Right Stuff
Rikky and Pete
Ring of Fire
Rio Grande
Rio Lobo
Rip van Winkle

Ripping

AWARDS INDEX

The **Awards Index** lists 1700 films honored by nine national and international award bodies, representing some 80 categories of competition. This information is also contained in the review following the credits. Only features available on video are listed in this index; movies not yet released on video are not covered. As award-winning films find their way to video, they will be added to the review section and covered in this index.

American Academy Awards
Australia Film Institute
British Academy of Film and Television Arts
Cannes Film Festival
Canadian Genie
Golden Globe
Directors Guild of America
Edgar Allan Poe Awards
MTV Awards

ACADEMY AWARDS

ACTOR

1928
Emil Jannings
Last Command

1931
Lionel Barrymore
A Free Soul

1932
Wallace Beery
The Champ
Fredric March
Dr. Jekyll and Mr. Hyde

1933
Charles Laughton
The Private Life of Henry VIII

1934
Clark Gable
It Happened One Night

1935
Victor McLaglen
The Informer

1936
Paul Muni
The Story of Louis Pasteur

1937
Spencer Tracy
Captains Courageous

1938
Spencer Tracy
Boys Town

1939
Robert Donat
Goodbye, Mr. Chips

1940
James Stewart
Philadelphia Story

1941
Gary Cooper
Sergeant York

1942
James Cagney
Yankee Doodle Dandy

1943
Paul Lukas
Watch on the Rhine

1944
Bing Crosby
Going My Way

1945
Ray Milland
The Lost Weekend

1946
Fredric March
The Best Years of Our Lives

1947
Ronald Colman
A Double Life

1948
Laurence Olivier
Hamlet

1949
Broderick Crawford
All the King's Men

1950
Jose Ferrer
Cyrano de Bergerac

1951
Humphrey Bogart
The African Queen

1952
Gary Cooper
High Noon

1953
William Holden
Stalag 17

1954
Marlon Brando
On the Waterfront

1955
Ernest Borgnine
Marty

1956
Yul Brynner
The King and I

1957
Alec Guinness
The Bridge on the River Kwai

1958
David Niven
Separate Tables

1959
Charlton Heston
Ben-Hur

1960
Burt Lancaster
Elmer Gantry

1961
Maximilian Schell
Judgment at Nuremberg

1962
Gregory Peck
To Kill a Mockingbird

1963
Sidney Poitier
Lilies of the Field

1964
Rex Harrison
My Fair Lady

1965
Lee Marvin
Cat Ballou

1966
Paul Scofield
A Man for All Seasons

1967
Rod Steiger
In the Heat of the Night

1968
Cliff Robertson
Charly

1969
John Wayne
True Grit

1970
George C. Scott
Patton

1971
Gene Hackman
The French Connection

1972
Marlon Brando
The Godfather

1973
Jack Lemmon
Save the Tiger

1974
Art Carney
Harry and Tonto

1975
Jack Nicholson
One Flew Over the Cuckoo's Nest

1976
Peter Finch
Network

1977
Richard Dreyfuss
Goodbye Girl

1978
Jon Voight
Coming Home

1979
Dustin Hoffman
Kramer vs. Kramer

1980
Robert De Niro
Raging Bull

1981
Henry Fonda
On Golden Pond

1982
Ben Kingsley
Gandhi

1983
Robert Duvall
Tender Mercies

1984
F. Murray Abraham
Amadeus

1985
William Hurt
Kiss of the Spider Woman

1906
Paul Newman
Color of Money

1987
Michael Douglas
Wall Street

1988
Dustin Hoffman
Rain Man

1989
Daniel Day Lewis
My Left Foot

1990
Jeremy Irons
Reversal of Fortune

1991
Anthony Hopkins
The Silence of the Lambs

ACTOR— SUPPORTING

1936
Walter Brennan
Come and Get It

1937
Joseph Schildkraut
The Life of Emile Zola

1939
Thomas Mitchell
Stagecoach

1940
Walter Brennan
The Westerner

1941
Donald Crisp
How Green was My Valley

1943
Charles Coburn
The More the Merrier

1944
Barry Fitzgerald
Going My Way

1945
James Dunn
A Tree Grows in Brooklyn

1946
Harold Russell
The Best Years of Our Lives

1947
Edmund Gwenn
Miracle on 34th Street

1948
Walter Huston
Treasure of the Sierra Madre

1949
Dean Jagger
Twelve O'Clock High

1950
George Sanders
All About Eve

1951
Karl Malden
A Streetcar Named Desire

1952
Anthony Quinn
Viva Zapata!

1953
Frank Sinatra
From Here to Eternity

1954
Edmond O'Brien
Barefoot Contessa

1955
Jack Lemmon
Mister Roberts

1956
Anthony Quinn
Lust for Life

1957
Red Buttons
Sayonara

1958
Burl Ives
Big Country

1959
Hugh Griffith
Ben-Hur

ACTOR—
SUPPORTING
▶cont.

1960
Peter Ustinov
Spartacus

1961
George Chakiris
West Side Story

1962
Ed Begley, Sr.
Sweet Bird of Youth

1963
Melvyn Douglas
Hud

1964
Peter Ustinov
Topkapi

1965
Martin Balsam
Thousand Clowns

1966
Walter Matthau
The Fortune Cookie

1967
George Kennedy
Cool Hand Luke

1969
Gig Young
They Shoot Horses, Don't They?

1970
John Mills
Ryan's Daughter

1971
Ben Johnson
The Last Picture Show

1972
Joel Grey
Cabaret

1973
John Houseman
The Paper Chase

1974
Robert De Niro
The Godfather: Part 2

1975
George Burns
The Sunshine Boys

1976
Jason Robards, Jr.
All the President's Men

1977
Jason Robards, Jr.
Julia

1978
Christopher Walken
The Deer Hunter

1980
Timothy Hutton
Ordinary People

1981
John Gielgud
Arthur

1982
Louis Gossett, Jr.
An Officer and a Gentleman

1983
Jack Nicholson
Terms of Endearment

1984
Haing S. Ngor
The Killing Fields

1985
Don Ameche
Cocoon

1986
Michael Caine
Hannah and Her Sisters

1987
Sean Connery
The Untouchables

1988
Kevin Kline
A Fish Called Wanda

1989
Denzel Washington
Glory

1990
Joe Pesci
Goodfellas

1991
Jack Palance
City Slickers

ACTRESS

1928
Janet Gaynor
Sunrise

1930
Norma Shearer
The Divorcee

1931
Marie Dressler
Min & Bill

1932
Helen Hayes
The Sin of Madelon Claudet

1933
Katharine Hepburn
Morning Glory

1934
Claudette Colbert
It Happened One Night

1935
Bette Davis
Dangerous

1936
Luise Rainer
The Great Ziegfeld

1937
Luise Rainer
The Good Earth

1938
Bette Davis
Jezebel

1939
Vivien Leigh
Gone with the Wind

1940
Ginger Rogers
Kitty Foyle

1941
Joan Fontaine
Suspicion

1942
Greer Garson
Mrs. Miniver

1943
Jennifer Jones
The Song of Bernadette

1944
Ingrid Bergman
Gaslight

1945
Joan Crawford
Mildred Pierce

1947
Loretta Young
The Farmer's Daughter

1948
Jane Wyman
Johnny Belinda

1949
Olivia de Havilland
The Heiress

1950
Judy Holliday
Born Yesterday

1951
Vivien Leigh
A Streetcar Named Desire

1952
Shirley Booth
Come Back, Little Sheba

1953
Audrey Hepburn
Roman Holiday

1954
Grace Kelly
Country Girl

1955
Anna Magnani
The Rose Tattoo

1956
Ingrid Bergman
Anastasia

1957
Joanne Woodward
The Three Faces of Eve

1958
Susan Hayward
I Want to Live!

1959
Simone Signoret
Room at the Top

1960
Elizabeth Taylor
Butterfield 8

1961
Sophia Loren
Two Women

1962
Anne Bancroft
The Miracle Worker

1963
Patricia Neal
Hud

1964
Julie Andrews
Mary Poppins

1965
Julie Christie
Darling

1966
Elizabeth Taylor
Who's Afraid of Virginia Woolf?

1967
Katharine Hepburn
Guess Who's Coming to Dinner

1968
Barbra Streisand
Funny Girl
Katharine Hepburn
The Lion in Winter

1969
Maggie Smith
The Prime of Miss Jean Brodie

1970
Glenda Jackson
Women in Love

1971
Jane Fonda
Klute

1972
Liza Minnelli
Cabaret

1973
Glenda Jackson
A Touch of Class

1974
Ellen Burstyn
Alice Doesn't Live Here Anymore

1975
Louise Fletcher
One Flew Over the Cuckoo's Nest

1976
Faye Dunaway
Network

1977
Diane Keaton
Annie Hall

1978
Jane Fonda
Coming Home

1979
Sally Field
Norma Rae

1980
Sissy Spacek
Coal Miner's Daughter

1981
Katharine Hepburn
On Golden Pond

1982
Meryl Streep
Sophie's Choice

1983
Shirley MacLaine
Terms of Endearment

1984
Sally Field
Places in the Heart

1985
Sally Field
Places in the Heart
Geraldine Page
Trip to Bountiful

1986
Marlee Matlin
Children of a Lesser God

1987
Cher
Moonstruck

1988
Jodie Foster
The Accused

1989
Jessica Tandy
Driving Miss Daisy

1990
Kathy Bates
Misery

1991
Jodie Foster
The Silence of the Lambs

ACTRESS—
SUPPORTING

1936
Gale Sondergaard
Anthony Adverse

1938
Fay Bainter
Jezebel

1939
Hattie McDaniel
Gone with the Wind

1940
Jane Darwell
The Grapes of Wrath

1941
Mary Astor
The Great Lie

1942
Teresa Wright
Mrs. Miniver

1944
Ethel Barrymore
None But the Lonely Heart

1945
Anne Revere
National Velvet

1946
Anne Baxter
The Razor's Edge

1948
Claire Trevor
Key Largo

1949
Mercedes McCambridge
All the King's Men

1950
Josephine Hull
Harvey

1951
Kim Hunter
A Streetcar Named Desire

1952
Gloria Grahame
The Bad and the Beautiful

1953
Donna Reed
From Here to Eternity

1954
Eva Marie Saint
On the Waterfront

1955
Jo Van Fleet
East of Eden

1956
Dorothy Malone
Written on the Wind

1957
Miyoshi Umeki
Sayonara

1958
Wendy Hiller
Separate Tables

1959
Shelley Winters
The Diary of Anne Frank

1960
Shirley Jones
Elmer Gantry

1961
Rita Moreno
West Side Story

1962
Patty Duke
The Miracle Worker

1964
Lila Kedrova
Zorba the Greek

1965
Shelley Winters
A Patch of Blue

1966
Sandy Dennis
Who's Afraid of Virginia Woolf?

1967
Estelle Parsons
Bonnie & Clyde

1968
Ruth Gordon
Rosemary's Baby

1969
Goldie Hawn
Cactus Flower

1970
Helen Hayes
Airport

1971
Cloris Leachman
The Last Picture Show

1972
Eileen Heckart
Butterflies Are Free

1973
Tatum O'Neal
Paper Moon

1974
Ingrid Bergman
Murder on the Orient Express

1975
Lee Grant
Shampoo

1976
Beatrice Straight
Network

1977
Vanessa Redgrave
Julia

1978
Maggie Smith
California Suite

1979
Meryl Streep
Kramer vs. Kramer

1980
Mary Steenburgen
Melvin and Howard

1981
Maureen Stapleton
Reds

1982
Jessica Lange
Tootsie

1983
Linda Hunt
The Year of Living Dangerously

1984
Peggy Ashcroft
A Passage to India

1985
Anjelica Huston
Prizzi's Honor

1986
Dianne Wiest
Hannah and Her Sisters

1987
Olympia Dukakis
Moonstruck

1988
Geena Davis
The Accidental Tourist

1989
Brenda Fricker
My Left Foot

1990
Whoopi Goldberg
Ghost

1991
Mercedes Ruehl
The Fisher King

ART DIRECTION

1934
Merry Widow

1937
Lost Horizon

1939
Gone with the Wind

1957
Sayonara

1958
Gigi

1967
Camelot

1968
Oliver!

1969
Hello, Dolly!

1970
Patton

1971
Nicholas and Alexandra

1972
Cabaret

1973
The Sting

1974
The Godfather: Part 2

1975
Barry Lyndon

1976
All the President's Men

1977
Star Wars

1978
Heaven Can Wait

1979
All That Jazz

1980
Tess

1981
Raiders of the Lost Ark

1982
Gandhi

1983
Fanny and Alexander

1984
Amadeus

1985
Out of Africa

1986
A Room with a View

1987
The Last Emperor

1990
Dick Tracy

1991
Bugsy

ART DIRECTION (B & W)

1940
Pride and Prejudice

1941
Blossoms in the Dust

1943
The Song of Bernadette

1944
Gaslight

1945
Blood on the Sun

1947
Great Expectations

1948
Hamlet

1949
The Heiress

1950
Sunset Boulevard

1951
A Streetcar Named Desire

1952
The Bad and the Beautiful

1953
Julius Caesar

1954
On the Waterfront

1955
The Rose Tattoo

1956
Somebody Up There Likes Me

1959
The Diary of Anne Frank

1960
The Apartment

1961
The Hustler

1962
To Kill a Mockingbird

1964
Zorba the Greek

1965
Ship of Fools

1966
Who's Afraid of Virginia Woolf?

ART DIRECTION (COLOR)

1943
Phantom of the Opera

1946
The Yearling

1947
Black Narcissus

1948
The Red Shoes

1949
Little Women

1950
Samson and Delilah

1951
An American in Paris

1952
Moulin Rouge

1953
The Robe

1954
20,000 Leagues Under the Sea

1955
Picnic

1956
The King and I

1959
Ben-Hur

1960
Spartacus

1961
West Side Story

1962
Lawrence of Arabia

1963
Cleopatra

1964
My Fair Lady

1965
Doctor Zhivago

1966
Fantastic Voyage

CINEMATOGRAPHY

1928
Sunrise

1931
Tabu

1932
A Farewell to Arms

1934
Cleopatra

1935
A Midsummer Night's Dream

1936
Anthony Adverse

1937
The Good Earth

1963
Cleopatra

1967
Bonnie & Clyde

1968
Romeo and Juliet

1969
Butch Cassidy and the Sundance Kid

1970
Ryan's Daughter

1971
Fiddler on the Roof

1972
Cabaret

1973
Cries and Whispers

1974
Towering Inferno

1975
Barry Lyndon

1976
Bound for Glory

1977
Close Encounters of the Third Kind

1978
Days of Heaven

1979
Apocalypse Now

1980
Tess

1981
Reds

1982
Gandhi

1983
Fanny and Alexander

1984
The Killing Fields

1985
Out of Africa

1986
The Mission

1987
The Last Emperor

1988
Mississippi Burning

1989
Glory

1990
Dances with Wolves

1991
JFK

CINEMATOGRAPHY (B & W)

1939
Wuthering Heights

1940
Rebecca

1941
How Green was My Valley

1942
Mrs. Miniver

1974
Francis Ford Coppola
The Godfather: Part 2

1975
Milos Forman
One Flew Over the Cuckoo's Nest

1976
John G. Avildsen
Rocky

1977
Woody Allen
Annie Hall

1978
Michael Cimino
The Deer Hunter

1979
Robert Benton
Kramer vs. Kramer

1980
Robert Redford
Ordinary People

1981
Warren Beatty
Reds

1982
Richard Attenborough
Gandhi

1983
James L. Brooks
Terms of Endearment

1984
Milos Forman
Amadeus

1985
Sydney Pollack
Out of Africa

1986
Oliver Stone
Platoon

1987
Bernardo Bertolucci
The Last Emperor

1988
Barry Levinson
Rain Man

1989
Oliver Stone
Born on the Fourth of July

1990
Kevin Costner
Dances with Wolves

1991
Jonathan Demme
The Silence of the Lambs

FILM

1927
Wings

1929
The Broadway Melody

1930
All Quiet on the Western Front

1931
Cimarron

1932
Grand Hotel

1934
It Happened One Night

1935
Mutiny on the Bounty

1936
The Great Ziegfeld

1937
The Life of Emile Zola

1938
You Can't Take It with You

1939
Gone with the Wind

1940
Rebecca

1941
How Green was My Valley

1942
Mrs. Miniver

1943
Casablanca

1944
Going My Way

1945
The Lost Weekend

1946
The Best Years of Our Lives

1948
Hamlet

1949
All the King's Men

1950
All About Eve

1951
An American in Paris

1952
Greatest Show on Earth

1953
From Here to Eternity

1954
On the Waterfront

1955
Marty

1956
Around the World in 80 Days

1957
The Bridge on the River Kwai

1958
Gigi

1959
Ben-Hur

1960
The Apartment

1961
West Side Story

1962
Lawrence of Arabia

1963
Tom Jones

1964
My Fair Lady

1965
The Sound of Music

1966
A Man for All Seasons

1967
In the Heat of the Night

1968
Oliver!

1969
Midnight Cowboy

1970
Patton

1971
The French Connection

1972
The Godfather

1973
The Sting

1974
The Godfather: Part 2

1975
One Flew Over the Cuckoo's Nest

1976
Rocky

1977
Annie Hall

1978
The Deer Hunter

1979
Kramer vs. Kramer

1980
Ordinary People

1981
Chariots of Fire

1982
Gandhi

1983
Terms of Endearment

1984
Amadeus

1985
Out of Africa

1986
Platoon

1987
The Last Emperor

1988
Rain Man

1989
Driving Miss Daisy

1990
Dances with Wolves

1991
The Silence of the Lambs

FILM— DOCUMENTARY

1952
The Sea Around Us

1953
Living Desert

1954
Vanishing Prairie

1961
The Sky Above, the Mud Below

1970
Woodstock

1971
Hellstrom Chronicle

1972
Marjoe

1973
Great American Cowboy

1975
The Man Who Skied Down Everest

1976
Harlan County, U.S.A.

1983
Times of Harvey Milk

1989
Common Threads: Stories from the Quilt

FILM EDITING

1935
A Midsummer Night's Dream

1936
Anthony Adverse

1937
Lost Horizon

1938
The Adventures of Robin Hood

1939
Gone with the Wind

1941
Sergeant York

1942
The Pride of the Yankees

1943
Air Force

1945
National Velvet

1946
The Best Years of Our Lives

1947
Body and Soul

1949
Champion

1950
King Solomon's Mines

1951
A Place in the Sun

1952
High Noon

1953
From Here to Eternity

1954
On the Waterfront

1955
Picnic

1956
Around the World in 80 Days

1957
The Bridge on the River Kwai

1958
Gigi

1960
The Apartment

1961
West Side Story

1962
Lawrence of Arabia

1963
How the West was Won

1964
Mary Poppins

1965
The Sound of Music

1966
Grand Prix

1968
Bullitt

1969
Z

1970
Patton

1971
The French Connection

1972
Cabaret

1973
The Sting

1974
Towering Inferno

1975
Jaws

1976
Rocky

1977
Star Wars

1978
The Deer Hunter

1979
All That Jazz

1980
Raging Bull

1981
Raiders of the Lost Ark

1982
Gandhi

1983
The Right Stuff

1984
The Killing Fields

1985
Witness

1986
Platoon

1987
The Last Emperor

1988
Who Framed Roger Rabbit?

1989
Born on the Fourth of July

1990
Dances with Wolves

1991
JFK

FILM—FOREIGN LANGUAGE

1947
Monsieur Vincent

1949
The Bicycle Thief

1951
Rashomon

1952
Forbidden Games

1954
Gate of Hell

FILM—FOREIGN LANGUAGE
▶cont.

1955
Samurai 1: Musashi Miyamoto

1956
La Strada

1957
Nights of Cabiria

1958
Mon Oncle

1959
Black Orpheus

1960
The Virgin Spring

1961
Through a Glass Darkly

1962
Sundays & Cybele

1963
8 1/2

1964
Yesterday, Today and Tomorrow

1965
Shop on Main Street

1966
A Man and a Woman

1967
Closely Watched Trains

1968
War and Peace

1969
Z

1971
Garden of the Finzi-Continis

1972
The Discreet Charm of the Bourgeoisie

1973
Day for Night

1974
Amarcord

1975
Dersu Uzala

1976
Black and White in Color

1977
Madame Rosa

1978
Get Out Your Handkerchiefs

1979
The Tin Drum

1980
Moscow Does Not Believe in Tears

1981
Mephisto

1983
Fanny and Alexander

1984
Dangerous Moves

1985
The Official Story

1986
The Assault

1987
Babette's Feast

1988
Pelle the Conqueror

1989
Cinema Paradiso

1990
Journey of Hope

MAKEUP

1964
Seven Faces of Dr. Lao

1968
Planet of the Apes

1981
An American Werewolf in London

1982
Quest for Fire

1984
Amadeus

1986
The Fly

1987
Harry and the Hendersons

1989
Driving Miss Daisy

1990
Dick Tracy

1991
Terminator 2: Judgement Day

MUSIC

1934
One Night of Love

1935
The Informer

1936
Anthony Adverse

1938
The Adventures of Robin Hood

1939
Stagecoach
The Wizard of Oz

1940
Pinocchio

1941
Dumbo

1942
Now, Voyager
Yankee Doodle Dandy

1943
The Song of Bernadette
This is the Army

1944
Cover Girl
Since You Went Away

1945
Anchors Aweigh
Spellbound

1946
The Best Years of Our Lives
The Jolson Story

1947
A Double Life

1948
Easter Parade
The Red Shoes

1949
The Heiress
On the Town

1950
Sunset Boulevard

1951
An American in Paris
A Place in the Sun

1952
High Noon

1953
Lili

1954
Seven Brides for Seven Brothers

1955
Love is a Many-Splendored Thing
Oklahoma!

1956
Around the World in 80 Days
The King and I

1957
The Bridge on the River Kwai

1958
Gigi

1959
Ben-Hur

1960
Exodus
Song Without End

1961
Breakfast at Tiffany's
West Side Story

1962
Lawrence of Arabia
The Music Man

1963
Irma La Douce
Tom Jones

1964
Mary Poppins
My Fair Lady

1965
Doctor Zhivago
The Sound of Music

1966
Born Free
A Funny Thing Happened on the Way to the Forum

1967
Camelot
Thoroughly Modern Millie

1968
The Lion in Winter
Oliver!

1969
Butch Cassidy and the Sundance Kid
Hello, Dolly!

1970
Let It Be
Love Story

1971
Fiddler on the Roof
Summer of '42

1972
Cabaret

1973
The Sting
The Way We Were

1974
Nino Rota
The Godfather: Part 2
The Great Gatsby

1975
Barry Lyndon
Jaws

1976
Bound for Glory
The Omen

1977
A Little Night Music
Star Wars

1978
Buddy Holly Story
Midnight Express

1979
All That Jazz
A Little Romance

1980
Fame

1981
Chariots of Fire

1982
E.T.: The Extra-Terrestrial
Victor/Victoria

1983
The Right Stuff
Yentl

1984
A Passage to India
Purple Rain

1985
Out of Africa

1986
Round Midnight

1987
The Last Emperor

1989
The Little Mermaid

1990
Dances with Wolves

1991
Beauty and the Beast

SONG

1934
"The Continental"
The Gay Divorcee

1936
"The Way You Look Tonight"
Swing Time

1939
"Over the Rainbow"
The Wizard of Oz

1940
"When You Wish Upon a Star"
Pinocchio

1941
"The Last Time I Saw Paris"
Lady Be Good

1942
"White Christmas"
Holiday Inn

1944
"Swinging on a Star"
Going My Way

1945
"It Might as Well Be Spring"
State Fair

1946
"On the Atchison, Topeka and Santa Fe"
The Harvey Girls

1948
"Buttons and Bows"
The Paleface

1949
"Baby It's Cold Outside"
Neptune's Daughter

1951
"In the Cool, Cool, Cool of the Evening"
Here Comes the Groom

1952
"High Noon (Do Not Forsake Me, Oh My Darlin)"
High Noon

1953
"Secret Love"
Calamity Jane

1955
"Love Is a Many-Splendored Thing"
Love is a Many-Splendored Thing

1956
"Whatever Will Be, Will Be (Que Sera, Sera)"
The Man Who Knew Too Much

1958
"Gigi"
Gigi

1959
"High Hopes"
Hole in the Head

1960
"Never on Sunday"
Never on Sunday

1961
"Moon River"
Breakfast at Tiffany's

1962
"Days of Wine and Roses"
Days of Wine and Roses

1963
"Call Me Irresponsible"
Papa's Delicate Condition

1964
"Chim Chim Cher-ee"
Mary Poppins

1965
"The Shadow of Your Smile"
The Sandpiper

1966
"Born Free"
Born Free

1967
"Talk to the Animals"
Doctor Doolittle

1968
"Windmills of Your Mind"
Thomas Crown Affair

1969
"Raindrops Keep Fallin'
on My Head"
Butch Cassidy and the
Sundance Kid

1970
"For All We Know"
Lovers and Other
Strangers

1971
Theme from "Shaft"
Shaft

1972
"The Morning After"
The Poseidon Adventure

1973
"The Way We Were"
The Way We Were

1974
"We May Never Love Like
this Again"
Towering Inferno

1975
"I'm Easy"
Nashville

1976
"Evergreen"
A Star is Born

1977
"You Light Up My Life"
You Light Up My Life

1978
"Last Dance"
Thank God It's Friday

1979
"It Goes Like It Goes"
Norma Rae

1980
"Fame"
Fame

1981
"Arthur's Theme"
Arthur

1982
"Up Where We Belong"
An Officer and a
Gentleman

1983
"Flashdance...What a
Feeling"
Flashdance

1984
"I Just Called to Say I
Love You"
The Woman in Red

1985
"Say You, Say Me"
White Nights

1986
"Take My Breath Away"
Top Gun

1987
"The Time of My Life"
Dirty Dancing

1988
"Let the River Run"
Working Girl

1989
"Under the Sea"
The Little Mermaid

1990
"Sooner or Later"
Dick Tracy

1991
"Beauty and the Beast"
Beauty and the Beast

SOUND

1932
A Farewell to Arms

1934
One Night of Love

1935
Naughty Marietta

1936
San Francisco

1937
Hurricane

1940
Strike Up the Band

1941
That Hamilton Woman

1942
Yankee Doodle Dandy

1943
This Land is Mine

1945
The Bells of St. Mary's

1946
The Jolson Story

1947
The Bishop's Wife

1949
Twelve O'Clock High

1950
All About Eve

1951
The Great Caruso

1953
From Here to Eternity

1954
The Glenn Miller Story

1955
Oklahoma!

1956
The King and I

1957
Sayonara

1958
South Pacific

1960
The Alamo

1961
West Side Story

1962
Lawrence of Arabia

1963
How the West was Won

1965
The Sound of Music

1966
Grand Prix

1968
Oliver!

1969
Hello, Dolly!

1970
Patton

1971
Fiddler on the Roof

1972
Cabaret

1973
The Exorcist

1974
Earthquake

1975
Jaws

1976
All the President's Men

1977
Star Wars

1978
The Deer Hunter

1979
Apocalypse Now

1981
Raiders of the Lost Ark

1982
E.T.: The Extra-Terrestrial

1983
The Right Stuff

1984
Amadeus

1985
Out of Africa

1986
Platoon

1987
The Last Emperor

1988
Bird

1989
Glory

1990
Dances with Wolves

1991
Terminator 2: Judgement
Day

WRITING

1931
Cimarron

1932
The Champ

1933
Little Women

1934
It Happened One Night

1935
The Informer

1936
The Story of Louis
Pasteur

1937
The Life of Emile Zola
A Star is Born

1938
Boys Town
Pygmalion

1939
Gone with the Wind
Mr. Smith Goes to
Washington

1940
The Great McGinty
Philadelphia Story

1941
Citizen Kane
Here Comes Mr. Jordan

1942
Mrs. Miniver
Woman of the Year

1943
Casablanca
The Human Comedy

1944
Going My Way

1945
The Lost Weekend

1946
The Best Years of Our
Lives
Seventh Veil

1947
The Bachelor and the
Bobby-Soxer
Miracle on 34th Street

1948
The Search
Treasure of the Sierra
Madre

1949
Battleground
A Letter to Three Wives

1950
All About Eve
Panic in the Streets
Sunset Boulevard

1951
An American in Paris
A Place in the Sun

1952
The Bad and the Beautiful
Greatest Show on Earth
The Lavender Hill Mob

1953
From Here to Eternity
Roman Holiday

1954
Broken Lance
Country Girl
On the Waterfront

1955
Love Me or Leave Me
Marty

1956
Around the World in 80
Days
The Brave One
The Red Balloon

1957
The Bridge on the River
Kwai
George Wells
Designing Woman

1958
Defiant Ones
Gigi

1959
Pillow Talk
Room at the Top

1960
The Apartment
Elmer Gantry

1961
Judgment at Nuremberg
Splendor in the Grass

1962
Divorce—Italian Style
To Kill a Mockingbird

1963
How the West was Won
Tom Jones

1964
Becket
Father Goose

1965
Darling
Doctor Zhivago

1966
A Man for All Seasons
A Man and a Woman

1967
Guess Who's Coming to
Dinner

1968
The Lion in Winter
The Producers

1969
Butch Cassidy and the
Sundance Kid
Midnight Cowboy

1970
M*A*S*H
Patton

1971
The French Connection
The Hospital

1972
The Candidate
The Godfather

1973
The Exorcist
The Sting

1974
Chinatown
The Godfather: Part 2

1975
Dog Day Afternoon
One Flew Over the
Cuckoo's Nest

1976
All the President's Men
Network

1977
Annie Hall
Julia

1978
Coming Home
Midnight Express

1979
Breaking Away
Kramer vs. Kramer

1980
Melvin and Howard
Ordinary People

1981
Chariots of Fire
On Golden Pond

1982
Gandhi
Missing

WRITING ▶cont.

1983
Tender Mercies
Terms of Endearment

1984
Amadeus
Places in the Heart

1985
Out of Africa
Witness

1986
Hannah and Her Sisters
A Room with a View

1987
The Last Emperor
Moonstruck

1988
Rain Man

1989
Driving Miss Daisy

1990
Dances with Wolves
Ghost

1991
The Silence of the Lambs
Thelma & Louise

AUSTRALIAN FILM INSTITUTE

ACTOR

1974
Jack Thompson
Sunday Too Far Away

1975
John Meillon
Fourth Wish

1976
Nick Tate
The Devil's Playground
Simon Burke
The Devil's Playground

1978
Bill Hunter
Newsfront

1979
Mel Gibson
Tim

1980
Jack Thompson
Breaker Morant

1981
Mel Gibson
Gallipoli

1983
Norman Kaye
Man of Flowers

1984
John Hargreaves
My First Wife

1986
Colin Friels
Malcolm

1987
Leo McKern
Travelling North

ACTRESS

1977
Pat Bishop
Don's Party

1980
Tracey Mann
Hard Knocks

1981
Judy Davis
The Winter of Our Dreams

1982
Noni Hazlehurst
Monkey Grip

1983
Wendy Hughes
Careful, He Might Hear You

1985
Noni Hazlehurst
Fran

1986
Judy Davis
Kangaroo

1987
Judy Davis
High Tide

BRITISH ACADEMY OF FILM AND TELEVISION ARTS

ACTOR

1953
Marlon Brando
Julius Caesar
John Gielgud
Julius Caesar

1954
Marlon Brando
On the Waterfront

1955
Ernest Borgnine
Marty
Laurence Olivier
Richard III

1956
Francois Perier
Gervaise

1957
Alec Guinness
The Bridge on the River Kwai
Henry Fonda
Twelve Angry Men

1958
Trevor Howard
The Key

1960
Jack Lemmon
The Apartment

1961
Paul Newman
The Hustler

1962
Marcello Mastroianni
Divorce—Italian Style

Peter O'Toole
Lawrence of Arabia

1963
Dirk Bogarde
The Servant
Richard Harris
This Sporting Life

1964
Richard Attenborough
Seance on a Wet Afternoon

1965
Dirk Bogarde
Darling

1966
Rod Steiger
Pawnbroker
Richard Burton
Who's Afraid of Virginia Woolf?

1967
Paul Scofield
A Man for All Seasons

1969
Dustin Hoffman
Midnight Cowboy

1971
Gene Hackman
The French Connection
Peter Finch
Sunday, Bloody Sunday

1973
Walter Matthau
Pete 'n' Tillie

1974
Jack Nicholson
Chinatown
Jack Nicholson
The Last Detail

1975
Al Pacino
Dog Day Afternoon

1976
Jack Nicholson
One Flew Over the Cuckoo's Nest

1979
Jack Lemmon
The China Syndrome

1980
John Hurt
Elephant Man
John Hurt
The Naked Civil Servant

1986
Bob Hoskins
Mona Lisa

1987
Sean Connery
The Name of the Rose

1989
Daniel Day Lewis
My Left Foot

ACTOR— SUPPORTING

1974
John Gielgud
Murder on the Orient Express

1976
Brad Dourif
One Flew Over the Cuckoo's Nest

1978
John Hurt
Midnight Express

1981
Ian Holm
Chariots of Fire

1986
Ray McAnally
The Mission

1987
Daniel Auteuil
Jean de Florette

ACTRESS

1952
Vivien Leigh
A Streetcar Named Desire

1953
Leslie Caron
Lili
Audrey Hepburn
Roman Holiday

1955
Katie Johnson
Ladykillers
Betsy Blair
Marty

1959
Audrey Hepburn
The Nun's Story

1960
Shirley MacLaine
The Apartment

1961
Sophia Loren
Two Women

1962
Anne Bancroft
The Miracle Worker

1965
Julie Christie
Darling

1966
Elizabeth Taylor
Who's Afraid of Virginia Woolf?

1971
Glenda Jackson
Sunday, Bloody Sunday

1972
Liza Minnelli
Cabaret

1973
Delphine Seyrig
The Discreet Charm of the Bourgeoisie

1975
Ellen Burstyn
Alice Doesn't Live Here Anymore

1976
Louise Fletcher
One Flew Over the Cuckoo's Nest

1977
Diane Keaton
Annie Hall

1978
Jane Fonda
Julia

1979
Jane Fonda
The China Syndrome

1980
Judy Davis
My Brilliant Career

1981
Meryl Streep
The French Lieutenant's Woman

1986
Maggie Smith
A Room with a View

1987
Anne Bancroft
84 Charing Cross Road

ACTRESS— SUPPORTING

1973
Valentina Cortese
Day for Night

1974
Ingrid Bergman
Murder on the Orient Express

1975
Diane Ladd
Alice Doesn't Live Here Anymore

1977
Jenny Agutter
Equus

1978
Julia

1986
Judy Dench
A Room with a View

1988
Michelle Pfeiffer
Dangerous Liaisons

ART DIRECTION

1968
2001: A Space Odyssey

1971
Death in Venice

1972
Cabaret

1978
Close Encounters of the Third Kind

1979
Alien

1980
Elephant Man

ART DIRECTION (B & W)

1965
Darling

ART DIRECTION (COLOR)

1964
Becket
1965
The Ipcress File
1967
A Man for All Seasons

CINEMATOGRAPHY

1968
2001: A Space Odyssey
1971
Death in Venice
1975
Barry Lyndon
1978
Julia
1979
The Deer Hunter
1980
All That Jazz
1986
A Room with a View

CINEMATOGRAPHY (B & W)

1963
The Servant

CINEMATOGRAPHY (COLOR)

1965
The Ipcress File
1967
A Man for All Seasons

COSTUME DESIGN

1971
Death in Venice
1972
Macbeth
1975
Day of the Locust
1978
Death on the Nile
1980
Kagemusha
1981
Chariots of Fire

COSTUME DESIGN (COLOR)

1964
Becket

DIRECTOR

1955
Laurence Olivier
Richard III

1968
Mike Nichols
The Graduate
1969
John Schlesinger
Midnight Cowboy
1971
John Schlesinger
Sunday, Bloody Sunday
1972
Bob Fosse
Cabaret
1973
Francois Truffaut
Day for Night
1974
Roman Polanski
Chinatown
1975
Stanley Kubrick
Barry Lyndon
1976
Milos Forman
One Flew Over the Cuckoo's Nest
1977
Woody Allen
Annie Hall
1978
Alan Parker
Midnight Express
1979
Francis Ford Coppola
Apocalypse Now
1980
Akira Kurosawa
Kagemusha
1987
John Boorman
Hope and Glory

FILM

1954
Wages of Fear
1959
Ben-Hur
1960
The Apartment
1964
Dr. Strangelove, or: How I Learned to Stop Worrying and Love the Bomb
1973
Day for Night
1979
Manhattan
1980
Elephant Man
1983
Educating Rita
1984
The Killing Fields
1986
A Room with a View
1987
Jean de Florette

FILM— DOCUMENTARY

1948
Louisiana Story

FILM EDITING

1968
The Graduate
1969
Midnight Cowboy
1971
The French Connection
1974
The Conversation
1975
Dog Day Afternoon
1976
One Flew Over the Cuckoo's Nest
1977
Annie Hall
1978
Midnight Express
1979
The Deer Hunter
1986
The Mission

MUSIC

1969
Z
1971
Death in Venice
1972
Cabaret
The Godfather
1974
The Conversation
Murder on the Orient Express
1975
Jaws
Nashville
1979
Days of Heaven
1981
The French Lieutenant's Woman
1982
E.T.: The Extra-Terrestrial
1986
The Mission
1987
The Untouchables

SOUND

1979
Alien
1981
The French Lieutenant's Woman
1987
Cry Freedom

WRITING

1955
Ladykillers
1957
The Bridge on the River Kwai
1962
Lawrence of Arabia
1965
Darling
1967
A Man for All Seasons
1968
The Graduate
1969
Midnight Cowboy
1972
tie
The Hospital
1973
The Discreet Charm of the Bourgeoisie
1974
Chinatown
1977
Annie Hall
1978
Julia
1979
Manhattan
1980
Being There

CANNES

ACTOR

1946
Ray Milland
The Lost Weekend
1953
Charles Vanel
Wages of Fear
1963
Richard Harris
This Sporting Life
1966
Per Oscarsson
Hunger
1969
Jean-Louis Trintignant
Z
1974
Jack Nicholson
The Last Detail
1978
Jon Voight
Coming Home
1979
Jack Lemmon
The China Syndrome
1982
Jack Lemmon
Missing
1983
Malou
1984
Alfredo Landa
The Holy Innocents

Francisco Rabal
The Holy Innocents
1985
William Hurt
Kiss of the Spider Woman
1986
Bob Hoskins
Mona Lisa
1987
Marcello Mastroianni
Dark Eyes
1988
Forest Whitaker
Bird
1989
James Spader
sex, lies and videotape
1991
John Turturro
Barton Fink

ACTOR— SUPPORTING

ACTRESS

1951
Bette Davis
All About Eve
1956
Susan Hayward
I'll Cry Tomorrow
1957
Giulietta Masina
Nights of Cabiria
1958
collective
Brink of Life
1959
Hiroshima, Mon Amour
1961
Sophia Loren
Two Women
1967
Pia Degermark
Elvira Madigan
1969
Vanessa Redgrave
Isadora
1975
Valerie Perrine
Lenny
1978
Jill Clayburgh
An Unmarried Woman
1979
Sally Field
Norma Rae
1982
Krystyna Janda
Interrogation
1984
Helen Mirren
Cal
1985
Cher
Mask
1988
Barbara Hershey
Jodhi May
A World Apart

ACTRESS ▶cont.

Linda Mvusi
A World Apart
1989
Meryl Streep
A Cry in the Dark
1991
Irene Jacob
The Double Life of Veronique

ACTRESS—
SUPPORTING

1978
Eva Mattes
Woyzeck

DIRECTOR

1951
Luis Bunuel
Los Olvidados
1957
Robert Bresson
Man Escaped
1959
Francois Truffaut
The 400 Blows
1969
Voltech Jasny
All My Good Countrymen
1976
Ettore Scola
Down & Dirty
1978
Nagisa Oshima
The Empire of Passion
1979
Terence Malick
Days of Heaven
1982
Werner Herzog
Fitzcarraldo
1984
Bertrand Tavernier
A Sunday in the Country
1986
Martin Scorsese
After Hours
1990
Emir Kusturica
Time of the Gypsies
1991
Joel Coen
Barton Fink

WRITING

1982
Moonlighting

GENIE

ACTOR

1980
Christopher Plummer
Murder by Decree

1982
Nick Mancuso
Ticket to Heaven
1983
Donald Sutherland
Threshold
1984
Eric Fryer
The Terry Fox Story
1986
John Wildman
My American Cousin
1987
Gordon Pinsent
John and the Missus
1988
Roger Le Bel
Night Zoo
1989
Jeremy Irons
Dead Ringers
1990
Remy Girard
Jesus of Montreal

ACTOR—
SUPPORTING

1981
German Houde
Les Bons Debarras
1982
Saul Rubinek
Ticket to Heaven
1983
R.H. Thomson
If You Could See What I Hear
1984
Michael Zelniker
The Terry Fox Story
1985
Alan Scarfe
The Bay Boy
1986
Alan Arkin
Joshua Then and Now
1987
Gabriel Arcand
The Decline of the American Empire
1988
German Houde
Night Zoo

ACTRESS

1980
Kate Lynch
Meatballs
1981
Marie Tifo
Les Bons Debarras
1982
Margot Kidder
Heartaches
1983
Rae Dawn Chong
Quest for Fire
1986
Margaret Langrick
My American Cousin

1987
Martha Henry
Dancing in the Dark
Sheila McCarthy
I've Heard the Mermaids Singing
1990
Rebecca Jenkins
Bye Bye Blues

ACTRESS—
SUPPORTING

1980
Genevieve Bujold
Murder by Decree
1981
Kate Reid
Atlantic City
1983
Jackie Burroughs
The Grey Fox
1986
Linda Sorensen
Joshua Then and Now
1987
Louise Portal
The Decline of the American Empire
Paule Baillargeon
I've Heard the Mermaids Singing
1989
Colleen Dewhurst
Obsessed
1990
Robyn Stevan
Bye Bye Blues

ART DIRECTION

1980
The Changeling
1981
Atlantic City
1983
The Grey Fox
1984
Maria Chapdelaine
1985
The Bay Boy
1986
Joshua Then and Now
1987
Dancing in the Dark
1988
Night Zoo
1989
Dead Ringers
1990
Jesus of Montreal

CINEMATOGRAPHY

1980
The Changeling
1981
Les Bons Debarras
1982
Silence of the North

1983
Threshold
1984
Maria Chapdelaine
1986
Joshua Then and Now
1988
Night Zoo
1989
Dead Ringers
1990
Jesus of Montreal

COSTUME DESIGN

1983
Quest for Fire
1984
Maria Chapdelaine
1985
The Bay Boy
1986
Joshua Then and Now
1987
Loyalties
1988
Night Zoo
1990
Jesus of Montreal

DIRECTOR

1980
Bob Clark
Murder by Decree
1981
Francis Mankiewicz
Les Bons Debarras
1983
Phillip Borsos
The Grey Fox
1984
Bob Clark
A Christmas Story
David Cronenberg
Videodrome
1986
Sandy Wilson
My American Cousin
1987
Denys Arcand
The Decline of the American Empire
1988
Jean-Claude Lauzon
Night Zoo
1989
David Cronenberg
Dead Ringers
1990
Denys Arcand
Jesus of Montreal

FILM

1980
The Changeling
1981
Les Bons Debarras

1982
Ticket to Heaven
1983
The Grey Fox
1984
The Terry Fox Story
1985
The Bay Boy
1986
My American Cousin
1987
The Decline of the American Empire
1988
Night Zoo
1989
Dead Ringers
1990
Jesus of Montreal

MUSIC

1980
Murder by Decree
1983
The Grey Fox
1984
Maria Chapdelaine
1987
John and the Missus
1988
Night Zoo
1989
Dead Ringers
1990
Jesus of Montreal

WRITING

1980
Meatballs
1981
The Changeling
Les Bons Debarras
1982
Heartaches
1983
The Grey Fox
Melanie
1984
A Christmas Story
1985
The Bay Boy
1986
My American Cousin
1987
Dancing in the Dark
The Decline of the American Empire
1988
Night Zoo
1989
Dead Ringers
1990
Jesus of Montreal

GOLDEN GLOBE AWARDS

ACTOR

1944
Paul Lukas
Watch on the Rhine

1946
Ray Milland
The Lost Weekend

1947
Gregory Peck
The Yearling

1948
Ronald Colman
A Double Life

1949
Laurence Olivier
Hamlet

1950
Broderick Crawford
All the King's Men

1951
Jose Ferrer
Cyrano de Bergerac

1953
Gary Cooper
High Noon

1956
Ernest Borgnine
Marty
Marlon Brando
On the Waterfront

1957
Kirk Douglas
Lust for Life

1958
Alec Guinness
The Bridge on the River Kwai

1959
David Niven
Separate Tables

1960
Anthony (Tony) Franciosa
Career

1961
Burt Lancaster
Elmer Gantry

1962
Marcello Mastroianni
Divorce—Italian Style
Maximilian Schell
Judgment at Nuremberg

1963
Gregory Peck
To Kill a Mockingbird

1964
Sidney Poitier
Lilies of the Field

1965
Peter O'Toole
Becket

1966
Omar Sharif
Doctor Zhivago

1967
Paul Scofield
A Man for All Seasons

1968
Rod Steiger
In the Heat of the Night

1969
Peter O'Toole
The Lion in Winter

1970
John Wayne
True Grit

1971
George C. Scott
Patton

1972
Gene Hackman
The French Connection

1973
Marlon Brando
The Godfather

1974
Al Pacino
Serpico

1975
Jack Nicholson
Chinatown

1976
Jack Nicholson
One Flew Over the Cuckoo's Nest

1977
Peter Finch
Network

1978
Richard Burton
Equus

1979
Jon Voight
Coming Home

1980
Dustin Hoffman
Kramer vs. Kramer

1981
Robert De Niro
Raging Bull

1982
Henry Fonda
On Golden Pond

1983
Ben Kingsley
Gandhi
Robert Duvall
Tender Mercies

1984
Tom Courtenay
Dresser

1985
F. Murray Abraham
Amadeus

1986
Jon Voight
Runaway Train

1987
Bob Hoskins
Mona Lisa

1988
Michael Douglas
Wall Street

1989
Dustin Hoffman
Rain Man

1990
Tom Cruise
Born on the Fourth of July

1991
Gerard Depardieu
Green Card
Jeremy Irons
Reversal of Fortune

1992
Robin Williams
The Fisher King
Nick Nolte
The Prince of Tides

ACTOR— SUPPORTING

1965
Edmond O'Brien
Seven Days in May

1966
Oskar Werner
The Spy Who Came in from the Cold

1967
Richard Attenborough
The Sand Pebbles

1968
David Attenborough
Doctor Doolittle

1970
Gig Young
They Shoot Horses, Don't They?

1971
John Mills
Ryan's Daughter

1972
Ben Johnson
The Last Picture Show

1973
Joel Grey
Cabaret

1974
John Houseman
The Paper Chase

1975
Fred Astaire
Towering Inferno

1976
Richard Benjamin
The Sunshine Boys

1977
Laurence Olivier
Marathon Man

1978
Peter Firth
Equus

1979
John Hurt
Midnight Express

1980
Robert Duvall
Apocalypse Now
Melvyn Douglas
Being There

1981
Timothy Hutton
Ordinary People

1982
John Gielgud
Arthur

1983
Louis Gossett, Jr.
An Officer and a Gentleman

1984
Jack Nicholson
Terms of Endearment

1985
Haing S. Ngor
The Killing Fields

1986
Klaus Maria Brandauer
Out of Africa

1987
Tom Berenger
Platoon

1988
Sean Connery
The Untouchables

1989
Martin Landau
Tucker: The Man and His Dream

1990
Denzel Washington
Glory

1991
Bruce Davison
Longtime Companion

1992
Jack Palance
City Slickers

ACTRESS

1944
Jennifer Jones
The Song of Bernadette

1945
Ingrid Bergman
Gaslight

1946
Ingrid Bergman
The Bells of St. Mary's

1947
Rosalind Russell
Sister Kenny

1949
Jane Wyman
Johnny Belinda

1950
Olivia de Havilland
The Heiress

1951
Gloria Swanson
Sunset Boulevard

1953
Shirley Booth
Come Back, Little Sheba

1954
Audrey Hepburn
Roman Holiday

1955
Grace Kelly
Country Girl

1956
Anna Magnani
The Rose Tattoo

1957
Ingrid Bergman
Anastasia

1958
Joanne Woodward
The Three Faces of Eve

1959
Susan Hayward
I Want to Live!

1960
Elizabeth Taylor
Suddenly, Last Summer

1961
Greer Garson
Sunrise at Campobello

1962
Geraldine Page
Summer and Smoke

1963
Geraldine Page
Sweet Bird of Youth

1966
Samantha Eggar
The Collector

1967
Anouk Aimee
A Man and a Woman

1969
Joanne Woodward
Rachel, Rachel

1970
Genevieve Bujold
Anne of the Thousand Days

1971
Ali MacGraw
Love Story

1972
Jane Fonda
Klute

1974
Marsha Mason
Cinderella Liberty

1975
Gena Rowlands
A Woman Under the Influence

1976
Louise Fletcher
One Flew Over the Cuckoo's Nest

1977
Faye Dunaway
Network

1978
Jane Fonda
Julia

1979
Jane Fonda
Coming Home

1980
Sally Field
Norma Rae

1981
Mary Tyler Moore
Ordinary People

1982
Meryl Streep
The French Lieutenant's Woman

1983
Meryl Streep
Sophie's Choice

1984
Shirley MacLaine
Terms of Endearment

ACTRESS ▶cont.

1986
Whoopi Goldberg
The Color Purple

1987
Marlee Matlin
Children of a Lesser God

1988
Sally Kirkland
Anna
Sigourney Weaver
Gorillas in the Mist

1989
Jodie Foster
The Accused
Shirley MacLaine
Madame Sousatzka
Julia Roberts
Pretty Woman

1990
Michelle Pfeiffer
The Fabulous Baker Boys

1991
Kathy Bates
Misery

1992
Bette Midler
For the Boys
Jodie Foster
The Silence of the Lambs

ACTRESS— SUPPORTING

1965
Agnes Moorehead
Hush, Hush, Sweet Charlotte

1966
Ruth Gordon
Inside Daisy Clover

1969
Ruth Gordon
Rosemary's Baby

1970
Goldie Hawn
Cactus Flower

1971
Maureen Stapleton
Airport
Karen Black
Five Easy Pieces

1972
Ann-Margret
Carnal Knowledge

1973
Shelley Winters
The Poseidon Adventure

1974
Linda Blair
The Exorcist

1975
Karen Black
The Great Gatsby

1976
Brenda Vaccaro
Once is Not Enough

1977
Katharine Ross
Voyage of the Damned

1978
Vanessa Redgrave
Julia

1979
Dyan Cannon
Heaven Can Wait

1980
Meryl Streep
Kramer vs. Kramer

1981
Mary Steenburgen
Melvin and Howard

1983
Joan Hackett
Only When I Laugh
Jessica Lange
Tootsie

1985
Meg Tilly
Agnes of God
Peggy Ashcroft
A Passage to India
Cher
Silkwood

1988
Olympia Dukakis
Moonstruck

1990
Whoopi Goldberg
Ghost
Julia Roberts
Steel Magnolias

1992
Mercedes Ruehl
The Fisher King

BEST FILM— DRAMA

1944
The Song of Bernadette

1945
Going My Way

1946
The Lost Weekend

1947
The Best Years of Our Lives

1949
Johnny Belinda
Treasure of the Sierra Madre

1950
All the King's Men

1951
Sunset Boulevard

1952
A Place in the Sun

1953
Greatest Show on Earth

1954
The Robe

1955
On the Waterfront

1956
East of Eden

1957
Around the World in 80 Days

1958
The Bridge on the River Kwai

1959
Defiant Ones

1960
Ben-Hur

1961
Spartacus

1962
The Guns of Navarone

1963
Lawrence of Arabia

1964
Cardinal

1965
Becket

1966
Doctor Zhivago

1967
A Man for All Seasons

1968
In the Heat of the Night

1969
The Lion in Winter

1970
Anne of the Thousand Days

1971
Love Story

1972
The French Connection

1973
The Godfather

1974
The Exorcist

1975
Chinatown

1976
One Flew Over the Cuckoo's Nest

1977
Rocky

1978
Turning Point

1979
Midnight Express

1980
Kramer vs. Kramer

1981
Ordinary People

1982
On Golden Pond

1983
E.T.: The Extra-Terrestrial

1984
Terms of Endearment

1985
Amadeus

1986
Out of Africa

1987
Platoon

1988
The Last Emperor

1989
Rain Man

1990
Born on the Fourth of July

1991
Dances with Wolves

1992
Bugsy

BEST FILM— MUSICAL/COMEDY

1952
An American in Paris

1955
Carmen Jones

1956
Guys & Dolls

1957
The King and I

1958
Les Girls

1959
Gigi

1960
Some Like It Hot

1961
The Apartment
Song Without End

1962
West Side Story

1963
The Music Man
That Touch of Mink

1964
Tom Jones

1965
My Fair Lady

1966
The Sound of Music

1967
The Russians are Coming, the Russians are Coming

1968
The Graduate

1969
Oliver!

1971
*M*A*S*H*

1972
Fiddler on the Roof

1973
Cabaret

1974
American Graffiti

1975
The Longest Yard

1976
The Sunshine Boys

1977
A Star is Born

1978
Goodbye Girl

1979
Heaven Can Wait

1980
Breaking Away

1981
Coal Miner's Daughter

1982
Arthur

1983
Tootsie

1984
Yentl

1985
Romancing the Stone

1986
Prizzi's Honor

1987
Hannah and Her Sisters

1988
Hope and Glory

1989
Working Girl

1990
Driving Miss Daisy

1991
Green Card

1992
Beauty and the Beast

DIRECTOR

1991
Kevin Costner
Dances with Wolves

1992
Oliver Stone
JFK

FILM—FOREIGN LANGUAGE

MUSIC

1990
The Sheltering Sky

WRITING

1991
Dances with Wolves

DIRECTORS GUILD OF AMERICA

DIRECTOR

1948
Joseph L. Mankiewicz
A Letter to Three Wives

1949
Carol Reed
The Third Man

1950
Joseph L. Mankiewicz
All About Eve

1951
George Stevens
A Place in the Sun

1952
John Ford
The Quiet Man

1953
Fred Zinneman
From Here to Eternity

1954
Elia Kazan
On the Waterfront

1955
Delbert Mann
Marty

1956
George Stevens
Giant

1957
David Lean
The Bridge on the River Kwai

1958
Vincente Minnelli
Gigi

1959
William Wyler
Ben-Hur

1960
Billy Wilder
The Apartment

1962
David Lean
Lawrence of Arabia

1964
George Cukor
My Fair Lady

1965
Robert Wise
The Sound of Music

1966
Fred Zinneman
A Man for All Seasons

1967
Mike Nichols
The Graduate

1969
John Schlesinger
Midnight Cowboy

1970
Francis Ford Coppola
Patton

1971
William Friedkin
The French Connection

1972
Francis Ford Coppola
The Godfather

1973
George Roy Hill
The Sting

1974
Francis Ford Coppola
The Godfather: Part 2

1975
Milos Forman
One Flew Over the Cuckoo's Nest

1977
Woody Allen
Annie Hall

1978
Michael Cimino
The Deer Hunter

1979
Robert Benton
Kramer vs. Kramer

1980
Robert Redford
Ordinary People

1981
Warren Beatty
Reds

1982
Richard Attenborough
Gandhi

1983
James L. Brooks
Terms of Endearment

1984
Milos Forman
Amadeus

1985
Steven Spielberg
The Color Purple

1986
Oliver Stone
Platoon

1987
Bernardo Bertolucci
The Last Emperor

1991
Jonathan Demme
The Silence of the Lambs

EDGAR ALLAN POE AWARDS

WRITING

1945
Murder My Sweet

1947
Crossfire

1949
Window

1950
Asphalt Jungle

1951
Detective Story

1953
Big Heat

1954
Rear Window

1955
Desperate Hours

1957
Twelve Angry Men

1958
Defiant Ones

1959
North by Northwest

1960
Psycho

1963
Charade

1964
Hush, Hush, Sweet Charlotte

1965
The Spy Who Came in from the Cold

1966
Harper

1967
In the Heat of the Night

1968
Bullitt

1969
Z

1971
The French Connection

1972
Sleuth

1973
The Last of Sheila

1974
Chinatown

1975
Three Days of the Condor

1976
Family Plot

1977
The Late Show

1981
Cutter's Way

1982
The Long Good Friday

1983
Gorky Park

1984
A Soldier's Story

1985
Witness

1986
Something Wild

1987
Stakeout

1988
The Thin Blue Line

1989
Heathers

MTV MOVIE AWARDS

ACTION SEQUENCE

1992
Terminator 2: Judgement Day

ACTOR

1992
Arnold Schwarzenegger
Terminator 2: Judgement Day

ACTRESS

1992
Linda Hamilton
Terminator 2: Judgement Day

BREAKTHROUGH PERFORMANCE

1992
Edward Furlong
Terminator 2: Judgement Day

COMEDIC PERFORMANCE

1992
Billy Crystal
City Slickers

KISS

1992
My Girl

MOST DESIRABLE FEMALE

1992
Linda Hamilton
Terminator 2: Judgement Day

MOST DESIRABLE MALE

1992
Keanu Reeves
Point Break

NEW FILMMAKER

1992
John Singleton
Boyz N the Hood

ON-SCREEN DUO

1992
Wayne's World

SONG

1992
''(Everything I Do) I Do for You''
Robin Hood: Prince of Thieves

VILLAIN

1992
Rebecca DeMornay
The Hand that Rocks the Cradle

FOREIGN FILM INDEX

The **Foreign Film Index** classifies movies by the country in which the feature was produced. Since the producers may hail from more than one country, some videos will be listed under more than one country. The codes appear in the review immediately following the color or black and white designation near the top of the review.

Key to Country Origination Tags

AR	Argentinean	*FR*	French	*NZ*	New Zealand
AU	Australian	*GE*	German	*NI*	Nicaraguan
BE	Belgian	*GR*	Greek	*NO*	Norwegian
BR	Brazilian	*HK*	Hong Kong	*PL*	Polish
GB	British	*HU*	Hungarian	*PT*	Portuguese
CA	Canadian	*IN*	Indian	*RU*	Russian
CH	Chinese	*IR*	Irish	*SA*	South African
CZ	Czech	*IS*	Israeli	*SP*	Spanish
DK	Danish	*IT*	Italian	*SW*	Swedish
NL	Dutch	*JP*	Japanese	*SI*	Swiss
PH	Filipino	*LI*	Lithuanian	*TU*	Turkish
FI	Finnish	*MX*	Mexican	*YU*	Yugoslavian

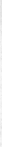

Argentinean

Camila
The Devil's Possessed
En Retirada (Bloody Retreat)
The Enchanted Forest
Eversmile New Jersey
Far Away and Long Ago
Fridays of Eternity
Funny, Dirty Little War
La Boca del Lobo
Made in Argentina
Man Facing Southeast
Miss Mary
The Official Story
The Stranger
Tango Bar
Time for Revenge
Times to Come
Two to Tango
Veronico Cruz

Australian

ABC's of Love & Sex,
    Australia Style
Air Hawk
Alice to Nowhere
Alison's Birthday
All the Rivers Run
...Almost
Alvin Purple
Alvin Rides Again
American Roulette
Anzacs: The War Down Under
Archer's Adventure
Around the World in 80 Ways
Attack Force Z
Backlash
Barry McKenzie Holds His
    Own
The Black Planet
Bliss
Blood Money
Bloodmoon
Blue Fin
Breaker Morant
Breakfast in Paris
Burke & Wills
Cactus
Camel Boy
Careful, He Might Hear You
Carry Me Back
Cars That Ate Paris
The City's Edge
The Clinic
The Club
The Coca-Cola Kid
Color Me Dead
Contagion
Crocodile Dundee
A Cry in the Dark
Dangerous Summer
Dark Age
Dark Forces

Dead Calm
Dead Easy
Dead End Drive-In
Deadly Possession
Demonstrator
The Devil's Playground
Dogs in Space
Don's Party
Dot & the Bunny
Dot and the Whale
The Dreaming
Dunera Boys
Dusty
Earthling
Echo Park
Echoes of Paradise
The Empty Beach
Endplay
Escape 2000
Fantasy Man
Fast Talking
Fatty Finn
The Fire in the Stone
Fists of Blood
For Love Alone
Fortress
Forty Thousand Horsemen
Fran
Frenchman's Farm
The Fringe Dwellers
Gallagher's Travels
Gallipoli
The Getting of Wisdom
The Good Wife
Grendel, Grendel, Grendel
Ground Zero
Hard Knocks
Heatwave
High Rolling in a Hot Corvette
High Tide
The Highest Honor
Howling 3
I Live with Me Dad
An Indecent Obsession
Inn of the Damned
Innocent Prey
The Irishman
Kangaroo
Killing of Angel Street
Kitty and the Bagman
Lady Stay Dead
The Last Bastion
The Last Wave
The Lighthorseman
Lonely Hearts
Long John Silver
Mad Dog Morgan
Mad Max Beyond
    Thunderdome
Magee and the Lady
Malcolm
Man of Flowers
The Mango Tree
Mesmerized
Midnight Dancer

Miracle Down Under
Money Movers
Monkey Grip
Moving Targets
My First Wife
Newsfront
Next of Kin
Norman Loves Rose
Now and Forever
The Odd Angry Shot
Overlanders
Palm Beach
Panic Station
Patrick
Phar Lap
Picnic at Hanging Rock
Plumber
Poor Girl, A Ghost Story
Prince and the Great Race
Puberty Blues
Puzzle
Queen of the Road
The Quest
Quigley Down Under
Race to the Yankee Zephyr
Razorback
Rebel
Return of Captain Invincible
Return to Snowy River
The Right Hand Man
Rikky and Pete
Road Games
Run, Rebecca, Run
Savage Attraction
Second Time Lucky
Shame
Silver City
Sky Pirates
Smuggler's Cove
Squizzy Taylor
Starstruck
Summer City
Sunday Too Far Away
Survivor
Sweet Talker
Sweetie
The Tale of Ruby Rose
Tearaway
Three's Trouble
Tim
Torn Allegiance
Touch & Go
A Town Like Alice
Travelling North
Typhoon Treasure
Under Capricorn
Vincent: The Life and Death of
    Vincent van Gogh
The Wacky World of Wills &
    Burke
Walk into Hell
Warm Nights on a Slow-
    Moving Train
Waterfront
Who Killed Baby Azaria?

The Wild Duck
Windrider
The Winter of Our Dreams
A Yank in Australia
The Year of Living
    Dangerously
The Year My Voice Broke
Young Einstein

Belgian

Daughters of Darkness
The Devil's Nightmare
Dust
Les Rendez-vous D'Anna
Mascara
The Music Teacher
One Sings, the Other Doesn't
Toute Une Nuit
A Wedding in Galilee

Brazilian

Amor Bandido
Black God & White Devil
Black Orpheus
Bye Bye Brazil
The Dolphin
Dona Flor & Her Two
    Husbands
Earth Entranced (Terra em
    Transe)
Gabriela
Hour of the Star
I Love You
Killer Fish
Kiss of the Spider Woman
Lady on the Bus
Opera do Malandro
Pixote
Quilombo
The Story of Fausta
Subway to the Stars
Women in Fury

British

Abdulla the Great
The Abominable Dr. Phibes
Above Us the Waves
Absolute Beginners
Absolution
Accident
Across the Bridge
Action for Slander
The Adventurers
Adventures of a Private Eye
The Adventures of Sherlock
    Holmes: Hound of the
    Baskervilles
The Adventures of Tartu
Adventures of a Taxi Driver
An Affair in Mind
The Affair of the Pink Pearl
After Darkness

After the Fox
After Pilkington
Afterward
Alamut Ambush
Alfie
Alias John Preston
Alice
Alice's Adventures in
    Wonderland
Alien Women
All Creatures Great and Small
American Roulette
And Nothing But the Truth
And Now the Screaming
    Starts
And Now for Something
    Completely Different
And Soon the Darkness
Another Country
Another Time, Another Place
Antonia and Jane
Apartment Zero
Appointment with Crime
Aria
The Arsenal Stadium Mystery
As You Like It
The Asphyx
The Assassination Run
Assassination of Trotsky
Assault
Assault on Agathon
Asylum
At the Earth's Core
The Atomic Man
The Baby and the Battleship
Baby Love
Bachelor of Hearts
Ballad in Blue
Ballet Shoes
Bananas Boat
Battle Beneath the Earth
Battle Hell
The Battle of the Sexes
The Bawdy Adventures of
    Tom Jones
Beachcomber
Beast in the Cellar
Beat Girl
Bedazzled
Belles of St. Trinian's
Bellman and True
The Belly of an Architect
Benny Hill's Crazy World
Best of the Benny Hill Show:
    Vol. 4
Best Enemies
Best of the Two Ronnies
Betrayal
Bewitched
Big Zapper
Biggles
Billy Budd
Billy Liar
The Bitch
Bitter Sweet

British

Black Adder 3
Black Adder The Third (Part 2)
Black Beauty
The Black Cat
Black Narcissus
Black Werewolf
Blackmail
Blackout
Blat
Bleak House
The Bliss of Mrs. Blossom
Blithe Spirit
Blockhouse
Blood Beast Terror
Blood Island
The Blood on Satan's Claw
Blood of the Vampire
Bloodbath at the House of
   Death
The Bloodsuckers
Blow-Up
The Blue Lamp
Blue Money
Blue Murder at St. Trinian's
Blueblood
Bobo
The Body in the Library
Bonnie Prince Charlie
The Boy Friend
The Brain
Brainwash
Brazil
Bread (BBC)
Breaking Glass
The Brides of Dracula
Brideshead Revisited
The Bridge on the River Kwai
A Bridge Too Far
Brief Encounter
Bright Smiler
Brighton Rock
Brimstone & Treacle
Britannia Hospital
Broken Melody
Browning Version
Bugsy Malone
Bulldog Jack
Bullshot
Burn Witch, Burn!
Burning Secret
Buster
Butler's Dilemma
Butterflies
Butterfly Ball
Caesar and Cleopatra
Call Him Mr. Shatter
Calling Paul Temple
Candide
Candles at Nine
A Canterbury Tale
Captain Horatio Hornblower
Captain Kronos: Vampire
   Hunter
Captain's Paradise
The Captain's Table
Captive Heart
Car Trouble
Caravaggio
Carlton Browne of the F.O.
Carrington, V.C.
Carry On Behind
Carry On Cleo
Carry On Cowboy
Carry On Dick
Carry On Doctor
Carry On Emmanuelle
Carry On England
Carry On Nurse
Carry On 'Round the Bend
Carry On Screaming
Carry On at Your
   Convenience
The Case of the Frightened
   Lady
The Case of the Missing Lady

The Case of the Mukkinese
   Battle Horn
Cash
The Cassandra Crossing
The Castle of Fu Manchu
The Cat and the Canary
Catch Me a Spy
The Chalk Garden
Chamber of Horrors
Champagne
Champions
Chariots of Fire
Charlie Grant's War
Chicago Joe & the Showgirl
Children of the Full Moon
Children of Rage
Chinese Boxes
Chitty Chitty Bang Bang
Choice of Weapons
A Chorus of Disapproval
Christabel
Christian the Lion
A Christmas Carol
Christopher Columbus
Chronicles of Narnia
Circle of Iron
Circus of Fear
Circus of Horrors
Citadel
Clash of the Titans
The Class of Miss MacMichael
The Clergyman's Daughter
Clockwise
A Clockwork Orange
Codename: Icarus
Cold War Killers
The Collector
The Comeback
Comfort and Joy
Coming Up Roses
Conduct Unbecoming
The Confessional
The Conqueror Worm
Consuming Passions
The Cook, the Thief, His Wife
   & Her Lover
Cool Mikado
Corridors of Blood
Cottage to Let
Count of Monte Cristo
Counterblast
The Courageous Mr. Penn
Courtesans of Bombay
The Courtney Affair
The Crackler
The Crawling Eye
The Creeping Flesh
Crimes at the Dark House
The Crimes of Stephen
   Hawke
Crimson Pirate
Cromwell
Crooks & Coronets
Crow Hollow
Crucible of Horror
Cruel Sea
Cry of the Banshee
Cry, the Beloved Country
A Cry from the Streets
Cry Terror
Curse of the Demon
The Curse of Frankenstein
The Curse of the Werewolf
Curtain Up
Daleks Invasion Earth 2150
   A.D.
Dam Busters
Damn the Defiant
Dance with a Stranger
Dandy in Aspic
Dangerous Moonlight
Dark Journey
Dark Places
Darling
A Day in the Death of Joe Egg
The Day Will Dawn

The Dead
Dead Lucky
Dead of Night
Deadly Recruits
Deadly Strangers
Dealers
Death of Adolf Hitler
Death Goes to School
Death on the Nile
Death Ship
Decameron Nights
The Deceivers
A Dedicated Man
Deep End
Deep in the Heart
Defense of the Realm
Deja Vu
The Demi-Paradise
Demon Barber of Fleet Street
Demons of the Mind
The Destructors
The Detective
Devil Doll
Devil Girl from Mars
Devil on Horseback
The Devils
The Devil's Undead
Diamonds are Forever
Dick Barton, Special Agent
Dick Barton Strikes Back
Die! Die! My Darling!
Die, Monster, Die!
Digby, the Biggest Dog in the
   World
Dinner at the Ritz
Disciple of Death
Distant Voices, Still Lives
The Divorce of Lady X
Doctor Blood's Coffin
The Doctor and the Devils
Doctor in Distress
Doctor in the House
Dr. Jekyll and Sister Hyde
Doctor at Large
Dr. No
Doctor Phibes Rises Again
Doctor at Sea
Dr. Strangelove, or: How I
   Learned to Stop Worrying
   and Love the Bomb
Dr. Syn
Dr. Terror's House of Horrors
Doctor Who and the Daleks
Doctor Who: Death to the
   Daleks
Doctor Who: Pyramids of
   Mars
Doctor Who: Spearhead from
   Space
Doctor Who: Terror of the
   Zygons
Doctor Who: The Ark in Space
Doctor Who: The Seeds of
   Death
Doctor Who: The Talons of
   Weng-Chiang
Doctor Who: The Time
   Warrior
The Dogs of War
A Doll's House
Don't Open Till Christmas
Doomwatch
Down Among the Z Men
The Draughtsman's Contract
Dreamchild
Dreaming Lips
Dressed for Death
Dressmaker
Drowning by Numbers
Drums
Dual Alibi
The Duellists
The Dummy Talks
Dust to Dust
Dutch Girls
Dynasty of Fear

Eagle's Wing
East of Elephant Rock
Easy Virtue
Eat the Rich
Ebony Tower
Edge of Darkness
Edge of Sanity
Edge of the World
Educating Rita
The Electronic Monster
Elephant Boy
11 Harrowhouse
The Elusive Pimpernel
Embassy
Empire State
Encore
The Endless Game
Endless Night
An Englishman Abroad
Enigma
The Entertainer
Erik the Viking
Escapade
Escape to Athena
Escape from El Diablo
Eureka!
Evergreen
The Evil of Frankenstein
The Evil Mind
Evil Under the Sun
The Executioner
Experience Preferred... But
   Not Essential
Eye Witness
The Fall of the House of Usher
Fall and Rise of Reginald
   Perrin 1
Fall and Rise of Reginald
   Perrin 2
The Fallen Idol
Family Life
Fanny Hill
Far from the Madding Crowd
Farewell, My Lovely
The Farmer's Wife
Fatal Confinement
The Fearless Vampire Killers
Feet Foremost
The Fiend
Fiendish Plot of Dr. Fu
   Manchu
The Fifth Musketeer
The Final Option
The Final Programme
Finessing the King
Fire Down Below
Fire Maidens from Outer
   Space
Fire Over England
Fire and Sword
First Men in the Moon
Five Golden Dragons
Five Miles to Midnight
Flame to the Phoenix
Flanagan Boy
Flesh and Blood Show
Flight from Singapore
Florence Chadwick: The
   Challenge
Follow That Camel
Footlight Frenzy
For Your Eyes Only
Forbidden Passion: The Oscar
   Wilde Movie
Foreign Body
The Foreman Went to France
Forever Young
Fortunes of War
The Forty-Ninth Parallel
The Four Feathers
Four Sided Triangle
The Fourth Protocol
The Franchise Affair
The Freakmaker
The French Lesson
Frenzy

Frieda
Fright
Frightmare
From Beyond the Grave
From Russia with Love
Frozen Alive
Funny Money
The Gambler & the Lady
Game for Vultures
The Gamma People
Gandhi
The Gay Lady
Genevieve
Georgy Girl
Getting It Right
The Ghost Goes West
Ghost in the Noonday Sun
Ghost Ship
The Ghost Train
The Ghoul
The Gift Horse
The Gilded Cage
A Girl in a Million
The Girl on a Motorcycle
The Girl in the Picture
Girly
Give My Regards to Broad
   Street
The Go-Between
The Golden Salamander
Goldfinger
The Good Father
Gorgo
The Gorgon
The Gospel According to Vic
Great Expectations
Great McGonagall
Great St. Trinian's Train
   Robbery
Great Train Robbery
The Greed of William Hart
Greek Street
Green for Danger
Green Grow the Rushes
Green Ice
The Green Man
Gregory's Girl
Gulag
Gulliver's Travels
Gumshoe
Guns at Batasi
Hail Mary
Hamlet
The Hand
A Handful of Dust
The Hands of Orlac
Hands of the Ripper
Hard Day's Night
Harry Black and the Tiger
The Haunted Strangler
Haunted: The Ferryman
The Haunting of Julia
Hawks
Hear My Song
Heart's Desire
Heat and Dust
Heat Wave
Heavens Above
Hedda
Hellbound: Hellraiser 2
Hellraiser
Hennessy
Henry IV, Part I
Henry IV, Part II
Henry V
Henry VIII
Hi-Di-Hi
The Hidden Room
High Command
High Hopes
High Season
The Hit
Hobson's Choice
The Holcroft Covenant
Hope and Glory
The Horror of Dracula

The Horror of Frankenstein
Horror Hospital
Horror Hotel
The Horse's Mouth
Hotel Du Lac
Hotel Reserve
The Hound of the Baskervilles
Hour of Decision
The House of 1,000 Dolls
The House of the Arrow
The House that Dripped Blood
The House of Lurking Death
The House in Marsh Road
House of Mystery
The House that Vanished
House of Whipcord
How to Get Ahead in
  Advertising
How I Learned to Love
  Women
How I Won the War
How Many Miles to Babylon?
Hue and Cry
Hullabaloo Over Georgie &
  Bonnie's Pictures
The Human Factor
The Human Monster
Hussy
Hysteria
I Am a Camera
I, Claudius, Vol. 1: A Touch of
  Murder/Family Affairs
I Don't Want to Be Born
I Know Where I'm Going
I Live in Grosvenor Square
I See a Dark Stranger
I Stand Condemned
I Want What I Want
If...
If I Were Rich
The Impossible Spy
Impulse
In Celebration
In Search of the Castaways
In the Shadow of Kilimanjaro
In Which We Serve
Ingrid
Inheritance
Innocents in Paris
Inserts
Inside Out
Insignificance
Inspector Hornleigh
Instant Justice
International Velvet
Internecine Project
Interrupted Journey
Intimate Contact
Invasion
Invasion of the Body Stealers
The Ipcress File
Island of the Burning Doomed
Island of Desire
Island of Terror
It Couldn't Happen Here
It Takes a Thief
The Italian Job
It's Love Again
Jabberwocky
Jack the Ripper
Jamaica Inn
Jane Eyre
Jane & the Lost City
Jason and the Argonauts
Java Head
Jericho
The Jewel in the Crown
The Jigsaw Man
Johnny Frenchman
Johnny Nobody
Joseph Andrews
Journey to the Far Side of the
  Sun
Jubilee
Juggernaut
Julius Caesar

Just William's Luck
The Key
The Key Man
Killer Fish
The Killing Fields
Killing Heat
Kind Hearts and Coronets
A Kind of Loving
King of Hearts
King Lear
A King in New York
King Solomon's Treasure
Kings and Desperate Men
The King's Rhapsody
Kipperbang
Kipps
The Kitchen Toto
Knight Without Armour
Knights & Emeralds
The Krays
Krull
Lady Caroline Lamb
Lady Chatterley's Lover
Lady in Distress
Lady Jane
The Lady Vanishes
Lady Vanishes
Ladykillers
The Lair of the White Worm
Land of Fury
Land of the Minotaur
Land That Time Forgot
The Last Days of Dolwyn
The Last of England
Last Holiday
Late Extra
The Lavender Hill Mob
Lawrence of Arabia
League of Gentlemen
Leather Boys
The Legacy
Legend
Legend of the Werewolf
Les Carabiniers
Let Him Have It
Let's Make Up
Letter to Brezhnev
Letters to an Unknown Lover
License to Kill
Lies
Life & Death of Colonel Blimp
Life is Sweet
Lily of Killarney
The Limping Man
Link
Lion of the Desert
Lisztomania
Little Dorrit, Film 1: Nobody's
  Fault
Little Dorrit, Film 2: Little
  Dorrit's Story
Little Match Girl
Little Prince
Live and Let Die
The Living Dead
Local Hero
Lola
Lolita
London Melody
The Lonely Passion of Judith
  Hearne
Long Ago Tomorrow
The Long Dark Hall
The Long Good Friday
Look Back in Anger
The Looking Glass War
Loophole
Loose Connections
Loot...Give Me Money, Honey!
Lord of the Flies
Love and Bullets
Love on the Dole
Love from a Stranger
Lucky Jim
Lust for a Vampire
Mad Death

Made in Heaven
Madhouse
Madhouse Mansion
The Magic Christian
Magical Mystery Tour
The Mahabharata
Maid Marian & Her Merry
  Men: How the Band Got
  Together
Maid Marian & Her Merry
  Men: The Miracle of St.
  Charlene
Major Barbara
Make Me an Offer
Make Mine Mink
A Man About the House
A Man for All Seasons
Man Bait
Man of Evil
Man Friday
The Man with the Golden Gun
The Man in Grey
The Man in the Mist
The Man Outside
Man on the Run
Man in the White Suit
The Man Who Fell to Earth
Man Who Had Power Over
  Women
The Man Who Haunted
  Himself
The Man Who Knew Too
  Much
The Man Who Lived Again
The Man Who Never Was
Mandy
Mania
Maniac
Mansfield Park
The Manxman
Marat/Sade (Persecution and
  Assassination...)
March or Die
The Mark
Mark of the Devil
Mark of the Devil 2
The Mark of the Hawk
A Married Man
Masque of the Red Death
A Matter of Who
Max Headroom, the Original
  Story
Max and Helen
The Maze
The McGuffin
McVicar
Measure for Measure
The Medusa Touch
Meet Sexton Blake
Melody
Men are not Gods
Men of Sherwood Forest
Men of Two Worlds
Merry Christmas, Mr.
  Lawrence
Mesmerized
The Mikado
The Mill on the Floss
Mimi
Mine Own Executioner
The Miracle
The Mirror Crack'd
The Misadventures of Mr. Wilt
The Missionary
Mister Arkadin
Mohammed: Messenger of
  God
Mona Lisa
The Monster Club
Month in the Country
Monty Python and the Holy
  Grail
Monty Python Live at the
  Hollywood Bowl
Monty Python Meets Beyond
  the Fringe

Monty Python's Life of Brian
Monty Python's The Meaning
  of Life
The Moon Stallion
Moonlight Sonata
Moonlighting
The Moonraker
Moonraker
Morgan!
Morons from Outer Space
The Mouse That Roared
Mr. Love
The Mummy
Muppet Movie
Murder
A Murder is Announced
Murder at the Baskervilles
Murder by the Book
Murder on the Campus
Murder of a Moderate Man
Murder in the Old Red Barn
Murder on the Orient Express
Murders at Lynch Cross
Murphy's War
My Son, the Vampire
Mysterious Island
The Naked Truth
Nancy Astor
Nasty Habits
Negatives
Never Let Go
Next Victim
Nice Girl Like Me
Nicholas Nickleby
Night Ambush
Night Beat
Night Caller from Outer Space
The Night Has Eyes
Night of the Laughing Dead
Night Train to Munich
The Night Visitor
Night Watch
The Nightcomers
Nine Days a Queen
1984
99 Women
No Comebacks
No Love for Johnnie
No Surrender
No Trace
Nothing But the Night
Number 1 of the Secret
  Service
O Lucky Man
The Oblong Box
The Obsessed
Octopussy
The Odd Job
Odd Man Out
The Odessa File
Of Human Bondage
Office Romances
Oh Alfie
Oh, Mr. Porter
The Old Curiosity Shop
Old Mother Riley's Ghosts
Old Spanish Custom
Oliver!
Oliver Twist
On Approval
On Her Majesty's Secret
  Service
One Day in the Life of Ivan
  Denisovich
One-Eyed Soldiers
One of Our Aircraft Is Missing
One Wish Too Many
Only Two Can Play
Open All Hours
Order of the Black Eagle
Othello
Out of Season
Overture
Paid to Kill
Paper Mask
Paper Tiger

Paperhouse
Party! Party!
Pascali's Island
A Passage to India
The Passing of the Third Floor
  Back
Passport to Pimlico
The People That Time Forgot
Performance
Permission To Kill
Personal Services
The Persuaders
Phantom Fiend
Philby, Burgess and MacLean:
  Spy Scandal of the Century
The Pickwick Papers
Pimpernel Smith
Pink Floyd: The Wall
The Pink Panther
The Pink Panther Strikes
  Again
Pink Strings and Sealing Wax
Playing Away
A Pocketful of Rye
The Pope Must Diet
Porridge
Portrait of a Lady
Postmark for Danger
Postmark for Danger/
  Quicksand
Power
Power Play
Prick Up Your Ears
Pride and Prejudice
Priest of Love
The Prime of Miss Jean
  Brodie
Prince and the Pauper
The Prisoner
A Private Function
Private Life of Don Juan
The Private Life of Henry VIII
The Private Life of Sherlock
  Holmes
Privates on Parade
Promise Her Anything
Promoter
Providence
Psychomania
Psychopath
Pulp
Puppet on a Chain
The Pure Hell of St. Trinian's
Pursuit of the Graf Spee
Pygmalion
Quadrophenia
Quartet
Quatermass 2
Quatermass Conclusion
The Quatermass Experiment
Queen of Hearts
The Queen of Spades
Quest for Love
Race for Life
Railway Children
The Rainbow
Rattle of a Simple Man
The Red Shoes
Redneck
Reilly: The Ace of Spies
Rentadick
Repulsion
Return of the Soldier
Revenge
Revolution
Rhodes
Rich and Strange
Richard II
Richard III
Richard's Things
The Riddle of the Sands
The Ring
Ring of Bright Water
Ripping Yarns
Rise of Catherine the Great
Rita, Sue & Bob Too

British

Road to 1984
Roadhouse Girl
Robbery
Robbery Under Arms
Robin Hood: The Movie
Robin Hood...The Legend
Robin Hood...The Legend:
    Herne's Son
Robin Hood...The Legend:
    The Swords of Wayland
Robin Hood...The Legend:
    The Time of the Wolf
Robin and Marian
Roboman
Rogue's Yarn
Romance with a Double Bass
Romantic Englishwoman
Romeo and Juliet
Romeo and Juliet (Gielgud)
Room to Let
Room at the Top
Room with a View
The Ruling Class
The Runaway Bus
Running Blind
Saigon: Year of the Cat
Sailing Along
The Sailor Who Fell from
    Grace with the Sea
The Saint
The Salamander
Sammy & Rosie Get Laid
Sanders of the River
Sapphire
The Satanic Rites of Dracula
Saturn 3
Say Hello to Yesterday
Scandal
The Scarlet Pimpernel
The Scars of Dracula
Scotland Yard Inspector
Scott of the Antarctic
Scream of Fear
Scream and Scream Again
Scrooge
Scrubbers
Scum
Sea Devils
Sea Wolves
Seance on a Wet Afternoon
Seaside Swingers
Sebastian
Second Best Secret Agent in
    the Whole Wide World
Second Chance
The Secret Adversary
The Secret Agent
The Secret Four
The Secret Garden
The Secret of the Loch
Secret Places
The Secret Policeman's Other
    Ball
Secret Policeman's Private
    Parts
Secrets
The Sender
A Sense of Freedom
The Servant
The 7 Brothers Meet Dracula
Seventh Veil
Sexton Blake and the Hooded
    Terror
Shades of Darkness
Shadey
The Shadow
The Shadow Man
Shadow on the Sun
Shaka Zulu
Shattered
Sherlock Holmes
Sherlock Holmes and the
    Incident at Victoria Falls
Shirley Valentine
The Shooting Party
A Shot in the Dark

The Shout
Shout at the Devil
Sidewalks of London
Silas Marner
The Silent Enemy
The Silent Mr. Sherlock
    Holmes
The Silent Passenger
Silent Scream
Simba
Sinbad and the Eye of the
    Tiger
Singleton's Pluck
Sister Dora
Skin Game
Slayground
Slipstream
The Smallest Show on Earth
The Snake Woman
Someone at the Door
Song of Freedom
The Spaniard's Curse
Spellbound
Spiral Staircase
Spitfire
Split Second
Spy in Black
The Spy Who Loved Me
Spy With a Cold Nose
Spymaker: The Secret Life of
    Ian Fleming
Stage Fright
The Stars Look Down
Starship
The Statue
Staying On
Stealing Heaven
Steaming
Stevie
A Stolen Face
Storm in a Teacup
Strange Awakening
Stranger in Town
Straw Dogs
The Stud
A Study in Terror
Subterfuge
Success Is the Best Revenge
Sudden Terror
A Summer Story
Sunday, Bloody Sunday
Supergirl
Supergrass
Suspended Alibi
Sweeney Todd: The Demon
    Barber of Fleet Street
Sweet William
Sword of Lancelot
Sword of Sherwood Forest
Sword of the Valiant
Sykes
A Tale of Two Cities
Tales of Beatrix Potter
Tales from the Crypt
A Taste of Honey
The Tell-Tale Heart
The Tempest
The Temptress
Ten Little Indians
10 Rillington Place
Terronauts
The Terrornauts
Tess
That Sinking Feeling
Theatre of Blood
There's a Girl in My Soup
Things to Come
Things Happen at Night
Think Dirty
Thirteenth Day of Christmas
30 Is a Dangerous Age,
    Cynthia
The 39 Steps
The Thirty-Nine Steps
This Happy Breed

This Lightning Always Strikes
    Twice
This Sporting Life
Those Daring Young Men in
    Their Jaunty Jalopies
Those Fantastic Flying Fools
Those Glory, Glory Days
Those Magnificent Men in
    Their Flying Machines
Threads
Three of a Kind
Three Stops to Murder
The Three Weird Sisters
Thunder in the City
Thunderball
Tiger Bay
Time Bandits
Time Lock
Time for Murder
Tintorera...Tiger Shark
To See Such Fun
To Sir, with Love
Tom Brown's School Days
Tom Jones
tom thumb
Tomb of Ligeia
Tomb of the Undead
Too Many Crooks
Torture Garden
Touch Me Not
Tough Guy
Tower of Evil
Tower of Terror
A Town Called Hell
Transatlantic Tunnel
Transmutations
Trap
Treasure Island
Treasure of Monte Cristo
Triple Cross
Triple Echo
Triumph of Sherlock Holmes
Truth About Women
Tunes of Glory
Turtle Diary
Twelfth Night
Twins of Evil
Twinsanity
2001: A Space Odyssey
Two-Way Stretch
The Unbreakable Alibi
The Uncanny
Under Capricorn
Under the Doctor
Under Milk Wood
Under the Red Robe
Under Suspicion
The Unearthly Stranger
Uneasy Terms
Unexplained Laughter
Unholy Four
Unicorn
Unmasked Part 25
Unmasking the Idol
The Unstoppable Man
Unsuitable Job for a Woman
Up the Creek
Valley of the Eagles
Vampire Circus
The Vampire Lovers
Vampyres
Vault of Horror
Vendetta for the Saint
Venom
Venus in Furs
Veronico Cruz
The Very Edge
The Vicious Circle
Victim
A View to a Kill
Village of the Damned
The Virgin and the Gypsy
The Virgin Soldiers
The Virgin Witch
The Vision
A Voyage Round My Father

Wagner: The Movie
Waltz of the Toreadors
War Lover
The Warriors
Water
Water Babies
Watership Down
We Dive at Dawn
We Think the World of You
Wellington: The Duel Scandal
Wetherby
What a Carve-Up!
When Dinosaurs Ruled the
    Earth
When Thief Meets Thief
When the Whales Came
When the Wind Blows
Where the Bullets Fly
While I Live
Whiskey Galore
The Whistle Blower
Whistle Down the Wind
White Fire
The Wicked Lady
Windom's Way
Winner Takes All
The Winslow Boy
Winter Flight
Wish You Were Here
Withnail and I
Without a Clue
Wombling Free
Women in Love
Wonderland
Wonderwall: The Movie
A World Apart
The World is Full of Married
    Men
The World of Suzie Wong
The Wrong Arm of the Law
The Wrong Box
Wuthering Heights
Xtro
Yellow Submarine
Yesterday's Hero
You Only Live Twice
Young and Innocent
The Young Ones
Young Winston
Yuri Nosenko, KGB
A Zed & Two Noughts

Canadian

Abducted
The Agency
American Boyfriends
The Amityville Curse
And Then You Die
Angela
Anne of Avonlea
Anne of Green Gables
The Apprenticeship of Duddy
    Kravitz
Atlantic City
The Bay Boy
The Big Crimewave
Big Meat Eater
Black Christmas
Blades of Courage
Blood Relations
Blood Relatives
Boy in Blue
Breaking All the Rules
The Brood
Buffalo Jump
Bullies
Busted Up
By Design
Bye Bye Blues
Candy Mountain
Cathy's Curse
The Changeling
Christina
A Christmas Gift

Circle of Two
City on Fire
Class of 1984
Close to Home
Cold Comfort
Concrete Angels
Confidential
Covergirl
Cross Country
Curtains
Dan Candy's Law
Dancing in the Dark
Daughter of Death
Dead Ringers
Death Ship
Death Weekend
Deathdream
The Decline of the American
    Empire
Deranged
Dirty Tricks
Disappearance
Distant Thunder
Dog Pound Shuffle
The Dogs Who Stopped the
    War
Double Identity
The Double Negative
Drying Up the Streets
Duel-Duo
Eddie & the Cruisers 2: Eddie
    Lives!
Eliza's Horoscope
En Passant
Family Viewing
Fatal Attraction
Fight for Your Life
Final Assignment
Find the Lady
Fish Hawk
Friends, Lovers & Lunatics
Funeral Home
The Funny Farm
Gas
The Girl in Blue
Girls on the Road
The Godsend
Going Berserk
Goldenrod
The Grey Fox
The Groundstar Conspiracy
The Gunfighters
The Gunrunner
Happy Birthday to Me
Harry Tracy
The Haunting of Julia
Heartaches
Heaven on Earth
Heavenly Bodies
High Ballin'
High Country
Highpoint
Hockey Night
Hog Wild
The Housekeeper
Humongous
Iceman
If You Could See What I Hear
In Celebration
In Praise of Older Women
In Trouble
Incredible Manitoba Animation
Incubus
Into the Fire
Iron Eagle 2
I've Heard the Mermaids
    Singing
Jesus of Montreal
John and the Missus
Joshua Then and Now
Journey
Journey into Fear
The Kidnapping of the
    President
Killing 'em Softly
King Solomon's Treasure

French

Head Over Heels
Heart of a Nation
Heat of Desire (Plein Sud)
Hercules Against the Moon Men
High Heels
Hiroshima, Mon Amour
The Holes (Les Gaspards)
Holiday Hotel
L'Homme Blesse
Honor Among Thieves
Horror Chamber of Dr. Faustus
The Horse of Pride
Hothead
Hothead (Coup de Tete)
Howling 2: Your Sister Is a Werewolf
The Hurried Man
I Killed Rasputin
I Love All of You
I Sent a Letter to My Love
Icy Breasts
In the Realm of the Senses
Infernal Trio
The Invaders
Investigation
Invitation au Voyage
Is Paris Burning?
Istanbul
It Happened in the Park
Italian Straw Hat
J'Accuse
Jacko & Lise
Jailbird's Vacation
je tu il elle
Jean de Florette
Jenny Lamour
Jesus of Montreal
Josepha
Jour de Fete
Journey Beneath the Desert
Journey to the Lost City
Judex
The Judge and the Assassin
Jules and Jim
Jupiter's Thigh
Just Another Pretty Face
Kapo
The Killing Game
King of Hearts
The Klutz
La Balance
La Barbiec de Seville
La Bete Humaine
La Boum
La Cage aux Folles
La Cage aux Folles 2
La Cage aux Folles 3: The Wedding
La Chevre
La Chienne
La Chute de la Maison Usher
La Femme Nikita
La Jetee/An Occurrence at Owl Creek Bridge
La Lectrice
La Marseillaise
La Nuit de Varennes
La Passante
La Petite Sirene
La Ronde
La Symphonie Pastorale
La Truite (The Trout)
The Lacemaker
L'Addition
Lady Chatterley's Lover
The Last Metro
Last Tango in Paris
The Last Winters
Last Year at Marienbad
L'Atalante
Le Bal
Le Beau Mariage
Le Beau Serge
Le Boucher

Le Bourgeois Gentilhomme
Le Cas du Dr. Laurent
Le Cavaleur (Practice Makes Perfect)
Le Chat
Le Complot (The Conspiracy)
Le Corbeau
Le Crabe Tambour
Le Dernier Combat
Le Doulos
Le Gentleman D'Epsom
Le Grand Chemin
Le Joli Mai
Le Jour Se Leve
Le Magnifique
Le Million
Le Petit Amour
Le Plaisir
Le Repos du Guerrier
Le Secret
Le Sex Shop
Leonor
Les Assassins de L'Ordre
Les Biches
Les Carabiniers
Les Choses de la Vie
Les Comperes
Les Enfants Terrible
Les Grandes Gueules
Les Miserables
Les Mistons
Les Rendez-vous D'Anna
Les Visiteurs du Soir
Letters from My Windmill
Letters to an Unknown Lover
Life and Nothing But
Light Years
Liliom
The Little Thief
Little World of Don Camillo
Lola
Lola Montes
Loulou
Love and the Frenchwoman
A Love in Germany
Love on the Run
Love Songs
Love Without Pity
The Lovers
Lovers Like Us
The Loves of Hercules
The Lower Depths
Lucky Luciano
Lumiere
Madame Bovary
Madame Rosa
The Mahabharata
Make Room for Tomorrow
Mama Dracula
Mama, There's a Man in Your Bed
Man on Fire
A Man in Love
Man from Nowhere
Man in the Raincoat
Man in the Silk Hat
The Man Who Loved Women
A Man and a Woman
A Man and a Woman: 20 Years Later
Manon of the Spring
Maria Chapdelaine
Marius
A Married Woman
Marry Me, Marry Me
Martin Luther
Masculine Feminine
May Fools
Melodie en Sous-Sol
The Milky Way
Mill of the Stone Women
Minnesota Clay
Mon Oncle
Mon Oncle D'Amerique
The Mongols
Monsieur Hire

Monsieur Vincent
Moon in the Gutter
Morgan the Pirate
Mouchette
Mr. Hulot's Holiday
Mr. Peek-A-Boo
Muriel
Murmur of the Heart
My Best Friend's Girl
My Life to Live
My New Partner
My Night at Maud's
My Other Husband
Mystery of Alexina
Nais
Nana
Napoleon
Next Summer
Next Year If All Goes Well
Night Flight from Moscow
1900
No Problem
A Nos Amours
A Nous le Liberte
Nous N'Irons Plus au Bois
Nudo di Donna
The Nun
Oasis of the Zombies
An Occurrence at Owl Creek Bridge
L'Odeur des Fauves
Oldest Profession
One Deadly Summer (L'Ete Meurtrier)
One Sings, the Other Doesn't
One Wild Moment
One Woman or Two
Overseas: Three Women with Man Trouble
Pain in the A—
Panique
Pantaloons
Parade
Pardon Mon Affaire
Pardon Mon Affaire, Too!
Paris Belongs to Us
Paris, Texas
Passion of Joan of Arc
Passion of Love
Passione d'Amore
Pattes Blanches
Pauline at the Beach
The Peking Blond
Pepe le Moko
Peril
The Perils of Gwendoline
Petit Con
Phantom of Liberty
Pickpocket
Picnic on the Grass
Pierrot le Fou
Planets Against Us
Playtime
The Plot Against Harry
Poil de Carotte
Police
Possession
Pouvoir Intime
Princess Academy
Princess Tam Tam
Prix de Beaute
Prize of Peril
A Propos de Nice
The Purple Taxi
Quartet
Quatorze Juliet
Quest for Fire
Qui Etes Vous, Mr. Sorge?
Ramparts of Clay
Rape of Love
Rape of the Sabines
The Rascals
The Raven
Ravishing Idiot
The Red Balloon

Red Balloon/Occurrence at Owl Creek Bridge
The Red and the Black
Red Kiss (Rouge Baiser)
Red Sun
Rendez-Moi Ma Peau
Rendez-vous
The Respectful Prostitute
The Return of Dr. Mabuse
Return of Martin Guerre
Return of the Tall Blond Man With One Black Shoe
Ricco
Rider on the Rain
Rififi
The Rise of Louis XIV
The River
Road to Salina
Roads to the South
Robert et Robert
Round Midnight
The Rules of the Game
Salut l'Artiste
Sand and Blood
Sauteuse de L'ANGE (The Angel Jump Girl)
Scene of the Crime
Scoumoune
Secret Agent Super Dragon
Seven Deadly Sins
The Sheep Has Five Legs
Sheer Madness
Shoot the Piano Player
Simple Story
Sincerely Charlotte
Six Fois Deux/Sur et Sous la Communication
The Slap
Small Change
Smiling Madame Beudet/The Seashell and the Clergyman
Soft and Hard
The Soft Skin
Sois Belle et Tais-Toi
Soldat Duroc...Ca Va Etre Ta Fete!
Son of Samson
The Sorrow and the Pity
Soviet Spy
Spirits of the Dead
Sputnik
State of Siege
Stay As You Are
Stolen Kisses
Stormy Waters
Story of Adele H.
The Story of a Love Story
The Story of a Three Day Pass
Story of Women
Subway
Sugar Cane Alley
Summer
A Sunday in the Country
Sundays & Cybele
Swann in Love
Swashbuckler
Sweet Ecstasy
Swimming Pool
Sylvia and the Phantom
Sympathy for the Devil
The Tall Blond Man with One Black Shoe
Tartuffe
Tatie Danielle
Tchao Pantin
Ten Days Wonder
The Tenant
Tendres Cousines
Terror of Rome Against the Son of Hercules
Tess
The Testament of Dr. Cordelier
The Testament of Orpheus

That Naughty Girl
Therese
36 Fillete
This Man Must Die
This Special Friendship
The 317th Platoon
Three Men and a Cradle
Tiger of the Seven Seas
Toni
Too Beautiful for You
Too Pretty to Be Honest
Too Shy to Try
Topaze
Tournament
Toute Une Nuit
The Trial
Triple Cross
Trop Jolie pour Etre Honette
12 Plus 1
Twist
Two English Girls
Two of Us
Ubu and the Great Gidouille
Umbrellas of Cherbourg
Un Chien Andalou
Uncle Tom's Cabin
Under the Roofs of Paris (Sous les Toits de Paris)
Under Satan's Sun
Une Jeunesse
Vagabond
The Vanishing
Venus on Fire
Very Curious Girl
A Very Private Affair
Vincent, Francois, Paul and the Others
Violette
Volpone
Voulez-Vous Danser Avec Moi?
Voyage en Balloon
Voyage Surprise
Wages of Fear
We Will Not Enter the Forest
Web of the Spider
Wedding in Blood
A Wedding in Galilee
Weekend
Well-Digger's Daughter
Where the Hot Wind Blows
Where the Sky Begins
Widow Couderc
Window Shopping
Without Warning
The Wolf at the Door
The Woman Next Door
Woman Times Seven
A Woman is a Woman
The Women
Wonders of Aladdin
Would-Be Gentleman
Yellow Fever
Yesterday, Today and Tomorrow
Z
Zazie dans le Metro
Zero for Conduct/Taris/A Propos de Nice
Zero for Conduct (Zero de Conduit)
Zombie Lake
Zou Zou

German

Aguirre, the Wrath of God
Ali: Fear Eats the Soul
American Friend
The American Soldier
Among the Cinders
Andy Warhol's Frankenstein
Angry Harvest
Anita, Dances of Vice
The Avenger

Italian

Blood Brothers
Blood Feud
Blood Ties
The Bloody Pit of Horror
Blow-Up
Boccaccio '70
Body Moves
Boomerang
Boot Hill
Bullet for the General
Bullet for Sandoval
Buried Alive
Burn!
Bye Bye Baby
Cabiria
Cafe Express
Caged Women
Caligula
Cambio de Cara (A Change of Face)
Camorra: The Naples Connection
The Canterbury Tales
Carmen
Carthage in Flames
Casanova '70
Castle of Blood
The Castle of Fu Manchu
Castle of the Living Dead
The Cat o' Nine Tails
Catch as Catch Can
Catherine & Co.
Chaste & Pure
The Children Are Watching Us
Chino
Christ Stopped at Eboli
The Christmas That Almost Wasn't
Cinema Paradiso
City of the Walking Dead
City of Women
The Clowns
Cobra
Code Name Alpha
Codename: Wildgeese
Cold Steel for Tortuga
Cold Sweat
Collector's Item
Colossus and the Amazon Queen
Commandos
The Con Artists
Confessions of a Police Captain
The Conformist
Conquest
Conquest of Mycene
Contempt
Contraband
Control
Conversation Piece
Coriolanus, Man without a Country
Corleone
Corrupt
Cosmos: War of the Planets
Could It Happen Here?
Count Dracula
Counter Punch
Creepers
Crime Busters
The Damned
Danger: Diabolik
Dangerous Liaisons
Dangerous Obsession
Dark Eyes
Daughters of Darkness
Day of the Cobra
Day and the Hour
The Day the Sky Exploded
Deadly Impact
Deadly Revenge
The Deadly Trap
Death Rage
Death in Venice
The Decameron

Delta Force Commando 2
Demons
Demons 2
Devil of the Desert Against the Son of Hercules
Devil in the Flesh
Devilfish
Devil's Crude
The Devil's Nightmare
Dial Help
Diary of Forbidden Dreams
Divine Nymph
Divorce—Italian Style
Doctor Butcher M.D.
Dr. Frankenstein's Castle of Freaks
Domino
Dorian Gray
Down & Dirty
Dracula vs. Frankenstein
Driver's Seat
Duel of Champions
The Easy Life
The Eclipse
Edipo Re (Oedipus Rex)
8 1/2
El Pirata Negro (The Black Pirate)
Emerald Jungle
Empty Canvas
Era Notte a Roma
Ernesto
Erotic Illusion
Everybody's Fine
Exterminators of the Year 3000
Eyeball
Eyes Behind the Stars
Eyes, the Mouth
The Family
Fangs of the Living Dead
Fear in the City
Fellini Satyricon
Fifth Day of Peace
A Fistful of Dollars
Fistful of Dynamite
Flatfoot
The Flowers of St. Francis
For a Few Dollars More
Forever Mary
Four Ways Out
Foxtrap
Frankenstein '80
Full Hearts & Empty Pockets
The Fury of Hercules
Gangsters
Garden of the Finzi-Continis
Generale Della Rovere
The Ghost
The Giant of Marathon
The Giant of Metropolis
Giants of Rome
The Giants of Thessaly
Ginger & Fred
Gladiator of Rome
Go for It
The Gold of Naples
Goliath Against the Giants
Goliath and the Barbarians
Goliath and the Dragon
Goliath and the Sins of Babylon
The Good, the Bad and the Ugly
Goodnight, Michelangelo
Gospel According to St. Matthew
Graveyard Shift
Great Adventure
The Great War
Grim Reaper
The Grim Reaper (La Commare Secca)
Gunfire
The Hanging Woman
Happy New Year

Happy Sex
Hatchet for the Honeymoon
Hawks & the Sparrows (Uccellacci e Uccellini)
Head of the Family
Hearts & Armour
Hell Commandos
Hellbenders
Henry IV
Hercules
Hercules 2
Hercules Against the Moon Men
Hercules and the Captive Women
Hercules in the Haunted World
Hercules and the Princess of Troy
Hercules, Prisoner of Evil
Hercules Unchained
Hero of Rome
Herod the Great
High Frequency
His Name was King
Homo Eroticus/Man of the Year
Honey
Honor Among Thieves
The Horrible Dr. Hichcock
Hotel Colonial
The House by the Cemetery
House on the Edge of the Park
How Funny Can Sex Be?
Howling 2: Your Sister Is a Werewolf
The Human Factor
Hundra
Husband Hunting
I Hate Blondes
I Killed Rasputin
I Vitelloni
The Icicle Thief
Immortal Bachelor
Indiscretion of an American Wife
Infernal Trio
Inferno
The Inheritance
The Innocent
The Inquiry
Intermezzo
The Invaders
Invasion of the Flesh Hunters
The Invincible Gladiator
The Invisible Dead
Island Monster
The Italian Connection
A Joke of Destiny, Lying in Wait Around the Corner Like a Bandit
Journey Beneath the Desert
Journey to the Lost City
Joyful Laughter
Judex
Juliet of the Spirits
Kapo
Kidnap Syndicate
Kill, Baby, Kill
La Cicada (The Cricket)
La Grande Bourgeoise
La Notte Brava (Lusty Night in Rome)
La Nuit de Varennes
La Signora di Tutti
La Strada
La Terra Trema
La Traviata
La Vie Continue
Lady Caroline Lamb
Lady of the Evening
Lady Frankenstein
The Lady Without Camelias
The Last Days of Pompeii
The Last Emperor
The Last Hunter

The Last Man on Earth
Last Tango in Paris
Last Year at Marienbad
L'Avventura
Le Bal
The Legend of Blood Castle
The Legend of the Wolf Woman
Les Carabiniers
Les Miserables
The Lie
Life of Verdi
Linchamiento (Lynching)
The Lion of Thebes
Lisa and the Devil
Little World of Don Camillo
Lobster for Breakfast
Lola
Love and Anarchy
Love Angels
Love in the City
Love Lessons
Lovers and Liars
The Loves of Hercules
Loves & Times of Scaramouche
Lucky Luciano
Lunatics & Lovers
Lure of the Sila
Macaroni
Maciste in Hell
The Mad Butcher
Malicious
Man from Cairo
A Man Called Rage
Man with a Cross
Man from Deep River
Man of Legend
The Man Who Wagged His Tail
Manhattan Baby
Manhunt
Marauder
Master Touch
Mean Frank and Crazy Tony
Medea
Medusa Against the Son of Hercules
Messalina vs. the Son of Hercules
The Messiah
Miami Cops
Miami Horror
Mill of the Stone Women
Minnesota Clay
A Minute to Pray, A Second to Die
The Miracle
Miracle in Milan
Mission Stardust
Mister Mean
Mole Men Against the Son of Hercules
Mondo Cane
Mondo Cane 2
The Mongols
Moon in the Gutter
Morgan the Pirate
Moving Target
Mr. Rossi's Vacation
Mr. Superinvisible
Muriel
Murri Affair
Nana
Nathalie Comes of Age
National Adultery
Navajo Joe
Nefertiti, Queen of the Nile
The New Gladiators
New Mafia Boss
New York Ripper
The Night After Halloween
The Night Evelyn Came Out of the Grave
Night Flight from Moscow
A Night Full of Rain

The Night Porter
Night of the Shooting Stars
Night of the Zombies
Nightmare Castle
Nights of Cabiria
1900
1990: The Bronx Warriors
99 Women
Nudo di Donna
Odds and Evens
Oedipus Rex
Oldest Profession
Once Upon a Time in the West
Open City
Open Doors (Porte Aperte)
The Opponent
Orgy of the Vampires
Ossessione
Otello
Other Hell
Padre Padrone
Pain in the A—
Paisan
Panic Button
Paranoia
Pardon My Trunk
The Passenger
The Passion of Evelyn
Passion of Love
Passionate Thief
Passione d'Amore
Peddlin' in Society
Phantom of Death
Phenomenal and the Treasure of Tutankamen
Pieces
Pierrot le Fou
Planet on the Prowl
Planet of the Vampires
Planets Against Us
Pool Hustlers
Prayer for World Peace
Private Affairs
Quo Vadis
Rape of the Sabines
The Red Desert
Red Sun
Redneck
Regina
The Return of Dr. Mabuse
Revenge of the Dead
Revolt of the Barbarians
Ricco
Rocco and His Brothers
Roland the Mighty
Romeo and Juliet
The Roof
Rough Justice
Rush
The Ruthless Four
Sacrilege
Salammbo
Salo, or the 120 Days of Sodom
Samson
Samson Against the Sheik
Samson and His Mighty Challenge
Santa Sangre
Sardinia Kidnapped
Satanik
Savage Island
The Scalawag Bunch
Scenes from a Murder
Secret Agent Super Dragon
Seduced and Abandoned
Seduction of Mimi
Senso
The Sensuous Nurse
Seven Beauties
Seven Deadly Sins
Seven Deaths in the Cat's Eye
Seven Doors of Death
Sex with a Smile
The She-Beast
Shoeshine

Shoot Loud, Louder, I Don't Understand!
Sicilian Connection
Sinbad of the Seven Seas
Sins of Rome
Slaughter of the Vampires
Slave of the Cannibal God
The Sleazy Uncle
Smugglers
Sodom and Gomorrah
Son of Captain Blood
Son of Samson
Sonny Boy
Sonny and Jed
Sotto, Sotto
Spaghetti House
A Special Day
The Spider's Stratagem
Spirits of the Dead
The Squeeze
Star Crash
Stateline Motel
The Story of Boys & Girls
Stranger in Paso Bravo
Stromboli
Submission
Summer Affair
Summer Night
Summer Night With Greek Profile, Almond Eyes . .
Super Fuzz
Superargo
Suspiria
Sweet Smell of Woman
Swept Away...
The Swindle
Taming of the Shrew
The Tempter
Tentacles
10th Victim
Teorema
Teresa Venerdi
Terror Creatures from the Grave
Terror of Rome Against the Son of Hercules
Terror of the Steppes
They Call Me Trinity
Thief of Baghdad
Third Solution
Three Brothers
Throne of Fire
Thunder Warrior 2
Tiger of the Seven Seas
Tigers in Lipstick
Till Marriage Do Us Part
Time of Indifference
'Tis a Pity She's a Whore
To Forget Venice
Together
Torso
Torture Chamber of Baron Blood
Tragedy of a Ridiculous Man
Tramplers
The Tree of Wooden Clogs
Trinity is Still My Name
Triumph of the Son of Hercules
12 Plus 1
Twitch of the Death Nerve
Two Nights with Cleopatra
2020 Texas Gladiators
Two Women
Tyrant of Lydia Against the Son of Hercules
L'Udienza
Ulysses
Umberto D
Uncle Tom's Cabin
Unsane
Ursus in the Valley of the Lions
Vanina Vanini
Variety Lights

Venus Against the Son of Hercules
Verdi
A Very Old Man with Enormous Wings
Viking Massacre
Violent Breed
Violent Professionals
The Virgin of Nuremberg
The Wanton Contessa
War Bus Commando
War Victims
The Warning
Warriors of the Wasteland
Watch Me When I Kill
Waterloo
We All Loved Each Other So Much
We the Living
Web of the Spider
Wedding in Blood
Weekend
Werewolf in a Girl's Dormitory
When Women Had Tails
When Women Lost Their Tails
Where the Hot Wind Blows
Where's Piccone
White Nights
Wifemistress
Wild, Wild Planet
The Witch
A Woman of Rome
Woman Times Seven
Women in Cell Block 7
Wonders of Aladdin
Yesterday, Today and Tomorrow
Yor, the Hunter from the Future
Zombie

Japanese

Akira
Akira Production Report
Antarctica
Attack of the Mushroom People
An Autumn Afternoon
The Bad Sleep Well
The Ballad of Narayama
Battle of the Japan Sea
Black Lizard
Black Magic Terror
Black Rain
The Burmese Harp
The Bushido Blade
Caledonian Dreams
The Crucified Lovers
Cruel Story of Youth
Crystal Triangle
Dersu Uzala
Destroy All Monsters
Dodes 'ka-den
Dog Soldier
Dominion: Tank Police, Act 1
Double Suicide
Drifting Weeds
Drunken Angel
Early Summer (Bakushu)
Eijanaika (Why Not?)
The Empire of Passion
Equinox Flower
Face of Another
The Family Game
Fires on the Plain
Floating Weeds
Forest of Little Bear
Forty-Seven Ronin, Part 1
Forty-Seven Ronin, Part 2
Fugitive Alien
The Funeral
Galaxy Express
The Gambling Samurai
Gamera

Gamera vs. Barugon
Gamera vs. Gaos
Gamera vs. Guiron
Gamera vs. Zigra
Gate of Hell
Ghidrah the Three Headed Monster
The Ghost of Yotsuya
The Go-Masters
Godzilla 1985
Godzilla, King of the Monsters
Godzilla on Monster Island
Godzilla Raids Again
Godzilla vs. the Cosmic Monster
Godzilla vs. Gigan
Godzilla vs. Mechagodzilla
Godzilla vs. Megalon
Godzilla vs. Monster Zero
Godzilla vs. Mothra
Godzilla vs. the Sea Monster
Godzilla vs. the Smog Monster
Godzilla's Revenge
Golden Demon
Gonza the Spearman
Gorath
Gunbuster 1
H-Man
Half Human
Heaven & Earth
Hidden Fortress
High & Low
Himatsuri (Fire Festival)
Hiroshima, Mon Amour
Human Beasts
The Human Condition: A Soldier's Prayer
The Human Condition: No Greater Love
The Human Condition: Road to Eternity
The Human Vapor
The Humanoid
Hunter in the Dark
I Give My All
Ikiru
The Imperial Japanese Empire
Imperial Navy
In the Realm of the Senses
Infra-Man
The Insect Woman
Irezumi (Spirit of Tattoo)
The Island
Kagemusha
Kamikaze
King Kong Versus Godzilla
Kojiro
Kwaidan
Late Spring
Life of O'Haru
Love and Faith
The Lower Depths
MacArthur's Children
Madox-01
The Manster
MD Geist
The Men Who Tread on the Tiger's Tail
Merry Christmas, Mr. Lawrence
Mistress (Wild Geese)
Monster from a Prehistoric Planet
Mother
Mothra
Murder in the Doll House
The Mysterians
Nanami, First Love
No Regrets for Our Youth
Nomugi Pass
Odd Obsession
One-Eyed Swordsman
Onibaba
Osaka Elegy
Panda and the Magic Serpent

Phoenix
The Pornographers
Port Arthur
Proof of the Man
Ran
Rashomon
Red Beard
Red Lion
Revenge of a Kabuki Actor
Riding Bean
Rikisha-Man
Rikyu
Rodan
Saga of the Vagabond
Samurai 1: Musashi Miyamoto
Samurai 2: Duel at Ichijoji Temple
Samurai 3: Duel at Ganryu Island
Samurai Reincarnation
Sandakan No. 8
Sanjuro
Sanshiro Sugata
Sansho the Bailiff
Sea Prince and the Fire Child
Seven Samurai
Shin Heike Monogatari
Shiokari Pass
Shogun Assassin
Sisters of Gion
Snow Country
Son of Godzilla
Station
The Story of the Late Chrysanthemum
Stray Dog
Street of Shame
Summer Vacation: 1999
Sword of Doom
Sword of Fury I
Sword of Fury II
Sword of Heaven
Swords of Death
Takanaka: World
Tampopo
Tattooed Hit Man
A Taxing Woman
A Taxing Woman Returns
Terror Beneath the Sea
Terror of Mechagodzilla
Throne of Blood
Tidal Wave
Time of the Apes
Tokyo Olympiad
Tokyo Story
Traitors of the Blue Castle
Twenty-Four Eyes
Ugetsu
The Ultimate Teacher
Varan the Unbelievable
Vengeance Is Mine
Violence at Noon
Virus
Wanna Be's
War of the Gargantuas
War in Space
Warning from Space
Warriors of the Wind
What's Up, Tiger Lily?
The White Sheik
The Wolves
Woman in the Dunes
X from Outer Space
Yog, Monster from Space
Yojimbo
Zatoichi, the Blind Samurai
Zatoichi vs. Yojimbo
Zero Pilot

Lithuanian

Velnio Nuotaka (The Devil's Bride)

Mexican

Amenaza Nuclear
Arizona
Attack of the Mayan Mummy
Boom in the Moon
The Brainiac
Bring Me the Vampire
By Hook or Crook
Caches de Oro
Calacan
Camelia
Children of Sanchez
Conexion Mexico
Conquest
The Craving
Creature of the Walking Dead
The Criminal Life of Archibaldo de la Cruz
Crossfire
The Curse of the Aztec Mummy
The Curse of the Crying Woman
Curse of the Devil
Curse of the Stone Hand
Dance of Death
Despedida de Casada
Diamante, Oro y Amor
Doctor of Doom
Dona Herlinda & Her Son
Dos Esposas en Mi Cama
Dos Gallos Y Dos Gallinas (Two Roosters For ...)
Dos Postolas Gemelas
El Ametralladora
El Aviador Fenomeno
El Aviso Inoportuno
El Bruto (The Brute)
El Ceniciento
El Ciclon
El Cuarto Chino (The Chinese Room)
El Derecho de Comer (The Right to Eat)
El Dia Que Me Quieras
El Fin del Tahur
El Forastero Vengador
El Fugitivo
El Gallo de Oro (The Golden Rooster)
El Gavilan Pollero
El Hombre Contra el Mundo
El Hombre de la Furia (Man of Fury)
El Hombre y el Monstruo
El Libro de Piedra
El Penon de las Animas
El Profesor Hippie
El Rey
El Robo al Tren Correo
El Sabor de la Venganza
El Sinaloense
El Tesoro de Chucho el Roto
El Zorro Blanco
El Zorro Vengador
Emiliano Zapata
Emilio Varela vs. Camelia La Texana
Erendira
The Exterminating Angel
Face of the Screaming Werewolf
The Fear Chamber
Fearmaker
Fever Mounts at El Pao
Flor Silvestre (Wild Flower)
Foxtrot
Frontera Sin Ley
Garbancito de la Mancha
Garringo
Genie of Darkness
Gente Violenta
Guitarras Lloren Guitarras
Guns for Dollars

Mexican

Hay Muertos Que No Hacen Ruido
House of Evil
The Illusion Travels by Streetcar
Invasion of the Vampires
Invasion of the Zombies
Jungle Warriors
Justicia para los Mexicanos
La Bella Lola
La Carcel de Laredo
La Casa del Amor
La Casa de Madame Lulu
La Casita del Pecado
La Chamuscada
La Choca
La Corona de un Campeon
La Diosa Arrodillada (The Kneeling Goddess)
La Fuerza Inutil
La Fuga del Rojo
La Generala
La Hifa Sin Padre
La Justicia Tiene Doce Anos
La Mansion de la Locura
La Marchanta
La Mentira
La Montana del Diablo (Devil's Mountain)
La Pachange (The Big Party)
La Sombra del Murcielago
La Trampa Mortal
La Trinchera
La Venganza del Kung Fu
La Venganza del Rojo
Lagrimas de Amor
Las Doce Patadas Mortales del Kung Fu
Las Munecas del King Kong
The Living Coffin
The Living Head
Los Amores de Marieta
Los Apuros de Dos Gallos (Troubles of Two Roosters)
Los Asesinos
Los Caciques
Los Cacos
Los Caifanes
Los Dos Hermanos
Los Hermanos Centella
Los Hermanos Diablo
Los Meses y los Dias
Los Olvidados
Los Sheriffs de la Frontera
Los Tres Amores de Losa
Los Tres Calaveras (The Three Skeletons)
Los Vampiros de Coyacan
Macario
The Man and the Monster
Manos Torpes
Marcados por el Destino
Maria Candelaria
Mariachi
Mexican Bus Ride
Mision Suicida
Misterios de Ultratumba
The Monster Demolisher
Murderer in the Hotel
Musico, Poeta y Loco
Narda or The Summer
Nazarin
The Neighborhood Thief
Neutron and the Black Mask
Neutron vs. the Amazing Dr. Caronte
Neutron vs. the Death Robots
Neutron vs. the Maniac
The New Invisible Man
Ni Chana Ni Juana
Nido de Aguilas (Eagle's Nest)
Night of the Bloody Apes
Night of a Thousand Cats
Nosotros (We)
Orlak, the Hell of Frankenstein
Pancho Villa Returns

Para Servir a Usted
Patsy, Mi Amor
The Pearl
The Robot vs. the Aztec Mummy
Samson vs. the Vampire Women
Samson in the Wax Museum
Santa Claus
Santa Sangre
Savage!
Semaforo en Rojo
Sexo Y Crimen
Siempre Hay una Primera Vez
Simon Blanco
Sin Salida
The Sinister Invasion
Sisters of Satan
The Snake People
Somos Novios (In Love and Engaged)
Soy un Golfo
Spiritism
Susana
Swamp of the Lost Monster
Tacos al Carbon
Target Eagle
Temporada Salvaje (Savage Season)
Thor el Conquistador
Tintorera...Tiger Shark
Torero (Bullfighter)
The Vampire
The Vampire's Coffin
Viridiana
Viva Chihuahua (Long Live Chihuahua)
Vivo o Muerto
Warlock Moon
Witches' Mountain
The Witch's Mirror
A Woman Without Love
World of the Vampires
Wrestling Women vs. the Aztec Ape
Wrestling Women vs. the Aztec Mummy
Wuthering Heights
Y Donde Esta el Muerto? (And... Where Is the Dead)
Yanco

New Zealand

Among the Cinders
An Angel at My Table
Bad Blood
Bad Taste
Beyond Reasonable Doubt
The Bridge to Nowhere
Came a Hot Friday
Dangerous Orphans
Dark of the Night
Goodbye Pork Pie
Heart of the Stag
Hot Target
Iris
Mesmerized
Nate and Hayes
The Navigator
Never Say Die
Nutcase
The Quiet Earth
Race to the Yankee Zephyr
Second Time Lucky
Shaker Run
The Silent One
Sleeping Dogs
Smash Palace
Starlight Hotel
Sylvia
Tearaway
Utu
Vigil
Wild Horses

Nicaraguan

Alsino and the Condor

Norwegian

One Day in the Life of Ivan Denisovich
Pathfinder

Polish

Ashes and Diamonds
Baritone
Border Street
A Brivele der Mamen (A Letter to Mother)
Casimir the Great
Colonel Wolodyjowski
Danton
Der Purimshpiler (The Jester)
Dr. Judym
The Dybbuk
Enigma Secret
Fever
First Spaceship on Venus
Foul Play
Greta
Hands Up
Hero of the Year
In Desert and Wilderness
In an Old Manor House
Interrogation
Kanal
Karate Polish Style
Knife in the Water
Mamele (Little Mother)
Man of Marble
Merry Christmas
Moonlighting
Mother of Kings
Nights and Days
The Peasants
Pharaoh
The Singing Blacksmith (Yankl der Shmid)
Train to Hollywood
The Unapproachable
W Hour
Without Love
Year of the Quiet Sun
Yidl Mitn Fidl (Yiddle with His Fiddle)

Portuguese

Black Orpheus
In the White City
Life is Beautiful
Return of the Evil Dead
Tombs of the Blind Dead

Russian

Aelita: Queen of Mars
Alexander Nevsky
The Amphibian Man
Andrei Roublev
Arsenal
Ballad of a Soldier
Baltic Deputy
The Battleship Potemkin
Bed and Sofa
Burglar
Chapayev
The Cigarette Girl of Mosselprom
Come and See
Commissar
The Cranes are Flying
Crime and Punishment
Dersu Uzala
Don Quixote
Earth
Enthusiasm

A Forgotten Tune for the Flute
Freeze-Die-Come to Life
The General Line
The Girl with the Hat Box
The Golem
Guerilla Brigade
Happiness
Incident at Map-Grid 36-80
The Inspector General
Ivan the Terrible
Ivan the Terrible, Part 1
Ivan the Terrible, Part 2
Kino Pravda/Enthusiasm
The Lady with the Dog
The Legend of Suram Fortress
Little Vera
The Magic Voyage of Sinbad
The Man with a Movie Camera
The Mirror
Moscow Does Not Believe in Tears
Mother
My Name Is Ivan
No Greater Love
Oblomov
October
The Overcoat
The Patriots
Peter the First: Part 1
Peter the First: Part 2
Planet Burg
Prince Igor
Private Life
Que Viva Mexico
Rasputin
Repentance
Shadows of Forgotten Ancestors
Sideburns
A Slave of Love
Solo Voyage: The Revenge
Storm Over Asia
Strike
A Summer to Remember
Sword & the Dragon
The Tsar's Bride
An Unfinished Piece for a Player Piano
War and Peace
Waterloo
Zvenigora

South African

City of Blood
Deadly Passion
Emissary
The Evil Below
Fatal Assassin
House of the Living Dead
Kill Slade
Place of Weeping
Stay Awake

Spanish

Anguish
Anita de Montemar
Any Gun Can Play
Ay, Carmela!
Bad Man's River
Benjamin Argumedo
Bestia Nocturna
Black Beauty
The Blood Spattered Bride
Blood Wedding (Bodas de Sangre)
Brandy Sheriff
Bromas S.A.
Brujo Luna
Bullet for Sandoval
Cada Noche un Amor (Every Night a New Lover)
Camila

Carmen
The Castle of Fu Manchu
Chimes at Midnight
The Christmas Kid
The City and the Dogs
City of the Walking Dead
Como Dos Gotas de Agua (Like Two Drops of Water)
Con Mi Mujer No Puedo
Conquest
Count Dracula
Cria
The Criminal Life of Archibaldo de la Cruz
Curse of the Devil
Dark Habits
Dawn of the Pirates (La Secta de los Tughs)
Demons in the Garden
Departamento Compartido
Desde el Abismo
Desperados
Details of a Duel: A Question of Honor
The Devil's Possessed
The Diabolical Dr. Z
Dias de Ilusion
Dr. Jekyll and the Wolfman
Dr. Orloff's Monster
Don Juan My Love
Don Quixote
Donde Duermen Dos...Duermen Tres
Dos Chicas de Revista
Dracula vs. Frankenstein
Dracula's Great Love
Duel of Champions
El Barbaro
El Cochecito
El Condor
El Coyote Emplumado
El Cristo de los Milagros
El Diputado (The Deputy)
El Muerto
El Norte
El Pirata Negro (The Black Pirate)
El Preprimido (The Timid Bachelor)
El Sheriff Terrible
El Super
El: This Strange Passion
Escape from El Diablo
Esposa y Amante
Esta Noche Cena Pancho
Expertos en Pinchazos
The Exterminating Angel
Exterminators of the Year 3000
The Fable of the Beautiful Pigeon Fancier
Fangs of the Living Dead
Fin de Fiesta
Fotografo de Senoras
Frida
From Hell to Victory
Frontera (The Border)
Fuerte Perdido
Furia Pasional
The Fury of the Wolfman
The Garden of Delights
Gemidos de Placer
Gendarme Desconocido
Goliath Against the Giants
Gran Valor en la Facultad de Medicina
Grandes Amigos (Great Friends)
Graveyard of Horror
Great Adventure
The Green Wall
Half of Heaven
The Hanging Woman
Hatchet for the Honeymoon
Hawk of the Caribbean
Heidi

CATEGORY INDEX

Please refer to the front of the book for the category definitions.

CATEGORY INDEX

Action Adventure

The Abductors
Above the Law
The Abyss
Ace of Aces
Aces: Iron Eagle 3
Across the Great Divide
Act of Piracy
Act of Vengeance
Action in Arabia
Action Jackson
Action U.S.A.
Adventure of the Action
  Hunters
The Adventurers
Adventures of Captain Fabian
Adventures of Captain Marvel
Adventures of Don Juan
The Adventures of Ford
  Fairlane
The Adventures of Frontier
  Fremont
The Adventures of Grizzly
  Adams at Beaver Dam
The Adventures of
  Huckleberry Finn
The Adventures of the
  Masked Phantom
The Adventures of Milo & Otis
Adventures of Red Ryder
The Adventures of Rex &
  Rinty
The Adventures of Rin Tin Tin
The Adventures of Robin
  Hood
The Adventures of Sinbad
Adventures of Smilin' Jack
The Adventures of Tartu
The Adventures of Tarzan
The Adventures of Tom
  Sawyer
The Adventures of Ultraman
The Adventures of the
  Wilderness Family
Africa Texas Style
African Rage
After the Fall of Saigon
Against All Flags
Against the Drunken Cat
  Paws
Ain't No Way Back
Air America
Air Force
Airport
Alamut Ambush
The Alaska Wilderness
  Adventure
Alice to Nowhere
Alien Private Eye
All the Lovin' Kinfolk
Allan Quartermain and the
  Lost City of Gold
Alley Cat
Alligator Alley
The Amateur

The Amazing Dobermans
The Amazing Spider-Man
The Amazing Transparent
  Man
Amazon
Amazon Jail
The Amazons
Amazons
American Autobahn
American Born
American Eagle
American Empire
American Justice
American Kickboxer 1
American Ninja
American Ninja 2: The
  Confrontation
American Ninja 3: Blood Hunt
American Ninja 4: The
  Annihilation
American Roulette
Americano
The Amsterdam Connection
The Amsterdam Kill
And I Alone Survived
And Then You Die
Android
Angel 3: The Final Chapter
Angel of Death
Angel Town
Angel's Brigade
Angels of the City
Angels Die Hard
Angels from Hell
Angels One Five
Angels' Wild Women
The Angry Dragon
The Annihilators
Another Pair of Aces: Three of
  a Kind
Antarctica
Apache Blood
Apocalypse Now
Appointment in Honduras
Archer: The Fugitive from the
  Empire
Archer's Adventure
Arena
Ark of the Sun God
Armed for Action
Armed Response
Around the World in 80 Days
Around the World Under the
  Sea
Arrest Bulldog Drummond
Ashanti, Land of No Mercy
Assassin
Assassination
Assault on Agathon
Assault with a Deadly Weapon
The Assault of Final Rival
Assault on Precinct 13
Assault on a Queen
Assignment Skybolt
At Gunpoint

At Sword's Point
Atlas in the Land of the
  Cyclops
Ator the Fighting Eagle
Attack Force Z
Attack of the Super Monsters
Avenged
The Avenger
Avenging Force
Avenging Ninja
The Aviator
Back to Back
Back to Bataan
Bad Guys
Badge 373
Bail Out
Balkan Express
The Ballad of Paul Bunyan
Ballbuster
Band of the Hand
Bandolero!
The Banker
Banzai Runner
Barbarian Queen
Barbarian Queen 2: The
  Empress Strikes Back
The Barbarians
Barbary Coast
The Barbary Coast
Bare Knuckles
Baron
Bat 211
Batman
Batman and Robin: Volume 1
Batman and Robin: Volume 2
The Battle of Britain
Battle of the Bulge
Battle of the Commandos
Battle of Neretva
Battle of Valiant
Battling Amazons
Bear Island
The Beast
Beastmaster
Beastmaster 2: Through the
  Portal of Time
Beasts
Beau Geste
Bedford Incident
Bells of Coronado
Bells of Death
The Belstone Fox
Ben-Hur
Beneath Arizona Skies/
  Paradise Canyon
Beneath the 12-Mile Reef
Benji
Benji the Hunted
Berlin Tunnel 21
Best of the Best
Best Revenge
Betrayal and Revenge
Beyond the Call of Duty
Beyond the Poseidon
  Adventure

Big Bad John
The Big Brawl
Big Bust Out
Big Chase
Big Country
The Big Doll House
Big Jim McLain
The Big Push
The Big Rascal
The Big Red One
The Big Scam
Big Showdown
The Big Slice
Big Steal
The Big Sweat
Big Trouble in Little China
The Big Wheel
Biggles
Billy Budd
Billy the Kid in Texas
Bimini Code
Bird of Paradise
Birds of Prey
The Black Arrow
Black Arrow
Black Beauty
Black Beauty/Courage of
  Black Beauty
Black Belt
Black Belt Fury
Black Belt Jones
Black Brigade
Black Caesar
The Black Cobra
Black Cobra 2
Black Cobra 3: The Manila
  Connection
The Black Coin
The Black Devils of Kali
Black Dragon's Revenge
Black Eagle
Black Eliminator
Black Force
Black Force 2
Black Gestapo
Black Magic
Black Market Rustlers
Black Moon Rising
Black Oak Conspiracy
The Black Pirate
Black Rain
The Black Raven
Black Samurai
Black Shampoo
The Black Stallion
The Black Stallion Returns
Black Sunday
Black Terrorist
Black Water Gold
Black & White Swordsmen
The Black Widow
Blackbeard the Pirate
Blackjack
Blackstar Volume 3
Blade

Blade in Hong Kong
Blade Master
The Blazing Ninja
Blind Fists of Bruce
Blind Fury
Blind Vengeance
Blonde in Black Leather
Blonde Savage
Blood Alley
Blood Avenger
Blood Brothers
Blood & Concrete: A Love
  Story
Blood of the Dragon
Blood of the Dragon Peril/
  Revenge of the Dragon
Blood and Guns
Blood Money
Blood Money - The Story of
  Clinton and Nadine
Blood for a Silver Dollar
Blood on the Sun
Bloodfight
Bloodfist
Bloodfist 2
Bloodfist 3: Forced to Fight
Bloodmatch
Bloodsport
The Bloody Fight
Bloody Fist
The Bloody Tattoo (The Loot)
Blue Fin
The Blue and the Grey
Blue Jeans and Dynamite
The Blue Lightning
Blue Steel
Blue Thunder
Blue Tornado
BMX Bandits
Bobbie Jo and the Outlaw
The Bodyguard
The Bold Caballero
Bolo
Boomerang
Border Radio
Border of Tong
Borderline
Born American
Born to Be Wild
Born Invincible
Born Killer
Born to Race
Bounty Hunters
Boy of Two Worlds
Braddock: Missing in Action 3
Brady's Escape
Branded
Brannigan
The Brave Bunch
Breaker! Breaker!
Breaking Loose
Breakout
Breathing Fire
A Breed Apart
The Bridge to Nowhere

Action

The Bridge at Remagen
A Bridge Too Far
The Bridges at Toko-Ri
Bring Me the Head of Alfredo
  Garcia
Bronze Buckaroo
Brothers Lionheart
Brothers of the West
Bruce Is Loose
Bruce Lee Fights Back From
  the Grave
Bruce Lee's Ways of Kung Fu
Bruce Le's Greatest Revenge
Bruce Li the Invincible
Bruce Li in New Guinea
Bruce and Shaolin Kung Fu
Bruce the Superhero
Bruce vs. Bill
Bruce's Deadly Fingers
Bruce's Fists of Vengeance
The Buccaneer
Buck Rogers Cliffhanger
  Serials, Vol. 2
Buckeye and Blue
Buddha Assassinator
Buffalo Rider
Buffalo Stampede
Bulldog Drummond
Bulldog Drummond Double
  Feature
Bulldog Drummond Escapes
Bulldog Drummond's Bride
Bulldog Drummond's Peril
Bulldog Drummond's Secret
  Police
Bulletproof
Bummer
Bunco
Burke & Wills
Burn 'Em Up Barnes
The Bushido Blade
The Bushwackers
The C-Man
The Cage
Caged Fury
Caged Heat
Caged Terror
Cahill: United States Marshal
Call Him Mr. Shatter
Call of the Wild
Camorra: The Naples
  Connection
Candleshoe
Cannonball
The Canterbury Tales
Cantonen Iron Kung Fu
Captain America
Captain America 2: Death Too
  Soon
Captain Blood
Captain Caution
Captain Gallant, Foreign
  Legionnaire
Captain Kidd
Captain Scarlett
Captive Rage
The Capture of Bigfoot
Captured in Chinatown
Car Crash
The Cariboo Trail
Cartel
The Cary Grant Collection
Casablanca Express
Casey Jones and Fury (Brave
  Stallion)
Castaway Cowboy
The Cat
Catch Me...If You Can
Chain Gang Killings
Chain Lightning
Chained Heat
Chains
The Challenge
Challenge to Be Free
Challenge of McKenna
Challenge to White Fang

Chance
Chandu on the Magic Island
Chandu the Magician
Charge of the Light Brigade
Charlie Bravo
Charlie Grant's War
Charlie and the Great Balloon
  Chase
Charlie, the Lonesome Cougar
The Cheaters
Cheetah
Child of Glass
Children of the Wild
Children's Island
China O'Brien
China Seas
Chinatown Connection
The Chinatown Kid
Chinese Boxes
Chinese Connection
Chinese Connection 2
Chinese Stuntman
Chinese Web
The Chisholms
Choke Canyon
Chrome Hearts
Chrome and Hot Leather
Circle of Fear
Circle Man
Clarence, the Cross-eyed Lion
Clash of the Ninja
Claws
Clayton County Line
The Climb
Cloak & Dagger
Cloud Waltzing
The Cobra
Cobra
Cobra Against Ninja
Code Name: Dancer
Code Name: Emerald
Code Name: Zebra
Code of Silence
Codename Kyril
Codename: Terminate
Codename: Vengeance
Codename: Wildgeese
Cold Blooded Murder
Cold Steel
Cold Steel for Tortuga
Cold Sweat
Cold War Killers
Coldfire
The Colditz Story
Cole Justice
The Colombian Connection
Color Adventures of
  Superman
Colorado
Colossus and the Amazon
  Queen
Combat Shock
Commando
Commando Cody
Commando Invasion
Commando Squad
Commandos
Commandos Strike at Dawn
Conan the Barbarian
Conan the Destroyer
The Concentratin' Kid
Conexion Mexico
Confidential
The Conqueror
Conqueror & the Empress
Conquest of the Normans
Contra Company
Contraband
Convoy
Cool As Ice
Cop in Blue Jeans
Cop-Out
Corleone
Coroner Creek
Corrupt Ones
Corsican Brothers

Could It Happen Here?
Counter Measures
Counterforce
Courage Mountain
Courage of Rin Tin Tin
Cover-Up
Covert Action
Crack House
Crazed Cop
Crime Killer
Crime Story
Crime Zone
Crimebusters
The Crimson Ghost
Crimson Pirate
Cross of Iron
Cross Mission
Crossfire
Crossing the Line
Crusoe
Cry of Battle
Cry of the Black Wolves
Cry of the Innocent
Crystalstone
Cuba
The Cut Throats
Cycle Vixens
Cyclone
Cyclone Cavalier
Cyclotrode ''X''
Daggers 8
Damn the Defiant
Damned River
Dance of Death
Dance with Death
Danger: Diabolik
Danger Zone
Danger Zone 2
Danger Zone 3: Steel Horse
  War
Dangerous Charter
Dangerous Orphans
Daredevils of the Red Circle
Daring Dobermans
Dark Age
Dark Mountain
Dark Rider
Darkman
The Darkside
Davy Crockett, King of the
  Wild Frontier
Davy Crockett and the River
  Pirates
Dawn of the Pirates (La Secta
  de los Tughs)
Day of the Assassin
Day of the Cobra
The Day of the Dolphin
Day of the Panther
Days of Hell
Days of Thunder
Dayton's Devils
Dead Aim
Dead Bang
Dead Easy
Dead End City
The Dead Pool
Dead Wrong
The Deadliest Art: The Best of
  the Martial Arts Films
Deadline Auto Theft
Deadly Alliance
The Deadly and the Beautiful
Deadly Bet
Deadly Breed
Deadly Commando
Deadly Diamonds
Deadly Encounter
Deadly Intent
Deadly Mission
Deadly Prey
Deadly Reactor
Deadly Revenge
Deadly Stranger
Deadly Surveillance
Deadly Thief

Deadly Vengeance
Deadly Weapon
Deadly Weapons
Death Before Dishonor
Death Chase
Death Duel of Mantis
Death Feud
Death Hunt
Death of the Incredible Hulk
Death Journey
The Death Merchant
Death Race 2000
Death Raiders
Death Ray
Death Shot
The Death Squad
Death Target
Death Warrant
Death Wish 2
Death Wish 3
Death Wish 4: The
  Crackdown
Deathstalker 4: Clash of the
  Titans
Defiant
Delos Adventure
Delta Force
Delta Force 2: Operation
  Stranglehold
Delta Force Commando
Delta Force Commando 2
Delta Fox
Demolition
The Demon Fighter
Demon with a Glass Hand
Dersu Uzala
Desert Commandos
The Desert of the Tartars
Desperate Cargo
Desperate Target
Destroyer
The Destructors
Detroit 9000
Devastator
Devil of the Desert Against the
  Son of Hercules
Devil Monster
Devil Wears White
Devil's Angels
Devil's Crude
Diamond Ninja Force
Diamonds are Forever
Diary of a Hitman
Diary of a Rebel
Dick Tracy
Dick Tracy Detective
Dick Tracy Double Feature #1
Dick Tracy Double Feature #2
Dick Tracy Meets Gruesome
Dick Tracy and the Oyster
  Caper
Dick Tracy Returns
Dick Tracy: The Original
  Serial, Vol. 1
Dick Tracy: The Original
  Serial, Vol. 2
Dick Tracy: The Spider Strikes
Dick Tracy vs. Crime Inc.
Dick Tracy vs. Cueball
Dick Tracy's Dilemma
Dick Turpin
Die Hard
Die Hard 2: Die Harder
The Dirty Dozen
The Dirty Dozen: The Next
  Mission
Dirty Mary Crazy Larry
Dirty Partners
The Divine Enforcer
Divine Nymph
Dixie Dynamite
Do or Die
The Doberman Gang
Doc Savage
Doctor Doom Conquers the
  World

Dr. No
Dr. Syn
Dr. Syn, Alias the Scarecrow
Dog Eat Dog
The Dogs of War
The Doll Squad
Dollman
Don Daredevil Rides Again
The Don is Dead
Don Juan
Don Q., Son of Zorro
Don Winslow of the Coast
  Guard
Don Winslow of the Navy
Don't Change My World
Double Impact
Double Revenge
The Dove
Down Twisted
The Dragon Fist
Dragon Force
Dragon Lee Vs. the Five
  Brothers
The Dragon Lives Again
Dragon Lord
Dragon the Odds
Dragon Princess
The Dragon Returns
Dragon Rider
Dragon from Shaolin
The Dragon Strikes Back
Dragon vs. Needles of Death
Dragonard
Dragon's Claw
The Dragon's Showdown
Dreadnaught
Dreadnaught Rivals
Driven to Kill
The Driver
Driving Force
Drums
Drums of Fu Manchu
Dual Flying Kicks
Dudes
Duel of the Brave Ones
Duel of Champions
Duel of Fists
Duel of the Iron Fist
Duel of the Masters
Duel of the Seven Tigers
Dune Warriors
Dust
Dynamite
Dynamo
Dynasty
Each Dawn I Die
The Eagle
Eagle Claws Champion
The Eagle Has Landed
Eagles Attack at Dawn
Eagle's Shadow
East of Kilimanjaro
East L.A. Warriors
Easy Kill
Edge of Fury
The Eiger Sanction
18 Bronzemen
Eighteen Jade Arhats
18 Weapons of Kung-Fu
El Barbaro
El Cid
El Hombre Contra el Mundo
El Pirata Negro (The Black
  Pirate)
El Rey
El Zorro Blanco
Electra Glide in Blue
Elephant Boy
Elsa & Her Cubs
The Elusive Pimpernel
Emerald of Aramata
Emil & the Detectives
Emilio Varela vs. Camelia La
  Texana
Emperor of the Bronx
Empire of the Dragon

Empire of Spiritual Ninja
Empire State
Empire of the Sun
The Enchanted Forest
Enemy Unseen
The Enforcer
Enforcer from Death Row
Enigma
Enter the Dragon
Enter the Game of Death
Enter the Ninja
Enter the Panther
Enter Three Dragons
Erik
Erik, the Viking
Erotic Touch of Hot Skin
Escapade in Japan
Escape from Alcatraz
Escape to Athena
Escape from the Bronx
Escape to Burma
Escape from Cell Block Three
Escape from Death Row
Escape from DS-3
Escape from El Diablo
Escape from the KGB
Escape from New York
Escape from Safehaven
Escape to the Sun
Eureka Stockade
Eve of Destruction
The Evil Below
The Evil That Men Do
Executioner of Venice
The Expendables
Exterminator
Exterminator 2
Extreme Prejudice
Eye of the Eagle
Eye of the Eagle 2
Eye of the Eagle 3
An Eye for an Eye
Eye of the Needle
Eye of the Octopus
Eye of the Tiger
F/X 2: The Deadly Art of
    Illusion
Fables of the Green Forest
Fair Game
Falcon
The Family
Famous Five Get Into Trouble
Fantastic Balloon Voyage
Fantasy Mission Force
Far East
Farewell to the King
Fass Black
Fast Company
Fast Fists
Fast Kill
Fast Money
Faster Pussycat! Kill! Kill!
Fatal Assassin
Fatal Beauty
Fatal Chase
Fatal Error
Fatal Mission
Fatal Skies
The Fate of Lee Khan
Fear in the City
Fearless Fighters
Fearless Hyena
Fearless Hyena: Part 2
The Fearless Young Boxer
Felix in Outer Space
The Female Bunch
Ferry to Hong Kong
Fever
ffolkes
Fight to the Death
Fight for Survival
Fight for Us
Fighting Ace
Fighting Black Kings
Fighting Devil Dogs
Fighting Duel of Death

Fighting Life
Fighting Mad
Fighting Marines
Fighting Prince of Donegal
The Fighting Rookie
Fighting Seabees
The Final Alliance
The Final Countdown
Final Impact
Final Justice
Final Mission
Final Sanction
Fire
Fire Birds
Fire and Ice
Fire, Ice and Dynamite
Fire in the Night
Fire and Rain
The Fire in the Stone
Fire and Sword
Fireback
Fireballs
Firecracker
Firefight
Firefox
Firehead
Firepower
Firewalker
The Firing Line
First Blood
First Yank into Tokyo
Fish Hawk
F.I.S.T.
Fist
Fist of Death
Fist of Fear, Touch of Death
Fist Fighter
A Fist Full of Talons
The Fist, the Kicks, and the
    Evils
Fist of Vengeance
Fists of Blood
Fists of Bruce Lee
Fists of Dragons
Fists of Fury
Fists of Fury 2
Fists of the White Lotus
Five Came Back
Five Fighters from Shaolin
Five Golden Dragons
Five for Hell
Five Miles to Midnight
Five Weeks in a Balloon
The Fix
The Flame & the Arrow
Flaming Signal
The Flash
Flash Challenger
Flash & Firecat
Flash Gordon Battles the
    Galactic Forces of Evil
Flash Gordon: Space
    Adventurer
Flashpoint Africa
Flat Top
Flatbed Annie and Sweetiepie:
    Lady Truckers
Flatfoot
Flesh and Blood
Fleshburn
The Flight
Flight of Black Angel
Flight of Dragons
Flight of the Eagle
Flight to Fury
Flight from Glory
Flight of the Intruder
Flight to Nowhere
The Flight of the Phoenix
Flight from Singapore
Flipper
Flipper's New Adventure
Flipper's Odyssey
Flood!
Flor Silvestre (Wild Flower)
The Florida Connection

Florida Straits
Flying Leathernecks
Follow Me
Follow That Car
For Your Eyes Only
Force Five
Force: Five
Force of One
Force 10 from Navarone
Forced Vengeance
Foreign Legionnaire: Court
    Martial
Forest Duel
The Forgotten
Forgotten Warrior
Fortune Dane
48 Hours to Live
The Four Feathers
Four Feathers
The Four Hands of Death:
    Wily Match
Four Robbers
Foxtrap
Foxy Brown
F.P. 1
Frame Up
Framed
The French Connection
French Connection 2
French Quarter Undercover
Friday Foster
From China with Death
From Hell to Borneo
From Russia with Love
Frontera Sin Ley
Fugitive Samurai
Full Metal Ninja
The Furious
The Furious Avenger
Furious Slaughter
Fury
Fury to Freedom: The Life
    Story of Raul Ries
The Fury of Hercules
Fury of King Boxer
Fury in the Shaolin Temple
Fury on Wheels
Future Cop
Future Force
Future Hunters
Future Kill
Future Zone
G-Men Never Forget
G-Men vs. the Black Dragon
Gallagher's Travels
The Galloping Ghost
Galyon
The Gamble
The Gambling Samurai
Game of Death
Game of Survival
Game for Vultures
Gang Wars
Gangsters
Gator
Gator Bait
Gator Bait 2: Cajun Justice
The Gauntlet
The General Died at Dawn
Gentle Giant
Get Christie Love!
The Getaway
Ghetto Blaster
Ghost of the Ninja
Ghostwarrior
G.I. Executioner
The Giant of Marathon
The Giant of Metropolis
The Gift Horse
Ginger
Girls Are for Loving
Girls Riot
The Gladiator
The Glass Jungle
Gloria

Glory Boys
The Glory Stompers
The Glove
Go for Broke
Go Kill and Come Back
Go Tell the Spartans
God's Bloody Acre
God's Country
Goetz von Berlichingen (Iron
    Hand)
Goin' Coconuts
Gold Raiders
Golden Destroyers
Golden Exterminators
Golden Lady
Golden Ninja Invasion
Golden Ninja Warrior
The Golden Salamander
Golden Sun
The Golden Triangle
Golden Voyage of Sinbad
Goldfinger
Goliath Against the Giants
Goliath and the Barbarians
Goliath and the Dragon
Goliath and the Sins of
    Babylon
Gone in 60 Seconds
Good Guys Wear Black
Goodbye Pork Pie
The Goonies
Gordon's War
Grand Canyon Trail
Grand Master of Shaolin Kung
    Fu
Grand Theft Auto
Gray Lady Down
Great Adventure
The Great Battle
The Great Escape
The Great Escape 2
The Great Hunter
Great K & A Train Robbery
The Great Locomotive Chase
The Great Los Angeles
    Earthquake
Great Massacre
Great Movie Stunts & the
    Making of Raiders of the
    Lost Ark
Great Ride
The Great Skycopter Rescue
Great Smokey Roadblock
The Great Texas Dynamite
    Chase
Great Train Robbery
The Great Waldo Pepper
The Greatest Fights of Martial
    Arts
The Greatest Flying Heroes
The Green Berets
Green Dragon Inn
The Green Hornet
Green Inferno
Greenstone
Greystoke: The Legend of
    Tarzan, Lord of the Apes
Grizzly Mountain
Ground Zero
The Groundstar Conspiracy
Gunbuster 1
Gung Ho!
Gunga Din
Guns
The Guns of Navarone
The Guy from Harlem
Gymkata
Gypsy
H-Bomb
Half a Loaf of Kung Fu
Hands of Death
Hangar 18
Hangfire
Hangmen
Hanna's War
Hard Drivin'

Hard to Kill
Hard Knocks
Hard Ticket to Hawaii
Hardcase and Fist
Hardware
Harley Davidson and the
    Marlboro Man
Harry Black and the Tiger
Hatari
Hawk of the Caribbean
Hawk the Slayer
Hawk of the Wilderness
Hawkeye
Heartbreak Ridge
Heartbreaker
Hearts & Armour
Heat
Heat of the Flame
Heat Street
Heated Vengeance
The Heist
Hell on the Battleground
Hell is for Heroes
Hell Hounds of Alaska
Hell Hunters
Hell in the Pacific
Hell Raiders
Hell Ship Mutiny
Hell Up in Harlem
Hell on Wheels
Hellbent
Hellcats
Hellriders
Hell's Angels '69
Hell's Angels Forever
Hell's Headquarters
Hell's Wind Staff
Hercules
Hercules 2
Hercules Against the Moon
    Men
Hercules and the Captive
    Women
Hercules in the Haunted World
Hercules in New York
Hercules, Prisoner of Evil
Hercules Unchained
Herculoids
Herculoids, Vol. 2
Here Come the Littles: The
    Movie
Hero Bunker
Hero of Rome
Hero and the Terror
Heroes in Hell
Heroes in the Ming Dynasty
Heroes Stand Alone
Hero's Blood
Hi-Riders
Hidden Fortress
High Ballin'
High Crime
High Ice
High Risk
High Rolling in a Hot Corvette
High Velocity
Highlander
Himatsuri (Fire Festival)
Hired to Kill
His Majesty O'Keefe
Hit Men
Hitler: Dead or Alive
The Hitman
The Hitter
Hollywood Cop
Holt of the Secret Service
Home Alone
Homeboy
Honeybaby
Hong Kong Nights
Honor Among Thieves
The Hooked Generation
Horsemen
Hostage
The Hostage Tower
Hostile Takeover

Action

Hot Blood
Hot Box
The Hot, the Cool, and the Vicious
Hot Pursuit
Hot Rod
Hot Rod Girl
Hour of the Assassin
Hula
The Human Shield
Hundra
The Hunt for Red October
Hunted
The Hunted Lady
Hunter
The Hunter
Hunter's Blood
Hunters of the Golden Cobra
Hunting the White Rhino
Hurricane
Hurricane Express
Hurricane Smith
Hurricane Sword
I Spit on Your Corpse
Ice Station Zebra
If Looks Could Kill
Image of Bruce Lee
Imperial Venus
In Deadly Heat
In Eagle Shadow Fist
In Gold We Trust
In Hot Pursuit
In Love and War
In Search of the Castaways
In Your Face
The Incredible Hulk
The Incredible Hulk Returns
The Incredible Journey
Incredible Master Beggars
Indiana Jones and the Last Crusade
Indiana Jones and the Temple of Doom
Indio
Indio 2: The Revolt
Inside Information
The Inside Man
Inside Out
Instant Justice
Instant Kung-Fu Man
The Invaders
Invaders from Mars
Invasion Force
Invasion U.S.A.
Invincible
Invincible Barbarian
Invincible Gladiators
Invincible Obsessed Fighter
The Invincible Six
Invincible Sword
Iron Angel
The Iron Crown
Iron Dragon Strikes Back
Iron Eagle
Iron Eagle 2
Iron Horsemen
The Iron Mask
Iron Thunder
Ironmaster
The Island
Island of Adventure
Island of the Blue Dolphins
Island of the Lost
Island at the Top of the World
Island Trader
Island Warriors
Istanbul
The Italian Connection
Ivanhoe
Jack the Giant Killer
Jackie Chan's Police Force
Jaguar
Jaguar Lives
Jail Bait
Jailbreakin'
Jakarta

Jamaica Inn
Japanese Connection
Jaws of Death
Jaws of the Dragon
Jaws of Justice
Jericho
Jessie's Girls
The Jesus Trip
Jet Attack/Paratroop Command
Jive Turkey
Johnny Handsome
Johnny Shiloh
Johnny Tremain & the Sons of Liberty
Jornada de Muerte
Journey
Journey to the Center of the Earth
Journey of Honor
Journey to the Lost City
Journey of Natty Gann
Joy Ride to Nowhere
Juggernaut
Jungle
Jungle Book
Jungle Bride
Jungle Drums of Africa
Jungle Goddess
Jungle Heat
Jungle Hell
Jungle Holocaust
Jungle Inferno
Jungle Master
Jungle Raiders
Jungle Siren
Jungle Warriors
Junior G-Men
Junior G-Men of the Air
Junkman
Juve Contre Fantomas
K-9000
Kagemusha
Kamikaze '89
The Karate Kid
The Karate Kid: Part 2
The Karate Kid: Part 3
Karate Killer
Karate Warrior
Karate Warriors
Kavik the Wolf Dog
Keeper of the City
Keeping Track
Kelly's Heroes
Kentucky Blue Streak
The Key to Rebecca
Key to Vengeance
Khartoum
Kick of Death: The Prodigal Boxer
The Kick Fighter
Kickboxer
Kickboxer the Champion
Kicks
Kid Colter
Kid and the Killers
Kidnapped
The Kill
Kill Alex Kill
Kill or Be Killed
Kill Factor
Kill the Golden Goose
Kill and Kill Again
Kill Line
Kill Slade
Kill Squad
Killcrazy
Killer Commandos
Killer Elephants
Killer Force
Killers
The Killer's Edge
The Killing Edge
The Killing Game
Killing at Hell's Gate
The Killing Machine

Killing Streets
Killing in the Sun
The Killing Zone
Killzone
Kilma, Queen of the Amazons
Kim
King Arthur, the Young Warlord
King Boxer
King Boxers
King of the Congo
King Dinosaur
King Kong
King Kong Lives
King Kong Versus Godzilla
King of the Kongo
King of Kung-Fu
King of the Mountain
King of the Rocketmen
King Solomon's Mines
King Solomon's Treasure
Kingfisher Caper
Kingfisher the Killer
King's Ransom
Kinjite: Forbidden Subjects
Klondike Fever
Knightriders
Knights of the City
Knock Outs
Kojiro
Koroshi
Kowloon Assignment
Kuffs
Kung Fu Avengers
Kung Fu Commandos
Kung Fu of Eight Drunkards
Kung Fu Genius
Kung Fu Hero
Kung Fu Kids
Kung Fu Massacre
Kung Fu Rebels
Kung Fu for Sale
Kung Fu Shadow
The Kung Fu Warrior
L.A. Bounty
La Choca
La Corona de un Campeon
La Fuga de Caro
La Fuga del Rojo
L.A. Gangs Rising
L.A. Heat
La Isla de Tesoro de los Pinos
La Montana del Diablo (Devil's Mountain)
La Tigeresa (The Tigress)
L.A. Vice
Lady in the Lake
Lady Scarface
Lady Street Fighter
Lady Terminator
Land of the Pharaohs
Las Doce Patadas Mortales del Kung Fu
Las Munecas del King Kong
Laser Mission
Laserblast
Lassie, Come Home
The Last Boy Scout
Last Challenge of the Dragon
Last Chase
The Last Contract
The Last Dragon
Last Flight of Noah's Ark
The Last Hour
Last Mercenary
Last of the Mohicans
Last Plane Out
The Last Valley
Last Warrior
Last Witness
Law of the Jungle
Law of the Sea
Law of the Wild
The Lawless Land
Le Mans
The Leatherneck

The Left Hand of God
Legend of Eight Samurai
The Legend of Hillbilly John
Legend of Lobo
Legend of the Lost
Legend of the Northwest
The Legend of Sea Wolf
Legion of Iron
Leopard Fist Ninja
Let 'Em Have It
Lethal Weapon
Lethal Weapon 2
Lethal Woman
Let's Get Harry
Liberty & Bash
License to Kill
Life & Times of Grizzly Adams
The Lifetaker
Light at the Edge of the World
Light in the Forest
Lightblast
Lightning Kung Fu
Lion of Africa
Lion Man
The Lion of Venice
Lionheart
The Lion's Share
Little Big Master
Little Caesar
Little Dragons
Little Heroes of Shaolin Temple
Little Laura & Big John
Little Treasure
Littlest Horse Thieves
Littlest Warrior
Live and Let Die
Live a Little, Steal a Lot
The Lives of a Bengal Lancer
The Living Daylights
Loaded Guns
The Lone Runner
Long John Silver
Long Shot
Loophole
Lords of Magick
Los Caciques
Los Caifanes
Lost
The Lost City
Lost City of the Jungle
The Lost Command
The Lost Continent
The Lost Empire
The Lost Idol/Shock Troop
The Lost Patrol
The Lost Samurai Sword
Lost Squadron
The Lost Tribe
Lost Zeppelin
Love and Bullets
The Loves of Hercules
Low Blow
The Lucifer Complex
Macao
Machoman
Maciste in Hell
The Mad Bomber
Mad Dog
Mad Max
Mad Max Beyond Thunderdome
Mad Wax: The Surf Movie
Madman
Mafia vs. Ninja
The Magic Sword
The Magic Voyage of Sinbad
The Magnificent
Magnificent Adventurer
Magnificent Duo
The Magnificent Kick
Magnum Force
Magnum Killers
Malibu Express
Malone
Mama's Dirty Girls

Man from Cairo
Man from Deep River
Man Escaped
The Man Inside
Man in the Iron Mask
Man of Legend
The Man Who Would Be King
A Man, a Woman and a Killer
Manfish
Manhunt
Manhunt in the African Jungle
Mankillers
Marauder
March or Die
Mark of the Beast
Mark of Zorro
Marked for Death
Marked Money
Marked for Murder
Martial Law
Martial Law 2: Undercover
Marvelous Stunts of Kung Fu
Mask of the Dragon
The Masked Marvel
Massacre in Dinosaur Valley
Massive Retaliation
The Master of Ballantrae
Master Blaster
Master of Dragonard Hill
Master Key
Master Killer (The Thirty-Sixth Chamber)
Master of Kung Fu
Master Ninja
Master Ninja 2
Master Ninja 3
Master Ninja 4
Master Ninja 5
Master's Revenge
Mata Hari
Maximum Breakout
McBain
The McConnell Story
McQ
Mean Frank and Crazy Tony
Mean Johnny Barrows
Mean Machine
Mean Streets of Kung Fu
The Mechanic
Melvin Purvis: G-Man
Memphis Belle
Men of Sherwood Forest
Men of Steel
Menace on the Mountain
The Mercenaries
Mercenary Fighters
Merchants of War
Merlin and the Sword
Messalina vs. the Son of Hercules
The Messenger
Messenger of Death
Meteor
Miami Cops
Miami Vendetta
Miami Vice 2: The Prodigal Son
Midnight Crossing
Midnight Warrior
Midnite Spares
Mighty Jungle
Militant Eagle
Mind Trap
Mines of Kilimanjaro
Ministry of Vengeance
Mision Suicida
Missing in Action
Missing in Action 2: The Beginning
Missing Link
Mission Batangas
Mission Corbari
Mission to Death
Mission Kiss and Kill
Mission Manila
Mission in Morocco

Action

Robbers of the Sacred
    Mountain
Robby
Robin Hood
Robin Hood, the Legend:
    Volume 2
Robin Hood: Prince of Thieves
Robin Hood...The Legend
Robin Hood...The Legend:
    Herne's Son
Robin Hood...The Legend:
    Robin Hood and the
    Sorcerer
Robin Hood...The Legend:
    The Swords of Wayland
Robin Hood...The Legend:
    The Time of the Wolf
Robinson Crusoe
Robinson Crusoe of Clipper
    Island
Robinson Crusoe of Mystery
    Island
Robinson Crusoe & the Tiger
RoboCop
RoboCop 2
Robot Ninja
Robotman & Friends 2: I Want
    to be Your Robotman
Rock House
The Rocketeer
Rocketship X-M
Rogue Lion
Roller Blade Warriors: Taken
    By Force
Rolling Vengeance
The Rookie
R.O.T.O.R.
Rough Cut
Roughnecks
The Rousters
Ruckus
Rulers of the City
Run, Angel, Run!
Run, Rebecca, Run
Runaway
The Runaway Barge
Runaway Island, Part 1
Runaway Island, Part 2
Runaway Island, Part 3
Runaway Island, Part 4
Runaway Nightmare
Runaway Train
Runaway Truck
Running Blind
Running Hot
The Running Man
Running Scared
Rush
The Russian Terminator
The Rutanga Tapes
Safari 3000
Saga of the Vagabond
Sahara
Sahara Cross (Extrana
    Aventura en el Sahara)
Saigon Commandos
Sakura Killers
Salty
Salvador
Samar
Samson
Samson Against the Sheik
The Samurai
Samurai 1: Musashi Miyamoto
Samurai 2: Duel at Ichijoji
    Temple
Samurai 3: Duel at Ganryu
    Island
Samurai Blood, Samurai Guts
Samurai Death Bells
Samurai Sword of Justice
Sands of Iwo Jima
Sanjuro
Sardinia Kidnapped
S.A.S. San Salvador
Satan's Satellites

Savage!
Savage Beach
Savage Dawn
Savage Drums
Savage Fury
The Savage Girl
Savage Hunger
Savage Island
Savage Justice
Savage Streets
Scaramouche
Scarface
The Scarlet & the Black
Scarlet Spear
Scorpion
Scott of the Antarctic
Sea Chase
Sea Devils
Sea Gypsies
Sea Hound
Sea Prince and the Fire Child
The Sea Shall Not Have Them
Sea Wolves
Search and Destroy
Second Chance
Secret Agent Super Dragon
Secret of El Zorro
Secret Executioners
Secret of the Ice Cave
The Secret of the Snake and
    Crane
Secret of the Sword
Secret of Tai Chi
Secret Weapon
Sell Out
Serpent Island
Sesame Street Presents:
    Follow That Bird
Seven Blows of the Dragon
Seven Indignant
Seven Magnificent Gladiators
Seven Sinners
The Seven-Ups
The Seventh Voyage of
    Sinbad
Shadow of Chikara
Shadow Killers
Shadow Man
Shadow World
Shakedown
Shaker Run
Shallow Grave
Shame
Shamus
Shanghai Massacre
The Shaolin: Blood Mission/
    Deadly Shaolin Long Fist
The Shaolin Brothers
Shaolin Challenges Ninja
Shaolin Devil/Shaolin Angel
Shaolin Disciple
Shaolin Drunk Fighter
The Shaolin Drunk Monkey
Shaolin Ex-Monk
Shaolin Executioner
Shaolin Fox Conspiracy
Shaolin Incredible Ten
The Shaolin Invincibles
Shaolin Kids
Shaolin Kung Fu Mystagogue
The Shaolin Kung Fu
    Mystagogue/Iron Ox, the
    Tiger's Killer
Shaolin Red Master/Two
    Crippled Heroes
Shaolin Temple Strikes Back
Shaolin Thief
Shaolin Vengence
Shaolin vs. Lama
Shaolin vs. Manchu
Shaolin Warrior
Shaolin Wooden Men
Shark Attack
Shark Hunter
Shark River
Sharks' Treasure

Sharky's Machine
Sharp Fists in Kung Fu
She
S*H*E
She Came to the Valley
She Devils in Chains
Sheba, Baby
Sheena
The Sheik
Shell Shock
Shep Comes Home
Sherlock's Rivals & Where's
    My Wife
Shinobi Ninja
Ships in the Night
Shipwrecked
Shogun Assassin
Shogun's Ninja
Shoot
Shoot to Kill
Shoot the Sun Down
Shotgun
Shout at the Devil
Showdown at the Equator
Showdown in Little Tokyo
Shredder Orpheus
Sidewinder One
Sign of Zorro
Silent Assassins
Silent Killers
Silent Rage
Silent Raiders
Silk 2
Silver Hermit from Shaolin
    Temple/The Devil's
    Assignment
Sinai Commandos
Sinbad and the Eye of the
    Tiger
Sinbad the Sailor
Sinbad of the Seven Seas
Single Fighter
Sins of Rome
The Sister-in-Law
Sister Street Fighters
Sixteen Fathoms Deep
Skeleton Coast
Ski Bum
Ski Troop Attack
Skull: A Night of Terror
Skull & Crown
Sky Hei$t
Sky High
Sky Pirates
Slate, Wyn & Me
Slaughter
Slaughter's Big Ripoff
Slave of the Cannibal God
Sloane
Slow Burn
A Small Killing
Smokey & the Hotwire Gang
Smuggler's Cove
Snake in the Eagle's Shadow
    2
Snake Fist of the Buddhist
    Dragon
Snake Fist Dynamo
Snake Fist Fighter
Snake in the Monkey's
    Shadow
SnakeEater
SnakeEater 3...His Law
Snuff-Bottle Connection
Sodom and Gomorrah
Solarman
Soldier of Fortune
Soldier's Fortune
Solo Voyage: The Revenge
Son of Captain Blood
Son of Hercules vs. Venus
Son of Kong
The Son of Monte Cristo
Son of Samson
Son of Sinbad
Son of Zorro

S.O.S. Coast Guard
Soul of Samurai
Soul Vengeance
Sourdough
South of Pago Pago
South Seas Massacre
Southern Comfort
Space Mutiny
Space Rage
Space Raiders
Space Riders
SpaceCamp
Spacehunter: Adventures in
    the Forbidden Zone
The Spanish Main
Spartacus
Special Forces
Speeding Up Time
Speedtrap
Spider-Woman
Spiderman: The Deadly Dust
Spiders
Spirit of the Eagle
The Spirit of Youth
Spitfire: The First of the Few
The Spoilers
The Spring
Spy
Spy in Black
Spy Smasher Returns
The Spy Who Loved Me
Squadron of Doom
Stalking Danger
Stamp Day for Superman
Stand and Deliver
Stanley and Livingstone
Star Quest
Starbird and Sweet William
Starchaser: The Legend of
    Orin
The Starfighters
Starflight One
Steel Arena
Steel Dawn
The Steel Helmet
Steele Justice
Steele's Law
Stickfighter
The Sting
Stone Cold
Stop That Train
Stormquest
Story of the Dragon
The Story of Drunken Master
Story of Robin Hood & His
    Merrie Men
Story in the Temple Red Lily
Street Asylum
Street Crimes
Street Fighter
The Street Fighter's Last
    Revenge
Street Fighters Part 2/Masters
    of Tiger Crane
Street Hawk
Street Hunter
Street Soldiers
Streets of Fire
Strike Back
Strike of Thunderkick Tiger
Striker
Striker's Mountain
Striking Back
Struggle Through Death
Stunt Pilot
Stunt Rock
Stunts
Submarine Seahawk
Subway
Sudden Death
Sudden Impact
Suicide Patrol
Summerdog
Sundown
Sunshine Run
Super Dragons/Mafia vs. Ninja

Super Dynamo
Super Gang
Super Ninja
Super Ninjas
Superargo
Superbug Super Agent
Superchick
Supergirl
Superman: The Movie
Superman 2
Superman 3
Superman 4: The Quest for
    Peace
Superman: The Cartoons
Supersonic Man
Supervixens
Surf Nazis Must Die
Surprise Attack
Survival of a Dragon
Survival Game
Survival Quest
The Survivalist
Survivor
Swamp Fire
Swamp Thing
Swamp Women
Swan Lake
Swashbuckler
Sweet Dirty Tony
Sweet Revenge
Sweet Sugar
Sweet Trash
Swift Justice
The Switch
Switchblade Sisters
Sword of Doom
Sword of Fury I
Sword of Fury II
Sword of Gideon
Sword of Heaven
Sword of Lancelot
Sword of Monte Cristo
Sword & the Rose
Sword of Sherwood Forest
Sword of Venus
Swords of Death
Taffin
The Take
The Taking of Beverly Hills
The Taking of Pelham One
    Two Three
Taking it to the Street
Tales of the Klondike: In a Far
    Country
Tales of the Klondike: Race
    for Number One
Tales of the Klondike: The
    Unexpected
Tales of Robin Hood
Tangier
Target Eagle
Target: Favorite Son
Tarka the Otter
Tarzan, the Ape Man
Tarzan of the Apes
Tarzan the Fearless
Tarzan and the Green
    Goddess
Tarzan and His Mate
Tarzan the Tiger
Tarzan and the Trappers
Tarzan's Revenge
Tattoo Connection
Tattooed Dragon
Tattooed Hit Man
Teenage Mutant Ninja Turtles:
    The Movie
Teenage Mutant Ninja Turtles
    2: The Secret of the Ooze
Temporada Salvaje (Savage
    Season)
Ten Brothers of Shao-lin
Ten to Midnight
The Ten Million Dollar
    Getaway
Ten Tigers of Shaolin

Adoption & Orphans

Adventure Drama

Adventure

Blood Debts
Blood Money
Blood and Sand
Bloody Che Contra
Blowing Wild
Border Heat
Botany Bay
The Bounty
Brighty of the Grand Canyon
Brotherhood of Death
A Bullet Is Waiting
Cadence
California Gold Rush
Camel Boy
Casualties of War
The Challenge
Challenge of a Lifetime
China Gate
The Coast Patrol
Cold Heat
Count of Monte Cristo
Courage of Black Beauty
Crimson Romance
Dances with Wolves
Daring Game
Dark Before Dawn
The Day Will Dawn
Deadly Sanctuary
Death Stalk
Death Wish
The Deceivers
Deliverance
Desert Warrior
Dillinger
Diplomatic Immunity
Dragonfly Squadron
East of Borneo
Edge of Honor
The Emerald Forest
En Retirada (Bloody Retreat)
Enchanted Island
The Enforcer
Evel Knievel
Fantastic Seven
55 Days at Peking
F.P. 1 Doesn't Answer
Gente Violenta
Gun Cargo
Hamlet
Harley
Hawk and Castile
Hell Squad
Hell's Angels
Heroes Three
High Country Calling
Hiroshima: Out of the Ashes
In Search of a Golden Sky
Inferno in Paradise
Islands
Isle of Forgotten Sins
Istanbul
It Rained All Night the Day I
  Left
K2
Kamikaze
Karate Polish Style
Kashmiri Run
Kidnap Syndicate
The Killer Elite
King of New York
Kiss and Kill
La Isla Encantada (Enchanted
  Island)
Lady Avenger
Lady Cocoa
Lady Scarface/The Threat
Lassiter
Last Man Standing
The Last Riders
The Last Season
Legend of Billie Jean
Legend of the Sea Wolf
Lethal Pursuit
Lion of the Desert
The Live Wire
Lockdown

Lost
Made for Love
The Man from Beyond
Manhunt for Claude Dallas
Meetings with Remarkable
  Men
Miami Vice
Mobsters
Moby Dick
Moon-Spinners
Night Crossing
No Greater Love
Obsessed
One Man Force
Outlaw Blues
Paco
The Palermo Connection
Papillon
Pathfinder
Payoff
Pharaoh
Platoon
Plunder Road
Plunge Into Darkness
Pray for the Wildcats
PT 109
Radio Patrol
Raiders in Action
Raw Courage
Reap the Wild Wind
Red Flag: The Ultimate Game
Report to the Commissioner
Requiem for a Heavyweight
Retreat, Hell!
Revenge
Riot in Cell Block 11
The Roaring Twenties
The St. Valentine's Day
  Massacre
The Sand Pebbles
Savage Sam
The Scarlet Car
The Sea Hawk
Sergeant York
Seven Alone
Shark!
Shipwreck Island
Shooting Stars
The Siege of Firebase Gloria
The Silence of the Lambs
The Silent Enemy
Simon Blanco
The Sinister Urge
Six Pack
Sky Riders
Slow Moves
Sonny Boy
Sorcerer
South of Hell Mountain
State of Grace
The Stone Killer
Strike Force
Surabaya Conspiracy
Surfacing
Survival
Survival Run
Swiss Family Robinson
Tank
Taps
Tenement
Thelma & Louise
Thomas Crown Affair
Those Calloways
Thunderground
Tides of War
A Tiger Walks
Tono Bicicleta
Toy Soldiers
True Blood
Truth or Die
Turnaround
Valley of Wanted Men
Viking Massacre
Wake Island
A Walk in the Sun
Walking Back

The War Boy
Warkill
What a Way to Die
Wheel of Fortune
Where the Spirit Lives
White Fury
White Hot
The White Tower
The Wicked Lady
The Wild Pair
Wild Thing
Winning
Wooden Horse
The Year of Living
  Dangerously
Yellowneck
Young Warlord
Young Warriors

Advertising

The Agency
Beer
Bliss
C.O.D.
Crazy People
The Ernest Film Festival
A Few Moments with Buster
  Keaton and Laurel and
  Hardy
The Girl Can't Help It
Her Husband's Affairs
The Horse in the Gray Flannel
  Suit
How to Get Ahead in
  Advertising
The Hucksters
I Married a Woman
Image of Passion
Kramer vs. Kramer
Lost in America
Lover Come Back
The Man in the Gray Flannel
  Suit
Mr. Blandings Builds His
  Dream House
Nothing in Common
Pray for the Wildcats
Rhino's Guide to Safe Sex
Super Bloopers #1
Super Bloopers #2
Tee Vee Treasures
Ten Speed
Think Dirty
The Thrill of It All!
Thunder in the City
Transformers: The Movie

Africa

Abdulla the Great
The Adventurers
Africa Screams
African Dream
African Journey
The African Queen
Albino
Algiers
Allan Quartermain and the
  Lost City of Gold
Beyond Obsession
Black Force
Black Terrorist
Black Vampire
Bo-Ru the Ape Boy
Cheetah
Chocolat
City of Blood
Cleopatra
Come Back, Africa
Country Lovers, City Lovers
Curse 3: Blood Sacrifice
The Desert Fox
Desert Tigers (Los Tigres del
  Desierto)

Dingaka
The Dogs of War
A Dry White Season
East of Kilimanjaro
The Egyptian
Elsa & Her Cubs
Fatal Assassin
Five Weeks in a Balloon
The Garden of Allah
The Gods Must Be Crazy
The Gods Must Be Crazy 2
The Golden Salamander
Guns at Batasi
Hatari
In Desert and Wilderness
In the Shadow of Kilimanjaro
It Rained All Night the Day I
  Left
Ivory Hunters
Jaguar
Jericho
Jewel of the Nile
Jungle Drums of Africa
Jungle Master
The Key to Rebecca
King Kong
King Solomon's Mines
The Kitchen Toto
The Light in the Jungle
Lion of Africa
The Lost Patrol
Made for Love
Mandela
Manhattan Baby
Manhunt in the African Jungle
March or Die
The Mark of the Hawk
Master Harold and the Boys
Men of Two Worlds
Mercenary Fighters
Mighty Joe Young
Mines of Kilimanjaro
Mister Johnson
Mogambo
Mondo Africana
Monster from Green Hell
Mountains of the Moon
Mr. Kingstreet's War
My African Adventure
Nairobi Affair
The Naked Prey
Night of the Sorcerers
Oddball Hall
Oh, Kojo! How Could You!
One Step to Hell
Options
Out of Africa
Overseas: Three Women with
  Man Trouble
The Passenger
Pepe le Moko
Place of Weeping
Red Scorpion
Return to Africa
Rhodes
Rogue Lion
Sanders of the River
Satan's Harvest
Secret Obsession
Shaka Zulu
Simba
Skeleton Coast
The Snows of Kilimanjaro
Something of Value
Song of Freedom
Sundown
Time to Kill
Tobruk
The Treasure of Bengal
Tuxedo Warrior
We are the Children
White Hunter, Black Heart
White Mischief
A World Apart
Zulu
Zulu Dawn

AIDS

Common Threads: Stories
  from the Quilt
An Early Frost
Intimate Contact
Longtime Companion
Love Without Fear
Men in Love
Parting Glances
Secret Passions

Airborne

Ace of Aces
Ace Drummond
Aces: Iron Eagle 3
Afterburn
Air Hawk
Airplane!
Airplane 2: The Sequel
Airport
Airport '75
Airport '77
Always
And I Alone Survived
Attack Squadron
The Aviator
Bail Out at 43,000
The Battle of Britain
Battle of the Eagles
Biggles
Birds of Prey
Black Box Affair
Blazing Stewardesses
Blue Angels: A Backstage
  Pass
The Blue Max
Blue Tornado
Brewster McCloud
Buddy Holly Story
Camouflage
Chain Lightning
Christopher Strong
The Concorde: Airport '79
The Crash of Flight 401
Crimson Romance
Dakota
Danger in the Skies
Dangerous Moonlight
Dawn Patrol
Deadly Encounter
Delta Force
Die Hard 2: Die Harder
The Doomsday Flight
Dumbo
Eddie Rickenbacker: Ace of
  Aces
Enola Gay: The Men, the
  Mission, the Atomic Bomb
Fighting Pilot
Final Approach
Fire Birds
Fire and Rain
Firefox
Five Came Back
Five Weeks in a Balloon
Flaming Signal
Flat Top
Flight of Black Angel
Flight to Fury
Flight from Glory
Flight of the Intruder
The Flight of the Phoenix
Flight from Singapore
Flights & Flyers
Flights & Flyers: Amelia
  Earhart
Flying Blind
The Flying Deuces
Flying Leathernecks
Flying Tigers
Ghosts of the Sky
The Great Skycopter Rescue
The Great Waldo Pepper
A Guy Named Joe

Hell's Angels
High Road to China
Hostage
Hot Shots!
The Hunters
Into the Sun
Iron Eagle
Iron Eagle 2
Island at the Top of the World
It Happened at the World's Fair
Jet Over the Atlantic
Jet Pilot
Journey Together
Kamikaze
Keep 'Em Flying
La Bamba
Lost Squadron
Malta Story
The Man with the Golden Gun
The McConnell Story
The Medal of Honor
Memphis Belle
Men of Steel
Millenium
Moon Pilot
Murder on Flight 502
Mystery Squadron
Night Flight
One of Our Aircraft Is Missing
Only Angels Have Wings
Out of Control
Party Plane
The Pilot
The Pursuit of D. B. Cooper
Reach for the Sky
Red Flag: The Ultimate Game
Return to Earth
The Right Stuff
Sabre Jet
Savage Hunger
Shadow on the Sun
Sky Hei$t
Sky Liner
Sky Riders
Spirit of St. Louis
Spitfire
Spitfire: The First of the Few
The Starfighters
Starflight One
Strategic Air Command
Stunt Pilot
Survivor
Tailspin: Behind the Korean Airline Tragedy
The Taking of Flight 847: The Uli Derickson Story
Target for Today: The 8th Air Force Story
Test Pilot
Those Endearing Young Charms
Those Magnificent Men in Their Flying Machines
Top Gun
Top Gun: The Real Story
Tragedy of Flight 103: The Inside Story
Treasure of the Yankee Zephyr
Twelve O'Clock High
War Lover
Warbirds
Where the Sky Begins
The Wild Aces
Winds of Kitty Hawk
A Wing and a Prayer
Wings
Wings of Danger
Wings of Eagles
Zero
Zero Pilot

Alien Beings-Benign

The Abyss
Aftershock
Alien from L.A.
Alien Nation
Alien Private Eye
Alien Women
The Aurora Encounter
*batteries not included
Beamship Meier Chronicles
Beamship Metal Analysis
Beamship Movie Footage
Big Meat Eater
The Brother from Another Planet
Cat from Outer Space
Close Encounters of the Third Kind
Cocoon
Communion
The Cosmic Eye
The Cosmic Man
The Day the Earth Stood Still
Doin' Time on Planet Earth
Earth Girls are Easy
Earth vs. the Flying Saucers
The Empire Strikes Back
Enemy Mine
Escape to Witch Mountain
E.T.: The Extra-Terrestrial
The Ewok Adventure
The Ewoks: Battle for Endor
Flight of the Navigator
Gremlins
Gremlins 2: The New Batch
Hello
The Hidden
Howard the Duck
Hyper-Sapien: People from Another Star
It Came From Outer Space
Labyrinth
The Last Starfighter
Mac and Me
The Man from Atlantis
Man Facing Southeast
The Man Who Fell to Earth
Martians Go Home!
Masters of Venus
Missile to the Moon
Morons from Outer Space
Munchie
Mysterious Planet
Mysterious Two
Pajama Party
The Phantom Planet
Purple People Eater
Real Men
Repo Man
Return of the Jedi
Simon
Slaughterhouse Five
Spaced Invaders
Star Trek 4: The Voyage Home
Star Wars
Stardust Memories
Starman
Suburban Commando
Superman: The Movie
Teenage Mutant Ninja Turtles 2: The Secret of the Ooze
Terronauts
The 27thth Day
UFO: Target Earth
Voyage of the Rock Aliens
War of the Robots
Warning from Space
Wavelength
Wee Wendy

Alien Beings-Vicious

The Adventures of Buckaroo Banzai
Alien
Alien Contamination
Alien Dead
The Alien Factor
Alien Massacre
Alien Predators
Alien Prey
Alien Seed
Alien Space Avenger
Alienator
Aliens
Aliens Are Coming
The Alpha Incident
The Ambushers
The Angry Red Planet
The Astounding She-Monster
The Astro-Zombies
Atomic Submarine
Bad Channels
Bad Taste
The Bamboo Saucer
Battle Beyond the Stars
Battle of the Worlds
Battlestar Galactica
The Blob
Bloodsuckers from Outer Space
Blue Monkey
The Borrower
The Brain
The Brain Eaters
The Brain from Planet Arous
Cat Women of the Moon
The Cosmic Monsters
Creature
Creeping Terror
The Cremators
Critters
Critters 2: The Main Course
Critters 3
Critters 4
The Curse
D-Day on Mars
The Dark
Dark Star
Day Time Ended
Dead Space
Deep Space
Destroy All Monsters
Devil Girl from Mars
Dr. Alien
Doctor Who and the Daleks
Doctor Who: Death to the Daleks
Doctor Who: Pyramids of Mars
Doctor Who: Spearhead from Space
Doctor Who: Terror of the Zygons
Doctor Who: The Ark in Space
Doctor Who: The Seeds of Death
Doctor Who: The Talons of Weng-Chiang
Doctor Who: The Time Warrior
Dollman
Dracula vs. Frankenstein
Earth vs. the Flying Saucers
Eat and Run
The Empire Strikes Back
Encounter at Raven's Gate
End of the World
Escape from Galaxy Three
Escapes
Evils of the Night
The Eye Creatures
Eyes Behind the Stars
The Fear Chamber

First Man into Space
Flight to Mars
Forbidden World
From Beyond
Fugitive Alien
Galaxy Invader
Galaxy of Terror
Gamera vs. Guiron
Gamera vs. Zigra
Ghidrah the Three Headed Monster
The Giant Claw
The Giant Spider Invasion
God Told Me To
Godzilla vs. the Cosmic Monster
Grampa's Sci-Fi Hits
Gremlins
Gremlins 2: The New Batch
Hands of Steel
The Hidden
High Desert Kill
Horror of the Blood Monsters
Horror Planet
The Human Duplicators
I Come in Peace
I Married a Monster from Outer Space
Interzone
Invaders from Mars
Invasion
Invasion of the Animal People
Invasion of the Bee Girls
Invasion of the Body Snatchers
Invasion of the Body Stealers
Invasion Earth: The Aliens Are Here!
Invasion of the Girl Snatchers
Invasion of the Space Preachers
Invisible Adversaries
Island of the Burning Doomed
It Came From Outer Space
It Conquered the World
J-Men Forever!
Killer Klowns from Outer Space
Killers from Space
Killings at Outpost Zeta
Kronos
Laboratory
Leviathan
Liquid Sky
The Little Colonel
Lords of the Deep
Mars Needs Women
The Martian Chronicles: Part 3
Masters of Venus
Metallica
Midnight Movie Massacre
Mindwarp
Mission Mars
The Monolith Monsters
Monster
Monster a Go-Go!
Monster High
Munchies
The Mysterians
Night Beast
Night of the Blood Beast
Night Caller from Outer Space
Night of the Creeps
No Survivors, Please
Peacemaker
The Phantom Planet
Phantom from Space
The Pink Chiquitas
Plan 9 from Outer Space
Planet of Blood
Planets Against Us
Predator
Predator 2
Quatermass 2
The Quatermass Experiment
Radar Men from the Moon

Reactor
Remote Control
The Return
Return of the Aliens: The Deadly Spawn
Return of the Jedi
Revenge of the Teenage Vixens from Outer Space
Robot Monster
Rock 'n' Roll Nightmare
Satan's Satellites
Scared to Death
The Sinister Invasion
Son of Blob
Split
Star Crystal
Star Trek: The Motion Picture
Star Trek 3: The Search for Spock
Star Wars
Strange Invaders
The Stranger from Venus
Superman: The Movie
A Taste for Flesh and Blood
Teen Alien
Teenagers from Outer Space
Tekkaman, the Space Knight: Vol. 2
Terror of Mechagodzilla
They
They Came from Beyond Space
The Thing
This Island Earth
Time Walker
Tremors
Tripods: The White Mountains
20 Million Miles to Earth
The Unearthly Stranger
Village of the Damned
War in Space
War of the Wizards
The War of the Worlds
Wild, Wild Planet
Women of the Prehistoric Planet
X from Outer Space
Xtro
Xtro 2: The Second Encounter
Zontar, the Thing from Venus

American South

The Adventures of Huckleberry Finn
Ain't No Way Back
Alligator Alley
Angel Baby
Angel City
As Summers Die
Autobiography of Miss Jane Pittman
Baby Doll
Bad Georgia Road
The Ballad of the Sad Cafe
The Beguiled
Between Heaven and Hell
The Big Easy
Blind Vengeance
Cape Fear
Cat on a Hot Tin Roof
Chiefs
Cold Sassy Tree
The Color Purple
The Concrete Cowboys
Dead Right
The Displaced Person
Dixiana
Driving Miss Daisy
The Drowning Pool
Drum
Final Cut
Flamingo Road
Fletch Lives
Follow That Car

A Connecticut Yankee in King Arthur's Court
The Cosmic Eye
Cruise
Crusader Rabbit vs. the Pirates
Crusader Rabbit vs. the State of Texas
Crystal Triangle
Cyberpunk
Daffy!
Daffy Duck Cartoon Festival: Ain't That Ducky
Daffy Duck Cartoon Festival: Daffy Duck & the Dinosaur
Daffy Duck & Company
Daffy Duck: The Nuttiness Continues...
Daffy Duck's Madcap Mania
Daffy Duck's Movie: Fantastic Island
Daffy Duck's Quackbusters
Dangaio
David Copperfield
David and Moses
Daydreamer
A Delicate Thread
Destination Moon
Dick Tracy and the Oyster Caper
Dino & Juliet
Disney Beginnings
A Disney Christmas Gift
Disney's American Heroes
Disney's Halloween Treat
Disney's Storybook Classics
Doctor Doom Conquers the World
Dr. Seuss' Grinch Grinches the Cat in the Hat/Pontoffel Pock
Dr. Seuss' Halloween Is Grinch Night
Doctor Snuggles
Dog Soldier
Dominion: Tank Police, Act 1
Donald Duck: The First 50 Years
Donald: Limited Gold Edition 1
Donald's Bee Pictures: Limited Gold Edition 2
A Doonesbury Special
Dorothy in the Land of Oz
Dot & the Bunny
Dot and the Whale
Dracula
The Dragon That Wasn't (Or Was He?)
DuckTales, the Movie
Dumbo
Early Warner Brothers Cartoons
Electric Light Voyage
Elmer Fudd Cartoon Festival
Elmer Fudd's Comedy Capers
En Passant
The Enchanted Journey
EPIC: Days of the Dinosaurs
Evil Toons
Fables of the Green Forest
The Fables of Harry Allard
Fabulous Fleischer Folio, Vol. 3
A Family Circus Christmas
Fantasia
Fantastic Animation Festival
Fantastic Planet
Farmer Gray Goes to the Dogs (Cats, Monkeys & Lions)
The Fat Albert Christmas Special
The Fat Albert Easter Special
The Fat Albert Halloween Special

Favorite Black Exploitation Cartoons
Favorite Celebrity Cartoons
Favorite Racist Cartoons
Felix the Cat: The Movie
Felix in Outer Space
Felix's Magic Bag of Tricks
Ferngully: The Last Rain Forest
50 of the Greatest Cartoons
The First Christmas
Flash Gordon Battles the Galactic Forces of Evil
Flash Gordon: Space Adventurer
Flash Gordon: The Beast Men's Prey
Fleischer Color Classics
Flight of Dragons
Flights of Fancy
Flintstones
The Flintstones Meet Rockula & Frankenstone
Flower Angel
Foghorn Leghorn's Fractured Funnies
Fraidy Cat
Frantic Antics
Fritz the Cat
From the Earth to the Moon
Frosty the Snowman
Fun & Fancy Free
Galaxy Express
Garbancito de la Mancha
Gentlemen of Titipu
George and the Christmas Star
George Pal Color Cartoon Carnival
George Pal Puppetoons #1
George Pal Puppetoons #2
Get Along Gang
G.I. Joe: The Movie
Gnomes
Goldie Gold & Action Jack
Goober and the Ghost Chasers
The Good, the Bad, and Huckleberry Hound
Good Grief, Charlie Brown
Goofy Over Sports
Great Expectations
The Great Mouse Detective
Grendel, Grendel, Grendel
Grizzly Golfer
Guido Manuli: Animator (Nightmare at the "Opera")
Gulliver's Travels
Gumby and the Moon Boggles
Gumby's Supporting Cast
Gunbuster 1
Halloween Monster Madness
Hansel & Gretel
Happy Face Cartoon Classics
Happy New Year, Charlie Brown
Heathcliff and the Catillac Cats
Heathcliff & Marmaduke
Heavy Traffic
Heckle y Jeckle
Heidi
Hello
Herculoids
Herculoids, Vol. 2
Here Come the Littles: The Movie
Here Comes Droopy
Here Comes Garfield
He's Your Dog, Charlie Brown!
Hey Good Lookin'
Hey There, It's Yogi Bear
The Hobbit
Hollywood Cartoons Go to War

Hollywood at War
Hoomania
Hoppity Goes to Town
How the Best was Won
Huckleberry Finn
Hugo the Hippo
The Humanoid
Humanoid Woman
Hurray for Betty Boop
I Give My All
Importance of Being Donald
In a Cartoon Studio
Inch High Private Eye
Incredible Detectives
Incredible Hulk, Volume 1
Incredible Hulk, Volume 2
Incredible Manitoba Animation
Inspector Clouseau Cartoon Festival
Interview
Is This Goodbye, Charlie Brown?
It Was a Short Summer, Charlie Brown
It's an Adventure, Charlie Brown!
It's Arbor Day, Charlie Brown
It's Flashbeagle, Charlie Brown/She's a Good Skate, Charlie Brown
It's the Great Pumpkin, Charlie Brown
It's Magic, Charlie Brown/ Charlie Brown's All Star
It's a Mystery, Charlie Brown
It's Three Strikes, Charlie Brown
It's Your First Kiss, Charlie Brown/Someday You'll Find Her Charlie Brown
Jesus: The Nativity
Jesus: The Resurrection
A Jetson Christmas Carol
The Jetsons Meet the Flintstones
The Jetsons: Movie
Jiminy Cricket's Christmas
Joseph & His Brothers
Journey Back to Oz
Journey to the Center of the Earth
Just Plain Daffy
Kids is Kids
King Solomon's Mines
Knick Knack
Koko Cartoons: No. 1
Koko the Clown, Vol. 1
The Land Before Time
The Last Unicorn
Laurel & Hardy Volume 1
Laurel & Hardy Volume 2
Laurel & Hardy Volume 3
Laurel & Hardy Volume 4
Lazer Tag Academy: The Champion's Biggest Challenge
The Legend of Hiawatha
The Legend of Robin Hood
The Legend of Sleepy Hollow
Les Miserables
Let Sleeping Minnows Lie
The Life and Adventures of Santa Claus
Life is a Circus, Charlie Brown/You're the Greatest, Charlie Brown
Life with Mickey
Light Years
The Lion, the Witch & the Wardrobe
The Little Drummer Boy
Little Lulu
The Little Mermaid
Littlest Warrior
Lone Ranger

Looney Looney Looney Bugs Bunny Movie
Looney Tunes & Merrie Melodies 1
Looney Tunes & Merrie Melodies 2
Looney Tunes & Merrie Melodies 3
Looney Tunes Video Show, Vol. 1
Looney Tunes Video Show, Vol. 2
Looney Tunes Video Show, Vol. 3
Looney Tunes Video Show, Vol. 4
Looney Tunes Video Show, Vol. 5
Looney Tunes Video Show, Vol. 6
Looney Tunes Video Show, Vol. 7
Lord of the Rings
Los 3 Reyes Magos
Lucky Luke: Daisy Town
Lucky Luke: The Ballad of the Daltons
Mad Monster Party
Madox-01
Magic Adventure
Magoo's Puddle Jumper
Man from Button Willow
Man in the Iron Mask
Man's Best Friend
Marco Polo, Jr.
Maria and Mirabella
Master of the World
Matinee at the Bijou, Vol. 1
Matinee at the Bijou, Vol. 2
Matinee at the Bijou, Vol. 3
Matinee at the Bijou, Vol. 4
Matinee at the Bijou, Vol. 5
Matinee at the Bijou, Vol. 6
Matinee at the Bijou, Vol. 7
Matinee at the Bijou, Vol. 8
Max Fleischer's Popeye Cartoons
MD Geist
Merry Christmas to You
Mickey & the Beanstalk
Mickey Commemorative Edition
Mickey Knows Best
Mickey Mouse Cartoon
Mickey's Christmas Carol
Miss Peach of the Kelly School
Moby Dick
Moby Dick & the Mighty Mightor
Moon Madness
Mouse and His Child
Mr. Bill Looks Back Featuring Sluggo's Greatest Hits
Mr. Magoo in the King's Service
Mr. Magoo: Man of Mystery
Mr. Magoo in Sherwood Forest
Mr. Magoo's Christmas Carol
Mr. Magoo's Storybook
Mr. Rossi's Dreams
Mr. Rossi's Vacation
My Friend Liberty
My Little Pony: The Movie
Mysterious Island
Nelvanamation
Nelvanamation II
Nicholas Nickleby
Nine Lives of Fritz the Cat
Ninja the Wonder Boy
Noel's Fantastic Trip
The Nutcracker Prince
Old Curiosity Shop
Oliver Twist
Once Upon a Time

1001 Arabian Nights
101 Dalmations
The Original Adventures of Betty Boop
Otto Messmer and Felix the Cat
Panda and the Magic Serpent
Panda's Adventures
Pepe Le Pew's Skunk Tales
Peter Pan
Peter and the Wolf
Pete's Dragon
Phantom Tollbooth
The Phantom Treehouse
Pink at First Sight
Pink Panther: Pink-a-Boo
Pinocchio in Outer Space
The Plague Dogs
Play It Again, Charlie Brown
Playful Little Audrey
Pogo for President: "I Go Pogo"
The Point
Popeye Assorted Cartoons
Popeye Cartoons
Popeye Color Festival
Popeye and Friends in Outer Space
Popeye and Friends in the South Seas
Popeye and Friends in the Wild West
Popeye Parade
Popeye the Sailor
Popeye: Travelin' on About Travel
Porky Pig Cartoon Festival
Porky Pig Cartoon Festival: Tom Turk and Daffy Duck
Porky Pig and Company
Porky Pig and Daffy Duck Cartoon Festival
Porky Pig Tales
Porky Pig's Screwball Comedies
Pound Puppies and the Legend of Big Paw
The Prince and the Pauper
Private Snafu Cartoon Festival
Project A-ko
The Puppetoon Movie
Race for Your Life, Charlie Brown
Raggedy Ann and Andy: A Musical Adventure
Rainbow Parade
Rainbow Parade 2
Rectangle & Rectangles
Red Barry
The Reluctant Dragon
Remembering Winsor McCay
The Rescuers
The Rescuers Down Under
Return of the Dinosaurs
Riding Bean
The Road Runner vs. Wile E. Coyote: Classic Chase
Robby the Rascal
Robin Hood
Robinson Crusoe
Robotix
Robotman & Friends 2: I Want to be Your Robotman
Rock & Rule
Roland and Rattfink
Rose
Rover Dangerfield
Rudolph & Frosty's Christmas in July
Rudolph the Red-Nosed Reindeer
Sabrina, the Teenage Witch
Salute to Chuck Jones
Salute to Friz Freleng
Salute to Mel Blanc
Sammy Bluejay

Animation

Avant-Garde

The Man Who Envied Women
A Man, a Woman and a Killer
Men in the Park
Mere Jeevan Saathi
Mixed Blood
My Neighborhood
News from Home
An Occurrence at Owl Creek Bridge
Offnight
Paradise Now
Paul Bartel's The Secret Cinema
Peril
Periphery
Permanent Vacation
Phantom of Liberty
Powaqqatsi: Life in Transformation
Rectangle & Rectangles
Red Grooms
Renoir Shorts
The Report on the Party and the Guests
Repulsion
Rome '78
Santa Sangre
Sauteuse de L'ANGE (The Angel Jump Girl)
Scenario du Film Passion
Scherzo
Seduction of Mimi
Shoot Loud, Louder, I Don't Understand!
Signals Through the Flames
Six Fois Deux/Sur et Sous la Communication
Smiling Madame Beudet/The Seashell and the Clergyman
Smothering Dreams
Soft and Hard
Sometimes It Works, Sometimes It Doesn't
Street of Crocodiles
Strong Medicine
The Testament of Dr. Cordelier
The Testament of Orpheus
The Thrill Killers
Towers Open Fire & Other Films
Two Men & a Wardrobe
Un Chien Andalou
Underground U.S.A.
Veronika Voss
Video Americana
WR: Mysteries of the Organism
You are Not I
A Zed & Two Noughts

Babysitting

Adventures in Babysitting
Are You in the House Alone?
The Babysitter
Don't Bother to Knock
Fright
The Guardian
The Hand that Rocks the Cradle
Laurel & Hardy: Brats
Look Who's Talking
Three's Trouble
Uncle Buck
Wanted: Babysitter
Weekend with the Babysitter
When a Stranger Calls

Ballet

Black Tights
Body Beat
Brain Donors

Carmen
The Children of Theatre Street
Cowboy & the Ballerina
Dance
Dancers
Ivan the Terrible
The Lady of the Camellias
Le Bourgeois Gentilhomme
Limelight
Little Ballerina
Mirrors
Nijinsky
Nutcracker Sweet
Nutcracker: The Motion Picture
The Red Shoes
Shall We Dance
Shattered Innocence
Spectre of the Rose
Suspiria
Tales of Beatrix Potter
Tango
Turning Point
Waterloo Bridge
White Nights

Ballooning

Around the World in 80 Days
The Balloonatic/One Week
Charlie and the Great Balloon Chase
Fantastic Balloon Voyage
Five Weeks in a Balloon
Flight of the Eagle
Frankenstein Island
Lost Zeppelin
Mysterious Island
Olly Olly Oxen Free
The Red Balloon
Trottie True
Voyage en Balloon

Baseball

Amazing Grace & Chuck
Amos
Arthur's Hallowed Ground
The Babe
Babe Ruth Story
The Bad News Bears
The Bad News Bears in Breaking Training
The Bad News Bears Go to Japan
Bang the Drum Slowly
Big Mo
Bingo Long Traveling All-Stars & Motor Kings
Blood Games
Blue Skies Again
Brewster's Millions
Bull Durham
The Busher
Charlie Brown's All-Stars
Chasing Dreams
Chu Chu & the Philly Flash
Comeback Kid
Damn Yankees
Don't Look Back: The Story of Leroy "Satchel" Paige
Eight Men Out
Ernie Banks: History of the Black Baseball Player
Fear Strikes Out
Field of Dreams
The Glory of Their Times
Hey! Hey!
Ironweed
It's Arbor Day, Charlie Brown
It's Good to Be Alive
It's My Turn
The Jackie Robinson Story
Jose Canseco's Baseball Camp

The Kid from Left Field
Long Gone
Lou Gehrig: King of Diamonds
Lou Gehrig Story
Love Affair: The Eleanor & Lou Gehrig Story
Major League
Mr. Destiny
The Natural
Night Game
One in a Million: The Ron LaFlore Story
Pastime
Pride of St. Louis
The Pride of the Yankees
The Slugger's Wife
Squeeze Play
Stealing Home
Strategic Air Command
Take Me Out to the Ball Game
Taking Care of Business
Talent for the Game
Tall Tales & Legends: Casey at the Bat
Tiger Town
Timothy and the Angel
Trading Hearts
Wait Until Spring, Bandini
The Winning Team
Woman of the Year
Workin' for Peanuts

Basketball

The Absent-Minded Professor
Coach
Cornbread, Earl & Me
Fast Break
The Fish that Saved Pittsburgh
For Heaven's Sake
Heaven is a Playground
Hoosiers
Inside Moves
One on One
Pistol: The Birth of a Legend
Scoring
Tall Story
Teen Wolf
That Championship Season
White Men Can't Jump

Beach Blanket Bingo

Back to the Beach
The Beach Girls and the Monster
Body Waves
A Swingin' Summer
Where the Boys Are

Behind the Scenes

Action
Actors and Sin
The Affairs of Annabel
Bela Lugosi Scrapbook
Bellissimo: Images of the Italian Cinema
Best Friends
Big Show
Bombay Talkie
Bugs Bunny Superstar
Buster Keaton Rides Again/ The Railroader
The Chaplin Review
A Chorus of Disapproval
A Chorus Line
Chuck Berry: Hail! Hail! Rock 'N' Roll
Contempt
Crossover Dreams
Dames
Dancing Lady

The Death Kiss
The Doors: Soft Parade, a Retrospective
Dracula: A Cinematic Scrapbook
Dresser
Elvis: That's the Way It Is
Elvis on Tour
Epic That Never Was
Falling for the Stars
Fangoria's Weekend of Horrors
Film Firsts
Filmmaker: A Diary by George Lucas
Footlight Parade
Forced March
Frankenstein: A Cinematic Scrapbook
The Freeway Maniac
The French Lieutenant's Woman
George Melies, Cinema Magician
The Ghost Walks
Give a Girl a Break
Goldwyn Follies
Good Morning, Babylon
Goodbye, Norma Jean
Goodnight, Sweet Marilyn
Grunt! The Wrestling Movie
The Hard Way
He Said, She Said
Hearts of Darkness: A Filmmaker's Apocalypse
Help!
Hollywood Boulevard
Hollywood Man
Hollywood Mystery
Hollywood Scandals and Tragedies
Hollywood Shuffle
Hooper
The Horror of It All
Inside Daisy Clover
Into the Sun
The Jayne Mansfield Story
Journey: Frontiers and Beyond
The Killing of Sister George
La Dolce Vita
Legs
The Lonely Lady
Major Bowes' Original Amateur Hour
Make a Wish
Making of "Gandhi"
Making of a Legend: "Gone With the Wind"
The Making of the Stooges
A Man in Love
Man of a Thousand Faces
Master Class
Miss All-American Beauty
Movie Struck
The Mozart Brothers
Muppets Take Manhattan
New Faces
A Night for Crime
On an Island with You
Pastime
Private Conversations: On the Set of "Death of a Salesman"
Purple Rose of Cairo
Roadie
Secret File of Hollywood
Sex and Buttered Popcorn
Sioux City Sue
Soap Opera Scandals
S.O.B.
Stage Door
Stand-In
Star Trek 25th Anniversary Special
State of Things

Steve Martin Live
Stunts
Summer Stock
Sunset
Sunset Boulevard
Sweet Liberty
Sweet Perfection
Teenager
Tootsie
Top Gun: The Real Story
Truth or Dare
25x5: The Continuing Adventures of the Rolling Stones
Under the Rainbow
Users
Velvet Touch
A Very Private Affair
The Wall: Live in Berlin
What Price Hollywood?
Who Framed Roger Rabbit?
Whodunit
Zou Zou

Bicycling

American Flyers
BMX Bandits
Breaking Away
Off the Mark
Pee-Wee's Big Adventure
Quicksilver
Rad
Rush It
Ten Speed

Big Battles

The Alamo
Alexander the Great
Alexander Nevsky
Aliens
Apocalypse Now
Arena
Back to Bataan
Barry Lyndon
Battle Beyond the Sun
The Battle of Britain
Battle of the Bulge
The Battle of El Alamein
Battle of the Worlds
Battleground
The Battleship Potemkin
The Big Red One
Birth of a Nation
The Blazing Ninja
A Bridge Too Far
Circus World
Cromwell
Cross of Iron
Das Boot
Diamonds are Forever
Die Hard
Die Hard 2: Die Harder
Dragonslayer
El Cid
The Empire Strikes Back
Enemy Below
Excalibur
The Fall of the Roman Empire
The Fighting Sullivans
Gallipoli
The Giant of Marathon
Giants of Rome
Glory
Godzilla vs. the Smog Monster
Goldfinger
The Guns of Navarone
Heaven & Earth
Henry V
Highlander
The King of the Kickboxers
The Last Command
The Lighthorseman

Buddies

The Color Purple
The Concrete Cowboys
Cops and Robbers
Country Gentlemen
Courage of Black Beauty
Crime Busters
Crossfire
Cry Freedom
Cycle Vixens
Dead Reckoning
December
The Deer Hunter
Defenseless
Defiant Ones
Departamento Compartido
Diner
Donovan's Reef
Down Under
Downtown
Dresser
Driving Miss Daisy
Drop Dead Fred
Drums in the Deep South
Easy Rider
Eat the Peach
El Dorado
Elmer
Enemy Mine
Entre-Nous (Between Us)
Every Which Way But Loose
Extreme Prejudice
F/X 2: The Deadly Art of
    Illusion
The Five Heartbeats
Forsaking All Others
48 Hrs.
Francis in the Navy
Freeze-Die-Come to Life
Gallipoli
Going Back
Goodfellas
Grandes Amigos (Great
    Friends)
Grizzly Mountain
Gumby and the Moon Boggles
Gunga Din
Hangin' with the Homeboys
The Hard Way
Harley Davidson and the
    Marlboro Man
The Hasty Heart
Hats Off
Hawaii Calls
Heartaches
Heartbreakers
The Highest Honor
Hired Hand
Hollywood or Bust
Home Free All
Hong Kong Nights
The Hunters
In-Laws
In the Navy
Inn of Temptation
Ishtar
Johnny Come Lately
Juice
Jules and Jim
Julia
Just Between Friends
Just Like Us
K-9000
K2
The Karate Kid
The Karate Kid: Part 2
The Karate Kid: Part 3
The Killing Fields
King of the Mountain
Kings of the Road (In the
    Course of Time)
Ladies on the Rocks
Last American Virgin
The Last Boy Scout
Last Summer
Le Beau Serge
Le Grand Chemin

Leaving Normal
The Lemon Sisters
Les Comperes
Lethal Weapon
Lethal Weapon 2
Let's Do It Again
Little Rascals, Book 1
Little Rascals, Book 2
Little Rascals, Book 3
Little Rascals, Book 14
Little Rascals, Book 16
Little Rascals: Choo Choo/
    Fishy Tales
Little Rascals Collector's
    Edition, Vol. 1-6
Little Rascals Comedy
    Classics, Vol. 2
Little Rascals: Fish Hooky/
    Spooky Hooky
Little Rascals: Honkey
    Donkey/Sprucin' Up
Little Rascals: Little Sinner/
    Two Too Young
Little Rascals: Mush and Milk/
    Three Men in a Tub
Little Rascals on Parade
Little Rascals: Pups is Pups/
    Three Smart Boys
Little Rascals: Readin' and
    Writin'/Mail and Female
Little Rascals: Reunion in
    Rhythm/Mike Fright
Little Rascals: Spanky/Feed
    'Em and Weep
Little Rascals: The Pinch
    Singer/Framing Youth
Little Rascals, Volume 1
Little Rascals, Volume 2
Littlest Outlaw
The Long Voyage Home
Looking for Miracles
The Lords of Flatbush
Los Chantas
Lost in the Barrens
Macaroni
The Man Who Would Be King
Manhattan Melodrama
March or Die
Me and Him
Mean Frank and Crazy Tony
The Men's Club
Merry Christmas
Midnight Cowboy
Midnight Run
Mikey & Nicky
The Mission
Misunderstood
Moon Over Miami
Mortal Thoughts
My Beautiful Laundrette
My Best Friend's Girl
My Dinner with Andre
My Girl
Mystic Pizza
Naked Youth
The Naughty Nineties
Night Beat
Nightforce
The Nightingale
The Odd Couple
Of Mice and Men
Old Enough
Once Upon a Time in America
One Crazy Summer
One Sings, the Other Doesn't
Pals
Papillon
Peck's Bad Boy with the
    Circus
The Penitent
Perfectly Normal
The Phantom of 42nd Street
Phone Call from a Stranger
P.K. and the Kid
Platoon
Playmates

A Pleasure Doing Business
Plunder Road
Portrait of a Showgirl
Racing with the Moon
Ramblin' Man
Rape
The Red Balloon
The Reivers
Reno and the Doc
The Return of Grey Wolf
Return of the Secaucus 7
Revenge of the Nerds
Revenge of the Nerds 2:
    Nerds in Paradise
Rich and Famous
Ride the High Country
Rio Conchos
The Road to Hong Kong
Road to Morocco
Robby
Robert et Robert
Rubin & Ed
Running Scared
Russkies
St. Elmo's Fire
St. Helen's, Killer Volcano
Salut l'Artiste
Seventh Veil
Shag: The Movie
Shaking the Tree
She Gods of Shark Reef
Ship of Fools
Showdown in Little Tokyo
Silence Like Glass
Snoopy's Getting Married,
    Charlie Brown
Snow White and the Seven
    Dwarfs
So This is Washington
Something of Value
Songwriter
Sophisticated Gents
Spies Like Us
Spittin' Image
Spotlight Scandals
Stand By Me
Star Trek 4: The Voyage
    Home
Starbird and Sweet William
Stealing Home
Steaming
The Steel Helmet
Stuckey's Last Stand
Sudden Death
Summer City
Sunny Skies
The Sunshine Boys
Sweet Hearts Dance
Tango and Cash
Ten Who Dared
Tendres Cousines
Tequila Sunrise
Terminal Bliss
That Sinking Feeling
Thelma & Louise
There's No Time for Love,
    Charlie Brown
Three Amigos
Three Legionnaires
Three Men and a Baby
Three Men and a Cradle
Three Musketeers
Thunder in the Pines
Thursday's Game
To Forget Venice
Top Gun
Tough Guys
Tour of Duty
True Colors
Turning Point
Two Weeks to Live
Ugetsu
Unholy Four
Veronico Cruz
Vincent, Francois, Paul and
    the Others

Wayne's World
We're in the Legion Now
What Price Glory?
Whatever Happened to Susan
    Jane?
Where the River Runs Black
White Christmas
White Mama
White Men Can't Jump
Why Me?
The Wild Life
Wild Rovers
Windy City
Winners Take All
Withnail and I
The Wrong Guys
The Wrong Road
The Yakuza
You Are Not Alone

Buddy Cops

Another 488 Hrs.
Arizona Heat
City Heat
Collision Course
48 Hrs.
The Hard Way
La Chevre
Lethal Weapon
Lethal Weapon 2
Liberty & Bash
Night Children
Nighthawks
The Presidio
Red Heat
Renegades
The Rookie
Shakedown
Shoot to Kill
Stakeout
Tango and Cash

Campus Capers

Accident
All-American Murder
American Tiger
Animal Behavior
Assault of the Party Nerds
Attack of the Killer
    Refrigerator
Baby Love
Bachelor of Hearts
Back to School
Beyond Dream's Door
Black Christmas
Blood Cult
Blood Sisters
Bloodmoon
The Campus Corpse
Campus Knights
Campus Man
Carnal Knowledge
Catch Me...If You Can
The Cheerleaders
A Chump at Oxford
Class
Class of '44
College
Computer Beach Party
The Computer Wore Tennis
    Shoes
The Cousins
Curse of the Alpha Stone
Deadly Obssession
Deadly Possession
Death Screams
Doctor in the House
Dorm That Dripped Blood
Escapade in Florence
Final Exam
Finishing School
First Affair
First Love

First Time
Fraternity Vacation
The Freshman
Fright Night 2
Ghoulies 3: Ghoulies Go to
    College
Girls Night Out
Good News
Gotcha!
Gross Anatomy
Hamburger... The Motion
    Picture
Harrad Summer
HauntedWeen
Heart of Dixie
Hell Night
Higher Education
Horse Feathers
H.O.T.S.
House Party 2: The Pajama
    Jam
The House on Sorority Row
How I Got Into College
Incoming Freshmen
The Initiation
Initiation of Sarah
Intimate Games
Jock Peterson
Kid with the 200 I.Q.
A Killer in Every Corner
Killer Party
Let's Go Collegiate
Little Sister
Lucky Jim
Making the Grade
Mazes and Monsters
Midnight Madness
Misadventures of Merlin
    Jones
Monkey's Uncle
Mortuary Academy
Mugsy's Girls
Murder on the Campus
National Lampoon's Animal
    House
Naughty Co-Eds
Night of the Creeps
A Night in the Life of Jimmy
    Reardon
Nightwish
Odd Jobs
The Paper Chase
Party Animal
Pieces
Pledge Night
Preppies
Revenge of the Nerds
Revenge of the Nerds 2:
    Nerds in Paradise
Ring of Terror
Rona Jaffe's Mazes &
    Monsters
Roommate
Rosalie
Rush Week
Scandal
School Daze
School Spirit
Scoring
The Seniors
Silent Madness
Silent Scream
Sky High
Sorority Babes in the Slimeball
    Bowl-A-Rama
Sorority House Massacre
Sorority House Massacre 2
Soul Man
Splatter University
Splitz
Spring Break
Star Spangled Girl
Stitches
Swinging Sorority Girls
Talking Walls
Tall Lie

Chases

It's a Mad, Mad, Mad, Mad World
Jackson County Jail
Jewel of the Nile
Junkman
King of the Rodeo
A Kink in the Picasso
The Lavender Hill Mob
Les Miserables
Lethal Weapon
Lethal Weapon 2
License to Drive
Live and Let Die
Mackintosh Man
Mad Max
The Man with the Golden Gun
A Man with a Maid
Man from Snowy River
Manhunt for Claude Dallas
Memoirs of an Invisible Man
Midnight Run
The Most Dangerous Game
Mr. Billion
Mr. Majestyk
Murder of a Moderate Man
Never Give a Sucker an Even Break
No Way Out
North by Northwest
Number Seventeen
Nuns on the Run
Octopussy
On Her Majesty's Secret Service
One Down, Two to Go!
One Man Jury
Out of Bounds
Outrageous Fortune
Paris Holiday
Paths to Paradise
Pee-Wee's Big Adventure
Planet of the Apes
Point Break
The Presidio
Pursuit
The Pursuit of D. B. Cooper
Race with the Devil
Raiders of the Lost Ark
Raising Arizona
Return of the Jedi
Return to Macon County
The Road Warrior
Romancing the Stone
Round Trip to Heaven
Running Scared
Seven Chances
The Shining
Slither
Smokey and the Bandit
Smokey and the Bandit, Part 2
Smokey and the Bandit, Part 3
Smokey Bites the Dust
Smokey & the Hotwire Gang
The Spy Who Loved Me
Stagecoach
Star Wars
Stephen King's Golden Years
The Sting
The Sugarland Express
Sweet Sweetback's Baadasssss Song
Tagget
The Taking of Pelham One Two Three
Tell Them Willie Boy Is Here
The Terminator
Terror of the Bloodhunters
Thelma & Louise
The Third Man
This Man Must Die
Thunder and Lightning
Thunder Road
Thunderball
Thunderbolt & Lightfoot

Time After Time
To Live & Die in L.A.
Top Gun
12 Plus 1
Valley of the Eagles
Vanishing Point
Watch the Birdie
What's Up, Doc?
White Lightning
Who Framed Roger Rabbit?
The Wilby Conspiracy
Wolfen
You Only Live Twice

Child Abuse

Adam
Blind Fools
Close to Home
A Dream of Passion
Fallen Angel
Fatal Confinement
Hush Little Baby, Don't You Cry
Impulse
Johnny Belinda
Judgment
Mommie Dearest
Psychopath
A Question of Guilt
Radio Flyer
South Bronx Heroes
Where the Spirit Lives
Who Killed Baby Azaria?

Childhood Visions

Alsino and the Condor
And Now Miguel
Army Brats
Au Revoir Les Enfants
Beethoven Lives Upstairs
The Beniker Gang
The Best of the Little Rascals
Big
Black Rain
The Black Stallion
Born to Run
But Where Is Daniel Wax?
Cameron's Closet
Captain January
Careful, He Might Hear You
Casey's Shadow
The Cat
The Cellar
Chapter in Her Life
Child in the Night
The Children Are Watching Us
Child's Play
Chocolat
A Christmas Story
Cria
David Copperfield
Days of Heaven
Demons in the Garden
Desert Bloom
Deutschland im Jahre Null
Diva
Emma's Shadow
Empire of the Sun
Escapade
E.T.: The Extra-Terrestrial
The Fallen Idol
Fanny and Alexander
Far Away and Long Ago
Father
Flight of the Grey Wolf
Girl from Hunan
A Girl of the Limberlost
The Golden Seal
Grazuole
Home Alone
Hook
Hope and Glory
Hue and Cry

Invaders from Mars
The Kid with the Broken Halo
The Kid from Left Field
The Kitchen Toto
The Lady in White
Lassie's Great Adventure
Le Grand Chemin
Lies My Father Told Me
Little Man Tate
Little Monsters
The Little Princess
Little Rascals, Book 17
Littlest Horse Thieves
The Long Days of Summer
Lord of the Flies
Louisiana Story
Magic of Lassie
The Mirror
Misty
Mosby's Marauders
My Friend Flicka
My Life As a Dog
My Name Is Ivan
National Velvet
New York Stories
Night of the Shooting Stars
Old Swimmin' Hole
Orphan Boy of Vienna
Orphans
Our Gang
Our Gang Comedy Festival
Our Gang Comedy Festival 2
Parents
Peck's Bad Boy
Pippi Longstocking
Pixote
Poltergeist
Poltergeist 2: The Other Side
The Poor Little Rich Girl
Psycho 4: The Beginning
Radio Flyer
Railway Children
The Red Balloon
The Reflecting Skin
Return from Witch Mountain
Roll of Thunder, Hear My Cry
Savage Sam
The Secret Garden
Shadow World
The Silence of the Lambs
Small Change
Sounds of Silence
Spirit of the Beehive
Square Shoulders
Stealing Home
Sudden Terror
A Summer to Remember
Sunday Daughters
Taking Care of Terrific
Taps
The Tin Drum
To Kill a Mockingbird
Tom Sawyer
Train to Hollywood
Treasure Island
Turf Boy
Unicorn
Voltus 5
Wait Until Spring, Bandini
The War Boy
When Father was Away on Business
Whistle Down the Wind
White Fang
The Wild Country
The Wizard of Loneliness
Wombling Free
A World Apart
The Yearling
Yes, Sir, Mr. Bones
Zazie dans le Metro
Zero for Conduct (Zero de Conduit)

China

Adventures of Smilin' Jack
Chinese Gods
Dragon Seed
Eat a Bowl of Tea
Empire of the Sun
A Fist Full of Talons
The General Died at Dawn
Girl from Hunan
The Good Earth
A Great Wall
The Inn of the Sixth Happiness
Iron & Silk
Kashmiri Run
The Keys of the Kingdom
The Last Emperor
The Leatherneck
The Mountain Road
Raiders of the Buddhist Kung Fu
The Rebellious Reign
Red Sorghum
The Sand Pebbles
Seven Indignant
The Shaolin Invincibles
The Silk Road
Single Fighter
Soldier of Fortune
Taipan
The World of Suzie Wong

Christmas

ABC Stage 67: Truman Capote's A Christmas Memory
An American Christmas Carol
Babes in Toyland
Benji's Very Own Christmas Story
The Best Christmas Pageant Ever
The Bishop's Wife
Black Christmas
Bloodbeat
Bugs Bunny's Looney Christmas Tales
A Charlie Brown Christmas
Children's Carol
Child's Christmas in Wales
A Christmas Carol
Christmas Carol
Christmas Coal Mine Miracle
Christmas Comes to Willow Creek
Christmas in Connecticut
A Christmas Gift
Christmas Lilies of the Field
The Christmas Messenger
Christmas Raccoons
A Christmas to Remember
A Christmas Story
The Christmas That Almost Wasn't
The Christmas Tree
The Christmas Wife
Christmas Without Snow
A Disney Christmas Gift
Don't Open Till Christmas
Dorm That Dripped Blood
Elves
Ernest Saves Christmas
A Family Circus Christmas
The Fat Albert Christmas Special
The First Christmas
Frosty the Snowman
The Gathering
The Gathering: Part 2
George and the Christmas Star
The Gift of Love
The Great Frost
Gremlins

Here Comes Santa Claus
Hobo's Christmas
Holiday Affair
Holiday Inn
Home Alone
Homecoming: A Christmas Story
It Came Upon a Midnight Clear
It's a Wonderful Life
Jack Frost
A Jetson Christmas Carol
Jiminy Cricket's Christmas
King's Christmas
La Pastorela
The Life and Adventures of Santa Claus
The Lion in Winter
The Little Drummer Boy
The Little Match Girl
Little Match Girl
Little Rascals Christmas Special
Magic Christmas Tree
The Man in the Santa Claus Suit
A Matter of Principle
Merry Christmas
Merry Christmas to You
Mickey's Christmas Carol
Miracle on 34th Street
Miracle Down Under
Mr. Corbett's Ghost
Mr. Magoo's Christmas Carol
National Lampoon's Christmas Vacation
Nativity
The Night They Saved Christmas
Nutcracker, A Fantasy on Ice
The Nutcracker Prince
Nutcracker: The Motion Picture
One Magic Christmas
Pee-Wee's Playhouse Christmas Special
Rich Little's Christmas Carol
Rudolph & Frosty's Christmas in July
Rudolph the Red-Nosed Reindeer
Rudolph's Shiny New Year
Santa Claus
Santa Claus is Coming to Town
Santa Claus Conquers the Martians
Santa Claus: The Movie
Scrooge
Scrooged
Silent Night, Deadly Night
Silent Night, Deadly Night 2
Silent Night, Deadly Night 3: Better Watch Out!
Silent Night, Deadly Night 5: The Toymaker
Silent Night, Lonely Night
The Simpsons Christmas Special
A Smoky Mountain Christmas
Stingiest Man in Town
Terror in Toyland
Thirteenth Day of Christmas
Three Godfathers
A Time to Remember
To All a Goodnight
Trail of Robin Hood
A Very Merry Cricket
A Walt Disney Christmas
The Waltons' Christmas Carol
When Wolves Cry
White Christmas
Who Slew Auntie Roo?
Year Without Santa Claus
Yogi's First Christmas

City Lights

Abbott and Costello in
    Hollywood
Across 110th Street
Adventures in Babysitting
After Hours
Against All Odds
Alone in the Neon Jungle
Berlin Alexanderplatz
    (episodes 1-2)
Beverly Hills Cop
Beverly Hills Cop 2
The Big Easy
Black Orpheus
Black Rain
The Blues Brothers
Breathless (A Bout de Souffle)
City of Hope
City That Never Sleeps
Cool Blue
The Cotton Club
Delightfully Dangerous
Dodes 'ka-den
Don't Look Now
Down and Out in Beverly Hills
The Dream Team
Fear, Anxiety and Depression
Ferris Bueller's Day Off
The Foreigner
Foul Play
Foxstyle
Frame Up
French Quarter
French Quarter Undercover
Gorky Park
Hangin' with the Homeboys
La Boum
L.A. Story
The Last Metro
Last Tango in Paris
A Little Sex
Long Island Four
Manhattan
Moscow on the Hudson
Nashville
A New Kind of Love
New Orleans After Dark
New York, New York
New York Stories
News from Home
Next Stop, Greenwich Village
1918
Once in Paris...
One More Saturday Night
The Out-of-Towners
Outrageous!
The Pope of Greenwich
    Village
A Rage in Harlem
The Rat Race
Risky Business
Running Scared
Scrooged
Slaves of New York
Tampopo
Times Square
To Live & Die in L.A.
Tokyo Pop
Tootsie
Urban Cowboy
V.I. Warshawski
Viva Las Vegas
Wall Street

Civil Rights

African Dream
As Summers Die
Blaxploitation Cartoons
Bucktown
The Bus Is Coming
The Color Purple
Come See the Paradise
The Conversation
Driving Miss Daisy

A Dry White Season
Favorite Racist Cartoons
Final Comedown
For Us, the Living
Free, White, and 21
Freedom Road
The Fringe Dwellers
Game for Vultures
Half Slave, Half Free
Hawmps!
Heart of Dixie
Hidden Agenda
Home of the Brave
The Kitchen Toto
Long Shadows
The Long Walk Home
The McMasters
Midnight Express
Mississippi Burning
A Nation Aflame
The Red Fury
The Ring
The Rodney King Case: What
    the Jury Saw in California
    vs. Powell
The Shadow Riders
A Soldier's Story
Two of Us
We of the Never Never

Civil War

Abraham Lincoln
Alvarez Kelly
The Andersonville Trial
Bad Company
The Beguiled
Birth of a Nation
Blood at Sundown
Bloody Trail
The Blue and the Grey
The Bushwackers
Cavalry
Charlotte Forten's Mission:
    Experiment in Freedom
The Civil War: Episode 1: The
    Cause: 1861
The Civil War: Episode 2: A
    Very Bloody Affair: 1862
The Civil War: Episode 3:
    Forever Free: 1862
The Civil War: Episode 4:
    Simply Murder: 1863
The Civil War: Episode 5: The
    Universe of Battle: 1863
The Civil War: Episode 6:
    Valley of the Shadow of
    Death: 1864
The Civil War: Episode 7:
    Most Hallowed Ground:
    1864
The Civil War: Episode 8: War
    is All Hell: 1865
The Civil War: Episode 9: The
    Better Angels of Our
    Nature: 1865
Colorado
Copper Canyon
Dark Command
Desperados
Drums in the Deep South
Finger on the Trigger
The Fool Killer
Friendly Persuasion
Frontier Pony Express
Frontier Scout
The General
General Spanky
Glory
Gone with the Wind
The Good, the Bad and the
    Ugly
Great Day in the Morning
The Great Locomotive Chase
Hangman's Knot

Hellbenders
The Horse Soldiers
How the West was Won
Ironclads
Johnny Shiloh
Kansas Pacific
Last Full Measure
Little Women
Littlest Rebel
Louisiana
Love Me Tender
Love's Savage Fury
Macho Callahan
The McMasters
Menace on the Mountain
Mosby's Marauders
Mysterious Island
No Drums, No Bugles
Old Maid
The Ordeal of Dr. Mudd
Outlaw Josey Wales
The Private History of a
    Campaign That Failed
Raintree County
Rebel Love
Rebel Vixens
The Red Badge of Courage
The Rose and the Jackal
Run of the Arrow
The Sacketts
Shadow of Chikara
Shenandoah
South of St. Louis
A Southern Yankee
Stephen Crane's Three
    Miraculous Soldiers
The Supernaturals
Tramplers
The Undefeated
Virginia City
War Brides
Yellowneck

Classic Horror

The Abominable Dr. Phibes
And Now the Screaming
    Starts
The Ape
The Ape Man
The Appointment
Asylum
Asylum of Satan
Atom Age Vampire
Attack of the Beast Creatures
Attack of the Mayan Mummy
Attack of the Swamp Creature
Avenging Conscience
Awakening
Barn of the Naked Dead
The Bat
The Bat People
The Beast from Haunted Cave
Bedlam
Before I Hang
The Birds
The Black Cat
The Black Room
Black Sabbath
The Black Scorpion
Black Sunday
Black Werewolf
Blood Beast Terror
Blood of Dracula
Blood of Dracula's Castle
Blood Island
Blood and Roses
Blood of the Vampire
The Bloodsuckers
Bluebeard
The Body Snatcher
The Bride
The Bride of Frankenstein
Bride of the Gorilla
Bride of the Monster

Bring Me the Vampire
The Brute Man
Burn Witch, Burn!
The Cabinet of Dr. Caligari
Carnival of Souls
Castle of Blood
Castle of Evil
Castle of the Living Dead
Cat in the Cage
Cat Girl
Cat People
Cauldron of Blood
Cave of the Living Dead
Chamber of Fear
Chamber of Horrors
The Changeling
Circus of Fear
Condemned to Live
The Conqueror Worm
Corridors of Blood
Count Dracula
Count Yorga, Vampire
The Crawling Eye
Creature from the Black
    Lagoon
Creature with the Blue Hand
Creature of the Walking Dead
The Creeper
The Creeping Flesh
Creeping Terror
Creepshow
Creepshow 2
The Crime of Dr. Crespi
Crimes at the Dark House
The Crimes of Stephen
    Hawke
Crucible of Horror
Cry of the Banshee
Crypt of the Living Dead
Curse of the Devil
The Curse of Frankenstein
Curse of the Swamp Creature
Cyclops
Dance of the Damned
Daughters of Darkness
Dementia 13
Demon Barber of Fleet Street
Demons of the Mind
The Devil Bats
The Devil Bat's Daughter
Devil Doll
Devil Kiss
Dr. Death, Seeker of Souls
Dr. Jekyll and Mr. Hyde
Doctor X
Doin' What the Crowd Does
Dolly Dearest
Dorian Gray
Dracula
Dracula/Garden of Eden
Dracula (Spanish Version)
Dracula vs. Frankenstein
Dracula's Daughter
Dracula's Great Love
Dracula's Last Rites
Dracula's Widow
El Hombre y el Monstruo
The Evil of Frankenstein
The Exorcist 3
Eye of the Demon
Eyes of Fire
Eyes of the Panther
The Fall of the House of Usher
Fangs of the Living Dead
The Fear Chamber
Fear No Evil
The Fog
Frankenstein
Frankenstein '80
Frankenstein Island
Frankenstein Meets the
    Wolfman
Frankenstein and the Monster
    from Hell
Frankenstein Unbound
Frankenstein's Daughter

Fright Night
Fright Night 2
The Fury
The Fury of the Wolfman
Gallery of Horror
Genie of Darkness
The Ghoul
The Gorgon
Gothic
Grampa's Monster Movies
Grave of the Vampire
Graveyard Shift
The Greed of William Hart
Growing Pains
Guardian of the Abyss
Halloween Night
The Hanging Woman
The Haunted
The Haunted Strangler
Haunted Summer
The Haunting
Haunting Fear
The Haunting of Morella
The Head
Horrible Double Feature
The Horror of Dracula
House of the Black Death
House of Dark Shadows
The House that Dripped Blood
House of Evil
House of Frankenstein
House on Haunted Hill
The House of Usher
House of Wax
How to Make a Monster
The Hunger
I Walked with a Zombie
I Was a Teenage Werewolf
Invasion of the Vampires
The Invisible Man
The Invisible Ray
Island of Dr. Moreau
Isle of the Dead
Jaws
Jaws 2
Jaws 3
Jaws: The Revenge
King of the Zombies
La Chute de la Maison Usher
Lady Frankenstein
The Lair of the White Worm
The Legend of Blood Castle
Legend of the Werewolf
The Legend of the Wolf
    Woman
Lemora, Lady Dracula
The Living Coffin
The Mad Monster
The Man They Could Not
    Hang
Mark of the Vampire
Martin
Masque of the Red Death
Masque of the Red Death/
    Premature Burial
Mirrors
Monsters on the March
The Mummy
Mummy & Curse of the Jackal
The Mummy's Hand
Murders in the Rue Morgue
Mystery of the Wax Museum
Night of Dark Shadows
Once Upon a Midnight Scary
Orlak, the Hell of Frankenstein
Pandora's Box
Phantom of the Opera
Picture of Dorian Gray
The Pit and the Pendulum
The Raven
Return of the Vampire
Revenge of the Zombies
Revolt of the Zombies
Scared to Death
The Snow Creature
Son of Dracula

Classic

Comedy

Comedy

Hothead (Coup de Tete)
H.O.T.S.
House Party
House Party 2: The Pajama Jam
Housemaster
How to Beat the High Cost of Living
How to Break Up a Happy Divorce
How Come Nobody's On Our Side?
How Funny Can Sex Be?
How I Got Into College
How I Learned to Love Women
How I Won the War
How to Kill 400 Duponts
How to Murder Your Wife
How to Pick Up Girls
How to Seduce a Woman
How to Stuff a Wild Bikini
How Sweet It Is!
Howdy Doody
Huckleberry Finn
Hullabaloo Over Georgie & Bonnie's Pictures
Hunk
Hurry, Charlie, Hurry
Hurry Up or I'll Be Thirty
I Give My All
I Love My...Wife
I Love You, Alice B. Toklas
I Married a Witch
I Wanna Be a Beauty Queen
I Wanna Hold Your Hand
I Was a Teenage TV Terrorist
I Will, I Will for Now
I Wonder Who's Killing Her Now?
The Icicle Thief
Identity Crisis
If I Were Rich
If You Don't Stop It...You'll Go Blind
Illegally Yours
I'm on My Way/The Non-Stop Kid
The Immigrant
Immortal Bachelor
Importance of Being Earnest
Impromptu
Improper Channels
In-Laws
In the Spirit
Incoming Freshmen
Inn of Temptation
Innerspace
The Innocents Abroad
Innocents in Paris
Inspector Clouseau Cartoon Festival
The Inspector General
Inspector Hornleigh
Instant Karma
Intimate Games
Into the Sun
Invasion Earth: The Aliens Are Here!
Invasion of the Space Preachers
The Invisible Kid
Invitation to the Wedding
Irish Luck
Irreconcilable Differences
Is This Goodbye, Charlie Brown?
It Could Happen to You
It Was a Short Summer, Charlie Brown
Italian Straw Hat
It's A Gift
It's a Joke, Son!
It's Not Easy Bein' Me
It's Not the Size That Counts
It's the Old Army Game

Izzy & Moe
Jabberwalk
Jack Benny Show
The Jackpot
Jackson & Jill
The January Man
Jean Sheperd on Route #1... and Other Thoroughfares
Jekyll & Hyde...Together Again
Jimmy the Kid
Jinxed
Jocks
Joe Bob Briggs Dead in Concert
Joe Piscopo: New Jersey Special
John & Julie
Johnny Be Good
Johnny Carson
Johnny Dangerously
Jokes My Folks Never Told Me
Jonah Who Will Be 25 in the Year 2000
Jonathan Winters on the Edge
Joy of Sex
Joy Sticks
Jumpin' Jack Flash
Jupiter's Thigh
Just Me & You
Just One of the Guys
Just Tell Me What You Want
Just the Way You Are
Just William's Luck
K-9
Keaton's Cop
Keep It Up, Jack!
Kid with the 200 I.Q.
Kid from Brooklyn
Kid 'n' Hollywood and Polly Tix in Washington
Kid Sister
Kidco
Kid's Auto Race/Mabel's Married Life
Kids from Candid Camera
Kill My Wife...Please!
Killer Dill
King of Jazz
King Ralph
Kings, Queens, Jokers
A Kink in the Picasso
A Kiss for Mary Pickford
Kiss Me, Stupid!
Kitty and the Bagman
The Knockout/Dough and Dynamite
Konrad
Kovacs!
Kung Fu Vampire Buster
La Boum
La Cage aux Folles
La Cage aux Folles 2
La Cage aux Folles 3: The Wedding
La Casa del Amor
La Casita del Pecado
La Criada Malcriada
La Guerra de los Sostenes
La Nuit de Varennes
La Pachanga (The Big Party)
L.A. Story
The Ladies' Man
Lady for a Day
Lady of the Evening
Lady Godiva Rides
Ladybugs
Las Vegas Hillbillys
Las Vegas Weekend
The Last American Hero
Last American Virgin
The Last Married Couple in America
Last Night at the Alamo
The Last Porno Flick

Last of the Red Hot Lovers
Last Resort
Lauderdale
Laugh with Linder!
Laurel & Hardy Comedy Classics Volume 1
Laurel & Hardy Comedy Classics Volume 2
Laurel & Hardy Comedy Classics Volume 3
Laurel & Hardy Comedy Classics Volume 4
Laurel & Hardy Comedy Classics Volume 5
Laurel & Hardy Comedy Classics Volume 6
Laurel & Hardy Comedy Classics Volume 8
Laurel & Hardy Comedy Classics Volume 9
Laurel & Hardy Special
The Lavender Hill Mob
Law of Desire
Le Bourgeois Gentilhomme
Le Cavaleur (Practice Makes Perfect)
Le Gentleman D'Epsom
Le Magnifique
Le Million
Le Sex Shop
Leader of the Band
League of Gentlemen
Leap Year
Leave It to the Marines
The Lemon Drop Kid
Lena's Holiday
Leningrad Cowboys Go America
Les Choses de la Vie
Les Comperes
Les Patterson Saves the World
Lesson in Love
Let It Ride
Let's Do It Again
Let's Get Tough
License to Drive
Life with Father
Life with Luigi
Life Stinks
The Life & Times of the Chocolate Killer
Lifeguard
Like Father, Like Son
Li'l Abner
Limit Up
Lip Service
Little Darlings
Little Miss Marker
Little Rascals
Little Rascals, Book 1
Little Rascals, Book 2
Little Rascals, Book 3
Little Rascals, Book 4
Little Rascals, Book 5
Little Rascals, Book 6
Little Rascals, Book 7
Little Rascals, Book 8
Little Rascals, Book 9
Little Rascals, Book 10
Little Rascals, Book 11
Little Rascals, Book 12
Little Rascals, Book 13
Little Rascals, Book 14
Little Rascals, Book 15
Little Rascals, Book 16
Little Rascals, Book 17
Little Rascals, Book 18
Little Rascals: Choo Choo/Fishy Tales
Little Rascals Christmas Special
Little Rascals Collector's Edition, Vol. 1-6
Little Rascals Comedy Classics, Vol. 1

Little Rascals Comedy Classics, Vol. 2
Little Rascals: Fish Hooky/Spooky Hooky
Little Rascals: Honkey Donkey/Sprucin' Up
Little Rascals: Little Sinner/Two Too Young
Little Rascals: Mush and Milk/Three Men in a Tub
Little Rascals on Parade
Little Rascals: Pups is Pups/Three Smart Boys
Little Rascals: Readin' and Writin'/Mail and Female
Little Rascals: Reunion in Rhythm/Mike Fright
Little Rascals: Spanky/Feed 'Em and Weep
Little Rascals: The Pinch Singer/Framing Youth
Little Rascals Two Reelers, Vol. 1
Little Rascals Two Reelers, Vol. 2
Little Rascals Two Reelers, Vol. 3
Little Rascals Two Reelers, Vol. 4
Little Rascals Two Reelers, Vol. 5
Little Rascals Two Reelers, Vol. 6
Little Rascals, Volume 1
Little Rascals, Volume 2
The Little Theatre of Jean Renoir
Little World of Don Camillo
Live from Washington it's Dennis Miller
Livin' Large
Livin' the Life
Living on Tokyo Time
The Lizzies of Mack Sennett
Lobster for Breakfast
Local Hero
Lonely Wives
The Longest Yard
The Longshot
Look Who's Laughing
Look Who's Talking, Too
Lookin' to Get Out
Looney Looney Looney Bugs Bunny Movie
Looney Tunes & Merrie Melodies 1
Looney Tunes & Merrie Melodies 2
Looney Tunes & Merrie Melodies 3
Looney Tunes Video Show, Vol. 4
Looney Tunes Video Show, Vol. 5
Looney Tunes Video Show, Vol. 6
Looney Tunes Video Show, Vol. 7
Loose Cannons
Loose Screws
Los Cacos
Los Chantas
Los Hombres Piensan Solo en Solo
Losin' It
Lost in America
Lost Diamond
Lost Honeymoon
Lots of Luck
Love by Appointment
The Love Bug
Love, Drugs, and Violence
Love at First Bite
Love at First Sight
Love or Money?
Love Nest

Love Notes
The Love Pill
Lovelines
Lovers and Liars
Lovers Like Us
Loves & Times of Scaramouche
Loving Couples
Lt. Robin Crusoe, U.S.N.
Lucky Devil
Lucky Jim
Lucky Partners
Luggage of the Gods
Lunatics & Lovers
Lunch Wagon
Lupo
Luv
Mabel & Fatty
Macaroni
Mack & Carole
Mack Sennett Comedies: Volume 1
Mack Sennett Comedies: Volume 2
Mad About Money
Mad About You
Mad Miss Manton
Mad Monster Party
Madame in Manhattan
Made in the USA
Madhouse
Madigan's Millions
The Magic Christian
Magic on Love Island
The Magnificent Dope
The Maid
Maid Marian & Her Merry Men: How the Band Got Together
Maid Marian & Her Merry Men: The Miracle of St. Charlene
Maid to Order
Major League
Make Me an Offer
Make a Million
Makin' It
Making the Grade
Making Mr. Right
Malcolm
Malibu Beach
Malibu Bikini Shop
Mama, There's a Man in Your Bed
The Man with Bogart's Face
A Man Called Sarge
Man from Clover Grove
Man of Destiny
A Man with a Maid
Man from Nowhere
The Man with One Red Shoe
Man in the Raincoat
Man in the Silk Hat
The Man That Corrupted Hadleyburg
The Man Who Loved Women
The Man Who Wasn't There
A Man, a Woman, & a Bank
The Mandarin Mystery
Manny's Orphans
Many Wonder
March of the Wooden Soldiers
Marihuana
Marriage is Alive and Well
The Marriage Circle
Married?
Martians Go Home!
M*A*S*H: Goodbye, Farewell & Amen
Master Mind
Masters of Comedy
The Matchmaker
Matilda
Matrimaniac
Matrimonio a la Argentina
A Matter of Degrees

A Matter of Who
Max
Maxie
May Fools
Mazurka
Me and Him
Meatballs
Meatballs 2
Meatballs 3
Meatballs 4
Medicine Man
Meet the Hollowheads
Meet the Parents
Mel Brooks: An Audience
Melvin and Howard
Memoirs of an Invisible Man
Men at Work
Merry Christmas
Messalina, Empress of Rome
Mexican Spitfire/Smartest Girl
  in Town
Miami Supercops
Mickey the Great
Mickey Rooney in the Classic
  Mickey McGuire Silly
  Comedies
Midnight Auto Supply
Midnight Madness
A Midsummer Night's Sex
  Comedy
Million Dollar Duck
Million Dollar Mystery
Milo Milo
Minsky's Follies
Minute Movie Masterpieces
The Misadventures of Mr. Wilt
Mischief
Miss Firecracker
Miss Grant Takes Richmond
Miss Right
The Missing Corpse
Mister Roberts
Mob Boss
Mob Story
Modern Girls
Modern Love
Modern Problems
Moliere
Mon Oncle
Mon Oncle D'Amerique
The Money
Monkey Hustle
Monkeys, Go Home!
Monkey's Uncle
Morgan Stewart's Coming
  Home
Moron Movies
Morons from Outer Space
Mother Goose Rock 'n'
  Rhyme
Movie Maker
Movie Struck
Moving
Moving Violations
Mr. Bill Looks Back Featuring
  Sluggo's Greatest Hits
The Mr. Bill Show
Mr. Bill's Real Life Adventures
Mr. Blandings Builds His
  Dream House
Mr. Hobbs Takes a Vacation
Mr. Hulot's Holiday
Mr. Magoo in the King's
  Service
Mr. Magoo: Man of Mystery
Mr. Mike's Mondo Video
Mr. Mom
Mr. Nice Guy
Mr. Peek-A-Boo
Mr. Skitch
Mr. Superinvisible
Mr. Walkie Talkie
Mr. Wise Guy
Mugsy's Girls
Munster's Revenge
Murder Ahoy

Murder at the Gallop
Murder Most Foul
Murder She Said
Music Box/Helpmates
Musico, Poeta y Loco
Mutant Video
Mutants In Paradise
My African Adventure
My American Cousin
My Best Friend Is a Vampire
My Blue Heaven
My Breakfast with Blassie
My Chauffeur
My Cousin Vinny
My Demon Lover
My Dog, the Thief
My Dream is Yours
My Favorite Wife
My Favorite Year
My Friend Liberty
My Geisha
My Girl
My Grandpa is a Vampire
My Lady of Whims
My Little Chickadee
My Mom's a Werewolf
My New Partner
My Other Husband
My Pleasure Is My Business
My Science Project
My Stepmother Is an Alien
My Tutor
Myra Breckinridge
Mystery Date
Mystery of the Leaping Fish/
  Chess Fever
The Mystery Man
Mystery Train
Naked Is Better
The Naked Truth
National Adultery
National Lampoon's Animal
  House
National Lampoon's
  Christmas Vacation
National Lampoon's Class
  Reunion
National Lampoon's European
  Vacation
National Lampoon's Vacation
Nature's Playmates
Naughty Co-Eds
Naughty Knights
Naughty Roommates
Near Misses
Necessary Roughness
The Neighborhood Thief
Neighbors
Never a Dull Moment
Never on Tuesday
Never Wave at a WAC
Never Weaken/Why Worry?
New Pastures
New York Stories
New York's Finest
Next Time I Marry
Next Year If All Goes Well
Ni Chana Ni Juana
Nice Girl Like Me
The Night Before
The Night Club
Night of the Comet
A Night in the Life of Jimmy
  Reardon
Night of the Living Babes
Night Partners
Night Patrol
Night Shift
The Night They Raided
  Minsky's
The Night They Robbed Big
  Bertha's
Nine Ages of Nakedness
9 to 5
Nine Lives of Fritz the Cat
1941

99 & 44/100 Dead
Ninja Academy
No Problem
No Time for Sergeants
Nobody's Perfect
Nobody's Perfekt
Norman Conquests, Part 1:
  Table Manners
Norman Conquests, Part 2:
  Living Together
Norman Conquests, Part 3:
  Round and Round the
  Garden
Norman Loves Rose
Norman's Awesome
  Experience
Now You See Him, Now You
  Don't
The Nude Bomb
Nudity Required
Nudo di Donna
The Nut House
Nutcase
O.C. and Stiggs
Ocean Drive Weekend
Ocean's 11
The Odd Couple
The Odd Job
The Odd Squad
Oddballs
Off the Mark
Off Your Rocker
Oh, God!
Oh, God! Book 2
Oh, God! You Devil
Oh, Heavenly Dog!
Oh, Mr. Porter
Oklahoma Bound
Old Spanish Custom
Oldest Profession
On the Air Live with Captain
  Midnight
On the Right Track
On Top of the Whale
Once Bitten
Once Upon a Crime
Once Upon a Honeymoon
Once Upon a Scoundrel
One Body Too Many
One Cooks, the Other Doesn't
One Crazy Summer
One More Saturday Night
The One and Only
One of Our Dinosaurs Is
  Missing
One-Punch O'Day
One Wild Moment
One Wish Too Many
Only Two Can Play
Open All Hours
Operation Dames
Operation Petticoat
Opportunity Knocks
Options
Oscar
Our Gang
Our Gang Comedies
Our Gang Comedy Festival
Our Gang Comedy Festival 2
Our Hospitality
Our Miss Brooks
The Out-of-Towners
Outrageous Fortune
Outside Chance of Maximillian
  Glick
Outtakes
Over the Brooklyn Bridge
Overboard
Pack Up Your Troubles
Packin' It In
Pajama Tops
Pals
Panama Lady
Panic Button
Pantaloons
Papa's Delicate Condition

Paper Lion
Paper Moon
Para Servir a Usted
Parade
Paradise Motel
Paradisio
Paramedics
Paramount Comedy Theater,
  Vol. 1: Well Developed
Paramount Comedy Theater,
  Vol. 2: Decent Exposures
Paramount Comedy Theater,
  Vol. 3: Hanging Party
Paramount Comedy Theater,
  Vol. 4: Delivery Man
Paramount Comedy Theater,
  Vol. 5: Cutting Up
Pardon Mon Affaire, Too!
Pardon Us
The Parent Trap
Paris Holiday
Parlor, Bedroom and Bath
Partners
Party Animal
Party Camp
Party Favors
Party Incorporated
Party! Party!
Party Plane
Passionate Thief
Passport to Pimlico
Paternity
Paths to Paradise
Paul Reiser: Out on a Whim
Paul Rodriguez Live! I Need
  the Couch
Pauline at the Beach
Peck's Bad Boy
Peck's Bad Boy with the
  Circus
The Pee-Wee Herman Show
Pee-Wee's Playhouse
  Christmas Special
Pee-Wee's Playhouse Festival
  of Fun
Penn and Teller's Cruel Tricks
  for Dear Friends
Pennywhistle Blues
People Are Funny
People's Choice
Pepper and His Wacky Taxi
Perfect Furlough
Perfectly Normal
Personal Exemptions
Petit Con
Phantom Empire
Phffft!
The Pick-Up Artist
Pick-Up Summer
The Pickwick Papers
Piece of the Action
The Pinchcliffe Grand Prix
The Pink Angels
The Pink Chiquitas
Pink Flamingos
Pink Motel
Pippi Longstocking
Plain Clothes
Planes, Trains & Automobiles
Playboy's Bedtime Stories
Playboy's Party Jokes
Playing Away
Playing for Keeps
Playmates
Plaza Suite
Please Don't Eat the Daisies
Please Don't Eat My Mother
A Pleasure Doing Business
Pleasure Resort
The Plot Against Harry
Poison Ivy
Poker Alice
Police Academy
Police Academy 2: Their First
  Assignment

Police Academy 3: Back in
  Training
Police Academy 4: Citizens on
  Patrol
Police Academy 5:
  Assignment Miami Beach
Police Academy 6: City Under
  Siege
A Polish Vampire in Burbank
Polyester
Pom Pom Girls
Pool Sharks
Pop Always Pays
Pop Goes the Cork
The Pope Must Diet
Popeye Assorted Cartoons
Porky's
Porky's 2: The Next Day
Porky's Revenge
Portrait of a White Marriage
Pot O' Gold
Pouvoir Intime
The Prairie Home Companion
  with Garrison Keillor
Prep School
Preppies
Pretty Smart
Pride of the Bowery
The Primitive Lover
Prince of Pennsylvania
The Prince and the Showgirl
Princess Academy
Private Affairs
Private Benjamin
Private Eyes
Private Lessons
Private Manoeuvres
Private Popsicle
Private Resort
Private School
Prize Fighter
Probation
Problem Child 2
Projectionist
Promises! Promises!
Promoter
Property
Protocol
Prurient Interest
The Pure Hell of St. Trinian's
Pure Luck
Purlie Victorious
Pygmalion
Quality Street
Queen for a Day
Quick Change
Rabbit Test
Race for Your Life, Charlie
  Brown
Radio Days
Rafferty & the Gold Dust
  Twins
The Rage of Paris
Ranch
Rascal Dazzle
Ratings Game
Ravishing Idiot
Real Genius
Real Life
Real Men
Really Weird Tales
Rebel High
Recruits
Red Desert Penitentiary
Red Skelton: A Comedy
  Scrapbook
Reg'lar Fellers
Rendez-Moi Ma Peau
Rented Lips
Repo Man
Return to Mayberry
Return of Our Gang
Return of the Tall Blond Man
  With One Black Shoe
Return of Video Yesterbloop

Comedy

Reunion: 10th Annual Young
  Comedians
Revenge of the Cheerleaders
Revenge of the Nerds
Revenge of the Nerds 2:
  Nerds in Paradise
Revenge of the Teenage
  Vixens from Outer Space
Revenge of TV Bloopers
Rhinestone
The Ribald Tales of Robin
  Hood
Rich Hall's Vanishing America
Richard Pryor: Here and Now
Richard Pryor: Live on the
  Sunset Strip
Riders of the Storm
Riding on Air
Rikky and Pete
The Rink
Rink/Immigrant
Ripping Yarns
Risky Business
Rita, Sue & Bob Too
Road Lawyers and Other
  Briefs
Robin! Tour de Face!
Rock & the Money-Hungry
  Party Girls
Rockin' Road Trip
Rockin' Ronnie
Rockula
Romance with a Double Bass
Ron Reagan Is the President's
  Son/The New
  Homeowner's Guide to
  Happiness
The Roseanne Barr Show
Rosebud Beach Hotel
Rosencrantz & Guildenstern
  Are Dead
Ruddigore
Rude Awakening
Ruggles
Ruggles of Red Gap
The Ruling Class
Run Virgin Run
The Runaway Bus
The Running Kind
Running Wild
The Russians are Coming, the
  Russians are Coming
Ruthless People
Ryder P.I.
Sabrina
The Sad Sack
Sailor-Made Man/Grandma's
  Boy
St. Benny the Dip
Sallah
Sally of the Sawdust
Salut l'Artiste
Sam Kinison Live!
Sammy & Rosie Get Laid
Sammy, the Way-Out Seal
Santa Claus Conquers the
  Martians
The Saphead
Saturday Night Live: Carrie
  Fisher
Saturday Night Live: Richard
  Pryor
Saturday Night Live: Steve
  Martin, Vol. 1
Saturday Night Live: Steve
  Martin, Vol. 2
Saturday Night Live, Vol. 1
Sawmill/The Dome Doctor
Say Yes!
Scandalous
Scared Stiff
Scavenger Hunt
Scenes from the Class
  Struggle in Beverly Hills
School Daze
School for Sex

School Spirit
Scoring
Screen Test
Screwball Academy
Screwball Hotel
Screwballs
Scrooged
Second Best Secret Agent in
  the Whole Wide World
Second City Comedy Special
Second City Insanity
Second Sight
Second Time Lucky
Secret Admirer
The Secret Diary of Sigmund
  Freud
Secret Life of Walter Mitty
The Secret War of Harry Frigg
Secrets of Sweet Sixteen
See No Evil, Hear No Evil
Semi-Tough
The Senator was Indiscreet
Senior Trip
Senior Week
The Seniors
Sensations
Sensuous Caterer
Serial
A Session with the Committee
Seven Chances
Seven Minutes in Heaven
Seven Thieves
Sex Adventures of the Three
  Musketeers
Sex Appeal
Sex and Buttered Popcorn
Sex in the Comics
Sex O'Clock News
Sex and the Other Woman
Sex Ray
Sex on the Run
Sex with a Smile
Sexpot
Shadow of the Thin Man
Shag: The Movie
The Shaggy D.A.
Shaggy Dog
Shake Hands with Murder
Shampoo
Shanghai Surprise
She Couldn't Say No
The Sheep Has Five Legs
Sherlock's Rivals & Where's
  My Wife
She's in the Army Now
She's Out of Control
Shirley Temple Festival
Shoot Loud, Louder, I Don't
  Understand!
Shore Leave
Short Circuit 2
Show People
The Shrimp on the Barbie
Sibling Rivalry
Side By Side
Silent Movie
Silent Movies: In Color
Silent Movies...Just For the
  Fun of It!
Silver Bears
Simon
The Sin of Harold Diddlebock
Singleton's Pluck
Sinners
Sitting Ducks
Ski School
Skinner's Dress Suit
Skyline
Slammer Girls
The Slap
Slapstick of Another Kind
Slaves of New York
The Sleazy Uncle
Sleepaway Camp 2: Unhappy
  Campers

Sleeper
A Slightly Pregnant Man
Slither
Slumber Party '57
Small Change
Small Hotel
The Smallest Show on Earth
Smart Money
Smile
Smiles of a Summer Night
Smokey Bites the Dust
Snapshot
Snowball Express
Snowballin'
Snowballing
So Fine
So This is Paris
So This is Washington
S.O.B.
Social Error
Soggy Bottom U.S.A.
Soldat Duroc...Ca Va Etre Ta
  Fete!
Some Like It Hot
Something Special
Son of Flubber
Sonar no Cuesta nada Joven
  (Dreams are Free)
Song of the Thin Man
Sorority House Scandals
Sorrowful Jones
Sotto, Sotto
Soup for One
A Southern Yankee
Soy un Golfo
Spaceballs
Spaced Invaders
Spaced Out
Spaceship
Spaghetti House
Spalding Gray: Terrors of
  Pleasure
Speed Zone
Speedy
Spies, Lies and Naked Thighs
Spike of Bensonhurst
Spirit of '76
Spite Marriage
Splitz
Spooks Run Wild
Sport Goofy
Sport Goofy's Vacation
Spring Break
Spring Fever
The Sprinter
Spy With a Cold Nose
S*P*Y*S
The Squeeze
Squeeze Play
Stallion
Stan Without Ollie, Volume 2
Stan Without Ollie, Volume 3
Stan Without Ollie, Volume 4
Stand-In
Star Spangled Girl
Star Trek 4: The Voyage
  Home
Starhops
Starring Bugs Bunny!
Stars and Bars
Start the Revolution Without
  Me
State of the Union
Steagle
Steamboat Bill, Jr.
Steelyard Blues
Stewardess School
The Stick-Up
Sticky Fingers
The Sting 2
Stir Crazy
Stitches
Stocks and Blondes
Stoogemania
The Stoogephile Trivia Movie
Stop! or My Mom Will Shoot

Storm in a Teacup
The Story of Fausta
Story of William S. Hart/Sad
  Clowns
Strange Brew
Strangers of the Evening
Strike It Rich
Stripes
Stroker Ace
Strong Man
Stuck on You
Stuckey's Last Stand
Student Affairs
The Student Body
The Student Body/Jailbait
  Babysitter
Student Confidential
The Student Teachers
Substitute Teacher
The Suitors
Sullivan's Travels
Summer Camp
Summer Job
Summer Night Fever
Summer Night With Greek
  Profile, Almond Eyes . .
Summer Rental
Summer School
Summer School Teachers
Summer Switch
Sundance
The Sunshine Boys
The Super
Super Fuzz
Super Kung Fu Kid
Super Seal
Super Soul Brother
Superdad
Supergrass
Suppose They Gave a War
  And Nobody Came?
The Sure Thing
Surf 2
Survivors
Susan Slept Here
Swap Meet
Sweater Girls
Swedenhielms
Swedish Wildcats
Sweet Hearts Dance
Sweet Liberty
Sweet Lies
Sweet Movie
Sweet Revenge
Sweet Young Thing
Sweetie
Swim Team
Swimming to Cambodia
Swimsuit: The Movie
A Swingin' Summer
The Swinging Cheerleaders
Swinging Wives
Switch
Switching Channels
Sykes
Sylvia and the Phantom
Table Settings
Tabloid!
Takin' It All Off
Takin' It Off
Taking Care of Business
Talkin' Dirty After Dark
Talking Walls
The Tall Blond Man with One
  Black Shoe
The Tall Guy
Tall Story
Tall, Tan and Terrific
Tammy and the Bachelor
Tammy and the Doctor
Tampopo
Tanner '88
Tartuffe
A Taste for Flesh and Blood
Tax Season
A Taxing Woman

A Taxing Woman Returns
Teachers
Teacher's Pet
Teddy at the Throttle
Teddy at the Throttle/
  Speeding Along
Teen Lust
Teen Vamp
Teen Witch
Teen Wolf
Teen Wolf Too
The Telephone
Television's Golden Age of
  Comedy
The Tender Trap
Tex Avery's Screwball
  Classics, Vol. 3
That Certain Thing
That Darn Cat
That Gang of Mine
That Man Is Pregnant
That Naughty Girl
That Sinking Feeling
That Uncertain Feeling
That's Adequate
That's My Baby!
There's a Girl in My Soup
There's Naked Bodies on My
  T.V.!
These Girls Won't Talk
They All Laughed
They Call Me Bruce
They Still Call Me Bruce
They Went That-a-Way &
  That-a-Way
The Thin Man Goes Home
Things Change
Things Happen at Night
Things We Did Last Summer
Think Dirty
Thinkin' Big
30 Is a Dangerous Age,
  Cynthia
This Happy Feeling
This Time I'll Make You Rich
Thomas Graal's Best Film
Thomas Graal's First Child
Those Daring Young Men in
  Their Jaunty Jalopies
Those Fantastic Flying Fools
Those Wild Bloopers
Thousand Clowns
Three Ages
Three for Bedroom C
Three Broadway Girls
Three in the Cellar
Three Day Weekend
Three Husbands
Three of a Kind
Three Kinds of Heat
Three Men and a Baby
Three Men and a Cradle
Three Men and a Little Lady
Three Nuts in Search of a Bolt
Three O'Clock High
The Three Stooges Meet
  Hercules
Three Way Weekend
Three's Trouble
The Thrill of It All!
Thunder & Mud
Thursday's Child
Thursday's Game
Tigers in Lipstick
A Tiger's Tale
Till Marriage Do Us Part
Tillie's Punctured Romance
The Time of Their Lives
Timepiece
The Tip-Off
To Be or Not to Be
To See Such Fun
Toast of the Town
Toga Party
Tom Sawyer
Tomboy

Tonight for Sure
Tony Draws a Horse
Too Many Crooks
Too Much Sun
Topper
Topper Returns
Topper Takes a Trip
Tough Guys
The Toy
Trading Places
Train to Hollywood
Tramp & a Woman
Transylvania 6-5000
Trash
Traveling Companions
Travels with My Aunt
Treasure of Fear
Trenchcoat
Tricks of the Trade
Triple Trouble/Easy Street
TripleCross
Troop Beverly Hills
Trottie True
The Trouble with Angels
Trouble in Paris
Trouble in Store
Truck Stop
True Identity
Truly Tasteless Jokes
Trust Me
Tune in Tomorrow
Tunnelvision
Turner and Hooch
Turumba
The Tuttles of Tahiti
Twelve Chairs
12 Plus 1
Twentieth Century
Twins
Twist
Twister
200 Motels
2 Idiots in Hollywood
Two Nights with Cleopatra
Two Reelers: Comedy
  Classics 1
Two Reelers: Comedy
  Classics 2
Two Reelers: Comedy
  Classics 3
Two Reelers: Comedy
  Classics 4
Two Reelers: Comedy
  Classics 5
Two Reelers: Comedy
  Classics 6
Two Reelers: Comedy
  Classics 7
Two Reelers: Comedy
  Classics 8
Two Reelers: Comedy
  Classics 9
2000 Year Old Man
Two Top Bananas
Two Weeks to Live
Two Women in Gold
Uforia
The Ugly Dachshund
UHF
Una Viuda Descocada
The Unbelievable Truth
Uncensored
Uncle Buck
Under the Doctor
Under the Rainbow
Under the Sheets
The Underachievers
Undergrads
Underground Aces
Unfaithfully Yours
Unidentified Flying Oddball
Unkissed Bride
The Unknown Comedy Show
Up the Academy
Up the Creek
Up from the Depths

Up Your Anchor
Uphill All the Way
Uptown Comedy Express
Uptown Saturday Night
Used Cars
The Vagabond
Valet Girls
Vals
The Van
Van Nuys Blvd.
Varieties on Parade
Vasectomy: A Delicate Matter
Very Close Quarters
Very Curious Girl
Vibes
Vice Academy
Vice Academy 2
Vice Academy 3
Vice Versa
The Victim
Victor/Victoria
Victorian Fantasies
Video Vixens
Video Yesterbloop
Virgin High
Virgin Queen of St. Francis
  High
Virgins of Purity House
Visitants
Viva Chihuahua (Long Live
  Chihuahua)
Voice of Hollywood
Volunteers
The Wacky World of Wills &
  Burke
Wait Till Your Mother Gets
  Home
Waitress
Watch the Birdie
Water
Wayne's World
W.C. Fields Comedy Bag
W.C. Fields: Flask of Fields
W.C. Fields Rediscovered
A Wedding
Wedding Band
Wedding Rehearsal
Weekend with the Babysitter
Weekend Pass
Weekend Warriors
Weird Science
Welcome to 18
Welcome Home, Roxy
  Carmichael
We're in the Legion Now
What About Bob?
What Do You Say to a Naked
  Lady?
Whatever Happened to Aunt
  Alice?
What's Up Front
What's Up, Tiger Lily?
When Comedy was King
When Knights were Bold
When Nature Calls
When Women Had Tails
When Women Lost Their Tails
When's Your Birthday?
Where the Boys Are '84
Where the Buffalo Roam
Where the Bullets Fly
Where's Willie?
Which Way to the Front?
Which Way Is Up?
Whiffs
Whiskey Galore
White Men Can't Jump
Who Framed Roger Rabbit?
Who Killed Doc Robbin?
Wholly Moses!
Whoops Apocalypse
Who's Harry Crumb?
Who's Minding the Mint?
Who's That Girl?
Wicked Stepmother
The Wild Life

Wild Women of Wongo
Wildcats
Wildest Dreams
Win, Place, or Steal
The Wizard of Speed and
  Time
Woman of Paris/Sunnyside
Woman Times Seven
Women's Club
Work/Police
Working Winnie/Oh So Simple
World's Greatest Athlete
World's Greatest Lover
The Worst of Hollywood: Vol.
  1
The Worst of Hollywood: Vol.
  2
The Worst of Hollywood: Vol.
  3
Would-Be Gentleman
The Wrong Guys
Y Donde Esta el Muerto?
  (And... Where Is the Dead)
You Can't Cheat an Honest
  Man
You Can't Hurry Love
You Can't Take It with You
You Talkin' to Me?
Young Doctors in Love
Young Einstein
The Young in Heart
Young Love - Lemon Popsicle
  7
The Young Ones
Young & Willing
Yours, Mine & Ours
You've Got to Have Heart
Zapped!
Zapped Again
Zazie dans le Metro
Zelig
Zero to Sixty
Zoo Radio
Zotz!

Comedy Drama

About Last Night...
Air America
Alfie
Alice Doesn't Live Here
  Anymore
Alice's Restaurant
All My Good Countrymen
An Almost Perfect Affair
Aloha Summer
Always
Amarcord
American Boyfriends
American Graffiti
And Baby Makes Six
Androcles and the Lion
Andy Hardy Gets Spring Fever
Andy Hardy's Double Life
Andy Hardy's Private
  Secretary
Angel in a Taxi
The Apprenticeship of Duddy
  Kravitz
Arabian Nights
Around the World in 80 Days
As You Like It
Autumn Marathon
Avalon
Awakenings
Ay, Carmela!
Babette's Feast
Baby It's You
Baby, Take a Bow
Back Roads
Bail Jumper
The Baker's Wife (La Femme
  du Boulanger)
The Ballad of Paul Bunyan
Band of Outsiders

Bartleby
Baxter
The Bells of St. Mary's
Benji
The Best Way
Big
The Big Chill
Big Wednesday
Billy Liar
The Bishop's Wife
Black Rain
Blat
Blueberry Hill
Blues Busters
Blume in Love
Boccaccio '70
Born to Win
The Breakfast Club
Breaking Away
Breaking In
Breaking Up Is Hard to Do
Bright Eyes
Brighton Beach Memoirs
Brimstone & Treacle
Broadcast News
Buck and the Preacher
Bud and Lou
The Buddy System
Buffalo Bill
Bulldog Drummond's
  Revenge
Busting
Butterflies Are Free
Caesar and Cleopatra
Cafe Express
California Casanova
Candy Mountain
Cannery Row
Captain Newman, M.D.
Carbon Copy
Carnal Knowledge
The Cary Grant Collection
Casey's Shadow
Cesar
Cesar & Rosalie
Charles et Lucie
Charlie and the Great Balloon
  Chase
Cheaper to Keep Her
Chimes at Midnight
Chloe in the Afternoon
Choose Me
A Chorus of Disapproval
The Christmas Wife
City Lights
City Slickers
City of Women
Claire's Knee
Clara's Heart
Class Reunion
Cleo from 5 to 7
Clipped Wings
Closely Watched Trains
Clown
Coast to Coast
Cocoon: The Return
Conspiracy of Terror
Copenhagen Nights
Corvette Summer
Coupe de Ville
Cousin, Cousine
Cousins
Crazy from the Heart
The Crime of Monsieur Lange
Crimes & Misdemeanors
Danny Boy
Death of the Rat
The Decameron
The Decline of the American
  Empire
Der Purimshpiler (The Jester)
Desperate Teenage Lovedolls
Details of a Duel: A Question
  of Honor
Detour to Danger
Devil & Daniel Webster

Devil on Horseback
The Devil's Eye
Diary of a Chambermaid
Diary of Forbidden Dreams
Diary of a Mad Housewife
Diary of a Young Comic
Dim Sum: A Little Bit of Heart
Diner
The Dining Room
Dinner at Eight
Dixie Lanes
Do the Right Thing
Dr. Christian Meets the
  Women
Doctor at Sea
Dog Day Afternoon
Dog Eat Dog
Don Quixote
Don't Drink the Water
Down by Law
Dream Chasers
The Dream Team
Drugstore Cowboy
Eat a Bowl of Tea
Eat My Dust
Eat the Peach
Educating Rita
El Bruto (The Brute)
El Cochecito
Eli Eli
Elizabeth of Ladymead
Emanon
End of the Line
Enemies, a Love Story
Enormous Changes
Extra Girl
Eyes Right!
The Fabulous Baker Boys
Family Business
Far North
Fast Talking
Fatso
The Fifth Monkey
First Name: Carmen
The Fisher King
Five Corners
The Five Heartbeats
The Flamingo Kid
Flesh
FM
For Keeps
For the Love of Benji
Four Adventures of Reinette
  and Mirabelle
Four Deuces
Four Friends
The Four Musketeers
The Four Seasons
Fried Green Tomatoes
Full Moon in Blue Water
Fuzz
Garbo Talks
Generation
George!
Georgy Girl
Getting Over
Getting Straight
Ghost Chasers
Ghosts on the Loose
The Gig
Gin Game
Ginger & Fred
The Girl with the Hat Box
Girlfriends
A Girl's Folly
The Gold of Naples
Gone are the Days
Good Morning, Vietnam
Good Sam
Goodbye Columbus
The Goodbye People
Goodnight, Michelangelo
The Gospel According to Vic
The Graduate
Grandes Amigos (Great
  Friends)

Comedy

Comedy Performance

Andy Kaufman: Sound Stage
Andy Kaufman Special
Benny Hill's Crazy World
The Best of Abbott & Costello Live
Best of the Benny Hill Show: Vol. 1
Best of Comic Relief
Best of Comic Relief '90
Best of Dan Aykroyd
Best of Eddie Murphy
Best of Gilda Radner
Bill Cosby: 49
Bill Cosby, Himself
Bill Murray Live from the Second City
Billy Crystal: A Comic's Line
Billy Crystal: Don't Get Me Started
Billy Crystal: Midnight Train to Moscow
Bloodhounds of Broadway
Bob Goldthwait: Is He Like That All the Time?
Carrott Gets Rowdie
Catch a Rising Star's 10th Anniversary
Charles Pierce at the Ballroom
Chris Elliott: FDR, A One-Man Show/Action Family
Clark & McCullough
Comedy Tonight
Comic Relief 2
Comic Relief 3
David Steinberg: In Concert
Did You Hear the One About ...?
Eddie Murphy "Delirious"
Eddie Murphy: Raw
Elayne Boosler: Broadway Baby
Elayne Boosler: Party of One
Ernie Kovacs/Peter Sellers
An Evening with Bobcat Goldthwait: Share the Warmth
Evening at the Improv
An Evening with Robin Williams
Father Guido Sarducci Goes to College
The Franken and Davis Special
Franken and Davis at Stockton State
Funhouse
Gabe Kaplan as Groucho
Gallagher in Concert
Gallagher: Melon Crazy
Gallagher: Over Your Head
Gallagher: Stuck in the 60's
Gallagher: The Bookkeeper
Gallagher: The Maddest
Gallagher's Overboard
Garry Shandling Show, 25th Anniversary Special
George Burns in Concert
George Burns: His Wit & Wisdom
George Carlin on Campus
George Carlin at Carnegie Hall
George Carlin - Live!: What am I Doing in New Jersey?
George Carlin: Playin' with Your Head
Gilda Live
Girls of the Comedy Store
Gross Jokes
Hollywood Goes to War
Howie From Maui - Live!
Howie Mandel's North American Watusi Tour
Hungry i Reunion
It's Not Easy Bein' Me
Jackie Mason on Broadway

Jo Jo Dancer, Your Life is Calling
Jock Jokes
Joe Bob Briggs Dead in Concert
Joe Piscopo
Joe Piscopo Live!
Joe Piscopo: New Jersey Special
Kovacs!
Laughing Room Only
Lenny
Lenny Bruce
Lenny Bruce Performance Film
Lily Tomlin
Live from Washington it's Dennis Miller
Madame in Manhattan
Mambo Mouth
Martin Mull Live: From Ridgeville, Ohio
Mel Brooks: An Audience
Melon Crazy
Monty Python Live at the Hollywood Bowl
Monty Python Meets Beyond the Fringe
More Laughing Room Only
Mr. Mike's Mondo Video
National Lampoon's Class of '86
New Wave Comedy
Paramount Comedy Theater, Vol. 1: Well Developed
Paramount Comedy Theater, Vol. 2: Decent Exposures
Paramount Comedy Theater, Vol. 3: Hanging Party
Paramount Comedy Theater, Vol. 4: Delivery Man
Paramount Comedy Theater, Vol. 5: Cutting Up
Paul Reiser: Out on a Whim
Paul Rodriguez Live! I Need the Couch
Penn and Teller Go Public
The Prairie Home Companion with Garrison Keillor
Punchline
Rap Master Ronnie
Red Skelton: A Comedy Scrapbook
Red Skelton: King of Laughter
Red Skelton's Funny Faces
Red Skelton's Funny Faces 3
Redd Foxx—Video in a Plain Brown Wrapper
Reunion: 10th Annual Young Comedians
Rich Little: One's a Crowd
Rich Little's Christmas Carol
Rich Little's Great Hollywood Trivia Game
Richard Lewis: I'm Exhausted
Richard Lewis: I'm in Pain
Richard Pryor: Here and Now
Richard Pryor: Live in Concert
Richard Pryor: Live and Smokin'
Richard Pryor: Live on the Sunset Strip
Rise and Fall of the Borscht Belt
Robert Klein on Broadway
Robert Klein: Child of the '50s, Man of the '80s
Robin! Tour de Face!
Robin Williams Live!
Rodney Dangerfield: Nothin' Goes Right
Sam Kinison Live!
Scrambled Feet
Second City Comedy Special
The Secret Policeman's Other Ball

Secret Policeman's Private Parts
Slow Men Working
Spalding Gray: Terrors of Pleasure
Stand Up Reagan
Steve Martin Live
Steven Wright Live
Things We Did Last Summer
Toast to Lenny
Truly Tasteless Jokes
Uptown Comedy Express
Without You I'm Nothing

Comic Adventure

The Adventures of Bullwhip Griffin
All the Way, Boys
The Ambushers
The Americano
Andy and the Airwave Rangers
Angel of H.E.A.T.
Angels Hard as They Come
Another 488 Hrs.
Avenging Disco Godfather
Back to the Future
Back to the Future, Part 2
Back to the Future, Part 3
Bad Guys
The Baltimore Bullet
Behind the Front
Beverly Hills Cop
Beverly Hills Cop 2
Big Zapper
Bird on a Wire
Bloodstone
Blue Iguana
Brink's Job
Burglar
California Straight Ahead
Cannibal Women in the Avocado Jungle of Death
Cannonball Run
Cannonball Run II
Captain Swagger
City Heat
Dangerous Relations
Doctor Detroit
Dog Pound Shuffle
Dolemite 2: Human Tornado
Down to Earth
El Profesor Hippie
Fast Getaway
Feds
The Fifth Musketeer
The Fighting American
Filthy Harry
Flirting with Fate
Flush It
48 Hrs.
Foul Play
Free Ride
The Further Adventures of Tennessee Buck
Ghostbusters
Ghostbusters 2
Go for It
The Golden Child
Hambone & Hillie
Having Wonderful Time/ Carnival Boat
Hawthorne of the USA
High Road to China
High Voltage
Highpoint
Hudson Hawk
Ice Pirates
In Like Flint
In 'n Out
It's in the Bag
Jake Speed
Jane & the Lost City
Jewel of the Nile

Just Tell Me You Love Me
Kindergarten Cop
King Kung Fu
Kiss My Grits
La Petite Bande
Landlord Blues
Leonard Part 6
Mademoiselle Striptease
The Masters of Menace
Midnight Run
Miracles
Mitchell
Mr. Billion
Mr. Robinson Crusoe
The Naked Sword of Zorro
No Deposit, No Return
Oddball Hall
Odds and Evens
Our Man Flint
Ramblin' Man
Red Lion
Reggie Mixes In
The Reivers
Return of the Rebels
Romancing the Stone
Round Trip to Heaven
Running Scared
Russkies
Seven
She's Back
Short Time
The Silver Streak
Ski Patrol
Sky High
Smokey and the Bandit
Smokey and the Bandit, Part 2
Smokey and the Bandit, Part 3
Smokey & the Judge
Special Delivery
Stakeout
Stingray
Stuntwoman
Suburban Commando
Suffering Bastards
Summer Night
Sunburn
Swashbuckler
Terminal Exposure
Thief Who Came to Dinner
Think Big
Those Magnificent Men in Their Flying Machines
Three Amigos
Three Fugitives
Three Musketeers
Thunder and Lightning
The Trouble with Spies
Wackiest Ship in the Army
We're No Angels
When the Clouds Roll By
The Wild and the Free
Wrong is Right

Coming of Age

The Abe Lincoln of Ninth Avenue
Across the Great Divide
The Affairs of Annabel
All Fall Down
All the Right Moves
Aloha Summer
Alpine Fire
Amarcord
American Boyfriends
American Graffiti
An American Summer
Among the Cinders
Angele
Angelo My Love
Anne of Green Gables
Anthony Adverse
Aparajito

The Apprenticeship of Duddy Kravitz
Baby It's You
Back Door to Heaven
Battling Bunyon
Big
The Big Bet
The Big Red One
Billy Bathgate
The Birch Interval
Blue Fin
Blue Jeans
The Blue Lagoon
Blue Velvet
Boss' Son
The Boys in Company C
Breaking Away
Breaking Home Ties
Brighton Beach Memoirs
Can You Feel Me Dancing?
Can't Buy Me Love
Career Opportunities
Carnal Knowledge
The Chicken Chronicles
Class
Class of '44
Closely Watched Trains
Cocktail
Cold Sassy Tree
Color of Money
Come and See
Courage Mountain
Courtship
Crazy Moon
Cria
Cry of Battle
Culpepper Cattle Co.
Dakota
Dead Poets Society
December
The Deer Hunter
Desert Bloom
The Devil's Playground
Diner
Dirty Dancing
Dogfight
Dragstrip Girl
Dutch Girls
Edward Scissorhands
Emanon
Ernesto
Experience Preferred... But Not Essential
Eyes Right!
The Eyes of Youth
Fandango
Fast Times at Ridgemont High
Father
Felicity
Fellini Satyricon
The Fire in the Stone
First Affair
First Love
The First Time
The Flamingo Kid
For Love Alone
Forever
Forever Young
The 400 Blows
Foxes
The Getting of Wisdom
Gidget
A Girl of the Limberlost
Goin' All the Way
The Gold & Glory
Goodbye Columbus
Goodbye, My Lady
The Graduate
The Grand Highway
The Great Santini
Gregory's Girl
Hairspray
Hangin' with the Homeboys
Hard Choices
Heathers
Heaven Help Us

Coming

Crime

Bugsy Malone
Bulldog Drummond's
  Revenge
Bullitt
Burglar
Busted Up
Buster
Busting
Butch and Sundance: The
  Early Days
Butterfly Affair
Buying Time
The C-Man
Caged Fury
Call Him Mr. Shatter
Call It Murder
Came a Hot Friday
Camorra: The Naples
  Connection
Cape Fear
Caper of the Golden Bulls
Captain Blackjack
Captain Swagger
Captive
Captured in Chinatown
Car Crash
Caribe
Cartier Affair
A Case of Deadly Force
Cash
The Cat
The Cat o' Nine Tails
The Catamount Killing
Catch the Heat
The Centerfold Girls
Challenge
Chameleon
Chan is Missing
Chance
Charade
Charles et Lucie
Charley Varrick
The Cheaters
Cheech and Chong: Things
  Are Tough All Over
Choice of Weapons
The Choirboys
Circle of Fear
The City
City in Fear
City in Panic
City of Shadows
Cleopatra Jones
Cleopatra Jones & the Casino
  of Gold
Clown Murders
Club Life
The Coast Patrol
Cobra
The Cobra Strikes
Cocaine Wars
Code Name: Zebra
Code of Silence
Coffy
Cohen and Tate
Cold Blood
Cold Eyes of Fear
Cold Feet
Cold Front
Cold Steel
Cold Sweat
Color Me Dead
Colors
Columbo: Murder By the Book
Commando Squad
Compromising Positions
Concrete Beat
Condemned to Hell
Confessions of Tom Harris
Confidentially Yours
Conflict
The Connection
Contraband
Convicts at Large
Coogan's Bluff
The Cool World

Cop
Cop in Blue Jeans
Cop & the Girl
Cop-Out
Cordelia
Corleone
Corrupt
Cosh Boy
Couples and Robbers
The Courier
Courier of Death
Covered Wagon Days
Covert Action
Crackers
Crashout
Crazy Mama
Creature with the Blue Hand
Crime Killer
Crime Story
Crime Story: The Complete
  Saga
Crimebusters
Crimes & Misdemeanors
Crimewave
Criminal Act
Criminal Code
Criminal Law
Criminals Within
Criss Cross
Crocodile Dundee 2
Crooks & Coronets
Cruisin' High
Daddy-O
Daddy's Boys
Dagger Eyes
Dance or Die
Dance Fools Dance
Dance with Me, Henry
Danger: Diabolik
Danger Zone
Dangerous Charter
Dangerous Love
Dangerous Mission
Dangerous Orphans
Dangerous Pursuit
Dangerously Close
Danny Boy
Daring Dobermans
Dark Mountain
Dark Passage
Dark Ride
Dark Side of Midnight
Darker Than Amber
Darktown Strutters
Das Testament des Dr.
  Mabuse
Daughter of the Tong
Day of the Assassin
Dayton's Devils
Dead Bang
Dead as a Doorman
Dead Easy
Dead End
Dead End City
Dead Heat
Dead Heat on a Merry-Go-
  Round
Dead Men Don't Wear Plaid
The Dead Pool
Dead Reckoning
Dead Wrong
Dead Wrong: The John Evans
  Story
Deadline Assault
Deadline at Dawn
Deadly Alliance
Deadly Business
Deadly Darling
Deadly Diamonds
Deadly Force
Deadly Game
Deadly Hero
Deadly Illusion
Deadly Intent
Deadly Mission
Deadly Revenge

Deadly Sting
Deadly Strangers
Deadly Sunday
Deadly Thief
Deadly Twins
Death in Deep Water
Death of a Hooker
The Death Merchant
Death of a Scoundrel
Death Scream
Death Sentence
Death in the Shadows
Death of a Soldier
The Death Squad
Death Stalk
Death Train
Death Wish
Death Wish 2
Death Wish 3
Death Wish 4: The
  Crackdown
Deathrow Gameshow
Deceptions
Dedee D'Anvers
The Defender
Defiance
Defiant
Demon Hunter
The Deputy Drummer
Desperate
Desperate Cargo
Desperate Hours
The Destructors
The Detective
Detective Sadie & Son
Detective School Dropouts
Detour
Detour to Danger
Detroit 9000
Devil at 4 O'Clock
Devil Doll
Devil & Leroy Basset
Devil Thumbs a Ride/Having
  Wonderful Crime
Devil's Angels
The Devil's Brother
The Devil's Commandment
The Devil's Sleep
Devlin Connection
Diamond Trail
Diamonds
Diamonds on Wheels
Diary of a Teenage Hitchhiker
Dick Tracy
Dick Tracy Detective
Dick Tracy, Detective/The
  Corpse Vanishes
Dick Tracy Double Feature #1
Dick Tracy Double Feature #2
Dick Tracy Meets Gruesome
Dick Tracy Returns
Dick Tracy: The Original
  Serial, Vol. 1
Dick Tracy: The Original
  Serial, Vol. 2
Dick Tracy: The Spider Strikes
Dick Tracy vs. Crime Inc.
Dick Tracy vs. Cueball
Dick Tracy's Dilemma
Dick Turpin
Die Hard
Die Hard 2: Die Harder
Dillinger
Diplomatic Immunity
Dirty Hands
Dirty Harry
Dirty Mary Crazy Larry
Disciple
Disorganized Crime
Distortions
Disturbance
Diva
Dixie Lanes
Do or Die
The Doberman Gang

Dr. Christian Meets the
  Women
Dr. Death, Seeker of Souls
The Doctor and the Devils
Dr. Mabuse the Gambler
Dr. Mabuse vs. Scotland Yard
Dr. Otto & the Riddle of the
  Gloom Beam
Dr. Syn, Alias the Scarecrow
Dog Day
Dog Day Afternoon
Dog Eat Dog
The Doll Squad
Dollars
Don Amigo/Stage to Chino
The Don is Dead
Don't Change My World
Doomed to Die
Door-to-Door Maniac
Double Identity
Double Indemnity
The Double Negative
Double Revenge
Double Trouble
Down & Dirty
Down the Drain
Down by Law
Down Twisted
Downtown
Dragnet
Dream No Evil
The Dream Team
Drive-In Massacre
The Driver
Drunken Angel
Dual Alibi
Dumb Waiter
Each Dawn I Die
East Side Kids
East Side Kids/The Lost City
Echo Murders
Echoes in the Darkness
Eddie Macon's Run
Edge of Darkness
Edge of Fury
Education of Sonny Carson
Edwin S. Porter/Edison
  Program
8 Million Ways to Die
El Callao (The Silent One)
El Coyote Emplumado
El Cristo de los Milagros
El Fego Baca: Six-Gun Law
El Muerto
El Tesoro de Chucho el Roto
Electra Glide in Blue
The Element of Crime
11 Harrowhouse
Emil & the Detectives
Emilio Varela vs. Camelia La
  Texana
Emperor of the Bronx
Emperor Jones
Emperor Jones/Paul
  Robeson: Tribute to an
  Artist
Endangered Species
Endplay
The Enforcer
Enforcer from Death Row
Enter the Dragon
Ernest Goes to Jail
Escapade in Florence
Escape 2000
Escape from Alcatraz
Escape from New York
Eternal Evil
Every Girl Should Have One
Everybody Wins
Evil Judgement
The Execution of Private
  Slovik
Execution of Raymond
  Graham
The Executioner

The Executioner, Part 2:
  Frozen Scream
The Executioner's Song
Exorcism
The Expendables
Experiment Perilous
Experiment in Terror
Explosion
Exquisite Corpses
Extreme Prejudice
An Eye for an Eye
Eye of the Tiger
Eye Witness
Eyeball
Eyewitness
The Face at the Window
Fake Out
The Falcon in Hollywood
The Falcon and the Snowman
The Family
Family Enforcer
A Family Matter (Vendetta)
Family Plot
Famous Five Get Into Trouble
The Fantasist
Fantastic Seven
Fast Getaway
Fast Money
Fast Talking
Fatal Beauty
Fatal Chase
Fatal Error
Fatal Exposure
The Fatal Hour
Fatal Skies
Fatal Vision
FBI Girl
The FBI Story
Fear
Fear in the City
Fear City
Fear in the Night
Female Trouble
Fever
Fiction Makers
Fiendish Plot of Dr. Fu
  Manchu
52 Pick-Up
Fight for Your Life
Fighting Back
The Fighting Rookie
Filthy Harry
Final Extra
Final Justice
Final Mission
Final Notice
Finders Keepers
A Fine Mess
The Finger Man
Fingerprints Don't Lie
Fingers
Fire
The Fire in the Stone
Firepower
The First Deadly Sin
First Name: Carmen
The First Power
A Fish Called Wanda
F.I.S.T.
Fists of Blood
Five Golden Dragons
Five Miles to Midnight
The Fix
Flame of the Islands
Flanagan Boy
Flash & Firecat
A Flash of Green
Flashpoint
Flesh and Blood
Flight to Fury
Flight to Nowhere
Flim-Flam Man
Flipper's New Adventure
The Florida Connection
Florida Straits
Flying Wild

Fog Island
For Your Love Only
Force of Evil
Force Five
Force of One
Formula for a Murder
Fort Apache, the Bronx
Fortune Dane
48 Hours to Live
48 Hrs.
Four Robbers
Four Ways Out
Foxy Brown
Framed
Frantic
Free to Love
Free Ride
A Free Soul
Freebie & the Bean
Freeway
The French Connection
French Connection 2
French Quarter Undercover
Frenchman's Farm
Frenzy
Fresh Kill
Friday Foster
Frontera Sin Ley
Fugitive Lovers
Fun with Dick and Jane
Future Force
The Galloping Ghost
Gambit
Game of Seduction
Gang Bullets
Gang Busters
Gangs, Inc.
Gangsters
Gator
The Gauntlet
Gemidos de Placer
Gentleman Bandit
Genuine Risk
Get Rita
Get That Man
The Getaway
Getting Even
The Gilded Cage
Ginger
The Girl
Girl in Black Stockings
Girl in Gold Boots
The Girl in Lover's Lane
The Girl Said No
The Girl from Tobacco Row
Girls Are for Loving
Girls in Chains
Girls in Prison
The Gladiator
The Glass Key
Glitch!
Glitz
The Glove
God Told Me To
The Godfather
Godfather 1902-1959—The
  Complete Epic
The Godfather: Part 2
Godson
Going in Style
The Golden Triangle
Gone in 60 Seconds
Gone to Ground
The Goonies
Gordon's War
Grace Quigley
Great Bank Hoax
Great Flamarion
Great Gold Swindle
The Great Lover
The Great Riviera Bank
  Robbery
The Great Texas Dynamite
  Chase
Great Train Robbery
The Greed of William Hart

The Green Hornet
Green Ice
The Grey Fox
The Grifters
The Grissom Gang
Ground Zero
The Guardian
Guilty as Charged
Gulliver
A Gun in the House
Gunblast
Gunpowder
The Gunrunner
Hallucination
Hands of a Stranger
Hangfire
Happy New Year
Hard Frame
The Hard Way
The Harder They Come
Harder They Fall
Harlem Nights
Harley Davidson and the
  Marlboro Man
The Haunted Strangler
Having Wonderful Crime
Hawkeye
He Walked by Night
Heart Condition
Heartbeat
Heat
Heat Street
The Heist
Held for Murder
Hell on Frisco Bay
Hell Harbor
Hell Up in Harlem
Hell on Wheels
Hellcats
Helter Skelter
Henry: Portrait of a Serial
  Killer
Her Alibi
Herbie Goes Bananas
Heroes in Blue
Hi-Jacked
Hide-Out
High Ballin'
High Command
High Crime
High & Low
High Risk
High Rolling in a Hot Corvette
The Hillside Strangler
Hired to Kill
Hit Lady
The Hit List
Hit Man (Il Sicario)
Hit Men
Hitz
Hollow Triumph
Hollywood Cop
Hollywood Crime Wave
Hollywood Vice Sqaud
Homer and Eddie
Homicide
Honeymoon
Honeymoon Killers
Hong Kong Nights
Honor Among Thieves
The Hooded Terror
The Hoodlum
Hook, Line and Sinker
The Hooked Generation
The Horror Show
The Horse Without a Head
The Hostage
The Hostage Tower
Hot Blood
Hot Child in the City
Hot Money
Hot Potato
Hot Pursuit
The Hot Rock
The Hot Spot
Hot Target

Hot Touch
House of Mystery
The House of Secrets
House of Terror
How to Beat the High Cost of
  Living
How Come Nobody's On Our
  Side?
How to Kill 400 Duponts
Hudson Hawk
Hue and Cry
Human Desire
The Human Factor
Hunt the Man Down
The Hunt for the Night Stalker
The Hunter
Hurricane Smith
Husbands, Wives, Money, and
  Murder
Hush Little Baby, Don't You
  Cry
Hustle
I Accuse My Parents
I Am the Law
I Can't Escape
I Died a Thousand Times
I Hate Blondes
I, the Jury
I Killed That Man
I Met a Murderer
I, Mobster
I Wake Up Screaming
I Want to Live!
I Wonder Who's Killing Her
  Now?
I'd Give My Life
I'll Get You
Illegal
Image of Death
Immortal Bachelor
Impact
Impulse
In Cold Blood
In Deadly Heat
In the Heat of the Night
In the King of Prussia
The Incident
The Inner Circle
An Innocent Man
Inside Information
Instant Justice
Internal Affairs
Interrupted Journey
Into Thin Air
The Invincible Six
The Invisible Man Returns
The Invisible Strangler
The Ipcress File
Iron Thunder
Island Monster
Island Trader
The Italian Connection
The Italian Job
It's Called Murder, Baby
Jack the Ripper
Jail Bait
Jailbird Rock
Jamaica Inn
Jane Doe
The January Man
Jesse James
Jessie's Girls
The Jesus Trip
Jigsaw
Johnnie Gibson F.B.I.
Johnny Angel
Johnny Barrows, El Malo
Johnny Come Lately
Johnny Dangerously
Johnny Handsome
Jornada de Muerte
Joyride
The Judge
Just William's Luck
Justice
Kamikaze '89

Kansas
Kansas City Confidential
Kansas City Massacre
Karate Warriors
Keaton's Cop
Keeper of the City
Keeping Track
Kemek
Kennel Murder Case
The Key Man
Kickboxer the Champion
Kid and the Killers
Kidnap Syndicate
Kidnapped
The Kill
Kill Alex Kill
Kill Me Again
The Kill Reflex
Killer
A Killer in Every Corner
Killer Force
Killer Inside Me
The Killer's Edge
Killer's Kiss
The Killing
Killing 'em Softly
Killing in the Sun
The Killing Time
The Killing Zone
King Creole
King of New York
King's Ransom
Kiss of Death
Kiss Tomorrow Goodbye
Kitty and the Bagman
Knights of the City
The Krays
Kuffs
L.A. Bounty
L.A. Crackdown
L.A. Crackdown 2
La Femme Nikita
La Tigeresa (The Tigress)
L'Addition
Ladies Club
Ladies of the Lotus
Lady in a Cage
Lady in Cement
The Lady Confesses
Lady in the Death House
Lady Killers
Lady from Nowhere
Lady in Red
Lady Scarface
Lady Scarface/The Threat
Lady Street Fighter
Lady Vanishes
Ladykillers
Laguna Heat
Las Vegas Serial Killer
Lassiter
The Last Detail
The Last of the Finest
Last House on the Left
Last Innocent Man
Last Mile
The Last of Mrs. Cheyney
Last Ride of the Dalton Gang
The Last Riders
Late Extra
Laughing Policeman
The Lavender Hill Mob
Law and Disorder
Law of the Underworld
Le Doulos
Le Jour Se Leve
League of Gentlemen
Left for Dead
The Leopard Woman
Lepke
Les Assassins de L'Ordre
Les Enfants Terrible
Les Grandes Gueules
Les Miserables
Let 'Em Have It
Let 'er Go Gallegher

Let Him Have It
Lethal Obsession
Lethal Weapon
Lethal Weapon 2
Let's Get Harry
The Letter
Letting the Birds Go Free
Liberators
License to Kill
The Life of Emile Zola
The Life & Times of the
  Chocolate Killer
Lights! Camera! Murder!
The Lightship
The Limping Man
The Lindbergh Kidnapping
  Case
The Lion's Share
Little Caesar
Little Murders
The Little Valentino
Little White Lies
Live a Little, Steal a Lot
Loaded Guns
Loan Shark
Lock Up
The Lone Rider in Ghost
  Town
Lone Star Law Men
The Long Good Friday
The Long Haul
Loophole
Loose Cannons
Loot...Give Me Money, Honey!
Los Cacos
Los Chantas
Los Drogadictos
Los Muchachos de Antes No
  Usaban Arsenico
The Love Flower
Love from a Stranger
Lucky Luciano
Lunch Wagon
M
Ma Barker's Killer Brood
Macao
The Mad Bomber
Mad Dog
Mad Dog Morgan
Mad Max
Mad Mission 3
Madame X
Made in the USA
Mademoiselle Striptease
Madigan's Millions
Magnificent Obsession
Magnum Force
Make Haste to Live
Malarek
Malcolm
Malone
Mama's Dirty Girls
Man on Fire
Man Inside
The Man Outside
Man in the Raincoat
Man on the Run
The Man They Could Not
  Hang
The Man Who Broke 1,000
  Chains
A Man, a Woman, & a Bank
The Manhandlers
Manhunt
Mania
Maniac
Mankillers
Manson
Marathon Man
Marbella
Marked for Death
Marked Money
Marked Woman
Martial Law
Master Touch
Maximum Security

Crime

Cult

Deep Blue

Demons & Wizards

Dental Mayhem

Detective Spoofs

Detectives

Detectives

Deathmask
The Detective
Detective Sadie & Son
Detective School Dropouts
Detective Story
The Devil's Cargo
Devlin Connection
Dick Barton, Special Agent
Dick Barton Strikes Back
Dick Tracy
Dick Tracy Detective
Dick Tracy, Detective/The Corpse Vanishes
Dick Tracy Double Feature #1
Dick Tracy Double Feature #2
Dick Tracy Meets Gruesome
Dick Tracy and the Oyster Caper
Dick Tracy Returns
Dick Tracy: The Original Serial, Vol. 1
Dick Tracy: The Original Serial, Vol. 2
Dick Tracy vs. Crime Inc.
Dick Tracy vs. Cueball
Dick Tracy's Dilemma
Dirkham Detective Agency
Dragnet
Dressed to Kill
The Drowning Pool
Edge of Darkness
El Coyote Emplumado
The Element of Crime
Emil & the Detectives
The Empty Beach
Endangered Species
The Enforcer
Everybody Wins
Evil Under the Sun
Ex-Mrs. Bradford
The Falcon in Hollywood
The Falcon in Mexico
Falcon Takes Over/Strange Bargain
Falcon's Adventure/Armored Car Robbery
The Falcon's Brother
Farewell, My Lovely
Fearless
Fiction Makers
Final Notice
Finessing the King
Follow Me Quietly
The Foreigner
Fourth Story
The French Detective
From Hollywood to Deadwood
Ginger
Girl Hunters
Girls Are for Loving
Gotham
Great K & A Train Robbery
The Great Mouse Detective
Green for Danger
Gumshoe
The Gumshoe Kid
Hammered: The Best of Sledge
Hands of a Stranger
Hard-Boiled Mahoney
Hat Box Mystery
Heart Condition
The Hollywood Detective
Hollywood Harry
Hollywood Stadium Mystery
The Hound of the Baskervilles
Hour of Decision
The House of the Arrow
House of Fear
The House of Lurking Death
House of Mystery
I, the Jury
I Killed That Man
In the Custody of Strangers
Infamous Crimes
The Inner Circle

Inspector Clouseau Cartoon Festival
Inspector Hornleigh
The Invisible Avenger
The Invisible Killer
Irish Luck
Jigsaw
Just Another Pretty Face
K-9
Kennel Murder Case
The Kill
Lady in Cement
Lady in the Lake
The Late Show
Laura
The Long Goodbye
Long Time Gone
Loose Cannons
The Love Flower
Love Happy
Love at Large
Lure of the Islands
Lying Lips
The Mad Bomber
Magnum Force
The Maltese Falcon
The Man with Bogart's Face
Man from Cairo
Man on the Eiffel Tower
The Man in the Mist
The Man Who Would Not Die
The Mandarin Mystery
Margin for Murder
Marlowe
Masks of Death
Meet Sexton Blake
Meeting at Midnight
Messenger of Death
Midnight Limited
Missing Pieces
Mister Arkadin
Moonlighting
More! Police Squad
Mr. Moto's Last Warning
Mr. Wong in Chinatown
Mr. Wong, Detective
Murder Ahoy
A Murder is Announced
Murder at the Baskervilles
Murder by the Book
Murder by Death
Murder by Decree
Murder in the Doll House
Murder at the Gallop
Murder Most Foul
Murder My Sweet
Murder on the Orient Express
Murder Over New York
Murder She Said
Murder at the Vanities
Murderous Vision
My Favorite Brunette
My New Partner
Mystery Island
Mystery Liner
Mystery of Mr. Wong
The Naked Gun: From the Files of Police Squad
Nature's Playmates
New Orleans After Dark
Newman's Law
Nick Knight
Night Moves
Night School
Night Stalker
No Way to Treat a Lady
Nocturne
Number Seventeen
One Body Too Many
One Good Cop
One of My Wives Is Missing
One Shoe Makes It Murder
Operation Julie
The Organization
Out of the Darkness
Out of the Past

The Panther's Claw
The Pearl of Death
Peter Gunn
Phantom of Chinatown
The Phantom of Soho
Pier 23
The Pink Chiquitas
The Pink Panther
The Pink Panther Strikes Again
Police Squad! Help Wanted!
The Presidio
Private Eyes
The Private Life of Sherlock Holmes
Pursuit to Algiers
Radioactive Dreams
The Red Raven Kiss-Off
Rentadick
The Return of Boston Blackie
The Return of Dr. Mabuse
Return of Frank Cannon
Return of the Pink Panther
Return of Superfly
Revenge of the Pink Panther
The Saint
Saint in London
The Saint in New York
The Saint Strikes Back
The Saint Takes Over
The Saint's Double Trouble
The Saint's Vacation/The Narrow Margin
Sapphire
Satan's Harvest
Scalp Merchant
Scarlet Claw
Sea of Love
Seven Doors to Death
The Seven Percent Solution
The Seven-Ups
Sexton Blake and the Hooded Terror
The Shadow Strikes
Shadow of the Thin Man
Shamus
Sherlock Holmes
Sherlock Holmes Faces Death
Sherlock Holmes and the Incident at Victoria Falls
Sherlock Holmes and the Secret Weapon
Sherlock Holmes: The Sign of Four
Sherlock Holmes: The Silver Blaze
Sherlock Holmes: The Voice of Terror
Sherlock Holmes in Washington
Sherlock's Rivals & Where's My Wife
Shoot to Kill
Shooting Stars
A Shot in the Dark
The Silent Mr. Sherlock Holmes
The Silent Passenger
Sir Arthur Conan Doyle
Slow Burn
Smart-Aleck Kill
Song of the Thin Man
The Speckled Band
Speedtrap
Spiderwoman
The Squeeze
Stakeout
Star of Midnight
Station
Stolen Kisses
Strange Shadows in an Empty Room
Stray Dog
A Study in Scarlet
A Study in Terror
Sweet Lies

Terror by Night
Terror by Night/Meeting at Midnight
They All Laughed
They Might Be Giants
The Thin Man
The Thin Man Goes Home
Third Degree Burn
The Third Key
13th Guest
Tony Rome
Trail of the Pink Panther
TripleCross
Triumph of Sherlock Holmes
The Unbreakable Alibi
Under Suspicion
Uneasy Terms
Unsuitable Job for a Woman
Vegas
Vertigo
V.I. Warshawski
Video Murders
Violated
Vortex
Waikiki
Web of Deception
Who Done It?
Who Framed Roger Rabbit?
Who Is the Black Dahlia?
Who's Harry Crumb?
Without a Clue
Witness
X Marks the Spot
Young Sherlock Holmes

Dinosaurs

Adventures in Dinosaur City
At the Earth's Core
Attack of the Super Monsters
Baby...Secret of the Lost Legend
Caveman
The Crater Lake Monster
Dino & Juliet
Dinosaurus!
EPIC: Days of the Dinosaurs
Ganjasaurus Rex
King Dinosaur
The Land Before Time
Land That Time Forgot
Legend of the Dinosaurs
The Lost Continent
The Lost World
Massacre in Dinosaur Valley
My Science Project
One Million B. C.
One of Our Dinosaurs Is Missing
The People That Time Forgot
Planet of the Dinosaurs
Return of the Dinosaurs
Sound of Horror
Teenage Caveman
Two Lost Worlds
Unknown Island
The Valley of Gwangi
When Dinosaurs Ruled the Earth

Disaster Flicks

Accident
After the Shock
Airport
Airport '75
Airport '77
Assignment Outer Space
Avalanche
The Cassandra Crossing
Catastrophe
Chain Reaction
The China Syndrome
City on Fire
The Concorde: Airport '79

The Crash of Flight 401
The Day the Sky Exploded
Die Hard
Die Hard 2: Die Harder
The Doomsday Flight
Earthquake
Eruption: St. Helens Explodes
Final Warning
Fire
Fire Alarm
Fire and Rain
Flood!
Goliath Awaits
Gray Lady Down
The Great Los Angeles Earthquake
Hurricane
Jet Over the Atlantic
Kameradschaft
Last Days of Pompeii
The Last Woman on Earth
The Lost Missile
Meteor
A Night to Remember
Planet on the Prowl
The Poseidon Adventure
St. Helen's, Killer Volcano
Snowblind
S.O.S. Titanic
Tailspin: Behind the Korean Airline Tragedy
Terror on the 40th Floor
Tidal Wave
Two Minute Warning
When Time Ran Out

Disease of the Week

The Affair
The Ann Jillian Story
Bang the Drum Slowly
Beaches
The Best Little Girl in the World
Between Two Women
Bobby Deerfield
Boy in the Plastic Bubble
Brian's Song
Camille
The Carrier
The Cassandra Crossing
Champions
Checking Out
C.H.U.D.
Cleo from 5 to 7
The Clinic
Coma
Cries and Whispers
Crystal Heart
The Curse
Damien: The Leper Priest
Dark River: A Father's Revenge
Dead Space
Dick Barton, Special Agent
The Doctor
A Dream of Kings
Duet for One
An Early Frost
The End
Eric
The First Deadly Sin
Fourth Wish
Gaby: A True Story
Germicide
Griffin and Phoenix: A Love Story
Ice Castles
Incredible Melting Man
Intimate Contact
Isle of the Dead
Killer on Board
L.A. Bad
The Last Man on Earth
Lightning Over Water

Documentary

Ciao Federico! Fellini Directs
   Satyricon
Cigarette Blues
Cinema Shrapnel
Citizen Welles
City of Gold/Drylanders
The Civil War: Episode 1: The
   Cause: 1861
The Civil War: Episode 2: A
   Very Bloody Affair: 1862
The Civil War: Episode 3:
   Forever Free: 1862
The Civil War: Episode 4:
   Simply Murder: 1863
The Civil War: Episode 5: The
   Universe of Battle: 1863
The Civil War: Episode 6:
   Valley of the Shadow of
   Death: 1864
The Civil War: Episode 7:
   Most Hallowed Ground:
   1864
The Civil War: Episode 8: War
   is All Hell: 1865
The Civil War: Episode 9: The
   Better Angels of Our
   Nature: 1865
The Clowns
Combat Bulletin
Commission
Common Threads: Stories
   from the Quilt
Cordell Hull: The Good
   Neighbor
Cult People
Cyberpunk
Dakota
Dance Hall Racket
Dario Argento's World of
   Horror
Dear America: Letters Home
   from Vietnam
The Decathalon
December 7th: The Movie
Declassified: The Plot to Kill
   President Kennedy
Decline of Western Civilization
   1
The Decline of Western
   Civilization 2: The Metal
   Years
Derby
Devil's Triangle
Divorce Hearing
D.O.A.: A Right of Passage
Dolphin Adventure
Don't Look Back
The Doors: Soft Parade, a
   Retrospective
Dorothy Stratten: The Untold
   Story
Dracula: A Cinematic
   Scrapbook
Dracula: The Great Undead
A Dream Called Walt Disney
   World
The Dream Continues
Ebony Dreams
Eddie Rickenbacker: Ace of
   Aces
Eisenhower: A Place in
   History
Eisenstein
Electric Boogie
Elsa & Her Cubs
The Elvis Files
Elvis Memories
Elvis in the Movies
Elvis Presley's Graceland
Elvis: Rare Moments with the
   King
Elvis: That's the Way It Is
Elvis: The Echo Will Never Die
Emperor Jones/Paul
   Robeson: Tribute to an
   Artist

Empire of the Air: The Men
   Who Made Radio
Encounter with Disaster
Encounter with the Unknown
The Endless Summer
Enemy of Women
Entertaining the Troops
Enthusiasm
Epic That Never Was
Ernie Banks: History of the
   Black Baseball Player
Eros, Love & Lies
Eruption: St. Helens Explodes
Faces of Death, Part 1
Faces of Death, Part 2
The Fall of the Berlin Wall
Falling for the Stars
Fangoria's Weekend of
   Horrors
The Fantasy Film Worlds of
   George Pal
Father Hubbard: The Glacier
   Priest
Film Firsts
Filmmaker: A Diary by George
   Lucas
Filmmakers: King Vidor
Finest Hours
Fiorello La Guardia: The
   Crusader
First Twenty Years: Part 1
   (Porter pre-1903)
First Twenty Years: Part 2
   (Porter 1903-1904)
First Twenty Years: Part 3
   (Porter 1904-1905)
First Twenty Years: Part 4
First Twenty Years: Part 5
   (Comedy 1903-1904)
First Twenty Years: Part 6
   (Comedy 1904-1907)
First Twenty Years: Part 7
   (Camera Effects)
First Twenty Years: Part 8
   (Camera Effects 1903-
   1904)
First Twenty Years: Part 9
   (Cameramen)
First Twenty Years: Part 10
   (Special Effects)
First Twenty Years: Part 11
   (Drama on Location)
First Twenty Years: Part 12
   (Direction & Scripts)
First Twenty Years: Part 13
   (D.W.Griffith)
First Twenty Years: Part 14
   (Griffith's Dramas)
First Twenty Years: Part 15
   (Griffith's Editing)
First Twenty Years: Part 16
   (Later Griffith)
First Twenty Years: Part 17
   (Make-up Effects)
First Twenty Years: Part 18
   (2-reelers)
First Twenty Years: Part 19 (A
   Temporary Truce)
First Twenty Years: Part 20
   (Lubin films)
First Twenty Years: Part 21 (3
   Independent Producers)
First Twenty Years: Part 22
   (Melies)
First Twenty Years: Part 23
   (British Comedies)
First Twenty Years: Part 26
   (Danish Superiority)
Flights & Flyers
Flights & Flyers: Amelia
   Earhart
Florence Chadwick: The
   Challenge
For All Mankind
Forever James Dean

France/Tour/Detour/Deux/
   Enfants
Francisco Oller
Frankenstein: A Cinematic
   Scrapbook
Franklin D. Roosevelt: F.D.R.
Frida
Fun with the Fab Four
Gap-Toothed Women
Gar Wood: The Silver Fox
Gates of Heaven
George Melies, Cinema
   Magician
George Stevens: A
   Filmmaker's Journey
Get 'Em Off
Ghosts of the Sky
The G.I. Road to Hell
Gimme Shelter
Girl Groups: The Story of a
   Sound
Gizmo!
The Glory of Their Times
Going Hollywood
Golden Age of Hollywood
Good Old Days of Radio
Gorilla Farming
Gotta Dance, Gotta Sing
Grant at His Best
The Grateful Dead Movie
Great American Cowboy
Great Battle of the Volga
Great White Death
Greed & Wildlife: Poaching in
   America
Grey Gardens
Guilty of Treason
Happy Anniversary 007: 25
   Years of James Bond
Harlan County, U.S.A.
Hearts of Darkness: A
   Filmmaker's Apocalypse
Heaven
Heavy Petting
Helen Hayes
Hell's Angels Forever
Hellstrom Chronicle
Henry Fonda: The Man and
   His Movies
Hepburn & Tracy
Hide and Seek
His Memory Lives On
Hollywood Clowns
Hollywood Home Movies
Hollywood My Hometown
Hollywood Scandals and
   Tragedies
Hollywood Uncensored
Hooker
The Horror of It All
The Illustrated Hitchcock
In Heaven There is No Beer?
Indomitable Teddy Roosevelt
Ingrid
Inside Hitchcock
Inside the Labyrinth
Iron Bodies
Irving Berlin's America
It Happened in the Park
James Cagney: That Yankee
   Doodle Dandy
James Dean
The James Dean 35th
   Anniversary Collection
James Dean Story
Jazz is Our Religion
The JFK Conspiracy
JFK Remembered
Jimi Hendrix: Live in
   Monterey, 1967
Jimi Hendrix: Story
John F. Kennedy
John Huston: The Man, the
   Movies, the Maverick
John Jacob Niles
John Lennon: Imagine

John Lennon/Yoko Ono: Then
   and Now
John Wayne: American Hero
   of the Movies
John Wayne: Bigger Than Life
Jungle Cat
Jungle Cavalcade
Kidco
Kino Pravda/Enthusiasm
Knute Rockne: The Rock of
   Notre Dame
The Lash of the Penitents
Last of the One Night Stands
The Last Waltz
The Late Great Planet Earth
LBJ: A Biography
Le Joli Mai
The Legend of Valentino
Let It Be
Let There Be Light
Let's Get Lost
Life in Camelot: The Kennedy
   Years
Lightning Over Water
Living Desert
Living Free
Lonely Boy/Satan's Choice
Long Shadows
Lou Gehrig: King of Diamonds
Louie Bluie
Louisiana Story
The Love Goddesses
Love Without Fear
Making of "Gandhi"
Making of a Legend: "Gone
   With the Wind"
Man of Aran
The Man with a Movie Camera
Man in the Silk Hat
The Man Who Saw Tomorrow
The Man Who Skied Down
   Everest
Manson
Many Faces of Sherlock
   Holmes
Maps to Stars' Homes Video
Marc and Ann
Marilyn
Mario Lanza: The American
   Caruso
Marjoe
Marlon Brando
Marvin Gaye
Max Fleischer's Documentary
   Features
Mein Kampf
The Mercenary Game
MGM: When the Lion Roars
Moana, a Romance of the
   Golden Age
Mondo Africana
Mondo Cane
Mondo Cane 2
Mondo Elvis (Rock 'n' Roll
   Disciples)
Mondo Lugosi: A Vampire's
   Scrapbook
Mondo Magic
Monsters, Madmen, Machines
Monterey Pop
Montgomery Clift
The Most Death-Defying
   Circus Acts of All Time
Movie Museum Series
Mysteries From Beyond Earth
The Naked Gershwin
Nashville Story
New Orleans: 'Til the Butcher
   Cuts Him Down
The 1930s: Music, Memories
   & Milestones
No, I Have No Regrets
Nostradamus
The Occult Experience
October
Of the Dead

On Any Sunday
On the Bowery
One Last Time: The Beatles'
   Final Concert
Ornette Coleman Trio
Orphans of the North
Oscar's Greatest Moments:
   1971-1991
The Other Side of Nashville
Paper Lion
Paris is Burning
Piano Players Rarely Ever
   Play Together
Pioneers of the Cinema, Vol. 2
Placido: A Year in the Life of
   Placido Domingo
Playboy Video Centerfold:
   35th Anniversary Playmate
Playboy's 1989 Playmate
   Video Calendar
Playboy's 1990 Playmate
   Video Calendar
Plow That Broke the Plains/
   River
Pot Shots
Private Conversations: On the
   Set of "Death of a
   Salesman"
Private Practices: The Story of
   a Sex Surrogate
Punk Rock Movie
A Quiet Hope
Rascal Dazzle
Rate It X
The Real Buddy Holly Story
Red Grooms
Remembering LIFE
Reminiscences of a Journey
   to Lithuania
Reporters
Revolution
Rhino's Guide to Safe Sex
Richard Lester
Rings Around the World
Rise and Fall of the Third
   Reich
The Roaring Twenties
Robert A. Taft: Mr.
   Republican
Rock 'n' Roll Heaven
Rodgers & Hammerstein: The
   Sound of American Music
The Rodney King Case: What
   the Jury Saw in California
   vs. Powell
Roger Corman: Hollywood's
   Wild Angel
Roger & Me
Ronald Reagan & Richard
   Nixon on Camera
Ronnie Dearest: The Lost
   Episodes
The Rose Parade: Through
   the Years
Salt of the Earth
Sarah Bailey
Say Amen, Somebody
Scenario du Film Passion
Scuba
The Sea Around Us
The Secret Life of Sergei
   Eisenstein
The Secret World of Reptiles
Secrets of Life
See It Now
The Sensual Taboo
September 1939
Sex and Buttered Popcorn
Sharks, Past & Present
Shocking Asia
Shocking Asia 2
Shout!: The Story of Johnny
   O'Keefe
Showbiz Goes to War
Signals Through the Flames
The Situation

Drama

Berlin Alexanderplatz
(episodes 1-2)
Berlin Alexanderplatz
(episodes 3-4)
Berlin Alexanderplatz
(episodes 5-6)
Berlin Alexanderplatz
(episodes 7-8)
The Berlin Conspiracy
Bernadette
The Best Christmas Pageant
Ever
Best Enemies
Best Kept Secrets
The Best Little Girl in the
World
The Best Man
The Best Years of Our Lives
Bethune
Betrayal
Betrayed
The Betsy
Between Friends
Between Men
Between Two Women
Between Wars
Beulah Land
Beverly Hills Call Girls
Beverly Hills Madam
Beware, My Lovely
Beyond the Doors
Beyond Erotica
Beyond Fear
Beyond the Forest
Beyond Innocence
Beyond the Limit
Beyond the Next Mountain
Beyond Reason
Beyond Reasonable Doubt
Beyond the Time Barrier
Beyond the Walls
Bible...In the Beginning
The Bicycle Thief
Big Bird Cage
The Big Blue
The Big Chill
The Big Man: Crossing the
Line
Big Mo
Big News
Big Red
Big Score
Big Town
Big Town After Dark
Big Trees
Bilitis
Bill
A Bill of Divorcement
Bill: On His Own
Billy Bathgate
Billy Galvin
Billy Jack
Billy: Portrait of a Street Kid
Billyboy
The Bionic Woman
The Birch Interval
Bird
Birdman of Alcatraz
Birds of Prey
Birdy
The Bitch
Bits and Pieces
Bitter Harvest
Bizarre
The Black Dragons
Black Emmanuelle
Black Fury
Black God & White Devil
Black Godfather
The Black Hand
Black Hand
The Black Klansman
Black Like Me
Black Lizard
Black Narcissus
Black Orpheus

Black Rain
Black Rainbow
Black Robe
Black Sister's Revenge
The Black Six
Black Starlet
Black Tower
A Black Veil for Lisa
Black Widow
Blackboard Jungle
Blacklash
Blackout
Blades of Courage
Blaise Pascal
Blake of Scotland Yard
The Blasphemer
Blastfighter
Blaze
Bless the Beasts and Children
Blind Ambition
Blind Fools
Blind Vengeance
Blinded by the Light
Blockhouse
Blood Brothers
Blood on the Mountain
Blood Red
Blood at Sundown
Blood Vows: The Story of a
Mafia Wife
Blood Wedding (Bodas de
Sangre)
Bloodbath
Bloodbrothers
Bloody Mama
The Blot
Blow-Up
Blown Away
The Blue Angel
Blue Canadian Rockies
Blue Collar
Blue Desert
Blue Fire Lady
Blue Heaven
Blue Jeans
The Blue Knight
The Blue Lagoon
The Blue Lamp
The Blue Max
The Blue Yonder
Bluebeard
Blueberry Hill
Bluffing It
The Blum Affair
Boat Is Full
Bob le Flambeur
Bobby Deerfield
Body Heat
Body and Soul
Body & Soul
Bogie
Bolero
Bombay Talkie
Bonjour Tristesse
Bonnie & Clyde
Bonnie Prince Charlie
Bonnie's Kids
The Boost
Boots Malone
The Border
Border Street
Born Beautiful
Born in Flames
Born on the Fourth of July
Born Free
Born to Kill
Born Losers
Born to Run
Boss of Big Town
Boss' Son
The Bostonians
Boulevard Nights
Bound for Glory
Bouquet of Barbed Wire
Boxcar Bertha
The Boxer and the Death

Boy in Blue
Boy with the Green Hair
Boy in the Plastic Bubble
Boy Who Could Fly
Boys in the Band
The Boys in Company C
The Boys Next Door
Boy's Reformatory
Boys Town
Boyz N the Hood
Brainwash
Brand New Life
Brass
The Brave One
Bread (BBC)
Break of Hearts
Breakdown
Breaking Home Ties
Breaking Up
Breakthrough
Breathless
Brian's Song
Bride & the Beast
The Bride Is Much Too
Beautiful
Brideshead Revisited
The Bridge on the River Kwai
Bridge of San Luis Rey
Bridge to Silence
The Brig
Bright Angel
Bright Lights, Big City
Brimstone & Treacle
Brink of Life
A Brivele der Mamen (A Letter
to Mother)
The Broadway Drifter
Broken Badge
Broken Blossoms
Broken Lance
The Broken Mask
Broken Melody
Broken Strings
Bronson's Revenge
The Bronx War
Brother Future
Brother John
Brother Orchid
Brother Sun, Sister Moon
The Brotherhood
Brotherhood of Justice
The Brothers Karamazov
Browning Version
Brubaker
Brussels Transit
Brutal Glory
Buck and the Preacher
Bucktown
Bugsy
Bulldog Edition
Bulldog Jack
Bullets or Ballots
Bullies
Bullitt
Burgess, Philby, and
MacLean: Spy Scandal of
the Century
Burglar
Buried Alive
Burn!
Burning Bed
Burning Rage
Burning Vengeance
Bury Me an Angel
The Bus Is Coming
Business as Usual
Busted Up
Buster
Busting
But Where Is Daniel Wax?
Butterfield 8
Butterflies
Butterfly
Butterfly Affair
Bye Bye Brazil
Caches de Oro

The Cage
Caged Heart
Caged in Paradiso
Caged Women
The Caine Mutiny
The Caine Mutiny Court
Martial
California Dreaming
Caligula
Call of the Forest
Call to Glory
Call Me
Callie and Son
Camp Double Feature
Can You Hear the Laughter?
The Story of Freddie Prinze
Cancion en el Alma (Song
from the Heart)
Candy Tangerine Man
Cape Fear
Captain Blackjack
Captain Horatio Hornblower
Captains Courageous
Captive
Captive Heart
The Capture of Grizzly Adams
Cardinal
Career
Careful, He Might Hear You
Carnal Crimes
Carnival in Flanders (La
Kermesse Heroique)
Carny
Caroline?
Carrington, V.C.
Carry Me Back
A Case of Deadly Force
The Case of the Frightened
Lady
A Case of Libel
Casimir the Great
Cass
Cassie
Cast the First Stone
Castle
Cat on a Hot Tin Roof
The Catered Affair
Catholic Hour
Catholics
Cease Fire
Celebrity
Celeste
Celia: Child Of Terror
Certain Fury
Certain Sacrifice
Chain Gang Women
Chain Reaction
Chained for Life
The Chalk Garden
Challenge
Challenge to Lassie
Chameleon
The Champ
Champagne
Champagne for Breakfast
Champion
Champions
Change of Habit
Changes
Chapayev
Chapter in Her Life
Chariots of Fire
Charlotte Forten's Mission:
Experiment in Freedom
The Chase
Chase
Chasing Dreams
Chaste & Pure
Chattahoochee
Cheaters
Cheers for Miss Bishop
Chicago Joe & the Showgirl
Chiefs
Child Bride of Short Creek
A Child is Waiting
The Children

Children of An Lac
The Children Are Watching Us
Children in the Crossfire
Children of Divorce
Children of Sanchez
The Children of Times Square
Children's Carol
The Children's Hour
China Girl
The China Lake Murders
China Sky
The China Syndrome
Chinese Roulette
Chocolat
The Chocolate War
The Choice
Choices
The Choppers
Christ Stopped at Eboli
Christabel
Christiane F.
A Christmas Carol
Christmas Coal Mine Miracle
Christmas Comes to Willow
Creek
Christmas Lilies of the Field
A Christmas to Remember
The Christmas Tree
Christmas Without Snow
Christopher Columbus
Cincinnati Kid
Cinema Paradiso
Citadel
Citizen Kane
The City
City for Conquest
The City and the Dogs
City in Fear
City of Hope
The City's Edge
Civilization
Clarence Darrow
Class of '63
Class of 1984
Class Action
The Class of Miss MacMichael
Claudia
Clean and Sober
Clearcut
Cleopatra
A Climate for Killing
The Clockmaker
Clodhopper
The Closer
Closet Land
Cloud Dancer
Clowning Around
The Club
Club Life
Coach
Cocaine Cowboys
Cocaine Fiends
Cockfighter
Code of Honor
Code Name Alpha
Cody
Coffy
Cold Front
Cold Justice
Cold River
Colonel Wolodyjowski
Color of Money
The Color Purple
Colors
Come Along with Me
Come Back to the 5 & Dime
Jimmy Dean, Jimmy Dean
Come Back, Africa
Come Back Little Reba
Come Back, Little Sheba
Come and Get It
Come and See
Comeback
The Comfort of Strangers
The Comic
Coming Home

Drama

Emperor Jones/Paul
  Robeson: Tribute to an
  Artist
The Empire of Passion
The Empty Beach
An Empty Bed
Encore
End of Desire
The End of Innocence
End of the Road
The Endless Game
Enemy Mine
Enemy Territory
An Englishman Abroad
Enigma Secret
Enola Gay: The Men, the
  Mission, the Atomic Bomb
The Entertainer
Entre-Nous (Between Us)
Equalizer 2000
Equinox Flower
Equus
Era Notte a Roma
Erendira
Eric
Ernesto
Erotic Images
Erotica: Fabulous Female
  Fantasies
Escape Artist
Escape from Hell
Escape from the Planet of the
  Apes
Escort Girls
Esposa y Amante
L'Etat Sauvage
The Eternal Waltz
Eternally Yours
Eternity
Eureka!
Europa '51
Europa, Europa
The Europeans
Even Angels Fall
Every Man for Himself & God
  Against All
Everybody's Fine
The Evil Mind
The Execution
The Execution of Private
  Slovik
Execution of Raymond
  Graham
The Executioner
The Executioner, Part 2:
  Frozen Scream
The Executioner's Song
Executive Action
Executive Suite
Exiled to Shanghai
Exodus
Extremities
Eye on the Sparrow
Eye Witness
F. Scott Fitzgerald in
  Hollywood
The Fabulous Dorseys
Face of Another
A Face in the Crowd
Face to Face
Face in the Mirror
Fail-Safe
Fall from Innocence
Fallen Angel
The Fallen Idol
The Fallen Sparrow
Falling from Grace
False Note
Fame is the Spur
The Family
Family Enforcer
Family Life
A Family Matter (Vendetta)
Family Secrets
Family Sins
Family Upside Down

Family Viewing
Fandango
Fanny and Alexander
Fanny Hill
Fanny Hill: Memoirs of a
  Woman of Pleasure
Fantasies
Fantasy in Blue
Far Away and Long Ago
Far Cry from Home
The Fast and the Furious
Fast Lane Fever
Fast Walking
Fat City
Fat Man and Little Boy
Fatal Confinement
Fatal Fix
Fatal Vision
Father
Father Figure
Faust
The FBI Story
Fear Strikes Out
Feelin' Up
Felicity
Fellow Traveler
Fever
Fever Mounts at El Pao
Fever Pitch
Fictitious Marriage
The Field
Field of Honor
Fifth Day of Peace
The Fig Tree
A Fight for Jenny
Fight for Your Life
The Fighter
Fighter Attack
Fighting Back
Fighting Father Dunne
Fighting Pilot
Final Comedown
Final Extra
Final Warning
Fine Gold
Finish Line
Fire Alarm
Firehouse
Fires on the Plain
First Born
The First Deadly Sin
First Olympics: Athens 1896
First Strike
Fitzcarraldo
Five Easy Pieces
Five Forty-Eight
The Five of Me
The $5.20 an Hour Dream
Flame of the Islands
Flame to the Phoenix
The Flaming Urge
A Flash of Green
Flatliners
Flesh & Blood
The Flesh and the Devil
Flight of the Grey Wolf
Flight from Vienna
The Flowers of St. Francis
Fly with the Hawk
Flying Blind
Flying Machine/Coup de
  Grace/Interlopers
The Flying Scotsman
A Fool There Was
Foolish Wives
Fools of Fortune
For the Boys
For Ladies Only
For Us, the Living
Forbidden Games
Forbidden Passion: The Oscar
  Wilde Movie
Force of Evil
Forced March
Ford Startime
Ford: The Man & the Machine

The Foreigner
The Foreman Went to France
Forest of Little Bear
Forever Emmanuelle
Forever Mary
Forever Young
Forgotten Prisoners
Fort Apache, the Bronx
Fortunes of War
The Forty-Ninth Parallel
Forty-Seven Ronin, Part 1
Forty-Seven Ronin, Part 2
Forty Thousand Horsemen
Foul Play
The Fountainhead
The Four Horsemen of the
  Apocalypse
Four Horsemen of the
  Apocalypse
Four Infernos to Cross
Four Ways Out
The 400 Blows
The Fourth War
Fourth Wise Man
Fourth Wish
Fox and His Friends
Foxes
Foxstyle
Foxtrot
Fran
Frances
Francis Gary Powers: The
  True Story of the U-2 Spy
Franck Goldberg Videotape
Free Amerika Broadcasting
Free to Love
A Free Soul
Free, White, and 21
Freedom
Freefall
Freeze-Die-Come to Life
French Can-Can
The French Woman
Frida
Frieda
Friendly Fire
Friendly Persuasion
The Fringe Dwellers
From Hell to Victory
From Here to Eternity
From the Life of the
  Marionettes
From the Terrace
Frontera (The Border)
The Fugitive
Full Hearts & Empty Pockets
Full Metal Jacket
Fun Down There
Fury to Freedom: The Life
  Story of Raul Ries
Fury is a Woman
Futz
Fyre
''G'' Men
Gaby: A True Story
The Gallant Hours
The Gambler
The Gambler & the Lady
The Game is Over
Gandhi
Gang Bullets
Gang Busters
The Gang's All Here
The Garden of Eden
Garden of the Finzi-Continis
Gardens of Stone
A Gathering of Eagles
The Gathering: Part 2
Gathering Storm
Gemini Affair
The Gene Krupa Story
Generale Della Rovere
Gentle Ben
Gentleman Bandit
Gentleman Jim
Genuine Risk

Germicide
Gertrud
Gerty, Gerty, Gerty Stein Is
  Back, Back, Back
Gervaise
Get Rita
Getting Physical
The Getting of Wisdom
Ghetto Revenge
The Ghost of H.L. Mencken
Ghostwriter
Giant
Gilda
The Gilded Cage
The Girl
Girl on a Chain Gang
Girl in Gold Boots
A Girl on Her Own
Girl from Hunan
A Girl of the Limberlost
The Girl on a Motorcycle
The Girl Said No
The Girl from Tobacco Row
Girl Who Spelled Freedom
Girls in Chains
Girls of Huntington House
Girls in Prison
Girls of the White Orchid
Git!
Give 'Em Hell, Harry!
Gladiator
Glass
Glass House
The Glass Menagerie
Glass Tomb
Gleaming the Cube
Glen or Glenda
Glitter Dome
Glitz
Glory
Go for Broke!
Go Down Death
Go for the Gold
The Go-Masters
The Goalie's Anxiety at the
  Penalty Kick
God, Man and Devil
Godard/Truffaut Shorts
The Goddess
The Godfather
Godfather 1902-1959—The
  Complete Epic
The Godfather: Part 2
The Godfather: Part 3
God's Little Acre
Gods of the Plague
Godson
Goin' All the Way
Going Back
Going for the Gold: The Bill
  Johnson Story
Going Home
The Gold & Glory
Golden Boy
The Golden Coach
The Golden Seal
Goldengirl
Goldenrod
Goldy 2: The Saga of the
  Golden Bear
Goldy: The Last of the Golden
  Bears
Goliath Awaits
The Good Earth
The Good Father
Good Morning, Babylon
The Good Mother
The Good Wife
Goodbye Emmanuelle
Goodbye Love
Goodbye, Norma Jean
Goodfellas
Goodnight, Sweet Marilyn
Gorilla
Gorillas in the Mist

Gospel According to St.
  Matthew
Grain of Sand
Grambling's White Tiger
Grand Canyon
The Grand Highway
Grand Hotel
Grand Prix
Grandma Didn't Wave Back
Grant at His Best
The Grapes of Wrath
Grazuole
The Great Adventure
The Great Armored Car
  Swindle
Great Dan Patch
Great Expectations
Great Flamarion
The Great Gabbo
The Great Gatsby
Great Guy
Great Kate: The Films of
  Katharine Hepburn
Great Leaders
Great Love Experiment
The Great Man Votes
Great Sadness of Zohara
The Great Santini
The Great Train Robbery:
  Cinema Begins
Great Wallendas
The Great White Hope
The Greatest
The Greatest Man in the
  World
Greatest Show on Earth
Greed
The Greek Tycoon
Green Eyes
The Green Glove
Green Horizon
The Green Promise
Green Room
The Green Wall
Greta
The Grey Fox
Greyfriars Bobby
The Grifters
The Grim Reaper (La
  Commare Secca)
The Grissom Gang
Grit of the Girl Telegrapher/In
  the Switch Tower
Grizzly Adams: The Legend
  Continues
The Group
Grown Ups
Guadalcanal Diary
The Guardian
Guilty of Innocence
Guilty by Suspicion
Guilty of Treason
Guitarras Lloren Guitarras
Gulag
Gun Crazy
A Gun in the House
The Gunrunner
Guns at Batasi
The Guns and the Fury
Guns of War
The Guyana Tragedy: The
  Story of Jim Jones
Gypsy Blood
The Gypsy Warriors
Hadley's Rebellion
Hail, Hero!
Hail Mary
The Hairy Ape
Half of Heaven
Half a Lifetime
Half Moon Street
Half Slave, Half Free
Hallmark Theater (Sometimes
  She's Sunday)
Halls of Montezuma
Hamburger Hill

Hamlet
Hamsin
A Handful of Dust
The Handmaid's Tale
Hands Up
Hansel & Gretel
Happily Ever After
Happiness Is...
The Happy Ending
Hard Choices
Hard Country
Hard Frame
Hard Knox
Hard Part Begins
Hard Times
Hard Traveling
Hard Way
Hardcore
Harder They Fall
Harrad Experiment
Harrad Summer
Harry & Son
Harry and Tonto
Harvest
The Hasty Heart
Hatfields & the McCoys
The Haunting Passion
Haunts of the Very Rich
Hawaii
Hawks & the Sparrows
  (Uccellacci e Uccellini)
Hazel's People
He is My Brother
He Who Walks Alone
Head of the Family
Healer
The Healing
Heart
Heart Beat
Heart of a Champion: The Ray
  Mancini Story
Heart of Dixie
Heart Like a Wheel
The Heart is a Lonely Hunter
Heart of Midnight
Heart of a Nation
Heart of the Stag
Heartbreak House
Hearts of Fire
Hearts of the World
Heat
Heat of Desire (Plein Sud)
Heat and Sunlight
Heat Wave
Heatwave
Heaven on Earth
Heaven & Earth
Heaven is a Playground
Heavenly Bodies
Heaven's Heroes
Hedda
Heidi
Held Hostage
Held for Murder
Hell on Frisco Bay
Hellcats of the Navy
Hellfighters
Hell's Angels on Wheels
Hell's House
Helltrain
Helter Skelter
The Henderson Monster
Hennessy
Henry IV
Henry & June
Henry: Portrait of a Serial
  Killer
Her Summer Vacation
Here Comes Kelly
Hero
Hero Ain't Nothin' But a
  Sandwich
Herod the Great
Heroes in Blue
The Heroes of Desert Storm

Heroes Die Young
Hester Street
Hidden Enemy
The Hiding Place
High Command
High Country
High Noon
High School Caesar
High School Confidential
High Sierra
High Tide
The Highest Honor
Hill 24 Doesn't Answer
Hill Number One
Hiroshima Maiden
Hiroshima, Mon Amour
His Kind of Woman
The Hit
Hit & Run
Hitchcock Collection
Hitchhikers
Hitler: The Last Ten Days
Hitler's Children
Hitler's Daughter
Hitler's Henchmen
Hobo's Christmas
Hockey Night
Hollow Triumph
Hollywood Confidential
Hollywood Heartbreak
Hollywood Man
Hollywood Mystery
Holocaust
The Holy Innocents
Holy Terror
Home of the Brave
Home Feeling
Home From the Hill
A Home of Our Own
Home Safe
Home to Stay
Home Town Story
Homeboy
Homeboys
Homecoming
Homecoming: A Christmas
  Story
Homeward Bound
Homework
Homicide
L'Homme Blesse
Honey
Honeyboy
Honeymoon Killers
Honkytonk Man
Hoosiers
Hope and Glory
The Horse
The Horse of Pride
The Horse Soldiers
The Horse Thief
The Hostage
Hot Money
The Hot Spot
Hot Target
Hotel
Hotel Du Lac
Houdini
Hour of the Star
Hour of Stars
Hour of the Wolf
The House of 1,000 Dolls
House Across the Bay
The House on Chelouche
  Street
The House on Garibaldi Street
House of Strangers
Housewife
How Green was My Valley
How Many Miles to Babylon?
The Hucksters
Hud
Hue and Cry
The Human Comedy
The Human Condition: A
  Soldier's Prayer

The Human Condition: No
  Greater Love
The Human Condition: Road
  to Eternity
The Human Factor
The Hunchback of Notre
  Dame
Hunger
The Hunt for the Night Stalker
Hunted!
The Hunters
The Hunting
The Hurried Man
Husband Hunting
Husbands and Lovers
Hush Little Baby, Don't You
  Cry
The Hustler
Hustling
Hypothesis of the Stolen
  Painting
I Accuse My Parents
I Am a Camera
I Am Curious (Yellow)
I Am a Fugitive from a Chain
  Gang
I Am the Law
I Am the Law & the Hunter
I Beheld His Glory
I, Claudius, Vol. 1: A Touch of
  Murder/Family Affairs
I, Claudius, Vol. 2: Waiting in
  the Wings/What Shall We
  Do With Claudius?
I, Claudius, Vol. 3: Poison is
  Queen/Some Justice
I, Claudius, Vol. 4: Queen of
  Heaven/Reign of Terror
I, Claudius, Vol. 5: Zeus, By
  Jove!/Hail Who?
I, Claudius, Vol. 6: Fool's
  Luck/A God in Colchester
I, Claudius, Vol. 7: Old King
  Log
I Cover the Waterfront
I Don't Give a Damn
I Heard the Owl Call My Name
I Killed Rasputin
I Know Why the Caged Bird
  Sings
I Love You...Goodbye
I Met a Murderer
I Posed for Playboy
I Remember Mama
I Wake Up Screaming
I Want to Live!
I Want What I Want
Ice Castles
Ice House
Ice Palace
Iceman
Icy Breasts
Identity Unknown
The Idiot
If...
If I Perish
If Looks Could Kill
If Things Were Different
Ike
Ikiru
The Illusion Travels by
  Streetcar
Ilsa, She-Wolf of the SS
Ilsa, the Tigress of Siberia
Ilsa, the Wicked Warden
I'm Dancing as Fast as I Can
I'm the One You're Looking
  For
The Image
Imagemaker
Imitation of Life
Immediate Family
Immortal Battalion
Immortal Sergeant
The Impossible Spy
The Imposter

Impulse
In Between
In Celebration
In Cold Blood
In Country
In the Custody of Strangers
In Desert and Wilderness
In a Glass Cage
In the Heat of the Night
In the Heat of Passion
In the King of Prussia
In Memoriam
In Name Only
In the Name of the Pope-King
In the Realm of the Senses
In the Region of Ice
In This House of Brede
In for Treatment
In Trouble
In the White City
The Incident
Incident at Map-Grid 36-80
An Inconvenient Woman
The Incredible Journey of Dr.
  Meg Laurel
The Incredible Sarah
Independence
The Indian Runner
Infamous Daughter of Fanny
  Hill
The Informer
Inga
Inherit the Wind
The Inheritance
The Inheritors
The Inn of the Sixth
  Happiness
The Inner Circle
Inner Sanctum
Innocence Unprotected
The Innocent
An Innocent Man
The Insect Woman
Inserts
Inside the Lines
Inside Moves
Inside Out
Interiors
International Velvet
The Interns
Interrogation
Intimate Contact
Intimate Moments
Intimate Power
Intimate Story
Intimate Strangers
Into the Homeland
Into Thin Air
Intolerance
Intruder in the Dust
Invincible Mr. Disraeli
Invitation au Voyage
Iphigenia
Iran: Days of Crisis
Irezumi (Spirit of Tattoo)
Iris
The Irishman
Iron Duke
Iron Maze
Iron & Silk
Ironclads
Ironweed
Is Paris Burning?
Isadora
The Island
Island of Lost Girls
Island in the Sun
The Islander
Islands in the Stream
Isn't Life Wonderful
It Came Upon a Midnight
  Clear
It Could Happen to You
It Happened in New Orleans
Italian
Italian Stallion

It's Called Murder, Baby
It's Good to Be Alive
Ivan the Terrible, Part 1
Ivan the Terrible, Part 2
Ivory Hunters
J. Edgar Hoover
Jack Frost
The Jack Knife Man
Jack the Ripper
The Jackie Robinson Story
Jacknife
Jackson County Jail
Jacob: The Man Who Fought
  with God
Jacob's Ladder
Jacqueline Bouvier Kennedy
Jailbait: Betrayed By
  Innocence
Jamaica Inn
James Dean
The James Dean 35th
  Anniversary Collection
James Joyce: A Portrait of the
  Artist as a Young Man
James Joyce's Women
Jane Austen in Manhattan
Java Head
The Jayne Mansfield Story
je tu il elle
Jealousy
Jean de Florette
The Jericho Mile
Jesse
The Jesse Owens Story
Jesus
Jesus of Montreal
Jesus of Nazareth
Jet Over the Atlantic
Jet Pilot
The Jeweller's Shop
Jezebel
JFK
Jigsaw
The Jilting of Granny
  Weatherall
Jim Thorpe: All American
Jimmy Valentine
Joan of Arc
Joe Louis Story
Joe Panther
Joey
John and the Missus
Johnny Apollo
Johnny Belinda
Johnny Come Lately
Johnny Firecloud
Johnny Got His Gun
Johnny Nobody
Johnny Tiger
Johnny We Hardly Knew Ye
Joni
Joseph Andrews
Joseph & His Brethren
Joseph & His Brothers
Journey of Hope
Journey Together
Journeys from Berlin/1971
Joy House
Joyful Laughter
Joyride
Juarez
Jud
Jud Suess
Judex
The Judge
The Judge and the Assassin
Judge Priest
Judgment
Judgment in Berlin
Judgment at Nuremberg
Juggernaut
Juggler of Notre Dame
Juice
Julia
Julius Caesar
July Group

Drama

June Night
Juno and the Paycock
Just a Gigolo
Just for the Hell of It
Just Hold Still
Just Like Us
Just Like Weather
Justice
Justin Morgan Had a Horse
Justine
Kameradschaft
Kandyland
Kangaroo
Kansas
Kansas City Confidential
Kansas City Massacre
Katherine
Keeping On
Keetje Tippei
Kennedy
Kenneth Anger, Vol. 1:
  Fireworks
Kenneth Anger, Vol. 2:
  Inauguration of the
  Pleasure Dome
Kenneth Anger, Vol. 4:
  Invocation of My Demon
  Brother
Kent State
Kerouac
Key Largo
The Key Man
The Keys of the Kingdom
Kid from Not-So-Big
A Kid for Two Farthings
Kidnapping of Baby John Doe
Kill Castro
Kill Me Again
Killer
Killer on Board
The Killers
Killer's Kiss
Killing of Angel Street
Killing 'em Softly
The Killing Fields
Killing Floor
Killing Hour
The Killing Kind
Killing of Randy Webster
Killing Stone
A Kind of Loving
King
King of America
King David
King of the Gypsies
King of Kings
King Lear
King Rat
Kings and Desperate Men
Kings of the Road (In the
  Course of Time)
Kings Row
Kipling's Women
Kipps
Kismet
The Kiss
Kiss of Death
Kiss the Night
Kiss of the Spider Woman
Kiss Tomorrow Goodbye
The Kissing Place
The Kitchen Toto
The Klansman
Knife in the Water
Knock on Any Door
Knute Rockne: All American
Kramer vs. Kramer
The Krays
Kung Fu
La Amante
L.A. Bad
La Balance
La Boca del Lobo
La Cicada (The Cricket)
L.A. Crackdown
L.A. Crackdown 2

La Dinastia de la Muerte
La Diosa Arrodillada (The
  Kneeling Goddess)
La Femme Nikita
La Forza del Destino
La Fuga
La Generala
La Grande Bourgeoise
La Hifa Sin Padre
La Hija del General
La Hora 24
La Justicia Tiene Doce Anos
L.A. Law
La Mentira
La Muerte del Che Guevara
La Notte Brava (Lusty Night in
  Rome)
La Primavera de los
  Escorpiones
La Puritaine
La Rebelion de las Muertas
La Signora di Tutti
La Symphonie Pastorale
La Terra Trema
La Truite (The Trout)
La Vida Sigue Igual
La Vie Continue
La Virgen de Guadalupe
The Lacemaker
L'Addition
Ladies Club
Ladies on the Rocks
Lady on the Bus
Lady Chatterley's Lover
The Lady Confesses
Lady in Distress
Lady Grey
Lady of the House
Lady Ice
Lady Killers
Lady for a Night
Lady from Nowhere
Lady in Red
The Lady Refuses
Lady Windermere's Fan
Lady from Yesterday
Lai Shi: China's Last Eunuch
Lamp at Midnight
Land of Fury
Lantern Hill
Larry
Las Barras Bravas
Las Vegas Lady
Las Vegas Story
Las Vegas Strip War
Lassie's Great Adventure
The Last American Hero
The Last Angry Man
The Last Bastion
Last Call
The Last Chance
Last Command
Last Cry for Help
The Last Days of Dolwyn
The Last Days of Patton
Last Days of Pompeii
The Last Emperor
The Last of England
Last Exit to Brooklyn
Last Fight
Last Flight Out: A True Story
Last Four Days
Last Game
The Last of His Tribe
Last Hurrah
Last Innocent Man
The Last Laugh
The Last Metro
The Last Movie
Last of Mrs. Lincoln
The Last Picture Show
Last Summer
The Last Supper
Last Tango in Paris
The Last Temptation of Christ
The Last Tycoon

The Last Wave
Last Word
Last Year at Marienbad
L'Atalante
Late Extra
Late Spring
Late Summer Blues
Lathe of Heaven
Laughing at Life
L'Avventura
Law of the Underworld
Lawrence of Arabia
Lazarus Syndrome
LBJ: The Early Years
Le Bal
Le Cas du Dr. Laurent
Le Chat
Le Complot (The Conspiracy)
Le Crabe Tambour
Le Dernier Combat
Le Doulos
Le Grand Chemin
Le Plaisir
Lean on Me
The Learning Tree
Leather Boys
Leaves from Satan's Book
Leaving Normal
Legal Tender
The Legend of Sleepy Hollow
Legend of Valentino
Legend of Walks Far Woman
The Legend of Young Robin
  Hood
Legs
Lenny
Leonor
The Leopard Woman
Lepke
Les Biches
Les Bons Debarras
Les Carabiniers
Les Enfants Terrible
Les Grandes Gueules
Les Miserables
Les Mistons
Lai Rendez-vous D'Anna
Less Than Zero
Let 'er Go Gallegher
Let Him Have It
Let It Rock
Lethal Obsession
Let's Get Married
The Letter
Letter of Introduction
A Letter to Three Wives
Letters to an Unknown Lover
Letting the Birds Go Free
Liar's Moon
Liberation of L.B. Jones
License to Kill
The Lie
Lies My Father Told Me
The Life and Assassination of
  the Kingfish
The Life of Christ
Life & Death of Colonel Blimp
The Life of Emile Zola
The Life & Loves of a Male
  Stripper
Life of O'Haru
Lifeboat
The Light Ahead
The Light in the Jungle
Lightning: The White Stallion
Lights! Camera! Murder!
The Lightship
Lilies of the Field
Liliom
Lily was Here
Limelight
The Lindbergh Kidnapping
  Case
The Line
The Lion in Winter
Lipstick

Lisa
Lisbon
Lisztomania
Little Ballerina
Little Boy Lost
Little Church Around the
  Corner
Little Dorrit, Film 1: Nobody's
  Fault
The Little Foxes
Little Girl...Big Tease
Little Gloria...Happy at Last
Little Heroes
Little Ladies of the Night
Little Lips
Little Lord Fauntleroy
Little Man Tate
Little Men
Little Miss Innocence
Little Orphan Annie
A Little Princess
A Little Sex
Little Sweetheart
The Little Theatre of Jean
  Renoir
Little Tough Guys
The Little Valentino
Little Vera
Little Women
Littlest Outlaw
Loan Shark
Lock Up
Lois Gibbs and the Love Canal
Lola Montes
Lolita
The Lone Wolf
The Loneliest Runner
The Lonely Lady
The Lonely Passion of Judith
  Hearne
Lonelyhearts
Loners
The Long Dark Hall
Long Day's Journey into Night
The Long Days of Summer
Long Gone
The Long Good Friday
The Long Gray Line
The Long Haul
Long Island Four
Long Journey Back
The Long Voyage Home
The Long Walk Home
A Long Way Home
The Long Weekend
The Longest Day
Longtime Companion
Look Back in Anger
Looking for Miracles
Looking for Mr. Goodbar
Lord of the Flies
Lord Jim
Lords of Discipline
The Lords of Flatbush
Los Chicos Crecen
Los Dos Hermanos
Los Meses y los Dias
Los Olvidados
The Losers
Lost Angels
Lost in the Barrens
The Lost Honor of Katharina
  Blum
Lost Horizon
Lost Legacy: A Girl Called
  Hatter Fox
Lost Moment
The Lost One
The Lost Weekend
The Lost World
Lou Gehrig Story
Love
Love Child
Love on the Dole
Love and Hate
Love Leads the Way

The Love Machine
Love Strange Love
Love from a Stranger
Love Under Pressure
Loveless
Lovely...But Deadly
The Loves of Edgar Allen Poe
Lovey: A Circle of Children 2
Loving
The Lower Depths
Loyalties
Loyola, the Soldier Saint
Lumiere
Lure of the Islands
Lust for Freedom
Lust for Life
Luther
M
Ma Barker's Killer Brood
Macao
Macario
MacArthur's Children
Machine Gun Kelly
Macho Dancer
Mack
Macon County Line
Macumba Love
Mad Bull
Mad Death
Mad Dog Morgan
The Mad Whirl
Mad Youth
Madame Bovary
Madame Curie
Madame Rosa
Madame X
M.A.D.D.: Mothers Against
  Drunk Driving
Made in Argentina
Mae West
Maedchen in Uniform
Mafia Princess
Magdalena Viraga
Magdalene
Magee and the Lady
Magic
Magic of Lassie
The Magician
The Magnificent Ambersons
Mahogany
Maid in Sweden
Maitresse
Major Barbara
Make Haste to Live
Make Me an Offer
Make-Up
Making Love
Malarek
Malibu High
Malibu Hot Summer
Malicious
Malou
The Maltese Falcon
A Man for All Seasons
A Man Called Adam
A Man Called Peter
Man with a Cross
Man Facing Southeast
Man of Flowers
Man Friday
The Man with the Golden Arm
The Man in the Gray Flannel
  Suit
Man Inside
A Man Like Eva
Man of Marble
The Man Outside
Man Outside
Man on the Run
Man of a Thousand Faces
Man of Violence
The Man Who Broke 1,000
  Chains
The Man Who Knew Too
  Much
The Man Who Never Was

Drama

One Last Run
One Man's War
One More Chance
One Night Stand
One on One
One Plus One
One Russian Summer
One Sings, the Other Doesn't
One Third of a Nation
Onibaba
The Onion Field
The Only Way Home
Open City
Open Doors (Porte Aperte)
The Oppermann Family
The Opponent
The Ordeal of Dr. Mudd
Ordet
Ordinary Heroes
Ordinary People
Oriane
Original Intent
Orphans
Orpheus Descending
Osaka Elegy
Oscar
Ossessione
Othello
The Other Side of Midnight
Our Daily Bread
Our Dancing Daughters
Our Family Business
Our Town
Out of Africa
Out of the Blue
Out of the Rain
The Outcasts
The Outlaw and His Wife
Outlaw Riders
Outrage
Outside Chance
Outside the Law
The Outsiders
Over the Edge
Over Indulgence
Over the Top
The Overcoat
The Overlanders
Overnight Sensation
Overseas: Three Women with
   Man Trouble
Overture to Glory (Der Vilner
   Shtot Khazn)
Pace That Kills
Padre Padrone
The Painted Hills
The Painted Veil
Paisan
Palm Beach
Panama Menace
Panama Patrol
Panic in Echo Park
Panic in Needle Park
Panic Station
Panic in the Year Zero!
Panique
Paper Man
Paper Mask
Paperback Hero
Paradise
Paradise Alley
Paris Belongs to Us
Paris, Texas
Paris Trout
The Park is Mine
Parole, Inc
Parting Glances
Party Girl
Party Girls
Pascali's Island
A Passage to India
Passage to Marseilles
The Passenger
The Passing of Evil
Passion
Passion for Life

Passion for Power
The Passover Plot
Pastime
Pather Panchali
Paths of Glory
The Patriots
Patterns
Patton
Patty Hearst
Paul Robeson
Paula
Paul's Case
Pawnbroker
Payday
Peacekillers
The Peacock Fan
The Pearl
Pearl of the South Pacific
Peddlin' in Society
The Pedestrian
The Peking Blond
Pelle the Conqueror
Penalty Phase
The Penitent
The People Next Door
The People vs. Jean Harris
Perfect
Perfect Killer
Perfect Strangers
Perfect Witness
Performance
Permanent Record
Permanent Vacation
Perry Mason Returns
Persona
Personal Best
Personals
Peter the Great
Peter Gunn
Petrified Forest
Petulia
Peyton Place
The Phantom in the House
Phar Lap
Phedre
Phone Call from a Stranger
A Piano for Mrs. Cimino
Pickpocket
Pickup on South Street
A Piece of Pleasure
Pierrot le Fou
Pigeon Feathers
Pilgrim Farewell
The Pilot
Pimpernel Smith
Pink Strings and Sealing Wax
Pipe Dreams
Pistol: The Birth of a Legend
Pixote
A Place Called Today
Place of Weeping
Places in the Heart
Play Murder for Me
Playboy of the Western World
Players
Playing with Fire
Playing for Time
Pleasure Palace
Plenty
The Ploughman's Lunch
Plutonium Incident
Poco
Poil de Carotte
Poison
Police
Police Court
Policewoman Centerfold
Poor Pretty Eddie
Poor White Trash 2
The Pope of Greenwich
   Village
Pope John Paul II
The Poppy Is Also a Flower
Porridge
Port of New York

Portfolio
Portrait of the Artist as a
   Young Man
Portrait of Grandpa Doc
Portrait of a Lady
Portrait of a Rebel: Margaret
   Sanger
Portrait of a Showgirl
Portrait of a Stripper
Portrait of Teresa
Portrait in Terror
Postal Inspector
The Postman Always Rings
   Twice
Postmark for Danger
P.O.W. Deathcamp
Power
Power, Passion & Murder
Prancer
Pray TV
Prayer for the Dying
Praying Mantis
Premonition
Presumed Innocent
Pretty Baby
Pretty Poison
Prick Up Your Ears
Pride of the Bowery
Pride of Jesse Hallum
Pride and the Passion
Pride and Prejudice
Pride of St. Louis
The Pride of the Yankees
Priest of Love
Prime Cut
Prime Suspect
Prince of Bel Air
The Prince of Central Park
Prince and the Great Race
Prince Igor
Prince Jack
Prince and the Pauper
Princess Daisy
Princess in Exile
The Principal
Prison Break
Prison Stories: Women on the
   Inside
Prison Train
The Prisoner
Prisoner of Honor
Prisoners of the Sun
The Private Affairs of Bel Ami
Private Contentment
Private Duty Nurses
The Private Files of J. Edgar
   Hoover
Private Hell 36
Private Life
The Private Life of Henry VIII
Prix de Beaute
Probe
Problem Child
The Prodigal
The Prodigal Planet
Project: Kill!
Promises in the Dark
Providence
Prudential Family Playhouse
Psych-Out
Puberty Blues
A Public Affair
Public Enemy
Pudd'nhead Wilson
Pueblo Affair
Pulsebeat
Pump Up the Volume
Purgatory
The Purple Taxi
The Pursuit of D. B. Cooper
The Pursuit of Happiness
Q Ships
QB VII
Quality Street
Quartet
Quatorze Juillet

Que Viva Mexico
Queen Kelly
The Queen of Mean
The Queen of Spades
Queen of the Stardust
   Ballroom
Queenie
Querelle
A Question of Guilt
Question of Honor
A Question of Love
Question of Silence
Quicksilver
Quiet Day In Belfast
Quilombo
Quo Vadis
Race for Glory
Race for Life
Rachel, Rachel
Rachel River
Rachel's Man
Racing with the Moon
The Racketeer
Rad
The Radical
Radio Flyer
Rafferty & the Gold Dust
   Twins
Rage
Rage of Angels
A Rage in Harlem
Raggedy Man
The Raggedy Rawney
Raging Bull
Ragtime
Raid on Entebbe
Railway Children
Rain
Rain Man
Rain People
The Rainbow
Rainbow Bridge
The Rainmaker
Rainy Day
Raise the Titanic
A Raisin in the Sun
Ramparts of Clay
Ransom
Rape of Love
Rape and Marriage: The
   Rideout Case
Rappaccini's Daughter
The Rapture
A Rare Breed
Rashomon
The Razor's Edge
Reach for the Sky
A Real American Hero
Rear Window
Rebecca
Rebel
Rebel Rousers
The Rebel Set
Rebel Vixens
Rebel Without a Cause
Reborn
Reckless Disregard
The Reckless Way
The Red Badge of Courage
Red Balloon/Occurrence at
   Owl Creek Bridge
Red Beard
The Red and the Black
The Red Desert
The Red Half-Breed
Red Kiss (Rouge Baiser)
Red Line 7000
Red Nightmare
Red Nights
The Red Pony
Red Shoe Diaries
The Red Stallion
Red Surf
Red, White & Busted
Redneck
Reds

Reefer Madness
The Reflecting Skin
A Reflection of Fear
Reflections: A Moment in
   Time
Reflections in a Golden Eye
Reform School Girls
Regarding Henry
Regina
The Reincarnation of Golden
   Lotus
Reincarnation of Peter Proud
Rembrandt
Rendez-vous
Requiem for Dominic
The Respectful Prostitute
Resurrection
Return to the Blue Lagoon
Return to Earth
Return Engagement
Return of Martin Guerre
Return to Peyton Place
Reunion
Reunion in France
Revealing of Elsie
Revenge of a Kabuki Actor
Reversal of Fortune
The Revolt of Job
The Revolt of Mother
Rhodes
Rich Kids
Rich and Strange
Richard II
Richard Petty Story
Richard's Things
Richie
Riddance
Ride in a Pink Car
Ride a Wild Pony
Ride the Wind
Riel
The Right Hand Man
The Right Stuff
Right of Way
Rikyu
The Ring
Ring of Bright Water
Ringside
Riot
Riot Squad
Rise of Catherine the Great
The Rise of Louis XIV
Rise & Rise of Daniel Rocket
Rising Son
Risky Business
Rita Hayworth: The Love
   Goddess
Rituals
Rivals
The River
River Beat
The River Niger
The River Rat
River of Unrest
Riverbend
River's Edge
Road to 1984
Road House
Road Racers
Roadhouse 66
Roadhouse Girl
Roanoak
Roaring Roads
Roaring Speedboats
Robert Kennedy and His
   Times
Robot Pilot
Rocco and His Brothers
Rock, Pretty Baby
Rocket Gibraltar
Rocket to the Moon
Rocky
Rocky 2
Rocky 3
Rocky 4
Rocky 5

Rodeo Champ
Rodrigo D: No Future
Roe vs. Wade
Roll of Thunder, Hear My Cry
Rolling Home
Rolling Thunder
Rollover
Roman Spring of Mrs. Stone
Romance in Manhattan
Rome '78
Romero
Rona Jaffe's Mazes & Monsters
The Roof
The Room
Room to Let
Rooster—Spurs of Death!
Roots, Vol. 1
The Rose Garden
The Rose and the Jackal
The Rose Tattoo
Roseland
Rosie: The Rosemary Clooney Story
Round Midnight
The Round Up
R.P.M.* (*Revolutions Per Minute)
R.S.V.P.
Ruby
Ruby Gentry
The Rules of the Game
Rumble Fish
The Rumor Mill
A Rumor of War
Run for the Roses
Runaway
The Runaways
The Runner Stumbles
Running Away
Running Brave
Running on Empty
Running Mates
Rush
The Russia House
The Rutherford County Line
Ryan's Daughter
The Sacrifice
Sacrilege
Saigon: Year of the Cat
The Sailor Who Fell from Grace with the Sea
St. Elmo's Fire
St. Helen's, Killer Volcano
St. Ives
Saint Jack
Saint Joan
Sakharov
Salaam Bombay!
Salammbo
Salo, or the 120 Days of Sodom
Salome
Salome/Queen Elizabeth
Salome, Where She Danced
Salt of the Earth
Samaritan: The Mitch Snyder Story
Same Time, Next Year
Sammy
Sam's Son
Samson and Delilah
Samurai Reincarnation
Sand and Blood
Sandakan No. 8
The Sandpiper
Sanshiro Sugata
Sansho the Bailiff
Santa
Santa Sangre
Sara Dane
Sarah, Plain and Tall
Satan's Harvest
Saturday Night Fever
Saul and David
Savage Attraction

Savage is Loose
The Savage Seven
Savannah Smiles
Save the Lady
Save the Tiger
Sawdust & Tinsel
Say Goodbye, Maggie Cole
Say Hello to Yesterday
Sayonara
Scalp Merchant
Scandal
Scandal in a Small Town
The Scar
Scarecrow
The Scarlet Letter
The Scarlet Pimpernel
Scarred
Scenes from the Goldmine
Scenes from a Marriage
Scheherazade
Scissors
Score
The Scorpion Woman
Scout's Honor
Scrooge
Scrubbers
Scruples
Scum
Sea Lion
Sea Racketeers
The Seagull
The Search
The Search for Bridey Murphy
Season for Assassins
Season of Fear
Sebastian
Sebastiane
Second Chance
Second Coming of Suzanne
Second Wind
Secret Ceremony
Secret Fantasy
Secret File of Hollywood
The Secret Four
Secret Games
The Secret Garden
Secret Honor
The Secret of the Loch
Secret Passions
A Secret Space
Secret Weapons
Secret of Yolanda
Secrets
Secrets of a Soul
Seduced
Seducers
The Seduction
The Seduction of Joe Tynan
See How She Runs
The Seekers
Seize the Day
Semaforo en Rojo
A Sense of Freedom
Sense & Sensibility
Sensual Partners
The Sensuous Teenager
Separacion Matrimonial
Separate But Equal
A Separate Peace
Separate Tables
Separate Ways
September
September 30, 1955
Sergeant Matlovich vs. the U.S. Air Force
Sergeant Ryker
Sergeant Sullivan Speaking
The Serpent's Egg
Serpico
The Servant
Sessions
The Set-Up
Seven Days in May
Seven Deadly Sins
Seven Miles from Alcatraz/ Flight from Glory

The Seventeenth Bride
The 7th Commandment
The Seventh Seal
Seventh Veil
Severance
Sewing Woman
Sex and the College Girl
Sex, Drugs, and Rock-n-Roll
sex, lies and videotape
Sex Madness
Sex and the Office Girl
Sex with the Stars
Sex Surrogate
Sex Through a Window
Shack Out on 101
The Shadow Box
Shadow Dancing
Shadow on the Sun
Shadows
Shadows of Forgotten Ancestors
Shadows of the Orient
Shadows Over Shanghai
Shadows on the Stairs
Shadows in the Storm
Shaka Zulu
Shakespeare Wallah
Shaking the Tree
Shame
The Shaming
The Shanghai Gesture
Sharma & Beyond
Shattered
Shattered Innocence
Shattered Silence
Shattered Spirits
Shattered Vows
She-Devils on Wheels
Sheer Madness
She'll Be Wearing Pink Pajamas
The Sheltering Sky
The Shepherd
Shin Heike Monogatari
Shining Through
Shiokari Pass
Ship of Fools
Shock Corridor
The Shoes of the Fisherman
Shoeshine
Shogun
Shoot the Moon
Shoot the Piano Player
Shooting
The Shooting Party
Shop on Main Street
Short Eyes
Short Fuse
The Shrieking
Shy People
The Sicilian
Sicilian Connection
Sid & Nancy
Side Show
Sidewalks of London
Signal 7
Silas Marner
The Silence
Silence
Silence of the Heart
Silence Like Glass
Silence of the North
The Silent One
The Silent Partner
Silent Rebellion
Silent Victory: The Kitty O'Neil Story
Silent Witness
Silhouette
Silkwood
Silver Chalice
Silver City
Silver Dream Racer
The Silver Streak
Simba
Simon Bolivar

Simple Justice
Simple Story
Sin of Adam & Eve
The Sin of Madelon Claudet
Sinbad
Sincerely Charlotte
The Sinful Bed
Sing Sing Nights
The Singing Blacksmith (Yankl der Shmid)
Singing the Blues in Red
Single Room Furnished
The Single Standard
Sink the Bismarck
Sinners in Paradise
Sins
Sins of Jezebel
Sister Dora
Sister Kenny
Sisters of Death
Sisters of Gion
Sixteen
'68
Sizzle
Skag
Skateboard
Skeezer
Skier's Dream
Skin Game
Skinheads: The Second Coming of Hate
Skullduggery
The Sky is Gray
The Sky Pilot
Slacker
Slash
Slashed Dreams
Slaughterhouse
A Slave of Love
Slavers
Slaves of Love
Sleeping Dogs
Sleeping with the Enemy
The Sleeping Tiger
The Slender Thread
Slipstream
A Small Circle of Friends
Small Sacrifices
Small Town Boy
A Small Town in Texas
Smash Palace
Smash-Up: The Story of a Woman
Smile, Jenny, You're Dead
Smiling Madame Beudet/The Seashell and the Clergyman
Smith!
Smithereens
Smoke
Smoke Screen
A Smoky Mountain Christmas
Smooth Talk
Smouldering Fires
Snake Dancer
Sniper
Snow Country
Snow Treasure
The Snows of Kilimanjaro
So Dear to My Heart
So Ends Our Night
Solaris
Soldier's Revenge
A Soldier's Story
The Solitary Man
Solo
Solomon and Sheba
Some Came Running
Somebody Has to Shoot the Picture
Somebody Up There Likes Me
Someone at the Door
Someone I Touched
Someone to Love
Something Special
Something of Value

Sometimes a Great Notion
Somewhere Tomorrow
The Song of Bernadette
Sons
Sooner or Later
Sophia Loren: Her Own Story
Sophie's Choice
Sophisticated Gents
Sorceress
Sorrows of Gin
S.O.S. Titanic
Soul Hustler
Soul Patrol
Soultaker
Sound of Murder
Sounder
South Bronx Heroes
South of Reno
The Southerner
Souvenir
Soviet Spy
Soylent Green
Spangles
The Spanish Gardener
Speaking Parts
Special Bulletin
A Special Day
Special Effects
The Specialist
Spellbound
Spetters
Spices
The Spider's Stratagem
Spirit of the Beehive
Spirit of the Eagle
Spirit of St. Louis
The Spirit of West Point
Spitfire
Spittin' Image
Splendor in the Grass
Split Decisions
Split Image
Split Second
The Spoilers
Spoon River Anthology
The Sporting Club
Spring Symphony
Springhill
Spy of Napoleon
Spymaker: The Secret Life of Ian Fleming
Square Dance
Square Shoulders
The Squeeze
S.S. Experiment Camp 2
S.S. Love Camp No. 27
Stacking
Stacy's Knights
Stage Fright
Stage Struck
Stalag 17
Stalk the Wild Child
Stand Alone
Stand By Me
Stand and Deliver
Stand Off
Stanley and Iris
Star 80
A Star is Born
The Star Chamber
Star Reporter
Star Trek 6: The Undiscovered Country
Starlight Hotel
The Stars Look Down
State of Siege
Stateline Motel
State's Attorney
Static
Station
Stavisky
Stay As You Are
Stay Hungry
Stay Tuned for Murder
Staying Alive
Staying On

Drama

Steal the Sky
Stealing Home
Steaming
Steel
Steel Cowboy
Stella
Stella Dallas
Stephen Crane's Three
   Miraculous Soldiers
The Stepmother
Steppenwolf
Stevie
The Stewardesses
Stigma
Still of the Night
The Stilts
A Stolen Life
The Stone Boy
Stone Fox
Storm
Storm Over Asia
Stormy Monday
Stormy Waters
Story of Adele H.
The Story of the Late
   Chrysanthemum
The Story of Louis Pasteur
The Story of a Love Story
The Story of O
The Story of O, Part 2
The Story of Ruth
The Story of a Three Day
   Pass
Story of Women
Storyville
Straight Out of Brooklyn
Straight Time
Straight Up
Strange Awakening
Strange Cargo
A Strange and Deadly
   Occurence
Strange Fruit
Strange Illusion
Strange Interlude
The Strange Love of Martha
   Ivers
The Strange Woman
A Stranger in My Forest
Stranger Who Looks Like Me
The Stranger Within
Strangers in the City
Strangers in Good Company
Strangers Kiss
Strangers: The Story of a
   Mother and Daughter
Strangers When We Meet
Strategic Air Command
Straw Dogs
The Strawberry Statement
Street
Street Hero
Street Justice
Street People
Street of Shame
Street Smart
Street Warriors
Street Warriors, Part 2
A Streetcar Named Desire
Streets
Streets of Gold
The Streets of L.A.
The Streets of San Francisco
Streetwalkin'
Strike
Strike Force
Strikeback
The Stripper
Strong Medicine
Stroszek
The Stud
The Student Nurses
Studs Lonigan
The Stunt Man
Submission
Subterfuge

Suburban Roulette
Suburbia
Subway to the Stars
Success Is the Best Revenge
Sudden Terror
Suddenly, Last Summer
Sugar Cane Alley
The Suicide Club
Summer of '42
Summer Affair
Summer City
Summer Heat
Summer Interlude
Summer Lovers
Summer Magic
The Summer of Miss Forbes
Summer of My German
   Soldier
A Summer to Remember
A Summer in St. Tropez
Summer Solstice
Summer Vacation: 1999
Summer Wishes, Winter
   Dreams
Summertime
Summertree
The Sun Shines Bright
Sunday, Bloody Sunday
A Sunday in the Country
Sunday Daughters
Sunday Too Far Away
Sundays & Cybele
The Sundowners
Sunny Side Up
Sunnyside
Sunrise
Sunset Boulevard
Sunset Strip
Super Bitch
Super Brother
Supercarrier
Superfly
Susan and God
Susana
Susannah of the Mounties
Suspended Alibi
Suzanne
Svengali
Swann in Love
The Swap
Swedenhielms
Sweet Beat
Sweet Bird of Youth
Sweet Country
Sweet Country Road
Sweet Creek County War
Sweet Ecstasy
Sweet 15
Sweet Hostage
Sweet Love, Bitter
Sweet Perfection
Sweet Poison
Sweet Smell of Success
Sweet Smell of Woman
Sweet Sweetback's
   Baadasssss Song
Sweet William
Sweetheart!
Sweetwater
Swimmer
Swimming Pool
The Swindle
Swing High, Swing Low
Swinging Ski Girls
Swinging Sorority Girls
Sword & the Cross
Sybil
Sylvester
Sylvia
Symphony of Living
T-Bird Gang
T-Men
Table for Five
Tabu
Tag: The Assassination Game
Tainted

Taking Care of Terrific
The Taking of Flight 847: The
   Uli Derickson Story
The Tale of Ruby Rose
A Tale of Two Cities
Tale of Two Sisters
Talent for the Game
Tales of the Klondike: The
   Scorn of Women
Tales of Ordinary Madness
Talk to Me
Talk Radio
Tall Lie
The Tamarind Seed
Tango
Tango Bar
Tanya's Island
Targets
A Taste of Honey
Tattoo
Taxi Driver
Tchao Pantin
Tea For Three
Tea and Sympathy
Teen Mothers
Teenage Bad Girl
Teenage Mother
Teenage Seductress
Teenager
Tell Me a Riddle
Tell Me Where It Hurts
The Tempest
Tempest
The Temptress
The Ten Commandments
Ten Nights in a Bar Room
Ten Nights in a Bar-Room
Ten North Frederick
Ten Speed
The Tenant
The Tender Age
Tender Is the Night
Tender Loving Care
The Tender Years
Tennessee Stallion
Tenth Month
Teorema
Terminal Bliss
Terminal Force
Termini Station
Terraces
The Terry Fox Story
Tess
A Test of Love
Testament
Tevye (Teyve der Milkhiker)
Tex
Texas Lightning
Thanksgiving Story
That Brennan Girl
That Championship Season
That Cold Day in the Park
That Forsyte Woman
That Hamilton Woman
That Long Night in '43
That Was Then...This Is Now
That'll Be the Day
Therese & Isabelle
These Three
They Came to Cordura
They Knew What They
   Wanted
They Live by Night
They Made Me a Criminal
They Meet Again
They Never Come Back
They Paid with Bullets:
   Chicago 1929
They Shoot Horses, Don't
   They?
They were So Young
They were Ten
They Won't Forget
Thief of Hearts
Third Solution
Third Walker

Thirteenth Day of Christmas
36 Fillete
This Happy Breed
This Land is Mine
This Property is Condemned
This Special Friendship
This Sporting Life
Thompson's Last Run
The Thorn Birds
Those Glory, Glory Days
Thou Shalt Not Kill...Except
Three Brothers
Three Came Home
The Three Faces of Eve
3:15—The Moment of Truth
Three Lives of Thomasina
Three on a Match
Three Sovereigns for Sarah
Three Stops to Murder
Three Strange Loves
Three Warriors
The Three Weird Sisters
Threshold
Throne of Blood
Through the Breakers
Thunder Alley
Thunder Warrior 3
Thunderheart
Ticket to Heaven
Tidal Wave
Tidy Endings
Tiefland
Tiger Town
Tiger Warsaw
Tigers Don't Cry
Till the End of Time
Tim
Timber Queen
Time of Destiny
Time of Indifference
Time to Kill
Time Lock
Time for Miracles
A Time to Remember
Time Stands Still
Time of Tears
Times Square
Timothy and the Angel
The Tin Drum
Tin Man
'Tis a Pity She's a Whore
To All My Friends on Shore
To Be a Rose
To Forget Venice
To Joy
To Kill a Mockingbird
To Kill a Priest
To the Lighthouse
To Race the Wind
To the Shores of Tripoli
To Sir, with Love
Toast of New York
Today I Hang
Tokyo Story
Tol'able David
The Tomboy
Tomboy & the Champ
Tomorrow is Forever
Tomorrow's Child
The Tongfather
Toni
Tonio Kroger
Too Hot to Handle
Too Late the Hero
Too Late for Tears
Too Wise Wives
Topkapi
Torchlight
Torment
Torn Allegiance
Tornado
Torpedo Alley
Torpedo Attack
Torpedo Run
Torrents of Spring
Tortilla Flat

Torture Ship
Touch and Go
Touched
Touched by Love
Tough Enough
Tough to Handle
Tough Kid
Toughlove
Tournament
Tower of Terror
Towering Inferno
Towers Open Fire & Other
   Films
A Town Like Alice
Toys in the Attic
Tracks
Traenen in Florenz (Tears in
   Florence)
Tragedy of Flight 103: The
   Inside Story
Tragedy of King Richard II
Tragedy of a Ridiculous Man
Train of Events
Train Station Pickups
Trained to Kill, U.S.A.
Tramp at the Door
Trapped
Trapped by the Mormons
Trapped in Silence
Traveling Man
Travelling North
Treasure of Arne
Treasure of the Lost Desert
A Tree Grows in Brooklyn
The Tree of Wooden Clogs
Trespasser
The Trial
Trial of Bernhard Goetz
Trial of Billy Jack
Trial of the Catonsville Nine
The Trial of Lee Harvey
   Oswald
Tribes
Tribute
Tricheurs
Trilby
Trio
The Trip
Trip to Bountiful
Triple Echo
Troma's War
Tromba, the Tiger Man
Trucker's Woman
True Colors
True Confessions
True West
Truman Capote's "The Glass
   House"
Trumps
Tucker: The Man and His
   Dream
Tuff Turf
Tunes of Glory
Turf Boy
Turning Point
Turtle Diary
Tuxedo Warrior
Twelve Angry Men
Twelve Months
Twelve O'Clock High
21 Hours at Munich
Twenty-Four Eyes
25 Fireman's Street
Twenty Dollar Star
Twice in a Lifetime
Twice a Woman
Twigs
Twin Peaks
Twirl
Two Can Play
The Two Jakes
Two of a Kind
Two Kinds of Love
Two Men & a Wardrobe
Two for the Road
The Two Sisters

Playmate of the Year Video
  Centerfold 1990
Posed For Murder
Priceless Beauty
Princess Daisy
Private Affairs
Private Collection
Queen of Outer Space
The Reckless Way
Revenge of the Cheerleaders
Rick's: Your Place for Fantasy
Round Trip to Heaven
Savage Beach
The Seven Year Itch
She Gods of Shark Reef
Sheer Heaven
She's Dressed to Kill
Sizzle Beach U.S.A.
Slaves of Love
The Slumber Party Massacre
Some Like It Hot
Sorceress
Sorority House Scandals
Soup for One
Spinout
The Spring
Star Crash
The Story of a Love Story
Strawberry Blonde
Strip Teasers
Strip Teasers 2
Stripper T's
Swimsuit: The Movie
The Swinging Cheerleaders
Swinging Ski Girls
Tammy and the Doctor
Tanya's Island
Tarzan, the Ape Man
10
Thieves of Fortune
Thunder & Mud
Too Shy to Try
Total Exposure
Ultimate Sampler Video
Vampire's Kiss
Vice Academy 3
Voyage to the Planet of
  Prehistoric Women
When Dinosaurs Ruled the
  Earth
The Woman in Red
You Only Live Twice
Young Nurses in Love

Ethics & Morals

All the King's Men
And Justice for All
Any Man's Death
The Assault
The Best Man
The Bicycle Thief
Blame It on Rio
The Border
Boy with the Green Hair
Brainwash
Casualties of War
Chinese Roulette
Chloe in the Afternoon
Citadel
The Conversation
Crimes & Misdemeanors
Cutter's Way
Deep Six
Defiant Ones
The Doctor
The Dog Soldier
Double Identity
Edipo Re (Oedipus Rex)
El Profesor Hippie
El: This Strange Passion
Elephant Man
Elmer Gantry
Fifth Day of Peace
Force of Evil

The Funeral
The Image
Jezebel
The Judge and the Assassin
Judgment at Nuremberg
Juggler of Notre Dame
Kings Row
La Ronde
La Truite (The Trout)
The Last of England
L'Avventura
Lazarus Syndrome
Les Choses de la Vie
License to Kill
The Littlest Angel
Love with the Proper Stranger
Malicious
Man Inside
The Man Who Shot Liberty
  Valance
Manhattan Melodrama
Marihuana
A Matter of Dignity
Midnight Express
Mindwalk
Mom & Dad
Mr. Smith Goes to
  Washington
Muriel
Network
News at Eleven
The Next Voice You Hear
No Man's Land
Oedipus Rex
One Flew Over the Cuckoo's
  Nest
One Man Jury
Other People's Money
Overture to Glory (Der Vilner
  Shtot Khazn)
The Ox-Bow Incident
Paris Trout
Penalty Phase
A Place in the Sun
Platoon
The Ploughman's Lunch
Private Practices: The Story of
  a Sex Surrogate
Ran
Riel
The Sacrifice
Scandal Man
The Scarlet Letter
Sermon on the Mount
Shampoo
The Shootist
Soul Man
S*P*Y*S
SS Girls
The Star Chamber
Story of Women
Strangers When We Meet
Survival of Spaceship Earth
Sweet Smell of Success
Test Tube Babies
To Live & Die in L.A.
A Touch of Class
The Toy
Tulsa
Twelve Angry Men
Twenty-One
Tycoon
Under Fire
Viridiana
Wall Street
Zabriskie Point

Etiquette

Born Yesterday
Educating Rita
My Fair Lady
Papa's Delicate Condition
Phantom of Liberty
Pleasure

Pygmalion
The Ruling Class
True Love

Exploitation

Alice in Wonderland
American Nightmare
Angel of H.E.A.T.
Angels of the City
Armed Response
Artist's Studio Secrets
Assassin of Youth
Bachelor Party
Bachelorette Party
Bad Girls from Mars
Bad Girls in the Movies
Barn of the Naked Dead
Basic Training
Battle of the Bombs
Beatrice
Beginner's Luck
Best of Everything Goes
Best of the Benny Hill Show:
  Vol. 2
Best of the Benny Hill Show:
  Vol. 4
Best of Everything Goes
Beverly Hills Call Girls
Beyond the Valley of the Dolls
Big Bust Out
Black Emmanuelle
The Black Maid
Black Shampoo
Blackenstein
Blood Games
Blue Movies
Blue Ribbon Babes
Boss
Buxom Boxers
The Cage
Caged Fury
Caged Terror
California Manhunt
Cartoon Collection, No. 5:
  Racial Cartoons
Centerfold Confidential
Certain Sacrifice
Chained for Life
Child Bride
Club Fed
Come Back, Africa
Confessions of a Vice Baron
Cool it Carol
The Cut Throats
Cycle Psycho
Cycle Vixens
Dance Hall Racket
Dangerous Obsession
Dark Side of Love
Deadly Sanctuary
Death of a Centerfold
Delinquent School Girls
Devil's Wedding Night
Erotica: Fabulous Female
  Fantasies
Escape from Safehaven
Escort Girls
Exterminator 2
Eyes of a Stranger
Faces of Death, Part 1
Faces of Death, Part 2
Fanny Hill
Fightin' Foxes
Film House Fever
For Ladies Only
Fox and His Friends
Foxy Brown
Ghoulies 3: Ghoulies Go to
  College
Girl on a Chain Gang
Great British Striptease
Hang Tough
Harlow
High School Caesar
Hitler's Children

Hollywood Centerfolds,
  Volume 1
Hollywood Centerfolds,
  Volume 2
Hollywood Centerfolds,
  Volume 3
Hollywood Centerfolds,
  Volume 4
Hollywood Confidential
The Hollywood Game
Hollywood Nights
The Hollywood Strangler
  Meets the Skid Row
  Slasher
House of Whipcord
I Accuse My Parents
I Spit on Your Grave
In Trouble
Intimate Lessons
Island of Lost Girls
Jock Peterson
Kidnapped
Kinjite: Forbidden Subjects
L.A. Heat
La Nueva Cigarra
La Primavera de los
  Escorpiones
L.A. Vice
The Lash of the Penitents
Last Call
Little Girl...Big Tease
Little Miss Innocence
Love Camp
The Love Pill
Mack
Malibu Express
Mandinga
M'Lady's Court
Mondo Africana
My Wonderful Life
Naked Cage
Naked Vengeance
Nathalie Comes of Age
Night of Evil
Night Friend
Nightmare in Red China
Nomugi Pass
Oh! Those Heavenly Bodies:
  The Miss Aerobics USA
  Competition
One Down, Two to Go!
One Night Only
One Plus One
Outrageous Strip Review
Overexposed
Paradise Motel
The Party
Party Incorporated
Penthouse Video, Vol. 1: The
  Girls of Penthouse
Playboy: Playmates of the
  Year-The '80s
Playboy's 1990 Playmate
  Video Calendar
Playboy's Farmers' Daughters
Playboy's Playmate of the
  Year Video Centerfold
Playboy's Sexy Lingerie
Playboy's Wet & Wild
Pleasure Resort
Policewoman Centerfold
Poor Pretty Eddie
Rate It X
Rebel Vixens
Reefer Madness
Reform School Girls
Return to the Blue Lagoon
Revealing of Elsie
The Ribald Tales of Robin
  Hood
The Road to Ruin
Satan's Cheerleaders
Saturday Night Sleazies, Vol.
  1
Savage Island
Secrets of Sweet Sixteen

Sensual Partners
Sex Adventures of the Three
  Musketeers
Sexcetera
Shocking Asia
Silent Witness
Single Room Furnished
Slaughter
Slaughter's Big Ripoff
Sleazemania
Sleazemania III: The Good,
  the Bad & the Sleazy
Sleazemania Strikes Back
Smile
Sneakin' and Peekin'
S.O.B.
Something Weird
SS Girls
Stag Party
Stand-In
Star 80
Starlet
Street Trash
Stripper T's
Striptease
The Student Body/Jailbait
  Babysitter
Student Confidential
Summer Heat
Summer School Teachers
Sunset Strip
Super Models
Swamp Women
Sweet Spirits
Sweet Sugar
The Swinging Cheerleaders
Swinging Ski Girls
Takin' It Off
The Teacher
Teen Lust
Teenage Seductress
Test Tube Babies
Therese & Isabelle
They're Playing with Fire
This Nude World
Three in the Attic
Three Day Weekend
Thunder & Mud
Tomorrow's Children
Topless Dancing Texas Style
Truck Stop
Up Your Anchor
Vanessa
Vendetta
Venus on Fire
Victims
Warrior Queen
Weekend with the Babysitter
Wild Rapture
Wild, Wild World of Jayne
  Mansfield
Witchdoctor in Tails
Woman Hunt
Women in Cell Block 7
Women in Fury
Women's Club
Women's Prison Massacre
Wrestling Racket Girls
Yellow Emanuelle
Yesterday's Hero
Young Lady Chatterly
Young Lady Chatterly 2

Fairy Tales

Adventures of Prince Achmed
Aladdin and His Wonderful
  Lamp
Alice in Wonderland
Alice's Adventures in
  Wonderland
Beauty and the Beast
The Boy Who Left Home to
  Find Out About the Shivers

Family Viewing

Family

Fantasy

Fantasy

The Tempest
Terror of Rome Against the Son of Hercules
Terror of the Steppes
The Testament of Dr. Cordelier
The Testament of Orpheus
Thief of Baghdad
Thor and the Amazon Women
Thor el Conquistador
The Three Worlds of Gulliver
Throne of Fire
Thumbelina
Time Bandits
Time Burst—The Final Alliance
Time Fighters
The Time of Their Lives
Timestalkers
Tobor the Great
tom thumb
Transformers: The Movie
The Trip
Triumph of the Son of Hercules
Troll
Trouble in Mind
The Trumpet and I
Tut & Tuttle
20,000 Leagues Under the Sea
The Two Worlds of Jenny Logan
Tyrant of Lydia Against the Son of Hercules
Ub Iwerks Cartoonfest 4
Ultra Flash
Unidentified Flying Oddball
Ursus in the Valley of the Lions
The Valley of Gwangi
Venus Against the Son of Hercules
A Very Old Man with Enormous Wings
Vibes
Vice Versa
Vulcan God of Fire
Walking on Air
Warrior Queen
The Warrior & the Sorceress
Warriors of the Wind
Watership Down
Waxworks
Welcome to Blood City
The White Sheik
Wild Women of Wongo
Wildest Dreams
Willow
Wings of Desire
A Witch Without a Broom
The Wizard of Oz
The Wizard of Speed and Time
Wizards
Wizards of the Lost Kingdom
Wizards of the Lost Kingdom 2
Wombling Free
Wonderful World of the Brothers Grimm
Yor, the Hunter from the Future
Zapped!
Zardoz
Zero for Conduct (Zero de Conduit)
Zone Troopers
Zvenigora

Fast Cars

Baffled
Banzai Runner
The Big Wheel

Bobby Deerfield
Born to Race
Burn 'Em Up Barnes
Cannonball
Cannonball Run
Cannonball Run II
Car Crash
Catch Me...If You Can
Daddy-O
Days of Thunder
Death Driver
Death Race 2000
Death Sport
The Devil on Wheels
Dorf Goes Auto Racing
Dragstrip Girl
Eat My Dust
Fast Company
The Fast and the Furious
Fast Lane Fever
Faster Pussycat! Kill! Kill!
Flash & Firecat
Fury on Wheels
Genevieve
Grand Prix
Grandview U.S.A.
Greased Lightning
Great Race
Gumball Rally
Hard Drivin'
The Heavenly Kid
Hell on Wheels
Herbie Goes Bananas
Herbie Goes to Monte Carlo
Hot Rod
Hot Rod Girl
King of the Mountain
The Last American Hero
Last Chase
Le Mans
A Man and a Woman: 20 Years Later
The Pinchcliffe Grand Prix
Private Road: No Trespassing
Race for Life
The Racers
Rebel Without a Cause
Red Line 7000
Return to Macon County
Richard Petty Story
Road Racers
Roaring Roads
Safari 3000
Sahara
Shaker Run
Six Pack
Smash Palace
Smokey and the Bandit
Speed Zone
Speedway
Spinout
Stroker Ace
Those Daring Young Men in Their Jaunty Jalopies
Van Nuys Blvd.
Vanishing Point
Viva Las Vegas
Winning
The Wraith

FBI Stories

Assassin
Betrayed
Chu Chu & the Philly Flash
Code Name Alpha
The Eiger Sanction
Exterminator
FBI Girl
Feds
Flashback
Flight to Nowhere
Follow That Car
''G'' Men
I Was a Zombie for the FBI

I'll Get You
J. Edgar Hoover
Johnnie Gibson F.B.I.
Killer Likes Candy
Let 'Em Have It
Manhunter
Mankillers
Married to the Mob
Melvin Purvis: G-Man
Mississippi Burning
No Safe Haven
Point Break
The Private Files of J. Edgar Hoover
S.A.S. San Salvador
The Silence of the Lambs
Sky High
Street Hawk
Target
That Darn Cat
357 Magnum
Thunderheart
White Ghost

Femme Fatale

The Accused
Algiers
Backstab
Best of Betty Boop
Beverly Hills Vamp
Black Magic Woman
Blood of Dracula
The Blood Spattered Bride
The Blue Angel
Body Heat
Cat on a Hot Tin Roof
Champagne for Caesar
Cleopatra
Conquest of Mycene
Dance Hall
Dangerous Liaisons
Daughter of the Tong
Dishonored Lady
Dr. Caligari
Dr. Jekyll and Sister Hyde
Double Indemnity
Drowning by Numbers
Emmanuelle, the Queen
Eve of Destruction
Fangs of the Living Dead
Far from the Madding Crowd
Faster Pussycat! Kill! Kill!
First Name: Carmen
Flame of the Islands
The Flesh and the Devil
A Fool There Was
Frankenhooker
Girl Hunters
A Girl to Kill for
The Girl from Tobacco Row
Girlfriend from Hell
Great Flamarion
Grievous Bodily Harm
Hellgate
I Spit on Your Corpse
Impulse
In a Lonely Place
Inner Sanctum
Into the Fire
Invasion of the Bee Girls
The Jayne Mansfield Story
Jezebel
Lady Frankenstein
Lady for a Night
Lady and the Tramp
The Lair of the White Worm
The Legend of the Wolf Woman
Lemora, Lady Dracula
The Leopard Woman
The Lifetaker
Mado
Mantis in Lace
Married to the Mob

Mata Hari
Medusa Against the Son of Hercules
Mesa of Lost Women
Mountaintop Motel Massacre
Ms. 45
Nana
Nanami, First Love
The Natural
Night Angel
Night of the Cobra Woman
The Night Evelyn Came Out of the Grave
Night Tide
No Way Out
Of Human Bondage
On Her Majesty's Secret Service
Out Cold
Out of the Past
Party Girls
Passione d'Amore
Pepe le Moko
Planet of Blood
Play Murder for Me
Possession: Until Death Do You Part
The Postman Always Rings Twice
Prey of the Chameleon
A Rage in Harlem
Roadhouse Girl
Salome
Samson and Delilah
Satanik
Scandal
Scarlet Street
Sea of Love
Secret Weapons
Seven Sinners
Sexpot
Sins of Jezebel
Sleep of Death
Some Came Running
Something Wild
The Strange Woman
Sunrise
Teenage Seductress
The Temptress
Third Degree Burn
Tomb
Too Late for Tears
Tootsie
Truth or Dare
Under Suspicion
The Untamable
Valmont
Vampire Princess Miyu
The Velvet Vampire
Venus Against the Son of Hercules
Venus in Furs
The Wasp Woman
Who Framed Roger Rabbit?
Witchcraft 3: The Kiss of Death
The Women

Film & Film History

Academy Award Winners Animated Short Films
Action
The Al Jolson Collection
Almonds & Raisins: A History of the Yiddish Cinema
An Almost Perfect Affair
America at the Movies
Assunta Spina
Bacall on Bogart
The Bad and the Beautiful
Bass on Titles
Battle of Elderbush Gulch
The Battleship Potemkin

Bellissimo: Images of the Italian Cinema
The Big Picture
Birth of a Legend
The Birth of Soviet Cinema
Bombshell
The Boys from Termite Terrace
Bruce Conner Films 1
Bruce Conner Films 2
Burden of Dreams
Buster Keaton Rides Again/ The Railroader
Camp Classics #1
Candid Hollywood
Chaplin: A Character Is Born
Chaplin: A Character Is Born/ Keaton: The Great Stone Face
Cinema Paradiso
Cinema Shrapnel
Citizen Welles
The Comic
Commission
Constructing Reality: A Film on Film
Contempt
Crimes & Misdemeanors
Custer's Last Fight
Dario Argento's World of Horror
Doctor X
Dracula: The Great Undead
Early Cinema Vol. 1
Early Films #1
Early Sound Films
Edwin S. Porter/Edison Program
8 1/2
Eisenstein
En Passant
Epic That Never Was
Falling for the Stars
Fangoria's Weekend of Horrors
The Fantasy Film Worlds of George Pal
Film Firsts
Filmmaker: A Diary by George Lucas
Filmmakers: King Vidor
First Twenty Years: Part 1 (Porter pre-1903)
First Twenty Years: Part 2 (Porter 1903-1904)
First Twenty Years: Part 3 (Porter 1904-1905)
First Twenty Years: Part 4
First Twenty Years: Part 5 (Comedy 1903-1904)
First Twenty Years: Part 6 (Comedy 1904-1907)
First Twenty Years: Part 7 (Camera Effects)
First Twenty Years: Part 8 (Camera Effects 1903-1904)
First Twenty Years: Part 9 (Cameramen)
First Twenty Years: Part 10 (Special Effects)
First Twenty Years: Part 11 (Drama on Location)
First Twenty Years: Part 12 (Direction & Scripts)
First Twenty Years: Part 13 (D.W.Griffith)
First Twenty Years: Part 14 (Griffith's Dramas)
First Twenty Years: Part 15 (Griffith's Editing)
First Twenty Years: Part 16 (Later Griffith)
First Twenty Years: Part 17 (Make-up Effects)

Folklore

Clash of the Titans
Colossus and the Amazon
  Queen
Crossroads
Darby O'Gill & the Little
  People
Disney's American Heroes
Disney's Storybook Classics
Donkey Skin (Peau d'Ane)
Down to Earth
Dreamwood
Erik the Viking
Faust
Field of Dreams
The Fool Killer
Force on Thunder Mountain
The Fury of Hercules
The Giant of Marathon
The Giant of Metropolis
The Giants of Thessaly
Golden Voyage of Sinbad
Goliath Against the Giants
Goliath and the Barbarians
Goliath and the Dragon
Goliath and the Sins of
  Babylon
The Gorgon
Grendel, Grendel, Grendel
Hercules
Hercules 2
Hercules Against the Moon
  Men
Hercules and the Captive
  Women
Hercules in the Haunted World
Hercules Unchained
Invincible Gladiators
Invitation to the Dance
Iphigenia
Jason and the Argonauts
Kenneth Anger, Vol. 1:
  Fireworks
Kismet
Kriemhilde's Revenge
The Legend of Suram
  Fortress
The Loves of Hercules
Maciste in Hell
Marc and Ann
Medea
Merlin and the Sword
Messalina vs. the Son of
  Hercules
Monty Python and the Holy
  Grail
The Natural
Oh, Kojo! How Could You!
Pancho Villa Returns
Pathfinder
Peter and the Wolf
Phedre
Phoenix
Robin Hood, the Legend:
  Volume 2
Rudyard Kipling's Classic
  Stories
The Secret of the Loch
The Secret of Navajo Cave
Siegfried
Son of Samson
Sons of Hercules in the Land
  of Darkness
Stories from Ireland
Tall Tales & Legends: Johnny
  Appleseed
Terror of Rome Against the
  Son of Hercules
Thor and the Amazon Women
Ulysses
Ursus in the Valley of the
  Lions
Winds of Change
Wonderful World of the
  Brothers Grimm
Young Aphrodites
Yum, Yum, Yum!

Zvenigora

Football

All the Right Moves
The Best of Times
Black Sunday
Brian's Song
Cheering Section
Cheerleaders' Beach Party
Cheerleaders' Wild Weekend
Diner
The Dropkick
Easy Living
Everybody's All American
First & Ten
First & Ten: The Team Scores
  Again
The Freshman
The Galloping Ghost
Good News
Grambling's White Tiger
Gus
Heaven Can Wait
Hold 'Em Jail
Horse Feathers
Johnny Be Good
Knute Rockne: All American
Knute Rockne: The Rock of
  Notre Dame
The Last Boy Scout
Last Game
The Longest Yard
Necessary Roughness
North Dallas Forty
One Night Only
Paper Lion
Quarterback Princess
Roberta
Semi-Tough
The Spirit of West Point
Trouble Along the Way
Two Minute Warning
Wildcats

Foreign Intrigue

Ambassador Bill
American Roulette
Battle Beneath the Earth
Chinese Boxes
Chinese Web
Cover-Up
The Destructors
Devil Wears White
An Englishman Abroad
The Executioner
The Fourth Protocol
Gunpowder
Jaguar Lives
Jumpin' Jack Flash
Puzzle
Reilly: The Ace of Spies
Saigon: Year of the Cat
Sebastian
Sweet Country
The Third Man
Torn Curtain

France

The Ambassador's Daughter
An American in Paris
And Soon the Darkness
Bedtime Story
The Blood of Others
Breathless (A Bout de Souffle)
Buffet Froid
Can-Can
Captain Scarlett
Carnival in Flanders (La
  Kermesse Heroique)
Carve Her Name with Pride
Catherine & Co.
Cesar

Charade
Cloud Waltzing
Danton
Desiree
Dirty Rotten Scoundrels
Dubarry
The Duellists
El Preprimido (The Timid
  Bachelor)
Elusive Corporal
The Fatal Image
The Foreman Went to France
Frantic
French Connection 2
French Intrigue
The French Lesson
French Postcards
The French Woman
Gentlemen Prefer Blondes
Gigi
Harvest
Invitation to Paris
Is Paris Burning?
Joan of Paris
La Marseillaise
La Passante
The Lacemaker
The Last Metro
Last Tango in Paris
The Last Train
Le Bal
Le Joli Mai
Le Million
Les Miserables
The Life of Emile Zola
The Longest Day
Lovely to Look At
Loves & Times of
  Scaramouche
Madame Bovary
Man on the Eiffel Tower
Man in the Iron Mask
Maniac
Marie Antoinette
Marius
May Fools
The Moderns
Moliere
Monkeys, Go Home!
Monsieur Beaucaire
Monsieur Verdoux
The Moonraker
Moulin Rouge
The Mountain
Mr. Klein
My Father's Glory
Napoleon
A New Kind of Love
Paths of Glory
Prisoner of Honor
The Raven
Reign of Terror
Reporters
Reunion in France
Round Midnight
The Rules of the Game
The Scarlet Pimpernel
Silk Stockings
Spy of Napoleon
A Tale of Two Cities
Theatre of Death
This Land is Mine
Those Magnificent Men in
  Their Flying Machines
To Catch a Thief
True Colors
Until September
Valmont
Waterloo
What Price Glory?

Front Page

Absence of Malice

Act of Passion: The Lost
  Honor of Kathryn Beck
Action in Arabia
The Adventures of Nellie Bly
Alice in the Cities
All Over Town
All the President's Men
American Autobahn
And Nothing But the Truth
AngKor: Cambodia Express
Anzio
The Average Woman
Barton Fink
The Beniker Gang
Between the Lines
Big News
Big Town After Dark
Broadcast News
Bulldog Edition
Cambio de Cara (A Change of
  Face)
Chill Factor
The China Syndrome
Citizen Kane
City in Fear
Concrete Beat
Conexion Oriental (East
  Connection)
Continental Divide
Criminal Act
Cry Freedom
Cut and Run
Deadline
Dear America: Letters Home
  from Vietnam
Defense of the Realm
Demonstone
The Devil's Commandment
Diamond Trail
Double Cross
Double Exposure
Double Exposure: The Story
  of Margaret Bourke-White
Eternity
Exiled to Shanghai
Eyes Behind the Stars
Eyes of a Stranger
A Face in the Fog
Far East
Fever Pitch
Final Extra
Finnegan Begin Again
Fit for a King
A Flash of Green
Fletch
Fletch Lives
Foreign Correspondent
The Front Page
The Ghost of H.L. Mencken
The Great Muppet Caper
Grown Ups
Headline Woman
Heat Wave
Her Life as a Man
His Girl Friday
Hit the Ice
Hour of Decision
Hue and Cry
Hustling
I Cover the Waterfront
The Image
An Inconvenient Woman
Istanbul
Johnny Come Lately
Just One of the Guys
Kid from Not-So-Big
Kill Slade
The Killing Fields
Killing Stone
La Dolce Vita
Lana in Love
Late Extra
Libeled Lady
Lip Service
Lonelyhearts
The Love Machine

Love is a Many-Splendored
  Thing
Malarek
The Man Inside
Mean Season
Meet John Doe
Midnight Warrior
Murder on the Campus
Murrow
Natas...The Reflection
News at Eleven
Newsfront
Newsies
The Nightmare Years
No Time to Die
Not for Publication
Nothing Sacred
The Odessa File
L'Odeur des Fauves
One Man Out
Paper Lion
Paralyzed
The Payoff
Perfect
Philadelphia Story
Platinum Blonde
The Ploughman's Lunch
Primary Motive
Private Investigations
Qui Etes Vous, Mr. Sorge?
Reckless Disregard
Resurrection of Zachary
  Wheeler
Revenge of the Radioactive
  Reporter
Riding on Air
Roman Holiday
Safari 3000
Salome, Where She Danced
Salvador
The Secrets of Wu Sin
Sex O'Clock News
Sex and the Single Girl
Sex Through a Window
Shock Corridor
Short Fuse
Shriek in the Night
Soul Patrol
Special Bulletin
Star Reporter
Stavisky
Straight Talk
Stranger on the Third Floor
Street Smart
Sucker Money
Sweet Smell of Success
Switching Channels
Teacher's Pet
That Lucky Touch
They Got Me Covered
The Thirteenth Man
Thirteenth Reunion
Those Glory, Glory Days
Three O'Clock High
Too Hot to Handle
Tough Assignment
Transylvania 6-5000
Under Fire
The Underworld Story
Unexplained Laughter
Up Your Alley
Where the Buffalo Roam
While the City Sleeps
Whispering City
Without Love
Woman Condemned
Woman of the Year
A World Apart
The World Gone Mad
The Year of Living
  Dangerously
Yesterday's Witness

Gambling

Aces and Eights
Action for Slander
Atlantic City
A Big Hand for the Little Lady
Big Town
A Billion for Boris
Birds & the Bees
Bitter Sweet
Bob le Flambeur
Boldest Job in the West
Casino
Catch Me...If You Can
The Cheaters
Cincinnati Kid
Cockfighter
Color of Money
Dead Lucky
Deadly Impact
Diamonds are Forever
Dona Flor & Her Two
    Husbands
Double Dynamite
Draw!
The Duchess and the
    Dirtwater Fox
Duke of the Derby
Easy Money
Eight Men Out
El Fin del Tahur
El Gallo de Oro (The Golden
    Rooster)
Fever Pitch
Five Card Stud
Flame of the Barbary Coast
Flame of the Islands
The Flamingo Kid
Force of Evil
Frankie and Johnny
The Gamble
Gamble on Love
The Gambler
The Gambler & the Lady
Gilda
Glory Years
Golden Rendezvous
Gun Crazy
Guys & Dolls
Half a Lifetime
Harder They Fall
Havana
Here Come the Marines
House of Games
In Like Flint
Inn of Temptation
Inside Out
Jinxed
Kenny Rogers as the Gambler
Kenny Rogers as the
    Gambler, Part 2: The
    Adventure Continues
Lady from Louisiana
Lady for a Night
Las Vegas Hillbillys
Las Vegas Lady
Las Vegas Story
Las Vegas Strip War
Las Vegas Weekend
Le Million
The Lemon Drop Kid
Let It Ride
Let's Do It Again
Let's Make it Legal
Little Miss Marker
Little Vegas
Living to Die
Lookin' to Get Out
Lost in America
Manhattan Melodrama
The Marrying Man
McCabe & Mrs. Miller
Meet Me in Las Vegas
Melodie en Sous-Sol
Mr. Lucky
Murderers' Row

My Darling Clementine
My Little Chickadee
Ninja, the Violent Sorcerer
No Man of Her Own
Oklahoma Kid
Order to Kill
Our Man Flint
Paradise Island
Pink Cadillac
Pleasure Palace
Poker Alice
The Queen of Spades
Quintet
Rain Man
Ramblin' Man
Riders of the Range
Rimfire
The Roaring Twenties
Robbery
Ruggles of Red Gap
Say Yes!
The Shadow Man
The Shanghai Gesture
Silver Queen
Sno-Line
Song of the Thin Man
Sorrowful Jones
The Squeeze
Stacy's Knights
The Sting
Strike It Rich
Support Your Local Sheriff
Swing Time
Tall Story
Tax Season
Tennessee's Partner
Things Change
This is a Hijack
Trading Places
Uphill All the Way
Uptown Saturday Night
The Vegas Strip Wars
Walking Tall
Wanda Nevada
Welcome to 18
The Westerner
Wheel of Fortune

Game Show

Best of Everything Goes
Champagne for Caesar
Do You Trust Your Wife/I've
    Got a Secret
Endurance
Family Game
For Love or Money
Gametime Vol. 1
Major Bowes' Original
    Amateur Hour
Prize of Peril
Queen for a Day
Rich Little's Great Hollywood
    Trivia Game
Shock Treatment

Gangs & Gangsters

Abduction of St. Anne
Al Capone
All Dogs Go to Heaven
Alley Cat
Alvin Rides Again
The American Gangster
American Me
The American Soldier
Angels Die Hard
Armed Response
Baby Face Morgan
Bad Bunch
Ballbuster
The Barber Shop
Big Bad Mama
Big Boss
The Big Brawl

Big Combo
Billy Bathgate
Black Bikers from Hell
Black Caesar
Black Gestapo
Black Godfather
Bloody Mama
Blue
Blues Busters
Bonnie & Clyde
Boss of Big Town
Bowery Blitzkrieg
Boyz N the Hood
Bring Me the Head of Alfredo
    Garcia
Broadway Danny Rose
The Bronx War
Brother Orchid
Buddy Buddy
Bugsy
Bulldog Edition
Bulldog Jack
Bullets or Ballots
Camorra: The Naples
    Connection
Chains
The Chase
China Girl
China O'Brien
Chrome Hearts
Chrome and Hot Leather
City Heat
Class of 1984
Class of 1999
Clipped Wings
Cobra
Code of Silence
Colors
Comfort and Joy
Commando Squad
Confidential
Contraband
Cookie
The Cool World
Cop in Blue Jeans
Corruption
Corsair
Cosh Boy
The Cross & the Switchblade
Cruisin' High
Cry of a Prostitute: Love Kills
Cry Terror
Danger Zone 2
Dangerous Mission
Dark Rider
Darktown Strutters
The Day It Came to Earth
Dead End City
Deadly Encounter
Deadly Revenge
Defiance
Defiant
Delta Fox
Designing Woman
Devil's Angels
Diamante, Oro y Amor
Diamonds on Wheels
Dick Tracy
Dick Tracy vs. Cueball
Dillinger
Dirt Gang
Dirty Laundry
Disorganized Crime
Diva
The Divine Enforcer
Dixie Jamboree
The Doberman Gang
The Dogs Who Stopped the
    War
The Don is Dead
Each Dawn I Die
East L.A. Warriors
East Side Kids
Education of Sonny Carson
El Callao (The Silent One)
Emil & the Detectives

Enemy Territory
The Enforcer
Escape from the Bronx
Every Which Way But Loose
Exterminator
Eye of the Tiger
The Family
The Final Alliance
A Fine Mess
Fireback
Force of Evil
Forced Vengeance
Forever, Lulu
Four Deuces
Four Jacks and a Jill
Foxy Brown
"G" War
The Gambler & the Lady
Game of Death
Gang Bullets
Gang Wars
Gangbusters
The Gangster
Get Rita
Ghetto Blaster
Ghost Chasers
The Ghost and the Guest
Gimme Shelter
Girls of the White Orchid
The Glory Stompers
The Golden Triangle
Gone are the Days
Goodfellas
The Great Skycopter Rescue
The Grissom Gang
The Gumshoe Kid
Gun Crazy
The Gunrunner
Hard-Boiled Mahoney
Harley
Heartbreaker
Heat Street
Hell on Frisco Bay
Hell Up in Harlem
High Sierra
Hit Men
Hitz
Hollywood Cop
Homeboy
Homeboys
Hoodlum Empire
I Died a Thousand Times
I, Mobster
I'd Give My Life
Impulse
The Jesus Trip
Jinxed
Johnny Angel
Johnny Apollo
Johnny Dangerously
Kansas City Massacre
Key Largo
Killer Dill
Killing in the Sun
Knights of the City
L.A. Bad
La Balance
L.A. Gangs Rising
Ladies of the Lotus
Lady from Nowhere
Lady in Red
Lady Scarface
The Last Riders
Law of the Underworld
Lethal Obsession
Little Caesar
Little Laura & Big John
Lonely Boy/Satan's Choice
The Long Good Friday
The Longshot
The Lost Capone
Love Me or Leave Me
Ma Barker's Killer Brood
Machine Gun Kelly
Madigan's Millions
The Manhandlers

Manhattan Merry-go-round
Manhunt
Marked Woman
The Masters of Menace
Master's Revenge
Maverick Queen
Mean Frank and Crazy Tony
Melvin Purvis: G-Man
Men of Respect
The Messenger
Midnight
Miller's Crossing
Million Dollar Kid
Mob Boss
Mob Story
Mob War
Mobsters
The Monster and the Stripper
Mr. Majestyk
Nashville Beat
'Neath Brooklyn Bridge
The Neon Empire
Never a Dull Moment
Never Let Go
Never Love a Stranger
Never Steal Anything Small
The New Kids
1931: Once Upon a Time in
    New York
99 & 44/100 Dead
Ninja Brothers of Blood
No Trace
Oscar
Paco
Paradise in Harlem
Parole, Inc
Party Girl
The Payoff
Performance
Petrified Forest
Pickup on South Street
Pierrot le Fou
Play Murder for Me
Postmark for Danger/
    Quicksand
The Power of the Ninjitsu
Prime Cut
Prize Fighter
Public Enemy
Pulp
Punk Vacation
Quatorze Juliet
The Racket
The Racketeer
Raw Deal
Renegades
Return of the Dragon
Return of the Rebels
Return of the Tiger
Revenge
Revenge of the Ninja
The Revenger
Revolt of the Dragon
Ricco
Riot Squad
The Rise and Fall of Legs
    Diamond
The Ritz
The Roaring Twenties
Robin and the Seven Hoods
Rooftops
Rulers of the City
Rumble Fish
The St. Valentine's Day
    Massacre
The Savage Seven
Scared Stiff
Scarface
The Sea Serpent
Second Chance
The Set-Up
Sexton Blake and the Hooded
    Terror
Shaft
Shark Hunter
She's Back

Horror

Horror

Bride of Re-Animator
A Bucket of Blood
C.H.U.D.
C.H.U.D. 2: Bud the Chud
The Comedy of Terrors
Creature from the Haunted
 Sea
Critters
Critters 2: The Main Course
Critters 3
Cutting Class
Dead Heat
Dr. Jekyll and Sister Hyde
Dolls
Doom Asylum
Dracula Blows His Cool
Dracula & Son
Dracula—Up in Harlem
Eat and Run
Elvira, Mistress of the Dark
Evil Spawn
Face of the Screaming
 Werewolf
The Fearless Vampire Killers
Frankenhooker
Frankenstein's Great Aunt
 Tillie
From Beyond
Ghoulies
Ghoulies 2
Ghoulies 3: Ghoulies Go to
 College
Gore-Met Zombie Chef From
 Hell
Gremlins
Gremlins 2: The New Batch
Highway to Hell
Hillbillys in a Haunted House
The Horror of Frankenstein
House
House 2: The Second Story
House of the Long Shadows
Human Beasts
Hysterical
I Married a Vampire
I Was a Teenage Zombie
Incredibly Strange Creatures
 Who Stopped Living and
 Became Mixed-Up Zombies
The Invisible Maniac
It Came from Hollywood
Killer Klowns from Outer
 Space
Killer Tomatoes Strike Back
Little Shop of Horrors
Mama Dracula
Microwave Massacre
Mom
Monster in the Closet
Motel Hell
Munchies
My Son, the Vampire
Night of the Creeps
Night of the Laughing Dead
Night of a Thousand Cats
Nocturna
Nothing But Trouble
Old Mother Riley's Ghosts
Out of the Dark
Phantom Brother
Piranha
Piranha 2: The Spawning
Psychos in Love
Q (The Winged Serpent)
The Re-Animator
Redneck Zombies
Return of the Killer Tomatoes
Return of the Living Dead
Return of the Living Dead Part
 2
Saturday the 14th
Saturday the 14th Strikes
 Back
Schlock
Son of Ingagi

Sorority Babes in the Slimeball
 Bowl-A-Rama
Speak of the Devil
Spider Baby
Student Bodies
The Stuff
Terrorvision
Theatre of Blood
The Thrill Killers
The Undertaker and His Pals
Vamp
The Vampire Happening
Zombies on Broadway

Horses

The Adventures of Gallant
 Bess
Appaloosa
Bells of San Angelo
Black Beauty
Black Beauty/Courage of
 Black Beauty
The Black Stallion
The Black Stallion Returns
Blue Fire Lady
Born to Run
Casey's Shadow
Chino
Come On Tarzan
Courage of Black Beauty
Courage of the North
Danny
Day the Bookies Wept
Devil Horse
Eagle's Wing
The Electric Horseman
Equus
Far North
The Fighting Stallion
The Flying Deuces
Forbidden Trails
A Girl in Every Port
Glory
The Golden Stallion
Heart of the Rockies
Hell's Hinges
Hit the Saddle
The Horse in the Gray Flannel
 Suit
Horsemasters
Horsemen
Hot to Trot!
Indian Paint
International Velvet
Italian Straw Hat
Justin Morgan Had a Horse
King of the Stallions
King of the Wild Horses
The Lighthorseman
Lightning: The White Stallion
Littlest Horse Thieves
Littlest Outlaw
Man from Snowy River
Marshal of Cedar Rock
Miracle of the White Stallions
The Misfits
Misty
My Friend Flicka
My Pal Trigger
National Lampoon's Animal
 House
Prince and the Great Race
The Quiet Man
A Rare Breed
The Red Fury
The Red Stallion
Return to Snowy River
Return of Wildfire
Ride a Wild Pony
Rodeo Champ
Rolling Home
Running Wild
Saddle Buster
Silver Stallion

Speedy
Sundown Saunders
Texas to Bataan
The Undefeated
Western Double Feature 16
When the West Was Young
White Mane
Wild Hearts Can't Be Broken
Wild Horse Hank
Wild Horses
Wild Pony

Hospitals &
Medicine

B.O.R.N.
Britannia Hospital
Carry On Nurse
A Child is Waiting
China Beach
The Clonus Horror
Coma
Critical Condition
Dead Ringers
Disorderlies
Disorderly Orderly
Dr. Kildare's Strange Case
Doctor at Large
Doctors' Wives
East of Kilimanjaro
The Egyptian
Frankenstein General Hospital
Green for Danger
Gross Anatomy
The Hasty Heart
Horror Hospital
The Hospital
Hospital Massacre
Hospital of Terror
In for Treatment
The Interns
Invasion
Lifespan
Naked Is Better
Nurse
One Flew Over the Cuckoo's
 Nest
Resurrection
A Stolen Face
Tender Loving Care
Threshold
Under the Sheets
Vital Signs
Where the Girls Are
Whose Life Is It Anyway?
Young Doctors in Love

Hunting

The Bear
The Belstone Fox
The Bridge to Nowhere
Caddyshack
Dan Candy's Law
Deadly Prey
Elmer Fudd Cartoon Festival
Elmer Fudd's Comedy Capers
First Blood
Forest of Little Bear
Gaiety
Git!
Greed & Wildlife: Poaching in
 America
Harry Black and the Tiger
Hunting the White Rhino
Mogambo
The Most Dangerous Game
Night of the Grizzly
Shoot
Shooting
Those Calloways
White Hunter, Black Heart

Identity

The Barber Shop
Biggles
Brainwaves
Captive Heart
Changes
The Crimes of Stephen
 Hawke
Dark Passage
Diabolically Yours
Dr. Heckyl and Mr. Hype
Dumbo
Far Out Man
The Girl Said No
The Great Gabbo
High Country
La Femme Nikita
Marnie
Middle Age Crazy
Montenegro
Moving Out
My Brilliant Career
My Dinner with Andre
Next Year If All Goes Well
Norma
The Nutty Professor
O Youth and Beauty
Old Boyfriends
Pocketful of Miracles
A Raisin in the Sun
Return of the Secaucus 7
The River Niger
Sorrows of Gin
Timothy and the Angel
Torment
Vincent, Francois, Paul and
 the Others
The Wrong Move
Zelig

Immigration

Arizona
Avalon
Blood Red
Borderline
Buck Privates Come Home
Does This Mean We're
 Married?
Drachenfutter (Dragon's
 Food)
El Super
Ellis Island
Green Card
Hester Street
I Cover the Waterfront
I Remember Mama
The Immigrant
Journey of Hope
A King in New York
The Manions of America
Marcados por el Destino
Maricela
Midnight Auto Supply
Monkey Business
Moscow on the Hudson
My Girl Tisa
Pelle the Conqueror
Skyline
Wait Until Spring, Bandini
Where is My Child?

India

The Black Devils of Kali
Black Narcissus
Conduct Unbecoming
Courtesans of Bombay
The Deceivers
Distant Thunder
Drums
The Far Pavilions
Gandhi
Gunga Din
Heat and Dust

The Home and the World
India's Master Musicians
The Jewel in the Crown
Journey to the Lost City
Jungle Hell
Kim
The Lives of a Bengal Lancer
The Mahabharata
The Man Who Would Be King
Northwest Frontier
Octopussy
A Passage to India
Pather Panchali
The Razor's Edge
The River
Sabaka
Shakespeare Wallah
Spices
Staying On
Two Daughters
Wee Willie Winkie

Insects

Beehive
Donald's Bee Pictures:
 Limited Gold Edition 2
Forever Darling
Hellstrom Chronicle
Hoppity Goes to Town
Woman in the Dunes

Interviews

Bela Lugosi: The Forgotten
 King
Born Famous
Bugs Bunny Superstar
Chuck Berry: Hail! Hail! Rock
 'N' Roll
Citizen Welles
The Compleat Beatles
Cult People
Cyberpunk
Drawing the Line: A Portrait of
 Keith Haring
Eros, Love & Lies
Falling for the Stars
Hearts of Darkness: A
 Filmmaker's Apocalypse
Hedda Hopper's Hollywood
High Adventure
Hollywood Ghost Stories
Ingrid
Inside Hitchcock
James Cagney: That Yankee
 Doodle Dandy
James Dean
The JFK Conspiracy
John Lennon: Interview with a
 Legend
John Lennon/Yoko Ono: Then
 and Now
Making of a Legend: ''Gone
 With the Wind''
Mondo Lugosi: A Vampire's
 Scrapbook
Person to Person
The Prince and Princess of
 Wales...Talking Personally
Private Conversations: On the
 Set of ''Death of a
 Salesman''
Problems, 1950s' Style
Remembering LIFE
The Return of Ruben Blades
Roger & Me
Sometimes It Works,
 Sometimes It Doesn't
Stork Club
Stripper
Tribute to Ricky Nelson

The Last Picture Show
MacArthur
The Manchurian Candidate
M*A*S*H
M*A*S*H: Goodbye, Farewell
  & Amen
Men in War
Mission Inferno
Mr. Walkie Talkie
One Minute to Zero
Operation Dames
Operation Inchon
Pork Chop Hill
Pueblo Affair
Retreat, Hell!
Sabre Jet
Sayonara
Sergeant Ryker
The Steel Helmet
Torpedo Alley
Wonsan Operation

Labor & Unions

Act of Vengeance
The Big Man: Crossing the
  Line
Black Fury
Border Heat
Boxcar Bertha
Brother John
Devil & Miss Jones
The Displaced Person
End of St. Petersburg
Fame is the Spur
F.I.S.T.
The $5.20 an Hour Dream
Gung Ho
I'm All Right Jack
Joyride
Keeping On
Killing Floor
Last Exit to Brooklyn
Man of Marble
Matewan
Molly Maguires
Moonlighting
Never Steal Anything Small
Newsies
Norma Rae
North Dallas Forty
On the Waterfront
Pajama Game
Salt of the Earth
Singleton's Pluck
Strike
Sweetheart!
Take This Job & Shove It
The Triangle Factory Fire
  Scandal
Waterfront

Late Bloomin' Love

Adorable Julia
The African Queen
Ali: Fear Eats the Soul
Atlantic City
Breakfast in Paris
Brief Encounter
Charly
The Christmas Wife
Circle of Two
Cold Sassy Tree
Crazy from the Heart
Devil in the Flesh
Doctor in Distress
End of August
Finnegan Begin Again
Forbidden Love
Forty Carats
Gal Young 'Un
Gin Game
Harold and Maude
If Ever I See You Again

La Symphonie Pastorale
The Last Winters
Lonely Hearts
The Lonely Passion of Judith
  Hearne
Love Among the Ruins
Macaroni
A Man and a Woman
The Night of the Iguana
Other People's Money
Players
Queen of the Stardust
  Ballroom
The Rainmaker
Roman Spring of Mrs. Stone
San Antonio
Shirley Valentine
The Shock
The Shop Around the Corner
Shore Leave
A Small Killing
Something in Common
Stay As You Are
Stormy Waters
Stroszek
To Love Again
Travelling North
24 Hours in a Woman's Life
Twice in a Lifetime
With Six You Get Eggroll

Law & Lawyers

Adam's Rib
Afterburn
And Justice for All
Angel on My Shoulder
Beyond a Reasonable Doubt
Beyond the Silhouette
Brain Donors
The Candidate
Cape Fear
A Case of Libel
Chase
Class Action
The Conqueror Worm
Criminal Law
Death of a Soldier
Dirty Mary Crazy Larry
Disorganized Crime
The Divorce of Lady X
Down the Drain
The Enforcer
Ernest Goes to Jail
Eye Witness
Final Verdict
First Monday in October
The Franchise Affair
A Free Soul
From the Hip
Gideon's Trumpet
Illegal
In the King of Prussia
The Incident
The Jagged Edge
Judge Priest
L.A. Law
Last Innocent Man
Legal Eagles
Legend of Billie Jean
Love Crimes
Man from Colorado
Manhattan Melodrama
Marked Woman
My Cousin Vinny
Narrow Margin
Ordeal by Innocence
Original Intent
The Paper Chase
Perfect Witness
Physical Evidence
Reet, Petite and Gone
Regarding Henry
Road Agent

The Rodney King Case: What
  the Jury Saw in California
  vs. Powell
Rustler's Valley
Slightly Honorable
The Sun Shines Bright
To Kill a Mockingbird
True Believer
True Colors
Twelve Angry Men
The Verdict
Victim
Violent Blood Bath
The Winslow Boy
The Wistful Widow of Wagon
  Gap
The World's Oldest Living
  Bridesmaid
The Young Philadelphians

Loner Cops

Above the Law
Action Jackson
The Blue Knight
Cop in Blue Jeans
Dead Bang
The Dead Pool
Deadly Dancer
Die Hard
Die Hard 2: Die Harder
Dirty Harry
The Enforcer
Force of One
Heat
In the Heat of the Night
Klute
Magnum Force
McQ
One Good Cop
Rent-A-Cop
Serpico
Sharky's Machine
Stick
Stone Cold
Sudden Impact
Ten to Midnight
Year of the Dragon

Macho Men

Abilene Town
Across the Pacific
Action in the North Atlantic
The Adventurers
Adventures of Captain Marvel
The Adventures of Ford
  Fairlane
Alone Against Rome
The Amazing Transplant
American Kickboxer 1
American Ninja
American Ninja 4: The
  Annihilation
Atlas
The Avenger
Backdraft
Bad Boys
Bail Out
Bail Out at 43,000
The Barbarians
Best of Popeye
Black Cobra 2
The Black Pirate
Bootleg
Brannigan
A Bridge Too Far
Brutal Glory
Buck Rogers in the 25th
  Century
Bulletproof
California Manhunt
Cannonball Run
Cannonball Run II
Captain America

Chino
The Cobra
Cobra
Codename: Terminate
Color Adventures of
  Superman
Colossus and the Amazon
  Queen
Conan the Barbarian
Conquest of Mycene
Convoy
Cool Hand Luke
Coriolanus, Man without a
  Country
Dead Man Walking
Deathstalker 4: Clash of the
  Titans
D.I.
Diamonds are Forever
Dick Tracy
Dick Tracy Detective
Dick Tracy, Detective/The
  Corpse Vanishes
Dick Tracy Meets Gruesome
Dick Tracy and the Oyster
  Caper
Dick Tracy Returns
Dick Tracy: The Original
  Serial, Vol. 1
Dick Tracy: The Original
  Serial, Vol. 2
Dick Tracy vs. Crime Inc.
Dick Tracy vs. Cueball
Dick Tracy's Dilemma
The Dirty Dozen
The Dirty Dozen: The Next
  Mission
Dirty Harry
Disney's American Heroes
Dr. No
The Dogs of War
The Dolls
Double Impact
Double Revenge
Duel of Champions
Dune Warriors
18 Bronzemen
Firewalker
First Blood
Fist
A Fistful of Dollars
For Your Eyes Only
Force of One
Forced Vengeance
Fury on Wheels
Gator
Giant
The Giant of Marathon
The Giant of Metropolis
The Giants of Thessaly
Gladiator of Rome
Goliath Against the Giants
Goliath and the Dragon
Gone with the Wind
Greystoke: The Legend of
  Tarzan, Lord of the Apes
Guns
Hangfire
Hard to Kill
Hard Knox
Hard Promises
Hawkeye
Hell in the Pacific
Hellfighters
Hercules
Hercules 2
Hercules Against the Moon
  Men
Hercules in New York
Hercules and the Princess of
  Troy
Hero of Rome
Heroes Three
Hired to Kill
The Hitman
Hooper

Horsemen
Hustle
I Come in Peace
Infra-Man
The Invaders
Invasion U.S.A.
The Iron Crown
Jackie Chan's Police Force
Judex
K2
Kill and Kill Again
The Killing Zone
Kindergarten Cop
Kinjite: Forbidden Subjects
Ladies Night Out
Last Warrior
Legend of the Lone Ranger
License to Kill
The Lion of Thebes
Machoman
Maciste in Hell
Mad Wax: The Surf Movie
The Magnificent Matador
Magnificent Seven
Magnum Force
Mambo Kings
A Man Called Rage
The Man with the Golden Gun
The Masked Marvel
Master Blaster
McBain
Medusa Against the Son of
  Hercules
Messalina vs. the Son of
  Hercules
Messenger of Death
Missing in Action
Missing in Action 2: The
  Beginning
Mobsters
Mole Men Against the Son of
  Hercules
Mr. Majestyk
Murphy's Law
Neutron and the Black Mask
Neutron vs. the Amazing Dr.
  Caronte
Neutron vs. the Death Robots
Neutron vs. the Maniac
Never Say Never Again
No Holds Barred
No Way Back
One Down, Two to Go!
One-Eyed Jacks
One Man Jury
Out for Justice
Partners
Party Animal
Paternity
Patton
Ragin' Cajun
Raging Bull
Rambo: First Blood, Part 2
Rambo: Part 3
Revolt of the Barbarians
The Road Warrior
Robin Hood
Rocky
Rocky 5
Roland the Mighty
The Rookie
The Runaway Barge
Samson Against the Sheik
Samson and His Mighty
  Challenge
Sands of Iwo Jima
The Searchers
Secrets in the Dark
Seven
The Seventh Voyage of
  Sinbad
Shaft
Sinbad of the Seven Seas
Sirocco
Soldier's Fortune
Son of Sinbad

Codename: Foxfire
Codename: Icarus
Cold Comfort
Cold Room
Columbo: Murder By the Book
Columbo: Prescription Murder
Comeback Kid
Comic Relief 2
Comic Relief 3
Conagher
The Concrete Cowboys
The Connection
Consenting Adults
Conspiracy of Terror
Control
Convicted
Convicted: A Mother's Story
Cotton Candy
Courage
Courtship
Coward of the County
Cowboy & the Ballerina
The Cracker Factory
The Crackler
The Cradle Will Fall
The Crash of Flight 401
Crime Story
Criminal Justice
Crisis at Central High
The Crucifer of Blood
Cruise Into Terror
A Cry for Love
Cry Panic
Cry Terror
Curse of the Black Widow
Curse of King Tut's Tomb
Damien: The Leper Priest
Dancing Princesses
Danger in the Skies
Dark Night of the Scarecrow
The Dark Secret of Harvest
   Home
Tho Day the Women Got Even
Days of Wine and Roses
The Dead Don't Die
Dead on the Money
Deadly Surveillance
Deadman's Curve
Dear America: Letters Home
   from Vietnam
Dear Detective
Death of Adolf Hitler
Death Be Not Proud
Death Cruise
Death of the Incredible Hulk
Death at Love House
Death of a Salesman
Death Scream
Death Sentence
The Death Squad
Death Stalk
Deathmoon
The Defender
The Defense Never Rests
Dempsey
Desperate Lives
Desperate Women
Destination Moonbase Alpha
Devil Dog: The Hound of Hell
The Devil's Messenger
Devlin Connection
Dick Powell Theatre: Price of
   Tomatoes
The Dining Room
Dinner at Eight
Directions '66
The Dirty Dozen: The Next
   Mission
The Disappearance of Aimee
The Displaced Person
Dixie: Changing Habits
Do You Trust Your Wife/I've
   Got a Secret
Dr. Strange
Dr. Syn, Alias the Scarecrow

Doctor Who and the Daleks
Doctor Who: Death to the
   Daleks
Doctor Who: Pyramids of
   Mars
Doctor Who: Spearhead from
   Space
Doctor Who: Terror of the
   Zygons
Doctor Who: The Ark in Space
Doctor Who: The Seeds of
   Death
Doctor Who: The Talons of
   Weng-Chiang
Doctor Who: The Time
   Warrior
Dollmaker
Don't Be Afraid of the Dark
Don't Go to Sleep
Don't Look Back: The Story of
   Leroy "Satchel" Paige
A Doonesbury Special
Door to Door
Double Exposure: The Story
   of Margaret Bourke-White
D.P.
Dracula
Draw!
A Dream for Christmas
Dreams Lost, Dreams Found
Drop-Out Mother
Duel
Duel of Hearts
Dumb Waiter
Dust to Dust
Dying Room Only
Dynamite and Gold
An Early Frost
East of Ninevah
Eat a Bowl of Tea
Ebony Tower
Echoes in the Darkness
Edge of Darkness
Edward and Mrs. Simpson
El Diablo
El Fego Baca: Six-Gun Law
Eleanor & Franklin
Electric Grandmother
Elizabeth, the Queen
Elvis and Me
Elvis: The Movie
Embassy
The Emperor's New Clothes
Endurance
An Englishman Abroad
Enola Gay: The Men, the
   Mission, the Atomic Bomb
Eric
Ernie Kovacs: Between the
   Laughter
Escape from Sobibor
The Execution
The Execution of Private
   Slovik
Execution of Raymond
   Graham
The Executioner's Song
Exiled to Shanghai
Experience Preferred... But
   Not Essential
Eye of the Demon
Eye on the Sparrow
F. Scott Fitzgerald in
   Hollywood
Fallen Angel
The Family Man
Family Secrets
Family Upside Down
The Fantastic World of D.C.
   Collins
Fantasy Island
The Far Pavilions
Fatal Chase
The Fatal Image
Fatal Vision
Father Figure

Fear
Feet Foremost
Fellow Traveler
Fer-De-Lance
A Few Moments with Buster
   Keaton and Laurel and
   Hardy
The Fig Tree
A Fight for Jenny
Final Verdict
Final Warning
Finessing the King
Finish Line
Finnegan Begin Again
Fire
Fire and Rain
Firehouse
Firesign Theatre Presents
   Nick Danger in the Case of
   the Missing Yolks
First Affair
First Olympics: Athens 1896
First & Ten
First & Ten: The Team Scores
   Again
The Five of Me
Five Miles to Midnight
The $5.20 an Hour Dream
Flame is Love
Flatbed Annie and Sweetiepie:
   Lady Truckers
Flood!
Florida Straits
For Heaven's Sake
For Ladies Only
For the Love of It
For Love or Money
For Us, the Living
For Your Love Only
Forbidden
Forbidden Love
Force of Evil
Force Five
Foreign Legionnaire: Court
   Martial
Forever
Forever Young
The Forgotten
Fortress
Four Feathers
France/Tour/Detour/Deux/
   Enfants
Francis Gary Powers: The
   True Story of the U-2 Spy
Frankenstein
Freedom Road
Friendly Fire
From Here to Eternity
From Here to Maternity
Fu Manchu
Funhouse
Funnier Side of Eastern
   Canada with Steve Martin
Future Cop
Gametime Vol. 1
Garry Shandling Show, 25th
   Anniversary Special
The Gathering
The Gathering: Part 2
Gentleman Bandit
Get Christie Love!
Getting Physical
Ghostwriter
Gideon's Trumpet
The Gift of Love
A Girl of the Limberlost
Girl Who Spelled Freedom
Girls of Huntington House
Girls of the White Orchid
The Gladiator
Glass House
Glitter Dome
Glitz
Gold of the Amazon Women
The Golden Honeymoon
Goldenrod

Goliath Awaits
Gone are the Days
The Good Father
Gore Vidal's Billy the Kid
Gotham
Grambling's White Tiger
Great American Traffic Jam
Great Expectations
The Great Los Angeles
   Earthquake
Great Wallendas
Green Eyes
Griffin and Phoenix: A Love
   Story
Gryphon
The Guardian
A Guide for the Married
   Woman
Guilty of Innocence
Gulag
A Gun in the House
The Gunfighters
The Guyana Tragedy: The
   Story of Jim Jones
The Gypsy Warriors
Half Slave, Half Free
Halloween with the Addams
   Family
Hamlet
Hammered: The Best of
   Sledge
Hands of a Stranger
The Hanged Man
Hansel & Gretel
Happily Ever After
Hard Frame
Hard Knox
Hard Rain-The Tet 1968
Hardhat & Legs
Harvey
Hatfields & the McCoys
Haunting of Harrington House
The Haunting Passion
Haunting of Sarah Hardy
Haunts of the Very Rich
Having It All
He Who Walks Alone
Heart of a Champion: The Ray
   Mancini Story
Heat Wave
Hector's Bunyip
Heidi
The Heist
Held Hostage
Help Wanted: Male
Helter Skelter
The Henderson Monster
Henry IV, Part I
Henry IV, Part II
Henry V
Henry VI, Part 1
Henry VIII
Henry VI, Part 3
Her Life as a Man
Hi-Di-Hi
High Desert Kill
High Noon: Part 2
High School USA
The Hill
Hill Number One
The Hillside Strangler
Hiroshima Maiden
The History of White People in
   America
The History of White People in
   America: Vol. 2
Hit Lady
Hitchhiker 1
Hitchhiker 2
Hitchhiker 3
Hitchhiker 4
The Hobbit
Hobo's Christmas
Hold the Dream
The Hollywood Detective
Holocaust

Holocaust Survivors...
   Remembrance of Love
Home for the Holidays
Home to Stay
Homecoming: A Christmas
   Story
Homeward Bound
Honeyboy
Honor Thy Father
The Hostage Tower
Hotel Du Lac
Hotline
Hour of Stars
The House of Dies Drear
The House on Garibaldi Street
The House of Lurking Death
How to Break Up a Happy
   Divorce
How Many Miles to Babylon?
How to Pick Up Girls
Huckleberry Finn
Hunchback of Notre Dame
The Hunted Lady
Hunter
Hurricane
Hustling
I, Claudius, Vol. 1: A Touch of
   Murder/Family Affairs
I, Claudius, Vol. 2: Waiting in
   the Wings/What Shall We
   Do With Claudius?
I, Claudius, Vol. 3: Poison is
   Queen/Some Justice
I, Claudius, Vol. 4: Queen of
   Heaven/Reign of Terror
I, Claudius, Vol. 5: Zeus, By
   Jove!/Hail Who?
I, Claudius, Vol. 6: Fool's
   Luck/A God in Colchester
I, Claudius, Vol. 7: Old King
   Log
I Heard the Owl Call My Name
I Know Why the Caged Bird
   Sings
I Married a Centerfold
I Posed for Playboy
I Will Fight No More Forever
I'm a Fool
The Image
The Impossible Spy
The Imposter
In the Custody of Strangers
In Love with an Older Woman
In This House of Brede
The Incredible Hulk
The Incredible Hulk Returns
The Incredible Journey of Dr.
   Meg Laurel
The Incredible Rocky
   Mountain Race
Indiscreet
Initiation of Sarah
Inside the Third Reich
Intimate Contact
Intimate Power
Intimate Stranger
Intimate Strangers
Into the Homeland
Into Thin Air
Intrigue
Intruder Within
Invincible Mr. Disraeli
Invitation to Hell
Ironclads
Islands
It Came Upon a Midnight
   Clear
It's Good to Be Alive
Ivanhoe
Ivory Hunters
Izzy & Moe
J. Edgar Hoover
Jack Frost
Jack the Ripper
Jacqueline Bouvier Kennedy

Mafia

Mafia

Godfather 1902-1959—The
    Complete Epic
The Godfather: Part 2
The Godfather: Part 3
Goodfellas
Gunblast
Hiding Out
High Stakes
Hired to Kill
Hit Lady
The Hit List
Hollywood Man
Honor Thy Father
Hooch
The Hunted Lady
I, Mobster
The Italian Connection
Keaton's Cop
Keeper of the City
The Kill Reflex
Lady of the Evening
The Last Contract
Last Exit to Brooklyn
Last Rites
Little Vegas
Lucky Luciano
Mafia Princess
Mafia vs. Ninja
Manhunt
Married to the Mob
Men of Respect
Mikey & Nicky
Mister Mean
Mobsters
Moving Target
Mr. Majestyk
New Mafia Boss
Nitti: The Enforcer
Odd Jobs
Once Upon a Time in America
Our Family Business
Out for Justice
The Palermo Connection
Panic Button
Payoff
Perfect Killer
Perfect Witness
The Pope Must Diet
Port of New York
Prime Target
Prizzi's Honor
The Red Light Sting
Ricco
The Rise and Fall of Legs
    Diamond
Rock, Baby, Rock It
Sicilian Connection
Sitting Ducks
Sizzle
Slaughter
Slaughter's Big Ripoff
Slow Burn
Smokescreen
Stark
State of Grace
Stiletto
The Stone Killer
Street Hero
Street People
Street War
Sweet Dirty Tony
Terminal Force
They Call Me Bruce
They Paid with Bullets:
    Chicago 1929
Thief
Things Change
Triad Savages
True Identity
Underworld, U.S.A.
White Hot: Mysterious Murder
    of Thelma Todd
Wise Guys
The Wrestler

Magic

Alabama's Ghost
The Alchemist
Bedknobs and Broomsticks
Black Magic Terror
Blackstone on Tour
The Butcher's Wife
Chandu on the Magic Island
Chandu the Magician
Cthulhu Mansion
Deathstalker 3
Devil Thumbs a Ride/Having
    Wonderful Crime
Doctor Mordrid: Master of the
    Unknown
Escape to Witch Mountain
Eternally Yours
Gryphon
Houdini
Lady in Distress
The Linguini Incident
Magic Adventure
Magic of Derek Dingle
The Magician
The Magician of Lublin
The Man in the Santa Claus
    Suit
Max Maven's Mindgames
Mondo Magic
My Chauffeur
The Mysterious Stranger
Mystery Magician
Penn and Teller Get Killed
Penn and Teller Go Public
Quicker Than the Eye
Stunt Rock
Sword of Heaven
Sword & the Sorcerer
Teen Witch
A Touch of Magic in Close-Up
Turnaround
Tut & Tuttle
Wedding Band
Wildest Dreams
Witchdoctor in Tails
The Wizard of Gore
Zotz!

Marriage

Across the Wide Missouri
The Adultress
An Affair in Mind
Alfredo, Alfredo
All This and Heaven Too
Almost You
Always
Amazons
Any Wednesday
Appassionata
Arnold
As You Desire Me
Assault and Matrimony
Autumn Leaves
Baby Doll
The Baby Maker
Bachelor Party
Backstreet Dreams
Barefoot in the Park
Barkleys of Broadway
The Bat People
Battle Shock
Bedroom Eyes 2
Best Enemies
Best Friends
Betrayal
Betsy's Wedding
Bill Cosby, Himself
A Black Veil for Lisa
The Bliss of Mrs. Blossom
Blonde in Black Leather
Blonde Venus
Blood Wedding (Bodas de
    Sangre)
Blue Heaven

Blume in Love
Body Chemistry
Brand New Life
Bride & the Beast
Bride Walks Out
The Brides Wore Blood
Bulldog Drummond's Bride
Bullwhip
Burn Witch, Burn!
Burning Bed
By Love Possessed
Captain's Paradise
Car Trouble
The Catered Affair
Cause for Alarm
A Change of Seasons
Charles and Diana: For Better
    or For Worse
Cheaters
Child Bride of Short Creek
Chinese Roulette
Chloe in the Afternoon
The Chocolate Soldier
Christina
Claire's Knee
Clash by Night
Coast to Coast
Come Back, Little Sheba
Conflict
Consolation Marriage
Conspiracy of Terror
Country Girl
Crime & Passion
Dancing in the Dark
Dangerous Liaisons
Date Bait
The Dead
Dead Mate
Dead in the Water
Death Kiss
Departamento Compartido
Designing Woman
Desire and Hell at Sunset
    Motel
Desire Under the Elms
Despedida de Casada
Desperate Hours
Devil In Silk
Diabolically Yours
Dial "M" for Murder
Diary of the Dead
Diary of a Mad Housewife
Dirty Hands
Divorce Hearing
Divorce—Italian Style
The Divorce of Lady X
Doctors' Wives
Dodsworth
Dollar
Doll's House
A Doll's House
Dominique is Dead
Dona Flor & Her Two
    Husbands
Donkey Skin (Peau d'Ane)
Don't Raise the Bridge, Lower
    the River
Double Deal
Double Indemnity
Double Standard
Doubting Thomas
Dressmaker
Drifting Souls
Drowning by Numbers
Drums Along the Mohawk
Easy Living
Eat a Bowl of Tea
Eating Raoul
Ecstasy
Ecstasy Inc.
Edward and Mrs. Simpson
Egg and I
El Preprimido (The Timid
    Bachelor)
El: This Strange Passion
Eleanor & Franklin

Elizabeth of Ladymead
Ellie
Elvis and Me
Emilienne
Emmanuelle
Emmanuelle, the Queen
Empty Canvas
Encore
End of Desire
End of the Road
Enemies, a Love Story
Enormous Changes
The Entertainer
Entre-Nous (Between Us)
Erotic Images
Escapade
Esposa y Amante
Estate of Insanity
Eternally Yours
Eureka!
Every Girl Should Be Married
Everybody's All American
Experiment Perilous
The Eyes of Youth
Falling from Grace
Family Upside Down
Far Cry from Home
The Farmer's Wife
Fatal Attraction
Father of the Bride
Fatty and Mabel Adrift/Mabel,
    Fatty and the Law
Fatty's Tin-Type Tangle/Our
    Congressman
Femme Fatale
Fever
Fictitious Marriage
First Affair
Fools
For Keeps
For Love or Money
For Pete's Sake
Forever Darling
Fortune's Fool
Fortunes of War
The Four Seasons
Fourth Story
Frieda
From the Terrace
Fun with Dick and Jane
Gaslight
A Gathering of Eagles
Gertrud
Get Out Your Handkerchiefs
The Ghost and the Guest
Giant
The Gift of Love
Girl from Hunan
The Golden Honeymoon
The Good Wife
Grand Theft Auto
The Grass is Greener
Greed
The Greek Tycoon
Grievous Bodily Harm
Group Marriage
Guess Who's Coming to
    Dinner
Guest Wife
A Guide for the Married Man
A Guide for the Married
    Woman
A Handful of Dust
Hard Promises
Heartburn
Here We Go Again!
High Heels
High Society
Homecoming
Honeymoon
Honeymoon Academy
The Horrible Dr. Hichcock
Hot Spell
House Across the Bay
Housewife

How to Break Up a Happy
    Divorce
How to Marry a Millionaire
How to Murder Your Wife
The Howards of Virginia
Hurry, Charlie, Hurry
Husbands and Lovers
I Do! I Do!
I Dood It
I Love My...Wife
I Love You to Death
I Love You Rosa
I Love You...Goodbye
I Married a Vampire
I Married a Woman
I Will, I Will for Now
I Wonder Who's Killing Her
    Now?
I'm on My Way/The Non-Stop
    Kid
Immortal Bachelor
In Celebration
In Name Only
The Innocent
Intimate Story
Intimate Strangers
The Invasion of Carol Enders
The Invisible Ghost
It Takes Two
Java Head
Jenny
The Jeweller's Shop
Josepha
Jubal
Jubilee Trail
Juliet of the Spirits
Just Like Weather
Killing Heat
A Kind of Loving
Kiss and Be Killed
La Bete Humaine
La Chienne
La Vie Continue
Lady Chatterley's Lover
Lady Windermere's Fan
The Last Married Couple in
    America
L'Atalante
Late Spring
Laurel & Hardy: Be Big
Le Beau Mariage
Le Chat
Lesson in Love
Let's Get Married
A Letter to Three Wives
Lianna
Liar's Moon
Life of O'Haru
A Little Sex
Lonely Wives
The Long Haul
The Long, Long Trailer
Loulou
Love from a Stranger
The Lovers
Lovers and Other Strangers
Loving Couples
Madame Bovary
Made for Each Other
Made in Heaven
Making Love
Man from Nowhere
Mannequin
Maria Chapdelaine
Marriage is Alive and Well
The Marriage Circle
The Marriage of Maria Braun
Married?
Married Too Young
A Married Woman
Marty
Matrimaniac
The Member of the Wedding
Memories of a Marriage
Memory of Us
Men...

Microwave Massacre
Mid-Channel
Midnight Crossing
Miracle of Morgan's Creek
Mistress (Wild Geese)
Modern Love
The Money Pit
Montana
Moon's Our Home
Moonstruck
Mr. Love
Mr. Mom
Mr. & Mrs. Smith
Mr. Skeffington
Mr. Sycamore
My Dear Secretary
My Favorite Wife
My First Wife
My Michael
My Other Husband
Naked Jungle
Niagara
Nightkill
Not Tonight Darling
Now and Forever
Nudo di Donna
Odd Obsession
O'Hara's Wife
On Approval
On Golden Pond
On Her Majesty's Secret
   Service
On Valentine's Day
One of My Wives Is Missing
One Trick Pony
Only With Married Men
Orchestra Wives
Overnight Sensation
Paint Your Wagon
The Painted Veil
Pajama Tops
The Palm Beach Story
A Paper Wedding
Paradise
Pardon Mon Affaire
Pardon Mon Affaire, Too!
Peggy Sue Got Married
The Penitent
Pennies From Heaven
Pete 'n' Tillie
Pete Townshend: White City
Petulia
Phffft!
Philadelphia Story
A Piece of Pleasure
Pipe Dreams
Platinum Blonde
Portrait of Teresa
The Postman Always Rings
   Twice
The Primitive Lover
Prisoner of Second Avenue
Prisoners of Inertia
Prowler
Quartet
Race for Life
Rachel and the Stranger
Raging Bull
Rape and Marriage: The
   Rideout Case
Rebecca
Return of the Soldier
Revenge of the Stepford
   Wives
Reversal of Fortune
Rich and Famous
Rich and Strange
Rider from Tucson
Risky Business
Rocky 2
Romantic Englishwoman
The Roof
Runaways
Ruthless People
Ryan's Daughter
Satan's Black Wedding

Say Yes!
Scenes from a Mall
Scenes from a Marriage
Scorchers
Season of Fear
Second Chance
Secret Beyond the Door
Secrets of Women
Seems Like Old Times
Send Me No Flowers
Separacion Matrimonial
Separate Vacations
Separate Ways
Seven Chances
Seven Days Ashore/Hurry,
   Charlie, Hurry
The Seven Year Itch
The Seventeenth Bride
Sex and the Single Girl
Shattered
She-Devil
She Waits
She's Having a Baby
Shoot the Moon
The Singing Blacksmith (Yankl
   der Shmid)
The Single Standard
The Sleeping Tiger
The Slugger's Wife
Smash Palace
Smash-Up: The Story of a
   Woman
Smouldering Fires
Snoopy's Getting Married,
   Charlie Brown
So This is Paris
The Soft Skin
The Story of Boys & Girls
Stromboli
Suspicion
Sweet Poison
Sweet Revenge
Sweethearts
Tell Me a Riddle
Tell Me That You Love Me
Tender Is the Night
The Tender Trap
Test Pilot
That Certain Thing
That Uncertain Feeling
Third Walker
30 Is a Dangerous Age,
   Cynthia
Thursday's Game
Till Death Do Us Part
Till Marriage Do Us Part
Tin Men
To Joy
Too Beautiful for You
Torchlight
The Torture of Silence
A Touch of Class
Tricks of the Trade
True Love
The Tunnel
Twice a Woman
The Two Mrs. Carrolls
Two for the Road
Ulysses
The Unchastened Woman
Under Capricorn
Under the Volcano
Une Jeunesse
Unfaithfully Yours
Unholy Wife
Unkissed Bride
Until September
Uptown New York
Vanishing Act
Vasectomy: A Delicate Matter
Violets Are Blue
The Virgin of Nuremberg
Vivacious Lady
Voulez-Vous Danser Avec
   Moi?
Voyage to Italy

W
Walk in the Spring Rain
Waltz of the Toreadors
War Brides
The Wash
Way Down East
A Wedding
Wedding in Blood
A Wedding in Galilee
The Wedding Party
We're Not Married
White Gold
Who's Afraid of Virginia
   Woolf?
Wifemistress
The Woman Next Door
A Woman Under the Influence
Woman of the Year
The Women
Women on the Verge of a
   Nervous Breakdown
The World is Full of Married
   Men
World's Greatest Lover
Worth Winning
X, Y & Zee
You Can't Fool Your Wife
Yours, Mine & Ours
You've Got to Have Heart

Martial Arts

Above the Law
Against the Drunken Cat
   Paws
Alley Cat
Aloha Summer
American Ninja
American Ninja 2: The
   Confrontation
American Ninja 3: Blood Hunt
American Ninja 4: The
   Annihilation
Angel Town
The Angry Dragon
Assassin
The Assault of Final Rival
Avenging Ninja
The Awaken Punch
Battling Amazons
Bells of Death
Best of the Best
Betrayal and Revenge
The Big Brawl
The Big Rascal
Big Showdown
Big Trouble in Little China
Black Belt
Black Belt Fury
Black Belt Jones
Black Dragon
Black Dragon's Revenge
Black Eagle
Black Samurai
Black & White Swordsmen
The Blazing Ninja
Blind Fists of Bruce
Blind Fury
Blind Rage
Blood Avenger
Blood of the Dragon
Blood of the Dragon Peril/
   Revenge of the Dragon
Blood on the Sun
Bloodfight
Bloodfist
Bloodfist 2
Bloodfist 3: Forced to Fight
Bloodmatch
Bloodsport
The Bloody Fight
Bloody Fist
The Bodyguard
Bolo
Border of Tong

Born Invincible
Born Losers
Braddock: Missing in Action 3
Breaker! Breaker!
Breathing Fire
Bronson Lee, Champion
The Bronx Executioner
Bruce Is Loose
The Bruce Lee Collection
Bruce Lee Fights Back From
   the Grave
Bruce Lee: The Legend
Bruce Lee: The Man/The Myth
Bruce Lee's Ways of Kung Fu
Bruce Le's Greatest Revenge
Bruce Li the Invincible
Bruce Li in New Guinea
Bruce and Shaolin Kung Fu
Bruce the Superhero
Bruce vs. Bill
Bruce's Deadly Fingers
Bruce's Fists of Vengeance
Buddha Assassinator
The Bushido Blade
Cantonen Iron Kung Fu
Catch the Heat
The Cavalier
Challenge of the Dragon
Challenge of the Masters
Challenge the Ninja
Champ Against Champ
Champion Operation
China O'Brien
China O'Brien 2
The Chinatown Kid
Chinese Connection
Chinese Connection 2
Chinese Stuntman
Circle of Iron
Clash of the Ninja
Cleopatra Jones
Cleopatra Jones & the Casino
   of Gold
Clones of Bruce Lee
The Cobra
Cobra
Cobra Against Ninja
Cold Blooded Murder
Counter Attack
Crack Shadow Boxers
Crazy Kung Fu Master
Daggers 8
Dance of Death
Day of the Panther
The Deadliest Art: The Best of
   the Martial Arts Films
Deadly Bet
Deadly Kick
Deadly Strike
Death Beach
Death Challenge
Death Code Ninja
Death Duel of Mantis
Death Machines
Death Mask of the Ninja
Death Raiders
Death Warrant
The Demon Fighter
Diamond Ninja Force
Dirty Partners
The Divine Enforcer
Dolemite
Double Impact
The Dragon Fist
Dragon Force
Dragon Lady Ninja/Devil Killer
Dragon Lee Vs. the Five
   Brothers
The Dragon Lives Again
Dragon Lord
Dragon the Odds
Dragon Princess
The Dragon Returns
Dragon Rider
Dragon from Shaolin
The Dragon Strikes Back

Dragon vs. Needles of Death
Dragon's Claw
The Dragon's Showdown
Dreadnaught
Dreadnaught Rivals
Dual Flying Kicks
Duel of the Brave Ones
Duel of Fists
Duel of the Iron Fist
Duel of the Masters
Duel of the Seven Tigers
Dynamo
Dynasty
Eagle Claws Champion
Eagle's Shadow
Edge of Fury
18 Bronzemen
Eighteen Jade Arhats
18 Weapons of Kung-Fu
The Eliminators
Empire of the Dragon
Empire of Spiritual Ninja
Enter the Dragon
Enter the Game of Death
Enter the Ninja
Enter the Panther
Enter Three Dragons
Exit the Dragon, Enter the
   Tiger
An Eye for an Eye
Fantasy Mission Force
Fast Fists
The Fate of Lee Khan
Fearless Fighters
Fearless Hyena
Fearless Hyena: Part 2
The Fearless Young Boxer
Ferocious Female Freedom
   Fighters
Fight to the Death
Fight for Survival
Fighting Ace
Fighting Black Kings
Fighting Duel of Death
Fighting Life
Final Impact
Fire in the Night
Firecracker
Fist of Death
Fist of Fear, Touch of Death
A Fist Full of Talons
The Fist, the Kicks, and the
   Evils
Fist of Vengeance
Fists of Blood
Fists of Bruce Lee
Fists of Dragons
Fists of Fury
Fists of Fury 2
Fists of the White Lotus
Five Fighters from Shaolin
Flash Challenger
Force of One
Forced Vengeance
Forest Duel
Forgotten Warrior
The Four Hands of Death:
   Wily Match
Four Infernos to Cross
Four Robbers
From China with Death
Fugitive Samurai
Full Metal Ninja
The Furious
Furious
The Furious Avenger
Furious Slaughter
Fury of King Boxer
Fury in the Shaolin Temple
The Gambling Samurai
Game of Death
Gang Wars
Ghost of the Ninja
Ghostwarrior
Golden Destroyers
Golden Exterminators

The Warrior
Warrior Within
Warriors of Fire
Warriors of the Sung Dynasty
Watch Out, Crimson Bat!
Way of the Black Dragon
Ways of Kung Fu
Weapons of Death
When Taekwondo Strikes
The White Butterfly Killer
White Phantom: Enemy of
 Darkness
The Wild Panther
Wolfen Ninja
Young Avengers/18 Jade
 Pearls
The Young Bruce Lee
Young Dragon
Young Dragons: The Kung Fu
 Kids
Young Dragons: The Kung Fu
 Kids 2
Young Hero
Young Hero of Shaolin
Young Hero of the Shaolin
 Part 2
Young Heroes
The Young Taoism Fighter
Young Tiger
Zatoichi, the Blind Samurai
Zatoichi vs. Yojimbo
Zombie vs. Ninja

Martyred Pop Icons

Beatlemania! The Movie
Blue Hawaii
Buddy Holly Story
Bus Stop
Clambake
Don't Look Back
The Doors
The Doors: A Tribute to Jim
 Morrison
The Doors: Live in Europe,
 1968
The Doors: Live at the
 Hollywood Bowl
The Doors: Soft Parade, a
 Retrospective
Elvis in the '50s
Elvis: Aloha from Hawaii
Elvis in Concert 1968
The Elvis Files
Elvis and Me
Elvis Memories
Elvis in the Movies
Elvis: One Night with You
Elvis Presley's Graceland
Elvis: Rare Moments with the
 King
Elvis Stories
Elvis: That's the Way It Is
Elvis: The Great
 Performances, Vol. 1:
 Center Stage
Elvis: The Great
 Performances, Vol. 2: Man
 & His Music
Elvis: The Movie
Elvis on Tour
Forever James Dean
Fun in Acapulco
Gabe Kaplan as Groucho
The James Dean 35th
 Anniversary Collection
James Dean Story
Jimi Hendrix: Rainbow Bridge
John Lennon: Imagine
John Lennon Live in New York
 City
John Lennon and the Plastic
 Ono Band: Live Peace in
 Toronto, 1969
John & Yoko: A Love Story

King Creole
Kissin' Cousins
La Bamba
Lenny
Live a Little, Love a Little
Love Me Tender
Marilyn
Marilyn: The Untold Story
Mondo Elvis (Rock 'n' Roll
 Disciples)
Paradise, Hawaiian Style
The Real Buddy Holly Story
Rebel Without a Cause
The Rose
Roustabout
Sgt. Pepper's Lonely Hearts
 Club Band
The Seven Year Itch
Speedway
Spinout
Stardust Memories
Stay Away, Joe
The Story of Elvis Presley
This is Elvis
Train to Hollywood
The Trouble with Girls

Mass Media

America
The Best of Not Necessarily
 the News
Breakfast in Hollywood
A Case of Libel
Chill Factor
The China Syndrome
A Cry in the Dark
Dead Men Don't Die
Death Watch
E. Nick: A Legend in His Own
 Mind
Eternity
Exiled to Shanghai
Eyewitness
France/Tour/Detour/Deux/
 Enfants
The Gladiators
The Groove Tube
Hollywood Scandals and
 Tragedies
I Was a Teenage TV Terrorist
The Image
International House
Livin' Large
The Lost Honor of Katharina
 Blum
Loving You
My Favorite Year
Network
News at Eleven
Nothing Sacred
The Passenger
Pray TV
Prime Suspect
Ratings Game
Remembering LIFE
Scrooged
Six Fois Deux/Sur et Sous la
 Communication
Soft and Hard
Superstar: The Life and Times
 of Andy Warhol
Tabloid!
The Thrill of It All!
Tunnelvision
Video Vixens
Zelig

Medieval Romps

The Adventures of Robin
 Hood
Becket
The Beloved Rogue
Camelot

Choice of Weapons
Conquest of the Normans
Cyrano de Bergerac
Deathstalker
Deathstalker 2: Duel of the
 Titans
The Decameron
Decameron Nights
Donkey Skin (Peau d'Ane)
Dragonslayer
El Cid
Excalibur
Fighting Prince of Donegal
Fire and Sword
The Flame & the Arrow
Flesh and Blood
Goetz von Berlichingen (Iron
 Hand)
Hawk and Castile
Hawk the Slayer
Inquisition
Ivanhoe
Jabberwocky
King Arthur, the Young
 Warlord
Knights of the Round Table
Kojiro
Ladyhawke
Legend
The Legend of Young Robin
 Hood
The Lion in Winter
Lionheart
Magdalene
The Magic Sword
Mark of the Devil
Mark of the Devil 2
Mark Twain's A Connecticut
 Yankee in King Arthur's
 Court
The Mongols
Monty Python and the Holy
 Grail
The Name of the Rose
Naughty Knights
The Navigator
Pippin
The Prince and the Pauper
The Prince of Thieves
The Ribald Tales of Robin
 Hood
Robin Hood
Robin Hood...The Legend
Robin Hood...The Legend:
 Herne's Son
Robin Hood...The Legend:
 The Swords of Wayland
Robin Hood...The Legend:
 The Time of the Wolf
Robin and Marian
The Scalawag Bunch
Secret of El Zorro
Secret of the Phantom Knight
Seven Samurai
The Seventh Seal
Skullduggery
Stealing Heaven
Sword & the Dragon
Sword of Lancelot
Sword of Sherwood Forest
Sword & the Sorcerer
Sword in the Stone
Sword of the Valiant
Tales of Robin Hood
Tharus Son of Attila
Three Musketeers
Time Trackers
Tor
The Undead
Unidentified Flying Oddball
The Vikings
The Warrior & the Sorceress
When Knights were Bold
Wizards of the Lost Kingdom
Young Warlord

The Zany Adventures of
 Robin Hood

Meltdown

Access Code
Amazing Grace & Chuck
Atoll K
The Atomic Cafe
The Atomic Kid
Big Bus
Black Rain
Bombs Away!
Burgess, Philby, and
 MacLean: Spy Scandal of
 the Century
Chain Reaction
Cyclotrode ''X''
Day After
The Day the Earth Caught Fire
The Death Merchant
Desert Bloom
Dick Barton Strikes Back
Dr. Strangelove, or: How I
 Learned to Stop Worrying
 and Love the Bomb
Enola Gay: The Men, the
 Mission, the Atomic Bomb
Fail-Safe
Fat Man and Little Boy
Final Sanction
First Spaceship on Venus
First Strike
First Yank into Tokyo
Flight of Black Angel
The Fourth Protocol
Full Fathom Five
Future Kill
Glen and Randa
Ground Zero
H-Man
Hiroshima Maiden
Hiroshima: Out of the Ashes
Incident at Map-Grid 36-80
A Man Called Rage
The Manhattan Project
Massive Retaliation
Miracle Mile
Modern Problems
Nuclear Conspiracy
Omega Man
On the Beach
One Night Stand
Panic in the Year Zero!
Patriot
Plutonium Incident
Prisoner in the Middle
Rats
Reactor
The Reflecting Skin
Refuge of Fear
Rocket Attack U.S.A.
The Sacrifice
Silkwood
Slugs
Special Bulletin
The Spy Who Loved Me
State of Things
Superman 4: The Quest for
 Peace
Testament
This is Not a Test
Threads
The 27thth Day
Unknown World
Voyage to the Bottom of the
 Sea
War Games
When the Wind Blows
Young Lady Chatterly 2

Men

Bachelor Apartment
Bachelorette Party

Carnal Knowledge
City Slickers
Esta Noche Cena Pancho
Evening at the Improv
The Great Escape 2
In a Lonely Place
The Incredible Shrinking Man
The Men's Club
Mr. Mom
90 Days
Patti Rocks
Planet of the Apes
The Private Life of Henry VIII
Private Practices: The Story of
 a Sex Surrogate
Raging Bull
Rate It X
Three Men and a Baby
Three Men and a Cradle
Thursday's Game
Truth or Die

Mental Retardation

Bill
Bill: On His Own
Charly
A Child is Waiting
Dark Night of the Scarecrow
A Day in the Death of Joe Egg
Dominick & Eugene
Homer and Eddie
Larry
Lovey: A Circle of Children 2
Of Mice and Men
Rain Man
Square Dance
Sweetie
Tim
Two of a Kind
The Wizard
The Woman Who Willed a
 Miracle

Mexico

Across the Bridge
Against All Odds
The Black Scorpion
The Border
Borderline
Breakout
Bullet for the General
Bullfighter & the Lady
Bullfighters
Commando Squad
Crazy from the Heart
Django
Dos Gallos Y Dos Gallinas
 (Two Roosters For ...)
Emiliano Zapata
Escape from Cell Block Three
The Falcon in Mexico
Five Giants from Texas
Found Alive
Fun in Acapulco
Hot Chili
In 'n Out
Interval
Juarez
La Guerrillera de Villa (The
 Warrior of Villa)
Los Olvidados
Love has Many Faces
Magnificent Seven
Mexican Hayride
The Night of the Iguana
Octaman
Old Gringo
On the Line
The Pearl
Professionals
Que Viva Mexico
Shoot the Living, Pray for the
 Dead

Movie

Teenage Confidential
Terror in the Aisles
Terror on Tape
To See Such Fun
True Duke: A Tribute to John
  Wayne

Music

All You Need is Cash
Allegro Non Troppo
Amadeus
America Live in Central Park
American Blue Note
Americathon
The Apple
Baja Oklahoma
Ballad in Blue
The Band Reunion
Basileus Quartet
Basin Street Revue
Beach Boys: An American
  Band
Beethoven
Beethoven Lives Upstairs
Beethoven's Nephew
Bessie Smith and Friends
The Best of Bing Crosby: Part
  1
The Best of the Kenny Everett
  Video Show
Beyond the Doors
Bill & Ted's Bogus Journey
Bird
Black & Tan/St. Louis Blues
Blame It on the Night
Bloodhounds of Broadway
Blue Angels: A Backstage
  Pass
Blueberry Hill
Born to Boogie
Bound for Glory
Breaking Glass
Buddy Holly Story
Bummer
Burglar
Caledonian Dreams
The California Raisins: Meet
  the Raisins
Captain January
Carmen
Carnegie Hall at 100: A Place
  of Dreams
Carnival of Animals
Cartoon Classics, No. 1:
  Looney Tunes & Merrie
  Melodies
Cassie
Cigarette Blues
The City
The Clash: Rude Boy
Coal Miner's Daughter
Cobham Meets Bellson
Comeback
The Commitments
Competition
The Compleat Beatles
Concert for Bangladesh
Concrete Angels
Cool As Ice
Cool Mikado
The Cotton Club
The Courier
Crossroads
Cruise
David Sanborn: Love and
  Happiness
Dead Girls
Death Drug
Death Games
Deception
Decline of Western Civilization
  1

The Decline of Western
  Civilization 2: The Metal
  Years
The Deputy Drummer
Diner
Disorderlies
Diva
Divine Madness
Doctor Duck's Super Secret
  All-Purpose Sauce
Don't Look Back
The Doors
The Doors: A Tribute to Jim
  Morrison
The Doors: Live in Europe,
  1968
The Doors: Live at the
  Hollywood Bowl
The Doors: Soft Parade, a
  Retrospective
The Dream Continues
Dreaming Lips
Dubeat-E-O
Duel-Duo
Duet for One
Duke Ellington & His
  Orchestra: 1929-1952
Duke Is Tops
Dutch Treat
Eddie & the Cruisers
Eddie & the Cruisers 2: Eddie
  Lives!
The Eddy Duchin Story
Edie in Ciao! Manhattan
Elephant Parts
Elvis in the '50s
Elvis: Aloha from Hawaii
Elvis in Concert 1968
Elvis and Me
Elvis Memories
Elvis: One Night with You
Elvis: Rare Moments with the
  King
Elvis Stories
Elvis: The Lost Performances
Elvis: The Movie
Elvis on Tour
Encounter at Raven's Gate
Eubie!
Evolutionary Spiral
The Fabulous Baker Boys
The Fabulous Dorseys
Fantasia
Fastest Guitar Alive
Fats Domino and Friends
Fats Waller and Friends
Fighting Mad
Fingers
Five Easy Pieces
The Five Heartbeats
The Five Pennies
The Fixx: Live in the USA
A Flock of Seagulls
FM
Follow Me, Boys!
Forget Mozart
Franck Goldberg Videotape
Frank Zappa's Does Humor
  Belong in Music?
Frankie and Johnny
Freddie Hubbard
Fun with the Fab Four
The Gay Ranchero
The Gene Krupa Story
George and the Christmas
  Star
Gerry Mulligan
Get Crazy
Getting Over
The Gig
Gimme Shelter
Give My Regards to Broad
  Street
The Glenn Miller Story
Go Go's: Wild at the Greek
Good Morning, Vietnam

Gotta Dance, Gotta Sing
The Grateful Dead Movie
Great Balls of Fire
The Great Waltz
Green Card
GRP All-Stars from the
  Record Plant
Guess Who: Together Again
Hard Day's Night
Hard Part Begins
Hear My Song
Heartbreak Hotel
Hearts of Fire
Help!
Hi-De-Ho
High Noon
His Memory Lives On
H.M.S. Pinafore
Hollywood Revels
Home Sweet Home
Honeysuckle Rose
Honkytonk Man
Honkytonk Nights
House Party
House Party 2: The Pajama
  Jam
Humoresque
I Wanna Hold Your Hand
Identity Crisis
In Heaven There is No Beer?
Incident at Channel Q
India's Master Musicians
Irving Berlin's America
It Couldn't Happen Here
Jazz Ball
Jazz in Exile
Jazz & Jive
Jazz is Our Religion
Jazz Singer
Jazz at the Smithsonian: Art
  Blakey
Jazz at the Smithsonian: Art
  Farmer
Jazz at the Smithsonian:
  Benny Carter
Jazz at the Smithsonian: Bob
  Wilber
Jazz at the Smithsonian: Joe
  Williams
Jazz at the Smithsonian: Mel
  Lewis
Jazz at the Smithsonian: Red
  Norvo
Jazz on a Summer's Day
Jet Sets
Jimi Hendrix: Rainbow Bridge
Jimi Hendrix: Story
Joe Cocker: Mad Dogs
Joey
John Jacob Niles
John Lennon: Imagine
John Lennon: Interview with a
  Legend
Johnny Guitar
The Jolson Story
Journey: Frontiers and
  Beyond
Kid from Gower Gulch
Killing 'em Softly
King of Jazz
Knights & Emeralds
Krush Groove
La Bamba
La Forza del Destino
Lady Be Good
Lady Grey
Lady Sings the Blues
Las Vegas Hillbillys
Las Vegas Story
Last of the One Night Stands
The Last Waltz
Led Zeppelin: The Song
  Remains the Same
Legendary Ladies of Rock &
  Roll
Legends of Rock & Roll

Leningrad Cowboys Go
  America
Let It Be
Let It Rock
Let's Get Lost
Life's a Beach
Lisztomania
Little River Band
Lonely Boy/Satan's Choice
Los Apuros de Dos Gallos
  (Troubles of Two Roosters)
Mad Dogs & Englishmen
Mahler
The Man and the Monster
Marc and Ann
Mario Lanza: The American
  Caruso
A Matter of Degrees
Matters of the Heart
Maxwell Street Blues
The Medium
Melody for Three
Mo' Better Blues
Monkey Grip
Monterey Pop
Moonlight Sonata
More Song City U.S.A.
Moscow on the Hudson
Mozart: A Childhood
  Chronicle
The Mozart Brothers
The Mozart Story
Mr. Charlie
Muscle Rock Madness
Music & the Spoken Word
The Music Teacher
Musicourt
My Pal Trigger
Mystery in Swing
Naked Eyes
Nashville
Nashville Girl
Nashville Goes International
Nashville Story
Natural States
Natural States Desert Dream
Netherworld
New Orleans: 'Til the Butcher
  Cuts Him Down
New York, New York
The Night the Lights Went Out
  in Georgia
Not Quite Love
Oklahoma Kid
On the Air Live with Captain
  Midnight
Oregon
Ornette Coleman Trio
Otello
Outlaw Blues
Paris Blues
The Paris Reunion Band
Party! Party!
Payday
Pete Kelly's Blues
Peter Pan
Phantom of the Paradise
Piano Players Rarely Ever
  Play Together
Pied Piper of Hamelin
Play Misty for Me
Play Murder for Me
Playing for Time
Point of Terror
Portfolio
Privates on Parade
Quadrophenia
RCA's All-Star Country Music
  Fair
Ready Steady Go, Vol. 1
The Real Buddy Holly Story
The Real Patsy Cline
Revolution
Rhinestone
Rich Girl
River of No Return

Road to Nashville
Rob McConnell & the Boss
  Brass
Rock Adventure
Rock, Baby, Rock It
Rock & the Money-Hungry
  Party Girls
Rock 'n' Roll Heaven
Rock 'n' Roll High School
Rock 'n' Roll High School
  Forever
Rock 'n' Roll History: English
  Invasion, the '60s
Rock 'n' Roll History: Girls of
  Rock 'n' Roll
Rock 'n' Roll History: Rock 'n'
  Soul Heaven
Rock 'n' Roll History:
  Rockabilly Rockers
Rock 'n' Roll History: Sixties
  Soul
Rock 'n' Roll: The Early Days
Rock & Roll Review
Rock & Rule
Rockin' the '50s
Rocktober Blood
Rodgers & Hammerstein: The
  Sound of American Music
Rolling Stone Presents
  Twenty Years of Rock &
  Roll
Rosanne Cash: Retrospective
Round Midnight
Roy Acuff's Open House: Vol.
  1
Roy Acuff's Open House: Vol.
  2
Rubber Rodeo
The Running Kind
Running Out of Luck
The Sacred Music of Duke
  Ellington
Satisfaction
Say Amen, Somebody
Scream Dream
Sepia Cinderella
Seventh Veil
Shock 'Em Dead
Shockwave
Shoot the Piano Player
Shout
Snow Queen
Soldier's Tale
Some Like It Hot
Sometimes It Works,
  Sometimes It Doesn't
Somos Novios (In Love and
  Engaged)
Son of Paleface
Song of Arizona
Song to Remember
Sophisticated Ladies
Spellcaster
The Spike Jones Story
Splitz
Spyro Gyra
A Star is Born
Stars on Parade/Boogie
  Woogie Dream
Sticky Fingers
Straight Talk
Stunt Rock
Sundae in New York
Sundance
Sunset Strip
Superdrumming
Sweet Country Road
Sweet Dreams
Sweet Love, Bitter
Swing High, Swing Low
Swing It, Professor
Symphony of Swing
Talking in Your Sleep
Talmage Farlow
The T.A.M.I. Show
Tapeheads

Tender Mercies
Terror on Tour
They Shall Have Music
This is Elvis
This is Spinal Tap
The Three Caballeros
Thunder Alley
Thunder & Mud
Tougher than Leather
Tribute to Ricky Nelson
Trick or Treat
The Trumpet and I
Trumpet Kings
25x5: The Continuing
   Adventures of the Rolling
   Stones
Twilight in the Sierras
Two-Gun Troubador
Underground U.S.A.
Unfaithfully Yours
United Front: Outside in Sight
URGH! A Music War
The Utopia Sampler
Vagabond Lover
Verdi
Voyage of the Rock Aliens
Wagner: The Movie
The Wall: Live in Berlin
Waltz King
Wild Orchid 2: Two Shades of
   Blue
Woodstock
World of Henry Orient
X: The Unheard Music
Yanco
Young Man with a Horn
Zachariah

Musical

Absolute Beginners
Actor: The Paul Muni Story
An American in Paris
Anchors Aweigh
The Apple
Babes in Arms
Babes in Toyland
Band Wagon
Barkleys of Broadway
Beat Street
The Belle of New York
Bells are Ringing
Black Tights
Blue Hawaii
Body Beat
Born to Dance
Breakin' Through
Breaking Glass
Breaking the Ice
The Broadway Melody
Broadway Melody of 1938
Broadway Melody of 1940
Cabaret
Calamity Jane
Camelot
Can-Can
Carmen Jones
Carousel
Charm of La Boheme
Cinderella
Congress Dances (Der
   Kongress taenzt)
Cover Girl
Daddy Long Legs
Damn Yankees
Damsel in Distress
Dancing Pirate
Dangerous When Wet
A Date with Judy
Deep in My Heart
Dogs in Space
Doll Face
Dos Gallos Y Dos Gallinas
   (Two Roosters For ...)
Dos Postolas Gemelas

Dream to Believe
Dynamite Chicken
Easter Parade
Easy Come, Easy Go
Easy to Love
Evergreen
Fabulous Fred Astaire
The Farmer Takes a Wife
The Five Pennies
Flower Drum Song
Flying Down to Rio
Follies in Concert
Follies Girl
Follow the Fleet
Follow That Dream
Follow That Rainbow
Footlight Parade
Footlight Serenade
For Me & My Gal
Frankie and Johnny
The French Line
The French Way
Fun in Acapulco
Funny Face
Gaiety
The Gay Divorcee
G.I. Blues
Gigi
Girl Crazy
Girl of the Golden West
Girls! Girls! Girls!
Give My Regards to Broad
   Street
Glorifying the American Girl
Go, Johnny Go!
Gold Diggers of 1933
Grease
Grease 2
Great Balls of Fire
The Great Caruso
The Great Waltz
Guys & Dolls
Hair
Half a Sixpence
Hands Across the Border
Hans Brinker
Hans Christian Andersen
Harlem on the Prairie
Harmony Lane
Harvest Melody
The Harvey Girls
Hats Off
Hawaii Calls
Hello, Dolly!
Hi-De-Ho
Higher and Higher
Hillbillys in a Haunted House
Hit the Deck
Holiday in Havana
Huckleberry Finn
I Do! I Do!
I Dream of Jeannie
Incident at Channel Q
Invitation to the Dance
It Happened in Brooklyn
Jailbird Rock
Johnny Lightning
Junction 88
Kid Galahad
Killer Diller
KISS Meets the Phantom of
   the Park
Krush Groove
The Last Radio Station
Les Girls
Let's Sing Again
Lili
Little Johnny Jones
The Little Match Girl
A Little Night Music
The Littlest Angel
London Melody
Los Tres Calaveras (The
   Three Skeletons)
Loving You
Main Street to Broadway

Man of La Mancha
Manhattan Merry-go-round
Marco
A Matter of Time
Maytime
Maytime in Mayfair
Meet Me in St. Louis
Meet the Navy
The Mikado
Million Dollar Mermaid
Moon Over Harlem
Moon Over Miami
Mrs. Brown, You've Got a
   Lovely Daughter
Murder at the Vanities
The Music Man
Music in My Heart
Music Shoppe
Musicals 2: Trailers on Tape
My Dream is Yours
My Fair Lady
Naughty Marietta
Newsies
Northwest Outpost
Oh! Calcutta!
On an Island with You
One from the Heart
One and Only, Genuine,
   Original Family Band
Pagan Love Song
Pajama Party
Paradise Island
Pinocchio
Pippin
Pirate Movie
Private Buckaroo
Reet, Petite and Gone
Rhapsody in Blue
Road to Hollywood
Roberta
Rosalie
Rumpelstiltskin
Running Out of Luck
Sailing Along
Sgt. Pepper's Lonely Hearts
   Club Band
Second Chorus
Sepia Cinderella
The Seven Little Foys
She Shall Have Music
Singin' in the Rain
Skirts Ahoy!
Small Town Girl
Song O' My Heart
South Pacific
Speedway
Spinout
Springtime in the Rockies
Stage Door Canteen
Stars on Parade/Boogie
   Woogie Dream
Stars and Stripes Forever
State Fair
Stormy Weather
The Story of Vernon and Irene
   Castle
Sweeney Todd: The Demon
   Barber of Fleet Street
Swing
Swing Parade of 1946
Take It Big
Taking My Turn
Terror of Tiny Town
Thank Your Lucky Stars
That's Dancing!
That's Entertainment
That's Entertainment, Part 2
That's Singing
Thousands Cheer
Three Little Words
The Three Penny Opera
The Threepenny Opera
Thrill of a Romance
Till the Clouds Roll By
The Toast of New Orleans
Tom Sawyer

Tonight and Every Night
The Trouble with Girls
Un Novio Para Dos Hermanas
   (One Boyfriend for Two
   Sisters)
Under the Cherry Moon
Vagabond Lover
Velnio Nuotaka (The Devil's
   Bride)
Viva Las Vegas
Wallaby Jim of the Islands
Whoopee!
Window Shopping
Words & Music
Yes, Sir, Mr. Bones

Musical Comedy

Almost Angels
Amateur Night
Andy Hardy Meets Debutante
April In Paris
Artists and Models
Babes on Broadway
Bathing Beauty
Beach Blanket Bingo
Beach Party
Bert Rigby, You're a Fool
Best Foot Forward
The Best Little Whorehouse in
   Texas
Big Store
Bikini Beach
Billy Rose's Jumbo
The Blues Brothers
The Boy Friend
Boy! What a Girl
Broadway Melody of 1936
Bugsy Malone
Bundle of Joy
Bye, Bye, Birdie
Call Out the Marines
Can't Stop the Music
Captain January
Carefree
The Chocolate Soldier
The Cocoanuts
Como Dos Gotas de Agua
   (Like Two Drops of Water)
Copacabana
Court Jester
Cry-Baby
Dames
Dance, Girl, Dance
Delightfully Dangerous
The Devil's Brother
Diplomaniacs
Disorderlies
Dixiana
DuBarry was a Lady
Earth Girls are Easy
The Eighties
Escape to Paradise
Fashions of 1934
Feel the Motion
Femmes de Paris
Fiesta
Finian's Rainbow
The First Nudie Musical
Four Jacks and a Jill
Frolics on Ice
A Funny Thing Happened on
   the Way to the Forum
Gentlemen Prefer Blondes
George White's Scandals
The Girl Can't Help It
Girl Happy
The Girl Most Likely
Girl Rush
Give a Girl a Break
Going My Way
Gold Diggers of 1935
Goldwyn Follies
Good News

The Great American
   Broadcast
Gypsy
Hairspray
Hallelujah, I'm a Bum
The Happiest Millionaire
Happy Go Lovely
Head
Here Comes the Groom
High Society
Hips, Hips, Hooray
Holiday Inn
Hollywood Canteen
How to Succeed in Business
   without Really Trying
I Dood It
I Dream Too Much
I Love Melvin
I Married an Angel
In the Navy
International House
It's a Date
It's a Great Feeling
It's Love Again
Jive Junction
Juke Joint
Jupiter's Darling
Keep 'Em Flying
Kid Millions
Kiss Me Kate
Kissin' Cousins
La Marchanta
Lady Be Good
Let's Dance
Let's Go Collegiate
Life Begins for Andy Hardy
Li'l Abner
Lily of Killarney
Little Shop of Horrors
Live a Little, Love a Little
Lovely to Look At
Lucky Me
Lullaby of Broadway
Make a Wish
Mame
Mamele (Little Mother)
A Matter of Degrees
Meet Me in Las Vegas
Melody Cruise
Merry Widow
Muppet Movie
Muppets Take Manhattan
Muscle Beach Party
My Song Goes Round the
   World
Nancy Goes to Rio
Neptune's Daughter
Never a Dull Moment
New Faces
No Time for Romance
Oklahoma!
On a Clear Day You Can See
   Forever
On the Town
One Touch of Venus
The Opposite Sex
Paint Your Wagon
Pajama Game
Pal Joey
Panama Hattie
The Perils of Pauline
Phantom of the Paradise
Pin-Up Girl
The Pirate
The Pirates of Penzance
Popeye
Presenting Lily Mars
Rebecca of Sunnybrook Farm
Road to Bali
Road to Lebanon
Road to Morocco
Road to Nashville
Road to Rio
Road to Singapore
Road to Utopia
Road to Zanzibar

Musical

Roadie
Roberta
Robin and the Seven Hoods
Rock 'n' Roll High School
Rock 'n' Roll High School
  Forever
Roman Scandals
Romance on the High Seas
Royal Wedding
Sensations of 1945
Seven Brides for Seven
  Brothers
Seven Days' Leave
The Seven Little Foys
1776
Sextette
Shall We Dance
Shock Treatment
Show Business
Silk Stockings
Sing Your Worries Away
Sioux City Sue
The Sky's the Limit
Song of the Islands
Spotlight Scandals
Starstruck
Step Lively
Strike Up the Band
Summer Holiday
Summer Stock
Sun Valley Serenade
Sunny
Sunny Skies
Sweethearts
Swing It, Professor
Swing Time
Take Me Out to the Ball Game
Tea for Two
Texas Carnival
Thank God It's Friday
That Girl from Paris
There's No Business Like
  Show Business
Thoroughly Modern Millie
Tickle Me
Tokyo Pop
Too Many Girls
Top Hat
23 1/2 Hours Leave
Two Girls & a Sailor
Two Tickets to Broadway
Two Weeks with Love
The Unsinkable Molly Brown
Up in Arms
Vogues of 1938
White Christmas
A Woman is a Woman
Wonder Man
Yes, Giorgio
Yidl Mitn Fidl (Yiddle with His
  Fiddle)
Yolanda & the Thief
You Were Never Lovelier
You'll Find Out
You'll Never Get Rich
Young Bing Crosby
Ziegfeld Follies
Zis Boom Bah

Musical Drama

Beatlemania! The Movie
Bitter Sweet
Blame It on the Night
Blonde Venus
Body Moves
Body Rock
Breakin'
Breakin' 2: Electric Boogaloo
Buddy Holly Story
The Cantor's Son (Dem
  Khann's Zindl)
Carnival Rock
Christmas Carol
Clambake

Coal Miner's Daughter
Concrete Angels
Cool Mikado
The Cotton Club
Crossover Dreams
Curly Top
Dance Hall
Dancing Lady
Deadman's Curve
Dimples
Dogs in Space
Double Trouble
Down Argentine Way
Duke Is Tops
El Dia Que Me Quieras
Fast Forward
Fiddler on the Roof
Fisherman's Wharf
Flashdance
Footloose
The Forbidden Dance
42nd Street
Frankie and Johnny
Funny Girl
Funny Lady
The Glenn Miller Story
Goodbye, Mr. Chips
Graffiti Bridge
The Great Ziegfeld
Greek Street
Hard to Hold
The Harder They Come
Harum Scarum
Heart's Desire
Honeysuckle Rose
Honkytonk Nights
I Could Go on Singing
Idolmaker
If You Knew Susie
I'll Cry Tomorrow
The In Crowd
It Happened at the World's
  Fair
It's Always Fair Weather
Jailhouse Rock
Jazz Singer
Jesus Christ Superstar
Jolson Sings Again
The Jolson Story
Jubilee
Jungle Patrol
Just Around the Corner
King Creole
The King and I
La Bamba
La Barbiec de Seville
La Cucaracha
Ladies of the Chorus
Lady Sings the Blues
Lambada
Life of Verdi
Light of Day
The Little Colonel
Little Miss Broadway
Littlest Rebel
Lottery Bride
Love Me or Leave Me
Love Me Tender
Mack the Knife
Mambo Kings
Melody Master
Miss Sadie Thompson
Motown's Mustang
Murder with Music
Musica Proibita
Never Steal Anything Small
New Moon
New York, New York
Night and Day
The Old Curiosity Shop
Oliver!
One Night of Love
One Trick Pony
Opera do Malandro
Orchestra Wives
Orphan Boy of Vienna

Paradise in Harlem
Paradise, Hawaiian Style
Paris Blues
Pennies From Heaven
Pete Kelly's Blues
Pete Townshend: White City
The Poor Little Rich Girl
Princess Tam Tam
Purple Rain
Quadrophenia
Rainbow
Rappin'
The Red Shoes
Rock, Baby, Rock It
Rock, Pretty Baby
Rock, Rock, Rock
Rooftops
The Rose
Rose Marie
Roustabout
Salsa
Satisfaction
Scrooge
Senora Tentacion
Shining Star
Shout
Show Boat
Sincerely Yours
Sing
Smilin' Through
Something to Sing About
Song of Freedom
Song of Norway
Song Without End
Songwriter
The Sound of Music
Sparkle
Stay Away, Joe
Stepping Out
Stowaway
Sunday in the Park with
  George
Sweet Charity
Sweet Dreams
Tap
They Shall Have Music
This is the Army
Till the Clouds Roll By/A Star
  Is Born
Tommy
Transatlantic Merry-Go-Round
Umbrellas of Cherbourg
Una Vez en la Vida
Waltz King
West Side Story
Yankee Doodle Dandy
Yentl
Young Caruso
Young at Heart
Young Man with a Horn
Ziegfeld Girl
Zoot Suit
Zou Zou

Musical Fantasy

Alice
Alice in Wonderland
Bedknobs and Broomsticks
Brigadoon
Butterfly Ball
Cabin in the Sky
Calacan
Chitty Chitty Bang Bang
Cinderella 2000
Comic Book Kids
A Connecticut Yankee in King
  Arthur's Court
Doctor Doolittle
Down to Earth
Fairytales
The 5000 Fingers of Dr. T
Kismet
La Pastorela
Labyrinth

Little Prince
Magical Mystery Tour
Marvelous Land of Oz
Mary Poppins
The New Adventures of Pippi
  Longstocking
Once Upon a Brothers Grimm
Peter Pan
The Rocky Horror Picture
  Show
Rumpelstiltskin
Willy Wonka & the Chocolate
  Factory
The Wiz
The Wizard of Oz
Wonderwall: The Movie
Xanadu

Mystery & Suspense

Abduction of St. Anne
Absolution
Accidents
Ace Crawford, Private Eye
Ace Drummond
Across the Bridge
Across the Pacific
Adventures of a Private Eye
The Adventures of Sherlock
  Holmes
The Adventures of Sherlock
  Holmes: Hound of the
  Baskervilles
An Affair in Mind
The Affair of the Pink Pearl
After Darkness
After Pilkington
After the Thin Man
Against A Crooked Sky
Against All Odds
Agatha
Alfred Hitchcock Presents
Alias John Preston
All-American Murder
All the Kind Strangers
All Through the Night
Alligator Eyes
Almost Partners
Aloha, Bobby and Rose
The Alphabet Murders
Alphaville
The Amateur
Amazing Mr. Blunden
The Amazing Mr. X
Amazing Stories, Book 1
The Ambassador
Ambition
American Friend
American Gigolo
Amos
Amsterdamned
And Soon the Darkness
And Then There Were None
The Anderson Tapes
Android
The Andromeda Strain
Angel Heart
Another Thin Man
Any Man's Death
Apology
Appointment with Death
Appointment with Fear
Apprentice to Murder
Arabesque
Are You in the House Alone?
Arizona Stagecoach
The Arousers
The Arsenal Stadium Mystery
The Art of Crime
Asphalt Jungle
Assassin
The Assassination Run
Assault
Assault on Precinct 13
At Close Range

Atomic Attack
Autopsy
The Babysitter
Back from Eternity
Back Track
Backfire
Backstab
Bad Boys
B.A.D. Cats
Bad Influence
Bad Ronald
The Bad Seed
Baffled
The Banker
The Barcelona Kill
The Bat Whispers
Battle Shock
Beat Girl
Bedford Incident
Bedroom Eyes
Bedroom Eyes 2
The Bedroom Window
Before Morning
Behind Locked Doors
Bell from Hell
The Berlin Affair
Berlin Express
Berserk!
Besame
Best Kept Secrets
Best Seller
Betrayal
Betrayed
Between Men
Beware, My Lovely
Beyond the Bermuda Triangle
Beyond a Reasonable Doubt
Beyond Reasonable Doubt
Beyond the Silhouette
Big Deadly Game
The Big Easy
The Big Fix
The Big Hurt
Big News
The Big Sleep
Big Switch
Big Town After Dark
Bikini Island
Bird with the Crystal Plumage
Birgitt Haas Must Be Killed
Bits and Pieces
Black Bird
Black Box Affair
The Black Cat
Black Cobra
Black Glove
Black Magic Woman
Black Panther
Black Rainbow
The Black Raven
Black Sunday
Black Tower
A Black Veil for Lisa
Black Widow
The Black Windmill
Blackmail
Blackout
Blind Date
Blind Fear
Blind Justice
Blind Man's Bluff
Blind Vision
Blindside
Blood Frenzy
Blood Relations
Blood Relatives
Blood Simple
Blood on the Sun
Bloody Avenger
Blow Out
Blue City
Blue Jeans and Dynamite
Blue Steel
Blue Tornado
Blue Velvet
Body Chemistry

Body Double
Body Heat
The Body in the Library
Body Parts
B.O.R.N.
Born to Kill
The Boston Strangler
Bounty Hunters
Bowery at Midnight/Dick
 Tracy vs. Cueball
The Boys from Brazil
Brass Target
Breakheart Pass
Breaking All the Rules
Bright Smiler
Brighton Strangler
Brighton Strangler Before
 Dawn
Brotherly Love
Bulldog Drummond
Bulldog Drummond in Africa/
 Arrest Bulldog Drummond
Bulldog Drummond Comes
 Back
Bulldog Drummond Double
 Feature
Bulldog Drummond's Bride
Bulldog Drummond's Peril
Bulldog Drummond's
 Revenge
Bulldog Drummond's
 Revenge/Bulldog
 Drummond's Peril
Bulldog Drummond's Secret
 Police
Burndown
Buying Time
By the Blood of Others
By Dawn's Early Light
By Hook or Crook
Cada Noche un Amor (Every
 Night a New Lover)
Call It Murder
Call Me
The Caller
Calling Paul Temple
Cambio de Cara (A Change of
 Face)
Candles at Nine
Cape Fear
Caper of the Golden Bulls
Cardiac Arrest
Caribe
Carnal Crimes
Carnival Lady
Case of the Baby Sitter
The Case of the Lucky Legs
The Case of the Missing Lady
Cast a Deadly Spell
Castle in the Desert
The Cat and the Canary
Cat Chaser
Cat and Mouse
The Cat o' Nine Tails
Catacombs
The Catamount Killing
Catch Me a Spy
Cause for Alarm
Center of the Web
The Centerfold Girls
Challenge of McKenna
Charade
Charlie Chan and the Curse of
 the Dragon Queen
Charlie Chan at the Opera
Charlie Chan in Paris
Charlie Chan in Rio
Charlie Chan at the Wax
 Museum
Charlie Chan's Secret
Child of Glass
Child in the Night
Chill Factor
Chinatown
Choice of Arms
Christina

City of Blood
City on Fire
City Killer
City in Panic
City of Shadows
The Clay Pigeon
The Clergyman's Daughter
A Climate for Killing
Cloak and Dagger
Clouds Over Europe
Clown Murders
Clue
The Clue According to
 Sherlock Holmes
The Clutching Hand
The Cobra Strikes
The Cobweb
Code of Honor
Code Name: Diamond Head
Codename: Foxfire
Codename: Icarus
Cohen and Tate
Cold Comfort
Cold Eyes of Fear
Cold Room
Color Me Dead
Columbo: Murder By the Book
Columbo: Prescription Murder
Coma
The Comfort of Strangers
Coming Out Alive
Committed
Compromising Positions
Concrete Beat
Confidentially Yours
Conflict
The Conversation
Conversation Piece
Convicted
The Corpse Vanishes
Counter Punch
Crack-Up
The Crackler
The Cradle Will Fall
Criminal Court
Criminal Law
Criminals Within
Criss Cross
Crow Hollow
The Crucifer of Blood
Cruise Into Terror
Cruise Missile
Cruising
Cry Danger
Cry Panic
Cry Terror
Curse of King Tut's Tomb
Curse of the Yellow Snake
Cutter's Way
Cycle Psycho
Dagger Eyes
The Dain Curse
Damned River
Dance with Death
Dance Macabre
Danger on the Air
Danger Man
Danger Zone
Dangerous Company
Dangerous Game
Dangerous Love
Dangerous Mission
Dangerous Obsession
Dangerous Passage
Dangerous Summer
Dark Corner
Dark Forces
Dark of the Night
Dark Night of the Scarecrow
Dark Passage
Dark Past
Dark Ride
Dark Side of Midnight
Darker Than Amber
Darkman
The Darkside

Daughter of Horror
The Day of the Jackal
Day of the Maniac
Dead Again
Dead as a Doorman
Dead Eyes of London
Dead Heat on a Merry-Go-
 Round
Dead Lucky
Dead on the Money
Dead Reckoning
Dead in the Water
Dead of Winter
Dead Women in Lingerie
Dead Zone
Deadline
Deadline Assault
Deadline at Dawn
Deadly Business
Deadly Dancer
Deadly Embrace
Deadly Fieldtrip
Deadly Force
Deadly Games
Deadly Illusion
Deadly Impact
Deadly Innocence
Deadly Intent
Deadly Passion
Deadly Possession
Deadly Sting
Deadly Strangers
Deadly Surveillance
Dear Dead Delilah
Dear Detective
Death Cruise
Death from a Distance
Death Games
Death Goes to School
Death of a Hooker
The Death Kiss
Death Kiss
Death at Love House
Death on the Nile
Death by Prescription
Death Rage
Death of a Scoundrel
Death Sentence
Death in the Shadows
Death Train
Death Valley
Death Warrant
Death Weekend
Deathmask
Deathtrap
Deceived
December Flower
Deceptions
The Deep
Deep Cover
Defense of the Realm
Defenseless
Delusion
The Deputy Drummer
Deranged
Descending Angel
Desire and Hell at Sunset
 Motel
Desperate
Desperate Hours
Desperately Seeking Susan
The Detective
Detective
Devil Thumbs a Ride
Devil Thumbs a Ride/Having
 Wonderful Crime
The Devil's Cargo
Devil's Party
The Devil's Undead
Diabolically Yours
Diabolique
Dial Help
Dial ''M'' for Murder
Diamond Run
Diamonds
Diamond's Edge

Diamonds on Wheels
Dias de Ilusion
Dick Tracy
Dick Tracy Detective
Dick Tracy, Detective/The
 Corpse Vanishes
Dick Tracy Double Feature #1
Dick Tracy Double Feature #2
Dick Tracy Meets Gruesome
Dick Tracy and the Oyster
 Caper
Dick Tracy Returns
Dick Tracy: The Original
 Serial, Vol. 1
Dick Tracy: The Original
 Serial, Vol. 2
Dick Tracy vs. Crime Inc.
Dick Tracy vs. Cueball
Dick Tracy's Dilemma
Die! Die! My Darling!
Die Screaming, Marianne
Die Sister, Die!
Dinner at the Ritz
Dirkham Detective Agency
Dirty Hands
Dirty Harry
Disappearance
Disconnected
Dishonored Lady
Distortions
Disturbance
Diva
The Dive
D.O.A.
Dr. Mabuse vs. Scotland Yard
Dr. Tarr's Torture Dungeon
Dog Day
Dollars
Dominique is Dead
The Domino Principle
Don't Bother to Knock
Don't Go in the House
Don't Look in the Attic
Don't Look Now
Don't Open the Door!
Doomed to Die
The Doomsday Flight
Double Agents
Double Cross
Double Exposure
Double Face
Double Identity
Double Indemnity
The Double McGuffin
The Double Negative
The Draughtsman's Contract
Dream Lover
Dream No Evil
Dressed for Death
Dressed to Kill
The Drifter
Driven to Kill
Driver's Seat
The Drowning Pool
Drums of Jeopardy
Duel
The Dummy Talks
Dust to Dust
Dying Room Only
The Dying Truth
Dynasty of Fear
Early Frost
Echo Murders
Echoes
Edge of Darkness
The Eiger Sanction
El Libro de Piedra
El Tesoro de Chucho el Roto
Embassy
Emergency
Emil & the Detectives
Emissary
The Empty Beach
Endangered Species
Endless Night

Enigma
Enrapture
Equus
Escapade in Florence
Escape
Escapist
Estate of Insanity
Even Angels Fall
Everybody Wins
Everybody's Dancin'
Evil Under the Sun
Ex-Mrs. Bradford
The Executioner
Executive Suite
Experiment Perilous
Experiment in Terror
Exposed
Exposure
Express to Terror
Exterminator
Extreme Close-Up
Eye of the Needle
Eyes of the Amaryllis
Eyes of Laura Mars
Eyewitness
F/X
A Face in the Fog
The Face at the Window
Fake Out
The Falcon in Hollywood
The Falcon in Mexico
The Falcon and the Snowman
Falcon Takes Over/Strange
 Bargain
Falcon's Adventure/Armored
 Car Robbery
The Falcon's Brother
False Identity
Family Plot
The Fan
The Fantasist
Far from Home
Farewell, My Lovely
Fatal Attraction
Fatal Exposure
Fatal Games
The Fatal Hour
The Fatal Image
Fatal Instinct
FBI Girl
Fear
Fear City
Fear in the Night
Fearless
Fearmaker
The Female Jungle
Femme Fatale
Fer-De-Lance
Fever
Fiction Makers
The Fifth Floor
52 Pick-Up
The Fighting Westerner
Fin de Fiesta
Final Analysis
Final Assignment
Final Cut
Final Notice
The Final Option
Finessing the King
Fingerprints Don't Lie
Flashpoint
Flatliners
Flight of Black Angel
Flowers in the Attic
Flying from the Hawk
Fog Island
Follow Me Quietly
The Fool Killer
Forbidden Sun
Force of Evil
Foreign Correspondent
Forget Mozart
The Forgotten One
The Formula
Formula for a Murder

Mystery

Murder on the Bayou (A Gathering of Old Men)
Murder by the Book
Murder on the Campus
Murder by Decree
Murder in the Doll House
Murder at the Gallop
Murder at Midnight
Murder on the Midnight Express
Murder Most Foul
Murder My Sweet
Murder by Natural Causes
Murder by Night
Murder by Numbers
Murder Once Removed
Murder Ordained
Murder on the Orient Express
Murder Over New York
Murder by Phone
Murder Rap
Murder She Said
Murder Story
Murder by Television
Murder: Ultimate Grounds for Divorce
Murderer in the Hotel
Murderers' Row
Murderlust
Murderous Vision
Murders at Lynch Cross
Murders in the Rue Morgue
The Murders in the Rue Morgue
Murri Affair
My Man Adam
My Nights With Susan, Sandra, Olga, and Julie
My Sister, My Love
Mysterious Doctor Satan
Mysterious Mr. Wong
Mystery at Castle House
Mystery at Fire Island
Mystery Liner
The Mystery Man
Mystery Mansion
The Mystery of the Mary Celeste
Mystery of Mr. Wong
Mystery Plane
Mystery of the Riverboat
Mystery in Swing
Mystery Theatre
The Naked Face
Naked Massacre
The Name of the Rose
Narrow Margin
Natas es Satan
Nature's Playmates
Neutron vs. the Maniac
Never Say Die
New Year's Evil
New York Ripper
Newman's Law
Next Victim
Niagara
Night Angel
Night of the Assassin
Night Cries
A Night for Crime
Night of the Cyclone
Night Eyes
Night of the Generals
The Night Has Eyes
Night of the Hunter
Night of the Juggler
Night Moves
Night of the Ninja
A Night to Remember
Night Rhythms
Night School
Night Shadow
Night Stalker
Night Terror
Night Train to Munich
The Night Visitor

Night Warning
Night Watch
The Nightcomers
Nighthawks
Nightkill
Nightmare in Badham County
Nightmare at Bittercreek
Nightwing
No Comebacks
No Mercy
No One Cries Forever
No Place to Hide
No Problem
No Secrets
No Way Out
No Way to Treat a Lady
Nocturne
Non-Stop New York
North by Northwest
Nothing Underneath
Notorious
Nowhere to Hide
Nuclear Conspiracy
Number Seventeen
The Oblong Box
The Obsessed
Obsession
Obsession: A Taste for Fear
Obsessive Love
The Occult Experience
Octagon
October Man
Octopussy
The Odessa File
On Dangerous Ground
On the Edge
One Frightened Night
One of My Wives Is Missing
One Shoe Makes It Murder
101 Dalmations
Open House
Operation Amsterdam
Ordeal by Innocence
Orion's Belt
The Osterman Weekend
Out on Bail
Out of Bounds
Out of the Darkness
Out of Order
Out of the Past
Out of the Rain
Out of Sight, Out of Her Mind
The Oval Portrait
Over Indulgence
Overexposed
Paid to Kill
Paint it Black
Panic on the 5:22
Panic in the Streets
Panique
The Panther's Claw
Paper Man
Paperhouse
Parallax View
Paranoia
Paris Belongs to Us
Paris Express
Parker
Party Line
Passion Flower
The Payoff
The Pearl of Death
Pendulum
Perfect Crime
Perfect Strangers
Perfect Victims
Peril
Perry Mason Returns
Perry Mason: The Case of the Lost Love
Personals
Peter Gunn
The Phantom of 42nd Street
The Phantom Broadcast
Phantom of Chinatown
The Phantom Creeps

Phantom Express
Phantom Fiend
The Phantom of Soho
Phobia
Phoenix Team
The Phone Call
Photographer
Physical Evidence
Picnic at Hanging Rock
Picture Mommy Dead
Pier 23
Plain Clothes
Play Misty for Me
Plumber
A Pocketful of Rye
Point Blank
Poor Girl, A Ghost Story
Posed For Murder
Positive I.D.
Postmark for Danger/Quicksand
Power Play
The Practice of Love
President's Mistress
The President's Mystery
The President's Plane is Missing
The Presidio
Prey of the Chameleon
Prime Suspect
Private Hell 36
The Private Life of Sherlock Holmes
The Prize
Probe
Project: Eliminator
Project X
Promise to Murder
Prowler
Psychic
Psycho
Psycho 2
Psycho 3
Psycho Cop
Psychomania
Pulp
Pursuit
Pursuit to Algiers
Puzzle
Quicker Than the Eye
Quiet Fire
The Quiller Memorandum
Rage
Railroaded
The Rain Killer
The Raven
Raw Nerve
Rear Window
Rebecca
Reckless Moment
Red Alert
The Red House
Red King, White Knight
The Red Raven Kiss-Off
Red Wind
Reflections of Murder
Regina
Rehearsal for Murder
Reilly: The Ace of Spies
Reincarnation of Peter Proud
Relentless
Remember Me
Repulsion
Requiem for Dominic
Return
The Return of Boston Blackie
Return of Chandu
The Return of Dr. Mabuse
Return of Frank Cannon
Return of Martin Guerre
Revenge
Revenge of the Stepford Wives
Reversal of Fortune
Rich and Strange
Rider on the Rain

Riders to the Seas
Rififi
Ring of Death
Road Games
Road to Salina
Roaring City
Roaring Fire
Robinson Crusoe of Mystery Island
Rogue's Gallery
Rogue's Tavern
Rogue's Yarn
Room 43
Roots of Evil
Rope
Rosary Murders
R.S.V.P.
Rude Awakening
Run If You Can
Runaways
Running Blind
Russian Roulette
Sabotage
Saboteur
The Saint
Saint in London
The Saint in New York
The Saint Strikes Back
Saint Strikes Back: Criminal Court
The Saint Takes Over
The Saint's Double Trouble
The Saint's Vacation/The Narrow Margin
The Salamander
The Salzburg Connection
Sanctuary of Fear
Sanders of the River
Sapphire
Savages
Scalp Merchant
Scandal Man
Scandalous
Scared to Death/Dick Tracy's Dilemma
Scarlet Claw
Scarlet Street
Scavengers
Scene of the Crime
Scenes from a Murder
Schnelles Geld
Scissors
The Scorpio Factor
Scotland Yard Inspector
Scream
Scream Bloody Murder
Scream and Die
Scream for Help
Screamer
Sea of Love
The Sea Shall Not Have Them
Seance on a Wet Afternoon
Season of Fear
Second Woman
The Secret Adversary
Secret Agent 00
Secret Beyond the Door
Seven Days in May
Seven Doors to Death
Seven Hours to Judgment
The Seven Percent Solution
The Seventh Sign
Sexo Y Crimen
Sexpot
Sexton Blake and the Hooded Terror
Sexual Response
Sexy Cat
Shades of Darkness
The Shadow
Shadow Dancing
Shadow of a Doubt
Shadow of the Eagle
The Shadow Man
Shadow Play
The Shadow Strikes

Shadow of the Thin Man
Shadows on the Stairs
Shadows in the Storm
Shaft
Shake Hands with Murder
Shallow Grave
Shattered
Shattered Silence
She Waits
Sherlock Holmes
Sherlock Holmes: A Study in Scarlet
Sherlock Holmes Faces Death
Sherlock Holmes and the Incident at Victoria Falls
Sherlock Holmes and the Secret Weapon
Sherlock Holmes: The Baskerville Curse
Sherlock Holmes: The Sign of Four
Sherlock Holmes: The Silver Blaze
Sherlock Holmes: The Valley of Fear
Sherlock Holmes: The Voice of Terror
Sherlock Holmes in Washington
She's Dressed to Kill
Shock
Shogun's Ninja
Shoot to Kill
A Shot in the Dark
The Shout
A Show of Force
Show Them No Mercy
Shriek in the Night
Side Show
Sidney Sheldon's Bloodline
Siesta
Silent Assassins
Silent Killers
The Silent Mr. Sherlock Holmes
The Silent Partner
The Silent Passenger
Silhouette
Sincerely Charlotte
Sing Sing Nights
The Sister-in-Law
Sister, Sister
Sisters of Death
Sketches of a Strangler
Sky Liner
Slamdance
The Slasher
Slayground
Sleeping Car to Trieste
Sleeping with the Enemy
Sleepwalk
Sleuth
Slightly Honorable
Slightly Scarlet
Slipping into Darkness
Slow Burn
Small Sacrifices
Smart-Aleck Kill
Smart Money
Smile, Jenny, You're Dead
Smooth Talker
Snatched
Snow Kill
Snowblind
Sois Belle et Tais-Toi
Somebody Has to Shoot the Picture
Someone Behind the Door
Someone at the Door
Someone to Watch Over Me
Song of the Thin Man
Sorry, Wrong Number
Sound of Murder
Spaceways
The Spaniard's Curse
Spare Parts

Occult

Kill, Baby, Kill
The Kiss
KISS Meets the Phantom of
 the Park
Lady Terminator
The Lady's Maid's Bell
Land of the Minotaur
The Legacy
Legend of Hell House
Leonor
The Living Head
Low Blow
The Magician
Making Contact
The Man Who Saw Tomorrow
Manos, the Hands of Fate
Mardi Gras Massacre
Maxim Xul
Medea
Men of Two Worlds
Midnight
Midnight Cabaret
Mirror of Death
Mirror, Mirror
Mirrors
The Monster Demolisher
Mrs. Amworth
Natas...The Reflection
Necromancer: Satan's
 Servant
Necropolis
The Night Stalker: Two Tales
 of Terror
Ninja, the Violent Sorcerer
The Occult Experience
The Occultist
Omoo Omoo, the Shark God
The Oracle
Orgy of the Dead
Phantasm
Phantasm II
Poltergeist
Poltergeist 2: The Other Side
Poltergeist 3
Possession: Until Death Do
 You Part
The Power
The Psychic
Psychic
Psychic Killer
Psychomania
Puppet Master
Race with the Devil
The Raven
A Reflection of Fear
The Reincarnate
Rendez-Moi Ma Peau
Repossessed
Rest In Pieces
Retribution
Return
Return of Chandu
Return from Witch Mountain
Rosemary's Baby
Ruby
Run If You Can
The Runestone
Sabaka
Satan's Princess
Satan's Touch
Scared Stiff
Season of the Witch
Second Sight
Seizure
The Sentinel
The Seventh Sign
The Seventh Victim
The Shaman
Shock 'Em Dead
Silent Night, Deadly Night 3:
 Better Watch Out!
Sorceress
The Spaniard's Curse
Speak of the Devil
The Spellbinder
Spellbound

Spiritism
Student of Prague
Summer of Fear
Supergirl
Teen Witch
The Tempter
Terror Beach
Terror Creatures from the
 Grave
That Darn Sorceress
Three Sovereigns for Sarah
To the Devil, A Daughter
Torture Chamber of Dr.
 Sadism
The Touch of Satan
Treasure of the Four Crowns
Tropic Heat
The Undead
Unknown Powers
Venus in Furs
Vibes
The Visitor
Warlock Moon
White Gorilla
The Wicker Man
Wild Man
Witch Who Came from the
 Sea
Witchboard
Witchcraft 2: The Temptress
Witchcraft 4: Virgin Heart
Witchcraft Through the Ages
Witches of Eastwick
The Witching
Witching Time
The Witchmaker
The Witch's Mirror
The Wizard of Oz
Wizards
The Woman Who Came Back
Zapped Again
The Zodiac Killer

Oldest Profession

Accatone!
Alexa: A Prostitute's Own
 Story
Alexander: The Other Side of
 Dawn
Alien Warrior
The Allnighter
American Nightmare
Amor Bandido
The Amsterdam Connection
Angel
Anna Christie
Auntie
Avenging Angel
Back Roads
The Balcony
The Best Little Whorehouse in
 Texas
Beverly Hills Call Girls
Beverly Hills Madam
Beverly Hills Vamp
Blood Money - The Story of
 Clinton and Nadine
Bordello
Butterfield 8
Candy Tangerine Man
Caravaggio
Catherine & Co.
The Cheyenne Social Club
Chicken Ranch
Christiane F.
Cinderella Liberty
Class of 1984
Club Life
Coffy
Confessions of a Vice Baron
Cool it Carol
Courtesans of Bombay
Crimes of Passion
Cross Country

Cry of a Prostitute: Love Kills
Dead End Street
Deadly Sanctuary
Death of a Hooker
Death Shot
Death Target
Dedee D'Anvers
Diary of a Lost Girl
Diva
Dixie: Changing Habits
Doctor Detroit
Dr. Jekyll and Sister Hyde
Drying Up the Streets
The Duchess and the
 Dirtwater Fox
East End Hustle
Education of Sonny Carson
8 Million Ways to Die
El Preprimido (The Timid
 Bachelor)
Elmer Gantry
Emily
Emmanuelle & Joanna
Erendira
Escort Girls
Everybody Wins
Fall from Innocence
Flesh
For Pete's Sake
Foxy Brown
Frankenhooker
Frankie and Johnny
French Quarter
Fyre
Get Rita
The Gift
Gigi
Girls of the White Orchid
Half Moon Street
Happy Hooker
Happy Hooker Goes
 Hollywood
Happy Hooker Goes to
 Washington
High Stakes
Hollywood Chainsaw Hookers
Hollywood Vice Sqaud
Hooker
Hustle
Hustler Squad
Hustling
I Want to Live!
Impulse
Intimate Moments
Irma La Douce
Jack's Back
Just a Gigolo
Justine
Kinjite: Forbidden Subjects
Kiss the Night
Kitty and the Bagman
Klute
La Chienne
L.A. Crackdown
L.A. Crackdown 2
Lady of the Evening
Lady of the House
Lady of the Rising Sun
The Last Prostitute
Life of O'Haru
Little Ladies of the Night
Lola Montes
Love Angels
Love by Appointment
Mack
Madame Rosa
Magdalena Viraga
Magdalene
Malibu High
Man of Flowers
Mayflower Madam
McCabe & Mrs. Miller
Miami Blues
Mid Knight Rider
Midnight Auto Supply
Midnight Cop

Midnight Cowboy
Miss Sadie Thompson
The Missionary
Mona Lisa
Mona's Place
Murderlust
Murphy's Law
My Life to Live
My Own Private Idaho
My Pleasure Is My Business
Naked Kiss
Nana
Nanami, First Love
Never on Sunday
New York's Finest
Night Friend
Night of the Generals
Night Shift
Night Slasher
The Night They Robbed Big
 Bertha's
Nights of Cabiria
The Ninja Empire
No One Cries Forever
No Prince for My Cinderella
The Nun
Nuts
Off Limits
Oldest Profession
Olivia
On the Block
One Night Only
Out of the Dark
The Owl and the Pussycat
Pandora's Box
Panic in Needle Park
The Passing of Evil
Pennies From Heaven
Personal Services
Poker Alice
Pretty Baby
Pretty Woman
Prettykill
Prime Cut
The Pyx
Queen Kelly
Quick, Let's Get Married
Rain
Reason to Die
Red Kimono
Risky Business
Room 43
Roots of Evil
Saint Jack
Sandakan No. 8
Sansho the Bailiff
Scarred
Schnelles Geld
Scorchers
Secret Games
Secrets of a Married Man
Sessions
Sex on the Brain
Single Room Furnished
Sisters of Gion
Sketches of a Strangler
SS Girls
S.S. Love Camp No. 27
Stone Cold Dead
Storyville
Street Girls
Street of Shame
Street Smart
Streetwalkin'
Sweet Charity
Sweet Revenge
Sweet Sweetback's
 Baadasssss Song
Sweet Trash
Sword & the Cross
Taxi Driver
Teen Lust
They were So Young
Thousand Pieces of Gold
Tightrope
To Be a Rose

Trading Places
Traffic in Souls
Train Station Pickups
Tricks of the Trade
Truck Stop Women
Twenty Dollar Star
Ultimate Desires
The Vampire Hookers
Vice Squad
Walk on the Wild Side
Warm Nights on a Slow-
 Moving Train
Warrior Queen
Waterloo Bridge
Where the Girls Are
Whore
Wild Orchid 2: Two Shades of
 Blue
The Winter of Our Dreams
A Woman in Flames
A Woman of Rome
Working Girls
The World of Suzie Wong
Wrestling Racket Girls

On the Rocks

Arthur 2: On the Rocks
Bad Georgia Road
Barfly
Blue Heaven
Break of Hearts
Clean and Sober
Cocktail
Come Back, Little Sheba
Country Girl
A Cry for Love
Danger in the Skies
Dangerous
Days of Wine and Roses
Desde el Abismo
Desert Bloom
Devil at 4 O'Clock
Devil Thumbs a Ride
Devil Thumbs a Ride/Having
 Wonderful Crime
Dona Flor & Her Two
 Husbands
Easy Virtue
Educating Rita
El Dorado
The Electric Horseman
The Entertainer
Fat City
Fish Hawk
Frances
A Free Soul
Gervaise
The Great Man Votes
The Green Man
Honkytonk Man
Hot Summer in Barefoot
 County
I Live with Me Dad
I'll Cry Tomorrow
The Informer
Ironweed
Izzy & Moe
The Last American Hero
Le Beau Serge
License to Kill
Long Day's Journey into Night
The Lost Weekend
M.A.D.D.: Mothers Against
 Drunk Driving
Marvin & Tige
Mean Dog Blues
Morning After
My Favorite Year
On the Bowery
Only When I Laugh
Papa's Delicate Condition
The Pilot
Police Court
Reuben, Reuben

The Robe
Shakes the Clown
The Sin of Harold Diddlebock
The Singing Blacksmith (Yankl der Shmid)
Skin Deep
Smash-Up: The Story of a Woman
A Star is Born
Stranded
Strange Brew
Tales of Ordinary Madness
Ten Nights in a Bar Room
Ten Nights in a Bar-Room
Termini Station
Under Capricorn
Under the Volcano
What Price Hollywood?
Whiskey Galore
White Lightning
Willa
The Winning Team

Only the Lonely

Abducted
The Abduction of Kari Swenson
An Autumn Afternoon
The Body Shop
Boys Night Out
Breakfast in Paris
Browning Version
Choose Me
A Chorus of Disapproval
Citizen Kane
Cotter
Cyrano de Bergerac
A Day in the Country
Desperately Seeking Susan
Dirty Dishes
Distant Thunder
Docks of New York
Doin' Time on Planet Earth
The Doll
Doll's House
A Doll's House
The Eclipse
El Super
An Empty Bed
Empty Suitcases
Enchanted Cottage
Escapade in Japan
E.T.: The Extra-Terrestrial
Every Man for Himself & God Against All
Eye of the Needle
The Final Alliance
Full Moon in Blue Water
Full Moon in Paris
The Game of Love
Harold and Maude
The Heart is a Lonely Hunter
Hell Hounds of Alaska
High Country
Hiroshima, Mon Amour
The Hunchback of Notre Dame
Husband Hunting
I Don't Give a Damn
I Sent a Letter to My Love
In the White City
La Petite Sirene
La Vie Continue
Lana in Love
The Last Horror Film
Les Bons Debarras
Lonely Hearts
The Lonely Passion of Judith Hearne
Lonelyhearts
Looking for Mr. Goodbar
Luther the Geek
Man of Flowers
Marianela

Marty
Men Don't Leave
Miles to Go Before I Sleep
Mouchette
Mr. Kingstreet's War
976-EVIL
Ninotchka
None But the Lonely Heart
Only the Lonely
Our Miss Brooks
Panic Station
Persona
Priest of Love
The Prime of Miss Jean Brodie
The Prince of Central Park
The Pursuit of Happiness
Queen of the Stardust Ballroom
The Rachel Papers
Rachel, Rachel
The Red Desert
Red Kimono
Restless
Return Engagement
Robert et Robert
Roseland
Samurai 2: Duel at Ichijoji Temple
Scream and Die
Secret Executioners
Secret Life of an American Wife
Seize the Day
Separate Tables
Single Bars, Single Women
Slipstream
The Solitary Man
Someone to Love
South of Reno
Strip Teasers
Strip Teasers 2
Summer
Sweetwater
Tales of the Klondike: The Scorn of Women
Tenth Month
Terror at the Opera
That Cold Day in the Park
Tomorrow
Too Shy to Try
Topless Dancing Texas Style
Turtle Diary
The Ultimate Ninja Challenge
Ultimate Sampler Video
Vagabond
When Your Lover Leaves
Wildest Dreams
Woman in the Shadows
Zelly & Me

Opera

Aria
Barber of Seville
Baritone
Broken Melody
Burlesque of Carmen
Carmen
Carmen Jones
The Chocolate Soldier
Cosi Fan Tutte
Don Carlo
L'Elisir D'Amore
Elmer Fudd's Comedy Capers
Faust
Fidelio
Fitzcarraldo
Gentlemen of Titipu
Gilbert & Sullivan Present Their Greatest Hits
Gondoliers
The Great Caruso
The Great Waltz
H.M.S. Pinafore

I Dream Too Much
Jesus Christ Superstar
Jonathan Miller's Production of Verdi's Rigoletto
Julius Caesar
La Barbiec de Seville
La Forza del Destino
La Perichole
La Traviata
Les Contes D'Hoffman
Life of Verdi
Lucia di Lammermoor
The Magic Flute
Mario Lanza: The American Caruso
Maytime
The Medium
Meeting Venus
Midnight Girl
Mignon
The Mozart Brothers
Musica Proibita
Naughty Marietta
A Night at the Opera
Otello
The Panther's Claw
Phantom of the Opera
Placido: A Year in the Life of Placido Domingo
Prince Igor
Rigoletto
Rose Marie
Simon Boccanegra
Terror at the Opera
The Toast of New Orleans
Tosca
The Tsar's Bride
Verdi
Wagner: The Movie

Order in the Court

Absence of Malice
The Accused
Action for Slander
Adam's Rib
The Adventures of Ichabod and Mr. Toad
All God's Children
The Ambush Murders
Anatomy of a Murder
And Justice for All
The Andersonville Trial
As Summers Die
Beyond a Reasonable Doubt
Breaker Morant
The Caine Mutiny Court Martial
Call It Murder
Carrington, V.C.
A Case of Libel
Clarence Darrow
Class Action
Clearcut
Conduct Unbecoming
Conspiracy: The Trial of the Chicago Eight
Convicted
Cordelia
The Court Martial of Billy Mitchell
Criminal Court
Cross Examination
A Cry in the Dark
Death Sentence
Death of a Soldier
The Defender
The Defense Never Rests
Defenseless
Delinquent Parents
Devil & Daniel Webster
The Disappearance of Aimee
Dishonored Lady
The Drake Case
A Dry White Season

Eureka!
Final Verdict
From the Hip
Gideon's Trumpet
The Good Father
The Good Mother
Guilty of Innocence
Hard Traveling
Hitz
How Many Miles to Babylon?
I Want to Live!
Illegally Yours
Inherit the Wind
It Could Happen to You
The Jagged Edge
Jailbait: Betrayed By Innocence
The Judge
The Judge and the Assassin
Judge Horton and the Scottsboro Boys
Judgment in Berlin
Judgment at Nuremberg
Justice
Killing in a Small Town
King of the Pecos
Knock on Any Door
L.A. Law
Last Innocent Man
The Last Wave
Les Assassins de L'Ordre
Little Gloria...Happy at Last
The Long Dark Hall
Madeleine
Midnight
Monsieur Beaucaire
Mrs. R's Daughter
Murder
Murder or Mercy
Music Box
Naked Lie
Never Forget
Nuts
Outrage
The Paradine Case
Penalty Phase
Perry Mason Returns
Perry Mason: The Case of the Lost Love
Presumed Innocent
Prisoners of the Sun
Question of Silence
Red Kimono
The Respectful Prostitute
Reversal of Fortune
Roe vs. Wade
The Rose Garden
The Runner Stumbles
Sacco & Vanzetti
The Scorpion Woman
Separate But Equal
Skullduggery
The Star Chamber
State's Attorney
Stuck on You
Suspect
They Won't Believe Me
To Kill a Mockingbird
The Trial
Trial of Bernhard Goetz
Trial of the Catonsville Nine
The Trial of Lee Harvey Oswald
Twelve Angry Men
The Verdict
When Every Day was the Fourth of July
When Thief Meets Thief
The Winslow Boy
Witness for the Prosecution
Wives Under Suspicion
Woman in Brown
A Woman's Face
The Young Philadelphians

Outtakes & Bloopers

Abbott and Costello Scrapbook
Bela Lugosi Scrapbook
Candid Candid Camera, Vol. 1
Candid Hollywood
Cinema Shrapnel
Hollywood Bloopers
Hollywood Outtakes & Rare Footage
Hollywood Uncensored
Presidential Blooper Reel
Return of Video Yesterbloop
Revenge of TV Bloopers
Rhino's Guide to Safe Sex
Rings Around the World
Salute to the "King of the Cowboys," Roy Rogers
Soap Opera Scandals
Son of Hollywood Bloopers
Star Bloopers
Star Trek Bloopers
Super Bloopers #1
Super Bloopers #2
Those Wild Bloopers
The Ultimate Stuntman: A Tribute to Dar Robinson
Video Yesterbloop
Voice of Hollywood
What Do You Say to a Naked Lady?
The World's Greatest Stunts
Zombiethon

Pacific Islands

Adventure Island
Aloha Summer
Balboa
Blue Hawaii
Borneo
Carlton Browne of the F.O.
Castaway
Dangerous Life
Enchanted Island
Ethan
Eye of the Octopus
Father Goose
Fighting Marines
Flaming Signal
Fog Island
Food of the Gods
From Hell to Borneo
From Here to Eternity
Hell in the Pacific
The Imperial Japanese Empire
The Island
Island of Dr. Moreau
Last Flight of Noah's Ark
Last Warrior
Lt. Robin Crusoe, U.S.N.
Mad Doctor of Blood Island
The Mermaids of Tiburon
Midway
Miss Sadie Thompson
Mission Manila
Moana, a Romance of the Golden Age
Mr. Robinson Crusoe
Mysterious Island
Noon Sunday
Paradise, Hawaiian Style
Pardon My Sarong
Rain
Return to the Blue Lagoon
Robinson Crusoe
South Pacific
Tabu
Till There was You
Tora! Tora! Tora!
Wackiest Ship in the Army
Windom's Way

The Year of Living
Dangerously

Parades & Festivals

Dream Boys Revue
Easter Parade
Ferris Bueller's Day Off
Miracle on 34th Street
National Lampoon's Animal
House
Rejoice
The Rose Parade: Through
the Years
Sweet Beat

Parenthood

Adam
Alice in the Cities
Aliens
All the Kind Strangers
Amazing Grace
An Autumn Afternoon
Baby Boom
Bachelor Mother
The Ballad of Narayama
Behind Prison Walls
Bellissima
Between Two Women
The Bicycle Thief
The Big Wheel
Blue City
Boys Town
Cahill: United States Marshal
Careful, He Might Hear You
Chapter in Her Life
The Children of Times Square
Cody
Cold River
Commissar
Crisscross
A Cry in the Dark
Da
Desde el Abismo
Distant Voices, Still Lives
Dona Herlinda & Her Son
Donde Duermen
Dos...Duermen Tres
Drop-Out Mother
East of Eden
El Monte de las Brujas
The Emerald Forest
Endless Love
Enormous Changes
Epitaph
Ernie Kovacs: Between the
Laughter
Esposa y Amante
Fatal Confinement
Father of the Bride
Field of Dreams
Flowers in the Attic
Follow That Rainbow
Forbidden Planet
Ghost Dad
The Good Father
The Good Mother
Growing Pains
Guess What We Learned in
School Today?
Gypsy
Harry & Son
Held for Murder
High Tide
Hole in the Head
Home From the Hill
Homeward Bound
Houseboat
I Live with Me Dad
I Never Sang For My Father

I Remember Mama
Immediate Family
Innocent Victim
Irreconcilable Differences
Johnny Belinda
Kiss Daddy Goodbye
Long Time Gone
Lost Angels
Love Under Pressure
Ma Barker's Killer Brood
Mad Youth
M.A.D.D.: Mothers Against
Drunk Driving
Max Dugan Returns
Memory of Us
Mirele Efros
Misunderstood
Mom
Mom & Dad
Monkey Grip
Mother & Daughter: A Loving
War
Mr. Mom
Mrs. R's Daughter
My Father's Nurse
My Mom's a Werewolf
National Lampoon's European
Vacation
Natural Enemies
Next of Kin
Night Cries
Nothing in Common
On Golden Pond
On the Third Day
One Man's War
Only the Lonely
Only When I Laugh
Padre Padrone
Parenthood
Parents
Paris, Texas
Poil de Carotte
Psycho 4: The Beginning
Raising Arizona
The Red House
Red River
The Reluctant Debutante
Revenge
The Revolt of Mother
Rising Son
The River Rat
Scream for Help
Sex and the Single Parent
The Shaggy D.A.
A Slightly Pregnant Man
Smoke
Son of Godzilla
Speeding Up Time
Stella
Stella Dallas
Superdad
Suzanne
Target
Teen Mothers
Teenage Bad Girl
Tequila Sunrise
Terms of Endearment
This Property is Condemned
Three Fugitives
Three Men and a Baby
Three Men and a Cradle
Throw Momma from the Train
Turning Point
Unfinished Business
The Unstoppable Man
Wanted: The Perfect Guy
When Wolves Cry
Where Are the Children?
Where's Poppa?
Without a Trace
The Woman Who Willed a
Miracle
Yours, Mine & Ours

Patriotism & Paranoia

Amber Waves
The Chase
The Commies are Coming, the
Commies are Coming
The Cosmic Man
Death of a Bureaucrat
Delta Force
F/X
The Falcon and the Snowman
Fellow Traveler
Firefox
First Blood
Guilty by Suspicion
Harry's War
Ivan the Terrible, Part 1
Ivan the Terrible, Part 2
Missing in Action
Mom & Dad
Murrow
Panama Patrol
Paris Belongs to Us
Patton
The Prisoner
Rambo: First Blood, Part 2
Rambo: Part 3
Red Dawn
The Sea Hawk
The Secret Four
Sky High
The Spider's Stratagem
Strictly G.I.
Taps
The 27thth Day
Wild Weed

Peace

The Day the Earth Stood Still
Deserters
Destry Rides Again
Diplomaniacs
Escapade
Firehead
Friendly Persuasion
Gandhi
War and Peace

Performing Arts

Aria
Caligari's Curse
Cosi Fan Tutte
L'Elisir D'Amore
Fame
Ivan the Terrible
Jonathan Miller's Production
of Verdi's Rigoletto
Ladies Night Out
The Lady of the Camellias
The Last Metro
Lucia di Lammermoor
Moliere
Nutcracker: The Motion
Picture
Resurrection at Masada
Rigoletto
Rolling Stone Presents
Twenty Years of Rock &
Roll
Shakespeare Wallah
Simon Boccanegra
Strip Teasers
Strip Teasers 2
Swimming to Cambodia
Time Flies When You're Alive
Topless Dancing Texas Style
Tosca
Two Can Play

Period Piece

Adventures of Don Juan

Al Capone
Alan & Naomi
Allonsanfan
Alone Against Rome
Amadeus
American Graffiti
The Amorous Adventures of
Moll Flanders
Anne of Avonlea
Anne of Green Gables
Anthony Adverse
At Sword's Point
The Babe
The Barretts of Wimpole
Street
Barry Lyndon
Barton Fink
Battle of Valiant
The Bay Boy
Belizaire the Cajun
Ben-Hur
Beulah Land
Black Robe
Blood Clan
Blood Red
The Bostonians
Botany Bay
Boxcar Bertha
Bronx Cheers
Brutal Glory
Carthage in Flames
Casanova's Big Night
Charge of the Light Brigade
Cheech and Chong's, The
Corsican Brothers
Chinatown
A Christmas Carol
Christmas Carol
Christopher Columbus
The Clan of the Cave Bear
Cleopatra
Cold Sassy Tree
Colonel Redl
Conduct Unbecoming
The Conqueror Worm
Conquest
Conquest of Cochise
The Covered Wagon
Cromwell
Cry of the Banshee
Daisy Miller
Damn the Defiant
Dangerous Liaisons
Danton
Daughters of the Dust
Decameron Nights
The Deceivers
Delusion
Delusions of Grandeur
Demons of the Mind
Desiree
The Devils
Diner
The Disappearance of Aimee
Doctor Zhivago
Dollmaker
Doll's House
A Doll's House
DuBarry was a Lady
Duel of Hearts
East of Elephant Rock
Eight Men Out
Eijanaika (Why Not?)
El Cid
Elephant Man
Elizabeth of Ladymead
Elizabeth, the Queen
Emma's Shadow
Empire of the Sun
End of Desire
Ernesto
The Europeans
The Fall of the Roman Empire
Fanny and Alexander
Fanny Hill: Memoirs of a
Woman of Pleasure

Far from the Madding Crowd
The Far Pavilions
Farewell, My Lovely
The Farmer Takes a Wife
Fiddler on the Roof
55 Days at Peking
The Fighting Kentuckian
Fire and Sword
First Olympics: Athens 1896
F.I.S.T.
Five Corners
Five Weeks in a Balloon
Flame is Love
Flight of the Eagle
Forest of Little Bear
Forget Mozart
Fortunes of War
The Four Feathers
Four Feathers
The Four Musketeers
Frankie and Johnny
Freedom Road
The French Lieutenant's
Woman
A Funny Thing Happened on
the Way to the Forum
Gaslight
Gate of Hell
Gervaise
The Getting of Wisdom
Glory
The Go-Between
Gold Diggers of 1933
Good Morning, Babylon
The Gorgeous Hussy
Gothic
Great Expectations
The Great Gatsby
The Great Moment
The Great Waldo Pepper
Green Dolphin Street
The Guns and the Fury
Hamlet
Haunted Summer
Hawaii
Hawk the Slayer
The Heiress
Henry V
Heroes in the Ming Dynasty
Hester Street
High Road to China
Hobson's Choice
The Horse Soldiers
The House on Carroll Street
I Remember Mama
Impromptu
The Inheritors
Inside Daisy Clover
The Iron Mask
Ivanhoe
Java Head
Jezebel
Joseph Andrews
Journey of Honor
Kagemusha
The King's Rhapsody
La Grande Bourgeoise
La Marseillaise
La Nuit de Varennes
La Ronde
Lady Caroline Lamb
Lady Jane
Lady from Louisiana
Lady for a Night
Land of the Pharaohs
The Last Days of Pompeii
The Last Emperor
The Last of His Tribe
Les Miserables
Life with Father
Light at the Edge of the World
The Lighthorseman
The Lion in Winter
Little Minister
The Little Princess
Little Women

Lethal Weapon
Let's Get Harry
Let's Get Lost
License to Kill
Live and Let Die
Long Day's Journey into Night
Los Drogadictos
Lost Angels
Lovely...But Deadly
The Man with the Golden Arm
Marihuana
Marked for Death
McBain
McQ
The Messenger
Midnight Cop
Midnight Express
Mike's Murder
Mindfield
Mission Manila
Mixed Blood
Mob War
Monkey Grip
More
Naked Lunch
Naked Youth
New Jack City
Newman's Law
Night Friend
The Night of the Iguana
Ninja Turf
Ninja in the U.S.A.
Ninja vs. Ninja
Nite Song
Not My Kid
On the Block
On the Edge
On the Edge: Survival of Dana
The Organization
Out of the Rain
Over the Edge
Pace That Kills
The Palermo Connection
Panic in Needle Park
The People Next Door
Performance
Pharmacist
Platoon the Warriors
Playing with Fire
Police
The Poppy Is Also a Flower
Postcards from the Edge
Pot Shots
Premonition
Pretty Smart
Prime Cut
Private Investigations
Professionals
Protector
Psych-Out
Puppet on a Chain
Question of Honor
Quiet Cool
Reckless Disregard
Red Blooded American Girl
Red Surf
Reefer Madness
Return of the Tiger
Revenge of the Ninja
Revenge of the Stepford
  Wives
Rip-Off
River's Edge
Rock House
Rodrigo D: No Future
The Rose
Running Scared
Rush
Saigon Commandos
Satan's Harvest
Scarface
Schnelles Geld
Scorchy
Scream Free!
Secret Agent Super Dragon
The Seven Percent Solution

Severance
Shotgun
Showdown in Little Tokyo
Sicilian Connection
Sid & Nancy
Silk
The Slender Thread
The Snake People
Sno-Line
Sorry, Wrong Number
Stand Alone
The Stewardesses
Straight Up
Street Girls
Street Hawk
Street People
Streetwalkin'
Strike Force
Sunset Strip
Super Bitch
Superfly
The Take
Target Eagle
Tchao Pantin
Teenage Devil Dolls
Tequila Sunrise
The Third Man
This Time I'll Make You Rich
Thunder Alley
Timothy and the Angel
Tongs: An American
  Nightmare
Torchlight
Toughlove
Toxic Zombies
Transmutations
Trash
The Trip
Tropical Snow
Twenty-One
Undercover
Vanishing Point
Wait Until Dark
Warlords from Hell
Watched
Weekend with the Babysitter
What Have I Done to Deserve
  This?
Who'll Stop the Rain?
The Wild Pair
Wild in the Streets
Wild Weed
The Winter of Our Dreams
Wired
Withnail and I
Woodstock

Poetry

The Barretts of Wimpole
  Street
Carnival of Animals
Just Hold Still
Kerouac
Lady Caroline Lamb
Tales of Ordinary Madness

Poisons

Arsenic and Old Lace
Barracuda
Color Me Dead
The Crazies
D.O.A.
Doomwatch
The Invisible Killer
Juggernaut
Kiss and Kill
Surf 2

Politics

The Act
The Americano
Americathon

Antonio Das Mortes
Beggars in Ermine
Being There
Burgess, Philby, and
  MacLean: Spy Scandal of
  the Century
Buried Alive
Carry On Emmanuelle
China Gate
The Colombian Connection
Conspiracy
Conspiracy: The Trial of the
  Chicago Eight
Country
Dear America: Letters Home
  from Vietnam
Death of a Bureaucrat
Defense of the Realm
Demonstone
The Devils
Diplomatic Immunity
Don's Party
Double Cross
Eijanaika (Why Not?)
Eminent Domain
En Retirada (Bloody Retreat)
Fame is the Spur
The Farmer's Daughter
Fever Mounts at El Pao
First Family
First Monday in October
A Flash of Green
Foreplay
The French Detective
The French Woman
Friday Foster
A Girl on Her Own
The Gorgeous Hussy
The Great McGinty
Guilty by Suspicion
The Home and the World
Hoodlum Empire
Hour of the Assassin
JFK
John and the Missus
A Joke of Destiny, Lying in
  Wait Around the Corner
  Like a Bandit
Just Like Weather
Killer Image
Larks on a String
Laurel & Hardy: Chickens
  Come Home
The Life and Assassination of
  the Kingfish
The Little Drummer Girl
Love, Drugs, and Violence
Mandela
Meeting Venus
Mein Kampf
Millhouse: A White Comedy
Nightbreaker
Open Doors (Porte Aperte)
Presumed Innocent
Quiet Fire
Red Kiss (Rouge Baiser)
Report to the Commissioner
Robert Kennedy and His
  Times
Salt of the Earth
Samaritan: The Mitch Snyder
  Story
Under Western Stars
Viva Zapata!
Washington Mistress
Werewolf of Washington
Whoops Apocalypse
Wild in the Streets

Pool

The Baltimore Bullet
Baron and the Kid
Color of Money
The Hustler

Kiss Shot
Pool Hustlers

Pornography

Body Double
Defenseless
Erotic Illusion
Fallen Angel
Glitter Dome
Hardcore
Heavy Traffic
Hi, Mom!
Hollywood Vice Sqaud
Inserts
It's Called Murder, Baby
The Last Porno Flick
M'Lady's Court
The Pornographers
Rented Lips
The Sinister Urge
South Bronx Heroes
Variety

Post Apocalypse

After the Fall of New York
Aftermath
Battle for the Planet of the
  Apes
Beneath the Planet of the
  Apes
The Blood of Heroes
A Boy and His Dog
Circuitry Man
City Limits
Class of 1999
The Creation of the
  Humanoids
Crime Zone
Cyborg
Damnation Alley
Day of the Dead
Dead Man Walking
Deadly Reactor
Def-Con 4
Driving Force
Dune Warriors
The Element of Crime
Endgame
Equalizer 2000
Escape from Safehaven
Exterminators of the Year
  3000
The Final Executioner
Final Sanction
Firefight
First Spaceship on Venus
Fortress of Amerikka
Future Kill
Glen and Randa
The Handmaid's Tale
Hell Comes to Frogtown
In the Aftermath
Interzone
The Killing Edge
Land of Doom
The Last Man on Earth
The Late Great Planet Earth
The Lawless Land
Le Dernier Combat
Lord of the Flies
Mad Max
Mad Max Beyond
  Thunderdome
A Man Called Rage
Neon City
Night of the Comet
984: Prisoner of the Future
1990: The Bronx Warriors
Omega Cop
Omega Man
On the Beach
Osa
Panic in the Year Zero!

People Who Own the Dark
Phoenix the Warrior
Planet of the Apes
Planet Earth
Prayer of the Rollerboys
The Quiet Earth
Quintet
Radioactive Dreams
Raiders of the Sun
Rats
Rebel Storm
Refuge of Fear
The Road Warrior
Robot Holocaust
Roller Blade
Rush
The Seventh Sign
She
Silent Running
The Sisterhood
The Slime People
Steel Dawn
Strange New World
Survival Earth
Survival Zone
Survivor
Teenage Caveman
Terminator 2: Judgement Day
The Terror Within 2
Testament
Things to Come
The Time Travelers
Time Troopers
2020 Texas Gladiators
Urban Warriors
Virus
WarCat
Warlords
Warlords of the 21st Century
Warriors of the Apocalypse
Warriors of the Wasteland
Wheels of Fire
Where Have All the People
  Gone?
World Gone Wild

Postwar

Americana
Baltic Deputy
Behold a Pale Horse
Berlin Express
Bloody Trail
The Blum Affair
Brussels Transit
The Burmese Harp
The Bushwackers
Cabo Blanco
Chattahoochee
Cinema Paradiso
Cruel Story of Youth
Demons in the Garden
Deutschland im Jahre Null
The Displaced Person
Distant Thunder
Distant Voices, Still Lives
D.P.
Early Summer (Bakushu)
Eat a Bowl of Tea
Ebony Dreams
Europa '51
Father
The Fighting Kentuckian
First Blood
Foolish Wives
Four in a Jeep
Frieda
Green Eyes
Gun Fury
Heroes
Hiroshima, Mon Amour
Hiroshima: Out of the Ashes
Horizons West
House

Religion

Goodnight God Bless
Gospel According to St.
   Matthew
Greaser's Palace
Great Sadness of Zohara
Greatest Heroes of the Bible
The Greatest Story Ever Told
Green Pastures
The Guyana Tragedy: The
   Story of Jim Jones
Hail Mary
Happiness Is...
Hawaii
Hazel's People
The Healing
Hear O Israel
Heavens Above
Hellfire
High Adventure
The Holy Father Loves You
A Home of Our Own
Home Safe
Hot Under the Collar
I Confess
I Heard the Owl Call My Name
If I Perish
Impure Thoughts
In Search of Historic Jesus
In This House of Brede
Inherit the Wind
The Inn of the Sixth
   Happiness
The Inquiry
Invasion of the Space
   Preachers
Jesus
Jesus Christ Superstar
Jesus of Montreal
Jesus of Nazareth
Jesus: The Nativity
Jesus: The Resurrection
The Jesus Trip
Joan of Arc
Joni
Judgment
The Keys of the Kingdom
King of Kings
The Last Temptation of Christ
The Late Great Planet Earth
Leaves from Satan's Book
The Left Hand of God
The Life of Christ
Lilies of the Field
Little Church Around the
   Corner
The Littlest Angel
Los 3 Reyes Magos
Love and Faith
Loyola, the Soldier Saint
Luther
Major Barbara
A Man for All Seasons
Marjoe
Martin Luther
Mary and Joseph: A Story of
   Faith
Mass Appeal
The Meeting: Eddie Finds a
   Home
The Messiah
The Milky Way
The Miracle
Miracle of Our Lady of Fatima
Miracle in Rome
The Mission
Mission to Glory
The Missionary
Mohammed: Messenger of
   God
Monsieur Vincent
Monsignor
Monty Python's Life of Brian
Mormon Maid
Moses
Most Wanted
Music & the Spoken Word

The Name of the Rose
Nasty Habits
Nativity
Nazarin
New Star Over Hollywood
The Next Voice You Hear
Night of the Assassin
The Night of the Iguana
Nite Song
Noah: The Deluge
The Nun
Nuns on the Run
The Nun's Story
Ordet
Pass the Ammo
Passion of Joan of Arc
The Passover Plot
The Penitent
Pier 23
Pope John Paul II
The Pope Must Diet
The Possessed
Power of the Resurrection
Pray TV
Prayer for the Dying
Prayer for World Peace
Prisoner of Rio
The Prodigal Planet
Quo Vadis
Rain
The Rapture
Resurrection
Ride the Wind
The Robe
Romero
Rosary Murders
The Rose Tattoo
The Runner Stumbles
St. Benny the Dip
Saint Joan
Salome
Salvation!
Sammy
Saving Grace
Second Sight
Sermon on the Mount
Shattered Vows
The Shepherd
Shiokari Pass
The Shoes of the Fisherman
Show My People
Sisters of Satan
The Song of Bernadette
Sorceress
Spirits
Split Image
Star Trek 5: The Final Frontier
Stealing Heaven
A Stranger in My Forest
Suicide Cult
Survival
Susan and God
Sword & the Cross
Therese
This Stuff'll Kill Ya!
Through a Glass Darkly
Ticket to Heaven
Time for Miracles
Tournament
Trapped by the Mormons
Trouble Along the Way
The Trouble with Angels
True Confessions
Under the Red Robe
Under Satan's Sun
The Unholy
Vatican Conspiracy
The Vision
We're No Angels
Where Jesus Walked
Whistle Down the Wind
Whitcomb's War
The White Rose
The White Sister
Wholly Moses!
The Wicker Man

The Winter Light
Wise Blood
Witness
The Word
Yentl

Religious Epics

The Agony and the Ecstasy
Bible...In the Beginning
Cleopatra
David and Bathsheba
David and Moses
Day of Triumph
Demetrius and the Gladiators
The Egyptian
El Cid
Exodus
Fourth Wise Man
Great Leaders
Herod the Great
In the Name of the Pope-King
The Inquiry
Jacob: The Man Who Fought
   with God
Jesus Christ Superstar
Joseph & His Brethren
Joseph & His Brothers
The Keys of the Kingdom
The Last Supper
A Man Called Peter
Moses
Nativity
Nefertiti, Queen of the Nile
Power of the Resurrection
Rachel's Man
The Robe
Salome
Samson and Delilah
Saul and David
Sebastian
Second Time Lucky
Silver Chalice
Sins of Jezebel
Sodom and Gomorrah
The Song of Bernadette
The Story of Ruth
Ten Commandments
The Ten Commandments

Restored Footage

Beau Revel
Crimes of Passion
Dr. Jekyll and Mr. Hyde
Foolish Wives
Frankenstein
Isadora
Lawrence of Arabia
Lost Horizon
New York, New York
Que Viva Mexico
Red River
The Rules of the Game
Sadie Thompson
Spartacus
Spiders
A Star is Born
Star Trek: The Motion Picture
Sunny Side Up
Tarzan and His Mate
Twelve O'Clock High
Two English Girls
Wages of Fear
When Time Ran Out

Revenge

Act of Aggression
Act of Passion: The Lost
   Honor of Kathryn Beck
Act of Vengeance
Adios, Hombre
The Adventurers
The Alchemist

Alien Seed
All Dogs Go to Heaven
American Commandos
American Kickboxer 1
Angel and the Badman
Angels from Hell
The Angry Dragon
Archer: The Fugitive from the
   Empire
Arizona Terror
Asylum
Avenging Force
Back to Back
Ballad of Cable Hogue
The Ballad of the Sad Cafe
Basket Case
The Beguiled
Bell from Hell
Big Jake
The Black Arrow
Black Arrow
Black Cobra
The Black Hand
The Black Raven
The Black Six
Black Sunday
Blind Vengeance
Blood and Guns
Blood on the Mountain
Blood Red
Blood for a Silver Dollar
Bloodfist
Bloodmatch
Bloody Birthday
The Blue Knight
The Blue Lamp
Border Rangers
Born Invincible
Bounty Hunters
Branded a Coward
Bronson's Revenge
Bronze Buckaroo
Brotherhood of Justice
Bruce Li the Invincible
The Brute Man
Bulldance
Bulldog Courage
Bullet for Sandoval
Bullies
The Burning
Bury Me an Angel
Cadillac Man
Caged Terror
Cape Fear
Carson City Kid
Chain Gang Women
The Chair
Challenge
The Chase
Chisum
Chrome and Hot Leather
Class of 1984
Clayton County Line
Cobra
Cobra Against Ninja
Code of Honor
Coffy
Cold Eyes of Fear
Cold Justice
Cold Steel
Cold Sweat
Cole Justice
Commando Invasion
Conqueror & the Empress
Cornered
Count of Monte Cristo
The Count of the Old Town
The Courier
The Cowboys
Crack House
Crazed Cop
Creepers
Crimebusters
Cry Danger
Cry of the Innocent
Cry Vengeance

Curse of the Devil
Daddy's Gone A-Hunting
Danger Zone 2
Dangerous Obsession
Dangerous Orphans
Dark August
Dark River: A Father's
   Revenge
Dawn Rider
The Day the Women Got Even
Dead on the Money
Deadly Darling
Deadly Game
Deadly Reactor
The Deadly Trackers
Deadly Twins
Deadly Vengeance
Deadly Weapon
Death Blow
Death Row Diner
Death Wish
Death Wish 2
Death Wish 3
Deep in the Heart
Demon Warrior
Demon Witch Child
The Demons
Demonwarp
Despedida de Casada
Devastator
Devil Times Five
Devil Woman
Devil's Canyon
The Devil's Mistress
The Devil's Possessed
Devonsville Terror
The Diabolical Dr. Z
Diamond Ninja Force
Die! Die! My Darling!
Dingaka
Dinner at the Ritz
Diplomatic Immunity
Dirty Harry
The Divine Enforcer
Dixie Dynamite
Do or Die
Dr. Terror's House of Horrors
Dolemite
The Double Negative
Double Revenge
The Dragon Fist
Drums of Jeopardy
Dudes
Duel in the Eclipse
Dynasty
Eagles Attack at Dawn
Easy Kill
The Eiger Sanction
El Barbaro
El Fugitivo
El Hombre de la Furia (Man of
   Fury)
El Sabor de la Venganza
El Zorro Vengador
The Eliminators
Emmanuelle, the Queen
The Empire of Passion
Escape
The Evil That Men Do
Evilspeak
The Execution
The Executioner, Part 2:
   Frozen Scream
Exterminator
Exterminator 2
Eye of the Eagle 2
An Eye for an Eye
Eye of the Tiger
Fair Game
The Fall of the House of Usher
Far Country
Fearless Hyena
Fight for Us
Fight for Your Life
The Fighter
Fighting Ace

Revenge

The Vernonia Incident
Vietnam, Texas
The Vigilantes Are Coming
Village on Fire
Viper
The Virgin Spring
Voodoo Black Exorcist
Walking the Edge
Walking Tall
Walking Tall: Part 2
The War Wagon
Water Rustlers
The Weirdo
The White Sheik
Wild Man
Wild Wheels
Wired to Kill
Without Honors
Wolf Lake
The Women
Wonder Man
Write to Kill
Young Warriors
Yuma
Zombie Lake
Zombie Nightmare

Revolutionary War

America/The Fall of Babylon
The Howards of Virginia
Johnny Tremain & the Sons of
    Liberty
The Rebels
Revolution
The Seekers
1776
Sweet Liberty
The Time of Their Lives
Yankee Doodle

Road Trip

Across the Great Divide
Alice Doesn't Live Here
    Anymore
Alligator Eyes
Aloha, Bobby and Rose
American Autobahn
Ariel
Armed for Action
Bad Company
The Big Crimewave
The Big Slice
Bingo
Blue De Ville
Blue Murder at St. Trinian's
Bolero
Bound for Glory
Bright Angel
Bustin' Loose
Butch Cassidy and the
    Sundance Kid
Bye Bye Brazil
Candy Mountain
Cannonball
Cannonball Run
Cannonball Run II
Changes
Cheech and Chong: Things
    Are Tough All Over
Clodhopper
Coast to Coast
The Concrete Cowboys
Coupe de Ville
Crazy Mama
Crossroads
Cycle Vixens
Delusion
Der Purimshpiler (The Jester)
Devil Thumbs a Ride
Devil Thumbs a Ride/Having
    Wonderful Crime
Diamonds on Wheels
The Easy Life

Easy Rider
End of the Line
Endplay
Every Which Way But Loose
Fandango
Far Out Man
Flim-Flam Man
Goodbye Pork Pie
Great Smokey Roadblock
The Great Texas Dynamite
    Chase
Greedy Terror
Harry and Tonto
Having Wonderful Crime
Heart Beat
Hell's Angels on Wheels
Highway 61
Hitcher
Hollywood or Bust
Homer and Eddie
Honkytonk Man
The Incredible Journey
Into the Night
It Takes Two
Jean Sheperd on Route #1...
    and Other Thoroughfares
Journey of Natty Gann
Joyride
The Judge Steps Out
Kings of the Road (In the
    Course of Time)
Kiss My Grits
La Petite Bande
La Strada
Ladies on the Rocks
Le Grand Chemin
Leaving Normal
Leningrad Cowboys Go
    America
Les Rendez-vous D'Anna
Lion of Africa
Loose Connections
Lost in America
Merry Christmas
Midnight Run
Mob Story
More
My Own Private Idaho
National Lampoon's Vacation
Neon City
Never on Tuesday
No Place to Run
Out
Paper Moon
The Passing of Evil
Patti Rocks
Pee-Wee's Big Adventure
Pierrot le Fou
Pink Cadillac
Planes, Trains & Automobiles
Powwow Highway
Prime Target
Rain Man
Rainbow Bridge
The Reivers
Rich Hall's Vanishing America
Ride in the Whirlwind
Road to Bali
Road Games
The Road to Hong Kong
Road to Lebanon
Road to Morocco
Road to Rio
Road to Singapore
Road to Utopia
The Road Warrior
Road to Zanzibar
Roadie
Rockin' Road Trip
Round Trip to Heaven
Rubin & Ed
The Runaway Bus
Running Hot
Sawdust & Tinsel
Scarecrow
Slither

Something Wild
Sorcerer
Spies Like Us
Starlight Hotel
Starman
Stay As You Are
Summer City
Thelma & Louise
Those Magnificent Men in
    Their Flying Machines
Three for the Road
Thumb Tripping
Tilt
Tokyo Pop
Traveling Man
Twentieth Century
Voyage Surprise
The Way West
Weekend
Where the Eagle Flies
White Line Fever
The Wizard of Oz

Robots & Androids

And You Thought Your
    Parents Were Weird!
Android
The Astro-Zombies
Bill & Ted's Bogus Journey
Blade Runner
Chopping Mall
Class of 1999
Cyborg
Daleks Invasion Earth 2150
    A.D.
The Day the Earth Stood Still
Devil Girl from Mars
Dr. Goldfoot and the Bikini
    Machine
Doctor Satan's Robot
Doctor Who and the Daleks
Doctor Who: Death to the
    Daleks
Doctor Who: Pyramids of
    Mars
Doctor Who: Spearhead from
    Space
Doctor Who: Terror of the
    Zygons
Doctor Who: The Ark in Space
Doctor Who: The Seeds of
    Death
Doctor Who: The Talons of
    Weng-Chiang
Doctor Who: The Time
    Warrior
The Eliminators
The Empire Strikes Back
Eve of Destruction
Forbidden Planet
Future Cop
The Human Duplicators
The Humanoid
Kronos
Megaville
Planets Against Us
Project: Shadowchaser
Prototype
Return of the Jedi
RoboCop
RoboCop 2
Roboman
Robot Holocaust
Robot Jox
Robot Monster
The Robot vs. the Aztec
    Mummy
R.O.T.O.R.
Runaway
Short Circuit
Short Circuit 2
Star Crash
Star Wars
Steel and Lace

Tekkaman, the Space Knight:
    Vol. 2
The Terminator
Terminator 2: Judgement Day
Terror of Mechagodzilla
Tobor the Great
War of the Robots
Westworld

Rock Stars on Film

All You Need is Cash
Blow-Up
Blue Hawaii
Breaking Glass
The Bride
Carny
Caveman
Charro!
Chuck Berry: Hail! Hail! Rock
    'N' Roll
Clambake
Double Trouble
Dune
Easy Come, Easy Go
Flaming Star
Follow That Dream
Frankie and Johnny
Freddy's Dead: The Final
    Nightmare
Fun in Acapulco
G.I. Blues
The Girl Can't Help It
Girl Happy
Girls! Girls! Girls!
Give My Regards to Broad
    Street
Go, Johnny Go!
Hard Day's Night
The Harder They Come
Harum Scarum
Head
Help!
How I Won the War
The Hunger
It Happened at the World's
    Fair
Jailhouse Rock
Jubilee
Kid Galahad
King Creole
KISS Meets the Phantom of
    the Park
Kissin' Cousins
La Bamba
The Last Temptation of Christ
The Legacy
Let It Be
Lisztomania
Live a Little, Love a Little
Love Me Tender
Loving You
The Magic Christian
Magical Mystery Tour
The Man Who Fell to Earth
Merry Christmas, Mr.
    Lawrence
Monster Dog
Mrs. Brown, You've Got a
    Lovely Daughter
Nomads
One Trick Pony
Paradise, Hawaiian Style
Pat Garrett & Billy the Kid
Performance
Pink Floyd: The Wall
Quadrophenia
Rock 'n' Roll High School
Roustabout
Speedway
Spinout
Stay Away, Joe
That'll Be the Day
This is Elvis
Tickle Me

Tommy
The Trouble with Girls
Viva Las Vegas
Wild in the Country
Woodstock
Yellow Submarine

Rodeos

The Adventures of Gallant
    Bess
Arizona Cowboy
Desert Trail
Feud of the West
Girl from Calgary
Goldenrod
Great American Cowboy
Junior Bonner
Kid from Gower Gulch
Lady Takes a Chance
Lights of Old Santa Fe
The Lusty Men
Man from Utah
The Misfits
My Heroes Have Always Been
    Cowboys
Never a Dull Moment
Oklahoma Cyclone
Rider from Tucson
Rodeo Champ
Rodeo Girl
Rolling Home
Saddle Buster
Trouble in Texas
Western Double Feature 5
Wild Horse
Wild Horse Rodeo

Role Reversal

All of Me
Anastasia
Angel
Angel on My Shoulder
Bad Manners
Big
Biggles
Blood of Dracula
Bloodlink
Captain from Koepenick
Cheech and Chong's, The
    Corsican Brothers
Christmas in Connecticut
Cinderfella
Clambake
Cleo/Leo
Coming to America
Como Dos Gotas de Agua
    (Like Two Drops of Water)
Condorman
The Couch Trip
Desperately Seeking Susan
Devil's Canyon
Dr. Black, Mr. Hyde
Doctor Detroit
Doctor Doolittle
Dr. Heckyl and Mr. Hype
Dr. Jekyll and Mr. Hyde
Doctors and Nurses
Dollars
Donde Duermen
    Dos...Duermen Tres
Down and Out in Beverly Hills
Dream a Little Dream
A Dream of Passion
18 Again!
Extremities
The $5.20 an Hour Dream
For Ladies Only
48 Hrs.
Frank and I
Freaky Friday
The Grand Duchess and the
    Waiter
The Great Gabbo

Romantic

The Wedding Party
We're Not Married
What's New Pussycat?
When Harry Met Sally
When Ladies Meet
Where the Boys Are
Who Am I This Time?
The Whoopee Boys
Wimps
With Six You Get Eggroll
Without Reservations
The Woman in Red
Woman of the Year
The Women
Working Girl
The World's Oldest Living
   Bridesmaid
Worth Winning
You Can't Fool Your Wife
You Can't Take It with You
Your Place or Mine

Romantic Drama

Aaron Loves Angela
The Accidental Tourist
The Adorable Cheat
The Affair
The African Queen
After Midnight
Algiers
Ali: Fear Eats the Soul
Alice
Alice Adams
All This and Heaven Too
Always
Anatomy of a Seduction
And Now My Love
Angel Square
The Animal Kingdom
Anna Karenina
Annapolis
Anne of Avonlea
Anthony Adverse
Arch of Triumph
The Atonement of Gosta
   Berling
The Average Woman
Back Street
Backstreet Dreams
The Ballad of the Sad Cafe
Barbarian and the Geisha
The Barretts of Wimpole
   Street
Bayou Romance
Beau Revel
Beloved Enemy
Berlin Blues
Betty Blue
Beware of Pity
The Big Parade
Big Street
The Bigamist
Birds of Paradise
Bitter Sweet
Black Orchid
Blind Husbands
Blood Feud
The Blood of Others
Blood and Sand
Blood and Sand/Son of the
   Sheik
Bolero
Born to Be Bad
Boy Takes Girl
Breakfast in Paris
Breathless (A Bout de Souffle)
The Bride Wore Red
Buffalo Jump
Bullfighter & the Lady
Burning Secret
Buster and Billie
By Love Possessed
Bye Bye Blues
Cabiria

Cabo Blanco
Cactus in the Snow
Caddie
Cafe Romeo
Camelia
Camila
Camille
Camille Claudel
Captive Hearts
Carmen
Carnival Story
The Carpetbaggers
Carthage in Flames
Casablanca
Casino
Castaway
Caught
Chanel Solitaire
Chapter Two
Children of a Lesser God
Christopher Strong
Cinderella Liberty
Circle of Love
Circle of Two
Circus World
City Girl
Clash by Night
Close My Eyes
Cobra
Cocktail
A Coeur Joie
Cold Sassy Tree
Collector's Item
The Coming of Amos
Competition
Conquest
Consolation Marriage
The Cousins
Cowboy & the Ballerina
The Cranes are Flying
Cyrano de Bergerac
Dance
Dancers
Dangerous Liaisons
Dark Eyes
Deception
A Dedicated Man
Deja Vu
Desert Hearts
Devil in the Flesh
Devil's Wanton
Devotion
Die Grosse Freiheit Nr. 7
Different Kind of Love
The Divorcee
Dogfight
The Dolls
The Dolphin
Dos Chicas de Revista
A Dream of Kings
Dressmaker
Dubarry
Duel of Hearts
Dying Young
Dynasty
Ecstasy
Edward and Mrs. Simpson
84 Charing Cross Road
El Diputado (The Deputy)
El Penon de las Animas
Elena and Her Men
Elephant Walk
Eliza's Horoscope
Empty Canvas
End of August
Endless Love
Escape to Love
Ethan
Every Time We Say Goodbye
Everybody's All American
Eyes, the Mouth
Falling in Love
Fanny
Far from the Madding Crowd
The Far Pavilions
The Farmer Takes a Wife

Fatal Attraction
Finishing School
Fire with Fire
Fire and Ice
First Affair
First Love
Five Days One Summer
Flame is Love
Flying Fool
Flying Tigers
Fool for Love
Fools
For Love Alone
For the Love of Angela
For Love or Money
For Your Love Only
Forbidden
Forever
Foxfire Light
Franz
The French Lieutenant's
   Woman
French Postcards
Fresh Horses
The Fugitive Kind
Full Moon in Paris
Furia Pasional
Gal Young 'Un
Gamble on Love
The Game of Love
Georgia, Georgia
Girl in Black
The Girl in Blue
The Go-Between
Golden Demon
Gone with the Wind
Goodbye Again
The Gorgeous Hussy
The Great Lie
Green Dolphin Street
Green Fields
Griffin and Phoenix: A Love
   Story
Harem
Havana
Head Over Heels
Heartbreakers
Heat and Dust
The Heiress
Heroes
Hide-Out
His Private Secretary
History is Made at Night
Hold the Dream
Holiday Affair
Holocaust Survivors...
   Remembrance of Love
Home Before Midnight
Home Sweet Home
The Home and the World
Honky
Humoresque
I Can't Escape
I Conquer the Sea
I Live in Grosvenor Square
I Love All of You
I Love N.Y.
I Love You
I Never Sang For My Father
Iguana
I'm a Fool
Image of Passion
In Dangerous Company
In Search of Anna
In a Shallow Grave
In Too Deep
Independence Day
Indiscretion of an American
   Wife
Interval
Iron Cowboy
Island of Desire
Isle of Secret Passion
Jacko & Lise
Jane Eyre
Jenny

Johnny Frenchman
Josepha
Jules and Jim
Jungle Fever
Just Between Friends
Katie's Passion
The Key
Killing Heat
The King's Rhapsody
Kipperbang
Kitty Foyle
Knight Without Armour
Knights of the Round Table
La Bella Lola
La Carcel de Laredo
La Fuerza Inutil
La Strada
La Traviata
Lady Chatterley's Lover
The Lady with the Dog
Lady from Louisiana
The Lady in Question
Lagrimas de Amor
Lana in Love
The Last Prostitute
Last Time I Saw Paris
The Last Train
The Last Winters
Le Repos du Guerrier
A Legacy for Leonette
Leopard in the Snow
Letters from the Park
Lianna
Lies of the Twins
The Life and Loves of Mozart
Life and Nothing But
Lilith
Listen to Me
The Little American
Little Dorrit, Film 2: Little
   Dorrit's Story
Little Minister
Lola
Long Ago Tomorrow
The Long, Hot Summer
Louisiana
Loulou
Love Affair: The Eleanor &
   Lou Gehrig Story
Love and Faith
Love and the Frenchwoman
A Love in Germany
Love Letters
Love is a Many-Splendored
   Thing
Love with a Perfect Stranger
Love with the Proper Stranger
Love on the Run
Love Streams
The Love of Three Queens
Love at the Top
Love Without Pity
The Lovers
The Loves of Carmen
Love's Savage Fury
Lovespell
Lydia
Macho y Hembra
Madame Bovary
The Magnificent Matador
Mambo
The Man in Grey
Man is Not a Bird
The Man Who Guards the
   Greenhouse
A Man and a Woman
Mannequin
Maria
Maria Chapdelaine
Marianela
Marjorie Morningstar
Marty
A Matter of Love
Mayerling
Melody
The Mill on the Floss

A Million to One
Mirrors
Mississippi Masala
Model Behavior
Moonlight Sonata
Morocco
Mrs. Soffel
My Champion
My Forbidden Past
Mysteries
Nais
Naked Sun
Narda or The Summer
Nativity
The Nest
Nevil Shute's The Far Country
New York Nights
Nicole
A Night Full of Rain
Night Games
A Night in Heaven
Night Is My Future
North Shore
Not Quite Paradise
Of Human Bondage
Oliver's Story
One Brief Summer
One Summer Love
Only Once in a Lifetime
Out of Season
Outcasts of the City
Over the Summer
Oxford Blues
A Paper Wedding
Passion of Love
Passione d'Amore
A Patch of Blue
The Peasants
Pepe le Moko
Picnic
The Playboys
Playing the Field
Pleasure
Poor White Trash
Port of Call
Portrait of Jennie
Possessed
Pride of the Clan
The Prince of Tides
The Private Lives of Elizabeth
   & Essex
Queen Christina
Red Headed Woman
Red Sorghum
The Road to Yesterday
Rocky
Romance on the Orient
   Express
Rome Adventure
Romeo and Juliet
Romeo and Juliet (Gielgud)
Room at the Top
Room with a View
Ryan's Daughter
Sadie Thompson
San Francisco
Say Hello to Yesterday
Second Wind
Secret Obsession
Secret Passions
September Affair
Seventh Veil
The Shining Hour
Shoot the Piano Player
Shop Angel
Shore Leave
Silver City
Sincerely Charlotte
Slipstream
Snow Country
The Soft Skin
A Soldier's Tale
Somewhere Tomorrow
Song to Remember
Sooner or Later
The Sorrows of Satan

Sea Gypsies
Sea Racketeers
Sharks' Treasure
Ship of Fools
Ships in the Night
Shipwreck Island
Shout at the Devil
Sinbad the Sailor
South Seas Massacre
Steamboat Bill, Jr.
Submarine Seahawk
Swashbuckler
They were Expendable
Till There was You
To Have & Have Not
Torpedo Run
Transatlantic Merry-Go-Round
The Treasure of Jamaica Reef
Two Lost Worlds
The Unsinkable Molly Brown
Up the Creek
Voyage of the Damned
Wake of the Red Witch
Wild Orchids
Windrider
Witch Who Came from the Sea
Wonderland Cove
Yankee Clipper

Sanity Check

Aguirre, the Wrath of God
Alone in the Dark
American Eagle
Amos
Anguish
Another Time, Another Place
Appointment with Fear
Arsenic and Old Lace
Asylum
Asylum of Satan
Bad Dreams
Bambi Meets Godzilla and Other Weird Cartoons
Bedlam
Behind Locked Doors
The Bell Jar
Bestia Nocturna
Beyond Therapy
A Blade in the Dark
Blood Frenzy
Blood Salvage
Blood Song
Bloody Wednesday
Body Chemistry
The Boston Strangler
Buried Alive
Camille Claudel
Captain Newman, M.D.
The Carpenter
The Case of the Frightened Lady
Cat in the Cage
Cause for Alarm
The Centerfold Girls
Chattahoochee
Clockwise
Coast to Coast
The Cobweb
The Collector
The Couch Trip
The Cracker Factory
Crazed
Crazy Fat Ethel II
The Crime of Dr. Crespi
Crime and Punishment
Criminally Insane
Crow Hollow
Cutting Class
Dancing in the Dark
Dark Obsession
Dark Places
Dark Sanity
Dark Waters

Darkroom
Das Testament des Dr. Mabuse
A Day at the Races
Dead as a Doorman
Dead Man Out
Dead of Night
Dead Pit
Deadly Innocence
Deadly Intruder
Deadly Strangers
The Deadly Trap
Death Kiss
Death Nurse
Dementia 13
Demons of the Mind
Deranged
Desde el Abismo
Despair
Desperate Living
Deutschland im Jahre Null
Devi
The Devil Bats
The Devil Bat's Daughter
Devil Doll
Devil in the Flesh
Devil In Silk
Devil Times Five
Devil Wears White
Diary of Forbidden Dreams
Diary of a Mad Housewife
Diary of a Madman
Diary of a Young Comic
Dick Barton, Special Agent
Die! Die! My Darling!
Die Screaming, Marianne
Disturbance
Disturbed
Doctor Blood's Coffin
Doctor Butcher M.D.
Dr. Caligari
The Doctor and the Devils
Doctor of Doom
Dr. Frankenstein's Castle of Freaks
Doctor Gore
Dr. Hackenstein
Dr. Jekyll and Mr. Hyde
Dr. Jekyll and the Wolfman
Dr. Mabuse the Gambler
Dr. Mabuse vs. Scotland Yard
Dr. Strangelove, or: How I Learned to Stop Worrying and Love the Bomb
Dr. Tarr's Torture Dungeon
Doctor X
Dog Day Afternoon
The Doll
Dominique is Dead
Don't Go in the Woods
Don't Look in the Basement
Don't Look Now
Don't Mess with My Sister!
Don't Open Till Christmas
Doom Asylum
A Double Life
Down the Drain
Dream Lover
Dream No Evil
The Dream Team
Easy Rider
Eaten Alive
Echoes
Edge of Sanity
El: This Strange Passion
The End of Innocence
End of the Road
Endless Love
Endless Night
Entr'acte
Equus
Escape from Safehaven
Estate of Insanity
Europa '51
Eyeball
Eyes of the Amaryllis

The Fall of the House of Usher
False Identity
Fat Guy Goes Nutzoid
Fear
Fear Strikes Out
The Fifth Floor
The Fisher King
The Five of Me
Follow Me Quietly
Forced March
The 4th Man
Frances
Fright
From the Life of the Marionettes
Frozen Terror
The Girl in Room 2A
The Goalie's Anxiety at the Penalty Kick
Hamlet
Hammersmith is Out
Harvey
The Haunting
Haunting of Sarah Hardy
Haunts
Heart of Midnight
Heat and Sunlight
Henry IV
Heroes
Hider in the House
The Horrible Dr. Hichcock
Horror Chamber of Dr. Faustus
Horror Hospital
Horror House on Highway 5
Hotline
House of Insane Women
The Housekeeper
How to Get Ahead in Advertising
How to Make a Monster
Hysteria
I Am the Cheese
I Dismember Mama
I Never Promised You a Rose Garden
Icy Breasts
If Things Were Different
Impulse
Innocent Victim
Insanity
Interiors
The Invisible Ghost
It's Alive!
Keeper
Killer's Moon
King of Hearts
Lassie from Lancashire
The Last of Philip Banter
Lethal Weapon
Lethal Weapon 2
Let's Scare Jessica to Death
Lilith
Lost Angels
Lower Level
Lunatics: A Love Story
Luther the Geek
Man Facing Southeast
Marat/Sade (Persecution and Assassination...)
Midnight Warning
Mind Games
Mine Own Executioner
Morgan!
Mother Kusters Goes to Heaven
My First Wife
Night of Bloody Horror
Night of Terror
The Night Visitor
Night Wars
Night Watch
The Ninth Configuration
Nobody's Perfekt

October Man
Offerings
One Flew Over the Cuckoo's Nest
One Summer Love
125 Rooms of Comfort
Out of Sight, Out of Her Mind
Overexposed
Pale Blood
Paperback Hero
Pay Off
Peeping Tom
Performance
A Piano for Mrs. Cimino
Picture Mommy Dead
Pigs
The Pink Panther Strikes Again
The Pit and the Pendulum
Poison
The President's Analyst
Prettykill
Prime Suspect
Psychic Killer
Psycho
Psycho 2
Psycho Cop
Psychos in Love
Question of Silence
Ransom
Rasputin
The Rejuvenator
Relentless
Remember Me
Return of the Ape Man
Return to Treasure Island, Vol. 2
Revolt of the Zombies
Richard III
Room to Let
The Ruling Class
Scissors
Scream Bloody Murder
Scream for Help
Screamer
Seance on a Wet Afternoon
Secret Beyond the Door
The Sender
Seven Days in May
Seven Hours to Judgment
Shadow Play
Shattered
Shattered Silence
Shattered Spirits
The Shining
Shock Corridor
The Silence of the Lambs
Silent Night, Bloody Night
Sisters
Sketches of a Strangler
Slaughter Hotel
Sledgehammer
Slipping into Darkness
Sniper
Solaris
Soldier's Revenge
South of Reno
Spectre of the Rose
Spellbound
Steven Wright Live
Story of Adele H.
Strait-Jacket
Strange Illusion
Street Asylum
A Streetcar Named Desire
Strong Medicine
Summer of Secrets
The Survivalist
Sweet Hostage
Sybil
Tainted Image
Teenage Frankenstein
The Tenant
The Testament of Dr. Cordelier
They Might Be Giants

Three Nuts in Search of a Bolt
Torture Ship
Torture Train
Touched
Tracks
Trapped in Silence
Twinsanity
Twister
Urge to Kill
Vampire's Kiss
Vengeance Is Mine
Watched
What About Bob?
What Ever Happened to Baby Jane?
Whatever Happened to Susan Jane?
While I Live
The Wind
Witchfire
A Woman Under the Influence
You are Not I
Young Hero of the Shaolin Part 2
The Young Ones

Satanism

Alison's Birthday
Amazon Jail
Angel on My Shoulder
Asylum of Satan
Bad Dreams
Beast of the Yellow Night
Beyond the Door 3
The Black Cat
Black Roses
Black Sunday
Blood Orgy of the She-Devils
The Blood on Satan's Claw
The Bloodsuckers
The Brotherhood of Satan
The Chosen
Damien: Omen 2
The Demon Lover
Demonoid, Messenger of Death
Devil & Daniel Webster
Devil & Max Devlin
Devil Rider
The Devils
The Devil's Eye
The Devil's Messenger
The Devil's Mistress
The Devil's Nightmare
The Devil's Partner
The Devil's Possessed
Devil's Rain
Devil's Son-in-Law
The Devil's Web
Disciple of Death
Doctor Faustus
El Monte de las Brujas
Equinox
Erotic Dreams
Eternal Evil
Evil Altar
Evilspeak
Exorcism
The Exorcist
Exorcist 2: The Heretic
Eye of the Demon
Fear No Evil
Final Conflict
Guardian of the Abyss
Hack O'Lantern
Halloween Night
Highway to Hell
House of the Black Death
I Don't Want to Be Born
I Drink Your Blood
Inferno
Invitation to Hell
Jaws of Satan

Kenneth Anger, Vol. 4:
  Invocation of My Demon
  Brother
Land of the Minotaur
Leaves from Satan's Book
The Legend of Hillbilly John
Leonor
The Man and the Monster
Masque of the Red Death/
  Premature Burial
Maxim Xul
The Mephisto Waltz
Midnight
Midnight Cabaret
Mr. Frost
Mysterious Doctor Satan
Necromancer: Satan's
  Servant
The Night Visitor
976-EVIL
976-EVIL 2: The Astral Factor
The Occultist
Oh, God! You Devil
The Omen
Other Hell
Prime Evil
Prince of Darkness
The Pyx
Race with the Devil
Rest In Pieces
Revenge
Rosemary's Baby
The Satanic Rites of Dracula
Satan's Black Wedding
Satan's Blade
Satan's Cheerleaders
Satan's Princess
Satan's School for Girls
Satan's Touch
Satanwar
The Sentinel
Servants of Twilight
The Seventh Victim
Shock 'Em Dead
Sisters of Satan
Speak of the Devil
Specters
The Spellbinder
Sugar Cookies
Suicide Cult
The Tempter
Terror Beach
To the Devil, A Daughter
Tombs of the Blind Dead
The Visitor
Warlock
Witchcraft
Witchcraft 3: The Kiss of
  Death
WitchTrap

Satire & Parody

ABC's of Love & Sex,
  Australia Style
The Act
Adventures of Hairbreadth
  Harry
The Adventures of Picasso
The Affairs of Annabel
L'Age D'Or
Airplane!
Airplane 2: The Sequel
Alex in Wonderland
All Through the Night
All You Need is Cash
Amazon Women On the Moon
America 3000
American Tickler
Americathon
And If I'm Elected
Andy Kaufman: Sound Stage
Andy Warhol's Dracula
Andy Warhol's Frankenstein
Angel City

Animators from Allegro Non
  Troppo
Anna Russell's Farewell
  Special
The Apartment
Bambi Meets Godzilla
Bananas
Beau Pere
Being There
The Best of Not Necessarily
  the News
Beyond Therapy
Beyond the Valley of the Dolls
Big Deal on Madonna Street
The Big Picture
Bingo Inferno
Black and White in Color
Blaze Glory
The Bonfire of the Vanities
Boob Tube
Boss
Bullshot
Bullshot Crummond
The Candidate
Candide
Carnal Knowledge
Carry On Cleo
Carry On Dick
Casino Royale
Caveman
The Cheap Detective
Chicken Thing
Clockwise
A Clockwork Orange
Closet Cases of the Nerd Kind
Cold Turkey
The Committee
Coup de Grace
Crainquebille
Crimewave
Curtain Up
Dark Habits
Dark Star
Darktown Strutters
Dead Men Don't Wear Plaid
Deal of the Century
Death of a Bureaucrat
Deathrow Gameshow
Diary of a Chambermaid
Dirty Dishes
The Discreet Charm of the
  Bourgeoisie
Dr. Strangelove, or: How I
  Learned to Stop Worrying
  and Love the Bomb
Doctors and Nurses
A Doonesbury Special
Dope Mania
The Dove
Down & Dirty
Down to Earth
Down and Out in Beverly Hills
Down Under
Dragnet
E. Nick: A Legend in His Own
  Mind
11 Harrowhouse
Ella Cinders
Erik the Viking
The Extraordinary Adventures
  of Mr. West in the Land of
  the Bolsheviks
The Family Game
Fat and the Lean
First Family
Flesh Gordon
Flicks
Footlight Frenzy
The Fortune Cookie
The French Touch
The Front
Funny Money
Galactic Gigolo
Galaxina
Ganjasaurus Rex
Gas

Geek Maggot Bingo
Gentlemen Prefer Blondes
Glory! Glory!
Going Places
Greaser's Palace
The Great Dictator
Great McGonagall
The Groove Tube
Gulliver
Gypsy
Hail
Hail the Conquering Hero
Hairspray
Half a Sixpence
Hammered: The Best of
  Sledge
Hardware Wars
Hardware Wars & Other Film
  Farces
Hell Comes to Frogtown
Hellfire
High Anxiety
High Hopes
High Season
The History of White People in
  America
The History of White People in
  America: Vol. 2
Hollywood Harry
Home Movies
The Horror of Frankenstein
Hot Shots!
The Hound of the Baskervilles
I Married a Vampire
I Was a Zombie for the FBI
I'm All Right Jack
I'm Gonna Git You Sucka
The Incredible Shrinking
  Woman
Incredibly Strange Creatures
  Who Stopped Living and
  Became Mixed-Up Zombies
The Inspector General
Is There Sex After Death?
It Couldn't Happen Here
J-Men Forever!
Jabberwocky
Jet Benny Show
A Joke of Destiny, Lying in
  Wait Around the Corner
  Like a Bandit
Kentucky Fried Movie
King of Hearts
La Casa de Madame Lulu
La Nueva Cigarra
The Last Polka
Lincoln County Incident
Loose Shoes
Love and Death
Love at Stake
Mad Mission 3
The Man with Two Brains
Man in the White Suit
The Man Who Came to Dinner
Man Who Had Power Over
  Women
Men...
Mickey
The Milky Way
Millhouse: A White Comedy
Mindkiller
The Missionary
Modern Times
Monty Python and the Holy
  Grail
Monty Python's Life of Brian
Monty Python's Parrot Sketch
  not Included
Monty Python's The Meaning
  of Life
More! Police Squad
The Mouse That Roared
Movers and Shakers
Movie, Movie
Murder by Death
Mutant on the Bounty

My Best Girl
The Naked Gun: From the
  Files of Police Squad
The Naked Gun 2 1/2: The
  Smell of Fear
Nasty Habits
Nocturna
The Nutty Professor
Overdrawn at the Memory
  Bank
Pandemonium
Pass the Ammo
Personal Services
Phantom of Liberty
Pickles
Police Squad! Help Wanted!
Pray TV
The President's Analyst
Prime Time
The Princess Bride
Prisoner of Rio
A Private Function
Prize of Peril
The Producers
A Propos de Nice
Putney Swope
The Ramrodder
Rat Pfink and Boo-Boo
Rentadick
Repentance
The Report on the Party and
  the Guests
Repossessed
Return of Captain Invincible
Revolution! A Red Comedy
Ricky 1
Road to Lebanon
The Rocky Horror Picture
  Show
Rosalie Goes Shopping
The Royal Bed
The Rules of the Game
The Ruling Class
Salvation!
Scenes from the Class
  Struggle in Beverly Hills
School for Scandal
Scrambled Feet
Scrooged
Second Best Secret Agent in
  the Whole Wide World
Second City Comedy Special
Second Time Lucky
Seduction of Mimi
Semi-Tough
The Senator was Indiscreet
Serial
A Session with the Committee
Sex Adventures of the Three
  Musketeers
Sex O'Clock News
Shame of the Jungle
Shampoo
A Shock to the System
Sideburns
Simon of the Desert
Slammer Girls
Sleeper
Smile
S.O.B.
Spaceballs
Spaced Out
Spaceship
The Spike Jones Story
Spirit of '76
State of Things
Static
Steven Wright Live
Straight to Hell
Strange Invaders
Streetfight
The Stunt Man
Sullivan's Travels
Sundae in New York
Sunset
Sunset Boulevard

Suppose They Gave a War
  And Nobody Came?
Surf Nazis Must Die
Susana
Sympathy for the Devil
The Tamer of Wild Horses
Tapeheads
Tartuffe
A Taste for Flesh and Blood
A Taxing Woman Returns
Teenager
That's My Hero!
There's Naked Bodies on My
  T.V.!
They Live
This is Spinal Tap
Thorn
Three Ages
Three of a Kind
The Thrill of It All!
To Be or Not to Be
Top Secret!
Trash
Traxx
True Stories
Trust Me
Tunnelvision
Turumba
Uforia
UHF
Under the Biltmore Clock
Unmasking the Idol
Very Curious Girl
Victor/Victoria
The Villain Still Pursued Her
Virgin High
Viridiana
Volpone
Voyage of the Rock Aliens
Walker
Watermelon Man
Way He Was
Where the Buffalo Roam
Where the Bullets Fly
White Mischief
The White Sheik
Whoops Apocalypse
Wild in the Streets
Wild, Wild West Revisited
Without a Clue
World Gone Wild
The Worm Eaters
The Worst of Hollywood: Vol.
  1
The Worst of Hollywood: Vol.
  2
The Worst of Hollywood: Vol.
  3
Would-Be Gentleman
Yin & Yang of Mr. Go
You'll Find Out
Young Doctors in Love
Young Frankenstein
Young Nurses in Love
Zachariah
The Zany Adventures of
  Robin Hood
Zelig
Zoo Radio
Zorro, the Gay Blade
Zvenigora

Savants

Being There
Champagne for Caesar
Charly
Doctor Doolittle
King Lear
Malcolm
Rain Man
Tony Draws a Horse

Close Encounters of the Third Kind
Club Extinction
Cocoon
Colossus: The Forbin Project
Commando Cody
Communion
Conquest of the Planet of the Apes
The Cosmic Eye
The Cosmic Man
The Cosmic Monsters
Cosmos: War of the Planets
Crash and Burn
The Crazies
The Creation of the Humanoids
Creature
Creatures the World Forgot
Crime Zone
Critters 4
Cyborg
D-Day on Mars
Dagora, the Space Monster
Daleks Invasion Earth 2150 A.D.
Damnation Alley
Dangaio
Dark Side of the Moon
D.A.R.Y.L.
The Day the Earth Caught Fire
The Day the Earth Stood Still
The Day It Came to Earth
The Day the Sky Exploded
Day Time Ended
Day of the Triffids
Dead End Drive-In
Dead Man Walking
Dead Space
Deadly Harvest
Deadly Weapon
Death Ray 2000
Death Sport
Death Watch
Deep Space
Deepstar Six
Def-Con 4
Deluge
Demon Seed
Destination Moon
Destination Moonbase Alpha
Destination Saturn
Destroy All Monsters
Devil Girl from Mars
Dinosaurus!
Doc Savage
Dr. Alien
Dr. Cyclops
Dr. Goldfoot and the Bikini Machine
Doctor Mordrid: Master of the Unknown
Doctor Satan's Robot
Doctor Who and the Daleks
Doctor Who: Death to the Daleks
Doctor Who: Pyramids of Mars
Doctor Who: Spearhead from Space
Doctor Who: Terror of the Zygons
Doctor Who: The Ark in Space
Doctor Who: The Seeds of Death
Doctor Who: The Talons of Weng-Chiang
Doctor Who: The Time Warrior
The Dog Soldier
Doin' Time on Planet Earth
Dominion: Tank Police, Act 1
Donovan's Brain
Dune
Dune Warriors
Earth vs. the Flying Saucers

Earth vs. the Spider
The Electronic Monster
The Element of Crime
The Eliminators
Embryo
Empire of the Ants
The Empire Strikes Back
Encounter at Raven's Gate
End of the World
Endgame
Endless Descent
Enemy Mine
Equalizer 2000
Escape 2000
Escape from the Bronx
Escape from DS-3
Escape from Galaxy Three
Escape from New York
Escape from the Planet of the Apes
Escape from Planet Earth
Escapes
E.S.P.
E.T.: The Extra-Terrestrial
Eve of Destruction
The Ewoks: Battle for Endor
Explorers
Exterminators of the Year 3000
The Eye Creatures
Eyes Behind the Stars
Fahrenheit 451
Fangs
Fantastic Planet
Fantastic Voyage
Far Out Space Nuts, Vol. 1
Fiend Without a Face
Final Approach
The Final Executioner
The Final Programme
Fire Maidens from Outer Space
Firebird 2015 A.D.
First Man into Space
First Men in the Moon
First Spaceship on Venus
Flash Gordon
Flash Gordon Battles the Galactic Forces of Evil
Flash Gordon Conquers the Universe
Flash Gordon: Mars Attacks the World
Flash Gordon: Rocketship
Flash Gordon: Space Adventurer
Flash Gordon: The Beast Men's Prey
Flash Gordon: Vol. 1
Flash Gordon: Vol. 2
Flash Gordon: Vol. 3
The Flesh Eaters
Flight to Mars
The Fly
The Fly 2
The Flying Saucer
Forbidden Planet
Forbidden World
Forbidden Zone
Force on Thunder Mountain
Fortress of Amerikka
Four Sided Triangle
The 4D Man
Frankenstein Meets the Space Monster
Freejack
From the Earth to the Moon
From ''Star Wars'' to ''Jedi'': The Making of a Saga
Frozen Alive
Fugitive Alien
Future Force
Future Hunters
Future Zone
Futureworld
Galactic Gigolo

Galaxy Express
Galaxy Invader
Galaxy of Terror
Game of Survival
Gamera
Gamera vs. Barugon
Gamera vs. Gaos
Gamera vs. Guiron
Gamera vs. Zigra
The Gamma People
Generations: Mind Control— From the Future
Germicide
Ghidrah the Three Headed Monster
Ghost Patrol
The Giant Claw
The Giant Gila Monster
The Giant Spider Invasion
The Gladiators
Glen and Randa
Godzilla 1985
Godzilla, King of the Monsters
Godzilla on Monster Island
Godzilla Raids Again
Godzilla vs. the Cosmic Monster
Godzilla vs. Mechagodzilla
Godzilla vs. Megalon
Godzilla vs. Monster Zero
Godzilla vs. Mothra
Godzilla vs. the Sea Monster
Godzilla vs. the Smog Monster
Godzilla's Revenge
Goliath Awaits
Gorath
Gorgo
Grampa's Sci-Fi Hits
The Greatest Flying Heroes
The Groundstar Conspiracy
Gunbuster 1
H-Man
Half Human
The Handmaid's Tale
Hands of Steel
Hangar 18
Hardware
Heartbeeps
The Hidden
Hide and Seek
Hideous Sun Demon
High Desert Kill
Highlander 2: The Quickening
Hollywood Boulevard 2
Horror Planet
Horrors of the Red Planet
Howard the Duck
The Human Duplicators
The Human Vapor
The Humanoid
Humanoid Defender
Humanoid Woman
Hyper-Sapien: People from Another Star
I Come in Peace
I Married a Monster from Outer Space
Iceman
Idaho Transfer
The Illustrated Man
In the Aftermath
In the Cold of the Night
Incredible Melting Man
The Incredible Petrified World
The Incredible Shrinking Man
Infra-Man
Innerspace
Interzone
Intruder Within
Invaders from Mars
Invasion
Invasion of the Animal People
Invasion of the Bee Girls
Invasion of the Body Snatchers

Invasion of the Body Stealers
Invasion Earth: The Aliens Are Here!
Invasion of the Girl Snatchers
Invasion of the Space Preachers
Invisible Adversaries
The Invisible Man Returns
Island of the Burning Doomed
Island of the Lost
Island of Terror
It Came from Beneath the Sea
It Came From Outer Space
It Conquered the World
Journey Beneath the Desert
Journey into the Beyond
Journey to the Center of the Earth
Journey to the Center of Time
Journey to the Far Side of the Sun
Killers from Space
The Killing Edge
Killings at Outpost Zeta
King Dinosaur
King Kong Versus Godzilla
The Kirlian Witness
Kronos
Krull
La Jetee/An Occurrence at Owl Creek Bridge
Laboratory
Land That Time Forgot
Laserblast
Last Chase
The Last Starfighter
Last War
The Last Woman on Earth
Lathe of Heaven
The Lawless Land
The Lawnmower Man
Leviathan
Lifeforce
Lifepod
Light Years
Liquid Sky
The Living Dead
Lobster Man from Mars
Lock and Load
Logan's Run
Looker
Lords of the Deep
The Lost Empire
The Lost Missile
Lost Planet Airmen
Lost in Space
The Lucifer Complex
Making Mr. Right
The Man from Atlantis
A Man Called Rage
The Man Who Fell to Earth
Manhunt of Mystery Island
The Manster
Marooned
Mars Needs Women
The Martian Chronicles: Part 1
The Martian Chronicles: Part 2
The Martian Chronicles: Part 3
Masters of Venus
MD Geist
Mechanical Crabs
Meet the Hollowheads
Megaforce
Megaville
Mesa of Lost Women
Metallica
Metalstorm
Metamorphosis
Meteor Monster
Miami Horror
Midnight Movie Massacre
Millenium
Mind Snatchers
Mind Trap
Mind Warp
Mindwarp

Misfits of Science
Missile to the Moon
Mission Galactica: The Cylon Attack
Mission Mars
Mission Stardust
Mistress of the World
Monkey Boy
The Monolith Monsters
The Monster that Challenged the World
Monster a Go-Go!
Monster from Green Hell
The Monster Maker
Monster from the Ocean Floor
The Monster of Piedras Blancas
Monster from a Prehistoric Planet
Monsters, Madmen, Machines
Moon 44
Moon Pilot
Mothra
Murder in Space
Mutant Hunt
Mutator
The Mysterians
Mysterious Island
Mysterious Planet
Mysterious Two
Navy vs. the Night Monsters
Neon City
The Nest
Neutron and the Black Mask
Neutron vs. the Amazing Dr. Caronte
Neutron vs. the Death Robots
The New Gladiators
The New Invisible Man
Next One
Night Beast
Night of the Blood Beast
Night Caller from Outer Space
Nightfall
Nightflyers
984: Prisoner of the Future
1984
No Survivors, Please
Norman's Awesome Experience
Not of This Earth
Nude on the Moon
Octaman
Omega Man
On the Comet
One Million B. C.
The Original Fabulous Adventures of Baron Munchausen
The Original Flash Gordon Collection
Outland
Parasite
Peacemaker
The People
The People That Time Forgot
People Who Own the Dark
The Phantom from 10,000 Leagues
The Phantom Empire
The Phantom Planet
Phantom from Space
Phase 4
Phenomenal and the Treasure of Tutankamen
The Philadelphia Experiment
Phoenix the Warrior
The Pink Chiquitas
Plan 9 from Outer Space
Planet of the Apes
Planet of Blood
Planet Burg
Planet of the Dinosaurs
Planet Earth
Planet on the Prowl
Planet of the Vampires

Science & Scientists

Screwball Comedy

Scuba

Flipper's New Adventure
Full Fathom Five
Great White Death
Isle of Forgotten Sins
Leviathan
The Man from Atlantis
Navy SEALS
Neptune Factor
Never Say Never Again
Night of the Sharks
Oceans of Fire
Scuba
The Sea Serpent
Sharks, Past & Present
Thunderball
The Treasure Seekers
Underwater

Sea Critter Attack

Alligator
Around the World Under the Sea
Attack of the Giant Leeches
Attack of the Swamp Creature
Barracuda
The Beach Girls and the Monster
The Beast from 20,000 Fathoms
Beyond Atlantis
Blood Beach
Bloodstalkers
Creature from the Black Lagoon
Creature from Black Lake
Creature from the Haunted Sea
Curse of the Swamp Creature
Deepstar Six
Demon of Paradise
Devil Monster
Devilfish
The Flesh Eaters
Godzilla vs. the Sea Monster
Gorgo
The Great Alligator
Horror of Party Beach
Island Claw
Jason and the Argonauts
Jaws
Jaws 2
Jaws 3
Jaws of Death
Jaws: The Revenge
Killer Fish
The Loch Ness Horror
The Man from Atlantis
Moby Dick
The Monster that Challenged the World
Monster from the Ocean Floor
The Monster of Piedras Blancas
Octaman
Orca
The Phantom from 10,000 Leagues
Piranha
Piranha 2: The Spawning
Reap the Wild Wind
Scorpion with Two Tails
Screams of a Winter Night
The Sea Serpent
Serpent Island
Shark!
Shark's Paradise
The Spawning
Tentacles
Terror in the Swamp
Tintorera...Tiger Shark
20,000 Leagues Under the Sea
Up from the Depths
What Waits Below

Serials

Ace Drummond
Adventures of Captain Marvel
The Adventures of the Flying Cadets
The Adventures of Grizzly Adams at Beaver Dam
The Adventures of the Masked Phantom
Adventures of Red Ryder
The Adventures of Rex & Rinty
The Adventures of Tarzan
Batman and Robin: Volume 1
Batman and Robin: Volume 2
Battling with Buffalo Bill
The Best of Dark Shadows
The Black Coin
The Black Widow
Blake of Scotland Yard
Bread (BBC)
Buck Rogers Cliffhanger Serials, Vol. 1
Buck Rogers Cliffhanger Serials, Vol. 2
Buck Rogers: Planet Outlaws
Bugs Bunny Superstar
Burn 'Em Up Barnes
Candid Candid Camera, Vol. 2 (More Candid Candid Camera)
Captain America
Captain Gallant, Foreign Legionnaire
Carmilla
Casey Jones and Fury (Brave Stallion)
Cheyenne Rides Again
Cliffhangers
The Clutching Hand
Commando Cody
The Crimson Ghost
Custer's Last Stand
Cyclotrode ''X''
D-Day on Mars
Daredevils of the Red Circle
Devil Horse
Devlin Connection
Dick Tracy
Dick Tracy Returns
Dick Tracy: The Original Serial, Vol. 1
Dick Tracy: The Original Serial, Vol. 2
Dick Tracy vs. Crime Inc.
Doctor Satan's Robot
Don Daredevil Rides Again
Don Winslow of the Coast Guard
Don Winslow of the Navy
Drums of Fu Manchu
Fighting Devil Dogs
Fighting with Kit Carson
Fighting Marines
Firesign Theatre's Hot Shorts
Flaming Frontiers
Flash Gordon Conquers the Universe
Flash Gordon: Mars Attacks the World
Flash Gordon: Vol. 1
Flash Gordon: Vol. 2
Flash Gordon: Vol. 3
G-Men Never Forget
G-Men vs. the Black Dragon
Genie of Darkness
Great Serial Prevues
Green Archer
The Green Hornet
Hawk of the Wilderness
Heroes of the West
Holt of the Secret Service
Hurricane Express
In Old New Mexico
Jesse James Rides Again

Jungle Drums of Africa
Junior G-Men
Junior G-Men of the Air
Juve Contre Fantomas
King of the Congo
King of the Kongo
King of the Rocketmen
King of the Texas Rangers
Last Frontier
Last of the Mohicans
Law of the Wild
A Lawman is Born
Lightning Warrior
Little Rascals, Book 4
Little Rascals, Book 5
Little Rascals, Book 6
Little Rascals, Book 7
Little Rascals Christmas Special
The Lone Defender
The Lone Ranger
Lone Ranger
The Lost City
Lost City of the Jungle
The Lost Jungle
Lost Planet Airmen
Manhunt in the African Jungle
Manhunt of Mystery Island
The Masked Marvel
Master Key
Masters of Venus
Miracle Rider: Vol. 1
Miracle Rider: Vol. 2
Mission: Monte Carlo
The Monster Demolisher
Mysterious Doctor Satan
Mystery Mountain
Mystery Squadron
Mystery Trooper
The New Adventures of Tarzan: Vol. 1
The New Adventures of Tarzan: Vol. 2
The New Adventures of Tarzan: Vol. 3
The New Adventures of Tarzan: Vol. 4
Nyoka and the Tigermen
Oregon Trail
The Original Flash Gordon Collection
Overland Mail
The Painted Stallion
Perils of the Darkest Jungle
Perils of Pauline
The Phantom Creeps
The Phantom Empire
Phantom Rider
Phantom of the West
Pirates of the High Seas
The Purple Monster Strikes
Queen of the Jungle
Radar Men from the Moon
Radio Patrol
Radio Ranch
Red Barry
Return of Chandu
Riders of Death Valley
Rise & Rise of Daniel Rocket
Robinson Crusoe of Clipper Island
Robinson Crusoe of Mystery Island
Rocket to the Moon
Rocky Jones, Space Ranger: Pirates of Prah
Rocky Jones, Space Ranger: Renegade Satellite
Rocky Jones, Space Ranger: The Cold Sun
Rocky Jones, Space Ranger: Trial of Rocky Jones
Rocky Jones, Space Ranger, Vol. 3: Blast Off
Roots, Vol. 1
Saturday Night Serials

Savage Fury
Sea Hound
Shadow of the Eagle
Sign of Zorro
Son of Zorro
S.O.S. Coast Guard
Space Soldiers Conquer the Universe
Sunset on the Desert
Superman: The Serial: Vol. 1
Suspense
Texas to Bataan
Three Musketeers
Tom Corbett, Space Cadet Volume 2
Tom Mix Compilation 1
Tom Mix Compilation 2
Trail Riders
Trail of the Royal Mounted
Trail of the Silver Spurs
Tumbledown Ranch in Arizona
Undersea Kingdom
Vanishing Legion
The Vigilantes Are Coming
Whispering Shadow
Winners of the West
Wolf Dog
Young Eagles
Zombies of the Stratosphere
Zorro Rides Again
Zorro's Black Whip
Zorro's Fighting Legion

Sex & Sexuality

ABC's of Love & Sex, Australia Style
Accident
The Adultress
Adventures of a Plumber's Helper
Adventures of a Private Eye
Adventures of a Taxi Driver
After School
Alexa: A Prostitute's Own Story
Alfie
Alice Goodbody
All the Way Down
The Allnighter
Almost Pregnant
Alpine Fire
Alvin Purple
Alvin Rides Again
The Amazing Transplant
American Drive-In
American Gigolo
The Amorous Adventures of Moll Flanders
And God Created Woman
And When She Was Bad...
Andy Warhol's Dracula
Andy Warhol's Frankenstein
Angel Heart
Angel of H.E.A.T.
L'Annee des Meduses
The Apartment
Aphrodite
Appassionata
The Arousers
Assault
Baby Doll
Baby Face
Baby Love
The Baby Maker
Bachelor Party
Bad Girls
Bad Girls from Mars
Ballyhoo Baby
Barbarella
Bare Necessities
Baring It All
Basic Training
Battling Amazons

The Bawdy Adventures of Tom Jones
Beach Girls
Beachballs
Beauties and the Beast
Bedroom Eyes 2
Beginner's Luck
The Bellboy and the Playgirls
Beneath the Valley of the Ultra-Vixens
The Berlin Affair
Best Buns on the Beach
The Best Little Whorehouse in Texas
Betrayal
Betty Blue
Beverly Hills Call Girls
Beverly Hills Madam
Beyond Erotica
Beyond the Silhouette
The Big Bet
Big Bust Out
Bikini Story
Bilitis
Birth Right
The Bitch
Black Emmanuelle
Black Narcissus
The Black Room
Black Starlet
Black Venus
Blind Vision
Blood and Sand
Bloodbath
The Blue Angel
Blue Jeans
The Blue Lagoon
Blue Movies
Blue Velvet
Boarding School
Bob & Carol & Ted & Alice
Boccaccio '70
Body Chemistry
Body Chemistry 2: Voice of a Stranger
Body Double
Body Waves
Bonjour Tristesse
Boob Tube
The Boss' Wife
Bottoms Up
A Boy and His Dog
Breathless
The Bride Is Much Too Beautiful
Brides of the Beast
Bus Station
Business as Usual
Butterfield 8
Butterfly
Bye Bye Baby
Cactus in the Snow
Caledonian Dreams
California Casanova
Caligula
Can I Do It...Till I Need Glasses?
Candid Candid Camera, Vol. 1
The Canterbury Tales
Carnal Crimes
Carnal Knowledge
Carry On Emmanuelle
Casanova '70
Casual Sex?
Cat on a Hot Tin Roof
Cat People
Cave Girl
Cesar & Rosalie
The Chambermaid's Dream
Champagne for Breakfast
Chaste & Pure
Chatterbox
Cheaters
Cheech and Chong's Up in Smoke
Cheerleaders' Beach Party

Cherry 2000
The Chicken Chronicles
Child Bride
Chinese Roulette
Choose Me
Cinderella
Cinderella 2000
Claire's Knee
Class
Class of '63
Class Reunion
Cleo/Leo
The Clinic
Close My Eyes
Closely Watched Trains
Collector's Item
The Comfort of Strangers
Compromising Positions
Con Mi Mujer No Puedo
Confessions of a Young
   American Housewife
Conspiracy
Contagion
The Cook, the Thief, His Wife
   & Her Lover
Cool it Carol
Copenhagen Nights
Corruption
Courtesans of Bombay
Crimes of Passion
Cruel Story of Youth
Cry Uncle
Dangerous Liaisons
The Dark Room
Dark Side of Love
Daughters of Satan
Death of a Centerfold
The Decameron
Decameron Nights
Deceptions
The Decline of the American
   Empire
Delinquent School Girls
Denial: The Dark Side of
   Passion
Desire and Hell at Sunset
   Motel
Desk Set
Devil in the Flesh
The Devils
The Devil's Eye
The Devil's Nightmare
The Devil's Playground
Devil's Wedding Night
Diary of a Chambermaid
Diary of a Mad Housewife
Diary of a Mad Old Man
Diner
Dirty Dancing
Divine Nymph
Divorce—Italian Style
Dr. Alien
Dr. Caligari
Dr. Minx
Dog Day Afternoon
Doll Face
Dollar
Domino
Dona Herlinda & Her Son
Don's Party
Don't Tell Daddy
Dorothy Stratten: The Untold
   Story
Double Suicide
Dracula Sucks
Dreams
Dreams Come True
Dreams of Desire
Driver's Seat
Drum
East End Hustle
Eating Raoul
Ebony Tower
Echoes of Paradise
The Eclipse
Ecstasy

Ecstasy Inc.
Eddie Murphy "Delirious"
Eddie Murphy: Raw
The Eerie Midnight Horror
   Show
El Preprimido (The Timid
   Bachelor)
El: This Strange Passion
Elmer Gantry
Elvira, Mistress of the Dark
Emilienne
Emily
Emmanuelle
Emmanuelle 4
Emmanuelle in Bangkok
Emmanuelle in the Country
Emmanuelle & Joanna
Emmanuelle, the Joys of a
   Woman
Emmanuelle, the Queen
Emmanuelle in Soho
Emmanuelle's Daughter
Emperor Caligula: The Untold
   Story
End of August
Endless Love
Enemies, a Love Story
Enormous Changes
Enrapture
Ernesto
Erotic Adventures of
   Pinocchio
Erotic Dreams
Erotic Escape
Erotic Illusion
Erotic Images
Erotic Taboo
The Erotic Three
Erotic Touch of Hot Skin
Erotica: Fabulous Female
   Fantasies
Escort Girls
Et La Tenoresse?...Bordel!
Everything You Always
   Wanted to Know About
   Sex (But Were Afraid to
   Ask)
Evils of the Night
Extreme Close-Up
Fairytales
Fanny Hill
Fanny Hill: Memoirs of a
   Woman of Pleasure
The Fantasist
Fantasy in Blue
Fantasy Masquerade
Faraway Fantasy
Fast Food
Fast Times at Ridgemont High
Fatal Attraction
Fatal Instinct
Feelin' Screwy
Felicity
Fellini Satyricon
Female Trouble
Firehouse
First Affair
The First Time
First Time
First Turn On
Five Forty-Fight
Flanagan Boy
Flesh
Flesh Gordon
A Flight of Rainbirds
Forbidden Impulse
Foreign Body
Foreplay
Forever Emmanuelle
The 4th Man
Foxes
Frank and I
Frankenstein Unbound
Frasier the Sensuous Lion
Fraternity Vacation
The French Woman

Frenzy
Fritz the Cat
The Fugitive Kind
Fun Down There
Gabriela
Galactic Gigolo
The Game is Over
Games Girls Play
Gap-Toothed Women
Gas Pump Girls
Gemidos de Placer
Gentlemen Prefer Blondes
Genuine Risk
Georgy Girl
Get 'Em Off
Get Out Your Handkerchiefs
Getting It On
Getting Lucky
Ghosts Can Do It
The Gift
Gilda
Ginger Ale Afternoon
The Girl
The Girl on a Motorcycle
Girl-Toy
Girls of Don Juan
The Girls of Malibu
Girls of the Moulin Rouge
Glen or Glenda
The Goddess
Goin' All the Way
The Good Mother
Goodbye Emmanuelle
Gorp
Gothic
The Graduate
Great British Striptease
The Handmaid's Tale
Hang Tough
Happily Ever After
Happy Hooker
Happy Hooker Goes to
   Washington
Happy Sex
Hardbodies 2
Harrad Experiment
Harrad Summer
Has Anyone Seen My Pants?
Heart of the Stag
Heaven Can Wait
Heavy Petting
Hell Comes to Frogtown
Henry & June
Her Summer Vacation
Higher Education
Hollywood Confidential
The Hollywood Game
Homework
Honey
Horseplayer
Hot Child in the City
Hot Chili
The Hot Spot
Hot Target
Hot Under the Collar
The Hotel New Hampshire
How Funny Can Sex Be?
How to Pick Up Girls
How to Seduce a Woman
The Hurried Man
Husbands and Lovers
I Am Curious (Yellow)
I Give My All
I Posed for Playboy
I Will, I Will for Now
Ilsa, Harem Keeper of the Oil
   Sheiks
Ilsa, She-Wolf of the SS
Ilsa, the Tigress of Siberia
Ilsa, the Wicked Warden
Impure Thoughts
In the Heat of Passion
In Praise of Older Women
In the Realm of the Senses
Incoming Freshmen
An Inconvenient Woman

Infamous Daughter of Fanny
   Hill
Inside Out
Intimate Games
Intimate Moments
Invasion of the Bee Girls
Irezumi (Spirit of Tattoo)
Is There Sex After Death?
Isadora
It
Italian Stallion
It's Not the Size That Counts
Jet Sets
John & Julie
Julia
Justine
Kanako
Kandyland
Keep It Up, Jack!
Kenneth Anger, Vol. 1:
   Fireworks
Kenneth Anger, Vol. 3:
   Scorpio Rising
Kenneth Anger, Vol. 4:
   Invocation of My Demon
   Brother
Kidnapped
Killer Workout
Kipling's Women
Kiss Me, Stupid!
La Amante
La Casita del Pecado
La Cicada (The Cricket)
La Ronde
L.A. Woman
Labyrinth of Passion
Ladies Night Out
Lady on the Bus
Lady Caroline Lamb
Lady Chatterley's Lover
Lady Frankenstein
Lady of the Rising Sun
Last American Virgin
Last Call
Last of the Red Hot Lovers
Last Resort
Last Summer
Last Tango in Paris
Law of Desire
Le Boucher
Le Cavaleur (Practice Makes
   Perfect)
Le Sex Shop
Legal Tender
Let's Do It!
The Lickerish Quartet
The Lie
Liebestraum
The Life & Loves of a Male
   Stripper
Lifeforce
The Lifetaker
Lights, Camera, Action, Love
Liquid Sky
Little Darlings
Little Lips
A Little Sex
Livin' the Life
Lolita
Looking for Mr. Goodbar
Loose Screws
Los Hombres Piensan Solo en
   Solo
Losin' It
Loulou
The Love Affair, or Case of
   the Missing Switchboard
   Operator
Love Angels
Love Circles
Love Crimes
Love, Drugs, and Violence
The Love Goddesses
Love Lessons
Love Letters
The Love Pill

Love Scenes
Loverboy
The Loves of Irina
Loving Couples
Lunch Wagon
Macho Dancer
Macho y Hembra
Madame Bovary
Mae West
Maid in Sweden
Maitresse
Makin' It
Making Love
A Man with a Maid
Man Who Had Power Over
   Women
The Man Who Loved Women
Maria's Lovers
Marilyn: The Untold Story
The Marquis de Sade's
   Justine
A Married Woman
Mars Needs Women
Mata Hari
Matrimonio a la Argentina
Mazurka
Me and Him
Meatballs 3
A Midsummer Night's Sex
   Comedy
Miss Julie
Miss Right
Miss Sadie Thompson
Mistress
Mo' Better Blues
Mona's Place
Mondo Trasho
Monsieur Hire
Montenegro
Monty Python's The Meaning
   of Life
Moon is Blue
Mortal Sins
Movers and Shakers
Murder Weapon
My Best Friend's Girl
My Chauffeur
My Demon Lover
My Father's Wife
My Life to Live
My Nights With Susan,
   Sandra, Olga, and Julie
My Wonderful Life
Myra Breckinridge
Mystery of Alexina
Naked Lie
Naked Lunch
Nanami, First Love
Nathalie Comes of Age
National Adultery
Naughty Co-Eds
Naughty Negligee Nights
Naughty Roommates
Nea
Negatives
New Look
New York Nights
Nice Girls Don't Explode
Night Eyes
Night Games
Night of the Living Babes
Night Stalker
Night Zoo
9 1/2 Weeks
No Secrets
Norma
Norman Loves Rose
Nudo di Donna
Oh! Calcutta!
On the Block
Once is Not Enough
One Plus One
Oriane
The Other Side of Midnight
Paddy
Paradisio

Paranoia
Pardon Mon Affaire, Too!
Party Animal
Party Incorporated
The Passion of Evelyn
Pauline at the Beach
Peeping Tom
The Penitent
Penthouse Love Stories
Penthouse Video, Vol. 1: The Girls of Penthouse
Penthouse on the Wild Side
Performance
The Perils of Gwendoline
Picnic
The Pink Angels
Playboy's 1988 Playmate Video Calendar
Playboy's 1990 Playmate Video Calendar
Playboy's Bedtime Stories
Playboy's Farmers' Daughters
Playboy's Inside Out
Playboy's Playmate of the Year Video Centerfold
Playboy's Wet & Wild 2
Playgirl Killer
Playing with Fire
Playmate Review
A Pleasure Doing Business
Pleasure Resort
Pleasure Unlimited
Point of Terror
Pom Pom Girls
Porky's
The Pornographers
Portnoy's Complaint
The Postman Always Rings Twice
Power, Passion & Murder
Prep School
Preppies
Pretty Baby
Pretty Smart
Pretty Woman
Private Lessons
Private Popsicle
Private Practices: The Story of a Sex Surrogate
Private Resort
Rabbit Test
Rain
The Rainbow
Rattle of a Simple Man
Red Kiss (Rouge Baiser)
Red Shoe Diaries
Redd Foxx—Video in a Plain Brown Wrapper
Reflections: A Moment in Time
The Reincarnation of Golden Lotus
Rendez-vous
Repo Jake
Repulsion
Restless
Revealing of Elsie
Rhino's Guide to Safe Sex
The Ribald Tales of Robin Hood
Rick's: Your Place for Fantasy
Rita, Sue & Bob Too
The Rocky Horror Picture Show
Romantic Encounters
R.S.V.P.
The Rue Morgue Massacres
Run Virgin Run
The Sailor Who Fell from Grace with the Sea
Salome
Sammy & Rosie Get Laid
Sandstone
Scandal
Scene of the Crime
School for Sex

Scorchy
Score
Scoring
Screwball Academy
Screwball Hotel
Scrubbers
Sea of Love
Secret Fantasy
Secret Games
Secret Life of an American Wife
Secrets
Secrets in the Dark
Secrets of Three Hungry Wives
The Seniors
Sensations
The Sensual Taboo
Sensuous Caterer
Separate Vacations
Sex
Sex Appeal
Sex on the Brain
Sex in the Comics
sex, lies and videotape
Sex Madness
Sex O'Clock News
Sex and the Office Girl
Sex and the Other Woman
Sex Ray
Sex on the Run
Sex with the Stars
Sex Surrogate
Sexcetera
Sextette
Sexual Response
The Shaming
Shampoo
Shattered Innocence
She's Gotta Have It
Shotgun
Sibling Rivalry
The Silence
Sinbad
The Sinful Bed
Single Bars, Single Women
The Sister-in-Law
Sisters of Satan
Ski School
Skin Deep
The Slap
Slaves of Love
Smooth Talker
Sneakin' and Peekin'
Sooner or Later
Sorority House Scandals
Spaced Out
Splendor in the Grass
Star 80
Starlet
The Statue
The Stewardesses
The Stilts
The Story of O
The Story of O, Part 2
Stripped to Kill
The Stripper
Stripper T's
The Stud
The Student Body
The Student Body/Jailbait Babysitter
Substitute Teacher
Sugarbaby
Summer Camp
Summer Heat
Summer Job
Summer Lovers
Summer Night Fever
A Summer in St. Tropez
Summer School Teachers
Sunday, Bloody Sunday
Super Bitch
Supervixens
The Sure Thing
Surfacing

Susana
Sweet Ecstasy
Sweet Lies
Sweet Poison
Sweet Spirits
Swimming Pool
The Swinging Cheerleaders
Swinging Ski Girls
Swinging Sorority Girls
Takin' It Off
Tale of Two Sisters
Talkin' Dirty After Dark
Talking Walls
Tall, Dark & Handsome
Tattoo
Taxi zum Klo
Tea For Three
Tea and Sympathy
Teen Lust
Teenage Seductress
Tendres Cousines
Teorema
The Terror Within
Test Tube Babies
Texasville
That Obscure Object of Desire
There's Naked Bodies on My T.V.!
Therese & Isabelle
Thief of Hearts
36 Fillete
This Property is Condemned
Three in the Attic
Three in the Cellar
Three Day Weekend
Tie Me Up! Tie Me Down!
Tigers in Lipstick
'Tis a Pity She's a Whore
To Be a Rose
Together
Tom Jones
Tonight for Sure
Traenen in Florenz (Tears in Florence)
Trash
Traveling Companions
Tropic of Desire
Tropic Heat
Trouble in Paris
Truck Stop
Truth About Women
Twenty-One
Twins of Evil
Two Women in Gold
The Ultimate Thrill
The Unbearable Lightness of Being
Under the Doctor
Under the Sheets
The Underachievers
Undercover Vixens
Up Your Anchor
Valet Girls
The Valley
Valmont
Vanessa
Venus on Fire
Very Curious Girl
Victorian Fantasies
Video Vixens
A View to a Kill
Virgin Queen of St. Francis High
The Virgin Soldiers
Virgins of Purity House
Wedding in Blood
Weekend Pass
Welcome to L.A.
What Would Your Mother Say?
What's Up Front
When Women Had Tails
Where the Boys Are '84
Where the Hot Wind Blows
A Whisper to a Scream
Whore

Wild Orchid
Wild, Wild World of Jayne Mansfield
Wish You Were Here
Witches of Eastwick
A Woman in Flames
A Woman, Her Men and Her Futon
The Woman Next Door
Woman Times Seven
Women in Fury
Women in Love
Women's Club
Working Girls
The World is Full of Married Men
X, Y & Zee
Yanks
Yellow Emanuelle
Yesterday, Today and Tomorrow
Young Aphrodites
Young Lady Chatterly
Young Lady Chatterly 2
Young Love, First Love
Young Nurses in Love
Your Ticket Is No Longer Valid
You've Got to Have Heart
The Yum-Yum Girls
Zabriskie Point
Zandalee
Zapped!
Zombie Island Massacre

Sexploitation

The Abductors
Alice Goodbody
Amazon Jail
And When She Was Bad...
Angel 3: The Final Chapter
Baby Love
Bad Girls Go to Hell
Barbarian Queen 2: The Empress Strikes Back
Bare Hunt
Bare Necessities
Baring It All
The Bawdy Adventures of Tom Jones
Bedroom Eyes
The Bellboy and the Playgirls
Beneath the Valley of the Ultra-Vixens
Best Buns on the Beach
Beverly Hills Knockouts
The Bikini Car Wash Company
Bikini Story
Black Emmanuelle
Black Venus
Blazing Stewardesses
Blue Ribbon Babes
Bolero
Bottoms Up
Boys Night Out
The Bride Is Much Too Beautiful
California Manhunt
Candy Stripe Nurses
Candy Tangerine Man
Casting Call
Chain Gang Women
The Chambermaid's Dream
Champagne for Breakfast
The Cheerleaders
Cheerleaders' Beach Party
Cheerleaders' Wild Weekend
Cherry Hill High
Chesty Anderson USN
Cinderella
Class Reunion
Copenhagen Nights
Corruption
Dandelions

Deadly Embrace
Delinquent School Girls
Devil's Wedding Night
Dirty Mind of Young Sally
Dr. Minx
Domino
Dracula Sucks
Ecstasy
Ecstasy Inc.
Emily
Emmanuelle
Emmanuelle 4
Emmanuelle 5
Emmanuelle in Bangkok
Emmanuelle in the Country
Emmanuelle & Joanna
Emmanuelle, the Joys of a Woman
Emmanuelle in Soho
Emmanuelle on Taboo Island
Erotic Adventures of Pinocchio
Erotic Dreams
Erotic Escape
Erotic Illusion
Erotic Taboo
Erotic Touch of Hot Skin
Erotica: Fabulous Female Fantasies
Escort Girls
Fanny Hill: Memoirs of a Woman of Pleasure
Feelin' Up
Firehouse
The First Nudie Musical
Flesh Gordon
Forbidden Impulse
Forever Emmanuelle
Frank and I
The French Woman
Ghosts Can't Do It
Gimme an F
Ginger
The Girls of Crazy Horse Saloon
Good Time with a Bad Girl
Goodbye Emmanuelle
Great British Striptease
The Happy Gigolo
Happy Hooker
Happy Hooker Goes to Washington
Hard Body Workout
Hardbodies
Hollywood Boulevard 2
Hollywood Erotic Film Festival
Hollywood High
Hollywood High Part 2
Hollywood Hot Tubs
Hollywood Hot Tubs 2: Educating Crystal
Honkytonk Nights
Hot Moves
Hot T-Shirts
H.O.T.S.
Infamous Daughter of Fanny Hill
Inhibition
Inn of Temptation
Intimate Games
Invasion of the Girl Snatchers
The Invisible Maniac
Joy of Sex
Kandyland
Kipling's Women
Knock Outs
Lady Godiva Rides
Lady of the Rising Sun
Last Dance
Let's Do It!
The Lickerish Quartet
Liquid Dreams
Little Miss Innocence
Love Angels
Love Circles
Love Desperados

Sexploitation

Love Lessons
Love Notes
Lust for Freedom
Mistress Pamela
Modern Girls
Mona's Place
My Father's Wife
My Pleasure Is My Business
My Therapist
Mysterious Jane
The Naked Sword of Zorro
Naked Venus
Nature's Playmates
Naughty Knights
Night Call Nurses
The Night Evelyn Came Out of the Grave
The Night Porter
Nine Ages of Nakedness
Nude on the Moon
Nurse on Call
Orgy of the Dead
Outrageous Strip Review
Paranoia
Party Favors
Party Plane
The Passion of Evelyn
Penthouse Love Stories
Penthouse on the Wild Side
Perfect Timing
The Perils of Gwendoline
Playboy Video Centerfold: 35th Anniversary Playmate
Playboy's 1989 Playmate Video Calendar
Playboy's Bedtime Stories
Playmate of the Year Video Centerfold 1990
Please Don't Eat My Mother
Pleasure Resort
Pleasure Unlimited
Preppies
Private Collection
Private Duty Nurses
Private Passions
Private School
Rebel Vixens
Revealing of Elsie
Revenge of the Cheerleaders
Revenge of the Virgins
Romantic Encounters
Round Trip to Heaven
Saturday Night Sleazies, Vol. 1
Screen Test
Sea of Dreams
Secrets in the Dark
Sensual Partners
The Sensual Taboo
Sensuous Caterer
The Sensuous Teenager
Sex Adventures of the Three Musketeers
Sex on the Brain
Sex and Buttered Popcorn
Sex Surrogate
She Devils in Chains
Sheer Heaven
Slammer Girls
Slumber Party Massacre 3
Snake Dancer
Sorority Babes in the Slimeball Bowl-A-Rama
Sorority House Massacre 2
Spaced Out
S.S. Love Camp No. 27
The Story of O, Part 2
Strip Teasers
Strip Teasers 2
Striptease
The Student Nurses
Submission
Suburban Roulette
Sugar Cookies
Supervixens
Sweet Georgia

The Swinging Cheerleaders
Swinging Ski Girls
Swinging Sorority Girls
Tender Loving Care
Texas Layover
Thinkin' Big
Train Station Pickups
Trouble in Paris
Truck Stop
Truck Stop Women
2069: A Sex Odyssey
Ultimate Desires
Ultimate Sampler Video
The Vampire Hookers
Vanessa
Venus on Fire
Venus in Furs
Vice Academy
Vice Academy 2
Vice Academy 3
Victorian Fantasies
Video Vixens
What Would Your Mother Say?
Wild Orchid 2: Two Shades of Blue
The Wildest Office Party
Witchcraft 2: The Temptress
Witchcraft 3: The Kiss of Death
The Young Nurses

Sexual Abuse

Demented
The Devils
Fall from Innocence
Heart of the Stag
Ilsa, Harem Keeper of the Oil Sheiks
Ilsa, She-Wolf of the SS
Ilsa, the Tigress of Siberia
Ilsa, the Wicked Warden
Mesmerized
The Night Porter

Shower & Bath Scenes

About Last Night...
American Gigolo
The Big Chill
Body Double
Breathless
Bull Durham
Club Paradise
Diabolique
Dressed to Kill
Fatal Attraction
Ghostbusters 2
High Anxiety
Hollywood Hot Tubs
Hollywood Hot Tubs 2: Educating Crystal
Jinxed
The Legacy
Married to the Mob
Mister Roberts
National Lampoon's Vacation
Porky's
The Postman Always Rings Twice
Psycho
Psycho 2
St. Elmo's Fire
Schizo
Silkwood
Snow White and the Seven Dwarfs
Stripes
Tequila Sunrise
Tie Me Up! Tie Me Down!
Valley Girl
Valmont

Shrinks

Agnes of God
Beauty on the Beach
Beyond Therapy
Blood Frenzy
Body Chemistry 2: Voice of a Stranger
The Brain
Captain Newman, M.D.
The Cobweb
The Couch Trip
Dark Mirror
Dead Man Out
The Dream Team
Fear
Final Analysis
Final Approach
A Fine Madness
High Anxiety
Hollow Triumph
Home Movies
House of Games
Human Experiments
I Never Promised You a Rose Garden
Lies of the Twins
Lovesick
Lovey: A Circle of Children 2
The Man Who Loved Women
Mine Own Executioner
Mr. Frost
Norma
On a Clear Day You Can See Forever
One Flew Over the Cuckoo's Nest
The Prince of Tides
Red Wind
Schizoid
Shock
The Sleeping Tiger
Spellbound
Still of the Night
Sybil
They Might Be Giants
The Three Faces of Eve
Under the Doctor
What's New Pussycat?
Wild in the Country

Shutterbugs

Bikini Island
Blow-Up
Centerfold
Corrupt Ones
Dagger Eyes
Darkroom
Dr. Mabuse vs. Scotland Yard
The Dolls
Don't Answer the Phone
Double Exposure
Double Exposure: The Story of Margaret Bourke-White
The Double Negative
Eyes Behind the Stars
Eyes of Laura Mars
Fatal Exposure
Fatal Images
The Favor, the Watch, & the Very Big Fish
Funny Face
High Season
I Love Melvin
Kanako
Killer Image
Love Crimes
Love & Murder
Model Behavior
Natural States
Not for Publication
Peeping Tom
Perfect Timing
Photographer
Remembering LIFE

Reporters
Return to Boggy Creek
Snapshot
A Time to Die
The Unbearable Lightness of Being
Watch the Birdie

Silent Films

The Adorable Cheat
The Adventurer
Adventures of Hairbreadth Harry
The Adventures of Tarzan
The Adventures of Walt Disney's Alice
Aelita: Queen of Mars
The Al Jolson Collection
All Night Long/Smile Please
America/The Fall of Babylon
American Aristocracy
American Pluck
The Americano
The Americano/Variety
Annapolis
Are Parents People?
Arsenal
Assunta Spina
The Atonement of Gosta Berling
Avenging Conscience
The Average Woman
Baby Face Harry Langdon
The Babylon Story from "Intolerance"
Backfire
Backstairs
Bakery/Grocery Clerk
The Balloonatic/One Week
The Bargain
Battle of Elderbush Gulch
The Battleship Potemkin
Battling Bunyon
Battling Butler
Beau Brummel
Beau Revel
Bed and Sofa
Behind the Front
Bellhop/The Noon Whistle
The Beloved Rogue
Ben-Hur
Ben Turpin: An Eye for an Eye
Best of Chaplin
The Big Parade
Birth of a Nation
The Black Pirate
The Blacksmith/The Balloonatic
The Blacksmith/The Cops
The Blasphemer
Blind Husbands
Blood and Sand
Blood and Sand/Son of the Sheik
The Blot
The Broadway Drifter
Broken Blossoms
The Broken Mask
Burlesque of Carmen
The Busher
Buster Keaton Festival: Vol. 1
Buster Keaton Festival: Vol. 2
Buster Keaton Festival: Vol. 3
Buster Keaton Rides Again/The Railroader
Buster Keaton: The Golden Years
The Cabinet of Dr. Caligari
Cabiria
California Straight Ahead
The Cameraman
Campus Knights
Cannon Ball/Eyes Have It

Cannonball/Dizzy Heights and Daring Hearts
Captain Swagger
Cartoonal Knowledge: Confessions of Farmer Gray
Cartoonal Knowledge: Farmer Gray Looks at Life
Cartoonal Knowledge: Farmer Gray and the Mice
The Cat and the Canary
The Cats and Mice of Paul Terry
Champagne
The Champion/His New Job
Chaplin Essanay Book 1
Chaplin Mutuals
The Chaplin Review
Chapter in Her Life
Charleston
Charley's Aunt
Charlie Chaplin Carnival, Four Early Shorts
Charlie Chaplin Cavalcade, Mutual Studios Shorts
Charlie Chaplin Centennial Collection
Charlie Chaplin Festival
Charlie Chaplin, the Mutual Studio Years
Charlie Chaplin: Night at the Show
Charlie Chaplin: Rare Chaplin
Charlie Chaplin: The Early Years, Volume 1
Charlie Chaplin: The Early Years, Volume 2
Charlie Chaplin: The Early Years, Volume 3
Charlie Chaplin: The Early Years, Volume 4
Charlie Chaplin...Our Hero!
Charlie Chaplin's Keystone Comedies #1
Charlie Chaplin's Keystone Comedies #2
Charlie Chaplin's Keystone Comedies #3
Charlie Chaplin's Keystone Comedies #4
Charlie Chaplin's Keystone Comedies #5
Charlie Chase and Ben Turpin
The Cheat
The Cheerful Fraud
Child of the Prairie
The Cigarette Girl of Mosselprom
The Circus
City Girl
City Lights
Civilization
Clodhopper
Cobra
College
Comedy Classics of Mack Sennett and Hal Roach
The Coming of Amos
Cops/One Week
Cord & Dischords/Naughty Nurse
The Count
Count/Adventurer
Count of Monte Cristo
The Country Kid
County Fair
The Covered Wagon
Crainquebille
The Crazy Ray
Cricket on the Hearth
The Crowd
Cure
Custer's Last Fight
Cyclone Cavalier
Dancing Mothers
Danger Ahead!

Tempest
Tempest/Eagle
Ten Commandments
Ten Days That Shook the World/October
Ten Nights in a Bar Room
That Certain Thing
These Girls Won't Talk
The Thief
Thief of Baghdad
Thomas Graal's Best Film
Thomas Graal's First Child
Three Ages
Three Charlies and One Phoney!
The Three Musketeers
The Three Musketeers (D'Artagnan)
Three Word Brand
Through the Breakers
Tillie's Punctured Romance
Tol'able David
Toll Gate
Tom Mix Compilation 1
Tom Mix Compilation 2
The Tomboy
Too Wise Wives
The Torture of Silence
Tournament
Traffic in Souls
Tramp & a Woman
Trapped by the Mormons
Treasure of Arne
Triple Trouble/Easy Street
True Heart Susie
20,000 Leagues Under the Sea
Two Men & a Wardrobe
Un Chien Andalou
The Unchastened Woman
Uncle Tom's Cabin
The Unholy Three
The Unknown
The Untamable
Until They Get Me
The Vagabond
Vanishing American
Variety
Virtue's Revolt
Walking Back
The Walloping Kid
Warning Shadows
Waxworks
Way Down East
Wedding March
West-Bound Limited
When the Clouds Roll By
White Gold
The White Rose
The White Sin
The White Sister
Wild Horse Canyon
Wild Orchids
The Wind
Wings
The Wishing Ring
Witchcraft Through the Ages
With Kit Carson Over the Great Divide
The Wizard of Oz
Wolfheart's Revenge
A Woman of Affairs
Woman in Grey
Woman in the Moon
Woman of Paris/Sunnyside
Work/Police
Working Winnie/Oh So Simple
Yankee Clipper
Yankee Doodle in Berlin
Zvenigora

Skateboarding

Gleaming the Cube
Skateboard

Skateboard Madness

Skating

Blades of Courage
Breaking the Ice
The Cutting Edge
Derby
Frolics on Ice
Hans Brinker
Ice Castles
Nutcracker, A Fantasy on Ice
Schizo
Sun Valley Serenade
Unholy Rollers
Xanadu

Skiing

All the Way Down
Avalanche
Better Off Dead
Club Med
Downhill Racer
Fire and Ice
Fire, Ice and Dynamite
For Your Eyes Only
Going for the Gold: The Bill Johnson Story
Hot Dog...The Movie!
Iced
Lost and Found
The Man Who Skied Down Everest
On Her Majesty's Secret Service
One Last Run
The Other Side of the Mountain
The Other Side of the Mountain, Part 2
Red Snow
Reno and the Doc
Ski Bum
Ski Patrol
Ski School
Ski Troop Attack
Skier's Dream
Snow: The Movie
Snowball Express
Snowballin'
Snowballing
Snowblind
Striker's Mountain
Swinging Ski Girls
Tales from the Snow Zone

Slapstick Comedy

Abbott and Costello in Hollywood
Abbott and Costello Meet Captain Kidd
Abbott and Costello Meet Dr. Jekyll and Mr. Hyde
Abbott and Costello Meet Frankenstein
Abbott & Costello Meet the Invisible Man
Abbott and Costello Meet the Killer, Boris Karloff
Abroad with Two Yanks
The Adventurer
The Adventures of Buckaroo Banzai
Africa Screams
All of Me
All Night Long/Smile Please
American Pluck
Animal Crackers
Assault and Matrimony
At War with the Army
The Bank Dick
The Bellboy
Big Mouth

Block-heads
Bohemian Girl
Bonnie Scotland
Boy, Did I Get a Wrong Number!
Brain Donors
Buck Privates
Buck Privates Come Home
Bullfighters
Buster Keaton Rides Again/The Railroader
The Caddy
Caddyshack
Caddyshack 2
Cannonball/Dizzy Heights and Daring Hearts
Charlie Chaplin: The Early Years, Volume 1
Charlie Chaplin: The Early Years, Volume 2
Charlie Chaplin: The Early Years, Volume 3
Charlie Chaplin: The Early Years, Volume 4
Cheech and Chong: Still Smokin'
Cheech and Chong: Things Are Tough All Over
Cheech and Chong's Next Movie
Cheech and Chong's, The Corsican Brothers
Cheech and Chong's Up in Smoke
Christmas in July
A Chump at Oxford
Cinderfella
Classic Comedy Video Sampler
Cockeyed Cavaliers
College
The Concrete Cowboys
Count/Adventurer
Cracking Up
Cure
Curse of the Pink Panther
Dance with Me, Henry
A Day at the Races
The Delicate Delinquent
The Dentist
Detective School Dropouts
Disorderly Orderly
Don't Raise the Bridge, Lower the River
Don't Shove/Two Gun Gussie
Dorf and the First Games of Mount Olympus
Dorf Goes Auto Racing
Dorf on Golf
Dorf's Golf Bible
Duck Soup
Ernest Goes to Camp
Ernest Goes to Jail
Errand Boy
Family Jewels
The Fatal Glass of Beer
The Feud
Finders Keepers
A Fine Mess
Flim-Flam Man
The Flying Deuces
Folks
For Auld Lang Syne: The Marx Brothers
Fuller Brush Man
Fun Factory/Clown Princes of Hollywood
Funny Farm
Ghost in the Noonday Sun
A Girl in Every Port
Going Bananas
The Gorilla
Great Guns
The Great Lover
A Guide for the Married Man
Gus

Hands Up
Hardly Working
A Hash House Fraud/The Sultan's Wife
Hold That Ghost
Hollywood or Bust
The Horn Blows at Midnight
Horse Feathers
Hot Lead & Cold Feet
I Hate Blondes
Ishtar
It's a Mad, Mad, Mad, Mad World
Jack & the Beanstalk
The Jerk
Jour de Fete
Kentucky Kernels
The Klutz
The Last Remake of Beau Geste
Laughfest
Laurel & Hardy: Another Fine Mess
Laurel & Hardy: At Work
Laurel & Hardy: Be Big
Laurel & Hardy: Below Zero
Laurel & Hardy: Berth Marks
Laurel & Hardy: Blotto
Laurel & Hardy: Brats
Laurel & Hardy: Chickens Come Home
Laurel & Hardy Comedy Classics Volume 7
Laurel & Hardy and the Family
Laurel & Hardy: Hog Wild
Laurel & Hardy: Laughing Gravy
Laurel & Hardy: Men O'War
Laurel & Hardy: Night Owls
Laurel & Hardy On the Lam
Laurel & Hardy: Perfect Day
Laurel & Hardy Spooktacular
Laurel & Hardy: Stan "Helps" Ollie
Laurel & Hardy: The Hoose-Gow
Let's Go!
The Lost Stooges
Love Happy
Mad Wednesday
Man's Favorite Sport?
Mexican Hayride
Milky Way
Monkey Business
The Naughty Nineties
The Navigator
Never Give a Sucker an Even Break
A Night in Casablanca
A Night at the Opera
The North Avenue Irregulars
Nuns on the Run
Odd Jobs
Off Limits
Off the Wall
One A.M.
Our Relations
Out of the Blue
Pardon My Sarong
Pardon My Trunk
The Party
The Patsy
Pawnshop
Pee-Wee's Big Adventure
Pharmacist
The Pink Panther
The Pink Panther Strikes Again
Playtime
Quick, Let's Get Married
Railroader
Return of the Pink Panther
Revenge of the Pink Panther
The Road to Hong Kong
Room Service
Saps at Sea

Seven Years Bad Luck
A Shot in the Dark
Silent Laugh Makers, No. 1
Silent Laugh Makers, No. 2
Silent Laugh Makers, No. 3
Slapstick
Snow White and the Three Stooges
Sons of the Desert
The Speed Spook
Spies Like Us
Stan Without Ollie, Volume 1
Stooges Shorts Festival
The Stooges Story
Swiss Miss
Take the Money and Run
30-Foot Bride of Candy Rock
Three Charlies and One Phoney!
Three Stooges Comedy Capers: Volume 1
Three Stooges Comedy Classics
Three Stooges: Cookoo Cavaliers and Other Nyuks
Three Stooges Festival
Three Stooges: From Nurse to Worse and Other Nyuks
Three Stooges: Half-Wit's Holiday and Other Nyuks
Three Stooges: Idiots Deluxe and Other Nyuks
Three Stooges: Nutty but Nice and Other Nyuks
Three Stooges: They Stooge to Conga and Other Nyuks
Too Shy to Try
Trail of the Pink Panther
A Tribute to the Boys
Two-Way Stretch
Up the Creek
Utopia
Walk Like a Man
Weekend at Bernie's
Who Done It?
Why Me?
A Woman of Distinction
Wonders of Aladdin
Yankee Doodle in Berlin
Yellowbeard

Slavery

The Abductors
The Adventures of Huckleberry Finn
Angel City
The Association
Big Bust Out
Brother Future
Buck and the Preacher
Burn!
The Civil War: Episode 1: The Cause: 1861
The Civil War: Episode 2: A Very Bloody Affair: 1862
The Civil War: Episode 3: Forever Free: 1862
The Civil War: Episode 4: Simply Murder: 1863
The Civil War: Episode 5: The Universe of Battle: 1863
The Civil War: Episode 6: Valley of the Shadow of Death: 1864
The Civil War: Episode 7: Most Hallowed Ground: 1864
The Civil War: Episode 8: War is All Hell: 1865
The Civil War: Episode 9: The Better Angels of Our Nature: 1865
The Colombian Connection
Dragonard

Furious Slaughter
Gone are the Days
Half Slave, Half Free
Hell's Wind Staff
The House of 1,000 Dolls
Intimate Power
Ladies of the Lotus
Lust for Freedom
Man from Deep River
Man Friday
Mandinga
Mandingo
Naked in the Sun
Naked Warriors
Phantom Empire
Pudd'nhead Wilson
Road to Morocco
The Robe
Roots, Vol. 1
Sansho the Bailiff
Skin Game
Slave Trade in the World
 Today
Slavers
Spartacus
Sunshine Run
Traffic in Souls
Uncle Tom's Cabin
White Slave
A Woman Called Moses

Slice of Life

The Adventures of
 Huckleberry Finn
Alice Doesn't Live Here
 Anymore
All Creatures Great and Small
Amarcord
American Graffiti
And Another Honky Tonk Girl
 Says She Will
Andy Hardy Gets Spring Fever
Andy Hardy's Double Life
Andy Hardy's Private
 Secretary
The Bicycle Thief
Billy Galvin
Bloodbrothers
Blueberry Hill
Breaking In
Bus Stop
Captains Courageous
Car Wash
Cesar
Citizens Band
Coming Up Roses
Courtship
Crisscross
Crossing Delancey
The Crowd
Days of Heaven
Dim Sum: A Little Bit of Heart
Diner
The Dining Room
Dodes 'ka-den
Early Summer (Bakushu)
Earthworm Tractors
Eat a Bowl of Tea
Echo Park
Edge of the World
Egg
El Super
An Empty Bed
End of the Line
Every Man for Himself & God
 Against All
Everybody's All American
The Family Game
Family Secrets
Fanny and Alexander
Five Corners
Flintstones
Fried Green Tomatoes
The Funeral

Girlfriends
God's Little Acre
The Graduate
Grand Canyon
The Grand Highway
Gregory's Girl
The Group
Hannah and Her Sisters
Heavy Traffic
Here Comes Kelly
Hester Street
High Hopes
Honor Thy Father
The Horse of Pride
I Remember Mama
I Wanna Hold Your Hand
The Inheritors
Ironweed
Jacknife
Joyless Street
Judge Priest
Key Exchange
Kings Row
La Terra Trema
The Last American Hero
The Last Days of Dolwyn
The Last Picture Show
Late Spring
Le Grand Chemin
Leather Boys
Life Begins for Andy Hardy
Life with Father
The Little Thief
Little Women
Living on Tokyo Time
Local Hero
Look Back in Anger
The Lords of Flatbush
Love on the Run
Lumiere
MacArthur's Children
Madame Sousatzka
The Mango Tree
Mansfield Park
Marius
Marty
Mean Streets
Meet Me in St. Louis
Memories of a Marriage
Mississippi Masala
Moonrise
Moonstruck
The Music Man
Mystic Pizza
Nashville
1918
90 Days
Offnight
Oklahoma!
Old Swimmin' Hole
One More Saturday Night
One Third of a Nation
Our Town
The Out-of-Towners
Over the Brooklyn Bridge
Papa's Delicate Condition
The Paper Chase
The Patriots
Pete 'n' Tillie
Pharmacist
The Playboys
Prisoner of Second Avenue
Queens Logic
Rachel River
Raggedy Man
Raging Bull
Rainy Day
Real Life
Return to Mayberry
Return of the Secaucus 7
Reuben, Reuben
Room at the Top
Rosalie Goes Shopping
The Rules of the Game
Sinners
Sinners in Paradise

The Smallest Show on Earth
Smithereens
Some Came Running
The Southerner
State Fair
Steamboat Bill, Jr.
Steel Magnolias
Stranger than Paradise
Strong Medicine
Stroszek
Sugarbaby
The Sun Shines Bright
Tex
That Sinking Feeling
Tin Men
To Kill a Mockingbird
Toni
Tonio Kroger
A Touch of Class
The Tree of Wooden Clogs
Trouble in the Glen
True Love
29thth Street
Twice in a Lifetime
Twist & Shout
The Two Sisters
Urban Cowboy
Walls of Glass
Wanderers
A Wedding
The Whole Shootin' Match
Who's That Knocking at My
 Door?
Why Shoot the Teacher
Wild Style
Wish You Were Here

Soccer

The Club
Hero
Hot Shot
Hothead
Hothead (Coup de Tete)
La Vida Sigue Igual
Ladybugs
Long Shot
Those Glory, Glory Days
Victory
Yesterday's Hero

South America

Aguirre, the Wrath of God
Amazon
The Americano
Americano
At Play in the Fields of the
 Lord
Bananas
Black Orpheus
Bloody Che Contra
Burden of Dreams
Bye Bye Brazil
Catch the Heat
The Colombian Connection
Diplomatic Immunity
El Muerto
The Emerald Forest
En Retirada (Bloody Retreat)
Fitzcarraldo
Five Came Back
The Fugitive
Fury
La Muerte del Che Guevara
The Last Movie
Let's Get Harry
Mighty Jungle
Miracles
Miss Mary
Missing
The Mission
Monster
Moon Over Parador
The Muthers

Naked Jungle
Nancy Goes to Rio
Night Gallery
One Man Out
Only Angels Have Wings
Quilombo
Rodrigo D: No Future
Romancing the Stone
Salvador
Simon Bolivar
Sweet Country
Terror in the Jungle
The Three Caballeros
To Kill a Stranger
Wild Orchid
You Were Never Lovelier

Space Exploration

Alien
Assignment Outer Space
Battle Beyond the Sun
Battlestar Galactica
Beamship Meier Chronicles
Black Hole
Buck Rogers Conquers the
 Universe
Capricorn One
Cat Women of the Moon
Countdown
Creature
Dark Side of the Moon
Dark Star
The Day the Sky Exploded
Destination Moon
Destination Moonbase Alpha
Destination Saturn
Escape from Planet Earth
Explorers
Far Out Space Nuts, Vol. 1
Fire Maidens from Outer
 Space
First Men in the Moon
First Spaceship on Venus
Flight to Mars
For All Mankind
Forbidden Planet
From the Earth to the Moon
Galaxy of Terror
Horror Planet
Horrors of the Red Planet
Journey to the Far Side of the
 Sun
Killings at Outpost Zeta
Laser Mission
Lifepod
Lost in Space
Marooned
Megaforce
Mission Galactica: The Cylon
 Attack
Mission Mars
Moonraker
Murder in Space
Mutant on the Bounty
Nude on the Moon
Planet of the Apes
Planet Burg
Planet of the Dinosaurs
Planet of the Vampires
Project Moon Base
PSI Factor
Return to Earth
The Right Stuff
Rocketship X-M
Rocky Jones, Space Ranger:
 Pirates of Prah
Rocky Jones, Space Ranger:
 Renegade Satellite
Rocky Jones, Space Ranger:
 The Cold Sun
Rocky Jones, Space Ranger:
 Trial of Rocky Jones
Rocky Jones, Space Ranger,
 Vol. 3: Blast Off

Silent Running
Skyrider
Space Movie
Space Mutiny
Space Raiders
Spaceballs
SpaceCamp
Star Crash
Star Slammer
Star Trek: The Motion Picture
Star Trek 2: The Wrath of
 Khan
Star Trek 3: The Search for
 Spock
Star Trek 4: The Voyage
 Home
Star Trek 5: The Final Frontier
Star Trek 6: The
 Undiscovered Country
Star Wars
Starbirds
Terronauts
Transformations
2001: A Space Odyssey
2010: The Year We Make
 Contact
Voyage to the Planet of
 Prehistoric Women
Voyage to the Prehistoric
 Planet
Walking on Air
Warlords of the 21st Century
Woman in the Moon
Women of the Prehistoric
 Planet

Spaghetti Western

Ace High
Beyond the Law
Boot Hill
Bullet for Sandoval
Chino
A Fistful of Dollars
Fistful of Dynamite
Five Giants from Texas
For a Few Dollars More
The Good, the Bad and the
 Ugly
Hang 'Em High
Hellbenders
Linchamiento (Lynching)
A Minute to Pray, A Second to
 Die
My Name Is Nobody
Once Upon a Time in the West
Rough Justice
The Ruthless Four
Spaghetti Western
Straight to Hell
The Stranger and the
 Gunfighter
Stranger in Paso Bravo
Take a Hard Ride
Tramplers
Trinity is Still My Name
You're Jinxed, Friend, You've
 Met Sacramento

Spain

Ay, Carmela!
Behold a Pale Horse
Blood and Sand
Caper of the Golden Bulls
I'm the One You're Looking
 For
Marbella
Pride and the Passion
Sand and Blood
Siesta
The Spanish Gardener
Surprise Attack
A Touch of Class

Speculation

Spies & Espionage

Spousal Abuse

Star Gazing

Supernatural

Necromancer: Satan's
  Servant
Necropolis
Neon Maniacs
The Nesting
Netherworld
Night of the Death Cult
Night of the Demons
The Night Evelyn Came Out of
  the Grave
Night of the Ghouls
Night of Horror
Night Life
Night Nurse
Night of the Sorcerers
Night Stalker
The Night Stalker: Two Tales
  of Terror
Night Vision
The Night Visitor
Nightlife
Nightmare in Blood
Nightmare Castle
Nightmare Sisters
Nightmare Weekend
Nightstalker
Nightwish
Nomads
Nothing But the Night
The Occultist
Of Unknown Origin
The Offspring
The Omen
The Oracle
Orgy of the Dead
Orgy of the Vampires
The Other
Other Hell
Out of the Body
The Outing
Patrick
Pet Sematary
Phantasm
Phantasm II
Phantom of the Ritz
Poltergeist
Poltergeist 2: The Other Side
Poltergeist 3
The Possessed
Possession of Joel Delaney
The Premonition
The Psychic
Psychomania
Pumpkinhead
Rawhead Rex
The Sleeping Car
Specters
The Supernaturals
Terror Creatures from the
  Grave
13 Ghosts
Turn of the Screw
The Visitors
The Wraith

Surfing

Adventures in Paradise
Aloha Summer
Beach Blanket Bingo
Big Wednesday
Bikini Beach
California Dreaming
Computer Beach Party
The Endless Summer
Gidget
Lauderdale
Life's a Beach
Mad Wax: The Surf Movie
Muscle Beach Party
North Shore
Point Break
Puberty Blues
Summer City
Sundance

Surf 2
Surf Nazis Must Die

Survival

Adventure Island
Adventures of Eliza Fraser
The Alaska Wilderness
  Adventure
Alpine Fire
And I Alone Survived
Antarctica
Attack of the Beast Creatures
The Aviator
Back from Eternity
Bat 211
Black Rain
Born Killer
The Bridge on the River Kwai
Buffalo Rider
A Bullet Is Waiting
Captain January
Challenge to Be Free
Clown
Cold River
Cool Hand Luke
Damned River
Das Boot
Dawson Patrol
Day After
Dead Calm
Deadly Harvest
Deadly Prey
Death Hunt
Def-Con 4
Deliverance
Delos Adventure
Delusion
Desert Tigers (Los Tigres del
  Desierto)
Desperate Target
The Devil
Devil at 4 O'Clock
Diamonds of the Night
The Dive
Earthling
Empire of the Sun
Endgame
Enemy Mine
Enemy Territory
Erendira
Eric
Escape from Hell
Escapist
Eye of the Eagle 2
Fight for Your Life
Five Came Back
Flight of the Eagle
Flight from Glory
The Flight of the Phoenix
Flight from Singapore
Fortress
Found Alive
Four Infernos to Cross
Gas-s-s-s!
High Ice
High Noon
Hunter's Blood
The Incredible Journey
Interrogation
The Island
Island of the Blue Dolphins
Island of the Lost
Island at the Top of the World
Jeremiah Johnson
Jungle Inferno
Just Before Dawn
Kameradschaft
Killing at Hell's Gate
King Rat
Land That Time Forgot
The Last Chance
Last Exit to Brooklyn
Last Flight of Noah's Ark
Le Dernier Combat

Legend of Alfred Packer
Life & Death of Colonel Blimp
Lifeboat
Lily was Here
Little Dorrit, Film 1: Nobody's
  Fault
Little Dorrit, Film 2: Little
  Dorrit's Story
Lord of the Flies
Lost
Love on the Dole
Male and Female
Melody
Merry Christmas, Mr.
  Lawrence
The Mosquito Coast
The Most Dangerous Game
Mr. Robinson Crusoe
My Side of the Mountain
Mysterious Island
The Naked Prey
Nightmare at Bittercreek
Nightmare at Noon
Oliver Twist
On the Beach
Osa
Out of Control
The Overlanders
Packin' It In
Panic in the Year Zero!
Paradise
Pioneer Woman
The Poseidon Adventure
Quest for Fire
The Return of Grey Wolf
Return of the Jedi
The Revolt of Job
Robby
Robinson Crusoe
Sahara
Savage Hunger
Sea Gypsies
Seven Days in May
Silence
Silence of the North
Skyline
Sourdough
SpaceCamp
Stagecoach
Starbird and Sweet William
Survival
Survival Quest
Survival Run
Survival of Spaceship Earth
The Survivalist
Sweet Sweetback's
  Baadasssss Song
Swiss Family Robinson
The Swiss Family Robinson
The Terror
Testament
Thief
Three Came Home
Two Dragons Fight Against
  the Tiger
Unsettled Land
The Voyage of the Yes
Wagon Master
Warriors of the Apocalypse
The White Dawn
White Fang and the Hunter
White Water Summer

Swashbucklers

Abbott and Costello Meet
  Captain Kidd
Adventures of Captain Fabian
Adventures of Don Juan
The Adventures of Robin
  Hood
Against All Flags
Anthony Adverse
At Sword's Point
The Black Pirate

Blackbeard the Pirate
Blackbeard's Ghost
Bluebeard
The Bold Caballero
The Buccaneer
Captain Blood
Captain Kidd
Captain Kronos: Vampire
  Hunter
The Challenge
Cheech and Chong's, The
  Corsican Brothers
China Seas
Conqueror & the Empress
Corsair
Corsican Brothers
Count of Monte Cristo
Court Jester
Crimson Pirate
Dancing Pirate
Dr. Syn
Dr. Syn, Alias the Scarecrow
Don Q., Son of Zorro
El Pirata Negro (The Black
  Pirate)
The Elusive Pimpernel
Executioner of Venice
The Fifth Musketeer
Fighting Prince of Donegal
A Fist Full of Talons
The Flame & the Arrow
The Four Musketeers
Ghost in the Noonday Sun
Gun Cargo
Gypsy
Hawk of the Caribbean
Hawk of the Wilderness
Hell Ship Mutiny
His Majesty O'Keefe
Hook
The Iron Mask
Island Trader
Jungle Raiders
Kojiro
La Isla de Tesoro de los Pinos
The Lamb
Long John Silver
Man in the Iron Mask
Marauder
Mark of Zorro
The Master of Ballantrae
Master of Dragonard Hill
Monty Python's The Meaning
  of Life
Mooncussers
Moonfleet
Morgan the Pirate
My Favorite Year
My Wicked, Wicked Ways
The Naked Sword of Zorro
Nate and Hayes
Naughty Marietta
New Moon
The Ninja Pirates
Old Ironsides
The Pirate
Pirate Movie
Pirate Warrior
Pirates
Pirates of the Coast
The Pirates of Penzance
Pirates of the Seven Seas
The Princess Bride
The Princess and the Pirate
Prisoner of Zenda
Return to Treasure Island,
  Vol. 1
Return to Treasure Island,
  Vol. 2
Return to Treasure Island,
  Vol. 3
Return to Treasure Island,
  Vol. 4
Return to Treasure Island,
  Vol. 5

The Ribald Tales of Robin
  Hood
Robin Hood
Robin Hood...The Legend
Romola
Scaramouche
The Scarlet Pimpernel
The Sea Hawk
Sea Hound
Secret of El Zorro
Shipwrecked
Sign of Zorro
Sinbad and the Eye of the
  Tiger
Sinbad the Sailor
Smuggler's Cove
Son of Captain Blood
The Son of Monte Cristo
Son of Zorro
South of Pago Pago
South Seas Massacre
The Spanish Main
Story of Robin Hood & His
  Merrie Men
Swashbuckler
Swiss Family Robinson
The Swiss Family Robinson
Sword of Sherwood Forest
Sword & the Sorcerer
Three Amigos
Three Musketeers
Tiger of the Seven Seas
Treasure of the Golden
  Condor
Treasure Island
Treasure of the Moon
  Goddess
Two Lost Worlds
Under the Red Robe
Wallaby Jim of the Islands
The Warriors
The Wicked Lady
Yellowbeard
Zorro, the Gay Blade

Swimming

Barracuda
Bathing Beauty
Cocoon
Dangerous When Wet
Dawn!
Easy to Love
Florence Chadwick: The
  Challenge
Gremlins
Jaws
Lifeguard
The Man from Atlantis
Million Dollar Mermaid
Neptune's Daughter
Off the Mark
Swim Team
Swimmer
Thrill of a Romance

Tearjerkers

An Affair to Remember
After Julius
All Fall Down
All Mine to Give
Ann Vickers
Baby Face
Balboa
Battle Circus
Beaches
Beyond Obsession
Big Bluff
Bittersweet Love
Carrie
The Champ
The Chase
The Cheat
Chinatown After Dark

The Christmas Tree
City That Never Sleeps
The Country Kid
The Courtney Affair
Craig's Wife
Dance Fools Dance
Dedee D'Anvers
Delinquent Parents
Easy Virtue
Fire Down Below
Flamingo Road
Fortune's Fool
Found Alive
Gangs, Inc.
The Garden of Allah
The Gathering
Goodbye, Mr. Chips
Goodbye, My Lady
Great Day
Harlow
Her Silent Sacrifice
Honky Tonk
Hot Spell
Human Desire
Human Hearts
I Sent a Letter to My Love
If Ever I See You Again
In This Our Life
Inside Daisy Clover
Intermezzo
The Kid with the Broken Halo
La Bete Humaine
La Chienne
Lady Caroline Lamb
The Lady Without Camelias
Le Jour Se Leve
Letter from an Unknown
    Woman
The Little Princess
The Love Flower
Love has Many Faces
Love in the Present Tense
Love Songs
Love Story
Lure of the Sila
Magnificent Obsession
Man of Evil
The Manxman
Matters of the Heart
Mildred Pierce
Miles to Go
Millie
Mimi
Miss Julie
Mistress
Mr. Skeffington
Night Life in Reno
Nomads of the North
Now, Voyager
Only Angels Have Wings
The Other Side of the
    Mountain
The Other Side of the
    Mountain, Part 2
The Passing of the Third Floor
    Back
Pattes Blanches
Penny Serenade
A Place in the Sun
The Purple Heart
Quicksand
The Racket
Random Harvest
Rebecca of Sunnybrook Farm
Red Kimono
Rikisha-Man
The Road to Ruin
Romola
Secrets of a Married Man
Secrets of Three Hungry
    Wives
Seizure: The Story of Kathy
    Morris
Sex
A Shining Season
The Shock

Silent Night, Lonely Night
Since You Went Away
Sparrows
Steel Magnolias
Stella Maris
Stromboli
Suds
A Summer Place
Summer and Smoke
Susan Lenox: Her Fall and
    Rise
Tell Me That You Love Me
Tender Comrade
Terms of Endearment
Test Tube Babies
They Drive by Night
Things in Their Season
Three Secrets
Time to Live
Tomorrow's Children
Torch Song
Toute Une Nuit
Traffic in Souls
Under the Roofs of Paris
    (Sous les Toits de Paris)
The Underdog
The Unholy Three
Uptown New York
Walk on the Wild Side
Walpurgis Night
Waterloo Bridge
Way Down East
The White Rose
Without a Trace
A Woman of Affairs
The World Accuses
The World is Full of Married
    Men
Written on the Wind
X, Y & Zee

Technology-
Rampant

Assassin
Attack of the Killer
    Refrigerator
Attack of the Robots
Back to the Future
Best Defense
The Bionic Woman
Black Cobra 3: The Manila
    Connection
Black Hole
Blade Runner
Blades
Chandu the Magician
Charly
Chopping Mall
Circuitry Man
Colossus: The Forbin Project
Computer Beach Party
Computer Dreams
Computer Wizard
Crash and Burn
The Creation of the
    Humanoids
Cyberpunk
Death of the Rat
Death Ray 2000
Defense Play
Demon Seed
Diamonds are Forever
Dr. Goldfoot and the Bikini
    Machine
Dungeonmaster
Electric Dreams
The Eliminators
Eve of Destruction
Fail-Safe
Family Viewing
The Final Programme
The Fly
The Fly 2
Future Cop

Futureworld
Geheimakte WB1 (Secret
    Paper WB1)
Godzilla vs. Mechagodzilla
Hide and Seek
Improper Channels
Interface
A Joke of Destiny, Lying in
    Wait Around the Corner
    Like a Bandit
A King in New York
The Lift
Lost Legacy: A Girl Called
    Hatter Fox
Maximum Overdrive
Microwave Massacre
Modern Times
Mon Oncle
Moonraker
Murder by Phone
Murder by Television
Mutant Hunt
Night of the Kickfighters
A Nous le Liberte
One Deadly Owner
Out of Order
The Philadelphia Experiment
The Power Within
Project: Eliminator
Project: Shadowchaser
The Quiet Earth
RoboCop 2
Robot Holocaust
Robot Jox
Robot Ninja
R.O.T.O.R.
Runaway
Search and Destroy
Shadowzone
Shaker Run
Shocker
Short Circuit
Short Circuit 2
Speaking Parts
The Spy Who Loved Me
Steel and Lace
Strange New World
Street of Crocodiles
Tekkaman, the Space Knight
Terminal Choice
Terminal Entry
The Terminal Man
The Terminator
Terrorvision
Things to Come
Thrillkill
Tron
Twisted
2001: A Space Odyssey
The Ultimate Imposter
The Unborn
The Vindicator
War Games
Weird Science
Westworld
Wired to Kill
Wrestling Women vs. the
    Aztec Ape
Zardoz

Teen Angst

Across the Tracks
Adoption
Age Isn't Everything
Almos' a Man
American Drive-In
American Graffiti
An American Summer
Amor Bandido
Andy Hardy Gets Spring Fever
Andy Hardy Meets Debutante
Andy Hardy's Double Life
Andy Hardy's Private
    Secretary

Angel
Angels with Dirty Faces
Baby It's You
The Bachelor and the Bobby-
    Soxer
Backstreet Dreams
Bad Boys
Bad Ronald
Battle of the Bullies
The Bay Boy
Beach House
The Beat
The Beniker Gang
Berserker
Better Off Dead
Beyond Innocence
The Big Bet
Bikini Beach
Bill & Ted's Excellent
    Adventure
Billyboy
Blackboard Jungle
BMX Bandits
Book of Love
Born Innocent
Boyz N the Hood
The Breakfast Club
Breaking All the Rules
Breaking Away
Brotherhood of Justice
Burglar
Buster and Billie
Buying Time
Can You Feel Me Dancing?
Can't Buy Me Love
Career Opportunities
Carrie
Cave Girl
The Chalk Garden
The Chicken Chronicles
Choices
The Choppers
Christine
Class
The Class of Miss MacMichael
Clayton County Line
The Cool World
Corvette Summer
Cotton Candy
Cruel Story of Youth
Cry-Baby
Dangerous Curves
Date Bait
A Date with Judy
Daughter of Death
Dead End
Dead Poets Society
Deadly Weapon
Defense Play
Delinquent Daughters
Delivery Boys
Desert Bloom
Desperate Lives
The Devil on Wheels
The Devil's Sleep
The Diary of Anne Frank
Diner
Dino
Directions '66
Diving In
Dogs in Space
Doin' Time on Planet Earth
Don't Tell Mom the
    Babysitter's Dead
Dream to Believe
Dream Machine
Drive-In
Dudes
Dutch Girls
East of Eden
East Side Kids
Endless Love
Escape Artist
Escape from El Diablo
E.T.: The Extra-Terrestrial
Evils of the Night

Eyes of the Panther
Eyes Right!
Face in the Mirror
Fame
Fast Talking
Fast Times at Ridgemont High
Fear No Evil
Ferris Bueller's Day Off
Fire with Fire
The First Time
First Turn On
The Flamingo Kid
Fly with the Hawk
Footloose
For Keeps
Forever
Foxes
The Franchise Affair
Getting Lucky
Girl from Hunan
A Girl of the Limberlost
Girlfriend from Hell
Girls Just Want to Have Fun
Gleaming the Cube
Grease
Grease 2
Hadley's Rebellion
Hairspray
Halloween
Hang Tough
Hard Knocks
Hardcore
Harley
Heathers
The Heavenly Kid
Heavy Petting
Her Summer Vacation
Hide and Go Shriek
High School Caesar
High School Confidential
High School USA
Hollywood High
Hollywood High Part 2
Hometown U.S.A.
Homework
Hoosiers
Hot Bubblegum
Hot Times
How I Got Into College
I Am the Cheese
I Never Promised You a Rose
    Garden
I Wanna Hold Your Hand
I Was a Teenage Werewolf
If...
If Looks Could Kill
The In Crowd
In the Land of the Owl Turds
Inga
The Invisible Kid
Jennifer
Junior High School
Just for the Hell of It
Just Like Us
Kansas
Kid Dynamite
King Creole
Knights & Emeralds
La Bamba
La Hija del General
La Notte Brava (Lusty Night in
    Rome)
La Petite Sirene
Lantern Hill
Last Summer
Lauderdale
Les Enfants Terrible
Les Mistons
License to Drive
Life Begins for Andy Hardy
Lisa
A Little Romance
Little Sweetheart
The Little Thief
The Loneliest Runner
Loners

Bonnie Prince Charlie
Born on the Fourth of July
Bound for Glory
Boy in Blue
Bruce Lee: The Legend
Bruce Lee: The Man/The Myth
Bud and Lou
Buddy Holly Story
Buster Keaton Rides Again/
  The Railroader
Caddie
Calamity Jane
Camille Claudel
Can You Hear the Laughter?
  The Story of Freddie Prinze
Caravaggio
Casimir the Great
Cast a Giant Shadow
Catholic Hour
Champions
Chanel Solitaire
Chariots of Fire
Christopher Columbus
Churchill and the Generals
Coal Miner's Daughter
Conrack
Cordell Hull: The Good
  Neighbor
The Courageous Mr. Penn
The Court Martial of Jackie
  Robinson
Cromwell
Cross Creek
Damien: The Leper Priest
Dance with a Stranger
Daniel Boone
Dawn!
Deep in My Heart
Dempsey
The Desert Fox
Dick Turpin
Dillinger
Don't Look Back: The Story of
  Leroy "Satchel" Paige
Dorothy Stratten: The Untold
  Story
Double Exposure: The Story
  of Margaret Bourke-White
Dreamchild
Eddie Rickenbacker: Ace of
  Aces
The Eddy Duchin Story
Edie in Ciao! Manhattan
Edith & Marcel
Education of Sonny Carson
Edward and Mrs. Simpson
Eisenhower: A Place in
  History
Eisenstein
Elephant Man
Elizabeth, the Queen
Elvis and Me
Elvis Memories
Elvis in the Movies
Elvis Presley's Graceland
Elvis: Rare Moments with the
  King
Elvis Stories
Elvis: The Echo Will Never Die
Elvis: The Movie
Enemy of Women
Enter Laughing
Ernie Banks: History of the
  Black Baseball Player
Ernie Kovacs: Between the
  Laughter
The Eternal Waltz
Eubie!
Europa, Europa
Evel Knievel
Exit the Dragon, Enter the
  Tiger
F. Scott Fitzgerald in
  Hollywood
The Fabulous Dorseys
Fanny and Alexander

Fast Company
Father Hubbard: The Glacier
  Priest
Finest Hours
Fiorello La Guardia: The
  Crusader
First Twenty Years: Part 2
  (Porter. 1903-1904)
The Five of Me
The Five Pennies
Flights & Flyers
Flights & Flyers: Amelia
  Earhart
Florence Chadwick: The
  Challenge
For Us, the Living
Ford: The Man & the Machine
Forever James Dean
The Fountainhead
Francisco Oller
Franklin D. Roosevelt: F.D.R.
Frida
Funny Girl
Funny Lady
The Gallant Hours
Gandhi
Gar Wood: The Silver Fox
Gathering Storm
The Gene Krupa Story
Gentleman Jim
George Burns: His Wit &
  Wisdom
George Melies, Cinema
  Magician
Gerty, Gerty, Gerty Stein Is
  Back, Back, Back
Give 'Em Hell, Harry!
The Glenn Miller Story
The Goddess
Going for the Gold: The Bill
  Johnson Story
Goodbye, Norma Jean
Goodnight, Sweet Marilyn
The Gorgeous Hussy
Gorillas in the Mist
Grambling's White Tiger
Greased Lightning
Great Balls of Fire
The Great Caruso
The Great Impostor
Great Missouri Raid
The Great Moment
Great Wallendas
The Great Waltz
The Great White Hope
The Great Ziegfeld
The Greatest
The Greatest Story Ever Told
Grey Gardens
Gypsy
Hammett
Happy Hooker Goes
  Hollywood
Harlow
Haunted Summer
Heart Beat
Heart of a Champion: The Ray
  Mancini Story
Heart Like a Wheel
Hedda Hopper's Hollywood
Helen Hayes
Helen Wills: Miss Poker Face
Henry Fonda: The Man and
  His Movies
Henry & June
Hepburn & Tracy
Hitler
Holy Terror
Home Sweet Home
Hoodoo Ann
Houdini
Hughes & Harlow: Angels in
  Hell
The Hunter
I Dream of Jeannie
I Shot Billy the Kid

Ike
I'll Cry Tomorrow
The Illustrated Hitchcock
Imperial Venus
In Search of Historic Jesus
The Incredible Sarah
Ingrid
The Inn of the Sixth
  Happiness
Inside Hitchcock
Inside the Third Reich
Invincible Mr. Disraeli
Iris
Iron Duke
Isadora
It's Good to Be Alive
J. Edgar Hoover
Jack Frost
Jack London
The Jackie Robinson Story
Jacqueline Bouvier Kennedy
James Cagney: That Yankee
  Doodle Dandy
James Dean
James Dean Story
James Dean: The First
  American Teenager
James Joyce: A Portrait of the
  Artist as a Young Man
James Joyce's Women
Janis: A Film
The Jayne Mansfield Story
Jazz is Our Religion
Jesse James Under the Black
  Flag
The Jesse Owens Story
Jesus
Jesus Christ Superstar
Jim Thorpe: All American
Joan of Arc
Joe Louis Story
John F. Kennedy
John Huston: The Man, the
  Movies, the Maverick
John Wayne: American Hero
  of the Movies
John Wayne: The Duke Lives
  On
John & Yoko: A Love Story
Johnny We Hardly Knew Ye
Jolson Sings Again
The Jolson Story
The Josephine Baker Story
Juarez
King
King of Kings
Knute Rockne: All American
Knute Rockne: The Rock of
  Notre Dame
La Bamba
La Signora di Tutti
Lady Jane
Lady Sings the Blues
Lamp at Midnight
Last Days of Frank & Jesse
  James
The Last Emperor
Last Four Days
Last of Mrs. Lincoln
Lawrence of Arabia
LBJ: The Early Years
The Learning Tree
The Left-Handed Gun
Legend of Valentino
The Legend of Valentino
Lenny
Lepke
Let's Get Lost
The Life and Assassination of
  the Kingfish
Life in Camelot: The Kennedy
  Years
The Life of Christ
The Life of Emile Zola
The Life and Loves of Mozart

Life & Times of Captain Lou
  Albano
Life of Verdi
The Light in the Jungle
Lightning Over Water
Lisztomania
The Loneliest Runner
Lonely Boy/Satan's Choice
The Long Gray Line
Long Journey Back
Lou Gehrig: King of Diamonds
Lou Gehrig Story
Love Affair: The Eleanor &
  Lou Gehrig Story
Love Me or Leave Me
The Loves of Edgar Allen Poe
Loyola, the Soldier Saint
Lucky Luciano
Lust for Life
MacArthur
Madame Curie
Madman
Mae West
Magnificent Adventurer
Mahler
A Man Called Peter
Man of a Thousand Faces
Marciano
Marco
Marie Antoinette
Marilyn
Marilyn: The Untold Story
Mario Lanza: The American
  Caruso
Marjoe
Marlon Brando
Martin Luther
Mary of Scotland
Mary White
Matter of Heart
A Matter of Life and Death
McVicar
Melody Master
The Miracle Worker
Mishima: A Life in Four
  Chapters
Mo' Better Blues
Moliere
Mommie Dearest
Mondo Elvis (Rock 'n' Roll
  Disciples)
Mondo Lugosi: A Vampire's
  Scrapbook
Money Madness
Monsieur Vincent
Montgomery Clift
Moulin Rouge
Mozart: A Childhood
  Chronicle
The Mozart Story
Murderers Among Us: The
  Simon Wiesenthal Story
Murrow
Mussolini & I
My Father's Glory
My Left Foot
My Wicked, Wicked Ways
Nadia
The Naked Civil Servant
Nancy Astor
Napoleon
Night and Day
Nijinsky
No, I Have No Regrets
Nostradamus
Nurse Edith Cavell
One in a Million: The Ron
  LaFlore Story
Otto Messmer and Felix the
  Cat
The Outlaw
Padre Padrone
Pancho Villa
Papa's Delicate Condition
Patton
Paul Robeson

Perfumed Nightmare
The Perils of Pauline
Pistol: The Birth of a Legend
Placido: A Year in the Life of
  Placido Domingo
Pope John Paul II
Portrait of a Rebel: Margaret
  Sanger
Prick Up Your Ears
Pride of St. Louis
The Pride of the Yankees
Priest of Love
Prince Jack
PT 109
Queenie
Raging Bull
Rainbow
Rasputin
A Real American Hero
The Real Buddy Holly Story
The Real Patsy Cline
Rembrandt
Remembering Winsor McCay
Return to Earth
Rhapsody in Blue
Rhodes
Richard Lester
Richard Petty Story
Riel
Rise of Catherine the Great
The Rise and Fall of Legs
  Diamond
The Rise of Louis XIV
Rita Hayworth: The Love
  Goddess
Road to 1984
Robert A. Taft: Mr.
  Republican
Robert Kennedy and His
  Times
Rodeo Girl
Roger Corman: Hollywood's
  Wild Angel
Romero
Ronnie Dearest: The Lost
  Episodes
Rosie: The Rosemary Clooney
  Story
Ruby
Sakharov
Sam's Son
San Francisco de Asis
The Secret Diary of Sigmund
  Freud
Secret Honor
The Secret Life of Sergei
  Eisenstein
A Sense of Freedom
The Seven Little Foys
Shadow on the Sun
The Sicilian
Silent Victory: The Kitty O'Neil
  Story
Simon Bolivar
Sir Arthur Conan Doyle
Sister Kenny
Soldier in Love
Somebody Up There Likes Me
Song of Norway
Song to Remember
Song Without End
Sophia Loren: Her Own Story
The Spectre of Edgar Allen
  Poe
Spirit of St. Louis
Spitfire
Spitfire: The First of the Few
Spring Symphony
Spymaker: The Secret Life of
  Ian Fleming
Squizzy Taylor
Stars of the Century
Stars and Stripes Forever
Stevie
Story of Adele H.
The Story of Elvis Presley

The Story of Louis Pasteur
The Story of Vernon and Irene Castle
Striker's Mountain
Sunrise at Campobello
Sweet Dreams
Sweet Love, Bitter
Sweetheart!
Sylvia
Tall Tales & Legends: Annie Oakley
Taxi zum Klo
The Ten Commandments
The Testament of Orpheus
That Hamilton Woman
Therese
This is Elvis
Three Little Words
Till the Clouds Roll By
Till the Clouds Roll By/A Star Is Born
Time for Miracles
To Hell and Back
To Race the Wind
Toast to Lenny
Tom Edison: The Boy Who Lit Up the World
A Tribute to Lucy
The Ultimate Stuntman: A Tribute to Dar Robinson
Veronika Voss
Victoria Regina
Vincent: The Life and Death of Vincent van Gogh
Vincent & Theo
The Virgin Queen
The Virginia Hill Story
Viva Zapata!
A Voyage Round My Father
Wagner: The Movie
Waiting for the Moon
Waltz King
Weeds
White Hunter, Black Heart
Whoever Says the Truth Shall Die
Who's That Knocking at My Door?
Wild, Wild World of Jayne Mansfield
Will: G. Gordon Liddy
Wilma
Winds of Kitty Hawk
Wings of Eagles
The Winning Team
Wired
Without Warning: The James Brady Story
The Wolf at the Door
A Woman Called Golda
A Woman Called Moses
Wonderful World of the Brothers Grimm
Words & Music
Yankee Doodle Dandy
Young Bill Hickok
Young Buffalo Bill
Young Caruso
Young Charlie Chaplin
Young Mr. Lincoln
Young Winston

3-D Flicks

Amityville 3: The Demon
Andy Warhol's Frankenstein
Cat Women of the Moon
Creature from the Black Lagoon
Dogs of Hell
Dynasty
Flesh and Blood Show
The French Line
Friday the 13th, Part 3
Gun Fury

House of Wax
It Came From Outer Space
Jaws 3
Kiss Me Kate
The Man Who Wasn't There
The Mask
Miss Sadie Thompson
Paradisio
Parasite
Robot Monster
Spacehunter: Adventures in the Forbidden Zone
Starchaser: The Legend of Orin

Time Travel

Amazing Mr. Blunden
The Arrival
Back to the Future
Back to the Future, Part 2
Back to the Future, Part 3
Beastmaster 2: Through the Portal of Time
Beyond the Time Barrier
Bill & Ted's Bogus Journey
Bill & Ted's Excellent Adventure
Brother Future
Buck Rogers in the 25th Century
Chronopolis
Doctor Mordrid: Master of the Unknown
Doctor Who and the Daleks
Doctor Who: Death to the Daleks
Doctor Who: Pyramids of Mars
Doctor Who: Spearhead from Space
Doctor Who: Terror of the Zygons
Doctor Who: The Ark in Space
Doctor Who: The Seeds of Death
Doctor Who: The Talons of Weng-Chiang
Doctor Who: The Time Warrior
The Eliminators
Escape from the Planet of the Apes
Escapes
Freejack
Hercules in New York
Highlander 2: The Quickening
Idaho Transfer
The Jetsons Meet the Flintstones
Lords of Magick
Mannequin 2: On the Move
Millenium
The Navigator
Next One
Night of Dark Shadows
Peggy Sue Got Married
The Philadelphia Experiment
Roman Scandals
Running Against Time
Spirit of '76
Terminator 2: Judgement Day
Time After Time
Time Bandits
The Time Guardian
The Time Machine
Time Machine
Time Trackers
The Time Travelers
Timestalkers
Trancers
Trancers 2: The Return of Jack Deth
Trancers 3: Deth Lives
Warlock

Waxwork 2: Lost in Time
A Witch Without a Broom

Torrid Love Scenes

The Adventurers
Alien Prey
Angel Heart
The Berlin Affair
Betty Blue
The Big Easy
Blue Velvet
Body Double
Body Heat
Breathless
Bull Durham
Cat Chaser
Cat on a Hot Tin Roof
Close My Eyes
The Cook, the Thief, His Wife & Her Lover
Crimes of Passion
Dark Obsession
The Dolphin
Dona Flor & Her Two Husbands
Don't Look Now
The Executioner's Song
The Fabulous Baker Boys
Fatal Attraction
Five Easy Pieces
From Here to Eternity
Heat of Desire (Plein Sud)
The Hunting
In the Realm of the Senses
Kiss Me a Killer
Lady Godiva Rides
Last Tango in Paris
The Lawnmower Man
Mambo Kings
Men in Love
Moon in the Gutter
Moonstruck
9 1/2 Weeks
1900
An Officer and a Gentleman
The Postman Always Rings Twice
Prizzi's Honor
Rage of Angels
Revenge
Risky Business
Sea of Love
Secrets
Sexual Response
Siesta
Something Wild
The Story of O
Swept Away...
Tattoo
Tequila Sunrise
Two Women in Gold
The Unbearable Lightness of Being
Wild Orchid
Wild Orchid 2: Two Shades of Blue
The Woman Next Door
Women in Love
Women & Men: Stories of Seduction, Part 2
Year of the Gun

Toys

Babes in Toyland
Child's Play 2
Dance of Death
Demonic Toys
Devil Doll
Dolls
Dolly Dearest
Silent Night, Deadly Night 5: The Toymaker
The Toy

Wired to Kill

Track & Field

Across the Tracks
Billie
College
Finish Line
Go for the Gold
The Jericho Mile
The Jesse Owens Story
Running Brave
Sam's Son
A Shining Season

Tragedy

Anna Karenina
Bitter Sweet
Camille
Charly
The Crucified Lovers
Deadman's Curve
Deep End
Different Kind of Love
The Discovery Program
Dying Young
East of Eden
Edipo Re (Oedipus Rex)
El Amor Brujo
End of the Road
Enemies, a Love Story
Ernie Kovacs: Between the Laughter
Everybody's All American
The Fighting Sullivans
Franz
The French Lieutenant's Woman
Gate of Hell
Girl in Black
Golden Boy
Grand Illusion
Hamlet
Henry IV
I, Claudius, Vol. 6: Fool's Luck/A God in Colchester
I Want to Live!
In Memoriam
Johnny Belinda
Jules and Jim
Julius Caesar
King Lear
La Traviata
Law of Desire
Love is a Many-Splendored Thing
Love Story
Lovespell
Macbeth
Madame Bovary
Maria Candelaria
Mary of Scotland
Medea
The Merchant of Four Seasons
The Mummy
The Nest
Nine Days a Queen
Oedipus Rex
Of Mice and Men
Othello
Pascali's Island
Pennies From Heaven
A Place in the Sun
Romeo and Juliet
Romeo and Juliet (Gielgud)
September 30, 1955
Shadows of Forgotten Ancestors
Shop on Main Street
The Story of the Late Chrysanthemum
Sundays & Cybele
Tess

Throne of Blood
Tragedy of Flight 103: The Inside Story
Trojan Women
Variety
A Very Private Affair
The Virgin Spring
Voyager
The White Sister
The Woman Next Door
The World According to Garp

Trains

Across the Bridge
Berlin Express
Boxcar Bertha
The Brain
Breakheart Pass
Cafe Express
Cajon Pass
The Cassandra Crossing
Chattanooga Choo Choo
Closely Watched Trains
Confessions of a Train Spotter
Dakota
Danger Ahead!
Danger Lights
Death Train
The Denver & Rio Grande
End of the Line
Express to Terror
Finders Keepers
Five Forty-Eight
The Flying Scotsman
From Russia with Love
The General
The Ghost Train
Go Kill and Come Back
Great K & A Train Robbery
The Great Locomotive Chase
Great Train Robbery
The Grey Fox
Grit of the Girl Telegrapher/In the Switch Tower
The Harvey Girls
Horror Express
Hot Lead & Cold Feet
Indiscretion of an American Wife
Inner Sanctum
Interrupted Journey
Jesse James at Bay
Jewel of the Nile
La Bete Humaine
The Lady Eve
The Lady Vanishes
Lady Vanishes
Man Who Loved Cat Dancing
Man Without a Star
Midnight Limited
Murder on the Midnight Express
Murder on the Orient Express
My Twentieth Century
Mystery Mountain
Night Train to Munich
Night Train to Terror
Number Seventeen
The Odyssey of the Pacific
Oh, Mr. Porter
Once Upon a Time in the West
Panic on the 5:22
Paris Express
Phantom Express
Plunder Road
Prison Train
Railroaded
Railroader
Red Signals
Risky Business
The Road to Yesterday
Robbery

Samson vs. the Vampire
   Women
The Satanic Rites of Dracula
The Scars of Dracula
Scream Blacula Scream
The 7 Brothers Meet Dracula
Slaughter of the Vampires
Son of Dracula
Subspecies
Sundown
Theatre of Death
Thirst
The Thirsty Dead
To Die for
To Die for 2
Track of the Vampire
Transylvania Twist
Twins of Evil
Understudy: The Graveyard
   Shift 2
Vamp
The Vampire
The Vampire Bat
Vampire Circus
Vampire Cop
The Vampire Hookers
The Vampire Lovers
Vampire at Midnight
Vampire Princess Miyu
The Vampire's Coffin
The Vampire's Ghost
Vampire's Kiss
Vampyr
Vampyres
The Velvet Vampire
The Werewolf vs. the Vampire
   Woman
The Wicked
World of the Vampires
Zoltan...Hound of Dracula

Variety

Bob & Ray, Jane, Laraine &
   Gilda
Bottoms Up '81
Debbie Does Las Vegas
Doctor Duck's Super Secret
   All-Purpose Sauce
Dream Boys Revue
Entertaining the Troops
Flying Karamazov Brothers
   Comedy Hour
Girls of the Moulin Rouge
Give a Girl a Break
Goldie & Kids
Good Old Days of Radio
Great British Striptease
Heller in Pink Tights
Here It is, Burlesque
Hollywood Goes to War
Hollywood Palace
Hollywood Revels
Hollywood Varieties
I Wanna Be a Beauty Queen
If You Don't Stop It...You'll Go
   Blind
International House
Invitation to Paris
Jabberwalk
Jack Benny Show
Johnny Carson
Major Bowes' Original
   Amateur Hour
Minsky's Follies
Motown 25: Yesterday,
   Today, Forever
Mr. Mike's Mondo Video
New Look
New Star Over Hollywood
Oh! Those Heavenly Bodies:
   The Miss Aerobics USA
   Competition
Parade
The Pee-Wee Herman Show

Pee-Wee's Playhouse
   Christmas Special
Pee-Wee's Playhouse Festival
   of Fun
Penn and Teller's Cruel Tricks
   for Dear Friends
Penthouse on the Wild Side
Perry Como Show
Playboy Playmates at Play
Playboy Video Centerfold
   Double Header
Playboy Video Magazine:
   Volumes 1-7
Playboy's Party Jokes
Playboy's Sexy Lingerie
Playboy's Wet & Wild
Playmate Review
Ripley's Believe It or Not!
Saturday Night Live: Richard
   Pryor
Saturday Night Live: Steve
   Martin, Vol. 1
Saturday Night Live, Vol. 1
Sleazemania
Sports Illustrated's 25th
   Anniversary Swimsuit
   Video
Stork Club
Strictly G.I.
Tall, Dark & Handsome
Tee Vee Treasures
Television's Golden Age of
   Comedy
That's Dancing!
That's Entertainment
That's Entertainment, Part 2
That's Singing
Toast of the Town
Two Top Bananas
Varieties on Parade
Ziegfeld Follies

Veterans

Alamo Bay
Anatomy of Terror
Bell Diamond
Billy Jack
Braddock: Missing in Action 3
Child Bride of Short Creek
Cry of the Innocent
Dark Before Dawn
Fear
First Blood
Fleshburn
Forced Vengeance
The Forgotten
Ghetto Revenge
The Great American
   Broadcast
Heroes
House
In Country
Jungle Assault
Killzone
Last Mercenary
Lethal Weapon
The Line
Moon in Scorpio
The Park is Mine
A Quiet Hope
Rambo: Part 3
Red, White & Busted
Riders of the Storm
Robbery
Savage Dawn
Stand Alone
Stanley
Tagget
To Kill a Clown
Tracks
The Woman Inside

Vietnam War

Above and Beyond
After the Fall of Saigon
Air America
American Commandos
Americana
The Annihilators
Apocalypse Now
Ashes and Embers
Autopsy
Backfire
Bat 211
Bell Diamond
Beyond the Call of Duty
Birdy
Born on the Fourth of July
The Boys in Company C
Braddock: Missing in Action 3
Cactus in the Snow
The Cage
Casualties of War
Cease Fire
Charlie Bravo
Children of An Lac
China Beach
China Gate
Combat Shock
Coming Home
Commando Invasion
Conspiracy: The Trial of the
   Chicago Eight
Crossfire
Dear America: Letters Home
   from Vietnam
Deathdream
The Deer Hunter
Deserters
Don't Cry, It's Only Thunder
84 Charlie Mopic
Explosion
Eye of the Eagle
Eye of the Eagle 2
Eye of the Eagle 3
Fatal Mission
Fear
Fighting Mad
Fireback
First Blood
Flight of the Intruder
For the Boys
The Forgotten
Forgotten Warrior
Four Friends
Friendly Fire
Full Metal Jacket
Gardens of Stone
Ghetto Revenge
Go Tell the Spartans
Good Morning, Vietnam
The Green Berets
Green Eyes
Hail, Hero!
Hair
Hamburger Hill
Hanoi Hilton
Hard Rain-The Tet 1968
Heated Vengeance
The Hill
In Country
In Gold We Trust
In Love and War
Invasion of the Flesh Hunters
The Iron Triangle
Jacknife
Jacob's Ladder
Jenny
Jud
Jungle Assault
Kent State
Lady from Yesterday
Last Flight Out: A True Story
The Last Hunter
Medal of Honor Rag
Missing in Action

Missing in Action 2: The
   Beginning
Nam Angels
Night Wars
1969
No Dead Heroes
The Odd Angry Shot
Off Limits
Once Upon a Time in Vietnam
Operation C.I.A.
Operation 'Nam
Operation War Zone
Ordinary Heroes
Platoon
Platoon Leader
P.O.W. Deathcamp
The P.O.W. Escape
Primary Target
Private War
Purple Hearts
A Quiet Hope
Rambo: First Blood, Part 2
Rambo: Part 3
Red, White & Busted
Return of the Secaucus 7
A Rumor of War
Running on Empty
Search and Destroy
The Shepherd
The Siege of Firebase Gloria
Soldier's Revenge
Some Kind of Hero
Streamers
Strike Commando
Summertree
Thou Shalt Not Kill...Except
To the Shores of Hell
Tornado
Tour of Duty
Trial of the Catonsville Nine
Twilight's Last Gleaming
Uncommon Valor
Unnatural Causes
Vietnam: Chronicle of a War
Warbus
Wardogs (The Assassination
   Team)
Welcome Home
When Hell Was in Session
Which Way Home
The Wild Aces
Winner Takes All
Woodstock

Volcanoes

Devil at 4 O'Clock
Eruption: St. Helens Explodes
Joe Versus the Volcano
Last Days of Pompeii
The Last Days of Pompeii
Nutcase
St. Helen's, Killer Volcano

War Between the Sexes

Adam's Rib
The Adventures of Robin
   Hood
The African Queen
All of Me
All's Fair
Alternative
Always
America 3000
Anita de Montemar
Annie Hall
Autumn Marathon
The Awful Truth
Ballyhoo Baby
Bare Essentials
Basic Training
Beware of Pity
The Bigamist

Blood and Sand
Blue Paradise
Bride Walks Out
Bringing Up Baby
Buffalo Jump
Bull Durham
Cannibal Women in the
   Avocado Jungle of Death
Carmen
Carmen Jones
Carrington, V.C.
Carry On Nurse
Casablanca
Casanova's Big Night
Castaway
Casual Sex?
Cold Heat
Concrete Beat
The Cutting Edge
Dangerous Liaisons
Dear Wife
Designing Woman
Divorce—Italian Style
Dogfight
Don Juan My Love
Easy Wheels
El Forastero Vengador
El: This Strange Passion
The Female Bunch
Fever
First Monday in October
Forever Darling
Frankie & Johnny
Fury is a Woman
A Girl in a Million
Gone with the Wind
The Good Father
Goodbye Love
The Happy Ending
Hardhat & Legs
Harrad Summer
Having It All
He Said, She Said
Heartburn
His Girl Friday
I, Claudius, Vol. 1: A Touch of
   Murder/Family Affairs
I Live My Life
I Love All of You
I Love My...Wife
I Love You
Intimate Story
Irreconcilable Differences
It Happened One Night
Jewel of the Nile
Juliet of the Spirits
Just Tell Me What You Want
Key Exchange
Kill My Wife...Please!
The King and I
Kiss Me Kate
La Diosa Arrodillada (The
   Kneeling Goddess)
L.A. Story
The Lady Says No
The Last Woman on Earth
Le Chat
Legion of Iron
Letters to an Unknown Lover
Lily in Love
The Lion in Winter
Love in the Afternoon
Lover Come Back
Loyalties
Lucky Partners
The Man Who Guards the
   Greenhouse
Manhattan
Many Wonder
Maria's Lovers
The Marriage of Maria Braun
The Marrying Man
A Matter of Love
Men...
A Midsummer Night's Dream
Moonlighting

Moonstruck
Mortal Thoughts
Murphy's Law
My Brilliant Career
Near Misses
A New Leaf
9 to 5
Nothing in Common
One Woman or Two
Pajama Tops
Pat and Mike
Patti Rocks
Pauline at the Beach
A Piece of Pleasure
Pillow Talk
Places in the Heart
Portrait of Teresa
The Practice of Love
Public Enemy
Queen of Outer Space
Question of Silence
The Quiet Man
Ramrod
Romancing the Stone
Rooster Cogburn
Sabrina
Scenes from a Mall
Seance on a Wet Afternoon
The Sensual Man
Sex and the College Girl
Shades of Love: Champagne
    for Two
She-Devil
Shore Leave
The Single Standard
The Social Secretary
Sotto, Sotto
Spies, Lies and Naked Thighs
Star 80
The Story of Fausta
Summer Night With Greek
    Profile, Almond Eyes . .
Switch
Switching Channels
Taming of the Shrew
A Taxing Woman
Teacher's Pet
That's My Baby!
Thelma & Louise
The Thing
Thor and the Amazon Women
Thunder in the Pines
Tiger and the Pussycat
To Have & Have Not
Tony Draws a Horse
True Love
Two-Faced Woman
Vasectomy: A Delicate Matter
War Between the Tates
The War of the Roses
When Ladies Meet
When Women Lost Their Tails
Who's Afraid of Virginia
    Woolf?
Witches of Eastwick
A Woman of Distinction
Woman of the Year
Women & Men: Stories of
    Seduction
Women & Men: Stories of
    Seduction, Part 2
Women on the Verge of a
    Nervous Breakdown
The World According to Garp
Yesterday, Today and
    Tomorrow
You're a Big Boy Now

War, General

Ace of Aces
The Adventures of the Flying
    Cadets
Against the Wind
Alexander Nevsky

All the Young Men
Alone Against Rome
Alsino and the Condor
American Eagle
Angels One Five
Anna Karenina
Arsenal
Atlas
The Avenger
Ay, Carmela!
Back to Bataan
Ballad of a Soldier
Battle of Algiers
The Battle of Austerlitz
Battle of Blood Island
Battle of the Commandos
Battle Hell
Battle of the Japan Sea
Battleforce
The Battleship Potemkin
The Beast
Beau Geste
Behind the Front
Behind the Rising Sun
Benjamin Argumedo
The Best of Bogart
Betrayed
Big Lift
The Boys in Company C
Breaker Morant
Bridge to Hell
A Brivele der Mamen (A Letter
    to Mother)
The Buccaneer
Bugles in the Afternoon
Bullet for the General
By Dawn's Early Light
Cambodia: Year Zero and
    Year One
A Canterbury Tale
Captain Caution
Captain Horatio Hornblower
Carthage in Flames
Cast a Giant Shadow
Cavalry Command
Chapayev
Charge of the Light Brigade
Civilization
Codename: Terminate
Command Decision
Commando Invasion
The Conqueror
Conquest of the Normans
Corregidor
Cottage to Let
Coup de Grace
The Court Martial of Billy
    Mitchell
Cromwell
Danton
Days of Hell
Deadline
December 7th: The Movie
The Desert of the Tartars
Deserters
Devil in the Flesh
Devil Wears White
Distant Drums
Distant Thunder
Dr. Strangelove, or: How I
    Learned to Stop Worrying
    and Love the Bomb
Doctor Zhivago
Donald: Limited Gold Edition 1
Dragon Seed
Duck Soup
Duke: The Films of John
    Wayne
The Eagle Has Landed
Eagles Attack at Dawn
Earth vs. the Flying Saucers
Eddie Rickenbacker: Ace of
    Aces
El Cid
El Paso Stampede
Elizabeth of Ladymead

Elusive Corporal
Emissary
Enchanted Cottage
End of St. Petersburg
Enemy Below
Enola Gay: The Men, the
    Mission, the Atomic Bomb
Escape
Escape from Sobibor
Exodus
The Expendables
Eye of the Eagle 2
Field of Honor
The Fighter
Fighting Marines
Fighting Pilot
Fighting Prince of Donegal
Fire Over England
Flor Silvestre (Wild Flower)
Forever and a Day
Fort Apache
Forty Days of Musa Dagh
The Four Feathers
Four Feathers
F.P. 1
Funny, Dirty Little War
Geheimakte WB1 (Secret
    Paper WB1)
The G.I. Road to Hell
Guerilla Brigade
Gunga Din
Guns at Batasi
Hail the Conquering Hero
Harry's War
Heart of a Nation
Heartbreak Ridge
Hearts & Armour
Hell on the Battleground
Hell to Eternity
Hell is for Heroes
Hell in the Pacific
Hell Raiders
Hell's Brigade: The Final
    Assault
Henry V
Henry VI, Part 1
Hidden Fortress
Hill 24 Doesn't Answer
Hitler: The Last Ten Days
Hollywood Canteen
Hollywood at War
Home of the Brave
How I Won the War
I Don't Give a Damn
In Gold We Trust
In the Name of the Pope-King
The Informer
The Invincible Gladiator
Iron Angel
Is Paris Burning?
J'Accuse
Johnny Got His Gun
Khartoum
Killers
The Killing Fields
Kolberg
La Marseillaise
The Last Command
Last of the Mohicans
The Last Reunion
The Last Valley
Last Warrior
The Last Winter
Late Summer Blues
Latino
Laughing at Life
Legion of Missing Men
Let There Be Light
Life is Beautiful
Life & Death of Colonel Blimp
Lion of the Desert
The Lost Command
The Lost Idol/Shock Troop
Man from the Alamo
Man Escaped
Mask of the Dragon

Massive Retaliation
Megaforce
Mein Kampf
Men in War
Mephisto
Mercenary Fighters
The Mercenary Game
Midway
Mission to Glory
The Mongols
The Moonraker
Mother
Mountain Fury
The Mouse That Roared
Napoleon
Noon Sunday
Old Gringo
On the Beach
One Minute to Zero
One Night Stand
One of Our Aircraft Is Missing
Operation Cross Eagles
Operation Thunderbolt
Orphans of the Storm
Overthrow
The Phantom Planet
Port Arthur
Pride and the Passion
Prisoners of the Sun
Project: Alien
Rage to Kill
A Reason to Live, A Reason
    to Die
The Rebellious Reign
The Red and the Black
Red Dawn
The Red and the White
Reds
Reign of Terror
Reilly: The Ace of Spies
Return of the Soldier
Revenge of the Barbarians
Revolution
Rhodes
Richard II
Ride to Glory
Running Away
Ryan's Daughter
Sabre Jet
Saga of the Vagabond
Saigon: Year of the Cat
Sanders of the River
Sands of Iwo Jima
The Sea Shall Not Have Them
Seven Indignant
Shark River
She Wore a Yellow Ribbon
Shell Shock
Simon Bolivar
Sinai Commandos
Sirocco
Sky High
The Small Back Room
Smugglers
Soldier of the Night
Something of Value
Sons
Split Second
Spy Story
Steele Justice
Strategic Air Command
Striker
Surprise Attack
Target for Today: The 8th Air
    Force Story
Tempest/Eagle
Thirty-Six Hours of Hell
This Island Earth
Three Came Home
The 317th Platoon
Tiger Fangs
Tiger Joe
Time to Kill
Torn Allegiance
Treasure of the Lost Desert
Troilus & Cressida

Troma's War
The Ugly American
Under Fire
The Uprising
Vera Cruz
Villa Rides
Walker
The Wanton Contessa
War Bus Commando
War and Peace
Warbirds
Waterloo
World War III
Yank in Libya
Young Winston
Zeppelin

Werewolves

An American Werewolf in
    London
Black Werewolf
Blood of Dracula's Castle
Bride of the Gorilla
Children of the Full Moon
The Company of Wolves
The Craving
Curse of the Devil
The Curse of the Werewolf
Deathmoon
Dr. Jekyll and the Wolfman
Dr. Terror's House of Horrors
Dracula vs. Frankenstein
Face of the Screaming
    Werewolf
Frankenstein Meets the
    Wolfman
The Fury of the Wolfman
Hercules, Prisoner of Evil
House of Frankenstein
The Howling
Howling 2: Your Sister Is a
    Werewolf
Howling 3
Howling 4: The Original
    Nightmare
Howling 6: The Freaks
I Was a Teenage Werewolf
Legend of the Werewolf
The Legend of the Wolf
    Woman
Lone Wolf
The Mad Monster
Meridian: Kiss of the Beast
Monster Dog
The Monster Squad
Moon of the Wolf
My Mom's a Werewolf
Night of the Howling Beast
Orgy of the Dead
Rats are Coming! The
    Werewolves are Here!
Return of the Vampire
Silver Bullet
Teen Wolf
Teen Wolf Too
The Undying Monster
Werewolf in a Girl's Dormitory
Werewolf of London
The Werewolf vs. the Vampire
    Woman
Werewolf of Washington
Werewolves on Wheels
The Wolf Man
Wolfen
The Wolfman

Western Comedy

Adios Amigo
Along Came Jones
Bad Man's River
Ballad of Cable Hogue
Bang Bang Kid

North of the Great Divide
Oath of Vengeance
Oh Susannah
Oklahoma Annie
Oklahoma Badlands
Oklahoma Cyclone
The Oklahoman
Old Barn Dance
Old Corral
On the Night Stage
On the Old Spanish Trail
On Top of Old Smoky
One-Eyed Jacks
One Little Indian
100 Rifles
Only the Valiant
Oregon Trail
Outcast
The Outlaw
Outlaw Express
Outlaw Fury
Outlaw Gang
Outlaw Josey Wales
Outlaw Justice
Outlaw of the Plains
Outlaw Roundup
Outlaw Rule
The Outlaw Tamer
Outlaw Trail
Outlaw Women
Outlaw's Paradise
Outlaws of the Range
Outlaws of Sonora
Overland Mail
The Ox-Bow Incident
Painted Desert
The Painted Stallion
Pale Rider
Pals of the Range
Panamint's Bad Man
Pancho Villa
Pancho Villa Returns
Paroled to Dio
Passion
Pat Garrett & Billy the Kid
Pecos Kid
Peter Lundy and the Medicine
    Hat Stallion
The Phantom Bullet
The Phantom Empire
Phantom Patrol
Phantom Rancher
Phantom Ranger
Phantom Rider
Phantom Thunderbolt
Phantom of the West
Pinto Canyon
Pinto Rustlers
Pioneer Marshal
Pioneer Woman
The Pioneers
The Plainsman
Pocatello Kid
Pony Express
Pony Express Rider
Posse
Powdersmoke Range
Prairie Badmen
The Prairie King
Prairie Moon
Prairie Pals
Professionals
The Proud and the Damned
Proud Rebel
Public Cowboy No. 1
Purple Vigilantes
Pursued
The Quick and the Dead
Quigley Down Under
Racketeers of the Range
Radio Ranch
Rage at Dawn
Raiders of the Border
Raiders of Red Gap
Raiders of Sunset Pass
Rainbow Ranch

Rainbow's End
Ramrod
The Ramrodder
Rancho Notorious
Randy Rides Alone
Range Busters
The Range Feud/Two Fisted
    Law
Range Law
Range Rider
Range Riders
Rangeland Empire
Rangeland Racket
Ranger and the Lady
Ranger's Roundup
Rangers Take Over
The Rare Breed
Rattler Kid
Rawhide
Rawhide Romance
A Reason to Live, A Reason
    to Die
Red Desert
Red Headed Stranger
Red River
Red River Valley
Red Rock Outlaw
The Red Rope
Red Sun
Renegade Girl
Renegade Ranger/Scarlet
    River
Renegade Trail
Renfrew on the Great White
    Trail
Renfrew of the Royal
    Mounted
The Restless Breed
Return of the Bad Men
Return of Draw Egan
Return of Frank James
Return of Jesse James
The Return of Josey Wales
Return of the Lash
The Return of a Man Called
    Horse
Return of the Seven
Return to Snowy River
Return of Wildfire
The Revenge Rider
Revenge of the Virgins
Ride 'Em Cowgirl
Ride to Glory
Ride the High Country
Ride the Man Down
Ride, Ranger, Ride
Ride in the Whirlwind
Ride the Wind
Rider of the Law
Rider from Tucson
Riders of Death Valley
Riders of the Desert
Riders of Destiny
Riders of the Range
Riders of the Range/Storm
    Over Wyoming
Riders of the Rockies
Riders of the Timberline
Riders of the West
Riders of the Whistling Pines
Riders of the Whistling Skull
Ridin' the California Trail
Ridin' the Lone Trail
Ridin' on a Rainbow
Ridin' Thru
Ridin' the Trail
The Riding Avenger
Riding On
Riding Speed
Rim of the Canyon
Rimfire
Rin Tin Tin, Hero of the West
Rio Bravo
Rio Conchos
Rio Grande
Rio Grande Raiders

Rio Hondo
Rio Lobo
Rio Rattler
River of No Return
Road Agent
The Roamin' Cowboy
Roamin' Wild
Roarin' Lead
Roaring Guns
Roaring Six Guns
Robbery Under Arms
Robin Hood of the Pecos
Robin Hood of Texas
Rock River Renegades
Rodeo Girl
Rogue of the Range
Rogue of the Rio Grande
Roll Along Cowboy
Roll on Texas Moon
Roll, Wagons, Roll
Rollin' Plains
Romance on the Range
Rooster Cogburn
Rootin' Tootin' Rhythm
Rose of Rio Grande
Rough Riders of Cheyenne
Rough Riders' Roundup
Rough Riding Rangers
Round-Up Time in Texas
Rounding Up the Law
Run of the Arrow
Running Wild
Rustler's Hideout
Rustler's Paradise
Rustler's Valley
The Sacketts
Sacred Ground
Saddle Buster
Saddle Mountain Round-Up
Saga of Death Valley
Sagebrush Trail
Salute to the ''King of the
    Cowboys,'' Roy Rogers
San Antonio
San Fernando Valley
Santa Fe Bound
Santa Fe Marshal
Santa Fe Trail
Santa Fe Uprising
Santee
Sartana's Here...Trade Your
    Pistol for a Coffin
Savage Guns
Savage Journey
Savage Run
Scalps
The Searchers
Seven Cities of Gold
Seven Sixgunners
Seventh Cavalry
The Shadow Riders
Shadows of Death
Shalako
Shane
She Wore a Yellow Ribbon
Sheriff of Tombstone
Shine on, Harvest Moon
Shoot the Living, Pray for the
    Dead
Shoot the Sun Down
The Shooting
The Shootist
Shotgun
The Showdown
Showdown
Showdown at Boot Hill
Showdown at Williams Creek
Silent Code
Silent Valley
Silver Bandit
Silver City Kid
Silver Lode
Silver Queen
Silver Spurs
Silver Stallion
Silver Star

Sing, Cowboy, Sing
Singing Buckaroo
Sinister Journey
Six Gun Gospel
Six Gun Rhythm
Six Shootin' Sheriff
Sky Bandits
Slaughter Trail
Smith!
Smoke in the Wind
Smokey Smith
Soldier Blue
The Sombrero Kid
Son of God's Country
Son of a Gun
Son of the Morning Star
Song of Arizona
Song of the Gringo
Song of Nevada
Song of Old Wyoming
Song of Texas
Song of the Trail
Songs and Bullets
Sonora Stagecoach
Sons of Katie Elder
Sons of the Pioneers
South of the Border
South of Monterey
South of the Rio Grande
South of St. Louis
South of Santa Fe
Southward Ho!
Spirit of the West
Springtime in the Rockies
Springtime in the Sierras
Spurs
Square Dance Jubilee
Stage to Mesa City
Stagecoach
Stagecoach to Denver
The Stalking Moon
Standing Tall
The Star Packer
Station West
Storm Over Wyoming
Stormy Trails
Straight Shooter
Straight Shootin'
Stranger's Gold
Strawberry Roan
Sudden Death
Sundance and the Kid
Sundown Fury
Sundown Kid
Sundown Riders
Sundown Saunders
Sunset
Sunset on the Desert
Sunset in El Dorado
Sunset Range
Sunset Serenade
Sunset Trail
Susanna Pass
Sweet Georgia
Swifty
Talion
The Tall Men
Tall in the Saddle
The Tall T
Tall Tales & Legends: Annie
    Oakley
Taste of Death
Tell Them Willie Boy Is Here
Ten Wanted Men
Tennessee's Partner
Tension at Table Rock
Terror of Tiny Town
Terror Trail
Test of Donald Norton
Tex Rides With the Boy
    Scouts
Texas to Bataan
Texas Cyclone
Texas Gunfighter
Texas Guns

Texas John Slaughter:
    Geronimo's Revenge
Texas John Slaughter:
    Stampede at Bitter Creek
Texas Justice
Texas Kid
Texas Lady
Texas Legionnaires
Texas Masquerade
Texas Pioneers
The Texas Rangers
Texas Terror
Texas Trouble
Texas Wildcats
They Died with Their Boots
    On
This Gun for Hire
Thousand Pieces of Gold
Three Bullets for a Long Gun
Three Desperate Men
Three Faces West
Three in the Saddle
3:10 to Yuma
Three Word Brand
Thunder in the Desert
Thunder in God's Country
Thunder Mountain
Thunder Over Texas
Thunder Pass
Thunder River Feud
Thundering Gunslingers
Thundering Trail
Timber Terrors
Tin Star
To the Last Man
Today We Kill, Tomorrow We
    Die
Toll of the Desert
Toll Gate
Tom Mix Compilation 1
Tom Mix Compilation 2
Tomahawk Territory
Tombstone Canyon
Tombstone Terror
The Tonto Kid
The Tracker
The Trackers
Trail Beyond
Trail Drive
Trail of the Hawk
Trail of the Mounties
Trail Riders
Trail of Robin Hood
Trail of the Royal Mounted
Trail of the Silver Spurs
Trail Street
Trail of Terror
Trailing Double Trouble
Trailing Trouble
Trails West
Train Robbers
Train to Tombstone
The Traitor
Treasure of Pancho Villa
Trigger, Jr.
Trigger Pals
Trigger Trio
Trouble Busters
Trouble in Texas
True Duke: A Tribute to John
    Wayne
True Grit
Tulsa
Tumbledown Ranch in
    Arizona
Tumbleweed Trail
Tumbleweeds
Twice a Judas
Twilight on the Rio Grande
Twilight in the Sierras
Twilight on the Trail
Two Fisted Justice
Two-Fisted Justice
Two Gun Man
Two-Gun Man from Harlem
Two-Gun Troubador

Westerns

Two Mules for Sister Sara
Two Rode Together
Ulzana's Raid
The Undefeated
Under California Stars
Under Nevada Skies
Under Western Stars
Undercover Man
Underground Rustlers
The Unforgiven
Up River
Utah
Utah Trail
Valdez is Coming
Valley of Fire
Valley of the Lawless
Valley of the Sun
Valley of Terror
Vanishing American
Vanishing Legion
The Vanishing Westerner
Vengeance Valley
Vigilantes of Boom Town
Villa Rides
Virginia City
The Virginian
Wagon Master
Wagon Trail
Wagon Wheels
Wall Street Cowboy
The Walloping Kid
Wanda Nevada
The War Wagon
War of the Wildcats
Warlock
Water Rustlers
Way of the West
The Way West
Welcome to Blood City
West of the Divide
West to Glory
West of the Law
West of Nevada
West of Pinto Basin
Western Double Feature 1
Western Double Feature 1:
    Wild Bill & Sunset Carson
Western Double Feature 2
Western Double Feature 2:
    Wayne & Hale
Western Double Feature 3
Western Double Feature 3:
    Rogers & Allen
Western Double Feature 4
Western Double Feature 4:
    Rogers & Evans
Western Double Feature 5
Western Double Feature 5:
    Rogers & Autry
Western Double Feature 6:
    Rogers & Trigger
Western Double Feature 7:
    Wild Bill & Crabbe
Western Double Feature 8:
    Hale & Carradine
Western Double Feature 9:
    LaRue & Beery
Western Double Feature 10:
    Roy, Dale, & Trigger
Western Double Feature 11
Western Double Feature 12
Western Double Feature 13
Western Double Feature 14
Western Double Feature 15
Western Double Feature 16
Western Double Feature 17
Western Double Feature 18
Western Double Feature 19
Western Double Feature 20
Western Double Feature 21
Western Double Feature 22
Western Double Feature 23
Western Double Feature 24
Western Double Feature 25
Western Frontier
Western Justice

Western Mail
Western Pacific Agent
Western Trails
Western Union
The Westerner
Westward Ho
Westward Ho, the 'Wagons!
When the Legends Die
When a Man Rides Alone
When the West Was Young
Where the North Begins
Where Trails End
Whirlwind Horseman
Whistlin' Dan
Whistling Bullets
White Apache
The White Buffalo
White Comanche
White Eagle
White Gold
The Wild Bunch
Wild Country
The Wild Country
Wild Horse
Wild Horse Canyon
Wild Horse Rodeo
Wild Mustang
Wild Times
Wild West
Wild, Wild West Revisited
Wild Women
Wildfire
Will Penny
Winchester '73
The Wind
Winds of the Wasteland
Winners of the West
Winning of the West
With Kit Carson Over the
    Great Divide
Without Honors
Wolf Call
Wolfheart's Revenge
The Woman of the Town
The Yellow Rose of Texas
Yodelin' Kid From Pine Ridge
Young Bill Hickok
Young Blood
Young Buffalo Bill
Young Guns
Young Guns 2
Yukon Flight
Yuma
Zachariah
Zorro Rides Again
Zorro's Black Whip
Zorro's Fighting Legion

Wild Kingdom

The Adventures of Grizzly
    Adams at Beaver Dam
The Adventures of Tarzan
Africa Texas Style
All Creatures Great and Small
Bambi
Barefoot Executive
The Bear
The Bears & I
Beastmaster
Beastmaster 2: Through the
    Portal of Time
Beasts
Bedtime for Bonzo
The Belstone Fox
Benji Takes a Dive at
    Marineland/Benji at Work
Black Cobra
Bless the Beasts and Children
Bo-Ru the Ape Boy
Bongo
Born Free
The Brave One
A Breed Apart
Bringing Up Baby

Buffalo Rider
Bugs Bunny Superstar
Call of the Wild
Carnival of Animals
The Carnivores
Catch as Catch Can
Charlie, the Lonesome Cougar
Cheetah
Christian the Lion
Chronicles of Narnia
Clarence, the Cross-eyed Lion
Courage of the North
A Cry in the Dark
Daffy Duck Cartoon Festival:
    Ain't That Ducky
Dark Age
Day of the Animals
Distant Thunder
Doctor Doolittle
Dolphin Adventure
Dot & the Bunny
Dumbo
Dusty
Eaten Alive
The Electric Horseman
Elephant Boy
Elmer Fudd Cartoon Festival
Elsa & Her Cubs
The Emmett Kelly Circus
The Enchanted Journey
Every Which Way But Loose
Eyes of the Panther
The Fifth Monkey
Flight of the Grey Wolf
Fraidy Cat
Francis in the Navy
Frasier the Sensuous Lion
Futz
Gaiety
Gates of Heaven
Gentle Ben
Gentle Giant
Going Ape!
Going Bananas
The Golden Seal
Goldy 2: The Saga of the
    Golden Bear
Goldy: The Last of the Golden
    Bears
Gorilla
Gorillas in the Mist
The Great Adventure
The Great Rupert
Greed & Wildlife: Poaching in
    America
Grizzly Adams: The Legend
    Continues
Gus
Harry and the Hendersons
Hatari
Hawmps!
High Country Calling
Himatsuri (Fire Festival)
Hunting the White Rhino
In the Shadow of Kilimanjaro
Island of the Blue Dolphins
Jungle
The Jungle Book
Jungle Cat
Jungle Cavalcade
Jungle Drums of Africa
Just Call Me Kitty
Kids is Kids
King of the Grizzlies
King Kung Fu
Lassie, Come Home
Lassie: The Miracle
Legend of Lobo
Leonard Part 6
Little Heroes
Living Free
Misadventures of Merlin
    Jones
Monkeys, Go Home!
Monkey's Uncle
Mountain Family Robinson

Mountain Man
Mouse and His Child
Mr. Kingstreet's War
Murders in the Rue Morgue
My Sister, My Love
Napoleon and Samantha
Never Cry Wolf
Night of the Grizzly
Nikki, the Wild Dog of the
    North
Operation Haylift
Panda and the Magic Serpent
Panda's Adventures
Pardon My Trunk
Planet of the Apes
Project X
Razorback
The Rescuers
The Rescuers Down Under
Ring of Bright Water
Rogue Lion
Rub a Dub Dub
The Runaways
Sammy, the Way-Out Seal
Secret of NIMH
The Secret World of Reptiles
Shakma
Shaolin Traitor
The Silver Streak
Stalk the Wild Child
A Summer to Remember
Super Seal
Survival of Spaceship Earth
Tarka the Otter
The Tender Warrior
Their Only Chance
Those Calloways
Thumbelina
A Tiger Walks
Time of the Apes
Tomboy & the Champ
Trap on Cougar Mountain
Un Elefante Color Ilusion
Vampire Circus
Vanishing Wilderness
Walk Like a Man
Warner Brothers Cartoon
    Festival 3
Watership Down
Whale of a Tale
When the North Wind Blows
When the Whales Came
The White Buffalo
White Gorilla
White Pongo
Wild Horses
Wolfen Ninja
The Yearling

Wilderness

Abducted
Across the Great Divide
The Adventures of Frontier
    Fremont
The Adventures of the
    Wilderness Family
The Alaska Wilderness
    Adventure
All Mine to Give
Backwoods
Black Robe
Blood Brothers
The Bridge to Nowhere
Bullies
Cajon Pass
The Capture of Grizzly Adams
Claws
Cold River
Continental Divide
Davy Crockett, King of the
    Wild Frontier
Davy Crockett and the River
    Pirates
Death Hunt

Earthling
Edge of Honor
Eyes of Fire
The Fatal Glass of Beer
Flight from Glory
God's Country
Goldy 2: The Saga of the
    Golden Bear
Great Adventure
The Great Outdoors
Grizzly
Grizzly Adams: The Legend
    Continues
Heaven on Earth
Hunter's Blood
Ice Palace
Jeremiah Johnson
Just Before Dawn
Life & Times of Grizzly Adams
Lost
Lost in the Barrens
Mountain Charlie
Mountain Family Robinson
Napoleon and Samantha
Natural States
Natural States Desert Dream
Northwest Passage
Orion's Belt
Orphans of the North
Quest for Fire
Secrets of Life
Shoot to Kill
Silence
Silence of the North
Slashed Dreams
Sourdough
Spirit of the Eagle
Starbird and Sweet William
Survival
The Tale of Ruby Rose
Tales of the Klondike: In a Far
    Country
Tales of the Klondike: Race
    for Number One
Tales of the Klondike: The
    Scorn of Women
Tales of the Klondike: The
    Unexpected
Timber Queen
Vanishing Prairie
Violent Zone
The Voyage of Tanai
White Fang
White Fang and the Hunter
The White Tower
White Water Summer
Wilderness Family, Part 2
Young & Free

Women

Alice Doesn't Live Here
    Anymore
An Angel at My Table
Anna
Another Way
Another Woman
Antonia and Jane
The Astounding She-Monster
Bachelor Mother
Beaches
Below the Belt
The Big Doll House
Blood Games
Blood and Sand
Born in Flames
The Bostonians
Business as Usual
Caged Fury
Caged Heat
Caged Women
Cave Girls
Challenge of a Lifetime
The Children's Hour
China Beach

Wrestling

Wrong Side of the Tracks

Yuppie Nightmares

Zombies

Zombies

SERIES INDEX

The **Series Index** provides 50 concise listings of major movie series, ranging from James Bond to National Lampoon to the Three Mesquiteers. It follows the arrangement of the Category Index, with series names arranged alphabetically.

Abbott & Costello
Andy Hardy
Billy the Kid
Bowery Boys
Bulldog Drummond
Charlie Brown and the Peanuts Gang
Charlie Chan
Cheech & Chong
Dr. Christian
Dr. Mabuse
Dracula
Elvisfilm
Faerie Tale Theatre
The Falcon
Frankenstein
Godzilla
Hallmark Hall of Fame
Hopalong Cassidy
Jack the Ripper
James Bond
Lassie
Laurel & Hardy
Lone Rider
Marx Brothers
Mr. Wong
Modern Shakespeare
Monty Python
Muppets
National Lampoon
Our Gang
Pink Panther
Planet of the Apes

Raiders of the Lost Ark
Rangebusters
Renfrew of the Mounties
Ritz Brothers
Rough Riders
The Saint
Sexton Blake
Sherlock Holmes
Star Wars
Tarzan
Texas Rangers
Thin Man
Three Mesquiteers
3 Stooges
Trail Blazers
Wonderworks Family Movies
Zucker/Abrahams/Zucker

SERIES INDEX

Abbott & Costello

Abbott and Costello in
  Hollywood
Abbott and Costello Meet
  Captain Kidd
Abbott and Costello Meet Dr.
  Jekyll and Mr. Hyde
Abbott and Costello Meet
  Frankenstein
Abbott & Costello Meet the
  Invisible Man
Abbott and Costello Meet the
  Killer, Boris Karloff
Africa Screams
Buck Privates
Buck Privates Come Home
Dance with Me, Henry
Hold That Ghost
In the Navy
Jack & the Beanstalk
Keep 'Em Flying
Mexican Hayride
The Naughty Nineties
Pardon My Sarong
Ride 'Em Cowboy
30-Foot Bride of Candy Rock
Who Done It?
The Wistful Widow of Wagon
  Gap

Andy Hardy

Andy Hardy Gets Spring Fever
Andy Hardy Meets Debutante
Andy Hardy's Double Life
Andy Hardy's Private
  Secretary
Life Begins for Andy Hardy
Love Finds Andy Hardy
Love Laughs at Andy Hardy

Billy the Kid

Bill & Ted's Excellent
  Adventure
Billy the Kid Returns
Billy the Kid in Texas
Billy the Kid Trapped
Billy the Kid Versus Dracula
Border Badmen
Deadwood
Gangster's Den
Gore Vidal's Billy the Kid
I Shot Billy the Kid
Law and Order
The Left-Handed Gun
Oath of Vengeance
Outlaw of the Plains
Pat Garrett & Billy the Kid
Prairie Badmen
Rustler's Hideout
Shadows of Death
Thundering Gunslingers
Wild Horse Phantom

Young Guns
Young Guns 2

Bowery Boys

Blues Busters
Bowery Blitzkrieg
Bowery Buckaroos
Boys of the City
Clipped Wings
Dead End
East Side Kids
East Side Kids/The Lost City
Flying Wild
Ghost Chasers
Ghosts on the Loose
Hard-Boiled Mahoney
Here Come the Marines
Kid Dynamite
Let's Get Tough
Little Tough Guys
Million Dollar Kid
Mr. Wise Guy
'Neath Brooklyn Bridge
Pride of the Bowery
Smugglers' Cove
Spook Busters
Spooks Run Wild
That Gang of Mine
They Made Me a Criminal

Bulldog Drummond

Arrest Bulldog Drummond
Bulldog Drummond
Bulldog Drummond in Africa/
  Arrest Bulldog Drummond
Bulldog Drummond Comes
  Back
Bulldog Drummond Double
  Feature
Bulldog Drummond Escapes
Bulldog Drummond's Bride
Bulldog Drummond's Peril
Bulldog Drummond's
  Revenge
Bulldog Drummond's
  Revenge/Bulldog
  Drummond's Peril
Bulldog Drummond's Secret
  Police
Bulldog Jack

Charlie Brown and
the Peanuts Gang

Be My Valentine, Charlie
  Brown
Bon Voyage, Charlie Brown
A Boy Named Charlie Brown
A Charlie Brown Celebration
A Charlie Brown Christmas
A Charlie Brown Thanksgiving
Charlie Brown's All-Stars

He's Your Dog, Charlie
  Brown!
It Was a Short Summer,
  Charlie Brown
It's Arbor Day, Charlie Brown
It's the Great Pumpkin,
  Charlie Brown
It's a Mystery, Charlie Brown
Play It Again, Charlie Brown
Race for Your Life, Charlie
  Brown
Snoopy, Come Home
Snoopy's Getting Married,
  Charlie Brown
There's No Time for Love,
  Charlie Brown
You're a Good Man, Charlie
  Brown

Charlie Chan

Castle in the Desert
Charlie Chan and the Curse of
  the Dragon Queen
Charlie Chan at the Opera
Charlie Chan in Paris
Charlie Chan in Rio
Charlie Chan at the Wax
  Museum
Charlie Chan's Secret
Meeting at Midnight
Murder Over New York
Terror by Night/Meeting at
  Midnight

Cheech & Chong

After Hours
Cheech and Chong: Still
  Smokin'
Cheech and Chong: Things
  Are Tough All Over
Cheech and Chong's Next
  Movie
Cheech and Chong's Nice
  Dreams
Cheech and Chong's, The
  Corsican Brothers
Cheech and Chong's Up in
  Smoke

Dr. Christian

Courageous Dr. Christian
Dr. Christian Meets the
  Women
Meet Dr. Christian
Melody for Three
Remedy for Riches
They Meet Again

Dr. Mabuse

Dr. Mabuse the Gambler
Dr. Mabuse vs. Scotland Yard

The Return of Dr. Mabuse
The Thousand Eyes of Dr.
  Mabuse

Dracula

Abbott and Costello Meet
  Frankenstein
Bela Lugosi Scrapbook
Best of the Two Ronnies
Blood of Dracula
Blood of Dracula's Castle
The Brides of Dracula
Count Dracula
Dracula
Dracula: A Cinematic
  Scrapbook
Dracula/Garden of Eden
Dracula & Son
Dracula (Spanish Version)
Dracula Sucks
Dracula: The Great Undead
Dracula—Up in Harlem
Dracula vs. Frankenstein
Dracula's Daughter
Dracula's Great Love
Dracula's Last Rites
Dracula's Widow
The Horror of Dracula
House of Frankenstein
Hysterical
Love at First Bite
The Monster Squad
Nocturna
Return of the Vampire
The Satanic Rites of Dracula
The Scars of Dracula
The Screaming Dead
Son of Dracula
To Die for
Transylvania 6-5000
Vampire Circus
Waxwork
Zoltan...Hound of Dracula

Elvisfilm

Blue Hawaii
Bye, Bye, Birdie
Change of Habit
Charro!
Clambake
Double Trouble
The Dream Continues
Early Elvis
Easy Come, Easy Go
Elvis in the '50s
Elvis: Aloha from Hawaii
Elvis in Concert 1968
The Elvis Files
Elvis and Me
Elvis Memories
Elvis in the Movies
Elvis: One Night with You
Elvis Presley's Graceland

Elvis: Rare Moments with the
  King
Elvis Stories
Elvis: That's the Way It Is
Elvis: The Echo Will Never Die
Elvis: The Great
  Performances, Vol. 1:
  Center Stage
Elvis: The Great
  Performances, Vol. 2: Man
  & His Music
Elvis: The Lost Performances
Elvis: The Movie
Elvis on Tour
Flaming Star
Follow That Dream
Frankie and Johnny
Fun in Acapulco
G.I. Blues
Girl Happy
Girls! Girls! Girls!
Harum Scarum
Heartbreak Hotel
His Memory Lives On
It Happened at the World's
  Fair
Jailhouse Rock
Kid Galahad
King Creole
Kissin' Cousins
Live a Little, Love a Little
Love Me Tender
Loving You
Mondo Elvis (Rock 'n' Roll
  Disciples)
Paradise, Hawaiian Style
Roustabout
Speedway
Spinout
Stay Away, Joe
The Story of Elvis Presley
This is Elvis
Tickle Me
Top Secret!
The Trouble with Girls
Viva Las Vegas
Wild in the Country

Faerie Tale Theatre

Aladdin and His Wonderful
  Lamp
Beauty and the Beast
The Boy Who Left Home to
  Find Out About the Shivers
Cinderella
Dancing Princesses
The Emperor's New Clothes
Goldilocks & the Three Bears
Hansel & Gretel
Jack & the Beanstalk
The Little Mermaid
Little Red Riding Hood
The Nightingale
Pied Piper of Hamelin

Faerie

Pinocchio
Princess and the Pea
The Princess Who Never
  Laughed
Puss 'n Boots
Rapunzel
Rip van Winkle
Rumpelstiltskin
Sleeping Beauty
Snow Queen
Snow White and the Seven
  Dwarfs
The Tale of the Frog Prince
The Three Little Pigs
Thumbelina

The Falcon

The Amityville Curse
The Devil's Cargo
The Falcon in Hollywood
The Falcon in Mexico
Falcon Takes Over/Strange
  Bargain
Falcon's Adventure/Armored
  Car Robbery
The Falcon's Brother

Frankenstein

Abbott and Costello Meet
  Frankenstein
Andy Warhol's Frankenstein
Blackenstein
The Bride
The Bride of Frankenstein
The Curse of Frankenstein
Dr. Frankenstein's Castle of
  Freaks
Dracula vs. Frankenstein
The Evil of Frankenstein
The Flintstones Meet Rockula
  & Frankenstone
Frankenstein
Frankenstein '80
Frankenstein: A Cinematic
  Scrapbook
Frankenstein General Hospital
Frankenstein Island
Frankenstein Meets the Space
  Monster
Frankenstein Meets the
  Wolfman
Frankenstein and the Monster
  from Hell
Frankenstein Unbound
Frankenstein's Daughter
Frankenstein's Great Aunt
  Tillie
Frankenweenie
Gothic
The Horror of Frankenstein
House of Frankenstein
Jesse James Meets
  Frankenstein's Daughter
Lady Frankenstein
Munster's Revenge
Orlak, the Hell of Frankenstein
Prototype
The Screaming Dead
Son of Frankenstein
Teenage Frankenstein
The Vindicator
Young Frankenstein

Godzilla

Destroy All Monsters
Ghidrah the Three Headed
  Monster
Godzilla 1985
Godzilla, King of the Monsters
Godzilla on Monster Island
Godzilla Raids Again

Godzilla vs. the Cosmic
  Monster
Godzilla vs. Gigan
Godzilla vs. Mechagodzilla
Godzilla vs. Megalon
Godzilla vs. Monster Zero
Godzilla vs. Mothra
Godzilla vs. the Sea Monster
Godzilla vs. the Smog
  Monster
Godzilla's Revenge
King Kong Versus Godzilla
The Search for Bridey Murphy
Son of Godzilla
Terror of Mechagodzilla

Hallmark Hall of Fame

Caroline?
O Pioneers!
Sarah, Plain and Tall
The Secret Garden

Hopalong Cassidy

Bar-20
Devil's Playground
False Colors
Hopalong Cassidy: Borrowed
  Trouble
Hopalong Cassidy: Dangerous
  Venture
Hopalong Cassidy: False
  Paradise
Hopalong Cassidy: Hoppy's
  Holiday
Hopalong Cassidy: Riders of
  the Deadline
Hopalong Cassidy: Silent
  Conflict
Hopalong Cassidy: Sinister
  Journey
Hopalong Cassidy: The Dead
  Don't Dream
Hopalong Cassidy: The Devil's
  Playground
Hopalong Cassidy: The
  Marauders
Hopalong Cassidy:
  Unexpected Guest
Renegade Trail
Riders of the Timberline
Rustler's Valley
Santa Fe Marshal
The Showdown
Sinister Journey
Texas Masquerade
Twilight on the Trail
Undercover Man

Jack the Ripper

Blade of the Ripper
Hands of the Ripper
He Kills Night After Night After
  Night
Jack the Ripper
Jack's Back
The Monster of London City
Murder by Decree
Night Ripper
Phantom Fiend
The Ripper
Room to Let
A Study in Terror
Terror in the Wax Museum
Time After Time

James Bond

Casino Royale
Diamonds are Forever
Dr. No
For Your Eyes Only

From Russia with Love
Goldfinger
License to Kill
Live and Let Die
The Living Daylights
The Man with the Golden Gun
Moonraker
Never Say Never Again
Octopussy
On Her Majesty's Secret
  Service
The Spy Who Loved Me
Thunderball
A View to a Kill
You Only Live Twice

Lassie

Challenge to Lassie
Lassie, Come Home
Lassie: The Miracle
Lassie's Great Adventure
The Painted Hills

Laurel & Hardy

Atoll K
Bellhop/The Noon Whistle
Block-heads
Bohemian Girl
Bonnie Scotland
Bullfighters
A Chump at Oxford
The Devil's Brother
A Few Moments with Buster
  Keaton and Laurel and
  Hardy
The Flying Deuces
Golden Age of Comedy
Great Guns
Laurel & Hardy: Another Fine
  Mess
Laurel & Hardy: At Work
Laurel & Hardy: Be Big
Laurel & Hardy: Below Zero
Laurel & Hardy: Berth Marks
Laurel & Hardy: Blotto
Laurel & Hardy: Brats
Laurel & Hardy: Chickens
  Come Home
Laurel & Hardy Comedy
  Classics Volume 1
Laurel & Hardy Comedy
  Classics Volume 2
Laurel & Hardy Comedy
  Classics Volume 3
Laurel & Hardy Comedy
  Classics Volume 4
Laurel & Hardy Comedy
  Classics Volume 5
Laurel & Hardy Comedy
  Classics Volume 6
Laurel & Hardy Comedy
  Classics Volume 7
Laurel & Hardy Comedy
  Classics Volume 8
Laurel & Hardy Comedy
  Classics Volume 9
Laurel & Hardy and the Family
Laurel & Hardy: Hog Wild
Laurel & Hardy: Laughing
  Gravy
Laurel & Hardy: Men O'War
Laurel & Hardy: Night Owls
Laurel & Hardy On the Lam
Laurel & Hardy: Perfect Day
Laurel & Hardy Special
Laurel & Hardy Spooktacular
Laurel & Hardy: Stan "Helps"
  Ollie
Laurel & Hardy: The Hoose-
  Gow
Laurel & Hardy Volume 1
Laurel & Hardy Volume 2
Laurel & Hardy Volume 3

Laurel & Hardy Volume 4
Masters of Comedy
Movie Struck
Our Relations
Pack Up Your Troubles
Pardon Us
Saps at Sea
Silent Laugh Makers, No. 1
Sons of the Desert
Swiss Miss
A Tribute to the Boys
Utopia
Way Out West

Lone Rider

Border Roundup
Death Rides the Plains
Law of the Saddle
The Lone Rider in Cheyenne
The Lone Rider in Ghost
  Town
Raiders of Red Gap
Rangeland Racket

Marx Brothers

Animal Crackers
At the Circus
Big Store
The Cocoanuts
A Day at the Races
Go West
Horse Feathers
Love Happy
Monkey Business
A Night in Casablanca
A Night at the Opera

Mr. Wong

Doomed to Die
The Fatal Hour
Mr. Wong in Chinatown
Mr. Wong, Detective
Mysterious Mr. Wong
Mystery of Mr. Wong
Phantom of Chinatown

Modern Shakespeare

Chimes at Midnight
A Double Life
Forbidden Planet
The Godfather: Part 3
Men are not Gods
Men of Respect
My Own Private Idaho
Ran
Rosencrantz & Guildenstern
  Are Dead
Throne of Blood
West Side Story

Monty Python

The Adventures of Baron
  Munchausen
All You Need is Cash
And Now for Something
  Completely Different
Brazil
Clockwise
Consuming Passions
Down Among the Z Men
Erik the Viking
A Fish Called Wanda
Jabberwocky
The Missionary
Monty Python and the Holy
  Grail
Monty Python Live at the
  Hollywood Bowl

Monty Python's Life of Brian
Monty Python's Parrot Sketch
  not Included
Monty Python's The Meaning
  of Life
Nuns on the Run
The Odd Job
Personal Services
A Private Function
Privates on Parade
Ripping Yarns
Romance with a Double Bass
The Secret Policeman's Other
  Ball
Secret Policeman's Private
  Parts
Time Bandits
Whoops Apocalypse
Yellowbeard

Muppets

The Great Muppet Caper
Labyrinth
Muppet Movie
Muppets Take Manhattan
Sesame Street Presents:
  Follow That Bird

National Lampoon

National Lampoon's Animal
  House
National Lampoon's
  Christmas Vacation
National Lampoon's Class of
  '86
National Lampoon's Class
  Reunion
National Lampoon's European
  Vacation
National Lampoon's Vacation

Our Gang

Little Rascals
Little Rascals, Book 1
Little Rascals, Book 2
Little Rascals, Book 3
Little Rascals, Book 4
Little Rascals, Book 5
Little Rascals, Book 6
Little Rascals, Book 7
Little Rascals, Book 8
Little Rascals, Book 9
Little Rascals, Book 10
Little Rascals, Book 11
Little Rascals, Book 12
Little Rascals, Book 13
Little Rascals, Book 14
Little Rascals, Book 15
Little Rascals, Book 16
Little Rascals, Book 17
Little Rascals, Book 18
Little Rascals: Choo Choo/
  Fishy Tales
Little Rascals Christmas
  Special
Little Rascals Collector's
  Edition, Vol. 1-6
Little Rascals Comedy
  Classics, Vol. 1
Little Rascals Comedy
  Classics, Vol. 2
Little Rascals: Fish Hooky/
  Spooky Hooky
Little Rascals: Honkey
  Donkey/Sprucin' Up
Little Rascals: Little Sinner/
  Two Too Young
Little Rascals: Mush and Milk/
  Three Men in a Tub
Little Rascals on Parade
Little Rascals: Pups is Pups/
  Three Smart Boys

Little Rascals: Readin' and
  Writin'/Mail and Female
Little Rascals: Reunion in
  Rhythm/Mike Fright
Little Rascals: The Pinch
  Singer/Framing Youth
Little Rascals Two Reelers,
  Vol. 1
Little Rascals Two Reelers,
  Vol. 2
Little Rascals Two Reelers,
  Vol. 3
Little Rascals Two Reelers,
  Vol. 4
Little Rascals Two Reelers,
  Vol. 5
Little Rascals Two Reelers,
  Vol. 6
Little Rascals, Volume 1
Little Rascals, Volume 2
Our Gang
Our Gang Comedies
Our Gang Comedy Festival
Our Gang Comedy Festival 2

Pink Panther

Curse of the Pink Panther
The Pink Panther
The Pink Panther Strikes
  Again
Return of the Pink Panther
Revenge of the Pink Panther
A Shot in the Dark
Trail of the Pink Panther

Planet of the Apes

Battle for the Planet of the
  Apes
Beneath the Planet of the
  Apes
Conquest of the Planet of the
  Apes
Escape from the Planet of the
  Apes
Planet of the Apes

Raiders of the Lost Ark

Indiana Jones and the Temple
  of Doom
Raiders of the Lost Ark

Rangebusters

Black Market Rustlers
Boot Hill Bandits
Haunted Ranch
Kid's Last Ride
Land of Hunted Men
Range Busters
Rock River Renegades
Saddle Mountain Round-Up
Texas to Bataan
Thunder River Feud
Trail Riders
Trail of the Silver Spurs
Trailing Double Trouble
Tumbledown Ranch in
  Arizona
Two Fisted Justice
Underground Rustlers
West of Pinto Basin

Renfrew of the Mounties

Danger Ahead
Fighting Mad
Murder on the Yukon
Renfrew on the Great White
  Trail

Renfrew of the Royal
  Mounted
Sky Bandits
Yukon Flight

Ritz Brothers

Goldwyn Follies
The Gorilla

Rough Riders

Arizona Bound
Below the Border
Down Texas Way
Fighting Caballero
Forbidden Trails
Ghost Town Law
Gunman from Bodie
Riders of the West

The Saint

Saint in London
The Saint in New York
The Saint Strikes Back
The Saint Takes Over
The Saint's Double Trouble
The Saint's Vacation/The
  Narrow Margin

Sexton Blake

Echo Murders
Meet Sexton Blake
Sexton Blake and the Hooded
  Terror

Sherlock Holmes

The Adventures of Sherlock
  Holmes
The Adventures of Sherlock
  Holmes: Hound of the
  Baskervilles
The Adventures of Sherlock
  Holmes' Smarter Brother
The Crucifer of Blood
Dressed to Kill
The Hound of the Baskervilles
House of Fear
Masks of Death
Murder at the Baskervilles
Murder by Decree
The Pearl of Death
The Private Life of Sherlock
  Holmes
Pursuit to Algiers
Scarlet Claw
The Seven Percent Solution
Sherlock Holmes
Sherlock Holmes: A Study in
  Scarlet
Sherlock Holmes Faces Death
Sherlock Holmes and the
  Incident at Victoria Falls
Sherlock Holmes and the
  Secret Weapon
Sherlock Holmes: The
  Baskerville Curse
Sherlock Holmes: The Sign of
  Four
Sherlock Holmes: The Silver
  Blaze
Sherlock Holmes: The Valley
  of Fear
Sherlock Holmes: The Voice
  of Terror
Sherlock Holmes in
  Washington
The Silent Mr. Sherlock
  Holmes
The Speckled Band
Spiderwoman
A Study in Scarlet

A Study in Terror
Terror by Night
Terror by Night/Meeting at
  Midnight
They Might Be Giants
Triumph of Sherlock Holmes
Without a Clue
Young Sherlock Holmes

Star Wars

The Empire Strikes Back
The Ewok Adventure
The Ewoks: Battle for Endor
Return of the Jedi
Star Wars

Tarzan

Greystoke: The Legend of
  Tarzan, Lord of the Apes
The New Adventures of
  Tarzan: Vol. 1
The New Adventures of
  Tarzan: Vol. 2
The New Adventures of
  Tarzan: Vol. 3
The New Adventures of
  Tarzan: Vol. 4
Tarzan, the Ape Man
Tarzan of the Apes
Tarzan Escapes
Tarzan the Fearless
Tarzan and the Green
  Goddess
Tarzan and His Mate
Tarzan the Tiger
Tarzan and the Trappers
Tarzan's Revenge
Walk Like a Man

Texas Rangers

Boss of Rawhide
Dead or Alive
Enemy of the Law
Gangsters of the Frontier
Marked for Murder
Outlaw Roundup
Ranger's Roundup
Rangers Take Over
Three in the Saddle
Wild West

Thin Man

After the Thin Man
Another Thin Man
Shadow of the Thin Man
Song of the Thin Man
The Thin Man
The Thin Man Goes Home

Three Mesquiteers

Come On, Cowboys
Covered Wagon Days
Cowboys from Texas
Frontier Horizon
Ghost Town Gold
Heart of the Rockies
Heroes of the Hills
Hit the Saddle
Law of the Saddle
Lonely Trail/Three Texas
  Steers
Outlaws of Sonora
Powdersmoke Range
Purple Vigilantes
Riders of the Whistling Skull
Roarin' Lead
Trigger Trio
Wild Horse Rodeo

3 Stooges

Dancing Lady
Gold Raiders
It's a Mad, Mad, Mad, Mad
  World
The Lost Stooges
The Making of the Stooges
Snow White and the Three
  Stooges
Stoogemania
The Stoogephile Trivia Movie
Stooges Shorts Festival
The Stooges Story
Three Stooges Comedy
  Capers: Volume 1
Three Stooges Comedy
  Classics
Three Stooges: Cookoo
  Cavaliers and Other Nyuks
Three Stooges Festival
Three Stooges: From Nurse to
  Worse and Other Nyuks
Three Stooges: Half-Wit's
  Holiday and Other Nyuks
Three Stooges: Idiots Deluxe
  and Other Nyuks
The Three Stooges Meet
  Hercules
Three Stooges: Nutty but Nice
  and Other Nyuks
Three Stooges: They Stooge
  to Conga and Other Nyuks

Trail Blazers

Arizona Whirlwind
Death Valley Rangers
The Law Rides Again
Outlaw Trail

Wonderworks Family Movies

African Journey
Anne of Avonlea
Anne of Green Gables
The Boy Who Loved Trolls
Bridge to Terabithia
Brother Future
Chronicles of Narnia
Clowning Around
The Fig Tree
A Girl of the Limberlost
Gryphon
Hector's Bunyip
Hiroshima Maiden
The House of Dies Drear
Konrad
Lantern Hill
A Little Princess
Looking for Miracles
Maricela
The Mighty Pawns
Miracle at Moreaux
Necessary Parties
Runaway
Sweet 15
Taking Care of Terrific
Walking on Air
Words by Heart
You Must Remember This

Zucker/Abrahams/ Zucker

Airplane!
Hot Shots!
Kentucky Fried Movie
More! Police Squad
The Naked Gun: From the
  Files of Police Squad
The Naked Gun 2 1/2: The
  Smell of Fear
Police Squad! Help Wanted!

Ruthless People
Top Secret!

KIBBLES INDEX

The **Kibbles Index** is a potpourri of movie trivia, with more than 100 lists of books-to-film, producers, special effects wizards, and composers such as John Barry and John Williams. It also provides lists of annual box-office winners (that are now on video) since 1939, classic movies, four-bone delights, and of course, the bottom of the barrel Woof! features.

KIBBLES INDEX

Adapted Screenplay

ABC Stage 67: Truman
  Capote's A Christmas
  Memory
Abduction
Abe Lincoln in Illinois
About Last Night...
Absolute Beginners
Accident
The Accidental Tourist
Acorn People
Across the Bridge
Across the Pacific
Act of Passion: The Lost
  Honor of Kathryn Beck
Adam Had Four Sons
Adorable Julia
The Adventurers
Adventures of Captain Marvel
The Adventures of
  Huckleberry Finn
The Adventures of Ichabod
  and Mr. Toad
The Adventures of Sherlock
  Holmes: Hound of the
  Baskervilles
The Adventures of Tom
  Sawyer
Advise and Consent
The Affair of the Pink Pearl
The African Queen
After the Promise
After the Thin Man
Age Old Friends
Agnes of God
The Agony and the Ecstacy
Airport
Aladdin
Alfie
Alice
Alice Adams
Alice Through the Looking
  Glass
Alice in Wonderland
Alien Women
All About Eve
All Creatures Great and Small
All the King's Men
All of Me
All My Sons
All This and Heaven Too
Allan Quartermain and the
  Lost City of Gold
Altered States
Amadeus
Amazing Mr. Blunden
Ambassador Bill
The Ambush Murders
An American Christmas Carol
The Americanization of Emily
The Amityville Horror
The Amorous Adventures of
  Moll Flanders
Anastasia

Anastasia: The Mystery of
  Anna
And Another Honky Tonk Girl
  Says She Will
And Then There Were None
The Anderson Tapes
The Andersonville Trial
Androcles and the Lion
The Andromeda Strain
Angel Heart
Angele
Animal Crackers
Animal Farm
The Animal Kingdom
Ann Vickers
Anna Christie
Anna to the Infinite Power
Anna Karenina
Anne of Avonlea
Anne of Green Gables
Annie
Another Country
Anthony Adverse
Any Wednesday
Aphrodite
Apocalypse Now
Appointment with Crime
Appointment with Death
Arch of Triumph
Are You in the House Alone?
Around the World in 80 Days
The Arrangement
Arrowsmith
Arsenic and Old Lace
As Is
As Summers Die
As You Desire Me
As You Like It
As Young As You Feel
Asphalt Jungle
Assassin
The Assault
At the Earth's Core
At Play in the Fields of the
  Lord
At Sword's Point
The Atonement of Gosta
  Berling
Attica
Audrey Rose
Autobiography of Miss Jane
  Pittman
Awakenings
The Awful Truth
Babes in Arms
Babette's Feast
Baby Doll
The Bacchantes
Back Street
Bad Day at Black Rock
Bad Ronald
The Bad Seed
The Balcony
The Ballad of the Sad Cafe
Bang the Drum Slowly

Bank Shot
Barabbas
Barefoot in the Park
Barry Lyndon
Bartleby
The Bat
The Bat Whispers
Battle Beyond the Stars
Battle Cry
The Battle of the Sexes
The Bawdy Adventures of
  Tom Jones
Beachcomber
Beaches
Bear/All the Troubles of the
  World
The Beast Within
Beau Brummel
Beau Geste
Becket
Becky Sharp
Being There
The Bell Jar
Belles of St. Trinian's
Bells are Ringing
Berlin Alexanderplatz
  (episodes 1-2)
Berlin Tunnel 21
Bernice Bobs Her Hair
The Best Man
The Best Years of Our Lives
Betrayal
The Betsy
Between Friends
Beyond the Limit
Beyond Therapy
The Big Fix
The Big Man: Crossing the
  Line
The Big Sky
The Big Sleep
Big Street
Big Zapper
Billy Budd
Biloxi Blues
The Birds
Birdy
Birth of a Nation
The Bitch
The Black Arrow
Black Like Me
Black Lizard
Black Narcissus
Black Robe
The Black Stallion
The Black Stallion Returns
Black Sunday
Blackboard Jungle
Blackmail
Blade Runner
Bleak House
Bless the Beasts and Children
Blithe Spirit
Blockhouse
Blood Island

Blood on the Moon
The Blood of Others
Blood and Roses
Blood and Sand
Blood Wedding (Bodas de
  Sangre)
Blue City
The Blue Max
The Body in the Library
The Body Snatcher
Bonjour Tristesse
Book of Love
Born on the Fourth of July
The Boston Strangler
The Bostonians
Botany Bay
A Boy and His Dog
Boys in the Band
The Boys from Brazil
Breakfast at Tiffany's
Breakheart Pass
Breath of Scandal
Brewster's Millions
Brideshead Revisited
The Bridge on the River Kwai
Bridge of San Luis Rey
The Bridges at Toko-Ri
Brief Encounter
The Brig
Brigadoon
Bright Lights, Big City
Brighton Beach Memoirs
Brighton Rock
Bringing Up Baby
The Brothers Grimm Fairy
  Tales
The Brothers Karamazov
Browning Version
Bulldog Drummond
Bulldog Drummond's Secret
  Police
Bullitt
The Burning Hills
Bus Stop
Butterfield 8
Butterflies Are Free
Butterfly Ball
Cabaret
Cactus Flower
Cadence
The Caine Mutiny
The Caine Mutiny Court
  Martial
Call of the Wild
Camille
Can-Can
Cancel My Reservation
Candide
Cannery Row
A Canterbury Tale
The Canterbury Tales
Cape Fear
Captain America
Captain Blood
Captain Horatio Hornblower

Captain from Koepenick
Captains Courageous
Cardinal
Carmen
Carmen Jones
Carmilla
Caroline?
The Carpetbaggers
Carrie
Casablanca
The Case of the Missing Lady
Casino Royale
Castle
Casual Sex?
Cat Chaser
Cat on a Hot Tin Roof
Catch-22
Catholics
Cat's Eye
Cease Fire
Celebrity
The Chalk Garden
Chamber of Horrors
Champion
Chapter Two
Charlotte's Web
Charly
The Chase
Cheech and Chong's, The
  Corsican Brothers
Cheyenne Autumn
Children of the Corn
Children of a Lesser God
Children of Sanchez
The Children's Hour
Chilly Scenes of Winter
Chimes at Midnight
China Seas
China Sky
The Chocolate Soldier
The Chocolate War
The Choirboys
A Chorus of Disapproval
A Chorus Line
Christ Stopped at Eboli
Christiane F.
Christine
A Christmas Carol
Christmas Carol
The Christmas Wife
Chronicles of Narnia
Churchill and the Generals
Cinderella
Circle of Love
Circle of Two
Citadel
The City and the Dogs
The Clan of the Cave Bear
Clarence, the Cross-eyed Lion
Clarence Darrow
Clash by Night
The Class of Miss MacMichael
The Clergyman's Daughter
A Clockwork Orange
Closely Watched Trains

Adapted

Cold Room
The Collector
The Color Purple
Colossus: The Forbin Project
Columbo: Prescription Murder
Coma
Come Back to the 5 & Dime
  Jimmy Dean, Jimmy Dean
Come Back, Little Sheba
Come Blow Your Horn
Come and Get It
Commandos Strike at Dawn
Communion
The Company of Wolves
Compromising Positions
Conagher
Conduct Unbecoming
Confidentially Yours
A Connecticut Yankee
A Connecticut Yankee in King
  Arthur's Court
The Connection
Conrack
Consenting Adults
Contempt
Cool Hand Luke
Cool Mikado
Corn is Green
Corsican Brothers
Cosi Fan Tutte
Count of Monte Cristo
Country Girl
Country Lovers, City Lovers
Coup de Grace
The Courtship of Eddie's
  Father
Coward of the County
The Crackler
The Cradle Will Fall
Craig's Wife
Crainquebille
Creature with the Blue Hand
Crime and Punishment
Crimes at the Dark House
Crimes of the Heart
Crisis at Central High
Crooked Hearts
Crossing Delancey
The Crucifer of Blood
Cry, the Beloved Country
A Cry for Love
Cthulhu Mansion
Cujo
Curse of the Demon
The Curse of Frankenstein
Curse of the Yellow Snake
Cyrano de Bergerac
Dad
Daddy Long Legs
Daddy's Dyin'...Who's Got the
  Will?
The Dain Curse
Daisy Miller
Damn Yankees
Dances with Wolves
Dandy in Aspic
Dangerous Liaisons
Dark Age
Dark Eyes
The Dark Secret of Harvest
  Home
Das Boot
David and Bathsheba
David Copperfield
David and Lisa
A Day in the Country
A Day in the Death of Joe Egg
The Day of the Jackal
Day of the Locust
Day of the Triffids
Days of Wine and Roses
The Dead
Dead Calm
Dead End
Dead Eyes of London
Dead Lucky

Dead Ringers
Dead Zone
Deadline at Dawn
Death Be Not Proud
Death on the Nile
Death Race 2000
Death of a Salesman
Death in Venice
Deathtrap
The Decameron
Decameron Nights
Deception
Deerslayer
Dempsey
Desert Bloom
Desert Gold
Desert Hearts
Desire Under the Elms
Desiree
Desk Set
Despair
Desperate Hours
The Detective
Detective
Detective Story
Devil & Daniel Webster
Devil in the Flesh
The Devils
The Devil's Eye
Dial "M" for Murder
Diamond Head
Diamonds are Forever
The Diary of Anne Frank
Diary of a Chambermaid
Diary of a Hitman
Diary of a Lost Girl
Diary of a Madman
Dick Tracy
Dick Tracy vs. Cueball
Dick Tracy's Dilemma
Die Hard
Die Hard 2: Die Harder
Die, Monster, Die!
Different Kind of Love
The Dining Room
Dinner at Eight
Dirty Gertie from Harlem
  U.S.A.
The Displaced Person
The Divorce of Lady X
Doc Savage
The Doctor
The Doctor and the Devils
Doctor Doolittle
Doctor Faustus
Dr. Jekyll and Mr. Hyde
Dr. No
Dr. Seuss' Grinch Grinches
  the Cat in the Hat/Pontoffel
  Pock
Dr. Seuss' Halloween Is
  Grinch Night
Dr. Strangelove, or: How I
  Learned to Stop Worrying
  and Love the Bomb
Doctor Zhivago
Doctors' Wives
Dodsworth
The Dogs of War
Doll Face
Dollmaker
Doll's House
A Doll's House
Don Quixote
Donovan's Brain
Don's Party
Don't Drink the Water
Don't Look Back: The Story of
  Leroy "Satchel" Paige
Don't Look Now
Don't Tell Her It's Me
Dorian Gray
Double Indemnity
The Double Negative
Down and Out in Beverly Hills
Dracula

Dracula (Spanish Version)
Dragnet
Dragon Seed
A Dream of Kings
Dressed to Kill
Dresser
Driver's Seat
Driving Miss Daisy
The Drowning Pool
Drugstore Cowboy
Drums Along the Mohawk
Dubarry
DuBarry was a Lady
Dude Ranger
Duel of Hearts
Duel in the Sun
The Duellists
Dune
The Dunwich Horror
The Dybbuk
Dynasty
East of Eden
Easy Living
Easy Virtue
Eat a Bowl of Tea
Ebony Tower
Echo Murders
Echoes in the Darkness
Edipo Re (Oedipus Rex)
Educating Rita
Education of Sonny Carson
Egg and I
The Eiger Sanction
Eight Men Out
8 Million Ways to Die
El Amor Brujo
Eleanor & Franklin
Electric Grandmother
Eleni
Elmer Gantry
The Elusive Pimpernel
Emil & the Detectives
Emperor Jones
Emperor Jones/Paul
  Robeson: Tribute to an
  Artist
The Emperor's New Clothes
Empire of the Ants
Empire of the Sun
Empty Canvas
Enchanted Cottage
Enchanted Island
End of August
End of Desire
End of the Road
Endless Love
Endless Night
Endplay
Enemies, a Love Story
Enola Gay: The Men, the
  Mission, the Atomic Bomb
Enormous Changes
Ensign Pulver
Enter Laughing
The Entertainer
Entertaining Mr. Sloane
Equus
Erendira
Eric
Escapade
Escape to Witch Mountain
L'Etat Sauvage
Eureka!
The Europeans
Everybody Wins
Everybody's All American
Everything You Always
  Wanted to Know About
  Sex (But Were Afraid to
  Ask)
The Evil That Men Do
Evil Under the Sun
The Execution of Private
  Slovik
The Executioner's Song
Exodus

Extremities
Eye of the Needle
Eyes of the Amaryllis
Eyes of the Panther
Eyes Right!
A Face in the Crowd
Face to Face
Fahrenheit 451
The Fall of the House of Usher
The Fallen Idol
Fanny
Far from the Madding Crowd
The Far Pavilions
A Farewell to Arms
Farewell, My Lovely
The Farmer's Wife
Fatal Vision
Father Figure
Fatty Finn
Faust
Fellini Satyricon
Fiddler on the Roof
Field of Dreams
52 Pick-Up
The Fighter
Fighting Caravans
Fighting Prince of Donegal
A Fine Madness
Finessing the King
Finian's Rainbow
The Fire in the Stone
Firestarter
The First Deadly Sin
Fitzcarraldo
Five Card Stud
Five Days One Summer
Five Forty-Eight
The Five of Me
Five Weeks in a Balloon
Flame is Love
Flash Gordon
Flash Gordon: Rocketship
A Flash of Green
Fletch
Fletch Lives
Flower Drum Song
Flowers in the Attic
Flying Machine/Coup de
  Grace/Interlopers
Follow That Dream
Food of the Gods: Part 2
Fool for Love
For Your Eyes Only
Forbidden Passion: The Oscar
  Wilde Movie
Forbidden Planet
Forever
Forlorn River
The Formula
Fortunes of War
Forty Carats
Forty Days of Musa Dagh
The Fountainhead
Four Daughters
The Four Feathers
The Fourth Protocol
The Franchise Affair
Frankenstein
Frankie & Johnny
Freaky Friday
Freedom Road
The French Lieutenant's
  Woman
Friday Foster
Fried Green Tomatoes
Friendly Persuasion
From Beyond
From the Earth to the Moon
From Here to Eternity
From the Terrace
The Front Page
The Fugitive
The Fugitive Kind
Full Metal Jacket
A Funny Thing Happened on
  the Way to the Forum

Fury is a Woman
Futz
Gabriela
Gal Young 'Un
Gaslight
Gathering Storm
The Gay Divorcee
Generation
Gentlemen Prefer Blondes
Georgia, Georgia
Georgy Girl
Gervaise
The Getaway
Getting It Right
The Getting of Wisdom
The Ghost and Mrs. Muir
Ghost Story
Giant
Gideon's Trumpet
Gidget
The Gift of Love
Gigi
Girl Hunters
A Girl of the Limberlost
The Girl in a Swing
Glass House
The Glass Key
The Glass Menagerie
Glitter Dome
Glitz
Glory
Gnome-Mobile
Gnomes
Go Tell the Spartans
The Goalie's Anxiety at the
  Penalty Kick
The Godfather
Godfather 1902-1959—The
  Complete Epic
God's Little Acre
Golden Boy
The Golden Coach
Golden Demon
The Golden Honeymoon
Golden Rendezvous
The Golden Salamander
Gone are the Days
Gone with the Wind
The Good Earth
The Good Mother
Goodbye Again
Goodbye Columbus
Goodbye Girl
Goodbye, Mr. Chips
Goodbye, My Lady
The Gorilla
Gorky Park
The Graduate
Grambling's White Tiger
Grand Hotel
The Grapes of Wrath
The Grass is Always Greener
  Over the Septic Tank
Graveyard Shift
Grease
Great Adventure
The Great Escape
Great Expectations
The Great Frost
The Great Gabbo
The Great Gatsby
The Great Mouse Detective
Great St. Trinian's Train
  Robbery
The Great Santini
Great Train Robbery
The Great White Hope
Greed
The Green Berets
Green Dolphin Street
Green Fields
The Green Man
Green Pastures
Green Room
The Grifters
The Group

Adapted

Mexican Hayride
Midnight
Midnight Cowboy
Midnight Lace
A Midsummer Night's Dream
The Mikado
Mildred Pierce
The Mill on the Floss
Mimi
Mindwalk
The Miracle of the Bells
Miracle at Moreaux
The Miracle Worker
The Mirror Crack'd
Misery
Miss Firecracker
Miss Julie
Miss Sadie Thompson
Missing Pieces
Mission Stardust
Missouri Traveler
Mister Johnson
Mister Roberts
Mistral's Daughter
Mistress Pamela
Misty
Moby Dick
Mommie Dearest
Money Movers
Monkey Boy
Monkey Grip
Monkey Shines
Monsignor
Monte Walsh
Month in the Country
The Moon and Sixpence
Moonfleet
Morning Glory
Morocco
The Mosquito Coast
The Most Dangerous Game
Mother
The Mountain
Mountains of the Moon
The Mouse That Roared
Mr. Corbett's Ghost
Mr. Horn
Mr. Magoo's Christmas Carol
Mr. & Mrs. Bridge
Mr. North
Mr. Peabody & the Mermaid
Mr. Reeder in Room 13
Mr. Skeffington
Mr. Winkle Goes to War
Mrs. Miniver
Murder
A Murder is Announced
Murder on the Bayou (A
   Gathering of Old Men)
Murder by the Book
Murder by Death
Murder My Sweet
Murder in Texas
Murders in the Rue Morgue
The Murders in the Rue
   Morgue
The Music Man
Mutiny on the Bounty
My African Adventure
My Fair Lady
My Father's Glory
My Girl Tisa
My Man Godfrey
My Michael
My Old Man
Myra Breckinridge
Mysteries
Mysterious Island
The Mysterious Stranger
The Naked and the Dead
The Naked Face
Naked Jungle
Naked Lunch
Nana
National Velvet
Native Son

The Natural
Neighbors
The Neon Empire
Never Cry Wolf
Never Love a Stranger
Never Say Never Again
Never So Few
Never Steal Anything Small
Never Too Late to Mend
NeverEnding Story
Nevil Shute's The Far Country
The New Adventures of Pippi
   Longstocking
New Centurions
The New Invisible Man
Nice Girl Like Me
Nicholas Nickleby
Nickel Mountain
Night Ambush
Night Breed
Night Flight
Night of the Generals
The Night of the Hunter
The Night of the Iguana
A Night in the Life of Jimmy
   Reardon
The Night the Lights Went Out
   in Georgia
'night, Mother
A Night to Remember
Night Train to Munich
The Nightcomers
Nightfall
Nightflyers
Nikki, the Wild Dog of the
   North
1918
1984
92 in the Shade
No Time for Sergeants
Noon Wine
North to Alaska
North Dallas Forty
Northern Pursuit
Northwest Passage
Now and Forever
Now, Voyager
Number Seventeen
The Nun's Story
The Nutcracker Prince
O Pioneers!
O Youth and Beauty
Oblomov
The Oblong Box
O.C. and Stiggs
An Occurrence at Owl Creek
   Bridge
Octopussy
The Odd Couple
The Odessa File
Oedipus Rex
Of Human Bondage
Of Mice and Men
Oh! Calcutta!
Oh Dad, Poor Dad (Momma's
   Hung You in the Closet &
   I'm Feeling So Sad)
The Old Curiosity Shop
Old Curiosity Shop
Old Gringo
Old Ironsides
Old Maid
Oliver!
Oliver Twist
Omega Man
The Omen
On Approval
On the Beach
On a Clear Day You Can See
   Forever
On the Comet
On Golden Pond
On the Town
On Wings of Eagles
Once is Not Enough

One Day in the Life of Ivan
   Denisovich
One Deadly Summer (L'Ete
   Meurtrier)
One Flew Over the Cuckoo's
   Nest
One in a Million: The Ron
   LaFlore Story
One of My Wives Is Missing
One Night Stand
One Russian Summer
One Shoe Makes It Murder
One Touch of Venus
The Onion Field
Ordeal by Innocence
Ordet
Orphan Train
The Osterman Weekend
Otello
Othello
The Other
Other People's Money
The Other Side of Midnight
Our Relations
Our Town
Out
Out of Africa
Out of Season
Outlaw of Gor
Outlaw Josey Wales
Outrage
The Oval Portrait
The Overcoat
Overnight Sensation
The Owl and the Pussycat
The Ox-Bow Incident
Paint Your Wagon
The Painted Veil
Pajama Game
Pal Joey
Panama Hattie
Panique
Papa's Delicate Condition
The Paper Chase
Paper Moon
The Paradine Case
Paradise
Paradise Now
Paris Express
Paris Trout
A Passage to India
The Passing of the Third Floor
   Back
Paths of Glory
Patterns
Patty Hearst
Paul's Case
Pawnbroker
The Pearl
Peck's Bad Boy
Pelle the Conqueror
Pennies From Heaven
The People
The People Next Door
Peril
The Perils of Gwendoline
Personal Services
Pet Sematary
Pete 'n' Tillie
Peter Pan
Peter and the Wolf and Other
   Tales
Petit Con
Petrified Forest
Phantom Fiend
Phantom of the Opera
Phantom Tollbooth
Philadelphia Story
The Pickwick Papers
Picnic at Hanging Rock
Picture of Dorian Gray
Pied Piper of Hamelin
Pigeon Feathers
Pippi Goes on Board
Pippi Longstocking
Pippi in the South Seas

The Pirates of Penzance
The Pit and the Pendulum
A Place in the Sun
The Plague Dogs
Planet of the Apes
Play It Again, Sam
Playboy of the Western World
Plaza Suite
Pocketful of Miracles
A Pocketful of Rye
Pollyanna
The Poppy Is Also a Flower
Portnoy's Complaint
Postcards from the Edge
The Postman Always Rings
   Twice
Power
Power of One
Premature Burial
Presenting Lily Mars
The President's Mystery
Pride and Prejudice
The Prince of Central Park
Prince and the Pauper
The Prince and the Pauper
Prince and the Pauper
The Prince of Thieves
The Prince of Tides
Prince Valiant
The Princess Bride
Princess Daisy
The Princess and the
   Swineherd
Princess Tam Tam
Prisoner of Second Avenue
Prisoner of Zenda
The Private History of a
   Campaign That Failed
Private War
Privates on Parade
The Prize
Prizzi's Honor
Prudential Family Playhouse
Psycho
Pudd'nhead Wilson
The Pure Hell of St. Trinian's
Purlie Victorious
The Purple Taxi
Puss 'n Boots
Pygmalion
QB VII
Quartet
The Queen of Spades
Queenie
Querelle
Quest for Love
Question of Honor
The Quick and the Dead
Quiet Day In Belfast
The Quiller Memorandum
Quo Vadis
Rachel, Rachel
A Rage in Harlem
Ragtime
Railway Children
Rain
Rainbow
The Rainbow
Rainbow Drive
The Rainmaker
Raintree County
A Raisin in the Sun
Ran
Rap Master Ronnie
Rappaccini's Daughter
Rawhead Rex
Ray Bradbury Theater,
   Volume 1
Ray Bradbury Theater,
   Volume 2
The Razor's Edge
Rebecca of Sunnybrook Farm
The Rebels
The Red Badge of Courage
Red Balloon/Occurrence at
   Owl Creek Bridge

The Red and the Black
The Red House
Red Kimono
The Red Pony
The Red Raven Kiss-Off
The Red Shoes
Reflections in a Golden Eye
The Reivers
Remo Williams: The
   Adventure Begins
The Resurrected
Return to Earth
Return to Oz
The Return of Peter Grimm
Return of the Soldier
The Return of Swamp Thing
Reuben, Reuben
Revenge
Reversal of Fortune
Rich Little's Christmas Carol
Richard II
Richard III
The Riddle of the Sands
Ride a Wild Pony
The Right Stuff
Rime of the Ancient Mariner
Ring of Bright Water
Rip van Winkle
The Ritz
The River Niger
The Robe
Robert Kennedy and His
   Times
Robin Hood
Robinson Crusoe
Rocket to the Moon
The Rocking Horse Winner
Roll Along Cowboy
Roll of Thunder, Hear My Cry
Roman Holiday
Roman Spring of Mrs. Stone
Romantic Comedy
Romeo and Juliet
Romeo and Juliet (Gielgud)
The Room
Room Service
Room with a View
Roommate
Roots, Vol. 1
Rope
Rosary Murders
Rosemary's Baby
Rosencrantz & Guildenstern
   Are Dead
Roxanne
Ruggles of Red Gap
Rumble Fish
A Rumor of War
The Runner Stumbles
Running Against Time
Rush
The Russia House
The Russians are Coming, the
   Russians are Coming
Sabotage
Sabrina
The Sacketts
The Sad Sack
Sadie Thompson
The Sailor Who Fell from
   Grace with the Sea
Saint Jack
Saint Joan
Salo, or the 120 Days of
   Sodom
Salome
The Salzburg Connection
Same Time, Next Year
The Saphead
Sarah, Plain and Tall
Saturday Night Fever
Sayonara
The Scarlet & the Black
The Scarlet Letter
The Scarlet Pimpernel
Scheherazade

Scorchers
Scrooge
Scruples
Scum
The Search for Bridey Murphy
The Secret Adversary
The Secret Four
The Secret Garden
Secret Life of Walter Mitty
Secret of NIMH
The Seekers
Seize the Day
A Sense of Freedom
Sense & Sensibility
A Separate Peace
Separate Tables
The Serpent and the Rainbow
Serpico
The Servant
Servants of Twilight
The Set-Up
Seven Blows of the Dragon
Seven Faces of Dr. Lao
The Seven Percent Solution
Seven Thieves
1776
Shades of Darkness
The Shadow Box
The Shadow Riders
The Shadow Strikes
Shadow of the Thin Man
Shalako
Shane
The Shanghai Gesture
Sharky's Machine
She Came to the Valley
She-Devil
She-Freak
The Sheltering Sky
Sherlock Holmes: The Silver
   Blaze
Sherlock Holmes: The Voice
   of Terror
Sherlock Hound: Tales of
   Mystery
Shinbone Alley
The Shining
Shining Through
Ship of Fools
Shirley Valentine
Shiver, Gobble & Snore
The Shoes of the Fisherman
Shogun
The Shooting Party
The Shootist
The Shop Around the Corner
The Shout
Shout at the Devil
Show Boat
The Sicilian
Sidney Sheldon's Bloodline
Silas Marner
The Silence of the Lambs
Silent Night, Lonely Night
Silent Victory: The Kitty O'Neil
   Story
The Silk Road
Silk Stockings
Silver Bears
Silver Bullet
Silver Chalice
Sinful Life
Sins of Dorian Gray
Skin Game
The Skull
The Sky is Gray
Slaughterhouse Five
Slaves of New York
Sleep of Death
Sleeping Car to Trieste
Sleuth
Slightly Honorable
Slow Burn
Smart-Aleck Kill
SnakeEater 3...His Law
Snow Country

Snow Queen
Snow Treasure
The Snows of Kilimanjaro
So Ends Our Night
Solaris
Soldier Blue
Soldier in the Rain
A Soldier's Story
Solomon and Sheba
Some Came Running
Something of Value
Something Wicked This Way
   Comes
Sometimes a Great Notion
Son of the Morning Star
Song of the Thin Man
Sophie's Choice
Sorrows of Gin
Sorry, Wrong Number
The Southerner
The Spectre of Edgar Allen
   Poe
Spirits of the Dead
The Spook Who Sat By the
   Door
Spoon River Anthology
Spy in Black
Spy Story
The Spy Who Came in from
   the Cold
The Squeaker
Stage Door
Stagecoach
Star Spangled Girl
Stars and Bars
The Stars Look Down
Station West
Staying On
Steaming
Stella
Stephen Crane's Three
   Miraculous Soldiers
Steppenwolf
Stick
Stiletto
Stopover Tokyo
The Story of O
Strange Case of Dr. Jekyll &
   Mr. Hyde
Strange Fruit
Strange Interlude
Strange Tales/Ray Bradbury
   Theatre
Strangers on a Train
Street Scene
A Streetcar Named Desire
The Stripper
Studs Lonigan
Suddenly, Last Summer
Summer Heat
Summer of My German
   Soldier
A Summer Place
A Summer Story
Summertime
Summertree
The Sundowners
Sunrise at Campobello
The Sunshine Boys
Superman: The Movie
Superman 2
Superman 3
Superman: The Complete
   Cartoon Collection
Svengali
Swann in Love
Sweet Bird of Youth
Sweet Charity
Sweet Hostage
Swimmer
Swiss Family Robinson
The Swiss Family Robinson
Sylvia and the Phantom
Taipan
A Tale of Two Cities
Tales from the Crypt

Tales from the Darkside: The
   Movie
Tales of the Klondike: In a Far
   Country
Tales of the Klondike: Race
   for Number One
Tales of the Klondike: The
   Scorn of Women
Tales of the Klondike: The
   Unexpected
Tales of Terror
Tales of Washington Irving
Talk Radio
The Talking Parcel
Tall Story
Tall Tales & Legends: Casey
   at the Bat
Taming of the Shrew
Taras Bulba
Tartuffe
Tarzan, the Ape Man
A Taste of Honey
Tea and Sympathy
Teenage Mutant Ninja Turtles:
   The Movie
The Tell-Tale Heart
Tempest
The Tempest
Tempest
Ten Days That Shook the
   World/October
Ten Days Wonder
Ten Little Indians
Ten North Frederick
The Tenant
Tender Is the Night
Tennessee's Partner
10th Victim
Tereza
The Terminal Man
Terror is a Man
Tess
That Championship Season
That Darn Cat
That Forsyte Woman
That Obscure Object of Desire
That Was Then...This Is Now
There's a Girl in My Soup
These Three
They Came to Cordura
They Knew What They
   Wanted
They Live by Night
They Won't Forget
Thief Who Came to Dinner
The Thin Man
The Thin Man Goes Home
The Thing
Things to Come
The Third Man
Third Man on the Mountain
This Happy Breed
This Property is Condemned
Those Calloways
Those Endearing Young
   Charms
Thousand Clowns
Three Brothers
Three Days of the Condor
The 317th Platoon
The Three Little Pigs
The Three Musketeers
Three Musketeers
The Three Musketeers
   (D'Artagnan)
The Three Penny Opera
3:10 to Yuma
The Three Worlds of Gulliver
The Threepenny Opera
Throne of Blood
Thumbelina
Thunder Mountain
A Tiger's Tale
Till Death Do Us Part
Tim
Time of Indifference

Time Lock
A Time to Love & a Time to
   Die
The Time Machine
Time Machine
Time for Murder
The Tin Drum
To the Camp & Back
To the Devil, A Daughter
To Have & Have Not
To Hell and Back
To Kill a Mockingbird
To the Last Man
To the Lighthouse
To Sir, with Love
Tom Brown's School Days
Tom Jones
Tom Sawyer
Tomb
Tomb of Ligeia
Topaz
Topaze
Torch Song Trilogy
Torn Apart
Tortilla Flat
Torture Garden
Tough Guy
Tough Guys Don't Dance
A Town Like Alice
Toys in the Attic
Tragedy of King Richard II
Transmutations
Travels with My Aunt
Treasure Island
Treasure of the Sierra Madre
A Tree Grows in Brooklyn
Trespasser
The Trial
Trio
Trip to Bountiful
Triple Cross
Triumph of Sherlock Holmes
Trojan Women
Tropic of Desire
True Grit
True West
Truman Capote's "The Glass
   House"
Tune in Tomorrow
Tunes of Glory
Tunes of Glory/Horse's Mouth
Turn of the Screw
Twelfth Night
Twelve Angry Men
29thth Street
20,000 Leagues Under the
   Sea
Twice-Told Tales
Twilight's Last Gleaming
Two Can Play
Two English Girls
Two Evil Eyes
Two by Forsyth
Two Kinds of Love
Two to Tango
2001: A Space Odyssey
2010: The Year We Make
   Contact
The Two Worlds of Jenny
   Logan
Ubu and the Great Gidouille
Ugetsu
The Ugly American
Ulysses
The Unapproachable
The Unbearable Lightness of
   Being
The Unbreakable Alibi
Uncle Tom's Cabin
Under the Biltmore Clock
Under Capricorn
Under Milk Wood
Under Satan's Sun
Under the Volcano
Underworld
Uneasy Terms

An Unfinished Piece for a
   Player Piano
The Unnamable
Unsuitable Job for a Woman
Up the Down Staircase
Users
Valmont
The Vampire Lovers
Vampyr
Vanina Vanini
Vault of Horror
Velnio Nuotaka (The Devil's
   Bride)
A Very Old Man with
   Enormous Wings
Village of the Damned
Village of the Giants
The Virgin and the Gypsy
The Virgin Soldiers
The Virginian
Vision Quest
Viva Zapata!
Volpone
Voyage of the Damned
A Voyage Round My Father
Voyager
Wackiest Wagon Train in the
   West
Wages of Fear
Wagon Wheels
Wait Until Dark
Wait Until Spring, Bandini
Walk on the Wild Side
Walking on Air
Wanderers
War Between the Tates
The War Lord
War & Love
War Lover
War and Peace
War & Remembrance, Part 1
War & Remembrance, Part 2
The War of the Worlds
Warlock
Warriors of the Sung Dynasty
Watch on the Rhine
Watchers
Watership Down
The Way We Were
The Way West
We the Living
We Think the World of You
Wee Willie Winkie
We're No Angels
Whale for the Killing
Whales of August
What Price Hollywood?
When the Bough Breaks
When the West Was Young
When the Wind Blows
Where Are the Children?
Where the Buffalo Roam
Where Time Began
Where's Poppa?
Whispers
Whistle Down the Wind
White Fang
White Fang and the Hunter
White Hot: Mysterious Murder
   of Thelma Todd
White Lie
White Nights
Who Am I This Time?
Who Framed Roger Rabbit?
Who is Killing the Great Chefs
   of Europe?
Who Slew Auntie Roo?
Who'll Stop the Rain?
Who's Afraid of Virginia
   Woolf?
Whose Life Is It Anyway?
Why Shoot the Teacher
The Wicked Lady
The Widow
The Wild Country
The Wild Duck

Adapted

Wild Party
Wild Times
Will: G. Gordon Liddy
The Wind in the Willows
Window
Winds of War
The Winslow Boy
Winter People
Winterset
Wise Blood
Witches of Eastwick
Without Warning: The James
    Brady Story
Witness for the Prosecution
The Wizard of Loneliness
The Wizard of Oz
Wolf Call
Wolfen
A Woman Rebels
Woman in the Shadows
A Woman of Substance,
    Episode 1: Nest of Vipers
A Woman of Substance,
    Episode 2: The Secret is
    Revealed
A Woman of Substance,
    Episode 3: Fighting for the
    Dream
A Woman Without Love
The Women
Women in Love
Women & Men: Stories of
    Seduction
Women & Men: Stories of
    Seduction, Part 2
Wonders of Aladdin
Wooden Horse
The World According to Garp
The World is Full of Married
    Men
World of Henry Orient
The Wrong Box
The Wrong Move
Wuthering Heights
The Yearling
Yentl
You Can't Take It with You
You Only Live Twice
The Young in Heart
Young at Heart
Young Lady Chatterly 2
Young & Willing
Young Winston
Your Ticket Is No Longer Valid
Zoot Suit
Zotz!

"Alien" & Rip-Offs

Alien
Alien Prey
Alien Seed
Alien Space Avenger
Aliens
Creature
Deep Space
Deepstar Six
Forbidden World
Galaxy of Terror
Horror Planet
Leviathan
Nightflyers
Predator
Star Crystal

BBC TV Productions

And Now for Something
    Completely Different
Bread (BBC)
Candide
Christabel
Chronicles of Narnia
Danger Man

Books to Film: Agatha Christie

The Affair of the Pink Pearl
The Alphabet Murders
And Then There Were None
Appointment with Death
The Body in the Library
The Case of the Missing Lady
The Clergyman's Daughter
The Crackler
Death on the Nile
Endless Night
Evil Under the Sun
Finessing the King
The House of Lurking Death
Love from a Stranger
The Man in the Mist
The Mirror Crack'd
Murder Ahoy
A Murder is Announced
Murder by the Book
Murder at the Gallop
Murder Most Foul
Murder on the Orient Express
Murder She Said
Ordeal by Innocence
A Pocketful of Rye
The Secret Adversary
Ten Little Indians
The Unbreakable Alibi
Witness for the Prosecution

Books to Film: Alexandre Dumas

Black Magic
Camille
Count of Monte Cristo
The Four Musketeers
Man in the Iron Mask
The Prince of Thieves
Sword of Monte Cristo
The Three Musketeers
Three Musketeers

Books to Film: Alistair MacLean

Bear Island
Breakheart Pass
Caravan to Vaccares
Force 10 from Navarone
Golden Rendezvous
The Guns of Navarone
Puppet on a Chain
River of Death
Where Eagles Dare

Books to Film: Charles Dickens

Bleak House
A Christmas Carol
Christmas Carol
David Copperfield
Great Expectations
Little Dorrit, Film 1: Nobody's
    Fault
Little Dorrit, Film 2: Little
    Dorrit's Story
Nicholas Nickleby
The Old Curiosity Shop
Old Curiosity Shop
Oliver Twist
The Pickwick Papers
Stingiest Man in Town
A Tale of Two Cities
Tale of Two Cities/In the
    Switch Tower

Books to Film: Edgar Allan Poe

The Black Cat
Dr. Tarr's Torture Dungeon
Doin' What the Crowd Does
The Fall of the House of Usher
The Haunted Palace
Haunting Fear
The Haunting of Morella
The House of Usher
Jaws of Justice
The Living Coffin
Manfish
Masque of the Red Death
Murders in the Rue Morgue
The Murders in the Rue
    Morgue
The Oblong Box
The Oval Portrait
The Pit and the Pendulum
The Raven
The Spectre of Edgar Allen
    Poe
Spirits of the Dead
Tales of Terror
The Tell-Tale Heart
Tomb of Ligeia
Torture Chamber of Dr.
    Sadism
Two Evil Eyes
Web of the Spider

Books to Film: Edgar Rice Burroughs

At the Earth's Core
Greystoke: The Legend of
    Tarzan, Lord of the Apes
Land That Time Forgot
The People That Time Forgot
Tarzan, the Ape Man
Tarzan Escapes
Tarzan the Fearless
Tarzan and the Green
    Goddess
Tarzan and His Mate
Tarzan and the Trappers
Tarzan's Revenge

Books to Film: F. Scott Fitzgerald

The Great Gatsby
Last Time I Saw Paris
The Last Tycoon
Marie Antoinette
Tender Is the Night

Books to Film: H.G. Wells

Empire of the Ants
First Men in the Moon
Food of the Gods
Food of the Gods: Part 2
Half a Sixpence
The Invisible Man
Island of Dr. Moreau
The Man Who Could Work
    Miracles
Things to Come
The Time Machine
Time Machine
Village of the Giants
The War of the Worlds

Books to Film: H.P. Lovecraft

Blood Island
Bride of Re-Animator

Cthulhu Mansion
The Curse
Die, Monster, Die!
The Dunwich Horror
From Beyond
The Re-Animator
The Sinister Invasion
The Unnamable

Books to Film: Ian Fleming

Casino Royale
Chitty Chitty Bang Bang
Diamonds are Forever
Dr. No
For Your Eyes Only
From Russia with Love
Goldfinger
License to Kill
Live and Let Die
The Living Daylights
The Man with the Golden Gun
Moonraker
Never Say Never Again
Octopussy
On Her Majesty's Secret
    Service
The Poppy Is Also a Flower
The Spy Who Loved Me
Thunderball
A View to a Kill
You Only Live Twice

Books to Film: John Steinbeck

Cannery Row
East of Eden
The Grapes of Wrath
Lifeboat
Of Mice and Men
The Pearl
The Red Pony

Books to Film: Jules Verne

Around the World in 80 Days
Five Weeks in a Balloon
From the Earth to the Moon
In Search of the Castaways
Journey to the Center of the
    Earth
Light at the Edge of the World
Master of the World
Mysterious Island
On the Comet
Those Fantastic Flying Fools
20,000 Leagues Under the
    Sea

Books to Film: Larry McMurtry

The Best Little Whorehouse in
    Texas
The Last Picture Show
Lonesome Dove
Terms of Endearment
Texasville

Books to Film: Mark Twain

The Adventures of
    Huckleberry Finn
The Adventures of Tom
    Sawyer
A Connecticut Yankee
A Connecticut Yankee in King
    Arthur's Court
Huckleberry Finn

The Innocents Abroad
Mark Twain's A Connecticut
    Yankee in King Arthur's
    Court
Prince and the Pauper
Tom Sawyer
Unidentified Flying Oddball

Books to Film: Neil Simon

Barefoot in the Park
Biloxi Blues
Brighton Beach Memoirs
California Suite
Chapter Two
The Cheap Detective
Come Blow Your Horn
Goodbye Girl
The Heartbreak Kid
Last of the Red Hot Lovers
The Marrying Man
Max Dugan Returns
Murder by Death
Only When I Laugh
Plaza Suite
Seems Like Old Times
The Slugger's Wife
Star Spangled Girl
The Sunshine Boys

Books to Film: Ray Bradbury

The Illustrated Man
It Came From Outer Space
The Martian Chronicles: Part 1
The Martian Chronicles: Part 2
The Martian Chronicles: Part 3

Books to Film: Robert Louis Stevenson

Abbott and Costello Meet Dr.
    Jekyll and Mr. Hyde
Adventure Island
The Black Arrow
The Body Snatcher
Dr. Jekyll and Mr. Hyde
Dr. Jekyll and Sister Hyde
Jekyll & Hyde...Together
    Again
Kidnapped
Man with Two Heads
Return to Treasure Island,
    Vol. 1
Return to Treasure Island,
    Vol. 2
Return to Treasure Island,
    Vol. 3
Return to Treasure Island,
    Vol. 4
Return to Treasure Island,
    Vol. 5
Strange Case of Dr. Jekyll &
    Mr. Hyde
Treasure Island
The Wrong Box

Books to Film: Stephen King

Carrie
Cat's Eye
Children of the Corn
Christine
Creepshow
Creepshow 2
Cujo
Dead Zone
Firestarter
Graveyard Shift

The Lawnmower Man
Maximum Overdrive
Misery
Pet Sematary
Return to Salem's Lot
Salem's Lot: The Movie
The Shining
Silver Bullet
Stand By Me
Stephen King's Golden Years
Stephen King's Nightshift
  Collection

Books to Film: Tennessee Williams

Baby Doll
Cat on a Hot Tin Roof
The Fugitive Kind
The Glass Menagerie
The Night of the Iguana
Roman Spring of Mrs. Stone
The Rose Tattoo
Senso
A Streetcar Named Desire
Suddenly, Last Summer
Summer and Smoke
Sweet Bird of Youth
This Property is Condemned

Books to Film: William Burroughs

Kerouac
Naked Lunch
Towers Open Fire & Other
  Films

Books to Film: William Shakespeare

Hamlet
Henry IV
Henry IV, Part I
Henry V
Henry VI, Part 1
Henry VIII
Henry VI, Part 3
Julius Caesar
King Lear
Macbeth
Measure for Measure
A Midsummer Night's Dream
Othello
Richard II
Richard III
Romeo and Juliet
Romeo and Juliet (Gielgud)
Taming of the Shrew
Tempest
Troilus & Cressida
Twelfth Night

Disney Animated Movies

Alice in Wonderland
Bambi
Beauty and the Beast
Cinderella
Donald Duck: The First 50
  Years
Donald: Limited Gold Edition 1
Donald's Bee Pictures:
  Limited Gold Edition 2
Dumbo
Fantasia
Fun & Fancy Free
The Great Mouse Detective
The Jungle Book
Lady and the Tramp
The Little Mermaid

101 Dalmations
Peter Pan
Pinocchio
The Reluctant Dragon
The Rescuers
The Rescuers Down Under
Robin Hood
Sleeping Beauty
So Dear to My Heart
Sword in the Stone
The Three Caballeros

Disney Family Movies

The Absent-Minded Professor
The Adventures of Bullwhip
  Griffin
Barefoot Executive
Candleshoe
Freaky Friday
Gus
Hot Lead & Cold Feet
In Search of the Castaways
The Love Bug
Misadventures of Merlin
  Jones
Monkeys, Go Home!
Monkey's Uncle
Moon-Spinners
No Deposit, No Return
The North Avenue Irregulars
The Parent Trap
Sammy, the Way-Out Seal
Savage Sam
Shaggy Dog
Story of Robin Hood & His
  Merrie Men
That Darn Cat
Those Calloways
A Tiger Walks
Treasure Island
20,000 Leagues Under the
  Sea

4 Bones

Abe Lincoln in Illinois
Adam's Rib
The Adventures of Robin
  Hood
The African Queen
L'Age D'Or
Aguirre, the Wrath of God
All About Eve
All the King's Men
All Quiet on the Western Front
American Friend
An American in Paris
Anatomy of a Murder
Andrei Roublev
An Angel at My Table
Angels with Dirty Faces
Animal Crackers
Annie Hall
Aparajito
Apocalypse Now
Ashes and Diamonds
Asphalt Jungle
Au Revoir Les Enfants
The Awful Truth
The Babylon Story from
  ''Intolerance''
The Ballad of Narayama
Bambi
The Bank Dick
Battle of Algiers
The Battleship Potemkin
Beauty and the Beast
Ben-Hur
Berlin Alexanderplatz
  (episodes 1-2)
Berlin Alexanderplatz
  (episodes 3-4)

Berlin Alexanderplatz
  (episodes 5-6)
Berlin Alexanderplatz
  (episodes 7-8)
The Best Years of Our Lives
The Bicycle Thief
The Big Parade
The Big Sleep
Birth of a Nation
Black Narcissus
Black Orpheus
Blood and Sand
The Blue Angel
Boyz N the Hood
Breathless (A Bout de Souffle)
Brian's Song
The Bride of Frankenstein
The Bridge on the River Kwai
Brief Encounter
Bringing Up Baby
The Burmese Harp
Butch Cassidy and the
  Sundance Kid
The Cabinet of Dr. Caligari
The Caine Mutiny
The Cameraman
Casablanca
Cesar
Children of Paradise (Les
  Enfants du Paradis)
Chimes at Midnight
Chinatown
Christ Stopped at Eboli
A Christmas Carol
Citizen Kane
City Lights
The Civil War: Episode 1: The
  Cause: 1861
The Civil War: Episode 2: A
  Very Bloody Affair: 1862
The Civil War: Episode 3:
  Forever Free: 1862
The Civil War: Episode 4:
  Simply Murder: 1863
The Civil War: Episode 5: The
  Universe of Battle: 1863
The Civil War: Episode 6:
  Valley of the Shadow of
  Death: 1864
The Civil War: Episode 7:
  Most Hallowed Ground:
  1864
The Civil War: Episode 8: War
  is All Hell: 1865
The Civil War: Episode 9: The
  Better Angels of Our
  Nature: 1865
A Clockwork Orange
Close Encounters of the Third
  Kind
Colonel Redl
Come and See
Commissar
The Conformist
Crime and Punishment
The Crowd
Cyrano de Bergerac
Danton
Das Boot
David Copperfield
Day for Night
Day of Wrath
Dead of Night
The Deer Hunter
Deliverance
Destry Rides Again
Diamonds of the Night
Dinner at Eight
The Discreet Charm of the
  Bourgeoisie
Dr. Strangelove, or: How I
  Learned to Stop Worrying
  and Love the Bomb
Double Indemnity
Duck Soup
Dumbo

Berlin Alexanderplatz
The Earrings of Madame De...
Earth
East of Eden
8 1/2
El Norte
Elephant Man
The Empire Strikes Back
E.T.: The Extra-Terrestrial
Every Man for Himself & God
  Against All
Fanny
Fanny and Alexander
Fantasia
Fitzcarraldo
Five Easy Pieces
Flight of Dragons
Floating Weeds
Forbidden Games
Foreign Correspondent
Forever and a Day
42nd Street
The Four Feathers
The 400 Blows
Frankenstein
The Freshman
From Here to Eternity
The Fugitive
Gallipoli
Garden of the Finzi-Continis
The General
The General Line
Gigi
The Godfather
Godfather 1902-1959—The
  Complete Epic
The Godfather: Part 2
The Gold Rush
Gone with the Wind
Goodfellas
Gospel According to St.
  Matthew
The Graduate
Grand Illusion
Great Expectations
Greed
Gunga Din
Hamlet
Harold and Maude
Harvest
Hearts of Darkness: A
  Filmmaker's Apocalypse
Henry: Portrait of a Serial
  Killer
Henry V
Here Comes Mr. Jordan
High Noon
Hiroshima, Mon Amour
His Girl Friday
Hollywood Canteen
How Green was My Valley
Hud
The Hunchback of Notre
  Dame
The Hustler
I Am a Fugitive from a Chain
  Gang
The Idiot
If...
Ikiru
The Informer
Intolerance
The Invisible Man
Isn't Life Wonderful
It Happened One Night
It's a Wonderful Life
Ivan the Terrible, Part 1
Jazz on a Summer's Day
JFK
Judgment at Nuremberg
Jules and Jim
The King and I
King Kong
Knife in the Water
Koyaanisqatsi
Kriemhilde's Revenge

Kwaidan
La Chienne
La Dolce Vita
La Strada
La Traviata
The Lady Eve
Lady and the Tramp
The Lady Vanishes
Last Command
The Last Detail
The Last Emperor
The Last Picture Show
Late Spring
Laura
Lawrence of Arabia
Les Miserables
The Letter
A Letter to Three Wives
Life & Death of Colonel Blimp
The Lion in Winter
Little Women
The Lives of a Bengal Lancer
Local Hero
Lola Montes
Long Day's Journey into Night
Lost Horizon
The Lost Weekend
M
The Magnificent Ambersons
Magnificent Seven
Male and Female
The Maltese Falcon
A Man for All Seasons
The Man Who Would Be King
The Manchurian Candidate
Manhattan
Marius
The Marriage of Maria Braun
M*A*S*H
McCabe & Mrs. Miller
Mean Streets
Miracle on 34th Street
Miracle of Morgan's Creek
Mister Roberts
Modern Times
Mon Oncle
Mr. & Mrs. Bridge
Mr. Smith Goes to
  Washington
Mrs. Miniver
The Music Man
Mutiny on the Bounty
My Left Foot
My Life As a Dog
My Life to Live
My Man Godfrey
Napoleon
Nashville
National Velvet
The Night of the Hunter
A Night at the Opera
Night of the Shooting Stars
North by Northwest
Notorious
A Nous le Liberte
O Lucky Man
Of Mice and Men
Oliver Twist
On the Waterfront
One Flew Over the Cuckoo's
  Nest
Only Angels Have Wings
Open City
Our Hospitality
The Ox-Bow Incident
Pandora's Box
Passion of Joan of Arc
Pather Panchali
Paths of Glory
Pelle the Conqueror
Persona
Petulia
Philadelphia Story
Pinocchio
Pixote
Poltergeist

You Only Live Twice
Zulu

Musical Score: John Williams

Angelo My Love
Black Sunday
Cinderella Liberty
Close Encounters of the Third Kind
Conrack
The Cowboys
Daddy's Gone A-Hunting
The Deer Hunter
Diamond Head
Dracula
Earthquake
The Eiger Sanction
The Empire Strikes Back
Empire of the Sun
E.T.: The Extra-Terrestrial
Family Plot
Fiddler on the Roof
The Fury
Gidget Goes to Rome
Goodbye, Mr. Chips
A Guide for the Married Man
Heartbeeps
Hook
Hot Lead & Cold Feet
Indiana Jones and the Last Crusade
Indiana Jones and the Temple of Doom
Jaws
Jaws 2
Jaws 3
The Killers
The Long Goodbye
Lost in Space
Man Who Loved Cat Dancing
Midway
Missouri Breaks
Monsignor
1941
No Deposit, No Return
None But the Brave
The Paper Chase
Pete 'n' Tillie
The Poseidon Adventure
Psychopath
Raiders of the Lost Ark
The Rare Breed
The Reivers
Return of the Jedi
The River
Sergeant Ryker
Star Wars
The Sugarland Express
Superman: The Movie
Superman 2
Superman 3
Superman 4: The Quest for Peace
The Swarm
Tom Sawyer
Towering Inferno
Yes, Giorgio

Producers: Andy Warhol

Andy Warhol's Bad
Cocaine Cowboys
Flesh
Heat
Trash

Producers: Coppola/American Zoetrope

American Graffiti

The Black Stallion
The Conversation
Escape Artist
The Godfather: Part 2
Hammett

Producers: George Lucas

The Empire Strikes Back
The Ewok Adventure
Raiders of the Lost Ark
Return of the Jedi

Producers: Robert Altman

Buffalo Bill & the Indians
James Dean Story
The Late Show
Nashville
Quintet
Rich Kids
A Wedding
Welcome to L.A.

Producers: Roger Corman/New World

Atlas
Attack of the Giant Leeches
Avalanche
Battle Beyond the Stars
Battle Beyond the Sun
Beyond the Call of Duty
Big Bad Mama
Big Bird Cage
The Big Doll House
Bloodfist
Bloodfist 2
Bloody Mama
Body Waves
Boxcar Bertha
A Bucket of Blood
Caged Heat
Candy Stripe Nurses
Carnival Rock
Cocaine Wars
Cockfighter
Daddy's Boys
Death Race 2000
Death Sport
Dementia 13
The Dunwich Horror
Eat My Dust
The Fall of the House of Usher
Forbidden World
Frankenstein Unbound
Galaxy of Terror
Gas-s-s-s!
Grand Theft Auto
The Gunslinger
The Haunted Palace
Humanoids from the Deep
I, Mobster
I Never Promised You a Rose Garden
In the Heat of Passion
It Conquered the World
Jackson County Jail
Lady in Red
The Last Woman on Earth
Little Shop of Horrors
Lords of the Deep
Lumiere
Machine Gun Kelly
Masque of the Red Death
Monster from the Ocean Floor
Munchies
Night of the Blood Beast
Piranha
The Pit and the Pendulum
Premature Burial
Primary Target

Private Duty Nurses
Raiders of the Sun
The Raven
Rock 'n' Roll High School Forever
The St. Valentine's Day Massacre
Ski Troop Attack
Small Change
Smokey Bites the Dust
The Student Nurses
T-Bird Gang
Tales of Terror
Targets
Teenage Caveman
The Terror
Thunder and Lightning
Time Trackers
Tomb of Ligeia
Too Hot to Handle
The Trip
The Undead
Unholy Rollers
Voyage to the Planet of Prehistoric Women
The Wasp Woman
The Wild Angels
The Wild Ride
X: The Man with X-Ray Eyes
The Young Nurses

Producers: Steven Spielberg

Amazing Stories, Book 1
Amazing Stories, Book 2
Amazing Stories, Book 3
Amazing Stories, Book 4
Amazing Stories, Book 5
An American Tail
An American Tail: Fievel Goes West
Back to the Future
The Color Purple
Empire of the Sun
The Goonies
Gremlins
Honey, I Shrunk the Kids
Innerspace
Poltergeist
Twilight Zone: The Movie
Who Framed Roger Rabbit?
Young Sherlock Holmes

Producers: Val Lewton

Bedlam
Cat People
Isle of the Dead

Producers: William Castle

Bug
House on Haunted Hill
Macabre
Riot
Rosemary's Baby
Strait-Jacket
13 Ghosts
Zotz!

"Road Warrior" & Rip-Offs

City Limits
Dead Man Walking
Driving Force
Dune Warriors
Endgame
Escape from Safehaven

Exterminators of the Year 3000
Firefight
Future Hunters
Future Kill
The Killing Edge
Never Too Young to Die
Phoenix the Warrior
Radioactive Dreams
The Road Warrior
Solarbabies
Steel Dawn
Survival Zone
2020 Texas Gladiators
Warriors of the Apocalypse
Warriors of the Wasteland
Wheels of Fire
World Gone Wild

Special FX Extravaganzas

The Abyss
Aliens
Batman
Black Hole
The Ewok Adventure
The Exorcist
Exorcist 2: The Heretic
Fantastic Voyage
Lifeforce
Return of the Jedi
Star Trek 2: The Wrath of Khan
Star Trek 4: The Voyage Home
Star Wars
Superman 2
The Terminator
Terminator 2: Judgement Day
The Thing
Total Recall
2001: A Space Odyssey
2010: The Year We Make Contact
Who Framed Roger Rabbit?
Wolfen

Special FX Extravaganzas: Make-Up

An American Werewolf in London
Battle for the Planet of the Apes
Beneath the Planet of the Apes
Conquest of the Planet of the Apes
Dawn of the Dead
Day of the Dead
Escape from the Planet of the Apes
The Exorcist
The Fly
The Fly 2
The Howling
Planet of the Apes
Scanners
The Thing

Special FX Wizards: Anton Furst

Alien
Batman
The Company of Wolves
Full Metal Jacket
Moonraker
Star Wars

Special FX Wizards: Dick Smith

Altered States
Amadeus
The Exorcist
Ghost Story
The Godfather
House of Dark Shadows
Little Big Man
Marathon Man
Midnight Cowboy
Scanners
The Sentinel
Spasms
The Sunshine Boys
Taxi Driver
World of Henry Orient

Special FX Wizards: Douglas Trumball

The Andromeda Strain
Blade Runner
Brainstorm
Close Encounters of the Third Kind
Silent Running
Star Trek: The Motion Picture
2001: A Space Odyssey

Special FX Wizards: Herschell Gordon Lewis

Blood Feast
Color Me Blood Red
Gruesome Twosome
Just for the Hell of It
The Psychic
She-Devils on Wheels
Suburban Roulette
This Stuff'll Kill Ya!
2000 Maniacs
The Wizard of Gore

Special FX Wizards: Ray Harryhausen

The Beast from 20,000 Fathoms
Clash of the Titans
Earth vs. the Flying Saucers
First Men in the Moon
Golden Voyage of Sinbad
It Came from Beneath the Sea
Jason and the Argonauts
Mighty Joe Young
Mysterious Island
The Seventh Voyage of Sinbad
Sinbad and the Eye of the Tiger
The Three Worlds of Gulliver
The Valley of Gwangi

Special FX Wizards: Rick Baker

An American Werewolf in London
Food of the Gods
The Fury
Greystoke: The Legend of Tarzan, Lord of the Apes
Incredible Melting Man
It's Alive
King Kong
Octaman
Squirm
Star Wars
Tanya's Island
Track of the Moonbeast

Special

Special FX Wizards: Rob Bottin

The Fog
The Howling
Humanoids from the Deep
RoboCop
Tanya's Island
The Thing
Twilight Zone: The Movie

Special FX Wizards: Tom Savini

Creepshow
Dawn of the Dead
Day of the Dead
Deathdream
Deranged
Friday the 13th
Friday the 13th, Part 4: The Final Chapter
Knightriders
Maniac
Martin
Midnight
Monkey Shines
The Ripper
Texas Chainsaw Massacre Part 2

Top Grossing Films of 1939

Gone with the Wind
The Hunchback of Notre Dame
Jesse James
Mr. Smith Goes to Washington
The Wizard of Oz

Top Grossing Films of 1940

Boom Town
Fantasia
Pinocchio
Rebecca
Santa Fe Trail

Top Grossing Films of 1941

Honky Tonk
Philadelphia Story
Sergeant York
A Yank in the R.A.F.

Top Grossing Films of 1942

Bambi
Casablanca
Mrs. Miniver
Random Harvest
Yankee Doodle Dandy

Top Grossing Films of 1943

The Outlaw
The Song of Bernadette
Stage Door Canteen
This is the Army

Top Grossing Films of 1944

Going My Way
Meet Me in St. Louis

Since You Went Away
Thirty Seconds Over Tokyo

Top Grossing Films of 1945

Anchors Aweigh
The Bells of St. Mary's
Spellbound

Top Grossing Films of 1946

The Best Years of Our Lives
Duel in the Sun
The Jolson Story

Top Grossing Films of 1947

Egg and I
Life with Father

Top Grossing Films of 1948

Johnny Belinda
On an Island with You
The Paleface
Red River
The Red Shoes
Three Musketeers

Top Grossing Films of 1949

Battleground
Jolson Sings Again
Samson and Delilah
Sands of Iwo Jima

Top Grossing Films of 1950

Cinderella
Father of the Bride
King Solomon's Mines

Top Grossing Films of 1951

Alice in Wonderland
David and Bathsheba
The Great Caruso
Quo Vadis
Show Boat

Top Grossing Films of 1952

Greatest Show on Earth
Hans Christian Andersen
Ivanhoe
The Snows of Kilimanjaro

Top Grossing Films of 1953

From Here to Eternity
How to Marry a Millionaire
Peter Pan
The Robe
Shane

Top Grossing Films of 1954

The Caine Mutiny
The Glenn Miller Story
Rear Window

20,000 Leagues Under the Sea
White Christmas

Top Grossing Films of 1955

Battle Cry
Lady and the Tramp
Mister Roberts
Oklahoma!

Top Grossing Films of 1956

Around the World in 80 Days
Giant
The King and I
The Ten Commandments

Top Grossing Films of 1957

The Bridge on the River Kwai
Old Yeller
Peyton Place
Raintree County
Sayonara

Top Grossing Films of 1958

Auntie Mame
Cat on a Hot Tin Roof
Gigi
No Time for Sergeants
South Pacific

Top Grossing Films of 1959

Ben-Hur
Darby O'Gill & the Little People
Operation Petticoat
Shaggy Dog
Sleeping Beauty

Top Grossing Films of 1960

The Alamo
Exodus
Psycho
Spartacus
Swiss Family Robinson

Top Grossing Films of 1961

The Absent-Minded Professor
El Cid
The Guns of Navarone
West Side Story

Top Grossing Films of 1962

How the West was Won
In Search of the Castaways
Lawrence of Arabia
The Longest Day
The Music Man

Top Grossing Films of 1963

Cleopatra
Irma La Douce

It's a Mad, Mad, Mad, Mad World
Sword in the Stone
Tom Jones

Top Grossing Films of 1964

The Carpetbaggers
From Russia with Love
Goldfinger
Mary Poppins
My Fair Lady

Top Grossing Films of 1965

Doctor Zhivago
The Sound of Music
That Darn Cat
Those Magnificent Men in Their Flying Machines
Thunderball

Top Grossing Films of 1966

Bible...In the Beginning
Hawaii
Lt. Robin Crusoe, U.S.N.
A Man for All Seasons
Who's Afraid of Virginia Woolf?

Top Grossing Films of 1967

Bonnie & Clyde
The Dirty Dozen
The Graduate
Guess Who's Coming to Dinner
The Jungle Book

Top Grossing Films of 1968

Bullitt
Funny Girl
The Odd Couple
Romeo and Juliet
2001: A Space Odyssey

Top Grossing Films of 1969

Butch Cassidy and the Sundance Kid
Easy Rider
Hello, Dolly!
The Love Bug
Midnight Cowboy

Top Grossing Films of 1970

Airport
Love Story
M*A*S*H
Patton

Top Grossing Films of 1971

Billy Jack
Diamonds are Forever
Fiddler on the Roof
The French Connection
Summer of '42

Top Grossing Films of 1972

Deliverance
The Godfather
Jeremiah Johnson
The Poseidon Adventure
What's Up, Doc?

Top Grossing Films of 1973

American Graffiti
The Exorcist
Papillon
The Sting
The Way We Were

Top Grossing Films of 1974

Blazing Saddles
Earthquake
Towering Inferno
Trial of Billy Jack
Young Frankenstein

Top Grossing Films of 1975

Dog Day Afternoon
Jaws
One Flew Over the Cuckoo's Nest

Top Grossing Films of 1976

All the President's Men
King Kong
Rocky
The Silver Streak
A Star is Born

Top Grossing Films of 1977

Close Encounters of the Third Kind
Goodbye Girl
Saturday Night Fever
Smokey and the Bandit
Star Wars

Top Grossing Films of 1978

Every Which Way But Loose
Grease
Jaws 2
National Lampoon's Animal House
Superman: The Movie

Top Grossing Films of 1979

Alien
The Jerk
Kramer vs. Kramer
Rocky 2
Star Trek: The Motion Picture

Top Grossing Films of 1980

Airplane!
Any Which Way You Can
The Empire Strikes Back
9 to 5
Stir Crazy

Woofs

Spare Parts
Sphinx
Splatter University
Splitz
Spookies
The Sporting Club
Spring Break
Starlet
Stepsisters
Stewardess School
Stocks and Blondes
Stone Cold
Stop! or My Mom Will Shoot
Stormquest
The Strangeness
The Stranger from Venus
Street Hero
Street Hunter
Street Law
Street People
Street Trash
Street Warriors
Street Warriors, Part 2
Strike Commando
Stroker Ace
Stuck on You
The Stud
Student Affairs
Sugar Cookies
Suicide Cult
Summer Job
Summer School Teachers
Sunnyside
Sunset Strip
Superchick
Supersonic Man
Surf 2
Surfacing
Survival Zone
Sweater Girls
Sweet Spirits
Sweet Trash
Switchblade Sisters
Symphony of Living
Takin' It Off
A Taste of Hell
Teenage Mother
Teenage Seductress
Teenage Strangler
Teenage Zombies
Teenagers from Outer Space
The Telephone
10 Violent Women
Terronauts
Terror in Beverly Hills
Terror in the Wax Museum
Terrorvision
Test Tube Babies
Texas Chainsaw Massacre
   Part 2
Texas Layover
They Saved Hitler's Brain
They Went That-a-Way &
   That-a-Way
They're Playing with Fire
Thieves of Fortune
The Thirsty Dead
This Stuff'll Kill Ya!
Thousand Mile Escort
Three on a Meathook
The Thrill Killers
Throne of Fire
Tintorera...Tiger Shark
TNT Jackson
To All a Goodnight
Toga Party
Tomboy
The Tormentors
Torso
Traveling Companions
Treasure of the Moon
   Goddess
Trial of Billy Jack
Trick or Treats
Troma's War
Tropic Heat

Twice Dead
Twilight People
Two Lost Worlds
2020 Texas Gladiators
Two Women in Gold
Under the Doctor
Undercover Vixens
Underground Aces
Unholy Rollers
The Uninvited
Unknown Island
Unknown Powers
Up Your Anchor
Vampire Raiders - Ninja
   Queen
Vasectomy: A Delicate Matter
Vengeance Is Mine
Venus on Fire
Vice Academy 3
The Video Dead
Video Vixens
Video Wars
Vigilante
Violated
Violent Ones
The Violent Years
Virgins of Purity House
Viva Knievel
Vulcan God of Fire
Waitress
Wanted: Babysitter
Warlords from Hell
Warrior of the Lost World
Warriors of the Apocalypse
Warriors of the Wasteland
Weekend with the Babysitter
Werewolves on Wheels
What Would Your Mother
   Say?
When Lightning Strikes
When Nature Calls
White Cannibal Queen
White Gorilla
White Pongo
The Wild Beasts
Wild Riders
Wild Wheels
The Wildest Office Party
Wired to Kill
Witchcraft
Woman Hunt
Women in Cell Block 7
Women in Fury
Women of the Prehistoric
   Planet
Women's Prison Massacre
Woodchipper Massacre
The World is Full of Married
   Men
The Worm Eaters
The Worst of Hollywood: Vol.
   1
The Worst of Hollywood: Vol.
   2
The Worst of Hollywood: Vol.
   3
Wrestling Racket Girls
Wrestling Women vs. the
   Aztec Ape
Wrestling Women vs. the
   Aztec Mummy
Xanadu
Xtro
The Yesterday Machine
Zandalee
Zero Boys
Zero Pilot
Zombie Nightmare
Zombies of Moratau

CAST & DIRECTOR INDEX

The **Cast & Director Index** includes 30,000 classifications by cast member or director in a straight alphabetical format by last name. More than 100,000 credits are cited. Every cast member and director credited in the main review section is indexed here, creating an intriguing array of movie-making lists and a sure-fire locator for those hard-to-place videos.

CAST & DIRECTOR INDEX

Beverly Aadland
Assault of the Rebel Girls '59
Raiders of Sunset Pass '43
South Pacific '58

Lee Aaker
Courage of Rin Tin Tin '83
Rin Tin Tin, Hero of the West '55

Angela Aames
Bachelor Party '84
Basic Training '86
The Lost Empire '83

Willie Aames
Cut and Run '85
Frankenstein '73
Paradise '82
Scavenger Hunt '79
Zapped! '82

Caroline Aaron
Crimes & Misdemeanors '89
Heartburn '86

Paul Aaron *(D)*
Deadly Force '83
Different Story '78
Force of One '79
Imperial Navy '80s
In Love and War '91
Maxie '85
The Miracle Worker '79

Judith Abarbanel
The Cantor's Son (Dem Khann's Zindl) '37

Angelo Abazoglou
Toute Une Nuit '82

Shelly Abblett
Blood Games '90

Bruce Abbott
Bad Dreams '88
Bride of Re-Animator '89
Interzone '88
The Re-Animator '85
Summer Heat '87
Trapped '89

Bud Abbott
Abbott and Costello in Hollywood '45
Abbott and Costello Meet Captain Kidd '52
Abbott and Costello Meet Dr. Jekyll and Mr. Hyde '52
Abbott and Costello Meet Frankenstein '48

Abbott & Costello Meet the Invisible Man '51
Abbott and Costello Meet the Killer, Boris Karloff '49
Abbott and Costello Scrapbook '70s
Africa Screams '49
The Best of Abbott & Costello Live '54
Buck Privates '41
Buck Privates Come Home '47
Classic Comedy Video Sampler '49
Dance with Me, Henry '56
Entertaining the Troops '89
Hey Abbott! '78
Hit the Ice '43
Hold That Ghost '41
Hollywood Clowns '85
Hollywood Goes to War '54
In the Navy '41
Jack & the Beanstalk '52
Keep 'Em Flying '41
Mexican Hayride '48
The Naughty Nineties '45
Pardon My Sarong '42
Ride 'Em Cowboy '42
Super Bloopers #1 '70s
The Time of Their Lives '46
Who Done It? '42
The Wistful Widow of Wagon Gap '47

Charles Abbott *(D)*
The Adventures of the Masked Phantom '38

Diahnne Abbott
Jo Jo Dancer, Your Life is Calling '86
King of Comedy '82
Love Streams '84
New York, New York '77

George Abbott *(D)*
Damn Yankees '58
Pajama Game '57
Too Many Girls '40

John Abbott
Adventure Island '47
Deception '46
The Falcon in Hollywood '44
Humoresque '47
Pursuit to Algiers '45
The Vampire's Ghost '45

Mike Abbott
Death Code Ninja '80s
Hands of Death '88

Phillip Abbott
The Fantastic World of D.C. Collins '84

Sparky Abbrams
Bottoms Up '87

Kareem Abdul-Jabbar
Airplane! '80
Chuck Berry: Hail! Hail! Rock 'N' Roll '87
The Fish that Saved Pittsburgh '79
Game of Death '79
Pee-Wee's Playhouse Christmas Special '88
Purple People Eater '88

Paula Abdul
Junior High School '81

Hakeem Abdul-Samad
Ernest Goes to Camp '87

Naoyuki Abe
Gamera vs. Gaos '67

Alan Abel
Is There Sex After Death? '71

Alan Abel *(D)*
Is There Sex After Death? '71

Alfred Abel
Dr. Mabuse the Gambler '22

Jeanne Abel *(D)*
Is There Sex After Death? '71

Robert Abel *(D)*
Elvis on Tour '72

Walter Abel
Curley '47
Fabulous Joe '47
Grace Quigley '84
Holiday Inn '42
Kid from Brooklyn '46
Law of the Underworld '38
Mirage '66
Mr. Skeffington '44
Prudential Family Playhouse '50
Quick, Let's Get Married '71
Raintree County '57
Silent Night, Bloody Night '73
13 Rue Madeleine '46
Three Musketeers '35
Wake Island '42

Jim Abele
Student Affairs '88
Wimps '87

Ridgely Abele
Master's Revenge '71

Jordan Abeles
South Bronx Heroes '85

Ian Abercrombie
Catacombs '89
Kicks '85
Puppet Master 3 '91

Keith Aberdein
Smash Palace '82
Wild Horses '82

Sivi Aberg
Dr. Death, Seeker of Souls '73

Lewis Abernathy *(D)*
House 4 '91

Ric Abernathy
The Squeeze '87

Michael Aberne
The Commitments '91

Abigail
Adventures of Eliza Fraser '76
Breaking Loose '90

William Abney
Flight from Singapore '62

F. Murray Abraham
All the President's Men '76
Amadeus '84
Beyond the Stars '89
The Big Fix '78
An Innocent Man '89
Intimate Power '89
Madman '79
Mobsters '91
The Name of the Rose '86
Prisoner of Second Avenue '74
The Ritz '76
Scarface '83
Serpico '73
Slipstream '89
The Sunshine Boys '75
They Might Be Giants '71
Third Solution '89

Ken Abraham
Creepozoids '87
Deadly Embrace '88

Marked for Murder '89
Vice Academy '88

Jim Abrahams
Kentucky Fried Movie '77

Jim Abrahams *(D)*
Airplane! '80
Big Business '88
Hot Shots! '91
Police Squad! Help Wanted! '82
Ruthless People '86
Top Secret! '84
Welcome Home, Roxy Carmichael '90

Klaus Abramowsky
Europa, Europa '91

Andrei Abrikosov
Sword & the Dragon '56

Al Abrikossov
Alexander Nevsky '38

Victoria Abril
After Darkness '85
Caged Heart '85
Esposa y Amante '70s
High Heels '91
L'Addition '85
Moon in the Gutter '84
On the Line '83
Tie Me Up! Tie Me Down! '90

Tenghiz Abuladze *(D)*
Repentance '87

Jay Acavone
Doctor Mordrid: Master of the Unknown '92

Goodman Ace
Two Reelers: Comedy Classics 6 '35

Anna Achdian
A Wedding in Galilee '87

Cristina Ache
Amor Bandido '79

Jovan Acin *(D)*
Hey, Babu Riba '88

Sharon Acker
The Hanged Man '74
Off Your Rocker '80
The Stranger '73

Georges Adel

Love and Death '75

Jan Adele

High Tide '87

Pierre Adidge (D)

Elvis on Tour '72
Mad Dogs & Englishmen '71

Vern Adix

Teen Alien '88

Isabelle Adjani

Camille Claudel '89
The Deadly Summer '83
The Driver '78
Ishtar '87
Next Year If All Goes Well '83
One Deadly Summer (L'Ete Meurtrier) '83
Possession '81
Quartet '81
The Slap '76
Story of Adele H. '75
Subway '85
The Tenant '76

Bill Adler

Pom Pom Girls '76
Van Nuys Blvd. '79

Celia Adler

Where is My Child? '37

Ethan Adler

HauntedWeen '91

Jay Adler

Grave of the Vampire '72

Joseph Adler (D)

Scream, Baby, Scream '69
Sex and the College Girl '64

Lou Adler (D)

Cheech and Chong's Up in Smoke '79

Luther Adler

Absence of Malice '81
The Brotherhood '68
Cornered '45
Crashout '55
The Desert Fox '51
D.O.A. '49
Hoodlum Empire '52
House of Strangers '49
Kiss Tomorrow Goodbye '50
The Last Angry Man '59
The Loves of Carmen '48
Mean Johnny Barrows '75
Murph the Surf '75
Voyage of the Damned '76
Wake of the Red Witch '49

Matt Adler

Diving In '90
Doin' Time on Planet Earth '88
Flight of the Navigator '86
North Shore '87
White Water Summer '87

Stella Adler

My Girl Tisa '48

Percy Adlon (D)

Bagdad Cafe '88
Celeste '81
Rosalie Goes Shopping '89
Sugarbaby '85

Ed Adlum (D)

Invasion of the Blood Farmers '72

Hank Adly

Boarding House '83

Edvin Adolphson

Boy of Two Worlds '70
Dollar '38
Only One Night '42

Renee Adoree

The Big Parade '25
The Michigan Kid '28

Mario Adorf

Hired to Kill '73
Hit Men '73
The Holcroft Covenant '85
Invitation au Voyage '83
The Italian Connection '73
The Lost Honor of Katharina Blum '75
Manhunt '73
Paralyzed '70s
The Tin Drum '79

Iris Adrian

Blue Hawaii '62
The Fast and the Furious '54
G.I. Jane '51
I Killed That Man '42
I'm from Arkansas '44
Infamous Crimes '47
The Lovable Cheat '49
Million Dollar Kid '44
The Paleface '48
Road to Zanzibar '41
Shake Hands with Murder '44
Stop That Cab '51
Varieties on Parade '52

Max Adrian

The Boy Friend '71
The Devils '71

Michael Adrian

The Life & Times of the Chocolate Killer '88

Patricia Adriani

I'm the One You're Looking For '88
The Nest '81

Frank Adu

Love and Death '75

Sade Adu

Absolute Beginners '86

Ljubica Adzovic

Time of the Gypsies '90

John Agar

Along the Great Divide '51
The Brain from Planet Arous '57
Cavalry Command '63
Curse of the Swamp Creature '66
The Daughter of Dr. Jekyll '57
Fear '90
Flesh and the Spur '57
Fort Apache '48
Hell Raiders '68
Jet Attack/Paratroop Command '58
Miracle Mile '89
Sands of Iwo Jima '49
She Wore a Yellow Ribbon '49

Zontar, the Thing from Venus '66

Tetchie Agbayani

The Dolls '83
Gymkata '85
Rikky and Pete '88

Suzanne Ager

Evil Toons '90

Robert Agnew

The Taxi Mystery '26

Carlos Agosti

El Zorro Blanco '87
Invasion of the Vampires '61

Olga Agostini

El Callao (The Silent One) '70s

Pierre Agostino

Hollywood Strangler '82
The Hollywood Strangler Meets the Skid Row Slasher '79
Las Vegas Serial Killer '86

Frank Agrama (D)

Dawn of the Mummy '82

Joseph Louis Agraz (D)

Treasure of the Moon Goddess '88

Jose Miguel Agrelot

La Criada Malcriada '60s

Janet Agren

Aladdin '86
Emerald Jungle '80
Gates of Hell '83
Hands of Steel '86
Magdalene '88
Night of the Sharks '89
Panic '83
Perfect Crime '78
The Uranium Conspiracy '78

David Agresta

Best of the Best '89

Antonio Aguilar

Benjamin Argumedo '78
El Rey '87
Emiliano Zapata '87
Simon Blanco '87
The Undefeated '69
Volver, Volver, Volver! '87

Kris Aguilar

Bloodfist '89

Luis Aguilar

Duelo en el Dorado
El Fugitivo '65
El Zorro Vengador '47
La Trampa Mortal
Los Tres Amores de Losa '47
Stallion '88

Luz Maria Aguilar

Soy un Golfo '65

Adriana Aguirre

Gran Valor en la Facultad de Medicina '80s

Beatriz Aguirre

Mariachi '65

Hilda Aguirre

Diamante, Oro y Amor '87
El Zorro Blanco '87
Stallion '88

Javier Aguirre (D)

Dracula's Great Love '72
The Rue Morgue Massacres '73

Agust Agustsson (D)

Maya '82

Jenny Agutter

An American Werewolf in London '81
Amy '81
Child's Play 2 '90
Dark Tower '87
Darkman '90
Dominique is Dead '79
The Eagle Has Landed '77
Equus '77
Gunfire '78
Logan's Run '76
Man in the Iron Mask '77
Railway Children '70
The Riddle of the Sands '84
Secret Places '85
Silas Marner '85
Survivor '80
Sweet William '79

Charlie Ahearn (D)

Twins '80
Wild Style '83

Brian Aherne

Beloved Enemy '36
A Bullet Is Waiting '54
Forever and a Day '43
I Confess '53
I Live My Life '35
The Lady in Question '40
A Night to Remember '42
Prince Valiant '54
Smilin' Through '41
The Swan '56
Sword of Lancelot '63
Sylvia Scarlett '35
Waltz King '63

Bjorje Ahistedt

Emma's Shadow '88

Mac Ahlberg (D)

Gangsters '79

Arthur Ahmbling

Greek Street '30

Philip Ahn

Back to Bataan '45
Betrayal from the East '44
China Sky '44
The General Died at Dawn '36
Hawaii Calls '38
His Majesty O'Keefe '53
Paradise, Hawaiian Style '66
Shock Corridor '63

Ahui

Robinson Crusoe & the Tiger '72

Kyoko Ai

Destroy All Monsters '68

Charles Aidman

Adam at 6 A.M. '70
The Barbary Coast '74
Countdown '68
Hour of the Gun '67

House of the Dead '80
The Invasion of Carol Enders '74
Menace on the Mountain '70
Picture of Dorian Gray '74
Prime Suspect '82
Zone of the Dead '78
Zoot Suit '81

Danny Aiello

Alone in the Neon Jungle '87
Bang the Drum Slowly '73
Bloodbrothers '78
Chu Chu and the Philly Flash '81
The Closer '91
Deathmask '69
Defiance '79
Do the Right Thing '89
Fingers '78
Fort Apache, the Bronx '81
The Front '76
The Godfather: Part 2 '74
Harlem Nights '89
Hide in Plain Sight '80
Hooch '76
Hudson Hawk '91
Jacob's Ladder '90
The January Man '89
Key Exchange '85
The Lost Idol/Shock Troop '80s
Man on Fire '87
Moonstruck '87
Old Enough '84
Once Around '91
Once Upon a Time in America '84
The Pick-Up Artist '87
Protector '85
Purple Rose of Cairo '85
Question of Honor '80
Radio Days '87
Ruby '92
The Stuff '85
Third Solution '89
29 Street '91
White Hot '88

Rick Aiello

The Closer '91
29 Street '91

Elaine Aiken

Lonely Man '57

Anouk Aimee

8 1/2 '63
Justine '69
La Dolce Vita '60
Lola '61
A Man and a Woman '66
A Man and a Woman: 20 Years Later '86
Paris Express '53
Sodom and Gomorrah '62
Success Is the Best Revenge '85
Tragedy of a Ridiculous Man '81

Henry Ainley

As You Like It '36

Richard Ainley

I Dood It '43

Mary Ainslee

Mad Youth '40

Holly Aird

Over Indulgence '87

Jane Aird

Death Goes to School '53

Christina Airoldi

Next Victim '71

Maria Aitken

A Fish Called Wanda '88

Spottiswoode Aitken

The Americano '17

Michael Aitkens

The Highest Honor '84
Moving Targets '87

Franklin Ajaye

Car Wash '76
Convoy '78
Fraternity Vacation '85
Jazz Singer '80
Jock Jokes '89
The Wrong Guys '88

Tarik Akan

Yol '82

John Akana

Nasty Rabbit '64

Chantal Akerman

Akermania, Vol. 1 '92

Chantal Akerman (D)

Akermania, Vol. 1 '92
je tu il elle '74
Les Rendez-vous D'Anna '78
News from Home '76
Toute Une Nuit '82

Andra Akers

Desert Hearts '86
E. Nick: A Legend in His Own Mind '84

Dona Akersten

Sleeping Dogs '82

Takejo Aki

The Ballad of Narayama '83

Boris Akimov

Ivan the Terrible '89

Claude Akins

Battle for the Planet of the Apes '73
The Big Push '75
The Caine Mutiny '54
The Concrete Cowboys '79
The Curse '87
The Death Squad '73
Eric '75
Falling from Grace '92
From Here to Eternity '53
Killer on Board '77
The Killers
A Man Called Sledge '71
Manhunt for Claude Dallas '86
Monster in the Closet '86
Ramblin' Man '80s
Return of the Seven '66
Rio Bravo '59
Sea Chase '55
Tarantulas: The Deadly Cargo '77
Waterhole #3 '67

Miyuki Akiyama

Gamera vs. Guiron '69

Moustapha Akkad (D)

Lion of the Desert '81
Mohammed: Messenger of God '77

Akkemay

Army Brats '84

Nazih Akleh

A Wedding in Galilee '87

Hiroshi Akutagawa

Mistress (Wild Geese) '53

Tomoko Al

Terror of Mechagodzilla '78

Sheeba Alahani

The Barbarians '87

Marc Alaimo

Archer: The Fugitive from the Empire '81
Arena '88
Tango and Cash '89

Steve Alaimo

Alligator Alley '72
The Hooked Generation '69
Wild Rebels '71

Gina Alajar

Fight for Us '89

Joe Alaskey

Lucky Stiff '88
Paramount Comedy Theater, Vol. 2: Decent Exposures '87

Benito Alazraki (D)

Spiritism '61

Maria Alba

Chandu on the Magic Island '34
Mr. Robinson Crusoe '32
Return of Chandu '34

Rafael Morena Alba (D)

House of Insane Women '74

Rose Alba

School for Sex '69

Captain Lou Albano

Body Slam '87
Life & Times of Captain Lou Albano '85
Wise Guys '86

Jose Albar

El Callao (The Silent One) '70s

John Albasiny

Kipperbang '82

Josh Albee

The Adventures of Tom Sawyer '73
The Runaways '75

Anna Maria Alberghetti

Cinderfella '60
Here Comes the Groom '51
The Medium '51

Luis Alberni

Dancing Pirate '36
Goodbye Love '34
The Great Man Votes '38
Harvest Melody '43
Hats Off '37
One Night of Love '34
The Sphinx '33

Sherry Alberoni

Barn of the Naked Dead '73

Hans Albers

Baron Munchausen '43
Die Grosse Freiheit Nr. 7 '45
F.P. 1 Doesn't Answer '33

Chava Alberstein

Intimate Story '81

Annie Albert

Tomb of Torture '65

Eddie Albert

The Act '82
Actors and Sin '52
Beulah Land '80
The Big Picture '89
The Birch Interval '78
Bombardier '43
Burning Rage '84
Captain Newman, M.D. '63
Carrie '52
The Concorde: Airport '79 '79
The Crash of Flight 401 '78
Devil's Rain '75
Dreamscape '84
Escape to Witch Mountain '75
Foolin' Around '80
Goliath Awaits '81
Head Office '86
The Heartbreak Kid '72
How to Beat the High Cost of Living '80
I'll Cry Tomorrow '55
The Longest Day '62
The Longest Yard '74
McQ '74
Miracle of the White Stallions '63
Moving Violation '76
Oklahoma! '55
Out of Sight, Out of Her Mind '89
Return of Video Yesterbloop '47
Roman Holiday '53
Smash-Up: The Story of a Woman '47
Stitches '85
Take This Job & Shove It '81
The Teahouse of the August Moon '56
Turnaround '87
The Victory '88
Whiffs '75
The Word '78
Yes, Giorgio '82

Edward Albert

Accidents '89
Butterflies Are Free '72
Butterfly '82
Death Cruise '74
Distortions '87
The Domino Principle '77
Ellie '84
Fist Fighter '88
The Fool Killer '65
Forty Carats '73
Galaxy of Terror '81
Getting Even '86
The Greek Tycoon '78
The Heist '88

House Where Evil Dwells '82
Mind Games '89
Night School '81
The Purple Taxi '77
The Rescue '88
The Rip Off '78
Silent Victory: The Kitty O'Neil Story '79
The Squeeze '80
Terminal Entry '87
Time to Die '83
The Underachievers '88
When Time Ran Out '80
Wild Zone '89

Jerry Albert

Bloodstalkers '76

Laura Albert

Blood Games '90
Death by Dialogue '88
Dr. Caligari '89

Maxine Albert

Home Remedy '88

Giorgia Albertazzi

Last Year at Marienbad '61

Guido Alberti

Casanova '70 '65
Ten Days Wonder '72

Coit Albertson

The Return of Boston Blackie '27

Frank Albertson

Hollywood Mystery '34
Killer Dill '47
Nightfall '56
Room Service '38
Way Back Home '32

Jack Albertson

Changes '69
Charlie and the Great Balloon Chase '82
Days of Wine and Roses '62
Dead and Buried '81
Flim-Flam Man '67
How to Murder Your Wife '64
Justine '69
Kissin' Cousins '63
Lover Come Back '61
Marriage is Alive and Well '80
Miracle on 34th Street '47
The Patsy '64
The Poseidon Adventure '72
Roustabout '64
Where the Eagle Flies '72
Willy Wonka & the Chocolate Factory '71

Elsy Albiin

Terror Street '54

Hans Albin (D)

No Survivors, Please '63

Marcy Albrecht

Hollywood High '77

Carlton J. Albright (D)

Luther the Geek '90

Hardie Albright

Champagne for Breakfast '79
Mom & Dad '47
The Scarlet Letter '34
Sing Sing Nights '35

Lola Albright

Champion '49
Joy House '64
Kid Galahad '62
The Monolith Monsters '57
The Tender Trap '55
The Way West '67

Wally Albright

Vengeance '37

Chris Alcaide

Jupiter's Darling '55

Angel Alcazar

I'm the One You're Looking For '88

Alan Alda

Betsy's Wedding '90
California Suite '78
Crimes & Misdemeanors '89
The Four Seasons '81
Glass House '72
Gone are the Days '63
Jenny '70
Lily Tomlin '80s
M*A*S*H: Goodbye, Farewell & Amen '83
The Mephisto Waltz '71
A New Life '88
Paper Lion '68
Playmates '72
Purlie Victorious '63
Same Time, Next Year '78
The Seduction of Joe Tynan '79
Sweet Liberty '86
To Kill a Clown '72
Truman Capote's "The Glass House" '73

Alan Alda (D)

Betsy's Wedding '90
The Four Seasons '81
A New Life '88
Sweet Liberty '86

Antony Alda

Homeboy '88
Movie Maker '86

Robert Alda

Bittersweet Love '76
Cloak and Dagger '46
The Devil's Hand '61
Every Girl Should Have One '78
Express to Terror '79
The Heist '88
Hollywood Varieties '50
I Will, I Will for Now '76
Lisa and the Devil '75
Love by Appointment '76
Night Flight from Moscow '73
Revenge of the Barbarians '64
Rhapsody in Blue '45
The Rip Off '80
The Squeeze '80

Rutanya Alda

Amityville 2: The Possession '82
The Deer Hunter '78
Girls Night Out '83
Mommie Dearest '81
Prancer '89
Racing with the Moon '84
Vigilante '83

Luis Aldas

Una Vez en la Vida '49

Ginger Alden

Lady Grey '82

Norman Alden

The Patsy '64
Sword in the Stone '63

Priscilla Alden

Crazy Fat Ethel II '85
Criminally Insane '75
Death Nurse '87

Richard Alden

The Sadist '63

Terry Alden

Last Game '80

John Alderman

The Alpha Incident '76
Boob Tube '75
Little Miss Innocence '73
The Pink Angels '71
Starlet '87

Thomas Alderman (D)

The Severed Arm '73

John Alderton

Zardoz '73

Will Aldis (D)

Stealing Home '88

Lynda Aldon

Mankillers '87

Mari Aldon

Distant Drums '51
Race for Life '55

Tom Aldredge

Gentleman Bandit '81
O Pioneers! '91
What About Bob? '91

Adell Aldrich (D)

The Kid from Left Field '79

John Aldrich

The Wild Beasts '85

Rhonda Aldrich

Jailbird Rock '88

Robert Aldrich (D)

All the Marbles '81
Apache '54
Autumn Leaves '56
The Choirboys '77
The Dirty Dozen '67
The Flight of the Phoenix '66
Four for Texas '63
The Frisco Kid '79
The Grissom Gang '71
Hush, Hush, Sweet Charlotte '65
Hustle '75
The Killing of Sister George '69
Kiss Me Deadly '55
The Longest Yard '74
Sodom and Gomorrah '62
Too Late the Hero '70
Twilight's Last Gleaming '77
Ulzana's Raid '72
Vera Cruz '53
What Ever Happened to Baby Jane? '62

Kay Aldridge

Nyoka and the Tigermen '42
The Phantom of 42nd Street '45

Kitty Aldridge

African Dream '90
American Roulette '88
Slipstream '89

Michael Aldridge

Bullshot '83
Footlight Frenzy '84

Thomas Gutierrez Alea

The Last Supper '76

Thomas Gutierrez Alea (D)

Death of a Bureaucrat '66
Memories of Underdevelopment '68

Norma Aleandro

Cousins '89
Gaby: A True Story '87
The Official Story '85
One Man's War '90
Vital Signs '90

Jimmy Aleck

One Last Run '89
Paramount Comedy Theater, Vol. 4: Delivery Man '87

Miguel Alejandro

Popi '69

Dragoljub Aleksic

Innocence Unprotected '68

Julio Aleman

Diamante, Oro y Amor '87
The Green Wall '70
La Guerrillera de Villa (The Warrior of Villa) '70s
La Hora 24 '87
La Trinchera '87
Marcados por el Destino '87
Neutron and the Black Mask '61
Neutron vs. the Amazing Dr. Caronte '61
Neutron vs. the Death Robots '62
Patsy, Mi Amor '87

Andre Aleme

Carnival in Flanders (La Kermesse Heroique) '35

Vera Alentova

Moscow Does Not Believe in Tears '80

Aki Aleong

Braddock: Missing in Action 3 '88

Toni Alessandra

Bachelor Party '84

Goffredo Alessandrini (D)

We the Living '42

Barbara Lee Alexander

Hired to Kill '91

Ben Alexander

Criminals Within '41
Dragnet '54
Hearts of the World '18
Legion of Missing Men '37

Don Alexander

The Wizard of Gore '70

Douglas Alexander

Deadly Sunday '82

Elizabeth Alexander

Scalp Merchant '77

Frank Alexander

Bimini Code '84

Jane Alexander

All the President's Men '76
The Betsy '78
Brubaker '80
Calamity Jane '82
City Heat '84
Death Be Not Proud '75
Eight Men Out '88
Eleanor & Franklin '76
Glory '89
The Great White Hope '70
A Gunfight '71
In the Custody of Strangers '82
In Love and War '91
Kramer vs. Kramer '79
Lovey: A Circle of Children 2 '82
New Centurions '72
Night Crossing '81
Playing for Time '80
A Question of Love '78
The Rumor Mill '86
Square Dance '87
Sweet Country '87
Testament '83

Jason Alexander

I Don't Buy Kisses Anymore '92
Pretty Woman '89
White Palace '90

Jeff Alexander

Curse of the Swamp Creature '66
Twisted Brain '74

Joan Alexander

Superman: The Cartoons '42

John Alexander

Arsenic and Old Lace '44
Fancy Pants '50
The Horn Blows at Midnight '45
The Jolson Story '46
Mr. Skeffington '44
A Tree Grows in Brooklyn '45

Katherine Alexander

The Barretts of Wimpole Street '34
Dance, Girl, Dance '40
The Great Man Votes '38

Liz Alexander

Killing of Angel Street '81

Max Alexander (D)

Flaming Lead '39

Newell Alexander

Home Safe '88

Nick Alexander

Love with the Proper Stranger '63

Richard Alexander

The Mysterious Lady '28
Renfrew on the Great White Trail '38

Spike Alexander

Brain Donors '92

Suzanne Alexander

Cat Women of the Moon '53

Terence Alexander

Waterloo '71

Terry Alexander

Day of the Dead '85

Hope Alexander-Willis

The Pack '77

Tiana Alexandra

Catch the Heat '87

Grigori Alexandrov

The Battleship Potemkin '25

Grigori Alexandrov (D)

Que Viva Mexico '32
Ten Days That Shook the World/October '27

Anna Alexiades

Fantasies '73

Dennis Alexio

Kickboxer '89

Alvin Alexis

Night of the Demons '88

Graciela Alfano

Departamento Compartido '85
Fotografo de Senoras '85
Los Drogadictos '80s

Richard Alferi

Echoes '83

Lidia Alfonsi

Black Sabbath '64

Omar Alfonso

Death of a Bureaucrat '66

Chuck Alford

Commando Squad '76
Hollywood Strangler '82

Phillip Alford

To Kill a Mockingbird '62

Hans Alfredson

The Adventures of Picasso '80

James Algar (D)

The Adventures of Ichabod and Mr. Toad '49
Jungle Cat '59
The Legend of Sleepy Hollow '49

Living Desert '53
Secrets of Life '56
Vanishing Prairie '54

Sidney Algier (D)

Wild Horse '31

Muhammad Ali

Body & Soul '81
Doin' Time '85
Freedom Road '79
The Greatest '77

Veronica Porsche Ali

Terror on Alcatraz '86

Grant Alianak

One Night Only '84

Sofia Aliberti

The Enchantress '88

Mary Alice

Charlotte Forten's Mission: Experiment in Freedom '85
He Who Walks Alone '78
Sparkle '76
To Sleep with Anger '90
The Women of Brewster Place '89

Ana Alicia

Coward of the County '81
Romero '89

Marta Alicia

Mindwarp '91

Lisa Aliff

Damned River '89
Playroom '90

Carlo Alighiero

The Cat o' Nine Tails '71

Slobodan Aligrudic

The Love Affair, or Case of the Missing Switchboard Operator '67

Jeff Alin

American Tickler '76
A Matter of Love '78

Marcello Aliprandi (D)

Vatican Conspiracy '81

N. Alisova

The Lady with the Dog '59

Howard Alk (D)

Janis: A Film '74

Andrea Allan

The House that Vanished '73
Scream and Die '74

Elizabeth Allan

Ace of Aces '33
Camille '36
Donovan's Reef '63
The Haunted Strangler '58
Java Head '35
Mark of the Vampire '35
Phantom Fiend '35
The Shadow '36
Star Spangled Girl '71
A Tale of Two Cities '35
A Woman Rebels '36

Hugh Allan

Annapolis '28

Jed Allan

Man from Clover Grove '78

Lloyd Allan

Gunblast '70s

Pamela Allan

Death Goes to School '53

Patrick Allan

Flight from Singapore '62

William Alland

Citizen Kane '41

Toni Allaylis

Fast Talking '86

Louise Allbritton

The Doolins of Oklahoma '49
Egg and I '47
Son of Dracula '43
Who Done It? '42

Marc Allegret

Blood and Roses '61

Marc Allegret (D)

Fanny '32
Lady Chatterley's Lover '55
The Love of Three Queens '54
Mademoiselle Striptease '57
Sois Belle et Tais-Toi '58
Zou Zou '34

Yves Allegret (D)

Dedee D'Anvers '49
Johnny Apollo '40
Seven Deadly Sins '53

A.K. Allen (D)

Ladies Club '86

Alicia Allen

California Girls '84

Bambi Allen

Outlaw Riders '72

Barbara Jo Allen

Sword in the Stone '63

Bill Allen

Rad '86

Chad Allen

Camp Cucamonga: How I
    Spent My Summer
    Vacation '90

Corey Allen

Party Girl '58

Corey Allen (D)

The Ann Jillian Story '88
Avalanche '78
Brass '85
Codename: Foxfire '85
Erotic Adventures of
    Pinocchio '71
The Last Fling '86
The Man in the Santa Claus
    Suit '79
Return of Frank Cannon '80
Thunder and Lightning '77

David Allen

Dungeonmaster '85

David Allen (D)

Puppet Master 2 '91

Debbie Allen

The Fish that Saved
    Pittsburgh '79
Jo Jo Dancer, Your Life is
    Calling '86
Kids from Fame '83
Ragtime '81

Don Allen

Suburbia '83

Fred Allen

It's in the Bag '45
We're Not Married '52

Fred Allen (D)

The Mysterious Rider '33
Saddle Buster '32

Ginger Lynn Allen

Buried Alive '89
Cleo/Leo '89
Hollywood Boulevard 2 '89
Vice Academy '88
Vice Academy 2 '90
Vice Academy 3 '91
Wild Man '89
Young Guns 2 '90

Gracie Allen

Damsel in Distress '37
International House '33
Two Girls & a Sailor '44

Irving Allen (D)

Slaughter Trail '51

Irwin Allen (D)

Beyond the Poseidon
    Adventure '79
Five Weeks in a Balloon '62
Lost in Space '65
The Swarm '78
Towering Inferno '74
Voyage to the Bottom of the
    Sea '61

James Allen (D)

Burndown '89

Jay Allen

Night of the Demon '80

Jeffrey Allen

The Life & Loves of a Male
    Stripper '87
This Stuff'll Kill Ya! '71
2000 Maniacs '64

Jo Harvey Allen

Checking Out '89

Joan Allen

All My Sons '86
Fat Guy Goes Nutzoid '86
In Country '89
Manhunter '86
Peggy Sue Got Married '86
Tucker: The Man and His
    Dream '88
Without Warning: The James
    Brady Story '91

John Allen

Roses Bloom Twice '77

Jonelle Allen

The Midnight Hour '86
Penalty Phase '86
The River Niger '76

Jordan Allen (D)

Terminal Bliss '91

Judith Allen

Boots & Saddles '37
Bright Eyes '34
Dancing Man '33
Git Along Little Dogies '37
Healer '36
The Port of Missing Girls '38
Tough Kid '39
Train to Tombstone '50

Karen Allen

Animal Behavior '89
Backfire '88
Cruising '80
East of Eden '80
The Glass Menagerie '87
Manhattan '79
National Lampoon's Animal
    House '78
Raiders of the Lost Ark '81
Scrooged '88
Secret Weapon '90
Shoot the Moon '82
A Small Circle of Friends '80
Split Image '82
Starman '84
Sweet Talker '91
Until September '84
Wanderers '79

Keith Allen

Chicago Joe & the Showgirl
    '90

Lee Taylor Allen

Pulsebeat '85

Lewis Allen (D)

Another Time, Another Place
    '58
At Sword's Point '51
Illegal '55
Suddenly '54
Those Endearing Young
    Charms '45
The Uninvited '44

Linsley Allen

Body Moves '90

Mark Allen

The Best of Dark Shadows
    '60s

Marty Allen

Ballad of Billie Blue '72
Harrad Summer '74
Whale of a Tale '76

Mike Allen

Death Driver '78
Hooch '76

Mikki Allen

Regarding Henry '91

Nancy Allen

Blow Out '81
The Buddy System '83
Carrie '76
Dressed to Kill '80
Forced Entry '75
The Gladiator '86
Home Movies '79
I Wanna Hold Your Hand '78
The Last Detail '73
Limit Up '89
Memories of Murder '90
1941 '79
Not for Publication '84

The Philadelphia Experiment
    '84
Poltergeist 3 '88
RoboCop '87
RoboCop 2 '90
Strange Invaders '83
Sweet Revenge '87
Terror in the Aisles '84

Pat Allen

Full Metal Ninja '89

Patrick Allen

Invasion of the Body Stealers
    '69
Island of the Burning Doomed
    '67
The Long Haul '57
Puppet on a Chain '72
Thirteenth Day of Christmas
    '85
When Dinosaurs Ruled the
    Earth '70

Rex Allen

Arizona Cowboy '49
Border Saddlemates '52
Charlie, the Lonesome Cougar
    '67
Colorado Sundown '52
I Dream of Jeannie '52
Legend of Lobo '62
The Secret of Navajo Cave
    '76
Thunder in God's Country '51
Trail of Robin Hood '50
Western Double Feature 3:
    Rogers & Allen '50

Richard Allen

The Snows of Kilimanjaro '52

Robert Allen

The Black Room '35
Perils of Pauline '34
The Revenge Rider '35

Ronald Allen

Supergrass '87

Rosalind Allen

To Die for 2 '91

Ruth Allen

Down to Earth '17

Shaun Allen

Pink Nights '87

Sheila Allen

The Alphabet Murders '65
Bouquet of Barbed Wire
Pascali's Island '88

Sian Barbara Allen

Eric '75

Steve Allen

Amazon Women On the Moon
    '87
The Benny Goodman Story
    '55
Best of Comic Relief '86
Good Old Days of Radio '70s
Heart Beat '80
Hey Abbott! '78
Hollywood Palace '68
Kerouac '84
The Making of the Stooges
    '85
Ratings Game '84
Toast to Lenny '84
Video Yesterbloop '70s

Todd Allen

Brothers in Arms '88
Witchboard '87

Tom Allen

Clash of the Ninja '86

Woody Allen

Annie Hall '77
Bananas '71
Broadway Danny Rose '84
Casino Royale '67
Crimes & Misdemeanors '89
Everything You Always
    Wanted to Know About
    Sex (But Were Afraid to
    Ask) '72
The Front '76
Hannah and Her Sisters '86
King Lear '87
Love and Death '75
Manhattan '79
A Midsummer Night's Sex
    Comedy '82
New York Stories '89
Play It Again, Sam '72
Radio Days '87
Scenes from a Mall '90
Shadows and Fog '92
Sleeper '73
Stardust Memories '80
Take the Money and Run '69
What's New Pussycat? '65
What's Up, Tiger Lily? '66
Zelig '83

Woody Allen (D)

Alice '90
Annie Hall '77
Another Woman '88
Bananas '71
Broadway Danny Rose '84
Crimes & Misdemeanors '89
Everything You Always
    Wanted to Know About
    Sex (But Were Afraid to
    Ask) '72
Hannah and Her Sisters '86
Interiors '78
Love and Death '75
Manhattan '79
A Midsummer Night's Sex
    Comedy '82
New York Stories '89
Purple Rose of Cairo '85
Radio Days '87
September '88
Shadows and Fog '92
Sleeper '73
Stardust Memories '80
Take the Money and Run '69
What's Up, Tiger Lily? '66
Zelig '83

Frank Allenby

The Flame & the Arrow '50

Fernando Allende

Frontera (The Border)
Heartbreaker '83
La Virgen de Guadalupe '87
Maria '68

Kirstie Alley

Blind Date '84
Champions '84
Look Who's Talking '89
Look Who's Talking, Too '90
Loverboy '89
Madhouse '90
One More Chance '90
Prince of Bel Air '87
Runaway '84
Shoot to Kill '88

Sibling Rivalry '90
Star Trek 2: The Wrath of
Khan '82
Summer School '87

Sara Allgood
The Accused '48
Blackmail '29
Challenge to Lassie '49
The Fabulous Dorseys '47
How Green was My Valley '41
Juno and the Paycock '30
The Keys of the Kingdom '44
The Spiral Staircase '46
Storm in a Teacup '37
The Strange Affair of Uncle
Harry '45
That Hamilton Woman '41

Louis Allibert
Le Million '31

Alix Allin
Smiling Madame Beudet/The
Seashell and the
Clergyman '28

Dorothy Allison
Rikky and Pete '88
See No Evil '71
The Third Key '57

Jean Allison
The Devil's Partner '58

Wayne Allison
Deadly Twins '85

Daniel Allman
Close to Home '86

Gregg Allman
Rush '91
Rush Week '88

Christopher Allport
Brainwash '82
News at Eleven '86
Savage Weekend '80
Special Bulletin '83

Sammy Allred
Fast Money '83

Pernilla Allwin
Fanny and Alexander '83

Astrid Allwyn
Beggars in Ermine '34
Charlie Chan's Secret '35
It Could Happen to You '37
Love Affair '39
Monte Carlo Nights '34
Mystery Liner '34
Stowaway '36

June Allyson
Battle Circus '53
Best Foot Forward '43
Blackout '78
Curse of the Black Widow '77
Executive Suite '54
Girl Crazy '43
The Glenn Miller Story '54
Good News '47
The Kid with the Broken Halo
'82
Little Women '49
The McConnell Story '55
The Opposite Sex '56
Strategic Air Command '55
Thousands Cheer '43
Three Musketeers '48
Till the Clouds Roll By '46

Two Girls & a Sailor '44
Vegas '78
Words & Music '48

Ferdinando Almada
All for Nothing '75
La Venganza del Rojo '87
Los Jinetes de la Bruja (The
Horseman...) '76

Maria Almada
Emilio Varela vs. Camelia La
Texana '87

Mario Almada
El Fin del Tahur '87
El Sabor de la Venganza '65
Emiliano Zapata '87
La Dinastia de la Muerte
La Fuga del Rojo '87
La Venganza del Rojo '87
Los Jinetes de la Bruja (The
Horseman...) '76
Nido de Aguilas (Eagle's Nest)
'76
Simon Blanco '87
Sin Salida '87

Gila Almagor
Hide and Seek '80
The House on Chelouche
Street '73
Sallah '63

James Almanzar
Weekend with the Babysitter
'70

**Michael
Almereyda** (D)
Twister '89

Susan Almgren
Deadly Surveillance '91

**Pedro
Almodovar** (D)
Dark Habits '84
High Heels '91
Labyrinth of Passion '82
Law of Desire '86
Matador '86
Tie Me Up! Tie Me Down! '90
What Have I Done to Deserve
This? '85
Women on the Verge of a
Nervous Breakdown '88

Paul Almond (D)
Captive Hearts '87
Final Assignment '80
Journey '77
Prep School '81

Brigittia Almsrom
ABC's of Love & Sex,
Australia Style '86

Rami Alon (D)
Choices '81

Chelo Alonso
Atlas in the Land of the
Cyclops '61
Girl Under the Sheet '60s
Son of Samson '62

Ernesto Alonso
The Criminal Life of
Archibaldo de la Cruz '55
Wuthering Heights '53

Jose Alonso
El Diputado (The Deputy) '78
Murderer in the Hotel '83

**Maria Conchita
Alonso**
Blood Ties '87
Colors '88
Extreme Prejudice '87
Fear City '85
A Fine Mess '86
McBain '91
Moscow on the Hudson '84
Predator 2 '90
The Running Man '87
Touch and Go '86
Vampire's Kiss '88

Anthony Alonzo
Trident Force '88

John Alonzo (D)
Belle Starr Story '70s
Blinded by the Light '82
Portrait of a Stripper '79

Corinne Alphen
C.O.D. '83
Hot T-Shirts '79
New York Nights '84
Screwball Hotel '88
Spring Break '83

Domenico Alpi
Before the Revolution '65

Catherine Alric
Associate '82

Emmett Alston
Demonwarp '87

Emmett Alston (D)
New Year's Evil '78
Nine Deaths of the Ninja '85
Three Day Weekend
Three Way Weekend '81
Tigershark '87

Carol Alt
Bye Bye Baby '88
A Family Matter (Vendetta) '91
My Wonderful Life '90
Portfolio '88

John Altamura
New York's Finest '87
Toxic Avenger, Part 3: The
Last Temptation of Toxie
'89

Juan C. Altavista
La Casa de Madame Lulu '65

Hector Alterio
Basileus Quartet '82
Camila '84
Fridays of Eternity '81
The Nest '81
The Official Story '85

Dirke Altevogt
Intimate Moments '82

Urs Althaus
Warbus '85

Jeff Altman
Doin' Time '85
In Love with an Older Woman
'82
It's Not Easy Bein' Me '87

Laughing Room Only '86

Robert Altman
Countdown '68

Robert Altman (D)
Aria '88
Beyond Therapy '86
Brewster McCloud '70
Buffalo Bill & the Indians '76
The Caine Mutiny Court
Martial '88
Come Back to the 5 & Dime
Jimmy Dean, Jimmy Dean
'82
Countdown '68
Dumb Waiter '87
Fool for Love '86
The Long Goodbye '73
M*A*S*H '70
McCabe & Mrs. Miller '71
Nashville '75
O.C. and Stiggs '87
Popeye '80
Quintet '79
The Room '87
Secret Honor '85
Streamers '83
Tanner '88 '88
That Cold Day in the Park '69
Vincent & Theo '90
A Wedding '78

Steve Altman
Transylvania Twist '89

Robert Alton (D)
Pagan Love Song '50

**Walter George
Alton**
Heavenly Bodies '84
Puma Man '80

Tony Alva
Skateboard '77

Dawn Alvan
Doom Asylum '88

Crox Alvarado
The Curse of the Aztec
Mummy '59
Curse of the Mummy/Robot
vs. the Aztec Mummy '60s
La Corona de un Campeon
'87

Magali Alvarado
Salsa '88

Trini Alvarado
American Blue Note '91
The Babe '91
The Chair '87
A Movie Star's Daughter '79
Mrs. Soffel '85
Nitti: The Enforcer '88
Private Contentment '83
Rich Kids '79
Satisfaction '88
Stella '89
Sweet Lorraine '87
Times Square '80

**Adolph "Oz"
Alvarez**
Pilot '84

Angel Alvarez
Django '68

**Enrique Garcia
Alvarez**
Invasion of the Vampires '61

E.R. Alvarez
Creature from the Haunted
Sea '60

Miguel A. Alvarez
Natas es Satan

Joe Alves (D)
Jaws 3 '83

**Luis Fernando
Alves**
Cthulhu Mansion '91

Kirk Alyn
Superman: The Serial: Vol. 1
'48

Lyle Alzado
Club Fed '91
Destroyer '88
Ernest Goes to Camp '87
The Lost Idol/Shock Troop
'80s
Neon City '91
Oceans of Fire '86
Tapeheads '89
Tough Guys '86
Zapped Again '89

Daniela Alzone
Domino '88

Benito Alzraki (D)
Invasion of the Zombies '61

Shigeru Amachi
The Ghost of Yotsuya '58

Zenat Aman
Deadly Thief '78

Vittorio Amandola
Queen of Hearts '89

Antogone Amanitou
The Enchantress '88

Betty Amann
Rich and Strange '32

Denis Amar (D)
Caged Heart '85
L'Addition '85

Leonora Amar
Captain Scarlett '53

Suzana Amaral (D)
Hour of the Star '85

Rod Amateau (D)
The Bushwackers '52
Drive-In '76
The Garbage Pail Kids Movie
'87
High School USA '84
Lovelines '84
The Seniors '78
The Statue '71

Julie Amato
Ghost Dance '83

Julie Amber
Sweet Beat '62

Ernie Kovacs: Between the Laughter '84
Final Notice '89
Firewalker '86
Flash Gordon '80
Hitler's Daughter '90
Policewoman Centerfold '83
Speed Zone '88

Paul Anstad

Snow Treasure '67

Norman Anstey

Scavengers '87

Adam Ant

Cold Steel '87
Motown 25: Yesterday,
   Today, Forever '83
Nomads '86
Slamdance '87
Spellcaster '91
Trust Me '89
World Gone Wild '88

Jerzy Antczak (D)

Nights and Days '76

Franz Antel

Fearless '78

Gerald Anthony

Secret of the Ice Cave '89
To Die Standing '91

Jane Anthony

Games Girls Play '75

Joseph Anthony

Shadow of the Thin Man '41

Joseph Anthony (D)

All in a Night's Work '61
Career '59
The Matchmaker '58
The Rainmaker '56
Tomorrow '83

Len Anthony (D)

Fright House '89

Lysette Anthony

Ivanhoe '82
Krull '83
Oliver Twist '82
Without a Clue '88

Paul Anthony

House Party '90

Ray Anthony

The Girl Can't Help It '56
High School Confidential '58

Tony Anthony

Treasure of the Four Crowns
   '82

Manuel Antin (D)

Far Away and Long Ago '74

Robin Antin

Jailbird Rock '88

Steve Antin

The Accused '88
Last American Virgin '82
Penitentiary 3 '87
Survival Quest '89

Michel Antoine

All the Way, Boys '73

Karl Anton (D)

The Avenger '60

Ronald Anton

Hurry Up or I'll Be Thirty '73

Susan Anton

The Boy Who Loved Trolls '84

Cannonball Run II '84
Goldengirl '79
Lena's Holiday '91
Options '88
Spring Fever '81

Laura Antonelli

Chaste & Pure '77
Collector's Item '89
Divine Nymph '71
High Heels '72
How Funny Can Sex Be? '76
The Innocent '76
Malicious '74
A Man Called Sledge '71
Passion of Love '82
Secret Fantasy '81
Swashbuckler '84
Tigers in Lipstick '80
Till Marriage Do Us Part '74
Wifemistress '79

Lou Antonio

Cool Hand Luke '67

Lou Antonio (D)

Between Friends '83
Breaking Up Is Hard to Do '79
The Gypsy Warriors '78
The Last Prostitute '91
Mayflower Madam '87
Pals '87
A Real American Hero '78
Silent Victory: The Kitty O'Neil
   Story '79
Someone I Touched '75

**Michelangelo
Antonioni** (D)

Blow-Up '66
The Eclipse '66
L'Avventura '60
Love in the City '53
The Passenger '75
The Red Desert '64

Alexander Antonov

The Battleship Potemkin '25

Omero Antonutti

Basileus Quartet '82
Good Morning, Babylon '87
Night of the Shooting Stars
   '82
Padre Padrone '77

Anulka

Vampyres '74

Gabrielle Anwar

If Looks Could Kill '91
Wild Hearts Can't Be Broken
   '91

Teruhiko Aoi

The Imperial Japanese Empire
   '85

Rocky Aoki

For Us, the Living '88
Midnight Faces '26
A Public Affair '62
That's Adequate '90

Kazuya Aoyama

Godzilla vs. the Cosmic
   Monster '74
Godzilla vs. Mechagodzilla '75

Oscar Apfel

Bulldog Edition '36
Rainbow's End '35

Apollonia

Back to Back '90
Black Magic Woman '91
Heartbreaker '83
Ministry of Vengeance '89
Purple Rain '84
Tricks of the Trade '88

Carmine Appice

Black Roses '88

Noel Appleby

My Grandpa is a Vampire '90s
The Navigator '88

Christina Applegate

Don't Tell Mom the
   Babysitter's Dead '91
Streets '90

Royce D. Applegate

Outside Chance '78

Mary Ann Appleseth

Slumber Party '57 '76

John Aprea

The Arousers '70
Idolmaker '80
Picasso Trigger '89
Savage Beach '89

Michael Apted

Spies Like Us '85

Michael Apted (D)

Agatha '79
Class Action '91
Coal Miner's Daughter '80
Continental Divide '81
Critical Condition '86
First Born '84
Gorillas in the Mist '88
Gorky Park '83
Kipperbang '82
Poor Girl, A Ghost Story '74
The Squeeze '77
Thunderheart '92
Triple Echo '77

Amy Aquino

Alan & Naomi '92
Descending Angel '90

Giacomo Aragall

Tosca '84

Art Aragon

The Ring '52

Frank Aragon

Angel Town '89

Jesse Aragon

Street Asylum '90

**Manuel Gutierrez
Aragon** (D)

Demons in the Garden '82
Half of Heaven '86

Sergio Aragones

To Kill a Stranger '70s

Hiro Arai

Captive '87

Oscar Araiz

Tango '83

Tomas Arana

The Church '90

Domino '88
The Last Temptation of Christ
   '88

Angel Aranda

Planet of the Vampires '65

Michael Aranda

Creepozoids '87

Vicente Aranda (D)

The Blood Spattered Bride '72

Manuel Aranguiz

A Paper Wedding '89

Romulo Arantes

Hell Hunters '87

Michiyo Aratama

The Gambling Samurai '60
The Human Condition: A
   Soldier's Prayer '61
The Human Condition: No
   Greater Love '58
The Human Condition: Road
   to Eternity '59
Kwaidan '64
Saga of the Vagabond '59

Alfonso Arau

Dynamite and Gold '88
Posse '75
Romancing the Stone '84
Scandalous John '71
Three Amigos '86

Araujo

Machoman '70s

Alexandre Arbatt

Dangerous Moves '84

**Jeffrey Allen
Arbaugh**

Darkroom '90

Richard Arbolino

Mirror Images '91

Fatty Arbuckle

Charlie Chaplin's Keystone
   Comedies #4 '14
Fatty and Mabel Adrift/Mabel,
   Fatty and the Law '16
Fatty's Tin-Type Tangle/Our
   Congressman '20s
General/Slapstick '26
The Knockout/Dough and
   Dynamite '14
Leap Year '21
The Lizzies of Mack Sennett
   '15
Mabel & Fatty '16
Mack Sennett Comedies:
   Volume 2 '16

Allan Arbus

Coffy '73
Greaser's Palace '72
The Young Nurses '73

Jonas Arby

Running on Empty '88

Kathleen Arc

Deadly Alliance '78

Denys Arcand (D)

The Decline of the American
   Empire '86
Jesus of Montreal '89

**George
Archainbaud** (D)

Blue Canadian Rockies '52
Devil's Playground '46
False Colors '43
Gene Autry Matinee Double
   Feature #1 '53
Gene Autry Matinee Double
   Feature #2 '53
Gene Autry Matinee Double
   Feature #3 '53
Hopalong Cassidy: Borrowed
   Trouble '48
Hopalong Cassidy: False
   Paradise '48
Hopalong Cassidy: Hoppy's
   Holiday '47
Hopalong Cassidy: Silent
   Conflict '48
Hopalong Cassidy: Sinister
   Journey '48
Hopalong Cassidy: The Dead
   Don't Dream '48
Hopalong Cassidy: The Devil's
   Playground '46
Hopalong Cassidy: The
   Marauders '47
Hopalong Cassidy:
   Unexpected Guest '47
Hunt the Man Down '50
Hunt the Man Down/
   Smashing the Rackets '50
The Kansan '43
Last of the Pony Riders '53
Lost Squadron '32
Night Stage to Galveston '52
On Top of Old Smoky '53
Sinister Journey '48
State's Attorney '31
Texas Masquerade '44
Winning of the West '53
The Woman of the Town '43

Anne Archer

Cancel My Reservation '72
Check is in the Mail '85
Eminent Domain '91
Fatal Attraction '87
Good Guys Wear Black '78
Green Ice '81
Hero at Large '80
The Last of His Tribe '92
Lifeguard '75
Love at Large '89
The Naked Face '84
Narrow Margin '90
Paradise Alley '78
Raise the Titanic '80
Too Scared to Scream '85
Waltz Across Texas '83

John Archer

Best of the Badmen '50
Blue Hawaii '62
Gangs, Inc. '41
King of the Zombies '41
Sherlock Holmes in
   Washington '43

Army Archerd

George Burns: His Wit &
   Wisdom '88

Irene Arcila

Macho y Hembra '85

Fanny Ardant

Confidentially Yours '83
The Family '88
Next Summer '84
Swann in Love '84
The Woman Next Door '81

Arianne Arden

Beyond the Time Barrier '60

Eve Arden

Anatomy of a Murder '59
At the Circus '39
Cinderella '84
Cover Girl '44
Eternally Yours '39
Grease '78
Grease 2 '82
A Guide for the Married
  Woman '78
Having Wonderful Time '38
Kid from Brooklyn '46
Letter of Introduction '38
Mildred Pierce '45
My Dream is Yours '49
Night and Day '46
One Touch of Venus '48
Our Miss Brooks '56
Pandemonium '82
Phone Call from a Stranger
  '52
Return of Video Yesterbloop
  '47
Slightly Honorable '40
Stage Door '37
Tea for Two '50
That Uncertain Feeling '41
Three Husbands '50
Under the Rainbow '81
We're Not Married '52
Ziegfeld Girl '41

Mary Arden

Blood and Black Lace '64

George Ardisson

Hercules in the Haunted World
  '64
The Invaders '63

Emile Ardolino (D)

Chances Are '89
Dirty Dancing '87
Rise & Rise of Daniel Rocket
  '86
Three Men and a Little Lady
  '90

Mats Arehn (D)

The Assignment '78
Istanbul '90

Dita Arel

Drifting '82

Gerardo Arellano

Miracle in Rome '88

James Arena

Love Desperados '86

Maurizio Arena

Boss Is Served '60s

Rosita Arenas

The Curse of the Aztec
  Mummy '59
The Curse of the Crying
  Woman '61
Curse of the Mummy/Robot
  vs. the Aztec Mummy '60s
El Bruto (The Brute) '52
Neutron and the Black Mask
  '61
Neutron vs. the Amazing Dr.
  Caronte '61
Neutron vs. the Death Robots
  '62
The Robot vs. the Aztec
  Mummy '59

The Witch's Mirror '60

Elke Arendt

The Fury of Hercules '61

Lois Areno

Pray TV '82

Eddi Arent

Circus of Fear '67
Dead Eyes of London '61
The Squeaker '65

Joey Aresco

Circle of Fear '89
Primary Target '89

Niels Arestrup

je tu il elle '74
Meeting Venus '91
Sincerely Charlotte '86

Patrick L'Argent

Key to Vengeance

Asia Argento

Demons 2 '87

Dario Argento (D)

Bird with the Crystal Plumage
  '70
The Cat o' Nine Tails '71
Creepers '85
Dario Argento's World of
  Horror '85
Deep Red: Hatchet Murders
  '75
Inferno '78
Suspiria '77
Terror at the Opera '88
Two Evil Eyes '91
Unsane '82

Fiore Argento

Creepers '85
Demons '86

Carmen Argenziano

Death Force '78
Red Scorpion '89

Alberto Argibay

Desde el Abismo

Allison Argo

Return of Frank Cannon '80

Victor Argo

King of New York '90
Quick Change '90

David Argue

Backlash '86
Blacklash '87
BMX Bandits '83
Gallipoli '81
Snow: The Movie '83

Tina Arhondis

A Test of Love '84

Ariane

Year of the Dragon '85

Imanol Arias

Camila '84
Demons in the Garden '82
Labyrinth of Passion '82

Ineko Arima

Equinox Flower '58

Adolfo

Aristarain (D)

The Stranger '87
Time for Revenge '82

Alida Arizmendy

Tono Bicicleta

Adam Arkin

Fourth Wise Man '85
Necessary Parties '88
Tom Edison: The Boy Who Lit
  Up the World '78

Alan Arkin

Bad Medicine '85
Big Trouble '86
Catch-22 '70
Chu Chu & the Philly Flash '81
Coupe de Ville '90
Edward Scissorhands '90
The Emperor's New Clothes
  '84
Escape from Sobibor '87
Fourth Wise Man '85
Freebie & the Bean '74
Havana '90
The Heart is a Lonely Hunter
  '68
Hearts of the West '75
Improper Channels '82
In-Laws '79
Joshua Then and Now '85
Last of the Red Hot Lovers
  '72
The Last Unicorn '82
Little Murders '71
The Magician of Lublin '79
A Matter of Principle '70s
Necessary Parties '88
Popi '69
Rafferty & the Gold Dust
  Twins '75
Return of Captain Invincible
  '83
The Rocketeer '91
The Russians are Coming, the
  Russians are Coming '66
The Seven Percent Solution
  '76
Simon '80
Wait Until Dark '67
Woman Times Seven '67

Alan Arkin (D)

Little Murders '71

David Arkin

Nashville '75

Steve Arkin

Lion Man '79

Tony Arkin

A Matter of Principle '70s

Robert Arkins

The Commitments '91

Robert Arkless (D)

The Man Who Would Not Die
  '75

Allan Arkush (D)

Caddyshack 2 '88
Death Sport '78
Get Crazy '83
Heartbeeps '81
Hollywood Boulevard '76
Rock 'n' Roll High School '79

Elizabeth Arlen

The First Power '89

Richard Arlen

Behind the Front '26
Buffalo Bill Rides Again '47
Cavalry Command '63
The Crawling Hand '63
Feel My Pulse '28
Flying Blind '41
The Human Duplicators '64
Identity Unknown '45
Let 'Em Have It '35
The Mountain '56
Return of Wildfire '48
Road to Nashville '67
Sex and the College Girl '64
Timber Queen '44
The Virginian '29
Warlock '59
Wildcat '42
Wings '27

Arletty

Children of Paradise (Les
  Enfants du Paradis) '44
Le Jour Se Leve '39
Les Visiteurs du Soir '42
The Longest Day '62

Dimitra Arliss

Eleni '85
Ski Bum '75

George Arliss

Dr. Syn '37
Iron Duke '34
Transatlantic Tunnel '35

Leslie Arliss (D)

A Man About the House '47
The Man in Grey '45
The Night Has Eyes '42

Ralph Arliss

The Asphyx '72

Brad Armacot

Backwoods '87

Kay Armen

Hit the Deck '55

Pedro Armendariz,
Jr.

All for Nothing '75
The Deadly Trackers '73
Don't Be Afraid of the Dark
  '73
La Chevre '81
Los Asesinos '68
Old Gringo '89

Pedro Armendariz

The Conqueror '56
El Bruto (The Brute) '52
Flor Silvestre (Wild Flower) '58
Fort Apache '48
From Russia with Love '63
The Fugitive '48
Littlest Outlaw '54
Los Caciques '87
Maria Candelaria '46
The Pearl '48
Three Godfathers '48

Gwen Arment

Mardi Gras Massacre '78

Henry Armetta

The Black Cat '34
Boss Foreman '39
Fisherman's Wharf '39
Let's Sing Again '36
Merry Widow '34

Armida

Border Romance '30
Fiesta '47
Gaiety '43
Jungle Goddess '49

George Armitage

Gas-s-s-s! '70

George
Armitage (D)

Hot Rod '79
Miami Blues '90
Private Duty Nurses '71

Gabi Armoni

Lupo '70

Russell Arms

Stage to Mesa City '48

Alun Armstrong

Split Second '92

Bekki Armstrong

The Immortalizer '89

Bess Armstrong

The Four Seasons '81
High Road to China '83
How to Pick Up Girls '78
Jaws 3 '83
Jekyll & Hyde...Together
  Again '82
Nothing in Common '86
Second Sight '89
Walking Through the Fire '80

Billy Armstrong

Charlie Chaplin: Rare Chaplin
  '15
Work/Police '16

Curtis Armstrong

Bad Medicine '85
Better Off Dead '85
One Crazy Summer '86
Revenge of the Nerds '84
Revenge of the Nerds 2:
  Nerds in Paradise '87
Risky Business '83

Gillian
Armstrong (D)

High Tide '87
Mrs. Soffel '85
My Brilliant Career '79
Starstruck '82

Howard Armstrong

Louie Bluie '85

Jerry Armstrong

Gator Bait 2: Cajun Justice
  '88

Katherine
Armstrong

Street Soldiers '91

Kerry Armstrong

The Hunting '92

Louis Armstrong

Betty Boop Special Collector's
  Edition: Volume 1 '35
Cabin in the Sky '43
The Five Pennies '59
The Glenn Miller Story '54
Hello, Dolly! '69
High Society '56

Margaret Avery

Blueberry Hill '88
The Color Purple '85
The Fish that Saved
    Pittsburgh '79
Heat Wave '90
Hell Up in Harlem '73
Lathe of Heaven '80
Return of Superfly '90
Riverbend '89
The Sky is Gray '80
Terror at Red Wolf Inn '72
Which Way Is Up? '77

Tex Avery (D)

Bugs Bunny & Elmer Fudd
    Cartoon Festival '44
Cartoon Moviestars: Daffy!
    '48
Cartoon Moviestars: Porky!
    '47
Daffy Duck: The Nuttiness
    Continues... '56
Tex Avery's Screwball
    Classics, Vol. 1 '40s
Tex Avery's Screwball
    Classics, Vol. 2 '40s

Val Avery

Black Caesar '73
Choices '81
Continental Divide '81
Courage '86
A Dream of Kings '69
Firehouse '72

John G. Avildsen (D)

Cry Uncle '71
For Keeps '88
Foreplay '75
The Formula '80
Guess What We Learned in
    School Today? '70
Happy New Year '87
Joe '70
The Karate Kid '84
The Karate Kid: Part 2 '86
The Karate Kid: Part 3 '89
Lean on Me '89
Neighbors '81
A Night in Heaven '83
Power of One '92
Rocky '76
Rocky 5 '90
Save the Tiger '73

Tom Avildsen (D)

Cheech and Chong: Things
    Are Tough All Over '82

Meiert Avis (D)

Far from Home '89

Jon Avnet (D)

Fried Green Tomatoes '91

Elvire Avoray

White Slave '86

Chris Avran

Bay of Blood '71
Twitch of the Death Nerve '72

Jane Avril

Women in Cell Block 7 '77

Nicole Avril

Summer Heat '73

Philippe Avron

Circus Angel '65

Keiko Awaji

The Bridges at Toko-Ri '55

Chikage Awashima

Early Summer (Bakushu) '51

Gabriel Axel (D)

Babette's Feast '87

George Axelrod (D)

Secret Life of an American
    Wife '68

Nina Axelrod

Cross Country '83
Motel Hell '80

Robert Axelrod

Repo Jake '90

Hoyt Axton

Act of Vengeance '86
Buried Alive '90
Christmas Comes to Willow
    Creek '87
Deadline Auto Theft '83
Disorganized Crime '89
Dixie Lanes '88
Endangered Species '82
Goldilocks & the Three Bears
    '83
Gremlins '84
Guilty of Innocence '87
Heart Like a Wheel '83
Junkman '82
Retribution '88
The Rousters '90
We're No Angels '89

Fernando Ayala (D)

El Profesor Hippie '69

Anne Ayars

Fiesta '47
Gaiety '43
Reunion in France '42

Dan Aykroyd

All You Need is Cash '78
Best of Chevy Chase '87
Best of Dan Aykroyd '86
Best of Gilda Radner '89
Best of John Belushi '85
The Blues Brothers '80
Caddyshack 2 '88
The Couch Trip '87
Doctor Detroit '83
Dragnet '87
Driving Miss Daisy '89
Ghostbusters '84
Ghostbusters 2 '89
The Great Outdoors '88
Into the Night '85
It Came from Hollywood '82
Loose Cannons '90
Love at First Sight '76
The Masters of Menace '90
Mr. Mike's Mondo Video '79
My Girl '91
My Stepmother Is an Alien '88
Neighbors '81
1941 '79
Nothing But Trouble '91
Saturday Night Live: Carrie
    Fisher '78
Saturday Night Live: Richard
    Pryor '75
Saturday Night Live: Steve
    Martin, Vol. 1 '78
Saturday Night Live: Steve
    Martin, Vol. 2 '78
Saturday Night Live, Vol. 1
    '70s

Spies Like Us '85
Things We Did Last Summer
    '78
This is My Life '91
Trading Places '83
Twilight Zone: The Movie '83

Dan Aykroyd (D)

Nothing But Trouble '91

Peter Aykroyd

The Funny Farm '82
Gas '81

Felix Aylmer

Action for Slander '38
Alice in Wonderland '50
As You Like It '36
The Case of the Frightened
    Lady '39
The Chalk Garden '64
Dreaming Lips '37
The Evil Mind '34
Exodus '60
Eye Witness '49
From the Terrace '60
Ghosts of Berkeley Square
    '47
Hamlet '48
The Hands of Orlac '60
Henry V '44
Ivanhoe '52
Knights of the Round Table
    '54
A Man About the House '47
The Master of Ballantrae '53
Mr. Emmanuel '44
The Mummy '59
Nine Days a Queen '34
October Man '48
Paris Express '53
Quo Vadis '51
Saint Joan '57
School for Scandal '65
Separate Tables '58
The Shadow '36
Spellbound '41
Trio '50
Victoria Regina '61
The Wicked Lady '45

Derek Aylward

The House in Marsh Road '60
School for Sex '69

Julian Aymes (D)

Jane Eyre '83

Alan Aynesworth

The Last Days of Dolwyn '49

Agnes Ayres

The Sheik '21
Son of the Sheik '26

Joann Ayres

Treasure of the Moon
    Goddess '88

Leah Ayres

Bloodsport '88
The Burning '82
Eddie Macon's Run '83

Lew Ayres

Advise and Consent '62
All Quiet on the Western Front
    '30
Battle for the Planet of the
    Apes '73
Big News '29
The Carpetbaggers '64
Cast the First Stone '89
Damien: Omen 2 '78

Dark Mirror '46
Dr. Kildare's Strange Case '40
Donovan's Brain '53
End of the World '77
Francis Gary Powers: The
    True Story of the U-2 Spy
    '76
Holiday '38
Johnny Belinda '48
The Kiss '29
Noah: The Deluge '79
Of Mice and Men '81
Salem's Lot: The Movie '79
The Stranger '73

Richard Ayres

Death Weekend '76

Robert Ayres

Battle Beneath the Earth '68
Cat Girl '57
River Beat '54
The Shepherd '88

Hank Azaria

Cool Blue '88

Annette Azcuy

Bill & Ted's Bogus Journey
    '91

Sabine Azema

Life and Nothing But '89
A Sunday in the Country '84

Ayu Azhari

Diamond Run '90

Tony Azito

Bloodhounds of Broadway '89
Chattanooga Choo Choo '84
Private Resort '84

Shabana Azmi

Madame Sousatzka '88

Charles Aznavour

The Adventurers '70
Blockhouse '73
Edith & Marcel '83
No, I Have No Regrets '73
Shoot the Piano Player '62
Sky Riders '76
Ten Little Indians '75
The Testament of Orpheus
    '59
Twist '76

Candice Azzara

Doin' Time on Planet Earth '88
Easy Money '83
Fatso '80
House Calls '78

Silvio Azzolini

Mafia vs. Ninja '84

Mario Azzopardi (D)

Deadline '82
Nowhere to Hide '87

Beth B (D)

Salvation! '87
Vortex '81

Jimi B, Jr.

Saigon Commandos '88

Scott B (D)

Vortex '81

Karin Baal

Dead Eyes of London '61

Thom Babbes

Deadly Dreams '88

Barbara Babcock

Christmas Coal Mine Miracle
    '78
Happy Together '89
Heart of Dixie '89
Lords of Discipline '83
News at Eleven '86
On the Edge: Survival of Dana
    '79
That Was Then...This Is Now
    '85

Fabienne Babe

Singing the Blues in Red '87

Hector Babenco (D)

At Play in the Fields of the
    Lord '91
Ironweed '87
Kiss of the Spider Woman '85
Pixote '81

Babita

Distant Thunder '73

Boris Babochkin

Chapayev '34

Tom Babson

Beasts '83

Baby Rose Marie

International House '33

Lauren Bacall

Appointment with Death '88
Bacall on Bogart '88
The Big Sleep '46
Blood Alley '55
The Cobweb '55
Dark Passage '47
Designing Woman '57
Dinner at Eight '89
The Fan '81
Harper '66
Hollywood Bloopers '50s
How to Marry a Millionaire '53
Innocent Victim '90
John Huston: The Man, the
    Movies, the Maverick '89
Key Largo '48
Misery '90
Mr. North '88
Murder on the Orient Express
    '74
Northwest Frontier '59
Return of Video Yesterbloop
    '47
Sex and the Single Girl '64
The Shootist '76
To Have & Have Not '44
Written on the Wind '56
Young Man with a Horn '50

Barbara Bach

Caveman '81
Force 10 from Navarone '78
Give My Regards to Broad
    Street '84
The Great Alligator '81
Jaguar Lives '79
The Legend of Sea Wolf '75
Princess Daisy '83
Screamers '80
The Spy Who Loved Me '77
Stateline Motel '75
Street Law '79

Conrad Bain

Bananas '71
Child Bride of Short Creek '81
C.H.O.M.P.S. '79
A Pleasure Doing Business '79
Postcards from the Edge '90
Who Killed Mary What's 'Er Name? '71

Cynthia Bain

Spontaneous Combustion '89

Ron Bain

Experience Preferred... But Not Essential '83

Sherry Bain

Pipe Dreams '76
Wild Riders '71
Wild & Wooly '78

Whitney Bain (D)

That Darn Sorceress '88

Fay Bainter

Babes on Broadway '41
Dark Waters '44
The Human Comedy '43
Jezebel '38
Kid from Brooklyn '46
Our Town '40
Presenting Lily Mars '43
Quality Street '37
Secret Life of Walter Mitty '47
The Shining Hour '38
State Fair '45
Woman of the Year '42

Jimmy Baio

The Bad News Bears in Breaking Training '77
Kiss and Be Killed '91
Playing for Keeps '86

Scott Baio

Bugsy Malone '76
Foxes '80
I Love N.Y. '87
Senior Trip '81
Zapped! '82

Alecs Baird

Children Shouldn't Play with Dead Things '72

Harry Baird

The Oblong Box '69
Tor '64

Jimmie Baird

Black Orchid '59
Brass '85

Sharon Baird

Ratboy '86

Curt Bais

Boat Is Full '81

Chieko Baisho

Station '81

Don Bajema

Heat and Sunlight '87

Richard Bakalayan

The Computer Wore Tennis Shoes '69
Jet Attack/Paratroop Command '58
The St. Valentine's Day Massacre '67

Up Periscope '59

Gary Bakeman

2000 Maniacs '64

Anthony Baker

What a Way to Die '70

Art Baker

A Southern Yankee '48
The Underworld Story '50

Betsy Baker

The Evil Dead '83

Betty Baker

My Lady of Whims '25

Blanche Baker

Awakening of Candra '81
Cold Feet '84
Embassy '85
French Postcards '79
The Handmaid's Tale '90
Livin' Large '91
Mary and Joseph: A Story of Faith '79
Shakedown '88
Sixteen Candles '84

Bob Baker

Outlaw Express '38
Western Trails '38

Bobby Baker

King of Kung-Fu '80s

Bonnie Baker

Spotlight Scandals '43

Carroll Baker

Andy Warhol's Bad '77
Baby Doll '56
Big Country '58
Bloodbath
But Not for Me '59
Captain Apache '71
The Carpetbaggers '64
Cheyenne Autumn '64
Easy to Love '53
Giant '56
The Greatest Story Ever Told '65
Harlow '65
How the West was Won '62
Ironweed '87
Kindergarten Cop '90
Kiss Me, Kill Me '69
My Father's Wife '80s
Native Son '86
Next Victim '74
Paranoia '69
Private Lessons
The Secret Diary of Sigmund Freud '84
Star 80 '83
Watcher in the Woods '81
The World is Full of Married Men '80
You've Got to Have Heart '77

Chet Baker

Let's Get Lost '89

David Baker

Air Hawk '84

David Baker (D)

Air Hawk '84
Best Enemies '80s

Dawn Baker

Miami Cops '89

Diane Baker

Baker's Hawk '76
The Closer '91
Danger in the Skies '79
The Diary of Anne Frank '59
The Horse in the Gray Flannel Suit '68
Journey to the Center of the Earth '59
Marnie '64
Mirage '66
The Pilot '82
The Prize '63
The Silence of the Lambs '91
Strait-Jacket '64

Dylan Baker

The Long Walk Home '89
The Wizard of Loneliness '88

Elsie Baker

The Ghosts of Hanley House '68

Frank Baker

The New Adventures of Tarzan: Vol. 1 '35
Tarzan and the Green Goddess '38

George Baker

Goodbye, Mr. Chips '69
The Moonraker '58
Robin Hood...The Legend: Herne's Son '85
Sword of Lancelot '63

Graham Baker (D)

Alien Nation '88
Final Conflict '81
Impulse '84

Henry Judd Baker

Clean and Sober '88

Janet Baker

Julius Caesar '84

Jay Baker

April Fool's Day '86

Jill Baker

Hope and Glory '87

Jim Baker

Tripods: The White Mountains '84

Joby Baker

Gidget '59

Joe Don Baker

The Abduction of Kari Swenson '87
Adam at 6 A.M. '70
Cape Fear '91
Charley Varrick '73
The Children '80
Cool Hand Luke '67
Criminal Law '89
Edge of Darkness '86
Final Justice '84
Fletch '85
Framed '75
Getting Even '86
Guns of the Magnificent Seven '69
Joy Sticks '83
Junior Bonner '72
The Killing Time '87
Leonard Part 6 '87
The Living Daylights '87
Mitchell '75

The Natural '84
The Pack '77
Shadow of Chikara '77
Speedtrap '78
Wacko '83
Walking Tall '73
Wild Rovers '71

Josephine Baker

The French Way '52
Princess Tam Tam '35
Zou Zou '34

Kai Baker

American Eagle '90
Stormquest '87

Kathy Baker

Clean and Sober '88
Dad '89
The Image '89
Jacknife '89
A Killing Affair '85
Mr. Frost '89
Permanent Record '88
The Right Stuff '83
Street Smart '87

Kelly Baker

Slaughter High '86

Kenny Baker

Elephant Man '80
The Empire Strikes Back '80
Return of the Jedi '83
Star Wars '77
Time Bandits '81

Kenny L. Baker

Amadeus '84
At the Circus '39
Cannon Movie Tales: Sleeping Beauty '89
The Harvey Girls '46
The Mikado '39

Kirsten Baker

Gas Pump Girls '79
Girls Next Door '79
Teen Lust '78

Leeanne Baker

Necropolis '87

Lenny Baker

Next Stop, Greenwich Village '76

Michael Baker (D)

An Affair in Mind '89

Nellie Bly Baker

Red Kimono '25

Pamela Baker

Bloody Wednesday '87

Penny Baker

Million Dollar Mystery '87

Phil Baker

Goldwyn Follies '38

Ray Baker

Everybody's All American '88
Heart Condition '90

Robert Baker

Chinese Connection '73

Robert S. Baker (D)

Blackout '50
Jack the Ripper '60

Roy Ward Baker (D)

And Now the Screaming Starts '73
Asylum '72
Dr. Jekyll and Sister Hyde '71
Don't Bother to Knock '52
Masks of Death '86
Mission: Monte Carlo '75
The Monster Club '85
A Night to Remember '58
October Man '48
One That Got Away '57
The Scars of Dracula '70
The 7 Brothers Meet Dracula '73
Switch '77
The Vampire Lovers '70
Vault of Horror '73

Ruth Baker

Marat/Sade (Persecution and Assassination...) '66

Sam Baker

Ninja of the Magnificence '89

Scott Thompson Baker

Cleo/Leo '89
New York's Finest '87
Rest In Pieces '87

Sharon Baker

Captive Planet '78
Metallica '85

Stanley Baker

Accident '67
Butterfly Affair '71
Cruel Sea '53
Dingaka '65
The Guns of Navarone '61
Knights of the Round Table '54
Queen of Diamonds '80s
Richard III '55
Robbery '67
Sodom and Gomorrah '62
Zulu '64

Timothy Baker

Bloodfist 2 '90

Tom Baker

Angels Die Hard '84
Candy Stripe Nurses '74
Chronicles of Narnia '89
Doctor Who: Pyramids of Mars '85
Doctor Who: Terror of the Zygons '78
Doctor Who: The Ark in Space '78
Doctor Who: The Talons of Weng-Chiang '88
The Freakmaker '73
Fyre '78
Golden Voyage of Sinbad '73
Nicholas and Alexandra '71
Vault of Horror '73
The Zany Adventures of Robin Hood '84

William L. Baker

Witchcraft 3: The Kiss of Death '90

William Bakewell

Annapolis '28
Dance Fools Dance '31
Dawn Express '42
Exiled to Shanghai '37
Gone with the Wind '39

Bakewell ►cont.

Lucky Me '54
Radar Men from the Moon '52
Roaring Speedboats '37
Romance on the High Seas '48

Brenda Bakke

Dangerous Love '87
Death Spa '87
Fist Fighter '88
Hardbodies 2 '86
Scavengers '87

Brigitte Bako

Red Shoe Diaries '92

Nebojsa Bakocevic

Hey, Babu Riba '88

Muhamad Bakri

Beyond the Walls '85
Hanna K. '83

Ralph Bakshi (D)

Fire and Ice '83
Fritz the Cat '72
Heavy Traffic '73
Hey Good Lookin' '82
Lord of the Rings '78
Streetfight '75
Wizards '77

Scott Bakula

The Last Fling '86
Necessary Roughness '91
Sibling Rivalry '90

Henry Bal

Angry Joe Bass '76

Walter Bal (D)

Jacko & Lise '82

Bob Balaban

Absence of Malice '81
Altered States '80
Bank Shot '74
Close Encounters of the Third Kind '80
Dead Bang '89
End of the Line '88
Girlfriends '78
Midnight Cowboy '69
The Strawberry Statement '70
2010: The Year We Make Contact '84
Whose Life Is It Anyway? '81

Bob Balaban (D)

Amazing Stories, Book 5 '85
Parents '89

Belinda Balaski

The Howling '81

Josiane Balasko

Too Beautiful for You '88

Lajos Balaszovits

Hungarian Rhapsody '78

Bill Balbridge

P O W Deathcamp '89

Anthony Balch

Towers Open Fire & Other Films '72

Anthony Balch (D)

Horror Hospital '73

Towers Open Fire & Other Films '72

Nick Baldasare

Beyond Dream's Door '88

Ferdinando Baldi (D)

Duel of Champions '61
Treasure of the Four Crowns '82

Rebecca Balding

Silent Scream '80

Renalto Baldini

Four Ways Out '57

Armenia Balducci (D)

Together '79

Richard Balducci (D)

L'Odeur des Fauves '66
Scandal Man '67
Too Pretty to Be Honest '72

Adam Baldwin

Bad Guys '86
Cohen and Tate '88
D.C. Cab '84
Full Metal Jacket '87
Hadley's Rebellion '84
Internal Affairs '90
Love on the Run '85
My Bodyguard '80
Next of Kin '89
Ordinary People '80
Predator 2 '90
Radio Flyer '91
Reckless '84
3:15—The Moment of Truth '86

Alec Baldwin

The Alamo: Thirteen Days to Glory '87
Alice '90
Beetlejuice '88
Forever, Lulu '87
Great Balls of Fire '89
The Hunt for Red October '90
Married to the Mob '88
The Marrying Man '91
Miami Blues '90
She's Having a Baby '88
Talk Radio '88
Working Girl '88

Daniel Baldwin

The Heroes of Desert Storm '91

Janet Baldwin

Gator Bait '73
Humongous '82
Ruby '77

Marty Baldwin

Ride the Wind '88

Mary Baldwin

Rock & the Money-Hungry Party Girls '89

Michael Baldwin

Phantasm '79

Peter Baldwin

The Ghost '63

Peter Baldwin (D)

Lots of Luck '85

Robert Baldwin

Courageous Dr. Christian '40
Meet Dr. Christian '39

Stephen Baldwin

The Beast '88
Born on the Fourth of July '89

William Baldwin

Backdraft '91
Flatliners '90

Christian Bale

Empire of the Sun '87
Henry V '89
The Land of Faraway '87
Newsies '92
Treasure Island '89

Rafael Baledon

La Fuerza Inutil '87
Los Hermanos Diablo '65

Rafael Baledon (D)

The Man and the Monster '65
Orlak, the Hell of Frankenstein '60
Swamp of the Lost Monster '65

John Balee

King Kung Fu '87

Carla Balenda

Outlaw Women '52

Betty Balfour

Champagne '28
Evergreen '34

Michael Balfour

The Canterbury Tales '71
Fiend Without a Face '58
Mania '60
Moulin Rouge '52
Quatermass 2 '57

Jennifer Balgobin

City Limits '85
Dr. Caligari '89
Repo Man '83

Denise Balik

Humanoids from the Deep '80

Ina Balin

Black Orchid '59
Charro! '69
Children of An Lac '80
Comancheros '61
The Don is Dead '73
From the Terrace '60
Galyon '77
Panic on the 5:22 '74
The Patsy '64
Projectionist '71
Vasectomy: A Delicate Matter '86

Mireille Balin

Pepe le Moko '37

Andras Balint

Father '67

Eszter Balint

Bail Jumper '89
The Linguini Incident '92
Stranger than Paradise '84

Fairuza Balk

Outside Chance of Maximillian Glick '88
Return to Oz '85
Valmont '89

Geza Balkay

A Hungarian Fairy Tale '87

Dave Balko

Strike Back '80

Angeline Ball

The Commitments '91

Frank Ball

Boothill Brigade '37
Gun Lords of Stirrup Basin '37
When a Man Rides Alone '33

Henry Ball

Ghost Dance '83

Lucille Ball

Abbott and Costello in Hollywood '45
The Affairs of Annabel '38
Annabel Takes a Tour/Maid's Night Out '38
Beauty for the Asking '39
Best Foot Forward '43
Big Street '42
Comedy Classics '35
Dance, Girl, Dance '40
Dark Corner '46
DuBarry was a Lady '43
Easy Living '49
Entertaining the Troops '89
Fancy Pants '50
Five Came Back '39
Follow the Fleet '36
Forever Darling '56
A Girl, a Guy and a Gob '41
A Guide for the Married Man '67
Having Wonderful Time '38
Having Wonderful Time/Carnival Boat '38
Hedda Hopper's Hollywood '42
Her Husband's Affairs '47
I Dream Too Much '35
Joy of Living '38
The Long, Long Trailer '54
Look Who's Laughing '41
Mame '74
Miss Grant Takes Richmond '49
Next Time I Marry '38
Panama Lady '39
Roberta '35
Roman Scandals '33
Room Service '38
Seven Days' Leave '42
Sorrowful Jones '49
Stage Door '37
That Girl from Paris '36
Thousands Cheer '43
Toast of the Town '56
Too Many Girls '40
Top Hat '35
A Tribute to Lucy '90
Valley of the Sun '42
A Woman of Distinction '50
You Can't Fool Your Wife '40
Yours, Mine & Ours '68
Ziegfeld Follies '46

Nicholas Ball

Claudia '85
The House that Bled to Death '81
Lifeforce '85

Vincent Ball

Blood of the Vampire '58
Dead Lucky '60
Demolition '77
A Matter of Who '62
Not Tonight Darling '72

William Ball

Suburban Commando '91

Carl Ballantine

Hollywood Chaos '89
Revenge of the Cheerleaders '76
Speedway '68
World's Greatest Lover '77

Maxine Ballantyne

Lemora, Lady Dracula '73

Carroll Ballard (D)

The Black Stallion '79
Never Cry Wolf '83
Nutcracker: The Motion Picture '86

Doug Ballard

Sex O'Clock News '85

John Ballard (D)

The Orphan '79

Kaye Ballard

Eternity '90
Fate '90
The Girl Most Likely '57
Modern Love '90
The Ritz '76
Tiger Warsaw '87

Smith Ballew

Gun Cargo '49
Hawaiian Buckaroo '38
Panamint's Bad Man '38
Rawhide '38
Roll Along Cowboy '37

Thomas Balltore

Once Bitten '85

Mario Balmasada

The Last Supper '76

Jean-Francois Balmer

Madame Bovary '91
Window Shopping '86

Joe Balogh

Visions '90

Haydee Balsa

La Nueva Cigarra '86

Martin Balsam

After the Fox '66
Al Capone '59
All the President's Men '76
The Anderson Tapes '71
Bedford Incident '65
Brand New Life '72
Breakfast at Tiffany's '61
Cape Fear '61
Cape Fear '91
The Carpetbaggers '64
Catch-22 '70
Confessions of a Police Captain '72
Death Rage '77
Death Wish 3 '85
The Defender '57
Delta Force '86

Eyes Behind the Stars '72
The Goodbye People '83
Grown Ups '86
Hard Frame '70
Harlow '65
Hombre '67
The House on Garibaldi Street '79
Innocent Prey '88
The Lindbergh Kidnapping Case '76
Little Big Man '70
Little Gloria...Happy at Last '84
Marjorie Morningstar '58
Miles to Go Before I Sleep '74
Mitchell '75
Murder on the Orient Express '74
Murder in Space '85
The People vs. Jean Harris '81
P.I. Private Investigations '87
Psycho '60
Queenie '87
Raid on Entebbe '77
St. Elmo's Fire '85
The Salamander '82
Season for Assassins '71
Silver Bears '78
Spaghetti Western '60s
The Stone Killer '73
Summer Wishes, Winter Dreams '73
The Taking of Pelham One Two Three '74
Thousand Clowns '65
Tora! Tora! Tora! '70
Twelve Angry Men '57
Two Minute Warning '76
The Warning '80

Talia Balsam

Calamity Jane '82
Consenting Adults '85
Crawlspace '86
In the Mood '87
Kent State '81
The Kindred '87
Mass Appeal '84
Nadia '84
On the Edge: Survival of Dana '79
Private Investigations '87
Sunnyside '79
Trust Me '89

Allison Balson

Best Seller '87

Donna Baltron

Bloodbath in Psycho Town '89

Agnes Baltsa

Les Contes D'Hoffman '84

Fernando Balzaretti

The Summer of Miss Forbes '88

Robert E. Balzer (D)

It's Three Strikes, Charlie Brown '86

Judy Bamber

The Atomic Brain '64

Gerry Bamman

Lightning Over Water '80
The Ten Million Dollar Getaway '91

Junzaburo Ban

Dodes 'ka-den '70

Lynn Banashek

Fatal Games '84

Anne Bancroft

Agnes of God '85
The Bell Jar '79
Bert Rigby, You're a Fool '89
Demetrius and the Gladiators '54
Don't Bother to Knock '52
84 Charing Cross Road '86
Elephant Man '80
Fatso '80
Garbo Talks '84
Girl in Black Stockings '57
The Graduate '67
Jesus of Nazareth '77
Lipstick '76
Mel Brooks: An Audience '84
The Miracle Worker '62
'night, Mother '86
Nightfall '56
Prisoner of Second Avenue '74
The Restless Breed '58
Silent Movie '76
The Slender Thread '65
To Be or Not to Be '83
Torch Song Trilogy '88
Treasure of the Golden Condor '53
Turning Point '79
Young Winston '72

Anne Bancroft (D)

Fatso '80

Bradford Bancroft

Damned River '89
A Time to Die '91

George Bancroft

Angels with Dirty Faces '38
Docks of New York '28
Each Dawn I Die '39
Little Men '40
Mr. Deeds Goes to Town '36
Old Ironsides '26
Stagecoach '39
Texas '41
White Gold '28

Roy Bancroft

Sunset on the Desert '42

Albert Band (D)

Doctor Mordrid: Master of the Unknown '92
Ghoulies 2 '88
I Bury the Living '58
Tramplers '66
Zoltan...Hound of Dracula '77

Charles Band

Dungeonmaster '85

Charles Band (D)

Crash and Burn '90
Doctor Mordrid: Master of the Unknown '92
Dungeonmaster '85
Meridian: Kiss of the Beast '90
Metalstorm '83
Parasite '81
Trancers '85
Trancers 2: The Return of Jack Deth '91

Antonio Banderas

Labyrinth of Passion '82
Law of Desire '86
Mambo Kings '92
Matador '86
Tie Me Up! Tie Me Down! '90

Yasosuke Bando

Mishima: A Life in Four Chapters '85

Kanu Banerjee

Aparajito '57
Pather Panchali '54

Karuna Banerjee

Aparajito '57
Pather Panchali '54

Victor Banerjee

Foreign Body '86
The Home and the World '84
Hullabaloo Over Georgie & Bonnie's Pictures '78
A Passage to India '84

Subir Banerji

Pather Panchali '54

Lisa Banes

Cocktail '88

Brenda Banet

Ballbuster

Joy Bang

Messiah of Evil '74
Night of the Cobra Woman '72
Play It Again, Sam '72

Peter Banicevic

Bomb at 10:10 '67

Donatas Banionis

Solaris '72

Mirra Bank (D)

Enormous Changes '83

Tallulah Bankhead

Daydreamer '66
Die! Die! My Darling! '65
Lifeboat '44
Main Street to Broadway '53
Making of a Legend: "Gone With the Wind" '89
Stage Door Canteen '43

Aaron Banks

The Bodyguard '76
Fist of Fear, Touch of Death '77

Ernie Banks

King '78

Jonathan Banks

Armed and Dangerous '86
Assassin '86
Beverly Hills Cop '84
Cold Steel '87
Nadia '84

Laura Banks

Demon of Paradise '87

Leslie Banks

The Arsenal Stadium Mystery '39
Chamber of Horrors '40
Cottage to Let '41
Eye Witness '49

Fire Over England '37
Henry V '44
Jamaica Inn '39
Madeleine '50
The Man Who Knew Too Much '34
The Most Dangerous Game '32
Sanders of the River '35
Transatlantic Tunnel '35

Montague Banks (D)

Great Guns '41

Monty Banks

Sherlock's Rivals & Where's My Wife '27

Vilma Banky

Blood and Sand/Son of the Sheik '26
The Eagle '25
Son of the Sheik '26

Ian Bannen

The Big Man: Crossing the Line '91
Bite the Bullet '75
Carlton Browne of the F.O. '59
The Courier '88
Defense of the Realm '85
Doomwatch '72
Driver's Seat '73
Eye of the Needle '81
The Flight of the Phoenix '66
Fright '71
From Beyond the Grave '73
Gathering Storm '74
George's Island '91
Ghost Dad '90
Hope and Glory '87
Mackintosh Man '73
The Prodigal '83
Ride to Glory '71
A Tale of Two Cities '58
The Third Key '57
Too Late the Hero '70
Watcher in the Woods '81

Jill Banner

Hard Frame '70

John Banner

The Fallen Sparrow '43
Guilty of Treason '50
Togetherness '70

Tony Banner

Image of Death '77

Walter Bannert (D)

The Inheritors '85

Reggie Bannister

Phantasm '79
Phantasm II '88
Silent Night, Deadly Night 4: Initiation '90
Survival Quest '89

Yoshimitu Banno (D)

Godzilla vs. the Smog Monster '72

Jack Bannon

Miracle of the Heart: A Boys Town Story '86
Take Your Best Shot '82

Jim Bannon

Man from Colorado '49

Unknown World '51

Ildiko Bansagi

Mephisto '81

Richard Bansbach (D)

Claws '77

Arell Banton

Wild Riders '71

Jennifer Baptist

Toxic Avenger '86

Maurice Baquet

Voyage en Balloon '59

Shlomo Bar-Aba

Fictitious Marriage '88

Nina Bara

Missile to the Moon '59

Theda Bara

A Fool There Was '14
The Love Goddesses '65
The Unchastened Woman '25

John Baragrey

Gamera '66

Darrel Baran

The Big Crimewave '86

Catherine Baranov

Metamorphosis '90

Vera Baranovskaya

Mother '26

Jadwiga Baranska

Nights and Days '76

Christine Baranski

Crackers '84

Cassie Barasch

Little Sweetheart '90

Olivia Barash

Child of Glass '78
Dr. Alien '88
Grave Secrets '89
Repo Man '83
Tuff Turf '85

Richie Barathy

Caged Fury '90

Luca Barbareschi

Bye Bye Baby '88
Private Affairs '89

Uri Barbash (D)

Beyond the Walls '85
Unsettled Land '88

Adrienne Barbeau

Back to School '86
Cannibal Women in the Avocado Jungle of Death '89
Cannonball Run '81
Charlie and the Great Balloon Chase '82
The Crash of Flight 401 '78
Creepshow '82
Double-Crossed '91
Escape from New York '81
The Fog '78
Magic on Love Island '80

Griff Barnett

Holiday Affair '49

Ivan Barnett *(D)*

The Fall of the House of Usher '52

Jeni Barnett

Indiscreet '88

Ken Barnett *(D)*

Dark Tower '87

Steve Barnett *(D)*

Emmanuelle 5 '87
Hollywood Boulevard 2 '89
Mindwarp '91

Vince Barnett

Baby Face Morgan '42
The Corpse Vanishes '42
Crimson Romance '34
The Death Kiss '33
East Side Kids/The Lost City '35
Gangs, Inc. '41
I Killed That Man '42
I Live My Life '35
Ride 'Em Cowgirl '41
Shoot to Kill '47
Thunder in the Pines '49
We're in the Legion Now '37
X Marks the Spot '42
Yellow Cargo '36

Lem Barney

The Black Six '74

Rick Barns

Death Wish Club '83

Walter Barns

Escape from the KGB '87

Allen Baron *(D)*

Foxfire Light '84

Bruce Baron

Fireback '78
Powerforce '83
The Ultimate Ninja '85

Lita Baron

Savage Drums '51

Lynda Baron

Open All Hours '83

Sandy Baron

Birdy '84
Broadway Danny Rose '84
The Out-of-Towners '70

Suzanne Baron

Run Virgin Run '90

Pierre Barouh

A Man and a Woman '66

Byron Barr

Double Indemnity '44
Down Dakota Way '49

Douglas Barr

Spaced Invaders '90

Edna Barr

Drums O'Voodoo '34

Jean-Marc Barr

The Big Blue '88
Hope and Glory '87

Jeanne Barr

Long Day's Journey into Night '62

Karen Barr

Spittin' Image '83

Leonard Barr

Diamonds are Forever '71

Patrick Barr

The Case of the Frightened Lady '39
Flesh and Blood Show '73
The Godsend '79
House of Whipcord '75
Sailing Along '38
The Satanic Rites of Dracula '73

Sharon Barr

Archer: The Fugitive from the Empire '81
Spittin' Image '83

Michael Barrack

Alone in the T-Shirt Zone '86

Maria Barranco

Don Juan My Love '90
Tie Me Up! Tie Me Down! '90
Women on the Verge of a Nervous Breakdown '88

Pierre Barrat

The Rise of Louis XIV '66

Robert Barratt

Bad Lands '39
Captain Caution '40
Distant Drums '51
Mary of Scotland '36
Northwest Passage '40
Shadows Over Shanghai '38
Strangler of the Swamp '46

Jean-Louis Barrault

Beethoven '36
Children of Paradise (Les Enfants du Paradis) '44
La Nuit de Varennes '82
La Ronde '51
The Longest Day '62
The Testament of Dr. Cordelier '59

Marie-Christine Barrault

Chloe in the Afternoon '72
Cousin, Cousine '76
The Daydreamer '77
L'Etat Sauvage '90
Jesus of Montreal '89
The Medusa Touch '78
My Night at Maud's '69
Stardust Memories '80
Swann in Love '84
Table for Five '83

Harry Barrd

The Story of a Three Day Pass '68

Gabriel Barre

Luggage of the Gods '87

Jorge Barreiro

Los Hijos de Lopez
Matrimonio a la Argentina '80s
Naked Is Better '73

Luis Barreiro

Boom in the Moon '46

Katherine Barrese

Jezebel's Kiss '90

Bruno Barreto *(D)*

Amor Bandido '79
Dona Flor & Her Two Husbands '78
Gabriela '84
Happily Ever After '86
A Show of Force '90
The Story of Fausta '92

Adrienne Barrett

Daughter of Horror '55

Anne Barrett

Follies Girl '43

Claudia Barrett

Robot Monster '53

Don Barrett

Slaughterhouse '87

Edith Barrett

I Walked with a Zombie '43
Lady for a Night '42

Jamie Barrett

Club Life '86

Jane Barrett

Eureka Stockade '49
Sword & the Rose '53

John Barrett

American Kickboxer 1 '91
Lights! Camera! Murder! '89

Judith Barrett

Road to Singapore '40
Yellowstone '36

Laurinda Barrett

The Heart is a Lonely Hunter '68

Lezli-Ann Barrett *(D)*

Business as Usual '88

Liz Barrett

Superargo '67

Louise Barrett

The Bloody Pit of Horror '65

Majel Barrett

Westworld '73

Nancy Barrett

House of Dark Shadows '70
Night of Dark Shadows '71

Nitchie Barrett

Preppies '82
A Time to Die '91

Raina Barrett

Oh! Calcutta! '72

Ray Barrett

Rebel '86
Revenge '71
Waterfront '83
Where the Green Ants Dream '84

Rona Barrett

Sextette '78

Roy Barrett

Contagion '87

Tim Barrett

Where the Bullets Fly '66

Victoria Barrett

Three Kinds of Heat '87

Mario Barri

The Steel Claw '61

Amanda Barrie

Carry On Cleo '65

Anthony Barrie

Girlfriend from Hell '89

Barbara Barrie

The Bell Jar '79
Breaking Away '79
Child of Glass '78
End of the Line '88
The Execution '85
Private Benjamin '80
Real Men '87
To Race the Wind '80
Two of a Kind '82

Colin Barrie

Melody '71

Judith Barrie

Party Girls '29

Mona Barrie

Dawn on the Great Divide '42
One Night of Love '34
Today I Hang '42

Wendy Barrie

Cash '34
Dead End '37
Five Came Back '39
Follies Girl '43
The Hound of the Baskervilles '39
I Am the Law '38
If I Were Rich '33
The Saint Strikes Back '39
Saint Strikes Back: Criminal Court '46
The Saint Takes Over '40
Vengeance '37
Wedding Rehearsal '32

Edgar Barrier

The Giant Claw '57
Phantom of the Opera '43

Pat Barringer

Orgy of the Dead '65

Christopher Barrington-Leigh

Loyalties '86

Pat Barrington

Mantis in Lace '68

Phyliss Barrington

Sucker Money '34

Rebecca Barrington

Dance or Die '87

George Barris

Smokey & the Hotwire Gang '79

Seth Barrish

Home Remedy '88

Bill Barron

Alone in the T-Shirt Zone '86

Bob Barron

Ballad of a Gunfighter '64
Song of Old Wyoming '45

Gary Barron *(D)*

Gorilla Farming '90

John Barron

Whoops Apocalypse '83

Keith Barron

Baby Love '69
Nothing But the Night '72

Robert Barron

Bill & Ted's Excellent Adventure '89
Sea Hound '47

Steven Barron *(D)*

Electric Dreams '84
Teenage Mutant Ninja Turtles: The Movie '90

Zelda Barron *(D)*

Bulldance '88
Forbidden Sun '89
Secret Places '85
Shag: The Movie '89

Diana Barrows

My Mom's a Werewolf '89

Bruce Barry

Plunge Into Darkness '77

Donald (Don "Red") Barry

Adventures of Red Ryder '40
Blazing Stewardesses '75
Border Rangers '50
California Joe '43
Frontier Vengeance '40
Gunfire '50
Hard Frame '70
I Shot Billy the Kid '50
Iron Angel '64
Outlaw Gang '49
Panama Patrol '39
The Purple Heart '44
Red Desert '50
Ringside '49
Saga of Death Valley '39
The Shakiest Gun in the West '68
Sinners in Paradise '38
The Sombrero Kid '42
Square Dance Jubilee '51
Sundown Fury '42
Sundown Kid '43
Texas Layover '75
Tough Assignment '49
Trail of the Mounties '47
Train to Tombstone '50
Western Double Feature 9: LaRue & Beery '40s
Western Double Feature 15 '50

Dorothy Barry

Sweetie '89

Duece Barry

Werewolves on Wheels '71

Hal Barwood (D)

Warning Sign '85

Nick Barwood (D)

Nasty Hero '89

Leon Bary

King of the Wild Horses '24

Mikhail Baryshnikov

Dancers '87
That's Dancing! '85
Turning Point '79
White Nights '85

Anat "Topol" Barzilai

Witchcraft '88

Gary Basaraba

One Magic Christmas '85

Richard Basehart

The Andersonville Trial '70
Being There '79
The Bounty Man '72
The Brothers Karamazov '57
Chato's Land '71
Flood! '76
Great Bank Hoax '78
Hans Brinker '69
He Walked by Night '48
Hitler '62
Island of Dr. Moreau '77
La Strada '54
Marilyn: The Untold Story '80
Mark Twain's A Connecticut
    Yankee in King Arthur's
    Court '78
Moby Dick '56
Rage '72
The Rebels '79
Reign of Terror '49
The Swindle '55

Tom Basham

The Pink Angels '71
Psychopath '73

Count Basie

Cinderfella '60
Stage Door Canteen '43

Toni Basil

Angel 3: The Final Chapter '88
Easy Rider '69
Mother, Jugs and Speed '76
Pajama Party '64
Rockula '90
Slaughterhouse Rock '88

Nadine Basile

Crazy for Love '52

The Basin Street Boys

Duke Is Tops '38

Kim Basinger

Batman '89
Blind Date '87
Final Analysis '92
Fool for Love '86
From Here to Eternity '79
Hard Country '81
Killjoy '81
The Man Who Loved Women
    '83
The Marrying Man '91
My Stepmother Is an Alien '88
Nadine '87

The Natural '84
Never Say Never Again '83
9 1/2 Weeks '86
No Mercy '86

A.W. Baskcomb

Phantom Fiend '35

Elya Baskin

Enemies, a Love Story '89
Moscow on the Hudson '84
The Name of the Rose '86

Richard Baskin (D)

Sing '89

Marianne Basler

Overseas: Three Women with
    Man Trouble '90
A Soldier's Tale '91

Lena Basquette

Hard Hombre '31
A Night for Crime '42
Pleasure '31
Rose of Rio Grande '38

Alfie Bass

The Fearless Vampire Killers
    '67
The Lavender Hill Mob '51
Man on the Run '49

Bobby Bass

The Squeeze '87

Harriet Bass

An Empty Bed '90

Jules Bass (D)

The Ballad of Paul Bunyan '72
Daydreamer '66
Flight of Dragons '82
Frosty the Snowman '69
The Hobbit '78
The Last Unicorn '82
Mad Monster Party '68
Santa Claus is Coming to
    Town '70

Mike Bass

Brotherhood of Death '76

Saul Bass (D)

Phase 4 '74

John Bassberger

Thorn '73

Albert Basserman

Foreign Correspondent '40
Madame Curie '43
Melody Master '41
The Moon and Sixpence '42
The Private Affairs of Bel Ami
    '47
The Red Shoes '48
Reunion in France '42
Rhapsody in Blue '45
The Shanghai Gesture '41
Since You Went Away '44
A Woman's Face '41

Heide Basset

First Turn On '83

Angela Bassett

City of Hope '91
Critters 4 '91

Angela Bassett

The Heroes of Desert Storm
    '91

Linda Bassett

Waiting for the Moon '87

Steve Bassett

Spring Break '83

Cynthia Bassinet

Last Dance '91

Paul Bassis

Ganjasaurus Rex '87

Bob Basso

Paradise Motel '84

Lawrence Bassoff (D)

Hunk '87
Weekend Pass '84

Christine Basson

Snake Dancer

William Bast

Forever James Dean '88

William Bast (D)

James Dean '76

Alexandra Bastedo

The Blood Spattered Bride '72
Draw! '81
Find the Lady '75
The Ghoul '75
Kashmiri Run '69

Billy Bastiani

Salvation! '87

Fanny Bastien

The Wolf at the Door '87

Hobart Basworth

Sea Lion '21

Michal Bat-Adam

Atalia '85
The House on Chelouche
    Street '73
I Love You Rosa '73

Michal Bat-Adam (D)

Boy Takes Girl '83

Sylvia Bataille

The Crime of Monsieur Lange
    '36
A Day in the Country '46

Alexei Batalov

The Cranes are Flying '57
The Lady with the Dog '59
Moscow Does Not Believe in
    Tears '80

Alexei Batalov (D)

The Overcoat '59

Nikolai Batalov

Bed and Sofa '27
Mother '26

Pierre Batcheff

Napoleon '27
Un Chien Andalou '28

Joy Batchelor (D)

Animal Farm '55

Anthony Bate

Madhouse Mansion '74
Philby, Burgess and MacLean:
    Spy Scandal of the Century
    '84

Natalie Bate

Ground Zero '88

Charles Bateman

The Brotherhood of Satan '71

Jason Bateman

Can You Feel Me Dancing?
    '85
Teen Wolf Too '87

Justine Bateman

Can You Feel Me Dancing?
    '85
The Closer '91
The Fatal Image '90
Primary Motive '92
Satisfaction '88

Kent Bateman (D)

Headless Eyes '83

Alan Bates

Britannia Hospital '82
Club Extinction '89
A Day in the Death of Joe Egg
    '71
Duet for One '86
An Englishman Abroad '83
The Entertainer '60
Far from the Madding Crowd
    '67
Georgy Girl '66
The Go-Between '71
Hamlet '90
In Celebration '75
A Kind of Loving '62
King of Hearts '66
Mr. Frost '89
Nijinsky '80
Prayer for the Dying '87
Quartet '81
Return of the Soldier '84
The Rose '79
Separate Tables '83
The Shout '78
The Story of a Love Story '73
Trespasser '85
An Unmarried Woman '78
A Voyage Round My Father
    '89
We Think the World of You
    '88
Whistle Down the Wind '62
The Wicked Lady '83
Women in Love '70
Zorba the Greek '64

Barbara Bates

The Caddy '53
The Inspector General '49
Let's Make it Legal '51

Bill Bates

Ghosts on the Loose '43

Caroline Bates

Grad Night '81

Florence Bates

The Chocolate Soldier '41
Diary of a Chambermaid '46
Heaven Can Wait '43
I Remember Mama '48
The Judge Steps Out '49
Lullaby of Broadway '51
The Moon and Sixpence '42

Mr. Lucky '43
My Dear Secretary '49
Portrait of Jennie '48
Rebecca '40
San Antonio '45
Secret Life of Walter Mitty '47
Since You Went Away '44
The Son of Monte Cristo '40
The Tuttles of Tahiti '42

Granville Bates

The Great Man Votes '38
Next Time I Marry '38

Jeanne Bates

Eraserhead '78
Mom '89
Sababa '55
The Strangler '64

Kathy Bates

At Play in the Fields of the
    Lord '91
Come Back to the 5 & Dime
    Jimmy Dean, Jimmy Dean
    '82
Dick Tracy '90
Fried Green Tomatoes '91
High Stakes '89
Men Don't Leave '89
Misery '90
Murder Ordained '87
Roe vs. Wade '89
Shadows and Fog '92
Summer Heat '87
White Palace '90

Michael Bates

Bedazzled '68
Patton '70

Ralph Bates

Dr. Jekyll and Sister Hyde '71
Dynasty of Fear '72
The Graveyard '74
The Horror of Frankenstein
    '70
I Don't Want to Be Born '75
Letters to an Unknown Lover
    '84
Lust for a Vampire '71
Murder Motel '74
Second Chance '80

William Bates

Orgy of the Dead '65

Randall Batinkoff

For Keeps '88
Streetwalkin' '85

Lloyd Batista

Last Plane Out '83

Stiv Bators

Polyester '81
Tapeheads '89

Rick Battaglia

Cold Steel for Tortuga '65
Roland the Mighty '58
'Tis a Pity She's a Whore '73

Bradley Battersby (D)

Blue Desert '91

Roy Battersby (D)

Mr. Love '86
Winter Flight '84

Jeanette Batti

Four Bags Full '56

Giacomo Battiato *(D)*

Blood Ties '87
Hearts & Armour '83

Carlo Battista

Umberto D '55

Miriam Battista

Blind Fools '40

Patrick Bauchau

Creepers '85
Double Identity '89
Emmanuelle 4 '84
La Amante '71
The Music Teacher '88
The Rapture '91

Georgette Baudry

Perfumed Nightmare '89

Belinda Bauer

Archer: The Fugitive from the
    Empire '81
Flashdance '83
The Game of Love '90
RoboCop 2 '90
Rosary Murders '87
Samson and Delilah '84
Servants of Twilight '91
Sins of Dorian Gray '82
Timerider '83
Winter Kills '79

Bruce Bauer

My Tutor '82

Charita Bauer

The Cradle Will Fall '83

Harry Bauer

Volpone '39

Irv Bauer

Sweet Beat '62

Jamie Lyn Bauer

Mysterious Island of Beautiful
    Women '79

Michelle Bauer

Deadly Embrace '88
Evil Toons '90
Hollywood Chainsaw Hookers
    '88
Phantom Empire '87
Wild Man '89

Steven Bauer

The Beast '88
A Climate for Killing '91
Gleaming the Cube '89
Running Scared '86
Scarface '83
Sweet Poison '91
Sword of Gideon '86
Thief of Hearts '84
Wildfire '88

Andre Bauge

La Barbiec de Seville '34

Sammy Baugh

King of the Texas Rangers '41

Ricardo Bauleo

Lost Diamond '74

Bruce Baum

Kandyland '87

L. Frank Baum *(D)*

His Majesty, the Scarecrow of
    Oz/The Magic Cloak of Oz
    '14
Patchwork Girl of Oz '14

Andy Bauman

Night of the Kickfighters '91

Erna Martha Bauman

Invasion of the Vampires '61
World of the Vampires '60

Anne Marie Baumann

The Last Days of Pompeii '60

Katherine Baumann

Chrome and Hot Leather '71
Slashed Dreams '74

Michele Baumgartner

The Woman Next Door '81

Monika Baumgartner

The Nasty Girl '90

Harry Baur

Beethoven '36
Crime and Punishment '35
Poil de Carotte '31

Marc Baur

Ultimate Desires '91

Lamberto Bava *(D)*

A Blade in the Dark '83
Blastfighter '85
Demons '86
Demons 2 '87
Devilfish '84
Frozen Terror '80

Mario Bava *(D)*

Bay of Blood '71
Beyond the Door 2 '79
Black Sabbath '64
Black Sunday '60
Blood and Black Lace '64
Danger: Diabolik '68
Hatchet for the Honeymoon
    '70
Hercules in the Haunted World
    '64
The Invaders '63
Kill, Baby, Kill '66
Lisa and the Devil '75
Planet of the Vampires '65
Torture Chamber of Baron
    Blood '72
Twitch of the Death Nerve '72
Wonders of Aladdin '61

Francis Bavier

The Day the Earth Stood Still
    '51
Horizons West '52
The Lady Says No '51

Jose Baviera

The Exterminating Angel '62

Conrad Baw

Death of a Hooker '71

Barbara Baxley

All Fall Down '62
Countdown '68

Tropic of Desire '80s

Craig R. Baxley *(D)*

Action Jackson '88
I Come in Peace '90
Stone Cold '91

David Baxt

Cloud Waltzing '87

Alan Baxter

Abe Lincoln in Illinois '40
Behind Prison Walls '43
Each Dawn I Die '39
It Could Happen to You '37
Saboteur '42
The Set-Up '49
Shadow of the Thin Man '41
Wild Weed '49

Anne Baxter

All About Eve '50
Angel on My Shoulder '46
Carnival Story '54
East of Eden '80
The Fighting Sullivans '42
Guest in the House '44
I Confess '53
Jane Austen in Manhattan '80
The Magnificent Ambersons
    '42
Masks of Death '86
The North Star '43
The Razor's Edge '46
The Ten Commandments '56
Ticket to Tomahawk '50
Walk on the Wild Side '62

Meredith Baxter Birney

Ben '72
Beulah Land '80
Bittersweet Love '76
Broken Badge '85
The Family Man '79
The Invasion of Carol Enders
    '74
Jezebel's Kiss '90
The Kissing Place '90
Little Women '78
Stranger Who Looks Like Me
    '74
Take Your Best Shot '82

Jane Baxter

The Evil Mind '34

John Baxter *(D)*

Love on the Dole '41
Old Mother Riley's Ghosts '41

Keith Baxter

Ash Wednesday '73
Berlin Blues '89
Chimes at Midnight '67

Lora Baxter

Before Morning '33

Lynsey Baxter

The French Lieutenant's
    Woman '81
The Girl in a Swing '89
To the Lighthouse '83

Warner Baxter

Adam Had Four Sons '41
Broadway Bill '34
42 Street '33
Stand Up and Cheer '34
Vogues of 1938 '37

Francis Bay

The Attic '80
Critters 3 '91

Sara (Rosalba Neri) Bay

Devil's Wedding Night '73
Lady Frankenstein '72
Naked Warriors '73

Susan Bay

Big Mouth '67

Nathalie Baye

Beau Pere '81
Beethoven's Nephew '88
Green Room '78
Honeymoon '87
La Balance '82
The Man Inside '90
Return of Martin Guerre '83

Gary Bayer

Will: G. Gordon Liddy '82

Rolf Bayer *(D)*

Pacific Inferno '85

Geoffrey Bayldon

Madame Sousatzka '88

Peter Baylis *(D)*

Finest Hours '64

Terence Baylor

Macbeth '71

Gene Baylos

The Love Machine '71

Stephen Bayly *(D)*

Coming Up Roses '87
Diamond's Edge '88

Lawrence Bayne

A Whisper to a Scream '88

Brenda Bazinet

Self-Defense '88

Andrew Be

Michael Prophet '87

Michael Beach

Cadence '89
In a Shallow Grave '88

Scott Beach

Out '88

Stephanie Beacham

And Now the Screaming
    Starts '73
The Confessional '75
The Dying Truth '91
Horror Planet '80
The Nightcomers '72
Super Bitch '77
Troop Beverly Hills '89

David Beaird *(D)*

It Takes Two '88
My Chauffeur '86
Octavia '82
Party Animal '83
Pass the Ammo '88
Scorchers '92

Rick Beairsto *(D)*

Close to Home '86

Cindy Beal

Slave Girls from Beyond
    Infinity '87

Crusty Beal

Sinner's Blood '70

John Beal

Amityville 3: The Demon '83
Break of Hearts '35
The Great Commandment '41
I Am the Law '38
Les Miserables '35
Little Minister '34
Ten Who Dared '60

Kay Beal

Torture Train '83

Jennifer Beals

Blood & Concrete: A Love
    Story '90
The Bride '85
Cinderella '84
Club Extinction '89
Flashdance '83
The Gamble '88
Split Decisions '88
That's Dancing! '85
Vampire's Kiss '88

Dovie Beams

Wild Wheels '75

Christopher F. Bean

Firepower '79

Orson Bean

Anatomy of a Murder '59
The Hobbit '78
Innerspace '87
Instant Karma '90
Lola '69
Movie Maker '86

Rick Bean

Whodunit '82

Robert Bean

Creature from the Haunted
    Sea '60
The Wild Ride '60

Sean Bean

Caravaggio '86
The Field '90
Stormy Monday '88

Hilary Beane

Zapped! '82

Harry Bear

Gods of the Plague '69

Matthew "Stymie" Beard

Broken Strings '40
East of Eden '80
Little Rascals, Book 1 '30s
Little Rascals, Book 2 '30s
Little Rascals, Book 3 '38
Little Rascals, Book 4 '30s
Little Rascals, Book 5 '30s
Little Rascals, Book 6 '30s
Little Rascals, Book 7 '30s
Little Rascals, Book 9 '30s
Little Rascals, Book 16 '30s
Little Rascals: Choo Choo/
    Fishy Tales '32
Little Rascals Christmas
    Special '79

Little Rascals Collector's
Edition, Vol. 1-6
Little Rascals Comedy
Classics, Vol. 1
Little Rascals Comedy
Classics, Vol. 2
Little Rascals: Fish Hooky/
Spooky Hooky '33
Little Rascals: Honkey
Donkey/Sprucin' Up '34
Little Rascals: Little Sinner/
Two Too Young '35
Little Rascals: Mush and Milk/
Three Men in a Tub '33
Little Rascals: Pups is Pups/
Three Smart Boys '30
Little Rascals: Readin' and
Writin'/Mail and Female '31
Little Rascals: Reunion in
Rhythm/Mike Fright '37
Little Rascals: Spanky/Feed
'Em and Weep '32
Little Rascals: The Pinch
Singer/Framing Youth '36
Little Rascals Two Reelers,
Vol. 1 '29
Little Rascals Two Reelers,
Vol. 2 '33
Little Rascals Two Reelers,
Vol. 3 '36
Little Rascals Two Reelers,
Vol. 4 '30
Little Rascals Two Reelers,
Vol. 5 '34
Little Rascals Two Reelers,
Vol. 6 '36
Little Rascals, Volume 1 '29
Two-Gun Man from Harlem
'38

Chris Bearde (D)

Hysterical '83

Nicholas Beardsley

Savage Island '85

Cody Bearpaw

Devil & Leroy Basset '73

Amanda Bearse

Fraternity Vacation '85
Fright Night '85
Protocol '84

Emmanuelle Beart

Date with an Angel '87
Manon of the Spring '87

Allyce Beasley

Moonlighting '85
Silent Night, Deadly Night 4:
Initiation '90

Linda Beatie

The Prodigal Planet '88

The Beatles

Rock 'n' Roll History: English
Invasion, the '60s '90
Rock 'n' Roll History: Rock 'n'
Soul Heaven '90

Norman Beaton

Growing Pains '82
Playing Away '87

Alan Beattie (D)

Delusion '84
Stand Alone '85

Clyde Beatty

Africa Screams '49
The Lost Jungle '34

Ned Beatty

All God's Children '80
Angel Square '92
Back to School '86
Big Bad John '90
Big Bus '76
The Big Easy '87
Blind Vision '91
Captain America '89
Charlotte Forten's Mission:
Experiment in Freedom '85
Chattahoochee '89
Deliverance '72
Dying Room Only '73
The Execution of Private
Slovik '74
Exorcist 2: The Heretic '77
The Fourth Protocol '87
Friendly Fire '79
Gray Lady Down '77
Great Bank Hoax '78
The Guyana Tragedy: The
Story of Jim Jones '80
Hear My Song '91
Hopscotch '80
The Incredible Shrinking
Woman '81
The Last American Hero '73
Midnight Crossing '87
Mikey & Nicky '76
Ministry of Vengeance '89
Network '76
1941 '79
Our Town '77
Physical Evidence '89
Pray TV '82
Promises in the Dark '79
Purple People Eater '88
A Question of Love '78
Repossessed '90
Robert Kennedy and His
Times '90
Rolling Vengeance '87
Rumpelstiltskin '82
Sex and Buttered Popcorn '91
Shadows in the Storm '88
The Silver Streak '76
Sniper '75
Spy '89
Stroker Ace '83
Superman: The Movie '78
Superman 2 '80
Switching Channels '88
Thief Who Came to Dinner '73
Time Trackers '88
Touched '82
The Toy '82
Tragedy of Flight 103: The
Inside Story '91
The Trouble with Spies '87
The Unholy '88
White Lightning '73
Wise Blood '79
A Woman Called Golda '82

Robert Beatty

Against the Wind '48
Appointment with Crime '45
Captain Horatio Hornblower
'51
Counterblast '48
The Love of Three Queens '54
Postmark for Danger '56
Postmark for Danger/
Quicksand '56
San Demetrio, London '47
Time Lock '57
Where Eagles Dare '68
Wings of Danger '52

Warren Beatty

All Fall Down '62
Bonnie & Clyde '67
Bugsy '91

Dick Tracy '90
Dollars '71
Heaven Can Wait '78
Ishtar '87
Lilith '64
McCabe & Mrs. Miller '71
Parallax View '74
Promise Her Anything '66
Reds '81
Roman Spring of Mrs. Stone
'61
Shampoo '75
Splendor in the Grass '61

Warren Beatty (D)

Dick Tracy '90
Heaven Can Wait '78
Reds '81

Carolyn Beauchamp

Troma's War '88

William
Beaudine (D)

The Ape Man '43
Bela Lugosi Meets a Brooklyn
Gorilla '52
Billy the Kid Versus Dracula
'66
Blues Busters '50
Bowery Buckaroos '47
The Country Kid '23
Ghost Chasers '51
Ghosts on the Loose '43
Hard-Boiled Mahoney '47
Here Come the Marines '52
Here Comes Kelly '43
Infamous Crimes '47
Jesse James Meets
Frankenstein's Daughter
'66
Little Annie Rooney '25
Mom & Dad '47
The Panther's Claw '42
Robot Pilot '41
Smugglers' Cove '48
Sparrows '26
Spook Busters '46
Spotlight Scandals '43
Ten Who Dared '60
Turf Boy '42
Westward Ho, the Wagons!
'56

Julie Beaulieu

Bridge to Terabithia '85

Gabrielle
Beaumont (D)

Carmilla '89
Death of a Centerfold '81
The Godsend '79
Gone are the Days '84
He's My Girl '87

Harry
Beaumont (D)

Beau Brummel '24
The Broadway Melody '29
Dance Fools Dance '31
Our Dancing Daughters '28
Our Dancing Daughters '20s
When's Your Birthday? '37

Hugh Beaumont

Danger Zone '51
The Fallen Sparrow '43
The Human Duplicators '64
The Lady Confesses '45
The Lost Continent '51
Panama Menace '41
Phone Call from a Stranger
'52

Pier 23 '51
Railroaded '47
Roaring City '51
The Seventh Victim '43
To the Shores of Tripoli '42

Kathryn Beaumont

Alice in Wonderland '51
Peter Pan '53

Lucy Beaumont

Condemned to Live '35
The Crowd '28
A Free Soul '31

Nicholas Beauvy

Rage '72
The Toolbox Murders '78

Ellen Beaven

Buster '88

Terry Beaver

Impure Thoughts '86

Louise Beavers

Big Street '42
Bullets or Ballots '38
Delightfully Dangerous '45
Dixie Jamboree '44
DuBarry was a Lady '43
General Spanky '36
Goodbye, My Lady '56
The Jackie Robinson Story
'50
Mr. Blandings Builds His
Dream House '48
Reap the Wild Wind '42
She Done Him Wrong '33
Tammy and the Bachelor '57

George Beban

Italian '15

Rodolfo Beban

Matrimonio a la Argentina '80s

Claudine Beccaire

Inhibition '86

Gorman
Bechard (D)

Assault of the Killer Bimbos
'87
Cemetery High '89
Galactic Gigolo '87
Psychos in Love '87

John C. Becher

Below the Belt '80

Sidney Bechet

Moon Over Harlem '39

Davide Bechini

The Story of Boys & Girls '91

Joan Bechtel

Split '90

William Bechtel

Spite Marriage '29

Bonnie Beck

Merchants of War '90

George Beck (D)

Behave Yourself! '52

Jackson Beck

Popeye and Friends in the
South Seas '61

Jeff Beck

The Secret Policeman's Other
Ball '82

Jennifer Beck

Tightrope '84

John Beck

Audrey Rose '77
Big Bus '76
A Climate for Killing '91
Deadly Illusion '87
Fire and Rain '89
Great American Traffic Jam
'80
In the Cold of the Night '89
King of the Pecos '36
The Other Side of Midnight
'77
Paperback Hero '73
Sky Riders '76
Sleeper '73
Three in the Attic '68
Time Machine '78

Josh Beck (D)

Lunatics: A Love Story '92

Julian Beck

Edipo Re (Oedipus Rex) '67
Emergency '68
Poltergeist 2: The Other Side
'86
Signals Through the Flames
'83

Kimberly Beck

Friday the 13th, Part 4: The
Final Chapter '84
Massacre at Central High '76
Nightmare at Noon '87
Private War '90

Martin Beck (D)

Challenge '74
Last Game '80

Michael Beck

Blackout '85
Celebrity '85
Deadly Game '91
The Golden Seal '83
Holocaust '78
Madman '79
Megaforce '82
Warlords of the 21st Century
'82
The Warriors '79
Xanadu '80

Reginald Beck (D)

The Long Dark Hall '51

Stanley Beck

Lenny '74

Thomas Beck

Charlie Chan at the Opera '36
Charlie Chan in Paris '35

Vincent Beck

Santa Claus Conquers the
Martians '64

Graham Beckel

Hazel's People '75
The Money '75
The Paper Chase '73
Rising Son '90

Desiree Becker

Good Morning, Babylon '87

Jeff Belker (D)

Last of the One Night Stands '83

Ann Bell

Champions '84
The Statue '71

Arnold Bell

The Temptress '49

Christopher Bell

Sarah, Plain and Tall '91

Dan Bell

Terror Eyes '87

Edward Bell

Image of Passion '86
The Premonition '75

Hank Bell

Valley of the Sun '42

James Bell

The Leopard Man '43
My Friend Flicka '43

Jeanne Bell

Fass Black '77
The Muthers '76
Policewomen '73
TNT Jackson '75

Keith Bell

Hands of the Ripper '71

Marie Bell

Phedre '68

Marshall Bell

Air America '90
The Heroes of Desert Storm '91
Johnny Be Good '88
No Way Out '87
Total Recall '90
Wildfire '88

Martin Bell (D)

Streetwise '85

Rex Bell

Battling with Buffalo Bill '31
Broadway to Cheyenne '32
Diamond Trail '33
Fighting Texans '33
From Broadway to Cheyenne '32
Law of the Sea '38
Rainbow Ranch '33
Stormy Trails '36
The Tonto Kid '35
West of Nevada '36

Sam Bell

Dangerous Relations '73

Tobin Bell

False Identity '90

Tom Bell

Ballad in Blue '66
Dressed for Death '74
Let Him Have It '91
Quest for Love '71
Wish You Were Here '87

Diana Bellamy

The Nest '88

Earl Bellamy (D)

Against A Crooked Sky '75
Desperate Women '78
Fire '77
Flood! '76
Justice of the West '61
Magic on Love Island '80
Seven Alone '75
Sidewinder One '77
Speedtrap '78
The Trackers '71
Walking Tall: Part 2 '75

Madge Bellamy

Soul of the Beast '23
The White Sin '24
The White Zombie '32

Ralph Bellamy

Ace of Aces '33
The Awful Truth '37
Boy in the Plastic Bubble '76
Brother Orchid '40
Cancel My Reservation '72
Carefree '38
The Court Martial of Billy Mitchell '55
Dance, Girl, Dance '40
The Defender '57
Delightfully Dangerous '45
Disorderlies '87
Doctors' Wives '70
Fourth Wise Man '85
The Good Mother '88
Guest in the House '44
Healer '36
His Girl Friday '40
Love Leads the Way '84
Missiles of October '74
Murder on Flight 502 '75
Nightmare in Badham County '76
Oh, God! '77
Pretty Woman '89
Professionals '66
Return to Earth '76
Spitfire '34
Sunrise at Campobello '60
Trading Places '83
War & Remembrance, Part 1 '88
War & Remembrance, Part 2 '89
Winds of War '83
The Wolf Man '41
Woman in the Shadows '34

Harry Bellaver

The Lemon Drop Kid '51

Annie Belle

House on the Edge of the Park '84
Naked Paradise '78

Kathleen Beller

Are You in the House Alone? '78
The Betsy '78
Cloud Waltzing '87
Fort Apache, the Bronx '81
The Godfather: Part 2 '74
The Manions of America '81
Mary White '77
Mother & Daughter: A Loving War '80
No Place to Hide '81
Promises in the Dark '79
Rappaccini's Daughter '80
Surfacing '84
Sword & the Sorcerer '82
Time Trackers '88
Touched '82

Agostina Belli

The Chosen '77
The Sex Machine '76

Laura Belli

Almost Human '79

Melvin Belli

Ground Zero '88

Olga Bellin

Tomorrow '83

Kylie Belling

The Fringe Dwellers '86

Warren Bellinger

I Will, I Will for Now '76

Cal Bellini

Waikiki '80

Roxana Bellini

Dos Esposas en Mi Cama '87
Samson in the Wax Museum '63

Donald P. Bellisario (D)

Last Rites '88

Marco Bellochio (D)

Devil in the Flesh '87
Eyes, the Mouth '83
Henry IV '85

Yannick Bellon (D)

Rape of Love '79

Saul Bellow

Zelig '83

Louis Bellson

Cobham Meets Bellson '83
Duke Ellington & His Orchestra: 1929-1952 '86
Superdrumming '88

Pamela Bellwood

Cellar Dweller '87
Cocaine: One Man's Seduction '83
Deadman's Curve '78
Double Standard '88
Hangar 18 '80

Jean-Paul Belmondo

The Brain '69
Breathless (A Bout de Souffle) '59
Casino Royale '67
High Heels '72
Is Paris Burning? '68
Le Doulos '61
Le Magnifique '76
Love and the Frenchwoman '60
Pierrot le Fou '65
Scoumoune '74
Sois Belle et Tais-Toi '58
Stavisky '74
Stuntwoman '81
Swashbuckler '84
Two Women '61
Un Singe en Hiver '62
A Woman is a Woman '60

Vera Belmont (D)

Red Kiss (Rouge Baiser) '85

Lara Belmonte

Lauderdale '89

Vicki Belmonte

The Grass is Always Greener Over the Septic Tank '78

Daisy Belmore

Sylvia Scarlett '35

Antoine Belpetre

The Raven '43

Hal Belsoe

The Ghastly Ones '68

Jerry Belson (D)

Jekyll & Hyde...Together Again '82
Surrender '87

Lola Beltran

Duelo en el Dorado

Robert Beltran

Eating Raoul '82
El Diablo '90
Kiss Me a Killer '91
La Pastorela '92
Latino '85
Night of the Comet '84
Scenes from the Class Struggle in Beverly Hills '89
To Die Standing '91

James Belushi

About Last Night... '86
Best Legs in the 8th Grade '84
Birthday Boy '85
Curly Sue '91
Diary of a Hitman '91
Homer and Eddie '89
Jumpin' Jack Flash '86
K-9 '89
Little Shop of Horrors '86
The Man with One Red Shoe '85
The Masters of Menace '90
Mr. Destiny '90
Mutant Video '76
Once Upon a Crime '91
Only the Lonely '91
The Palermo Connection '91
The Principal '87
Real Men '87
Red Heat '88
Salvador '86
Taking Care of Business '90
Thief '81
Trading Places '83
Wedding Band '89

John Belushi

All You Need is Cash '78
Best of Chevy Chase '87
Best of Dan Aykroyd '86
Best of Gilda Radner '89
Best of John Belushi '85
The Blues Brothers '80
Continental Divide '81
Goin' South '78
National Lampoon's Animal House '78
Neighbors '81
1941 '79
Old Boyfriends '79
Saturday Night Live: Carrie Fisher '78
Saturday Night Live: Richard Pryor '75
Saturday Night Live: Steve Martin, Vol. 1 '78

Saturday Night Live: Steve Martin, Vol. 2 '78
Saturday Night Live, Vol. 1 '70s
Shame of the Jungle '90
Things We Did Last Summer '78

Richard Belzer

America '86
The Big Picture '89
Fame '80
Flicks '85
Freeway '88
The Groove Tube '72
Reunion: 10th Annual Young Comedians '87
The Wrong Guys '88

Maria-Luisa Bemberg (D)

Camila '84
Miss Mary '86

Maggie Bemby

Angels' Wild Women '72

Cliff Bemis

Modern Love '90

Jacob Ben-Ami (D)

Green Fields '37

Bea Benadaret

Flintstones '60

Steven Benally, Jr.

The Secret of Navajo Cave '76

Pat Benatar

Catch a Rising Star's 10th Anniversary '83
Union City '81

Gloriella Ruben Benavides

El Fin del Tahur '87

Brian Benben

Clean and Sober '88
I Come in Peace '90
Mortal Sins '90

Robert Benchley

China Seas '35
Foreign Correspondent '40
It's in the Bag '45
Road to Utopia '46
The Sky's the Limit '43
You'll Never Get Rich '41
Young & Willing '42

Andrew Bendarski

Family Sins '87

Don Bendell

The Instructor '83

Don Bendell (D)

The Instructor '83

Jack Bender (D)

Child's Play 3 '91
In Love with an Older Woman '82
Letting Go '85
The Midnight Hour '86
Shattered Vows '84
Side By Side '88
Tricks of the Trade '88
Two Kinds of Love '85

Jack Benny

Broadway Melody of 1936 '35
Daffy Duck & Company '50
Entertaining the Troops '89
A Guide for the Married Man '67
Hollywood Canteen '44
The Horn Blows at Midnight '45
It's in the Bag '45
It's a Mad, Mad, Mad, Mad World '63
Jack Benny Show '58
Medicine Man '30
Television's Golden Age of Comedy '50s
To Be or Not to Be '42
Transatlantic Merry-Go-Round '34

Oliver Benny

High Frequency '88

Martin Benrath

From the Life of the Marionettes '80

Deborah Benson

Just Before Dawn '80
Mutant on the Bounty '89
September 30, 1955 '77

Elaine Benson

The Old Curiosity Shop '35

George Benson

Convoy '40
The Creeping Flesh '72

Ivy Benson

The Dummy Talks '43

Jodi Benson

The Little Mermaid '89

Leon Benson (D)

Flipper's New Adventure '64

Lucille Benson

Cactus in the Snow '72
Duel '71
Huckleberry Finn '74
Reflections of Murder '74
Tom Sawyer '73

Lyric Benson

Modern Love '90

Martin Benson

The Cosmic Monsters '58
The King and I '56
A Matter of Who '62
Tiffany Jones '75

Peter Benson

Henry VI, Part 1 '82
Henry VI,Part 3 '82

Richard Benson (D)

The Day the Sky Exploded '57
Werewolf in a Girl's Dormitory '61

Robby Benson

All the Kind Strangers '74
City Limits '85
Death Be Not Proud '75
Death of Richie '76
Die Laughing '80
The End '78
Harry & Son '84
Ice Castles '79
Jory '72

Last of Mrs. Lincoln '76
Modern Love '90
Ode to Billy Joe '76
One on One '77
Our Town '77
Rent-A-Cop '88
Richie '77
Running Brave '83
Tribute '80
Two of a Kind '82
The Virginia Hill Story '76
White Hot '88

Robby Benson (D)

Modern Love '90
White Hot '88

Steve Benson (D)

Endgame '85

Vickie Benson

Bimini Code '84

W.E. Benson

Redneck Zombies '88

Michael Bentine

Down Among the Z Men '52
Rentadick '72

Fabrizio Bentivoglio

Apartment Zero '88

Dana Bentley

Sorority House Massacre 2 '92

John Bentley

Calling Paul Temple '48
The Chair '87
Flight from Vienna '58
River Beat '54
Scarlet Spear '54
Submarine Seahawk '59
White Huntress '57

Thomas Bentley (D)

Murder at the Baskervilles '37
The Old Curiosity Shop '35

Barbi Benton

Deathstalker '84
For the Love of It '80
Hospital Massacre '81

Chris Benton

20,000 Leagues Under the Sea '16

Eddie Benton

Dr. Strange '78

Helen Benton

Bloodbeat '80s

Jerome Benton

Graffiti Bridge '90
Under the Cherry Moon '86

Robert Benton (D)

Bad Company '72
Billy Bathgate '91
Kramer vs. Kramer '79
The Late Show '77
Nadine '87
Places in the Heart '84
Still of the Night '82

Suzanne Benton

A Boy and His Dog '75
That Cold Day in the Park '69

Yakov Bentsvi (D)

Legion of Iron '90

Femi Benussi

Hawks & the Sparrows (Uccellacci e Uccellini) '67
The Passion of Evelyn '80s

Marina Benvenga

Vice Academy 2 '90

Michael Benveniste

Flesh Gordon '72

Donna Kei Benz

Pray for Death '85

Obie Benz (D)

Heavy Petting '89

Daniel Benzali

Messenger of Death '88

John Beradino

Don't Look Back: The Story of Leroy "Satchel" Paige '81
Moon of the Wolf '72

Augustine Beral

The Killing Zone '90

George Beranger

Flirting with Fate '16

Luc Beraud (D)

Heat of Desire (Plein Sud) '84

Iris Berben

Tea For Three '84

Ady Berber

Dead Eyes of London '61

Luca Bercovici

Clean and Sober '88
Frightmare '83
K2 '92
Mortal Passions '90
Pacific Heights '90
Parasite '81

Luca Bercovici (D)

Ghoulies '85
Rockula '90

Geza Beremenyi (D)

The Midas Touch '89

Eric Berenger

Monsieur Hire '89

Tom Berenger

At Play in the Fields of the Lord '91
Betrayed '88
Beyond Obsession '89
The Big Chill '83
Born on the Fourth of July '89
Butch and Sundance: The Early Days '79
The Dogs of War '81
Eddie & the Cruisers '83
Fear City '85
The Field '90
In Praise of Older Women '78
Johnny We Hardly Knew Ye '77
Last Rites '88
Looking for Mr. Goodbar '77
Love at Large '89
Major League '89
Platoon '86

Rush It '77
Rustler's Rhapsody '85
The Sentinel '76
Shattered '91
Shoot to Kill '88
Someone to Watch Over Me '87

Harold Berens

Dual Alibi '47

Marisa Berenson

Barry Lyndon '75
Cabaret '72
Death in Venice '71
Killer Fish '79
Night of the Cyclone '90
The Secret Diary of Sigmund Freud '84
Sex on the Run '78
S.O.B. '81
White Hunter, Black Heart '90

Al Beresford (D)

Screamtime '83

Bruce Beresford (D)

Aria '88
Barry McKenzie Holds His Own '74
Black Robe '91
Breaker Morant '80
The Club '81
Crimes of the Heart '86
Don's Party '76
Driving Miss Daisy '89
The Fringe Dwellers '86
The Getting of Wisdom '77
Her Alibi '88
King David '85
Mister Johnson '91
Money Movers '78
Puberty Blues '81
Tender Mercies '83

Carmen Berg

Playboy's 1989 Playmate Video Calendar '89

Joanna Berg

The Visitors '89

Paul Berg

Fatal Confinement '20s

Peter Berg

Crooked Hearts '91
Genuine Risk '89
Late for Dinner '91
Never on Tuesday '88
Race for Glory '89
Shocker '89

Phyllis Berg-Pigorsch (D)

The Islander '88

Judith-Marie Bergan

Abduction '75
Never Pick Up a Stranger '79

Francine Berge

Circle of Love '64
Judex '64
The Nun '66

Candice Bergen

The Adventurers '70
Best of Chevy Chase '87
Best of Gilda Radner '89
Bite the Bullet '75
Carnal Knowledge '71

The Domino Principle '77
11 Harrowhouse '74
Gandhi '82
Getting Straight '70
The Group '66
March or Die '77
Mayflower Madam '87
Merlin and the Sword '85
A Night Full of Rain '78
Oliver's Story '78
Rich and Famous '81
The Sand Pebbles '66
Soldier Blue '70
Starting Over '79
Stick '85
The Wind and the Lion '75

Edgar Bergen

Do You Trust Your Wife/I've Got a Secret '56
Fun & Fancy Free '47
Goldwyn Follies '38
Here We Go Again! '42
Homecoming: A Christmas Story '71
I Remember Mama '48
Letter of Introduction '38
Look Who's Laughing '41
Muppet Movie '79
Stage Door Canteen '43
You Can't Cheat an Honest Man '39

Polly Bergen

Anatomy of Terror '74
At War with the Army '50
Born Beautiful '82
Cape Fear '61
Cry-Baby '90
Death Cruise '74
A Guide for the Married Man '67
Haunting of Sarah Hardy '89
How to Pick Up Girls '78
Making Mr. Right '86
Murder on Flight 502 '75
War & Remembrance, Part 1 '88
War & Remembrance, Part 2 '89
Winds of War '83

Grete Berger

Student of Prague '13

Gustav Berger

God, Man and Devil '49

Harris Berger

East Side Kids '40

Harvey Berger

Rebel High '88

Helmut Berger

Ash Wednesday '73
Battleforce '78
Code Name: Emerald '85
Conversation Piece '75
The Damned '69
Deadly Game '83
Dorian Gray '70
Fatal Fix '80s
Garden of the Finzi-Continis '71
The Great Battle '79
Mad Dog
Order to Kill '73
Romantic Englishwoman '75

Joachim Berger

Curse of the Yellow Snake '63

Katya Berger

Little Lips '70s
Nana '82
Tales of Ordinary Madness '83

Ludwig Berger (D)

Thief of Baghdad '40

Michelle Berger

Midnight Warrior '89

Nicole Berger

Shoot the Piano Player '62
The Story of a Three Day Pass '68

Sarah Berger

The Green Man '91

Senta Berger

The Ambushers '67
Boss Is Served '60s
Cast a Giant Shadow '66
Cobra '71
Cross of Iron '76
Diabolically Yours '67
Full Hearts & Empty Pockets '63
Killing Cars '80s
Mother Doesn't Always Know Best '60s
The Quiller Memorandum '66
The Scarlet Letter '73
Smugglers '75
Swiss Conspiracy '77
Waltz King '63
When Women Had Tails '70
When Women Lost Their Tails '75

Sidney Berger

Carnival of Souls '62

Toni Berger

Sugarbaby '85

William Berger

Day of the Cobra '84
Hell Hunters '87
Hercules '83
Hercules 2 '85
Oil '78
Slaughterday '77
Today We Kill, Tomorrow We Die '71
What a Way to Die '70

Wolfram Berger

Quicker Than the Eye '88

Teddy Bergeron

Paramount Comedy Theater, Vol. 5: Cutting Up '87

Micha Bergese

The Company of Wolves '85

Thommy Berggren

The Adventurers '70
Elvira Madigan '67

Herbert Berghof

Master Mind '73

Patrick Bergin

The Courier '88
Highway to Hell '92
Love Crimes '91
Mountains of the Moon '90
Robin Hood '91
Sleeping with the Enemy '91

Bjorn Berglund

Swedenhielms '35

Bullen Berglund

Intermezzo '36

Erik Berglund

Only One Night '42
Walpurgis Night '41

Andrew Bergman (D)

The Freshman '90
So Fine '81

Henry Bergman

Charlie Chaplin Festival '17
Charlie Chaplin: The Early Years, Volume 2 '17
City Lights '31
Modern Times '36

Ingmar Bergman (D)

After the Rehearsal '84
Autumn Sonata '78
Brink of Life '57
Cries and Whispers '72
The Devil's Eye '60
Devil's Wanton '49
Dreams '55
Fanny and Alexander '83
From the Life of the Marionettes '80
Hour of the Wolf '68
Lesson in Love '54
The Magic Flute '73
The Magician '58
Monika '52
Night Is My Future '47
Persona '48
Port of Call '48
Sawdust & Tinsel '53
Scenes from a Marriage '73
Secrets of Women '52
The Serpent's Egg '78
The Seventh Seal '56
The Silence '63
Smiles of a Summer Night '55
Summer Interlude '50
Three Strange Loves '49
Through a Glass Darkly '61
To Joy '50
The Touch '71
The Virgin Spring '59
Wild Strawberries '57
The Winter Light '62

Ingrid Bergman

Adam Had Four Sons '41
Anastasia '56
Arch of Triumph '48
Autumn Sonata '78
The Bells of St. Mary's '45
The Best of Bogart '43
Cactus Flower '69
The Cary Grant Collection '90
Casablanca '42
The Count of the Old Town '34
Dr. Jekyll and Mr. Hyde '41
Dollar '38
Elena and Her Men '56
Europa '51 '52
Gaslight '44
Goodbye Again '61
The Hideaways '73
Indiscreet '58
The Inn of the Sixth Happiness '58
Intermezzo '36
Intermezzo '39
Joan of Arc '48

June Night '40
A Matter of Time '76
Murder on the Orient Express '74
Notorious '46
Only One Night '42
Spellbound '45
Stromboli '50
Swedenhielms '35
24 Hours in a Woman's Life '61
Under Capricorn '49
Voyage to Italy '53
Walk in the Spring Rain '70
Walpurgis Night '41
A Woman Called Golda '82
A Woman's Face '38

Peter Bergman

Firesign Theatre Presents Nick Danger in the Case of the Missing Yolks '83
Hot Shorts '84
J-Men Forever! '79
Phantom of the Ritz '88

Robert Bergman (D)

Skull: A Night of Terror '88
A Whisper to a Scream '88

Sandahl Bergman

Airplane 2: The Sequel '82
All That Jazz '79
Conan the Barbarian '82
Getting Physical '84
Hell Comes to Frogtown '88
Kandyland '87
Programmed to Kill '86
Raw Nerve '91
Red Sonja '85
She '83
Stewardess School '86
Xanadu '80

Art Bergmann

Highway 61 '91

Elisabeth Bergner

As You Like It '36
Cry of the Banshee '70
Dreaming Lips '37
The Pedestrian '74
Rise of Catherine the Great '34

Ary Beri

Stand Off '89

Luis Beristain

El: This Strange Passion '52
The Exterminating Angel '62

William Berke (D)

Bandit Queen '51
Betrayal from the East '44
Border Rangers '50
Dangerous Passage '44
Dick Tracy, Detective/The Corpse Vanishes '42
The Falcon in Mexico '44
Falcon's Adventure/Armored Car Robbery '50
FBI Girl '52
Gunfire '50
Highway 13 '48
I Shot Billy the Kid '50
Jungle '52
The Lost Missile '58
The Marshal's Daughter '53
Operation Haylift '50
Roaring City '51
Savage Drums '51
Shoot to Kill '47

Toll of the Desert '35
Treasure of Monte Cristo '50

Busby Berkeley

Babes on Broadway '41
Fashions of 1934 '34
42 Street '33
Million Dollar Mermaid '52
Roman Scandals '33

Busby Berkeley (D)

Babes in Arms '39
Babes on Broadway '41
For Me & My Gal '42
Gold Diggers of 1935 '35
Strike Up the Band '40
Take Me Out to the Ball Game '49
They Made Me a Criminal '39

Keith Berkeley

The Visitors '89

Leon Berkeley

The Lawless Land '88

Xander Berkeley

Assassin '89
Deadly Dreams '88
The Lawless Land '88
Short Time '90

Steven Berkoff

Beverly Hills Cop '84
The Krays '90
Prisoner of Rio '89
Rambo: First Blood, Part 2 '85
Revolution '85
Transmutations '86
Underworld '85
War & Remembrance, Part 1 '88

Terri Berland

The Strangeness '85

Milton Berle

Amazing Stories, Book 5 '85
The Bellboy '60
Broadway Danny Rose '84
Cracking Up '83
Hey Abbott! '78
It's a Mad, Mad, Mad, Mad World '63
Jonathan Winters on the Edge '86
Journey Back to Oz '71
Legend of Valentino '75
Lepke '75
Let's Make Love '60
The Loved One '65
Muppet Movie '79
Off Your Rocker '80
Oscar '66
Side By Side '88
Sun Valley Serenade '41
Who's Minding the Mint? '67

Francois Berleand

Au Revoir Les Enfants '88

Abby Berlin (D)

Double Deal '50

Irving Berlin

This is the Army '43

Jeannie Berlin

The Baby Maker '70
Getting Straight '70
The Heartbreak Kid '72
Housewife '72
In the Spirit '90
Portnoy's Complaint '72

The Strawberry Statement '70

Tom Berlin

Empire of Spiritual Ninja

Warren Berlinger

Billie '65
Four Deuces '75
Sex and the Single Parent '82
Spinout '66
Take Two '87

Harvey Berman (D)

The Wild Ride '60

Monty Berman (D)

Jack the Ripper '60

Pandro S. Berman (D)

Romance in Manhattan '34

Shelley Berman

The Best Man '64
More Laughing Room Only Rented Lips '88
Son of Blob '72
Think Dirty '70

Susan Berman

Smithereens '84

Patrick Bermel

Ladies of the Lotus '86

Tonia Bern

Glass Tomb '55

Armand Bernard

Napoleon '27

Chris Bernard (D)

Letter to Brezhnev '86

Crystal Bernard

Slumber Party Massacre 2 '87

Ed Bernard

Across 110th Street '72
Reflections of Murder '74

John Bernard (D)

Stay Awake '87

Michael Bernard (D)

Nights in White Satin '87

Paul Bernard

Pattes Blanches '49

Susan Bernard

Faster Pussycat! Kill! Kill! '66
Teenager '74

Thelonious Bernard

A Little Romance '79

Herschel Bernardi

Actor: The Paul Muni Story '78
The Front '76
Green Fields '37
Irma La Douce '63
Love with the Proper Stranger '63
No Place to Run '72

Edward L. Bernds (D)

Clipped Wings '53

Zena Bethune
Who's That Knocking at My Door? '68

Francesca Bett
Duel of Champions '61

John Bett
Gregory's Girl '80

Lyle Bettger
Carnival Story '54
Lone Ranger '56
Sea Chase '55
Union Station '50

Laura Betti
Allonsanfan '73
The Canterbury Tales '71
Hatchet for the Honeymoon '70
Lovers and Liars '81
1900 '76
Sonny and Jed '73
Teorema '68

Valerie Bettis
Affair in Trinidad '52

Robert Bettles
Fourth Wish '75
Ride a Wild Pony '75

Gil Bettman (D)
Crystal Heart '87
Never Too Young to Die '86

Franca Bettoya
The Last Man on Earth '64
Marauder '65

Jack Betts
Dead Men Don't Die '91

Jonathan Betuel (D)
My Science Project '85

Carl Betz
Deadly Encounter '78
The Meal '75
Spinout '66

Matthew Betz
From Broadway to Cheyenne '32
Tarzan the Fearless '33
The Unholy Three '25
Wedding March '28

Francois Beukelaers
Istanbul '85

Jack Beutel
The Outlaw '43

Billy Bevan
Comedy Classics '35
Girl of the Golden West '38
High Voltage '29
The Lizzies of Mack Sennett '15
Silent Laugh Makers, No. 2 '20s
Two Reelers: Comedy Classics 8 '33
The White Sin '24
The Wrong Road '37

Stewart Bevan
The Ghoul '75

Clem Bevans
Abe Lincoln in Illinois '40
Gold Raiders '51
Highway 13 '48
Rim of the Canyon '49

Carol Bever
Waitress '81

Eddie Beverly, Jr. (D)
Ballbuster
Escapist '83

Helen Beverly
Green Fields '37
The Light Ahead '39
Overture to Glory (Der Vilner Shtot Khazn) '40

Victor Bevine
Princess and the Call Girl '84

Rodney Bewes
Unidentified Flying Oddball '79

Leighton Bewley
The Ann Jillian Story '88

Turhan Bey
The Amazing Mr. X '48
Dragon Seed '44
Out of the Blue '47
Parole, Inc '49
Shadows on the Stairs '41

Troy Beyer
Rooftops '89

Richard Beymer
Cross Country '83
The Diary of Anne Frank '59
Indiscretion of an American Wife '54
Johnny Tremain & the Sons of Liberty '58
The Longest Day '62
Scream Free! '71
Silent Night, Deadly Night 3: Better Watch Out! '89
The Stripper '63
Twin Peaks '90
West Side Story '61

Didier Bezace
The Little Thief '89

Sandra Bezic
Snow Queen '82

Kim Bhang
Ninja Enforcer '70s

Andrew Bianchi (D)
Burial Ground '85
Cry of a Prostitute: Love Kills '72

Daniela Bianchi
Dirty Heroes '71
From Russia with Love '63
Secret Agent 00 '90

Edward Bianchi (D)
The Fan '81

Eleanora Bianchi
Goliath and the Sins of Babylon '64

Marta Bianchi
Made in Argentina '86

Roberto Bianco
Fight for Gold

Charles K. Bibby
Order of the Black Eagle '87
Unmasking the Idol '86

Abner Biberman
Betrayal from the East '44
Elephant Walk '54
His Girl Friday '40
The Keys of the Kingdom '44
Panama Patrol '39

Herbert Biberman (D)
Master Race '44
Salt of the Earth '54

Charles Bickford
Anna Christie '30
Babe Ruth Story '48
A Big Hand for the Little Lady '66
Branded '50
Command Decision '48
The Court Martial of Billy Mitchell '55
Days of Wine and Roses '62
Duel in the Sun '46
East of Borneo '31
The Farmer Takes a Wife '35
The Farmer's Daughter '47
Fatal Confinement '20s
Four Faces West '48
Guilty of Treason '50
Jim Thorpe: All American '51
Johnny Belinda '48
Little Miss Marker '34
Mr. Lucky '43
Mutiny in the Big House '39
Of Mice and Men '39
The Plainsman '37
Reap the Wild Wind '42
The Song of Bernadette '43
A Star is Born '54
Thunder Pass '37
The Unforgiven '60
A Wing and a Prayer '44

Joan Bickhill
Follow That Rainbow '66

Jean-Luc Bideau
Jonah Who Will Be 25 in the Year 2000 '76

Robert Bideman
Skull: A Night of Terror '88

Anthony Bido
Watch Me When I Kill '81

Michael Biehn
The Abyss '89
Aliens '86
Coach '78
The Fan '81
Hog Wild '80
In a Shallow Grave '88
K2 '92
Lords of Discipline '83
Navy SEALS '90
The Seventh Sign '88
The Terminator '84
Timebomb '91

Ann Christine Biel
Cosi Fan Tutte '85

Dick Biel
Slime City '89
Splatter University '84

Josef Bierbichler
Woyzeck '78

Ramon Bieri
The Frisco Kid '79
Grandview U.S.A. '84
Love Affair: The Eleanor & Lou Gehrig Story '77
A Matter of Life and Death '81
Nicole '72
Panic in Echo Park '77
The Passing of Evil '70
Sorcerer '77

Robert Bierman (D)
Apology '86
Vampire's Kiss '88

Kathryn Bigelow (D)
Blue Steel '90
Loveless '83
Near Dark '87
Point Break '91

Scott "Bam Bam" Bigelow
SnakeEater 3...His Law '92

Susan Bigelow
Wild & Wooly '78

Ven Bigelow
Soul Vengeance '75

Roxann Biggs
Guilty by Suspicion '91

Joseph Bigwood (D)
Never Pick Up a Stranger '79

Theodore Bikel
Above Us the Waves '56
The African Queen '51
Dark Tower '87
Darker Than Amber '70
Defiant Ones '58
Dog of Flanders '59
Enemy Below '57
Flight from Vienna '58
The Four Sons
I Bury the Living '58
I Want to Live! '58
Murder on Flight 502 '75
My Fair Lady '64
My Side of the Mountain '69
The Russians are Coming, the Russians are Coming '66
See You in the Morning '89
Stingiest Man in Town '78
T.G.I.S.(Thank Goodness It's Shabbat)
Very Close Quarters '84

Fernando Bilbao
Apocalipsis Joe '65
Fangs of the Living Dead '68
The Screaming Dead '72

David Bilcock (D)
Alvin Rides Again '74

Paul Bildt
The Blum Affair '48

John Bill
Red, White & Busted '75

Tony Bill
Are You in the House Alone? '78
Come Blow Your Horn '63

Haunts of the Very Rich '72
Heart Beat '80
Ice Station Zebra '68
Initiation of Sarah '78
The Killing Mind '90
Little Dragons '80
Never a Dull Moment '68
None But the Brave '65
Shampoo '75
Soldier in the Rain '63
Washington Mistress '81
You're a Big Boy Now '66

Tony Bill (D)
Crazy People '90
Five Corners '88
My Bodyguard '80
Princess and the Pea '83
Six Weeks '82

Raoul Billerey
The Grand Highway '88
The Little Thief '89

Don Billett
Prince of the City '81

Tom Billett
Lurkers '88

Dawn Ann Billings
Trancers 3: Deth Lives '92

Barbara Billingsley
Airplane! '80
Back to the Beach '87
Eye of the Demon '87

Jennifer Billingsley
Chrome Hearts '70
Hollywood Man '76
Lady in a Cage '64
The Thirsty Dead '74
White Lightning '73

Peter Billingsley
Beverly Hills Brats '89
A Christmas Story '83
Death Valley '81
Dirt Bike Kid '86
Russkies '87

Richard Billingsley
After the Promise '87

Sherman Billingsley
Stork Club '52

Kevin Billington (D)
Light at the Edge of the World '71

Michael Billington
The KGB: The Secret War

Bruce Bilson (D)
Chattanooga Choo Choo '84
The North Avenue Irregulars '79

Danny Bilson (D)
The Wrong Guys '88
Zone Troopers '85

Jiang Bin
Black & White Swordsmen '85

Jerzy Binczycki
Nights and Days '76

John Binder (D)
Uforia '81

Black ▶cont.

Blood Money '91
Born to Win '71
Burnt Offerings '76
Can She Bake a Cherry Pie? '83
Capricorn One '78
Chanel Solitaire '81
The Children '91
Club Fed '91
Come Back to the 5 & Dime Jimmy Dean, Jimmy Dean '82
Crime & Passion '75
Cut and Run '85
Day of the Locust '75
Dixie Lanes '88
Easy Rider '69
Eternal Evil '87
Evil Spirits '91
Family Plot '76
Five Easy Pieces '70
The Great Gatsby '74
A Gunfight '71
Haunting Fear '91
The Heist '88
Hitchhiker 1 '85
Hitz '92
Homer and Eddie '89
Hostage '87
In Praise of Older Women '78
Invaders from Mars '86
The Invisible Kid '88
It's Alive 3: Island of the Alive '87
Killer Fish '79
The Killer's Edge '90
Killing Heat '84
Last Word '80
Law and Disorder '74
Little Laura & Big John '73
The Little Mermaid '84
Love Under Pressure '80s
Martin's Day '85
Mirror, Mirror '90
Miss Right '81
Mr. Horn '79
Nashville '75
Night Angel '90
Out of the Dark '88
Overexposed '90
Portnoy's Complaint '72
The Pyx '73
Quiet Fire '91
The Rip Off '78
Rubin & Ed '92
Savage Dawn '84
Separate Ways '82
The Squeeze '80
Trilogy of Terror '75
Twisted Justice '89
You're a Big Boy Now '66
Zapped Again '89

Larry Black

Web of Deceit '90

Lindsay Armstrong Black

The Worm Eaters '77

Mary Black

A Song for Ireland '88

Maurice Black

Marked Money '28
Sixteen Fathoms Deep '34

Noel Black (D)

A Conspiracy of Love '87
Electric Grandmother '81
Eyes of the Panther '90

I'm a Fool '77
A Man, a Woman, & a Bank '79
Mirrors '78
Pretty Poison '68
Prime Suspect '82
Private School '83
Quarterback Princess '85

Royana Black

Almost Partners '87

Black Uhuru

Island Reggae Greats '85

Greta Blackburn

Party Line '88

Richard Blackburn

Lemora, Lady Dracula '73

Richard Blackburn (D)

Lemora, Lady Dracula '73

Gene Blackey

Sweet Trash '89

Bond Blackman

Trauma '62

Honor Blackman

The Cat and the Canary '79
Fright '71
Glass Tomb '55
Goldfinger '64
Green Grow the Rushes '51
Lola '69
A Matter of Who '62
A Night to Remember '58
Suspended Alibi '56
To the Devil, A Daughter '76
The Virgin and the Gypsy '70

Jack Blackman

Shalako '68

Joan Blackman

Blue Hawaii '62
Career '59
Daring Game '68
Kid Galahad '62
Macon County Line '74

Sidney Blackmer

Beyond a Reasonable Doubt '56
Buffalo Bill '44
Count of Monte Cristo '34
Deluge '33
Goodbye Love '34
High Society '56
The House of Secrets '37
How to Murder Your Wife '64
Little Caesar '30
The Little Colonel '35
The Panther's Claw '42
The President's Mystery '36
Rosemary's Baby '68
Tammy and the Bachelor '57
Transatlantic Merry-Go-Round '34
War of the Wildcats '43

Harry Blackstone, Jr.

Blackstone on Tour '84

Ron Blackstone

Two Wrongs Make a Right '89

Carlyle Blackwell

She '21

David Blackwell

China O'Brien '88

Evelyn Blackwell

Ashes and Embers '82

Monika Blackwell

The Wizard of Gore '70

Christian Blackwood (D)

Roger Corman: Hollywood's Wild Angel '78

Taurean Blacque

Deepstar Six '89
The $5.20 an Hour Dream '80

Richard Blade

Spellcaster '91

Steven Blade

Alien Seed '89

Ruben Blades

Crazy from the Heart '91
Critical Condition '86
Crossover Dreams '85
Dead Man Out '89
Disorganized Crime '89
Fatal Beauty '87
Homeboy '88
The Josephine Baker Story '90
Last Fight '82
The Lemon Sisters '90
The Milagro Beanfield War '88
Mo' Better Blues '90
One Man's War '90
Predator 2 '90
The Return of Ruben Blades '87
The Super '91
The Two Jakes '90

Dasha Blahova

Howling 3 '87

Estella Blain

The Diabolical Dr. Z '65
Pirates of the Coast '61

Gerard Blain

The Cousins '59
Hatari '62
Le Beau Serge '58

Cullen Blaine (D)

R.O.T.O.R. '88

Jerry Blaine

Blood of Dracula '57

Sally Blane

Cross Examination '32
Law of the Sea '38
When the West Was Young '32

Vivian Blaine

The Cracker Factory '79
The Dark '79
Doll Face '45
Guys & Dolls '55
Parasite '81
Skirts Ahoy! '52
State Fair '45

Betsy Blair

Marty '55

Eleanor Blair

Violent Women '59

Janet Blair

The Black Arrow '48
Burn Witch, Burn! '62
The Fabulous Dorseys '47
Fuller Brush Man '48
One and Only, Genuine, Original Family Band '68
Tonight and Every Night '45

Kevin Blair

Friday the 13th, Part 7: The New Blood '88
Hills Have Eyes, Part 2 '84

Linda Blair

Airport '75 '75
Bail Out '90
Bedroom Eyes 2 '89
Born Innocent '74
Chained Heat '83
The Chilling '89
Dead Sleep '91
The Exorcist '73
Exorcist 2: The Heretic '77
Grotesque '87
Hell Night '81
Moving Target '89
Night Patrol '85
Nightforce '86
Red Heat '85
Repossessed '90
Ruckus '81
Savage Island '85
Savage Streets '83
Silent Assassins '88
The Sporting Club '72
Summer of Fear '78
Sweet Hostage '75
Up Your Alley '89
Wild Horse Hank '79
Witchery '88
A Woman Obsessed '88
Zapped Again '89

Nicky Blair

Viva Las Vegas '63

Pamela Blair

Svengali '83

Reno Blair

Gentleman from Texas '46

Tom Blair

The Game '89

Deborah Blaisdell

Student Affairs '88
Wildest Dreams '90
Wimps '87

Paul Blaisdell

It Conquered the World '56

Alfonso Corona Blake (D)

Samson vs. the Vampire Women '61
Samson in the Wax Museum '63
World of the Vampires '60

Amanda Blake

Betrayal
The Boost '88
B.O.R.N. '88
Duchess of Idaho '50
The Glass Slipper '55
Sabre Jet '53

Eubie Blake

Bessie Smith and Friends '86

Frank Blake

Return of the Evil Dead '75

Geoffrey Blake

The Abduction of Kari Swenson '87
Fatal Exposure '91

Jon Blake

Anzacs: The War Down Under '85
Early Frost '84
Freedom '81
The Lighthorseman '87

Julia Blake

Lonely Hearts '82
Man of Flowers '84
Travelling North '87
Under Capricorn '82

Karen Blake

Sundance and the Kid '76

Larry J. Blake

Holiday Affair '49

Marie Blake

Love Finds Andy Hardy '38
Sensations of 1945 '44

Pamela Blake

Border Rangers '50
Case of the Baby Sitter '47
Gunfire '50
Hat Box Mystery '47
Highway 13 '48
Kid Dynamite '43
Rolling Home '48
Sea Hound '47
Sky Liner '49
Son of God's Country '48

Peter Blake

Murder on Line One '90

Robert (Bobby) Blake

The Adventures of Rin Tin Tin '47
Andy Hardy's Double Life '42
Busting '74
Coast to Coast '80
Counter Punch '71
Electra Glide in Blue '73
Heart of a Champion: The Ray Mancini Story '85
Homesteaders of Paradise Valley '47
Humoresque '47
In Cold Blood '67
Of Mice and Men '81
Our Gang Comedies '40s
Pork Chop Hill '59
PT 109 '63
Santa Fe Uprising '46
Stagecoach to Denver '46
Tell Them Willie Boy Is Here '69
This Property is Condemned '66
Treasure of the Golden Condor '53
Treasure of the Sierra Madre '48
Vigilantes of Boom Town '46
Western Double Feature 12 '44
Western Double Feature 13 '40s

Western Double Feature 21
'40s

Sarah Blake

Mom & Dad '47

T.C. Blake (D)

Nightflyers '87

Teresa Blake

Payback '90

Valerie Blake

Blastfighter '85

Whitney Blake

Stranger Who Looks Like Me
'74

Brien Blakely

HauntedWeen '91

Colin Blakely

The Big Sleep '78
The Dogs of War '81
Equus '77
Evil Under the Sun '82
Little Lord Fauntleroy '80
Loophole '83
Love Among the Ruins '75
Murder on the Orient Express
'74
Nijinsky '80
Operation Julie '85
The Pink Panther Strikes
Again '76
The Private Life of Sherlock
Holmes '70
Shattered '72
This Sporting Life '63

Donald Blakely

In the Shadow of Kilimanjaro
'86
Short Eyes '79
Strike Force '75

**James "Doc"
Blakely**

Small Town Boy '37

Rubel Blakely

Sepia Cinderella '47

Susan Blakely

Blackmail '91
The Concorde: Airport '79 '79
A Cry for Love '80
Dead Reckoning '89
Dreamer '79
Hiroshima Maiden '88
The Incident '89
The Lords of Flatbush '74
Make Me an Offer '80
My Mom's a Werewolf '89
Out of Sight, Out of Her Mind
'89
Over the Top '86
Report to the Commissioner
'74
Savages '72
Secrets '77
The Survivalist '87
Towering Inferno '74

**Michael
Blakemore** (D)

Privates on Parade '84

Art Blakey

Jazz is Our Religion '72

Ronee Blakley

The Baltimore Bullet '80
Desperate Women '78
The Driver '78
Lightning Over Water '80
Mannikin '77
Murder by Numbers '89
Nashville '75
A Nightmare on Elm Street '84
The Private Files of J. Edgar
Hoover '77
Return to Salem's Lot '87
She Came to the Valley '77
Someone to Love '87
Student Confidential '87

Leslie Blalock

Incoming Freshmen '79

Erika Blanc

The Devil's Nightmare '71
Kill, Baby, Kill '66
Mark of the Devil 2 '72
The Night Evelyn Came Out of
the Grave '71
Sartana's Here...Trade Your
Pistol for a Coffin '70
Special Forces '80s
Sweet Spirits '80s

Mel Blanc

Best of Warner Brothers:
Volume 2 '30s
Betty Boop '30s
Bugs! '88
Bugs Bunny: Hollywood
Legend '90
Bugs Bunny in King Arthur's
Court '89
Bugs Bunny Superstar '75
Bugs Bunny's Comedy
Classics '88
Bugs Bunny's Hare-Raising
Tales '89
Bugs Bunny's Looney
Christmas Tales '79
Bugs Bunny's Wacky
Adventures '57
Bugs & Daffy's Carnival of the
Animals '89
Daffy Duck's Quackbusters
'89
Dino & Juliet '89
Flintstones '60
Gay Purr-ee '62
Hey There, It's Yogi Bear '64
A Jetson Christmas Carol '89
The Jetsons Meet the
Flintstones '88
The Jetsons: Movie '90
Kiss Me, Stupid! '64
Looney Looney Looney Bugs
Bunny Movie '81
Neptune's Daughter '49
Phantom Tollbooth '69
Porky Pig Tales '89
Salute to Chuck Jones '60
Salute to Friz Freleng '58
Salute to Mel Blanc '58
Sylvester & Tweety's Crazy
Capers '61
Who Framed Roger Rabbit?
'88

Michel Blanc

The Favor, the Watch, & the
Very Big Fish '92
Monsieur Hire '89

Ken Blancato (D)

Stewardess School '86

Jewel Blanch

Against A Crooked Sky '75

**Dominique
Blanchar**

L'Avventura '60

Pierre Blanchar

Crime and Punishment '35
La Symphonie Pastorale '46
Man from Nowhere

Alan Blanchard

Slithis '78

**Felix "Doc"
Blanchard**

The Spirit of West Point '47

Francois Blanchard

Venus on Fire '84

John Blanchard (D)

The Last Polka '84

Keith Blanchard

Portrait of Grandpa Doc '77

Mari Blanchard

Son of Sinbad '55

Susan Blanchard

President's Mistress '78
She's in the Army Now '81

Roland Blanche

Too Beautiful for You '88

Dorothee Blanck

Cleo from 5 to 7 '61
Umbrellas of Cherbourg '64

Joyce Bland

Dreaming Lips '37

Peter Bland

Came a Hot Friday '85
Dangerous Orphans '80s
Tearaway '87

Clara Blandick

The Adventures of Tom
Sawyer '38
Dreaming Out Loud '40
Infamous Crimes '47
Murder at Midnight '31
Small Town Boy '37
The Wizard of Oz '39

Sally Blane

Fighting Mad '39
Heritage of the Desert '33
Phantom Express '32
Probation '32
The Silver Streak '34
Vagabond Lover '29

Karen Blanguernon

No Survivors, Please '63

Harrod Blank (D)

In the Land of the Owl Turds
'89

Les Blank (D)

Burden of Dreams '82
Cigarette Blues '89
Gap-Toothed Women '89
Marc and Ann '91
Yum, Yum, Yum! '91

Mark Blankfield

Angel 3: The Final Chapter '88

Frankenstein General Hospital
'88
Jack & the Beanstalk '83
Jekyll & Hyde...Together
Again '82
The Midnight Hour '86

Billy Blanks

Bloodfist '89
The King of the Kickboxers
'91

Arell Blanton

House of Terror '87

**Alexandro
Blasetti** (D)

The Iron Crown '41

Freddie Blassie

My Breakfast with Blassie '83

**William Peter
Blatty** (D)

The Ninth Configuration '79

Charles Blavette

Toni '34

**Hans-Christian
Blech**

The Blum Affair '48
Dirty Hands '76
The Scarlet Letter '73
The Wrong Move '78

Jeff Bleckner (D)

Brotherly Love '85
When Your Lover Leaves '83
White Water Summer '87

William Bledsoe

Dark Side of the Moon '90

Debra Blee

Beach Girls '82
Malibu Bikini Shop '86
Savage Streets '83
Sloane '84

Brian Blessed

Flash Gordon '80
Henry V '89
High Road to China '83
I, Claudius, Vol. 1: A Touch of
Murder/Family Affairs '91
I, Claudius, Vol. 2: Waiting in
the Wings/What Shall We
Do With Claudius? '80
I, Claudius, Vol. 3: Poison is
Queen/Some Justice '80
King Arthur, the Young
Warlord '75
Man of La Mancha '72
Prisoner of Honor '91
Return to Treasure Island,
Vol. 1 '85
Return to Treasure Island,
Vol. 2 '85
Return to Treasure Island,
Vol. 3 '85
Return to Treasure Island,
Vol. 4 '85
Return to Treasure Island,
Vol. 5 '85
Robin Hood: Prince of Thieves
'91
Trojan Women '71
Young Warlord '75

Brenda Blethyn

The Witches '90

Michael Blevins

A Chorus Line '85

Jason Blicker

African Journey '91
American Boyfriends '89

Georg Blickingberg

Walpurgis Night '41

Bernard Blier

Buffet Froid '79
Casanova '70 '65
Catch Me a Spy '71
The Daydreamer '77
Dedee D'Anvers '49
Jenny Lamour '47
Les Miserables '57
Man in the Raincoat '57
Passion for Life '48
The Tall Blond Man with One
Black Shoe '72

Bertrand Blier (D)

Beau Pere '81
Buffet Froid '79
Get Out Your Handkerchiefs
'78
Going Places '74
My Best Friend's Girl '84
Too Beautiful for You '88

Maggie Blinco

The Wicked '89

Bob Blizz (D)

Vengeance '80

Bruce Block (D)

Princess Academy '87

Hunt Block

Secret Weapons '85

Larry Block

Dead Man Out '89

Dan Blocker

Errand Boy '61
Lady in Cement '68
Ride the Wind '66

Michael Blodgett

The Ultimate Thrill '74
The Velvet Vampire '71

Joan Blondell

Angel Baby '61
Battered '78
Brother Orchid '40
Bullets or Ballots '38
Cincinnati Kid '65
Dames '34
The Dead Don't Die '75
Death at Love House '75
Desk Set '57
Footlight Parade '33
The Glove '78
Gold Diggers of 1933 '33
Hollywood Bloopers '50s
Lady for a Night '42
Millie '31
The Opposite Sex '56
Public Enemy '31
The Rebels '79
Sergeant Sullivan Speaking
'53
Son of Hollywood Bloopers
'40s
Stand-In '37
Stay Away, Joe '68
Three Broadway Girls '32
Three on a Match '32

Martin Boddeg

Valley of the Eagles '51

Tain Bodkin

At Gunpoint '90

Karl-Heinz Boehm

Fox and His Friends '75
Peeping Tom '63
Unnatural '52
Wonderful World of the
  Brothers Grimm '62

Earl Boen

Terminator 2: Judgement Day
  '91

Herbert Boenne

Mill of the Stone Women '60

Beatrice Boepple

Quarantine '89

Kathryn Boese

Revenge of the Radioactive
  Reporter '91

Budd Boetticher (D)

Behind Locked Doors '48
Bullfighter & the Lady '50
Horizons West '52
The Magnificent Matador '55
Man from the Alamo '53
The Rise and Fall of Legs
  Diamond '60
The Tall T '57

Sandra Bogan

Punk Vacation '90

Ted Bogan

Louie Bluie '85

Dirk Bogarde

Accident '67
The Blue Lamp '49
A Bridge Too Far '77
Damn the Defiant '62
The Damned '69
Darling '65
Death in Venice '71
Despair '79
Doctor in Distress '63
Doctor in the House '53
Doctor at Large '57
Doctor at Sea '56
Epic That Never Was '65
I Could Go on Singing '63
Ingrid '89
Justine '69
Night Ambush '57
Night Flight from Moscow '73
The Night Porter '74
Permission To Kill '75
Providence '76
The Sea Shall Not Have Them
  '55
Sebastian '68
The Servant '63
Simba '55
The Sleeping Tiger '54
Song Without End '60
The Spanish Gardener '57
A Tale of Two Cities '58
To See Such Fun '81
Victim '62
The Vision '87

Humphrey Bogart

Across the Pacific '42
Action in the North Atlantic '43
The African Queen '51
All Through the Night '42

Angels with Dirty Faces '38
Bacall on Bogart '88
Barefoot Contessa '54
Battle Circus '53
Beat the Devil '53
The Best of Bogart '43
The Big Sleep '46
Brother Orchid '40
Bullets or Ballots '38
The Caine Mutiny '54
Call It Murder '34
Casablanca '42
Chain Lightning '50
Conflict '45
Dark Passage '47
Dark Victory '39
Dead End '37
Dead Reckoning '47
Desperate Hours '55
The Enforcer '51
Great Movie Trailers '70s
Harder They Fall '56
High Sierra '41
Hollywood Outtakes & Rare
  Footage '83
In a Lonely Place '50
Key Largo '48
Knock on Any Door '49
The Left Hand of God '55
The Maltese Falcon '41
Marked Woman '37
Midnight '34
Oklahoma Kid '39
Passage to Marseilles '44
Petrified Forest '36
Return of Video Yesterbloop
  '47
The Roaring Twenties '39
Sabrina '54
Sahara '43
Sirocco '51
Stand In '37
Thank Your Lucky Stars '43
They Drive by Night '40
Three on a Match '32
To Have & Have Not '44
Tokyo Joe '49
Treasure of the Sierra Madre
  '48
The Two Mrs. Carrolls '47
Virginia City '40
We're No Angels '55

Paul Bogart (D)

Cancel My Reservation '72
Class of '44 '73
Marlowe '69
Oh, God! You Devil '84
Power, Passion & Murder '83
Skin Game '71
Tell Me Where It Hurts '74
Torch Song Trilogy '88

Yuri Bogatyrev

An Unfinished Piece for a
  Player Piano '77

**Yurek
Bogayevicz** (D)

Anna '87

Peter Bogdanovich

The Trip '67

**Peter
Bogdanovich** (D)

Daisy Miller '74
Illegally Yours '87
The Last Picture Show '71
Mask '85
Paper Moon '73
Saint Jack '79
Targets '69
Texasville '90

They All Laughed '81
Voyage to the Planet of
  Prehistoric Women '68
What's Up, Doc? '72

Willy Bogner (D)

Fire and Ice '87
Fire, Ice and Dynamite '91

**Lucic Bogoljub
Benny**

SS Girls '78

Eric Bogosian

The Caine Mutiny Court
  Martial '88
Funhouse '87
Last Flight Out: A True Story
  '90
Special Effects '85
Suffering Bastards '90
Talk Radio '88

Kelley Bohanan

Idaho Transfer '73

John Bohn

The Crime of Dr. Crespi '35

Roman Bohnen

The Best Years of Our Lives
  '46
The Hairy Ape '44

Jose Bohr

Rogue of the Rio Grande '30

Corinne Bohrer

Stewardess School '86
Vice Versa '88

Richard Bohringer

Caged Heart '85
Diva '82
The Grand Highway '88
L'Addition '85
Le Grand Chemin '87
Peril '85

Jean Boht

The Girl in a Swing '89

Elaine Boies

Legacy of Horror '78

Curt Bois

Bitter Sweet '41
Boom Town '40
Caught '49
The Great Waltz '38
The Lady in Question '40
The Tuttles of Tahiti '42
Wings of Desire '88

Michel Boisrond (D)

Catherine & Co. '76
That Naughty Girl '58
Voulez-Vous Danser Avec
  Moi? '59

Jocelyne Boisseau

Girls Riot '88

Yves Boisset (D)

Cobra '71
Dog Day '83
Double Identity '89
Prize of Peril '84
The Purple Taxi '77

Christine Boisson

Sorceress '88
Unsettled Land '88

Jerome Boivin (D)

Baxter '89

**Alberto
Bojorquez** (D)

Los Meses y los Dias '70

Hal Bokar

The Last Reunion '80

James Bolam

Crucible of Terror '72
Dressed for Death '74
In Celebration '75
The Maze '85

Marc Bolan

Born to Boogie '72

Mary Boland

New Moon '40
Pride and Prejudice '40
Ruggles of Red Gap '35
The Women '39

**Jose Antonio
Bolanos** (D)

Born in America '90

Cal Bolder

Jesse James Meets
  Frankenstein's Daughter
  '65

Buddy Boles

Reg'lar Fellers '41

John Boles

Craig's Wife '36
Curly Top '35
Frankenstein '31
King of Jazz '30
Littlest Rebel '35
Sinners in Paradise '38
Stand Up and Cheer '34
Stella Dallas '37
Thousands Cheer '43

Steve Boles

Rockin' Road Trip '85

**Richard
Boleslawski** (D)

The Garden of Allah '36
The Last of Mrs. Cheyney '37
Les Miserables '35
The Painted Veil '34

John Bolger

Parting Glances '86

Ray Bolger

April In Paris '52
Babes in Toyland '61
Daydreamer '66
For Heaven's Sake '79
Four Jacks and a Jill '41
The Great Ziegfeld '36
The Harvey Girls '46
Rosalie '37
The Runner Stumbles '79
Stage Door Canteen '43
Sunny '41
Sweethearts '38
That's Dancing! '85
The Wizard of Oz '39

Elena Bolkan

Acqua e Sapone '83

Florinda Bolkan

Acqua e Sapone '83
Collector's Item '89
Day that Shook the World '78
The Last Valley '71
Master Touch '74
Ring of Death '72
Some Girls '88
The Word '78

Tiffany Bolling

Bonnie's Kids '73
The Centerfold Girls '74
Ecstasy '84
Kingdom of the Spiders '77
Love Scenes '84
Wild Party '74

Michael Bollner

Willy Wonka & the Chocolate
  Factory '71

Joseph Bologna

Alligator 2: The Mutation '90
Big Bus '76
Blame It on Rio '84
Chapter Two '79
Cops and Robbers '73
Coupe de Ville '90
Honor Thy Father '73
My Favorite Year '82
One Cooks, the Other Doesn't
  '83
Rags to Riches '87
Torn Between Two Lovers '79
Transylvania 6-5000 '85
The Woman in Red '84

**Mauro
Bolognini** (D)

Husbands and Lovers '92
The Inheritance '76
La Grande Bourgeoise '74
Oldest Profession '67

Ben Bolt (D)

Big Town '87

Faye Bolt

Star Crystal '85

Jean Bolt

Arthur's Hallowed Ground '84

Jonathan Bolt

Eyes of the Amaryllis '82

Pat Bolt

Natas...The Reflection '83

Robert Bolt (D)

Lady Caroline Lamb '73

Emily Bolton

Moonraker '79

Heather Bolton

Dark of the Night '85

Bon Jovi

Incident at Channel Q '86

Jon Bon Jovi

Young Guns 2 '90

Sheila Bon

Curse of the Stone Hand '64

Paolo Bonacelli

Christ Stopped at Eboli '79
Henry IV '85

Wade Boteler
Big News '29
The Green Hornet '40
Let 'er Go Gallegher '28
The Mandarin Mystery '37

Sara Botsford
By Design '82
Deadly Eyes '82
The Gunrunner '84
Murder by Phone '82
Still of the Night '82

Zeinab Botsvadze
Repentance '87

Benjamin Bottoms
A Shining Season '79
Stalk the Wild Child '76

Gaye Bottoms
The Abomination '88

Joseph Bottoms
Black Hole '79
Blind Date '84
Born to Race '88
Celebrity '85
Cloud Dancer '80
Crime & Passion '75
The Dove '74
High Rolling in a Hot Corvette '77
Holocaust '78
Inner Sanctum '91
Intruder Within '81
King of the Mountain '81
Open House '86
Return Engagement '78
Sins of Dorian Gray '82
Stalk the Wild Child '76
Surfacing '84

Sam Bottoms
After School '88
Apocalypse Now '79
Bronco Billy '80
Class of '44 '73
Desperate Lives '82
Dolly Dearest '92
East of Eden '80
Gardens of Stone '87
Hearts of Darkness: A Filmmaker's Apocalypse '91
Hunter's Blood '87
In 'n Out '86
Joseph & His Brothers '79
Outlaw Josey Wales '76
Prime Risk '84
Ragin' Cajun '90
Return to Eden '89
Savages '75
Up from the Depths '79

Timothy Bottoms
The Drifter '88
East of Eden '80
The Fantasist '89
The Gift of Love '90
Hambone & Hillie '84
High Country '81
Hurricane '79
Husbands, Wives, Money, and Murder '86
In the Shadow of Kilimanjaro '86
Invaders from Mars '86
Istanbul '90
Johnny Got His Gun '71
The Land of Faraway '87
The Last Picture Show '71
Love Leads the Way '84

The Other Side of the Mountain, Part 2 '78
The Paper Chase '73
Rollercoaster '77
The Sea Serpent '85
A Shining Season '79
A Small Town in Texas '76
Texasville '90
Tin Man '83
What Waits Below '83
The White Dawn '75

Shiao Bou-Lo
Dance of Death '84

Pegi Boucher
Private Duty Nurses '71

Sherry Boucher
Sisters of Death '76

Barbara Bouchet
Cry of a Prostitute: Love Kills '72
In Harm's Way '65
Mean Machine '73
Rogue '76
Surabaya Conspiracy '75

Patrick Bouchitey
The Best Way '82

Jetta Boudal
Road to Yesterday/The Yankee Clipper '27

Fred Boudin
Summer Job '88

Geoffery Boues
Kidnapping of Baby John Doe '88

Jean-Claude Bouillon
Julia '74

Jean Bouise
Le Dernier Combat '84
Old Gun '76

Evelyne Bouix
Edith & Marcel '83

Michel Boujenah
Three Men and a Cradle '85

Jonathan Bould
Ninja Powerforce '90

Ingrid Boulting
Deadly Passion '85
The Last Tycoon '77

John Boulting (D)
Brighton Rock '47
Heavens Above '63
I'm All Right Jack '69
Lucky Jim '58

Roy Boulting (D)
Carlton Browne of the F.O. '59
Fame is the Spur '47
Heavens Above '63
There's a Girl in My Soup '70

Carol Bouquet
Buffet Froid '79
Dagger Eyes '80s
For Your Eyes Only '81
That Obscure Object of Desire '77

Too Beautiful for You '88

Michel Bouquet
Le Complot (The Conspiracy) '73
Pattes Blanches '49
This Special Friendship '67

Serge Bourguignon (D)
A Coeur Joie '67
Head Over Heels '67
Sundays & Cybele '62

Whitney Bourne
Flight from Glory '37

Philip Bourneuf
Beyond a Reasonable Doubt '56

Antoine Bourseillor
Cleo from 5 to 7 '61

Bourvil
The Brain '69
Crazy for Love '52
Four Bags Full '56
Les Grandes Gueules '65
Les Miserables '57
Mr. Peek-A-Boo '50
When Wolves Cry '69

Andre Bourvil
The Christmas Tree '69

Kathleen Boutall
The Golden Salamander '51

Genee Boutell
Rawhide Romance '34

Dennis Boutsikaris
The Dream Team '89

Robert Bouvier (D)
City in Panic '86

Haim Bouzaglo (D)
Fictitious Marriage '88

Brunella Bova
Miracle in Milan '51
The White Sheik '52

Joe Bova
Roboman '75

Julie Bovasso
Gentleman Bandit '81
Moonstruck '87
Wise Guys '86

Clara Bow
Dancing Mothers '26
Down to the Sea in Ships '22
Free to Love '25
Hula '28
It '27
Mantrap '26
My Lady of Whims '25
Wings '27

Simmy Bow
The Doberman Gang '72

Tui Bow
Frenchman's Farm '87

Dorris Bowdon
Drums Along the Mohawk '39
The Grapes of Wrath '40

Dennis Bowen
Gas Pump Girls '79
Van Nuys Blvd. '79

Jenny Bowen (D)
Street Music '81
The Wizard of Loneliness '88

John Bowen (D)
Knock Outs '92

Michael Bowen
The Abduction of Kari Swenson '87
Echo Park '86
Mortal Passions '90
Night of the Comet '84
Season of Fear '89

Roger Bowen
M*A*S*H '70
What About Bob? '91
Zapped! '82

Sean Bowen
Troma's War '88

Malick Bowens
The Believers '87
Out of Africa '85

Antoinette Bower
Blood Song '82
Die Sister, Die! '74
The Evil That Men Do '84

Dallas Bower (D)
Alice in Wonderland '50

Tom Bower
Ballad of Gregorio Cortez '83
Family Sins '87
The Lightship '86
Massive Retaliation '85
True Believer '89
Wildrose '85

Anthony Bowers (D)
Shame, Shame on the Bixby Boys '92

Geoffrey G. Bowers (D)
Danger Zone 2 '89

George Bowers (D)
Body & Soul '81
The Hearse '80
My Tutor '82
Private Resort '84

John Bowers
The Sky Pilot '21
The White Sin '24

John R. Bowey (D)
Mutator '90

David Bowie
Absolute Beginners '86
Christiane F. '82
The Hunger '83
Inside the Labyrinth '86
Into the Night '85
Just a Gigolo '79
Labyrinth '86
The Last Temptation of Christ '88
The Linguini Incident '92
The Man Who Fell to Earth '76
Merry Christmas, Mr. Lawrence '83

UHF '89
Yellowbeard '83

Rosemary Bowie
Big Bluff '55

Judi Bowker
Brother Sun, Sister Moon '73
Clash of the Titans '81
East of Elephant Rock '76
The Shooting Party '85

Peter Bowles
A Day in the Death of Joe Egg '71
Endless Night '71

Dave Bowling
Dark Side of Midnight '86

F.E. Bowling
Oklahoma Bound '81

Don Bowman
Hillbillys in a Haunted House '67

Laura Bowman
Birth Right '39
Drums O'Voodoo '34
Son of Ingagi '40

Lee Bowman
Bataan '43
Cover Girl '44
Love Affair '39
My Dream is Yours '49
Next Time I Marry '38
Smash-Up: The Story of a Woman '47
Tonight and Every Night '45

Loretta L. Bowman
Offerings '89

Ralph Bowman
Flaming Frontiers '38

Riley Bowman
Bloodfist '89

Tina Bowmann
Kill Alex Kill '83

John Bown (D)
Nudity Required '90

Kenneth Bowser (D)
In a Shallow Grave '88

Muriel Box (D)
Rattle of a Simple Man '64
Truth About Women '58

Bruce Boxleitner
The Babe '91
The Baltimore Bullet '80
Diplomatic Immunity '91
East of Eden '80
Kenny Rogers as the Gambler '80
Kenny Rogers as the Gambler, Part 2: The Adventure Continues '83
Kuffs '91
Murderous Vision '91
Passion Flower '86
Tron '82

Sully Boyar
Car Wash '76

Alan Boyce

Permanent Record '88

William Boyce

The Slime People '63

Beverly Boyd

Ghost Rider '43
Jive Junction '43

Blake Boyd

Dune Warriors '91
Raiders of the Sun '92

Daniel Boyd (D)

Chillers '88
Invasion of the Space
Preachers '90

Don Boyd (D)

East of Elephant Rock '76
Twenty-One '91

Dorothy Boyd

A Shot in the Dark '33

Guy Boyd

Body Double '84
Eyes of the Amaryllis '82
Eyes of Fire '84
Keeping On '81
Kiss Me a Killer '91
The Last of the Finest '90
Murder Ordained '87

Ian Boyd

Skier's Dream '88

Janette Boyd

Stripper '86

Joe Boyd (D)

Jimi Hendrix: Story '73

Julianne Boyd (D)

Eubie! '82

Karin Boyd

Mephisto '81

Rick Boyd

Guns for Dollars '73

Sarah Boyd

Old Enough '84

Stephen Boyd

Ben-Hur '59
Bible...In the Beginning '66
Billy Rose's Jumbo '62
Black Brigade
The Bravados '58
Caper of the Golden Bulls '67
The Fall of the Roman Empire
'64
Fantastic Voyage '66
Hannie Caulder '72
Imperial Venus '71
The Man Who Never Was '55
Oscar '66
Shalako '68
The Squeeze '77
The Treasure of Jamaica Reef
'74

Tanya Boyd

Black Shampoo '76

William Boyd

Along the Sundown Trail '42
Bar-20 '43
Devil's Playground '46

East Side Kids/The Lost City
'35
False Colors '43
Flying Fool '29
Having Wonderful Time/
Carnival Boat '38
High Voltage '29
His First Command '29
Hopalong Cassidy: Borrowed
Trouble '48
Hopalong Cassidy: Dangerous
Venture '48
Hopalong Cassidy: False
Paradise '48
Hopalong Cassidy: Hoppy's
Holiday '47
Hopalong Cassidy: Riders of
the Deadline '43
Hopalong Cassidy: Silent
Conflict '48
Hopalong Cassidy: Sinister
Journey '48
Hopalong Cassidy: The Dead
Don't Dream '48
Hopalong Cassidy: The Devil's
Playground '46
Hopalong Cassidy: The
Marauders '47
Hopalong Cassidy:
Unexpected Guest '47
King of Kings '27
Laughing at Life '33
The Leatherneck '28
The Lost City '34
Midnight Warning '32
Oliver Twist '33
Painted Desert '31
Power '28
Prairie Pals '42
Renegade Trail '39
Riders of the Timberline '41
The Road to Yesterday '25
Road to Yesterday/The
Yankee Clipper '27
Rustler's Valley '37
Santa Fe Marshal '40
The Showdown '40
Sinister Journey '48
State's Attorney '31
Texas Masquerade '44
Tumbleweed Trail '42
Twilight on the Trail '41
Yankee Clipper '27

Sally Boyden

Barnaby and Me '77
Little Dragons '80

Chance Boyer

One Man Force '89

Charles Boyer

Adorable Julia '62
Algiers '38
All This and Heaven Too '40
April Fools '69
Arch of Triumph '48
Around the World in 80 Days
'56
Barefoot in the Park '67
Break of Hearts '35
The Buccaneer '58
Casino Royale '67
The Cobweb '55
Conquest '32
The Earrings of Madame De...
'54
Fanny '61
Four Horsemen of the
Apocalypse '62
The Garden of Allah '36
Gaslight '44
Heart of a Nation '43
History is Made at Night '37
Hot Line '69

Is Paris Burning? '68
Liliom '35
Love Affair '39
A Matter of Time '76
Mayerling '36
Nana '55
Red Headed Woman '32
Stavisky '74

Jean Boyer (D)

Dressmaker '56
Fernandel the Dressmaker '57
The French Touch '54
Mr. Peek-A-Boo '50

Katy Boyer

Long Gone '87
Tapeheads '89
Trapped '89

Miriam Boyer

Jonah Who Will Be 25 in the
Year 2000 '76
Window Shopping '86

Sully Boyer

The Manhattan Project '86

Pat Boyette

Dungeon of Harrow '64

Pat Boyette (D)

Dungeon of Harrow '64

Tom Boylan

Night Shadow '90

Lara Flynn Boyle

Eye of the Storm '91
How I Got Into College '89
May Wine '90
Mobsters '91
Poltergeist 3 '88
The Rookie '90
Twin Peaks '90
Wayne's World '92

Peter Boyle

Beyond the Poseidon
Adventure '79
Brink's Job '78
The Candidate '72
Conspiracy: The Trial of the
Chicago Eight '87
The Dream Team '89
Echoes in the Darkness '87
F.I.S.T. '78
From Here to Eternity '79
Ghost in the Noonday Sun '74
Hammett '82
Hardcore '79
Joe '70
Johnny Dangerously '84
Men of Respect '91
Outland '81
Red Heat '88
Slither '73
Speed Zone '88
Steelyard Blues '73
Surrender '87
Swashbuckler '76
Taxi Driver '76
Tragedy of Flight 103: The
Inside Story '91
Turk 182! '85
Walker '87
Where the Buffalo Roam '80
Yellowbeard '83
Young Frankenstein '74

Ray Boyle

Satan's Satellites '58

Richard Boyle

The Situation '87

William Boyle

Hawk of the Wilderness '38

Max Bozyk

The Jolly Paupers '38
Mamele (Little Mother) '38
Yidl Mitn Fidl (Yiddle with His
Fiddle) '36

Reizl Bozyk

Crossing Delancey '88

Bruno Bozzetto (D)

Allegro Non Troppo '76
Mr. Rossi's Dreams '83
Mr. Rossi's Vacation '83
Vip, My Brother Superman '90
West and Soda '90

Marcel Bozzuffi

The French Connection '71
La Cage aux Folles 2 '81

Pearl Braaten

Survival '88

Charles Brabant (D)

The Respectful Prostitute '52

Teda Bracci

Big Bird Cage '72
Chrome Hearts '70

Lorraine Bracco

Camorra: The Naples
Connection '85
The Dream Team '89
Goodfellas '90
Medicine Man '92
The Pick-Up Artist '87
Radio Flyer '91
Sing '89
Someone to Watch Over Me
'87
Switch '91
Talent for the Game '91

John Brace

Truth or Dare? '86

Alejandro Bracho

Romero '89

Eddie Bracken

Hail the Conquering Hero '44
Hour of Stars '58
Miracle of Morgan's Creek '44
National Lampoon's Vacation
'83
Shinbone Alley '70
Summer Stock '50
Too Many Girls '40
We're Not Married '52
Young & Willing '42

Susan Bracken

Don't Open the Door! '74

Sidney Bracy

The Cameraman '28

Kitty Bradbury

Our Hospitality '23

Lane Bradbury

The Ultimate Warrior '75

**Robert N.
Bradbury** (D)

Alias John Law '35
Between Men '35
Big Calibre '35
Blue Steel '34
Brand of the Outlaws '36
Breed of the Border '33
Cavalry '36
Dawn Rider '35
Dawn Rider/Frontier Horizon
'39
Desert Trail '35
Forbidden Trails '41
Lawless Frontier '35
Lucky Texan '34
Man from Hell's Edges '32
Riders of Destiny '33
Riders of the Rockies '37
Smokey Smith '36
The Star Packer '34
Trail Beyond '34
Trouble in Texas '37
Valley of the Lawless '36
West of the Divide '34
Western Double Feature 2 '33
Western Double Feature 5 '35
Western Justice '35
Westward Ho '35

**Robert
Bradbury** (D)

Headin' for the Rio Grande '36
Hittin' the Trail '37
John Wayne Anthology: Star
Packer/Lawless Range '34
Lawless Range '35
Man from Utah '34

Greg Braddock

Satan's Black Wedding '75

Kim Braden

Billyboy '79
Bloodsuckers from Outer
Space '83

Greg Bradford

Let's Do It! '84
Lovelines '84
Zapped! '82

Richard Bradford

Ambition '91
Badge of the Assassin '85
Goin' South '78
Heart of Dixie '89
Legend of Billie Jean '85
Little Nikita '88
Mean Season '85
The Milagro Beanfield War '88
A Rumor of War '80
Running Hot '83
Servants of Twilight '91
Sunset '88
Trip to Bountiful '85
The Untouchables '87

**Samuel
Bradford** (D)

Teen Vamp '88

Virginia Bradford

Marked Money '28

Jean Bradin

Prix de Beaute '30

Kelly Bradish

Clayton County Line

Hollywood Strangler '82
The Hollywood Strangler
  Meets the Skid Row
  Slasher '79
Incredibly Strange Creatures
  Who Stopped Living and
  Became Mixed-Up Zombies
  '63
Rat Pfink and Boo-Boo '66
The Thrill Killers '65

Thordis Brandt

The Witchmaker '69

Victor Elliot Brandt

Neon Maniacs '86

Craig Branham

One Last Run '89

Laura Branigan

Mugsy's Girls '85

Paul Branney

Cobra Against Ninja

Fred Brannon (D)

The Crimson Ghost '46
Cyclotrode "X" '46
D-Day on Mars '45
G-Men Never Forget '48
Jesse James Rides Again '47
Jungle Drums of Africa '53
King of the Rocketmen '49
Lost Planet Airmen '49
Radar Men from the Moon '52
Satan's Satellites '58
Zombies of the Stratosphere
  '52

Marjorie Bransfield

Easy Wheels '89

Betsy Brantley

Another Country '84
Dreams Lost, Dreams Found
  '87
Five Days One Summer '82
I Come in Peace '90

Timothy Brantley

Brutal Glory '89

Albert Bras

Napoleon '27

Dominick Brascia

Evil Laugh '86

**Dominick
Brascia** (D)

Evil Laugh '86

Enrique Braso (D)

In Memoriam '76

Tinto Brass (D)

Caligula '80

Keefe Brasselle

Skirts Ahoy! '52

Keefe Brasselle (D)

If You Don't Stop It...You'll Go
  Blind '77

George Brassens

French Singers '50s

Claude Brasseur

Act of Aggression '73
Band of Outsiders '64
Elusive Corporal '62

L'Etat Sauvage '90
Josepha '82
La Boum '81
Lobster for Breakfast '82
Pardon Mon Affaire, Too! '77
Simple Story '80

Pierre Brasseur

Carthage in Flames '60
Children of Paradise (Les
  Enfants du Paradis) '44
Horror Chamber of Dr.
  Faustus '59

Andre Braugher

The Court Martial of Jackie
  Robinson '90
Glory '89
Somebody Has to Shoot the
  Picture '90

Michel Brault (D)

A Paper Wedding '89

Gunther Braun

Terror Beneath the Sea '66

John Braun

S.S. Experiment Camp 2 '86

Pinkas Braun

The Man Outside '68
Mission Stardust '68
Praying Mantis '83

Asher Brauner

American Eagle '90
B.A.D. Cats '80
Living to Die '91
Merchants of War '90
Treasure of the Moon
  Goddess '88

Jeff Braunstein

One Night Only '84

**Joseph
Braunstein** (D)

Edge of the Axe '89
Rest In Pieces '87

Arthur Brauss

The Goalie's Anxiety at the
  Penalty Kick '71

Bart Braverman

The Great Texas Dynamite
  Chase '76
Hollywood Hot Tubs 2:
  Educating Crystal '89

**Charles
Braverman** (D)

Brotherhood of Justice '86
Hit & Run '82
Prince of Bel Air '87

Bob Bravler (D)

Rush Week '88

John Bravman (D)

Zombie Nightmare '86

Charlie Bravo

Scalps '83

Tony Bravo

Caches de Oro
East L.A. Warriors '89

Hilda Brawner

One Plus One '61

Kriss Braxton

Danger Zone '87

Robert Bray

Never Love a Stranger '58

William Brayne (D)

Cold War Killers '86
Flame to the Phoenix '85

Edward Brayshaw

Unmasked Part 25 '88

Lidia Brazzi

The Christmas That Almost
  Wasn't '66

Rossano Brazzi

The Adventurers '70
Barefoot Contessa '54
Bobo '67
The Christmas That Almost
  Wasn't '66
Dr. Frankenstein's Castle of
  Freaks '74
The Far Pavilions '84
Fear City '85
Final Conflict '81
Final Justice '84
Formula for a Murder '85
The Italian Job '69
Legend of the Lost '57
Mr. Kingstreet's War '71
One Step to Hell '68
Rome Adventure '62
South Pacific '58
Summertime '55
Third Solution '89
Three Coins in the Fountain
  '54
Time for Miracles '80
We the Living '42
Woman Times Seven '67

Rossano Brazzi (D)

The Christmas That Almost
  Wasn't '66

Sebastian Breaks

Big Switch '70

George Breakston

The Return of Peter Grimm
  '35

**George
Breakston** (D)

Scarlet Spear '54
White Huntress '57

Gil Brealey (D)

A Test of Love '84

Peter Breck

The Crawling Hand '63
Highway 61 '91
Shock Corridor '63

Candice Brecker

Bridge to Silence '89

**Guillermo
Bredeston**

Con Mi Mujer No Puedo '80s

Barley Bree

Let's Have an Irish Party '83

Larry Breeding

A Matter of Life and Death '81
Street Music '81

Bobby Breen

Breaking the Ice '38
Escape to Paradise '39
Fisherman's Wharf '39
Hawaii Calls '38
It Happened in New Orleans
  '36
Let's Sing Again '36
Make a Wish '37

Danny Breen

The Best of Not Necessarily
  the News '88
Bill Murray Live from the
  Second City '70s

Mary Breen

Honeymoon Killers '70

Patrick Breen

Nobody's Perfect '90

Paulette Breen

The Clonus Horror '79
The Wizard of Speed and
  Time '88

**Richard L.
Breen** (D)

Stopover Tokyo '57

Thomas E. Breen

The River '51

Bobbie Breese

Ghoulies '85
Mausoleum '83

Tracy Bregman

Concrete Jungle '82
Happy Birthday to Me '81

**Catherine
Breillat** (D)

36 Fillete '88

Marie Breillat

Dracula & Son '76

Valerie Breiman

Tale of Two Sisters '89

Jana Brejchova

Loves of a Blonde '66
The Original Fabulous
  Adventures of Baron
  Munchausen '61

Peter Brek

Benji '73

Jacques Brel

Franz '72

Jacques Brel (D)

Franz '72

Lucille Bremer

Behind Locked Doors '48
Human Gorilla '48
Meet Me in St. Louis '44
Till the Clouds Roll By '46
Till the Clouds Roll By/A Star
  Is Born '46
Yolanda & the Thief '45
Ziegfeld Follies '46

Sylvia Bremer

Narrow Trail '17

Milton Bren (D)

Three for Bedroom C '52

El Brendel

The Beautiful Blonde from
  Bashful Bend '49
Big Trail '30
I'm from Arkansas '44

Tom Breneman

Breakfast in Hollywood '46

Claire Brennan

She-Freak '67

Eileen Brennan

Babes in Toyland '86
Blood Vows: The Story of a
  Mafia Wife '87
The Cheap Detective '78
Clue '85
Daisy Miller '74
Death of Richie '76
FM '78
Fourth Wise Man '85
The Funny Farm '82
Great Smokey Roadblock '76
I Don't Buy Kisses Anymore
  '92
The Last Picture Show '71
Murder by Death '76
My Father's House '75
My Old Man '79
The New Adventures of Pippi
  Longstocking '88
Pandemonium '82
Playmates '72
Private Benjamin '80
Rented Lips '88
Richie '77
Scarecrow '73
Stella '90
Sticky Fingers '88
The Sting '73
Texasville '90
White Palace '90

Kerry Brennan

Terror Squad '87

Michael Brennan

No Trace '50

Stephen Brennan

Eat the Peach '86

Walter Brennan

The Adventures of Tom
  Sawyer '38
Along the Great Divide '51
At Gunpoint '55
Bad Day at Black Rock '54
Barbary Coast '35
Best of the Badmen '50
Blood on the Moon '48
Come and Get It '36
Dakota '45
Far Country '55
Glory '56
Gnome-Mobile '67
Goodbye, My Lady '56
The Green Promise '49
Home for the Holidays '72
How the West was Won '62
Meet John Doe '41
Moon's Our Home '36
My Darling Clementine '46
The North Star '43
Northwest Passage '40
One and Only, Genuine,
  Original Family Band '68
Oscar '66
The Pride of the Yankees '42

Silent Night, Lonely Night '69
Starman '84
Stay Hungry '76
Texasville '90
Thunderbolt & Lightfoot '74
Tron '82
Tucker: The Man and His Dream '88
Winter Kills '79
Yin & Yang of Mr. Go '71

Lloyd Bridges

Abilene Town '46
Airplane! '80
Airplane 2: The Sequel '82
Around the World Under the Sea '65
Bear Island '80
Big Deadly Game '54
The Blue and the Grey '85
Cousins '89
Daring Game '68
East of Eden '80
Force of Evil '77
The Goddess '58
Great Movie Trailers '70s
Great Wallendas '78
The Happy Ending '69
Haunts of the Very Rich '72
High Noon '52
Home of the Brave '49
Hot Shots! '91
Joe Versus the Volcano '90
Last of the Comanches '52
The Limping Man '53
Little Big Horn '51
Loving '84
Master Race '44
Mission Galactica: The Cylon Attack '79
Moonrise '48
Mutual Respect '77
The Queen of Mean '90
The Rainmaker '56
Ramrod '47
Rocketship X-M '50
Roots, Vol. 1 '77
Running Wild '73
Sahara '43
Scuba '72
Silent Night, Lonely Night '69
Tall Texan '53
The Tattered Web '71
Trapped '49
Try and Get Me '50
Tucker: The Man and His Dream '88
A Walk in the Sun '46
War in the Sky '82
Weekend Warriors '86
The White Tower '50
The Wild Pair '87
Winter People '89

Todd Bridges

Homeboys '92
Twice Dead '88

Dee Dee Bridgewater

The Brother from Another Planet '84

David Brierly

Threads '85

Richard Briers

A Chorus of Disapproval '89
Henry V '89
A Matter of Who '62
Norman Conquests, Part 1: Table Manners '80
Rentadick '72

Bernard Brieux

Petit Con '84

Gary Briggle

Alice in Wonderland '82

Julie Briggs

Bowery Buckaroos '47

Richard Bright

Idolmaker '80
Panic in Needle Park '75
Red Heat '88

Omar Brignoll

The Tree of Wooden Clogs '78

Lilla Brignone

Coriolanus, Man without a Country '64

Charles Brill

Bail Out '90

Charlie Brill

Bloodstone '88

Fran Brill

Look Back in Anger '80
Old Enough '84
What About Bob? '91

Jason Brill

Hell High '86

Nancy Brill

Demons 2 '87

Patti Brill

Girl Rush '44
Hard-Boiled Mahoney '47

Steven Brill

sex, lies and videotape '89

Cynthia Brimhall

Hard Ticket to Hawaii '87

Wilford Brimley

Absence of Malice '81
Act of Vengeance '86
American Justice '86
Borderline '80
The China Syndrome '79
Cocoon '85
Cocoon: The Return '88
Country '84
Death Valley '81
The Electric Horseman '79
End of the Line '88
Eternity '90
The Ewoks: Battle for Endor '85
Gore Vidal's Billy the Kid '89
Harry & Son '84
High Road to China '83
The Hotel New Hampshire '84
Murder in Space '85
The Natural '84
Remo Williams: The Adventure Begins '85
Rodeo Girl '80
Roughnecks '80
The Stone Boy '84
Ten to Midnight '83
Tender Mercies '83
The Thing '82
Thompson's Last Run '90
Tough Enough '87

Michele Brin

Secret Games '90s

Burt Brinckerhoff (D)

Can You Hear the Laughter? The Story of Freddie Prinze '79
The Day the Women Got Even '80
Mother & Daughter: A Loving War '80

Adrian Brine

Lily was Here '89

Paul Brinegar

Charro! '69
How to Make a Monster '58

Christie Brinkley

National Lampoon's Vacation '83
Sports Illustrated's 25th Anniversary Swimsuit Video '89

John Brinkley

T-Bird Gang '59

Ritch Brinkley

Silhouette '91

Bo Brinkman

Ice House '89

Charles Brinley

Moran of the Lady Letty '22

Bill Brinsfield

Warbirds '88

Breezy Brisbane

Little Rascals, Book 5 '30s
Little Rascals, Book 6 '30s
Little Rascals, Book 7 '30s

William Brisbane

Maid's Night Out '38

David Brisbin

Kiss Daddy Goodnight '87

Danielle Brisbois

Mom, the Wolfman and Me '80

Gwen Brisco

Getting Over '81

Jimmy Briscoe

Spaced Invaders '90

Danielle Brisebois

Big Bad Mama 2 '87
Killcrazy '89
The Premonition '75

Irving Briskin (D)

Texas Cyclone '32

Jean Brismee (D)

The Devil's Nightmare '71

Carl Brisson

The Manxman '29
Murder at the Vanities '34
The Ring '27

Donald Britain (D)

Accident '83

May Britt

Haunts '77

The Hunters '58

Morgan Brittany

Initiation of Sarah '78
LBJ: The Early Years '88
The Prodigal '83
Sundown '91

Tally Brittany

Sex Appeal '86
Slammer Girls '87

Lawrence Britten (D)

Whose Child Am I? '75

Barbara Britton

Bandit Queen '51
Captain Kidd '45
Dragonfly Squadron '54
I Shot Jesse James '49

James Britton

The Yesterday Machine '63

Nicol Britton

South of Hell Mountain '70

Pamela Britton

Anchors Aweigh '45
D.O.A. '49

Tony Britton

Agatha '79
The Day of the Jackal '73
Dr. Syn, Alias the Scarecrow '64
Horsemasters '61
Operation Amsterdam '60

Tracy Lynch Britton (D)

Maximum Breakout '91

Herman Brix

Daredevils of the Red Circle '38
Hawk of the Wilderness '38
A Million to One '37
The New Adventures of Tarzan: Vol. 1 '35
Shadow of Chinatown '36
Tarzan and the Green Goddess '38

Ian Broadbent

ABC's of Love & Sex, Australia Style '86

Jim Broadbent

The Good Father '87
Life is Sweet '91

Bernd Broaderup

Taxi zum Klo '81

David Broadnax

Zombie Island Massacre '84

Curt Brober

Ninja Mission '84

Lily Broberg

Famous Five Get Into Trouble '70s

Peter Brocco

The Balcony '63
Homebodies '74

Martine Brochard

Eyeball '78

Anne Brochet

Cyrano de Bergerac '89

Alan Brock

Shriek of the Mutilated '74

Anita Brock

People Who Own the Dark '75

Deborah Brock (D)

Andy and the Airwave Rangers '89
Rock 'n' Roll High School Forever '91
Slumber Party Massacre 2 '87

Stan Brock

Galyon '77
Return to Africa '89

Terry Ten Brock

Delirium '77

Gerd Brockmann

The Wannsee Conference '84

Roy Brocksmith

Arachnophobia '90
Big Business '88
Killer Fish '79
King of the Gypsies '78
Tales of Ordinary Madness '83
Total Recall '90

Gladys Brockwell

The Drake Case '29
Oliver Twist '22

Beth Broderick

Are You Lonesome Tonight '92
Thousand Pieces of Gold '91

Chris Broderick

Legacy of Horror '78

Helen Broderick

Bride Walks Out '36
The Rage of Paris '38
Swing Time '36
Top Hat '35

James Broderick

Alice's Restaurant '69
Dog Day Afternoon '75
The Group '66
Keeping On '81
The Shadow Box '80
The Taking of Pelham One Two Three '74

John Broderick (D)

Bad Georgia Road '77
The Warrior & the Sorceress '84

Matthew Broderick

Biloxi Blues '88
Cinderella '84
Family Business '89
Ferris Bueller's Day Off '86
The Freshman '90
Glory '89
Ladyhawke '85
Master Harold and the Boys '84
Max Dugan Returns '83
1918 '85
On Valentine's Day '86
Project X '87
Torch Song Trilogy '88
War Games '83

Greer Brodie

Terror Squad '87

Kevin Brodie (D)

Mugsy's Girls '85

Steve Brodie

The Admiral was a Lady '50
Armored Car Robbery '50
Badman's Territory '46
The Beast from 20,000
  Fathoms '53
Blue Hawaii '62
Criminal Court '46
Desperate '47
Donovan's Brain '53
The Giant Spider Invasion '75
Kiss Tomorrow Goodbye '50
Out of the Past '47
Roustabout '64
Saint Strikes Back: Criminal
  Court '46
The Steel Helmet '51
Three for Bedroom C '52
Tough Assignment '49
Treasure of Monte Cristo '49
The Wild World of Batwoman
  '66
The Wizard of Speed and
  Time '88

Harry Brogan

Broth of a Boy '59

Giulio Brogi

The Spider's Stratagem '70

James Brolin

The Ambush Murders '82
The Amityville Horror '79
Backstab '90
Bad Jim '89
The Boston Strangler '68
Capricorn One '78
Class of '63 '73
Fantastic Voyage '66
Finish Line '89
High Risk '76
Hold the Dream '85
Mae West '84
Night of the Juggler '80
Nightmare on the 13th Floor
  '90
Our Man Flint '66
Steel Cowboy '78
Von Ryan's Express '65
Westworld '73

Josh Brolin

Finish Line '89
The Goonies '85
Thrashin' '86

J. Edward Bromberg

Cloak and Dagger '46
The Missing Corpse '45
Phantom of the Opera '43
Queen of the Amazons '47
Return of Frank James '40
Reunion in France '42
Son of Dracula '43
Strange Cargo '40

John Bromfield

Big Bluff '55
Easy to Love '53
Flat Top '52
Manfish '56

Rex Bromfield (D)

Cafe Romeo '91
Home is Where the Hart is '88

Love at First Sight '76

Melanie '82

Valri Bromfield

Home is Where the Hart is '88
Mr. Mom '83
Nothing But Trouble '91

Sheila Bromley

Torture Ship '39

Sydney Bromley

Crystalstone '88
The NeverEnding Story '84

Eleanor Bron

Bedazzled '68
Help! '65
Turtle Diary '86
Two for the Road '67
Women in Love '70

Douglas Bronco (D)

Danger Zone 3: Steel Horse
  War '90

William Bronder

Flush '81

Henry Bronett

The Mozart Brothers '86

Fritz Bronner

Tax Season '90

Albert Bronski

The 13th Mission '91

Brick Bronsky

Class of Nuke 'Em High Part
  2: Subhumanoid Meltdown
  '91

Anthony Bronson

Bruce Lee Fights Back From
  the Grave '76

Betty Bronson

Are Parents People? '25
Ben-Hur '26
Medicine Man '30
Naked Kiss '64
Yodelin' Kid From Pine Ridge
  '37

Charles Bronson

Act of Vengeance '86
Assassination '87
Borderline '80
Breakheart Pass '76
Breakout '75
Bull of the West '89
Cabo Blanco '81
Chato's Land '71
Chino '75
Cold Sweat '74
Death Hunt '81
Death Wish '74
Death Wish 2 '82
Death Wish 3 '85
Death Wish 4: The
  Crackdown '87
The Dirty Dozen '67
Drum Beat '54
The Evil That Men Do '84
The Family '73
Four for Texas '63
The Great Escape '63
Guns of Diablo '64
Hard Times '75
Honor Among Thieves '68
House of Wax '53
The Indian Runner '91
Jubal '56

Kid Galahad '62

Kinjite: Forbidden Subjects
  '89
Lola '69
Love and Bullets '79
Machine Gun Kelly '58
Magnificent Seven '60
Master of the World '61
The Meanest Men in the West
  '60s
The Mechanic '72
Messenger of Death '88
Miss Sadie Thompson '54
Mr. Majestyk '74
Murphy's Law '86
Never So Few '59
Once Upon a Time in the West
  '68
Pat and Mike '52
Raid on Entebbe '77
Red Sun '71
Rider on the Rain '70
Run of the Arrow '56
St. Ives '76
The Sandpiper '65
Showdown at Boot Hill '58
Someone Behind the Door '71
The Stone Killer '73
Telefon '77
Ten to Midnight '83
This Property is Condemned
  '66
Vera Cruz '53
Villa Rides '68
The White Buffalo '77
Wild West '87

Harry Bronson

The Thief '52

Lillian Bronson

The Next Voice You Hear '50

Claudio Brook

The Bees '78
Daniel Boone: Trail Blazer '56
Dr. Tarr's Torture Dungeon
  '75
La Mansion de la Locura '87
Neutron and the Black Mask
  '61
The Peking Blond '68
Samson in the Wax Museum
  '63
Sisters of Satan '75

Clive Brook

Action for Slander '38
Convoy '40
Hula '28
On Approval '44

Clive Brook (D)

On Approval '44

Doris Brook

Lone Bandit '33

Faith Brook

After Julius
Bloodbath
To Sir, with Love '67
Uneasy Terms '48

Irina Brook

Captive '87
The Girl in the Picture '86

Elwyn Brook-Jones

Bonnie Prince Charlie '48
The Gilded Cage '54
Rogue's Yarn '56

Lyndon Brook

Above Us the Waves '56
Invasion '65
The Spanish Gardener '57

Peter Brook (D)

King Lear '71
Lord of the Flies '63
The Mahabharata '89
Marat/Sade (Persecution and
  Assassination...) '66
Meetings with Remarkable
  Men '79

Bunny Brooke

Alison's Birthday '79

Hillary Brooke

Abbott and Costello Meet
  Captain Kidd '52
Africa Screams '49
Big Town After Dark '47
Enchanted Cottage '45
Heat Wave '54
The Lost Continent '51
The Man Who Knew Too
  Much '56
Never Wave at a WAC '52
Rangeland Racket '41
Road to Utopia '46
Sherlock Holmes Faces Death
  '43
Sherlock Holmes: The Voice
  of Terror '42
The Strange Woman '46
To the Shores of Tripoli '42

Paul Brooke

The Lair of the White Worm
  '88

Rebecca Brooke

Confessions of a Young
  American Housewife '78
Little Girl...Big Tease '75

Sandy Brooke

Crazed Cop '88
Miami Vendetta '87
Star Slammer '87
Terror on Alcatraz '86

Sorrell Brooke

Up the Down Staircase '67

Ted Brooke

Ninja Demon's Massacre '80s

Tim Brooke

12 Plus 1 '70

Tyler Brooke

The Divorcee '30

Walter Brooke

Joseph & His Brothers '79

Richard Brooker

Deathstalker '84
Friday the 13th, Part 3 '82

Jacqueline Brookes

Hardhat & Legs '80
Rodeo Girl '80
Silent Witness '85

Howard Brookner (D)

Bloodhounds of Broadway '89

Adam Brooks (D)

Almost You '85

Cannon Movie Tales: Red
  Riding Hood '89

Aimee Brooks

Critters 3 '91

Albert Brooks

Broadcast News '87
Defending Your Life '91
Lost in America '85
Modern Romance '81
Private Benjamin '80
Real Life '79
Taxi Driver '76
Twilight Zone: The Movie '83
Unfaithfully Yours '84

Albert Brooks (D)

Defending Your Life '91
Lost in America '85
Modern Romance '81
Real Life '79

Avery Brooks

Half Slave, Half Free '85

Bob Brooks (D)

Tattoo '81

Christopher Brooks

Alabama's Ghost '72

Clarence Brooks

Bronze Buckaroo '39
Harlem Rides the Range '39

David Brooks

Scream for Help '86

David Allan Brooks

The Kindred '87

Dusty Brooks

Fats Waller and Friends '86

Elisabeth Brooks

The Forgotten One '89
The Howling '81

Foster Brooks

Cracking Up '83
More Laughing Room Only
Oddballs '84
Super Seal '77

Geraldine Brooks

Challenge to Lassie '49
The Green Glove '52
Johnny Tiger '66
Reckless Moment '49

Hazel Brooks

Body and Soul '47

Hildy Brooks

Forbidden Love '82

Iris Brooks

I Drink Your Blood '71

James L. Brooks

Modern Romance '81

James L. Brooks (D)

Broadcast News '87
Terms of Endearment '83
Thursday's Game '74

Jan Brooks

Naked Youth '59

Jean Brooks

Boot Hill Bandits '42

Boss of Big Town '43
The Leopard Man '43
The Seventh Victim '43

Jennifer Brooks

The Abductors '72

Jeremy Brooks

Torture Dungeon '70

Jim Brooks

Ninja Holocaust '80s

Joel Brooks

Are You Lonesome Tonight '92

Joseph Brooks

If Ever I See You Again '78

Joseph Brooks (D)

If Ever I See You Again '78
Invitation to the Wedding '73
You Light Up My Life '77

Leslie Brooks

The Cobra Strikes '48
Hollow Triumph '48
Romance on the High Seas '48
The Scar '48
Tonight and Every Night '45
You Were Never Lovelier '42

Lola Brooks

The Sundowners '60

Louise Brooks

Diary of a Lost Girl '29
It's the Old Army Game '26
Love 'Em and Leave 'Em '27
Pandora's Box '28
Prix de Beaute '30

Lucius Brooks

Harlem Rides the Range '39

Mel Brooks

Blazing Saddles '74
High Anxiety '77
History of the World: Part 1 '81
Life Stinks '91
Look Who's Talking, Too '90
Mel Brooks: An Audience '84
Muppet Movie '79
Putney Swope '69
Silent Movie '76
Spaceballs '87
To Be or Not to Be '83
Twelve Chairs '70

Mel Brooks (D)

Blazing Saddles '74
High Anxiety '77
History of the World: Part 1 '81
Life Stinks '91
The Producers '68
Silent Movie '76
Spaceballs '87
Twelve Chairs '70
Young Frankenstein '74

Michael Alan Brooks

The Heroes of Desert Storm '91

Pauline Brooks

Make a Million '35

Peter Brooks

Gidget Goes to Rome '63

Phyllis Brooks

Dangerous Passage '44
Little Miss Broadway '38
Rebecca of Sunnybrook Farm '38
The Shanghai Gesture '41
Silver Spurs '43
Slightly Honorable '40

Rand Brooks

Devil's Playground '46
Gone with the Wind '39
Hopalong Cassidy: Borrowed Trouble '48
Hopalong Cassidy: Dangerous Venture '48
Hopalong Cassidy: False Paradise '48
Hopalong Cassidy: Hoppy's Holiday '47
Hopalong Cassidy: Silent Conflict '48
Hopalong Cassidy: Sinister Journey '48
Hopalong Cassidy: The Dead Don't Dream '48
Hopalong Cassidy: The Devil's Playground '46
Hopalong Cassidy: The Marauders '47
Hopalong Cassidy: Unexpected Guest '47
Ladies of the Chorus '49
Sinister Journey '48
The Sombrero Kid '42
The Vanishing Westerner '50

Randi Brooks

Colors '88
Cop '88
Forbidden Love '82
Hamburger... The Motion Picture '86
Tightrope '84

Ray Brooks

Daleks Invasion Earth 2150 A.D. '66
Flesh and Blood Show '73
House of Whipcord '75
Office Romances '86
Tiffany Jones '75

Richard Brooks

84 Charlie Mopic '89
Memphis '91
Shakedown '88
Short Fuse '88
To Sleep with Anger '90

Richard Brooks (D)

Battle Circus '53
Bite the Bullet '75
Blackboard Jungle '55
The Brothers Karamazov '57
Cat on a Hot Tin Roof '58
The Catered Affair '56
Dollars '71
Elmer Gantry '60
Fever Pitch '85
The Happy Ending '69
In Cold Blood '67
Last Time I Saw Paris '54
Looking for Mr. Goodbar '77
Lord Jim '65
Professionals '66
Something of Value '57
Sweet Bird of Youth '62
Wrong is Right '82

Van Brooks

Trespasses '86

Gladys Brookwell

Spangles '26

Lois Broomfield

Paramount Comedy Theater, Vol. 5: Cutting Up '87

Nick Broomfield

Chicken Ranch '83

Nick Broomfield (D)

Dark Obsession '90

Brian Brophy

Skinheads: The Second Coming of Hate '88

Edward Brophy

The Cameraman '28
The Champ '31
Cover Girl '44
Dance, Girl, Dance '40
Destroyer '43
Dumbo '41
Great Guy '36
Hide-Out '34
Last Hurrah '58
Renegade Girl '46
Roaring City '51
Show Them No Mercy '35
Swing Parade of 1946 '46
The Thin Man Goes Home '44

Kevin Brophy

Delos Adventure '86
Hell Night '81
The Seduction '82

Pierce Brosnan

The Carpathian Eagle '81
The Deceivers '88
The Fourth Protocol '87
The Heist '89
The Lawnmower Man '92
The Long Good Friday '79
The Manions of America '81
The Mirror Crack'd '80
Mister Johnson '91
Murder 101 '91
Nomads '86
Taffin '88

Claude Brosset

My New Partner '84

Collette Brosset

Femmes de Paris '53

Gudrun Brost

Hour of the Wolf '68

John Brotherton

Murder: No Apparent Motive '84

James Broughton (D)

Dreamwood '72

Liliane Brousse

Maniac '63

Lee De Broux

Hangfire '91

Otto Brower (D)

Devil Horse '32
Fighting Caravans '32
Hard Hombre '31

I Can't Escape '34
Postal Inspector '36
Spirit of the West '32

Amelda Brown

Little Dorrit, Film 1: Nobody's Fault '88
Little Dorrit, Film 2: Little Dorrit's Story '88

Andre Brown

The Bronx War '91

Angela Brown

Teen Vamp '88

Arvin Brown (D)

Diary of the Dead '80s

Barbara Brown

The Fighting Sullivans '42
Hollywood Canteen '44
Home Town Story '51

Barry Brown

Bad Company '72
The Disappearance of Aimee '76
He Who Walks Alone '78
The Ultimate Thrill '74

Barry Brown (D)

Cloud Dancer '80

Blair Brown

Altered States '80
And I Alone Survived '78
The Bad Seed '85
The Choirboys '77
Continental Divide '81
A Flash of Green '85
Hands of a Stranger '87
Kennedy '83
One Trick Pony '80
Stealing Home '88
Strapless '90

Bruce Brown (D)

The Endless Summer '66
On Any Sunday '71

Bryan Brown

Blame It on the Bellboy '91
Blood Money '80
Breaker Morant '80
Cocktail '88
Dead in the Water '91
The Empty Beach '85
F/X '86
F/X 2: The Deadly Art of Illusion '91
Far East '85
Give My Regards to Broad Street '84
The Good Wife '86
Gorillas in the Mist '88
The Irishman '78
Money Movers '78
Newsfront '78
Palm Beach '79
Parker '84
Prisoners of the Sun '91
Rebel '86
Sweet Talker '91
Taipan '86
The Thorn Birds '83
A Town Like Alice '85
Wild Thing '87
The Winter of Our Dreams '82

Charles D. Brown

Brother Orchid '40
Follow Me Quietly '49
Old Swimmin' Hole '40

Clancy Brown

Ambition '91
Bad Boys '83
Blue Steel '90
The Bride '85
Cast a Deadly Spell '91
Highlander '86
Season of Fear '89
Shoot to Kill '88

Clarence Brown (D)

Anna Christie '30
Anna Karenina '35
Conquest '32
The Eagle '25
The Flesh and the Devil '27
A Free Soul '31
The Gorgeous Hussy '36
The Human Comedy '43
Idiot's Delight '39
Intruder in the Dust '49
National Velvet '44
Possessed '31
Sadie McKee '34
Smouldering Fires '25
Tempest/Eagle '27
A Woman of Affairs '28
The Yearling '46

David G. Brown

Chasing Dreams '81
Deadly Harvest '72

Dennis Brown

The Warrior

Dewey Brown

Jazz & Jive '30s

D.W. Brown

Mischief '85
Weekend Pass '84

Dwier Brown

The Guardian '90

Dyann Brown

Lone Wolf '88

Earle Brown

Mr. Robinson Crusoe '32

Edwin Scott Brown (D)

The Prey '80

Eleonora Brown

Two Women '61

Eric Brown

Private Lessons '81
They're Playing with Fire '84
Video Murders '87

Ewing Miles Brown

The Astounding She-Monster '58

Ewing Miles Brown (D)

Killers '88

Gary Brown

Glory Boys '84
Invasion of the Space Preachers '90

Gaye Brown

Mata Hari '85

Unknown Powers '80
World's Greatest Athlete '73

Suzanne Browne

The Bikini Car Wash
Company '92

Cara Brownell (D)

Cave Girls '80s

Ricou Browning

Creature from the Black
Lagoon '54
Flipper's New Adventure '64

Ricou Browning (D)

Salty '73

Rod Browning

The Life & Times of the
Chocolate Killer '88

Tod Browning (D)

Devil Doll '36
Dracula '31
Freaks '32
Mark of the Vampire '35
Outside the Law '21
The Unholy Three '25
The Unknown '27

S.F. Brownrigg (D)

Don't Look in the Basement
'73
Don't Open the Door! '74
Keep My Grave Open '80
Poor White Trash 2 '75
Thinkin' Big '87

**Radoslav
Brozobohaty**

All My Good Countrymen '68

Angela Bruce

Charlie Boy '81

Betty Bruce

Gypsy '62

Brenda Bruce

Antonia and Jane '91

Brenda Bruce

Peeping Tom '63

Cheryl Lynn Bruce

Daughters of the Dust '91
Music Box '89

Clifford Bruce

A Fool There Was '14

Colin Bruce

Chicago Joe & the Showgirl
'90
Gotham '88

David Bruce

Jungle Hell '55
Pier 23 '51
Salome, Where She Danced
'45

Eve Bruce

The Love Machine '71

James Bruce (D)

The Suicide Club '88

Kate Bruce

Short Films of D.W. Griffith,
Volume 2 '09

Kitty Bruce

Switchblade Sisters '75

Lenny Bruce

Dance Hall Racket '58
Hungry i Reunion '81
Lenny Bruce '67
Lenny Bruce Performance
Film '68

Nigel Bruce

The Adventures of Sherlock
Holmes '39
Becky Sharp '35
The Blue Bird '40
Charge of the Light Brigade
'36
The Chocolate Soldier '41
Corn is Green '45
Dressed to Kill '46
The Hound of the Baskervilles
'39
House of Fear '45
Lassie, Come Home '43
The Last of Mrs. Cheyney '37
Limelight '52
The Pearl of Death '44
Pursuit to Algiers '45
Rebecca '40
Scarlet Claw '44
Sherlock Holmes Faces Death
'43
Sherlock Holmes and the
Secret Weapon '42
Sherlock Holmes: The Voice
of Terror '42
Sherlock Holmes in
Washington '43
Spiderwoman '44
Stand Up and Cheer '34
Suean and God '40
Suspicion '41
Terror by Night '46
Terror by Night/Meeting at
Midnight '46
Thunder in the City '37
Treasure Island '34
The Two Mrs. Carrolls '47

Virginia Bruce

Action in Arabia '44
Born to Dance '36
The Great Ziegfeld '36
Jane Eyre '34
Let 'Em Have It '35
Pardon My Sarong '42
State Department File 649 '49
Strangers When We Meet '60

**Clyde
Bruckman** (D)

Feet First '30
The General '27

Patrick Bruel

Secret Obsession '88

**Christopher
Brugger**

Bloody Moon '83

Eddie Brugman

Katie's Passion '75

Werner Bruhns

1900 '76

Beau Brummel

Three Bullets for a Long Gun
'73
Village of the Giants '65

Bo Brundin

Headless Eyes '83
Russian Roulette '75
Shoot the Sun Down '81

Adrian Brunel

Spitfire '42

Adrian Brunel (D)

Old Spanish Custom '36

Robert Bruning

The Night After Halloween '83

Natja Brunkhorst

Christiane F. '82

Fernando Bruno

Two Nights with Cleopatra '54

Paul Bruno

Mankillers '87

Philip Bruns

Dead Men Don't Die '91

Franco Brusati (D)

The Sleazy Uncle '90
To Forget Venice '70s

Dora Bryan

Great St. Trinian's Train
Robbery '66
My Son, the Vampire '52
No Trace '50
Screamtime '83
Small Hotel '57
A Taste of Honey '61

Jane Bryan

Each Dawn I Die '39
Marked Woman '37
Old Maid '39

Tonia Bryan

Hot Summer in Barefoot
County '74

Walter Bryan

The Respectful Prostitute '52

Betty Bryant

Forty Thousand Horsemen
'41

Bill Bryant

King Dinosaur '55

Charles Bryant (D)

Salome '22
Salome/Queen Elizabeth '23

James Bryant (D)

The Executioner, Part 2:
Frozen Scream '84

John Bryant

Courage of Black Beauty '57

Joyce Bryant

Across the Plains '39
East Side Kids '40

Marie Bryant

Duke Is Tops '38

Michael Bryant

Girly '70
Goodbye, Mr. Chips '69
Sakharov '84

Nana Bryant

Eyes of Texas '48
Harvey '50
Inner Sanctum '48
Ladies of the Chorus '49
Sinners in Paradise '38

Pamela Bryant

Lunch Wagon '81
Tigershark '87

Virginia Bryant

The Barbarians '87
Demons 2 '87

William Bryant

Corvette Summer '79
Mountain Family Robinson '79

Claudia Bryar

Green Eyes '76

Bill Bryden (D)

Aria '88

Jon Bryden

Deserters '80s

Yul Brynner

Anastasia '56
Battle of Neretva '71
The Brothers Karamazov '57
The Buccaneer '58
Cast a Giant Shadow '66
Death Rage '77
Futureworld '76
Fuzz '72
Invitation to a Gunfighter '64
The King and I '56
Light at the Edge of the World
'71
The Magic Christian '69
Magnificent Seven '60
Morituri '65
Night Flight from Moscow '73
The Poppy Is Also a Flower
'66
Port of New York '49
Return of the Seven '66
Romance of a Horsethief '71
Solomon and Sheba '59
Taras Bulba '62
The Ten Commandments '56
The Testament of Orpheus
'59
Triple Cross '67
The Ultimate Warrior '75
Villa Rides '68
Westworld '73

Zbynek Brynych (D)

Transport from Paradise '65

Kathleen Bryon

Twins of Evil '71

Leesa Bryte

Senior Week '88

Barry Buchanan

The Loch Ness Horror '82

Cheryl Buchanan

Escape from Galaxy Three '76

Claud Buchanan

Running Wild '27

Edgar Buchanan

Abilene Town '46
Big Trees '52
The Black Arrow '48
Buffalo Bill '44

Coroner Creek '48
Destroyer '43
The Devil's Partner '58
Human Desire '54
Lust for Gold '49
Make Haste to Live '54
Man from Button Willow '65
Man from Colorado '49
Penny Serenade '41
Rage at Dawn '55
Rawhide '50
Ride the High Country '62
Shane '53
She Couldn't Say No '52
Silver Star '55
Talk of the Town '42
Texas '41
Yuma '70

Jack Buchanan

Band Wagon '53
When Knights were Bold '36

Larry Buchanan (D)

Beyond the Doors '83
Curse of the Swamp Creature
'66
The Eye Creatures '65
Free, White, and 21 '62
Goodbye, Norma Jean '75
Goodnight, Sweet Marilyn '89
Hell Raiders '68
Hughes & Harlow: Angels in
Hell '77
It's Alive! '68
The Loch Ness Horror '82
Mars Needs Women '66
Mistress of the Apes '79
Zontar, the Thing from Venus
'66

Miles Buchanan

Dangerous Game '90
Runaway Island, Part 1
Runaway Island, Part 2
Runaway Island, Part 3
Runaway Island, Part 4

Robert Buchanan

That Sinking Feeling '79

Sherry Buchanan

Eyes Behind the Stars '72

Simone Buchanan

Run, Rebecca, Run '81
Runaway Island, Part 1
Runaway Island, Part 2
Runaway Island, Part 3
Runaway Island, Part 4
Shame '87

West Buchanan

Cosmos: War of the Planets
'80

Sherry Buchanna

Emmanuelle & Joanna '86

**Christine
Buchegger**

From the Life of the
Marionettes '80

Horst Buchholz

Aces: Iron Eagle 3 '92
Berlin Tunnel 21 '81
The Catamount Killing '74
Code Name: Emerald '85
Empty Canvas '64
Fanny '61
From Hell to Victory '79
Magnificent Seven '60
One, Two, Three '61

Burstyn ▶cont.

The Exorcist '73
Hanna's War '88
Harry and Tonto '74
Into Thin Air '85
The Last Picture Show '71
The People vs. Jean Harris '81
Providence '80
Resurrection '80
Same Time, Next Year '78
Silence of the North '81
Something in Common '86
Thursday's Game '74
Twice in a Lifetime '85

Benny Burt

Hawaiian Buckaroo '38
Sea Racketeers '37

Clarissa Burt

The NeverEnding Story 2:
Next Chapter '91

Chad Burton

In the Spirit '90

Clarence Burton

The Navigator '24

Corey Burton

Galaxy Express '82

David Burton (D)

Fighting Caravans '32
Lady by Choice '34

Devera Burton

Omoo Omoo, the Shark God
'49

Jennifer Burton

Truck Stop Women '74

Julian Burton

A Bucket of Blood '59

LeVar Burton

Acorn People '82
Almos' a Man '78
Battered '78
Billy: Portrait of a Street Kid
'77
Grambling's White Tiger '81
The Guyana Tragedy: The
Story of Jim Jones '80
The Hunter '80
The Jesse Owens Story '84
The Midnight Hour '86
One in a Million: The Ron
LaFlore Story '78
Roots, Vol. 1 '77
The Supernaturals '86

Maggie Burton

Keep It Up, Jack! '75

Norman Burton

Attack of the Mayan Mummy
'63
Bad Guys '86
Bloodsport '88
Crimes of Passion '84
Diamonds are Forever '71
Fright '56
Jud '71
Mausoleum '83
Save the Tiger '73
Simon, King of the Witches
'71
The Terminal Man '74

Pierre Burton

Love Circles '85

Richard Burton

Absolution '81
Alexander the Great '55
Anne of the Thousand Days
'69
Assassination of Trotsky '72
Becket '64
Bluebeard '72
Breakthrough '78
Circle of Two '80
Cleopatra '63
The Desert Rats '53
Divorce His, Divorce Hers '72
Doctor Faustus '68
Equus '77
Exorcist 2: The Heretic '77
Gathering Storm '74
Green Grow the Rushes '51
Hammersmith is Out '72
Ice Palace '60
The Klansman '74
The Last Days of Dolwyn '49
The Longest Day '62
Look Back in Anger '58
Lovespell '79
Massacre in Rome '73
The Medusa Touch '78
The Night of the Iguana '64
1984 '84
Raid on Rommel '71
The Robe '53
The Sandpiper '65
The Spy Who Came in from
the Cold '65
Taming of the Shrew '67
Tempest '63
Under Milk Wood '73
Wagner: The Movie '85
Where Eagles Dare '68
Who's Afraid of Virginia
Woolf? '66

Richard Burton (D)

Doctor Faustus '68

Robert Burton

The Gallant Hours '60
Invasion of the Animal People
'62
A Man Called Peter '55
A Reason to Live, A Reason
to Die '73
The Slime People '63
The Tall T '57
Teenage Frankenstein '58
Trilogy of Terror '75

Tim Burton (D)

Aladdin and His Wonderful
Lamp '84
Batman '89
Beetlejuice '88
Edward Scissorhands '90
Frankenweenie '80s
Pee-Wee's Big Adventure '85

Tony Burton

Armed and Dangerous '86
Blackjack '78
Oceans of Fire '86

Tracy Burton

Zombiethon '86

Wendell Burton

East of Eden '80
The Sterile Cuckoo '69

Steve Buscemi

Heart '87
Parting Glances '86

Dennis Busch

Faster Pussycat! Kill! Kill! '66

Ernst Busch

Kameradschaft '31

Mae Busch

Bohemian Girl '36
Foolish Wives '22
Laurel & Hardy Comedy
Classics Volume 1 '33
Laurel & Hardy Comedy
Classics Volume 3 '34
Laurel & Hardy Comedy
Classics Volume 6 '35
Laurel & Hardy Comedy
Classics Volume 7 '30s
The Mad Monster '42
Marie Antoinette '38
Sons of the Desert '33
Sucker Money '34
The Unholy Three '25
Without Honors '32
Ziegfeld Girl '41

Gary Busey

Act of Piracy '89
Angels Hard as They Come
'71
Barbarosa '82
Big Wednesday '78
Buddy Holly Story '78
Bulletproof '88
Carny '80
Dangerous Life '89
D.C. Cab '84
Didn't You Hear? '83
The Execution of Private
Slovik '74
Eye of the Tiger '86
Foolin' Around '80
Gumball Rally '76
Half a Lifetime '86
Hider in the House '90
Hitchhiker 1 '85
Insignificance '85
The Last American Hero '73
Lethal Weapon '87
Let's Get Harry '87
My Heroes Have Always Been
Cowboys '91
The Neon Empire '89
Point Break '91
Predator 2 '90
The Shrieking '73
Silver Bullet '85
A Star is Born '76
Straight Time '78
Thunderbolt & Lightfoot '74

Timothy Busfield

Field of Dreams '89
Revenge of the Nerds '84
Revenge of the Nerds 2:
Nerds in Paradise '87

Billy "Green" Bush

Alice Doesn't Live Here
Anymore '74
Amazing Stories, Book 5 '85
Critters '86
Culpepper Cattle Co. '72
Electra Glide in Blue '73
Elvis and Me '88
Five Easy Pieces '70
The River '84

Chuck Bush

Fandango '85

Grand Bush

Blind Vengeance '90
Colors '88
Hang Tough '89

Streets of Fire '84

James Bush

Beggars in Ermine '34
Crimson Romance '34
King of the Cowboys '43
The Return of Peter Grimm
'35
A Shot in the Dark '35

Jovita Bush

The Cheerleaders '72
Foxstyle '73

Rebecca Bush

Hunk '87

Anthony Bushell

The Arsenal Stadium Mystery
'39
Dark Journey '37
The Ghoul '34
A Night to Remember '58
The Royal Bed '31
The Scarlet Pimpernel '34
The Small Back Room '49

Anthony Bushell (D)

The Long Dark Hall '51

John Bushelman (D)

L.A. Gangs Rising '89

Joe Bushkin

The Rat Race '60

Francis X. Bushman

Ben-Hur '26
David and Bathsheba '51
Dick Tracy: The Original
Serial, Vol. 1 '37
Dick Tracy: The Original
Serial, Vol. 2 '37
Dick Tracy: The Spider Strikes
'37
Eyes Right! '26
Honky Tonk '41
Last Frontier '32
Midnight Faces '26
The Phantom Planet '61
Sabrina '54
When Lightning Strikes '34

Christel
Bushmann (D)

Comeback '83

William H.
Bushnell, Jr. (D)

Four Deuces '75

Akosua Busia

The Color Purple '85
The Final Terror '83
Native Son '86
The Seventh Sign '88

Marion Busia

Deadline Auto Theft '83
Gone in 60 Seconds '74

Ricky Busker

Big Shots '87

Narciso Busquets

El Gallo de Oro (The Golden
Rooster) '65
El Hombre Contra el Mundo
'87

Raymond Bussieres

Jonah Who Will Be 25 in the
Year 2000 '76

Paris When It Sizzles '64

Maria Bustamante

Yanco '64

Sergio Bustamante

Conexion Mexico '87
Lagrimas de Amor '87

Budd Buster

Border Badmen '45
The Lone Rider in Ghost
Town '41
Outlaw of the Plains '46
Pinto Canyon '40
Raiders of Sunset Pass '43
Songs and Bullets '38
Thunder in the Desert '38

Range Busters

Texas to Bataan '42

Kim Butcher

Frightmare '74

Dick Butkus

Cracking Up '83
Deadly Games '80
Hamburger... The Motion
Picture '86
Johnny Dangerously '84
The Legend of Sleepy Hollow
'79
Mother, Jugs and Speed '76

Calvin Butler

Drying Up the Streets '76
July Group '70s

Cindy Butler

Boggy Creek II '83
Grayeagle '77

Daniel Butler

Ernest Goes to Camp '87

David Butler

County Fair '20
The Sky Pilot '21

David Butler (D)

April In Paris '52
Bright Eyes '34
Calamity Jane '53
Captain January '36
A Connecticut Yankee '31
Doubting Thomas '35
It's a Great Feeling '49
The Little Colonel '35
Littlest Rebel '35
Lullaby of Broadway '51
Playmates '41
The Princess and the Pirate
'44
Road to Morocco '42
San Antonio '45
Tea for Two '50
Thank Your Lucky Stars '43
They Got Me Covered '43
You'll Find Out '40

Daws Butler

The Good, the Bad, and
Huckleberry Hound '88
Hey There, It's Yogi Bear '64
A Jetson Christmas Carol '89
The Jetsons Meet the
Flintstones '88

Dean Butler

Desert Hearts '86
Forever '78
Kid with the 200 I.Q. '83

Dick Butler *(D)*

Glory '56

Frank Butler

King of the Wild Horses '24
Made for Love '26

Jerry Butler

Rock 'n' Roll History: Sixties
Soul '90

Jimmy Butler

Manhattan Melodrama '34

Lois Butler

Mickey '48

Paul Butler

Crime Story: The Complete
Saga '86
To Sleep with Anger '90

Robert Butler *(D)*

Barefoot Executive '71
The Computer Wore Tennis
Shoes '69
Concrete Beat '84
Hot Lead & Cold Feet '78
James Dean '76
Long Time Gone '86
Moonlighting '85
Night of the Juggler '80
Now You See Him, Now You
Don't '72
A Question of Guilt '78
Scandalous John '71
Strange New World '75
The Ultimate Thrill '74
Underground Aces '80
Up the Creek '04

Tom Butler

Confidential '86
Scanners 2: The New Order
'91
Tales of the Klondike: The
Scorn of Women '81

William Butler

Inner Sanctum '91
Leatherface: The Texas
Chainsaw Massacre 3 '89

Hendel Butoy *(D)*

The Rescuers Down Under
'90

Merritt Butrick

Death Spa '87
Shy People '87
Star Trek 2: The Wrath of
Khan '82
Star Trek 3: The Search for
Spock '84
When the Bough Breaks '86
Wired to Kill '86
Zapped! '82

Johnny Butt

Q Ships '28

Lawson Butt

Dante's Inferno '24

**Charles
Butterworth**

Dixie Jamboree '44
Forsaking All Others '35
Moon's Our Home '36
Second Chorus '40
Swing High, Swing Low '37

Donna Butterworth

Family Jewels '65
Paradise, Hawaiian Style '66

Tyler Butterworth

Consuming Passions '88

Red Buttons

The Ambulance '90
Chaplin: A Character Is Born/
Keaton: The Great Stone
Face
C.H.O.M.P.S. '79
Death of a Hooker '71
18 Again! '88
Five Weeks in a Balloon '62
Gay Purr-ee '62
George Burns: His Wit &
Wisdom '88
Harlow '65
Hatari '62
Leave 'Em Laughing '81
The Longest Day '62
Movie, Movie '78
Off Your Rocker '80
Pete's Dragon '77
The Poseidon Adventure '72
Rudolph & Frosty's Christmas
in July '82
Sayonara '57
Side Show '84
They Shoot Horses, Don't
They? '69
13 Rue Madeleine '46
Users '78
Viva Knievel '77
When Time Ran Out '80
Who Killed Mary What's 'Er
Name? '71

Pat Buttram

Back to the Future, Part 3 '90
Blue Canadian Rockies '52
The Gatling Gun '72
Hills of Utah '51
Night Stage to Galveston '52
Robin Hood '73
Roustabout '64
Valley of Fire '51

Ulrike Butz

Secrets in the Dark '83

Sarah Buxton

Primal Rage '90
Rock 'n' Roll High School
Forever '91

Bernard Buys *(D)*

Demon Lust

George Buza

Meatballs 3 '87

Johannes Buzalski

Mark of the Devil '69

Zane Buzby

Cracking Up '83

Zane Buzby *(D)*

Last Resort '86

Lando Buzzanca

Homo Eroticus/Man of the
Year '71

Edward Buzzell *(D)*

At the Circus '39
Best Foot Forward '43
Go West '40
Neptune's Daughter '49
Song of the Thin Man '47

A Woman of Distinction '50

Ruth Buzzi

Bad Guys '86
The Being '83
Chu Chu & the Philly Flash '81
Digging Up Business '91
Dixie Lanes '88
Dream Boys Revue '87
Freaky Friday '76
My Mom's a Werewolf '89
The North Avenue Irregulars
'79
Once Upon a Brothers Grimm
'77
Pogo for President: ''I Go
Pogo'' '84
Pound Puppies and the
Legend of Big Paw '88
Scavenger Hunt '79
Surf 2 '84
Up Your Alley '89

Bobbie Byers

Wild Rebels '71

Mark Byers *(D)*

Criminal Act '88
Digging Up Business '91

Max Bygraves

A Cry from the Streets '57

Spring Byington

The Big Wheel '49
The Blue Bird '40
Devil & Miss Jones '41
Enchanted Cottage '45
Heaven Can Wait '43
In the Good Old Summertime
'49
Jezebel '38
Little Women '33
Lucky Partners '40
Meet John Doe '41
Mutiny on the Bounty '35
Please Don't Eat the Daisies
'60
Thrill of a Romance '45
Walk Softly, Stranger '50
Werewolf of London '35
When Ladies Meet '41
You Can't Take It with You
'38

Rolan Bykov

Commissar '67

John Byner

Great Smokey Roadblock '76
The Man in the Santa Claus
Suit '79
A Pleasure Doing Business
'79
Transylvania 6-5000 '85

Caruth C. Byrd *(D)*

Hollywood High Part 2 '81

Ralph Byrd

Blake of Scotland Yard '36
Born to Be Wild '38
Desperate Cargo '41
Dick Tracy '37
Dick Tracy Double Feature #1
'45
Dick Tracy Double Feature #2
'47
Dick Tracy Meets Gruesome
'47
Dick Tracy Returns '38
Dick Tracy: The Original
Serial, Vol. 1 '37

Dick Tracy: The Original
Serial, Vol. 2 '37
Dick Tracy: The Spider Strikes
'37
Dick Tracy vs. Crime Inc. '41
Dick Tracy's Dilemma '47
Guadalcanal Diary '43
The Howards of Virginia '40
Jungle Book '42
Jungle Goddess '49
Life Begins for Andy Hardy
'41
Radar Secret Service '50
Scared to Death/Dick Tracy's
Dilemma '47
The Son of Monte Cristo '40
S.O.S. Coast Guard '37
Thunder in the Pines '49
Trigger Trio '37

Tom Byrd

Wet Gold '84

The Byrds

Rock 'n' Roll History:
American Rock, the '60s
'89

Bill Byrge

Ernest Goes to Jail '90

Annie Byrne

Manhattan '79
A Night Full of Rain '78

Barbara Byrne

Mystery at Fire Island '81
Sunday in the Park with
George '86
Svengali '83

Catherine Byrne

Eat the Peach '86

David Byrne

Stop Making Sense '84
True Stories '86

David Byrne *(D)*

True Stories '86

Debbie Byrne

Rebel '86

Gabriel Byrne

Christopher Columbus '91
The Courier '88
Dark Obsession '90
Defense of the Realm '85
Excalibur '81
Gothic '87
Hanna K. '83
Hello Again '87
Julia and Julia '87
The Keep '83
Lionheart '87
Miller's Crossing '90
Shipwrecked '90
Siesta '87
A Soldier's Tale '91

Josh Byrne

Wild Pony '83

Martha Byrne

Anna to the Infinite Power '84
Eyes of the Amaryllis '82

Michael Byrne

Over Indulgence '87

Niall Byrne

The Miracle '91

Patricia T. Byrne

Night Call Nurses '72

Paula Byrne

The Key Man '57

Bara Byrnes

Extreme Close-Up '73
Sex Through a Window '72

Burke Byrnes

Witchboard '87

Edd Byrnes

Any Gun Can Play '67
Back to the Beach '87
Erotic Images '85
Go Kill and Come Back '68
Mankillers '87
Twirl '81

Maureen Byrnes

Hurry Up or I'll Be Thirty '73

Annie Byron

Fran '85

Bruce Byron

Kenneth Anger, Vol. 3:
Scorpio Rising '65

Delma Byron

Dimples '36

Jack Byron

Hard Hombre '31

Jeffrey Byron

Dungeonmaster '85
Metalstorm '83
The Seniors '78

Kathleen Byron

Burn Witch, Burn! '62
The Gambler & the Lady '52
The Small Back Room '49

Marion Byron

Breed of the Border '33
Steamboat Bill, Jr. '28

Walter Byron

Queen Kelly '29
The Savage Girl '32

John Byrum *(D)*

Heart Beat '80
Inserts '76
The Razor's Edge '84
The Whoopee Boys '86

James Caan

Alien Nation '88
Another Man, Another Chance
'77
Bolero '82
Brian's Song '71
A Bridge Too Far '77
Chapter Two '79
Cinderella Liberty '73
Comes a Horseman '78
Countdown '68
Dick Tracy '90
El Dorado '67
For the Boys '91
Freebie & the Bean '74
Funny Lady '75
The Gambler '74
Gardens of Stone '87
The Godfather '72
Godfather 1902-1959—The
Complete Epic '81
The Godfather: Part 2 '74

Camp ▶*cont.*

The Swinging Cheerleaders
'74
They All Laughed '81
Track 29 '88
Valley Girl '83
Walk Like a Man '87
Wicked Stepmother '89

Hamilton Camp

Arena '88
Meatballs 2 '84
Rosebud Beach Hotel '85
Tall Tales & Legends: Casey
at the Bat '85
Under Fire '83

Joe Camp (D)

Benji '73
Benji the Hunted '87
The Double McGuffin '79
For the Love of Benji '77
Hawmps! '76
Oh, Heavenly Dog! '80

Philip Campanaro

Sex Appeal '86

Frank Campanella

Blood Red '88
Chesty Anderson USN '76

Joe Campanella

Original Intent '92

Joseph Campanella

Ben '72
Body Chemistry '90
Club Fed '90
Comic Book Kids '82
Down the Drain '89
Hangar 18 '80
Last Call '90
Meteor '79
Murder Once Removed '71
No Retreat, No Surrender 3:
Blood Brothers '91
Plutonium Incident '82
The President's Plane is
Missing '71
Return to Fantasy Island '77
The St. Valentine's Day
Massacre '67
Sky Hei$t '75
Steele Justice '87
Terror on the 40th Floor '74

**Pasquale Festa
Campanile** (D)

When Women Had Tails '70
When Women Lost Their Tails
'75

Beatrice Campbell

Last Holiday '50
The Master of Ballantrae '53

Bill Campbell

Crime Story: The Complete
Saga '86
The Rocketeer '91

Brent Campbell

Happiness Is... '88

Bruce Campbell

Crimewave '86
The Evil Dead '83
Evil Dead 2: Dead by Dawn
'87
Going Back '83

Lunatics: A Love Story '92
Maniac Cop '88
Maniac Cop 2 '90
Mindwarp '91
Sundown '91
Waxwork 2: Lost in Time '91

C. Jutson Campbell

Star Crystal '85

Charlene Campbell

Dance '90

Cheryl Campbell

Chariots of Fire '81
Greystoke: The Legend of
Tarzan, Lord of the Apes
'84
McVicar '80
The Shooting Party '85

Chip Campbell

Driven to Kill '90

Colin Campbell

Leather Boys '63

Colin Campbell (D)

The Spoilers '14

Daphne Campbell

Overlanders '46

**David James
Campbell**

Killzone '85

**Donald
Campbell** (D)

White of the Eye '88

Douglas Campbell

Oedipus Rex '57

**Douglas
Campbell** (D)

Season of Fear '89
Zapped Again '89

Elizabeth Campbell

Doctor of Doom '62
Wrestling Women vs. the
Aztec Ape '62

Eric Campbell

Charlie Chaplin Festival '17
Charlie Chaplin: The Early
Years, Volume 1 '17
Charlie Chaplin: The Early
Years, Volume 2 '17
Charlie Chaplin: The Early
Years, Volume 3 '16
Charlie Chaplin: The Early
Years, Volume 4 '16
Count/Adventurer '17
The Floorwalker/By the Sea
'15
Triple Trouble/Easy Street '18

Glen Campbell

Rock-a-Doodle '92
True Grit '69
Uphill All the Way '85

Graeme Campbell

And Then You Die '88

**Graeme
Campbell** (D)

Blood Relations '87
Into the Fire '88
Murder One '88

**Anna Campbell-
Jones**

Secrets '82

Judy Campbell

Bonnie Prince Charlie '48
Convoy '40
Dust to Dust '85

Julia Campbell

Livin' Large '91
Opportunity Knocks '90

Kate Campbell

Monte Carlo Nights '34

Ken Campbell

Letter to Brezhnev '86
Smart Money '88

Louise Campbell

Bulldog Drummond Comes
Back '37
Bulldog Drummond's Peril '38
Bulldog Drummond's
Revenge '37
Bulldog Drummond's
Revenge/Bulldog
Drummond's Peril '38

**Martin
Campbell** (D)

Cast a Deadly Spell '91
Criminal Law '89
Defenseless '91
Edge of Darkness '86

Naomi Campbell

Cool As Ice '91

Nicholas Campbell

The Big Slice '91
Certain Fury '85
Fast Company '78
Going Home '86
July Group '70s
Knights of the City '86
Shades of Love: Champagne
for Two '87
The Victory '88

Patrick Campbell

Smokey Bites the Dust '81

Peggy Campbell

Big Calibre '35

Ron Campbell

Silhouette '91

Tisha Campbell

House Party '90
House Party 2: The Pajama
Jam '91
Rags to Riches '87
Rooftops '89
School Daze '88

Torguil Campbell

The Golden Seal '83
Heaven on Earth '89

William Campbell

Battle Circus '53
Dementia 13 '63
Man Without a Star '55
Night of Evil '62
Portrait in Terror '62
Track of the Vampire '66

Frank Campeau

When the Clouds Roll By '19

**Anna Campbell-
Jones**

Cris Campion

Field of Honor '87
Pirates '86

Jane Campion (D)

An Angel at My Table '89
Sweetie '89

Liz Campion

Those Glory, Glory Days '83

Wally Campo

The Beast from Haunted Cave
'60
Hell Squad '58
Machine Gun Kelly '58

**Carlo
Campogalliani** (D)

Goliath and the Barbarians '60
Son of Samson '62
Ursus in the Valley of the
Lions '62

Rafael Campos

The Astro-Zombies '67
The Hanged Man '74
Lady in a Cage '64
The Return of Josey Wales
'86
Slumber Party '57 '76

Susana Campos

Los Chicos Crecen

Victor Campos

Archer: The Fugitive from the
Empire '81

Michael Campus (D)

Education of Sonny Carson
'74
Mack '73
The Passover Plot '75

Rocky Camron

Sonora Stagecoach '44

Marcel Camus (D)

Black Orpheus '58
Where the Sky Begins '84

Mario Camus (D)

The Holy Innocents '84

Robert Canada

Danger Zone '87

Giana Maria Canale

The Devil's Commandment
'60
Go for Broke! '51
Hercules '58
Man from Cairo '54
Sins of Rome '54
Tiger of the Seven Seas '62

Maria Canale

Marauder '65

Lee Canalito

The Glass Jungle '88
Paradise Alley '78

David Canary

Johnny Firecloud '75
Melvin Purvis: G-Man '74
Posse '54
Sharks' Treasure '75

Stelio Candelli

A Man Called Rage '84

Candy Candido

The Great Mouse Detective
'86
Peter Pan '53

John Candy

Armed and Dangerous '86
Best of Comic Relief '86
The Blues Brothers '80
Brewster's Millions '85
Clown Murders '83
Find the Lady '75
Going Berserk '83
The Great Outdoors '88
Home Alone '90
Hot to Trot! '88
It Came from Hollywood '82
JFK '91
The Last Polka '84
Little Shop of Horrors '86
Lost and Found '79
The Masters of Menace '90
National Lampoon's Vacation
'83
Nothing But Trouble '91
Once Upon a Crime '91
Only the Lonely '91
Planes, Trains & Automobiles
'87
Really Weird Tales '86
The Rescuers Down Under
'90
Second City Comedy Special
'79
Second City Insanity '81
Sesame Street Presents:
Follow That Bird '85
The Silent Partner '78
Spaceballs '87
Speed Zone '88
Splash '84
Stripes '81
Summer Rental '85
Tales of the Klondike: The
Unexpected '83
Tunnelvision '76
Uncle Buck '89
Volunteers '85
Who's Harry Crumb? '89

Charles Cane

Dead Reckoning '47
Lucky Me '54
Native Son '51

Joseph Cane (D)

Oh Susannah '38

Roberto Canedo

Bestia Nocturna '86
Doctor of Doom '62
El Sinaloense '84
Wrestling Women vs. the
Aztec Ape '62

Lina Canelajas

The Garden of Delights '70

Jeff Canfield

Night of Horror '80s

Lee Canfield

Escapes '86

**Mary Grace
Canfield**

South of Reno '87

Enzo Cannavale

Flatfoot '78

James Canning

The Boys in Company C '77

Amanda Cannings

Forbidden '85

Richard Cannistraro (D)

Violated '84

David Cannon

The Severed Arm '73

Dyan Cannon

The Anderson Tapes '71
Author! Author! '82
Bob & Carol & Ted & Alice '69
Caddyshack 2 '88
Coast to Coast '80
Deathtrap '82
Doctors' Wives '70
The End of Innocence '90
Having It All '82
Heaven Can Wait '78
Honeysuckle Rose '80
Lady of the House '78
The Last of Sheila '73
The Love Machine '71
Merlin and the Sword '85
Revenge of the Pink Panther '78
Shamus '73
The Virginia Hill Story '76

Dyan Cannon (D)

The End of Innocence '90

Ed Cannon

Vampire Cop '90

Esma Cannon

Crow Hollow '52

Freddy Cannon

Village of the Giants '65

J.D. Cannon

The Adventures of Nellie Bly '81
Cool Hand Luke '67
Death Wish 2 '82
Killing Stone '78
Pleasure Palace '80
Raise the Titanic '80
Street Justice '89
Walking Through the Fire '80

Katherine Cannon

High Noon: Part 2 '80
Matters of the Heart '90
Private Duty Nurses '71
The Red Fury '70s
Will: G. Gordon Liddy '82

Mary Cannon

Trucker's Woman '83

Pomeroy Cannon

The Four Horsemen of the Apocalypse '21

Vince Cannon

The Manhandlers '73

Gaspar Cano

Hot Blood '89

Anne Canoras

Revenge of the Dead '84

Diana Canova

The First Nudie Musical '75
Night Partners '83

Judy Canova

The Adventures of Huckleberry Finn '60
Oklahoma Annie '51

Anne Canovos

High Frequency '88

Antonio Cantafora

Gabriela '84

Luis Aceves Cantaneda

Curse of the Mummy/Robot vs. the Aztec Mummy '60s

Elena Cantarone

Evil Clutch '89

Kieran Canter

Buried Alive '81

Jo Canterbury

Teenage Strangler '64

Cantinflas

Around the World in 80 Days '56

Toni Canto

I'm the One You're Looking For '88

Eddie Cantor

Entertaining the Troops '89
Glorifying the American Girl '30
Hollywood Goes to War '54
If You Knew Susie '48
Kid Millions '34
Roman Scandals '33
Show Business '44
Thank Your Lucky Stars '43
Whoopee! '30

Yakima Canutt

Branded a Coward '35
Circle of Death '36
Cyclone of the Saddle '35
Dawn Rider '35
The Fighting Stallion '26
Gone with the Wind '39
Heart of the Rockies '37
Hit the Saddle '37
King of the Pecos '36
Last Frontier '32
Lucky Texan '34
The Man from Hell '34
Outlaw Rule '36
The Painted Stallion '37
Paradise Canyon '35
Riders of the Rockies '37
Riders of the Whistling Skull '37
Sagebrush Trail '33
The Star Packer '34
Trouble in Texas '37
West of the Divide '34
Wild Horse Canyon '25
Winds of the Wasteland '36

Yakima Canutt (D)

G-Men Never Forget '48
Oklahoma Badlands '48

Peter Capaldi

Dangerous Liaisons '89
John & Yoko: A Love Story '85
The Lair of the White Worm '88
Local Hero '83

Louis Capauno (D)

Executioner of Venice '63

Jad Capelja

Freedom '81
Puberty Blues '81

Barbara Capell

The Werewolf vs. the Vampire Woman '70

Hedge Capers

The Legend of Hillbilly John '73

Virginia Capers

The North Avenue Irregulars '79
White Mama '80

Manuel Capetillo

Passion for Power '85
Vivo o Muerto '87

Raymundo Capetillo

Bestia Nocturna '86

Giorgio Capitani (D)

I Hate Blondes '83
Lobster for Breakfast '82
The Ruthless Four '70

Frank Capizzi

Desert Snow '89

Katie Caple

Shotgun '89

Carmine Capobianco

Galactic Gigolo '87
Psychos in Love '87

Lino Capolicchio

Garden of the Finzi-Continis '71

Nadia Capone

Skull: A Night of Terror '88
A Whisper to a Scream '88

Truman Capote

ABC Stage 67: Truman Capote's A Christmas Memory '66
Murder by Death '76

Carl Capotorto

American Blue Note '91

Patrizia Capparelli

The Decameron '70

Pier Paola Capponi

Man of Legend '71

Piero Cappucilli

Otello '82

Bernt Capra (D)

Mindwalk '91

Frank Capra (D)

Arsenic and Old Lace '44
Broadway Bill '34
Here Comes the Groom '51
Hole in the Head '59
It Happened One Night '34
It's a Wonderful Life '46
Lady for a Day '33

Lost Horizon '37
Meet John Doe '41
Mr. Deeds Goes to Town '36
Mr. Smith Goes to Washington '39
Platinum Blonde '31
Pocketful of Miracles '61
State of the Union '48
Strong Man '26
That Certain Thing '28
You Can't Take It with You '38

Jordanna Capra

Hired to Kill '91

Tony Caprari

River of Diamonds '90

Anna Capri

The Brotherhood of Satan '71
Darker Than Amber '70
Payday '72
The Specialist '75

Kate Capshaw

Best Defense '84
Black Rain '89
Code Name: Dancer '87
Dreamscape '84
Indiana Jones and the Temple of Doom '84
A Little Sex '82
Love at Large '89
My Heroes Have Always Been Cowboys '91
Power '86
Private Affairs '89
The Quick and the Dead '87
SpaceCamp '86
Windy City '84

Luigi Capuano (D)

Cold Steel for Tortuga '65
Marauder '65
The Snake Hunter Strangler '66
Tiger of the Seven Seas '62

Capucine

The Con Artists '80
Curse of the Pink Panther '83
Fellini Satyricon '69
From Hell to Victory '79
The Honey Pot '67
Jaguar Lives '70s
North to Alaska '60
The Pink Panther '64
Scandalous '88
Song Without End '60
The Switch '89
Trail of the Pink Panther '82
Walk on the Wild Side '62
What's New Pussycat? '65

Irene Cara

Aaron Loves Angela '75
Busted Up '86
Caged in Paradiso '89
Certain Fury '85
City Heat '84
D.C. Cab '84
Fame '80
For Us, the Living '88
Killing 'em Softly '85
Sparkle '76

Flora Carabella

A Night Full of Rain '78

Paul Carafotes

Choices '81
Journey to the Center of the Earth '88

Lilli Carati

Copenhagen Nights '80s

Dacosta Carayan (D)

The Brave Bunch '70

Costa Carayiannis (D)

Land of the Minotaur '77

Richard Carballo

Guess What We Learned in School Today? '70

Joseph Carberry

The Ten Million Dollar Getaway '91

Christopher Carbis

The Psychotronic Man '91

Antony Carbone

A Bucket of Blood '59
Creature from the Haunted Sea '60
Extreme Close-Up '73
The Last Porno Flick '74
The Last Woman on Earth '61
Sex Through a Window '72
Skateboard '77

Larry Carby

The Dying Truth '91

Lamar Card (D)

The Clones '73

Christine Cardan

Gunblast '70s

Richard Cardella

The Crater Lake Monster '77

Elsa Cardenas

Fun in Acapulco '63
Los Asesinos '68
Mision Suicida '87
Rio Hondo '47

Hernan Cardenas (D)

Island Claw '80

Jane Carder

Sex and the Other Woman '79

Pat Cardi

And Now Miguel '66
Twisted Brain '74

Jack Cardiff (D)

The Freakmaker '73
The Girl on a Motorcycle '68
My Geisha '62

Lori Cardillo

Day of the Dead '85

Maria Cardinal

La Casita del Pecado
Mouchette '60

Roger Cardinal (D)

Malarek '89

Tantoo Cardinal

Dances with Wolves '90
Loyalties '86

Claudia Cardinale

The Battle of Austerlitz '60
Big Deal on Madonna Street
  '60
Blonde in Black Leather '77
Blood Brothers '74
Butterfly Affair '71
Circus World '64
Conversation Piece '75
Corleone '79
8 1/2 '63
Escape to Athena '79
Fitzcarraldo '82
The Gift '82
Henry IV '85
Immortal Bachelor '80
Jesus of Nazareth '77
Legend of Frenchie King '71
The Lost Command '66
A Man in Love '87
Next Summer '84
Once Upon a Time in the West
  '68
One Russian Summer '73
The Pink Panther '64
Princess Daisy '83
Professionals '66
Queen of Diamonds '80s
Red Tent '69
Rocco and His Brothers '60
The Salamander '82
Scoumoune '74
The Sniper '87
Time of Indifference '64
L'Udienza '71

Annette Cardona

Latino '85

Rene Cardona, Sr. (D)

Night of the Bloody Apes '68
Night of a Thousand Cats '72
Santa Claus '59
Wrestling Women vs. the
  Aztec Mummy '59

Rene Cardona, Jr.

Spiritism '61

Rene Cardona, Jr. (D)

Beaks: The Movie '87
Doctor of Doom '62
Robinson Crusoe & the Tiger
  '72
Tintorera...Tiger Shark '78
Wrestling Women vs. the
  Aztec Ape '62

J.S. Cardone (D)

A Climate for Killing '91
Shadowzone '90
The Slayer '82
The Slayer/The Scalps '82
Thunder Alley '85

Nathalie Cardone

The Little Thief '89

John Cardos

Blood of Dracula's Castle '69

John Cardos (D)

The Dark '79
Day Time Ended '80
The Female Bunch '69
Kingdom of the Spiders '77
Mutant '83
Outlaw of Gor '87
Skeleton Coast '89

Benny Cardosa

Scalps '83

James B. Cardwell

The Devil on Wheels '47
Down Dakota Way '49
Fear '46
The Fighting Sullivans '42

David Cardy

Xtro '83

Roger Carel

Soldat Duroc...Ca Va Etre Ta
  Fete! '60s
Two of Us '68

Annette Carell

The Tell-Tale Heart '62
Vulture '67

Lianella Carell

The Bicycle Thief '48

Julien Carette

Grand Illusion '37
La Bete Humaine '38
The Rules of the Game '39

James Carew

Midnight at the Wax Museum
  '36

Topper Carew (D)

Talkin' Dirty After Dark '91

Christopher Carey

Captain America 2: Death Too
  Soon '79

George E. Carey

Weekend with the Babysitter
  '70

Harry Carey, Jr.

Alvarez Kelly '66
Back to the Future, Part 3 '90
Bad Jim '89
Bandolero! '68
Billy the Kid Versus Dracula
  '66
Cahill: United States Marshal
  '73
Challenge to White Fang '86
Copper Canyon '50
Endangered Species '82
The Long Riders '80
Lou Gehrig Story '56
Mister Roberts '55
Moonrise '48
A Public Affair '62
Pursued '47
Rio Grande '50
The Searchers '56
Seventh Cavalry '56
She Wore a Yellow Ribbon '49
Three Godfathers '48
Trinity is Still My Name '75
The Undefeated '69
Wagon Master '50
Whales of August '87

Harry Carey

Aces Wild '37
Air Force '43
Angel and the Badman '47
Beyond Tomorrow '40
Border Devils '32
Breaking In '89
Buffalo Stampede '33
Cavalier of the West '31
Devil Horse '32
Duel in the Sun '46

Ghost Town '37
The Great Moment '44
Last of the Clintons '35
Last of the Mohicans '32
Last Outlaw '36
Law West of Tombstone '38
Man of the Forest '33
Mr. Smith Goes to
  Washington '39
Night Rider '32
The Port of Missing Girls '38
Powdersmoke Range '35
Rustler's Paradise '35
Short Films of D.W. Griffith,
  Volume 2 '09
So Dear to My Heart '49
The Spoilers '42
Straight Shootin' '17
Sundown '41
They Knew What They
  Wanted '40
Vanishing Legion '31
Wagon Trail '35
Wild Mustang '35
Without Honors '32

Joyce Carey

Blithe Spirit '45
Cry, the Beloved Country '51
October Man '48

Leonard Carey

The Snows of Kilimanjaro '52

MacDonald Carey

Access Code '84
Copper Canyon '50
End of the World '77
Great Missouri Raid '51
It's Alive 3: Island of the Alive
  '87
Let's Make it Legal '51
The Rebels '79
Shadow of a Doubt '42
Tammy and the Doctor '63
Wake Island '42
Who Is the Black Dahlia? '75

Michele Carey

Changes '69
El Dorado '67
In the Shadow of Kilimanjaro
  '86
Live a Little, Love a Little '68
Scandalous John '71

Phil Carey

Gun Fury '53
The Long Gray Line '55
Monster '78
The Time Travelers '64
Tonka '58

Roland Carey

The Giants of Thessaly '60
Revolt of the Barbarians '64

Ron Carey

Fatso '80
High Anxiety '77
History of the World: Part 1
  '81
Who Killed Mary What's 'Er
  Name? '71

Sandy Carey

Mysterious Jane '80s

Timothy Carey

Echo Park '86
Fast Walking '82
The Finger Man '55
The Killing '56
The Mermaids of Tiburon '62

Paths of Glory '57
Poor White Trash '57
Speedtrap '78

Tino Carey

The Legend of the Wolf
  Woman '77

Timothy Carhart

Party Animal '83
Pink Cadillac '89
Thelma & Louise '91

Gia Carides

Backlash '86
Blacklash '87
Midnite Spares '85

Zoe Carides

Stones of Death '88

Carmine Caridi

Split Decisions '88

Leo Carillo

Riders of Death Valley '41

Len Cariou

Drying Up the Streets '76
The Four Seasons '81
The Lady in White '88
A Little Night Music '77
Louisiana '87
Who'll Save Our Children? '82

Adam Carl

Summer Camp Nightmare '86

Cynthia Carle

Alamo Bay '85

Gilles Carle (D)

In Trouble '67
Maria Chapdelaine '84
The Red Half-Breed '70

Richard Carle

The Ghost Walks '34

Sophie Carle

Quicker Than the Eye '88

John Carlen

House of Dark Shadows '70

Claire Carleton

The Devil's Partner '58
The Fighter '52

George Carlin

Best of Comic Relief '86
Bill & Ted's Bogus Journey
  '91
Bill & Ted's Excellent
  Adventure '89
Car Wash '76
George Carlin on Campus '84
George Carlin at Carnegie Hall
  '83
George Carlin - Live!: What
  am I Doing in New Jersey?
  '89
George Carlin: Playin' with
  Your Head '88
Outrageous Fortune '87

Gloria Carlin

Goldenrod '77

Lynn Carlin

Deathdream '72
Forbidden Love '82
Silent Night, Lonely Night '69

Superstition '82
Wild Rovers '71

Paolo Carlini

It Started in Naples '60

Lewis John Carlino (D)

Class '83
The Great Santini '80
The Sailor Who Fell from
  Grace with the Sea '76

Anne Carlisle

Liquid Sky '83
Perfect Strangers '84
The Suicide Club '88

Belinda Carlisle

Go Go's: Wild at the Greek
  '84
Legendary Ladies of Rock &
  Roll '86
Swing Shift '84

Bruce Carlisle

The Fast and the Furious '54

Kitty Carlisle

Murder at the Vanities '34
A Night at the Opera '35
Radio Days '87

Lucille Carlisle

Bakery/Grocery Clerk '21

Mary Carlisle

Baby Face Morgan '42
Champagne for Breakfast '79
Dance, Girl, Dance '40
Dead Men Walk '43
Held for Murder '32
Kentucky Kernels '34
One Frightened Night '35

Spencer Carlisle

The Devil's Partner '58

Steve Carlisle

Sonny Boy '87

Ismael Carlo

One Man Out '89

Johann Carlo

Nadia '84

Margit Carlquist

Smiles of a Summer Night '55

Henning Carlsen (D)

The Wolf at the Door '87

Carl Carlson

Juggler of Notre Dame '84

Gene Carlson

Psychopath '73

Joel Carlson

Communion '89

June Carlson

Delinquent Daughters '44
Mom & Dad '47

Karen Carlson

American Dream '81
Black Oak Conspiracy '77
Brotherly Love '85
The Candidate '72

Fleshburn '84
In Love with an Older Woman '82
Octagon '80
On Wings of Eagles '86
The Student Nurses '70
Teen Vamp '88
Wild Horses '84

Leslie Carlson

Deranged '74

Moose Carlson

Can I Do It...Till I Need Glasses? '77

Richard Carlson

The Amazing Mr. X '48
Behind Locked Doors '48
Beyond Tomorrow '40
Creature from the Black Lagoon '54
Flat Top '52
Hold That Ghost '41
The Howards of Virginia '40
Human Gorilla '48
Hunted '88
It Came From Outer Space '53
King Solomon's Mines '50
The Last Command '55
The Little Foxes '41
Presenting Lily Mars '43
Retreat, Hell! '52
Saturday Night Shockers, Vol. 2 '58
Too Many Girls '40
Tormented '60
Try and Get Me '50
The Valley of Gwangi '69
The Young in Heart '38

Steve Carlson

Brothers O'Toole '73

Susie Carlson

Swinging Sorority Girls '70s

Karin Carlsson

Walpurgis Night '41
A Woman's Face '38

Hope Marie Carlton

Bloodmatch '91
Hard Ticket to Hawaii '87
Picasso Trigger '89
Playmate Playoffs '86
Round Numbers '91
Savage Beach '89
Slaughterhouse Rock '88
Slumber Party Massacre 3 '90
Terminal Exposure '89

Milly Carlucci

Hercules 2 '85

Pat Carlyle

Marihuana '36

Roger C. Carmel

Gambit '66
Hardly Working '81
My Dog, the Thief '69
Skullduggery '70
Thunder and Lightning '77

Jeanne Carmen

In Old Montana '39
The Monster of Piedras Blancas '57

Jewel Carmen

American Aristocracy '17
Flirting with Fate '16

Julie Carmen

Blue City '86
Can You Hear the Laughter? The Story of Freddie Prinze '79
Fright Night 2 '88
Gloria '80
Gore Vidal's Billy the Kid '89
The Hunt for the Night Stalker '91
Kiss Me a Killer '91
Last Plane Out '83
The Milagro Beanfield War '88
The Neon Empire '89
Night of the Juggler '80
Paint it Black '89
The Penitent '88
She's in the Army Now '81

Loene Carmen

The Year My Voice Broke '87

Francesca Carmeno

Other Hell '85

Jean Carmet

Black and White in Color '76
Buffet Froid '79
The Little Theatre of Jean Renoir '71
Return of the Tall Blond Man With One Black Shoe '74
Secret Obsession '88
Sorceress '88
The Tall Blond Man with One Black Shoe '72
Violette '78

Hoagy Carmichael

The Best Years of Our Lives '46
Johnny Angel '45
Las Vegas Story '52
To Have & Have Not '44
Young Man with a Horn '50

Ian Carmichael

Betrayed '54
The Colditz Story '55
Heavens Above '63
I'm All Right Jack '69
Lucky Jim '58

Patricia Carmichael

Dear Dead Delilah '72

Tullio Carminati

London Melody '37
One Night of Love '34
Roman Holiday '53

Michael Carmine

Band of the Hand '86
*batteries not included '87
Leviathan '89

Don Carmody (D)

Surrogate '88

Stella Carnachia

The Eerie Midnight Horror Show '82

Judy Carne

The Americanization of Emily '64
Only With Married Men '74

Marcel Carne (D)

Bizarre Bizarre '39
Children of Paradise (Les Enfants du Paradis) '44
Le Jour Se Leve '39

Les Visiteurs du Soir '42

Primo Carnera

Hercules Unchained '59

Alan Carney

Girl Rush '44
Roadie '80
Zombies on Broadway '44

Art Carney

Better Late Than Never '83
Bitter Harvest '81
The Blue Yonder '86
Death Scream '75
Defiance '79
The Emperor's New Clothes '84
Firestarter '84
Going in Style '79
A Guide for the Married Man '67
Harry and Tonto '74
House Calls '78
Izzy & Moe '85
Katherine '75
The Late Show '77
Miracle of the Heart: A Boys Town Story '86
Movie, Movie '78
Muppets Take Manhattan '84
The Naked Face '84
Night Friend '87
The Night They Saved Christmas '87
Pot O' Gold '41
The Radical '75
Roadie '80
St. Helen's, Killer Volcano '82
Sunburn '79
Take This Job & Shove It '81
Undergrads '85

Flint Carney

Desert Snow '89

George Carney

Brighton Rock '47
Love on the Dole '41

Uncle Don Carney

Robinson Crusoe '36

Darcia Carnie

Ladies of the Lotus '86

Angela Carnon

Pleasure Unlimited '86

Morris Carnovsky

Cyrano de Bergerac '50
Dead Reckoning '47
Gun Crazy '50
Joe's Bed-Stuy Barbershop: We Cut Heads '83
Rhapsody in Blue '45

Alicia Caro

El Ceniciento '47
Maria '68

Cathia Caro

Triumph of the Son of Hercules '63

Cindy Carol

Dear Brigitte '65
Gidget Goes to Rome '63

Jean Carol

Payback '89

Jeannie Carol

Vice Academy '88

Linda Carol

Carnal Crimes '91
Future Hunters '88
Reform School Girls '86

Martine Carol

Around the World in 80 Days '56
The Battle of Austerlitz '60
Lola Montes '55
Nana '55

Sheila Carol

The Beast from Haunted Cave '60
Ski Troop Attack '60

Sue Carol

The Lone Star Ranger '39
Walking Back '26

Glenn Gordon Caron (D)

Clean and Sober '88

Leslie Caron

An American in Paris '51
The Battle of Austerlitz '60
The Cary Grant Collection '90
Courage Mountain '89
Daddy Long Legs '55
Dangerous Moves '84
Fanny '61
Father Goose '64
Gigi '58
The Glass Slipper '55
Goldengirl '79
Head of the Family '71
Is Paris Burning? '68
Lili '53
Madron '70
The Man Who Loved Women '77
Nicole '72
Promise Her Anything '66
QB VII '74
The Unapproachable '82

Howard Carpendale

No One Cries Forever '85

Bethany Carpenter

Tiger Town '83

Carleton Carpenter

Summer Stock '50
Two Weeks with Love '50
Up Periscope '59

Horace Carpenter

Maniac '34
The Worst of Hollywood: Vol. 1 '90

Horace Carpenter (D)

Arizona Kid '39

John Carpenter

Night of the Ghouls '59
Song of Old Wyoming '45

John Carpenter (D)

Assault on Precinct 13 '76
Big Trouble in Little China '86
Christine '84
Dark Star '74
Elvis: The Movie '79
Escape from New York '81
The Fog '78
Halloween '78

Memoirs of an Invisible Man '92
Prince of Darkness '87
Starman '84
They Live '88
The Thing '82

Paul Carpenter

Fire Maidens from Outer Space '56
Heat Wave '54
Paid to Kill '54
Unholy Four '54

Peter Carpenter

Blood Mania '70
Point of Terror '71

Stephen Carpenter (D)

Dorm That Dripped Blood '82
The Kindred '87
The Power '80

Thelma Carpenter

The Wiz '78

Tony Carpenter

Night Vision '87

Fabio Carpi (D)

Basileus Quartet '82

Adrian Carr (D)

Now and Forever '82

Bernard Carr

Who Killed Doc Robbin? '48

Bernard Carr (D)

Curley '47
Fabulous Joe '47

Betty Ann Carr

Inferno in Paradise '88

Camilla Carr

Don't Look in the Basement '73
Keep My Grave Open '80
Logan's Run '76
Making Love '82
Poor White Trash 2 '75

Carole Carr

Down Among the Z Men '52
Scream Dream '89

Charmian Carr

The Sound of Music '65

Darlene Carr

The Jungle Book '67

Jane Carr

The Prime of Miss Jean Brodie '69

John Carr (D)

Death Wish Club '83

Lindsay Carr

Glitch! '88

Louis Carr, Jr.

Motown's Mustang '86

Marian Carr

The Indestructible Man '56

Mary Carr

Gun Law '33

Val Chapman

The Legend of Jedediah Carver '70s

Anna Chappell

Mountaintop Motel Massacre '86

Robert Chappell (D)

Diamond Run '90

Corynne Charbit

La Chevre '81

Patricia Charbonneau

Brain Dead '89
Call Me '88
Desert Hearts '86
K2 '92
RoboCop 2 '90
Shakedown '88
Stalking Danger '88

Jon Chardiet

Beat Street '84

Erik Charell (D)

Congress Dances (Der Kongress taenzt) '31

Cyd Charisse

Band Wagon '53
Brigadoon '54
Deep in My Heart '54
The Harvey Girls '46
It's Always Fair Weather '55
Meet Me in Las Vegas '56
On an Island with You '48
Party Girl '58
Silk Stockings '57
Singin' in the Rain '52
Stars of the Century '88
Swimsuit: The Movie '89
That's Dancing! '85
Till the Clouds Roll By/A Star Is Born '46
Two Weeks in Another Town '62
Words & Music '48

Alok Charkravarty

World of Apu '59

Bobby H. Charles

Toga Party '79

David Charles

Julia has Two Lovers '91

Emile Charles

Wonderland '88

Josh Charles

Dead Poets Society '89
Don't Tell Mom the Babysitter's Dead '91

Leon Charles

Whitcomb's War '88

Lynne Charles

The Lady of the Camellias '87

Ray Charles

Ballad in Blue '66
The Blues Brothers '80
Carnegie Hall at 100: A Place of Dreams '91
Limit Up '89
Quincy Jones: A Celebration '83

R.K. Charles

Girl on a Chain Gang '65

Ian Charleson

Car Trouble '86
Chariots of Fire '81
Codename Kyril '91
Gandhi '82
Greystoke: The Legend of Tarzan, Lord of the Apes '84
Louisiana '87
Terror at the Opera '88

Leslie Charleson

Most Wanted '76

George Charlia

Prix de Beaute '30

Robert Charlton (D)

No Big Deal '83

Charo

The Concorde: Airport '79 '79
Moon Over Parador '88
Pee-Wee's Playhouse Christmas Special '88

Charpin

The Baker's Wife (La Femme du Boulanger) '33
Cesar '36
Marius '31
Well-Digger's Daughter '46

Antonio Charriello

La Bella Lola '87

Henri Charriere

Butterfly Affair '71
Queen of Diamonds '80s

Spencer Charters

The Bat Whispers '30
The Ghost Walks '34
Lady from Nowhere '36

Melanie Chartoff

Doin' Time '85
Having It All '82
Stoogemania '85

Alden Chase

Buried Alive '39
Cowboy Millionaire '35

Annazette Chase

Fist '79
Truck Turner '74

Barbara Chase

The Ghosts of Hanley House '68

Barrie Chase

Fabulous Fred Astaire '58

Charley Chase

Block-heads '38
Charley Chase Festival: Vol. 1 '20s
Charlie Chase and Ben Turpin '21
Fun Factory/Clown Princes of Hollywood '20s
King of the Wild Horses '24
Silent Laugh Makers, No. 1 '20s
Silent Laugh Makers, No. 2 '20s
Slapstick '30s
Sons of the Desert '33

Three Charlies and One Phoney! '18

Chevy Chase

Best of Chevy Chase '87
Best of Dan Aykroyd '86
Best of Gilda Radner '89
Best of John Belushi '85
Caddyshack '80
Caddyshack 2 '88
Deal of the Century '83
Fletch '85
Fletch Lives '89
Foul Play '78
Funny Farm '88
The Groove Tube '72
Memoirs of an Invisible Man '92
Modern Problems '81
National Lampoon's Christmas Vacation '89
National Lampoon's European Vacation '85
National Lampoon's Vacation '83
Nothing But Trouble '91
Oh, Heavenly Dog! '80
Saturday Night Live: Richard Pryor '75
Saturday Night Live, Vol. 1 '70s
Seems Like Old Times '80
Sesame Street Presents: Follow That Bird '85
Spies Like Us '85
Three Amigos '86
Tunnelvision '76
Under the Rainbow '81

Ilka Chase

The Animal Kingdom '32

James Chase

The Virgin Witch '70

Jeffrey Chase

Scream of the Demon Lover '71

Jennifer Chase

Balboa '82
Death Screams '83
House of Death '82

Karen Chase

Vendetta '85

Richard Chase (D)

Hell's Angels Forever '83

Stephan Chase

Black Arrow '84
The Dying Truth '91

Nathan Lee Chasing His Horse

Dances with Wolves '90

David Chaskin

The Curse '87

Don Chastain

Black Godfather '74

Corinne Chateau

Witchfire '86

Peter Chatel

Fox and His Friends '75

Soumitra Chaterjee

The Home and the World '84

Jack Chatham

The Mutilator '85

Etienne Chatiliez (D)

Tatie Danielle '91

Gregory Allen Chatman

Beyond the Doors '83

Sorapong Chatri

The Ultimate Ninja '85

Soumitra Chatterjee

Devi '60
Distant Thunder '73
World of Apu '59

Ruth Chatterton

Dodsworth '36
Prudential Family Playhouse '50

Tom Chatterton

Drums of Fu Manchu '40

Daniel Chatto

Little Dorrit, Film 1: Nobody's Fault '88
Little Dorrit, Film 2: Little Dorrit's Story '88

Francois Chau

Teenage Mutant Ninja Turtles 2: The Secret of the Ooze '91

Amin Q. Chaudhri (D)

Tiger Warsaw '87
An Unremarkable Life '89

Emmanuelle Chaulet

Boyfriends & Girlfriends '88

Lilyan Chauvin

Silent Night, Deadly Night '84

Jacqueline Chauveau

Murmur of the Heart '71

Elizabeth Chauvet

Rattlers '76

Jaime Chavarri (D)

I'm the One You're Looking For '88

Kotti Chave

Dollar '38

Ingrid Chavez

Graffiti Bridge '90

Julio Chavez

The Lion's Share '79

Marcelo Chavez

Viva Chihuahua (Long Live Chihuahua) '47

Oscar Chavez

Los Caifanes '87

Jeff Chayette

Cave Girl '85

Maury Chaykin

Breaking In '89
Cold Comfort '90
Dances with Wolves '90
Def-Con 4 '85
George's Island '91
Higher Education '88
Iron Eagle 2 '88
July Group '70s
Leaving Normal '92
Mr. Destiny '90
Stars and Bars '88
Sweetheart!

Mariann Chazel

Next Year If All Goes Well '83

Don Cheadle

Colors '88
Hamburger Hill '87

Andrea Checchi

The Invaders '63

Al Checco

Extreme Close-Up '73
Sex Through a Window '72

Jeremiah S. Chechik (D)

National Lampoon's Christmas Vacation '89

Douglas Cheek (D)

C.H.U.D. '84

Micheline Cheirel

Carnival in Flanders (La Kermesse Heroique) '35
Cornered '45

Michael Chekhov

Spectre of the Rose '46
Spellbound '45

Ara Chekmayan (D)

Forever James Dean '88

Peter Chelsom (D)

Hear My Song '91

Tsilla Chelton

Tatie Danielle '91

Wayne Chema

Nomad Riders '81

Sophie Chemineau

A Girl on Her Own '76

Bruce Chen

Ninja Holocaust '80s

Chen Chen

Vengeance of the Snowmaid '70s

Cheung Ying Chen

Return of the Chinese Boxer '74

Tang Chen Da

Little Heroes of Shaolin Temple '72

Jacky Chen

Of Cooks and Kung-Fu '80s

Joan Chen

The Blood of Heroes '89
Dim Sum: A Little Bit of Heart '85

Chen ▶cont.

The Last Emperor '87
Night Stalker '87
Taipan '86
Twin Peaks '90

Jon Chen

Death Mask of the Ninja '80s

Lei Chen

Kung Fu Strongman

Wong Chen-Li

Secret Executioners '70s

Lu Pi Chen

Fist of Vengeance '70s

Moira Chen

Endgame '85

Tina Chen

Lady from Yesterday '85
Paper Man '71

Tsao Chen

Enter the Panther '79

Yeung Ching Chen (D)

The Two Great Cavaliers '73

Pierre Chenal (D)

Crime and Punishment '35
Man from Nowhere
Native Son '51

Jacky Cheng

Ninja Exterminators '81

Kent Cheng

Crazy Kung Fu Master

Tommy Cheng (D)

Ninja: American Warrior '90
Ninja Death Squad '87

Chan Kin Cheong

War of the Shaolin Temple

Bill Chepil

Street Trash '87

Cher

Come Back to the 5 & Dime
   Jimmy Dean, Jimmy Dean
   '82
Mask '85
Mermaids '90
Moonstruck '87
Silkwood '83
Suspect '87
Witches of Eastwick '87

Patrice Chereau

Danton '82

Patrice Chereau (D)

L'Homme Blesse '83

Nikolai Cherkassov

Alexander Nevsky '38
Baltic Deputy '37
Don Quixote '57
Ivan the Terrible, Part 1 '44
Ivan the Terrible, Part 2 '46

Kaethe Cherney

Cthulhu Mansion '91

Virginia Cherrill

City Lights '31
Late Extra '35

Byron Cherry

The Fix '84

John R. Cherry, III (D)

Dr. Otto & the Riddle of the
   Gloom Beam '86
The Ernest Film Festival '86
Ernest Goes to Camp '87
Ernest Goes to Jail '90
Ernest Saves Christmas '88

John Cherry (D)

Ernest Scared Stupid '92

Karen Cheryl

Here Comes Santa Claus '84

George Cheseboro

Cheyenne Takes Over '47
In Old Santa Fe '34
Laramie Kid '35
Red River Valley '36
Return of the Lash '47
Roamin' Wild '36
Stage to Mesa City '48

Elizabeth Cheshire

Melvin and Howard '80

Arthur Chesney

The Lodger '26

Alma Chester

When a Man Rides Alone '33

Ed Chester

City in Panic '86

Holly Chester

SnakeEater 3...His Law '92

Morris Chestnut

Boyz N the Hood '91

Cynthia Cheston

Angels of the City '89

Lionel Chetwynd (D)

Hanoi Hilton '87

Alton Cheung (D)

Kickboxer the Champion '91

Maggie Cheung

Jackie Chan's Police Force
   '85

Pearl Cheung

Venus the Ninja '80s

Philip Cheung

Bruce Lee's Ways of Kung Fu
   '80s

Yuen Cheung Yan

Taoism Drunkard

Kevin Timothy Chevalia

Folks '92

Maurice Chevalier

Breath of Scandal '60
Can-Can '60
Fanny '61

French Singers '50s
Gigi '58
In Search of the Castaways
   '62
Invitation to Paris '60
Love in the Afternoon '57
Merry Widow '34
Monkeys, Go Home! '66
A New Kind of Love '63
Panic Button '62

Kim Chew

Dim Sum: A Little Bit of Heart
   '85

Laureen Chew

Chan is Missing '82
Dim Sum: A Little Bit of Heart
   '85

Sam Chew

Rattlers '76

Ceclie Cheyreau

Spaceways '53

Cheng Tien Chi

Super Ninjas '80s
The Warrior

Lee Chi-Keung

Just Like Weather '86

Lam Chun Chi

King of Kung-Fu '80s

Yuen Man Chi

Bruce's Deadly Fingers '80s

Yung Man Chi

Young Dragon '77

Liu Chia-Hui

Challenge of the Masters '89

Lang Shih Chia

Renegade Monk '82

Lu Chia-Liang (D)

Challenge of the Masters '89
Shaolin Executioner '80s

Sun Chia-Lin

Eighteen Jade Arhats '84
The Hot, the Cool, and the
   Vicious '80s

Liang Chia-Yen

Sleeping Fist '80s

Liu Chia-Yung

Challenge of the Masters '89
Two Graves to Kung-Fu '82

Minoru Chiaki

Godzilla Raids Again '59
Hidden Fortress '58
Throne of Blood '57

Charlie Chiang

Tough Guy '70s

Chen Chiang

Shadow Killers '80s

David Chiang

Duel of Fists
Duel of the Iron Fist '72
Seven Blows of the Dragon
   '73

Hui Hsiao Chiang

Ninja vs. Ninja '90

Tu Chiang

Single Fighter '78

Roy Chiao

Bloodsport '88

Walter Chiari

Bonjour Tristesse '57
Girl Under the Sheet '60s
Husband Hunting '60s

Sonny Chiba

Aces: Iron Eagle 3 '92
Assassin '79
The Bodyguard '76
Dragon Princess '70s
Hunter in the Dark '80
Karate Warriors '70s
The Killing Machine '76
Kowloon Assignment '77
Roaring Fire '82
Shogun's Ninja '83
Sister Street Fighters '76
Street Fighter '75
The Street Fighter's Last
   Revenge '79
Terror Beneath the Sea '66
Virus '82

Marissa Chibas

Cold Feet '84

Michele Chicione

Explosion '69

Etienne Chicot

Osa '85
36 Fillete '88

Don Chien

Dragon vs. Needles of Death
   '82

Yu Chien

Killer Elephants '76

Michael Chiklis

Wired '89

Linden Chiles

Deadline Assault '90
Death Be Not Proud '75
Forbidden World '82
Red Flag: The Ultimate Game
   '81
Who Is the Black Dahlia? '75

Lois Chiles

Broadcast News '87
Coma '78
Creepshow 2 '87
Death on the Nile '78
Diary of a Hitman '91
Moonraker '79
Raw Courage '84
Twister '89

Simon Chilvers

The Big Hurt '87
Dunera Boys '85
Ground Zero '88
Windrider '86

Andre Chimene

Computer Beach Party '88

Lee Chin-Chuan (D)

Single Fighter '78

Jackie Chin

Enter Three Dragons '81

Joey Chin

China Girl '87
King of New York '90

Kieu Chin

Operation C.I.A. '65

Tan Chin

The Bloody Fight '70s

Tsai Chin

The Castle of Fu Manchu '68
Invasion '65
Kiss and Kill '68
The Virgin Soldiers '69
You Only Live Twice '67

Chen Ching

Shadow Killers '80s

William Ching

Give a Girl a Break '53
Never Wave at a WAC '52
Pat and Mike '52
Terror in the Haunted House
   '58
The Wistful Widow of Wagon
   Gap '47

Wong Ching

Crazy Kung Fu Master
Young Dragon '77

Kieu Chinh

Children of An Lac '80

Jade Chinn

Girl Who Spelled Freedom '86

Michael Chiodo

Carnage '84

Stephen Chiodo (D)

Killer Klowns from Outer
   Space '88

Carlos Chionetti

The Red Desert '64

Erik Chitty

First Men in the Moon '64

Amy Chludzinski

Cannibal Campout '88

Anna Chlumsky

My Girl '91

Tim Choate

Def-Con 4 '85
First Time '82
Spy '89

Han Kwok Choi

The Young Bruce Lee

Marvin J. Chomsky (D)

Anastasia: The Mystery of
   Anna '86
Attica '80
Evel Knievel '72
Holocaust '78
Inside the Third Reich '82
Little Ladies of the Night '77
Live a Little, Steal a Lot '75
Murph the Surf '75
Nairobi Affair '88
Robert Kennedy and His
   Times '90
The Shaming '71
Tank '83

Billy Chong

Kung Fu from Beyond the
  Grave
Kung Fu Zombie '82

Elton Chong

Invincible Obsessed Fighter
  '80s
Shaolin Incredible Ten '70s

Ida F.O. Chong

Dim Sum: A Little Bit of Heart
  '85

Jun Chong

Ninja Turf '86
Silent Assassins '88
Street Soldiers '91

Michael Chong

Death Machines '76

Mona Chong

On Her Majesty's Secret
  Service '69

Paris Chong

Far Out Man '89

Rae Dawn Chong

Amazon '90
American Flyers '85
Badge of the Assassin '85
Beat Street '84
The Borrower '89
Cheech and Chong's, The
  Corsican Brothers '84
Choose Me '84
City Limits '85
The Color Purple '85
Commando '85
Common Bonds '91
Denial: The Dark Side of
  Passion '91
Far Out Man '89
Fear City '85
The Principal '87
Prison Stories: Women on the
  Inside '91
Quest for Fire '82
Running Out of Luck '86
Soul Man '86
The Squeeze '87
Tales from the Darkside: The
  Movie '90

Shelby Chong

Far Out Man '89

Thomas Chong

After Hours '85
Cheech and Chong: Still
  Smokin' '83
Cheech and Chong: Things
  Are Tough All Over '82
Cheech and Chong's Next
  Movie '80
Cheech and Chong's Nice
  Dreams '81
Cheech and Chong's, The
  Corsican Brothers '84
Cheech and Chong's Up in
  Smoke '79
Far Out Man '89
It Came from Hollywood '82
Spirit of '76 '91
Tripwire '89
Yellowbeard '83

Thomas Chong *(D)*

Cheech and Chong: Still
  Smokin' '83

Cheech and Chong's Next
  Movie '80
Cheech and Chong's Nice
  Dreams '81
Cheech and Chong's, The
  Corsican Brothers '84
Far Out Man '89

Farid Chopel

Caged Heart '85

Joyce Chopra *(D)*

The Lemon Sisters '90
Smooth Talk '85

Yuen Tat Chor

Taoism Drunkard
The Young Taoism Fighter

Ala Chostakova

The Lady with the Dog '59

Wang Mo Chou

The Furious Avenger '76

Sarita Choudhury

Mississippi Masala '92

Michel Choupon

Petit Con '84

Elie Chouraqui *(D)*

Love Songs '85
Man on Fire '87

Raymond Chow *(D)*

When Taekwondo Strikes '83

Marilyn Chris

Honeymoon Killers '70

Keith Christensen

Little Heroes '91

Stacy Christensen

Virgin Queen of St. Francis
  High '88

A. Christiakov

Storm Over Asia '28

Claudia Christian

Arena '88
Clean and Sober '88
Mad About You '90
Maniac Cop 2 '90
Mom '89
Never on Tuesday '88
Tale of Two Sisters '89

Helen Christian

Zorro Rides Again '37

John Christian

Covert Action '88
Mob War '88

**Linda Christian-
Jones**

Severance '88

Keely Christian

Slumber Party Massacre 3 '90

Linda Christian

The Devil's Hand '61
Full Hearts & Empty Pockets
  '63

Michael Christian

Hard Knocks

The Legend of Frank Woods
  '77
Mid Knight Rider '84
Poor Pretty Eddie '73

**Nathaniel
Christian** *(D)*

California Casanova '91
Club Fed '91

Paul Christian

The Beast from 20,000
  Fathoms '53
Journey to the Lost City '58

Robert Christian

Bustin' Loose '81

Roger Christian *(D)*

The Sender '82
Starship '87

Susanne Christian

Paths of Glory '57

Mady Christians

Letter from an Unknown
  Woman '48
Tender Comrade '43

**Benjamin
Christiansen**

Only Way '70

Audrey Christie

Frankie and Johnny '65
Splendor in the Grass '61

Helen Christie

Escort Girls '74

Julie Christie

Billy Liar '63
Darling '65
Demon Seed '77
Doctor Zhivago '65
Don't Look Now '73
Fahrenheit 451 '66
Far from the Madding Crowd
  '67
Fools of Fortune '91
The Go-Between '71
Heat and Dust '82
Heaven Can Wait '78
McCabe & Mrs. Miller '71
Miss Mary '86
Petulia '68
Power '86
Return of the Soldier '84
Secret Obsession '88
Separate Tables '83
Shampoo '75

Katia Christine

Cosmos: War of the Planets
  '80
Spirits of the Dead '68

Virginia Christine

Billy the Kid Versus Dracula
  '66
The Cobweb '55
The Inner Circle '46

Lindsay Christman

Twister '89

Eric Christmas

Child of Darkness, Child of
  Light '91
Home is Where the Hart is '88
Middle Age Crazy '80

The Philadelphia Experiment
  '84

Leonard Christmas

Harlem Rides the Range '39

**Debra
Christofferson**

Round Numbers '91

**Francoise
Christophe**

The Invaders '63
King of Hearts '66
Walk into Hell '57

Jean Christophe *(D)*

One-Eyed Soldiers '67

Bojesse Christopher

Meatballs 4 '92
Point Break '91

Dennis Christopher

Alien Predators '80
Breaking Away '79
California Dreaming '79
Chariots of Fire '81
Circuitry Man '90
Didn't You Hear? '83
Don't Cry, It's Only Thunder
  '81
Elvis: The Movie '79
Fade to Black '80
Jack & the Beanstalk '83
Jake Speed '86
Last Word '80
September 30, 1955 '77
Sinful Life '89
A Wedding '78
The Young Graduates '71

Jean Christopher

Playgirl Killer '66

Kathy Christopher

Beasts '83

Kay Christopher

Dick Tracy's Dilemma '47
Scared to Death/Dick Tracy's
  Dilemma '47

Robin Christopher

Equinox '71

Sharon Christopher

Girl School Screamers '86
Girls School Screamers '86

Thom Christopher

Deathstalker 3 '89
Wizards of the Lost Kingdom
  '85

**Tony
Christopher** *(D)*

Sorority House Scandals '86

**William
Christopher**

For the Love of It '80
M*A*S*H: Goodbye, Farewell
  & Amen '83

Ann Christy

Speedy '28

Dorothy Christy

Convicted '32
Night Life in Reno '31

June Christy

Symphony of Swing '86

Kao Tien Chu

Two Dragons Fight Against
  the Tiger

Yeh Yung Chu *(D)*

Fists of Dragons '69

Paul Chubb

Sweet Talker '91

Simon Chuckster

Sweet Sweetback's
  Baadasssss Song '71

Art Chudabala

Gleaming the Cube '89

**Byron Ross
Chudnow** *(D)*

The Amazing Dobermans '76
Daring Dobermans '73
The Doberman Gang '72

George Kee Chueng

Opposing Force '87

Richard Chui

Fighting Duel of Death

Grigori Chukrai *(D)*

Ballad of a Soldier '60

Chang Ying Chun

Ninja Swords of Death '90

Dennis Chun

Inferno in Paradise '88

Lung Chun-Eng

Shaolin Traitor '82
Two Assassins in the Dark
  '80s

Lau Chun

Four Robbers '70s

William Chun

The Steel Helmet '51

Lung Chunem

To Subdue the Evil

George Chung

Hawkeye '88

Nam Chung-Im

Yongary, Monster of the Deep
  '67

Kuo Shu Chung

Fury of King Boxer '83

Kwan Chung

The Young Taoism Fighter

Ou-Yang Chung

Golden Sun '80s

Tong Chun Chung

Champion Operation '70s

Wang Chung

Kung Fu Rebels '80s

Cyril Elaine Church

The Killing of Sister George
  '69

Samuel Claxton

The Last Supper '76

Andrew Dice Clay

The Adventures of Ford
    Fairlane '90
Amazon Women On the Moon
    '87
Casual Sex? '88
Making the Grade '84
Night Patrol '85
Pretty in Pink '86
Private Resort '84
Rodney Dangerfield: Nothin'
    Goes Right '88
Truly Tasteless Jokes '85
Wacko '83

Jennifer Clay

Suburbia '83

Nicholas Clay

Cannon Movie Tales: Sleeping
    Beauty '89
Evil Under the Sun '82
Excalibur '81
Lady Chatterley's Lover '81
Lionheart '87
Lovespell '79

Noland Clay

The Stalking Moon '69

Philip Clay

Deadly Sting '82

Jill Clayburgh

The Art of Crime '75
Best of Chevy Chase '87
First Monday in October '81
Griffin and Phoenix: A Love
    Story '76
Hanna K. '83
Hustling '75
I'm Dancing as Fast as I Can
    '82
It's My Turn '80
Miles to Go '86
Portnoy's Complaint '72
Semi-Tough '77
Shy People '87
The Silver Streak '76
Starting Over '79
The Terminal Man '74
Thief Who Came to Dinner '73
An Unmarried Woman '78
The Wedding Party '69
Where Are the Children? '85

Bob Clayton

The Bellboy '60

Ethey Clayton

Risky Business '28

Jack Clayton (D)

The Great Gatsby '74
The Lonely Passion of Judith
    Hearne '87
Room at the Top '59
Something Wicked This Way
    Comes '83

Jan Clayton

The Showdown '40

John Clayton

Freedom '81
Warm Nights on a Slow-
    Moving Train '87

John Clayton (D)

Summer Dog '78

Summerdog '78

Melissa Clayton

House 4 '91

Merry Clayton

Blame It on the Night '84

June Clayworth

Criminal Court '46

Corinne Cleary

Insanity '82

Eddie Cleary

Encounter at Raven's Gate
    '88

Jerry Cleary

Hollywood in Trouble '87

John Cleese

An American Tail: Fievel Goes
    West '91
And Now for Something
    Completely Different '72
The Big Picture '89
The Bliss of Mrs. Blossom '68
Bullseye! '90
Clockwise '86
Erik the Viking '89
A Fish Called Wanda '88
The Great Muppet Caper '81
The Magic Christian '69
Monty Python and the Holy
    Grail '75
Monty Python Live at the
    Hollywood Bowl '82
Monty Python Meets Beyond
    the Fringe
Monty Python's Life of Brian
    '79
Monty Python's Parrot Sketch
    not Included '90
Monty Python's The Meaning
    of Life '83
Privates on Parade '84
Romance with a Double Bass
    '74
The Secret Policeman's Other
    Ball '82
Secret Policeman's Private
    Parts '81
Silverado '85
The Statue '71
Time Bandits '81
Whoops Apocalypse '83
Yellowbeard '83

Tom Clegg (D)

Any Man's Death '90
A Casualty of War '90
Children of the Full Moon '84
The Inside Man '84
McVicar '80
Stroke of Midnight '90

Brian Clemens (D)

Captain Kronos: Vampire
    Hunter '74

Paul Clemens

The Beast Within '82
The Family Man '79
Promises in the Dark '79
They're Playing with Fire '84

Christian Clemenson

Bad Influence '90

Andree Clement

La Symphonie Pastorale '46

Aurore Clement

The Eighties '83
Invitation au Voyage '83
Le Crabe Tambour '77
Les Rendez-vous D'Anna '78
Paris, Texas '83
Toute Une Nuit '82

Clay Clement

Star Reporter '39

Dick Clement (D)

Bullshot '83
Catch Me a Spy '71
Porridge '91
Water '85

Rene Clement (D)

And Hope to Die '72
Day and the Hour '63
The Deadly Trap '71
Forbidden Games '52
Gervaise '57
Is Paris Burning? '68
Joy House '64
Rider on the Rain '70
Wanted: Babysitter '75

Pierre Clementi

The Conformist '71

Edward Clements

Metropolitan '90

Hal Clements

Grit of the Girl Telegrapher/In
    the Switch Tower '15
Let's Go! '23

John Clements

Convoy '40
The Four Feathers '39
The Silent Enemy '58
Train of Events '49

Ron Clements

The Great Mouse Detective
    '86

Ron Clements (D)

The Little Mermaid '89

Stanley Clements

Ghosts on the Loose '43

Clarence Clemmons

Legendary Ladies of Rock &
    Roll '86

David Clendenning

Death Games '82

David Clennon

Conspiracy: The Trial of the
    Chicago Eight '87
The Couch Trip '87
Falling in Love '84
Go Tell the Spartans '78
Hanna K. '83
Legal Eagles '86
Missing '82
The Right Stuff '83
Star 80 '83
Sweet Dreams '85
The Thing '82

Judy Cler

The Wizard of Gore '70

Corinne Clery

The Con Artists '80
Dangerous Obsession '88
Fatal Fix '80s

Love by Appointment '76
Moonraker '79
The Story of O '75
The Switch '89
Yor, the Hunter from the
    Future '83

Amanda Cleveland

Let's Do It! '84

Carol Cleveland

Monty Python and the Holy
    Grail '75
Monty Python's The Meaning
    of Life '83

George Cleveland

The Abe Lincoln of Ninth
    Avenue '39
Abroad with Two Yanks '44
Drums of Fu Manchu '40
Hidden Enemy '40
I Conquer the Sea '36
Midnight Limited '40
Miss Grant Takes Richmond
    '49
Monte Carlo Nights '34
Mutiny in the Big House '39
Playmates '41
Revolt of the Zombies '36
Texas Legionnaires '43
Two-Faced Woman '41
Valley of the Sun '42
The Wistful Widow of Wagon
    Gap '47

Max Cleven (D)

Night Stalker '87

Van Cliburn

Carnegie Hall at 100: A Place
    of Dreams '91

Jimmy Cliff

Club Paradise '86
The Harder They Come '72

John Cliff

Teenage Frankenstein '58

Clare Clifford

Wish You Were Here '87

Colleen Clifford

Where the Green Ants Dream
    '84

Graeme Clifford (D)

The Boy Who Left Home to
    Find Out About the Shivers
    '81
Burke & Wills '85
Frances '82
Gleaming the Cube '89
Little Red Riding Hood '83
Turn of the Screw '89

Jack Clifford

King of the Pecos '36
The Man from Gun Town '36
The Revenge Rider '35
Sweet Adeline '26

Kathleen Clifford

When the Clouds Roll By '19

Tommy Clifford

Song O' My Heart '29

Faith Clift

Cataclysm '81
Savage Journey '83

Jeanette Clift

The Hiding Place '75

Montgomery Clift

Big Lift '50
From Here to Eternity '53
The Heiress '49
I Confess '53
Indiscretion of an American
    Wife '54
Judgment at Nuremberg '61
Lonelyhearts '58
The Misfits '61
Montgomery Clift '85
A Place in the Sun '51
Raintree County '57
Red River '48
The Search '48
Suddenly, Last Summer '59
The Young Lions '58

Elmer Clifton

America/The Fall of Babylon
    '24
Intolerance '16

Elmer Clifton (D)

Assassin of Youth '35
The Black Coin '36
Captured in Chinatown '35
Custer's Last Stand '36
Cyclone of the Saddle '35
Down to the Sea in Ships '22
Fighting Caballero '35
Gangsters of the Frontier '44
The Judge '49
Let 'er Go Gallegher '28
Marked for Murder '45
Roaring Speedboats '37
Seven Doors to Death '44

Jane Clifton

A Kink in the Picasso '90

Cynthia Cline

UFO: Target Earth '74

Eddie Cline (D)

The Bank Dick '40
Cops/One Week '22
Cowboy Millionaire '35
Dude Ranger '34
Hawaii Calls '38
Hook, Line and Sinker '30
My Little Chickadee '40
Never Give a Sucker an Even
    Break '41
Peck's Bad Boy with the
    Circus '38
Private Buckaroo '42
Three Ages '23
The Villain Still Pursued Her
    '41

Patsy Cline

The Real Patsy Cline '90

Renee Cline

Fatal Exposure '90
Invasion Force '90

Marcia Clingan

Deadly Impact '84

Debra Clinger

Midnight Madness '80

George Clinton

Graffiti Bridge '90

Colin Clive

The Bride of Frankenstein '35
Christopher Strong '33

Frankenstein '31
History is Made at Night '37
Jane Eyre '34

E.E. Clive

The Adventures of Sherlock
  Holmes '39
Arrest Bulldog Drummond '38
The Bride of Frankenstein '35
Bulldog Drummond Escapes
  '37
Bulldog Drummond's Bride
  '39
Bulldog Drummond's Peril '38
Bulldog Drummond's
  Revenge/Bulldog
  Drummond's Peril '38
Bulldog Drummond's Secret
  Police '39
Charge of the Light Brigade
  '36
The Dark Hour '36
The Hound of the Baskervilles
  '39
Libeled Lady '36
The Little Princess '39

Henry Clive

Her Silent Sacrifice '18

Teagan Clive

Alienator '89
Sinbad of the Seven Seas '89
Vice Academy 2 '90

Al Cliver

Big Boss '77
The Black Cat '81
Forever Emmanuelle '82
Mr. Scarface '77
Rulers of the City '71
2020 Texas Gladiators '85
Zombie '80

Maurice Cloche (D)

Monsieur Vincent '47

George Clooney

Red Surf '90
Return of the Killer Tomatoes
  '88

Rosemary Clooney

The Best of Bing Crosby: Part
  1 '91
Deep in My Heart '54
White Christmas '54

Glenn Close

The Big Chill '83
Dangerous Liaisons '89
Fatal Attraction '87
Hamlet '90
Immediate Family '89
The Jagged Edge '85
Light Years '88
Maxie '85
Meeting Venus '91
The Natural '84
Orphan Train '79
Reversal of Fortune '90
Sarah, Plain and Tall '91
The Stone Boy '84
The World According to Garp
  '82

John Scott Clough

Fast Forward '84
Gross Anatomy '89

Michael Clough

To the Lighthouse '83

Dave Clouse

Summer Job '88

Robert Clouse (D)

The Amsterdam Kill '78
The Big Brawl '80
Black Belt Jones '74
China O'Brien '88
China O'Brien 2 '89
Darker Than Amber '70
Deadly Eyes '82
Enter the Dragon '73
Force: Five '81
Game of Death '79
Gymkata '85
The Pack '77
The Ultimate Warrior '75

Raymond Cloutier

Riel '79

Henri-Georges Clouzot (D)

Diabolique '55
Jenny Lamour '47
Le Corbeau '43
The Raven '43
Wages of Fear '55

Vera Clouzot

Diabolique '55

Brian Clover

A Midsummer Night's Dream
  '82

David Clover

Ten Speed '76

Jason Clow

Bikini Summer '91

D.E.P. Clucher (D)

The Odd Squad '86

E.B. Clucher (D)

Crime Busters '78
Go for It '83
They Call Me Trinity '72
Trinity is Still My Name '75

Rick Cluchey

The Cage '90

Rick Cluchey (D)

The Cage '90

Richard Cluck

Nomad Riders '81

Jennifer Cluff

Death Games '82

H. M. Clugston

The Navigator '24

Genevieve Cluny

The Cousins '59

Harold Clurman (D)

Deadline at Dawn '46

Francois Cluzet

Chocolat '88
The Horse of Pride '80
Round Midnight '86
Story of Women '88

Andy Clyde

Annie Oakley '35
Bad Lands '39
Bar-20 '43

Comedy Festival #1 '30s
Comedy Festival #2 '30s
Comedy Festival #4 '30s
Devil's Playground '46
False Colors '43
Hopalong Cassidy: Borrowed
  Trouble '48
Hopalong Cassidy: Dangerous
  Venture '48
Hopalong Cassidy: False
  Paradise '48
Hopalong Cassidy: Hoppy's
  Holiday '47
Hopalong Cassidy: Riders of
  the Deadline '43
Hopalong Cassidy: Silent
  Conflict '48
Hopalong Cassidy: Sinister
  Journey '48
Hopalong Cassidy: The Dead
  Don't Dream '48
Hopalong Cassidy: The Devil's
  Playground '46
Hopalong Cassidy: The
  Marauders '47
Hopalong Cassidy:
  Unexpected Guest '47
Riders of the Timberline '41
Ships in the Night '28
Shirley Temple Festival '33
Sinister Journey '48
Sundown Riders '48
Texas Masquerade '44
Twilight on the Trail '41

Craig Clyde (D)

Little Heroes '91

June Clyde

Hollywood Mystery '34
Make-Up '37
Morals for Women '31
Seven Doors to Death '44
She Shall Have Music '36
A Study in Scarlet '33
Wayne Murder Case '38

Terry Coady

Ilsa, the Tigress of Siberia '79

Michael Coard

Jigsaw '90

The Coasters

Phantom of the Ritz '88

Franklin Coates

Jesse James Under the Black
  Flag '21

Franklin Coates (D)

Jesse James Under the Black
  Flag '21

Kim Coates

The Amityville Curse '90
Blind Fear '89
Cold Front '89
Red Blooded American Girl
  '90
Smokescreen '90

Lewis (Luigi Cozzi) Coates (D)

Alien Contamination '81
Black Cat '90
Hercules '83
Hercules 2 '85
Star Crash '78

Phyllis Coates

Blues Busters '50
El Paso Stampede '53

The Incredible Petrified World
  '58
Jungle Drums of Africa '53
Marshal of Cedar Rock '53
Teenage Frankenstein '58

Sammantha Coates

Water Babies '79

Shirley "Muggsy" Coates

Our Gang Comedies '40s

Cindy Coatman

Game of Survival '89

Necmettin Cobanoglu

Journey of Hope '90

Dita Cobb

Dunera Boys '85

Edmund Cobb

Buffalo Bill Rides Again '47
Danger Trails '35
Gun Law '33
Gunners & Guns '34
Riders of the Range '24
Rustler's Paradise '35
Wild Horse '31

Irvin S. Cobb

Hawaii Calls '38

Jerry Cobb

Hawk and Castile

Joe Cobb

Little Rascals '24
Little Rascals, Book 1 '30s
Little Rascals, Book 18 '29
Our Gang '26
Our Gang Comedy Festival
  '88
Our Gang Comedy Festival 2
  '89
Return of Our Gang '25

Julie Cobb

Lisa '90

Kacey Cobb

The Crater Lake Monster '77

Lee J. Cobb

The Brothers Karamazov '57
Buckskin Frontier '43
Bull of the West '89
But Not for Me '59
Come Blow Your Horn '63
Coogan's Bluff '68
Danger on the Air '38
Dark Past '49
Day of Triumph '54
Exodus '60
The Exorcist '73
The Fighter '52
Four Horsemen of the
  Apocalypse '62
Golden Boy '39
How the West was Won '62
In Like Flint '67
The Left Hand of God '55
Liberation of L.B. Jones '70
Macho Callahan '70
MacKenna's Gold '69
The Man in the Gray Flannel
  Suit '56
Man of the West '58
Man Who Loved Cat Dancing
  '73

The Meanest Men in the West
  '60s
Men of Boys Town '41
The Miracle of the Bells '48
On the Waterfront '54
Our Man Flint '66
Party Girl '58
The Racers '55
Rustler's Valley '37
Sirocco '51
The Song of Bernadette '43
Tall Texan '53
That Lucky Touch '75
The Three Faces of Eve '57
Trap '59
Twelve Angry Men '57

Randall "Tex" Cobb

Blind Fury '90
Buy & Cell '89
Ernest Goes to Jail '90
The Golden Child '86
Raising Arizona '87
Raw Nerve '91
Uncommon Valor '83

Brian Cobby

The Great Armored Car
  Swindle '64

Billy Cobham

Cobham Meets Bellson '83

David Cobham (D)

Tales of the Klondike: Race
  for Number One '81

Eva Cobo

Matador '86

Roberto Cobo

Los Olvidados '50

German Cobos

Fuerte Perdido '78

Glen Cobum

Tabloid!

Charles Coburn

Around the World in 80 Days
  '56
Bachelor Mother '39
Colonel Effingham's Raid '45
Devil & Miss Jones '41
Gentlemen Prefer Blondes '53
Heaven Can Wait '43
In This Our Life '42
Kings Row '41
The Lady Eve '41
Made for Each Other '39
Monkey Business '52
The More the Merrier '43
The Paradine Case '47
Rhapsody in Blue '45
Road to Singapore '40
Stanley and Livingstone '39
Three Faces West '40
Trouble Along the Way '53
Vivacious Lady '38

David Coburn

Born American '86

Glenn Coburn (D)

Bloodsuckers from Outer
  Space '83

Harrison Coburn

Deadly Passion '85

James Coburn

The Americanization of Emily '64
The Baltimore Bullet '80
Bite the Bullet '75
Bruce Lee: The Legend '84
Charade '63
Cowboys of the Saturday Matinee '84
Cross of Iron '76
Crossover '82
The Dain Curse '78
Dead Heat on a Merry-Go-Round '66
Death of a Soldier '85
Draw! '81
Firepower '79
Fistful of Dynamite '72
Goldengirl '79
The Great Escape '63
Hard Times '75
Hell is for Heroes '62
High Risk '76
Hudson Hawk '91
In Like Flint '67
Internecine Project '73
The Last of Sheila '73
Looker '81
The Loved One '65
Loving Couples '80
Magnificent Seven '60
Major Dundee '65
Martin's Day '85
Midway '76
Muppet Movie '79
Our Man Flint '66
Pat Garrett & Billy the Kid '73
Pinocchio '83
The President's Analyst '67
A Reason to Live, A Reason to Die '73
Revenge of TV Bloopers '60s
Sky Riders '76
Waterhole #3 '67
Young Guns 2 '90

Michael Coby

The Bitch '78
Supersonic Man '78

Imogene Coca

National Lampoon's Vacation '83
Rabbit Test '78

Philip Coccioletti

Dagger Eyes '80s

Kean-Laurent Cochot

An Affair in Mind '89

Eddie Cochran

Go, Johnny Go! '59
Rock 'n' Roll Heaven '84

Linda Cochran

2000 Maniacs '64

Robert Cochran

I Stand Condemned '36
Sanders of the River '35
Scrooge '35

Steve Cochran

Carnival Story '54
The Chase '46
Copacabana '47
Deadly Companions '61
I, Mobster '58
Jim Thorpe: All American '51
Private Hell 36 '54
Shark River '53

White Heat '49
Wonder Man '45

Daisy Cockburn

Secrets '82

Joe Cocker

Mad Dogs & Englishmen '71

Gary Cockrell

War Lover '62

James Coco

The Chair '87
The Cheap Detective '78
End of the Road '70
Ensign Pulver '64
Generation '69
Hunk '87
Man of La Mancha '72
Muppets Take Manhattan '84
A New Leaf '71
Only When I Laugh '81
Ray Bradbury Theater, Volume 1 '85
Scavenger Hunt '79
The Strawberry Statement '70
That's Adequate '90
Wholly Moses! '80
Wild Party '74
You Can't Take It with You '84

Jean Cocteau

Les Enfants Terrible '50
The Testament of Orpheus '59

Jean Cocteau (D)

Beauty and the Beast '46
Orpheus '49
The Testament of Orpheus '59

Pablo Luis Codevilla

Un Elefante Color Ilusion

Claude Codgen

Follow Me '69

Camille Coduri

Hawks '89
King Ralph '91
Nuns on the Run '90
Strapless '90

Bill Cody

Frontier Days '34
Ghost City '32
Mason of the Mounted '32
Outlaws of the Range '36
Texas Pioneers '32

Iron Eyes Cody

Bowery Buckaroos '47
El Condor '70
Ernest Goes to Camp '87
Grayeagle '77
King of the Stallions '42
The Longest Drive '76
The Road to Yesterday '25
Son of Paleface '52
Spirit of '76 '91
Under Nevada Skies '46

Kathleen Cody

Charley and the Angel '73
Girls on the Road '73
Superdad '73

Lew Cody

Mickey '17

Barry Coe

The Bravados '58
But Not for Me '59
The Cat '66
Dr. Death, Seeker of Souls '73
Love Me Tender '56
The Oval Portrait '88
Peyton Place '57

David Allen Coe

Lady Grey '82

Fred Coe (D)

Thousand Clowns '65

George Coe

Blind Date '87
Bustin' Loose '81
The End of Innocence '90
A Flash of Green '85
The Hollywood Detective '89
Rage of Angels '85
Red Flag: The Ultimate Game '81
Sessions '83

Peter Coe

House of Frankenstein '44

Wayne Coe (D)

Grim Prairie Tales '89

Susie Coelho

Mysterious Island of Beautiful Women '79

Joel Coen (D)

Barton Fink '91
Blood Simple '85
Miller's Crossing '90
Raising Arizona '87

Susan Coetzer

City of Blood '88

Scott Coffey

Amazing Stories, Book 2 '86
Montana '90
Satisfaction '88
Shag: The Movie '89

Peter Coffield

Times Square '80

Frederick Coffin

Hard to Kill '89
Manhunt for Claude Dallas '86
One Summer Love '76

Tristram Coffin

The Corpse Vanishes '42
Dick Tracy, Detective/The Corpse Vanishes '42
Jesse James Rides Again '47
King of the Rocketmen '49
Land of the Lawless '47
Lost Planet Airmen '49
Ma Barker's Killer Brood '60
Pirates of the High Seas '50
Queen for a Day '51
Spy Smasher Returns '42
Under Nevada Skies '46

Nikki Coghill

Dark Age '88
The Time Guardian '87

Frank Coghlan, Jr.

Adventures of Captain Marvel '41
Little Red Schoolhouse '36
Yankee Clipper '27

Junior Coghlan

Hell's House '32
Last of the Mohicans '32
Let 'er Go Gallegher '28
Marked Money '28
Square Shoulders '29

Alain Cohen

Two of Us '68

Dan Cohen (D)

Madman '79

David Cohen (D)

Hollywood Zap '86

Emma Cohen

Bronson's Revenge '79
Horror Rises from the Tomb '72

Gary P. Cohen (D)

Video Violence '87
Video Violence Part 2...The Exploitation! '87

Herman Cohen (D)

Crocodile '81

Howard R. Cohen (D)

Deathstalker 4: Clash of the Titans '91
Saturday the 14th '81
Saturday the 14th Strikes Back '88
Space Raiders '83
Time Trackers '88

Jeff B. Cohen

The Goonies '85

Larry Cohen (D)

The Ambulance '90
Black Caesar '73
Deadly Illusion '87
God Told Me To '76
Hell Up in Harlem '73
Housewife '72
It's Alive '74
It's Alive 2: It Lives Again '78
It's Alive 3: Island of the Alive '87
Perfect Strangers '84
The Private Files of J. Edgar Hoover '77
Q (The Winged Serpent) '82
Return to Salem's Lot '87
Special Effects '85
The Stuff '85
Wicked Stepmother '89

Martin B. Cohen (D)

Rebel Rousers '69

Mitchell Cohen

Toxic Avenger '86

Rob Cohen (D)

Scandalous '84
A Small Circle of Friends '80

S.E. Cohen (D)

Martial Law '90

Saskia Cohen Tanugi

Never Say Never Again '83

Thomas A. Cohen (D)

Massive Retaliation '85

Randolph Cohlan (D)

Night Shadow '90

Cyril Coke (D)

Pride and Prejudice '85

Harley Cokliss (D)

Black Moon Rising '86
Malone '87
Warlords of the 21st Century '82

Stefano Colagrande

Misunderstood '87

Nicholas Colasanto

Fat City '72

Angel Colbert

Women Unchained '70s

Catero Colbert

Sweet Perfection '90
Up Against the Wall '91

Claudette Colbert

Boom Town '40
Cleopatra '34
Drums Along the Mohawk '39
Egg and I '47
Guest Wife '45
I Cover the Waterfront '33
It Happened One Night '34
Let's Make it Legal '51
The Love Goddesses '65
The Palm Beach Story '42
Since You Went Away '44
Texas Lady '56
Three Came Home '50
Tomorrow is Forever '46
Without Reservations '46

Curt Colbert

The Psychotronic Man '91

Anita Colby

Cover Girl '44

Barbara Colby

Memory of Us '74

Albert Cole

The Incredible Two-Headed Transplant '71

Alexandra Cole

Doctor Butcher M.D. '80

Ben Cole

Howling 5: The Rebirth '89

Celea Ann Cole

Bloodstalkers '76

Dallas Cole

Glitch! '88

Dennis Cole

Amateur Night '85
The Barbary Coast '74
The Connection '73
Dead End City '88
Powder Keg '70
Pretty Smart '87

Conrad

Patrick Conrad (D)

Mascara '87

Robert Conrad

Assassin '86
Bandits '73
Breaking Up Is Hard to Do '79
The Commies are Coming, the Commies are Coming '57
Crossfire '67
Falling for the Stars '85
Hard Knox '83
Lady in Red '79
Live a Little, Steal a Lot '75
More Wild, Wild West '80
Murph the Surf '75
Palm Springs Weekend '63
Sudden Death '77
Two Fathers' Justice '85
Wild, Wild West Revisited '79
Will: G. Gordon Liddy '82
Wrong is Right '82

Robert Conrad (D)

Bandits '73
Crossfire '67

Sid Conrad

The Annihilators '85

William Conrad

The Adventures of Rocky & Bullwinkle: Birth of Bullwinkle '91
The Adventures of Rocky & Bullwinkle: Blue Moose '91
The Adventures of Rocky & Bullwinkle: Canadian Gothic '91
The Adventures of Rocky & Bullwinkle: La Grande Moose '91
The Adventures of Rocky & Bullwinkle: Mona Moose '91
The Adventures of Rocky & Bullwinkle: Vincent Van Moose '91
Body and Soul '47
The Conqueror '56
Cowboy '54
Cry Danger '51
Killing Cars '80s
Moonshine County Express '77
Naked Jungle '54
Night Cries '78
The Racket '51
The Rebels '79
Return of Frank Cannon '80
Vengeance '89

William Conrad (D)

Side Show '84

Hans Conried

Behave Yourself! '52
Birds & the Bees '56
Brothers O'Toole '73
Bus Stop '56
Davy Crockett, King of the Wild Frontier '55
Falcon Takes Over/Strange Bargain '49
The 5000 Fingers of Dr. T '53
Hitler's Children '43
Jet Pilot '57
The Monster that Challenged the World '57
Nancy Goes to Rio '50
Oh, God! Book 2 '80
The Patsy '64
Peter Pan '53
Phantom Tollbooth '69

Scruffy '80
The Senator was Indiscreet '47
Summer Stock '50
Three for Bedroom C '52
Tut & Tuttle '82

Frank Conroy

Ace of Aces '33
I Live My Life '35
Manhattan Melodrama '34
The Ox-Bow Incident '43

Jarleth Conroy

Day of the Dead '85

John Considine

Brainwash '82
Choose Me '84
Dixie: Changing Habits '85
Dr. Death, Seeker of Souls '73
Endangered Species '82
Forbidden Love '82
The Late Show '77
Opposing Force '87
Reunion in France '42
Rita Hayworth: The Love Goddess '83
See How She Runs '78
The Shadow Box '80
The Thirsty Dead '74
A Wedding '78
Welcome to L.A. '77

Tim Considine

Clown '53
Daring Dobermans '73
Patton '70
Shaggy Dog '59

Barbara Constable

Lady Terminator '89

Michelle Constant

Return of the Family Man '89

Yvonne Constant

Monkeys, Go Home! '66

Eddie Constantine

Alphaville '65
Attack of the Robots '66
Cleo from 5 to 7 '61
It's Alive 2: It Lives Again '78
The Long Good Friday '79
The Long Goodbye '73
No, I Have No Regrets '73
Room 43 '58

Michael Constantine

Billy: Portrait of a Street Kid '77
Conspiracy of Terror '75
Death Cruise '74
Dirty Heroes '71
Don't Drink the Water '69
Fear in the City '81
Forty Days of Musa Dagh '85
Justine '69
Killing in the Sun
The North Avenue Irregulars '79
Prancer '89
The Reivers '69
Say Goodbye, Maggie Cole '72
Silent Rebellion '82

John Conte

The Man with the Golden Arm '55
Trauma '62

Michael Conte

Necropolis '87

Richard Conte

Assault on a Queen '66
Big Combo '55
Circus World '64
Explosion '69
The Fighter '52
The Godfather '72
Godfather 1902-1959—The Complete Epic '81
Guadalcanal Diary '43
Hotel '67
House of Strangers '49
I'll Cry Tomorrow '55
Lady in Cement '68
1931: Once Upon a Time in New York '72
Operation Cross Eagles '69
The Purple Heart '44
Race for Life '55
They Came to Cordura '59
13 Rue Madeleine '46
Tony Rome '67
Violent Professionals '82
A Walk in the Sun '46

Richard Conte (D)

Operation Cross Eagles '69

Steve Conte

Attack of the Mayan Mummy '63
Terror of the Bloodhunters '62

Therese Conte (D)

Chasing Dreams '81

John Content

Death Row Diner '88

Catherine Conti

Cry of the Black Wolves '70s
2069: A Sex Odyssey '78

Tom Conti

American Dreamer '84
Beyond Therapy '86
Deep Cover '88
The Duellists '78
Dumb Waiter '87
The Gospel According to Vic '87
The Haunting of Julia '77
Merry Christmas, Mr. Lawrence '83
Miracles '86
Norman Conquests, Part 1: Table Manners '80
Norman Conquests, Part 2: Living Together '80
Norman Conquests, Part 3: Round and Round the Garden '80
Princess and the Pea '83
The Quick and the Dead '87
Reuben, Reuben '83
Saving Grace '86
Shirley Valentine '89
That Summer of White Roses '90

Vincent Conti

Deadly Impact '84

Dick Contino

Daddy-O '59

James A. Contner (D)

Hitler's Daughter '90

The Ten Million Dollar Getaway '91

Chantal Contouri

The Night After Halloween '83
Thirst '87

Maria Contrera

Brandy Sheriff '78

Patricio Contreras

The Official Story '85
Old Gringo '89

Frank Converse

Alone in the Neon Jungle '87
Anne of Avonlea '87
Brother Future '91
The Bushido Blade '80
Cruise Into Terror '78
Danger in the Skies '79
Everybody Wins '90
Killer on Board '77
Marilyn: The Untold Story '80
A Movie Star's Daughter '79
Mystery at Fire Island '81
The Pilot '82
Sergeant Matlovich vs. the U.S. Air Force '78
The Tattered Web '71

William Converse-Roberts

Courtship '87
The Fig Tree '87
On Valentine's Day '86

Bert Convy

A Bucket of Blood '59
Cannonball Run '81
Help Wanted: Male '82
Hero at Large '80
Jennifer '78
The Love Bug '68
The Man in the Santa Claus Suit '79
Racquet '79
Semi-Tough '77

Bert Convy (D)

Weekend Warriors '86

Blake Conway

Ghost Town '88

Sarah Conway Ciminera

Beyond the Door 3 '91

Gary Conway

American Ninja 2: The Confrontation '87
How to Make a Monster '58
Liberty & Bash '90
Teenage Frankenstein '58

Jack Conway (D)

Boom Town '40
Dragon Seed '44
Honky Tonk '41
The Hucksters '47
Libeled Lady '36
Red Headed Woman '32
A Tale of Two Cities '35
Tarzan and His Mate '34
Too Hot to Handle '38

James L. Conway (D)

Donner Pass: The Road to Survival '84
The Fall of the House of Usher '80

Hangar 18 '80
The Incredible Rocky Mountain Race '77
Last of the Mohicans '85

Joseph Conway

Ring of Terror '62

Kevin Conway

F.I.S.T. '78
Homeboy '88
Johnny We Hardly Knew Ye '77
Lathe of Heaven '80
One Good Cop '91
Paradise Alley '78
Rage of Angels '85

Morgan Conway

Bowery at Midnight/Dick Tracy vs. Cueball '42
Dick Tracy Detective '45
Dick Tracy, Detective/The Corpse Vanishes '42
Dick Tracy Double Feature #1 '45
Dick Tracy Double Feature #2 '47
Dick Tracy: The Spider Strikes -'37
Dick Tracy vs. Cueball '46
The Saint Takes Over '40
Sinners in Paradise '38

Pat Conway

Brighty of the Grand Canyon '67

Russ Conway

The Screaming Skull '58

Tim Conway

Ace Crawford, Private Eye
The Apple Dumpling Gang '75
The Apple Dumpling Gang Rides Again '79
The Billion Dollar Hobo '78
Cannonball Run II '84
Dorf and the First Games of Mount Olympus '87
Dorf Goes Auto Racing '89
Dorf on Golf '86
Dorf's Golf Bible '89
Gus '76
The Longshot '86
Private Eyes '80
Prize Fighter '79
Revenge of TV Bloopers '60s
The Shaggy D.A. '76
They Went That-a-Way & That-a-Way '78
World's Greatest Athlete '73

Tom Conway

Bride of the Gorilla '51
Cat People '42
Criminal Court '46
The Falcon in Hollywood '44
The Falcon in Mexico '44
Falcon's Adventure/Armored Car Robbery '50
The Falcon's Brother '42
I Walked with a Zombie '43
Lady Be Good '41
Mrs. Miniver '42
Mystery Theatre '51
One Touch of Venus '48
Peter Pan '53
Saint Strikes Back: Criminal Court '46
The Seventh Victim '43
Three Stops to Murder '53
Voodoo Woman '57

The Rainmaker '56
Rear Window '54
The Search '48
Sorry, Wrong Number '48
Women of the Prehistoric
Planet '66

Robert Corff

Gas-s-s-s! '70

Ann Corio

Here It is, Burlesque '79
Jungle Siren '42

Catherine Corkill

Two Fathers' Justice '85

Danny Corkill

Between Two Women '86

Anthony Corlan

Something for Everyone '70
Vampire Circus '71

Al Corley

Incident at Channel Q '86
Squeeze Play '79
Torchlight '85

Pat Corley

Mr. Destiny '90
Of Mice and Men '81
Poker Alice '87
Silent Witness '85
The Two Worlds of Jenny
Logan '79

Maddie Corman

The Adventures of Ford
Fairlane '90
Seven Minutes in Heaven '86

Roger Corman

Cannonball '76
Godfather 1902-1959—The
Complete Epic '81
The Godfather: Part 2 '74
Roger Corman: Hollywood's
Wild Angel '78
The Silence of the Lambs '91
State of Things '82
Swing Shift '84

Roger Corman (D)

Atlas '60
Bloody Mama '70
A Bucket of Blood '59
Carnival Rock '57
Creature from the Haunted
Sea '60
The Fall of the House of Usher
'60
Frankenstein Unbound '90
Gas-s-s-s! '70
The Gunslinger '56
The Haunted Palace '63
I, Mobster '58
It Conquered the World '56
The Last Woman on Earth '61
Little Shop of Horrors '60
Machine Gun Kelly '58
Masque of the Red Death '65
Masque of the Red Death/
Premature Burial '65
The Pit and the Pendulum '61
Premature Burial '62
The Raven '63
The St. Valentine's Day
Massacre '67
Shame '61
She Gods of Shark Reef '56
Swamp Women '55
Tales of Terror '62
Teenage Caveman '58

The Terror '63
Tomb of Ligeia '64
Tower of London '62
The Trip '67
The Undead '57
The Wasp Woman '49
The Wild Angels '66
X: The Man with X-Ray Eyes
'63

Al Cormier

Barn of the Naked Dead '73

Alain Corneau (D)

Choice of Arms '83

Don Cornelius

No Way Back '74
Tapeheads '89

Henry Cornelius (D)

Genevieve '53
I Am a Camera '55
Passport to Pimlico '49

Ellie Cornell

Halloween 4: The Return of
Michael Myers '88
Halloween 5: The Revenge of
Michael Myers '89

John Cornell (D)

Almost an Angel '90
Crocodile Dundee 2 '88

**Hubert
Cornfield** (D)

Angel Baby '61
Plunder Road '57

Robert Cornthwaite

The War of the Worlds '53

Aurora Cornu

Claire's Knee '71

A. Cornwall

College '27

Charlotte Cornwell

The Krays '90
White Hunter, Black Heart '90

Judy Cornwell

Santa Claus: The Movie '85
Think Dirty '70
Wuthering Heights '70

**Stephen
Cornwell** (D)

Killing Streets '91

Alfonso Corona (D)

Deathstalker 3 '89

Isabela Corona

The Witch's Mirror '60

Eugene Corr (D)

Desert Bloom '86

Georges Corraface

Impromptu '90

Arturo Correa

El Derecho de Comer (The
Right to Eat) '87

Charles Correll

Check & Double Check '30

Mady Correll

Texas Masquerade '44

Richard Correll (D)

Ski Patrol '89

Adrienne Corri

Blat
A Clockwork Orange '71
Corridors of Blood '58
Devil Girl from Mars '54
Madhouse '74
Make Me an Offer '55
The River '51
The Tell-Tale Heart '62
Vampire Circus '71

Nick Corri

Gotcha! '85
In the Heat of Passion '91
The Lawless Land '88
A Nightmare on Elm Street '84
Slaves of New York '89
Tropical Snow '88

Sergio Corrieri

Memories of
Underdevelopment '68

James Corrigan

The Jack Knife Man '20

Lloyd Corrigan

The Chase '46
Ghost Chasers '51
A Girl, a Guy and a Gob '41
Hitler's Children '43
The Lady in Question '40
Since You Went Away '44
The Thin Man Goes Home '44

Lloyd Corrigan (D)

Dancing Pirate '36

Ray Corrigan

Arizona Stagecoach '42
Black Market Rustlers '43
Boot Hill Bandits '42
Come On, Cowboys '37
Country Gentlemen '36
Frontier Horizon '39
Ghost Town Gold '36
Heart of the Rockies '37
Heroes of the Hills '38
Hit the Saddle '37
Kid's Last Ride '41
Land of Hunted Men '43
Lonely Trail/Three Texas
Steers '39
Outlaws of Sonora '38
The Painted Stallion '37
Purple Vigilantes '38
Range Busters '40
Renegade Girl '46
Riders of the Whistling Skull
'37
Roarin' Lead '37
Rock River Renegades '42
Saddle Mountain Round-Up
'41
Saturday Night Serials '40
Thunder River Feud '42
Trail of Robin Hood '50
Trail of the Silver Spurs '41
Trailing Double Trouble '40
Trigger Trio '37
Tumbledown Ranch in
Arizona '41
Underground Rustlers '41
Undersea Kingdom '36
West of Pinto Basin '40
White Gorilla '47
Wild Horse Rodeo '37

Shirley Corrigan

The Devil's Nightmare '71
Dr. Jekyll and the Wolfman
'71

Aneta Corseaut

The Blob '58
The Toolbox Murders '78

Silvana Corsini

Accatone! '61

Bill Cort

Sammy '88

Bud Cort

Bernice Bobs Her Hair '76
Brain Dead '89
Brewster McCloud '70
The Chocolate War '88
Die Laughing '80
Electric Dreams '84
Gas-s-s-s! '70
Harold and Maude '71
Hysterical '83
Invaders from Mars '86
Love Letters '83
Love at Stake '87
Maria's Lovers '84
M*A*S*H '70
The Nightingale '83
Out of the Dark '88
Rumpelstiltskin '82
The Secret Diary of Sigmund
Freud '84
The Strawberry Statement '70
Why Shoot the Teacher '79

Michael Cort (D)

Alien Woman '60

Elaine Cortadellas

Vagabond '85

Frederic Corte (D)

Creature of the Walking Dead
'65

Carlos Cortes

Patsy, Mi Amor '87

**Fernando
Cortes** (D)

Los Tres Calaveras (The
Three Skeletons) '64

Mapita Cortes

Misterios de Ultratumba '47

Margarita Cortes

Maria Candelaria '46

Mary Cortes

Seven Days' Leave '42

Rowena Cortes

Heroes Three '84
To Live and Die in Hong Kong
'89

Joe Cortese

Assault and Matrimony '87
The Closer '91
Deadly Illusion '87
Evilspeak '82
Family Enforcer '77
Letting Go '85
Python Wolf '88
Stalking Danger '88
Windows '80

Valentina Cortese

The Adventures of Baron
Munchausen '89
Appassionata '74
Assassination of Trotsky '72
Barefoot Contessa '54
Brother Sun, Sister Moon '73
Day for Night '73
First Love '70
Juliet of the Spirits '65
Kidnap Syndicate '76
Les Miserables '52
When Time Ran Out '80
Widow's Nest '77

Bella Cortez

The Giant of Metropolis '61
Tor '64
Vulcan God of Fire

Carlos Cortez

El Robo al Tren Correo '47
Postal Inspector '36
Rio Hondo '47

Ricardo Cortez

Gentleman from California '37
I Killed That Man '42
The Inner Circle '46
Last Hurrah '58
Lost Zeppelin '29
The Maltese Falcon '31
Mr. Moto's Last Warning '39
Murder Over New York '40
The Phantom in the House '29
The Sorrows of Satan '26
The Swan '25

Jerome Cortland

Man from Colorado '49

Jerry Cortwright

The Giant Gila Monster '59

Henry Cory

Hot Times '74

Ernest Cosart

Love from a Stranger '47

Bill Cosby

Bill Cosby: 49 '89
Bill Cosby, Himself '81
California Suite '78
Devil & Max Devlin '81
The Fat Albert Christmas
Special '77
The Fat Albert Easter Special
'82
The Fat Albert Halloween
Special '77
Ghost Dad '90
Hungry i Reunion '81
Leonard Part 6 '87
Let's Do It Again '75
Man & Boy '71
Mother, Jugs and Speed '76
Piece of the Action '77
To All My Friends on Shore
'71
Toast to Lenny '84
Uptown Saturday Night '74

Bill Cosby (D)

Bill Cosby, Himself '81

**Don A.
Coscarelli** (D)

Beastmaster '82
Phantasm '79
Phantasm II '88
Survival Quest '89

Howard Cosell

Bananas '71
Broadway Danny Rose '84
Sleeper '73
Tall Tales & Legends: Casey
at the Bat '85

George P. Cosmatos (D)

The Cassandra Crossing '76
Cobra '86
Escape to Athena '79
Leviathan '89
Massacre in Rome '73
Of Unknown Origin '83
Rambo: First Blood, Part 2 '85
Restless '72

Luis Cosme

El Derecho de Comer (The
Right to Eat) '87

John Cosola

Beach House '82

Ernest Cossart

The Great Ziegfeld '36
Kings Row '41
Kitty Foyle '40

James Cossins

Dynasty of Fear '72
Wuthering Heights '70

Pierre Cosso

My Wonderful Life '90

Brigete Cossu

Blood Sisters '86

Michele Cossu

Bandits of Orgosolo '61

Cosie Costa

Missing in Action 2: The
Beginning '85

Constantin Costa-Gavras

Spies Like Us '85

Constantin Costa-Gavras (D)

Betrayed '88
Hanna K. '83
Missing '82
Music Box '89
State of Siege '73
Z '69

Marina Costa

Jungle Raiders '85

Mario Costa (D)

Gladiator of Rome '63
Rough Justice '87

Mary Costa (D)

Pirate Warrior '64

Sara Costa

Stripper '86

Suzanne Costallos

True Love '89

Robert Costanzo

Delusion '91
Honeyboy '82
The Vegas Strip Wars '84

Dolores Costello

Breaking the Ice '38
Little Lord Fauntleroy '36
The Magnificent Ambersons
'42

Don Costello

Texas Masquerade '44

Elvis Costello

Americathon '79
No Surrender '86
Straight to Hell '87

Helene Costello

Don Juan '26

Lou Costello

Abbott and Costello in
Hollywood '45
Abbott and Costello Meet
Captain Kidd '52
Abbott and Costello Meet Dr.
Jekyll and Mr. Hyde '52
Abbott and Costello Meet
Frankenstein '48
Abbott and Costello Meet the
Invisible Man '51
Abbott and Costello Meet the
Killer, Boris Karloff '49
Abbott and Costello
Scrapbook '70s
Africa Screams '49
The Best of Abbott & Costello
Live '54
Buck Privates '41
Buck Privates Come Home
'47
Classic Comedy Video
Sampler '49
Dance with Me, Henry '56
Entertaining the Troops '89
Hey Abbott! '78
Hit the Ice '43
Hold That Ghost '41
Hollywood Clowns '85
Hollywood Goes to War '54
In the Navy '41
Jack & the Beanstalk '52
Keep 'Em Flying '41
Mexican Hayride '48
The Naughty Nineties '45
Pardon My Sarong '42
Ride 'Em Cowboy '42
Super Bloopers #1 '70s
30-Foot Bride of Candy Rock
'59
The Time of Their Lives '46
Who Done It? '42
The Wistful Widow of Wagon
Gap '47

Mariclare Costello

All God's Children '80
Conspiracy of Terror '75
Coward of the County '81
The Execution of Private
Slovik '74
Heart of a Champion: The Ray
Mancini Story '85
Skeezer '82

Maurice Costello

Tale of Two Cities/In the
Switch Tower '15

Pat Costello

Mexican Hayride '48

Ward Costello

The City '76
The Gallant Hours '60
MacArthur '77

Nicholas Coster

The Electric Horseman '79
MacArthur '77
The Solitary Man '82
The Sporting Club '72
Stir Crazy '80

Mary Costi

Terror Beach '80s

George Costigan

Rita, Sue & Bob Too '87

Kevin Costner

Amazing Stories, Book 1 '85
American Flyers '85
Bull Durham '88
Chasing Dreams '81
Dances with Wolves '90
Fandango '85
Field of Dreams '89
The Gunrunner '84
JFK '91
Night Shift '82
No Way Out '87
Revenge '90
Robin Hood: Prince of Thieves
'91
Shadows Run Black '84
Silverado '85
Sizzle Beach U.S.A. '86
Stacy's Knights '83
Table for Five '83
Testament '83
The Untouchables '87

Kevin Costner (D)

Dances with Wolves '90

Marina Costo

The Final Executioner '83

Peter Cotes (D)

The Young and the Guilty '58

Kami Cotler

Children's Carol '80

Manny Coto (D)

Cover-Up '91
Playroom '90

Ileana Cotrubas

Les Contes D'Hoffman '84

Vittorio Cottafavi (D)

Goliath and the Dragon '61
Hercules and the Captive
Women '63

Joseph Cotten

The Abominable Dr. Phibes
'71
Airport '77 '77
Beyond the Forest '49
The Big Push '75
Brighty of the Grand Canyon
'67
Casino '80
Churchill and the Generals '81
Citizen Kane '41
Delusion '84
Duel in the Sun '46
The Farmer's Daughter '47
From the Earth to the Moon
'58
Gaslight '44
The Hearse '80
Heaven's Gate '81
Hellbenders '67
Hush, Hush, Sweet Charlotte
'65

Nicole Courcel

Journey into Fear '42
Lady Frankenstein '72
The Lindbergh Kidnapping
Case '76
Lydia '41
The Magnificent Ambersons
'42
Niagara '52
Oscar '66
The Passing of Evil '70
Perfect Crime '78
Petulia '68
Portrait of Jennie '48
Return to Fantasy Island '77
Screamers '80
September Affair '50
Shadow of a Doubt '42
Since You Went Away '44
Soylent Green '73
Survivor '80
The Third Man '49
Tiger '87
Tora! Tora! Tora! '70
Torture Chamber of Baron
Blood '72
Touch of Evil '58
Tramplers '66
Twilight's Last Gleaming '77
Under Capricorn '49
Walk Softly, Stranger '50
White Comanche '67

Fanny Cottencon

Window Shopping '86

Catherine Cotter

Outlaws of the Range '36
Pinto Rustlers '36

John Cotter (D)

Mountain Family Robinson '79

Chrissie Cotterill

Scrubbers '82

Ralph Cotterill

Howling 3 '87
Starship '87

John Cotton

Personal Exemptions '88

Oliver Cotton

Eleni '85
Robin Hood...The Legend:
Herne's Son '85

Jack Couffer (D)

Living Free '72
Nikki, the Wild Dog of the
North '61

Paul Coufos

Busted Up '86
City of Shadows '86
Food of the Gods: Part 2 '88
The Lost Empire '83
Thunderground '89

Ian Coughlan (D)

Alison's Birthday '79

David Coughlin

Cemetery High '89

Junior Coughlin

Drum Taps '33
Kentucky Blue Streak '35

Yvonne Coulette

A Married Man '84

George Coulouris

Arabesque '66

Citizen Kane '41

Citizen Kane '41
Doctor in the House '53
The Lady in Question '40
The Long Good Friday '79
Master Race '44
Mr. Skeffington '44
Murder on the Orient Express
'74
None But the Lonely Heart '44
Papillon '73
Race for Life '55
The Runaway Bus '54
The Skull '65
Song to Remember '45
A Southern Yankee '48
The Stranger '73
The Tempter '74
This Land is Mine '43
Watch on the Rhine '43

Bernie Coulson

The Accused '88
Eddie & the Cruisers 2: Eddie
Lives! '89

Catherine Coulson

Twin Peaks '90

Roy Coulson

Robin Hood '22

Richard Council

The Manhattan Project '86

Mary Count

Naked Warriors '73

Barbara Couper

The Last Days of Dolwyn '49

Carolyn Courage

The Terror '79

Nicole Courcel

Le Cas du Dr. Laurent '57
Sundays & Cybele '62

Hazel Court

The Curse of Frankenstein '57
Devil Girl from Mars '54
Doctor Blood's Coffin '62
Ghost Ship '53
Hour of Decision '57
Masque of the Red Death '65
Masque of the Red Death/
Premature Burial '65
Premature Burial '62
The Raven '63

Jason Court

Grandview U.S.A. '84
A Night in the Life of Jimmy
Reardon '88

Nigel Court

Third Solution '89

Tom Courtenay

Billy Liar '63
Catch Me a Spy '71
Dandy in Aspic '68
Doctor Zhivago '65
Dresser '83
Happy New Year '87
I Heard the Owl Call My Name
'73
King Rat '65
Leonard Part 6 '87
Let Him Have It '91
Night of the Generals '67
One Day in the Life of Ivan
Denisovich '71

Scatman Crothers

Between Heaven and Hell '56
Black Belt Jones '74
Bronco Billy '80
Chesty Anderson USN '76
Deadly Eyes '82
Detroit 9000 '73
The Great White Hope '70
Journey of Natty Gann '85
Lady in a Cage '64
Mean Dog Blues '78
One Flew Over the Cuckoo's Nest '75
The Patsy '64
Scavenger Hunt '79
The Shining '80
The Shootist '76
The Silver Streak '76
Stay Hungry '76
Streetfight '75
Truck Turner '74
Twilight Zone: The Movie '83
Two of a Kind '83
Yes, Sir, Mr. Bones '51
Zapped! '82

David Crotto

Oriane '85

William Forest Crouch (D)

Reet, Petite and Gone '47

Brian Croucher

The House that Bled to Death '81

Avery Crounse (D)

Eyes of Fire '84
The Invisible Kid '88

Lindsay Crouse

All the President's Men '76
Between the Lines '77
Communion '89
Daniel '83
Desperate Hours '90
Eleanor & Franklin '76
House of Games '87
Iceman '84
Paul's Case '80
Places in the Heart '84
Prince of the City '81
Slap Shot '77
Summer Solstice '81
The Verdict '82

Roger Crouzet

Letters from My Windmill '54

Luciano Crovato

Evil Clutch '89

Carl Crow

Premonition '71

Dean Crow (D)

Backwoods '87

Graham Crowden

Britannia Hospital '82
Code Name: Emerald '85
The Company of Wolves '85
Naughty Knights '71
Romance with a Double Bass '74

Cameron Crowe (D)

Say Anything '89

Christopher Crowe (D)

Off Limits '87

Tanya Crowe

Dark Night of the Scarecrow '81

Frank Crowell

Attack of the Swamp Creature '75

Josephine Crowell

Hearts of the World '18

Ed Crowley

Running on Empty '88

Jeananne Crowley

Reilly: The Ace of Spies '87

Kathleen Crowley

The Female Jungle '56
The Rebel Set '59
Westward Ho, the Wagons! '56

Pat Crowley

Force of Evil '77
Hollywood or Bust '56
Return to Fantasy Island '77
Wild Women of Wongo '59

Patricia Crowley

Menace on the Mountain '70

Susan Crowley

Born of Fire '87

Dick Crown

Platoon the Warriors '88

Laura Cruickshank

Buying Time '89

Su Cruickshank

Young Einstein '89

Tom Cruise

All the Right Moves '83
Born on the Fourth of July '89
Cocktail '88
Color of Money '86
Days of Thunder '90
Endless Love '81
Legend '86
Losin' It '82
The Outsiders '83
Rain Man '88
Risky Business '83
Taps '81
Top Gun '86

Liza Crusat

Sweet Perfection '90

Rosalie Crutchley

Creatures the World Forgot '70
Eleni '85
Hunchback of Notre Dame '82
Make Me an Offer '55
Night of the Laughing Dead '73
Wuthering Heights '70

Abigail Cruttenden

Intimate Contact '87
Kipperbang '82

Alexis Cruz

Gryphon '88

Andres Jose Cruz

Teorema '68

Carlina Cruz

Maricela '88

Celia Cruz

Mambo Kings '92

Charmaine Cruz

The Bronx War '91

Ernesto Cruz

El Norte '83

James Cruze (D)

The Covered Wagon '23
The Great Gabbo '29
Hawthorne of the USA '19
I Cover the Waterfront '33
Mr. Skitch '33
Old Ironsides '26
The Roaring Road '19
The Wrong Road '37

Gretchen Cryer

Hiding Out '87

Jon Cryer

Dudes '87
Hiding Out '87
Hot Shots! '91
Morgan Stewart's Coming Home '87
No Small Affair '84
Noon Wine '84
O.C. and Stiggs '87
Pretty in Pink '86
Rap Master Ronnie '88
Superman 4: The Quest for Peace '87

Billy Crystal

Best of Comic Relief '86
Best of Comic Relief '90 '90
Billy Crystal: A Comic's Line '83
Billy Crystal: Don't Get Me Started '86
Billy Crystal: Midnight Train to Moscow '89
Breaking Up Is Hard to Do '79
Catch a Rising Star's 10th Anniversary '83
City Slickers '91
Comic Relief 2 '87
Comic Relief 3 '89
Enola Gay: The Men, the Mission, the Atomic Bomb '80
Evening at the Improv '86
Memories of Me '88
The Princess Bride '87
Rabbit Test '78
Running Scared '86
This is Spinal Tap '84
The Three Little Pigs '84
Throw Momma from the Train '87
When Harry Met Sally '89

Eszter Csakanyi

A Hungarian Fairy Tale '87

Vince Csapos

Clayton County Line

Gyorgy Cserhalmi

Hungarian Rhapsody '78
Mephisto '81

Larry Csonka

SnakeEater '89

Peppeddu Cuccu

Bandits of Orgosolo '61

Enrique Cuenca

El Aviso Inoportuno '87
Hijazo De Mi Vidaza (Son of My Life) '70s

John Haslett Cuff

Psycho Girls '86

Xavier Cugat

A Date with Judy '48
Neptune's Daughter '49
Stage Door Canteen '43
Two Girls & a Sailor '44
You Were Never Lovelier '42

George Cukor

Making of a Legend: ''Gone With the Wind'' '89

George Cukor (D)

Adam's Rib '49
A Bill of Divorcement '32
Born Yesterday '50
Camille '36
David Copperfield '35
Dinner at Eight '33
A Double Life '47
Gaslight '44
Heller in Pink Tights '60
Holiday '38
It Should Happen to You '53
Justine '69
Les Girls '57
Let's Make Love '60
Little Women '33
Love Among the Ruins '75
My Fair Lady '64
Pat and Mike '52
Philadelphia Story '40
Rich and Famous '81
Romeo and Juliet '36
Song Without End '60
A Star is Born '54
Susan and God '40
Sylvia Scarlett '35
Travels with My Aunt '72
Two-Faced Woman '41
What Price Hollywood? '32
A Woman's Face '41
The Women '39

Norris Culf

Robot Holocaust '87

Macaulay Culkin

Home Alone '90
Jacob's Ladder '90
My Girl '91
Rocket Gibraltar '88
See You in the Morning '89
Uncle Buck '89

Brett Cullen

Leaving Normal '92

Edward Cullen

Catholic Hour '60

Max Cullen

Hard Knocks '80

David Culliname

Play Dead '81

Mark Cullingham (D)

Cinderella '84
Dead on the Money '91
Gryphon '88

Dana Culliver

Blood Freak '72

John Cullum

Glory '89
Marie '85
Morgan Stewart's Coming Home '87
The Prodigal '83
Sweet Country '87

Jason Culp

Skinheads: The Second Coming of Hate '88

Joseph Culp

The Arrival '90
Dream Lover '85

Robert Culp

Big Bad Mama 2 '87
The Blue Lightning '86
Bob & Carol & Ted & Alice '69
Castaway Cowboy '74
Demon with a Glass Hand '64
Flood! '76
The Gladiator '86
Goldengirl '79
Great Scout & Cathouse Thursday '76
Hannie Caulder '72
Her Life as a Man '83
Hot Rod '79
Inside Out '75
The Key to Rebecca '85
Killjoy '81
Murderous Vision '91
A Name for Evil '70
PT 109 '63
Pucker Up and Bark Like a Dog '89
Sammy, the Way-Out Seal '62
Silent Night, Deadly Night 3: Better Watch Out! '89
Sky Riders '76
That's Action! '90
Timebomb '91
Turk 182! '85

Calvin Culver

Score '73

Howard Culver

Home Safe '88

Jeffrey Culver

Hobgoblins '80s

Michael Culver

Alamut Ambush '86
Conduct Unbecoming '75
Deadly Recruits '86
The Devil's Web '74
Fast Kill '73
Philby, Burgess and MacLean: Spy Scandal of the Century '84

Roland Culver

Betrayed '54
The Day Will Dawn '42
Dead of Night '45
Down to Earth '47
Encore '52
The Great Lover '49
The Obsessed '51
On Approval '44
Spitfire: The First of the Few '43

George Culzon

Sexton Blake and the Hooded Terror '38

Jill Cumer

Deranged '87

Dorothy Cumming

Applause '29
King of Kings '27
The Wind '28

Leslie Cumming

Witchery '88

Burton Cummings

Melanie '82

Constance Cummings

The Battle of the Sexes '60
Blithe Spirit '45
Criminal Code '31
The Foreman Went to France '42

Eli Cummings

Harley '90

Gregory Cummings

Blood Games '90

Howard Cummings (D)

Courtship '87

Irving Cummings

The Saphead '21
Sex '20
Uncle Tom's Cabin '14

Irving Cummings (D)

Curly Top '35
Double Dynamite '51
Down Argentine Way '40
Flesh & Blood '22
Just Around the Corner '38
Little Miss Broadway '38
The Poor Little Rich Girl '36
Springtime in the Rockies '42
Vogues of 1938 '37

Kristin Cummings

Don't Go to Sleep '82

Quinn Cummings

The Babysitter '80
Goodbye Girl '77
Night Terror '76

Robert Cummings

The Accused '48
Beach Party '63
The Carpetbaggers '64
The Chase '46
Desert Gold '36
Devil & Miss Jones '41
Dial "M" for Murder '54
Five Golden Dragons '67
Forever and a Day '43
Kings Row '41
Lost Moment '47
Lucky Me '54
Moon Over Miami '41
My Geisha '62
Promise Her Anything '66
Reign of Terror '49
Saboteur '42

Sandy Cummings

The Messenger '87

Susan Cummings

Swamp Women '55

Verboten! '59

Eli Cummins

Dakota '88

Greg Cummins

Action U.S.A. '89
Caged Fury '90
Dead End City '88

James Cummins (D)

The Bone Yard '90

Juliette Cummins

Deadly Dreams '88
Friday the 13th, Part 5: A New Beginning '85
Running Hot '83
Slumber Party Massacre 2 '87

Peggy Cummins

The Captain's Table '60
Curse of the Demon '57
Gun Crazy '50
Salute John Citizen '42

Peter Cummins

Blue Fire Lady '78

Quinn Cummins

Intimate Strangers '77

Brett Cumo

The Wicked '89

Liam Cundill

Reason to Die '90
Return of the Family Man '89

Roger Cundy

Barbarian Queen 2: The Empress Strikes Back '90s

Richard Cunha (D)

Frankenstein's Daughter '58
Giant from the Unknown '58
Missile to the Moon '59
She Demons '58

Cecil Cunningham

The Awful Truth '37

Curtis Cunningham (D)

Mata Hari '85

Dennis Cunningham

Stuff Stephanie in the Incinerator '89

James Cunningham

Gunblast '70s

John Cunningham

Lost and Found '79
Twisted '86

Margo Cunningham

The Sailor Who Fell from Grace with the Sea '76

Peter "Sugarfoot" Cunningham

Bloodfist 3: Forced to Fight '91
Bloodmatch '91

Robert Cunningham

No Survivors, Please '63

Sean S. Cunningham (D)

Deepstar Six '89
Friday the 13th '80
Manny's Orphans '78
The New Kids '85
Spring Break '83
A Stranger is Watching '82

Alain Cuny

Basileus Quartet '82
Camille Claudel '89
Christ Stopped at Eboli '79
Emmanuelle '74
La Dolce Vita '60
Les Visiteurs du Soir '42
The Lovers '59
The Milky Way '68

Barbara Cupisti

Stagefright '87
Terror at the Opera '88

E.J. Curcio

Hard Rock Zombies '85

Frederick Curiel (D)

Genie of Darkness '60
The Monster Demolisher '60
Neutron and the Black Mask '61
Neutron vs. the Amazing Dr. Caronte '61
Neutron vs. the Death Robots '62

Juanita Curiel

Smokey & the Judge '80

Pauline Curley (D)

The Walloping Kid '26

Richard Curnock

Paradise '82

Bill Curran

Graveyard of Horror '71

Eileen Curran

Men of Ireland '38

Lynette Curran

Bliss '85

Lee Curreri

Crystal Heart '87
Kids from Fame '83

Anthony Currie (D)

The Pink Chiquitas '86

Cherie Currie

Foxes '80
Parasite '81
Rosebud Beach Hotel '85
Twilight Zone: The Movie '83
Wavelength '83

Finlay Currie

The Adventures of Huckleberry Finn '60
Big Deadly Game '54
Billy Liar '63
Bonnie Prince Charlie '48
Corridors of Blood '58
The Day Will Dawn '42
Edge of the World '37
The Forty-Ninth Parallel '41
Great Expectations '46
I Know Where I'm Going '47
Ivanhoe '52

The King's Rhapsody '55
Make Me an Offer '55
Quo Vadis '51
Sleeping Car to Trieste '45
Stars and Stripes Forever '52
Treasure Island '50
Trio '50

Louise Currie

Adventures of Captain Marvel '41
The Masked Marvel '43
Million Dollar Kid '44
Wild West '46

Margeret Currie

Rockin' Road Trip '85

Michael Currie

The Dead Pool '88
Pray TV '82

Sondra Currie

Concrete Jungle '82
Fugitive Lovers '70s
Jessie's Girls '75
Mama's Dirty Girls '74
Policewomen '73
Runaways '84
Teenage Seductress

Mary Currier

Return of the Ape Man '44

Anne Curry

Murphy's Fault '88

Christopher Curry

C.H.U.D. '84

John Curry

Snow Queen '82

Steven Curry

Glen and Randa '71

Tim Curry

Annie '81
Blue Money '84
Clue '85
The Hunt for Red October '90
Legend '86
Oliver Twist '82
Oscar '91
Pass the Ammo '88
The Ploughman's Lunch '83
The Rocky Horror Picture Show '75
The Shout '78
Times Square '80

Jane Curtin

Best of Chevy Chase '87
Best of Dan Aykroyd '86
Best of Gilda Radner '89
Best of John Belushi '85
Bob & Ray, Jane, Laraine & Gilda '83
How to Beat the High Cost of Living '80
Mr. Mike's Mondo Video '79
O.C. and Stiggs '87
Saturday Night Live: Carrie Fisher '78
Saturday Night Live: Richard Pryor '75
Saturday Night Live: Steve Martin, Vol. 1 '78
Saturday Night Live: Steve Martin, Vol. 2 '78
Saturday Night Live, Vol. 1 '70s
Suspicion '87

Valerie Curtin

Big Trouble '86
Christmas Without Snow '80
Different Story '78
Maxie '85

Alan Curtis

Apache Chief '50
Buck Privates '41
Flight to Nowhere '46
Gung Ho! '43
Last Outlaw '36
Mannequin '37
Melody Master '41
The Naughty Nineties '45
Renegade Girl '46

Bill Curtis

Terror of Tiny Town '33

Carolyn Curtis

Mesquite Buckaroo '39

Dan Curtis (D)

Burnt Offerings '76
Curse of the Black Widow '77
Dead of Night '77
Dracula '73
Express to Terror '79
House of Dark Shadows '70
Kansas City Massacre '75
Last Ride of the Dalton Gang '79
The Long Days of Summer '80
Melvin Purvis: G-Man '74
Mrs. R's Daughter '79
Night of Dark Shadows '71
Trilogy of Terror '75
Turn of the Screw '74
War & Remembrance, Part 1 '88
War & Remembrance, Part 2 '89
When Every Day was the Fourth of July '78
Winds of War '83

Dick Curtis

Boothill Brigade '37
A Lawman is Born '37
Motel Hell '80
Valley of Terror '38

Donald Curtis

The Amazing Mr. X '48
Criminals Within '41
Earth vs. the Flying Saucers '56
It Came from Beneath the Sea '55
Seventh Cavalry '56

Douglas Curtis (D)

The Campus Corpse '77
The Sleeping Car '90

Jack Curtis

Lawless Range '35

Jack Curtis (D)

The Flesh Eaters '64

Jackie Curtis

WR: Mysteries of the Organism '71

Jamie Lee Curtis

The Adventures of Buckaroo Banzai '84
Amazing Grace & Chuck '87
As Summers Die '86
Blue Steel '90

Augusta Dabney

Running on Empty '88

Maryam D'Abo

Immortal Sins '91
The Living Daylights '87
Nightlife '90
Xtro '83

Olivia D'Abo

Beyond the Stars '89
Bolero '84
Conan the Destroyer '84
Dream to Believe '80s
Into the Fire '88
Mission to Kill '85
Spirit of '76 '91

Jesse Dabson

Platoon Leader '87

Morton DaCosta (D)

Auntie Mame '58
The Music Man '62

Andrew Daddo

A Kink in the Picasso '90

Daddy Life

Live & Red Hot '87

Dimis Dadiras (D)

The Mediterranean in Flames '72

Willem Dafoe

Born on the Fourth of July '89
Cry-Baby '90
Flight of the Intruder '90
Hitchhiker 3 '87
The Hunger '83
The Last Temptation of Christ '88
Loveless '83
Mississippi Burning '88
New York Nights '84
Off Limits '87
Platoon '86
Roadhouse 66 '84
Streets of Fire '84
To Live & Die in L.A. '85
Triumph of the Spirit '89
White Sands '92

Jensen Daggett

Friday the 13th, Part 8: Jason Takes Manhattan '89

Angie Daglas

Sinners '89

Lil Dagover

The Cabinet of Dr. Caligari '19
Dr. Mabuse the Gambler '22
The Pedestrian '74
Spiders '18

Arlene Dahl

Journey to the Center of the Earth '59
Land Raiders '69
Night of the Warrior '91
Reign of Terror '49
Slightly Scarlet '56
A Southern Yankee '48
Three Little Words '50
Watch the Birdie '50

John Dahl (D)

Kill Me Again '89

Lisbet Dahl

Topsy Turvy '84

Eva Dahlbeck

Brink of Life '57
The Counterfeit Traitor '62
Dreams '55
Lesson in Love '54
Secrets of Women '52
Smiles of a Summer Night '55

Bob Dahlin (D)

Monster in the Closet '86

Paul Dahlke

The Head '59

Dan Dailey

It's Always Fair Weather '55
Lady Be Good '41
Meet Me in Las Vegas '56
Panama Hattie '42
Pride of St. Louis '52
The Private Files of J. Edgar Hoover '77
There's No Business Like Show Business '54
What Price Glory? '52
Wings of Eagles '57
Ziegfeld Girl '41

Bill Daily

Alligator 2: The Mutation '90
Magic on Love Island '80

E.G. Daily

Dogfight '91

Elizabeth Daily

Bad Dreams '88
Fandango '85
Funny Money '82
No Small Affair '84
Pee-Wee's Big Adventure '85
Street Music '81
Streets of Fire '84

Pete Daily

Yes, Sir, Mr. Bones '51

Masaaki Daimon

Godzilla vs. the Cosmic Monster '74
Godzilla vs. Mechagodzilla '75

Patricia Dainton

The House in Marsh Road '60

Vickie Dakil

The Personals '83

Patricia Dal

Virgins of Purity House '79

Alberto Dalbes

The Screaming Dead '72

Lynn Dalby

Legend of the Werewolf '75

Badgett Dale

Lord of the Flies '90

Carlo Dale

Two Nights with Cleopatra '54

Cynthia Dale

Heavenly Bodies '84

Dick Dale

Back to the Beach '87

Dolly Dale

Riders of the Range '24

Esther Dale

Curly Top '35
Holiday Affair '49
A Stolen Life '46

Grover Dale

Half a Sixpence '67

Janet Dale

Prick Up Your Ears '87

Jennifer Dale

Of Unknown Origin '83
Separate Vacations '86
When Angels Fly '70s

Jim Dale

The Adventures of Huckleberry Finn '85
Carry On Cleo '65
Carry On Cowboy '66
Carry On Doctor '68
Carry On Screaming '66
Digby, the Biggest Dog in the World '73
Follow That Camel '67
Hot Lead & Cold Feet '78
Joseph Andrews '77
Pete's Dragon '77
Scandalous '84
Unidentified Flying Oddball '79

Richard Dale

Ladies of the Lotus '86

Virginia Dale

Danger Zone '51
Holiday Inn '42

James Dalesandro

The Killing Zone '90

James Daley

The Court Martial of Billy Mitchell '55

Tom Daley (D)

The Outing '87

Fabienne Dali

Erotic Touch of Hot Skin '65
Kill, Baby, Kill '66

Salvador Dali (D)

Un Chien Andalou '28

Tracy Dali

Virgin High '90

David Dalie

Mighty Jungle '64
Virgin Sacrifice '59

Marcel Dalio

Beethoven '36
Can-Can '60
Captain Blackjack '51
Casablanca '42
China Gate '57
Dedee D'Anvers '49
Grand Illusion '37
How Sweet It Is! '68
Lovely to Look At '52
Lucky Me '54
The Rules of the Game '39
Sabrina '54
The Snows of Kilimanjaro '52
To Have & Have Not '44

John Dall

Corn is Green '45
Gun Crazy '50
Rope '48

Massimo Dallamano (D)

A Black Veil for Lisa '68
Dorian Gray '70

Charlene Dallas

Criminal Act '88

Beatrice Dalle

Betty Blue '86

Joe Dallesandro

Andy Warhol's Dracula '74
Andy Warhol's Frankenstein '74
The Cotton Club '84
Double Revenge '89
Flesh '68
Heat '72
The Hollywood Detective '89
Private War '90
Season for Assassins '71
Seeds of Evil '76
Sunset '88
Trash '70
Wild Orchid 2: Two Shades of Blue '92

Maurice Dallimore

The Collector '65

Neville D'Almedia (D)

Lady on the Bus '78

Abby Dalton

Roller Blade Warriors: Taken By Force '90
Whale of a Tale '76

Audrey Dalton

The Monster that Challenged the World '57
The Prodigal '55

Darren Dalton

Brotherhood of Justice '86

Dorothy Dalton

Disciple '15
Moran of the Lady Letty '22
The Three Musketeers (D'Artagnan) '16

Timothy Dalton

Agatha '79
Chanel Solitaire '81
Cromwell '70
The Doctor and the Devils '85
The Emperor's New Clothes '84
Flame is Love '79
Flash Gordon '80
Happy Anniversary 007: 25 Years of James Bond '87
Hawks '89
Jane Eyre '83
License to Kill '89
The Lion in Winter '68
The Living Daylights '87
Mistral's Daughter '84
Permission To Kill '75
The Rocketeer '91
Sextette '78
Sins '85
Wuthering Heights '70

Roger Daltrey

Cold Justice '89
Forgotten Prisoners '90
The Legacy '79
Lisztomania '75

Little Match Girl '87
Mack the Knife '89
McVicar '80
Monterey Pop '68
Murder: Ultimate Grounds for Divorce '85
Ready Steady Go, Vol. 1 '83
Ready Steady Go, Vol. 2 '85
Tommy '75
Woodstock '70

Robert Dalva (D)

The Black Stallion Returns '83

Candice Daly

Hell Hunters '87
Liquid Dreams '92

James Daly

Planet of the Apes '68
Resurrection of Zachary Wheeler '71
Tender Is the Night '55

Jane Daly

Children Shouldn't Play with Dead Things '72

Mark Daly

Lassie from Lancashire '38

Timothy Daly

Diner '82
For Love or Money '88
I Married a Centerfold '84
Just the Way You Are '84
Love or Money? '88
Made in Heaven '87
Mirrors '85
Rise & Rise of Daniel Rocket '86
The Spellbinder '88

Tyne Daly

The Adultress '77
The Aviator '85
Better Late Than Never '79
The Enforcer '76
Intimate Strangers '77
Larry '74
A Matter of Life and Death '81
Movers and Shakers '85
Speedtrap '78
Telefon '77
Your Place or Mine '83
Zoot Suit '81

William Robert Daly (D)

Uncle Tom's Cabin '14

Jacqueline Dalya

Behind Prison Walls '43
Miss Melody Jones '73

Stefania D'Amario

Zombie '80

Jill Damas

Games Girls Play '75

Joe D'Amato (D)

Buried Alive '81
Grim Reaper '81

Charlotte d'Amboise

American Blue Note '91

Jacques D'Amboise

Off Beat '86

Salsa '88

Davey Davidson

When the Line Goes Through '71

Diana Davidson

Around the World in 80 Ways '86
Scared to Death '80

Eduardo Davidson

Sonar no Cuesta nada Joven (Dreams are Free) '70s

Eileen Davidson

Easy Wheels '89
Eternity '90
The House on Sorority Row '83

Gordon Davidson (D)

Trial of the Catonsville Nine '72
The Trial of Lee Harvey Oswald '77

Jack Davidson

Baby It's You '82

James Davidson

Parasite '81

John Davidson

The Concorde: Airport '79 '79
The Happiest Millionaire '67
Monsieur Beaucaire '24
The Squeeze '87

John Davidson (D)

The Hanging Woman '72

Martin Davidson (D)

Eddie & the Cruisers '83
Hard Promises '92
Heart of Dixie '89
Hero at Large '80
Long Gone '87
The Lords of Flatbush '74

Max Davidson

Roamin' Wild '36

Richard Davidson

Variety '83

Tommy Davidson

Strictly Business '91

William B. Davidson

Held for Murder '32
Vice Squad '31

Arthur Davies

Rigoletto '82

Betty Ann Davies

Alias John Preston '56
The Blue Lamp '49

Gary Davies

Deadly Harvest '72

Jackson Davies

Dead Wrong '83
Jane Doe '83

James Davies

Wildest Dreams '90
Young Nurses in Love '89

John Davies

Interface '84
Positive I.D. '87

John Howard Davies

Oliver Twist '48
Tom Brown's School Days '51

John Davies (D)

A Married Man '84
Sherlock Holmes: The Silver Blaze '77

Lane Davies

Impure Thoughts '86

Lindy Davies

Malcolm '86

Marion Davies

Show People '28

Rachel Davies

The House that Bled to Death '81

Ray Davies

Absolute Beginners '86

Rudi Davies

The Lonely Passion of Judith Hearne '87

Rupert Davies

The Conqueror Worm '68
Frightmare '74
The Spy Who Came in from the Cold '65
Waterloo '71
Zeppelin '71

Sian Leisa Davies

Heaven on Earth '89

Stephen Davies

The Berlin Conspiracy '91
The Nest '88

Terence Davies (D)

Distant Voices, Still Lives '88

Valentine Davies (D)

The Benny Goodman Story '55

Luis Davila

Como Dos Gotas de Agua (Like Two Drops of Water) '70s
Pancho Villa '72
The Scalawag Bunch '75

Allyson Davis

Deadly Reactor '89

Amy Davis

All-American Murder '91

Andrew Davis

The Light in the Jungle '91

Andrew Davis (D)

Above the Law '88
The Final Terror '83
The Package '89

Andy Davis (D)

Code of Silence '85

Ann B. Davis

Lover Come Back '61

Art Davis

Along the Sundown Trail '42
Tumbleweed Trail '42

Arthur Davis (D)

Brutes and Savages '77
Cartoon Moviestars: Daffy! '48
Cartoon Moviestars: Elmer! '48

Bette Davis

All About Eve '50
All This and Heaven Too '40
As Summers Die '86
Beyond the Forest '49
The Bride Came C.O.D. '41
Burnt Offerings '76
The Catered Affair '56
Corn is Green '45
Dangerous '35
The Dark Secret of Harvest Home '78
Dark Victory '39
Death on the Nile '78
Deception '46
The Disappearance of Aimee '76
Empty Canvas '64
Fashions of 1934 '34
The Great Lie '41
Hell's House '32
Hollywood Canteen '44
Hollywood Outtakes & Rare Footage '83
Hush, Hush, Sweet Charlotte '65
In This Our Life '42
Jezebel '38
Juarez '39
The Letter '40
The Little Foxes '41
Little Gloria...Happy at Last '84
Madame Sin '71
The Man Who Came to Dinner '41
Marked Woman '37
Mr. Skeffington '44
Now, Voyager '42
Of Human Bondage '34
Old Maid '39
Petrified Forest '36
Phone Call from a Stranger '52
A Piano for Mrs. Cimino '82
Pocketful of Miracles '61
The Private Lives of Elizabeth & Essex '39
Return of Video Yesterbloop '47
Return from Witch Mountain '78
Right of Way '84
A Stolen Life '46
Strangers: The Story of a Mother and Daughter '79
Thank Your Lucky Stars '43
Three on a Match '32
The Virgin Queen '55
Watch on the Rhine '43
Watcher in the Woods '81
Way Back Home '32
Whales of August '87
What Ever Happened to Baby Jane? '62
White Mama '80
Wicked Stepmother '89

Betty Ann Davis

Blackout '54

B.J. Davis (D)

White Ghost '88

Brad Davis

Blood Ties '87
The Caine Mutiny Court Martial '88
The Campus Corpse '77
Chariots of Fire '81
Chiefs '83
Child of Darkness, Child of Light '91
Cold Steel '87
The Greatest Man in the World '80
Hangfire '91
Heart '87
Midnight Express '78
Querelle '83
Robert Kennedy and His Times '90
Rosalie Goes Shopping '89
A Rumor of War '80
A Small Circle of Friends '80
Sybil '76
Vengeance '89
When the Time Comes '91

Brownlee Davis

Day of Judgment '81

Bud Davis

House of the Rising Sun '87

Carole Davis

C.O.D. '83
The Flamingo Kid '84
Mannequin '87
Princess Academy '87
The Shrimp on the Barbie '90

Cathy Davis

The Guy from Harlem '77

Chet Davis

The Eye Creatures '65
Yes, Sir, Mr. Bones '51

Clifton Davis

Don't Look Back: The Story of Leroy "Satchel" Paige '81
Little Ladies of the Night '77

Catherine Davis Cox

Dr. Hackenstein '88

Desmond Davis (D)

Clash of the Titans '81
Nice Girl Like Me '69
Ordeal by Innocence '84

Diane Davis

The Meateater '79

Doug Davis

Future Kill '85

Eddie Davis

Fighting Pilot '35
Gunners & Guns '34

Eddie Davis (D)

Color Me Dead '69

Elaine Davis

The Atomic Kid '54

Elizabeth Davis

Gruesome Twosome '67

Frances Davis

Devil's Wedding Night '73

Frank Davis

Return of the Killer Tomatoes '88

Gail Davis

Blue Canadian Rockies '52
Cow Town '50
On Top of Old Smoky '53
Valley of Fire '51
Winning of the West '53

Geena Davis

The Accidental Tourist '88
Beetlejuice '88
Earth Girls are Easy '89
Fletch '85
The Fly '86
Quick Change '90
Secret Weapons '85
Thelma & Louise '91
Tootsie '82
Transylvania 6-5000 '85

Gene Davis

Messenger of Death '88
Night Games '80

Glenn Davis

The Spirit of West Point '47

Guy Davis

Beat Street '84

Humphrey Davis

Fright '56

Jack Davis

Endgame '85
Return of Our Gang '25

Jim Davis

Bad Company '72
Big Chase '54
The Big Sky '52
The Choirboys '77
Comes a Horseman '78
Day Time Ended '80
Don't Look Back: The Story of Leroy "Satchel" Paige '81
Dracula vs. Frankenstein '71
El Dorado '67
Gun Riders '69
Hellfire '48
Hi-Jacked '48
Inferno in Paradise '88
Iron Angel '64
Jesse James Meets Frankenstein's Daughter '65
Killing Stone '78
Law of the Land '76
Little Big Horn '51
Maverick Queen '55
Monster from Green Hell '58
Monte Walsh '70
Ninja Force of Assassins '70s
Outcast '54
The Restless Breed '58
The Runaway Barge '75
Three Desperate Men '50
The Trackers '71

Joan Davis

George White's Scandals '45
Hold That Ghost '41
If You Knew Susie '48
Just Around the Corner '38
Outcast '54
Show Business '44
Sun Valley Serenade '41

John Henry Davis

Maxwell Street Blues '89

Wild West '46

Felicity Dean

The Whistle Blower '87

Isabel Dean

Weather in the Streets '84

Ivor Dean

Theatre of Death '67

James Dean

America at the Movies '76
East of Eden '55
Forever James Dean '88
Giant '56
Hill Number One '51
Hollywood Outtakes & Rare
    Footage '83
James Dean '85
James Dean Story '57
James Dean: The First
    American Teenager '76
Rebel Without a Cause '55

Jeanne Dean

Clipped Wings '53

Jimmy Dean

Big Bad John '90
The City '76
Diamonds are Forever '71

Laura Dean

Almost You '85

Loren Dean

Billy Bathgate '91

Margia Dean

The Quatermass Experiment
    '56
Rimfire '49

Max Dean

Suicide Patrol '70s

Priscilla Dean

Law of the Sea '38
Outside the Law '21

Rick Dean

Bloodfist 3: Forced to Fight
    '91
Heroes Stand Alone '89
Raiders of the Sun '92

Ron Dean

Above the Law '88
Big Score '83
Cocktail '88
Cold Justice '89

Marcie Deane-Johns

The Winter of Our Dreams '82

Leslie Deane

Freddy's Dead: The Final
    Nightmare '91
Girlfriend from Hell '89
976-EVIL '88

Gina DeAngelis

Radio Days '87

**Emile
DeAntonio** *(D)*

Millhouse: A White Comedy
    '71

William Dear *(D)*

Harry and the Hendersons '87
If Looks Could Kill '91

Timerider '83

Basil Dearden *(D)*

The Blue Lamp '49
Captive Heart '47
Dead of Night '45
Frieda '47
Khartoum '66
League of Gentlemen '60
The Man Who Haunted
    Himself '70
Mission: Monte Carlo '75
The Persuaders '71
Sapphire '59
The Smallest Show on Earth
    '57
Train of Events '49
Victim '62

James Dearden *(D)*

Cold Room '84
A Kiss Before Dying '91
Pascali's Island '88

Edgar Dearing

Seven Doors to Death '44

JoAnn Dearing

Suburban Commando '91

Max Dearly

Madame Bovary '34

Jill Dearman

Gap-Toothed Women '89

Justin Deas

Dream Lover '85
Montana '90

Frank Deasy *(D)*

The Courier '88

Donna Death

Geek Maggot Bingo '83

George Deaton

E.S.P. '83

Moreno D'EBartolli

When Father was Away on
    Business '85

Kristine DeBell

Alice in Wonderland '77
Cheerleaders' Wild Weekend
    '85
Lifepod '80
Rooster—Spurs of Death! '83
Tag: The Assassination Game
    '82

John DeBello *(D)*

Attack of the Killer Tomatoes
    '77
Happy Hour '87
Killer Tomatoes Strike Back
    '90
Return of the Killer Tomatoes
    '88

Nikki DeBoer

Prom Night 4: Deliver Us From
    Evil '91

Dorothy DeBorba

Little Rascals, Book 6 '30s

**Gianfranco
DeBosio** *(D)*

Moses '76

**Phillipe
DeBroca** *(D)*

Jupiter's Thigh '81
King of Hearts '66
Louisiana '87

Lee DeBroux

Sweet Hostage '75

Jean Debucourt

The Earrings of Madame De...
    '54
La Chute de la Maison Usher
    '28
Seven Deadly Sins '53

Rosemary DeCamp

Cheers for Miss Bishop '41
Commandos Strike at Dawn
    '42
Yankee Doodle Dandy '42

William W. Decker

Redneck Zombies '88

Eugene Deckers

The Golden Salamander '51

Linda DeCoff

Hurry Up or I'll Be Thirty '73

Guy Decomble

The 400 Blows '59
Jour de Fete '48

Fred DeCordova *(D)*

Bedtime for Bonzo '51
Frankie and Johnny '65

Ted DeCorsia

Slightly Scarlet '56

Kutira Decosterd

Men in Love '90

David DeCoteau *(D)*

Creepozoids '87
Dr. Alien '88
Nightmare Sisters '87
Puppet Master 3 '91
Sorority Babes in the Slimeball
    Bowl-A-Rama '87

Blue Dedeort

Lady from Yesterday '85

Yann Dedet

Under Satan's Sun '87

Buffy Dee

Jaws of Death '76

Candy Dee

Psycho from Texas '83

Catherine Dee

In Deadly Heat '87
Stuff Stephanie in the
    Incinerator '89

Frances Dee

Becky Sharp '35
Finishing School '33
Four Faces West '48
I Walked with a Zombie '43
Little Women '33
Of Human Bondage '34
The Private Affairs of Bel Ami
    '47
So Ends Our Night '41
Wheel of Fortune '41

Gloria Dee

King of the Congo '52

Ruby Dee

All God's Children '80
The Balcony '63
Buck and the Preacher '72
Cat People '82
The Court Martial of Jackie
    Robinson '90
Do the Right Thing '89
Gone are the Days '63
I Know Why the Caged Bird
    Sings '79
The Incident '67
It's Good to Be Alive '74
The Jackie Robinson Story
    '50
Jungle Fever '91
Love at Large '89
Purlie Victorious '63
A Raisin in the Sun '61
Zora Is My Name! '90

Sandra Dee

The Dunwich Horror '70
Fantasy Island '76
Gidget '59
Imitation of Life '59
Lost '83
The Reluctant Debutante '58
A Summer Place '59
Tammy and the Doctor '63

Miles Deem *(D)*

Jungle Master '56

Deep Purple

Incident at Channel Q '86

Olive Deering

Samson and Delilah '49

Kevin Dees

Jet Benny Show '86

Rick Dees

The Jetsons: Movie '90

Eddie Deezen

Beverly Hills Vamp '88
Desperate Moves '86
Hollywood Boulevard 2 '89
I Wanna Hold Your Hand '78
Midnight Madness '80
Million Dollar Mystery '87
Mugsy's Girls '85
A Polish Vampire in Burbank
    '80
Rosebud Beach Hotel '85
Surf 2 '84

Frank DeFelitta *(D)*

Dark Night of the Scarecrow
    '81

**Eduardo
DeFilippo** *(D)*

Shoot Loud, Louder, I Don't
    Understand! '66

Don DeFore

A Girl in Every Port '52
Ramrod '47
A Rare Breed '81
Romance on the High Seas
    '48
Too Late for Tears '49
Without Reservations '46

Calvert Deforest

Heaven Help Us '85
Leader of the Band '87

Bob DeFrank

Driller Killer '74

Isabelle DeFunes

Kiss Me, Kill Me '69

**Allessandro
DeGaetano** *(D)*

Bloodbath in Psycho Town
    '89

**Michael
DeGaetano** *(D)*

The Haunted '79

Andre Degas *(D)*

American Autobahn '84

Pia Degermark

Elvira Madigan '67
The Looking Glass War '69

**Joseph
DeGrasse** *(D)*

The Scarlet Car '17

Sam DeGrasse

Robin Hood '22
The Scarlet Car '17

Philip DeGuere *(D)*

Dr. Strange '78
Misfits of Science '85

Gloria DeHaven

Best Foot Forward '43
Bog '84
Summer Holiday '48
The Thin Man Goes Home '44

Lisa DeHaven

Redneck Zombies '88

Jamil Dehlavi *(D)*

Born of Fire '87

John Dehner

California Gold Rush '81
Cowboy '54
Creator '85
The Day of the Dolphin '73
Girl in Black Stockings '57
The Hallelujah Trail '65
The Jagged Edge '85
Killer Inside Me '76
The Left-Handed Gun '58
Man of the West '58
Mountain Man '77
Nothing Personal '80
The Prodigal '55
Vigilantes of Boom Town '46

Mark Deimel *(D)*

Perfect Match '88

Edward Dein *(D)*

Shack Out on 101 '55

Donna Deitch *(D)*

Desert Hearts '86
Prison Stories: Women on the
    Inside '91
The Women of Brewster
    Place '89

Steve DeJarnatt *(D)*

Cherry 2000 '88
Miracle Mile '89

Christine Dejoux

One Wild Moment '78

Dennehy ►cont.

The River Rat '84
A Rumor of War '80
Semi-Tough '77
Silent Victory: The Kitty O'Neil Story '79
Silverado '85
Split Image '82
Tall Tales & Legends: Annie Oakley '85
10 '79
To Catch a Killer '92
To Love Again '80
Twice in a Lifetime '85

Barbara Dennek

Playtime '67

Barry Dennen

Liquid Dreams '92

Charles Denner

Bluebeard '63
The Holes (Les Gaspards) '72
Les Assassins de L'Ordre '71
Mado '76
The Man Who Loved Women '77
Robert et Robert '79
Two of Us '68
Window Shopping '86
Z '69

Dodo Denney

Willy Wonka & the Chocolate Factory '71

Richard Denning

Adam Had Four Sons '41
An Affair to Remember '57
Alice Through the Looking Glass '66
Black Beauty '46
The Black Scorpion '57
Creature from the Black Lagoon '54
Double Deal '50
Girls in Prison '56
The Glass Key '42
Hangman's Knot '52
Unknown Island '48

Charles Dennis (D)

Reno and the Doc '84

John Dennis

Tomb of the Undead '72

Sandy Dennis

Another Woman '88
Come Back to the 5 & Dime Jimmy Dean, Jimmy Dean '82
The Execution '85
The Four Seasons '81
God Told Me To '76
The Indian Runner '91
Mr. Sycamore '74
Nasty Habits '77
976-EVIL '88
The Out-of-Towners '70
Parents '89
Splendor in the Grass '61
That Cold Day in the Park '69
Up the Down Staircase '67
Who's Afraid of Virginia Woolf? '66

Winifred Dennis

Die! Die! My Darling! '65

Sabrina Dennison

Santa Sangre '90

Leslie Denniston

Blue Heaven '84

Reginald Denny

Abbott and Costello Meet Dr. Jekyll and Mr. Hyde '52
Anna Karenina '35
Around the World in 80 Days '56
Arrest Bulldog Drummond '38
Assault on a Queen '66
Bulldog Drummond Comes Back '37
Bulldog Drummond Double Feature '30s
Bulldog Drummond Escapes '37
Bulldog Drummond's Bride '39
Bulldog Drummond's Peril '38
Bulldog Drummond's Revenge '37
Bulldog Drummond's Revenge/Bulldog Drummond's Peril '38
Bulldog Drummond's Secret Police '39
California Straight Ahead '25
Cat Ballou '65
The Cheerful Fraud '27
Dancing Man '33
Escape to Burma '55
Hour of Stars '58
The Lost Patrol '34
Mr. Blandings Builds His Dream House '48
My Favorite Brunette '47
Of Human Bondage '34
Parlor, Bedroom and Bath '31
Rebecca '40
Sabaka '55
Sherlock Holmes: The Voice of Terror '42
Skinner's Dress Suit '26
We're in the Legion Now '37

Pen Densham (D)

The Kiss '88
The Zoo Gang '85

John Densmore

The Doors: A Tribute to Jim Morrison '81
The Doors: Live in Europe, 1968 '68
The Doors: Live at the Hollywood Bowl '68

John Densmore (D)

The Doors: Live in Europe, 1968 '68

Vernon Dent

Baby Face Harry Langdon '26
The Cameraman '28
Extra Girl '23
His First Flame '26
Texas Cyclone '32

Christa Denton

Convicted: A Mother's Story '87
The Gate '87
Scandal in a Small Town '88

Donna Denton

Death Blow '87
Glory Years '87
Outlaw of Gor '87
Slaughterhouse Rock '88

Paul Dentzer

They '80s

Bob Denver

Back to the Beach '87
Far Out Space Nuts, Vol. 1 '75
High School USA '84
Wackiest Wagon Train in the West '84
Who's Minding the Mint? '67

John Denver

Fire and Ice '87
Oh, God! '77
Rock & Roll Call '84

Manuela Denz

Sugarbaby '85

Ruggero Deodato (D)

The Barbarians '87
Dial Help '88
House on the Edge of the Park '84
The Lone Runner '88
Phantom of Death '87

Brian DePalma (D)

Blow Out '81
Body Double '84
The Bonfire of the Vanities '90
Carrie '76
Casualties of War '89
Dressed to Kill '80
The Fury '78
Greetings '68
Hi, Mom! '70
Home Movies '79
Obsession '76
Phantom of the Paradise '74
Scarface '83
Sisters '73
The Untouchables '87
The Wedding Party '69
Wise Guys '86

Elisabeth Depardieu

Jean de Florette '87
Tartuffe '84

Gerard Depardieu

Buffet Froid '79
Camille Claudel '89
Choice of Arms '83
Cyrano de Bergerac '89
Danton '82
Get Out Your Handkerchiefs '78
Going Places '74
Green Card '90
The Holes (Les Gaspards) '72
I Love All of You '83
Jean de Florette '87
La Chevre '81
The Last Metro '80
Les Comperes '83
Loulou '80
Maitresse '76
Mon Oncle D'Amerique '80
Moon in the Gutter '84
1900 '76
One Woman or Two '86
Police '85
Return of Martin Guerre '83
Stavisky '74
Tartuffe '84
Too Beautiful for You '88
Under Satan's Sun '87
Vincent, Francois, Paul and the Others '76

The Woman Next Door '81

Gerard Depardieu (D)

Tartuffe '84

Brian DePersia

Deadtime Stories '86

Johnny Depp

Cry-Baby '90
Edward Scissorhands '90
Freddy's Dead: The Final Nightmare '91
A Nightmare on Elm Street '84
Platoon '86
Private Resort '84

Theo Depuay

Gore-Met Zombie Chef From Hell '87

Jacques Deray (D)

Swimming Pool '70

Bo Derek

Bolero '84
A Change of Seasons '80
Fantasies '81
Ghosts Can't Do It '90
Orca '77
Tarzan, the Ape Man '81
10 '79

John Derek

All the King's Men '49
Ambush at Tomahawk Gap '53
The Annapolis Story '55
Exodus '60
Knock on Any Door '49
Omar Khayyam '57
Once Before I Die '65
Outcast '54
The Ten Commandments '56

John Derek (D)

Bolero '84
Confessions of Tom Harris '72
Fantasies '73
Ghosts Can't Do It '90
Once Before I Die '65
Tarzan, the Ape Man '81

Joe DeRita

It's a Mad, Mad, Mad, Mad World '63
The Three Stooges Meet Hercules '61

Edouard Dermithe

Les Enfants Terrible '50
The Testament of Orpheus '59

Anton Dermota

The Life and Loves of Mozart '59

Bruce Dern

Big Town '87
Black Sunday '77
Bloody Mama '70
The 'Burbs '89
Coming Home '78
The Court Martial of Jackie Robinson '90
The Cowboys '72
The Driver '78
Family Plot '76
The Great Gatsby '74
Harry Tracy '83

Hush, Hush, Sweet Charlotte '65
The Incredible Two-Headed Transplant '71
Into the Badlands '92
Laughing Policeman '74
Marnie '64
Middle Age Crazy '80
1969 '89
On the Edge '86
Posse '75
Psych-Out '68
Rebel Rousers '69
The St. Valentine's Day Massacre '67
Silent Running '71
Smile '75
Support Your Local Sheriff '69
Tattoo '81
That Championship Season '82
They Shoot Horses, Don't They? '69
Thumb Tripping '72
Toughlove '85
The Trip '67
Twist '76
The War Wagon '67
Waterhole #3 '67
The Wild Angels '66
Will Penny '67
World Gone Wild '88

Laura Dern

Afterburn '92
Blue Velvet '86
Fat Man and Little Boy '89
Foxes '80
Haunted Summer '88
Mask '85
Smooth Talk '85
Strange Case of Dr. Jekyll & Mr. Hyde '89
Teachers '84

Derna-Hazell

The Dummy Talks '43

Richard Derr

Adam at 6 A.M. '70
Castle in the Desert '42
Guilty of Treason '50
The Invisible Avenger '58
Terror is a Man '59
When Worlds Collide '51

Cleavant Derricks

Bluffing It '87
Moscow on the Hudson '84
Off Beat '86

Joe Derrig

Maximum Thrust '87

Rick Derringer

Rick Derringer '82

Dominique Deruddere

Istanbul '85

Dominique Deruddere (D)

Wait Until Spring, Bandini '90

Jacqueline Derval

Battle of the Worlds '61

Lamya Derval

Howling 4: The Original Nightmare '88

Dewhurst ▶cont.

Boy Who Could Fly '86
The Cowboys '72
Dead Zone '83
Dying Young '91
Final Assignment '80
A Fine Madness '66
Glitter Dome '84
The Guyana Tragedy: The
    Story of Jim Jones '80
Ice Castles '79
Lantern Hill '90
Mary and Joseph: A Story of
    Faith '79
McQ '74
Medea '59
The Nun's Story '59
Obsessed '88
Silent Victory: The Kitty O'Neil
    Story '79
Sword of Gideon '86
Termini Station '91
Third Walker '79
When a Stranger Calls '79
You Can't Take It with You
    '84

Tom Dewier (D)

Death by Dialogue '88

Johana DeWinter

Planet Earth '74

Billy DeWolfe

Billie '65
Dear Wife '49
Frosty the Snowman '69
Lullaby of Broadway '51
The Perils of Pauline '47
Tea for Two '50

Peter Dews (D)

Taming of the Shrew '81

Anthony Dexter

Fire Maidens from Outer
    Space '56
Married Too Young '62

Brad Dexter

Between Heaven and Hell '56
Las Vegas Story '52
Last Train from Gun Hill '59
Magnificent Seven '60
None But the Brave '65
The Oklahoman '56
Run Silent, Run Deep '58
Von Ryan's Express '65
Winter Kills '79

John Dexter (D)

I Want What I Want '72
The Virgin Soldiers '69

Maury Dexter (D)

Surf Party '64

Tony Dexter

The Phantom Planet '61

Susan Dey

Comeback Kid '80
Echo Park '86
First Love '77
L.A. Law '86
Little Women '78
Looker '81
Love Leads the Way '84
Sunset Limousine '83
That's Adequate '90
The Trouble with Dick '88

Cliff DeYoung

Awakening of Candra '81
Bulldance '88
Code Name: Dancer '87
Dangerous Company '88
F/X '86
Fear '88
Flight of the Navigator '86
Forbidden Sun '89
Fourth Story '90
The Hunger '83
Immortal Sins '91
In Dangerous Company '88
Independence Day '83
King '78
The Lindbergh Kidnapping
    Case '76
Protocol '84
Pulse '88
Reckless '84
Robert Kennedy and His
    Times '90
Rude Awakening '89
Secret Admirer '85
Shock Treatment '81
The Survivalist '87
Tall Tales & Legends: Annie
    Oakley '85
To Die Standing '91

Bernard Deyries (D)

Here Come the Littles: The
    Movie '85

Eddie Dezen

Mob Boss '90

Dharmendra

Deadly Thief '78

Bernard Dheran

Bernadette '90

Robert Dhery

Femmes de Paris '53

Danica D'Hondt

Unkissed Bride '66

Maria Di Aragon

The Cremators '72

Dalia di Lazzaro

Andy Warhol's Frankenstein
    '74
Creepers '85

Lea di Lea

The Earrings of Madame De...
    '54

Fernando Di Leo (D)

The Italian Connection '73
Kidnap Syndicate '76
Loaded Guns '75
Mr. Scarface '77
Slaughter Hotel '71
Violent Breed '83

Mario Di Leo (D)

The Final Alliance '89

Victor Di Mello (D)

Her Summer Vacation

Rino Di Silvestro (D)

Women in Cell Block 7 '77

Alice Diabo

Strangers in Good Company
    '91

Rika Dialina

Black Sabbath '64

Catherine Diamant

The Nun '66

George Diamond

Video Wars '80s

Joel Diamond

Crossover Dreams '85

Marcia Diamond

Deranged '74

Neil Diamond

Jazz Singer '80
The Last Waltz '78
Rock & Roll Call '84

Reed Edward Diamond

Ironclads '90
O Pioneers! '91

Selma Diamond

All of Me '84
Bang the Drum Slowly '73
Twilight Zone: The Movie '83

John DiAquino

Pumpkinhead '88
Slipping into Darkness '88

Chico Diaz

The Fable of the Beautiful
    Pigeon Fancier '88

Justino Diaz

Otello '86

Maria Diaz

The Heroes of Desert Storm
    '91

Jose Diaz Morales (D)

Loyola, the Soldier Saint '52

Sully Diaz

Gryphon '88

Tony Diaz

Hay Muertos Que No Hacen
    Ruido '47

Vic Diaz

Beast of the Yellow Night '70
Children of An Lac '80
The Deadly and the Beautiful
    '73
Deathhead Virgin '74
Flight to Fury '66
Night of the Cobra Woman '72
Savage! '73
Too Hot to Handle '76

Kem Dibbs

Daniel Boone: Trail Blazer '56

Tony DiBenedetto

Someone to Watch Over Me
    '87

Luigi Diberti

All Screwed Up '74

Louie Dibianco

Mafia Princess '86

Leonardo DiCaprio

Critters 3 '91

George DiCenzo

The Exorcist 3 '90
The Frisco Kid '79
Helter Skelter '76
The New Adventures of Pippi
    Longstocking '88
Omega Syndrome '87
Sing '89

Bobby DiCicco

Frame Up '91
I Wanna Hold Your Hand '78
The Last Hour '91
The Philadelphia Experiment
    '84
Splash '84
Tiger Warsaw '87

Danny Dick

Dungeonmaster '85

Douglas Dick

The Accused '48
The Oklahoman '56
The Red Badge of Courage
    '51

Kirby Dick (D)

Private Practices: The Story of
    a Sex Surrogate '86

Nigel Dick (D)

Deadly Intent '88
Private Investigations '87

Beach Dickerson

Creature from the Haunted
    Sea '60
T-Bird Gang '59

Ernest R. Dickerson

Juice '92

Ernest R. Dickerson (D)

Juice '92

George Dickerson

Blue Velvet '86
Cutter's Way '81
Death Warrant '90

James Dickey

Deliverance '72

Lucinda Dickey

Breakin' '84
Breakin' 2: Electric Boogaloo
    '84
Cheerleader Camp '88
Ninja 3: The Domination '84

Paul Dickey

Robin Hood '22

Angie Dickinson

Big Bad Mama '74
Big Bad Mama 2 '87
Captain Newman, M.D. '63
Cast a Giant Shadow '66
Charlie Chan and the Curse of
    the Dragon Queen '81
The Chase '66
China Gate '57
Death Hunt '81
Dressed to Kill '80

Fire and Rain '89
Jealousy '84
The Killers '64
Klondike Fever '79
Lucky Me '54
Ocean's 11 '60
One Shoe Makes It Murder
    '82
Point Blank '67
The Poppy Is Also a Flower
    '66
Pray for the Wildcats '74
Resurrection of Zachary
    Wheeler '71
Rio Bravo '59
Rome Adventure '62
Texas Guns '90
Thief '71

Thorold Dickinson (D)

The Arsenal Stadium Mystery
    '39
High Command '37
Hill 24 Doesn't Answer '55
Men of Two Worlds '46
The Queen of Spades '49

George Dickson

Ninja Phantom Heroes '87

Lance Dickson

Project: Nightmare

Neil Dickson

Biggles '85
It Couldn't Happen Here '88
The Murders in the Rue
    Morgue '86

Marc Didden (D)

Istanbul '85

Bo Diddley

Chuck Berry: Hail! Hail! Rock
    'N' Roll '87
Hell's Angels Forever '83
Rockula '90

Arlette Didier

Bernadette '90

Samuel Diege (D)

Ride 'Em Cowgirl '41
Water Rustlers '39

Dan Diego

Three Day Weekend
Three Way Weekend '81

Gabino Diego

Ay, Carmela! '90

Carlos Diegues (D)

Bye Bye Brazil '79
Quilombo '84
Subway to the Stars '87

John Diehl

Angel '84
Cool Blue '88
Dark Side of the Moon '90
Glitz '88
Madhouse '90
Miami Vice '84
Mikey '92

Baruch Dienar (D)

They were Ten '61

John Dierkes

The Daughter of Dr. Jekyll '57
The Haunted Palace '63

Charles Dierkop

Banzai Runner '86
The Fix '84
Roots of Evil '91

Gustav Diesl

Das Testament des Dr.
  Mabuse '33
Westfront 1918 '30

William Dieterle

Backstairs '21
Waxworks '24

William Dieterle (D)

The Accused '48
Boots Malone '52
Devil & Daniel Webster '41
Elephant Walk '54
Fashions of 1934 '34
The Hunchback of Notre
  Dame '39
Juarez '39
The Life of Emile Zola '37
A Midsummer Night's Dream
  '35
Mistress of the World '59
Omar Khayyam '57
Portrait of Jennie '48
Quick, Let's Get Married '71
Salome '53
September Affair '50
The Story of Louis Pasteur
  '36

Marsha Dietlein

Return of the Living Dead Part
  2 '88

Daniel Dietrich

Crossfire '89

Marlene Dietrich

Around the World in 80 Days
  '56
Blonde Venus '32
The Blue Angel '30
Destry Rides Again '39
Funny Guys & Gals of the
  Talkies '30s
The Garden of Allah '36
Judgment at Nuremberg '61
Just a Gigolo '79
Knight Without Armour '37
The Love Goddesses '65
Morocco '30
Paris When It Sizzles '64
Rancho Notorious '52
Seven Sinners '40
The Spoilers '42
Stage Fright '50
Touch of Evil '58
Witness for the Prosecution
  '57

Frank Dietz

The Jitters '88
Zombie Nightmare '86

Albert Dieudonne

Napoleon '27

Anton Diffring

Black Werewolf '75
Call Him Mr. Shatter '74
Circus of Horrors '60
The Colditz Story '55
Fahrenheit 451 '66
Mark of the Devil 2 '72
Seven Deaths in the Cat's Eye
  '72
Where Eagles Dare '68
Winds of War '83

Zeppelin '71

Jenny Diggers

The Body Shop

Dudley Digges

China Seas '35
Devotion '31
Emperor Jones '33
Emperor Jones/Paul
  Robeson: Tribute to an
  Artist '33
The General Died at Dawn '36
The Invisible Man '33
The Maltese Falcon '31
Mutiny on the Bounty '35

Arthur Dignam

The Devil's Playground '76
Grendel, Grendel, Grendel '82
King Solomon's Mines '86
The Right Hand Man '87
Strange Behavior '81
We of the Never Never '82
The Wild Duck '84

Eric Dignam (D)

Denial: The Dark Side of
  Passion '91

Alan Dijon

Assignment Outer Space '61

Irasema Dilian

Wuthering Heights '53

Matt Dill

The Boy Who Loved Trolls '84

Denise Dillaway

The Cheerleaders '72

Phyllis Diller

The Bone Yard '90
Boy, Did I Get a Wrong
  Number! '66
Hungry i Reunion '81
Jonathan Winters on the Edge
  '86
Laughing Room Only '86
Mad Monster Party '68
Maniac '34
Minsky's Follies '83
The Nutcracker Prince '91
Pink Motel '82
A Pleasure Doing Business
  '79
Splendor in the Grass '61
The Worst of Hollywood: Vol.
  1 '90

Bradford Dillman

The Amsterdam Kill '78
Black Water Gold '69
The Bridge at Remagen '69
Brother John '70
Bug '75
Death in Deep Water '74
The Enforcer '76
Escape from the Planet of the
  Apes '71
Force Five '75
Heroes Stand Alone '89
Legend of Walks Far Woman
  '82
Lords of the Deep '89
Love and Bullets '79
Man Outside '88
Master Mind '73
The Mephisto Waltz '71
Moon of the Wolf '72
Murder or Mercy '74
99 & 44/100 Dead '74
One Away '76

Piranha '78
Resurrection of Zachary
  Wheeler '71
Running Scared '79
Sergeant Ryker '68
Sudden Impact '83
Suppose They Gave a War
  And Nobody Came? '70
The Swarm '78
The Way We Were '73
The Widow '76

Brandan Dillon

The Killing of Sister George
  '69

Brendan Dillon, Jr.

Lords of Magick '88

C.J. Dillon

Return of the Killer Tomatoes
  '88

Edward Dillon

The Broadway Melody '29

Jack Dillon (D)

Suds '20

John Francis Dillon (D)

Millie '31

Kevin Dillon

The Blob '88
The Doors '91
Heaven Help Us '85
Immediate Family '89
No Big Deal '83
Platoon '86
Remote Control '88
The Rescue '88
War Party '89

Matt Dillon

Big Town '87
Bloodhounds of Broadway '89
Drugstore Cowboy '89
The Flamingo Kid '84
Kansas '88
A Kiss Before Dying '91
Liar's Moon '82
Little Darlings '80
My Bodyguard '80
Native Son '86
The Outsiders '83
Over the Edge '79
Rebel '86
Rumble Fish '83
Target '85
Tex '82
Women & Men: Stories of
  Seduction, Part 2 '91

Melinda Dillon

Absence of Malice '81
Bound for Glory '76
Captain America '89
A Christmas Story '83
Close Encounters of the Third
  Kind '80
Fallen Angel '81
F.I.S.T. '78
Harry and the Hendersons '87
Juggler of Notre Dame '84
Marriage is Alive and Well '80
Nightbreaker '89
The Prince of Tides '91
Right of Way '84
The Shadow Box '80
Shattered Spirits '91
Slap Shot '77
Songwriter '84
Spontaneous Combustion '89

Staying Together '89

Mick Dillon

Champions '84

Paul Dillon

The Beat '88
Kiss Daddy Goodnight '87

Stephen Dillon

An Affair in Mind '89
Christabel '89

Tom Dillon

Dressed to Kill '46
Night Tide '63

Joe DiMaggio

Manhattan Merry-go-round
  '37

Joseph Dimambro

Concrete Angels '87

James Dime

Stand and Deliver '28

Richard Dimitri

Johnny Dangerously '84

Dennis Dimster

Olly Olly Oxen Free '78

Dennis Dimster (D)

Mikey '92

Ayub Khan Din

Sammy & Rosie Get Laid '87

Alan Dinehart

After Midnight '33
Baby, Take a Bow '34
A Study in Scarlet '33

Arthur Dingham

Between Wars '74

Mark Dingham

Hamlet '69

Charles Dingle

The Court Martial of Billy
  Mitchell '55
Guest Wife '45
If You Knew Susie '48
The Little Foxes '41
A Southern Yankee '48

Ernie Dingo

The Fringe Dwellers '86
A Waltz Through the Hills '88

Kelly Dingwall

Around the World in 80 Ways
  '86

Michael Dinner (D)

Heaven Help Us '85
Hot to Trot! '88
Off Beat '86

Jerry Dinome

Tomboy '85

Lou Dinos

Paramount Comedy Theater,
  Vol. 4: Delivery Man '87

Reece Dinsdale

Threads '85
Winter Flight '84
Young Catherine '91

Rose Dione

Salome '22
Suds '20

Silvia Dionisio

The Scalawag Bunch '75

Carlo DiPalma (D)

Blonde in Black Leather '77

Dire Straits

Nelson Mandela 70th Birthday
  Tribute '88

John Dirlam (D)

Fatal Instinct '92

Dyanne DiRosario

Dr. Hackenstein '88

Mark DiSalle (D)

Kickboxer '89
The Perfect Weapon '91

John Disanti

Eyes of a Stranger '81

Alesandro DiSanzo

Forever Mary '91

Bob Dishy

Brighton Beach Memoirs '86
I Wonder Who's Killing Her
  Now? '76
Kill My Wife...Please!
Lovers and Other Strangers
  '70

Walt Disney

Mickey & the Beanstalk '47
The Rose Parade: Through
  the Years '88

Walt Disney (D)

Canine Commando '45
Cartoon Classics, Vol. 1:
  Here's Mickey! '41
Cartoon Classics, Vol. 2:
  Here's Donald! '39
Cartoon Classics, Vol. 3:
  Here's Goofy! '39
Cartoon Classics, Vol. 5:
  Here's Pluto! '39
Chip 'n' Dale & Donald Duck
  '39
Donald Duck: The First 50
  Years '53
Donald: Limited Gold Edition 1
  '53
Donald's Bee Pictures:
  Limited Gold Edition 2 '52
Fun & Fancy Free '47
How the Best was Won '60
Kids is Kids '53
Life with Mickey '51

Joe DiSue

Blackenstein '73

Divine

Divine '70s
Female Trouble '75
Hairspray '88
I Wanna Be a Beauty Queen
  '85
Lust in the Dust '85
Mondo Trasho '70
Multiple Maniacs '70
Out of the Dark '88
Pink Flamingos '73
Polyester '81
Trouble in Mind '86

Andrew Divoff
Toy Soldiers '91

Dorothy Dix
Drum Taps '33

Richard Dix
Ace of Aces '33
American Empire '42
Archer: The Fugitive from the Empire '81
Buckskin Frontier '43
Cimarron '30
The Kansan '43
Lost Squadron '32
Lucky Devil '25
Ten Commandments '23
Transatlantic Tunnel '35
Vanishing American '26

Robert Dix
Blood of Dracula's Castle '69
Cain's Cutthroats '71
Deadwood '65
Horror of the Blood Monsters '70
Killers '88
Wild Wheels '75

Tommy Dix
Best Foot Forward '43

Phyllis Dixey
Dual Alibi '47

Carter Dixon
Young Eagles '34

David Dixon
Escort Girls '74

Denver Dixon (D)
Fighting Cowboy '33

Donna Dixon
Beverly Hills Madam '86
The Couch Trip '87
Doctor Detroit '83
Lucky Stiff '87
Margin for Murder '81
Speed Zone '88
Spies Like Us '85

Ernest Dixon
Freedom Road '79

Ivan Dixon
Car Wash '76
Fer-De-Lance '74
A Patch of Blue '65
A Raisin in the Sun '61
Suppose They Gave a War And Nobody Came? '70

Ivan Dixon (D)
The Spook Who Sat By the Door '73

James Dixon
It's Alive '74
It's Alive 3: Island of the Alive '87
Return to Salem's Lot '87

Jean Dixon
Joy of Living '38

Joan Dixon
Gunplay '51

Ken Dixon (D)
Slave Girls from Beyond Infinity '87

MacIntyre Dixon
Comedy Tonight '77
Funny Farm '88
Ghostwriter '84

Michael Dixon
Chains '89

Pamela Dixon
Chance '89
Hollywood in Trouble '87
L.A. Crackdown '87
L.A. Crackdown 2 '88
Mayhem '87

Steve Dixon
The Carrier '87

Willie Dixon
Rich Girl '91

Gregg D'Jah
Cheering Section '73

Les Djinns
Invitation to Paris '60

Badja Djola
Mississippi Burning '88
The Serpent and the Rainbow '87

Edward Dmytryk (D)
Alvarez Kelly '66
Anzio '68
Back to Bataan '45
Behind the Rising Sun '43
Bluebeard '72
Broken Lance '54
The Caine Mutiny '54
The Carpetbaggers '64
Cornered '45
Crossfire '47
He is My Brother '75
The Hidden Room '49
Hitler's Children '43
The Human Factor '75
The Left Hand of God '55
Mirage '66
The Mountain '56
Murder My Sweet '44
Mutiny '52
Raintree County '57
Seven Miles from Alcatraz/ Flight from Glory '43
Shalako '68
Soldier of Fortune '55
Tender Comrade '43
Till the End of Time '46
Trail of the Hawk '37
Walk on the Wild Side '62
Warlock '59
The Young Lions '58

Ho Chung Do
Fists of Fury 2 '80

Mauricio Do Valle
Amazon Jail '85
Antonio Das Mortes '68

Wong Do
Moonlight Sword & Jade Lion '70s

Tony B. Dobb (D)
Blue Tornado '90

Frank Q. Dobbs (D)
Uphill All the Way '85

Heidi Dobbs
Death Journey '76
Jornada de Muerte '80s

James Dobbs
Impulse '74

Valeria D'Obici
Escape from the Bronx '85
Passion of Love '82

Alan Dobie
White Mischief '88

Lawrence Dobkin
Patton '70

Lawrence Dobkin (D)
Children's Carol '80
Sixteen '72

Gosia Dobrowolska
Around the World in 80 Ways '86
Silver City '84

James Dobson
Jet Attack/Paratroop Command '58

Kevin Dobson
All Night Long '81
Hardhat & Legs '80
Margin for Murder '81
Midway '76
Orphan Train '79

Kevin Dobson (D)
Demolition '77
Image of Death '77
The Mango Tree '77
Squizzy Taylor '84

Peter Dobson
Sing '89

Tamara Dobson
The Amazons '84
Chained Heat '83
Cleopatra Jones '73
Cleopatra Jones & the Casino of Gold '75

Vernon Dobtcheff
Nijinsky '80
Pascali's Island '88

Trudi Dochtermann
Down Twisted '89

Leslie Dockery
Eubie! '82

Carol Doda
Honkytonk Nights '70s

Marilyn Dodd-Frank
Sweet Lies '88

Lorenzo Dodo
The Executioner '78

Jack Dodson
The Getaway '72

John Doe
Great Balls of Fire '89
A Matter of Degrees '90
Slamdance '87
Without You I'm Nothing '90
X: The Unheard Music '86

Tatiana Dogileva
A Forgotten Tune for the Flute '88

Shannen Doherty
Girls Just Want to Have Fun '85
Heathers '89

Donald M. Dohler (D)
The Alien Factor '78
Fiend '83

Lou Doillon
Le Petit Amour '87

Lou Doillon (D)
La Puritaine '86

Katie Dolan
The Vampire Hookers '78

Michael Dolan
Hamburger Hill '87

Rainbow Dolan
In the Aftermath '87

Trent Dolan
A Stranger in My Forest '88

Thomas Dolby
Rockula '90
The Wall: Live in Berlin '90

Guy Doleman
The Ipcress File '65
Thunderball '65

Jim Dolen
The Virgin of Nuremberg '65

Mickey Dolenz
Head '68
Night of the Strangler '73

Birgit Doll
Nuclear Conspiracy '85

Dora Doll
Black and White in Color '76
Boomerang '76

Patrick Dollaghan
Circle of Fear '89
Murphy's Fault '88

Martin Dolman (D)
After the Fall of New York '85
American Tiger '89
Hands of Steel '86

Neige Dolsky
Tatie Danielle '91

Larry Domasin
Fun in Acapulco '63
Island of the Blue Dolphins '64

Arielle Dombasle
The Boss' Wife '86
Le Beau Mariage '82
Pauline at the Beach '83

Twisted Obsession '90

Andrea Domburg
Katie's Passion '75

Faith Domergue
The Atomic Man '56
Blood Legacy '73
The House of Seven Corpses '73
It Came from Beneath the Sea '55
The Sibling '72
This Island Earth '55
Voyage to the Prehistoric Planet '65

Placido Domingo
Les Contes D'Hoffman '84
Mario Lanza: The American Caruso '83
Otello '86
Placido: A Year in the Life of Placido Domingo '84
Tosca '85

Berta Dominguez
Maya '82

Arturo Dominici
Black Sunday '60
Hercules '58

Fats Domino
Fats Domino and Friends '88
Rock, Rock, Rock '56

Solveig Dommartin
Wings of Desire '88

Angelica Domrose
Girls Riot '88
The Scorpion Woman '89

Elinor Donahue
Freddy's Dead: The Final Nightmare '91
High School USA '84

Mitzi Donahue
Lethal Pursuit '88

Troy Donahue
Assault of the Party Nerds '89
The Chilling '89
Cockfighter '74
Cry-Baby '90
Cyclone '87
Deadly Diamonds '91
Deadly Prey '87
Dr. Alien '88
Godfather 1902-1959—The Complete Epic '81
The Godfather: Part 2 '74
Grandview U.S.A. '84
Hawkeye '88
Hollywood Cop '87
Imitation of Life '59
The Legend of Frank Woods '77
Love Thrill Murders '71
Nudity Required '90
Omega Cop '90
Palm Springs Weekend '63
Rome Adventure '62
Sexpot '88
Shock 'Em Dead '90
Sounds of Silence '91
South Seas Massacre
A Summer Place '59
Terminal Force '88
Those Fantastic Flying Fools '67
Tin Man '83

Axel Duborg

The Virgin Spring '59

Paulette Dubost

The Last Metro '80
The Sheep Has Five Legs '54

Paul Dubov

China Gate '57
Ma Barker's Killer Brood '60
Underworld, U.S.A. '60
Verboten! '59

Arthur Dubs (D)

Vanishing Wilderness '73

Cecile Ducasse

Chocolat '88

Kristie Ducati

The Bikini Car Wash
Company '92

Jean Duceppe

Mon Oncle Antoine '71

Michael Duchaussoy

The Killing Game '67
May Fools '90
Return of the Tall Blond Man
With One Black Shoe '74
Road to Ruin '91
This Man Must Die '70

Roger Duchesne

Bob le Flambeur '55

**Roger
Duchonwny** (D)

Camp Cucamonga: How I
Spent My Summer
Vacation '90

David Duchovny

Beethoven '92
Julia has Two Lovers '91
The Rapture '91
Red Shoe Diaries '92

Dortha Duckworth

Honeymoon Killers '70

Rick Ducommun

The 'Burbs '89
Little Monsters '89
No Small Affair '84

Louis Ducreux

The Double Life of Veronique
'91
A Sunday in the Country '84

John Joseph Duda

Prancer '89

Michael Dudikoff

American Ninja '85
American Ninja 2: The
Confrontation '87
American Ninja 4: The
Annihilation '91
Avenging Force '86
Bachelor Party '84
Black Marble '79
Bloody Birthday '80
The Human Shield '92
I Ought to Be in Pictures '82
Making Love '82
Platoon Leader '87
Radioactive Dreams '86
River of Death '90

Doris Dudley

The Moon and Sixpence '42

Lesley Dudley

John & Julie '77

Peter Duel

Generation '69

William Duell

Grace Quigley '84
Police Squad! Help Wanted!
'82

Eppy Dueno

Ye-Yo '70s

**Annemarie
Duerringer**

The Eternal Waltz '59

Howard Duff

Battered '78
The Double Negative '80
East of Eden '80
Flame of the Islands '55
Kramer vs. Kramer '79
Love on the Run '85
Magic on Love Island '80
Monster in the Closet '86
No Way Out '87
Oh, God! Book 2 '80
Private Hell 36 '54
Snatched '72
Snowblind '78
Spaceways '53
A Wedding '78
While the City Sleeps '56

Malcolm Duff

Last Mercenary '84

Peter Duffell (D)

Experience Preferred... But
Not Essential '83
The Far Pavilions '84
The House that Dripped Blood
'71
Inside Out '75
Letters to an Unknown Lover
'84

Kevin Duffis

The Expendables '89
Nam Angels '88

Dee Duffy

Hellcats '68

Julia Duffy

Children in the Crossfire '84
Night Warning '82

Patrick Duffy

Enola Gay: The Men, the
Mission, the Atomic Bomb
'80
Last of Mrs. Lincoln '76
The Man from Atlantis '77
Vamping '84

Jacques Dufilho

Black and White in Color '76
The Horse of Pride '80
Le Crabe Tambour '77

Dennis Dugan

Can't Buy Me Love '87
The Howling '81
The New Adventures of Pippi
Longstocking '88
Night Call Nurses '72

Night Moves '75
She's Having a Baby '88
Unidentified Flying Oddball '79

Dennis Dugan (D)

Brain Donors '92
Problem Child '90

Gerry Dugan

The Devil's Playground '76

John Dugan

Texas Chainsaw Massacre
'74

Michael Dugan (D)

Mausoleum '83
Super Seal '77

Tom Dugan

Medicine Man '30
Perfect Victims '87
Take Me Out to the Ball Game
'49
To Be or Not to Be '42

George Dugdale (D)

Slaughter High '86

Sandra Dugdale

Ruddigore '70s

Andrew Duggan

The Bears & I '74
The Commies are Coming, the
Commies are Coming '57
Doctor Detroit '83
Firehouse '72
Frankenstein Island '81
Housewife '72
The Hunted Lady '77
In Like Flint '67
The Incredible Mr. Limpet '64
It's Alive '74
It's Alive 2: It Lives Again '78
J. Edgar Hoover '87
Jigsaw '71
The Long Days of Summer
'80
Palm Springs Weekend '63
Pueblo Affair '73
Return to Salem's Lot '87
The Secret War of Harry Frigg
'68
Skin Game '71
The Streets of San Francisco
'72
Time Machine '78
Winds of War '83

Richard Duggan

Beach House '82

**Christopher
Duguay** (D)

Scanners 2: The New Order
'91

John Duigan (D)

Far East '85
One Night Stand '84
Romero '89
The Winter of Our Dreams '82
The Year My Voice Broke '87

Davor Dujmovic

Time of the Gypsies '90

John Dukakis

Delusion '84

Olympia Dukakis

Dad '89

Death Wish '74
Idolmaker '80
In the Spirit '90
Look Who's Talking '89
Look Who's Talking, Too '90
Moonstruck '87
Rich Kids '79
Steel Magnolias '89
Walls of Glass '85
Working Girl '88

Bill Duke

Bird on a Wire '90
Killing Floor '85
Predator '87

Bill Duke (D)

Johnnie Gibson F.B.I. '87
Maximum Security '87
A Rage in Harlem '91
A Raisin in the Sun '89

Daryl Duke (D)

Griffin and Phoenix: A Love
Story '76
Hang Tough '89
I Heard the Owl Call My Name
'73
Payday '72
The President's Plane is
Missing '71
The Silent Partner '78
Taipan '86
The Thorn Birds '83

Forrest Duke

Hollywood Strangler '82

Patty Duke

Amityville 4 '89
The Babysitter '80
Best Kept Secrets '88
Billie '65
By Design '82
Curse of the Black Widow '77
Family Upside Down '78
Fire '77
The 4D Man '59
The Goddess '58
Killer on Board '77
The Miracle Worker '62
The Miracle Worker '79
Mom, the Wolfman and Me
'80
My Sweet Charlie '70
September Gun '83
She Waits '71
Something Special '86
The Swarm '78

Robin Duke

Best of Eddie Murphy '89
The Last Polka '84

William Duke (D)

Killing Floor '85

David Dukes

Cat on a Hot Tin Roof '84
Catch the Heat '87
Date with an Angel '87
The First Deadly Sin '80
Held Hostage '91
The Josephine Baker Story
'90
The Men's Club '86
Miss All-American Beauty '82
Only When I Laugh '81
Portrait of a Rebel: Margaret
Sanger '80
Rawhead Rex '87
The Rutanga Tapes '90
See You in the Morning '89
Snow Kill '90

Strange Interlude '90
The Strawberry Statement '70
The Triangle Factory Fire
Scandal '79
War & Remembrance, Part 1
'88
War & Remembrance, Part 2
'89
Wild Party '74
Winds of War '83
Without a Trace '83

Germaine Dulac (D)

Smiling Madame Beudet/The
Seashell and the
Clergyman '28

Paul Dullac

La Marseillaise '37

Keir Dullea

Black Christmas '75
Black Water Gold '69
Blind Date '84
Brainwaves '82
David and Lisa '63
The Haunting of Julia '77
The Hostage Tower '80
Leopard in the Snow '78
Love Under Pressure '80s
Madame X '66
Mannikin '77
Next One '84
No Place to Hide '81
Paperback Hero '73
2001: A Space Odyssey '68
2010: The Year We Make
Contact '84
Welcome to Blood City '77

Sandra Dumas

Twice a Woman '79

Sandrine Dumas

The Double Life of Veronique
'91

Wade Dumas

Harlem Rides the Range '39

Douglas Dumbrille

Baby Face '33
Blonde Savage '47
Castle in the Desert '42
A Day at the Races '37
I Married an Angel '42
Jupiter's Darling '55
Mr. Deeds Goes to Town '36
The Mysterious Rider '38
Naughty Marietta '35
Ride 'Em Cowboy '42
Road to Utopia '46
Spook Busters '46
The Ten Commandments '56
Treasure Island '34
Under Nevada Skies '46
Virginia City '40

Margaret Dumont

Animal Crackers '30
At the Circus '39
Big Store '41
The Cocoanuts '29
A Day at the Races '37
Duck Soup '33
The Horn Blows at Midnight
'45
Never Give a Sucker an Even
Break '41
A Night at the Opera '35
Seven Days Ashore/Hurry,
Charlie, Hurry '44
Sing Your Worries Away '42
Three for Bedroom C '52

Thom Eberhardt (D)

Gross Anatomy '89
The Night Before '88
Night of the Comet '84
Sole Survivor '84
Without a Clue '88

Christine Ebersole

Acceptable Risks '86
Amadeus '84
Folks '92
Ghost Dad '90
Mac and Me '88
Thief of Hearts '84

David Eberts

Burning Secret '89

Ghasem Ebrahimian (D)

The Suitors '88

Buddy Ebsen

The Adventures of Tom Sawyer '73
The Andersonville Trial '70
Between Heaven and Hell '56
Born to Dance '36
Breakfast at Tiffany's '61
Broadway Melody of 1936 '35
Broadway Melody of 1938 '37
Captain January '36
Davy Crockett, King of the Wild Frontier '55
Davy Crockett and the River Pirates '56
Falling for the Stars '85
Girl of the Golden West '38
The Interns '62
One and Only, Genuine, Original Family Band '68
The President's Plane is Missing '71
Sing Your Worries Away '42
Stone Fox '87

Maude Eburne

Courageous Dr. Christian '40
Doughnuts & Society '36
Ruggles of Red Gap '35
The Vampire Bat '32

Aimee Eccles

Concrete Jungle '82
Group Marriage '72
Humanoid Defender '85
Little Big Man '70
Lovelines '84

Teddy Eccles

The Little Drummer Boy '68
My Side of the Mountain '69

Christopher Eccleston

Let Him Have It '91

Gisela Echevarria

In a Glass Cage '86

Guy Ecker

Devil Wears White '86
Night Terror '89

James Eckhouse

Blue Heaven '84

Billy Eckstine

Jo Jo Dancer, Your Life is Calling '86
Skirts Ahoy! '52

Barbara Eda-Young

Serpico '73

Helen Jerome Eddy

The Country Kid '23
County Fair '20
Rebecca of Sunnybrook Farm '17
A Shot in the Dark '35

Nelson Eddy

Bitter Sweet '41
The Chocolate Soldier '41
Dancing Lady '33
Girl of the Golden West '38
I Married an Angel '42
Maytime '37
Naughty Marietta '35
New Moon '40
Northwest Outpost '47
Phantom of the Opera '43
Rosalie '38
Rose Marie '36
Sweethearts '38

Uli Edel (D)

Christiane F. '82
Last Exit to Brooklyn '90

Gregg Edelman

Green Card '90

Herb Edelman

The Charge of the Model T's '76
Cracking Up '83
A Cry for Love '80
Goin' Coconuts '78
Marathon '80
On the Right Track '81
A Strange and Deadly Occurence '74
Strike Force '81
The Way We Were '73
The Yakuza '75

Barbara Eden

The Amazing Dobermans '76
Chattanooga Choo Choo '84
Five Weeks in a Balloon '62
Flaming Star '60
From the Terrace '60
Harper Valley P.T.A. '78
How to Break Up a Happy Divorce '76
Quick, Let's Get Married '71
Return of the Rebels '81
Seven Faces of Dr. Lao '63
The Stranger Within '74
Voyage to the Bottom of the Sea '61
The Woman Hunter '72
Wonderful World of the Brothers Grimm '62

Chana Eden

Holocaust Survivors... Remembrance of Love '83

Elana Eden

The Story of Ruth '60

Jerome Eden

2000 Maniacs '64

Mark Eden

Seance on a Wet Afternoon '64

Richard Eden

Liberty & Bash '90

Robert Edeson

On the Night Stage '15
Walking Back '26

George Edgely

Free, White, and 21 '62

Earle Edgerton

Carnival of Blood '80s

Gustaf Edgren (D)

Walpurgis Night '41

Walker Edmiston

The Beach Girls and the Monster '65

Jeff Edmond

Getting It On '83

Dale Edmonds

Paisan '46

Don Edmonds (D)

Bare Knuckles '77
Tender Loving Care '73
Terror on Tour '83

Elizabeth Edmonds

Experience Preferred... But Not Essential '83
Scrubbers '82

Louis Edmonds

The Best of Dark Shadows '60s

Ade Edmondson

The Young Ones '88

Adrian Edmondson

Supergrass '87

Lada Edmund

Savage! '73

Marie Edmund

Assault of the Rebel Girls '59

Don Edmunds (D)

Ilsa, She-Wolf of the SS '74

Beatie Edney

Diary of a Mad Old Man '88
Mister Johnson '91

Nekohachi Edoya

The Funeral '84

Richard Edson

Do the Right Thing '89
Eight Men Out '88
Good Morning, Vietnam '87
Let It Ride '89
Platoon '86
Stranger than Paradise '84
Tougher than Leather '88

Allan Edwall

Brothers Lionheart '85
Fanny and Alexander '83
The Sacrifice '86

Anthony Edwards

Downtown '89
El Diablo '90
Going for the Gold: The Bill Johnson Story '85
Gotcha! '85
Hawks '89
Heart Like a Wheel '83
How I Got Into College '89

Miracle Mile '89
Mr. North '88
Revenge of the Nerds '84
Revenge of the Nerds 2: Nerds in Paradise '87
Summer Heat '87
The Sure Thing '85
Top Gun '86

Barbara Edwards

Malibu Express '85

Bill Edwards

First Man into Space '59
Ladies of the Chorus '49
War Lover '62

Blake Edwards

Strangler of the Swamp '46

Blake Edwards (D)

Blind Date '87
Breakfast at Tiffany's '61
The Cary Grant Collection '90
Curse of the Pink Panther '83
Days of Wine and Roses '62
Experiment in Terror '62
A Fine Mess '86
Great Race '65
The Man Who Loved Women '83
Micki & Maude '84
Operation Petticoat '59
The Party '68
Perfect Furlough '59
Peter Gunn '89
The Pink Panther '64
The Pink Panther Strikes Again '76
Return of the Pink Panther '74
Revenge of the Pink Panther '78
A Shot in the Dark '64
Skin Deep '89
S.O.B. '81
Sunset '88
Switch '91
The Tamarind Seed '74
10 '79
That's Life! '86
This Happy Feeling '58
Trail of the Pink Panther '82
Victor/Victoria '82
Wild Rovers '71

Bruce Edwards

The Black Widow '47
The Fallen Sparrow '43
Hitler: Dead or Alive '43
The Iron Major '43

Cassandra Edwards

Vasectomy: A Delicate Matter '86

Cliff Edwards

Dance Fools Dance '31
Dumbo '41
Girl of the Golden West '38
His Girl Friday '40
Parlor, Bedroom and Bath '31
Pinocchio '40
The Sin of Madelon Claudet '31

Don Edwards (D)

Black Cobra 3: The Manila Connection '90

Elizabeth Edwards

The Pink Chiquitas '86

Ella Edwards

Sweet Sugar '72

Gail Edwards

Get Crazy '83

Gene Edwards

Grizzly Adams: The Legend Continues '90

George Edwards (D)

The Attic '80

Gerald Edwards

The Fat Albert Christmas Special '77
The Fat Albert Easter Special '82
The Fat Albert Halloween Special '77

Harry Edwards

Baby Face Harry Langdon '26

Harry Edwards (D)

Baby Face Harry Langdon '26

Henry Edwards

The Golden Salamander '51

Henry Edwards (D)

Scrooge '35

Hugh Edwards

Lord of the Flies '63

James Edwards

Home of the Brave '49
Patton '70
Pork Chop Hill '59
The Set-Up '49
The Steel Helmet '51

Jennifer Edwards

All's Fair '89
Heidi '67
The Man Who Loved Women '83
Overexposed '90
Perfect Match '88
Peter Gunn '89
S.O.B. '81
Sunset '88
That's Life! '86

Jimmy Edwards

Innocents in Paris '53

Judy Edwards

The Mermaids of Tiburon '62

Kyle Edwards

Death Weekend '76

Lance Edwards

Peacemaker '90
A Woman, Her Men and Her Futon '92

Luke Edwards

The Wizard '89

Mark Edwards

Boldest Job in the West '70s

Meredith Edwards

The Electronic Monster '57

Patty Edwards

Torture Train '83

Paul Edwards

Combat Killers '80

Elliott ▸cont.

Beverly Hills Cop '84
Can You Hear the Laughter?
  The Story of Freddie Prinze
  '79
Cutter's Way '81
Days of Hell '70s
Jacqueline Bouvier Kennedy
  '81
Mrs. R's Daughter '79
Perry Mason: The Case of the
  Lost Love '87
Report to the Commissioner
  '74
Roadhouse 66 '84
Sergeant Matlovich vs. the
  U.S. Air Force '78
Taking Care of Business '90

Wild Bill Elliott
Western Double Feature 1:
  Wild Bill & Sunset Carson
  '40s
Western Double Feature 7:
  Wild Bill & Crabbe '45
Western Double Feature 12
  '44

William Elliott
Hellfire '48

Bob Ellis
Man of Flowers '84

Bob Ellis (D)
Warm Nights on a Slow-
  Moving Train '87

Brent Ellis
La Traviata '88

Chuck Ellis
Blood Cult '85

Desmond Walter Ellis
Great St. Trinian's Train
  Robbery '66

Diane Ellis
The Leatherneck '28

Edward Ellis
The Return of Peter Grimm
  '35
The Thin Man '34
Wheel of Fortune '41

George Ellis
Demon Hunter '88

James Ellis
No Surrender '86

Laura Ellis
Bloodsuckers from Outer
  Space '83

Mary Ellis
The Three Worlds of Gulliver
  '59

Patricia Ellis
Back Door to Heaven '39
Block-heads '38
The Case of the Lucky Legs
  '35
Postal Inspector '36
Three on a Match '32

Paul Ellis
Camp Double Feature '39

Robert Ellis
Captured in Chinatown '35
From Broadway to Cheyenne
  '32
White Eagle '32

Robin Ellis
Curse of King Tut's Tomb '80
The Europeans '79

Art Ellison
Carnival of Souls '62

Christopher Ellison
Buster '88

Gwen Ellison
Luggage of the Gods '87

James Ellison
Fifth Avenue Girl '39
Guns of Justice '50
I Walked with a Zombie '43
Kentucky Jubilee '51
Last Bullet '50
Last of the Wild Horses '49
Marshal of Heldorado '50
Next Time I Marry '38
Outlaw Fury '50
Rangeland Empire '50
Sudden Death '50
23 1/2 Hours Leave '37
The Undying Monster '42
Vivacious Lady '38
You Can't Fool Your Wife '40

Joseph Ellison (D)
Don't Go in the House '80

Max Elloy
Atoll K '51
Utopia '52

Maura Ellyn
Home Free All '84

Rebekah Elmaloglou
In Too Deep '90

Raymond Elmendorf
Bloody Wednesday '87

Iron "Amp" Elmore
Iron Thunder '89

Nicoletta Elmi
Demons '86

Javier Elorrieta (D)
Blood and Sand '89

Michael Elphick
Arthur's Hallowed Ground '84
The Element of Crime '85
Let Him Have It '91
Little Dorrit, Film 1: Nobody's
  Fault '88
Little Dorrit, Film 2: Little
  Dorrit's Story '88
Ordeal by Innocence '84

Tom Elpin
Sunset Strip '85

Hannelore Elsner
Time Troopers '89

Isobel Elsom
The Ghost and Mrs. Muir '47
Love is a Many-Splendored
  Thing '55
Love from a Stranger '47
Monsieur Verdoux '47

Timothy Elston
Bellamy '80

Paul Eluard
L'Age D'Or '30

Maurice Elvey (D)
Beware of Pity '46
The Evil Mind '34
Lily of Killarney '34
The Obsessed '51
Phantom Fiend '35
Transatlantic Tunnel '35

June Elvidge
A Girl's Folly '17

Cary Elwes
Another Country '84
Days of Thunder '90
Glory '89
Hot Shots! '91
Lady Jane '85
The Princess Bride '87

Roger Elwin
Lord of the Flies '63

Dennis Ely
Deadly Sunday '82

John Ely
Slaughter Hotel '71

Ron Ely
Cry of the Black Wolves '70s
Doc Savage '75
Once Before I Die '65
Slavers '77

Sallee Elyse
Demented '80
Home Sweet Home '80

Ruby Elzy
Emperor Jones '33

Kelly Emberg
Portfolio '88

Heidi Emerich
Tainted Image '91

Faye Emerson
Destination Tokyo '43

John Emerson (D)
The Americano '17
The Americano/Variety '25
Down to Earth '17
His Picture in the Papers '16
Mystery of the Leaping Fish/
  Chess Fever '25
Reaching For the Moon '17
The Social Secretary '16

Jonathon Emerson
84 Charlie Mopic '89

Alain Emery
White Mane '52

Dick Emery
The Case of the Mukkinese
  Battle Horn '56

Dick Emery (D)
Yellow Submarine '68

Gilbert Emery
The Royal Bed '31

James H. Emery
Phoenix the Warrior '88

Jesse Emery
Chillers '88

John Emery
Forever Darling '56
Kronos '57
Rocketship X-M '50
The Spanish Main '45

Katherine Emery
The Private Affairs of Bel Ami
  '47

Ralph Emery
The Girl from Tobacco Row
  '66

Robert Emery (D)
Ride in a Pink Car '74
Scream Bloody Murder '72

Ian Emes (D)
Knights & Emeralds '87

David Emgee
Dawn of the Dead '78

Robert Emhardt
Die Sister, Die! '74
Noah: The Deluge '79
Shame '61
Underworld, U.S.A. '60

Michael Emil
Always '85
Can She Bake a Cherry Pie?
  '83
Insignificance '85
New Year's Day '89
Sitting Ducks '80
Someone to Love '87

Daniel Emilfork
The Devil's Nightmare '71

Mari Emlyn
Coming Up Roses '87

Alphonsia Emmanuel
Under Suspicion '91

Takis Emmanuel
Kostas '79

Colette Emmanuelle
The Devil's Nightmare '71

Roland Emmerich (D)
Ghost Chase '88
Making Contact '86
Moon 44 '90

Michael Emmet
Night of the Blood Beast '58

Robert Emmett (D)
Way of the West '35

Fred Emney
Naughty Knights '71

Maria Emo
Hitler '62

Richard Emory
Mask of the Dragon '51

Kyoko Enami
Gamera vs. Barugon '66

Alicia Encinas
The Bees '78

Robert Enders (D)
Stevie '78

Cy Endfield (D)
Mysterious Island '61
Try and Get Me '50
The Underworld Story '50
Zulu '64

Lena Endre
The Visitors '89

Erich Engel (D)
The Blum Affair '48

Georgia Engel
The Care Bears Movie '84
The Day the Women Got Even
  '80

Thomas E. Engel (D)
Rich Little: One's a Crowd '88

Tina Engel
Boat Is Full '81

Tobias Engel
'38: Vienna Before the Fall '88

David Engelbach (D)
America 3000 '86

Jim Engelhardt
Comic Book Kids '82

Roy Engelman
Children Shouldn't Play with
  Dead Things '72

Heinz Engelmann
More '69

Wera Engels
Hong Kong Nights '35

Randall England
Dream Machine '91

Harrison Engle (D)
Indomitable Teddy Roosevelt
  '83

Alex English
Amazing Grace & Chuck '87

Cameron English
A Chorus Line '85

Doug English
Big Bad John '90

Kenny Everett

The Best of the Kenny Everett Video Show '81
Bloodbath at the House of Death '85

Rupert Everett

Another Country '84
The Comfort of Strangers '91
Dance with a Stranger '85
Hearts of Fire '87
Livin' the Life '84
The Right Hand Man '87

Tom Everett

Best of the Best '89
Friday the 13th, Part 4: The Final Chapter '84
Prison '88

Nancy Everhard

The China Lake Murders '90
Deepstar Six '89
Demonstone '89
Double Revenge '89
Trial of the Incredible Hulk '89

Rex Everhart

Family Business '89

Trish Everly

Madhouse '87

Herb Evers

The Brain That Wouldn't Die '63

Jason Evers

Barracuda '78
Dasket Case 2 '90
Claws '77
Fer-De-Lance '74

Cory Everson

Double Impact '91

John L. Eves

Chains '89

Greg Evigan

Deepstar Six '89
Private Road: No Trespassing '87
Shades of Love: Echoes in Crimson '87
Stripped to Kill '87

Pat Evison

The Clinic '83
Starstruck '82
Tim '79

John Ewart

Island Trader '71
Newsfront '78
Prince and the Great Race '83
The Quest '86
Which Way Home '90

Tom Ewell

Adam's Rib '49
Easy Money '83
The Girl Can't Help It '56
The Seven Year Itch '55
State Fair '62
Tropic of Desire '80s

Bill Ewens

Second Time Lucky '84

Barbara Ewing

Guardian of the Abyss '82

Stephane Excoffier

The Pointsman '88

Valie Export (D)

Invisible Adversaries '77
The Practice of Love '84

Richard Eyer

Desperate Hours '55
The Seventh Voyage of Sinbad '58

Frank Eyman

Riot '69

Peter Eyre

Diamond's Edge '88
Hedda '75

Richard Eyre (D)

Loose Connections '87
The Ploughman's Lunch '83
Singleton's Pluck '84

John Eyres (D)

Project: Shadowchaser '92

William Eythe

Colonel Effingham's Raid '45
The Ox-Bow Incident '43
A Wing and a Prayer '44

Bessie Eyton

The Spoilers '14

Maynard Eziashi

Mister Johnson '91
Twenty-One '91

Ka Sa Fa

Duel of the Seven Tigers '80

Fa'amgase

Moana, a Romance of the Golden Age '26

Shelley Fabares

Brian's Song '71
Clambake '67
Girl Happy '65
Great American Traffic Jam '80
Love or Money? '88
Sky Hei$t '75
Spinout '66

Jacques Fabbri

Diva '82

Christian Faber (D)

Bail Jumper '89

Julliette Faber

Passion for Life '48

Peter Faber

Army Brats '84

Fabian

Dear Brigitte '65
Five Weeks in a Balloon '62
The Longest Day '62
Mr. Hobbs Takes a Vacation '62
North to Alaska '60
Soul Hustler '76

Ava Fabian

Ski School '91
To Die for '89
Welcome Home, Roxy Carmichael '90

Francoise Fabian

Dressmaker '56
Fernandel the Dressmaker '57
The French Woman '79
Happy New Year '74
Love by Appointment '76
My Night at Maud's '69
Salut l'Artiste '74
That Naughty Girl '58

Janice Fabian

Invasion Earth: The Aliens Are Here! '87

Joel Fabiani

President's Mistress '78
Reuben, Reuben '83

Andria Fabio

Stryker '83

Luce Fabiole

Two of Us '68

Nanette Fabray

Alice Through the Looking Glass '66
Amy '81
Band Wagon '53
The Happy Ending '69
Harper Valley P.T.A. '78
The Man in the Santa Claus Suit '79
Personal Exemptions '88
The Private Lives of Elizabeth & Essex '39

Dominique Fabre

Hot Line '69

Pierre Fabre

The 317th Platoon '65

Manolo Fabregas

Captain Scarlett '53

Aldo Fabrizi

Open City '45

Franco Fabrizi

Besame '87
Ginger & Fred '86
I Vitelloni '53
The Swindle '55
A Woman of Rome '56

Valeria Fabrizi

Beauty on the Beach '60s
Women in Cell Block 7 '77

Giannina Facio

Delta Force Commando 2 '90

Robert Factor

Fear '88
Ninja Academy '90

Louise Fadenza

The Night Club '25

Roberto Faenza (D)

Corrupt '84

Suzanna Fagan

Grad Night '81

Jeff Fahey

Backfire '88
Body Parts '91
Execution of Raymond Graham '85
Impulse '90

Iran: Days of Crisis '91
Iron Maze '91
The Last of the Finest '90
The Lawnmower Man '92
Psycho 3 '86
Silverado '85
Split Decisions '88
True Blood '89
White Hunter, Black Heart '90

Mary-Anne Fahey

Celia: Child Of Terror '89
Dunera Boys '85
Mutant Hunt '87

Myrna Fahey

The Fall of the House of Usher '60

Zulma Faiad

La Casa del Amor '87
La Flor de la Mafia
Night of a Thousand Cats '72

J.W. Fails

No Retreat, No Surrender '86

Peter Faiman (D)

Crocodile Dundee '86

Matty Fain

The Thirteenth Man '37

Samson Fainsilber

Charles et Lucie '79

Elinor Fair

Let 'er Go Gallegher '28
Road to Yesterday/The Yankee Clipper '27
Yankee Clipper '27

Jody Fair

The Brain Eaters '58
High School Confidential '58

Bruce Fairbairn

Cyclone '87
Nightstick '87
The Vampire Hookers '78

Christopher Fairbank

How Many Miles to Babylon? '82

Douglas Fairbanks, Sr.

American Aristocracy '17
The Americano '17
The Americano/Variety '25
Birth of a Legend '84
The Black Pirate '26
Camp Classics #1 '50s
Days of Thrills and Laughter '61
Don Q., Son of Zorro '25
Down to Earth '17
Flirting with Fate '16
His Picture in the Papers '16
The Iron Mask '29
A Kiss for Mary Pickford '27
The Lamb '15
Mark of Zorro '20
Matrimaniac '16
Mr. Robinson Crusoe '32
Mystery of the Leaping Fish/Chess Fever '25
Private Life of Don Juan '34
Reaching for the Moon '17
Reaching for the Moon '31
Reggie Mixes In '16
Robin Hood '22

Taming of the Shrew '29
Thief of Baghdad '24
The Three Musketeers '21
When the Clouds Roll By '19

Douglas Fairbanks, Jr.

Angels Over Broadway '40
Corsican Brothers '41
Ghost Story '81
Great Kate: The Films of Katharine Hepburn
Gunga Din '39
Having Wonderful Time '38
Having Wonderful Time/Carnival Boat '38
Hollywood Uncensored '87
The Hostage Tower '80
Joy of Living '38
Little Caesar '30
Mimi '35
Morning Glory '33
Party Girls '29
Prisoner of Zenda '37
The Rage of Paris '38
Rise of Catherine the Great '34
The Sacred Music of Duke Ellington '82
Sinbad the Sailor '47
When Thief Meets Thief '37
A Woman of Affairs '28
The Young in Heart '38

Gladys Fairbanks

A Poor Little Rich Girl '17

Ann Fairchild

Death Wish Club '83

Max Fairchild

Howling 3 '87

Morgan Fairchild

Campus Man '87
Cannon Movie Tales: Sleeping Beauty '89
The Concrete Cowboys '79
Deadly Illusion '87
Even Angels Fall '90
Haunting of Sarah Hardy '89
Honeyboy '82
Initiation of Sarah '78
Midnight Cop '88
Mob Boss '90
Phantom of the Mall: Eric's Revenge '89
Ramblin' Man '80s
Red Headed Stranger '87
The Seduction '82
The Zany Adventures of Robin Hood '84

William Fairchild (D)

Horsemasters '61
The Silent Enemy '58

Virginia Brown Faire

Cricket on the Hearth '23

Deborah Fairfax

Frightmare '74

Ferdinand Fairfax (D)

Nate and Hayes '83
The Rescue '88
Spymaker: The Secret Life of Ian Fleming '90

Fei ▶cont.

Mission Kiss and Kill
Return of the Chinese Boxer '74
Shaolin Temple Strikes Back '72
Shaolin Traitor '82
Snake in the Eagle's Shadow 2 '83

Meng Fei

The Dragon Returns
Prodigal Boxer: The Kick of Death '73
The Unbeaten 28 '80s
Venus the Ninja '80s

Xie Fei (D)

Girl from Hunan '86

Paul Feig

Paramount Comedy Theater, Vol. 2: Decent Exposures '87

Seita Kathleen Feigny

Murder Rap '87

Beda Ocampo Feijoo (D)

Under the Earth '86

Esther Feild

Eli Eli '40

Alan Feinstein

Bunco '83
The Hunt for the Night Stalker '91
The Hunted Lady '77
Joe Panther '76
The Two Worlds of Jenny Logan '79

Felix Feist (D)

Big Trees '52
Deluge '33
Devil Thumbs a Ride '47
Devil Thumbs a Ride/Having Wonderful Crime '47
Donovan's Brain '53
George White's Scandals '45
Guilty of Treason '50
Threat '49

Frances Feist

Carnival of Souls '62

Raphael Fejto

Au Revoir Les Enfants '88

Fritz Feld

The Affairs of Annabel '38
At the Circus '39
Bringing Up Baby '38
Errand Boy '61
Herbie Goes Bananas '80
History of the World: Part 1 '81
Kentucky Jubilee '51
The Lovable Cheat '49
Mexican Hayride '48
Promises! Promises! '63
World's Greatest Lover '77

Clarence Felder

Killing Floor '85

Corey Feldman

The 'Burbs '89
Dream a Little Dream '89
Edge of Honor '91
Friday the 13th, Part 4: The Final Chapter '84
Friday the 13th, Part 5: A New Beginning '85
The Goonies '85
Gremlins '84
License to Drive '88
The Lost Boys '87
Meatballs 4 '92
Rock 'n' Roll High School Forever '91
Round Trip to Heaven '92
Stand By Me '86
Teenage Mutant Ninja Turtles: The Movie '90

Dennis Feldman (D)

Real Men '87

Gene Feldman (D)

Danny '79

John Feldman (D)

Alligator Eyes '90

Marty Feldman

The Adventures of Sherlock Holmes' Smarter Brother '78
The Last Remake of Beau Geste '77
Sex with a Smile '76
Silent Movie '76
Slapstick of Another Kind '84
Think Dirty '70
To See Such Fun '81
Yellowbeard '83
Young Frankenstein '74

Marty Feldman (D)

The Last Remake of Beau Geste '77

Tibor Feldman

Fat Guy Goes Nutzoid '86

Barbara Feldon

Children of Divorce '80
A Guide for the Married Woman '78
No Deposit, No Return '76
Playmates '72
Smile '75

Tovah Feldshuh

Amazing Howard Hughes '77
Blue Iguana '88
Brewster's Millions '85
Cheaper to Keep Her '80
Holocaust '78
Idolmaker '80
The Triangle Factory Fire Scandal '79

Mario Feliciani

Revolt of the Barbarians '64

Jose Feliciano

Aaron Loves Angela '75
La Bamba Party '80s

Emeterio Y. Felipes

Amenaza Nuclear

Enrique Alvarez Felix

Los Caifanes '87
Narda or The Summer '76

Maria Felix

Camelia
El Penon de las Animas '47
French Can-Can '55
La Diosa Arrodillada (The Kneeling Goddess) '47
La Generala '87

Barbara Felker

King of the Stallions '42

Norman Fell

The Boatniks '70
The Bone Yard '90
Bullitt '68
Charley Varrick '73
Cleopatra Jones & the Casino of Gold '75
Death Stalk '74
The End '78
For the Love of It '80
The Graduate '67
The Killers '64
On the Right Track '81
Paternity '81
Pork Chop Hill '59
The Stone Killer '73
Stripped to Kill '87
Transylvania 6-5000 '85

Catherine Feller

The Curse of the Werewolf '61

Federico Fellini

Alex in Wonderland '71
Bellissimo: Images of the Italian Cinema '87
Ciao Federico! Fellini Directs Satyricon '69
The Flowers of St. Francis '50
The Miracle '48
We All Loved Each Other So Much '77

Federico Fellini (D)

Amarcord '74
And the Ship Sails On '84
Boccaccio '70 '62
City of Women '81
The Clowns '71
8 1/2 '63
Fellini Satyricon '69
Ginger & Fred '86
I Vitelloni '53
Juliet of the Spirits '65
La Dolce Vita '60
La Strada '54
Love in the City '53
Nights of Cabiria '57
Spirits of the Dead '68
The Swindle '55
Variety Lights '51
The White Sheik '52

Julian Fellowes

Baby...Secret of the Lost Legend '85

Edith Fellows

Heart of the Rio Grande '42
Music in My Heart '40

Hansjorg Felmy

The Monster of London City '64

Verna Felton

Dumbo '41
The Jungle Book '67
The Oklahoman '56
Picnic '55

John Femia

Almost Royal Family '84

Georg Fenady (D)

Arnold '73
Terror in the Wax Museum '73

Freddy Fender

La Pastorela '92
The Milagro Beanfield War '88
She Came to the Valley '77

Edwige Fenech

Blade of the Ripper '84
Phantom of Death '87
Sex with a Smile '76

Hsu Feng

The Great Hunter '75

Huang Feng (D)

Sting of the Dragon Masters '74

Ku Feng

Crack Shadow Boxers '70s
Incredible Master Beggars '82
Kung Fu Avengers '85
Shaolin Disciple

Wang Hsiao Feng

Kung Fu Vampire Buster

Tanya Fenmore

Lisa '90
My Stepmother Is an Alien '88

Sherilyn Fenn

Backstreet Dreams '90
Crime Zone '88
Desire and Hell at Sunset Motel '92
Diary of a Hitman '91
Just One of the Guys '85
Meridian: Kiss of the Beast '90
Ruby '92
True Blood '89
Twin Peaks '90
Two Moon Junction '88
The Wild Life '84
The Wraith '87
Zombie High '87

Sylvie Fennec

The Music Teacher '88

George Fenneman

How to Succeed in Business without Really Trying '67

Frank Fenton

The Golden Stallion '49
Isle of Forgotten Sins '43
Mexican Hayride '48
Naked Hills '56

Lance Fenton

Heathers '89
Night of the Demons '88

Leslie Fenton

F.P. 1 '33
The House of Secrets '37
Marie Galante '34
Murder at Midnight '31
Public Enemy '31

Colm Feore

Blades of Courage '88
Personals '90

Scott Feraco

Nasty Hero '89

Maurice Feraudy

Crainquebille '23

Pamelyn Ferdin

Sealab 2020 '72
The Toolbox Murders '78

Tawny Fere

Angel 3: The Final Chapter '88
Night Children '89
Rockula '90

Rene Feret (D)

Mystery of Alexina '86

Al Ferguson

Laramie Kid '35
Roamin' Wild '36
Rose Marie '54
Tarzan the Tiger '29

Andrew Ferguson

Miracle Down Under '87

Casson Ferguson

Cobra '25

J. Don Ferguson

Loveless '83
Running Mates '86

Jane Ferguson

Mystery Mansion '83

Kathleen Ferguson

Nightmare '82

Marlow Ferguson

American Tickler '76

Michael Ferguson (D)

Doctor Who: The Seeds of Death '85
Glory Boys '84

Karen Fergusson

An Angel at My Table '89

John Ferita

Video Murders '87

Lawrence Ferlinghetti

Kerouac '84

Maria Fernanda

Hijazo De Mi Vidaza (Son of My Life) '70s

Fernandel

Angele '34
Around the World in 80 Days '56
Dressmaker '56
Fernandel the Dressmaker '57
The French Touch '54
Harvest '37
Invitation to Paris '60
Little World of Don Camillo '51
Man in the Raincoat '57
Nais '45
Pantaloons '57
Paris Holiday '57
The Sheep Has Five Legs '54
Topaze '51
Well-Digger's Daughter '46

Abel Fernandez
Rose Marie '54

Angel Fernandez
Street Warriors, Part 2 '87

Arturo Fernandez
Sound of Horror '64

Emilio Fernandez
Bring Me the Head of Alfredo
Garcia '74
Duelo en el Dorado

Emilio Fernandez (D)
Flor Silvestre (Wild Flower) '58
La Choca '73
Maria Candelaria '46
The Pearl '48

Esther Fernandez
Pancho Villa Returns '50

Evelina Fernandez
American Me '92

Fernando Fernandez
El Sinaloense '84

Jaime Fernandez
El Forastero Vengador '65
Emiliano Zapata '87
Los Hermanos Centella '65
Samson vs. the Vampire
Women '61

Jesus Fernandez
Nazarin '58

Juan Fernandez
The Amazing Transplant '70
Cat Chaser '90
Kinjite: Forbidden Subjects '89
Liquid Dreams '92

Lucia Gil Fernandez
La Ley del Revolver (The Law of the Gun) '78

Maribel Fernandez
Bus Station '88

Miguel Fernandez
The Kidnapping of the President '80
Sorry, Wrong Number '89
True Believer '89

Ricardo Fernandez
Y Donde Esta el Muerto? (And... Where Is the Dead) '87

Vicente Fernandez
Tacos al Carbon '87

Wilhelmenia Wiggins Fernandez
Diva '82

Rachid Ferrache
My Other Husband '85

Catherine Ferran
Baxter '89

Sabrina Ferrand
Sinners '89

Tony Ferrandis
Heat of the Flame '82

Tim Ferrante
A Taste for Flesh and Blood '90

Abel Ferrara (D)
Cat Chaser '90
China Girl '87
Crime Story '86
Driller Killer '74
Fear City '85
The Gladiator '86
King of New York '90
Ms. 45 '81

Juan Ferrara
La Hija del General '84

Romano Ferrara (D)
Planets Against Us '61

Ashley Ferrare
Cyclone '87

Christina Ferrare
Mary, Mary, Bloody Mary '76

Antonio Ferrari
The Tree of Wooden Clogs '78

Doug Ferrari
Paramount Comedy Theater, Vol. 2: Decent Exposures '87

Marco Ferrari (D)
Tales of Ordinary Madness '83

Nick Ferrari
Baby It's You '82

Sarah Ferrati
The Witch '66

Rebecca Ferratti
Cheerleader Camp '88
Gor '88
Outlaw of Gor '87
Playboy's 1989 Playmate Video Calendar '89

Francis Ferre
Bridge to Hell '87

Andrea Ferreal
Despair '79

Jamie Ferreira
Danger Zone '87

Taumaturgo Ferreira
Subway to the Stars '87

Conchata Ferrell
Deadly Hero '75
Eye on the Sparrow '91
Heartland '81
Lost Legacy: A Girl Called Hatter Fox '77
Portrait of a White Marriage '88
Rape and Marriage: The Rideout Case '80
To Love Again '80

Where the River Runs Black '86

Jeff Ferrell (D)
Revenge of the Teenage Vixens from Outer Space '86

Tyra Ferrell
White Men Can't Jump '92

Andrea Ferreol
Control '87
The Last Metro '80
Stroke of Midnight '90
Three Brothers '80
A Zed & Two Noughts '88

Jose Ferrer
The Art of Crime '75
The Being '83
Berlin Tunnel 21 '81
The Big Brawl '80
Big Bus '76
Blood Tide '82
Bloody Birthday '80
The Caine Mutiny '54
Christopher Columbus '91
Cyrano de Bergerac '50
Deep in My Heart '54
Dune '84
Enter Laughing '67
The Evil That Men Do '84
The Fifth Musketeer '79
Gideon's Trumpet '80
The Greatest Story Ever Told '65
Hired to Kill '91
The Horror of It All '91
Ingrid '89
Joan of Arc '48
Lawrence of Arabia '62
The Little Drummer Boy '68
A Midsummer Night's Sex Comedy '82
Miss Sadie Thompson '54
Moulin Rouge '52
Natural Enemies '79
Old Explorers '90
Order to Kill '73
Paco '75
Pleasure Palace '80
The Private Files of J. Edgar Hoover '77
Samson and Delilah '84
Seduced '85
The Sentinel '76
Ship of Fools '65
Strange Interlude '90
The Swarm '78
To Be or Not to Be '83
Voyage of the Damned '76
Who Has Seen the Wind? '77
Zoltan...Hound of Dracula '77

Jose Ferrer (D)
Return to Peyton Place '61
State Fair '62

Lupita Ferrer
Balboa '82
Children of Sanchez '79

Mel Ferrer
Blood and Roses '61
Born to Be Bad '50
Brannigan '75
City of the Walking Dead '83
Deadly Game '83
Eaten Alive '76
Elena and Her Men '56
Emerald Jungle '80
The Fall of the Roman Empire '64
The Fifth Floor '80

The Great Alligator '81
The Hands of Orlac '60
Hi-Riders '77
Knights of the Round Table '54
Lili '53
The Longest Day '62
Norseman '78
One Shoe Makes It Murder '82
Outrage '86
Paris When It Sizzles '64
Rancho Notorious '52
Scaramouche '52
Sex and the Single Girl '64
The Tempter '74
The Visitor '80
War and Peace '56

Miguel Ferrer
Deepstar Six '89
Flashpoint '84
Lovelines '84
Python Wolf '88
Revenge '90
RoboCop '87
Star Trek 3: The Search for Spock '84

Tony Ferrer
Blind Rage '78

Marco Ferreri
Bellissimo: Images of the Italian Cinema '87

Marco Ferreri (D)
El Cochecito '60
L'Udienza '71

Martin Ferrero
High Spirits '88
Planes, Trains & Automobiles '87

Eve Ferret
Foreign Body '86

Diana Ferreti
La Fuga de Caro
La Hora 24 '87

Robert A. Ferretti (D)
Fear '88

Benoit Ferreux
Murmur of the Heart '71

Alessandra Ferri
Dancers '87

Mark Alan Ferri
Getting It On '83

Peter Ferri
Fly with the Hawk

Babette Ferrier
Chloe in the Afternoon '72

Noel Ferrier
Adventures of Eliza Fraser '76
Return of Captain Invincible '83

Carla Ferrigno
Black Roses '88
Seven Magnificent Gladiators '84

Lou Ferrigno
All's Fair '89

The Cage '89
Death of the Incredible Hulk '90
Desert Warrior '88
Hercules '83
Hercules 2 '85
The Incredible Hulk '77
The Incredible Hulk Returns '88
Liberty & Bash '90
Seven Magnificent Gladiators '84
Sinbad of the Seven Seas '89
Trial of the Incredible Hulk '89

Frank Ferrin (D)
Sabaka '55

Barbara Ferris
A Chorus of Disapproval '89
Nice Girl Like Me '69

Irena Ferris
Covergirl '83

John Ferris
Monique '83
Torpedo Attack '72

Loraine Ferris
Cycle Vixens '79

Stan Ferris (D)
Tulips '81

Turi Ferro
Malicious '74

George Ferron (D)
The Scalawag Bunch '75

Giorgio Ferroni (D)
The Bacchantes '63
Conquest of Mycene '63
Mill of the Stone Women '60

Frank Ferrucci
The Hitman '91

Ferrusquilla
Mariachi '65

David Ferry
Men of Steel '88
Tales of the Klondike: Race for Number One '81

Gabriel Ferzetti
Angel in a Taxi '59
Appassionata '74
Basileus Quartet '82
Counter Punch '71
Julia and Julia '87
L'Avventura '60
The Night Porter '74
On Her Majesty's Secret Service '69
The Psychic '78

Stepin Fetchit
Amazing Grace '74
Judge Priest '34
Stand Up and Cheer '34
The Sun Shines Bright '53
Wild Horse '31

Darrell Fetty
Stunts '77

Debra Feuer
Homeboy '88
Night Angel '90

Travis Fine
Child's Play 3 '91

John P. Finegan (D)
Girl School Screamers '86

B.P. Fineman (D)
Beauty for the Asking '39

Anthony Finetti
The Expendables '89

Elizabeth Fink
Alice in Wonderland '82

Janace Fink (D)
Velvet Smooth '76

Michael Fink (D)
Black Force '75

F.G. Finkbinder
The Redeemer '76

T.K. Finkbinder
Class Reunion Massacre '77

Abe Finkel
Sergeant York '41

Ken Finkleman
Airplane 2: The Sequel '82

Ken Finkleman (D)
Airplane 2: The Sequel '82
Head Office '86

Frank Finlay
Assault '70
Bouquet of Barbed Wire
Candide '76
Cromwell '70
Cthulhu Mansion '91
Death of Adolf Hitler '84
Enigma '82
The Four Musketeers '75
Gumshoe '72
In the Secret State
Lifeforce '85
Molly Maguires '69
Murder by Decree '79
The Ploughman's Lunch '83
Return of the Soldier '84
Robbery '67
A Study in Terror '66
Thief of Baghdad '78
Three Musketeers '74

James Finlayson
All Over Town '37
Block-heads '38
Bohemian Girl '36
Bonnie Scotland '35
A Chump at Oxford '40
The Devil's Brother '33
Eyes of Turpin Are Upon You!
'21
Fatty's Tin-Type Tangle/Our
Congressman '20s
The Flying Deuces '39
Laurel & Hardy: Chickens
Come Home '20s
Laurel & Hardy Comedy
Classics Volume 4 '32
Laurel & Hardy Comedy
Classics Volume 5 '31
Laurel & Hardy Comedy
Classics Volume 6 '35
Laurel & Hardy Comedy
Classics Volume 7 '30s
Our Relations '36
Pack Up Your Troubles '32
Pardon Us '31

Saps at Sea '40
Stan Without Ollie, Volume 3
'23
Stan Without Ollie, Volume 4
'25
Way Out West '37

Jon Finlayson
Girl-Toy '84
A Kink in the Picasso '90
Lonely Hearts '82

Evelyn Finley
Trail Riders '42

Joe Finley (D)
Barbarian Queen 2: The
Empress Strikes Back '90s

William Finley
Phantom of the Paradise '74
The Wedding Party '69

George Finn
Youngblood '86

John Finn
Cover-Up '91
Glory '89

Mickey Finn
Earth vs. the Spider '58

Robyn Finn
Comic Book Kids '82

Dave Finnegan
The Commitments '91

John Finnegan
Finish Line '89

Siobhan Finneran
Rita, Sue & Bob Too '87

Warren Finnerty
The Brig '64
The Connection '61
Panic in Needle Park '75
Scream Free! '71

Alan Finney
Alvin Rides Again '74

Albert Finney
Alpha Beta '73
Annie '81
Dresser '83
The Duellists '78
The Endless Game '89
The Entertainer '60
The Green Man '91
Gumshoe '72
The Image '89
Looker '81
Loophole '83
Miller's Crossing '90
Murder on the Orient Express
'74
Orphans '87
The Playboys '92
Pope John Paul II '84
Scrooge '70
Shoot the Moon '82
Tom Jones '63
Two for the Road '67
Under the Volcano '84
The Wall: Live in Berlin '90
Wolfen '81

Edward Finney (D)
King of the Stallions '42
Riot Squad '41

Silver Stallion '41

Michelle Finney
Hot Money '80s

Shirley Jo Finney
Echo Park '86
Wilma '77

Allen Finzat
For the Love of Benji '77

Fionna
Hearts of Fire '87

Maria Fiore
Love Angels '87

Nada Fiorelli
The Golden Coach '52

Linda Fiorentino
After Hours '85
Gotcha! '85
The Moderns '88
The Neon Empire '89
Queens Logic '91
Shout '91
Vision Quest '85
Wildfire '88

Ann Firbank
Flame to the Phoenix '85

Eddie Firestone
Duel '71

Rex Firlin (D)
Death of Adolf Hitler '04

Sam Firstenberg (D)
American Ninja '85
American Ninja 2: The
Confrontation '87
Avenging Force '86
Breakin' 2: Electric Boogaloo
'84
My African Adventure '87
Ninja 3: The Domination '84
One More Chance '90
Revenge of the Ninja '83
Riverbend '89

Colin Firth
Another Country '84
Apartment Zero '88
Dutch Girls '87
Femme Fatale '90
Month in the Country '87
The Secret Garden '87
Valmont '89

Michael Firth (D)
Heart of the Stag '84
Sylvia '86

Peter Firth
Born of Fire '87
Burndown '89
Equus '77
Fire and Sword '82
The Hunt for Red October '90
The Incident '89
Innocent Victim '90
Joseph Andrews '77
King Arthur, the Young
Warlord '75
Letter to Brezhnev '86
Lifeforce '85
Prisoner of Honor '91
Tess '80

Michael Fischa (D)
Crack House '89
Death Spa '87
My Mom's a Werewolf '89

Mario Fischel
David '79

Kai Fischer
The Goalie's Anxiety at the
Penalty Kick '71

Madeleine Fischer
The Day the Sky Exploded '57

Max Fischer (D)
Killing 'em Softly '85

O.W. Fischer
Uncle Tom's Cabin '69

Vera Fischer
Love Strange Love '82
Quilombo '84

Margie Fisco
The Atomic Brain '64

Nancy Fish
The Exorcist 3 '90

Tony Fish
Madman '82

Joe Fishback
The Night Brings Charlie '90

Larry Fishburne
Apocalypse Now '79
Boyz N the Hood '91
Cadence '89
Gardens of Stone '87
Hearts of Darkness: A
Filmmaker's Apocalypse
'91
King of New York '90
A Nightmare on Elm Street 3:
Dream Warriors '87
Red Heat '88
School Daze '88

Alan Fisher
Young Nurses in Love '89

Arem Fisher
Alice Goodbody '74

Brad Fisher
Cthulhu Mansion '91

Carrie Fisher
Amazon Women On the Moon
'87
Appointment with Death '88
The Blues Brothers '80
The 'Burbs '89
Drop Dead Fred '91
The Empire Strikes Back '80
Frankenstein '82
From Here to Maternity '85
From "Star Wars" to "Jedi":
The Making of a Saga '83
Garbo Talks '84
Hannah and Her Sisters '86
Hollywood Vice Squad '86
Loverboy '89
The Man with One Red Shoe
'85
Mr. Mike's Mondo Video '79
Paul Reiser: Out on a Whim
'88
Return of the Jedi '83

Saturday Night Live: Carrie
Fisher '78
Shampoo '75
She's Back '88
Sibling Rivalry '90
Star Wars '77
Sweet Revenge '90
This is My Life '91
Thumbelina '82
The Time Guardian '87
Under the Rainbow '81
When Harry Met Sally '89

Cindy Fisher
Hometown U.S.A. '79
Liar's Moon '82

David Fisher (D)
Liar's Moon '82
Toy Soldiers '84

Eddie Fisher
Bundle of Joy '56
Butterfield 8 '60
Revenge of TV Bloopers '60s

Frances Fisher
Can She Bake a Cherry Pie?
'83
Cold Sassy Tree '89
Frame Up '91
Heart '87
Patty Hearst '88

Gail Fisher
Mankillers '87

Gregory Fisher
1984 '84

Jack Fisher (D)
Torn Apart '89

Jasen Fisher
The Witches '90

Mary Ann Fisher (D)
Lords of the Deep '89

Shug Fisher
The Giant Gila Monster '59
Huckleberry Finn '75

Terence Fisher (D)
Blackout '54
The Brides of Dracula '60
The Curse of Frankenstein '57
The Curse of the Werewolf
'61
Four Sided Triangle '53
Frankenstein and the Monster
from Hell '74
The Gorgon '64
The Horror of Dracula '58
The Hound of the Baskervilles
'59
Island of the Burning Doomed
'67
Island of Terror '66
Man Bait '52
The Mummy '59
Race for Life '55
Robin Hood: The Movie '91
Spaceways '53
A Stolen Face '52
Sword of Sherwood Forest
'60
Three Stops to Murder '53
Unholy Four '54

Tom Fisher (D)
Cannibal Campout '88

Tricia Leigh Fisher

Book of Love '91
C.H.U.D. 2: Bud the Chud '89
Hollywood Chaos '89
Pretty Smart '87

Vera Fisher

I Love You '81

Bill Fishman (D)

Tapeheads '89

Jack Fisk

Eraserhead '78

Jack Fisk (D)

Daddy's Dyin'...Who's Got the Will? '90
Final Verdict '91
Raggedy Man '81
Violets Are Blue '86

Robert Fiske

Green Archer '40

Alan Fisler

If Looks Could Kill '86
New York's Finest '87
Student Affairs '88

Louise Fitch

Starbird and Sweet William '73

Christopher Fitchett (D)

Blood Money '80
Fair Game '85

Rick Fitts

Platoon Leader '87

Peter Fitz

Au Revoir Les Enfants '88

Barry Fitzgerald

And Then There Were None '45
Bringing Up Baby '38
Broth of a Boy '59
The Catered Affair '56
Dawn Patrol '38
Going My Way '44
How Green was My Valley '41
Juno and the Paycock '30
The Long Voyage Home '40
Marie Antoinette '38
None But the Lonely Heart '44
The Quiet Man '52
The Saint Strikes Back '39
Saint Strikes Back: Criminal Court '46
Union Station '50

Ella Fitzgerald

Pete Kelly's Blues '55
Ride 'Em Cowboy '42

Fern Fitzgerald

A Cry for Love '80

Geraldine Fitzgerald

Arthur '81
Arthur 2: On the Rocks '88
Bloodlink '86
Dark Victory '39
Diary of the Dead '80s
Dixie: Changing Habits '85
Easy Money '83
Harry and Tonto '74

The Jilting of Granny Weatherall '80
Kennedy '83
The Last American Hero '73
Lovespell '79
The Mango Tree '77
The Mill on the Floss '37
The Obsessed '51
Pawnbroker '65
Poltergeist 2: The Other Side '86
Rachel, Rachel '68
The Strange Affair of Uncle Harry '45
Ten North Frederick '58
Watch on the Rhine '43
Wuthering Heights '39

Lewis Fitzgerald

A Cry in the Dark '88

Tara Fitzgerald

Hear My Song '91

Walter Fitzgerald

Great Day '46
San Demetrio, London '47
Treasure Island '50

George Fitzmaurice (D)

As You Desire Me '32
The Last of Mrs. Cheyney '37
Mata Hari '32
Son of the Sheik '26

Colleen Fitzpatrick

Hairspray '88

Kate Fitzpatrick

Fantasy Man '84
Night Nurse '77
Return of Captain Invincible '83

Michael Fitzpatrick

Blood Tracks '86
Pay Off '89

Neil Fitzpatrick

A Cry in the Dark '88
Ground Zero '88

Emily Fitzroy

The Cheerful Fraud '27

Marjorie Fitzsimmons

Chillers '88

Robert S. Fiveson (D)

The Clonus Horror '79

Paul Fix

The Bad Seed '56
Blood Alley '55
Desert Trail '35
El Dorado '67
Fargo Express '32
Grayeagle '77
Gun Law '33
Hitler: Dead or Alive '43
Jet Pilot '57
Mannequin '37
Phantom Patrol '36
Sherlock Holmes and the Secret Weapon '42
South of Santa Fe '42
Star Reporter '39
Wanda Nevada '79
Zabriskie Point '70

Gary Fjellgaard

Ranch '88

Cash Flagg

Incredibly Strange Creatures Who Stopped Living and Became Mixed-Up Zombies '63
The Thrill Killers '65

Fannie Flagg

Five Easy Pieces '70
My Best Friend Is a Vampire '88

Joe Flaherty

Back to the Future, Part 2 '89
Blue Monkey '84
Going Berserk '83
Looking for Miracles '90
Really Weird Tales '86
Sesame Street Presents: Follow That Bird '85
Speed Zone '88
Stripes '81
Tunnelvision '76
Used Cars '80
Wedding Band '89

Pat Flaherty

Midnight Limited '40

Paul Flaherty (D)

Billy Crystal: Midnight Train to Moscow '89
18 Again! '88
Who's Harry Crumb? '89

Robert Flaherty (D)

Elephant Boy '37
Louisiana Story '48
Man of Aran '34
Moana, a Romance of the Golden Age '26
Tabu '31

Snow Flake

Twentieth Century '34

Flame

My Dog Shep '46

Georges Flament

La Chienne '31

Fionnula Flanagan

The Ewok Adventure '84
In the Region of Ice '76
James Joyce's Women '85
Mary White '77
Picture of Dorian Gray '74
Voyager from the Unknown '83
Young Love, First Love '79
Youngblood '86

Tommy Flanagan

Talmage Farlow '81

Ed Flanders

Amazing Howard Hughes '77
Eleanor & Franklin '76
The Exorcist 3 '90
Mary White '77
The Ninth Configuration '79
The Passing of Evil '70
The Pursuit of D. B. Cooper '81
Sophia Loren: Her Own Story '80
Special Bulletin '83
Things in Their Season '74
Tomorrow's Child '82

Trial of the Catonsville Nine '72
True Confessions '81

Ian Geer Flanders

Silence '73

Erin Flannery

Incubus '82

Susan Flannery

Anatomy of a Seduction '79
Gumball Rally '76

Rob Flash

Vulcan God of Fire

John Flaus

Blood Money '80
In Too Deep '90

James Flavin

Abbott and Costello Meet the Killer, Boris Karloff '49
Abroad with Two Yanks '44
Brand of Hate '34
Cloak and Dagger '46
Hollywood Canteen '44

Alex Flavorsham

The Ghost of Rashmon Hall '47

Flea

Back to the Future, Part 2 '89
Dudes '87
My Own Private Idaho '91

John Fleck

Mutant on the Bounty '89

Sharry Fleet

War Brides '80s

Mick Fleetwood

The Running Man '87

Susan Fleetwood

The Krays '90
The Sacrifice '86
White Mischief '88

Charles Fleischer

Back to the Future, Part 2 '89
Bad Dreams '88
Honey, I Shrunk the Kids '89
A Nightmare on Elm Street '84
Who Framed Roger Rabbit? '88

Dave Fleischer (D)

Best of Betty Boop '35
Betty Boop Special Collector's Edition: Volume 1 '35
Fabulous Fleischer Folio, Vol. 3 '30s
Gulliver's Travels '39
Hoppity Goes to Town '41

Max Fleischer

Betty Boop Festival: No. 3 '30s

Max Fleischer (D)

Betty Boop Cartoon Festival '30s
Betty Boop Festival: No. 3 '30s
Betty Boop Special Collector's Edition: Volume 1 '35
Cartoon Collection, No. 1: Porky in Wackyland '40s
Koko Cartoons: No. 1 '27

Max Fleischer's Documentary Features '23
Max Fleischer's Popeye Cartoons '39

Richard Fleischer (D)

Amityville 3: The Demon '83
Armored Car Robbery '50
Ashanti, Land of No Mercy '79
Barabbas '62
Between Heaven and Hell '56
The Boston Strangler '68
The Clay Pigeon '49
Conan the Destroyer '84
Doctor Doolittle '67
The Don is Dead '73
Fantastic Voyage '66
Follow Me Quietly '49
The Incredible Sarah '76
Jazz Singer '80
Mandingo '75
Million Dollar Mystery '87
Mr. Majestyk '74
New Centurions '72
Prince and the Pauper '78
Red Sonja '85
See No Evil '71
Soylent Green '73
10 Rillington Place '71
Tora! Tora! Tora! '70
Tough Enough '87
Trapped '49
20,000 Leagues Under the Sea '54
The Vikings '58

Jane Fleiss

Kent State '81

Andrew Fleming (D)

Bad Dreams '88

Cythia Fleming

Invasion of the Blood Farmers '72

Edward Fleming (D)

Topsy Turvy '84

Eric Fleming

Fright '56
The Glass Bottom Boat '66
Queen of Outer Space '58

Erin Fleming

Demon Hunter '88

Ian Fleming

Butler's Dilemma '43
Ingrid '89
Land of Fury '55
Murder at the Baskervilles '37
Sherlock Holmes: The Silver Blaze '41
Triumph of Sherlock Holmes '35
When Thief Meets Thief '37

Lone Fleming

Tombs of the Blind Dead '72

Rhonda Fleming

Abilene Town '46
Adventure Island '47
Bullwhip '58
A Connecticut Yankee in King Arthur's Court '49
Cry Danger '51
The Great Lover '49
Gunfight at the O.K. Corral '57
The Nude Bomb '80
Out of the Past '47
Pony Express '53

Since You Went Away '44
Slightly Scarlet '56
Spellbound '45
The Spiral Staircase '46
Tennessee's Partner '55
While the City Sleeps '56

Susan Fleming

The Range Feud/Two Fisted
    Law '31

Victor Fleming

Making of a Legend: "Gone
    With the Wind" '89

Victor Fleming (D)

Bombshell '33
Captains Courageous '37
Dr. Jekyll and Mr. Hyde '41
The Farmer Takes a Wife '35
Gone with the Wind '39
A Guy Named Joe '44
Hula '28
Joan of Arc '48
Mantrap '26
Red Dust '32
Test Pilot '39
Tortilla Flat '42
Treasure Island '34
The Virginian '29
When the Clouds Roll By '19
The Wizard of Oz '39

Gordon Flemyng (D)

Cloud Waltzing '87
Daleks Invasion Earth 2150
    A.D. '66
Doctor Who and the Daleks
    '65
Philby, Burgess and MacLean:
    Spy Scandal of the Century
    '84

Robert Flemyng

Blood Beast Terror '67
The Blue Lamp '49
The Horrible Dr. Hichcock '62
The Man Who Never Was '55

Rodman Flender (D)

In the Heat of Passion '91
The Unborn '91

Alan Fletcher

Fran '85

Bramwell Fletcher

Random Harvest '42
The Undying Monster '42

Dexter Fletcher

Lionheart '87
The Rachel Papers '89
The Raggedy Rawney '90
Twisted Obsession '90

Dusty Fletcher

Killer Diller '48

Hal Fletcher

Stepsisters '85

Jay Fletcher

Born to Win '71

Juanita Fletcher

Mesquite Buckaroo '39

Louise Fletcher

Best of the Best '89
Blind Vision '91

Blue Steel '90
Boy Who Could Fly '86
Brainstorm '83
The Cheap Detective '78
Exorcist 2: The Heretic '77
Final Notice '89
Firestarter '84
Flowers in the Attic '87
Invaders from Mars '86
Islands '87
J. Edgar Hoover '87
Lady in Red '79
The Magician of Lublin '79
Mama Dracula '80
Natural Enemies '79
Nightmare on the 13th Floor
    '90
Nobody's Fool '86
One Flew Over the Cuckoo's
    Nest '75
Overnight Sensation '83
Russian Roulette '75
Shadowzone '90
Strange Behavior '81
Strange Invaders '83
A Summer to Remember '84
Talk to Me '82
Two Moon Junction '88

Mandie Fletcher (D)

Black Adder The Third (Part 2)
    '89

Page Fletcher

Friends, Lovers & Lunatics '89

Suzanne Fletcher

Bloodsucking Pharoahs of
    Pittsburgh '90
Sleepwalk '88

Tex Fletcher

Six Gun Rhythm '39

Theodore J. Flicker (D)

Playmates '72
The President's Analyst '67
Soggy Bottom U.S.A. '84
Three in the Cellar '70

Sam Flint

A Face in the Fog '36
The Lawless Nineties '36
Red River Valley '36
Spy Smasher Returns '42

Jay C. Flippen

Devil's Canyon '53
Far Country '55
Hellfighters '68
Jet Pilot '57
The Killing '56
The King and Four Queens
    '56
The Lemon Drop Kid '51
Marie Galante '34
Oklahoma! '55
The Restless Breed '58
Seventh Cavalry '56
They Live by Night '49
The Wild One '54
Winchester '73 '50
A Woman's Secret '49

Lucy Lee Flippin

Summer School '87

James T. Flocker (D)

Ghosts That Still Walk '77
Ground Zero '88

The Secret of Navajo Cave
    '76

Charles Flohe

Rappin' '85

Suzanne Flon

One Deadly Summer (L'Ete
    Meurtrier) '83
The Train '65
Un Singe en Hiver '62

Joe Flood

Student Bodies '81

Michael Flood

Crazy Fat Ethel II '85
Criminally Insane '75
Death Nurse '87

John Florea (D)

Computer Wizard '77
The Invisible Strangler '76
Island of the Lost '68
Where the Eagle Flies '72
Where's Willie? '77

Florelle

The Crime of Monsieur Lange
    '36
Liliom '35

Sheila Florence

Cactus '86

John Flores (D)

Hot Child in the City '87

Laura Flores

Bestia Nocturna '86

Lola Flores

Los Tres Amores de Losa '47

Stefano Satta Flores

We All Loved Each Other So
    Much '77

Robert Florey (D)

The Cocoanuts '29
Murders in the Rue Morgue
    '32
Outpost in Morocco '49

Holly Floria

Bikini Island '91
Netherworld '91

Aldo Florio (D)

Five Giants from Texas '66

Agata Flory

Guns for Dollars '73

Don Flourney

The Giant Gila Monster '59

George Flower

Across the Great Divide '76
Candy Tangerine Man '75
The Capture of Bigfoot '79
Cheerleader Camp '88
Devil & Leroy Basset '73
In Search of a Golden Sky '84
Mountain Family Robinson '79
Norma '89
Relentless '89
They Live '88
Wilderness Family, Part 2 '77

Kim Flowers

Nobody's Perfect '90

Calvin Floyd (D)

Sleep of Death '79

Charles R. Floyd

The P.O.W. Escape '86

Christopher Floyd

Nurse on Call '88

Drago Floyd (D)

Troll 2 '92

Joey Floyd

Honeysuckle Rose '80

Darlanne Fluegel

Battle Beyond the Stars '80
Border Heat '88
Bulletproof '88
Concrete Beat '84
Crime Story '86
Crime Story: The Complete
    Saga '86
Deadly Stranger '88
Eyes of Laura Mars '78
Freeway '88
Last Fight '82
Lock Up '89
Once Upon a Time in America
    '84
Project: Alien '89
Running Scared '86
To Live & Die in L.A. '85

Joel Fluellen

A Dream for Christmas '73
The Great White Hope '70
The Jackie Robinson Story
    '50
Man Friday '76

The Flying Karamazov Brothers

Flying Karamazov Brothers
    Comedy Hour '83
Jewel of the Nile '85

Bill Flynn

Kill and Kill Again '81

Eric Flynn

A Killer in Every Corner '74

Errol Flynn

Adventures of Captain Fabian
    '51
Adventures of Don Juan '49
The Adventures of Robin
    Hood '38
Against All Flags '52
Assault of the Rebel Girls '59
Captain Blood '35
Charge of the Light Brigade
    '36
Dawn Patrol '38
Dodge City '39
Gentleman Jim '42
Hollywood Bloopers '50s
It's a Great Feeling '49
Kim '50
The King's Rhapsody '55
Let's Make Up '55
The Master of Ballantrae '53
Northern Pursuit '43
Prince and the Pauper '37
The Private Lives of Elizabeth
    & Essex '39
Return of Video Yesterbloop
    '47
San Antonio '45
Santa Fe Trail '40

The Sea Hawk '40
Son of Hollywood Bloopers
    '40s
Thank Your Lucky Stars '43
That Forsyte Woman '49
They Died with Their Boots
    On '41
Virginia City '40
The Warriors '55

Joe Flynn

The Computer Wore Tennis
    Shoes '69
Gentle Savage '73
Lover Come Back '61
Million Dollar Duck '71
My Dog, the Thief '69
Now You See Him, Now You
    Don't '72
Superdad '73

John Flynn

Split '90

John Flynn (D)

Best Seller '87
Defiance '79
Lock Up '89
Marilyn: The Untold Story '80
Rolling Thunder '77
Touched '82

Kelly Flynn

A Patch of Blue '65

Kimberly Flynn

Revolution! A Red Comedy
    '91

Miriam Flynn

18 Again! '88
National Lampoon's
    Christmas Vacation '89

Sean Flynn

Dos Postolas Gemelas '65
Son of Captain Blood '62

Spiros Focas

Jewel of the Nile '85
Rambo: Part 3 '88

Nina Foch

An American in Paris '51
Child of Glass '78
Columbo: Prescription Murder
    '67
Dark Past '49
Dixie Lanes '88
Executive Suite '54
Illegal '55
Jennifer '78
Mahogany '76
Return of the Vampire '43
St. Benny the Dip '51
Salty '73
Scaramouche '52
Skin Deep '89
Song to Remember '45
Spartacus '60
The Ten Commandments '56

Helmut Foernbacher (D)

What a Way to Die '70

Brenda Fogarty

Deadly Fieldtrip '74
Fairytales '79

Vladimar Fogel

Bed and Sofa '27
The Girl with the Hat Box '27

Foldes

Lawrence Foldes (D)

Curse of the Living Dead '89
The Great Skycopter Rescue '82
Nightforce '86
Young Warriors '83

Peter Foleg (D)

The Unseen '80

Ellen Foley

Fatal Attraction '87
Married to the Mob '88
Star Shorts '80s

James Foley (D)

At Close Range '86
Reckless '84
Who's That Girl? '87

Macka Foley

Las Vegas Weekend '85
WarCat '88

Michael Foley

The Divine Enforcer '91

Sandra Foley

Assault with a Deadly Weapon '82

Randy Follet

John and the Missus '87

Megan Follows

Anne of Avonlea '87
Anne of Green Gables '85
Hockey Night '84
The Nutcracker Prince '91
Silver Bullet '85
Stacking '87
Termini Station '91
Time of Destiny '88

Megan Folson

Heartland '81

Bridget Fonda

Aria '88
Doc Hollywood '91
Frankenstein Unbound '90
The Godfather: Part 3 '90
Iron Maze '91
Light Years '88
Out of the Rain '90
Scandal '89
Shag: The Movie '89
Strapless '90
You Can't Hurry Love '88

Henry Fonda

Advise and Consent '62
American Film Institutes Life Achievement Awards: Jimmy Stewart '89
Ash Wednesday '73
Battle of the Bulge '65
Battleforce '78
Bernice Bobs Her Hair '76
The Best Man '64
A Big Hand for the Little Lady '66
Big Street '42
The Boston Strangler '68
The Cheyenne Social Club '70
City on Fire '78
Clarence Darrow '78
The Displaced Person '76
Drums Along the Mohawk '39
Fail-Safe '64
The Farmer Takes a Wife '35
Fort Apache '48

The Fugitive '48
Gideon's Trumpet '80
Golden Age of Hollywood '77
The Grapes of Wrath '40
The Great Battle '79
Great Smokey Roadblock '76
Henry Fonda: The Man and His Movies '84
Home to Stay '79
How the West was Won '62
I Dream Too Much '35
Immortal Sergeant '43
In Harm's Way '65
Jesse James '39
Jezebel '38
The Jilting of Granny Weatherall '80
The Lady Eve '41
Last Four Days '77
The Longest Day '62
Mad Miss Manton '38
Madigan '68
The Magnificent Dope '42
Main Street to Broadway '53
The Man That Corrupted Hadleyburg '80
Meteor '79
Midway '76
Mister Roberts '55
Moon's Our Home '36
My Darling Clementine '46
My Name Is Nobody '74
Night Flight from Moscow '73
Oldest Living Graduate '82
On Golden Pond '81
Once Upon a Time in the West '68
The Ox-Bow Incident '43
Perry Como Show '56
Rappaccini's Daughter '80
The Red Pony '76
Return of Frank James '40
Rollercoaster '77
Sex and the Single Girl '64
Sometimes a Great Notion '71
Stage Struck '57
Summer Solstice '81
The Swarm '78
Tentacles '77
There was a Crooked Man '70
Tin Star '57
Too Late the Hero '70
Twelve Angry Men '57
Wanda Nevada '79
War and Peace '56
War Years '77
Warlock '59
The Wrong Man '56
You Only Live Once '37
Young Mr. Lincoln '39
Yours, Mine & Ours '68

Jane Fonda

Agnes of God '85
Any Wednesday '66
Barbarella '68
Barefoot in the Park '67
California Suite '78
Cat Ballou '64
The Chase '66
The China Syndrome '79
Circle of Love '64
Comes a Horseman '78
Coming Home '78
Dollmaker '84
A Doll's House '73
The Electric Horseman '79
Fun with Dick and Jane '77
The Game is Over '66
Joy House '64
Julia '77
Klute '71
Morning After '86
9 to 5 '80
Old Gringo '89

On Golden Pond '81
Rollover '81
Spirits of the Dead '68
Stanley and Iris '90
Steelyard Blues '73
Tall Story '60
They Shoot Horses, Don't They? '69
Walk on the Wild Side '62

Peter Fonda

Cannonball Run '81
Certain Fury '85
Dirty Mary Crazy Larry '74
Easy Rider '69
Fatal Mission '89
Futureworld '76
Hawken's Breed '87
High Ballin' '78
Hired Hand '71
Hollywood Uncensored '87
The Hostage Tower '80
Jungle Heat '84
Killer Force '75
The Last Movie '71
Lilith '64
Mercenary Fighters '88
92 in the Shade '76
Outlaw Blues '77
Race with the Devil '75
Roger Corman: Hollywood's Wild Angel '78
The Rose Garden '89
Spasms '82
Spirits of the Dead '68
Split Image '82
Tammy and the Doctor '63
The Trip '67
Wanda Nevada '79
The Wild Angels '66

Peter Fonda (D)

Hired Hand '71
Idaho Transfer '73
Wanda Nevada '79

Phil Fondacaro

Ghoulies 2 '88
Meridian: Kiss of the Beast '90

Marcello Fondato (D)

Immortal Bachelor '80

Stephanie Fondue

The Cheerleaders '72

Allen Fong

Just Like Weather '86

Allen Fong (D)

Just Like Weather '86

Au Fong

The White Butterfly Killer

Benson Fong

Deception '46
Girls! Girls! Girls! '62
His Majesty O'Keefe '53
Our Man Flint '66
The Purple Heart '44

Chee Fong

The White Butterfly Killer

Leo Fong

Blind Rage '78
Enforcer from Death Row '78
Killpoint '84
The Last Reunion '80
Low Blow '85
Ninja Nightmare '88

Leo Fong (D)

Hawkeye '88
24 Hours to Midnight '92

Angelino Fons (D)

Marianela '70s

Jorge Fons (D)

Jory '72

Alisha Fontaine

French Quarter '78

Jean Fontaine

The Sinister Urge '60

Joan Fontaine

Annabel Takes a Tour/Maid's Night Out '38
Beyond a Reasonable Doubt '56
The Bigamist '53
Born to Be Bad '50
Casanova's Big Night '54
Damsel in Distress '37
Decameron Nights '53
Gunga Din '39
Island in the Sun '57
Ivanhoe '52
Letter from an Unknown Woman '48
Maid's Night Out '38
A Million to One '37
Rebecca '40
September Affair '50
Suspicion '41
Voyage to the Bottom of the Sea '61
The Women '39

Genevieve Fontanel

Grain of Sand '84

Catherine Fontenoy

Poil de Carotte '31

Guilherme Fontes

Subway to the Stars '87

Lloyd Fonvielle (D)

Gotham '88

Hallie Foote

Courtship '87
1918 '85
On Valentine's Day '86

Horton Foote, Jr.

Blood Red '88
1918 '85

Richard Foote

Dangerous Charter '62

Dick Foran

Atomic Submarine '59
Brighty of the Grand Canyon '67
Dangerous '35
Deputy Marshall '50
Earthworm Tractors '36
Four Daughters '38
Guest Wife '45
The Mummy's Hand '40
My Little Chickadee '40
Petrified Forest '36
Private Buckaroo '42
Return of Video Yesterbloop '47
Ride 'Em Cowboy '42
Riders of Death Valley '41
Winners of the West '40

Leo Fong (D)

June Foray

The Adventures of Rocky & Bullwinkle: Birth of Bullwinkle '91
The Adventures of Rocky & Bullwinkle: Blue Moose '91
The Adventures of Rocky & Bullwinkle: Canadian Gothic '91
The Adventures of Rocky & Bullwinkle: La Grande Moose '91
The Adventures of Rocky & Bullwinkle: Mona Moose '91
The Adventures of Rocky & Bullwinkle: Vincent Van Moose '91
Looney Looney Looney Bugs Bunny Movie '81
Sabaka '55
Scruffy '80
Who Framed Roger Rabbit? '88

Bryan Forbes

The Baby and the Battleship '56
The Colditz Story '55
December Flower
League of Gentlemen '60
Quatermass 2 '57
A Shot in the Dark '64
Wooden Horse '50

Bryan Forbes (D)

Better Late Than Never '83
International Velvet '78
King Rat '65
Long Ago Tomorrow '71
The Naked Face '84
Seance on a Wet Afternoon '64
Whistle Down the Wind '62
The Wrong Box '66

Francine Forbes

Splatter University '84

Mary Forbes

The Awful Truth '37
Tender Comrade '43
You Can't Cheat an Honest Man '39
You Can't Take It with You '38

Ralph Forbes

Convicts at Large '38
Inside the Lines '30
Legion of Missing Men '37
Make a Wish '37
The Phantom Broadcast '33
Twentieth Century '34

John Force

The Visitors '89

Lewis J. Force (D)

Night After Night After Night '70

Anitra Ford

Big Bird Cage '72
Invasion of the Bee Girls '73
Stacey '73

Ann Ford

Logan's Run '76
The Love Machine '71

Cecil Ford

Men of Ireland '38

Constance Ford

All Fall Down '62
Home From the Hill '60
Rome Adventure '62
A Summer Place '59

Faith Ford

You Talkin' to Me? '87

Francis Ford (D)

Custer's Last Fight '12

Geoffrey Ford

Dick Barton, Special Agent '48

Glenn Ford

Affair in Trinidad '52
Americano '55
Appointment in Honduras '53
Big Heat '53
Blackboard Jungle '55
Border Shootout '90
Casablanca Express '89
The Courtship of Eddie's Father '81
Day of the Assassin '81
Destroyer '43
Experiment in Terror '62
Final Verdict '91
Four Horsemen of the Apocalypse '62
Ghosts of the Sky '70s
Gilda '46
Great White Death '81
The Green Glove '52
Happy Birthday to Me '81
Human Desire '54
Is Paris Burning? '68
Jubal '56
The Lady in Question '40
The Loves of Carmen '48
Lust for Gold '49
Man from the Alamo '53
Man from Colorado '49
Midway '76
Pocketful of Miracles '61
Raw Nerve '91
The Sacketts '79
Santee '73
Smith! '69
So Ends Our Night '41
A Stolen Life '46
Superman: The Movie '78
The Teahouse of the August Moon '56
Texas '41
3:10 to Yuma '57
Torpedo Run '58
Virus '82
The Visitor '80
The White Tower '50

Greg Ford (D)

Daffy Duck's Quackbusters '89

Harrison Ford

American Graffiti '73
Apocalypse Now '79
The Average Woman '24
Blade Runner '82
The Conversation '74
Dead Heat on a Merry-Go-Round '66
The Empire Strikes Back '80
Force 10 from Navarone '78
Frantic '88
The Frisco Kid '79
From ''Star Wars'' to ''Jedi'': The Making of a Saga '83
Getting Straight '70

Great Movie Stunts & the Making of Raiders of the Lost Ark '81

Hanover Street '79
Hawthorne of the USA '19
Heroes '77
Indiana Jones and the Last Crusade '89
Indiana Jones and the Temple of Doom '84
The Mosquito Coast '86
The Nervous Wreck '26
Presumed Innocent '90
Raiders of the Lost Ark '81
Regarding Henry '91
Return of the Jedi '83
Star Wars '77
Witness '85
Working Girl '88
Zabriskie Point '70

Jan Ford

The Devil on Wheels '47

John Ford

The Long Gray Line '55

John Ford (D)

Arrowsmith '31
Cheyenne Autumn '64
December 7th: The Movie '91
Donovan's Reef '63
Drums Along the Mohawk '39
Fort Apache '48
The Fugitive '48
The Grapes of Wrath '40
The Horse Soldiers '59
How Green was My Valley '41
How the West was Won '62
Hurricane '37
The Informer '35
Judge Priest '34
Last Hurrah '58
The Long Voyage Home '40
The Lost Patrol '34
The Man Who Shot Liberty Valance '62
Mary of Scotland '36
Mister Roberts '55
Mogambo '53
My Darling Clementine '46
The Quiet Man '52
Rio Grande '50
The Searchers '56
She Wore a Yellow Ribbon '49
Stagecoach '39
Straight Shootin' '17
The Sun Shines Bright '53
They were Expendable '45
Three Godfathers '48
Two Rode Together '61
Wagon Master '50
Wee Willie Winkie '37
What Price Glory? '52
Wings of Eagles '57
Young Mr. Lincoln '39

Maria Ford

Deathstalker 4: Clash of the Titans '91
The Haunting of Morella '91
Masque of the Red Death '89
The Rain Killer '90
Ring of Fire '91
Slumber Party Massacre 3 '90
Stripped to Kill II: Live Girls '89

Melissa Ford

Roommate '84

Michael Ford

To Die Standing '91

Mick Ford

How to Get Ahead in Advertising '89
Scum '79

Montgomery Ford

Today We Kill, Tomorrow We Die '71

Paul Ford

Advise and Consent '62
Lola '69
Lust for Gold '49
The Matchmaker '58
Missouri Traveler '58
The Music Man '62
The Russians are Coming, the Russians are Coming '66
The Teahouse of the August Moon '56

Philip Ford (D)

The Inner Circle '46
The Vanishing Westerner '50

Ross Ford

Blue Canadian Rockies '52
Jungle Patrol '48

Ruth Ford

Eyes of the Amaryllis '82
The Woman Who Came Back '45

Steven Ford

Body Count '87
Dungeonmaster '85
When Harry Met Sally '89

Wallace Ford

The Ape Man '43
Back Door to Heaven '39
Blood on the Sun '45
Coroner Creek '48
Crack-Up '46
Dead Reckoning '47
Exiled to Shanghai '37
Freaks '32
Get That Man '35
Great Jesse James Raid '49
Harvey '50
The Informer '35
Jericho '38
Last Hurrah '58
The Lost Patrol '34
Man from Laramie '55
The Matchmaker '58
Maverick Queen '55
The Mummy's Hand '40
Mysterious Mr. Wong '35
One Frightened Night '35
A Patch of Blue '65
Possessed '31
The Rainmaker '56
Rogue's Tavern '36
The Set-Up '49
Seven Days' Leave '42
Shadow of a Doubt '42
She Couldn't Say No '52
Spellbound '45
Swing It, Sailor! '37
T-Men '47
Warlock '59

Eugene Forde (D)

Inspector Hornleigh '39

Jessica Forde

Four Adventures of Reinette and Mirabelle '89

Walter Forde (D)

Bulldog Jack '35

The Ghost Train '41
The Secret Four '40

Ken Foree

Dawn of the Dead '78
Glitz '88
Leatherface: The Texas Chainsaw Massacre 3 '89

Deborah Foreman

April Fool's Day '86
Destroyer '88
The Experts '89
Friends, Lovers & Lunatics '89
Lobster Man from Mars '89
Lunatics: A Love Story '92
My Chauffeur '86
Sundown '91
3:15—The Moment of Truth '86
Valley Girl '83
Waxwork '88

Michelle Foreman

Sunset Strip '91

Richard Foreman (D)

Sudden Death '77

Delphine Forest

Europa, Europa '91

Mark Forest

Goliath and the Dragon '61
Goliath and the Sins of Babylon '64
The Lion of Thebes '64
Mole Men Against the Son of Hercules '61
Son of Samson '62
Terror of Rome Against the Son of Hercules '64

Michael Forest

The Beast from Haunted Cave '60
Dirt Gang '71
Mohammed: Messenger of God '77
Ski Troop Attack '60

Robert Forest (D)

The Final Programme '73

Carol Forman

The Black Widow '47

Joey Forman

The Nude Bomb '80

Milos Forman

Heartburn '86
James Cagney: That Yankee Doodle Dandy '86
New Year's Day '89

Milos Forman (D)

Amadeus '84
Firemen's Ball '68
Hair '79
Loves of a Blonde '66
One Flew Over the Cuckoo's Nest '75
Ragtime '81
Valmont '89

Tom Forman (D)

The Fighting American '24

Richard Foronjy

Prince of the City '81

Andy Forrest

Bridge to Hell '87

Brett Forrest

Murder on Line One '90

Christine Forrest

Monkey Shines '88

Frederic Forrest

The Adventures of Huckleberry Finn '85
Apocalypse Now '79
Best Kept Secrets '88
Calamity Jane '82
Cat Chaser '90
The Conversation '74
The Don is Dead '73
Double Exposure: The Story of Margaret Bourke-White '89
Gotham '88
Hammett '82
Hearts of Darkness: A Filmmaker's Apocalypse '91
It's Alive 2: It Lives Again '78
Larry '74
Missouri Breaks '76
Music Box '89
One from the Heart '82
Permission To Kill '75
Quo Vadis '85
Return '88
The Rose '79
Saigon: Year of the Cat '87
Shadow on the Sun '88
Stacking '87
The Stone Boy '84
Tucker: The Man and His Dream '88
The Two Jakes '90
Valentino Returns '88
Valley Girl '83
When the Legends Die '72
Where Are the Children? '85

Helen Forrest

Bathing Beauty '44
Symphony of Swing '86

Irene Forrest

Sitting Ducks '80

Michael Forrest

Atlas '60

Mike Forrest

Shark Hunter '84

Sally Forrest

Son of Sinbad '55
Vengeance Valley '51

Steve Forrest

Captain America '79
Clown '53
Deerslayer '78
Flaming Star '60
The Hanged Man '74
Hatfields and the McCoys '75
Heller in Pink Tights '60
Hotline '82
Last of the Mohicans '85
Mommie Dearest '81
North Dallas Forty '79
Roughnecks '80
A Rumor of War '80
Sahara '83
Spies Like Us '85
The Wild Country '71

William Forrest

Jailhouse Rock '57
The Masked Marvel '43

John Forristal

Osa '85

Rolf Forsberg

Beyond the Next Mountain
'70s

Constance Forslund

Dear Detective '79
A Shining Season '79

Kathrine Forster

Hollywood Harry '86

Robert Forster

Alligator '80
Avalanche '78
The Banker '89
Black Hole '79
The City '76
Committed '91
The Death Squad '73
Delta Force '86
Diplomatic Immunity '91
The Don is Dead '73
Goliath Awaits '81
Hollywood Harry '86
In Between '90s
Journey Through Rosebud
'72
Medium Cool '69
Once a Hero '88
Peacemaker '90
Satan's Princess '90
The Stalking Moon '69
Standing Tall '78
Stunts '77
Vigilante '83
Walking the Edge '83

Robert Forster (D)

Hollywood Harry '86

Rudolph Forster

The Return of Dr. Mabuse '61
The Threepenny Opera '31

Bill Forsyth (D)

Breaking In '89
Comfort and Joy '84
Gregory's Girl '80
Housekeeping '87
Local Hero '83
That Sinking Feeling '79

Bruce Forsyth

Bedknobs and Broomsticks
'71

Ed Forsyth (D)

Chesty Anderson USN '76
Inferno in Paradise '88
Superchick '71

Rosemary Forsyth

Columbo: Murder By the Book
'71
The Gladiator '86
Gray Lady Down '77
My Father's House '75
The War Lord '65
Whatever Happened to Aunt
Alice? '69

Tony Forsyth

Wonderland '88

Drew Forsythe

Doctors and Nurses '82

A Test of Love '84

**Henderson
Forsythe**

Crisis at Central High '80
Sessions '83

John Forsythe

The Ambassador's Daughter
'56
And Justice for All '79
Cruise Into Terror '78
Cry Panic '74
Destination Tokyo '43
The Happy Ending '69
In Cold Blood '67
Madame X '66
Murder Once Removed '71
Mysterious Two '82
Scrooged '88
Sizzle '81
Sniper '75
Terror on the 40th Floor '74
Time for Miracles '80
Topaz '69
The Trouble with Harry '55
Users '78

Rosemary Forsythe

Shenandoah '65

Stephen Forsythe

Hatchet for the Honeymoon
'70

William Forsythe

American Me '92
Dead Bang '89
Dick Tracy '90
The Lightship '86
The Long, Hot Summer '86
Out for Justice '91
Patty Hearst '88
Raising Arizona '87
Savage Dawn '84
Stone Cold '91
Torrents of Spring '90
Weeds '87

Fabian Forte

Kiss Daddy Goodbye '81
Little Laura & Big John '73

Fabrizio Forte

Padre Padrone '77

Marlene Forte

The Bronx War '91

Nick Apollo Forte

Broadway Danny Rose '84

Albert Fortell

Nuclear Conspiracy '85
Scandalous '88
Time Troopers '89

Gregory Fortescue

The Carrier '87

Jane Forth

Trash '70

Nurla Forway

Hawk and Castile

Bob Fosse

Give a Girl a Break '53
Kiss Me Kate '53
Little Prince '74

Bob Fosse (D)

All That Jazz '79

Cabaret '72
Lenny '74
Star 80 '83
Sweet Charity '68

Nicole Fosse

A Chorus Line '85

Ernie Fosselius

Heat and Sunlight '87

Brigitte Fossey

Blue Country '78
Chanel Solitaire '81
Enigma '82
Forbidden Games '52
Going Places '74
Honor Among Thieves '68
La Boum '81
The Man Who Loved Women
'77
Quintet '79

Simche Fostel

Yidl Mitn Fidl (Yiddle with His
Fiddle) '36

Barry Foster

Frenzy '72
Quiet Day In Belfast '74
Robbery '67

Charles Foster

Emmanuelle 5 '87

Clayton Foster

Hospital of Terror '78

Dianne Foster

Kentuckian '55
Last Hurrah '58

Edward Foster

Stop That Train

Eric Foster

Grandma's House '88

Frances Foster

Enemy Territory '87

Giles Foster (D)

Consuming Passions '88
Dutch Girls '87
Hotel Du Lac '86
Innocent Victim '90
Silas Marner '85

Gloria Foster

City of Hope '91
The Cool World '63
The House of Dies Drear '88
Leonard Part 6 '87
Man & Boy '71
To All My Friends on Shore
'71

Helen Foster

Ghost City '32
The Road to Ruin '28
Saddle Buster '32
Young Blood '33

Jodie Foster

The Accused '88
Alice Doesn't Live Here
Anymore '74
Back Track '91
The Blood of Others '84
Bugsy Malone '76
Candleshoe '78
Carny '80
Five Corners '88

Foxes '80
Freaky Friday '76
The Hotel New Hampshire '84
The Little Girl Who Lives
Down the Lane '76
Little Man Tate '91
Mesmerized '84
Napoleon and Samantha '72
O'Hara's Wife '82
One Little Indian '73
Shadows and Fog '92
Siesta '87
The Silence of the Lambs '91
Smile, Jenny, You're Dead '74
Stealing Home '88
Svengali '83
Taxi Driver '76
Tom Sawyer '73

Jodie Foster (D)

Little Man Tate '91

Julia Foster

F. Scott Fitzgerald in
Hollywood '76
Great McGonagall '75
Half a Sixpence '67
Henry VI, Part 3 '82
Thirteenth Reunion '81

Karen Foster

Playboy Video Centerfold
Double Header '90

Kimberly Foster

It Takes Two '88

Lewis R. Foster

Laurel & Hardy Comedy
Classics Volume 8 '30s

Lewis R. Foster (D)

Crashout '55
Dakota Incident '56
Sign of Zorro '60
Tonka '58

Lisa Foster

Blade Master '84

Marvin Foster

Mad Wax: The Surf Movie '90

Meg Foster

Adam at 6 A.M. '70
Backstab '90
Blind Fury '90
Carny '80
Different Story '78
Diplomatic Immunity '91
The Emerald Forest '85
James Dean '76
Jezebel's Kiss '90
The Legend of Sleepy Hollow
'79
Leviathan '89
Masters of the Universe '87
The Osterman Weekend '83
Project: Shadowchaser '92
Relentless '89
Relentless 2: Dead On '91
Stepfather 2: Make Room for
Daddy '89
They Live '88
Things in Their Season '74
Thumb Tripping '72
Ticket to Heaven '81
To Catch a Killer '92
Tripwire '89
The Wind '87

Norman Foster

Journey into Fear '42

Norman Foster (D)

Brighty of the Grand Canyon
'67
Davy Crockett, King of the
Wild Frontier '55
Davy Crockett and the River
Pirates '56
Indian Paint '64
Journey into Fear '42
Mr. Moto's Last Warning '39
Nine Lives of Elfego Baca '58
Rachel and the Stranger '48
Sign of Zorro '60

Peter Foster

Dangerous Charter '62

Phil Foster

Bang the Drum Slowly '73
Hail '73

Preston Foster

American Empire '42
Annie Oakley '35
The Big Cat '49
Doctor X '32
Guadalcanal Diary '43
The Harvey Girls '46
I Am a Fugitive from a Chain
Gang '32
I Shot Jesse James '49
The Informer '35
Kansas City Confidential '52
Last Days of Pompeii '35
Last Mile '32
The Marshal's Daughter '53
My Friend Flicka '43
Ramrod '47
Sea Devils '37
Three Desperate Men '50
The Time Travelers '64

Robert Foster

Counterforce '87

**Steffen Gregory
Foster**

Dead Pit '89

Susanna Foster

Phantom of the Opera '43

Andre Fouche

Cesar '36

Byron Foulger

Bells of San Fernando '47
The Devil's Partner '58
The Panther's Claw '42
Prisoner of Zenda '37
Sullivan's Travels '41

Alain Foures

Rape of Love '79

Claude Fournier (D)

Dan Candy's Law '73
Tales of the Klondike: The
Scorn of Women '81

Noel Fournier

Fatal Chase '77

Silvia Fournier

World of the Vampires '60

Susannah Fowle

The Getting of Wisdom '77

Blaque Fowler

Tainted '88

Noel Francis

Bachelor Apartment '31
Fire Alarm '38

Robert Francis

The Long Gray Line '55

Wilma Francis

Guest Wife '45

Pietro Francisci (D)

Hercules '58
Hercules Unchained '59

James Franciscus

The Amazing Dobermans '76
Beneath the Planet of the
    Apes '70
Butterfly '82
The Cat o' Nine Tails '71
Dario Argento's World of
    Horror '85
Good Guys Wear Black '78
The Greek Tycoon '78
Hunter '77
Jacqueline Bouvier Kennedy
    '81
Jonathan Livingston Seagull
    '73
Killer Fish '79
Man Inside '76
Marooned '69
Nightkill '80
One of My Wives Is Missing
    '76
Puzzle '78
Secret Weapons '85
Secrets of Three Hungry
    Wives '78
Snow Treasure '67
The Valley of Gwangi '69

Don Francks

The Christmas Wife '88
Drying Up the Streets '76
Fast Company '78
Finian's Rainbow '68
984: Prisoner of the Future
    '84
Phoenix Team '80

Jess (Jesus) Franco

Ilsa, the Wicked Warden '78

**Jess (Jesus)
Franco** (D)

Angel of Death '86
Attack of the Robots '66
Bloody Moon '83
The Castle of Fu Manchu '68
Count Dracula '71
Deadly Sanctuary '68
The Diabolical Dr. Z '65
Dr. Orloff's Monster '64
Ilsa, the Wicked Warden '78
The Invisible Dead
Jack the Ripper '80
Kiss and Kill '68
The Loves of Irina '80s
99 Women '69
Revenge in the House of
    Usher '82
The Screaming Dead '72
Venus in Furs '70
Virgin Among the Living Dead
    '71

Ramon Franco

The Hill '88
Kiss Me a Killer '91
Tour of Duty '87

**Richardo
Franco** (D)

Berlin Blues '89

Jacques Francois

Barkleys of Broadway '49
The Gift '82

Jean Francoise

Violette '78

Georges Franju (D)

Horror Chamber of Dr.
    Faustus '59
Judex '64

Adam Frank

The Power of the Ninjitsu '88

A.M. Frank (D)

Oasis of the Zombies '82

Ben Frank

Don't Answer the Phone '80
Hollywood Vice Sqaud '86
Hollywood Zap '86

Billy Frank

Nudity Required '90

Carol Frank (D)

Sorority House Massacre '86

Charles Frank

The Chisholms '79
A Guide for the Married
    Woman '78
LBJ: The Early Years '88
The Right Stuff '83
Snowblind '78
Tarantulas: The Deadly Cargo
    '77

Charles Frank (D)

Inheritance '47

**Christopher
Frank** (D)

Josepha '82

Consuelo Frank

Volver, Volver, Volver! '87

Diana Frank

The Glass Jungle '88
Monster High '89
Pale Blood '91

Gary Frank

Deadly Weapon '88
Enemy Territory '87
Enola Gay: The Men, the
    Mission, the Atomic Bomb
    '80

Herbert Frank

The Social Secretary '16

Horst Frank

Albino '76
Code Name Alpha '67
Desert Commandos '60s
The Head '59

Hubert Frank (D)

The Dolls '83
Melody in Love '78

Ian Frank

Ninja Fantasy '86

Joanna Frank

Always '85
The Savage Seven '68

Melvin Frank (D)

Court Jester '56
The Duchess and the
    Dirtwater Fox '76
The Jayhawkers '59
Li'l Abner '59
Lost and Found '79
Prisoner of Second Avenue
    '74
A Touch of Class '73
Walk Like a Man '87

Robert Frank (D)

Candy Mountain '87

T.C. Frank (D)

Billy Jack '71
Born Losers '67

Tony Frank

Riverbend '89

Ronald Frankau

Dual Alibi '47

Cyril Frankel (D)

Devil on Horseback '54
Make Me an Offer '55
Permission To Kill '75
The Very Edge '63

Mark Frankel

Young Catherine '91

Al Franken

Franken and Davis at
    Stockton State '84
One More Saturday Night '86
Tunnelvision '76

Steve Franken

Ants '77
Freeway '88
Which Way to the Front? '70

**John
Frankenheimer** (D)

All Fall Down '62
Birdman of Alcatraz '62
Black Sunday '77
The Challenge '82
Days of Wine and Roses '58
Dead Bang '89
52 Pick-Up '86
The Fourth War '90
French Connection 2 '75
Grand Prix '66
The Holcroft Covenant '85
Horsemen '70
The Manchurian Candidate
    '62
99 & 44/100 Dead '74
Prophecy '79
Seven Days in May '64
The Story of a Love Story '73
The Train '65
Year of the Gun '91

Paul Frankeur

Le Gentleman D'Epsom '62
The Milky Way '68
Un Singe en Hiver '62

**William
Frankfather**

Defense Play '88

David Frankham

Return of the Fly '59

Aretha Franklin

The Blues Brothers '80

Bonnie Franklin

Breaking Up Is Hard to Do '79
Portrait of a Rebel: Margaret
    Sanger '82
Your Place or Mine '83

Carl Franklin

Eye of the Eagle 3 '91

Carl Franklin (D)

Eye of the Eagle 2 '89
Full Fathom Five '90
Nowhere to Run '88

David Franklin

Early Frost '84

Diane Franklin

Better Off Dead '85
Last American Virgin '82
Second Time Lucky '84
Terrorvision '86

Don Franklin

Fast Forward '84

Gloria Franklin

Drums of Fu Manchu '40
Without Warning '52

Jimmy Franklin (D)

Ripping Yarns '75

Joe Franklin

Ghoul School '90

John Franklin

Children of the Corn '84

Pamela Franklin

And Soon the Darkness '70
David Copperfield '70
Eleanor & Franklin '76
Flipper's New Adventure '64
Food of the Gods '76
The Horse Without a Head '63
Legend of Hell House '73
The Prime of Miss Jean
    Brodie '69
Satan's School for Girls '73
Screamer '74
A Tiger Walks '64
The Witching '72

**Richard
Franklin** (D)

Cloak & Dagger '84
F/X 2: The Deadly Art of
    Illusion '91
Link '86
Patrick '78
Psycho 2 '83
Road Games '81

Roger Franklin (D)

Raiders of Atlantis '70s

Sidney Franklin

The Three Musketeers '21

Sidney Franklin (D)

The Barretts of Wimpole
    Street '34
The Good Earth '37
The Primitive Lover '16
Wild Orchids '28

William Franklyn

Quatermass 2 '57
The Satanic Rites of Dracula
    '73

Chloe Franks

The House that Dripped Blood
    '71
Ivanhoe '82

Philip Franks

Bleak House '85

Mary Frann

I'm Dangerous Tonight '90

Fely Franquelli

Back to Bataan '45

Arthur Franz

Abbott & Costello Meet the
    Invisible Man '51
Atomic Submarine '59
Beyond a Reasonable Doubt
    '56
Bogie '80
Dream No Evil '75
Flight to Mars '52
Hellcats of the Navy '57
The Human Factor '75
Jungle Patrol '48
The Member of the Wedding
    '52
Sisters of Death '76

Chris Franz

Stop Making Sense '84

Dennis Franz

Blow Out '81
Body Double '84
Die Hard 2: Die Harder '90
Dressed to Kill '80
Kiss Shot '90s
The Package '89
Psycho 2 '83

Eduard Franz

The Burning Hills '56
Hollow Triumph '48
Outpost in Morocco '49
The Scar '48
Sins of Jezebel '54
The Story of Ruth '60

Filippa Franzen

The Sacrifice '86

Bill Fraser

Alias John Preston '56
Little Dorrit, Film 1: Nobody's
    Fault '88
Little Dorrit, Film 2: Little
    Dorrit's Story '88
Naughty Knights '71
Pirates '86
Ripping Yarns '75

Brent Fraser

Wild Orchid 2: Two Shades of
    Blue '92

Christopher Fraser

Summer City '77

Elisabeth Fraser

The Man Who Came to Dinner
    '41
A Patch of Blue '65
Young at Heart '54

Harry Fraser (D)

Aces Wild '37

Broadway to Cheyenne '32
Cavalcade of the West '36
Chained for Life '51
Diamond Trail '33
Enemy of the Law '45
Fighting Parson '35
From Broadway to Cheyenne '35
Galloping Dynamite '37
Ghost City '32
Ghost Town '37
Heroes of the Alamo '37
Outlaw Roundup '44
Rainbow Ranch '33
Rustler's Paradise '35
The Savage Girl '32
Six Shootin' Sheriff '38
Three in the Saddle '45
The Tonto Kid '35
Wagon Trail '35
Wild Mustang '35
Wolf Dog '33

Hugh Fraser

The Draughtsman's Contract '82

John Fraser

Horsemasters '61
Isadora '68
Repulsion '65

Liz Fraser

Adventures of a Taxi Driver '76
The Americanization of Emily '64
Carry On Cruising '62
Chicago Joe & the Showgirl '90
Two-Way Stretch '60

Moyra Fraser

The Boy Friend '71

Phyllis Fraser

Tough to Handle '37
Western Double Feature 3 '36
Winds of the Wasteland '36

Richard Fraser

Bedlam '45
The Cobra Strikes '48
The Private Affairs of Bel Ami '47
White Pongo '45

Ronald Fraser

The Killing of Sister George '69
Rentadick '72

Sally Fraser

Dangerous Charter '62
Giant from the Unknown '58
It Conquered the World '56

Stuart Fratkin

Dr. Alien '88
Ski School '91

Charles Frawley

The Devil's Playground '76

James Frawley (D)

Assault and Matrimony '87
Big Bus '76
Fraternity Vacation '85
Great American Traffic Jam '80
Hansel & Gretel '82
Muppet Movie '79
Spies, Lies and Naked Thighs '91

William Frawley

Abbott & Costello Meet the Invisible Man '51
The Adventures of Huckleberry Finn '39
Babe Ruth Story '48
The Bride Came C.O.D. '41
Down to Earth '47
Fighting Seabees '44
Flame of the Barbary Coast '45
The General Died at Dawn '36
Gentleman Jim '42
Harmony Lane '35
The Inner Circle '46
The Lemon Drop Kid '51
Miracle on 34th Street '47
Monsieur Verdoux '47
Rancho Notorious '52
Something to Sing About '36
Toast of the Town '56
Wildcat '42

Jane Frazee

The Gay Ranchero '42
Grand Canyon Trail '48
Last of the Wild Horses '49
On the Old Spanish Trail '47
Springtime in the Sierras '47
Under California Stars '48

Henry Frazer (D)

'Neath the Arizona Skies '34
Randy Rides Alone '34

Quincy Frazer

Forgotten Warrior '86

Robert Frazer

Dawn Express '42
Death from a Distance '36
The Drake Case '29
Fighting Parson '35
Fighting Pilot '35
Found Alive '34
Monte Carlo Nights '34
Mystery Trooper '32
Renfrew on the Great White Trail '38
Saddle Buster '32
Ten Nights in a Bar-Room '31
Trail of the Royal Mounted '34

Rupert Frazer

The Girl in a Swing '89

Sheila Frazier

Firehouse '72
The Hitter '79
Superfly '72
Three the Hard Way '74

Sterling Frazier

Starhops '78

Patrick Frbezar

Ninja of the Magnificence '89

Stephen Frears (D)

Dangerous Liaisons '89
The Grifters '90
Gumshoe '72
The Hit '85
My Beautiful Laundrette '86
Prick Up Your Ears '87
Saigon: Year of the Cat '87
Sammy & Rosie Get Laid '87

Stan Freberg

Amazing Stories, Book 2 '86
Lady and the Tramp '55
Pogo for President: "I Go Pogo" '84

Mark Frechette

Zabriskie Point '70

Peter Frechette

No Small Affair '84
Paint it Black '89

Riccardo Freda (D)

Devil of the Desert Against the Son of Hercules '62
The Devil's Commandment '60
The Giants of Thessaly '60
The Horrible Dr. Hichcock '62
Les Miserables '52
Maciste in Hell '60
The Mongols '60
Sins of Rome '54

Freddie & the Dreamers

Seaside Swingers '65

Freddie Redd Quartet

The Connection '61

Blanche Frederici

A Farewell to Arms '32
Sadie Thompson '28

Lynne Frederick

Amazing Mr. Blunden '72
Phase 4 '74
Prisoner of Zenda '79
Schizo '77
Vampire Circus '71
Voyage of the Damned '76

Pauline Frederick

Smouldering Fires '25

Vicki Frederick

All the Marbles '81
Body Rock '84
Chopper Chicks in Zombietown '91

Dean Fredericks

The Phantom Planet '61

Nina Fredric

2069: A Sex Odyssey '78

Vicki Fredrick

A Chorus Line '85

Christa Free

Revealing of Elsie '86

Alan Freed

Go, Johnny Go! '59
Rock, Rock, Rock '56

Bert Freed

The Cobweb '55
Detective Story '51
Evel Knievel '72
Halls of Montezuma '50

Herb Freed (D)

Beyond Evil '80
Graduation Day '81
Haunts '77
Survival Game '87
Tomboy '85

Mark Freed (D)

Shock 'Em Dead '90

Rona Freed

The Penitent '88

Sam Freed

Call Me '88

Harry Freedman

Storm '87

Jerrold Freedman (D)

Best Kept Secrets '88
Borderline '80
Family Sins '87
He Who Walks Alone '78
Legs '83
Native Son '86
Seduced '85
The Streets of L.A. '79
Thompson's Last Run '90

Robert Freedman (D)

Goin' All the Way '82

Thornton Freeland (D)

Flying Down to Rio '33
Jericho '38
Whoopee! '30

Al Freeman, Jr.

The Defense Never Rests '90
Detective '68
Ensign Pulver '64
Finian's Rainbow '68
My Sweet Charlie '70
Perry Mason Returns '85
Seven Hours to Judgment '88

Eric Freeman

Silent Night, Deadly Night 2 '87

Hal Freeman (D)

Blood Frenzy '87

Howard Freeman

Double Dynamite '51
Million Dollar Mermaid '52
Scaramouche '52

J.E. Freeman

Aces: Iron Eagle 3 '92
Hard Traveling '85
Memphis '91
Miller's Crossing '90

Joan Freeman

Fastest Guitar Alive '68
Panic in the Year Zero! '62
Roustabout '64
Tower of London '62

Joan Freeman (D)

Satisfaction '88
Streetwalkin' '85

John Freeman

Buffalo Rider '78

Kathleen Freeman

Disorderly Orderly '64
Errand Boy '61
Hollywood Chaos '89
Malibu Bikini Shop '86
The Nutty Professor '63
The Willies '90

Mona Freeman

Battle Cry '55

Black Beauty '46
Black Beauty/Courage of Black Beauty '57
Branded '50
Copper Canyon '50
Dear Wife '49
That Brennan Girl '46

Morgan Freeman

Blood Money - The Story of Clinton and Nadine '88
The Bonfire of the Vanities '90
Brubaker '80
Clean and Sober '88
Driving Miss Daisy '89
Execution of Raymond Graham '85
Eyewitness '81
Glory '89
Harry & Son '84
Johnny Handsome '89
Lean on Me '89
Marie '85
Power of One '92
Robin Hood: Prince of Thieves '91
Roll of Thunder, Hear My Cry '78
Street Smart '87
Teachers '84
That Was Then...This Is Now '85

Paul Freeman

Aces: Iron Eagle 3 '92
The Dogs of War '81
Eminent Domain '91
Murderers Among Us: The Simon Wiesenthal Story '89
Prisoner of Rio '89
Raiders of the Lost Ark '81
The Sender '82
Shanghai Surprise '86
Unsuitable Job for a Woman '82
Whose Child Am I? '75
Without a Clue '88
A World Apart '88

Tony Freeman

Invincible Gladiators '64
Ocean Drive Weekend '85

Warwick Freeman (D)

Demonstrator '71

Willie Freeman

Leopard Fist Ninja '80s

Paul Frees

Hunt the Man Down/Smashing the Rackets '50
The Little Drummer Boy '68
The Milpitas Monster '75
Santa Claus is Coming to Town '78
Stingiest Man in Town '78
Suddenly '54

Hugo Fregonese (D)

Blowing Wild '54
Decameron Nights '53
Dracula vs. Frankenstein '69
Harry Black and the Tiger '58

Ace Frehley

KISS Meets the Phantom of the Park '78

Sally Frei

The Undertaker and His Pals '67

Edward Furlong

Terminator 2: Judgement Day '91

Yvonne Furneaux

Frankenstein's Great Aunt Tillie '83
The House of the Arrow '53
La Dolce Vita '60
The Lion of Thebes '64
The Master of Ballantrae '53
The Mummy '59
Repulsion '65
The Warriors '55

Betty Furness

Beggars in Ermine '34
The President's Mystery '36
Renegade Ranger/Scarlet River '38
Swing Time '36

Deborra-Lee Furness

The Last of the Finest '90
Shame '87

Chris Furrh

Lord of the Flies '90

Julia Fursich

Thundersquad '85

Esther Furst

Ring of Terror '62

Joseph Furst

Diamonds are Forever '71
Dunera Boys '85

Stephen Furst

The Dream Team '89
Getting Wasted '80
Midnight Madness '80
National Lampoon's Animal House '78
Swim Team '79
The Unseen '80
Up the Creek '84

Ira Furstenberg

The Battle of El Alamein '68

George Furth

Doctor Detroit '83

Yasuo Furuhata (D)

Station '81

Carl Fury

Posed For Murder '89

Ed Fury

Colossus and the Amazon Queen '64
Samson Against the Sheik '62
Ursus in the Valley of the Lions '62
Wild Women of Wongo '59

Susumo Fusuita

Sanshiro Sugata '43

Herbert Fux

Mark of the Devil '69

Robert Fyfe

Xtro '83

Victoria Fyodorova

Crime and Punishment '75

Marianne Gaba

The Choppers '61

Richard Gabai

Assault of the Party Nerds '89
Hot Under the Collar '91
Virgin High '90

Richard Gabai (D)

Assault of the Party Nerds '89
Hot Under the Collar '91
Virgin High '90

Martin Gabel

James Dean Story '57
Lady in Cement '68
Smile, Jenny, You're Dead '74
The Thief '52

Martin Gabel (D)

Lost Moment '47

Scilla Gabel

Mill of the Stone Women '60

Jean Gabin

The Big Grab '63
Duke of the Derby '62
Four Bags Full '56
French Can-Can '55
Grand Illusion '37
La Bete Humaine '38
Le Cas du Dr. Laurent '57
Le Chat '75
Le Gentleman D'Epsom '62
Le Jour Se Leve '39
Le Plaisir '52
Les Miserables '57
The Lower Depths '36
Maria Chapdelaine '34
Melodie en Sous-Sol '63
Napoleon '55
Pepe le Moko '37
Stormy Waters '41
Un Singe en Hiver '62
Zou Zou '34

Christopher Gable

The Boy Friend '71
The Rainbow '89

Clark Gable

Across the Wide Missouri '51
Betrayed '54
Boom Town '40
But Not for Me '59
Candid Hollywood '62
China Seas '35
Command Decision '48
Dance Fools Dance '31
Dancing Lady '33
Forsaking All Others '35
A Free Soul '31
Gone with the Wind '39
Honky Tonk '41
The Hucksters '47
Idiot's Delight '39
It Happened One Night '34
It Started in Naples '60
The King and Four Queens '56
The Lost Stooges '33
Making of a Legend: "Gone With the Wind" '89
Manhattan Melodrama '34
The Misfits '61
Mogambo '53
Mutiny on the Bounty '35
1939, The Movies' Vintage Year: Trailers on Tape '39
No Man of Her Own '32
Painted Desert '31
Possessed '31
Red Dust '32

Eva Gabor

Artists and Models '55
Gigi '58
Last Time I Saw Paris '54
A New Kind of Love '63
Princess Academy '87
The Rescuers Down Under '90
Tales of the Klondike: The Scorn of Women '81
Truth About Women '58

Miklos Gabor

Father '67

Pal Gabor (D)

Brady's Escape '84

Zsa Zsa Gabor

Death of a Scoundrel '56
Every Girl Should Have One '78
Frankenstein's Great Aunt Tillie '83
Lili '53
Lovely to Look At '52
Moulin Rouge '52
Movie Maker '86
Pee-Wee's Playhouse Christmas Special '88
Picture Mommy Dead '66
Queen of Outer Space '58
Stars of the Century '88
Touch of Evil '58
We're Not Married '52

Mitchell Gabourie (D)

Buying Time '89

Richard Gabourie

Title Shot '81

Radu Gabrea (D)

A Man Like Eva '83
Secret of the Ice Cave '89

Mike Gabriel (D)

The Rescuers Down Under '90

Peter Gabriel

Nelson Mandela 70th Birthday Tribute '88
New York Stories '89

Roman Gabriel

The Undefeated '69

Monique Gabrielle

Amazon Women On the Moon '87
Bachelor Party '84
Deathstalker 2: Duel of the Titans '87
Emmanuelle 5 '87
Evil Toons '90
Hard to Hold '84
Hot Moves '84

Run Silent, Run Deep '58

Run Silent, Run Deep '58
San Francisco '36
Soldier of Fortune '55
Strange Cargo '40
Susan Lenox: Her Fall and Rise '31
The Tall Men '55
Teacher's Pet '58
Test Pilot '39
Too Hot to Handle '38

John Clark Gable

Bad Jim '89

Eva Gabor

Silk 2 '89

Silk 2 '89

Gabriel Gabrio

Harvest '37
Pepe le Moko '37

Leo Gabrowski

Deadly Business

Amelia Gade

House of Insane Women '74

Analia Gade

Murder Mansion '70

Antonio Gades

Blood Wedding (Bodas de Sangre) '81
Carmen '83
El Amor Brujo '86

Frederic Gadette (D)

This is Not a Test '62

Jon Gadsby

Second Time Lucky '84

Anna Gael

Alien Women '69
Therese & Isabelle '68

Eddie M. Gaerlan

Trident Force '88

John Gaffari

Hundra '85

Nathy Gaffney

Contagion '87

Robert Gaffney (D)

Frankenstein Meets the Space Monster '65

Erford Gage

The Seventh Victim '43

George Gage (D)

Fleshburn '84

John Gage (D)

Velvet Touch '48

Verne Gagne

The Wrestler '73

Holly Gagnier

Girls Just Want to Have Fun '85

Jenny Gago

Old Gringo '89
Sweet 15 '90

Blake Gahner

Lethal Pursuit '88

Michael Gahr

The Nasty Girl '90

Jane Gail

Traffic in Souls '13

Jeanne Gail

The Woman Who Came Back '45

June Gail

20,000 Leagues Under the Sea '16

Max Gail

Aliens Are Coming '80
Amazing Stories, Book 4 '85
Cardiac Arrest '74
D.C. Cab '84
The Game of Love '90
Heartbreakers '84
Judgment in Berlin '88
Night Moves '75
Street Crimes '92
Where Are the Children? '85

Tim Gail

If Looks Could Kill '86

Ambrose Gaines, III (D)

Scuba '72

Boyd Gaines

Call Me '88
The Sure Thing '85

Jim Gaines

Blood Debts '83
Codename: Terminate '90
Commando Invasion '87

Richard Gaines

Enchanted Cottage '45
Humoresque '47
The More the Merrier '43

Will Gaines

Masters of Tap '88

Courtney Gains

Back to the Future '85
The 'Burbs '89
Can't Buy Me Love '87
Children of the Corn '84
Hardbodies '84
Lust in the Dust '85

Charlotte Gainsbourg

Le Petit Amour '87
The Little Thief '89

Serge Gainsbourg

The Fury of Hercules '61
I Love All of You '83
Samson '61
Trop Jolie pour Etre Honette '60s

Rene Gainville (D)

Associate '82
Le Complot (The Conspiracy) '73

Cristina Gajoni

Ursus in the Valley of the Lions '62

Yoram Gal

Holocaust Survivors... Remembrance of Love '83

Michel Galabru

The Judge and the Assassin '75
La Cage aux Folles '78
La Cage aux Folles 2 '81
La Cage aux Folles 3: The Wedding '86
Soldat Duroc...Ca Va Etre Ta Fete! '60s

Tony Galati

The Darkside '87

Richard Galbreath (D)

Night of Evil '62

David Gale

The Brain '88
Bride of Re-Animator '89
The Re-Animator '85
Syngenor '90

Eddra Gale

Revenge of the Cheerleaders '76

Joan Gale

Miracle Rider: Vol. 1 '35
Miracle Rider: Vol. 2 '35

John Gale (D)

Commando Invasion '87
The Firing Line '91
Slash '87

June Gale

It Could Happen to You '39
Rainbow's End '35

Genevieva Galea

Les Carabiniers '63

Johnny Galecki

National Lampoon's Christmas Vacation '89

Timothy Galfas (D)

Sunnyside '79

Peter Galfy

Time Stands Still '82

Katie Galian

Marie Galante '34

Juan Luis Galiardo

Don Juan My Love '90

Kelly Galindo

Angels of the City '89

Frank Gallacher

Waterfront '83

Gallagher

Comedy Tonight '77
Gallagher in Concert '88
Gallagher: Melon Crazy '85
Gallagher: Over Your Head '86
Gallagher's Overboard '87
Melon Crazy '85

Bronagh Gallagher

The Commitments '91

Francis Gallagher

Hurry Up or I'll Be Thirty '73

John A. Gallagher

Beach House '82

John A. Gallagher (D)

Street Hunter '90

Megan Gallagher

The Ambulance '90

Patti Gallagher

King Dinosaur '55

Peter Gallagher

The Caine Mutiny Court Martial '88
Dreamchild '85
High Spirits '88
Idolmaker '80
An Inconvenient Woman '91
Late for Dinner '91
Long Day's Journey into Night '88
The Murder of Mary Phagan '87
My Little Girl '87
Private Contentment '83
sex, lies and videotape '89
Skag '79
Summer Lovers '82
Tune in Tomorrow '90

Skeets Gallagher

Bachelor Bait '34
Danger on the Air '38
Hats Off '37
Possessed '31

Vera Gallagher

Girl School Screamers '86
Girls School Screamers '86

Jean Galland

La Barbiec de Seville '34

Philippe Galland

Overseas: Three Women with Man Trouble '90

Juan Gallardo

El Cristo de los Milagros '80s
Inquisition '76
Los Sheriffs de la Frontera '47

Rosa Maria Gallardo

The Brainiac '61

Silvania Gallardo

Out of the Dark '88

Ely Galleani

Redneck '73

Gina Gallego

Lust in the Dust '85
Personals '90

Mariquita Gallegos

La Guerra de los Sostenes

Ira Gallen (D)

Ronnie Dearest: The Lost Episodes '88

Georges Galley

Gorilla '56

Rosina Galli

Escape to Paradise '39
Fisherman's Wharf '39

Zach Galligan

Gremlins '84
Gremlins 2: The New Batch '90
Mortal Passions '90
Psychic '91
Rebel Storm '90
Round Trip to Heaven '92
Waxwork '88
Waxwork 2: Lost in Time '91

Ely Gallo

Kiss Me, Kill Me '69

Fred Gallo (D)

Dead Space '90

George Gallo (D)

29 Street '91

Maresa Gallo

Love in the City '53

Robert Gallo

Mayhem '87
Sinners '89

William Gallo

Night of the Demons '88

Carmine Gallone (D)

Carthage in Flames '60

Frank Gallop

Great Chase '63

Dada Galloti

Frankenstein '80 '79

Dick Galloway

Meet the Parents '91

Don Galloway

Demon Rage '82
Snowblind '78
Two Moon Junction '88

Jack Galloway

Codename: Icarus '85

Scott Galloway

Red Snow '91

Samuel Gallu (D)

The Man Outside '68
Theatre of Death '67

Agi Gallus

Psycho Girls '86

Dave Galluzzo

Terror on Tour '83

Jose Galvez

Benjamin Argumedo '78
Semaforo en Rojo '47

W. Randolph Galvin

Ballbuster

Milan Galvonic

Rogue '76

Rita Gam

Distortions '87
King of Kings '61
Klute '71
Midnight '89
Mohawk '56
Noah: The Deluge '79
Seeds of Evil '76
The Thief '52

Larry Gamal

Feelin' Screwy '90

Gloria Gamata

Kashmiri Run '69

Sally Gamble

10 Violent Women '79

Michael Gambon

The Cook, the Thief, His Wife & Her Lover '90
Forbidden Passion: The Oscar Wilde Movie '87
Missing Link '88
Mobsters '91
The Rachel Papers '89
Turtle Diary '86

Tom Gamen

Lord of the Flies '63

Lee Games (D)

Actors and Sin '52

Robin Gammell

Guilty by Suspicion '91
The Haunting of Julia '77
Missing Pieces '83
Panic in Echo Park '77

James Gammon

Ballad of Gregorio Cortez '83
Coupe de Ville '90
Crisscross '92
Deadly Encounter '72
Hard Traveling '85
Leaving Normal '92
Major League '89
The Milagro Beanfield War '88
Revenge '90
Roe vs. Wade '89

Ken Gampu

Chain Gang Killings '85
Dingaka '65
Enemy Unseen '91
Kill and Kill Again '81
The Naked Prey '65
Scavengers '87
Soul Patrol '80

Yvonne Gamy

Letters from My Windmill '54

Abel Gance

Napoleon '27

Abel Gance (D)

The Battle of Austerlitz '60
J'Accuse '37
Napoleon '27
The Torture of Silence '17

Marguerite Gance

La Chute de la Maison Usher '28

Jin Gang

Dynasty '77

Ray Gange

The Clash: Rude Boy '80

Tony Ganios

Continental Divide '81
Porky's '82
Porky's 2: The Next Day '83
Porky's Revenge '85

Gail Ganley

Blood of Dracula '57

Albert C. Gannaway (D)

Daniel Boone: Trail Blazer '56

Dennis Gannon

Murderlust '86

Richard Ganoung

Parting Glances '86

Richard Gant

The Freshman '90

Paul Ganus

Crash and Burn '90

Bruno Ganz

American Friend '77
In the White City '83
Strapless '90
Wings of Desire '88

Teresa Ganzel

C.O.D. '83

Yehoram Gaon

Dead End Street '83
Operation Thunderbolt '77

Angel Garasa

Boom in the Moon '46
Un Novio Para Dos Hermanas (One Boyfriend for Two Sisters) '70s

Kaz Garaz

Final Mission '84
Naked Vengeance '85

Matthew Garber

Gnome-Mobile '67
Three Lives of Thomasina '63

Greta Garbo

Anna Christie '30
Anna Karenina '35
As You Desire Me '32
The Atonement of Gosta Berling '24
Camille '36
Conquest '32
The Flesh and the Devil '27
Grand Hotel '32
Joyless Street '25
The Kiss '29
The Love Goddesses '65
Mata Hari '32
The Mysterious Lady '28
Ninotchka '39
The Painted Veil '34
Queen Christina '33
The Single Standard '29
Susan Lenox: Her Fall and Rise '31
Two-Faced Woman '41
Wild Orchids '28
A Woman of Affairs '28

Ingrid Garbo

Touch Me Not '74

Delia Garces

El: This Strange Passion '52

Isabel Garces

Como Dos Gotas de Agua (Like Two Drops of Water) '70s
El Preprimido (The Timid Bachelor) '70s

John Garces

Robby '68

Mauricio Garces

The Brainiac '61
Bromas S.A. '87
Despedida de Casada '87
The Living Head '59
Los Hermanos Diablo '65

Gavin ▶*cont.*

House of Shadows '83
Imitation of Life '59
Jennifer '78
Midnight Lace '60
Murder for Sale '70
Psycho '60
Sophia Loren: Her Own Story '80
Spartacus '60
Thoroughly Modern Millie '67
A Time to Love & a Time to Die '58

Cassandra Gaviola

The Black Room '82
Conan the Barbarian '82

Victor Gaviria (D)

Rodrigo D: No Future '91

Peter Gawthorne

Phantom Fiend '35

William Gaxton

Best Foot Forward '43
It's the Old Army Game '26

Arnette Gay

Zombie '80

Gregory Gay

Seven Doors to Death '44

Ramon Gay

The Curse of the Aztec Mummy '59
Curse of the Mummy/Robot vs. the Aztec Mummy '60s
The Robot vs. the Aztec Mummy '59

Annie Gaybis

Hollywood Zap '86

Gregory Gaye

Charlie Chan at the Opera '36

Howard Gaye

Dante's Inferno '24
Flirting with Fate '16

Lisa Gaye

Castle of Evil '66
Class of Nuke 'Em High Part 2: Subhumanoid Meltdown '91
Night of Evil '62
Sign of Zorro '60
Toxic Avenger, Part 3: The Last Temptation of Toxie '89

Marvin Gaye

Chrome and Hot Leather '71
Motown 25: Yesterday, Today, Forever '83
Ready Steady Go, Vol. 2 '85
Ready Steady Go, Vol. 3 '84
Rock 'n' Roll History: Rock 'n' Soul Heaven '90

Nora Gaye

Dubeat-E-O '84

Jackie Gayle

Bert Rigby, You're a Fool '89
Pepper and His Wacky Taxi '72
Tin Men '87

Monica Gayle

Nashville Girl '76
Switchblade Sisters '75
Undercover Vixens '87

Anna Gaylor

The Killing Game '67

Gerry Gaylor

Queen of Outer Space '58

Mitch Gaylord

American Anthem '86
American Tiger '89

Raymond Gaylord

Face of the Screaming Werewolf '59

George Gaynes

Breaking Up Is Hard to Do '79
Dead Men Don't Wear Plaid '82
It Came Upon a Midnight Clear '84
Micki & Maude '84
Police Academy '84
Police Academy 2: Their First Assignment '85
Police Academy 3: Back in Training '86
Police Academy 4: Citizens on Patrol '87
Police Academy 5: Assignment Miami Beach '88
Police Academy 6: City Under Siege '89

Janet Gaynor

The Farmer Takes a Wife '35
A Star is Born '37
Sunrise '27
Till the Clouds Roll By/A Star Is Born '46
The Young in Heart '38

Jock Gaynor

Deathhead Virgin '74

Mitzi Gaynor

Birds & the Bees '56
For Love or Money '63
Les Girls '57
South Pacific '58
There's No Business Like Show Business '54
We're Not Married '52

Eunice Gayson

Dr. No '62
From Russia with Love '63

Joe Gayton (D)

Warm Summer Rain '89

Armand Gazarian (D)

Game of Survival '89

John Gazarian

Real Bullets '90

Gyula Gazdag (D)

A Hungarian Fairy Tale '87
Stand Off '89

Wendy Gazelle

Hot Pursuit '87
Sammy & Rosie Get Laid '87
Triumph of the Spirit '89

Understudy: The Graveyard Shift 2 '88

Ben Gazzara

Anatomy of a Murder '59
The Bridge at Remagen '69
Control '87
Death of Richie '76
An Early Frost '85
High Velocity '76
Neptune Factor '73
Passionate Thief '60
Pursuit '72
QB VII '74
Question of Honor '80
Quicker Than the Eye '88
Richie '77
Road House '89
Saint Jack '79
Secret Obsession '88
Shattered Silence '71
Sicilian Connection '74
Sidney Sheldon's Bloodline '79
Tales of Ordinary Madness '83
They All Laughed '81
The Trial of Lee Harvey Oswald '77
Voyage of the Damned '76

Michael Gazzo

Black Sunday '77
Cookie '89
Fingers '78
Gangsters '79
Godfather 1902-1959—The Complete Epic '81
The Godfather: Part 2 '74
Kill Castro '80
King of the Gypsies '78
Love and Bullets '79
The Mercenaries '80
Sweet Dirty Tony '81

Luella Gear

Phffft! '54

Valerie Gearon

Invasion '65

Anthony Geary

Crack House '89
Dangerous Love '87
High Desert Kill '90
The Imposter '84
Kicks '85
Night of the Warrior '91
Pass the Ammo '88
Penitentiary 3 '87
Scorchers '92
UHF '89
You Can't Hurry Love '88

Bud Geary

D-Day on Mars '45

Maine Geary

Robin Hood '22

Gunther Gebel-Williams

Rings Around the World '67

Gordon Gebert

Narrow Margin '52

Glenn Gebhard (D)

One Last Run '89

Nicholas Gecks

Forever Young '85

Jason Gedrick

The Heavenly Kid '85
Iron Eagle '85
Massive Retaliation '85
Promised Land '88
Rooftops '89
Stacking '87
The Zoo Gang '85

Prunella Gee

Never Say Never Again '83
The Wilby Conspiracy '75
Witching Time '84

Ellen Geer

Babe! '75
Bloody Birthday '80
Hard Traveling '85
Harold and Maude '71
Memory of Us '74
A Shining Season '79
Silence '73

Roman Geer

After the Fall of New York '85

Will Geer

Bandolero! '68
The Billion Dollar Hobo '78
Black Like Me '64
Broken Arrow '50
Brother John '70
Bunco '83
Dear Dead Delilah '72
Executive Action '73
The Hanged Man '74
Hurricane '74
Jeremiah Johnson '72
Lust for Gold '49
Memory of Us '74
Moving Violation '76
My Sister, My Love '78
Napoleon and Samantha '72
The President's Analyst '67
The Reivers '69
Salt of the Earth '54
Silence '73
Thanksgiving Story '73
Unknown Powers '80
The Waltons' Christmas Carol '80s
Winchester '73 '50
A Woman Called Moses '78

Judy Geeson

Adventures of a Taxi Driver '76
Berserk! '67
Carry On England '76
Doomwatch '72
Dynasty of Fear '72
The Executioner '70
Horror Planet '80
Murder on the Midnight Express '74
The Plague Dogs '82
10 Rillington Place '71
To Sir, with Love '67
Twinsanity '70

Sally Geeson

Cry of the Banshee '70
The Oblong Box '69

Deborah Geffner

Exterminator 2 '84

Martha Gehman

The Flamingo Kid '84

Pleasant Gehman

The Runnin' Kind '89
The Running Kind '89

Lou Gehrig

Rawhide '38

William Geiger

Beyond the Door 3 '91

Joe "Corky" Geil

Our Gang Comedies '40s

Dieter Geissler (D)

Deadly Game '83

Bob Geldof

Pink Floyd: The Wall '82
The Secret Policeman's Other Ball '82

Ruth Geler

Hamsin '83

Daniel Gelin

Is Paris Burning? '68
La Ronde '51
Mademoiselle Striptease '57
The Man Who Knew Too Much '56
Murmur of the Heart '71
The Testament of Orpheus '59
A Woman of Rome '56

Manuel Gelin

Oasis of the Zombies '82

Gratien Gelinas

Agnes of God '85

Sarah Gellar

High Stakes '89

Bruce Geller (D)

The Savage Bees '76

Uri Geller

Beyond Belief '76

Larry Gelman

Chatterbox '76
Slumber Party '57 '76

Rhoda Gemignani

Concrete Beat '84

Giuliano Gemma

Battleforce '78
Corleone '79
Erik, the Viking '72
Goliath and the Sins of Babylon '64
Master Touch '74
Smugglers '75
Unsane '82
The Warning '80

Julian Gemma

The Opponent '89

Laura Gemser

Ator the Fighting Eagle '83
Black Cobra '83
Caged Women '84
Crime Busters '78
Emmanuelle in the Country '78
Emmanuelle, the Queen '75
Emmanuelle on Taboo Island '76
Emmanuelle's Daughter '79
Fury '80s
Love Camp '76
Naked Paradise '78

Women's Prison Massacre '85

Denise Gence

Buffet Froid '79

Francois-Eric Gendron

Boyfriends & Girlfriends '88
Cloud Waltzing '87

Bryan Genesse

California Casanova '91
Loose Screws '85

Marcelle Geniat

Crime and Punishment '35

Augusto Genina (D)

Prix de Beaute '30

Leo Genn

Circus of Fear '67
Green for Danger '47
Henry V '44
Lady Chatterley's Lover '55
The Longest Day '62
Mackintosh Man '73
Moby Dick '56
Strange Case of Dr. Jekyll &
    Mr. Hyde '68
Velvet Touch '48
When Thief Meets Thief '37
Wooden Horse '50

Lina Gennari

Umberto D '55

Arnaldo Genoino (D)

Herod the Great '60

Dick Genola

The Psychic '68

Claude Gensac

Cloud Waltzing '87

John Gentil (D)

Battle of Valiant '63

Denise Gentile

Netherworld '91

Linda Gentile

To All a Goodnight '80

Nikki Gentile

Traveling Companions '86
Truck Stop

Robert Gentile

Strangers in the City '62

Giacomo Gentilomo (D)

Hercules Against the Moon
    Men '64

John Gentry

Hard Drivin' '60

Minnie Gentry

Def by Temptation '90
Georgia, Georgia '72

Roger Gentry

Alien Massacre '67
Horrors of the Red Planet '64

Lisa Geoffreion

Disturbance '89

Paul Geoffrey

Excalibur '81
Flame to the Phoenix '85

Stephen Geoffreys

The Chair '87
Fraternity Vacation '85
Fright Night '85
Moon 44 '90
976-EVIL '88

Brian George

Smokescreen '90

Chief Dan George

Americathon '79
The Bears & I '74
Cancel My Reservation '72
Dan Candy's Law '73
Harry and Tonto '74
Little Big Man '70
Outlaw Josey Wales '76

Christopher George

AngKor: Cambodia Express
    '81
Chisum '70
Cruise Into Terror '78
Day of the Animals '77
Dixie Dynamite '76
El Dorado '67
Enter the Ninja '81
Exterminator '80
Gates of Hell '83
Graduation Day '81
Grizzly '76
Midway '76
Mortuary '81
Pieces '83
Train Robbers '73
Whiskey Mountain '77

George W. George (D)

James Dean Story '57

Gladys George

The Best Years of Our Lives
    '46
Detective Story '51
Flamingo Road '49
House Across the Bay '40
Lullaby of Broadway '51
The Maltese Falcon '41
Marie Antoinette '38
The Roaring Twenties '39

Goetz George

Out of Order '84

Grace George

Johnny Come Lately '43

Heinrich George

Jud Suess '40
Kolberg '45

Jan George

The American Soldier '70

Lynda Day George

Aliens from Spaceship Earth
    '77
Ants '77
The Barbary Coast '74
Beyond Evil '80
Casino '80
Cruise Into Terror '78
Day of the Animals '77
Junkman '82

Mortuary '81
Panic on the 5:22 '74
Pieces '83
Racquet '79
Roots, Vol. 1 '77
Young Warriors '83

Maude George

Foolish Wives '22
Wedding March '28

Nathan George

Short Eyes '79

Pat George

T-Bird Gang '59

Peter George (D)

Surf Nazis Must Die '87

Rita George

Hollywood Boulevard '76
On the Run '85

Susan George

Die Screaming, Marianne '73
Dirty Mary Crazy Larry '74
Dr. Jekyll and Mr. Hyde '73
Enter the Ninja '81
Fright '71
House Where Evil Dwells '82
Jack the Ripper '88
The Jigsaw Man '84
Kiss My Grits '82
Lightning: The White Stallion
    '86
The Looking Glass War '69
Mandingo '75
Mission: Monte Carlo '75
Out of Season '75
Pajama Tops '83
A Small Town in Texas '76
Sonny and Jed '73
Straw Dogs '72
Sudden Terror '70
That Summer of White Roses
    '90
Tintorera...Tiger Shark '78
Tomorrow Never Comes '77
Venom '82

Olga Georges-Picot

Children of Rage '75
The Day of the Jackal '73
The Graveyard '74
Honor Among Thieves '68
The Man Who Haunted
    Himself '70

Tom Georgeson

A Fish Called Wanda '88
No Surrender '86

Aviva Ger

Secret of Yolanda '82

Carmelita Geraghty

Flaming Signal '33
My Lady of Whims '25
Night Life in Reno '31
Rogue of the Rio Grande '30

Maurice Geraghty (D)

Sword of Monte Cristo '51

Bernard Gerard (D)

Gone with the West '72

Charles Gerard

Happy New Year '74

Danny Gerard

Drop-Out Mother '88

Gil Gerard

Buck Rogers in the 25th
    Century '79
Final Notice '89
For Love or Money '84
Fury to Freedom: The Life
    Story of Raul Ries '85
Help Wanted: Male '82
Hooch '76
Killing Stone '78
Monsters, Madmen, Machines
    '84
Soldier's Fortune '91

Harriet Gerard

Vampyr '31

Steve Geray

Affair in Trinidad '52
The Big Sky '52
Inspector Hornleigh '39
Jesse James Meets
    Frankenstein's Daughter
    '65
Ladies of the Chorus '49
The Moon and Sixpence '42
Tobor the Great '54

Emil Gerber

The Last Chance '45

Joan Gerber

Tobor the Great '54

Neva Gerber

The Fighting Stallion '26

Steve Gerbson (D)

Elayne Boosler: Broadway
    Baby '87

Richard Gere

American Gigolo '79
Beyond the Limit '83
Bloodbrothers '78
Breathless '83
The Cotton Club '84
Days of Heaven '78
Final Analysis '92
Internal Affairs '90
King David '85
Looking for Mr. Goodbar '77
Miles from Home '88
No Mercy '86
An Officer and a Gentleman
    '82
Power '86
Pretty Woman '89
Report to the Commissioner
    '74
Reporters '81
Strike Force '75
Yanks '79

Georges Geret

Diary of a Chambermaid '64
Z '69

Lucy Gerhman

A Brivele der Mamen (A Letter
    to Mother) '38

Misha Gerhman

A Brivele der Mamen (A Letter
    to Mother) '38

Haile Gerima (D)

Ashes and Embers '82

Marion Gering (D)

Thunder in the City '37

Richard Gering

Date Bait '60

Sean Gerlis

A Summer to Remember '84

Gretchen German

A Man Called Sarge '90

Gala Germani

Castle of the Living Dead '64

Pietro Germi (D)

Alfredo, Alfredo '73
Divorce—Italian Style '62
Four Ways Out '51
Seduced and Abandoned '64

Clyde Geronimi (D)

The Adventures of Ichabod
    and Mr. Toad '49
Alice in Wonderland '51
The Legend of Sleepy Hollow
    '49
Sleeping Beauty '59

Charles Gerrard

California Straight Ahead '25
Down to Earth '17
The Nervous Wreck '26

Peter Gerretsen

Kidnapping of Baby John Doe
    '88

Peter Gerretsen (D)

Night Friend '87

Flo Gerrish

Don't Answer the Phone '80

Daniel Gerroll

Big Business '88
Drop Dead Fred '91

Kurt Gerron

The Blue Angel '30

Alex Gerry

The Bellboy '60

Gerry & the Pacemakers

The T.A.M.I. Show '64

Savina Gersak

Beyond the Door 3 '91
Curse 2: The Bite '88
Iron Warrior '87
Sonny Boy '87
War Bus Commando '89

Sandy Gershman

Angels of the City '89

Gina Gershon

Cocktail '88
Pretty in Pink '86
Voodoo Dawn '89

Theodore Gershuny (D)

Silent Night, Bloody Night '73

George Gershwin

Clark & McCullough '30s

Betty Lou Gerson

101 Dalmations '61

Jeanne Gerson

She Gods of Shark Reef '56
The Touch of Satan '70

Paul Gerson

Cricket on the Hearth '23

Berta Gersten

A Brivele der Mamen (A Letter
   to Mother) '38
Mirele Efros '38

Frank Gerstle

The Atomic Brain '64
Gang Busters '55
Hell on Wheels '67
Submarine Seahawk '59
The Wasp Woman '49

Valeska Gert

The Threepenny Opera '31

Jami Gertz

Alphabet City '84
Crossroads '86
Don't Tell Her It's Me '90
Endless Love '81
Less Than Zero '87
Listen to Me '89
The Lost Boys '87
Mischief '85
Quicksilver '86
Renegades '89
Sibling Rivalry '90
Silence Like Glass '90
Sixteen Candles '84
Solarbabies '86

Bruno Gerussi

The Hitman '91

Nicolas Gessner (D)

It Rained All Night the Day I
   Left '78
The Little Girl Who Lives
   Down the Lane '76
The Peking Blond '68
Quicker Than the Eye '88
Someone Behind the Door '71
12 Plus 1 '70

Richard Gesswein

R.O.T.O.R. '88

Inna Gest

Six Gun Gospel '43

Steven Gethers

Amazing Stories, Book 3 '86

Steven Gethers (D)

Billy: Portrait of a Street Kid
   '77
Damien: The Leper Priest '80
The Hillside Strangler '89
Jacqueline Bouvier Kennedy
   '81

Balthazar Getty

December '91
Lord of the Flies '90
My Heroes Have Always Been
   Cowboys '91
The Pope Must Diet '91
Young Guns 2 '90

Estelle Getty

Mannequin '87
Mask '85

Stop! or My Mom Will Shoot '91

Paul Getty, III

State of Things '82

John Getz

Blood Simple '85
Born on the Fourth of July '89
Concrete Beat '84
Curly Sue '91
Don't Tell Mom the
   Babysitter's Dead '91
The Fly '86
The Fly 2 '89
Kent State '81
Tattoo '81
Thief of Hearts '84

Robert Getz

Heroes Die Young '60

Stan Getz

The Benny Goodman Story
   '55

Stuart Getz

The Van '77

Bernard Geurts

Thundering Ninja '80s

Gina Gevshon

Red Heat '88

Steve Geyer

Slow Men Working '90

Alle Ghadban

Prom Night 4: Deliver Us From
   Evil '91

Alice Ghostley

Blue Sunshine '78
Gator '76
The Graduate '67
New Faces '54
Not for Publication '84
To Kill a Mockingbird '62
With Six You Get Eggroll '68

Fosco Giachetti

We the Living '42

Francis Giacobetti (D)

Emmanuelle 4 '84
Emmanuelle, the Joys of a
   Woman '76

Louis Giambalvo

The Ambush Murders '82
Devlin Connection '82

Joseph Gian

Blackout '88
Mad About You '90
Night Stalker '87

Rick Gianasi

Escape from Safehaven '88
Maximum Thrust '87
Mutant Hunt '87
The Occultist '89
Posed For Murder '89
Robot Holocaust '87

Alfredo Gianelli (D)

Mother Doesn't Always Know
   Best '60s

Ettore Giannini

Europa '51 '52

Giancarlo Giannini

American Dreamer '84
Bellissimo: Images of the
   Italian Cinema '87
Blood Feud '79
Blood Red '88
Fever Pitch '85
Goodnight, Michelangelo '89
How Funny Can Sex Be? '76
Immortal Bachelor '80
The Innocent '76
La Grande Bourgeoise '74
Life is Beautiful '79
Love and Anarchy '73
Lovers and Liars '81
Murri Affair '74
New York Stories '89
A Night Full of Rain '78
Once Upon a Crime '91
Saving Grace '86
Seduction of Mimi '74
The Sensual Man '74
Seven Beauties '76
The Sleazy Uncle '90
Swept Away... '75
Time to Kill '89
Where's Piccone '84

Simone Gianozzi

Misunderstood '87

Ian Giatti

Long Shot '81

Andy Gibb

Olivia '80

Cynthia Gibb

Death Warrant '90
Jack's Back '87
Malone '87
Modern Girls '86
Salvador '86
Short Circuit 2 '88
Stardust Memories '80
Youngblood '86

Donald Gibb

Bloodsport '88

Robyn Gibbes

Gone to Ground '76

Duncan Gibbins (D)

Eve of Destruction '90
Fire with Fire '86

Harry Gibbon

The Tomboy '24

Pamela Gibbons (D)

Midnight Dancer '87

Angela Gibbs

The Young Nurses '73

Judi Gibbs

Deadline Auto Theft '83

Rick Gibbs

Buckeye and Blue '87

Robyn Gibbs

Wild Horses '82

Timothy Gibbs

The Kindred '87

Josianne Gibert

The Screaming Dead '72

Belinda Giblin

Demolition '77

Rebecca Gibney

Among the Cinders '83
Jigsaw '90

Alan Gibson (D)

Martin's Day '85
The Satanic Rites of Dracula
   '73
Silent Scream '84
Twinsanity '70
The Two Faces of Evil '82
A Woman Called Golda '82

Brian Gibson (D)

Breaking Glass '80
The Josephine Baker Story
   '90
Murderers Among Us: The
   Simon Wiesenthal Story '89
Poltergeist 2: The Other Side
   '86

Colin Gibson

John & Julie '77

Deborah Gibson

Heat Street '87

Ethlyn Gibson

Working Winnie/Oh So Simple
   '22

Felicity Gibson

One Brief Summer '70

Gerry Gibson

Rage of Honor '87

Henry Gibson

The Blues Brothers '80
Charlotte's Web '79
For the Love of It '80
The Incredible Shrinking
   Woman '81
Innerspace '87
Kentucky Fried Movie '77
Kiss Me, Stupid! '64
Long Gone '87
The Long Goodbye '73
Monster in the Closet '86
Nashville '75
The Night Visitor '89
Switching Channels '88
Tulips '81
Tune in Tomorrow '90

Hoot Gibson

Arizona Whirlwind '44
Boiling Point '32
Cavalcade of the West '36
Clearing the Range '31
The Concentratin' Kid '30
Cowboy Counselor '33
Death Valley Rangers '44
Dude Bandit '32
Feud of the West '35
Fighting Parson '35
Frontier Justice '35
Hard Hombre '31
The Horse Soldiers '59
King of the Rodeo '28
Last Outlaw '36
The Law Rides Again '43
Local Badman '32
Lucky Terror '36
Man's Land '32
Marked Trails '44
The Marshal's Daughter '53
Outlaw Trail '44
The Painted Stallion '37

The Phantom Bullet '26
Powdersmoke Range '35
The Prairie King '27
Rainbow's End '35
The Riding Avenger '36
Sonora Stagecoach '44
Spirit of the West '32
Spurs '30
Straight Shootin' '17
Sunset Range '35
Swifty '36
Wild Engine/Mistaken Orders
   '23
Wild Horse '31

Hoot Gibson (D)

Spurs '30

Jody Gibson

Evil Laugh '86

Judith Gibson

Return of the Ape Man '44

Kevin Gibson

What Would Your Mother
   Say? '81

Kitty Gibson

Cal '84

Marcus Gibson

Ninja Brothers of Blood '89

Margaret Gibson

Last Days of Frank & Jesse
   James '86
Never a Dull Moment '50

Mel Gibson

Air America '90
Attack Force Z '84
Bird on a Wire '90
The Bounty '84
Gallipoli '81
Hamlet '90
Lethal Weapon '87
Lethal Weapon 2 '89
Mad Max '80
Mad Max Beyond
   Thunderdome '85
Mrs. Soffel '85
The River '84
The Road Warrior '82
Summer City '77
Tequila Sunrise '88
Tim '79
The Year of Living
   Dangerously '82

Mimi Gibson

Courage of Black Beauty '57
The Monster that Challenged
   the World '57

Robert Gibson (D)

Snow: The Movie '83

Sonny Gibson

Dark Before Dawn '89

Tom Gibson (D)

Singing Buckaroo '37

Virginia Gibson

Seven Brides for Seven
   Brothers '54
Tea for Two '50

Bond Gideon

Gold of the Amazon Women
   '79
Stepsisters '85
Storyville '74

Pamela Gidley

Blue Iguana '88
Disturbed '90
Highway to Hell '92
Liebestraum '91
Permanent Record '88
Thrashin' '86

John Gielgud

Appointment with Death '88
Around the World in 80 Days '56
Arthur '81
Arthur 2: On the Rocks '88
Becket '64
Brideshead Revisited '81
Caligula '80
Chariots of Fire '81
Chimes at Midnight '67
Elephant Man '80
11 Harrowhouse '74
The Formula '80
Frankenstein '82
Gandhi '82
Getting It Right '89
Hamlet '48
Hunchback of Notre Dame '82
Inside the Third Reich '82
Invitation to the Wedding '73
James Joyce: A Portrait of the Artist as a Young Man '77
Julius Caesar '53
Julius Caesar '70
Les Miserables '78
Lion of the Desert '81
The Loved One '65
A Man for All Seasons '88
Murder by Decree '79
Murder on the Orient Express '74
Plenty '85
Portrait of the Artist as a Young Man '78
Power of One '92
Priest of Love '81
Probe '72
Providence '76
QB VII '74
Richard II '79
Richard III '55
Romance on the Orient Express '89
Romeo and Juliet '54
Romeo and Juliet (Gielgud) '79
Saint Joan '57
Scandalous '84
The Scarlet & the Black '83
Sebastian '68
The Secret Agent '36
Shining Through '91
The Shoes of the Fisherman '68
The Shooting Party '85
Sphinx '81
Strike It Rich '90
Wagner: The Movie '85
War & Remembrance, Part 1 '88
War & Remembrance, Part 2 '89
The Whistle Blower '87
The Wicked Lady '83

Gary Giem

Hush Little Baby, Don't You Cry '86

Frances Gifford

American Empire '42
Thrill of a Romance '45

Frank Gifford

Up Periscope '59

Gloria Gifford

D.C. Cab '84
Vice Versa '88

Donn Gift

Fighting Father Dunne '48

Roland Gift

Sammy & Rosie Get Laid '87
Scandal '89

Elaine Giftos

Angel '84
Gas-s-s-s! '70
The Student Nurses '70
The Wrestler '73

John Gigante

Phantom Brother '88

Sandro Giglio

The War of the Worlds '53

Mark Gil

Trident Force '88

Rosemarie Gil

Devil Woman '76

Leslie Gilb

Lemora, Lady Dracula '73

Darrel Gilbeau

Lauderdale '89

Billy Gilbert

Block-heads '38
Breaking the Ice '38
Comedy Festival #1 '30s
Comedy Festival #2 '30s
Destry Rides Again '39
The Great Dictator '40
His Girl Friday '40
Joy of Living '38
Laurel & Hardy Comedy Classics Volume 1 '33
Laurel & Hardy Comedy Classics Volume 2 '30
Laurel & Hardy Comedy Classics Volume 3 '34
Laurel & Hardy Comedy Classics Volume 6 '35
Little Rascals, Book 6 '30s
Maid's Night Out '38
Melody Master '41
Mr. Wise Guy '42
Peck's Bad Boy with the Circus '38
Seven Sinners '40
Song of the Islands '42
Spotlight Scandals '43
Toast of New York '37
Two Reelers: Comedy Classics 2 '40s
Valley of the Sun '42

Brian Gilbert (D)

The French Lesson '86
Not Without My Daughter '90
Sharma & Beyond '84
Vice Versa '88

Catherine Gilbert

The Hanging Woman '72

Eugenia Gilbert

Test of Donald Norton '26

Florence Gilbert

Backfire '22

Helen Gilbert

Andy Hardy Gets Spring Fever '39
Death Valley '46
Girls in Prison '56
God's Country '46

John Gilbert

The Big Parade '25
The Busher '19
The Flesh and the Devil '27
Queen Christina '33
A Woman of Affairs '28

Lewis Gilbert (D)

The Adventurers '70
Alfie '66
A Cry from the Streets '57
Damn the Defiant '62
Educating Rita '83
Ferry to Hong Kong '59
Moonraker '79
Not Quite Paradise '86
Reach for the Sky '57
The Sea Shall Not Have Them '55
Shirley Valentine '89
Sink the Bismarck '60
The Spy Who Loved Me '77
Stepping Out '91
Tough Guy '53
You Only Live Twice '67

Lou Gilbert

The Great White Hope '70

Melissa Gilbert

Blood Vows: The Story of a Mafia Wife '87
Christmas Coal Mine Miracle '78
Family Secrets '84
Ice House '89
The Miracle Worker '79
Penalty Phase '86
Snow Queen '83
Sylvester '85

Paul Gilbert

Three Nuts in Search of a Bolt '64

Taylor Gilbert

Alone in the T-Shirt Zone '86
Torment '85

Yvonne Gilbert

2000 Maniacs '64

Bruce Gilchrist

Demented '80

Connie Gilchrist

The Half-Breed '51
Here Comes the Groom '51
Houdini '53
Long John Silver '53
Misadventures of Merlin Jones '63
Ticket to Tomahawk '50
Tortilla Flat '42

Richard Gilden

The Black Klansman '66

David Giler (D)

Black Bird '75

David Giles (D)

Mansfield Park '85
A Murder is Announced '87

Sandra Giles

Daddy-O '59

Tom Giles (D)

Journey Through Rosebud '72

Jack Gilford

Arthur 2: On the Rocks '88
Catch-22 '70
Caveman '81
Cheaper to Keep Her '80
Cocoon '85
Cocoon: The Return '88
Daydreamer '66
Enter Laughing '67
A Funny Thing Happened on the Way to the Forum '66
Harry & Walter Go to New York '76
The Incident '67
Max '79
Save the Tiger '73
They Might Be Giants '71
Wholly Moses! '80

Joseph Gilford (D)

Max '79

Mark Gilhuis (D)

Bloody Wednesday '87

Stuart Gillard

Virus '82

Stuart Gillard (D)

A Man Called Sarge '90
Paradise '82

Andre Gille

Voyage en Balloon '59

Jeff Gillen

Children Shouldn't Play with Dead Things '72

Jeff Gillen (D)

Deranged '74

Walter Giller

Fanny Hill: Memoirs of a Woman of Pleasure '64

Dana Gillespie

The People That Time Forgot '77
Scrubbers '82

Dizzy Gillespie

A Delicate Thread '86
Jazz is Our Religion '72

Anita Gillette

Ants '77
Brass '85
Marathon '80
Moonstruck '87

Mickey Gilley

Off the Wall '82

David Gilliam

Frogs '72
Gunpowder '84
Sharks' Treasure '75

Stu Gilliam

Dr. Black, Mr. Hyde '76

Terry Gilliam

And Now for Something Completely Different '72
Monty Python and the Holy Grail '75
Monty Python Live at the Hollywood Bowl '82
Monty Python Meets Beyond the Fringe
Monty Python's Life of Brian '79
Monty Python's Parrot Sketch not Included '90
Monty Python's The Meaning of Life '83
Spies Like Us '85

Terry Gilliam (D)

The Adventures of Baron Munchausen '89
Brazil '86
The Fisher King '91
Jabberwocky '77
Monty Python and the Holy Grail '75
Time Bandits '81

Ian Gillian

Butterfly Ball '76

Lawrence Gilliard

Straight Out of Brooklyn '91

Sidney Gilliat (D)

Endless Night '71
Great St. Trinian's Train Robbery '66
Green for Danger '47
Only Two Can Play '62

Andrew Gillies

Big Meat Eater '85

Isabel Gillies

Metropolitan '90

Richard Gilliland

Acceptable Risks '86
Challenge of a Lifetime '85
Embassy '85
Happy Hour '87
Killing in a Small Town '90

Linda Gillin

Terror at Red Wolf Inn '72
Windows '80

John Gilling (D)

The Gamma People '56
The Gilded Cage '54
It Takes a Thief '59
Mania '60
My Son, the Vampire '52
Night Caller from Outer Space '66
No Trace '50
Where the Bullets Fly '66
White Fire '53

Rebecca Gilling

The Blue Lightning '86

Anne Gillis

The Adventures of Tom Sawyer '38
Big Town After Dark '47
'Neath Brooklyn Bridge '42
Peck's Bad Boy with the Circus '38
Texas Legionnaires '43

Barry Gillis

Things '80s

James Gillis

Deranged '87
Dracula Sucks '79

Paul Michael Glaser (D)

Amazing Stories, Book 4 '85
The Amazons '84
Band of the Hand '86
The Running Man '87

Ned Glass

Fright '56
Kid Galahad '62
The Rebel Set '59
Requiem for a Heavyweight '56
Street Music '81
West Side Story '61

Ron Glass

The Crash of Flight 401 '78
Deep Space '87

Seamon Glass

This is Not a Test '62

Phillip Glasser

An American Tail: Fievel Goes West '91

Lesli Linka Glatter (D)

Amazing Stories, Book 3 '86
Amazing Stories, Book 5 '85
Into the Homeland '87

Bob Glaudini

The Alchemist '81
Angel City '84
Chameleon '81

Louise Glaum

The Leopard Woman '20
Return of Draw Egan '16
Sex '20
The Three Musketeers (D'Artagnan) '16

Eugene Glazer

Dollman '91

Jackie Gleason

All Through the Night '42
Don't Drink the Water '69
Fun Shows '50
The Hustler '61
Izzy & Moe '85
Mister Halpren & Mister Johnson '83
Mr. Billion '77
Nothing in Common '86
Orchestra Wives '42
Papa's Delicate Condition '63
Smokey and the Bandit '77
Smokey and the Bandit, Part 2 '80
Smokey and the Bandit, Part 3 '83
Soldier in the Rain '63
Springtime in the Rockies '42
The Sting 2 '83
The Toy '82

James Gleason

Arsenic and Old Lace '44
The Bishop's Wife '47
The Clock '45
Down to Earth '47
Ex-Mrs. Bradford '36
Falcon Takes Over/Strange Bargain '49
Footlight Serenade '42
A Free Soul '31
A Guy Named Joe '44
Here Comes Mr. Jordan '41

The Jackpot '50
The Keys of the Kingdom '44
Last Hurrah '58
Loving You '57
Meet John Doe '41
Miss Grant Takes Richmond '49
Movie Stunt Man '53
The Night of the Hunter '55
Suddenly '54
A Tree Grows in Brooklyn '45
Tycoon '47
We're Not Married '52
What Price Glory? '52

Joanna Gleason

Crimes & Misdemeanors '89
F/X 2: The Deadly Art of Illusion '91
For Richer, For Poorer '92
Heartburn '86

Michie Gleason (D)

Summer Heat '87

Pat Gleason

Criminal Court '46
I Killed That Man '42
Rogue's Gallery '44
Swing It, Professor '37

Paul Gleason

Challenge of a Lifetime '85
Die Hard '88
The Ewoks: Battle for Endor '85
Forever, Lulu '87
Fourth Story '90
Ghost Chase '88
He Knows You're Alone '80
Johnny Be Good '88
Miami Blues '90
Morgan Stewart's Coming Home '87
Night Game '89
Private Duty Nurses '71
Rich Girl '91
She's Having a Baby '88
Trading Places '83

Russell Gleason

All Quiet on the Western Front '30
Condemned to Live '35
Flying Fool '29
I Can't Escape '34

Nicholas Gledhill

Careful, He Might Hear You '84

Michele Gleizer

Europa, Europa '91

Iain Glen

Fools of Fortune '91
Mountains of the Moon '90
Rosencrantz & Guildenstern Are Dead '90

John Glen (D)

Aces: Iron Eagle 3 '92
For Your Eyes Only '81
License to Kill '89
The Living Daylights '87
Octopussy '83
A View to a Kill '85

Candace Glendenning

Flesh and Blood Show '73

Cody Glenn

Border Shootout '90

Montgomery Glenn

Castle of Blood '64

Raymond Glenn

The Return of Boston Blackie '27

Roy Glenn

Carmen Jones '54
Jungle Drums of Africa '53

Scott Glenn

Angels Hard as They Come '71
Apocalypse Now '79
As Summers Die '86
The Baby Maker '70
Backdraft '91
The Challenge '82
The Hunt for Red October '90
Intrigue '90
The Keep '83
Man on Fire '87
Miss Firecracker '89
My Heroes Have Always Been Cowboys '91
Off Limits '87
Personal Best '82
The Right Stuff '83
The River '84
She Came to the Valley '77
The Shrieking '73
The Silence of the Lambs '91
Silverado '85
Urban Cowboy '80
Verne Miller '88
Women & Men: Stories of Seduction, Part 2 '91

Bert Glennon (D)

Paradise Island '30

Frank Glennon

Aces and Eights '36

Peter Glenville (D)

Becket '64
The Prisoner '55
Summer and Smoke '61

Sharon Gless

The Crash of Flight 401 '78
Hardhat & Legs '80
Hobson's Choice '83
Letting Go '85
Revenge of the Stepford Wives '80
The Star Chamber '83

Stacey Glick

Three O'Clock High '87

James Glickenhaus (D)

Exterminator '80
McBain '91
Protector '85
Shakedown '88
The Soldier '82

Paul Glickler (D)

The Cheerleaders '72
Running Scared '79

Robert Gligorov

Stagefright '87

Arne Glimcher (D)

Mambo Kings '92

Greg Glionna

Meet the Parents '91

Greg Glionna (D)

Meet the Parents '91

Brian Glover

An American Werewolf in London '81
The Company of Wolves '85
The McGuffin '85

Bruce Glover

Chrome Hearts '70
Diamonds are Forever '71
Ghost Town '88
Killcrazy '89
Stunts '77

Crispin Glover

At Close Range '86
Back to the Future '85
Friday the 13th, Part 4: The Final Chapter '84
Little Noises '91
My Tutor '82
Racing with the Moon '84
River's Edge '87
Rubin & Ed '92
Teachers '84
Twister '89
Where the Heart Is '90

Danny Glover

Bat 211 '88
Chu Chu & the Philly Flash '81
The Color Purple '85
Dead Man Out '89
Escape from Alcatraz '79
Flight of the Intruder '90
Grand Canyon '91
Iceman '84
Kooping On '81
Lethal Weapon '87
Lethal Weapon 2 '89
Lonesome Dove '89
Mandela '87
Out '88
Places in the Heart '84
Predator 2 '90
Pure Luck '91
A Rage in Harlem '91
A Raisin in the Sun '89
Silverado '85
Stand-In '85
To Sleep with Anger '90
Witness '85

John Glover

Apology '86
The Chocolate War '88
Dead on the Money '91
An Early Frost '85
El Diablo '90
Ernie Kovacs: Between the Laughter '84
The Evil That Men Do '84
52 Pick-Up '86
Gremlins 2: The New Batch '90
A Killing Affair '85
The Last Embrace '79
A Little Sex '82
Masquerade '88
Meet the Hollowheads '89
Mountain Men '80
Rocket Gibraltar '88
Scrooged '88
Something Special '86
Traveling Man '89

Julian Glover

For Your Eyes Only '81
The Fourth Protocol '87
Hearts of Fire '87
Henry VIII '79

Indiana Jones and the Last Crusade '89
Ivanhoe '82
Mandela '87
The Secret Garden '87
Theatre of Death '67
Tom Jones '63
Treasure Island '89
Tusks '89

Kara Glover

The Beat '88
Caribe '87

Wolfgang Gluck (D)

'38: Vienna Before the Fall '88

Carlin Glynn

Blood Red '88
Continental Divide '81
Night Game '89
Sixteen Candles '84
Trip to Bountiful '85

Mary Glynne

Scrooge '35

Mathias Gnaedinger

Boat Is Full '81
Journey of Hope '90

Tom Gniazdowski (D)

Death Collector '89

Hiromi Go

MacArthur's Children '85

Harry Goaz

Twin Peaks '90

Tito Gobbi

Musica Proibita '43

George Gobel

Better Late Than Never '79
Birds & the Bees '56
Ellie '84
The Fantastic World of D.C. Collins '84
I Married a Woman '56
More Laughing Room Only
Rabbit Test '78
Those Crazy Americans '67

Boy Gobert

Kamikaze '89 '83

Constantine S. Gochis (D)

The Redeemer '76

Justin Gocke

My Grandpa is a Vampire '90s

Jean-Luc Godard

Cleo from 5 to 7 '61
First Name: Carmen '83
King Lear '87

Jean-Luc Godard (D)

Alphaville '65
Aria '88
Band of Outsiders '64
Breathless (A Bout de Souffle) '59
Contempt '64
First Name: Carmen '83
France/Tour/Detour/Deux/ Enfants '78
Godard/Truffaut Shorts '58

Eve Gordon
Paradise '91

Gale Gordon
The 'Burbs '89
Here We Go Again! '42
Our Miss Brooks '56
Speedway '68
30-Foot Bride of Candy Rock '59

Gavin Gordon
The Bat '59
His First Command '29
I Killed That Man '42
The Matchmaker '58
Turf Boy '42

Gerald Gordon
Ants '77
Force Five '75
Hell Up in Harlem '73

Glenn Gordon
The Finger Man '55
Fu Manchu '56

Harold Gordon
Yellowneck '55

Hayes Gordon
Return of Captain Invincible '83

Hilary Gordon
The Mosquito Coast '86

James Gordon
Mark of the Beast

Julius Gordon
D.P. '85

Keith Gordon
Back to School '86
Christine '84
Combat Academy '86
Dressed to Kill '80
Home Movies '79
Kent State '81
Legend of Billie Jean '85
Silent Rebellion '82
Single Bars, Single Women '84
Static '87

Keith Gordon (D)
The Chocolate War '88

Leo Gordon
Alienator '89
Bog '84
Devil's Angels '67
The Haunted Palace '63
The Jayhawkers '59
Riot in Cell Block 11 '54
Seventh Cavalry '56
Shame '61
Soldier of Fortune '55
Ten Wanted Men '54

Lisa Gordon
Kipling's Women '63

Marjorie Gordon
Danger Trails '35

Marrianne Gordon
Demon Hunter '88

Mary Gordon
The Adventures of Sherlock Holmes '39
Boss of Big Town '43
Million Dollar Kid '44
Riot Squad '41
Sherlock Holmes and the Secret Weapon '42
Texas Cyclone '32

Michael Gordon (D)
Cyrano de Bergerac '50
For Love or Money '63
Pillow Talk '59

Mitchell Gordon
Atlas in the Land of the Cyclops '61

Muriel Gordon
The Lone Avenger '33

Philip Gordon
The Bridge to Nowhere '86
Came a Hot Friday '85

Rachel Gordon
I Married a Vampire '87

Rebecca Gordon
The Mosquito Coast '86

Robert Gordon
Loveless '83

Robert Gordon (D)
The Gatling Gun '72
It Came from Beneath the Sea '55

Roy Gordon
Attack of the 50 Foot Woman '58
The Great Man Votes '38

Russell Gordon
Rounding Up the Law '22

Ruth Gordon
Abe Lincoln in Illinois '40
Action in the North Atlantic '43
Any Which Way You Can '80
Big Bus '76
Don't Go to Sleep '82
Every Which Way But Loose '78
Harold and Maude '71
Inside Daisy Clover '65
Jimmy the Kid '82
Maxie '85
Mugsy's Girls '85
My Bodyguard '80
The Prince of Central Park '77
Rosemary's Baby '68
The Trouble with Spies '87
Two-Faced Woman '41
Voyage of the Rock Aliens '87
Whatever Happened to Aunt Alice? '69
Where's Poppa? '70

John Gordon-Sinclair
The Girl in the Picture '86

Steve Gordon (D)
Arthur '81

Stuart Gordon (D)
Dolls '87
From Beyond '86
The Re-Animator '85

Robot Jox '89

Susan Gordon
Picture Mommy Dead '66
Tormented '60

Tanya Gordon
Stay Awake '87

Virginia Gordon
Love Desperados '86

Berry Gordy (D)
Mahogany '76

John Gordy
Paper Lion '68

Laura Gore
Peddlin' in Society '47

Lesley Gore
Legendary Ladies of Rock & Roll '86

Sandy Gore
Street Hero '84

Serif Goren (D)
Yol '82

Claude Goretta (D)
The Lacemaker '77

Galyn Gorg
Body Beat '88
Malibu Bikini Shop '86
RoboCop 2 '90

Gabriel Gori
The Bronx Executioner '86

Gianni Gori
Operation Orient

Vladimir Gorikker (D)
The Tsar's Bride '66

Owen Gorin
Pace That Kills '28

Marius Goring
The Case of the Frightened Lady '39
Charlie Boy '81
Holocaust '78
The Moonraker '58
Night Ambush '57
Paris Express '53
The Red Shoes '48
Spy in Black '39
The Unstoppable Man '59
Zeppelin '71

Ruth Goring
Yentl '83

Buddy Gorman
Ghost Chasers '51

Cliff Gorman
All That Jazz '79
Angel '84
Boys in the Band '70
Class of '63 '73
Cops and Robbers '73
Night of the Juggler '80
Strike Force '75
An Unmarried Woman '78

Jonathan Gorman
Street Soldiers '91

Charles Gormley (D)
The Gospel According to Vic '87

Felim Gormley
The Commitments '91

Karen Gorney
Saturday Night Fever '77

Michael Gornick (D)
Creepshow 2 '87

Morleen Gorris (D)
Question of Silence '83

Frank Gorshin
Batman '66
Between Heaven and Hell '56
Beverly Hills Bodysnatchers '89
Dragstrip Girl '57
The Great Impostor '61
Hollywood Vice Sqaud '86
Hot Resort '85
Midnight '89
Rudolph's Shiny New Year '79
Sky Hei$t '75
Star Trek Bloopers '69
Studs Lonigan '60
That Darn Cat '65
Underground Aces '80
Upper Crust '88

Marjoe Gortner
American Ninja 3: Blood Hunt '89
Bobbie Jo and the Outlaw '76
Earthquake '74
Fire, Ice and Dynamite '91
Food of the Gods '76
Hellhole '85
Jungle Warriors '84
Marjoe '72
Mausoleum '83
Pray for the Wildcats '74
Sidewinder One '77
Star Crash '78
The Survivalist '87
Viva Knievel '77

Christopher Gosch
The Last Season '87

Freeman Gosden
Check & Double Check '30

Hideo Gosha (D)
Hunter in the Dark '80
The Wolves '82

Jurgen Goslar (D)
Albino '76
Slavers '77

Maarten Goslins
Miami Vendetta '87

David Goss
Hollywood Cop '87
She '83

Mark Paul Gosselaar
Necessary Parties '88

Louis Gossett, Jr.
Aces: Iron Eagle 3 '92
The Choirboys '77
Cover-Up '91
The Deep '77
Don't Look Back: The Story of Leroy "Satchel" Paige '81
El Diablo '90
Enemy Mine '85
Finders Keepers '84
Firewalker '86
The Guardian '84
He Who Walks Alone '78
Iron Eagle '85
Iron Eagle 2 '88
It Rained All Night the Day I Left '78
It's Good to Be Alive '74
Jaws 3 '83
J.D.'s Revenge '76
The Josephine Baker Story '90
Keeper of the City '92
Laughing Policeman '74
Lazarus Syndrome '79
Little Ladies of the Night '77
Murder on the Bayou (A Gathering of Old Men) '91
An Officer and a Gentleman '82
The Principal '87
The Punisher '90
A Raisin in the Sun '61
The River Niger '76
Skin Game '71
Straight Up '90
Toy Soldiers '91
Travels with My Aunt '72
The White Dawn '75
Zora Is My Name! '90

Roland Got
Across the Pacific '42
G-Men vs. the Black Dragon '43

Walter Gotell
The African Queen '51
Basic Training '86
For Your Eyes Only '81
From Russia with Love '63
Lord Jim '65
Moonraker '79
Sleepaway Camp 2: Unhappy Campers '88
The Spy Who Loved Me '77

Staffan Gotestam
Brothers Lionheart '85

Peter Gothar (D)
Time Stands Still '82

Michael Gothard
For Your Eyes Only '81
King Arthur, the Young Warlord '75
Lifeforce '85
Young Warlord '75

James T. Goto
The Gallant Hours '60

Toshio Goto (D)
Forest of Little Bear '79

Gilbert Gottfried
The Adventures of Ford Fairlane '90
Bad Medicine '85
Problem Child '90
Problem Child 2 '91

John Gottfried

Berlin Alexanderplatz
(episodes 1-2) '80
Berlin Alexanderplatz
(episodes 3-4) '83
Berlin Alexanderplatz
(episodes 5-6) '83
Berlin Alexanderplatz
(episodes 7-8) '83

Carl Gottlieb

The Committee '68
Jaws '75
A Session with the Committee
'69

Carl Gottlieb (D)

Amazon Women On the Moon
'87
Caveman '81
Paul Reiser: Out on a Whim
'88

Franz Gottlieb (D)

Curse of the Yellow Snake '63
The Phantom of Soho '64

Lisa Gottlieb (D)

Just One of the Guys '85

Michael Gottlieb (D)

Mannequin '87

Stanley Gottlieb

Putney Swope '69

John Gottowt

Waxworks '24

**Ferdinand
Gottschalk**

Sing Sing Nights '35

**Robert
Gottschalk** (D)

Dangerous Charter '62

Siggi Gotz (D)

Summer Night Fever '79

Jetta Goudal

The Coming of Amos '25
The Road to Yesterday '25
White Gold '28

Ingrid Goude

The Killer Shrews '59

Lloyd Gough

A Southern Yankee '48

Michael Gough

Batman '89
Crucible of Horror '69
Dresser '83
The Fourth Protocol '87
The Go-Between '71
The Horror of Dracula '58
Horror Hospital '73
The Horse's Mouth '58
Out of Africa '85
Oxford Blues '84
Richard III '55
Rob Roy—The Highland
Rogue '53
The Skull '65
The Small Back Room '49
Strapless '90
Sword & the Rose '53
They Came from Beyond
Space '67
Top Secret! '84

Venom '82
What a Carve-Up! '62
Women in Love '70

Elliot Gould

Inside Out '91

Elliott Gould

Best of John Belushi '85
Bob & Carol & Ted & Alice '69
A Bridge Too Far '77
Busting '74
Capricorn One '78
Conspiracy: The Trial of the
Chicago Eight '87
Dangerous Love '87
Dead Men Don't Die '91
Devil & Max Devlin '81
Dirty Tricks '81
Escape to Athena '79
Falling in Love Again '80
Frog '89
Getting Straight '70
Harry & Walter Go to New
York '76
Hitz '92
I Love My...Wife '70
I Will, I Will for Now '76
Jack & the Beanstalk '83
Lady Vanishes '79
Last Flight of Noah's Ark '80
The Lemon Sisters '90
Lethal Obsession '87
Little Murders '71
The Long Goodbye '73
M*A*S*H '70
Matilda '78
Mean Johnny Barrows '75
Muppet Movie '79
My Wonderful Life '90
The Naked Face '84
The Night They Raided
Minsky's '69
The Night Visitor '89
Over the Brooklyn Bridge '83
Paul Reiser: Out on a Whim
'88
Quick, Let's Get Married '71
Roboman '75
The Silent Partner '78
S*P*Y*S '74
Tall Tales & Legends: Casey
at the Bat '85
The Telephone '87
The Touch '71
Vanishing Act '88
Whiffs '75

Harold Gould

Actor: The Paul Muni Story
'78
Better Late Than Never '79
Dream Chasers '84
Fourth Wise Man '85
Help Wanted: Male '82
How to Break Up a Happy
Divorce '76
Kenny Rogers as the Gambler
'80
Kenny Rogers as the
Gambler, Part 2: The
Adventure Continues '83
The One and Only '78
Playing for Keeps '86
The Red Light Sting '84
Romero '89
Seems Like Old Times '80
The Sting '73

Heywood Gould (D)

One Good Cop '91

Jason Gould

The Prince of Tides '91

William Gould

Phantom Thunderbolt '33
Rio Rattler '35

Alfred Goulding (D)

A Chump at Oxford '40
Dick Barton, Special Agent
'48

**Edmund
Goulding** (D)

Dark Victory '39
Dawn Patrol '38
Forever and a Day '43
Grand Hotel '32
The Great Lie '41
Old Maid '39
The Razor's Edge '46
Reaching for the Moon '31
We're Not Married '52

Robert Goulet

Gay Purr-ee '62
The Naked Gun 2 1/2: The
Smell of Fear '91
Scrooged '88

Rosalind Gourgey

One Wish Too Many '55

Andre Govan

The Monster Squad '87

Alan Govenar (D)

Cigarette Blues '89

**Richard
Governor** (D)

Ghost Town '88

Bruce Gowers (D)

Eddie Murphy "Delirious" '83

Gene Gowing

Drifting Souls '32

Gibson Gowland

Greed '24
Gun Cargo '49
Hell Harbor '30
Phantom of the Opera '25
The Secret of the Loch '34
Without Honors '32

Chantal Goya

Masculine Feminine '66

Mona Goya

Juggernaut '37

Emmanuel Goyet

La Vie Continue '82

Sergio Goyri

Conexion Mexico '87

Harry Goz

Bill '81
Bill: On His Own '83
Mommie Dearest '81
Rappin' '85

Paolo Gozlino

Guns for Dollars '73

Fernand Graavey

Bitter Sweet '33

Betty Grable

The Beautiful Blonde from
Bashful Bend '49

Day the Bookies Wept '39
Down Argentine Way '40
The Farmer Takes a Wife '53
Follow the Fleet '36
Footlight Serenade '42
The Gay Divorcee '34
Gotta Dance, Gotta Sing '84
Hold 'Em Jail '32
Hollywood Goes to War '54
How to Marry a Millionaire '53
I Wake Up Screaming '41
The Love Goddesses '65
Moon Over Miami '41
Pin-Up Girl '44
Probation '32
Song of the Islands '42
Springtime in the Rockies '42
Strictly G.I. '44
Whoopee! '30
The Worst of Hollywood: Vol.
3 '90
A Yank in the R.A.F. '41

Sofie Graboel

The Wolf at the Door '87

Janusz Grabowski

Greta '86

Hale Grace

Goodbye Love '34

Nickolas Grace

Diamond's Edge '88
The Green Man '91
Heat and Dust '82
Max Headroom, the Original
Story '86
Robin Hood...The Legend '83
Robin Hood...The Legend:
Herne's Son '85
Robin Hood...The Legend:
Robin Hood and the
Sorcerer '83
Robin Hood...The Legend:
The Swords of Wayland '83
Robin Hood...The Legend:
The Time of the Wolf '85
Salome's Last Dance '88

Sally Grace

Madhouse Mansion '74

Wayne Grace

Heroes Stand Alone '89

William Grace (D)

Tomb of Torture '65

Elizabeth Gracen

Death of the Incredible Hulk
'90
Lower Level '91

Marc Gracie (D)

A Kink in the Picasso '90

Paulo Gracindo

Amor Bandido '79
Earth Entranced (Terra em
Transe) '66

Ed L. Grady

Last Game '80

Roger Graef (D)

Secret Policeman's Private
Parts '81

Tom Graeff

Teenagers from Outer Space
'59

Tom Graeff (D)

Teenagers from Outer Space
'59

Paul Graetz

Bulldog Jack '35
Heart's Desire '37
Mimi '35
Power '34

Paul Graetz (D)

Heart of a Nation '43

David Graf

Police Academy 2: Their First
Assignment '85
Police Academy 3: Back in
Training '86
Police Academy 4: Citizens on
Patrol '87
Police Academy 5:
Assignment Miami Beach
'88
Police Academy 6: City Under
Siege '89

Peter Graf

Sex Adventures of the Three
Musketeers '80s

Ilene Graff

Ladybugs '92

Todd Graff

The Abyss '89
City of Hope '91
Five Corners '88
Framed '90
Opportunity Knocks '90

Wilton Graff

The Mozart Story '48

Sondra Graffi

Amazon Jail '85

**Anders
Grafstrom** (D)

Long Island Four '80

Bill Graham

Gardens of Stone '87

Billy Graham

The Prodigal '83

Bob Graham (D)

End of August '82

**David C.
Graham** (D)

The Undertaker and His Pals
'67

Fred Graham

Heart of the Rockies '51

Gary Graham

Last Warrior '89
No Place to Hide '81
Robot Jox '89

Gerrit Graham

The Annihilators '85
Cannonball '76
Child's Play 2 '90
C.H.U.D. 2: Bud the Chud '89
Greetings '68
Hi, Mom! '70
Home Movies '79
It's Alive 3: Island of the Alive
'87

Richard E. Grant

Codename Kyril '91
Henry & June '90
How to Get Ahead in
Advertising '89
Hudson Hawk '91
L.A. Story '91
Mountains of the Moon '90
Warlock '91
Withnail and I '87

Rodney Grant

Dances with Wolves '90
Son of the Morning Star '91

Salim Grant

The Hitman '91

Schuyler Grant

Anne of Avonlea '87
Anne of Green Gables '85

Shelby Grant

Our Man Flint '66
The Witchmaker '69

Tony Grant

Hush Little Baby, Don't You
Cry '86

Graziella Granta

Slaughter of the Vampires '62

Lucy Grantham

Last House on the Left '72
Legend of Boggy Creek '75

Jean-Pierre Granval

Picnic on the Grass '59

Allan Granville

Black Terrorist '85

Bonita Granville

Breakfast in Hollywood '46
The Glass Key '42
Guilty of Treason '50
Hitler's Children '43
Lone Ranger '56
Love Laughs at Andy Hardy
'46
Now, Voyager '42
Quality Street '37
Seven Miles from Alcatraz/
Flight from Glory '43
These Three '36

Sydney Granville

The Mikado '39

Marcus J. Grapes

Mind Warp '72

Charley Grapewin

Captains Courageous '37
The Good Earth '37
The Grapes of Wrath '40
Hell's House '32
Johnny Apollo '40
Libeled Lady '36
One Frightened Night '35
Petrified Forest '36
They Died with Their Boots
On '41
The Wizard of Oz '39

Christiane Graskoff

Mephisto '81

Louis Grasnier (D)

Murder on the Yukon '40

Alex Grasshof (D)

Pepper and His Wacky Taxi
'72

Karen Grassle

Battered '78
The Best Christmas Pageant
Ever '86
Cocaine: One Man's
Seduction '83
Harry's War '84
President's Mistress '78

Bill Gratton

Syngenor '90

Jorge Grau (D)

The Legend of Blood Castle
'72

Walter Grauman (D)

Are You in the House Alone?
'78
Force Five '75
Lady in a Cage '64
Manhunter '74
Most Wanted '76
Nightmare on the 13th Floor
'90
Outrage '86
Paper Man '71
Pleasure Palace '80
Scene of the Crime '85
The Streets of San Francisco
'72
To Race the Wind '80

Lou Gravance

Evilspeak '82

Gary Graver (D)

Crossing the Line '90
Evil Spirits '91
Moon in Scorpio '86
Party Camp '87
Roots of Evil '91
Trick or Treats '82

Abraham Graves

Green Pastures '36

Ernest Graves

Hercules in New York '70
One Plus One '61

Karron Graves

The Fig Tree '87

Peter Graves

The Adventurers '70
Airplane! '80
Airplane 2: The Sequel '82
Beginning of the End '57
Beneath the 12-Mile Reef '53
Casey Jones and Fury (Brave
Stallion) '57
The Clonus Horror '79
The Court Martial of Billy
Mitchell '55
Cruise Missile '78
Encore '52
The Guns and the Fury '83
It Conquered the World '56
Killers from Space '54
Legend of the Sea Wolf '58
Let's Make Up '55
The Magic Christian '69
Maytime in Mayfair '52
The Night of the Hunter '55
Number One With a Bullet '87
Poor White Trash '57

The President's Plane is Missing '71

The President's Plane is
Missing '71
The Rebels '79
Sergeant Ryker '68
Stalag 17 '53
Super Bloopers #2 '70s
Survival Run '80
Where Have All the People
Gone? '74
Winds of War '83

Ralph Graves

The Black Coin '36
Dream Street '21
Extra Girl '23
That Certain Thing '28

Robert Graves

Epic That Never Was '65

Rupert Graves

The Children '91
Fortunes of War '90
A Handful of Dust '88
Maurice '87
A Room with a View '86

Teresa Graves

Get Christie Love! '74

Fernand Gravet

The Great Waltz '38
La Ronde '51

Carla Gravina

And Now My Love '74
Boomerang '76
Salut l'Artiste '74
The Tempter '74

Cesare Gravina

Foolish Wives '22

Andee Gray

9 1/2 Ninjas '90

Billy Gray

Werewolves on Wheels '71

Blind Arvella Gray

Maxwell Street Blues '89

Bobby Gray

The Day the Earth Stood Still
'51

Carole Gray

Island of Terror '66

Charles Gray

Black Werewolf '75
Diamonds are Forever '71
Dreams Lost, Dreams Found
'87
An Englishman Abroad '83
The Executioner '70
The House on Garibaldi Street
'79
The Jigsaw Man '84
Julius Caesar '79
The Man Outside '68
The Mirror Crack'd '80
Murder on the Midnight
Express '74
Nine Ages of Nakedness '69
The Seven Percent Solution
'76
Shock Treatment '81
Silver Bears '78
You Only Live Twice '67

Coleen Gray

Kansas City Confidential '52
The Killing '56

Kiss of Death '47

Kiss of Death '47
The Phantom Planet '61

Denise Gray

Sputnik '61

Dolores Gray

Designing Woman '57
Kismet '55
The Opposite Sex '56

Donald Gray

Flight from Vienna '58
The Four Feathers '39
Island of Desire '52
Murder on the Campus '52

Dorian Gray

Colossus and the Amazon
Queen '64

Dulcie Gray

Angels One Five '54
The Franchise Affair '52
A Man About the House '47
Mine Own Executioner '47

Elspet Gray

The Girl in a Swing '89

Erin Gray

Born Beautiful '82
Breaking Home Ties '87
Buck Rogers in the 25th
Century '79
Six Pack '82
The Ultimate Imposter '79

Gary Gray

The Next Voice You Hear '50
The Painted Hills '66
Rachel and the Stranger '48
Two Weeks with Love '50

Gilda Gray

The Great Ziegfeld '36

Gordon Gray

Kenneth Anger, Vol. 1:
Fireworks '47

James Gray

Ninja's Extreme Weapons '90

John Gray (D)

Billy Galvin '86
The Lost Capone '90

Lawrence Gray

A Face in the Fog '36
Love 'Em and Leave 'Em '27

Linda Gray

The Grass is Always Greener
Over the Septic Tank '78
Shadows Over Shanghai '38
The Two Worlds of Jenny
Logan '79
The Wild and the Free '80

Lorna Gray

The Man They Could Not
Hang '39

Lydia Gray

Derby '71

Maralou Gray

Secret File of Hollywood '62

Marc Gray

Which Way Home '90

Maxine Gray

King Kung Fu '87

Mike Gray (D)

Wavelength '83

Nadia Gray

The Captain's Table '60
La Dolce Vita '60
Maniac '63
Two for the Road '67
Valley of the Eagles '51

Rhonda Gray

Twisted Nightmare '87

Sally Gray

Dangerous Moonlight '41
The Hidden Room '49
I'll Get You '53
Lady in Distress '39
Saint in London '39
The Saint's Vacation/The
Narrow Margin '52

Sam Gray

Heart '87

Spalding Gray

Almost You '85
Beaches '88
Clara's Heart '88
Hard Choices '84
The Killing Fields '84
Spalding Gray: Terrors of
Pleasure '88
Stars and Bars '88
Straight Talk '92
Swimming to Cambodia '87
True Stories '86

Vernon Gray

To Paris with Love '55

Willoughby Gray

Absolution '81

Godfrey Grayson (D)

Dick Barton Strikes Back '48
Room to Let '49

Kathryn Grayson

Anchors Aweigh '45
Andy Hardy's Private
Secretary '41
It Happened in Brooklyn '47
Kiss Me Kate '53
Lovely to Look At '52
Show Boat '51
Thousands Cheer '43
Till the Clouds Roll By '46
Till the Clouds Roll By/A Star
Is Born '46
The Toast of New Orleans '50

Jose Greco

Goliath and the Sins of
Babylon '64
Ship of Fools '65

Juliette Greco

Elena and Her Men '56
The Green Glove '52
Uncle Tom's Cabin '69

Janet Greek (D)

The Spellbinder '88

Adolph Green

Follies in Concert '85
Lily in Love '85

David Greenwalt (D)

Rude Awakening '89
Secret Admirer '85

Bruce Greenwood

The Climb '87
Malibu Bikini Shop '86
Servants of Twilight '91
Spy '89
Striker's Mountain '80s
Wild Orchid '90

Carolyn Greenwood

Play Dead '81

Charlotte Greenwood

Dangerous When Wet '53
Down Argentine Way '40
Funny Guys & Gals of the Talkies '30s
Glory '56
Moon Over Miami '41
Oklahoma! '55
Parlor, Bedroom and Bath '31
Springtime in the Rockies '42
Two Reelers: Comedy Classics 5 '38

Joan Greenwood

The Detective '54
Frenzy '46
A Girl in a Million '46
The Hound of the Baskervilles '77
Importance of Being Earnest '52
Kind Hearts and Coronets '49
Little Dorrit, Film 1: Nobody's Fault '88
Little Dorrit, Film 2: Little Dorrit's Story '88
Man in the White Suit '51
Moonfleet '55
Mr. Peek-A-Boo '50
Mysterious Island '61
October Man '48
Stage Struck '57
Tom Jones '63
Whiskey Galore '48

Monty Greenwood

Place in Hell '65

Dabbs Greer

Rage '72
Rose Marie '54
Roustabout '64
Trouble Along the Way '53
Two Moon Junction '88

Germaine Greer

Riding On '37

Jane Greer

Against All Odds '84
Big Steal '49
Billie '65
Clown '53
Dick Tracy Detective '45
Dick Tracy, Detective/The Corpse Vanishes '42
George White's Scandals '45
Just Between Friends '86
Man of a Thousand Faces '57
Out of the Past '47
Prisoner of Zenda '52
Sinbad the Sailor '47
Station West '48
They Won't Believe Me '47

Leighton Greer

Skeezer '82

Michael Greer

Messiah of Evil '74

Timothy Greeson

Disturbance '89

William Grefe

The Death Curse of Tartu '66

William Grefe (D)

Alligator Alley '72
Impulse '74
Jaws of Death '76
Stanley '72
Whiskey Mountain '77
Wild Rebels '71

Bradley Gregg

Class of 1999 '90
Eye of the Storm '91
Madhouse '90

Clark Gregg

Lana in Love '92

Colin Gregg (D)

We Think the World of You '88

John Gregg

Heatwave '83

Julie Gregg

From Hell to Borneo '64
The Seekers '79

Ron Gregg

Nomad Riders '81

Virginia Gregg

The Amazing Mr. X '48
Columbo: Prescription Murder '67
D.I. '57
Love is a Many-Splendored Thing '55
No Way Back '74
Walk in the Spring Rain '70

Pascal Greggory

Pauline at the Beach '83

Nora Gregor

The Rules of the Game '39

Rose Gregorio

Five Corners '88
Last Innocent Man '87

Jerry Gregoris

Rana: The Legend of Shadow Lake '75

Andre Gregory

Always '85
Author! Author! '82
The Bonfire of the Vanities '90
Follies in Concert '85
The Last Temptation of Christ '88
The Linguini Incident '92
The Mosquito Coast '86
My Dinner with Andre '81
Protocol '84
Soldier's Tale '84
Some Girls '88
Street Smart '87

Celia Gregory

Agatha '79
Children of the Full Moon '84

David Gregory

Dead Mate '88

Dick Gregory

Chicago Blues '72
Sweet Love, Bitter '67

Frank Gregory (D)

The Invincible Gladiator '63

Iola Gregory

Coming Up Roses '87

James Gregory

Al Capone '59
The Ambushers '67
Beneath the Planet of the Apes '70
Captain Newman, M.D. '63
Clambake '67
Great American Traffic Jam '80
Lou Gehrig Story '56
The Manchurian Candidate '62
Murderers' Row '66
Nightfall '56
PT 109 '63
Two Weeks in Another Town '62
Wait Till Your Mother Gets Home '83

Jennifer Gregory

God's Bloody Acre '75

Mark Gregory

Blue Paradise '86
Escape from the Bronx '85
Thunder Warrior '85
Thunder Warrior 2 '85
Thunder Warrior 3 '88
War Bus Commando '89

Paul Gregory

Whoopee! '30

Robert Gregory

The Devil's Mistress '68

Sebastian Gregory

Mona's Place '89
Sweet Trash '89

Sharee Gregory

Mass Appeal '84

Thea Gregory

Paid to Kill '54

John Gregson

Above Us the Waves '56
Angels One Five '54
The Captain's Table '60
Fright '71
Genevieve '53
Hans Brinker '69
The Lavender Hill Mob '51
Night of the Generals '67

Stephen Greif

The Great Riviera Bank Robbery '79

Robert Greig

The Great Ziegfeld '36
Sullivan's Travels '41

Kim Greist

Brazil '86
C.H.U.D. '84
Manhunter '86
Punchline '88
Why Me? '90

Jean Gremillon (D)

Pattes Blanches '49
Stormy Waters '41

Wolf Gremm (D)

Kamikaze '89 '83

Joyce Grenfell

The Americanization of Emily '64
Belles of St. Trinian's '53
Blue Murder at St. Trinian's '56
The Pure Hell of St. Trinian's '61

Googy Gress

Babes in Toyland '86
Bloodhounds of Broadway '89
First Turn On '83
Vibes '88

Heinrich Gretler

Heidi '52

Laurent Grevill

Camille Claudel '89

Edmond T. Greville (D)

Beat Girl '60
The Hands of Orlac '60
Princess Tam Tam '35

Anne Grey

Kiss of Death '47
Number Seventeen '32

Denise Grey

Devil in the Flesh '46
La Boum '81

Jennifer Grey

American Flyers '85
Bloodhounds of Broadway '89
The Cotton Club '84
Criminal Justice '90
Dirty Dancing '87
Ferris Bueller's Day Off '86
Light Years '88
Reckless '84
Red Dawn '84
Stroke of Midnight '90

Joel Grey

Buffalo Bill & the Indians '76
Cabaret '72
Queenie '87
Remo Williams: The Adventure Begins '85
The Seven Percent Solution '76

Lita Grey

The Kid/Idle Class '21

Nan Grey

Danger on the Air '38
Dracula's Daughter '36
The Invisible Man Returns '40
The Tower of London '39

Reatha Grey

Soul Vengeance '75

Samantha Grey

Night of the Zombies '81

Shirley Grey

Drifting Souls '32
Hurricane Express '32
The Mystery of the Mary Celeste '37
The Phantom Ship '37
Texas Cyclone '32
Uptown New York '32

Virginia Grey

Another Thin Man '39
Big Store '41
Bullfighter & the Lady '50
Idaho '43
Lady Scarface/The Threat '49
The Last Command '55
Love has Many Faces '65
Madame X '66
Mexican Hayride '48
The Michigan Kid '28
Naked Kiss '64
The Rose Tattoo '55
Slaughter Trail '51
Swamp Fire '46
Threat '49
Three Desperate Men '50
Unknown Island '48
Who Killed Doc Robbin? '48

Clinton Greyn

The Love Machine '71
Raid on Rommel '71

Bill Gribble

Dogs of Hell '83
The Wild and the Free '80

Eddie Gribbon

Rio Rattler '35

Harry Gribbon

Ben Turpin: An Eye for an Eye '24
The Cameraman '28
Two Reelers: Comedy Classics 1 '33
Two Reelers: Comedy Classics 7 '38

Lucas Gridoux

Pepe le Moko '37

Richard Grieco

If Looks Could Kill '91
Mobsters '91

Sergio Grieco (D)

Mad Dog '
Man of Legend '71

Stephen Grief

The Dying Truth '91

Helmut Griem

Children of Rage '75
The Damned '69
Escape '90
La Passante '83
Les Rendez-vous D'Anna '78
Malou '83

David Alan Grier

Beer '85
Loose Cannons '90
Off Limits '87
A Soldier's Story '84

Pam Grier

Above the Law '88
Badge of the Assassin '85

Grier ▶cont.

Big Bird Cage '72
The Big Doll House '71
Bill & Ted's Bogus Journey
  '91
Bucktown '75
Class of 1999 '90
Coffy '73
Drum '76
Fort Apache, the Bronx '81
Foxy Brown '74
Friday Foster '75
Greased Lightning '77
Naked Warriors '73
On the Edge '86
The Package '89
Scream Blacula Scream '73
Sheba, Baby '75
Something Wicked This Way
  Comes '83
Stand Alone '85
Tough Enough '87
Twilight People '72
The Vindicator '85

Roosevelt Grier

The Big Push '75
The Glove '78
The Seekers '79
Sophisticated Gents '81
The Treasure of Jamaica Reef
  '74

Charles Gries

Chance '89

Jonathan Gries

Joy Sticks '83
Kill Me Again '89
Pucker Up and Bark Like a
  Dog '89
Real Genius '85
Swap Meet '79

Tom Gries (D)

Breakheart Pass '76
Breakout '75
The Connection '73
Fools '70
Glass House '72
The Greatest '77
Helter Skelter '76
Lady Ice '73
100 Rifles '69
QB VII '74
Truman Capote's "The Glass
  House" '73
Will Penny '67

Kim Griest

Payoff '91

Andrew Grieve (D)

Suspicion '87

Ken Grieve (D)

Alamut Ambush '86

Russ Grieve

The Hills Have Eyes '77

Joe Grifasi

Bad Medicine '85
Chances Are '89
The Feud '90
Gentleman Bandit '81
On the Yard '79
Still of the Night '82

Simone Griffeth

Death Race 2000 '75
Delusion '84

Patriot '86
Sixteen '72

**Giuseppe Patroni
Griffi** (D)

Collector's Item '89
Divine Nymph '71
Driver's Seat '73
'Tis a Pity She's a Whore '73

Hugh Griffin

Craze '74

Johnny Griffin

Jazz is Our Religion '72

Josephine Griffin

The Man Who Never Was '55
Postmark for Danger '56
The Spanish Gardener '57

Lorie Griffin

Aloha Summer '88
Cheerleader Camp '88

Lynne Griffin

Obsessed '88

Merv Griffin

The Lonely Guy '84

Michael Griffin

Sonny Boy '87

Robert Griffin

I Was a Teenage Werewolf
  '57
Monster from Green Hell '58

Terry Griffin

Survival '88

Tod Griffin

She Demons '58

Tony Griffin

Evil Laugh '86

Andy Griffith

A Face in the Crowd '57
Fatal Vision '84
From Here to Eternity '79
Hearts of the West '75
Murder in Coweta County '83
Murder in Texas '81
No Time for Sergeants '58
Pray for the Wildcats '74
Return to Mayberry '85
Rustler's Rhapsody '85
Savages '75

Bill Griffith

Whirlwind Horseman '38

Charles B. Griffith

It Conquered the World '56

**Charles B.
Griffith** (D)

Dr. Heckyl and Mr. Hype '80
Eat My Dust '76
Smokey Bites the Dust '81
Up from the Depths '79
Wizards of the Lost Kingdom
  2 '89

Corrine Griffith

Dracula/Garden of Eden '28
The Garden of Eden '28

D.W. Griffith (D)

Abraham Lincoln '30

Avenging Conscience '14
The Babylon Story from
  "Intolerance" '16
Battle of Elderbush Gulch '13
Birth of a Nation '15
Broken Blossoms '19
Dream Street '21
D.W. Griffith Triple Feature
  '13
First Twenty Years: Part 13
  (D.W.Griffith) '12
First Twenty Years: Part 14
  (Griffith's Dramas) '09
First Twenty Years: Part 15
  (Griffith's Editing) '09
First Twenty Years: Part 16
  (Later Griffith) '09
First Twenty Years: Part 17
  (Make-up Effects) '11
First Twenty Years: Part 18
  (2-reelers) '11
First Twenty Years: Part 19 (A
  Temporary Truce) '12
Griffith Biograph Program
  '10s
Hearts of the World '18
Home Sweet Home '14
Intolerance '16
Isn't Life Wonderful '24
The Love Flower '20
Orphans of the Storm '21
Sally of the Sawdust '25
Short Films of D.W. Griffith,
  Volume 1 '12
The Sorrows of Satan '26
True Heart Susie '19
Way Down East '20
The White Rose '23

**Edward H.
Griffith** (D)

The Animal Kingdom '32
Bahama Passage '42
My Love For Yours '39
The Sky's the Limit '43
Young & Willing '42

Eva Griffith

Ride a Wild Pony '75
Turn of the Screw '74

Geraldine Griffith

Experience Preferred... But
  Not Essential '83

Gordon Griffith

Little Annie Rooney '25
Outlaws of the Range '36

Hugh Griffith

The Abominable Dr. Phibes
  '71
Ben-Hur '59
The Big Scam '79
The Canterbury Tales '71
The Counterfeit Traitor '62
Cry of the Banshee '70
Diary of Forbidden Dreams
  '76
Doctor Phibes Rises Again '72
Exodus '60
The Final Programme '73
The Last Days of Dolwyn '49
Legend of the Werewolf '75
Lucky Jim '58
Luther '74
Mutiny on the Bounty '62
Oh Dad, Poor Dad (Momma's
  Hung You in the Closet &
  I'm Feeling So Sad) '67
Oliver! '68
The Passover Plot '75
The Sleeping Tiger '54

Start the Revolution Without
  Me '70
Tom Jones '63
Who Slew Auntie Roo? '71
Wuthering Heights '70

James Griffith

The Amazing Transparent
  Man '60
Bullwhip '58
The Legend of Sleepy Hollow
  '79

Jeff Griffith

The 13th Mission '91

Kenneth Griffith

Circus of Horrors '60
The Final Option '82
Koroshi '67
Night of the Laughing Dead
  '73
Revenge '71

Kristin Griffith

Interiors '78

Melanie Griffith

Body Double '84
The Bonfire of the Vanities '90
Cherry 2000 '88
The Drowning Pool '75
Fear City '85
Harrad Experiment '73
In the Spirit '90
Joyride '77
The Milagro Beanfield War '88
Night Moves '75
One on One '77
Pacific Heights '90
Paradise '91
She's in the Army Now '81
Shining Through '91
Smile '75
Something Wild '86
Steel Cowboy '78
Stormy Monday '88
Underground Aces '80
Women & Men: Stories of
  Seduction '90
Working Girl '88

Raymond Griffith

All Quiet on the Western Front
  '30
Hands Up '26
The Night Club '25
Paths to Paradise '25

Simone Griffith

Hot Target '85

Tom Griffith

The Alien Factor '78
Hawk and Castile
Night Beast '83

Tracy Griffith

Fast Food '89
The First Power '89
Sleepaway Camp 3: Teenage
  Wasteland '89

Linda Griffiths

Lianna '83
Overdrawn at the Memory
  Bank '83
Reno and the Doc '84

Mark Griffiths (D)

Hardbodies '84
Hardbodies 2 '86
Heroes Stand Alone '89
Running Hot '83

Richard Griffiths

Blame It on the Bellboy '91
Shanghai Surprise '86
Whoops Apocalypse '83
Withnail and I '87

Trevor Griffiths

Singing the Blues in Red '87

Ele Grigsby

Invasion of the Girl Snatchers
  '73

Johanna Grika

Vice Academy 3 '91
Visitants '87

Gary Grillo (D)

American Justice '86

John Grillo

Afterward '85

Lucky Grills

Money Movers '78

Dan Grimaldi

Don't Go in the House '80

Eva Grimaldi

The Black Cobra '87

Hugo Grimaldi

Godzilla Raids Again '59

Hugo Grimaldi (D)

The Human Duplicators '64

Frank Grimes

Crystalstone '88
The Dive '89

Gary Grimes

Cahill: United States Marshal
  '73
Class of '44 '73
Culpepper Cattle Co. '72
Summer of '42 '71

Rebecca Grimes

Rebel '73

Scott Grimes

Critters '86
Critters 2: The Main Course
  '88
Frog '89
It Came Upon a Midnight
  Clear '84
Night Life '90

Tammy Grimes

America '86
The Last Unicorn '82
Mr. North '88
No Big Deal '83
The Runner Stumbles '79

Tiny Grimes

Fats Waller and Friends '86

Daniel Grimm

Osa '85

Herbert Grimwood

Romola '25
When the Clouds Roll By '19

Nick Grinde (D)

Before I Hang '40
Delinquent Parents '38
Hitler: Dead or Alive '43

The Man They Could Not
Hang '39

Nikolai Grinko

Andrei Roublev '66

Alan Grint

The Secret Garden '87

Brad Grinter

Scream, Baby, Scream '69

Brad Grinter (D)

Blood Freak '72

Randy Grinter, Jr.

Blood Freak '72

Harry Grippe

Great K & A Train Robbery
'26

Christina Grisanti

A Woman Under the Influence
'74

Jerry Grisham

Escapes '86

Wallace Grissell (D)

King of the Congo '52

John Grissmer (D)

Blood Rage '87
Scalpel '76

Stephen Grives

Horror Planet '80

George Grizzard

Advise and Consent '62
Attica '80
Bachelor Party '84
Caroline? '90
Comes a Horseman '78
Embassy '85
From the Terrace '60
Iran: Days of Crisis '91
Oldest Living Graduate '82
Pueblo Affair '73
Seems Like Old Times '80
The Stranger Within '74
Wrong is Right '82

Rajko Grlic (D)

That Summer of White Roses
'90

Kenna Grob

Deadly Diamonds '91

David Grocey

Pride of the Bowery '41

Tomasz Grochoczki

Greta '86

Charles Grodin

Beethoven '92
Catch-22 '70
The Couch Trip '87
11 Harrowhouse '74
The Grass is Always Greener
Over the Septic Tank '78
The Great Muppet Caper '81
Grown Ups '86
The Heartbreak Kid '72
Heaven Can Wait '78
The Incredible Shrinking
Woman '81
Ishtar '87
It's My Turn '80
Just Me & You '78

King Kong '76
Last Resort '86
The Lonely Guy '84
The Meanest Men in the West
'60s
Midnight Run '88
Movers and Shakers '85
Real Life '79
Rosemary's Baby '68
Seems Like Old Times '80
Sex and the College Girl '64
Sunburn '79
Taking Care of Business '90
The Woman in Red '84
You Can't Hurry Love '88

Kathryn Grody

The Lemon Sisters '90
Quick Change '90

Ferde Grofe, Jr. (D)

Judgment Day '88
The Proud and the Damned
'72
Satan's Harvest '65
War Kill '68
Warkill '65

Claire Grogan

Gregory's Girl '80

C.P. Grogan

Comfort and Joy '84

David Groh

Hero Ain't Nothin' But a
Sandwich '78
Hot Shot '86
Zertigo Diamond Caper '82

Ake Gronberg

Monika '52
Sawdust & Tinsel '53

**Herbert
Gronemeyer**

Das Boot '81
Spring Symphony '86

Sam Groom

The Baby Maker '70
Betrayal
Beyond the Bermuda Triangle
'75
Deadly Eyes '82
Deadly Games '80
Run for the Roses '78

Malcolm Groome

Feelin' Up '76

Gary Groomes

Wired '89

Frank Groothof

In for Treatment '82

Ulu Grosbard (D)

Falling in Love '84
Straight Time '78
True Confessions '81

Arye Gross

The Couch Trip '87
Coupe de Ville '90
The Experts '89
House 2: The Second Story
'87
A Matter of Degrees '90
Shaking the Tree '91
Soul Man '86
Tequila Sunrise '88

Ed Gross

Black Gestapo '75

Edan Gross

And You Thought Your
Parents Were Weird! '91

Eva Gross

The Happy Gigolo '70s

Jerry Gross (D)

Girl on a Chain Gang '65
Teenage Mother '67

Larry Gross (D)

3:15—The Moment of Truth
'86

Mary Gross

Best of Eddie Murphy '89
Big Business '88
Bill Murray Live from the
Second City '70s
Casual Sex? '88
Club Paradise '86
The Couch Trip '87
Feds '88
Troop Beverly Hills '89

Michael Gross

Alan & Naomi '92
Big Business '88
Cool As Ice '91
Summer Fantasy '84
Tremors '89

Paul Gross

Buffalo Jump '90

Willard Gross

Creature of the Walking Dead
'65

Yoram Gross (D)

Camel Boy '84
Dot & the Bunny '83
EPIC: Days of the Dinosaurs
'87

David Grossman

Frog '89

**Douglas
Grossman** (D)

Hell High '86

Sam Grossman (D)

The Van '77

Scott Grossman

Body Beat '88

Sonny Grosso

The French Connection '71

Dennis Grosvenor

Gone to Ground '76

James Grout

Sister Dora '77

Sybil Grove

Campus Knights '29

Deborah Grover

The Christmas Wife '88
Mania '80s

Ed Grover

Roboman '75

Robin Groves

The Nesting '80

Robert Grubb

Mad Max Beyond
Thunderdome '85
Remember Me '85

Franz Gruber

Terror Beneath the Sea '66
X from Outer Space '67

**Krzysztof
Gruber** (D)

Greta '86

Franz Gruger

Deutschland im Jahre Null '47

Klaus Grunberg

More '69

Gustav Grundgens

M '31

Karl Grune (D)

Street '23

Olivier Gruner

Angel Town '89

**Allan
Grunewald** (D)

Nightmare Castle '65

Ilka Gruning

Desperate '47

**Halina
Gryglaszewska**

The Double Life of Veronique
'91

Shen Guan-Chu

The Go-Masters '82

Christopher Guard

Lord of the Rings '78
Return to Treasure Island,
Vol. 1 '85
Return to Treasure Island,
Vol. 2 '85
Return to Treasure Island,
Vol. 3 '85
Return to Treasure Island,
Vol. 4 '85
Return to Treasure Island,
Vol. 5 '85

Dominic Guard

Absolution '81
The Go-Between '71
Picnic at Hanging Rock '75

Kit Guard

El Diablo Rides '39

Pippa Guard

Unsuitable Job for a Woman
'82

Maribel Guardia

By Hook or Crook '85

Harry Guardino

The Adventures of Bullwhip
Griffin '66
Any Which Way You Can '80
Dirty Harry '71
The Enforcer '76
The Five Pennies '59

Get Christie Love! '74
Goldengirl '79
Hell is for Heroes '62
Houseboat '58
Lovers and Other Strangers
'70
Madigan '68
Matilda '78
The Neon Empire '89
Octaman '71
Pork Chop Hill '59
Rollercoaster '77
St. Ives '76
Whiffs '75

Paulo Guarniero

Amor Bandido '79

Vincent Guastaferro

Nitti: The Enforcer '88

Enrico Guazzoni (D)

Quo Vadis '12

Irina Gubanova

Private Life '83
War and Peace '68

**Nikolai
Gubenko** (D)

Orphans '83

Hans Gudegast

Dayton's Devils '68

Ann Gudrun

Terror Street '54

Vanessa Guedj

The Grand Highway '88
Le Grand Chemin '87

**James W.
Guercio** (D)

Electra Glide in Blue '73

Bruce Guerin

The Country Kid '23

Florence Guerin

Bizarre '87

Francois Guerin

Horror Chamber of Dr.
Faustus '59

Blanca Guerra

Murderer in the Hotel '83
Santa Sangre '90

Castulo Guerra

Where the River Runs Black
'86

Danny Guerra

Plutonium Baby '87

Rogelio Guerra

La Corona de un Campeon
'87

Ruy Guerra

Aguirre, the Wrath of God '72

Ruy Guerra (D)

Erendira '83
The Fable of the Beautiful
Pigeon Fancier '88
Opera do Malandro '87

Juan Guerreo (D)

Narda or The Summer '76

Trilok Gurtu

Oregon '88

Louis Guss

Moonstruck '87
Nitti: The Enforcer '88

Arlo Guthrie

Alice's Restaurant '69

Catherine Guthrie

The Girl in the Picture '86

Tani Phelps Guthrie

Daughters of Satan '72
The Thirsty Dead '74

Tyrone Guthrie

Beachcomber '38

Tyrone Guthrie (D)

Oedipus Rex '57

Antonio Gutierrez

Virgin Sacrifice '59

Zaide Silvia Gutierrez

El Norte '83

Nathaniel Gutman (D)

Deadline '87

Steve Guttenberg

Amazon Women On the Moon '87
Bad Medicine '85
The Bedroom Window '87
The Boys from Brazil '78
Can't Stop the Music '80
The Chicken Chronicles '77
Cocoon '85
Cocoon: The Return '88
Day After '83
Diner '82
Don't Tell Her It's Me '90
High Spirits '88
The Man Who Wasn't There '83
Miracle on Ice '81
Police Academy '84
Police Academy 2: Their First Assignment '85
Police Academy 3: Back in Training '86
Police Academy 4: Citizens on Patrol '87
Short Circuit '86
Surrender '87
Three Men and a Baby '87
Three Men and a Little Lady '90
To Race the Wind '80

Lucy Gutteridge

The Secret Garden '87
Top Secret! '84
The Trouble with Spies '87
Tusks '89

Andre Guttfreund (D)

Femme Fatale '90

Amos Guttman (D)

Drifting '82

Bertil Guve

Fanny and Alexander '83

Felicia Guy

The Naked Angels '69

J. Scott Guy

Armed for Action

Jasmine Guy

Harlem Nights '89
Runaway '89

Sheila Guyse

Sepia Cinderella '47

Joe Guzaldo

Smooth Talker '90

Sergio Guzik

La Justicia Tiene Doce Anos '87

Claudio Guzman (D)

Antonio '73
The Hostage Tower '80
Willa '79

Enrique Guzman

The Sinister Invasion '68

Roberto "Flaco" Guzman

Arizona '86
La Hifa Sin Padre '87
Marcados por el Destino '87

Teri Guzman

Escape from Cell Block Three '78
Women Unchained '70s

Margherita Guzzinati

The Witch '66

Jack Gwaltney

Vital Signs '90

Robert Gwaltney

Trading Hearts '87

Edmund Gwenn

The Bigamist '53
Cash '34
Challenge to Lassie '49
Cheers for Miss Bishop '41
Devil & Miss Jones '41
Doctor Takes a Wife '40
Forever and a Day '43
Green Dolphin Street '47
If I Were Rich '33
Java Head '35
The Keys of the Kingdom '44
Lassie, Come Home '43
Life with Father '47
Miracle on 34th Street '47
Pride and Prejudice '40
Skin Game '31
Sylvia Scarlett '35
Them! '54
The Trouble with Harry '55
A Woman of Distinction '50

David Gwillim

Henry IV, Part I '80
Henry IV, Part II '80
Island at the Top of the World '74

Jack Gwillim

Sword of Sherwood Forest '60

Andy Gwyn

Dark Sanity '82

Michael Gwynn

Jason and the Argonauts '63
The Scars of Dracula '70

Anne Gwynne

Adam at 6 A.M. '70
Black Tower '50
Breakdown '53
Fear '46
House of Frankenstein '44
Killer Dill '47
King of the Bullwhip '51
Meteor Monster '57
Ride 'Em Cowboy '42

Fred Gwynne

Boy Who Could Fly '86
The Cotton Club '84
Day with Conrad Green '80
Disorganized Crime '89
Fatal Attraction '87
Ironweed '87
The Littlest Angel '69
The Man That Corrupted Hadleyburg '80
Munster's Revenge '81
Murder by the Book '87
My Cousin Vinny '92
The Mysterious Stranger '82
Off Beat '86
Pet Sematary '89
Simon '80
Vanishing Act '88

Michael C. Gwynne

Cherry 2000 '88
Harry Tracy '83
The Last of the Finest '90
Payday '72
Seduced '85
Special Delivery '76
The Streets of L.A. '79

Peter Gwynne

Puzzle '78
Tim '79

Tibor Gyapjas

Moving Out '83

Agi Gyenes

Just for the Hell of It '68

Stephen Gyllenhaal (D)

The Abduction of Kari Swenson '87
Certain Fury '85
Killing in a Small Town '90
Paris Trout '91

Kim Gyngell

The Wacky World of Wills & Burke '85

Greta Gynt

The Arsenal Stadium Mystery '39
The Hooded Terror '38
The Human Monster '39
Mr. Emmanuel '44
Sexton Blake and the Hooded Terror '38

Imre Gyongyossy (D)

The Revolt of Job '84

Brion Gysin

Towers Open Fire & Other Films '72

Deanna Haas

Devonsville Terror '83

Hugo Haas

Days of Glory '43
The Fighting Kentuckian '49
King Solomon's Mines '50
My Girl Tisa '48
Northwest Outpost '47
The Private Affairs of Bel Ami '47

Lukas Haas

Alan & Naomi '92
The Lady in White '88
Music Box '89
See You in the Morning '89
Shattered Spirits '91
Solarbabies '86
Witness '85
The Wizard of Loneliness '88

Susanna Haavisto

Ariel '89

Sylvie Habault

Therese '86

Karen Haber

Girls of the Comedy Store '86

Matthias Habich

Coup de Grace '78

Cox Habrema

Question of Silence '83

Herve Hachuel (D)

The Last of Philip Banter '87

Shelley Hack

Blind Fear '89
A Casualty of War '90
If Ever I See You Again '78
Kicks '85
King of Comedy '82
Single Bars, Single Women '84
Stepfather '87
Troll '85

Joseph Hacker

Little Treasure '85
The Ultimate Imposter '79

Buddy Hackett

Bud and Lou '78
God's Little Acre '58
It's a Mad, Mad, Mad, Mad World '63
Jack Frost '79
The Little Mermaid '89
Loose Shoes '80
Muscle Beach Party '64
The Music Man '62
Scrooged '88
Wonderful World of the Brothers Grimm '62

Carl Hackett

Border Phantom '37

Jay Hackett

One Night Stand '84

Joan Hackett

Class of '63 '73
Dead of Night '77
Deadly Rivals '72

Escape Artist '82
Flicks '85
The Group '66
The Last of Sheila '73
The Long Days of Summer '80
One Trick Pony '80
Only When I Laugh '81
The Possessed '77
Reflections of Murder '74
Rivals '72
Support Your Local Sheriff '69
The Terminal Man '74
Treasure of Matecumbe '76
Will Penny '67

Karl Hackett

Cavalry '36
Gun Lords of Stirrup Basin '37
His Brother's Ghost '45
Outlaw of the Plains '46
Phantom Ranger '38
The Red Rope '37
Songs and Bullets '38
Sonora Stagecoach '44
Take Me Back to Oklahoma '40

Sandy Hackett

Hamburger... The Motion Picture '86

Taylor Hackford (D)

Against All Odds '84
Chuck Berry: Hail! Hail! Rock 'N' Roll '87
Everybody's All American '88
Idolmaker '80
An Officer and a Gentleman '82
Tribute to Ricky Nelson '86
White Nights '85

Pennie Hackforth-Jones

Image of Death '77

Gene Hackman

All Night Long '81
America at the Movies '76
Another Woman '88
Bat 211 '88
Bite the Bullet '75
Bonnie & Clyde '67
A Bridge Too Far '77
Class Action '91
The Conversation '74
Doctors' Wives '70
The Domino Principle '77
Downhill Racer '69
Eureka! '81
The French Connection '71
French Connection 2 '75
Full Moon in Blue Water '88
Hawaii '66
Hoosiers '86
I Never Sang For My Father '70
Lilith '64
Loose Cannons '90
March or Die '77
Marooned '69
Mississippi Burning '88
Misunderstood '84
Narrow Margin '90
Night Moves '75
No Way Out '87
The Package '89
The Poseidon Adventure '72
Postcards from the Edge '90
Power '86
Prime Cut '72
Reds '81

Nathaniel "Afrika" Hall

Livin' Large '91

Peter Hall (D)

La Traviata '88

Philip Baker Hall

The Last Reunion '80
Secret Honor '85

Porter Hall

The Beautiful Blonde from
   Bashful Bend '49
Bulldog Drummond Escapes
   '37
Bulldog Drummond's Peril '38
Bulldog Drummond's
   Revenge/Bulldog
   Drummond's Peril '38
The Case of the Lucky Legs
   '35
Double Indemnity '44
The General Died at Dawn '36
Going My Way '44
The Great Moment '44
The Half-Breed '51
His Girl Friday '40
Intruder in the Dust '49
Miracle on 34th Street '47
Miracle of Morgan's Creek '44
Mr. Smith Goes to
   Washington '39
Petrified Forest '36
The Plainsman '37
The Story of Louis Pasteur
   '36
Sullivan's Travels '41
They Shall Have Music '39
The Thin Man '34
The Woman of the Town '43

Rich Hall

Million Dollar Mystery '87
Rich Hall's Vanishing America
   '86

Ruth Hall

Between Fighting Men '32
Monkey Business '31
Strawberry Roan '33

Sam Hall

South of Hell Mountain '70

Shannah Hall

Princess and the Call Girl '84

Thurston Hall

The Black Room '35
Call of the Canyon '42
Each Dawn I Die '39
The Great Gildersleeve '43
I Dood It '43
Lady from Nowhere '36
Night Stage to Galveston '52
Rim of the Canyon '49
You Can't Cheat an Honest
   Man '39

Zooey Hall

I Dismember Mama '74

Charles Hallahan

Margin for Murder '81
Pale Rider '85
The Thing '82
True Believer '89
Vision Quest '85

John Hallam

Dragonslayer '81
Under Capricorn '82

Jane Hallaren

Lianna '83
A Night in the Life of Jimmy
   Reardon '88

Daniel Haller (D)

Buck Rogers in the 25th
   Century '79
Devil's Angels '67
Die, Monster, Die! '65
The Dunwich Horror '70
Follow That Car '80
Margin for Murder '81
Paddy '70

Neil Hallett

The Virgin Witch '70

Tom Hallick

Hangar 18 '80
A Rare Breed '81

Bryant Halliday

Tower of Evil '72

Heather Halliday

Peter Pan '60

John Halliday

Consolation Marriage '31
Finishing School '33

Peter Halliday

Fast Kill '73
Millie '31

Lori Hallier

Blindside '88
The Gunfighters '87
Higher Education '88
My Bloody Valentine '81

William Halligan

Jive Junction '43
Robot Pilot '41
Turf Boy '42

Todd Hallowell (D)

Love or Money? '88

Lasse Hallstrom (D)

My Life As a Dog '87
Once Around '91

David Hallyday

He's My Girl '87

Martin Halm

Ernesto '79

Billy Halop

Angels with Dirty Faces '38
Junior G-Men '40
Junior G-Men of the Air '42
Little Tough Guys '38
Pride of the Bowery '41

Victor Halperin (D)

Buried Alive '39
I Conquer the Sea '36
Party Girls '29
Revolt of the Zombies '36
Torture Ship '39
The White Zombie '32

Dina Halpern

The Dybbuk '37

Luke Halpin

Flipper '63
Flipper's New Adventure '64
Flipper's Odyssey '66

Island of the Lost '68
Peter Pan '60

Daria Halprin

Zabriskie Point '70

Luke Halprin

Shock Waves '77

Brett Halsey

Atomic Submarine '59
Black Cat '90
The Crash of Flight 401 '78
Dangerous Obsession '88
The Girl in Lover's Lane '60
Jet Over the Atlantic '59
Magnificent Adventurer '76
Return of the Fly '59
Return to Peyton Place '61
Submarine Seahawk '59
Twice-Told Tales '63

Michael Halsey

Dollman '91
Treasure Island '89
Under the Gun '88

Charles Halton

Across the Pacific '42
Dr. Cyclops '40
Rhapsody in Blue '45

Peter Ham

The Happy Gigolo '70s

Mie Hama

What's Up, Tiger Lily? '66
You Only Live Twice '67

Torahiko Hamada

X from Outer Space '67

Yuko Hamada

Gamera vs. Guiron '69

John Hamblin

Who Killed Baby Azaria? '83

Glenda Hambly (D)

Fran '85

Veronica Hamel

Cannonball '76
The Gathering '77
The Gathering: Part 2 '79
A New Life '88
Sessions '83
Snowblind '78
Taking Care of Business '90

Gerald Hamer

Scarlet Claw '44

Robert Hamer (D)

Dead of Night '45
The Detective '54
Kind Hearts and Coronets '49
Pink Strings and Sealing Wax
   '45
To Paris with Love '55

Rusty Hamer

Dance with Me, Henry '56

Jayne Hamil

Vice Academy '88
Vice Academy 2 '90

Dorothy Hamill

Nutcracker, A Fantasy on Ice
   '83
Snow Queen '82

Mark Hamill

The Big Red One '80
Black Magic Woman '91
Britannia Hospital '82
The City '76
Corvette Summer '79
The Empire Strikes Back '80
Eric '75
From "Star Wars" to "Jedi":
   The Making of a Saga '83
The Night the Lights Went Out
   in Georgia '81
Return of the Jedi '83
Slipstream '89
Star Wars '77

Anthony Hamilton

Howling 4: The Original
   Nightmare '88
Samson and Delilah '84

Antony Hamilton

Mirrors '85

Bernie Hamilton

Bucktown '75
The Losers '70
Scream Blacula Scream '73

Carrie Hamilton

Shag: The Movie '89
Tokyo Pop '88

Dan Hamilton

Opposing Force '87

David Hamilton (D)

Bilitis '77

Dean Hamilton

Rush Week '88

George Hamilton

Angel Baby '61
By Love Possessed '61
The Dead Don't Die '75
Doc Hollywood '91
Evel Knievel '72
Express to Terror '79
From Hell to Victory '79
The Godfather: Part 3 '90
Happy Hooker Goes to
   Washington '77
Home From the Hill '60
Killer on Board '77
Love at First Bite '79
Medusa '74
Monte Carlo '86
Once is Not Enough '75
Once Upon a Crime '91
Poker Alice '87
The Seekers '79
Sextette '78
Togetherness '70
Two Fathers' Justice '85
Two Weeks in Another Town
   '62
Users '78
Where the Boys Are '60
Zorro, the Gay Blade '81

Guy Hamilton

Barry Lyndon '75

Guy Hamilton (D)

The Battle of Britain '69
The Colditz Story '55
Diamonds are Forever '71
Evil Under the Sun '82
Force 10 from Navarone '78
Funeral in Berlin '66
Goldfinger '64
Live and Let Die '73

The Man with the Golden Gun
   '74
The Mirror Crack'd '80
Remo Williams: The
   Adventure Begins '85

Hale Hamilton

The Champ '31
Murder at Midnight '31
Susan Lenox: Her Fall and
   Rise '31

Jane Hamilton

Bedroom Eyes 2 '89
Bloodsucking Pharoahs of
   Pittsburgh '90
Cleo/Leo '89
Deranged '87
New York's Finest '87
Slammer Girls '87
Wildest Dreams '90

John Hamilton

Fangs of the Living Dead '68
The Fatal Hour '40
I Killed That Man '42
Outcasts of the City '58
The Saint's Double Trouble
   '40

Josh Hamilton

O Pioneers! '91

Judd Hamilton

Gun Crazy '69
The Last Horror Film '82
Star Crash '78

Kipp Hamilton

War of the Gargantuas '70

Ellen Hamilton-Latzen

Fatal Attraction '87

Linda Hamilton

Beauty and the Beast '87
Beauty and the Beast: Though
   Lovers Be Lost '89
Black Moon Rising '86
Children of the Corn '84
King Kong Lives '86
Mr. Destiny '90
Rape and Marriage: The
   Rideout Case '80
Secret Weapons '85
The Stone Boy '84
Tag: The Assassination Game
   '82
The Terminator '84
Terminator 2: Judgement Day
   '91

Lois Hamilton

Armed Response '86

Margaret Hamilton

The Adventures of Tom
   Sawyer '38
The Anderson Tapes '71
Breaking the Ice '38
Brewster McCloud '70
Broadway Bill '34
Daydreamer '66
George White's Scandals '45
Guest in the House '44
Mad Wednesday '51
Moon's Our Home '36
My Little Chickadee '40
Nothing Sacred '37
The Red Pony '49
The Sin of Harold Diddlebock
   '47
These Three '36

13 Ghosts '60
The Wizard of Oz '39
You Only Live Once '37

Murray Hamilton

The Amityville Horror '79
The Boston Strangler '68
The Brotherhood '68
Brubaker '80
Casey's Shadow '78
The Drowning Pool '75
The FBI Story '59
The Graduate '67
Houseboat '58
Jaws '75
Jaws 2 '78
Last Cry for Help '79
The Last Days of Patton '86
Mazes and Monsters '82
1941 '79
No Time for Sergeants '58
No Way to Treat a Lady '68
Papa's Delicate Condition '63
Rona Jaffe's Mazes &
   Monsters '82
Sergeant Ryker '68
Spirit of St. Louis '57
The Tattered Web '71
Too Scared to Scream '85
The Way We Were '73

Neil Hamilton

America/The Fall of Babylon
   '24
The Animal Kingdom '32
The Devil's Hand '61
Hollywood Stadium Mystery
   '38
Isn't Life Wonderful '24
King of the Texas Rangers '41
The Patsy '64
The Saint Strikes Back '39
The Sin of Madelon Claudet
   '31
Tarzan, the Ape Man '32
Tarzan and His Mate '34
What Price Hollywood? '32
The White Rose '23
X Marks the Spot '42

Patricia Hamilton

Anne of Avonlea '87
Anne of Green Gables '85
The Christmas Wife '88

Rex Hamilton

More! Police Squad '82

Richard Hamilton

In Country '89

Rusty Hamilton

Rock 'n' Roll Nightmare '85

Shorty Hamilton

On the Night Stage '15

Strathford Hamilton *(D)*

Blueberry Hill '88
Diving In '90

Suzanna Hamilton

Brimstone & Treacle '82
1984 '84
Out of Africa '85
Wetherby '85

Suzette Hamilton

Scream Bloody Murder '72

Ted Hamilton

Pirate Movie '82

Tony Hamilton

Fatal Instinct '92
Nocturna '79

Wendy Hamilton

The Scars of Dracula '70

Dilys Hamlett

Bananas Boat '78

Harry Hamlin

Blue Skies Again '83
Clash of the Titans '81
Deceptions '90
Dinner at Eight '89
Hitchhiker 1 '85
King of the Mountain '81
L.A. Law '86
Laguna Heat '87
Making Love '82
Maxie '85
Movie, Movie '78
Target: Favorite Son '87

Dan Hamlon

20,000 Leagues Under the
   Sea '16

Eric Hamm

Summer '86

Warren Hammack

The Eye Creatures '65

Fritz Hammel

The Scorpion Woman '89

Achim Hammer

Sex Adventures of the Three
   Musketeers '80s

Ben Hammer

Crazy People '90
The Execution of Private
   Slovik '74
Invasion of the Bee Girls '73
Survival Quest '89

Heidi Hammer

2069: A Sex Odyssey '78

Jon Hammer

Phantom Brother '88

Robert Hammer *(D)*

Don't Answer the Phone '80

Oscar Hammerstein

Toast of the Town '56

John Hammil

Crossover Dreams '85

Alicia Hammond

Killers '88

John Hammond

The Blue and the Grey '85
Enforcer from Death Row '78
The Prodigal '83

Kay Hammond

Abraham Lincoln '30
Blithe Spirit '45

Nicholas Hammond

The Amazing Spider-Man '77
Chinese Web '78
The Martian Chronicles: Part 1
   '79
The Martian Chronicles: Part 3
   '79

Spiderman: The Deadly Dust
   '78

Patricia Lee Hammond

Dracula's Last Rites '79

Peter Hammond

The Adventurers '52

Chastity Hammons

White Fury '90

Olivia Hamnett

Deadly Possession '88
Earthling '80
The Last Wave '77
Plunge Into Darkness '77

Walter Hampden

The Prodigal '55
Reap the Wild Wind '42
Sabrina '54

Susan Hampshire

Baffled '72
Cry Terror '74
David Copperfield '70
Fighting Prince of Donegal '66
Living Free '72
Those Daring Young Men in
   Their Jaunty Jalopies '69
Three Lives of Thomasina '63

James Hampton

Amazing Howard Hughes '77
The China Syndrome '79
Condorman '81
Force Five '75
Hangar 18 '80
Hawmps! '76
Teen Wolf '85

Jane Hampton

Iron Cowboy '68

Lionel Hampton

Basin Street Revue '55
The Benny Goodman Story
   '55

Paul Hampton

Private Duty Nurses '71
They Came from Within '75

Robert Hampton *(D)*

The Ghost '63
The Horrible Dr. Hichcock '62

Susie Hampton

Fast Kill '73

Charles Han

Strike of Thunderkick Tiger
   '80s

Eagle Han

Shaolin Incredible Ten '70s

Maggie Han

The Last Emperor '87

Master Bong Soo Han

Force: Five '81
Kill the Golden Goose '79

Shotaro Hanayagi

The Story of the Late
   Chrysanthemum '39

Yoshiaki Hanayagi

Sansho the Bailiff '54

Victor Hanbury *(D)*

Hotel Reserve '44

Herbie Hancock

Round Midnight '86

John Hancock

Archer: The Fugitive from the
   Empire '81
In the Custody of Strangers
   '82
Traxx '87
Why Me? '90

John Hancock *(D)*

Bang the Drum Slowly '73
California Dreaming '79
Let's Scare Jessica to Death
   '71
Prancer '89
Steal the Sky '88
Weeds '87

Lou Hancock

Miracle Mile '89

Sheila Hancock

Buster '88
Hawks '89

Susan Hancock

Finian's Rainbow '68

David Hand *(D)*

Bambi '42

Richard Handford *(D)*

Second Chance '80

Irene Handl

Adventures of a Private Eye
   '87
Morgan! '66
The Private Life of Sherlock
   Holmes '70
Riding High '78
Small Hotel '57
Wonderwall: The Movie '69

Alan Handley *(D)*

Alice Through the Looking
   Glass '66

Ken Handley *(D)*

Delivery Boys '84

Rene Handren-Seals

Powwow Highway '89

James Handy

Bird '88

Carol Haney

Pajama Game '57

Daryl Haney

Daddy's Boys '87
Lords of the Deep '89

David Haney

Hootch Country Boys '75

Dorothy Haney

Terror of the Bloodhunters '62

Helen Hanft

Stardust Memories '80

Susumu Hani *(D)*

Nanami, First Love '68

Roger Hanin

My Other Husband '85
Rocco and His Brothers '60

Barry Hankerson

Pipe Dreams '76

Tom Hankerson

Candy Tangerine Man '75

Larry Hankin

Armed and Dangerous '86

Brad Hanks

HauntedWeen '91

Steve Hanks

B.A.D. Cats '80

Tom Hanks

Bachelor Party '84
Big '88
The Bonfire of the Vanities '90
The 'Burbs '89
Dragnet '87
Every Time We Say Goodbye
   '86
He Knows You're Alone '80
Joe Versus the Volcano '90
The Man with One Red Shoe
   '85
Mazes and Monsters '82
The Money Pit '86
Nothing in Common '86
Punchline '88
Rona Jaffe's Mazes &
   Monsters '82
Splash '84
Turner and Hooch '89
Volunteers '85

Bridget Hanley

Chattanooga Choo Choo '84

Jenny Hanley

Flesh and Blood Show '73
On Her Majesty's Secret
   Service '69
The Scars of Dracula '70

Jimmy Hanley

The Blue Lamp '49
Captive Heart '47
Housemaster '38
Room to Let '49
Salute John Citizen '42

Johnny Hanley

Let's Have an Irish Party '83

Peter Hanley

Best of Ernie Kovacs '56

Robert Hanley

L.A. Bounty '89

Julie Hanlon

I Was a Teenage TV Terrorist
   '87

Adam Hann-Byrd

Little Man Tate '91

William Hanna *(D)*

Hey There, It's Yogi Bear '64
The Jetsons: Movie '90

Darryl Hannah

At Play in the Fields of the Lord '91
Blade Runner '82
The Clan of the Cave Bear '86
Crazy People '90
Crimes & Misdemeanors '89
The Final Terror '83
The Fury '78
Hard Country '81
High Spirits '88
Legal Eagles '86
Memoirs of an Invisible Man '92
The Pope of Greenwich Village '84
Reckless '84
Roxanne '87
Splash '84
Steel Magnolias '89
Summer Lovers '82
Wall Street '87

Page Hannah

After School '88
Creepshow 2 '87
My Man Adam '86
Racing with the Moon '84
Shag: The Movie '89

Will Hannah

Buckeye and Blue '87

Ken Hannam (D)

The Assassination Run '80
Sunday Too Far Away '74

Brian Hannant (D)

The Time Guardian '87

Alyson Hannigan

My Stepmother Is an Alien '88

Marilyn Hanold

Frankenstein Meets the Space Monster '65

Izhak Hanooka (D)

Red Nights '87

Wladyslaw Hanoza

Casimir the Great

Lawrence Hanray

Mimi '35

Wuk Ma No Hans

Bruce's Deadly Fingers '80s

Kali Hansa

Night of the Sorcerers '70

Glen Hansard

The Commitments '91

Marion Hansel (D)

Dust '85

Cheryl Hansen

Thou Shalt Not Kill...Except '87

Ed Hansen (D)

The Bikini Car Wash Company '92
Party Favors '89
Party Plane '90
Takin' It All Off '87

Gale Hansen

Dead Poets Society '89

The Deadly and the Beautiful '73

Shaking the Tree '91

Gunnar Hansen

The Demon Lover '77
Hollywood Chainsaw Hookers '88
Texas Chainsaw Massacre '74

Heidi Hansen

Superbug Super Agent '76

James Hansen

Night Ripper '86

Joachim Hansen

Frozen Alive '64

Juanita Hansen

Charlie Chase and Ben Turpin '21

Lory Hansen

Point of Terror '71

Patti Hansen

Hard to Hold '84
They All Laughed '81

Rolf Hansen (D)

Devil In Silk '56

William Hansen

Homebodies '74

Rhonda Hansome

Feelin' Up '76

Curtis Hanson (D)

The Arousers '70
Bad Influence '90
The Bedroom Window '87
The Children of Times Square '86
The Hand that Rocks the Cradle '91
Little Dragons '80
Losin' It '82

Eleanor Hanson

Flaming Frontiers '38

Jody Hanson

Felicity '83

John Hanson (D)

Northern Lights '79
Wildrose '85

Kristina Hanson

Dinosaurus! '60

Lars Hanson

The Atonement of Gosta Berling '24
The Flesh and the Devil '27
Homecoming '28
Walpurgis Night '41
The Wind '28

Peter Hanson

When Worlds Collide '51

Tom Hanson (D)

The Zodiac Killer '71

Tom Haoff

Summer's Children '84

Setsuko Hara

Early Summer (Bakushu) '51

The Idiot '51

Late Spring '49
No Regrets for Our Youth '46
Tokyo Story '53

Ernest Harada

Volunteers '85

Itoko Harada

X from Outer Space '67

Kinako Harada

Days of Hell '70s

Meiko Harada

Nomugi Pass '79
Ran '85

Haya Harareet

Ben-Hur '59
Hill 24 Doesn't Answer '55
The Interns '62
Journey Beneath the Desert '61

Clement Harari

Flight of the Eagle '82
Monkeys, Go Home! '66

Richard Harbinger (D)

T-Bird Gang '59

Christine Harbort

Mephisto '81

Martin Harburg

Hockey Night '84

Ernest Harden, Jr.

White Men Can't Jump '92

Ernest Harden

The Final Terror '83
White Mama '80

Jacques Harden

Lola '61

Marcia Gay Harden

Fever '91
Late for Dinner '91
Miller's Crossing '90

Crofton Hardester

Devastator '85

Kate Hardie

Conspiracy '89
Mona Lisa '86

Raymond Hardie

Look Back in Anger '80

Jerry Hardin

Blaze '89
Wanted Dead or Alive '86
Wolf Lake '79

Melora Hardin

Big Man on Campus '89
Lambada '89
The North Avenue Irregulars '79

Rellie Hardin

Bronze Buckaroo '39

Sherry Hardin

Hollywood High '77

Ty Hardin

Bad Jim '89
Berserk! '67
I Married a Monster from Outer Space '58
One Step to Hell '68
Palm Springs Weekend '63
PT 109 '63
Rooster—Spurs of Death! '83
You're Jinxed, Friend, You've Met Sacramento '70

Ann Harding

The Animal Kingdom '32
Devotion '31
Love from a Stranger '37
The Man in the Gray Flannel Suit '56
Promise to Murder '56
Two Weeks with Love '50

Jeff Harding

Blood Tracks '86

June Harding

The Trouble with Angels '66

Kay Harding

Scarlet Claw '44

Lyn Harding

The Man Who Lived Again '36
Murder at the Baskervilles '39
The Mutiny of the Elsinore '39
Old Spanish Custom '36
Sherlock Holmes: The Silver Blaze '41
The Speckled Band '31

Kadeem Hardison

Beat Street '84
Def by Temptation '90
White Men Can't Jump '92

Karl Hardman

Night of the Living Dead '68

Joe Hardt

Blood Cult '85

Derek Hardwick

Among the Cinders '83

Edward Hardwick

The Adventures of Sherlock Holmes: Hound of the Baskervilles '89

Cedric Hardwicke

Around the World in 80 Days '56
Becky Sharp '35
Beware of Pity '46
Botany Bay '53
Commandos Strike at Dawn '42
A Connecticut Yankee in King Arthur's Court '49
The Desert Fox '51
Five Weeks in a Balloon '62
The Ghoul '33
The Green Glove '52
The Howards of Virginia '40
The Hunchback of Notre Dame '39
I Remember Mama '48
The Invisible Man Returns '40
The Keys of the Kingdom '44
King Solomon's Mines '37
Les Miserables '35
Nicholas Nickleby '46
Nine Days a Queen '34
Power '34

Richard III '55
Rope '48
Salome '53
Stanley and Livingstone '39
Sundown '41
Suspicion '41
The Ten Commandments '56
Things to Come '36
Tom Brown's School Days '40
Tycoon '47
Valley of the Sun '42
The White Tower '50
A Wing and a Prayer '44
The Winslow Boy '48

Cedric Hardwicke (D)

Forever and a Day '43

Jonathon Hardy

The Devil's Playground '76
Lonely Hearts '82
The Wacky World of Wills & Burke '85

Joseph Hardy (D)

Love's Savage Fury '79
Return Engagement '78
To Love Again '80
Users '78

Oliver Hardy

Atoll K '51
Bakery/Grocery Clerk '21
Bellhop/The Noon Whistle '22
Block-heads '38
Bohemian Girl '36
Bonnie Scotland '35
Bullfighters '45
A Chump at Oxford '40
Comedy Classics '35
Days of Thrills and Laughter '61
The Devil's Brother '33
A Few Moments with Buster Keaton and Laurel and Hardy '63
The Fighting Kentuckian '49
The Flying Deuces '39
Golden Age of Comedy '58
Great Guns '41
Hollywood Clowns '85
Laurel & Hardy: Another Fine Mess '30
Laurel & Hardy: At Work '32
Laurel & Hardy: Be Big '20s
Laurel & Hardy: Below Zero '20s
Laurel & Hardy: Berth Marks '29
Laurel & Hardy: Blotto '30
Laurel & Hardy: Brats '20s
Laurel & Hardy: Chickens Come Home '20s
Laurel & Hardy Comedy Classics Volume 1 '33
Laurel & Hardy Comedy Classics Volume 2 '30
Laurel & Hardy Comedy Classics Volume 3 '34
Laurel & Hardy Comedy Classics Volume 4 '32
Laurel & Hardy Comedy Classics Volume 5 '31
Laurel & Hardy Comedy Classics Volume 6 '35
Laurel & Hardy Comedy Classics Volume 7 '30s
Laurel & Hardy Comedy Classics Volume 8 '30s
Laurel & Hardy Comedy Classics Volume 9 '30s
Laurel & Hardy and the Family '33
Laurel & Hardy: Hog Wild '20s

Phyllis Haver

The Balloonatic/One Week '20s
The Blacksmith/The Balloonatic '23
The Nervous Wreck '26

Nigel Havers

Burke & Wills '85
Chariots of Fire '81
Empire of the Sun '87
Farewell to the King '89
A Little Princess '87
A Passage to India '84
The Whistle Blower '87
Who is Killing the Great Chefs of Europe? '78

Thom Haverstock

Fall from Innocence '88
Skullduggery '79

Allen Havey

Checking Out '89
Love or Money? '88

June Havoc

Brewster's Millions '45
Four Jacks and a Jill '41
Return to Salem's Lot '87
Sing Your Worries Away '42
Timber Queen '44

Robin Hawdon

Alien Women '69
When Dinosaurs Ruled the Earth '70

Ethan Hawke

Dad '89
Dead Poets Society '89
Explorers '85
Mystery Date '91
White Fang '90

Ian Hawkes

Queen of Hearts '89

John Hawkes

Murder Rap '87

Steve Hawkes

Blood Freak '72

Steve Hawkes (D)

Blood Freak '72

Caroline Hawkins

Alien Women '69

Jack Hawkins

The Adventurers '52
Angels One Five '54
Ben-Hur '59
Bonnie Prince Charlie '48
The Bridge on the River Kwai '57
Cruel Sea '53
The Elusive Pimpernel '50
Escape to the Sun '72
The Fallen Idol '49
Guns at Batasi '64
Land of Fury '55
Land of the Pharaohs '55
Lawrence of Arabia '62
League of Gentlemen '60
Lola '69
Lord Jim '65
Malta Story '53
Mandy '53
Nicholas and Alexandra '71
Phantom Fiend '35
The Prisoner '55

Restless '72
A Shot in the Dark '33
The Small Back Room '49
The Third Key '57
Those Daring Young Men in Their Jaunty Jalopies '69
Waterloo '71
Young Winston '72
Zulu '64

Jay Hawkins

Mystery Train '89
Two Moon Junction '88

Loye Hawkins

The Guy from Harlem '77

Michael Hawkins

Trucker's Woman '83

Tara Hawkins Moore (D)

Tusks '89

Patricia Hawkins

Oh! Calcutta! '72

Ronnie Hawkins

Club Med '83
The Last Waltz '78

Don Hawks (D)

Hush Little Baby, Don't You Cry '86

Howard Hawks (D)

Air Force '43
Ball of Fire '42
Barbary Coast '35
The Big Sky '52
The Big Sleep '46
Bringing Up Baby '38
Come and Get It '36
Criminal Code '31
El Dorado '67
Gentlemen Prefer Blondes '53
Hatari '62
His Girl Friday '40
Land of the Pharaohs '55
Man's Favorite Sport? '63
Monkey Business '52
Monkey Business '52
Only Angels Have Wings '39
Red Line 7000 '65
Red River '48
Rio Bravo '59
Rio Lobo '70
Scarface '31
Sergeant York '41
The Thing '51
To Have & Have Not '44
Twentieth Century '34

Pamela Hawksford

Contagion '87

Jean Hawkshaw

Wild Women of Wongo '59

Monte Hawley

Mystery in Swing '40
Tall, Tan and Terrific '46

Wanda Hawley

American Pluck '25

Goldie Hawn

Best Friends '82
Bird on a Wire '90
Butterflies Are Free '72
Cactus Flower '69
Crisscross '92
Deceived '91

Dollars '71
The Duchess and the Dirtwater Fox '76
Foul Play '78
The Girl from Petrovka '74
Goldie & Kids '82
Lovers and Liars '81
One and Only, Genuine, Original Family Band '68
Overboard '87
Private Benjamin '80
Protocol '84
Revenge of TV Bloopers '60s
Seems Like Old Times '80
Shampoo '75
The Sugarland Express '74
Swing Shift '84
There's a Girl in My Soup '70
Wildcats '86

Jill Haworth

Exodus '60
The Freakmaker '73
In Harm's Way '65
Tower of Evil '72

Nigel Hawthorne

Firefox '82
Pope John Paul II '84
S*P*Y*S '74
Tartuffe '90

Charles Hawtrey

Alien Women '69
A Canterbury Tale '44
Carry On Cleo '65
Carry On Screaming '66
Carry On at Your Convenience '71
Room to Let '49
Terronauts '67
The Terrornauts '67

Kay Hawtry

Funeral Home '82

Alexandra Hay

How Come Nobody's On Our Side? '73
The Love Machine '71

Christian Hay

You're Jinxed, Friend, You've Met Sacramento '70

George Hay

Two Reelers: Comedy Classics 9 '34

Rod Hay (D)

Breaking Loose '90

Will Hay

Oh, Mr. Porter '37

William Hay

Doom Asylum '88

Sessue Hayakawa

The Bridge on the River Kwai '57
The Cheat '15
The Geisha Boy '58
Hell to Eternity '60
Swiss Family Robinson '60
Three Came Home '50
Tokyo Joe '49

Marc Hayashi

Chan is Missing '82

Yutaka Hayashi

Godzilla vs. Megalon '76

Marcia Haydee

The Lady of the Camellias '87

Harry Hayden

Double Dynamite '51

Linda Hayden

Baby Love '69
The Barcelona Kill '70s
The Blood on Satan's Claw '71
The House on Straw Hill '76
Madhouse '74

Richard Hayden

Pride of St. Louis '52

Russell Hayden

Apache Chief '50
Blazing Guns '50
Guns of Justice '50
Knights of the Range '40
Last Bullet '50
The Light of Western Stars '40
Lost City of the Jungle '45
Marshal of Heldorado '50
The Mysterious Rider '38
'Neath Canadian Skies '46
North of the Border '46
Outlaw Fury '50
Rangeland Empire '50
Renegade Trail '39
Rolling Home '48
Santa Fe Marshal '40
The Showdown '40
Sudden Death '50
Trail of the Mounties '47
Where the North Begins '47

Sterling Hayden

Asphalt Jungle '50
Bahama Passage '42
The Blue and the Grey '85
Cobra '71
Deadly Strangers '82
The Denver & Rio Grande '51
Dr. Strangelove, or: How I Learned to Stop Worrying and Love the Bomb '64
Fighter Attack '53
The Final Programme '73
Flat Top '52
Gas '81
The Godfather '72
Godfather 1902-1959—The Complete Epic '81
Johnny Guitar '53
Kansas Pacific '53
The Killing '56
King of the Gypsies '78
The Last Command '55
The Long Goodbye '73
9 to 5 '80
1900 '76
Prince Valiant '54
Shotgun '55
Spaghetti Western '60s
Suddenly '54
Venom '82
Winter Kills '79

Richard Haydn

Clarence, the Cross-eyed Lion '65
Jupiter's Darling '55
Please Don't Eat the Daisies '60

Richard Haydn (D)

Dear Wife '49

Helen Haye

The Case of the Frightened Lady '39
Man of Evil '48

Jeff Hayenga

Prince of Pennsylvania '88
The Unborn '91

Sidney Hayers (D)

Assault '70
Bananas Boat '78
Burn Witch, Burn! '62
Circus of Horrors '60
Deadly Strangers '82
King Arthur, the Young Warlord '75
Revenge '71
The Seekers '79
Trap '66

Allan Hayes

Neon Maniacs '86

Allison Hayes

Attack of the 50 Foot Woman '58
The Crawling Hand '63
The Gunslinger '56
The Undead '57
The Unearthly '57
Zombies of Moratau '57

Bernadene Hayes

Heroes in Blue '39
Santa Fe Marshal '40

Bill Hayes (D)

King Kung Fu '87

Billie Hayes

Li'l Abner '59

Frank Hayes

Fatty's Tin-Type Tangle/Our Congressman '20s

George "Gabby" Hayes

Bad Man of Deadwood '41
Badman's Territory '46
Beggars in Ermine '34
Bells of Rosarita '45
Billy the Kid Returns '38
Blue Steel '34
Border Devils '32
Brand of Hate '34
Breed of the Border '33
Broadway to Cheyenne '32
The Cariboo Trail '50
Cavalier of the West '31
Colorado '40
Come On Rangers '38
Dark Command '40
Death Valley Manhunt '43
From Broadway to Cheyenne '32
Heart of the Golden West '42
Helldorado '46
Home in Oklahoma '47
House of Mystery '34
In Old Caliente '39
In Old Cheyenne '41
In Old Santa Fe '34
Jesse James at Bay '41
Lawless Frontier '35
The Lawless Nineties '36
Lights of Old Santa Fe '47
Lonely Trail/Three Texas Steers '39
The Lost City '34
Lucky Texan '34
Man from Cheyenne '42

Pulse '88

David Healy

Supergirl '84

Katherine Healy

Six Weeks '82

Mary Healy

The 5000 Fingers of Dr. T '53

Myron Healy

Cavalry Command '63
Smoke in the Wind '75
The Unearthly '57
Varan the Unbelievable '61

Patricia Healy

Sweet Poison '91

Susan Healy

Paramount Comedy Theater,
    Vol. 3: Hanging Party '87

Ted Healy

Dancing Lady '33
Three Stooges Comedy
    Classics '49

Howard Heard (D)

Shadows Run Black '84

John Heard

After Hours '85
Awakenings '90
Beaches '88
Best Revenge '83
Betrayed '88
Between the Lines '77
Big '88
Blown Away '90
Cat People '82
Chilly Scenes of Winter '79
C.H.U.D. '84
Cutter's Way '81
Deceived '91
The End of Innocence '90
First Love '77
Gladiator '92
Heart Beat '80
Heaven Help Us '85
Home Alone '90
Legs '83
The Milagro Beanfield War '88
Mindwalk '91
On the Yard '79
The Package '89
Radio Flyer '91
The Seventh Sign '88
The Telephone '87
Too Scared to Scream '85
Trip to Bountiful '85
Violated '84

Susan Heard

The Body Beneath '70

George Hearn

A Piano for Mrs. Cimino '82
Sanctuary of Fear '79
See You in the Morning '89
Sweeney Todd: The Demon
    Barber of Fleet Street '84

Mary Ann Hearn

Death Driver '78

Richard Hearne

Butler's Dilemma '43

Patty Hearst

Cry-Baby '90

Rick Hearst

Crossing the Line '90

Marla Heasly

Born to Race '88

Thomas Heathcote

Above Us the Waves '56

Jean Heather

Double Indemnity '44

Clifford Heatherly

Bitter Sweet '33
Champagne '28
If I Were Rich '33

Joey Heatherton

Bluebeard '72
Happy Hooker Goes to
    Washington '77
Johnny Carson '70s

Patricia Heaton

Beethoven '92

David Heavener

Border of Tong '90
Deadly Reactor '89
Killcrazy '89
Outlaw Force '87
Prime Target '91
Ragin' Cajun '90
Twisted Justice '89

David Heavener (D)

Deadly Reactor '89
Killcrazy '89
Outlaw Force '87
Prime Target '91
Twisted Justice '89

Paul Hebert

Mouchette '60

Anne Heche

O Pioneers! '91

Ben Hecht (D)

Actors and Sin '52
Angels Over Broadway '40
Spectre of the Rose '46

Donatella Hecht

Flesh Eating Mothers '89

Paul Hecht

Mary and Joseph: A Story of
    Faith '79
The Savage Bees '76

Eileen Heckart

The Bad Seed '56
Bus Stop '56
Butterflies Are Free '72
A Doll's House '59
The Hiding Place '75
Hot Spell '58
No Way to Treat a Lady '68
Somebody Up There Likes Me
    '56
Sorrows of Gin '79
Table Settings '84
Up the Down Staircase '67
White Mama '80

Amy Heckerling

Into the Night '85

Amy Heckerling (D)

Fast Times at Ridgemont High
    '82

Johnny Dangerously '84
Look Who's Talking '89
Look Who's Talking, Too '90
National Lampoon's European
    Vacation '85

Dan Hedaya

The Addams Family '91
The Adventures of Buckaroo
    Banzai '84
Blood Simple '85
Commando '85
Courage '86
Endangered Species '82
The Hunger '83
Joe Versus the Volcano '90
Pacific Heights '90
The Prince of Central Park '77
Reckless '84
Running Scared '86
Slow Burn '86
A Smoky Mountain Christmas
    '86
Tightrope '84
Tune in Tomorrow '90
Wise Guys '86

Rob Hedden (D)

Friday the 13th, Part 8: Jason
    Takes Manhattan '89

Peter Hedges

Sammy '88

Serene Hedin

Boggy Creek II '83
Hawken's Breed '87

David Hedison

The Art of Crime '75
Enemy Below '57
ffolkes '80
The Fly '58
Kemek '88
Kenny Rogers as the
    Gambler, Part 2: The
    Adventure Continues '83
License to Kill '89
Live and Let Die '73
The Naked Face '84
The Power Within '79
Undeclared War '91

Jack Hedley

For Your Eyes Only '81
Goodbye, Mr. Chips '69

Trine Hedman (D)

Famous Five Get Into Trouble
    '70s

Tippi Hedren

The Birds '63
Foxfire Light '84
Harrad Experiment '73
In the Cold of the Night '89
Marnie '64
Mr. Kingstreet's War '71
Pacific Heights '90
Satan's Harvest '65

Deborah Hedwall

Sessions '83

Astrid Heeren

Silent Night, Bloody Night '73

Herb Heesel

2069: A Sex Odyssey '78

Honor Heffernan

Danny Boy '84

Wayne Heffley

Submarine Seahawk '59

Avram Heffner (D)

But Where Is Daniel Wax? '74

Kyle T. Heffner

Mutant on the Bounty '89

**Richard T.
Heffron** (D)

Convicted: A Mother's Story
    '87
Death Scream '75
Foolin' Around '80
Futureworld '76
Guilty of Innocence '87
I, the Jury '82
I Will Fight No More Forever
    '75
Newman's Law '74
Outlaw Blues '77
A Rumor of War '80
Samaritan: The Mitch Snyder
    Story '86
See How She Runs '78
Tagget '90
Whale for the Killing '81

Van Heflin

Airport '70
Back Door to Heaven '39
Battle Cry '55
Cry of Battle '63
Flight from Glory '37
The Greatest Story Ever Told
    '65
Green Dolphin Street '47
Madame Bovary '49
The Man Outside '68
Patterns '56
Presenting Lily Mars '43
Prowler '51
The Ruthless Four '70
Santa Fe Trail '40
Seven Miles from Alcatraz/
    Flight from Glory '43
Shane '53
The Strange Love of Martha
    Ivers '46
They Came to Cordura '59
Three Musketeers '48
3:10 to Yuma '57
Till the Clouds Roll By '46
Till the Clouds Roll By/A Star
    Is Born '46
A Woman Rebels '36

Hugh Hefner

Playboy Video Centerfold:
    35th Anniversary Playmate
    '88

Chris Hegedus (D)

Jimi Hendrix: Live in
    Monterey, 1967 '67

O.P. Heggie

Anne of Green Gables '34
The Bride of Frankenstein '35
Call It Murder '34
Count of Monte Cristo '34
Devotion '31
Midnight '34

Robert Hegyes

Just Tell Me You Love Me '80
Underground Aces '80

Robert Hegyes (D)

E. Nick: A Legend in His Own
    Mind '84

Peter Hehir

Fast Talking '86
I Live with Me Dad '86
Sweet Talker '91

Sasha Hehn

Melody in Love '78

Yim Nam Hei

Murder Masters of Kung Fu
    '85

Yu Ka Hei

Champion Operation '70s

Janice Heiden

The Student Body '76

Jascha Heifetz

They Shall Have Music '39

Yosif Heifitz (D)

Baltic Deputy '37
The Lady with the Dog '59

Lorna Heilbron

The Creeping Flesh '72

Elayne Heilveil

The Adventures of Nellie Bly
    '81
Birds of Prey '72

Laurie Heineman

Save the Tiger '73

Michael Heinz

Clayton County Line '

Stuart Heisler (D)

Along Came Jones '45
The Burning Hills '56
Chain Lightning '50
The Glass Key '42
Hitler '62
I Died a Thousand Times '55
Island of Desire '52
Lone Ranger '56
Smash-Up: The Story of a
    Woman '47
Tokyo Joe '49
Tulsa '49

Carol Heiss

Snow White and the Three
    Stooges '61

Michael Heit

Bare Knuckles '77

Martin Held

Captain from Koepenick '56

Brit Helfer

Alley Cat '84
Let's Do It! '84

Mats Helge (D)

Ninja Mission '84
The Russian Terminator '90

Marg Helgenberger

After Midnight '89
Always '89
Blind Vengeance '90
China Beach '88

Anne Marie Helger

Ladies on the Rocks '83

Jenny Helia

Toni '34

Richard Hell

Desperately Seeking Susan '85
Geek Maggot Bingo '83
Smithereens '84

Thomas Hellberg

The Assignment '78

Olle Hellbron (D)

Brothers Lionheart '85
Pippi Goes on Board '75
Pippi in the South Seas '74

Marjorie Hellen

Missile to the Moon '59

Jordan Heller

Ninja Fantasy '86

Randee Heller

Can You Hear the Laughter?
The Story of Freddie Prinze
'79
Fast Break '79
The Karate Kid '84

Gerome Hellman (D)

Promises in the Dark '79

Jacqueline Hellman

Flight to Fury '66

Monte Hellman

Someone to Love '87

Monte Hellman (D)

The Beast from Haunted Cave
'60
Cockfighter '74
Creature from the Haunted
Sea '60
Flight to Fury '66
Gunfire '78
Iguana '89
Ride in the Whirlwind '67
The Shooting '66
Silent Night, Deadly Night 3:
Better Watch Out! '89

Oliver Hellman (D)

Beyond the Door '75
Desperate Moves '86
Tentacles '77

Anne Helm

Follow That Dream '61
Nightmare in Wax '69
Unkissed Bride '66

Harry Helm

Unnatural '52

Levon Helm

Best Revenge '83
Coal Miner's Daughter '80
Dollmaker '84
End of the Line '88
Man Outside '88
The Right Stuff '83
Smooth Talk '85

Tiffany Helm

The Zoo Gang '85

Heidi Helmer

Beachballs '88

Paul Helmick (D)

Hard Drivin' '60

Charlotte J. Helmkamp

Frankenhooker '90
Posed For Murder '89

Katherine Helmond

Autobiography of Miss Jane
Pittman '74
Brazil '86
Family Plot '76
Jack & the Beanstalk '83
The Lady in White '88
Larry '74
Overboard '87
Rosie: The Rosemary Clooney
Story '82
Scout's Honor '80
Shadey '87
Time Bandits '81
World War III '86

Tom Helmore

Designing Woman '57
Flipper's New Adventure '64
The Tender Trap '55
Trouble Along the Way '53
Vertigo '58

Frits Helmuth

Memories of a Marriage '90

Tony Helou

Dogs in Space '87

David Helpern (D)

Something Short of Paradise
'79

Robert Helpmann

The Mango Tree '77
Patrick '78
Puzzle '78
The Red Shoes '48

Sheila Helpmann

The Getting of Wisdom '77
Image of Death '77

George Helton

Go Kill and Come Back '68

David Hemblen

The Room '87
Short Circuit 2 '88
Speaking Parts '89

Mark Hembrow

Out of the Body '88
Return to Snowy River '88

Margaux Hemingway

Inner Sanctum '91
Killer Fish '79
Lipstick '76
Over the Brooklyn Bridge '83
They Call Me Bruce '82

Mariel Hemingway

Creator '85
Falling from Grace '92
Into the Badlands '92
Lipstick '76
Manhattan '79
Mean Season '85
Personal Best '82
Star 80 '83
Steal the Sky '88
The Suicide Club '88
Sunset '88
Superman 4: The Quest for
Peace '87

Richard Hemingway

Woman Condemned '33

David Hemmings

Barbarella '68
Beyond Erotica '79
Beyond Reasonable Doubt
'80
Blood Relatives '78
Blow-Up '66
Calamity Jane '82
Camelot '67
Dark Forces '83
Deep Red: Hatchet Murders
'75
Disappearance '81
Islands in the Stream '77
Juggernaut '74
Just a Gigolo '79
The Key to Rebecca '85
The Love Machine '71
Man, Woman & Child '83
Murder by Decree '79
The Old Curiosity Shop '75
Power Play '81
The Rainbow '89
Snow Queen '83
The Squeeze '77
Thirst '87
Turn of the Screw '89

David Hemmings (D)

Just a Gigolo '79
The Key to Rebecca '85
Race to the Yankee Zephyr
'81
Survivor '80
Treasure of the Yankee
Zephyr '83

Myra D. Hemmings

Go Down Death '41

Anouska Hempel

On Her Majesty's Secret
Service '69
The Scars of Dracula '70
Tiffany Jones '75

Shirley Hemphill

Girls of the Comedy Store '86

Sherman Hemsley

Camp Cucamonga: How I
Spent My Summer
Vacation '90
Club Fed '91
Combat Academy '86
Ghost Dance '83
Ghost Fever '87
Love at First Bite '79

Joseph Henabery (D)

Cobra '25

George Henare

The Silent One '86

Richard Hench

Bio Hazard '85

John Hend (D)

Jimi Hendrix: Story '73

Frederique Hender

The Devil's Nightmare '71

Albert Henderson

Greaser's Palace '72

Bill Henderson

City Slickers '91
Get Crazy '83
Movie Maker '86
Murphy's Law '86

Chuck Henderson

Operation Dames '59

Clark Henderson (D)

Circle of Fear '89
Primary Target '89
Saigon Commandos '88
Warlords from Hell '87

Dell Henderson

The Crowd '28
Show People '28

Don Henderson

The Ghoul '75
The Island '80

Don Henderson (D)

The Touch of Satan '70
Weekend with the Babysitter
'70

Douglas Henderson

King Dinosaur '55

Ena Henderson

Fatal Exposure '90

Florence Henderson

Shakes the Clown '92
Song of Norway '70

Jeff Henderson

Last Witness '88

Jo Henderson

Lianna '83

Joe Henderson

The Paris Reunion Band '88

Maggie Henderson

Indiscreet '88

Marcia Henderson

Naked Hills '56
Thunder Bay '53

Scott A. Henderson

The Milpitas Monster '75

Ty Henderson

Competition '80
Happy Hour '87

Lauri Hendler

High School USA '84

Martin Hendler

Forbidden Impulse '85

Tony Hendra

This is Spinal Tap '84

Evelyn Hendricks

Night of Bloody Horror '76

Nancy Hendrickson

Mother's Day '80

Elaine Hendrix

Last Dance '91

Jimi Hendrix

Jimi Hendrix: Berkeley May
1970 '73
Jimi Hendrix: Johnny B.
Goode '67
Jimi Hendrix: Live in
Monterey, 1967 '67
Jimi Hendrix: Rainbow Bridge
'89
Monterey Pop '68
Woodstock '70

Leah Ayres Hendrix

Hot Child in the City '87

Wanda Hendrix

The Admiral was a Lady '50
My Outlaw Brother '51
The Oval Portrait '88

Gloria Hendry

Bare Knuckles '77
Black Belt Jones '74
Black Caesar '73
Hell Up in Harlem '73
Live and Let Die '73
Slaughter's Big Ripoff '73

Ian Hendry

The Bitch '78
Captain Kronos: Vampire
Hunter '74
Captain Kronos: Vampire
Hunter '74
Journey to the Far Side of the
Sun '69
Killer with Two Faces '74
The Passenger '75
Repulsion '65
The Saint '68
Tales from the Crypt '72
Theatre of Blood '73

Frank Henenlotter (D)

Basket Case '82
Basket Case 2 '90
Basket Case 3: The Progeny
'91
Brain Damage '88
Frankenhooker '90

Sonja Henie

Sun Valley Serenade '41

Anna Henkel

1900 '76

Peter Henkel (D)

Three Bullets for a Long Gun
'73

Drewe Henley

When Dinosaurs Ruled the
Earth '70

Carrie Henn

Aliens '86

Dermot Hennelly

Deserters '80s

Marilu Henner

Between the Lines '77
Bloodbrothers '78
Cannonball Run II '84
Grown Ups '86
Hammett '82
Johnny Dangerously '84
L.A. Story '91
Love with a Perfect Stranger
'86

The Man Who Loved Women '83
Perfect '85
Rustler's Rhapsody '85
Stark '85

Mark Hennessy
Terminal Exposure '89

Eva Henning
Devil's Wanton '49
Three Strange Loves '49

Astrid Henning-Jensen
The Element of Crime '85

Sam Hennings
Seedpeople '92

Paul Henreid
Battle Shock '56
Casablanca '42
Deception '46
Deep in My Heart '54
Four Horsemen of the Apocalypse '62
Goodbye, Mr. Chips '39
Hollow Triumph '48
Joan of Paris '41
Never So Few '59
Night Train to Munich '40
Now, Voyager '42
The Scar '48
The Spanish Main '45
A Stolen Face '52
Tall Lie '53

Paul Henreid (D)
Ballad in Blue '66
Battle Shock '56
Tall Lie '53

Bobby Henrey
The Fallen Idol '49

Lance Henriksen
Aliens '86
Choke Canyon '86
Damien: Omen 2 '78
Deadly Intent '88
Dog Day Afternoon '75
The Hit List '88
The Horror Show '89
The Jagged Edge '85
Johnny Handsome '89
Near Dark '87
Nightmares '83
Piranha 2: The Spawning '82
Prince of the City '81
Pumpkinhead '88
The Right Stuff '83
Savage Dawn '84
Stone Cold '91
Survival Quest '89
The Terminator '84
The Visitor '80

Anders Henrikson
Miss Julie '50
A Woman's Face '38

Buck Henry
Aria '88
Best of Gilda Radner '89
Best of John Belushi '85
Catch-22 '70
Dark Before Dawn '89
Defending Your Life '91
Eating Raoul '82
Gloria '80
The Graduate '67
Is There Sex After Death? '71
The Linguini Incident '92

The Man Who Fell to Earth '76
Old Boyfriends '79
Rude Awakening '89
Steve Martin Live '85
Tune in Tomorrow '90

Buck Henry (D)
First Family '80
Heaven Can Wait '78

Buss Henry
Rolling Home '48

Charlotte Henry
Bowery Blitzkrieg '41
Charlie Chan at the Opera '36
The Mandarin Mystery '37
March of the Wooden Soldiers '34

Gloria Henry
Miss Grant Takes Richmond '49

Greg Henry
Body Double '84
Fair Game '89
Funny Money '82
Hitchhiker 4 '87
Hot Rod '79
Just Before Dawn '80
The Last of Philip Banter '87
Mean Dog Blues '78
Patriot '86

Justin Henry
Kramer vs. Kramer '79
Martin's Day '85
Sixteen Candles '84
Sweet Hearts Dance '88
Tiger Town '83

Laura Henry
Heavenly Bodies '84

Lenny Henry
The Suicide Club '88
Three of a Kind '80s
True Identity '91

Martha Henry
Dancing in the Dark '86
White Light '90

Mike Henry
Adios Amigo '75
Smokey and the Bandit '77
Smokey and the Bandit, Part 2 '80
Smokey and the Bandit, Part 3 '83

Pat Henry
Lady in Cement '68

Robert "Buzzy" Henry
Turf Boy '42
Western Frontier '35
Wild West '46

Terence Henry
The Berlin Conspiracy '91

Tim Henry
Dawson Patrol '78

William Henry
Dance Hall '41
Marshal of Cedar Rock '53
Movie Stunt Man '53
Tarzan Escapes '36

Jim Henshaw
Snapshot '77

Craig Hensley
Natas...The Reflection '83

Lisa Hensley
The 13th Floor '88

Pamela Hensley
Buck Rogers in the 25th Century '79
Doc Savage '75
Double Exposure '82
The Nude Bomb '80
The Rebels '79

Basil Henson
Anatomy of Terror '74

Brad Henson
Warlords from Hell '87

Gladys Henson
Train of Events '49

Jim Henson
Inside the Labyrinth '86
Into the Night '85
Muppet Movie '79
Muppets Take Manhattan '84
Timepiece '65

Jim Henson (D)
The Dark Crystal '82
The Great Muppet Caper '81
Labyrinth '86

Nicky Henson
Number 1 of the Secret Service '77
Psychomania '73

Torsten Hentes
David '79

Perry Henzell (D)
The Harder They Come '72

Audrey Hepburn
Always '89
Breakfast at Tiffany's '61
Charade '63
The Children's Hour '62
Funny Face '57
The Lavender Hill Mob '51
Love in the Afternoon '57
My Fair Lady '64
The Nun's Story '59
Paris When It Sizzles '64
Robin and Marian '76
Roman Holiday '53
Sabrina '54
Sidney Sheldon's Bloodline '79
They All Laughed '81
Two for the Road '67
The Unforgiven '60
Wait Until Dark '67
War and Peace '56

Dee Hepburn
Gregory's Girl '80

Katharine Hepburn
Adam's Rib '49
The African Queen '51
Alice Adams '35
Bacall on Bogart '88
A Bill of Divorcement '32
Break of Hearts '35
Bringing Up Baby '38
Christopher Strong '33

Desk Set '57
Dragon Seed '44
Grace Quigley '84
Great Kate: The Films of Katharine Hepburn
Great Movie Trailers '70s
Guess Who's Coming to Dinner '67
Hepburn & Tracy '84
Holiday '38
The Lion in Winter '68
Little Minister '34
Little Women '33
Long Day's Journey into Night '62
Love Among the Ruins '75
Mary of Scotland '36
Morning Glory '33
Olly Olly Oxen Free '78
On Golden Pond '81
Pat and Mike '52
Philadelphia Story '40
Quality Street '37
The Rainmaker '56
Rooster Cogburn '75
Spitfire '34
Stage Door '37
Stage Door Canteen '43
State of the Union '48
Storytime Classics '83
Suddenly, Last Summer '59
Summertime '55
Sylvia Scarlett '35
Trojan Women '71
A Woman Rebels '36
Woman of the Year '42

Bernard Hepton
Eminent Domain '91
Mansfield Park '85

Douglas Herald
Typhoon Treasure '39

Bill Herbert (D)
Warlock Moon '73

Charles Herbert
13 Ghosts '60

Henry Herbert (D)
Emily '77

Holmes Herbert
Bad Boy '39
British Intelligence '40
Dr. Jekyll and Mr. Hyde '31
The House of Secrets '37
The Kiss '29
Sherlock Holmes and the Secret Weapon '42
Shop Angel '32
Stanley and Livingstone '39
Through the Breakers '28

Hugh Herbert
The Beautiful Blonde from Bashful Bend '49
Diplomaniacs '33
Eternally Yours '39
Fashions of 1934 '34
Gold Diggers of 1935 '35
The Great Waltz '38
Hook, Line and Sinker '30
A Midsummer Night's Dream '35
One Rainy Afternoon '36
Sherlock Holmes and the Secret Weapon '42
The Villain Still Pursued Her '41

James Herbert
Ricky 1 '88

Martin Herbert (D)
Miami Horror '87
Strange Shadows in an Empty Room '76

Percy Herbert
The Fiend '71
Mysterious Island '61

Rick Herbst
Brain Damage '88

Carla Herd
Deathstalker 3 '89
Wild Zone '89

Richard Herd
Lovely...But Deadly '82
Wolf Lake '79

Lisa Heredia
Summer '86

Stephen Herek (D)
Bill & Ted's Excellent Adventure '89
Critters '86
Don't Tell Mom the Babysitter's Dead '91

James Leo Herlihy
Four Friends '81

Jacques Herlin
Torrents of Spring '90

Roberto Herlitzka
Summer Night '87
Summer Night With Greek Profile, Almond Eyes... '87

Al Herman (D)
Man from Texas '39
Outlaws of the Range '36
Renfrew on the Great White Trail '38
Renfrew of the Royal Mounted '37
Roll, Wagons, Roll '39
Rollin' Plains '38
Take Me Back to Oklahoma '40
Utah Trail '38
Valley of Terror '38
Western Frontier '35
Whispering Shadow '33

Albert Herman (D)
Cowboy & the Bandit '35
Dawn Express '42
Delinquent Daughters '44
Gun Play '36
The Missing Corpse '45
Rogue's Gallery '44
Shake Hands with Murder '44

Jack Herman
The Yesterday Machine '63

Jean Herman (D)
Butterfly Affair '71
Honor Among Thieves '68

Jimmy Herman
Dances with Wolves '90

Mark Herman (D)
Blame It on the Bellboy '91

Paul Herman
The Last Temptation of Christ '88

Carnal Crimes '91
Endless Love '81
Killer Party '86
Night Rhythms '90s
Out of Control '85
Private War '90
Secret Games '90s
White Ghost '88
Yellowbeard '83

Pete Hewitt *(D)*
Bill & Ted's Bogus Journey
'91

Rob Hewitt *(D)*
Verne Miller '88

Sean Hewitt
Big Zapper '73
The Sender '82

Philip Hewland
Q Ships '28

Bentley Hewlett
Born to Be Wild '38

David Hewlett
Desire and Hell at Sunset
Motel '92
Pin '88
Scanners 2: The New Order
'91

Donald Hewlett
Saving Grace '86

Jon-Erik Hexum
Voyager from the Unknown
'83

Christian Hey
Autopsy '70s

Virginia Hey
Obsession: A Taste for Fear
'89

Weldon Heyburn
Criminals Within '41
Panama Patrol '39
Sea Racketeers '37
The Thirteenth Man '37

Yvette Heyden
Tides of War '90

Louis Jean Heydt
The Great McGinty '40
Thirty Seconds Over Tokyo
'44
Zombies on Broadway '44

Douglas Heyes *(D)*
Powder Keg '70

Herbert Heyes
Behind Locked Doors '48
The Cobra Strikes '48

John Heyl
A Separate Peace '73

Barton Heyman
Billy Galvin '86
He Who Walks Alone '78
Let's Scare Jessica to Death
'71
Valdez is Coming '71

Sean Heyman
The Lost Platoon '89

**Laurent
Heynemann** *(D)*
Birgitt Haas Must Be Killed '83

Anne Heywood
The Brain '62
Carthage in Flames '60
Doctor at Large '57
I Want What I Want '72
Scenes from a Murder '72
The Shaming '71
The Very Edge '63

Pat Heywood
Girly '70
Rude Awakening '81
Wish You Were Here '87

Ruth Hiatt
Ridin' Thru '35

Jesse Hibbs *(D)*
To Hell and Back '55

Winston Hibler *(D)*
Charlie, the Lonesome Cougar
'67

**George
Hickenlooper** *(D)*
Hearts of Darkness: A
Filmmaker's Apocalypse
'91

Barry Hickey
Las Vegas Weekend '85

Brendan Hickey
I Married a Vampire '07

Bruce Hickey *(D)*
Necropolis '87

Marguerite Hickey
Mirrors '85

Tom Hickey
Nuns on the Run '90

William Hickey
Any Man's Death '90
The Boston Strangler '68
Bright Lights, Big City '88
Hobo's Christmas '87
Invitation to a Gunfighter '64
Mikey & Nicky '76
Mob Boss '90
My Blue Heaven '90
The Name of the Rose '86
National Lampoon's
Christmas Vacation '89
One Crazy Summer '86
Pink Cadillac '89
Prizzi's Honor '85
Puppet Master '89
The Runestone '91
Sea of Love '89
Tales from the Darkside: The
Movie '90
Walls of Glass '85

Catherine Hickland
Ghost Town '88
Witchery '88

Darryl Hickman
The Devil on Wheels '47
Fighting Father Dunne '48
Geronimo's Revenge '60
Johnny Shiloh '63
Men of Boys Town '41
Tea and Sympathy '56

Texas John Slaughter:
Geronimo's Revenge '60

Dwayne Hickman
Cat Ballou '65
Dr. Goldfoot and the Bikini
Machine '66
High School USA '84
How to Stuff a Wild Bikini '65
My Dog, the Thief '69

Howard Hickman
Civilization '16

Anthony Hickox *(D)*
Sundown '91
Waxwork '88
Waxwork 2: Lost in Time '91

Douglas Hickox *(D)*
Blackout '85
Brannigan '75
Entertaining Mr. Sloane '70
Mistral's Daughter '84
Sins '85
Sky Riders '76
Theatre of Blood '73
Zulu Dawn '79

Bill Hicks
Rodney Dangerfield: Nothin'
Goes Right '88

Catherine Hicks
Better Late Than Never '83
Child's Play '88
Death Valley '81
Fever Pitch '85
Garbo Talks '84
Laguna Heat '87
Like Father, Like Son '87
Marilyn: The Untold Story '80
Peggy Sue Got Married '86
The Razor's Edge '84
Running Against Time '90
She's Out of Control '89
Souvenir '89
Spy '89
Star Trek 4: The Voyage
Home '86

Dan Hicks
Evil Dead 2: Dead by Dawn
'87

John Hicks
Free, White, and 21 '62

Kevin Hicks
Blood Relations '87
Final Notice '89
Higher Education '88

Leonard Hicks
Santa Claus Conquers the
Martians '64

Russell Hicks
Blind Fools '40
Hitler: Dead or Alive '43
Seventh Cavalry '56
Stanley and Livingstone '39

Seymour Hicks
Fame is the Spur '47
Scrooge '35
The Secret of the Loch '34

**Tommy Redmond
Hicks**
She's Gotta Have It '86

William T. Hicks
Challenge '74
Day of Judgment '81
Death Screams '83
House of Death '82
Order of the Black Eagle '87
Unmasking the Idol '86

Joan Hickson
The Body in the Library '87
A Day in the Death of Joe Egg
'71
Doctor in the House '53
Great Expectations '81
A Murder is Announced '87
A Pocketful of Rye '87

**Raymundo Hidalgo-
Gato**
El Super '79

Bokuzen Hidari
The Lower Depths '57

Sachiko Hidari
The Insect Woman '63

Tonpei Hidari
The Ballad of Narayama '83

**Chieko
Higashiyama**
Tokyo Story '53

Mary Jane Higby
Honeymoon Killers '70

Wilbur Higby
Reggie Mixes In '16

Howard Higgin *(D)*
Hell's House '32
High Voltage '29
Painted Desert '31
The Racketeer '29

Anthony Higgins
The Bride '85
Cold Room '84
The Draughtsman's Contract
'82
Quartet '81
She'll Be Wearing Pink
Pajamas '84

Clare Higgins
Hellbound: Hellraiser 2 '88
Hellraiser '87
Wonderland '88

Colin Higgins *(D)*
The Best Little Whorehouse in
Texas '82
Foul Play '78
9 to 5 '80

Joe Higgins
Flipper's New Adventure '64
Milo Milo '60s

Joel Higgins
First Affair '83

**John Michael
Higgins**
National Lampoon's Class of
'86 '86

Michael Higgins
Born Beautiful '82
The Conversation '74

Courtship '87
Dead Bang '89
1918 '85
On Valentine's Day '86
Paul's Case '80

Jane Higginson
Devil Wears White '86
Silent Night, Deadly Night 5:
The Toymaker '91

Mark High
The Glass Jungle '88

Jennifer Hilary
Miss A & Miss M '86
One Brief Summer '70

Lise Hilboldt
A Married Man '84
Noon Wine '84

Hilde Hildebrand
Die Grosse Freiheit Nr. 7 '45

**Charles George
Hildebrandt**
Return of the Aliens: The
Deadly Spawn '83

Mark Hildreth
After the Promise '87

Joel Hile
Visitants '87

Arthur Hill
The Amateur '82
The Andromeda Strain '71
A Bridge Too Far '77
Churchill and the Generals '81
Death Be Not Proud '75
Dirty Tricks '81
Futureworld '76
The Guardian '84
Harper '66
Henry Fonda: The Man and
His Movies '84
Judge Horton and the
Scottsboro Boys '76
The Killer Elite '75
Love Leads the Way '84
Making Love '82
One Magic Christmas '85
The Ordeal of Dr. Mudd '80
Petulia '68
The Pursuit of Happiness '70
Return of Frank Cannon '80
Revenge of the Stepford
Wives '80
Riel '79
Tomorrow's Child '82
The Ugly American '63

Benny Hill
Benny Hill's Crazy World '74
Best of the Benny Hill Show:
Vol. 2 '81
Best of the Benny Hill Show:
Vol. 3 '83
Chitty Chitty Bang Bang '68
The Italian Job '69
Those Magnificent Men in
Their Flying Machines '65
To See Such Fun '81

Bernard Hill
Bellman and True '88
The Bounty '84
Drowning by Numbers '87
Henry VI, Part 3 '82
Mountains of the Moon '90

Patricia Hodge

Betrayal '83
Diamond's Edge '88
Dust to Dust '85
Spymaker: The Secret Life of
    Ian Fleming '90
Sunset '88

Horace Hodgers

London Melody '37

Eddie Hodges

The Adventures of
    Huckleberry Finn '60
Hole in the Head '59
Live a Little, Love a Little '68
Summer Magic '63

Johnny Hodges

Duke Ellington & His
    Orchestra: 1929-1952 '86

Mike Hodges (D)

Black Rainbow '91
Flash Gordon '80
Florida Straits '87
Missing Pieces '83
Morons from Outer Space '85
Prayer for the Dying '87
Pulp '72
The Terminal Man '74

Ralph Hodges

Sea Hound '47

Runa Hodges

A Fool There Was '14

Thomas E. Hodges

Lucas '86

Tom Hodges

Amazing Stories, Book 3 '86

Earle Hodgins

Aces and Eights '36
Heroes of the Alamo '37
A Lawman is Born '37
Santa Fe Marshal '40

John Hodiak

Across the Wide Missouri '51
Ambush at Tomahawk Gap
    '53
Battleground '49
Command Decision '48
Conquest of Cochise '53
Dragonfly Squadron '54
The Harvey Girls '46
Lifeboat '44
Love from a Stranger '47

Katrina Hodiak

Jane Austen in Manhattan '80

Jeno Hodl (D)

Deadly Obssession '88

Mark Hoeger (D)

The Little Match Girl '84

Tobias Hoels

Mines of Kilimanjaro '87

Devin Hoelscher

Wired to Kill '86

Jeremy
Hoenack (D)

Dark Ride '80s

Heinz Hoenig

Judgment in Berlin '88

Paul Hoerbiger

The Mozart Story '48

Arthur Hoerl (D)

Drums O'Voodoo '34

Dennis Hoey

David and Bathsheba '51
House of Fear '45
The Pearl of Death '44
Power '34
Roll on Texas Moon '46
Sherlock Holmes and the
    Secret Weapon '42
Spiderwoman '44

Michael Hoey (D)

Navy vs. the Night Monsters
    '66

Gray Hof-meyr (D)

The Light in the Jungle '91

Ernest Hofbauer (D)

Code Name Alpha '67

John Hofeus

Return to Boggy Creek '77

Abbie Hoffman

Born on the Fourth of July '89

Alfred Hoffman

Antonia and Jane '91

Basil Hoffman

Communion '89
Lambada '89

Connie Hoffman

Texas Layover '75

David Hoffman

Children in the Crossfire '84
Wolf Lake '79

David Hoffman (D)

Remembering LIFE '85

Dustin Hoffman

Agatha '79
Alfredo, Alfredo '73
All the President's Men '76
Billy Bathgate '91
Death of a Salesman '86
Dick Tracy '90
Family Business '89
The Graduate '67
Hook '91
Ishtar '87
Kramer vs. Kramer '79
Lenny '74
Little Big Man '70
Madigan's Millions '67
Marathon Man '76
Midnight Cowboy '69
Papillon '73
Private Conversations: On the
    Set of "Death of a
    Salesman" '85
Rain Man '88
Straight Time '78
Straw Dogs '72
Tootsie '82

Elizabeth Hoffman

Fear No Evil '80

Gaby Hoffman

Field of Dreams '89
This is My Life '91

Gertrude Hoffman

The Ape '40

Isabella Hoffman

Tripwire '89

Jane Hoffman

Senior Trip '81
Sybil '76
Up the Sandbox '72

Jerzy Hoffman (D)

Colonel Wolodyjowski '69

Jon Hoffman

Night Terror '89

Joshua Hoffman

The Legend of Jedediah
    Carver '70s

Michael
Hoffman (D)

Promised Land '88
Soapdish '91
Some Girls '88

Peter Hoffman (D)

Valentino Returns '88

Robert Hoffman

A Black Veil for Lisa '68
Eyes Behind the Stars '72
Joe Panther '76

Roy Hoffman

Skyline '84

Shawn Hoffman

Adventures in Dinosaur City
    '92

Susan Lee Hoffman

Wizards of the Lost Kingdom
    2 '89

Thom Hoffman

The 4th Man '79
Lily was Here '89
Nite Song '88

Thurn Hoffman

In the Spirit '90

Susanna Hoffs

The Allnighter '87

Tamar Simon
Hoffs (D)

The Allnighter '87

Marco
Hofschneider

Europa, Europa '91

Rene Hofschneider

Europa, Europa '91

Jack Hofsiss (D)

Cat on a Hot Tin Roof '84
Family Secrets '84
I'm Dancing as Fast as I Can
    '82

Halvart Hoft

Leaves from Satan's Book '19

Bosco Hogan

James Joyce: A Portrait of the
    Artist as a Young Man '77

Helen Hogan

Dungeon of Harrow '64

Hulk Hogan

No Holds Barred '89
Suburban Commando '91

Jack Hogan

Jet Attack/Paratroop
    Command '58

James Hogan

Bulldog Drummond Double
    Feature '30s

James Hogan (D)

Arrest Bulldog Drummond '38
The Broken Mask '28
Bulldog Drummond Escapes
    '37
Bulldog Drummond's Bride
    '39
Bulldog Drummond's Peril '38
Bulldog Drummond's Secret
    Police '39

Michael Hogan

Clearcut '92
Lost '86
Smokescreen '90

Pat Hogan

Davy Crockett, King of the
    Wild Frontier '55
Indian Paint '64
Seventh Cavalry '56

Paul Hogan

Almost an Angel '90
Anzacs: The War Down Under
    '85
Crocodile Dundee '86
Crocodile Dundee 2 '88

Robert Hogan

Gone are the Days '84
Memory of Us '74

Susan Hogan

Phobia '80
Title Shot '81
White Fang '90

Ian Hogg

The Legacy '79

Michael Lindsey
Hogg (D)

Brideshead Revisited '81

Maria Hoglind

Cosi Fan Tutte '85

James Hogue

The Rejuvenator '88

Michael
Hohensee (D)

Mad Wax: The Surf Movie '90

Kong Hoi

Seven Indignant

Wang Hoi

Twister Kicker '85

Fred Holbert

Scream Bloody Murder '72

Gregory Holbit (D)

L.A. Law '86

David Holbrook

Creepshow 2 '87
Return to Salem's Lot '87

Hal Holbrook

All the President's Men '76
Behind Enemy Lines '85
Capricorn One '78
Creepshow '82
Fletch Lives '89
The Fog '78
The Girl from Petrovka '74
Girls Night Out '83
The Great White Hope '70
The Group '66
Julia '77
The Kidnapping of the
    President '80
Killing of Randy Webster '81
Killing in a Small Town '90
Magnum Force '73
Midway '76
Murder by Natural Causes '79
Natural Enemies '79
Our Town '77
The People Next Door '70
Pueblo Affair '73
Rituals '79
Sorry, Wrong Number '89
The Star Chamber '83
The Unholy '88
Wall Street '87
When Hell Was in Session '82
Wild in the Streets '68

Tami Holbrook

Hot Moves '84

Victor Holchak

Hughes & Harlow: Angels in
    Hell '77

Kathryn Holcomb

Skag '79

Kimberly Holcomb

Stripper '86

Rod Holcomb (D)

Blind Justice '86
Captain America '79
Cartier Affair '84
Chase '85
China Beach '88
The Red Light Sting '84
Stark '85
Two Fathers' Justice '85

Sarah Holcomb

Caddyshack '80

John Hold (D)

Viking Massacre '70s

Diane Holden

Black Starlet '74
Grave of the Vampire '72

Fay Holden

Andy Hardy Gets Spring Fever
    '39
Andy Hardy Meets Debutante
    '40
Andy Hardy's Double Life '42
Andy Hardy's Private
    Secretary '41
Bitter Sweet '41

Blossoms in the Dust '41
Life Begins for Andy Hardy '41
Love Finds Andy Hardy '38
Love Laughs at Andy Hardy '46
Samson and Delilah '49
Ziegfeld Girl '41

Gloria Holden

Behind the Rising Sun '43
Dracula's Daughter '36
Hawaii Calls '38
The Life of Emile Zola '37

Jan Holden

Fire Maidens from Outer Space '56
One Brief Summer '70

Marjean Holden

Stripped to Kill II: Live Girls '89

Mark Holden

Blue Fire Lady '78

Peter Holden

The Great Man Votes '38

Rebecca Holden

The Sisterhood '88
Twenty Dollar Star '91

Tommy Holden

What's Up Front '63

William Holden

Alvarez Kelly '66
Ashanti, Land of No Mercy '79
Boots Malone '52
Born Yesterday '50
The Bridge on the River Kwai '57
The Bridges at Toko-Ri '55
Casino Royale '67
The Christmas Tree '69
The Counterfeit Traitor '62
Country Girl '54
Damien: Omen 2 '78
Dance Fools Dance '31
Dark Past '49
Dear Wife '49
Earthling '80
Escape to Athena '79
Executive Suite '54
Golden Boy '39
The Horse Soldiers '59
The Key '58
Love is a Many-Splendored Thing '55
Man from Colorado '49
Miss Grant Takes Richmond '49
Moon is Blue '53
Network '76
Our Town '40
Paris When It Sizzles '64
Picnic '55
Rachel and the Stranger '48
Sabrina '54
S.O.B. '81
Stalag 17 '53
Sunset Boulevard '50
Texas '41
Towering Inferno '74
21 Hours at Munich '76
Union Station '50
When Time Ran Out '80
When Wolves Cry '69
The Wild Bunch '69
Wild Rovers '71
The World of Suzie Wong '60
Young & Willing '42

Winifred Holden

Strangers in Good Company '91

Chris Holder

Deadly Intruder '84

Geoffrey Holder

Doctor Doolittle '67
Everything You Always Wanted to Know About Sex (But Were Afraid to Ask) '72
Live and Let Die '73
Swashbuckler '76

Roy Holder

Loot...Give Me Money, Honey! '70

Thomas Holding

The Three Musketeers '21

Judd Holdren

Satan's Satellites '58
Zombies of the Stratosphere '52

William Hole, Jr. (D)

The Devil's Hand '61

Adam Holender (D)

Twisted '86

Ronald Holgate

1776 '72

Bill Holiday

French Quarter Undercover '85

Billie Holiday

Duke Ellington & His Orchestra: 1929-1952 '86

Bryant Holiday

Devil Doll '64

Hope Holiday

The Apartment '60

Agnieszka Holland (D)

Angry Harvest '85
Europa, Europa '91
Fever '81
To Kill a Priest '89

Cully Holland

Warbirds '88

Jeffrey Holland

Hi-Di-Hi '88

John Holland

Girl in Black Stockings '57
Hell Harbor '30
Morals for Women '31

Kristina Holland

The Strawberry Statement '70

Nicholas Holland

Dusty '85

Pamela Holland

Dorm That Dripped Blood '82

Ray Holland

Hot T-Shirts '79

Steve Holland

Flash Gordon: Vol. 1 '53
Flash Gordon: Vol. 2 '53

Steve Holland (D)

Better Off Dead '85
How I Got Into College '89
One Crazy Summer '86

Todd Holland (D)

The Wizard '89

Tom Holland (D)

Child's Play '88
Fatal Beauty '87
Fright Night '85

Eli Hollander (D)

Out '88

Xaviera Hollander

My Pleasure Is My Business '74
Penthouse Love Stories '86

Lloyd Hollar

The Crazies '76

Allan Holleb (D)

Candy Stripe Nurses '74
School Spirit '85

Billy Holliday

Terror in the Swamp '85

David Holliday

Cannon Movie Tales: Sleeping Beauty '89

Judy Holliday

Adam's Rib '49
Bells are Ringing '60
Born Yesterday '50
It Should Happen to You '53
Phffft! '54

Martha Holliday

George White's Scandals '45

Polly Holliday

Amazing Stories, Book 5 '85
Gremlins '84
Lots of Luck '85
Moon Over Parador '88

Richard Holliday

Assault with a Deadly Weapon '82

Earl Holliman

Alexander: The Other Side of Dawn '77
Anzio '68
Armored Command '61
Big Combo '55
The Bridges at Toko-Ri '55
Broken Lance '54
Cry Panic '74
Forbidden Planet '56
Giant '56
Gunfight at the O.K. Corral '57
Hot Spell '58
I Died a Thousand Times '55
I Love You...Goodbye '73
Last Train from Gun Hill '59
The Rainmaker '56
Sharky's Machine '81
Smoke '70
The Solitary Man '82
Sons of Katie Elder '65
Summer and Smoke '61
Trap '59
Tribes '70

Laura Hollingsworth

The Pit '81

Bridget Hollman

Evils of the Night '85
Slumber Party '57 '76

Carol Holloway

The Saphead '21

Doug Holloway (D)

Fast Money '83

Stanley Holloway

Brief Encounter '46
Hamlet '48
Immortal Battalion '44
In Harm's Way '65
The Lavender Hill Mob '51
Lily of Killarney '34
Mrs. Brown, You've Got a Lovely Daughter '68
My Fair Lady '64
Nicholas Nickleby '46
No Love for Johnnie '60
Passport to Pimlico '49
The Private Life of Sherlock Holmes '70
Salute John Citizen '42
This Happy Breed '47

Sterling Holloway

The Adventures of Huckleberry Finn '60
Alice in Wonderland '51
The Beautiful Blonde from Bashful Bend '49
Cheers for Miss Bishop '41
Death Valley '46
Disney's Storybook Classics '82
Doubting Thomas '35
Dumbo '41
I Live My Life '35
International House '33
Kentucky Rifle '56
Live a Little, Love a Little '68
Melody Master '41
Merry Widow '34
Robin Hood of Texas '47
Sioux City Sue '46
Super Seal '77
The Three Caballeros '45
Thunder and Lightning '77
Tomorrow's Children '34
Twilight on the Rio Grande '41
A Walk in the Sun '46
Wildfire '45

W.E. Holloway

Elephant Boy '37

Buddy Holly

Rock 'n' Roll Heaven '84
Rock 'n' Roll History: Rock 'n' Soul Heaven '90
Rock 'n' Roll History: Rockabilly Rockers '90

Ellen Holly

Cops and Robbers '73
School Daze '88

Lauren Holly

The Adventures of Ford Fairlane '90
Band of the Hand '86

Fred Hollyday

Edge of the Axe '89

Barnaby Holm

Final Conflict '81

Celeste Holm

All About Eve '50
Bittersweet Love '76
Champagne for Caesar '50
Cinderella '64
Death Cruise '74
High Society '56
Murder by the Book '87
Road House '48
The Tender Trap '55
Three Men and a Baby '87
Tom Sawyer '73

Christopher Holm

Thor el Conquistador '86

Claus Holm

Journey to the Lost City '58

Eleanor Holm

Tarzan's Revenge '38

Henrietta Holm

Italian Stallion '73

Ian Holm

Alien '79
Another Woman '88
Brazil '86
Chariots of Fire '81
Dance with a Stranger '85
Dreamchild '85
The Endless Game '89
Greystoke: The Legend of Tarzan, Lord of the Apes '84
Hamlet '90
Henry V '89
Holocaust '78
Inside the Third Reich '82
Juggernaut '74
Man in the Iron Mask '77
March or Die '77
Murder by the Book '87
Naked Lunch '91
Return of the Soldier '84
Shout at the Devil '76
Singleton's Pluck '84
S.O.S. Titanic '79
Thief of Baghdad '78
Time Bandits '81
Wetherby '85
Young Winston '72

Clare Holman

Let Him Have It '91

Rex Holman

The Choppers '61
Panic in the Year Zero! '62

Ben Holmes (D)

Maid's Night Out '38
The Saint in New York '38

Fred Holmes (D)

Dakota '88
Harley '90

Helen Holmes

Wild Engine/Mistaken Orders '23

Jennifer Holmes

The Demon '81

Luree Holmes

Pajama Party '64

Michelle Holmes

Rita, Sue & Bob Too '87

Phillips Holmes

Criminal Code '31
Dinner at Eight '33
General Spanky '36
Housemaster '38

Taylor Holmes

Before Morning '33
Beware, My Lovely '52
Copper Canyon '50
Double Deal '50
Hoodlum Empire '52
Tobor the Great '54

Rosie Holotik

Don't Look in the Basement '73
Encounter with the Unknown '75
Twisted Brain '74

Max Holsboer

The Blue Light '32

Betsy Holt

Thunder Road '58

Bobby Holt

Dark Sanity '82

Charlene Holt

El Dorado '67
Man's Favorite Sport? '63
Red Line 7000 '65

David Holt

Last Days of Pompeii '35

Hans Holt

Almost Angels '62
The Mozart Story '48

Jack Holt

Across the Wide Missouri '51
Cat People '42
The Chase '46
Flight to Nowhere '46
Holt of the Secret Service '42
King of the Bullwhip '51
The Little American '17
Littlest Rebel '35
Red Desert '50
San Francisco '36
They were Expendable '45
Trail of Robin Hood '50
Western Double Feature 14 '52

Jennifer Holt

Ghost Town Renegades '47
Outlaw Trail '44
Raiders of Sunset Pass '43
Song of Old Wyoming '45
Stage to Mesa City '48
Where the North Begins '47

Jim Holt

Fever '88

Nick Holt

They '80s

Patrick Holt

Alias John Preston '56
Flight from Singapore '62
Men of Sherwood Forest '57
Psychomania '73
Suspended Alibi '56
Unholy Four '54
Vulture '67

When Dinosaurs Ruled the Earth '70

Seth Holt (D)

Scream of Fear '61

Steven Holt

Preppies '82

Tim Holt

Dynamite Pass '50
Fifth Avenue Girl '39
Gun Smugglers/Hot Lead '51
Gunplay '51
His Kind of Woman '51
Hitler's Children '43
Law West of Tombstone '38
The Magnificent Ambersons '42
The Monster that Challenged the World '57
My Darling Clementine '46
Mysterious Desperado '49
Mysterious Desperado/Rider From Tucson '49
Renegade Ranger/Scarlet River '38
Rider from Tucson '50
Riders of the Range '49
Riders of the Range/Storm Over Wyoming '50
Storm Over Wyoming '50
This Stuff'll Kill Ya! '71
Treasure of the Sierra Madre '48
Western Double Feature 14 '52
The Yesterday Machine '63

Ula Holt

The New Adventures of Tarzan: Vol. 1 '35
The New Adventures of Tarzan: Vol. 2 '35
The New Adventures of Tarzan: Vol. 3 '35
The New Adventures of Tarzan: Vol. 4 '35
Tarzan and the Green Goddess '38

Mark Holton

Easy Wheels '89
Pee-Wee's Big Adventure '85

Sean Holton

White Fury '90

Thomas Holtzman

Qui Etes Vous, Mr. Sorge? '61

Allen Holubar

20,000 Leagues Under the Sea '16

Maria Holvoe

Last Warrior '89
Worth Winning '89

Jennifer Holy

Buffalo Bill Rides Again '47

Evander Holyfield

Blood Salvage '90

Roger Holzberg (D)

Midnight Crossing '87

Arabella Holzbog

Stone Cold '91

Baby Jane Holzer

Edie in Ciao! Manhattan '72

Allan Holzman (D)

Forbidden World '82
Grunt! The Wrestling Movie '85
Intimate Stranger '91
Out of Control '85
Programmed to Kill '86

Robert E. Homans

The Concentratin' Kid '30
Man from Hell's Edges '32
Rogue's Gallery '44
Spurs '30
X Marks the Spot '42

David Homb

The Channeler '89
Street Soldiers '91

Paul Home

The 13th Mission '91

Skip Homeier

The Burning Hills '56
Cry Vengeance '54
Gunfighter '50
Halls of Montezuma '50
Johnny Shiloh '63
Starbird and Sweet William '73
The Tall T '57
Ten Wanted Men '54
The Voyage of the Yes '72

Charles Homet

Beginner's Luck '84

Oscar Homolka

Ball of Fire '42
The Executioner '70
Funeral in Berlin '66
The House of the Arrow '53
I Remember Mama '48
The Key '58
Mooncussers '62
Rhodes '36
Sabotage '36
Seven Sinners '40
The Seven Year Itch '55
Song of Norway '70
Strange Case of Dr. Jekyll & Mr. Hyde '68
War and Peace '56
The White Tower '50
Wonderful World of the Brothers Grimm '62

Karl Arne Homsten

Secrets of Women '52

Jean-Marie Hon

Blade in Hong Kong '85

Kam Hon

Samurai Sword of Justice

Lau Shing Hon (D)

To Live and Die in Hong Kong '89

Shek Hon

Four Robbers '70s

Wah Hon

The White Butterfly Killer

Inoshiro Honda

King Kong Versus Godzilla '63

Inoshiro Honda (D)

Attack of the Mushroom People '63
Destroy All Monsters '68

Ghidrah the Three Headed Monster '65
Godzilla vs. Monster Zero '68
Godzilla vs. Mothra '64
Godzilla's Revenge '69
Gorath '67
H-Man '59
Half Human '58
Mothra '62
The Mysterians '58
Rodan '56
Varan the Unbelievable '61
Yog, Monster from Space '71

Germane Honde

Prettykill '87

Mikhael Honesseau

Anita, Dances of Vice '87

Elliot Hong (D)

Kill the Golden Goose '79
They Call Me Bruce '82

James Hong

Bethune '77
Caged Fury '90
The Golden Child '86
The Jitters '88
Revenge of the Nerds 2: Nerds in Paradise '87
Shadowzone '90
Tango and Cash '89
Tax Season '90
The Vineyard '89
Yes, Giorgio '82

James Hong (D)

Girls Next Door '79
The Vineyard '89

Kojiro Hongo

Gamera vs. Barugon '66
Gamera vs. Gaos '67

Misako Honno

The Go-Masters '82

Andre Honore

The Slumber Party Massacre '82

Ron Honthaner (D)

The House on Skull Mountain '74

Darla Hood

The Bat '59
Bohemian Girl '36
Little Rascals, Book 3 '38
Little Rascals, Book 4 '30s
Little Rascals, Book 5 '30s
Little Rascals, Book 6 '30s
Little Rascals, Book 7 '30s
Little Rascals, Book 8 '30s
Little Rascals, Book 9 '30s
Little Rascals, Book 10 '30s
Little Rascals, Book 11 '38
Little Rascals, Book 12 '37
Little Rascals, Book 14 '30s
Little Rascals, Book 16 '30s
Little Rascals, Book 17 '31
Little Rascals: Choo Choo/ Fishy Tales '32
Little Rascals Christmas Special '79
Little Rascals Collector's Edition, Vol. 1-6
Little Rascals Comedy Classics, Vol. 1
Little Rascals Comedy Classics, Vol. 2
Little Rascals: Fish Hooky/ Spooky Hooky '33

Little Rascals: Honkey Donkey/Sprucin' Up '34
Little Rascals: Little Sinner/ Two Too Young '35
Little Rascals: Mush and Milk/ Three Men in a Tub '33
Little Rascals on Parade '37
Little Rascals: Pups is Pups/ Three Smart Boys '30
Little Rascals: Readin' and Writin'/Mail and Female '31
Little Rascals: Reunion in Rhythm/Mike Fright '37
Little Rascals: Spanky/Feed 'Em and Weep '32
Little Rascals: The Pinch Singer/Framing Youth '36
Little Rascals Two Reelers, Vol. 1 '29
Little Rascals Two Reelers, Vol. 2 '33
Little Rascals Two Reelers, Vol. 3 '36
Little Rascals Two Reelers, Vol. 4 '30
Little Rascals Two Reelers, Vol. 5 '34
Little Rascals Two Reelers, Vol. 6 '36
Little Rascals, Volume 1 '29
Our Gang Comedies '40s

Don Hood

Absence of Malice '81
Blind Vengeance '90
Marie '85

Miki Hood

Inspector Hornleigh '39

Randall Hood (D)

Die Sister, Die! '74

Carla Hoogeveen

Alternative '76

Harry Hook (D)

The Kitchen Toto '87
The Last of His Tribe '92
Lord of the Flies '90

Kevin Hooks

Aaron Loves Angela '75
Can You Hear the Laughter? The Story of Freddie Prinze '79
Sounder '72

Kevin Hooks (D)

Heat Wave '90
Strictly Business '91

Robert Hooks

Aaron Loves Angela '75
The Execution '85
Fast Walking '82
Supercarrier '88
A Woman Called Moses '78
Words by Heart '84

Brett Hool

Steel Dawn '87

Lance Hool (D)

Missing in Action 2: The Beginning '85
Steel Dawn '87

Barbara Hooper

The Sisterhood '88

Geraldine Hooper

Cambio de Cara (A Change of Face) '71

Tobe Hooper

Fangoria's Weekend of
  Horrors '86

Tobe Hooper (D)

Eaten Alive '76
The Funhouse '81
I'm Dangerous Tonight '90
Invaders from Mars '86
Lifeforce '85
Poltergeist '82
Salem's Lot: The Movie '79
Spontaneous Combustion '89
Texas Chainsaw Massacre
  '74
Texas Chainsaw Massacre
  Part 2 '86

Peter Hooten

Deadly Mission '78
Dr. Strange '78
Fantasies '73
Slashed Dreams '74
2020 Texas Gladiators '85

William Hootkins

Hear My Song '91

Hoover the Dog

Little Heroes '91

Herbert Hoover

The Rose Parade: Through
  the Years '88

Phil Hoover

Black Gestapo '75
Polioowomon '73

Bob Hope

Boy, Did I Get a Wrong
  Number! '66
Cancel My Reservation '72
Candid Hollywood '62
Casanova's Big Night '54
Entertaining the Troops '89
Fancy Pants '50
The Great Lover '49
Hollywood Clowns '85
Hollywood Goes to War '54
Hollywood at War '40s
The Lemon Drop Kid '51
Masters of Comedy '30s
Muppet Movie '79
My Favorite Brunette '47
Off Limits '53
Oscar '66
The Paleface '48
Paris Holiday '57
The Princess and the Pirate
  '44
Road to Bali '53
The Road to Hong Kong '62
Road to Lebanon '60
Road to Morocco '42
Road to Rio '47
Road to Singapore '40
Road to Utopia '46
Road to Zanzibar '41
The Seven Little Foys '55
Son of Paleface '52
Sorrowful Jones '49
Strictly G.I. '44
Television's Golden Age of
  Comedy '50s
They Got Me Covered '43
Two Reelers: Comedy
  Classics 9 '34

Gary Hope

Big Zapper '73

Harry Hope

Escape from Planet Earth '67

Harry Hope (D)

Swift Justice '88

Leslie Hope

The Abduction of Allison Tate
  '92
The Big Slice '91
It Takes Two '88
Kansas '88
Men at Work '90
Prep School '81
Sword of Gideon '86
Talk Radio '88

Richard Hope

Antonia and Jane '91

Richard Hope

Bellman and True '88
A Casualty of War '90
Singleton's Pluck '84

Vida Hope

Angels One Five '54
Roadhouse Girl '53

William Hope

Hellbound: Hellraiser 2 '88

Jason Hopely

The War Boy '85

Anthony Hopkins

All Creatures Great and Small
  '74
Audrey Rose '77
The Bounty '84
A Bridge Too Far '77
A Change of Seasons '80
A Chorus of Disapproval '89
Desperate Hours '90
A Doll's House '89
84 Charing Cross Road '86
Elephant Man '80
Freejack '92
The Girl from Petrovka '74
The Good Father '87
Great Expectations '89
Hamlet '69
Hunchback of Notre Dame '82
International Velvet '78
Juggernaut '74
The Lindbergh Kidnapping
  Case '76
The Lion in Winter '68
The Looking Glass War '69
Magic '78
A Married Man '84
Mussolini & I '85
One Man's War '90
Othello '82
QB VII '74
The Silence of the Lambs '91
Young Winston '72

Arthur Hopkins (D)

His Double Life '33

Barrett Hopkins

Firehouse '87

Bo Hopkins

American Graffiti '73
Big Bad John '90
Bounty Hunters '89
Casino '80
Center of the Web '92
Culpepper Cattle Co. '72
Day of the Locust '75
The Fifth Floor '80

The Final Alliance '89
The Getaway '72
Kansas City Massacre '75
The Killer Elite '75
Midnight Express '78
Mutant '83
Nightmare at Noon '87
The Only Way Home '72
Plutonium Incident '82
Posse '75
Rodeo Girl '80
The Runaway Barge '75
A Small Town in Texas '76
A Smoky Mountain Christmas
  '86
Sweet 16 '81
Tentacles '77
Trapper County War '89
What Comes Around '85
White Lightning '73
The Wild Bunch '69

Harold Hopkins

Demonstrator '71
Fantasy Man '84
Monkey Grip '82
Sara Dane '81

Jermaine Hopkins

Juice '92

Joan Hopkins

Man on the Run '49

Katherine Hopkins

The Capture of Bigfoot '79
Legend of Big Foot '82

Mary Hopkins

Getting Over '81

Miriam Hopkins

Barbary Coast '35
Becky Sharp '35
Carrie '52
The Chase '66
The Children's Hour '62
Dr. Jekyll and Mr. Hyde '31
Fanny Hill: Memoirs of a
  Woman of Pleasure '64
The Heiress '49
Men are not Gods '37
Old Maid '39
Savage Intruder '68
These Three '36
Virginia City '40

**Rhonda Leigh
Hopkins**

Summer School Teachers '75
Tidal Wave '75

**Shirley Knight
Hopkins**

Secrets '71

**Stephen
Hopkins** (D)

Dangerous Game '90
A Nightmare on Elm Street 5:
  Dream Child '89
Predator 2 '90

Telma Hopkins

The Kid with the Broken Halo
  '82
Trancers '85
Trancers 3: Deth Lives '92
Vital Signs '90

Maja Hoppe

M'Lady's Court '80s

Rolf Hoppe

Mephisto '81
Spring Symphony '86

Dennis Hopper

American Friend '77
Apocalypse Now '79
Back Track '91
Black Widow '87
Blood Red '88
Bloodbath
Blue Velvet '86
Chattahoochee '89
Cool Hand Luke '67
Double-Crossed '91
Easy Rider '69
Eye of the Storm '91
Flashback '89
Giant '56
The Glory Stompers '67
Hearts of Darkness: A
  Filmmaker's Apocalypse
  '91
Hoosiers '86
The Indian Runner '91
The Inside Man '84
King of the Mountain '81
The Last Movie '71
Let It Rock '86
Mad Dog Morgan '76
My Science Project '85
Night Tide '63
O.C. and Stiggs '87
The Osterman Weekend '83
Out of the Blue '82
Paris Trout '91
The Pick-Up Artist '87
Planet of Blood '66
Rebel Without a Cause '55
Reborn '78
Riders of the Storm '88
River's Edge '87
Rolling Stone Presents
  Twenty Years of Rock &
  Roll '87
Rumble Fish '83
Running Out of Luck '86
Stark '85
Straight to Hell '87
Texas Chainsaw Massacre
  Part 2 '86
Tracks '76
The Trip '67
Wild Times '79

Dennis Hopper (D)

Back Track '91
Colors '88
Easy Rider '69
The Hot Spot '90
The Last Movie '71
Out of the Blue '82

**E. Mason
Hopper** (D)

Held for Murder '32
Hong Kong Nights '35
Square Shoulders '29

Hedda Hopper

Annabel Takes a Tour/Maid's
  Night Out '38
As You Desire Me '32
Breakfast in Hollywood '46
Dangerous Holiday '37
The Dark Hour '36
Don Juan '26
Doughnuts & Society '36
Dracula's Daughter '36
The Dropkick '27
Hedda Hopper's Hollywood
  '42
I Live My Life '35
Maid's Night Out '38

One Frightened Night '35
The Racketeer '29
Reap the Wild Wind '42
Skinner's Dress Suit '26
Tarzan's Revenge '38
Vogues of 1938 '37

Jerry Hopper (D)

Madron '70
Pony Express '53

Mason Hopper (D)

Shop Angel '32

Victoria Hopper

The Mill on the Floss '37

William Hopper

20 Million Miles to Earth '57

Russell Hopton

Death from a Distance '36
Last Outlaw '36
Renegade Trail '39
The World Accuses '35
Zombies on Broadway '44

Russell Hopton (D)

Song of the Trail '36

John Hora

Innerspace '87

Horace Heidt Band

Pot O' Gold '41

Barbara Horan

Malibu Bikini Shop '86

Gerard Horan

Look Back in Anger '89

Hilary Horan

Satan's Cheerleaders '77

James Horan

Image of Passion '86

Roy Horan

Eagle's Shadow '84
Snuff-Bottle Connection '82
Story of the Dragon '82

Michael Hordern

Anne of the Thousand Days
  '69
The Baby and the Battleship
  '56
A Christmas Carol '51
Cleopatra '63
Dark Obsession '90
Demons of the Mind '72
Dr. Syn, Alias the Scarecrow
  '64
A Funny Thing Happened on
  the Way to the Forum '66
The Green Man '91
How I Won the War '67
Ivanhoe '82
Joseph Andrews '77
Khartoum '66
Lady Jane '85
Lamp at Midnight '66
Mackintosh Man '73
The Man Who Never Was '55
The Medusa Touch '78
The Missionary '82
The Old Curiosity Shop '75
Oliver Twist '82
Possession of Joel Delaney
  '72
Promoter '52
The Secret Garden '87
Sink the Bismarck '60

Thelma Houston

And God Created Woman '88
The Hollywood Game '89

Tony Houston (D)

Outlaw Riders '72

Whitney Houston

Nelson Mandela 70th Birthday
Tribute '88

Anders Hove

Subspecies '91

Adrian Hoven

Castle of the Creeping Flesh
Cave of the Living Dead '65
Fox and His Friends '75
Tromba, the Tiger Man '52

Adrian Hoven (D)

Dandelions '74
Mark of the Devil 2 '72

Helen Hovey

The Sadist '63

Natasha Hovey

Acqua e Sapone '83
Demons '86

Adam Coleman Howard

No Secrets '91
Quiet Cool '86
Slaves of New York '89

Alan Howard

A Casualty of War '90
The Cook, the Thief, His Wife
& Her Lover '90
Strapless '90

Anne Howard

The Ghost of Rashmon Hall
'47

Arliss Howard

Crisscross '92
Door to Door '84
Full Metal Jacket '87
Hands of a Stranger '87
Iran: Days of Crisis '91
The Lightship '86
Men Don't Leave '89
Plain Clothes '88
The Prodigal '83
Ruby '92
Somebody Has to Shoot the
Picture '90
Tequila Sunrise '88
Till Death Do Us Part '92

Arthur Howard

Paradisio '61
The Reckless Way '36

Barbara Howard

Running Mates '86

Bob Howard

Fats Waller and Friends '86
Junction 88 '40
Murder with Music '45
Stars on Parade/Boogie
Woogie Dream '46

Booth Howard

Mystery Liner '34
Oh Susannah '38
Red River Valley '36

Brie Howard

Android '82
The Runnin' Kind '89
The Running Kind '89

Cathy Howard

School for Sex '69

Clint Howard

B.O.R.N. '88
Cotton Candy '82
Death of Richie '76
Disturbed '90
Eat My Dust '76
End of the Line '88
Evilspeak '82
Freeway '88
Gentle Ben '70s
Gentle Giant '67
Grand Theft Auto '77
Gung Ho '85
Huckleberry Finn '75
Rock 'n' Roll High School '79
Salty '73
Silent Night, Deadly Night 4:
Initiation '90
The Wild Country '71
The Wraith '87

Curly Howard

Classic Comedy Video
Sampler '49
Comedy Festival #1 '30s
The Lost Stooges '33
The Making of the Stooges
'85
Snow White and the Three
Stooges '61
The Stooges Story '90
Three Stooges Comedy
Capers: Volume 1 '40s
Three Stooges Comedy
Classics '49
Three Stooges: Cookoo
Cavaliers and Other Nyuks
'40
Three Stooges Festival '49
Three Stooges: From Nurse to
Worse and Other Nyuks
'45
Three Stooges: Half-Wit's
Holiday and Other Nyuks
'35
Three Stooges: Idiots Deluxe
and Other Nyuks '40
Three Stooges: Nutty but Nice
and Other Nyuks '36
Three Stooges: They Stooge
to Conga and Other Nyuks
'40

Cy Howard (D)

Lovers and Other Strangers
'70

David Howard

The Demon Lover '77

David Howard (D)

Crimson Romance '34
Daniel Boone '34
Hollywood Stadium Mystery
'38
In Old Santa Fe '34
Mystery Ranch '34
Renegade Ranger/Scarlet
River '38

Esther Howard

Vice Squad '31

Frances Howard

The Swan '25

Frank Howard

Hard Knox '83
That Was Then...This Is Now
'85

Henry Howard

The Brig '64

Jean Howard

Huckleberry Finn '75

John Howard

The Club '81
The Fighting Kentuckian '49
Lost Horizon '37
Love from a Stranger '47
Make Haste to Live '54
The Undying Monster '42
Young Einstein '89

Joyce Howard

Appointment with Crime '45
The Night Has Eyes '42

Kathleen Howard

Blossoms in the Dust '41
It's A Gift '34

Ken Howard

Country Girl '82
Damien: The Leper Priest '80
Manhunter '74
Pudd'nhead Wilson '87
Rage of Angels '85
Second Thoughts '83
1776 '72
Strange Interlude '90
The Thorn Birds '83

Kevyn Major Howard

Alien Nation '88
Full Metal Jacket '87

Leslie Howard

The Animal Kingdom '32
Devotion '31
The Forty-Ninth Parallel '41
A Free Soul '31
Gone with the Wind '39
Intermezzo '39
Making of a Legend: "Gone
With the Wind" '89
Of Human Bondage '34
Petrified Forest '36
Pimpernel Smith '42
Pygmalion '38
Romeo and Juliet '36
The Scarlet Pimpernel '34
Spitfire '42
Spitfire: The First of the Few
'43
Stand-In '37

Leslie Howard (D)

Pimpernel Smith '42
Pygmalion '38
Spitfire '42
Spitfire: The First of the Few
'43

Lisa Howard

Rolling Vengeance '87
Sabaka '55

Marvin Howard

The Cremators '72

Mary Howard

Abe Lincoln in Illinois '40
All Over Town '37
The Loves of Edgar Allen Poe
'42

Mel Howard

Hester Street '75

Moe Howard

Classic Comedy Video
Sampler '49
Comedy Festival #1 '30s
Comedy Reel, Vol. 2
Dancing Lady '33
Dr. Death, Seeker of Souls '73
Gold Raiders '51
It's a Mad, Mad, Mad, Mad
World '63
The Lost Stooges '33
The Making of the Stooges
'85
Revenge of TV Bloopers '60s
Snow White and the Three
Stooges '61
The Stooges Story '90
Swing Parade of 1946 '46
Three Stooges Comedy
Capers: Volume 1 '40s
Three Stooges Comedy
Classics '49
Three Stooges: Cookoo
Cavaliers and Other Nyuks
'40
Three Stooges Festival '49
Three Stooges: From Nurse to
Worse and Other Nyuks
'45
Three Stooges: Half-Wit's
Holiday and Other Nyuks
'35
Three Stooges: Idiots Deluxe
and Other Nyuks '40
The Three Stooges Meet
Hercules '61
Three Stooges: Nutty but Nice
and Other Nyuks '36
Three Stooges: They Stooge
to Conga and Other Nyuks
'40

Paul Howard (D)

Night Terror '89

Rance Howard

B.O.R.N. '88
Dark Before Dawn '89
Eat My Dust '76
Gentle Ben '70s
Huckleberry Finn '75

Ron Howard

American Graffiti '73
Bitter Harvest '81
The Courtship of Eddie's
Father '62
Door-to-Door Maniac '61
Eat My Dust '76
Grand Theft Auto '77
Huckleberry Finn '75
I'm a Fool '77
The Music Man '62
Return to Mayberry '85
Roger Corman: Hollywood's
Wild Angel '78
Run, Stranger, Run '73
The Shootist '76
Smoke '70
Video Yesterbloop '70s
Village of the Giants '65
The Wild Country '71

Ron Howard (D)

Backdraft '91
Cotton Candy '82
Grand Theft Auto '77
Gung Ho '85
Night Shift '82
Parenthood '89
Splash '84

Willow '88

Ronald Howard

Night Beat '48
The Queen of Spades '49

Sandy Howard

One Step to Hell '68

Shemp Howard

Africa Screams '49
The Bank Dick '40
Buck Privates '41
Dancing Lady '33
Gold Raiders '51
The Making of the Stooges
'85
Private Buckaroo '42
Revenge of TV Bloopers '60s
The Stooges Story '90
Three Stooges Comedy
Capers: Volume 1 '40s
Three Stooges Comedy
Classics '49
Three Stooges Festival '49

Stanford Howard

A Man Called Horse '70

Susan Howard

Moonshine County Express
'77
The Power Within '79
Sidewinder One '77

Trevor Howard

Adventures of Eliza Fraser '76
Albino '76
Around the World in 80 Days
'56
The Battle of Britain '69
The Bawdy Adventures of
Tom Jones '76
Brief Encounter '46
The Cary Grant Collection '90
Catch Me a Spy '71
Catholics '73
Conduct Unbecoming '75
Count of Monte Cristo '74
Craze '74
Deadly Game '82
A Doll's House '73
Dust '85
11 Harrowhouse '74
Father Goose '64
Flashpoint Africa '84
Foreign Body '86
Gandhi '82
The Gift Horse '52
The Golden Salamander '51
The Graveyard '74
Green for Danger '47
Hennessy '75
Hurricane '79
I See a Dark Stranger '47
Inside the Third Reich '82
Invincible Mr. Disraeli '63
The Key '58
Meteor '79
The Missionary '82
Morituri '65
Mutiny on the Bounty '62
Night Flight '79
The Night Visitor '70
The Poppy Is Also a Flower
'66
Roboman '75
Ryan's Daughter '70
Sea Wolves '81
Shaka Zulu '83
Slavers '77
Staying On '80
Stevie '78
Superman: The Movie '78
Sword of the Valiant '83

Josephine Hutchinson

The Adventures of Huckleberry Finn '60
Son of Frankenstein '39
The Story of Louis Pasteur '36
The Tender Years '47

Ken Hutchison

Deadly Strangers '82

Harold Huth (D)

Night Beat '48

J.B. Hutto

Chicago Blues '72

Betty Hutton

Greatest Show on Earth '52
Hollywood Clowns '85
Jazz Ball '56
Let's Dance '50
Miracle of Morgan's Creek '44
The Perils of Pauline '47

Brian Hutton

Carnival Rock '57
Last Train from Gun Hill '59

Brian G. Hutton (D)

The First Deadly Sin '80
High Road to China '83
Kelly's Heroes '70
Night Watch '72
Where Eagles Dare '68
X, Y & Zee '72

Dana J. Hutton

A Time to Love & a Time to Die '58

Jayne Hutton

Stay Awake '87

Jim Hutton

Don't Be Afraid of the Dark '73
The Green Berets '68
The Hallelujah Trail '65
Hellfighters '68
Major Dundee '65
Nightmare at 43 Hillcrest '74
Psychic Killer '75
A Time to Love & a Time to Die '58
Walk, Don't Run '66
Where the Boys Are '60
Who's Minding the Mint? '67

Lauren Hutton

American Gigolo '79
Bulldance '88
The Cradle Will Fall '83
Fear '90
Forbidden Sun '89
From Here to Maternity '85
The Gambler '74
Gap-Toothed Women '89
Gator '76
Guilty as Charged '91
Lassiter '84
Malone '87
Monte Carlo '86
Nashville '75
Once Bitten '85
Paper Lion '68
Paternity '81
Scandalous '88
Snow Queen '83
Starflight One '83
Timestalkers '87
Viva Knievel '77

A Wedding '78
Welcome to L.A. '77
Zorro, the Gay Blade '81

Marion Hutton

Love Happy '50

Robert Hutton

Big Bluff '55
Cinderfella '60
Destination Tokyo '43
Hollywood Canteen '44
Naked Youth '59
Outcasts of the City '58
The Persuaders '71
Showdown at Boot Hill '58
Slaughter Trail '51
The Slime People '63
The Steel Helmet '51
They Came from Beyond Space '67
Vulture '67

Robert Hutton (D)

The Slime People '63

Timothy Hutton

And Baby Makes Six '79
Daniel '83
Everybody's All American '88
The Falcon and the Snowman '85
Father Figure '80
Friendly Fire '79
Iceman '84
A Long Way Home '81
Made in Heaven '87
Oldest Living Graduate '82
Ordinary People '80
Q & A '90
Taps '81
Time of Destiny '88
Torrents of Spring '90
Turk 182! '85
Young Love, First Love '79

Tom Hutton

The Demon Lover '77

Richard Huw

Getting It Right '89

Amos Huxley

Preacherman '83

Sam Huxley

Ninja the Battalion '90

Judy Huxtable

Scream and Scream Again '70

Willard Huyck (D)

Best Defense '84
French Postcards '79
Howard the Duck '86
Messiah of Evil '74

Pelle Hvenegaard

Pelle the Conqueror '88

Hw-Luen

Little Heroes of Shaolin Temple '72

Chen Che Hwa (D)

Young Dragons: The Kung Fu Kids 2 '89

Li Shuen Hwa

Little Heroes of Shaolin Temple '72

Shih Bao Hwa

Young Hero of Shaolin

Yue Hwa

Green Dragon Inn '82
The Ninja Wolves '90

Jacky Hwong (D)

Ninja vs. the Shaolin '84

Leila Hyams

Freaks '32
Red Headed Woman '32
Ruggles of Red Gap '35
Spite Marriage '29

Nessa Hyams (D)

Leader of the Band '87

Peter Hyams (D)

Amazing Stories, Book 3 '86
Busting '74
Capricorn One '78
Death Target '83
Hanover Street '79
Narrow Margin '90
Outland '81
The Presidio '88
Running Scared '86
The Star Chamber '83
2010: The Year We Make Contact '84

Gigi Hyatt

The Lady of the Camellias '87

Jacquelyn Hyde

House of Terror '87

Jonathan Hyde

Fellow Traveler '89

Kimberly Hyde

Candy Stripe Nurses '74

Alex Hyde-White

Biggles '85
Echoes in the Darkness '87
First Olympics: Athens 1896 '84
Ironclads '90
Phantom of the Opera '89
The Seekers '79

Wilfrid Hyde-White

Betrayed '54
Brand New Life '72
Browning Version '51
Carry On Nurse '59
The Cat and the Canary '79
Damien: The Leper Priest '80
The Demi-Paradise '43
Fanny Hill '83
Ghosts of Berkeley Square '47
The Golden Salamander '51
Heartburn '86
In Search of the Castaways '62
King Solomon's Treasure '76
Last Holiday '50
The Magic Christian '69
My Fair Lady '64
Oh, God! Book 2 '80
The Rebels '79
Scout's Honor '80
Skullduggery '70
Teenage Bad Girl '59
The Third Man '49
The Toy '82
Truth About Women '58
Two-Way Stretch '60
Up the Creek '58

The Vicious Circle '57

Martha Hyer

Bikini Beach '64
The Carpetbaggers '64
Catch as Catch Can '68
The Chase '66
Cry Vengeance '54
The Delicate Delinquent '56
First Men in the Moon '64
Francis in the Navy '55
Gun Smugglers/Hot Lead '51
The House of 1,000 Dolls '67
Houseboat '58
Ice Palace '60
Lucky Me '54
Mistress of the World '59
Night of the Grizzly '66
Paris Holiday '57
Picture Mommy Dead '66
Sabrina '54
Scarlet Spear '54
Some Came Running '58
Sons of Katie Elder '65

Catherine Hyland

Vamping '84

Diana Hyland

Boy in the Plastic Bubble '76
Hercules and the Princess of Troy '65

Frances Hyland

Home to Stay '79

Scott Hylands

Coming Out Alive '84
Daddy's Gone A-Hunting '69
Fools '70
Savage Hunger '84
Tales of the Klondike: In a Far Country '81
Winds of Kitty Hawk '78

Jack Hylton

She Shall Have Music '36

Jane Hylton

Circus of Horrors '60
The Manster '59

Richard Hylton

Halls of Montezuma '50

Bob Hyman

The Crater Lake Monster '77

Earle Hyman

The Life and Adventures of Santa Claus '85

Phyllis Hyman

The Kill Reflex '89
The Sacred Music of Duke Ellington '82

Warren Hymer

Baby Face Morgan '42
Confidential '35
Hitler: Dead or Alive '43
Hong Kong Nights '35
Joy of Living '38
The Lone Star Ranger '39
Lure of the Islands '42
Mr. Wise Guy '42
The Mysterious Rider '33
Sea Racketeers '37
Show Them No Mercy '35
Tango '36

Mike Hynson

The Endless Summer '66

David Hyry

Northville Cemetery Massacre '76

Joyce Hyser

Just One of the Guys '85
Wedding Band '89

Dorothy Hyson

The Ghoul '34

Dafydd Hywel

Coming Up Roses '87

Chaing I

Return of the Tiger '78
They Shoot Horses, Don't They? '69

Chang Shinn I (D)

Snake in the Eagle's Shadow 2 '83

Dean Iandoli

Monster High '89

Bonaventure Ibanez

Romola '25

Juan Ibanez (D)

Chamber of Fear '68
Dance of Death '69
The Fear Chamber '68
House of Evil '68
The Sinister Invasion '68

Goro Ibuki

Watch Out, Crimson Bat! '69

Ice Cube

Boyz N the Hood '91

Ice-T

New Jack City '91
Ricochet '91

Leon Ichaso (D)

Crossover Dreams '85
El Super '79
Power, Passion & Murder '83
The Take '90

Kon Ichikawa (D)

The Burmese Harp '56
Fires on the Plain '59
Odd Obsession '60
Revenge of a Kabuki Actor '63
Tokyo Olympiad '66

Raizo Ichikawa

Shin Heike Monogatari '55

Kurt Ida

The Adventures of Huckleberry Finn '78

Eric Idle

The Adventures of Baron Munchausen '89
All You Need is Cash '78
And Now for Something Completely Different '72
Jabberwocky '77
Monty Python and the Holy Grail '75
Monty Python Live at the Hollywood Bowl '82
Monty Python Meets Beyond the Fringe
Monty Python's Life of Brian '79

Monty Python's Parrot Sketch
    not Included '90
Monty Python's The Meaning
    of Life '83
Nuns on the Run '90
Pied Piper of Hamelin '84
To See Such Fun '81
Too Much Sun '90
Yellowbeard '83

Eric Idle (D)

All You Need is Cash '78
The Tale of the Frog Prince
    '83

Cinnamon Idles

Kidco '83
Sixteen Candles '84

Hisashi Igawa

Ran '85

James Iglehardt

Death Force '78
Fighting Mad '77
Savage! '73

Kyunna Ignatova

Planet Burg '62

Robin Ignico

Don't Go to Sleep '82

Darrow Igus

Makin' It

Steve Ihnat

Countdown '68
Hour of the Gun '67
Hunter '73

Choko Iida

Drunken Angel '48

Ryo Ikebe

Snow Country '57
War in Space '77

Oh Young Il

Yongary, Monster of the Deep
    '67

Toby Iland

Tomboy '85

Igor Ilinsky

A Kiss for Mary Pickford '27

Pola Illery

Under the Roofs of Paris
    (Sous les Toits de Paris) '29

Rolf Illig

Alpine Fire '89

Peter Illing

Against the Wind '48
The Electronic Monster '57
Eureka Stockade '49

Shohei Imamura (D)

The Ballad of Narayama '83
Black Rain '88
Eijanaika (Why Not?) '81
The Insect Woman '63
The Pornographers '66
Vengeance Is Mine '79

Iman

House Party 2: The Pajama
    Jam '91
L.A. Story '91
Lies of the Twins '91

No Way Out '87
Star Trek 6: The
    Undiscovered Country '91
Surrender '87

**Shmuel
Imberman** (D)

I Don't Give a Damn '88

Gary Imhoff

The Seniors '78

Markus Imhoof (D)

Boat Is Full '81

Carlo Imperato

Kids from Fame '83

Rino Imperio

'Tis a Pity She's a Whore '73

Sayo Inaba

Space Riders '83

Yoshio Inaba

Seven Samurai '54

Hiroshi Inagaki (D)

Kojiro '67
Rikisha-Man '58
Samurai 1: Musashi Miyamoto
    '55
Samurai 2: Duel at Ichijoji
    Temple '55
Samurai 3: Duel at Ganryu
    Island '56

Ada Ince

Frontier Days '34
Rainbow's End '35

Ralph Ince

Law of the Sea '38

Thomas Ince (D)

Civilization '16
The Three Musketeers
    (D'Artagnan) '16

Jennifer Inch

Frank and I '83

Rafael Inclan

Bellas de Noche
By Hook or Crook '85

**Annabella
Incontrera**

Double Face '70

Sam Incorvia

Desert Snow '89

Franco Indovina (D)

Catch as Catch Can '68
Oldest Profession '67

Frieda Inescort

Beauty for the Asking '39
The Judge Steps Out '49
Pride and Prejudice '40
Return of the Vampire '43
Shadows on the Stairs '41
Sunny '41
The Underworld Story '50

Irene Inescort

Demonstrator '71

Cruz Infante

El Sinaloense '84

Eddie Infante

Cavalry Command '63
Ethan '71
Warkill '65

Pedro Infante

El Ametralladora '65
El Gavilan Pollero '47

Angelo Infanti

Black Emmanuelle '76
The Inquiry '87

Joyce Ingalls

Deadly Force '83

Marty Ingels

How to Seduce a Woman '74
Instant Karma '90

Amy Ingersoll

Knightriders '81

Barrie Ingham

The Great Mouse Detective
    '86
Invasion '65

Robert Ingham

Mutants In Paradise '88

Doran Inghram

Steele's Law '91

James Inglehart

Angels Hard as They Come
    '71

Marty Ingles

Round Numbers '91

Sarah Inglis

Battle of the Bullies '85

Lloyd Ingraham

Red River Valley '36
Undercover Man '36

Lloyd Ingraham (D)

American Aristocracy '17

Jack Ingram

The Adventures of the
    Masked Phantom '38
Arizona Roundup '42
Oath of Vengeance '44
Outlaw Roundup '44
Raiders of Sunset Pass '43
Ridin' the Trail '40
Valley of Terror '38
Whistling Bullets '36

James Ingram

Quincy Jones: A Celebration
    '83

Kate Ingram

Scrubbers '82

Rex Ingram

The Adventures of
    Huckleberry Finn '39
Cabin in the Sky '43
Dark Waters '44
Elmer Gantry '60
Green Pastures '36
Moonrise '48
Sahara '43
Talk of the Town '42
Thief of Baghdad '40

Rex Ingram (D)

The Four Horsemen of the
    Apocalypse '21

Boris Ingster (D)

The Judge Steps Out '49
Stranger on the Third Floor
    '40

**J. Christian
Ingvordsen** (D)

Covert Action '88
Firehouse '87
Hangmen '87
Mob War '88
The Roaring Twenties '20s
Search and Destroy '88

I. Inkizhinov

Storm Over Asia '28

Valeri Inkizhinov

Journey to the Lost City '58
Storm Over Asia '28

Frank Inn

Benji the Hunted '87

George Innes

Archer: The Fugitive from the
    Empire '81
Ivanhoe '82

Neil Innes

All You Need is Cash '78
Monty Python and the Holy
    Grail '75

Harold Innocent

Robin Hood: Prince of Thieves
    '91

**Markus
Innocenti** (D)

Murder Story '89

Dan Inosanto

Chinese Stuntman '70s
Counter Attack '84
Warrior Within '80

Tino Insana

The Masters of Menace '90
Wedding Band '89

Charles Insley

Work/Police '16

Franco Interlenghi

I Vitelloni '53
Shoeshine '46

Steve Inwood

The Human Shield '92
Hurry Up or I'll Be Thirty '73

Jeffrey R. Iorio

Deadly Obssession '88

Aharon Ipale

One Man Out '89
Too Hot to Handle '76

Enrique Irazoqui

Gospel According to St.
    Matthew '64

Jill Ireland

Assassination '87
Breakheart Pass '76
Breakout '75

Chino '75
Cold Sweat '74
Death Wish 2 '82
The Family '73
Hard Times '75
Love and Bullets '79
The Mechanic '72
Rider on the Rain '70
Someone Behind the Door '71
Villa Rides '68

John Ireland

The Adventurers '70
All the King's Men '49
Bloody Che Contra
Bordello '79
The Bushwackers '52
Challenge of McKenna '83
Crossbar '79
Dead for a Dollar '70s
Delta Fox '77
Diary of a Rebel '84
Dirty Heroes '71
The Doolins of Oklahoma '49
Escape to the Sun '72
The Fall of the Roman Empire
    '64
Farewell, My Lovely '75
The Fast and the Furious '54
55 Days at Peking '63
Flying from the Hawk '86
Glass Tomb '55
Gunfight at the O.K. Corral '57
The Gunslinger '56
The House of Seven Corpses
    '73
I Shot Jesse James '49
Incubus '82
Joan of Arc '48
Kavik the Wolf Dog '84
Little Big Horn '51
The Mad Dutcher '72
Marilyn: The Untold Story '80
Martin's Day '85
Messenger of Death '88
Miami Horror '87
Midnight Auto Supply '78
Mission to Glory '80
My Darling Clementine '46
Northeast of Seoul '72
Party Girl '58
Perfect Killer '77
Railroaded '47
Ransom '84
Return of Jesse James '50
Satan's Cheerleaders '77
A Southern Yankee '48
Spartacus '60
Tales of the Klondike: Race
    for Number One '81
Taste of Death '77
Thunder Run '86
Tomorrow Never Comes '77
Vengeance Valley '51
A Walk in the Sun '46
Wild in the Country '61

John Ireland (D)

The Fast and the Furious '54

Kathy Ireland

Alien from L.A. '87
Mr. Destiny '90
Necessary Roughness '91
Side Out '90
Sports Illustrated's 25th
    Anniversary Swimsuit
    Video '89

O'Dale Ireland (D)

Date Bait '60
High School Caesar '56

Tonto Irie

Rockers '87 '87

Doug Irk

Crazed Cop '88

Jeremy Irons

Betrayal '83
Brideshead Revisited '81
A Chorus of Disapproval '89
Dead Ringers '88
The French Lieutenant's
  Woman '81
The Mission '86
Moonlighting '82
Nijinsky '80
Reversal of Fortune '90
Swann in Love '84
The Wild Duck '84

Michael Ironside

American Nightmare '81
Coming Out Alive '84
Common Bonds '91
Cross Country '83
Deadly Surveillance '91
Extreme Prejudice '87
The Falcon and the Snowman
  '85
The Family Man '79
Ford: The Man & the Machine
  '87
Hello Mary Lou: Prom Night 2
  '87
Highlander 2: The Quickening
  '91
Hostile Takeover '88
Killer Image '92
McBain '91
Mindfield '89
Murder by Night '89
Murder in Space '85
Neon City '91
Nowhere to Hide '87
Payback '90
Spacehunter: Adventures in
  the Forbidden Zone '83
Surrogate '88
Top Gun '86
Total Recall '90
Visiting Hours '82
Watchers '88

Dom Irrera

Rodney Dangerfield: Nothin'
  Goes Right '88

John Irvin (D)

Champions '84
The Dogs of War '81
Eminent Domain '91
Ghost Story '81
Hamburger Hill '87
Haunted: The Ferryman '74
Next of Kin '89
Raw Deal '86
Robin Hood '91
Turtle Diary '86

Sam Irvin (D)

Guilty as Charged '91

Kathleen Irvine

Greenstone '85

Kevin Irvine (D)

Greenstone '85

Robin Irvine

Easy Virtue '27

Amy Irving

An American Tail: Fievel Goes
  West '91
Anastasia: The Mystery of
  Anna '86

Carrie '76
Competition '80
Crossing Delancey '88
The Far Pavilions '84
The Fury '78
Heartbreak House '86
Honeysuckle Rose '80
I'm a Fool '77
James Dean '76
Micki & Maude '84
Rumpelstiltskin '86
A Show of Force '90
Turn of the Screw '89
Who Framed Roger Rabbit?
  '88
Yentl '83

David Irving (D)

Cannon Movie Tales: Sleeping
  Beauty '89
Cannon Movie Tales: The
  Emperor's New Clothes '89
C.H.U.D. 2: Bud the Chud '89
Goodbye Cruel World '82
Night of the Cyclone '90
Rumpelstiltskin '82
Rumpelstiltskin '86

George Irving

The Abe Lincoln of Ninth
  Avenue '39
Captain January '36
The Divorcee '30
Maid's Night Out '38
The Mandarin Mystery '37

George S. Irving

Deadly Hero '75

Penny Irving

Big Zapper '73
House of Whipcord '75

Richard Irving (D)

The Art of Crime '75
Columbo: Prescription Murder
  '67
The Jesse Owens Story '84

William Irving

The Cameraman '28

Bill Irwin

Eight Men Out '88
Hot Shots! '91
My Blue Heaven '90
Popeye '80
Stepping Out '91

Boyd Irwin

The Three Musketeers '21

Alberto Isaac

Beyond the Next Mountain
  '70s

James Isaac (D)

The Horror Show '89

Chris Isaak

Married to the Mob '88
The Silence of the Lambs '91

Peter Isacksen

Rich Hall's Vanishing America
  '86

Antonio Isasi (D)

Ricco '74
Vengeance '86

Tom Isbell

Behind Enemy Lines '85
A Case of Deadly Force '86

Jose Isbert

El Cochecito '60

Tony Isbert

The Saga of the Draculas '72

Robert Iscove (D)

The Hill '88
The Little Mermaid '84
Puss 'n Boots '84

Tor Isedal

The Doll '62

John Isenbarger

The Black Six '74

Gerald I. Isenberg (D)

Seizure: The Story of Kathy
  Morris '82

Tatsuya Ishiguro

Shin Heike Monogatari '55

Akira Ishihama

The Human Condition: No
  Greater Love '58

Hiroshi Ishikawa

Godzilla on Monster Island '72
Godzilla vs. Gigan '72

Claudia Islas

La Pachange (The Big Party)
  '87
Para Servir a Usted '87
Volver, Volver, Volver! '87

Neal Israel

It's Alive 3: Island of the Alive
  '87

Neal Israel (D)

Americathon '79
Bachelor Party '84
Combat Academy '86
Moving Violations '85
Tunnelvision '76

Peter Israelson (D)

Side Out '90

Brad Issac

Shock! Shock! Shock! '87

Gregory Issacs

Gregory Issacs Live '87
The Moistro Meet the Cool
  Ruler '87

Agustin Isunza

The Illusion Travels by
  Streetcar '53

Juzo Itami

The Family Game '83
MacArthur's Children '85

Juzo Itami (D)

The Funeral '84
Tampopo '86
A Taxing Woman '87
A Taxing Woman Returns '88

Emi Ito

Mothra '62

Robert Ito

The Adventures of Buckaroo
  Banzai '84
Pray for Death '85

Jose Isbert

Jose Isbert

Women of the Prehistoric
  Planet '66

Toshiya Ito

Space Riders '83

Yumi Ito

Mothra '62

Yunosuke Ito

The Burmese Harp '56

Jose Iturbi

Anchors Aweigh '45
Two Girls & a Sailor '44

Aniko Ivan

Time Stands Still '82

Zeljko Ivanek

Echoes in the Darkness '87
Mass Appeal '84
The Sender '82

Hira Ivanov-Golarko

War and Peace '68

Stan Ivar

Creature '85

Vladimir Ivashov

Ballad of a Soldier '60

Rosalind Iven

Corn is Green '45
Pursuit to Algiers '45

Daniel Ivernel

Diary of a Chambermaid '64
High Heels '72

Robert Ivers

The Delicate Delinquent '56
G.I. Blues '60

Burl Ives

Baker's Hawk '76
Big Country '58
Cat on a Hot Tin Roof '58
Daydreamer '66
Desire Under the Elms '58
East of Eden '55
Ensign Pulver '64
The Ewok Adventure '84
Hugo the Hippo '76
The McMasters '70
Roots, Vol. 1 '77
Rudolph the Red-Nosed
  Reindeer '64
So Dear to My Heart '49
Station West '48
Summer Magic '63
Those Fantastic Flying Fools
  '67
Two Moon Junction '88
Uphill All the Way '85

Judith Ivey

Brighton Beach Memoirs '86
Compromising Positions '85
Dixie: Changing Habits '85
Everybody Wins '90
Harry & Son '84
Hello Again '87
In Country '89
The Lonely Guy '84
The Long, Hot Summer '86
Love Hurts '91
Miles from Home '88
Sister, Sister '87
We are the Children '87
The Woman in Red '84

James Ivory (D)

Bombay Talkie '70
The Bostonians '84
The Europeans '79
Heat and Dust '82
Hullabaloo Over Georgie &
  Bonnie's Pictures '78
Jane Austen in Manhattan '80
Maurice '87
Mr. & Mrs. Bridge '91
Quartet '81
A Room with a View '86
Roseland '77
Savages '72
Shakespeare Wallah '65
Slaves of New York '89
Wild Party '74

Sharon Iwai

A Great Wall '86

Shima Iwashita

An Autumn Afternoon '62
Double Suicide '69
Gonza the Spearman '86
MacArthur's Children '85
Red Lion '69

Victor Izay

Premonition '71
Trial of Billy Jack '74

David Izenon

Ornette Coleman Trio '66

Teresa Izewska

Kanal '56

Shigeru Izumiya

Eijanaika (Why Not?) '81

Arnaldo Jabor (D)

I Love You '81

Arch Jaboulian

The Meateater '79

Silvana Jachino

Princess Cinderella '30s

Wolfman Jack

American Graffiti '73
The Committee '68
Deadman's Curve '78
Hanging on a Star '78
Midnight '89
Mortuary Academy '91
Motel Hell '80

Jackee

Ladybugs '92
The Women of Brewster
  Place '89

Jackie Jackler

Assault of the Rebel Girls '59

Anne Jackson

The Bell Jar '79
Blinded by the Light '82
Blood Debts '83
The Family Man '79
Folks '92
Funny About Love '90
Independence '76
Leave 'Em Laughing '81
Lovers and Other Strangers
  '70
Nasty Habits '77
Sam's Son '84
Secret Life of an American
  Wife '68
Tall Story '60

Samuel H. James

Go Down Death '41

Sid James

Carry On Behind '75
Carry On Dick '75
Carry On 'Round the Bend '72

Sidney James

Carry On Cleo '65
Carry On Cowboy '66
Carry On Cruising '62
Carry On at Your
Convenience '71
Glass Tomb '55
Heat Wave '54
The Lavender Hill Mob '51
Quatermass 2 '57
The Silent Enemy '58
What a Carve-Up! '62

Sonny James

Las Vegas Hillbillys '66
Rock 'n' Roll History:
Rockabilly Rockers '90

Steve James

American Ninja '85
American Ninja 2: The
Confrontation '87
American Ninja 3: Blood Hunt
'89
Avenging Force '86
The Brother from Another
Planet '84
Hero and the Terror '88
I'm Gonna Git You Sucka '88
Johnny Be Good '88
McBain '91
The P O W Escape '86
Python Wolf '88
Riverbend '89
Stalking Danger '88
Street Hunter '90

Tina James

In Your Face '77

Walter James

Battling Butler '26
The Kid Brother '27
Little Annie Rooney '25

Conrad Jameson

Pinocchio in Outer Space '64

Jerry Jameson (D)

The Bat People '74
Cowboy & the Ballerina '84
Dirt Gang '71
Fire and Rain '89
High Noon: Part 2 '80
Hotline '82
Hurricane '74
Killing at Hell's Gate '81
Raise the Titanic '80
Sniper '75
Starflight One '83
Terror on the 40th Floor '74

Joyce Jameson

The Balcony '63
The Comedy of Terrors '64
Good Neighbor Sam '64
Hardbodies '84
Pray TV '80
Savage Run '70
Tales of Terror '62

Louise Jameson

Doctor Who: The Talons of
Weng-Chiang '88

Malcom Jamieson

Meridian: Kiss of the Beast
'90

Georges Jamin

Daughters of Darkness '71

Bud Jamison

Charlie Chaplin: Rare Chaplin
'15
The Dentist '32
W.C. Fields: Flask of Fields
'30

Mikki Jamison-Olsen

Sea Gypsies '78

Jan & Dean

The T.A.M.I. Show '64

DeWitt Jan

Harley '90

Lawrence Jan

To Live and Die in Hong Kong
'89

Adam Janas

Bottoms Up '87

Miklos Jancso (D)

Hungarian Rhapsody '78
The Red and the White '68
The Round Up '66

Tadeusz Janczar

Kanal '56

Gerhard Janda (D)

Tea For Three '84

Krystyna Janda

Interrogation '82
Man of Marble '76
Mephisto '81

Ivan Jandl

The Search '48

Randy Jandt

Thinkin' Big '87

Conrad Janis

Buddy Holly Story '78
The Duchess and the
Dirtwater Fox '76
Oh, God! Book 2 '80
Sonny Boy '87

Zoran Janjic (D)

A Connecticut Yankee in King
Arthur's Court '70

Annabel Jankel

Max Headroom, the Original
Story '86

Annabel Jankel (D)

D.O.A. '88

Stole Jankovic (D)

Hell River '75

Oleg Jankowski

My Twentieth Century '90

Leon Janney

Charly '68
Police Court '37

Rosanna Janni

Sonny and Jed '73

Emil Jannings

The Americano/Variety '25
The Blue Angel '30
Faust '26
Fortune's Fool '21
Last Command '28
The Last Laugh '24
Othello '22
Passion '19
Variety '25
Waxworks '24

Robert Jannucci

Exterminators of the Year
3000 '83

Victor Janos (D)

Last House on Dead End
Street '77

Ellen Janov

The Horse in the Gray Flannel
Suit '68

Walter Janovitz

Billy the Kid Versus Dracula
'66

Tama Janowitz

Slaves of New York '89

Andre Jansen

A Waltz Through the Hills '88

Gina Jansen

Sensual Partners '87

Horst Janson

Captain Kronos: Vampire
Hunter '74
To Catch a King '84

David Janssen

Birds of Prey '72
City in Fear '80
Fer-De-Lance '74
Francis in the Navy '55
Generation '69
Golden Rendezvous '77
The Green Berets '68
Hell to Eternity '60
High Ice '80
Macho Callahan '70
Marooned '69
Moon of the Wolf '72
Once is Not Enough '75
Pioneer Woman '73
Prisoner in the Middle '74
The Shoes of the Fisherman
'68
Smile, Jenny, You're Dead '74
S.O.S. Titanic '79
Stalk the Wild Child '76
Swamp Fire '46
Swiss Conspiracy '77
Two Minute Warning '76
The Word '78

Eilene Janssen

Curley '47

Walther Jansson

The Mozart Story '48

Lois January

Bulldog Courage '35
Cocaine Fiends '36
Lightning Bill Crandall '37
The Red Rope '37
The Roamin' Cowboy '37

Christian Jaque (D)

Legend of Frenchie King '71

Al Jardine

Beach Boys: An American
Band '85

Paully Jardine

Summer's Children '84

Claude Jarman, Jr.

Hangman's Knot '52
Intruder in the Dust '49
Rio Grande '50
The Yearling '46

Derek Jarman (D)

Aria '88
Caravaggio '86
The Last of England '87
Sebastiane '79

Jim Jarmusch

American Autobahn '84
Leningrad Cowboys Go
America '89
Straight to Hell '87

Jim Jarmusch (D)

Down by Law '86
Mystery Train '89
Stranger than Paradise '84

Marthe Jarnias

Vagabond '85

John Jarratt

Dark Age '88
Next of Kin '82
Sound of Love '77
Summer City '77

Kevin Jarre

Gotham '88

Andy Jarrell

Inferno in Paradise '88

Gabe Jarret

Real Genius '85

Jerry Jarret

Killer's Kiss '55

Art Jarrett

Trigger Pals '39

Catherine Jarrett

Quicker Than the Eye '88
S.A.S. San Salvador '84

Charles Jarrott (D)

The Amateur '82
Boy in Blue '86
Condorman '81
The Dove '74
Last Flight of Noah's Ark '80
Littlest Horse Thieves '76
Night of the Fox '90
The Other Side of Midnight
'77
Strange Case of Dr. Jekyll &
Mr. Hyde '68

Graham Jarvis

Cold Turkey '71
Doin' Time '85
Draw! '81
Middle Age Crazy '80
Mischief '85
Misery '90
Vanishing Act '88

Vojta Jasny (D)

The Great Land of Small '86

Voltech Jasny (D)

All My Good Countrymen '68

David Jason

Open All Hours '83

Harvey Jason

Dr. Minx '75
Joseph & His Brothers '79
Oklahoma Crude '73

Leigh Jason (D)

Bride Walks Out '36
The Choppers '61
Lady for a Night '42
Lost Honeymoon '47
Mad Miss Manton '38
Out of the Blue '47
That Girl from Paris '36

Neville Jason

Mohammed: Messenger of
God '77

Peter Jason

Alien Nation '88
Hyper-Sapien: People from
Another Star '86
Texas Lightning '81
They Live '88

Rick Jason

Color Me Dead '69
Eagles Attack at Dawn '74
Partners '82
Witch Who Came from the
Sea '76

Ron Jason

Las Vegas Serial Killer '86

Sybil Jason

The Blue Bird '40
The Little Princess '39

Will Jason (D)

Everybody's Dancin' '50

Star Jasper

True Love '89

Terry Jastrow

Waltz Across Texas '83

Ricky Jay

Things Change '88

Jennifer Jayne

The Crawling Eye '58

Michael Jayston

Alice's Adventures in
Wonderland '72
Craze '74
Dust to Dust '85
Macbeth '90
Nicholas and Alexandra '71

Elizabeth Jeager

Escape '90

Gloria Jean

Copacabana '47
The Ladies' Man '61
Never Give a Sucker an Even
Break '41

Isabel Jean

Easy Virtue '27

Lorin Jean
Rest In Pieces '87

Sonia Jeanine
M'Lady's Court '80s

Zizi Jeanmarie
Hans Christian Andersen '52

Anais Jeanneret
Peril '85

Isabel Jeans
Elizabeth of Ladymead '48
Gigi '58
Great Day '46
Heavens Above '63
The Magic Christian '69

Ursula Jeans
Dam Busters '55
Dark Journey '37
Life & Death of Colonel Blimp '43

Colin Jeavons
Bartleby '70

Kalina Jedrusik
The Double Life of Veronique '91

Barbara Jefferd
Murders at Lynch Cross '85

Lang Jefferies
Deadline Auto Theft '83

Lionel Jefferies
A Chorus of Disapproval '89

Herbert Jefferson, Jr.
Detroit 9000 '73
Private Duty Nurses '71

Roy Jefferson
Brotherhood of Death '76

Barbara Jefford
And the Ship Sails On '84
Lust for a Vampire '71
Ulysses '67

Peter Jeffrey
Horsemen '70
Ring of Bright Water '69
Twinsanity '70

Tom Jeffrey (D)
The Odd Angry Shot '79

Anne Jeffreys
Billy the Kid Trapped '42
Bordertown Gunfighters '43
Death Valley Manhunt '43
Dick Tracy Detective '45
Dick Tracy Double Feature #1 '45
Dick Tracy Double Feature #2 '47
Dick Tracy vs. Cueball '46
Dillinger '45
I Married an Angel '42
Man from Thunder River '43
Return of the Bad Men '48
Riff-Raff '47
Step Lively '44
Those Endearing Young Charms '45
Trail Street '47
X Marks the Spot '42

Chuck Jeffreys
Aftershock '88
Hawkeye '88

Fran Jeffries
Harum Scarum '65
Sex and the Single Girl '64

Herbert Jeffries
Bronze Buckaroo '39
Chrome and Hot Leather '71
Harlem on the Prairie '38
Harlem Rides the Range '39
Two-Gun Man from Harlem '38

Lang Jeffries
Alone Against Rome '62
Mission Stardust '68

Lionel Jeffries
The Baby and the Battleship '56
Bananas Boat '78
Blue Murder at St. Trinian's '56
Camelot '67
Chitty Chitty Bang Bang '68
The Colditz Story '55
Fanny '61
First Men in the Moon '64
Hour of Decision '57
Letting the Birds Go Free '86
Lola '69
Lust for Life '56
Murder Ahoy '64
Oh Dad, Poor Dad (Momma's Hung You in the Closet & I'm Feeling So Sad) '67
Prisoner of Zenda '79
Spy With a Cold Nose '66
Sudden Terror '70
Those Fantastic Flying Fools '67
Up the Creek '58
The Vicious Circle '57
Who Slew Auntie Roo? '71
The Wrong Arm of the Law '63

Lionel Jeffries (D)
Amazing Mr. Blunden '72
Railway Children '70
Water Babies '79
Wombling Free '77

Deanne Jeffs
Midnight Dancer '87

Andras Jeles (D)
The Little Valentino '79

Tom Jelfry (D)
Weekend of Shadows '77

Rudolph Jelinek
The Original Fabulous Adventures of Baron Munchausen '61

Anna Jemison
Heatwave '83
My First Wife '84
Smash Palace '82

Roger Jendly
Jonah Who Will Be 25 in the Year 2000 '76

Devon Jenkin
Slammer Girls '87

Allen Jenkins
The Big Wheel '49
Brother Orchid '40
Case of the Baby Sitter '47
The Case of the Lucky Legs '35
Dead End '37
Falcon Takes Over/Strange Bargain '49
Hat Box Mystery '47
Marked Woman '37
Oklahoma Annie '51
Robin and the Seven Hoods '64
Tomorrow at Seven '33

Anthony Jenkins
Bloodspell '87

Daniel H. Jenkins
Florida Straits '87
O.C. and Stiggs '87

David Jenkins
Invincible Barbarian '83

Elin Jenkins
Waterfront '83

Jackie "Butch" Jenkins
Summer Holiday '48

John Jenkins
Patti Rocks '88

Ken Jenkins
Edge of Honor '91

Megs Jenkins
Murder Most Foul '65
Trouble in Store '53

Michael Jenkins (D)
Rebel '86
Sweet Talker '91

Patrick Jenkins (D)
The Gambler & the Lady '52

Polly Jenkins
The Man from Music Mountain '38

Rebecca Jenkins
Bye Bye Blues '91
Cowboys Don't Cry '88
Family Reunion '88
Till Death Do Us Part '92

Richard Jenkins
Blaze '89
Rising Son '90
Sea of Love '89
The Tender Age '84
Witches of Eastwick '87

Terence Jenkins
Vampire Cop '90

Frank Jenks
Blonde Savage '47
Corregidor '43
The Falcon in Hollywood '44
Kid Sister '45
The Missing Corpse '45
The Navy Comes Through '42
The Phantom of 42nd Street '45
Rogue's Gallery '44
Shake Hands with Murder '44
That Brennan Girl '46

That Girl from Paris '36
Zombies on Broadway '44

Si Jenks
Captain January '36
Rawhide Romance '34

Michael Jenn
Another Country '84

Bruce Jenner
Can't Stop the Music '80
Grambling's White Tiger '81

Lucinda Jenney
Rain Man '88
Verne Miller '88
Wired '89

Claudia Jennings
Death Sport '78
Fast Company '78
Gator Bait '73
The Great Texas Dynamite Chase '76
Group Marriage '72
Jud '71
The Love Machine '71
Moonshine County Express '77
Sisters of Death '76
The Stepmother '71
Truck Stop Women '74
Unholy Rollers '72

DeWitt Jennings
The Bat Whispers '30
Criminal Code '31

Jane Jennings
The Magnificent

Joseph Jennings
Rock House '88

Maxine Jennings
Two Reelers: Comedy Classics 1 '33

Paul Craig Jennings
Invasion of the Blood Farmers '72

Peter Jennings
JFK Remembered '88

Tom Jennings
Night Master '87
Stones of Death '88

Waylon Jennings
Sesame Street Presents: Follow That Bird '85
Stagecoach '86

Rita Jenrette
Malibu Bikini Shop '86
Zombie Island Massacre '84

Salome Jens
Angel Baby '61
Diary of the Dead '80s
The Fool Killer '65
A Matter of Life and Death '81
Savages '72
Tomorrow's Child '82

Astrid Henning Jensen (D)
Boy of Two Worlds '70

Jan Jensen
The Last Slumber Party '87

Karen Jensen
Battlestar Galactica '78
Deadly Blessing '81
The Salzburg Connection '72

Kenneth Jensen
Buckeye and Blue '87

Maren Jensen
Deadly Blessing '81

Sasha Jensen
Dream Trap '90
A Girl to Kill for '90

Jerry Jenson
Ghosts That Still Walk '77

Linda Jenson
Cheerleaders' Beach Party '77

Roy Jenson
Bandits '73
Crossfire '67

Roy Cameron Jenson
Demonoid, Messenger of Death '81

Dominic Jephcott
African Dream '90

Helen Jepson
Goldwyn Follies '38

Adele Jergens
Abbott & Costello Meet the Invisible Man '51
Armored Car Robbery '50
Big Chase '54
Blues Busters '50
The Cobweb '55
Dark Past '49
Down to Earth '47
Falcon's Adventure/Armored Car Robbery '50
Fuller Brush Man '48
Girls in Prison '56
Ladies of the Chorus '49
The Prince of Thieves '48
Radar Secret Service '50
Treasure of Monte Cristo '50

Diane Jergens
High School Confidential '58

Mary Jerrold
The Queen of Spades '49

Anita Jesse
Home Safe '88

Dan Jesse
Angels with Dirty Faces '38

George Jessel
Diary of a Young Comic '79

Patricia Jessel
Horror Hotel '60

Alain Jessua (D)
The Killing Game '67

Michael Jeter
Dead Bang '89

Joan Jett
Dubeat-E-O '84
Light of Day '87

Mark V. Jevicky

The Majorettes '87

Betty Jewel

The Last Outlaw '27

Jimmy Jewel

Arthur's Hallowed Ground '84

Geri Jewell

Two of a Kind '82

Isabel Jewell

The Arousers '70
The Leopard Man '43
Lost Horizon '37
Manhattan Melodrama '34
Marked Woman '37
The Seventh Victim '43
Swing It, Sailor! '37
A Tale of Two Cities '35

Norman Jewison (D)

Agnes of God '85
And Justice for All '79
Best Friends '82
Cincinnati Kid '65
Fiddler on the Roof '71
F.I.S.T. '78
In Country '89
In the Heat of the Night '67
Jesus Christ Superstar '73
Moonstruck '87
Other People's Money '91
Rollerball '75
The Russians are Coming, the
   Russians are Coming '66
Send Me No Flowers '64
A Soldier's Story '84
Thomas Crown Affair '68
The Thrill of It All! '63

Takeo Jii

Nomugi Pass '79

Penn Jillette

Comic Relief 2 '87
Light Years '88
Miami Vice 2: The Prodigal
   Son '85
Penn and Teller Get Killed '90
Penn and Teller Go Public '89
Penn and Teller's Cruel Tricks
   for Dear Friends '88

Ann Jillian

The Ann Jillian Story '88
Convicted: A Mother's Story
   '87
Girls of the White Orchid '85
Gypsy '62
Little White Lies '89
Mae West '84
Mr. Mom '83
Sealab 2020 '72

Joyce Jillson

Slumber Party '57 '76
Superchick '71

Yolanda Jilot

Diving In '90

Chrysti Jimenez

Omega Cop '90

Falco Jimenez

La Pastorela '92

Jose Luis Jimenez

San Francisco de Asis
Spiritism '61
The Vampire '57

Sergio Jimenez

La Pachange (The Big Party)
   '87
Los Caifanes '87
Los Dos Hermanos '87

Orlando Jiminez-Leal

El Super '79

Maria Rosa Jiminez

Loyola, the Soldier Saint '52

Robert Jiras (D)

I Am the Cheese '83

Duan Jishun (D)

The Go-Masters '82

Mike Jittlov

The Wizard of Speed and
   Time '88

Mike Jittlov (D)

The Wizard of Speed and
   Time '88

Jo Stafford & the Pied Pipers

DuBarry was a Lady '43

Suzy Joachim

Ultimate Desires '91

Keith Joakum

Witches of Eastwick '87

Leo Joannon (D)

Atoll K '51
Utopia '52

Phil Joanou (D)

Final Analysis '92
State of Grace '90
Three O'Clock High '87

Marlene Jobert

Catch Me a Spy '71
Masculine Feminine '66
Rider on the Rain '70
Swashbuckler '84
Ten Days Wonder '72

Dickie Jobson (D)

Countryman '83

Rene Jodoin (D)

Rectangle & Rectangles '87

Adan Jodorowsky

Santa Sangre '90

Alejandro Jodorowsky (D)

Santa Sangre '90

Axel Jodorowsky

Santa Sangre '90

Steve Jodrell (D)

Shame '87

Michael Joens (D)

My Little Pony: The Movie '86

Alfred Joffe (D)

Harem '85

Mark Joffe (D)

Grievous Bodily Harm '89

Night Master '87

Roland Joffe (D)

Fat Man and Little Boy '89
The Killing Fields '84
The Mission '86

Zita Johann

The Mummy '32

Ingemar Johannson

48 Hours to Live '60

David Johansen

Candy Mountain '87
Desire and Hell at Sunset
   Motel '87
Freejack '92
Let It Ride '89
Light Years '88
Scrooged '88
Tales from the Darkside: The
   Movie '90

Paul Johansson

Martial Law 2: Undercover '91

Domenick John

Creepshow 2 '87

Elton John

Born to Boogie '72
Tommy '75

Gottfried John

Mata Hari '85
Mother Kusters Goes to
   Heaven '76

Karl John

The Lost One '51

The John Kirby Sextet

Sepia Cinderella '47

Robert John

Creatures the World Forgot
   '70

Rosamund John

Fame is the Spur '47
Spitfire '42
Spitfire: The First of the Few
   '43

Alexandra Johnes

Zelly & Me '88

Tom Johnigam

Bad Bunch '76

Johnny Bee & Tuffi

Michael Prophet '87

Anne Johns

Family Enforcer '77

Glynis Johns

The Adventures of Tartu '43
All Mine to Give '56
Another Time, Another Place
   '58
Around the World in 80 Days
   '56
Court Jester '56
Dear Brigitte '65
Encore '52
The Forty-Ninth Parallel '41
Frieda '47
Land of Fury '55
Mary Poppins '64
Mrs. Amworth '77

Papa's Delicate Condition '63
Promoter '52
Rob Roy—The Highland
   Rogue '53
Spragque '84
The Sundowners '60
Sword & the Rose '53
Under Milk Wood '73
Vault of Horror '73
Zelly & Me '88

Marion Johns

Gone to Ground '76

Mervyn Johns

Captive Heart '47
A Christmas Carol '51
Counterblast '48
Day of the Triffids '63
Dead of Night '45
Ingrid '89
The Master of Ballantrae '53
No Love for Johnnie '60
Pink Strings and Sealing Wax
   '45
Romeo and Juliet '54
San Demetrio, London '47
Tony Draws a Horse '51

Stratford Johns

Great Expectations '81
The Lair of the White Worm
   '88
Salome's Last Dance '88

Tracy C. Johns

New Jack City '91

A.J. Johnson

Double Trouble '91
House Party '90

Alan Johnson

Iced '88

Alan Johnson (D)

Solarbabies '86
To Be or Not to Be '83

Anne-Marie Johnson

Hollywood Shuffle '87
I'm Gonna Git You Sucka '88
Robot Jox '89
True Identity '91

Arnold Johnson

My Demon Lover '87
Putney Swope '69

Arte Johnson

Bud and Lou '78
Bunco '83
The Charge of the Model T's
   '76
Evil Spirits '91
Evil Toons '90
If Things Were Different '79
Love at First Bite '79
Munchie '92
The Red Raven Kiss-Off '90
Tax Season '90
What Comes Around '85

Ben Johnson

Back to Back '90
Bite the Bullet '75
Breakheart Pass '76
Champions '84
Cherry 2000 '88
Chisum '70
Dark Before Dawn '89
Dillinger '73
The Getaway '72

Grayeagle '77
Junior Bonner '72
The Last Picture Show '71
Let's Get Harry '87
Major Dundee '65
Mighty Joe Young '49
My Heroes Have Always Been
   Cowboys '91
One-Eyed Jacks '61
Radio Flyer '91
The Rare Breed '66
Red Dawn '84
Rio Grande '50
Ruckus '81
The Sacketts '79
The Savage Bees '76
The Shadow Riders '82
Shane '53
She Wore a Yellow Ribbon '49
Soggy Bottom U.S.A. '84
The Sugarland Express '74
The Swarm '78
Terror Train '80
Tex '82
Three Godfathers '48
Tomboy & the Champ '58
Town that Dreaded Sundown
   '76
Train Robbers '73
Trespasses '86
The Undefeated '69
Wagon Master '50
Wild Horses '84
Wild Times '79
Will Penny '67

Beverly Johnson

Ashanti, Land of No Mercy '79

Bill Johnson

Texas Chainsaw Massacre
   Part 2 '86

Bobby Johnson

Hollywood's New Blood '88

Brad Johnson

Always '89
Flight of the Intruder '90
The Losers '70
Nam Angels '88

Bruce Johnson

Beach Boys: An American
   Band '85

Candy Johnson

Muscle Beach Party '64
Pajama Party '64

Ceepee Johnson

Mystery in Swing '40

Celia Johnson

Brief Encounter '46
Captain's Paradise '53
The Hostage Tower '80
In Which We Serve '42
Les Miserables '78
The Prime of Miss Jean
   Brodie '69
Staying On '80
This Happy Breed '47
Unicorn '83

Chic Johnson

All Over Town '37
Country Gentlemen '36

Clark Johnson

Personals '90

Denny Johnson

Truly Tasteless Jokes '85

Don Johnson

Beulah Land '80
A Boy and His Dog '75
Cease Fire '85
The City '76
Dead Bang '89
G.I. Joe: The Movie '86
Harley Davidson and the Marlboro Man '91
Harrad Experiment '73
The Hot Spot '90
Law of the Land '76
The Long, Hot Summer '86
Melanie '82
Miami Vice '84
Miami Vice 2: The Prodigal Son '85
Paradise '91
The Rebels '79
Return to Macon County '75
Revenge of the Stepford Wives '80
Snowblind '78
Soggy Bottom U.S.A. '84
Sweet Hearts Dance '88
Zachariah '70

Dots Johnson

Paisan '46
Tall, Tan and Terrific '46

Earle Johnson

Pee-Wee's Playhouse Christmas Special '88

Edith Johnson

The Scarlet Car '17

Emory Johnson (D)

Phantom Express '32

Floch Johnson (D)

Maid in Sweden '83

Hall Johnson

Black & Tan/St. Louis Blues '29

Helen Johnson

The Divorcee '30
Vice Squad '31

Jed Johnson (D)

Andy Warhol's Bad '77

Jenny Johnson

First Turn On '83

Jerry Johnson

Kill Squad '81

Jesse Johnson

Invasion of the Space Preachers '90

Jill Johnson

Party Plane '90
Wildest Dreams '90

Joanna Johnson

Killer Party '86

Joey Johnson

Courier of Death '84
Grad Night '81

John H. Johnson (D)

Curse of the Blue Lights '88

Johnny Johnson

Two Reelers: Comedy Classics 7 '38

Joseph Alan Johnson

Berserker '87

Joshua Johnson

Death Machines '76

Jude Johnson

The Abomination '88

Julie Johnson

The Islander '88

Kathleen Johnson

Bottoms Up '87

Katie Johnson

Ladykillers '55

Kelly Johnson

Carry Me Back '82
Goodbye Pork Pie '81
Utu '83

Kenneth Johnson (D)

The Incredible Hulk '77
Senior Trip '81
Short Circuit 2 '88

Kent Johnson

Forever Evil '87

Kurt Johnson

Jane Austen in Manhattan '80
Sole Survivor '84

Kyle Johnson

The Learning Tree '69
Man on the Run '74

Lamont Johnson

Unnatural Causes '86

Lamont Johnson (D)

Crisis at Central High '80
Ernie Kovacs: Between the Laughter '84
Escape '90
The Execution of Private Slovik '74
FM '78
The Groundstar Conspiracy '72
A Gunfight '71
Jack & the Beanstalk '83
The Last American Hero '73
Lipstick '76
My Sweet Charlie '70
One on One '77
Paul's Case '80
Spacehunter: Adventures in the Forbidden Zone '83

Laura Johnson

Beyond Reason '77
Fatal Instinct '92
Murderous Vision '91
Nick Knight '80s

Lynn-Holly Johnson

Alien Predators '80
Digging Up Business '91
For Your Eyes Only '81
Ice Castles '79
The Sisterhood '88

Watcher in the Woods '81
Where the Boys Are '84 '84

Martin Johnson

Borneo '37

Mel Johnson, Jr.

Eubie! '82
Total Recall '90

Michael Johnson

Lust for a Vampire '71

Michelle Johnson

Beaks: The Movie '87
Blame It on Rio '84
Genuine Risk '89
Gung Ho '85
Slipping into Darkness '88
Waxwork '88

Myra Johnson

Fats Waller and Friends '86

Noble Johnson

East of Borneo '31

Noble Johnson

King Kong '33
The Navigator '24
North of the Great Divide '50

Nunnally Johnson (D)

The Man in the Gray Flannel Suit '56
The Three Faces of Eve '57

Orin Johnson

The Three Musketeers (D'Artagnan) '16

Osa Johnson

Borneo '37

Patrick Read Johnson (D)

Spaced Invaders '90

Pauline Johnson

The Flying Scotsman '29

Pete Johnson

Bessie Smith and Friends '86

Raymond K. Johnson (D)

Code of the Fearless '39
Daughter of the Tong '39
In Old Montana '39
Pinto Canyon '40
The Reckless Way '36
Ridin' the Trail '40

Reggie Johnson

Platoon '86
Seven Hours to Judgment '88

Richard Johnson

The Amorous Adventures of Moll Flanders '65
Beyond the Door '75
The Big Scam '79
The Comeback '77
The Crucifer of Blood '91
Diving '90
Duel of Hearts '92
Fifth Day of Peace '72
The Four Feathers '78
The Great Alligator '81
The Haunting '63
Hennessy '75

Lady Jane '85
A Man for All Seasons '88
Never So Few '59
Restless '72
Screamers '80
Spymaker: The Secret Life of Ian Fleming '90
Tomb of Ligeia '64
Treasure Island '89
Turtle Diary '86
The Witch '66
Zombie '80

Rita Johnson

My Friend Flicka '43
The Naughty Nineties '45
They Won't Believe Me '47

Robin Johnson

Splitz '84
Times Square '80

Russell Johnson

Fatal Chase '77
The Great Skycopter Rescue '82
It Came From Outer Space '53
This Island Earth '55

Sandy Johnson

Jokes My Folks Never Told Me '77

Steve Johnson

Angel of H.E.A.T. '82
Lemora, Lady Dracula '73

Sunny Johnson

Dr. Heckyl and Mr. Hype '80
The Red Light Sting '84

Terry Johnson

Class Reunion '87
Pleasure Unlimited '86

Toni Johnson

The Screaming Skull '58

Tor Johnson

The Beast of Yucca Flats '61
Behind Locked Doors '48
Bride of the Monster '56
Carousel '56
Houdini '53
Human Gorilla '48
Night of the Ghouls '59
Plan 9 from Outer Space '56
The Unearthly '57

Tucker Johnson (D)

Blood Salvage '90

Van Johnson

Battleground '49
Brigadoon '54
The Caine Mutiny '54
Command Decision '48
Delta Force Commando 2 '90
The Doomsday Flight '66
Duchess of Idaho '50
Easy to Love '53
Go for Broke! '51
A Guy Named Joe '44
The Human Comedy '43
In the Good Old Summertime '49
The Kidnapping of the President '80
Last Time I Saw Paris '54
Madame Curie '43
Pied Piper of Hamelin '57
Purple Rose of Cairo '85
Scorpion with Two Tails '82

State of the Union '48
Thirty Seconds Over Tokyo '44
Three Guys Named Mike '51
Thrill of a Romance '45
Too Many Girls '40
Two Girls & a Sailor '44
Yours, Mine & Ours '68

Victoria (Vicki) Johnson

Grizzly '76

Welton Benjamin Johnson

The Passing '88

Bruce Johnston

The Compleat Beatles '82

Grace Johnston

One Good Cop '91

James Johnston

Never Pick Up a Stranger '79

J.J. Johnston

Fatal Attraction '87
Things Change '88

Jo Johnston

The Swinging Cheerleaders '74

Joe Johnston (D)

Honey, I Shrunk the Kids '89
The Rocketeer '91

John Dennis Johnston

A Breed Apart '84
Dear Detective '79
KISS Meets the Phantom of the Park '78
Streets of Fire '84

John R. Johnston

Possession: Until Death Do You Part '90

Julanne Johnston

Thief of Baghdad '24

Margaret Johnston

Burn Witch, Burn! '62
A Man About the House '47
Sebastian '68

Michelle Johnston

A Chorus Line '85

Oliver Johnston

A King in New York '57
Tomb of Ligeia '64

Patrick Johnston

Buckeye and Blue '87

Jane Anne Johnstone

Dixie Dynamite '76

Marilyn Joi

Hospital of Terror '78

Mirjana Jokovic

Eversmile New Jersey '89

Pierre Jolivet

Le Dernier Combat '84

I. Stanford Jolley

The Black Raven '43
The Crimson Ghost '46
Cyclotrode "X" '46
King of the Rocketmen '49
Midnight Limited '40
Outlaw Roundup '44
The Rebel Set '59
Trail of the Silver Spurs '41
Wild Country '47

Mike Jolly

Bad Guys '86

Nick Jolly

Asylum of Satan '72

Al Jolson

The Al Jolson Collection '92
Hallelujah, I'm a Bum '33
Jazz Singer '27

Jonah

Jonah Who Will Be 25 in the Year 2000 '76

Allan Jones

A Day at the Races '37
A Night at the Opera '35
Rose Marie '36
Show Boat '36
A Swingin' Summer '65

Amanda Jones

Among the Cinders '83

Amy Jones (D)

Love Letters '83
Maid to Order '87
The Slumber Party Massacre '82

Andras Jones

Far from Home '89
A Nightmare on Elm Street 4: Dream Master '88
Sorority Babes in the Slimeball Bowl-A-Rama '87

Annie Jones

Moving Targets '87

Barbara Alyn Jones

Dance with Death '91

Barbara Ann Jones

The Carpenter '89

Barry Jones

Demetrius and the Gladiators '54
The Glass Slipper '55
Prince Valiant '54
Uneasy Terms '48

Ben Jones

Don't Change My World '83

Billy Jones

Ninja Death Squad '87

Bob Jones

The Zodiac Killer '71

Brian Jones

25x5: The Continuing Adventures of the Rolling Stones '90

Brian Thomas Jones (D)

Escape from Safehaven '88

Posed For Murder '89
The Rejuvenator '88

Bryan Jones (D)

Ocean Drive Weekend '85

Buck Jones

Arizona Bound '41
Below the Border '42
Dawn on the Great Divide '42
Down Texas Way '42
Forbidden Trails '41
Ghost Town Law '42
Gunman from Bodie '41
Phantom Rider '37
The Range Feud/Two Fisted Law '31
Riders of Death Valley '41
Riders of the West '42
West of the Law '42
White Eagle '32

Carolyn Jones

Big Heat '53
Career '59
Color Me Dead '69
Eaten Alive '76
Halloween with the Addams Family '79
Hole in the Head '59
House of Wax '53
How the West was Won '62
Ice Palace '60
Invasion of the Body Snatchers '56
King Creole '58
Last Train from Gun Hill '59
Little Ladies of the Night '77
The Man Who Knew Too Much '56
Marjorie Morningstar '58
Road to Bali '53
The Seven Year Itch '55
The Shaming '71
The Tender Trap '55

Charlie Jones

Return of the Killer Tomatoes '88

Chris Jones

Wild in the Streets '68

Christopher Jones

The Looking Glass War '69
Ryan's Daughter '70
Three in the Attic '68

Chuck Jones

Chuck Amuck: The Movie '91
Innerspace '87
Looney Tunes & Merrie Melodies 1 '33

Chuck Jones (D)

Best of Warner Brothers: Volume 1 '40s
Best of Warner Brothers: Volume 2 '30s
Bugs Bunny Cartoon Festival '44
Bugs Bunny Cartoon Festival: Little Red Riding Rabbit '43
Bugs Bunny & Elmer Fudd Cartoon Festival '44
Bugs Bunny in King Arthur's Court '79
Bugs Bunny/Road Runner Movie '79
Bugs Bunny's 3rd Movie: 1,001 Rabbit Tales '82
Bugs Bunny's Hare-Raising Tales '89

Bugs Bunny's Looney Christmas Tales '79
Bugs Bunny's Wacky Adventures '57
Bugs & Daffy: The Wartime Cartoons '45
Bugs & Daffy's Carnival of the Animals '89
Bugs vs. Elmer '53
Cartoon Classics in Color, No. 2 '40s
Cartoon Classics in Color, No. 3 '43
Cartoon Classics in Color, No. 4 '48
Cartoon Collection, No. 1: Porky in Wackyland '40s
Cartoon Collection, No. 2: Classic Warner Brothers '40s
Cartoon Collection, No. 3: Coal Black & de Sebben Dwarfs '40s
Cartoon Collection, No. 5: Racial Cartoons '40s
Cartoon Moviestars: Bugs! '48
Cartoon Moviestars: Elmer! '48
Cartoon Moviestars: Porky! '47
Christmas Carol '84
Daffy Duck Cartoon Festival: Ain't That Ducky '40s
Daffy Duck Cartoon Festival: Daffy Duck & the Dinosaur '45
Daffy Duck & Company '50
Daffy Duck: The Nuttiness Continues... '56
Daffy Duck's Madcap Mania '89
Elmer Fudd Cartoon Festival '40s
Elmer Fudd's Comedy Capers '55
Foghorn Leghorn's Fractured Funnies '52
Looney Looney Looney Bugs Bunny Movie '81
Looney Tunes & Merrie Melodies 2 '40s
Looney Tunes & Merrie Melodies 3 '40s
Looney Tunes Video Show, Vol. 1 '50s
Looney Tunes Video Show, Vol. 2 '50s
Looney Tunes Video Show, Vol. 3 '50s
Looney Tunes Video Show, Vol. 4 '50s
Looney Tunes Video Show, Vol. 5 '50s
Looney Tunes Video Show, Vol. 6 '50s
Looney Tunes Video Show, Vol. 7 '50s
Pepe Le Pew's Skunk Tales '53
Phantom Tollbooth '69
Porky Pig and Company '48
Porky Pig and Daffy Duck Cartoon Festival '44
Porky Pig's Screwball Comedies '52
The Road Runner vs. Wile E. Coyote: Classic Chase '63
Salute to Chuck Jones '60
Salute to Mel Blanc '58
Starring Bugs Bunny! '48
Sylvester & Tweety's Crazy Capers '61
Tweety & Sylvester '48
A Very Merry Cricket '73

Warner Brothers Cartoon Festival 1 '40s

Warner Brothers Cartoon Festival 2 '40s
Warner Brothers Cartoon Festival 3 '40s
Warner Brothers Cartoon Festival 4 '40s
Warner Brothers Cartoons '50s

Claude Earl Jones

Bride of Re-Animator '89
Evilspeak '82
Impulse '84

Clyde Jones

Motown's Mustang '86

Clyde R. Jones

Crack House '89

Dan Jones

Oklahoma Bound '81

Darby Jones

I Walked with a Zombie '43
Zombies on Broadway '44

David Jones (D)

Betrayal '83
The Christmas Wife '88
84 Charing Cross Road '86
Jacknife '89
Look Back in Anger '89

Davy Jones

Head '68

Dean Jones

Any Wednesday '66
Beethoven '92
Blackbeard's Ghost '67
Fire and Rain '89
Herbie Goes to Monte Carlo '77
High Adventure
The Horse in the Gray Flannel Suit '68
Jailhouse Rock '57
The Long Days of Summer '80
The Love Bug '68
Million Dollar Duck '71
Monkeys, Go Home! '66
Mr. Superinvisible '73
New Star Over Hollywood '70s
Once Upon a Brothers Grimm '77
Other People's Money '91
The Shaggy D.A. '76
Snowball Express '72
Tea and Sympathy '56
That Darn Cat '65
Torpedo Run '58
The Ugly Dachshund '65
When Every Day was the Fourth of July '78

Della Jones

Julius Caesar '84

Desmond Jones

Romance with a Double Bass '74

Dick Jones

Pinocchio '40
Trail of the Hawk '37

Don Jones (D)

The Forest '83
Love Butcher '82

Project: Nightmare

Sweater Girls '78

Donald M. Jones

Fatal Error '83

Donald M. Jones (D)

Deadly Sunday '82
Lethal Pursuit '88
Murderlust '86

Duane Jones

Beat Street '84
Black Vampire '73
Fright House '89
Night of the Living Dead '68
To Die for '89

Earl Jones

The Displaced Person '76
Lying Lips '39

Ed "Too Tall" Jones

The Double McGuffin '79

Eddie Jones

Apprentice to Murder '88
Mr. Charlie '40s

Eugene S. Jones (D)

High Ice '80

F. Richard Jones (D)

Bulldog Drummond '29
Extra Girl '23

Floyd Jones

Chicago Blues '72
Maxwell Street Blues '89

Frankie Jones

Michael Prophet '87

Freddie Jones

Accident '67
All Creatures Great and Small '74
And the Ship Sails On '84
Assault '70
The Bliss of Mrs. Blossom '68
Consuming Passions '88
Dune '84
Elephant Man '80
Erik the Viking '89
Firefox '82
Firestarter '84
Juggernaut '74
Krull '83
Romance with a Double Bass '74
The Satanic Rites of Dracula '73
Twinsanity '70
Young Sherlock Holmes '85
Zulu Dawn '79

Gary Jones (D)

Stay Tuned for Murder '88

Gemma Jones

The Devils '71

George Jones

A Stranger in My Forest '88

Gillian Jones

Echoes of Paradise '86
Shame '87

Jesus Juarez
Santa Sangre '90

Ruben Juarez
Tango Bar '88

Carolyn Judd
Escape from Cell Block Three '78
Women Unchained '70s

Edward Judd
The Day the Earth Caught Fire '62
First Men in the Moon '64
Invasion '65
Island of Terror '66
The Kitchen Toto '87
Murder Motel '74
Whose Child Am I? '75

Robert Judd
Fight for Your Life '79

Charles Judels
Baby Face Morgan '42
Two Reelers: Comedy Classics 8 '33

Arlene Judge
The Crawling Hand '63
Girls in Chains '43
Law of the Jungle '42
Mysterious Mr. Wong '35
Song of Texas '43
Take It Big '44
Wildcat '42

Naomi Judge
Terror Trail '33

Gordon Judges
Busted Up '86

Hernadi Judit
Another Way '82

Arno Juergling
Andy Warhol's Dracula '74

Harald Juhnke
Code Name Alpha '67

Raul Julia
The Addams Family '91
The Alamo: Thirteen Days to Glory '87
Compromising Positions '85
Death Scream '75
Escape Artist '82
Eyes of Laura Mars '78
Florida Straits '87
Frankenstein Unbound '90
Gumball Rally '76
Havana '90
Kiss of the Spider Woman '85
Mack the Knife '89
Moon Over Parador '88
Morning After '86
One from the Heart '82
The Organization '71
Overdrawn at the Memory Bank '83
Panic in Needle Park '75
The Penitent '88
Presumed Innocent '90
Romero '89
The Rookie '90
Strong Medicine '84
Tango Bar '88
Tempest '82
Tequila Sunrise '88
Trading Hearts '87

Janet Julian
Ghostwarrior '86
Humongous '82
King of New York '90
Smokey Bites the Dust '81

Rupert Julian (D)
Phantom of the Opera '25
Walking Back '26
Yankee Clipper '27

Lenny Juliano
Not of This Earth '88

Jorge Juliao
Pixote '81

Max Julien
Getting Straight '70
Mack '73

Montserrat Julio
The Blood Spattered Bride '72

Julissa
Chamber of Fear '68
Dance of Death '69
The Fear Chamber '68

Ami Julius
Malibu Bikini Shop '86

Sandra Jullien
The Sensuous Teenager '70

Gordon Jump
Dirkham Detective Agency '83
Making the Grade '84
Ransom Money '88

Chang Mei Jun (D)
Young Dragons: The Kung Fu Kids '87

Tito Junco
El Hombre Contra el Mundo '87
Invasion of the Vampires '61
A Woman Without Love '51

Victor Junco
Conexion Mexico '87
Gente Violenta '87

Calvin Jung
American Ninja 3: Blood Hunt '89
In the Shadow of Kilimanjaro '86

Sun Jung-Chi
Monkey Kung Fu '80s

Fabio Junior
Bye Bye Brazil '79

Walter Lima Junior (D)
The Dolphin '87

Bernie Junker
Clash of the Ninja '86

Katy Jurado
Arrowhead '53
The Badlanders '58
Barabbas '62
Broken Lance '54
Bullfighter & the Lady '50
Children of Sanchez '79
El Bruto (The Brute) '52
Fearmaker '89

High Noon '52
Once Upon a Scoundrel '73
One-Eyed Jacks '61
Pat Garrett & Billy the Kid '73
Stay Away, Joe '68
Trapeze '56
Under the Volcano '84

Nathan Juran (D)
The Brain from Planet Arous '57
First Men in the Moon '64
Hellcats of the Navy '57
Jack the Giant Killer '62
Land Raiders '69
The Seventh Voyage of Sinbad '58
20 Million Miles to Earth '57

Peter Jurasik
Peter Gunn '89
Problem Child '90

Curt Jurgens
And God Created Woman '57
The Battle of Britain '69
Battle of the Commandos '71
Battle of Neretva '71
Breakthrough '78
Cruise Missile '78
Devil In Silk '56
Dirty Heroes '71
Enemy Below '57
Ferry to Hong Kong '59
French Intrigue
Goldengirl '79
The Inn of the Sixth Happiness '58
The Invincible Six '68
Just a Gigolo '79
The Longest Day '62
Lord Jim '65
The Mephisto Waltz '71
Miracle of the White Stallions '63
The Mozart Story '48
Murder for Sale '70
Nicholas and Alexandra '71
Sleep of Death '79
The Spy Who Loved Me '77
This Happy Feeling '58
The Three Penny Opera '62
Twist '76
Vault of Horror '73

Deana Jurgens
Code Name: Zebra '84
Tin Man '83

Melita Jurisic
The Tale of Ruby Rose '87

Don Jurwich (D)
G.I. Joe: The Movie '86

Rich Jury
Blood on the Mountain '88

Juan Jose Jusid (D)
Made in Argentina '86

Ed Justice
The Vernonia Incident '89

James Robertson Justice
Above Us the Waves '56
Alien Women '69
David and Bathsheba '51
Doctor in Distress '63
Doctor in the House '53
Doctor at Large '57

James Robertson Justice
Doctor at Sea '56

James Robertson Justice
The Lady Says No '51
Land of the Pharaohs '55
Le Repos du Guerrier '62
Murder She Said '62

James Robertson Justice
Rob Roy—The Highland Rogue '53

James Robertson Justice
Spirits of the Dead '68
Sword & the Rose '53
Whiskey Galore '48

James Robertson Justice (D)
Superstition '82

Katherine Justice
Captain America 2: Death Too Soon '79
Columbo: Prescription Murder '67
Five Card Stud '68
Frasier the Sensuous Lion '73
The Stepmother '71

William Justine
Bride & the Beast '58
Trauma '62

Paul Justman (D)
The Doors: Live in Europe, 1968 '68
Gimme an F '85

Claude Jutra
Till Death Do Us Part '72

Claude Jutra (D)
By Design '82
Mon Oncle Antoine '71
Surfacing '84

Kaethe Kaack
Baron Munchausen '43

Jeff Kaake
Border Shootout '90

Suzanne Kaaren
The Devil Bats '41
Phantom Ranger '38

Barna Kabay (D)
The Revolt of Job '84

Roger Kabler
Alligator Eyes '90

George Kaczender
The Agency '81

George Kaczender (D)
The Agency '81
Chanel Solitaire '81
The Girl in Blue '74
In Praise of Older Women '78
Prettykill '87
Your Ticket Is No Longer Valid '84

Jane Kaczmarek
All's Fair '89
D.O.A. '88
Door to Door '84
Falling in Love '84
The Heavenly Kid '85
Vice Versa '88

Jan Kadar (D)
Death Is Called Engelchen '63
Freedom Road '79
Lies My Father Told Me '75
Shop on Main Street '65

Nicholas Kadi
Navy SEALS '90

Ellis Kadisan (D)
The Cat '66
Git! '65

Karen Kadler
The Devil's Messenger '62
It Conquered the World '56

Piotr Kadochnikov
Ivan the Terrible, Part 1 '44
Ivan the Terrible, Part 2 '46

Larisa Kadochnikova
Shadows of Forgotten Ancestors '64

Haruki Kadowawa (D)
Heaven & Earth '90

Helmut Kaeutner (D)
Captain from Koepenick '56

David Kagan
Body Chemistry '90
Friday the 13th, Part 6: Jason Lives '86

Diane Kagan
Barn Burning '80
The Life and Assassination of the Kingfish '76
Mr. & Mrs. Bridge '91

Jeremy Paul Kagan (D)
The Big Fix '78
Big Man on Campus '89
Conspiracy: The Trial of the Chicago Eight '87
Courage '86
Descending Angel '90
Heroes '77
Journey of Natty Gann '85
Katherine '75
The Sting 2 '83

Marilyn Kagan
Foxes '80

Kyoko Kagawa
The Crucified Lovers '54
The Lower Depths '57
Mother '52
Sansho the Bailiff '54

David Kagen
Conspiracy: The Trial of the Chicago Eight '87

Saul Kahan
Banana Monster '72

Takeshi Kato

None But the Brave '65

Rosanne Katon

Bachelor Party '84
Chesty Anderson USN '76
Lunch Wagon '81
The Muthers '76
She Devils in Chains
Zapped! '82

Milton Katselas *(D)*

Butterflies Are Free '72
Forty Carats '73
Report to the Commissioner
'74
Strangers: The Story of a
Mother and Daughter '79

Shintaro Katsu

Zatoichi, the Blind Samurai
Zatoichi vs. Yojimbo '70

Andreas Katsulas

Blame It on the Bellboy '91
Communion '89
Next of Kin '89
Someone to Watch Over Me
'87

Kokinji Katsura

The Human Condition: Road
to Eternity '59

William Katt

Baby...Secret of the Lost
Legend '85
Big Wednesday '78
Butch and Sundance: The
Early Days '79
Carrie '76
The Defense Never Rests '90
First Love '77
House '86
House 4 '91
Last Call '90
Perry Mason Returns '85
Perry Mason: The Case of the
Lost Love '87
Pippin '81
Swimsuit: The Movie '89
Thumbelina '82
Wedding Band '89
White Ghost '88

Allan Katz

Big Man on Campus '89

Gloria Katz *(D)*

Messiah of Evil '74

Omri Katz

Adventures in Dinosaur City
'92

Chuck Katzakian

Omega Cop '90

Lee H. Katzin

The Salzburg Connection '72

Lee H. Katzin *(D)*

Death Ray 2000 '81
Le Mans '71
The Longest Drive '76
The Man from Atlantis '77
The Salzburg Connection '72
Savages '75
Sky Hei$t '75
The Stranger '73
Terror Out of the Sky '78
The Voyage of the Yes '72

**Whatever Happened to Aunt
Alice?** '69
World Gone Wild '88

Yftach Katzur

Atalia '85
Hot Bubblegum '81
Private Popsicle '82
Up Your Anchor '85
Young Love - Lemon Popsicle
7 '87

**Jonathan
Kaufer** *(D)*

Soup for One '82

Cristen Kauffman

Jailbait: Betrayed By
Innocence '86

Daniel Kauffman

Alice Goodbody '74

John Kauffman

Didn't You Hear? '83

Andy Kaufman

Andy Kaufman Special '85
Comedy Tonight '77
Heartbeeps '81
My Breakfast with Blassie '83

**Charles
Kaufman** *(D)*

Jakarta '88
Mother's Day '80
When Nature Calls '85

David Kaufman

The Last Prostitute '91

**George S.
Kaufman** *(D)*

The Senator was Indiscreet
'47

Gunther Kaufman

Kamikaze '89 '83

Jim Kaufman *(D)*

Backstab '90

Lloyd Kaufman *(D)*

Toxic Avenger, Part 2 '89
Toxic Avenger, Part 3: The
Last Temptation of Toxie
'89

Maurice Kaufman

Die! Die! My Darling! '65
Hero '72
Next Victim '74

Philip Kaufman *(D)*

The Great Northfield
Minnesota Raid '72
Henry & June '90
Invasion of the Body
Snatchers '78
Outlaw Josey Wales '76
The Right Stuff '83
The Unbearable Lightness of
Being '88
Wanderers '79
The White Dawn '75

Christine Kaufmann

The Last Days of Pompeii '60
Murders in the Rue Morgue
'71
Taras Bulba '62
Welcome to 18 '87

Joseph Kaufmann

Jud '71
Private Duty Nurses '71

Aki Kaurismaki *(D)*

Ariel '89
Leningrad Cowboys Go
America '89

**Mika
Kaurismaki** *(D)*

Amazon '90

Caroline Kava

Born on the Fourth of July '89
Little Nikita '88

Brian Kavanagh *(D)*

Double Deal '84

Christine Kavanagh

Monkey Boy '90

John Kavanagh

Bellman and True '88
Cal '84

Julie Kavner

Awakenings '90
Bad Medicine '85
Hannah and Her Sisters '86
Katherine '75
New York Stories '89
The Radical '75
Radio Days '87
Revenge of the Stepford
Wives '80
The Simpsons Christmas
Special '89
Surrender '87
This is My Life '91

Anwar Kawadri *(D)*

Claudia '85
Nutcracker Sweet '84
Sex with the Stars '81

Saeda Kawaguchi

Violence at Noon

**Jerzy
Kawalerowicz** *(D)*

Pharaoh '66

**Gonjuro
Kawarazaki**

The Story of the Late
Chrysanthemum '39

Keizo Kawasaki

Warning from Space '56

Hiroyuki Kawase

Godzilla vs. Megalon '76
Godzilla vs. the Smog
Monster '72

Takuzo Kawatani

The Empire of Passion '76

Yusuke Kawazu

Cruel Story of Youth '60
The Human Condition: A
Soldier's Prayer '61

Beatrice Kay

Underworld, U.S.A. '60

Cheryl Kay

Crack House '89

Dianne Kay

Andy and the Airwave
Rangers '89
Portrait of a Showgirl '82

Fiona Kay

Vigil '84

Gilbert Kay *(D)*

White Comanche '67

James H. Kay *(D)*

Seeds of Evil '76

Jody Kay

Death Screams '83
House of Death '82
One Armed Executioner '80

Mary Ellen Kay

Colorado Sundown '52
Thunder in God's Country '51

Ron Kay

The Immortalizer '89

Walter Kay

Speak of the Devil '90

Yuzo Kayama

Red Beard '65
Zero '84

Caren Kaye

Help Wanted: Male '82
Kill Castro '80
The Mercenaries '80
My Tutor '82
Satan's Princess '90
Sweet Dirty Tony '81

Celia Kaye

Island of the Blue Dolphins '64

Claudelle Kaye

Manhattan Melodrama '34

Danny Kaye

Chasing Those Depression
Blues '35
Court Jester '56
The Five Pennies '59
Hans Christian Andersen '52
Hollywood Clowns '85
The Inspector General '49
Kid from Brooklyn '46
Masters of Comedy '30s
Pinocchio '76
Secret Life of Walter Mitty '47
Up in Arms '44
White Christmas '54
Wonder Man '45

Lila Kaye

Nuns on the Run '90

Norman Kaye

Frenchman's Farm '87
Lonely Hearts '82
Man of Flowers '84
Warm Nights on a Slow-
Moving Train '87
Where the Green Ants Dream
'84

Richard Kaye

The Wizard of Speed and
Time '88

Stubby Kaye

The Big Push '75
Cat Ballou '65

Cool it Carol '68
Cool Mikado '63
Guys & Dolls '55
Li'l Abner '59
Minsky's Follies '83
The Way West '67
Who Framed Roger Rabbit?
'88

Robert Kaylor *(D)*

Carny '80
Derby '71
Nobody's Perfect '90

Jan Kayne

Ghost Chasers '51

Allan J. Kayser

Hot Chili '85

Costa Kazakos

Iphigenia '77

Elia Kazan

The Arrangement '69
City for Conquest '40

Elia Kazan *(D)*

The Arrangement '69
Baby Doll '56
East of Eden '55
A Face in the Crowd '57
The Last Tycoon '77
On the Waterfront '54
Panic in the Streets '50
Splendor in the Grass '61
A Streetcar Named Desire '51
A Tree Grows in Brooklyn '45
Viva Zapata! '52

Lainie Kazan

Beaches '88
A Cry for Love '80
Dayton's Devils '68
Delta Force '86
Eternity '90
I Don't Buy Kisses Anymore
'92
Journey of Natty Gann '85
Lady in Cement '68
Love Affair: The Eleanor &
Lou Gehrig Story '77
Lust in the Dust '85
My Favorite Year '82
Obsessive Love '84
One from the Heart '82
Out of the Dark '88
Pinocchio '83
Romance of a Horsethief '71
Sunset Limousine '83
29 Street '91

Maria Kazan

Seance on a Wet Afternoon
'64

Sidney Kazan

Torpedo Attack '72

**Fran Rubel
Kazui** *(D)*

Tokyo Pop '88

Tim Kazurinsky

Best of Eddie Murphy '89
Bill Murray Live from the
Second City '70s
A Billion for Boris '90
Continental Divide '81
Dinner at Eight '89
Hot to Trot! '88
Police Academy 3: Back in
Training '86

Donald Kenney

Clayton County Line

James Kenney

Above Us the Waves '56
Tough Guy '53

Kathy Kenney

Clayton County Line

Sean Kenney

The Corpse Grinders '71
Savage Abduction '73

Grant Kenny

The Gold & Glory '88

June Kenny

Murderer's Wife

Tom Kenny

Shakes the Clown '92

Patsy Kensit

Absolute Beginners '86
Blame It on the Bellboy '91
Blue Tornado '90
Bullseye! '90
Chicago Joe & the Showgirl '90
A Chorus of Disapproval '89
Does This Mean We're Married? '90
The Great Gatsby '74
Lethal Weapon 2 '89
Monty Python and the Holy Grail '75
Silas Marner '85
Timebomb '91
Twenty-One '91

April Kent

The Incredible Shrinking Man '57

Arnold Kent

Hula '28

Barbara Kent

Chinatown After Dark '31
The Dropkick '27
Feet First '30
The Flesh and the Devil '27
Indiscreet '31
Oliver Twist '33

Crauford Kent

Tea for Two '50
Virtue's Revolt '24

Dorothea Kent

Danger Ahead '40

Elizabeth Kent

Mindwarp '91

Jean Kent

Bonjour Tristesse '57
Browning Version '51
The Gay Lady '49
Man of Evil '48
Sleeping Car to Trieste '45
Trottie True '49

Julie Kent

Dancers '87

Kenneth Kent

Dangerous Moonlight '41
House of Mystery '41

Larry Kent

Viva Las Vegas '63

Larry Kent (D)

The Victory '88

Robert Kent

Dimples '36
Gang Bullets '38
The Phantom Creeps '39

Willis Kent (D)

Mad Youth '40

Erle C. Kenton (D)

Escape to Paradise '39
House of Frankenstein '44
Melody for Three '41
Pardon My Sarong '42
Remedy for Riches '40
They Meet Again '41
Who Done It? '42

Linda Kenton

Hot Resort '85

Stan Kenton

Symphony of Swing '86

Doris Kenyon

A Girl's Folly '17
Monsieur Beaucaire '24

Gwen Kenyon

In Old New Mexico '45

Sandy Kenyon

Beyond the Doors '83
The Loch Ness Horror '82

Alexia Keogh

An Angel at My Table '89

Danny Keogh

Kill Slade '89
Playing with Fire '70s

Shell Kepler

Homework '82

Brad Kepnick

Gator Bait 2: Cajun Justice '88

Nicholas Kepros

Identity Crisis '90

Evelyne Ker

A Nos Amours '84

Marian Kerby

Tumbledown Ranch in Arizona '41

Steven Kerby

China O'Brien '88

Ken Kercheval

Calamity Jane '82
California Casanova '91
Judge Horton and the Scottsboro Boys '76

Jean-Pierre Kerien

Muriel '63

Warren Kerigan

The Covered Wagon '23

Jackie Kerin

Next of Kin '82

Max Kerlow

Frida '84

Robert Kerman

Emerald Jungle '80

Dan Kern

Rocket Attack U.S.A. '58

James V. Kern (D)

Second Woman '51
Two Tickets to Broadway '51

Russell Kern (D)

Spittin' Image '83

Diana Kerner

Creature with the Blue Hand '70

Karin Kernke

The Head '59

Sarah Kernochan (D)

Marjoe '72

Joanna Kerns

An American Summer '90
Broken Badge '85
Cross My Heart '88
The Great Los Angeles Earthquake '91
Street Justice '89

Roger Kerns

Delos Adventure '86

Bill Kerr

The Coca-Cola Kid '84
Deadline '81
Dusty '85
Gallipoli '81
The Lighthorseman '87
Miracle Down Under '87
Pirate Movie '82
Razorback '84
Sweet Talker '91
Vigil '84

Bruce Kerr

Man from Snowy River '82

Deborah Kerr

An Affair to Remember '57
America at the Movies '76
The Arrangement '69
Black Narcissus '47
Bonjour Tristesse '57
Casino Royale '67
The Chalk Garden '64
The Courageous Mr. Penn '41
The Day Will Dawn '42
From Here to Eternity '53
The Grass is Greener '61
Hold the Dream '85
The Hucksters '47
I See a Dark Stranger '47
Julius Caesar '53
The King and I '56
King Solomon's Mines '50
Life & Death of Colonel Blimp '43
Love on the Dole '41
Major Barbara '41
The Night of the Iguana '64
Prisoner of Zenda '52
Quo Vadis '51
Separate Tables '58
The Sundowners '60
Tea and Sympathy '56

A Woman of Substance, Episode 1: Nest of Vipers '84
A Woman of Substance, Episode 2: The Secret is Revealed '84
A Woman of Substance, Episode 3: Fighting for the Dream '84

E. Katherine Kerr

Children of a Lesser God '86
Power '86
Reuben, Reuben '83
Suspect '87

Geoffrey Kerr

Just Suppose '26

J. Herbert Kerr, Jr.

A Place Called Today '72

John Kerr

The Cobweb '55
The Pit and the Pendulum '61
Tea and Sympathy '56

Kelle Kerr

The Boost '88

Ken Kerr

Beneath the Valley of the Ultra-Vixens '79
Thundering Ninja '80s

Larry Kerr

The Lost Missile '58

Linda Kerridge

Alien from L.A. '87
Down Twisted '89
Fade to Black '80
Mixed Blood '84
Strangers Kiss '83
Surf 2 '84

J.M. Kerrigan

The Great Man Votes '38
Song O' My Heart '29

John Kerry

Bad Georgia Road '77
Memorial Valley Massacre '88

Norman Kerry

Bachelor Apartment '31
The Hunchback of Notre Dame '23
Phantom of the Opera '25
The Unknown '27

Kathy Kersh

Gemini Affair '80s

Doug Kershaw

Music Shoppe '82

Whitney Kershaw

Two Fathers' Justice '85

Irvin Kershner

The Last Temptation of Christ '88

Irvin Kershner (D)

The Empire Strikes Back '80
Eyes of Laura Mars '78
A Fine Madness '66
Flim-Flam Man '67
Never Say Never Again '83
Raid on Entebbe '77
The Return of a Man Called Horse '76

RoboCop 2 '90
S*P*Y*S '74
Traveling Man '89
Up the Sandbox '72

Corinne Kersten

Murmur of the Heart '71

Brian Kerwin

Getting Wasted '80
Hard Promises '92
Hometown U.S.A. '79
King Kong Lives '86
Miss All-American Beauty '82
Murphy's Romance '85
Nickel Mountain '85
A Real American Hero '78
Torch Song Trilogy '88
Wet Gold '84

Harry Kerwin (D)

Barracuda '78
Cheering Section '73
God's Bloody Acre '75

Lance Kerwin

Children of Divorce '80
Enemy Mine '85
Fourth Wise Man '85
The Loneliest Runner '76
The Mysterious Stranger '82
Reflections of Murder '74
Salem's Lot: The Movie '79
Shooting '82
Side Show '84
Snow Queen '83

Maureen Kerwin

The Destructors '74

William Kerwin

Playgirl Killer '66

D.J. Kerzner

Monster High '89

Alek Keshishian (D)

Truth or Dare '91

Rick Kesler

Beyond Dream's Door '88

Jillian Kesner

Firecracker '71
Roots of Evil '91
Starhops '78
The Student Body '76

Bruce Kessler (D)

Angels from Hell '68
Cruise Into Terror '78
Deathmoon '78
Simon, King of the Witches '71

Quin Kessler

She '83

Wulf Kessler

The White Rose '83

Magnus Kesster

Only One Night '42

Sara Kestelman

Lady Jane '85
Lisztomania '75
Zardoz '73

Bryan Kestner

Fire Birds '90

Kliff Keuhl *(D)*
Murder Rap '87

David Keung
The Bloody Tattoo (The Loot)

Tsui Siu Keung
The Bloody Tattoo (The Loot)
The Rebellious Reign

Michael Keusch *(D)*
Lena's Holiday '91

Lotis Key
Return of the Dragon '84

Evelyn Keyes
Around the World in 80 Days
   '56
Before I Hang '40
Gone with the Wind '39
Here Comes Mr. Jordan '41
The Jolson Story '46
The Lady in Question '40
Making of a Legend: ''Gone
   With the Wind'' '89
Prowler '51
Return to Salem's Lot '87
The Seven Year Itch '55
Slightly Honorable '40
Wicked Stepmother '89

Irwin Keyes
Frankenstein General Hospital
   '88

Mark Keyloun
Gimme an F '85
Mike's Murder '84
Separate Vacations '86

Lulu Keyser-Korff
The Haunted Castle (Schloss
   Vogelod) '21

The Keystone Cops
The Knockout/Dough and
   Dynamite '14

Glenn Kezer
Play Dead '81

Persis Khambatta
Conduct Unbecoming '75
Deadly Intent '88
First Strike '85
Megaforce '82
Nighthawks '81
Phoenix the Warrior '88
Star Trek: The Motion Picture
   '80
Warrior of the Lost World '84
The Wilby Conspiracy '75

Riffat A. Khan *(D)*
Feelin' Screwy '90

Arsinee Khanjian
Speaking Parts '89

Michel Khleifi *(D)*
A Wedding in Galilee '87

Kim Ki-dak *(D)*
Yongary, Monster of the Deep
   '67

Guy Kibbee
Babes in Arms '39
Captain Blood '35
Captain January '36
Dames '34

Dixie Jamboree '44
Earthworm Tractors '36
Footlight Parade '33
42 Street '33
Gold Diggers of 1933 '33
The Horn Blows at Midnight
   '45
Joy of Living '38
Lady for a Day '33
Little Lord Fauntleroy '36
Miss Annie Rooney '42
Mr. Smith Goes to
   Washington '39
Our Town '40
Rain '32
Three Godfathers '48

Milton Kibbee
Jungle Siren '42

Kid 'N' Play
House Party '90

Eddie Kidd
Riding High '78

Jonathan Kidd
The 7th Commandment '61

Michael Kidd
It's Always Fair Weather '55
Smile '75

Margot Kidder
The Amityville Horror '79
Black Christmas '75
The Bounty Man '72
Glitter Dome '84
The Great Waldo Pepper '75
Heartaches '82
Hitchhiker 2 '85
Keeping Track '86
Little Treasure '85
Louisiana '87
Miss Right '81
Mob Story '90
Mr. Mike's Mondo Video '79
92 in the Shade '76
Quackser Fortune Has a
   Cousin in the Bronx '70
Quiet Day In Belfast '74
Reincarnation of Peter Proud
   '75
Shoot the Sun Down '81
Sisters '73
Some Kind of Hero '82
Superman: The Movie '78
Superman 2 '80
Superman 3 '83
Superman 4: The Quest for
   Peace '87
To Catch a Killer '92
Trenchcoat '83
Vanishing Act '88

Nicole Kidman
Billy Bathgate '91
BMX Bandits '83
Days of Thunder '90
Dead Calm '89
Night Master '87
Prince and the Great Race '83
Windrider '86

Beeban Kidron *(D)*
Antonia and Jane '91

Elizabeth Kiefer
Rebel Storm '90

Ray Kieffer
Dance or Die '87

Rose Kiekens
Virgin Among the Living Dead
   '71

Richard Kiel
Cannonball Run II '84
Eegah! '62
Flash & Firecat '75
Force 10 from Navarone '78
The Human Duplicators '64
Hysterical '83
The Longest Yard '74
Mad Mission 3 '84
Moonraker '79
Pale Rider '85
The Phantom Planet '61
Roustabout '64
The Silver Streak '76
So Fine '81
The Spy Who Loved Me '77
They Went That-a-Way &
   That-a-Way '78
Think Big '90
War of the Wizards '81

Sue Kiel
Survivor '87

Wolfgang Kieling
Out of Order '84

Udo Kier
Andy Warhol's Dracula '74
Andy Warhol's Frankenstein
   '74
The House on Straw Hill '76
Mark of the Devil '69
My Own Private Idaho '91
Suspiria '77

Fritz Kiersch *(D)*
Children of the Corn '84
Gor '88
Into the Sun '91
Tuff Turf '85
Under the Boardwalk '89
Winners Take All '87

**Krzysztof
Kieslowski** *(D)*
The Double Life of Veronique
   '91

Robbie Kiger
Children of the Corn '84
Table for Five '83
Welcome Home, Roxy
   Carmichael '90

Susan Kiger
Death Screams '83
H.O.T.S. '79
House of Death '82
Seven '79

Jorgen Kiil
Parallel Corpse '83

Gerard Kikione *(D)*
Buried Alive '89
Dragonard '88
Edge of Sanity '89
Frank and I '83

Nicholas Kilbertus
The Kiss '88

Percy Kilbride
Egg and I '47
Riff-Raff '47
The Southerner '45
State Fair '45
The Woman of the Town '43

Terence Kilburn
A Christmas Carol '38
Fiend Without a Face '58
They Shall Have Music '39

Richard Kiley
The Adventures of
   Huckleberry Finn '85
Angel on My Shoulder '80
The Bad Seed '85
Blackboard Jungle '55
Endless Love '81
Jigsaw '71
Little Prince '74
Looking for Mr. Goodbar '77
Murder Once Removed '71
Pendulum '69
Pickup on South Street '53
Pray TV '82
Separate But Equal '90
The Thorn Birds '83

Dorothy Kilgallen
Stork Club '52

Victor Kilian
The Adventures of Tom
   Sawyer '38
Dr. Cyclops '40
Lady from Nowhere '36
Rimfire '49
Unknown World '51

Paul Killiam
Movie Museum Series '80

Cynthia Killion
The Killing Game '87

Jon Killough *(D)*
Skinned Alive '89

Edward Killy *(D)*
Don Amigo/Stage to Chino
   '40s

Jean-Claude Killy
Club Med '83

Peter Kilman
Olly Olly Oxen Free '78

Val Kilmer
The Doors '91
Gore Vidal's Billy the Kid '89
Kill Me Again '89
The Man Who Broke 1,000
   Chains '87
The Murders in the Rue
   Morgue '86
Real Genius '85
Thunderheart '92
Top Gun '86
Top Secret! '84
Willow '88

Val Kilmer *(D)*
Jazz is Our Religion '72

Lincoln Kilpatrick
Deadly Force '83
Prison '88

Patrick Kilpatrick
The Cellar '90
Class of 1999 '90
Death Warrant '90
Roanoak '86

Rose Kilpatrick
To Be a Rose '74

Shirley Kilpatrick
The Astounding She-Monster
   '58

Bobby Kim
Kill Line '91
Manchurian Avenger '84

Chris Kim
Magnificent Duo '70s

Evan C. Kim
The Dead Pool '88

Joon Kim
Street Soldiers '91

Joy Kim
Blood Alley '55

Kyehee Kim
Four Infernos to Cross '70s

Miki Kim
Primary Target '89

Richard H. Kim *(D)*
Kill Line '91

Valentine Kim
Harley '90

Anne Kimball
Monster from the Ocean Floor
   '54

Sharron Kimberly
The Party '68

Charles Kimbrough
Sunday in the Park with
   George '86

Clinton Kimbrough
Bloody Mama '70
Night Call Nurses '72

**Clinton
Kimbrough** *(D)*
The Young Nurses '73

Bruce Kimmel
The First Nudie Musical '75
Spaceship '81

Bruce Kimmel *(D)*
Spaceship '81

Dana Kimmell
Friday the 13th, Part 3 '82
Sweet 16 '81

**Anthony
Kimmins** *(D)*
Bonnie Prince Charlie '48
Captain's Paradise '53
Mine Own Executioner '47

Kenneth Kimmins
My Best Friend Is a Vampire
   '88

Isao Kimura
Black Lizard '68

Shek Kin
Two Graves to Kung-Fu '82

Spence Kinard
Music & the Spoken Word '74

Return to Mayberry '85
The Shakiest Gun in the West
'68

Kathryn Knotts

Tax Season '90

Bernard Knowles (D)

Frozen Alive '64

Elizabeth Knowles

Wild Riders '71

Patric Knowles

The Adventures of Robin
Hood '38
Auntie Mame '58
Beauty for the Asking '39
Big Steal '49
Charge of the Light Brigade
'36
Chisum '70
Five Came Back '39
Frankenstein Meets the
Wolfman '42
Hit the Ice '43
How Green was My Valley '41
Mutiny '52
Three Came Home '50
Who Done It? '42
The Wolf Man '41

Alexander Knox

Accident '67
Alias John Preston '56
Commandos Strike at Dawn
'42
Cry of the Innocent '80
Europa '51 '52
Holocaust 2000 '83
The Judge Steps Out '49
Khartoum '66
The Longest Day '62
Nicholas and Alexandra '71
Operation Amsterdam '60
Puppet on a Chain '72
Reach for the Sky '57
Shalako '68
Sister Kenny '46
The Sleeping Tiger '54
Tokyo Joe '49
The Vikings '58

Buddy Knox

Sweet Country Road '81

Elyse Knox

Don Winslow of the Coast
Guard '43
Hit the Ice '43
Joe's Bed-Stuy Barbershop:
We Cut Heads '83

Gary Knox

Exquisite Corpses '88

Mickey Knox

Western Pacific Agent '51

Robert Knox

Scream Bloody Murder '72

Terence Knox

City Killer '87
Distortions '87
Hard Rain-The Tet 1968 '89
The Hill '88
Humanoid Defender '85
The Mighty Pawns '87
Murder Ordained '87
Rebel Love '85
Snow Kill '90
Tour of Duty '87

Tripwire '89

Werner Knox (D)

Scalps '83

Peggy Knudsen

Copper Canyon '50
Humoresque '47
A Stolen Life '46

Gustav Knuth

Heidi '65
Tromba, the Tiger Man '52

Kent Ko

Dreadnaught Rivals '80s

Philip Ko

Return of the Scorpion '80s

Philip Ko (D)

Platoon the Warriors '88

Keiju Kobayashi

Godzilla 1985 '85
Tidal Wave '75

Masaki Kobayashi (D)

The Human Condition: A
Soldier's Prayer '61
The Human Condition: No
Greater Love '58
The Human Condition: Road
to Eternity '59
Kwaidan '64

Toshiko Kobayashi

Twenty-Four Eyes '54

Tsuruko Kobayashi

Varan the Unbelievable '61

Yukiko Kobayashi

Destroy All Monsters '68

Jeff Kober

Alien Nation '88
China Beach '88
The First Power '89
Out of Bounds '86

Otar Koberidze

The Tsar's Bride '66

Howard W. Koch (D)

Badge 373 '73
Girl in Black Stockings '57

Marianne Koch

A Fistful of Dollars '64
Frozen Alive '64
The Monster of London City
'64

Pete Koch

Adventures in Dinosaur City
'92

Philip Koch (D)

Pink Nights '87

Edda Kochi

Alice in the Cities '74

Oja Kodar

Someone to Love '87

James Kodl

The Female Jungle '56

Kuninori Kodo

Seven Samurai '54

Lothar Koemer

Student of Prague '13

Tommy Koenig

National Lampoon's Class of
'86 '86

Walter Koenig

Star Trek: The Motion Picture
'80
Star Trek 2: The Wrath of
Khan '82
Star Trek 3: The Search for
Spock '84
Star Trek 4: The Voyage
Home '86
Star Trek 5: The Final Frontier
'89
Star Trek 6: The
Undiscovered Country '91
Star Trek Bloopers '69

Michiyo Kogure

Shin Heike Monogatari '55

Christopher Kohlberg

Ninja Mission '84

Fred Kohler, Jr.

Daniel Boone: Trail Blazer '56
Deluge '33
Fighting Caravans '32
Honor of the Range '34
The Man from Hell '34
Pecos Kid '35
Toll of the Desert '35
Western Mail '42

Lee Kohlmar

Death from a Distance '36

Pancho Kohner (D)

Mr. Sycamore '74

Susan Kohner

Dino '57
The Gene Krupa Story '59
Imitation of Life '59

Mary Kohnert

Beyond the Door 3 '91
Valet Girls '86

Kristina Kohoutova

Alice '88

Hiroshi Koizumi

Attack of the Mushroom
People '63
Godzilla vs. Mothra '64

Marja Kok (D)

In for Treatment '82

Lily Kokodi

The Enchantress '88

Henry Kolker

Baby Face '33
Charlie Chan in Paris '35
The Ghost Walks '34
The Mystery Man '35

Amos Kollek

Goodbye, New York '85

Amos Kollek (D)

Forever, Lulu '87

Goodbye, New York '85
High Stakes '89

Xavier Koller (D)

Journey of Hope '90

Nicole Kolman

Body Moves '90

Ron Kologie

Iced '88

Monika Kolpek

Sexy Cat '60s

James Komack (D)

Porky's Revenge '85

Tetsu Komai

Hong Kong Nights '35

Sergei Komarov (D)

A Kiss for Mary Pickford '27

Hosei Komatsu

Double Suicide '69
A Taxing Woman Returns '88

Kazuo Komatsubara (D)

Warriors of the Wind '85

Rich Komenich

Two Wrongs Make a Right '89

Maja Komorowska

Year of the Quiet Sun '84

Andrei Konchalovsky (D)

Duet for One '86
Homer and Eddie '89
The Inner Circle '91
Maria's Lovers '84
Runaway Train '85
Shy People '87
Tango and Cash '89

Gabor Koncz

Hungarian Rhapsody '78

Massaomi Konda

The Pornographers '66

Jackie Kong (D)

The Being '83
Night Patrol '85
The Underachievers '88

Johnny Kong Kong

Steel Fisted Dragon '82

Joseph Kong (D)

Clones of Bruce Lee '80
Thundering Ninja '80s

Kam Kong

Kung Fu Strongman
Shaolin Warrior

Queen Kong

Deathstalker 2: Duel of the
Titans '87
Slashdance '89

Venice Kong

Playboy Playmates at Play '90

Young Kong

Counter Attack '84

Magda Konopka

Satanik '69
When Dinosaurs Ruled the
Earth '70

Phyllis Konstam

Murder '30

Madame Konstantin

Notorious '46

Guich Koock

American Ninja '85
Picasso Trigger '89
Seven '79

Frances Koon

Thunder Road '58

Simonie Kopapik

The White Dawn '75

Milos Kopecky

The Original Fabulous
Adventures of Baron
Munchausen '61

Bernie Kopell

Combat Academy '86
Joseph & His Brothers '79
The Loved One '65

Karen Kopins

Creator '85
Jake Speed '86
Once Bitten '85

Sharry Koponski

Playboy's 1989 Playmate
Video Calendar '89

Maria Koppenhofer

Tiefland '44

Barbara Kopple (D)

Harlan County, U.S.A. '76
Keeping On '81

Keystone Kops

Golden Age of Comedy '58

Michael Korb

2000 Maniacs '64

Alexander Korda (D)

Marius '31
Private Life of Don Juan '34
The Private Life of Henry VIII
'33
Rembrandt '36
That Hamilton Woman '41
Wedding Rehearsal '32

Zoltan Korda (D)

Cash '34
Cry, the Beloved Country '51
Drums '38
Elephant Boy '37
The Four Feathers '39
Jungle Book '42
Sahara '43
Sanders of the River '35

Arnold Korff

The Haunted Castle (Schloss
Vogelod) '21

K. Korieniev

The Amphibian Man '61

Barbara Laage
The Respectful Prostitute '52
Therese & Isabelle '68

Andre S. Labarthe
My Life to Live '62

Patti LaBelle
Fire and Rain '89
Sing '89
A Soldier's Story '84
Unnatural Causes '86

Cateria Sylos Labini
The Icicle Thief '89

Matthew Laborteaux
Aliens Are Coming '80
Amazing Stories, Book 5 '85
Deadly Friend '86
Shattered Spirits '91
A Woman Under the Influence '74

Patrick Laborteaux
Heathers '89
Prince of Bel Air '87

Elina Labourdette
Lola '61

Dominique Labourier
Jonah Who Will Be 25 in the Year 2000 '76

Jeanne Labrune (D)
Sand and Blood '87

Adele Lacey
When a Man Rides Alone '33

Catherine Lacey
Cottage to Let '41
Pink Strings and Sealing Wax '45

Leslee Lacey
HauntedWeen '91

Margaret Lacey
Diamonds are Forever '71
Seance on a Wet Afternoon '64

Ronald Lacey
Disciple of Death '72
Firefox '82
Into the Darkness '86
Jailbird Rock '88
The Lone Runner '88
Making the Grade '84
Next Victim '74
Nijinsky '80

Julie LaChapelle
In Trouble '67

Harry Lachman (D)
Baby, Take a Bow '34
Castle in the Desert '42
Charlie Chan in Rio '41
The Loves of Edgar Allen Poe '42
Murder Over New York '40
Our Relations '36

Stephen Lack
Fatal Attraction '80
Perfect Strangers '84
Scanners '81

Wilton Lackaye
Trilby '17

Mike Lackey
Street Trash '87

Skip Lackey
Once Bitten '85

Frank Lackteen
Red Barry '38

Andre Lacombe
The Bear '89

Georges Lacombe (D)
Seven Deadly Sins '53

Francis Lacombrade
This Special Friendship '67

Cathy Lacommaro
Splatter University '84

Philippe Lacoste
Hail Mary '85

Denis Lacroix
Running Brave '83

Beatriz Lacy
Graveyard of Horror '71

Jerry Lacy
Play It Again, Sam '72

Joe Lacy (D)
A Witch Without a Broom '68

John Ladalski
Chinese Stuntman '70s
Counter Attack '84

Andrea Ladanyi
Gate 2 '92

Alan Ladd
All the Young Men '60
The Badlanders '58
Botany Bay '53
Branded '50
Captain Caution '40
The Carpetbaggers '64
Citizen Kane '41
Deep Six '58
Drum Beat '54
Duel of Champions '61
Gangs, Inc. '41
The Glass Key '42
Hell on Frisco Bay '55
The Howards of Virginia '40
Joan of Paris '41
The McConnell Story '55
My Favorite Brunette '47
Proud Rebel '58
Shane '53
This Gun for Hire '42

Cheryl Ladd
The Hasty Heart '86
Lisa '90
Millenium '89
Now and Forever '82
Purple Hearts '84
Romance on the Orient Express '89
Satan's School for Girls '73
The Treasure of Jamaica Reef '74

David Ladd
Dog of Flanders '59
Misty '61
The Treasure of Jamaica Reef '74

Diane Ladd
Alice Doesn't Live Here Anymore '74
All Night Long '81
Chinatown '74
Desperate Lives '82
Embryo '76
I Married a Centerfold '84
A Kiss Before Dying '91
National Lampoon's Christmas Vacation '89
Plain Clothes '88
Rebel Rousers '69
Something Wicked This Way Comes '83
The Wild Angels '66
Willa '79

Nicole Ladmiral
Diary of a Country Priest '50

Aldo Lado (D)
Paralyzed '70s

Gabriele Lafari
A Woman in Flames '84

John Lafayette
Full Fathom Five '90
White Sands '92

Patricia Laffan
Devil Girl from Mars '54
Quo Vadis '51

John Lafia (D)
Blue Iguana '88
Child's Play 2 '90

Art LaFleur
Trancers '85
Zone Troopers '85

Jean LaFleur (D)
Ilsa, the Tigress of Siberia '79
Mystery of the Million Dollar Hockey Puck '80s

C.D. LaFleure
Alice Goodbody '74

Bernadette LaFont
Catch Me a Spy '71
Le Beau Serge '58
Too Pretty to Be Honest '72
Trop Jolie pour Etre Honette '60s
Violette '78

Jean-Philippe LaFont
Babette's Feast '87

Frederic Lagache
Emmanuelle, the Joys of a Woman '76

Sarah Lagenfeld
The Act '82

Sture Lagerwall
Walpurgis Night '41

Ron Lagomarsino (D)
Dinner at Eight '89

Valerie Lagrange
A Man and a Woman '66
Morgan the Pirate '60
Weekend '67

Craig Lahiff (D)
Deadly Possession '88
Fever '88

Bert Lahr
Just Around the Corner '38
The Night They Raided Minsky's '69
Rose Marie '54
Sing Your Worries Away '42
Two Reelers: Comedy Classics 4 '36
The Wizard of Oz '39

Christine Lahti
And Justice for All '79
Crazy from the Heart '91
The Doctor '91
The Executioner's Song '82
Funny About Love '90
Gross Anatomy '89
The Henderson Monster '80
Housekeeping '87
Just Between Friends '86
Leaving Normal '92
Miss Firecracker '89
Running on Empty '88
Single Bars, Single Women '84
Stacking '87
Swing Shift '84
Whose Life Is It Anyway? '81

Gary Lahti
Knightriders '81

Bruce Lai
Clones of Bruce Lee '80
The Magnificent

Christina Lai
Savage Island '85

Joseph Lai (D)
Cobra Against Ninja
Ninja Commandments '87
Ninja Operation: Licensed to Terminate '87
Ninja Powerforce '90
Ninja Showdown '90
Ninja Strike Force '88
The Power of the Ninjitsu '88

Lo Lai
Young Dragon '77

Harvey Laidman (D)
The Boy Who Loved Trolls '84
Steel Cowboy '78

Leah Lail
Body Waves '91

Frankie Laine
Meet Me in Las Vegas '56

Jeff Laine
Cheering Section '73

Jennifer Laine
Caged Fury '80

Jimmy Laine
Driller Killer '74
Ms. 45 '81

Ray Laine
Season of the Witch '73

John Laing (D)
Beyond Reasonable Doubt '80
Dangerous Orphans '80s
The Lost Tribe '89

R.D. Laing
Eros, Love & Lies '90

Marvin Laird
Beyond Reason '77

Oze Lajos
Oh, Bloody Life! '88

Alan Lake
Don't Open Till Christmas '84

Arthur Lake
Exiled to Shanghai '37
Silent Laugh Makers, No. 1 '20s
The Silver Streak '34
Skinner's Dress Suit '26
23 1/2 Hours Leave '37

Florence Lake
Edgar Kennedy Slow Burn Festival '38
Going to Town '44
Next Time I Marry '38
Quality Street '37
Savage Intruder '68

Michael Lake
Sweetie '89

Ricki Lake
Cookie '89
Cry-Baby '90
Hairspray '88
Last Exit to Brooklyn '90
Working Girl '88

Veronica Lake
Flesh Feast '69
The Glass Key '42
I Married a Witch '42
Ramrod '47
Sullivan's Travels '41
This Gun for Hire '42

Elaine Lakeman
Death Target '83

Joe Lala
The Take '90

Michael Lally
The Nesting '80

Frank Laloggia
The Wizard of Speed and Time '88

Frank Laloggia (D)
Fear No Evil '80
The Lady in White '88

Robert Lalonde
Men of Steel '88

Rene Laloux (D)
Fantastic Planet '73

Barry Lam
Rivals of the Silver Fox '80s

Debbie Lam
Snake Fist of the Buddhist Dragon '70s

Paula Lane

Goodnight, Sweet Marilyn '89

Priscilla Lane

Arsenic and Old Lace '44
Four Daughters '38
The Roaring Twenties '39
Saboteur '42
Silver Queen '42

Richard Lane

Bullfighters '45
Day the Bookies Wept '39
It Could Happen to You '39
The Jackie Robinson Story '50
Mr. Winkle Goes to War '44
Ride 'Em Cowboy '42
Sunny '41
Take Me Out to the Ball Game '49

Rocky Lane (D)

All's Fair '89

Rosemary Lane

Four Daughters '38
Harvest Melody '43
Oklahoma Kid '39

Tim Lane

Iron Warrior '87

Tracy Lane

Robinson Crusoe of Clipper Island '36

Yancey Lane

Trail of the Hawk '37

Eric Laneuville (D)

The Mighty Pawns '87

Sidney Lanfield (D)

The Hound of the Baskervilles '39
The Lemon Drop Kid '51
Skirts Ahoy! '52
Sorrowful Jones '49
Station West '48
You'll Never Get Rich '41

Laura Lanfranchi

Don't Mess with My Sister! '85

Ben Lang

The Plot Against Harry '69

Carl Lang

Creature with the Blue Hand '70

Chang Wu Lang

Duel of the Brave Ones '80
Image of Bruce Lee '70s
Shadow Killers '80s

Doreen Lang

Almost an Angel '90

Fritz Lang

Contempt '64

Fritz Lang (D)

Beyond a Reasonable Doubt '56
Big Heat '53
Clash by Night '52
Cloak and Dagger '46
Das Testament des Dr. Mabuse '33

Dr. Mabuse the Gambler '22
Human Desire '54
Journey to the Lost City '58
Kriemhilde's Revenge '24
Liliom '35
M '31
Moonfleet '55
Rancho Notorious '52
Return of Frank James '40
Scarlet Street '45
Secret Beyond the Door '48
Siegfried '24
Spiders '18
Spies '28
The Thousand Eyes of Dr. Mabuse '60
Western Union '41
While the City Sleeps '56
Woman in the Moon '29
You Only Live Once '37

Helen Lang

Revenge of the Cheerleaders '76

Jeanne Lang

Mother & Daughter: A Loving War '80

Judith Lang

Count Yorga, Vampire '70

June Lang

Bonnie Scotland '35
Captain January '36
Wee Willie Winkie '37

k.d. lang

Pee-Wee's Playhouse Christmas Special '88

Malisa Lang

Thor el Conquistador '86

Michael Lang (D)

The Gift '82
Holiday Hotel '77

Perry Lang

Eight Men Out '88
Girls Next Door '79
Great Ride '78
Greedy Terror '78
Jocks '87
Mortuary Academy '91
O'Hara's Wife '82
A Rumor of War '80
Sahara '83
Spring Break '83
Tag: The Assassination Game '82
Teen Lust '78

Perry Lang (D)

Little Vegas '90

Richard Lang (D)

A Change of Seasons '80
Christmas Comes to Willow Creek '87
Don't Go to Sleep '82
Fantasy Island '76
Force of Evil '77
The Hunted Lady '77
Mountain Men '80
Night Cries '78
Shooting Stars '85
Strike Force '81
Vegas '78

Rocky Lang (D)

Race for Glory '89

Stephen Lang

Band of the Hand '86
Crime Story '86
Crime Story: The Complete Saga '86
Death of a Salesman '86
Finish Line '89
The Hard Way '91
Last Exit to Brooklyn '90
Manhunter '86

Veronica Lang

The Clinic '83
Don's Party '76

Walter Lang (D)

The Blue Bird '40
But Not for Me '59
Can-Can '60
Desk Set '57
The Jackpot '50
The King and I '56
The Little Princess '39
The Magnificent Dope '42
Moon Over Miami '41
Red Kimono '25
Snow White and the Three Stooges '61
Song of the Islands '42
State Fair '45
There's No Business Like Show Business '54

Declan Langan (D)

Erotic Images '85

Glenn Langan

Big Chase '54
Hangman's Knot '52
Treasure of Monte Cristo '50

Harry Langdon

All Night Long/Smile Please '24
Baby Face Harry Langdon '26
Comedy Festival #3 '30s
Ella Cinders '26
Golden Age of Comedy '58
His First Flame '26
Mad About Money '30
Silent Laugh Makers, No. 3 '20s
Story of William S. Hart/Sad Clowns '20s
Strong Man '26
When Comedy was King '59

Lillian Langdon

The Americano '17
Flirting with Fate '16
The Lamb '15

Sue Ann Langdon

The Cheyenne Social Club '70
The Evictors '79
Frankie and Johnny '65
Frankie and Johnny '65
Roustabout '64
Zapped! '82

Terrence Langdon

Another Time, Another Place '58

Barbara Lange

Paris Blues '61

Carl Lange

Torture Chamber of Dr. Sadism '69

Claudia Lange

Invincible Gladiators '64

Hope Lange

Beulah Land '80
Blue Velvet '86
Bus Stop '56
Death Wish '74
Fer-De-Lance '74
Ford: The Man & the Machine '87
I Am the Cheese '83
I Love You...Goodbye '73
The Love Bug '68
A Nightmare on Elm Street 2: Freddy's Revenge '85
Peyton Place '57
Pleasure Palace '80
Pocketful of Miracles '61
The Prodigal '83
Tune in Tomorrow '90
Wild in the Country '61
The Young Lions '58

Jessica Lange

All That Jazz '79
Cape Fear '91
Cat on a Hot Tin Roof '84
Country '84
Crimes of the Heart '86
Everybody's All American '88
Far North '88
Frances '82
How to Beat the High Cost of Living '80
King Kong '76
Men Don't Leave '89
Music Box '89
O Pioneers! '91
The Postman Always Rings Twice '81
Sweet Dreams '85
Tootsie '82

Ted Lange

Glitch! '88
Terminal Exposure '89

Viktor Lange

Secrets in the Dark '83

Jack Langedijk

Blind Fear '89
Evil Judgement '85

Frank Langella

And God Created Woman '88
The Deadly Trap '71
Diary of a Mad Housewife '70
Dracula '79
Masters of the Universe '87
The Men's Club '86
Sphinx '81
Those Lips, Those Eyes '80
True Identity '91
Twelve Chairs '70

Sarah Langenfeld

Bloodlink '86
Gold Raiders '84

Heather Langenkamp

Nickel Mountain '85
A Nightmare on Elm Street '84
A Nightmare on Elm Street 3: Dream Warriors '87

Bonnie Langford

Wombling Free '77

Frances Langford

Born to Dance '36
Deputy Marshall '50
Dixie Jamboree '44
Dreaming Out Loud '40

Girl Rush '44
Too Many Girls '40
Yankee Doodle Dandy '42

Amanda Langlet

Pauline at the Beach '83

Noel Langley (D)

The Pickwick Papers '54
The Search for Bridey Murphy '56

Lisa Langlois

Blood Relatives '78
Deadly Eyes '82
Hang Tough '89
Joy of Sex '84
The Man Who Wasn't There '83
Mindfield '89
The Nest '88
Truth or Die '86

Chris Langman (D)

Moving Targets '87

Margaret Langrick

American Boyfriends '89
Cold Comfort '90
My American Cousin '85
Thunderground '89

Caroline Langrishe

Eagle's Wing '79

Murray Langston

Digging Up Business '91
Night Patrol '85
Two Top Bananas '87
Uncensored '85
The Unknown Comedy Show '87
Up Your Alley '89

David Langton

Quintet '79

Jeff Langton

Final Impact '91

Robert Langton-Lloyd

The Mahabharata '89

Paul Langton

The Cosmic Man '59
The Incredible Shrinking Man '57
The Snow Creature '54
Thirty Seconds Over Tokyo '44

Simon Langton (D)

Act of Passion: The Lost Honor of Kathryn Beck '83
Anna Karenina '85
Laguna Heat '87
The Whistle Blower '87

Robert Langyel

Attack of the Beast Creatures '85

Pua Lani

Hawaii Calls '38

Jaron Lanier

Cyberpunk '91

Susan Lanier

The Hills Have Eyes '77

Marie Laurin

Talking Walls '85

George Lauris (D)

Buffalo Rider '78

Pricilla Lauris

Buffalo Rider '78

Michelle Laurita

Vals '85

Ed Lauter

The Amateur '82
Big Score '83
Breakheart Pass '76
Cartier Affair '84
The Chicken Chronicles '77
Class of '63 '73
Cujo '83
Death Hunt '81
Death Wish 3 '85
Eureka! '81
Family Plot '76
Finders Keepers '84
Girls Just Want to Have Fun '85
In the Custody of Strangers '82
King Kong '76
Lassiter '84
The Last Days of Patton '86
The Longest Yard '74
Love's Savage Fury '79
Magic '78
Nickel Mountain '85
Noah: The Deluge '79
Rage '72
Revenge of the Nerds 2: Nerds in Paradise '87
Stephen King's Golden Years '91
Timerider '83
Youngblood '86
Yuri Nosenko, KGB '86

Heiner Lauterbach

Men... '85

Fernand Lauterier (D)

Nais '45

Georges Lautner (D)

Icy Breasts '75
La Cage aux Folles 3: The Wedding '86
My Other Husband '85
Road to Salina '68

Kathrin Lautner

Final Impact '91
The Last Riders '90
Night of the Wilding '90
Spirits '90

Linda Lautrec (D)

My Breakfast with Blassie '83

Gerard Lauzier (D)

Petit Con '84

Jean-Claude Lauzon (D)

Night Zoo '87

Arnold Laven (D)

The Monster that Challenged the World '57
Without Warning '52

June Laverick

Mania '60

Lucille LaVerne

Kentucky Kernels '34
The White Rose '23

Jean Laverty

Campus Knights '29

Maureen Lavette

Hardcase and Fist '89

Daliah Lavi

Lord Jim '65
The Return of Dr. Mabuse '61
Spy With a Cold Nose '66
Those Fantastic Flying Fools '67
Two Weeks in Another Town '62

Gabriele Lavia

Revenge of the Dead '84

Efrat Lavie

My Michael '75

Linda Lavin

The $5.20 an Hour Dream '80
Maricela '88
A Matter of Life and Death '81
Muppets Take Manhattan '84
See You in the Morning '89

Martin Lavut

The Mask '61

Martin Lavut (D)

Smokescreen '90

Albert Law (D)

Cold Blooded Murder

Clara Law (D)

The Reincarnation of Golden Lotus '89

George Law (D)

Torpedo Attack '72

John Phillip Law

African Rage '70s
Alienator '89
American Commandos '84
Attack Force Z '84
Barbarella '68
The Cassandra Crossing '76
Cold Heat '90
Danger: Diabolik '68
Fatal Assassin '78
Golden Voyage of Sinbad '73
L.A. Bad '85
The Last Movie '71
The Love Machine '71
Moon in Scorpio '86
Night Train to Terror '84
No Time to Die '78
Overthrow '82
The Russians are Coming, the Russians are Coming '66
Space Mutiny '88
Spiral Staircase '75
Tarzan, the Ape Man '81
Thunder Warrior 3 '88

Tom Law

Shallow Grave '87

Tom Law (D)

Tax Season '90

Peter Lawford

Advise and Consent '62
Angel's Brigade '79
April Fools '69
Body & Soul '81
The Canterville Ghost '44
Easter Parade '48
Exodus '60
Fantasy Island '76
Good News '47
Harlow '65
It Happened in Brooklyn '47
It Should Happen to You '53
Little Women '49
The Longest Day '62
A Man Called Adam '66
Mrs. Miniver '42
Mysterious Island of Beautiful Women '79
Never So Few '59
Ocean's 11 '60
On an Island with You '48
Oscar '66
Picture of Dorian Gray '45
Royal Wedding '51
Togetherness '70
War in the Sky '82

James Lawless

Satan's Touch '84

Yvonne Lawley

Among the Cinders '83

Stephanie Lawlor

Cherry Hill High '76
Hot T-Shirts '79

Adam Lawrence

Drive-In Massacre '74

Barbara Lawrence

Kronos '57
Oklahoma! '55
The Street With No Name '48

Bruno Lawrence

The Bridge to Nowhere '86
Grievous Bodily Harm '89
Heart of the Stag '84
An Indecent Obsession '85
The Quiet Earth '85
Race to the Yankee Zephyr '81
Rikky and Pete '88
Smash Palace '82
Utu '83

Bryan Lawrence

She Shall Have Music '36

Carol Lawrence

Just Like Us '83
New Faces '54
New Star Over Hollywood '70s

Daphne Lawrence

Cycle Vixens '79

Delphi Lawrence

Frozen Alive '64
Three Stops to Murder '53

Denny Lawrence (D)

Archer's Adventure '85

Elizabeth Lawrence

Sleeping with the Enemy '91

Gail Lawrence

Maniac '80

Gertrude Lawrence

Men are not Gods '37
Mimi '35
Rembrandt '36
Stage Door Canteen '43

Ingrid Lawrence

Thunder Warrior 3 '88

Jim Lawrence

High Country '81

Joey Lawrence

Pulse '88
Wait Till Your Mother Gets Home '83

John Lawrence

The Asphyx '72
They Live '88

John Lawrence (D)

Savage Abduction '73

Kiva Lawrence

To the Shores of Hell '65

Linda Lawrence

Gas Pump Girls '79

Marc Lawrence

Asphalt Jungle '50
The Black Hand '50
Cataclysm '81
Charlie Chan at the Wax Museum '40
Cloak and Dagger '46
Dream No Evil '75
Frasier the Sensuous Lion '73
Goin' Coconuts '78
I Am the Law '38
Jigsaw '49
King of Kong Island '78
The Man with the Golden Gun '74
'Neath Brooklyn Bridge '42
Night Train to Terror '84
Pigs '73
Vengeance '37

Marc Lawrence (D)

Pigs '73

Martin Lawrence

House Party 2: The Pajama Jam '91
Talkin' Dirty After Dark '91

Mittie Lawrence

Night Call Nurses '72

Muriel Lawrence

I Dream of Jeannie '52

Peter Lee Lawrence

Garringo '65
Manos Torpes '65
Special Forces '80s
They Paid with Bullets: Chicago 1929 '70s

Quentin Lawrence (D)

The Crawling Eye '58

Ray Lawrence (D)

Bliss '85

Robert Lawrence

Ten Nights in a Bar Room '13

Ronald Lawrence

Eye of the Eagle 2 '89

Rosina Lawrence

Charlie Chan's Secret '35
General Spanky '36
Movie Struck '37
Way Out West '37

Scott Lawrence

God's Bloody Acre '75

Sheldon Lawrence

Sweet Beat '62

Sid Lawrence

Naked Warriors '73

Stephanie Lawrence

Buster '88
Phantom of the Opera '89

Steve Lawrence

Express to Terror '79
The Lonely Guy '84

Toni Lawrence

Pigs '73

William E. Lawrence

Flirting with Fate '16

Dean Lawrie

Bad Taste '88

Denis Lawson

Local Hero '83
Return of the Jedi '83

Leigh Lawson

Brother Sun, Sister Moon '73
Charlie Boy '81
Fire and Sword '82
It's Not the Size That Counts '74
Love Among the Ruins '75
Madame Sousatzka '88
Madhouse Mansion '74
O Pioneers! '91
Sword of the Valiant '83
Tess '80

Louise Lawson

Creeping Terror '64

Priscilla Lawson

Flash Gordon: Rocketship '36
Girl of the Golden West '38

Richard Lawson

Fist '79
The Forgotten '89
Foxstyle '73
Johnnie Gibson F.B.I. '87
Poltergeist '82
Scream Blacula Scream '73
Streets of Fire '84

Sarah Lawson

Island of the Burning Doomed '67

Wilfred Lawson

Man of Evil '48
The Night Has Eyes '42
The Prisoner '55
Pygmalion '38
Tower of Terror '42

Meme Lei

The Element of Crime '85

Fritz Leiber

Equinox '71
The Story of Louis Pasteur '36

Paul Leiber

Bill: On His Own '83

Ron Leibman

The Art of Crime '75
Door to Door '84
The Hot Rock '70
Norma Rae '79
Phar Lap '84
A Question of Guilt '78
Rhinestone '84
Romantic Comedy '83
Seven Hours to Judgment '88
Slaughterhouse Five '72
Up the Academy '80
Where's Poppa? '70
Zorro, the Gay Blade '81

Arnold Leibovit (D)

The Puppetoon Movie '87

Carol Leifer

Rodney Dangerfield: Nothin'
  Goes Right '88

Neil Leifer (D)

Trading Hearts '87
Yesterday's Hero '79

Don Leifert

The Alien Factor '78

Barbara Leigh

Boss '74
Junior Bonner '72
Mistress of the Apes '79
Seven '79
Smile, Jenny, You're Dead '74
The Student Nurses '70

Barbara Leigh-Hunt

Frenzy '72
Paper Mask '91
The Plague Dogs '82

Ronald Leigh-Hunt

Le Mans '71

Janet Leigh

Bye, Bye, Birdie '63
The Fog '78
Harper '66
Holiday Affair '49
Houdini '53
Jet Pilot '57
Little Women '49
The Manchurian Candidate '62
The Naked Spur '52
Perfect Furlough '59
Pete Kelly's Blues '55
Prince Valiant '54
Psycho '60
Scaramouche '52
That Forsyte Woman '49
Touch of Evil '58
Two Tickets to Broadway '51
The Vikings '58
Words & Music '48

Jennifer Jason Leigh

Angel City '80
Backdraft '91

The Best Little Girl in the World '81
The Big Picture '89
Buried Alive '90
Crooked Hearts '91
Easy Money '83
Eyes of a Stranger '81
Fast Times at Ridgemont High '82
Flesh and Blood '85
Girls of the White Orchid '85
Grandview U.S.A. '84
Heart of Midnight '89
Hitcher '85
Just Like Us '83
Killing of Randy Webster '81
Last Exit to Brooklyn '90
The Men's Club '86
Miami Blues '90
Rush '91
Sister, Sister '87
Undercover '87
Wrong is Right '82

Mike Leigh (D)

High Hopes '89
Life is Sweet '91

Nelson Leigh

Ma Barker's Killer Brood '60
Texas Masquerade '44

Spencer Leigh

The Last of England '87
Smart Money '88

Steven Leigh

Deadly Bet '91

Suzanna Leigh

The Fiend '71
Lust for a Vampire '71
Paradise, Hawaiian Style '66

Vivien Leigh

Anna Karenina '48
Caesar and Cleopatra '45
Dark Journey '37
Fire Over England '37
Gone with the Wind '39
Hollywood Outtakes & Rare Footage '83
Making of a Legend: "Gone With the Wind" '89
1939, The Movies' Vintage Year: Trailers on Tape '39
Roman Spring of Mrs. Stone '61
Ship of Fools '65
Sidewalks of London '40
Storm in a Teacup '37
A Streetcar Named Desire '51
That Hamilton Woman '41
Waterloo Bridge '40

Lillian Leighton

The Jack Knife Man '20
Peck's Bad Boy '21

Margaret Leighton

The Best Man '64
Bonnie Prince Charlie '48
Carrington, V.C. '55
Choice of Weapons '76
The Elusive Pimpernel '50
From Beyond the Grave '73
The Go-Between '71
The Loved One '65
Under Capricorn '49
Waltz of the Toreadors '62
The Winslow Boy '48
X, Y & Zee '72

Roberta Leighton

Barracuda '78
Covergirl '83

Chang Leih

Bruce's Deadly Fingers '80s

Leila

From Russia with Love '63

Jefferson Leinberger

Maxim Xul '91

Tom Leindecker

Cheering Section '73

Harald Leipnitz

Fight for Gold

Mitchell Leisen (D)

The Girl Most Likely '57
Murder at the Vanities '34
Swing High, Swing Low '37

John Leisenring

Mondo Trasho '70

Erwin Leiser (D)

Mein Kampf '60

Frederick Leister

Green Grow the Rushes '51
Spellbound '41

David Leisure

You Can't Hurry Love '88

Christopher Leitch (D)

Courage Mountain '89
The Hitter '79
Teen Wolf Too '87

Donovan Leitch

And God Created Woman '88
The Blob '88
Cutting Class '89
Glory '89
The In Crowd '88

Virginia Leith

The Brain That Wouldn't Die '63

David Leivick (D)

Gospel '82

Beau Leland

Nail Gun Massacre '86

Chloe Leland

Wish You Were Here '87

David Leland (D)

The Big Man: Crossing the Line '91
Checking Out '89
Wish You Were Here '87

Claude Lelouch (D)

And Now My Love '74
Another Man, Another Chance '77
Bolero '82
Cat and Mouse '78
Edith & Marcel '83
Happy New Year '74
A Man and a Woman '66
A Man and a Woman: 20 Years Later '86

Robert et Robert '79

Lucienne Lemarchand

Crime and Punishment '35

Paul LeMat

Aloha, Bobby and Rose '74
American Graffiti '73
Burning Bed '85
Citizens Band '77
Death Valley '81
Easy Wheels '89
Firehouse '72
Grave Secrets '89
Hanoi Hilton '87
Into the Homeland '87
Jimmy the Kid '82
Long Time Gone '86
Melvin and Howard '80
The Night They Saved Christmas '87
On Wings of Eagles '86
P.K. and the Kid '85
Private Investigations '87
Puppet Master '89
Rock & Rule '83
Strange Invaders '83

Jeannine Lemay

Don't Mess with My Sister! '85

Harvey Lembeck

Beach Blanket Bingo '65
Beach Party '63
Between Heaven and Hell '56
Bikini Beach '64
How to Stuff a Wild Bikini '65
Love with the Proper Stranger '63
Pajama Party '64
Stalag 17 '53

Michael Lembeck

The Boys in Company C '77
Conspiracy: The Trial of the Chicago Eight '87
Gorp '80
On the Right Track '81

John LeMesurier

A Married Man '84

Marie-Adele Lemieux

Enemies, a Love Story '89

James Lemmo (D)

Heart '87
Tripwire '89

Chris Lemmon

C.O.D. '83
Dad '89
Firehead '90
Going Undercover '88
Happy Hooker Goes Hollywood '80
Just Before Dawn '80
Lena's Holiday '91
Swing Shift '84
That's Life! '86
Weekend Warriors '86

Jack Lemmon

Airport '77 '77
The Apartment '60
April Fools '69
Bell, Book and Candle '59
Buddy Buddy '81
The China Syndrome '79
Dad '89

Days of Wine and Roses '62
Fire Down Below '57
For Richer, For Poorer '92
The Fortune Cookie '66
The Front Page '74
Good Neighbor Sam '64
Great Race '65
Heaven for Betsy/Jackson & Jill
How to Murder Your Wife '64
Irma La Douce '63
It Should Happen to You '53
JFK '91
Long Day's Journey into Night '88
Luv '67
Macaroni '85
Mass Appeal '84
Missing '82
Mister Roberts '55
The Murder of Mary Phagan '87
The Odd Couple '68
The Out-of-Towners '70
Phffft! '54
Prisoner of Second Avenue '74
Save the Tiger '73
Some Like It Hot '59
That's Life! '86
Tribute '80
Wackiest Ship in the Army '61

Jack Lemmon (D)

Kotch '71

Kasi Lemmons

Vampire's Kiss '88

Michel Lemoine

The Chambermaid's Dream '86
Planets Against Us '61
The Sensuous Teenager '70

Lisa Lemole

Drive-In '76

Genevieve Lemon

Sweetie '89

Rusty Lemorande (D)

Journey to the Center of the Earth '88

Peter Lempert

In the Region of Ice '76

Mark Lenard

Noon Sunday '71
Star Trek 3: The Search for Spock '84
Star Trek 4: The Voyage Home '86

Paul Leni (D)

The Cat and the Canary '27
Waxworks '24

Jan Lenica (D)

Ubu and the Great Gidouille '79

Christine Lenier

The Nun '66

Harry J. Lennix

The Five Heartbeats '91

John Lennon

Chuck Berry: Hail! Hail! Rock 'N' Roll '87

Jose Lewgoy

Earth Entranced (Terra em Transe) '66
Fitzcarraldo '82
Kiss of the Spider Woman '85

Stephen Lewicki (D)

Certain Sacrifice '80

Albert Lewin (D)

The Moon and Sixpence '42
Picture of Dorian Gray '45
The Private Affairs of Bel Ami '47

Ben Lewin (D)

The Favor, the Watch, & the Very Big Fish '92

Sam Lewin (D)

Dunera Boys '85

Al Lewis

Comic Cabby '87
Fright House '89
Grampa's Monster Movies '90
Grampa's Sci-Fi Hits '90
Munster's Revenge '81
My Grandpa is a Vampire '90s
Way He Was '76

Al Lewis (D)

Our Miss Brooks '56

Alun Lewis

Experience Preferred... But Not Essential '83

Brittney Lewis

Dream Machine '91

Charlotte Lewis

Bare Essentials '91
Dial Help '88
The Golden Child '86
Pirates '86
Tripwire '89

Christopher Lewis (D)

The Red Raven Kiss-Off '90
Revenge '86
The Ripper '86

David Lewis

The Apartment '60

David Lewis (D)

Dangerous Curves '88

Diana Lewis

Andy Hardy Meets Debutante '40
Go West '40

Fiona Lewis

Blueblood '80s
Dracula '73
Drum '76
Innerspace '87
Lisztomania '75
Strange Behavior '81
Strange Invaders '83
Stunts '77
Tintorera...Tiger Shark '78
Wanda Nevada '79

Forrest Lewis

The Monster of Piedras Blancas '57

Gary Lewis

Hardly Working '81

Geoffrey Lewis

Catch Me...If You Can '89
Culpepper Cattle Co. '72
Disturbed '90
Double Impact '91
Every Which Way But Loose '78
The Geisha Boy '58
The Great Waldo Pepper '75
I, the Jury '82
The Lawnmower Man '92
Lust in the Dust '85
Macon County Line '74
Matters of the Heart '90
Maximum Security '87
Night of the Comet '84
Out of the Dark '88
The Return of a Man Called Horse '76
Return of the Man from U.N.C.L.E. '83
September Gun '83
Shoot the Sun Down '81
Smile '75
Stitches '85
Thunderbolt & Lightfoot '74
Tilt '78
When Every Day was the Fourth of July '78

George Lewis

The Big Sombrero '49
Indian Paint '64
King of the Bullwhip '51
Perils of the Darkest Jungle '44
Sign of Zorro '60
Zorro's Black Whip '44

Gilbert Lewis

Touched '82

Henry Lewis

Night of Terror '87

Herschell Gordon Lewis (D)

Blood Feast '63
Color Me Blood Red '64
Gruesome Twosome '67
Just for the Hell of It '68
Monster a Go-Go! '65
Nature's Playmates '62
The Psychic '68
She-Devils on Wheels '68
Something Weird '68
Suburban Roulette '67
This Stuff'll Kill Ya! '71
2000 Maniacs '64
The Wizard of Gore '70

Ida Lewis

Sweet Adeline '26

Janelle Lewis

Zombiethon '86

Jay Lewis (D)

The Baby and the Battleship '56

Jenny Lewis

The Wizard '89

Jerry Lewis

Artists and Models '55
At War with the Army '50
The Bellboy '60
Best of Comic Relief '86
Big Mouth '67
The Caddy '53
Cinderfella '60
Cookie '89
Cracking Up '83
The Delicate Delinquent '56
Disorderly Orderly '64
Don't Raise the Bridge, Lower the River '68
Errand Boy '61
Family Jewels '65
The Geisha Boy '58
Hardly Working '81
Hollywood or Bust '56
Hollywood Clowns '85
It's a Mad, Mad, Mad, Mad World '63
King of Comedy '82
The Ladies' Man '61
The Nutty Professor '63
The Patsy '64
Rascal Dazzle '81
Road to Bali '53
The Sad Sack '57
Scared Stiff '53
Slapstick of Another Kind '84
Television's Golden Age of Comedy '50s
Trading Hearts '87
Which Way to the Front? '70

Jerry Lee Lewis

Chuck Berry: Hail! Hail! Rock 'N' Roll '87
High School Confidential '58
Ready Steady Go, Vol. 2 '85

Jerry Lewis (D)

The Bellboy '60
Big Mouth '67
Cracking Up '83
Errand Boy '61
Family Jewels '65
Hardly Working '81
The Ladies' Man '61
The Nutty Professor '63
The Patsy '64
Which Way to the Front? '70

Joe Lewis

Force: Five '81
Jaguar Lives '70s

Joe E. Lewis

Lady in Cement '68
Private Buckaroo '42

Joseph H. Lewis (D)

Big Combo '55
Boys of the City '40
Criminals Within '41
Gun Crazy '50
The Invisible Ghost '41
Pride of the Bowery '41
Retreat, Hell! '52
Seventh Cavalry '56
That Gang of Mine '40

Juliette Lewis

Cape Fear '91
Meet the Hollowheads '89
My Stepmother Is an Alien '88
National Lampoon's Christmas Vacation '89

Linda Lewis

Alien Dead '85

Lori Lewis

Nightmare Weekend '86

Louise Lewis

Blood of Dracula '57
I Was a Teenage Werewolf '57

Martin Lewis

Greek Street '30

Mitchell Lewis

Salome '22
Salome/Queen Elizabeth '23
The Wizard of Oz '39

Monica Lewis

The Concorde: Airport '79 '79

Nicholas Lewis

Fass Black '77

Phill Lewis

Brother Future '91
City Slickers '91

Ralph Lewis

Dante's Inferno '24
Outside the Law '21
Sucker Money '34
When the Clouds Roll By '19

Richard Lewis

Diary of a Young Comic '79
Once Upon a Crime '91
Richard Lewis: I'm Exhausted '89
Richard Lewis: I'm in Pain '85
That's Adequate '90
The Wrong Guys '88

Robert Lewis

Monsieur Verdoux '47

Robert Lewis (D)

Child Bride of Short Creek '81
City Killer '87
Dead Reckoning '89
Desperate Lives '82
Embassy '85
Fallen Angel '81
If Things Were Different '79
Lady Killers '80s
Memories of Murder '90
No Room to Run '78
Pray for the Wildcats '74
S*H*E '79
A Summer to Remember '84

Ronald Lewis

Bachelor of Hearts '58
Billy Budd '62
Conspiracy of Hearts '60
Robbery Under Arms '57
Scream of Fear '61

Sheldon Lewis

Tombstone Canyon '35
With Kit Carson Over the Great Divide '25

Stanley Lewis (D)

Punk Vacation '90

Sybil Lewis

Broken Strings '40

Ted Lewis

Manhattan Merry-go-round '37

Terry Lewis (D)

Gorilla Farming '90

Tom Lewis

Steamboat Bill, Jr. '28

Denis Lewiston (D)

Hot Target '85

Perieles Lewnes

Redneck Zombies '88

John Ley

BMX Bandits '83

Pierre Leymarie

Pickpocket '59

Juan Leyrado

Times to Come '81

Johan Leysen

Egg '88
The Music Teacher '88

John Leyton

Guns at Batasi '64
Seaside Swingers '65
Von Ryan's Express '65

Thierry Lhermitte

My Best Friend's Girl '84
My New Partner '84
Next Year If All Goes Well '83
Until September '84

Bruce Li

Blind Fists of Bruce '70s
Bruce Li the Invincible '80
Bruce Li in New Guinea '80
Chinese Connection 2 '84
Chinese Stuntman '70s
Counter Attack '84
Edge of Fury '80s
Enter the Panther '79
Enter Three Dragons '81
Fists of Fury 2 '80
Image of Bruce Lee '70s
Iron Dragon Strikes Back '84
Kung Fu Avengers '85
Return of the Tiger '78
Story of the Dragon '82

Bruce Li (D)

Counter Attack '84

Chang Li

Black Belt Fury '81
Enter Three Dragons '81
Fight to the Death '83

Gong Li

Red Sorghum '87

Hwang Cheng Li

Dragon's Claw '80s
Young Heroes '69

John Li

Lightning Kung Fu '70s

Pi Li Lee

Kung Fu Massacre '82

Chou Li Lung

Crack Shadow Boxers '70s

Man Li Peng

Little Big Master '70s

Wang Li

Zombie vs. Ninja

Bruce Liang

The Fist, the Kicks, and the Evils '80
Showdown at the Equator '83

Guo Liang

Young Hero of Shaolin

Young Hero of the Shaolin
    Part 2

Ko Shou Liang

Ninja Exterminators '81
Renegade Monk '82
Shaolin Ex-Monk '82

Liu Chang Liang

Fighting Ace '80s
Renegade Monk '82

Liu Chia Liang (D)

Master Killer (The Thirty-Sixth
    Chamber) '84

Tan Tao Liang

Dual Flying Kicks '82
Incredible Master Beggars '82

Yu Lianqi (D)

Kung Fu Hero 2 '90

Peter Paul Liapis

Starhops '78

Brian Libby

Platoon Leader '87
Silent Rage '82

Liberace

The Loved One '65
Sincerely Yours '56

Anne Libert

The Demons '74
Girls of Don Juan '70s
Virgin Among the Living Dead
    '71

Richard Libertini

All of Me '84
Animal Behavior '89
Big Trouble '86
The Bonfire of the Vanities '90
Comedy Tonight '77
Deal of the Century '83
Fletch '85
Fletch Lives '89
Fourth Wise Man '85
Going Berserk '83
Popeye '80
Sharky's Machine '81
Soup for One '82
Unfaithfully Yours '84

Mitchell Lichenstein

Streamers '83

Jeremy Licht

Father Figure '80
Lois Gibbs and the Love Canal
    '82
Next One '84
The Seekers '79
Twilight Zone: The Movie '83

Chung Lick

Super Kung Fu Kid

Joe Lickshot

Raw and Rough Pinchers '87

Carlo Liconti (D)

Goodnight, Michelangelo '89

Gabriella Licudi

The Unearthly Stranger '64

Laura Liddell

Shakespeare Wallah '65

G. Gordon Liddy

Camp Cucamonga: How I
    Spent My Summer
    Vacation '90
Street Asylum '90
When Nature Calls '85

Anki Liden

My Life As a Dog '87

Jeff Lieberman (D)

Blue Sunshine '78
Just Before Dawn '80
Remote Control '88
Squirm '76

**Robert
Lieberman** (D)

Table for Five '83
Will: G. Gordon Liddy '82

Leon Liebgold

Yidl Mitn Fidl (Yiddle with His
    Fiddle) '36

Harry Liedtke

Passion '19

Don Liefert

Fiend '83
Galaxy Invader '85

Chen Hung Lieh

To Subdue the Evil

Lo Lieh

Born Invincible '76
Bruce's Deadly Fingers '80s
The Dragon Returns
Fists of Bruce Lee '80s
The Golden Triangle '80
King Boxer '90
Kung Fu from Beyond the
    Grave
Ninja Massacre '84
The Ninja Pirates '90
The Noble Ninja '90
Shaolin Executioner '80s
Showdown at the Equator '83
The Stranger and the
    Gunfighter '76

Lo Lieh (D)

Fists of the White Lotus '80s

Robert Liensol

Time to Kill '89

Johanna Lier

Alpine Fire '89

Albert Lieven

Beware of Pity '46
Frieda '47
Life & Death of Colonel Blimp
    '43
Seventh Veil '46
Sleeping Car to Trieste '45

Enyaw Liew

Machoman '70s

Judith Light

Stamp of a Killer '87

Mike Light (D)

Flesh Gordon '72

Gordon Lightfoot

Harry Tracy '83

Marilyn Lightstone

Lies My Father Told Me '75
Surrogate '88
Wild Pony '83

Tom Ligon

Fury on Wheels '71
Judge Horton and the
    Scottsboro Boys '76

Mitsos Liguisos

Never on Sunday '60

Ralph M. Like (D)

Chinatown After Dark '31

Lili Liliana

The Dybbuk '37

Peter Lilienthal (D)

David '79
The Uprising '81

Dennis Lill

Bad Blood '87

Tom Lillard

The Brig '64

Anna Lillian

Where is My Child? '37

Beatrice Lillie

On Approval '44
Thoroughly Modern Millie '67

Sandy Lillingston

Dangerous Game '90

Silvia Lillo

Love in the City '53

Yvonne Lime

I Was a Teenage Werewolf
    '57

Dina Limon

Boy Takes Girl '83

Bridget Lin

Jackie Chan's Police Force
    '85

Pearl Lin

Bruce Lee's Ways of Kung Fu
    '80s

Sinon Lin

Dreadnaught Rivals '80s

Traci Lin

Class of 1999 '90
Fright Night 2 '88
Spellcaster '91
Survival Quest '89
A Tiger's Tale '87

Kay Linaker

Hidden Enemy '40

Peter Linari

Fat Guy Goes Nutzoid '86

Abbey Lincoln

For Love of Ivy '68

Elmo Lincoln

The Adventures of Tarzan '21
Tarzan of the Apes '17

Fred J. Lincoln

Last House on the Left '72

Fred J. Lincoln (D)

Wild Man '89

Lar Park Lincoln

Friday the 13th, Part 7: The
    New Blood '88
Princess Academy '87

Pamela Lincoln

Anatomy of a Psycho '61

Richard Lincoln

Manny's Orphans '78

Warren Lincoln

The Power '80

Della Lind

Swiss Miss '38

Jenny Lind

Barnum '86

Traci Lind

No Secrets '91

Hala Linda

Legion of Missing Men '37

Kristina Lindberg

Maid in Sweden '83

Per Lindberg (D)

June Night '40

Gunnel Lindblom

The Seventh Seal '56
The Virgin Spring '59

Jainine Linde

Moving Target '89

Carrie Lindell

Cannibal Campout '88

Eric Linden

Criminals Within '41
Let 'Em Have It '35

Hal Linden

Beethoven Cycle '88
Father Figure '80
How to Break Up a Happy
    Divorce '76
I Do! I Do! '84
Mr. Inside, Mr. Outside '74
My Wicked, Wicked Ways '84
A New Life '88
Starflight One '83

Jennie Linden

Hedda '75
Women in Love '70

Liane Linden

The Arsenal Stadium Mystery
    '39

Cec Linder

Deadly Eyes '82
The Devil's Web '74
Goldfinger '64
Heavenly Bodies '84
A Touch of Class '73

Christa Linder

Night of a Thousand Cats '72

Maud Linder (D)

Man in the Silk Hat '15

Max Linder

Early Cinema Vol. 1 '19
Laugh with Linder! '13
Man in the Silk Hat '15
Man in the Silk Hat '83
Max '22
Pop Goes the Cork '22
Seven Years Bad Luck '20

Max Linder (D)

Early Cinema Vol. 1 '19

Viveca Lindfors

Adventures of Don Juan '49
The Ann Jillian Story '88
Bell from Hell '74
The Best Little Girl in the
    World '81
Cauldron of Blood '67
Child of Darkness, Child of
    Light '91
Creepshow '82
The Exorcist 3 '90
For Ladies Only '81
Four in a Jeep '51
Girlfriends '78
The Hand '81
Inside the Third Reich '82
King of Kings '61
La Campana del Infierno
Marilyn: The Untold Story '80
Mom, the Wolfman and Me
    '80
Moonfleet '55
Natural Enemies '79
A Question of Guilt '78
Rachel River '89
Secret Weapons '85
Silent Madness '84
The Story of Ruth '60
The Sure Thing '85
The Way We Were '73
A Wedding '78
Welcome to L.A. '77
Zandalee '91

Peter Lindgren

I Am Curious (Yellow) '67

Kristen Lindholm

Twins of Evil '71

Audra Lindley

Best Friends '82
Cannery Row '82
Desert Hearts '86
The Heartbreak Kid '72
Revenge of the Stepford
    Wives '80
The Spellbinder '88
Stamp of a Killer '87

Olga Lindo

The Hidden Room '49
Things Happen at Night '48

Betty H. Lindon

I Eat Your Skin '64

Vincent Lindon

A Man in Love '87

Carol Lindsay

She Gods of Shark Reef '56

Helen Lindsay

Secrets '82

**Michael Lindsay-
Hogg** (D)

As Is '85
Let It Be '70
Little Match Girl '87

Waterfront '83

Michelle Little

Appointment with Fear '85
Blood Clan '91
Bluffing It '87
My Demon Lover '87
Radioactive Dreams '86

Rich Little

Christmas Raccoons '84
Dirty Tricks '81
Happy Hour '87
Rich Little: One's a Crowd '88
Rich Little's Christmas Carol '90
Rich Little's Great Hollywood Trivia Game '84

Little Richard

Down and Out in Beverly Hills '86
Pee-Wee's Playhouse Christmas Special '88
Purple People Eater '88

Eddie Little Sky

Journey Through Rosebud '72
Savage Run '70

Lucien Littlefield

Bulldog Drummond's Revenge '37
Charley's Aunt '25
Hollywood Stadium Mystery '38
Let's Sing Again '36
The Sheik '21
Strangers of the Evening '32
Susanna Pass '49
Tom Sawyer '30
Tumbleweeds '25
Zorro's Black Whip '44

Gary Littlejohn

Angels Hard as They Come '71
Bury Me an Angel '71

Lynne Littman (D)

Testament '83

Mario Litto

The Avenger '60

Anatole Litvak (D)

All This and Heaven Too '40
Anastasia '56
City for Conquest '40
Goodbye Again '61
Mayerling '36
Night of the Generals '67
Sorry, Wrong Number '48

Carolyn Liu

Do or Die '91

Jimmy Liu

Dragon's Claw '80s

John Liu

Avenging Ninja '80s
Instant Kung-Fu Man '84
Ninja in the Claws of the C.I.A. '83
Shaolin Ex-Monk '82
Snuff-Bottle Connection '82

Boris Livanov

Ten Days That Shook the World/October '27

Gerry Lively (D)

Body Moves '90

Jason Lively

Ghost Chase '88
National Lampoon's European Vacation '85
Night of the Creeps '86

Robin Lively

Buckeye and Blue '87
Crazy from the Heart '91
Teen Witch '89

Jack Livesey

Never Too Late to Mend '37

Roger Livesey

Drums '38
The Entertainer '60
Green Grow the Rushes '51
I Know Where I'm Going '47
League of Gentlemen '60
Life & Death of Colonel Blimp '43
The Master of Ballantrae '53
Of Human Bondage '64

Robert Livesy

Devil Wears White '86

Living Theatre Company

Signals Through the Flames '83

Barry Livingston

Easy Wheels '89

Bob Livingston

Blazing Stewardesses '75
The Bold Caballero '36
Come On, Cowboys '37
Covered Wagon Days '40
Cowboys from Texas '39
Death Rides the Plains '44
Ghost Town Gold '36
Heart of the Rockies '37
Heroes of the Hills '38
Hit the Saddle '37
Law of the Saddle '45
Outlaws of Sonora '38
Purple Vigilantes '38
Raiders of Red Gap '43
Riders of the Whistling Skull '37
Roarin' Lead '37
Texas Layover '75
The Vigilantes Are Coming '36
Western Double Feature 6: Rogers & Trigger '50
Western Double Feature 12 '44
Western Double Feature 13 '40s

Jennie Livingston (D)

Paris is Burning '91

Margaret Livingston

The Lady Refuses '31
Sunrise '27
Through the Breakers '28

Robert Livingston

Grand Canyon Trail '48

Shelby Livingston

2000 Maniacs '64

Stanley Livingston

Smokey & the Hotwire Gang '79

Enrique Lizalde

La Mentira '87
Sexo Y Crimen '87

Kari Lizer

Private School '83
Smokey Bites the Dust '81

Carlo Lizzani (D)

House of the Yellow Carpet '84
Last Four Days '77
Love in the City '53

Julio Ruiz Llaneza (D)

By Hook or Crook '85

Aldolfo Llaurado

Portrait of Teresa '79

Barbara Llewellyn

Great Gold Swindle '84

Desmond Llewelyn

Diamonds are Forever '71
For Your Eyes Only '81
From Russia with Love '63
Goldfinger '64
License to Kill '89
The Living Daylights '87
The Man with the Golden Gun '74
Moonraker '79
On Her Majesty's Secret Service '69
The Spy Who Loved Me '77
Thunderball '65
A View to a Kill '85
You Only Live Twice '67

Luis Llosa (D)

Crime Zone '88
Hour of the Assassin '86

Christopher Lloyd

The Addams Family '91
The Adventures of Buckaroo Banzai '84
Amazing Stories, Book 2 '86
Back to the Future '85
Back to the Future, Part 2 '89
Back to the Future, Part 3 '90
Clue '85
Cowboy & the Ballerina '84
The Dream Team '89
Eight Men Out '88
Goin' South '78
Joy of Sex '84
Lady in Red '79
Legend of the Lone Ranger '81
Miracles '86
Mr. Mom '83
One Flew Over the Cuckoo's Nest '75
Pilgrim Farewell '82
The Postman Always Rings Twice '81
September Gun '83
Star Trek 3: The Search for Spock '84
Street Hawk '84
Suburban Commando '91
Three Warriors '77
To Be or Not to Be '83
Track 29 '88
Walk Like a Man '87

Who Framed Roger Rabbit? '88
Why Me? '90

Danny Lloyd

The Shining '80
Will: G. Gordon Liddy '82

Doris Lloyd

The Drake Case '29
A Shot in the Dark '35

Emily Lloyd

Chicago Joe & the Showgirl '90
Cookie '89
In Country '89
Scorchers '92
Wish You Were Here '87

Frank Lloyd (D)

Blood on the Sun '45
Forever and a Day '43
The Howards of Virginia '40
The Last Command '55
Mutiny on the Bounty '35
Oliver Twist '22

Harold Lloyd, Jr.

The Flaming Urge '53
Married Too Young '62

Harold Lloyd

Days of Thrills and Laughter '61
Dr. Jack/For Heaven's Sake '26
Don't Shove/Two Gun Gussie '19
Early Cinema Vol. 1 '19
Feet First '30
For Heaven's Sake '26
The Freshman '25
Funstuff '30s
Girl Shy '24
His Royal Slyness/Haunted Spooks '20
Hollywood Clowns '85
Hot Water/Safety Last '24
I'm on My Way/The Non-Stop Kid '10s
The Kid Brother '27
Kings, Queens, Jokers '30s
Mad Wednesday '51
Milky Way '36
Never Weaken/Why Worry? '23
Sailor-Made Man/Grandma's Boy '20s
The Freshman '25
Silent Laugh Makers, No. 3 '20s
The Sin of Harold Diddlebock '47
Slapstick '30s
Speedy '28

Harold Lloyd (D)

Early Cinema Vol. 1 '19
His Royal Slyness/Haunted Spooks '20
I'm on My Way/The Non-Stop Kid '10s

Jimmy Lloyd

G.I. Jane '51
The Jolson Story '46
Miss Grant Takes Richmond '49
Riders of the Whistling Pines '49
Sea Hound '47

John Bedford Lloyd

The Abyss '89

Sweet Lorraine '87
Tough Guys Don't Dance '87
Waiting for the Light '90

Kathleen Lloyd

It's Alive 2: It Lives Again '78
The Jayne Mansfield Story '80
Missouri Breaks '76
Skateboard '77
Take Down '79

Norman Lloyd

Amityville 4 '89
Jaws of Satan '81
Journey of Honor '91
The Nude Bomb '80
Saboteur '42
The Southerner '45

Richard Lloyd

Invincible Gladiators '64

Sue Lloyd

The Bitch '78
The Ipcress File '65
Number 1 of the Secret Service '77

Susan Lloyd

Hysteria '64

Ken Loach (D)

Family Life '72
Hidden Agenda '90
Singing the Blues in Red '87

Bruni Lobel

Big Lift '50

Tony LoBianco

The Ann Jillian Story '88
Blood Ties '87
Bloodbrothers '78
City Heat '84
City of Hope '91
F.I.S.T. '78
The French Connection '71
God Told Me To '76
Goldenrod '77
Honeymoon Killers '70
Last Cry for Help '79
Magee and the Lady '78
Marciano '79
Mean Frank and Crazy Tony '75
Mr. Inside, Mr. Outside '74
Separate Ways '82
The Seven-Ups '73
The Ten Million Dollar Getaway '91

Tony LoBianco (D)

Too Scared to Scream '85

Amy Locane

Cry-Baby '90
Lost Angels '89
No Secrets '91

Carol Locatell

Sammy '88

David Lochary

Female Trouble '75
Mondo Trasho '70
Multiple Maniacs '70
Pink Flamingos '73

Donna Locke

Neon Maniacs '86

Ingold Locke

Feel the Motion '86

Masque of the Red Death '90
Murders in the Rue Morgue '71
Mysterious Island '61
99 Women '69
Northwest Frontier '59
Paris Express '53
The Pink Panther Strikes Again '76
The Pope Must Diet '91
Return of the Pink Panther '74
Revenge of the Pink Panther '78
River of Death '90
Room 43 '58
Seventh Veil '46
A Shot in the Dark '64
Skeleton Coast '89
Spartacus '60
Ten Little Indians '75
Ten Little Indians '89
Third Man on the Mountain '59
Tiara Tahiti '62
Trail of the Pink Panther '82
Uncle Tom's Cabin '69
Villa Rides '68
War and Peace '56

D.J. Anthony Loma (D)

Counterforce '87
Target Eagle '84

Joseph Loman (D)

Boldest Job in the West '70s
Surprise Attack '70s

Carole Lombard

Big News '29
Golden Age of Comedy '58
High Voltage '29
Hollywood Outtakes & Rare Footage '83
In Name Only '39
Lady by Choice '34
Mack & Carole '28
Made for Each Other '39
Making of a Legend: "Gone With the Wind" '89
Mr. & Mrs. Smith '41
My Man Godfrey '36
No Man of Her Own '32
Nothing Sacred '37
Power '28
The Racketeer '29
Swing High, Swing Low '37
These Girls Won't Talk '20s
They Knew What They Wanted '40
To Be or Not to Be '42
Twentieth Century '34

Francisco J. Lombardi (D)

La Boca del Lobo '89

Ralph Lombardi

Raw Force '81

Guy Lombardo

Stage Door Canteen '43

Lou Lombardo (D)

P.K. and the Kid '85
Russian Roulette '75

Seline Lomez

Snowballin' '85

Ulli Lommel

Boogey Man 2 '83
Chinese Roulette '86

Ulli Lommel (D)

The Big Sweat '91
The Boogey Man '80
Brainwaves '82
Cocaine Cowboys '79
Cold Heat '90
Devonsville Terror '83
Olivia '83
Overkill '86
Strangers in Paradise '86
Warbirds '88

Jacek Lomnicki

Man of Marble '76

Tadeusz Lomnicki

Man of Marble '76

Britt Lomond

Sign of Zorro '60

Richard Loncraine (D)

Bellman and True '88
Brimstone & Treacle '82
Deep Cover '88
The Haunting of Julia '77
The Missionary '82

Debbie Lee London

Unmasked Part 25 '88

Jason London

December '91

Jerry London (D)

Chiefs '83
Father Figure '80
Haunting of Sarah Hardy '89
Kiss Shot '90s
Manhunt for Claude Dallas '86
Rent-A-Cop '88
The Scarlet & the Black '83
Shogun '80

Julie London

The Girl Can't Help It '56
Man of the West '58
Nabonga '44

Laurene London

Hundra '85

Lisa London

H.O.T.S. '79
Savage Beach '89

Rosalyn London

Amazing Mr. Blunden '72

Roy London (D)

Diary of a Hitman '91

Tom London

Blue Canadian Rockies '52
Courage of the North '35
Flaming Lead '39
Outlaw Justice '32
Red Desert '50
Rio Rattler '35
Toll of the Desert '35
Two Gun Man '31
Zorro's Black Whip '44

John Lone

Echoes of Paradise '86
Iceman '84
The Last Emperor '87
The Moderns '88
Year of the Dragon '85

Audrey Long

The Adventures of Gallant Bess '48
Desperate '47
Indian Uprising '51

Chin Long

Tiger's Claw '70s

Jack Long

Ninja Checkmate '90

Jodi Long

Patty Hearst '88

Joseph Long

Queen of Hearts '89

Lotus Long

Mr. Wong in Chinatown '39
Mystery of Mr. Wong '39
Phantom of Chinatown '40

Martin Long

A Man with a Maid '73

Melissa Long

Reactor '85
Scandal '80s
War of the Robots '78

Richard Long

Big Valley '68
Criss Cross '48
Death Cruise '74
Egg and I '47
Egg and I '47
House on Haunted Hill '58
The Stranger '46
Tomorrow is Forever '40

Richard Long (D)

The Word '78

Ricky Long

Ghostriders '87

Shelley Long

Caveman '81
Comic Relief 2 '87
Comic Relief 3 '89
The Cracker Factory '79
Don't Tell Her It's Me '90
Hello Again '87
Irreconcilable Differences '84
Losin' It '82
The Money Pit '86
Night Shift '82
Outrageous Fortune '87
A Small Circle of Friends '80
Troop Beverly Hills '89

Stanley Long (D)

Adventures of a Private Eye '87
Adventures of a Taxi Driver '76

Tarng Long

Challenge of the Dragon '82

Walter Long

Blood and Sand '22
Flaming Lead '39
The Little American '17
Little Church Around the Corner '23
Man's Country '38
Moran of the Lady Letty '22
Sea Devils '31
The Sheik '21
Silver Stallion '41
Six Shootin' Sheriff '38

Soul-Fire '25
Yankee Clipper '27

Humphrey Longan (D)

The Great Battle '79

Beba Longcar

Don't Look in the Attic '81

John Longden

Alias John Preston '56
Juno and the Paycock '30
Quatermass 2 '57
Sherlock Holmes '51
Tower of Terror '42

Terence Longden

Murder on the Campus '52

John Longdon

Blackmail '29

Claudine Longet

The Party '68

Stephen Longhurst

Invitation to Hell '84

Sue Longhurst

Keep It Up, Jack! '75
A Man with a Maid '73
Victorian Fantasies '86

Victoria Longley

Celia: Child Of Terror '89

Alix Longman

The Getting of Wisdom '77

Tony Longo

Bloodhounds of Broadway '89
Suburban Commando '91
Worth Winning '89

Emily Longstreth

American Drive-In '88
The Big Picture '89
Private Resort '84
Rising Son '90
Wired to Kill '86

Tadeusz Loniski

Colonel Wolodyjowski '69

Ray Lonnen

Belfast Assassin '84
Guardian of the Abyss '82

Michael Lonsdale

Enigma '82
Erendira '83
The Holcroft Covenant '85
Is Paris Burning? '68
Les Assassins de L'Ordre '71
Moonraker '79
Murmur of the Heart '71
The Name of the Rose '86
Souvenir '88

Richard Loo

Across the Pacific '42
Back to Bataan '45
Betrayal from the East '44
China Sky '44
The Clay Pigeon '49
First Yank into Tokyo '45
Love is a Many-Splendored Thing '55
The Man with the Golden Gun '74
Mr. Wong in Chinatown '39
The Purple Heart '44

The Steel Helmet '51

Christopher Loomis

The Nesting '80

Deborah Loomis

Hercules in New York '70

Nancy Loomis

Halloween '78

Rod Loomis

Jack's Back '87

Larry Loonin

The Man Who Envied Women '85

Friedl Loor

The Eternal Waltz '59

Silvia Lopel

Hercules Unchained '59

Tanya Lopert

Navajo Joe '67
Tales of Ordinary Madness '83

Charo Lopez

La Vida Sigue Igual

Estrellita Lopez

Only Once in a Lifetime '80s

Ivonne Lopez

Letters from the Park '88

Kamala Lopez

Born in East L.A. '87
Crazy from the Heart '91
Dollman '91

Margo Lopez

El Libro de Piedra '68
La Muneca Perversa
Nazarin '58

Perry Lopez

Chinatown '74
Taras Bulba '62

Priscilla Lopez

Jesse '88

Temistocles Lopez (D)

Exquisite Corpses '88

Trini Lopez

Antonio '73
The Dirty Dozen '67
La Bamba Party '80s
The Poppy Is Also a Flower '66

Laura Lopinski

Dorm That Dripped Blood '82

Gail Lorber

Hot Times '74

Isabel Lorca

Lightning: The White Stallion '86
She's Having a Baby '88

Del Lord (D)

Vengeance '37

Lenita Love

The Devil on Wheels '47

Lucretia Love

Dr. Heckyl and Mr. Hype '80
The Eerie Midnight Horror
Show '82
Naked Warriors '73
Phenomenal and the Treasure
of Tutankamen '70s
The She-Beast '65

Mike Love

Beach Boys: An American
Band '85

Montagu Love

Bulldog Drummond '29
Damsel in Distress '37
Don Juan '26
Fighting Devil Dogs '43
Gunga Din '39
Hands Up '26
His Double Life '33
Inside the Lines '30
Lady for a Night '42
Prince and the Pauper '37
Prisoner of Zenda '37
Sherlock Holmes: The Voice
of Terror '42
The Son of Monte Cristo '40
The Wind '28

Suzanna Love

The Boogey Man '80
Boogey Man 2 '83
Brainwaves '82
Cocaine Cowboys '79
Devonsville Terror '83
Olivia '83

Victor Love

Guilty of Innocence '87
Heaven is a Playground '91
Native Son '86

Alec Lovejoy

Birth Right '39
Moon Over Harlem '39

Frank Lovejoy

Americano '55
The Finger Man '55
The Hitch-Hiker '53
Home of the Brave '49
House of Wax '53
In a Lonely Place '50
Retreat, Hell! '52
Shack Out on 101 '55
Strategic Air Command '55
Three Secrets '50
Try and Get Me '50
The Winning Team '52

Harry Lovejoy

The Black Klansman '66

Nita Loveless

Married Too Young '62

Alan Lovell

Glass '90

Raymond Lovell

Appointment with Crime '45
Hotel Reserve '44

Ray Lovelock

Autopsy '78
From Hell to Victory '79
One Russian Summer '73

**Charles
Loventhal** (D)

First Time '82
My Demon Lover '87

William Lover

To All a Goodnight '80

Dorothy Lovett

Courageous Dr. Christian '40
Dr. Christian Meets the
Women '40
Remedy for Riches '40
They Meet Again '41

Bert Lovitt (D)

Prince Jack '83

Jon Lovitz

An American Tail: Fievel Goes
West '91
Big '88
Comic Relief 2 '87
Jumpin' Jack Flash '86
Last Resort '86
Mr. Destiny '90
My Stepmother Is an Alien '88
Three Amigos '86

Celia Lovsky

I, Mobster '58

Steven Lovy (D)

Circuitry Man '90

Arthur Lowe

The Bawdy Adventures of
Tom Jones '76
O Lucky Man '73
The Ruling Class '72
Sweet William '79
This Sporting Life '63

Chad Lowe

Apprentice to Murder '88
Highway to Hell '92
An Inconvenient Woman '91
Nobody's Perfect '90
Silence of the Heart '84
True Blood '89

Chris Lowe

It Couldn't Happen Here '88

Debbie Lowe

The Cheerleaders '72

Edmund Lowe

Call Out the Marines '42
Chandu the Magician '32
Dillinger '45
The Enchanted Forest '45
The Eyes of Youth '19
Good Sam '48
Heller in Pink Tights '60
Wings of Eagles '57

Lucas Lowe (D)

The King of the Kickboxers
'91

Patrick Lowe

Primal Rage '90
Slumber Party Massacre 2 '87

Rob Lowe

About Last Night... '86
Bad Influence '90
Class '83
The Hotel New Hampshire '84
Illegally Yours '87
Masquerade '88

The Outsiders '83
Oxford Blues '84
St. Elmo's Fire '85
Square Dance '87
Stroke of Midnight '90
Wayne's World '92
Youngblood '86

Susan Lowe

Desperate Living '77

William Lowe (D)

Slaughter in San Francisco '81

Bob Lowell

Mom & Dad '47

Carey Lowell

Club Paradise '86
Dangerously Close '86
Down Twisted '89
The Guardian '90
License to Kill '89
Me and Him '89
Road to Ruin '91

Robert Lowell

I Accuse My Parents '45

Curt Lowens

Werewolf in a Girl's Dormitory
'61

**Richard
Lowenstein** (D)

Dogs in Space '87
Pete Townshend: White City
'85

H. Wayne Lowery

Kill Line '91

Robert Lowery

Border Rangers '50
Call of the Forest '49
Dangerous Passage '44
Dark Mountain '44
Dawn on the Great Divide '42
Death Valley '46
Drums Along the Mohawk '39
God's Country '46
Gunfire '50
Highway 13 '48
I Shot Billy the Kid '50
Lure of the Islands '42
Murder Over New York '40
Mystery of the Riverboat '44
Navy Way '44
Outlaw Gang '49
Queen of the Amazons '47
Revenge of the Zombies '43
The Rise and Fall of Legs
Diamond '60
Shep Comes Home '49
Trail of the Mounties '47
Train to Tombstone '50
The Undertaker and His Pals
'67
Western Pacific Agent '51

William E. Lowery

The Lamb '15
Reggie Mixes In '16
Robin Hood '22

Klaus Lowitsch

Despair '79
The Marriage of Maria Braun
'79

Dick Lowry (D)

Coward of the County '81
The Jayne Mansfield Story '80

Kenny Rogers as the Gambler
'80
Kenny Rogers as the
Gambler, Part 2: The
Adventure Continues '83
Smokey and the Bandit, Part
3 '83
Wet Gold '84
Wild Horses '84

Ed Lowry

House of Mystery '34

Jennifer Lowry

Brain Damage '88

Lynn Lowry

The Crazies '76
Score '73
Sugar Cookies '77
They Came from Within '75

Morton Lowry

Immortal Sergeant '43

Scooter Lowry

Our Gang '26
Our Gang Comedy Festival
'88
Our Gang Comedy Festival 2
'89

David Loxton (D)

Lathe of Heaven '80

Mino Loy

Emerald Jungle '80

Myrna Loy

After the Thin Man '36
Airport '75 '75
The Ambassador's Daughter
'56
The Animal Kingdom '32
Another Thin Man '39
Ants '77
April Fools '69
Arrowsmith '31
The Bachelor and the Bobby-
Soxer '47
The Best Years of Our Lives
'46
Broadway Bill '34
A Connecticut Yankee '31
Consolation Marriage '31
Don Juan '26
The End '78
From the Terrace '60
The Great Ziegfeld '36
Jazz Singer '27
Just Tell Me What You Want
'80
Libeled Lady '36
Lonelyhearts '58
Manhattan Melodrama '34
Midnight Lace '60
Mr. Blandings Builds His
Dream House '48
The Red Pony '49
Renegade Ranger/Scarlet
River '38
Rogue of the Rio Grande '30
Shadow of the Thin Man '41
So This is Paris '26
Song of the Thin Man '47
Summer Solstice '81
Test Pilot '39
The Thin Man '34
The Thin Man Goes Home '44
Too Hot to Handle '38
Topaze '33

Nanni Loy (D)

Cafe Express '83

Head of the Family '71
Where's Piccone '84

Irma Lozano

All for Nothing '75
Benjamin Argumedo '78

Margarita Lozano

Half of Heaven '86
Night of the Shooting Stars
'82
Viridiana '61

John Lozier

Sixteen '72

Antonella Lualdi

Mafia la Ley que non Perdona
'79
The Mongols '60

Arthur Lubin (D)

Buck Privates '41
Delightfully Dangerous '45
Escapade in Japan '57
Francis in the Navy '55
Hold That Ghost '41
Impact '49
In the Navy '41
The Incredible Mr. Limpet '64
Keep 'Em Flying '41
Phantom of the Opera '43
Queen for a Day '51
Ride 'Em Cowboy '42
Thief of Baghdad '61
Yellowstone '36

Ernst Lubitsch

One Arabian Night '21

Ernst Lubitsch (D)

Gypsy Blood '18
Heaven Can Wait '43
Lady Windermere's Fan '26
The Marriage Circle '24
Merry Widow '34
Ninotchka '39
One Arabian Night '21
Passion '19
The Shop Around the Corner
'40
So This is Paris '26
The Student Prince in Old
Heidelberg '27
That Uncertain Feeling '41
To Be or Not to Be '42

S. Roy Luby (D)

Arizona Badman '42
Arizona Stagecoach '42
Black Market Rustlers '43
Boot Hill Bandits '42
Border Phantom '37
Kid's Last Ride '42
Outlaw Rule '36
The Red Rope '37
Tough to Handle '37
Trail of the Silver Spurs '41
Tumbledown Ranch in
Arizona '41
Underground Rustlers '41

Arthur Lucan

My Son, the Vampire '52
Old Mother Riley's Ghosts '41

Jack Lucarelli

American Justice '86

George Lucas

Hearts of Darkness: A
Filmmaker's Apocalypse
'91

Elves '89
Hardly Working '81
Red Wind '91

Huand Lund

Young Hero

Jana Lund

Married Too Young '62

John Lund

Dakota Incident '56
Duchess of Idaho '50
The Perils of Pauline '47
Wackiest Ship in the Army '61

Lucille Lund

The Black Cat '34

Zoe Tamerlaine Lund

Exquisite Corpses '88

Christine Lunde

Dead End City '88
Masque of the Red Death '90

Christopher Lunde

Mankillers '87

Gerda Lundequist

The Atonement of Gosta Berling '24

Dolph Lundgren

Cover-Up '91
I Come in Peace '90
Masters of the Universe '87
The Punisher '90
Red Scorpion '89
Rocky 4 '85
Showdown in Little Tokyo '91
A View to a Kill '85

William Lundigan

Andy Hardy's Double Life '42
Danger on the Air '38
Dishonored Lady '47
Follow Me Quietly '49
Love Nest '51
State Department File 649 '49
The White Orchid '54
Wives Under Suspicion '38

Vic Lundin

Ma Barker's Killer Brood '60

Steve Lundquist

Killer Tomatoes Strike Back '90
Return of the Killer Tomatoes '88

Randy Lundsford

Night Screams '87

Jeff Lundy

The American Angels: Baptism of Blood '89

Jessica Lundy

Caddyshack 2 '88
Madhouse '90

Alex Lung

Thunderfist '73

Huang I Lung

Fists of Dragons '69
Little Big Master '70s
Two Dragons Fight Against the Tiger

Lei Hsiao Lung

Golden Sun '80s

Liang Hsian Lung

Las Doce Patadas Mortales del Kung Fu '87

Lu I Lung

Mission Kiss and Kill

Shangkuan Lung

Kung Fu Commandos '80

Shou Lung

The True Game of Death '70s

Suma Wah Lung

Kick of Death: The Prodigal Boxer '80s
Prodigal Boxer: The Kick of Death '73

Thomas Lung

Sakura Killers '86

Ti Lung

Duel of Fists
Duel of the Iron Fist '72

Tong Lung

Fist of Death '87

Wen Ching Lung

Kung Fu Rebels '80s

Yuan Lung

Mission Kiss and Kill

Cherie Lunghi

Excalibur '81
King David '85
Letters to an Unknown Lover '84
The Mission '86
Oliver Twist '82
Praying Mantis '83
Tales of the Klondike: The Unexpected '83
To Kill a Priest '89

Dan Lunham

The Strangeness '85

Ann Lunn

Estate of Insanity '70

Lian Lunson

The Big Hurt '87

Alfred Lunt

Sally of the Sawdust '25

Min Luong

Gleaming the Cube '89

Leticia Lupersio

Dona Herlinda & Her Son '86

Roldano Lupi

The Giant of Metropolis '61
The Mongols '60

Ida Lupino

The Adventures of Sherlock Holmes '39
Beware, My Lovely '52
The Bigamist '53
Devil's Rain '75
Food of the Gods '76
Forever and a Day '43
High Sierra '41
Junior Bonner '72

Lust for Gold '49
My Boys are Good Boys '78
On Dangerous Ground '51
One Rainy Afternoon '36
Private Hell 36 '54
Road House '48
Sea Devils '37
Thank Your Lucky Stars '43
They Drive by Night '40
While the City Sleeps '56

Ida Lupino (D)

The Bigamist '53
The Hitch-Hiker '53
The Trouble with Angels '66

Wallace Lupino

The Deputy Drummer '35

Alberto Lupo

The Agony and the Ecstacy '65
Atom Age Vampire '61
The Bacchantes '63
Coriolanus, Man without a Country '64
Django Shoots First '74
The Giant of Marathon '60
Herod the Great '60

Michele Lupo (D)

Goliath and the Sins of Babylon '64
Master Touch '74
Mean Frank and Crazy Tony '75

Patti LuPone

Driving Miss Daisy '89
Fighting Back '82
LBJ: The Early Years '88
1941 '79
Wise Guys '86
Witness '85

Robert LuPone

High Stakes '89

Federico Luppi

Funny, Dirty Little War '83
Mayalunta '86
Time for Revenge '82

John Lupton

Jesse James Meets Frankenstein's Daughter '65
The Rebel Set '59

Peter Lupus

Escapist '83
More! Police Squad '82
Muscle Beach Party '64
Pulsebeat '85
Think Big '90

Esther Luquin

Nosotros (We) '77

Evan Lurie

Martial Law 2: Undercover '91

John Lurie

Down by Law '86
The Last Temptation of Christ '88
Permanent Vacation '84
Stranger than Paradise '84

Hamilton Luske

Alice in Wonderland '51
Bongo '47

Hamilton Luske (D)

Lady and the Tramp '55
101 Dalmations '61
Peter Pan '53

Jaques Lussier

Norman's Awesome Experience '88
Pouvoir Intime '87

Kathryn Luster

Cop-Out '91

Marlena Lustic

My Brother has Bad Dreams '88

Aaron Lustig

Bad Channels '92

William Lustig (D)

The Hit List '88
Maniac '80
Maniac '80
Maniac Cop '88
Maniac Cop 2 '90
Relentless '89
Vigilante '83

Salvo Luther (D)

Forget Mozart '86

Paul Luty

Water Babies '79

Joleen Lutz

Outtakes '85

Nola Luxford

Sylvia Scarlett '35

Franc Luz

Ecstasy '84
Ghost Town '88
The Nest '88

Maria Pia Luzi

Planets Against Us '61

Bruce Ly

Chinatown Connection '90

Georges Lycan

Triple Cross '67

Tom Lycos

Sweetie '89

Mona Lyden

Million Dollar Mystery '87

Robert Lyden

Rocky Jones, Space Ranger, Vol. 7: Forbidden Moon '53

Bob Lydiard

The Paper Chase '73

Jimmy Lydon

Back Door to Heaven '39
Life with Father '47
Little Men '40
Strange Illusion '45
Tom Brown's School Days '40

John (Johnny Rotten) Lydon

Corrupt '84

Reg Lye

Freedom '81

Dorothy Lyman

Camp Cucamonga: How I Spent My Summer Vacation '90
Summer Fantasy '84

Jacquie Lyn

Pack Up Your Troubles '32

Sandy Lyn

Violent Women '59

Jeffery Lynas

Lies My Father Told Me '75

Alfred Lynch

Bewitched '85
Blockhouse '73

Catherine Lynch

Under Capricorn '82

David Lynch

Zelly & Me '88

David Lynch (D)

Blue Velvet '86
Dune '84
Elephant Man '80
Eraserhead '78
Twin Peaks '90

Helen Lynch

The Return of Grey Wolf '22

Jimmy Lynch

Avenging Disco Godfather '76

John Lynch

Cal '84
Hardware '90
Monkey Boy '90

Kate Lynch

Def-Con 4 '85
Meatballs '79
Reckless Disregard '84
Summer's Children '84

Kelly Lynch

Cocktail '88
Curly Sue '91
Desperate Hours '90
Drugstore Cowboy '89
Osa '85
Road House '89
Warm Summer Rain '89

Ken Lynch

Jet Attack/Paratroop Command '58
My Pleasure Is My Business '74

Paul Lynch (D)

At Sword's Point '51
Blindside '88
Bullies '86
Cross Country '83
Humongous '82
Murder by Night '89
Prom Night '80

Ray Lynch

The Zodiac Killer '71

Richard Lynch

Aftershock '88
Bad Dreams '88
The Barbarians '87
Baron
Death Sport '78

Susannah of the Mounties '39
Texas Masquerade '44
13th Guest '32

Jeanette MacDonald

Bitter Sweet '41
Girl of the Golden West '38
I Married an Angel '42
Lottery Bride '30
Maytime '37
Merry Widow '34
Naughty Marietta '35
New Moon '40
Rose Marie '36
San Francisco '36
Smilin' Through '41
Sweethearts '38

Jessica Wright MacDonald

Lianna '83

Kenneth MacDonald

Six Gun Gospel '43

Mary Ann MacDonald

Love at First Sight '76

Michael MacDonald

Mr. Nice Guy '86
Mystery of the Million Dollar Hockey Puck '80s
The Nutcracker Prince '91

Ray Macdonald

Babes on Broadway '41

Wallace MacDonald

Between Fighting Men '32
Red Signals '27
Texas Cyclone '32

Wendy MacDonald

Blood Frenzy '87
Dark Side of the Moon '90
Heat Street '87
Mayhem '87

Andie MacDowell

Green Card '90
Greystoke: The Legend of Tarzan, Lord of the Apes '84
Hudson Hawk '91
St. Elmo's Fire '85
sex, lies and videotape '89
Women & Men: Stories of Seduction, Part 2 '91

Paul Mace

The Lords of Flatbush '74

Terry Mace

Bury Me an Angel '71

Victor Mace

Fighting Pilot '35

Rita Macedo

The Criminal Life of Archibaldo de la Cruz '55
The Curse of the Crying Woman '61
El Fugitivo '65
Nazarin '58

Mike MacFarland (D)

Hanging on a Star '78
Pink Motel '82

Moyna MacGill

The Strange Affair of Uncle Harry '45

Niall MacGinnis

Betrayed '54
Curse of the Demon '57
The Day Will Dawn '42
Edge of the World '37
Island of Terror '66
Lust for Life '56
Martin Luther '53
Sword of Sherwood Forest '60
The War Lord '65

Tara MacGowan

Las Vegas Serial Killer '86
Secret Places '85

Jack MacGowran

The Brain '62
The Exorcist '73
The Fearless Vampire Killers '67
King Lear '71
Lord Jim '65
The Quiet Man '52
Start the Revolution Without Me '70
Wonderwall: The Movie '69

Ali MacGraw

Convoy '78
The Getaway '72
Goodbye Columbus '69
Just Tell Me What You Want '80
Love Story '70
Murder Elite '86
Players '79
Winds of War '83

Brian MacGregor

Free Ride '86

Scottie MacGregor

The Student Nurses '70

Philip Machale

Slugs '87

Marla MacHart

Blood Sisters '86

Gustav Machaty (D)

Ecstasy '33

Ignacy Machowski

First Spaceship on Venus '60

Stephen Macht

American Dream '81
Amityville 1992: It's About Time '92
The Choirboys '77
Enola Gay: The Men, the Mission, the Atomic Bomb '80
Galaxina '80
Graveyard Shift '90
Killjoy '81
The Monster Squad '87
Trancers 3: Deth Lives '92

Ian MacInnes

Half Moon Street '86

Polly MacIntyre

Jet Benny Show '86

Betty Mack

Fighting Texans '33

Outlaw Rule '36
Toll of the Desert '35

Brice Mack (D)

Jennifer '78
Swap Meet '79

Charles Mack

Dream Street '21

Earle Mack (D)

The Children of Theatre Street '78

Floyd Mack

Rehearsal '47

Helen Mack

Fargo Express '32
Fit for a King '37
Melody Cruise '32
Milky Way '36
The Return of Peter Grimm '35
Son of Kong '33
The Wrong Road '37

Kerry Mack

Fair Game '85
Fantasy Man '84
Savage Attraction '83

Marion Mack

The General '27
General/Slapstick '26

Michael Mack

Star Quest '89

Russell Mack (D)

Lonely Wives '31

Wayne Mack

Crypt of Dark Secrets '76
Mardi Gras Massacre '78
Storyville '74

Dorothy Mackaill

Ranson's Folly '26
Shore Leave '25

Barry Mackay

Sailing Along '38

Fulton Mackay

Defense of the Realm '85
Nothing But the Night '72

John MacKay

Alligator Eyes '90
Fat Guy Goes Nutzoid '86
The Rejuvenator '88

Bronwyn MacKay-Payne

Dawn! '83

Yvonne Mackay (D)

The Silent One '86

Helen MacKellar

Bad Boy '39
Delinquent Parents '38

Alexander MacKendrick (D)

Ladykillers '55
Man in the White Suit '51
Mandy '53
Sweet Smell of Success '57
Whiskey Galore '48

Ella MacKenzie

Cord & Dischords/Naughty Nurse '28

Evan Mackenzie

Ghoulies 3: Ghoulies Go to College '91

Giselle MacKenzie

Music Shoppe '82
One Minute Before Death '88
The Oval Portrait '88

Jan MacKenzie

The American Angels: Baptism of Blood '89
Gator Bait 2: Cajun Justice '88

John MacKenzie (D)

Act of Vengeance '86
Beyond the Limit '83
The Fourth Protocol '87
The Last of the Finest '90
The Long Good Friday '79
One Brief Summer '70
Ruby '92
A Sense of Freedom '78

Patch MacKenzie

Defense Play '88
Goodbye, Norma Jean '75
Isle of Secret Passion '85

Peter M. MacKenzie (D)

Merchants of War '90

Philip Charles MacKenzie

Blind Justice '86

Will MacKenzie (D)

Hobo's Christmas '87
Worth Winning '89

Scott MacKenzsie

Hootch Country Boys '75

Vivian Mackerell

Madhouse Mansion '74

Barry Mackey

Evergreen '34

Paul Mackey

People Who Own the Dark '75

Mary Mackie

Alternative '76

Gilles Mackinnon (D)

The Playboys '92

Albert Macklin

Date with an Angel '87

Jim Mackrell

Just Between Friends '86

Duncan Maclachlan (D)

Scavengers '87

Janet MacLachlan

Big Mo '73
Darker Than Amber '70
Murphy's Law '86

Roll of Thunder, Hear My Cry '78
She's in the Army Now '81
Sounder '72

Kyle MacLachlan

Blue Velvet '86
Don't Tell Her It's Me '90
The Doors '91
Dune '84
The Hidden '87
Twin Peaks '90

Shirley MacLaine

All in a Night's Work '61
The Apartment '60
Around the World in 80 Days '56
Artists and Models '55
Being There '79
The Bliss of Mrs. Blossom '68
Can-Can '60
Cannonball Run II '84
Career '59
A Change of Seasons '80
The Children's Hour '62
Gambit '66
Golden Age of Hollywood '77
Hot Spell '58
Irma La Douce '63
Loving Couples '80
Madame Sousatzka '88
The Matchmaker '58
My Geisha '62
Possession of Joel Delaney '72
Postcards from the Edge '90
Some Came Running '58
Steel Magnolias '89
Sweet Charity '68
Terms of Endearment '83
That's Dancing! '85
The Trouble with Harry '55
Turning Point '79
Two Mules for Sister Sara '70
Waiting for the Light '90
War Years '77
Woman Times Seven '67

Barton MacLane

All Through the Night '42
Big Street '42
Black Fury '35
Bombardier '43
Bugles in the Afternoon '52
Bullets or Ballots '38
The Case of the Lucky Legs '35
Drums in the Deep South '51
"G" Men '35
The Geisha Boy '58
The Half-Breed '51
Kansas Pacific '53
Let's Dance '50
The Maltese Falcon '41
Man of the Forest '33
Marine Raiders '43
Melody Ranch '40
Mutiny in the Big House '39
Nabonga '44
Naked in the Sun '57
Prince and the Pauper '37
Prison Break '38
Song of Texas '43
The Spanish Main '45
To the Last Man '33
Treasure of Fear '45
Treasure of the Sierra Madre '48
The Underdog '43
Unknown Island '48
Western Union '41
You Only Live Once '37

Mansfield ▶cont.

Rocky Jones, Space Ranger: Trial of Rocky Jones '53
Rocky Jones, Space Ranger, Vol. 3: Blast Off '53
Rocky Jones, Space Ranger, Vol. 7: Forbidden Moon '53

Scott Mansfield (D)
Deadly Games '80

Leonor Manso
Far Away and Long Ago '74
House of Shadows '83
Made in Argentina '86

Alan Manson
Let's Scare Jessica to Death '71
Whiffs '75

Jean Manson
The Young Nurses '73

Laura Manszky
Sweet Lies '88

Joe Mantegna
Alice '90
Bugsy '91
Compromising Positions '85
Critical Condition '86
Elvis: The Movie '79
The Godfather: Part 3 '90
Homicide '91
House of Games '87
The Money Pit '86
Off Beat '86
Queens Logic '91
Second Thoughts '83
Suspect '87
Things Change '88
Three Amigos '86
Wait Until Spring, Bandini '90
Weeds '87

Fernando Soto Mantequilla
Guitarras Lloren Guitarras '47

Clive Mantle
Robin Hood...The Legend: Robin Hood and the Sorcerer '83

Randolph Mantooth
The Seekers '79
Terror at London Bridge '85

Paul Mantu
Wolf Lake '79

Mike Manty
Blood Debts '83

Robert Manuel
Rififi '54

Guido Manuli (D)
Guido Manuli: Animator (Nightmare at the "Opera") '90

Chantal Manz
The 13th Mission '91

Linda Manz
Days of Heaven '78
Orphan Train '79
Out of the Blue '82
Snow Queen '83

Wanderers '79

Armando Manzanero
Somos Novios (In Love and Engaged) '68

Eduardo Manzano
El Aviso Inoportuno '87
Hijazo De Mi Vidaza (Son of My Life) '70s

Miguel Manzano
The Illusion Travels by Streetcar '53

Ray Manzarek
The Doors: A Tribute to Jim Morrison '81
The Doors: Live in Europe, 1968 '68
The Doors: Live at the Hollywood Bowl '68

Ray Manzarek (D)
The Doors: Live in Europe, 1968 '68
The Doors: Soft Parade, a Retrospective '69

David Manzy
The Baby '72

Angela Mao
The Association '80s
The Fate of Lee Khan '73
Return of the Tiger '78
Sting of the Dragon Masters '74
When Taekwondo Strikes '83

Perla Mar
Sonar no Cuesta nada Joven (Dreams are Free) '70s

Adele Mara
Bells of Rosarita '45
The Inner Circle '46
Night Time in Nevada '48
Robin Hood of Texas '47
Twilight on the Rio Grande '41

La Mara
Rings Around the World '67

Steve Marachuk
Hot Target '85
Piranha 2: The Spawning '82
The Spawning '82
Waikiki '80

Jean Marais
The Battle of Austerlitz '60
Beauty and the Beast '46
Donkey Skin (Peau d'Ane) '70
Dos Cruces en Danger Pass '71
Elena and Her Men '56
Eternal Return '43
Orpheus '49
Rape of the Sabines '61
The Testament of Orpheus '59
White Nights '57

Maria Marais
Power of One '92

Sam Marais
Playing with Fire '70s

Andree Maranda
Toxic Avenger '86

Evi Marandi
Planet of the Vampires '65

Janna Marangosoff
Men... '85

Cindy Maranne
Slashdance '89

Richard Marant
On the Third Day '83

Mario Maranzana
Day of the Cobra '84

Pete Maravich
Scoring '80

Peter Marc
Dangerous Love '87

Joss Marcano
Delivery Boys '84

Robert Marcarelli (D)
I Don't Buy Kisses Anymore '92
Original Intent '92

Marcel Marceau
Barbarella '68
Red Skelton's Funny Faces '79
Silent Movie '76

Sophie Marceau
La Boum '81
Police '85

Terry Marcel (D)
Hawk the Slayer '81
Jane & the Lost City '87
Prisoners of the Lost Universe '84

Su Carnal Marcelo
Hay Muertos Que No Hacen Ruido '47

Alex March (D)
Firehouse '72
Master Mind '73
Paper Lion '68

Barbara March
Deserters '80s

Fredric March
Alexander the Great '55
Anna Karenina '35
Anthony Adverse '36
The Barretts of Wimpole Street '34
The Best Years of Our Lives '46
The Bridges at Toko-Ri '55
Christmas Carol '54
Christopher Columbus '49
Desperate Hours '55
Dr. Jekyll and Mr. Hyde '31
Executive Suite '54
Hombre '67
I Married a Witch '42
Inherit the Wind '60
Les Miserables '35
The Man in the Gray Flannel Suit '56
Mary of Scotland '36
Nothing Sacred '37
Seven Days in May '64
So Ends Our Night '41

A Star is Born '37
Susan and God '40
Till the Clouds Roll By/A Star Is Born '46

Tony March
Shallow Grave '87

Georges Marchal
The French Way '52

Colette Marchand
Moulin Rouge '52

Corinne Marchand
Cleo from 5 to 7 '61

Guy Marchand
Coup de Torchon '81
Cousin, Cousine '76
Entre-Nous (Between Us) '83
Heat of Desire (Plein Sud) '84
Loulou '80
May Wine '90
Petit Con '84

Henri Marchand
A Nous la Liberte '31

Nancy Marchand
The Bostonians '84
Brain Donors '92
Directions '66 '66
The Hospital '71
Killjoy '81
Marty '53
The Naked Gun: From the Files of Police Squad '88
Soldier's Home '77

Ron Marchini
Death Machines '76
Forgotten Warrior '86
Ninja Warriors '85
Omega Cop '90
Return Fire '91
Wolf '86

David Marciano
Hellbent '90
Kiss Shot '90s

Paul Marco
Night of the Ghouls '59

Silvia Marco
The Man Who Wagged His Tail '57

Saverio Marconi
Padre Padrone '77

Andrea Marcovicci
The Concorde: Airport '79 '79
The Devil's Web '74
The Front '76
The Hand '81
Kings and Desperate Men '83
Packin' It In '83
Someone to Love '87
Spacehunter: Adventures in the Forbidden Zone '83
Spragque '84
The Stuff '85

Deborah Marcus
Cycle Vixens '79

James Marcus
The Broken Mask '28
The Eagle '25
The Lone Avenger '33
Sadie Thompson '28

James Marcus (D)
Double Cross '90s

Philip Marcus (D)
Terror on Alcatraz '86

Richard Marcus
Cannibal Campout '88
Deadly Friend '86
Enemy Mine '85

Theo Marcuse
Harum Scarum '65

Elio Marcuzzo
Ossessione '42

Jordan Marder
Walking on Air '87

Heinz Marecek
Tea For Three '84

Lynn Maree
Deadly Passion '85

Vera Maretskaya
No Greater Love '43

Simone Marevil
Un Chien Andalou '28

Janie Mareze
La Chienne '31

Andreas Marfori (D)
Evil Clutch '89

Arthur Margetson
Juggernaut '37
Sherlock Holmes Faces Death '43

Antonio Margheriti (D)
Assignment Outer Space '61
Wild, Wild Planet '65

Lukacs Margit
Oh, Bloody Life! '88

Agi Margittay
Nice Neighbor '79

Margo
Behind the Rising Sun '43
I'll Cry Tomorrow '55
The Leopard Man '43
Lost Horizon '37
Viva Zapata! '52
Winterset '37

Bill Margold
What Would Your Mother Say? '81

Janet Margolin
Annie Hall '77
David and Lisa '63
Distant Thunder '88
Enter Laughing '67
The Game of Love '90
The Last Embrace '79
Morituri '65
Planet Earth '74
Plutonium Incident '82
Pray for the Wildcats '74
Take the Money and Run '69
The Triangle Factory Fire Scandal '79

Stuart Margolin

Big Bus '76
Bye Bye Blues '91
Class '83
Days of Heaven '78
Death Wish '74
A Fine Mess '86
Futureworld '76
Glitter Dome '84
Guilty by Suspicion '91
Iron Eagle 2 '88
Kelly's Heroes '70
Running Hot '83
S.O.B. '81

Stuart Margolin (D)

Glitter Dome '84
Paramedics '88
A Shining Season '79

Jeff Margolis (D)

Oscar's Greatest Moments:
  1971-1991 '92
Richard Pryor: Live in Concert
  '79

Mark Margolis

Descending Angel '90

Miriam Margolyes

Little Dorrit, Film 1: Nobody's
  Fault '88
Little Dorrit, Film 2: Little
  Dorrit's Story '88
Orpheus Descending '91

Michael Margotta

The Blue Knight '75
Can She Bake a Cherry Pie?
  '83

David Margulies

Dressed to Kill '80
Funny About Love '90
Ghostbusters 2 '89
9 1/2 Weeks '86
Running on Empty '88

Fiorella Mari

The Day the Sky Exploded '57

Gina Mari

The Slumber Party Massacre
  '82

Angelica Maria

Somos Novios (In Love and
  Engaged) '68
To Kill a Stranger '70s

Ferdinand Marian

Jud Suess '40

Mark Marian

Tomb of Torture '65

Ann Marie

Beneath the Valley of the
  Ultra-Vixens '79

Jeanne Marie

If Looks Could Kill '86
Young Nurses in Love '89

Nadia Marie

Quiet Fire '91

Jean-Pierre Marielle

Coup de Torchon '81
One Wild Moment '78

Marietto

Angel in a Taxi '59
It Started in Naples '60

Roy Marika

Where the Green Ants Dream
  '84

Wandjuk Marika

Where the Green Ants Dream
  '84

Cheech Marin

After Hours '85
Born in East L.A. '87
Cheech and Chong: Still
  Smokin' '83
Cheech and Chong: Things
  Are Tough All Over '82
Cheech and Chong's Next
  Movie '80
Cheech and Chong's Nice
  Dreams '81
Cheech and Chong's, The
  Corsican Brothers '84
Cheech and Chong's Up in
  Smoke '79
Echo Park '86
Far Out Man '89
Fatal Beauty '87
It Came from Hollywood '82
La Pastorela '92
Rude Awakening '89
The Shrimp on the Barbie '90
Yellowbeard '83

Cheech Marin (D)

Born in East L.A. '87

Edwin L. Marin (D)

Abilene Town '46
The Cariboo Trail '50
A Christmas Carol '38
The Death Kiss '33
I'd Give My Life '36
Johnny Angel '45
Miss Annie Rooney '42
Mr. Ace '46
Nocturne '46
Show Business '44
A Study in Scarlet '33
Tall in the Saddle '44

Gloria Marin

Bromas S.A. '87
Historia de un Gran Amor
  (History of a Great Love)
  '57

Jacques Marin

Island at the Top of the World
  '74

Jason Marin

The Little Mermaid '89

Luis Marin

Great Treasure Hunt '70s

Rikki Marin

Cheech and Chong: Things
  Are Tough All Over '82
Cheech and Chong's, The
  Corsican Brothers '84

Ed Marinaro

Born Beautiful '82
Dead Aim '87
The Game of Love '90
Policewoman Centerfold '83

Barbara Marineau

When Nature Calls '85

Alex Marino

Sex with a Smile '76

Kenny Marino

Alphabet City '84
Prince of the City '81

Lex Marinos (D)

An Indecent Obsession '85
Remember Me '85

Beth Marion

Between Men '35
Rip Roarin' Buckaroo '36

George F. Marion

Anna Christie '30
Death from a Distance '36
Hook, Line and Sinker '30

Salverio Marioni

Padre Padrone '77

Mona Maris

The Falcon in Mexico '44
Heartbeat '46
I Married an Angel '42

Peter Maris (D)

Delirium '77
Diplomatic Immunity '91
Hangfire '91
Land of Doom '84
Terror Squad '87
True Blood '89
Viper '88

Richard Maris

Thrilled to Death '88

Silvia Mariscal

La Fuerza Inutil '87

Marisol

Blood Wedding (Bodas de
  Sangre) '81

Kensaku Marita

Renegade Ninja '83
War in Space '77

Sari Maritza

Crimson Romance '34
Greek Street '30
International House '33

Robert Marius

The 13th Mission '91

Mary Mark

Voyage to the Planet of
  Prehistoric Women '68

Jane Marken

Crazy for Love '52
A Day in the Country '46
Dedee D'Anvers '49

Chris Marker (D)

La Jetee/An Occurrence at
  Owl Creek Bridge '64
Le Joli Mai '62

Russ Marker (D)

The Yesterday Machine '63

Anthony Markes (D)

Last Dance '91

Tony Markes

In the Aftermath '87

Tony Markes (D)

Bikini Island '91

David Markey (D)

Desperate Teenage Lovedolls
  '84

Enid Markey

Civilization '16
The Fugitive: Taking of Luke
  McVane '15
Tarzan of the Apes '17

Barbara Markham

House of Whipcord '75

Kika Markham

Deep Cover '88
Miss A & Miss M '86
Two English Girls '72

Monte Markham

Counter Measures
Defense Play '88
Ginger in the Morning '73
Guns of the Magnificent
  Seven '69
Hotline '82
Hustling '75
Judgment Day '88
Off the Wall '82
Shame, Shame on the Bixby
  Boys '82

Monte Markham (D)

Defense Play '88
Neon City '91

Kelly Markin (D)

Tiger Claws '91

Gabriel Markiw (D)

Mob Story '90

Jancarlo Markiw (D)

Mob Story '90

Ted Markland

Angels from Hell '68
Wanda Nevada '79

Fletcher Markle (D)

The Incredible Journey '63
Jigsaw '49

Peter Markle (D)

Bat 211 '88
El Diablo '90
Hot Dog...The Movie! '83
Nightbreaker '89
The Personals '83
Youngblood '86

Stephen Markle

984: Prisoner of the Future
  '84
Perfect Timing '84

Monica Marko

Possession: Until Death Do
  You Part '90

Margaret Markov

Hot Box '72
Naked Warriors '73
Run, Angel, Run! '69

Olivera Markovic

Fury is a Woman '61

Murray Markowitz (D)

Left for Dead '78

Robert Markowitz (D)

Afterburn '92
The Belarus File '85
A Long Way Home '81
Pray TV '82

Alfred Marks

Mission: Monte Carlo '75
Scream and Scream Again '70

Arthur Marks (D)

Bonnie's Kids '73
Bucktown '75
Detroit 9000 '73
Friday Foster '75
J.D.'s Revenge '76
Monkey Hustle '77
Togetherness '70

Barbara Marks

Teenage Devil Dolls '53

Harrison Marks

Nine Ages of Nakedness '69

Joe E. Marks

Li'l Abner '59

Logena Marks

Secrets in the Dark '83

Michael Marks

The Wasp Woman '49

Sherry Marks

Satan's Cheerleaders '77

Slash Marks

Night of the Cobra Woman '72

Willis Marks

The Jack Knife Man '20

Zoli Marks

No One Cries Forever '85

Morley Markson (D)

Off Your Rocker '80

Winnie Markus

Devil In Silk '56
The Mozart Story '48

Arnold Marle

Men of Two Worlds '46
The Snake Woman '61

Bob Marley

Island Reggae Greats '85

John Marley

Deathdream '72
Framed '75
Glitter Dome '84
The Godfather '72
Godfather 1902-1959—The
  Complete Epic '81
The Greatest '77
Hooper '78
It's Alive 2: It Lives Again '78
Jory '72
Kid Vengeance '75
Love Story '70
A Man Called Sledge '71
On the Edge '86

Marley ▶cont.

Robbers of the Sacred
Mountain '83
Threshold '83
Utilities '83

Anne Marlie

Swinging Sorority Girls '70s

Carla Marlier

The Avenger '62
The Big Grab '63
Spirits of the Dead '68
Zazie dans le Metro '61

Alan Marlowe

Deadly Vengeance '81

Hugh Marlowe

Birdman of Alcatraz '62
Bugles in the Afternoon '52
Castle of Evil '66
The Day the Earth Stood Still
'51
Earth vs. the Flying Saucers
'56
Elmer Gantry '60
Illegal '55
Monkey Business '52
Rawhide '50
Twelve O'Clock High '49

June Marlowe

Don Juan '26
The Lone Defender '32
The Night Cry '26
Pardon Us '31

Linda Marlowe

Big Zapper '73
The Green Man '91
He Kills Night After Night After
Night '70
Night After Night After Night
'70
Night Slasher '84

Florence Marly

Dr. Death, Seeker of Souls '73
Planet of Blood '66
Tokyo Joe '49

Percy Marmont

Four Sided Triangle '53
Lisbon '56
Mantrap '26
Rich and Strange '32
Young and Innocent '37

**Malcolm
Marmorstein** (D)

Dead Men Don't Die '91

**Elizabeth Marner-
Brooks**

I Drink Your Blood '71

Kelli Maroney

Fast Times at Ridgemont High
'82
Night of the Comet '84
Servants of Twilight '91
Slayground '84
Zero Boys '86

Carl Marotte

Breaking All the Rules '85
Hang Tough '89
Pick-Up Summer '79
Workin' for Peanuts '85

Christian Marquand

And God Created Woman '57
Corrupt Ones '67
End of Desire '62

**Richard
Marquand** (D)

Eye of the Needle '81
Hearts of Fire '87
The Jagged Edge '85
The Legacy '79
Return of the Jedi '83
Until September '84

**Maria Elena
Marques**

Across the Wide Missouri '51
Ambush at Tomahawk Gap
'53
The Pearl '48

**Jacques
Marquette** (D)

Meteor Monster '57

Esteban Marquez

Mexican Bus Ride '51

Evarist Marquez

Burn! '70

Andre Marquis

Paco '75

Charles Marquis (D)

Tension at Table Rock '56

Margaret Marquis

Brand of the Outlaws '36

Eddie Marr

Hollywood Canteen '44
I Was a Teenage Werewolf
'57

Leon Marr (D)

Dancing in the Dark '86

Pons Marr

The American Scream '80s

Moore Marriot

The Flying Scotsman '29
Oh, Mr. Porter '37

David Marriott

Monster High '89

John Marriott

Black Like Me '64

Sylvia Marriott

Crimes at the Dark House '39
Two English Girls '72

Dina Marrone

Quest for the Mighty Sword
'90

Jose Marrone

Una Viuda Descocada

**Roberto
Marrtero** (D)

Thirty-Six Hours of Hell '77

Kenneth Mars

The Apple Dumpling Gang
Rides Again '79
Beer '85

**Butch Cassidy and the
Sundance Kid** '69
Fletch '85
For Keeps '88
Goin' Coconuts '78
Illegally Yours '87
The Little Mermaid '89
Misfits of Science '85
Night Moves '75
Parallax View '74
The Producers '68
Protocol '84
Radio Days '87
Rented Lips '88
Shadows and Fog '92
What's Up, Doc? '72
Young Frankenstein '74

Maurice Marsac

Clarence, the Cross-eyed Lion
'65
Tarzan and the Trappers '58
Werewolf in a Girl's Dormitory
'61

Branford Marsalis

School Daze '88

Marita Marschall

Deadline '87

Beatrice Marsden

The Ghost of Rashmon Hall
'47

Carol Marsh

Alice in Wonderland '50
Brighton Rock '47
A Christmas Carol '51
The Horror of Dracula '58

David Marsh (D)

Lords of Magick '88

Garry Marsh

A Girl in a Million '46
Just William's Luck '47
Pink Strings and Sealing Wax
'45
Someone at the Door '50

Jean Marsh

The Changeling '80
Frenzy '72
Goliath Awaits '81
Horsemasters '61
Return to Oz '85
The Unearthly Stranger '64
Willow '88

Joan Marsh

Champagne for Breakfast '79
Manhunt in the African Jungle
'54

Linda Marsh

Homebodies '74

Mae Marsh

Birth of a Nation '15
D.W. Griffith Triple Feature
'13
The Fighting Sullivans '42
From the Terrace '60
Gunfighter '86
Home Sweet Home '14
Hoodoo Ann '16
Intolerance '16
Short Films of D.W. Griffith,
Volume 2 '09
Story of the Silent Serials Girls
in Danger
The Tall Men '55
Three Godfathers '48

While the City Sleeps '56
The White Rose '23

Marian Marsh

The Black Room '35
Svengali '31
When's Your Birthday? '37
Youth on Parole '37

Matthew Marsh

An Affair in Mind '89

Michele Marsh

Deadly Alliance '78
Evil Town '70s

Ray Marsh (D)

The Last Porno Flick '74

Yvonne Marsh

Little Ballerina '47

Philip Marshak (D)

Cataclysm '81
Dracula Sucks '79

Alan Marshal

After the Thin Man '36
Conquest '32
House on Haunted Hill '58
The Howards of Virginia '40
Lydia '41
Tom, Dick, and Harry '41

Brenda Marshall

The Sea Hawk '40

Bryan Marshall

Because of the Cats '74
The Long Good Friday '79
Return to Snowy River '88

Darrah Marshall

Teenage Caveman '58

**David Anthony
Marshall**

Another 488 Hrs. '90

Dodie Marshall

Easy Come, Easy Go '67
Spinout '66

Don Marshall

Terminal Island '73

E.G. Marshall

Abduction of St. Anne '75
The Bridge at Remagen '69
Broken Lance '54
The Buccaneer '58
The Caine Mutiny '54
The Chase '66
Creepshow '82
Independence '76
Interiors '78
Ironclads '90
Is Paris Burning? '68
Kennedy '83
Lazarus Syndrome '79
The Left Hand of God '55
The Littlest Angel '69
The Mountain '56
My Chauffeur '86
National Lampoon's
Christmas Vacation '89
The Poppy Is Also a Flower
'66
Power '86
Pursuit '72
The Pursuit of Happiness '70
Revenge of TV Bloopers '60s
Saigon: Year of the Cat '87

Silver Chalice '54
Superman 2 '80
Thank You Mr. President '84
13 Rue Madeleine '46
Tora! Tora! Tora! '70
Twelve Angry Men '57

Frank Marshall (D)

Arachnophobia '90

Garry Marshall

Lost in America '85

Garry Marshall (D)

Beaches '88
The Flamingo Kid '84
Frankie & Johnny '91
Nothing in Common '86
Overboard '87
Pretty Woman '89
Young Doctors in Love '82

George Marshall (D)

Boy, Did I Get a Wrong
Number! '66
Destry Rides Again '39
Fancy Pants '50
Goldwyn Follies '38
Houdini '53
How the West was Won '62
Never a Dull Moment '50
Off Limits '53
Pack Up Your Troubles '32
The Perils of Pauline '47
Pot O' Gold '41
The Sad Sack '57
Scared Stiff '53
Show Them No Mercy '35
Texas '41
Valley of the Sun '42
You Can't Cheat an Honest
Man '39

Herbert Marshall

Blonde Venus '32
Captain Blackjack '51
Crack-Up '46
Duel in the Sun '46
Enchanted Cottage '45
The Fly '58
Foreign Correspondent '40
Forever and a Day '43
The Letter '40
The Little Foxes '41
The Moon and Sixpence '42
Murder '30
The Painted Veil '34
The Razor's Edge '46
The Secret Garden '49
Stage Struck '57
The Underworld Story '50
The Virgin Queen '55
When Ladies Meet '41
A Woman Rebels '36

James Marshall

Gladiator '92

Ken Marshall

Double Exposure: The Story
of Margaret Bourke-White
'89
Feds '88
Krull '83
Tilt '78

Lee Marshall

The American Angels:
Baptism of Blood '89

Meri D. Marshall

Valet Girls '86

Nancy Marshall

Frankenstein Meets the Space Monster '65

Paul Marshall

Ninja Operation: Licensed to Terminate '87

Penny Marshall

Challenge of a Lifetime '85
The Hard Way '91
How Come Nobody's On Our Side? '73
Movers and Shakers '85

Penny Marshall (D)

Awakenings '90
Big '88
Jumpin' Jack Flash '86

Peter Marshall

A Guide for the Married Woman '78
Return of Jesse James '50

Sarah Marshall

The People vs. Jean Harris '81

Sean Marshall

Pete's Dragon '77

Steve Marshall

Night of the Creeps '86

Tony Marshall

I Was a Teenage Werewolf '57

Trudy Marshall

The Fighting Sullivans '42
Joe's Bed-Stuy Barbershop: We Cut Heads '83
Married Too Young '62

Tully Marshall

Behind Prison Walls '43
Big Trail '30
The Cat and the Canary '27
Fighting Caravans '32
Hawthorne of the USA '19
Let's Go! '23
Queen Kelly '29
Smouldering Fires '25
Strangers of the Evening '32
Tom Sawyer '30

William Marshall

Blacula '72
The Boston Strangler '68
Demetrius and the Gladiators '54
The Great Skycopter Rescue '82
Scream Blacula Scream '73
State Fair '45
That Brennan Girl '46
Vasectomy: A Delicate Matter '86

William Marshall (D)

Adventures of Captain Fabian '51
The Phantom Planet '61

Zena Marshall

Dr. No '62
Terronauts '67
The Terrornauts '67

Christina Marsillach

Every Time We Say Goodbye '86
Terror at the Opera '88

Tony Marsina

Last Mercenary '84
Tornado '83

Lynn Marta

Joe Kidd '72
Richard Petty Story '72

Arlene Martel

Angels from Hell '68
Demon with a Glass Hand '64

Jeanne Martel

Santa Fe Bound '37

June Martel

Forlorn River '37

K.C. Martel

Munster's Revenge '81

Wendy Martel

Sorority House Massacre '86

Chris Martell

Flesh Feast '69
Gruesome Twosome '67
Scream, Baby, Scream '69

Donna Martell

Abbott and Costello Meet the Killer, Boris Karloff '49
Give a Girl a Break '53
Project Moon Base '53
Ten Wanted Men '54

Gillian Martell

Oliver Twist '85

Lisa Repo Martell

American Boyfriends '89

Peter Martell

Mission Phantom '79
Planet on the Prowl '65

Norma Martelli

Everybody's Fine '91

Max Volkert Martens

Judgment in Berlin '88

Mona Martenson

The Atonement of Gosta Berling '24

Ian Marter

Doctor Faustus '68

Frank Marth

Fright '56

Virgilio Marti

Crossover Dreams '85

Alberto Martin

Los Hijos de Lopez

Ana Martin

Fin de Fiesta '71
Siempre Hay una Primera Vez '87
Tacos al Carbon '87

Andra Martin

Up Periscope '59

Andrea Martin

Club Paradise '86
Comic Relief 2 '87
Rude Awakening '89
Soup for One '82
Stepping Out '91
Too Much Sun '90
Worth Winning '89

Anne-Marie Martin

Hammered: The Best of Sledge '88

Aubrey Martin

This is Not a Test '62

Barney Martin

Arthur 2: On the Rocks '88

Billy Martin

One in a Million: The Ron LaFlore Story '78

Bob Martin

Hide and Seek '77

Charles Martin

Fighting Black Kings '76

Charles Martin (D)

Death of a Scoundrel '56
My Dear Secretary '49
One Man Jury '78

Christopher Martin

House Party '90
House Party 2: The Pajama Jam '91

Claude Martin

Inn of Temptation '86

Culture Martin

Raw and Rough Pinchers '87

Damon Martin

Amityville 1992: It's About Time '92
Ghoulies 2 '88

Danny Martin

Surprise Attack '70s

Dean Martin

Airport '70
All in a Night's Work '61
The Ambushers '67
Artists and Models '55
At War with the Army '50
Bandolero! '68
Bells are Ringing '60
The Caddy '53
Cannonball Run '81
Cannonball Run II '84
Career '59
Five Card Stud '68
Four for Texas '63
Hollywood or Bust '56
Hollywood Clowns '85
Kiss Me, Stupid! '64
Murderers' Row '66
Ocean's 11 '60
Rio Bravo '59
Road to Bali '53
Robin and the Seven Hoods '64
Scared Stiff '53
Showdown '73
Some Came Running '58
Sons of Katie Elder '65

Television's Golden Age of Comedy '50s
That's Dancing! '85
Toys in the Attic '63
The Young Lions '58

Dean Paul Martin

Backfire '88
Misfits of Science '85
Players '79

Debbie Martin

The Game '89

Dewey Martin

The Big Sky '52
Desperate Hours '55
Flight to Fury '66
Land of the Pharaohs '55
Savage Sam '63
Seven Alone '75
The Thing '51

Diana Martin

Minnesota Clay '65

Dick Martin

Carbon Copy '81
The Glass Bottom Boat '66

Don Martin

Punk Vacation '90

Duke Martin

Lost Zeppelin '29

D'Urville Martin

Big Score '83
Blind Rage '78
Boss '74
Death Journey '76
Sheba, Baby '75

D'Urville Martin (D)

Dolemite '75
Fass Black '77

Eugenio Martin (D)

Pancho Villa '72

Frank Martin (D)

Doctor Butcher M.D. '80
John Huston: The Man, the Movies, the Maverick '89

Gene Martin (D)

Bad Man's River '72
Horror Express '72

George Martin

The Compleat Beatles '82
Crossing Delancey '88
Falling in Love '84
Psychopath '68

Helen Martin

Doc Hollywood '91
Hero Ain't Nothin' But a Sandwich '78
Hollywood Shuffle '87
Night Angel '90

James Aviles Martin (D)

Flesh Eating Mothers '89

Jared Martin

Karate Warrior
The New Gladiators '87
Quiet Cool '86
The Sea Serpent '85

Jean Martin

Battle of Algiers '66

John Martin

Black Roses '88
Dark Before Dawn '89
Fire in the Night '85
The Tell-Tale Heart '62

Kiel Martin

Child Bride of Short Creek '81
Convicted: A Mother's Story '87
Panic in Needle Park '75

Lawrence Martin

House Party '90

Lewis Martin

The War of the Worlds '53

Lock Martin

The Day the Earth Stood Still '51

Luis Martin

Black Box Affair '66

Maribel Martin

The Blood Spattered Bride '72

Marion Martin

Man in the Iron Mask '39
Sinners in Paradise '38

Mary Martin

Irving Berlin's America '86
Main Street to Broadway '53
Night and Day '46
Peter Pan '60
Rodgers & Hammerstein: The Sound of American Music '85

Mary Catherine Martin

Clean and Sober '88

Michael Martin

Colt is My Law (Mi Revolver es la Ley)

Millicent Martin

Alfie '66
Horsemasters '61
Those Magnificent Men in Their Flying Machines '65

Nan Martin

For Love of Ivy '68
Goodbye Columbus '69
Matters of the Heart '90
Toys in the Attic '63
The Young Nurses '73

Pamela Sue Martin

Buster and Billie '74
Eye of the Demon '87
Flicks '85
Girls of Huntington House '73
Lady in Red '79
Torchlight '85

Pepper Martin

Angels from Hell '68
Evil Altar '89
Scream '83
The Shepherd '88

Ricci Martin

Just Tell Me You Love Me '80

Richard Martin

Dynamite Pass '50
Gun Smugglers/Hot Lead '51
Gunplay '51
Mysterious Desperado '49
Mysterious Desperado/Rider
From Tucson '49
Rider from Tucson '50
Riders of the Range '49
Riders of the Range/Storm
Over Wyoming '50
Storm Over Wyoming '50
Tender Comrade '43

Ross Martin

Dying Room Only '73
Experiment in Terror '62
More Wild, Wild West '80
Sealab 2020 '72
The Seekers '79
Wild, Wild West Revisited '79
Wild & Wooly '78

Sallie Martin

Say Amen, Somebody '80

Sandy Martin

Vendetta '85

Sharlene Martin

Possession: Until Death Do
You Part '90

Skip Martin

Horror Hospital '73

Sonja Martin

Young Love - Lemon Popsicle
7 '87

Steve Martin

All of Me '84
Best of Dan Aykroyd '86
Best of Gilda Radner '89
Dead Men Don't Wear Plaid
'82
Dirty Rotten Scoundrels '88
Father of the Bride '91
Grand Canyon '91
The Jerk '79
L.A. Story '91
Little Shop of Horrors '86
The Lonely Guy '84
The Man with Two Brains '83
Movers and Shakers '85
Muppet Movie '79
My Blue Heaven '90
Parenthood '89
Pennies From Heaven '81
Planes, Trains & Automobiles
'87
Roxanne '87
Sgt. Pepper's Lonely Hearts
Club Band '78
Saturday Night Live: Steve
Martin, Vol. 1 '78
Saturday Night Live: Steve
Martin, Vol. 2 '78
Steve Martin Live '85
Three Amigos '86

Strother Martin

Ballad of Cable Hogue '70
Better Late Than Never '79
The Brotherhood of Satan '71
Butch Cassidy and the
Sundance Kid '69
Cheech and Chong's Up in
Smoke '79
Cool Hand Luke '67
Hannie Caulder '72
Hard Times '75
The Horse Soldiers '59
Hot Wire '70s

Invitation to a Gunfighter '64
Kiss Me Deadly '55
Love and Bullets '79
Nightwing '79
Pocket Money '72
Rooster Cogburn '75
Slap Shot '77
Steel Cowboy '78
The Wild Bunch '69

Todd Martin

Finger on the Trigger '65

Tony Martin

Big Store '41
Easy to Love '53
Hit the Deck '55
Music in My Heart '40
Till the Clouds Roll By '46
Till the Clouds Roll By/A Star
Is Born '46
Two Tickets to Broadway '51
Ziegfeld Girl '41

Vera Martin

Loyalties '86

Vince Martin

Breaking Loose '90
Night Master '87

Vivian Martin

The Wishing Ring '14

W.T. Martin

Hardhat & Legs '80
High Stakes '89

Elsa Martinelli

Belle Starr Story '70s
Blood and Roses '61
Hatari '62
Madigan's Millions '67
Oldest Profession '67
10th Victim '65
The Trial '63
Woman Times Seven '67

A. Martinez

The Cowboys '72
The Hunt for the Night Stalker
'91
Joe Panther '76
Once Upon a Scoundrel '73
Powwow Highway '89
She-Devil '89
Starbird and Sweet William
'73

Adalberto Martinez

El Aviador Fenomeno '47
Soy un Golfo '65

Chuck Martinez (D)

Nice Girls Don't Explode '87

Claudio Martinez

Daring Dobermans '73

James Martinez

Leader of the Band '87

Jennie Martinez

Nurse on Call '88

Jorge Martinez

Virgins of Purity House '79

Leo Martinez

She Devils in Chains

Nacho Martinez

Matador '86

Patrice Martinez

Three Amigos '86

Rene Martinez, Jr. (D)

The Guy from Harlem '77

Waldo Martinez

Un Elefante Color Ilusion

Richard Martini (D)

Limit Up '89
You Can't Hurry Love '88

Al Martino

The Godfather: Part 3 '90

John Martino

Truck Stop Women '74

Luciano Martino

Emerald Jungle '80

Raymond Martino

Mayhem '87

Sergio Martino (D)

Casablanca Express '89
The Cheaters '80s
Day of the Maniac '77
The Great Alligator '81
Next Victim '71
The Opponent '89
Screamers '80
Sex with a Smile '76
Slave of the Cannibal God '79
Torso '73
Violent Professionals '82

Danny Martins

The Hostage '67

Ellie Martins

Night Vision '87

Orlando Martins

Men of Two Worlds '46
Simba '55

Henry Martinson

Northern Lights '79

Leslie Martinson (D)

The Atomic Kid '54
Batman '66
The Fantastic World of D.C.
Collins '84
Hot Rod Girl '56
Kid with the 200 I.Q. '83
The Kid with the Broken Halo
'82
PT 109 '63
Return to Africa '89

Andrew Marton (D)

Africa Texas Style '67
Around the World Under the
Sea '65
Clarence, the Cross-eyed Lion
'65

Eva Marton

Tosca '84
Tosca '89

Akihiro Maruyama

Black Lizard '68

Tom Maruzzi

Man Beast '55

Paul Marvel

The Psychotronic Man '91

The Marvelettes

Girl Groups: The Story of a
Sound '83

Lee Marvin

Bad Day at Black Rock '54
Big Heat '53
The Big Red One '80
The Caine Mutiny '54
Cat Ballou '65
Comancheros '61
Death Hunt '81
Delta Force '86
The Dirty Dozen '67
The Dirty Dozen: The Next
Mission '85
Dog Day '83
Donovan's Reef '63
Gorky Park '83
Great Scout & Cathouse
Thursday '76
Gun Fury '53
Hangman's Knot '52
Hell in the Pacific '69
I Died a Thousand Times '55
The Killers '64
The Klansman '74
The Man Who Shot Liberty
Valance '62
The Meanest Men in the West
'60s
Missouri Traveler '58
Monte Walsh '70
Paint Your Wagon '70
Pete Kelly's Blues '55
Pocket Money '72
Point Blank '67
Prime Cut '72
Professionals '66
Raintree County '57
Sergeant Ryker '68
Shack Out on 101 '55
Ship of Fools '65
Shout at the Devil '76
Union Station '50
The Wild One '54

Mike Marvin (D)

Hamburger... The Motion
Picture '86
The Wraith '87

Alan Marx

For Auld Lang Syne: The
Marx Brothers '32
Gore-Met Zombie Chef From
Hell '87

Chico Marx

Animal Crackers '30
At the Circus '39
Big Store '41
The Cocoanuts '29
A Day at the Races '37
Duck Soup '33
For Auld Lang Syne: The
Marx Brothers '32
Go West '40
Hollywood Clowns '85
Horse Feathers '32
Love Happy '50
Monkey Business '31
A Night in Casablanca '46
A Night at the Opera '35
Revenge of TV Bloopers '60s
Room Service '38

Groucho Marx

Animal Crackers '30
At the Circus '39
Big Store '41

The Cocoanuts '29
Copacabana '47
A Day at the Races '37
Double Dynamite '51
Duck Soup '33
For Auld Lang Syne: The
Marx Brothers '32
Funny Guys & Gals of the
Talkies '30s
A Girl in Every Port '52
Go West '40
Groucho/Howdy Doody Show
Hollywood Clowns '85
Horse Feathers '32
Love Happy '50
Monkey Business '31
A Night in Casablanca '46
A Night at the Opera '35
Revenge of TV Bloopers '60s
Room Service '38
Television's Golden Age of
Comedy '50s

Harpo Marx

Animal Crackers '30
At the Circus '39
Big Store '41
The Cocoanuts '29
A Day at the Races '37
Duck Soup '33
For Auld Lang Syne: The
Marx Brothers '32
Go West '40
Hollywood Clowns '85
Hollywood Goes to War '54
Horse Feathers '32
Love Happy '50
Monkey Business '31
A Night in Casablanca '46
A Night at the Opera '35
Revenge of TV Bloopers '60s
Room Service '38
Stage Door Canteen '43
Strictly G.I. '44

Zeppo Marx

Animal Crackers '30
The Cocoanuts '29
Duck Soup '33
Horse Feathers '32
Monkey Business '31
Revenge of TV Bloopers '60s

Carolyn Marz

Driller Killer '74

Franco Marzi

Island Monster '53

Ron Masak

Man from Clover Grove '78

Shin-Ichi Masaki (D)

The Humanoid '91

Mil Mascaras

Frontera Sin Ley
Los Vampiros de Coyacan '87

Pierrino Mascarino

Summer Rental '85

Joseph Mascelli (D)

The Atomic Brain '64

Laurence E. Mascott (D)

Brave Rifles '66
Those Crazy Americans '67

Marino Mase

Les Carabiniers '63

Martin Mase
Alien Contamination '81

Francesco Maselli (D)
Love in the City '53
Time of Indifference '64

Claudio Masenza (D)
James Dean '85
Marlon Brando '85
Montgomery Clift '85

Daniel Masey
The Cat and the Canary '79

Nelson Mashita
Darkman '90

Giulietta Masina
Europa '51 '52
Ginger & Fred '86
Juliet of the Spirits '65
La Strada '54
Nights of Cabiria '57
The Swindle '55
Variety Lights '51
The White Sheik '52

Giuseppe Masini (D)
Journey Beneath the Desert '61

Ace Mask
Transylvania Twist '89

Virginia Maskell
Only Two Can Play '82

Walter Maslow
Atlas '60

Bill Mason
Yellowneck '55

Connie Mason
Blood Feast '63
2000 Maniacs '64

Dana Mason
Death Row Diner '88

Eric Mason
Black Starlet '74
Kiss of the Tarantula '75

Herbert Mason (D)
Lady in Distress '39

Hilary Mason
Dolls '87
Don't Look Now '73
Meridian: Kiss of the Beast '90
Robot Jox '89

Ingrid Mason
Death Train '79

Jackie Mason
Caddyshack 2 '88
History of the World: Part 1 '81
Jackie Mason on Broadway '89
The Jerk '79

James Mason
The Assisi Underground '84
Bad Man's River '72
The Blue Max '66
Botany Bay '53
The Boys from Brazil '78
Caught '49
Cold Sweat '74
The Concentratin' Kid '30
Cross of Iron '76
Dangerous Summer '82
The Desert Fox '51
The Desert Rats '53
The Destructors '74
11 Harrowhouse '74
Evil Under the Sun '82
Face to Face '52
The Fall of the Roman Empire '64
ffolkes '80
Fire Over England '37
For Heaven's Sake '26
Forever Darling '56
Georgy Girl '66
Heaven Can Wait '78
High Command '37
Hotel Reserve '44
I Met a Murderer '39
Inside Out '75
Island in the Sun '57
Ivanhoe '82
Jesus of Nazareth '77
Journey to the Center of the Earth '59
Julius Caesar '53
Kidnap Syndicate '76
The Last of Sheila '73
Late Extra '35
Lolita '62
Lord Jim '65
Mackintosh Man '73
Madame Bovary '49
Man of Evil '48
The Man in Grey '45
Mandingo '75
The Mill on the Floss '37
Murder by Decree '79
The Night Has Eyes '42
North by Northwest '59
Odd Man Out '47
Prince Valiant '54
Prisoner of Zenda '52
Reckless Moment '49
Salem's Lot: The Movie '79
Seventh Veil '46
The Shooting Party '85
Sidney Sheldon's Bloodline '79
A Star is Born '54
Street War '70s
Tiara Tahiti '62
20,000 Leagues Under the Sea '54
The Verdict '82
Voyage of the Damned '76
Water Babies '79
The Wicked Lady '45
Yellowbeard '83
Yin & Yang of Mr. Go '71

Larry Mason
The Adventures of the Masked Phantom '38

Leroy Mason
The Mystery Man '35
Outlaw Express '38
The Painted Stallion '37
Silver Stallion '41
Vigilantes of Boom Town '46
Western Mail '42

Madison Mason
Glitz '88

Marlyn Mason
The Trouble with Girls '69

Marsha Mason
Audrey Rose '77
Blume in Love '73
Chapter Two '79
The Cheap Detective '78
Cinderella Liberty '73
Dinner at Eight '89
Drop Dead Fred '91
Goodbye Girl '77
Heartbreak Ridge '86
The Image '89
Lois Gibbs and the Love Canal '82
Max Dugan Returns '83
Only When I Laugh '81
Promises in the Dark '79
Stella '89
Trapped in Silence '90

Mary Mason
Cheyenne Kid '33

Morgan Mason
The Sandpiper '65

Noel Mason (D)
Fighting Pilot '35

Pamela Mason
Navy vs. the Night Monsters '66

Robert Mason
Codename: Terminate '90

Sharon Mason
Primal Scream '87

Syd Mason
Secret File of Hollywood '62

Thomas Boyd Mason
In a Shallow Grave '88

Tom Mason
Aliens Are Coming '80
Kicks '85
Mississippi Burning '88
Walking Through the Fire '80

Walter Mason
Black Like Me '64

John Massari
The Wizard of Speed and Time '88

Lea Massari
Allonsanfan '73
Christ Stopped at Eboli '79
L'Avventura '60
Les Choses de la Vie '70
Les Rendez-vous D'Anna '78
Murmur of the Heart '71
The Story of a Love Story '73
Vengeance '86

Francesco Massaro (D)
Private Affairs '89

Osa Massen
Master Race '44
Outcasts of the City '58
Rocketship X-M '50
A Woman's Face '41

Richard Massery
Killzone '85

Ivana Massetti (D)
Domino '88

Anna Massey
Another Country '84
A Doll's House '89
Foreign Body '86
Frenzy '72
Hotel Du Lac '86
The Little Drummer Girl '84
Mansfield Park '85
The McGuffin '85
Mountains of the Moon '90
Peeping Tom '63
Sweet William '79
Vault of Horror '73

Daniel Massey
The Entertainer '60
The Incredible Sarah '76
Intimate Contact '87
Love with a Perfect Stranger '86
Vault of Horror '73
Victory '81

Dick Massey
The Commitments '91

Edith Massey
Desperate Living '77
Female Trouble '75
Multiple Maniacs '70
Mutants In Paradise '88
Pink Flamingos '73
Polyester '81

Gina Massey
Thrillkill '84

Ilona Massey
Love Happy '50
Melody Master '41
Northwest Outpost '47
Rosalie '38

Raymond Massey
Abe Lincoln in Illinois '40
Action in the North Atlantic '43
Arsenic and Old Lace '44
Battle Cry '55
Chain Lightning '50
David and Bathsheba '51
Dreaming Lips '37
Drums '38
East of Eden '55
Fire Over England '37
The Forty-Ninth Parallel '41
The Fountainhead '49
The Great Impostor '61
How the West was Won '62
Hurricane '37
MacKenna's Gold '69
The Naked and the Dead '58
Omar Khayyam '57
The President's Plane is Missing '71
Prisoner of Zenda '37
Reap the Wild Wind '42
Santa Fe Trail '40
The Scarlet Pimpernel '34
The Speckled Band '31
Things to Come '36
Under the Red Robe '36

Paul Massie
Sapphire '59

Leonide Massine
The Red Shoes '48

Joe Massot (D)
Space Riders '83

Wonderwall: The Movie '69

Ben Masters
Celebrity '85
Dream Lover '85
Key Exchange '85
Making Mr. Right '86
Mandingo '75
The Shadow Box '80

James Masters
Rip-Off '77

Marie Masters
Scream for Help '86

Quentin Masters (D)
Dangerous Summer '82
PSI Factor '80
The Stud '78
Thumb Tripping '72

Sharon Masters
Deadline '82

Mary Stuart Masterson
Amazing Stories, Book 2 '86
At Close Range '86
Chances Are '89
Fried Green Tomatoes '91
Funny About Love '90
Gardens of Stone '87
Heaven Help Us '85
Immediate Family '89
Mr. North '88
My Little Girl '87
Some Kind of Wonderful '87

Peter Masterson
A Question of Guilt '78

Peter Masterson (D)
Blood Red '88
Full Moon in Blue Water '88
Night Game '89
Trip to Bountiful '85

Sean Masterson
Fatal Games '84

Valerie Masterson
Julius Caesar '84

Nico Mastorakis (D)
Blind Date '84
Glitch! '88
Hired to Kill '91
In the Cold of the Night '89
Next One '84
Nightmare at Noon '87
Ninja Academy '90
Sky High '84
Terminal Exposure '89
The Wind '87
Zero Boys '86

Mary Elizabeth Mastrantonio
The Abyss '89
Class Action '91
Color of Money '86
Fools of Fortune '91
The January Man '89
Robin Hood: Prince of Thieves '91
Scarface '83
Slamdance '87
White Sands '92

Camillo Mastrocinque (D)

Full Hearts & Empty Pockets '63

Gina Mastrogiacomo

Alien Space Avenger '91

Armand Mastroianni (D)

Cameron's Closet '89
Distortions '87
Double Revenge '89
He Knows You're Alone '80
Killing Hour '84
The Supernaturals '86

Marcello Mastroianni

Allonsanfan '73
Bellissimo: Images of the
 Italian Cinema '87
Beyond Obsession '89
Big Deal on Madonna Street
 '60
Blood Feud '79
Casanova '70 '65
City of Women '81
Dark Eyes '87
Diary of Forbidden Dreams
 '76
Divine Nymph '71
Divorce—Italian Style '62
8 1/2 '63
Everybody's Fine '91
Gabriela '84
Get Rita
Ginger & Fred '86
Henry IV '85
La Dolce Vita '60
La Nuit de Varennes '82
Lady of the Evening '79
Lunatics & Lovers '76
Macaroni '85
Massacre in Rome '73
The Poppy Is Also a Flower
 '66
Salut l'Artiste '74
Shoot Loud, Louder, I Don't
 Understand! '66
A Slightly Pregnant Man '79
A Special Day '77
Stay As You Are '78
10th Victim '65
A Very Private Affair '62
We All Loved Each Other So
 Much '77
Where the Hot Wind Blows
 '59
White Nights '57
Wifemistress '79
Yesterday, Today and
 Tomorrow '64

Toshio Masuda (D)

Tora! Tora! Tora! '70
Zero '84

Toru Masuoka

A Taxing Woman Returns '88

Richard Masur

Adam '83
Amazing Stories, Book 3 '86
The Believers '87
Betrayal '89
Burning Bed '85
Cast the First Stone '89
East of Eden '80
Embassy '85
Fallen Angel '81

Far from Home '89
Heartburn '86
Heaven's Gate '81
Hiroshima Maiden '88
License to Drive '88
Mean Season '85
My Science Project '85
Nightmares '83
Rent-A-Cop '88
Risky Business '83
Semi-Tough '77
Shoot to Kill '88
The Thing '82
Third Degree Burn '89
Timerider '83
Under Fire '83
Walker '87
Walking Through the Fire '80
When the Bough Breaks '86
Who'll Stop the Rain? '78
Wild Horses '84

Orestes Matacena (D)

Tainted '88

Eddy Matalon (D)

Blackout '78
Cathy's Curse '89

Vivian Matalon (D)

Private Contentment '83

Gabor Mate

My Twentieth Century '90

Rudolph Mate (D)

Branded '50
Dark Past '49
Deep Six '58
D.O.A. '49
The Green Glove '52
Second Chance '53
Union Station '50
When Worlds Collide '51

Steve Mateo

Vice Academy 3 '91

Julian Mateos

Four Rode Out '69
Hellbenders '67
Kashmiri Run '69

Aubrey Mather

Adventures of Don Juan '49
House of Fear '45

Aubrey Mather

Mrs. Miniver '42

Aubrey Mather

The Silent Passenger '35
That Forsyte Woman '49
The Undying Monster '42

George Mather

Ring of Terror '62

James Mathers

Dr. Jekyll's Dungeon of Death
 '82

Jerry Mathers

Back to the Beach '87
Down the Drain '89
The Trouble with Harry '55

Marissa Mathes

Track of the Vampire '66

Judy Matheson

The House that Vanished '73

Scream and Die '74
Twins of Evil '71

Michelle Matheson

Howling 6: The Freaks '90

Tim Matheson

The Apple Dumpling Gang
 Rides Again '79
Best Legs in the 8th Grade
 '84
Blind Justice '86
Buried Alive '90
Dreamer '79
Drop Dead Fred '91
Eye of the Demon '87
Fletch '85
Impulse '84
Listen to Your Heart '83
A Little Sex '82
Little White Lies '89
The Longest Drive '76
Magnum Force '73
Mary White '77
National Lampoon's Animal
 House '78
1941 '79
The Runaway Barge '75
Speed Zone '88
To Be or Not to Be '83
Up the Creek '84
Yours, Mine & Ours '68

Carmin Mathews

Sounder '72

Carole Mathews

Strange Awakening '58

Christopher Mathews

The Seducer '82

George Mathews

Last of the Comanches '52
The Man with the Golden Arm
 '55

Jessie Mathews

Sailing Along '38

Kerwin Mathews

Battle Beneath the Earth '68
Devil at 4 O'Clock '61
Jack the Giant Killer '62
Killer Likes Candy '78
Maniac '63
Nightmare in Blood '75
Octaman '71
The Seventh Voyage of
 Sinbad '58
The Three Worlds of Gulliver
 '59
Waltz King '63

Oliver Mathews

Devil Kiss '77

Thom Mathews

Alien from L.A. '87
Bloodmatch '91
Down Twisted '89
Friday the 13th, Part 6: Jason
 Lives '86
Return of the Living Dead Part
 2 '88

Darien Mathias

Blue Movies '88

Samantha Mathis

Pump Up the Volume '90
This is My Life '91

Steve Matilla

Last Night at the Alamo '83

Doug Matley

Sledgehammer '83

Marlee Matlin

Bridge to Silence '89
Children of a Lesser God '86
The Linguini Incident '92
Walker '87

John Matshikiza

Dust '85
Mandela '87

Shue Matsubayashi (D)

Last War '68

Eiko Matsuda

In the Realm of the Senses
 '76

Yusaku Matsuda

The Family Game '83
Murder in the Doll House '79

Yoko Matsuyama

Watch Out, Crimson Bat! '69

Sharon Matt

Sweet Trash '89

Dietrich Mattausch

The Wannsee Conference '84

Bruno Mattei (D)

Night of the Zombies '83
Seven Magnificent Gladiators
 '84

Marius Mattei (D)

Moving Target '89

Eva Mattes

Celeste '81
David '79
A Man Like Eva '83
Stroszek '77
Woyzeck '78

Pam Matteson

Girls of the Comedy Store '86
Punchline '88

Charles Matthau (D)

Doin' Time on Planet Earth '88

Walter Matthau

Atomic Attack '50
The Bad News Bears '76
Buddy Buddy '81
Cactus Flower '69
California Suite '78
Casey's Shadow '78
Charade '63
Charley Varrick '73
The Couch Trip '87
Earthquake '74
Ensign Pulver '64
A Face in the Crowd '57
Fail-Safe '64
First Monday in October '81
The Fortune Cookie '66
The Front Page '74
A Guide for the Married Man
 '67
Hello, Dolly! '69
Hopscotch '80
House Calls '78
I Ought to Be in Pictures '82

The Incident '89
JFK '91
Kentuckian '55
King Creole '58
Kotch '71
Laughing Policeman '74
Little Miss Marker '80
Lonely are the Brave '62
Mirage '66
Movers and Shakers '85
A New Leaf '71
The Odd Couple '68
Pete 'n' Tillie '72
Pirates '86
Plaza Suite '71
Secret Life of an American
 Wife '68
Strangers When We Meet '60
The Sunshine Boys '75
Survivors '83
Suspense '53
The Taking of Pelham One
 Two Three '74

A.E. Matthews

Made in Heaven '52

Al Matthews

American Roulette '88

Brian Matthews

The Burning '82
Red Nights '87

Carole Matthews

Shark River '53
Swamp Women '55

Dakin Matthews

Child's Play 3 '91
Naked Lie '89

Francis Matthews

Corridors of Blood '58
The McGuffin '85
Murder Ahoy '64
Small Hotel '57

Fritz Matthews

Born Killer '89
Deadly Prey '87
Hell on the Battleground '88

Jessie Matthews

Candles at Nine '44
Evergreen '34
Forever and a Day '43
The Hound of the Baskervilles
 '77
It's Love Again '36

John Matthews

A Secret Space '88

Junius Matthews

Sword in the Stone '63

Lee Matthews

Fu Manchu '56

Lester Matthews

The Raven '35
Werewolf of London '35

Peter Matthey

New York Nights '84

Darlene Mattingly

Women Unchained '70s

Burny Mattinson

The Great Mouse Detective
 '86

Luisa Mattiol

La Bella Lola '87

Frank S. Mattison (D)

With Kit Carson Over the Great Divide '25

Sally Mattison (D)

Slumber Party Massacre 3 '90

Mario Mattoli (D)

Two Nights with Cleopatra '54

Walt Mattox (D)

Scared to Death '47

Arne Mattson (D)

The Doll '62
The Girl '86

Robin Mattson

Bonnie's Kids '73
Candy Stripe Nurses '74
Captain America '79
Hot Rod '79
In Between '90s
Return to Macon County '75
Take Two '87
Wolf Lake '79

Veronika Mattson

Pay Off '89

Peter Matulavich (D)

Jupiter Menace '82

Victor Mature

After the Fox '66
Androcles and the Lion '52
Betrayed '54
Captain Caution '40
Dangerous Mission '54
Demetrius and the Gladiators '54
Easy Living '49
The Egyptian '54
Footlight Serenade '42
I Wake Up Screaming '41
Kiss of Death '47
Las Vegas Story '52
The Long Haul '57
Million Dollar Mermaid '52
My Darling Clementine '46
One Million B. C. '40
The Robe '53
Samson and Delilah '49
Samson and Delilah '84
Seven Days' Leave '42
The Shanghai Gesture '41
Song of the Islands '42

John Matuszak

Caveman '81
Down the Drain '89
Ghost Writer '89
The Goonies '85
Ice Pirates '84
North Dallas Forty '79
One Man Force '89
Semi-Tough '77

Johanna Matz

The Life and Loves of Mozart '59
They were So Young '55

Billy Mauch

Prince and the Pauper '37

Jack Mauck

Greenstone '85

Arthur Maude

The Man from Beyond '22

Joan Maude

Power '34
The Temptress '49

Mary Maude

Crucible of Terror '72

Howard Mauer

Ilsa, the Tigress of Siberia '79

Monica Maughan

Cactus '86

Bill Mauldin

The Red Badge of Courage '51

Wayne Maunder

Crazy Horse and Custer: "The Untold Story" '90

Sarah Maur-Thorp

Edge of Sanity '89
River of Death '90
Ten Little Indians '89

Carmen Maura

Ay, Carmela! '90
Dark Habits '84
Law of Desire '86
Matador '86
What Have I Done to Deserve This? '85
Women on the Verge of a Nervous Breakdown '88

Joshua Maurer

The Hill '88
Tour of Duty '87

Peggy Maurer

I Bury the Living '58

Nicole Maurey

Day of the Triffids '63
The Jayhawkers '59
Rogue's Yarn '56

Robert Mauri (D)

Invincible Gladiators '64
Slaughter of the Vampires '62

Jon Maurice

Primal Scream '87

Giovanni Mauriello

Another Time, Another Place '83

Claire Maurier

The 400 Blows '59
La Cage aux Folles '78
Sweet Ecstasy '62

Toralv Maurstad

Song of Norway '70

Gerda Maurus

Spies '28
Woman in the Moon '29

Derrel Maury

Massacre at Central High '76

Jean Max

J'Accuse '37

Ron Max

Forced Entry '75

Edwin Maxwell

Behind Prison Walls '43
Mystery Liner '34
Swamp Fire '46
Taming of the Shrew '29
Waterfront '44
Woman in Brown '48

Frank Maxwell

The Haunted Palace '63
Lonelyhearts '58
Shame '61

James Maxwell

One Day in the Life of Ivan Denisovich '71

Jane Maxwell

The Black Devils of Kali '55

John Maxwell

Boss of Big Town '43
Honky Tonk '41
The Payoff '43
Prowler '51

Larry Maxwell

Poison '91

Lois Maxwell

Diamonds are Forever '71
Dr. No '62
Eternal Evil '87
For Your Eyes Only '81
From Russia with Love '63
Goldfinger '64
Live and Let Die '73
The Man with the Golden Gun '74
Moonraker '79
On Her Majesty's Secret Service '69
The Spy Who Loved Me '77
Thunderball '65
The Unstoppable Man '59
A View to a Kill '85
You Only Live Twice '67

Marilyn Maxwell

Champion '49
The Lemon Drop Kid '51
Off Limits '53
Summer Holiday '48
Wild Women '70

Paul Maxwell

Cloud Waltzing '87
Going Home '86

Paul Maxwell (D)

Superargo '67

Peter Maxwell (D)

The Highest Honor '84
Platypus Cove '86
Plunge Into Darkness '77
Run, Rebecca, Run '81
Touch & Go '80

Roberta Maxwell

Lois Gibbs and the Love Canal '82
Psycho 3 '86

Ronald F. Maxwell (D)

Kidco '83
Little Darlings '80

The Night the Lights Went Out in Georgia '81

Angela May

Blackjack '78

April May

Hollywood High Part 2 '81

Deborah May

Johnny Be Good '88
Rage of Angels '85

Donald May

Bogie '80

Elaine May

California Suite '78
Enter Laughing '67
In the Spirit '90
Luv '67
A New Leaf '71

Elaine May (D)

The Heartbreak Kid '72
Ishtar '87
Mikey & Nicky '76
A New Leaf '71

Jack May

He Kills Night After Night After Night '70
Night After Night After Night '70
Night Slasher '84

Jodhi May

Eminent Domain '91
Max and Helen '90
A World Apart '88

Joe May (D)

Homecoming '28
The Invisible Man Returns '40

Lola May

Civilization '16

Lyn May

Las Munecas del King Kong '80s

Mathilda May

Letters to an Unknown Lover '84
Lifeforce '85

Paul May (D)

Dr. Mabuse vs. Scotland Yard '64

Rita May

Threads '85

John Mayall

Sunset Strip '85

Rik Mayall

Drop Dead Fred '91
Little Noises '91
The Young Ones '88

Anthony Mayans

Heat of the Flame '82
Zombie Lake '84

Big Maybelle

Jazz on a Summer's Day '59

Gale Mayberrie

Swinging Wives '79

Russ Mayberry (D)

Challenge of a Lifetime '85
Fer-De-Lance '74
The $5.20 an Hour Dream '80
The Jesus Trip '71
Marriage is Alive and Well '80
A Matter of Life and Death '81
Probe '72
The Rebels '79
Unidentified Flying Oddball '79

Eddie Mayehoff

Artists and Models '55
How to Murder Your Wife '64

Chip Mayer

Our Family Business '81
Survivor '87

Frederic Mayer

Field of Honor '87

Gerald Mayer (D)

Man Inside '76

Ray Mayer

Swing It, Sailor! '37

Val Mayerick

The Demon Lover '77

Paul Mayersberg (D)

Captive '87
Nightfall '88

Curtis Mayfield

Short Eyes '79

Peter Mayhew

The Empire Strikes Back '80
Return of the Jedi '83

Tony Maylam (D)

The Burning '82
The Riddle of the Sands '84
Sins of Dorian Gray '82
Split Second '92

Ken Maynard

Arizona Terror '31
Arizona Whirlwind '44
Between Fighting Men '32
Come On Tarzan '32
Death Rides the Range '40
Death Valley Rangers '44
Drum Taps '33
Dynamite Ranch '32
False Faces '32
Fargo Express '32
$50,000 Dollar Reward '25
Fightin' Ranch '30
Fighting Thru '30
Flaming Lead '39
Flamingo Lead '39
Harmony Trail '44
Hell Fire Austin '32
Honor of the Range '34
In Old Santa Fe '34
The Law Rides Again '43
Lightning Strikes West '40
The Lone Avenger '33
Mystery Mountain '34
Phantom Rancher '39
Phantom Thunderbolt '33
Pocatello Kid '31
Range Law '31
Six Shootin' Sheriff '38
Strawberry Roan '33
Sunset Trail '32
Texas Gunfighter '32
Tombstone Canyon '35
Trail Drive '35

Amazing Stories, Book 2 '86
Angel Baby '61
The Concorde: Airport '79 '79
Deadly Sanctuary '68
Echoes '83
Giant '56
Girls of Huntington House '73
Johnny Guitar '53
99 Women '69
The President's Plane is
  Missing '71
The Sacketts '79
Sixteen '72
Suddenly, Last Summer '59
Tender Is the Night '55
Touch of Evil '58

Tom McCamus

Norman's Awesome
  Experience '88

Paul McCandless

Oregon '88

Chuck McCann

Cameron's Closet '89
C.H.O.M.P.S. '79
Far Out Space Nuts, Vol. 1
  '75
Foul Play '78
The Heart is a Lonely Hunter
  '68
If Things Were Different '79
Projectionist '71
Rosebud Beach Hotel '85
That's Adequate '90
Thrashin' '86

Donal McCann

Cal '84
Danny Boy '84
The Dead '87
Hard Way '80
The Miracle '91
Screamer '74

Frances McCann

The Creation of the
  Humanoids '62

Henry McCann

Submarine Seahawk '59

Jim McCann

Jim McCann & the Morrisseys
  '88

John McCann

Erotic Images '85

Sean McCann

Mindfield '89

Leo McCarey (D)

The Awful Truth '37
The Bells of St. Mary's '45
Duck Soup '33
Going My Way '44
Good Sam '48
Indiscreet '31
Love Affair '39
Milky Way '36
Once Upon a Honeymoon '42
Ruggles of Red Gap '35

Ray McCarey (D)

Devil's Party '38
The Mystery Man '35
You Can't Fool Your Wife '40

Rod McCarey

No Drums, No Bugles '71

Fred McCarren

How to Pick Up Girls '78
Marriage is Alive and Well '80
Red Flag: The Ultimate Game
  '81

Andrew McCarthy

The Beniker Gang '83
Class '83
Club Extinction '89
Fresh Horses '88
Heaven Help Us '85
Kansas '88
Less Than Zero '87
Mannequin '87
Pretty in Pink '86
St. Elmo's Fire '85
Waiting for the Moon '87
Weekend at Bernie's '89
Year of the Gun '91

Annette McCarthy

Creature '85

Charlie McCarthy

Fun & Fancy Free '47
Here We Go Again! '42

Earl McCarthy

Sucker Money '34

Frank McCarthy

A Case of Deadly Force '86
Dead Men Don't Wear Plaid
  '82

John P. McCarthy

The Snake Woman '61

**John P.
McCarthy** (D)

Cavalier of the West '31
Oklahoma Cyclone '30
Song of the Gringo '36

Kevin McCarthy

Ace High '68
The Annapolis Story '55
The Best Man '64
A Big Hand for the Little Lady
  '66
Buffalo Bill & the Indians '76
Dan Candy's Law '73
Dark Tower '87
Dead on the Money '91
Dead Right '88
Eve of Destruction '90
Fast Food '89
Final Approach '92
For Love or Money '88
A Gathering of Eagles '63
Ghoulies 3: Ghoulies Go to
  College '91
Hero at Large '80
Hostage '87
Hotel '67
The Howling '81
Innerspace '87
Invasion of the Body
  Snatchers '56
Invasion of the Body
  Snatchers '78
Invitation to Hell '84
LBJ: The Early Years '88
Love or Money? '88
The Midnight Hour '86
Mirage '66
My Tutor '82
Order to Kill '73
Piranha '78
Ratings Game '84
The Rose and the Jackal '90

Rosie: The Rosemary Clooney
  Story '82
The Sleeping Car '90
Texas Guns '90
Those Lips, Those Eyes '80
Twilight Zone: The Movie '83
UHF '89

Lin McCarthy

D.I. '57
Yellowneck '55

**Michael
McCarthy** (D)

Crow Hollow '52
Operation Amsterdam '60

Molly McCarthy

The Flamingo Kid '84

Nobu McCarthy

The Geisha Boy '58
Pacific Heights '90
The Wash '88

Patti McCarthy

Outlaw of the Plains '46

Sheila McCarthy

Beethoven Lives Upstairs '92
Friends, Lovers & Lunatics '89
George's Island '91
I've Heard the Mermaids
  Singing '87
Pacific Heights '90
Paradise '91
Stepping Out '91

Terence McCarthy

This Stuff'll Kill Ya! '71

Linda McCartney

Give My Regards to Broad
  Street '84

Mike McCartney

The Compleat Beatles '82

Paul McCartney

The Compleat Beatles '82
Eat the Rich '87
Fun with the Fab Four '90
Give My Regards to Broad
  Street '84
Hard Day's Night '64
Help! '65
Let It Be '70
Magical Mystery Tour '67
One Last Time: The Beatles'
  Final Concert '66
Ready Steady Go, Vol. 1 '83
Ready Steady Go, Vol. 2 '85
Ready Steady Go, Vol. 3 '84
The Real Buddy Holly Story
  '87
Rock 'n' Roll Heaven '84
Yellow Submarine '68

Conan McCarty

An Empty Bed '90

Mary McCarty

The Fighting Sullivans '42

Rod McCary

A Girl to Kill for '90
Rebel Storm '90

**Constance
McCashin**

Love on the Run '85
Nightmare at Bittercreek '91
Obsessive Love '84

The Two Worlds of Jenny
  Logan '79

Sky McCatskill

The Maze '85

Charles McCaughan

Impulse '90
Slaves of New York '89

Vera McCaughan

Shredder Orpheus '89

John McCauley (D)

Rattlers '76

Candy McClain

Simple Justice '89

Joedida McClain

Season of the Witch '73

Saundra McClain

Mr. & Mrs. Bridge '91

Rue McClanahan

After the Shock '90
Little Match Girl '87
Modern Love '90
Sergeant Matlovich vs. the
  U.S. Air Force '78

**Gregory
McClatchy** (D)

Vampire at Midnight '88

Brian McClave

Ninja Brothers of Blood '89

Gary McCleery

Hard Choices '84

Michelle McClellan

Death Row Diner '88
Night of the Living Babes '87
Nightmare Sisters '87

Catherine McClenny

Creature from Black Lake '76

Don McCleod

Tanya's Island '81

Mike McClerie

Alien Space Avenger '91

Leigh McCloskey

Alexander: The Other Side of
  Dawn '77
Cameron's Closet '89
Dirty Laundry '87
Double Revenge '89
Fraternity Vacation '85
Hamburger... The Motion
  Picture '86
Hearts & Armour '83
Inferno '78
Just One of the Guys '85

Doug McClure

At the Earth's Core '76
Bananas Boat '78
Bull of the West '89
Dark Before Dawn '89
Enemy Below '57
52 Pick-Up '86
Fight for Gold
Firebird 2015 A.D. '81
Hell Hounds of Alaska '70s
House Where Evil Dwells '82
Humanoids from the Deep '80
Land That Time Forgot '75
Omega Syndrome '87

The People That Time Forgot
  '77
Playmates '72
Prime Suspect '88
The Rebels '79
Shenandoah '65
Tapeheads '89
The Unforgiven '60
Wild & Wooly '78

Marc McClure

After Midnight '89
Back to the Future '85
Grim Prairie Tales '89
I Wanna Hold Your Hand '78
Pandemonium '82
Perfect Match '88
Strange Behavior '81
Superman: The Movie '78
Superman 2 '80
Superman 3 '83
Superman 4: The Quest for
  Peace '87

Tane McClure

Hot Under the Collar '91

Edie McClurg

Carrie '76
Cheech and Chong's Next
  Movie '80
Cheech and Chong's, The
  Corsican Brothers '84
Eating Raoul '82
Elvira, Mistress of the Dark
  '88
Ferris Bueller's Day Off '86
The Little Mermaid '89
Mr. Mom '83
Pandemonium '82
Planes, Trains & Automobiles
  '87
Secret of NIMH '82
She's Having a Baby '88

Ken McCluskey

The Commitments '91

Mark McCollum

Paramount Comedy Theater,
  Vol. 3: Hanging Party '87

Warren McCollum

Boy's Reformatory '39
Reefer Madness '38

Katie McCombs

Silent Witness '85

Arn McConnell (D)

Shock! Shock! Shock! '87

Judith McConnell

The Thirsty Dead '74

Keith McConnell

Alice Goodbody '74

Rob McConnell

Rob McConnell & the Boss
  Brass '83

Marilyn McCoo

The Fantastic World of D.C.
  Collins '84
My Mom's a Werewolf '89

John McCook

Codename: Foxfire '85

Kent McCord

For Heaven's Sake '79
Illicit Behavior '91
Nashville Beat '89

Mississippi Burning '88
Raising Arizona '87

Donald McDougall (D)

Chinese Web '78

Betty McDowall

Ballad in Blue '66
Dead Lucky '60
Jack the Ripper '60
Time Lock '57

Roddy McDowall

The Adventures of Bullwhip Griffin '66
Arnold '73
Battle for the Planet of the Apes '73
Bedknobs and Broomsticks '71
The Big Picture '89
Carmilla '89
Cat from Outer Space '78
Charlie Chan and the Curse of the Dragon Queen '81
Circle of Iron '78
Class of 1984 '82
Cleopatra '63
Conquest of the Planet of the Apes '72
Cutting Class '89
Dead of Winter '87
Deadly Game '91
Dirty Mary Crazy Larry '74
Doin' Time on Planet Earth '88
Double Trouble '91
Embryo '76
Escape from the Planet of the Apes '71
Evil Under the Sun '82
Five Card Stud '68
Flood! '76
Fright Night '85
Fright Night 2 '88
Funny Lady '75
The Greatest Story Ever Told '65
Hill Number One '51
How Green was My Valley '41
An Inconvenient Woman '91
Inside Daisy Clover '65
The Keys of the Kingdom '44
Laserblast '78
Lassie, Come Home '43
Legend of Hell House '73
Life & Times of Judge Roy Bean '72
The Longest Day '62
The Loved One '65
Macbeth '48
Mae West '84
The Martian Chronicles: Part 2 '79
The Martian Chronicles: Part 3 '79
Mean Johnny Barrows '75
Midnight Lace '60
My Flicka '43
Night Gallery '69
Overboard '87
Planet of the Apes '68
Rabbit Test '78
Revenge of TV Bloopers '60s
Scavenger Hunt '79
Shakma '89
Tempest '63
Thief of Baghdad '78
The Zany Adventures of Robin Hood '84

Claire McDowall

Ben-Hur '26
The Big Parade '25

Short Films of D.W. Griffith, Volume 1 '12
West-Bound Limited '23

Malcolm McDowell

Blue Thunder '83
Britannia Hospital '82
Buy & Cell '89
Caligula '80
The Caller '87
Cat People '82
Class of 1999 '90
A Clockwork Orange '71
The Compleat Beatles '82
Cross Creek '83
Disturbed '90
Get Crazy '83
Gulag '85
If... '69
Jezebel's Kiss '90
The Light in the Jungle '91
Little Red Riding Hood '83
Long Ago Tomorrow '71
Look Back in Anger '80
Merlin and the Sword '85
Monte Carlo '86
Moon 44 '90
O Lucky Man '73
Sunset '88
Time After Time '79
Voyage of the Damned '76

Rick McDowell

Getting Lucky '89

James McEachin

The Dead Don't Die '75
The Groundstar Conspiracy '72
He Who Walks Alone '78
Honeyboy '82

Ellen McElduff

Imposters '84
Working Girls '87

John McEnery

Bartleby '70
Land That Time Forgot '75
One Russian Summer '73
Pope John Paul II '84

Peter McEnery

Entertaining Mr. Sloane '70
Fighting Prince of Donegal '66
The Game is Over '66
I Killed Rasputin '67
A Midsummer Night's Dream '82
Moon-Spinners '64
Negatives '68
Victim '62

Jamie McEnnan

Munchie '92

Annie McEnroe

Cop '88
The Hand '81
Howling 2: Your Sister Is a Werewolf '85
Purple Hearts '84
Running Scared '79
True Stories '86
Warlords of the 21st Century '82

Reba McEntire

Reba McEntyre '90
Tremors '89

Bernard McEveety

One Little Indian '73

Bernard McEveety (D)

The Bears & I '74
The Broadway Drifter '27
The Brotherhood of Satan '71
Napoleon and Samantha '72
One Little Indian '73
Roughnecks '80

Vincent McEveety (D)

Amy '81
The Apple Dumpling Gang Rides Again '79
Castaway Cowboy '74
Charley and the Angel '73
Gus '76
Herbie Goes Bananas '80
Herbie Goes to Monte Carlo '77
Million Dollar Duck '71
Superdad '73
Treasure of Matecumbe '76

Geraldine McEwan

Foreign Body '86
Robin Hood: Prince of Thieves '91

Rob McEwan

Fall from Innocence '88

Hamibh McEwen

Zombie Nightmare '86

Gates McFadden

Taking Care of Business '90

Hamilton McFadden (D)

Sea Racketeers '37
Stand Up and Cheer '34

T.J. McFadden

Slipping into Darkness '88

John McFadyen

At Sword's Point '51

Connie McFarland

Troll 2 '92

George "Spanky" McFarland

The Aurora Encounter '85
General Spanky '36
Kentucky Kernels '34
Little Rascals, Book 1 '30s
Little Rascals, Book 2 '30s
Little Rascals, Book 3 '38
Little Rascals, Book 4 '30s
Little Rascals, Book 5 '30s
Little Rascals, Book 6 '30s
Little Rascals, Book 7 '30s
Little Rascals, Book 8 '30s
Little Rascals, Book 9 '30s
Little Rascals, Book 10 '30s
Little Rascals, Book 11 '38
Little Rascals, Book 12 '37
Little Rascals, Book 13 '37
Little Rascals, Book 14 '30s
Little Rascals, Book 15 '30s
Little Rascals, Book 16 '30s
Little Rascals, Book 17 '31
Little Rascals: Choo Choo/Fishy Tales '32
Little Rascals Collector's Edition, Vol. 1-6
Little Rascals Comedy Classics, Vol. 1
Little Rascals Comedy Classics, Vol. 2

Little Rascals: Fish Hooky/Spooky Hooky '33
Little Rascals: Honkey Donkey/Sprucin' Up '34
Little Rascals: Little Sinner/Two Too Young '35
Little Rascals: Mush and Milk/Three Men in a Tub '33
Little Rascals on Parade '37
Little Rascals: Pups is Pups/Three Smart Boys '30
Little Rascals: Readin' and Writin'/Mail and Female '31
Little Rascals: Reunion in Rhythm/Mike Fright '37
Little Rascals: Spanky/Feed 'Em and Weep '32
Little Rascals: The Pinch Singer/Framing Youth '36
Little Rascals Two Reelers, Vol. 1 '29
Little Rascals Two Reelers, Vol. 2 '33
Little Rascals Two Reelers, Vol. 3 '36
Little Rascals Two Reelers, Vol. 4 '30
Little Rascals Two Reelers, Vol. 5 '34
Little Rascals Two Reelers, Vol. 6 '36
Little Rascals, Volume 1 '29
Little Rascals, Volume 2 '30
Our Gang Comedies '40s
Peck's Bad Boy with the Circus '38

Bruce McFarlane (D)

Kill Slade '89

Hamish McFarlane

The Navigator '88

William F. McGaha

Iron Horsemen '71

William F. McGaha (D)

Iron Horsemen '71

Celia McGann

Maniac '34

Mark McGann

John & Yoko: A Love Story '85
Let Him Have It '91

Paul McGann

Dealers '89
Innocent Victim '90
Paper Mask '91
The Rainbow '89
Withnail and I '87

William McGann (D)

American Empire '42
Dr. Christian Meets the Women '40
In Old California '42

Darren McGavin

Airport '77 '77
Baron and the Kid '84
Blood & Concrete: A Love Story '90
By Dawn's Early Light '89
Captain America '89
Child in the Night '90
A Christmas Story '83

The Court Martial of Billy Mitchell '55
Dead Heat '88
The Delicate Delinquent '56
Firebird 2015 A.D. '81
Hangar 18 '80
Hitchhiker 2 '85
Ike '79
The Man with the Golden Arm '55
The Martian Chronicles: Part 1 '79
The Martian Chronicles: Part 2 '79
The Martian Chronicles: Part 3 '79
Mission Mars '67
My Wicked, Wicked Ways '84
The Natural '84
The Night Stalker: Two Tales of Terror '74
No Deposit, No Return '76
Power, Passion & Murder '83
Queen for a Day '51
Raw Deal '86
Richard Petty Story '72
Say Goodbye, Maggie Cole '72
Sunset '88
Tribes '70
Turk 182! '85
Waikiki '80
Zero to Sixty '78

Darren McGavin (D)

Run, Stranger, Run '73

Cindy McGee

Fast Forward '84

Mary McGee

Corruption '89

Vic McGee

Horrors of the Red Planet '64

Vonetta McGee

Big Bust Out '73
Blacula '72
The Eiger Sanction '75
Repo Man '83
To Sleep with Anger '90

William Bill McGhee

Don't Look in the Basement '73

Bruce McGill

Charlotte Forten's Mission: Experiment in Freedom '85
Citizens Band '77
End of the Line '88
The Hand '81
Into the Night '85
Last Innocent Man '87
Little Vegas '90
National Lampoon's Animal House '78
No Mercy '86
Out Cold '89
Silkwood '83
Tough Enough '87
Whale for the Killing '81
Wildcats '86

Everett McGill

Brubaker '80
Dune '84
Field of Honor '86
Heartbreak Ridge '86
Iguana '89
Jezebel's Kiss '90
License to Kill '89

Summer of My German
  Soldier '78
Two Moon Junction '88
Women of Valor '86
You Can't Hurry Love '88

Glynis McNicoll

Roses Bloom Twice '77

Bill McNulty

Dungeon of Harrow '64

Dorothy McNulty

Sea Racketeers '37

Maggie McOmie

THX 1138 '71

Sandy McPeak

The Flight '89
Inside Out '91
Ode to Billy Joe '76
The Taking of Flight 847: The
  Uli Derickson Story '88
Tarantulas: The Deadly Cargo
  '77

Jack McPeat

Ninja Operation: Licensed to
  Terminate '87

**Murdock
McQuarrie**

Laramie Kid '35
The Tonto Kid '35

Butterfly McQueen

The Adventures of
  Huckleberry Finn '85
Amazing Grace '74
Duel in the Sun '46
Gone with the Wind '39
I Dood It '43
Killer Diller '48
Making of a Legend: "Gone
  With the Wind" '89
The Mosquito Coast '86

Chad McQueen

Fever Pitch '85
The Karate Kid '84
Martial Law '90
Nightforce '86
Skateboard '77

Steve McQueen

Baby, the Rain Must Fall '64
The Blob '58
Bruce Lee: The Legend '84
Bullitt '68
Cincinnati Kid '65
The Defender '57
The Getaway '72
The Great Escape '63
Hell is for Heroes '62
The Hunter '80
Junior Bonner '72
Le Mans '71
Love with the Proper Stranger
  '63
Magnificent Seven '60
Nevada Smith '66
Never Love a Stranger '58
Never So Few '59
On Any Sunday '71
Papillon '73
The Reivers '69
The Sand Pebbles '66
Soldier in the Rain '63
Somebody Up There Likes Me
  '56
Thomas Crown Affair '68
Towering Inferno '74
War Lover '62

Wild West '87

Carmen McRae

Jo Jo Dancer, Your Life is
  Calling '86

Elizabeth McRae

The Conversation '74
The Incredible Mr. Limpet '64

Frank McRae

*batteries not included '87
Cannery Row '82
Farewell to the King '89
License to Kill '89
Lock Up '89

Leslie McRae

Blood Orgy of the She-Devils
  '74
Girl in Gold Boots '69

Meredith McRae

Husbands, Wives, Money, and
  Murder '86
Sketches of a Strangler '78
Vultures '84

Gerald McRaney

American Justice '86
Blind Vengeance '90
The Brain Machine '72
City Killer '87
Dynamite and Gold '88
The Haunting Passion '83
Hobo's Christmas '87
Mind Warp '72
The NeverEnding Story '84
Night of Bloody Horror '76

Ian McShane

Code Name: Diamond Head
  '77
Exposed '83
The Fifth Musketeer '79
The Great Escape 2 '88
The Great Riviera Bank
  Robbery '79
Journey into Fear '74
The Last of Sheila '73
The Murders in the Rue
  Morgue '86
Ordeal by Innocence '84
The Terrorists '74
Too Scared to Scream '85
Torchlight '85
War & Remembrance, Part 1
  '88
Yesterday's Hero '79
Young Charlie Chaplin '88

Kitty McShane

Old Mother Riley's Ghosts '41

Monica McSwain

The Little Match Girl '84

Bud McTaggart

Billy the Kid Trapped '42

Janet McTeer

Hawks '89

John McTiernan *(D)*

Die Hard '88
The Hunt for Red October '90
Medicine Man '92
Nomads '86
Predator '87

Paul McVey

Buried Alive '39
Phantom of Chinatown '40

Margaret McWade

Mr. Deeds Goes to Town '36

Robert McWade

Feet First '30
Healer '36

Jillian McWhirter

After Midnight '89
Beyond the Call of Duty '92
Dune Warriors '91

**Caroline
McWilliams**

Pigeon Feathers '88
Rage '83
Shattered Vows '84

**Susanne Kevin
Mead**

Hollywood High '77

Julia Meade

Zotz! '62

Mary Meade

T-Men '47

Audrey Meadows

Jack Benny Show '58
That Touch of Mink '62

Jayne Meadows

City Slickers '91
David and Bathsheba '51
Hollywood Palace '68
James Dean '76
Lady in the Lake '46
Miss All-American Beauty '82
Murder by Numbers '89
Ratings Game '84
Song of the Thin Man '47
Suspense '53

Joyce Meadows

The Brain from Planet Arous
  '57
Flesh and the Spur '57
The Girl in Lover's Lane '60

Stanley Meadows

Terronauts '67

John Meagher *(D)*

Fantasy Man '84

Karen Meagher

Experience Preferred... But
  Not Essential '83
Threads '85

Ray Meagher

Bootleg '89
Dark Age '88

Anne Meara

Awakenings '90
Fame '80
The Longshot '86
Lovers and Other Strangers
  '70
My Little Girl '87
Nasty Habits '77
The Out-of-Towners '70
That's Adequate '90

DeAnn Mears

The Loneliest Runner '76

Mark Mears

Memorial Valley Massacre '88

Meatloaf

The Rocky Horror Picture
  Show '75
The Squeeze '87

Susan Mechsner

Lovely...But Deadly '82

Peter Meda *(D)*

The Rocking Horse Winner
  '50

Karen Medak

A Girl to Kill for '90

Peter Medak *(D)*

The Babysitter '80
Breakin' Through '84
The Changeling '80
Dancing Princesses '84
A Day in the Death of Joe Egg
  '71
The Emperor's New Clothes
  '84
Ghost in the Noonday Sun '74
The Krays '90
Let Him Have It '91
The Men's Club '86
Negatives '68
The Odd Job '78
Pinocchio '83
The Ruling Class '72
Snow Queen '83
Snow White and the Seven
  Dwarfs '83
Zorro, the Gay Blade '81

Vladimir Medar

Torture Chamber of Dr.
  Sadism '69

Cissy Meddings

Strangers in Good Company
  '91

Don Medford *(D)*

The Organization '71
Sizzle '81

Judy Medford

Little Miss Innocence '73

Kay Medford

Butterfield 8 '60
Funny Girl '68
Lola '69
The Rat Race '60
Windows '80

Benny Medina

Music Shoppe '82

Jesus Medina

Yanco '64

Ofelia Medina

Frida '84
Patsy, Mi Amor '87

Patricia Medina

The Big Push '75
Botany Bay '53
The Killing of Sister George
  '69
Mister Arkadin '55
Snow White and the Three
  Stooges '61

Raynaldo Medina

El Super '79

Martha Meding

Secrets in the Dark '83

Victoria Medlin

Vanishing Point '71

Cary Medoway *(D)*

The Heavenly Kid '85
Paradise Motel '84

**Alexander
Medvedkin** *(D)*

Happiness '34

Michael Medwin

Above Us the Waves '56
The Courtney Affair '47
Four in a Jeep '51
The Gay Lady '49
The Jigsaw Man '84
Rattle of a Simple Man '64
Someone at the Door '50

Lew Meehan

Backfire '22
Ridin' the Lone Trail '37
Ridin' Thru '35
Thunder in the Desert '38

Donald Meek

The Adventures of Tom
  Sawyer '38
Barbary Coast '35
Bathing Beauty '44
Colonel Effingham's Raid '45
DuBarry was a Lady '43
Fabulous Joe '47
Little Miss Broadway '38
Magic Town '47
Merry Widow '34
Mrs. Wiggs of the Cabbage
  Patch '34
My Little Chickadee '40
One Rainy Afternoon '36
Return of Frank James '40
The Return of Peter Grimm
  '35
Stagecoach '39
State Fair '45
They Got Me Covered '43
The Thin Man Goes Home '44
Three Legionnaires '37
Toast of New York '37
Tortilla Flat '42
Two Girls & a Sailor '44
A Woman's Face '41
You Can't Take It with You
  '38
Young Mr. Lincoln '39

Jeffrey Meek

Night of the Cyclone '90
Winter People '89

George Meeker

Apache Rose '47
Danger on the Air '38
Murder by Television '35
Omoo Omoo, the Shark God
  '49
People's Choice '46
Seven Doors to Death '44
Tango '36
Twilight in the Sierras '50

Ralph Meeker

The Alpha Incident '76
The Anderson Tapes '71
Battle Shock '56
Birds of Prey '72
Brannigan '75
Cry Panic '74
The Dead Don't Die '75
Detective '68
The Dirty Dozen '67
Food of the Gods '76

Meeker ▶cont.

Four in a Jeep '51
Gentle Giant '67
Hi-Riders '77
Johnny Firecloud '75
Kiss Me Deadly '55
Mind Snatchers '72
My Boys are Good Boys '78
Paths of Glory '57
Run of the Arrow '56
The St. Valentine's Day
  Massacre '67
Winter Kills '79

Edith Meeks
Poison '91

Jeanine Meerapfel (D)
Malou '83

Armand Meffre
Here Comes Santa Claus '84

Pomme Meffre (D)
Grain of Sand '84

Francis Megahy (D)
The Carpathian Eagle '81
The Great Riviera Bank
  Robbery '79
Livin' the Life '84
Taffin '88

Don Megowan
The Creation of the
  Humanoids '62
The Jayhawkers '59

Blanche Mehaffey
Devil Monster '46

Roy Mehaffey
Doctor Gore '75

Ketan Mehta (D)
Spices '86

Chang Mei
Kung Fu Rebels '80s

Lou Mei
Super Ninja

Armin Meier
Mother Kusters Goes to
  Heaven '76

Thomas Meighan
Male and Female '19
Peck's Bad Boy '21

Mary Meigs
Strangers in Good Company
  '91

Zhang Meijun (D)
Dynasty '77

John Meillon
Billy Budd '62
The Blue Lightning '86
Cars That Ate Paris '74
Crocodile Dundee '86
Dunera Boys '85
Fourth Wish '75
Frenchman's Farm '87
Inn of the Damned '74
Ride a Wild Pony '75

Michael Mein
2069: A Sex Odyssey '78

Gus Meins (D)
Gentleman from California '37
March of the Wooden Soldiers
  '34
Roll Along Cowboy '37

Norbert Meisel (D)
The Adultress '77
Night Children '89
Walking the Edge '83

Kathryn Meisle
Basket Case 2 '90

Gunter Meisner
Between Wars '74
In a Glass Cage '86
Magdalene '88

Karl Meixner
Das Testament des Dr.
  Mabuse '33

Alfonso Mejia
Los Olvidados '50

Miguel Aceves Mejia
Dos Gallos Y Dos Gallinas
  (Two Roosters For ...) '70s
El Ciclon '47
Los Apuros de Dos Gallos
  (Troubles of Two Roosters)
  '69
Viva Chihuahua (Long Live
  Chihuahua) '47

Alfonso Mejias
Rio Hondo '47

Isabelle Mejias
The Bay Boy '85
Daughter of Death '82
Fall from Innocence '88
Higher Education '88
Meatballs 3 '87
Scanners 2: The New Order
  '91
Unfinished Business '89

Adolfas Mekas
The Brig '64

John Mekas (D)
The Brig '64

Mariangela Melato
By the Blood of Others '73
Dancers '87
Love and Anarchy '73
Seduction of Mimi '74
So Fine '81
Summer Night '87
Summer Night With Greek
  Profile, Almond Eyes .. '87
Swept Away... '75
To Forget Venice '70s

Georges Melchior
Juve Contre Fantomas '14

Lauritz Melchior
Thrill of a Romance '45

Ib Melchoir (D)
The Angry Red Planet '59
The Time Travelers '64

Max Meldrum
Death Train '79

Annielo Mele
Shoeshine '46

Arthur N. Mele (D)
Soldier's Fortune '91

Nicholas Mele
Impulse '90

Asdrubal Melendez
A Very Old Man with
  Enormous Wings '88

Bill Melendez (D)
Bon Voyage, Charlie Brown
  '80
A Boy Named Charlie Brown
  '69
A Charlie Brown Celebration
  '84
A Charlie Brown Christmas
  '65
A Charlie Brown Thanksgiving
  '81
Charlie Brown's All-Stars '66
Good Grief, Charlie Brown '83
Happy New Year, Charlie
  Brown '82
It Was a Short Summer,
  Charlie Brown '69
It's an Adventure, Charlie
  Brown! '83
It's Flashbeagle, Charlie
  Brown/She's a Good
  Skate, Charlie Brown '84
It's Three Strikes, Charlie
  Brown '86
The Lion, the Witch & the
  Wardrobe '79
Snoopy, Come Home '72

George Melford (D)
Dracula (Spanish Version) '31
East of Borneo '31
Moran of the Lady Letty '22
The Sheik '21

Jack Melford
When Thief Meets Thief '37

Lilly Melgar
Midnight Warrior '89

Joe Melia
Privates on Parade '84

Georges Melies
George Melies, Cinema
  Magician '78

Georges Melies (D)
Early Cinema Vol. 1 '19
Early Films #1 '03
Film Firsts '60
First Twenty Years: Part 22
  (Melies)

Harley Melin
The Russian Terminator '90

Anna Melita
Naked Warriors '73

Joseph Mell
I Was a Teenage Werewolf
  '57

Marisa Mell
Danger: Diabolik '68
Hostages '79
Mad Dog
Secret Agent Super Dragon
  '66

John Cougar Mellencamp
Falling from Grace '92

John Cougar Mellencamp (D)
Falling from Grace '92

Sunnyi Melles
'38: Vienna Before the Fall '88

Fuller Mellish, Jr.
Applause '29

Breno Mello
Black Orpheus '58

Douglas Mellor
The Beast of Yucca Flats '61

Larry "Bud" Melman
Elayne Boosler: Party of One
  '89

Courtney Melody
Michael Prophet '87

Robin Meloy
The House on Sorority Row
  '83

Frank Melton
Stand Up and Cheer '34

Sid Melton
Leave It to the Marines '51
The Lost Continent '51
Sky High '51
The Steel Helmet '51
Stop That Cab '51

Jean-Pierre Melville
Breathless (A Bout de Souffle)
  '59

Jean-Pierre Melville (D)
Bob le Flambeur '55
Godson '72
Le Doulos '61
Les Enfants Terrible '50

Pauline Melville
Mona Lisa '86

Sam Melville
Roughnecks '80
Twice Dead '88

Winifred Melville
The Greed of William Hart '48

Donnie Melvin
ABC Stage 67: Truman
  Capote's A Christmas
  Memory '66

Murray Melvin
The Bawdy Adventures of
  Tom Jones '76
The Devils '71
Madhouse Mansion '74
A Taste of Honey '61

Steven Memel
Savage Justice '88

George Memmoli
Hot Potato '76

Men at Work
Australia Now '84

Ben Mendelsohn
Quigley Down Under '90
The Year My Voice Broke '87

George Mendeluk (D)
Doin' Time '85
The Kidnapping of the
  President '80
Meatballs 3 '87
Stone Cold Dead '80

David Mendenhall
Going Bananas '88
Over the Top '86
Secret of the Ice Cave '89
Space Raiders '83
Streets '90
Witchfire '86

Lothar Mendes (D)
The Man Who Could Work
  Miracles '37
Moonlight Sonata '38
Power '34

Fernando Mendez (D)
The Living Coffin '58
Misterios de Ultratumba '47
The Vampire '57
The Vampire's Coffin '58

Marcy Mendham
The Groove Tube '72

Bob Mendlesohn
Devil's Gift '84

Mauro Mendonca
Calacan
Dona Flor & Her Two
  Husbands '78
Love Strange Love '82

Victor Manuel Mendoza
Susana '51

Ramon Menendez (D)
Stand and Deliver '87

Ramiro Meneses
Rodrigo D: No Future '91

Bernard Menez
Dracula & Son '76

John Mengatti
Knights of the City '86
Meatballs 2 '84

Chris Menges (D)
Crisscross '92
A World Apart '88

Jon Menick
Easy Wheels '89

Adolphe Menjou
Across the Wide Missouri '51
The Ambassador's Daughter
  '56
Are Parents People? '25
Bundle of Joy '56
A Farewell to Arms '32
The Front Page '31

Gold Diggers of 1935 '35
Golden Boy '39
Goldwyn Follies '38
The Grand Duchess and the Waiter '26
Heartbeat '46
Hi Diddle Diddle '43
The Hucksters '47
I Married a Woman '56
The King on Main Street '25
Letter of Introduction '38
Little Miss Marker '34
The Marriage Circle '24
Milky Way '36
Morning Glory '33
Morocco '30
My Dream is Yours '49
Paths of Glory '57
Pollyanna '60
The Sheik '21
The Sorrows of Satan '26
Stage Door '37
A Star is Born '37
State of the Union '48
Step Lively '44
The Swan '25
Till the Clouds Roll By/A Star Is Born '46
Woman of Paris/Sunnyside '23
You Were Never Lovelier '42

Nina Menkes (D)

Great Sadness of Zohara '83
Magdalena Viraga '86

Gian-Carlo Menotti (D)

The Medium '51

Barnard Menoud (D)

Scenario du Film Passion '82

Vladimir Menshov (D)

Moscow Does Not Believe in Tears '80

Vladimir Mensik

All My Good Countrymen '68

Narciso Ibanez Menta

Los Muchachos de Antes No Usaban Arsenico '80s
The Saga of the Draculas '72

Jiri Menzel (D)

Closely Watched Trains '66
Larks on a String '68

Heather Menzies

Captain America '79
Piranha '78
Red, White & Busted '75
The Sound of Music '65

Robert Menzies

Cactus '86

William Cameron Menzies (D)

Chandu the Magician '32
Drums in the Deep South '51
Invaders from Mars '53
Things to Come '36

Doro Merande

Kiss Me, Stupid! '64
The Man with the Golden Arm '55

The Russians are Coming, the Russians are Coming '66
The Seven Year Itch '55

Maria Mercader

Musica Proibita '43

Louis Mercanton (D)

Salome/Queen Elizabeth '23

Beryl Mercer

Jane Eyre '34

Frances Mercer

Beauty for the Asking '39
Hunt the Man Down/Smashing the Rackets '50

Jack Mercer

Popeye and Friends in the South Seas '61

Mae Mercer

Frogs '72

Nicholas Mercer

Deadly Diamonds '91

Antonio Mercero (D)

Don Juan My Love '90

Ismail Merchant (D)

Courtesans of Bombay '85

Vivien Merchant

Accident '67
Alfie '66
Frenzy '72
Under Milk Wood '73

Chantal Mercier

Small Change '76

Michele Mercier

Black Sabbath '64
Call of the Wild '72
Casanova '70 '65
Oldest Profession '67
Web of the Spider '70

Melina Mercouri

A Dream of Passion '78
Nasty Habits '77
Never on Sunday '60
Once is Not Enough '75
Topkapi '64
Where the Hot Wind Blows '59

Jean Mercure

Baxter '89

Marthe Mercure

The Victory '88

Monique Mercure

The Odyssey of the Pacific '82
Tramp at the Door '87
Two Women in Gold '70

Gus Mercurio

All the Rivers Run '84

Burgess Meredith

Advise and Consent '62
Batman '66
A Big Hand for the Little Lady '66
Burnt Offerings '76
Cardinal '63

Clash of the Titans '81
Day of the Locust '75
Diary of a Chambermaid '46
Final Assignment '80
Foul Play '78
Full Moon in Blue Water '88
G.I. Joe: The Movie '86
Golden Rendezvous '77
Great Bank Hoax '78
Idiot's Delight '39
In Harm's Way '65
Johnny We Hardly Knew Ye '77
King Lear '87
Last Chase '81
MacKenna's Gold '69
Madame X '66
Magic '78
Man on the Eiffel Tower '48
The Manitou '78
Mine Own Executioner '47
Mr. Corbett's Ghost '90
92 in the Shade '76
Oddball Hall '91
Of Mice and Men '39
Outrage '86
Probe '72
Rocky '76
Rocky 2 '79
Rocky 3 '82
Rocky 5 '90
Santa Claus: The Movie '85
Second Chorus '40
The Sentinel '76
Son of Blob '72
State of Grace '90
Stay Away, Joe '68
That Uncertain Feeling '41
There was a Crooked Man '70
Thumbelina '82
Tom, Dick, and Harry '41
Torture Garden '67
True Confessions '81
Twilight Zone: The Movie '83
Wet Gold '84
When Time Ran Out '80
Winterset '37
Yin & Yang of Mr. Go '71

Burgess Meredith (D)

Man on the Eiffel Tower '48
Yin & Yang of Mr. Go '71

Don Meredith

Express to Terror '79
Sky Hei$t '75
Terror on the 40th Floor '74

Iris Meredith

Green Archer '40
A Lawman is Born '37
Mystery of the Hooded Horseman '37

Judi Meredith

Jack the Giant Killer '62
Planet of Blood '66

Lee Meredith

The Sunshine Boys '75

Madge Meredith

Falcon's Adventure/Armored Car Robbery '50
Trail Street '47

Tita Merello

Las Barras Bravas

Luc Merenda

The Cheaters '80s
Could It Happen Here? '80s
Le Mans '71

Violent Professionals '82

James Merendino (D)

Witchcraft 4: Virgin Heart '92

Jalal Merhi

Tiger Claws '91

Joseph Merhi (D)

Epitaph '87
Final Impact '91
Fresh Kill '87
The Glass Jungle '88
Heat Street '87
The Killing Game '87
L.A. Crackdown '87
L.A. Crackdown 2 '88
L.A. Heat '88
L.A. Vice '89
The Last Riders '90
Midnight Warrior '89
The Newlydeads '87
Night of the Wilding '90

Tarcisio Meria

I Love You '81

Macha Meril

A Married Woman '65
Vagabond '85

Eda Reiss Merin

Don't Tell Mom the Babysitter's Dead '91

J.L. Merino (D)

Hell Commandos '69
Scream of the Demon Lover '71

Ricardo Merino

Esposa y Amante '70s

Michele Meritz

Le Beau Serge '58

Philip Merivale

Lady for a Night '42
Sister Kenny '46

Lee Meriwether

Batman '66
Brothers O'Toole '73
The 4D Man '59
The Undefeated '69

Ron Merk (D)

Pinocchio's Storybook Adventures '79

Una Merkel

Abraham Lincoln '30
The Bank Dick '40
The Bat Whispers '30
Bombshell '33
Born to Dance '36
Broadway Melody of 1936 '35
Destry Rides Again '39
42 Street '33
The Girl Most Likely '57
I Love Melvin '53
It's a Joke, Son! '47
Kentuckian '55
The Maltese Falcon '31
Merry Widow '34
The Parent Trap '61
Red Headed Woman '32
Road to Zanzibar '41
Spinout '66
Summer Magic '63
Summer and Smoke '61
A Tiger Walks '64

Vasily Merkuryev

The Cranes are Flying '57

Franco Merli

Arabian Nights '74

Maurizio Merli

Fearless '78

Jan Merlin

Silk 2 '89
Tom Corbett, Space Cadet Volume 1
Twilight People '72

Joanna Merlin

Baby It's You '82
Class Action '91
Fame '80

Maria Luisa Merlo

Ursus in the Valley of the Lions '62

Ethel Merman

Airplane! '80
It's a Mad, Mad, Mad, Mad World '63
Kid Millions '34
Rudolph & Frosty's Christmas in July '82
Stage Door Canteen '43
That's Singing '84
There's No Business Like Show Business '54

Mary Merrall

Nicholas Nickleby '46
The Obsessed '51
Pink Strings and Sealing Wax '45

Charlotte Merriam

One-Punch O'Day '26

Ian Merrick (D)

Black Panther '77

Lawrence Merrick (D)

Black Bikers from Hell '70
The Outlaw Bikers - Gang Wars '70

Lynn Merrick

California Joe '43
Down to Earth '47
The Sombrero Kid '42
Western Double Feature 15 '50

T.J. Merrick

Slime City '89

B.D. Merrill

What Ever Happened to Baby Jane? '62

Dina Merrill

Anna to the Infinite Power '84
Butterfield 8 '60
Caddyshack 2 '88
The Cary Grant Collection '90
The Courtship of Eddie's Father '62
Deadly Encounter '78
Desk Set '57
Fear '90
The Greatest '77
Hot Pursuit '84
The Meal '75
Operation Petticoat '59

Merrill ▶cont.

Running Wild '73
The Sundowners '60
Tenth Month '79
Twisted '86
A Wedding '78

Frank Merrill

Tarzan the Tiger '29

Gary Merrill

All About Eve '50
Around the World Under the
  Sea '65
Clambake '67
The Great Impostor '61
Huckleberry Finn '74
The Incident '67
Missouri Traveler '58
Mysterious Island '61
Phone Call from a Stranger
  '52
Pueblo Affair '73
The Seekers '79
Twelve O'Clock High '49

Julie Merrill

Mirror of Death '87

Keith Merrill (D)

Great American Cowboy '73
Harry's War '84
Take Down '79
Windwalker '81

Louis D. Merrill

The Giant Claw '57

Michael Merrins

Posed For Murder '89

Lindsay Merrithew

Hitler's Daughter '90

Theresa Merritt

The Serpent and the Rainbow
  '87
Voodoo Dawn '89

Nicholas Merriwether (D)

Eegah! '62

Jane Merrow

The Appointment '82
Hands of the Ripper '71
Island of the Burning Doomed
  '67
The Lion in Winter '68

Molly Mershon

Angry Joe Bass '76

Mary Mertens

The Alien Factor '78

John Merton

Boot Hill Bandits '42
Border Bandits '46
Code of the Fearless '39
Covered Wagon Days '40
Gang Bullets '38
In Old Montana '39
Phantom Ranger '38
Roaring Six Guns '37
Rustler's Hideout '44
Two-Gun Troubador '37
Valley of Terror '38
Wild Horse Phantom '44

William Mervyn

Murder Ahoy '64
Railway Children '70

Riba Meryl

Beyond the Doors '83

Ayse Emel Mesci

Wall '83

Michael Mesmer

Deadtime Stories '86

Don Messick

The Jetsons: Movie '90

Philip Frank Messina (D)

Spy '89

Steve Messina

Body Moves '90

Gertrude Messinger

Aces Wild '37
The Adventurous Knights '35
Fighting Pilot '35
Roaring Roads '35
Rustler's Paradise '35
Social Error '35

Armand Mestral

Gervaise '57

Marta Meszaros (D)

Adoption '75
The Girl '68
Riddance '73

Sambat Metanen

The Golden Triangle '80

George Metaxa

Doctor Takes a Wife '40

Laurie Metcalf

Candy Mountain '87
Desperately Seeking Susan
  '85
Execution of Raymond
  Graham '85
Internal Affairs '90
JFK '91
Making Mr. Right '86
Miles from Home '88
Pacific Heights '90
Stars and Bars '88

Mark Metcalf

The Final Terror '83
National Lampoon's Animal
  House '78
One Crazy Summer '86

Arthur Metcalfe

Soul-Fire '25

Earl Metcalfe

With Kit Carson Over the
  Great Divide '25

Ken Metcalfe

Warriors of the Apocalypse
  '85

Sombat Methance

Magnum Killers '76

Mayo Methot

Corsair '31
Goodbye Love '34
Marked Woman '37

Art Metrano

Beverly Hills Bodysnatchers
  '89
Breathless '83
Going Ape! '81
The Heartbreak Kid '72
Malibu Express '85
Norma '89
Police Academy 2: Their First
  Assignment '85
Police Academy 3: Back in
  Training '86
Prisoner in the Middle '74
Seven '79
Slaughter's Big Ripoff '73

Peter Metro

Live & Red Hot '87

Tonto Metro

Live & Red Hot '87

Nancy Mette

Meet the Hollowheads '89

Alan Metter (D)

Back to School '86
Girls Just Want to Have Fun
  '85
Moving '88

Alan Metzger (D)

The China Lake Murders '90
Fatal Exposure '91
Red Wind '91

Radley Metzger (D)

The Cat and the Canary '79
The Lickerish Quartet '70
Princess and the Call Girl '84
Score '73
Therese & Isabelle '68

Jim Metzler

Circuitry Man '90
Delusion '91
Four Friends '81
Hot to Trot! '88
Love Kills '91
Murder by Night '89
976-EVIL '88
On Wings of Eagles '86
Squeeze Play '79
Tex '82

Paul Meurisse

Diabolique '55
Gypsy '75
Picnic on the Grass '59

Harry Meyen

Unnatural '52

Andrew Meyer (D)

Night of the Cobra Woman '72
Tidal Wave '75

Frank Meyer-Brockman

Feel the Motion '86

Emile Meyer

Riot in Cell Block 11 '54

Eve Meyer

Operation Dames '59

Jean Meyer

Would-Be Gentleman '58

Joe Meyer

Forgotten Warrior '86

Michael Meyer

Crossfire '89

Michelle Meyer

Nail Gun Massacre '86

Nancy Meyer

The Last Slumber Party '87

Nicholas Meyer (D)

Day After '83
The Deceivers '88
Pied Piper of Hamelin '84
Star Trek 2: The Wrath of
  Khan '82
Star Trek 6: The
  Undiscovered Country '91
Time After Time '79
Volunteers '85

Russ Meyer

Cult People '89

Russ Meyer (D)

Beneath the Valley of the
  Ultra-Vixens '79
Beyond the Valley of the Dolls
  '70
Fanny Hill: Memoirs of a
  Woman of Pleasure '64
Faster Pussycat! Kill! Kill! '66
Motor Psycho '65
Supervixens '75

Scott Meyer

Quick, Let's Get Married '71

Suzy Meyer

The Abomination '88

Thom Meyer

Bloodsuckers from Outer
  Space '83

Torben Meyer

The Matchmaker '58
Sullivan's Travels '41

Ari Meyers

License to Kill '84
Shakma '89
Think Big '90

Fredricka Meyers

The Thirsty Dead '74

Michelle Meyrink

Joy of Sex '84
Nice Girls Don't Explode '87
One Magic Christmas '85
Permanent Record '88
Real Genius '85
Revenge of the Nerds '84

Arturo Meza

Dona Herlinda & Her Son '86

Myriam Meziere

Jonah Who Will Be 25 in the
  Year 2000 '76

Vittorio Mezzogiorno

Cafe Express '83
L'Homme Blesse '83
Moon in the Gutter '84
Three Brothers '80

Geini Mhlophe

Place of Weeping '86

Buntaro Miake

Warning from Space '56

Cora Miao

Dim Sum: A Little Bit of Heart
  '85
Eat a Bowl of Tea '89

Nora Miao

Bruce's Deadly Fingers '80s
The Kung Fu Kid
Return of the Dragon '73
Young Dragon '77

Dominic Micelli

The Executioner '78

Bill Michael

The Farmer's Other Daughter
  '65

George Michael

Nelson Mandela 70th Birthday
  Tribute '88

Gertrude Michael

Behind Prison Walls '43
Cleopatra '34
Flamingo Road '49

Olive Michael

Coming Up Roses '87

Ralph Michael

Diary of a Mad Old Man '88
Johnny Frenchman '46
Murder Most Foul '65
San Demetrio, London '47

David Michael-Standing

Back to Back '90

Tim Michael

Ninja of the Magnificence '89

Athena Michaelidou

A Matter of Dignity '57

Dario Michaelis

The Day the Sky Exploded '57
The Devil's Commandment
  '60

Beverly Michaels

Crashout '55

Corinne Michaels

Consenting Adults '85
Laboratory '80

Greg Michaels

Jupiter Menace '82

Johnathan Jamcovic Michaels

Pink Nights '87

Johnny Michaels

Jive Junction '43

Jordan Michaels

Lifepod '80

Julie Michaels

Road House '89

Ken Michaels

Homeboys '92

Matt Michaels

Summer Camp '78

Michele Michaels

The Slumber Party Massacre '82

Pat Michaels

The Cut Throats '89

Richard Michaels (D)

Berlin Tunnel 21 '81
Blue Skies Again '83
Heart of a Champion: The Ray Mancini Story '85
Homeward Bound '80
Indiscreet '88
One Cooks, the Other Doesn't '83
Plutonium Incident '82
The Queen of Mean '90
Silence of the Heart '84

Roxanna Michaels

Caged Fury '90

T.J. Michaels

Igor & the Lunatics '85

Helena Michaelson

The Russian Terminator '90

Karli Michaelson

Just Like Us '83
Kid with the 200 I.Q. '83

John E. Michalakis (D)

I Was a Teenage Zombie '87

Claude Michaud

The Klutz '73

Francoise Michaud

Eminent Domain '91

Michael Michaud

Ricky 1 '88

Oscar Micheaux (D)

Lying Lips '39

Delaune Michel

A Woman, Her Men and Her Futon '92

Dominique Michel

The Decline of the American Empire '86

Lore Lee Michel

Tokyo Joe '49

Marc Michel

Lola '61
Umbrellas of Cherbourg '64

Odile Michel

Venus on Fire '84

Marcella Michelangeli

Could It Happen Here? '80s
Padre Padrone '77

Keith Michell

The Deceivers '88
Gondoliers
Grendel, Grendel, Grendel '82
Ruddigore '70s

Soldier in Love '67

Anne Michelle

French Quarter '78
The Haunted '79
House of Whipcord '75
Mistress Pamela '74
The Virgin Witch '70

Charlotte Michelle

Kill or Be Killed '80

Janee Michelle

The House on Skull Mountain '74
Scream Blacula Scream '73

Patricia Michelle

Cheering Section '73

Shelley Michelle

Bikini Summer '91
Sunset Strip '91

Vicki Michelle

The Virgin Witch '70

Dave Michener

The Great Mouse Detective '86

Maria Michi

Paisan '46
Redneck '73

Ivan Micholaichuk

Shadows of Forgotten Ancestors '64

Terry Michos

The Great Skycopter Rescue '82

Jered Mickey

The Swap '71

Frank Middlemass

The Island '80
Oliver Twist '85

Charles Middleton

Dick Tracy Returns '38
Flash Gordon Conquers the Universe '40
Flash Gordon: Mars Attacks the World '38
Flash Gordon: Rocketship '36
Miracle Rider: Vol. 1 '35
Miracle Rider: Vol. 2 '35
Mystery Ranch '34
Oklahoma Kid '39
Space Soldiers Conquer the Universe '40
Spook Busters '46
Strangler of the Swamp '46
Western Union '41
Yodelin' Kid From Pine Ridge '37

Guy Middleton

Dangerous Moonlight '41
A Man About the House '47

Kathleen Middleton

Wild Man '89

Noelle Middleton

Carrington, V.C. '55
The Vicious Circle '57

Ray Middleton

Christmas Carol '54
I Dream of Jeannie '52

Jubilee Trail '54
Lady from Louisiana '42
Lady for a Night '42

Robert Middleton

Career '59
Desperate Hours '55
Harrad Experiment '73

Robert Middletown

Court Jester '56

Alain Midgette

Before the Revolution '65

Fanny Midgley

Italian '15

Dale Midkiff

Blackmail '91
Elvis and Me '88
Nightmare Weekend '86
Pet Sematary '89
Streetwalkin' '85

Bette Midler

Beaches '88
Bette Midler Show '76
Big Business '88
Divine Madness '80
Down and Out in Beverly Hills '86
For the Boys '91
Hawaii '66
Jinxed '82
Outrageous Fortune '87
The Rose '79
Ruthless People '86
Scenes from a Mall '90
Stella '89
Thorn '73

Midnight Oil

Australia Now '84

Midori

Blood Voyage '77

Paulmichel Miekhe (D)

Murderer's Keep '88

Anne-Marie Mieville (D)

France/Tour/Detour/Deux/Enfants '78
Scenario du Film Passion '82
Six Fois Deux/Sur et Sous la Communication '76
Soft and Hard '85

Toshiro Mifune

The Bad Sleep Well '60
Battle of the Japan Sea '70s
The Challenge '82
Drunken Angel '48
The Gambling Samurai '60
Grand Prix '66
Hell in the Pacific '69
Hidden Fortress '58
High & Low '62
The Idiot '51
Journey of Honor '91
Life of O'Haru '52
Love and Faith '78
The Lower Depths '57
Midway '76
1941 '79
Paper Tiger '74
Proof of the Man '84
Rashomon '51
Red Beard '65
Red Lion '69

Red Sun '71
Rikisha-Man '58
Saga of the Vagabond '59
Samurai 1: Musashi Miyamoto '55
Samurai 2: Duel at Ichijoji Temple '55
Samurai 3: Duel at Ganryu Island '56
Sanjuro '62
Seven Samurai '54
Shogun '80
Stray Dog '49
Sword of Doom '67
Throne of Blood '57
Winter Kills '79
Yojimbo '62
Zatoichi vs. Yojimbo '70

Julia Migenes-Johnson

Berlin Blues '89
Mack the Knife '89

Darlene Mignacco

Psycho Girls '86

Cynthia Miguel

L.A. Crackdown 2 '88

R.A. Mihailoff

Leatherface: The Texas Chainsaw Massacre 3 '89

George Mihalka (D)

Eternal Evil '87
Hostile Takeover '88
My Bloody Valentine '81
Pick-Up Summer '79

Tatsuya Mihashi

The Burmese Harp '56
High & Low '62
None But the Brave '65
Tora! Tora! Tora! '70
What's Up, Tiger Lily? '66

Judy Mihei

Chan is Missing '82

Humberto Martinez Mijares (D)

Bestia Nocturna '86

Shin-Ichiro Mikami

An Autumn Afternoon '62

Ted V. Mikels (D)

Aftermath '85
The Astro-Zombies '67
The Black Klansman '66
The Corpse Grinders '71
Cruise Missile '78
The Doll Squad '73
Girl in Gold Boots '69
Hustler Squad '76
10 Violent Women '79
WarCat '88

Nikita Mikhailkov

An Unfinished Piece for a Player Piano '77

Nikita Mikhailkov (D)

Dark Eyes '87
Oblomov '81
A Slave of Love '78
An Unfinished Piece for a Player Piano '77

Mikhal Mikhalesko

God, Man and Devil '49

George Miki

Go for Broke! '51

Mark Mikulski

Stuck on You '84

Rentaro Mikuni

The Burmese Harp '56
The Go-Masters '82
Kwaidan '64
Nomugi Pass '79
Rikyu '90
Swords of Death '71
A Taxing Woman Returns '88

Frank Milan

Joy of Living '38

Lita Milan

I, Mobster '58
The Left-Handed Gun '58
Naked in the Sun '57
Never Love a Stranger '58
Poor White Trash '57

Alyssa Milano

Commando '85
Little Sister '92
Old Enough '84

Lisa Milano

Satan's Black Wedding '75

Mario Milano

Beyond Evil '80

Adolph Milar

The Michigan Kid '28
The Savage Girl '32

David Milbern

Sorceress '82

Bernard Miles

Fame is the Spur '47
Great Expectations '46
In Which We Serve '42
The Man Who Knew Too Much '56
Nicholas Nickleby '46
Sapphire '59
The Smallest Show on Earth '57

Betty Miles

The Law Rides Again '43
Lone Star Law Men '42
Sonora Stagecoach '44

Christopher Miles (D)

Priest of Love '81
That Lucky Touch '75
The Virgin and the Gypsy '70

Joanna Miles

Blackout '88
Born Innocent '74
Bug '75
The Orphan '79
Sound of Murder '82
The Ultimate Warrior '75

Kevin Miles

Cars That Ate Paris '74
Endplay '75

Lillian Miles

Get That Man '35

F.E. Miller

Bronze Buckaroo '39
Harlem on the Prairie '38
Harlem Rides the Range '39
Mystery in Swing '40
Yes, Sir, Mr. Bones '51

Garry Miller

Amazing Mr. Blunden '72

Geoffrey Lewis Miller

Truth or Dare? '86

George Miller

More Laughing Room Only

George Miller *(D)*

All the Rivers Run '84
Anzacs: The War Down Under '85
The Aviator '85
Les Patterson Saves the World '90
Mad Max '80
Mad Max Beyond Thunderdome '85
Man from Snowy River '82
Miracle Down Under '87
The NeverEnding Story 2: Next Chapter '91
The Road Warrior '82
Twilight Zone: The Movie '83
Witches of Eastwick '87

Geri Miller

Trash '70

Glenn Miller

Orchestra Wives '42
Sun Valley Serenade '41

Harvey Miller *(D)*

Bad Medicine '85

Ira Miller *(D)*

Loose Shoes '80

James Miller

Platoon the Warriors '88

Jason Miller

The Best Little Girl in the World '81
The Dain Curse '78
The Exorcist '73
The Exorcist 3 '90
F. Scott Fitzgerald in Hollywood '76
The Henderson Monster '80
A Home of Our Own '75
Light of Day '87
Marilyn: The Untold Story '80
Monsignor '82
The Ninth Configuration '79
That Championship Season '82
Toy Soldiers '84
Vengeance '86

Jason Miller *(D)*

That Championship Season '82

J.C. Miller *(D)*

No Dead Heroes '87

Jean Miller

Silent Night, Deadly Night 2 '87
Speak of the Devil '90

Jeremy Miller

Emanon '86
The Willies '90

John Miller

The Brave Bunch '70
Hero Bunker '71

Jonathan Miller *(D)*

Long Day's Journey into Night '88

Joshua Miller

And You Thought Your Parents Were Weird! '91
Class of 1999 '90
Near Dark '87
River's Edge '87
Teen Witch '89

Julie Miller

Primal Scream '87

Kathleen Miller

Strange New World '75

Ken Miller

I Was a Teenage Werewolf '57

Kenny Miller

Bloodstalkers '76
Surf Party '64

Kid Punch Miller

New Orleans: 'Til the Butcher Cuts Him Down '71

Kristine Miller

Jungle Patrol '48
Too Late for Tears '49

Larry Miller

Suburban Commando '91

Lee Miller

Video Violence Part 2...The Exploitation! '87

Linda Miller

Alice Sweet Alice '76
Elvis and Me '88
One Summer Love '76

Lorraine Miller

Border Badmen '45
Lonesome Trail '55
White Gorilla '47

Lydia Miller

Backlash '86
Blacklash '87

Lynn Miller

Poor Girl, A Ghost Story '74

Mandy Miller

Mandy '53

Marjorie Miller

Party Favors '89

Mark Miller

Ginger in the Morning '73
Savannah Smiles '82

Mark Thomas Miller

Blue De Ville '86
Misfits of Science '85
Mom '89

Mark Miller *(D)*

Sister Dora '77

Martin Miller

Bergonzi Hand '70
Frenzy '46
Hotel Reserve '44
Man on the Run '49

Marvin Miller

Dead Reckoning '47
Kiss Daddy Goodbye '81
Off Limits '53
Panda and the Magic Serpent '61

Mary Louise Miller

The Night Cry '26

Michael B. Miller

One Summer Love '76

Michael Miller *(D)*

Blown Away '90
A Case of Deadly Force '86
Jackson County Jail '76
National Lampoon's Class Reunion '82
Outside Chance '78
Silent Rage '82
Silent Witness '85
Street Girls '75

Mindi Miller

Sacred Ground '83

Mirta Miller

Icebox Murders '80s
La Rebelion de las Muertas '72

Miss Miller

Best of Everything Goes '83

Neal Miller *(D)*

Under the Biltmore Clock '85

Patsy Ruth Miller

Headwinds '28
The Hunchback of Notre Dame '23
Lonely Wives '31
The Sheik '21
So This is Paris '26

Penelope Ann Miller

Adventures in Babysitting '87
Awakenings '90
Big Top Pee Wee '88
Biloxi Blues '88
Dead Bang '89
Downtown '89
The Freshman '90
Kindergarten Cop '90
Miles from Home '88
Other People's Money '91

Rebecca Miller

The Murder of Mary Phagan '87

Robert Ellis Miller *(D)*

Any Wednesday '66
The Baltimore Bullet '80
The Girl from Petrovka '74
Hawks '89
The Heart is a Lonely Hunter '68
Her Life as a Man '83
Reuben, Reuben '83

Roger Miller

Robin Hood '73

Roosevelt Miller, Jr.

Game of Survival '89

Sharron Miller *(D)*

The Woman Who Willed a Miracle '83

Sherie Miller

Goin' All the Way '82

Sidney Miller

Boys Town '38
Men of Boys Town '41

Sidney Miller *(D)*

30-Foot Bride of Candy Rock '59

Stephen E. Miller

Home is Where the Hart is '88
Jane Doe '83

Stephen E. Miller *(D)*

Visions '90

Tony Miller

T-Bird Gang '59

Walter Miller

Ghost Patrol '36
The Lone Defender '32

Andra Millian

Nightfall '88
Stacy's Knights '83

James Millican

The Adventures of Gallant Bess '48
Man from Colorado '49
Rimfire '49
The Winning Team '52

Andy Milligan *(D)*

The Body Beneath '70
Carnage '84
The Ghastly Ones '68
Legacy of Horror '78
Man with Two Heads '72
Rats are Coming! The Werewolves are Here! '72
Torture Dungeon '70
The Weirdo '89

Spike Milligan

The Case of the Mukkinese Battle Horn '56
Digby, the Biggest Dog in the World '73
Down Among the Z Men '52
Ghost in the Noonday Sun '74
Goon Movie (Stand Easy) '53
Great McGonagall '75
The Hound of the Baskervilles '77
The Magic Christian '69
Rentadick '72
To See Such Fun '81

Bill Milling *(D)*

Caged Fury '90
Lauderdale '89

Alec Mills *(D)*

Bloodmoon '91
Dead Sleep '91

Alley Mills

Rape and Marriage: The Rideout Case '80

Barbara Mills

Chain Gang Women '72

Brooke Mills

The Big Doll House '71
Dream No Evil '75
The Student Teachers '73

Danny Mills

Female Trouble '75
Pink Flamingos '73

Donna Mills

Beyond the Bermuda Triangle '75
Bunco '83
Curse of the Black Widow '77
Fire '77
Haunts of the Very Rich '72
The Hunted Lady '77
The Incident '67
Killer with Two Faces '74
Live a Little, Steal a Lot '75
Murph the Surf '75
One Deadly Owner '74
Play Misty for Me '71
Waikiki '80
The World's Oldest Living Bridesmaid '92

Grace Mills

Night of the Howling Beast '75

Hayley Mills

Appointment with Death '88
Bananas Boat '70
The Chalk Garden '64
Deadly Strangers '82
Endless Night '71
In Search of the Castaways '62
Kingfisher Caper '76
Moon-Spinners '64
The Parent Trap '61
Pollyanna '60
Summer Magic '63
That Darn Cat '65
Tiger Bay '59
The Trouble with Angels '66
Whistle Down the Wind '62

Jacob Mills

The Islander '88

Jed Mills

New Year's Evil '78

Jenny Mills

Ninja Fantasy '86

John Mills

Above Us the Waves '56
Africa Texas Style '67
Around the World in 80 Days '56
The Baby and the Battleship '56
The Big Sleep '78
A Black Veil for Lisa '68
The Chalk Garden '64
Choice of Weapons '76
Chuka '67
The Colditz Story '55
Cottage to Let '41
Dr. Strange '78
Escapade '55
Gandhi '82
Goodbye, Mr. Chips '39
Great Expectations '46

John Mitchum

Big Foot '72
Bloody Trail
The Devil's Sleep '51
The Enforcer '76
Escapes '86

Robert Mitchum

The Agency '81
The Ambassador '84
The Amsterdam Kill '78
Anzio '68
Bar-20 '43
The Big Sleep '78
Big Steal '49
Blood on the Moon '48
Breakthrough '78
Cape Fear '61
Cape Fear '91
Crossfire '47
El Dorado '67
Enemy Below '57
False Colors '43
Farewell, My Lovely '75
Fire Down Below '57
Five Card Stud '68
Girl Rush '44
The Grass is Greener '61
Gung Ho! '43
His Kind of Woman '51
Holiday Affair '49
Hollywood Home Movies '87
Home From the Hill '60
Hopalong Cassidy: Riders of
    the Deadline '43
The Human Comedy '43
The Hunters '58
John Huston: The Man, the
    Movies, the Maverick '89
The Last Tycoon '77
The List of Adrian Messenger
    '63
The Longest Day '62
The Lusty Men '52
Macao '52
Maria's Lovers '84
Matilda '78
Midway '76
Mr. North '88
My Forbidden Past '51
The Night of the Hunter '55
Nightkill '80
One Minute to Zero '52
One Shoe Makes It Murder
    '82
Out of the Past '47
Pursued '47
Rachel and the Stranger '48
The Racket '51
The Red Pony '49
River of No Return '54
Ryan's Daughter '70
Scrooged '88
Second Chance '53
Secret Ceremony '69
She Couldn't Say No '52
The Sundowners '60
That Championship Season
    '82
Thirty Seconds Over Tokyo
    '44
Thompson's Last Run '90
Thunder Road '58
Till the End of Time '46
Villa Rides '68
War & Remembrance, Part 1
    '88
War & Remembrance, Part 2
    '89
The Way West '67
Winds of War '83
The Yakuza '75

Roberto Mitrotti (D)

Little Girl...Big Tease '75

Felix Mitterer

Requiem for Dominic '91

Tomokazu Miura

The Imperial Japanese Empire
    '85

Art Mix

Border Devils '32
Powdersmoke Range '35
Way of the West '35
Yodelin' Kid From Pine Ridge
    '37

Ruth Mix

The Black Coin '36
Custer's Last Stand '36
The Riding Avenger '36
The Tonto Kid '35

Tom Mix

Child of the Prairie '18
Dick Turpin '25
Early Cinema Vol. 1 '19
Great K & A Train Robbery
    '26
Heart of Texas Ryan '16
Hidden Gold '33
In the Days of the Thundering
    Herd & the Law & the
    Outlaw '14
Justice Rides Again '30s
Miracle Rider: Vol. 1 '35
Miracle Rider: Vol. 2 '35
Terror Trail '33
Tom Mix Compilation 1 '16
Tom Mix Compilation 2 '16

Tom Mix (D)

In the Days of the Thundering
    Herd & the Law & the
    Outlaw '14

Ichirota Miyagawa

The Family Game '83

Eri Miyagima

Summer Vacation: 1999 '88

Nobuko Miyamoto

The Funeral '84
Tampopo '86
A Taxing Woman '87
A Taxing Woman Returns '88

Kim Miyori

The Punisher '90
When the Bough Breaks '86

Kenji Mizoguchi (D)

The Crucified Lovers '54
Forty-Seven Ronin, Part 1 '42
Forty-Seven Ronin, Part 2 '42
Life of O'Haru '52
Osaka Elegy '36
Sansho the Bailiff '54
Shin Heike Monogatari '55
Sisters of Gion '36
The Story of the Late
    Chrysanthemum '39
Street of Shame '56
Ugetsu '53

Moshe Mizrahi (D)

Every Time We Say Goodbye
    '86
The House on Chelouche
    Street '73
I Love You Rosa '73
I Sent a Letter to My Love '81

La Vie Continue '82
Madame Rosa '77
Rachel's Man '75
War & Love '84

Rie Mizuhara

Summer Vacation: 1999 '88

Kimi Mizuno

War of the Gargantuas '70

Moana

Moana, a Romance of the
    Golden Age '26

Noah Moazezi

Tiger Town '83

Luke Moberly (D)

Little Laura & Big John '73

Mary Ann Mobley

Crazy Horse and Custer: "The
    Untold Story" '90
Girl Happy '65
Harum Scarum '65
My Dog, the Thief '69

Roger Mobley

Emil & the Detectives '64
Jack the Giant Killer '62

Peter Mochrie

The Winter of Our Dreams '82

W.D. Mochtar

Black Magic Terror '79

**Carlos Lopez
Moctezuma**

The Curse of the Crying
    Woman '61
El Aviso Inoportuno '87
Lagrimas de Amor '87
Night of the Bloody Apes '68

**Carlos Lopez
Moctezuma** (D)

Mary, Mary, Bloody Mary '76

Mara Modair

Blood Voyage '77

Jayne Modean

Spring Break '83
Street Hawk '84

Richard Moder (D)

The Bionic Woman '75

Matthew Modine

Baby It's You '82
Birdy '84
Full Metal Jacket '87
The Gamble '88
Gross Anatomy '89
The Hotel New Hampshire '84
Married to the Mob '88
Memphis Belle '90
Mrs. Soffel '85
Orphans '87
Pacific Heights '90
Private School '83
Streamers '83
Vision Quest '85

Gaston Modot

L'Age D'Or '30
Grand Illusion '37
The Rules of the Game '39
Under the Roofs of Paris
    (Sous les Toits de Paris) '29

Marcello Modugno

Dial Help '88

**Hans Joachim
Moebis**

Westfront 1918 '30

Titus Moede

Rat Pfink and Boo-Boo '66

Phillip Moeller (D)

Break of Hearts '35

Marianna Moen

Dance Macabre '91

Donald Moffatt

Alamo Bay '85
The Best of Times '86
The Bonfire of the Vanities '90
Far North '88
The Great Northfield
    Minnesota Raid '72
Jacqueline Bouvier Kennedy
    '81
License to Kill '84
The Long Days of Summer
    '80
Mary White '77
Mrs. R's Daughter '79
Music Box '89
Necessary Parties '88
On the Nickel '80
Promises in the Dark '79
Rachel, Rachel '68
The Right Stuff '83
Showdown '73
Strangers: The Story of a
    Mother and Daughter '79
The Thing '82
The Unbearable Lightness of
    Being '88
When the Time Comes '91
Winter Kills '79

Graham Moffatt

Oh, Mr. Porter '37

Charles Moffett

Ornette Coleman Trio '66

D.W. Moffett

Lisa '90
The Misfit Brigade '87

Gregory Moffett

Let's Dance '50
Robot Monster '53

Michelle Moffett

Deathstalker 4: Clash of the
    Titans '91
Hired to Kill '91

Sharyn Moffett

The Judge Steps Out '49

John Moffitt (D)

Love at Stake '87

Anna Moffo

The Adventurers '70

Otae Mofo

Eighteen Jade Arhats '84

**Flavio
Mogherini** (D)

Lunatics & Lovers '76

Leonide Moguy (D)

Action in Arabia '44

Whistle Stop '46

Victor Mohica

Johnny Firecloud '75

Carl Mohner

Cave of the Living Dead '65
Cave of the Living Dead '65
It Takes a Thief '59
Rififi '54
Sink the Bismarck '60
Wanted: Babysitter '75

Gerald Mohr

The Angry Red Planet '59
Hunt the Man Down '50
Hunt the Man Down/
    Smashing the Rackets '50
The Ring '52
Sirocco '51
Terror in the Haunted House
    '58

Hanro Mohr (D)

Hostage '87

Zia Mohyeddi

Bombay Talkie '70
We are the Children '87

Richard Moir

Heatwave '83
In Search of Anna '79
Panic Station '82
Remember Me '85

Zakes Mokae

Dad '89
A Dry White Season '89
Gross Anatomy '89
A Rage in Harlem '91
The Serpent and the Rainbow
    '87

Gustaf Molander (D)

Dollar '38
Intermezzo '36
Only One Night '42
Swedenhielms '35
A Woman's Face '38

Karin Molander

Thomas Graal's Best Film '17
Thomas Graal's First Child '18

Jeff Moldovan

Master Blaster '85

Jerry Molen

Rain Man '88

Alberto Molina

Portrait of Teresa '79

Alfred Molina

Blat
Not Without My Daughter '90
Prick Up Your Ears '87

Angela Molina

Camorra: The Naples
    Connection '85
Demons in the Garden '82
Eyes, the Mouth '83
Half of Heaven '86
That Obscure Object of Desire
    '77

Jack Molina (D)

The Craving '80

Vidal Molina

Sexy Cat '60s

Stefano Molinari

Evil Clutch '89

Edouard Molinaro (D)

Dracula & Son '76
Just the Way You Are '84
La Cage aux Folles '78
La Cage aux Folles 2 '81
Pain in the A— '77
Ravishing Idiot '64

Richard Molinos

Strange Awakening '58

Charles Moll

Cataclysm '81
Combat Academy '86
Night Train to Terror '84

Georgia Moll

Contempt '64
Thief of Baghdad '61
What a Way to Die '70

Richard Moll

Dungeonmaster '85
Hard Country '81
House '86
Metalstorm '83
Savage Journey '83
Survivor '87
Sword & the Sorcerer '82
Think Big '90
Wicked Stepmother '89

Flemming Quist Moller

Ladies on the Rocks '83

Clifford Mollison

A Christmas Carol '51

Darbnia Molloy

Paddy '70

Patrick Molloy

Phantom Brother '88
Plutonium Baby '87

Steve Molone

The Big Sweat '91

Alexandre Molthe

The Best of Dark Shadows '60s

Alessandro Momo

Malicious '74

Ghislaine Mona

Therese '86

Dan Monahan

Porky's '82
Porky's 2: The Next Day '83
Porky's Revenge '85

Maria Monay

Flesh and the Spur '57

Dick Monda

Body Parts '90

Jorge Mondragon

The New Invisible Man '58
Spiritism '61

Pierre Mondy

The Gift '82

Paul Mones

Tuff Turf '85

Peter Mones (D)

The Beat '88

Richard Monette

Big Zapper '73
Far Cry from Home '70s
Hello Mary Lou: Prom Night 2 '87
Higher Education '88
Murder by Night '89

Eddie Money

Money Madness '79

Monique Monge

What Would Your Mother Say? '81

Christopher Monger (D)

Waiting for the Light '90

Corbett Monica

The Passing of Evil '70

Mario Monicelli (D)

Big Deal on Madonna Street '60
Casanova '70 '65
The Great War '59
Joyful Laughter '54
Lovers and Liars '81
Passionate Thief '60

Alan Monk

La Traviata '83

Egon Monk (D)

The Oppermann Family '82

Louisa Monk

Dreadnaught Rivals '80s

Thelonious Monk

Jazz on a Summer's Day '59

Noel Monkman (D)

Typhoon Treasure '39

Yvonne Monlaur

The Brides of Dracula '60
Circus of Horrors '60
Hawk of the Caribbean '70s

Valentine Monnier

After the Fall of New York '85

Lawrence Monoson

Dangerous Love '87
Last American Virgin '82

Cinzia Monreale

Buried Alive '81

Louise Monroe

Alien Contamination '81

Marilyn Monroe

All About Eve '50
As Young As You Feel '51
Asphalt Jungle '50
Bus Stop '56
Clash by Night '52
Don't Bother to Knock '52
Gentlemen Prefer Blondes '53
Hollywood Home Movies '87
Hollywood Outtakes & Rare Footage '83
Home Town Story '51

How to Marry a Millionaire '53
Ladies of the Chorus '49
Let's Make it Legal '51
Let's Make Love '60
The Love Goddesses '65
Love Happy '50
Love Nest '51
The Misfits '61
Monkey Business '52
Niagara '52
The Prince and the Showgirl '57
River of No Return '54
The Seven Year Itch '55
Some Like It Hot '59
There's No Business Like Show Business '54
We're Not Married '52

Phil Monroe (D)

Bugs Bunny/Road Runner Movie '79

Jacques Monseau

The Devil's Nightmare '71

Carl Monson (D)

Blood Legacy '73
Please Don't Eat My Mother '72

Renzo Montagnani

A Joke of Destiny, Lying in Wait Around the Corner Like a Bandit '84

Edward Montagne (D)

They Went That-a-Way & That-a-Way '78

Bruce Montague

Whoops Apocalypse '83

Lee Montague

Brother Sun, Sister Moon '73
Mahler '74

Monte Montague

King of the Rodeo '28
Radio Patrol '37

Carlos Montalban

Bananas '71

Ricardo Montalban

Across the Wide Missouri '51
Alice Through the Looking Glass '66
Battleground '49
Black Water Gold '69
Blue '68
Cannonball Run II '84
Cheyenne Autumn '64
Conquest of the Planet of the Apes '72
Escape from the Planet of the Apes '71
Fantasy Island '76
Iron Cowboy '68
Joe Panther '76
Madame X '66
Mission to Glory '80
The Naked Gun: From the Files of Police Squad '88
Neptune's Daughter '49
Nosotros (We) '77
On an Island with You '48
Pirate Warrior '64
Return to Fantasy Island '77
Ride to Glory '71
Sayonara '57
Star Trek 2: The Wrath of Khan '82

Sweet Charity '68
Train Robbers '73
Two Weeks with Love '50

Guiliano Montaldo (D)

Control '87
Fifth Day of Peace '72
Sacco & Vanzetti '71
Time to Kill '89

Celia Montalvan

Toni '34

Karla Montana

La Pastorela '92
Sweet 15 '90

Monte Montana

Circle of Death '36
Down Dakota Way '49

Silvia Montanari

Deadly Revenge '83

Yves Montand

Cesar & Rosalie '72
Choice of Arms '83
Delusions of Grandeur '76
Goodbye Again '61
Grand Prix '66
Is Paris Burning? '68
Jean de Florette '87
Let's Make Love '60
Lovers Like Us '75
Manon of the Spring '87
My Geisha '62
Napoleon '55
On a Clear Day You Can See Forever '70
Resurrection at Masada '89
Roads to the South '78
State of Siege '73
Vincent, Francois, Paul and the Others '76
Wages of Fear '55
Where the Hot Wind Blows '59
Z '69

Lew Montania

Wonsan Operation '78

Arlene Montano

Furious '83

Lane Montano

Ricky 1 '88

Marlo Monte

Soul Vengeance '75

Juan Montezuma (D)

Sisters of Satan '75

Kelly Monteith

Hollywood Boulevard 2 '89
Screwball Hotel '88

Carmen Montejo

El Rey '87
La Dinastia de la Muerte
Ni Chana Ni Juana '87
The Vampire '57

Lisa Montell

Nine Lives of Elfego Baca '58
She Gods of Shark Reef '56

Mario Montenegro

Brides of the Beast '68

Sasha Montenegro

Los Vampiros de Coyacan '87
Passion for Power '85

Elsa Montero

Death of a Bureaucrat '66

John Montero

Blood Frenzy '87

Robert Montero (D)

Island Monster '53
Tharus Son of Attila '70s

Zully Montero

El Super '79

Rosenda Monteros

Cauldron of Blood '67
Tiara Tahiti '62
The White Orchid '54

Enrico Monterrano

Sex with a Smile '76

Fernando Montes

The Diabolical Dr. Z '65

Enrico Montesano

Lobster for Breakfast '82

Jorge Montesi

Death Target '83

Liliane Montevecchi

Meet Me in Las Vegas '56

Belinda J. Montgomery

Blackout '78
The Man from Atlantis '77
Silent Madness '84
Stone Cold Dead '80
Stone Fox '87
Tell Me That You Love Me '84
The Todd Killings '71

Douglass Montgomery

Harmony Lane '35

Elizabeth Montgomery

Amos '85
The Court Martial of Billy Mitchell '55
Deadline Assault '90
Missing Pieces '83

George Montgomery

Bomb at 10:10 '67
From Hell to Borneo '64
Hallucination '67
Hostile Guns '67
Indian Uprising '51
Orchestra Wives '42
Rough Riders' Roundup '34
Samar '62
Satan's Harvest '65
The Steel Claw '61
Sword of Monte Cristo '51
The Texas Rangers '51
War Kill '68
Warkill '65

George Montgomery (D)

From Hell to Borneo '64
Samar '62

Montgomery ▶cont.

The Steel Claw '61

Jeff Montgomery

Oasis of the Zombies '82

Julia Montgomery

Girls Night Out '83
Revenge of the Nerds '84
Savage Justice '88
South of Reno '87
Stewardess School '86
Up the Creek '84

Lee Montgomery

Ben '72
Girls Just Want to Have Fun '85
Into the Fire '88
The Midnight Hour '86
Mutant '83
Pete 'n' Tillie '72
Prime Risk '84

Patrick Montgomery (D)

The Compleat Beatles '82

Richard Montgomery

The Legend of Jedediah Carver '70s

Robert Montgomery

The Divorcee '30
Eye Witness '49
Forsaking All Others '35
Here Comes Mr. Jordan '41
Hide-Out '34
Lady in the Lake '46
The Last of Mrs. Cheyney '37
The Lost Stooges '33
Mr. & Mrs. Smith '41
They were Expendable '45

Robert Montgomery (D)

Eye Witness '49
The Gallant Hours '60
Lady in the Lake '46

Thomas Montgomery (D)

King Kong Versus Godzilla '63

Aldo Monti

Vivo o Muerto '87

Ivana Monti

Contraband '86

Silvia Monti

The Brain '69
Sicilian Connection '74

Anna Maria Monticelli

The Empty Beach '85
Nomads '86

Dora Montiel

Calacan

Sara Montiel

Besame '87
Cada Noche un Amor (Every Night a New Lover) '75
La Bella Lola '87

La Violetera (The Violet) '71

Sarita Montiel

La Amante '71

Lugi Montini

Satanik '69

Vincent Monton (D)

Windrider '86

Alex Montoya

West to Glory '47

The Montreal Canadians

Mystery of the Million Dollar Hockey Puck '80s

Dave Montresor

Assignment Outer Space '61

Michael Monty

Slash '87
The 13th Mission '91

Jim Moody

Bad Boys '83

Lynne Moody

A Fight for Jenny '90
Nightmare in Badham County '76
Scream Blacula Scream '73
Some Kind of Hero '82
Wait Till Your Mother Gets Home '83

Ralph Moody

Road to Bali '53

Ron Moody

David Copperfield '70
Dog Pound Shuffle '75
Dominique is Dead '79
Legend of the Werewolf '75
Murder Most Foul '65
Oliver! '68
Twelve Chairs '70
Unidentified Flying Oddball '79
Wrong is Right '82

De Sacia Mooers

The Average Woman '24

Randy Mooers

Blood Sisters '86

Heinz Moog

Senso '68
The Wanton Contessa '55

Elina Moon

Don't Tell Daddy '90

Keith Moon

Monterey Pop '68
Ready Steady Go, Vol. 1 '83
Ready Steady Go, Vol. 2 '85
Sextette '78
That'll Be the Day '73
Tommy '75
Woodstock '70

Kwon Young Moon

Young Heroes '69

Dennis Mooney

Death Row Diner '88

Martin Mooney (D)

The Phantom of 42nd Street '45

Saba Moor

Cave Girl '85

Alice Moore

Pulsebeat '85

Alvy Moore

A Boy and His Dog '75
The Brotherhood of Satan '71
Designing Woman '57
Dr. Minx '75
Intruder '88
Scream '83
Smokey & the Hotwire Gang '79
The Specialist '75
They're Playing with Fire '84
Three Nuts in Search of a Bolt '64
The Wild One '54
The Witchmaker '69

Archie Moore

The Adventures of Huckleberry Finn '60
Breakheart Pass '76
My Sweet Charlie '70

Barbara Moore

Escape from Death Row '76

Ben Moore

2000 Maniacs '64

Bobby Moore

Victory '81

Candy Moore

Lunch Wagon '81
Tomboy & the Champ '58

Charles Philip Moore (D)

Dance with Death '91
Demon Wind '90

Chris Moore

Punch the Clock '90

Christine Moore

Alexa: A Prostitute's Own Story '88
Lurkers '88
Prime Evil '88
Thrilled to Death '88

Clayton Moore

The Black Dragons '42
The Crimson Ghost '46
Cyclotrode ''X'' '46
Far Frontier '48
G-Men Never Forget '48
Jesse James Rides Again '47
Jungle Drums of Africa '53
Justice of the West '61
Lone Ranger '56
Lone Ranger and the Lost City of Gold '58
Nyoka and the Tigermen '42
Radar Men from the Moon '52
The Son of Monte Cristo '40
Tomahawk Territory '52

Cleo Moore

On Dangerous Ground '51

Colleen Moore

The Busher '19

Ella Cinders '26
The Scarlet Letter '34
The Sky Pilot '21
These Girls Won't Talk '20s

Constance Moore

Buck Rogers Conquers the Universe '39
Buck Rogers: Planet Outlaws '38
Delightfully Dangerous '45
Destination Saturn '39
Wives Under Suspicion '38
You Can't Cheat an Honest Man '39

Deborah Maria Moore

Into the Sun '91

Debrah Moore

Warriors of the Apocalypse '85

Del Moore

The Patsy '64

Demi Moore

About Last Night... '86
Blame It on Rio '84
The Butcher's Wife '91
Choices '81
Ghost '90
Mortal Thoughts '91
No Small Affair '84
Nothing But Trouble '91
One Crazy Summer '86
Parasite '81
Ron Reagan Is the President's Son/The New Homeowner's Guide to Happiness '85
St. Elmo's Fire '85
The Seventh Sign '88
We're No Angels '89
Wisdom '87
Young Doctors in Love '82

Dennis Moore

Across the Plains '39
Arizona Bound '41
Black Market Rustlers '43
Border Roundup '41
D-Day on Mars '45
East Side Kids '40
East Side Kids/The Lost City '35
Fast Talking '86
King of the Bullwhip '51
Land of Hunted Men '43
Master Key '45
Mutiny in the Big House '39
The Purple Monster Strikes '45
Spooks Run Wild '41
Sylvia Scarlett '35
Texas Justice '42
Wild Horse Canyon '39

Dickie Moore

Blonde Venus '32
Heaven Can Wait '43
Jive Junction '43
Little Rascals, Book 7 '30s
Little Red Schoolhouse '36
Miss Annie Rooney '42
Oliver Twist '33
Out of the Past '47
Sergeant York '41
The World Accuses '35

Dorothy Moore

Death Kiss '77

Dudley Moore

The Adventures of Milo & Otis '89
Arthur '81
Arthur 2: On the Rocks '88
Bedazzled '68
Best Defense '84
Blame It on the Bellboy '91
Comic Relief 2 '87
Crazy People '90
A Delicate Thread '86
Foul Play '78
The Hound of the Baskervilles '77
Like Father, Like Son '87
Lovesick '83
Micki & Maude '84
Romantic Comedy '83
Santa Claus: The Movie '85
Six Weeks '82
10 '79
30 Is a Dangerous Age, Cynthia '68
Those Daring Young Men in Their Jaunty Jalopies '69
Unfaithfully Yours '84
Wholly Moses! '80
The Wrong Box '66

Duke Moore

Night of the Ghouls '59
The Sinister Urge '60

Eileen Moore

Men of Sherwood Forest '57

Ellen Moore

New Orleans After Dark '58

Erin O'Brien Moore

Destination Moon '50

Frank Moore

Kings and Desperate Men '83
Rabid '77

Frank Moore (D)

Beehive '85

Gar Moore

Abbott and Costello Meet the Killer, Boris Karloff '49
Paisan '46
The Underworld Story '50

Glenn Moore

Oregon '88

Grace Moore

One Night of Love '34

Gwen Moore

High Voltage '29

James Moore

Dark Side of Midnight '86

Jeanie Moore

Dream Trap '90
The Final Alliance '89
Vampire at Midnight '88

Joanna Moore

Countdown '68
The Dunwich Horror '70
Follow That Dream '61
Iron Horsemen '71
Never a Dull Moment '68
Scout's Honor '80
Son of Flubber '63

Juanita Moore

A Dream for Christmas '73

Shoot It Black, Shoot It Blue
'74
Sound of Murder '82
The Stuff '85
Tailspin: Behind the Korean
Airline Tragedy '89
Troll '85
Who'll Stop the Rain? '78
Winds of Kitty Hawk '78

Philippe Morier-Genoud

Au Revoir Les Enfants '88
Confidentially Yours '83
Cyrano de Bergerac '89

Francesca Moriggi

The Tree of Wooden Clogs
'78

Michel Morin

Ilsa, the Tigress of Siberia '79

Patricia Morison

Dressed to Kill '46
The Fallen Sparrow '43
The Prince of Thieves '48
Queen of the Amazons '47
Return of Wildfire '48
Song of the Thin Man '47
Song Without End '60

David Morisson

The 13th Mission '91

Miki Morita

Border Phantom '37

Pat Morita

Amos '85
Babes in Toyland '86
Cancel My Reservation '72
Captive Hearts '87
Collision Course '89
Do or Die '91
For the Love of It '80
Hiroshima: Out of the Ashes
'90
The Karate Kid '84
The Karate Kid: Part 2 '86
The Karate Kid: Part 3 '89
Las Vegas Strip War '84
Lena's Holiday '91
Night Patrol '85
Slapstick of Another Kind '84
Thoroughly Modern Millie '67
The Vegas Strip Wars '84
When Time Ran Out '80

Yoshimitsu Morita (D)

The Family Game '83

Dorothea Moritz

Alpine Fire '89

Louisa Moritz

Chained Heat '83
Hot Chili '85
Last American Virgin '82
New Year's Evil '78
True Confessions '81

Henning Moritzen

Cries and Whispers '72
Memories of a Marriage '90

Mary Morlass

This is Not a Test '62

Karen Morley

Beloved Enemy '36
Healer '36

Littlest Rebel '35
Mata Hari '32
Our Daily Bread '34
Scarface '31
The Sin of Madelon Claudet
'31

Malcolm Morley

The Trouble with Spies '87

Rita Morley

The Flesh Eaters '64

Robert Morley

The African Queen '51
The Alphabet Murders '65
Around the World in 80 Days
'56
The Battle of the Sexes '60
Beat the Devil '53
Cromwell '70
Curtain Up '53
Deadly Game '82
The Foreman Went to France
'42
Ghosts of Berkeley Square
'47
The Great Muppet Caper '81
High Road to China '83
Hugo the Hippo '76
I Live in Grosvenor Square '46
Istanbul '90
Joseph & His Brethren '60
Lady Killers '80s
Lola '69
Loophole '83
The Loved One '65
Major Barbara '41
Marie Antoinette '38
Murder at the Gallop '63
Of Human Bondage '64
Oh, Heavenly Dog! '80
Scavenger Hunt '79
Second Time Lucky '84
The Small Back Room '49
Song of Norway '70
A Study in Terror '66
Theatre of Blood '73
Those Magnificent Men in
Their Flying Machines '65
Topkapi '64
The Trouble with Spies '87
War & Remembrance, Part 1
'88
Who is Killing the Great Chefs
of Europe? '78
The Wind '87
Woman Times Seven '67

Mormon Tabernacle Choir

Music & the Spoken Word '74

Louis Morneau (D)

To Die Standing '91

Stanley Morner

I Conquer the Sea '36

Alicia Moro

Exterminators of the Year
3000 '83
Hot Blood '89

Frank Moro

La Fuga de Caro

Mike Moroff

Angel Town '89
The Cage '89

Joshua Morrell

Making Contact '86

Leo Morrell

Crime Killer '85

Priscilla Morrill

The People vs. Jean Harris '81
Right of Way '84

Adrian Morris

Radio Patrol '37

Anita Morris

Absolute Beginners '86
Aria '88
Bloodhounds of Broadway '89
Blue City '86
18 Again! '88
The Hotel New Hampshire '84
Maria's Lovers '84
Martians Go Home! '90
Ruthless People '86
Sinful Life '89
A Smoky Mountain Christmas
'86

Aubrey Morris

Lifeforce '85
Night Caller from Outer Space
'66

Barboura Morris

Atlas '60
A Bucket of Blood '59
Machine Gun Kelly '58
The Wasp Woman '49

Brad Morris

American Kickboxer 1 '91

Chester Morris

The Bat Whispers '30
Corsair '31
The Divorcee '30
Five Came Back '39
Flight from Glory '37
Frankie and Johnny '36
The Great White Hope '70
Hunt the Man Down/
Smashing the Rackets '50
Law of the Underworld '38
Red Headed Woman '32
Seven Miles from Alcatraz/
Flight from Glory '43
Tomorrow at Seven '33

Colleen Morris

Crossing the Line '90

David Burton Morris (D)

Patti Rocks '88

Ernest Morris (D)

The Tell-Tale Heart '62

Errol Morris (D)

Gates of Heaven '78
The Thin Blue Line '88

Frances Morris

Boss Cowboy '35

Garrett Morris

The Anderson Tapes '71
Best of Chevy Chase '87
Best of Gilda Radner '89
Best of John Belushi '85
Husbands, Wives, Money, and
Murder '86
Saturday Night Live: Carrie
Fisher '78
Saturday Night Live: Richard
Pryor '75

Saturday Night Live: Steve
Martin, Vol. 1 '78
Saturday Night Live: Steve
Martin, Vol. 2 '78
Saturday Night Live, Vol. 1
'70s
Severed Ties '92
The Stuff '85
Things We Did Last Summer
'78
The Underachievers '88

Glenn Morris

Tarzan's Revenge '38

Greg Morris

Vegas '78

Haviland Morris

For Love or Money '88
Love or Money? '88
Sixteen Candles '84
Who's That Girl? '87

Howard Morris

End of the Line '88
High Anxiety '77
History of the World: Part 1
'81
Life Stinks '91
Munster's Revenge '81
The Nutty Professor '63
Portrait of a Showgirl '82
Splash '84

Howard Morris (D)

Don't Drink the Water '69
Goin' Coconuts '78
Who's Minding the Mint? '67
With Six You Get Eggroll '68

Jeff Morris

Payday '72

Jim Morris

Rap Master Ronnie '88

John Morris

Beyond Innocence '87

Judy Morris

Best Enemies '80s
Between Wars '74
Cass '78
In Search of Anna '79
Phar Lap '84
Plumber '79
Razorback '84

Kim Morris

Playboy's 1988 Playmate
Video Calendar '87

Kirk Morris

Devil of the Desert Against the
Son of Hercules '62
Maciste in Hell '60
Terror of the Steppes '64
Triumph of the Son of
Hercules '63

Lana Morris

The Gay Lady '49
Trottie True '49

Marianne Morris

Vampyres '74

Mary Morris

Pimpernel Smith '42
Train of Events '49

Mercury Morris

The Black Six '74

Robert Morris

One Deadly Owner '74

Robert Morris (D)

King of Kong Island '78

Wayne Morris

The Bushwackers '52
Paths of Glory '57
Plunder Road '57

Bruce Morrison (D)

Shaker Run '85
Tearaway '87

E.G. Morrison

The Last Chance '45

Ernie Morrison

Return of Our Gang '25

Jim Morrison

The Doors: A Tribute to Jim
Morrison '81
The Doors: Live in Europe,
1968 '68
The Doors: Live at the
Hollywood Bowl '68

Kenny Morrison

The NeverEnding Story 2:
Next Chapter '91
The Quick and the Dead '87

Sammy Morrison

Flying Wild '41
Ghosts on the Loose '43
Let's Get Tough '42
'Neath Brooklyn Bridge '42
Spooks Run Wild '41

Shelly Morrison

Castle of Evil '66
Devil Times Five '82

Steven Morrison

Last House on Dead End
Street '77

Tom Morrison

Rocky 5 '90

Van Morrison

The Last Waltz '78
The Wall: Live in Berlin '90

Billy Morrissette

Severed Ties '92

Eamon Morrissey

Eat the Peach '86

Nani Morrissey

The Mermaids of Tiburon '62

Paul Morrissey (D)

Andy Warhol's Dracula '74
Andy Warhol's Frankenstein
'74
Beethoven's Nephew '88
Flesh '68
Heat '72
The Hound of the Baskervilles
'77
Mixed Blood '84
Spike of Bensonhurst '88
Trash '70

W.K. Moward (D)

Let's Go! '23

Alan Mowbray

Abbott and Costello Meet the
  Killer, Boris Karloff '49
I Wake Up Screaming '41
In Person '35
It Happened in New Orleans
  '36
The King and I '56
The Lovable Cheat '49
The Man Who Knew Too
  Much '56
Mary of Scotland '36
Music in My Heart '40
My Dear Secretary '49
Panama Hattie '42
The Phantom of 42nd Street
  '45
The Prince of Thieves '48
So This is Washington '42
Stand-In '37
A Study in Scarlet '33
That Hamilton Woman '41
That Uncertain Feeling '41
Topper Takes a Trip '39
The Villain Still Pursued Her
  '41
Vogues of 1938 '37

**Malcolm
Mowbray** (D)

Don't Tell Her It's Me '90
Out Cold '89
A Private Function '84

Jack Mower

Mystery Squadron '33
Ships in the Night '20

Patrick Mower

The Bloodsuckers '70
Carry On England '76
Catch Me a Spy '71
Cry of the Banshee '70

John Mowod

Revenge of the Living
  Zombies '88

**John Llewellyn
Moxey**

No Place to Hide '81

**John Llewellyn
Moxey** (D)

The Bounty Man '72
Circus of Fear '67
Conspiracy of Terror '75
The Cradle Will Fall '83
Home for the Holidays '72
Horror Hotel '60
Intimate Strangers '77
Killjoy '81
Mating Season '81
Nightmare in Badham County
  '76
No Place to Hide '81
Panic in Echo Park '77
The Power Within '79
President's Mistress '78
Sanctuary of Fear '79
The Solitary Man '82
A Strange and Deadly
  Occurence '74
Through Naked Eyes '87
Where Have All the People
  Gone? '74

Wood Moy

Chan is Missing '82

Allan Moyle

Outrageous! '77

Allan Moyle (D)

Pump Up the Volume '90
Times Square '80

Rodolpho Moyos

The Brave One '56

Sandra Mozar

Terror Beach '80s

Sandra Mozarowsky

Night of the Death Cult '75

George Mozart

Song of Freedom '38

Mr. Stubbs

Toby Tyler '59

Mr. T

D.C. Cab '84
Penitentiary 2 '82
Rocky 3 '82

Mr. Wizard

Starlight Hotel '90

Chen Mu-Chuan

Monkey Kung Fu '80s

Carlo Mucari

Obsession: A Taste for Fear
  '89

Leonard Mudie

British Intelligence '40

Cookie Mueller

Female Trouble '75

Julia Mueller

The Unbelievable Truth '90

Maureen Mueller

In a Shallow Grave '88

**Armin Mueller-
Stahl**

Angry Harvest '85
Avalon '90
Colonel Redl '84
Forget Mozart '86
Lethal Obsession '87
A Love in Germany '84
Midnight Cop '88
Music Box '89
Power of One '92

Lola Muethel

From the Life of the
  Marionettes '80

**Amanda
Muggleston**

Queen of the Road '84

Alfonso Muguia

Furia Pasional

Idries Muhammed

The Paris Reunion Band '88

Nino Muhlach

The Boy God '86

David Muir

Dr. Hackenstein '88

Donald Muir

Ninja's Extreme Weapons '90

Gavin Muir

Abbott & Costello Meet the
  Invisible Man '51
House of Fear '45
Mary of Scotland '36
Night Tide '63
Sherlock Holmes Faces Death
  '43

Geraldine Muir

Hope and Glory '87

Barbara Mujica

Los Muchachos de Antes No
  Usaban Arsenico '80s
Mayalunta '86

Karl Muka

Mad Mission 3 '84

Romesh Mukerji

Distant Thunder '73

Russell Mulcahy (D)

Highlander '86
Highlander 2: The Quickening
  '91
Razorback '84
Ricochet '91

Diana Muldaur

Beyond Reason '77
Master Ninja 3 '83
McQ '74
The Other '72
Planet Earth '74
Return of Frank Cannon '80

**Deborah
Muldowney**

Cellar Dweller '87

Gary Muledeer

Diary of a Young Comic '79

Matt Mulern

Biloxi Blues '88

Nancy Mulford

Skeleton Coast '89

Kate Mulgrew

Lovespell '79
The Manions of America '81
Remo Williams: The
  Adventure Begins '85
Round Numbers '91
A Stranger is Watching '82
Throw Momma from the Train
  '87
Time for Miracles '80
The Word '78

Jack Mulhall

Burn 'Em Up Barnes '34
The Clutching Hand '36
Custer's Last Stand '36
Dawn Express '42
Desperate Cargo '41
A Face in the Fog '36
Hell's Headquarters '32
I Killed That Man '42
Mr. Wise Guy '42
Mystery Squadron '33
'Neath Canadian Skies '46
Outlaws of Sonora '38
Skull & Crown '38

Edward Mulhare

Hill 24 Doesn't Answer '55
Megaforce '82
Our Man Flint '66
Von Ryan's Express '65

Mark Mulholland

No Surrender '86

Chris Mulkey

Dangerous Company '88
Heartbreak Hotel '88
In Dangerous Company '88
K-9000 '89
Patti Rocks '88
Roe vs. Wade '89
Tricks of the Trade '88
Write to Kill '91

Randy Mulkey

Natas...The Reflection '83

Martin Mull

Bad Manners '84
The Boss' Wife '86
Clue '85
Cutting Class '89
Dance with Death '91
Far Out Man '89
Flicks '85
FM '78
The History of White People in
  America '85
The History of White People in
  America: Vol. 2 '85
Home is Where the Hart is '88
Jonathan Winters on the Edge
  '86
Lots of Luck '85
Martin Mull Live: From
  Ridgeville, Ohio '88
Mr. Mom '83
My Bodyguard '80
O.C. and Stiggs '87
Portrait of a White Marriage
  '88
Rented Lips '88
Serial '80
Ski Patrol '89
Take This Job & Shove It '81
Think Big '90

Jack Mullaney

George! '70
Spinout '66
Tickle Me '65

Greg Mullavey

Body Count '87
Chrome Hearts '70
Husbands, Wives, Money, and
  Murder '86
I Dismember Mama '74
The Love Machine '71

Patty Mullen

Doom Asylum '88
Frankenhooker '90

Edward Muller (D)

Escape from Hell '80s
Savage Island '85

Harrison Muller

The Final Executioner '83
Miami Cops '89
She '83
2020 Texas Gladiators '85
Violent Breed '83

Katrin Muller

Perfumed Nightmare '89

Paul Muller

The Black Devils of Kali '55
The Devil's Commandment
  '60
Fangs of the Living Dead '68
Four Ways Out '57
Goliath and the Sins of
  Babylon '64
Lady Frankenstein '72
Naked Warriors '73
Nightmare Castle '65
Two Nights with Cleopatra '54
Virgin Among the Living Dead
  '71

Peter Muller

25 Fireman's Street '73

Gerry Mulligan

Gerry Mulligan '81

Richard Mulligan

Babes in Toyland '86
Big Bus '76
Doin' Time '85
A Fine Mess '86
The Group '66
The Heavenly Kid '85
The Hideaways '73
Little Big Man '70
Meatballs 2 '84
Micki & Maude '84
Poker Alice '87
Pueblo Affair '73
Scavenger Hunt '79
S.O.B. '81
Teachers '84

Robert Mulligan (D)

Baby, the Rain Must Fall '64
Bloodbrothers '78
Clara's Heart '88
Fear Strikes Out '57
The Great Impostor '61
Inside Daisy Clover '65
Kiss Me Goodbye '82
Love with the Proper Stranger
  '63
The Other '72
The Pursuit of Happiness '70
The Rat Race '60
Same Time, Next Year '78
The Stalking Moon '69
Summer of '42 '71
To Kill a Mockingbird '62
Up the Down Staircase '67

Mark Mullin (D)

Cool Blue '88

Rod Mullinar

Breakfast in Paris '81
Echoes of Paradise '86
Patrick '78

Michael Mullins

Pom Pom Girls '76

Claude Mulot (D)

Black Venus '83

Dermot Mulroney

Bright Angel '91
Career Opportunities '91
Longtime Companion '90
Staying Together '89
Sunset '88
Survival Quest '89
Young Guns '88

Kieran Mulroney

Nowhere to Run '88

Chuck Norris

Braddock: Missing in Action 3 '88
Breaker! Breaker! '77
Code of Silence '85
Delta Force '86
Delta Force 2: Operation Stranglehold '90
An Eye for an Eye '81
Firewalker '86
Force of One '79
Forced Vengeance '82
Game of Death '79
Good Guys Wear Black '78
Hero and the Terror '88
The Hitman '91
Invasion U.S.A. '85
Lone Wolf McQuade '83
Missing in Action '84
Missing in Action 2: The Beginning '85
Octagon '80
Return of the Dragon '73
Silent Rage '82
Slaughter in San Francisco '81
The Student Teachers '73
The Ultimate Stuntman: A Tribute to Dar Robinson '90
Warrior Within '80

Dean Norris

Hard to Kill '89

Edward Norris

Heartaches '47
The Lady in Question '40
Show Them No Mercy '35
They Won't Forget '37

Jim Norris

Secret Executioners '70s

Mike Norris

Born American '86
Survival Game '87

Rufus Norris

Demon Wind '90

Alan North

Billy Galvin '86
Lean on Me '89
More! Police Squad '82
Police Squad! Help Wanted! '82

Christopher North

Apology '86

Heather North

Barefoot Executive '71
Git! '65

Jay North

Scout's Honor '80
The Teacher '74

Noelle North

Slumber Party '57 '76
Sweater Girls '78

Robert North

Blades '89

Sheree North

Breakout '75
Charley Varrick '73
Defenseless '91
Legs '83
Maniac Cop '88
Marilyn: The Untold Story '80
Most Wanted '76
Only Once in a Lifetime '80s

The Organization '71
Portrait of a Stripper '79
The Shootist '76
Snatched '72
Telefon '77
The Trouble with Girls '69

Ted North

Charlie Chan in Rio '41
Devil Thumbs a Ride '47

Virginia North

The Abominable Dr. Phibes '71

Wayne Northrop

Going for the Gold: The Bill Johnson Story '85

Nadine Nortier

Mouchette '60

Alex Norton

Comfort and Joy '84
Gregory's Girl '80
Hidden City '87
A Sense of Freedom '78

Barry Norton

Devil Monster '46
Dracula (Spanish Version) '31

Bill L. Norton *(D)*

Hard Rain-The Tet 1968 '89
The Hill '88
Tour of Duty '87

B.W.L. Norton *(D)*

Baby...Secret of the Lost Legend '85
Three for the Road '87

Dee Dee Norton

Dakota '88

Edgar Norton

Dr. Jekyll and Mr. Hyde '31
Son of Frankenstein '39

Jack Norton

The Bank Dick '40
Finishing School '33
Two Reelers: Comedy Classics 1 '33

Jim Norton

Hidden Agenda '90

Ken Norton

Drum '76
Kiss and Be Killed '91
Mandingo '75
Oceans of Fire '86

Randy Norton

Fire and Ice '83

Richard Norton

China O'Brien '88
China O'Brien 2 '89
Crossfire '89
Equalizer 2000 '86
Gymkata '85
The Kick Fighter '91
Raiders of the Sun '92

Judy Norton-Taylor

Children's Carol '80
The Waltons' Christmas Carol '80s

Terry Norton

American Kickboxer 1 '91

Emissary '89

Tony Norton

Special Forces '80s

Ellie Norwood

The Silent Mr. Sherlock Holmes '12

Jaime Nos

Skyline '84

Yoko Nosaki

Murder in the Doll House '79

Raphael Nossbaum *(D)*

Private Road: No Trespassing '87

Max Nosseck *(D)*

Black Beauty '46
Black Beauty/Courage of Black Beauty '57
Brighton Strangler '45
Dillinger '45
The Hoodlum '51

Noel Nosseck *(D)*

Dreamer '79
King of the Mountain '81
Las Vegas Lady '76
Night Partners '83
Return of the Rebels '81
Summer Fantasy '84

Ralph Nossek

Chicago Joe & the Showgirl '90

Christopher Noth

Jakarta '88

Vic Noto

Street Trash '87

Thierry Notz *(D)*

The Terror Within '88
Watchers 2 '90

Michael Nouri

Between Two Women '86
Flashdance '83
The Hidden '87
Imagemaker '86
Little Vegas '90
Project: Alien '89
Psychic '91
Spraggue '84
Thieves of Fortune '89
Total Exposure '91

Lou Nova

Double Dynamite '51

Shelly Novack

Most Wanted '76

Blaine Novak

Strangers Kiss '83

Blaine Novak *(D)*

Short Fuse '88

Eva Novak

Red Signals '27

Ivana Novak

The Devil's Nightmare '71

Kim Novak

The Amorous Adventures of Moll Flanders '65
Bell, Book and Candle '59
The Children '91
The Eddy Duchin Story '56
The French Line '54
Just a Gigolo '79
Kiss Me, Stupid! '64
Liebestraum '91
The Man with the Golden Arm '55
The Mirror Crack'd '80
Of Human Bondage '64
Pal Joey '57
Phffft! '54
Picnic '55
Strangers When We Meet '60
Vertigo '58
The White Buffalo '77

Klaus Novak

The Inheritors '85

Lenka Novak

The Vampire Hookers '78

Mel Novak

Cat in the Cage '68
Lovely...But Deadly '82

Ramon Novarro

Ben-Hur '26
Big Steal '49
Eyes of Turpin Are Upon You! '21
Mata Hari '32
The Student Prince in Old Heidelberg '27

Don Novello

Father Guido Sarducci Goes to College '85
Gilda Live '80
The Godfather: Part 3 '90
La Pastorela '92
New York Stories '89
Paul Rodriguez Live! I Need the Couch '89
Spirit of '76 '91

Ivor Novello

The Lodger '26
Phantom Fiend '35
The White Rose '23

Jay Novello

Harum Scarum '65
Miracle of Our Lady of Fatima '52
The Prodigal '55
Robin Hood of the Pecos '41
Sabaka '55

Ruben Zepeda Novelo

Los Tres Calaveras (The Three Skeletons) '64

Jarmila Novotna

The Search '48

Alfred Nowke

Kill Slade '89

Philomena Nowlin

Ebony Dreams '80
Miss Melody Jones '73

Zachi Noy

Hot Bubblegum '81
Private Manoeuvres '83
Private Popsicle '82

Young Love - Lemon Popsicle 7 '87

Nippy Noya

Superdrumming '88

Phillip Noyce *(D)*

Blind Fury '90
Dead Calm '89
Echoes of Paradise '86
Heatwave '83
Newsfront '78

Winston Ntshona

A Dry White Season '89
Night of the Cyclone '90

Alan Nuary

Black Magic Terror '79

Danny Nucci

The Children of Times Square '86

Simon Nuchtern *(D)*

New York Nights '84
Savage Dawn '84
Silent Madness '84

Eddie Nugent

Doughnuts & Society '36
Kentucky Blue Streak '35

Elliott Nugent *(D)*

My Favorite Brunette '47
My Girl Tisa '48
My Outlaw Brother '51
Up in Arms '44

Richard Nugent

Sahara '43

Mizan Nunes

Maximum Thrust '87

Victor Nunez *(D)*

A Flash of Green '85
Gal Young 'Un '79

Alice Nunn

Fangs '75

Bill Nunn

Def by Temptation '90
Do the Right Thing '89
Mo' Better Blues '90
Regarding Henry '91
White Lie '91

Teri Nunn

Follow That Car '80
Thank God It's Friday '78

Trevor Nunn *(D)*

Hedda '75
Lady Jane '85

Rudolf Nureyev

Exposed '83

Loredana Nusciak

Django '68

Deland Nuse *(D)*

The Chilling '89

Raphael Nussbaum *(D)*

Death Blow '87
Sinai Commandos '68

Bombardier '43
Bombshell '33
Boy with the Green Hair '48
Consolation Marriage '31
Crack-Up '46
Devil Thumbs a Ride/Having
    Wonderful Crime '47
The End '78
Fighting Father Dunne '48
The Front Page '31
Having Wonderful Crime '45
Hawaiian Buckaroo '38
Hell's House '32
The Iron Major '43
James Cagney: That Yankee
    Doodle Dandy '86
Jubilee Trail '54
Knute Rockne: All American
    '40
Last Hurrah '58
Marine Raiders '43
The Navy Comes Through '42
Ragtime '81
Return of Video Yesterbloop
    '47
Riff-Raff '47
Scout's Honor '80
Slightly Honorable '40
Some Like It Hot '59
The World Gone Mad '33

Peter O'Brien

A Kink in the Picasso '90

Richard O'Brien

Robin Hood...The Legend:
    The Time of the Wolf '85
The Rocky Horror Picture
    Show '75
Shock Treatment '81

Tex O'Brien

Outlaw Roundup '44

**Timothy Eric
O'Brien**

Suburbia '83

Virginia O'Brien

Big Store '41
DuBarry was a Lady '43
The Harvey Girls '46
Lady Be Good '41
Panama Hattie '42
Thousands Cheer '43
Till the Clouds Roll By '46
Till the Clouds Roll By/A Star
    Is Born '46
Two Girls & a Sailor '44

Jeffrey Obrow

The Power '80

Jeffrey Obrow (D)

Dorm That Dripped Blood '82
Servants of Twilight '91

Karyn O'Bryan

Private Resort '84

Melody O'Bryan

Melody in Love '78

Pat O'Bryan

976-EVIL '88
Relentless '89

Patrick O'Bryan

976-EVIL 2: The Astral Factor
    '91

Ric Ocasek

Made in Heaven '87

Andrea Occhipinti

Bolero '84
Conquest '83
Control '87
El Barbaro '84

Kevin Och

Fun Down There '88

Andrea Ochhipinti

A Blade in the Dark '83

**Manuel Lopez
Ochoa**

Crossfire '67
El Tesoro de Chucho el Roto
    '47
Los Tres Calaveras (The
    Three Skeletons) '64

Uwe Ochsenknecht

Fire, Ice and Dynamite '91
Men... '85

Arthur O'Connell

Anatomy of a Murder '59
Ben '72
Bus Stop '56
Fantastic Voyage '66
Follow That Dream '61
Gidget '59
The Great Impostor '61
The Hiding Place '75
Huckleberry Finn '74
Kissin' Cousins '63
The Last Valley '71
Law of the Jungle '42
The Man in the Gray Flannel
    Suit '56
Man of the West '58
Misty '61
Monkey's Uncle '65
Operation Petticoat '59
Picnic '55
Pocketful of Miracles '61
Seven Faces of Dr. Lao '63
There was a Crooked Man '70

Charlie O'Connell

Derby '71

Deidre O'Connell

Falling from Grace '92
Pastime '91
Straight Talk '92

Eddie O'Connell

Absolute Beginners '86

Jack O'Connell (D)

Revolution '69

Jerry O'Connell

Stand By Me '86

Taffee O'Connell

Caged Fury '80

**Christopher
O'Conner**

Tongs: An American
    Nightmare '88

John O'Conner (D)

Prisoner in the Middle '74

Nancy O'Conner

Whale of a Tale '76

**James
O'Connolly** (D)

Berserk! '67
Crooks & Coronets '69
Tower of Evil '72
The Valley of Gwangi '69
Vendetta for the Saint '68

Brian O'Connor

Night Wars '88

Carroll O'Connor

Brass '85
Cleopatra '63
Convicted '86
Death of a Gunfighter '69
Doctors' Wives '70
For Love of Ivy '68
Hawaii '66
In Harm's Way '65
Johnny Frenchman '46
Kelly's Heroes '70
Law and Disorder '74
Lonely are the Brave '62
Marlowe '69
Point Blank '67
Waterhole #3 '67

Derrick O'Connor

Dealers '89
Hope and Glory '87
Lethal Weapon 2 '89

Donald O'Connor

Beau Geste '39
Francis in the Navy '55
I Love Melvin '53
James Cagney: That Yankee
    Doodle Dandy '86
Pandemonium '82
Private Buckaroo '42
Ragtime '81
Singin' in the Rain '52
There's No Business Like
    Show Business '54
A Time to Remember '90
Wonders of Aladdin '61

**Frank
O'Connor** (D)

Free to Love '25
Spangles '26

Glynnis O'Connor

Boy in the Plastic Bubble '76
California Dreaming '79
A Conspiracy of Love '87
Dark Side of Love '79
Johnny Dangerously '84
Kid Vengeance '75
Melanie '82
Night Crossing '81
Ode to Billy Joe '76
Our Town '77
Someone I Touched '75
Those Lips, Those Eyes '80

Hazel O'Connor

Breaking Glass '80

Jim O'Connor (D)

Mistress Pamela '74

Kenneth O'Connor

Carry On Emmanuelle '78

Kevin J. O'Connor

Bogie '80
Candy Mountain '87
Let's Scare Jessica to Death
    '71
Love at Large '89

The Moderns '88
Peggy Sue Got Married '86
Signs of Life '89
Steel Magnolias '89

Pat O'Connor (D)

Cal '84
Fools of Fortune '91
Month in the Country '87
Stars and Bars '88

**Sandra Day
O'Connor**

Gap-Toothed Women '89

Sinead O'Connor

The Wall: Live in Berlin '90

Terry O'Connor (D)

Early Frost '84

Tim O'Connor

Across 110th Street '72
The Groundstar Conspiracy
    '72
Manhunter '74

Una O'Connor

The Adventures of Robin
    Hood '38
The Barretts of Wimpole
    Street '34
The Bride of Frankenstein '35
The Canterville Ghost '44
Christmas in Connecticut '45
Fighting Father Dunne '48
Hopalong Cassidy:
    Unexpected Guest '47
The Informer '35
The Invisible Man '33
The Sea Hawk '40
Strawberry Blonde '41
This Land is Mine '43
Witness for the Prosecution
    '57

Hugh O'Conor

My Left Foot '89

Anita O'Day

The Gene Krupa Story '59
Jazz on a Summer's Day '59

Denis O'Dea

Captain Horatio Hornblower
    '51
The Long Dark Hall '51
Treasure Island '50
Under Capricorn '49

Judith O'Dea

Night of the Living Dead '68

David Odell (D)

Martians Go Home! '90

Georgia O'Dell

Big Calibre '35
Cannon Ball/Eyes Have It '20s

Sean Odell

Full Metal Ninja '89

Christophe Odent

First Name: Carmen '83

Clifford Odets

The General Died at Dawn '36

Clifford Odets (D)

None But the Lonely Heart '44

Mary Odette

She '21

George T. Odis

Straight Out of Brooklyn '91

Cathy O'Donnell

The Amazing Mr. X '48
Ben-Hur '59
Detective Story '51
The Love of Three Queens '54
Man from Laramie '55
Terror in the Haunted House
    '58
They Live by Night '49

Chris O'Donnell

Fried Green Tomatoes '91
Men Don't Leave '89

"Spec" O'Donnell

The Country Kid '23

**Michael
O'Donoghue**

Mr. Mike's Mondo Video '79
The Suicide Club '88

Ross O'Donovan

Starstruck '82

**Luciano
Odorisio** (D)

Sacrilege '86

Ron Odriozola

Homeboys '92

Martha O'Driscoll

Criminal Court '46
The Fallen Sparrow '43
Reap the Wild Wind '42
Young & Willing '42

**Bernadette
O'Farrell**

Robin Hood, the Legend:
    Volume 2 '52
Robin Hood: The Movie '91
Scotland Yard Inspector '52

Frederick Offrein

The Russian Terminator '90

Damian O'Flynn

Crack-Up '46
Daniel Boone: Trail Blazer '56
The Devil on Wheels '47
X Marks the Spot '42

Ken Ogata

The Ballad of Narayama '83
Eijanaika (Why Not?) '81
Mishima: A Life in Four
    Chapters '85
Vengeance Is Mine '79
Virus '82

Sammy Ogg

Miracle of Our Lady of Fatima
    '52

Bulle Ogier

Candy Mountain '87
The Discreet Charm of the
    Bourgeoisie '72

Pascale Ogier

Full Moon in Paris '84

Stranger in Town '57

Frank Packard *(D)*

In Your Face '77

David Packer

The Runnin' Kind '89
The Running Kind '89
Trust Me '89
You Can't Hurry Love '88

Maria Pacome

The Daydreamer '77

Joanna Pacula

Death Before Dishonor '87
Escape from Sobibor '87
Gorky Park '83
Husbands and Lovers '92
The Kiss '88
Marked for Death '90
Not Quite Paradise '86
Options '88
Sweet Lies '88

Sarah Padden

Cross Examination '32
Exiled to Shanghai '37
Girl Rush '44
The Mad Monster '42
Reg'lar Fellers '41
Riders of the West '42
Song of Old Wyoming '45
Tomorrow's Children '34
Wild West '46

Hugh Paddick

The Killing of Sister George '69
Naughty Knights '71

Ignace Jan Paderewski

Moonlight Sonata '38

Calvin Jackson Padget *(D)*

The Battle of El Alamein '68
Secret Agent Super Dragon '66

Haydee Padilla

Time for Revenge '82

Robert Padilla

The Great Gundown '75
Twisted Nightmare '87

Ernesto Pagani

Salammbo '14

Giulia Pagano

Masada '81

Anita Page

After Midnight '33
The Broadway Melody '29
Jungle Bride '33
Our Dancing Daughters '28
Our Dancing Daughters '20s

Ann Page

Stranger in Town '57

Anthony Page

Rebel '73

Anthony Page *(D)*

Absolution '81
Alpha Beta '73
Bill '81
Bill: On His Own '83

F. Scott Fitzgerald in Hollywood '76
Final Warning '90
Forbidden '85
Heartbreak House '86
I Never Promised You a Rose Garden '77
Johnny Belinda '82
Lady Vanishes '79
Missiles of October '74
Monte Carlo '86
The Nightmare Years '89
Pueblo Affair '73
Scandal in a Small Town '88

Betty Page

That Darn Sorceress '88

Bradley Page

The Affairs of Annabel '38
Champagne for Breakfast '79

Dorothy Page

Ride 'Em Cowgirl '41
Water Rustlers '39

Elaine Page

Unexplained Laughter '89

Gale Page

Four Daughters '38
Knute Rockne: All American '40
They Drive by Night '40

Genevieve Page

Buffet Froid '79
Day and the Hour '63
El Cid '61
Gun Crazy '69
The Private Life of Sherlock Holmes '70
Song Without End '60

Geraldine Page

ABC Stage 67: Truman Capote's A Christmas Memory '66
The Adventures of Huckleberry Finn '85
Barefoot in Athens '66
The Beguiled '70
The Bride '85
Day of the Locust '75
Dollmaker '84
The Happiest Millionaire '67
Harry's War '84
Hazel's People '75
Hitchhiker 1 '85
Honky Tonk Freeway '81
I'm Dancing as Fast as I Can '82
Interiors '78
Loving '84
My Little Girl '87
Nasty Habits '77
Native Son '86
Pete 'n' Tillie '72
The Pope of Greenwich Village '84
Riders to the Seas '88
Summer and Smoke '61
Sweet Bird of Youth '62
Toys in the Attic '63
Trip to Bountiful '85
Walls of Glass '85
Whatever Happened to Aunt Alice? '69
White Nights '85
You're a Big Boy Now '66

Harrison Page

Hammered: The Best of Sledge '88

Lionheart '90

Ilse Page

Creature with the Blue Hand '70

Jimmy Page

Led Zeppelin: The Song Remains the Same '73

Joy Page

Conquest of Cochise '53
Fighter Attack '53

LaWanda Page

B.A.D. Cats '80
Mausoleum '83
Shakes the Clown '92
Zapped! '82

Patti Page

Elmer Gantry '60

Rebecca Page

Danny '79

Teddy Page *(D)*

Blood Debts '83

Tony Page

Gangsters '79

Alfred Paget

America/The Fall of Babylon '24
The Babylon Story from "Intolerance" '16
The Lamb '15

Debra Paget

Broken Arrow '50
Demetrius and the Gladiators '54
From the Earth to the Moon '58
The Haunted Palace '63
House of Strangers '49
Journey to the Lost City '58
Love Me Tender '56
The Mercenaries '65
Omar Khayyam '57
Prince Valiant '54
Stars and Stripes Forever '52
Tales of Terror '62
The Ten Commandments '56

Robert Paget

The Choppers '61

Nicola Pagetti

Oliver's Story '78
Privates on Parade '84

Marcel Pagliero

Dedee D'Anvers '49

Marcel Pagliero *(D)*

The Respectful Prostitute '52

Aldo Paglisi

Seduced and Abandoned '64

Marcel Pagnol *(D)*

Angele '34
The Baker's Wife (La Femme du Boulanger) '33
Cesar '36
Harvest '37
Letters from My Windmill '54
Topaze '51
Well-Digger's Daughter '46

Sue Francis Pai

Jakarta '88

Suzze Pai

Big Trouble in Little China '86

Ian Paice

Superdrumming '88

Janice Paige

Angel on My Shoulder '80
Hollywood Canteen '44
Magic on Love Island '80
Please Don't Eat the Daisies '60
Romance on the High Seas '48
Silk Stockings '57

Kymberly Paige

Playboy's 1988 Playmate Video Calendar '87

Mabel Paige

Nocturne '46
The Scar '48

Raymond Paige

Hawaii Calls '38

Robert Paige

The Green Promise '49
Pardon My Sarong '42
The Red Stallion '47
Son of Dracula '43

Didier Pain

My Father's Glory '91

Brian Painchaud

Who Has Seen the Wind? '77

Bonnie Paine

Ballbuster
Bare Necessities '85

Cathy Paine

Image of Death '77

Heidi Paine

Alien Seed '89
New York's Finest '87

Nestor Paiva

Ballad of a Gunfighter '64
Creature from the Black Lagoon '54
Double Dynamite '51
The Falcon in Mexico '44
Fear '46
Follow Me Quietly '49
Jesse James Meets Frankenstein's Daughter '65
Nine Lives of Elfego Baca '58
Outcasts of the City '58
Prison Train '38
The Purple Heart '44
Shoot to Kill '47

John Paizs

The Big Crimewave '86

John Paizs *(D)*

The Big Crimewave '86

Turo Pajala

Ariel '89

Andres Pajares

Ay, Carmela! '90

Christine Pak

90 Days '86

Lisa Pak

Attack of the Beast Creatures '85

John Pakos

Futz '69

Alan J. Pakula *(D)*

All the President's Men '76
Comes a Horseman '78
Dream Lover '85
Klute '71
Orphans '87
Parallax View '74
Presumed Innocent '90
Rollover '81
See You in the Morning '89
Sophie's Choice '82
Starting Over '79
The Sterile Cuckoo '69

George Pal *(D)*

George Pal Color Cartoon Carnival '44
Seven Faces of Dr. Lao '63
The Time Machine '60
tom thumb '58
Wonderful World of the Brothers Grimm '62

Riccardo Palacio

Ark of the Sun God '82

Holly Palance

The Best of Times '86
Cast the First Stone '89
The Omen '76
Tuxedo Warrior '82

Jack Palance

Alice Through the Looking Glass '66
Alone in the Dark '82
Angel's Brigade '79
Arrowhead '53
Bagdad Cafe '88
Barabbas '62
Batman '89
The Battle of Austerlitz '60
Battle of the Commandos '71
Big Boss '77
Black Cobra '83
Bloody Avenger '80s
Chato's Land '71
City Slickers '91
Cocaine Cowboys '79
Contempt '64
Cop in Blue Jeans '78
Craze '74
Deadly Sanctuary '68
Desperados '69
Dr. Jekyll and Mr. Hyde '68
Dracula '73
Four Deuces '75
God's Gun '75
Gor '88
Great Adventure '75
Halls of Montezuma '50
Hatfields & the McCoys '75
Hawk the Slayer '81
Hell's Brigade: The Final Assault '80
Horsemen '70
I Died a Thousand Times '55
The Last Contract '86
Last Ride of the Dalton Gang '79
Lonely Man '57
The McMasters '70
The Mongols '60
Monte Walsh '70

Palance ▶cont.

Mr. Scarface '77
Oklahoma Crude '73
One Man Jury '78
Outlaw of Gor '87
Panic in the Streets '50
Portrait of a Hitman '77
Professionals '66
Requiem for a Heavyweight '56
Revenge of TV Bloopers '60s
Rulers of the City '71
Second Chance '53
The Sensuous Nurse '76
Shane '53
Silver Chalice '54
Sting of the West '76
Strange Case of Dr. Jekyll & Mr. Hyde '68
Tango and Cash '89
Torture Garden '67
Unknown Powers '80
Welcome to Blood City '77
Young Guns '88

Tom Palazzolo (D)

Caligari's Curse '74
Sneakin' and Peekin' '76

Euzhan Palcy (D)

A Dry White Season '89
Sugar Cane Alley '83

Joe Palese

Sinners '89

Natalie Paley

Sylvia Scarlett '35

Phillip Paley

Beachballs '88

David Palfy

Storm '87

Ron Palillo

Committed '91
Doin' Time '85
Hellgate '89
SnakeEater '89

Michael Palin

And Now for Something Completely Different '72
Around the World in 80 Days '90
Brazil '86
A Fish Called Wanda '88
Jabberwocky '77
The Missionary '82
Monty Python and the Holy Grail '75
Monty Python Live at the Hollywood Bowl '82
Monty Python Meets Beyond the Fringe
Monty Python's Life of Brian '79
Monty Python's Parrot Sketch not Included '90
Monty Python's The Meaning of Life '83
A Private Function '84
Ripping Yarns '75
The Secret Policeman's Other Ball '82
Secret Policeman's Private Parts '81
Time Bandits '81

Larry Pall (D)

Off Your Rocker '80

Jean Marie Pallardy (D)

Truck Stop

Anita Pallenberg

Barbarella '68
Performance '70

Rospo Pallenberg (D)

Cutting Class '89

Eugene Pallette

The Adventures of Robin Hood '38
Big Street '42
Birth of a Nation '15
The Bride Came C.O.D. '41
The Ghost Goes West '36
Heaven Can Wait '43
Intolerance '16
Kennel Murder Case '33
The Lady Eve '41
Mantrap '26
Mr. Smith Goes to Washington '39
My Man Godfrey '36
Pin-Up Girl '44
Sensations of 1945 '44
Silver Queen '42
Stowaway '36
Strangers of the Evening '32
The Three Musketeers '21
Topper '37
The Virginian '29

Gabriella Pallotti

Hero of Rome '63
The Mongols '60
The Roof '56

Anders Palm (D)

Murder on Line One '90
Unmasked Part 25 '88

Andrea Palma

The Criminal Life of Archibaldo de la Cruz '55

Loretta Palma

Sensations '87

Mimmo Palmara

The Last Days of Pompeii '60

Ulf Palme

Dreams '55
Miss Julie '50

Betsy Palmer

Friday the 13th '80
Friday the 13th, Part 2 '81
The Last Angry Man '59
The Long Gray Line '55
Mister Roberts '55
Tin Star '57

Corliss Palmer

The Return of Boston Blackie '27

Geoffrey Palmer

Christabel '89

Gregg Palmer

The Rebel Set '59
Zombies of Moratau '57

Joseph Palmer

The Legend of Sea Wolf '75

Leland Palmer

All That Jazz '79

Lilli Palmer

Adorable Julia '62
The Amorous Adventures of Moll Flanders '65
Beware of Pity '46
Body and Soul '47
The Boys from Brazil '78
But Not for Me '59
Chamber of Horrors '40
Cloak and Dagger '46
Conspiracy of Hearts '60
The Counterfeit Traitor '62
Devil In Silk '56
The Holcroft Covenant '85
The Long Dark Hall '51
Main Street to Broadway '53
Miracle of the White Stallions '63
Murders in the Rue Morgue '71
My Girl Tisa '48
Sebastian '68
The Secret Agent '36
What the Peeper Saw '72

Maria Palmer

Days of Glory '43
Outcasts of the City '58

Patricia Palmer

Rounding Up the Law '22

Peter Palmer

Li'l Abner '59

Reno Palmer

Street Law '79

Shirley Palmer

Campus Knights '29

Tony Palmer (D)

The Children '91
Space Movie '80
Wagner: The Movie '85

Conrad Palmisano (D)

Busted Up '86
Space Rage '86

Carlos Palomino

Nasty Hero '89
Strangers Kiss '83

Matilde Palou

Susana '51

Bruce Paltrow (D)

A Little Sex '82

Lucianna Paluzzi

A Black Veil for Lisa '68
Carlton Browne of the F.O. '59
Chuka '67
Conexion Oriental (East Connection) '70
The Italian Connection '73
Journey to the Lost City '58
The Klansman '74
Medusa '74
Muscle Beach Party '64
99 Women '69
One-Eyed Soldiers '67
Return to Peyton Place '61
Thunderball '65

Giuseppe Pambiere

Yellow Emanuelle '70s

Rick Pamplin (D)

Provoked '89

George Pan-Andreas

Crime Killer '85

George Pan-Andreas (D)

Crime Killer '85

Nam Chun Pan

King of Kung-Fu '80s

Yang Pan Pan

Duel of the Seven Tigers '80
Return of the Scorpion '80s
The Story of Drunken Master

Norman Panama (D)

Barnaby and Me '77
Court Jester '56
I Will, I Will for Now '76
The Road to Hong Kong '62
Trap '59

Alessandra Panaro

The Bacchantes '63

Sandra Panaro

Executioner of Venice '63

Heidi Pane

Wildest Dreams '90

Richard Panebianco

China Girl '87
Dogfight '91

Miguel Paneque

Letters from the Park '88

Tien Pang

Murder Masters of Kung Fu '85
The Samurai
To Subdue the Evil

Franklin Pangborn

All Over Town '37
The Bank Dick '40
Call Out the Marines '42
Carefree '38
Christmas in July '40
Comedy Reel, Vol. 2
Dangerous Holiday '37
Doughnuts & Society '36
Fifth Avenue Girl '39
Flying Down to Rio '33
A Girl, a Guy and a Gob '41
The Great Moment '44
Hail the Conquering Hero '44
Hats Off '37
Just Around the Corner '38
Mad Wednesday '51
The Mandarin Mystery '37
Never Give a Sucker an Even Break '41
The Palm Beach Story '42
Romance on the High Seas '48
The Sin of Harold Diddlebock '47
A Star Is Born '37
Sullivan's Travels '41
Tango '36
Till the Clouds Roll By/A Star Is Born '46
Topper Takes a Trip '39
Two Weeks to Live '43
Vivacious Lady '38

Patrick Pankhurst

Whitcomb's War '88

Stuart Pankin

The Best of Not Necessarily the News '88
Dirt Bike Kid '86
Fatal Attraction '87
Life Stinks '91
Love at Stake '87
Mannequin 2: On the Move '91
Second Sight '89
That's Adequate '90

John Pankow

Johnny Be Good '88
Monkey Shines '88
Mortal Thoughts '91
Talk Radio '88
To Live & Die in L.A. '85
Year of the Gun '91

Gerulf Pannach

Singing the Blues in Red '87

Joe Pantoliano

Downtown '89
Eddie & the Cruisers '83
Empire of the Sun '87
The Final Terror '83
Idolmaker '80
The In Crowd '88
La Bamba '87
The Last of the Finest '90
Mean Season '85
Midnight Run '88
Risky Business '83
Running Scared '86
Scenes from the Goldmine '87
Short Time '90
The Squeeze '87
Tales from the Crypt '89
Zandalee '91

Domenico Paolella (D)

Pirates of the Coast '61

Deno Paoli (D)

The Legend of Frank Woods '77

Anny Papa

A Blade in the Dark '83

Nicols Papaconstantinou

The Enchantress '88

Tatiana Papamoskou

Iphigenia '77

Helen Papas

Graveyard Shift '87

Irene Papas

Anne of the Thousand Days '69
The Assisi Underground '84
The Brotherhood '68
Christ Stopped at Eboli '79
A Dream of Kings '69
Erendira '83
The Guns of Navarone '61
High Season '88

The Bat People '74
Dead and Buried '81
Delinquent School Girls '84
Dirt Gang '71
Grave of the Vampire '72
Halloween 4: The Return of Michael Myers '88
The Last Porno Flick '74
On the Edge: Survival of Dana '79
One More Chance '90
The Pink Angels '71
Rocky 4 '85
The Underachievers '88
When Every Day was the Fourth of July '78
Zoltan...Hound of Dracula '77

Michael Pataki (D)

Cinderella '77

Wally Patch

Calling Paul Temple '48
Cottage to Let '41
Inspector Hornleigh '39

Tom Patchett (D)

Best Legs in the 8th Grade '84

Christopher Pate

The Mango Tree '77

Michael Pate

Face to Face '52
Houdini '53
Howling 3 '87
Julius Caesar '53
The Oklahoman '56
Return of Captain Invincible '83
Seventh Cavalry '56
Tower of London '62

Michael Pate (D)

Tim '79

Nana Patekar

Salaam Bombay! '88

Raju Patel (D)

In the Shadow of Kilimanjaro '86

Sharad Patel (D)

Amin: The Rise and Fall '82

Bill Paterson

The Adventures of Baron Munchausen '89
Comfort and Joy '84
Coming Up Roses '87
Defense of the Realm '85
Dutch Girls '87
Hidden City '87
The Killing Fields '84
The Ploughman's Lunch '83
A Private Function '84
The Rachel Papers '89
Truly, Madly, Deeply '91
The Witches '90

Smita Patil

Spices '86

Mandy Patinkin

Alien Nation '88
The Big Fix '78
Daniel '83
Dick Tracy '90
The Doctor '91
Follies in Concert '85
French Postcards '79

The House on Carroll Street '88
Impromptu '90
The Last Embrace '79
Maxie '85
Night of the Juggler '80
The Princess Bride '87
Ragtime '81
Sunday in the Park with George '86
True Colors '91
Yentl '83

Jorge Patino

La Venganza del Rojo '87

Charles Paton

Blackmail '29

Laurie Paton

Norman's Awesome Experience '88

Stuart Paton (D)

20,000 Leagues Under the Sea '16

Jason Patric

The Beast '88
Denial: The Dark Side of Passion '91
Frankenstein Unbound '90
The Lost Boys '87
Rush '91
Solarbabies '86
Toughlove '85

David Patrick

Hide and Seek '77

Dennis Patrick

Choices '81
Dear Dead Delilah '72
Joe '70
The Time Travelers '64

Dorothy Patrick

Follow Me Quietly '49
Torch Song '53

Gail Patrick

Badmen of Nevada '33
Brewster's Millions '45
Doctor Takes a Wife '40
My Favorite Wife '40
My Man Godfrey '36
The Mysterious Rider '33
The Phantom Broadcast '33
Stage Door '37
To the Last Man '33
Wagon Wheels '34
Wives Under Suspicion '38

Gregory Patrick

Sexpot '88
A Woman Obsessed '88

Joan Patrick

The Astro-Zombies '67

Lee Patrick

Black Bird '75
Fisherman's Wharf '39
In This Our Life '42
Inner Sanctum '48
Law of the Underworld '38
The Maltese Falcon '41
Pillow Talk '59
Seven Faces of Dr. Lao '63
Summer and Smoke '61

Michael Patrick (D)

Hider in the House '90

Nigel Patrick

The Battle of Britain '69
Browning Version '51
Encore '52
The Executioner '70
Johnny Nobody '61
League of Gentlemen '60
Mackintosh Man '73
The Pickwick Papers '54
Raintree County '57
Sapphire '59
Trio '50
The Virgin Soldiers '69

Nigel Patrick (D)

Johnny Nobody '61

Randal Patrick

Fast Food '89

Robert Patrick

Behind Enemy Lines '85
Eye of the Eagle '87
Future Hunters '88
Terminator 2: Judgement Day '91

Robert Patrick (D)

Road to Nashville '67

Tom Patricola

Two Reelers: Comedy Classics 8 '33

Stefano Patrizi

The Hurried Man '77

Jean-Marie Patte

The Rise of Louis XIV '66

Luanna Patten

Follow Me, Boys! '66
Johnny Tremain & the Sons of Liberty '58
Rock, Pretty Baby '56
So Dear to My Heart '49

Chuck Patterson

The Five Heartbeats '91

David Patterson

The Secret Garden '75
The Secret Garden '84

Elizabeth Patterson

A Bill of Divorcement '32
Hide-Out '34
Intruder in the Dust '49
No Man of Her Own '32
The Sky's the Limit '43

Frank Patterson

The Dead '87

George Patterson

I Drink Your Blood '71

Jay Patterson

Double Exposure: The Story of Margaret Bourke-White '89
Heaven Help Us '85
McBain '91

J.G. Patterson

Doctor Gore '75

John D. Patterson

Forlorn River '37

John D. Patterson (D)

Deadly Innocence '88
Legend of Earl Durand '74
The Spring '89

Lee Patterson

Above Us the Waves '56
Chato's Land '71
He Lives: The Search for the Evil One '88
Jack the Ripper '60
The Key Man '57
The Spaniard's Curse '58
Time Lock '57

Lorna Patterson

The Imposter '84

Melody Patterson

The Immortalizer '89

Neva Patterson

An Affair to Remember '57
David and Lisa '63
Desk Set '57
The Runaways '75
Women of Valor '86

Pat Patterson

Gas-s-s-s! '70

Pat Patterson (D)

Doctor Gore '75

Ray Patterson (D)

The Good, the Bad, and Huckleberry Hound '88
Yogi's First Christmas '80

Richard Patterson (D)

J-Men Forever! '79

Robert Patterson

Nuns on the Run '90

Rocky Patterson

Armed for Action
Nail Gun Massacre '86

Sarah Patterson

The Company of Wolves '85

Willi Patterson (D)

Dreams Lost, Dreams Found '87

Brigitta Pattersson

The Virgin Spring '59

Michael Pattinson (D)

...Almost '90
Ground Zero '88
Moving Out '83

Jeremy Pattnosh

Certain Sacrifice '80

Bart Patton

Dementia 13 '63

Bill Patton

Fangs of Fate '25

Mark Patton

Anna to the Infinite Power '84
A Nightmare on Elm Street 2: Freddy's Revenge '85

Will Patton

Belizaire the Cajun '86
Chinese Boxes '84
Everybody Wins '90
No Way Out '87
The Rapture '91
A Shock to the System '90
Signs of Life '89
Stars and Bars '88
Variety '83
Wildfire '88

Peter Patzak (D)

Lethal Obsession '87
Midnight Cop '88
Slaughterday '77

Henrik Pauer

Time Stands Still '82

Adrian Paul

Masque of the Red Death '89

Alexandra Paul

American Flyers '85
American Nightmare '81
Christine '84
8 Million Ways to Die '85
Getting Physical '84
In Between '90s
Just the Way You Are '84
Prey of the Chameleon '91

Byron Paul (D)

Lt. Robin Crusoe, U.S.N. '66

David Paul

The Barbarians '87
Double Trouble '91
Ghost Writer '89
Think Big '90

Don Michael Paul

Alien from L.A. '87
Aloha Summer '88
Brotherhood of Justice '86
Heart of Dixie '89
Lovelines '84
Rich Girl '91
Rolling Vengeance '87
Winners Take All '87

Frankie Paul

Prince Jammy '87

Jean Paul

Full Metal Ninja '89

John Paul

Doomwatch '72

Lee Paul

Children of An Lac '80

Nancy Paul

Lifeforce '85

Peter Paul

The Barbarians '87
Double Trouble '91
Ghost Writer '89
Think Big '90

Rene Paul

The C-Man '49

Richard Paul

Bloodfist 3: Forced to Fight '91
Not for Publication '84
Princess Academy '87

Richard Joseph Paul
Under the Boardwalk '89

Ricky Paul
Piranha 2: The Spawning '82

Rosemary Paul
Dead Easy '82

Steven Paul (D)
Eternity '90
Falling in Love Again '80
Slapstick of Another Kind '84

Stuart Paul
Emanon '86
Fate '90

Stuart Paul (D)
Emanon '86
Fate '90

Talia Paul
Anguish '88

Scott Paulin
Captain America '89
From Hollywood to Deadwood '89
Grim Prairie Tales '89
The Last of Philip Banter '87
Pump Up the Volume '90
The Right Stuff '83
A Soldier's Story '84
Teen Wolf '85
Tricks of the Trade '88
Turner and Hooch '89

Morgan Paull
Surf 2 '84

David Paulsen (D)
Schizoid '80

Pat Paulsen
Ellie '84
Foreplay '75
Harper Valley P.T.A. '78
Night Patrol '85
They Still Call Me Bruce '86

Rob Paulsen
Eyes of Fire '84
Perfect Match '88

George Paulsin
Simon, King of the Witches '71

Marisa Pavan
Drum Beat '54
The Man in the Gray Flannel Suit '56
The Rose Tattoo '55
Solomon and Sheba '59
What Price Glory? '52

Anna Pavane
Portrait in Terror '62

Luciano Pavarotti
Yes, Giorgio '82

Lucy Pavey
The Fall of the House of Usher '52

Johanna Pavlis
Until September '84

George Pavlou (D)
Rawhead Rex '87
Transmutations '86
Underworld '85

Nadia Pavlova
Murder She Said '62

Drina Pavlovic
Games Girls Play '75

Pavel Pavlovsky
All My Good Countrymen '68

Muriel Pavlow
Doctor in the House '53
Doctor at Large '57
Malta Story '53

Lennox Pawle
Sylvia Scarlett '35

Adam Pawlikowski
Ashes and Diamonds '59

James Pax
In Love and War '91

Katina Paxinou
Inheritance '47
Rocco and His Brothers '60

Bill Paxton
Aliens '86
Back to Back '90
Brain Dead '89
Impulse '84
The Last of the Finest '90
Navy SEALS '90
Near Dark '87
Next of Kin '89
Pass the Ammo '88
Predator 2 '90
Slipstream '89
Streets of Fire '84
The Terminator '84

Lesley Paxton
Pilgrim Farewell '82

Tom Paxton
The Folk Music Reunion '88

Chien Pay
Dance of Death '84

Gilles Payant
Big Red '62

Johnny Paycheck
Hell's Angels Forever '83
Sweet Country Road '81

Alain Payet (D)
Helltrain '80

Clark Paylow (D)
Ring of Terror '62

Michelle Paymar (D)
And Another Honky Tonk Girl Says She Will '90

Allen Payne
New Jack City '91

Bruce Payne
Wonderland '88

George Pavlou (D)

Bruce Martin Payne
Howling 6: The Freaks '90

John Payne
Bail Out at 43,000 '57
Dodsworth '36
Footlight Serenade '42
The Great American Broadcast '41
Hats Off '37
Kansas City Confidential '52
Miracle on 34th Street '47
The Razor's Edge '46
Silver Lode '54
Slightly Scarlet '56
Springtime in the Rockies '42
Sun Valley Serenade '41
Tennessee's Partner '55
To the Shores of Tripoli '42

Julie Payne
Fraternity Vacation '85

Laurence Payne
The Crawling Eye '58
One Deadly Owner '74
The Tell-Tale Heart '62
Train of Events '49
Vampire Circus '71

Sally Payne
Jesse James at Bay '41
Man from Cheyenne '42
Romance on the Range '42

Amanda Pays
Cold Room '84
Dead on the Money '91
Exposure '91
The Flash '90
The Kindred '87
Leviathan '89
Max Headroom, the Original Story '86
Off Limits '87
Oxford Blues '84

Blanche Payson
Laurel & Hardy Comedy Classics Volume 5 '31
Stan Without Ollie, Volume 4 '25

Barbara Payton
Bride of the Gorilla '51
Drums in the Deep South '51
Flanagan Boy '53
Four Sided Triangle '53
Great Jesse James Raid '49
Kiss Tomorrow Goodbye '50
Only the Valiant '50
Trapped '49

Pamela Payton-Wright
Resurrection '80

Dixie Peabody
Bury Me an Angel '71

Rock Peace
Return of the Killer Tomatoes '88

Mary Peach
Ballad in Blue '66
A Gathering of Eagles '63
No Love for Johnnie '60

Kim Peacock
Midnight at the Wax Museum '36

Robert Peacock
Ocean Drive Weekend '85

Trevor Peacock
Henry VI, Part 1 '82
Twelfth Night '80

Barry Peak (D)
The Big Hurt '87

Teri Peake
Boys Night Out '88

E.J. Peaker
The Banker '89
Hello, Dolly! '69

Adele Pearce
Pop Always Pays '40

Alice Pearce
The Belle of New York '52
Dear Brigitte '65

Craig Pearce
Vicious '88

Jacqueline Pearce
Don't Raise the Bridge, Lower the River '68

Joanne Pearce
Morons from Outer Space '85

Mary Vivian Pearce
Desperate Living '77
Female Trouble '75
Mondo Trasho '70
Multiple Maniacs '70
Pink Flamingos '73

Michael Pearce (D)
James Joyce's Women '85

Peggy Pearce
Sex '20

Richard Pearce (D)
Country '84
Dead Man Out '89
Heartland '81
The Long Walk Home '89
No Mercy '86
Sessions '83
Threshold '83

Patricia Pearcy
Delusion '84
Squirm '76

Barry Pearl
Avenging Angel '85

Minnie Pearl
Nashville Story '86

Randy Pearlstein
Revenge of the Radioactive Reporter '91

Alfred Pears
Cobra Against Ninja

Robert E. Pearso (D)
Devil & Leroy Basset '73

Christopher Pearson
Frank and I '83

Clive Pearson
Witchcraft 4: Virgin Heart '92

Drew Pearson
Betrayal from the East '44

Fort Pearson
Queen for a Day '51

George Pearson (D)
Midnight at the Wax Museum '36
A Shot in the Dark '33

Jake Pearson
Dr. Jekyll's Dungeon of Death '82

Karen Pearson
Shattered Silence '71

Peter Pearson (D)
Paperback Hero '73
Tales of the Klondike: The Unexpected '83

Richard Pearson
Pirates '86
Thirteenth Reunion '81

Virginia Pearson
The Taxi Mystery '26
The Wizard of Oz '25

Harold Peary
Clambake '67
The Great Gildersleeve '43
Here We Go Again! '42
Look Who's Laughing '41
Seven Days' Leave '42

Patsy Pease
He Knows You're Alone '80

Max Pecas (D)
Erotic Touch of Hot Skin '65
The Sensuous Teenager '70
Sweet Ecstasy '62

William F. Pecchi
Honeymoon Horror '70s

Andrew Pece
Suburbia '83

Bruce Pecheur
Trash '70

Maria Pechukas
Spookies '85

Bill Peck
The Eye Creatures '65

Bob Peck
After Pilkington '88
Edge of Darkness '86
The Kitchen Toto '87
Slipstream '89

Brian Peck (D)
The Willies '90

Cecilia Peck
Ambition '91
My Best Friend Is a Vampire '88
Torn Apart '89

David Peck
Gal Young 'Un '79

Penn ▶cont.

Fast Times at Ridgemont High '82
Judgment in Berlin '88
Killing of Randy Webster '81
Racing with the Moon '84
Shanghai Surprise '86
State of Grace '90
Taps '81
We're No Angels '89

Sean Penn (D)

The Indian Runner '91

Malvina Penne

Chloe in the Afternoon '72

D.A. Pennebaker (D)

Don't Look Back '67
Jimi Hendrix: Live in Monterey, 1967 '67
John Lennon and the Plastic Ono Band: Live Peace in Toronto, 1969 '69
Otis Redding: Live in Monterey, 1967 '67

Judy Pennebaker

The Farmer's Other Daughter '65

Eagle Pennell (D)

Ice House '89
Last Night at the Alamo '83
The Whole Shootin' Match '79

Larry Pennell

The FBI Story '59

Joe Penner

Day the Bookies Wept '39

Alan Penney

Around the World in 80 Ways '86
Gone to Ground '76

Jack Pennick

Lady from Louisiana '42

Laurie Pennington

Chillers '88

Chris Pennock

The Great Texas Dynamite Chase '76

Hannibal Penny

Iron Horsemen '71

Joe Penny

Blood Vows: The Story of a Mafia Wife '87
Bloody Birthday '80
Deathmoon '78
Lifepod '80

Sydney Penny

Bernadette '90
Child of Darkness, Child of Light '91
Hyper-Sapien: People from Another Star '86
News at Eleven '86
Pale Rider '85

John Penrose

The Shadow Man '53

Lonnie Pense

Cycle Vixens '79

Virginia Penta

Stuck on You '84

Arthur Pentelow

The Gladiators '70

Anthony Penya

Humanoids from the Deep '80

Alena Penz

2069: A Sex Odyssey '78

David Peoples (D)

The Blood of Heroes '89

Clare Peploe (D)

High Season '88

George Peppard

Bang the Drum Slowly '56
Battle Beyond the Stars '80
The Blue Max '66
Breakfast at Tiffany's '61
The Carpetbaggers '64
Damnation Alley '77
The Executioner '70
From Hell to Victory '79
The Groundstar Conspiracy '72
Home From the Hill '60
How the West was Won '62
Newman's Law '74
Night of the Fox '90
Pendulum '69
Pork Chop Hill '59
Race to the Yankee Zephyr '81
Silence Like Glass '90
Target Eagle '84
Tobruk '66
Torn Between Two Lovers '79
Treasure of the Yankee Zephyr '84
Your Ticket Is No Longer Valid '84

Barbara Pepper

Girls in Chains '43
Hollywood Stadium Mystery '38
Kiss Me, Stupid! '64
Last Outlaw '36
Our Daily Bread '34

Cynthia Pepper

Kissin' Cousins '63

John Pepper

Specters '87

Marilia Pera

Mixed Blood '84
Pixote '81

Stacy Peralta

Skateboard Madness '80

Joe Perce

Don't Mess with My Sister! '85

Eileen Percy

Down to Earth '17
Let's Go! '23
Reaching For the Moon '17

Esme Percy

Bitter Sweet '33
Old Spanish Custom '36

Leticia Perdigon

Stallion '88

Helen Perdriere

Topaze '51

Paulo Cesar Pereio

I Love You '81

Miguel Pereira (D)

Veronico Cruz '87

Zeni Pereira

Women in Fury '84

Hope Perello (D)

Howling 6: The Freaks '90

Antonio Perez (D)

The 13th Mission '91

Conchita Perez

Cria '76

Jose Perez

Courage '86
Miami Blues '90
One Shoe Makes It Murder '82
Short Eyes '79

Beatriz Perez-Porro

Skyline '84

Rosie Perez

Criminal Justice '90
Do the Right Thing '89
White Men Can't Jump '92

Vincent Perez

Cyrano de Bergerac '89

Etienne Perier (D)

Hot Line '69
Investigation '79
Zeppelin '71

Francois Perier

Gervaise '57
Godson '72
Orpheus '49
Stavisky '74
Sylvia and the Phantom '45
Tartuffe '84
The Testament of Orpheus '59

Camila Perisse

Departamento Compartido '85

Anthony Perkins

Black Hole '79
Catch-22 '70
Crimes of Passion '84
Desire Under the Elms '58
Destroyer '88
The Double Negative '80
Edge of Sanity '89
Fear Strikes Out '57
ffolkes '80
The Fool Killer '65
Friendly Persuasion '56
Glory Boys '84
Goodbye Again '61
I'm Dangerous Tonight '90
Is Paris Burning? '68
Les Miserables '78
Life & Times of Judge Roy Bean '72
Lonely Man '57
Mahogany '76
The Matchmaker '58

Murder on the Orient Express '74
On the Beach '59
Pretty Poison '68
Psycho '60
Psycho 2 '83
Psycho 3 '86
Psycho 4: The Beginning '91
Ravishing Idiot '64
Sins of Dorian Gray '82
Someone Behind the Door '71
Tall Story '60
Ten Days Wonder '72
Tin Star '57
The Trial '63
Twice a Woman '79
Winter Kills '79

Anthony Perkins (D)

Lucky Stiff '88
Psycho 3 '86

Carl Perkins

Into the Night '85
The Other Side of Nashville '84
The Real Patsy Cline '90
Rock 'n' Roll History: Rockabilly Rockers '90

Elizabeth Perkins

About Last Night... '86
Avalon '90
Big '88
The Doctor '91
Enid is Sleeping '90
From the Hip '86
He Said, She Said '90
Love at Large '89
Sweet Hearts Dance '88

Millie Perkins

Cockfighter '74
The Diary of Anne Frank '59
Ensign Pulver '64
The Haunting Passion '83
License to Kill '84
Pistol: The Birth of a Legend '90
Ride in the Whirlwind '67
Shattered Vows '84
The Shooting '66
Table for Five '83
Two Moon Junction '88
Wild in the Country '61
Wild in the Streets '68
Witch Who Came from the Sea '76

Osgood Perkins

Scarface '31

Rebecca Perle

Not of This Earth '88

Max Perlich

Drugstore Cowboy '89
Rush '91

Rhea Perlman

Amazing Stories, Book 1 '85
Intimate Strangers '77
Ratings Game '84
Stamp of a Killer '87

Ron Perlman

Beauty and the Beast '87
Beauty and the Beast: Though Lovers Be Lost '89
Blind Man's Bluff '91
Ice Pirates '84
The Name of the Rose '86
Quest for Fire '82

Gigi Perreau

Dance with Me, Henry '56
Hell on Wheels '67
Journey to the Center of Time '67
The Man in the Gray Flannel Suit '56
Never a Dull Moment '50

Mireille Perrey

Umbrellas of Cherbourg '64

Paul Perri

Delta Force 2: Operation Stranglehold '90
Hit & Run '82

Greg Perrie

Scoring '80

Mireille Perrier

Chocolat '88
Love Without Pity '91

Jack Perrin

Apache Kid's Escape '30
Bandit Queen '51
The Court Martial of Billy Mitchell '55
Hell Fire Austin '32
I Shot Billy the Kid '50
Midnight Faces '26
New Moon '40
The North Star '43
The Story of Vernon and Irene Castle '39
Sunrise at Campobello '60
Ten Wanted Men '54
Them! '54
West of Pinto Basin '40
When Gangland Strikes '56

Jacques Perrin

Black and White in Color '76
Cinema Paradiso '88
The Desert of the Tartars '82
Donkey Skin (Peau d'Ane) '70
Dos Cruces en Danger Pass '71
Le Crabe Tambour '77
State of Siege '73
The 317th Platoon '65

Valerie Perrine

The Agency '81
The Border '82
Bright Angel '91
Can't Stop the Music '80
The Electric Horseman '79
The Last American Hero '73
Lenny '74
The Magician of Lublin '79
Maid to Order '87
Mr. Billion '77
Slaughterhouse Five '72
Superman: The Movie '78
Superman 2 '80
The Three Little Pigs '84
Water '85
When Your Lover Leaves '83

Nicola Perring

Nothing Underneath '85

Leslie Perrins

Man on the Run '49
Nine Days a Queen '34
The Silent Passenger '35

Francoise Perrot

Life and Nothing But '89
My Best Friend's Girl '84

Women's Prison Massacre '85

Carol Perry

Demons of Ludlow '75

Edd Perry

Battle Beyond the Sun '63

Felton Perry

Checking Out '89
Night Call Nurses '72
RoboCop 2 '90
Sudden Death '77
Walking Tall '73

Frank Perry

Accident '83

Frank Perry (D)

ABC Stage 67: Truman
 Capote's A Christmas
 Memory '66
Compromising Positions '85
David and Lisa '63
Diary of a Mad Housewife '70
Hello Again '87
Last Summer '69
Mommie Dearest '81
Monsignor '82
Skag '79
Swimmer '68

Gil Perry

Sourdough '77

Harvey Perry

Falling for the Stars '85

Jeffery Perry

Chronicles of Narnia '89

John Bennett Perry

The Last Fling '86
A Matter of Life and Death '81
Only When I Laugh '81

Joseph Perry

Iron Cowboy '68

Paul Perry Jr. (D)

Revenge of the Virgins '66

Lou Perry

Last Night at the Alamo '83

Luke Perry

Sweet Trash '89
Terminal Bliss '91

Matthew L. Perry

A Night in the Life of Jimmy
 Reardon '88
The Whole Shootin' Match '79

Navarre Perry

Bloody Wednesday '87

Peter Perry (D)

Cycle Vixens '79

Rod Perry

Black Gestapo '75
Black Godfather '74

Roger Perry

The Cat '66
Conspiracy of Terror '75
Count Yorga, Vampire '70

Clara Perryman

Escape to Love '82

Jill Perryman

Windrider '86

Maria Perschy

The Castle of Fu Manchu '68
Exorcism '74
Horror of the Zombies '74
House of Psychotic Women
 '73
Last Day of the War '69
Man's Favorite Sport? '63
No Survivors, Please '63
People Who Own the Dark '75
The Rue Morgue Massacres
 '73
A Witch Without a Broom '68

Bill Persky (D)

How to Pick Up Girls '78
Serial '80
Wait Till Your Mother Gets
 Home '83

Lisa Jane Persky

The Cotton Club '84
The Great Santini '80
The Last of the Finest '90
Peggy Sue Got Married '86
Shattered Vows '84
Vital Signs '90
When Harry Met Sally '89

Nehemiah Persoff

Al Capone '59
An American Tail: Fievel Goes
 West '91
The Badlanders '58
Comancheros '61
Deadly Harvest '72
Eric '75
Francis Gary Powers: The
 True Story of the U-2 Spy
 '76
The Henderson Monster '80
Killing Stone '78
The Last Temptation of Christ
 '88
The People Next Door '70
The Rebels '79
Some Like It Hot '59
The Stranger Within '74
Voyage of the Damned '76
The Wrong Man '56
Yentl '83

Essy Persson

Cry of the Banshee '70
Mission Stardust '68
Therese & Isabelle '68

Gene Persson

Earth vs. the Spider '58

Pippa Perthree

The Dining Room '86

Theo Pertsindis

The Fire in the Stone '85

Jon Pertwee

Carry On Screaming '66
Doctor Who: Death to the
 Daleks '87
Doctor Who: Spearhead from
 Space
Doctor Who: The Time
 Warrior
The House that Dripped Blood
 '71

Dana Peru

Something Wild '86

**Francisco J.
Lombardi Pery** (D)

The City and the Dogs '85

Frank Pesce

29 Street '91

P.J. Pesce (D)

Body Waves '91

Joe Pesci

Back Track '91
Betsy's Wedding '90
Easy Money '83
Eureka! '81
Family Enforcer '77
Goodfellas '90
Home Alone '90
I'm Dancing as Fast as I Can
 '82
JFK '91
Lethal Weapon 2 '89
Man on Fire '87
My Cousin Vinny '92
Once Upon a Time in America
 '84
Raging Bull '80
The Super '91

Lisa Pescia

Body Chemistry '90
Body Chemistry 2: Voice of a
 Stranger '91

Donna Pescow

Glory Years '87
Policewoman Centerfold '83
Saturday Night Fever '77

Andorai Peter

Another Way '82

Bernadette Peters

Alice '90
Annie '81
Heartbeeps '81
Impromptu '90
The Jerk '79
The Longest Yard '74
The Martian Chronicles: Part 2
 '79
The Martian Chronicles: Part 3
 '79
Pennies From Heaven '81
Pink Cadillac '89
Silent Movie '76
Slaves of New York '89
Sleeping Beauty '83
Sunday in the Park with
 George '86
Tulips '81

Bonny Peters (D)

Sandstone '74

Brock Peters

Abe Lincoln: Freedom Fighter
 '78
Ace High '68
The Adventures of
 Huckleberry Finn '78
Alligator 2: The Mutation '90
Carmen Jones '54
Daring Game '68
Framed '75
The Incident '67
The McMasters '70
Pawnbroker '65
Slaughter's Big Ripoff '73
Soylent Green '73
To Kill a Mockingbird '62

**Brooke L.
Peters** (D)

Anatomy of a Psycho '61
The Unearthly '57

Clarke Peters

A Casualty of War '90
Mona Lisa '86

Elizabeth Peters

She-Devil '89

Erika Peters

The Atomic Brain '64

George Peters

Teenage Mother '67

House Peters

Headwinds '28
Human Hearts '22

Jean Peters

Apache '54
As Young As You Feel '51
Broken Lance '54
A Man Called Peter '55
Niagara '52
Pickup on South Street '53
Three Coins in the Fountain
 '54
Viva Zapata! '52

Jim Peters

Ghostriders '87

Krika Peters

Heroes Die Young '60

Luan Peters

Flesh and Blood Show '73
Land of the Minotaur '77
Man of Violence '71
Not Tonight Darling '72
Twins of Evil '71

Lyn Peters

Grave of the Vampire '72

Molly Peters

Thunderball '65

Noel Peters

The Invisible Maniac '90

Ralph Peters

Death Rides the Range '40
Man's Country '38

Scott Peters

Girl Hunters '63

Susan Peters

Andy Hardy's Double Life '42
Random Harvest '42

Tiffany Peters

Soul Vengeance '75

Werner Peters

Corrupt Ones '67
The Counterfeit Traitor '62
Curse of the Yellow Snake '63
The Phantom of Soho '64

Chris Petersen

Little Dragons '80

Colin Petersen

A Cry from the Streets '57

Kent Petersen

Ride the Wind '88

Pat Petersen

Cold River '81
Little Dragons '80

William L. Petersen

Amazing Grace & Chuck '87
Cousins '89
Hard Promises '92
Long Gone '87
Manhunter '86
To Live & Die in L.A. '85
Young Guns 2 '90

**Wolfgang
Petersen** (D)

Das Boot '81
Enemy Mine '85
For Your Love Only '79
The NeverEnding Story '84
Shattered '91

Amanda Peterson

Can't Buy Me Love '87
Explorers '85
I Posed for Playboy '91
The Lawless Land '88
Listen to Me '89

Cassandra Peterson

Balboa '82
Echo Park '86
Elvira, Mistress of the Dark
 '88
Uncensored '85
Working Girls '75

Chris Peterson

Ninja Masters of Death '85
Vampire Raiders - Ninja
 Queen '89
When Every Day was the
 Fourth of July '78

Clifford Peterson

Charlie, the Lonesome Cougar
 '67

**Daniel M.
Peterson** (D)

Girlfriend from Hell '89

David Peterson

Deadly Business

Eric Peterson

Tramp at the Door '87

Gil Peterson

The Brain Machine '72
Mind Warp '72

Julie Peterson

Playboy's 1988 Playmate
 Video Calendar '87

**Kristine
Peterson** (D)

Body Chemistry '90
Critters 3 '91
Deadly Dreams '88
Lower Level '91

Nan Peterson

Hideous Sun Demon '59

Sarah Peterson

Cannon Movie Tales: Snow
 White '89

Shelly Peterson

The Housekeeper '86

Stewart Peterson

Against A Crooked Sky '75
Pony Express Rider '76
Seven Alone '75
Where the Red Fern Grows '74

Vidal Peterson

Wizards of the Lost Kingdom '85

Edward Petherbridge

Lovers of Their Time '85
Strange Interlude '90

Christopher Petit (D)

Chinese Boxes '84
Unsuitable Job for a Woman '82

Pascale Petit

End of Desire '62

Victor Petit

Night of the Death Cult '75
Street Warriors '87
Terror Beach '80s

Russ Petranto (D)

Uptown Comedy Express '89

Gio Petre

The Doll '62
Gorilla '56

Ian Petrella

A Christmas Story '83

Alexei Petrenko

Rasputin '85

Elio Petri (D)

10th Victim '65

Marcella Petri

Nathalie Comes of Age '86

Mario Petri

Conqueror & the Empress '64

Anne Petrie

Close to Home '86

Daniel Petrie (D)

The Bay Boy '85
The Betsy '78
Buster and Billie '74
Cocoon: The Return '88
Dollmaker '84
Eleanor & Franklin '76
Execution of Raymond Graham '85
Fort Apache, the Bronx '81
Lifeguard '75
Moon of the Wolf '72
Mystic Pizza '88
Neptune Factor '73
A Raisin in the Sun '61
Resurrection '80
Rocket Gibraltar '88
Silent Night, Lonely Night '69
Six Pack '82
Spy With a Cold Nose '66
Square Dance '87
Sybil '76
Toy Soldiers '91

Donald Petrie (D)

Amazing Stories, Book 4 '85
Opportunity Knocks '90

Hay Petrie

A Canterbury Tale '44
Crimes at the Dark House '39
The Old Curiosity Shop '35
Spellbound '41
Spy in Black '39

Susan Petrie

Snapshot '77

Sammy Petrillo

Bela Lugosi Meets a Brooklyn Gorilla '52

Boris L. Petroff (D)

Hats Off '37
Outcasts of the City '58

Giulio Petroni (D)

Blood and Guns '79

Vladimir Petrov (D)

The Inspector General '54
Peter the First: Part 1 '37
Peter the First: Part 2 '38

Sonia Petrovna

The Fatal Image '90

Joanna Pettet

Blue '68
Casino Royale '67
Cry of the Innocent '80
Double Exposure '82
The Evil '78
The Group '66
A Killer in Every Corner '74
Night of the Generals '67
Pioneer Woman '73
Return of Frank Cannon '80
Robbery '67
Sweet Country '87

Valarie Pettiford

Street Hunter '90

Lori Petty

Cadillac Man '90
Point Break '91

Richard Petty

Richard Petty Story '72

Ross Petty

The Housekeeper '86

Tom Petty

Made in Heaven '87

Angelique Pettyjohn

Bio Hazard '85
Confessions of Tom Harris '72
G.I. Executioner '71
The Lost Empire '83
Repo Man '83
Revenge of Doctor X '80s
Takin' It Off '84
The Wizard of Speed and Time '88

Joseph Pevney

Nocturne '46
The Street With No Name '48

Joseph Pevney (D)

Away All Boats '56

Man of a Thousand Faces '57
Mysterious Island of Beautiful Women '79
Night of the Grizzly '66
Tammy and the Bachelor '57
Torpedo Run '58
Who Is the Black Dahlia? '75

John Peyser (D)

The Centerfold Girls '74
Four Rode Out '69
Kashmiri Run '69

Penny Peyser

The Frisco Kid '79
Messenger of Death '88

Claudia Peyton

Bloodbeat '80s

Christian Pezy

Sweet Ecstasy '62

Ben Pfeifer

Kill Line '91

Dedee Pfeiffer

Brothers in Arms '88
The Horror Show '89
King's Ransom '91
The Midnight Hour '86
Toughlove '85

James Pfeiffer

Firefight '87

Michelle Pfeiffer

Amazon Women On the Moon '87
B.A.D. Cats '80
Callie and Son '81
Charlie Chan and the Curse of the Dragon Queen '81
Dangerous Liaisons '89
The Fabulous Baker Boys '89
Falling in Love Again '80
Frankie & Johnny '91
Grease 2 '82
Into the Night '85
Ladyhawke '85
Married to the Mob '88
Power, Passion & Murder '83
The Russia House '90
Scarface '83
The Solitary Man '82
Sweet Liberty '86
Tequila Sunrise '88
Witches of Eastwick '87

Scott Pfeiffer (D)

Firefight '87

Welsey Pfenning

Circle of Fear '89

Dedee Pfieffer

Red Surf '90

Michael Pfleghar (D)

Oldest Profession '67

JoAnn Pflug

The Day the Women Got Even '80
M*A*S*H '70

Joe Phelan

Fatal Pulse '88
South of Reno '87
Terminal Exposure '89
Zero Boys '86

Buster Phelps

Little Orphan Annie '32

Peter Phelps

Breaking Loose '90
The Lighthorseman '87
Starlight Hotel '90

Will Phelps (D)

North Shore '87

Michael Phenicie

9 1/2 Ninjas '90

John Philbin

Martians Go Home! '90
North Shore '87
Shy People '87

Mary Philbin

Human Hearts '22
Phantom of the Opera '25

Phil Philbin

Flesh Feast '69

James Philbrook

Finger on the Trigger '65
Last Day of the War '69
Sound of Horror '64

Robert Philip

Children Shouldn't Play with Dead Things '72

Gerard Philippe

Dangerous Liaisons '60
Devil in the Flesh '46
Fever Mounts at El Pao '59
La Ronde '51
The Red and the Black '57

Lee Philips

Peyton Place '57
Psychomania '63

Lee Philips (D)

Barnum '86
Blind Vengeance '90
The Blue Lightning '86
Hardhat & Legs '80
Mae West '84
On the Right Track '81
Samson and Delilah '84
The Spell '77
The Stranger Within '74
Sweet Hostage '75
War Between the Tates '76

Mary Philips

A Farewell to Arms '32
Prince Valiant '54

Robert Philips

Adios Amigo '75

Sian Philips

Valmont '89

Harald Phillipe (D)

Escape from the KGB '87

Barney Phillips

I Was a Teenage Werewolf '57
Savage Run '70

Barry Phillips

Dr. Caligari '89

Britta Phillips

Satisfaction '88

Chynna Phillips

Caddyshack 2 '88
The Invisible Kid '88

Conrad Phillips

Murder She Said '62

E.D. Phillips

Welcome Back Mr. Fox '83

Eddie Phillips

Phantom Patrol '36
Probation '32

Emo Phillips

Meet the Parents '91

Ethan Phillips

Bloodhounds of Broadway '89

Frank Phillips

Battle of the Eagles '81

Gary Phillips

Whodunit '82

Howard Phillips

Last Mile '32

John Phillips

Max and Helen '90

Joseph C. Phillips

Strictly Business '91

Julianne Phillips

Fletch Lives '89
Seven Hours to Judgment '88
Skin Deep '89
Summer Fantasy '84
Sweet Lies '88

Lawrence King Phillips

Abducted '86

Lee Phillips

The Hunters '58

Leo Phillips

Wedding in White '72

Leslie Phillips

The Limping Man '53
Mountains of the Moon '90
The Smallest Show on Earth '57
Train of Events '49

Lou Diamond Phillips

Ambition '91
Dakota '88
Disorganized Crime '89
The First Power '89
Harley '90
La Bamba '87
Renegades '89
A Show of Force '90
Stand and Deliver '87
Trespasses '86
Young Guns '88
Young Guns 2 '90

MacKenzie Phillips

American Graffiti '73
Eleanor & Franklin '76
Love Child '82
Miles to Go Before I Sleep '74
Rafferty & the Gold Dust Twins '75

Plummer ►cont.

Somewhere in Time '80
The Sound of Music '65
Souvenir '88
Spiral Staircase '75
Stage Struck '57
Star Crash '78
Star Trek 6: The
   Undiscovered Country '91
The Thorn Birds '83
Triple Cross '67
Waterloo '71
Where the Heart Is '90
Young Catherine '91

Glenn Plummer

Pastime '91

Lo Chine Po

Young Heroes '69

Shum Shim Po

Fists of Fury 2 '80

Werner Pochat

Flatfoot '78
Thunder Warrior 3 '88

Rick Podell

Lunch Wagon '81
World Gone Wild '88

Rossana Podesta

Alone Against Rome '62
Homo Eroticus/Man of the
   Year '71
The Sensual Man '74
Sodom and Gomorrah '62
Ulysses '55
The Virgin of Nuremberg '65

Cathy Podewell

Night of the Demons '88

Alexander Podorozhny

Zvenigora '28

Amos Poe (D)

Alphabet City '84
The Foreigner '78

Harlan Cary Poe

Stigma '73

Rudiger Poe (D)

Monster High '89

Judith Pogany

A Hungarian Fairy Tale '87

Kathryn Pogson

Over Indulgence '87

Ken Pogue

Blind Man's Bluff '91
Crazy Moon '87
The Grey Fox '83
The Hitman '91
Keeping Track '86
Miracle at Moreaux '86

Klaus Pohl

Woman in the Moon '29

Eric Pohlmann

Belles of St. Trinian's '53
Blackout '50
Glass Tomb '55
Moulin Rouge '52

Terror Street '54
Three Stops to Murder '53
Tiffany Jones '75

Larry Poindexter

American Ninja 2: The
   Confrontation '87
Blue Movies '88

Priscilla Pointer

Archer: The Fugitive from the
   Empire '81
Competition '80
Disturbed '90
Honeysuckle Rose '80
Mysterious Two '82
A Nightmare on Elm Street 3:
   Dream Warriors '87
Rumpelstiltskin '86

Jean Poiret

The Last Metro '80

Sidney Poitier

All the Young Men '60
Bedford Incident '65
Blackboard Jungle '55
Brother John '70
Buck and the Preacher '72
Cry, the Beloved Country '51
Defiant Ones '58
Duel at Diablo '66
Emperor Jones/Paul
   Robeson: Tribute to an
   Artist '33
For Love of Ivy '68
Goodbye, My Lady '56
The Greatest Story Ever Told
   '65
Guess Who's Coming to
   Dinner '67
In the Heat of the Night '67
Let's Do It Again '75
Lilies of the Field '63
Little Nikita '88
The Mark of the Hawk '57
The Organization '71
Paris Blues '61
A Patch of Blue '65
Piece of the Action '77
A Raisin in the Sun '61
Separate But Equal '90
Shoot to Kill '88
The Slender Thread '65
Something of Value '57
They Call Me Mr. Tibbs! '70
To Sir, with Love '67
Uptown Saturday Night '74
The Wilby Conspiracy '75

Sidney Poitier (D)

Buck and the Preacher '72
Fast Forward '84
Ghost Dad '90
Hanky Panky '82
Let's Do It Again '75
Piece of the Action '77
Stir Crazy '80
Uptown Saturday Night '74

Hanna Pola

Ninja Mission '84

Isa Pola

The Children Are Watching Us
   '44

James Polakof (D)

Balboa '82
Demon Rage '82
Midnight Auto Supply '78
Slashed Dreams '74
Swim Team '79
Vals '85

Diana Polakov

Supersonic Man '78

Anna Maria Polani

Hercules Against the Moon
   Men '64

Roman Polanski

Andy Warhol's Dracula '74
Back in the USSR '91
Ciao Federico! Fellini Directs
   Satyricon '69
Diary of Forbidden Dreams
   '76
The Fearless Vampire Killers
   '67
The Magic Christian '69
The Tenant '76

Roman Polanski (D)

Chinatown '74
Diary of Forbidden Dreams
   '76
Fat and the Lean '61
The Fearless Vampire Killers
   '67
Frantic '88
Knife in the Water '62
Macbeth '71
Pirates '86
Repulsion '65
Rosemary's Baby '68
The Tenant '76
Tess '80
Two Men & a Wardrobe '58

Stephen Poliakoff (D)

Close My Eyes '91
Hidden City '87

Gian Polidoro (D)

Loose in New York '80s

Jon Polito

Barton Fink '91
Fire with Fire '86
The Freshman '90
Miller's Crossing '90

Lina Polito

All Screwed Up '74
Blood Brothers '74

Bridgit Polk

Watched '73

Oscar Polk

Green Pastures '36

Ben Pollack

The Benny Goodman Story
   '55
The Glenn Miller Story '54

Kevin Pollack

L.A. Story '91

Sydney Pollack (D)

Absence of Malice '81
Bobby Deerfield '77
The Electric Horseman '79
Havana '90
Jeremiah Johnson '72
Out of Africa '85
Scalphunters, The '68
The Slender Thread '65
They Shoot Horses, Don't
   They? '69
This Property is Condemned
   '66
Three Days of the Condor '75
Tootsie '82

The Way We Were '73
The Yakuza '75

Cheryl Pollak

My Best Friend Is a Vampire
   '88
Night Life '90
Pump Up the Volume '90
Swimsuit: The Movie '89

Kevin Pollak

Ricochet '91

Tracy Pollan

The Abduction of Kari
   Swenson '87
Baby It's You '82
Bright Lights, Big City '88
Great Love Experiment '84
Promised Land '88
The Tender Age '84

Bud Pollard (D)

Tall, Tan and Terrific '46

Daphne Pollard

Bonnie Scotland '35
Laurel & Hardy Comedy
   Classics Volume 7 '30s
Our Relations '36

Harry Pollard (D)

California Straight Ahead '25

Michael J. Pollard

America '86
American Gothic '88
The Arrival '90
The Art of Dying '90
Between the Lines '77
Bonnie & Clyde '67
Dick Tracy '90
Enter Laughing '67
Fast Food '89
Heated Vengeance '87
I Come in Peace '90
Legend of Frenchie King '71
Melvin and Howard '80
Next of Kin '89
The Night Visitor '89
Patriot '86
Paul Reiser: Out on a Whim
   '88
Riders of the Storm '88
Roxanne '87
Scrooged '88
Season of Fear '89
Sleepaway Camp 3: Teenage
   Wasteland '89
Split Second '92
Tango and Cash '89
Vengeance Is Mine '84
Why Me? '90
The Wild Angels '66

Snub Pollard

Arizona Days '37
Comedy Classics '35
Don't Shove/Two Gun Gussie
   '19
Errand Boy '61
Funstuff '30s
I'm on My Way/The Non-Stop
   Kid '10s
Laughfest '10s
A Nation Aflame '37
Riders of the Rockies '37
Silent Laugh Makers, No. 3
   '20s
The White Legion '36

Albert Pollet

The Mysterious Lady '28

Jack Pollexfen (D)

The Indestructible Man '56

Sarah Polley

The Adventures of Baron
   Munchausen '89
Babar: The Movie '88
Lantern Hill '90

Channing Pollock

Judex '64

Dee Pollock

Beware, My Lovely '52

Ellen Pollock

Horror Hospital '73

George Pollock (D)

Broth of a Boy '59
Murder Ahoy '64
Murder at the Gallop '63
Murder Most Foul '65
Murder She Said '62
Stranger in Town '57

Polo Polo

La Casita del Pecado

Teri Polo

Mystery Date '91

Rui Polonah

The Emerald Forest '85

Abraham Polonsky (D)

Force of Evil '49
Romance of a Horsethief '71
Tell Them Willie Boy Is Here
   '69

Francisco Lara Polop (D)

Murder Mansion '70

John Polson

Dangerous Game '90

Jeff Pomerantz

Group Marriage '72

Erich Pommer (D)

Beachcomber '38

Elena Pompei

Mines of Kilimanjaro '87

Antonio Pompeo

Quilombo '84

Paul Pompian

Street Girls '75

Eusebio Poncela

Law of Desire '86

Marcel Poncin

The Golden Salamander '51

Mary Paz Pondal

Apocalipsis Joe '65

Lily Pons

I Dream Too Much '35
That Girl from Paris '36

Martina Pons

Mummy & Curse of the Jackal
   '67

Gillo Pontecorvo (D)

Battle of Algiers '66
Burn! '70
Kapo '59

Chris Ponti

Kill Alex Kill '83

Enrique Ponton

Semaforo en Rojo '47

Yvan Ponton

Men of Steel '88
Pouvoir Intime '87
Scanners 2: The New Order '91

Clara Pontoppidan

Witchcraft Through the Ages '22

David Pontremoli

To Forget Venice '70s

Maurizio Ponzi (D)

Pool Hustlers '83

John Poole

Fass Black '77

Patrick C. Poole (D)

Oklahoma Bound '81

Iggy Pop

Hardware '90

Julie Pop

Purgatory '89

Petru Popescu (D)

Death of an Angel '86

Leo Popkin (D)

The Well '51

N. Popov

Ten Days That Shook the World/October '27

Rada Popovic

Bomb at 10:10 '67

Nils Poppe

The Seventh Seal '56

Marc Poppel

Damned River '89

Johnny Popwell

Challenge '74
Manhunter '83

Paul Porcasi

Morocco '30

Jorge Porcel

Expertos en Pinchazos '79
Fotografo de Senoras '85
La Casa del Amor '87
Los Hombres Piensan Solo en Solo '76
Naked Is Better '73

Marc Porel

The Psychic '78

Paulina Porizkova

Anna '87
Her Alibi '88
Portfolio '88

Porky Pig

Daffy Duck's Quackbusters '89

Jean-Francois Poron

Charlie Bravo '80s

Frank Porretta

Song of Norway '70

Sverre Porsanger

Pathfinder '88

Louise Portal

The Decline of the American Empire '86
The Klutz '73

Javier Portales

Con Mi Mujer No Puedo '80s

Henny Porten

Backstairs '21

Alison Porter

Curly Sue '91

Ashley Porter

The Young Nurses '73

Brett Porter

Firehead '90

Don Porter

Buck Privates Come Home '47
The Candidate '72
Live a Little, Love a Little '68
Our Miss Brooks '56
The Racket '51
Who Done It? '42

Edwin S. Porter (D)

Early Films #1 '03
Film Firsts '60
First Twenty Years: Part 3 (Porter 1904-1905) '05
First Twenty Years: Part 4 '05
The Great Train Robbery: Cinema Begins '03

Eric Porter

Antony and Cleopatra '73
The Belstone Fox '73
Churchill and the Generals '81
The Fall of the Roman Empire '64
Hamlet '79
Hands of the Ripper '71
Hennessy '75
Little Lord Fauntleroy '80
Macbeth '85
Oliver Twist '85
The Thirty-Nine Steps '79

Eric Porter (D)

Marco Polo, Jr. '72

Heath Porter

Stay Awake '87

Jean Porter

Bathing Beauty '44
G.I. Jane '51
Kentucky Jubilee '51

Robert Porter

Deadly Fieldtrip '74
The Jesus Trip '71

Rafael Portillo (D)

The Curse of the Aztec Mummy '59
Curse of the Mummy/Robot vs. the Aztec Mummy '60s
The Robot vs. the Aztec Mummy '59

Eric Portman

Bedford Incident '65
A Canterbury Tale '44
The Colditz Story '55
The Crimes of Stephen Hawke '36
The Forty-Ninth Parallel '41
Great Day '46
Men of Two Worlds '46
Moonlight Sonata '38
Murder in the Old Red Barn '36
One of Our Aircraft Is Missing '41
Saturday Night Shockers, Vol. 2 '58
We Dive at Dawn '43

Richard Portnow

In Dangerous Company '88
Meet the Hollowheads '89
Say Anything '89

Bill Posner (D)

Teenage Strangler '64

Geoffrey Posner (D)

The Young Ones '88

Markie Post

Glitz '88
Scene of the Crime '85
Tricks of the Trade '88
TripleCross '85

Saskia Post

Dogs in Space '87
One Night Stand '84

Ted Post (D)

The Baby '72
Beneath the Planet of the Apes '70
Diary of a Teenage Hitchhiker '82
Go Tell the Spartans '78
Good Guys Wear Black '78
Hang 'Em High '67
Harrad Experiment '73
The Human Shield '92
Magnum Force '73
Nightkill '80
Stagecoach '86
Whiffs '75
Yuma '70

William Post, Jr.

Sherlock Holmes and the Secret Weapon '42

Adrienne Posta

Adventures of a Taxi Driver '76

Peter Postlethwaite

Distant Voices, Still Lives '88
Dressmaker '89
Split Second '92
To Kill a Priest '89

Laurens C. Postma

Heroes Three '84
To Live and Die in Hong Kong '89

Michal Postnikow

First Spaceship on Venus '60

Pat Poston

She-Devils on Wheels '68

Tiffanie Poston

Adventures in Dinosaur City '92

Tom Poston

Carbon Copy '81
Cold Turkey '71
Soldier in the Rain '63
Tempest '63
Up the Academy '80
Zotz! '62

Victor Potel

Miracle of Morgan's Creek '44
Paradise Island '30
Thunder Over Texas '34

Charlie Potter

The Witches '90

George Potter (D)

Seven Sixgunners '87

H.C. Potter (D)

Beloved Enemy '36
The Farmer's Daughter '47
Mr. Blandings Builds His Dream House '48
Mr. Lucky '43
Second Chorus '40
The Story of Vernon and Irene Castle '39

Madeleine Potter

The Bostonians '84
Slaves of New York '89
The Suicide Club '88
Two Evil Eyes '91

Martin Potter

Ciao Federico! Fellini Directs Satyricon '69
Fellini Satyricon '69
Gunpowder '84
The Marquis de Sade's Justine '70
Only Way '70
Twinsanity '70

Steve Potter

The Meal '75

Terry Potter

Bad Taste '88

Vern Potter

Beasts '83

Gerald Potterton (D)

George and the Christmas Star '85

Richard Pottier (D)

Rape of the Sabines '61

Annie Potts

Amazing Stories, Book 2 '86
Corvette Summer '79
Crimes of Passion '84
Flatbed Annie and Sweetiepie: Lady Truckers '79
Ghostbusters '84
Ghostbusters 2 '89
Heartaches '82

It Came Upon a Midnight Clear '84
Jumpin' Jack Flash '86
King of the Gypsies '78
Pass the Ammo '88
Pretty in Pink '86
Texasville '90
Who's Harry Crumb? '89

Cliff Potts

The Groundstar Conspiracy '72
Last Ride of the Dalton Gang '79
Little Women '78
Silent Running '71

Wallace Potts (D)

Psycho Cop '88

Ken Pouge

Dead of Winter '87

Ely Pouget

Cool Blue '88
Endless Descent '90

Georges Poujouly

Forbidden Games '52
Frantic '58

Gerry Poulson (D)

Under the Doctor '76

CCH Pounder

Bagdad Cafe '88

Andre Pousse

Deadly Sting '82

Phyllis Povah

Pat and Mike '52

Jason Rai Pow

Black Dragon '74
Black Dragon's Revenge '75

Alisa Powell

Satan's Cheerleaders '77

Arla Powell

Battle Beyond the Sun '63

Cari Powell

Ghostriders '87

Cozy Powell

Superdrumming '88

David Powell

The Average Woman '24

Dick Powell

The Bad and the Beautiful '52
Christmas in July '40
Cornered '45
Cry Danger '51
Dames '34
Footlight Parade '33
42 Street '33
Gold Diggers of 1933 '33
Gold Diggers of 1935 '35
In the Navy '41
A Midsummer Night's Dream '35
Murder My Sweet '44
Station West '48
Susan Slept Here '54

Dick Powell (D)

The Conqueror '56
Enemy Below '57
The Hunters '58

Mel Brooks: An Audience '84
The Ploughman's Lunch '83
Praying Mantis '83
The Rachel Papers '89
Something Wicked This Way
  Comes '83

Nicholas Pryor

Brain Dead '89
East of Eden '80
The Fish that Saved
  Pittsburgh '79
The $5.20 an Hour Dream '80
Force Five '75
Gumball Rally '76
Happy Hooker '75
Last Song '80
The Life and Assassination of
  the Kingfish '76
Morgan Stewart's Coming
  Home '87
Night Terror '76
Nightbreaker '89
Smile '75

Richard Pryor

Adios Amigo '75
Best of Chevy Chase '87
Bingo Long Traveling All-Stars
  & Motor Kings '76
Black Brigade
Blue Collar '78
Brewster's Millions '85
Bustin' Loose '81
California Suite '78
Car Wash '76
Critical Condition '86
Greased Lightning '77
Harlem Nights '89
Jo Jo Dancer, Your Life is
  Calling '86
Lady Sings the Blues '72
Lily Tomlin '80s
Mack '73
Motown 25: Yesterday,
  Today, Forever '83
Moving '88
Richard Pryor: Here and Now
  '83
Richard Pryor: Live in Concert
  '79
Richard Pryor: Live and
  Smokin' '71
Richard Pryor: Live on the
  Sunset Strip '82
Saturday Night Live: Richard
  Pryor '75
See No Evil, Hear No Evil '89
The Silver Streak '76
Some Call It Loving '73
Some Kind of Hero '82
Stir Crazy '80
Superman 3 '83
The Toy '82
Uptown Saturday Night '74
Which Way Is Up? '77
Wholly Moses! '80
Wild in the Streets '68
The Wiz '78

Richard Pryor (D)

Jo Jo Dancer, Your Life is
  Calling '86
Richard Pryor: Here and Now
  '83

Roger Pryor

Headline Woman '35
Identity Unknown '45
Kid Sister '45
Lady by Choice '34
The Man They Could Not
  Hang '39
Panama Menace '41

Geof Pryssir

Hot Child in the City '87

Lenita Psillakis

Girl-Toy '84

Wojciech Pszoniak

Angry Harvest '85
Danton '82

Alexander Ptushko (D)

The Magic Voyage of Sinbad
  '52
Sword & the Dragon '56

Robert Pucci

The Last Hour '91

Vladimir Pucholt

Loves of a Blonde '66

Vsevolod Pudovkin

The Extraordinary Adventures
  of Mr. West in the Land of
  the Bolsheviks '24

Vsevolod Pudovkin (D)

End of St. Petersburg '27
Mother '26
Mystery of the Leaping Fish/
  Chess Fever '25
Storm Over Asia '28

Tito Puente

Mambo Kings '92

Luis Puenzo (D)

The Official Story '85
Old Gringo '89

David Pugh

The Love Pill '71

Willard Pugh

Ambition '91
The Color Purple '85

William E. Pugh

Traxx '87

Frank Puglia

The Black Hand '50
Bulldog Drummond's
  Revenge '37
The Fatal Hour '40
Romola '25
20 Million Miles to Earth '57
Without Reservations '46

Bill Pullman

The Accidental Tourist '88
Brain Dead '89
Bright Angel '91
Cold Feet '89
Liebestraum '91
Newsies '92
Rocket Gibraltar '88
Ruthless People '86
The Serpent and the Rainbow
  '87
Sibling Rivalry '90
Spaceballs '87

Lilo Pulver

A Time to Love & a Time to
  Die '58

Liselotte Pulver

The Nun '66

Kong Pun

Ninja Swords of Death '90
Samurai Blood, Samurai Guts
Soul of Samurai

Ng Pun

Samurai Blood, Samurai Guts

Angela Punch-McGregor

Double Deal '84
Newsfront '78

Henri Puopon

Angele '34

Dick Purcell

The Abe Lincoln of Ninth
  Avenue '39
Captain America '44
Heroes in Blue '39
Irish Luck '39
King of the Zombies '41
Timber Queen '44
Tough Kid '39
X Marks the Spot '42

Evelyn Purcell (D)

Nobody's Fool '86

James Purcell

Playroom '90
White Light '90

Joseph Purcell (D)

Delos Adventure '86

Lee Purcell

Adam at 6 A.M. '70
Amazing Howard Hughes '77
Big Wednesday '78
Eddie Macon's Run '83
Homework '82
The Incredible Hulk Returns
  '88
Jailbait: Betrayed By
  Innocence '86
Kenny Rogers as the Gambler
  '80
Killing at Hell's Gate '81
Mr. Majestyk '74
Space Rage '86
Stir Crazy '80
The Witching '72

Noel Purcell

Encore '52
Land of Fury '55
Lust for Life '56

Una Richard Purcell

The Bank Dick '40

Nathan Purdee

Return of Superfly '90

Reginald Purdell

Brighton Rock '47
Candles at Nine '44
The Old Curiosity Shop '35

Edmund Purdom

After the Fall of New York '85
Ator the Fighting Eagle '83
Big Boss '77
Dr. Frankenstein's Castle of
  Freaks '74
Don't Open Till Christmas '84
The Egyptian '54
Herod the Great '60
Mr. Scarface '77

Nefertiti, Queen of the Nile '64
The Prodigal '55
Rulers of the City '71

Edmund Purdom (D)

Don't Open Till Christmas '84

Carolyn Purdy-Gordon

Dolls '87

Amrish Puri

Indiana Jones and the Temple
  of Doom '84

Linda Purl

The Adventures of Nellie Bly
  '81
Crazy Mama '75
Eleanor & Franklin '76
Flame is Love '79
High Country '81
Last Cry for Help '79
Little Ladies of the Night '77
The Manions of America '81
Outrage '73
Spies, Lies and Naked Thighs
  '91
Viper '88
Visiting Hours '82
Web of Deceit '90

Purple Sage

Western Double Feature 11
  '50

Bob Purvey

Follow Me '69

Edna Purviance

Burlesque of Carmen '16
The Champion/His New Job
  '15
Chaplin Essanay Book 1 '15
Chaplin Mutuals '16
The Chaplin Review '58
Charlie Chaplin Festival '17
Charlie Chaplin: Night at the
  Show '15
Charlie Chaplin: Rare Chaplin
  '15
Charlie Chaplin: The Early
  Years, Volume 1 '17
Charlie Chaplin: The Early
  Years, Volume 2 '17
Charlie Chaplin: The Early
  Years, Volume 3 '16
Charlie Chaplin: The Early
  Years, Volume 4 '16
Charlie Chaplin...Our Hero!
  '15
Count/Adventurer '17
The Floorwalker/By the Sea
  '15
His Prehistoric Past/The Bank
  '15
The Kid '21
The Kid/Idle Class '21
Rink/Immigrant '10s
Three Charlies and One
  Phoney! '18
Tramp & a Woman '15
Triple Trouble/Easy Street '18
Woman of Paris/Sunnyside
  '23
Work/Police '16

Manfred Purzer (D)

Girls Riot '88

Anne Puskin

His Name was King '85

John Putch

Curfew '88
Impure Thoughts '86

Nat Puvanai

Crocodile '81

Dorothy Ann Puzo (D)

Cold Steel '87

Hy Pyke

Halloween Night '90

Monty Pyke

Lemora, Lady Dracula '73

Jean Vander Pyl

Flintstones '60

Denver Pyle

The Adventures of Frontier
  Fremont '75
Dynamite Pass '50
Hawmps! '76
Hills of Utah '51
The Legend of Hillbilly John
  '73
Legend of the Northwest '78
Life & Times of Grizzly Adams
  '74
Mountain Man '77
Murder or Mercy '74
Naked Hills '56
Oklahoma Annie '51
Rim of the Canyon '49
Seventh Cavalry '56
Too Late for Tears '49

Natasha Pyne

Madhouse '74

John Pyper-Ferguson

Killer Image '92

Joe Pytka (D)

Let It Ride '89

Albert Pyun (D)

Alien from L.A. '87
Bloodmatch '91
Captain America '89
Cyborg '89
Dangerously Close '86
Dollman '91
Down Twisted '89
Radioactive Dreams '86
Sword & the Sorcerer '82

Stacey Q

One Man Force '89

Wang Qiang

Super Ninja

Pan Qingfu

Iron & Silk '91

Lin Qinqin

A Great Wall '86

Michel Qissi

Bloodmatch '91

John Quade

Dirkham Detective Agency '83
Fury to Freedom: The Life
  Story of Raul Ries '85
Night Terror '76
Peter Lundy and the Medicine
  Hat Stallion '77

The Purple Taxi '77
Sardinia Kidnapped '75
Ski Bum '75
Stardust Memories '80
'Tis a Pity She's a Whore '73
The Verdict '82
Zardoz '73

Remak Ramsay

The Dining Room '86
Mr. & Mrs. Bridge '91

John Ramsbottom

Kill and Kill Again '81

Frances Ramsden

The Sin of Harold Diddlebock '47

Anne Ramsey

Boy in the Plastic Bubble '76
Deadly Friend '86
Dr. Hackenstein '88
The Goonies '85
Marilyn: The Untold Story '80
Meet the Hollowheads '89
A Small Killing '81
Throw Momma from the Train '87
Weeds '87
White Mama '80

Logan Ramsey

Confessions of Tom Harris '72
Conspiracy of Terror '75
Dr. Hackenstein '88
Fury on Wheels '71
Say Yes! '86

Marion Ramsey

Police Academy 3: Back in Training '86
Police Academy 5: Assignment Miami Beach '88
Police Academy 6: City Under Siege '89

Thea Ramsey

Maniac '34

Ward Ramsey

Dinosaurus! '60

Peter Ramwa

After the Fall of Saigon '80s

Frederica Ranchi

Son of Samson '62

Renee Rancourt

Beyond the Door 3 '91

John Rand

Charlie Chaplin: The Early Years, Volume 1 '17
Charlie Chaplin: The Early Years, Volume 2 '17

Patrick Rand (D)

Mom '89

Addison Randall

Chance '89
Deadly Breed '89
Hollow Gate '88

Addison Randall (D)

Chance '89
East L.A. Warriors '89
The Killing Zone '90

Anne Randall

Stacey '73

Jack Randall

Across the Plains '39
Gunsmoke Trail '38
Man's Country '38
Wild Horse Canyon '39

Kelly Randall

Ninja Academy '90

Lexi Randall

Sarah, Plain and Tall '91

Monica Randall

Commando Attack '67
Witches' Mountain '71

Monique Randall

Five Giants from Texas '66

Richard Randall

Cross Mission '89

Suze Randall

Chloe in the Afternoon '72

Tony Randall

The Adventures of Huckleberry Finn '60
The Alphabet Murders '65
Everything You Always Wanted to Know About Sex (But Were Afraid to Ask) '72
Foolin' Around '80
King of Comedy '82
Let's Make Love '60
The Littlest Angel '69
Lover Come Back '61
My Little Pony: The Movie '86
Pillow Talk '59
Scavenger Hunt '79
Send Me No Flowers '64
Seven Faces of Dr. Lao '63
Super Bloopers #2 '70s
That's Adequate '90

Tony Randel (D)

Amityville 1992: It's About Time '92
Hellbound: Hellraiser 2 '88

Ron Randell

Girl in Black Stockings '57
The Loves of Carmen '48
Omoo Omoo, the Shark God '49

Ermanno Randi

Young Caruso '51

Ric Randing

Splatter University '84

Anders Randolf

The Love Flower '20
Ranson's Folly '26

Amanda Randolph

Lying Lips '39

Anders Randolph

The Kiss '29

Elsie Randolph

Rich and Strange '32

Isabelle Randolph

The Missing Corpse '45

Jane Randolph

Abbott and Costello Meet Frankenstein '48
Cat People '42
Curse of the Cat People '44
The Falcon's Brother '42

John Randolph

The Adventures of Nellie Bly '81
As Summers Die '86
Iron Maze '91
King Kong '76
Lovely...But Deadly '82
National Lampoon's Christmas Vacation '89
Pretty Poison '68
Prizzi's Honor '85
The Runaways '75
Secrets '77
Serpico '73
There was a Crooked Man '70
Winds of Kitty Hawk '78
The Wizard of Loneliness '88

Lillian Randolph

How to Seduce a Woman '74

Ty Randolph

Deadly Embrace '88
Nudity Required '90

Windsor Taylor Randolph

Amazons '86

Robert Random

Danger Zone 2 '89
Danger Zone 3: Steel Horse War '90

Salvo Randone

Fellini Satyricon '69
Spirits of the Dead '68
10th Victim '65

Eduardo Ranez

Erotic Adventures of Pinocchio '71

The Range Ramblers

Hittin' the Trail '37

Maria Rangel

Full Fathom Five '90

Dan Ranger

Cop-Out '91

Mike Ranger

What Would Your Mother Say? '81

Arthur Rankin, Jr.

Walking Back '26

Arthur Rankin, Jr. (D)

The Ballad of Paul Bunyan '72
Flight of Dragons '82
Frosty the Snowman '69
The Hobbit '78
Santa Claus is Coming to Town '70

Arthur Rankin

Terror Trail '33

Shaba Ranks

Prince Jammy '87

Juanita Ranney

Danger Zone 3: Steel Horse War '90

Rodolfo Ranni

Deadly Revenge '83
En Retirada (Bloody Retreat) '84

Les Rannow

Summer Affair '79

Lazlo Ranody (D)

Nobody's Daughter '

Joan Ranquet

Murder: No Apparent Motive '84

Tim Ransom

Dressmaker '89
Vital Signs '90

Prunella Ransome

Far from the Madding Crowd '67

Ryan Rao

Evil Altar '89

Anna Rapagna

Future Force '89

Frederic Raphael (D)

Women & Men: Stories of Seduction '90

Anthony Rapp

Adventures in Babysitting '87
Far from Home '89

Joel Rapp (D)

Battle of Blood Island '60

David Rappaport

The Bride '85
Mysteries '84
Peter Gunn '89

Jean-Paul Rappeneau (D)

Cyrano de Bergerac '89
Lovers Like Us '75
Swashbuckler '84

Irving Rapper (D)

The Brave One '56
Corn is Green '45
Deception '46
Joseph & His Brethren '60
Marjorie Morningstar '58
Now, Voyager '42
Rhapsody in Blue '45

Michael Rapport

Hardbodies '84

Eva Ras

The Love Affair, or Case of the Missing Switchboard Operator '67
Man is Not a Bird '65

David Rasche

Bingo '91
Hammered: The Best of Sledge '88
An Innocent Man '89
The Masters of Menace '90
Silhouette '91

Wedding Band '89
Wicked Stepmother '89

Stephanie Rascoe

Positive I.D. '87

Steve Rash (D)

Buddy Holly Story '78
Can't Buy Me Love '87
Queens Logic '91
Rich Hall's Vanishing America '86
Under the Rainbow '81

Jay Raskin (D)

I Married a Vampire '87

Daniel Raskov (D)

The Masters of Menace '90
Wedding Band '89

Mark Rasmussen

Deadly Encounter '78

Fritz Rasp

Diary of a Lost Girl '29
Spies '28
The Threepenny Opera '31

Ivan Rassimov

Beyond the Door 2 '79
Man from Deep River '77
Next Victim '71

Rada Rassimov

Torture Chamber of Baron Blood '72

Thalmus Rasulala

Adios Amigo '75
Blacula '72
Blind Vengeance '90
Born American '86
Bucktown '75
Bulletproof '88
Friday Foster '75
Sophisticated Gents '81

Wayne Ratay

The Wizard of Gore '70

Sandy Ratcliff

Family Life '72
Yesterday's Hero '79

Robin Rates

Puppet Master '89

Basil Rathbone

The Adventures of Robin Hood '38
The Adventures of Sherlock Holmes '39
Anna Karenina '35
Bathing Beauty '44
Captain Blood '35
Casanova's Big Night '54
Christmas Carol '54
The Comedy of Terrors '64
Court Jester '56
David Copperfield '35
Dawn Patrol '38
Dressed to Kill '46
The Garden of Allah '36
Heartbeat '46
Hillbillys in a Haunted House '67
The Hound of the Baskervilles '39
House of Fear '45
Last Days of Pompeii '35
Last Hurrah '58
Love from a Stranger '37
The Magic Sword '62

Julian Reyes

Point Break '91

Kamar Reyes

Coldfire '90
East L.A. Warriors '89

Ricardo Reyes

Ye-Yo '70s

Rina Reyes

Bloodfist 2 '90

Madeleine Reynal

Dr. Caligari '89

Janine Reynaud

The Sensuous Teenager '70

David Reyne

Frenchman's Farm '87

Burt Reynolds

All Dogs Go to Heaven '89
Angel Baby '61
Armored Command '61
Best Friends '82
The Best Little Whorehouse in
    Texas '82
Breaking In '89
Cannonball Run '81
Cannonball Run II '84
City Heat '84
Deliverance '72
The End '78
Everything You Always
    Wanted to Know About
    Sex (But Were Afraid to
    Ask) '72
Fuzz '72
Gator '76
Gentle Ben '70s
Hard Frame '70
Heat '87
Hooper '78
Hustle '75
Iron Cowboy '68
The Longest Yard '74
Malone '87
Man Who Loved Cat Dancing
    '73
The Man Who Loved Women
    '83
Modern Love '90
Navajo Joe '67
100 Rifles '69
Operation C.I.A. '65
Paternity '81
Physical Evidence '89
Rent-A-Cop '88
Rough Cut '80
Savage Run '70
Semi-Tough '77
Shamus '73
Shark! '68
Sharky's Machine '81
Silent Movie '76
Skulduggery '70
Smokey and the Bandit '77
Smokey and the Bandit, Part
    2 '80
Smokey and the Bandit, Part
    3 '83
Starting Over '79
Stick '85
Stroker Ace '83
Switching Channels '88
Uphill All the Way '85
White Lightning '73

Burt Reynolds (D)

The End '78
Gator '76
Sharky's Machine '81

Stick '85

**C.D.H.
Reynolds** (D)

Day of Judgment '81

**Christopher
Reynolds** (D)

Offerings '89

Dave Reynolds

One More Saturday Night '86

Debbie Reynolds

Bundle of Joy '56
The Catered Affair '56
Charlotte's Web '79
Detective Sadie & Son '84
Give a Girl a Break '53
Hit the Deck '55
How Sweet It Is! '68
How the West was Won '62
I Love Melvin '53
Meet Me in Las Vegas '56
The Rat Race '60
Singin' in the Rain '52
Skirts Ahoy! '52
Susan Slept Here '54
Tammy and the Bachelor '57
The Tender Trap '55
That's Singing '84
This Happy Feeling '58
Three Little Words '50
Two Weeks with Love '50
The Unsinkable Molly Brown
    '64

Don Reynolds (D)

His Name was King '85

Gene Reynolds

Andy Hardy's Private
    Secretary '41
Boys Town '38
Country Girl '54
Jungle Patrol '48
Life Begins for Andy Hardy
    '41
Love Finds Andy Hardy '38
They Shall Have Music '39
The Tuttles of Tahiti '42

Gene Reynolds (D)

Truth or Die '86

Jay Reynolds

The Reincarnate '71

John Reynolds

Manos, the Hands of Fate '66

Kevin Reynolds (D)

The Beast '88
Fandango '85
Robin Hood: Prince of Thieves
    '91

Kristina Reynolds

The Wolfman '82

Marjorie Reynolds

The Abe Lincoln of Ninth
    Avenue '39
Doomed to Die '40
The Fatal Hour '40
Gone with the Wind '39
Holiday Inn '42
Home Town Story '51
Man's Country '38
Midnight Limited '40
Mr. Wong in Chinatown '39
Mystery Plane '39
Racketeers of the Range '39

Robin Hood of the Pecos '41
Six Shootin' Sheriff '38
Stunt Pilot '39
The Time of Their Lives '46
Western Trails '38

**Norman
Reynolds** (D)

Amazing Stories, Book 5 '85

Ollie Reynolds

The Legend of the Wolf
    Woman '77

Patrick Reynolds

The Eliminators '86

Paul Reynolds

Let Him Have It '91

Peter Reynolds

Devil Girl from Mars '54
The Great Armored Car
    Swindle '64
It Takes a Thief '59

Quincy Reynolds

Feelin' Screwy '90

Simon Reynolds

Gate 2 '92

Ursi Reynolds (D)

Ganjasaurus Rex '87

Vera Reynolds

The Monster Walks '32
The Night Club '25
Risky Business '28
The Road to Yesterday '25
Road to Yesterday/The
    Yankee Clipper '27
Sunny Side Up '28

William Reynolds

Carrie '52

David Reynoso

Invasion of the Vampires '61
La Trinchera '87
Orlak, the Hell of Frankenstein
    '60

Mark Rezyka (D)

South of Reno '87

Ving Rhames

Homicide '91
The Long Walk Home '89
Patty Hearst '88
Rising Son '90

Allison Rhea

Last Dance '91

Jhoon Rhee

When Taekwondo Strikes '83

Phillip Rhee

Best of the Best '89
Ninja Turf '86
Silent Assassins '88

Simon Rhee

Best of the Best '89
Furious '83

Thoon Rhee

Sting of the Dragon Masters
    '74

Rhema

Voyage of the Rock Aliens '87

Barbara Rhoades

The Choirboys '77
Conspiracy of Terror '75
The Day the Women Got Even
    '80
Goodbye Girl '77
Hunter '73
Scream Blacula Scream '73
Serial '80
Sex and the Single Parent '82
The Shakiest Gun in the West
    '68
Up the Sandbox '72

Bobby Rhodes

Demons 2 '87
Endgame '85
The Last Hunter '80

Christopher Rhodes

Gorgo '61

Cynthia Rhodes

Dirty Dancing '87
Flashdance '83
Runaway '84
Staying Alive '83

Donnelly Rhodes

After the Promise '87
Goldenrod '77
Hard Part Begins '80s
Showdown at Williams Creek
    '91

Erik Rhodes

Charlie Chan in Paris '35
The Gay Divorcee '34
Top Hat '35

Hari Rhodes

Conquest of the Planet of the
    Apes '72
Detroit 9000 '73
A Dream for Christmas '73
Shock Corridor '63
A Woman Called Moses '78

Harry Rhodes

Coma '78

Kenneth Rhodes

Death Rides the Range '40

Marjorie Rhodes

Decameron Nights '53
Great Day '46

**Michael Ray
Rhodes** (D)

The Killing Mind '90

Michael Rhodes (D)

Matters of the Heart '90

Sonny Rhodes

Cigarette Blues '89

Madlyn Rhue

The Cary Grant Collection '90

John Rhys-Davies

Best Revenge '83
Great Expectations '89
In the Shadow of Kilimanjaro
    '86
Indiana Jones and the Last
    Crusade '89
Ivanhoe '82

Journey of Honor '91
King Solomon's Mines '85
Little Match Girl '87
The Living Daylights '87
Nairobi Affair '88
Nativity '78
Raiders of the Lost Ark '81
Rebel Storm '90
Sahara '83
Secret Weapon '90
Shogun '80
Sphinx '81
Sword of the Valiant '83
Trial of the Incredible Hulk '89
Tusks '89
Victor/Victoria '82
War & Remembrance, Part 1
    '88

Paul Rhys

Vincent & Theo '90

Candice Rialson

Candy Stripe Nurses '74
Chatterbox '76
Hollywood Boulevard '76
Mama's Dirty Girls '74
Moonshine County Express
    '77
Summer School Teachers '75

Diego Riba

Evil Clutch '89

Billy Riback

Paramount Comedy Theater,
    Vol. 3: Hanging Party '87

Alfonso Ribeiro

The Mighty Pawns '87

**Carlos Alberto
Riccelli**

The Dolphin '87

Christina Ricci

The Addams Family '91
Mermaids '90

Tonino Ricci (D)

Great Treasure Hunt '70s
Rush '84

Katia Ricciarelli

Otello '86
Placido: A Year in the Life of
    Placido Domingo '84

Frederick Riccio

Teenage Mother '67

Bill Rice

Doomed Love '83
Landlord Blues '87
Vortex '81

Bill Rice (D)

The Vineyard '89

Brett Rice

Final Cut '88

David Rice (D)

Quiet Thunder '87

Florence Rice

At the Circus '39
Boss of Big Town '43
The Ghost and the Guest '43
Sweethearts '38

Frank Rice

Phantom Thunderbolt '33

Dead Man Walking '88
Fairytales '79
Repo Man '83
Straight to Hell '87
Street Asylum '90
They Live '88

Tony Richardson (D)

The Border '82
The Entertainer '60
Hamlet '69
The Hotel New Hampshire '84
Joseph Andrews '77
Look Back in Anger '58
The Loved One '65
Penalty Phase '86
Shadow on the Sun '88
A Taste of Honey '61
Tom Jones '63
Women & Men: Stories of Seduction '90

William Richert

My Own Private Idaho '91

William Richert (D)

A Night in the Life of Jimmy Reardon '88
Winter Kills '79

Clint Richie

Against A Crooked Sky '75

Peter Mark Richman

City Killer '87
Dempsey '83
Friday the 13th, Part 8: Jason Takes Manhattan '89
Judgment Day '88
PSI Factor '80

Anthony Richmond

Devil Girl from Mars '54
One Wish Too Many '55

Anthony Richmond (D)

Deja Vu '84
A Man Called Rage '84
Night of the Sharks '89

Fiona Richmond

The House on Straw Hill '76

June Richmond

Reet, Petite and Gone '47

Kane Richmond

The Adventures of Rex & Rinty '35
Anything for a Thrill '37
Cavalier of the West '31
Confidential '35
East Side Kids/The Lost City '35
The Lost City '34
The Reckless Way '36
Spy Smasher Returns '42
Tough to Handle '37

Mark Richmond

Black Orchid '59

Warner Richmond

Big News '29
Headin' for the Rio Grande '36
A Lawman is Born '37
Rainbow's End '35
Song of the Gringo '36
The Speed Spook '24
Wild Horse Canyon '39

Wild West '46

Dan Richter

2001: A Space Odyssey '68

Debi Richter

Hot Moves '84
The Rebels '79
Swap Meet '79
Twirl '81

Deborah Richter

The Banker '89
Cyborg '89
Promised Land '88

Ota Richter (D)

Skullduggery '79

Paul Richter

Dr. Mabuse the Gambler '22
Kriemhilde's Revenge '24
Siegfried '24

W.D. Richter (D)

The Adventures of Buckaroo Banzai '84
Late for Dinner '91

Slick Rick

Tougher than Leather '88

Tony Rickards

Castaway '87
Great Gold Swindle '84
Waterfront '83

James Ricketson (D)

Kiss the Night '87

Don Rickles

Beach Blanket Bingo '65
Bikini Beach '64
Enter Laughing '67
For the Love of It '80
Johnny Carson '70s
Keaton's Cop '90
Kelly's Heroes '70
Muscle Beach Party '64
Pajama Party '64
The Rat Race '60
Revenge of TV Bloopers '60s
Run Silent, Run Deep '58
Two Top Bananas '87
X: The Man with X-Ray Eyes '63

Alan Rickman

Close My Eyes '91
Closet Land '90
Die Hard '88
The January Man '89
Quigley Down Under '90
Robin Hood: Prince of Thieves '91
Truly, Madly, Deeply '91

Allen Rickman

I Was a Teenage Zombie '87
Shock! Shock! Shock! '87

Robert Rickman

Thou Shalt Not Kill...Except '87

Tom Rickman (D)

The River Rat '84

Maria Ricossa

Dead Man Out '89

Rachel Ricotin

Critical Condition '86

Vincenzo Ricotta

Hanna's War '88

Bob Ridgely

Hootch Country Boys '75

John Ridgely

Air Force '43
Arsenic and Old Lace '44
Destination Tokyo '43

Robert Ridgely

Rosie: The Rosemary Clooney Story '82
Who Am I This Time? '82

Stanley Ridges

Each Dawn I Die '39
Master Race '44
Mr. Ace '46

Linda Ridgeway

The Mechanic '72

Emma Ridley

The House that Bled to Death '81

Philip Ridley (D)

The Reflecting Skin '91

Giovanni Ridolfi

Yesterday, Today and Tomorrow '64

Leni Riefenstahl

The Blue Light '32
Sacred Mountain/White Flame '31
Tiefland '44

Leni Riefenstahl (D)

The Blue Light '32
Tiefland '44
Triumph of the Will '34

Peter Riegert

Americathon '79
Chilly Scenes of Winter '79
Crossing Delancey '88
Local Hero '83
A Man in Love '87
National Lampoon's Animal House '78
News at Eleven '86
Oscar '91
The Runestone '91
A Shock to the System '90
The Stranger '87
That's Adequate '90

Virgilio Riento

Peddlin' in Society '47

Charles Riesner (D)

Big Store '41
Manhattan Merry-go-round '37
Steamboat Bill, Jr. '28

Dean Riesner (D)

Bill and Coo '47

Vladmir Rif (D)

Very Close Quarters '84

Adam Rifkin (D)

Never on Tuesday '88

Richard Rifkin

Assault of the Party Nerds '89

Ron Rifkin

Courage '86
Mrs. R's Daughter '79
Silent Running '71

Anabella Rigaud

Quatorze Juliet '32

Jorge Rigaud

Black Box Affair '66
Dos Postolas Gemelas '65
Quatorze Juliet '32

Edward Rigby

Convoy '40

Jean Rigby

Rigoletto '82

Carl Rigg

The Oblong Box '69

Diana Rigg

Bleak House '85
Cannon Movie Tales: Snow White '89
Evil Under the Sun '82
The Great Muppet Caper '81
The Hospital '71
In This House of Brede '75
Julius Caesar '70
A Little Night Music '77
On Her Majesty's Secret Service '69
Theatre of Blood '73
Unexplained Laughter '89

Rebecca Rigg

Doctors and Nurses '82
Fortress '85
The Hunting '92

Nikki Riggins

Scream Dream '89

Bobby Riggs

Racquet '79

John Riggs

Invasion of the Space Preachers '90

Gennaro Righelli (D)

Peddlin' in Society '47

Roger Rignack

True Love '89

Caterina Rigoglioso

Love in the City '53

Brad Rijn

Perfect Strangers '84
Special Effects '85

Robin Riker

Alligator '80
Body Chemistry 2: Voice of a Stranger '91

Elaine Riley

Clipped Wings '53
Devil's Playground '46
Hopalong Cassidy: Borrowed Trouble '48
Hopalong Cassidy: The Devil's Playground '46
Rider from Tucson '50

Sinister Journey '48

Gary Riley

Amazing Stories, Book 5 '85

Jack Riley

Martin Mull Live: From Ridgeville, Ohio '88
Night Patrol '85
Portrait of a White Marriage '88

Jeannine Riley

Big Mouth '67
The Comic '69
Electra Glide in Blue '73
Wackiest Wagon Train in the West '77

John Riley

Greenstone '85

Larry Riley

Crackers '84
A Soldier's Story '84

Michael Riley

Perfectly Normal '91
To Catch a Killer '92

Rob Riley

Mutant Video '76

Eli Rill

Slipstream '72

Walter Rilla

Frozen Alive '64
The Gamma People '56
The Golden Salamander '51
House of Mystery '41
Mr. Emmanuel '44

Wolf Rilla (D)

Bachelor of Hearts '58
Roadhouse Girl '53
Village of the Damned '60

Carlo Rim (D)

Seven Deadly Sins '53

Shane Rimmer

The People That Time Forgot '77
S*P*Y*S '74

Taro Rin (D)

Galaxy Express '82

Rin Tin Tin Jr

The Adventures of Rex & Rinty '35
Lightning Warrior '31

Susan Rinell

Pals '87

Blanche Ring

It's the Old Army Game '26

Erica Ringston

Last Dance '91

Molly Ringwald

Betsy's Wedding '90
The Breakfast Club '85
For Keeps '88
Fresh Horses '88
King Lear '87
Packin' It In '83
The Pick-Up Artist '87
P.K. and the Kid '85
Pretty in Pink '86

Ringwald ►cont.

Sixteen Candles '84
Spacehunter: Adventures in the Forbidden Zone '83
Strike It Rich '90
Tall Tales & Legends: Johnny Appleseed '86
Tempest '82
Women & Men: Stories of Seduction '90

Brad Rinn

Smithereens '84

Dante Rino

The Golden Coach '52

David Rintoul

Legend of the Werewolf '75
Pride and Prejudice '85

Nicole Rio

Sorority House Massacre '86
Visitants '87
Zero Boys '86

Marjorie Riordan

The Hoodlum '51
Pursuit to Algiers '45
South of Monterey '47

Guillermo Rios

Old Gringo '89

Lalo Rios

The Ring '52

Genevieve Rioux

The Decline of the American Empire '86

Arthur Ripley (D)

The Chase '46
Thunder Road '58

Michael Ripper

The Curse of the Werewolf '61
Dead Lucky '60
Eye Witness '49
Girly '70
Quatermass 2 '57
The Scars of Dracula '70
Where the Bullets Fly '66

Frank Ripploh

Taxi zum Klo '81

Frank Ripploh (D)

Taxi zum Klo '81

Leon Rippy

Eye of the Storm '91
Jesse '88
Kuffs '91
Track 29 '88

Arturo Ripstein (D)

Foxtrot '76

Carlos Riqueline

The Milagro Beanfield War '88

Elisabeth Risdon

Down Dakota Way '49
The Great Man Votes '38
Guilty of Treason '50
Mannequin '37
Reap the Wild Wind '42
Scaramouche '52

Miriam Riselle

The Singing Blacksmith (Yankl der Shmid) '38
Tevye (Teyve der Milkhiker) '39

Dino Risi (D)

The Easy Life '63
How Funny Can Sex Be? '76
Love in the City '53
Sweet Smell of Woman '75
Tiger and the Pussycat '67

Marco Risi (D)

Forever Mary '91

Jeff Risk

Kill Squad '81

Jacques Rispal

Le Chat '75

Michael Rissi (D)

Soultaker '90

Roberto Risso

Gladiator of Rome '63

Gary Rist

The Young Graduates '71

Robbie Rist

He is My Brother '75
Teenage Mutant Ninja Turtles: The Movie '90

Little Risto

Prince Jammy '87

Cyril Ritchard

Blackmail '29
The First Christmas '75
Half a Sixpence '67
Hans Brinker '69
Peter Pan '60

Bruce Ritchey

A Child is Waiting '63

Clint Ritchie

Peacekillers '71
The St. Valentine's Day Massacre '67

June Ritchie

A Kind of Loving '62
The Three Penny Opera '62

Michael Ritchie (D)

An Almost Perfect Affair '79
The Bad News Bears '76
The Candidate '72
The Couch Trip '87
Divine Madness '80
Downhill Racer '69
Fletch '85
Fletch Lives '89
The Golden Child '86
The Island '80
Prime Cut '72
Semi-Tough '77
Smile '75
Survivors '83
Wildcats '86

Viktors Ritelis (D)

Crucible of Horror '69

Kristen Riter

Student Bodies '81

Jennifer Ritt

The Young Graduates '71

Martin Ritt

Conrack '74
The Slugger's Wife '85

Martin Ritt (D)

Back Roads '81
Black Orchid '59
The Brotherhood '68
Casey's Shadow '78
Conrack '74
Cross Creek '83
The Front '76
The Great White Hope '70
Hombre '67
Hud '63
The Long, Hot Summer '58
Molly Maguires '69
Murphy's Romance '85
Norma Rae '79
Nuts '87
Paris Blues '61
Pete 'n' Tillie '72
Sounder '72
The Spy Who Came in from the Cold '65
Stanley and Iris '90

Brent Ritter

Curse of the Blue Lights '88

John Ritter

Americathon '79
Barefoot Executive '71
Comeback Kid '80
Flight of Dragons '82
Hero at Large '80
In Love with an Older Woman '82
The Last Fling '86
Letting Go '85
The Other '72
Pray TV '82
Problem Child '90
Problem Child 2 '91
Real Men '87
Skin Deep '89
A Smoky Mountain Christmas '86
The Stone Killer '73
Sunset Limousine '83
They All Laughed '81
Tricks of the Trade '88
Unnatural Causes '86
Wholly Moses! '80

Tex Ritter

Arizona Days '37
Cowboy '54
Cowboys of the Saturday Matinee '84
Dead or Alive '44
Enemy of the Law '45
Gangsters of the Frontier '44
The Girl from Tobacco Row '66
Headin' for the Rio Grande '36
Hittin' the Trail '37
Holiday Rhythm '50
Man from Texas '39
Marked for Murder '45
The Marshal's Daughter '53
Mystery of the Hooded Horseman '37
The Pioneers '41
Riders of the Rockies '37
Roll, Wagons, Roll '39
Rollin' Plains '38
Sing, Cowboy, Sing '37
Song of the Gringo '36
Take Me Back to Oklahoma '40

Tex Rides With the Boy Scouts '37
Three in the Saddle '45
Trouble in Texas '37
Utah Trail '38

Thelma Ritter

All About Eve '50
As Young As You Feel '51
Birdman of Alcatraz '62
Daddy Long Legs '55
The Farmer Takes a Wife '53
For Love or Money '63
Ford Startime '60
Hole in the Head '59
How the West was Won '62
The Incident '67
A Letter to Three Wives '49
Miracle on 34th Street '47
The Misfits '61
A New Kind of Love '63
Pickup on South Street '53
Pillow Talk '59
Rear Window '54

The Ritz Brothers

Blazing Stewardesses '75
Goldwyn Follies '38

Harry Ritz

Silent Movie '76
Texas Layover '75

Jimmy Ritz

Texas Layover '75

Emmanuelle Riva

Hiroshima, Mon Amour '59
Kapo '59

Sandra Riva

The Green Wall '70

Giorgio Rivalta (D)

The Avenger '62

Carlos Rivas

The Black Scorpion '57
El Cuarto Chino (The Chinese Room) '71
The King and I '56

Guillermo Rivas

Bus Station '88
Los Hermanos Centella '65

Sonia Rivas

Lost Diamond '74

Amparo Rivelles

Anita de Montemar '60s
El Fugitivo '65

Gabriel River

Ecstasy Inc. '88

Cecilia Rivera

Aguirre, the Wrath of God '72

Chita Rivera

Mayflower Madam '87
Once Upon a Brothers Grimm '77
Pippin '81
Sweet Charity '68
That's Singing '84

Kirk Rivera

Body Moves '90

Patricia Rivera

La Fuga del Rojo '87

George Rivero

Bordello '79
Counterforce '87
Fist Fighter '88
Target Eagle '84

Jorge Rivero

Bellas de Noche
Conquest '83
Day of the Assassin '81
El Hombre Contra el Mundo '87
La Montana del Diablo (Devil's Mountain) '87
Los Caciques '87
Los Dos Hermanos '87
Priest of Love '81
Rio Lobo '70
Sin of Adam & Eve '72
Sin Salida '87
Soldier Blue '70
Volver, Volver, Volver! '87

Julian Rivero

Arizona Gangbusters '40
Heroes of the Alamo '37
Man from Hell's Edges '32
Phantom Patrol '36
Ridin' the Lone Trail '37

Joan Rivers

Les Patterson Saves the World '90
Muppets Take Manhattan '84
Pee-Wee's Playhouse Christmas Special '88
Spaceballs '87

Joan Rivers (D)

Rabbit Test '78

Michael Rivers

La Ley del Revolver (The Law of the Gun) '78

Catherine Rivet

Emmanuelle, the Joys of a Woman '76

Jacques Rivette (D)

The Nun '66
Paris Belongs to Us '60

Jean Riveyre

Diary of a Country Priest '50

George Riviere

Castle of Blood '64
Journey Beneath the Desert '61
The Virgin of Nuremberg '65

Isabelle Riviere

Summer '86

Marie Riviere

Summer '86

Lucille Rivim

Home Free All '84

Cheryl Rixon

Swap Meet '79

Gianni Rizzo

Desert Commandos '60s
Mission Stardust '68

Bert Roach

The Crowd '28
The Great Waltz '38

Bruce Roach

Things '80s

Chris Roach (D)

The 13th Floor '88

Frank Roach (D)

Nomad Riders '81

Hal Roach, Jr. (D)

One Million B. C. '40

Hal Roach (D)

The Devil's Brother '33
One Million B. C. '40

Pat Roach

Willow '88

Rickey Roach

Black Devil Doll from Hell '84

Adam Roarke

Beach Girls '82
Dirty Mary Crazy Larry '74
Four Deuces '75
Frogs '72
Hell's Angels on Wheels '67
How Come Nobody's On Our Side? '73
Hughes & Harlow: Angels in Hell '77
The Losers '70
The Savage Seven '68
This is a Hijack '75
Trespasses '86

John Roarke

Mutant on the Bounty '89

Jason Robards, Sr.

Bedlam '45
Black Rainbow '91
Crimson Romance '34
Desperate '47
The Fatal Hour '40
Fighting Marines '36
Reunion '88
Riff-Raff '47
Rimfire '49
Stunt Pilot '39
Wayne Murder Case '38
Western Pacific Agent '51
White Eagle '32

Jason Robards, Jr.

All the President's Men '76
Any Wednesday '66
Ballad of Cable Hogue '70
A Big Hand for the Little Lady '66
A Boy and His Dog '75
Breaking Home Ties '87
Bright Lights, Big City '88
Burden of Dreams '82
By Love Possessed '61
Cabo Blanco '81
A Christmas to Remember '78
The Christmas Wife '88
Comes a Horseman '78
Day After '83
A Doll's House '59
Dream a Little Dream '89
Final Warning '90
Fools '70
The Good Mother '88
Hour of the Gun '67
Hurricane '79
An Inconvenient Woman '91
Isadora '77
Johnny Got His Gun '71
Julia '77
Julius Caesar '70

Laguna Heat '87
Legend of the Lone Ranger '81
Long Day's Journey into Night '62
The Long, Hot Summer '86
Mark Twain and Me '91
Max Dugan Returns '83
Melvin and Howard '80
Mr. Sycamore '74
Murders in the Rue Morgue '71
The Night They Raided Minsky's '69
Once Upon a Time in the West '68
Parenthood '89
Quick Change '90
Raise the Titanic '80
Return of the Bad Men '48
The St. Valentine's Day Massacre '67
Sakharov '84
Something Wicked This Way Comes '83
Square Dance '87
Thousand Clowns '65
Tora! Tora! Tora! '70
Trail Street '47
You Can't Take It with You '84

Sam Robards

Fandango '85
Into Thin Air '85
Not Quite Paradise '86
Tempest '82

Willis Robards

The Three Musketeers '21

Barbra Robb

Young Nurses in Love '89

Seymour Robbie (D)

Chrome Hearts '70
Marco '73

Michael Robbin

Rock House '88

Brian Robbins

Cellar Dweller '87
C.H.U.D. 2: Bud the Chud '89

Christmas Robbins

The Demon Lover '77

Cindy Robbins

I Was a Teenage Werewolf '57

Gale Robbins

Barkleys of Broadway '49
The Belle of New York '52
My Girl Tisa '48
Parasite '81
Three Little Words '50

Jerome Robbins (D)

West Side Story '61

Jessie Robbins

The Fearless Vampire Killers '67

Marty Robbins

Ballad of a Gunfighter '64
Hell on Wheels '67
Road to Nashville '67

Matthew Robbins (D)

*batteries not included '87
Bingo '91
Corvette Summer '79
Dragonslayer '81
Legend of Billie Jean '85

Oliver Robbins

Poltergeist '82
Poltergeist 2: The Other Side '86

Skeeter Bill Robbins

Fighting Parson '35
Hard Hombre '31
Man's Land '32

Tim Robbins

Bull Durham '88
Cadillac Man '90
Erik the Viking '89
Five Corners '88
Fraternity Vacation '85
Howard the Duck '86
Jacob's Ladder '90
Made in Heaven '87
Miss Firecracker '89
No Small Affair '84
The Sure Thing '85
Tapeheads '89
Top Gun '86
Toy Soldiers '84

Mike Robe (D)

Child in the Night '90
Murder Ordained '87
News at Eleven '86
Son of the Morning Star '91
Urge to Kill '84

Miguel Robelo

Only Once in a Lifetime '80s

Richard Rober

Jet Pilot '57
The Well '51

Jim Roberson (D)

Legend of Alfred Packer '80

Syd Roberson (D)

Robin Hood...The Legend: The Time of the Wolf '85

Genevieve Robert (D)

Casual Sex? '88

Mark Robert

Superbug Super Agent '76

Yves Robert (D)

My Father's Glory '91
Pardon Mon Affaire '77
Pardon Mon Affaire, Too! '77
Return of the Tall Blond Man With One Black Shoe '74
Salut l'Artiste '74
The Tall Blond Man with One Black Shoe '72

Alan Roberts (D)

Happy Hooker Goes Hollywood '80
Round Trip to Heaven '92
Young Lady Chatterly 2 '85

Allene Roberts

The Hoodlum '51
The Red House '47

Union Station '50

Arthur Roberts

Deadly Vengeance '81
Not of This Earth '88

Beverly Roberts

Buried Alive '39

Bill Roberts

Bongo '47

Carol Roberts

Commando Invasion '87

Christian Roberts

The Adventurers '70
Desperados '69
To Sir, with Love '67

Conrad Roberts

The Mosquito Coast '86
The Serpent and the Rainbow '87

Deborah Roberts (D)

Frankenstein General Hospital '88

Des Roberts

Black Bikers from Hell '70
The Outlaw Bikers - Gang Wars '70

Doris Roberts

The Fig Tree '87
Hester Street '75
Honeymoon Killers '70
National Lampoon's Christmas Vacation '89
Ordinary Heroes '85
Simple Justice '89

Edith Roberts

The Taxi Mystery '26

Edward Roberts (D)

The Adventurous Knights '35

Eric Roberts

The Ambulance '90
Best of the Best '89
Blood Red '88
The Coca-Cola Kid '84
Descending Angel '90
A Family Matter (Vendetta) '91
Final Analysis '92
King of the Gypsies '78
The Lost Capone '90
Nobody's Fool '86
Options '88
Paul's Case '80
The Pope of Greenwich Village '84
Raggedy Man '81
Rude Awakening '89
Runaway Train '85
Slow Burn '86
Star 80 '83

Francesca Roberts

Heart of Dixie '89

Glenn Roberts

The Crater Lake Monster '77

Jason Roberts

Claws '85

Jay Roberts, Jr.

Aftershock '88

White Phantom: Enemy of Darkness '87

J.H. Roberts

The Courageous Mr. Penn '41

Joe Roberts

Our Hospitality '23

Julia Roberts

Baja Oklahoma '87
Blood Red '88
Dying Young '91
Flatliners '90
Hook '91
Mystic Pizza '88
Pretty Woman '89
Satisfaction '88
Sleeping with the Enemy '91
Steel Magnolias '89

Ken Roberts

Fatal Pulse '88
The Great Land of Small '86

Larry Roberts

Lady and the Tramp '55

Lee Roberts

Battling Marshal '48

Luanne Roberts

Weekend with the Babysitter '70

Lynn Roberts

Heart of the Rockies '37
Hollywood Stadium Mystery '38
Hunt the Man Down '50
Robin Hood of Texas '47

Lynne Roberts

Billy the Kid Returns '38
Eyes of Texas '48
Frolics on Ice '40
Hunt the Man Down/ Smashing the Rackets '50
Sioux City Sue '46

Mariwin Roberts

Enforcer from Death Row '78
Jokes My Folks Never Told Me '77

Mark Roberts

Posse '75

Michael D. Roberts

Heartbreaker '83
Ice Pirates '84
Rain Man '88

Pascale Roberts

The Grand Highway '88
The Peking Blond '68

Pernell Roberts

Errand Boy '61
Four Rode Out '69
High Noon: Part 2 '80
Hot Rod '79
Kashmiri Run '69
Magic of Lassie '78
The Night Train to Kathmandu '88
Paco '75
Sniper '75

Rachel Roberts

Alpha Beta '73
Baffled '72
The Belstone Fox '73

Going to Congress/Don't Park
    There '24
Golden Age of Comedy '58
Judge Priest '34
Mack Sennett Comedies:
    Volume 2 '16
Mr. Skitch '33

Lionel Rogosin (D)

On the Bowery '56

Jaromir Rogoz

Ecstasy '33

Cliff Roguemore (D)

Devil's Son-in-Law '77

Little Rohan

Raw and Rough Pinchers '87

Maria Rohm

Count Dracula '71
The House of 1,000 Dolls '67
Kiss and Kill '68
Venus in Furs '70

Eric Rohmer (D)

Boyfriends & Girlfriends '88
Chloe in the Afternoon '72
Claire's Knee '71
Four Adventures of Reinette
    and Mirabelle '89
Full Moon in Paris '84
Le Beau Mariage '82
My Night at Maud's '69
Pauline at the Beach '83
Summer '86

Patrice Rohmer

Revenge of the Cheerleaders
    '76

Clayton Rohner

Bat 211 '88
I, Madman '89
Just One of the Guys '85
Private Investigations '87

Alberto Rojas

Esta Noche Cena Pancho '87
The Neighborhood Thief '83
Stallion '88

Esther Rojo

Fuerte Perdido '78

Gustavo Rojo

The Christmas Kid '68
No Survivors, Please '63
A Witch Without a Broom '68

Helena Rojo

Aguirre, the Wrath of God '72
Mary, Mary, Bloody Mary '76
Siempre Hay una Primera Vez
    '87

Maria Rojo

Candy Stripe Nurses '74

Rueben Rojo

Cauldron of Blood '67
Samson in the Wax Museum
    '63

Gilbert Roland

Any Gun Can Play '67
Around the World in 80 Days
    '56
The Bad and the Beautiful '52
Barbarosa '82
Beneath the 12-Mile Reef '53

Between God, the Devil & a
    Winchester '70s
Bullfighter & the Lady '50
Cabo Blanco '81
Captain Kidd '45
Cheyenne Autumn '64
The French Line '54
Go Kill and Come Back '68
Islands in the Stream '77
Miracle of Our Lady of Fatima
    '52
The Poppy Is Also a Flower
    '66
The Racers '55
Ridin' the California Trail '47
Running Wild '73
The Ruthless Four '70
The Sacketts '79
Samar '62
The Sea Hawk '40
She Done Him Wrong '33
South of Monterey '47
Thunder Bay '53
Thunder Pass '37
Treasure of Pancho Villa '55
Underwater '55

Gylian Roland

Barn of the Naked Dead '73
Deadly Sunday '82

Sutton Roley (D)

Loners '72
Snatched '72

Tutta Rolf

Dollar '38
Swedenhielms '35

Guy Rolfe

The Alphabet Murders '65
And Now the Screaming
    Starts '73
Dolls '87
Puppet Master 3 '91

Tom Rolfing

He Knows You're Alone '80

Michelle Rolia

Mr. Hulot's Holiday '53

Judi Rolin

Alice Through the Looking
    Glass '66

Henri Rollan

The Crazy Ray '22

Esther Rolle

Age Old Friends '89
Driving Miss Daisy '89
I Know Why the Caged Bird
    Sings '79
Journey Together '82
The Mighty Quinn '89
P.K. and the Kid '85
A Raisin in the Sun '89
Summer of My German
    Soldier '78

Jean Rollin

Zombie Lake '84

Jean Marie Rollin

Love Without Pity '91

Rolling Stones

Ready Steady Go, Vol. 2 '85
Sympathy for the Devil '70

Bernie Rollins (D)

Getting Over '81

**Howard E. Rollins,
Jr.**

The Children of Times Square
    '86
For Us, the Living '88
The House of Dies Drear '88
Johnnie Gibson F.B.I. '87
King '78
The Member of the Wedding
    '83
On the Block '91
Ragtime '81
A Soldier's Story '84

Mark Rolston

Prancer '89
Sinful Life '89
Survival Quest '89
Weeds '87

Yvonne Romain

The Curse of the Werewolf
    '61
Double Trouble '67

Alex Roman

Erotic Adventures of
    Pinocchio '71

Candice Roman

Big Bird Cage '72
Unholy Rollers '72

Freddie Roman

Sweet Lorraine '87

Leticia Roman

Fanny Hill: Memoirs of a
    Woman of Pleasure '64
G.I. Blues '60

Phil Roman (D)

Be My Valentine, Charlie
    Brown '75
A Charlie Brown Thanksgiving
    '81
Good Grief, Charlie Brown '83
It's Magic, Charlie Brown/
    Charlie Brown's All Star '81
It's a Mystery, Charlie Brown
    '87
It's Three Strikes, Charlie
    Brown '86
It's Your First Kiss, Charlie
    Brown/Someday You'll
    Find Her Charlie Brown '77
Life is a Circus, Charlie
    Brown/You're the Greatest,
    Charlie Brown '80
She's a Good Skate, Charlie
    Brown '80

Ruth Roman

The Baby '72
Beyond the Forest '49
Blowing Wild '54
Champion '49
Day of the Animals '77
Echoes '83
Far Country '55
Good Sam '48
Great Day in the Morning '56
Impulse '74
The Killing Kind '73
Love has Many Faces '65
The Sacketts '79
Since You Went Away '44
Strangers on a Train '51
Three Secrets '50
Window '49

Susan Roman

Weekend with the Babysitter
    '70

Viviane Romance

The Big Grab '63
Melodie en Sous-Sol '63
Panique '47
Princess Tam Tam '35

Beatrice Romand

Claire's Knee '71
House of the Yellow Carpet
    '84
Le Beau Mariage '82
Summer '86

Gina Romand

Los Amores de Marieta '87

Mark Romanek (D)

Static '87

Andy Romano

Get Christie Love! '74

Maria Romano

Caged Women '84
Thor el Conquistador '86

Ara Romanoff

Troma's War '88

Richard Romanus

The Couch Trip '87
Hollywood Heartbreak '89
Mean Streets '73
Murphy's Law '86
Night Terror '76
Protocol '84
Sitting Ducks '80
Strangers Kiss '83

Robert Romanus

Bad Medicine '85
Fast Times at Ridgemont High
    '82
Pulse '88

Anatoly Romashin

Rasputin '85

Lina Romay

Gemidos de Placer '70s
Ilsa, the Wicked Warden '78
Love Laughs at Andy Hardy
    '46
The Loves of Irina '80s

Luis Romay

Death of a Bureaucrat '66

Cindy Rome

Swift Justice '88

Sindi Rome

Knock Outs '92

Sydne Rome

Diary of Forbidden Dreams
    '76
Just a Gigolo '79
Order to Kill '73
Puma Man '80
That Lucky Touch '75
Twist '76
Wanted: Babysitter '75

Rachel Romen

The Girl from Tobacco Row
    '66
Ransom Money '88

Fred Romer

Great Adventure '75

Cesar Romero

Americano '55
Around the World in 80 Days
    '56
Batman '66
The Beautiful Blonde from
    Bashful Bend '49
The Big Push '75
The Computer Wore Tennis
    Shoes '69
Crooks & Coronets '69
Dance Hall '41
Doin' What the Crowd Does
    '73
Donovan's Reef '63
FBI Girl '52
The Great American
    Broadcast '41
Gun Crazy '69
Happy Go Lovely '51
Judgment Day '88
Jungle '52
The Little Princess '39
The Lost Continent '51
Lust in the Dust '85
Madigan's Millions '67
Mission to Glory '80
Mortuary Academy '91
Now You See Him, Now You
    Don't '72
Orchestra Wives '42
The Proud and the Damned
    '72
The Racers '55
Scotland Yard Inspector '52
The Shadow Man '53
Show Them No Mercy '35
Simple Justice '89
The Spectre of Edgar Allen
    Poe '73
Springtime in the Rockies '42
The Thin Man '34
Vera Cruz '53
Wee Willie Winkie '37

Chanda Romero

The Last Reunion '80
Olongape: The American
    Dream '89

David Romero

Over the Summer '85

Eddie Romero (D)

Beast of the Yellow Night '70
Beyond Atlantis '73
Brides of the Beast '68
Cavalry Command '63
Twilight People '72
WhiteForce '88
Woman Hunt '72

**George A.
Romero** (D)

The Crazies '76
Creepshow '82
Dawn of the Dead '78
Day of the Dead '85
Knightriders '81
Martin '77
Monkey Shines '88
Night of the Living Dead '68
Season of the Witch '73

George Romero (D)

Two Evil Eyes '91

Joanelle Romero

Lost Legacy: A Girl Called
    Hatter Fox '77

Joanelle Romero (D)
Powwow Highway '89

Marta Romero
La Sombra del Murcielago

Ned Romero
Deerslayer '78
House 4 '91
I Will Fight No More Forever '75
Last of the Mohicans '85

Tina Romero
Sisters of Satan '75

Daniela Romo
Frontera (The Border)

John Romo
Dead Women in Lingerie '90

Danny Ronan
Nightmare '82

Carla Ronanelli
Fighting Fists of Shanghai Joe '65

Charles R. Rondeau (D)
The Devil's Partner '58
The Girl in Lover's Lane '60

Ronnie Rondell (D)
No Safe Haven '87

Clelia Rondinelli
Where's Piccone '84

Maurice Ronet
Beau Pere '81
Besame '87
The Deadly Trap '71
Frantic '58
How Sweet It Is! '68
The Lost Command '66
L'Odeur des Fauves '66
Scandal Man '67
Sidney Sheldon's Bloodline '79
Sphinx '81
Swimming Pool '70
Without Warning '52
The Women '69

Ronettes
Girl Groups: The Story of a Sound '83
Rock 'n' Roll History: Girls of Rock 'n' Roll '90

John Roney
Barnum '86

Chito Rono (D)
Olongape: The American Dream '89

Linda Ronstadt
Chuck Berry: Hail! Hail! Rock 'N' Roll '87
FM '78
La Pastorela '92
The Pirates of Penzance '83
The Return of Ruben Blades '87

Darrell Roodt (D)
City of Blood '88
Place of Weeping '86

Michael Rooker
Afterburn '92
Days of Thunder '90
Eight Men Out '88
Henry: Portrait of a Serial Killer '90
JFK '91
Mississippi Burning '88
Music Box '89
Sea of Love '89

Abram Room (D)
Bed and Sofa '27

Charles Rooner
The Pearl '48

Mickey Rooney
Ace of Hearts '70s
The Adventures of Huckleberry Finn '39
Andy Hardy Gets Spring Fever '39
Andy Hardy Meets Debutante '40
Andy Hardy's Double Life '42
Andy Hardy's Private Secretary '41
The Atomic Kid '54
Babes in Arms '39
Babes on Broadway '41
The Big Wheel '49
Bill '81
Bill: On His Own '83
The Black Stallion '79
Boys Town '38
Breakfast at Tiffany's '61
The Bridges at Toko-Ri '55
Captains Courageous '37
The Care Bears Movie '84
The Comic '69
Dark Side of Love '79
The Domino Principle '77
Erik the Viking '89
Find the Lady '75
Girl Crazy '43
Healer '36
Hide-Out '34
How to Stuff a Wild Bikini '65
The Human Comedy '43
It Came Upon a Midnight Clear '84
It's a Mad, Mad, Mad, Mad World '63
Leave 'Em Laughing '81
Life Begins for Andy Hardy '41
Lightning: The White Stallion '86
Little Lord Fauntleroy '36
Love Finds Andy Hardy '38
Love Laughs at Andy Hardy '46
Magic of Lassie '78
Manhattan Melodrama '34
Manipulator '71
Men of Boys Town '41
Mickey the Great '39
Mickey Rooney in the Classic Mickey McGuire Silly Comedies
A Midsummer Night's Dream '35
My Heroes Have Always Been Cowboys '91
My Outlaw Brother '51
National Velvet '44
The Odyssey of the Pacific '82
Off Limits '53
Pete's Dragon '77
Postmark for Danger/Quicksand '56
Pulp '72
Quicksand '50

Rachel's Man '75
Ronnie Dearest: The Lost Episodes '88
Rudolph & Frosty's Christmas in July '82
Santa Claus is Coming to Town '70
Senior Trip '81
Silent Night, Deadly Night 5: The Toymaker '91
Strike Up the Band '40
Summer Holiday '48
Thousands Cheer '43
Thunder County '74
Women's Prison Escape
Words & Music '48

Pat Rooney, Jr.
Two Reelers: Comedy Classics 7 '38

Tim Rooney
Storyville '74

Thorkild Roose
Day of Wrath '43

Buddy Roosevelt
Boss Cowboy '35
Circle Canyon '33
Lightning Range '33
Powdersmoke Range '35
Range Riders '35

Stephen Root
Monkey Shines '88

Wills Root (D)
The Bold Caballero '36

Tom Ropelewski (D)
Madhouse '90

Brian Roper
The Secret Garden '49

Gil Roper
Basket Case 3: The Progeny '91

Noel Roquevort
Le Corbeau '43
The Raven '43

Robby Rosa
Salsa '88

Donna Rosae
Master Blaster '85

Rosa Rosal
Ethan '71

Tom Rosales
Bail Out '90

Marion Rosamond
I Married an Angel '42

Bert Rosario
Cold Justice '89

Faliero Rosati (D)
High Frequency '88

Tony Rosato
Busted Up '86
City of Shadows '86
Hog Wild '80

Nello Rosatti (D)
The Sensuous Nurse '76

Francoise Rosay
Bizarre Bizarre '39
Carnival in Flanders (La Kermesse Heroique) '35
Johnny Frenchman '46
The Pedestrian '74
September Affair '50
Seven Deadly Sins '53

Patty Rosborough
New Wave Comedy '85

Bernard Rose (D)
Chicago Joe & the Showgirl '90
Paperhouse '89

Bill Rose
Hollywood Revels '47

Clifford Rose
Marat/Sade (Persecution and Assassination...) '66

Cristine Rose
Singing the Blues in Red '87

Deborah Rose
The Bone Yard '90

Felissa Rose
Sleepaway Camp '83

Gabrielle Rose
Family Viewing '88

George Rose
The Hideaways '73
Jack the Ripper '60
Mania '60
A New Leaf '71
A Night to Remember '58
The Pirates of Penzance '83
You Can't Take It with You '84

Jamie Rose
Chopper Chicks in Zombietown '91
Heartbreakers '84
In Love with an Older Man '82
Playroom '90
Rebel Love '85
Tightrope '84
To Die Standing '91
Twirl '81

Laurie Rose
Blood Voyage '77
Hot Box '72
Woman Hunt '72
Working Girls '75

Les Rose (D)
Gas '81
Hog Wild '80
Title Shot '81

Rose Marie
Bruce's Deadly Fingers '80s
Cheaper to Keep Her '80
Dead Heat on a Merry-Go-Round '66
Ghetto Blaster '89
Man from Clover Grove '78
Memory of Us '74
Witchboard '87

Mickey Rose (D)
Student Bodies '81

Pauline Rose
Hot T-Shirts '79

Pierre Rose (D)
The Klutz '73

Reuben Rose (D)
Screwball Academy '86

Roger Rose
Ski Patrol '89

Sherrie Rose
Martial Law 2: Undercover '91
Summer Job '88

Steve Rose
Summer School '77

Will Rose
Red Desert Penitentiary '83

Ben Roseman
Daughter of Horror '55

Barry Rosen (D)
The Yum-Yum Girls '78

Daniela Rosen
Sin Salida '87

Eric Rosen
Dollar '38

Martin Rosen (D)
The Plague Dogs '82
Stacking '87
Watership Down '78

Neal Rosen
Flesh Eating Mothers '89

Phil Rosen (D)
Beggars in Ermine '34
Gangs, Inc. '41
I Killed That Man '42
It Could Happen to You '37
Man's Land '32
Meeting at Midnight '44
The Phantom Broadcast '33
Phantom of Chinatown '40
The Phantom in the House '29
The President's Mystery '36
Return of the Ape Man '44
The Sphinx '33
Spooks Run Wild '41
Tango '36
Two Gun Man '31
Whistlin' Dan '32
Young Blood '33

Robert L. Rosen (D)
Raw Courage '84

Alan Rosenberg
After Midnight '89
White of the Eye '88

Michael Rosenberg
His Wife's Lover (Zayn Vaybs Lubovnik) '31
Mirele Efros '38
The Two Sisters '37

Rick Rosenberg (D)
Distant Thunder '88

Kaspar Rostrup (D)

Memories of a Marriage '90

Maggie Roswell

Midnight Madness '80
The Simpsons Christmas Special '89

Andrea Roth

Seedpeople '92

Bobby Roth (D)

Baja Oklahoma '87
Brainwash '82
The Game of Love '90
Heartbreakers '84
Keeper of the City '92
The Man Inside '90
Rainbow Drive '90

Celia Roth

Labyrinth of Passion '82
The Stranger '87

Cy Roth (D)

Fire Maidens from Outer Space '56

Gene Roth

Attack of the Giant Leeches '59
Earth vs. the Spider '58
Pirates of the High Seas '50

Ivan E. Roth

Hollywood Zap '86

Joan Roth

Luther the Geek '90

Joe Roth (D)

Coupe de Ville '90
Revenge of the Nerds 2: Nerds in Paradise '87
Streets of Gold '86

Lillian Roth

Animal Crackers '30

Louis Roth

Clash of the Ninja '86
Ninja Connection '90

Martha Roth

El Hombre y el Monstruo '47
The Man and the Monster '65

Phillip J. Roth (D)

Red Snow '91

Tim Roth

The Cook, the Thief, His Wife & Her Lover '90
The Hit '85
Rosencrantz & Guildenstern Are Dead '90
To Kill a Priest '89
Vincent & Theo '90
A World Apart '88

Wolf Roth

Spare Parts '85

Bendt Rothe

Gertrud '64

John Rothman

The Boost '88

Stephanie Rothman (D)

Group Marriage '72

The Student Nurses '70
Terminal Island '73
Track of the Vampire '66
The Velvet Vampire '71
Working Girls '75

Cynthia Rothrock

China O'Brien '88
China O'Brien 2 '89
Fast Getaway '91
Martial Law '90
Martial Law 2: Undercover '91
No Retreat, No Surrender 2 '89
Tiger Claws '91
24 Hours to Midnight '92

William Rotsler

Norma '89

William Rotsler (D)

Mantis in Lace '68
Norma '89

Yella Rottlaender

Alice in the Cities '74
The Scarlet Letter '73

Carlos Rotzinger

La Hora 24 '87

Brigitte Rouan

Overseas: Three Women with Man Trouble '90

Brigitte Rouan (D)

Overseas: Three Women with Man Trouble '90

Theodore Roubanis

Medusa '74

Jean Rouch (D)

Jaguar '56

Jacques Rouffio (D)

La Passante '83

Michael Rougas

I Was a Teenage Werewolf '57

Jean Rougerie

The Perils of Gwendoline '84

Philippe Rouleau

Oriane '85

Richard Roundtree

Angel 3: The Final Chapter '88
Bad Jim '89
The Banker '89
Baron and the Kid '84
Big Score '83
Bloodfist 3: Forced to Fight '91
City Heat '84
Crack House '89
Day of the Assassin '81
Diamonds '72
Earthquake '74
Embassy '72
Escape to Athena '79
An Eye for an Eye '81
Firehouse '72
Game for Vultures '86
Jocks '87
Killpoint '84
The Last Contract '86
Man Friday '76
Maniac Cop '88
Miami Cops '89
The Night Visitor '89
One Down, Two to Go! '82

Opposing Force '87
Party Line '88
Portrait of a Hitman '77
Q (The Winged Serpent) '82
Shaft '71
A Time to Die '91
Young Warriors '83

Robert Rounseville

Carousel '56

Hayden Rourke

Project Moon Base '53

Mickey Rourke

Angel Heart '87
Barfly '87
Body Heat '81
City in Fear '80
Desperate Hours '90
Diner '82
Eureka! '81
Fade to Black '80
Harley Davidson and the Marlboro Man '91
Heaven's Gate '81
Homeboy '88
Johnny Handsome '89
9 1/2 Weeks '86
1941 '79
The Pope of Greenwich Village '84
Prayer for the Dying '87
Rape and Marriage: The Rideout Case '80
Rumble Fish '83
White Sands '92
Wild Orchid '90
Year of the Dragon '85

Graham Rouse

Ride a Wild Pony '75
Weekend of Shadows '77

Russel Rouse (D)

Caper of the Golden Bulls '67
Oscar '66
The Thief '52
The Well '51

Deborah Roush

Assault of the Party Nerds '89

Anne Roussel

The Music Teacher '88

Gilbert Roussel (D)

Women's Prison Massacre '85

Myriem Roussel

First Name: Carmen '83
Hail Mary '85
Sacrilege '86

Nathalie Roussel

My Father's Glory '91

Jean-Paul Roussillon

Baxter '89

Alison Routledge

The Bridge to Nowhere '86
The Quiet Earth '85

Catherine Rouvel

Black and White in Color '76
Les Assassins de L'Ordre '71
Picnic on the Grass '59
Sand and Blood '87

Josef Rovensky

Diary of a Lost Girl '29

Gay Rowan

The Girl in Blue '74

Earl Rowe

The Blob '58

George Rowe

The Mermaids of Tiburon '62

Greg Rowe

Blue Fin '78

Misty Rowe

Goodbye, Norma Jean '75
Goodnight, Sweet Marilyn '89
Hitchhikers '72
Loose Shoes '80
The Man with Bogart's Face '80
Meatballs 2 '84
A Pleasure Doing Business '79

Nevan Rowe

Nutcase '83
Sleeping Dogs '82

Nicholas Rowe

Young Sherlock Holmes '85

Peter Rowe (D)

Lost '86
Personal Exemptions '88
Take Two '87

Tom Rowe (D)

Get Rita

Leesa Rowland

Class of Nuke 'Em High Part 2: Subhumanoid Meltdown '91

Roy Rowland (D)

Bugles in the Afternoon '52
The 5000 Fingers of Dr. T '53
Girl Hunters '63
Hit the Deck '55
Meet Me in Las Vegas '56
Two Weeks with Love '50

Steve Rowland

Naked Youth '59

William Rowland (D)

Flight to Nowhere '46
Follies Girl '43

Gena Rowlands

Another Woman '88
Brink's Job '78
A Child is Waiting '63
An Early Frost '85
Gloria '80
Light of Day '87
Lonely are the Brave '62
Love Streams '84
Montana '90
Once Around '91
A Question of Love '78
Rapunzel '82
Strangers: The Story of a Mother and Daughter '79
Tempest '82
Tony Rome '67
Two Minute Warning '76
A Woman Under the Influence '74

Polly Rowles

Power '86
Springtime in the Rockies '37

Christopher Rowley (D)

Soul Patrol '80

Deep Roy

Starship '87

Diana Roy

Invincible Barbarian '83

Esther Roy

Return of the Evil Dead '75

Jean-Claude Roy (D)

Deadly Sting '82

Sandhya Roy

Distant Thunder '73

Allan Royal

Men of Steel '88

Selena Royale

The Harvey Girls '46

Christiane Royce

Dr. Frankenstein's Castle of Freaks '74

Lionel Royce

Panama Menace '41
Road to Zanzibar '41
White Pongo '45

Roselyn Royce

Malibu Hot Summer '81
Retrievers '82
Sizzle Beach U.S.A. '86

Virginia Roye

Pace That Kills '28

Carol Royle

Tuxedo Warrior '82

Selena Royle

A Date with Judy '48
The Fighting Sullivans '42
Moonrise '48
My Dream is Yours '49
Robot Monster '53
Summer Holiday '48

William Royle

Mr. Wong in Chinatown '39
Mutiny in the Big House '39

Gregory Rozakis

Abduction '75
Five Corners '88

Patricia Rozema (D)

I've Heard the Mermaids Singing '87

Janosz Rozsa (D)

Sunday Daughters '80

Rubber Neck Boys

Duke Is Tops '38

Michael Rubbo (D)

The Peanut Butter Solution '85

Bad Manners '84
Eye of the Demon '87
Fire and Rain '89
Funny About Love '90
L.A. Law '86
Sweet 15 '90

Barbara Rutting

The Phantom of Soho '64
The Squeaker '65

Wolf Ruvinskis

Neutron and the Black Mask '61
Neutron vs. the Amazing Dr. Caronte '61
Neutron vs. the Death Robots '62
Neutron vs. the Maniac '62

Richard Ruxton

Galaxy Invader '85

Basil Ruysdael

Broken Arrow '50
Carrie '52

Anne Ryan

All the Lovin' Kinfolk '89
Three O'Clock High '87

Chet Ryan

Rounding Up the Law '22

Christopher Ryan

The Young Ones '88

Deborah Ryan

KISS Meets the Phantom of the Park '78

Eddie Ryan

Breakfast in Hollywood '46
The Fighting Sullivans '42
Hollywood Varieties '50

Fran Ryan

Eyes of Fire '84
Johnny Belinda '82
Private School '83
Quiet Cool '86

Frank Ryan

Call Out the Marines '42

Frank Ryan (D)

Call Out the Marines '42

Fred Ryan

Rebel Love '85

Hilary Ryan

The Getting of Wisdom '77

Irene Ryan

Heading for Heaven '47

James Ryan

Go for the Gold '84
Kill or Be Killed '80
Kill and Kill Again '81
Pursuit '90
Rage to Kill '88
Space Mutiny '88

John P. Ryan

Avenging Force '86
Best of the Best '89
Breathless '83
City of Shadows '86
Class of 1999 '90
Cops and Robbers '73
Death Scream '75

Death Wish 4: The Crackdown '87
Delta Force 2: Operation Stranglehold '90
Eternity '90
It's Alive '74
It's Alive 2: It Lives Again '78
Paramedics '88
The Postman Always Rings Twice '81
Rent-A-Cop '88
The Right Stuff '83
Runaway Train '85
Shamus '73

Kathleen Ryan

Odd Man Out '47
Try and Get Me '50

Marlo Ryan

Date Bait '60

Meg Ryan

Amityville 3: The Demon '83
Armed and Dangerous '86
D.O.A. '88
The Doors '91
Innerspace '87
Joe Versus the Volcano '90
The Presidio '88
Promised Land '88
Rich and Famous '81
Top Gun '86
When Harry Met Sally '89

Mitchell Ryan

Aces: Iron Eagle 3 '92
Angel City '80
The Best of Dark Shadows '60s
The Choice '01
Christmas Coal Mine Miracle '78
Death of a Centerfold '81
Double Exposure: The Story of Margaret Bourke-White '89
Electra Glide in Blue '73
The Five of Me '81
High Plains Drifter '73
Lethal Weapon '87
Magnum Force '73
Monte Walsh '70
My Old Man's Place '71
Of Mice and Men '81
Peter Lundy and the Medicine Hat Stallion '77
Sergeant Matlovich vs. the U.S. Air Force '78
Winter People '89

Paul Ryan

Starhops '78

Peggy Ryan

Miss Annie Rooney '42

R.L. Ryan

Eat and Run '86
Street Trash '87

Robert Ryan

And Hope to Die '72
Anzio '68
Back from Eternity '56
Bad Day at Black Rock '54
Battle of the Bulge '65
Behind the Rising Sun '43
Berlin Express '48
Best of the Badmen '50
Beware, My Lovely '52
Billy Budd '62
Bombardier '43
Born to Be Bad '50

Boy with the Green Hair '48
Caught '49
Clash by Night '52
Crossfire '47
The Dirty Dozen '67
Escape to Burma '55
Executive Action '73
Flying Leathernecks '51
God's Little Acre '58
Horizons West '52
Hour of the Gun '67
Ice Palace '60
The Iron Major '43
King of Kings '61
Lonelyhearts '58
The Longest Day '62
The Love Machine '71
Marine Raiders '43
Men in War '57
A Minute to Pray, A Second to Die '67
The Naked Spur '52
On Dangerous Ground '51
Professionals '66
The Racket '51
Return of the Bad Men '48
The Set-Up '49
The Sky's the Limit '43
The Tall Men '55
Tender Comrade '43
Trail Street '47
The Wild Bunch '69

Sheila Ryan

The Cobra Strikes '48
Fingerprints Don't Lie '51
Gold Raiders '51
Great Guns '41
Heartaches '47
Mask of the Dragon '51
On Top of Old Smoky '53
Railroaded '47
Ringside '49
Western Pacific Agent '51

Steve Ryan

Crime Story: The Complete Saga '86

Thomas Ryan

Body Count '87

Tim Ryan

China Beach '88
Here Come the Marines '52
The Runestone '91

Will Ryan

The Land Before Time '88

Melody Ryane

Lords of the Deep '89

Edgar Ryazanov (D)

A Forgotten Tune for the Flute '88
Private Life '83

Raisa Ryazanova

Moscow Does Not Believe in Tears '80

Georg Rydeberg

Dollar '38
Walpurgis Night '41
A Woman's Face '38

Bobby Rydell

Bye, Bye, Birdie '63
Marco Polo, Jr. '72

Christopher Rydell

Blood and Sand '89
How I Got Into College '89

Side Out '90

Derek Rydell

The Night Visitor '89
Phantom of the Mall: Eric's Revenge '89
Popcorn '89

Mark Rydell

The Long Goodbye '73
Punchline '88

Mark Rydell (D)

Cinderella Liberty '73
The Cowboys '72
For the Boys '91
Harry & Walter Go to New York '76
On Golden Pond '81
The Reivers '69
The River '84
Rose '79
The Rose '79

Alfred Ryder

Bogie '80
Probe '72
T-Men '47

Irene Ryder

Rage of Wind '82

Michael Ryder

Troma's War '88

Scott Ryder

Terror on Alcatraz '86

Winona Ryder

Beetlejuice '88
Edward Scissorhands '90
Great Balls of Fire '89
Heathers '89
Lucas '86
Mermaids '90
1969 '89
Square Dance '87
Welcome Home, Roxy Carmichael '90

Isabella Rye

S*H*E '79

Stellan Rye (D)

Student of Prague '13

Patrick Ryecart

Romeo and Juliet (Gielgud) '79
Silas Marner '85
True Colors '87
Twenty-One '91

Rolan Rykov

The Overcoat '59

Judy Rymer (D)

Who Killed Baby Azaria? '83

Rex Ryon

Jack's Back '87

Helle Ryslinge

Ladies on the Rocks '83

Chishu Ryu

An Autumn Afternoon '62
Equinox Flower '58
The Human Condition: A Soldier's Prayer '61
Late Spring '49
Tokyo Story '53
Twenty-Four Eyes '54

Bruno S

Every Man for Himself & God Against All '75
Stroszek '77

Yacef Saadi

Battle of Algiers '66

Bob Saal

The Instructor '83

Saban

Wall '83

William Sabatier

The Clockmaker '76

Anthony Sabato

Crimebusters '80s
Thundersquad '85

Antonio Sabato

Beyond the Law '68
Black Lemons '80s
New Mafia Boss '72
War of the Robots '78

Bo Sabato

L.A. Crackdown 2 '88
Neon Maniacs '86

Tony Sabato

This Time I'll Make You Rich '75

Simon Sabela

African Rage '70s
Fatal Assassin '78

Stefano Sabelli

Bizarre '87

Mark Sabin

Code of Honor

Robert C. Sabin

Slime City '89

Andrew Sabiston

Prep School '81

Jean Sablon

Invitation to Paris '60

Marcel Sabourin

The Hitman '91
Two Women in Gold '70

Sabu

Black Narcissus '47
Drums '38
Elephant Boy '37
Jungle Book '42
Jungle Hell '55
Mistress of the World '59
Pardon My Trunk '52
Savage Drums '51
Thief of Baghdad '40
A Tiger Walks '64
The Treasure of Bengal '53

Shin Saburi

Equinox Flower '58

Robert Sacchi

Cold Heat '90
The Man with Bogart's Face '80

Chuck Sacci

The Choirboys '77

Renato Sala

Full Metal Ninja '89
Ninja of the Magnificence '89

Chloe Salaman

Dragonslayer '81

Abel Salazar

The Brainiac '61
The Curse of the Crying
   Woman '61
El Hombre y el Monstruo '47
The Living Head '59
Los Hermanos Diablo '65
Los Tres Amores de Losa '47
The Man and the Monster '65
Paula
The Vampire '57
The Vampire's Coffin '58

Theresa Saldana

Angel Town '89
Defiance '79
Double Revenge '89
The Evil That Men Do '84
I Wanna Hold Your Hand '78
The Night Before '88

Charles Sale

Two Reelers: Comedy
   Classics 6 '35

Chic Sale

The Fighting Westerner '35
Treasure Island '34
Two Reelers: Comedy
   Classics 9 '34

Richard Sale (D)

Let's Make it Legal '51
Ticket to Tomahawk '50

El Hedi Ben Salem

Ali: Fear Eats the Soul '74

Murray Salem

Riding High '78

Pamela Salem

After Darkness '85
Never Say Never Again '83
Salome '85

Meredith Salenger

Dream a Little Dream '89
Edge of Honor '91
Journey of Natty Gann '85
The Kiss '88
A Night in the Life of Jimmy
   Reardon '88

Carmen Salinas

Esta Noche Cena Pancho '87

Chucho Salinas

Diamante, Oro y Amor '87

Diane Salinger

Creature '85
Pee-Wee's Big Adventure '85

Matt Salinger

Captain America '89
Manhunt for Claude Dallas '86
Options '88
Power '86

Monroe Salisbury

The Lamb '15

Pascale Salkin

Akermania, Vol. 1 '92

The Eighties '83
Window Shopping '86

Sidney Salkow (D)

The Last Man on Earth '64
Twice-Told Tales '63

Albert Salmi

The Ambushers '67
Bang the Drum Slowly '56
Black Oak Conspiracy '77
Born American '86
The Bravados '58
Breaking In '89
Dragonslayer '81
Empire of the Ants '77
Gore Vidal's Billy the Kid '89
The Guns and the Fury '83
Hard to Hold '84
Hour of the Gun '67
Jesse '88
Joseph & His Brothers '79
Kill Castro '80
Love Child '82
The Meanest Men in the West
   '60s
Menace on the Mountain '70
The Mercenaries '80
Moonshine County Express
   '77
St. Helen's, Killer Volcano '82
Superstition '82
Sweet Creek County War '82
The Unforgiven '60
Viva Knievel '77

Syda Salmonava

The Golem '20

Joanasie Salomonie

The White Dawn '75

Frank S. Salsedo

Creepshow 2 '87

Jennifer Salt

Hi, Mom! '70
Play It Again, Sam '72

Victor Salva (D)

Clown House '88

Jaime Salvador (D)

Boom in the Moon '46

Phillip Salvador

Fight for Us '89

Guilia Salvatori

La Vie Continue '82

Renato Salvatori

Burn! '70
La Cicada (The Cricket) '83
Rocco and His Brothers '60
State of Siege '73

Catherine Salviat

Mama, There's a Man in Your
   Bed '89

Julie Saly

The Craving '80

Mark Salzman

Iron & Silk '91

Kareem Samar

Courtesans of Bombay '85

Gloria Samara

Battle of the Eagles '81

Yelena Samarina

The Scarlet Letter '73

Valerie Samarine

The Werewolf vs. the Vampire
   Woman '70

Isaac Samberg

The Dybbuk '37

Aldo Sambrel

Hot Blood '89
Navajo Joe '67
Voodoo Black Exorcist '89

Barbara Sammeth

Murder in Texas '81

Emma Samms

George Burns: His Wit &
   Wisdom '88
Goliath Awaits '81
The Shrimp on the Barbie '90

Tatyana Samoilova

The Cranes are Flying '57

**Salvatore
Samperi** (D)

Ernesto '79
Malicious '74
Submission '77

Candy Samples

Flesh Gordon '72

**Edwards
Sampson** (D)

The Fast and the Furious '54

Robert Sampson

The Arrival '90
Dark Side of the Moon '90
Ethan '71
Gates of Hell '83
The Grass is Always Greener
   Over the Septic Tank '78
The Re-Animator '85
Robot Jox '89

Tim Sampson

War Party '89

Will Sampson

Buffalo Bill & the Indians '76
Fish Hawk '79
From Here to Eternity '79
The Hunted Lady '77
Insignificance '85
One Flew Over the Cuckoo's
   Nest '75
Orca '77
Poltergeist 2: The Other Side
   '86
Roanoak '86
Standing Tall '78
Vegas '78
The White Buffalo '77

David Samson

Polyester '81

**Samson &
Hopkins** (D)

Torment '85

Joanne Samuel

Alison's Birthday '79
Mad Max '80
Queen of the Road '84

Andy Samuels

Return of Our Gang '25

Haydon Samuels

I Live with Me Dad '86

Joanne Samuels

Gallagher's Travels '80s

Laura San Giacomo

Miles from Home '88
Once Around '91
Pretty Woman '89
Quigley Down Under '90
sex, lies and videotape '89
Under Suspicion '91
Vital Signs '90

**Maria Luisa San
Jose**

El Diputado (The Deputy) '78

Olga San Juan

The Beautiful Blonde from
   Bashful Bend '49
One Touch of Venus '48

**Maria Eugenia San
Martin**

Bring Me the Vampire '61
El Aviador Fenomeno '47
El Zorro Vengador '47

Henry Sanada

Shogun's Ninja '83

Alina Sanchez

Portrait of Teresa '79

Blanca Sanchez

La Mentira '87
Los Jinetes de la Bruja (The
   Horseman...) '76

Cuco Sanchez

Guitarras Lloren Guitarras '47

Jaime Sanchez

Pawnbroker '65

Paul Sanchez

Navy SEALS '90

Pedro Sanchez

Any Gun Can Play '67
White Fang and the Hunter
   '85

Fernando Sancho

Duel in the Eclipse
Voodoo Black Exorcist '89

Frank Sancho

Boldest Job in the West '70s

Jose Sancho

Ay, Carmela! '90

Paul Sand

Can't Stop the Music '80
Great Bank Hoax '78
The Hot Rock '70
The Last Fling '86
The Legend of Sleepy Hollow
   '79
Once Upon a Brothers Grimm
   '77
Teen Wolf Too '87
Wholly Moses! '80

Dominique Sanda

Cabo Blanco '81
The Conformist '71
Damnation Alley '77
First Love '70
Garden of the Finzi-Continis
   '71
The Inheritance '76
Mackintosh Man '73
1900 '76
Steppenwolf '74
The Story of a Love Story '73

Olof Sandborg

Only One Night '42

Walter Sande

Nocturne '46
Rim of the Canyon '49

Otto Sander

Wings of Desire '88

Stephen Sander

Trained to Kill, U.S.A. '83

John Sanderford

The Alchemist '81

Ann D. Sanders

Straight Out of Brooklyn '91

Byron Sanders

The Flesh Eaters '64

Denis Sanders (D)

Elvis: That's the Way It Is '70
Invasion of the Bee Girls '73

Dirk Sanders

Pierrot le Fou '65

George Sanders

Action in Arabia '44
All About Eve '50
Allegheny Uprising '39
The Amorous Adventures of
   Moll Flanders '65
Bitter Sweet '41
Captain Blackjack '51
Death of a Scoundrel '56
Doomwatch '72
Endless Night '71
Falcon Takes Over/Strange
   Bargain '49
The Falcon's Brother '42
Foreign Correspondent '40
From the Earth to the Moon
   '58
The Ghost and Mrs. Muir '47
In Search of the Castaways
   '62
Invasion of the Body Stealers
   '69
Ivanhoe '52
The Jungle Book '67
Jupiter's Darling '55
The Moon and Sixpence '42
Moonfleet '55
Mr. Moto's Last Warning '39
Nurse Edith Cavell '39
One Step to Hell '68
Outcasts of the City '58
Picture of Dorian Gray '45
The Private Affairs of Bel Ami
   '47
Psychomania '73
The Quiller Memorandum '66
Rebecca '40
Saint in London '39
The Saint Strikes Back '39
Saint Strikes Back: Criminal
   Court '46

Cuba '79
Dog Day Afternoon '75
Forced March '90
Fright Night '85
Lipstick '76
Mayflower Madam '87
The Osterman Weekend '83
The Princess Bride '87
Protocol '84
The Resurrected '91
The Sentinel '76
Slaves of New York '89
Tailspin: Behind the Korean
Airline Tragedy '89
Whispers '89

Susan Sarandon

Atlantic City '81
Beauty and the Beast '83
The Buddy System '83
Bull Durham '88
Compromising Positions '85
A Dry White Season '89
The Front Page '74
Great Smokey Roadblock '76
The Great Waldo Pepper '75
The Hunger '83
The January Man '89
Joe '70
King of the Gypsies '78
Loving Couples '80
Mussolini & I '85
One Summer Love '76
The Other Side of Midnight
'77
Pretty Baby '78
The Rocky Horror Picture
Show '75
Something Short of Paradise
'79
Sweet Hearts Dance '88
Tempest '82
Thelma & Louise '91
White Palace '90
Who Am I This Time? '82
Witches of Eastwick '87
Women of Valor '86

Lane Sarasohn

The Groove Tube '72

Martine Sarcey

One Wild Moment '78

Kate Sarchet

Sweater Girls '78

**Frank Sarcinello,
Sr.**

Death Row Diner '88

Fernand Sardou

The Little Theatre of Jean
Renoir '71
Picnic on the Grass '59

Barry Sarecky (D)

Buck Rogers Cliffhanger
Serials, Vol. 1 '39
Buck Rogers Cliffhanger
Serials, Vol. 2 '39

Lelani Sarelle

Neon Maniacs '86

Isabel Sarfi

Una Viuda Descocada

Dick Sargent

Billie '65
Body Count '87
The Clonus Horror '79
Fantasy Island '76
Frame Up '91

Hardcore '79
Live a Little, Love a Little '68
Melvin Purvis: G-Man '74
Murder by Numbers '89
Teen Witch '89
That Touch of Mink '62
Twenty Dollar Star '91

Joseph Sargent (D)

Amber Waves '82
Caroline? '90
Coast to Coast '80
Colossus: The Forbin Project
'70
Freedom '81
Goldengirl '79
Hustling '75
Ivory Hunters '90
Jaws: The Revenge '87
MacArthur '77
The Manions of America '81
Never Forget '91
Nightmares '83
Passion Flower '86
The Taking of Pelham One
Two Three '74
Tomorrow's Child '82
Tribes '70
White Lightning '73

Kenny Sargent

Thinkin' Big '87

Lewis Sargent

The New Adventures of
Tarzan: Vol. 1 '35

Richard Sargent

For Love or Money '63
Tanya's Island '81

Marina Sargenti (D)

Child of Darkness, Child of
Light '91
Mirror, Mirror '90

Dick Sarin (D)

Family Reunion '88

Vic Sarin (D)

Cold Comfort '90

Maurice Sarli (D)

The Bodyguard '76

Michael Sarne

Cult People '89
Head Over Heels '67
Seaside Swingers '65

Michael Sarne (D)

Myra Breckinridge '70

Jonathan Sarno (D)

The Kirlian Witness '78

Michael Sarrazin

Beulah Land '80
Captive Hearts '87
The Double Negative '80
Fighting Back '82
Flim-Flam Man '67
For Pete's Sake '74
The Groundstar Conspiracy
'72
Gumball Rally '76
Joshua Then and Now '85
Keeping Track '86
Lena's Holiday '91
Loves & Times of
Scaramouche '76
Malarek '89
Mascara '87
The Phone Call '89

The Pursuit of Happiness '70
Reincarnation of Peter Proud
'75
The Seduction '82
Sometimes a Great Notion '71
They Shoot Horses, Don't
They? '69
The Train Killer '83

Gailard Sartain

Blaze '89
Ernest Goes to Camp '87
Ernest Goes to Jail '90
The Grifters '90
Guilty by Suspicion '91
Hard Country '81
Leader of the Band '87
Mississippi Burning '88
Roadie '80

Katsuhiko Sasaki

Godzilla vs. Megalon '76
Terror of Mechagodzilla '78

Sumie Sasaki

The Insect Woman '63

Peter Sasdy (D)

The Devil's Undead '75
Doomwatch '72
Hands of the Ripper '71
I Don't Want to Be Born '75
King Arthur, the Young
Warlord '75
The Lonely Lady '83
Nothing But the Night '72
Rude Awakening '81
Thirteenth Reunion '81
The Two Faces of Evil '82
Young Warlord '75

Catya Sasoon

Dance with Death '91

Barbara Sass (D)

Without Love '80

Jacqueline Sassard

Accident '67
Bad Girls '69
Les Biches '68
Pirates of the Seven Seas '62

Oley Sassone

Storyville '74

Catya Sassoon

Tuff Turf '85

Lina Sastri

Goodnight, Michelangelo '89
The Inquiry '87
Where's Piccone '84

Tura Satana

The Astro-Zombies '67
The Doll Squad '73
Faster Pussycat! Kill! Kill! '66

Ron Satlof (D)

Perry Mason Returns '85
Perry Mason: The Case of the
Lost Love '87
Spiderman: The Deadly Dust
'78
Waikiki '80

Hajime Sato (D)

Terror Beneath the Sea '66

Junya Sato (D)

The Go-Masters '82
The Silk Road '92

Kei Sato

Onibaba '64

Koichi Sato

The Silk Road '92

Jaroslav Satoransky

Larks on a String '68

Barry Sattels

Dawn of the Mummy '82

Paul Satterfield

Arena '88

Ludwig Satz

His Wife's Lover (Zayn Vaybs
Lubovnik) '31

Mo Man Sau

The Four Hands of Death:
Wily Match

Rick Saucedo

Elvis Stories '89

Jennifer Saunders

Supergrass '87

Justine Saunders

The Fringe Dwellers '86

Lori Saunders

Track of the Vampire '66

Mary Jane Saunders

Sorrowful Jones '49

Pamela Saunders

Alien Warrior '85

Rebecca Saunders

Tearaway '87

Terry Saunders

The King and I '56

Tom Saunders

Tainted Image '91

Claude Sautet (D)

Cesar & Rosalie '72
Les Choses de la Vie '70
Mado '76
Simple Story '80
Vincent, Francois, Paul and
the Others '76

Ann Savage

Detour '46
The More the Merrier '43
Pier 23 '51
Renegade Girl '46
Treasure of Fear '45

Archie Savage

Assignment Outer Space '61

Ben Savage

Little Monsters '89

Derek Savage (D)

The Meateater '79

Fred Savage

Boy Who Could Fly '86
Convicted: A Mother's Story
'87
Little Monsters '89
The Princess Bride '87
Vice Versa '88
The Wizard '89

Houston Savage

The Losers '70

John Savage

All the Kind Strangers '74
The Amateur '82
Any Man's Death '90
Bad Company '72
The Beat '88
Brady's Escape '84
Caribe '87
Coming Out of the Ice '82
The Deer Hunter '78
Do the Right Thing '89
Eric '75
The Godfather: Part 3 '90
Hair '79
Hotel Colonial '88
The Hunting '92
Inside Moves '80
The Killing Kind '73
Maria's Lovers '84
Nairobi Affair '88
The Onion Field '79
Primary Motive '92
Salvador '86
Silent Witness '85
The Sister-in-Law '74
Soldier's Revenge '70s
Steelyard Blues '73
The Tender Age '84

Rick Savage

Sensations '87

Vic Savage

Creeping Terror '64

Dany Saval

Moon Pilot '62

George Savalas

The Belarus File '85

Telly Savalas

The Belarus File '85
Beyond the Poseidon
Adventure '79
Beyond Reason '77
Birdman of Alcatraz '62
Cannonball Run II '84
Cape Fear '61
Capricorn One '78
Cartier Affair '84
Crooks & Coronets '69
The Dirty Dozen '67
Escape to Athena '79
Fake Out '82
The Family '73
The Greatest Story Ever Told
'65
The Hollywood Detective '89
Horror Express '72
Inside Out '75
The Interns '62
Kelly's Heroes '70
Killer Force '75
Land Raiders '69
Lisa and the Devil '75
MacKenna's Gold '69
Muppet Movie '79
New Mafia Boss '72
On Her Majesty's Secret
Service '69
Pancho Villa '72
A Reason to Live, A Reason
to Die '73
Redneck '73
Scalphunters, The '68
Scenes from a Murder '72
Silent Rebellion '82
The Slender Thread '65
Sonny and Jed '73
A Town Called Hell '72

Cynthia Scott (D)

Strangers in Good Company '91

Dan Scott

Spookies '85

Deborah Scott

Witchcraft '88

Debralee Scott

Deathmoon '78
Incoming Freshmen '79
Just Tell Me You Love Me '80
Police Academy '84

Donovan Scott

Goldilocks & the Three Bears '83
Savannah Smiles '82
Sheena '84

Eric Scott

Children's Carol '80

Evelyn Scott

Devil Kiss '77

Ewing Scott (D)

Windjammer '31

Fred Scott

Code of the Fearless '39
In Old Montana '39
Ranger's Roundup '38
Ridin' the Trail '40
The Roamin' Cowboy '37
Singing Buckaroo '37
Songs and Bullets '38
Two-Gun Troubador '37

Geoffrey Scott

First & Ten '85
First & Ten: The Team Scores Again '85

George C. Scott

Anatomy of a Murder '59
Bank Shot '74
Bible...In the Beginning '66
The Changeling '80
The Day of the Dolphin '73
Descending Angel '90
Dr. Strangelove, or: How I Learned to Stop Worrying and Love the Bomb '64
The Exorcist 3 '90
Firestarter '84
Flim-Flam Man '67
The Formula '80
Hardcore '79
The Hospital '71
The Hustler '61
Islands in the Stream '77
The Last Days of Patton '86
The List of Adrian Messenger '63
Movie, Movie '78
The Murders in the Rue Morgue '86
New Centurions '72
Oklahoma Crude '73
Oliver Twist '82
Pals '87
Patton '70
Petulia '68
Prince and the Pauper '78
Rage '72
The Rescuers Down Under '90
Savage is Loose '74
Taps '81
They Might Be Giants '71

George C. Scott (D)

The Andersonville Trial '70
Descending Angel '90
Don't Look Back: The Story of Leroy ''Satchel'' Paige '81
Rage '72
Savage is Loose '74

Gordon Scott

Conquest of Mycene '63
Coriolanus, Man without a Country '64
Death Ray
Gladiator of Rome '63
Hercules and the Princess of Troy '65
Hero of Rome '63
Marauder '65
Tarzan and the Trappers '58
Tramplers '66
Tyrant of Lydia Against the Son of Hercules '63

Gregg Scott

Ghosthouse '88

Gil Scott-Heron

Saturday Night Live: Richard Pryor '75

Hilary Scott

Bare Necessities '85

Howard Scott

Revenge of the Teenage Vixens from Outer Space '86

Ian Scott

Never Pick Up a Stranger '79

Imogen Millais Scott

Salome's Last Dance '88

Jacqueline Scott

Empire of the Ants '77

James Scott (D)

Strike It Rich '90

Janet Scott

Day of the Triffids '63
Day of the Triffids '63

Jason Scott

Ultimate Desires '91

Jay Scott

All the Lovin' Kinfolk '89
The Cut Throats '89
Grave of the Vampire '72

Jesse Scott

The Champ '31

John Scott

Horror of Party Beach '64
The Tell-Tale Heart '62

Jonathan Scott

Chronicles of Narnia '89

Kathryn Leigh Scott

House of Dark Shadows '70
The Last Days of Patton '86
Visitor from the Grave '81
Witches' Brew '79

Kay Scott

Fear in the Night '47

Ken Scott

Stopover Tokyo '57

Kimberly Scott

The Abyss '89
Flatliners '90

Kirk Scott

Cinderella '77
End of the World '77

Larry B. Scott

The Children of Times Square '86
Hero Ain't Nothin' But a Sandwich '78
My Man Adam '86
Revenge of the Nerds '84
Revenge of the Nerds 2: Nerds in Paradise '87
SpaceCamp '86
That Was Then...This Is Now '85

Lizabeth Scott

Dead Reckoning '47
Easy Living '49
Loving You '57
Pulp '72
The Racket '51
Scared Stiff '53
Silver Lode '54
A Stolen Face '52
The Strange Love of Martha Ivers '46
Too Late for Tears '49

Margaretta Scott

Action for Slander '38
Calling Paul Temple '48
Counterblast '48
Man of Evil '48

Martha Scott

Adam '83
Ben-Hur '59
Cheers for Miss Bishop '41
Desperate Hours '55
Falcon Takes Over/Strange Bargain '49
Father Figure '80
Hi Diddle Diddle '43
The Howards of Virginia '40
Our Town '40
Sayonara '57
The Ten Commandments '56
Turning Point '77
War of the Wildcats '43

Nina Scott

Invitation au Voyage '83

Oz Scott (D)

Bustin' Loose '81

Patricia Scott

Kentucky Blue Streak '35

Peter Graham Scott (D)

Subterfuge '68

Pippa Scott

Bad Ronald '74
Cold Turkey '71
Petulia '68

Randolph Scott

Abilene Town '46
Badman's Territory '46
Bombardier '43
Buffalo Stampede '33
Captain Kidd '45
The Cariboo Trail '50
China Sky '44
Coroner Creek '48
The Doolins of Oklahoma '49
The Fighting Westerner '35
Follow the Fleet '36
Gung Ho! '43
Hangman's Knot '52
Heritage of the Desert '33
Jesse James '39
Last of the Mohicans '36
Man of the Forest '33
My Favorite Wife '40
The Nevadan '50
Rage at Dawn '55
Rebecca of Sunnybrook Farm '38
Return of the Bad Men '48
Ride the High Country '62
Roberta '35
Seventh Cavalry '56
The Spoilers '42
Susannah of the Mounties '39
The Tall T '57
Ten Wanted Men '54
To the Last Man '33
To the Shores of Tripoli '42
Trail Street '47
Virginia City '40
Wagon Wheels '34
Western Union '41
When the West Was Young '32

Ridley Scott (D)

Alien '79
Black Rain '89
Blade Runner '82
The Duellists '78
Legend '86
Someone to Watch Over Me '87
Thelma & Louise '91

Robert Scott (D)

The Video Dead '87

Sherman Scott (D)

Billy the Kid Trapped '42

Sherry Scott

Demon Rage '82

Susan Scott

Day of the Maniac '77
The Slasher '74
Student Confidential '87
Trap Them & Kill Them '84

Thomas Scott

Motor Psycho '65

Timothy Scott

Inside Out '91
Love Me Deadly '76

Tony Scott (D)

Beverly Hills Cop 2 '87
Days of Thunder '90
The Hunger '83
The Last Boy Scout '91
Revenge '90
Top Gun '86

Victoria Scott

Blades '89

Zachary Scott

Appointment in Honduras '53
Born to Be Bad '50
Flame of the Islands '55
Flamingo Road '49
Let's Make it Legal '51
Medic & the Star & the Story '55
Mildred Pierce '45
Shotgun '55
South of St. Louis '48
The Southerner '45
Wings of Danger '52

Andrea Scotti

Love Angels '87

Vito Scotti

Caper of the Golden Bulls '67
I Wonder Who's Killing Her Now? '76
Life with Luigi '53
Two Weeks in Another Town '62
Von Ryan's Express '65

Aubrey Scotto (D)

Uncle Moses '32

Angela Scoular

The Adventurers '70

Alexander Scourby

Affair in Trinidad '52
Big Heat '53

Derf Scratch

Dubeat-E-O '84

Don Scribner

Wild Man '89

Angus Scrimm

The Lost Empire '83
Mindwarp '91
Phantasm '79
Phantasm II '88
Subspecies '91
Transylvania Twist '89

Earl Scruggs

Nashville Story '86
Scruggs '70

Simon Scuddamore

Slaughter High '86

Samantha Scully

Silent Night, Deadly Night 3: Better Watch Out! '89

Sean Scully

Sara Dane '81

Sandra Seacat

Promised Land '88

Sandra Seacat (D)

In the Spirit '90

Susan Seaforth Hayes

Billie '65
Dream Machine '91

Steven Seagal

Above the Law '88
Hard to Kill '89
Marked for Death '90
Out for Justice '91

Jenny Seagrove

Appointment with Death '88
Bullseye! '90
A Chorus of Disapproval '89
Deadly Game '91
The Guardian '90
Hold the Dream '85
Local Hero '83

Nate and Hayes '83
Sherlock Holmes and the Incident at Victoria Falls '91
A Woman of Substance, Episode 1: Nest of Vipers '84
A Woman of Substance, Episode 2: The Secret is Revealed '84
A Woman of Substance, Episode 3: Fighting for the Dream '84

Elizabeth Seal

Vampire Circus '71

Douglas Seale

Ernest Saves Christmas '88
Mr. Destiny '90

John Seale (D)

Till There was You '91

Franklyn Seales

The Onion Field '79
Southern Comfort '81

Jackie Searl

Topaze '33

Francis Searle (D)

A Girl in a Million '46
Someone at the Door '50
Things Happen at Night '48

Fred F. Sears

Down to Earth '47

Fred F. Sears (D)

Ambush at Tomahawk Gap '53
Earth vs. the Flying Saucers '56
The Giant Claw '57

Heather Sears

Estate of Insanity '70
Room at the Top '59

Victor Sears (D)

Ninja the Battalion '90
Ninja Force of Assassins '70s
Ninja's Extreme Weapons '90

George Seaton (D)

Airport '70
Big Lift '50
The Counterfeit Traitor '62
Country Girl '54
Miracle on 34th Street '47
Showdown '73
Teacher's Pet '58

James Seay

Beginning of the End '57
Heartaches '47
Phantom from Space '53
Turf Boy '42

Josef Sebanek

Loves of a Blonde '66

Ben Sebastian

Gator Bait 2: Cajun Justice '88

Beverly Sebastian

Rocktober Blood '85

Beverly Sebastian (D)

The American Angels: Baptism of Blood '89

Delta Fox '77
Gator Bait '73
Gator Bait 2: Cajun Justice '88

Dorothy Sebastian

His First Command '29
Our Dancing Daughters '28
Rough Riders' Roundup '34
The Single Standard '29
Spite Marriage '29
They Never Come Back '32
Two Reelers: Comedy Classics 8 '33
A Woman of Affairs '28

Ferd Sebastian (D)

The American Angels: Baptism of Blood '89
Delta Fox '77
Flash & Firecat '75
Gator Bait '73
Gator Bait 2: Cajun Justice '88
Hitchhikers '72
On the Air Live with Captain Midnight '70s
Rocktober Blood '85

John Sebastian (D)

Voyage to the Prehistoric Planet '65

Tracy Sebastian

On the Air Live with Captain Midnight '70s

Jean Seberg

Airport '70
Bonjour Tristesse '57
Breathless (A Bout de Souffle) '59
A Fine Madness '66
Lilith '64
Macho Callahan '70
The Mouse That Roared '59
Paint Your Wagon '70
Pendulum '69
Saint Joan '57

Douta Seck

Sugar Cane Alley '83

Harry Secombe

Down Among the Z Men '52
Goon Movie (Stand Easy) '53
Song of Norway '70

Kyle Secor

City Slickers '91
Delusion '91
Heart of Dixie '89
Sleeping with the Enemy '91

David Secter (D)

Feelin' Up '76

Neil Sedaka

Playgirl Killer '66

Margaret Seddon

Little Church Around the Corner '23
Mr. Deeds Goes to Town '36

Edie Sedgwick

Edie in Ciao! Manhattan '72

Edna Sedgwick

Red Barry '38

Edward Sedgwick (D)

The Cameraman '28
Fit for a King '37
Movie Struck '37
Parlor, Bedroom and Bath '31
Riding on Air '37
A Southern Yankee '48
Spite Marriage '29

Kyra Sedgwick

Born on the Fourth of July '89
Kansas '88
Mr. & Mrs. Bridge '91
Taipan '86
War & Love '84
Women & Men: Stories of Seduction, Part 2 '91

Robert Sedgwick

Morgan Stewart's Coming Home '87
Nasty Hero '89
Tune in Tomorrow '90

Elena Sedova

Virgins of Purity House '79

Miriam Seegar

Strangers of the Evening '32

Mindy Seeger

Relentless '89

Cari Seel

Tripods: The White Mountains '84

S.K. Seeley (D)

Blonde Savage '47

Charles Seeling (D)

Rounding Up the Law '22

Sybil Seely

The Balloonatic/One Week '20s

Steven Seemayer (D)

Miami Vendetta '87

Carol Seflinger

Sweater Girls '78

George Segal

All's Fair '89
Black Bird '75
Blume in Love '73
Born to Win '71
The Bridge at Remagen '69
Carbon Copy '81
Cold Room '84
Deadly Game '82
The Duchess and the Dirtwater Fox '76
The Endless Game '89
For the Boys '91
Fun with Dick and Jane '77
The Hot Rock '70
Invitation to a Gunfighter '64
Killing 'em Softly '85
King Rat '65
The Last Married Couple in America '80
Look Who's Talking '89
The Lost Command '66
Lost and Found '79
No Way to Treat a Lady '68
Not My Kid '85
The Owl and the Pussycat '70
The Quiller Memorandum '66

Rollercoaster '77
Russian Roulette '75
The St. Valentine's Day Massacre '67
Ship of Fools '65
Stick '85
The Terminal Man '74
A Touch of Class '73
Where's Poppa? '70
Who is Killing the Great Chefs of Europe? '78
Who's Afraid of Virginia Woolf? '66
The Zany Adventures of Robin Hood '84

Howard Segal

Last Game '80

Kerry Segal

Grown Ups '86

Shahar Segal

Late Summer Blues '87

Zohra Segal

Courtesans of Bombay '85

Jonathan Segall

Hot Bubblegum '81
Private Popsicle '82
Young Love - Lemon Popsicle 7 '87

Pamela Segall

After Midnight '89
Bad Manners '84
Gate 2 '92
Something Special '86

Dortha Segda

My Twentieth Century '90

Lavinia Segurini

Control '87

Mang Sei

Kick of Death: The Prodigal Boxer '80s

Charles Seibert

Wild & Wooly '78

Ileana Seidel

Beach House '82

Tim Seidel

Ghost Rider '43

Arthur Seidelman (D)

The Caller '87
Children of Rage '75
Echoes '83
Glory Years '87
Hercules in New York '70
Poker Alice '87

Susan Seidelman (D)

Cookie '89
Desperately Seeking Susan '85
Making Mr. Right '86
She-Devil '89
Smithereens '84

Joseph Seiden (D)

Eli Eli '40
God, Man and Devil '49
Paradise in Harlem '40

Emmanuelle Seigner

Frantic '88

Louis Seigner

The Eclipse '66
Seven Deadly Sins '53
This Special Friendship '67
Would-Be Gentleman '58

Lewis Seiler (D)

Charlie Chan in Paris '35
Doll Face '46
Great K & A Train Robbery '26
Guadalcanal Diary '43
The Winning Team '52

Jerry Seinfeld

It's Not Easy Bein' Me '87

John Seinpolis

The Untamable '23

William A. Seiter (D)

Allegheny Uprising '39
The Cheerful Fraud '27
Destroyer '43
Dimples '36
Diplomaniacs '33
In Person '35
It's a Date '40
Lady Takes a Chance '43
Little Church Around the Corner '23
Make Haste to Live '54
Moon's Our Home '36
One Touch of Venus '48
Roberta '35
Room Service '38
Skinner's Dress Suit '26
Sons of the Desert '33
Stowaway '36
Susannah of the Mounties '39
Way Back Home '32
The White Sin '24
You Were Never Lovelier '42

George B. Seitz (D)

Andy Hardy Meets Debutante '40
Andy Hardy's Double Life '42
Andy Hardy's Private Secretary '41
Danger Lights '30
Drums of Jeopardy '31
The Ice Flood '26
Kit Carson '40
Last of the Mohicans '36
Life Begins for Andy Hardy '41
Love Finds Andy Hardy '38
Vanishing American '26

John Seitz

Forced March '90
Hard Choices '84
Out of the Rain '90

Steve Sekely (D)

Behind Prison Walls '43
Day of the Triffids '63
Hollow Triumph '48
Revenge of the Zombies '43
The Scar '48
Waterfront '44

Kyoko Seki

Ikiru '52

Jacques Serres

Blue Country '78

Virginia Serret

Boom in the Moon '46

Daguey Servaes

Fortune's Fool '21

Jean Servais

The Devil's Nightmare '71
Fever Mounts at El Pao '59
Rififi '54

Juan Carlos Sesanzo (D)

Deadly Revenge '83

Alex Sessa (D)

Amazons '86
Stormquest '87

Almira Sessions

Diary of a Chambermaid '46
Miracle of Morgan's Creek '44
Oklahoma Annie '51

John Sessions

Sweet Revenge '90

Roshan Seth

Gandhi '82
Little Dorrit, Film 1: Nobody's Fault '88
Little Dorrit, Film 2: Little Dorrit's Story '88
Mississippi Masala '92
My Beautiful Laundrette '86

Bruce Seton

The Blue Lamp '49
Demon Barber of Fleet Street '36
Gorgo '61
Love from a Stranger '37

Joan Seton

Frenzy '46

Brian Setzer

La Bamba '87

Lynn Seus

Mountain Charlie '80

Eric Sevareid

Churchill and the Generals '81

Rex Sevenoaks

Trap '66

Joan Severance

Almost Pregnant '91
Another Pair of Aces: Three of a Kind '91
Illicit Behavior '91
No Holds Barred '89
The Runestone '91
See No Evil, Hear No Evil '89
Write to Kill '91

Susanne Severeid

Howling 4: The Original Nightmare '88

Doc Severinsen

Best of Comic Relief '86

Carmen Sevilla

Boldest Job in the West '70s
La Guerrillera de Villa (The Warrior of Villa) '70s

Pantaloons '57

Ninon Sevilla

Senora Tentacion '49

Philip Seville (D)

Those Glory, Glory Days '83

Billie Seward

Branded a Coward '35
The Man from Gun Town '36
The Revenge Rider '35

Kathleen Seward

Feelin' Up '76

Richard Seward

The Dolls '83

Rufus Sewell

Twenty-One '91

Vernon Sewell (D)

Blood Beast Terror '67
Frenzy '46
Ghost Keeper '80
Ghost Ship '53
Ghosts of Berkeley Square '47
Horrors of Burke & Hare '71
Rogue's Yarn '56
Uneasy Terms '48

Denise Sexton

Forbidden Impulse '85

Susan Sexton

The American Angels: Baptism of Blood '89

Athene Seyler

The Franchise Affair '52
Private Life of Don Juan '34
Sailing Along '38

Anne Seymour

All the King's Men '49
Gemini Affair '80s
Portrait of Grandpa Doc '77

Carolyn Seymour

The Assignment '78
The Bitch '78

Dan Seymour

Hard-Boiled Mahoney '47
A Night in Casablanca '46
Return of the Fly '59
To Have & Have Not '44

Jane Seymour

Are You Lonesome Tonight '92
Battlestar Galactica '78
East of Eden '80
The Four Feathers '78
The Haunting Passion '83
Head Office '86
Jack the Ripper '88
Jamaica Inn '82
Killer on Board '77
Lassiter '84
Live and Let Die '73
Matters of the Heart '90
Oh, Heavenly Dog! '80
Only Way '70
The Scarlet Pimpernel '82
Sinbad and the Eye of the Tiger '77
Somewhere in Time '80
The Tunnel '89
War & Remembrance, Part 1 '88

War & Remembrance, Part 2 '89
Young Winston '72

Lynn Seymour

Dancers '87

Ralph Seymour

Amazing Stories, Book 3 '86
Killer Party '86
Meatballs 2 '84
Rain Man '88

Sheldon Seymour

Monster a Go-Go! '65

Stephanie Seymour

Sports Illustrated's 25th Anniversary Swimsuit Video '89

Coralie Seyrig

Grain of Sand '84

Delphine Seyrig

Accident '67
The Black Windmill '74
Daughters of Darkness '71
The Discreet Charm of the Bourgeoisie '72
A Doll's House '73
Donkey Skin (Peau d'Ane) '70
Dos Cruces en Danger Pass '71
Grain of Sand '84
I Sent a Letter to My Love '81
Last Year at Marienbad '61
The Milky Way '68
Muriel '63
Stolen Kisses '69
Window Shopping '86

Serif Sezer

Yol '82

Nicholas Sgarro (D)

Fortune Dane '86
Happy Hooker '75

Sha-Na-Na

Grease '78

Ted Shackleford

Summer Fantasy '84
Sweet Revenge '87

Michael Shackleton (D)

Survivor '87

John Shackley

Tripods: The White Mountains '84

Susan Shadburne (D)

Shadow Play '86

Glenn Shadix

Heathers '89

Rebecca Shaeffer

The End of Innocence '90

Bobby Ray Shafer

Psycho Cop '88

Matt Shaffen (D)

Tabloid!

Josef Shaftel (D)

Naked Hills '56

Jamal Shah

K2 '92

Krishna Shah (D)

American Drive-In '88
Deadly Rivals '72
Deadly Thief '78
Hard Rock Zombies '85
Rivals '72
The River Niger '76

Martin Shaker

The Children '80

Matt Shakman

Meet the Hollowheads '89

Shakti

Three Kinds of Heat '87

Tupac Shakur

Juice '92

William Shallert

Twilight Zone: The Movie '83

Chuck Shamata

Death Weekend '76
Joshua Then and Now '85
Mafia Princess '86
Night Friend '87
Princess in Exile '90

Jeremy Shamos

Kid Colter '85

Chang Shan

Empire of the Dragon '80s
Shaolin Temple Strikes Back '72

Kuan Hai Shan

The Kung Fu Warrior '80s

Garry Shandling

Garry Shandling Show, 25th Anniversary Special '85

Forman Shane

Lady Godiva Rides '68

Gene Shane

The Velvet Vampire '71

Harvey Shane

Pleasure Unlimited '86
Saturday Night Sleazies, Vol. 1 '66

Jim Shane

Chasing Dreams '81

Maxwell Shane (D)

Fear in the Night '47

Paul Shane

Hi-Di-Hi '88

Sara Shane

The King and Four Queens '56

Michael Shaner

Bloodfist '89

Shangri-Las

Girl Groups: The Story of a Sound '83

Ravi Shankar

Concert for Bangladesh '72

Amelia Shankley

A Little Princess '87

Doug Shanklin

Dark Rider '91

Don Shanks

Halloween 5: The Revenge of Michael Myers '89
Last of the Mohicans '85
Life & Times of Grizzly Adams '74
Spirit of the Eagle '90

Debi Shanley

Delirium '77

John Patrick Shanley (D)

Joe Versus the Volcano '90

Al Shannon

The Drifter '88
Out of the Rain '90

Ethel Shannon

Charley's Aunt '25

Frank Shannon

Flash Gordon Conquers the Universe '40
Flash Gordon: Rocketship '36

Harry Shannon

Citizen Kane '41
Cow Town '50
Fighting Father Dunne '48
Idaho '43
The Jackie Robinson Story '50
Song of Texas '43
Summer and Smoke '61
The Underworld Story '50
Written on the Wind '56
The Yellow Rose of Texas '44

Joy Shannon (D)

Uptown Angel '90

Marc Shannon

Voyage to the Prehistoric Planet '65

Michael J. Shannon

Afterward '85
Future Cop '76
Smart-Aleck Kill '85

Peggy Shannon

The Case of the Lucky Legs '35
Deluge '33
False Faces '32

Susan Shantall

Romeo and Juliet '54

Alan Shapiro (D)

Tiger Town '83

Ken Shapiro (D)

The Groove Tube '72
Modern Problems '81

Mike Shapiro

Fireballs '90

Mike Shapiro (D)

Fireballs '90

Paul Shapiro (D)

Hockey Night '84

Gordon Shields

Vigil '84

Nicholas Shields

Lost in the Barrens '91
Princess in Exile '90

Pat Shields (D)

Frasier the Sensuous Lion '73

Dhiu Min Shien

Five Fighters from Shaolin '70s

James Shigeta

The Cage '89
Enola Gay: The Men, the Mission, the Atomic Bomb '80
Flower Drum Song '61
Paradise, Hawaiian Style '66
Tomorrow's Child '82

Marion Shilling

Captured in Chinatown '35
Cavalcade of the West '36
Gun Play '36
Inside Information '34
Man's Land '32
Rio Rattler '35
Shop Angel '32
A Shot in the Dark '35

Shmuel Shilo

Goodbye, New York '85
Noa at Seventeen '82

Yoseph Shiloah

I Love You Rosa '73
Private Manoeuvres '83

Joseph Shiloal

Eagles Attack at Dawn '74
Lion of Africa '87

Koji Shima (D)

Golden Demon '53
Warning from Space '56

Yoko Shimada

My Champion '84
Shogun '80

Armin Shimerman

Arena '88

Dalia Shimko

Noa at Seventeen '82

Joanna Shimkus

The Virgin and the Gypsy '70

Sab Shimono

Come See the Paradise '90
Gung Ho '85
The Wash '88

Takashi Shimura

The Bad Sleep Well '60
Drunken Angel '48
Godzilla, King of the Monsters '56
Ikiru '52
Love and Faith '78
The Mysterians '58
Sanshiro Sugata '43
Seven Samurai '54
Stray Dog '49
Throne of Blood '57

Nelson Shin (D)

Transformers: The Movie '86

Eitaro Shindo

The Crucified Lovers '54
Sansho the Bailiff '54

Kaneto Shindo (D)

The Island '61
Onibaba '64

Chen Shing

The Amsterdam Connection '70s
The Two Great Cavaliers '73

Cheng Shing

Snake in the Eagle's Shadow 2 '83

Li Tai Shing

Hurricane Sword '70s

Mo Si Shing

Samurai Sword of Justice

Sung Kam Shing

Kung Fu from Beyond the Grave

Tony Wai Shing

Duel of the Brave Ones '80

Christine Shinn

White Fury '90

Hiroko Shino

Murder in the Doll House '79

Masahiro Shinoda (D)

Double Suicide '69
Gonza the Spearman '86
MacArthur's Children '85

Saburo Shinoda

The Imperial Japanese Empire '85

Sue Shiomi

Dragon Princess '70s
Shogun's Ninja '83
The Street Fighter's Last Revenge '79

Toshi Shioya

Prisoners of the Sun '91

John Wesley Shipp

The Flash '90
The NeverEnding Story 2: Next Chapter '91

Yumi Shirakawa

H-Man '59
The Mysterians '58
Rodan '56

Talia Shire

Blood Vows: The Story of a Mafia Wife '87
For Richer, For Poorer '92
Gas-s-s-s! '70
The Godfather '72
Godfather 1902-1959—The Complete Epic '81
The Godfather: Part 2 '74
The Godfather: Part 3 '90
Hyper-Sapien: People from Another Star '86
Lionheart '87
Mark Twain and Me '91

Murderer's Keep '88
New York Stories '89
Old Boyfriends '79
Prophecy '79
Rad '86
Rip van Winkle '85
Rocky '76
Rocky 2 '79
Rocky 3 '82
Rocky 4 '85
Rocky 5 '90
Windows '80

Bill Shirk

Ballbuster
Escapist '83

Anne Shirley

Anne of Green Gables '34
Bombardier '43
Devil & Daniel Webster '41
Four Jacks and a Jill '41
Law of the Underworld '38
Murder My Sweet '44
Stella Dallas '37

Bill Shirley

Abbott and Costello Meet Captain Kidd '52
I Dream of Jeannie '52

Tom Shirley

Rehearsal '47

Cathie Shirriff

Covergirl '83
Star Trek 3: The Search for Spock '84

Tiki Shirtee

Ninja Massacre '84

Stephan Shkurat

Earth '30

Sallie Shockley

The Tattered Web '71

William Shockley

Howling 5: The Rebirth '89

Pierre Shoendoerffer (D)

The 317th Platoon '65

Ingrid Sholder

Mission Phantom '79

Jack Sholder (D)

By Dawn's Early Light '89
The Hidden '87
A Nightmare on Elm Street 2: Freddy's Revenge '85
Renegades '89

Lee Sholem (D)

Escape from Planet Earth '67
Hell Ship Mutiny '57
Tobor the Great '54

Roger Sholes (D)

The Tale of Ruby Rose '87

Lindsay Shonteff (D)

Big Zapper '73
Devil Doll '64
Fast Kill '73
Number 1 of the Secret Service '77
Second Best Secret Agent in the Whole Wide World '65

Pamela Shoop

One Man Jury '78

Dan Shor

Bill & Ted's Excellent Adventure '89
Daddy's Boys '87
Mesmerized '84
Mike's Murder '84
A Rumor of War '80
Strange Behavior '81
Strangers Kiss '83

Dinah Shore

Fun & Fancy Free '47
Hollywood Goes to War '54
Oh, God! '77
Pee-Wee's Playhouse Christmas Special '88
Thank Your Lucky Stars '43
Till the Clouds Roll By '46
Till the Clouds Roll By/A Star Is Born '46
Up in Arms '44

Sig Shore (D)

The Act '82
Shining Star '75
Sudden Death '85
The Survivalist '87

Lynn Shores

Charlie Chan at the Wax Museum '40
A Million to One '37

Richard Shorr (D)

Witches' Brew '79

Bobby Short

Hardhat & Legs '80

Dorothy Short

Brothers of the West '37
Camp Double Feature '39
Daughter of the Tong '39
Reefer Madness '38
Savage Fury '33
Trail of the Silver Spurs '41
Wild Horse Canyon '39

Florence Short

The Love Flower '20

Martin Short

Best of Comic Relief '86
The Big Picture '89
Cross My Heart '88
The Family Man '79
Father of the Bride '91
Innerspace '87
Lost and Found '79
Pure Luck '91
Really Weird Tales '86
Sunset Limousine '83
Tall Tales & Legends: Johnny Appleseed '86
Three Amigos '86
Three Fugitives '89

Joe Lick Shot

Live & Red Hot '87

Jack Shoulder (D)

Alone in the Dark '82

Grant Show

A Woman, Her Men and Her Futon '92

Max Showalter

Racing with the Moon '84
Sixteen Candles '84

Kathy Shower

Bedroom Eyes 2 '89
Commando Squad '87
Frankenstein General Hospital '88
The Further Adventures of Tennessee Buck '88
Out on Bail '89
Playboy: Playmates of the Year-The '80s '89
Robo-Chic '89

Kin Shriner

Angel 3: The Final Chapter '88
Escape '90
Obsessive Love '84
Vendetta '85

Will Shriner

Time Trackers '88

Ko Shu-How (D)

Five Fighters from Shaolin '70s

Irene Shubik (D)

Staying On '80

Elizabeth Shue

Adventures in Babysitting '87
Back to the Future, Part 2 '89
Back to the Future, Part 3 '90
Call to Glory '84
Cocktail '88
The Karate Kid '84
Link '86
The Marrying Man '91

Lee Shue

Samurai Blood, Samurai Guts
Samurai Sword of Justice

Stephanie Shuford

Dreams Come True '84

Richard B. Shull

Big Bus '76
Cockfighter '74
Dreamer '79
Hail '73
The Pack '77
Slither '73
Splash '84
Unfaithfully Yours '84

Yisroel Shumacher

The Jolly Paupers '38

Herman Shumlin (D)

Watch on the Rhine '43

Yuen Shun-Yee

Dreadnaught

Antonina Shuranova

An Unfinished Piece for a Player Piano '77
War and Peace '68

Ethel Shutta

Whoopee! '30

A. Shvorin

The Cranes are Flying '57

Ritch Shyder

Paramount Comedy Theater, Vol. 3: Hanging Party '87

Charles Shyer (D)

Baby Boom '87
Father of the Bride '91
Irreconcilable Differences '84

James Shyman (D)

Hollywood's New Blood '88
Slashdance '89

Ting Shan Si

La Venganza del Kung Fu '87

Ting Shan Si (D)

Fury of King Boxer '83

Yeung Si

Super Kung Fu Kid

Sabrina Siami

Throne of Fire '82

Ryp Siani

Robby '68

Sabrina Siani

Ator the Fighting Eagle '83
Conquest '83

Stefano Sibaldi

Mondo Cane '63

Mussef Sibay (D)

A Woman, Her Men and Her
Futon '92

Jane Sibbet

The Resurrected '91

John Sibbit

Love Circles '85

Nadia Sibirskaia

The Crime of Monsieur Lange
'36

Joseph R. Sicari

Night School '81

Sid Vicious

Mr. Mike's Mondo Video '79

Andy Sidaris (D)

Do or Die '91
Guns '90
Hard Ticket to Hawaii '87
Malibu Express '85
Picasso Trigger '89
Savage Beach '89
Seven '79
Stacey '73

David Siddon (D)

Their Only Chance '78

George Sidney

Manhattan Melodrama '34

George Sidney (D)

Anchors Aweigh '45
Bathing Beauty '44
Bye, Bye, Birdie '63
The Eddy Duchin Story '56
Half a Sixpence '67
The Harvey Girls '46
Jupiter's Darling '55
Kiss Me Kate '53
Pal Joey '57
Scaramouche '52
Show Boat '51
Thousands Cheer '43
Three Musketeers '48
Viva Las Vegas '63

Scott Sidney (D)

Charley's Aunt '25
The Nervous Wreck '26
Tarzan of the Apes '17

Sylvia Sidney

Corrupt '84
Damien: Omen 2 '78
Dead End '37
God Told Me To '76
Hammett '82
Having It All '82
I Never Promised You a Rose
Garden '77
Love from a Stranger '47
Mr. Ace '46
One Third of a Nation '39
Pals '87
Sabotage '36
The Shadow Box '80
A Small Killing '81
Snowbeast '77
Summer Wishes, Winter
Dreams '73
You Only Live Once '37

Charles Siebert

Blue Sunshine '78
A Cry for Love '80
Tarantulas: The Deadly Cargo
'77

Oliver Siebert

The White Rose '83

Paul Siederman

Deranged '87

Jim Siedow

Texas Chainsaw Massacre
'74
Texas Chainsaw Massacre
Part 2 '86

Don Siegel

Invasion of the Body
Snatchers '56
Invasion of the Body
Snatchers '78
Play Misty for Me '71

Don Siegel (D)

The Annapolis Story '55
The Beguiled '70
Big Steal '49
The Black Windmill '74
Charley Varrick '73
Coogan's Bluff '68
Death of a Gunfighter '69
Dirty Harry '71
Escape from Alcatraz '79
Flaming Star '60
Hell is for Heroes '62
Invasion of the Body
Snatchers '56
Jinxed '82
The Killers '64
Madigan '68
Private Hell 36 '54
Riot in Cell Block 11 '54
Rough Cut '80
The Shootist '76
Telefon '77
Two Mules for Sister Sara '70

Harvey Siegel

Enrapture '90

Mark Siegel

The Crater Lake Monster '77

Robert Siegel (D)

The Line '80

George Siegmann

Oliver Twist '22
The Three Musketeers '21

Casey Siemaszko

Amazing Stories, Book 1 '85
Back to the Future '85
Back to the Future, Part 2 '89
The Big Slice '91
Biloxi Blues '88
Breaking In '89
Class '83
Gardens of Stone '87
Miracle of the Heart: A Boys
Town Story '86
Near Misses '91
Secret Admirer '85
Stand By Me '86
Three O'Clock High '87
Young Guns '88

Nina Siemaszko

Wild Orchid 2: Two Shades of
Blue '92

Tom Sierchio

Delivery Boys '84

Gregory Sierra

The Clones '73
Code Name: Dancer '87
Dynamite and Gold '88
Let's Get Harry '87
Mean Dog Blues '78
Miami Vice '84
Pocket Money '72
Thief Who Came to Dinner '73

Miguel Sierra

The Bronx War '91

Everett Sifuentes

The Return of Josey Wales
'86

Elsbeth Sigmund

Heidi '52

James Signorelli (D)

Easy Money '83
Elvira, Mistress of the Dark
'88

Tom Signorelli

Crossover Dreams '85
The Last Porno Flick '74

Simone Signoret

Against the Wind '48
Day and the Hour '63
Dedee d'Anvers '49
Diabolique '55
I Sent a Letter to My Love '81
Is Paris Burning? '68
La Ronde '51
Le Chat '75
Madame Rosa '77
Room at the Top '59
Ship of Fools '65
Widow Couderc '74

Caroline Sihol

Confidentially Yours '83

Huang Ing Sik

When Taekwondo Strikes '83

Lung Sikar

The Unbeaten 28 '80s

Cynthia Sikes

Goodbye Cruel World '82
Love Hurts '91

The Man Who Loved Women
'83
Oceans of Fire '86

James B. Sikking

Black Force 2 '78
Charro! '69
Competition '80
Final Approach '92
Man on the Run '74
Morons from Outer Space '85
Narrow Margin '90
The Night God Screamed '80s
Ordinary People '80
Outland '81
Star Trek 3: The Search for
Spock '84
The Terminal Man '74
Up the Creek '84

Tom Silardi

Fury to Freedom: The Life
Story of Raul Ries '85

Vic Silayan

The Last Reunion '80

Adam Silbar

Hot Moves '84

Joel Silberg (D)

Bad Guys '86
Breakin' '84
Catch the Heat '87
Lambada '89
Rappin' '85

Tusse Silberg

The Company of Wolves '85
Hidden City '87

Vira Silenti

Atlas in the Land of the
Cyclops '61
Maciste in Hell '60
Son of Samson '62

Allison Silva

Bridge to Silence '89

David Silva

El Aviador Fenomeno '47
Senora Tentacion '49
Sisters of Satan '75

Franco Silva

The Mongols '60

Henry Silva

Above the Law '88
Alligator '80
Almost Human '79
The Bravados '58
Buck Rogers in the 25th
Century '79
Bulletproof '88
Cannonball Run II '84
Chained Heat '83
Cinderfella '60
Code of Silence '85
The Colombian Connection
'91
Crimebusters '80s
Cry of a Prostitute: Love Kills
'72
Day of the Assassin '81
Deadly Sting '82
Escape from the Bronx '85
A Gathering of Eagles '63
Hired to Kill '73
Hit Man '73
The Italian Connection '73
The Jayhawkers '59
Killer '70s

Killing in the Sun

Love and Bullets '79
Lust in the Dust '85
Man & Boy '71
The Manchurian Candidate
'62
Manhunt '73
Megaforce '82
Never a Dull Moment '68
Sharky's Machine '81
Shoot '85
The Tall T '57
Thirst '87
Violent Breed '83
Wrong is Right '82

Ivan Silva

Francisco Oller '84

Maria Silva

Curse of the Devil '73
The Mummy's Revenge '73

Rebeca Silva

Gente Violenta '87

Simone Silva

Big Deadly Game '54
The Shadow Man '53

Trinidad Silva

Colors '88
Crackers '84
The Night Before '88

Silvagni

The Rise of Louis XIV '66

Aldo Silvani

La Strada '54

Alvin Silver

Nudity Required '90

Andrew Silver (D)

Return '88

Borah Silver

Elves '89

Christine Silver

Room to Let '49

Cindy Silver

Hardbodies '84

Fawn Silver

Orgy of the Dead '65
Terror in the Jungle '68

Jeanne Silver

Sex, Drugs, and Rock-n-Roll
'84

Joan Micklin
Silver (D)

Bernice Bobs Her Hair '76
Between the Lines '77
Chilly Scenes of Winter '79
Crossing Delancey '88
Finnegan Begin Again '84
Hester Street '75
Loverboy '89
Prison Stories: Women on the
Inside '91

Joe Silver

Almost You '85
The Apprenticeship of Duddy
Kravitz '74
The Gig '85
Mr. Nice Guy '86
Rabid '77

Rambo: Part 3 '88
RoboCop '87
True Believer '89

Lane Smith

Air America '90
Blind Vengeance '90
The Displaced Person '76
My Cousin Vinny '92
Night Game '89
Prime Suspect '82
Prison '88
The Solitary Man '82
Weeds '87

Lewis Smith

The Adventures of Buckaroo
　Banzai '84
Diary of a Hitman '91
The Final Terror '83
The Heavenly Kid '85

Linda Smith

The Phone Call '89

Liz Smith

Apartment Zero '88
We Think the World of You
　'88

Lois Smith

Five Easy Pieces '70
The Jilting of Granny
　Weatherall '80
Reckless '84
Reuben, Reuben '83
Twisted '86

Loring Smith

Shadow of the Thin Man '41

Louise Smith

Working Girls '87

Madeleine Smith

Frankenstein and the Monster
　from Hell '74
Live and Let Die '73
The Vampire Lovers '70

Madolyn Smith

All of Me '84
The Caller '87
Ernie Kovacs: Between the
　Laughter '84
Final Approach '92
Funny Farm '88
Pray TV '82
Rehearsal for Murder '82
The Rose and the Jackal '90
The Super '91
2010: The Year We Make
　Contact '84
Urban Cowboy '80

Maggie Smith

Better Late Than Never '83
California Suite '78
Clash of the Titans '81
Death on the Nile '78
Evil Under the Sun '82
The Honey Pot '67
Hook '91
Lily in Love '85
The Lonely Passion of Judith
　Hearne '87
The Missionary '82
Mrs. Silly '85
Murder by Death '76
The Prime of Miss Jean
　Brodie '69
A Private Function '84
Quartet '81
A Room with a View '86

Travels with My Aunt '72

Mamie Smith

Paradise in Harlem '40

Mark Clifford Smith

An Empty Bed '90

Maurice Smith

Bloodfist 2 '90

Mel Smith

Brain Donors '92
The Misadventures of Mr. Wilt
　'90
Morons from Outer Space '85
The Princess Bride '87
Slayground '84

Mel Smith (D)

The Tall Guy '89

Melanie Smith

Trancers 3: Deth Lives '92

Michael Smith

Twins '80

Oliver Smith

Hellbound: Hellraiser 2 '88
Hellraiser '87

Pam Smith

Misty '61

Patricia Smith

Save the Tiger '73
Spirit of St. Louis '57

Paul Smith

Caged Fury '90
Crimewave '86
Crossing the Line '90
Death Chase '87
Dune '84
The Fire in the Stone '85
Gor '84
Haunted Honeymoon '86
Jungle Warriors '84
Madron '70
Midnight Express '78
Mortuary '81
Outlaw Force '87
Pieces '83
Platypus Cove '86
Popeye '80
Raiders in Action '71
Red Sonja '85
Return of the Tiger '78
The Salamander '82
Sno-Line '85
Sonny Boy '87

Pete Smith

A Few Moments with Buster
　Keaton and Laurel and
　Hardy '63

Peter Smith

The Quiet Earth '85

Peter Smith (D)

No Surrender '86

Putter Smith

Diamonds are Forever '71

Queenie Smith

The Great Rupert '50

Reid Smith

Teenager '74

Rex Smith

The Pirates of Penzance '83
Snow White and the Seven
　Dwarfs '83
Sooner or Later '78
Street Hawk '84
Transformations '80s
Trial of the Incredible Hulk '89

Richard Smith (D)

Trident Force '88

Robert Smith

Call Out the Marines '42

Roger Smith

Auntie Mame '58
Man of a Thousand Faces '57

Savannah Smith

Everybody's All American '88
The Long Riders '80

Sharon Smith

Nightmare '82

Shawn Smith

The River Rat '84

Shawnee Smith

The Blob '88
Summer School '87
Who's Harry Crumb? '89

Shelley Smith

The Fantastic World of D.C.
　Collins '84
National Lampoon's Class
　Reunion '82

Sinjin Smith

Side Out '90

Stuart Smith

Ninja Kill '87
The Ultimate Ninja '85

Terri Susan Smith

Basket Case '82

Tracy N. Smith

Hot Dog...The Movie! '83

Verne Smith

Mystery Theatre '51

William Smith

Action U.S.A. '89
Angels Die Hard '84
At Sword's Point '51
Blackjack '78
B.O.R.N. '88
Boss '74
Chrome Hearts '70
Chrome and Hot Leather '71
Commando Squad '87
Conan the Barbarian '82
Darker Than Amber '70
Deadly Breed '89
The Deadly Trackers '73
Dr. Minx '75
East L.A. Warriors '89
Emperor of the Bronx '89
Evil Altar '89
Eye of the Tiger '86
Fast Company '78
Fever Pitch '85
The Frisco Kid '79
Gentle Savage '73
Grave of the Vampire '72
Hell on the Battleground '88
Hollywood Man '76
Invasion of the Bee Girls '73

Jungle Assault '89
L.A. Vice '89
The Last Riders '90
The Losers '70
Maniac Cop '88
Mean Season '85
Memorial Valley Massacre '88
Moon in Scorpio '86
Platoon Leader '87
The Rebels '79
Red Dawn '84
Run, Angel, Run! '69
Scorchy '76
Seven '79
Slow Burn '90
A Taste of Hell '73
Terror in Beverly Hills '90
The Ultimate Warrior '75

**Willie May Ford
Smith**

Say Amen, Somebody '80

Yeardley Smith

City Slickers '91
Ginger Ale Afternoon '89
Heaven Help Us '85
Legend of Billie Jean '85
Maximum Overdrive '86
The Simpsons Christmas
　Special '89

Alan Smithee (D)

Appointment with Fear '85
Bloodsucking Pharoahs of
　Pittsburgh '90
City in Fear '80
Ghost Fever '87
I Love N.Y. '87
Iron Cowboy '68
Let's Get Harry '87
Morgan Stewart's Coming
　Home '87
The Shrimp on the Barbie '90
Stitches '85

Jan Smithers

Mr. Nice Guy '86

William Smithers

Death Sport '78

Bill Smitrovich

Crime Story: The Complete
　Saga '86
A Killing Affair '85
Renegades '89

Jimmy Smits

The Believers '87
Glitz '88
L.A. Law '86
Old Gringo '89
Running Scared '86
Stamp of a Killer '87
Switch '91
Vital Signs '90

Sonja Smits

Hitchhiker 4 '87
The Pit '81
That's My Baby! '88
Videodrome '83
War Brides '80s

Stephen Smoke (D)

Final Impact '91
Street Crimes '92

Phil Smoot (D)

The Dark Power '85

Rinaldo Smordoni

Shoeshine '46

Dick Smothers

And If I'm Elected '84
Rap Master Ronnie '88
Speed Zone '88

Tom Smothers

Alice Through the Looking
　Glass '66
And If I'm Elected '84
Hurray for Betty Boop '80
Pandemonium '82
A Pleasure Doing Business
　'79
Rap Master Ronnie '88
Serial '80
Silver Bears '78
Speed Zone '88

Josephine Smulders

The Gold & Glory '88

David Smulker

Mark of the Beast

Ann Smyrner

Beyond the Law '68
The House of 1,000 Dolls '67

Brian Sneagle

Deranged '74

Susan Sneers

Tiffany Jones '75

Gerti Sneider

2069: A Sex Odyssey '78

Mike Snell

Derby '71

Mortimer Snerd

Here We Go Again! '42

Helmut Snider

Kemek '88

Wesley Snipes

Jungle Fever '91
King of New York '90
Major League '89
Mo' Better Blues '90
New Jack City '91
Streets of Gold '86
White Men Can't Jump '92

Carrie Snodgress

Across the Tracks '89
The Attic '80
Blueberry Hill '88
Chill Factor '90
Diary of a Mad Housewife '70
The Fury '78
L.A. Bad '85
Murphy's Law '86
Nadia '84
A Night in Heaven '83
Pale Rider '85
The Rose and the Jackal '90
Silent Night, Lonely Night '69
The Solitary Man '82
Trick or Treats '82

Peter Snook

Fly with the Hawk

Carrie Snow

Girls of the Comedy Store '86

Marguerite Snow

With Kit Carson Over the
　Great Divide '25

Park Jong Soo

Search and Destroy '81
Striking Back '81

Jonathan Soper

Ninja Brothers of Blood '89

Mark Soper

Understudy: The Graveyard
    Shift 2 '88

Mike Soper

Blood Rage '87

Michael Sopkiw

After the Fall of New York '85
Blastfighter '85
Massacre in Dinosaur Valley
    '85

Mario Soqui (D)

Tramplers '66

Agnes Soral

Killing Cars '80s
One Wild Moment '78

Elga Sorbas

The American Soldier '70

Alberto Sordi

The Great War '59
I Vitelloni '53
Those Magnificent Men in
    Their Flying Machines '65
Two Nights with Cleopatra '54
The White Sheik '52

Agnes Sorel

Tchao Pantin '85

Guy Sorel

Honeymoon Killers '70

Jean Sorel

Paralyzed '70s

Louise Sorel

Get Christie Love! '74
Mazes and Monsters '82
Plaza Suite '71
The President's Plane is
    Missing '71
Rona Jaffe's Mazes &
    Monsters '82
When Every Day was the
    Fourth of July '78

Ted Sorel

Basket Case 2 '90
From Beyond '86

Linda Sorensen

Family Reunion '88
Kavik the Wolf Dog '84
Stone Cold Dead '80

Ricky Sorenson

Sword in the Stone '63
Tarzan and the Trappers '58

Sylvia Sorente

Castle of Blood '64

Charo Soriano

Dracula's Great Love '72
The Garden of Delights '70
Orgy of the Vampires '73

Judie Soriano

Black Force '75

Jim Soriero

Delivery Boys '84

Arleen Sorkin

From Here to Maternity '85

Janet Sorley

Last House on Dead End
    Street '77

Jason Sorokin

Oddballs '84

Paul Sorvino

Age Isn't Everything '91
Almost Partners '87
Bloodbrothers '78
Brink's Job '78
Chiefs '83
Cruising '80
Cry Uncle '71
The Day of the Dolphin '73
Dick Tracy '90
A Fine Mess '86
Fury on Wheels '71
The Gambler '74
Goodfellas '90
I, the Jury '82
I Will, I Will for Now '76
Jailbait: Betrayed By
    Innocence '86
Lost and Found '79
Melanie '82
Off the Wall '82
Oh, God! '77
Panic in Needle Park '75
Question of Honor '80
Reds '81
The Rocketeer '91
The Stuff '85
Tell Me Where It Hurts '74
That Championship Season
    '82
A Touch of Class '73
Turk 182! '85
Vasectomy: A Delicate Matter
    '86
Very Close Quarters '84

Sander Soth

Time Stands Still '82

Ann Sothern

The Best Man '64
Crazy Mama '75
The Judge Steps Out '49
Kid Millions '34
The Killing Kind '73
Lady Be Good '41
Lady in a Cage '64
A Letter to Three Wives '49
Little Dragons '80
The Manitou '78
Mexican Spitfire/Smartest Girl
    in Town '40
Nancy Goes to Rio '50
Panama Hattie '42
Whales of August '87
Words & Music '48

Hugh Sothern

Fighting Devil Dogs '43

Dimitri Sotirakis (D)

Beverly Hills Brats '89

Fernando Soto

Boom in the Moon '46
El Hombre de la Furia (Man of
    Fury) '65
The Illusion Travels by
    Streetcar '53

Invasion of the Vampires '61

Hugo Soto

Man Facing Southeast '86
Times to Come '81

Luchy Soto

The Garden of Delights '70

Talisa Soto

License to Kill '89
Mambo Kings '92
Prison Stories: Women on the
    Inside '91
Silhouette '91
Spike of Bensonhurst '88

Jim Sotos (D)

Forced Entry '75
Hot Moves '84
Sweet 16 '81

Miao Ko Sou

Samurai Death Bells

Alain Souchon

One Deadly Summer (L'Ete
    Meurtrier) '83

Christine Souder

Street Girls '75

Russell S. Soughten (D)

The Healing '88

David Soul

Appointment with Death '88
Dog Pound Shuffle '75
Hanoi Hilton '87
Homeward Bound '80
In the Cold of the Night '89
The Key to Rebecca '85
Little Ladies of the Night '77
Magnum Force '73
The Manions of America '81
Rage '80
Salem's Lot: The Movie '79
The Stick-Up '77
Through Naked Eyes '87
Tides of War '90
World War III '86

Stephen Soul

Silent Killers

Renee Soutendijk

Cold Room '84
Eve of Destruction '90
Forced March '90
The 4th Man '79
Grave Secrets '89
Inside the Third Reich '82
Keeper of the City '92
Murderers Among Us: The
    Simon Wiesenthal Story '89
Out of Order '84
Spetters '80

Colin South (D)

In Too Deep '90

Eve Southern

Morocco '30

Linda Southern

The Amazing Transplant '70

Tom Southern

Harlem Rides the Range '39

Charles Southwood

Guns for Dollars '73

Sartana's Here...Trade Your
    Pistol for a Coffin '70

Ben Soutten

The Crimes of Stephen
    Hawke '36
The Mutiny of the Elsinore '39

George Sowards

Backfire '22

Agnes Spaak

Dr. Orloff's Monster '64

Catherine Spaak

The Cat o' Nine Tails '71
Circle of Love '64
Counter Punch '71
The Easy Life '63
Empty Canvas '64
Honey '81
Hotel '67

Arthur Space

Miss Grant Takes Richmond
    '49
Riot Squad '41
A Southern Yankee '48
Terror at Red Wolf Inn '72
The Vanishing Westerner '50

Sissy Spacek

Badlands '74
Carrie '76
Coal Miner's Daughter '80
Crimes of the Heart '86
Ginger in the Morning '73
Girls of Huntington House '73
Hard Promises '92
Heart Beat '80
JFK '91
Katherine '75
The Long Walk Home '89
The Man with Two Brains '83
Marie '85
Missing '82
'night, Mother '86
Prime Cut '72
The Radical '75
Raggedy Man '81
The River '84
Violets Are Blue '86
Welcome to L.A. '77

Kevin Spacey

Dad '89
Henry & June '90
Long Day's Journey into Night
    '88
The Murder of Mary Phagan
    '87

Odoardo Spadaro

The Golden Coach '52

James Spader

Baby Boom '87
Bad Influence '90
Endless Love '81
Family Secrets '84
Jack's Back '87
Less Than Zero '87
Mannequin '87
The New Kids '85
Pretty in Pink '86
The Rachel Papers '89
sex, lies and videotape '89
True Colors '91
Tuff Turf '85
Wall Street '87
White Palace '90

Merrie Spaeth

World of Henry Orient '64

Fay Spain

Al Capone '59
Dragstrip Girl '57
Flight to Fury '66
The Godfather: Part 2 '74
Hercules and the Captive
    Women '63

Roda Spain

Rebel Vixens '85

B.J. Spalding

Bail Jumper '89

Timothy Spall

Crusoe '89
Dutch Girls '87
Gothic '87
Life is Sweet '91
The Sheltering Sky '90
White Hunter, Black Heart '90

Laurette Spang

Battlestar Galactica '78

Larry Spangler (D)

Joshua '76

Joe Spano

American Graffiti '73
Brotherhood of Justice '86
Brujo Luna
Cast the First Stone '89
Dunera Boys '85
Fever '91
Northern Lights '79
Terminal Choice '85
Warlock Moon '73

Vincent Spano

Afterburn '92
Alphabet City '84
And God Created Woman '88
Baby It's You '82
The Black Stallion Returns '83
Blood Ties '87
City of Hope '91
Creator '85
The Double McGuffin '79
Good Morning, Babylon '87
High Frequency '88
Maria's Lovers '84
Over the Edge '79
Rumble Fish '83
Senior Trip '81

Hans A. Spanuth (D)

The Great Train Robbery:
    Cinema Begins '03

Adrian Sparks

My Stepmother Is an Alien '88
Roanoak '86

Don Sparks

Fairytales '79

Ned Sparks

Bride Walks Out '36
Corsair '31
Gold Diggers of 1933 '33
Hawaii Calls '38
Lady for a Day '33
Marie Galante '34

Teresa Sparks (D)

Over the Summer '85

Robert Sparr (D)

A Swingin' Summer '65

Krao Spartacus

After the Fall of Saigon '80s

Malta Story '53
Massacre in Rome '73
The Master of Ballantrae '53
Perfect Crime '78
The Story of O '75
Tiger of the Seven Seas '62
Wooden Horse '50

Don Steel

Into the Night '85

Adrian Steele

The Mosquito Coast '86

Anthony Steele

Revenge of the Barbarians '64

Barbara Steele

Black Sunday '60
Caged Heat '74
Castle of Blood '64
8 1/2 '63
The Ghost '63
The Horrible Dr. Hichcock '62
La Dolce Vita '60
Nightmare Castle '65
Piranha '78
The Pit and the Pendulum '61
Pretty Baby '78
The She-Beast '65
Silent Scream '80
Terror Creatures from the
    Grave '66
They Came from Within '75
Winds of War '83

Bob Steele

Alias John Law '35
Arizona Gunfighter '37
Arizona Whirlwind '44
Atomic Submarine '59
Battling Outlaws '40s
Big Calibre '35
The Big Sleep '46
Billy the Kid in Texas '40
Border Phantom '37
Brand of Hate '34
Brand of the Outlaws '36
Breed of the Border '33
Cavalry '36
City for Conquest '40
Death Valley Rangers '44
Demon for Trouble '34
Durango Valley Raiders '38
El Diablo Rides '39
The Enforcer '51
Giant from the Unknown '58
Gun Lords of Stirrup Basin '37
The Gun Ranger '34
Kid Courageous '35
Kid Ranger '36
Last of the Warrens '36
The Law Rides '36
Lightning Bill Crandall '37
Man from Hell's Edges '32
Marked Trails '44
Mesquite Buckaroo '39
Mystery Squadron '33
No Man's Range '35
Northwest Trail '46
Of Mice and Men '39
Oklahoma Cyclone '30
Outlaw Trail '44
Paroled to Die '37
Pinto Canyon '40
Pork Chop Hill '59
Powdersmoke Range '35
The Red Rope '37
Rider of the Law '35
Riders of the Desert '32
Ridin' the Lone Trail '37
Rio Bravo '59
Rio Grande Raiders '46
Smokey Smith '36
Sonora Stagecoach '44

Sundown Saunders '36
Texas Trouble '30s
Thunder in the Desert '38
Tombstone Terror '34
Trail of Terror '35
Trails West '40s
Twilight on the Rio Grande '41
Western Double Feature 6:
    Rogers & Trigger '50
Western Justice '35
Wildfire '45
Young Blood '33

Fred Steele

Demolition '77

George Steele

Tall Texan '53

Henry Steele

Ninja Connection '90

Jadrien Steele

The Secret Garden '87

Karen Steele

Bail Out at 43,000 '57
Marty '55
The Rise and Fall of Legs
    Diamond '60
Trap on Cougar Mountain '72

Marjorie Steele

Face to Face '52
Tough Assignment '49

Pippa Steele

Lust for a Vampire '71
The Vampire Lovers '70

Tommy Steele

Finian's Rainbow '68
G-Men Never Forget '48
Half a Sixpence '67
The Happiest Millionaire '67

Irving Steen

The Fall of the House of Usher
    '52

Jessica Steen

John and the Missus '87
Sing '89

Stuart Steen

Ninja in Action '80s
Ninja Connection '90

Mary Steenburgen

Back to the Future, Part 3 '90
The Butcher's Wife '91
Cross Creek '83
Dead of Winter '87
End of the Line '88
Goin' South '78
Little Red Riding Hood '83
Melvin and Howard '80
A Midsummer Night's Sex
    Comedy '82
Miss Firecracker '89
One Magic Christmas '85
Parenthood '89
Ragtime '81
Romantic Comedy '83
Time After Time '79
Whales of August '87

**David
Steensland** (D)

Escapes '86

Jessica Steer

Workin' for Peanuts '85

Yvette Stefens

Carnal Crimes '91

Anthony Steffen

Escape from Hell '80s
Garringo '65
The Gentleman Killer '78
The Night Evelyn Came Out of
    the Grave '71
On the Edge '82
Savage Island '85
Stranger in Paso Bravo '73

Michael Stefoni

Trancers '85

Giorgio Stegani (D)

Beyond the Law '68

Bernice Stegers

City of Women '81
Frozen Terror '80
The Girl '86
Xtro '83

Renate Steiger

Boat Is Full '81

Rod Steiger

Across the Bridge '57
Al Capone '59
American Gothic '88
The Amityville Horror '79
Back from Eternity '56
The Ballad of the Sad Cafe
    '91
Breakthrough '78
Catch the Heat '87
The Court Martial of Billy
    Mitchell '55
Dirty Hands '76
Doctor Zhivago '65
F.I.S.T. '78
Fistful of Dynamite '72
Glory Boys '84
Guilty as Charged '91
Harder They Fall '56
Hennessy '75
The Illustrated Man '69
In the Heat of the Night '67
The January Man '89
Jubal '56
The Kindred '87
Klondike Fever '79
The Last Contract '86
Last Four Days '77
Lion of the Desert '81
The Longest Day '62
Love and Bullets '79
The Loved One '65
Lucky Luciano '74
The Mark '61
Marty '53
The Naked Face '84
No Way to Treat a Lady '68
Oklahoma! '55
On the Waterfront '54
Pawnbroker '65
Portrait of a Hitman '77
Run of the Arrow '56
The Sacred Music of Duke
    Ellington '82
Seven Thieves '60
Sword of Gideon '86
That Summer of White Roses
    '90
Time of Indifference '64
Unholy Wife '57
Waterloo '71
Wolf Lake '79

Chris Stein

Men of Respect '91

David Stein

The Amityville Curse '90

Herbert Stein (D)

Escape to Love '82

Ken Stein (D)

The Rain Killer '90

**Margaret Sophie
Stein**

Enemies, a Love Story '89
Sarah, Plain and Tall '91

Paul Stein (D)

Counterblast '48
Heart's Desire '37
Lottery Bride '30
Mimi '35

Sandor Stein (D)

Glitz '88

Ingeborg Steinbach

Train Station Pickups '70s

Victor Steinbach

Quiet Thunder '87

David Steinberg

Comedy Tonight '77
The End '78
Showbiz Goes to War '82
Something Short of Paradise
    '79

**David Max
Steinberg** (D)

Severance '88

David Steinberg (D)

Going Berserk '83
Paternity '81
Tall Tales & Legends: Casey
    at the Bat '85

Ziggy Steinberg (D)

The Boss' Wife '86

John Steiner

Ark of the Sun God '82
Beyond the Door 2 '79
Blood and Guns '79
Dagger Eyes '80s
Hunters of the Golden Cobra
    '82
The Last Hunter '80
The Lone Runner '88
Massacre in Rome '73
Operation 'Nam '86
Sinbad of the Seven Seas '89
Yor, the Hunter from the
    Future '83

Riley Steiner

Beginner's Luck '84

Sherry Steiner

Asylum of Satan '72
Heaven Help Us '85
Three on a Meathook '72

Sigfrit Steiner

The Last Chance '45

Jake Steinfeld

Home Sweet Home '80

**Danny
Steinmann** (D)

Friday the 13th, Part 5: A New
    Beginning '85
Savage Streets '83

Richard Steinmetz

Liquid Dreams '92

Albert Steinruck

The Golem '20

William Steis

Demon of Paradise '87
Equalizer 2000 '86
Eye of the Eagle '87
Raiders of the Sun '92

Shirley Stelfox

Personal Services '87

Anna Sten

The Girl with the Hat Box '27
So Ends Our Night '41
Soldier of Fortune '55

Brigitta Stenberg

Raiders of the Sun '92

Susan Stenmark

Innocent Prey '88

Steno (D)

Flatfoot '78

Yutte Stensgaard

Alien Women '69
Lust for a Vampire '71
Scream and Scream Again '70

Karel Stepanek

Counterblast '48
Second Best Secret Agent in
    the Whole Wide World '65
Sink the Bismarck '60

Nicole Stephane

Les Enfants Terrible '50

**Frederick
Stephani** (D)

Flash Gordon: Rocketship '36

A.C. Stephen (D)

Lady Godiva Rides '68
Orgy of the Dead '65
Saturday Night Sleazies, Vol.
    1 '66

Daniel Stephen

2020 Texas Gladiators '85
Warbus '85

Susan Stephen

Heat Wave '54
A Stolen Face '52
White Huntress '57

Ann Stephens

The Franchise Affair '52

Barbara Stephens

Hector's Bunyip '86

Harvey Stephens

Abe Lincoln in Illinois '40
Forlorn River '37
Let 'Em Have It '35
Oklahoma Kid '39

Peter Strauss

Angel on My Shoulder '80
Flight of Black Angel '91
Hail, Hero! '69
The Jericho Mile '79
The Last Tycoon '77
Man of Legend '71
Masada '81
Penalty Phase '86
Peter Gunn '89
Secret of NIMH '82
Soldier Blue '70
Spacehunter: Adventures in the Forbidden Zone '83
Trial of the Catonsville Nine '72
Whale for the Killing '81

Robert Strauss

The Atomic Kid '54
The Bridges at Toko-Ri '55
Family Jewels '65
The 4D Man '59
Frankie and Johnny '65
Girls! Girls! Girls! '62
I, Mobster '58
The Man with the Golden Arm '55
The Seven Year Itch '55
Stalag 17 '53

Frank Strayer (D)

Condemned to Live '35
Death from a Distance '36
The Ghost Walks '34
The Monster Walks '32
Murder at Midnight '31
The Vampire Bat '32

Meryl Streep

A Cry in the Dark '88
The Deer Hunter '78
Defending Your Life '91
Falling in Love '84
The French Lieutenant's Woman '81
Heartburn '86
Holocaust '78
Ironweed '87
Julia '77
Kramer vs. Kramer '79
Manhattan '79
Out of Africa '85
Plenty '85
Postcards from the Edge '90
The Seduction of Joe Tynan '79
She-Devil '89
Silkwood '83
Sophie's Choice '82
Still of the Night '82

David Street

Holiday Rhythm '50

Elliot Street

Harrad Experiment '73
Melvin Purvis: G-Man '74
Paper Man '71

Russell Streiner

Night of the Living Dead '68

Barbra Streisand

All Night Long '81
For Pete's Sake '74
Funny Girl '68
Funny Lady '75
Hello, Dolly! '69
The Main Event '79
Nuts '87
On a Clear Day You Can See Forever '70
The Owl and the Pussycat '70

The Prince of Tides '91
A Star is Born '76
Up the Sandbox '72
The Way We Were '73
What's Up, Doc? '72
Yentl '83

Barbra Streisand (D)

The Prince of Tides '91
Yentl '83

Melissa Stribling

Crow Hollow '52
The Horror of Dracula '58

Elaine Strich

Perfect Furlough '59

Joseph Strick (D)

The Balcony '63
Ulysses '67

Connie Strickland

Act of Vengeance '74
Bummer '73

Gail Strickland

The Gathering '77
The Gathering: Part 2 '79
Hyper-Sapien: People from Another Star '86
Lies '83
A Matter of Life and Death '81
One on One '77
Oxford Blues '84
Protocol '84
Rape and Marriage: The Rideout Case '80
Snowblind '78
Who'll Stop the Rain? '78

John Stride

Henry VIII '79

Karen Stride

The Vampire Hookers '78

Hal Strieb

Maxim Xul '91

Anita Strinberg

The Tempter '74
Women in Cell Block 7 '77

Elaine Stritch

Follies in Concert '85
Providence '77
September '88

Axel Strobye

Pelle the Conqueror '88

Herbert L. Strock (D)

Blood of Dracula '57
The Crawling Hand '63
The Devil's Messenger '62
How to Make a Monster '58
Man on the Run '74
Monster '78
Teenage Frankenstein '58
Witches' Brew '79

Woody Strode

Boot Hill '69
The Final Executioner '83
The Gatling Gun '72
Hired to Kill '73
Hit Men '73
The Italian Connection '73
Jaguar Lives '70s
Jungle Warriors '84

Kill Castro '80
Kingdom of the Spiders '77
Loaded Guns '75
Lust in the Dust '85
The Man Who Shot Liberty Valance '62
Manhunt '73
The Mercenaries '80
Oil '78
Once Upon a Time in the West '68
Pork Chop Hill '59
Professionals '66
Ride to Glory '71
Scream '83
Shalako '68
Spartacus '60
Super Brother '90
Sweet Dirty Tony '81
Violent Breed '83

Valeri Stroh

Mystery of Alexina '86

Tara Strohmeier

The Great Texas Dynamite Chase '76

William R. Stromberg (D)

The Crater Lake Monster '77

Tami Stronach

The NeverEnding Story '84

Andrew Strong

The Commitments '91

Eugene Strong

The Dropkick '27

Gwyneth Strong

Nothing But the Night '72

John Strong

The Executioner '78

Michael Strong

Patton '70
Point Blank '67
Queen of the Stardust Ballroom '75

James Strother

Heroes Die Young '60

Claude Stroud

Promises! Promises! '63

Don Stroud

The Amityville Horror '79
Armed and Dangerous '86
Bloody Mama '70
Buddy Holly Story '78
The Choirboys '77
Coogan's Bluff '68
Death Weekend '76
The Divine Enforcer '91
Down the Drain '89
Explosion '69
Express to Terror '79
Hollywood Man '76
Joe Kidd '72
Killer Inside Me '76
The King of the Kickboxers '91
Live a Little, Steal a Lot '75
Madigan '68
Murph the Surf '75
The Night the Lights Went Out in Georgia '81
Prime Target '91
Search and Destroy '81

Slaughter's Big Ripoff '73
Striking Back '81
Sudden Death '77
Sweet 16 '81
Twisted Justice '89
Two to Tango '88

Sheppard Strudwick

All the King's Men '49
Beyond a Reasonable Doubt '56
Cops and Robbers '73
Let's Dance '50
The Loves of Edgar Allen Poe '42
Psychomania '63
The Red Pony '49
Three Husbands '50

Joe Strummer

Candy Mountain '87
Mystery Train '89
Straight to Hell '87

Sally Struthers

The Getaway '72
A Gun in the House '81
Intimate Strangers '77

Carel Struycken

The Addams Family '91
Servants of Twilight '91
Witches of Eastwick '87

Hans Strydom

The Gods Must Be Crazy 2 '89

Amy Stryker

Impulse '84

Christopher Stryker

Hell High '86

Jonathan Stryker (D)

Curtains '83

Brian Stuart (D)

Sorceress '82

Cassie Stuart

Dolls '87
Hidden City '87

Eleanor Stuart

Oedipus Rex '57

Elizabeth Stuart

Policewomen '73

Giacomo Rossi Stuart

The Avenger '62
Kill, Baby, Kill '66

Gloria Stuart

Gold Diggers of 1935 '35
The Invisible Man '33
It Could Happen to You '39
My Favorite Year '82
The Poor Little Rich Girl '36
Rebecca of Sunnybrook Farm '38
Roman Scandals '33
The Two Worlds of Jenny Logan '79

Jack Stuart

Macabre '77
Planet on the Prowl '65

James R. Stuart

Reactor '85
War of the Robots '78

Jeanne Stuart

The Shadow '36

John Stuart

Alias John Preston '56
Candles at Nine '44
The Gilded Cage '54
Man on the Run '49
Number Seventeen '32
Old Mother Riley's Ghosts '41
The Temptress '49

Maxine Stuart

Coast to Coast '80
The Rousters '90

Mel Stuart (D)

The Chisholms '79
Happy Anniversary 007: 25 Years of James Bond '87
I Love My...Wife '70
Mean Dog Blues '78
Sophia Loren: Her Own Story '80
The Triangle Factory Fire Scandal '79
Willy Wonka & the Chocolate Factory '71

Randy Stuart

The Incredible Shrinking Man '57

Suzanne Stuart

Scream, Baby, Scream '69

Imogen Stubbs

Erik the Viking '89
Fellow Traveler '89
A Summer Story '88
True Colors '91

Stephen Stucker

Airplane! '80
Airplane 2: The Sequel '82
Delinquent School Girls '84

Johnny Stumper

Covert Action '88
Mob War '88

Melvin Sturdy

A Session with the Committee '69

John Sturges (D)

Alice Goodbody '74
Bad Day at Black Rock '54
By Love Possessed '61
Chino '75
The Eagle Has Landed '77
The Great Escape '63
Gunfight at the O.K. Corral '57
The Hallelujah Trail '65
Hour of the Gun '67
Ice Station Zebra '68
Joe Kidd '72
Last Train from Gun Hill '59
Magnificent Seven '60
Marooned '69
McQ '74
Never So Few '59
Underwater '55

Preston Sturges

Paris Holiday '57

Preston Sturges (D)

The Beautiful Blonde from
 Bashful Bend '49
Christmas in July '40
The Great McGinty '40
The Great Moment '44
Hail the Conquering Hero '44
Hail the Conquering Hero '44
The Lady Eve '41
Mad Wednesday '51
Miracle of Morgan's Creek '44
The Palm Beach Story '42
The Sin of Harold Diddlebock
 '47
Sullivan's Travels '41
Unfaithfully Yours '48

**Charles
Sturridge** (D)

Aria '88
Brideshead Revisited '81
A Handful of Dust '88

Darlene Stuto

Ms. 45 '81

Trudie Styler

Fair Game '89

Richard Styles

Terror on Tour '83

Richard Styles (D)

Escape '90
Shallow Grave '87

Bobby Suarez (D)

American Commandos '84
Pay or Dio '83
Warriors of the Apocalypse
 '85

Hector Suarez

Despedida de Casada '87
Diamante, Oro y Amor '87
Fin de Fiesta '71
Passion for Power '85

Jose Suarez

Carthage in Flames '60
Maria '68

Ricardo Suarez

Death of a Bureaucrat '66

Eliseo Subiela (D)

Man Facing Southeast '86

Barne Suboski

Danger Zone 3: Steel Horse
 War '90

David Suchet

Crime of Honor '80s
Gulag '85
Harry and the Hendersons '87
To Kill a Priest '89
When the Whales Came '89
A World Apart '88

Arne Sucksdorff

The Great Adventure '53

Arne Sucksdorff (D)

The Great Adventure '53

Kjell Sucksdorff

The Great Adventure '53

L. Sudjio (D)

Black Magic Terror '79

Susie Sudlow

Ark of the Sun God '82

Pak Ying-Mai Suet

Thousand Mile Escort

Ichiro Sugai

Early Summer (Bakushu) '51
Sansho the Bailiff '54

Kin Sugai

The Funeral '84

**Andrew
Sugarman** (D)

Basic Training '86

Bunta Sugawara

Tattooed Hit Man '76

Michael Sugich

The Night God Screamed '80s

Toshio Sugie (D)

Saga of the Vagabond '59

Haruko Sugimura

Drifting Weeds '59
Floating Weeds '59
Late Spring '49
Tokyo Story '53

**Tsuneharu
Sugiyama**

Mad Mission 3 '84

Barbara Sukowa

Berlin Alexanderplatz
 (episodes 1-2) '80
Berlin Alexanderplatz
 (episodes 3-4) '83
Berlin Alexanderplatz
 (episodes 5-6) '83
Berlin Alexanderplatz
 (episodes 7-8) '83
Deadly Game '83
The Sicilian '87
Voyager '92

Terrence Sul (D)

Wonsan Operation '78

Margaret Sullavan

Moon's Our Home '36
The Shining Hour '38
The Shop Around the Corner
 '40
So Ends Our Night '41

Barry Sullivan

Another Time, Another Place
 '58
The Bad and the Beautiful '52
Casino '80
Cause for Alarm '51
Earthquake '74
A Gathering of Eagles '63
The Human Factor '75
Hurricane '74
Kung Fu '72
Legend of the Sea Wolf '58
Maverick Queen '55
Nancy Goes to Rio '50
Night Gallery '69
No Room to Run '78
Oh, God! '77
Planet of the Vampires '65
Shark! '68
Skirts Ahoy! '52
Strategic Air Command '55
Take a Hard Ride '75

Tell Them Willie Boy Is Here
 '69
Texas Lady '56
Three Guys Named Mike '51
Washington Affair '77
The Woman of the Town '43
Yuma '70

Billy Sullivan

One-Punch O'Day '26
Stamp of a Killer '87

Bob Sullivan

Commando Attack '67

Brad Sullivan

Funny Farm '88
Guilty by Suspicion '91
Orpheus Descending '91

Don Sullivan

The Giant Gila Monster '59
The Rebel Set '59
Teenage Zombies '58

Ed Sullivan

Bye, Bye, Birdie '63
Early Elvis '56
A Few Moments with Buster
 Keaton and Laurel and
 Hardy '63
Toast of the Town '56

Francis L. Sullivan

Action for Slander '38
Behave Yourself! '52
Butler's Dilemma '43
Caesar and Cleopatra '45
Christopher Columbus '49
Tho Day Will Dawn '42
Great Expectations '46
Non-Stop New York '37
Oliver Twist '48
Pimpernel Smith '42
Power '34
The Prodigal '55
The Secret Four '40
Spy of Napoleon '36

Fred G. Sullivan (D)

Cold River '81

Jack Sullivan (D)

Charge of the Light Brigade
 '36

Jean Sullivan

Squirm '76

Jenny Sullivan

The Radical '75

Kevin Sullivan (D)

Anne of Avonlea '87
Anne of Green Gables '85
Lantern Hill '90
Wild Pony '83

Maureen Sullivan

Private Practices: The Story of
 a Sex Surrogate '86

Michael Sullivan

Great Ride '78

**Sean Gregory
Sullivan**

Howling 6: The Freaks '90

Sheila Sullivan

A Name for Evil '70

Susan Sullivan

City in Fear '80
Dark Ride '80s
Deadman's Curve '78
The Incredible Hulk '77
Marriage is Alive and Well '80
The Ordeal of Dr. Mudd '80

Tom Sullivan

Cocaine Cowboys '79

**William Bell
Sullivan**

Diamond Run '90

Frank Sully

The Tender Trap '55

Yma Sumac

Omar Khayyam '57

Bart Summer

Video Violence '87

Donna Summer

Hot Summer Night...with
 Donna '84

Diane Summerfield

Black Godfather '74
Blackjack '78

**Eleanor
Summerfield**

Black Glove '54
Man on the Run '49

**Jeremy
Summers** (D)

Five Golden Dragons '67
The House of 1,000 Dolls '67

**Walter
Summers** (D)

House of Mystery '41
The Human Monster '39

Roy Summerset

Cold Heat '90
Overkill '86

Slim Summerville

All Quiet on the Western Front
 '30
The Beloved Rogue '27
Captain January '36
Charlie Chaplin's Keystone
 Comedies #4 '14
The Farmer Takes a Wife '35
I'm from Arkansas '44
King of the Rodeo '28
Rebecca of Sunnybrook Farm
 '38
Under Montana Skies '30
Western Union '41

Bart Sumner

Video Violence Part 2...The
 Exploitation! '87

Joan Sumner

American Tickler '76

Donald Sumpter

Black Panther '77

Shirley Sun (D)

Iron & Silk '91

Wong Sun

Dragon Rider '85

Clinton Sundberg

As Young As You Feel '51
The Belle of New York '52
In the Good Old Summertime
 '49
Two Weeks with Love '50

Bjorn Sundquist

The Dive '89

Gerry Sundquist

Don't Open Till Christmas '84
Great Expectations '81

**Cedric
Sundstrom** (D)

American Ninja 3: Blood Hunt
 '89
Captive Rage '88
The Revenger '90

Neal Sundstrom (D)

Howling 5: The Rebirth '89

Asao Suno

Fires on the Plain '59

Jack A. Sunseri (D)

The Chilling '89

Bunny Sunshine

The Southerner '45

Superzan

Los Vampiros de Coyacan '87

The Supremes

Rock 'n' Roll History: Girls of
 Rock 'n' Roll '90

Nur Surer

Journey of Hope '90

Cristina Suriani

The Saga of the Draculas '72

Fred Surin (D)

The Bride Is Much Too
 Beautiful '58

Nicolas Surovy

The Act '82
Anastasia: The Mystery of
 Anna '86
Stark '85

Allan Surtees

Murder on Line One '90

Almanta Suska

Hunters of the Golden Cobra
 '82

Todd Susman

Star Spangled Girl '71

Henry Suso (D)

Death Sport '78

**Jean-Claude
Sussfeld** (D)

Elle Voit des Nains Partout '83

David Sust

In a Glass Cage '86

Eric Suter

The Fat Albert Christmas
 Special '77
The Fat Albert Halloween
 Special '77

Chief Ted Thin Elk

Thunderheart '92

Roy Thinnes

Code Name: Diamond Head
  '77
From Here to Eternity '79
Journey to the Far Side of the
  Sun '69
Rush Week '88
Satan's School for Girls '73
Secrets '77
Sizzle '81

Third World

Island Reggae Greats '85

Albie Thomas (D)

Palm Beach '79

Betty Thomas

Falling for the Stars '85
Homework '82
Troop Beverly Hills '89
When Your Lover Leaves '83

Billie "Buckwheat" Thomas

General Spanky '36
Our Gang Comedies '40s

B.J. Thomas

Jory '72

Damien Thomas

Mohammed: Messenger of
  God '77
Pirates '86
Shogun '80
Twins of Evil '71

Danny Thomas

Road to Lebanon '60
Side By Side '88

Dave Thomas

Love at Stake '87
Moving '88
My Man Adam '86
Sesame Street Presents:
  Follow That Bird '85
Strange Brew '83
Stripes '81

Dave Thomas (D)

The Experts '89
Strange Brew '83

Doris Thomas

The Demons '74

Doug Thomas

Hangmen '87

Frank M. Thomas

Law of the Underworld '38
Sunset on the Desert '42

Frankie Thomas

Boys Town '38
Tom Corbett, Space Cadet
  Volume 1

Gareth Thomas

Super Bitch '77
Visitor from the Grave '81

Gerald Thomas (D)

The Cariboo Trail '50
Carry On Behind '75
Carry On Cleo '65
Carry On Cruising '62

Carry On Dick '75
Carry On Doctor '68
Carry On Emmanuelle '78
Carry On England '76
Carry On Nurse '59
Carry On 'Round the Bend '72
Carry On Screaming '66
Carry On at Your
  Convenience '71
Follow That Camel '67
Time Lock '57
The Vicious Circle '57

Heather Thomas

Cyclone '87
Ford: The Man & the Machine
  '87
Red Blooded American Girl
  '90
Zapped! '82

Henry Thomas

Cloak & Dagger '84
E.T.: The Extra-Terrestrial '82
Misunderstood '84
Murder One '88
Psycho 4: The Beginning '91
The Quest '86
Raggedy Man '81
Valmont '89

Hugh Thomas

Last of the One Night Stands
  '83

Jameson Thomas

Beggars in Ermine '34
Convicted '32
The Farmer's Wife '28
Jane Eyre '34
Night Life in Reno '31
Sing Sing Nights '35

John G. Thomas (D)

Tin Man '83

John Thomas (D)

Arizona Heat '87
Banzai Runner '86

Kevin Thomas

Cleo/Leo '89

Kristin Scott Thomas

The Endless Game '89
Framed '90
A Handful of Dust '88
Spymaker: The Secret Life of
  Ian Fleming '90
Under the Cherry Moon '86

Kurt Thomas

Gymkata '85

Larry Thomas

Night Ripper '86

Lowell Thomas

Borneo '37

Mark Thomas

Working Girls '75

Marlo Thomas

Act of Passion: The Lost
  Honor of Kathryn Beck '83
Consenting Adults '85
Held Hostage '91
In the Spirit '90
Jenny '70

Philip Michael Thomas

Death Drug '83
A Fight for Jenny '90
Homeboy '80s
Miami Vice '84
Miami Vice 2: The Prodigal
  Son '85
Sparkle '76
Stigma '73
Streetfight '75
The Wizard of Speed and
  Time '88

Ralph L. Thomas (D)

Apprentice to Murder '88
The Terry Fox Story '83
Ticket to Heaven '81

Ralph Thomas (D)

Above Us the Waves '56
The Big Scam '79
Conspiracy of Hearts '60
Doctor in Distress '63
Doctor in the House '53
Doctor at Large '57
Doctor at Sea '56
It's Not the Size That Counts
  '74
No Love for Johnnie '60
Quest for Love '71
A Tale of Two Cities '58

Richard Thomas

All Quiet on the Western Front
  '79
Andy and the Airwave
  Rangers '89
Battle Beyond the Stars '80
Berlin Tunnel 21 '81
Cactus in the Snow '72
A Doll's House '59
Glory! Glory! '90
Hobson's Choice '83
Homecoming: A Christmas
  Story '71
Johnny Belinda '82
Last Summer '69
September 30, 1955 '77
Thanksgiving Story '73
The Todd Killings '71
The Waltons' Christmas Carol
  '80s
Winning '69

Robin Thomas

Memories of Murder '90
Personals '90

Ron Thomas

The Big Bet '85
Night Screams '87

Scott Thomas

Guns of the Magnificent
  Seven '69

Scott Thomas (D)

Silent Assassins '88

Terry Thomas

Danger: Diabolik '68

Tressa Thomas

The Five Heartbeats '91

William C. Thomas (D)

Big Town After Dark '47
Dark Mountain '44
Navy Way '44

Howard Thomashefsky

America '86

Harry Z. Thomason (D)

The Day It Came to Earth '77
Encounter with the Unknown
  '75
Hootch Country Boys '75

William Thomason

The Devil's Sleep '51

Lee Thomburg

Hollywood High Part 2 '81

Tim Thomerson

Air America '90
Cherry 2000 '88
Dollman '91
Fade to Black '80
Glory Years '87
Intimate Stranger '91
Iron Eagle '85
Jekyll & Hyde...Together
  Again '82
Metalstorm '83
Near Dark '87
Rhinestone '84
Terraces '77
A Tiger's Tale '87
Trancers '85
Trancers 2: The Return of
  Jack Deth '91
Trancers 3: Deth Lives '92
Uncommon Valor '83
Vietnam, Texas '90
Volunteers '85
Who's Harry Crumb? '90
The Wrong Guys '88
Zone Troopers '85

Beatrix Thompson

The Old Curiosity Shop '35

Blue Thompson

The Abomination '88

Brett Thompson (D)

Adventures in Dinosaur City
  '92

Brian Thompson

Commando Squad '87
Doctor Mordrid: Master of the
  Unknown '92
Hired to Kill '91
Lionheart '90

Carolyn Thompson

Three on a Meathook '72

Cindy Ann Thompson

Cave Girl '85

Derek Thompson

Belfast Assassin '84
The Long Good Friday '79

Don Thompson (D)

Ugly Little Boy '77

Donald W. Thompson (D)

Blood on the Mountain '88
Heaven's Heroes '88
Home Safe '88
Mannikin '77
The Prodigal Planet '88

The Shepherd '88
A Stranger in My Forest '88
Survival '88

Dorri Thompson

Chesty Anderson USN '76

Elizabeth Thompson

Metropolitan '90

Emma Thompson

Dead Again '91
Fortunes of War '90
Henry V '89
Impromptu '90
Look Back in Anger '89
The Tall Guy '89

Eric Thompson

One Day in the Life of Ivan
  Denisovich '71

Ernest Thompson (D)

1969 '89

Fred Dalton Thompson

Aces: Iron Eagle 3 '92
Curly Sue '91
Die Hard 2: Die Harder '90
Feds '88
The Hunt for Red October '90
Marie '85

Hal Thompson

Lassie from Lancashire '38

J. Lee Thompson (D)

The Ambassador '84
Battle for the Planet of the
  Apes '73
The Blue Knight '75
Cabo Blanco '81
Cape Fear '61
Conquest of the Planet of the
  Apes '72
Death Wish 4: The
  Crackdown '87
The Evil That Men Do '84
Firewalker '86
The Greek Tycoon '78
The Guns of Navarone '61
Happy Birthday to Me '81
Huckleberry Finn '74
King Solomon's Mines '85
MacKenna's Gold '69
Messenger of Death '88
Murphy's Law '86
Northwest Frontier '59
Reincarnation of Peter Proud
  '75
St. Ives '76
Taras Bulba '62
Ten to Midnight '83
Tiger Bay '59
The White Buffalo '77
The Widow '76

Jack Thompson

Bad Blood '87
Breaker Morant '80
Burke & Wills '85
Caddie '81
The Club '81
Earthling '80
Ground Zero '88
Jock Peterson '74
Man from Snowy River '82
Merry Christmas, Mr.
  Lawrence '83

Gun Crazy '69
The House on Garibaldi Street '79
Queenie '87
Sallah '63
War & Remembrance, Part 1 '88
Winds of War '83

Tom Topor (D)
Judgment '90

Ralph Toporoff (D)
American Blue Note '91

Peta Toppano
Echoes of Paradise '86
Which Way Home '90

Burt Topper (D)
Hell Squad '58
Soul Hustler '76
The Strangler '64

Paul Torcha
Clash of the Ninja '86

Terry Torday
Julia '74
Tower of Screaming Virgins '68

Dan Toren
Atalia '85

Marta Toren
Paris Express '53

Lyllah Torena
And When She Was Bad... '84

Sarah Torgov
American Gothic '88
If You Could See What I Hear '82

Peter Tork
Head '68

Mel Torme
Good News '47
A Man Called Adam '66
Words & Music '48

Rip Torn
Airplane 2: The Sequel '82
Another Pair of Aces: Three of a Kind '91
Baby Doll '56
Beastmaster '82
Beer '85
Betrayal '78
The Birch Interval '78
Blind Ambition '82
By Dawn's Early Light '89
Cat on a Hot Tin Roof '84
Cincinnati Kid '65
City Heat '84
Cold Feet '89
Coma '78
Cotter '72
Cross Creek '83
Defending Your Life '91
Dolly Dearest '92
The Execution '85
Extreme Prejudice '87
Flashpoint '84
Heartland '81
The Hit List '88
J. Edgar Hoover '87
Jinxed '82
King of Kings '61
Laguna Heat '87
The Man Who Fell to Earth '76

Manhunt for Claude Dallas '86
Misunderstood '84
Nadine '87
Nasty Habits '77
One Trick Pony '80
Payday '72
Pork Chop Hill '59
The President's Plane is Missing '71
The Private Files of J. Edgar Hoover '77
Rape and Marriage: The Rideout Case '80
The Seduction of Joe Tynan '79
A Shining Season '79
Silence Like Glass '90
Slaughter '72
Songwriter '84
Steel Cowboy '78
A Stranger is Watching '82
Summer Rental '85
Sweet Bird of Youth '61
24 Hours in a Woman's Life '61
You're a Big Boy Now '66

Rip Torn (D)
The Telephone '87

Pierre Tornade
Soldat Duroc...Ca Va Etre Ta Fete! '60s

Toni Tornado
Quilombo '84

Giuseppe Tornatore (D)
Cinema Paradiso '88
Everybody's Fine '91

Joe Tornatore (D)
Code Name: Zebra '84
Grotesque '87
Zebra Force '76

Regina Torne
El Cuarto Chino (The Chinese Room) '71
Los Asesinos '68

Elizabeth Toro
Sheena '84

Mari Torocsik
Love '72

Sarah Torov
Drying Up the Streets '76

Danil Torppe
Hanging on a Star '78

Ernest Torrance
Fighting Caravans '32
The Hunchback of Notre Dame '23
I Cover the Waterfront '33
Mantrap '26
Tol'able David '21

Robert Torrance (D)
Mutant on the Bounty '89

Joel Torre
Olongape: The American Dream '89

David Torrence
Jane Eyre '34

Ernest Torrence
The Covered Wagon '23
King of Kings '27
Steamboat Bill, Jr. '28

Ana Torrent
Blood and Sand '89
Cria '76
The Nest '81
Spirit of the Beehive '73

Diana Torres
Las Munecas del King Kong '80s

Fernanda Torres
One Man's War '90

Gabe Torres (D)
December '91

Liz Torres
America '86
Lena's Holiday '91

Miguel Contreras Torres (D)
Pancho Villa Returns '50

Cinzia Torrini (D)
Hotel Colonial '88

David Tors
Return to Africa '89

Ivan Tors
Return to Africa '89

Peter Tors
Return to Africa '89

Steven Tors
Return to Africa '89

Robert Torti
Alley Cat '84

Conrad Tortosa
Street Warriors, Part 2 '87

Sylvia Tortosa
When the Screaming Stops '73

Maria Tortuga
What Would Your Mother Say? '81

Jose Torvay
The Hitch-Hiker '53
My Outlaw Brother '51

Hal Tory
Earth vs. the Spider '58

Bruce Toscano (D)
The Jar '84

Luigi Tosi
The Black Devils of Kali '55

Stacey Toten
Lovelines '84

Leslie Toth
Buying Time '89
Unfinished Business '89

Szilvia Toth
A Hungarian Fairy Tale '87

Toto
Bellissimo: Images of the Italian Cinema '87
Big Deal on Madonna Street '60
The Gold of Naples '54
Hawks & the Sparrows (Uccellacci e Uccellini) '67
Passionate Thief '60

Robert Totten (D)
Dark Before Dawn '89
Huckleberry Finn '75
Pony Express Rider '76
The Red Pony '76
The Sacketts '79
The Wild Country '71

Merle Tottenham
Room to Let '49

Audrey Totter
FBI Girl '52
Jet Attack/Paratroop Command '58
Lady in the Lake '46
Murderer's Wife
The Postman Always Rings Twice '46
The Set-Up '49

Lorraine Touissant
A Case of Deadly Force '86

George Touliatos
Firebird 2015 A.D. '81
Robbers of the Sacred Mountain '83

Mikhail Toumanichulli (D)
Solo Voyage: The Revenge '90

Tamara Toumanova
Days of Glory '43
Invitation to the Dance '56

Jacques Tourneur (D)
Appointment in Honduras '53
Berlin Express '48
Cat People '42
The Comedy of Terrors '64
Curse of the Demon '57
Days of Glory '43
Easy Living '49
Experiment Perilous '44
The Flame & the Arrow '50
The Giant of Marathon '60
Great Day in the Morning '56
I Walked with a Zombie '43
The Leopard Man '43
Nightfall '56
Out of the Past '47

Maurice Tourneur (D)
A Girl's Folly '17
A Poor Little Rich Girl '17
Pride of the Clan '18
Trilby '17
Volpone '39
The Wishing Ring '14

Sheila Tousey
Thunderheart '92

Allen Toussaint
Piano Players Rarely Ever Play Together '89

Beth Toussaint
Blackmail '91

Lorraine Toussaint
Breaking In '89

Roland Toutain
Liliom '35
The Rules of the Game '39

Lupita Tova
Dracula (Spanish Version) '31
Old Spanish Custom '36

Wade Tower
The Ripper '86

Constance Towers
Naked Kiss '64
On Wings of Eagles '86
Shock Corridor '63
Sylvester '85

Tom Towles
The Borrower '89
Henry: Portrait of a Serial Killer '90
Night of the Living Dead '90

Aline Towne
Radar Men from the Moon '52
The Vanishing Westerner '50
Zombies of the Stratosphere '52

Robert Towne
The Pick-Up Artist '87
Villa Rides '68

Robert Towne (D)
Personal Best '82
Tequila Sunrise '88

Ralph Towner
Oregon '88

Harry Townes
The Warrior & the Sorceress '84

Bud Townsend (D)
Alice in Wonderland '77
Coach '78
Nightmare in Wax '69
Terror at Red Wolf Inn '72

Jill Townsend
Oh Alfie '75

Patrice Townsend
Always '85
Sitting Ducks '80

Patrice Townsend (D)
Beach Girls '82

Pete Townsend
Jimi Hendrix: Story '73
Monterey Pop '68
Pete Townshend: White City '85
Ready Steady Go, Vol. 1 '83
Ready Steady Go, Vol. 2 '85
The Secret Policeman's Other Ball '82
Secret Policeman's Private Parts '81

Tommy '75
Woodstock '70

Robert Townsend

American Flyers '85
The Five Heartbeats '91
It's Not Easy Bein' Me '87
The Mighty Quinn '89
Odd Jobs '85
Ratboy '86
A Soldier's Story '84
Streets of Fire '84
That's Adequate '90
Uptown Comedy Express '89

Robert Townsend (D)

Eddie Murphy: Raw '87
The Five Heartbeats '91
Hollywood Shuffle '87

Ian Toynton (D)

The Maid '90

Shiro Toyoda (D)

Mistress (Wild Geese) '53
Snow Country '57

Fausto Tozzi

East of Kilimanjaro '59
The Return of Dr. Mabuse '61
Sicilian Connection '74

Giorgio Tozzi

Torn Between Two Lovers '79

Benno Trachtmann (D)

Jumper '85

Brooke Tracy

Twisted '86

Emerson Tracy

Champagne for Breakfast '79

Lee Tracy

The Best Man '64
Betrayal from the East '44
Bombshell '33
Doctor X '32
The Payoff '43

Marlene Tracy

I Dismember Mama '74

Spencer Tracy

Adam's Rib '49
Bad Day at Black Rock '54
Boom Town '40
Boys Town '38
Broken Lance '54
Captains Courageous '37
Desk Set '57
Devil at 4 O'Clock '61
Dr. Jekyll and Mr. Hyde '41
Father of the Bride '50
Father's Little Dividend '51
Guess Who's Coming to
    Dinner '67
A Guy Named Joe '44
Hepburn & Tracy '84
How the West was Won '62
Inherit the Wind '60
It's a Mad, Mad, Mad, Mad
    World '63
Judgment at Nuremberg '61
Last Hurrah '58
Libeled Lady '36

Mannequin '37
Marie Galante '34
Men of Boys Town '41
The Mountain '56
Northwest Passage '40
Pat and Mike '52
San Francisco '36
Stanley and Livingstone '39
State of the Union '48
Test Pilot '39
Thirty Seconds Over Tokyo
    '44
Tortilla Flat '42
Woman of the Year '42

Steve Tracy

Desperate Moves '86

William Tracy

Angels with Dirty Faces '38
As You Were '51
Mr. Walkie Talkie '52
Strike Up the Band '40
To the Shores of Tripoli '42

Marie Trado

Ten Nights in a Bar Room '13

Mary Ellen Trainer

The Goonies '85
Tales from the Crypt '89

Katheryn Trainor

Walking on Air '87

Eugene Trammel

Mafia vs. Ninja '84

Jean-Claude Tramont (D)

All Night Long '81
As Summers Die '86

Deborah Tranelli

Naked Vengeance '85

Cordula Trantow

Castle '68
Hitler '62

Gordy Trapp

Sweet Country Road '81

Ami Traub

Drifting '82

Helen Traubel

Deep in My Heart '54
The Ladies' Man '61

Daniel J. Travanti

Adam '83
A Case of Libel '83
Fellow Traveler '89
Megaville '91
Midnight Crossing '87
Millenium '89
Murrow '85
St. Ives '76
Tagget '90

Alfred Travers (D)

Dual Alibi '47

Bill Travers

Born Free '66
Browning Version '51
Christian the Lion '76
Duel at Diablo '66
Gorgo '61
Ring of Bright Water '69
Romeo and Juliet '54

The Smallest Show on Earth
    '57
Trio '50

Bill Travers (D)

Christian the Lion '76

Henry Travers

Ball of Fire '42
The Bells of St. Mary's '45
Dark Victory '39
Dodge City '39
Dragon Seed '44
A Girl, a Guy and a Gob '41
High Sierra '41
It's a Wonderful Life '46
Madame Curie '43
Mrs. Miniver '42
The Naughty Nineties '45
Primrose Path '40
Random Harvest '42
Shadow of a Doubt '42
Stanley and Livingstone '39
Thrill of a Romance '45
The Yearling '46

Marisa Travers

Frankenstein '80 '79

Roy Travers

Q Ships '28

Susan Travers

The Abominable Dr. Phibes
    '71
Peeping Tom '63
The Snake Woman '61

Madeline Traverse

A Poor Little Rich Girl '17

Susana Traverso

Barbarian Queen '85

Greg Travis

Paradise '91

Henry Travis

The Brain from Planet Arous
    '57

June Travis

Earthworm Tractors '36
Exiled to Shanghai '37
Monster a Go-Go! '65

Nancy Travis

Air America '90
Internal Affairs '90
Loose Cannons '90
Three Men and a Baby '87
Three Men and a Little Lady
    '90

Richard Travis

Big Town After Dark '47
Danger Zone '51
Fingerprints Don't Lie '51
The Man Who Came to Dinner
    '41
Mask of the Dragon '51
Mesa of Lost Women '52
Missile to the Moon '59
Roaring City '51
Sky Liner '49

Stacy Travis

Dr. Hackenstein '88
Hardware '90

Joey Travolta

American Born '89
Car Crash
Ghost Writer '89

The Prodigal '83
Rosie: The Rosemary Clooney
    Story '82
Sinners '89
Sunnyside '79
Wilding '90

John Travolta

Blow Out '81
Boy in the Plastic Bubble '76
Carrie '76
Devil's Rain '75
Dumb Waiter '87
The Experts '89
Grease '78
Look Who's Talking '89
Look Who's Talking, Too '90
Olivia: Twist of Fate '84
Perfect '85
Saturday Night Fever '77
Shout '91
Staying Alive '83
Two of a Kind '83
Urban Cowboy '80

Stephen Traxler (D)

Slithis '78

William Traylor

One Plus One '61

Peter S. Traynor (D)

Death Game '76
Seducers '77

Arthur Treacher

Curly Top '35
Delightfully Dangerous '45
Heidi '37
I Live My Life '35
The Little Princess '39
A Midsummer Night's Dream
    '35
National Velvet '44
Stowaway '36

Emerson Treacy

A Star is Born '54

Terri Treas

Deathstalker 3 '89
House 4 '91
The Nest '88
The Terror Within '88

C. Court Treatt (D)

Bo-Ru the Ape Boy '30s

Tree

Clown House '88

David Tree

Drums '38

Dorothy Tree

Abe Lincoln in Illinois '40
Hitler: Dead or Alive '43
Mystery of Mr. Wong '39

Mary Treen

Clipped Wings '53
Hands Across the Border '43
I Married a Monster from
    Outer Space '58
Paradise, Hawaiian Style '66
The Sad Sack '57
Swing It, Sailor! '37

Lisa Trego

It's Called Murder, Baby '82

Norma Trelvar (D)

A Nation Aflame '37

Les Tremayne

The Angry Red Planet '59
Fangs '75
The Gallant Hours '60
I Love Melvin '53
A Man Called Peter '55
The Monolith Monsters '57
The Monster of Piedras
    Blancas '57
The Slime People '63
The War of the Worlds '53

Johanne-Marie Tremblay

Jesus of Montreal '89

Joseph Tremice

Autobiography of Miss Jane
    Pittman '74

Brian Trenchard-Smith (D)

BMX Bandits '83
Day of the Assassin '81
Day of the Panther '88
Dead End Drive-In '86
Escape 2000 '81
Fists of Blood '87
Out of the Body '88
The Quest '86
The Siege of Firebase Gloria
    '89

Kim Trengove

Fair Game '85

Luis Trenker

The Challenge '38

Jean Trent

Western Mail '42

John Trent

Mystery Plane '39
Stunt Pilot '39

John Trent (D)

Best Revenge '83
Crossbar '79
Find the Lady '75
Middle Age Crazy '80
Vengeance Is Mine '84

Cheryl Trepton

Mantis in Lace '68

James Treuer

Walking on Air '87

Noel Trevarthen

Dusty '85

Frederick Treves

Flame to the Phoenix '85
Paper Mask '91

Leonardo Treviglio

Sebastiane '79

Roger Treville

The Green Glove '52
Mr. Peek-A-Boo '50

Dale Trevillion (D)

Las Vegas Weekend '85
One Man Force '89

Witches' Brew '79
Ziegfeld Girl '41

Tierre Turner

Cornbread, Earl & Me '75

Tina Turner

Mad Max Beyond
    Thunderdome '85
Tommy '75

Nadyne Turney

The Arousers '70

Ben Turpin

Ben Turpin: An Eye for an Eye
    '24
Burlesque of Carmen '16
Cannon Ball/Eyes Have It '20s
The Champion/His New Job
    '15
Chaplin Essanay Book 1 '15
Charlie Chase and Ben Turpin
    '21
Chasing Those Depression
    Blues '35
Desperate Scoundrel/The
    Pride of Pikeville '22
Eyes of Turpin Are Upon You!
    '21
Fun Factory/Clown Princes of
    Hollywood '20s
Golden Age of Comedy '58
Laughfest '10s
Laurel & Hardy Comedy
    Classics Volume 6 '35
Mack Sennett Comedies:
    Volume 1 '21
Saps at Sea '40
Silent Laugh Makers, No. 1
    '20s
Slapstick '30s
When Comedy was King '59
Yankee Doodle in Berlin '19

Aida Turturro

True Love '89

John Turturro

Back Track '91
Barton Fink '91
Brain Donors '92
Color of Money '86
Desperately Seeking Susan
    '85
Do the Right Thing '89
Five Corners '88
Gung Ho '85
Hannah and Her Sisters '86
Jungle Fever '91
Men of Respect '91
Miller's Crossing '90
Mo' Better Blues '90
Off Beat '86
The Sicilian '87
State of Grace '90
To Live & Die in L.A. '85

Rita Tushingham

Doctor Zhivago '65
Dream to Believe '80s
Dressed for Death '74
Green Eyes '76
The Housekeeper '86
The Human Factor '75
Leather Boys '63
Loose in New York '80s
Mysteries '84
Rachel's Man '75
Slaughterday '77
Spaghetti House '82
A Taste of Honey '61
Trap '66

Dorothy Tutin

The Shooting Party '85
Sister Dora '77
A Tale of Two Cities '58

Frank Tuttle *(D)*

Hell on Frisco Bay '55
Love 'Em and Leave 'Em '27
Lucky Devil '25
Roman Scandals '33
This Gun for Hire '42

Lurene Tuttle

Don't Bother to Knock '52
The Fortune Cookie '66
The Glass Slipper '55
Ma Barker's Killer Brood '60
White Mama '80

Michael Twaine

Cheap Shots '91

Shannon Tweed

Cannibal Women in the
    Avocado Jungle of Death
    '89
Codename: Vengeance '87
The Firing Line '91
Hitchhiker 4 '87
Hot Dog...The Movie! '83
In the Cold of the Night '89
Last Call '90
The Last Fling '86
The Last Hour '91
Lethal Woman '88
Meatballs 3 '87
The Night Visitor '89
Of Unknown Origin '83
Sexual Response '92
Surrogate '88
Twisted Justice '89

Terry Tweed

Night Rhythms '90s
The Reincarnate '71

Helen Twelvetrees

Millie '31
Painted Desert '31
State's Attorney '31

Jason Twelvetrees

Not Tonight Darling '72

Twiggy

The Blues Brothers '80
The Boy Friend '71
Butterfly Ball '76
Club Paradise '86
The Doctor and the Devils '85
Istanbul '90
Little Match Girl '87
Madame Sousatzka '88
W '74
Young Charlie Chaplin '88

**Twinkie Clark & the
Clark Sisters**

Gospel '82

Derek Twist *(D)*

Green Grow the Rushes '51

Archie Twitchell

Thundering Trail '51

Pat Twohill

Forty Thousand Horsemen
    '41

Anne Twomey

Behind Enemy Lines '85

Deadly Friend '86
Last Rites '88
Orpheus Descending '91

Dominique Tyawa

River of Diamonds '90

Jean Tych *(D)*

Great Expectations '83

Jeff Tyler

The Adventures of Tom
    Sawyer '73

Judy Tyler

Jailhouse Rock '57

Tom Tyler

Adventures of Captain Marvel
    '41
Battling with Buffalo Bill '31
Brothers of the West '37
Cheyenne Rides Again '38
Coyote Trails '35
Fast Bullets '44
Feud of the Trail '37
Guns of Justice '50
Laramie Kid '35
Last Outlaw '36
The Mummy's Hand '40
Phantom of the West '31
Pinto Rustlers '36
Powdersmoke Range '35
Riders of the Timberline '41
Ridin' Thru '35
Riding On '37
Rio Rattler '35
Rip Roarin' Buckaroo '36
Roamin' Wild '36
Santa Fe Bound '37
Silent Code '35
Silent Valley '35
Two Fisted Justice '31
Valley of the Sun '42
When a Man Rides Alone '33

Charles Tyner

Harold and Maude '71
Planes, Trains & Automobiles
    '87
Pulse '88

Glenn Tyron *(D)*

Law West of Tombstone '38

Susan Tyrrell

Andy Warhol's Bad '77
Angel '84
Another Man, Another Chance
    '77
Avenging Angel '85
Big Top Pee Wee '88
Cry-Baby '90
Far from Home '89
Fast Walking '82
Fat City '72
Fire and Ice '83
Flesh and Blood '85
Forbidden Zone '80
Islands in the Stream '77
Killer Inside Me '76
Lady of the House '78
Liar's Moon '82
Loose Shoes '80
Night Warning '82
The Offspring '87
Poker Alice '87
Rockula '90
September 30, 1955 '77
Tales of Ordinary Madness
    '83
Tapeheads '89
Thompson's Last Run '90
The Underachievers '88

Cathy Tyson

Business as Usual '88
Mona Lisa '86
The Serpent and the Rainbow
    '87

Cicely Tyson

Acceptable Risks '86
Autobiography of Miss Jane
    Pittman '74
Bustin' Loose '81
The Concorde: Airport '79 '79
Fried Green Tomatoes '91
The Heart is a Lonely Hunter
    '68
Heat Wave '90
Hero Ain't Nothin' But a
    Sandwich '78
King '78
A Man Called Adam '66
The River Niger '76
Roots, Vol. 1 '77
Samaritan: The Mitch Snyder
    Story '86
Sounder '72
Wilma '77
A Woman Called Moses '78
The Women of Brewster
    Place '89

Richard Tyson

The Babe '91
Kindergarten Cop '90
Three O'Clock High '87
Two Moon Junction '88

Margaret Tyzack

Mr. Love '86
Quatermass Conclusion '79

UB40

Nelson Mandela 70th Birthday
    Tribute '88

Tomu Uchida *(D)*

Swords of Death '71

**Seiichiro
Uchikawa** *(D)*

One-Eyed Swordsman '63

Fabiana Udenio

Bride of Re-Animator '89
Hardbodies 2 '86

Chris Udvarnoky

The Other '72

Martin Udvarnoky

The Other '72

Claudia Udy

Captive Rage '88
Out of Control '85
The Pink Chiquitas '86

Helene Udy

The Hollywood Detective '89

Bob Uecker

Major League '89

Kichijiro Ueda

Gamera vs. Gaos '67

Julian Ugarte

Fangs of the Living Dead '68

Leslie Uggams

Poor Pretty Eddie '73
Sizzle '81

Two Weeks in Another Town
    '62

Peter Ujaer

The Power of the Ninjitsu '88

Liv Ullman

Autumn Sonata '78
The Bay Boy '85
A Bridge Too Far '77
Cold Sweat '74
Cries and Whispers '72
Dangerous Moves '84
Forty Carats '73
Gaby: A True Story '87
Hour of the Wolf '68
Leonor '75
Mindwalk '91
The Night Visitor '70
Persona '48
Richard's Things '80
The Rose Garden '89
Scenes from a Marriage '73
The Serpent's Egg '78
The Wild Duck '84

Tracy Ullman

Give My Regards to Broad
    Street '84
I Love You to Death '90
Jumpin' Jack Flash '86
Plenty '85
Three of a Kind '80s

Edgar G. Ulmer *(D)*

The Amazing Transparent
    Man '60
Beyond the Time Barrier '60
The Black Cat '34
Bluebeard '44
The Daughter of Dr. Jekyll '57
Detour '46
Girls in Chains '43
Isle of Forgotten Sins '43
Jive Junction '43
Journey Beneath the Desert
    '61
The Light Ahead '39
Moon Over Harlem '39
Naked Venus '58
St. Benny the Dip '51
The Singing Blacksmith (Yankl
    der Shmid) '38
Strange Illusion '45
The Strange Woman '46
Thunder Over Texas '34

George Ulmer

Invitation to Paris '60

Lenore Ulric

Camille '36
Northwest Outpost '47

Mikhail Ulyanov

Private Life '83

Martin Umbach

The NeverEnding Story 2:
    Next Chapter '91

Margaret Umbers

The Bridge to Nowhere '86
Death Warmed Up '85

Jolanta Umecka

Knife in the Water '62

Tomoko Umeda

Godzilla on Monster Island '72
Godzilla vs. Gigan '72

Miyoshi Umeki

Flower Drum Song '61

Jeanne Valeri

Dangerous Liaisons '60
Desert Commandos '60s

Joan Valerie

Charlie Chan at the Wax
Museum '40

Tonino Valerii (D)

My Name Is Nobody '74
A Reason to Live, A Reason
to Die '73

Pilar Valesquez

Sex on the Brain

Rosa Valetti

The Blue Angel '30

Frederick Valk

Dangerous Moonlight '41
Dead of Night '45
Frenzy '46
Hotel Reserve '44

Kitty Vallacher

Grave of the Vampire '72
The Legend of Frank Woods
'77

Louise Vallance

Robbers of the Sacred
Mountain '83

Vic Vallard

Hollywood in Trouble '87

Marcel Vallee

The Crazy Ray '22
Topazo '33

Rudy Vallee

The Admiral was a Lady '50
The Bachelor and the Bobby-
Soxer '47
The Beautiful Blonde from
Bashful Bend '49
Glorifying the American Girl
'30
How to Succeed in Business
without Really Trying '67
I Remember Mama '48
International House '33
Live a Little, Love a Little '68
Mad Wednesday '51
My Dear Secretary '49
The Palm Beach Story '42
People Are Funny '46
The Sin of Harold Diddlebock
'47
Slashed Dreams '74
Unfaithfully Yours '48
Vagabond Lover '29

Al Valletta

Runaway Nightmare '84

Alida Valli

Beyond Erotica '79
The Cassandra Crossing '76
Edipo Re (Oedipus Rex) '67
Horror Chamber of Dr.
Faustus '59
The Miracle of the Bells '48
1900 '76
Oedipus Rex '67
The Paradine Case '47
Senso '68
The Spider's Stratagem '70
Suspiria '77
The Tempter '74
The Third Man '49
Walk Softly, Stranger '50

The Wanton Contessa '55
We the Living '42

Frankie Valli

Dirty Laundry '87
Eternity '90
Modern Love '90

Joe Valli

Typhoon Treasure '39

Romolo Valli

Bobby Deerfield '77
Fistful of Dynamite '72

Virginia Valli

Lost Zeppelin '29
Night Life in Reno '31
The Shock '23

Rick Vallin

Corregidor '43
King of the Stallions '42
The Panther's Claw '42

Elenora Vallone

Green Horizon '83
La Fuga

Raf Vallone

An Almost Perfect Affair '79
Christopher Columbus '91
El Cid '61
The Girl in Room 2A '76
The Greek Tycoon '78
A Gunfight '71
Harlow '65
Honor Thy Father '73
The Human Factor '75
The Italian Job '69
La Violetera (The Violet) '71
Lion of the Desert '81
Nevada Smith '66
The Other Side of Midnight
'77
Ricco '74
The Scarlet & the Black '83
That Lucky Touch '75
Time to Die '83
Two Women '61

Jean Valmont

Erotic Touch of Hot Skin '65

Fernand Valois

The Brain '69

Maximo Valverde

The Barcelona Kill '70s

Vampira

Plan 9 from Outer Space '56

**Robert Van
Ackeren**

A Woman in Flames '84

**Robert Van
Ackeren** (D)

A Woman in Flames '84

**Willeke Van
Ammelrooy**

Fatal Error '83
The Lift '85
My Nights With Susan,
Sandra, Olga, and Julie

Lee Van Atta

Dick Tracy: The Original
Serial, Vol. 1 '37

Dick Tracy: The Original
Serial, Vol. 2 '37

Ingrid van Bergen

The Avenger '60

Lewis Van Bergen

Overthrow '82
Rage of Honor '87
South of Reno '87

Bobby Van

Escape from Planet Earth '67
Kiss Me Kate '53
Navy vs. the Night Monsters
'66
Small Town Girl '53

**Tomas Van
Bromssen**

My Life As a Dog '87

Mabel van Buren

The Four Horsemen of the
Apocalypse '21

Lee Van Cleef

Armed Response '86
Bad Man's River '72
The Beast from 20,000
Fathoms '53
Beyond the Law '68
Big Combo '55
The Bravados '58
Captain Apache '71
China Gate '57
Codename: Wildgeese '84
Commandos '73
The Conqueror '56
El Condor '70
Escape from Death Row '76
Escape from New York '81
Fatal Chase '77
For a Few Dollars More '65
God's Gun '75
The Good, the Bad and the
Ugly '67
Gunfight at the O.K. Corral '57
Hard Way '80
The Heist '88
High Noon '52
It Conquered the World '56
Jungle Raiders '85
Kansas City Confidential '52
Kid Vengeance '75
A Man Alone '55
Master Ninja '78
Master Ninja 2 '83
Master Ninja 3 '83
Master Ninja 4 '83
Master Ninja 5 '83
Mean Frank and Crazy Tony
'75
New Mafia Boss '72
Octagon '80
Perfect Killer '77
The Rip Off '78
Speed Zone '88
The Squeeze '80
The Stranger and the
Gunfighter '76
Take a Hard Ride '75
Thieves of Fortune '89
Tin Star '57
The Young Lions '58

Ron Van Cliff

Black Dragon '74
Black Dragon's Revenge '75
Fight to the Death '83
Fist of Fear, Touch of Death
'77
Super Weapon '75
Way of the Black Dragon '81

Jose Van Dam

The Music Teacher '88

**Jean-Claude Van
Damme**

Black Eagle '88
Bloodsport '88
Cyborg '89
Death Warrant '90
Double Impact '91
Kickboxer '89
Lionheart '90
No Retreat, No Surrender '86

David Van Day

Screamtime '83

**Monique Van De
Ven**

Amsterdamned '88
Katie's Passion '75
Keetje Tippei '75
The Man Inside '90
Paint it Black '89

**Gertrude Van Der
Berger**

The Naked Prey '65

**Nadine Van Der
Velde**

After Midnight '89
Munchies '87
Shadow Dancing '88

**Kelley Van Der
Velden**

I Love N Y '87

Diana Van Der Vlis

Girl in Black Stockings '57
Lovespell '79
X: The Man with X-Ray Eyes
'63

Keith Van Der Wat

Three Bullets for a Long Gun
'73

Jim Van Der Woude

The Pointsman '88

**Richard Van Der
Wyk**

Skateboard '77

Trish Van Devere

All God's Children '80
The Changeling '80
The Day of the Dolphin '73
The Hearse '80
Hollywood Vice Sqaud '86
Messenger of Death '88
Movie, Movie '78
Savage is Loose '74
Stalk the Wild Child '76
Uphill All the Way '85
Where's Poppa? '70

Peter Van Dissel

Funeral for an Assassin '77

Mamie Van Doren

Girl in Black Stockings '57
High School Confidential '58
Las Vegas Hillbillys '66
Navy vs. the Night Monsters
'66
Revenge of TV Bloopers '60s
Teacher's Pet '58

Three Nuts in Search of a Bolt
'64
Voyage to the Planet of
Prehistoric Women '68

Ton Van Dort

The Pointsman '88

**Bruce Van
Dusen** (D)

Cold Feet '84

**Granville Van
Dusen**

Breaking Up '78
Hotline '82
The Rose and the Jackal '90
War Between the Tates '76
The Wild and the Free '80

Barry Van Dyke

Ants '77
Foxfire Light '84

Conny Van Dyke

Framed '75
Hell's Angels '69 '69

Dick Van Dyke

Bye, Bye, Birdie '63
Chitty Chitty Bang Bang '68
Cold Turkey '71
The Comic '69
Country Girl '82
Dick Tracy '90
Lt. Robin Crusoe, U.S.N. '66
Mary Poppins '64
Never a Dull Moment '68
Revenge of TV Bloopers '60s
The Runner Stumbles '79

Jerry Van Dyke

The Courtship of Eddie's
Father '62
Death Blow '87
Johnny Carson '70s
Run If You Can '87

W.S. Van Dyke (D)

After the Thin Man '36
Andy Hardy Gets Spring Fever
'39
Another Thin Man '39
Bitter Sweet '41
Forsaking All Others '35
Hide-Out '34
I Live My Life '35
I Married an Angel '42
Manhattan Melodrama '34
Marie Antoinette '38
Naughty Marietta '35
Rosalie '38
Rose Marie '36
San Francisco '36
Shadow of the Thin Man '41
Sweethearts '38
Tarzan, the Ape Man '32
The Thin Man '34

Peter Van Eyck

The Brain '62
Dr. Mabuse vs. Scotland Yard
'64
The Spy Who Came in from
the Cold '65
The Thousand Eyes of Dr.
Mabuse '60
Wages of Fear '55

John Van Eyssen

Four Sided Triangle '53
The Horror of Dracula '58
Men of Sherwood Forest '57

The Ernest Film Festival '86
Ernest Goes to Camp '87
Ernest Goes to Jail '90
Ernest Saves Christmas '88
Ernest Scared Stupid '92
Fast Food '89
Hey Vern! It's My Family
  Album '85
The Rousters '90

Reg Varney

Great St. Trinian's Train
  Robbery '66

Roland Varno

Return of the Vampire '43

Kenneth Varnum

Dangerous Relations '73

Diane Varsi

Bloody Mama '70
I Never Promised You a Rose
  Garden '77
Johnny Got His Gun '71
The People '71
Peyton Place '57
Sweet Love, Bitter '67
Ten North Frederick '58
Wild in the Streets '68

Henri Vart (D)

Secret Obsession '88

Vagelis Varton

Emmanuelle's Daughter '79

Magda Vasarykova

On the Comet '68

**Gregory
Vasiliev** (D)

Chapayev '34

**Jose Luis Lopez
Vasquez**

The Garden of Delights '70

Joseph P. Vasquez

The Bronx War '91

**Joseph P.
Vasquez** (D)

The Bronx War '91
Hangin' with the Homeboys
  '91

Roberta Vasquez

Do or Die '91
Easy Wheels '89
The Rookie '90
Street Asylum '90

Queenie Vasser

Primrose Path '40

Woody Vasulka (D)

Commission '70s

Robert Vattier

The Baker's Wife (La Femme
  du Boulanger) '33
Letters from My Windmill '54

Alberta Vaughan

The Dropkick '27
Laramie Kid '35
Randy Rides Alone '34

Frankie Vaughan

Let's Make Love '60

Peter Vaughan

Bleak House '85
Blockhouse '73
Brazil '86
Coming Out of the Ice '82
Die! Die! My Darling! '65
Forbidden '85
Haunted Honeymoon '86
Mackintosh Man '73
The Man Outside '68
Mountains of the Moon '90
The Razor's Edge '84
Straw Dogs '72
Time Bandits '81

Sarah Vaughan

Basin Street Revue '55

Vanessa Vaughan

Crazy Moon '87

Alberta Vaughn

All Night Long/Smile Please
  '24
The Live Wire '34
Wild Horse '31

Ashley Vaughn

Incoming Freshmen '79

Heidi Vaughn

Heaven's Heroes '88
Trained to Kill, U.S.A. '83

Robert Vaughn

Battle Beyond the Stars '80
Black Moon Rising '86
Blind Vision '91
Blue Jeans and Dynamite
Brass Target '78
The Bridge at Remagen '69
Brutal Glory '89
Bullitt '68
Buried Alive '89
Captive Rage '88
C.H.U.D. 2: Bud the Chud '89
City in Fear '80
Delta Force '86
Demon Seed '77
Emissary '89
The Glass Bottom Boat '66
Hangar 18 '80
Hitchhiker 4 '87
Hour of the Assassin '86
Inside the Third Reich '82
Julius Caesar '70
Kill Castro '80
The Last Bastion '84
The Lucifer Complex '78
Magnificent Seven '60
The Mercenaries '80
Nightstick '87
Nobody's Perfect '90
Prince of Bel Air '87
Question of Honor '80
The Rebels '79
Return of the Man from
  U.N.C.L.E. '83
River of Death '90
The Shaming '71
Skeleton Coast '89
S.O.B. '81
The Statue '71
Superman 3 '83
Sweet Dirty Tony '81
Teenage Caveman '58
That's Adequate '90
Towering Inferno '74
Transylvania Twist '89
Virus '82
Wanted: Babysitter '75
The Woman Hunter '72
The Young Philadelphians '59

Marcus Vaughter

Visitants '87

Natalie Vavilova

Moscow Does Not Believe in
  Tears '80

Ron Vawter

Internal Affairs '90
sex, lies and videotape '89
Sudden Death '77

Selma Vaz Dias

The Tell-Tale Heart '62

Bruno Ve Sota

Attack of the Giant Leeches
  '59
The Choppers '61
Creature of the Walking Dead
  '65
Daughter of Horror '55
The Gunslinger '56
The Wasp Woman '49

Bruno Ve Sota (D)

The Brain Eaters '58
The Female Jungle '56

Francis Veber (D)

La Chevre '81
Les Comperes '83
Three Fugitives '89

Edward Vedder

The Mozart Story '48

Deeann Veeder

Carnago '84

Isela Vega

Barbarosa '82
Bordello '79
Bring Me the Head of Alfredo
  Garcia '74
Chamber of Fear '68
Drum '76
El Sabor de la Venganza '65
Fin de Fiesta '71
Joshua '76
La Primavera de los
  Escorpiones
Temporada Salvaje (Savage
  Season) '87

Pastor Vega (D)

Portrait of Teresa '79

Tommy Vegas

Tono Bicicleta

Misa Vehara

Hidden Fortress '58

Conrad Veidt

All Through the Night '42
The Beloved Rogue '27
The Cabinet of Dr. Caligari '19
Casablanca '42
Congress Dances (Der
  Kongress taenzt) '31
Dark Journey '37
F.P. 1 '33
The Passing of the Third Floor
  Back '36
Power '34
Spy in Black '39
Thief of Baghdad '40
Under the Red Robe '36
Waxworks '24
A Woman's Face '41

Ofra Veingarten

Fictitious Marriage '88

Harry Vejar

West to Glory '47

**Reginald Vel
Johnson**

Die Hard '88
Die Hard 2: Die Harder '90

Paco Vela

The Return of Josey Wales
  '86

Maria Elena Velasco

El Coyote Emplumado '83
Ni Chana Ni Juana '87

Andres Velasquez

Littlest Outlaw '54

Jaime Velasquez

Semaforo en Rojo '47

Lorena Velasquez

Doctor of Doom '62
Mision Suicida '87
Samson vs. the Vampire
  Women '61
Wrestling Women vs. the
  Aztec Ape '62
Wrestling Women vs. the
  Aztec Mummy '59

Pilar Velasquez

Ace of Hearts '70s
Manos Torpes '65

Teresa Velasquez

Dos Esposas en Mi Cama '87
Night of a Thousand Cats '72

Henri Velbert

Letters from My Windmill '54

Hans Man Int Veld

In for Treatment '82

Eddie Velez

Doin' Time '85
Romero '89
Rooftops '89
Split Decisions '88
Women's Club '87

Lupe Velez

Hell Harbor '30
Hell's Harbor '34
Mad About Money '30
Mexican Spitfire/Smartest Girl
  in Town '40
Palooka '34
Playmates '41
Stand and Deliver '28

John Vella

Wild Rebels '71

Carlos Velo (D)

Torero (Bullfighter) '56

Charles Veltman

Revenge of the Virgins '66

Evelyn Venable

Alice Adams '35
Harmony Lane '35
Hollywood Stadium Mystery
  '38
The Little Colonel '35

Mrs. Wiggs of the Cabbage
  Patch '34
Pinocchio '40

Luca Venantini

Aladdin '86

Shirley Venard

Satan's Touch '84

Amy Veness

Man of Evil '48

Marijke Vengelers

Egg '88

Ingrid Veninger

Hide and Seek '77

Chick Vennera

Kidnapped '87
Last Rites '88
The Milagro Beanfield War '88
The Terror Within 2 '91
Thank God It's Friday '78
Yanks '79

Diane Venora

Bird '88
The Cotton Club '84
F/X '86
Ironweed '87
Terminal Choice '85
Wolfen '81

Playboy Venson

Maxwell Street Blues '89

Wanda Ventham

Blood Beast Terror '67
Captain Kronos: Vampire
  Hunter '74

Itsvan Ventilla (D)

Nicole '72

Clyde Ventura

Bury Me an Angel '71
Gator Bait '73

Jesse Ventura

Predator '87
Repossessed '90
Ricochet '91
The Running Man '87
Thunderground '89

Lino Ventura

The French Detective '75
Happy New Year '74
Jailbird's Vacation '65
Les Grandes Gueules '65
The Medusa Touch '78
Mistress of the World '59
Pain in the A— '77
The Slap '76
Sword of Gideon '86
The Three Penny Opera '62

Viviane Ventura

Battle Beneath the Earth '68

Richard Venture

Missing '82
Street Hawk '84

Silvana Venturelli

Erotic Illusion '86
The Lickerish Quartet '70

**Massimo
Venturiello**

My Wonderful Life '90

Billy Vera

Finish Line '89

Vera-Ellen

The Belle of New York '52
Happy Go Lovely '51
Love Happy '50
On the Town '49
Perry Como Show '56
Three Little Words '50
White Christmas '54
Wonder Man '45
Words & Music '48

Victoria Vera

Monster Dog '82

Ben Verbong (D)

Lily was Here '89

Gwen Verdon

Alice '90
Cocoon '85
Cocoon: The Return '88
The Cotton Club '84
Damn Yankees '58
Legs '83
Nadine '87

Carlo Verdone

Acqua e Sapone '83

Carlo Verdone (D)

Acqua e Sapone '83

Elena Verdugo

The Big Sombrero '49
Cyrano de Bergerac '50
House of Frankenstein '44
The Moon and Sixpence '42

Ben Vereen

All That Jazz '79
Breakin' Through '84
Buy & Cell '89
Funny Lady '75
Gas-s-s-s! '70
The Jesse Owens Story '84
Pippin '81
Puss 'n Boots '84
Roots, Vol. 1 '77
The Zoo Gang '85

Enrique Vergara (D)

The Snake People '68

Alicia Vergel

Cavalry Command '63

Betty Verges

Dracula Blows His Cool '82
Summer Night Fever '79

Michael Verhoeven (D)

Killing Cars '80s
The Nasty Girl '90
The White Rose '83

Paul Verhoeven (D)

The Eternal Waltz '59
Flesh and Blood '85
The 4th Man '79
Katie's Passion '75
Keetje Tippei '75
RoboCop '87
Soldier of Orange '78
Spetters '80
Total Recall '90
Turkish Delights '74

Karen Veriler

Charlie Bravo '80s

Bernard Verley

Chloe in the Afternoon '72
The Milky Way '68

Francoise Verley

Chloe in the Afternoon '72

Renaud Verley

Bell from Hell '74
La Campana del Infierno

Melanie Verliin

Midnight '81

Arpad Vermes

A Hungarian Fairy Tale '87

Harold Vermilyea

Sorry, Wrong Number '48

Denise Vernanc

Unnatural '52

Karen Verne

All Through the Night '42
Kings Row '41
Sherlock Holmes and the
   Secret Weapon '42

Ines Vernengo

Man Facing Southeast '86

Henri Verneuil (D)

The Big Grab '63
Just Another Pretty Face '58
Melodie en Sous-Sol '63
Night Flight from Moscow '73
The Sheep Has Five Legs '54
Un Singe en Hiver '62

Guy Verney

Fame is the Spur '47
Martin Luther '53

Pierre Vernier

Mama, There's a Man in Your
   Bed '89

Anne Vernon

Therese & Isabelle '68
Umbrellas of Cherbourg '64

Bobby Vernon

A Hash House Fraud/The
   Sultan's Wife '15
Teddy at the Throttle '16
Teddy at the Throttle/
   Speeding Along '17

Glenn Vernon

Bedlam '45
Hollywood Varieties '50

Henry Vernon (D)

Danger Zone '87

Howard Vernon

The Diabolical Dr. Z '65
The Invisible Dead
Revenge in the House of
   Usher '82
The Screaming Dead '72
Triple Cross '67
Virgin Among the Living Dead
   '71
Zombie Lake '84

Jackie Vernon

Frosty the Snowman '69

Hungry i Reunion '81
Microwave Massacre '83
Rudolph & Frosty's Christmas
   in July '82

John Vernon

Angela '77
Bail Out '90
The Barbary Coast '74
The Black Windmill '74
The Blood of Others '84
Blue Monkey '84
Border Heat '88
Brannigan '75
Chained Heat '83
Charley Varrick '73
Curtains '83
Deadly Stranger '88
Dirty Harry '71
Dixie Lanes '88
Doin' Time '85
Ernest Goes to Camp '87
Fraternity Vacation '85
Golden Rendezvous '77
Hammered: The Best of
   Sledge '88
Herbie Goes Bananas '80
Hunter '73
Journey '77
Jungle Warriors '84
Justine '69
Killer Klowns from Outer
   Space '88
Mob Story '90
National Lampoon's Animal
   House '78
Outlaw Josey Wales '76
Savage Streets '83
Tell Them Willie Boy Is Here
   '69
Terminal Exposure '89
The Virginia Hill Story '76
W '74
War Bus Commando '89

Kate Vernon

Alphabet City '84
Hostile Takeover '88
The Last of Philip Banter '87
Mob Story '90
Roadhouse 66 '84

Richard Vernon

The Satanic Rites of Dracula
   '73

Richard Vernon (D)

The Shadow Man '53

Sherri Vernon

10 Violent Women '79

Wally Vernon

Gunfire '50
Square Dance Jubilee '51

Gennadi Vernov

Planet Burg '62

Stephen Verona (D)

The Lords of Flatbush '74
Pipe Dreams '76
Talking Walls '85

Christine Veronica

Love Notes '88
Party Incorporated '89

Nancy Veronica

Black Dragon '74

Cec Verrell

Eye of the Eagle '87
Silk '86

Marie Versini

Escape from the KGB '87

Wim Verstappen (D)

Fatal Error '83

Odile Versuis

Le Crabe Tambour '77
To Paris with Love '55

Dziga Vertov (D)

Enthusiasm '31
Kino Pravda/Enthusiasm '31
The Man with a Movie Camera
   '29

Veruschka

Blow-Up '66
The Bride '85

Louw Verwey

The Gods Must Be Crazy '84

Virginia Vestoff

1776 '72

Victoria Vetri

Group Marriage '72
Invasion of the Bee Girls '73
When Dinosaurs Ruled the
   Earth '70

Nora Veyran

Spiritism '61

Venetia Vianello

Born in America '90

Henry Vibart

Just Suppose '26

Marco Vicario (D)

The Sensual Man '74
Wifemistress '79

Renata Vicario

Island Monster '53

Carmen Vicarte

La Montana del Diablo (Devil's
   Mountain) '87
Los Caciques '87

Lindsey C. Vickers (D)

The Appointment '82

Martha Vickers

Big Bluff '55
The Big Sleep '46

Yvette Vickers

Attack of the Giant Leeches
   '59
Evil Spirits '91

Charles Victor

Motor Patrol '50
San Demetrio, London '47
Seven Days' Leave '42

Edward Victor (D)

Alley Cat '84

Henry Victor

The Beloved Rogue '27
King of the Zombies '41
Sherlock Holmes and the
   Secret Weapon '42

Katherine Victor

Curse of the Stone Hand '64
House of the Black Death '65
Mesa of Lost Women '52
Teenage Zombies '58
The Wild World of Batwoman
   '66

Piero Vida

Dead for a Dollar '70s

Gil Vidal

Night of the Howling Beast '75
Rape '80s

Henri Vidal

Just Another Pretty Face '58
Sois Belle et Tais-Toi '58
Voulez-Vous Danser Avec
   Moi? '59

Mario Vidal

Fuerte Perdido '78

Gala Videnovic

Hey, Babu Riba '88

Steven Vidler

Dunera Boys '85
Three's Trouble '85

Charles Vidor (D)

Cover Girl '44
Gilda '46
Hans Christian Andersen '52
The Lady in Question '40
Love Me or Leave Me '55
The Loves of Carmen '48
Song to Remember '45
Song Without End '60
The Swan '56
The Tuttles of Tahiti '42

Florence Vidor

Are Parents People? '25
Beau Revel '21
The Coming of Amos '25
The Grand Duchess and the
   Waiter '26
The Marriage Circle '24

King Vidor

Filmmakers: King Vidor '65

King Vidor (D)

Beyond the Forest '49
The Big Parade '25
Bird of Paradise '32
The Champ '31
Citadel '38
The Crowd '28
Duel in the Sun '46
The Fountainhead '49
The Jack Knife Man '20
Man Without a Star '55
Northwest Passage '40
Our Daily Bread '34
Ruby Gentry '52
Show People '28
The Sky Pilot '21
Solomon and Sheba '59
Stella Dallas '37
Street Scene '31
War and Peace '56

Steve Viedor

Keep It Up, Jack! '75

Eric Viellard

Boyfriends & Girlfriends '88

Enzo Viena

La Casa de Madame Lulu '65

Michel Vitold
Basileus Quartet '82

Ray Vitte
Grambling's White Tiger '81
Sky Hei$t '75

Monica Vitti
An Almost Perfect Affair '79
Bellissimo: Images of the
Italian Cinema '87
Blonde in Black Leather '77
The Eclipse '66
Girls of Don Juan '70s
Immortal Bachelor '80
L'Avventura '60
Phantom of Liberty '74
The Red Desert '64
Tigers in Lipstick '80

Viva
Edie in Ciao! Manhattan '72
Forbidden Zone '80
Play It Again, Sam '72
State of Things '82

Giana Vivaldi
Kill, Baby, Kill '66

Piero Vivarelli (D)
Satanik '69

Ruth Vivian
The Man Who Came to Dinner
'41

Sonia Viviani
Hercules 2 '85

Sal Viviano
Black Roses '88
The Jitters '88

Floyd Vivino
Crazy People '90

Richard Vlacia
Mixed Blood '84

Yuri Vladimirov
Ivan the Terrible '89

Marina Vlady
Chimes at Midnight '67
Double Agents '59

Emmett Vogan
Down Dakota Way '49
The Lady Confesses '45
Robot Pilot '41

Jack Vogel
Lock and Load '90

Mitch Vogel
Texas Detour '77

Nikolas Vogel
Dangerous Liaisons '60
The Inheritors '85

Peter Vogel
The Phantom of Soho '64

Virgil W. Vogel (D)
Invasion of the Animal People
'62
Portrait of a Rebel: Margaret
Sanger '82

Ruediger Vogler
Alice in the Cities '74

Kings of the Road (In the
Course of Time) '76
The Wrong Move '78

Alfred Vohrer (D)
Creature with the Blue Hand
'70
Dead Eyes of London '61
The Squeaker '65

Joan Vohs
As You Were '51
Cry Vengeance '54

Jon Voight
Catch-22 '70
The Champ '79
Coming Home '78
Conrack '74
Deliverance '72
Desert Bloom '86
Eternity '90
Final Warning '90
Hour of the Gun '67
The Last of His Tribe '92
Lookin' to Get Out '82
Midnight Cowboy '69
The Odessa File '74
Runaway Train '85
Table for Five '83

Vicki Volante
Angels' Wild Women '72
Blood of Dracula's Castle '69
Brain of Blood '71
Horror of the Blood Monsters
'70

Merete Voldstedlund
The Wolf at the Door '87

Paul G. Volk (D)
Sunset Strip '91

Claudio Volonte
Bay of Blood '71

Gian Marie Volonte
Bullet for the General '68
Christ Stopped at Eboli '79
A Fistful of Dollars '64
For a Few Dollars More '65
Journey Beneath the Desert
'61
Lucky Luciano '74
Open Doors (Porte Aperte) '89
Sacco & Vanzetti '71
The Witch '66

Julian Voloshin
Aladdin '86

Phillipe Volter
Cyrano de Bergerac '89
The Double Life of Veronique
'91
The Music Teacher '88

Ferdinand von Alten
Champagne '28

Josef von Baky (D)
Baron Munchausen '43

Brad von Beltz
Kill the Golden Goose '79

Christina von Blanc
Virgin Among the Living Dead
'71

William von Brinken
Hidden Enemy '40

Michael von der Goltz
American Autobahn '84

Lenny Von Dohlen
Billy Galvin '86
Blind Vision '91
Dracula's Widow '88
Electric Dreams '84
Leaving Normal '92
Love Kills '91
Under the Biltmore Clock '85

Sholto Von Douglas
Boogey Man 2 '83

Theodore von Eltz
Confidential '35
Drifting Souls '32
Strangers of the Evening '32

Loni von Friedl
Journey to the Far Side of the
Sun '69

Gunther Von Fristch (D)
Curse of the Cat People '44

Veith von Furstenberg (D)
Fire and Sword '82

Paul von Hauser
Demons of Ludlow '75

Wilhelm von Homburg
Ghostbusters 2 '89

Hunter von Leer
Under the Boardwalk '89

Irene von Meyendorf
The Mozart Story '48

Rik von Nutter
Assignment Outer Space '61
Tharus Son of Attila '70s

Joel von Ornsteiner
Robot Holocaust '87
Slashdance '89

Heidi von Palleske
Blind Fear '89
Dead Ringers '88
White Light '90

Rosa von Praunheim (D)
Anita, Dances of Vice '87

Geza von Radvanyi (D)
Uncle Tom's Cabin '69

Akos Von Rathony (D)
Cave of the Living Dead '65

Catherine von Schell
On Her Majesty's Secret
Service '69

Hans von Schlettow
Isn't Life Wonderful '24

Gustav von Seyffertitz
Don Juan '26
Down to Earth '17
The Mysterious Lady '28
Mystery Liner '34
Sparrows '26

Josef von Sternberg
Epic That Never Was '65

Josef von Sternberg (D)
Blonde Venus '32
The Blue Angel '30
Docks of New York '28
Jet Pilot '57
Last Command '28
Macao '52
Morocco '30
The Shanghai Gesture '41

Erich von Stroheim
As You Desire Me '32
Birth of a Nation '15
Blind Husbands '19
The Crime of Dr. Crespi '35
Crimson Romance '34
Foolish Wives '22
Grand Illusion '37
Great Flamarion '45
The Great Gabbo '29
Hearts of the World '18
Intolerance '16
Lost Squadron '32
Napoleon '55
The North Star '43
So Ends Our Night '41
Sunset Boulevard '50
Unnatural '52
Wedding March '28
The Worst of Hollywood: Vol.
2 '90

Erich von Stroheim (D)
Blind Husbands '19
Foolish Wives '22
Greed '24
Queen Kelly '29
Wedding March '28
The Worst of Hollywood: Vol.
2 '90

Max von Sydow
Awakenings '90
The Belarus File '85
Brass Target '78
Brink of Life '57
Christopher Columbus '91
Code Name: Emerald '85
Conan the Barbarian '82
Death Watch '80
Dreamscape '84
Duet for One '86
Dune '84
Embassy '72
The Exorcist '73
Exorcist 2: The Heretic '77
Flash Gordon '80
Flight of the Eagle '82
Foxtrot '76
The Greatest Story Ever Told
'65

Hannah and Her Sisters '86
Hawaii '66
Hiroshima: Out of the Ashes
'90
Hour of the Wolf '68
Hurricane '79
A Kiss Before Dying '91
The Magician '58
March or Die '77
Never Say Never Again '83
The Night Visitor '70
Pelle the Conqueror '88
The Quiller Memorandum '66
Quo Vadis '85
Red King, White Knight '89
Samson and Delilah '84
The Seventh Seal '56
Soldier's Tale '84
Steppenwolf '74
Strange Brew '83
Target Eagle '84
Three Days of the Condor '75
Through a Glass Darkly '61
The Touch '71
The Ultimate Warrior '75
Victory '81
The Virgin Spring '59
Voyage of the Damned '76
Wild Strawberries '57
The Winter Light '62
The Wolf at the Door '87

Hanz von Teuffen
The Flying Saucer '50

Ernst R. von Theumer (D)
Jungle Warriors '84

Lars von Trier (D)
The Element of Crime '85

Margarethe von Trotta
Coup de Grace '78

Margarethe von Trotta (D)
The Lost Honor of Katharina
Blum '75
Sheer Madness '84

Hans von Twardowski
Dawn Express '42

Gustav von Wagenheim
Woman in the Moon '29

Harry von Zell
Son of Paleface '52
The Strange Affair of Uncle
Harry '45

Danielle von Zernaeck
Dangerous Curves '88
La Bamba '87
My Science Project '85
Under the Boardwalk '89

Kurt Vonnegut, Jr.
Back to School '86

Donald Voorhees
Rehearsal '47

Klaus Voorman

John Lennon and the Plastic Ono Band: Live Peace in Toronto, 1969 '69

Bernard Vorhaus (D)

The Amazing Mr. X '48
Broken Melody '34
Courageous Dr. Christian '40
Fisherman's Wharf '39
Lady from Louisiana '42
Meet Dr. Christian '39
Three Faces West '40

Zandor Vorkov

Brain of Blood '71
Dracula vs. Frankenstein '71

George Voskovec

The Boston Strangler '68
The Bravados '58
Twelve Angry Men '57
The 27 Day '57

Arnold Vosloo

Living to Die '91
Reason to Die '90
The Rutanga Tapes '90

Frank Vosper

Power '34

Kurt Voss (D)

Genuine Risk '89
Horseplayer '91

Andreas Voutsinas

A Dream of Passion '78

Yorgo Voyagis

The Adventurers '70
Courage Mountain '89
Frantic '88
The Little Drummer Girl '84

Vlasta Vrana

Deadly Surveillance '91
Scanners 2: The New Order '91

Irene Vrkijan

David '79

Frederick Vroom

The Navigator '24

Eric Vu-An

The Sheltering Sky '90

Olivera Vuco

Mark of the Devil '69

Vuillemin

Mystery of Alexina '86

Tito Vuolo

The Enforcer '51

Murvyn Vye

Al Capone '59
A Connecticut Yankee in King Arthur's Court '49
Pickup on South Street '53
River of No Return '54
Road to Bali '53

Ivan Vyskocil

The Report on the Party and the Guests '66

Yeung Oi Wa

Twister Kicker '85

Robert Wachs

Eddie Murphy "Delirious" '83

Daniel Wachsmann (D)

Hamsin '83

Jonathan Wacks (D)

Mystery Date '91
Powwow Highway '89

Christopher Wade

Mindkiller '87

John Wade

Sundance and the Kid '76

Russell Wade

The Body Snatcher '45
The Iron Major '43
Shoot to Kill '47
Sundown Riders '48

Sara Lee Wade

Darkroom '90

Stuart Wade

Meteor Monster '57
Monster from the Ocean Floor '54

Michael Wadleigh (D)

Wolfen '81
Woodstock '70

Henry Wadsworth

Applause '29

Eva Waegner

Paradisio '61

Anthony Wager

Great Expectations '46

Michael Wager

Hill 24 Doesn't Answer '55
Jane Austen in Manhattan '80

George Waggner

The Sheik '21

George Waggner (D)

The Commies are Coming, the Commies are Coming '57
The Fighting Kentuckian '49
Mystery Plane '39
Outlaw Express '38
Stunt Pilot '39
Western Trails '38
Wolf Call '39
The Wolf Man '41

Lyle Waggoner

Dead Women in Lingerie '90
Dream Boys Revue '87
Journey to the Center of Time '67
Love Me Deadly '76
Mind Trap '91
Robo-Chic '89
Surf 2 '84

Bob Wagner

Honeymoon Horror '70s

Christie Wagner

She-Devils on Wheels '68

Chuck Wagner

America 3000 '86
The Sisterhood '88

Fernando Wagner

The Pearl '48
Virgin Sacrifice '59

Fernando Wagner (D)

Virgin Sacrifice '59

Jack Wagner

Jive Junction '43
Play Murder for Me '91
Swimsuit: The Movie '89

Lindsay Wagner

The Bionic Woman '75
Callie and Son '81
Convicted '86
The Flight '89
High Risk '76
The Incredible Journey of Dr. Meg Laurel '79
Martin's Day '85
Nighthawks '81
Nightmare at Bittercreek '91
The Paper Chase '73
Princess Daisy '83
Scruples '80
Second Wind '90
The Taking of Flight 847: The Uli Derickson Story '88
Two Kinds of Love '83
The Two Worlds of Jenny Logan '79

Robert Wagner

Abduction of St. Anne '75
The Affair '73
Beneath the 12-Mile Reef '53
Between Heaven and Hell '56
Broken Lance '54
The Concorde: Airport '79 '79
Curse of the Pink Panther '83
Death at Love House '75
Halls of Montezuma '50
Harper '66
The Hunters '58
I Am the Cheese '83
Indiscreet '88
Let's Make it Legal '51
The Longest Day '62
Madame Sin '71
Midway '76
The Mountain '56
The Ox-Bow Incident '55
The Pink Panther '64
Prince Valiant '54
Stars and Stripes Forever '52
Stopover Tokyo '57
The Streets of San Francisco '72
To Catch a King '84
Towering Inferno '74
Trail of the Pink Panther '82
War Lover '62
What Price Glory? '52
Winning '69

Yueh Wah

Ninja Supremo '90

Adam Wahl

Dead Mate '88

Corinne Wahl

Equalizer 2000 '86

Evelyn Wahl

Jungle Siren '42

Ken Wahl

The Dirty Dozen: The Next Mission '85
Fort Apache, the Bronx '81
The Gladiator '86
Jinxed '82
Omega Syndrome '87
Purple Hearts '84
Race to the Yankee Zephyr '81
Running Scared '79
The Soldier '82
The Taking of Beverly Hills '91
Treasure of the Yankee Zephyr '83
Wanderers '79

Lam Wai

Triad Savages

Chan Wai-Man

Ninja Blacklist '90

Samuel Wail (D)

Stuck on You '84

Edward Wain

Creature from the Haunted Sea '60
The Last Woman on Earth '61

James Wainwright

Jigsaw '71
Joe Kidd '72
The President's Plane is Missing '71
Warlords of the 21st Century '82
A Woman Called Moses '78

Loudon Wainwright, III

Jacknife '89
The Slugger's Wife '85

Rupert Wainwright

Another Country '84

Michael Waite

Fun Down There '88

Ralph Waite

Angel City '80
Chato's Land '71
Cool Hand Luke '67
Crash and Burn '90
Five Easy Pieces '70
Gentleman Bandit '81
Girls on the Road '73
The Grissom Gang '71
On the Nickel '80
Red Alert '77
The Sporting Club '72
The Stone Killer '73
Thanksgiving Story '73
The Waltons' Christmas Carol '80s

Ralph Waite (D)

On the Nickel '80

Thomas G. Waites

And Justice for All '79
The Clan of the Cave Bear '86
McBain '91
On the Yard '79
The Thing '82
Verne Miller '88
The Warriors '79

Tom Waits

At Play in the Fields of the Lord '91
Candy Mountain '87
Cold Feet '89
The Cotton Club '84
Down by Law '86
Ironweed '87
Mystery Train '89
The Outsiders '83
Queens Logic '91
Rumble Fish '83
Shakedown '88

Christopher Waitz

Fire and Sword '82

Andrzej Wajda (D)

Ashes and Diamonds '59
Danton '82
Fury is a Woman '61
Kanal '56
A Love in Germany '84
Man of Marble '76

Alexander Wajnberg

Mama Dracula '80

Marc-Henri Wajnberg

Mama Dracula '80

Akiko Wakabayashi

What's Up, Tiger Lily? '66
You Only Live Twice '67

Ayako Wakao

Drifting Weeds '59
Floating Weeds '59

Tomisaburo Wakayama

Phoenix '78
Shogun Assassin '80

Hugh Wakefield

Blithe Spirit '45

Deborah Wakeham

Covergirl '83
House of the Rising Sun '87
Middle Age Crazy '80

Jimmy Wakely

Lonesome Trail '55
Silver Star '55

Chris Walas (D)

The Fly 2 '89

Garry Walberg

Rage '80

Anton Walbrook

Dangerous Moonlight '41
The Forty-Ninth Parallel '41
La Ronde '51
Life & Death of Colonel Blimp '43
Lola Montes '69
The Queen of Spades '49
The Red Shoes '48
Saint Joan '57

Raymond Walburn

Born to Dance '36
Christmas in July '40
Count of Monte Cristo '34
Hail the Conquering Hero '44
It Could Happen to You '39

A Night to Remember '42
Sinbad the Sailor '47
Tycoon '47
The Young in Heart '38

Rick Wallace (D)

Acceptable Risks '86
California Girls '84
Time to Live '85

Royce Wallace

Green Eyes '76

Stephen Wallace (D)

For Love Alone '86
Prisoners of the Sun '91

Sue Wallace

Experience Preferred... But
Not Essential '83

**Tommy Lee
Wallace** (D)

Aloha Summer '88
Fright Night 2 '88
Halloween 3: Season of the
Witch '82

Eli Wallach

Ace High '68
Baby Doll '56
The Brain '69
Christopher Columbus '91
Cinderella Liberty '73
Circle of Iron '78
The Deep '77
The Domino Principle '77
Embassy '85
The Executioner's Song '82
A Family Matter (Vendetta) '91
Girlfriendo '78
The Godfather: Part 3 '90
The Good, the Bad and the
Ugly '67
How the West was Won '62
The Hunter '80
The Impossible Spy '87
Independence '76
Lord Jim '65
MacKenna's Gold '69
Magnificent Seven '60
The Misfits '61
Moon-Spinners '64
Movie, Movie '78
Nasty Habits '77
Nuts '87
The People Next Door '70
The Poppy Is Also a Flower
'66
Pride of Jesse Hallum '81
Rocket to the Moon '86
Romance of a Horsethief '71
The Salamander '82
Sam's Son '84
Seven Thieves '60
Something in Common '86
Stateline Motel '75
Tough Guys '86
The Two Jakes '90
Winter Kills '79

Katherine Wallach

Gap-Toothed Women '89

Sigurd Wallen

Swedenhielms '35

Eddy Waller

El Paso Stampede '53
Marshal of Cedar Rock '53

Fats Waller

Fats Waller and Friends '86
Stormy Weather '43

**Herb
Wallerstein** (D)

Snowbeast '77

Deborah Walley

Benji '73
Bon Voyage! '62
Dr. Goldfoot and the Bikini
Machine '66
Gidget Goes Hawaiian '61
The Severed Arm '73
Spinout '66
Summer Magic '63

Richard Walling

Walking Back '26

William Walling

Great K & A Train Robbery
'26

Jimmy Wallington

Hollywood Stadium Mystery
'38

Hal B. Wallis (D)

Anne of the Thousand Days
'69

Shani Wallis

Oliver! '68
Round Numbers '91

Frika Wallner

Esposa y Amante '70s

Samuel Walls

Enter Three Dragons '81

Tom Walls, Jr.

Maytime in Mayfair '52

Tom Walls

Johnny Frenchman '46
Maytime in Mayfair '52
While I Live '47

Jon Walmsley

Children's Carol '80

Louis Walser

Last Mercenary '84

Angela Walsh

Distant Voices, Still Lives '88

Dale Walsh

The New Adventures of
Tarzan: Vol. 1 '35

Dermot Walsh

Ghost Ship '53
The Great Armored Car
Swindle '64
Mania '60
Murder on the Campus '52
The Tell-Tale Heart '62

Ed Walsh

Let It Ride '89

George Walsh

American Pluck '25
The Broadway Drifter '27
The Live Wire '34
Test of Donald Norton '26

Gwynyth Walsh

Blue Monkey '84

Joey Walsh

Hans Christian Andersen '52

Johnny Walsh

Wild Women of Wongo '59

J.T. Walsh

Backdraft '91
The Big Picture '89
Crazy People '90
Dad '89
Defenseless '91
The Grifters '90
Hard Choices '84
House of Games '87
Iron Maze '91
Narrow Margin '90
The Russia House '90
Tequila Sunrise '88
Why Me? '90
Wired '89

Kay Walsh

Encore '52
Greyfriars Bobby '61
The Horse's Mouth '58
In Which We Serve '42
Last Holiday '50
October Man '48
Oliver Twist '48
Stage Fright '50
This Happy Breed '47
Tunes of Glory '60
Tunes of Glory/Horse's Mouth
'60

Ken Walsh

Love and Hate '90
Reno and the Doc '84

Kevin T. Walsh

Legion of Iron '90

M. Emmet Walsh

The Abduction of Kari
Swenson '87
Back to School '86
The Best of Times '86
Blade Runner '82
Blood Simple '85
Brubaker '80
Cannery Row '82
Catch Me...If You Can '89
Clean and Sober '88
Critters '86
Dear Detective '79
East of Eden '80
Fast Walking '82
Fletch '85
Fourth Story '90
Grandview U.S.A. '84
Killer Image '92
The Mighty Quinn '89
The Milagro Beanfield War '88
Missing in Action '84
Murder Ordained '87
Narrow Margin '90
Night Partners '83
Ordinary People '80
The Pope of Greenwich
Village '84
Raising Arizona '87
Raw Courage '84
Red Scorpion '89
Reds '81
Rich Hall's Vanishing America
'86
Scandalous '84
Stiletto '69
Straight Time '78
Sunset '88
They Might Be Giants '71
Thunderground '89
War Party '89
White Sands '92
Wildcats '86

Percy Walsh

The Golden Salamander '51

Philip Walsh

John & Yoko: A Love Story
'85

Raoul Walsh

Birth of a Nation '15
Sadie Thompson '28

Raoul Walsh (D)

Along the Great Divide '51
Battle Cry '55
Big Trail '30
Blackbeard the Pirate '52
Captain Horatio Hornblower
'51
Dark Command '40
Distant Drums '51
The Enforcer '51
Gentleman Jim '42
Gun Fury '53
High Sierra '41
The Horn Blows at Midnight
'45
The King and Four Queens
'56
The Naked and the Dead '58
Northern Pursuit '43
Pursued '47
The Roaring Twenties '39
Sadie Thompson '28
Sea Devils '53
Strawberry Blonde '41
The Tall Men '55
They Died with Their Boots
On '41
They Drive by Night '40
Thief of Baghdad '24
When Thief Meets Thief '37
White Heat '49

Sharon Walsh

Incredibly Strange Creatures
Who Stopped Living and
Became Mixed-Up Zombies
'63

Susan Walsh

Female Trouble '75

Sydney Walsh

To Die for '89

Ray Walston

Amos '85
The Apartment '60
Blood Relations '87
Damn Yankees '58
The Fall of the House of Usher
'80
Fast Times at Ridgemont High
'82
Fine Gold
For Love or Money '84
From the Hip '86
Galaxy of Terror '81
Happy Hooker Goes to
Washington '77
Johnny Dangerously '84
The Kid with the Broken Halo
'82
Kiss Me, Stupid! '64
O.C. and Stiggs '87
O'Hara's Wife '82
Paramedics '88
Popcorn '89
Popeye '80
Private School '83
Saturday the 14th Strikes
Back '88
The Silver Streak '76
Ski Patrol '89

South Pacific '58
The Sting '73
Tall Story '60
That's Singing '84

Roy Walston

Blood Salvage '90

Harriet Walter

The Good Father '87

Jerry Walter

Nightmare in Blood '75

Jessica Walter

Dr. Strange '78
The Execution '85
The Flamingo Kid '84
Going Ape! '81
Goldengirl '79
The Group '66
Home for the Holidays '72
Hurricane '74
Miracle on Ice '81
Play Misty for Me '71
Secrets of Three Hungry
Wives '78
She's Dressed to Kill '79
Tapeheads '89
Wild & Wooly '78

Tracy Walter

At Close Range '86
Batman '89
City Slickers '91
Delusion '91
Liquid Dreams '92
Mad Bull '77
Mortuary Academy '91
Out of the Dark '88
Raggedy Man '81
Repo Man '83
Something Wild '86
Timerider '83
The Two Jakes '90
Under the Boardwalk '89
Young Guns 2 '90

Anrae Walterhouse

The Grass is Always Greener
Over the Septic Tank '78

Charles Walters (D)

Barkleys of Broadway '49
The Belle of New York '52
Billy Rose's Jumbo '62
Dangerous When Wet '53
Easter Parade '48
Easy to Love '53
The Glass Slipper '55
Good News '47
High Society '56
Lili '53
Please Don't Eat the Daisies
'60
Summer Stock '50
The Tender Trap '55
Texas Carnival '51
Three Guys Named Mike '51
Torch Song '53
The Unsinkable Molly Brown
'64
Walk, Don't Run '66

James Walters

Shout '91

Julie Walters

Buster '88
Car Trouble '86
Educating Rita '83
Mack the Knife '89
Personal Services '87
Prick Up Your Ears '87

Mona Washbourne

The Brides of Dracula '60
Brideshead Revisited '81
The Collector '65
December Flower
Driver's Seat '73
Mrs. Brown, You've Got a
    Lovely Daughter '68
My Fair Lady '64
O Lucky Man '73
Stevie '78
Stranger in Town '57

Beverly Washburn

Old Yeller '57
When the Line Goes Through
    '71

Bryant Washburn

Drifting Souls '32
The Wizard of Oz '25

Rick Washburne

Covert Action '88
Hangmen '87

Debbie Washington

In Hot Pursuit '82

Denzel Washington

Carbon Copy '81
Cry Freedom '87
Glory '89
Heart Condition '90
License to Kill '84
The Mighty Quinn '89
Mississippi Masala '92
Mo' Better Blues '90
Power '86
Ricochet '91
A Soldier's Story '84

Fredi Washington

Emperor Jones '33

Gene Washington

The Black Six '74
Lady Cocoa '75

Judy Washington

The Kill '73

Ken Washington

She Devils in Chains

Shirley Washington

Darktown Strutters '74
The Deadly and the Beautiful
    '73

William Washington

Broken Strings '40

Ted Wass

Curse of the Pink Panther '83
Fine Gold
The Longshot '86
Oh, God! You Devil '84
Sheena '84
The Triangle Factory Fire
    Scandal '79
TripleCross '85

Jerry Wasserman

Quarantine '89

Craig Wasson

Body Double '84
The Boys in Company C '77
Carny '80
Four Friends '81

Ghost Story '81
Go Tell the Spartans '78
The Innocents Abroad '84
The Men's Club '86
A Nightmare on Elm Street 3:
    Dream Warriors '87
Schizoid '80
Second Thoughts '83
Skag '79

Michal Waszynski (D)

The Dybbuk '37

Gedde Watanabe

Gung Ho '85
Sixteen Candles '84
The Spring '89
Vamp '86
Volunteers '85

Ken Watanabe

Bruce's Fists of Vengeance
    '84
Commando Invasion '87
Karate Warrior
MacArthur's Children '85
Tampopo '86

Tsunehiko Watase

The Silk Road '92
Tattooed Hit Man '76

Loeta Waterdown

Bounty Hunters '89

Nick Wateres

The Lighthorseman '87

Dennis Waterman

Cold Justice '89
Fright '71
The Scars of Dracula '70

Felicity Waterman

Lena's Holiday '91

Ida Waterman

Stella Maris '18

Willard Waterman

Fun Shows '50
Hail '73
Hollywood or Bust '56

Cheryl Waters

Didn't You Hear? '83
Image of Death '77
Macon County Line '74

Ethel Waters

Cabin in the Sky '43
The Member of the Wedding
    '52
Stage Door Canteen '43

Harry Waters, Jr.

Back to the Future, Part 2 '89

John Waters

Alice to Nowhere '86
All the Rivers Run '84
Attack Force Z '84
Breaker Morant '80
Cass '78
Demolition '77
Endplay '75
The Getting of Wisdom '77
Grievous Bodily Harm '89
Miracle Down Under '87
Scalp Merchant '77
Something Wild '86
Three's Trouble '85

True Colors '87
Weekend of Shadows '77
Which Way Home '90

John Waters (D)

Cry-Baby '90
Desperate Living '77
Divine '70s
Female Trouble '75
Hairspray '88
Mondo Trasho '70
Multiple Maniacs '70
Pink Flamingos '73
Polyester '81

Muddy Waters

Chicago Blues '72
The Last Waltz '78

Nick Waters

The Big Hurt '87

Roger Waters

The Wall: Live in Berlin '90

James Waterson

Dead Poets Society '89

Sam Waterston

Amazing Stories, Book 4 '85
The Boy Who Loved Trolls '84
Capricorn One '78
Crimes & Misdemeanors '89
Death of a Hooker '71
Dempsey '83
Eagle's Wing '79
Finnegan Begin Again '84
Friendly Fire '79
Generation '69
The Great Gatsby '74
Hannah and Her Sisters '86
Heaven's Gate '81
Hopscotch '80
Interiors '78
Journey into Fear '74
Just Between Friends '86
The Killing Fields '84
Lantern Hill '90
Mindwalk '91
The Nightmare Years '89
Reflections of Murder '74
Savages '72
September '88
Sweet William '79
Warning Sign '85
Welcome Home '89
Who Killed Mary What's 'Er
    Name? '71

Gwen Watford

The Fall of the House of Usher
    '52
Ripping Yarns '75

Ian Watkin

Nutcase '83

Pierre Watkin

Country Gentlemen '36
The Magnificent Dope '42
Road to Singapore '40
Superman: The Serial: Vol. 1
    '48
Swamp Fire '46

Charlene Watkins

Tough Enough '87

Gary Watkins

Wheels of Fire '84

Joanne Watkins

The Big Sweat '91
Cold Heat '90

Miles Watkins

The Great Texas Dynamite
    Chase '76

Peter Watkins (D)

The Gladiators '70

Jack Watling

The Courtney Affair '47
Under Capricorn '49

Alberta Watson

Best Revenge '83
The Hitman '91
The Keep '83
White of the Eye '88
Women of Valor '86

Bob Watson

Dreaming Out Loud '40
Hitler: Dead or Alive '43
In Hot Pursuit '82

Doc Watson

Scruggs '70

Don Watson

In Hot Pursuit '82

Douglas Watson

Trial of the Catonsville Nine
    '72

Harry Watson

The Barber Shop '33

Jack Watson

King Arthur, the Young
    Warlord '75
Night Caller from Outer Space
    '66
Peeping Tom '63
Young Warlord '75

James A. Watson, Jr.

Extreme Close-Up '73
Killings at Outpost Zeta '80
Sex Through a Window '72

James C. Watson (D)

Night of the Demon '80

John Watson (D)

Deathstalker '84
The Zoo Gang '85

Lucile Watson

The Great Lie '41
Let's Dance '50
Made for Each Other '39
Mr. & Mrs. Smith '41
The Thin Man Goes Home '44
Tomorrow is Forever '46
Waterloo Bridge '40
The Women '39

Michael Watson

Subspecies '91

Mills Watson

Heated Vengeance '87
Kansas City Massacre '75

Minor Watson

Abe Lincoln in Illinois '40
The Ambassador's Daughter
    '56
As Young As You Feel '51
Beyond the Forest '49
Charlie Chan in Paris '35

Face to Face '52
Guadalcanal Diary '43
The Jackie Robinson Story
    '50
A Southern Yankee '48
To the Shores of Tripoli '42
Western Union '41
Woman of the Year '42

Mitch Watson

Primal Rage '90

Owen Watson

Black Force '75

Robert Watson

The Paleface '48

Tom Watson

Another Time, Another Place
    '83

Vernee Watson

Death Drug '83

William Watson

Force of Evil '77
Girl on a Chain Gang '65

William Watson (D)

Heroes in Blue '39

Woody Watson

All-American Murder '91
The Last Prostitute '91

Wylie Watson

Brighton Rock '47
A Girl in a Million '46
The Sundowners '60
Things Happen at Night '48

Harry Watt (D)

Eureka Stockade '49
Overlanders '46

Marty Watt

Almost You '85

Nate Watt

Frontier Vengeance '40

Nate Watt (D)

Rustler's Valley '37

Richard C. Watt

Death Shot '73

William Watters

What's Up Front '63

Richard Wattis

Doctor in the House '53
My Son, the Vampire '52
Sex and the Other Woman '79
Three Stops to Murder '53

Charlie Watts

Gimme Shelter '70
Ready Steady Go, Vol. 1 '83
Ready Steady Go, Vol. 2 '85
Ready Steady Go, Vol. 3 '84
25x5: The Continuing
    Adventures of the Rolling
    Stones '90

Chick Watts

Yes, Sir, Mr. Bones '51

Cotton Watts

Yes, Sir, Mr. Bones '51

Hugo Weaving

For Love Alone '86
The Right Hand Man '87

Alan Webb

Challenge to Lassie '49
Deadly Game '82
Entertaining Mr. Sloane '70
Great Train Robbery '79
King Lear '71

Cassandra Webb

Starship '87

Chloe Webb

The Belly of an Architect '90
China Beach '88
Heart Condition '90
Queens Logic '91
Sid & Nancy '86
Twins '88

Clifton Webb

Dark Corner '46
Laura '44
The Man Who Never Was '55
The Razor's Edge '46
Stars and Stripes Forever '52
Three Coins in the Fountain
  '54

Daniel Webb

The Unapproachable '82

Greg Webb

Puppet Master 2 '91
Running Mates '86

Harry S. Webb (D)

Laramie Kid '35
The Live Wire '34
Mesquite Buckaroo '39
Ridin' Thru '35
Riding On '37
Santa Fe Bound '37

Jack Webb

The Commies are Coming, the
  Commies are Coming '57
D.I. '57
Dragnet '54
Halls of Montezuma '50
He Walked by Night '48
The Men '50
Pete Kelly's Blues '55
Red Nightmare '62
Sunset Boulevard '50

Jack Webb (D)

D.I. '57
Dragnet '54
Pete Kelly's Blues '55

Jane Webb

Sabrina, the Teenage Witch
  '69

Kenneth Webb (D)

Just Suppose '26

Millard Webb

Reaching For the Moon '17

Millard Webb (D)

The Dropkick '27
Glorifying the American Girl
  '30

Peter Webb (D)

Give My Regards to Broad
  Street '84

Richard Webb

Attack of the Mayan Mummy
  '63
Distant Drums '51
Git! '65
Hell Raiders '68

Rita Webb

Alien Women '69

Robert D. Webb (D)

Beneath the 12-Mile Reef '53
Love Me Tender '56
Seven Cities of Gold '55

Tamara Webb

What Would Your Mother
  Say? '81

William Webb (D)

The Banker '89
Dirty Laundry '87
Party Line '88

Beth Webber

Strangers in Good Company
  '91

Diane Webber

The Mermaids of Tiburon '62

Peggy Webber

The Screaming Skull '58

Robert Webber

Assassin '86
Bring Me the Head of Alfredo
  Garcia '74
Casey's Shadow '78
The Choirboys '77
Dead Heat on a Merry-Go-
  Round '66
Death Stalk '74
Dollars '71
Don't Go to Sleep '82
The Final Option '82
The Great White Hope '70
Hysteria '64
Nuts '87
Private Benjamin '80
Revenge of the Pink Panther
  '78
The Sandpiper '65
S.O.B. '81
Soldat Duroc...Ca Va Etre Ta
  Fete! '60s
Starflight One '83
The Streets of L.A. '79
The Stripper '63
10 '79
Thief '71
Twelve Angry Men '57
Wrong is Right '82

Timothy Webber

The Grey Fox '83
That's My Baby! '88
Toby McTeague '87

Bruce Weber (D)

Let's Get Lost '89

Jacques Weber

Cyrano de Bergerac '89

Lois Weber (D)

The Blot '21
Chapter in Her Life '23
Discontent '16
Too Wise Wives '21

Sharon Weber

Little Dragons '80

Suzanne Weber

Cold River '81

Jeff Weborg

The Islander '88

Ben Webster

The Old Curiosity Shop '35

D.J. Webster (D)

Dark Side of the Moon '90

Nicholas Webster (D)

Gone are the Days '63
Mission Mars '67
Purlie Victorious '63
Santa Claus Conquers the
  Martians '64

Paddy Webster

Cat Girl '57

David Wechter (D)

Malibu Bikini Shop '86
Midnight Madness '80

Peter Weck

Almost Angels '62

Ann Wedgeworth

Bang the Drum Slowly '73
The Birch Interval '78
Bogie '80
The Catamount Killing '74
Citizens Band '77
Far North '88
Killjoy '81
Law and Disorder '74
Made in Heaven '87
Miss Firecracker '89
My Science Project '85
No Small Affair '84
One Summer Love '76
Scarecrow '73
Soggy Bottom U.S.A. '84
Steel Magnolias '89
Sweet Dreams '85
A Tiger's Tale '87

Barbara Weeks

Hell's Headquarters '32
The Violent Years '56
White Eagle '32

Christopher Weeks

Valet Girls '86
Violent Zone '89

Jimmie Ray Weeks

Apology '86
Frantic '88

Stephen Weeks (D)

Madhouse Mansion '74
Sword of the Valiant '83

Paul Wegener

The Golem '20
Kolberg '45
One Arabian Night '21
Student of Prague '13

Paul Wegener (D)

The Golem '20
Student of Prague '13

Lam Men Wei

Duel of the Seven Tigers '80
Return of the Scorpion '80s

Lo Wei (D)

Chinese Connection '73
Fists of Fury '73
Shaolin Wooden Men
Tattooed Dragon '82

Tang Wei

The Cavalier '80s
Dynasty '77

Yang Wei

Japanese Connection '82

Peter Weibel

Invisible Adversaries '77

Blair Weickgenant

Smooth Talker '90

Alfred Weidenmann (D)

Adorable Julia '62

Ken Weiderhorn (D)

Return of the Living Dead Part
  2 '88

Virginia Weidler

All This and Heaven Too '40
Babes on Broadway '41
Best Foot Forward '43
The Great Man Votes '38
Too Hot to Handle '38

Paul Weigel

For Heaven's Sake '26

Teri Weigel

The Banker '89
Cheerleader Camp '88
Glitch! '88
Marked for Death '90
Predator 2 '90
Return of the Killer Tomatoes
  '88
Savage Beach '89

Samuel Weil (D)

First Turn On '83
Squeeze Play '79
Troma's War '88
Waitress '81

Paul Weiland (D)

Leonard Part 6 '87

Claudia Weill (D)

Girlfriends '78
It's My Turn '80
Once a Hero '88

Kate Weiman

Sex O'Clock News '85

Chuck Wein (D)

Jimi Hendrix: Rainbow Bridge
  '89
Rainbow Bridge '71

Marc Weiner

New Wave Comedy '85

Isabelle Weingarten

State of Things '82

Bob Weinstein (D)

Playing for Keeps '86

Harvey Weinstein (D)

Light Years '88

Carl Weintraub

Sorry, Wrong Number '89

Cindy Weintraub

Humanoids from the Deep '80

Sandra Weintraub (D)

The Deadliest Art: The Best of
  the Martial Arts Films '90
Women's Club '87

Helen Weir

The Social Secretary '16

Peter Weir

Cat Chaser '90

Peter Weir (D)

Cars That Ate Paris '74
Dead Poets Society '89
Gallipoli '81
Green Card '90
The Last Wave '77
The Mosquito Coast '86
Picnic at Hanging Rock '75
Plumber '79
Witness '85
The Year of Living
  Dangerously '82

Bob Weis (D)

The Wacky World of Wills &
  Burke '85

Don Weis (D)

The Gene Krupa Story '59
I Love Melvin '53
Munster's Revenge '81
Pajama Party '64
Zero to Sixty '78

Gary Weis (D)

Chris Elliott: FDR, A One-Man
  Show/Action Family '89
Diary of a Young Comic '79
Jimi Hendrix: Story '73
Things We Did Last Summer
  '78
Wholly Moses! '80

Heidelinde Weis

The Man Outside '68
Something for Everyone '70

Jack Weis (D)

Crypt of Dark Secrets '76
Mardi Gras Massacre '78
Storyville '74

Grethe Weiser

Tromba, the Tiger Man '52

Robin Weisman

Three Men and a Little Lady
  '90

Straw Weisman (D)

Dead Mate '88

Adrian Weiss (D)

Bride & the Beast '58

Don Weiss (D)

Billie '65

Florence Weiss

The Cantor's Son (Dem Khann's Zindl) '37
Overture to Glory (Der Vilner Shtot Khazn) '40
The Singing Blacksmith (Yankl der Shmid) '38

Gary Weiss

All You Need is Cash '78

Helmut Weiss (D)

Tromba, the Tiger Man '52

Joel Weiss

Street Soldiers '91

Michael T. Weiss

Howling 4: The Original Nightmare '88

Robert Weiss

Hitchhiker 3 '87

Robert Weiss (D)

Amazon Women On the Moon '87

Roberta Weiss

Abducted '86

Norbert Weisser

Secret of the Ice Cave '89
Sweet Lies '88

Jeffrey Weissman

Back to the Future, Part 2 '89

Mitch Weissman

Beatlemania! The Movie '81

Johnny Weissmuller, Jr.

Shame of the Jungle '90

Johnny Weissmuller

Stage Door Canteen '43
Swamp Fire '46
Tarzan, the Ape Man '32
Tarzan Escapes '36
Tarzan and His Mate '34

Hilde Weissner

Tromba, the Tiger Man '52

Bruce Weitz

Death of a Centerfold '81
The Queen of Mean '90
Rainbow Drive '90

Elizabeth Welch

Song of Freedom '38

Raquel Welch

Bandolero! '68
Bedazzled '68
Bluebeard '72
Fantastic Voyage '66
The Four Musketeers '75
Fuzz '72
Hannie Caulder '72
Lady in Cement '68
The Last of Sheila '73
Legend of Walks Far Woman '82
The Magic Christian '69
Mother, Jugs and Speed '76
Myra Breckinridge '70
Oldest Profession '67
100 Rifles '69
Prince and the Pauper '78
Restless '72

Roustabout '64

Roustabout '64
Scandal in a Small Town '88
Shoot Loud, Louder, I Don't Understand! '66
Stuntwoman '81
A Swingin' Summer '65
Three Musketeers '74
Wild Party '74

Tahnee Welch

Cocoon '85
Cocoon: The Return '88
Lethal Obsession '87

Tracy Welch

Cannon Movie Tales: Sleeping Beauty '89

Gertrude Welcker

Dr. Mabuse the Gambler '22

Tuesday Weld

Author! Author! '82
Cincinnati Kid '65
F. Scott Fitzgerald in Hollywood '76
The Five Pennies '59
Heartbreak Hotel '88
Looking for Mr. Goodbar '77
Mother & Daughter: A Loving War '80
Once Upon a Time in America '84
Pretty Poison '68
A Question of Guilt '78
Reflections of Murder '74
Return to Peyton Place '61
Rock, Rock, Rock '56
Serial '80
Soldier in the Rain '63
Something in Common '86
Thief '81
Who'll Stop the Rain? '78
Wild in the Country '61

Nancy Welford

The Phantom in the House '29

Philip Welford

Paramount Comedy Theater, Vol. 1: Well Developed '87

Karen Well

Burial Ground '85

Mary Louise Weller

Forced Vengeance '82
National Lampoon's Animal House '78

Peter Weller

The Adventures of Buckaroo Banzai '84
Apology '86
Butch and Sundance: The Early Days '79
Dancing Princesses '84
First Born '84
A Killing Affair '85
Leviathan '89
Naked Lunch '91
Of Unknown Origin '83
Rainbow Drive '90
Road to Ruin '91
RoboCop '87
RoboCop 2 '90
Shakedown '88
Shoot the Moon '82
The Tunnel '89
Two Kinds of Love '85
Women & Men: Stories of Seduction '90

Gwen Welles

Desert Hearts '86
The Men's Club '86
New Year's Day '89
Sticky Fingers '88

Mel Welles

Commando Squad '87
Dr. Heckyl and Mr. Hype '80
Joy Ride to Nowhere '78
Little Shop of Horrors '60
Rented Lips '88
The Undead '57
Wizards of the Lost Kingdom 2 '89

Mel Welles (D)

Lady Frankenstein '72

Orson Welles

America at the Movies '76
The Battle of Austerlitz '60
Battle of Neretva '71
Battleforce '78
Black Magic '49
Blood and Guns '79
Bugs Bunny Superstar '75
Butterfly '82
Casino Royale '67
Catch-22 '70
Chimes at Midnight '67
Citizen Kane '41
The Double McGuffin '79
The Enchanted Journey '70s
Ferry to Hong Kong '59
Finest Hours '64
The Great Battle '79
History of the World: Part 1 '81
Hot Money '80s
Inspiration of Mr. Budd '75
Is Paris Burning? '68
Journey into Fear '42
King of Kings '61
The Lady from Shanghai '48
The Late Great Planet Earth '80s
Little World of Don Camillo '51
The Long, Hot Summer '58
Macbeth '48
A Man for All Seasons '66
The Man Who Saw Tomorrow '81
Mister Arkadin '55
Moby Dick '56
Muppet Movie '79
Napoleon '55
Revenge of TV Bloopers '60s
Scene of the Crime '85
Shogun '81
Slapstick of Another Kind '84
Someone to Love '87
Start the Revolution Without Me '70
The Stranger '46
Tales of the Klondike: In a Far Country '81
Tales of the Klondike: Race for Number One '81
Tales of the Klondike: The Scorn of Women '81
Tales of the Klondike: The Unexpected '83
Ten Days Wonder '72
Ten Little Indians '75
The Third Man '49
Toast of the Town '56
Tomorrow is Forever '46
Touch of Evil '58
Treasure Island '72
The Trial '63
Trouble in the Glen '54
12 Plus 1 '70
Voyage of the Damned '76

Waterloo '71
The Witching '72

Orson Welles (D)

Chimes at Midnight '67
Citizen Kane '41
The Lady from Shanghai '48
Macbeth '48
The Magnificent Ambersons '42
Mister Arkadin '55
The Stranger '46
Touch of Evil '58
The Trial '63

Sabra Welles

Tropic Heat '70s

Steve Welles

Puppet Master 2 '91

Virginia Welles

Dynamite '49

Charles Wellesley

A Poor Little Rich Girl '17

David Wellington (D)

The Carpenter '89

Cathy Wellman

The Prodigal Planet '88

William Wellman, Jr.

Born Losers '67
High School Confidential '58
Macumba Love '60
The Prodigal Planet '88
A Swingin' Summer '65
Trial of Billy Jack '74

William A. Wellman

Curfew '88

William A. Wellman (D)

Across the Wide Missouri '51
Battleground '49
Beau Geste '39
Blood Alley '55
Buffalo Bill '44
Goodbye, My Lady '56
Lady of Burlesque '43
Magic Town '47
The Next Voice You Hear '50
Nothing Sacred '37
The Ox-Bow Incident '43
Public Enemy '31
A Star is Born '37
Till the Clouds Roll By/A Star Is Born '46
Wings '27

Aarika Wells

Walking the Edge '83

Ann Wells

Creature of the Walking Dead '65

Carole Wells

Funny Lady '75
The House of Seven Corpses '73

Carrie Wells

The Bad Seed '85

Christine Wells

Ninja Phantom Heroes '87

Daniel Wells

Ninja Masters of Death '85

Dawn Wells

Return to Boggy Creek '77
Town that Dreaded Sundown '76

Dolores Wells

The Time Travelers '64

Doris Wells

Oriane '85

Jacqueline Wells

The Black Cat '34
Bohemian Girl '36
Laurel & Hardy Comedy Classics Volume 4 '32
Riders of the Range/Storm Over Wyoming '50
Tarzan the Fearless '33
Torture Ship '39

Jennifer Wells

Confessions of a Young American Housewife '78

Jerold Wells

The Element of Crime '85

Junior Wells

Chicago Blues '72

Mary Wells

Legendary Ladies of Rock & Roll '86
Rock 'n' Roll History: Girls of Rock 'n' Roll '90

Mel Wells

The She-Beast '65

Shani Wells

Arnold '73

Simon Wells (D)

An American Tail: Fievel Goes West '91

Vernon Wells

American Eagle '90
Circle of Fear '89
Circuitry Man '90
Commando '85
Crossing the Line '90
Enemy Unseen '91
Last Man Standing '87
The Losers '70
Nam Angels '88
The Road Warrior '82
Sexual Response '92
The Shrimp on the Barbie '90
Undeclared War '91

Ken Welsh

And Then You Die '88
The Big Slice '91
The Climb '87
Crocodile Dundee 2 '88
Lost '86
Loyalties '86
Of Unknown Origin '83
Perfectly Normal '91
Screwball Academy '86
The War Boy '85

Margaret Welsh

Mr. & Mrs. Bridge '91

William Welsh

Traffic in Souls '13

Stuart Wilson

The Highest Honor '84
Romance on the Orient
Express '89
Running Blind '70s
Secret Weapon '90

Sue Wilson

Creatures the World Forgot
'70

Teddy Wilson

The Ambush Murders '82
The Benny Goodman Story
'55
Bessie Smith and Friends '86
Kiss Shot '90s
Life Stinks '91
The Mighty Pawns '87
Stars on Parade/Boogie
Woogie Dream '46

Thick Wilson

Adventures Beyond Belief '87

Thomas F. Wilson

Back to the Future '85
Back to the Future, Part 2 '89
Back to the Future, Part 3 '90

Tom Wilson

The Americano '17
Battling Butler '26
California Straight Ahead '25
Let's Get Harry '87

Trey Wilson

Bull Durham '88
End of the Line '88
Great Balls of Fire '89
Miss Firecracker '89
Raising Arizona '87
The Vampire Hookers '78
Welcome Home '89

**Wendy Dawn
Wilson**

The Scorpio Factor '90

Yale Wilson (D)

Truth or Dare? '86

Yvonne Wilson

The Touch of Satan '70

Ana Beatriz Wiltgen

Subway to the Stars '87

Ann Wilton

Man of Evil '48

Cindy Wilton

Swinging Ski Girls '70s

Penelope Wilton

Blame It on the Bellboy '91
Clockwise '86
Cry Freedom '87
Othello '82
Singleton's Pluck '84

George Wiltshire

Killer Diller '48

Simeon Wiltsie

The Wishing Ring '14

Roderick Wimberly

Charlotte Forten's Mission:
Experiment in Freedom '85

Brian Wimmer

China Beach '88
Dangerous Pursuit '89
Late for Dinner '91
The World's Oldest Living
Bridesmaid '92

Cie Cie Win

Marco '73

Jim Winburn (D)

The Death Merchant '91
Evil Altar '89

Simon Wincer (D)

Dark Forces '83
D.A.R.Y.L. '85
Girl Who Spelled Freedom '86
Harley Davidson and the
Marlboro Man '91
The Lighthorseman '87
Phar Lap '84
Quigley Down Under '90

April Winchell

Honey, I Shrunk the Kids '89

Paul Winchell

Perry Como Show '56

**Anna-Maria
Winchester**

Deadly Possession '88

Jeff Winchester

Olivia '83

Jeff Wincott

Deadly Bet '91
Martial Law 2: Undercover '91

Michael Wincott

Robin Hood: Prince of Thieves
'91
Talk Radio '88
Tragedy of Flight 103: The
Inside Story '91
Wild Horse Hank '79

Marc Wincourt

Murmur of the Heart '71

Nadia Windell

Street Warriors '87

**Fred
Windermere** (D)

The Taxi Mystery '26

Marek Windheim

I Married an Angel '42

William Windom

Abduction of St. Anne '75
Brewster McCloud '70
Columbo: Prescription Murder
'67
Committed '91
Desperate Lives '82
Detective '68
Escape from the Planet of the
Apes '71
For Love or Money '63
Funland '89
Girls of Huntington House '73
Grandview U.S.A. '84
Last Plane Out '83
Leave 'Em Laughing '81
Mean Dog Blues '78
The Mephisto Waltz '71
Planes, Trains & Automobiles
'87

Pursuit '72
She's Having a Baby '88
Street Justice '89
To Kill a Mockingbird '62

Allen Windsor

The Incredible Petrified World
'58

Barbara Windsor

Carry On Doctor '68

Chris Windsor (D)

Big Meat Eater '85

Claire Windsor

The Blot '21
Too Wise Wives '21

Marie Windsor

Bedtime Story '63
Cahill: United States Marshal
'73
Cat Women of the Moon '53
City That Never Sleeps '53
Commando Squad '87
Double Deal '50
The Fighting Kentuckian '49
Force of Evil '49
Girl in Black Stockings '57
Hellfire '48
Jungle '52
The Killing '56
Little Big Horn '51
Narrow Margin '52
Outlaw Women '52
Outpost in Morocco '49
The Saint's Vacation/The
Narrow Margin '52
Salem's Lot: The Movie '79
Silver Star '55
Swamp Women '55
Tall Texan '53
Trouble Along the Way '53
Unholy Wife '57
Wild Women '70

Romy Windsor

Big Bad John '90
The House of Usher '88
Howling 4: The Original
Nightmare '88

Simon Windsor (D)

Lonesome Dove '89

Bretaigne Windust

Face to Face '52

**Bretaigne
Windust** (D)

The Enforcer '51
Pied Piper of Hamelin '57

Penelope Windust

Ghost Town '88

Janu Wine

Mantis in Lace '68

Harry Winer (D)

Mirrors '85
Single Bars, Single Women
'84
SpaceCamp '86

Lucy Winer (D)

Rate It X '86

Gil Winfield

Fiend Without a Face '58

Paul Winfield

Angel City '80
Big Shots '87
Blue City '86
Brother John '70
Carbon Copy '81
Conrack '74
Damnation Alley '77
Death Before Dishonor '87
For Us, the Living '88
Gordon's War '73
The Greatest '77
Green Eyes '76
Guilty of Innocence '87
Hero Ain't Nothin' But a
Sandwich '78
High Velocity '76
Huckleberry Finn '74
It's Good to Be Alive '74
King '78
The Mighty Pawns '87
Mike's Murder '84
Presumed Innocent '90
R.P.M.* (*Revolutions Per
Minute) '70
The Serpent and the Rainbow
'87
Sophisticated Gents '81
Sounder '72
Star Trek 2: The Wrath of
Khan '82
The Terminator '84
Twilight's Last Gleaming '77
The Women of Brewster
Place '89

Oprah Winfrey

The Color Purple '85
Native Son '86
Pee-Wee's Playhouse
Christmas Special '88
The Women of Brewster
Place '89

Anna Wing

Xtro '83

Leslie Wing

Cowboy & the Ballerina '84
Dungeonmaster '85
Retribution '88

Mickey Lotus Wing

Hollywood Revels '47

Lee Wing Shan

Four Robbers '70s

Eugenie Wingate

Scream, Baby, Scream '69

Debra Winger

Betrayed '88
Black Widow '87
Cannery Row '82
Everybody Wins '90
French Postcards '79
Legal Eagles '86
Made in Heaven '87
Mike's Murder '84
An Officer and a Gentleman
'82
The Sheltering Sky '90
Slumber Party '57 '76
Terms of Endearment '83
Thank God It's Friday '78
Urban Cowboy '80

Mark Wingett

Quadrophenia '79

Gary Winick (D)

Out of the Rain '90

Angela Winkler

Danton '82
The Lost Honor of Katharina
Blum '75
Sheer Madness '84
The Tin Drum '79

Charles Winkler (D)

Disturbed '90
You Talkin' to Me? '87

Chris Winkler

You Talkin' to Me? '87

Henry Winkler

An American Christmas Carol
'79
Heroes '77
Katherine '75
The Lords of Flatbush '74
Night Shift '82
The One and Only '78
The Radical '75
Video Yesterbloop '70s

Henry Winkler (D)

Memories of Me '88
A Smoky Mountain Christmas
'86

Irwin Winkler (D)

Guilty by Suspicion '91

**Terence H.
Winkless** (D)

The Berlin Conspiracy '91
Bloodfist '89
The Nest '88

Kitty Winn

The Exorcist '73
Exorcist 2: The Heretic '77
Miles to Go Before I Sleep '74
Panic in Needle Park '75
They Might Be Giants '71

Michael Winner (D)

Appointment with Death '88
The Big Sleep '78
Bullseye! '90
Chato's Land '71
A Chorus of Disapproval '89
Cool Mikado '63
Death Wish '74
Death Wish 2 '82
Death Wish 3 '85
Firepower '79
The Mechanic '72
Murder on the Campus '52
The Nightcomers '72
Scream for Help '86
The Sentinel '76
The Stone Killer '73
The Wicked Lady '83

Vic Winner

Horror Rises from the Tomb
'72
The Rue Morgue Massacres
'73

Olof Winnerstrand

Night Is My Future '47

Gary Winnick (D)

Curfew '88

Lucyna Winnicka

First Spaceship on Venus '60
25 Fireman's Street '73

Edward Woodward

The Appointment '82
The Bloodsuckers '70
Breaker Morant '80
Champions '84
Codename Kyril '91
The Final Option '82
King David '85
Merlin and the Sword '85
Mister Johnson '91
The Wicker Man '75
Young Winston '72

Joanne Woodward

A Big Hand for the Little Lady '66
A Christmas to Remember '78
Crisis at Central High '80
The Drowning Pool '75
The End '78
A Fine Madness '66
From the Terrace '60
The Fugitive Kind '60
The Glass Menagerie '87
Harry & Son '84
The Long, Hot Summer '58
Mr. & Mrs. Bridge '91
A New Kind of Love '63
Paris Blues '61
Rachel, Rachel '68
See How She Runs '78
The Shadow Box '80
The Streets of L.A. '79
The Stripper '63
Summer Wishes, Winter Dreams '73
Sybil '76
They Might Be Giants '71
The Three Faces of Eve '57
Winning '69

Joanne Woodward (D)

Come Along with Me '84

John Woodward (D)

Stephen King's Nightshift Collection '80s

Morgan Woodward

Dark Before Dawn '89
Girls Just Want to Have Fun '85
The Longest Drive '76
Moonshine County Express '77
Ride in a Pink Car '74
Running Wild '73
Speedtrap '78
Walking Tall: The Final Chapter '77
Which Way Is Up? '77

Tim Woodward

The Europeans '79

Norman Wooland

Teenage Bad Girl '59

Susan Wooldridge

How to Get Ahead in Advertising '89
Loyalties '86

Chuck Woolery

A Guide for the Married Woman '78

Gerry Woolery (D)

Jokes My Folks Never Told Me '77

Charles Woolf

No Way Back '74

Henry Woolf

The Love Pill '71

Monty Woolley

As Young As You Feel '51
The Bishop's Wife '47
The Man Who Came to Dinner '41
Night and Day '46
Since You Went Away '44

Susan Woolridge

Hope and Glory '87
The Jewel in the Crown '84
Twenty-One '91

Robert Woolsey

Cockeyed Cavaliers '34
Diplomaniacs '33
Dixiana '30
Half-Shot at Sunrise '30
Hips, Hips, Hooray '34
Hold 'Em Jail '32
Hook, Line and Sinker '30
Kentucky Kernels '34

Tom Wopat

Burning Rage '84
Christmas Comes to Willow Creek '87

Smith Wordes

Deadly Dancer '90

Richard Wordsworth

The Quatermass Experiment '56

Carl Workman (D)

The Money '75

Chuck Workman (D)

Kill Castro '80
The Mercenaries '80
Stoogemania '85
Sweet Dirty Tony '81

Jimmy Workman

The Addams Family '91

Nanette Workman

Evil Judgement '85

Christina World

Golden Lady '79

Frederick Worlock

101 Dalmations '61
Strange Cargo '40

Mary Woronov

Angel of H.E.A.T. '82
Challenge of a Lifetime '85
Chopping Mall '86
Club Fed '91
Death Race 2000 '75
Eating Raoul '82
Hellhole '85
Hollywood Boulevard '76
Hollywood Man '76
Jackson County Jail '76
Kemek '88
Let It Ride '89
Mortuary Academy '91
The Movie House Massacre '78
Night of the Comet '84

The Princess Who Never Laughed '84

Rock 'n' Roll High School '79
Rock 'n' Roll High School Forever '91
Scenes from the Class Struggle in Beverly Hills '89
Silent Night, Bloody Night '73
Sugar Cookies '77
Terrorvision '86
Warlock '91

Mary Woronov (D)

Eating Raoul '82

Wallace Worsley (D)

The Hunchback of Notre Dame '23

Clark Worswick (D)

Agent on Ice '86

Aaron Worth (D)

9 1/2 Ninjas '90

Barbara Worth

The Prairie King '27

Constance Worth

Criminals Within '41
Dawn Express '42
G-Men vs. the Black Dragon '43
Kid Sister '45
Windjammer '31

David Worth (D)

Kickboxer '89
Warrior of the Lost World '84

Harry Worth

Tough to Handle '37

Irene Worth

Deathtrap '82
The Displaced Person '76
Eyewitness '81
Forbidden '85
King Lear '71
Separate Tables '83

Lilian Worth

Tarzan the Tiger '29

Mike Worth

Final Impact '91
Street Crimes '92

Nancy Worth

Raiders of Sunset Pass '43

Nicholas Worth

Dirty Laundry '87
Doin' Time '85
The Red Raven Kiss-Off '90

J.T. Wotton

Virgin Queen of St. Francis High '88

Helmut Woudenberg

In for Treatment '82

Martin Wragge (D)

Last Warrior '89

Fay Wray

Adam Had Four Sons '41
Blind Husbands '19
Bulldog Jack '35

The Coast Patrol '25

The Cobweb '55
Doctor X '32
The Evil Mind '34
Gideon's Trumpet '80
Hell on Frisco Bay '55
King Kong '33
Melody for Three '41
The Most Dangerous Game '32
Mystery of the Wax Museum '33
Rock, Pretty Baby '56
Small Town Girl '53
Tammy and the Bachelor '57
Treasure of the Golden Condor '53
The Vampire Bat '32
Wedding March '28
When Knights were Bold '36
Woman in the Shadows '34

John Wray

All Quiet on the Western Front '30

John Wray (D)

Beau Revel '21

Caspar Wrede (D)

One Day in the Life of Ivan Denisovich '71
The Terrorists '74

Clare Wren

Season of Fear '89
Steel and Lace '90

Doug Wren

Bad Taste '88

Michael Wren

Dynamite and Gold '88

Amy Wright

The Accidental Tourist '88
The Amityville Horror '79
Breaking Away '79
Daddy's Dyin'...Who's Got the Will? '90
The Deer Hunter '78
Final Verdict '91
Girlfriends '78
Inside Moves '80
Love Hurts '91
Off Beat '86
Stardust Memories '80
The Telephone '87
Trapped in Silence '90
Wise Blood '79

Ben Wright

101 Dalmations '61

Cobina Wright, Jr.

Charlie Chan in Rio '41

Jenny Wright

The Chocolate War '88
The Executioner's Song '82
I, Madman '89
The Lawnmower Man '92
Near Dark '87
Out of Bounds '86
St. Elmo's Fire '85
A Shock to the System '90
Twister '89
Valentino Returns '88
The Wild Life '84
Young Guns 2 '90

Ken Wright

Eye of the Eagle 3 '91
Hollywood Boulevard 2 '89

Opposing Force '87

Mack V. Wright (D)

Big Show '37
Hit the Saddle '37
Riders of the Whistling Skull '37
Robinson Crusoe of Clipper Island '36
Robinson Crusoe of Mystery Island '36
Rootin' Tootin' Rhythm '38
Sea Hound '47
The Vigilantes Are Coming '36
Western Double Feature 3 '36
Winds of the Wasteland '36

Maggie Wright

Sex and the Other Woman '79
Twins of Evil '71

Marcella Wright

The Hanging Woman '72

Max Wright

Fraternity Vacation '85
Touch and Go '86

Max Wright (D)

Western Double Feature 5: Rogers & Autry '37

Michael Wright

The Five Heartbeats '91
Streamers '83

Michael David Wright

Malibu Bikini Shop '86

Nory Wright

Hustler Squad '76

Patrick Wright

The Abductors '72
If You Don't Stop It...You'll Go Blind '77

Patrick Wright (D)

Hollywood High '77

Richard Wright

Native Son '51

Robin Wright

Denial: The Dark Side of Passion '91
Hollywood Vice Sqaud '86
The Playboys '92
The Princess Bride '87
State of Grace '90

Samuel E. Wright

Bird '88
The Little Mermaid '89

Steven Wright

Comic Relief 3 '89
Desperately Seeking Susan '85
Evening at the Improv '86
Men of Respect '91
Reunion: 10th Annual Young Comedians '87
Sledgehammer '83

Teresa Wright

The Best Years of Our Lives '46
Bill: On His Own '83
Escapade in Japan '57
The Fig Tree '87
Flood! '76

Wright ▶cont.

The Good Mother '88
Hail, Hero! '69
The Happy Ending '69
The Little Foxes '41
The Men '50
Mrs. Miniver '42
The Pride of the Yankees '42
Pursued '47
Roseland '77
The Search for Bridey Murphy '56
Shadow of a Doubt '42
Somewhere in Time '80

Thomas J. Wright (D)

The Fatal Image '90
No Holds Barred '89
Snow Kill '90
Torchlight '85

Tom Wright

Marked for Death '90

Tom Wright (D)

Deadly Game '91

Tony Wright

Broth of a Boy '59
Flanagan Boy '53
The House in Marsh Road '60
The Spaniard's Curse '58

Willaim Wright

American Gothic '88
Infamous Crimes '47
Miss Grant Takes Richmond '49
A Night to Remember '42

Maris Wrixon

The Ape '40
British Intelligence '40
Highway 13 '48
Sons of the Pioneers '42
Waterfront '44
White Pongo '45

Donald Wrye (D)

Born Innocent '74
Death Be Not Proud '75
Ice Castles '79

James Wu (D)

Ninja Condors '87

John Wu

Ninja vs. the Shaolin '84

Vivian Wu

Iron & Silk '91

Robert Wuhl

Batman '89
Blaze '89
Bull Durham '88
Good Morning, Vietnam '87
Tales from the Crypt '89

Kari Wuhrer

The Adventures of Ford Fairlane '90
Beastmaster 2: Through the Portal of Time '91

Kai Wulff

Firefox '82
Jungle Warriors '84

Richard Wulicher (D)

House of Shadows '83

Shelley Wyant

Home Free All '84
Pilgrim Farewell '82

Jane Wyatt

The Adventures of Tom Sawyer '73
Amityville 4 '89
Buckskin Frontier '43
Katherine '75
Lost Horizon '37
Nativity '78
The Navy Comes Through '42
None But the Lonely Heart '44
The Radical '75

Margaret Wycherly

Call It Murder '34
Midnight '34
Random Harvest '42
White Heat '49
The Yearling '46

Alix Wyeth

Jud '71

Katya Wyeth

Dressed for Death '74
Twins of Evil '71

Paul Wyett

A Soldier's Tale '91

Michael Wyle

Appointment with Fear '85

Link Wyler

Grizzly Adams: The Legend Continues '90

Richard Wyler

Rattler Kid '70s

William Wyler (D)

Ben-Hur '59
The Best Years of Our Lives '46
Big Country '58
Carrie '52
The Children's Hour '62
The Collector '65
Come and Get It '36
Dead End '37
Desperate Hours '55
Detective Story '51
Dodsworth '36
Friendly Persuasion '56
Funny Girl '68
The Heiress '49
Jezebel '38
The Letter '40
Liberation of L.B. Jones '70
The Little Foxes '41
Mrs. Miniver '42
Roman Holiday '53
These Three '36
War in the Sky '82
The Westerner '40
Wuthering Heights '39

John Wylie

An Empty Bed '90

Bill Wyman

Eat the Rich '87
Gimme Shelter '70
Ready Steady Go, Vol. 1 '83
Ready Steady Go, Vol. 2 '85
Ready Steady Go, Vol. 3 '84

25x5: The Continuing Adventures of the Rolling Stones '90

Jane Wyman

Bon Voyage! '62
Footlight Serenade '42
Here Comes the Groom '51
The Incredible Journey of Dr. Meg Laurel '79
It's a Great Feeling '49
Johnny Belinda '48
The Lost Weekend '45
Magic Town '47
Magnificent Obsession '54
Night and Day '46
Pollyanna '60
Stage Fright '50
Three Guys Named Mike '51
The Yearling '46

John Wyman

For Your Eyes Only '81
Tuxedo Warrior '82

Patrick Wymark

The Blood on Satan's Claw '71
The Conqueror Worm '68
Journey to the Far Side of the Sun '69
Repulsion '65
The Skull '65
Where Eagles Dare '68
Woman Times Seven '67

Patricia Wymer

The Young Graduates '71

Patrice Wymore

Big Trees '52
The King's Rhapsody '55
Tea for Two '50

Carol Wyndham

Roamin' Wild '36

George Wyner

The Bad News Bears Go to Japan '78
Spaceballs '87

Peter Wyngarde

Burn Witch, Burn! '62
Double Cross '90s

Bob Wynn (D)

Resurrection of Zachary Wheeler '71

Ed Wynn

The Absent-Minded Professor '61
Alice in Wonderland '51
Babes in Toyland '61
Cinderfella '60
Dear Brigitte '65
The Diary of Anne Frank '59
Gnome-Mobile '67
The Greatest Story Ever Told '65
Marjorie Morningstar '58
Mary Poppins '64
Requiem for a Heavyweight '56
Those Calloways '65

Keenan Wynn

The Absent-Minded Professor '61
The Americanization of Emily '64
Around the World Under the Sea '65

Battle Circus '53
The Belle of New York '52
Best Friends '82
Call to Glory '84
Cancel My Reservation '72
The Capture of Grizzly Adams '82
Chaplin: A Character Is Born '79
Chaplin: A Character Is Born/ Keaton: The Great Stone Face
The Clock '45
The Clonus Horror '79
Coach '78
The Dark '79
Deep Six '58
Devil's Rain '75
Dr. Strangelove, or: How I Learned to Stop Worrying and Love the Bomb '64
Finian's Rainbow '68
For Me & My Gal '42
The Glass Slipper '55
Great Race '65
Hard Knocks
He is My Brother '75
Herbie Rides Again '74
High Velocity '76
Hit Lady '74
Hole in the Head '59
The Hucksters '47
Hyper-Sapien: People from Another Star '86
Internecine Project '73
Just Tell Me What You Want '80
Killer Inside Me '76
Kiss Me Kate '53
Laserblast '78
The Last Unicorn '82
Legend of Earl Durand '74
The Long, Long Trailer '54
The Longest Drive '76
The Lucifer Complex '78
MacKenna's Gold '69
The Man in the Gray Flannel Suit '56
The Man Who Would Not Die '75
Manipulator '71
The Mechanic '72
Mid Knight Rider '84
Mission to Glory '80
Mom, the Wolfman and Me '80
Monster '78
My Dear Secretary '49
Naked Hills '56
Nashville '75
Neptune's Daughter '49
Once Upon a Time in the West '68
Orca '77
The Patsy '64
Perfect Furlough '59
Phone Call from a Stranger '52
A Piano for Mrs. Cimino '82
Piranha '78
Point Blank '67
Promise Her Anything '66
Requiem for a Heavyweight '56
Royal Wedding '51
Santa Claus is Coming to Town '70
Shack Out on 101 '55
The Shaggy D.A. '76
Since You Went Away '44
Smith! '69
Snowball Express '72
Son of Flubber '63
Song of the Thin Man '47
Texas Carnival '51
Three Little Words '50

Three Musketeers '48
A Time to Love & a Time to Die '58
The Treasure Seekers '79
Viva Max '69
The War Wagon '67
Wavelength '83

Kitty Wynn

Mirrors '78

Peggy Wynn

Wild Country '47

Tracy Keenan Wynn (D)

Hit Lady '74

Christopher Wynne

Remote Control '88

Greg Wynne

Mystery Mansion '83

Jim Wynorski (D)

Big Bad Mama 2 '87
Chopping Mall '86
Deathstalker 2: Duel of the Titans '87
The Lost Empire '83
Munchie '92
976-EVIL 2: The Astral Factor '91
Not of This Earth '88
The Return of Swamp Thing '89
Sorority House Massacre 2 '92
Transylvania Twist '89

Dana Wynter

The Connection '73
D-Day, the Sixth of June '56
Dead Right '88
Invasion of the Body Snatchers '56
The List of Adrian Messenger '63
Lovers Like Us '75
Santee '73
Sink the Bismarck '60
Something of Value '57

Charlotte Wynters

Harvest Melody '43
Panama Patrol '39
Renegade Trail '39
Sinners in Paradise '38

Amanda Wyss

Better Off Dead '85
Black Magic Woman '91
Deadly Innocence '88
Fast Times at Ridgemont High '82
A Nightmare on Elm Street '84
Powwow Highway '89
Shakma '89
To Die for '89
To Die for 2 '91

Megan Wyss

Night Screams '87

Nelson Xavier

Gabriela '84

Maria Xenia

Hero Bunker '71

Deng Xiaotuang

Girl from Hunan '86

Richard Wulicher (D)

Yabo Yablonsky *(D)*

Manipulator '71

Frank Yaconelli

The Barber Shop '33
Escape to Paradise '39
Gun Play '36
Lone Star Law Men '42
South of Monterey '47
Western Frontier '35
Western Mail '42
Wild Horse Canyon '39

Joseph Yadin

Four in a Jeep '51

Yossi Yadin

Lies My Father Told Me '75

Kelly Yaegermann

Cruisin' High '75

Lior Yaeni

But Where Is Daniel Wax? '74

Jeff Yagher

Lower Level '91

Karen Yahng

Almost You '85
Birdy '84
Criminal Law '89
Deep in the Heart '84
Heat '87
Jaws: The Revenge '87
Little Sweetheart '90
Maria's Lovers '84
Night Game '89
9 1/2 Weeks '86
The Ten Million Dollar
   Getaway '91
Torch Song Trilogy '88

Karen Yahng *(D)*

Deadly Darling '85

Yuri Yakovlev

The Seagull '71

Michael Yakub *(D)*

The Shaman '87

Koji Yakusho

Tampopo '86

Simon Yam

Tongs: An American
   Nightmare '88

Isuzu Yamada

The Lower Depths '57
Osaka Elegy '36
Sisters of Gion '36
Throne of Blood '57

Isao Yamagata

Warning from Space '56

**Kazuhiko
Yamaguchi** *(D)*

Sister Street Fighters '76

**Takaya
Yamamauchi**

MacArthur's Children '85

Fujiko Yamamoto

Golden Demon '53

So Yamamura

Barbarian and the Geisha '58

Gung Ho '85
The Human Condition: No
   Greater Love '58
Tokyo Story '53

Al Yamanouchi

The Lone Runner '88
2020 Texas Gladiators '85

Susan W. Yamasaki

Going Back '83

Tadashi Yamashita

Bronson Lee, Champion '78
Shinobi Ninja '84
Sword of Heaven '81

Akira Yamauchi

Godzilla vs. the Smog
   Monster '72

Tsutomu Yamazaki

The Funeral '84
Kagemusha '80
Rikyu '90
Tampopo '86
A Taxing Woman '87

Suuzu Yamda

Yojimbo '62

Nancy Yan

Born Invincible '76

Eijiro Yanagi

Shin Heike Monogatari '55

**Mitsuo
Yanagimachi** *(D)*

Himatsuri (Fire Festival) '85

Shinichi Yanagisawa

X from Outer Space '67

Ernesto Yanez

La Montana del Diablo (Devil's
   Mountain) '87

Chiang Yang

Challenge of the Masters '89

C.Y. Yang *(D)*

Bruce Li in New Guinea '80

Laura Yang

Key to Vengeance

Tiger Yang

Rage of the Master '70s

Yeh Fei Yang

Instant Kung-Fu Man '84

Yangzi

Kung Fu Massacre '82

Weird Al Yankovic

The Naked Gun: From the
   Files of Police Squad '88
UHF '89

Oleg Yankovsky

The Mirror '76

Philip Yankovsky

The Mirror '76

Jean Yanne

Cobra '71
Hanna K. '83
Le Boucher '69
Madame Bovary '91

Quicker Than the Eye '88
This Man Must Die '70
Weekend '67
The Wolf at the Door '87

Rossana Yanni

Fangs of the Living Dead '68
The Rue Morgue Massacres
   '73
White Comanche '67

Lillian Yarbo

You Can't Take It with You
   '38

**Jean
Yarborough** *(D)*

The Brute Man '46
The Creeper '48
The Devil Bats '41
The Gang's All Here '41
Hillbillys in a Haunted House
   '67
Holiday in Havana '49
Jack & the Beanstalk '52
King of the Zombies '41
Law of the Jungle '42
Let's Go Collegiate '41
Lure of the Islands '42
The Naughty Nineties '45
Panama Menace '41
Under Western Stars '45

Margaret Yarde

The Deputy Drummer '35

Bob Yari *(D)*

Mind Games '89

Celeste Yarnall

The Velvet Vampire '71

Lorene Yarnell

Spaceballs '87

Trever Yarrish

Spirit of the Eagle '90

Amy Yasbeck

Problem Child '90
Problem Child 2 '91

Shoji Yasui

The Burmese Harp '56

Rikiya Yasuoka

Tampopo '86

Cassie Yates

Convoy '78
The Evil '78
Father Figure '80
F.I.S.T. '78
FM '78
Of Mice and Men '81
Perry Mason Returns '85
St. Helen's, Killer Volcano '82
Unfaithfully Yours '84

Peter Yates *(D)*

Breaking Away '79
Bullitt '68
The Deep '77
Dresser '83
Eleni '85
Eyewitness '81
For Pete's Sake '74
The Hot Rock '70
The House on Carroll Street
   '88
An Innocent Man '89
Koroshi '67
Krull '83

Mother, Jugs and Speed '76
Murphy's War '71
Robbery '67
Suspect '87

Tomonori Yazaki

Godzilla's Revenge '69

Biff Yeager

Girls Just Want to Have Fun
   '85

**Irvin S. Yeawarth,
Jr.** *(D)*

The Blob '58
Dinosaurus! '60
The 4D Man '59

Eric Yee

Snake Fist Dynamo '84

Kelvin Han Yee

A Great Wall '86

Kong Yeh

Super Kung Fu Kid

Tien Yeh

The Furious Avenger '76

Gaye Yellen

Night of Bloody Horror '76

Peter Yellen

Ms. 45 '81

Yellowman

Live & Red Hot '87

Alan Yen

Killer Elephants '76

Chao Hsin Yen

Bells of Death '90

Lam Chen Yen

The Reincarnation of Golden
   Lotus '89

Nancy Yen

The Ninja Pirates '90

Isaac Yeshurun *(D)*

Noa at Seventeen '82

John Yesno

King of the Grizzlies '69

Richard Yesteran

Green Inferno '80s

Bolo Yeung

Bloodfight '90
Bloodsport '88
Breathing Fire
Tiger Claws '91

Chang Yi

Shaolin Traitor '82
Ten Brothers of Shao-lin '80s
Two Assassins in the Dark
   '80s

Chiang Yi

Bells of Death '90

Liu Hao Yi

Kung Fu Terminator
The Young Taoism Fighter

Maria Yi

Fists of Fury '73

Zhang Yimou *(D)*

Red Sorghum '87

Mao Yin

Moonlight Sword & Jade Lion
   '70s

Yu In Yin

The Bloody Fight '70s

Chen Ying

Tough Guy '70s

Cheung Ying

Lightning Kung Fu '70s

Mao Ying

Dance of Death '84
La Venganza del Kung Fu '87
The Two Great Cavaliers '73

Pai Ying

Dynasty '77
Militant Eagle
Ninja Massacre '84

David Yip

Ping Pong '87

Richard Yniguez

Boulevard Nights '79
Sniper '75

Stephen Yoakam

Patti Rocks '88

Erica Yohn

An American Tail: Fievel Goes
   West '91

Edie Yolles *(D)*

That's My Baby! '88

Soh Yomamura

Tora! Tora! Tora! '70

Kevin Yon

Blind Faith '89

Wladimir Yordanoff

Vincent & Theo '90

Amanda York

Scrubbers '82

Brittany York

I Posed for Playboy '91

Dick York

Inherit the Wind '60
They Came to Cordura '59

Francine York

The Centerfold Girls '74
Curse of the Swamp Creature
   '66
The Doll Squad '73
Flood! '76
Secret File of Hollywood '62

Jeff York

Davy Crockett and the River
   Pirates '56
The Lady Says No '51
Old Yeller '57
Savage Sam '63
Westward Ho, the Wagons!
   '56
The Yearling '46

John York

House of the Rising Sun '87

Wedding Rehearsal '32
The Young in Heart '38

Roxie Young

In Your Face '77

Sean Young

Baby...Secret of the Lost
Legend '85
Blade Runner '82
The Boost '88
Cousins '89
Dune '84
Fire Birds '90
Jane Austen in Manhattan '80
A Kiss Before Dying '91
Love Crimes '91
No Way Out '87
Once Upon a Crime '91
Stripes '81
Under the Biltmore Clock '85
Wall Street '87
Young Doctors in Love '82

Stephen Young

Deadline '82
Lifeguard '75
Patton '70

Steve Young

Crossfire '89

Terence Young (D)

The Amorous Adventures of
Moll Flanders '65
The Christmas Tree '69
Cold Sweat '74
Dr. No '62
From Russia with Love '63
The Jigsaw Man '84
The Klansman '74
The Poppy Is Also a Flower
'66
Red Sun '71
Sidney Sheldon's Bloodline
'79
Thunderball '65
Triple Cross '67
Valley of the Eagles '51
Wait Until Dark '67
When Wolves Cry '69

Tony Young

Chrome and Hot Leather '71
Policewomen '73

Trudy Young

The Reincarnate '71

Victor Sen Young

Castle in the Desert '42

**William Allen
Young**

Johnnie Gibson F.B.I. '87
Outrage '86

Jack Youngblood

Python Wolf '88
Stalking Danger '88

Barrie Youngfellow

It Came Upon a Midnight
Clear '84
Lady from Yesterday '85
Nightmare in Blood '75

Gary Youngman (D)

Dead as a Doorman '85

Henny Youngman

Amazon Women On the Moon
'87

Laughing Room Only '86
Unkissed Bride '66

Gail Youngs

Belizaire the Cajun '86
Hockey Night '84
Last Days of Frank & Jesse
James '86
Timestalkers '87

Jim Youngs

Hot Shot '86
Nobody's Fool '86
You Talkin' to Me? '87

**Robert
Youngson** (D)

Days of Thrills and Laughter
'61
When Comedy was King '59

Jerry Younkins (D)

The Demon Lover '77

Igor Youskevitch

Invitation to the Dance '56

Chief Yowlachie

Bowery Buckaroos '47
King of the Stallions '42
Rose Marie '54
Wild West '46

Frank Ysconelli

Across the Plains '39

Albert Yu (D)

Devil Woman '76

Jimmy Wang Yu

Furious Slaughter
Return of the Chinese Boxer
'74
Tattooed Dragon '82
Thundering Ninja '80s

Steve Yu

Thunderfist '73

Wang Yu

Blood of the Dragon '73
Crazy Kung Fu Master
Fury of King Boxer '83
The Great Hunter '75
Invincible Sword '78

Yu Chan Yuan

Master of Kung Fu '77

Noriaki Yuasa (D)

Gamera '66
Gamera vs. Gaos '67
Gamera vs. Guiron '69
Gamera vs. Zigra '71

Choi Yue

Militant Eagle

Sun Yueh

Ninja Supremo '90

Corey Yuen (D)

No Retreat, No Surrender '86
No Retreat, No Surrender 2
'89

Lee Fat Yuen

Rivals of the Silver Fox '80s

Simon Yuen

Eagle's Shadow '84

Yeh Yuen

The Cavalier '80s

Jimmy Yuill

Paper Mask '91

Saori Yuki

The Family Game '83

Ian Yule

Chain Gang Killings '85
City of Blood '88

Joe Yule

New Moon '40

Harris Yulin

Another Woman '88
Bad Dreams '88
The Believers '87
Candy Mountain '87
Dynasty '82
End of the Road '70
Fatal Beauty '87
Ghostbusters 2 '89
Judgment in Berlin '88
Melvin Purvis: G-Man '74
Night Moves '75
St. Ives '76
Scarface '83
Short Fuse '88
Tailspin: Behind the Korean
Airline Tragedy '89
Watched '73
When Every Day was the
Fourth of July '78

Raoru Yumi

Phoenix '78

Ling Yun

Marvelous Stunts of Kung Fu
'83

Johnny Yune

They Call Me Bruce '82
They Still Call Me Bruce '86

Johnny Yune (D)

They Still Call Me Bruce '86

Sen Yung

Betrayal from the East '44
Charlie Chan in Rio '41
Charlie Chan at the Wax
Museum '40
Murder Over New York '40

Victor Sen Yung

Across the Pacific '42
The Letter '40
She Demons '58
Ticket to Tomahawk '50

Blanche Yurka

Bridge of San Luis Rey '44
City for Conquest '40
Lady for a Night '42
A Night to Remember '42
The Southerner '45

Larry Yust (D)

Homebodies '74
Say Yes! '86

Peter Yuval (D)

Dead End City '88
Firehead '90

Brian Yuzna (D)

Bride of Re-Animator '89
Silent Night, Deadly Night 4:
Initiation '90

Society '92

**Jose Maria
Zabalza** (D)

The Fury of the Wolfman '70

Lila Zaborin

Blood Orgy of the She-Devils
'74

Zabou

The Perils of Gwendoline '84

Grace Zabriskie

Ambition '91
Body Rock '84
The Boost '88
Child's Play 2 '90
Drugstore Cowboy '89
East of Eden '80
Intimate Stranger '91
Megaville '91
Nickel Mountain '85
Prison Stories: Women on the
Inside '91
Servants of Twilight '91

Oliver Zabriskie

Summerdog '78

Tavia Zabriskie

Summerdog '78

**Alfredo
Zacharias** (D)

Bandits '73
Crossfire '67
Demonoid, Messenger of
Death '81

Ann Zacharias

Nea '78

Steffen Zacharias

They Call Me Trinity '72

Michael Zachary (D)

Sin of Adam & Eve '72

John Zacherle

Geek Maggot Bingo '83

**Krystyna
Zachwatowicz**

Man of Marble '76

Ramy Zada

After Midnight '89
Two Evil Eyes '91

Arnon Zadok

Beyond the Walls '85
Torn Apart '89
Unsettled Land '88

Pia Zadora

Butterfly '82
Fake Out '82
Hairspray '88
The Lonely Lady '83
Pajama Tops '83
Santa Claus Conquers the
Martians '64
Voyage of the Rock Aliens '87

Yoav Zafir

Late Summer Blues '87

Eleni Zafirou

A Matter of Dignity '57

Anita Zagaria

Queen of Hearts '89

Frank Zagarino

Barbarian Queen '85
Project: Eliminator '91
Project: Shadowchaser '92
The Revenger '90
Striker '88

Robert Zagone (D)

Stand-In '85

Raja Zahr (D)

The Last Season '87

Ludmila Zaisova

Little Vera '88

Ichiro Zaitsu

The Funeral '84

Jerry Zaks

Gentleman Bandit '81

Roxana Zal

River's Edge '87
Shattered Spirits '91
Table for Five '83
Testament '83
Under the Boardwalk '89

Nancy Zala (D)

Round Numbers '91

Halina Zalewska

Planet on the Prowl '65

Zaira Zambello

Bye Bye Brazil '79

Del Zamora

Heat Street '87

Lydia Zamora

Amenaza Nuclear

Luigi Zampa (D)

Tigers in Lipstick '80
A Woman of Rome '56

Mario Zampi (D)

The Naked Truth '58
Too Many Crooks '59

Elliot Zamuto

The Legend of the Wolf
Woman '77

Philip Zanden

The Mozart Brothers '86

Billy Zane

Back to the Future '85
Blood & Concrete: A Love
Story '91
Brotherhood of Justice '86
Dead Calm '89
Femme Fatale '90
The Hillside Strangler '89
Megaville '91
Memphis Belle '90

Lisa Zane

Bad Influence '90
Femme Fatale '90
Freddy's Dead: The Final
Nightmare '91
Pucker Up and Bark Like a
Dog '89

Giancarlo Zanetti

The Warning '80

Bruno Zanin

Amarcord '74
Boss Is Served '60s

Susanne Zanke (D)

The Scorpion Woman '89

Lenore Zann

American Nightmare '81
Mania '80s
One Night Only '84

Angelo Zanolli

Maciste in Hell '60
Son of Samson '62

Ramon Zanora

Return of the Dragon '84

Lili Fini Zanuck (D)

Rush '91

Krzysztof Zanussi (D)

The Catamount Killing '74
The Unapproachable '82
Year of the Quiet Sun '84

Bob Zany

Up Your Alley '89

Janusz Zaorski (D)

Baritone '85

Vanessa Zaoui

Alan & Naomi '92

Zbigniew Zapassiewicz

Baritone '85
Stand Off '89

Carmen Zapata

Boulevard Nights '79
The Last Porno Flick '74

Dweezil Zappa

Pretty in Pink '86
The Running Man '87

Frank Zappa

Baby Snakes: The Complete
    Version '79
Frank Zappa's Does Humor
    Belong in Music? '86
Head '68
200 Motels '71

Frank Zappa (D)

Baby Snakes: The Complete
    Version '79
200 Motels '71

Moon Zappa

The Boys Next Door '85
Nightmares '83
Spirit of '76 '91

Alfredo Zarcharias (D)

The Bees '78

Mier Zarchi (D)

Don't Mess with My Sister!
    '85
I Spit on Your Grave '77

Joan Zaremba

20 Million Miles to Earth '57

Asher Zarfati

Secret of Yolanda '82

Tony Zarindast

Kill Alex Kill '83

Tony Zarindast (D)

The Guns and the Fury '83
Hardcase and Fist '89
Treasure of the Lost Desert
    '83

Janet Zarish

Danny '79

Manuel Zarzo

The 317th Platoon '65

Michael Zaslow

You Light Up My Life '77

Cesare Zavattini (D)

Love in the City '53

Magdalena Zawadzka

Colonel Wolodyjowski '69

Alfonso Zayas

Esta Noche Cena Pancho '87
The Neighborhood Thief '83

Edmund Zayenda

A Brivele der Mamen (A Letter
    to Mother) '38
Mamele (Little Mother) '38

Edwin Zbonek (D)

The Monster of London City
    '64

Alexandre Zbruev

The Inner Circle '91

Robert Z'Dar

The Big Sweat '91
Blood Money '91
Dead End City '88
The Divine Enforcer '91
Evil Altar '89
Final Sanction '89
The Killer's Edge '90
Maniac Cop '88
Maniac Cop 2 '90
Night Stalker '87
Quiet Fire '91
Soultaker '90
Tango and Cash '89

Kristi Zea (D)

Women & Men: Stories of
    Seduction, Part 2 '91

Nick Zedd

Franck Goldberg Videotape

Nick Zedd (D)

Geek Maggot Bingo '83

Young Zee

Bruce's Deadly Fingers '80s

Franco Zeffirelli

Bellissimo: Images of the
    Italian Cinema '87

Franco Zeffirelli (D)

Brother Sun, Sister Moon '73
The Champ '79

Endless Love '81
Hamlet '90
Jesus of Nazareth '77
Otello '86
Romeo and Juliet '68
Taming of the Shrew '67

Primo Zeglio (D)

Mission Stardust '68

Rafal Zeilinski (D)

Screwballs '83

Alfred Zeisler (D)

Amazing Adventure '37
Enemy of Women '44
Fear '46
Make-Up '37
Parole, Inc '49

Michelle Zeitlin

Dance Macabre '91

Jitka Zelenohorska

Larks on a String '68

Andrew Zeller

American Born '89
Screwball Hotel '88

Jerzy Zelnick

Pharaoh '66

Michael Zelniker

Bird '88
Pick-Up Summer '79

Yuri Zeltser (D)

Eye of the Storm '91

Karel Zeman (D)

On the Comet '68
The Original Fabulous
    Adventures of Baron
    Munchausen '61

Jaimie Zemarel

Night Beast '83

Robert Zemeckis (D)

Amazing Stories, Book 2 '86
Back to the Future '85
Back to the Future, Part 2 '89
Back to the Future, Part 3 '90
I Wanna Hold Your Hand '78
Romancing the Stone '84
Tales from the Crypt '89
Used Cars '80
Who Framed Roger Rabbit?
    '88

Suzanne Zenor

The Baby '72

Will Zens (D)

The Fix '84
Hell on Wheels '67
Hot Summer in Barefoot
    County '74
Trucker's Woman '83

Edward Zentara

Ay, Carmela! '90

Fereno Zenthe

The Revolt of Job '84

Delphine Zentout

36 Fillete '88

Eracio Zepeda

El Norte '83

Gerard Zepeda

Night of the Bloody Apes '68

Anthony Zerbe

Attica '80
Child of Glass '78
Cool Hand Luke '67
Dead Zone '83
Farewell, My Lovely '75
KISS Meets the Phantom of
    the Park '78
Liberation of L.B. Jones '70
License to Kill '89
Omega Man '71
Opposing Force '87
Papillon '73
Parallax View '74
Private Investigations '87
Question of Honor '80
Rooster Cogburn '75
See No Evil, Hear No Evil '89
Soggy Bottom U.S.A. '84
Steel Dawn '87
They Call Me Mr. Tibbs! '70
Who'll Stop the Rain? '78
Will Penny '67

Alvaro Zermeno

Caches de Oro

Gene Zerna

Dr. Caligari '89

Maria Louise Zetha

Love Lessons '85

Mai Zetterling

Frieda '47
Hidden Agenda '90
Ingrid '89
Night Is My Future '47
Only Two Can Play '62
Truth About Women '58
The Witches '90

Mai Zetterling (D)

Scrubbers '82

Esther Zevko

But Where Is Daniel Wax? '74

Warren Zevon

Warren Zevon '82

Yu Zhang

Girl from Hunan '86

Ye Zharikov

My Name Is Ivan '62

Arunas Zhebrunas (D)

Grazuole '69

Yuri Zhelyabuzhsky (D)

The Cigarette Girl of
    Mosselprom '24

Stacia Zhivago

Sorority House Massacre 2
    '92

Tian Zhuangzhuang (D)

The Horse Thief '87

Nick Zickefoose

Cole Justice '89

Claude Zidi (D)

My New Partner '84
Stuntwoman '81

Howard Zieff

Flesh Gordon '72

Howard Zieff (D)

The Dream Team '89
Hearts of the West '75
House Calls '78
The Main Event '79
My Girl '91
Private Benjamin '80
Slither '73
Unfaithfully Yours '84

Rafal Zielinski (D)

Ginger Ale Afternoon '89
Loose Screws '85
Recruits '86
Screwball Hotel '88
Spellcaster '91
Valet Girls '86

Sonja Ziemann

Made in Heaven '52
A Matter of Who '62

Chip Zien

Grace Quigley '84

Hanns Zieschler

Les Rendez-vous D'Anna '78

Paul Ziller

Pledge Night '90

Paul Ziller (D)

Deadly Surveillance '91

Madeline Zima

The Hand that Rocks the
    Cradle '91

Efrem Zimbalist, Jr.

Airport '75 '75
By Love Possessed '61
Deep Six '58
Family Upside Down '78
The Gathering: Part 2 '79
Hot Shots! '91
House of Strangers '49
Scruples '80
Shooting Stars '85
Terror Out of the Sky '78
Wait Until Dark '67
Who Is the Black Dahlia? '75

Stephanie Zimbalist

Awakening '80
The Babysitter '80
Caroline? '90
Forever '78
The Gathering '77
The Killing Mind '90
Long Journey Back '78
Love on the Run '85
Magic of Lassie '78
Personals '90
Tomorrow's Child '82
The Triangle Factory Fire
    Scandal '79

Marlene Zimmerman

Kill Line '91

Patric Zimmerman

The Jetsons: Movie '90

MUSIC VIDEO GUIDE

The **Music Video Guide** alphabetically lists the titles of more than 3,000 music videos and music performance recordings. These videos are typically prefaced by the name of the artist for quick identification. Due to space limitations and the remarkable sameness of most of the reviews (Music star sings, dances, songs include...), these videos are not covered elsewhere in the book.

MUSIC VIDEO GUIDE

Music Video

ABBA
ABBA: Again
ABBA: In Concert
ABBA: Story (1974-1982)
ABBA: Waterloo
ABC: Absolutely ABC
ABC: Mantrap
Abrasive Wheels
AC/DC: Clipped
AC/DC: Fly on the Wall
AC/DC: Let There Be Rock
AC/DC: The Interview
  Sessions
AC/DC: Who Made Who
A.C. Jobim/Gal Costa: Rio
  Revisited
Acoustic Sounds from Africa
Adam Ant: Antics in the
  Forbidden Zone
Adam and the Ants
Adam and the Ants: Prince
  Charming Revue
Aerosmith: 3 x 5
Aerosmith: The Making of
  ''Pump''
Aerosmith: Things That Go
  Pump in the Night
Aerosmith: Video Scrapbook
Afrika Bambaataa & Family:
  Electric Dance Hop
Agent Steel: Mad Locust
  Rising
Agony
Aida
Aim High America
Air Supply: Live in Hawaii
Al Green: Gospel According to
  Al Green
Al Green: On Fire in Tokyo
Al Jarreau in London
Alabama: Greatest Hits Video
Alabama: Pass It on Down
Alan Holdsworth: Tokyo
  Dream
Alannah Myles
Alarm: Spirit of '86
Alarm: Standards
Alcatraz: Metallic Live '84 in
  Japan
Alcatraz: Power Live
Alexander O'Neal: Live in
  London
Alice in Chains: Live Facelift
Alice Cooper & Friends
Alice Cooper: The Nightmare
  Returns
Alice Cooper: Trash These
  Videos
Alice Cooper Trashes the
  World
Alice Cooper: Welcome to My
  Nightmare
Alien Sex Fiend: A Purple
  Glistener

Alien Sex Fiend: Edit
Alien Sex Fiend: Overdose
All Aboard: A Collection of
  Music Videos for Children
All Soul's Day Concert 1983
All Soul's Day Concert 1985
All-Star Country Music Fair
All-Star Gospel Session
The All-Star Reggae Session
All-Star Swing Festival
All Strings Attached
Allman Brothers Band:
  Brothers of the Road
Alyssa Milano: Look in My
  Heart
Amazing Grace with Bill
  Moyers
Amazon Dreams
America Live in Central Park
American Patchwork: Jazz
  Parades
The American Patchwork:
  Land Where Blues Began
American Patchwork: Songs
  & Stories About America
American Suite
America's Music: Chicago and
  All That Jazz
America's Music, Volume 2:
  Country & Western 2
America's Music, Volume 3:
  Blues 1
America's Music, Volume 4:
  Blues 2
America's Music, Volume 5:
  Rhythm and Blues 1
America's Music, Volume 6:
  Rhythm and Blues 2
America's Music, Volume 7:
  Folk 1
America's Music, Volume 8:
  Folk 2
America's Music, Volume 9:
  Jazz Then Dixieland 1
America's Music, Volume 10:
  Dixieland 2
America's Music, Volume 11:
  Soul 1
America's Music, Volume 12:
  Soul 2
America's Music, Volume 13:
  Gospel 1
America's Music, Volume 14:
  Gospel 2
Amy Grant: Age to Age
Amy Grant: Find a Way
Amy Grant: Heart in Motion
Anasa Briggs: Gospel Festival
Anderson, Bruford, Wakeman
  and Howe: In the Big
  Dream
Andre Walton: Arise Skates
Andrew Lloyd Webber:
  Requiem

Angelic Gospel Singers:
  Gospel in Motion
Animotion: Video EP
Anita Baker: One Night of
  Rapture
Anita O'Day: Live at Ronnie
  Scotts
Anna Russell: The (First)
  Farewell Concert
Anna Russell's Farewell
  Special
Annie Lennox: Diva
Anthrax: Attack of the Killer
  Videos
Anthrax: Oidivnikufesin
  (N.F.V.)
Anthrax: Through Time P.O.V.
Any Child is My Child
April Wine
April Wine: Live in London
Arcadia
''Are My Ears on Wrong?''—
  A Profile of Charles Ives
Aretha Franklin: Ridin' on the
  Freeway
Aretha Franklin: The Queen of
  Soul
Arlen Roth Band: Live in
  England
Armored Saint: A Trip Thru
  Red Times
A.R.M.S. Concert: Complete
A.R.M.S. Concert: Part 1
A.R.M.S. Concert: Part 2
Around the World
Art Blakey: The Jazz Life
The Art of Noise: In Visible
  Silence
Artur Rubinstein
Ashford and Simpson
Asia in Asia
Asia: Live in Moscow, 1990
Asia: Live in the UK, 1990
Aswad: Always Wicked
Aswad: Distant Thunder Live
Aswad: Live at Light House
  Studio
Aswad & Pablo Moses:
  Reggae in the Hills
Athens, Georgia: Inside/Out
Atlantic Starr: As the Band
  Turns...The Video
Auf der schwaeb'sche
  Eise'bahne
Aurex Jazz Festival Fusion
  Super Jam
Australia Now
Avengers
Average White Band Shine
B-52s: 1979-1989
Babalu Music! I Love Lucy's
  Greatest Hits
Baby Snakes: The Complete
  Version
Babyface: Tender Lover

Bach: All That Bach - A
  Celebration
Back to the 50's, Vol. 1
Bad English
Bad News Tour
B.A.L.L.
Bananarama: And That's Not
  All...
Bananarama: The Greatest
  Hits Collection
Bananarama: True
  Confessions
The Band Reunion
Bangles Greatest Hits
Banjo Feedback
The Barber of Seville
Barbra Streisand: A
  Happening in Central Park
Barbra Streisand: Color Me
  Barbra
Barbra Streisand: My Name is
  Barbra
Barbra Streisand: One Voice
Barbra Streisand: Putting It
  Together
Barenboim and the Berlin
  Philharmonic: Mozart
Barenboim/Schiff/Solti and
  the English Chamber
  Orchestra: Mozart
Barnes and Barnes:
  Zabagabee
Barrington Levy
Barry Gibb: Now Voyager
Barry Manilow: Because It's
  Christmas
Barry Manilow: First Special
Barry Manilow: Live on
  Broadway
Barry Manilow: Live at the
  Greek
Barry Manilow: Making of 2
  A.M. Paradise Cafe
Barry Manilow: The Concert
  at Blenheim Palace
Baryshnikov: The Dance and
  the Dancer
Basia: Prime Time TV
Batouka: 1st International
  Festival of Percussion
Bauhaus: Archive
Bauhaus/Chrome:
  Compilation
Bauhaus: Shadow of Light
Bay City Rollers: Japan Tour,
  1976
B.B. King and Friends: A Night
  of Red Hot Blues
B.B. King: Live in Africa
B.B. King: Live at Nick's
B.B. King: Memories of
  Greatness Live
Beast of I.R.S., Vol. 1
The Beastie Boys

The Beastie Boys: The Skills
  to Pay the Bills
Beat of the Live Drum
Beatlemania! The Movie
The Beatles: Live at Budokan
Beatles Live: Ready, Steady,
  Go
Beats Go On: Percussion from
  Pleistocene to Paradiddle
Bee Gees: One for All Tour,
  Vol. 1
Bee Gees: One for All Tour,
  Vol. 2
Beethoven
Beethoven By Barenboim
Beethoven, the Cello, and
  Paul Tortelier
Beethoven Cycle
Beethoven and His Music
Beethoven: Ordeal and
  Triumph
Beethoven Piano Concertos
Beethoven: Symphony No. 5
Beethoven: Symphony No. 5 -
  Concert Aid
Beggar's Opera
Behind the Scenes
Bela Fleck and the
  Flecktones: Flight of the
  Cosmic Hippo
Belinda
Belinda Carlisle: Belinda Live
Belinda Carlisle: Runaway
  Live
Belinda Carlisle: Runaway
  Videos
Bell Biv Devoe: Mental Videos
Bellini: La Sonnambula
The Beloved: Happiness
Ben Webster: Big Ben - In
  Europe
Ben Webster/Dexter Gordon:
  Top Tenors
Benefit Concert 1
Benefit Concert 2
Benko of Hungary Jazz Band
Benny Golson/Tubby Hayes
Benny Goodman: 1980 Aurex
  Jazz Festival
Benny Goodman: Live
Berlin
Berlin Dream
Bernadette Peters in Concert
Bernstein on Beethoven: A
  Tribute
Bernstein on Beethoven:
  Fidelio
Bernstein on Beethoven: Ode
  to Joy
Bernstein on Beethoven
  Series
Bernstein Conducts
  Beethoven: Program 1
Bernstein Conducts
  Beethoven: Program 2

Music

Bernstein Conducts Beethoven: Program 3
Bernstein Conducts Beethoven: Program 4
Bernstein Conducts Beethoven: Program 5
Bernstein Conducts Beethoven: Program 6
Bernstein Conducts Beethoven: Program 7
Bernstein Conducts Beethoven: Program 8
Bernstein Conducts Beethoven: Program 9
Bernstein Conducts Beethoven: Program 10
Bernstein In Berlin: Ode to Freedom, Beethoven's Symphony No. 9
Bernstein: West Side Story
Bessie Smith and Friends
Best of Berlin Independence Days
Best of Blondie
Best of the Cutting Edge, Vol. 1
Best of the Cutting Edge, Vol. 2
Best of Elvis Costello and the Attractions
The Best of the Fest: New Orleans Jazz & Heritage Festival
Best of Judy Garland
The Best of MTV's 120 Minutes
Best of New Faces
The Best of New Wave Theatre, Vol. 1
The Best of New Wave Theatre, Vol. 2
Best of Reggae Sunsplash, Part 1
Best of Reggae Sunsplash, Part 2
Best of Spike Jones, Vol. 1
Best of Spike Jones, Vol. 2
Best of West Coast Rock, Vol. 1
Best of West Coast Rock, Vol. 2
Best of West Coast Rock, Vol. 3
Bette Midler: Art or Bust
Bette Midler: Mondo Beyondo
Bette Midler Show
Big Audio Dynamite: BAD 1 & 2
Big Bands at Disneyland (3-volume set)
Big Bands, Vol. 101
Big Bands, Vol. 102
Big Bands, Vol. 103
Big Bands, Vol. 104
Big Bands, Vol. 105
Big Bands, Vol. 106
Big Bands, Vol. 107
Big Bands, Vol. 108
Big Bands, Vol. 109
Big Bands, Vol. 110
Big Bands, Vol. 111
Big Black: Live Video
Big Black: The Last Blast
Big Brother and the Holding Company: Ball and Chain
Big Country Live
Big Country: Live in New York
Big Daddy Kane: Chocolate City
Big Pig—Bonk: The Videos
Bill Bruford: Bruford and the Beat
Bill Evans: Universal Mind
Bill Watrous
Bill Wyman
Bill Wyman: Digital Dreams

Billy Bragg Goes to Moscow & Norton, Virginia Too
Billy Idol: The Charmed Life Videos
Billy Idol: Vital Idol
Billy Joe Royal
Billy Joel: Eye of the Storm
Billy Joel Live from Long Island
Billy Joel Live at Yankee Stadium
Billy Joel: Stormfront
Billy Joel: The Video Album, Vol. 1
Billy Joel: The Video Album, Vol. 2
Billy Ocean: In London
Billy Ocean: Tear Down These Hits
Billy Squier
Bingoboys: How To Dance (The Extended Dance Version)
Birthday Party: Pleasure Heads Must Burn
Bizarre Music Television
Black Box: Video Dreams 1991
Black Diamond Jazz Band
Black Eagle Jazz Band
Black Flag: Live in San Francisco
Black Flag: Six Pack
Black Flag: T.V. Party
Black Jazz & Blues
Black Music in America: The Seventies
Black Sabbath Live: Never Say Die
Black & Tan/St. Louis Blues
Black Uhuru and Steel Pulse: Tribute to Bob Marley
Black Uhuru: Tear It Up Live
The Blanchebom in Opera and Song
Blancmange
Blondie: Eat to the Beat
Blondie Live!
Blood on the Cats III...For a Few Pussies More
Blotto
Blow Monkeys: Digging Your Video
Blow Monkeys: Video LP
Blue Note I
Blue Suede Shoes: A Rockabilly Session with Carl Perkins and Friends
The Blues
The Blues Accordin' to Lightnin' Hopkins
Blues Alive
Blues Like Showers of Rain
The Bluffers: Music Video
Bo Diddley and the All Star Jam Show
Bob Dylan: Hard to Handle
Bob Dylan, Tom Petty & the Heartbreakers: Hard to Handle
Bob James Live
Bob Marley: Legend
The Bob Marley Story: Caribbean Nights
Bob Marley and the Wailers: Live at the Rainbow
Bob Marley and the Wailers: Live from the Santa Barbara
Bobby Brown: His Prerogative
Bobby Darin: The Darin Invasion
Bobby McFerrin: Spontaneous Inventions
Bobby and the Midnites

Bobby Short at the Cafe Carlyle
Bobby Short & Friends
Bobby Vinton
Body Music
Bolshoi Ballet: Les Sylphides
Bon Jovi: Access All Areas
Bon Jovi: Breakout
Bon Jovi: New Jersey - The Videos
Bon Jovi: Slippery When Wet
Bon Jovi: The Interview Sessions
Bon Jovi: Tokyo Road
Bones
Bonnie Raitt: The Video Collection
Boogie-Doodle
Boogie Down Productions: Edutainment
Boogie Down Productions Live
Born to Boogie
Born to Rock: The T.A.M.I.-T.N.T. Show
Born to Swing: The Count Basie Alumni
Boulez on Varese
Boulez X 3
Boy George: Sold
The Boys: Crazy
The Boys: Video Messages
Brahms: Concerto in ''A'' for Violin & Cello
Brahms and His Music
Branford Marsalis: Steep
The Brass Menagerie
Brazilian Knights & A Lady
Breakin' Metal
Breakin' Metal Special
Breathe: All That Jazz
Brenda Lee: An Evening with Brenda Lee
Brian Eno: Excerpt from the Khumba Nela
Brian Eno: Imaginary Landscapes
Brian Eno: Thursday Afternoon
British Big Beat: The Invasion
British Rock: The First Wave
British Rock: The Legends of Punk & New Wave
Britny Fox: Year of the Fox
Britten, the Voice, and Peter Pears
Broken Bones: Live at Leeds
Bruce Dickinson: Dive! Dive! Live
Bruce Hornsby and the Range: A Night on the Town
Bruce Springsteen and Friends
Bruce Springsteen Video Anthology: 1978-88
Bruce Willis: Return of Bruno
Bryan Adams: Reckless
Bryan Ferry: New Town
Buck Creek Jazz Band No. 1
Buck Creek Jazz Band No. 2
Buckwheat Zydeco: Taking it Home
Buddy Barnes: Live from Studio B
Buddy Rich
Buddy Rich Memorial Scholarship Concert
Buffalo Jazz Fest
BulletBoys: Pigs in Mud
Buried Treasures, Volume 1: The Directors
Buried Treasures, Volume 2: Reggae Classics
Burning Spear: Live in Paris
Buzzcocks: Auf Wiedersehen

Cab Calloway and His Orchestra
Cabaret Voltaire: Gasoline in Your Eye
Cajun Visits
California to New York Island
The California Raisins 2: Raisins, Sold Out
Camel: Pressure Points Live in Concert
Cameo: Video EP
Cameo: Word Up
Canadian Reggae Music Awards '88
Cancion Romantica
C&C Music Factory: Everbody Dance Now
Canned Heat: Boogie Assault
Cannonball Adderly: 1983 Live
Caribbean Music and Dance
Carla Bley Band: Joyful Noise
Carly Simon: Live from Martha's Vineyard
Carly Simon: My Romance in Concert
Carman: Coming on Strong (Live in Concert)
Carmen
Carmen McRae Live
Carmina Burana
Carnegie Hall at 100: A Place of Dreams
The Carnival of the Animals
Carol Jo Brown: Turn It Up
Carole King: One to One
Carols for Christmas
Carreras, Domingo, Pavarotti: Three Tenors in Concert
The Cars: 1984-1985—Live
The Cars: Heartbeat City
Casey Kasem's Rock 'n Roll Goldmine: The San Francisco Sound
Casey Kasem's Rock 'n Roll Goldmine: The Sixties
Casey Kasem's Rock 'n Roll Goldmine: The Soul Years
Catch a Rising Star's 10th Anniversary
Cavalleria Rusticana
Cecil Taylor: Burning Poles
A Celebration
Celine Dion: Unison
Celtic Frost: Live at the Hammersmith Odeon
Central Ohio Hot Jazz Fest
Chaka Khan Live
Chaka Khan: My Night
Chamber Music: The String Quartet
The Chameleons: Live at Camden Palace
Chant a Capella
Charles Aznavour: Memories of Greatness Live
Charles Haughey's Ireland
Charles Johnson & the Revivers: One Night Revival
Charles Mingus: Live in Norway, 1964
The Charlie Daniels Band: Saratoga Concert
Charlie Daniels: Homefolks and Highways
Charlie Peacock
Charlie Pride: An Evening with Charlie Pride
Chartbusters from Kids Incorporated
Chase the Devil: Religious Music of the Appalachians
Cheap Trick: Every Trick in the Book
Cheech and Chong: Get Out of My Room

Cheetachrome: M1S (Live)
Chemical People: Live
Chen and China's Symphony
Cher: Extravaganza: Live at the Mirage
Cherry Bombz
Cheryl Ladd: Fascinated
Chess Moves
Chet Atkins and Friends: Music from the Heart
Chet Baker: Candy
Chet Baker/Elvis Costello: Live at Ronnie Scott's
Chet Baker: My Funny Valentine
Chicago Basketball: A Musical Celebration
Chicago Bears: Super Bowl Shuffle
Chicago Beat: Rock Lobster
Chicago Blues
Chicago: Chicago 17
Chicago Rhythm
Chicago Six Jazz Band
Chick Corea
Chick Corea: A Very Special Concert
Chick Corea Elektric Band: Inside Out
Chick Corea & Gary Burton: Live in Tokyo
Chick Corea: In Madrid
Chico Hamilton: The Jazz Life
The Chieftains: An Irish Evening
The Chieftains in China
The Chieftains: The Bells of Dublin
Children's Songs Around the World
Children's Songs & Stories with the Muppets
Choices for the Future
Chopi Music of Mozambique/ Banguza Timbila
Chopin/Schubert Recital: Zimerman (laserdisc)
Chris Isaak: Wicked Game
Christians: The Collection
Christine McVie: The Video Album
The Christmas Carol Video
Christmas Carols from England
Christmas Classics 3
Christmas with Eleanor Steber
Christmas with Flicka
Christmas with Luciano Pavarotti
The Christmas Oratorio
Christmas at Ripon Cathedral
Christmas from St. Patrick's Cathedral
The Christmas Tree Video
Christmas with the Westminster Choir
Christopher Tree
Chronos
Chrysanthemum Ragtime Band
Chuck Berry: Hail! Hail! Rock 'N' Roll
Chuck Berry: Rock & Roll Music
Chuck's Choice Cuts
The Church: Goldfish (Jokes, Magic & Souvenirs)
Cinderella: Heartbreak Station Video Collection
Cinderella: Night Songs
Cinderella: Tales from the Gypsy Road
Circuit Rider With Noel Paul Stookey
Cissy Houston: Sweet Inspiration

Music

Edinburgh Military Tattoo 1988
Edith Piaf: La Vie en Rose
Edvard Grieg: The Man and His Music
Egberto Gismonti
1812 Overture
El Maestro: Primitivo Santos
Eleanor Steber in Opera and Oratorio
Eleanor Steber in Opera and Song
Eleanor Steber in Opera and Song, Vol. 2
Electric Light Orchestra Live at Wembly
Elektra
11th & B
L'Elisir d'Amore
Ella Fitzgerald: J.A.T.P.
Elton John
Elton John: Breaking Hearts Tour
Elton John Live in Australia
Elton John Live in Central Park
Elton John: Night & Day - The Nighttime Concert
Elton John: The Very Best
Elton John: To Russia...With Elton
Elton John: Visions
Elvin Jones: A Different Drummer
Elvis
Elvis in the '50s
Elvis: '68 Comeback Special
Elvis: Aloha from Hawaii
Elvis in Concert 1968
Elvis and Marilyn
Elvis: One Night with You
Elvis Presley: 27 Songs That Shook the World
Elvis on Television
Elvis: That's the Way It Is
Elvis: The Great Performances, Vol. 1: Center Stage
Elvis: The Great Performances, Vol. 2: Man & His Music
Elvis: The Lost Elvis
Elvis: The Lost Performances
Elvis on Tour
Emerson, Lake and Palmer: Pictures at an Exhibition
EMI Video Series, Vol. 1: Between Us
EMI Video Series, Vol. 2: Breakthrough Video
Emmylou Harris and the Nash Ramblers at the Ryman
An Emotional Fish
An Emperor's Music
En Vogue: Born to Sing
Engelbert Humperdinck in Concert
Engelbert Humperdinck Live
The Enid
Entertaining the Troops
Enuff Z'Nuff
Enya: Moon Shadows
Erasure: Live Wild
Eric B. & Rakim: Let the Rhythm Hit 'em
Eric Clapton: 24 Nights
Eric Clapton & Friends
Eric Clapton: Live '85
Eric Clapton: On the Whistle Test
Erick Friedman Plays Fritz Kreisler
Ernani
Ernest Tubb: "Thanks Troubadour, Thanks"

Erroll Garner: London/ Copenhagen
The Escape Club: Dollars, Sex & Wild, Wild West
Eugene Onegin: Tchaikovsky
Europe in America #1
Europe in America #2
Europe: Live
Europe: The Final Countdown World Tour
Eurythmics: Greatest Hits
Eurythmics Live
Eurythmics: Savage
The Eurythmics: Sweet Dreams, the Video Album
Eurythmics: We Two Are One Too
Eva Marton in Concert
The Eve
An Evening with Danny Kaye
An Evening with Kiri Te Kanawa
An Evening with Liza Minnelli
An Evening with Marlene Dietrich
An Evening with Paul Anka
An Evening with Ray Charles
An Evening with Sammy Davis Jr. and Jerry Lewis
An Evening with Sister Sledge
Everly Brothers Album: Flash
The Everly Brothers Reunion Concert
The Everly Brothers: Rock 'n' Roll Odyssey
Evolutionary Spiral
Evviva Belcanto
Exile in Concert
Expanding Visions: An Introduction to the New Age
The Exploited: Live at Palm Grove
The Exploited: Sexual Favours
Expose: Video Exposure
Extreme Art, Vol. 1
Extreme Art, Vol. 2
Extreme Noise Terror: From One Extreme to Another
Extreme: Photograffiti
Fabulous Thunderbirds: Live from London
Fabulous Thunderbirds: Play It Tough & Play It Live
Fact Shorts
A Factory Video
Factrix: Night of the Succubus
The Facts of Life
Fairport Convention: It All Comes Around Again
Faith No More: Live at the Brixton Academy
Falco: Rock Me Falco
The Fall: Perverted by Language
Falstaff
Fantastic All-Electric Music Movie
Fast Forward, Vol. 1
Fast Forward, Vol. 2
Faster than Lightning
Fat Boys: 3x3
Fat Boys on Video: Brrr, Watch 'Em!
Fats Domino: 15 Greatest Hits Live
Fats Domino and Friends
Fats Domino Live!
Fats Waller and Friends
Fela: A Concert
Fela Anikulapo Kuti: Teacher Don't Teach Me Nonsense
Ferruccio Tagliavini in Opera and Song
Fidelio

Fields of the Nephilim
Fields of the Nephilim: Morphic Fields
The 5th Dimension Travelling Sunshine Show
Fine Young Cannibals: Live at the Paramount
Firehouse: Rock on the Road
First Ladies of Opera
The Fishbone: Reality of My Surroundings
Five Star: Luxury of Life
The Fixx: Live in the USA
Flaming Lips: Live
Fleetwood Mac in Concert: Mirage Tour 1982
Fleetwood Mac: Documentary & Live Concert
Fleetwood Mac: Tango in the Night
Flesh for Lulu: Live from London
Flipper
Flipside Video, Vol. 1
Flipside Video, Vol. 2
Flipside Video, Vol. 3
Flipside Video, Vol. 5
Flipside Video, Vol. 6
Flipside Video, Vol. 7
Flipside Video, Vol. 8
Flipside Video, Vol. 9
Flipside Video, Vol. 10
The Floating World
A Flock of Seagulls
Flora Purim & Airto: Latin Jazz All-Stars
Florida Mass Choir: Higher Hope
A Flower Out of Place
Folk City 25th Anniversary
The Folk Music Reunion
Fools: World Dance Party
Ford Star Jubilee Salute to Cole Porter
Ford Star Jubilee: "Together with Music"
The Foreigner Story: Feels Like the Very First Time
Foreigner: Super Rock '85
The Forester Sisters: Talkin' 'Bout Men
Foster & Lloyd: Version of the Truth
Four American Composers
The Four Seasons
The Four Tops Live
Frank Cappelli: All Aboard the Train and Other Favorites
Frank Patterson: Ireland's Golden Voice
Frank Sinatra: A Man and His Music
Frank Sinatra: A Man and His Music Ella Jobim
Frank Sinatra in Concert at Royal Festival Hall
Frank Sinatra: Ol' Blue Eyes Is Back
Frank Sinatra: Reprise Collection, Vol. 1
Frank Sinatra: Reprise Collection, Vol. 2
Frank Sinatra: Reprise Collection, Vol. 3
Frank Sinatra Show
Frank Sinatra: Sinatra
Frank Sinatra: Sinatra, the Main Event
Frank Zappa Presents: The Amazing Mr. Bickford
Frank Zappa: Video from Hell
Frank Zappa's Does Humor Belong in Music?
Frankie Goes to Hollywood: Wasteland...Paradise

Frankie Valli: Hits From the 60's
Freddie Hubbard
Freddie Jackson: You Are My Lady
Freddie King: Blues Band Live
Frederica von Stade: Christmas
Free: The Best of Free
Freedom Beat: The Video
Frehley's Comet: Live 4
French Singers
Fresh 89, Part 1
Fresh 89, Part 2
Fresh 89, Part 2: Vol. 1
Fresh 89, Part 2: Vol. 2
Fresh 89, Part 3
Frightwig: Strip for Frightwig
From Mao to Mozart: Isaac Stern in China
From the New World
Front Room
Front Row with Charlie Peacock
Front Row with Margaret Becker
Front Row with Michael Card
Front Row with Steve Curtis Chapman
Frontier Records' Picture Book
Frontline Music Festival
Frozen Ghosts
Fullerton College Jazz Bands
Fulton Street #1
Fulton Street #2
Further on Down the Road
Futurama 6
G. Lewis and D. Ewart Play Music by D. Ewart and G. Lewis
The Gadd Gang Live
Gala Concert
Gala Lirica
A Galway Christmas
Gap Band Video Train
Garden Avenue Seven #2
Gary Glitter's Gangshow: The Gang, the Band, the Leader
Gary Moore and the Midnight Blues Band: An Evening of the Blues
Gary Numan: Berserker
Gary Numan: The Touring Principal '79
Gatemouth Brown
G.B.H.: Brit Boys Attacked by Rats
G.B.H.: Live
Genesis: Invisible Touch Tour
Genesis Live: The Mama Tour
Genesis: Three Sides Live
Genesis: Videos, Vol. 1
Genesis: Videos, Vol. 2
Genesis: Visible Touch
George Clinton: Parliament/ Funkadelic
George Jones: Living Legend in Concert
George Jones/Tammy Wynette: In Concert
George Michael
George Michael: Faith
George Strait Live!
George Thorogood: Born to Be Bad
George Thorogood & the Destroyers: Live
Georgia Mass Choir: Hold On, Help is on the Way
Gerardo: Rico Suave
Gerry Mulligan
Gerry Mulligan: Live at Eric's
Gibson Jazz Concert
Gil Bailey's 19th Anniversary Radio Show

Gil Evans
Gil Evans and His Orchestra
Gil Scott-Heron: Black Wax
Gil Scott-Heron: Tales of Gil
Gilbert & Sullivan Present Their Greatest Hits
Gimme Shelter
Gino Vannelli
Giorno Video Pak 3
Girl Groups: The Story of a Sound
Girlschool
Give Peace a Chance
Gladys Knight & the Pips & Ray Charles in Concert
Glam Rock
Glen Campbell: Live in London
Glenn Branca: Symphony No. 4
Glenn Gould: A Portrait
Glenn Medeiros, Vol. 1
Glenn Miller: A Moonlight Serenade
Gloria Estefan: Coming Out of the Dark
Gloria Estefan: Into the Light World Tour
Gloria Estefan & the Miami Sound Machine: Evolution
Gloria Estefan & the Miami Sound Machine: Homecoming
The Glory of Spain
The Go-Betweens: Video Singles
Go Go Big Beat
Go Go's: Live at the Capitol Center
Go Go's: Prime Time
Go Go's: Wild at the Greek
Godley & Creme: History Mix
Golden Earring
Golden Earring: Live from the Twilight Zone
Good Mornin' Blues
Goodtime Rock 'n' Roll
Goodyear Jazz Concert with Bobby Hackett
Goodyear Jazz Concert with Duke Ellington
Goodyear Jazz Concert with Eddie Condon
Goodyear Jazz Concert with Louis Armstrong
Gospel
Gospel from the Holy Land
Gospel Keynotes: Live at Jackson State University
Gotta Make This Journey: Sweet Honey in the Rock
Grace Jones: One Man Show
Grace Jones: State of Grace
Graham Parker
Grand Dominion Jazz Band
Grand Funk: The American Band Live
Grand Ole Opry Country Music Celebration
Grand Ole Opry Time
Grand Ole Opry Time: Vol. 02
Grand Ole Opry's Classic Country Tributes
Grateful Dead: Dead Ahead
The Grateful Dead: Making of the "Touch of Grey" Video
The Grateful Dead Movie
Grateful Dead: So Far
Great American Gospel Sound, Vol. 1
Great American Gospel Sound, Vol. 2
Great American Gospel Sound, Vol. 3
Great American Gospel Sound, Vol. 4
Great American Jazz Band

Great Jazz Bands of the '30s
Great Music from Chicago
Great Video Hits, Vol. 1
Great Video Hits, Vol. 2
Great Video Hits, Vol. 3
Great White: My...My...My...
Great White: Videos
The Greatest Sound in Gospel: Vol. 1 - 4
The Greatest Week in Gospel
Greatest Week in Gospel, Part 1
Greatest Week in Gospel, Part 2
Greatest Week in Gospel, Part 3
Greatest Week in Gospel, Part 4
The Greatest Years in Rock
Greenpeace Non-Toxic Video Hits
Greetings from Eden Alley
Greg Giuffria: Live in Japan '85
Gregg Allman: One Way Out
Gregory Isaacs: Maestro Meets the Cool Ruler
Gregory Issacs Live
Grok Gazer
Grover Washington, Jr. in Concert
GRP Christmas Collection
GRP: Live in Session
GRP Superlive in Concert
GRP Video Collection
GTR: The Making of GTR
Guarneri String Quartet
The Guess Who Reunion
Guess Who: Together Again
Guitar Men
Guitar from Three Centuries
Gun: Taking on the World
Guys Next Door
Gwar: Live from Antarctica
Haggadah: A Search for Freedom
Hail to the Chieftains
Hal Smith Trio
Hal Smith's Rhythmakers
Hallelujah: A Gospel Celebration
Hampton Hawes: All Stars
Hank Williams, Jr.
Hank Williams, Jr.: A Star-Spangled Country Party
Hank Williams, Jr.: Full Access
Hank Williams Jr.: Greatest Video Hits
Hank Williams, Jr.: Pure Hank
Hanoi Rocks: The Nottingham Tapes
Hanoi Rocks: Video LP
Hansel & Gretel
Happy Hour with the Humans
Hard 'n' Heavy: Thrash Metal Speed
Hard 'n' Heavy, Vol. 1
Hard 'n' Heavy, Vol. 2
Hard 'n' Heavy, Vol. 3
Hard 'n' Heavy, Vol. 4
Hard 'n' Heavy, Vol. 5
Hard 'n' Heavy, Vol. 6
Hard 'n' Heavy, Vol. 7
Hard 'n' Heavy, Vol. 8
Hard 'n' Heavy, Vol. 9
Hard 'n' Heavy, Vol. 10
Hard 'n' Heavy, Vol. 11
Hard 'n' Heavy, Vol. 12
Hard Time '77
Hardcore
Hardcore, Volume 1
Hardcore, Volume 2
Hardcore, Volume 3
Harlem Harmonies, Vol. 1
Harlem Harmonies, Vol. 2
Harmony

Harry Belafonte: Don't Stop the Carnival
Harry Belafonte's Global Carnival
Harry Chapin: The Final Concert
Harry Connick, Jr.: Do You Know What It Means to Miss New Orleans?
Harry Connick, Jr.: Singin' & Swingin'
Harry Connick, Jr.: Swingin' Out Live
Harry James: One Night Stand
Harry Owens and His Royal Hawaiians
Harvest Jazz Series
Harvest Jazz Series: Alto Madness
Hawkwind: Chronicle of the Black Sword
Hawkwind: Live
Hawkwind: Night of the Hawks
HBTV Music Videos
Healing Yourself
Hear 'N' Aid—The Sessions/Concert for Famine Relief
Heart: If Looks Could Kill
Heart & Soul
Heart: Video 45
Heartland Reggae
Heartstrings: Peter, Paul and Mary in Central America
Heavy D. & the Boyz: We Got Our Own Thang
Heavy Metal
Heifetz & Piatigorsky
Heimat, deine Lieder 2 (Songs We Remember)
Helix
Henry Mancini and Friends with Robert Goulet and Vicki Carr
Herbie Hancock: Jazz Africa
Herbie Hancock and the Rockit Band
Herbie Hancock: Rockit Band Live in Tokyo
Herbie Hancock Trio: Hurricane!
Here It Is, Burlesque
Heroes and Villains
He's Right on Time-Live from Los Angeles
Hi-Five: The Video Hits
High Fidelity
High School Yearbook: The '50s & the Early '60s
High Society Jazz Band #1
High Society Jazz Band #2
Highway 101: Greatest Hits (1987-1990)
Highwaymen Live
His Legacy for Home Video Series
History Never Repeats—The Best of Split Enz
Hits Live from London
The Hoffnung Festival Concert
Holiday Sing Along With Mitch
Holly Dunn: Milestones - Greatest Hits
Hollywood Goes to War
Homage to Verdi
Homemade American Music
Hooters: Nervous Night
Horowitz in London
Horowitz in Moscow
Horowitz in Vienna
Hot Antic Jazz Band
Hot Frogs Jumping Jazz Band #1
Hot Frogs Jumping Jazz Band #2

Hot 'n' Heavy
Hot Rock Videos, Volume 1
Hot Rock Videos, Volume 2
Hot Shop '89
Hot Summer Night...with Donna
Hothouse Flowers: Take a Last Look at the Sun
How Many Fingers?
Howard Jones: Last World Dream
Howard Jones: Like to Get to Know You Well
Howard Jones: Live in Japan
The Huberman Festival
Huey Lewis & the News
Huey Lewis & the News: All the Way Live
Huey Lewis & the News: Fore and More
Huey Lewis & the News: Live
Huey Lewis & the News: The Heart of Rock n' Roll
Huey Lewis & the News: Video Hits
Hugh Masakela: Notice to Quit (The Lion Never Sleeps)
Hugh Shannon: Saloon Singer
Hulk Hogan Real American
Hullabaloo, Vol. 1
Human League: Greatest Hits
Humans: Happy Hour
Hungry i Reunion
Hurricane Irene
Husker Du: Makes No Sense
The Hutchinson Family Singers
Hymn of the Nations
I Lombardi Alla Prima Crociata
Ian Gillian: Live at the Rainbow
Ian and Sylvia
Ice-T: Iceberg
Ice-T: O.G. The Original Gangster Video
I.D.F.: Dark Arts (Live Treatment)
Idomeneo
Iggy Pop: "Live in San Francisco, 81"
Illumination
Im Land der Lieder
Imagine the Sound
Impact Video Magazine
Imperials: 20th Anniversary Concert
Implosions
The Impressions Series
In Celebration of the Piano
In Concert
In a Dance All Style: New Yorker, New Yorker in Jamaica
In mir klingt ein Lied
In Love with These Times
In Our Hands
In Performance: David Shifrin, Clarinet
In the Silence of the Night
Incantation: The Best of Incantation
Indecent Obsession
Indie Top Video, Take 4
Indie Top Video, Take 5
Indigo Girls: Live at the Uptown Lounge
Indoor Cycling
The Industrial Symphony No. 1: Dream of the Broken Hearted
Infantasia
Infantastic Lullabyes on Video
Inside Country Music, Vol. 1: Edition 2

The International Choral Festival Film
Invitation to Paris
INXS: Flick the Video Kick
INXS: Greatest Video Hits (1980-1990)
INXS: In Search of Excellence
INXS: Live Baby Live
INXS: Living
INXS: What You Need
Irene Cara: Live in Japan
Irish Magic, Irish Music
Iron Maiden: 12 Wasted Years
Iron Maiden: Behind the Iron Curtain
Iron Maiden: Live After Death
Iron Maiden: Live in London
Iron Maiden: Maiden England
Iron Maiden: Video Pieces
Isaac Hayes: Black Moses of Soul
Isaac Stern in Jerusalem: A Musical Celebration
Island Reggae Greats
Israel on Parade: A Variety Show
Israel Philharmonic Orchestra Welcomes Berliner Philharmoniker
It's Clean, It Just Looks Dirty
It's in Every One of Us
Itzhak Perlman
Ivo Perelman: Live in New York
Ivo Pogorelich
Ivo Pogorelich in Villa Contarini
J. Geils Band
Jack Jones: An Evening with Jack Jones
Jack Ruby's 25th Anniversary Dance
Jack Sheldon in New Orleans
Jackie Patterson: Hip Hop Dance
The Jackson Southernaires: Live at Jackson State University
The Jacksons
Jacqueline Du Pre and the Elgar Cello Concerto
J'ai Ete au Bal (I Went to the Dance)
Jaki Byard: Anything for Jazz
Jam: The Trans-Global Unity Express
Jam: The Video Snap!
James Brown: Cold Sweat - In Japan
James Brown & Friends: Set Fire to the Soul
James Brown Live in America
James Brown Live in Berlin
James Brown Live in London
James Brown: Soulful Night in L.A.
James Brown: Story 1956 - 1976
James Cleveland: Down Memory Lane
James Cleveland & the Northern and Southern California Choirs: Breathe on Me
James Galway & the Chieftains in Ireland
James Galway in Concert
James Galway's Christmas Carol
James Moore & the Mississippi Mass Choir
James Taylor in Concert
Jan Peerce Anniversary
Jan Peerce: If I Were a Rich Man

Jan Pietrzak: Mr. Censor
Jane's Addiction: The Fan's Video - Soul Kiss
Janet Jackson: Control—The Videos
Janet Jackson: Control—The Videos, Part II
Janet Jackson: The Rhythm Nation Compilation
Janet Jackson's Rhythm Nation
Janis: A Film
Jansen/Barbieri: The World's in a Small Room
Jason Donovan: The Videos, Vol. 2
Jazz in America
Jazz Legends, Part 1
Jazz Legends, Part 2
Jazz Legends, Part 3
Jazz Legends, Part 4
Jazz at Ronnie's
Jazz Shorts
Jazz at the Smithsonian: Alberta Hunter
Jazz at the Smithsonian: Bob Wilber
Jazz at the Smithsonian: Joe Lewis
Jazz at the Smithsonian: Joe Williams
Jazz at the Smithsonian: Mel Lewis
Jazz at the Smithsonian: Red Norvo
Jazz from Studio 61
Jazz Summit
Jazz Vocal Special
Jazzvisions: Implosions
Jeff Healy Band: See the Light
Jefferson Starship: Live
Jellyfish: Gone Jellyfishin'
Jermaine Jackson: Dynamite Videos
Jermaine Stewart: Frantic Romantic
Jerome Hines in Opera and Song
Jerry Lee Lewis & Friends
Jerry Lee Lewis: I Am What I Am
Jerry Lee Lewis: Live at the Arena
Jerry Lee Lewis Live in London
Jerusalem: A Musical Celebration
Jesse Rae: Rusha/D.E.S.I.R.E.
Jesus and Mary Chain: Chain Reaction
Jesus & Mary Chain: Videos: (1985-1989)
Jethro Tull: Slipstream
Jethro Tull: The First Twenty Years
The Jets: Airplay
The Jickets: Direct Art
Jim & Jesse: West Virginia Boys
Jim McCann & the Morrisseys
Jimi Hendrix: At the Isle of Wight
Jimi Hendrix: Berkeley May 1970
Jimi Hendrix Concerts Videogram
Jimi Hendrix: Experience
Jimi Hendrix: Johnny B. Goode
Jimi Hendrix: Live in Monterey, 1967
Jimi Hendrix: Rainbow Bridge
Jimi Hendrix: Rarities
Jimi Hendrix: Story

Music

Jimmy Buffett: Live by the Bay
Jimmy Cliff: Bongo Man
Jimmy Somerville: Video Collection 1984-90
Jive's Greatest Rap Hits
Jivin' in Bebop
J.J. Johnson & Kai Windig All Stars
Joan Armatrading: Track Record
Jodeci
Jody Watley: Video Classics, Vol. 1
Joe Cocker: Live from Tokyo
Joe Cocker: Mad Dogs
Joe Cocker: Shelter Me
Joe Cool Live
Joe Ely: Live From Texas
Joe Jackson Live in Tokyo
Joe Jackson: The Big World Sessions
John Abercrombie: Live at the Village Vanguard
John Carter & Bobby Bradford: The New Music
John Cougar Mellencamp: Ain't That America
John Hammond: Bessie Smith to Bruce Springsteen
John Hartford: Learning to Smile All Over Again
John Lee Hooker: Blue Monday Party
John Lee Hooker & Charlie Musselwhite
John Lennon Live in New York City
John Lennon and the Plastic Ono Band: Live Peace in Toronto, 1969
John Lennon/Yoko Ono: Then and Now
John Scofield Live
John Scofield: Live Three Ways
John Waite: No Brakes Live
Johnnie Taylor: Live in Dallas
Johnny Cash: Live in London
Johnny Cash: Ridin' the Rails
Johnny Gill: Video Hits
Johnny Griffin: The Jazz Life
Johnny Maddox Plays Ragtime
Johnny Mathis: 25th Anniversary Concert
Johnny Mathis: Chances Are
Johnny Mathis: Greatest Hits
Johnny Mathis: Home for Christmas
Johnny McEvoy: In Concert
Johnny Winter Live
Jonathan Butler: Heal Our Land
Joni Mitchell
Joni Mitchell: Refuge of Roads
Joni Mitchell: Shadows & Light
Jose Carreras: Silent Night
Jose Greco in Performance
Jose' Jose' en Acapulco
Jose Serebrier
Journey: Frontiers and Beyond
Journey into Space
The Joy of Bach
Joy Division: Here are the Young Men
Joyful Gospel Christmas
Joyous Sounds of Christmas
Ju Ju
Jubilee U.S.A.
Judas Priest: Fuel for Life
Judas Priest Live
Judas Priest: Pain Killer

The Judds: Across the Heartland
The Judds: Greatest Video Hits
The Judds: Love Can Build a Bridge
The Judds: Their Final Concert
Judy Collins: Live
Judy Garland in Concert, Vol. 1
Judy Garland in Concert, Vol. 2
Judy Garland (General Electric Theatre)
Judy Garland: Judy Garland & Friends
Judy Garland Scrapbook
The Judy Garland Show
Juke Box Saturday Night: All Star Golden Classics
Julia Fordham: Porcelain
Julian Lennon: Live in Japan, 1986
Julian Lennon: Stand By Me
Julie Andrews: In Concert
Julio Iglesias in Spain
Julio Iglesias: Starry Night
Jump the Blues Away
A Jumpin' Night in the Garden on Eden
Just Say Yes, the Video: Volume 1
Just Say Yes, the Video: Volume 2
Justice League: Hyp Existance
Kajagoogoo
Kansas
Kansas City Jazz
KAOMA Worldbeat: The Lambada Videos
Karen Akers: On Stage at Wolf Trap
Kate Bush: Live at Hammersmith
Kate Bush: Single File
Kate Bush: The Sensual World (The Videos)
Kate Bush: The Whole Story
Kathy Mattea: From the Heart Hits
Katrina and the Waves
K.C. and the Sunshine Band in Concert
k.d. lang: Harvest of Seven Years (Cropped and Chronicled)
Keith Jarrett: Last Solo
Keith Jarrett: Standards
Keith Jarrett: Standards Vol. 2
Keith Jarrett: Tokyo Live '85
Keith Richards & the X-Pensive Winos: Live at the Hollywood Palladium
Keith Whitley: I Wonder, Do You Think of Me?
Kenny G Live
Kenny Loggins Alive
Kenny Loggins: Live from the Grand Canyon
Kenny Rogers/Dolly Parton: Real Love
Kenny Rogers and the First Edition
Kenny Rogers: Great Video Hits
Kentucky Headhunters: Pickin' on Nashville
Kerrang!
The Kid Creole & Coconuts: Leisure Tour 1985
Kid Creole and the Coconuts Live
Kid Creole & the Coconuts: Live at the Ritz

Kid Creole & Coconuts: The Lifeboat
Kid 'n Play: The Video
Kid "Punch" Miller: 'Til the Butcher Cuts Him Down
The Kids are Alright
Kids from Fame
Killdozer
Kim Carnes
Kim Wilde
Kim Wilde: You Keep Me Hangin' On
King Crimson: The Noise
King Crimson: Three of a Perfect Pair
King Sunny Ade: Concert in the Park
King Sunny Ade: Juju Music
King Sunny Ade: With His African Beats
King's Christmas
Kings of Independence
The King's Singers Christmas
The King's Singers in Concert
King's X: Chronicles
The Kingston Trio: An Evening with the Kingston Trio
Kingston Trio & Friends: Reunion
Kinks: Come Dancing
Kinks: One for the Road
Kiri Te Kanawa at Christmas
Kirkpatrick Plays Bach
KISS: Animalized Live & Uncensored
KISS: Crazy Nights
KISS: Exposed
KISS Meets the Phantom of the Park
KISS: The Interview Sessions
Kitaro: Light of the Spirit
Kitaro: Live
Kix: Blow My Fuse - The Videos
KLF: The Stadium House Trilogy
KMFDM: Live
Knack: Live at Carnegie Hall
Koko Taylor: Queen of the Blues
Komm mit ins Land der Lieder
Kooky Classics
Kool and the Gang in Concert
Kool & the Gang: Tonight!
Kool Moe Dee: Funke, Funke Wisdom
Krokus: Screaming in the Night
Krokus: The Video Blitz
K.T. Oslin: Love in a Small Town
Kylie Minogue: On the Go - Live in Japan
Kylie Minogue: The Video
Kyoto Vivaldi: The Four Seasons
La Bamba Party
L.A. Guns: Love, Peace & Geese
L.A. Guns: One More Reason
La Toya Jackson
Labor Songs
The Ladies Sing the Blues
Lady Day: The Many Faces of Billie Holiday
Laibach: Live
Lambada Dance Party
Larry Carlton: Live
The LaSalle String Quartet
The Last of the Blue Devils
The Last Rally
Last Romantic
The Last Waltz
Late Night Romantics
Latina Familia

LaTour: People are Still Having Sex
Laura Branigan: Self Control
Laurie Anderson: Collected Videos
Laurie Anderson: Home of the Brave
Lawrence Welk: 1985 Christmas Reunion Special/ New Year's Party
Lawrence Welk: America, What it Used to Be/Armed Forces Day
Lawrence Welk: Carnival/ Fashions & Hits Through the Years
Lawrence Welk: Country and Western Show/Salute to New York City
Lawrence Welk: Halloween Party/Thanksgiving
Lawrence Welk: Love Songs/ History of American Musical Entertainment
Lawrence Welk On Tour, Volumes 1 and 2
Lawrence Welk: Riverboat Show/The Norma Zimmer Show
Lawrence Welk: St. Patrick's Day Show/Easter Show
Lawrence Welk: Salute to the Ladies/Spotlight on Our Musical Family
Lawrence Welk: Salute to Swing Bands/Tribute to George Gershwin
Lawrence Welk: Songs from the Movies/South of the Border
Lawrence Welk: Summer/ Vaudeville
Lawrence Welk: The Early Years Collector Series, Volume 1
Lawrence Welk: The Roaring Twenties/Sights and Sounds of L.A.
Le 'El Palm: Pull Back the Covers
Le Grandi Primadonne
Le Nozze di Figaro
Le Piano Vivant
The Leaders: Jazz in Paris 1988
Led Zeppelin: The Song Remains the Same
Lee Aaron Live
Lee Greenwood: God Bless the U.S.A.
Lee Konitz: Portrait of the Artist as Saxophonist
Lee Ritenour & Dave Grusin: Live from the Record Plant
Lee Ritenour & Friends: Live from the Coconut Grove, Vol. 1
Lee Ritenour & Friends: Live from the Coconut Grove, Vol. 2
Lee Ritenour: Rit Special
Lee Ritenour: Village, Live
Legendary Ladies of Rock & Roll
Legendary Pink Dots: Live
Legends of Rock & Roll
Leise flehen meine Lieder
Lena Horne: The Lady & Her Music
Lennon: A Tribute
Leo Kottke: Home and Away
Leo Sayer: The Very Best of Leo Sayer
Leon Russell: Best of the Festivals Live

Leon Russell/Edgar Winter: Main Street Cafe
Leonid Kogan: Interpretations
Les Blues de Balfa
Les Contes D'Hoffman
Les McCann Trio
Les Paul & Friends: He Changed the Music
Lester Young & Billie Holiday
Let's Cotton Together
Let's Have an Irish Party
Lettermen: Christmas Reflections
Level 42: Family of Five
Level 42: Level Best
Level 42: Live and More
Level 42: Live at Wembley
Liberace
Liberace in Las Vegas
Liberace Live
The Liberace Show
Liberace Show Vol. 2
Lieder, die vom Herzen Kommen
Lietuviskos Dainos Suente
Lietuvos Dainu Svente (Lithuanian Song Festival)
Lifers Group: World Tour "Rahway Prison, That's It"
Lifestreams Video
Light of the Spirit
Lightnin' Hopkins: Live in the Late '60s
Lightning Seeds
Lincoln Center's Mostly Mozart
Linda Ronstadt/Nelson Riddle "What's New"
Lindisfarne: Live
Lionel Hampton
Lionel Hampton: Gypsy in My Soul
Lionel Hampton's One Night Stand
Lionel Richie: All Night Long
Lionel Richie: Making of Dancing on the Ceiling
Lionel Richie: Super Live 1987
Lisa Stansfield: Real Life
Liszt in Mid-Life at Mid-Century
Lita Ford: A Midnight Snack
Lita Ford: Dangerous Videos, Volume 1
Lita Ford: Lita
Lita Ford: Wembley
The Lithuanian Ethnographic Ensemble
Little Richard: Keep on Rockin'
Little River Band
Live at the Biltmore Ballroom
Live from the Met Highlights, Vol. 1
Live from New York City
Live & Red Hot
Live Skull: Live
Live Skull: Live from Ohio
Live at Target
Live at the Village Vanguard, Vol. 1
Live at the Village Vanguard, Vol. 2
Live at the Village Vanguard, Vol. 3
Live at the Village Vanguard, Vol. 4
Live at the Village Vanguard, Vol. 5
Live at the Village Vanguard, Vol. 6
Living Colour: Primer
Living Colour: Time Tunnel
Liza in Concert
Liza Minnelli: Visible Results

Lizzy Borden: Live Murderous Row Show
L.L. Cool J: Future of the Funk
Lohengrin
London Beat: Quick Pix
Lone Justice: Live
Lonely is an Eyesore
Lords of the New Church: Live from London
Loretta Lynn: An Evening with Loretta Lynn
Loretta Lynn in Concert
Lorrie Morgan: Great Video Hits
Los Angeles Raiders: The Silver & Black Attack
Los Joao
Lou Rawls: An Evening with Lou Rawls
Lou Rawls with Duke Ellington
Lou Reed: A Night with Lou Reed
Lou Reed: Coney Island Baby
Lou Reed & John Cale: Songs for Drella
Lou Reed: The New York Album
Loudness: Eurobunds
Loudness: In Tokyo
Loudness: Live in Tokyo/Lightning Strikes
Loudness: Thunder in the East (Complete Version)
Louie Bellson & His Big Band
Louie Bluie
Louis Armstrong: 1963
Louis Armstrong & His Orchestra: 1942-1965
Louis Armstrong: Satchmo
Louis Bellson and His Big Band
Louis Bellson Meets Billy Cobham
Louis Jordan & Friends
Louis Jordan and His Tympany Five
Love of Destiny
Love Lion
Love and Rockets: The Haunted Fish Tank
Loverboy: Any Way You Look At It
Loverboy: Rock & Roll
Love's Awakening
Lowell Fulson/Percy Mayfield: Mark Naftalin's Blue Monday Party
Lucia il Lammermoor
Luciano Pavarotti Gala Concert
Luciano Pavarotti: The Event
The Lute and the Sword
Luther Vandross: Live at Wembley
Luther Vandross: The Best of Luther Vandross
Lynyrd Skynyrd: Tribute Tour
M/W/F Music Video One
Ma Vlast (My Fatherland)
Mabel Mercer
Mabel Mercer: A Singer's Singer
Mabel Mercer: Cabaret Artist "Forever and Always"
Macbeth
Mad Dogs & Englishmen
Madama Butterfly
Madonna: Blond Ambition
Madonna: Ciao Italia
Madonna and Dance Music
Madonna: Immaculate Collection
Madonna: Justify My Love
Madonna: Like a Virgin

Madonna Live: The Virgin Tour
Madonna Live: Who's That Girl?
Maestros in Moscow
Maggie Staton Peebles: Born Again
Magnolia Jazz Band
Magnum: Live
Magnum: Live at Hammersmith Odeon, 1988
Mahalia Jackson: Give God the Glory
Mahler: Symphony No. 1
The Making of the Coming Out of Their Shells Tour
Making Michael Jackson's Thriller
Making of "Teenage Mutant Ninja Turtles: Coming Out of Their Shell Tour"
The Making of "Will the Circle Be Unbroken," Volume 2
Mal Waldron/Jackie MacLean: Left Alone '86
Malcolm McLaren: Duck Rock
Man: Bananas
Manhattan Jazz Quartet
Manhattan Jazz Quartet: Live in Japan
Manhattan Project
Manhattan Transfer in Concert
The Manhattan Transfer: Vocalese
Manhattan Transfer: Vocalese Live 1986
Mannheim Steamroller Video Sampler 1 & 2
Manon Lescaut
Manu Dibango: King Makossa
Many Faces of Bird
Marc Bolan: Story, Vol. 1 (1968-1973)
Marc Bolan: Story, Vol. 2 (1973-1977)
Marcia Griffiths/Yellowman: Caribbean Night
Marcus Roberts: Deep in the Shed
Margaret Becker
Maria Callas 1962 Hamburg Concert
Mariah Carey: The First Vision
Marian Anderson
Marianne Faithful: Blazing Away
Marillion: From Stoke Row to Ipanema
Marillion: Live from Loreley
Marky Mark & the Funky Bunch: Music for the People
The Marriage of Figaro
The Marshall Tucker Band: Live
Martin Briley: Dangerous Moments
Marvin Gaye
Marvin Gaye: Greatest Hits Live
Mary Lou's Mass
Mary's Danish: Live Foxey Lady
Max Roach: In Concert/In Session
Max Roach: In Washington D.C.
Max Roach: Jazz in America - Video 45
Maynard Ferguson: Big, Bold, & Brassy
Maze
Maze Featuring Frankie Beverly
M.C. Hammer: 2 Legit 2 Quit

M.C. Hammer: Addams Groove
M.C. Hammer: Hammer Time
M.C. Hammer: Here Comes the Hammer
M.C. Hammer: Please Hammer Don't Hurt 'Em
McCoy Tyner
The McDonald's Gospelfest: Vol. 1
The McDonald's Gospelfest: Vol. 2
Meat Loaf: Hits Out of Hell
Meat Loaf Live!
Meatloaf: Hits Out of Hell
Media: Zbig Rybczynski, a Collection
Medieval Music Performed on Antique Instruments
Mefistofele
Megadeth: Rusted Pieces
Mel Lewis & His Big Band
Mel Torme
Mel Torme & Della Reese in Concert
Mel Torme Special
Melody Magic
The Memories: Music from the Yank Years, Vol. 1
The Memories: Music from the Yank Years, Volume 2
Memphis Slim at Ronnie Scott's
The Men in the Blue Suits
Men at Work: Live in San Francisco...or Was It Berkeley?
Menudo: Live in Concert
Menudo: Video Explosion
Merengue: Music from the Dominican Republic
Merle Haggard: Poet of the Common Man
Messiah
The Messiah
Messiah
Messin' with the Blues
Metal Edge: Heavy Metal Superstars
Metal Mania
Metal Masters
Metal Variations: Enigma Metal/Hard Rock Videos
Metalhead Video Magazine, Vol. 1
Metallica: Cliff 'em All
Metallica: Two of One
The Meteors: Video Nasty
Metro Media vs. Inna City, Part 1
Metro Media vs. Inna City, Part 2
Mica Paris
Michael Bolton: Soul and Passion
Michael Bolton: Soul Provider
Michael Damian: Rock On
Michael Jackson: Moonwalker
Michael Jackson's Thriller
Michael Monroe: Rock Like XXXX—Live at the Whiskey A Go Go 2
Michael Nesmith: Nezmuzic
Michael Nesmith: Rio/Cruisin'
Michael Prophet
Michael Schenker Group: Rock Will Never Die
Michael Stanley Band
Michael W. Smith: Big Picture Tour
Michael W. Smith: Concert Video
Michel Beroff Plays Olivier Messiaen
Michel Petrucciani: Live at the Village Vanguard

Michel Petrucciani: Power of Three Video
Michel'le
Michelle Shocked: Captain Swing Revue
Michigan Nighthawks Jazz Band #1
Michigan Nighthawks Jazz Band #2
Michigan Nighthawks Jazz Band #3
Midnight Oil: Black Rain Falls
Midori: Live at Carnegie Hall
A Midsummer Night's Dream
Mighty Clouds of Joy in Concert
The Mighty Sparrow: Live
The Mikado
Mike Mainieri: The Jazz Life
Mike & the Mechanics: A Closer Look
Mildred Dilling: Memoirs of a Harp Virtuoso
Miles Davis/John Coltrane: Miles & Trane
Miles Davis: Live '85
Miles Davis: The Sound of Miles Davis
Milli Vanilli: In Motion
Milton Berle, the Second Time Around: Legends
Milton Brunson & the Thompson Community Singers: Rise Up and Walk
Mingus: Charlie Mingus
Ministry: In Case You Didn't Feel Like Showing Up
Ministry: Live 1-28-90
Minor Detail
Minor Threat: Live
Misbehavin' Jazz Band
Missa Luba
Missing Persons
Mission: Crusade
Mission: Dawn to Dusk
Mission U.K.: Waves Upon the Sand
Mississippi Mass Choir: Live in Jackson, MS
Mississippi Six Dixieland Band
Mister Mister: Videos from the Real World
Mit Musik durch Stadt und Land
Mitch Ryder: Live in Ann Arbor
The Modern Jazz Quartet
Modern Jazz Quartet and the Juillard String Quartet
The Moistro Meet the Cool Ruler
Moliendo Vidrio
Money Madness
Monkees: Heart & Soul
Monkees, Volume 1
Monkees, Volume 2
Monkees, Volume 3
Monkees, Volume 4
Monkees, Volume 5
Monkees, Volume 6
Monster T.V. Rap Hits
Monterey Jazz Festival
Monterey Pop
More Baby Songs
More of Jive's Greatest Rap Video Hits
More Than the Music
More Women in Rock
The Mormon Tabernacle Choir Christmas
The Mormon Tabernacle Choir Greatest Hits
Morrissey: Hulmerist
Morrissey: Live in Dallas
The Moscow Sax Quintet: Jazznost Tour

Motels: Shock
Mother Mallard's Portable Masterpiece Company
Motley Crue: Dr. Feelgood—The Videos
Motley Crue: Uncensored
Motorhead: 1916 Live...And It Serves You Right!
Motorhead: Another Perfect Day
Motorhead: Deaf Not Blind
Motorhead: The Birthday Party
Motown 25: Yesterday, Today, Forever
Motown Returns to the Apollo
Motown Sounds
Motown Time Capsule: The '60s
Motown Time Capsule: The '70s
Moussorgsky: Pictures at an Exhibition
Mouthful of Sweat Compilation
Movie Memories Series
Mozart
Mozart No End and the Paradise Band
Mozart/Smetana/Dvorak
Mozart Under a Microscope
Mozart's Requiem with Colin Davis
Mr. Big: Lean into It
Mr. Charlie
Mr. Mister: Videos from the Real World
MTV Closet Classics
MTV Collection: 3rd Annual Video Music Awards
Muse Concert: No Nukes
Music in the 12th Century
Music of the Baroque
Music Classics
Music and Comedy Masters Series
Music Court
Music and Dance of Pakistan
Music? or Else
Music at Large—The Manhattan Percussion Ensemble
The Music of Melissa Manchester
Music in the Midnight Sun
Music in Time
Music Video From "Streets of Fire"
Music for Wilderness Lake
Musica Latina: "Hot"
Musica Tipica
Musical Personalities No. 1
Musicourt
Musik kennt keine Grenzen #1
Musik kennt keine Grenzen #2
Musik kennt keine Grenzen #3
Musik aus allen Himmelsrichtungen
Musik Laden, Part 1: Beat Club
Musik Laden, Part 2: Beat Club
Musik liegt in der Luft
Musik geht um die Welt, Part 1
Musik geht um die Welt, Part 2
Mutabaruka: Live
Mylon & Broken Heart: Crank It Up!
Myrna Summers & the Myrna Summers Workshop Choir of Washington D.C.

Myrna Summers: We're Going to Make It Live
Nabucco
Naked Eyes
The Naked Gershwin
Nana Mouskouri: Live at Herod Atticus
Nanci Griffith: One Fair Summer Evening
Nancy Wilson
Nancy Wilson at Carnegie Hall
Napalm Death: Live Corruption
Napoli
Nashville Bluegrass Band
Nashville Goes International
Nashville Story
The Nat "King" Cole Story
Nat King Cole: The Incomparable Nat "King" Cole
Nat "King" Cole/The Mills Brothers/The Delta Rhythm
Natalie Cole: Everlasting
Natasha: A Dance Entertainment
Natural Light: Windance
Nazareth: Live!
Ned's Atomic Dustbin: Nothing Is Cool
Neil Diamond: Greatest Hits Live
Neil Diamond: I'm Glad You're Here With Me Tonight
Neil Diamond: Love at the Greek
Neil Sedaka in Concert
Neil Sedaka: Timeless In Concert
Neil Young in Berlin
Neil Young and Crazy Horse: Ragged Glory
Neil Young: Freedom
Neil Young: Rust Never Sleeps
Neil Young: Solo Trans
Neil Young: Weld
Nelson Mandela 70th Birthday Tribute
The Neneh Cherry: Rise of Neneh Cherry
Neptune's Garden
Neville Brothers & Friends: Tell It Like It Is
Neville Marriner
New Born King
New Edition: Past and Present
New From London, Volume 1
New Kids on the Block: Hangin' Tough
New Kids on the Block: Hangin' Tough Live
New Kids on the Block: Step by Step
New Order: Substance
New Sousa Band: On Stage at Wolf Trap
New Stars on Video
New York City: Performance
New York Dolls: Live in a Doll's House
New York Giants: We're the New York Giants
New Yorker, New Yorker
Nicolai Ghiaurov: Tribute to the Great Basso
Night of the Guitar, Vol. 1
Night of the Guitar, Vol. 2
Night of the Long Knives
Night Ranger: 7 Wishes Tour
Night Ranger: Japan in Motion
Night Ranger: Japan Tour
Nightblooming Jazzmen 1
Nightblooming Jazzmen 2
Nightblooming Jazzmen 3

Nina Simone: Live at Ronnie Scotts
999: Feelin' Alright with the Crew
Nitty Gritty Dirt Band: Twenty Years of Dirt
Nocturna
Nocturnal Emissions: Bleed
Norman and Nancy Blake: Planet Riders
Nostalgia for the '90s
Now That's What I Call Music
Nuclear Assault: Handle with Care
Nugget Hotel Jazz Fest
Nugget Jazz Fest (Hymn-a-Long)
Odyssey: Ron Hays' Music Image
Of Men and Music
Ofra Haza: From Sunset till Dawn
Oh Happy Day
Oingo Boingo: Skeletons in the Closet
Oktoberfest 1 - 2 - g'suffa
Oktoberfest Stimmungsvideo
Olatunji and His Drums of Passion
Old-Fashioned Christmas
Olivia
Olivia in Concert
Olivia: Physical
Olivia: Soul Kiss
Olivia: Twist of Fate
One Last Time: The Beatles' Final Concert
One Little Indian Compilation
One Night With Blue Note Preserved
One Voice, One Guitar
Onegin
Opening Concert Berlin: 750 Years
The Operas of Thomas Pasatieri
Operation Rock 'n' Roll: The Home Video
Orchestral Manoeuvres in the Dark: Crush, the Movie
An Orchestral Tribute to the Beatles
Oregon
Oregon: Live at 87 Freiburg
Ornette Coleman: Ornette Made in America
Oscar Peterson: Quartet Live
Osibisa: Warrior
The Other Side of Nashville
Otis Day and the Knights: Otis My Man!
Otis Redding: Live in Monterey, 1967
Otis Redding: Ready, Steady, Go!
Outlaws
Ozzy Osbourne: Bark at the Moon
Ozzy Osbourne: Don't Blame Me
Ozzy Osbourne: The Ultimate Ozzy
Ozzy Osbourne: Wicked Videos
Paddy Reilly Live
Pain Teens: Live
Palm Springs Yacht Club
Pantera: Cowboys from Hell
The Paris Reunion Band
Party Tillyadrop Tour Video
Pat Benatar in Concert
Pat Benatar Hit Videos
Pat Benatar: The Visual Music Collection
Pat Travers: Just Another Killer Day

Patti La Belle: Live at the Apollo Theater
Patti La Belle: Look to the Rainbow Tour
Patti Page Video Songbook
Paul Dresher Ensemble: "Slowfire"
Paul Dresher: Was Are/Will Be
Paul McCartney: Once Upon a Video
Paul McCartney: Put It There
Paul McCartney: Rupert and the Frog Song
Paul McCartney's Get Back
Paul Revere and the Raiders: The Last Mad Man of Rock and Roll
Paul Simon in Concert
Paul Simon: Concert in Central Park
Paul Simon: Graceland
Paul Simon: Graceland, the African Concert
Paul Simon Special
Paul Winter: Canyon Consort
Paul Young: The Video Singles
Paula Abdul: Straight Up
Pavarotti
Pavarotti & Levine In Recital
Pavarotti in London
P.D.Q. Bach: The Abduction of Figaro
Peace/Allelujah
The Peace Concert
The Peace Tapes
Peerce, Anderson and Segovia
Peggy Lee: The Quintessential
Penelope Houston: Dash-30-Dash
Pentecostal Community Choir with Minister Keith Pringle: No Greater Love
Perfect Gentlemen: Rated PG
Perfect Harmony: The Whiffenpoofs in China
The Perfumed Handkerchief
Personal Property
Pet Shop Boys: Actually
Pet Shop Boys: Highlights
Pet Shop Boys: Television
Pete Townshend
Pete Townshend: Deep End Live
Pete Townshend: Video EP
Peter Allen and the Rockettes
Peter Gabriel: Compilation Video
Peter Gabriel: POV
Peter Metro: Live & Hot
Peter, Paul and Mary: 25th Anniversary Concert
Peter, Paul and Mary: Holiday Concert
Peter Tosh Band with Andrew Tosh: Live
Peter Tosh: Live
Peter Tosh: Live in Africa
Peter and the Wolf
Peter and the Wolf: Natalia Sats
Petra: Beyond Belief Video Album
Phantom Empire
Phil Collins
Phil Collins: Live at Perkins Palace
Phil Collins: No Jacket Required
Phil Collins: No Ticket Required
Phil Collins: Seriously Live
Phil Collins: The Singles Collection

Phil Woods in Concert
Photonos
Pia Zadora's American Songbook
Piano Players Rarely Ever Play Together
Picture Music
P.I.L.: '83
P.I.L.: Live in 1985 - The Anarchy Tour
Piledriver: The Wrestling Album 2
Pilgrim Jubilees: Live in Jackson
Pinchers Live: Jumbo
Pink Floyd in Concert: Delicate Sound of Thunder
Pink Floyd: La Carrera Panamericana
Pink Floyd at Pompeii
Pink Floyd's David Gilmour
Pirates of Panasonic
Placido: A Year in the Life of Placido Domingo
Placido Domingo: An Evening with Placido
Placido Domingo: Songs of Mexico, Vol. 1
The Platters/The Coasters: In Concert
Playboy Jazz Festival, Vol. 1
Playboy Jazz Festival, Vol. 2
Playboy's Girls of Rock & Roll
P.M. Dawn: Of the Heart, Of the Soul, and Of the Cross
P.M. Dawn: Of the Heart, of the Soul and of the Cross
The Pogues: Live at the Town and Country
Pointer Sisters: Live in Africa
Pointer Sisters: So Excited
Poison Idea: Mating Walruses
Poison: Sight for Sore Eyes
Police: Around the World
Police: Every Breath You Take (The Videos)
The Police: Synchronicity Concert
Polish Chamber Orchestra, Conducted by Jerzy Maksy
Polish Christmas Carols
A Portrait of Beverly Sills
Power Station: Video EP
Praisin' His Name: The Gospel Soul Children of New Orleans
Presenting Joan Sutherland
The Pretenders: The Singles
Pretty Woman: The Music Videos
Priest Live
Primal Sounds from N.Y.C.
Prime Cuts
Prime Cuts: Heavy Metal
Prime Cuts: Jazz and Beyond
Prime Cuts: Red Hots
Prince Jammy
Prince: Nothing Can Stop...Prince and the New Power Generation: Gett Off
Prince and the Revolution: Live
Prince: Sign O' the Times
Prince's Trust All-Star Rock Concert
Prince's Trust Rock Gala
Private Music Video Collection
Professor Don Burns
Professor Plum's Jazz
Psychedelic Furs: All of This and Nothing
Psychedelic Rock
Psychic T.V.: 8Transmissions8
Psychic T.V.: Joy

Public Enemy: The Fight Power—Live
Puerto Rican Danza
Puerto Rican Plena
Punk in London
Punk Pioneers
Punk Rock Movie
Punk Special
Pussy Galore: Live
Pussy Galore: Maximum Penetration
Queen: Greatest Hits
Queen: Live at Budapest
Queen: Live in Japan
Queen: Live in Rio
Queen of Spades: Bolshoi Opera
Queen: The Magic Years, Vol. 1
Queen: The Magic Years, Vol. 2
Queen: The Magic Years, Vol. 3
Queen: The Works
Queen: We Will Rock You
Queensryche: Live in Tokyo
Queensryche: Operation Livecrime
Queensryche: Video: Mindcrime
Quiet Riot: Bang Thy Head
Quiet Riot: Bang Thy Head (long play version)
Quincy Jones: A Celebration
Quincy Jones: Live in Hawaii
Quincy Jones: Reflections
Raccoons: Let's Dance
Radiance/Celebration
Rafael en Raphael
Raffi in Concert with the Rise & Shine Band
Raffi: Young Children's Concert with Raffi
Raggamuffin Tek' Over
Rahsaan Roland Kirk/John Cage: Music
Rainbow Bridge
Rainbow Goblins Story
Rainbow: Japan Tour '84
Rainbow: Live Between the Eyes
Rainbow: The Final Cut
Rainier Jazz Band
The Rake's Progress
Ralph Kirkpatrick Plays Bach
Ramones: Lifestyles of the Rich and Ramones
Randy Newman Live at the Odeon
Randy Stonehill: One Night in Twenty Years
Randy Travis: Forever and Ever
Rank & File: Live
Rap from Atlantic Street
Rap Classics
Ratt: Detonator: Videoaction 1991
Ratt: The Video
Ravi Shankar: Raga
Raw and Rough Pinchers
Raw Talent
Ray Charles Live 1991: A Romantic Evening at the McCallum Theatre
Ray Davies: Return to Waterloo
RCA's All-Star Country Music Fair
Reader's Digest: Nature's Symphony
Ready Steady Go, Vol. 2
Ready Steady Go, Vol. 3
Real Country
Reba McEntyre
Rebel

Red, Hot & Blue
Red Hot Chili Peppers: Funky Monks
Red Hot Chili Peppers: Positive Mental Octopus
Red Hot Chili Peppers: Psychedelic Sexfunk Live from Heaven
Red Hot Chili Peppers: Red Hot Skate Rock
Red Hot Dance
Red Hot Rock
Red Rose Ragtime Band
Reggae Dance Hall Clash Video
Reggae Stars in Concert
Reggae Sting '85
Reggae Sting '89: Vol. 1
Reggae Sting '89: Vol. 2
Reggae Sting '89: Vol. 3
Reggae Sting '89: Vol. 4
Reggae Sunsplash 1, 2
Reggae Sunsplash '90: Dance Hall Night
Reggae Sunsplash '90: Singer's Night
Reggae Sunsplash '90: Vintage Night
Reggae Sunsplash: Dance Hall 1987
Reggae Sunsplash: Dance Hall 1988
Reggae Sunsplash: Dance Hall 1989
Reggae Sunsplash: Variety Night
Reggae Superstars in Concert
Reggae Tribute: A Tribute to Bob Marley
Rehearsal
Reinaldo Jorge: Parts 1 & 2
R F M · Pop Screen
R.E.M.: Selected Videos
R.E.M.: Succumbs
R.E.M.: This Film Is On
R.E.M.: Tourfilm
Renata Scotto: Prima Donna in Recital
Renee & Angela: The Video Singles
Rent Party Revellers
REO Speedwagon: Live Infidelity
REO Speedwagon: Wheels Are Turnin'
Requiem
The Residents: Moleshow/ Whatever Happened to Vileness Fats
The Residents: Video Voodoo
Restless Heart: Big Dreams in a Small Town
Restless Heart: Fast Moving Train
Return of the Goddess for the New Millenium
The Return of Ruben Blades
Return of Ulysses
Revival Temple Mass Choir
Revolting Cocks: You Goddam Sonofabitch
Rhythm & Blues Review
Rhythms of the World Anthology
Richard Clayderman Live In Concert
Richard Marx: Hold On to the Nights
Richard Marx, Vol. 1
Richard Smallwood Singers: Video Celebration
Richard Thompson
Rick Astley: Best Video Hits
Rick Derringer
Rick Springfield: Platinum Videos

Rick Springfield: The Beat of a Live Drum
The Ricky Grundy Chorale
Ricky Nelson in Concert
Ricky Nelson: Special Memories
Ricky van Shelton: RVS...To Be Continued
Ricky Skaggs: Live in London
The Right Stuff
Rigoletto
Ringo Starr and His All-Star Band
Rise Stevens in Opera and Song, Vol. 2
Rob McConnell & the Boss Brass
Robert Cray: Collection
Robert Lockwood, Jr.: Annie's Boogie
Robert Palmer: Riptide
Robert Palmer: Super Nova
Robert Palmer: Video Addictions
Robert Plant: Mumbo Jumbo
Robert (Ragtime Bob) Darch
Roberta Flack: An Evening with Roberta Flack
Robyn Hitchcock: Brenda of the Lightbulb Eyes
Robyn Hitchcock: Gotta Let this Hen Out
The Roches: Live Nude Review
Rocio Banquells, Video 1
Rock Aid Armenia: The Earth Hour Video
Rock Aid Armenia: The Making of "Smoke on the Water"
Rock Classix
Rock Classix: Heavy Metal
Rock Encyclopedia: 1973
Rock Encyclopedia: 1974
Rock Encyclopedia: 1975
Rock Music with the Muppets
Rock 'n' Roll #3: Sexy Girls & Sexy Guns
Rock 'N Roll Classics, Vol. 1
Rock 'N Roll Classics, Vol. 2
Rock 'N Roll Classics, Vol. 3
Rock 'N Roll Classics, Vol. 4
Rock 'N Roll Classics, Vol. 5
Rock 'N Roll Classics, Vol. 6
Rock 'n' Roll Heaven
Rock 'n' Roll History: American Rock, the '60s
Rock 'n' Roll History: English Invasion, the '60s
Rock 'n' Roll History: Fabulous Fifties
Rock 'n' Roll History: Girls of Rock 'n' Roll
Rock 'n' Roll History: Rock 'n' Soul Heaven
Rock 'n' Roll History: Rockabilly Rockers
Rock 'n' Roll History: Sixties Soul
Rock 'n Roll Meltdown
Rock 'n' Roll: The Early Days
Rock & Read
Rock & Roll Call
Rock & Roll Review
Rock & Roll Wrestling Music Television
Rock & Roll Wrestling Music Television: The Special Edition
Rockers
Rockers '87
Rockin' the '50s
Rockin' the Night Away
Rockin' Reindeer Christmas
Rockshow
Rod Stewart

Rod Stewart Concert Video
Rod Stewart & the Faces: Videography 1969-1974
Rod Stewart Live at the L.A. Forum
Rod Stewart: Storyteller 1984-1991
Rod Stewart: Tonight He's Yours
Rodeo: The Music Video
Roger Daltrey: Ride a Rock Horse
Roger McGuinn/Nitty Gritty Dirt Band: Live!
Roger Waters: Radio K.A.O.S.
Roger Whittaker
Roll Over Alice
Rolling Stone Presents Twenty Years of Rock & Roll
Rolling Stone: The First 20 Years
Rolling Stones: Let's Spend the Night Together
Rolling Stones: Live In Hyde Park
The Rolling Stones: Rolling On
Rolling Stones: Video Rewind-Great Video Hits
Romeo and Juliet (Bolshoi Anniversary)
Romeo and Juliet in Kansas City
Ron Carter & Art Farmer: Live at Sweet Basil
Ron Carter: Live Double Bass
Ronnie Milsap: Golden Video Hits
Ronnie Milsap: Greatest Video Hits
The Roots of Gospel
The Roots of Gospel: Part 2
Roots, Rock and Reggae: Inside the Jamaican Music Scene
Rosanne Cash: Retrospective
Rosanne Cash: The Interiors Tour
Roxette: Look Sharp Live
Roxy Music: The High Road
Roxy Music: Total Recall
Roy Acuff's Open House: Vol. 1
Roy Acuff's Open House: Vol. 2
Roy Ayers: Live at Ronnie Scott's
Roy Brown: Aires Bucaneros
Roy Orbison
Roy Orbison & Friends: A Black and White Night
Royal Society Jazz Orchestra
RRRecords Compilation Testament
Rubber Rodeo
Rudolf Schock: Auf Schusters Rappen
Run D.M.C.
Run D.M.C.: The Video
Running Wild: Death or Glory Tour Live
Rush: A Show of Hands
Rush: Chronicles
Rush: Exit...Stage Left
Rush: Grace Under Pressure Tour
Rush: Through the Camera Eye
Russ Carlyle: Story
Russian Folk Song & Dance
Ryuichi Sakamoto: Beauty
Ryuichi Sakamoto: Media Ban Live
Ryuichi Sakamoto: Neo Geo
Ryuichi Sakamoto: Tokyo Melody

Sabbat: End of the Beginning
The Sacred Music of Duke Ellington
Sadao Watanabe: Mysha
Sade: Diamond Life Video
The St. John Passion
St. Louis Blues
St. Louis Ragtimers
St. Matthew Passion
Salif Keita: Destiny of a Noble Outcast
Salute to the Edinburgh Tattoo
Sam Kinison: Banned
Samantha Fox: Just One Night
Samantha Fox: Making Music
Samantha Fox: The Music Video
San Francisco Blues Festival
Sandi Patti: Let There Be Praise: Concert Video
Sandi Patti... Live
Santana: Viva Santana (A Conversation with Carlos)
Sarah Vaughan: Live from Monterey
Sarah Vaughan: Live in Tokyo
Sarah Vaughan/Mel Torme: The Nearness of You
Sass & Brass: A Jazz Session
Savage Republic: Disarmament
Sawyer Brown: Greatest Hits
Sawyer Brown: Shakin'
Saxon: Live
Scarlatti/Debussy/Ravel
Scenic Harvest from the Kingdom of Pain
Scorpions
Scorpions: Crazy World Tour Live - Berlin 1991
Scorpions: First Sting
Scorpions: To Russia With Love
Scorpions: World Wide Live
Scott Joplin, 1868-1917
Screamers: Live in San Francisco, 1978
Scritti Politti
Scruggs
Seal
Secret Policeman's Other Ball
Secret Policeman's Private Parts
Sensational Nightingales: Ministry in Song
Sermon on the Mount
Seventeenth Annual Telluride Bluegrass Festival
'70s Music Video Album
'70s Revival
Severed Heads: Overhead
Sex Pistols: Buried Alive
Sex Pistols: Decade
Sex Pistols: Filth & the Fury
Sex Pistols: Last Winterland Concert
Sex Pistols Live
Sex Pistols: The Great Rock & Roll Swindle
Sha Na Na Rock 'n' Roll Concert and Party Video
Shabba Ranks: Talk of the Town
Shabba Ranks vs. Ninja Man: Super Clash Round
Shadows on the Grass
Sharing... a Visual Musical Experience
Sheena Easton
Sheena Easton: Act One
Sheena Easton: Live at the Palace, Hollywood
Sheena Easton: Private Heaven

Sheila E.: Live Romance 1600
Sheila Walsh: "Shadowlands" in Concert
Shelley Duvall's Rock 'n' Rhymeland
Shelly Manne Quartet
Shenandoah: Greatest Video Hits
Sherrill Milnes: Homage to Verdi
Shimmy-Disc Compilation, Vol. 1
Shimmy-Disc Compilation, Vol. 2
Shindig! Presents Frat Party
Shindig! Presents Groovy Gals
Shindig! Presents Jackie Wilson
Shindig! Presents Motor City Magic
Shindig! Presents the Righteous Brothers: Unchained Melody
Shindig! Presents Soul
Shirley Bassey: You Ain't Heard Nothin' Yet - Live
Shirley Caesar: Hold My Mule (Live in Memphis)
Shirley Caesar: I Remember Mama
Shmenges: The Last Polka
Shockwave
Short Pieces
Shorty Rogers: West Coast Giants
Shout!: The Story of Johnny O'Keefe
Show My People
Showdown 1
Showdown at the Hoedown
Siam Odyssey
Simon Boccanegra
Simon and Garfunkel: The Concert in Central Park
Simply Red: Let Me Take You Home
Simultaneous
Sinead O'Connor: Nothing Compares 2 U
Sinead O'Connor: The Value of Ignorance
Sinead O'Connor: Year of the Horse
Sing Along with Israel
Sing mit Heino
Sing mir das Lied noch einmal
Sing mit Mir, Tanz mit Mir
Singer Presents "Elvis" (The 1968 Comeback Special)
Singing French Songs for Children
Siouxsie and the Banshees
Siouxsie & the Banshees: Nocturne
Siouxsie & the Banshees: Once Upon a Time
Sir George Solti Conducts the Chicago Symphony Orchestra in Tokyo, 1990
'60s Music Video Album
Ska Beatz
Ska Explosion
Skid Row: Oh Say, Can You Scream?
Skin Deep: The Men
Skin Deep: The Women
Skinny Puppy: Ain't It Dead Yet?
Slammin' Rap Video Magazine, Vol. 1
Slammin' Rap Video Magazine, Vol. 2
Slammin' Rap Video Magazine, Vol. 3

Music

Slammin' Rap Video Magazine, Vol. 4
Slaughter: From the Beginning
Slipping Through the Cracks
Smashin' the Palace '84
Smithereens: Smithereens 10
Snap: World Power - The Videos
Snub TV, Vol. 2
So Proudly We Hail: A Salute to American Patriotism
So Schoen wie Heut, so Muesst es Bleiben (Schlager, die Man nie Vergisst)
Soft Cell
Soft Cell/Marc Almond: Memorabilia - Video Singles
Soldiers of Music: Rostropovich Returns to Russia
Solo Tribute to Keith Jarrett
A Song for Ireland
Songs from Papua New Guinea
Songs from the Soul: The Negro Spiritual
Sonic Youth: Goo
Sonny & Cher Nitty Gritty Hour
Sonny Rollins: Live
Sonny Rollins: Montreux Jazz Festival
Sonny Rollins: Saxaphone Colossus
Sonny Terry: Shoutin' the Blues
Sophisticated Ladies
Soul Classix
Soul Experience
The Soul of R & B
Soul to Soul
Sound?
Sound of Jazz
Soundgarden: Louder Than Live
The Soundies, Vol. 1: 1940s' Music Machine
Soundies, Vol. 2: Singing Stars of Swing Era
Soundies, Vol. 3: Big Band Swing
Soundies, Vol. 4: Harlem Highlights
Sounds of Motown
South Market Street Jazz Band: 1
South Market Street Jazz Band: 2
South Pacific: The London Sessions
Southern Comfort Jazz Band #1
Southern Comfort Jazz Band #2
Southern Comfort Jazz Band #3
Soviet Army Chorus, Band, and Dance Ensemble
Spandau Ballet: Live Hits
Spandau Ballet: Over Britain
Spartacus
Spend It All
Spike Lee & Co.: Do It A Cappella
Spirituals in Concert
S.P.K.: 2 Autopsy Films
S.P.K.: Despair
Splashin' the Palace '84
Split Enz: History Never Repeats - The Best of Split Enz
Split Enz: Live
Spring Symphony
Sprout Wings and Fly
Spyro Gyra

Spyro Gyra: Graffiti
Squeeze: A Round & a Bout
Squeeze Play: The Videos 1978-1987
Stage Show
Stage Show with the Dorsey Brothers
Stamping Ground
Stan Getz with Richie Cole
Stan Kenton & His Orchestra: London, 1970
Stan Kenton: In the USA
Stand By Me: AIDS Day Benefit Concert
Standards II with Keith Jarrett, Gary Peacock and Jack DeJohnette
Stanley Jordan: Cornucopia
Stanley Jordan: Magic Touch
Stanley Jordan: The Blue Note Concert
Stanley Turrentine: In Concert
Star Flight
Star Licks
Star Spangled Banner 1
Star Spangled Video: Operation Desert Storm Welcome Home
Stars on 45
Starship: Greatest Hits (Ten Years and Change 1979-1991)
Starship: Video Hoopla
State of the Arts Performance Special: A Festival
The Statler Brothers: Brothers in Song
Status Quo: Rockin' All Over the Years
Steeleye Span: A 20th Anniversary Celebration
Stephan Kates, Cellist
Stephane Grappelli: Live in San Francisco
Stephane Grappelli in New Orleans
Stephanie Mills: Home is Where the Heart is
Stephanie Mills: Television Medicine
Steps Ahead: Live in Japan
Steve and Annie Chapman: The Greatest Gift
Steve Gadd II: In Session
Steve Miller Band
Stevie Nicks in Concert
Stevie Nicks & Friends
Stevie Nicks: I Can't Wait
Stevie Nicks: Live at Red Rocks
Stevie Ray Vaughn and Double Trouble: Live at the El Mocombo 1983
Stevie Ray Vaughn & Double Trouble: Pride & Joy
Stewart Copeland: Rhythmatist
Sting: Bring on the Night
Sting: Nothing Like the Sun: The Videos
Sting: The Soul Cages Concert
Sting: The Soul Cages Video
Sting: The Videos Part 1
Stonehenge '84
Stop Making Sense
Stop the Violence Movement: Self Destruction
The Story of the Clancy Brothers and Tommy Makem
The Story of Creation
Story of Hands Across America
The Stranglers

Stratasphere: Portrait of Teresa Stratas
Stray Cats
Stray Cats: Bring It Back Again
Stray Cats Rock Tokyo
Streisand Collection
Strictly for the Beat
String Trio of New York: Built by Hand
Structures from Silence
Stryper: Against the Law
Stryper: In the Beginning
Stryper: Live In Japan
The Style Council: Far East & Far Out
Style Council: Jerusalem
Style Council: Live at Full House
Style Council: Showbiz
Style Council: What We Did On Our Holiday
Styx: Caught in the Act
Subbulakshmi: Queen of Song
The Sugar Cubes: Live Zabor
Sugar Minott: Dance Hall Video
Sugarcubes: Live Zabor
Suicidal Tendencies: Lights...Camera...Suicidal
Summerwind
Sun City
Sun, Moon & Feather
Sun Ra: Mystery, Mr. Ra
Sunday Mornin' Country
Super Drumming
Super Jam 3: Stone Love vs. Metromedia
Super Jam '87
Super Rock '85
Superdrumming
Supershow
Superstars in Concert
Supertramp: Brother Where You Bound
Supertramp: The Story So Far
Surfing Beach Party
Survivor: Live in Japan
Survivors: The Blues Today
Suzanne Vega
Suzuki Violin Concert Series
Sweet Baby: Live
Sweet Sensation: Sweet Sensation
Swing Out Sister: And Why Not?
Swing Out Sister: Kaleidoscope World
Swing: The Best of the Big Bands
Swinging Singing Years
Swinging U.K.
The Swingle Singers in Concert
The Sylvia Plath—Part 2—Getting There
Sympathy for the Devil
Symphony of Swing
T. Rex: Solid Gold
Taj Mahal: Live at Ronnie Scott's
Takanaka: World
Take 6: All Access
The Tales of Hoffman
Talk Talk
Talk Talk: Natural History
Talking Heads: Storytelling Giants
Tam Tam: Do It Tam Tam
The T.A.M.I. Show
Tammy Wynette in Concert
Tanita Tikaram: Ancient Heart
Tankard: Open All Night - Live at the Berlin Wall
Tannhaeuser
Tanya Tucker: Live

Taylor Dayne: Twists of Fate
Tchaikovsky Competition: Violin & Piano
Tchaikovsky Gala in Leningrad
Tchaikovsky: Piano Concerto #1
Tear In My Beer & Other Great Country Videos
Tears for Fears
Tears for Fears: Going to California
Tears for Fears: In My Mind's Eye
Tears for Fears: Scenes From the Big Chair
Tears for Fears: Sowing the Seeds
Tears for Fears: Tears Roll Down (Greatest Hits 82-92)
Teatro Alla Scala: Adrianna Lecouvrer
Technotronic: Trip on This - The Videos
Ted Nugent's New Year's Eve Whiplash Bash
Teddy Pendergrass: Live in London
Teenage Mutant Ninja Turtles: Coming Out of Their Shells Tour
Teenage Mutant Ninja Turtles: Music Videos
TeenVid Video Magazine, Vol. 1
TeenVid Video Magazine, Vol. 2
The Temps and the Tops
Temptations & Four Tops
Temptations: Get Ready
Temptations: Live in Concert
10,000 Maniacs: Time Capsule 1982-1990
Ten Years After: Going Home
10CC: Changing Faces
Terence Trent D'Arby: Introducing the Hardline...Live
Tesla: Five Man Video Band
Test Department: Program for Progress
Testament: Seen Between the Lines
Tex Wyndham's Red Lion Jazz Band
Texaco's Swing into Spring
Texas: Southside
Texas Style
That Teen Show #1: Rock Concert Violence
That Was Rock (The TAMI/TNT Show)
That'll Be the Day
The The: Infected
The The: Versus the World
Thelonious Monk: '63 in Japan
Thelonious Monk: Monk in Europe
Thelonious Monk: Music in Monk's Time
Thelonious Monk: Straight, No Chaser
Thelonious Sphere Monk: Celebrating a Jazz Master
There But for Fortune
There was No Room for You
There's a Meetin' Here Tonight
They Might Be Giants: The Videos 1986-1989
Thin Lizzy: Dedication
Thin Lizzy: Live & Dangerous
3rd Bass: The Cactus Vidie/Yo
Third World: Live in Japan '85

Third World: Prisoner in the Street
38 Special: Wild Eyed and Live
Thomas Dolby: Live Wireless
Thomas Dolby: The Golden Age of Video
Thomas Dolby: Video 45
Thompson Twins: Into the Gap Live
Thompson Twins: Live at Liverpool
Thompson Twins: Single Vision
3-Way Thrash
The Thrill of the Orchestra
Throbbing Gristle: Destiny
Throbbing Gristle: Live at Kezar
Thunders, Kane, Nolan: Live on the Sunset Strip
Tiffany: Live in Japan
Tiger
Tiger: Spectacular
Tigertailz: Bezerk Live '90
'Til Tuesday
Timbuk 3
Time Groove
Timex Watch Hour
Timmy T.
Tina Turner: Break Every Rule
Tina Turner: Do You Want Some Action!
Tina Turner: Foreign Affair
Tina Turner Live: Nice 'n' Rough
Tina Turner: Live, Private Dancer Tour
Tina Turner: Live in Rio
Tina Turner: Private Dancer
Tina Turner: Queen of Rock & Roll
Tina Turner: Simply the Best
Tino Rossi: Thanks to Dear Friends - Live
Tiny and Ruby: Hell Divin' Women
Todd Rundgren: 2nd Wind Live Recording Sessions
Todd Rundgren: Grok Gazer, Visual Concert Hall
Todd Rundgren Live in Japan
Todd Rundgren: Nearly Human - Live in Japan
Todd Rundgren Videosyncracy
Tom Jones: Live in Las Vegas
Tom Jones: Live at This Moment
Tom Petty: A Bunch of Videos & Some Other Stuff
Tom Petty: Full Moon Fever, the Videos
Tom Petty & the Heartbreakers: Pack Up the Plantation Live!
Tom Waits: Big Time
Tommy
Tommy Page: I'll Be Your Everything
Tommy Puett: Heart Attack
Tonadillas of Granados
Toni Basil: Word of Mouth
Toni Childs: The Videos
Tony Bennett Live: Watch What Happens
Tony Bennett Sings
The Tony Marshall Show
Tony Powers
Tony! Toni! Tone!
Tony Williams: New York Live Video
Too Short: Short Dog's in the House
Toots and the Maytals Live from New Orleans

Toots Thielemans in New Orleans
Tosca
Toscanini: The Maestro/Hymn of the Nations
Toscanini: The Television Concerts
Toshiko Akiyoshi: Jazz is My Native Language
Totally Go-Go's
Toto: Past to Present
Tout Ecartile
Toxic Reasons
Toy Dolls
Toy Dolls: Idle Gossip
Traffic: Live at Santa Monica
Tramaine Hawkins: Bring it Home, Live
Transformation is Our Secret
Travis Tritt: It's All About to Change
T.Rex: T-Rexmas
A Tribe Called Quest: The Art of Moving Butts
Tribute to Bechet
Tribute to John Coltrane
Tribute to Ricky Nelson
The Triplets: Video Triple Single
Triumph: A Night of Triumph
The Triumph of Charlie Parker: Celebrating Bird
Triumph: Live at the US Festival
TROOP: Attitude
The Tropicana Music Bowl 3
Troy Boys
True Story of 200 Motels
Truth or Dare
Tubes: Live at the Greek
The Tubes Video
Turandot
Turn Up the Volume, 3
Turn Up the Volume, 4
Turnpike Cruisers: Cable T.V. Show
The Turtles: Happy Together
Tuxedomoon: Four Major Events
TV/ARM Video Magazine
Twelve O'Clock High Video Compilation
Twisted Sister: Come Out and Play
Twisted Sister's Stay Hungry
Two Fiddles, No Waiting
200 Motels
2-Kut: Rock That
2 Live Crew: Banned in the USA
Two Rooms: Celebrating the Songs of Elton John & Bernie Taupin
2 Sound Clash Dance Hall
Two X 4
U2: Live at Red Rocks "Under A Blood Red Sky"
U2: Rattle and Hum
U2: The Unforgettable Fire Collection
UB40: CCCP (The Video Mix)
UB40: Labour of Love 2
UB40: The Best of UB40
U.F.O. Live
U.K. Subs: Grossout U.K.
Ultimate Revenge 2
Ultimate Rock
Ultravox: The Collection
Una Aventura Llamada Menudo
Uncle Floyd's Nostalgic Music Special
Uncle Meat
Und weiter geht's
Underground Forces Volumes 1-8

Underground U.S.A. Vol. 1: Music Magazine
Underground U.S.A. Vol. 2: Hardcore
Underground U.S.A. Vol. 3: Heavy Metal
Underground U.S.A. Vol. 4: New Music
Underground U.S.A. Vol. 5: New Music
Underground U.S.A. Vol. 6: Heavy Metal 2
Underground U.S.A. Vol. 7
Underground U.S.A. Vol. 8
Underground U.S.A. Vol. 9: Son of Heavy Metal
Underground U.S.A. Vol. 10: Rap Rap Rap
Underground U.S.A. Vol. 11: Rap Rap Rap
Underground U.S.A. Vol. 12: Rap Rap Rap
Underground U.S.A. Vol. 13: Heavy Metal
Underground U.S.A. Vol. 14
Underground U.S.A. Vol. 15
Underground U.S.A. Vol. 16
Underground U.S.A. Vol. 17
Underground U.S.A. Vol. 18
Universal Congress Of: Live
Unser kleines Platzkonzert 1
Unser kleines Platzkonzert 2
Unsere schoensten volkstuemlichen Lieder
Upper Dallas Jazz Band
Urban Dance Squad: Mental Floss for the Globe
URGH! A Music War
Uriah Heep: Gypsy
Utopia: An Evening with Utopia
Utopia: Live at the Royal Oak
The Utopia Sampler
Van Halen: Live Without a Net
Van Morrison in Ireland
Van Morrison: The Concert
Vancouver New Music 1978-80
Vandenburg: Live in Japan
Vanessa Williams: De Right Stuff Collection
Vanilla Ice: Play That Funky Music, White Boy
Vassar Clements: Dobro Summit
Venom: Alive in '85!
Venom: Live '90
Venom: Live at Hammersmith Odeon
The Ventures: Live in Japan '84
The Ventures: Live in Japan '90
The Ventures: Live in L.A. '81
Verdi: Requiem
Vespers of the Blessed Virgin
VH-1 '60s Music Video Album
VH-1 '70s Music Video Album
Vico Torriani's internationale Schlager-Revue
Victor Borge in London
Video Aid
Video Band: War Dance
A Video Christmas Card
Video a Go-Go: Volume 1
Video a Go-Go: Volume 2
Video Love Songs
Video Meltdown
Video Rap Pack
Video Sheet Metal
Video Sheet Metal, Volume 2
Video Void
Videotomy, Vol. 1
Videotomy, Vol. 2
Vietnam Experience
Vietnam, Volume 1

Virgin Prunes: The Sons Find Devils
Vision Shared: A Tribute to Woody Guthrie & Leadbelly
Vixen: Revved Up
Vocomotion: Best of the '50s & '60s
Vocomotion: Classic Country
Vocomotion: Famous Star Classic Series
Vocomotion: Favorite Standards
Vocomotion: Holiday Favorites
Vocomotion: Spotlighting the '70s & '80s
A Voice from Soviet Russia
Voice of the Whale
Voices that Care
Voices of Firestone Classic Performances
Voices of Sarafina
Volkstuemliches Europa
Vow Wow: Live
Wagner: Concert in Leipzig
The Wall: Live in Berlin
Walt Whitman & the Soul Children: Live and Blessed
Walter Hawkins & the Love Alive Choir 4
Warfare: A Concept of Hatred
Warlock: Live from London
Warrant: Blood, Sweat & Beers
Warrant: Cherry Pie
Warrant: Dirty Rotten Filthy Stinking Rich Live
Warren Zevon
W.A.S.P.
W.A.S.P.: Videos in the Raw
Waylon Jennings: America
Wayne Newton at the London Palladium
We Like the Blues, Vol. 1
We Like the Blues, Vol. 2
We are the World: The Video Event
Weather Report: Domino Theory
Weavers: Wasn't That a Time!
Weird Al Yankovic: This is Life
Welt der Oper
Wendy O. Williams: Live
We're All Devo
West End Jazz Band #1
West End Jazz Band #2
West End Jazz Band #3
West End Jazz Band #4
Wet, Wet, Wet
Wha Dat Disco
Whalesong
Wham! in China: Foreign Skies
Wham! The Final
Wham! The Video
When the Music's Over
Whitesnake
Whitesnake: Live Commandos
Whitesnake: Live at the Monsters of Rock Festival
Whitesnake: Trilogy
Whitney Houston: #1 Video Hits
Whitney Houston: The Star Spangled Banner
Whitney Houston: Welcome Home Heroes
The Who: Rocks America, 1982 American Tour
The Who: The Who Live!
Tommy
The Who: Who's Better, Who's Best
Whodini: Back in Black
Whodini: Greatest Rap Hits

The Williams Brothers: I'm Just a Nobody
Willie Banks & the Messengers: In Concert
Willie Nelson & Family in Concert
Willie Nelson: Live at Budokan
Willie Nelson: Some Enchanted Evening
Willie Nelson Special with Ray Charles
Willie Nelson's Greatest Hits: Live
Willie & the Poor Boys
Wilson Phillips
The Winans: Live in Concert
The Winans: Return
Windham Hill in Concert
Windham Hill Music Video Series
Winger: In the Heart of the Young
Winger: In the Heart of the Young, Part 2
Winger: The Videos, Volume One
Wir lassen uns das Singen nicht verbieten
Wishbone Ash: Phoenix
Wojceich Mlynarski: Lyrical Evening
The Wolfe Tones: 25th Anniversary Concert
Women in Rock
Women in Rock: The Women of Heavy Metal
The Wonderful Planet
Woodstock
Woodstock: The Lost Performances
Woody Guthrie: Hard Travelin'
Woody Herman: 1982 Aurex Jazz Festival
Woody Herman: Live Thunder
Word Rap Video Compilation, Vol. 1
Word Up: The Best
The Work of Sefan Wolpe
World Drums
The World of Music
World War II: The Music Video, Vol. 1
World War II: The Music Video, Vol. 2
The Worship of Music: The Dagar Singers of India
Wozzeck
Wrathchild: Live from London
Wrestling Music Television: Born to Bleed
Wynton Marsalis: Blues & Swing
Y & T Live at the San Francisco Civic
Y & T: Summertime Girls & All American Boys - The Videos
Yakety Yak: Take It Back
Yankee Air Pirates
Yankee Doodle's Odyssey
Yehudi Menuhin: Brahms Violin Concerto in D, Opus 7
Yehudi Menuhin: Concert for the Pope
Yehudi Menuhin: Tribute to J. S. Bach
Yellow Magic Orchestra: Computer Game
Yellow Magic Orchestra: Live '81
Yellow Magic Orchestra: Propaganda
Yellow Magic Orchestra: YMO Live in L.A.
Yellowman: Live at Woodbury

Yellowman: Raw & Rough
Yes
Yes: 9012 Live
Yes: Greatest Video Hits
Yessongs
Yngwie Malmsteen: Rising Force - Live '85
Yngwie Malmsteen: Trial By Fire/Live in Leningrad
A Young Children's Concert with Raffi
The Young Composer
Young Country
Young M.C.: Bustin' Moves
You're the Top: The Cole Porter Story
Youri Vamos: Der Nussknacker - Eine Weihnachtsgeschichte
Yukihiro Takahashi: Boys Will Be Boys
Yuletide TV Treats 2
Yumi Matsutoya: Train of Thought
Zbigniew Rybezynski: A Collection
Zenon Laskowik's Benefit: By the Back Door
Zev: Six Examples
Ziggy Marley and the Melody Makers: Conscious Party Live at the Palladium
Zildjian Day in New York
Zoot Sims
Zoot Sims Quartet
Zydeco
Zydeco: Night 'n' Day

DISTRIBUTOR INDEX

Each video review has at least one and as many as three distributor codes. In the reviews, these are located immediately before the critical rating at the bottom of the review. This **Distributor Guide** lists the full address and phone, toll-free, and fax numbers of these organizations. The key below interprets the codes. Those reviews annotated with the code *OM* are on moratorium (distributed at one time, though not currently). Since a title enjoying such status was once distributed, it may well linger on your local video store shelf. When the distributor is not known, the code *NO* appears in the review.

Distributor Code Key:

A&M—A & M Video
AAE—A & E Home Video
AAI—Arts America, Inc.
ABA—ABA Consortium for Professional Education
ABC—ABC Distribution Company
ACA—Academy Home Entertainment
ACD—Academy Videos
ACE—Ace Video
ACF—Action Films & Video
ACN—Acorn Video
AEF—American Educational Films
AFE—Audiofidelity Enterprises
AFF—American Film Foundation
AFR—Afro-Am Distributing Company
AHV—Active Home Video
AIM—AIMS Media
AIP—A.I.P. Home Video, Inc.
ALD—Alden Films
ALE—Allison Entertainment
ALL   Allied Artists
ALS—Allinson Educational Service
AMB—Ambrose Video Publishing, Inc.
AMI—AMI Entertainment
AMV—Amvest Video
AOV—Admit One Video
APD—Applause Productions, Inc.
API—Abacus Productions, Inc.
APL—Appalshop Films
APP—Applause Theatre and Cinema
ARP—ARP Videos
ART—Artec
ASE—All Seasons Entertainment
ATL—Atlas Entertainment Corporation
ATP—Apple Tree Productions
ATV—Atlantic Video
AUP—Aura Productions
AVC—Allied Video Corp.
AVD—American Video Corporation
AVE—A*Vision
AVL—Aviation A.V. Library
AVP—American Video Productions
AVT—American Video Tape
AWF—Award Films
AXV—Axon Video Corporation
BBF—Bonnie Business Forms
BBV—Bleak Beauty Video
BEI—Bandera Enterprises, Inc.
BFA—Phoenix/BFA Films
BFS— BFS Video
BFV—Best Film & Video Corporation
BLC—Blacast Entertainment
BMF—Benchmark Films, Inc.

BMG—BMG
BMV—Bennett Marine Video
BOK—Bookpeople
BPG—Bridgestone Production Group
BRS—Barry Raymond Steiner
BRT—Britannica Films
BST—Best Video
BTV—Baker & Taylor Video
BUL—Bullfrog Films, Inc.
BUP—Bob Jones University Press
BUR—Burbank Video
BVG—Bennett Video Group
BVV—Buena Vista Home Video
CAB—Cable Films
CAF—Cabin Fever Entertainment
CAL—California Newsreel
CAM—Camp Motion Pictures
CAN—Cannon Video
CAP—Capitol Video Communications, Inc.
CAR—Rosalie & Jerry Carter
CCR—Critics' Choice Video, Inc.
CCE—Classic Cinema Entertainment
CCF—CC Films
CCI—Cinema Concepts, Inc.
CCN—Cinevista
CCV—Concord Video
CEG—Cinecom Entertainment Group
CEL—Celebrity Home Entertainment
CEN—Centre Communications
CFV—Carousel Film & Video
CGH—Cinema Group Home Video
CGI—Continental Group, Inc.
CGV—Cineglobe Video Inc.
CHA—Charter Entertainment
CHF—Churchill Films
CHI—Center for Humanities, Inc.
CHN—Channel 13/Madhouse Video
CHV—Coyote Home Video
CIC—Cinema International Canada
CIG—Cinema Guild
CIN—Cinema Home Video
CKV—Cocktail Video, Ltd.
CLV—City Lights Home Video
CMR—Camera 3
CNM—Cinemacabre Video
CNT—Canterbury Distribution
COL—Columbia Tristar Home Video
COM—Commtron
CON—Continental Video
COV—Covenant Video
CPB—Captain Bijou
CPM—Central Park Media/U.S. Manga Corps

CPT—Capital
CRC—Criterion Collection
CRO—Crocus Entertainment, Inc.
CRP—Creative Programming, Inc.
CRR—Career Track
CRV—Crown Video
CRY—Crystal Productions
CSB—Communication Skill Builders
CSM—Coliseum Video
CST—Castelli Sonnabend Tapes & Films
CTC—Coast to Coast Video
CTH—Corinth Video
CTL—Capitol Home Video
CTP—Century Productions
CTY—Century Home Video, Inc.
CUN—Cunningham Dance Foundation, Inc.
CVA—Children's Video of America
CVC—Connoisseur Video Collection
CVD—Consumer Video Distributers
CVL—Children's Video Library
CVS—Condor Video CVW Cine-Video West
DAR—Darryl L. Sink & Associates, Inc.
DAY—Day Films
DCH—Doris Chase
DCL—Direct Cinema Limited
DER—Documentary Educational Resources
DIS—Walt Disney Home Video
DMD—Diamond Entertainment, Inc.
DMP—Donna Michelle Productions
DNB—Danny Burk
DSN—Disney Educational Productions
DSP—Daughters of Saint Paul
DVE—Digital Vision Entertainment
DVT—Discount Video Tapes, Inc.
DWE—Dusty Woods Entertainment
EAG—Eagle Productions
EAI—Electronic Arts Intermix
EBE—Earl Blair Enterprises
EDV—EDDE Entertainment
EJB—Enrique J. Bouchard
EKC—Eastman Kodak Company
ELE—Electric Productions
ELV—Electric Video
EMB—Embassy Home Entertainment
EME—Emerald Video Productions
EMI—EMI Video
ERG—Ergo Media Inc.
ESP—ESPN Home Video
EVD—European Video Distributor
EVG—Evergreen Video
EXP—Expanded Entertainment
FCT—Facets Multimedia, Inc.
FFM—Flower Films

FGF—Ferde Grofe Films
FHE—Family Home Entertainment
FHF—Finley-Holiday Film Corporation
FHS—Films for the Humanities & Sciences
FLI—Films, Inc.
FLL—Filmakers Library, Inc.
FMP—Fire Mountain Pictures
FMT—Format International
FOR—Force Video
FOX—CBS/Fox Video
FOX—CBS/Fox Video
FRG—Fright Video
FRH—Fries Home Video
FRI—First Run/Icarus Films
FRV—First Run Video
FST—Festival Films
FXL—Fox/Lorber Home Video
FXV—FoxVideo
GEM—Video Gems
GEO—Geovision Limited
GHA—GHA Communications
GHV—Genesis Home Video
GKK—Goodtimes/Kids Klassics
    Distribution Corporation
GLV—German Language Video Center
GNS—21st Genesis Home Video
GPV—Grapevine Video
GRG—Gorgon Video
GRP—GRP Video
GVV—Glenn Video Vistas, Ltd.
HAR—HarmonyVision
HBO—HBO Home Video
HCC—Home Cinema Corporation
HEG—Horizon Entertainment
HFE—Hollywood Family Entertainment
HHE—Hollywood Home Entertainment
HHT—Hollywood Home Theatre
HIL—Hilltop Productions
HMD—Hemdale Home Video
HMV—Home Vision Cinema
HNB—Hanna-Barbera Home Video
HPH—Hollywood Pictures Home Video
HRS—Hal Roach Studios
HSE—High/Scope Educational Research
    Foundation
HSR—HSR Sales
HTV—Hen's Tooth Video
HVL—Hollywood Video Library
ICA—Icarus Films
IEC—Interamerican Entertainment Corp.
IFB—International Film Bureau, Inc.
IFE—International Film Exchange, Ltd.
IFF—IFEX Films
IGP—Ignatius Press
IGV—Interglobal Video
IHF—International Historic Films, Inc.
    (IHF)
IMA—IMA
IMD—Impact Distributors/Cornerstone
    Films
IME—Image Entertainment
IMG—Image Associates
IMP—Imperial Entertainment Corporation
INC—Increase Video
IND—Independent United Distributors
INF—International Film Foundation
ING—Ingram Video Inc.
INT—Interama, Inc.
INU—Indiana University Center for Media
    & Teaching Resources
INV—International Home Video
IPI—Impulse Productions
IRN—Ironwood Productions
IRP—Instant Replay Videocassette
    Magazine

ISS—Image Magnetic Associates
IUF—Interurban Films
IVA—Island Visual Arts
IVC—Ivanhoe Communications, Inc.
IVE—International Video Entertainment
    (IVE)
IVY—Ivy Classics Video
JAF—James Agee Film Project
JCF—Judaica Captioned Film Center, Inc.
JCI—JCI Video
JEF—JEF Films
JER—Jeremiah Films (JF)
JFK—Just for Kids Home Video
JLT—JLT Films, Inc.
JOU—Journal Films, Inc.
JTC—J2 Communications
JWD—J World
KAR—Karol Media/Karol Video
KBE—K-Beech Video, Inc.
KBV—King Bee Video
KEP—Keep the Faith Inc.
KET—KET, The Kentucky Network
    Enterprise Division
KID—Kids Klassics, Inc.
KIT—Kit Parker Video
KIV—Kino on Video
KOV—King of Video
KRT—Barr Entertainment
KRV—Keeling's Records & Videos
KTC—KTCA-TV Video Services
KUI—Knowledge Unlimited, Inc.
KUL—Kultur Video
KVI—K Video
KWI—KnoWhutImean? Home Video
LBD—LB Distributors, Inc.
LCA—Learning Corporation of America
LDC—Pioneer LDCA, Inc.
LDV—LD Video
LEG—Liberty Entertainment Group, Inc.
LGC—Legacy Home Video
LHV—Lorimar Home Video
LIV—Live Home Video
LME—Lucerne Media
LOO—Loonic Video
LPC—Liberty Publishing Company, Inc.
LTG—Lightning Video
LUM—Lumivision Corporation
LUN—Luna Video
MAA—Malin & Associates
MAD—Madera Cinevideo
MAG—Magnum Entertainment Inc.
MAS—Mastervision, Inc.
MAV—Master Arts Video
MBP—Moonbeam Publications
MCA—MCA/Universal Home Video
MCG—Management Company
    Entertainment Group (MCEG), Inc.
MDP—Media Project, Inc.
MDS—Modern Signs Press, Inc.
MED—Media Home Entertainment
MET—Metro Packaging
MFI—Mark V International
MFV—Mystic Fire Video
MGL—Mogul Communications
MGM—MGM Home Video, Inc.
MGS—Magna Systems, Inc.
MHV—Mexican Home Video
MID—Midnight Video
MIL—Milestone Film & Video
MIR—Miramar Productions
MIV—Mark Four Pictures Film & Video
MLB—Mike LeBell's Video
MNC—Monarch Home Video
MNE—Monogram Entertainment
MNP—Monticello Productions

MON—Monterey Home Video
MOR—Morris Video
MOV—Movies Unlimited
MPI—MPI Home Video
MRC—Miracle Video, Inc.
MRV—Moore Video
MSM—Music Media
MSP—Modern Sound Pictures, Inc.
MTH—MTI Home Video
MTI—Coronet/MTI Film & Video
MTP—Maryland Public Television
MTT—MTI Teleprograms, Inc.
MTX—MNTEX Entertainment, Inc.
MUS—Museum of Modern Art of Latin
    America
MVC—Moviecraft, Inc.
MVD—Music Video Distributors
MVI—Major Video Concepts, Inc.
MWF—Monday/Wednesday/Friday Video
MWP—Mossman Williams Productions
NAC—NAC Home Video
NAM—North American Home Video
NAV—North American Video, Inc.
NBI—Newsbank, Inc.
NCJ—National Center for Jewish Film
NEG—National Entertainment Group, Inc.
NEW—New Video Center
NFC—National Film Board of Canada
NHO—New Horizons Home Video
NJN—New Jersey Network
NLC—New Line Home Video
NOR—Norman Beerger Productions
NOS—Nostalgia Family Video
NRT—Northeast Video, Ltd.
NSV—New Star Video
NUV—New & Unique Videos
NVH—Nova Home Video
NWV—New World Entertainment
NYF—New Yorker Video
NYS—New York State Education Department
OCE—Ocean Video
OHC—Ohlmeyer Communications
ORI—Orion Home Video
OWM—One West Media
PAR—Paramount Home Video
PAU—Paulist Productions
PAV—Pacific Arts Video
PBC—Princeton Book Company Publishers
PBS—PBS Video
PCC—Plexus Communications Corporation
PCV—Pacific Coast Video
PEN—Pentrex Video
PEV—Palisades Entertainment
PFI—Public Films, Inc.
PGN—Paragon Video Productions
PGV—Polygram Music Video (PMV)
PHX—Phoenix Distributers
PIA—Pioneer Artists, Inc.
PIC—Picture Start
PII—Peerless Films, Inc.
PLV—Planet Video
PME—Public Media Video
PMH—PM Entertainment Group, Inc.
PMR—Premiere Home Video
PMV—Passport Music Video
PNR—Pioneer Video Imports
PPI—Peter Pan Industries
PPM—Pied Piper Media
PRE—Premier Promotions
PRM—Primate Productions
PRS—Proscenium Entertainment
PRV—Prime Video
PSM—Prism Entertainment
PSV—Powersports
PTB—Proud To Be...A Black Video Collection

PYR—Pyramid Film & Video
QCM—Quantum Communications
QHV—Questar Home Video
QNE—Quintex Entertainment
QUE—Quest Entertainment
QVD—Quality Video
RAE—Raedon Entertainment
RAP—RAP Entertainment Inc.
RCA—RCA Video
RDG—Reader's Digest Home Video
REA—Real Productions
REB—Club Reba Video
REG—Regal Video
REP—Republic Pictures Home Video
RGO—Rego: Irish Records & Tapes, Inc.
RHI—Rhino Video
RHP—Rhapsody Films
RHV—Regency Home Video/Banana
        Republic
RKO—RKO Pictures
RLX—Relax Video
RMF—RM Films, Inc.
RRP—Ready Reference Press
RXM—Rex Miller
SAL—Salenger Films, Inc.
SAT—Saturn Productions, Inc.
SCB—Schwartz Brothers, Inc.
SCL—Scholastic Lorimar
SDS—San Diego State University
SEC—Sutton Entertainment Corporation
SED—Studio Entertainment
SEL—Select Home Video
SFF—Strike Force Films
SFR—San Francisco Rush Video
SGE—SGE Home Video
SHE—Shine Home Entertainment
SHG—Shapiro Glickenhaus Home Video
SHO—Shokus Video
SHP—Shapiro Arts
SHV—Snoopy's Home Video Library
SIM—Simitar Entertainment
SLC—Select-A-Tape
SLF—S-L Film Productions
SLP—Sell Pictures
SMV—Sony Music Video Enterprises
SMW—Something Weird Video
SNC—Sinister Cinema
SOF—Sportsmen on Film
SOU—Southgate Entertainment
SPI—Stevenson Productions, Inc.
SPV—Sports on Video
SPW—Sparrow Distribution
SSI—Super Sitters, Inc.
STC—Star Classics
STE—Starmaker Entertainment, Inc.
STP—Streamline Pictures
STV—Studio Video Productions
SUE—Sultan Entertainment
SUM—Summit Media Company
SUN—Sun Video
SUP—Super Source
SVA—Send Video Arts
SVC—Sony Video Communications
SVD—Specialty Video Distributors
SVI—Strand/VCI Entertainment
SVN—Surf Video Network
SVS—Sony Video Software (SVS, Inc.)
SVT—SVS/Triumph
SYB—SyberVision Systems, Inc.
TAF—Trade-A-Flick
TAM—Tamarelle's
TAV—10th Avenue Video
TCF—20th Century Fox Film Corporation
TEL—Televideos
TEX—Texture Films, Inc.

TGV—Target Video
THA—3M Audio Visual Division
THM—3M/Leisure Time Products
THR—Thriller Video
TIM—Timeless Video Inc.
TLF—Time-Life Video and Television
TMG—The Maier Group
TOU—Touchstone Home Video
TPI—Thomson Productions, Inc.
TPV—Tapeworm Video Distributors
TRI—Triboro Entertainment Group
TRV—Trans-Atlantic Video
TRY—Trylon Video, Inc.
TSF—Two Star Films, Inc.
TSV—Telecine Spanish Video
TTC—Turner Home Entertainment
        Company
TTE—Twin Tower Enterprises
TVS—TV Sports Scene, Inc. (TVSS)
TVT—Tournament Video Tapes TWE
        Trans-World Entertainment
UAV—UAV Corporation
UCV—Urban Classics Video
UHV—Video Communications, Inc.
UND—Uni Distribution
UNI—Unicorn Video, Inc.
UNR—Unreel Productions
USA—USA Home Video
UTM—United Training Media
VAI—Video Artists International, Inc.
VAL—Value Video
VAN—Vantage Communications, Inc.
VBL—Video Bible Library, Inc.
VCD—Video City Productions
VCI—VCI, Inc.
VCL—VCL Home Video
VCN—Video Connection
VDA—Video Action
VDC—Vidcrest
VDE—Video Video
VDM—Video Dimensions
VDY—Video Dynamics
VEC—Valencia Entertainment Corporation
VES—Vestron Video
VGD—Vanguard Video
VGR—Videograf
VHE—VCII Home Entertainment, Inc.
VHM—Video Home Library
VHV—Vista Home Video
VIC—Victory Video
VID—VidAmerica, Inc.
VIL—Vidmag International, Ltd.
VIP—VIP Video
VIR—Virgin Vision
VLA—Video Latino
VMK—Vidmark Entertainment
VMV—Virgin Music Video
VOI—Video Outreach, Inc.
VPH—Videograph
VPI—VPI/AC Video
VPL—Video Placement International
VPR—Video Presentations, Inc.
VRL—Video Rails
VSI—Valley Studios VSV Visionsmiths,
        Inc.
VTI—Video Trend, Inc.
VTK—Videotakes
VTO—Video Tours, Inc.
VTP—Vidtape/Duravision
VTR—Video Treasures
VVC—Video Visa
VVI—VideoVision, Inc.
VVP—Vineyard Video Productions
VWV—V.I.E.W. Video
VYG—Voyager Company

VYY—Video Yesteryear
WAC—World Artists Cinema Corporation
WAR—Warner Home Video, Inc.
WAX—Waxworks
WBF—Water Bearer Films
WBV—WB Video Productions
WCV—West Coast Video Distributers
WEA—Warner/Elektra/Atlantic Corporation
WEM—Worldwide Entertainment Marketing
WES—Western World
WFV—Western Film & Video, Inc.
WGE—Weiss Global Enterprises
WIF—Women in Focus
WKV—Wood Knapp & Company, Inc.
WLA—Western L.A. Film Productions, Inc.
WMB—Wombat Productions
WNE—WNET/Thirteen Non-Broadcast
WOM—Wombat Film and Video
WOR—World Artists Releasing
WOV—Worldvision Home Video, Inc.
WPB—Western Publishing
WPM—World Premiere
WRD—World Wide Video
WRV—Warner Reprise Video
WSH—Wishing Well Distributing
WST—White Star
WTA—Whole Toon Access
WVE—World Video Enterprises, Inc.
WWV—Westernworld Video, Inc.
XVC—Xenon
YES—YES Entertainment
YOC—Yankee-Oriole Company
ZBI—Zeitgeist Films

DISTRIBUTOR GUIDE

ABA CONSORTIUM FOR PROFESSIONAL EDUCATION *(ABA)*
750 N. Lake Shore Dr.
Chicago, IL 60611
312-988-6191
800-621-8986

ABACUS PRODUCTIONS, INC. *(API)*
PO Box 10298
Canoga Park, CA 91309

ABC DISTRIBUTION COMPANY *(ABC)*
Capitol Cities/ABC Video Enterprises
825 7th Ave.
New York, NY 10019-6001
212-887-1725

ACADEMY HOME ENTERTAINMENT *(ACA)*
1 Pine Haven Shore Rd.
PO Box 788
Shelburne, VT 05482
800-972-0001
Fax: 802-985-3403

ACADEMY VIDEOS *(ACD)*
Box 5224
Sherman Oaks, CA 91423-5224
818-788-6662
800-423-2397

ACE VIDEO *(ACE)*
19749 Dearborn St.
Chatsworth, CA 91311
818-718-1116
Fax: 818-718-9109

ACORN VIDEO *(ACN)*
c/o Academy Entertainment
1 Pine Haven Shore Rd.
PO Box 788
Shelburne, VT 05482
802-985-2060
800-972-0001
Fax: 802-985-3403

ACTION FILMS & VIDEO *(ACF)*
PO Box 1160
Salt Lake City, UT 84110

ACTIVE HOME VIDEO *(AHV)*
12121 Wilshire Blvd.
Los Angeles, CA 90025
310-447-6131
800-824-6109
Fax: 310-274-2621

ADMIT ONE VIDEO *(AOV)*
PO Box 66, Sta. O
Toronto, ON, Canada M4A 2M8
416-463-5714

AFRO-AM DISTRIBUTING COMPANY *(AFR)*
407 E. 25th St., Ste. 600
Chicago, IL 60616
312-791-1611

AIMS MEDIA *(AIM)*
9710 DeSoto Ave.
Chatsworth, CA 91311-9734
818-773-4300
800-367-2467
Fax: 818-341-6700

A.I.P. HOME VIDEO, INC. *(AIP)*
10726 McCune Ave.
Los Angeles, CA 90034
800-456-2471
Fax: 213-559-8849

ALDEN FILMS *(ALD)*
PO Box 449
Clarksburg, NJ 08510
201-462-3522

ALL SEASONS ENTERTAINMENT *(ASE)*
18121 Napa St.
Northridge, CA 91325
818-886-8680
800-423-5599
Fax: 818-886-8972

ALLIED ARTISTS *(ALL)*
3415 S. Sepulveda Blvd., Ste. 630
Los Angeles, CA 90034
310-390-6161

ALLIED VIDEO CORP. *(AVC)*
PO Box 702618
Tulsa, OK 74170
918-496-3332

ALLINSON EDUCATIONAL SERVICE *(ALS)*
Vanwell Rd.
Maidenhead, Berks., England

ALLISON ENTERTAINMENT *(ALE)*
8700 Reseda Blvd., Ste. 214
Northridge, CA 91325

AMBROSE VIDEO PUBLISHING, INC. *(AMB)*
1290 Avenue of the Americas, Ste. 2245
New York, NY 10104
212-696-4545
800-526-4663

AMERICAN EDUCATIONAL FILMS *(AEF)*
3807 Dickerson Rd.
Nashville, TN 37207
800-822-5678

AMERICAN FILM FOUNDATION *(AFF)*
PO Box 2000
Santa Monica, CA 90406
310-394-5689

AMERICAN VIDEO CORPORATION *(AVD)*
844 Reseda Blvd., Ste. M
Northridge, CA 91324
818-407-0590
800-323-7979
Fax: 818-907-0598

AMERICAN VIDEO PRODUCTIONS *(AVP)*
26541 Espolter Dr.
Mission Viego, CA 92691
714-583-9789

AMERICAN VIDEO TAPE *(AVT)*
1116 Edgewater Ave.
Ridgefield, NJ 07657

AMI ENTERTAINMENT *(AMI)*
12500 Riverside Dr., Ste. 207
North Hollywood, CA 91607
213-969-1709

AMVEST VIDEO *(AMV)*
937 E. Hazelwood Ave.
Rahway, NJ 07065

APPALSHOP FILMS *(APL)*
306 Madison St.
Whitesburg, KY 41858
606-633-0108
800-545-7467

APPLAUSE PRODUCTIONS, INC. *(APD)*
85 Longview Rd.
Port Washington, NY 11050
516-883-2825
Fax: 516-883-7460

APPLAUSE THEATRE AND CINEMA *(APP)*
211 W. 71st St.
New York, NY 10023
212-496-7511

APPLE TREE PRODUCTIONS *(ATP)*
5200 N. Sawyer Ave.
Boise, ID 83714-1491
208-322-6155

ARP VIDEOS *(ARP)*
PO Box 4617
North Hollywood, CA 91617
213-877-4406
800-843-3672

ARTEC *(ART)*
1 Pine Haven Shore Rd.
Shelburne, VT 05482

802-985-9411
800-451-5180
Fax: 802-985-3403

ARTS AMERICA, INC. *(AAI)*
12 Havermeyer Pl.
Greenwich, CT 06830
203-869-4693
Fax: 203-661-1174

ATLANTIC VIDEO *(ATV)*
150 S. Gordon St.
Alexandria, VA 22304
703-823-2800
Fax: 703-370-6748

ATLAS ENTERTAINMENT CORPORATION *(ATL)*
1 Jocama Blvd., Ste. 2E
Old Bridge, NJ 08857
908-591-1155
Fax: 908-591-0660

AUDIOFIDELITY ENTERPRISES *(AFE)*
PO Box 86
Rahway, NJ 07065

AURA PRODUCTIONS *(AUP)*
14849 Lull St.
Van Nuys, CA 91405

AVIATION A.V. LIBRARY *(AVL)*
PO Box 399
Uvoon, WA 98592
206-893-8080
800-626-6095
Fax: 206-858-8081

A*VISION *(AVE)*
Div. of Atlantic Recording Corp.
1290 6th Ave.
New York, NY 10019
212-484-7697
Fax: 212-484-7551

AWARD FILMS *(AWF)*
c/o Facets Video
1517 W. Fullerton Ave.
Chicago, IL 60614
312-281-9075
Fax: 312-929-5437

AXON VIDEO CORPORATION *(AXV)*
1900 Broadway
New York, NY 10022
212-787-8228

BAKER & TAYLOR VIDEO *(BTV)*
501 S. Gladiolus
Momence, IL 60954
800-435-5111

BANDERA ENTERPRISES, INC. *(BEI)*
Box 1107

Studio City, CA 91604
818-985-5050

BARR ENTERTAINMENT *(KRT)*
8500 Keystone Crossing, Ste. 540
Indianapolis, IN 46240
317-297-1888
800-582-2000
Fax: 317-254-4455

BARRY RAYMOND STEINER *(BRS)*
1211 W. Farwell Ave.
Chicago, IL 60626
312-274-1053

BENCHMARK FILMS, INC. *(BMF)*
145 Scarborough Rd.
Briarcliff Manor, NY 10510
914-762-0030
Fax: 914-762-3895

BENNETT MARINE VIDEO *(BMV)*
730 Washington St.
Marina del Rey, CA 90292
213-821-3329
800-262-8862
Fax: 213-306-3162

BENNETT VIDEO GROUP *(BVG)*
730 Washington St.
Marina del Rey, CA 90292
310-821-3329
800-262-8862
Fax: 310-306-3162

BEST FILM & VIDEO CORPORATION *(BFV)*
98 Cutter Mill Rd.
Great Neck, NY 11021
516-487-4515
800-527-2189
Fax: 516-487-4834

BEST VIDEO *(BST)*
10580 Newkirk, Ste. 100
Dallas, TX 75220
214-869-9641
800-541-1008
Fax: 214-556-1686

BFS VIDEO *(BFS)*
350 Newkirk Rd., N.
Richmond Hill, ON, Canada L4C 367
416-884-2323
Fax: 416-884-8292

BLACAST ENTERTAINMENT *(BLC)*
199-19 Linden Blvd.
St. Albans, NY 11412

718-712-2300
800-955-9285
Fax: 718-712-5345

**BLEAK BEAUTY
VIDEO** *(BBV)*
67 Barclay Rd.
Clintondale, NY 12515
914-883-6077

BMG *(BMG)*
6550 E. 30th St.
Indianapolis, IN 46219
317-542-0414

**BOB JONES UNIVERSITY
PRESS** *(BUP)*
Greenville, SC 29614
803-242-5100

**BONNIE BUSINESS
FORMS** *(BBF)*
2691 Freewood
Dallas, TX 75220
214-357-3956

BOOKPEOPLE *(BOK)*
7900 Edgewater Dr.
Oakland, CA 94621
510-632-4700

**BRIDGESTONE
PRODUCTION GROUP** *(BPG)*
2091 Las Palmas Dr.
Carlsbad, CA 92009
619-431-9888
800-523-0988

BRITANNICA FILMS *(BRT)*
310 S. Michigan Ave.
Chicago, IL 60604
312-347-7958
Fax: 312-347-7966

**BUENA VISTA HOME
VIDEO** *(BVV)*
500 S. Buena Vista St.
Burbank, CA 91521
818-562-3560
Fax: 818-567-6464

**BULLFROG FILMS,
INC.** *(BUL)*
PO Box 149
Oley, PA 19547
215-779-8226
800-543-3764
Fax: 215-370-1978

BURBANK VIDEO *(BUR)*
PO Box 10069
Burbank, CA 91505

**CABIN FEVER
ENTERTAINMENT** *(CAF)*
100 W. Putnam Ave.
Greenwich, CT 06830
203-863-5200
Fax: 203-863-5258

CABLE FILMS *(CAB)*
Country Club Sta.
PO Box 7171
Kansas City, MO 64113
816-362-2804
Fax: 816-381-0000

**CALIFORNIA
NEWSREEL** *(CAL)*
149 Ninth St., Ste. 420
San Francisco, CA 94103
415-621-6196
Fax: 415-621-6522

CAMERA 3 *(CMR)*
PO Box 994
Kent, CT 06757
203-927-1964
Fax: 203-927-1965

**CAMP MOTION
PICTURES** *(CAM)*
2412 S. Thurman Ave.
Los Angeles, CA 90016
213-935-8650
800-826-9763

CANNON VIDEO *(CAN)*
8200 Wilshire Blvd., 3rd Fl.
Beverly Hills, CA 90211
213-966-5600

**CANTERBURY
DISTRIBUTION** *(CNT)*
9925 Horn Rd.
Sacramento, CA 95827
916-364-5084
800-878-5084
Fax: 916-364-0216

CAPITAL *(CPT)*
4818 Yuma St. NW
Washington, DC 20016
202-363-8800
Fax: 202-363-4680

CAPITOL HOME VIDEO *(CTL)*
4818 Yuma St. NW
Washington, DC 20016
202-363-8800
Fax: 202-363-4680

**CAPITOL VIDEO
COMMUNICATIONS,
INC.** *(CAP)*
2121 Wisconsin Ave. NW
Washington, DC 20007
202-965-7800

CAPTAIN BIJOU *(CPB)*
PO Box 87
Toney, AL 35773
205-852-0198

CAREER TRACK *(CRR)*
1800 38th St.
Boulder, CO 80301
303-447-2323

**CAROUSEL FILM &
VIDEO** *(CFV)*
260 5th Ave., Rm. 705
New York, NY 10001
212-683-1660

**CASTELLI SONNABEND
TAPES & FILMS** *(CST)*
420 W. Broadway
New York, NY 10012
212-431-6279

CBS/FOX VIDEO *(FOX)*
1330 Avenue of the Americas,
5th Fl.
New York, NY 10019
212-373-4800
800-800-2369
Fax: 212-373-4802

CBS/FOX VIDEO *(FOX)*
7930 Alabama
Cogna Park, CA 91304

CC FILMS *(CCF)*
National Council of Churches
475 Riverside Dr., Rm. 860
New York, NY 10115-0050
212-870-2575
Fax: 212-870-2511

**CELEBRITY HOME
ENTERTAINMENT** *(CEL)*
22025 Ventura Blvd.
PO Box 4112
Woodland Hills, CA 91365-
4112
818-595-0666
Fax: 818-716-0168

**CENTER FOR HUMANITIES,
INC.** *(CHI)*
Communications Park

Box 1000
Mount Kisco, NY 10549
914-666-4100
800-431-1242
Fax: 914-666-5319

**CENTRAL PARK MEDIA/U.S.
MANGA CORPS** *(CPM)*
250 W. 57th St., Ste. 831
New York, NY 10107
212-977-7456
Fax: 212-977-8709

**CENTRE
COMMUNICATIONS** *(CEN)*
1800 30th St., Ste. 207
Boulder, CO 80301
303-444-1166
Fax: 303-444-1168

**CENTURY HOME VIDEO,
INC.** *(CTY)*
1509 N. Crescent Heights
Blvd., Ste. 3
Los Angeles, CA 90046
213-848-2345
Fax: 213-848-2343

**CENTURY
PRODUCTIONS** *(CTP)*
PO Box 1156
Sandy, UT 84091-1156
801-571-5161
Fax: 801-571-5161

**CHANNEL 13/MADHOUSE
VIDEO** *(CHN)*
PO Box 15602
North Hollywood, CA 91615
800-TVC-ALLS

**CHARTER
ENTERTAINMENT** *(CHA)*
335 N. Maple Dr., Ste. 350
Beverly Hills, CA 90210-3899
310-285-6000
Fax: 310-285-6190

**CHILDREN'S VIDEO OF
AMERICA** *(CVA)*
28231 Ave. Crocker, Ste. 120
Valencia, CA 91355
805-257-6054
800-323-2061
Fax: 805-257-6921

**CHILDREN'S VIDEO
LIBRARY** *(CVL)*
60 Long Ridge Rd.
PO Box 4995
Stamford, CT 06907
203-968-0100

CHURCHILL FILMS *(CHF)*
12210 Nebraska Ave.
Los Angeles, CA 90025
310-207-6600
Fax: 310-207-1330

CINE-VIDEO WEST *(CVW)*
PO Box 391
Pasedena, CA 91102
818-792-0842

**CINECOM ENTERTAINMENT
GROUP** *(CEG)*
85 3rd Ave.
New York, NY 10022
212-819-5000

**CINEGLOBE VIDEO
INC.** *(CGV)*
5250 Finch Ave., E., Unit 2
Scarborough, ON, Canada
M1S 5A4
416-609-1891
Fax: 416-609-1894

**CINEMA CONCEPTS,
INC.** *(CCI)*
2461 Berlin Tpke.

Newington, CT 06111
203-667-1251

**CINEMA GROUP HOME
VIDEO** *(CGH)*
2320 Cotner
Los Angeles, CA 90064

CINEMA GUILD *(CIG)*
1697 Broadway
New York, NY 10019
212-246-5522
Fax: 212-246-5525

CINEMA HOME VIDEO *(CIN)*
7095 Hollywood Blvd., Ste.
104
Hollywood, CA 90028
310-288-1569

**CINEMA INTERNATIONAL
CANADA** *(CIC)*
8255 Mountain Sights, Ste.
507
Montreal, PQ, Canada H4P
2B5
514-737-3363
Fax: 514-737-6165

**CINEMACABRE
VIDEO** *(CNM)*
PO Box 10005-D
Baltimore, MD 21285-0005

CINEVISTA *(CCN)*
560 W. 43rd, No. 8J
New York, NY 10036
212-947-4373
Fax: 212-947-0644

**CITY LIGHTS HOME
VIDEO** *(CLV)*
8981 Sunset Blvd.
West Hollywood, CA 90069
800-537-4552

**CLASSIC CINEMA
ENTERTAINMENT** *(CCE)*
19425-B Soledad Canyon
Rd., Ste. 225
Canyon Country, CA 91351
805-252-3262

CLUB REBA VIDEO *(REB)*
40 N. Moore St.
New York, NY 10013

**COAST TO COAST
VIDEO** *(CTC)*
148 Lafayette St., 10th Fl.
New York, NY 10013
212-431-3890
800-223-3420

**COCKTAIL VIDEO,
LTD.** *(CKV)*
1501 Broadway, Rm. 1603
New York, NY 10129
212-840-0010

COLISEUM VIDEO *(CSM)*
430 W. 54th St.
New York, NY 10019
212-489-8130
800-288-8130
Fax: 212-582-5690

**COLUMBIA TRISTAR HOME
VIDEO** *(COL)*
3400 Riverside Dr.
Burbank, CA 91505-4627
818-972-8193
Fax: 818-972-0937

COMMTRON *(COM)*
7900 Hickman Rd.
Des Moines, IA 50322
515-226-3000

**COMMUNICATION SKILL
BUILDERS** *(CSB)*
3830 E. Bellevue

PO Box 42050
Tucson, AZ 85733
602-323-7500

CONCORD VIDEO *(CCV)*
7506 N. Broadway Ext., Ste.
505
Oklahoma City, OK 73116
405-840-6031
800-222-2811
Fax: 405-848-3960

CONDOR VIDEO *(CVS)*
c/o Media Home
Entertainment
5730 Buckingham Pkwy.
Culver City, CA 90230-6508
213-216-7900

**CONNOISSEUR VIDEO
COLLECTION** *(CVC)*
8436 W. 3rd St., Ste. 600
Los Angeles, CA 90048
213-653-8873
Fax: 213-651-0055

**CONSUMER VIDEO
DISTRIBUTERS** *(CVD)*
1427 Woodhurst St.
Bowling Green, KY 42104
502-843-8300

**CONTINENTAL GROUP,
INC.** *(CGI)*
1 Harbor Plaza
Stamford, CT 06904
203-263-1900

CONTINENTAL VIDEO *(CON)*
2320 Cotner
Los Angeles, CA 90064
800-821-3427

CORINTH VIDEO *(CTH)*
34 Gansevoort St.
New York, NY 10014
212-463-0305
800-221-4720

**CORONET/MTI FILM &
VIDEO** *(MTI)*
108 Wilmot Rd.
Deerfield, IL 60015
708-940-1260
800-621-2131
Fax: 708-940-3640

COVENANT VIDEO *(COV)*
3200 W. Foster Ave.
Chicago, IL 60625
312-478-4676
800-621-1290
Fax: 312-478-2622

COYOTE HOME VIDEO *(CHV)*
c/o Hemdale Home Video
7966 Beverly Blvd.
Los Angeles, CA 90048
213-966-3158
Fax: 213-651-3107

**CREATIVE PROGRAMMING,
INC.** *(CRP)*
30 E. 60th St.
New York, NY 10022
Fax: 212-355-6221

**CRITERION
COLLECTION** *(CRC)*
2139 Manning Ave.
Los Angeles, CA 90025-6315
800-446-2001

**CRITICS' CHOICE VIDEO,
INC.** *(CCB)*
PO Box 549
Oak Grove Village, IL 60009
800-367-7765
Fax: 800-544-9852

CROCUS ENTERTAINMENT, INC. *(CRO)*
762 12 Oaks Center
15500 Wayzata Blvd.
Wayzata, MN 55391
612-473-9002
800-942-2992
Fax: 612-473-0648

CROWN VIDEO *(CRV)*
225 Park Ave., S.
New York, NY 10003
212-254-1600

CRYSTAL PRODUCTIONS *(CRY)*
1812 Johns Dr.
Box 2159
Glenview, IL 60025
708-657-8144
800-255-8629
Fax: 708-657-8149

CUNNINGHAM DANCE FOUNDATION, INC. *(CUN)*
463 West St.
New York, NY 10014
212-633-2453

DANNY BURK *(DNB)*
2316 Mishawaka Ave.
South Bend, IN 46615

DARRYL L. SINK & ASSOCIATES, INC. *(DAR)*
1155 N. Capitol Ave., Ste. 200
San Jose, CA 95132
408-272-8384

DAUGHTERS OF SAINT PAUL *(DSP)*
50 St. Paul's Ave.
Jamaica Plain
Boston, MA 02130
617-522-8911
Fax: 617-524-8035

DAY FILMS *(DAY)*
121 W. 27th St., Ste. 902
New York, NY 10001

DIAMOND ENTERTAINMENT, INC. *(DMD)*
PO Box 695
New Canaan, CT 06840

DIGITAL VISION ENTERTAINMENT *(DVE)*
7080 Hollywood Blvd., Ste. 901
Los Angeles, CA 90028
213-462-3790
Fax: 213-462-4408

DIRECT CINEMA LIMITED *(DCL)*
PO Box 69799
Los Angeles, CA 90069-9976
213-652-8000
800-345-6748
Fax: 213-652-2346

DISCOUNT VIDEO TAPES, INC. *(DVT)*
833-A N. Hollywood Way
PO Box 7122
Burbank, CA 91510
818-843-3366
Fax: 818-843-3821

DISNEY EDUCATIONAL PRODUCTIONS *(DSN)*
500 S. Buena Vista St.
Burbank, CA 91521
818-972-3410
800-621-2131

DOCUMENTARY EDUCATIONAL RESOURCES *(DER)*
101 Morse St.
Watertown, MA 02172
617-926-0491
Fax: 617-926-9519

DONNA MICHELLE PRODUCTIONS *(DMP)*
6253 Hollywood Blvd., Ste. 818
Hollywood, CA 90028
213-469-0672
800-356-4386

DORIS CHASE *(DCH)*
Chelsea Hotel
222 W. 23rd St.
New York, NY 10011
212-243-3700

DUSTY WOODS ENTERTAINMENT *(DWE)*
6003 Castor Ave.
Philadelphia, PA 19149

A & E HOME VIDEO *(AAE)*
c/o New Video Group
419 Park Ave., S., 20th Fl.
New York, NY 10016
800-229-9994
Fax: 212-685-2625

EAGLE PRODUCTIONS *(EAG)*
7860 Mission Center Ct., Ste. 106
San Diego, CA 92108
800-621-0852

EARL BLAIR ENTERPRISES *(EBE)*
PO Box 87
Toney, AL 35773
205-852-0198

EASTMAN KODAK COMPANY *(EKC)*
c/o Wood Knapp
Knapp Press
5900 Wilshire Blvd.
Los Angeles, CA 90036
213-937-5486

EDDE ENTERTAINMENT *(EDV)*
19749 Dearborn St.
Chatsworth, CA 91311
800-727-2229

ELECTRIC PRODUCTIONS *(ELE)*
8700 Reseda Blvd.
Ste. 214
Northridge, CA 91325
818-993-1787

ELECTRIC VIDEO *(ELV)*
PO Box 784
Tiffin, OH 44883
419-447-0110

ELECTRONIC ARTS INTERMIX *(EAI)*
536 Broadway, 9th Fl.
New York, NY 10012
212-966-4605

EMBASSY HOME ENTERTAINMENT *(EMB)*
335 N. Maple Dr., Ste. 350
Beverly Hills, CA 90210
310-285-6000

EMERALD VIDEO PRODUCTIONS *(EME)*
15 Ridge Ave.
San Rafael, CA 94901
415-331-5185

EMI VIDEO *(EMI)*
1290 Ave. of the Americas

New York, NY 10104

ENRIQUE J. BOUCHARD *(EJB)*
Charcas 3 8 5 8
6 Piso "A"
1425 Buenos Aires, Argentina

ERGO MEDIA INC. *(ERG)*
668 Front St.
PO Box 2037
Teaneck, NJ 07666
201-692-0404
800-695-3746

ESPN HOME VIDEO *(ESP)*
PO Box 3390
Department B
Wallingford, CT 06494
203-661-6040

EUROPEAN VIDEO DISTRIBUTOR *(EVD)*
2321 W. Olive Ave., Ste. A
Burbank, CA 91506
818-848-5902
800-423-6752
Fax: 818-848-1965

EVERGREEN VIDEO *(EVG)*
228 W. Houston St.
New York, NY 10014
212-691-7362
800-225-7783

EXPANDED ENTERTAINMENT *(EXP)*
2222 S. Barrington Ave.
Los Angeles, CA 90064
310-473-6701
Fax: 310-444-9850

FACETS MULTIMEDIA, INC. *(FCT)*
1517 W. Fullerton Ave.
Chicago, IL 60614
312-281-9075

FAMILY HOME ENTERTAINMENT *(FHE)*
A Division of LIVE Home Video
15400 Sherman Way
Ste. 500
Van Nuys, CA 91410-0124
818-499-5827

FERDE GROFE FILMS *(FGF)*
PO Box 399
Union Washington
Union, WA 98592
206-898-8080
Fax: 206-898-8081

FESTIVAL FILMS *(FST)*
2841 Irving Ave., S.
Minneapolis, MN 55408
612-870-4744

FILMAKERS LIBRARY, INC. *(FLL)*
124 E. 40th
New York, NY 10016
212-808-4980

FILMS FOR THE HUMANITIES & SCIENCES *(FHS)*
743 Alexander Rd.
Princeton, NJ 08540
609-452-1128
800-257-5126
Fax: 609-452-1602

FILMS, INC. *(FLI)*
5547 N. Ravenswood Ave.
Chicago, IL 60640-1199
312-878-2600
800-323-4222

FINLEY-HOLIDAY FILM CORPORATION *(FHF)*
PO Box 619
Whittier, CA 90601
310-945-3325
800-345-6707
Fax: 310-693-4756

FIRE MOUNTAIN PICTURES *(FMP)*
6455 Metrowest Blvd., Ste. 300
Orlando, FL 32811

FIRST RUN/ICARUS FILMS *(FRI)*
153 Waverly Pl., 6th Fl.
New York, NY 10014
212-727-1711

FIRST RUN VIDEO *(FRV)*
3620 Overland Ave.
Los Angeles, CA 90034
213-838-2111
Fax: 213-838-5212

FLOWER FILMS *(FFM)*
10341 San Pablo Ave.
El Cerrito, CA 94530
510-525-0942

FORCE VIDEO *(FOR)*
60 Long Ridge Rd.
PO Box 4000
Stamford, CT 06907
203-359-3537

FORMAT INTERNATIONAL *(FMT)*
3921 N. Meridian St.
Indianapolis, IN 46208-4011
317-924-5163

FOX/LORBER HOME VIDEO *(FXL)*
419 Park Ave., S., 20th Fl.
New York, NY 10016
800-229-9994
Fax: 212-685-2625

FOXVIDEO *(FXV)*
2121 Avenue of the Stars, 25th Fl.
Los Angeles, CA 90067
213-203-3900
800-800-2FOX
Fax: 213-774-5811

FRIES HOME VIDEO *(FRH)*
6922 Hollywood Blvd.
Ste. 415
Hollywood, CA 90028
213-466-2266
800-248-1113
Fax: 213-466-2126

FRIGHT VIDEO *(FRG)*
PO Box 179
Billerica, MA 01821
508-663-2510

GENESIS HOME VIDEO *(GHV)*
15820 Arminta St.
Van Nuys, CA 91406
818-787-0660
800-344-1060

GEOVISION LIMITED *(GEO)*
PO Box 60297
Los Angeles, CA 90060

GERMAN LANGUAGE VIDEO CENTER *(GLV)*
7625 Pendleton Pke.
Indianapolis, IN 46226-5298
317-547-1257
800-252-0957
Fax: 317-547-1263

GHA COMMUNICATIONS *(GHA)*
73 Main St.
Tuckahoe, NY 10707
914-961-3832

GLENN VIDEO VISTAS, LTD. *(GVV)*
6924 Canby Ave., Ste. 103
Reseda, CA 91335
818-981-5506
Fax: 818-981-5506

GOODTIMES/KIDS KLASSICS DISTRIBUTION CORPORATION *(GKK)*
401 5th Ave.
New York, NY 10016
212-889-0044
Fax: 212-213-9319

GORGON VIDEO *(GRG)*
6525 Sunset Blvd., Ste. 706
Hollywood, CA 90028
818-505-1981
800-367-3845

GRAPEVINE VIDEO *(GPV)*
PO Box 46161
Phoenix, AZ 85063
602-245-0210

GRP VIDEO *(GRP)*
555 W. 57th St., Ste. 1228
New York, NY 10019
212-245-7033

HAL ROACH STUDIOS *(HRS)*
345 N. Maple Dr.
Beverly Hills, CA 90210
310-281-2600

HANNA-BARBERA HOME VIDEO *(HNB)*
1 CNN Center
N. Tower, 5th Fl.
Atlanta, GA 30303
404-827-3261

HARMONYVISION *(HAR)*
116 N. Robertson Blvd., Ste. 701
Los Angeles, CA 90048
213-652-8844

HBO HOME VIDEO *(HBO)*
1114 6th Ave.
New York, NY 10036
212-512-7400

HEMDALE HOME VIDEO *(HMD)*
7966 Beverly Blvd.
Los Angeles, CA 90048
213-966-3758
Fax: 213-651-3107

HEN'S TOOTH VIDEO *(HTV)*
1124 S. Solano Dr.
Las Cruces, NM 88001
505-525-8233

HIGH/SCOPE EDUCATIONAL RESEARCH FOUNDATION *(HSE)*
600 N. River St.
Ypsilanti, MI 48198
313-485-2000

HILLTOP PRODUCTIONS *(HIL)*
PO Box 853
Warwick, NY 10990

HOLLYWOOD FAMILY ENTERTAINMENT *(HFE)*
7959 Deering Ave.
Canoga Park, CA 91304
800-426-9491

HOLLYWOOD HOME ENTERTAINMENT *(HHE)*
1747 Van Buren St., Ste. 700
Hollywood, FL 33020
305-925-2709
800-333-2709
Fax: 305-925-5655

HOLLYWOOD HOME THEATRE *(HHT)*
1540 N. Highland Ave., Ste. 110
Hollywood, CA 90028
213-466-0127

HOLLYWOOD PICTURES HOME VIDEO *(HPH)*
Fairmont Bldg. 526
500 S. Buena Vista St.
Burbank, CA 91505-9842

HOLLYWOOD VIDEO LIBRARY *(HVL)*
1831 Hyperion Ave.
Hollywood, CA 90027
213-664-7234

HOME CINEMA CORPORATION *(HCC)*
PO Box 8032
Tucson, AZ 85725
602-791-9531

HOME VISION CINEMA *(HMV)*
5547 N. Ravenswood Ave.
Chicago, IL 60640-1199
312-878-2600
800-826-3456
Fax: 312-878-8648

HORIZON ENTERTAINMENT *(HEG)*
45030 Trevor Ave.
Lancaster, CA 93534
805-940-1040
800-323-2061
Fax: 805-940-8511

HSR SALES *(HSR)*
c/o Henry Rosenberg Associates
550 Sylvan Ave.
Englewood Cliffs, NJ 07632
201-569-6560

ICARUS FILMS *(ICA)*
153 Waverly Pl.
New York, NY 10014
212-727-1711

IFEX FILMS *(IFF)*
201 W. 52nd St.
New York, NY 10019
212-582-4318

IGNATIUS PRESS *(IGP)*
15 Oakland Ave.
Harrison, NY 10528-9974
914-835-4216

IMA *(IMA)*
5460 White Oak Ave., Ste. H303
Encino, CA 91316
818-501-4327

IMAGE ASSOCIATES *(IMG)*
352 Conejo Rd.
PO Box 40106
Santa Barbara, CA 93103
805-962-6009

IMAGE ENTERTAINMENT *(IME)*
9333 Oso Ave.
Chatsworth, CA 91311
818-407-9100

IMAGE MAGNETIC ASSOCIATES *(ISS)*
1525 Saltair, No. 107

West Los Angeles, CA 90025

IMPACT DISTRIBUTERS/ CORNERSTONE FILMS *(IMD)*
400 Coe Ave.
East Haven, CT 06512
800-766-1234

IMPERIAL ENTERTAINMENT CORPORATION *(IMP)*
4640 Lankershim Blvd., 4th Fl.
North Hollywood, CA 91602

IMPULSE PRODUCTIONS *(IPI)*
16882 Stagg St.
Van Nuys, CA 91406
800-635-0167
Fax: 818-780-2584

INCREASE VIDEO *(INC)*
6860 Canby Ave., Ste. 118
Reseda, CA 91335
818-342-2880
800-233-2880
Fax: 818-342-4029

INDEPENDENT UNITED DISTRIBUTORS *(IND)*
430 W. 54th St.
New York, NY 10019
212-489-8130
800-457-0056

INDIANA UNIVERSITY CENTER FOR MEDIA & TEACHING RESOURCES *(INU)*
Bloomington, IN 47405-5901
812-855-8087
Fax: 812-855-8404

INGRAM VIDEO INC. *(ING)*
2259 Merritt Dr.
Garland, TX 75041
214-840-6621

INSTANT REPLAY VIDEOCASSETTE MAGAZINE *(IRP)*
2601 S. Bayshore Dr., Ste. 1050
Coconut Grove, FL 33133
305-854-8777

INTERAMA, INC. *(INT)*
301 W. 53rd St., Ste. 19E
New York, NY 10019
212-977-4830

INTERAMERICAN ENTERTAINMENT CORP. *(IEC)*
500 Mill Rd.
Bensalem, PA 19080
215-638-4222
800-447-3399

INTERGLOBAL VIDEO *(IGV)*
18-22 Spadina Ave.
Toronto, ON, Canada M5V 2H6
Fax: 416-585-9927

INTERNATIONAL FILM BUREAU, INC. *(IFB)*
332 S. Michigan Ave.
Chicago, IL 60604-4382
312-427-4545

INTERNATIONAL FILM EXCHANGE, LTD. *(IFE)*
201 W. 52nd St.
New York, NY 10019
212-582-4318

INTERNATIONAL FILM FOUNDATION *(INF)*
155 W. 72nd St.
New York, NY 10023
212-580-1111

INTERNATIONAL HISTORIC FILMS, INC. (IHF) *(IHF)*
PO Box 29035
Chicago, IL 60629
312-927-2900
Fax: 312-927-9211

INTERNATIONAL HOME VIDEO *(INV)*
431 N. Figueroa St.
Wilmington, CA 90744
800-243-4835

INTERNATIONAL VIDEO ENTERTAINMENT (IVE) *(IVE)*
15400 Sherman Way, Ste. 500
Van Nuys, CA 91406
818-908-0303
800-423-7455
Fax: 818-778-3259

INTERURBAN FILMS *(IUF)*
PO Box 6444
Glendale, CA 91225-0444
818-240-9130

IRONWOOD PRODUCTIONS *(IRN)*
Ste. 255
Los Angeles, CA 90026
213-664-6069

ISLAND VISUAL ARTS *(IVA)*
8920 Sunset Blvd., 2nd Fl.
Los Angeles, CA 90069
213-288-5382
Fax: 213-276-5476

IVANHOE COMMUNICATIONS, INC. *(IVC)*
401 S. Rosalind Ave.
Orlando, FL 32801
407-423-8045

IVY CLASSICS VIDEO *(IVY)*
725 Providence Rd., Ste. 204
Charlotte, NC 28207
704-333-3991
Fax: 704-335-0672

J WORLD *(JWD)*
PO Box 1500
Newport, RI 02840
401-849-5492

J2 COMMUNICATIONS *(JTC)*
10850 Wilshire Blvd., Ste. 1000
Los Angeles, CA 90024
213-474-5252

JAMES AGEE FILM PROJECT *(JAF)*
316 1/2 E. Main St.
Johnson City, TN 37601
615-926-8637
800-352-5111

JCI VIDEO *(JCI)*
21550 Oxnard St., Ste. 920
Woodland Hills, CA 91367
818-593-3600
800-223-7479
Fax: 818-593-3610

JEF FILMS *(JEF)*
Film House
143 Hickory Hill Circle
Osterville, MA 02655
617-428-7198

JEREMIAH FILMS (JF) *(JER)*
PO Box 1710
Hemet, CA 92546
714-652-1006
800-828-2290

JLT FILMS, INC. *(JLT)*
7000 N. Austin Ave.
Niles, IL 60648

708-647-9513

JOURNAL FILMS, INC. *(JOU)*
930 Pitner Ave.
Evanston, IL 60202
708-328-6700
800-323-5448
Fax: 708-328-6706

JUDAICA CAPTIONED FILM CENTER, INC. *(JCF)*
PO Box 21439
Baltimore, MD 21208-0439

JUST FOR KIDS HOME VIDEO *(JFK)*
c/o Celebrity Home Entertainment
6320 Canoga Ave., Penthouse Ste.
PO Box 4112
Woodland Hills, CA 91365-4112
818-715-1980
800-445-8210
Fax: 818-716-0168

K-BEECH VIDEO, INC. *(KBE)*
8847 Wilbur
Northridge, CA 91324
818-772-4201
800-255-5923

K VIDEO *(KVI)*
157 Wiltshire Rd.
Claymont, DE 19703
302-798-2229
Fax: 302-798-7359

KAROL MEDIA/KAROL VIDEO *(KAR)*
22 Riverview Dr.
Wayne, NJ 07470
201-628-9111
800-526-4773

KEELING'S RECORDS & VIDEOS *(KRV)*
190 W. 135th St.
New York, NY 10031
212-283-5825

KEEP THE FAITH INC. *(KEP)*
PO Box 1069
141 Main
Clifton, NJ 07014-1065
201-471-7494
800-221-1564

KET, THE KENTUCKY NETWORK ENTERPRISE DIVISION *(KET)*
2230 Richmond Rd., Ste. 213
Lexington, KY 40502
606-233-3000
800-354-9067
Fax: 606-266-3562

KIDS KLASSICS, INC. *(KID)*
401 5th Ave.
New York, NY 10016
212-889-0044

KING BEE VIDEO *(KBV)*
PO Box 2521
Newbury Park, CA 91320
800-344-6680

KING OF VIDEO *(KOV)*
3529 S. Valley View Blvd.
Las Vegas, NV 89103
800-634-6143

KINO ON VIDEO *(KIV)*
333 W. 39th St., Ste. 503
New York, NY 10018
212-629-6880

KIT PARKER VIDEO *(KIT)*
c/o Central Park Media
301 W. 53rd St., 13th Fl.
New York, NY 10019

800-833-7456

KNOWHUTIMEAN? HOME VIDEO *(KWI)*
c/o The Ernest Fan Club
PO Box 23325
Nashville, TN 37202
615-255-6694

KNOWLEDGE UNLIMITED, INC. *(KUI)*
Box 52
Madison, WI 53701-0052
608-836-6660
800-356-2303
Fax: 608-831-1570

KTCA-TV VIDEO SERVICES *(KTC)*
Twin Cities Public TV
The Minnesota Telecenter
172 E. 4th St.
St. Paul, MN 55101
612-222-1717

KULTUR VIDEO *(KUL)*
121 Hwy. No. 36
West Long Branch, NJ 07764
908-229-2343
800-458-5887
Fax: 908-229-0066

LB DISTRIBUTORS, INC. *(LBD)*
3186 Doolittle Dr.
Northbrook, IL 60062
800-950-0255
Fax: 800-564-8656

LD VIDEO *(LDV)*
6911 Topanga Canyon Blvd.
Canoga Park, CA 91303
800-238-2302
Fax: 818-999-2731

LEARNING CORPORATION OF AMERICA *(LCA)*
108 Wilmot Rd.
Deerfield, IL 60015-9990
708-940-1260
800-621-2131
Fax: 708-940-3600

LEGACY HOME VIDEO *(LGC)*
9301 Wilshire Blvd., Ste. 507
Beverly Hills, CA 90210
310-273-2353
Fax: 310-273-2418

LIBERTY ENTERTAINMENT GROUP, INC. *(LEG)*
1019 E. 53rd
Davenport, IA 52807

LIBERTY PUBLISHING COMPANY, INC. *(LPC)*
50 Scott Adam Rd.
Cockeysville, MD 21030
410-667-6680

LIGHTNING VIDEO *(LTG)*
60 Long Ridge Rd.
PO Box 4000
Stamford, CT 06907

LIVE HOME VIDEO *(LIV)*
15400 Sherman Way, Ste. 500
Van Nuys, CA 91410-0124
818-988-5060

LOONIC VIDEO *(LOO)*
2022 Taraval St., Ste. 6427
San Francisco, CA 94116
510-526-5681

LORIMAR HOME VIDEO *(LHV)*
15838 N. 62nd St., Ste. 100
Scottsdale, AZ 85254
800-345-1441

LUCERNE MEDIA *(LME)*
37 Ground Pine Rd.
Morris Plains, NJ 07950
201-538-1401
800-341-2293
Fax: 201-538-0855

**LUMIVISION
CORPORATION** *(LUM)*
1490 Lafayette St., Ste. 305
Denver, CO 80218
303-860-0400
800-776-5864

LUNA VIDEO *(LUN)*
c/o Media Home
   Entertainment
5730 Buckingham Pkwy.
Culver City, CA 90230
213-216-7900
800-421-4509

A & M VIDEO *(A&M)*
1416 N. LaBrea Ave.
Hollywood, CA 90028
213-469-2411

MADERA CINEVIDEO *(MAD)*
525 E. Yosemite Ave.
Madera, CA 93638
209-661-6000
800-624-2204

**MAGNA SYSTEMS,
INC.** *(MGS)*
W. Countyline 95
Barrington, IL 60010
708-382-6477
Fax: 708-250-0038

**MAGNUM ENTERTAINMENT
INC.** *(MAG)*
9650 De Soto Ave., Unit M
Chateworth, CA 91311-5012
818-700-2822
800-624-6868
Fax: 818-700-2835

THE MAIER GROUP *(TMG)*
235 E. 95th St.
New York, NY 10128
212-534-4100
Fax: 212-410-2145

**MAJOR VIDEO CONCEPTS,
INC.** *(MVI)*
6103 Johns Rd., Ste. 5-8
Tampa, FL 33634
813-884-4050
800-274-0682
Fax: 813-888-5936

**MALIN &
ASSOCIATES** *(MAA)*
24352 Highlander Rd.
West Hills, CA 91307
818-884-5363

**MANAGEMENT COMPANY
ENTERTAINMENT GROUP
(MCEG), INC.** *(MCG)*
2121 Avenue of the Stars,
   Ste. 2630
Los Angeles, CA 90067
310-282-0871
Fax: 310-315-7850

**MARK FOUR PICTURES
FILM & VIDEO** *(MIV)*
Box 3810
Urbandale Sta.
5907 Meredith Dr.
Des Moines, IA 50322
515-278-4737
800-247-3456

**MARK V
INTERNATIONAL** *(MFI)*
19770 Bahama St.
Northridge, CA 91324

818-407-3800
800-433-9753

**MARYLAND PUBLIC
TELEVISION** *(MTP)*
11767 Owings Mills Blvd.
Owings Mills, MD 21117-1499
410-356-5600
Fax: 410-581-4338

MASTER ARTS VIDEO *(MAV)*
11549 Amigo Ave.
Northridge, CA 91326
818-368-9220

MASTERVISION, INC. *(MAS)*
969 Park Ave.
New York, NY 10028
212-879-0448

**MCA/UNIVERSAL HOME
VIDEO** *(MCA)*
70 Universal City Plaza
Universal City, CA 91608-
   9955
818-777-4300
Fax: 818-777-6419

**MEDIA HOME
ENTERTAINMENT** *(MED)*
5730 Buckingham Pkwy.
Culver City, CA 90230
310-216-7900
800-421-4509

MEDIA PROJECT, INC. *(MDP)*
PO Box 4093
Portland, OR 97208

METRO PACKAGING *(MET)*
3501 Biway
Ft. Worth, TX 76114
817-429-8230

**MEXICAN HOME
VIDEO** *(MHV)*
8455 Beverly Blvd., Ste. 303
Los Angeles, CA 90048
213-655-4040

**MGM HOME VIDEO,
INC.** *(MGM)*
10000 W. Washington Blvd.
Culver City, CA 90232
310-280-6212

MIDNIGHT VIDEO *(MID)*
3960 Laurel Canyon Blvd.,
   Ste. 275
Studio City, CA 91604
818-509-3948

MIKE LEBELL'S VIDEO *(MLB)*
75 Freemont Pl.
Los Angeles, CA 90005
213-938-3333

**MILESTONE FILM &
VIDEO** *(MIL)*
275 W. 96th St., Ste 28C
New York, NY 10025
212-865-7449
Fax: 212-222-8952

MIRACLE VIDEO, INC. *(MRC)*
PO Box 470484
Tulsa, OK 74147-0484
918-252-9346
800-548-1803

**MIRAMAR
PRODUCTIONS** *(MIR)*
200 2nd Ave., W.
Seattle, WA 98119
206-284-4700
800-245-6472
Fax: 206-286-4433

**MNTEX ENTERTAINMENT,
INC.** *(MTX)*
PO Box 667
Prior Lake, MN 55372-0667

612-440-6028
Fax: 612-447-8173

**MODERN SIGNS PRESS,
INC.** *(MDS)*
PO Box 1181
Los Alamitos, CA 90720
213-596-8548

**MODERN SOUND PICTURES,
INC.** *(MSP)*
1402 Howard St.
Omaha, NE 68102
402-341-8476
Fax: 402-341-8487

**MOGUL
COMMUNICATIONS** *(MGL)*
1311 N. Mansfield Ave.
Hollywood, CA 90028
213-650-2122

**MONARCH HOME
VIDEO** *(MNC)*
7900 Hickman Rd.
Des Moines, IA 50322
515-226-3000
800-247-8032
Fax: 515-226-3592

**MONDAY/WEDNESDAY/
FRIDAY VIDEO** *(MWF)*
c/o Moore
73 E. Houston St.
New York, NY 10012
212-219-0765

**MONOGRAM
ENTERTAINMENT** *(MNE)*
5710 E. Sheila St., Bldg. T
Commerce, CA 90040
213-726-0452

**MONTEREY HOME
VIDEO** *(MON)*
28038 Dorothy Dr., Ste. 1
Agoura Hills, CA 91301
818-597-0047
800-424-2593
Fax: 818-597-0105

**MONTICELLO
PRODUCTIONS** *(MNP)*
1822 Easterly Ter.
Los Angeles, CA 90026
213-662-2938

**MOONBEAM
PUBLICATIONS** *(MBP)*
18530 Mack Ave.
Grosse Pointe, MI 48236
313-884-5255
800-445-2391
Fax: 313-884-5166

MOORE VIDEO *(MRV)*
PO Box 5703
Richmond, VA 23220
804-745-9785
Fax: 804-745-9785

MORRIS VIDEO *(MOR)*
2730 Monterey St., No. 105
Monterey Business Park
Torrance, CA 90503
310-533-4800
800-843-3606

**MOSSMAN WILLIAMS
PRODUCTIONS** *(MWP)*
9840 Wornall Rd.
Kansas City, MO 64114-3907
816-241-6684

MOVIECRAFT, INC. *(MVC)*
13916 Charleston
PO Box 438
Orland Park, IL 60462
708-460-9082
Fax: 708-460-9099

MOVIES UNLIMITED *(MOV)*
6736 Castor Ave.
Philadelphia, PA 19149
215-722-8298
800-523-0823

MPI HOME VIDEO *(MPI)*
15825 Rob Roy Dr.
Oak Forest, IL 60452
708-687-7881
Fax: 708-687-3797

MTI HOME VIDEO *(MTH)*
14216 SW 136th St.
Miami, FL 33186
305-255-8684
800-821-7461
Fax: 305-233-6943

**MTI TELEPROGRAMS,
INC.** *(MTT)*
108 Wilmot Rd.
Deerfield, IL 60015-9990
708-940-1260
800-621-2131

**MUSEUM OF MODERN ART
OF LATIN AMERICA** *(MUS)*
Audio-Visual Program/
   Organization of American
   States
1889 F St., NW
Washington, DC 20006
202-458-6016

MUSIC MEDIA *(MSM)*
5730 Buckingham Pkwy
Culver City, CA 90230
310-216-7900
800-421-4509

**MUSIC VIDEO
DISTRIBUTORS** *(MVD)*
Equivest Industrial Center
Norristown, PA 19401
215-272-7771
800-888-0486
Fax: 215-272-6074

MYSTIC FIRE VIDEO *(MFV)*
PO Box 1092
Cooper Sta.
New York, NY 10276
212-941-0999
Fax: 212-941-1443

NAC HOME VIDEO *(NAC)*
1300 Quail, Ste. 201
Newport Beach, CA 92660

**NATIONAL CENTER FOR
JEWISH FILM** *(NCJ)*
Brandeis University
Lown Bldg. 102
Waltham, MA 02254-9110
617-899-7044
Fax: 617-736-2070

**NATIONAL
ENTERTAINMENT GROUP,
INC.** *(NEG)*
Congress Video Division
1776 Broadway, Ste. 1010
New York, NY 10019
212-581-4880
800-847-8273
Fax: 212-581-4962

**NATIONAL FILM BOARD OF
CANADA** *(NFC)*
1251 Avenue of the Americas,
   16th Fl.
New York, NY 10020-1173
212-586-5131

**NEW HORIZONS HOME
VIDEO** *(NHO)*
2951 Flowers Rd., S., Ste.
   237
Atlanta, GA 30341
404-458-3488

**NEW JERSEY
NETWORK** *(NJN)*
1573 Parkside Ave.
Trenton, NJ 08625
609-530-5252

**NEW LINE HOME
VIDEO** *(NLC)*
116 N. Robertson Blvd.
Los Angeles, CA 90048
213-967-6670
Fax: 213-854-0602

NEW STAR VIDEO *(NSV)*
260 S. Beverly Dr.
Beverly Hills, CA 90212

**NEW & UNIQUE
VIDEOS** *(NUV)*
2336 Sumac Dr.
San Diego, CA 92105
619-282-6126
Fax: 619-283-8264

NEW VIDEO CENTER *(NEW)*
90 University Pl.
New York, NY 10003
212-243-0400

**NEW WORLD
ENTERTAINMENT** *(NWV)*
1440 S. Sepulveda Blvd.
Los Angeles, CA 90025
213-444-8100

**NEW YORK STATE
EDUCATION
DEPARTMENT** *(NYS)*
Center for Learning
   Technologies
Media Distribution Network,
   Rm. C-7, Concourse Level
Albany, NY 12230
518-474-3852

NEW YORKER VIDEO *(NYF)*
16 W. 61st St.
New York, NY 10023
212-247-6110
800-447-0196
Fax: 212-307-7855

NEWSBANK, INC. *(NBI)*
58 Pine St.
New Canaan, CT 06840-5408
203-966-1100
800-243-7694

**NORMAN BEERGER
PRODUCTIONS** *(NOR)*
3217 S. Arville St.
Las Vegas, NV 89102-7612
702-876-2328

**NORTH AMERICAN HOME
VIDEO** *(NAM)*
95 Commerce Rd.
Stamford, CT 06902-4505
800-458-4051

**NORTH AMERICAN VIDEO,
INC.** *(NAV)*
385 Pine Grove Rd.
Roswell, GA 30075
404-992-1301

**NORTHEAST VIDEO,
LTD.** *(NRT)*
102 Madison Ave.
New York, NY 10016
212-661-8830

**NOSTALGIA FAMILY
VIDEO** *(NOS)*
PO Box 606
Baker City, OR 97814
503-523-9034

NOVA HOME VIDEO *(NVH)*
c/o Canadian Video Factory
3807 9th St. SE

Calgary, AB, Canada T2G 3C7
403-287-9070

OCEAN VIDEO (OCE)
2550 Corporate Pl., Ste. C-103
Monterey Park, CA 91754
213-881-0732

OHLMEYER COMMUNICATIONS (OHC)
962 N. La Cienega Blvd.
Los Angeles, CA 90069
213-659-8557

ONE WEST MEDIA (OWM)
PO Box 5766
559 Onate Pl.
Santa Fe, NM 87501
505-983-8685

ORION HOME VIDEO (ORI)
1325 Avenue of the Americas
New York, NY 10019
212-956-3800
Fax: 212-956-7449

PACIFIC ARTS VIDEO (PAV)
11858 La Grange
Los Angeles, CA 90025
310-820-0991
800-538-5856
Fax: 310-826-4779

PACIFIC COAST VIDEO (PCV)
635 Chapala St.
Santa Barbara, CA 93101
805-965-5015

PALISADES ENTERTAINMENT (PEV)
1875 Century Park, E., 3rd Fl.
Los Angeles, CA 90067

PARAGON VIDEO PRODUCTIONS (PGN)
PO Box 3478
San Mateo, CA 94403
415-362-2520

PARAMOUNT HOME VIDEO (PAR)
5555 Melrose Ave.
Los Angeles, CA 90038-3197
213-956-5000

PASSPORT MUSIC VIDEO (PMV)
3619 Kennedy Rd.
South Plainfield, NJ 07080
908-753-6100

PAULIST PRODUCTIONS (PAU)
PO Box 1057
Pacific Palisades, CA 90272
213-454-0688
800-624-8613

PBS VIDEO (PBS)
11858 La Grange Ave.
Los Angeles, CA 90025
213-820-0991
Fax: 213-826-4779

PEERLESS FILMS, INC. (PII)
1810 Johns Dr.
Glenview, IL 60025
708-729-8160

PENTREX VIDEO (PEN)
PO Box 94911
Pasadena, CA 91109
800-950-9333
Fax: 818-793-3797

PETER PAN INDUSTRIES (PPI)
88 Saint Frances St.
Newark, NJ 07105

201-344-4214
Fax: 201-344-0465

PHOENIX/BFA FILMS (BFA)
468 Park Ave., S.
New York, NY 10016
212-684-5910
800-221-1274

PHOENIX DISTRIBUTERS (PHX)
6253 Hollywood Blvd., No. 818
Hollywood, CA 90028
800-356-4386
Fax: 213-469-7041

PICTURE START (PIC)
1727 W. Catalpa
Chicago, IL 60640
312-769-2489

PIED PIPER MEDIA (PPM)
9710 De Soto Ave.
Chatsworth, CA 91311-4409
818-773-4300
800-367-2467

PIONEER ARTISTS, INC. (PIA)
200 W. Grand Ave.
Montvale, NJ 07645

PIONEER LDCA, INC. (LDC)
2265 E. 220th St.
PO Box 22782
Long Beach, CA 90801-5782
310-835-6117

PIONEER VIDEO IMPORTS (PNR)
200 W. Grand Ave.
Montvale, NJ 07645

PLANET VIDEO (PLV)
1560 Broadway, Rm. 1101
New York, NY 10036-1525

PLEXUS COMMUNICATIONS CORPORATION (PCC)
15760 Ventura Blvd., Ste. 532
Encino, CA 91436
818-995-1947
800-423-3061

PM ENTERTAINMENT GROUP, INC. (PMH)
16780 Schoenborn St.
Sepulveda, CA 91343
818-891-1288
Fax: 818-892-8391

POLYGRAM MUSIC VIDEO (PMV) (PGV)
825 8th Ave.
New York, NY 10019
212-333-8000
800-825-7781

POWERSPORTS (PSV)
15840 Ventura Blvd., Ste. 213
Encino, CA 91436
800-323-7979
Fax: 818-907-0598

PREMIER PROMOTIONS (PRE)
PO Box 19022
Charlotte, NC 28219
704-334-1111

PREMIERE HOME VIDEO (PMR)
755 N. Highland
Hollywood, CA 90038
213-934-8903
Fax: 213-934-8910

PRIMATE PRODUCTIONS (PRM)
PO Box 91436

Santa Barbara, CA 93190-1536

PRIME VIDEO (PRV)
7325 Fulton Ave.
Van Nuys, CA 91605
818-718-1116
800-727-2229

PRINCETON BOOK COMPANY PUBLISHERS (PBC)
PO Box 57
Pennington, NJ 08534
609-737-8177
800-326-7149

PRISM ENTERTAINMENT (PSM)
1888 Century Park, E., Ste. 1000
Los Angeles, CA 90067
213-277-3270
Fax: 213-203-8036

PROSCENIUM ENTERTAINMENT (PRS)
PO Box 909
Highstown, NJ 08520
800-222-6260

PROUD TO BE**A BLACK VIDEO COLLECTION** (PTB)
1235-E East Blvd., Ste. 209
Charlotte, NC 28203
704-523-2227

PUBLIC FILMS, INC. (PFI)
PO Box 1689
Wimberley, TX 78676
713-880-2604
800-628-4987

PUBLIC MEDIA VIDEO (PME)
5547 N. Ravenswood Ave.
Chicago, IL 60640
312-878-2600
800-323-4222
Fax: 312-878-8648

PYRAMID FILM & VIDEO (PYR)
Box 1048
2801 Colorado Ave.
Santa Monica, CA 90404
213-828-7577
800-421-2304
Fax: 213-453-9083

QUALITY VIDEO (QVD)
7399 Bush Lake Rd.
Minneapolis, MN 55439-2027
612-893-0903
Fax: 612-893-1585

QUANTUM COMMUNICATIONS (QCM)
3301 W. Hampden Ave., Ste. N.
Englewood, CO 80110
303-781-0679
Fax: 303-761-8556

QUEST ENTERTAINMENT (QUE)
1000 Universal Studios Plaza
Orlando, FL 32819
407-363-8440
Fax: 407-363-8449

QUESTAR HOME VIDEO (QHV)
PO Box 11345
Chicago, IL 60611
312-266-9400
800-544-8422
Fax: 312-266-9523

QUINTEX ENTERTAINMENT (QNE)
345 N. Maple Dr.

Beverly Hills, CA 90210
310-281-2600
Fax: 310-273-9453

RAEDON ENTERTAINMENT (RAE)
8707-D Lindley Ave., Ste. 173
Northridge, CA 91325
805-582-2550
Fax: 818-773-9770

RAP ENTERTAINMENT INC. (RAP)
932-17 Ave. SW, Ste. 310
Calgary, AB, Canada T2T 0A2
403-244-7000
800-565-7777
Fax: 403-245-0511

RCA VIDEO (RCA)
c/o Columbia TriStar Home Video
3400 Riverside Dr.
Burbank, CA 91505
818-972-8686

READER'S DIGEST HOME VIDEO (RDG)
Reader's Digest Rd.
Pleasantville, NY 10570
800-776-6868

READY REFERENCE PRESS (RRP)
PO Box 5249
Santa Monica, CA 90405

REAL PRODUCTIONS (REA)
1821 University Ave., Ste. N-153
St. Paul, MN 55104
612-646-9472

REGAL VIDEO (REG)
PO Box 674
Bowling Green Sta.
New York, NY 10274
212-732-3515

REGENCY HOME VIDEO/ BANANA REPUBLIC (RHV)
9911 W. Pico Blvd.
Los Angeles, CA 90035
310-552-2431
Fax: 310-552-9039

REGO: IRISH RECORDS & TAPES, INC. (RGO)
64 New Hyde Park Rd.
Garden City, NY 11530
516-328-7800
800-854-3746

RELAX VIDEO (RLX)
419 W. 119th St., Ste. 8I
New York, NY 10027
212-496-4400

REPUBLIC PICTURES HOME VIDEO (REP)
12636 Beatrice St.
Los Angeles, CA 90066-0930
310-306-4040

REX MILLER (RXM)
Rte. 1, Box 457-D
East Prairie, MO 63845
314-649-5048

RHAPSODY FILMS (RHP)
PO Box 179
New York, NY 10014
212-243-0152

RHINO VIDEO (RHI)
2225 Colorado Ave.
Santa Monica, CA 90404
310-828-1980

RKO PICTURES (RKO)
1900 Avenue of the Stars, Ste. 1562

Los Angeles, CA 90067
213-277-3133

RM FILMS, INC. (RMF)
Box 3748
Hollywood, CA 90078
213-466-7791

ROSALIE & JERRY CARTER (CAR)
RR 9, Box 64
Frankfort, IN 46041
317-324-2182

S-L FILM PRODUCTIONS (SLF)
PO Box 41108
Los Angeles, CA 90041
213-254-8528

SALENGER FILMS, INC. (SAL)
1635 12th St.
Santa Monica, CA 90404-9988
310-450-1300
Fax: 310-450-1010

SAN DIEGO STATE UNIVERSITY (SDS)
Learning Resource Center
San Diego, CA 92182
619-594-5200

SAN FRANCISCO RUSH VIDEO (SFR)
1574 Fell St., Ste. 1
San Francisco, CA 94117
415-921-8273

SATURN PRODUCTIONS, INC. (SAT)
1697 Broadway, Ste. 1105
New York, NY 10019
212-489-2460
800-228-1717
Fax: 212-397-0665

SCHOLASTIC LORIMAR (SCL)
17942 Cowan Ave.
Irvine, CA 92714

SCHWARTZ BROTHERS, INC. (SCB)
4901 Forbes Blvd.
Lanham, MD 20706
301-459-8000

SELECT-A-TAPE (SLC)
3960 Laurel Canyon Blvd., No. 275
Studio City, CA 91604

SELECT HOME VIDEO (SEL)
c/o Hemdale Home Video
7966 Beverly Blvd.
Los Angeles, CA 90048
213-966-3700
Fax: 213-651-3107

SELL PICTURES (SLP)
210 5th Ave.
New York, NY 10010

SEND VIDEO ARTS (SVA)
650 Missouri St.
San Francisco, CA 94107

SGE HOME VIDEO (SGE)
Div. of Shapiro Glickenhaus Entertainment
12001 Ventura Pl., 4th Fl.
Studio City, CA 91604
818-766-8500

SHAPIRO ARTS (SHP)
1335 Astor St., No. 14A
Chicago, IL 60610
312-761-6432

SHAPIRO GLICKENHAUS HOME VIDEO *(SHG)*
12001 Ventura Pl., 4th Fl.
Studio City, CA 91604

SHINE HOME ENTERTAINMENT *(SHE)*
5170 Sepulveda Blvd., Ste. 290
Sherman Oaks, CA 91409
818-784-1095
800-422-6484
Fax: 818-784-1389

SHOKUS VIDEO *(SHO)*
PO Box 8434
Van Nuys, CA 91409
818-704-0400
800-325-6800

SIMITAR ENTERTAINMENT *(SIM)*
3850 Annapolis Ln.
Plymouth, MN 55447
612-559-6660
Fax: 612-559-0210

SINISTER CINEMA *(SNC)*
PO Box 4369
Medford, OR 97501-0168
503-773-6860

SNOOPY'S HOME VIDEO LIBRARY *(SHV)*
c/o Media Home Entertainment
5730 Buckingham Pkwy.
Culver City, CA 90230
310-216-7900
800-421-4509

SOMETHING WEIRD VIDEO *(SMW)*
c/o Mike Vraney
PO Box 33664
Seattle, WA 98133
206-361-3759

SONY MUSIC VIDEO ENTERPRISES *(SMV)*
51 W. 52nd St.
New York, NY 10019
212-975-4321

SONY VIDEO COMMUNICATIONS *(SVC)*
Tape Production Department
700 W. Artesia Blvd.
Compton, CA 90220
213-537-4300

SONY VIDEO SOFTWARE (SVS, INC.) *(SVS)*
1700 Broadway, 16th Fl.
New York, NY 10019
518-972-8870
Fax: 518-972-0907

SOUTHGATE ENTERTAINMENT *(SOU)*
c/o Weissman/Angellotti
3855 Lankershim Blvd.
North Hollywood, CA 91604
818-763-2975
800-445-6887

SPARROW DISTRIBUTION *(SPW)*
101 Winners Circle
Brentwood, TN 37027
800-877-4443
Fax: 615-371-6909

SPECIALTY VIDEO DISTRIBUTORS *(SVD)*
371 Commercial, Ste. 13
Northbrook, IL 60062
708-205-4410

SPORTS ON VIDEO *(SPV)*
4129 Rosario Rd.
Woodland Hills, CA 91364
818-715-0583
Fax: 818-715-9863

SPORTSMEN ON FILM *(SOF)*
5038 N. Pkwy
Calabasas, CA 91302
818-713-1888
800-533-8111
Fax: 818-713-8572

STAR CLASSICS *(STC)*
4301 Glenwood Rd.
Brooklyn, NY 11210
718-434-1100
800-626-2723
Fax: 718-859-0759

STARMAKER ENTERTAINMENT, INC. *(STE)*
151 Industrial Way, E.
Eatontown, NJ 07724
908-389-1020
800-233-3738
Fax: 908-389-1021

STEVENSON PRODUCTIONS, INC. *(SPI)*
3227 Banks St.
New Orleans, LA 70119
504-822-7678
Fax: 504-822-5459

STRAND/VCI ENTERTAINMENT *(SVI)*
3350 Ocean Park Blvd., Ste. 205
Santa Monica, CA 90405
310-396-7011
Fax: 310-392-2472

STREAMLINE PICTURES *(STP)*
971 N. La Cienega, Ste. 209
Los Angeles, CA 90069
310-657-8559
Fax: 310-657-8642

STRIKE FORCE FILMS *(SFF)*
PO Box 954
Evanston, IL 60204

STUDIO ENTERTAINMENT *(SED)*
386 Park Ave., S., Ste. 900
New York, NY 10016
800-247-7004

STUDIO VIDEO PRODUCTIONS *(STV)*
200 Suburban, Ste. E.
San Luis Obispo, CA 93401
805-543-0707
800-541-1843
Fax: 805-549-0352

SULTAN ENTERTAINMENT *(SUE)*
335 N. Maple Dr., Ste. 351
Beverly Hills, CA 90210-3899
310-285-6000

SUMMIT MEDIA COMPANY *(SUM)*
27811 Hopkins Ave., Unit 1
Valencia, CA 91355
805-295-0675
800-777-8668
Fax: 805-295-0680

SUN VIDEO *(SUN)*
15 Donnybrook
Demarest, NJ 07627
201-784-0662

SUPER SITTERS, INC. *(SSI)*
10503 N. Cedarburg Rd.
PO Box 218
Mequon, WI 53092
800-323-3431

SUPER SOURCE *(SUP)*
PO Box 410777
San Francisco, CA 94141
415-777-1964
800-331-6304
Fax: 415-777-0187

SURF VIDEO NETWORK *(SVN)*
PO Box 6671
Santa Barbara, CA 93160
805-683-3501

SUTTON ENTERTAINMENT CORPORATION *(SEC)*
PO Box 7032
Edison, NJ 08837
908-417-1959

SVS/TRIUMPH *(SVT)*
3500 W. Olive Ave.
Burbank, CA 91505
818-953-7900

SYBERVISION SYSTEMS, INC. *(SYB)*
1 Sansome St., Ste. 1610
San Francisco, CA 94104
510-846-2244
800-777-4994

TAMARELLE'S *(TAM)*
7900 Hickman RD.
Des Moines, IA 50322
515-254-7253
800-356-3577
Fax: 515-254-7021

TAPEWORM VIDEO DISTRIBUTORS *(TPV)*
12420 Montague St., Ste. B
Arleta, CA 91331
818-896-8899
800-367-8437
Fax: 818-896-3855

TARGET VIDEO *(TGV)*
678 S. Van Ness
San Francisco, CA 94110

TELECINE SPANISH VIDEO *(TSV)*
2151 Belmont Ave.
Bronx, NY 10457

TELEVIDEOS *(TEL)*
PO Box 22
Lorane, OR 97451
800-284-3367
800-2VI-DEOS

10TH AVENUE VIDEO *(TAV)*
9333 Oso Ave.
Chatsworth, CA 91311
213-772-1034

TEXTURE FILMS, INC. *(TEX)*
5547 N. Ravenswood Ave.
Chicago, IL 60640
312-878-7300
800-323-4222

THOMSON PRODUCTIONS, INC. *(TPI)*
PO Box 1225
Orem, UT 84059
801-226-0155
800-228-8491

3M AUDIO VISUAL DIVISION *(THA)*
Bldg. 223-15-02
St. Paul, MN 55144
612-736-2964
800-328-1371
Fax: 612-736-7479

3M/LEISURE TIME PRODUCTS *(THM)*
3M Center
Bldg. 225-3N-04
St. Paul, MN 55144-1000

612-733-2665

THRILLER VIDEO *(THR)*
21800 Burbank Blvd., No. 300
PO Box 4062
Woodland Hills, CA 91365-4062

TIME-LIFE VIDEO AND TELEVISION *(TLF)*
1450 E. Parham Rd.
Richmond, VA 23280
804-266-6330
800-621-7026

TIMELESS VIDEO INC. *(TIM)*
10010 Canoga Ave., Ste. B2
Chatsworth, CA 91311
818-773-0284

TOUCHSTONE HOME VIDEO *(TOU)*
500 S. Buena Vista St.
Burbank, CA 91521
818-562-3883

TOURNAMENT VIDEO TAPES *(TVT)*
1615 W. Burbank Blvd.
Burbank, CA 91506
213-843-0373

TRADE-A-FLICK *(TAF)*
4125 Transit Rd.
Williamsville, NY 14221

TRANS-ATLANTIC VIDEO *(TRV)*
607 Montrose
South Plainfield, NJ 07080
201-756-9800

TRANS-WORLD ENTERTAINMENT *(TWE)*
3330 Cahuenga Blvd., W., Ste. 500
Los Angeles, CA 90068
213-969-2800

TRIBORO ENTERTAINMENT GROUP *(TRI)*
101 W. 57th St.
New York, NY 10019
212-262-5555

TRYLON VIDEO, INC. *(TRY)*
645 Madison Ave.
New York, NY 10022
212-836-4423

TURNER HOME ENTERTAINMENT COMPANY *(TTC)*
PO Box 105366
Atlanta, GA 30348-5366
404-827-3066

TV SPORTS SCENE, INC. (TVSS) *(TVS)*
5804 Ayrshire Blvd.
Minneapolis, MN 55436
612-925-9661

20TH CENTURY FOX FILM CORPORATION *(TCF)*
PO Box 900
Beverly Hills, CA 90213
310-277-2211

21ST GENESIS HOME VIDEO *(GNS)*
15820 Arminta St.
Van Nuys, CA 91406
818-787-0660
800-344-1060

TWIN TOWER ENTERPRISES *(TTE)*
18720 Oxnard St., Ste. 101
Tarzana, CA 91356

818-344-8424
800-553-4321
Fax: 818-344-8474

TWO STAR FILMS, INC. *(TSF)*
Box 495
Saint James, NY 11780
516-584-7283

UAV CORPORATION *(UAV)*
PO Box 7647
Charlotte, NC 28241
919-548-7300
Fax: 919-548-3335

UNI DISTRIBUTION *(UND)*
PO Box 3966
North Hollywood, CA 91609

UNICORN VIDEO, INC. *(UNI)*
9811 Independence Ave.
Chatsworth, CA 91311
818-407-1333
800-528-4336
Fax: 818-407-8246

UNITED TRAINING MEDIA *(UTM)*
6633 W. Howard St.
Niles, IL 60648-3305
800-759-0364
Fax: 708-647-0918

UNREEL PRODUCTIONS *(UNR)*
1012 Brioso Ave., Ste. 106
Costa Mesa, CA 92627

URBAN CLASSICS VIDEO *(UCV)*
Empire Entertainment
1551 N. La Brea Ave.
Los Angeles, CA 90028

USA HOME VIDEO *(USA)*
21800 Burbank Blvd.
PO Box 4062
Woodlands Hills, CA 91316

VALENCIA ENTERTAINMENT CORPORATION *(VEC)*
28231 Ave. Crocker, Ste. 120
Valencia, CA 91355
805-257-6054
800-323-2061
Fax: 805-949-3400

VALLEY STUDIOS *(VSI)*
292 Gibraltar Ave., Bldg. A-1
Sunnyvale, CA 94089
408-747-1491

VALUE VIDEO *(VAL)*
2700 Matheson Blvd., E., Ste. 304
Mississauga, ON, Canada L4W 4V9
416-624-2021
Fax: 416-624-0776

VANGUARD VIDEO *(VGD)*
436 Creamery Way, Ste. D
Exton, PA 19341
800-323-7432

VANTAGE COMMUNICATIONS, INC. *(VAN)*
PO Box 546-G
78 Main St.
Nyack, NY 10960
914-358-0147
800-872-0068
Fax: 914-358-0359

VCI, INC. *(VCI)*
7601 Washington Ave., S.
Minneapolis, MN 55435

...OME
...ERTAINMENT,
...C. *(VHE)*
13418 Wyandotte St.
North Hollywood, CA 91605
818-764-1777
800-350-1931
Fax: 818-764-0231

VCL HOME VIDEO *(VCL)*
Glen Professional Centre
2980 Beverly Glen Circle, Ste. 302
Los Angeles, CA 90077

VESTRON VIDEO *(VES)*
c/o LIVE Entertainment
15400 Sherman Way, Ste. 500
Van Nuys, CA 91410-0124
818-988-5060

VICTORY VIDEO *(VIC)*
PO Box 87
Toney, AL 35773
205-852-0198

VIDAMERICA, INC. *(VID)*
60 Madison Ave., 12th Fl.
New York, NY 10010
212-685-1300

VIDCREST *(VDC)*
8561 Cole Crest Dr.
PO Box 69642
Los Angeles, CA 90046
213-650-7310

VIDEO ACTION *(VDA)*
708 W. 1st St.
Los Angeles, CA 90012
213-687-8262
800-422-2241

VIDEO ARTISTS INTERNATIONAL, INC. *(VAI)*
158 Linwood Plaza, Ste. 301
Fort Lee, NJ 07024
201-944-0099
Fax: 201-947-8850

VIDEO BIBLE LIBRARY, INC. *(VBL)*
Box 17515
Portland, OR 97213
503-892-7707

VIDEO CITY PRODUCTIONS *(VCD)*
4266 Broadway
Oakland, CA 94611
510-428-0202
800-847-8400
Fax: 510-654-7802

VIDEO COMMUNICATIONS, INC. *(UHV)*
6535 E. Skelley Dr.
Tulsa, OK 74145
918-622-6460
800-331-4077
Fax: 918-665-6256

VIDEO CONNECTION *(VCN)*
3123 Sylvania Ave.
Toledo, OH 43613
419-472-7727

VIDEO DIMENSIONS *(VDM)*
530 W. 23rd St., Ste. 317
New York, NY 10011
212-929-6135

VIDEO DYNAMICS *(VDY)*
Box 9550
Jackson, MS 39286

VIDEO GEMS *(GEM)*
731 N. La Brea Ave.
PO Box 38188
Los Angeles, CA 90038

213-938-2385
800-421-3252

VIDEO HOME LIBRARY *(VHM)*
75 Spring St.
New York, NY 10012
212-925-7744
800-221-5480

VIDEO LATINO *(VLA)*
431 N. Figueroa
Wilmington, CA 90744

VIDEO OUTREACH, INC. *(VOI)*
14536 Roscoe Blvd.
Panorama City, CA 91402
818-891-5893

VIDEO PLACEMENT INTERNATIONAL *(VPL)*
240 E. 27th St., Ste. 81
New York, NY 10016
212-696-9207

VIDEO PRESENTATIONS, INC. *(VPR)*
6th & Battery Bldg.
2326 6th Ave., Ste. 230
Seattle, WA 98121
206-728-9241
800-458-5335

VIDEO RAILS *(VRL)*
PO Box 80001
San Diego, CA 92138
619-581-6226
800-262-2776
Fax: 619-581-0131

VIDEO TOURS, INC. *(VTO)*
300 Winding Brook Dr.
Glastonbury, CT 06033
203-659-8687
800-869-6789
Fax: 203-657-3334

VIDEO TREASURES *(VTR)*
2001 Glenn Pkwy.
Batavia, OH 45103
513-732-2790
Fax: 513-732-2611

VIDEO TREND, INC. *(VTI)*
1011 E. Touhy Ave., Ste. 500
Des Plaines, IL 60018-2806

VIDEO VIDEO *(VDE)*
1376 Soquel Ave.
Santa Cruz, CA 95062
408-427-2282

VIDEO VISA *(VVC)*
12901 Coral Tree Pl.
Los Angeles, CA 90066
213-827-7222

VIDEO YESTERYEAR *(VYY)*
Box C
Sandy Hook, CT 06482
203-426-2574
800-243-0987
Fax: 800-422-3460

VIDEOGRAF *(VGR)*
144 W. 27th St.
New York, NY 10001
212-242-7871

VIDEOGRAPH *(VPH)*
2833 25th St.
San Francisco, CA 94110

VIDEOTAKES *(VTK)*
187 Parker Ave., Ste. 71
Manasquan, NJ 08736

VIDEOVISION, INC. *(VVI)*
37 E. Washington St.
Hagerstown, MD 21740
301-791-3496

VIDMAG INTERNATIONAL, LTD. *(VIL)*
1572 Bitterroot Ct.
PO Box 1346
San Marcos, CA 92069

VIDMARK ENTERTAINMENT *(VMK)*
2644 30th St.
Santa Monica, CA 90405-3009
310-314-2000

VIDTAPE/DURAVISION *(VTP)*
Jefferson Business Park
600 S. Jefferson, Unit J
Placentia, CA 92670

V.I.E.W. VIDEO *(VWV)*
34 E. 23rd St.
New York, NY 10010
212-674-5550
Fax: 212-979-0266

VINEYARD VIDEO PRODUCTIONS *(VVP)*
PO Box 370
West Tisbury, MA 02575
508-693-3584

VIP VIDEO *(VIP)*
c/o JEF Films
Film House
143 Hickory Hill Circle
Osterville, MA 02655-1322
508-428-7198

VIRGIN MUSIC VIDEO *(VMV)*
111 N. Hollywood Way
Burbank, CA 91505-4356
818-840-6344

VIRGIN VISION *(VIR)*
6100 Wilshire Blvd., 16th Fl.
Los Angeles, CA 90048

VISIONSMITHS, INC. *(VSV)*
96 Spadina Ave., 3rd Fl.
Toronto, ON, Canada M5V 2J6

VISTA HOME VIDEO *(VHV)*
1370 Avenue of the Americas
New York, NY 10019

VOYAGER COMPANY *(VYG)*
1351 Pacific Coast Hwy.
Santa Monica, CA 90401
310-451-1383
800-446-2001
Fax: 310-394-2156

VPI/AC VIDEO *(VPI)*
381 Park Ave. S., Ste. 621
New York, NY 10016
212-685-5522
Fax: 212-615-5486

WALT DISNEY HOME VIDEO *(DIS)*
4111 W. Alameda
Burbank, CA 91505
818-562-3560

WARNER/ELEKTRA/ ATLANTIC CORPORATION *(WEA)*
9451 LBJ Fwy., Ste. 107
Dallas, TX 75243
214-234-6200

WARNER HOME VIDEO, INC. *(WAR)*
4000 Warner Blvd.
Burbank, CA 91522
818-954-6000

WARNER REPRISE VIDEO *(WRV)*
Div. of Warner Brothers Records
3300 Warner Blvd.

Burbank, CA 91505
818-846-9090

WATER BEARER FILMS *(WBF)*
205 West End Ave., Ste. 24H
New York, NY 10023
212-580-8185
Fax: 212-787-5455

WAXWORKS *(WAX)*
325 E. 3rd St.
Owensboro, KY 42303
502-926-0008
800-825-8558

WB VIDEO PRODUCTIONS *(WBV)*
6447 S. Heritage Pl., W.
Englewood, CO 80111

WEISS GLOBAL ENTERPRISES *(WGE)*
2055 Saviers Rd., Ste. 12
Oxnard, CA 93030
805-486-4495

WEST COAST VIDEO DISTRIBUTERS *(WCV)*
5750 E. Shields Ave., Ste. 101
Fresno, CA 93727
209-292-2013

WESTERN FILM & VIDEO, INC. *(WFV)*
30941 Agoura Rd., Ste. 302
Westlake Village, CA 91361
818-889-7350

WESTERN L.A. FILM PRODUCTIONS, INC. *(WLA)*
1604 Vista Del Mar
Hollywood, CA 90028
213-957-2110
Fax: 213-957-1050

WESTERN PUBLISHING *(WPB)*
12301 Wilshire Blvd. Ste. 611
Los Angeles, CA 90025
310-207-3156

WESTERN WORLD *(WES)*
104095 Santa Monica Blvd.
Los Angeles, CA 90025
310-475-5500

WESTERNWORLD VIDEO, INC. *(WWV)*
10523-45 Burbank Blvd.
North Hollywood, CA 91601
818-753-3000
800-347-8213

WHITE STAR *(WST)*
121 Hwy. 36
West Long Branch, NJ 07764
908-229-2343
800-458-5887
Fax: 908-229-0066

WHOLE TOON ACCESS *(WTA)*
PO Box 369
Issaquah, WA 98027-0369
206-391-8747
Fax: 206-391-9064

WISHING WELL DISTRIBUTING *(WSH)*
PO Box 2
Wilmot, WI 53192
414-864-2395
800-888-9355
Fax: 414-862-2398

WNET/THIRTEEN NON-BROADCAST *(WNE)*
356 W. 58th St.
New York, NY 10019
212-560-3045

WOMBAT FILM AND VIDEO *(WOM)*
930 Pitner
Evanston, IL 60202
708-328-6700
800-323-5448
Fax: 708-328-6706

WOMBAT PRODUCTIONS *(WMB)*
250 W. 57th St., Ste. 2421
New York, NY 10019
212-315-2502
Fax: 212-582-0585

WOMEN IN FOCUS *(WIF)*
849 Beatty St.
Vancouver, BC, Canada V6B 2M6
604-682-5848

WOOD KNAPP & COMPANY, INC. *(WKV)*
Knapp Press
5900 Wilshire Blvd.
Los Angeles, CA 90036
213-937-5486
800-521-2666

WORLD ARTISTS CINEMA CORPORATION *(WAC)*
135 N. Detroit St.
PO Box 36788
Los Angeles, CA 90036
213-933-7057
Fax: 213-933-2356

WORLD ARTISTS RELEASING *(WOR)*
256 S. Robertson Blvd., Ste. 3779
Beverly Hills, CA 90211-2832
213-650-7631

WORLD PREMIERE *(WPM)*
20611 Plummer St.
Chatsworth, CA 91311-5112
800-437-9063

WORLD VIDEO ENTERPRISES, INC. *(WVE)*
1447 Powell St.
San Francisco, CA 94133
415-543-4819

WORLD WIDE VIDEO *(WRD)*
7505 Foothill Blvd.
Tujunga, CA 91042
818-352-8735

WORLDVISION HOME VIDEO, INC. *(WOV)*
660 Madison Ave.
New York, NY 10021
212-261-2700
800-346-3000

WORLDWIDE ENTERTAINMENT MARKETING *(WEM)*
1133 Avenue of the Americas
New York, NY 10036
212-930-4500

XENON *(XVC)*
211 Arizona Ave.
Santa Monica, CA 90401
800-468-1913

YANKEE-ORIOLE COMPANY *(YOC)*
21-12 33rd Ave.
Astoria, NY 11106
718-721-0724

YES ENTERTAINMENT *(YES)*
2122 SE 43rd Ave.
Portland, OR 97215
503-234-6028

ZEITGEIST FILMS *(ZBI)*
200 Waverly Pl., Ste. 1
New York, NY 10014
212-274-1989